THE INDUSTRIAL ELECTRONICS HANDBOOK

The Electrical Engineering Handbook Series

Series Editor
Richard C. Dorf
University of California, Davis

Titles Included in the Series

The Electrical Engineering Handbook, Richard C. Dorf
The Biomedical Engineering Handbook, Joseph D. Bronzino
The Circuits and Filters Handbook, Wai-Kai Chen
The Transforms and Applications Handbook, Alexander D. Poularikas
The Control Handbook, William S. Levine
The Electronics Handbook, Jerry C. Whitaker
The Industrial Electronics Handbook, J. David Irwin
The Communications Handbook, Jerry D. Gibson
The Mobile Communications Handbook, Jerry D. Gibson

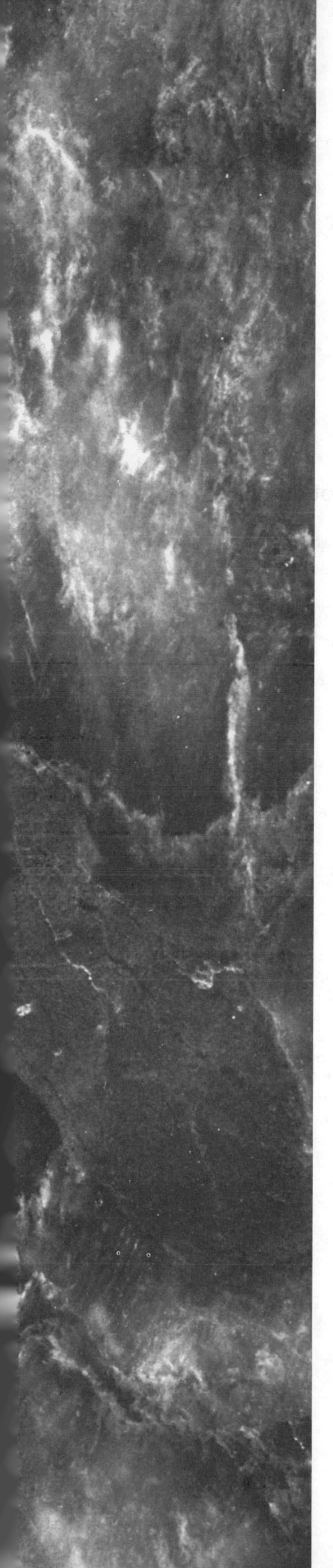

THE INDUSTRIAL ELECTRONICS HANDBOOK

Editor-in-Chief

J. DAVID IRWIN

A CRC Handbook Published in Cooperation with IEEE Press

Library of Congress Cataloging-in-Publication Data

The industrial electronics handbook/edited by J. David Irwin.
p. cm.--(The electrical engineering handbook series)
Includes bibliographical references and index.
ISBN 0-8493-8343-9 (alk. paper)
1. Industrial electronics—Handbooks, manuals, etc. I. Irwin, J. David, 1939– . 11. Series.
TK7881.I52 1996
621.3-dc20

96-3070
CIP

International Standard Book Number 0-8493-8343-9
Library of Congress Card Number 96-3070
Printed in the United States of America 1 2 3 4 5 6 7 8 9 0
Printed on acid-free paper

Preface

Introduction

The purpose of *The Industrial Electronics Handbook* is to provide a reference that is both concise and useful for individuals who range from students in engineering to experienced practicing professionals. The Handbook is designed to cover a very wide range of topics that comprise the subject of industrial electronics in a well-organized and highly informative manner. This volume is a careful blend of the traditional topics with the new and innovative technologies that are at the vanguard of the advancements being made in this subject. Special emphasis is placed on the practical application of the technologies discussed. Thus this Handbook is not a theoretical tome, but an enlightening presentation of the usefulness of the variety of technological entities that encompass the field. In an effort to further enhance the value of the book to the reader and foster a clear understanding of the material, the presentation is tutorial in nature, and items, such as examples of the practical use of the technology described, have been deliberately included.

The contributors to this Handbook span the globe (about one-fourth of them are from outside the U.S.) and are some of the leading authorities in their areas of expertise. Approximately one-third of them are from industry and government. The remaining contributors are from academe. All of them were chosen because of their intimate knowledge of their subjects as well as their ability to present them in an easily understandable manner.

Organization

The book is organized into two parts. Part 1 presents the traditional topics that have long been fundamental core areas of this technology. Part 2 introduces the new emerging and exciting areas that will play a vital and increasingly important role in the future of industrial electronics. The material not only stresses the key concepts, models, parameters, and equations that describe the technology, but in addition special emphasis is placed on the application of this technology to real-world problems. A unique mix of the basic principles and their application has been provided by authors who are eminently qualified and very experienced to address the important aspects of this technology.

Part 1 begins with a tutorial discussion of some of the topics which are basic to industrial electronics and provides the foundation for the development of the remaining sections. The section on Data Acquisition and Measurement Systems addresses the issues surrounding the manner in which data in an industrial process are obtained for analysis and subsequent use in such areas as communication, control, and automation. The remaining sections in Part 1 present the elements and methodologies of Power Electronics, Factory Communications, System Control, and Factory Automation which form the backbone of the industrial processes employed on the factory floor.

Part 2, which is entitled Intelligent Electronics and Emerging Technologies, presents the new modern techniques that are having, and will continue to have, a profound effect on the development and optimization of industrial systems. The fundamental characteristic, which permeates the new industrial electronics technologies, is the implied or specific use of some form of intelligence. These technologies include expert systems, neutral networks, fuzzy systems, soft computing, evolutionary systems, computational intelligence, and hybrid systems. The emerging technologies, which exhibit great promise for use in industrial processes, include virtual reality, asynchronous transfer mode technology, micro-electro-mechanical systems, and multisensor fusion and integration.

Locating Your Topic

Numerous avenues of access to information contained in the Handbook are provided. A complete table of contents is presented at the front of the book. In addition, an individual table of contents precedes each of the ten sections. Finally, each chapter begins with its own table of contents. The reader should look over these tables of contents to become familiar with the structure, organization, and content of the book.

Two alphabetical indexes have been compiled to provide access to information. The first is an index of contributing authors and the second a detailed subject index.

The Industrial Electronics Handbook is designed to provide both the young engineer and the experienced professional with answers to questions involving the wide spectrum of industrial electronics technology covered in this book. The topical coverage, as well as the numerous avenues to its access, hopefully will effectively satisfy the reader's needs.

Acknowledgments

This Handbook was made possible through the expertise and dedication of the International Advisory Board under the direction of Dr. James C. Hung, the outstanding authors from throughout the world, and my editorial associates. I gratefully acknowledge the personnel at CRC Press who produced the book. They are Felicia Shapiro, Associate Editor; Kristen Maus, Developmental Editor; and Carol Whitehead, Senior Project Editor. In addition, special thanks are due to Mrs. Peggy Turnquist, the Managing Editor for this book. Finally, I express my deep appreciation to my loving wife, Edie, who graciously puts up with my seemingly endless publication activities.

J. David Irwin
Editor-in-Chief

Preface

Editor-in-Chief

J. David Irwin was born in Minneapolis, Minnesota, on August 9, 1939. He received his B.E.E. degree from Auburn University, Auburn, Alabama, in 1961, and his M.S. and Ph.D. degrees from the University of Tennessee, Knoxville, in 1962 and 1967, respectively.

In 1967 he joined Bell Telephone Laboratories, Inc., Holmdel, New Jersey, as a member of the technical staff and was made a supervisor in 1968. He joined Auburn University in 1969 as an Assistant Professor of Electrical Engineering. He was made an Associate Professor in 1972, Associate Professor and Head of the Department in 1973, and Professor and Head in 1976. In 1993, he was named Earle C. Williams Eminent Scholar and Head.

Dr. Irwin has served the IEEE Computer Society as a member of the Education Committee and as Education Editor of *Computer.* He has served as Chairman of the Southeastern Association of Electrical Engineering Department Heads and the National Association of Electrical Engineering Department Heads and is past president of both the IEEE Industrial Electronics Society and the IEEE Education Society. He is a life member of the IEEE Industrial Electronics Society AdCom and has served as a member of the Oceanic Engineering Society AdCom. He served for two years as Editor of the *IEEE Transactions on Industrial Electronics.* He has served on the Executive Committee of the Southeastern Center for Electrical Engineering Education, Inc. and was president of the organization in 1983–84. He has served as an IEEE Adhoc Visitor for ABET Accreditation teams. He has served as a member of the IEEE Educational Activities Board, and was the Accreditation Coordinator for IEEE in 1989. He has served as a member of numerous IEEE committees including the Lamme Medal Award Committee, Fellow Committee, and the Nominations and Appointments Committee. He has served as a member of the Board of Directors of the IEEE Press. He has also served as a member of the Secretary of the Army's Advisory Panel for ROTC Affairs and Nominations Chairman for the National Electrical Engineering Department Heads Association. He has been and continues to be involved in the management of several international conferences sponsored by the IEEE Industrial Electronics Society.

He is the author/co-author of more than 50 publications, papers, and presentations including *Basic Engineering Circuit Analysis,* first, second, third and fourth editions, published by Macmillan Publishing Co. and the fifth edition published by Prentice Hall Book Co., Inc. He is the author of *On Becoming An Engineer: A Guide to Career Paths* published by the IEEE Press. He is also co-author of four other textbooks, *An Introduction to Computer Logic, Industrial Noise and Vibration Control, Introduction to Electrical Engineering,* and *Digital Logic Circuit Analysis and Design,* which are published by Prentice Hall Book Company, Inc.

He was made a Fellow of the Institute of Electrical and Electronics Engineers in 1982 and received an IEEE Centennial Medal in 1984. He was awarded the Bliss Medal by the Society of American Military Engineers in 1985. He received the IEEE Industrial Electronics Society's Anthony J. Hornfeck Outstanding Service Award in 1986, was named IEEE Region III (southeast) Outstanding Engineering Educator in 1989, and in 1991 received a Meritorious Service Citation from the IEEE Educational Activities Board, the 1991 Eugene Mittelmann Achievement Award from the IEEE Industrial Electronics Society, and the 1991 Achievement Award from the IEEE Education Society. In 1992 he was named a Distinguished Auburn Engineer. He is a member of the American Society for Engineering Education. In addition, he is a member of Sigma Xi, Phi Kappa Phi, Tau Beta Pi, Eta Kappa Nu, Pi Mu Epsilon, and Omicron Delta Kappa.

International Advisory Board

Contributors

Kenneth Agehed
Royal Institute of Technology
Stockholm, Sweden

Marcelo H. Ang, Jr.
National University of Singapore
Singapore, Singapore

Fumihito Arai
Nagoya University
Nagoya, Japan

Tamio Arai
University of Tokyo
Tokyo, Japan

Tom Baginski
Auburn University
Auburn, Alabama

Vrej Barkhordarian
International Rectifier
El Segundo, California

Antal K. Bejczy
Jet Propulsion Laboratory
California Institute of Technology
Pasadena, California

Upendra Belhe
The University of Iowa
Iowa City, Iowa

Michael Benard
Hewlett Packard
European Headquarters
Geneva, Switzerland

Janos Bencze
EKA Factory/El Appliance and Materials SC
Budapest, Hungary

Richard A. Blade
University of Colorado
Colorado Springs, Colorado

Jozsef Borka
Hungarian Academy of Sciences
Budapest, Hungary

A. John Boye
University of Nebraska
Lincoln, Nebraska

William Boyer
Sandia National Laboratories
Albuquerque, New Mexico

William L. Brogan
University of Nevada
Las Vegas, Nevada

Ronald H. Brown
Marquette University
Milwaukee, Wisconsin

Janusz Bryzek
Intelligent MicroSensor Technology (IMST)
Fremont, California

Anna L. Buczak
Allied Signal
Morristown, New Jersey

Dan Bugajski
Honeywell Technology Center
Minneapolis, Minnesota

Denis Calvet
CEA DSM/DAPNIA
Gif-Sur-Yvette, France

Gail A. Carpenter
Boston University
Boston, Massachusetts

John Carson
Irvine Sensors Corporation
Costa Mesa, California

Thomas Caudell
University of New Mexico
Albuquerque, New Mexico

Chen-Yu Chi
QualComm, Inc.
San Diego, California

Mo-yuen Chow
North Carolina State University
Raleigh, North Carolina

Robert W. Christie
University of Virginia
Arllington, Virginia

WooGon Chung
Sung Kyun Kwan University
SunWon City, KyungGi-Do
South Korea

Michel Combacau
L.A.A.S.—C.N.R.S.
Toulouse, France

J. Arlin Cooper
Sandia National Laboratories
Albuquerque, New Mexico

Mark G. Cooper
University of California
Los Angeles, California

Michele Costa
CERN
Geneva, Switzerland

Marc Courvoisier
I.N.S.A. de Toulouse
Toulouse, France

Timothy J. Dascy
MIT Lincoln Labs
Lexington, Massachusetts

Taher Daud
Jet Propulsion Laboratory
California Institute of Technology
Pasadena, California

J. Alexis De Abreu-Garcia
University of Akron
Akron, Ohio

Martin DePrycker
Alcatel Bell Telephone
Antwerp, Belgium

Jean-Dominique Decotignie
EPFL-LIT
Lausanne, Switzerland

Bert J. Dempsey
University of North Carolina
Chapel Hill, North Carolina

Thomas S. Denney, Jr.
Auburn University
Auburn, Alabama

Al Diy
International Rectifier Corporation
El Segundo, California

Kamel Djidi
CEA DSM/DAPNIA
Gif-Sur-Yvette, France

Rhonda F. Drayton
University of Illinois
Chicago, Illinois

Jean-Pierre Dufey
CERN
Geneva, Switzerland

Ernst Dummermuth
Rockwell Automation
Mayfield Heights, Ohio

Tuan A. Duong
Jet Propulsion Laboratory
California Institute of Technology
Pasadena, California

Ranadeep Dutta
International Rectifier Corporation
El Segundo, California

Åge Eide
Ostfold College
Halden, Norway

Charles W. Einolf, Jr.
Westinghouse Science and Technology Center
Pittsburgh, Pennsylvania

Prasad Enjeti
Texas A&M University
College Station, Texas

Dale Enns
Honeywell Technology Center
Minneapolis, Minnesota

Arthuro de la Escalera
Universidad Carlos III de Madrid
Madrid, Spain

Johnny Evers
USAF Armament Directorate
WL/MNAG
Eglin AFB, Florida

Faiq A. Fazal
AT&T Network Systems
Warren, New Jersey

Richard Francis
International Rectifier Corporation
El Segundo, California

A. Bruno Frazier
University of Utah
Salt Lake City, Utah

Craig R. Friedrich
Louisiana Tech University
Ruston, Louisiana

Hiroyuki Fujita
The University of Tokyo
Tokyo, Japan

Toshio Fukuda
Nagoya University
Nagoya, Japan

Kunihiko Fukushima
Osaka University
Toyonaka, Osaka, Japan

Diego Gachet
Universidad Carlos III de Madrid
Madrid, Spain

Fathi Ghorbel
Rice University
Houston, Texas

Michael Greene
Auburn University
Auburn, Alabama

Stephen Grossberg
Boston University
Boston, Massachusetts

Leif Gustafsson
University of Uppsala
Uppsala, Sweden

Sándor Halász
Technical University of Budapest
Budapest, Hungary

C. C. Hang
National University of Singapore
Singapore, Singapore

Royce D. Harbor
University of West Florida
Pensacola, Florida

Simon Haykin
McMaster University
Hamilton, Ontario, Canada

John Hecklesmiller
Best Power Technology, Inc.
Necedah, Wisconsin

James A. Heinen
Marquette University
Milwaukee, Wisconsin

James P. Helferty
Pennsylvania State University
State College, Pennsylvania

Sunderesh Heragu
Rensselaer Polytechnic Institute
Troy, New York

Gerry Heydt
Arizona State University
Tempe, Arizona

Robert M. Hines
University of Virginia
Charlottesville, Virginia

Isao Hirano
Hitachi Ltd.
Asao-ku, Kawasaki, Japan

A. S. Hodel
Auburn University
Auburn, Alabama

Mitsuyuki Hombu
Hitachi Ltd.
Ibaraki-ken, Japan

Takamasa Hori
Mie University
Tsu-Shi, Mie-Ken, Japan

Guan-Chyun Hsieh
National Taiwan Institute of Technology
Taipei, Taiwan, Republic of China

King-Lung Huang
Industrial Technology Research Institute
Hsinchu, Taiwan, Republic of China

Solve Hultberg
Royal Institute of Technology
Stockholm, Sweden

James C. Hung
University of Tennessee, Knoxville
Knoxville, Tennessee

John Y. Hung
Auburn University
Auburn, Alabama

Stephen T. Hung
Ford Motor Company
Dearborn, Michigan

Imre Ipsits
Technical University of Budapest
Budapest, Hungary

Rokuya Ishii
Yokohama National University
Yokohama, Japan

Mike Jackson
Honeywell Technology Center
Minneapolis, Minnesota

Yashvant Jani
Hitachi America Ltd.
Brisbane, California

James Jara-Almonte
SSI Technologies, Inc.
Janesville, Wisconsin

Ray Jarvis
Monash University
Clayton, Victoria, Australia

Bernard C. Jiang
Yuan-Ze Institute of Technology
Taiwan, Republic of China

Nicolaos B. Karayiannis
University of Houston
Houston, Texas

Attila Karpati
Technical University of Budapest
Budapest, Hungary

Linda P. B. Katehi
University of Michigan
Ann Arbor, Michigan

Michael G. Kay
North Carolina State University
Raleigh, North Carolina

Okyay Kaynak
Bogazici, University
Bebeck, Istanbul, Turkey

Sabrina Kemeny
Jet Propulsion Laboratory
California Institute of Technology
Pasadena, California

Chang-Jin Kim
University of California
Los Angeles, California

Lindsay Kleeman
Monash University
Clayton, Victoria, Australia

Karl Kluge
University of Michigan, AI Lab
Ann Arbor, Michigan

Klaudiusz Kobylecki
University of Uppsala
Uppsala, Sweden

Teuvo Kohonen
Helsinki University of Technology
Rakentajanaukio, Finland

Mieczyslaw M. Kokar
Northeastern University
Boston, Massachusetts

Zbigniew Korona
Northeastern University
Boston, Massachusetts

Mark E. Kotanchek
Pennsylvania State University
State College, Pennsylvania

Satoru Kotsu
Tokyo Institute of Technology
Yokohama, Japan

Karoly Kurutz
Technical University of Budapest
Budapest, Hungary

Andrew Kusiak
The University of Iowa
Iowa City, Iowa

Clifford Lau
Office of Naval Research
Arlington, Virginia

Dave Layden
Best Power Technology, Inc
Necedah, Wisconsin

Tawfik Lazrak
Royal Institute of Technology
Stockholm, Sweden

Robert N. Lea
Ortech Engineering, Inc
Houston, Texas

P. Le Dû
CEA DSM/DAPNIA
Gif-Sur-Yvette, France

Jay H. Lee
Auburn University
Auburn, Alabama

Jeong B. Lee
Georgia Institute of Technology
Pettit Microelectronics Research Center
Atlanta, Georgia

T. H. Lee
National University of Singapore
Singapore, Singapore

Michael Lehr
Stanford University
Stanford, California

Mike Letheren
CERN
Geneva, Switzerland

Khiang-Wee Lim
University of New South Wales
Sydney, Australia

Thomas Lindblad
Royal Institute of Technology
Stockholm, Sweden

Clark S. Lindsey
Royal Institute of Technology
Stockholm, Sweden

Christopher M. Lucarelli
Rensselaer Polytechnic Institute
Troy, New York

Ren C. Luo
North Carolina State University
Raleigh, North Carolina

Noel C. MacDonald
Cornell University
Ithaca, New York

Neeraj Magotra
University of New Mexico
Albuquerque, New Mexico

Atsushi Manabe
CERN
Ibaraki, Japan

I. Mandjavidze
CEA DSM/DAPNIA
Gif-Sur-Yvette, France

Giridhar Mandyam
University of New Mexico
Albuquerque, New Mexico

David J. Marchette
Naval Surface Warfare Center
Dahlgren, Virginia

Alessandro Marchioro
CERN
Geneva, Switzerland

Walt Maslowski
Allen-Bradley Company
Mequon, Wisconsin

Ryosuke Masuda
Tokai University
Kanagawa, Japan

Wes McCoy
Motorola Advanced Messaging Group
Ft. Worth, Texas

Joseph A. Mica
NASA/Goddard Space Flight Center
Green Belt, Maryland

Evangelia Micheli-Tzanakou
Rutgers University
Piscataway, New Jersey

S. A. Miller
Cornell University
Ithaca, New York

Nadine Miner
Sandia National Laboratories
Albuquerque, New Mexico

Curtis L. Moffit
Newbridge Networks, Inc.
Herndon, Virginia

Luis Moreno
Universidad Carlos III de Madrid
Madrid, Spain

Blaise Morton
Honeywell Technology Center
Minneapolis, Minnesota

W. Bosseau Murray
Pennsylvania State University
Hershey, Pennsylvania

Herschell Murry
Polhemus, Inc.
Colchester, Vermont

Hiroshi Nagase
Mito Works/Hitachi, Ltd.
Ibaraki-ken, Japan

Rakesh Nagi
State University of New York
Buffalo, New York

István Nagy
Technical University of Budapest
Budapest, Hungary

Victor P. Nelson
Auburn University
Auburn, Alabama

Russell J. Niederjohn
Marquette University
Milwaukee, Wisconsin

Kouhei Ohnishi
Keio University
Yokohama, Japan

Mary Lou Padgett
Auburn University
Auburn, Alabama

Christian Paillard
CERN
Geneva, Switzerland

Charles Palmer
Pennsylvania State University
Hershey, Pennsylvania

Mario Paludetto
L.A.A.S.—C.N.R.S.
Toulouse, France

Carmen M. Pancerella
Sandia National Laboratories
Livermore, California

John Parr
University of Evansville
Evansville, Indiana

B. Pauwels
Alcatel Bell Telephone
Antwerp, Belgium

Witold Pedrycz
University of Manitoba
Winnipeg, Manitoba, Canada

Sameer Pendharkar
Texas Instruments
Dallas, Texas

Guido Petit
Alcatel Bell Telephone
Antwerp, Belgium

Richard Pettit
Sandia National Laboratories
Albuquerque, New Mexico

Charles L. Phillips
Auburn University
Auburn, Alabama

Juan R. Pimentel
GMI Engineering and Management Institute
Flint, Michigan

Patrick Pleinevaux
EPFL-LIT
Lausanne, Switzerland

Timothy Poston
National University of Singapore
Institute of Systems Science
Singapore, Singapore

Carey E. Priebe
The Johns Hopkins University
Baltimore, Maryland

Jean-Marie Proth
INRIA Lorraine
Metz, France

M. F. Rahman
University of New South Wales
New South Wales, Australia

Gabriel M. Rebeiz
University of Michigan
Ann Arbor, Michigan

Ray P. Reed
Proteun Services
Albuquerque, New Mexico

Douglas L. Reilly
Nestor, Inc.
Providence, Rhode Island

Antonio J. Ricco
Sandia National Laboratories
Albuquerque, New Mexico

Gary Riley
NASA/Johnson Space Center
Houston, Texas

Anthony D. Robbi
New Jersey Institute of Technology
Newark, New Jersey

Stephen V. Robertson
University of Michigan
Ann Arbor, Michigan

Michael Robinson
International Rectifier Corporation
El Segundo, California

George W. Rogers
Naval Surface Warfare Center
Dahlgren, Virginia

Steven K. Rogers
Air Force Institute of Technology
Wright-Patterson AFB, Ohio

Thaddeus Roppel
Auburn University
Auburn, Alabama

James R. Rowland
University of Kansas
Lawrence, Kansas

Dennis W. Ruck
Air Force Institute of Technology
Wright-Patterson AFB, Ohio

David E. Rumelhart
Stanford University
Stanford, California

R. Andrew Russell
Monash University
Melbourne, Victoria, Australia

David Ryerson
Sandia National Laboratories
Albuquerque, New Mexico

Richard Saeks
Accurate Automation Corporation
Chattanooga, Tennessee

M.T.A. Saif
Cornell University
Ithaca, New York

Miguel A. Salichs
Universidad Carlos III de Madrid
Madrid, Spain

Debapriya Sarkar
University of Virginia
Charlottesville, Virginia

Michio Sasaki
Tokai University
Kanagawa, Japan

Chris Saunders
Irvine Sensors Corporation
Costa Mesa, California

Robert Shelton
NASA/Johnson Space Center
Houston, Texas

Krishna Shenai
University of Illinois
Chicago, Illinois

Bing J. Sheu
University of Southern California
Los Angeles, California

Takanori Shibata
Ministry of International Trade and Industry (MITI)
Tsukuba, Japan
and
Massachusetts Institute of Technology (MIT)
Cambridge, Massachusetts

N. K. Sinha
McMaster University
Hamilton, Ontario, Canada

Tarek M. Sobh
University of Bridgeport
Bridgeport, Connecticut

Jeffrey L. Solka
Naval Surface Warfare Center
Dahlgren, Virginia

Otis Solomon
Sandia National Laboratories
Albuquerque, New Mexico

Donald F. Specht
Lockheed Martin Missiles & Space
Palo Alto, California

P. Sphicas
MIT
Cambridge, Massachusetts

James Stanislawski
National Power Laboratory
Necedah, Wisconsin

Samuel D. Stearns
Sandia National Laboratories
Albuquerque, New Mexico

Laura Steffek
Best Power Technology, Inc.
Necedah, Wisconsin

Gunter Stein
Honeywell Technology Center
Minneapolis, Minnesota

W. Timothy Strayer
Sandia National Laboratories
Livermore, California

John W. Sublett
University of Virginia
Charlottesville, Virginia

Michio Sugeno
Tokyo Institute of Technology
Midori-ku, Yokohama, Japan

Konstanty Sumorok
MIT
Cambridge, Massachusetts

Harold H. Szu
Naval Surface Warfare Center
Dahlgren, Virginia

Yu-Chong Tai
California Institute of Technology
Pasadena, California

Hideyuki Takagi
Kyushyu Institute of Design
Fukuoka, Japan

Kota Takahashi
The University of Electro-Communications
Tokyo, Japan

K. K. Tan
National University of Singapore
Singapore, Singapore

Li-Zhe Tan
Iterated Systems
Atlanta, Georgia

Kazuo Tanie
Ministry of International Trade and Industry (MITI)
Tsukuba, Japan

Brian A. Telfer
Naval Surface Warfare Center
Dahlgren, Virginia

Hannu Tenhunen
Royal Institute of Technology
Stockholm, Sweden

S. Tether
MIT
Cambridge, Massachusetts

Anil Thakoor
Jet Propulsion Laboratory
California Institute of Technology
Pasadena, California

Victor Trent
Auburn University
Auburn, Alabama

Malay Trivedi
University of Wisconsin
Madison, Wisconsin

Lefteri H. Tsoukalas
Purdue University
West Lafayette, Indiana

Robert E. Uhrig
University of Tennessee
Knoxville, Tennessee
and
Oak Ridge National Laboratory
Oak Ridge, Tennessee

Kazunori Umeda
Chuo University
Tokyo, Japan

Belle Upadhyaya
University of Tennessee
Knoxville, Tennessee

Vadim Utkin
Ohio State University
Columbus, Ohio

Michael J. Vasile
Louisiana Tech University
Ruston, Louisiana

Robert J. Veillette
University of Akron
Akron, Ohio

V. Rao Vemuri
University of California at Davis
Livermore, California

H. Verhille
Alcatel Bell Telelphone
Antwerp, Belgium

Darrell Vines
Texas Tech University
Lubbock, Texas

Ljubisa Vlacic
Griffith University
Queensland, Australia

Patrick L. Walter
Texas Christian University
Fort Worth, Texas

Keith O. Warren
Litton Guidance and Control Systems
Woodland Hills, California

Alfred C. Weaver
University of Virginia
Charlottesville, Virginia

Tom M. Weller
University of South Florida
Tampa, Florida

Paul J. Werbos
National Science Foundation
Arlington, Virginia

Charles R. White
Auburn University
Auburn, Alabama

Bernard Widrow
Stanford University
Stanford, California

Robert G. Wilhelm
University of North Carolina
Charlotte, North Carolina

Theodore J. Williams
Purdue University
West Lafayette, Indiana

Howard A. Winston
United Technologies Research Center
East Hartford, Connecticut

Seth Wolpert
Penn State University at Harrisburg
Middletown, Pennsylvania

Walter Wong
Queensland University of Technology
Brisbane, Queensland, Australia

Tony H. Wu
University of Southern California
Los Angeles, California

Hiro Yamasaki
Yokogawa Electric Corporation
Tokyo, Japan

Choon-seng Yee
National University of Singapore
Singapore, Singapore

Gary G. Yen
Oklahoma State University
Oklahoma City, Oklahoma

Brian Young
Best Power Technology, Inc.
Necedah, Wisconsin

Lotfi A. Zadeh
University of California, Berkeley
Berkeley, California

Daniel A. Zahner
Rutgers University
Pompton Plains, New Jersey

Dian-cheng Zhang
Hefei University of Technology
Hefei, Anhwei, People's Republic of China

MengChu Zhou
New Jersey Institute of Technology
Newark, New Jersey

Richard Zurawski
Swinburne University of Technology
Melbourne, Victoria, Australia

To my loving wife, Edie.

Contents

Part 1 FUNDAMENTALS OF INDUSTRIAL ELECTRONICS

SECTION I Supporting Technologies

1 Electronics *Darrell Vines and Tom Baginski* . . . 5
- 1.1 Introduction . . . 5
- 1.2 Diodes . . . 5
- 1.3 Trasistors as Switches . . . 10
- 1.4 Models for Transistors . . . 15
- 1.5 Analog and Digital Circuits . . . 19

2 Digital Control Circuits *Marc Courvoisier, Michel Combacau, and Mario Paludetto* . . . 22
- 2.1 Logic Control . . . 22
- 2.2 Sequence Control . . . 28
- 2.3 Implementation Techniques . . . 41

3 Computer Architecture *Victor P. Nelson* . . . 48
- 3.1 Hardware Organization . . . 48
- 3.2 Computer Software . . . 50
- 3.3 Information Representation in Digital Computers . . . 51
- 3.4 Specifying Instruction Operands . . . 53
- 3.5 CPU Registers . . . 54
- 3.6 Memory Organization . . . 56
- 3.7 Computer Instruction Types . . . 58
- 3.8 Interrupts and Exceptions . . . 60
- 3.9 Evaluating Instruction Set Architectures . . . 61
- 3.10 Computer System Design . . . 62
- 3.11 Input/Output Device Interfaces . . . 67
- 3.12 Microcontroller Architectures . . . 67
- 3.13 Multiple Processor Architectures . . . 69

4 Signal Processing *James A. Heinen and Russell J. Niederjohn* . . . 73
- 4.1 Introduction . . . 73
- 4.2 Continuous-Time Signals . . . 74
- 4.3 Time-Domain Analysis of Continuous-Time Signals . . . 74
- 4.4 Frequency-Domain Analysis of Continuous-Time Signals . . . 75
- 4.5 Continuous-Time Signal Processors . . . 79
- 4.6 Time-Domain Analysis of Continuous-Time Signal Processors . . . 79
- 4.7 Frequency-Domain Analysis of Continuous-Time Signal Processors . . . 81
- 4.8 Continuous-Time (Analog) Filters . . . 80
- 4.9 Sampling . . . 81
- 4.10 Discrete-Time Signals . . . 83
- 4.11 Time-Domain Analysis of Discrete-Time Signals . . . 84

4.12 Frequency-Domain Analysis of Discrete-Time Signals 84
4.13 Discrete-Time Signal Processors 89
4.14 Time-Domain Analysis of Discrete-Time Signal Processors 89
4.15 Frequency-Domain Analysis of Discrete-Time Signal Processors 91
4.16 Discrete-Time (Digital) Filters 91
4.17 Discrete-Time Analysis of Continuous-Time Signals 93
4.18 Discrete-Time Processing of Continuous-Time Signals 94

SECTION II Data Aquisition and Measurement Systems

5 Sensors *Charles W. Einolf, Jr.* 97
5.1 Introduction 97
5.2 Passive Sensors 98
5.3 Active Sensors 98

6 Measurement System Architecture 103
6.1 Introduction *Patrick L. Walter* 103
6.2 System Considerations *Patrick L. Walter* 104
6.3 Signal Conditioning and Filtering *David Ryerson* 105
6.4 Signal/Data Transmission Components *Otis Solomon and William Boyer* 119
6.5 Software Data Correction *William Boyer and David Ryerson* 122
6.6 Computers in Instrumentation Systems *William Boyer* 126
6.7 Software for Instrumentation Systems *William Boyer* 129
6.8 Calibration and Testing *Richard Pettit* 132
6.9 Digital Signal Processing *Belle Upadhyaya* 138
6.10 Signal Pick-up and Interface Circuitry *Thaddeus Roppel* 146
6.11 Thermal Effects in Industrial Electronic Circuits *Ray P. Reed* 151
6.12 Lossless Waveform Compression *Giridhar Mandyam, Neeraj Magotra, Samuel D. Stearns, Li-Zhe Tan, and Wes McCoy* 164
6.13 3-D Measurement Techniques *Bernard C. Jiang* 174

SECTION III Power Electronics

7 Introduction to Power Electronics *Janos Bencze* 187
7.1 Introduction 187
7.2 Power Supplies 189
7.3 Electric Drives 190
7.4 Application Examples 191
7.5 Future Trends 194

8 Overview: Devices and Components *Malay Trivedi, Sameer Pendharkar, and Krishna Shenai* 195
8.1 Introduction 195
8.2 Diode 195
8.3 Thyristor 196
8.4 Transistors 197
8.5 New Devices 199

9 Devices and Components 203
9.1 Power Diodes *Imre Ipsits* 203
9.2 Power Bipolar Junction Transistors (BJTs) *Imre Ipsits* 211
9.3 Passive Networks *Karoly Kurutz* 215

10 Power MOSFETs *Vrej Barkhordarian* 218
10.1 Introduction 218
10.2 Static Characteristics 220
10.3 Dynamic Characteristics 224
10.4 Applications 227

11 Insulated Gate Bipolar Transistors *Michael Robinson, Richard Francis, Ranadeep Dutta, and Al Diy* 229
11.1 Introduction 229
11.2 Basic Structure and Operation 230
11.3 Design Considerations 232
11.4 Requirement for Anti-parallel Diode 236
11.5 Comparison Between the Power MOSFET, IGBT, and MCT 236
11.6 IGBT Data Sheet Parameters 237
11.7 Appendix: Typical IGBT Data Sheet 238

12 Conversion 244
12.1 AC-DC Converters *Attila Karpati* 244
12.2 DC-DC Converters *István Nagy* 253
12.3 DC-AC Conversion *Attila Karpati* 263
12.4 AC-AC Conversion *Sándor Halász* 273
12.5 Resonant Converters *István Nagy* 276

13 Motor Drives 288
13.1 Control Systems and Applications *Takamasa Hori* 288
13.2 DC Motor Control Systems *Takamasa Hori* 289
13.3 Induction Motor Control Systems *Takamasa Hori, Hiroshi Nagase, and Mitsuyuki Hombu* 294
13.4 Synchronous Motor Control Systems *Takamasa Hori* 315
13.5 PM Synchronous Motor Control *M. F. Rahman and Khiang-Wee Lim* 319
13.6 Step Motor Drives *Ronald H. Brown* 331
13.7 Servo Drives *Sándor Halász* 341
13.8 Switched Reluctance Motor Drives *Jozsef Borka* 344

14 Main Disturbances 349
14.1 Power Quality *James Stanislawski* 349
14.2 Reactive Power and Harmonics Compensation *Gerry Heydt* 352
14.3 New Power Converters *Prasad Enjeti* 363
14.4 Uninterruptible Power Supplies (UPS) *Laura Steffek, John Hecklesmiller, Dave Layden, and Brian Young* 367

15 Electromagnetic Compatibility for Drives *Walt Maslowski* 377
15.1 Compatibility: Emissions and Immunity 377

SECTION IV Factory Communications

16 Evolution of Factory Communication *W. Timothy Strayer and Carmen M. Pancerella* 385
16.1 Point-to-Point Communications 385
16.2 Network Communications 386
16.3 Advantages of Network Interconnection 387
16.4 Communications Requirements for Distributed Systems 388

17 Open Systems Interconnection Basic Reference Model *Robert M. Hines* 389
17.1 Introduction 389
17.2 Physical Layer 389
17.3 Datalink Layer 390
17.4 Network Layer 390

17.5 Transport Layer . . . 391
17.6 Session Layer . . . 392
17.7 Presentation Layer . . . 392
17.8 Application Layer . . . 392

18 Local Area Networks . . . 394
18.1 Ethernet and IEEE 802.3 Contention Bus *Alfred C. Weaver* . . . 394
18.2 IEEE 802.5 Token Ring *John W. Sublett* . . . 396
18.3 IEEE 802.4 Token Bus *Alfred C. Weaver* . . . 400
18.4 Fieldbus *Jean-Dominique Decotignie* . . . 403
18.5 Fiber Distributed Data Interface (FDDI) *Robert W. Christie* . . . 408
18.6 Asynchronous Transfer Mode *Curtis L. Moffit* . . . 412

19 Manufacturing Automation Protocol (MAP) *Juan R. Pimentel* . . . 417
19.1 History . . . 417
19.2 Purpose . . . 417
19.3 Description . . . 418
19.4 Standards Used . . . 420
19.5 Example of Use . . . 426

20 Essential Communications Protocols . . . 427
20.1 Datalink Protocols *Bert J. Dempsey* . . . 427
20.2 Network Protocols *Debapriya Sarkar* . . . 429
20.3 Transport Layer Protocols *Bert J. Dempsey* . . . 434

SECTION V System Control

21 Control System Fundamentals *A. S. Hodel* . . . 443
21.1 Modeling . . . 443
21.2 Controller Design . . . 444
21.3 Intelligent Control . . . 445
21.4 Other Control Approaches . . . 445

22 Modeling for System Control *A. John Boye and William L. Brogan* . . . 447
22.1 Introduction . . . 447
22.2 Analytical Modeling . . . 447
22.3 Defining the Problem . . . 448
22.4 Determining the System Components . . . 448
22.5 Writing the System Equations . . . 449
22.6 Verifying the Model . . . 450
22.7 Empirical or Experimental Modeling . . . 451

23 Basic Feedback Concept *T. H. Lee, C. C. Hang, and K. K. Tan* . . . 453
23.1 Beneficial Effects of Feedback . . . 454
23.2 Analysis of Design of Feedback Control Systems . . . 455
23.3 Implementation of Feedback Control Systems . . . 455

24 Stability Analysis *N. K. Sinha* . . . 456
24.1 Stability Analysis for Linear Systems . . . 456
24.2 Stability of Linear Time-Invariant Continuous-Time Systems . . . 456
24.3 Stability of Linear Time-Invariant Discrete-Time Systems . . . 463
24.4 Nonlinear Systems . . . 466

25 PID Control *James C. Hung* 470
25.1 Introduction 470
25.2 Classical PID Control (Ziegler-Nichols Tuning) 470
25.3 Remarks 472

26 Bode Diagram Method *John Parr* 474
26.1 Bode Diagram Analysis 474
26.2 Mathematical Model Determination 478
26.3 Correlation of Frequency Response and Time Response 480
26.4 Shaping the Cutoff Response 481
26.5 Compensator Design 482
26.6 Design for Digital Systems 486

27 The Root Locus Method *Robert J. Veillette and J. Alexis De Abreu-Garcia* 490
27.1 Motivation and Background 490
27.2 Root Locus Analysis 490
27.3 Compensator Design by Root Locus Method 495
27.4 Examples 497

28 Pole Placement Design *Michael Greene and Victor Trent* 504
28.1 Pole Placement 504
28.2 State Observation 506
28.3 Discrete Implementation 509

29 The Smith Predictor Technique *John Y. Hung* 511
29.1 Background—Control of Processes Having Time Delay 511
29.2 Basic Principle of the Smith Predictor 511
29.3 A Smith Predictor Design Example 512

30 Internal Model Control *James C. Hung* 513
30.1 Basic IMC Structures 513
30.2 IMC Design 514
30.3 Discussion 514

31 Model Predictive Control *Jay H. Lee* 515
31.1 Overview 515
31.2 Applications 516

32 Dynamic Matrix Control *James C. Hung* 522
32.1 The Dynamic Matrix 522
32.2 Output Projection 522
32.3 Control Computation 523
32.4 Remarks 523

33 Disturbance Observation-Cancellation Technique *Kouhei Ohnishi* 524
33.1 Why Estimate Disturbance? 524
33.2 Plant and Disturbance 524
33.3 Higher-Order Disturbance Approximation 526
33.4 Disturbance Observation 526
33.5 Disturbance Cancellation 526
33.6 Examples of Application 527
33.7 Conclusions 528

34 Phase-Locked Loop-Based Control *Guan-Chyun Hsieh* 529
34.1 Introduction 529
34.2 Configurations of PLL Applications 532

34.3 Analog, Digital, and Hybrid PLLs . . . 533
34.4 Popular PLL Integrated Circuits (ICs) . . . 533

35 Variable Structure Control Technique *Vadim Utkin* . . . 535
35.1 Introduction . . . 535
35.2 Mathematical Aspects . . . 536
35.3 Sliding Mode Control Design . . . 538
35.4 Chattering Problem . . . 540
35.5 Control of Manipulators . . . 540
35.6 Control of Mobile Robots . . . 541
35.7 Control of Railway Wheelset . . . 541
35.8 Control of Torsion Oscillations of a Flexible Shaft . . . 542
35.9 DC Motors . . . 542
35.10 Control of DC Motors Based on a Reduced-Order Model . . . 543
35.11 Conclusion . . . 544

36 Digital Computation *James R. Rowland* . . . 545
36.1 System Response . . . 545
36.2 Numerical Integration Formulas . . . 548
36.3 Exact Difference Equations for Linear Systems . . . 551
36.4 Summary . . . 552

37 Digital Control *John Y. Hung and Victor Trent* . . . 553
37.1 Introduction . . . 553
37.2 Discretization of Continuous-Time Systems . . . 553
37.3 Discretization of the Servomotor System . . . 554
37.4 Frequency Domain Design through the w-Transform . . . 555
37.5 Root Locus Design on the Unit Circle . . . 556
37.6 Simulation Comparisons . . . 557

38 Estimation and Identification *Thomas S. Denney, Jr.* . . . 559
38.1 Kalman Filters . . . 559
38.2 Other Types of Kalman Filters . . . 561
38.3 Identification . . . 561

39 Fuzzy Logic-Based Control *Mo-yuen Chow* . . . 564
39.1 Introduction to Intelligent Control . . . 564
39.2 DC Motor Dynamics . . . 565
39.3 Fuzzy Control . . . 566
39.4 Conclusion and Future Direction . . . 570

40 Neural Network-Based Control *Dian-cheng Zhang* . . . 572
40.1 Control Configuration . . . 572
40.2 Design Procedure . . . 580

41 Programmable Logic Control (PLC) *Ernst Dummermuth* . . . 587
41.1 Basic Concepts . . . 587
41.2 Hardware Components . . . 588
41.3 PLC Real-Time Operating Systems . . . 588
41.4 Software Components . . . 590
41.5 PLC Communications . . . 590
41.6 Selecting the Right PLC . . . 591

42 Adaptive Control *Stephen T. Hung* 593
42.1 Introduction 593
42.2 Update Strategies 595
42.3 Direct Adaptive Control 599
42.4 Indirect Adaptive Control 604
42.5 Adaptive/Self-Tuning Behavior 606
42.6 Summary 607

43 Hardware Compensating Networks *Royce D. Harbor and Charles L. Phillips* 609
43.1 Continuous Compensation 609
43.2 Other Compensation Procedures 611

44 μ-Synthesis and Analysis *Dan Bugajski, Dale Enns, Mike Jackson, Blaise Morton, and Gunter Stein* 613
44.1 Defining the Interconnection Structure 614
44.2 H_∞-Synthesis 615
44.3 μ-Analysis and D Scales 617
44.4 D-K Iteration 618
44.5 Changing Weights 619
44.6 Compensator Model Reduction 620
44.7 Summary 620

SECTION VI Factory Automation

45 An Overview of Factory Automation *Richard Zurawski* 625
45.1 Introduction 625
45.2 New Technologies for Factory Automation 626

46 Types of Automated Manufacturing Systems *Ljubisa Vlacic, Walter Wong, and Theodore J. Williams* 629
46.1 The Hierarchical Model Presentation of Manufacturing Activities 629
46.2 Enterprise/Factory Integration 632
46.3 The Methodology for CIE/CIM 634
46.4 Architectures of Automated Manufacturing Systems 638
46.5 Implementations of Factory Automation Systems 641
46.6 Flexible Manufacturing Systems (FMS) 642

47 Production Management Architecture *Rakesh Nagi and Jean-Marie Proth* 653
47.1 Introduction 653
47.2 Production Management in the Sixties and Beyond 654
47.3 Components of the Hierarchical Production Management System 654
47.4 Long-Term Production Plan (LTPP) 655
47.5 Master Production Scheduling (MPS) 656
47.6 Capacity Requirement Planning (CRP) 658
47.4 MRP Philosophy 658
47.8 Application of the MRP 662
47.9 Conclusion 662

48 Production Management Techniques *Upendra Belhe and Andrew Kusiak* 663
48.1 Material Requirements Planning (MRP) 663
48.2 Manufacturing Resource Planning (MRPII) 664
48.3 Optimized Production Technology (OPT) 665
48.4 Toyota System and Just-in-Time 666
48.5 The Kanban Concept 667

49 Automated Manufacturing System Development Methodology 669
49.1 Analysis of Functional Properties of Specification and Design Models of Industrial Automated Systems *Richard Zurawski and MengChu Zhou* 669
49.2 Automated Manufacturing System Design Using Analytical Techniques *Sunderesh S. Heragu and Christopher M. Lucarelli* 677
49.3 Discrete Event Simulation *MengChu Zhou, Anthony D. Robbi, and Richard Zurawski* 694

50 Hybrid Systems and Control *Tarek M. Sobh* 706
50.1 Introduction 706
50.2 Discrete Event and Hybrid Observation under Uncertainty 707
50.3 Conclusions 714

51 Virtual Manufacturing Environment *Robert G. Wilhelm* 718
51.1 Introduction 718
51.2 Scope for Virtual Manufacturing 718
51.3 Typical Applications 718
51.4 Emerging Technology 720

52 Signal Processing for Factory Production Lines *Rokuya Ishii* 723
52.1 Introduction 723
52.2 Examples of Signal Processing Systems 724

53 Robots 730
53.1 Robots: Qualities and Capabilities *Ray Jarvis* 730
53.2 Robot Vision *Ray Jarvis* 732
53.3 Ultrasonic Sensors *Lindsay Kleeman* 738
53.4 Robot Tactile Sensing *R. Andrew Russell* 745
53.5 A Robotic Sense of Smell *R. Andrew Russell* 749
53.6 Actuators in Robotics and Automation Systems *Marcelo H. Ang, Jr. and Choon-seng Yee* 750
53.7 Control *Fathi Ghorbel* 760
53.8 Mobile Robots *Miguel A. Salichs, Luis Moreno, Diego Gachet, Arthuro de la Escalera, and Juan R. Pimentel* 773
53.9 Teleoperators *Antal K. Bejczy* 784

PART 2 INTELLIGENT ELECTRONICS AND EMERGING TECHNOLOGIES

SECTION VII Expert Systems and Neural Networks

Expert Systems

54 Current Applications of Expert Systems in Industrial Electronics *Mary Lou Padgett and Robert Shelton* 805
54.1 Emerging Trends for Expert Systems in Industrial Electronics 805
54.2 Defining Terms 805
54.3 Resources 807

55 Expert Systems Methodology *Gary Riley* 808
55.1 Capturing Human Expertise in a Program 808
55.2 Rule-Based Programming 809
55.3 Truth Table Simplification Program 811

56 Expert Systems and Their Use in Complex Engineering Systems *Robert E. Uhrig and Lefteri H. Tsoukalas* 824
56.1 Introduction 824
56.2 Definition of Expert Systems 824

56.3 Characteristics of Expert Systems 825
56.4 Components of an Expert System 825
56.5 Knowledge Representation and Inference 826
56.6 Uncertainty Management 828
56.7 State of the Art of Expert Systems 830
56.8 Use of Expert Systems 830
56.9 Potential Implementation Issues for Expert Systems 831
56.10 Legal Aspects of Expert Systems 832
56.11 Use of Expert Systems in Nuclear Power Plants 833

Neural Networks

57 Strategies and Tactics for the Application of Neural Networks to Industrial Electronics *Mary Lou Padgett, Paul J. Werbos, and Teuvo Kohonen* 835
57.1 Computational Intelligence Connections and Future 835
57.2 Engineering Intelligent Electronics Applications 836
57.3 Summary of Basic Modeling Concepts 846
57.4 Applications 846
57.5 Future 846
57.6 Defining Terms 847
57.7 Resources 851

58 The Basic Ideas in Neural Networks *David E. Rumelhart, Bernard Widrow, and Michael Lehr* 853
58.1 Introduction 853
58.2 Learning By Example 855
58.3 Generalization 856
58.4 Hints for Successful Applications 857

59 Neural Networks on a Chip *Clifford Lau* 858
59.1 Artificial Neural Network Technology Compared with Conventional 858
59.2 Examples of Chips 858
59.3 Comparisons of NN VLSI Microchips 864
59.4 Applications of Neural Network Technology 864
59.5 BMDO/IST Demonstration Project: 3-D ANN Silicon Neuron Seeker 864

60 Commercially Available Artificial Neural Network Chips *Seth Wolpert* 867
60.1 Introduction 867
60.2 Analog ANN Products 867
60.3 Digital ANN Products 869
60.4 Hybrid ANN Producrts 871
60.5 Discussion 872

61 Implementing Neural Networks in Silicon *Seth Wolpert and Evangelia Micheli-Tzanakou* 874
61.1 Introduction 874
61.2 The Living Neuron 874
61.3 Neuromorphic Models 875
61.4 Neurological Process Modeling 881

62 An Avionics Application: MIMD Neural Network Processor *Richard Saeks* 885
62.1 NNP Architecture 885
62.2 Summary 887

63 Backpropagation to Neurocontrol *Paul J. Werbos* 888
63.1 Neurocontrol: Where It Is Going and Why It Is Crucial 888

64 CMAC Neural Networks and Color Correction *King-Lung Huang* 906
64.1 Introduction 906
64.2 High-Order CMAC Neural Networks for Color Correction 907
64.3 Experimental Result 907
64.4 Conclusion 908

65 Temporal Signal Processing *Simon Haykin* 910
65.1 Introduction 910
65.2 Temporal Neural Networks with Observable States 910
65.3 Temporal Neural Networks with Hidden States 912
65.4 Conclusions 914

66 Feature Selection for Pattern Recognition Using Multilayer Perceptrons *Dennis W. Ruck and Steven K. Rogers* . . 916
66.1 Introduction 916
66.2 Background 918
66.3 Methodology 918
66.4 Applications 920
66.5 Conclusions 921

67 Wavelets for Pattern Recognition *George W. Rogers, David J. Marchette, and Jeffrey L. Solka* 923
67.1 Wavelet-Based Segmentation 923
67.2 Resistive Grid Local Averaging 925
67.3 Examples 928

68 Fractals for Pattern Recognition *George W. Rogers, Carey E. Priebe, and Jeffrey L. Solka* 933
68.1 A PDP Approach to Localized Fractal Dimension Computation with Segmentation Boundaries 933

69 Multilayer Perceptrons with ALOPEX and Backpropagation *Daniel A. Zahner and Evangelia Micheli-Tzanakou* . . 942
69.1 Introduction 942
69.2 The Backpropagation Algorithm 943
69.3 The ALOPEX Algorithm 944
69.4 Miltilayer Perceptron Network 945
69.5 ALOPEX in VLSI 947
69.6 Discussion 949

70 Supervised Neural Networks for Handwritten Digit Recognition in Industrial Processing *WooGon Chung and Evangelia Micheli-Tzanakou* 951
70.1 Introduction 951
70.2 Preprocessing of Handwritten Digit Images 951
70.3 Zernike Moments (ZM) to Characterize Image Patterns 955
70.4 Dimensionality Reduction 960
70.5 Analysis of Prediction Error Rates from Bootstrapping Assessment 962
70.6 Summary 964

71 Neocognitron *Kunihiko Fukushima* 966
71.1 Neocognitron 966
71.2 Selective Attention Model (SAM) 969

72 Studies of Pattern Recognition with Self-Learning Layered Neural Networks *Faiq A. Fazal and Evangelia Micheli-Tzanakou* 975
72.1 Abstract 975
72.2 Introduction 975
72.3 Neocognitron and Pattern Classification 976
72.4 Objectives 978
72.5 Methods 978

72.6 Study A . . . 979
72.7 Study B . . . 985
72.8 Summary and Discussion . . . 989

73 Analog 3-D Neuroprocessor for Fast Frame Focal Plane Image Processing *Tuan A. Duong, Sabrina Kemeny, Taher Daud, Anil Thakoor, Chris Saunders, and John Carson* . . . 990
73.1 Introduction . . . 990
73.2 Neural Network Architecture . . . 991
73.3 Neural Network Design and Operation . . . 991
73.4 Experimental Results . . . 994
73.5 Cascade-Backpropagation (CBP) . . . 995
73.6 Six-Bit Parity Problem . . . 999
73.7 Conclusions . . . 999

74 Simulated Annealing, Boltzmann Machine, and Hardware Annealing *Tony H. Wu and Bing J. Sheu* . . . 1003
74.1 Simulated Annealing . . . 1003
74.2 Boltzmann Machine . . . 1004
74.3 Hardware Annealing on Hopfield Networks for Optimization . . . 1005
74.4 Hardware Annealing on Cellular Neural Networks . . . 1007

75 Radial Basis Function (RBF) Neural Networks *Thomas Lindblad, Clark S. Lindsey, and Åge Eide* . . . 1014
75.1 Introduction . . . 1014
75.2 Topology . . . 1014
75.3 Operation . . . 1015
75.4 Training . . . 1015
75.5 Summary . . . 1017
75.6 Defining Terms . . . 1017

76 Hardware Implemented Radial Basis Function (RBF): The IBM Zero Instruction Set Computer *Thomas Lindblad, Clark S. Lindsey, and Åge Eide* . . . 1019
76.1 Introduction . . . 1019
76.2 The ZISC036 VLSI Chip . . . 1019
76.3 Processing and Training . . . 1020
76.4 Implementing the Chip . . . 1021
76.5 Summary and Extrapolations . . . 1022

77 The RCE Neural Network *Douglas L. Reilly* . . . 1025
77.1 Introduction . . . 1025
77.2 Training the RCE Network . . . 1027
77.3 RCE Network Responses . . . 1032
77.4 Practical Guides to RCE Network Training and Use . . . 1033
77.5 Applications of RCE to Pattern Recognition . . . 1034
77.6 RCE Network on a Commercially Available Neural Network Chip . . . 1035

78 Probabilistic Neural Networks Model *Donald F. Specht* . . . 1038
78.1 Basic PNN . . . 1038
78.2 Adaptive PNN . . . 1041
78.3 High-Speed Classification . . . 1042
78.4 Other Considerations . . . 1044
78.5 Summary . . . 1046

79 General Regression Neural Network Model *Donald F. Specht* . . . 1047
79.1 GRNN . . . 1047
79.2 Adaptive GRNN . . . 1052
79.3 Summary . . . 1053

80 Classifiers: An Overview *WooGon Chung and Evangelia Micheli-Tzanakou* . . . 1055
80.1 Introduction . . . 1055
80.2 Criteria for Optimal Classifier Design . . . 1055
80.3 Categorizing the Classifiers . . . 1056
80.4 Classifiers . . . 1057
80.5 Neural Networks . . . 1062
80.6 Comparison of Experimental Results . . . 1075
80.7 System Performance Assessment . . . 1076
80.8 Analysis of Prediction Rates from Bootstrapping Assessment . . . 1080

SECTION VIII Fuzzy Systems and Soft Computing

81 Applications of Fuzzy Systems and Soft Computing in Industrial Electronics *Mary Lou Padgett* . . . 1087
81.1 Introduction . . . 1087
81.2 From Basic Implementations to New Research . . . 1087

82 Fuzzy Numbers: The Application of Fuzzy Algebra to Safety and Risk Analysis *J. Arlin Cooper* . . . 1091
82.1 Background . . . 1091
82.2 Analytical Processing of Input Data . . . 1091
82.3 Fuzzy-Algebra Background . . . 1091
82.4 Fuzzy-Algebra Depiction of Uncertainty . . . 1092
82.5 Example Applications . . . 1093

83 Fuzzy Systems *Mo-yuen Chow* . . . 1096
83.1 Brief Description of Fuzzy Logic . . . 1096
83.2 Qualitative (Linquistic) to Quantitative Description . . . 1097
83.3 Fuzzy Operations . . . 1098
83.4 Fuzzy Rules, Inference . . . 1100
83.5 Fuzzy Control . . . 1101

84 Fuzzy Hardware *Mary Lou Padgett* . . . 1103
84.1 Introduction . . . 1103
84.2 Challenges and Rewards . . . 1103
84.3 Approaches . . . 1103
84.4 Futures . . . 1110
84.5 Defining Terms . . . 1110

85 Fuzzy Modeling and Applications: Controls, Visions, Decisions *Mary Lou Padgett* . . . 1112
85.1 Introduction . . . 1112
85.2 Engineering Approaches . . . 1112
85.3 Futures . . . 1115

86 Fuzzy Logic Control: Basics and Applications *Robert N. Lea, Yashvant Jani, and Joseph A. Mica* . . . 1116
86.1 Introduction . . . 1116
86.2 A Simple Example of Fuzzy Logic Control . . . 1117
86.3 The Example of the Inverted Pendulum . . . 1118
86.4 Remote Manipulator System . . . 1122
86.5 Collision Avoidance . . . 1123
86.6 Summary . . . 1124

87 Development of an Intelligent Unmanned Helicopter Based on Fuzzy Systems *Michio Sugeno, Howard A. Winston, Isao Hirano, and Satoru Kotsu* . . . 1127
87.1 Introduction . . . 1127
87.2 Helicopter Hardware System . . . 1129

87.3 Software System for Helicopter Control 1131
87.4 Results 1135
87.5 Conclusions 1136

88 Fuzzy and Neural Modeling *Mary Lou Padgett* 1139
88.1 Introduction 1139
88.2 Engineering Approaches and Applications 1139
88.3 Futures 1141

89 NeuFuz: A Combined Neural Net/Fuzzy Logic Tool *Thomas Lindblad and Clark S. Lindsey* 1143
89.1 Introduction 1143
89.2 Working with the Neural Network of NeuFuz4 1143
89.3 Working with the Fuzzy Logic Part of NeuFuz4 1145
89.4 Working with the Code Generator Part of NeuFuz4 1145
89.5 Summary 1146

90 Neural Network Learning in Fuzzy Systems *Yashvant Jani and Robert N. Lea* 1147
90.1 Introduction 1147
90.2 Reinforcement Learning 1147
90.3 Architecture of ARIC 1147
90.4 ARIC and 6 DOF Space Operations 1149
90.5 GARIC and Attitude Control 1150
90.6 Six Degree-of-Freedom Proximity Operations Trajectory Controller 1154

91 Neurocontrol and Elastic Fuzzy Logic: Capabilities, Concepts, and Applications *Paul J. Werbos* 1157
91.1 Introduction 1157
91.2 Neurocontrol in General 1158
91.3 Basic Principles of Design 1159
91.4 Supervised Learning for Neurocontrol 1160
91.5 Elastic Fuzzy Logic: Principle and Subroutines 1162
91.6 Current Designs in Neurocontrol: A Roadmap 1165
91.7 Appendix (Tutorial Level Background Information): Neurocontrol and Fuzzy Logic 1166

92 Integrated Health Monitoring and Control in Rotorcraft Machines *Gary G. Yen* 1182
92.1 Introduction 1182
92.2 Artificial Neural Networks 1184
92.3 Fuzzy-Based Feedforward Neural Network 1185
92.4 FDIA Architecture 1187
92.5 Simulation Study 1189
92.6 Conclusions 1190

93 Autonomous Neural Control in Flexible Space Structures *Gary G. Yen* 1192
93.1 Learning Control System 1192
93.2 Adaptive Time-Delay Radial Basis Function Network 1194
93.3 Eigenstructure Bidirectional Associative Memory 1195
93.4 Fault Detection and Identification 1198
93.5 Reconfigurable Control 1199
93.6 Simulation Studies 1202
93.7 Conclusion 1205

94 Fuzzy Pattern Recognition *Witold Pedrycz* 1207
94.1 Introductory Remarks—Pattern Recognition in the Framework of Fuzzy Sets 1207
94.2 The General Methodological Structure of Fuzzy Modeling 1208
94.3 Formation of the Feature Space 1209
94.4 Implicit and Explicit Knowledge Representation in Pattern Recognition 1212
94.5 From Supervised to Unsupervised Pattern Recognition—A Continuum of Classification Models 1213
94.6 Fuzzy Neural Structures 1213

94.7 Supervised Learning . . . 1218
94.8 Implicitly Supervised Pattern Recognition . . . 1223
94.9 Unsupervised Learning . . . 1225

95 Neural Fuzzy Systems in Handwritten Digit Recognition *Timothy J. Dasey and Evangelia Micheli-Tzanakou* . . . 1231
95.1 Introduction . . . 1231
95.2 System Design . . . 1240
95.3 Application to Handwritten Digits . . . 1248
95.4 Discussion . . . 1256
95.5 Summary . . . 1258

96 Fuzzy Algorithms for Learning Vector Quantization *Nicolaos B. Karayiannis* . . . 1264
96.1 Introduction . . . 1264
96.2 Learning Vector Quantization . . . 1265
96.3 Generalized Learning Vector Quantization . . . 1266
96.4 Fuzzy Learning Vector Quantization Algorithms . . . 1268
96.5 GLVQ-F and FLVQ Algorithms . . . 1269
96.6 Fuzzy Algorithms for Learning Vector Quantization . . . 1270
96.7 The FALVQ 1 Family of Algorithms . . . 1272
96.8 The FALVQ 2 Family of Algorithms . . . 1274
96.9 The FALVQ 3 Family of Algorithms . . . 1275
96.10 Competition Measures . . . 1277
96.11 Alternative FALVQ Algorithms . . . 1280
96.12 Experimental Results . . . 1282
96.13 Discussion and Concluding Remarks . . . 1284

97 Adaptive Resonance Theory *Gail A. Carpenter and Stephen Grossberg* . . . 1286
97.1 Match-Based Learning and Error-Based Learning . . . 1287
97.2 ART and Fuzzy Logic . . . 1288
97.3 ART Dynamics . . . 1288
97.4 Fuzzy ART . . . 1290
97.5 Fuzzy ARTMAP . . . 1290
97.6 Fuzzy ART Algorithm . . . 1292
97.7 Fuzzy ARTMAP Algorithm . . . 1294
97.8 ART Applications . . . 1296

98 Future Directions for Fuzzy Systems and Soft Computing in Industrial Electronics *Mary Lou Padgett and Lotfi A. Zadeh* . . . 1299

SECTION IX Evolutionary Systems, Computational Intelligence, and Hybrid Systems Applications

Evolutionary Systems

99 Applications of Evolutionary Systems in Industrial Electronics *Mary Lou Padgett and V. Rao Vemuri* . . . 1303
99.1 Introduction . . . 1303
99.2 From Basic Implementations to New Research . . . 1303
99.3 Defining Terms . . . 1304

100 Evolutionary Computation *Mary Lou Padgett* . . . 1307
100.1 Introduction . . . 1307
100.2 Design of Evolutionary Systems . . . 1307
100.3 Applications . . . 1313
100.4 Summary . . . 1315

101 Genetic Algorithms *Mark G. Cooper and V. Rao Vemuri* 1316
101.1 Introduction 1316
101.2 The Basic Genetic Algorithm 1316
101.3 String Encoding 1317
101.4 Evaluation 1317
101.5 Test Fitness Functions 1317
101.6 Premature Convergence 1318
101.7 Selection 1318
101.8 Replacement 1320
101.9 Genetic Parameters 1320

102 Fuzzy Evolutionary and GA Systems *Mary Lou Padgett* 1321
102.1 Introduction 1321
102.2 Combining Evolutionary Systems and Fuzzy Systems 1321
102.3 Summary 1323

103 Information Fusion by Fuzzy Set Operations and Genetic Algorithms *Anna L. Buczak and Robert E. Uhrig* 1325
103.1 Information Fusion 1325
103.2 Fuzzy Aggregation Connectives 1326
103.3 Genetic Algorithms 1328
103.4 Two Fuzzy-Genetic Fusion Techniques 1329
103.5 Information Fusion for Object Classification 1331
103.6 Vibration Monitoring 1332
103.7 Results 1332
103.8 Conclusions 1335

104 Neural Evolutionary and GA Systems and Applications *Mary Lou Padgett* 1338
104.1 Introduction 1338
104.2 Combining Evolutionary Systems and Neural Systems 1338
104.3 Summary 1341

Computational Intelligence and Hybrid Systems Applications

105 Computational Intelligence Applications in Industrial Electronics *Mary Lou Padgett and Robert Shelton* 1343
105.1 Introduction 1343
105.2 Aerospace Applications of Computational Intelligence 1343
105.3 From Basic Implementations to New Research 1344

106 Hybrid Artificial Intelligence Systems *Lefteri H. Tsoukalas and Robert E. Uhrig* 1346
106.1 Introduction 1346
106.2 Expert Systems and Fuzzy Logic Systems 1347
106.3 Neural Networks and Expert Systems 1347
106.4 Neural Networks and Fuzzy Logic Systems 1347
106.5 Genetic Algorithms and Neural Networks 1356
106.6 Genetic Algorithms and Fuzzy Systems 1357
106.7 Discussion and Conclusions 1357

107 Application Techniques: Combining Fuzzy Logic, Artificial Neural Networks, and Probabilistic Reasoning—Soft Computing *Okyay Kaynak* 1360
107.1 Combining Soft Computing Methodologies 1361
107.2 Neurofuzzy Control 1361
107.3 The Use of NNs in Consumer Products 1361
107.4 The Fusion of GA and FS 1362

108 Synthesis of Fuzzy, Artificial Intelligence, Neural Networks, and Genetic Algorithm for Hierarchical Intelligent Control *Takanori Shibata, Toshio Fukuda, and Kazuo Tanie* 1364
108.1 Introduction 1364

108.2 Artificial Intelligence, Fuzzy, Neural Network, and Genetic Algorithm 1364
108.3 Hierarchical Intelligent Control of Robotic Motion 1366
108.4 Conclusions 1367

109 Advanced Tools for Adaptive Nonlinear Modeling and Control of Power in Large Systems *Harold H. Szu and Brian A. Telfer* 1369
109.1 Introduction 1369
109.2 Modeling, Control, and Neural Networks 1369
109.3 Wavelet and Adaptive Space-Frequency Techniques for Modeling and Control 1370
109.4 Summary and Conclusions 1371

110 Application of Model Reference Adaptive Control and Adaptive Time-Delay RBF Networks *Gary G. Yen* 1372
110.1 Introduction 1372
110.2 Dynamic Modeling of Flexible Multibody 1374
110.3 Adaptive Time-Delay Radial Basis Function Netowrk 1376
110.4 Pace Simulation Study 1377
110.5 Conclusions 1379

SECTION X Emerging Technologies

Virtual Reality

111 Virtual Reality 1383
111.1 Current Applications in Virtual Reality *Richard A. Blade and Mary Lou Padgett* 1383
111.2 The Virtual Workbench—A Path to Use for VR *Timothy Poston* 1390
111.3 Motion Tracking for Virtual Reality *Herschell Murry* 1393
111.4 Virtual Sound *Nadine Miner and Thomas Caudell* 1397
111.5 Virtual Reality Systems *Mary Lou Padgett, Richard A. Blade, Johnny Evers, and Charles R. White* 1404
111.6 Fuzzy Logic Applications in Image Processing Equipment: Intelligent VR Futures *Hideyuki Takagi* . . . 1426

Asynchronous Transfer Mode for High-Speed Communication

112 Asynchronous Transfer Mode Technology *Thomas Lindblad* 1438
112.1 What is ATM Offering? 1438
112.2 Why ATM? 1438
112.3 What is ATM? 1439
112.4 ATM Applications 1439
112.5 The NEBULAS Project 1440
112.6 Summary 1442

113 NEBULAS: High Performance Data-Driven Event Building Architectures Based on Asynchronous Self-Routing Packet-Switching Networks *M. Costa, J.-P. Dufey, M. Letheren, A. Manabe, A. Marchioro, C. Paillard, D. Calvet, K. Djidi, P. Le Dû, I. Mandjavidze, P. Sphicas, K. Sumorok, S. Tether, L. Gustafsson, K. Kobylecki, K. Agehed, S. Hultberg, T. Lazrak, T. Lindblad, C. S. Lindsey, H. Tenhunen, M. DePrycker, B. Pauwels, G. Petit, H. Verhille, and M. Benard* 1444
113.1 Introduction 1445
113.2 Technical Background 1446
113.3 Computer Modeling 1447
113.4 Event Building Protocols and Related Software Development 1454
113.5 Hardware Development 1460
113.6 Integration of Event Builder Demonstrators 1464
113.7 Plan of Work 1466

114 Microelectromechanical Systems (MEMS) *Yu-Chong Tai and Chang-Jin Kim* . . . 1468
- 114.1 Introduction . . . 1468
- 114.2 Bulk Micromachining . . . 1468
- 114.3 Surface Micromachining . . . 1469
- 114.4 First Applications . . . 1469

115 Micromachines *Hiroyuki Fujita* . . . 1472
- 115.1 Micromachines and the Scaling Effect . . . 1472
- 115.2 Difficulties in Miniaturization and Proposed Solutions . . . 1473
- 115.3 Microactuators . . . 1474
- 115.4 Architectures for MEMS: Autonomous Distributed Micromachines . . . 1479
- 115.5 Applications . . . 1483
- 115.6 Conclusion . . . 1487

116 Selected Micromachining Fabrication Technologies . . . 1489
- 116.1 Precision Metallic Micro Structures and Micro Molding Technologies *A. Bruno Frazier and James Jara-Almonte* . . . 1489
- 116.2 Nanotechnology *Noel C. MacDonald, M. T. A. Saif, and S. A. Miller* . . . 1500
- 116.3 Precision Micromachining Technologies *Craig R. Friedrich and Michael J. Vasile* . . . 1505

117 Microsensors . . . 1515
- 117.1 Pressure Sensors and Accelerometers *Keith O. Warren* . . . 1515
- 117.2 Acoustic Wave-Based Chemical Sensors *Antonio J. Ricco* . . . 1519

118 Micro Actuators and Energy Supply *Toshio Fukuda and Fumihito Arai* . . . 1526
- 118.1 Micro Actuators . . . 1526
- 118.2 Energy Supply Methods and Non-Contact Manipulation . . . 1533

119 On-Board Power Supply and Remote Driving Mechanisms for Microelectromechanical Systems *Jeong B. Lee* . . . 1538
- 119.1 Power Requirements of Microelectromechanical Systems . . . 1538
- 119.2 On-Board Power Supply: Solar Cell Array . . . 1540
- 119.3 On-Board Power Supply: Microbattery . . . 1542
- 119.4 Remote Driving Mechanisms . . . 1544
- 119.5 Conclusions . . . 1545

120 Si Micromachining in High-Frequency Applications *Linda P. B. Katehi, Gabriel M. Rebeiz, Tom M. Weller, Rhonda F. Drayton, Stephen V. Robertson, and Chen-Yu Chi* . . . 1547
- 120.1 Introduction . . . 1547
- 120.2 Applications . . . 1548
- 120.3 Fabrication Methodology . . . 1551
- 120.4 Membrane Supported Distributed Circuits . . . 1556
- 120.5 Conformal Micromachined Packaging . . . 1562
- 120.6 Micromachined Lumped Elements . . . 1567
- 120.7 Conclusions . . . 1572

121 MEMS Integration—Technical and Economic Considerations *Janusz Bryzek* . . . 1576
- 121.1 Introduction . . . 1576
- 121.2 Why MEMS Focus on Silicon . . . 1577
- 121.3 Market Growth Analogy: Transistors, Integrated Circuits and MEMS . . . 1578
- 121.4 Integrated MEMS Market Overview . . . 1580
- 121.5 To Integrate Or Not To Integrate . . . 1582
- 121.6 Mechanical On-Sensor-Chip Integration . . . 1585
- 121.7 Monolithic or Hybrid . . . 1585
- 121.8 Case Study: Lucas NovaSensor . . . 1586

121.9 Conclusions 1590

Multisensor Fusion and Integration for Intelligent Systems

122 Multisensor Fusion and Integration for Intelligent Systems 1592
122.1 Introduction *Ren C. Luo* 1593
122.2 Issues and Approaches of Multisensor Fusion and Integration *Ren C. Luo and Michael G. Kay* 1593
122.3 Audio-Visual Sensor Fusion System for Intelligent Sound Sensing *Kota Takahashi and Hiro Yamasaki* . . 1609
122.4 Industrial Vision System by Fusing Range Image and Intensity Image *Kazunori Umeda and Tamio Arai* 1615
122.5 Application of Data Fusion to Neonate Oxygenation Control *Mark E. Kotanchek, James P. Helferty, W. Bosseau Murray, and Charles Palmer* 1622
122.6 Multiresolution Multisensor Target Identification *Zbigniew Korona and Mieczyslaw M. Kokar* 1627
122.7 Shaping Control of Plastic Object by Robot Hand with Sensor Fusion Processing *Ryosuke Masuda and Michio Sasaki* 1632
122.8 Multisensor System Integration for Autonomous Navigation Tasks *Karl Kluge* 1639
122.9 Future Trends for the Further Development in Multisensor Fusion and Integration *Ren C. Luo* 1657

INDEXES

Author Index 1663
Subject Index 1669

THE

INDUSTRIAL ELECTRONICS

HANDBOOK

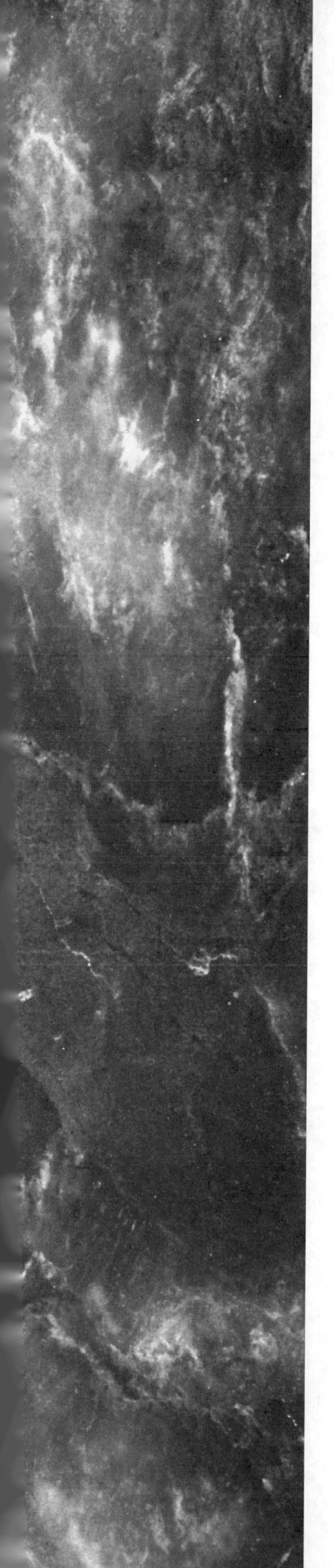

PART 1

FUNDAMENTALS of INDUSTRIAL ELECTRONICS

I

Supporting Technologies

1 **Electronics** *Darrell Vines and Tom Baginski* 5
Introduction • Diodes • Transistors as Switches • Models for Transistors • Analog and Digital Circuits

2 **Digital Control Circuits** *Marc Courvoisier, Michel Combacau, and Mario Paludetto* 22
Logic Control • Sequence Control • Implementation Techniques

3 **Computer Architecture** *Victor P. Nelson* 48
Hardware Organization • Computer Software • Information Representation in Digital Computers • Specifying Instruction Operands • CPU Registers • Memory Organization • Computer Instruction Types • Interrupts and Exceptions • Evaluating Instruction Set Architectures • Computer System Design • Input/Output Device Interfaces • Microcontroller Architectures • Multiple Processor Architectures

4 **Signal Processing** *James A. Heinen and Russell J. Niederjohn* 73
Introduction • Continuous-Time Signals • Time-Domain Analysis of Continuous-Time Signals • Frequency-Domain Analysis of Continuous-Time Signals • Continuous-Time Signal Processors • Time-Domain Analysis of Continuous-Time Signal Processors • Frequency-Domain Analysis of Continuous-Time Signal Processors • Continuous-Time (Analog) Filters • Sampling • Discrete-Time Signals • Time-Domain Analysis of Discrete-Time Signals • Frequency-Domain Analysis of Discrete-Time Signals • Discrete-Time Signal Processors • Time-Domain Analysis of Discrete-Time Signal Processors • Frequency-Domain Analysis of Discrete-Time Signal Processors • Discrete-Time (Digital) Filters • Discrete-Time Analysis of Continuous-Time Signals • Discrete-Time Processing of Continuous-Time Signals

1
Electronics

Darrell Vines
Texas Tech University

Tom Baginski
Auburn University

1.1 Introduction 5
1.2 Diodes 5
Device Behavior • Summary and Extension
1.3 Transistors as Switches 10
Transistor Device Behavior • Summary and Extension
1.4 Models for Transistors 15
The DC Bias Model • AC Model or Small Signal Model • High-Frequency or Transient Model • Foundations for Transistor Models • Summary and Extension
1.5 Analog and Digital Circuits 19
Device Behavior • Summary and Extension

1.1 Introduction

This discussion will concentrate on the terminal characteristics of devices and their applications. Comprehension of the terminal characteristics of components leads to strong analytic techniques as well as a basic understanding of the functionality of the devices. Initially, the load line analysis will be applied to diode circuits as their characteristics are explained. Next, the load line analysis will be applied to transistor characteristic curves to illustrate the bias behavior of three-terminal devices. Introduction of a small signal model provides an analytical approach which can confirm the load line analysis or, in turn, be supported by an understanding of load lines. With the circuit model, applications are presented for bias, frequency response, and transient response. The final application will be to use the concepts of diodes and transistors to analyze a basic logic circuit.

This discussion of electronics will be concerned with discrete devices which include semiconductor diodes, as well as transistors. A few applications will then illustrate how devices are used in circuits. The following discussion is designed to enable the reader to understand the significant issues involved in analyzing electronic circuits, regardless of whether the specific component is discussed herein. The format for the presentation is as follows:

- Discussion of terminal characteristics and models
- Examples using the terminal characteristics and models
- Explanations of device behavior
- Summary

1.2 Diodes

A discussion of the diode, which is the smallest discrete component to be examined, precedes the discussion of transistors. Terminal characteristics for components are sometimes referred to as the current-voltage, I-V characteristics. When the I-V of a resistor is plotted, the slope of the line will, of course, be inversely proportional to the value of the resistance. The I-V characteristic of a diode is shown in Figure 1.1 when the diode is represented with voltage and current references as indicated. Notice that as the voltage VD is increased, the current increases exponentially near $VD = 0.7$ volts. However, if the voltage VD is reversed, the current remains negligible. One interpretation of diode behavior recognizes that current flows easily in the forward direction and opposes current flow in the reverse direction. The forward and reverse directions may be implied by the suggested direction of the arrow in the diode. For historical reasons, the head of the diode is referred to as the anode and the other side becomes the cathode. Clearly, the diode is a nonlinear element as contrasted with a resistor.

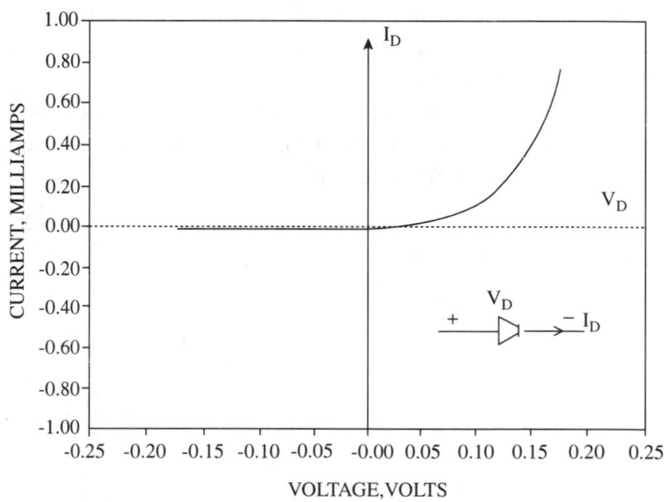

Figure 1.1 The I-V characteristic of a diode with reference directions indicated.

0-8493-8343-9/97/$0.00+$.50

The circuit of Figure 1.2 will be used to determine the currents and voltages in a circuit that contains the nonlinear circuit element, the diode. In the series circuit, the current that flows must be the same in both the resistor and the nonlinear device. Writing the Kirchhoff equation around the loop yields

$$V_{in} = ID*R1 + VD \tag{1.1}$$

The diode voltage depends on the actual value of the current in the circuit, yielding an indeterminacy. Thus, the linear circuit that the diode "sees" can be regarded as Thevenin's equivalent circuit (also known as the load line) and can be plotted on the characteristic curves as follows: replace the nonlinear circuit with a short circuit, then determine the maximum current in the circuit and plot it on the horizontal axis. Connect the two points to determine the load line.

A graphical process which determines the value of the circuit current locates the intersection of the nonlinear I-V curve and the load line. Since only one current can exist in the series components, the intersection of the curves yields both the current in the circuit as well as the voltage across the diode. Note that the voltage from 0 volts to the intersection is *VD*, while the voltage from *VD* to the supply voltage V_{in} represents the voltage across the resistor.

EXAMPLE 1.1 Forward Biased Diode

In the circuit of Figure 1.2, if $V_{in} = 0.175$ volts, and $R = 220$ ohms, let us determine the value of the series current if the I-V characteristic is that shown in Figure 1.3a.

The load line should intersect the voltage axis at 0.175 volts, and the current axis intersection should be approximately 0.78 milliamps. The intersection of the diode curve and the load line indicates that the current will be about 0.22 milliamps and the voltage across the diode will be 0.125 volts. In addition, the voltage across the resistor will be 0.050 volts.

For an undemonstrated example the reader may change R1 to 20 ohms; the intersection would yield a current of 0.4 milliamps and the voltage across the diode would be approximately equal to the supply voltage. In reality, a forward biased diode will almost always exhibit about 0.7 volts when it is forward biased with reasonably sized voltages over 1.0 volt.

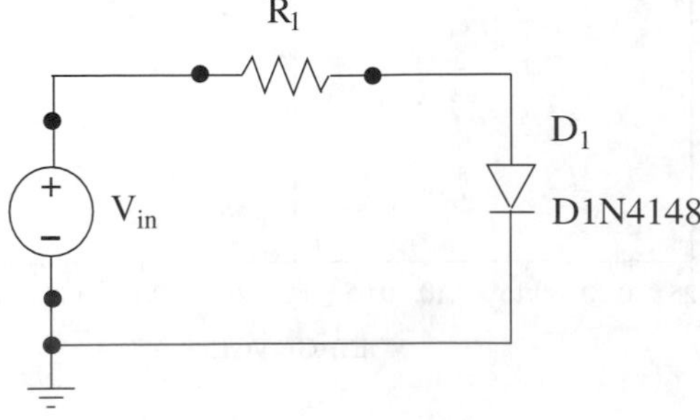

Figure 1.2 A simple nonlinear circuit.

EXAMPLE 1.2 Reverse Biased Diode

If the direction of the voltage supply in the circuit of Figure 1.2 is reversed in the example above, the value of current, as indicated in the third quadrant of Figure 1.3b, will always be the same negligible value regardless of the value of the supply voltage.

EXAMPLE 1.3 Rectifying Circuit

Suppose that a sinusoidal voltage were applied to the series circuit of Figure 1.4a in which a resistor and diode are connected. The I-V characteristic for the diode is shown in Figure 1.4b. Using the techniques described above, one may determine the diode current and the output voltage across the resistor at specific instants of time during the sinusoid cycle. The circuit will be analyzed three different ways to illustrate the validity of the load line procedure. First, using the load line technique, the value of the current and the output voltage will be determined when the peak positive voltage appears across the supply, and the process

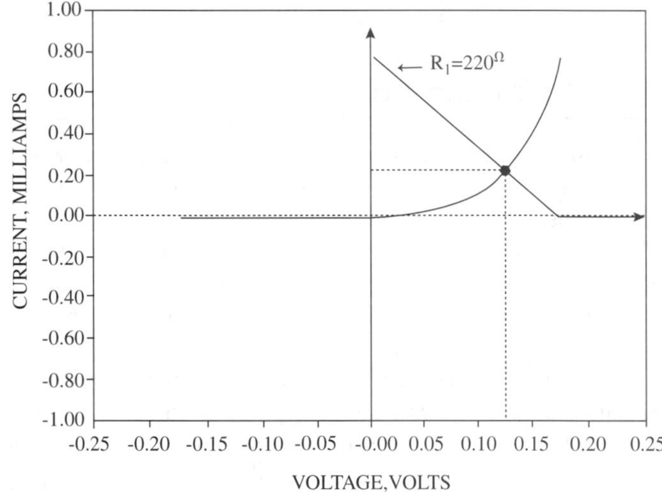

Figure 1.3a I-V curve for diode with a load line.

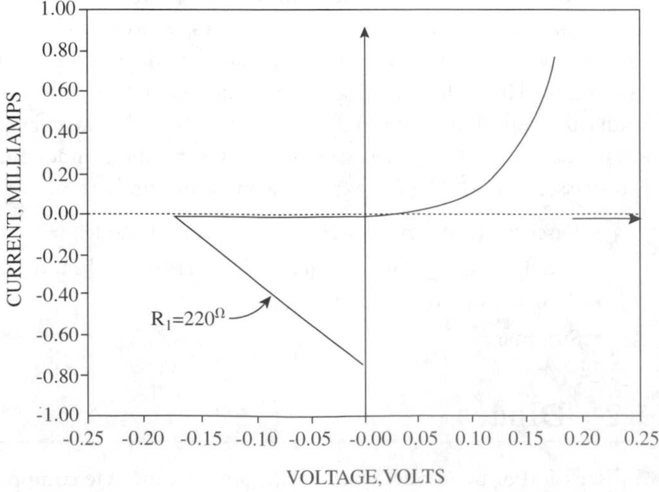

Figure 1.3b Analysis of reverse biased diode.

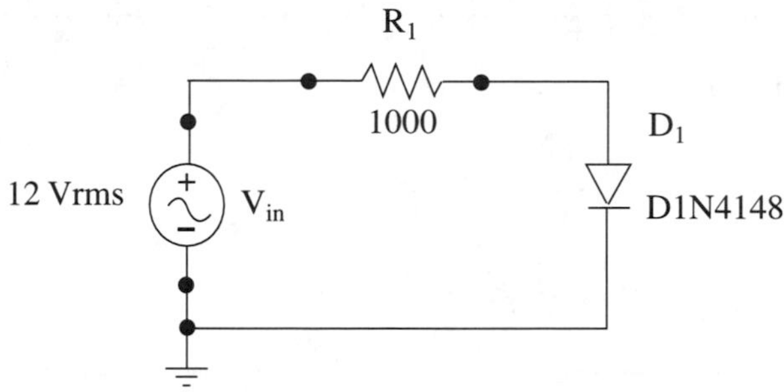

Figure 1.4a A 1N4148 used to half-wave rectify a sine wave.

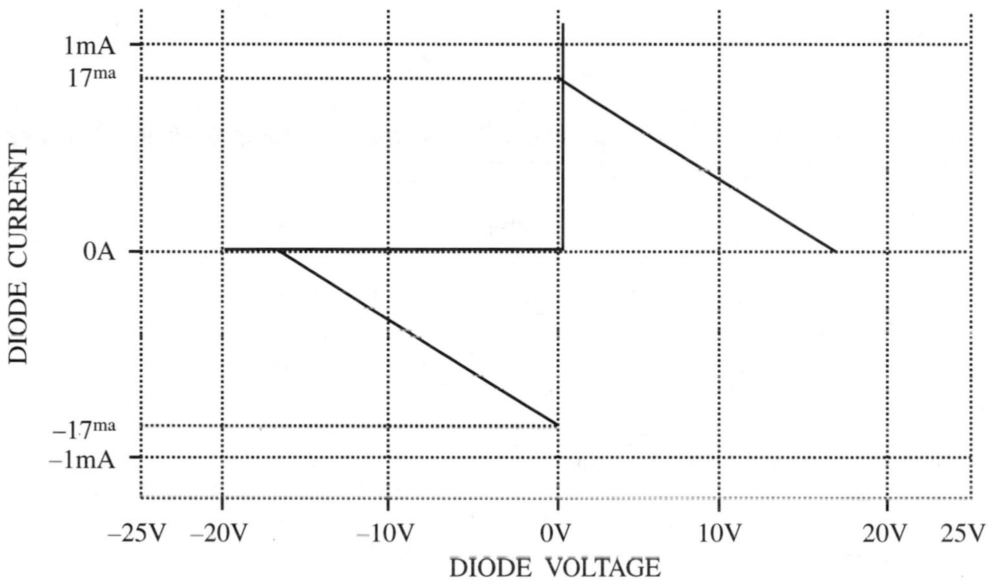

Figure 1.4b I-V characteristic of a diode, 1N4148.

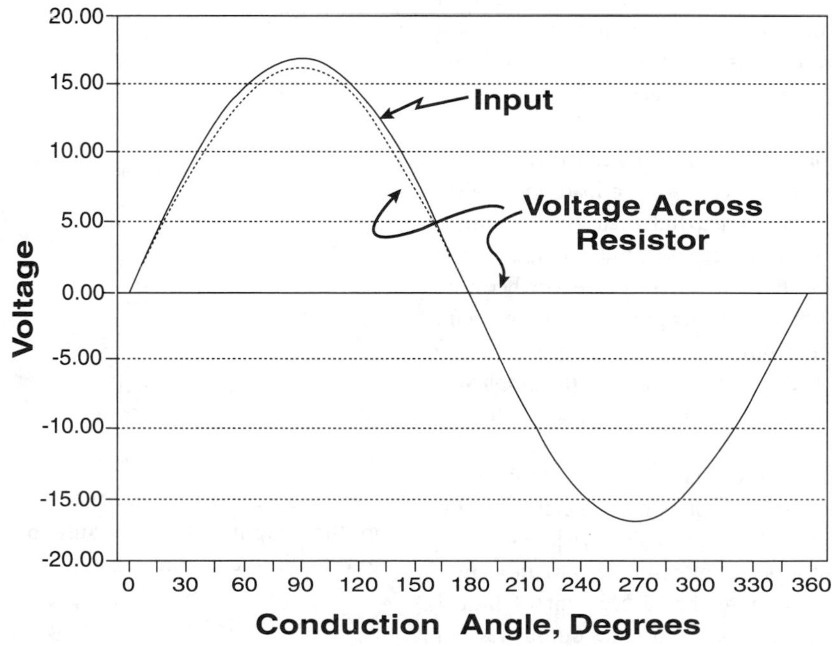

Figure 1.4c Graph of data taken from a load line analysis of rectifier circuit and from Table 1.1.

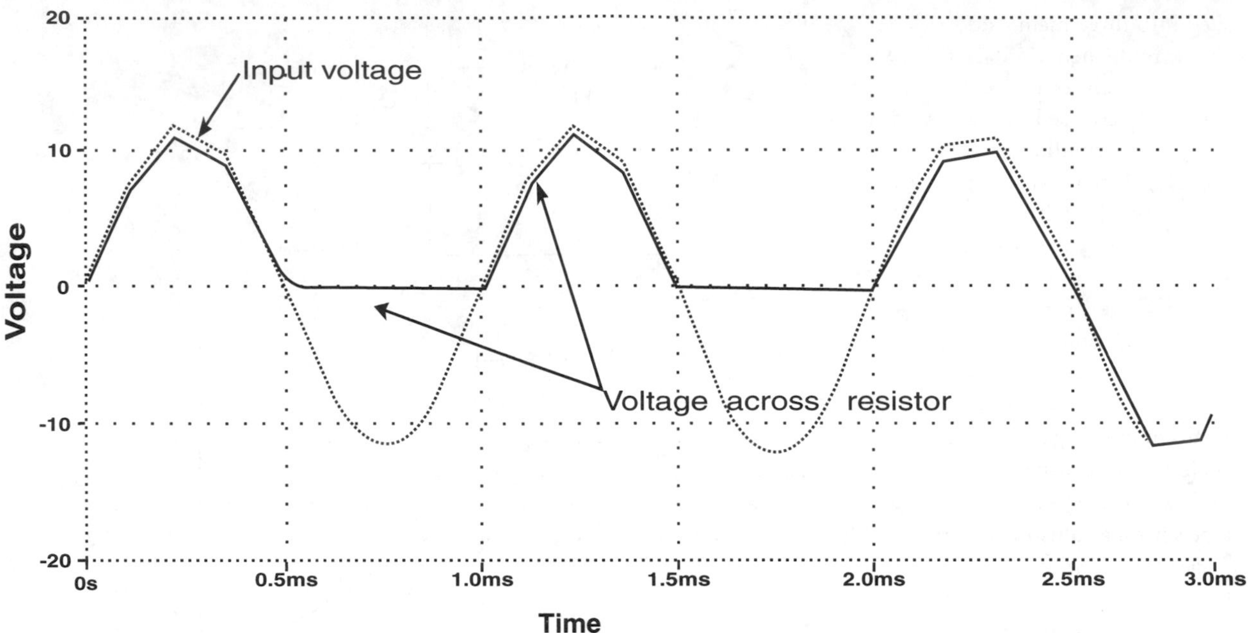

Figure 1.4d Graph of data taken from PSPICE analysis of rectifier circuit.

repeated for the peak negative voltage. Second, the current and voltage at 15-degree increments through a full cycle will be determined. Finally, the circuit will be analyzed using PSPICE, a computer-aided simulation package. The results of the three analyses will then be compared.

Since the peak voltages are ± 17 volts (i.e., $V_{peak} = \sqrt{2}V_{rms}$), then one load line will connect the +17 volt intercept on the voltage scale with +17 milliamps on the current axis. During the negative half-cycle, the maximum voltage of −17 volts will appear on the voltage scale and the −17 milliamps on the negative current axis, as shown in Figure 1.4b. The actual peak current that flows will be 16.3 milliamps during the positive half-cycle because the diode voltage drop is about 0.7^V, and zero during the negative half-cycle. The peak voltage across the load resistor will be 16.3 volts.

For a more accurate and detailed analysis, consider Table 1.1 in which the data have been extracted from a load line analysis for 15-degree increments through a full cycle. Each value of current for a particular instant corresponds to that which occurs at the intersection of the load line and the I-V curve. Furthermore, current and voltage data from the listing are plotted in the graph of Figure 1.4c. The results from the graphical analysis can be confirmed by examining the specific instants on the graph with the data calculated from the I-V curve for peak values of applied voltage.

Finally, for a considerably more detailed analysis, PSPICE was used, and the graph of Figure 1.4d illustrates the similarity of the results, confirming that the load line analysis yields essentially the same results as that of a powerful computer circuit analysis program. It is instructive to note that a segment of time is required before a response develops across the output resistor. Such a delay is illustrated in Figure 1.4d. The input voltage begins to rise, but the output voltage is delayed while the voltage is developing across the diode. Thus, a nonlinearity is introduced in the output voltage because of the nonlinear component inserted in the circuit.

Table 1.1 Data Extracted from Load Line Analysis

Full Wave Rectified Sine Wave			
Degrees	V in	V out	I, ma
0	0.0	0.0	0.0
15	4.4	3.7	3.7
30	8.5	7.8	7.8
45	12.1	11.4	11.4
60	14.8	14.1	14.1
75	16.5	15.8	15.8
90	17.0	16.3	16.3
105	16.4	15.7	15.7
120	14.6	13.9	13.9
135	11.9	11.2	11.2
150	8.3	7.6	7.6
165	4.2	3.5	3.5
180	−0.3	0.0	0.0
195	−4.7	0.0	0.0
210	−8.8	0.0	0.0
225	−12.3	0.0	0.0
240	−14.9	0.0	0.0
255	−16.5	0.0	0.0
270	−17.0	0.0	0.0
285	−16.3	0.0	0.0
300	−14.5	0.0	0.0
315	−11.7	0.0	0.0
330	−8.1	0.0	0.0
345	−3.9	0.0	0.0
360	0.0	0.0	0.0

Device Behavior

To explain how modern electronics operate, a brief discussion of some fundamental nomenclature is required. Materials are separated into three main categories based upon their ability to conduct current when a voltage potential is applied across them. A conductor, such as copper, freely transmits current and is therefore routinely used in many applications such as the wires in the power distribution system of a house. An insulator does not conduct current when a potential is applied. The plastic insulation around a wire is one such example.

A third group of materials is referred to as semiconductors. The ability of such a material as silicon to carry current is determined by how much mobile charge is available in the material. The addition of some impurities such as boron to silicon adds excess mobile positive charges which are referred to as holes (i.e., the material is *p*-type). The addition of phosphorous to silicon adds excess mobile negative charge (electrons) whereby the material is referred to as *n*-type. An interesting phenomenon occurs when a region of *p*-type silicon is brought in contact with a region of *n*-type silicon as illustrated in Figure 1.5a. The intersection of the *p*-type material with the *n*-type is referred to as a *pn* junction or diode.

Just as an ink drop disperses in a glass of clear water (i.e., the amount of ink per volume of water eventually becomes constant) so the excess concentration of holes on the *p*-side of the junction and electrons on the *n* side of the junction try to reach an equilibrium concentration. Devices which rely on holes (+ charge) and electrons (− charge) to make up current flowing through them are referred to as bipolar. The physics of how equilibrium is reached is very complicated. The end result is the formation of an energy barrier with a large concentration of holes on one side and a large concentration of electrons on the other. One can think simply of the energy barrier as a dam and current flow as a river of water.

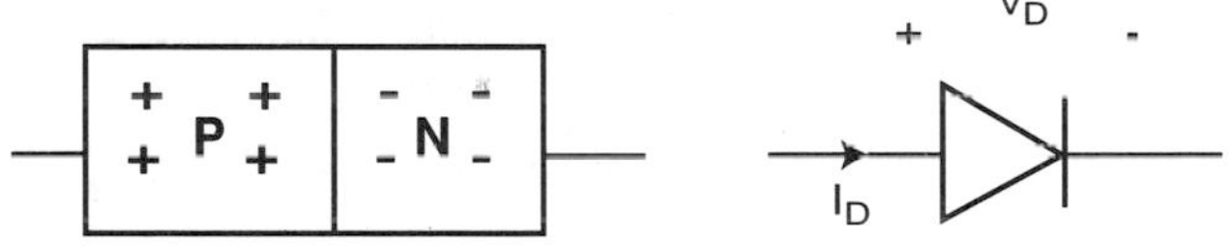

Figure 1.5a Doped silicon material exhibiting *p*- and *n*-type impurities to comprise a diode.

The application of an electrical potential to lower the energy barrier results in a torrent of current flowing from one side of the junction to the other. However, if the polarity of the potential is reversed, essentially no current flows. A *pn* junction only allows current to flow one way, from the *p*-side to the *n*-side, just like a dam only allows water to flow one way. The application of an energy source to lower the floodgates of a dam results in water flowing. The application of an energy source to raise the floodgates even further does not. The application of a voltage across a junction is referred to as forward bias when the potential on the *p*-side exceeds the potential on the *n*-side. The opposite is called reverse bias.

The current vs. voltage characteristic of a *pn* junction (i.e., diode) is shown in Figures 1.1 and 1.5b. The current is very small for reverse-bias conditions until the breakdown voltage of the junction is reached and increases exponentially for forward bias. Breakdown of the junction occurs when the applied potential is large enough to induce a large and unregulated flow of carriers across the junction during reverse bias. If the current flow is not limited externally the junction can be irreversibly damaged during such an event. It has been empirically observed that the equation which relates current to voltage (commonly called the diode equation) is

$$I_D = I_0[\exp(V_D/nV_t) - 1] \tag{1.2}$$

where I_D is current flow through the junction, I_o is a constant value (typically 10^{-12} A) and n is a value determined empirically as a value between 1 and 2. For simplicity it will be assumed $n = 1$. V_t is known as the thermal voltage which is Boltzman's constant multiplied by temperature and divided by unit charge.

$$V_t = kT/q \tag{1.3}$$

V_t at room temperature is 26 mV. It is evident that the current depends exponentially on biasing voltage. Any circuit which contains a diode will therefore be nonlinear. There are a variety of

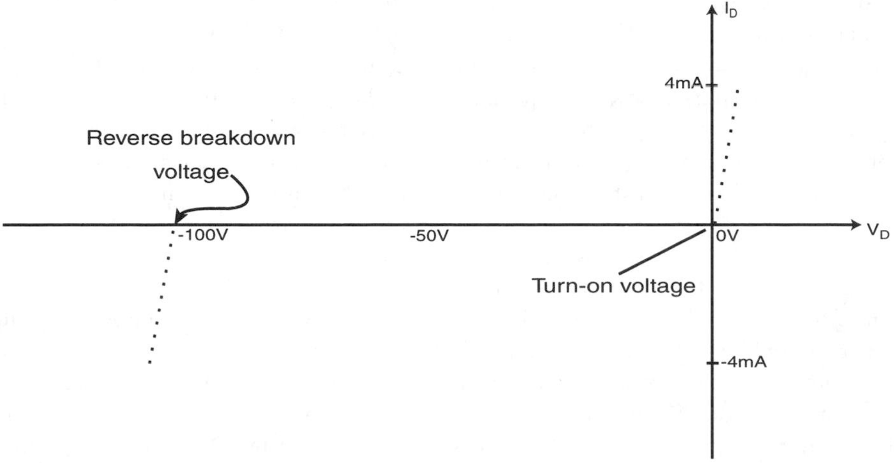

Figure 1.5b Full I-V characteristic of a diode.

techniques which can be used to solve nonlinear circuits. Load line analysis is one such technique shown earlier.

Summary and Extension

- Diodes permit current to flow in only one direction.
- Diodes are nonlinear components.
- A load line can be used to determine the currents and voltages in a circuit.
- Separate analyses confirm the usefulness of the load line for interpretation of operating conditions.
- A vacuum tube device called a diode probably has similar characteristics when compared to a semiconductor diode, differing in magnitudes of voltages or current.
- Diodes may conduct considerable current, or small values, depending on applications.
- Diodes may withstand large voltages if properly designed.
- Diodes exhibit a time-delay response if operated at high speeds.

1.3 Transistors as Switches

Transistors, generally made of silicon, as are diodes, are known by the variety of processes by which they are made. A few of the many types of transistors include bipolar junction transistors (BJT), junction field effect transistors (JFET), metal oxide semiconductor field effect transistor (MOSFET) and insulated-gate-bipolar-junction-transistors (IGBJT). Even the historical vacuum tube component behaves in much the same way as transistors. While this discussion will focus on the BJT, the Summary and Extension section will discuss its similarities with the other types of components. Thus, the following description of transistor behavior and analysis will be applicable, in principle, for any other transistor or active component.

Having examined the operation and function of diodes with well-defined terminal characteristics, the reader will review the behavior of transistors on the basis of their terminal characteristics, referred to as "characteristic curves." The characteristic curves in Figure 1.6a show the collector current on the ordinate and the collector-to-emitter voltage on the abscissa. The parameter that selects the separate curves in the display represents the input to the transistor and is named the base current in this example. The definition of the terminals of the transistor can be inferred from the Figure 1.6b.

Current gain, a figure of merit of a transistor, is determined by the ratio of the output current to the input current in the linear or active portion of the curves. The current gain is often referred to as the beta, β, or hfe, or hFE, or h21, which are symbols used in an h-parameter representation of two-port networks. In addition, current gain transistors are selected on the basis of several conditions which may represent limitations. Transistors may conduct large currents accompanied by low voltage, or conduct small currents and withstand large voltages, or pass extremely small currents at moderate voltages, or a variety of other combinations.

A biased transistor will have a defined collector current, a collector-to-emitter voltage, and an input base current. Thus, the identification of the three quantities will determine one bias point on the set of characteristic curves. For example, one may want to bias the transistor at 3 milliamps of collector current, with a collector-to-emitter voltage of 5 volts and a base current of 30 microamps. It would also be possible to bias the transistor at 3 milliamps collector current with a collector-to-emitter voltage of 10 volts, provided the 30 microamps input were available. However, if the base current were less than 30 microamps, there is no value of collector-to-emitter voltage that will cause the 3 milliamps of current to flow. The design of the bias circuitry can be accomplished by using the concepts of the load line previously described.

EXAMPLE 1.4 Transistor Load Line

As an example, the circuit illustrated in Figure 1.7a represents a biased 2N2222 and Figure 1.7b represents the characteristic curves of the output of the device. If $V_{CC} = 10$ volts and $V_{BB} = 3.7$ volts, the actual currents and voltage at which the transistor is biased can be determined.

First, the input of the transistor behaves like the forward biased diode that was discussed earlier. The base-to-emitter voltage of the transistor will be limited to 0.7 volts when it is forward biased, and the direction of the arrow on the emitter represents the direction of the base-to-emitter diode. Thus, the base current can be determined from Kirchhoff's equation,

$$V_{BB} - I_B \cdot R_B - 0.7 = 0 \tag{1.4}$$

and thus

$$I_B = 30 \text{ microamps}$$

The supply voltage identifies the intersection of the load line on the abscissa. The short circuit identifies the intersection of the load line on the ordinate. Thus, the load line, which is also the negative reciprocal of the load resistor, can be drawn on the curves as shown in Figure 1.7b. The current and voltage combination at the output of the transistor will occur ONLY at the intersection of the load line and the base bias curve. Thus, since the base current was determined to be 30 microamps, the collector current is constrained to be about 3 milliamps and the collector-to-emitter voltage VCE = 7.5 volts. The intersection of the base current curve and the load line identifies the bias condition referred to as the operating or quiescent point. The operating point, located approximately in the linear portion of the family of curves of the 2N2222, suggests that the collector current may be increased or decreased by adjusting the base current. For example, if the base current were increased by 10%, the collector current increases by 10%. Likewise, a decrease of base current by 10% results in a decrease in collector current of 10%. The analogy of the transistor and a valve seems appropriate at this point, for the "valve" is partially open and by

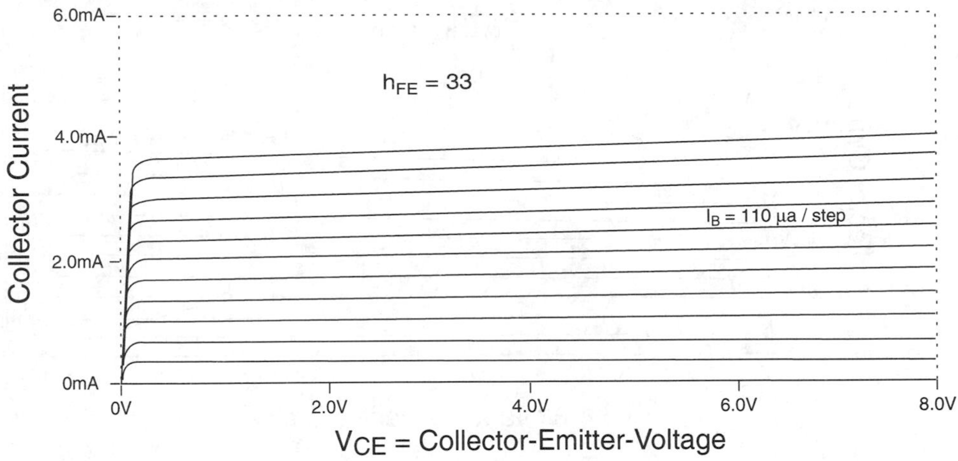

Figure 1.6a Characteristic curves for 2N2222.

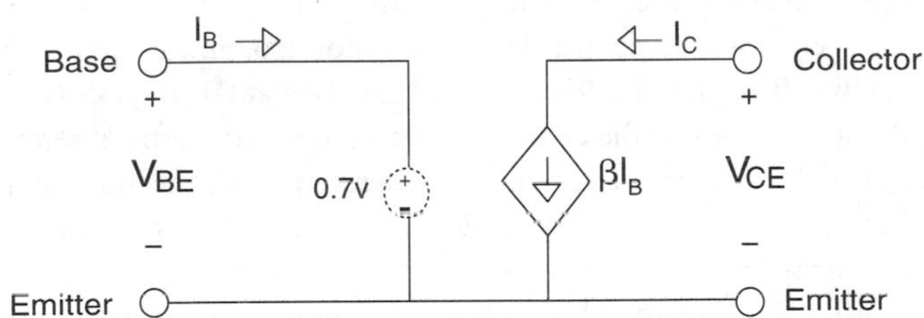

Figure 1.6b A DC bias model for a transistor.

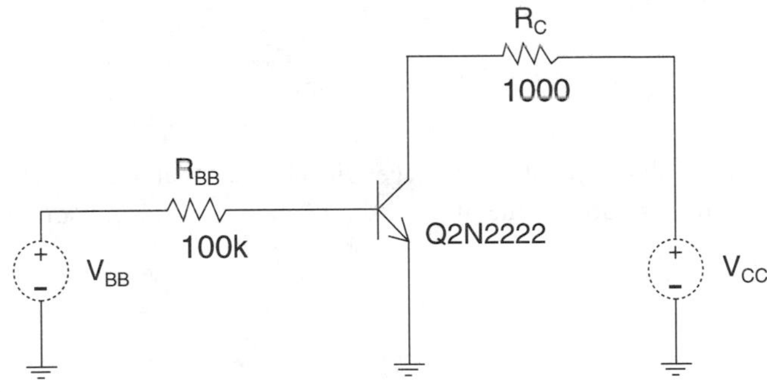

Figure 1.7a Biased circuit with 2N2222.

adjusting the controller, fluid flow can be increased or decreased.

EXAMPLE 1.5 Transistor Switch

Suppose that the circuit of Figure 1.8a is to be used as a switch to control the current in the 100 ohm load. The characteristic curves of Figure 1.8b already have the load line identified. The base voltage pulse alternates between zero and 10 volts. When the voltage is zero, there is no base current and the transistor is turned off. When the voltage is 10 volts, the base current can be calculated to be 9.3 milliamps or essentially, 10 milliamps. The corresponding collector current is limited to just less than 100 milliamps because of the load line. Clearly, the transistor switch is not functioning in a linear region. When the transistor is turned on with a base-to-emitter potential significantly greater than 0.7v, maximum current flows and a minimum voltage of VCE are observed. Even if the base current is doubled at this point, the current in the collector remains constant because the transistor is said to be in a "saturated" condition. Figure 1.8c illustrates the switching relationship between the input voltage pulse and the output current and output voltage, VCE, for this example.

In summary, once a transistor is biased, power from the DC supply can be controlled to make the circuit function as a switch by causing the base current to become zero, in which case the transistor does not allow current to flow; or causing the base current to become so large that maximum collector current flows and a minimum collector-to-emitter voltage is observed. Extensive use is made of transistors in this configuration as digital logic circuits. Another application of the transistor switch may be found in switching power to a motor or inductive device, as discussed in the next example.

EXAMPLE 1.6 Inductive Load

A transistor can be used to switch current through an inductive load, such as a motor or transformer circuit. Consider the circuit in Figure 1.9a where the curves for the transistor are given in Figure 1.9b. The input voltage is switched between 0 to 12 volts with a 50% duty cycle at 100 pulses per second as shown in Figure 1.9c. The time constant for the inductor will be approximately L_C/R_C = 1 millisecond while the inductor is charging or discharging. The diode across the inductive load protects the transistor when the transistor shuts off the current flow. The diode is reverse biased when the transistor is conducting and provides a path for current after the switch is opened and the inductor is discharging its energy.

For the examples considered thus far, the operating point always resides on the load line. Likewise, in this example, when the transistor is turned off, the collector current is zero, and the

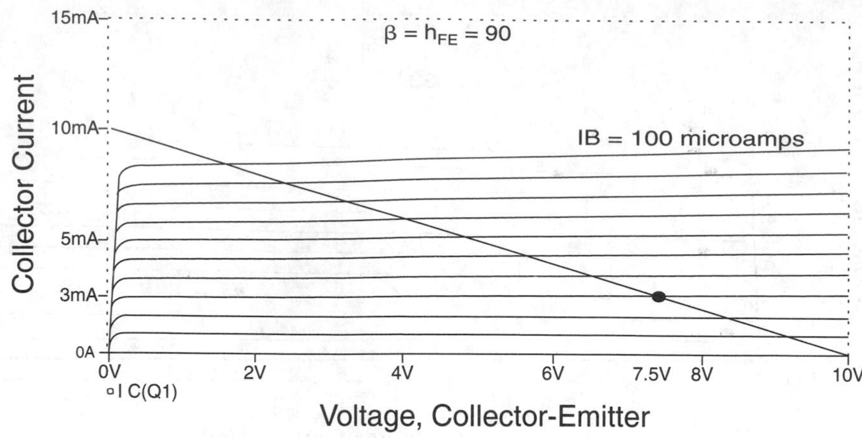

Figure 1.7b Characteristic curves for 2N2222 with a load line.

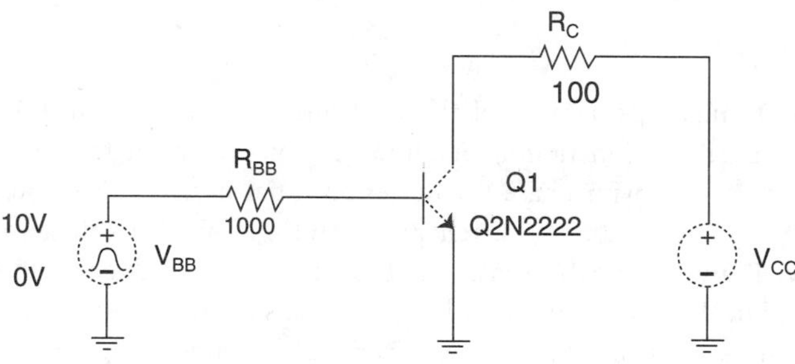

Figure 1.8a A transistor switching circuit.

voltage across the transistor is equal to the power supply voltage. When the transistor is turned on, the current is high (almost V_{CC}/R_C) and the voltage across the transistor is low, or about 0.2 volts. As the transistor moves from one operating condition to the other, the operating point for an inductive load does not follow the load line, as shown in Figure 1.9b. The transient response is shown in Figure 1.9c.

Suppose the transistor has been in the off condition for longer than 10 time constants, at which time it is turned on by the input voltage pulse. Since the transistor turns on instantly, the voltage across the device should drop to zero, and current can begin to flow into the inductor. The time constant, L_C/R_C, controls the speed at which the current increases to a maximum as determined by the load line and the intersection of the transistor curves. The transistor has no excessive voltage or current during the turn-on times.

After the transistor has been on for more than 10 time constants or sufficient time for all transients to subside, suppose that it is turned off by the input pulse voltage. Initially, since a steady-state current exists in the transistor, opening the circuit causes the inductor to develop a voltage determined by the equation

$$VL = L\, di/dt \tag{1.5}$$

and the direction of the voltage will be such that the current continues to flow in the original direction. Since current cannot

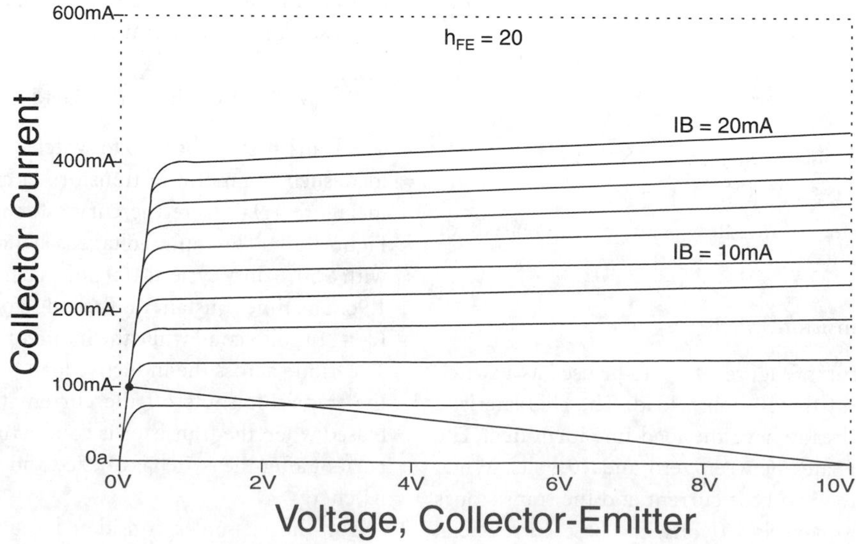

Figure 1.8b Characteristic curves with 100 ohm load line.

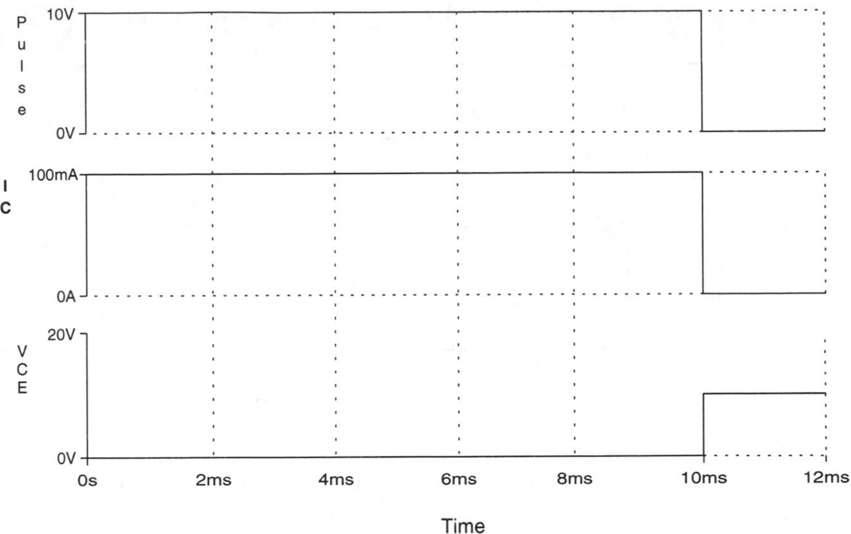

Figure 1.8c Waveforms indicating input voltage and output responses.

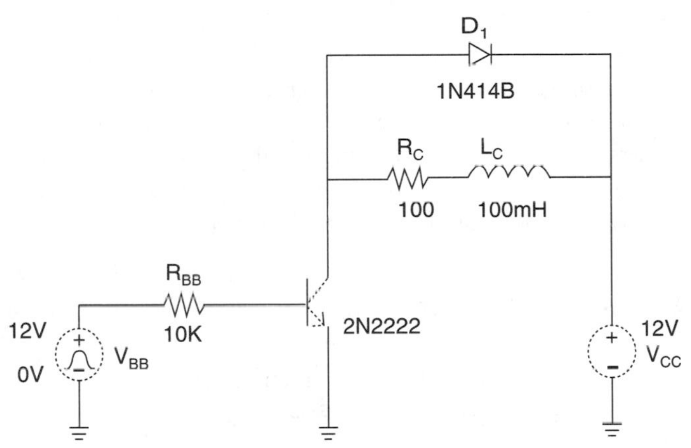

Figure 1.9a Transistor switch and inductive load.

flow into the transistor, it follows the path through the diode until the energy stored in the inductor has been dissipated. The time constant of the circuit will be L_C/R_C as before, because the forward biased diode is essentially a short circuit to the current.

Transistor Device Behavior

Transistors are made basically on a silicon substrate by physically implanting various elements into the silicon to make the devices function as desired. In Figure 1.10, the three terminals of the transistor are defined as collector, base, and emitter. In the block model with the *n-p-n* device representing the physical construction of the device, one can see which parts are labeled collector, base, and emitter.

A transistor is indeed only a current controlling device as evidenced by the information shown in the characteristic curves.

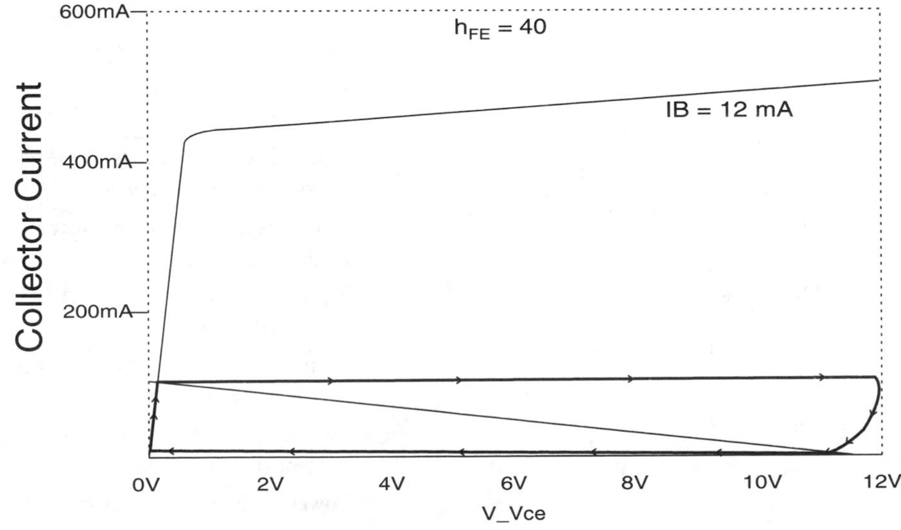

Figure 1.9b Locus of operating points for R/L network.

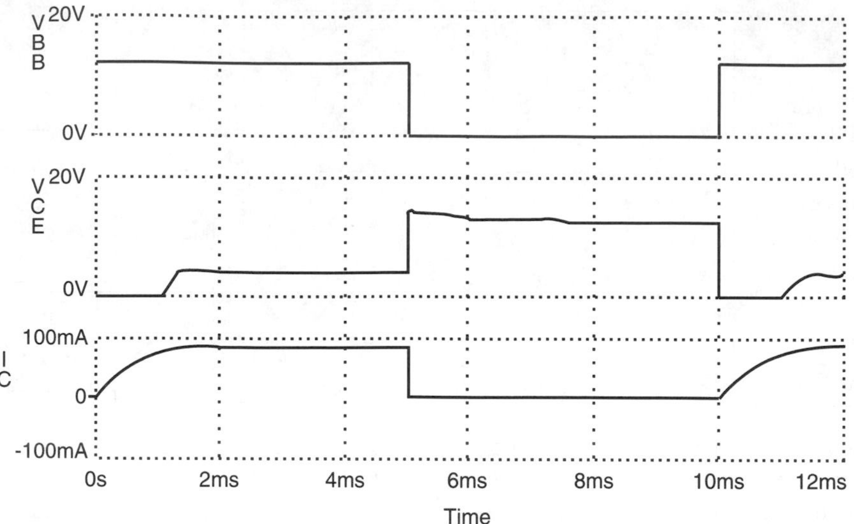

Figure 1.9c Transient response due to input voltage change.

Power is extracted from a power supply and channeled into a load in a manner controlled by the transistor. The transistor "valve" is opened more or less according to the value of the base current, which controls the absolute value of the collector current. One may well think of the transistor as a valve that opens to permit a flow of water. In fact, vacuum tubes were initially called "valves" for the same reason.

The different transistors which may be labeled *p-n-p* are similar to the *n-p-n* devices except that the bias voltages are all reversed. Likewise, as field effect transistors (FETs) are introduced, similar bias conditions must be observed. However, a different mechanism within the FET causes currents to flow, but the characteristic curves remain the same. A field is created between the drain and the source (collector and emitter analogy) and a gate (base equivalent) voltage influences the field in such a way that more or less current flows, depending on the polarity of the gate voltage. Observance of the data sheet information and characteristic curves will describe the device behavior.

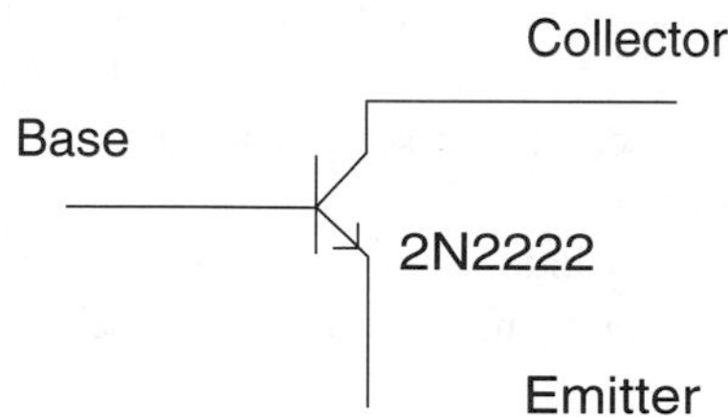

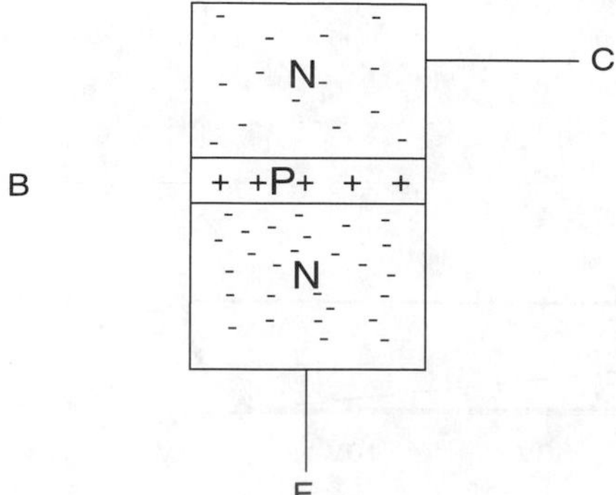

Figure 1.10 An *npn* transistor and a schematic symbol for the device.

Summary and Extension

- The *n-p-n* transistors were discussed extensively.
- Transistors can switch resistive loads.
- Transistors can switch inductive loads, if the designer is careful.
- Transistors can switch capacitive loads.
- Characteristic curves illustrate the current-voltage relationships for the transistor.
- Load lines can be used to visualize the location of operating point.
- Operating point always resides on the load line except for energy storage elements.
- The linear region of a transistor corresponds to the portion of the curves that are all horizontal.
- The saturation region generally is from zero volts to a few volts at the most.
- Transistors are limited by either maximum voltage or maximum current.
- A base-to-emitter voltage of less than 0.7 volts will not turn on the transistor.
- Power is dissipated near the collector.
- Power is generally calculated by the product of collector current and collector-emitter voltage.

- Heat sinks can extend the power-handling capability of transistors.
- Collectors correspond to drains.
- Emitters correspond to sources.
- Bases correspond to gates.
- Insulated gate bipolar transistors are crosses between BJT and FET transistors.
- Photo transistors allow light to activate the transistor rather than current or voltage.

1.4 Models for Transistors

In this section, three models will be discussed: direct current (DC) bias model, alternating current small signal (AC) model, and high-frequency or transient model. The models are useful in determining specific currents and voltages in more complex circuits than those that can be analyzed with the graphical scheme.

The DC Bias Model

The DC model consists of only two parts in the four terminal network. As shown in Figure 1.11, a voltage source of 0.7v, representing the voltage drop across a forward-biased diode, is in the base-emitter portion of the circuit. The output circuit contains a current source whose value is dependent on the magnitude of the base current that flows in the input circuit and the current gain, beta, or βI_B.

In an earlier discussion, the process of biasing a transistor to operate in its linear region was presented in addition to biasing the transistor to operate as a switch. Certainly, the circuit outside the transistor will determine where the actual bias point is located. One should always check the KVL or KCL, or check the operating point with a load line after an analysis using the linear models presented herein.

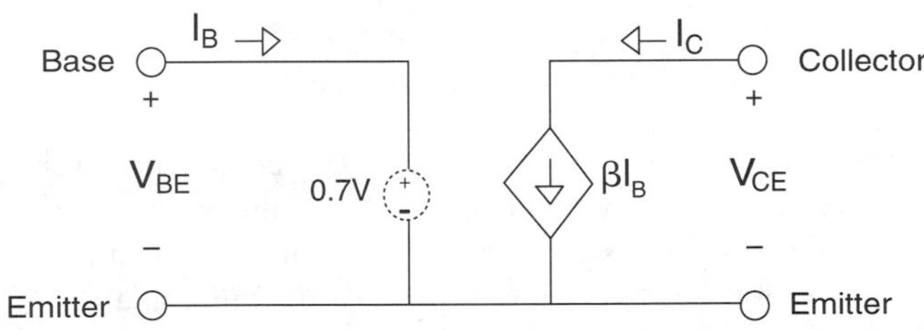

Figure 1.11 A DC model for a transistor.

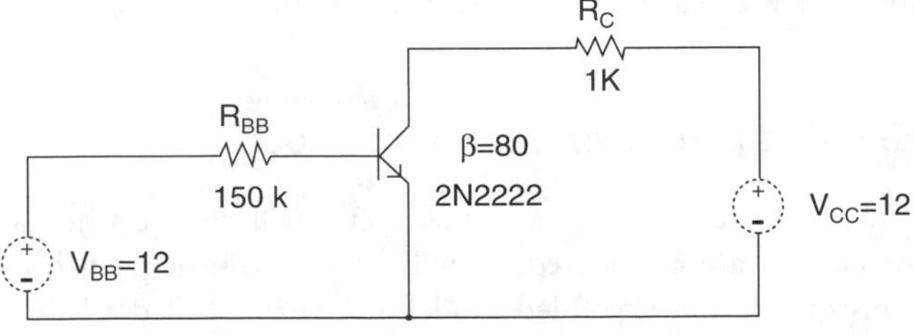

Figure 1.12 A biased transistor circuit.

EXAMPLE 1.7 Analytical Bias Calculations

The DC circuit in Figure 1.12 will be analyzed to determine the location of the bias point. First, the network in Figure 1.13 was created by replacing the transistor with the DC model in Figure 1.11.

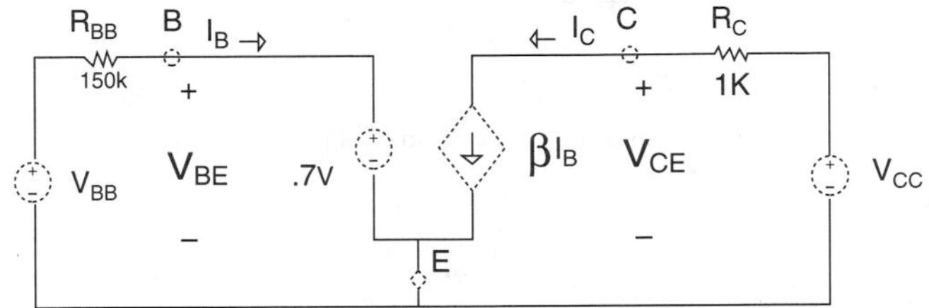

Figure 1.13 Biased circuit with model inserted.

The base current can be determined from the expression

$$V_{BB} = I_B{}^*R_{BB} + V_{BE}$$

and

$$I_B = 75\ \mu A$$

Since the circuit should be operating in the linear range, the collector current can be derived from the base current as βI_B. The collector-to-emitter voltage from Kirchhoff's voltage law using the output loop as

$$V_{CC} = \beta I_B{}^*RC + V_{CE}$$

and

$$V_{CE} = 6V$$

This value could also be obtained by utilizing the load line procedure presented earlier.

AC Model or Small Signal Model

The AC model or the small signal model implies that the transistor will operate with magnitudes of current and voltage that are well controlled. Many engineers regard the small signal model to be one in which the input voltage can be doubled or halved and the output signal would follow linearly. The strong implication, then, is that the transistor must be biased, as shown in Example 1.7, so that the device is in the active region of the transistor characteristics and so that the magnitudes of input voltage or current will not drive the device out of its properly biased condition.

The circuit model of Figure 1.14 can be regarded as one of several small signal models. Note that the series resistance in the base of the model replaces the 0.7v drop of the forward biased diode in the DC bias model. Typically the series resistor, $r\pi$, is in the range of 1000 ohms. Secondly, the output circuit is represented again by a current-dependent current source which is

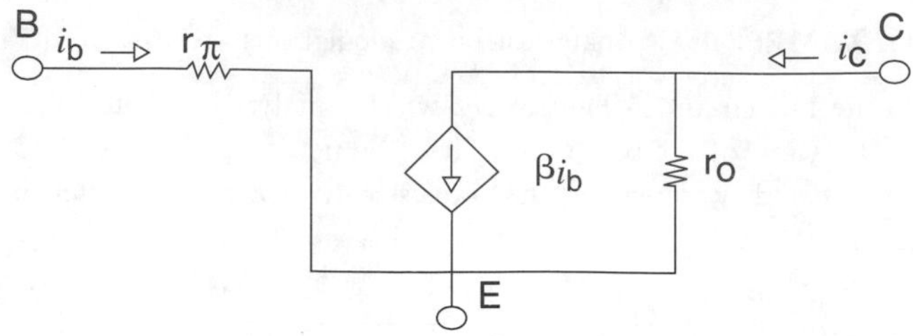

Figure 1.14 The transistor used in a small signal application.

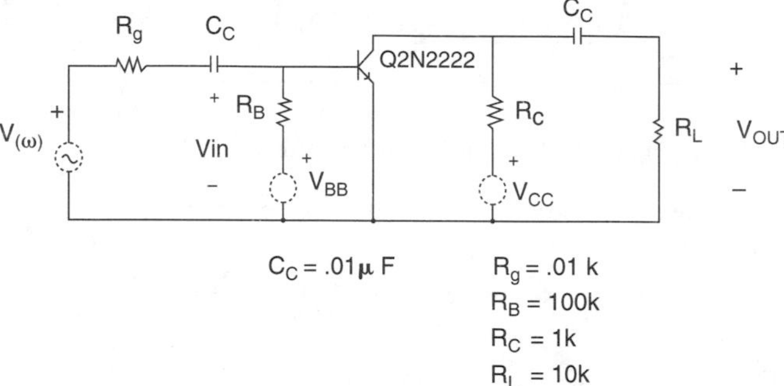

Figure 1.15 A transistor amplifier with low frequency considerations.

shown as βI_B. A resistor in parallel with the current source is also shown in the model. The resistor is usually in the range of 0.5 mΩ, so that it rarely impacts the first order approximations in a circuit analysis or design. Consequently, the parallel resistor is often ignored. A designer will determine the transistor current gain from the data sheets and use these approximations for the resistors.

EXAMPLE 1.8 An Audio Amplifier

The amplifier represented in Figure 1.15 illustrates a biased transistor which is separated from the AC input voltage by a coupling capacitor Cc which blocks DC. The output voltage across the load resistor RL is separated from the bias circuit by another Cc. Proper biasing will fix the operating point in the linear portion of the characteristic curves.

Figure 1.16 shows a small signal representation of the circuit of Figure 1.15 because the model of Figure 1.14 has been inserted in place of the transistor, and the two DC supplies have been replaced by their internal resistance, assumed to be zero in each case. The coupling capacitors have been replaced by short circuits with the knowledge that the circuit will operate at audio frequencies. It is not necessary to make any particular assumptions at the beginning of the analysis, but the engineer should recognize that some components are so small in comparison with others that many terms can be canceled or ignored later.

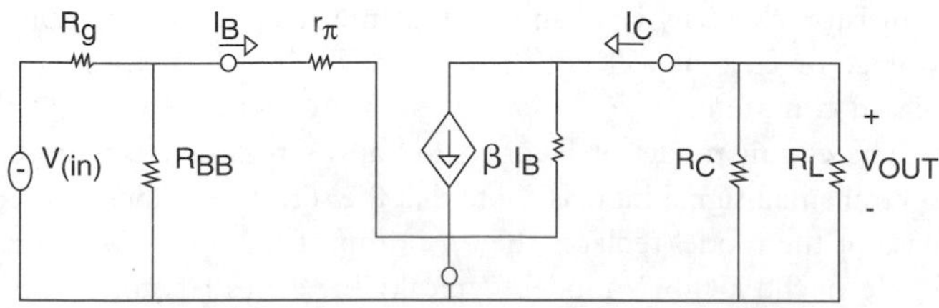

Figure 1.16 A biased small signal amplifying circuit.

To analyze the circuit, the KVL equations can be written to determine the value of current, i_b

$$i_b = \frac{V_{in}(r\pi IIR_{BB})}{r\pi(R_g + r\pi IIR_{BB})}$$

The voltage divider will provide a value for voltage from base to emitter, from which the base current can be easily determined. Alternatively, one could use a current divider procedure to determine the same value.

$$V_{OUT} = -\beta i_b\,[r_o IIR_C IIR_L]$$

The output voltage can be determined when the current source applies current in the direction shown to the three resistors in parallel.

One can substitute the value for i_b into the equation for the output voltage and divide both sides of the equation by the input voltage to arrive at the voltage transfer function, which is dependent only on circuit and component values.

$$\frac{V_{out}}{V_{in}} = \frac{-\beta(r\pi IIR_{BB})(r\pi IIR_C IIR_L)}{r\pi(R_g + r\pi IIR_{BB})}$$

Using the values of components listed in Figure 1.15, one can reduce the complexity of the voltage transfer function to the following equation, because R_c is small compared to its parallel counterparts. Likewise the value of $r\pi$ is small compared to the base bias resistor.

$$\frac{V_{out}}{V_{in}} = \frac{-\beta\, R_C}{(R_g + r\pi)}$$

A few observations are appropriate at this point. The transfer function will generally never be as large as this value suggests. Further, the negative sign in the equations indicates that a 180-degree phase shift has occurred between the input and output because as the input current was increasing in the transistor, the output voltage was decreasing. Finally, the coupling capacitors will cause a deterioration of the low-frequency response of an amplifier with such components present.

High-Frequency or Transient Model

A model for a transistor and the circuit will change when all frequencies are considered, as will be in a transient analysis. Consequently, an expanded model, sometimes called a hybrid model, includes internal capacitance which makes the transistor model seem more complex. In particular, capacitance exists

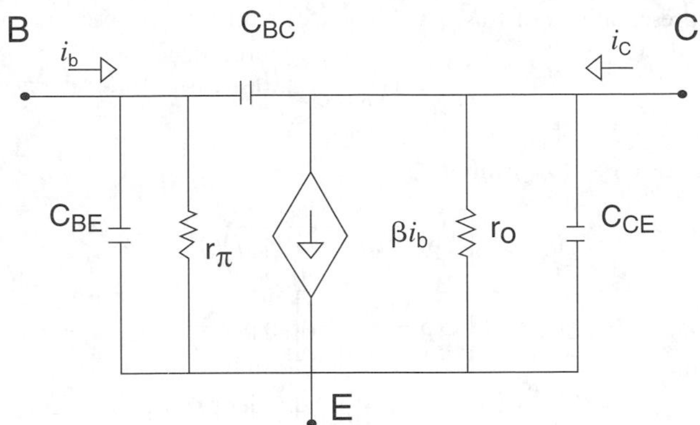

Figure 1.17 An all-frequency hybrid model for a transistor.

between all terminals as shown in Figure 1.17. Although the values of the capacitance are small, the frequency ranges of 10 MHz will create significant impedances, whereas, with the audio amplifier, the capacitances did not introduce significant impedances.

Consider the network of Figure 1.18 which represents an amplifier that could be analyzed with a frequency response or by analyzing its transient response. Notice that the DC base bias supply has been replaced with a voltage divider operating from the collector supply. It will be assumed that the transistor is appropriately biased according to admonitions read earlier. Thus, the small signal model can be placed into the figure to yield Figure 1.19, which shows all the significant components. Note that R_{BB} is the result of the parallel combination of the two resistors used in the voltage divider. Additionally, the low-frequency effects require that the coupling capacitors be included in the circuit analysis. The actual transfer function for this circuit will be composed of numerous terms. However, as an alternative to the development of the actual transfer function, a PSPICE simulation will provide the frequency response of the circuit. In this simulation the input voltage is constant in magnitude, but the frequency is varied from low to high values. Figure 1.20 is a plot of the magnitude of the voltage transfer function as a function of frequency. The low frequencies do not get amplified because of the coupling capacitors. The very high frequencies are not amplified because of internal capacitance. In the mid-range of frequencies, however, the output voltage is consistent with that calculated in the previous example.

Foundations for Transistor Models

The models presented thus far have been interpreted from the characteristic curves. To understand how the curves came to be in the first place, it may be helpful to review basic semiconductor physics and develop a model which can be used in high frequency analyses.

The solid-state physics which describes a transistor requires detailed understanding of material beyond the scope of the handbook. However, a transistor user is usually more concerned with terminal characteristics than internal details. Figure 1.10 portrays a schematic and pictorial side view of a bipolar transistor. It is evident that two junctions exist in the device. The intersection of the *p*-type material with *n*-type material on one side is referred to as the base-emitter (B-E) junction. The intersection of *n*-type material with the *p*-type material on the other side is referred to as the base collector junction (B C).

To understand qualitatively how the device works, let us apply a positive voltage on the base region with respect to the emitter

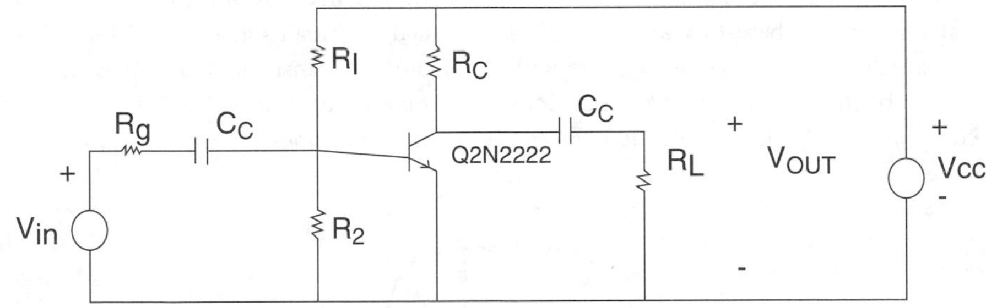

Figure 1.18 A high-frequency amplifier.

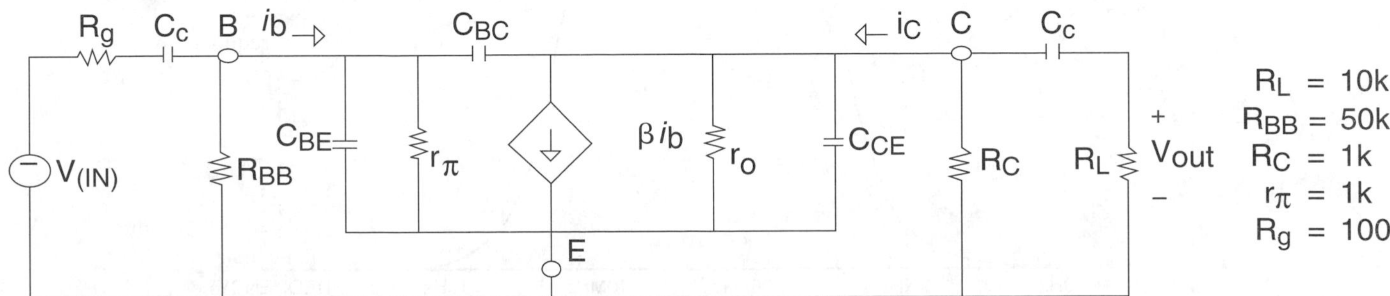

Figure 1.19 A biased small signal amplifier with a high-frequency transistor model included.

and leave the collector unconnected. Since the bias acts to lower the energy barrier of the junction, holes will diffuse from the high concentration in the base toward the emitter and electrons will flow from the emitter to the base. The net effect is a positive current flow from base to emitter since the two current components add. (Remember that positive current flow is opposite the direction of electron flow.) The hole and electron populations are continuously replenished via an internal generation mechanism in the transistor. The device as connected is operating as a forward biased diode.

Let us now apply another biasing voltage to the collector such that the collector-base junction is reverse biased. Also, for simplicity let us restrict our discussion to only electrons. The positive bias on the collector acts to attract the electrons which were injected from the emitter across the base-emitter junction. The result of this is that the electrons now begin their path in the emitter, diffuse across the base, and most are swept up and congregate at the collector where they now flow out of the device. Due to the geometrical arrangement of the device some electrons will flow to the base junction. The ratio of electrons which flow to the collector to those that flow into the base is commonly called *Beta* and is represented by β. Beta is also the ratio of I_C/I_B and is defined as the current gain.

This phenomenon is the essence of how a bipolar transistor operates. A small voltage difference across the base-emitter junction controls the current flow between the collector and emitter. Recall the diode equation of Equation 1.2.

$$I_{CE} = I_o[\exp(V_{BE}/nV_t) - 1]$$

The current which flows from the collector to emitter, I_{CE}, depends exponentially on the forward bias voltage of the base-emitter junction. We will assume the exponential term is much larger than 1. The voltage across the base-to-emitter junction need not be restricted to a value which does not change with time (i.e., DC value), but may be the summation of a DC component and time varying component (AC). Usually the base-emitter bias consists of both. The following equations are quantitative representations of this.

$$V_{BE} = \{V_{BE(DC)} + v_{be}\}$$

where $v_{be} = (v_{AC})\sin(\omega t)$

$$I_{CE} = I_o[\exp(\{V_{BE(DC)} + v_{be}\}/V_t)]$$

$$I_{CE} = I_o\left[\exp\frac{\{V_{BE(DC)}\}}{V_t}\right]\exp\left(\frac{v_{be}/V_t}{V_t}\right)$$

$$= I_o\left[\exp\frac{\{V_{BE(DC)}\}}{V_t}\right]\left[\exp\frac{\{V_{be}\}}{V_t}\right]$$

which can be rewritten as

$$I_{CE} = I_{DC} * \exp\{v_{be}\}/V_t)$$

where

$$I_{DC} = I_o\left[\exp\frac{\{V_{BE(DC)}\}}{V_t}\right]$$

which is simply the time invariant component of the collector current I_{DC} multiplied by an exponential term, restating again

$$I_{CE} = I_{DC}{}^*[\exp\{v_{be}\}/V_t].$$

The exponential is now expanded realizing that exp (x) is equal to $1 + x + \ldots$ for small values of x. Realizing that V_t is approximately 26 mV at room temperature leads us to what is commonly called the small signal model for the terminal characteristics of a bipolar transistor. For values of v_{be}, which are small when compared to 26 mV, the total current through the collector to emitter becomes

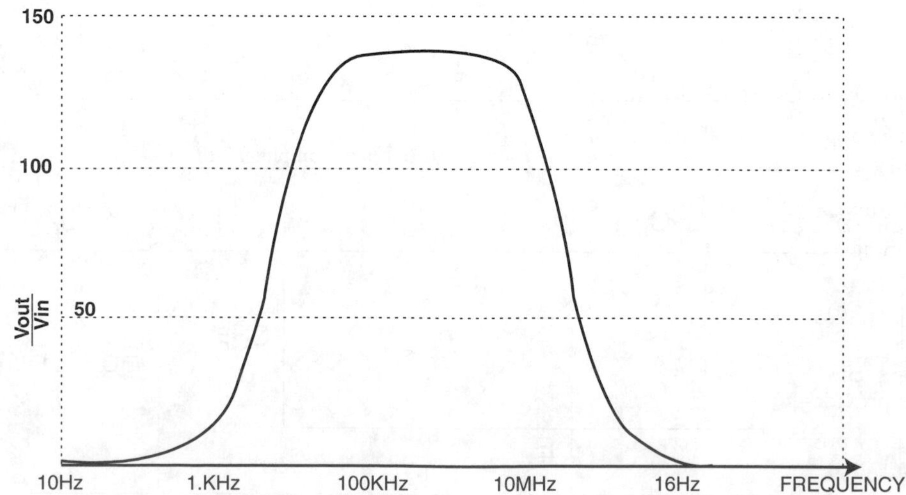

Figure 1.20 Magnitude of voltage transfer function for the circuit of Figure 1.19.

$$I_{CE} = I_{DC}{}^{*}(1 + v_{be}/V_t) \quad \text{or}$$

$$I_{CE} = I_{DC} + \{I_{DC}/V_t\}^{*} v_{be}$$

The quantity $\{I_{DC}/V_t\}$ is referred to as the transconductance and is usually written g_m. The expression illustrates that for small values of v_{be} the current flowing through the collector to the base consists of a constant value plus a time-varying component. The usual methodology involved in analyzing a bipolar transistor involves solving for the components separately since both components of current are now linearly dependent on the base-emitter bias.

A simple model can now be utilized to describe the small signal time-varying terminal characteristics for a bipolar transistor. It is observed that a dependent current source can model the connection between the collector to the emitter. The ratio of the voltage from the base to emitter is simply v_{be}/I_B which is $\{v_{be}{}^{*}\beta/I_C\}$. This parameter has dimensions of ohms and is an equivalent resistance which is called r_π. In order to further enhance the accuracy of the model, some additional parasitic elements must be included.

Internal to the transistor some charge is stored across the base-emitter junction. This charge storage phenomenon is represented by a parasitic capacitance connected between the base and the emitter called C_{BE}. A similar capacitance exists between the base-collector region called C_{BC}. A third capacitor called C_{CE} is also placed between the collector and emitter.

It is observed in Figure 1.6a that the collector current is not constant for all values of collector-emitter voltage but increases in a linear fashion. The change in collector current with respect to V_{CE} is a constant referred to as the output resistance and is called r_o. Thus, the complete small signal model is illustrated in Figure 1.17.

Summary and Extension

- Transistor models are a result of graphical and analytical analyses.
- Small signal models are useful for AC analyses.
- Transistor models represent a current or voltage-dependent relationship.
- The DC model should be used for biasing a transistor.
- All DC bias conditions must be met before choosing an AC model.
- High-frequency models are useful in transient analyses.
- PSPICE analyses are useful in AC or transient analyses.
- Simple models provide quick estimates for solutions.
- Elaborate models provide accuracy which should be checked by the "quick solution".
- Computer models account for temperature and parameter variations.

1.5 Analog and Digital Circuits

Transistors are found in numerous analog integrated circuits. These integrated circuits are specially built devices to perform designated functions, such as filtering and amplification.

Transistor applications are also found in digital circuits in which most of the devices operate as switches. Integrated circuits (IC) of this type include hundreds of applications that range from small scale logic circuits to large scale circuits, such as inverters, counters, memory (RAM), and microprocessors.

The basic concepts for transistors introduced thus far will be sufficient to expand the reader's understanding of the specialized circuits. The circuits generally consist of transistor biasing elements, capacitors, and diodes. The purpose of the additional elements is to ensure the circuit remains stable over a broad range of temperature and biasing points. Two examples will be provided to illustrate the procedures to analyze circuits that are more complex than those presented thus far. The first example establishes a stable operating point for an analog amplifier. The other example analyzes the behavior of a NAND circuit.

EXAMPLE 1.9 Bias Stability for Transistor

The circuit of Figure 1.21 looks different from some of the earlier circuits because the base bias is created with the power supply and a voltage divider, consisting of R_{B1} and R_{B2}. Secondly, any circuit used in an audio frequency amplifier must also have connections to the input and output circuits. The circuit presented in this example provides one illustration of a typical task that circuit designers accomplish as they develop circuits for audio, instrumentation, or power applications. It can be assumed that the transistor has been properly biased using the techniques described earlier in which an operating point is chosen within the linear portion of the transistor characteristic curves. A transistor circuit model for the audio amplifier can be inserted into the circuit, yielding the circuit shown in Figure 1.22.

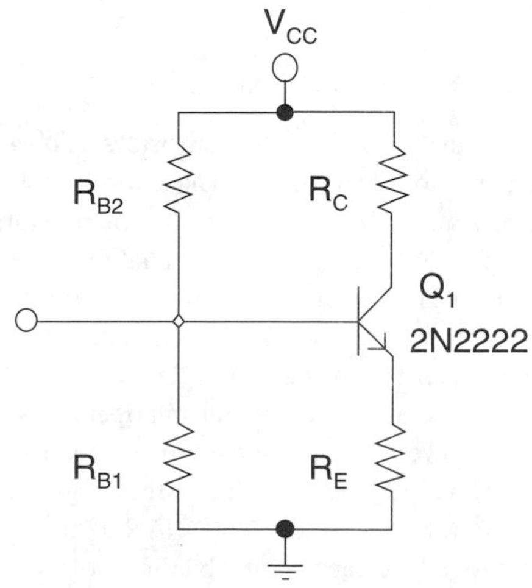

Figure 1.21 A four-resistor bias to obtain operating point stability.

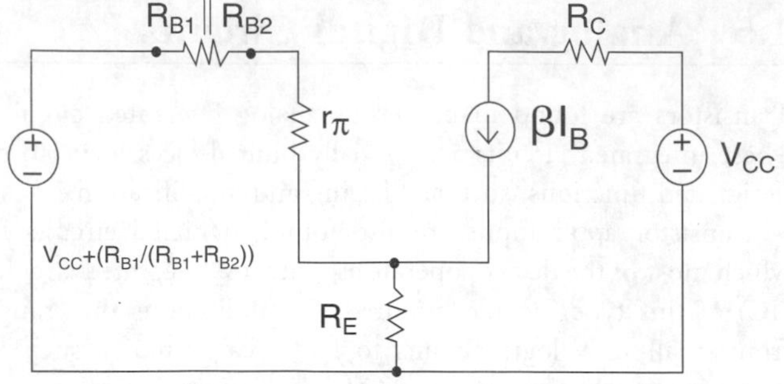

Figure 1.22 Decomposition of a bias circuit with model for transistor.

The analysis of the circuit will show that the transistor bias point and consequently the collector current will be independent of the temperature-dependent and device-dependent current gain, beta. Writing KVL around the input loop yields the following equations:

$$V_{BB} - I_B(R_{BB} + r_\pi) - (I_B + I_C)R_E = 0$$

$$V_{BB} = I_B(R_{BB} + r_\pi + (1 + \beta)R_E)$$

$$I_B = V_{BB}/(r_\pi + R_{BB} + (1 + \beta)R_E)$$

When the designer chooses RE such that

$$(1 + \beta)\ R_E = r\pi + R_{BB}$$

$$I_B \approx \frac{V_{BB}}{(1 + \beta)R_E}$$

$$I_C = \frac{\beta V_{BB}}{(1 + \beta)R_E}$$

$$\approx \frac{V_{BB}}{R_E}$$

We see that I_C depends only upon the value of V_{BB} and R_E and is not strongly related to other external components or device parameters such as β.

EXAMPLE 1.10 Logic Circuit Analysis

A combination of diodes and transistors allows the digital circuit designer an opportunity to make complex logic functions with transistors and a minimum number of resistors, either of which reduces power dissipation. An analysis of a transistor-transistor logic (TTL) circuit will provide opportunities to discuss diodes, transistors, and functional characteristics. Specifically, the manufacturers determined that multiple emitters can be placed on transistors and that the device will still operate as a transistor. The analysis, however, requires one to utilize the diode characteristic of the emitters. Figure 1.23 illustrates a discrete equivalent component circuit for an SN7400 NAND circuit, which is assumed to drive a load represented by the 5 volt supply in series with a 400 ohm resistor. Functionally, the output will be described by the following truth table:

Truth Table for a NAND Circuit

Input A	Input B	Output
5v	5v	0v
5v	0v	5v
0v	5v	5v
0v	0v	5v

When both input voltages are at 5 volts, the "diodes," composed of the base-emitter connections of transistor Q_1 will not let current flow, as can be seen by the circuit equations which include the V_{CC} in opposition to the 5v input voltages. Current may flow through the base-collector diode and through the low-resistance path to ground via the base-emitter diodes of both Q_2 and Q_3. As recognized earlier in the discussion of switching circuits, when base current flows, then the transistor may well be turned on. Thus one may conclude that Q_2 and Q_3 are turned on and may well be in saturation, which will cause the collector-emitter voltage of Q_3 to be close to zero volts. It is instructive to note at this point that current flows from the load into the saturated transistor, Q_3. When a load provides current to the driving circuit, the driving circuit is said to "sink" current. In most TTL logic circuits, the driving circuit sinks current when it presents a low voltage to its load.

As the base current flows through the base-emitter diode of Q_2, the transistor action occurs and causes collector current to flow, which adds to the base current, creating a relatively large emitter current. The resulting collector current into Q_2 is such that no current is available to flow into the base of Q_4, which causes it to turn off. By substituting the DC bias models in place of the transistors, the reader can then write circuit equations to verify the conditions stated.

Considering the other conditions in which the input voltages may be zero or 5 volts, one recognizes immediately that one of the base-emitter diodes will be conducting, which leaves no current to flow through the collector-base of Q_1. Q_1 will not function as a transistor because its collector is not attached to a DC bias supply. Thus, the transistors Q_2 and Q_3 must be turned off because there is no current from Q_1 to turn them on. The base-emitter of Q_4 now has current available because the current is not diverted to Q_2. The primary question at this point will be whether Q_4 is turned on and into saturation, or whether it is turned off.

The NAND circuit may drive loads that are like the one shown, or drive another TTL circuit, which could be modeled as a series resistor and source, similar to the one shown. The reader may conclude that because base current is available to Q_4, then the transistor will be in a saturated condition IF there is a path for the emitter current to go to ground. Otherwise, the output terminal (collector of Q_3) will float to a high level, as needed. Clearly, if the current load were replaced with only a 400 ohm resistor to ground, the path for the emitter current of Q_4 would allow Q_4 to be in saturation and a nominal 0.7v drop would be measured across the output diode. In any event, when either or both of the input voltages to the NAND circuit is zero, the output of the circuit will be high, or near 5 volts.

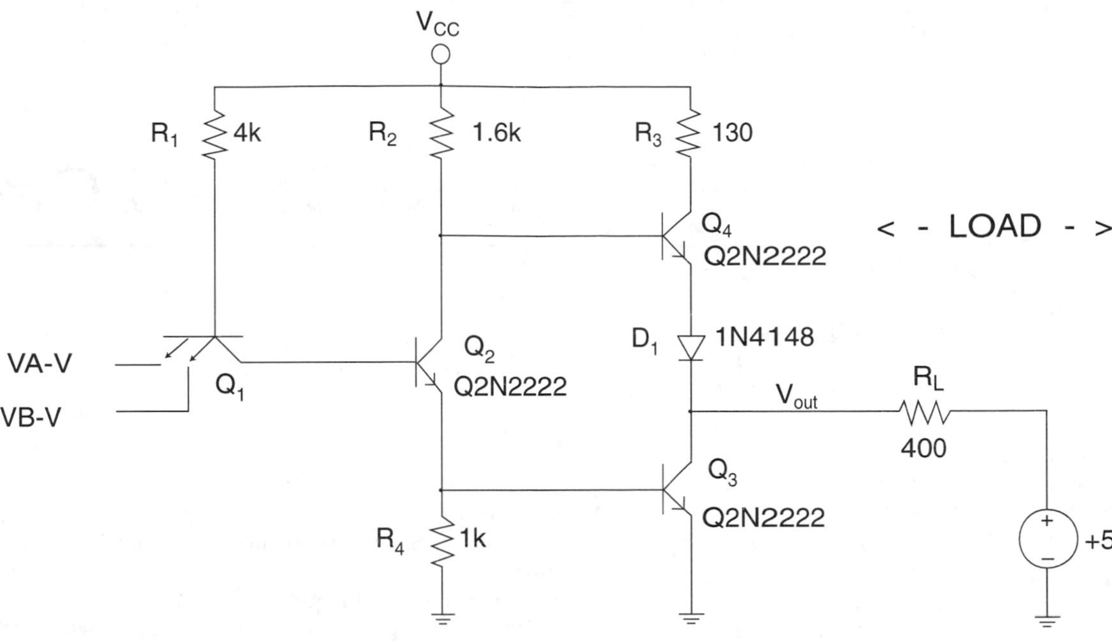

Figure 1.23 Discrete model of a two-input TTL NAND circuit.

Device Behavior

In many respects the linear integrated and digital integrated circuits behave alike because the transistors, diodes, capacitors, and resistors are all embedded into a silicon substrate. One may well envision that a printed circuit board, such as a "mother board" in a personal computer, could be transformed into smaller and smaller components until it fits on a board whose dimensions will be a quarter inch on a side. With a circuit so small, the engineers could implant the circuit components onto a slab of silicon, resulting in an integrated circuit. Thus, with all the circuit complexity described in a data sheet, the user may have great confidence that the circuit will behave as described. As printed circuit boards are tested extensively, the integrated circuits are tested also. Further, the integrated circuits may even be more reliable in a variety of applications because they are even temperature compensated, whereas, some of the printed circuit boards may not be compensated. In any event, the data sheets will describe the essential functions of the circuit.

Summary and Extension

- Linear integrated circuits contain many amplifiers and control circuits.
- Digital integrated circuits contain many switching circuits.
- Detailed circuit analysis without a computer is tedious.
- Computer analysis packages are helpful.
- Data sheets describe terminal conditions and behavior of the devices.
- Application notes provide helpful information and additional constraints of the circuits.
- Data sheets of complex devices may actually result in handbooks to describe the features.

Fortunately, the manufacturers of the complex components provide extensive literature to describe the function of the component, so that the circuits applications engineer can readily use the device.

References

Belanger, Adler, and Rumin, 1985. *Introduction to Circuits with Electronics,* Holt, Rinehart and Winston, New York, NY.

Cogdell, 1996. *Foundations of Electrical Engineering,* 2d ed., Prentice-Hall, Englewood Cliffs, NJ.

Demassa and Ciccone, 1996. *Digital Integrated Circuits,* John Wiley & Sons, New York, NY.

Millman, 1979. *Micro-Electronics,* McGraw-Hill, New York, NY.

Nelson, Nagle, Carroll, and Irwin, 1995. *Digital Logic Circuit Analysis and Design,* Prentice-Hall, Englewood Cliffs, NJ.

Sedra and Smith, 1987. *Microelectronic Circuits,* Holt, Rinehart and Winston, New York, NY.

Savant, Roden, and Carpenter, 1987. *Electronic Circuit Design,* Benjamin/Cummings, Menlo Park, CA.

2 Digital Control Circuits

Marc Courvoisier
Université Paul Sabatier, Institu National des Sciences Appliquées de Toulouse, Laboratoire d'Architecture et d'Analyse des Système, Centre National de la Recherche Scientifique

Michel Combacau
Université Paul Sabatier, Laboratoire d'Architecture et d'Analyse des Système, Centre National de la Recherche Scientifique

Mario Paludetto
Université Paul Sabatier, Laboratoire d'Architecture et d'Analyse des Système, Centre National de la Recherche Scientifique

2.1 Logic Control 22
Introduction
2.2 Sequence Control 28
State Machines and Algorithmic State Machines • Petri Nets • Design Methods
2.3 Implementation Techniques 41
Relay Ladder Diagrams • Implementation of Discrete Event Systems via Electronic Integrated Circuits • Software Implementation

2.1 Logic Control

Introduction

Discrete events systems deal with processes in which the state variables, the inputs, and the outputs are not continuous. They change from one value to another. In most of these systems, each input, output, or state variable can be constrained to take only two values (very often for implementation purposes). Computers, programmable logic controllers, logic automatisms, etc. belong to this class of systems. Some models have been developed to express formally the solutions of discrete systems control. This section deals with these models and begins with a description of the basic concept on which they have been built: Boolean algebra.

Boolean Algebra

Definitions. Definition domain: the Boolean set $B_2 = \{0,1\} = \{\text{False, True}\}$

Boolean variable: any symbol taking its value in the Boolean set with the two following axioms:

$$x = 0 \Leftrightarrow x \neq 1$$

$$x = 1 \Leftrightarrow x \neq 0$$

Operators. Unary: complement

Notation: complement (x) is noted $\bar{x}$

Definition:

$$x = 0 \Leftrightarrow \bar{x} = 1$$

$$x = 1 \Leftrightarrow \bar{x} = 0$$

Binary:

OR (logical sum)

Notation: OR $(x1, x2)$ is noted $x_1 + x_2$

Definition: $(x_1 + x_2 = 0) \Leftrightarrow (x_1 = 0)$ and $(x_2 = 0)$

AND (logical product)

Notation: AND (x_1, x_2) is noted x_1, x_2

Definition: $(x_1 \cdot x_2 = 1) \Leftrightarrow (x_1 = 1)$ and $(x_2 = 1)$

Properties of Boolean Algebra. x_1, x_2 and x_3 being three Boolean variables, we have:

	OR	AND
Involution	$\bar{\bar{x}}_1 = x_1$	
Idempotence	$x_1 + x_1 = x_1$	$x_1 \cdot x_1 = x_1$
Commutativity	$x_1 + x_2 = x_2 + x_1$	$x_1 \cdot x_2 = x_2 \cdot x_1$
Associativity	$(x_2 + x_1) + x_0 = x_2 + (x_1 + x_0)$	$(x_2 \cdot x_1) \cdot x_0 = x_2 \cdot (x_1 \cdot x_0)$
Complementarity	$x_1 + \bar{x}_1 = 1$	$x_1 \cdot \bar{x}_1 = 0$
Neutral element	$x_1 + 0 = x_1$	$x_i \cdot 1 = x_1$
Absorptive element	$x_1 + 1 = 1$	$x_1 \cdot 0 = 0$
Distributivity		
Product on sum	$x_2 \cdot (x_1 + x_0) = x_2 \cdot x_1 + x_2 \cdot x_0$	
Sum on product	$x_2 \cdot x_1 + x_0 = (x_2 + x_0) \cdot (x_1 + x_0)$	

0-8493-8343-9/97/$0.00+$.50

All of these basic properties are easily demonstrated by enumeration.

De Morgan's Laws. Let $\{x_0, \ldots, x_k\}$ be a set of Boolean variables

$$\overline{\sum_{i=0}^{k} x_i} = \prod_{i=0}^{k} \overline{x_i} \quad \text{and} \quad \overline{\prod_{i=0}^{k} x_i} = \sum_{i=0}^{k} \overline{x_i}$$

Inference rules:

$$x_2 \cdot x_1 + \overline{x_2} \cdot x_0 = x_2 \cdot x_1 + \overline{x_2} \cdot x_0 + x_1 \cdot x_0$$

$$(x_2 + x_1) \cdot (x_2 + x_0) = (x_2 + x_1) \cdot (\overline{x_2} + x_0) \cdot (x_1 + x_0)$$

Logic Expressions

Definitions. A logic expression is a combination of variables and operators of Boolean algebra. The variables appearing in a logic expression are the input variables of this expression. A logic expression has a Boolean value.

0, 1, x are logic expressions (x is a Boolean variable).

If E_1 and E_2 are logic expressions then $\bar{E}_1$, $E_1 \cdot E_2$, $E_1 + E_2$ are logic expressions also.

The Boolean value of a logic expression for a given combination of the input variables is obtained by replacing in the expression the variables by the Boolean value they have in this combination.

EXAMPLE 2.1:

$$E(x_1, x_2, x_3) = x_1 \cdot x_2 + x_3$$

$$E(0, 0, 1) = 0 \cdot 0 + 1 = 1$$

Two logic expressions E_1 and E_2 are equal if, and only if, the values of E_1 and E_2 are equal for any combination of their input variables.

Analogy With the Set Theory

Let F and G two subsets of a set X. If we define two functions f and g from X to B_2 by

$$\forall x \in X \left| \begin{array}{l} f(x) = 1 \Leftrightarrow x \in F \text{ and } f(x) = 0 \Leftrightarrow x \notin F \\ g(x) = 1 \Leftrightarrow x \in G \text{ and } g(x) = 0 \Leftrightarrow x \notin G \end{array} \right.$$

we have,

$$\left| \begin{array}{ll} \overline{f(x)} = 1 & x \in \overline{F} \\ f(x) + g(x) = 1 & x \in F \cup G \\ f(x) \cdot g(x) = 1 & x \in F \cap G \end{array} \right.$$

This analogy models many of the problems expressed by natural languages.

Logic Functions

Definition. A logic function of n Boolean variables is a function from B_2^n to B_2 associating a Boolean value to each combination of n Boolean variables (named the input variables). Example: OR (x_1, x_2) is a logic function of the variables x_1 and x_2.

Complementary Function. A logic function f of n variables splits up the set B_2^n into two subsets of combinations

- a subset F from which the function takes the value 1.
- a subset $\bar{F}$ from which the function takes the value 0.

The complementary function of f noted $\bar{f}$ takes the value

- 1 for each combination of $\overline{F}$.
- 0 for each combination of F.

Representation of Logic Functions

The goal of the representation of a logic function is to define the value of the function for each combination of its input variables. The most useful ones are algebraic expressions, truth tables, and Karnaugh maps. A complete description of some other representations can be found in (Givone, 1970, and Nussbaumer).

Algebraic Expressions. In this representation, the function is associated with a logic expression on the input variables. The value of the logic expression for each combination of the input variables defines the value taken by the logic function for this combination.

EXAMPLE 2.2:

$E_0 = x_1 + x_2$ is an algebraic representation of the logic function OR (x_1, x_2).

The Boolean value taken by OR (x_1, x_2) when $x_1 = 0$ and $x_2 = 1$ can be calculated through the logic expression $E_0 = x_1 + x_2$.

So, we have OR(0, 1) = 0 + 1 = 1.

A function has more than one algebraic expression. In fact, if the logic expression E_0 is an algebraic representation of a function f, any logic expression E_j equal to E_0 is also an algebraic representation of f.

EXAMPLE 2.3:

We saw that $x_1 + x_2$ is a representation of OR (x_1, x_2),

$$\left. \begin{array}{l} E_1 = x_1 + x_2 + x_2 \cdot x_1 \\ E_2 = \overline{\overline{x_1} \cdot \overline{x_2}} \\ E_3 = \overline{x_2} \cdot x_1 + x_2 \cdot \overline{x_1} + x_2 \cdot x_1 \end{array} \right| \text{are algebraic representations of OR } (x_1, x_2) \text{ also.}$$

The logic expression E_3 is called the first canonical algebraic representation of the function OR (x_1, x_2).

It consists of the logical OR between other logic expressions $E_3 = E_{31} + E_{32} + E_{33}$ in which each E_{ij} expression takes the value 1 for only a single combination of the input variables.

Truth Tables. The representation seen in Figure 2.1 relies on the enumeration of the values the function takes for all possible combinations of the input variables. A logic function f of n input variables is defined in a table with $n + 1$ columns and 2^n rows. Each of the n first columns is dedicated to an input variable while the last column receives the value of the logic function f. So, each row contains a combination of the input variables and the corresponding value of the function f.

The combinations of the input variables always appear according to the natural binary order. So, the first row of a truth table always contains the combination

$$\prod_{i=n}^{1} \overline{x_i}$$

and the last row the combination

$$\prod_{i=n}^{1} x_i$$

EXAMPLE 2.4:

OR (x_1, x_2) (Figure 2.2)

The representation seen in Figure 2.2 is the most common for functions with a small number of input variables. Indeed, the size of the table exponentially grows according to the number of input variables.

Karnaugh Maps. As with the truth table technique, this representation relies on the enumeration of the value of the function for each combination of the input variables.

A Karnaugh map is a double entry table. For a logic function of n variables, the combinations of m $(m < n)$ variables are associated with the rows of the table whereas the combinations of the other $n - m$ variables are associated with the columns. In such a table, the intersection of a row and a column defines a combination of all the input variables of the function and is associated with the value the function takes for this combination.

On both rows and columns the combinations are arranged according to the Gray code. This constitutes the specificity of Karnaugh maps.

EXAMPLE 2.5:

$f(x_4, x_3, x_2, x_1)$ (Figure 2.3)

The size of a Karnaugh map is theoretically not limited. However, representing a function with more than four to five input variables in only one table must be avoided. Indeed, the use of the table becomes difficult with more than four variables. It is better to represent a function of n variables $(n > 4)$ by means of 2^{n-4} tables, each one being associated with a combination of the variables $x_n, x_{n-1}, \ldots, x_5$. Like the truth table method, this technique is only suitable for functions having at most six input variables. For more complex functions, the map entered variable method, an extension of Karnaugh maps, is more efficient (see next paragraph).

From Tabular Representation to Algebraic Representation. Generally, to solve a combinational problem, the designer has to obtain an algebraic expression of the function that the control system must perform. We saw that a function has many algebraic representations and the main problem is to choose one of them.

Shannon's theorem says that any function of n input variables may be written as the sum of the products of each combination of the input variable and of the value of the function for this combination.

$$\begin{aligned} f(x_n, x_{n-1}, \ldots, x_1) = {} & \overline{x_n}\cdot\overline{x_{n-1}}\cdots\overline{x_1}\cdot f(0, 0, \ldots, 0) \\ & + \overline{x_n}\cdot\overline{x_{n-1}}\cdots x_1\cdot f(0, 0, \ldots, 1) \\ & + \cdots \\ & + x_n\cdot x_{n-1}\cdots x_1\cdot f(1, 1, \ldots, 1) \end{aligned}$$

Applying Shannon's theorem to a truth table or to a Karnaugh map gives us the first canonical form of the logic function. In this form, for a combination C_k of the input variables, the function f can take:

- The value 1 and $C_k \cdot 1 = C_k$; C_k belongs to the algebraic expression of the function.
- The value 0 and $C_k \cdot 0 = 0$; C_k does not belong to the algebraic representation.

xn	xn-1	...	x2	x1	f(xn,xn-1,...,x2,x1)
0	0	...	0	0	f(0,0,...,0,0)
0	0	...	0	1	f(0,0,...,0,1)
1	1	...	1	0	f(1,1,...,1,0)
1	1	...	1	1	f(1,1,...,1,1)

Figure 2.1 A truth table.

x2	x1	OR(x2 , x1) = x1 + x2
0	0	OR (0 , 0) = 0
0	1	OR (0 , 1) = 1
1	0	OR (1 , 0) = 1
1	1	OR (1 , 1) = 1

Figure 2.2 Truth table of OR function.

		Combinations of (x2,x1)			
	x4 x3 \ x2 x1	00	01	11	10
Combinations of (x4,x3)	00	f(0,0,0,0)	f(0,0,0,1)	f(0,0,1,1)	f(0,0,1,0)
	01	f(0,1,0,0)	f(0,1,0,1)	f(0,1,1,1)	f(0,1,1,0)
	11	f(1,1,0,0)	f(1,1,0,1)	f(1,1,1,1)	f(1,1,1,0)
	10	f(1,0,0,0)	f(1,0,0,1)	f(1,0,1,1)	f(1,0,1,0)

Figure 2.3 A Karnaugh map with four variables.

x_2	x_1	$OR(x_2, x_1) = x_1 + x_2$
0	0	OR (0 , 0) = 0
0	1	OR (0 , 1) = 1
1	0	OR (1 , 0) = 1
1	1	OR (1 , 1) = 1

Figure 2.4 Truth table of the example.

EXAMPLE 2.6:

OR (x_2, x_1) (Figure 2.4)

$$\begin{aligned} \text{OR}(x_2, x_1) &= \overline{x_2}\cdot\overline{x_1}\cdot\text{OR}(0,0) + \overline{x_2}\cdot x_1\cdot\text{OR}(0,1) \\ &\quad + x_2\cdot\overline{x_1}\cdot\text{OR}(1,0) + x_2\cdot x_1\cdot\text{OR}(1,1) \\ &= \overline{x_2}\cdot x_1 + x_2\cdot\overline{x_1} + x_2\cdot x_1 \end{aligned}$$

This expression is not optimal according to the number of operators and to the occurrence number of input variables. A lot of methods have been developed to optimize logic expressions. Their details can be found in Nussbaumer. Here we will focus on the Karnaugh map method only.

Function Simplification by the Karnaugh Map Method. This method is based on the complementarity property of the Boolean algebra $x + \bar{x} = 1$.

Principle:

Let C_0 and C_1 be two combinations of n variables such that: $C_0 = x_j \cdot C_{01}$ and $C_1 = \bar{x}_j \cdot C_{01}$ in which C_{01} is a combination of $\{x_n, x_{n-1}, \ldots, x_{j+1}, x_{j-1}, \ldots, x_1\}$ (C_0 and C_1 are called adjacent combinations).

Their logical sum $C_0 + C_1 = C_{01} \cdot (\bar{x}_j + x_j) = C_{01} \cdot 1 = C_{01}$ does not contain the variable x_j.

As a consequence, every time a logic function is true for two adjacent combinations of its input variables, an algebraic expression simpler than the first canonical one can be obtained.

As the combinations of the input variables of a Karnaugh map are arranged according to the Gray code at the rows and columns inputs, two combinations are found in two adjacent squares (squares sharing a side).

This property of Karnaugh maps allows a graphical application of the complementarity property of the Boolean algebra. By grouping together two 1s of the function in two adjacent squares, a variable can be removed between two combinations, so that the algebraic expression of the function is simplified.

EXAMPLE 2.7:

OR (x_2, x_1) (Figure 2.5)
Finally, we obtain

$$\text{OR}(x_2, x_1) = x_2 + x_1$$

For a function with more than three input variables, we can group four combinations together to remove two input variables.

These four combinations C_4, C_3, C_2, C_1 must have the following property:

$$C_1 = C_0\cdot\overline{x_i}\cdot\overline{x_j}$$
$$C_2 = C_0\cdot\overline{x_i}\cdot x_j$$
$$C_3 = C_0\cdot x_i\cdot\overline{x_j}$$
$$C_4 = C_0\cdot x_i\cdot x_j$$

Indeed,

$$\begin{aligned} C_1 + C_2 + C_3 + C_4 &= C_0\cdot\overline{x_i}(\overline{x_j} + x_j) + C_0\cdot x_i(\overline{x_j} + x_j) \\ &= C_0\cdot\overline{x_i} + C_0\cdot x_i = C_0 \end{aligned}$$

In a Karnaugh map, this corresponds to four 1s located:

- In four squares constituting a row or a column.
- In four squares sharing the same vertex.

Because the first row (resp. column) is adjacent to the last one, some groupings are not obvious in Karnaugh maps (Figure 2.6).

If two adjacent columns (resp. rows) contain eight 1s, the same reasoning leads to replacement of eight combinations of four variables by only one variable (the only variable whose value is the same in the eight combinations).

For functions with more than four variables, we saw that more than one Karnaugh map may be used to simplify the algebraic expression of the function.

Some simplifications can be done when the same group of 1s appears in two tables associated with adjacent combinations of the variables which are not inputs of the tables. For example, let f be the function of $(x_6, x_5, x_4, x_3, x_2, x_1)$, defined in Figure 2.7.

Applying the rules of Karnaugh maps, we obtain:

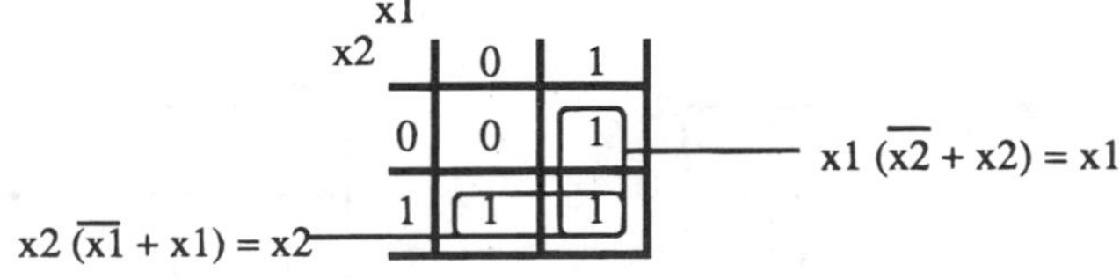

Figure 2.5 Karnaugh map of the example.

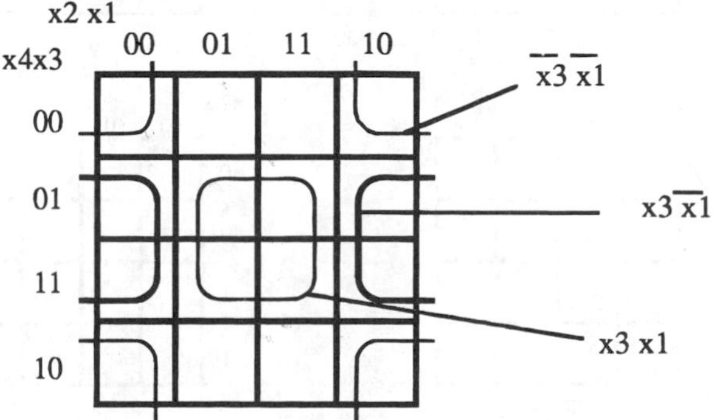

Figure 2.6 Example of grouping in a Karnaugh map.

$$F(x_6, x_5, x_4, x_3, x_2, x_1) = \overline{x_6} \cdot x_3 \cdot x_1 + \overline{x_6} \cdot \overline{x_5} \cdot x_4 \cdot \overline{x_3} \cdot x_2 \cdot \overline{x_1} + x_6 \cdot x_5 \cdot x_4 \cdot \overline{x_3} \cdot x_2 \cdot \overline{x_1}$$

in which the two last combinations cannot be grouped together because they are not adjacent even if they are at the same location in two of the four tables.

Note that, sometimes, it is easier to group the 0s of the table, in order to obtain an algebraic expression of the complementarity function. The complement of this expression is obviously an algebraic expression of the function.

Incompletely Specified Functions. In most cases, the problem concerns a function in which all the input variable combinations are not to be taken into account, for example, when the input variables correspond to logic sensors that cannot be simultaneously on. Such combinations are called unspecified combinations or "don't care" terms. For these combinations the function can take the value "0" or "1" indifferently. In tabular representations, the value of the function is noted "*" or "-". In Karnaugh maps, "*" can be considered like 1s if they allow simplifying the algebraic expression (they are grouped with 1s) and are considered like 0s otherwise.

EXAMPLE 2.8: (Figure 2.8)

Giving the value "1" to the unspecified combination (0,1,0,1) and the value "0" to the other ones simplifies the algebraic expression of the function:

$$f = x_3 \cdot \overline{x_2}$$

In this part, we described the basic tools needed to solve combinational problems. Let us now give an example to illustrate this.

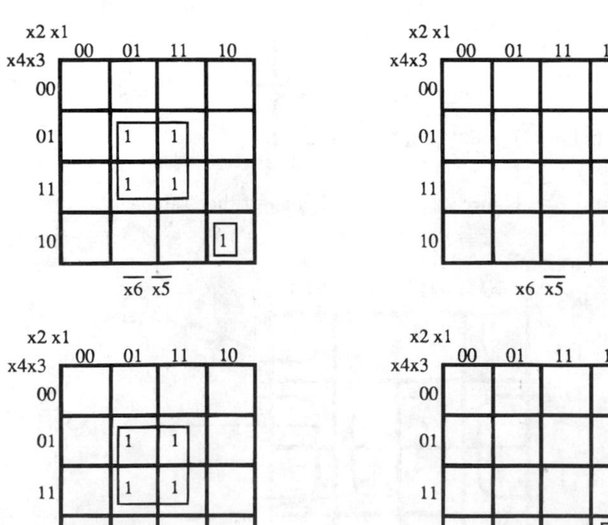

Figure 2.7 Grouping in a Karnaugh map—a few configurations.

EXAMPLE 2.9:

We have to design an encoder with four input lines (e_3 to e_0) and three output lines c_1, c_0, and K (keypressed).

The inputs are connected to keys of a keyboard and the outputs must give the following information.

K indicates that at least a key is pressed on the keyboard; the combinations (c_1, c_0) must take the binary value of the index of the key that is pressed.

If keys are simultaneously pressed, the encoder gives the value of the highest index of the keys pressed.

To solve this problem, three Karnaugh maps are necessary (see Figure 2.9).

We obtain:

$$\overline{K} = \overline{e_3} \cdot \overline{e_2} \cdot \overline{e_1} \cdot \overline{e_0}$$

$$K = e_3 + e_2 + e_1 + e_0$$

$$c_1 = e_3 + e_2$$

$$c_0 = e_3 + \overline{e_2} \cdot e_1$$

The combination (0,0,0,0) is unspecified for functions c_1 and c_0 because the signal K indicates that no key is pressed on the keyboard. Because of the value we give to c_1 and c_0 in the Karnaugh map, we can foresee that when no key is pressed on the keyboard, the encoder will give the values (0,0) to the outputs (c_1, c_0).

Complex Systems—Functions of Functions

We have seen that the size of tabular representations of logic functions exponentially grows with the number of input variables. For instance, if we want to design a similar encoder with eight inputs (Keys e_7 to e_1), we have to foresee $2^8 = 256$ combinations of the keys. Another way is to consider that this

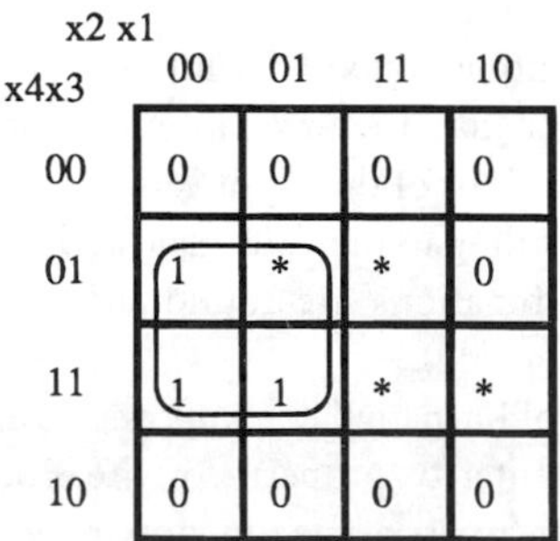

Figure 2.8 Grouping including unspecified combinations.

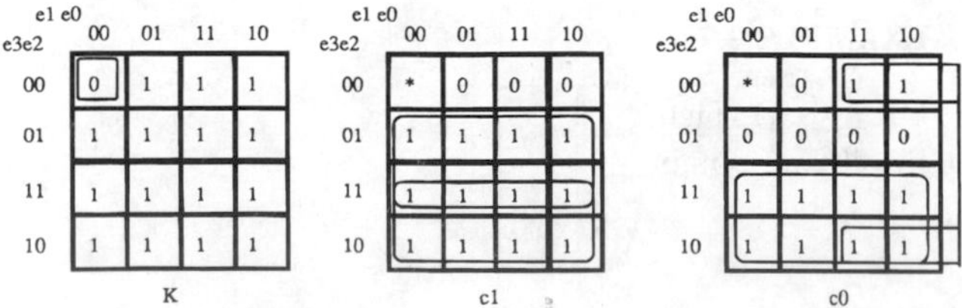

Figure 2.9 Karnaugh maps of an encoder.

encoder can be built by means of two four-input encoders as shown in Figure 2.10.

In this way, the system to be designed (? box) has six inputs only. So, the size of the problem has been divided by four.

Moreover, a quick analysis of the "?" box shows that:

K takes the value "1" when $K_1 = 1$ or $K_2 = 1$.

(C_2, C_1, C_0) takes the value $(0, C_{1a}, C_{0a})$ when simultaneously $K_1 = 1$ and $K_2 = 0$.

(C_2, C_1, C_0) takes the value $(1, C_{1b}, C_{0b})$ when $K_2 = 1$.

We can represent these implications in a table which is an extension of truth tables (Figure 2.11). And using Shannon's theorem, we obtain the logic expressions:

$$K = K_2 + K_1$$

$$C_2 = K_2$$

$$C_1 = \overline{K_2} \cdot C_{1a} + K_2 \cdot C_{1b} \text{ (after simplifications)}$$

$$C_0 = \overline{K_2} \cdot C_{0a} + K_2 \cdot C_{0b}$$

Finally, replacing the functions K_1, K_2, C_{1a}, C_{0a}, C_{1b}, and C_{0b} by their logic expressions we obtain the logic expressions of the four functions of the encoder:

$$K = e_7 + e_6 + e_5 + e_4 + e_3 + e_2 + e_1$$

$$C_2 = e_7 + e_6 + e_5 + e_4$$

$$C_1 = (e_7 + e_6 + e_5 + e_4) \cdot (e_7 + e_6) + \overline{(e_7 + e_6 + e_5 + {}_4)}(e_3 + e_2)$$

$$C_0 = (e_7 + e_6 + e_5 + e_4) \cdot (e_7 + e_6 \cdot e_5) + \overline{(e_7 + e_6 + e_5 + e_4)}(e_3 + \overline{e_2} \cdot e_1)$$

This method can be used each time the system to be designed can be structured in subsystems with a small number of inputs. The resulting expressions are not optimal in regard to the number of operators and to the number of occurrences of the input variables. Some algorithmic methods of optimization, like the Quine-McCluskey method, can be applied to simplify the expressions. These methods are not described here, but can be found in (Givone, 1970).

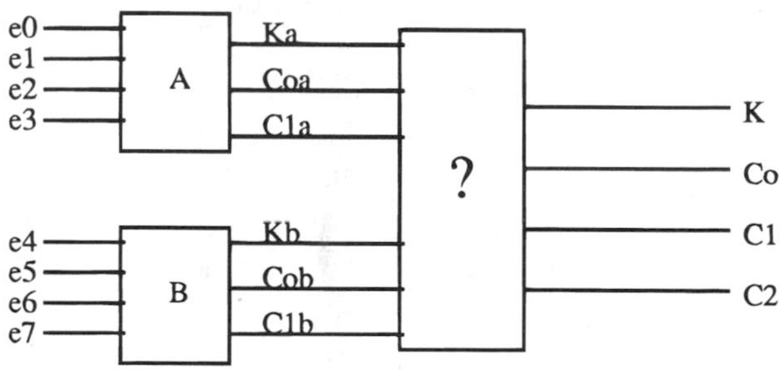

Figure 2.10 Eight inputs encoder.

K_2	K_1	K	C_2	C_1	C_0
0	0	0	*	*	*
0	1	1	0	C_{1a}	C_{0a}
1	0	1	1	C_{1b}	C_{0b}
1	1	1	1	C_{1b}	C_{0b}

Figure 2.11 Intermediary table of the eight inputs encoder.

In the previous example, we solved the problem by using truth tables. A similar method exists with Karnaugh maps and allows finding already simplified expressions of functions of functions.

Let f be the function of $(x_3, x_2, x_1, x_0, a, b)$ represented in Figure 2.12.

To obtain an expression of the function f we must:

1. Replace all the expressions in the table by "0", then group together the "1" (using eventually the "∗") just like in a classic Karnaugh map to obtain fl.
2. For each expression Ei replace:
   2.1 All the occurrences of "1" in the table by "∗"
   2.2 All the occurrences of the expression Ei by "1",
   2.3 All the other expressions by "0",
   2.4 Then, in the resulting table, group together the "1" (using eventually the "∗") to obtain Ri.
3. Finally, write $f = fl + \sum_i Ei \cdot Ri$.

Application to the example (see Figure 2.13):

Table 1 in Figure 2.13 corresponds to step 1. We get

$$fl = \overline{x_4} \cdot x_3 + \overline{x_4} \cdot x_1$$

Table 2 in Figure 2.13 corresponds to step 2 with $a = 1$ and $b = 0$. We get

$$fa = x_3 \cdot \overline{x_2}$$

Table 3 in Figure 2.13 corresponds to step 2 with $a = 0$ and $b = 1$. We get

$$fb = x_2 \cdot x_1 + \overline{x_3} \cdot x_1$$

And finally

$$f(x_4, x_3, x_2, x_1, a, b) = \overline{x_4} \cdot x_3 + \overline{x_4} \cdot x_1 + a \cdot x_3 \cdot \overline{x_2} + b \cdot (x_2 \cdot x_1 + \overline{x_3} \cdot x_1)$$

x4x3 \ x2 x1	00	01	11	10
00	*	*	1	0
01	1	*	1	1
11	*	a	b	0
10	0	b	*	0

Figure 2.12 Example of map entered variables.

These techniques can be used whatever the problem. We will see in the following section that it is widely used to determine logical expressions of finite state machines specified by means of a reduced state diagram in which only the events called significant events appear.

Two Universal Functions: NAND and NOR

Definitions. NOR and NAND are two logic functions of n variables $(x_m, \ldots, x_1)$ defined by:

$$\text{NOR}\ (x_m, x_{n-1}, \cdots, x_1) = 1 \Leftrightarrow x_i = 0\ \forall\ i \in [1, n]$$

$$\text{NAND}\ (x_m, x_{n-1}, \cdots, x_1) = 1 \Leftrightarrow x_i = 1\ \forall\ i \in [1, n]$$

Remarks.

- Limited to two input variables:

  NAND (x_2,x_1) is the complementary function of AND(x_2,x_1).
  NOR (x_2,x_1) is the complementary function of OR (x_2,x_1).

- Both NOR and NAND functions are nonassociative.
- NAND $(1,1, \ldots, x_2,x_1)$ = NAND (x_2,x_1).
- NOR $(0,0, \ldots, x_2,x_1)$ = NOR (x_2,x_1).
- NAND (x_2,x_1) is a binary operator denoted | : NAND (x_2,x_1) $= x_2 \mid x_1 = \overline{x_2 \cdot x_1}$.
- NOR (x_2,x_1) is a binary operator denoted $\downarrow$: NOR (x_2,x_1) $= x_2 \downarrow x_1 = \overline{x_2 \cdot x_1}$.

The NAND (resp. NOR) operator is universal (complete): any logic function has one algebraic representation in which only the operator NAND (resp. NOR) appears. In fact, the three basic Boolean operators NOT, AND, and OR can be written with NAND and NOR.

$$\overline{x_1} = \overline{x_1 \cdot x_1} = x_1 \mid x_1$$

$$x_2 + x_1 = \overline{\overline{x_2 + x_1}} = \overline{\overline{x_2} \cdot \overline{x_1}} = \overline{\overline{x_2 \cdot x_2} \cdot \overline{x_1 \cdot x_1}} = (x_2 \mid x_2) \mid (x_1 \mid x_1)$$

$$x_2 \cdot x_1 = \overline{\overline{x_2 \cdot x_1}} = \overline{\overline{x_2 \cdot x_1} \cdot \overline{x_2 \cdot x_1}} = (x_2 \mid x_1) \mid (x_2 \mid x_1)$$

The same demonstration can be done for the NOR operator.

The translation of a general logic expression with $^-$, $\cdot$ and + operators in a "NAND" expression can be done using De Morgan's theorem.

EXAMPLE 2.10:

$$E_0 = x_3 \cdot (x_2 + \overline{x_1}) + \overline{x_3} \cdot x_1$$
$$= \overline{\overline{x_3 \cdot (x_2 + \overline{x_1}) + \overline{x_3} \cdot x_1}} = \overline{\overline{x_3 \cdot (x_2 + \overline{x_1})} \cdot \overline{\overline{x_3} \cdot x_1}}$$
$$= (x_3 \mid (x_2 + \overline{x_1})) \mid (\overline{x_3} \cdot x_1)$$
$$= (x_3 \mid (\overline{\overline{x_2} \cdot x_1})) \mid (\overline{\overline{\overline{x_3} \cdot x_1}})$$
$$= (x_3 \mid ((x_2 \mid x_2) \mid x_1)) \mid ((x_3 \mid x_3) \mid x_1) \mid ((x_3 \mid x_3) \mid x_1)$$

Remarks. A $\Sigma\Pi$ form of a logic expression can be written in a NAND form directly by replacing OR and AND operators by the NAND one.

$$\Sigma\Pi = \Pi 1 + \Pi 2 + \cdots + \Pi K$$
$$= \overline{\overline{\Pi 1 + \Pi 2 + \cdots + \Pi k}} = \overline{\overline{\Pi 1} \cdot \overline{\Pi 2} \cdots \overline{\Pi k}}$$
$$= \text{NAND}(\overline{\Pi 1}, \overline{\Pi 2}, \ldots, \overline{\Pi k})$$

And obviously each term $\overline{\Pi i}$ is a NAND form on the input variables.

EXAMPLE 2.11:

$$E(x_3, x_2, x_1) = x_3 \cdot x_2 \cdot x_1 + x_2 \cdot \overline{x_1} + x_3 \cdot \overline{x_2}$$
$$= \text{NAND}(\text{NAND}(x_3, x_2, x_1),$$
$$\text{NAND}(x_2, \overline{x_1}), \text{NAND}(x_3, \overline{x_2}))$$

These two functions are of main importance because they are easily implemented by electronic components. NAND (7400) and NOR (7402) gates are the cheapest ones of the TTL family and they can be used to implement any logic function.

2.2 Sequence Control

State Machines and Algorithmic State Machines

State Machine: Definitions

Industrial processes are often characterized by repetitive and sequential behaviors. Specifying the control of such systems

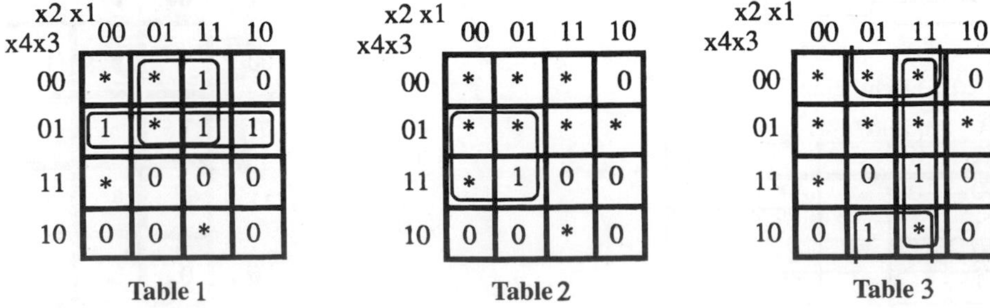

x4x3 \ x2 x1	00	01	11	10
00	*	*	1	0
01	1	*	1	1
11	*	0	0	0
10	0	0	*	0

Table 1

x4x3 \ x2 x1	00	01	11	10
00	*	*	*	0
01	*	*	*	*
11	*	1	0	0
10	0	0	*	0

Table 2

x4x3 \ x2 x1	00	01	11	10
00	*	*	*	0
01	*	*	*	*
11	*	0	1	0
10	0	1	*	0

Table 3

Figure 2.13 Steps of logic expressions extraction.

necessitates having adapted tools. To understand this, let us consider a very simple example in which the only problem is to identify particular sequences of events. The circuit to be designed, called sequence discriminator, has two pulsed type inputs x_1 and x_2 and one output z. Sequences recognized by this sequence discriminator SD are shown in Figure 2.14. A pulse on output z coincides with the second pulse on input x_2 after at least one pulse on x_1.

Consider the behavior of SD at two successive times t_1 and t_2: with the same input configuration (also called input vector or input combination) the output z is different. The consequence is obvious: SD cannot be a combinational circuit; its output z depends not only on the input vector applied at a given moment but also of the input vectors applied before, in the past.

Then, a very important concept appears: the state concept giving memory capability to the control system in order to manage sequences of events.

In this example, it is easy to understand that SD can be modeled with three states corresponding to the three instants t_0, t_1, t_2 represented in Figure 2.14. A way to illustrate the relations between these three states is to use a diagram as shown in Figure 2.15.

This diagram is called a state diagram and all systems that can be represented in this way are called state machines. It necessitates giving a more formal definition of state machines.

Let us consider a system with

a set of inputs
$e = (e_1, \ldots, e_n)$ $e_i \in \{0,1\}$ $\forall_i = 1, \ldots, n$ and
a set of outputs
$s = (s_1, \ldots, s_p)$ $s_i = \{0,1\}$ $\forall_i = 1, \ldots, p$.

Combinations over e and s are respectively called input and output vectors or input and output states written $I = \{i_1, \ldots, i_l\}$ and $Z = \{z_1, \ldots, z_q\}$. Each element of I, e.g., i_p, is one of the 2^n combinations of e (idem for Z).

Definition. A state machine is a five-tuple $\langle I,Q,Z,\delta,\omega\rangle$ where

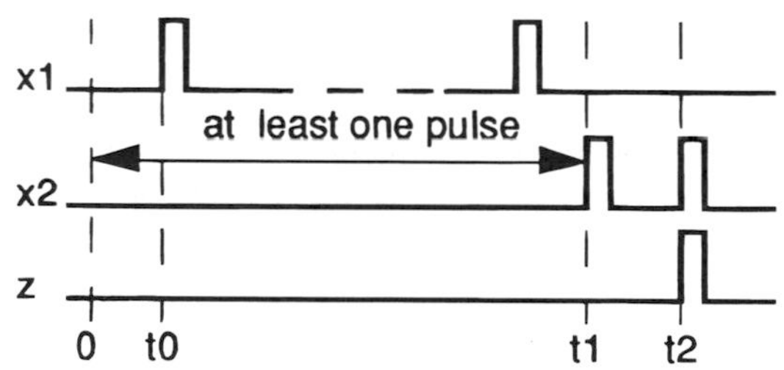

Figure 2.14 Sequence discriminator (SD).

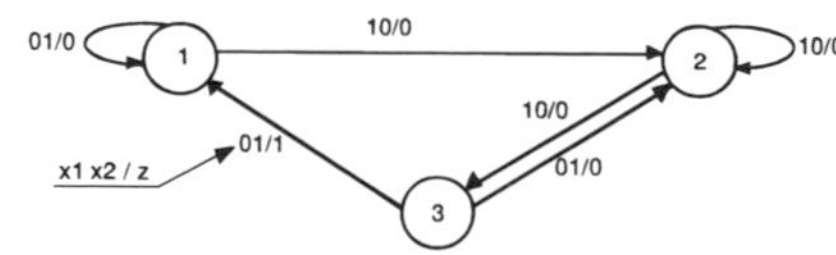

Figure 2.15 State diagram of the SD.

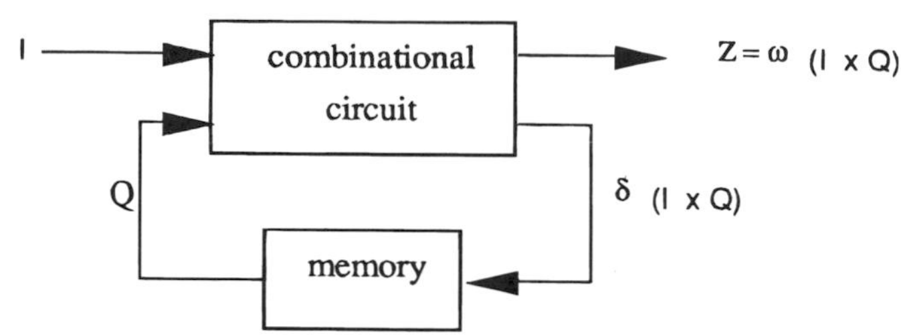

Figure 2.16 The state machine model.

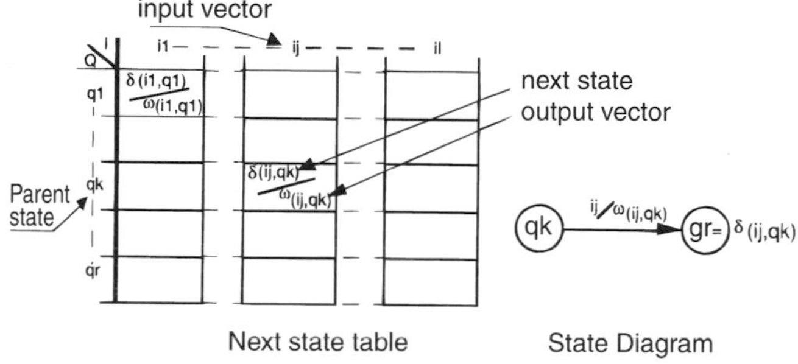

Figure 2.17 Representation tools of a state machine.

$I = \{i_1, \ldots, i_l\}$ is the set of input vectors the machine can receive.

$Z = \{z_1, \ldots z_q\}$ is the set of output vectors the machine generates.

$Q = \{q_1, \ldots, q_r\}$ is the set of states of the machine.

If Q is a finite set, the machine is a finite state machine.

δ is the next state function. In fact, it is an application defined by δ: $I \times Q \overset{\delta}{\Rightarrow} Q$ which models the behavior of the machine and its evolution from state to state according to the input vectors applied.

ω is the output function. Being a subjective application, ω: $I \times Q \overset{\omega}{\Rightarrow} Z$, it shows the external evolution of the machine (output changes) resulting from state changes.

According to this definition, it is now possible to see that a state machine is a closed loop system inside the black box model having inputs I and outputs Z (see Figure 2.16). This loop gives to a state machine its capacity to keep track of the past.

The input of the memory box is the next state of the machine and its output is the present state of the machine. It is important to notice that any change in the state and/or the output vector of a state machine results from a change of the input vector.

Representing a state machine makes use of one of two models: the state table or the state diagram. They give exactly the same information—the first in a tabular form the second in a graphical form.

The state table is a two-entry table: rows represent states and columns the input vectors. Each box of the table contains the next state and the output vector.

The state diagram is a graph in which a state is represented by a circle. State changes are shown as arcs labeled by a couple *input/output*.

Figure 2.17 depicts the two representation tools according to the definition.

Figure 2.18 represents the state table of machine SD also described by the state diagram of Figure 2.15.

Remarks. Two remarks come from the interpretation of Figure 2.18. The first is that x_1 and x_2 are pulsed inputs and cannot occur at the same time; then in column $x_1\ x_2 = (1,1)$, the next states and outputs are not specified. This is given by a dash. The second is that in column $x_1\ x_2 = (0,0)$ the next states are always identical to the corresponding present state and the output vector is always nil. This comes from the nature of the input signals which are pulses in machine SD. More generally, if all the pulsed inputs in an input vector are 0 the corresponding vector is called nil input vector and written as ℕ. The reason is that ℕ cannot make a machine changes its state as we shall see later. In machine SD ℕ = {00}.

Classification of State Machines

The evolution of a state machine depends on its inputs as well as its behavior. A classification generally admitted is based on the nature of the input signals: levels or pulses (Givone, 1970).

Fundamental Mode. A state machine operates in fundamental mode if, after an input change, no input can be changed until a new state is reached.

Two consequences result from this definition: the inputs of a state machine operating in fundamental mode are levels; the internal evolution of such a machine is not controlled, as shown in Figure 2.19 on the fragment of a state machine where the feedback loop is activated twice before the state reached is stable.

Such machines are also called *asynchronous state machines.* They correspond to the so-called Huffman model.

Q \ x1x2	00	01	11	10
1	1 / 0	1 / 0	-	2 / 0
2	2 / 0	3 / 0	-	2 / 0
3	3 / 0	1 / 1	-	2 / 0

Figure 2.18 State table of SD.

Pulsed Mode. A state machine behaves according to a pulsed mode if:

1. One of the inputs at least is a pulsed one.
2. The evolution of the machine results from a pulse signal at one of the pulsed inputs.
3. Resulting from an evolution, only one state change is allowed.

Figure 2.20 explains the state evolution of a pulsed state machine. The single state change results from the fact that the input stimulus is a pulse whose duration is shorter than the propagation time of the signals in the feedback loop so that, when the new present state is reached, the input vector is returned to ℕ. So the internal evolution is controlled by the pulsed inputs. These machines are also called *synchronous state machines.* Concerning the outputs, two classes of pulsed state machines can be defined:

- Mealy state machines: the outputs are pulsed signals which appear during the transitions between states. Machine SD (Figure 2.15 and Figure 2.18) is a Mealy state machine.
- Moore state machines: the outputs are level signals. They are defined on the states and depend only on the states, not on the inputs.

Figure 2.21 illustrates the behavior of a Moore state machine called SD2 having two inputs x (pulsed type) and p (level type) and one output z (level type). The output z changes each time an even number of pulses is applied to x and p is set. If p is reset to 0, pulses on x are ignored.

Two remarks are worth noting on this example:

- Columns ℕ do not bring any information on the evolution of the machine excepted on the output behavior. So columns ℕ are often omitted and replaced by an output column which describes the values of the outputs in the states.
- Very often pulsed state machines have one pulsed input only. In this case this input is called clock and the machines are clock synchronized state machines. The clock is applied to the memory box and becomes an implicit input which gives rhythm to the evolution of the feedback loop, as shown in Figure 2.22.

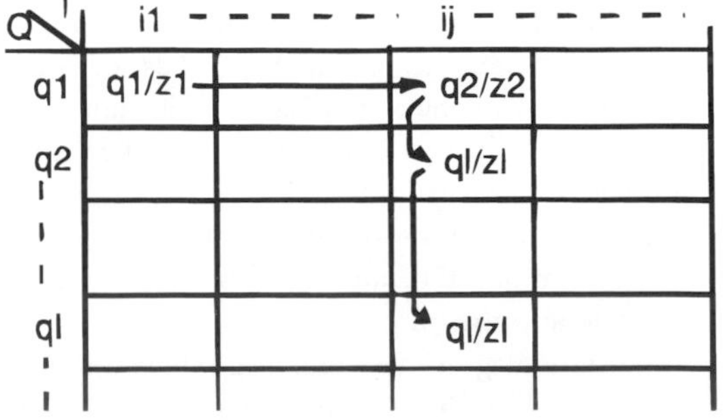

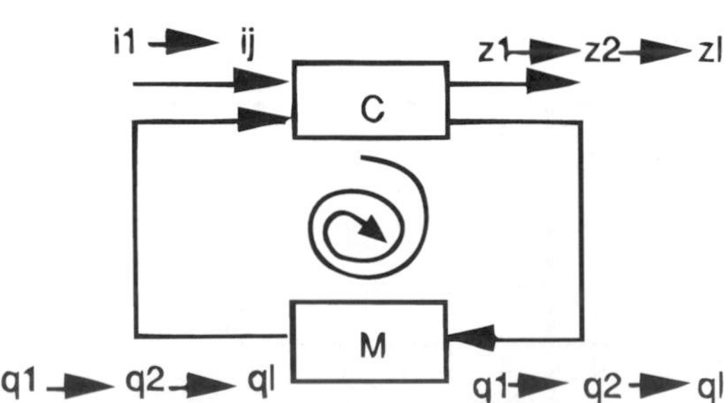

Figure 2.19 Fundamental mode evolution.

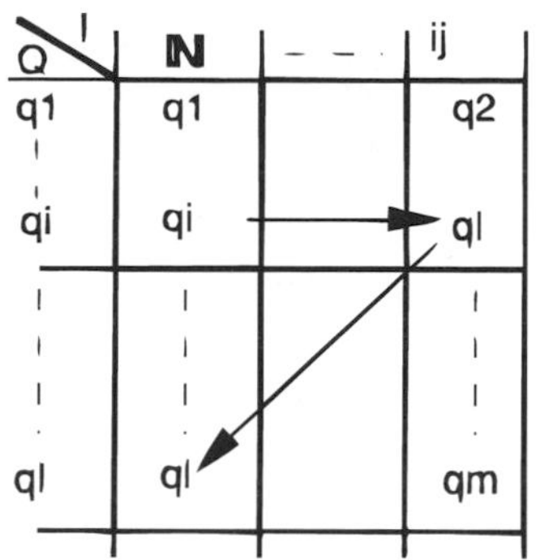

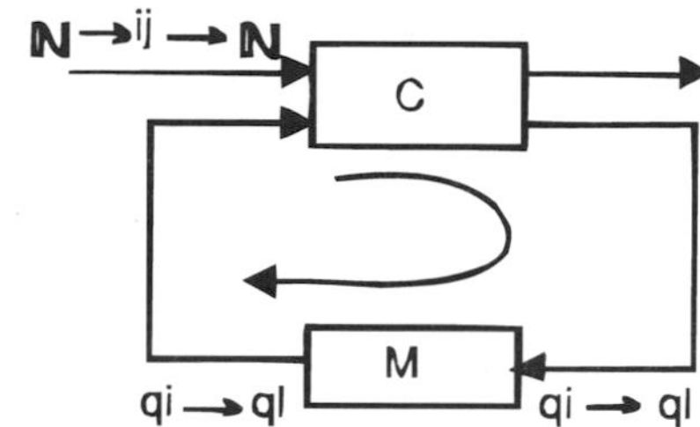

Figure 2.20 Pulsed mode evolution.

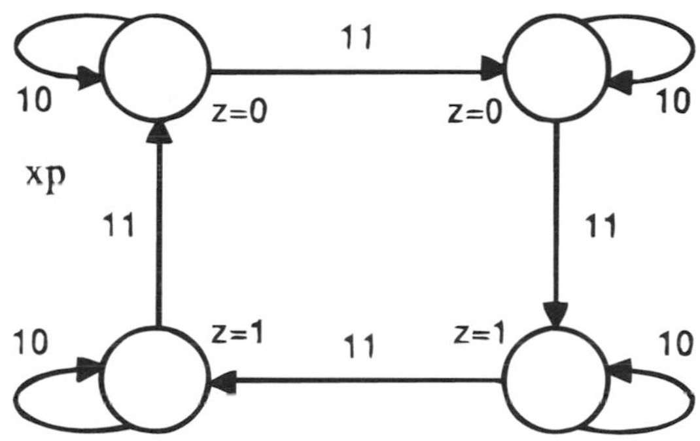

Q \ xp	N: 00	01	11	10
1	1/0	1/0	2	1
2	2/0	2/0	3	2
3	3/1	3/1	4	3
4	4/1	4/1	1	4

Figure 2.21 A Moore state machine SD2.

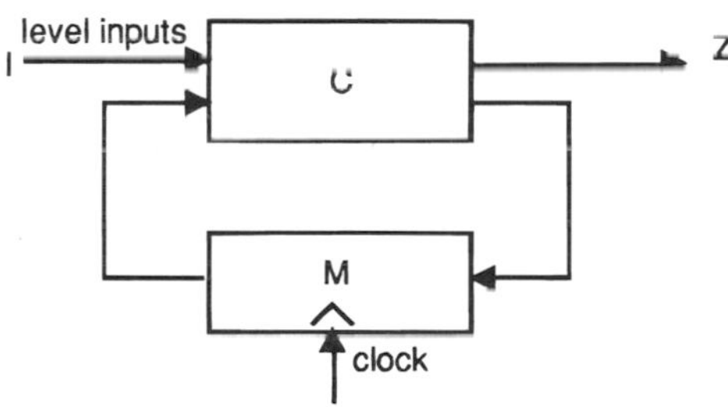

Figure 2.22 A clock synchronized state machine.

According to these two remarks, machine SD2 can be defined by the simplified stable state given in Figure 2.23.

A Special Case: The Primitive Flow Table

Consider a fundamental mode state machine such that each state is stable for one input combination only. It means

Q \ p	0	1	z
1	1	2	0
2	2	3	0
3	3	4	1
4	4	1	1

Figure 2.23 Machine SD2 simplified stable state.

that all the arcs reaching a state q_i have the same input vector as the labels in Figure 2.24a illustrate.

It is possible to have a simplified description by removing all the labels and inscribing the name of the input vector into the state as shown in Figure 2.24b. This model is called a primitive flow table. It has been defined by Huffman. If we consider the state table of a primitive flow table it is characterized by the presence of one stable state per row only. The following example will illustrate this model.

EXAMPLE 2.12: An electrical lock with combination

Consider a lock, shown in Figure 2.25, made up of a magnet bar moved inside a solenoid according to an excitation current.

To open the door a particular sequence (temporal combination) must be applied to push buttons A and B from a state where the door is closed. Initially, A and B are released and the solenoid is not excited ($S = 0$). The sequences can be

Push A, push B, release B, release A, then $S = 1$ and the door opens, or

Push B, push A, release A, release B, then $S = 1$ and the door opens.

To bolt the lock when the door is closed it is sufficient to push on A or B from a state where A and B are released and $S = 1$. Then $S = 0$ immediately.

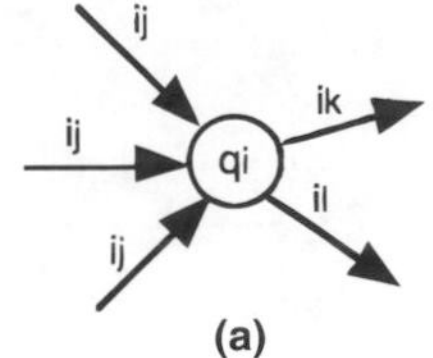

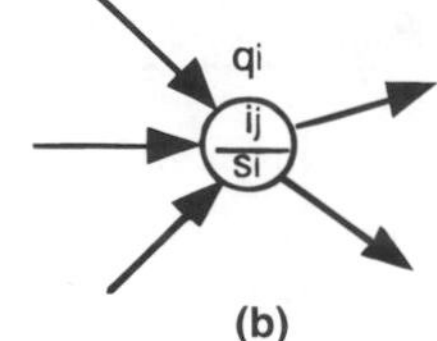

Figure 2.24 A simplified notation of a state in fundamental mode. (a) A special case. (b) Its simplified representation.

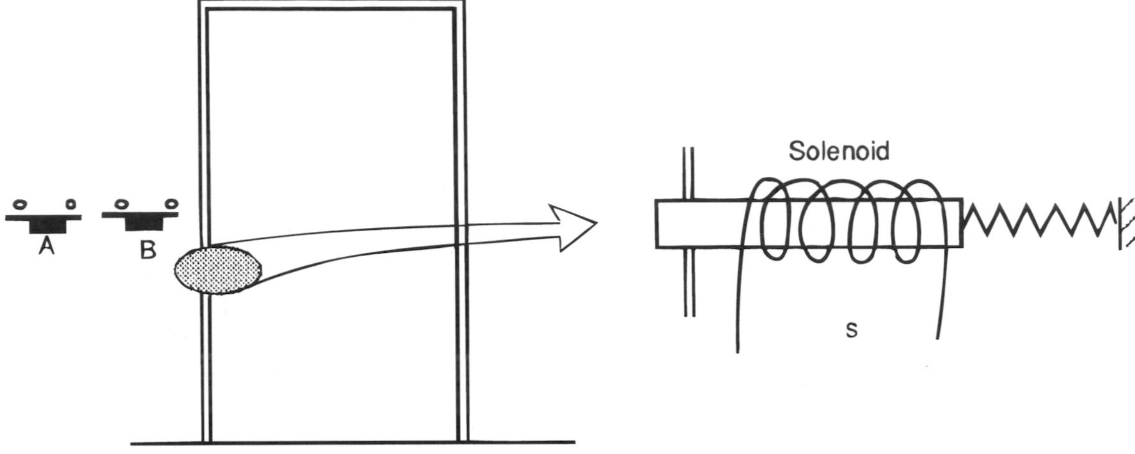

Figure 2.25 An electrical lock.

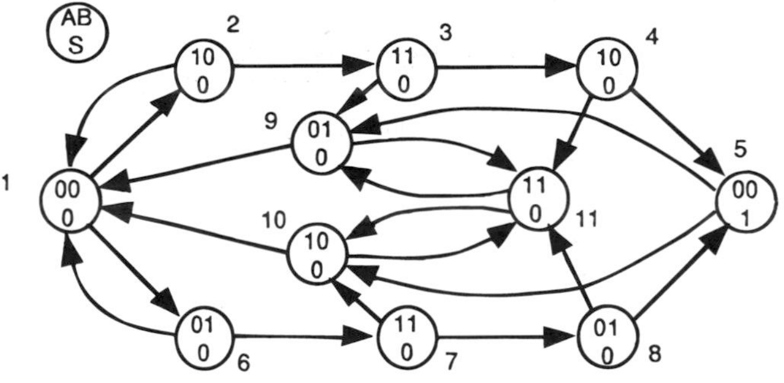

Figure 2.26 The primitive flow diagram of an electrical lock with time combination.

Figure 2.26 is a state machine of this system represented by a primitive flow diagram and a primitive flow table is shown in Figure 2.27.

States \ AB	00	01	11	10
1	(1)/0	6	-	2
2	1	-	3	(2)/0
3	-	9	(3)/0	4
4	5	-	11	(4)/0
5	(5)/1	9	-	10
6	1	(6)/0	7	-
7	-	8	(7)/0	10
8	5	(8)/0	11	-
9	1	(9)/0	11	-
10	1	-	11	(10)/0
11	-	9	(11)/0	10

Figure 2.27 A primitive flow table of an electrical lock.

Remarks.

1. Looking at Figure 2.27, it is easy to understand that one hypothesis has been made: only one input is changed at any time. This hypothesis is realistic for many reasons. First, the reaction of the operator is very slow compared to the system itself, so that a simultaneous push of buttons A and B will always be interpreted by the control system as sequential. A second reason, which is a consequence of the first one, is that a nonperfect simultaneous change induces races between inputs. Due to the fact that the feedback loop of a fundamental mode sequential machine is not controlled, these races can introduce hazards.

2. A primitive flow diagram is the most precise model one can use. As a matter of fact, any input change is taken into account even if it has no influence on the system evolution. This precision has a counterpart: using a primitive flow diagram is practicable with systems having a small number of inputs (up to five) only.

Reduced State Diagrams or Algorithmic State Machines: A Model to Manage Industrial Systems

In industry, discrete systems to be controlled generally have a lot of inputs (and outputs): dozens or hundreds are common. On the other hand, sequences are often simple (but long) and if we observe the evolution of such a system, it is sensible to some events only, not to all the input changes.

This consideration leads to introducing new concepts like receptivity which will allow filtering of input events. Let us define these concepts.

Event: is the occurrence of a logical function defined on the inputs of a control system. An input vector is a particular case in which all inputs are defined.

A significative event: is an event which leads to an evolution of the state machine modeling the control system: change of a state or an output.

Receptivity: is the capacity for a state machine to be sensitive to significative events only. If the receptivity concept is introduced in state machines, these machines will exhibit as minimal information those given by the input changes which affect the behavior of the system. Then, reduced state diagrams are obtained (but not inevitably minimal ones).

Let us consider an example: the control of a drill. This system is designed to drill two kinds of parts: thin parts and thick parts. The variable p allows the operator to program the kind of part ($p = 0$ thin, $p = 1$ thick) he has introduced. The drilling cycle is then started by the operator when he pushes button S, as shown in Figure 2.28 depending on the value of p. Switches A, B, C, and D allow calibration of the drill movement.

The right part of Figure 2.28 shows a reduced state diagram of the automatic drill.

Labels of the arcs make use of the receptivity concept: they are focused on the information which is necessary and sufficient to model the evolution phases of this system. Note also that in state 1 the output u is activated as long as switch A is not set.

The same use of the receptivity concept with a different graph inspired by flow charts leads to the so-called algorithmic state machines (ASM) introduced by Clare in 1972.

In an ASM chart, a state is represented as a box, a condition as a decision box, and a conditional output is specified in a conditional output box.

Figure 2.29 represents an ASM chart of the automatic drilling

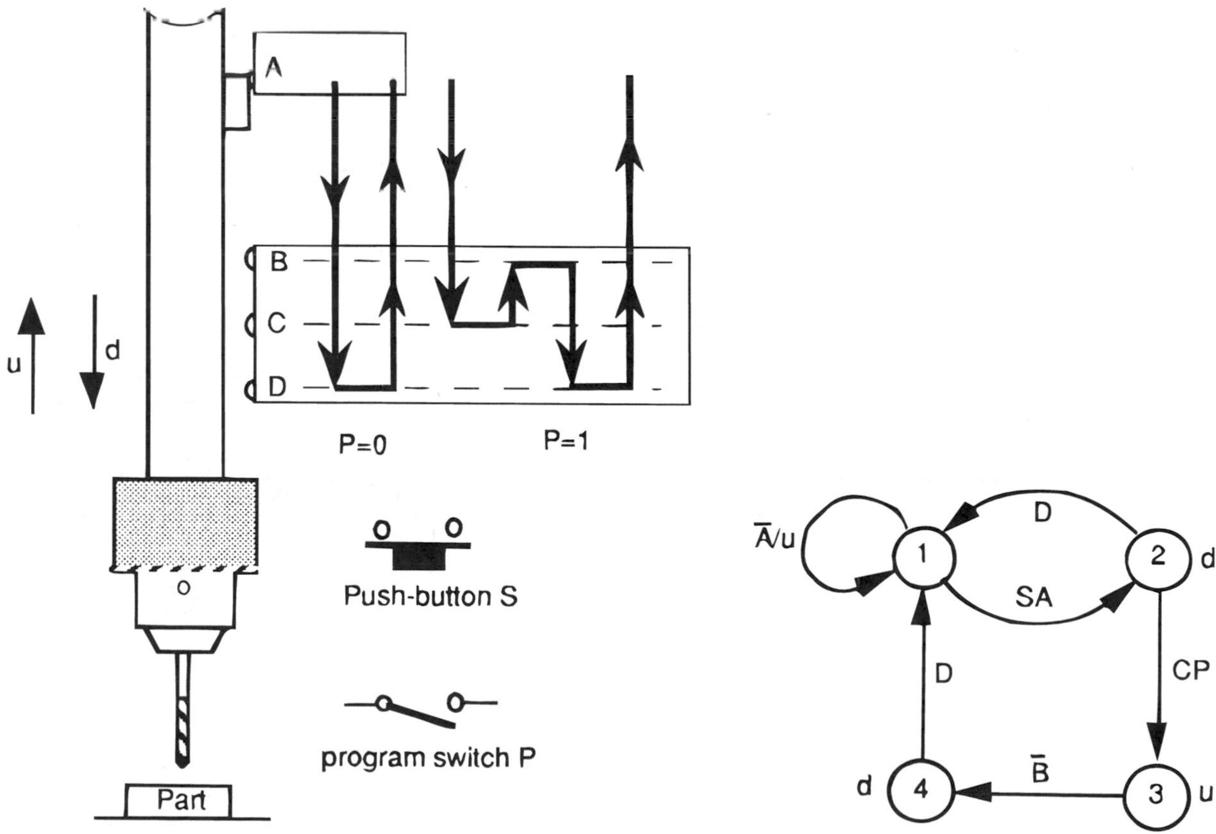

Figure 2.28 An automatic drill and its reduced state diagram.

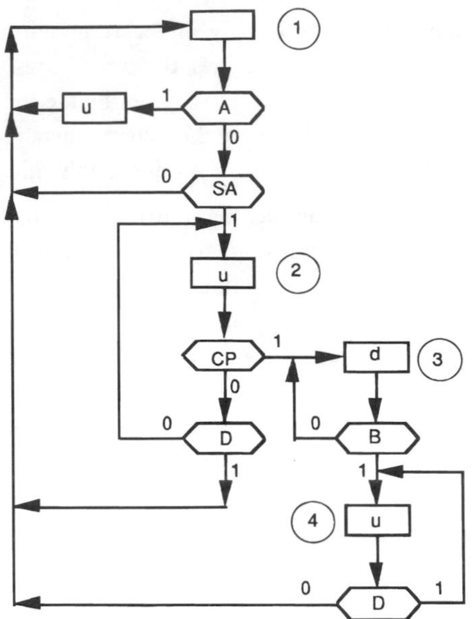

Figure 2.29 ASM chart of the automatic drilling machine.

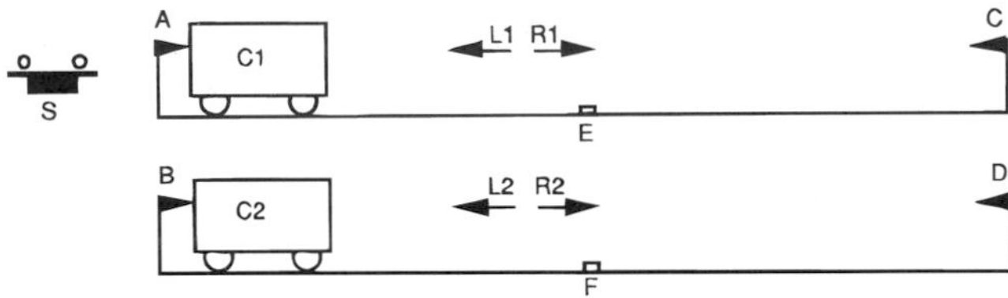

Figure 2.30 A two carts problem.

machine described above. In our opinion, this formalism is no improvement over the state diagram. It is graphically heavier and does not provide a better readability.

Petri Nets

In spite of the qualities offered by state machines such as rigorous formalism and ability to model complex systems by using receptivity, this model suffers from severe limitations. These limitations appear when a discrete system is composed of several subsystems which can operate simultaneously (or in parallel). Modeling the control of all the subsystems can lead to a combinational increase of states which is exponential. To master this kind of system a new model must be introduced. Let us first introduce the combinational state blowup by using an example. Petri nets and their interpretation will be presented next.

Modeling Parallelism: An Exponential State Blowup

Let us consider the following problem: two carts *C1* and *C2* can move on independent rails as Figure 2.30 shows.

Distances *AC* and *BD* are not known. Positions of sensors *E* and *F* are not known either. Speeds of carts *C1* and *C2* are not given.

As soon as the operator pushes button S, both carts move to the right. When cart *C1* (*C2*) arrives on sensor *E* (*F*) it stops. A timer t_1 (t_2) is triggered (*TT1*) (*TT2*), and after time *T1* (*T2*) cart *C1* (*C2*) resumes its running on the right. When cart *C1* reaches sensor *C*, it moves back (*L1*) to *A* without stopping on *E*. When cart *C2* reaches *D*, it moves back (*L2*) to *B* without stopping on *F*, if cart *C1* has reached *A*. Otherwise it awaits this event on *D* and resumes its left motion to *B*.

The state diagram in Figure 2.31 represents all the possible evolutions of carts *C1* and *C2*. The only cases not represented are simultaneous changes: for instance, events *E* and *F* in state 1 are not supposed to happen at the same time. Otherwise an arc should be added between state 1 and state 5.

Figure 2.31 illustrates what "exponential state increase" means: 20 states (initial state 0 not included) are necessary to model this simple problem. Imagine that a third cart *C3* were added with the same specification. It would need 100 states to model the problem. We can conclude that state machines are no longer adapted to model systems in which parallelism exists.

Modeling Parallelism, a Solution: State Machine Decomposition

A (partial) solution can be proposed to avoid such a state blowup. It consists of decomposing a system into subsystems, each of these subsystems being modeled by a state machine. The former example would be modeled by two state machines as shown in Figure 2.32.

Remarks.

1. This state diagram shows that the composition of state machines is a state machine in which the number of states is the product of the number of states of the composed machines (4 × 5 states in this example).
2. State machines modeling subsystems are often dependent on each other. They are synchronized. Synchronization is more or less strong according to the problem. In the two carts problem it is very weak. As a matter

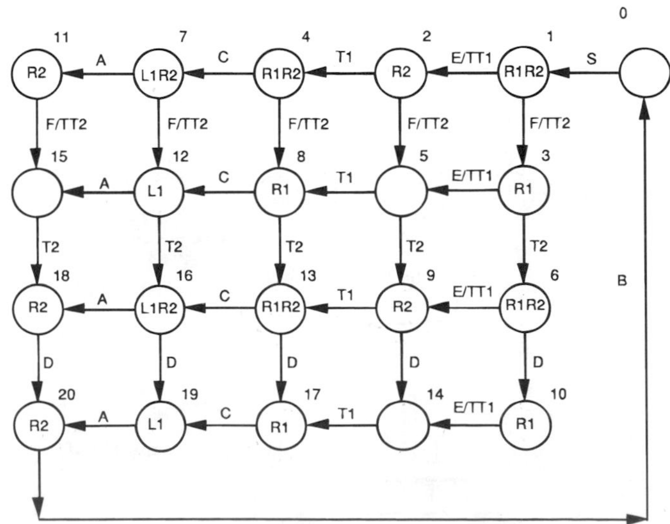

Figure 2.31 State diagram of the two carts problem.

of fact, the only necessary synchronization is the presence of sensor *B* in the event $S \cdot B$ outgoing state *i*1. It prevents cart *C1* from moving right before cart *C2* is back on *B*, even if the operator pushes button *S*. Generally synchronization constraints between machines are strong and, if not cautiously specified, can introduce deadlocks again. The reason is that state machines are not adapted to model synchronization problems.

Definition of Petri Nets

Let us give a formal definition of Petri nets.

A marked Petri net is a five-tuple $P = \langle P,T,I,O,M_o \rangle$

$P = \{p_1, \ldots, p_m\}$ is a finite set of elements called places.
$T = \{t_1, \ldots, t_n\}$ is a finite set of elements called transitions.
I: $P \times T \rightarrow N$ is the input place function, where N is the set of non-negative integers.
O: $P \times T \rightarrow N$ is the output place function.
M_o: $P \rightarrow N$ is the initial marking of the net.

Graphically, a place p_i is represented by a circle (as a state in a state machine) a transition t_j is represented by a bar; relations *I* and *O* are represented by arcs connecting, respectively, places to transitions and transitions to places.

Figure 2.33 shows a Petri net with three places $P = \{p_1, p_2, p_3\}$, four transitions $T = \{t_1, t_2, t_3, t_4\}$ and arcs whose weight are written as labels. Functions *I* and *O* can be represented also as matrices.

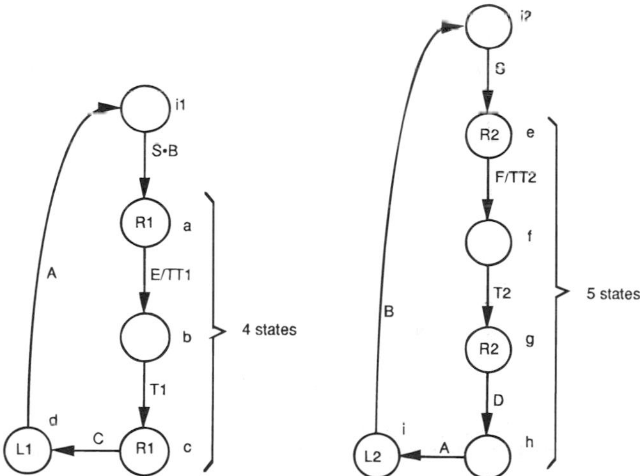

Figure 2.32 The two carts problem decomposed.

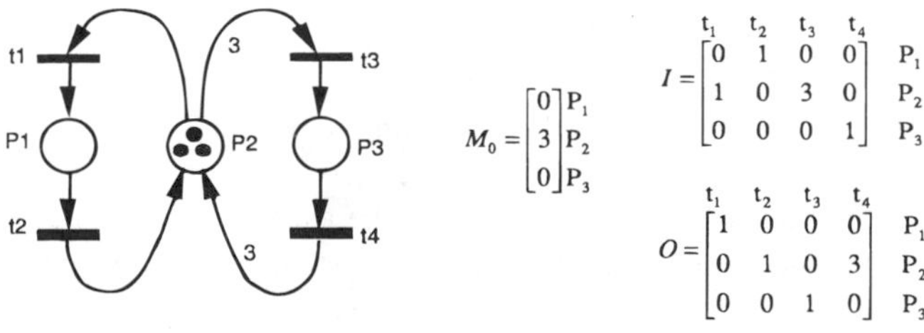

Figure 2.33 A marked Petri net.

A marked Petri net receives tokens in its place. The initial distribution of tokens constitutes the initial marking M_o. *I* and *O* functions define the structure of a Petri net whereas tokens and their positions at a given moment, called the marking of the net (its global state), represent its dynamic behavior.

Evolution rule: A transition is enabled when each of its input place has a number of tokens greater than or equal to the weight of the corresponding arc.

In Figure 2.34 t_1 is enabled, t_2 is not enabled. It would be enabled if p_5 had at least one token and p_6 had at least two tokens.

An enabled transition may fire. When a transition is fired, two indivisible operations are performed:

- each input place has a number of tokens removed from it, equal to the weight of the corresponding incoming arc and,
- each output place receives a number of tokens equal to the weight of the corresponding outgoing arc.

Firing of a transition changes the marking of a Petri net. Firing transition t_1 in Figure 2.34 leads to the marking given in Figure 2.35.

Each marking constitutes a state of the net and, generally, from each state more than one transition can be fired. So, in the Petri net of Figure 2.33, two transitions (t_1 and t_3) are enabled from the initial marking M_o.

But, note that if t_1 is fired, t_3 cannot be fired any longer and it is the same if t_3 is fired first. These two transitions are said to conflict. It is impossible to decide which one can fire first. So the situation is nondeterministic. It is always the case if a place is shared by two transitions or more. Solving this problem necessitates adding an interpretation of the transitions, as we shall see later.

With their structure and marking Petri nets are well suited to

Figure 2.34 An enabled and a not enabled transition.

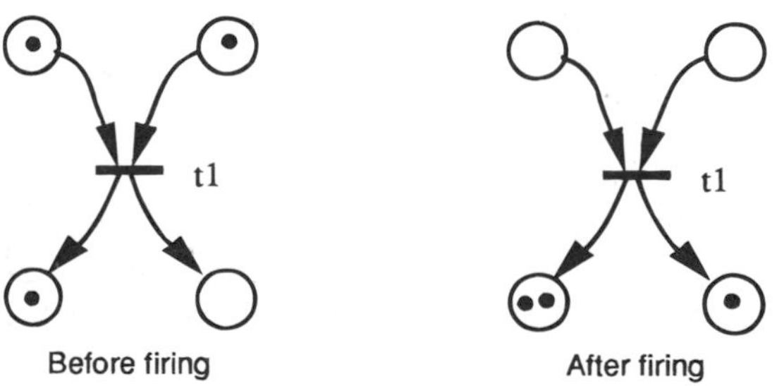

Figure 2.35 A transition before and after firing.

model situations like choices (conflicts), parallelism, and synchronization. Figure 2.36 illustrates these configurations.

Remark. A particular case: safe Petri nets.

A safe Petri net is such that all its arcs have a weight equal to one and its places can have at most one token. Safe Petri nets are well adapted to model discrete events control systems. Other properties not presented here allow avoidance of deadlocks (Peterson).

Interpreted Petri Nets

To model a control system, when a specification tool must offer a way to represent interactions of the control system with the process, it means inputs and outputs. Inputs are the events which rhythm the evolution of the control system, outputs are the actions sent to the process.

Taking this into account in a Petri net is easy:

- Events can be introduced on the transitions as conditions, and when these latter are true they will allow the transitions to fire if enabled.
- Actions can be introduced in two ways (similarly to Mealy and Moore state machines)
  - on the transitions if an action is a fugitive one (as a pulse)
  - on the places if an action must be maintained in a given state.
- Nets are supposed to be safe.

To illustrate these concepts, let us go back to the two carts problem described in Figure 2.30. Modeling this problem with an interpreted Petri net is represented in Figure 2.37.

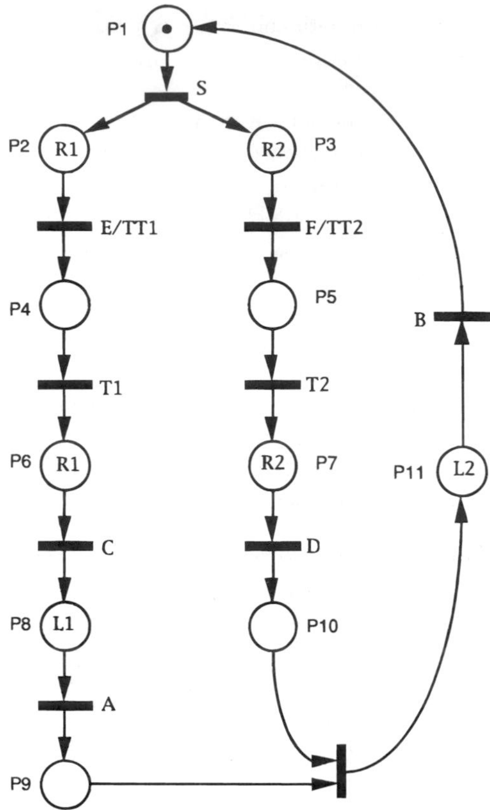

Figure 2.37 An interpreted Petri net for the two carts problem.

Design Methods

Traditional Design Methods

The design process of a sequential system starts from an FSM (finite state machine) and finishes in a list of Boolean functions. The logic diagram is drawn by means of specifications. The implementation can be made with a given or chosen technology.

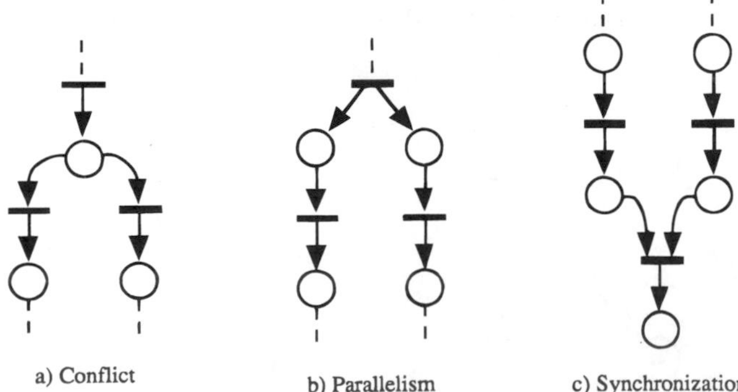

Figure 2.36 Conflict, parallelism, and synchronization Petri net structures.

This is equivalent to defining the combinational block and the memory block of our sequential block diagram model (see subsection Sequence Control). Such a model can be described by a set of state equations which give next state values on one hand and output values on the other hand (also named output state). The logical equations are Boolean combinational functions from which logic diagrams based on AND, OR, INVERTER gates and FLIP-FLOPS can be made. So, we talk about an equivalent sequential circuit of the specified FSM.

The design procedure of a sequential system may be summarized by a list of consecutive steps as follows:

1. The requirements of the system to be designed are given in textual mode accompanied by timing diagrams or other pertinent information.
2. From the given information, obtain the (primitive) state diagram and deduce the (primitive) state table.
3. Reduce the (primitive) state table by means of state reduction methods and obtain a reduced state table.
4. Determine the number of internal variables needed and determine a code such that a combination is assigned to each state. This is called the state-assignment problem. In some implementations, like some software implementation types, only a symbolic state assignment is necessary because at a given time a state may be represented by a software type variable.

5. Let us go on with conventional design. By now, there are two means:
   a. Direct feedback (for fundamental mode only). From the state transition table derive the logic equation of each next state variable, and from the output table derive the logic equation of each logic output.
   b. Flip-flop feedback (for both fundamental and pulsed modes). Determine the flip-flop type to be used. From the flip-flop truth table derive the flip-flop evolution table. Assign a flip-flop to each internal state variable. From the state transition table and the flip-flop evolution table derive the flip-flop excitation tables. Finally, from the flip-flop excitation tables derive the equations of the flip-flop excitations and from the output table derive the logic equation of each logic output.

To perform steps 1 and 2 correctly, the designer has to possess good intuition and experience in translating the system requirements into the (primitive) state diagram, because written specifications may be subject to incorrect facts, omissions, inconsistencies, and ambiguities. However, once the requirements are converted into the (primitive) state diagram and subsequently into the (primitive) state table, it is possible to make use of some formal methods to perform steps 3, 4, and 5.

Such methods are now well known by the community of discrete event systems and are usually computed. Moreover, they are used by a few specialists to design small or very small systems. The main reason is because they are not so convenient for designing systems of reasonable size. The reader may find the detailed theories and principles of such methods in (Givone, 1970; Miller, 1965; Paull and Unger, 1959). For now, it will be more interesting to consider a more user-friendly method.

Method of Map-Entered Variables

This design method can be applied to all kinds of FSM and especially to an ASM. ASM were discussed in the subsection on sequence control and the main principles of map-entered variables in the subsection on Logic Control.

The map-entered variables method applied to FSM results from a classical Karnaugh map, which has been folded in such a way that its size is only dependent on the number of states of the machine. This is equivalent to expressing the present state dependency of the internal variables parameterized by input variables. The main benefit of this technique is the small size of the Karnaugh maps required.

We will demonstrate how to fill up map-entered variables and how to deduce the logical expressions of the next state variables and outputs. We will give some general rules and illustrate them by means of the drill example seen in the section on sequence control.

We have to work with the FSM diagram as a result of an ASM specification technique. First, the state-assignment problem must be solved.

If a fundamental operating mode is chosen, the adjacency of the binary codes assigned to the adjacent states must be done

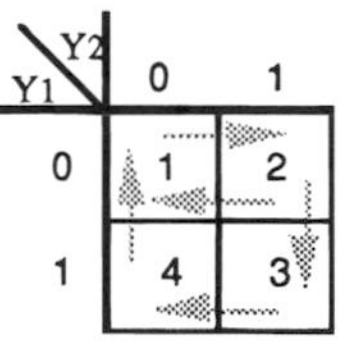

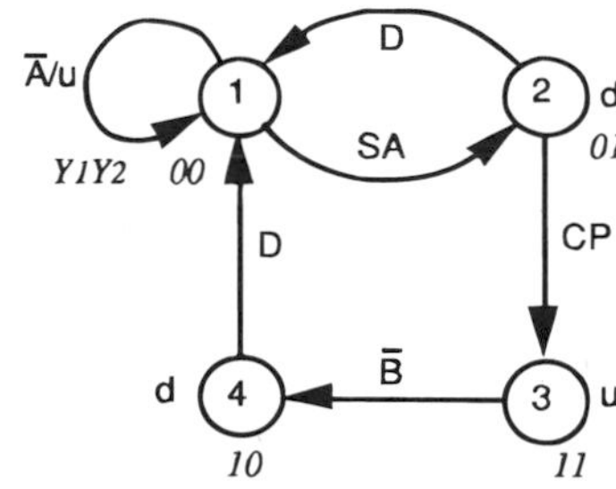

Figure 2.38 The FSM and state map of the drill example.

to avoid any race. The simplest method is an empirical one which may necessitate using additional states and/or additional internal variables. A good practice is to draw a state map which only depends on the present states. Note the states in it. This allows the designer to solve the adjacency problem and to locate the state concerned when he is working within a box of the next states map.

In pulsed mode the adjacency of the state codes does not matter. The problem is limited to fixing the number of internal variables to be used and to assigning a binary code to each state at random.

Once the state-assignment problem has been solved, write on the state diagram each binary code near the associated state. Let us illustrate this first step with the drill example (see Figure 2.38).

At this point, we have to provide the logical expression of the system. The procedure is as follows.

Make a map-entered function map for each internal variable and output. With a flip-flop based implementation, the maps will be for the excitation inputs of the flip-flops.

Now, consider one map-entered function map and one box of it. By looking simultaneously at the state map, note the associated state and the binary value of the corresponding present state variable. Three basic cases may be encountered:

1. The value of the state variable does not change when the system state changes.

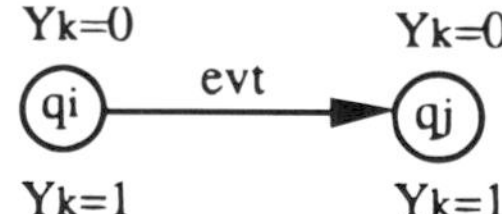

The variable state does not change. Obviously, it keeps its value. Then in the considered box write down this logical value (or the input flip-flop value deduced from the flip-flop evolution table in the flip-flop feedback loop. The next subsection gives details on flip-flop implementation).

2. The variable state is set when system state changes.

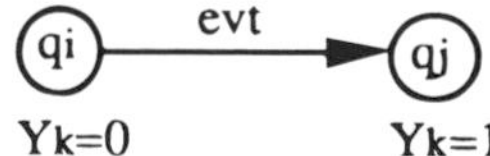

Note that the variable state changes from "0" to "1" when event *evt* also goes from "0" to "1". So, this is identical to the clause: *in state qi the internal variable*

behaves like event evt. The event *evt* represents the condition which sets the internal variable. Thus, in the considered box write down event *evt* in the direct feedback loop or associate it with the right input flip-flop deduced from the flip-flop conditional evolution table in the flip-flop feedback loop.

3. The variable state is reset when system state changes.

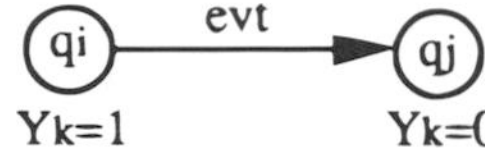

This time, the variable state changes from "1" to "0" when event *evt* goes from "0" to "1". So, this is identical to the clause: *in state qi, the internal variable behaves like the complement of event evt. Event evt* represents the condition which resets the internal variable. Thus, in the considered box write down:

- $\overline{evt}$ in the direct feedback loop because a dedicated reset function does not exist in this implementation mode.
- The event *evt* to the input flip-flop deduced from the flip-flop conditional evolution table in the flip-flop feedback loop. (Take care of the flip-flops having only one excitation input. For some of them, like *D* flip-flops, the negative value of the event is necessary. See conditional evolution table below.)

Fundamental Mode and Direct Feedback Loop. Let us apply the previous rules in the drill example. Figure 2.39 shows the resulting map-entered function maps and the resulting Boolean equations of the next state variables.

The last problem to solve is how to deduce the Boolean equations from the map-entered function maps.

Keep in mind that we look for the true points of the Boolean function to write. So, we have to consider the logical "1s" separately because we don't know the logical values of the other entered functions just now. Obviously, we may group several "1s" according to the Karnaugh map rules. Then, consider an entered function. This means we assume this function has the logical value "1". This time, we may group it with some "1s" of the table (when it is possible).

Output Equations Deduced from Map-Entered Function Maps. To fill up such a map is very simple. Indeed, the entered variables are the output conditional variables (if any). The latter ones may be at the level type or at the pulsed type. This depends on the machine type. For the drill example, the output equations are given by Figures 2.39 and 2.40.

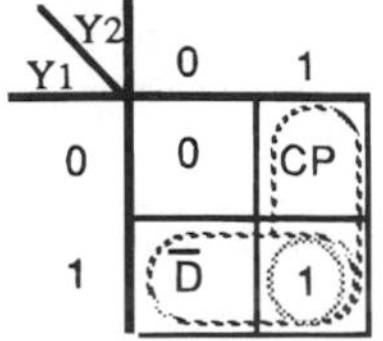

$y1 = Y1Y2 + \overline{D}\,Y1 + CPY2$

Y1 \ Y2	0	1
0	SA	$\overline{D}$
1	0	B

$y2 = BY1Y2 + \overline{D}\,\overline{Y1}Y2 + SA\overline{Y1}\,\overline{Y2}$

Figure 2.39 Map-entered function maps for the next state variables of the drill example.

Flip-Flop Feedback Loop. Like in classical implementation, the output flip-flops are used as the present state variables. Thus, when the machine commutes from a present state to a next state, we need to know how to connect the flip-flop excitations in order that its outputs behave like the corresponding next-state code. As in classical methods, when the flip-flop evolution depends on known logical values, the evolution table of the specified flip-flop gives us the logical values to apply to the flip-flop excitations. But with this method, an additional question comes to the designer's mind: when the flip-flops have to be set or reset by a conditional event, how are the equations of its excitation inputs found? The answer is given by an evolution table completed by the conditional set or reset map.

The connections of the conditional set or reset map are found as follows. Looking for the reset of the flip-flop output means that it is at present state "1." If the event is in state "0," the machine does not commute, thus the flip-flop output has to remain in state "1". On the other hand, if the event goes to "1", the flip-flop output has to go to state "0" (reset function). The event contributes as a reset signal. Considering flip-flop functioning, and looking at its evolution table, we determine to which excitation input and in which way the event has to be connected. With like reasoning we can find the part of the conditional set map. Figure 2.41 gives the complete evolution table for the flip-flops commonly used.

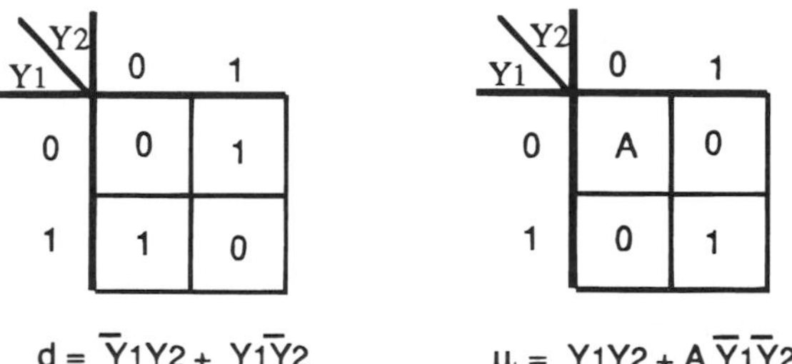

$d = \overline{Y1}Y2 + Y1\overline{Y2}$ $\mu = Y1Y2 + A\,\overline{Y1}\,\overline{Y2}$

Figure 2.40 The output equations of the drill example.

condit.	Q_n	Q_{n+1}	R	S	D	T	J	K
set	0	0	-	0	0	0	0	-
or	0	1	0	1	1	1	1	-
reset	1	0	1	0	0	1	-	1
↓	1	1	0	-	1	0	-	0
event								
0	0	0	0	evt	evt	evt	evt	-
1	0	1						
event								
0	1	1	evt	0	$\overline{evt}$	evt	-	evt
1	1	0						

Figure 2.41 Evolution and conditional evolution tables for RS, D, T, and JK flip-flops.

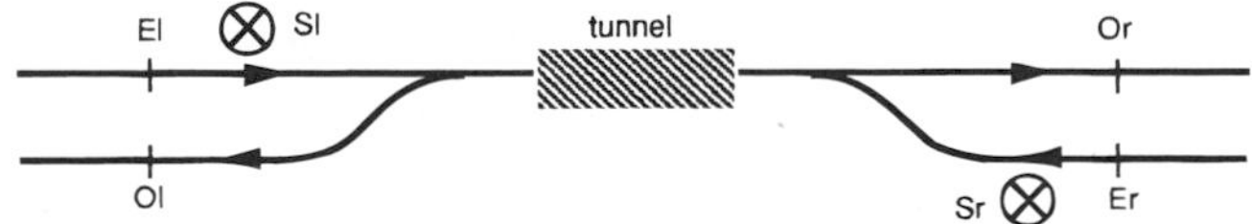

Figure 2.42 Schematic example of the two-railways tunnel.

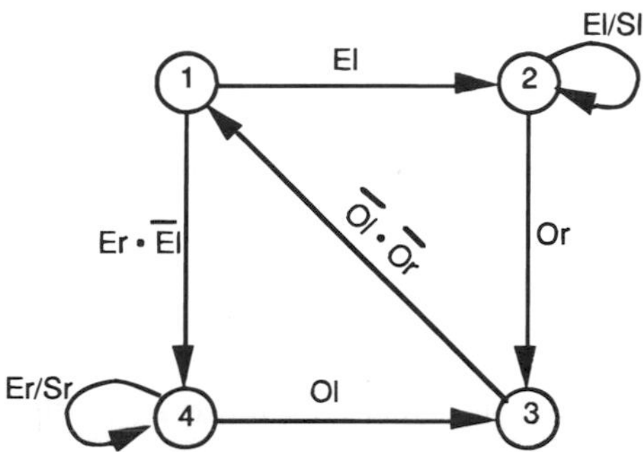

Figure 2.43 Reduced state diagram of the two-railways tunnel.

EXAMPLE 2.13: Two-railways and a single-way tunnel (Figure 2.42)

Sr, Sl: right and left semaphores.
Er, El: right and left tunnel input sensors.
Or, Ol: right and left tunnel output sensors.

In the following explanation, the tunnel area includes the tunnel itself and the railway structure located between right and left sensors (Figure 2.43).

Semaphores are always red when no train drives along the railway (logic level "0"). When a train reaches sensor *El*, semaphore *Sl* goes green (logic level to "1") if the tunnel area is free. While the train is above sensor *El*, the latter one gives a logical "1". Semaphore *Sl* goes back to red immediately after the tail of the train passes over the sensor *El*. The tunnel is assumed free again when the queue of the train passes over the sensor *Or*. The specification is symmetric for a train coming from the opposite direction.

When the tunnel area is occupied, another train may arrive and it must wait in position *El* and/or *Er*. Then, priority is given to the train waiting in position *El*.

First we make the ASM of the tunnel system.

State assignment: In the fundamental mode at least three internal variables would be necessary because states 1 and 3 are adjacent to all others, whereas a synchronous implementation needs two variables only as shown in Figure 2.44.

Let us implement this example by means of two *JK* flip-flops, the outputs Q_1 and Q_2 which represent the present state variables Y_1 and Y_2, respectively. The excitation inputs of the *JK* flip-flops are (see Figure 2.45)

$$\begin{cases} J_1 = E_r \cdot \overline{E_l} \cdot \overline{Q_2} + O_r \cdot Q_2 \\ K_1 = \overline{O_r} \cdot \overline{O_l} \cdot Q_2 \end{cases} \qquad \begin{cases} J_2 = E_r \cdot \overline{E_1} \cdot \overline{Q_2} + O_r \cdot Q_2 \\ K_2 = \overline{O_r} \cdot \overline{O_l} \cdot Q_2 \end{cases}$$

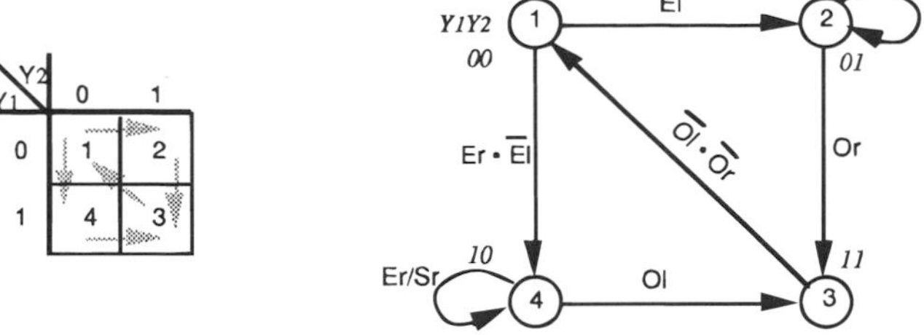

Figure 2.44 State assignment of the two-railways tunnel.

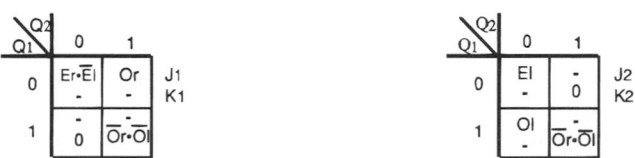

Figure 2.45 The two-railways tunnel: logic expressions extraction.

In this example the output equations can be directly written from the state diagram:

$$\begin{cases} S_r = E_r \cdot Q_1 \cdot \overline{Q_2} \\ S_l = E_l \cdot Q_1 \cdot Q_2 \end{cases}$$

One Out of *n* Coding and Its Extension to Petri Nets

To avoid state coding problems many authors have proposed systematic coding techniques, also called universal codes. One of them is particularly easy to apply. It is based on the following principle: one state variable is associated with each state of the machine; its value is "true" when the corresponding state is activated, "false" otherwise.

Let us consider an example: a level crossing control system represented in Figure 2.46.

A traffic light *F* is made red as soon as train is detected by sensor *A* or sensor *C*. It stays red until the queue of the train passes over *B*. Trains cannot maneuver in the *AC* zone; their size may be shorter or greater than *AB* or *BC* sections which are themselves not necessarily equal, but it is always shorter than the *AC* section. Figure 2.47 gives a state diagram of this control system.

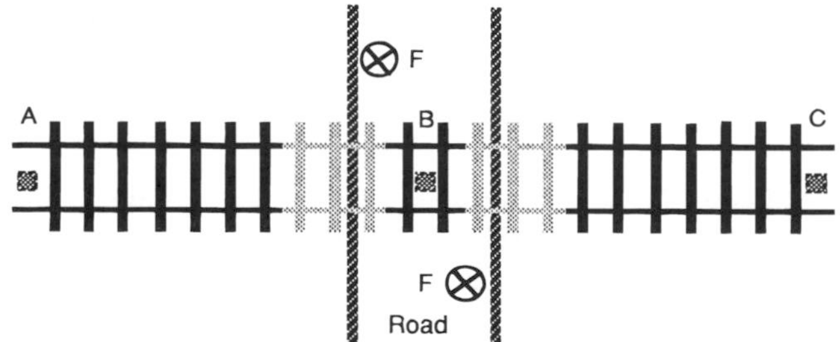

Figure 2.46 The level crossing example.

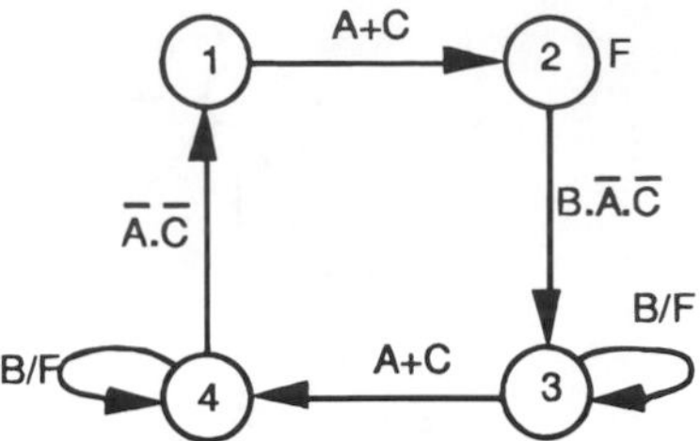

Figure 2.47 State diagram for level crossing control system.

This state diagram may be coded with a 1 out of n code having four variables.

States	Codes			
	Y_1	Y_2	Y_3	Y_4
1	1	0	0	0
2	0	1	0	0
3	0	0	1	0
4	0	0	0	1

Each code word is such that only one variable is true. So, the distance between code words is always two. A transition between two states consists of setting the variable representing the next state and resetting the variable representing the present state. This consideration shows a perfect similarity with the operating mode of a flip-flop.

So, let us associate a flip-flop with each variable and let us consider the synchronous case only (the asynchronous one could be considered in the same way). We associate a *JK* flip-flop (J_i, K_i; Q_i) with (Y_i).

Setting flip-flop (J_i, K_i; Q_i) corresponds to the activation of state q_i. Indeed, this is a result of the transition from one of its previous state to state q_i. Resetting flip-flop (J_i, K_i; Q_i) must coincide with a transition from state q_i to one of its successor states.

Let us apply such principles to the level crossing control system. The input excitations of the *JK* flip-flops are

$$J_1 = \overline{A} \cdot \overline{C} \cdot Q_4 \qquad K_1 = A + C$$

$$J_2 = (A + C)Q_1 \qquad K_2 = B \cdot \overline{A} \cdot \overline{C}.$$

$$J_3 = B \cdot \overline{A} \cdot \overline{C} \cdot Q_2 \qquad K_3 = A + C$$

$$J_4 = (A + C)Q_3 \qquad K_4 = \overline{A} \cdot \overline{C}$$

and the output

$$R = Q_2 + Q_3 \cdot F + Q_4 \cdot F$$

More generally, if we consider a state q_i and its environment (Figure 2.48), the input excitations of the corresponding *JK* flip-flop may be written as follows:

$$J_i = \sum_{l=1}^{k} e_{pl} \cdot Q_{pl}$$

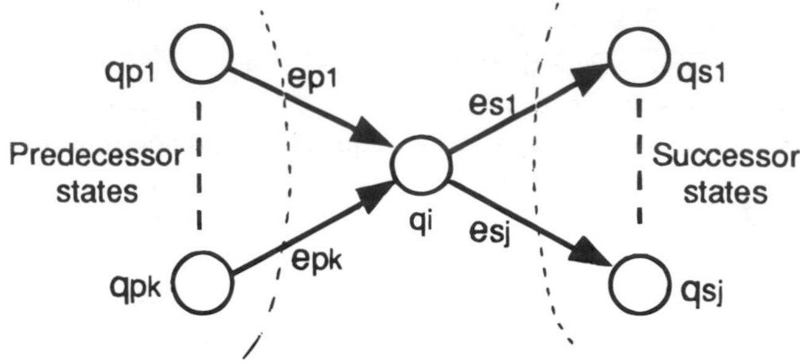

Figure 2.48 Generic state q_i and its environment.

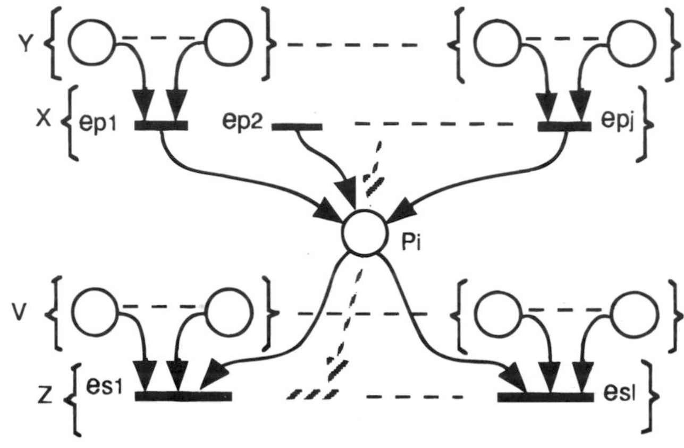

Figure 2.49 A generic place and its environment.

$$K_i = \sum_{l=1}^{j} e_{sl}$$

where p stands for the pth predecessor state and ℓ for the ℓth successor state of q_i.

Extending this concept to safe Petri nets is relatively easy. Keeping in mind the same principle, we associate a *JK* flip-flop with each place of a net (synchronous implementation).

Setting flip-flop (J, K_i; Q_i) corresponds to a token arriving in place p_i. This event results from the firing of an input transition of p_i (Figure 2.49). Then, looking for all the input transitions of p_i, the excitation of the J_i input of the *JK* flip-flop associated to p_i is given by:

$$J_i = \sum_{x=1}^{j} e_{p_x}\left(\prod_{Y} Qy\right)$$

$x = 1, \ldots, j$ indexes the events associated with the input transitions of p_i such that, ep_1, ep_2, ... Y defines the subset of places connected to transition t_x.

Resetting flip-flop (J_i, K_i; Q_i) results from firing one of the output transitions of p_i which removes the token from p_i. So, this can also be written:

$$K_i = \sum_{z=1}^{l} e_{s_z}\left(\prod_{V} Qv\right)$$

with $z = 1, \ldots, \ell$ indexes the output transitions of p_i and V defines for each transition t_z the set of its input places, p_i excluded.

Let us apply these equations to the two carts example modeled in the second paragraph in the Sequence Control section by a Petri net represented in Figure 2.37. We obtain the following equations:

$$J_1 = B \cdot Q_{11} \qquad K_1 = S$$
$$J_2 = S \cdot Q_1 \qquad K_2 = E$$
$$J_3 = S \cdot Q_1 \qquad K_3 = F$$
$$J_4 = E \cdot Q_2 \qquad K_4 = T_1$$
$$J_5 = F \cdot Q_3 \qquad K_5 = T_2$$
$$J_6 = T_1 \cdot Q_4 \qquad K_6 = C$$
$$J_7 = T_2 \cdot Q_5 \qquad K_7 = D$$
$$J_8 = C \cdot Q_6 \qquad K_8 = A$$
$$J_9 = A \cdot Q_8 \qquad K_9 = Q_{10}$$
$$J_{10} = D \cdot Q_7 \qquad K_{10} = Q_9$$
$$J_{11} = Q_9 \cdot Q_{10} \qquad K_{11} = B$$

and for the outputs:

$$R_1 = Q_2 + Q_6$$
$$R_2 = Q_3 + Q_7$$
$$TT1 = E \cdot Q_2$$
$$TT2 = F \cdot Q_3$$
$$L_1 = Q_8$$
$$L_2 = Q_{11}$$

Obtaining this set of equations directly from the model is very easy. For this reason this technique is also called the direct synthesis method.

2.3 Implementation Techniques

Relay Ladder Diagrams

In the field of electromechanical engineering, relay ladder diagrams have been widely used in the past. Now, semiconductors and integrated circuits have replaced this technology, but there are two domains where relay ladder diagrams are still used:

- Power systems because their structure ensures a high degree of insulation between the controlled and control circuits.
- Programming languages of programmable logic controllers (PLC), to make the transition from relay technology to PLC easier (due to the fact that in some countries like the U.S., relay ladder diagrams are still used to model switching circuits).

Principle

A relay diagram is built by using a symbolic system similar to a wiring diagram of relay technology. Many symbols have been proposed, but we will describe the most commonly used here (see Figure 2.50).

These basic symbols are sufficient to represent any logic function. A logic function is always associated with a relay solenoid.

EXAMPLE 2.14: (Figure 2.51)

$$f(x3, x2, x1) = x_2 \cdot \overline{x_1} + \overline{x_3} \cdot \overline{x_2} + \overline{x_3} \cdot \overline{x_1}$$

The analogy with electrical wiring is obvious:

- The left vertical line represents a supply wire (+Vcc).
- The right vertical line is the ground wire.
- xi is a normally open switch.
- $\overline{\text{xi}}$ is a normally closed switch.

So, f (the solenoid of the relay) receives power if and only if the function $f = 1$.

Notice that ΣΠ and ΠΣ expressions of logic functions are easier to translate into relay diagrams than any other forms of logic expressions.

Implementation of sequential systems by means of relay ladder diagrams does not raise any specific problem. However, for relays operating in asynchronous mode, the coding of a finite state machine implemented by this method must guarantee that the codes of any successive states differ by the value of one variable only (see the Sequence Control section state assignment problem).

The next state of variables are identified with relay solenoid excitations and the present state variables are the relay switches.

EXAMPLE 2.15: (Figure 2.52)

$$y = m + \overline{a} \cdot Y$$
$$S = Y$$

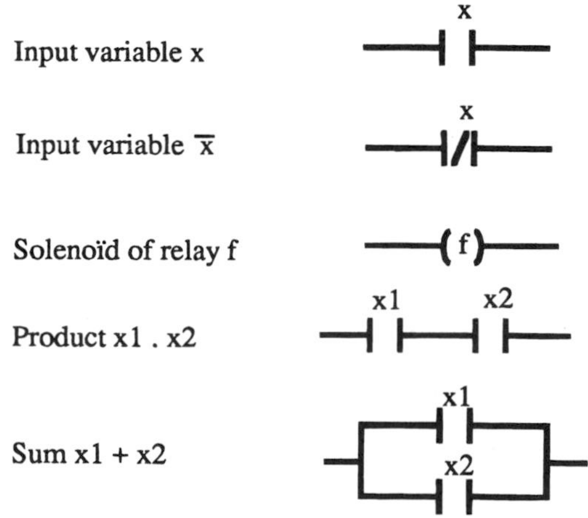

Figure 2.50 Basic symbols of relay ladder diagram.

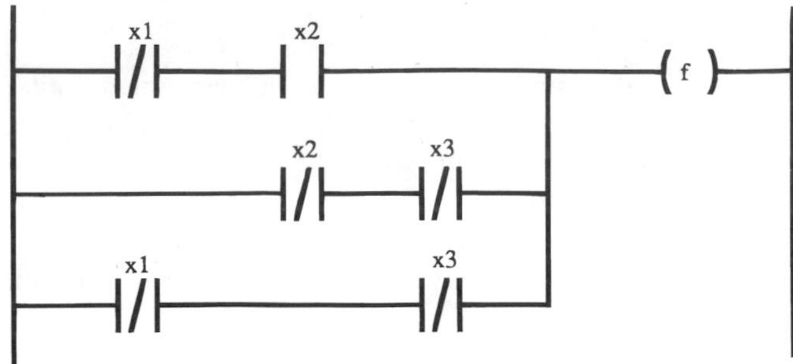

Figure 2.51 An example of relay diagram.

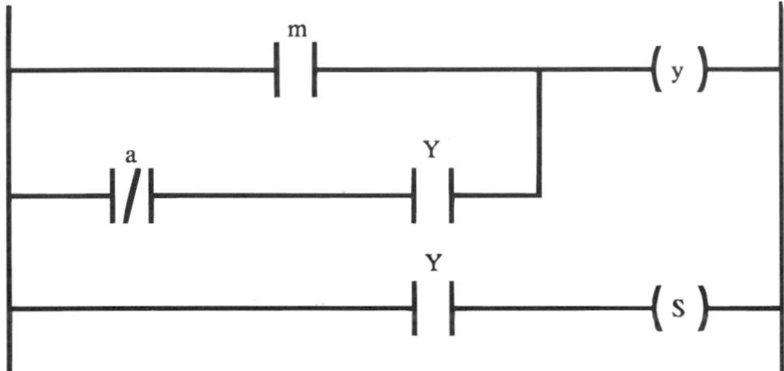

Figure 2.52 A state machine implementation.

Implementation of Discrete Event Systems via Electronic Integrated Circuits

During the last two decades, the implementation of large discrete event systems required a lot of integrated circuits. Designers used standard integrated functions. According to the number of integrated gates in one component, the classification was the following:

- SSI (small scale integration).
- MSI (medium scale integration, from 12 to 100 gates).
- LSI (large scale integration, from 100 to 3000 gates).
- VLSI (very large scale integration, more than 3000 gates).

The integrated circuits are not programmable. The basic functions provided by the manufacturer were gates, flip-flops, counters, multiplexers, etc. So, if the designer had to design a complex function, he had to build it with these basic functions in one or several printed circuit boards.

PLD Technology (Programmable Logic Devices)

The introduction of PLDs was a true evolution in the hardware design community. PLDs are digital, user-configurable, integrated circuits used to implement custom logic functions. The main benefits offered by PLDs are

- Reduction of package number. One PLD may replace several MSI/LSI functions. So the power consumption is also reduced.
- PCB (printed circuit board) area reduced because of higher integration.
- Introduction of flexibility by programmability feature.
- Improved reliability due to the significant reduction of the connections.
- Shorter and more flexible design cycle. Compared to the PCB, standard cells, or gate arrays, custom function can be implemented faster with PLDs because of fewer wired connections, software design, and simulation tools.
- Proprietary design protection ("fuse" protection). Circuit design can be protected against malicious readers by blowing the security "fuse."
- Versatility of the circuit. Functions are custom designed.

Until now, PLDs were a good alternative to discrete logic and custom or semicustom devices such as ASICs (application-specific integrated circuits) and gate arrays. Because of the improvement of electronic technologies and the benefits once realized, PLDs became the preferred choice of most designers.

PLDs encompass all digital logic circuits configured by the user, including PAL/GAL devices, field programmable gate arrays (FPGAs), function-specific PLDs or complex PLDs and erasable programmable logic devices (EPLD).

PLD Architecture

Most PLD architecture is based on two arrays, the AND array and the OR array. So, a sum of product terms (p-terms) may be easily implemented. According to the circuit type, both AND and OR arrays or only AND arrays are programmable. For instance, the two arrays of PROMs, EPROMs (erasable PROMs), and FPLAs (field programmable logic arrays) are programmable, whereas in complex PLDs, EPLDs (erasable PLDs), PALs, and PLSs (programmable logic sequencers) only the AND array is programmable.

A classical PLD block diagram is shown in Figure 2.53. The internal architecture of the device has a sum of p-terms (AND/OR structure). Inputs to the programmable AND array come from the true and complement signals of input pins, and the true and complement forms of the feedback signals from the I/O macrocells.

The outputs of the product terms are ORed together. Then the output of the OR gate is fed as an input of an XOR gate. The XOR gate allows the designer to specify the polarity (true form or complement form) of the output signal by blowing the EPROM fuse connected to the second entry of the XOR gate (Figure 2.54).

The XOR output then feeds the I/O macrocell, in which the output may be configured for registered or combinational operations. In registered mode, the output is registered via an edge-triggered flip-flop. The feedback signal comes from the output of the flip-flop. Some PLDs allow programming the flip-flop type, SR, D, T, or JK flip-flops. In combinational mode, the output is not registered, and the feedback signal comes directly from the I/O pin.

An output enable p-term determines whether the output signal of the I/O macrocell will propagate to the output pin. When it does not, the output buffer takes a high impedance and allows the pin to be used as an input. In some PLDs, the macrocells may be triggered by its global clock input signal(s) from clock input

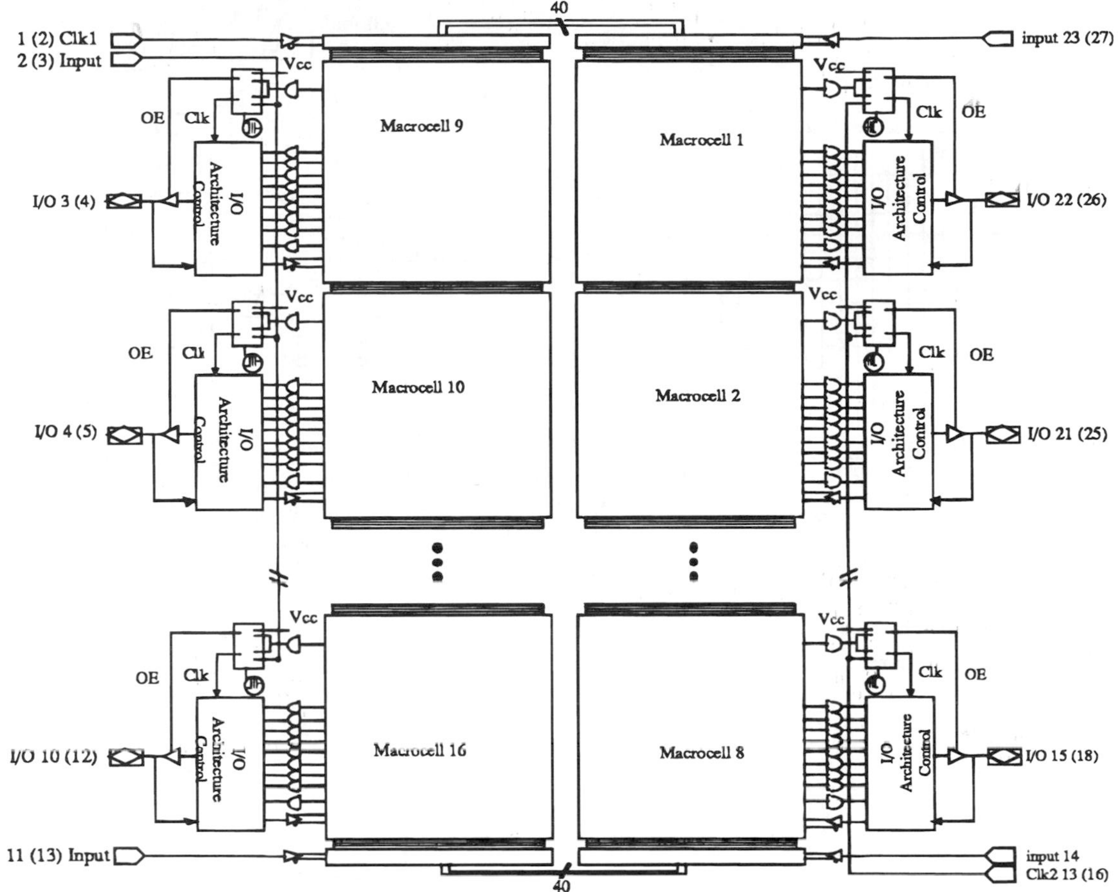

Figure 2.53 Example of an EPLD architecture—EP610 (Altera, 1992).

pin(s) or from an array clock signal generated by a p-term. Figures 2.54b and 2.54c show two modes of output enable and clock selection. This is done by a multiplexer controlled by a single EPROM bit. It can be individually configured for each I/O pin.

PLDs and EPLDs Design Development Cycle

Designers can use familiar CAE (computer aided environment) tools for design entry and simulation. Software interfaces to other design tools are provided by translators and via industry-standard EDIF (editor files) netlists. Usually, a manufacturer development system includes a standard netlist format that provides a bridge between syntactic and schematic editors or simulators and the manufacturer software for design implementation and real-time design verification. The manufacturer software works on the most popular engineering workstations.

The design entry software consists of libraries and netlist interfaces for standard CAE tools such as Cadence, or CAD, Viewlogic, PALASM, etc. PLD libraries allow design entry with standard TTL functions, Boolean equations, FSM description, and user-defined macros.

The simulation software includes models and netlists to standard simulator software that is used for logic and timing simulations.

The PLD development cycle is summarized in Figure 2.55.

Using PLDs for FSM Implementation

The programmable AND/OR structure, feedback paths, and flip-flops make PLDS very convenient for FSM implementation. These features allow building a generic FSM structure as shown in Figure 2.56.

F and G are combinational programmable blocks. The programmable I/O macrocells allow making combinational or registered outputs and feedback loops. In classical PLDs, the true and complement signals of the inputs are directly connected to the AND array, whereas some complex PLDs offer an I/O macrocell for each pin, allowing custom-programmable, combinational, or registered inputs.

So, the three machine classes, Huffman, Mealy, or Moore, can be implemented. The general descriptions are written below. In the Moore machine, we notice two subclasses.

When neither inputs/outputs nor feedback loops are registered, the machine is fully asynchronous. This is a direct-feedback-loop implementation of an FSM operating in the fundamental mode (Huffman machine).

Next states $y_{(t)} = F_{(i(t),\, Y(t))}$
Output states $z_{(t)} = G_{(i(t),\, Y(t))}$
Combinational I/O and feedback loops (Huffman machine).

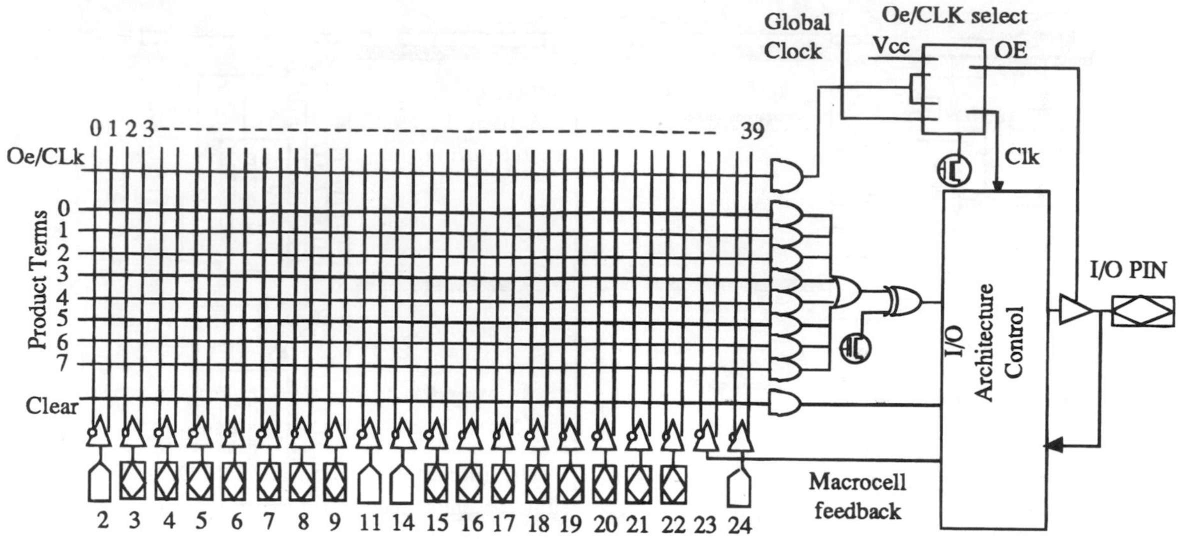

Figure 2.54a Altera EP610 macrocell.

Mode 0 operation :
The register is clocked by the global Clock Signal, which can be connected to the other macrocells. The output is enabled by the logic from the product term.

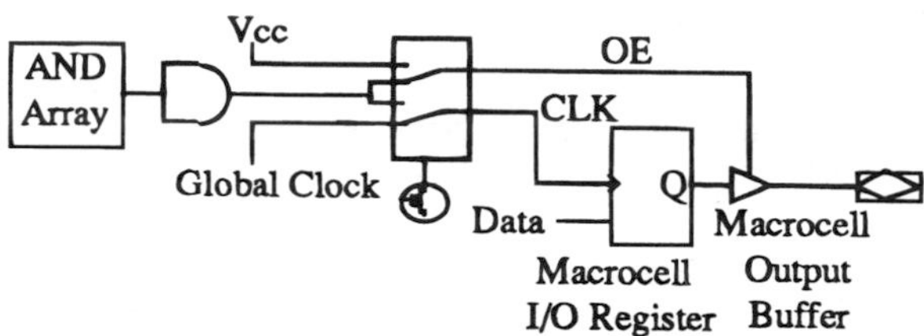

Figure 2.54b OE/CLK select multiplexer of Altera EP610 in mode 0.

Mode 1 operation :
The output is permanently enabled and the register is clocked by the product term, which allows gated clocks to be generated in EP610 EPLDs.

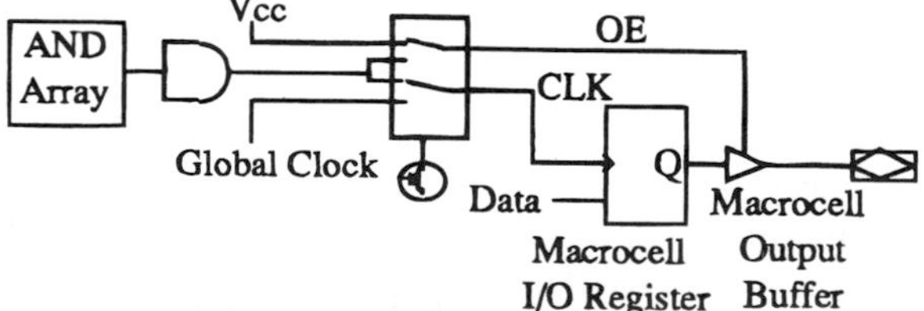

Figure 2.54c OE/CLK select multiplexer of Altera EP610 in mode 1.

If the next states $Y_{(t)} = F_{(i(t),\ Y(t))}$, the output states $z_{(t)} = G_{(i(t),\ Y(t))}$, and the I/O and the feedback loop are registered, the machine is a Mealy one.

Next states $y_{(t)} = F_{(i(t),\ Y(t))}$
Output states $z_{(t)} = G_{(i(t),\ Y(t))}$
Registered outputs and feedback loops (Mealy machine).

Moore machines can be of two classes:

Next states $y_{(t)} = F_{(i(t),\ Y(t))}$
Output states $z_{(t)} = G_{(Y(t))}$
Registered outputs and feedback loops.
(Class A Moore machine)

Next states $y_{(t)} = F_{(i(t),\ Y(t))}$
Output states $z_{(t)} = Y_{(t)}$
Registered outputs and feedback loops.
(Class B Moore machine)

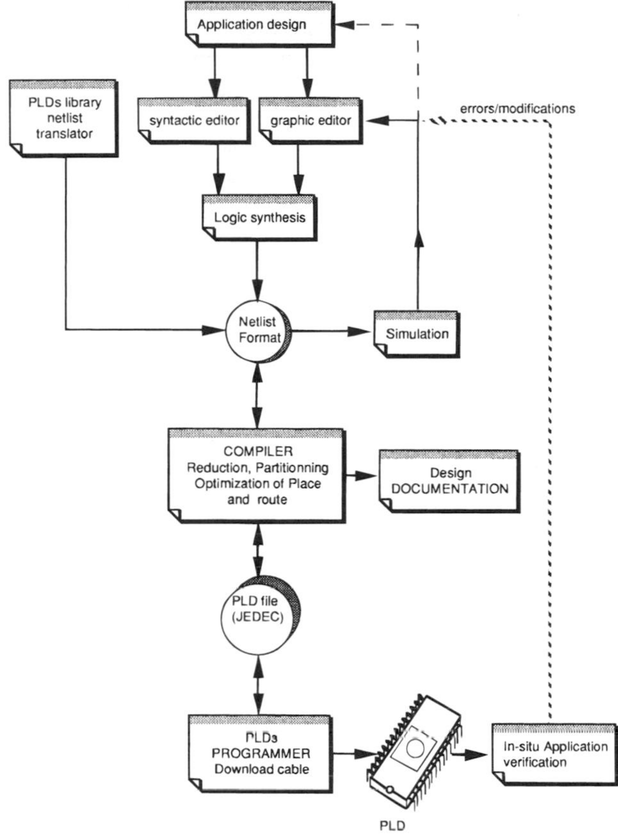

Figure 2.55 The development cycle for PLDs.

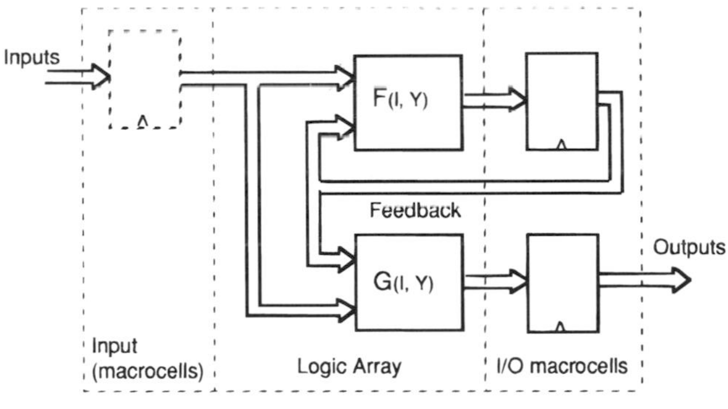

Figure 2.56 FSM generic implementation structure using PLDs.

To use the PLD area efficiently, the design has to provide the Boolean expressions of the FSM to be implemented. The design software tool sold by the PLD manufacturer allows implementing an FSM from a textual or graphical entry directly. In this case an optimal implementation is not guaranteed. For instance, when the custom is to use this kind of entry, what about the state-assignment principle used by the software tool? Unfortunately, the software documentation rarely gives a satisfactory answer to this question.

Memory Based Implementation

There is another way to implement FSMs by means of electronic devices. It consists of implementing classical devices such as memory, register, multiplexer, etc., and a special kind of PLD named a PLS (programmable logic sequencer) (Altera, 1992, Lattice, 1992).

A basic PLS is a PLD in which registered feedback loops are set during the manufacturing process. Most complex PLSs integrate a memory plan and counters, multiplexers, encoders, and any functional blocks. Following are some manufacturer references of PLSs:

Texas TIBPLS5xx series field programmable PLS for Mealy state machines.
AMD 29PL1xx series.
Altera SAM (stand alone microsequencer).
Xilinx XC3000-70 microsequencer.

The simplest PLSs are advanced PLDs, and the Boolean expressions or state diagram descriptions are needed as entries for programming.

The PLSs equipped with a memory plan may be directly programmed from the state diagram. Two basic principles used are given.

The PLSs equipped with a memory plan may be directly programmed from the state table. The memory stores the next states and the outputs of the FSM. The memory address bus is connected to the input and the present states. If the memory plan has a small number of address lines or data bits, an encoder and a decoder can be used as input stage and output stage, respectively. The feedback loops have to be registered in order to avoid any race (Figure 2.57).

The memory data field is specified from the state and output table described in Figure 2.58.

The implementation of a Moore machine saves a share of the data field because the outputs directly depend on present states.

The main disadvantage of this solution is the lack of address lines and/or data bits when the FSM to be implemented exceeds about ten inputs/outputs (Figure 2.59).

Software Implementation

Programmable Logic Controllers

A programmable logic controller is a specialized microcomputer that can operate in various and extreme environments (dust, vibrations, humidity, etc.). It is built around a microprocessor (or microcontroller), a random access memory in which user

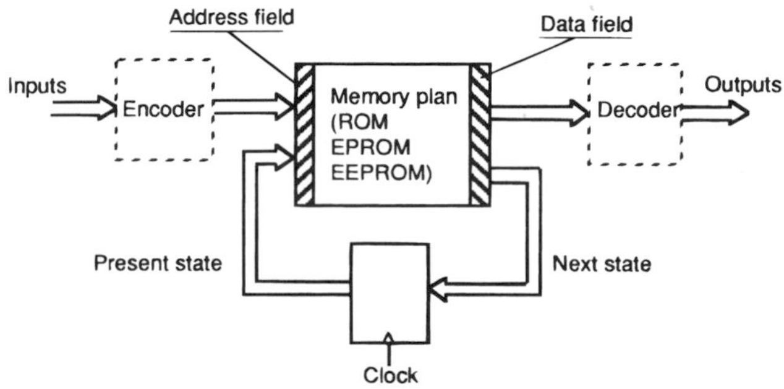

Figure 2.57 A simple memory implementation of FSM.

programs and data are stored, and a read-only memory that contains the monitor. A PLC has logic input and logic output lines that can be individually addressed by the processor. The user program must be written in order to implement a given specification (combinational expressions, finite state machines, Petri net, etc.) by means of an instruction set specific to the PLC. In this text, we want to give some basic programming concepts that could be used whatever the PLC. So we first define a standard PLC with a generic input/output configuration and a generic set of instructions.

To define a generic configuration, we assume the following:

- n input lines labeled *I1* to *In*.
- m output lines labeled *O1* to *Om*.
- k internal bits of RAM labeled *B1* to *Bk*.
- The following set of instructions to access to these bits.

LD xx	LoaD the accumulator with the Boolean value of bit at address xx.
LD/xx	LoaD the accumulator with the complement of the Boolean value of bit xx.
AND xx	Logical AND between the accumulator and bit xx; the result is stored in the accumulator.
OR xx	Logical OR between the accumulator and bit xx; the result is stored in the accumulator.
ST xx	STore the accumulator value into the bit at address xx.
PUT xx 0 (1)	Store the value 0 (1) into the bit at address xx; this instruction is conditional: it is executed only if the accumulator contains 1 and does not affect the value in the accumulator.

We will also assume that the program (list of instructions) is executed sequentially and that, when the last program instruction has been executed, the monitor reads the input lines, updates the output lines then starts the execution again at the first instruction of the program (synchronous operation).

With these hypotheses, a combinational problem modeled by the expressions

$$S_1 = x_3 + x_2 \cdot x_1$$

$$S_2 = \overline{x_2} \cdot x_1 + x_3 \cdot \overline{x_1}$$

with the following assignments of the inputs and outputs,

$$S_1 \leftrightarrow O1 \qquad S_2 \leftrightarrow O2$$

$$x_3 \leftrightarrow I3 \qquad x_2 \leftrightarrow I2 \qquad x_1 \leftrightarrow I1$$

can be implemented by

LD	*I2*	; acc ← x_2
AND	*I1*	; acc ← $x_2 \cdot x_1$
OR	*I3*	; acc ← $x_2 \cdot x_1 + x_3$
ST	*O1*	; S1 ← $x_2 \cdot x_1 + x_3$

LD	*/I2*	; acc ← $\overline{x_2}$
AND	*I1*	; acc ← $\overline{x_2} \cdot x_1$
ST	*B1*	; *B1* ← $\overline{x_2} \cdot x_1$ (store an intermediate result)
LD	*/I1*	; acc ← $\overline{x_1}$
AND	*I3*	; acc ← $\overline{x_1} \cdot x_3$
OR	*B1*	; acc ← $\overline{x_1} \cdot x_3 + \overline{x_2} \cdot x_1$
ST	*O2*	; *S2* ← $\overline{x_1} \cdot x_3 + \overline{x_2} \cdot x_1$

Q \ i_1i_2	00	01	11	10
q1	q1/00	q2/01	q2/01	q1/00
q2	q3/11	q2/01	q2/01	q4/10
q3	q3/11	q4/10	q1/00	q2/01
q4	q2/01	q4/10	q1/00	q4/10

y_1y_2/s_1s_2

Y_1Y_2 \ i_1i_2	00	01	11	10
00	00/00	01/01	01/01	00/00
01	11/11	01/01	01/01	10/10
11	11/11	10/10	00/00	01/01
10	01/01	10/10	00/00	10/10

Figure 2.58 State and output tables of the example.

Memory map :

Address field				Data field			
A_3 Y_1	A_2 Y_2	A_1 i_1	A_0 i2	D3 y_1	D_2 y_2	D_1 s_1	D0 s2
0	0	0	0	0	0	0	0
0	0	0	1	0	1	0	1
0	0	1	0	0	0	0	0
0	0	1	1	0	1	0	1
0	1	0	0	1	1	1	1
0	1	0	1	0	1	0	1
0	1	1	0	1	0	1	0
...	...	...	...	...	...	...	...

Figure 2.59 Memory map.

The implementation of a finite state machine is based on a similar principle. We can program the expressions of the next state variables and of the outputs, and, at the end of the program, update the value of the present state. The program must begin with a sequence of instructions which initialize the state variables.

An efficient method for avoiding implementation errors consists of using a 1 out of n coding of the finite state machine because the state diagram can be implemented directly.

For instance, to program the part of FSM shown in Figure 2.60 with the following assignments:

State i ↔ *Bi* State j ↔ *Bj*

A1 ↔ *I1* A2 ↔ *I2*

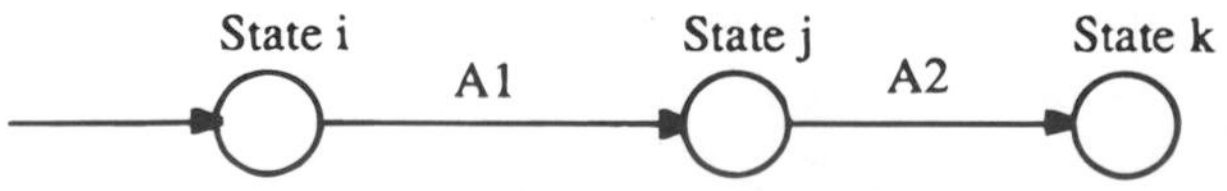

State i A1 State j A2 State k

Figure 2.60 A finite state machine.

an implementation can be

```
...
LD    Bi      ; acc ← State i (1 if the FSM is in state i)
AND   I1      ; acc ← State i·A1 (1 when the transition
                    between State i and State j is possible)
PUT   Bi 0    ; if acc = W1 State i ← 0
PUT   Bj 1    ; if acc = 1 State j ← 1 (state transition
                    executed)
LD    Bj      ; idem for state j
AND   I2
PUT   Bj 0
PUT   Bk 1
...
```

The same principle can be applied to implement safe Petri nets. A one place-one variable coding is necessary and the only difference is that all the input places of the transitions must be checked, while in a finite state machine only one state has to be checked. For general Petri nets' implementation, the binary information of internal bits is not sufficient. Fortunately, most PLCs have word oriented instructions and word internal variables, so that nonbinary place marking can be represented. However, the implementation of such models is more complex and will not be described here.

Microcontrollers

When a PLC solution is not realistic (cost, weight, size) and if the size of the solution forbids the use of PLDs, microcontrollers can be used to implement control models like finite state machines, Petri nets, etc. A microcontroller is a single-chip microcomputer containing processor and memory, as well as logic input and output lines, called ports, that can drive power interfaces and receive signals from logic sensors. Some of them also have digital-to-analog and analog-to-digital converters. Special instructions are used to access these parts.

Finite state machines, Petri nets, etc. are implemented by writing programs in the native assembly language of the microcontroller similar to those described for PLCs above. Alternatively, a high-level programming language can be used, such as C, which is widely used in microcontroller applications.

Microprocessor and microcontroller architectures will be discussed in the next chapter.

References

Altera Data Book, 1992.

Givone, D. D., 1970. *Introduction to Switching Theory.* McGraw-Hill, New York. Computer Science Series.

Grasselli, A. and Lucio, F., 1966. A method for combined row-column reduction of flow tables. *Conf. Record Seventh Annu. Symp. Switching and Automata Theory.* No. 3, October 1996.

Lattice, 1992. *pLSI and isPLSI Data Book*, Data Book *Supplement* and *Handbook.*

Miller, R. E., 1965. *Switching Theory.* Volume II: Sequential circuits and machines. John Wiley & Sons, New York.

Paull, M. C. and Unger, S. H., 1959. Minimizing the number of states in incompletely specified sequential switching functions. *IRE Trans. Electro. Computers*, Vol. EC-8, No. 3, September 1959.

Peterson, J. L., *Petri Net Theory and The Modeling of Systems.* Prentice-Hall, Englewood Cliffs, New Jersey, Edition 81.

3

Computer Architecture

3.1 Hardware Organization 48
3.2 Computer Software 50
Programming Languages • Operating Systems
3.3 Information Representation in Digital Computers 51
3.4 Specifying Instruction Operands 53
3.5 CPU Registers 54
3.6 Memory Organization 56
Memory Address Generation
3.7 Computer Instruction Types 58
Data Transfer • Arithmetic Instructions • Logical • Shift and Rotate • Control Transfer • Processor Control
3.8 Interrupts and Exceptions 60
3.9 Evaluating Instruction Set Architectures 61
3.10 Computer System Design 62
Memory Systems • Semiconductor Memory Technologies • Memory System Organization • Cache Memory • Virtual Memory Management
3.11 Input/Output Device Interfaces 67
3.12 Microcontroller Architectures 67
3.13 Multiple Processor Architectures 69

Victor P. Nelson
Auburn University

A digital computer is a device capable of solving problems and manipulating information under the direction of a given program of instructions. The hardware of a digital computer is a set of digital logic circuits that receives information from one or more sources, processes that information, and sends the results to one or more destinations. Digital computers allow the automation of arithmetic operations and provide an inexpensive way to solve complex numeric problems; store, retrieve, and communicate information; and control robots, appliances, automobiles, games, manufacturing plants, and a variety of other processes and machines.

In this section, we introduce the basic hardware and software elements of a digital computer and examine different computer architectures constructed from these elements. As examples, we will consider the architectures of two general-purpose microprocessors, the Intel 8086 and Motorola 68000, two microcontrollers, the Intel 8051 and Motorola 6805, and a reduced instruction set computer (RISC), the SUN SPARC.

3.1 Hardware Organization

The primary hardware elements of a digital computer are a central processing unit (CPU), memory, and assorted input and output devices, as illustrated in Figure 3.1. The CPU comprises a control unit, which coordinates the actions of the other elements in the computer, an *arithmetic and logic unit (ALU),* which is a digital logic circuit that manipulates data as instructed by the control unit, and a set of *registers,* which are high-speed storage locations used to temporarily store data, addresses, and other information within the CPU. The ALU, registers, memory, and input/output devices make up the *datapath* of the computer.

Each ALU is unique in the type of data that it can manipulate and the set of operations that it can perform on those data. Most ALUs support operations on binary integers of various sizes.

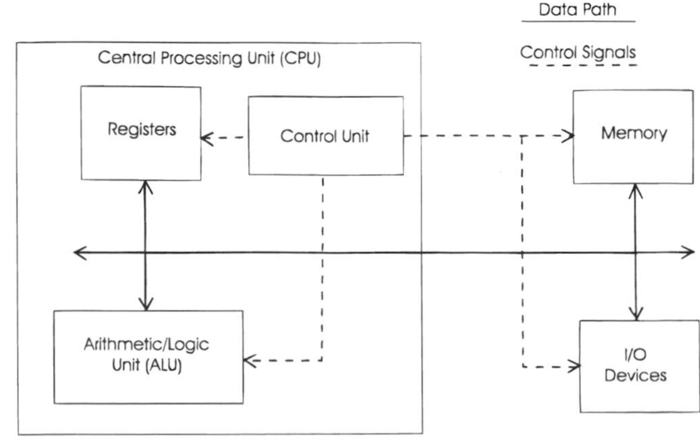

Figure 3.1 Computer hardware organization.

0-8493-8343-9/97/$0.00+$.50

Some also include operations to manipulate nonbinary and floating-point numbers, and various nonnumeric data. Typical ALU operations include:

- Arithmetic: add, subtract, multiply, divide, compare.
- Logical: AND, OR, exclusive-OR, complement, bit test.
- Shift and rotate data.

The control unit of a CPU is responsible for fetching program instructions from memory, interpreting or *decoding* the instruction codes, and executing instructions by issuing control signals to the elements of the data path. The control unit coordinates all operations of the ALU, memory, and I/O devices by continuously cycling through a set of operations that cause instructions to be fetched from memory and executed. This sequence of events is called the *instruction cycle* of the computer, and is illustrated in Figure 3.2. An instruction cycle includes five basic steps:

1. An instruction is fetched from the memory into the control unit of the CPU.
2. The control unit decodes the instruction, i.e., determines from the instruction code what operations to perform.
3. Any data, called *operands,* needed to perform these operations are read from input devices indicated by the instruction, retrieved from memory, or accessed from CPU storage registers.
4. The operation is performed on these operands.
5. The result is written to a register, a memory location, or an output device.

Program instructions and data are stored and retrieved from the memory of the computer. If a single memory is used for both, as is the case in most general-purpose computers and illustrated in Figure 3.3a, the computer is said to have a *Von Neumann* architecture, after Jon Von Neumann who is credited with developing the first stored program computer. A computer that uses one memory for instructions and a separate memory for data as shown in Figure 3.3b is referred to as having a *Harvard* architecture. Many microcontrollers fall into this category. In addition, a number of high-performance CPUs use Harvard architectures so that instruction and data memories can be accessed concurrently.

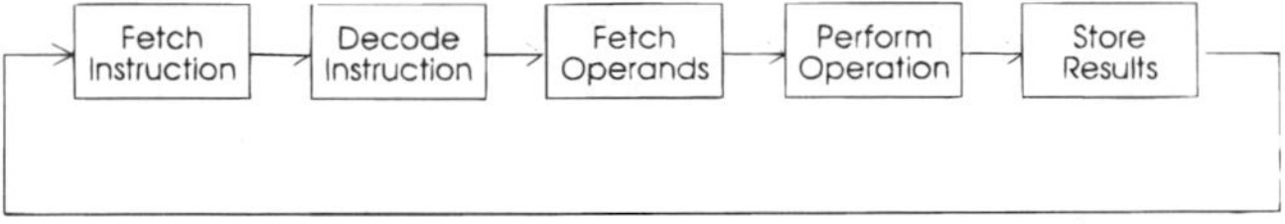

Figure 3.2 Instruction fetch and execute cycle.

The *instruction set architecture* of a computer refers to the organization of a computer instruction set as seen from a programmer's point of view. Every instruction set architecture is unique in how it supports different data types, operations on data, and access to information in registers, memories, and input/output devices.

Information is transferred between a computer and the outside world through various input/output devices. Programs are usually transferred into the memory of a computer from such peripheral equipment as magnetic or optical peripheral storage devices. Data to be used by a program can likewise be transferred into memory from keyboards, scanners, magnetic disks, analog to digital converters, and other input devices. A program may output data to several types of peripherals. Cathode-ray tubes (CRTs) and liquid crystal display (LCD) panels are often used to display the results of a program's calculations, and various types of printers are used to produce permanent results. Digital-to-analog converters, plotters, magnetic disks and other recording equipment are a few commonly used output devices.

Computers are often classified according to levels of integration. A *mainframe* computer is a large machine whose circuitry is typically contained on several circuit boards or cabinets of circuit boards. A *microprocessor* is an integrated circuit (IC) chip containing a complete CPU. A *minicomputer* falls in between mainframe and microprocessors. *Personal computers (PCs)* are typically built around microprocessors and include a video display system, a keyboard for data entry, and disk drives for information storage. Common PC add-ons include printers, pointing devices such as mice, track balls, and joysticks, CDROMs and tape drives for mass storage of information, sound generators for multimedia applications, and modems and network interface hardware for communication with other computers. Engineering *workstations* are similar to PCs, although workstations are oriented more toward intensive graphics applications and networking.

To improve performance, some processors incorporate high-speed memory, called *cache memory,* within the CPU itself.

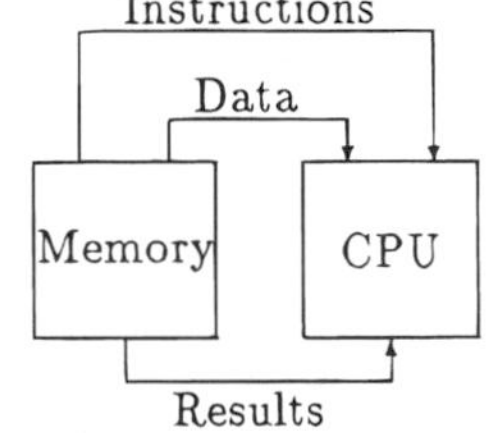

(a) Von-Neumann Architecure

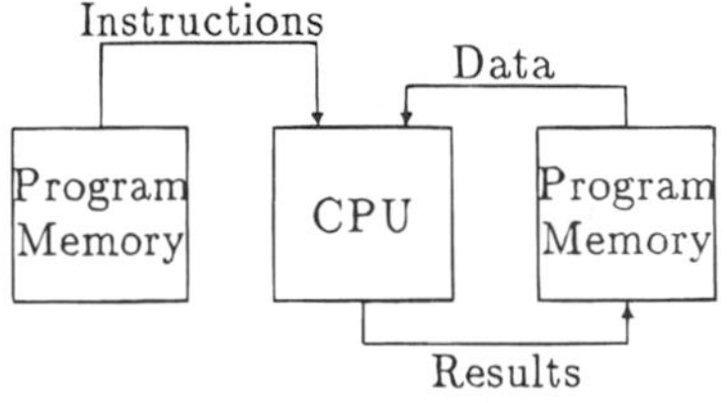

(b) Harvard Architecure

Figure 3.3 Computer memory architectures.

Superscalar processors further improve performance by integrating multiple ALUs and other functional units within a single CPU to allow multiple instructions to be executed concurrently.

A *microcontroller* is a complete computer on a single IC chip, comprising CPU, memory, and various I/O devices and interfaces to external sensors, actuators, and other devices. To make room for these elements on chip, most microcontrollers CPUs have fewer capabilities than general-purpose microprocessors. Microcontrollers are primarily used in embedded control systems, in which the computer is embedded within the hardware of such products as an automobile engines, kitchen appliances, communication equipment, and industrial control systems.

3.2 Computer Software

Computer software comprises programs of instructions that specify how the computer hardware is to be utilized to manipulate data. Programs can be classified as application or system programs. An *application program* is a set of instructions designed to perform a given task according to a specified algorithm. *System programs* comprise all of the software provided on a computer system to aid programmers in developing and executing application programs.

Programming Languages

The individual steps of an algorithm or task to be performed by a program must be expressed using the statements of a programming language. Every computer has a unique native *machine language* that is recognized by its hardware. A machine language is a set of binary codes that indicate to the CPU operations to be performed and operands (data items) to be used in those operations. All digital computer instructions must be represented by these binary codes before the computer can interpret them.

Rather than writing programs directly as sequences of binary codes, a symbolic representation of a CPU's machine language called *assembly language* is often used to develop programs, especially for applications that require very small or very efficient programs, such as control systems embedded into such products as home appliances and automobiles. Assembly language allows a programmer to symbolically specify operations to be performed on data stored in the internal registers and memory of a processor without becoming bogged down in creating binary code sequences. A system program called an *assembler* translates the symbolic assembly language instructions into machine language so that the program code can be interpreted by the CPU.

The following are a Motorola 6805 machine language instruction and its equivalent assembly language instruction.

```
Machine language:    10001011  00011001
Assembly Language:   ADDA      #25
```

This instruction tells the 6805 CPU to add 25 to the current contents of its *A* register and to replace the contents of the *A* register with the computed sum. The first 8 bits of the machine language code indicate that the operation is ADD, that the *A* register is to be used, and that an 8-bit data value follows as the second byte of the instruction code.

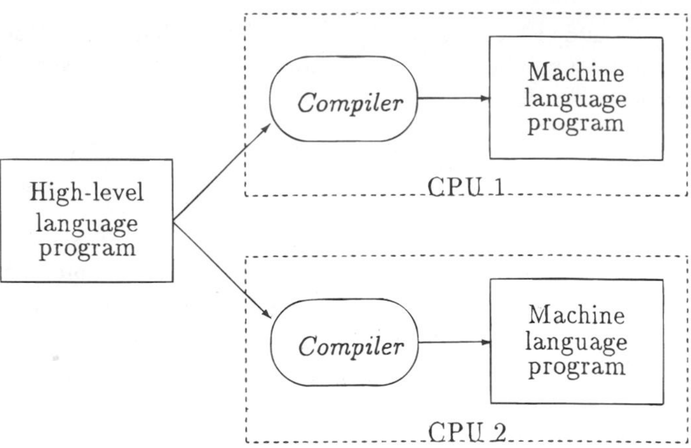

Figure 3.4 Compilation of a high-level language program for two different CPU architectures.

An assembly language programmer must be intimately familiar with the architecture of the specific CPU being programmed to efficiently represent algorithms for a given application in the assembly language of that CPU. In contrast, *high-level languages* allow a programmer to express an algorithm for a given application in a more natural way, independent of any particular computer architecture. Many different high-level languages have been developed, most tailored to specific applications. FORTRAN (FORmula TRANslation) was developed for numeric applications, COBOL (COmputer Business Oriented Language) for business applications and PROLOG and LISP to support artificial intelligence and expert system applications. Some languages, such as BASIC, PASCAL, ADA, C, and C++ are more general purpose, supporting a wide variety of different applications.

To execute a high-level language program on a computer, the program must be translated to the native machine language of that computer by a system program called a *compiler.* One of the primary benefits of using a high-level language to write a program is that the program can be targeted at a different computer architecture by simply recompiling it into the machine language of that architecture as illustrated in Figure 3.4. Thus, a high-level language program is portable across different computer architectures. The process of recompiling a program for a new computer architecture is referred to as *porting* the program to the new architecture.

As an example, the following C language statement is a natural way to indicate that the value of variable *a* is to be set to the sum of the values of variables *b* and *c*.

$$a = b + c$$

This statement might be translated by a C compiler to the following sequence of assembly language instructions for a Motorola 6805 microcontroller:

```
ldaa  b  ;load variable b into accumulator a (ACCA)
adda  c  ;add variable c to the value in ACCA
staa  a  ;store the value in ACCA into variable a
```

The same C instruction might be translated to the following for a SUN SPARC processor:

```
ld    b,r0       ;load variable b into register r0
ld    c,r1       ;load variable c into register r1
add   r0,r1,r2   ;write the sum r0 + r1 into register r2
st    r2,a       ;store the sum in variable a
```

In addition to assemblers and compilers, other system programs are usually available to assist in the development of programs. These include text editors to create and alter the text of a program, linkers to link together multiple program segments, including programs from libraries of routines, and program debugging tools.

Operating Systems

When a computer is dedicated to one specific task, the application program can be stored permanently in the memory of the computer and executed with no other support software. General-purpose computing systems, however, execute many different programs, which change from day to day and from user to user. In addition, multiple users may need to share a single system, or a single user may wish to concurrently perform multiple tasks on a single system. In such cases, a control program called an *operating system* is used to coordinate usage of the facilities of the computer.

An operating system is a program that interprets user commands typed at a keyboard, selected by clicking on an icon with a mouse, or read from a file. Some commands are executed by programs built into the operating system, while other commands correspond to programs that reside on a disk or are otherwise supplied by a user. An operating system also manages the file system on the computer, which comprises a directory of files stored on a disk or tape and programs that locate and access these files when requested by a program. An operating system also coordinates access to printers, networks, and other input/output devices, and manages CPU time and memory space.

Most PCs are controlled by single-user operating systems such as MSDOS or the Macintosh System 7. A single-user operating system interprets one command at a time from the user, executes the corresponding program, and then waits for another command to be entered. No other program may be executed until the current one is finished. Once an individual has control of the computer, that user may modify and reexecute programs or execute several different programs before turning the machine over to the next user.

In the single-user environment, much time can be spent idling while waiting for user inputs or data transfers involving slow input/output devices. To exploit this idle time, multitasking and multiuser operating systems allow CPU time to be shared by multiple programs. The operating system passes control of the CPU from program to program, with each program allowed to execute for a small allotment of time or until it becomes stalled waiting for input/output. In this manner, the execution of a program is interleaved with execution of other programs until it has completed. The end result is that programs execute concurrently with each program appearing to have exclusive control of the CPU. UNIX is an example of a multiuser, multitasking operating system and is used on a wide variety of personal computers, workstations, and larger machines. Multiple users can issue commands to the operating system of one computer from different terminals, with each user running several programs at the same time.

Process control and many other applications require *real-time* operation, in which the computer must respond to various events and perform designated actions within given time constraints as the events occur. In such applications, special real-time operating systems are used to coordinate the execution of processes in response to these events.

3.3 Information Representation in Digital Computers

All information in a computer must be represented by patterns of 1's and 0's. The assignment of a meaning to a bit pattern is called *coding.* In general, an *n*-bit pattern of information can represent 2^n unique items. CPUs generally support a limited number of pattern lengths. In most general-purpose computers the smallest pattern size is 8 bits, referred to as a *byte,* although some smaller microcontrollers work primarily with 4-bit *nibbles* of information. In general, the primary pattern length supported by a CPU is referred to as its *word size,* which is usually an integral number of bytes. For example, the word size in the Intel 8086 and Motorola 68000 CPUs is 16 bits, while the word size is 32 bits in the Intel 80486, Motorola 68040, and SUN SPARC CPUs. Some more advanced processors support 64-bit and larger information patterns.

Figure 3.5 shows a taxonomy of different information types found in a computer. Addresses are pointers to storage locations in memory or input/output device interfaces. Instruction codes tell the CPU what operation to perform in a given program step. Data are items to be manipulated by computer instructions.

Data can be classified as numeric or nonnumeric. Different numeric data formats are used in different applications. In most

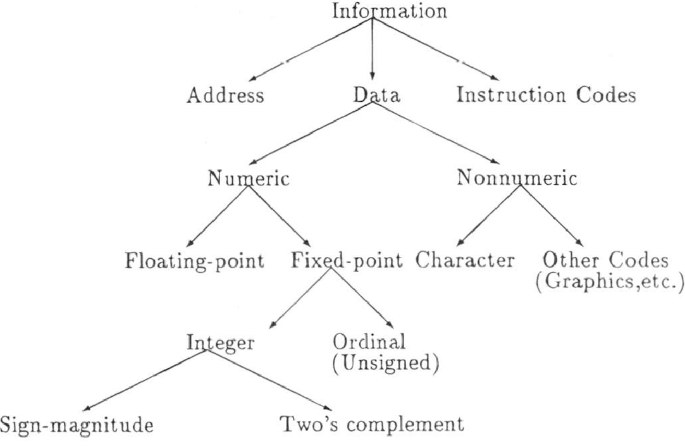

Figure 3.5 Taxonomy of information representation.

situations, simple binary integers are sufficient. Binary integers can be unsigned or signed. An n-bit unsigned (*ordinal*) number can represent values in the range $[0 \cdots (2^n - 1)]$.

Signed integers can be represented in several ways. An n-bit *sign-magnitude* number format uses the leftmost bit to represent the sign of the number and the remaining $n - 1$ bits to represent the magnitude or absolute value of the number. A sign bit of 0 indicates a positive value and 1 indicates a negative value. An n-bit sign-magnitude number can represent numbers in the range $[-(2^{n-1} - 1) \cdots + (2^{n-1} - 1)]$.

Digital logic circuits that add and subtract sign-magnitude format numbers are difficult to design. For this reason, most computers use the *two's complement* number system to represent signed numbers. In an n-bit two's complement number system, positive numbers are represented as they are in sign-magnitude; the most significant bit is 0 and the remaining $n - 1$ bits indicate the magnitude. A negative value A is represented by its two's complement, defined as $2^n - A$. This simplifies arithmetic hardware by allowing one to compute $A - B$ by forming the two's complement of B and then adding it to A, i.e., $A - B = A + (-B)$. The two's complement is fairly easy to compute; one method is to complement all of the bits and then add 1 to the result.

For example, the 8-bit two's complement number system representation of the value -5_{10} would be computed as follows:

1. Write the 8-bit binary code for $+5_{10} = 00000101_2$.
2. Complement the bits: $\overline{00000101} = 11111010$.
3. Add one to the result: $11111010 + 1 = 11111011$.

Therefore, the 8-bit code 11111011 represents the value -5_{10}. Note that the sign-magnitude code for the the value -5_{10} is 10000101, which is the code for $+5_{10}$ with the leftmost bit set to 1 to indicate a negative value. The reader is referred to Nelson et al. (1995) for additional algorithms and examples involving signed numbers.

Financial and other applications require manipulation of decimal rather than binary numbers. In these cases, *binary-coded decimal* (*BCD*) coding is used. In BCD coding, each decimal digit is represented independently by its 4-bit binary equivalent. Generally two BCD digits are packed into an 8-bit byte; this is referred to as *packed BCD*. For example, the packed BCD representation of 25_{10} is coded by packing into 8 bits the binary code for 2 (0010) and the binary code for 5 (0101). The result is the packed BCD code 00100101. Many general-purpose CPUs provide special arithmetic hardware and instructions to assist in manipulating BCD numbers without requiring conversions to and from binary.

Table 3.1 Integer Number Ranges for Different Coding Methods

Bits	Ordinal	Sign-magnitude	Two's complement
4	0 . . 15	−7 . . +7	−8 . . +7
8	0 . . 255	−255 . .+255	−256 . . +255
16	0 . . 65535	−32767 . . +32767	−32768 . . +32767
32	0 . . 4,294,967,295	−2,147,483,647 . . +2,147,483,647	−2,147,483,648 . . +2,147,483,647

If both integer and fractional numbers are needed, a fixed-point or floating-point number format is used. Fixed-point notation partitions the n bits used to represent a number into two fixed-length parts: k bits to represent the integer part and $n - k$ bits to represent the fractional part, as listed in Figure 3.6. A binary point is assumed to be at a fixed position between the two parts. An n-bit fixed-point number is said to have n bits of *precision*, and can accurately represent a value to within $2^{-(n-k)}$, the value of the least significant (leftmost) bit. The *range* of numbers is the span between largest and smallest magnitudes. The largest value is approximately 2^k, as determined by the number of integer bits, and the smallest value is $2^{-(n-k)}$, which is a function of the number of fraction bits. Integers are special cases of fixed-point numbers with the binary point to the immediate right of the least significant bit.

Scientific and a number of other applications require very wide ranges of numeric values. In these cases *floating-point* number formats are used. A floating-point number is generally written as:

$$\pm m \times r^e$$

where m is the *mantissa*, r the *radix*, and e the *exponent* of the number. Floating-point numbers are represented in a computer system by packing the codes for e and m into a single storage location, as illustrated in Figure 3.7. The code for the radix r need not be stored since it is always known. The most common radix value is $r = 2$, corresponding to binary numbers. The IBM 360 mainframe architecture used a floating-point format based on $r = 16$.

Prior to the mid-1980s, each computer manufacturer developed its own scheme for encoding floating point numbers, making it difficult to transfer data from one computer architecture to another. In 1985, ANSI (American National Standards Institute) and the IEEE developed the *IEEE Standard for Binary Floating Point Arithmetic*, ANSI/IEEE Standard 754–1985, which

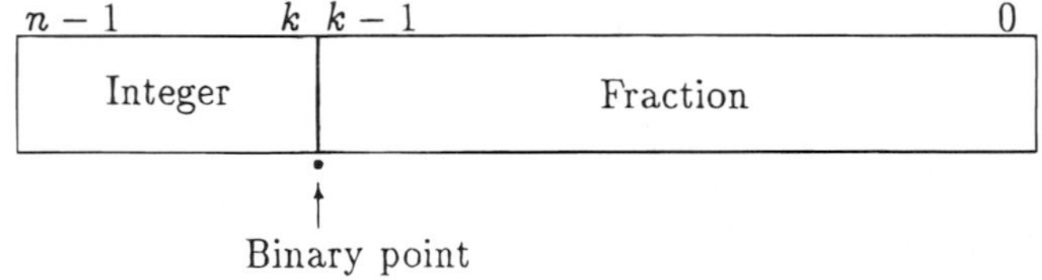

Figure 3.6 Fixed-point number format.

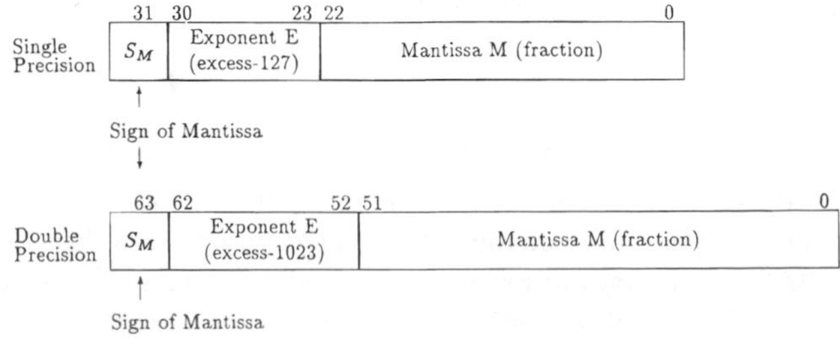

Figure 3.7 IEEE Standard 754–1985 floating-point formats.

defines standard single-precision (32-bit) and double-precision (64-bit) floating-point formats and operations that are now used in most computer systems. These are illustrated in Figure 3.7.

The mantissa is stored in normalized sign-magnitude format. The sign bit is placed in the leftmost bit of the 32-bit code, allowing a number to be easily identified as positive or negative. The magnitude of the mantissa is of the form 1.*F*, where *F* is a 23-bit binary fraction stored in bits 22–0 for the single-precision format, and a 53-bit fraction stored in bits 52–0 for the double-precision format. *Normalized* forms are characterized by their most significant digits being nonzero, and are used to ensure a single unique representation for each floating-point number. For example, the following are all representations of the same number:

$$10.11 \times 2^{9}$$

$$1.011 \times 2^{10}$$

$$0.1011 \times 2^{11}$$

$$0.01011 \times 2^{12}$$

Requiring the mantissa to be normalized to the form 1.*F* forces the unique code 1.011×2^{10} to represent this number.

The 8-bit exponent is stored in biased, *excess-127* format. In excess-*N* number format, a bias value of *N* is added to each value to force numbers in the range $[-N \ldots +(N-1)]$ to be represented by a linearly increasing number sequence $[0 \ldots (N-1)]$. In the IEEE Standard 754–1985 format, a bias value of 127 is added to each exponent value, with the result stored in bits 30–23. Exponents in the range $[-127 \ldots +126]$ are therefore represented by increasing binary codes from [00000001 . . . 11111110]. The exponent codes 00000000 and 11111111 are reserved to indicate special conditions. For example, the constant zero is represented by the all-0's word, and +/− infinity by an exponent of all 1's and a mantissa of all 0's.

Nonnumeric data are represented in a computer by designing a coding scheme that assigns a unique binary code to each data item. Alphanumeric and special character information is commonly represented by the *ASCII* (American Standard Code for Information Interchange) code, listed in Table 3.2. Each printable character has a unique 7-bit code that is recognized by most devices that send or receive alphanumeric information, such as printers, terminals, etc. For example, suppose we want to encode the message "ADD 1". This message has five characters, the fourth being a space or blank. In the ASCII code, our message becomes

A	D	D	space	1
1000001	1000100	1000100	0100000	0110001

ASCII characters are often padded with an extra zero on the left to allow each code to fit exactly into one 8-bit byte of memory.

Many other codes have been developed to represent graphical information, special symbols, and a wide variety of other information.

Table 3.2 7-Bit ASCII Character Codes ($c_6c_5c_4c_3c_2c_1c_0$)

<table>
<tr><th></th><th colspan="8">$c_6c_5c_4$</th></tr>
<tr><th>$c_3c_2c_1c_0$</th><th>000</th><th>001</th><th>010</th><th>011</th><th>100</th><th>101</th><th>110</th><th>111</th></tr>
<tr><td>0000</td><td>NUL</td><td>DLE</td><td>SP</td><td>0</td><td>@</td><td>P</td><td>'</td><td>p</td></tr>
<tr><td>0001</td><td>SOH</td><td>DC1</td><td>!</td><td>1</td><td>A</td><td>Q</td><td>a</td><td>q</td></tr>
<tr><td>0010</td><td>STX</td><td>DC2</td><td>"</td><td>2</td><td>B</td><td>R</td><td>b</td><td>r</td></tr>
<tr><td>0011</td><td>ETX</td><td>DC3</td><td>#</td><td>3</td><td>C</td><td>S</td><td>c</td><td>s</td></tr>
<tr><td>0100</td><td>EOT</td><td>DC4</td><td>$</td><td>4</td><td>D</td><td>T</td><td>d</td><td>t</td></tr>
<tr><td>0101</td><td>ENQ</td><td>NAK</td><td>%</td><td>5</td><td>E</td><td>U</td><td>e</td><td>u</td></tr>
<tr><td>0110</td><td>ACK</td><td>SYN</td><td>&</td><td>6</td><td>F</td><td>V</td><td>f</td><td>v</td></tr>
<tr><td>0111</td><td>BEL</td><td>ETB</td><td>'</td><td>7</td><td>G</td><td>W</td><td>g</td><td>w</td></tr>
<tr><td>1000</td><td>BS</td><td>CAN</td><td>(</td><td>8</td><td>H</td><td>X</td><td>h</td><td>x</td></tr>
<tr><td>1001</td><td>HT</td><td>EM</td><td>)</td><td>9</td><td>I</td><td>Y</td><td>i</td><td>y</td></tr>
<tr><td>1010</td><td>LF</td><td>SUB</td><td>*</td><td>;</td><td>J</td><td>Z</td><td>j</td><td>z</td></tr>
<tr><td>1011</td><td>VT</td><td>ESC</td><td>+</td><td>;</td><td>K</td><td>[</td><td>k</td><td>{</td></tr>
<tr><td>1100</td><td>FF</td><td>FS</td><td>'</td><td><</td><td>L</td><td>\</td><td>l</td><td>|</td></tr>
<tr><td>1101</td><td>CR</td><td>GS</td><td>−</td><td>=</td><td>M</td><td>]</td><td>m</td><td>}</td></tr>
<tr><td>1110</td><td>SO</td><td>RS</td><td>.</td><td>></td><td>N</td><td></td><td>n</td><td>~</td></tr>
<tr><td>1111</td><td>SI</td><td>US</td><td>/</td><td>?</td><td>O</td><td></td><td>o</td><td>DEL</td></tr>
</table>

3.4 Specifying Instruction Operands

A computer instruction must specify to the control unit what operation is to be performed, where to obtain operands for the operation, and where to store the result of the operation. As with other information, each instruction must be encoded into patterns of ones and zeros. Instruction codes are generally subdivided into separately coded fields as illustrated in Figure 3.8.

Operation Code	Operand 1 Specifier	Operand 2 Specifier	. . .

Figure 3.8 Instruction code format.

These fields include the *operation code* (opcode), which specifies what the instruction is to do, and one or more *operand specifiers*, which indicate operands to be used for the instruction.

A data value to be used as an instruction operand may be embedded within the instruction code, retrieved from register within the CPU, or read from an external memory location. Some CPUs also recognize special I/O device addresses. Those that do not access I/O devices via their memory address space.

When a constant is used as an operand for an instruction, the value of that constant is encoded within the instruction as the operand specifier. Such a data value is referred to as an *immediate* operand, because it is immediately available from the fetched instruction without having to access additional storage locations. The following instructions add the immediate value 5 to a designated CPU register.

```
8051:   ADD     5         ;A + 5 → A
6805:   ADDA    #5        ;ACCA + 5 → ACCA
8086:   ADD     AX,5      ;AX + 5 → AX
68000:  ADD.W   #5,D1     ;5 + D1 → D1
SPARC:  ADD     R2,5,R1   ;R2 + 5 → R1
```

In general, immediate values are encoded with the same number of bits as the second operand. The instruction codes for these machines include enough bits to represent these values. In the SUN SPARC (*The SPARC Architecture Manual,* 1987), all instructions are limited to 32-bits, which must include the operation code and three operand specifiers. Hence, the SPARC limits an immediate operand to a 13-bit two's complement value. This value will be extended by the CPU to 32 bits before doing the operation.

3.5 CPU Registers

A *register* is a set of high-speed binary storage elements that can be accessed concurrently. Registers are used within a CPU to temporarily hold data and memory address values that might be needed in the near future. Being located within the CPU, registers can be accessed more quickly and more efficiently than external memory. In addition, since there is typically only a small number of registers, operand specifiers can be several bits, as compared to memory addresses that require considerably more bits to represent. For example, SUN SPARC instruction codes include three 5-bit operand specifiers, each of which identifies one of 32 registers to be used as operands.

Every instruction set architecture has its own distinctive set of program-accessible registers that may be used to store data, addresses, and control or status information. Figure 3.9 shows the register sets of four popular microprocessors.

Every CPU contains one register called a *program counter* (*PC*) or *instruction pointer* (*IP*) that always contains the address in memory of the next program instruction to be executed. The PC register is updated automatically as each instruction is executed so that it points to the next instruction to be fetched from memory.

Most CPUs have one or more registers that can be used to hold memory addresses or information used to compute addresses. The 68000 (Motorola Inc., 1990) has eight such addressing registers, labelled A0–A7. The 8086 (Brey, 1994) registers BX, SI, DI, and BP may be used for both addresses and data. In contrast, microcontrollers generally have very limited memory addressing capabilities. For example, the 6805 (Motorola Inc., 1983) has a single register, X, that can be used in memory addressing.

The 8051 (Stewart, 1993) and 6805 microcontrollers are similar to many older computers in that arithmetic and logic operations are centered around a single *accumulator* register, labelled *A* in Figure 3.9. Arithmetic operations such as addition and subtraction combine the number in the accumulator with a second data value and write the resulting value back to the accumulator, overwriting the original accumulator contents. The second operand can be an immediate value or the contents of another register or memory location. To combine two data values from memory, one of them must be moved to the accumulator prior to the operation. After the operation, the result can be moved to a memory location if desired. The Motorola 68HC11 microcontroller register set is similar to that of the 6805, except that a second accumulator, B, is provided (Spasov, 1993). Either the A or B accumulator may be used by most instructions. In addition, the A and B accumulators can be used together as a single 16-bit accumulator referred to as register D.

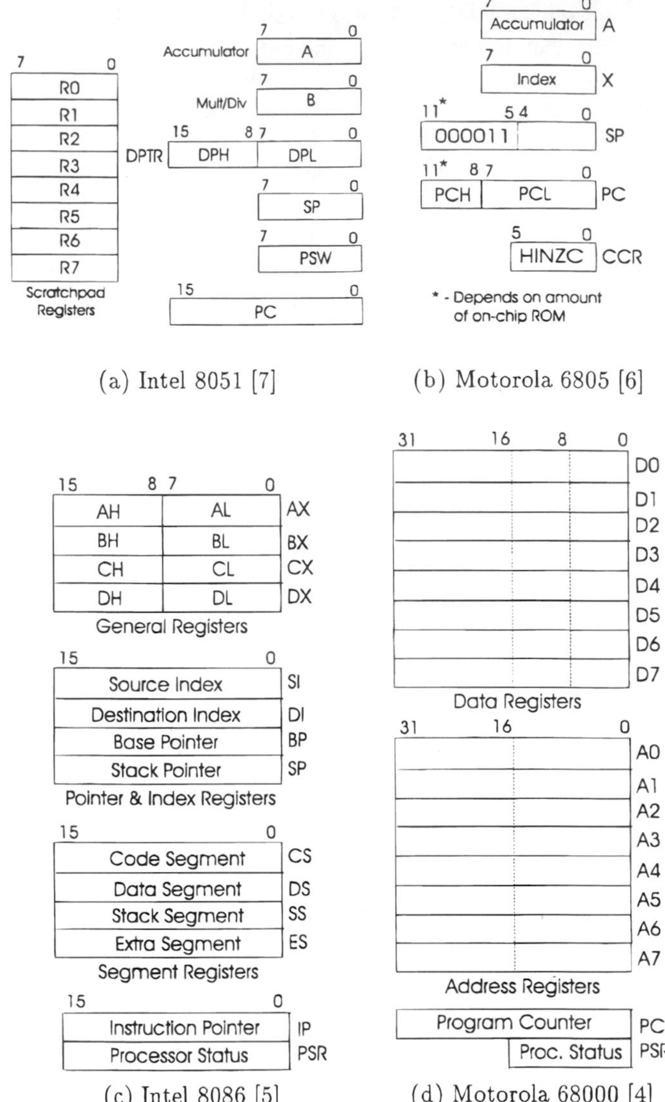

Figure 3.9 Register sets of four common CPUs. (a) *Source*: Stewart, 1993. *The 8051 Microcontroller: Hardware, Software and Interfacing,* Regents/Prentice-Hall, Englewood Cliffs, NJ. With permission. (b) *Source*: Motorola Inc., 1983. *M6805 HMOS/M146805 CMOS Family Users Manual,* Prentice-Hall, Englewood Cliffs, NJ. With permission. (c) *Source*: Brey, 1994. The *Intel Microprocessors: 8086/8088, 80286, 80386, and 80486: Architecture, Programming, and Interfacing,* 3d ed., Macmillan, New York, NY. With permission. (d) *Source*: Motorola Inc., 1990. *MC68000 8-/16-/32-Bit Microprocessors User's Manual,* 8th ed., Prentice-Hall, Englewood Cliffs, NJ. With permission.

The 8086 and 68000 CPUs give more flexibility to a programmer by providing a number of general purpose registers, any of which can supply operands for or receive the results of arithmetic and logical operations. The 68000 provides eight data registers named D0–D7. Any data register may be a source and/or a destination in any operation. 8-bit operations use the lowest 8 bits of the data register, 16-bit operations use the lower 16 bits of the register, and 32-bit operations use the entire register. The following illustrates 8-, 16-, and 32-bit addition operations; the suffix on the ADD mnemonic indicates the data size.

```
ADD.B   D0,D1   ;D0 + D1 → D1 (8-bit bytes)
ADD.W   D2,D3   ;D2 + D3 → D3 (16-bit words)
ADD.L   D4,D5   ;D4 + D5 → D5 (32-bit long words)
```

As in the 68000, the 8086 also provides eight registers that can be used in most arithmetic and logical instructions. However, most of these registers also have special functions and are named accordingly. The four general-purpose registers AX, BX, CX, and DX, are used by some instructions as an accumulator, a base address register, a count register, and an I/O addressing register, respectively. The four index and pointer registers SI, DI, SP, and BP are used for memory addressing. For byte operations, each half of the four general registers can be used independently, and hence the high and low parts of these registers are labeled AH/AL, BH/BL, CH/CL, and DH/DL. The following illustrates 8- and 16-bit addition operations; data sizes are deduced from the register sizes.

```
ADD   AX,DI   ;AX + DI → AX (16-bit words)
ADD   DH,CL   ;DH + CL → DH (8-bit bytes)
```

Most reduced instruction set computer (RISC) architectures, such as those of the SUN SPARC (The SPARC Architecture Manual, 1987) and MIPS R4000 (Patterson and Hennessy, 1994) allow three different registers to be used in arithmetic and logical operations: two to supply operands and a third to receive the result. This provides even more flexibility than the two-operand formats of the 8086 and 68000. For example, the SPARC architectures provide 32 registers that can be accessed at any given time, designated *r*0 ··· *r*31. The following is a SPARC add instruction.

```
ADD   R1,R2,R3   ;R1 + R2 → R3 (32-bit words)
```

Unlike the CPUs examined above, RISC architectures restrict operands for arithmetic and logical operations to register and immediate values only. Data in memory may only be accessed by special load and store instructions that move data between registers and memory.

To simplify the passing of parameters from one procedure to another, the SPARC actually includes more than 32 registers; the exact number depends on the specific implementation. At any given time the programmer only has access to 32 of these registers, as shown in Figure 3.10. The 8 global registers, *r*0 − *r*7, are accessible at all times. The remaining 24 registers, *r*8 − *r*31, are accessed through a sliding window. The window slides down 16 registers whenever a new procedure is called, overlapping the window of the calling procedure by 8 registers. The window of a subsequently called procedure is likewise overlapped by 8 registers. Note that registers *r*24 − *r*31 of one procedure refer to the same storage locations as registers *r*8 − *r*15 of the procedure that it calls. In this manner, a procedure can access parameters passed to it by reading registers *r*8 − *r*31, and can pass parameters to a new procedure by storing them in registers *r*24 − *r*31. Registers *r*16–*r*23 are local to each procedure, i.e., they can only be accessed by the currently-active procedure. Because the sliding register window is used to pass parameters to and from procedures and to save procedure return addresses, the SPARC does not directly support a stack data structure.

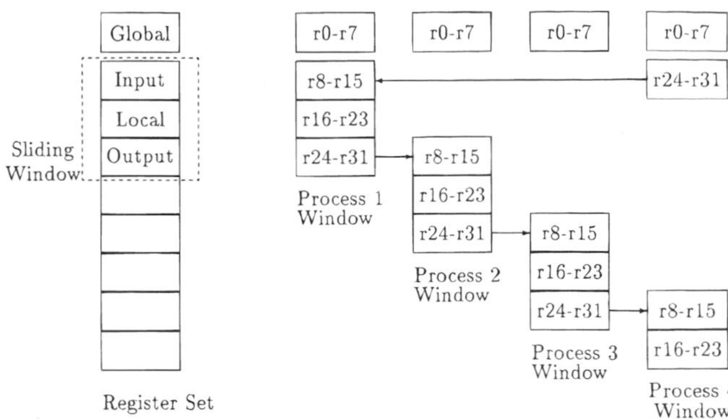

Figure 3.10 Process context changing with SPARC sliding register windows. *Source: The SPARC Architecture Manual,* 1987.

Table 3.3 6805 Condition Code Flags

Flag	Status of last result if flag = 1
Z (Zero)	Result zero
N (Sign)	Result negative
C (Carry)	Carry out of the most significant bit of the result
V (Overflow)	Result out of range for the given number of bits
H (Half Carry)	Carry from bit 3 to bit 4 of result

Source: Motorola, Inc., *M6805 HMOS/M146805 CMOS Family Users Manual,* 1983. Prentice-Hall, Englewood Cliffs, N.J.

Most CPUs contain a *processor status register* (*PSR*), sometimes called a *condition code register,* that contains information about internal CPU conditions and about operations that have been performed. PSRs usually contain one or more condition code bits, called *flags,* that characterize the result of a previous arithmetic or logical operation performed in the ALU. These allow decisions to be made based on the outcomes of these operations. Table 3.3 lists the condition code flags of the 6805 and 68000, which are typical of those found in most CPUs. The half carry flag is primarily used to support operations on BCD operations, representing a carry from one decimal digit to the next within a byte.

Many CPUs support a special last-in/first-out data structure in memory called a push-down *stack.* A stack is a convenient place to temporarily save information and subsequently restore it. For example, many CPUs allow a running program to be temporarily interrupted to execute some other program. The state of the running program can be saved temporarily on a stack while the other program is executed, and then be restored from the stack when that program is finished, allowing the original program to continue where it left off.

A special register called a *stack pointer* (*SP*) contains the address of the top element on the stack. An operation called PUSH adds an element to the stack, and an operation called POP or PULL removes an element from the stack, as illustrated in Figure 3.11. The SP automatically increments and decrements as elements are added to and removed from the stack.

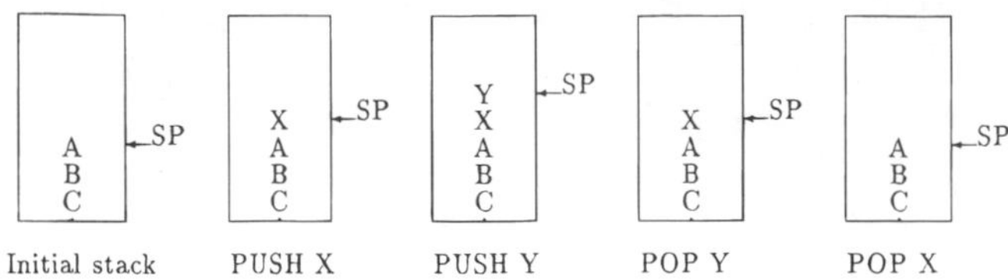

Figure 3.11 Push-down stack PUSH and POP operations.

D_{7-0}
0
1
⋮
$2^N - 1$

(a) Byte-wide

D_{15-8}	D_{7-0}
1	0
3	2
⋮	⋮
$2^N - 1$	$2^N - 2$

Little endian format

D_{15-8}	D_{7-0}
0	1
2	3
⋮	⋮
$2^N - 2$	$2^N - 1$

Big endian format

(b) Word-wide organizations

D_{31-24}	D_{23-16}	D_{15-8}	D_{7-0}
3	2	1	0
7	6	5	4
⋮	⋮	⋮	⋮
$2^N - 1$	$2^N - 2$	$2^N - 3$	$2^N - 4$

Little endian format

D_{31-24}	D_{23-16}	D_{15-8}	D_{7-0}
0	1	2	3
4	5	6	7
⋮	⋮	⋮	⋮
$2^N - 4$	$2^N - 3$	$2^N - 2$	$2^N - 1$

Big endian format

(c) Double-word-wide organizations

Figure 3.12 Byte-addressable memory organizations (2^N bytes).

3.6 Memory Organization

A computer system utilizes memory elements for storing program instructions, data, and other information. From the viewpoint of the instruction set, memory is a set of *words,* each identified by a unique *address* that indicates its location within the memory. The concept of a memory address is equivalent to that of a telephone number. Every telephone is assigned a unique number comprising an area code, exchange, and a number within that exchange. Similarly, each memory location is assigned a unique address that identifies a memory module and a specific storage location within that module.

Each memory word contains one or more addressable bytes, as illustrated in Figure 3.12. The number of bits in the data path of the CPU determine the number of bytes per word of memory. For example, the 8051 and 6805 have 8-bit data paths and thus have byte-wide memory organizations as in Figure 3.12a. The 8086 and 68000 have 16-bit data paths and support operations on both bytes and words. Therefore, memory must be *byte-addressable,* i.e., each byte of memory must have a unique address. Figure 3.12 shows two 16-bit memory formats. The 8086 uses the *little endian* format, i.e., the least significant byte of data is placed in the lower numbered address. The 68000 uses the *big endian* format in which the least significant byte of data is placed in the higher numbered address. The 80486 and 68040 CPUs have 32-bit data paths and support operations on 32-bit words, 16-bit halfwords, and 8-bit bytes. Memory for these CPUs is organized as four bytes per word as shown in Figure 3.12c. Both CPUs can access one, two, three, or all four bytes of a memory word with a single read or write operation. As do their predecessors, the 80486 uses the little endian format and the 68040 uses the big endian format.

The number of addressable memory locations in a computer is a function of the number of bits used by the CPU to represent memory addresses. An N-bit address can address 2^N locations. For example, the 8051 uses a 16-bit address, allowing it to address $2^{16} = 64$K bytes of memory. The 68000 uses 24 address bits, and can address $2^{24} = 16$M bytes of memory, organized as two bytes per word for a total of $2^{23} = 8$M words.

Memory Address Generation

To retrieve an operand from a memory location or write a result to a memory location, the address of that location must be specified by the instruction being executed. In most CPUs, memory addresses may be specified directly or indirectly.

Direct Addressing

The location of the operand is explicitly specified either numerically or symbolically when writing the instruction. The operand address is embedded within the assembled instruction code as illustrated in Figure 3.13. The following are examples of direct addresses specified by assembly language instructions. The last example uses the symbolic label BOB to represent a variable stored in memory. The actual address of BOB is determined and inserted into the instruction code when the instruction is assembled. Note that 68000 instructions specify the source as the left operand and the destination as the right operand, while the 8086 and 8051 do the opposite.

Indirect Addressing

It is often the case that a programmer will not know where a particular operand will be located at the time an instruction is executed; the operand address will be computed while the program is running. Examples include arrays of numbers that are accessed using a starting address of the array and an index into the array, and data accessed using pointers. In such cases the address is specified indirectly. The CPU is either told where to find the address, in a register or memory location, or how to calculate the address.

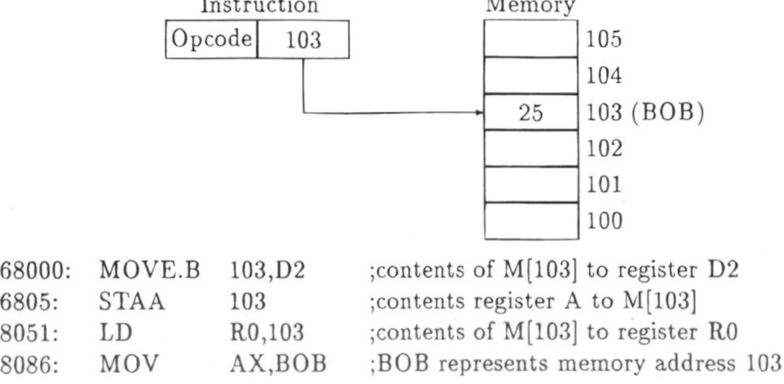


```
68000:  MOVE.B  103,D2    ;contents of M[103] to register D2
6805:   STAA    103       ;contents register A to M[103]
8051:   LD      R0,103    ;contents of M[103] to register R0
8086:   MOV     AX,BOB    ;BOB represents memory address 103
```

Figure 3.13 Direct memory addressing.

Figure 3.14 illustrates *register indirect* addressing of the data in memory location 103. The operand specifier in the instruction code indicates that the operand address is contained in register *R*. Since there are relatively few registers in a CPU, as compared to the number of addressable memory locations, the operand specifier is only a few bits.

The 68000 and some other CPUs support special autoincrement and autodecrement modes of register indirect addressing, in which the address register is automatically incremented after, or decremented before, each memory access. This simplifies access to tables of data in which the elements are to be accessed sequentially, as illustrated in the following example which computes the sum of a list of numbers in a table.

```
    MOVE.W   SIZE,D1     ;Load size of TABLE into D1
    LEA      TABLE,A0    ;Load address of TABLE into A0
    CLR.W    D0          ;Initialize SUM to 0
L:  ADD.W    (A0) +,D0   ;SUM = SUM + TABLE(I)
    DBR      D1,L        ;Repeat summation until finished
```

In the ADD instruction, two is automatically added to the address in register A0 after reading each two-byte data word, leaving A0 pointing to the next element of the table.

The Motorola 68040 and the Digital Equipment Corporation VAX CPUs also support *memory indirect* addressing modes, in which the address of an operand is contained in a designated memory location. This requires two memory accesses: one to fetch the operand address and another to fetch the operand itself.

Base/Indexed Addressing

Data are often stored in tables, lists, records, or other data structures in which addresses are specified in two parts: a *base* (beginning) address of the data structure and an offset, or *index*, from the beginning. The operand specifier either directly or indirectly identifies the base address and the index.

The base address, the index, or both is usually retrieved from registers. The example instructions in Figure 3.15 designate an address register for the base address and specify a constant index of 3. This form of base-indexed addressing is useful for accessing records and similar data structures in which each element is at a known offset from the beginning of the structure.

For accessing linear arrays of data, it may be more convenient to explicitly specify a base address and identify a register containing an index as shown in Figure 3.16. Alternatively, a number of CPUs allow a base address to be in one register and an index in another as shown in Figure 3.17.

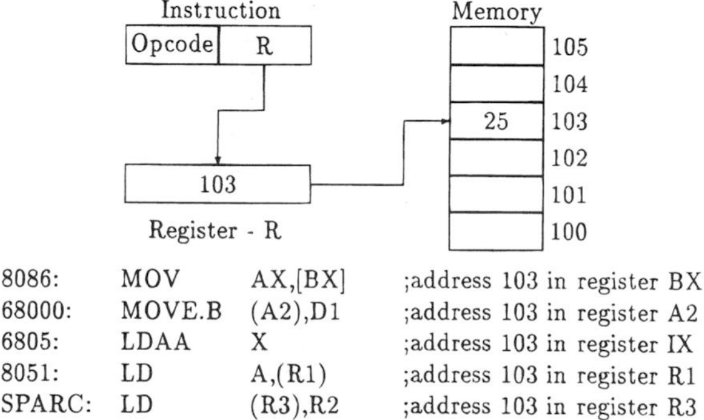

Figure 3.14 Indirect memory addressing.

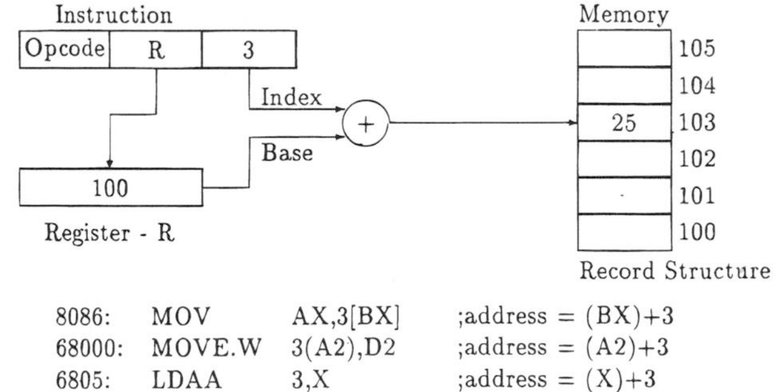

Figure 3.15 Addressing a record element with variable base and constant offset.

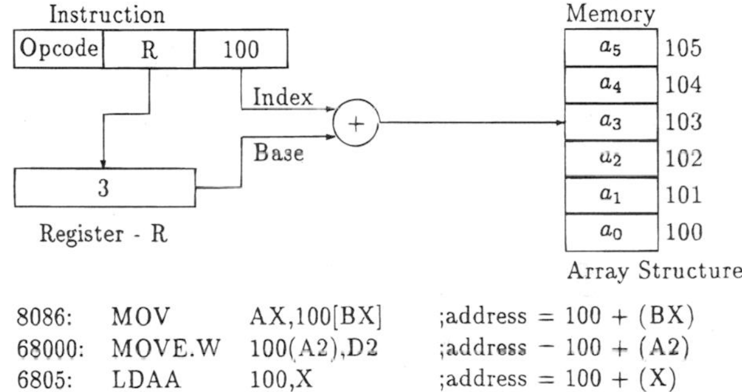

Figure 3.16 Addressing an array with constant base and variable offset.

The following program shows how to access an indexed array variable, TABLE(I), in the 68000.

```
LEA      TABLE,A0        ;Let A0 point to TABLE
MOVE.W   I,D0            ;Load index I into D0
MOVE.B   (A0,D0.W),D1    ;Load TABLE(I) into D1
```

The 68040 and 80486 support a *scaled index* addressing mode, in which the index is multiplied by a scale factor corresponding to the number of bytes in the accessed data item. This allows a simple index to be automatically converted into a displacement from the beginning of a table of values as illustrated in Figure 3.18.

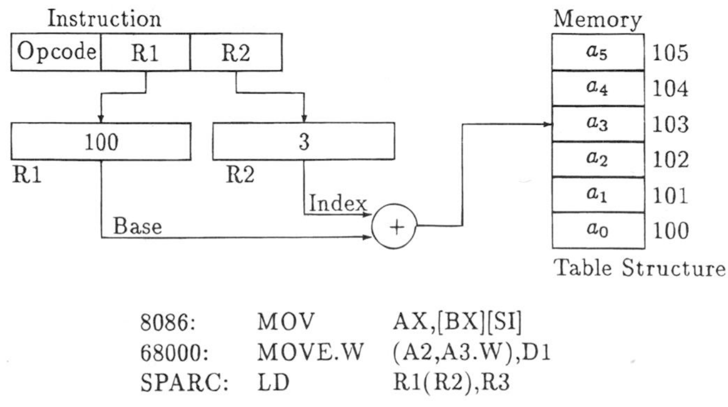

Figure 3.17 Addressing an array using both base and index registers.

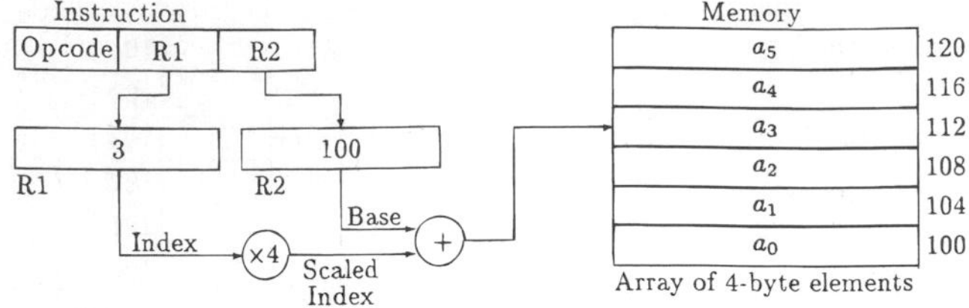

Figure 3.18 Scaled index for an array of 4-byte data values.

Program Counter Relative Addressing

It is often necessary to jump or branch from one point in a program to another. Instead of specifying the target address within the branch instruction, most CPUs compute the target address by adding to the program counter a displacement from the current instruction to the target instruction. This is referred to as program counter relative addressing. The advantage of doing this is that code can be made position-independent and relocatable. This means that the program can be loaded in any location in memory without reassembling or recompiling it, since branch instructions are only dependent on the distance to each target address and not the absolute value of the address.

3.7 Computer Instruction Types

Digital computer instructions can be organized into six basic categories: data transfer, arithmetic, logical, shift/rotate, control transfer, and processor control.

Data Transfer

Data transfer instructions load a CPU register with the contents of a memory location, store the contents of a CPU register into a memory location, move data from one CPU register to another, or from one memory location to another. The following are examples of data transfer instructions for the example CPUs.

```
8051:    MOV      A,R0       ;register R0 to accumulator A
         MOV      R0,A       ;accumulator A to register R0

6805:    LDAA     MEMY       ;memory to accumulator A
         STAA     MEMY       ;accumulator A to memory
         TBA                 ;accumulator B to accumulator A

8086:    MOV      AX,MEMY    ;memory to register AX
         MOV      MEMY,AX    ;register AX to memory
         MOV      AX,BX      ;register BX to register AX

68000:   MOVE.W   MEMY,D0    ;memory to register DO
         MOVE.W   D0,MEMY    ;register DO to memory
         MOVE.W   D0,D1      ;register DO to register D1

SPARC:   LD       MEMY,R0    ;load R0 from memory
         ST       R0,MEMY    ;store R0 into memory
```

Other examples of data transfer instructions are PUSH and POP operations using the stack pointer register, as described earlier, and instructions to load address registers with the computed operand addresses. The following instructions compute the sum of a base address and an index register and place the result into an address register to use as a pointer.

```
8086:    LEA   DI,TABLE[SI]    ;DI points to TABLE(I)
68000:   LEA   TABLE(A0),A1    ;A1 points to TABLE(I)
```

Arithmetic Instructions

Instructions such as add, subtract, multiply, and divide perform binary arithmetic on integer operands. Not all CPUs support all four functions. For example, many microcontrollers provide only add and subtract instructions and leave it to the programmer to write short programs to perform multiplication or division. Some CPUs include additional instructions to increment and decrement binary numbers to facilitate counting operations and modification of memory addresses.

CPUs that support decimal number formats provide special instructions to perform binary arithmetic on BCD numbers, either directly or indirectly. The 68000 has special BCD add and subtract instructions, while in most other CPUs BCD numbers are added by using the normal binary ADD instruction followed by a special *decimal adjust* instruction to correct the result. The following illustrate examples of adding two packed BCD numbers.

```
68000:   ABCD   D0,D1    ;BCD sum to D1

8086:    ADD    AL,DL    ;binary sum in AL
         DAA             ;decimal adjust result in AL
```

Often one must do arithmetic on multi-precision numbers, i.e., numbers with more bits than the word size of the CPU. This is done as with pencil and paper, where one adds the two least significant digits of a pair of numbers, producing a digit and possibly a carry. If there is a carry, 1 is added to the next pair of digits and so on. An add or subtract operation will set the carry flag in the PSR to 0 or 1 to indicate that a carry or borrow was produced. Using an add-with-carry or subtract-with-borrow instruction allows the next pair of digits to be adjusted as needed. The following example illustrates the computation of $H = H + G$ on an 8086, where H and G are 32-bit numbers and CF is the carry flag.

$$\begin{array}{rr} & CF \\ & G_1 \quad G_0 \\ + & H_1 \quad H_0 \\ \hline & H_1^* \quad H_0^* \end{array}$$

$H_0^* = G_0 + H_0$

$H_1^* = G_1 + H_1 + CF$

Check:

```
MOV   AX,G       ;Get low word of G
ADD   H,AX       ;Add to low word of H
MOV   AX,G+1     ;Get high word of G
ADC   H+1,AX     ;Add-with-carry to high
                  word of H
```

Logical

Logical instructions apply the Boolean AND, OR, and exclusive-OR (XOR) operations to corresponding bits of two operands. This gives the computer the ability to set, clear, complement, or test individual bits or groups of bits within a memory location

Table 3.4 Boolean Operations on Bit Variable b

AND	OR	XOR
$b \cdot 0 = 0$	$b + 0 = b$	$b \oplus 0 = b$
$b \cdot 1 = b$	$b + 1 = 1$	$b \oplus 1 = \bar{b}$

$$\begin{array}{rcccc} & b_3 & b_2 & b_1 & b_0 \\ \wedge & 1 & 1 & 0 & 1 \\ \hline & b_3 & b_2 & 0 & b_0 \end{array} \qquad \begin{array}{rcccc} & b_3 & b_2 & b_1 & b_0 \\ \vee & 0 & 0 & 1 & 0 \\ \hline & b_3 & b_2 & 1 & b_0 \end{array} \qquad \begin{array}{rcccc} & b_3 & b_2 & b_1 & b_0 \\ \oplus & 0 & 0 & 1 & 0 \\ \hline & b_3 & b_2 & \bar{b}_1 & b_0 \end{array}$$

(a) Clear b_1 (b) Set b_1 (c) Toggle b_1

Figure 3.19 Logical operations used to alter a selected bit.

or I/O device register. Table 3.4 summarizes the three Boolean operators applied to a one-bit Boolean variable.

The AND operator can be used to force selected bits of a word to 0, as illustrated in Figure 3.19a. The second operand is a bit pattern called a *mask* that contains a 0 in each bit position that is to be forced to 0, and a 1 in each bit position that is to be left unchanged. Similar masks can be created for the OR operator to force selected bits to 1, and for the XOR operator to force selected bits to be complemented. These are illustrated in Figures 3.19b and 3.19c, respectively.

Many input/output devices contain a status register whose bits reflect the readiness of the device to perform an operation. The AND operator can be used to isolate a selected bit of a byte read from a status register to determine if that bit is 0 or 1. This is illustrated in Figure 3.20. Here the mask is used to force all bits to 0 except for bit b_1. If the zero flag of the CPU's processor status register is set, indicating a result of 0000, then it follows that $b_1 = 0$; if the zero flag is not set, the result is nonzero which implies $b_1 = 1$.

For example, assume that a printer interface contains a status register in which the rightmost bit indicates whether the printer is ready to accept another character to print. The following 8086 program loop will be continuously executed as long as the "printer ready" bit is 0. The CPU will exit the loop and continue as soon as the ready bit becomes 1.

```
Check:  IN    AL, PrintStatus   ;read printer status register
        AND   AL,0000 0001      ;isolated "printer ready" bit
        JZ    Check             ;go back to Check if printer not ready
```

Shift and Rotate

Shift and rotate instructions slide bits right or left within a register or memory location as illustrated in Figure 3.21. These can be used for extracting or combining bit fields within an

$$\begin{array}{rcccc} & b_3 & b_2 & b_1 & b_0 \\ \wedge & 0 & 0 & 1 & 0 \\ \hline & 0 & 0 & b_1 & 0 \end{array}$$

Figure 3.20 Using AND to isolate one bit.

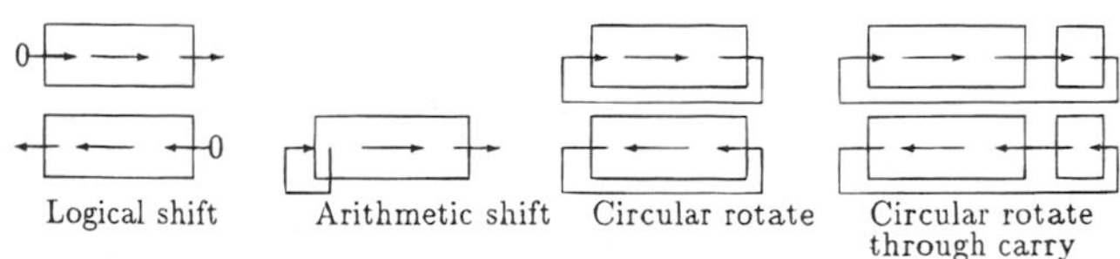

Figure 3.21 Shift and rotate operations.

operand, to convert data between parallel and serial form, and to perform multiplication and division by powers of 2.

In a *logical shift* operation, the bits are shifted right or left by one bit position, with the vacated bit replaced by a 0. For unsigned numbers, this is equivalent to dividing or multiplying the number by 2. An *arithmetic right shift* implements a divide by 2 operation on a two's complement number by preserving the sign bit as the operand is shifted. Some CPUs allow an operand to be shifted by more than one bit position with a single instruction. The following 68000 example packs two BCD digits into a single byte by shifting one digit four bits to the left and then combining the two digits.

```
Check:  SHL.B   #4,D0    ;shift BCD digit to upper nibble of D0
        OR.B    D1,D0    ;combine two BCD digits in D1 and D0
```

Circular rotate instructions perform a shift operation while replacing the vacated bit with the bit shifted out of the other end of the operand. A second rotate operation is often provided that rotates the number through the carry flag of the processor status register. In most CPUs, the bit shifted out of an operand is copied to the carry flag of the processor status register where it can be tested or used to support multi-precision shift operations. A multi-precision number can be shifted by using the carry flag as a link between parts of the number, allowing a bit shifted out of one part to be shifted into the other using a rotate-through-carry operation. The following 8086 example multiplies a 32-bit number by 2 by shifting one byte at a time one bit to the left.

```
SHL  NUMBER      ;shift memory byte 1 bit left
RLC  NUMBER+1    ;shift carry and 2nd byte 1 bit left
RLC  NUMBER+2    ;shift carry and 3rd byte 1 bit left
RLC  NUMBER+3    ;shift carry and 4th byte 1 bit left
```

Control Transfer

The normal flow of a program is to execute instructions in order from sequential memory addresses. To control this flow, the program counter increments automatically after each instruction. Jump, branch, and subroutine call instructions interrupt the normal flow by transferring control of the program to some instruction other than the next one in sequence. This allow looping and decision-making programs to be written, as well as supporting procedure and function calls. The following are examples of instructions that unconditionally transfer control of a program to location X within the current program:

```
8051/8086:      JMP X
6805/68000:     JMP X or BR X
SPARC:          BRA X
```

Decision making and looping require conditional branch instructions that jump only if a given condition is true and continue with the next sequential instruction if the condition is false.

Conditional branch instructions typically test selected bits of the processor status register, which reflect the result of a previous arithmetic or logical operation. The following 8086 program loop adds a list of four numbers in memory, decrementing the SI register at the end of each iteration and repeating the loop as long as SI is greater than or equal to 0.

```
        MOV   SI,3            ;set counter to 3
        MOV   AL,0            ;clear accumulator
Start:  ADD   AL,TABLES[SI]   ;add next element of TABLE
        DEC   SI              ;subtract 1 from SI
        JGE   Start           ;repeat if SI ≥ 0
```

The relationship between two operands can be tested by subtracting them and then testing the resulting condition codes according to Table 3.5. Many CPUs provide a compare instruction (CMP) that performs the subtraction and sets the condition code flags without altering either operand. The following 6805 program branches to location RICK if the unsigned number in accumulator *A* is less than or equal to 10, using the "branch if less or same" instruction to test the result of a compare instruction.

```
Check:  CMP   #10     ;subtract 10 from A
        BLS   RICK    ;go to RICK if A lower than or same as 10
```

Modular programming requires the ability to partition software into separate subroutines, such as procedures and functions, that can be invoked as needed. This is supported by special subroutine call instructions instructions that jump from a main program to the start of a subroutine after saving a pointer to the next instruction in the main program, allowing a return to the main program after completing the subroutine.

A subroutine call (CALL) or jump to subroutine (JSR) instruction typically pushes the current program counter onto the system stack to save the address of the next instruction in the main

Table 3.5 Condition Codes for Relational Operators

Condition	Symbol	Relation	Number type	Boolean condition
Zero	Z	$A = B$	Both	Z
Not zero	NZ	$A \neq B$	Both	$\bar{Z}$
Greater than	G	$A > B$	Signed	$\overline{(N \oplus V) + Z}$
Greater than or equal	GE	$A \geq B$	Signed	$\overline{N \oplus V}$
Less than	L	$A < B$	Signed	$N \oplus V$
Less than or equal	LE	$A \leq B$	Signed	$(N \oplus V) + Z$
Above	A	$A > B$	Unsigned	$\overline{C + Z}$
Above or equal	AE	$A \geq B$	Unsigned	$\bar{C}$
Below	B	$A < B$	Unsigned	C
Below or equal	BE	$A \leq B$	Unsigned	$C + Z$

program. A return (RET) or return from subroutine (RTS) is executed as the last instruction of the subroutine to pop the program counter from the stack and thus return to the main program. The SPARC does not support a system stack; subroutines are called with a jump and link (JMPL) instruction, which saves the program counter in register *r*31 of the current register window, and then slides the window down 16 registers as was illustrated in Figure 3.10. the subroutine returns to the main program by retrieving the return address from register *r*7 of its register window, which corresponds to *r*31 of the calling program.

Input and Output

Some CPUs utilize separate address spaces for memory and for input/output devices. In these cases, special instructions are provided to read information into the CPU from an input device and to write information from the CPU to an output device. The Intel CPUs support an isolated I/O address space that can be accessed only by the two special instructions IN and OUT as follows:

```
IN    AL,25   ;data from IO address 25 to AL register
OUT   25,AL   ;data from AL register to IO address 25
```

Processor Control

These instructions manipulate various hardware elements within the CPU and are therefore CPU-specific. The reader is referred to *The SPARC Architecture Manual, Ver. 7* (1983, 1987), Motorola Inc. (1990), Brey (1994), and Stewart (1993) for descriptions of processor control instructions for specific CPUs.

3.8 Interrupts and Exceptions

Events often occur that require interruption of normal instruction processing to perform some special action. Such exceptional events, or simply *exceptions*, can be triggered by condition signaled by devices external to the CPU, or by conditions detected within the CPU.

For example, desktop PCs often use a timer to interrupt the CPU once per second to make it update an image of a clock displayed on the screen. PCs used in process control are typically interrupted by sensors that detect various conditions in the plant that require immediate attention. An example of an internally detected condition is an attempt to divide a number by 0, which cannot produce a valid result. This type of exceptional condition should suspend normal processing to abort the operation and send a warning message to the user.

A primary advantage of external interrupt is that a CPU may work in parallel with one or more external processes, such as printing a document, and be interrupted only when the process requires attention. The alternative is to continuously monitor the process by checking a status register in the device to detect when the device requires attention. Such monitoring would prevent the CPU from doing other work while waiting for the device to be ready, whereas with interrupts the CPU can do other jobs until interrupted.

When an exception condition is detected, a CPU typically responds as follows.

1. Complete the instruction currently in progress to reach a convenient stopping point.
2. Save the current program counter on the system stack or in a designated register, preserving a pointer to the next instruction that would have been executed had the program not been interrupted.
3. Determine the condition that requested the interrupt. Many CPUs execute a special interrupt acknowledgement operation to allow an external interrupting device to identify itself with a unique number called an *interrupt vector.*
4. Load the program counter with the starting address of a program to perform the service requested by the interrupting device. This program is called an *interrupt service routine* (ISR). Where used, an interrupt vector points to a memory location containing the starting address of the ISR. Some CPUs restrict the ISR starting address to a fixed address in memory, while others require that the ISR address be stored in one specific memory location.
5. Fetch and execute the instructions of the ISR.
6. Upon completion of the ISR, execute an interrupt return instruction to restore the original program counter from the stack, allowing the CPU to return to the point in the original program at which it was interrupted.

Since interrupts may occur at any time, the ISR should begin by saving any registers that will be used within the ISR, and then restore them before returning to the main program. This allows the main program can be continued as if the interrupt had not occurred.

3.9 Evaluating Instruction Set Architectures

Many different metrics are used as indicators of computer performance. Perhaps the most commonly cited is *MIPS,* which stands for millions of instructions per second. Unfortunately, MIPS figures do not indicate how much work is done by each instruction in a particular CPU. In some machines instructions perform very primitive operations while in others each instruction may do a considerable amount of work. Therefore, simply knowing how many instructions a CPU can execute per second provides only a partial picture of how fast a computer can perform.

As computer architecture evolved from the first computers in the 1940s through the machines of the 1970s and 80s, the sizes of the instructions sets grew as designers became able to incorporate more circuit devices on a single integrated circuit chip. High-level languages became widespread, and emphasis was placed on making compiled programs as efficient as possible. For this reason, CPU instruction sets were expanded to include any instruction that might be needed to implement a high-level language statement. Thus instructions became more powerful, with compilers producing fewer instructions per program. However, this growth in the number and power of instructions required larger and more complex CPU control units, slowing down the entire processor.

In 1970, researchers at IBM designed the model 801 computer, in which the instruction set was reduced to only those that were used frequently. Later, researchers at Stanford and Berkeley observed that only a small core of computer instruction sets were executed the majority of the time. They developed reduced instruction set computers (RISC) that could execute programs faster than complex instruction set computers (CISC) by streamlining CPU designs. These efforts evolved to the SPARC, MIPS, PowerPC, and other RISC processors.

The true performance of a CPU can be measured by the time, T_{exec}, that it takes to execute a given program.

$$T_{exec} = (I/P) \times (C/I) \times (T/C)$$

where I/P is the number of instructions in the program, C/I is the average number of CPU clock cycles required to execute an instruction, and T/C is the time per CPU clock cycle, which is the inverse of the clock frequency.

CISC designs strive to minimize T_{exec} by reducing the number of machine instructions needed to execute a high-level language program, minimizing I/P. Designers of CPUs like the Digital Equipment Corporation VAX minicomputer and the Intel 80x86 and Motorola MC680x0 microprocessors attempted to anticipate the needs of compiler writers and provided as many instructions as might be needed. However, the cost of providing a large number of instructions is increased hardware complexity, which increases the factors C/I and T/C.

RISC designs target the C/I and T/C factors at the expense of increased I/P. Simplifying the instruction set can reduce the number of clock cycles required to execute each instruction. The target of most RISC processors is a single clock cycle per instruction. In addition, simplifying the hardware often enables the clock period to be shortened. However, more instructions are required to perform a given task than for an equivalent CISC machine.

With transistor switching time improvements slowing, computer architects have turned to parallelism to obtain additional improvements. One method of utilizing parallelism is *superscalar* design, in which multiple functional units are contained in the CPU and multiple instructions fetched and executed in parallel. For example, the Intel Pentium includes two integer and one floating point ALU that can be used concurrently. The Pentium control unit fetches multiple instruction codes at one time from memory, and can simultaneously initiate operation in one, two, or all three of these units, instead of executing the instructions sequentially. Likewise, the SUN Super Sparc and the Motorola PowerPC 601 CPUs each contain an integer unit, a floating-point unit, and a branch processing unit that can be used to execute multiple instructions concurrently.

Hennessy and Patterson (1995), Patterson and Hennessy (1994), and Stone (1993) provide thorough discussions of computer performance and the effects of architectural features on program execution times.

3.10 Computer System Design

Figure 3.22 illustrates the basic interconnection of memory and I/O devices to a CPU. Information is transferred between a CPU and selected memory or I/O devices via a *data bus,* which is a set of parallel signal lines, each carrying one data bit. The CPU selects one memory location or I/O device to receive or provide data by broadcasting its address to all devices in the system over an *address bus.* Logic circuits in each device interface compare the address on the bus to its assigned value to determine if it is the one being addressed. This is called *decoding* the address.

Data transfers are coordinated by the CPU with one or more control signals that indicate the type, direction, and timing of each data transfer. The direction of a data transfer is either device-to-CPU (a CPU *read cycle*) or CPU-to-device (a CPU *write cycle*). Motorola processors and many other CPUs issue a control signal $R/\overline{W}$ at the beginning of the cycle, setting it high to designate a read cycle and low to indicate a write cycle. Some CPUs, such as the Intel processors, use two separate control lines to indicate the type of cycle: $\overline{RD}$ to signify a read cycle and $\overline{WR}$ to signify a write cycle. (The overbar indicates that these signals are active-low, i.e., a logic 0 signals that the indicated operation is active.)

The timing of a data transfer is critical. The CPU must signal a device when it is time to begin and when it is time to end each data transfer. The Motorola processors issue a *data strobe* signal ($\overline{DS}$) that goes low when the transfer is to begin and returns high when the transfer is finished. In the Intel processors, $\overline{RD}$ and $\overline{WR}$ act as data strobes for read and write cycles, respectively.

Figure 3.23 illustrates bus timing for read and write cycles for CPUs that use control signals compatible with those of the Motorola CPUs, and for CPUs that use Intel-compatible control signals.

The number of bits in a data bus may or may not match the width of the internal data path of the CPU. For example, the 8086 and 68000 have internal and external 16-bit data buses. However, the Intel 8088 and Motorola 68008 processors have 16-bit wide internal data paths and 8-bit external data buses, while the Intel Pentium has a 32-bit internal data path and a 64-bit external data bus. The rate at which data can be transferred between a CPU and an external device is referred to as the *bandwidth* (BW) of the bus, and is a function of the bus speed (number of data transfers per second) and the number of bits per transfer.

$$BW = (transfers/second) \times (bits/transfer)$$

Given identical bus speeds, the bus bandwidth of an 8088 would be half that of an 8086, even though they have identical internal data paths.

For CPUs that support byte-addressable memory with 16-bit and wider buses, control signals are provided to indicate which byte or bytes of the bus are to be used to transfer data. For the 16-bit 68000 data bus, two control lines, $\overline{UDS}$ and $\overline{LDS}$ (upper data strobe and lower data strobe), indicate whether the upper byte, the lower byte, or both bytes are to be used for the data transfer. The 8086 activates a special control line, $\overline{BHE}$ (byte high enable), when the high byte of the data bus is to be used for a data transfer. Address line A0 indicates an offset from the low byte of the bus. A0 = 0 indicates that the low byte is to be used and and A0 = 1 indicates that only the high byte is to be used. For a 16-bit data transfer, both A0 = 0 and $\overline{BHE}$ = 0. This is summarized in Table 3.6.

This can be readily extended to 32-bit and larger data buses. The 32-bit data buses of the Intel 80386 and 80486 have four control lines: $\overline{BE_3}$, $\overline{BE_2}$, $\overline{BE_1}$, and $\overline{BE_0}$, corresponding to data bus lines D_{31-24}, D_{23-16}, D_{15-8}, and D_{7-0}, respectively. The Pentium has eight such control lines for its 64-bit bus.

Reliable data transfers require synchronization of the CPU and the device being accessed. A *synchronous bus* implies that the data transfer is synchronized to the clock signal that drives the CPU, and must be completed within a designated number of clock periods. A designer must select memory and I/O devices whose access times are short enough to fit within within this constraint.

In contrast, an *asynchronous bus* does not synchronize data transfers to a reference clock. Instead, the CPU signals the device to begin the transfer and the device signals the CPU when it has completed the data transfer. This allows each data transfer to take as much or as little time as required for the device being accessed, and therefore allows slower devices to be used. The 68000 has an input signal called $\overline{DTACK}$ (data transfer acknowledge) that must be activated by each accessed device to signal the CPU that the transfer has been completed.

A *semisynchronous bus* synchronizes data transfers to the CPU clock as in the synchronous case, but allows a slow device to signal the CPU that it needs more time to complete a data transfer. For example, the 8086 has an input called *READY* that would normally be held high, but can be pulled low by a device until it is ready to complete a data transfer. When this happens the CPU enters a special wait state for one clock cycle and checks the *READY* signal again, repeating this wait state indefinitely until *READY* has been activated by the device.

On most microcontrollers and some older microprocessors chip package sizes have a limited number of signal pins. In these cases it is impractical to allocate a separate pin for each address,

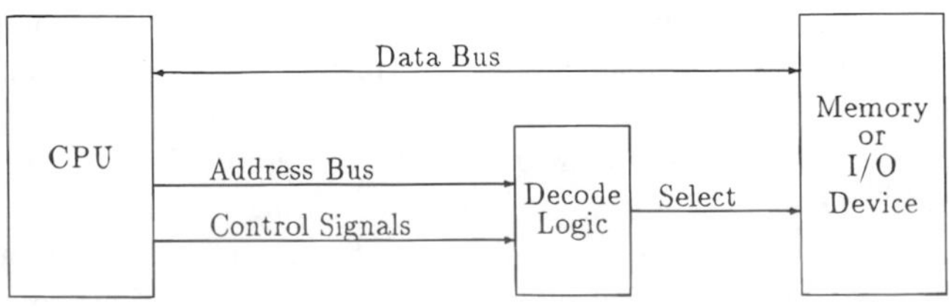

Figure 3.22 Computer system address and data buses.

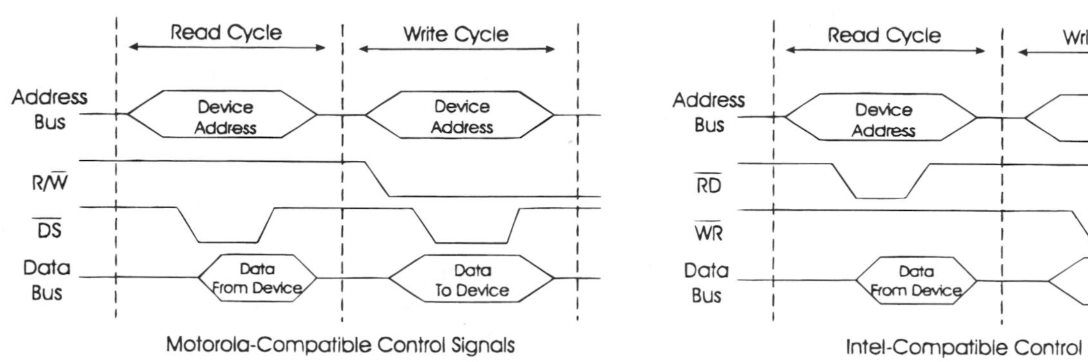

Figure 3.23 Data transfer cycles between CPU and memory.

Table 3.6 Byte and Word Transfers on a 16-bit Bus

	68000		8086	
Data transfer type	$\overline{UDS}$	$\overline{LDS}$	$\overline{BHE}$	A0
Byte: D_{7-0}	1	0	1	0
Byte: D_{15-8}	0	1	0	1
Word: D_{15-0}	0	0	0	0

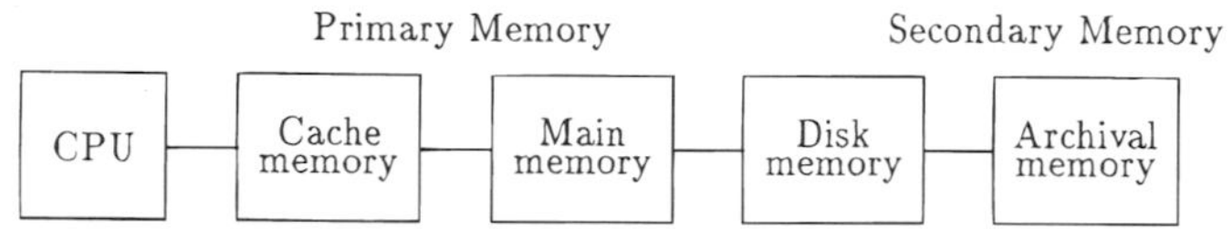

Figure 3.25 Computer system memory hierarchy.

data, and control signal. To reduce pin count, multiple signals are forced to share a single pin, i.e., two or more signals are time-multiplexed over one signal line. It is common for expandable microcontrollers to use one set of signal pins to transmit an address to external memory and then use the same pins to transfer data. A special control signal indicates when each type of information is on the pins, allowing the bus to be demultiplexed.

For example, the 8086 multiplexes its 16 bit data bus, D_{15-0}, with address bits A_{15-0} on signal pins AD_{15-0} as illustrated in Figure 3.24a. For each data transfer, the CPU first broadcasts an address on AD_{15-0} and then uses the same pins to transfer the data, as illustrated in the timing diagram of Figure 3.24b. A signal called *ALE* (Address Latch Enable) is set high by the CPU whenever an address is present on AD_{15-0}, and low otherwise. Since the CPU removes the address before the data is transferred, a latch must be provided external to the CPU chip, as shown in Figure 3.24a, to save the address while ALE=1 and hold it for the duration of the data transfer.

Memory Systems

General-purpose computer systems often utilize a hierarchy of memory devices, as shown in Figure 3.25. Memory devices differ in how they are accessed, their storage capacity, volatility, cost per bit, and access time.

Computer memory units are classified as *primary memory* if any storage location within the memory can be accessed directly by the CPU; otherwise they are classified as *secondary memory.* Direct or random access memories comprise a set of numbered storage locations that are accessed by supplying the address of the location to be accessed, along with one or more control signals to indicate whether information is to be read from memory or written to memory.

Secondary memory devices are used for bulk or mass storage of programs and data, and include rotating magnetic devices, such as floppy and hard disks, magnetic tapes, magnetic bubble memories, optical devices such as CDROMs (compact disk read-only memory), and a variety of other devices. In contrast to primary memory, information in secondary memory devices is not accessed directly. Instead, a special controller searches the device to locate the block of information containing the desired item. When found, the entire block is usually transferred into primary memory, where the desired items can be accessed in a more convenient fashion.

Volatility refers to the permanence of data stored in a memory. A read-only memory preserves information permanently and cannot be rewritten. This is useful for storing programs and data

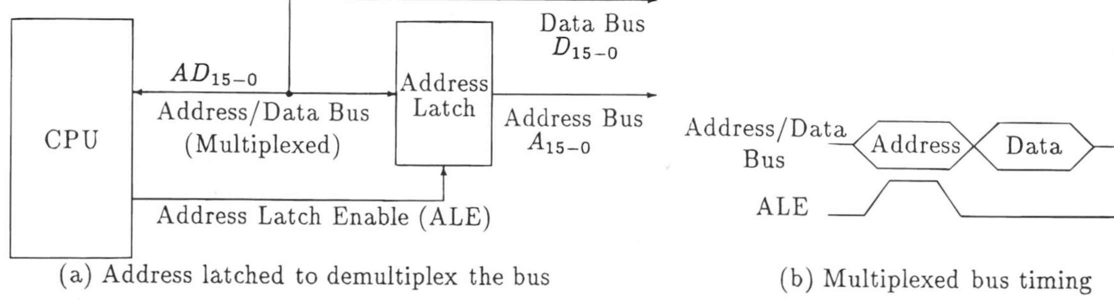

Figure 3.24 Multiplexed address and data buses.

values that will not change. Read-write memories can be erased and/or rewritten. Some read-write memories retain information only while powered up, while others can retain data indefinitely until erased or rewritten.

The *capacity* of a memory refers to the total number of bits of information that can be stored. Capacity is a function of the mechanism used to access the memory. A direct access memory is limited in size by the number of address bits provided by the CPU. The capacity of a disk drive is determined partially by its physical characteristics and control circuitry and the ability of operating system software to keep track of the information on the disk. Archival secondary storage devices with removable media provide virtually unlimited capacity.

The *cost per bit* of storage decreases as one moves farther from the CPU in Figure 3.25. High-speed cache memories are more expensive than slower main memories, while storage on most disk drives is often orders of magnitude less expensive per bit. Archival storage allows large quantities of information to be saved on inexpensive tape or diskettes with minimal cost.

As capacity increases and cost per bit goes down in the memory hierarchy, the performance parameters, or access and cycle times, of the memory become longer. Memory *access time* is the length of time required to retrieve (read) a word from the memory, and memory *cycle time* is the minimum interval of time required between successive memory operations. The access time of a memory determines how quickly information can be obtained by the CPU, while the cycle time determines the rate at which successive memory accesses may be made. In general, direct access devices can be accessed more quickly and more often than secondary devices.

Semiconductor Memory Technologies

Most computers built prior to 1970, some of which are still in operation today, utilized arrays of magnetic cores as their primary memory elements, while a few specialized systems, particularly in space vehicles, utilized plated wire as a replacement for magnetic core in applications where radiation hardness was required. In today's digital computers, primary memories are constructed of semiconductor integrated circuit chips.

Semiconductor memories are available as read-only or read-write devices. Read-only memory (ROM) is used to store programs, tables, and other data that will not be modified while the computer is operating. Table 3.7 lists several ROM technologies, which differ in how the devices are programmed and/or erased. ROMs are programmed when they are manufactured, while PROMs are field-programmable, i.e., they are programmed by the system designers that use them. Neither ROMs nor PROMs are alterable; they must be discarded when their contents are no longer valid.

Table 3.7 Read-Only Memory Technologies

Acronym	Device type
ROM	Read-Only Memory
PROM	Programmable Read-Only Memory
EPROM	Erasable Programmable Read-Only Memory
EEPROM	Electrically-Erasable Programmable Read-Only Memory
FLASH	Flash Memory—EEPROM that is erased by erasing the entire memory array

EPROM, EEPROM, and FLASH memories are field-programmable devices that can also be erased and reprogrammed. EPROMs are erased by exposing the storage cells to ultraviolet light to free electrical charge trapped in the memory cells, while EEPROM and FLASH memories are erased electrically. Selected locations can be erased and reprogrammed in an EEPROM, while with a FLASH memory the entire device must be erased before any location can be reprogrammed.

Read-write memory is commonly referred to as *RAM,* which is an acronym for *random access memory.* In reality, both ROM and RAM are random access memories since their contents can be accessed in any random order. Nonetheless, the term RAM has become associated with read and write memory.

RAM chips are further classified as static or dynamic. Once written, a *static RAM* retains information until its power is removed; the RAM cells will be in random states when power is restored. Each memory cell contains the equivalent of a digital flip-flop circuit to achieve this static storage. In contrast, a *dynamic RAM* retains information for only a short time as charge stored on a capacitor associated with a single transistor. This charge slowly leaks away over a short period of time. Therefore, to retain information the contents of a dynamic RAM must be periodically *refreshed,* that is, read and rewritten.

Because more transistors are used for each storage cell, static RAM devices have a higher cost per bit than dynamic RAMs, and can hold fewer bits than a dynamic RAM chip fabricated with a comparable technology. However, extra control circuitry is required for dynamic RAMs to refresh the memory, and dynamic RAMs are typically slower than comparable static devices. For small systems, the complexity and cost of the extra control circuitry for dynamic RAMs make them less attractive than static RAMs. However, for systems that use large amounts of RAM, the extra cost of the refresh circuitry is negligible compared to the lower cost per bit of dynamic RAM devices. Therefore, most PCs, workstations, and larger computers use dynamic RAMs for their main memories.

Memory System Organization

The organization of a primary memory system is determined primarily by the address and data buses of the host CPU. A memory chip is organized as $2^k \times n$, which means that there are k address input lines and n data input/output lines, and thus a total of 2^k addressable n-bit locations. An address decoder within the chip selects one location corresponding to each k-bit address and either sends the information from that location off the chip or writes new information into that location.

Using memory devices organized as $2^k \times n$ in a system that has j total address lines, where $j \leq k$, means that the system can address $2^j/2^k = 2^{j-k}$ devices. These devices are usually organized hierarchically into banks as illustrated in Figure 3.26, with the memory address partitioned into a bank number, a chip number, and an on-chip address as shown. Each part of the address is

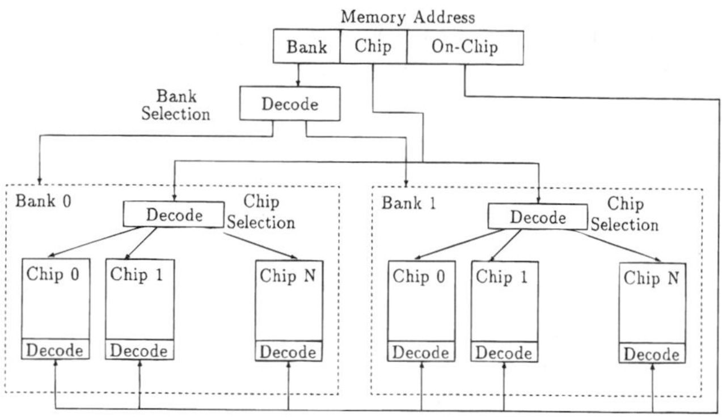

Figure 3.26 Hierarchical decoding of memory addresses.

decoded at a different level of the hierarchy. One level of address decoding selects one of the banks, another selects a chip within the bank, and the on-chip decoder selects one location within the selected chip.

If the system data bus width m is wider than the number of data pins n on the memory chip, then m/n chips must be used to create one addressable block of memory, with each chip connected to a different n-bit section of the data bus.

For example, the 68000 CPU bus interface comprises 16 data lines and 24 address lines. If a 16M byte memory system is to be constructed of 1 M-byte RAM chips ($2^{20} \times 8$), then chips must be connected to the data bus in pairs, one connected to bits D_{7-0} and containing odd-numbered bytes, and one to bits D_{15-8} and containing even numbered bytes for a total of 2M bytes of RAM. 16M bytes of RAM requires $16/2 = 8$ pairs of chips. Three address bits A_{23-21} must be decoded to select a pair of chips, with 20 address bits A_{20-1} decoded within each chip. The remaining address bit A_0 is used within the CPU to determine whether the even or odd numbered byte or both are to take part in the data transfer.

If only 4 pairs of chips will fit on one circuit board, then two boards would be used, with address bit A_{23} selecting one of the boards and A_{22-21} selecting a pair of chips within the board, as illustrated in Figure 3.26.

Cache Memory

CPU speeds have increased at dramatic rates from year to year. As a result, the design of memory systems that can keep information moving in and out of the CPU without making it wait is becoming more difficult. Memory chips with short access times are expensive. An alternative to building a large memory out of expensive high-speed RAM chips is to use a small high-speed memory to hold the information most likely to be used by a CPU, with the remaining information kept in the main memory. The main memory can be built with lower-speed, and therefore less expensive, RAM chips.

Memory accesses often exhibit a property called locality of reference. *Spatial locality of reference* means that if a memory location is accessed then there is a high probability of accessing the next sequential location in memory. Programs tend to exhibit high degrees of spatial locality of reference because instructions are fetched from sequential memory locations until a jump or branch is executed. *Temporal locality of reference* means that if an address has been accessed then there is a high likelihood that it will be accessed again in the near future. Data variables often exhibit good temporal locality of reference, as do instructions that are contained in loops.

A *cache memory* exploits locality of reference by holding in high-speed memory a subset of the main memory locations that are most likely to be needed by the CPU. Each memory reference is sent first to the cache. If the requested information is there, a *cache hit* is said to occur and the information is accessed and passed quickly to the CPU. If a *cache miss* occurs, i.e., if the information is not found in the cache, then the slower main memory must be accessed. The average access time is given by:

$$T_{access} = HT_{cache} + (1 - H)(T_{cache} + T_{main})$$

where H is the *hit ratio* of the cache, i.e., the percentage of cache accesses that result in hits, and T_{cache} and T_{main} are the access times of the cache and main memories, respectively. For example, a memory system with a 90% hit ratio with $T_{cache} = 20ns$ and $T_{main} = 100ns$, would have an average access time of $T_{access} = 30ns$, which is much closer to that of the higher-speed cache memory than the slower main memory.

Writes to memory can be handled in two different ways in systems with cache memory. In a *write through* strategy, each item is written directly to the main memory on every write cycle, with the cache updated concurrently. In a *write back* strategy, all writes are done only to the cache, leaving the cache and main memory contents temporarily inconsistent. Later, when information must be replaced in the cache, all modified cache entries are copied back to the main memory. This reduces the total number of main memory accesses when multiple updates are done to a single data item.

Cache memory designs and performance are examined extensively in Hennessy and Patterson (1995), Patterson and Hennessy (1994), and Stone (1993).

Virtual Memory Management

Increasing program sizes and workload demands for PCs, workstations, and larger computers have made it impractical to provide enough primary memory to store a user's entire program and data. This is especially true in multiuser and multitasking environments in which memory and CPU time are shared by multiple programs. Referring to the locality of reference principle described in the previous section, it is usually the case that a CPU does not need immediate access to all instructions and data of a program at any given time. Therefore it is sufficient to make a subset of this information available in memory, with the rest held on a disk or other secondary storage device until needed. This allows a program's address space to be much larger than the available or allocated physical memory.

This is achieved by using two different address spaces: a *virtual* or *logical address space* that specifies locations within a program, and a *physical address space* that identifies physical storage locations in the main memory. Since different portions of a program are placed in memory at different times, each location from the logical address space must be mapped to the current physical address containing that information as illustrated in Figure 3.27. This mapping is handled by a *memory management unit* (*MMU*), which is a hardware element that translates each logical address to the corresponding physical address.

A common method for translating logical to physical addresses is *paging*, in which a user's program is partitioned into fixed-length blocks called pages. Physical memory is likewise partitioned into fixed-length blocks, usually of the same length as the logical pages so that each logical page fits exactly into one physical page of memory. Each logical address is divided into an n-bit page number and a k-bit offset within the page as illustrated in Figure 3.28. The n-bit logical page number is mapped to an m-bit page number in the physical address space. Since logical and physical page sizes are the same, the offset portion of the logical address is used directly as the offset portion of the physical address without going through the translation process. Only the address bits corresponding to the page number need to be translated.

As logical pages are placed in the physical memory, a *page table* is created and updated. As illustrated in Figure 3.28, the page table contains a *descriptor* for each logical page containing a *present* bit (P) to indicate whether that page is currently resident in memory and, if so, the corresponding physical page number (PPN). The page descriptor may also contain a *modified* bit (M) to indicate whether the page has been modified since being loaded into memory, an *accessed* bit (A) to indicate if the page has been accessed, and possible restrictions on how the page may be used, such as a *writeable* bit (W) to indicate that the page may be altered. These bits are used by the operating system to help manage the pages in memory.

To perform an address mapping, the MMU uses the logical page number as an index into the page table, as shown in Figure 3.28, to fetch the page descriptor. If the page is present in memory the physical page number is retrieved from the page descriptor and concatenated with the page offset to produce the physical address. If the page is not present in memory, a *page fault* is said to occur. The CPU is interrupted and the operating system called to find the requested page on disk and load it into memory. Then the page table is updated and the original memory request is repeated.

In a multiuser or multitasking system, each process has its own page table. This is illustrated in Figure 3.29, which shows two processes sharing CPU time. Pages A and B of Process 1 are currently in memory along with pages L and N of Process 2. The page table of each process keeps track of which pages are in memory. When Process 1 is running, logical address 1, corresponding to page B, is mapped to page 0 of physical memory. When Process 2 is running, page L is accessed by mapping logical address 1 into page 1 of physical memory. If Process 1 attempts to access page C, which is not currently in memory, it will be suspended to allow operating system to retrieve page D from disk and replace one of the current pages in memory. Then Process 1 will be allowed to repeat the access to page D and continue processing.

An alternative method of address translation is *segmentation*,

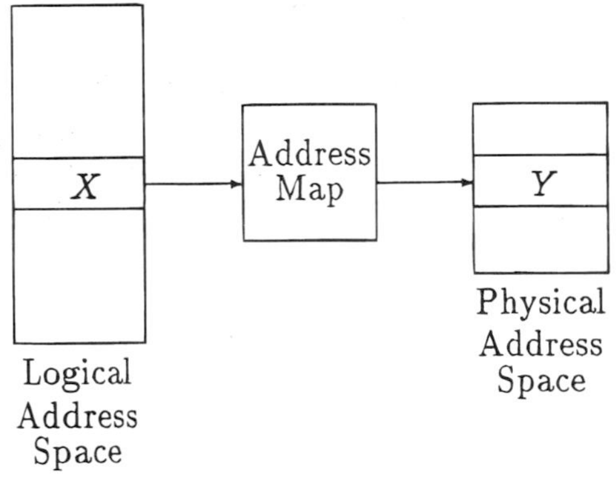

Figure 3.27 Mapping logical address X to physical address Y.

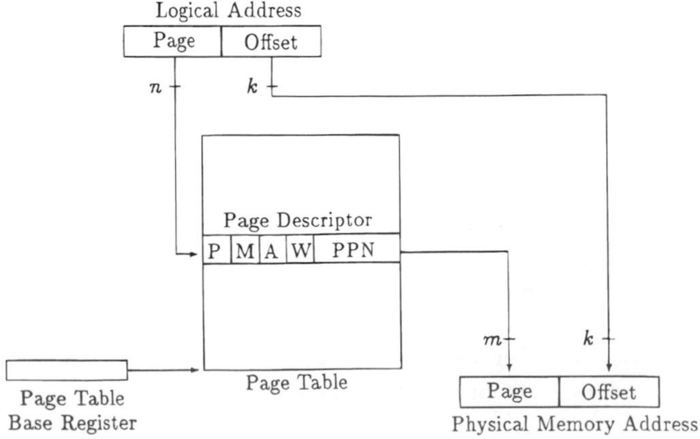

Figure 3.28 Address mapping through a page table.

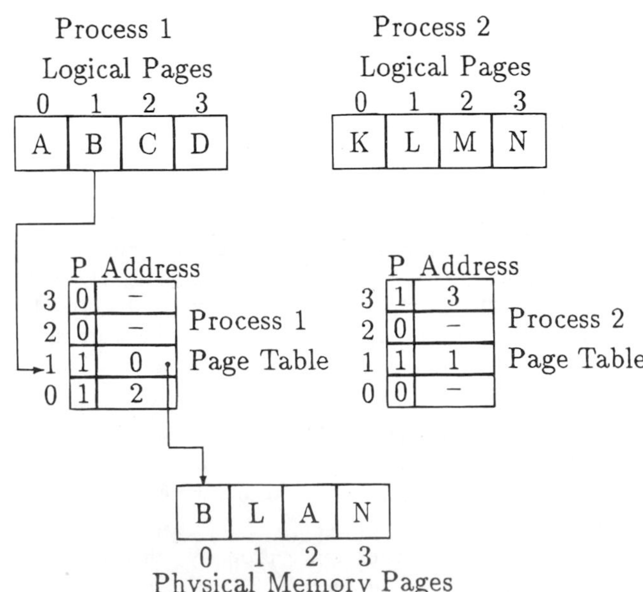

Figure 3.29 Address mapping in a multitasking environment.

in which information is organized by the programmer into segments of items that share common characteristics. Each segment has a segment number and an offset within the segment. Segment numbers are translated in the same manner as page numbers, indexing into a segment table to retrieve a segment descriptor that points to the beginning of a physical memory area. Unlike pages, segments can be of arbitrary size, and can be loaded at any memory address. The segment size is stored in the segment descriptor and compared to each segment offset to ensure that the requested information is not outside the bounds of the segment. The segment offset is then added to the starting memory address to form the complete physical address. The use of arbitrary segment sizes allows a programmer to use only as much memory as needed, although memory allocation is difficult since the operating system is not working with fixed sized blocks, and therefore free memory may become fragmented as segments are moved in and out of memory. There may be a sufficient amount of free memory to accomodate a new segment, but if the free memory is not contiguous, the segment cannot be loaded.

Occasionally a combination of segmentation and paging is used. In this case information is organized by the programmer into segments, with protection information placed in the segment descriptor. Then each segment is partitioned, transparent to the programmer, into fixed sized pages, simplifying the memory allocation process since logical and physical pages will be the same size, preventing fragmentation. Each logical address comprises a segment number, page number, and page offset. The segment number is used to access a segment descriptor in a segment table. The segment descriptor points to a page table containing descriptors of the pages of that segment. The page number is then used to access a page descriptor from the selected page table, which provides the physical page number. Finally, this page number is concatenated with the original page offset to form the physical memory address.

3.11 Input/Output Device Interfaces

An external device must be interfaced to the CPU's data bus so that data can be transferred by program instructions between it and the CPU. As with memory, each I/O device interface must be addressable and respond to bus control signals issued by the CPU. The most common approach is to use one or more registers in the I/O device interface as buffers between the device and the CPU as illustrated in Figure 3.30. Data is sent to a device by writing it to a register in the device interface, from where it is sent to the external device. Likewise, data from an external device is accessed by the CPU by reading a register in the device interface.

An external device may require data in some format other than the parallel binary digits provided by the CPU. For example, transmission of data over a telephone line via a modem requires that each byte be converted to a serial stream of bits, with additional control bits prepended and appended to each transmitted byte. Data from temperature, pressure, and other sensors is often produced as continuous analog voltages or currents. To be processed by a digital computer, analog values must be sampled and represented by digital values that can be read by the CPU. A similar function must be provided by digital to analog converters to send information from a CPU to an analog device. There are many different device interface chips that automatically convert data from one format to another. In most of these, the CPU simply transfers parallel digital data to and from registers in the device interface. Conversion and transmission of data between these registers and the external devices is handled automatically by circuits in the interface.

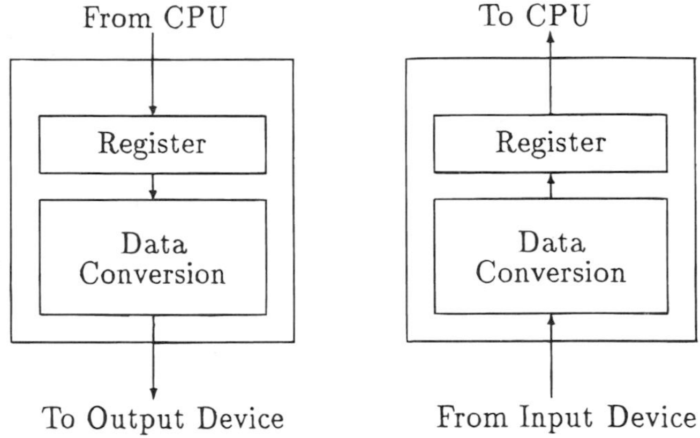

Figure 3.30 External input/output device interfaces.

Interfaces to disk drives, graphics display terminals, and other complex I/O devices operate in a similar manner. Despite the significant differences between the characteristics of different types of peripheral devices, most of them are viewed by the CPU as a set of addressable registers. These device interfaces incorporate intelligent control circuits that respond to commands sent by a CPU to special command registers within the device interface. Parameters and data are likewise transferred to and from registers in these interface circuits.

The operation of these circuits is monitored by reading status information from registers in the device interfaces. For example, the following 68000 program segment reads and tests the rightmost bit of a device status register. If that bit is 1, the device is known to be busy and the program remains in the loop. If the bit is 0, the device is not busy, and new data can be sent to a data register the device interface.

```
Check:  MOVE.B  Status,D0        ;read device status register
        AND.B   #00000001,D0     ;isolate BUSY-bit 0
        BNE     Check            ;stay in loop if device busy
        MOVE.B  D1,Data          ;send new data to device if not busy
```

3.12 Microcontroller Architectures

Microcontrollers are used in applications that require low cost and chip count, combining on a single chip a CPU, ROM, RAM, and various peripheral functions and I/O interfaces. Typical applications include embedded controllers for kitchen appliances, automotive electronics, cellular phones, home electronics (TVs, VCRs, etc.), process control, and many other applications. In general, it has proven cost-effective to use a single microcontroller chip to replace circuitry that would otherwise require several digital and/or analog ICs. In addition, the programmability of microcontrollers allows features to be changed or added with very little extra cost.

Microcontroller CPUs are available with data path sizes from 4 to 32 bits. On-chip RAM is generally small compared to on-chip ROM, usually 64 to 256 bytes. In embedded applications, RAM is used primarily to store a few temporary variables and to perform "scratch" work; hence, on-chip RAM is commonly referred to as *scratchpad* memory. Most microcontrollers provide a number of parallel I/O pins. These are used to send control signals to actuators, receive signals from sensors, and pass data to/from other external devices. Many microcontrollers also offer programmable serial communication interfaces. Other special I/O interfaces may be found in selected chips. I/O interfaces and other special functions are accessed by the instruction set via control, status, and data registers with permanently assigned addresses.

Programmable timers/counters are common functions included in most microcontrollers. These are used to provide timing for digital alarm clocks, cooking cycle times in microwave ovens, automobile engine control timing, timing of bit transmissions for serial I/O, the measurement of time periods between signal changes detected on external I/O lines events, and counting of such signal changes.

Figure 3.31 shows a block diagram of the 8051 microcontroller (Stewart, 1993). In addition to the CPU, 128 bytes of RAM, 4K bytes of ROM, and four 8-bit parallel I/O ports, the 8051 includes two programmable timers, a serial communication port, interrupt control logic, and an external bus interface. When activated, parallel ports P0 and P2 are disabled and these pins become the external data and address buses. Data is multiplexed with the low address byte on the 8 pins of port P0, with the remaining address bus bits on the eight pins of port P2. Additional memory and I/O function interface chips can be accessed via this external bus to expand the capabilities of the microcontroller. The Motorola 68HC11 microcontroller can be expanded off-chip in a similar manner.

The basic architecture of the Motorola 6805 is shown in Figure 3.32 (Motorola Inc., 1983). As in the 8051, the 6805 includes CPU, RAM, ROM, parallel I/O pins, a programmable timer, and interrupt support logic. Unlike the 8051 and the Motorola 68HC11, the 6805 is not expandable; the data bus cannot be accessed external to the chip. Therefore, no additional memory may be used other than what is provided on-chip, and all signals to and from the outside world must go through the provided parallel or other special I/O pins. Because applications differ in their memory and I/O requirements, dozens of different configurations of the 6805 are offered. Table 3.8 summarizes six members of the 6805 family. As can be seen in this table, members of the 6805 family differ in the amounts of on-chip RAM and ROM, number and types of I/O pins, package sizes, number of external interrupt pins, and availability of A/D converters, phase-lock loops, and other special functions. In addition, several different technologies are available, including HMOS and CMOS. A system designer must select a device from the 6805 family that most closely matches the needs of his or her application. Many other manufactures also supply single-chip microcontrollers as families of non-expandable devices.

Note in Table 3.8 that the 68705P3 microcontroller contains EPROM instead of ROM. This part is more expensive than an equivalent ROM-based part. However, the EPROM can be erased and reprogrammed many times, and therefore the 68705P3 can be used during development of the product prototype, when the program is subject to continuous changes. When the program is frozen in its final form, the less expensive ROM version can be ordered for production. This allows one to develop a product using a pin-compatible version of the microcontroller and then replace it with the less expensive ROM-based version in the final product. Many other suppliers offer EPROM-based versions of their microcontrollers for this purpose.

Microcontroller instruction sets generally support smaller

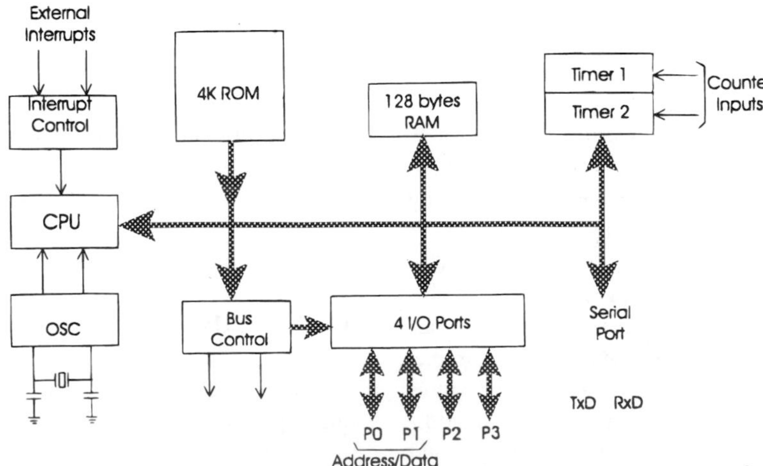

Figure 3.31 Intel 8051 microcontroller block diagram. *Source:* Stewart, 1993. *The 8051 Microcontroller: Hardware, Software and Interfacing,* Regents/Prentice-Hall, Englewood Cliffs, NJ. With permission.

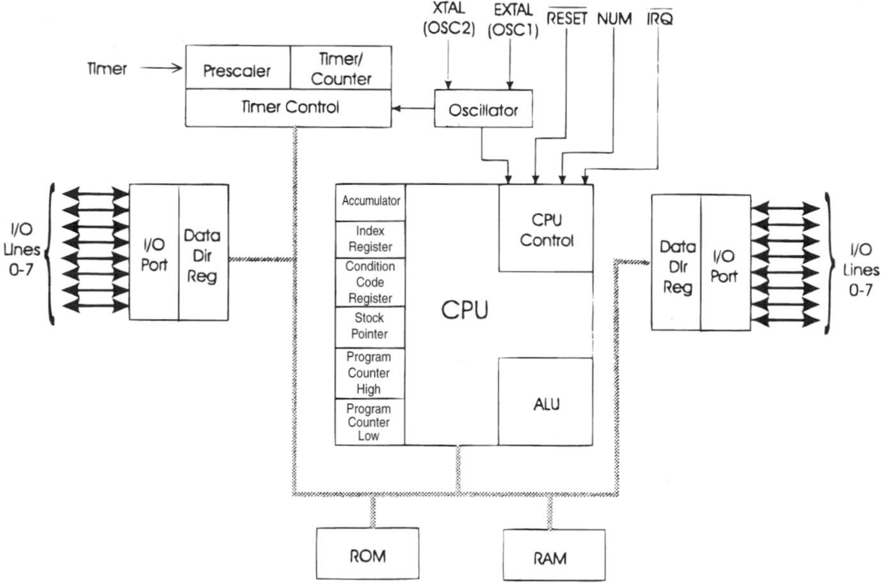

Figure 3.32 Motorola 6805 microcontroller block diagram. *Source*: Motorola Inc., 1983. *M6805 HMOS/M146805 CMOS Family Users Manual*, Prentice-Hall, Englewood Cliffs, NJ. With permission.

Table 3.8 Motorola 6805 Family Devices Configurations

	6805P2	6805R2	6805R3	6805T2	68705P3	146805F2
Technology	HMOS	HMOS	HMOS	HMOS	HMOS	CMOS
Pins	28	40	40	28	28	40
RAM	64	64	112	64	112	64
ROM	1.1K	2K	3.8K	2.5K	1.8K (EPROM)	1K
I/O pins	20	24	24	19	20	16
Input pins	0	6	6	0	0	4
Timer	Yes	Yes	Yes	Yes	Yes	Yes
Interrupt pins	1	2	0	2	1	1
Other I/O	—	A/D	A/D	PLL	—	—

Source: Motorola Inc., 1983. *M6805 HMOS/M146805, CMOS Family Users Manual*, Prentice-Hall, Englewood Cliffs, N.J. With permission.

memory spaces and provide fewer arithmetic and other high-level functions than general-purpose CPUs. However, many microcontroller CPUs include special instructions to facilitate I/O operations and to manipulate individual bits of input and output data, since these operations typically dominate microcontroller applications more so than arithmetic computations.

For designers needing more computation power in their embedded applications, or who wish to take advantage of the vast quantity of system and application software available for popular general-purpose CPUs, several IC manufacturers have developed microcontrollers around general-purpose 16 and 32-bit CPUs. One such device is the Motorola 68300 microcontroller family, which is built around the basic 68000 CPU and typical microcontroller memory and I/O device interfaces.

A number of microcontrollers are also available with instruction sets tailored to specific applications. For example, there are several microcontrollers available whose instructions sets have been designed specifically for digital signal processing, which relies heavily on multiply-accumulate and other mathematical computations.

A new trend in microcontroller design is to create custom chips for customers from libraries of standard cells. Standard cells are predesigned and characterized layouts of such functions as CPUs, RAMs, ROMs, I/O interface ports, and other special functions. A customer orders a microcontroller by specifying the desired CPU and exactly the types and amounts of memory I/O devices needed for the intended application. The chip layout is then created by the manufacturer by placing and interconnecting cells from the library, and the chip is fabricated for the customer in a relatively short time.

3.13 Multiple Processor Architectures

System designers have long been faced with applications requiring more computing power than can be provided by a single computer. Computing throughput requirements can be orders of magnitude greater than can be provided by even the fastest computers, despite the fact that CPU performances have increased dramatically over the years. Such high throughput requirements can only be met by the use of parallel processing, in which applications are partitioned into multiple tasks that are executed concurrently by multiple processors.

Multiple processor system architectures vary widely in the numbers of processors used, methods in which applications are partitioned and mapped onto an architecture, and methods for interconnecting processors to communicate and share information. Flynn (1966) proposed a commonly used method for classifying computer architectures, based on the number of instruction and data streams that can be processed concurrently. The Von-Neumann and Harvard architectures described previously in this chapter are examples of *single-instruction stream, single-data stream* (*SISD*) architectures. One stream of instructions is fetched from memory by the CPU, which also fetches a stream of data from memory as needed.

Throughputs in some computationally intensive applications can be increased by performing a single operation concurrently on an entire set of data. This is especially true for computations involving vectors and matrices. This type of parallel processing can be performed on a *single-instruction stream, multiple-data stream (SIMD)* architecture, as illustrated in Figure 3.33. In a SIMD architecture, a control processor fetches program instructions and identifies those instructions that involve computations on sets of numbers. As shown in Figure 3.33, each such instruction, I, is broadcast to N processing elements (PE_1, PE_2, . . . , PE_N), which perform that operation concurrently on N data items (D_1, D_2, . . . , D_N) accessed from a shared memory. SIMD architectures are commonly referred to as *array processors.* The reader is referred to Hwang's textbook (1983) which discusses the architectural features of a number of SIMD machines.

A *multiple-instruction stream, single-data stream (MISD)* architecture comprises N processing elements, each of which performs a different operation (I_1, I_2, . . . , I_N) on a single data item D in assembly-line fashion, as shown in Figure 3.34. MISD principles have found their way into single processor architectures in the form of *pipelining.* In a pipelined CPU, the steps needed to process each instruction are performed by separate hardware modules. As each module completes its part, the module passes the instruction on to the next module and begins processing the next instruction. If there are N such modules, then up to N instructions can be processed at one time within a CPU, producing one new result every clock period rather than one every N clock periods. Pipelined CPU designs are discussed extensively in Hennessy and Patterson (1995), Patterson and Hennessy (1994), and Stone (1993).

MISD principles have also been applied to experimental *data flow computers,* in which data is passed from processor to processor. Whenever a processor receives all the data items required for an operation, it performs that operation and passes the results

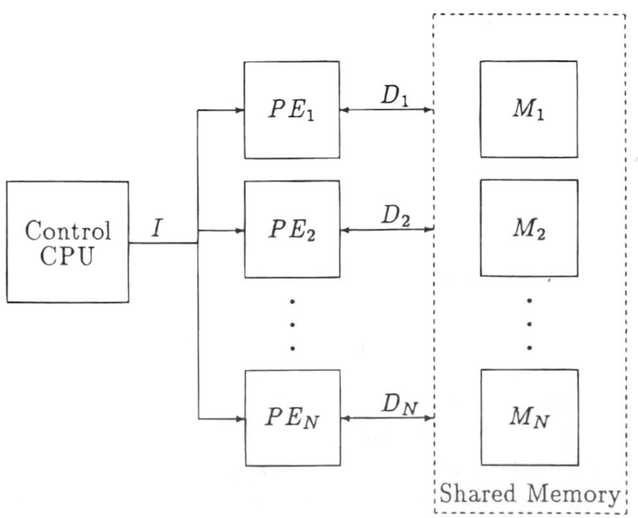

Figure 3.33 SIMD architecture: instruction I performed on N data items.

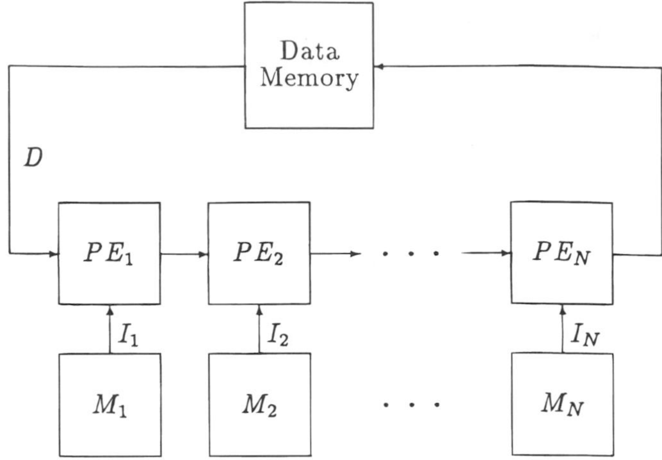

Figure 3.34 MISD architecture: each PE performs a different operation on one data item.

on to other processors. In this manner, computations are triggered by the flow of data through the system.

The most widely used multiple processor architectures are *multiple-instruction stream, multiple-data stream (MIMD)* configurations. In an MIMD system, each processor performs its own assigned tasks, and accesses its own stream of data. Consequently, MIMD architectures can be applied to a much broader range of problems than the more specialized SIMD and MISD configurations.

There are numerous ways to configure an MIMD system. The number of processors can range from as few as two to hundreds or even thousands of elements. The most significant differences between MIMD architectures are related to the manner in which processors cooperate in solving problems. In general, MIMD architectures can be classified according to the degree of coupling between processors. In a *loosely coupled* system, the processes are autonomous and communicate primarily by exchanging messages through a communication network. In a *tightly coupled* system, the processors are closely synchronized and work closely with each other in solving problems. The required degree of coupling will determine how the processors should be interconnected.

The simplest MIMD interconnection method is the shared bus, as illustrated in Figure 3.35. Multiple CPUs can cooperate in solving problems by sharing information in a global memory that is accessed via a shared bus. Because available bus bandwidth is limited, each processing element typically uses a private local memory as a cache for nonshared programs and data. The majority of each processor's memory accesses are to this local memory, with traffic on the system bus limited to accessing data that must be shared between processes. Many commercial microprocessors contain special bus interface signals and functions to support connections to shared buses. Therefore many multiprocessors systems have been developed using off-the-shelf CPU and memory boards.

If a single bus is unable to provide sufficient bandwidth to handle all of the shared memory accesses, processors will be

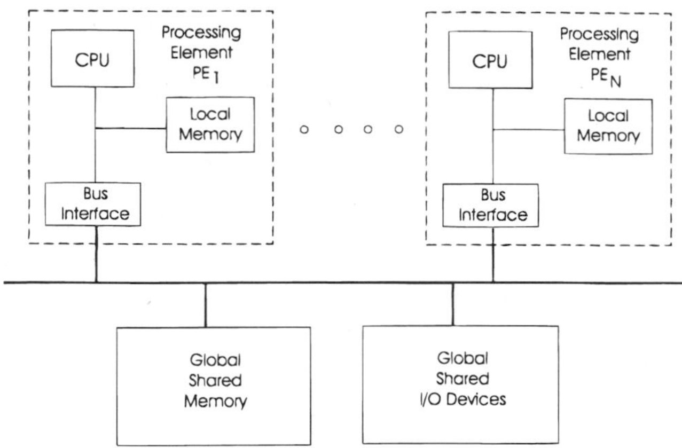

Figure 3.35 Shared-bus, shared-memory multiprocessor.

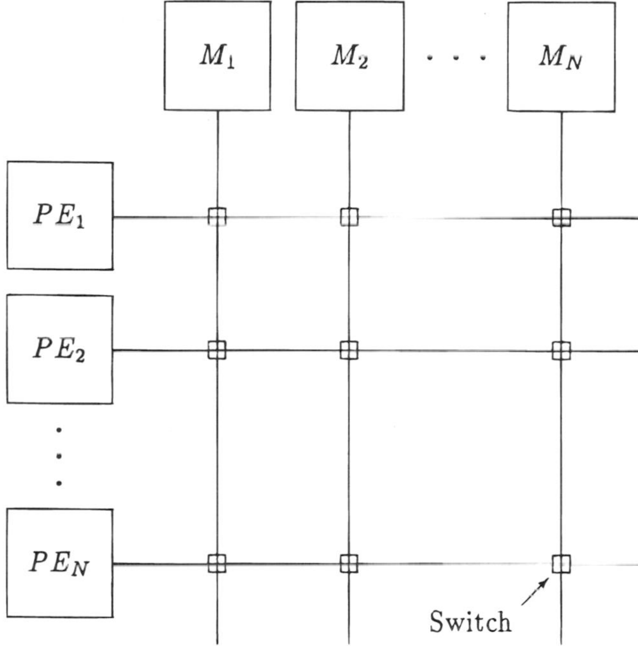

Figure 3.36 MIMD architecture based on a crossbar switch.

forced to wait for access to the bus, degrading system performance. In such cases, bandwidth can be improved by using multiple buses. In the extreme, maximum memory bandwidth can be achieved by interconnecting processors and memories with a *crossbar switch,* as shown in Figure 3.36. A separate switch and bus are provided between each processing element and each shared memory module. Therefore, any permutation of N processing elements concurrently accessing N shared memories can be achieved. Conflicts occur only when two processors must access the same shared memory.

To reduce the number of switches and buses, many other connection networks have been proposed for shared memory multiprocessors (Wu and Feng, 1984). Many of these networks allow a limited number of permutations of concurrent connections between N processing elements and N shared memory modules. However, since there are fewer switches, conflicts often occur in the connection network that block the progress of one processor while the other accesses memory, even though the processors may be attempting to access different memory modules. The tutorial by Wu and Feng (1984) provides an excellent overview of interconnection networks for multiprocessor systems.

Networks of PCs, workstations, and industrial computers are typical examples of loosely coupled systems. Loosely coupled systems contain autonomous processing elements with no shared resources. The processors share information by exchanging messages through a communication network. The most common interconnection methods for computer networks are buses and rings.

Ethernet is one of the most widely used interconnection buses for computer networks. Similar to the shared-bus configuration of Figure 3.35, the Ethernet bus is accessed using a *carrier-sense, multiple-access with collision detection (CSMA/CD)* protocol. All processors continuously "listen" to the bus. When any processor wishing to transmit a message senses that the bus is free, it may begin transmitting preliminary information, followed by the message. If multiple processors begin transmitting at the same time a *collision* is said to occur and the information on the bus becomes garbled, allowing the collision to be detected by all processors on the bus. When a collision is detected, all processors back off and wait a random amount of time before attempting to transmit again. Each processor makes its own decisions regarding access to the bus. Consequently, processors can be added to and removed from the bus without disturbing the rest of the network.

Another commonly used method of interconnecting PCs and other computers is a *token ring,* illustrated in Figure 3.37. Each processing element is connected to two neighboring processors to form a closed ring. Messages are passed from processor to processor in one direction around the ring. When a processor receives a message, it moves the message to its private memory if its address is contained in the message header, effectively removing the message from the ring. Otherwise the message is forwarded to the next processor in the ring. At any given time, one processor is in possession of a logical *token* that allows it to initiate the transmission of messages. The other processors may only receive and forward messages during this time. When a processor in possession of the token has finished, the token is

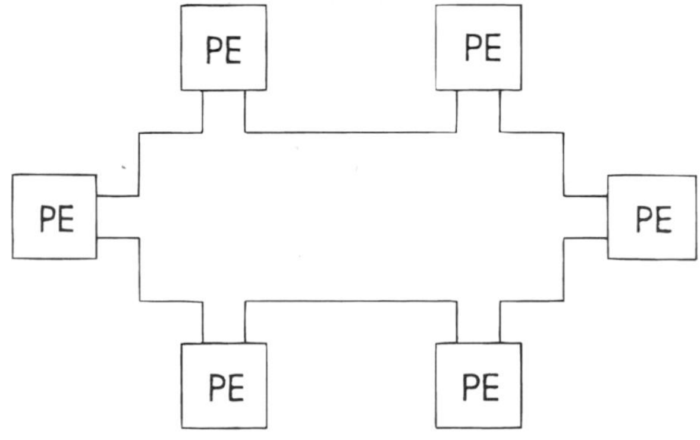

Figure 3.37 Processors connected by a token ring network.

passed to the next processor on the ring, which may then transmit its messages, and so on.

Many other communication network topologies and protocols have been developed for multiple processor systems. The interested reader is referred to Wu and Feng (1984) and Hwang (1993) for further information.

References

Brey, B. B. 1994. *The Intel Microprocessors: 8086/8088, 80286, 80386, and 80486: Architecture, Programming, and Interfacing,* 3d ed., Macmillan, New York, NY.

Flynn, M. J. 1966. Very high-speed computing systems, *Proc. IEEE* 54(12):1901–1909.

Hennessy, J. L. and Patterson, D. A. 1995. *Computer Architecture: A Quantitative Approach,* 2d ed., Morgan Kaufmann, San Mateo, CA.

Hwang, K. 1993. *Advanced Computer Architecture: Parallelism, Scalability, Programmability,* McGraw-Hill, New York, NY.

IEEE Standard for Binary Floating-Point Arithmetic, ANSI/IEEE Std. 754–1985, IEEE, Inc., 345 East 47th St., New York, NY, 10017.

Motorola, Inc., 1990. *MC68000 8-/16-/32-Bit Microprocessors User's Manual,* 8th ed., Prentice-Hall, Englewood Cliffs, NJ.

Motorola, Inc., 1983. *M6805 HMOS/H146805 CMOS Family Users Manual,*, Prentice-Hall, Englewood Cliffs, NJ.

Nelson, V. P. et al., 1995. *Digital Logic Circuit Analysis and Design,* Prentice-Hall, Englewood Cliffs, NJ.

Patterson, D. A. and Hennessy, J. L. 1994. *Computer Organization and Design: the Hardware/Software Interface, Morgan Kaufmann,* San Mateo, CA.

The SPARC Architecture Manual, Version 7, SUN Microsystems, 2550 Garcia Avenue, Mountain View, CA, 94043, 1987.

Spasov, Peter 1993. *Microcontroller Technology: The 68HC11,* Regents/Prentice-Hall, Englewood Cliffs, NJ.

Stewart, J. W. 1993. *The 8051 Microcontroller: Hardware, Software and Interfacing,* Regents/Prentice-Hall, Englewood Cliffs, NJ.

Stone, H. S. 1993. *High-Performance Computer Architecture,* 3d. ed., Addison-Wesley, Reading, MA.

Wu, C. and Feng, T. 1984. *Tutorial: Interconnection Networks for Parallel and Distributed Processing,* IEEE Computer Society Press, Los Alamitos, CA.

4

Signal Processing

James A. Heinen
Marquette University

Russell J. Niederjohn
Marquette University

4.1 Introduction 73
4.2 Continuous-Time Signals 74
Common Signals • Periodic Signals
4.3 Time-Domain Analysis of Continuous-Time Signals 74
Basic Operations on Signals • Convolution
4.4 Frequency-Domain Analysis of Continuous-Time Signals 75
Fourier Series • Fourier Transforms • Laplace Transforms
4.5 Continuous-Time Signal Processors 79
4.6 Time-Domain Analysis of Continuous-Time Signal Processors. 79
4.7 Frequency-Domain Analysis of Continuous-Time Signal Processors 81
4.8 Continuous-Time (Analog) Filters 80
Common Filter Types • Filter Design
4.9 Sampling 81
4.10 Discrete-Time Signals 83
Common Signals • Periodic Signals • Finite-Duration Signals
4.11 Time-Domain Analysis of Discrete-Time Signals 84
Basic Operations on Signals • Convolution • Periodic Convolution
4.12 Frequency-Domain Analysis of Discrete-Time Signals 84
Discrete Fourier Series • Discrete Fourier Transforms • Fast Fourier Transforms • Discrete-Time Fourier Transforms • z-Transforms
4.13 Discrete-Time Signal Processors 89
4.14 Time-Domain Analysis of Discrete-Time Signal Processors 89
4.15 Frequency-Domain Analysis of Discrete-Time Signal Processors 91
4.16 Discrete-Time (Digital) Filters 91
Common Filter Types • FIR Filter Design • IIR Filter Design
4.17 Discrete-Time Analysis of Continuous-Time Signals 93
4.18 Discrete-Time Processing of Continuous-Time Signals 94

4.1 Introduction

Signal processing is a very broad field. Examples include filtering of electrical signals to remove 60 Hz interference, equalization of audio signals, processing of electrocardiograms to determine features of the heartbeat, enhancement of noisy speech signals to improve intelligibility, and enhancement of images to emphasize edges.

A signal is a function of one or more independent variables (usually time in the case of one-dimensional signals). Only real-valued, one-dimensional signals will be considered here. This rules out, for example, images which are two-dimensional. Furthermore, only deterministic signals will be considered, ruling out random signals, which may be described only in a probabilistic sense. Both continuous-time (analog) and discrete-time (digital) signals will be described, as will the sampling (analog-to-digital conversion) process. These signals will be studied directly in the time domain and indirectly in the frequency (transform) domain. Both continuous-time and discrete-time systems (signal processors) will be discussed as well.

Because of the vast amount of information available on signal processing, only that deemed most fundamental will be discussed here. Several excellent texts on signal processing (Ambardar, 1995; Baher, 1990; Oppenheim and Schafer, 1989; Proakis and Manolakis, 1988; Roberts and Mullis, 1987), linear systems (Phillips and Parr, 1995; Taylor, 1994; Ziemer et al., 1993; Lathi, 1992; Jackson, 1991; Gabel and Roberts, 1987), and filters (Antoniou, 1993; Jackson, 1989; Ghausi and Laker, 1981; Lam, 1979; Johnson, 1976) are listed in the References at the end of this discussion. The reader is referred to these for more detailed information regarding the topics discussed and those omitted.

0-8493-8343-9/97/$0.00+$.50

4.2 Continuous-Time Signals

Continuous-time signals are functions of the form $x(t)$, $-\infty < t < \infty$, where t is an independent variable, normally time. Only physically meaningful signals, or those reasonably approximated as such, will be considered in this discussion. Signals with infinite discontinuities will not be allowed. Also disallowed will be signals with an infinite number of finite discontinuities and/or maxima and minima in a finite time range. Impulses (defined below) and signals with finite discontinuities will be considered, even though they are not physically meaningful.

Common Signals

Some continuous-time signals often encountered in practice, and/or which may be easily described mathematically, are described below and illustrated in Figure 4.1.

The unit step is the signal

$$u(t) = \begin{cases} 1, & t \geq 0 \\ 0, & t < 0 \end{cases}.$$

Its integral is the unit ramp

$$r(t) = \left\{\begin{matrix} t, & t \geq 0 \\ 0, & t < 0 \end{matrix}\right\} = \int_{-\infty}^{t} u(\tau)d\tau.$$

The derivative (appropriately defined) of the unit step is the unit impulse

$$\delta(t) = \left\{\begin{matrix} \infty, & t = 0 \\ 0, & t \neq 0 \end{matrix}\right\} = \frac{du(t)}{dt}.$$

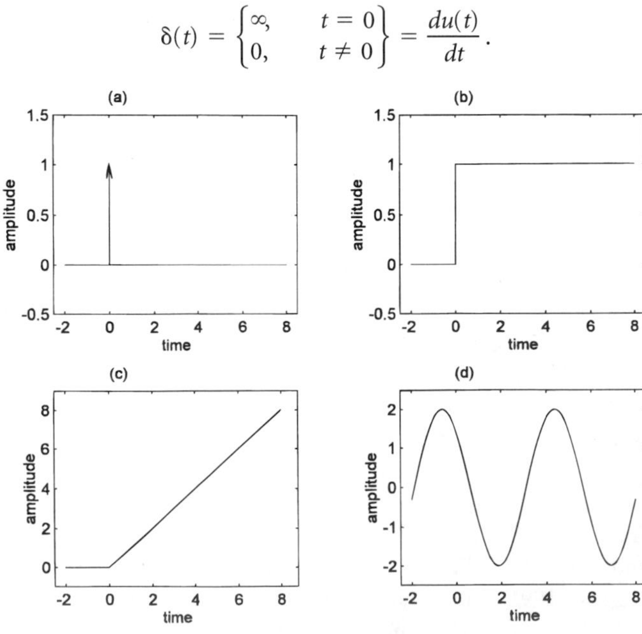

Figure 4.1 Illustration of four common continuous-time signals: (a) $\delta(t)$, (b) $u(t)$, (c) $r(t)$, and (d) $2\cos(2\pi t + \pi/4)$.

To completely define the unit impulse it is necessary to impose the condition that

$$\int_{-\infty}^{\infty} \delta(t)dt = 1.$$

This leads to the so-called sampling property of the impulse, namely,

$$\int_{-\infty}^{\infty} x(t)\delta(t - t_o)dt = x(t_o),$$

provided that $x(t)$ is continuous at t_o.

Sinusoids may be written in the form

$$x(t) = A\cos(\omega t + \theta),$$

where A is the amplitude, $\omega = 2\pi f$ is the angular frequency (f being the frequency), and θ is the phase. A very general signal that is important physically is

$$x(t) = At^p e^{-\alpha t}\cos(\omega t + \theta)u(t),$$

where $p \geq 0$. This signal includes steps, ramps, sinusoids, and damped sinusoids as special cases.

Periodic Signals

Periodic continuous-time signals are those satisfying the condition

$$x(t + T) = x(t), \qquad -\infty < t < \infty,$$

for some fixed $T > 0$. It is noted that if this condition is satisfied for a given T, it is also satisfied for all integer multiples of T. The smallest T for which it is satisfied is called the period. The frequency of a periodic signal with period T is given by $f = 1/T$. Its angular frequency is $\omega = 2\pi f = 2\pi/T$. Normally, T is specified in units of seconds, ω in radians/second and f in Hz. Sinusoids are simple but fundamental examples of periodic signals. Signals that are not periodic are said to be aperiodic.

4.3 Time-Domain Analysis of Continuous-Time Signals

Basic Operations on Signals

Various mathematical operations may be performed on a signal or combination of signals. Obvious ones include differentiating or integrating a signal, and combining two or more signals using addition, subtraction, multiplication, or division. Other operations are also important in the context of signal analysis. Amplitude scaling a signal $x(t)$ produces a modified signal

$$y(t) = Ax(t),$$

where $A \neq 0$ is a constant. Time shifting produces

$$y(t) = x(t - t_o),$$

where t_o is fixed. Time reversal results in

$$y(t) = x(-t),$$

and time scaling leads to

$$y(t) = x(at),$$

where $a > 0$ is a constant.

Convolution

As will be seen later, a very important operation used to combine two continuous-time signals is convolution. The convolution of the signals $x_1(t)$ and $x_2(t)$ is defined as

$$y(t) = x_1(t) * x_2(t) = \int_{-\infty}^{\infty} x_1(\tau)x_2(t - \tau)d\tau, \qquad -\infty < t < \infty.$$

This involves time reversal and time shifting of the signal $x_2(\tau)$ to produce $x_2(t - \tau)$. (It is noted that unless interpreted carefully the notation $y(t) = x_1(t)*x_2(t)$ can lead to confusion. For instance, $y(t - t_o) = x_1(t)*x_2(t - t_o)$, not $x_1(t - t_o)*x_2(t - t_o)$, as might be expected.)

A positive-time signal is one satisfying the condition

$$x(t) = 0, \qquad t < 0.$$

Steps and ramps are clearly examples of positive-time signals. If the signals $x_1(t)$ and $x_2(t)$ are positive-time, their convolution is also positive-time and may be simplified to the form

$$y(t) = x_1(t) * x_2(t) = \begin{cases} \int_0^t x_1(\tau)x_2(t - \tau)d\tau, & t \geq 0 \\ 0, & t < 0 \end{cases}.$$

4.4 Frequency-Domain Analysis of Continuous-Time Signals

It is often both convenient and informative to study continuous-time signals in the frequency (or transform) domain. The original signal is transformed by some appropriate mathematical operation to result in a frequency-domain quantity (often complex-valued in nature). This frequency-domain quantity allows for interpretation of the signal in terms of basic constituent components of various frequencies. The particular frequency-domain representation used will, at least in part, depend on the characteristics of the signal being studied.

Fourier Series

Suppose $x(t)$ is a periodic signal with period T_o and angular frequency $\omega_o = 2\pi/T_o$. Then $x(t)$ may be considered as consisting of a (possibly infinite) linear combination of sinusoids with angular frequencies at integer multiples of ω_o. Specifically

$$x(t) = a_o + \sum_{k=1}^{\infty} (a_k \cos(k\omega_o t) + b_k \sin(k\omega_o t)).$$

Written this way, $x(t)$ is called a (trigonometric) Fourier series with coefficients, a_o, a_k, b_k, $k = 1, 2, \ldots$. These coefficients may be calculated from the original signal using the formulas

$$a_o = \frac{1}{T_o}\int_0^{T_o} x(t)dt,$$

$$a_k = \frac{2}{T_o}\int_0^{T_o} x(t)\cos(k\omega_o t)dt, \qquad k = 1, 2, \ldots,$$

$$b_k = \frac{2}{T_o}\int_0^{T_o} x(t)\sin(k\omega_o t)dt, \qquad k = 1, 2, \ldots.$$

It is noted that, because the integrands are periodic, these integrals may be calculated over any range of t values of length T_o, e.g., $-T_o/2$ to $T_o/2$. It is also noted that a_o is the average (or "dc") value of the signal $x(t)$, as may be seen from its definition. The kth term in the series is referred to as the kth harmonic, with the first harmonic also called the fundamental.

If $x(t)$ is an even signal (i.e., one satisfying $x(-t) = x(t)$, $-\infty < t < \infty$), then $b_k = 0$, $k = 1, 2, \ldots$. On the other hand, if $x(t)$ is an odd signal (i.e., one satisfying $x(-t) = -x(t)$, $-\infty < t < \infty$), then $a_k = 0$, $k = 0, 1, 2, \ldots$. Thus an even signal consists only of cosine components (the cosine is even) and an odd signal consists only of sine components (the sine is odd).

Using trigonometric identities, the trigonometric Fourier series may be rewritten as an exponential (or complex) Fourier series. This takes the form

$$x(t) = \sum_{k=-\infty}^{\infty} X_k e^{jk\omega_o t},$$

where

$$X_k = \frac{1}{T_o}\int_0^{T_o} x(t)e^{-jk\omega_o t}dt, \qquad k = \ldots, -1, 0, 1, 2, \ldots.$$

It is observed that X_k is complex-valued. The coefficients of these two Fourier series representations are related by the formulas

$$a_o = X_o,$$
$$a_k = 2\mathrm{Re}\{X_k\}, \qquad k = 1, 2, \ldots,$$
$$b_k = -2\mathrm{Im}\{X_k\}, \qquad k = 1, 2, \ldots.$$

While it is not always expressed this way, it is instructive to think of X_k and $x(t)$ as forming a transform pair, i.e., a representation in which X_k is uniquely determined by $x(t)$ and vice versa. Thus

$$X_k = FS\{x(t)\}$$

and

$$x(t) = FS^{-1}\{X_k\}.$$

In this context, it is clear that X_k and $x(t)$ contain exactly the same information, but in a different form.

As an example consider the signal $x(t)$ which is assumed to be periodic with period T_o, and which is defined over one period by

$$x(t) = \begin{cases} 1, & 0 \le t < \dfrac{T_o}{2} \\ 0, & \dfrac{T_o}{2} \le t < T_o \end{cases}.$$

Using the defining relation, it is easily determined that the exponential Fourier series coefficients are

$$X_o = 1/2; \qquad X_k = j\frac{(-1)^k - 1}{2\pi k}, \; k \neq 0.$$

The trigonometric Fourier series coefficients are likewise easily found to be

$$a_o = 1/2; \qquad a_k = 0, \; k > 0; \qquad b_k = \frac{1 - (-1)^k}{\pi k}, \; k > 0.$$

Being complex-valued, X_k may be written as

$$X_k = |X_k| \angle \phi_k = |X_k| e^{j\phi_k},$$

where $\phi_k = \arg\{X_k\}$. Under our assumption that $x(t)$ is real-valued, $|X_k|$ is an even function of k and ϕ_k is an odd function of k, i.e., respectively,

$$|X_{-k}| = |X_k|, \qquad \phi_{-k} = -\phi_k.$$

Thus, the information contained in X_k, $k < 0$, is redundant. When plotted vs. k (the frequency index), $|X_k|$ is called the magnitude spectrum and ϕ_k the phase spectrum of $x(t)$. These spectra aid in visualizing the frequency-domain characteristics of the periodic signal under consideration. The spectra for the example discussed above are illustrated in Figure 4.2.

Some important properties of the exponential Fourier series are summarized in Table 4.1.

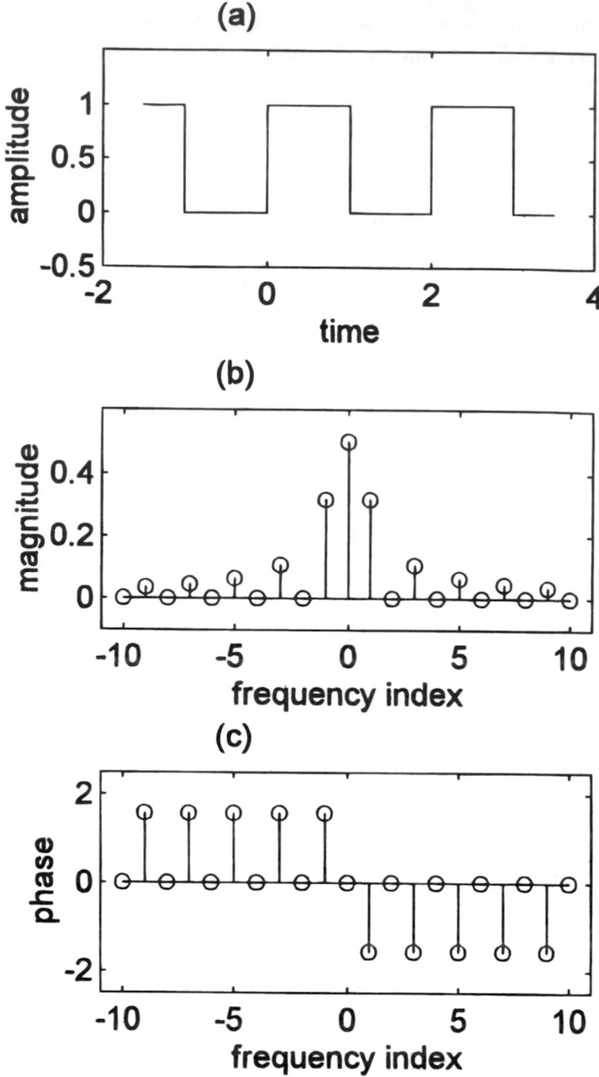

Figure 4.2 Illustration of (a) a periodic signal ($T_o = 2$) and its (b) magnitude spectrum and (c) phase spectrum.

Table 4.1 Properties of Exponential Fourier Series

Periodic signals	Fourier series coefficients
$Ax_1(t) + Bx_2(t)$	$AX_{1,k} + BX_{2,k}$
$x(t - t_o)$	$X_k e^{-jk\omega_o t_o}$
$x(-t)$	X_{-k}
$x(at)$,† $a > 0$	X_k
$\dfrac{dx(t)}{dt}$	$jk\omega_o X_k$

Note: $x(t)$, $x_1(t)$, $x_2(t)$ are arbitrary periodic signals with period T_o and angular frequency ω_o, and with Fourier series coefficients X_k, $X_{1,k}$, $X_{2,k}$, respectively. A, B, a, t_o are arbitrary constants.
† $x(at)$ has period T_o/a and angular frequency $a\omega_o$.

The Fourier series of a signal having finite discontinuities will necessarily contain an infinite number of terms. It is often necessary to approximate such a Fourier series by a truncated series containing only a finite number of terms. Of course, this truncation may take place at any point in the series. Fortunately, including an additional term will never increase the mean-squared error between the truncated series and the true signal. On the other hand, it is very difficult for a Fourier series to represent a discontinuity. Regardless of the number of terms included, the truncated Fourier series exhibits oscillations near each discontinuity. As additional terms are included, these oscillations move closer to the discontinuity, but they never disappear. The existence of these oscillations is known as Gibb's phenomenon.

Fourier Transforms

Fourier series are useful only for analyzing periodic signals. In addition to periodic signals, certain aperiodic signals may be studied using the Fourier transform. This representation may loosely be considered as arising from the Fourier series representation when the fundamental period is allowed to approach infinity.

A continuous-time signal $x(t)$ will have a Fourier transform if it is absolutely integrable, i.e., if

$$\int_{-\infty}^{\infty} |x(t)|\,dt < \infty.$$

With proper interpretation, certain signals not satisfying this condition will also have Fourier transforms. These include steps, constants, and all periodic signals.

The Fourier transform of $x(t)$ is defined as

$$X(j\omega) = FT\{x(t)\} = \int_{-\infty}^{\infty} x(t)e^{-j\omega t}dt, \qquad -\infty < \omega < \infty.$$

$x(t)$ may be recovered uniquely from $X(j\omega)$ using the inverse Fourier transform given by

$$x(t) = FT^{-1}\{X(j\omega)\} = \frac{1}{2\pi}\int_{-\infty}^{\infty} X(j\omega)e^{j\omega t}d\omega.$$

Note that while $X(j\omega)$ is considered as a function of ω, for convenience it is often written, as we have done, as a function of $j\omega$. While periodic signals have all their energy concentrated at only certain values of frequency (multiples of the fundamental frequency), other signals allowed by the Fourier transform have their energy spread out over, in general, all frequencies. Some common Fourier transforms are listed in Table 4.2.

As is true for the exponential Fourier series coefficients X_k, the Fourier transform $X(j\omega)$ is, in general, complex-valued, i.e.,

$$X(j\omega) = |X(j\omega)|\angle\phi(j\omega) = |X(j\omega)|e^{j\phi(j\omega)},$$

Table 4.2 Common Fourier Transforms

$x(t)$	$X(j\omega)$
$\delta(t)$	1
1	$2\pi\delta(\omega)$
$u(t)$	$\pi\delta(\omega) + \frac{1}{j\omega}$
$e^{-at}u(t),\ a > 0$	$\frac{1}{a + j\omega}$
$te^{-at}u(t),\ a > 0$	$\frac{1}{(a + j\omega)^2}$
$\sin(\omega_o t)$	$j\pi[\delta(\omega + \omega_o) - \delta(\omega - \omega_o)]$
$\cos(\omega_o t)$	$\pi[\delta(\omega + \omega_o) + \delta(\omega - \omega_o)]$

where $\phi(j\omega) = \arg\{X(j\omega)\}$. For our assumed real-valued $x(t)$, $|X(j\omega)|$ is an even function of ω, and $\phi(j\omega)$ is an odd function of ω, i.e., respectively

$$|X(-j\omega)| = |X(j\omega)|, \qquad \phi(-j\omega) = -\phi(j\omega).$$

Thus the information contained in $X(j\omega)$, $\omega < 0$, is redundant. When plotted versus ω, $|X(j\omega)|$ is called the magnitude spectrum and $\phi(j\omega)$ the phase spectrum of $x(t)$. An example is shown in Figure 4.3. While the spectra based on the Fourier series exist only at integer values of the frequency index k, the Fourier transform spectra exist, in general, at all frequencies.

Signals not satisfying the absolute integrability condition stated above, will, if they are Fourier transformable, normally contain frequency-domain impulses in their Fourier transforms. As a specific case, if $x(t)$ is a periodic signal with fundamental angular frequency ω_o and with Fourier series $X_k = FS\{x(t)\}$, then its Fourier transform is given by

$$X(j\omega) = \sum_{k=-\infty}^{\infty} X_k\delta(\omega - k\omega_o),$$

with all energy clearly still concentrated (now, mathematically, by means of impulse functions) at integer multiples of ω_o.

Some important properties of the Fourier transform are summarized in Table 4.3.

Laplace Transforms

As seen above, the Fourier transform is a generalization of the Fourier series, in that it allows for the analysis of a broader range of continuous-time signals than does the Fourier series. The (single-sided) Laplace transform is likewise a generalization of the Fourier transform, but with the restriction that the signal must be positive-time, i.e., zero for $t < 0$.

A positive-time signal $x(t)$ will have a Laplace transform if

$$\int_{0-}^{\infty} |x(t)|e^{-\sigma t}dt < \infty$$

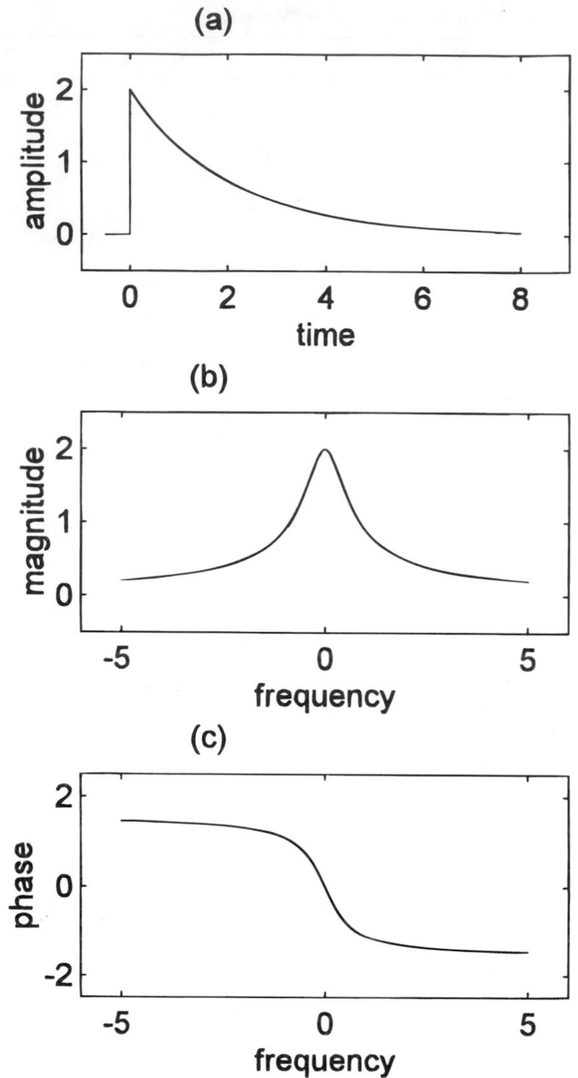

Figure 4.3 Illustration of (a) a signal ($x(t) = 2e^{-0.5t}u(t)$) and its (b) magnitude spectrum and (c) phase spectrum.

Table 4.3 Important Properties of Fourier Transforms

Signals	Fourier transforms
$Ax_1(t) + Bx_2(t)$	$AX_1(j\omega) + BX_2(j\omega)$
$x(t - t_o)$	$X(j\omega)e^{-j\omega t_o}$
$x(-t)$	$X(-j\omega)$
$x(at),\ a > 0$	$\frac{1}{a}X\left(j\frac{\omega}{a}\right)$
$\frac{dx(t)}{dt}$	$j\omega X(j\omega)$
$\int_{-\infty}^{t} x(\tau)d\tau$	$\frac{X(j\omega)}{j\omega} + \pi X(0)\delta(\omega)$
$x_1(t)*x_2(t)$	$X_1(j\omega)X_2(j\omega)$

Note: $x(t)$, $x_1(t)$, $x_2(t)$ are arbitrary signals with Fourier transforms $X(j\omega)$, $X_1(j\omega)$, $X_2(j\omega)$, respectively. A, B, a, t_o are arbitrary constants.

Table 4.4 Common Laplace Transforms

$x(t)$	$X(s)$
$\delta(t)$	1
$u(t)$	$\frac{1}{s},\ Re\{s\} > 0$
$e^{-at}u(t)$	$\frac{1}{s + a},\ Re\{s\} > -a$
$te^{-at}u(t)$	$\frac{1}{(s + a)^2},\ Re\{s\} > -a$
$\sin(\omega_o t)u(t)$	$\frac{\omega_o}{s^2 + \omega_o^2},\ Re\{s\} > 0$
$\cos(\omega_o t)u(t)$	$\frac{s}{s^2 + \omega_o^2},\ Re\{s\} > 0$

for some real σ. (The notation $0-$ is used to indicate that the lower limit of the integral is taken "just to the left of 0." In practice, this means that impulses at $t = 0$ are to be included entirely in the integral.) Since $\sigma = 0$ is allowed in this condition, if $x(t)$ is a positive-time signal with a Fourier transform, it will also have a Laplace transform, but the converse is not true. For example, a unit ramp has a Laplace transform but not a Fourier transform. On the other hand, nontrivial periodic signals and constants (which are necessarily nonzero for $t < 0$) do not have Laplace transforms even though they have Fourier transforms.

The Laplace transform of the positive-time signal $x(t)$ is defined as

$$X(s) = LT\{x(t)\} = \int_{0-}^{\infty} x(t)e^{-st}dt,$$

where $s = \sigma + j\omega$ is a complex variable. In general, this integral will converge only for certain choices of σ. $x(t)$ may be recovered uniquely from $X(s)$ using the inverse Laplace transform given by

$$x(t) = LT^{-1}\{X(s)\} = \frac{1}{2\pi j}\int_{\sigma_1 - j\infty}^{\sigma_1 + j\infty} X(s)e^{st}ds,$$

where σ_1 is appropriately chosen. (This formula is rarely directly used in practice because of its complexity. Instead, tables of Laplace transform pairs are used, with complicated transforms first being decomposed into simpler ones using partial fraction expansion techniques.) Some common Laplace transforms are listed in Table 4.4.

For a positive-time signal $x(t)$ having a Fourier transform $X(j\omega)$, it is the case that

$$FT\{x(t)\} = LT\{x(t)\}|_{\sigma=0}$$

or

$$X(j\omega) = X(s)|_{s=j\omega},$$

where $X(s)$ is the Laplace transform of $x(t)$. (This explains our choice of notation for the Fourier transform.) Thus, in this case, the Fourier transform is simply the Laplace transform evaluated on the $j\omega$ (imaginary) axis in the s (complex) plane.

Some important properties of the Laplace transform are summarized in Table 4.5.

4.5 Continuous-Time Signal Processors

A continuous-time signal processor (system) is a device (Figure 4.4) that acts on an input signal $x(t)$, modifying it in some manner to produce an output signal $y(t)$. This may be represented abstractly as

$$y(t) = \mathcal{H}\{x(t)\}.$$

(Here $x(t)$ and $y(t)$ should be thought of in their totality, rather than at specific instants t. Just as in the case of convolution, this notation can lead to confusion. It does not imply that $y(t)$ is a function of $x(t)$ at only the same instant t. $y(t)$ could, for example, depend on all past values $x(\tau)$, $\tau \leq t$.)

This abstract definition is very broad. In practice, because of mathematical tractability, only certain relatively simple classes of signal processors are usually considered. In this treatment, we will discuss only those continuous-time signal processors that can be described by linear constant-coefficient ordinary differential equations. Such signal processors are of necessity linear, time-invariant, and causal. (In the familiar case of electrical circuits, they can be built using resistors, capacitors, inductors, and operational amplifiers.) A system is linear if

$$\mathcal{H}\{Ax_1(t) + Bx_2(t)\} = A\mathcal{H}\{x_1(t)\} + B\mathcal{H}\{x_2(t)\}, \quad -\infty < t < \infty,$$

for all signals $x_1(t)$, $x_2(t)$, and all constants A, B. It is time-invariant if

$$y(t - t_o) = \mathcal{H}\{x(t - t_o)\}, \qquad -\infty < t < \infty,$$

for all input signals $x(t)$ and corresponding output signals $y(t)$ and all constants t_o. A system is causal if $\mathcal{H}\{x(t)\}$ depends only on $x(\tau)$, $\tau \leq t$, for all signals $x(t)$.

Table 4.5 Important Properties of Laplace Transforms

Signals	Laplace transforms
$Ax_1(t) + Bx_2(t)$	$AX_1(s) + BX_2(s)$
$x(t - t_o)$, $t_o \geq 0$	$X(s)e^{-st_o}$
$x(at)$, $a > 0$	$\frac{1}{a}X\left(\frac{s}{a}\right)$
$\frac{dx(t)}{dt}$	$sX(s) - x(0-)$
$\int_{-\infty}^{t} x(\tau)d\tau$	$\frac{X(s)}{s}$
$x_1(t) * x_2(t)$	$X_1(s)X_2(s)$

Note: $x(t)$, $x_1(t)$, $x_2(t)$ are arbitrary signals with Laplace transforms $X(s)$, $X_1(s)$, $X_2(s)$, respectively. A, B, a, t_o are arbitrary constants.

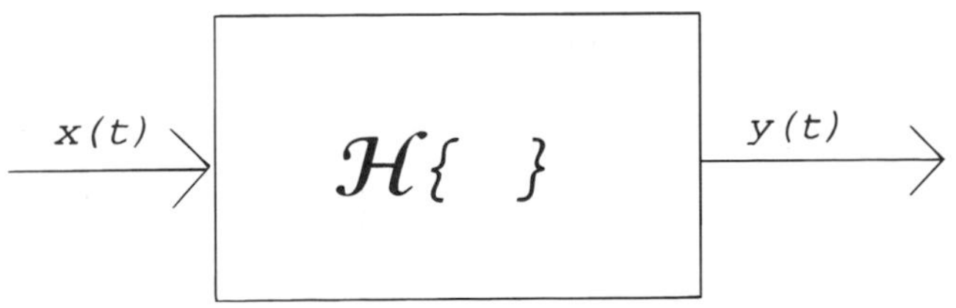

Figure 4.4 Continuous-time signal processing system.

4.6 Time-Domain Analysis of Continuous-Time Signal Processors

Under our assumptions, a continuous-time signal processor may be represented by a linear constant-coefficient ordinary differential equation. That is,

$$\sum_{r=0}^{N} a_r \frac{d^r y(t)}{dt^r} = \sum_{r=0}^{M} b_r \frac{d^r x(t)}{dt^r}.$$

Ordinarily $M \leq N$ and, often, this equation is normalized so that $a_N = 1$. This describes an implicit relationship between the input $x(t)$ and output $y(t)$. In specific cases, for a known input $x(t)$ and initial conditions on $y(t)$ and its derivatives at $t = 0$, it may be solved for $x(t)$, $t \geq 0$. Laplace transforms may be used for this purpose, as may other techniques.

An alternative means for describing our signal processor, in the form of an explicit relationship between input and output, is based on the use of its impulse response. The impulse response is defined as

$$h(t) = \mathcal{H}\{\delta(t)\},$$

i.e., the output when the input is a unit impulse. Since we have assumed causality, it may be shown that $h(t)$ is a positive-time signal, i.e., that $h(t) = 0$, $t < 0$. Because the signal processor is linear and time-invariant, it may additionally be shown that for any input $x(t)$ the corresponding output is given by the convolution of $h(t)$ and $x(t)$, namely

$$y(t) = \mathcal{H}\{x(t)\} = h(t) * x(t).$$

This is a very significant result, since it shows that all information regarding the signal processor (at least from an input-output viewpoint) is contained in its impulse response.

A very desirable property of signal processors is stability. A system is said to be (bounded-input bounded-output) stable if whenever $x(t)$ is bounded (i.e., $|x(t)| \leq K_x < \infty$, $-\infty < t < \infty$, for some K_x) then the corresponding $y(t)$ is also bounded (i.e., $|y(t)| \leq K_y < \infty$, $-\infty < t < \infty$, for some K_y). (The alternative would normally be undesirable.) It is possible to easily determine stability directly from $h(t)$. Specifically, the signal processor is stable if and only if $h(t)$ is absolutely integrable, i.e., if and only if

$$\int_{-\infty}^{\infty} |h(t)|\, dt < \infty.$$

4.7 Frequency-Domain Analysis of Continuous-Time Signal Processors

In the frequency (transform) domain, a signal processor is characterized by its transfer function $H(s)$. $H(s)$ may be determined by Laplace transforming the convolution relationship to obtain

$$Y(s) = H(s)X(s),$$

where

$$H(s) = LT\{h(t)\}.$$

Thus the transfer function is the Laplace transform of the impulse response, well defined since $h(t)$ is a positive-time signal.

Alternatively, the input-output differential equation may be Laplace transformed (assuming zero initial conditions) to obtain

$$\sum_{r=0}^{N} a_r s^r Y(s) = \sum_{r=0}^{M} b_r s^r X(s),$$

from which

$$H(s) = \frac{\sum_{r=0}^{M} b_r s^r}{\sum_{r=0}^{N} a_r s^r}.$$

Thus $H(s)$ is a rational function (ratio of polynomials) in s. The impulse response $h(t)$, being the inverse Laplace transform of a rational function, will necessarily consist only of terms of the form

$$At^p e^{-\alpha t} \cos(\omega t + \theta) u(t),$$

where $p \geq 0$.

The roots of the numerator polynomial are called the zeros of $H(s)$ and the roots of the denominator polynomial are called the poles of $H(s)$. It may be shown that a signal processor is stable if and only if $M \leq N$ and all poles lie strictly in the left half of the complex s plane (i.e., have strictly negative real parts).

Fourier transforming the convolution relationship leads to

$$Y(j\omega) = H(j\omega)X(j\omega),$$

where

$$H(j\omega) = FT\{h(t)\}.$$

$H(j\omega)$ is called the frequency response of the signal processor, and is the Fourier transform of the impulse response. The magnitude of $H(j\omega)$ is the magnitude response and its angle is the phase response. It is seen that

$$H(j\omega) = H(s)|_{s=j\omega},$$

where $H(s)$ is the transfer function. That is, the frequency response is the transfer function evaluated on the $j\omega$ (imaginary) axis in the complex s plane. Thus we may also write

$$H(j\omega) = \frac{\sum_{r=0}^{M} b_r (j\omega)^r}{\sum_{r=0}^{N} a_r (j\omega)^r}.$$

The frequency response of a signal processor is also useful for determining its output in the case of a periodic input. If $x(t)$ is a periodic input signal with Fourier series coefficients X_k, then the output signal $y(t)$ will also be periodic (with the same fundamental angular frequency ω_o) and will have Fourier series coefficients given by

$$Y_k = H(jk\omega_o)X_k.$$

This relationship may also be used for the steady-state analysis of systems, i.e., systems that have been operating for a sufficiently long time that all transient terms may be neglected.

It is thus clear that the way a signal processor processes an input signal (either aperiodic or periodic) is determined by its frequency response. It is thus very informative to have this quantity available, particularly in visual form as embodied in its magnitude and phase spectra.

4.8 Continuous-Time (Analog) Filters

A continuous-time (analog) filter is a signal processor designed to allow input signal components of certain frequencies to pass through to the output while preventing input signal components of other frequencies from doing so. Ideally, the frequency response $H(j\omega)$ of a filter should have magnitude one at those frequencies we wish to allow to pass through to the output and magnitude zero at those frequencies disallowed.

Common Filter Types

The most common types of filters are lowpass, highpass, bandpass, and bandstop. In the ideal case, these are described, respectively, by the following relationships (Figure 4.5):

$$H_{lp}(j\omega) = \begin{cases} 1, & |\omega| \leq \omega_c \\ 0, & \text{otherwise} \end{cases},$$

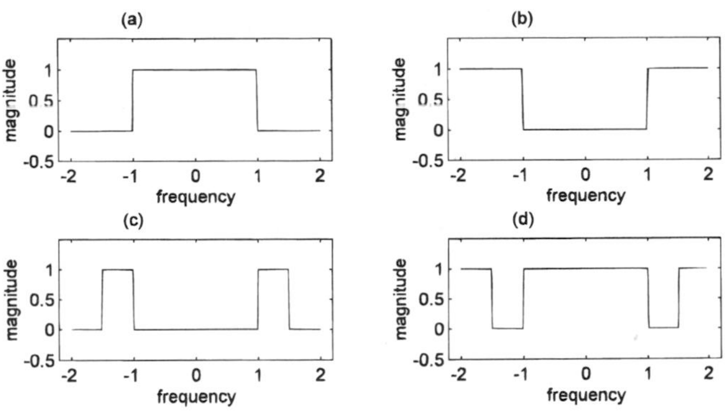

Figure 4.5 Illustration of four ideal filter types: (a) lowpass filter, (b) highpass filter, (c) bandpass filter, and (d) bandstop filter.

$$H_{hp}(j\omega) = \begin{cases} 0, & |\omega| < \omega_c \\ 1, & \text{otherwise} \end{cases},$$

$$H_{bp}(j\omega) = \begin{cases} 1, & \omega_l \leq |\omega| \leq \omega_u \\ 0, & \text{otherwise} \end{cases},$$

$$H_{bs}(j\omega) = \begin{cases} 0, & \omega_l < |\omega| < \omega_u \\ 1, & \text{otherwise} \end{cases}.$$

The values ω_c, ω_l, and ω_u are the cutoff frequencies. Those ranges of frequencies for which the frequency response is one are the passbands and those for which it is zero are the stopbands. The bandwidth is defined as ω_c for a lowpass filter and $\omega_u - \omega_l$ for a bandpass filter.

Unfortunately, ideal filters are noncausal and are thus not physically realizable. They may, however, be quite closely approximated in practice. In the nonideal (practical) case, the frequency response values are not identically one in the passbands and not identically zero in the stopbands. Furthermore, the response does not abruptly change from one to zero between passbands and stopbands. Instead, it gradually changes in transition bands which separate the passbands and stopbands. While cutoff frequencies are uniquely defined for ideal filters, in the case of practical filters any frequency within a transition band may appropriately be called a cutoff frequency. Often those frequencies at which the magnitude response is $1/\sqrt{2}$ (approximately -3db) are chosen as the cutoff frequencies.

In many instances, the phase response of a filter is of little consequence. In certain applications, however, it is required that the phase response be linear, i.e., that the phase be linearly related to frequency in the passbands. Again, this may only be approximated in practice. Linear phase ensures that the shape of a signal within the passband of a filter, rather than simply its frequency content, is preserved in the filtering process.

Filter Design

The design of a filter normally begins by specifying the locations of the passbands and stopbands, the allowable variations from the ideal values in these bands (tolerances), and the locations and widths of the transition bands. Generally, the tighter the tolerances and the narrower the transition bands, the higher the filter order will have to be to meet the specifications. Normally our goal will be to meet the specifications with the lowest order filter possible.

Despite the type of filter desired, an appropriate prototype lowpass filter $H_p(s)$, with a cutoff frequency of one, is often designed first. Design formulas exist for translating the original specifications to the prototype filter. Various standard lowpass prototypes are available. The simplest of these is perhaps the Butterworth filter. The transfer function of an Nth-order Butterworth filter is given by

$$H_{B_N}(s) = \frac{1}{B_N(s)},$$

where $B_N(s)$ is a Butterworth polynomial. Table 4.6 lists expressions for the first four of these. The poles of $H_{B_N}(s)$ are evenly spaced on the unit circle in the left half of the complex plane. The magnitude response of a Butterworth filter monotonically decreases from one to zero and has a value of $1/\sqrt{2}$ (-3db) at $\omega = 1$. Figure 4.6 shows the magnitude and phase responses of a typical lowpass Butterworth filter.

Chebyshev filters are equiripple in the passband (i.e., their magnitude response oscillates between certain tolerance limits) and monotonic in the stopband. These characteristics are interchanged in the case of inverse Chebyshev filters. Elliptic (Cauer) filters are equiripple in both the passband and stopband. Bessel filters achieve very nearly linear phase in the passband. In each case, various design formulas exist for choosing the appropriate filter order and other filter parameters.

Once the prototype lowpass filter $H_p(s)$ (with cutoff frequency of one) is determined, an appropriate frequency transformation is employed to result in the final filter design $H(s)$ as follows:

$$H(s) = H_p(s)|_{s=T(s)}$$

That is, s is replaced by $T(s)$ in the expression for $H_p(s)$. These transformations are summarized in Table 4.7 for cutoff frequencies of ω_c, ω_l, and ω_u.

4.9 Sampling

Sampling is the process of converting a continuous-time signal into a discrete-time signal. This may be represented abstractly by

$$x[n] = S\{x(t)\},$$

Table 4.6 Butterworth Polynomial Expressions

N	$B_N(s)$
1	$s + 1$
2	$s^2 + 1.414s + 1$
3	$s^3 + 2s^2 + 2s + 1$
4	$s^4 + 2.613s^3 + 3.414s^2 + 2.613s + 1$

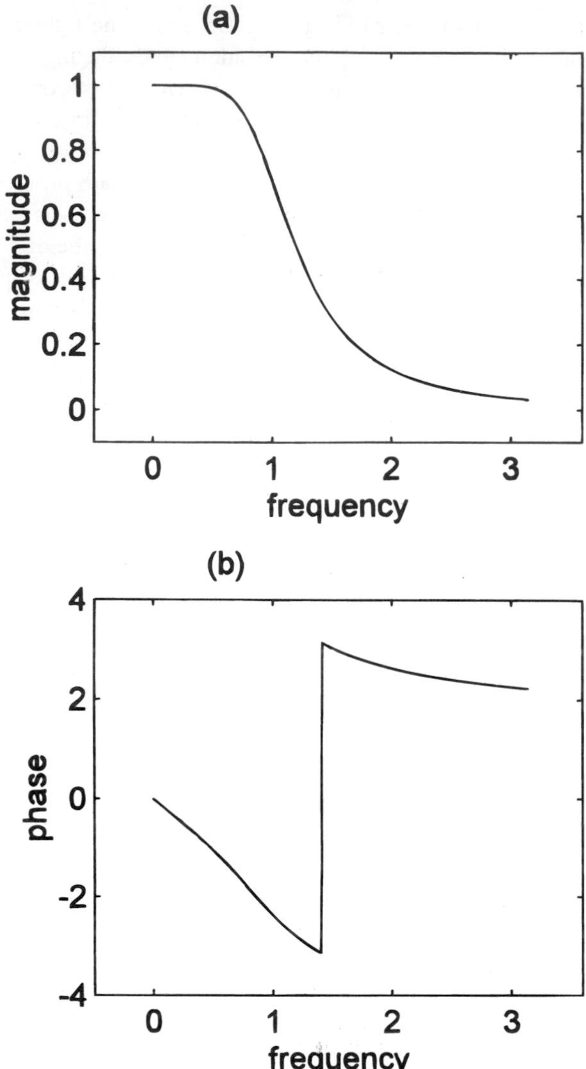

Figure 4.6 Third-order Butterworth filter (a) magnitude response and (b) phase response.

Table 4.7 Frequencies Transformations

Filter type	T(s)
Lowpass	$\frac{s}{\omega_c}$
Highpass	$\frac{\omega_c}{s}$
Bandpass	$\frac{(\omega_u - \omega_l)s}{s^2 + \omega_u\omega_l}$
Bandstop	$\frac{s^2 + \omega_u\omega_l}{(\omega_u - \omega_l)s}$

where $x(t)$, $-\infty < t < \infty$, is the original continuous-time signal and $x[n]$, $-\infty < n < \infty$, n being integer-valued, is a discrete-time signal consisting of samples of $x(t)$ (Figure 4.7). More precisely,

$$x[n] = x(nT_s) = x(t)|_{t=nT_s},$$

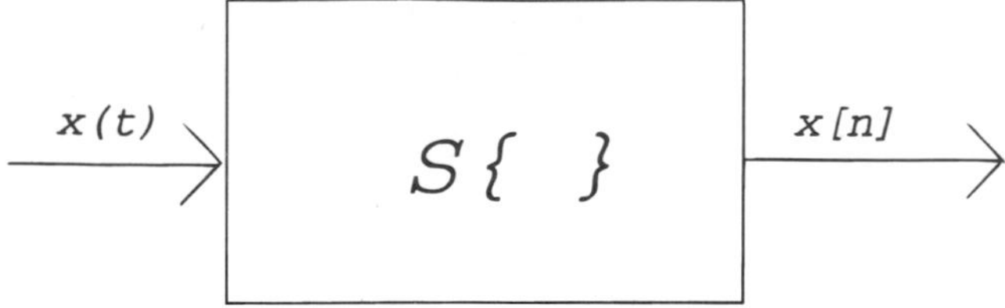

Figure 4.7 Sampling a continuous-time signal to produce a discrete-time signal.

where $T_s > 0$ is called the sampling period, assumed here to be fixed. The sampling frequency is $f_s = 1/T_s$ and the angular sampling frequency is $\omega_s = 2\pi f_s = 2\pi/T_s$. It is noted that n is really a time index, but will usually be referred to as time. $x[n]$ is thus seen to be simply a sequence of numbers. (The square-bracket notation, which will be used throughout to distinguish discrete-time from continuous-time quantities, follows that used by Oppenheim and Schafer, 1989.)

The actual process of sampling a physical signal is considerably more complicated than what is implied by this abstract description. The final result, namely a sequence of numbers, is, however, quite meaningful if, say, these numbers are fed to a digital computer for further processing. The physical sampling (or analog-to-digital conversion) process may be modeled by a switch that closes instantaneously, or more realistically, very briefly, at instants separated by T_s units of time. This produces narrow pulses of height $x(nT_s)$, which may further be considered as impulses. This leads to the "impulse-sampled" signal

$$x_s(t) = \sum_{n=-\infty}^{\infty} x(nT_s)\delta(t - nT_s) = \sum_{n=-\infty}^{\infty} x[n]\delta(t - nT_s).$$

Oddly enough, this impulse-sampled signal $x_s(t)$ is in fact a continuous-time signal. Clearly, from an information content viewpoint, the signal $x_s(t)$ and the sequence $x[n]$ are equivalent. Depending on the task at hand, one or the other of these representations may be the most convenient to consider.

In practice, when we sample a signal we wish to do it in such a manner that the samples comprise a reasonably accurate representation of the original continuous-time signal. Intuitively this suggests that we take a great many samples (T_s small, f_s large), especially if the signal is changing rapidly, so as not to lose any significant information. On the other hand, efficiency (both in how fast we must sample and perhaps how many samples we must store) dictates that we not take too large a number of samples. Trade-offs must therefore be made in choosing an appropriate sampling frequency for a given signal.

It is instructive to consider the sampling process in the frequency domain. If $x(t)$ is a continuous-time signal with Fourier transform $X(j\omega)$, then its impulse-sampled counterpart $x_s(t)$ will have Fourier transform

$$X_s(j\omega) = \frac{1}{T_s}\sum_{m=-\infty}^{\infty} X(j(\omega - m\omega_s)).$$

That is, $X_s(j\omega)$ consists of copies of $X(j\omega)$, shifted by all possible multiples of the angular sampling frequency ω_s, added together and scaled by $1/T_s$. $X_s(j\omega)$ is thus seen to be periodic with period ω_s.

In general, the various copies of $X(j\omega)$ will overlap each other, a phenomenon called aliasing. This prevents the recovery of $X(j\omega)$ from $X_s(j\omega)$, implying that information is lost in the sampling process. On the other hand, if $x(t)$ is bandlimited to half the sampling frequency, i.e., if

$$X(j\omega) = 0, \qquad |\omega| \geq \frac{\omega_s}{2},$$

then no aliasing or overlap occurs, and the original signal may be recovered from its samples. This may be summarized in the form of a theorem, the *sampling theorem*, which states that no information is lost in the sampling process if the sampling frequency is at least twice that of the highest frequency component present in the original signal. This minimum sampling frequency is called the Nyquist rate. The sampling theorem provides a quantitative basis for the choice of sampling frequency, and validates our intuition that rapidly varying signals must be sampled more frequently than slowly varying ones.

4.10 Discrete-Time Signals

Discrete-time signals are functions (sequences) of the form $x[n]$, $-\infty < n < \infty$, where n is an independent integer-valued variable, normally a time index (usually simply referred to as "time"). For our purposes, we will assume $x[n]$ to be real-valued.

Common Signals

Some discrete-time signals often encountered in practice, and/or which may be easily described mathematically, are described below and illustrated in Figure 4.8.

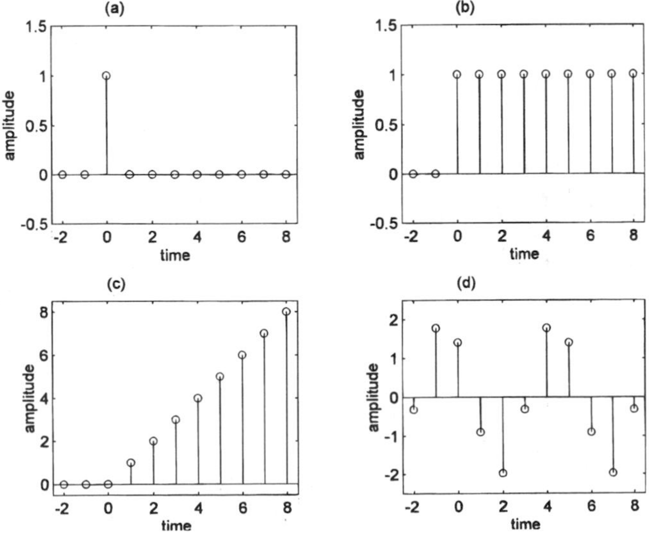

Figure 4.8 Illustration of four common discrete-time signals: (a) $\delta[n]$, (b) $u[n]$, (c) $r[n]$, and (d) $2\cos(2\pi n + \pi/4)$.

The unit step is the signal

$$u[n] = \begin{cases} 1, & n \geq 0 \\ 0, & n < 0 \end{cases}.$$

Its first difference is the unit impulse

$$\delta[n] = \begin{Bmatrix} 1, & n = 0 \\ 0, & n \neq 0 \end{Bmatrix} = u[n] - u[n-1].$$

The unit ramp is given by

$$r[n] = \begin{cases} n, & n \geq 0 \\ 0, & n < 0 \end{cases}.$$

Discrete-time sinusoids may be written in the form

$$x[n] = A\cos(\omega n + \theta).$$

A very general signal that is important physically is

$$x[n] = An^p a^n \cos(\omega n + \theta)u[n],$$

where $p \geq 0$.

Periodic Signals

Periodic discrete-time signals are those satisfying the condition

$$x[n + N] = x[n], \qquad -\infty < n < \infty,$$

for some fixed integer $N > 0$. As is the case for continuous-time periodic signals, if this condition is satisfied for a given N, it is also satisfied for all integer multiples of N. The smallest such N is called the period. Its frequency is $1/N$ and its angular frequency is $2\pi/N$.

Oddly enough, not all discrete-time sinusoids are periodic. The sinusoid $A\cos(\omega n + \theta)$ is periodic if and only if $\omega = 2\pi f$ where f is a rational number (ratio of integers). In this case, the period N is the smallest integer that is an integer multiple of $1/f$.

Finite-Duration Signals

As the name implies, finite-duration signals are signals that are nonzero for only a finite range of time values. We will be somewhat restrictive and define $x[n]$ to be a finite-duration signal if it satisfies the condition

$$x[n] = 0, \qquad n < 0 \text{ and } n > N - 1.$$

This signal thus has, at most, N nonzero values and will be said to be of length N.

There is a close relationship between finite-duration signals and periodic signals. Specifically, if $x[n]$ is periodic with period N, then

$$y[n] = \begin{cases} x[n], 0 \le n \le N-1 \\ 0, \text{otherwise} \end{cases}$$

is a finite-duration signal of length N. On the other hand, if $x[n]$ is a length-N finite-duration signal then

$$y[n] = \sum_{m=-\infty}^{\infty} x[n - mN]$$

is periodic with period N. Each therefore contains the same information, but in a different form.

4.11 Time-Domain Analysis of Discrete-Time Signals

Basic Operations on Signals

Just as in the case of continuous-time signals, various mathematical operations may be performed on a discrete-time signal or combination of signals. Basic ones include combining two or more signals using addition, subtraction, multiplication, or division.

The (backward) difference of a signal $x[n]$ is given by

$$y[n] = \Delta x[n] = x[n] - x[n-1].$$

The accumulation of a signal $x[n]$ is given by

$$y[n] = \sum_{m=-\infty}^{n} x[m].$$

The difference and accumulation operations are inverses of each other, and may be thought of as discrete-time analogs of differentiation and integration, respectively.

Other operations important in signal analysis include amplitude scaling a signal $x[n]$ to produce

$$y[n] = Ax[n],$$

where $A \neq 0$ is a constant. Time shifting produces

$$y[n] = x[n - n_o]$$

where n_o is a fixed integer. Finally, time reversal results in

$$y[n] = x[-n].$$

Convolution

The discrete-time convolution of the two signals $x_1[n]$ and $x_2[n]$ is defined as

$$\begin{aligned} y[n] &= x_1[n] * x_2[n] \\ &= \sum_{m=-\infty}^{\infty} x_1[m]x_2[n-m], \qquad -\infty < n < \infty. \end{aligned}$$

This involves time reversal and time shifting of the signal $x_2[m]$ to produce $x_2[n - m]$. (It is once again noted that caution must be exercised in interpreting the notation $y[n] = x_1[n]*x_2[n]$. For instance $y[n - n_o] = x_1[n]*x_2[n - n_o]$.)

A positive-time signal is one satisfying the condition

$$x[n] = 0, \qquad n < 0.$$

Unit steps, ramps, and impulses are positive-time. If the signals $x_1[n]$ and $x_2[n]$ are both positive-time then their convolution is also positive-time and may be written in the simplified form

$$y[n] = x_1[n] * x_2[n] = \begin{cases} \sum_{m=0}^{n} x_1[m]x_2[n-m], & n \ge 0 \\ 0, & n < 0 \end{cases}.$$

Periodic Convolution

If $x_1[n]$ and $x_2[n]$ are two periodic signals with period N, their convolution would clearly not be meaningful. Such signals may be combined, instead, using the operation of periodic convolution, defined by

$$\begin{aligned} y[n] &= x_1[n] \tilde{*} x_2[n] \\ &= \sum_{m=0}^{N-1} x_1[m]x_2[n-m], \qquad -\infty < n < \infty. \end{aligned}$$

The resulting signal $y[n]$ is also periodic with period N. It should be noted that periodic convolution is not ordinarily something we would wish to do—rather, it is an operation forced upon us by the mathematics.

4.12 Frequency-Domain Analysis of Discrete-Time Signals

As is also true for continuous-time signals, much can be gained by studying discrete-time signals in the frequency (or transform) domain. Again, various transforms are used, depending on the nature of the signal and the task at hand.

Discrete Fourier Series

Suppose $x[n]$ is a periodic discrete-time signal with period N. Then $x[n]$ may be considered as consisting of a finite linear combination of discrete-time complex exponentials. Specifically, it takes N such signals to represent $x[n]$, which can be written as

$$x[n] = \frac{1}{N}\sum_{k=0}^{N-1} X[k]e^{j(2\pi/N)kn}.$$

The coefficients $X[k]$ of the discrete Fourier series may be obtained from the original signal using the formula

$$X[k] = \sum_{n=0}^{N-1} x[n]e^{-j(2\pi/N)kn}, \qquad k = 0, 1, 2, \ldots, N-1.$$

It is noted that $X[k]$ is itself a (generally complex) periodic signal with period N, and as such is defined for all integers k. However, only N values of $X[k]$ are required to determine (or represent) $x[n]$.

$X[k]$ and $x[n]$ form a transform pair with $X[k]$ uniquely determined by $x[n]$ and vice versa. Thus we write

$$X[k] = DFS\{x[n]\}$$

and

$$x[n] = DFS^{-1}\{X[k]\}.$$

$X[k]$ and $x[n]$ thus contain exactly the same information, but in a different form.

As an example, consider the signal $x[n]$ which is assumed to be periodic with period N (where N is even), and which is defined over one period by

$$x[n] = \begin{cases} 1, & 0 \le n < \dfrac{N}{2} \\ 0, & \dfrac{N}{2} \le n < N \end{cases}.$$

Using the defining relation, it is easily determined that the discrete Fourier series coefficients are

$$X[k] = \frac{1-(-1)^k}{1-e^{-j(2\pi/N)k}}$$

for all integers k.

Being complex-valued, $X[k]$ may be written as

$$X[k] = |X[k]|\angle\phi[k] = |X[k]|e^{j\phi[k]},$$

where $\phi[k] = \arg\{X[k]\}$. Under our assumption that $x[n]$ is real-valued, $|X[k]|$ is an even function of k and $\phi[k]$ is an odd function of k, i.e., respectively,

$$|X[-k]| = |X[k]|, \qquad \phi[-k] = -\phi[k].$$

Thus the information contained in $X[k]$, $k < 0$, is redundant. In addition, since $X[k]$ is periodic, all but N consecutive values of $X[k]$ are redundant. Combining these two results leads to the conclusion that $X[k]$ is completely determined by $X[k]$, $0 \le k \le M$, where $M = N/2$ for N even and $M = (N-1)/2$ for N odd. When plotted versus k (the frequency index), $|X[k]|$ is called the magnitude spectrum and $\phi[k]$ the phase spectrum. The spectra for the example discussed above are illustrated in Figure 4.9.

Some important properties of the discrete Fourier series are summarized in Table 4.8.

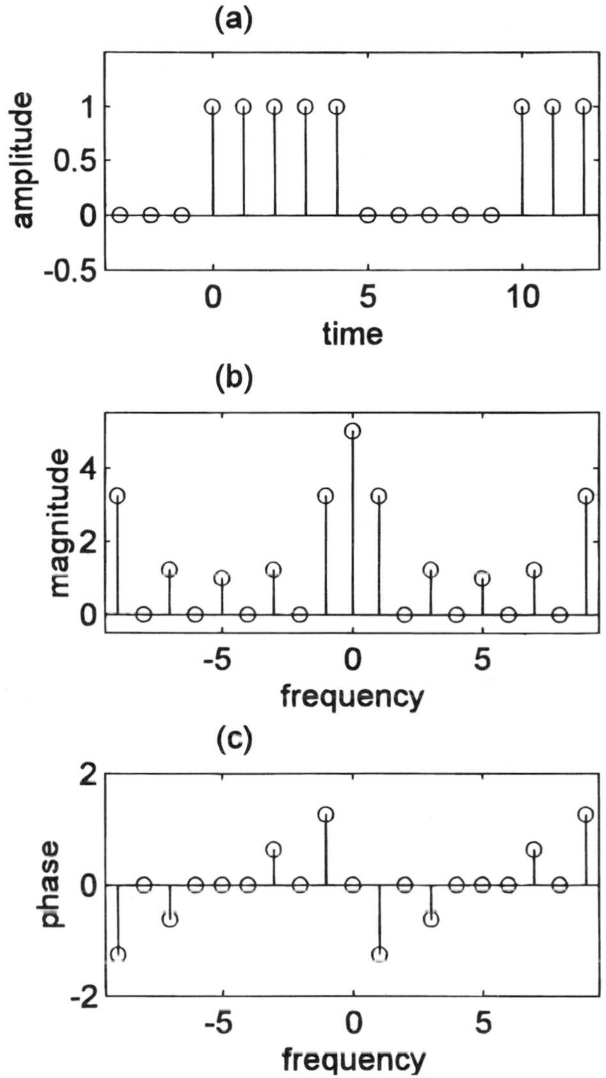

Figure 4.9 Discrete Fourier series of a periodic signal: (a) signal, (b) magnitude spectrum, and (c) phase spectrum.

Table 4.8 Important Properties of Discrete Fourier Series

Discrete-time periodic signals	Discrete fourier series coefficients
$Ax_1[n] + Bx_2[n]$	$AX_1[k] + BX_2[k]$
$x[n - n_o]$	$X[k]e^{-j(2\pi/N)kn_o}$
$x[-n]$	$X[-k]$
$x_1[n] \tilde{*} x_2[n]$	$X_1[k]X_2[k]$

Note: $x[n]$, $x_1[n]$, $x_2[n]$ are arbitrary discrete-time periodic signals with period N and with discrete Fourier series coefficients $X[k]$, $X_1[k]$, $X_2[k]$, respectively. A, B are arbitrary constants and n_o is an arbitrary integer constant.

Discrete Fourier Transforms

As seen earlier, from an information content viewpoint, there is no essential difference between a periodic signal and a finite duration signal. The discrete Fourier series is the frequency-domain representation of a periodic signal. The frequency-domain representation of a length-N finite-duration signal $x[n]$ is the discrete Fourier transform, given by

$$X[k] = DFT\{x[n]\}$$
$$= \begin{cases} \sum_{n=0}^{N-1} x[n]e^{-j(2\pi/N)kn}, & k = 0, 1, 2, \ldots, N-1 \\ 0, & \text{otherwise} \end{cases}.$$

The inverse discrete Fourier transform is given by

$$x[n] = DFT^{-1}\{X[k]\}$$
$$= \begin{cases} \frac{1}{N}\sum_{k=0}^{N-1} X[k]e^{j(2\pi/N)kn}, & n = 0, 1, 2, \ldots, N-1 \\ 0, & \text{otherwise} \end{cases}.$$

As an example consider the finite-duration signal $x[n]$ which is defined by

$$x[n] = \begin{cases} 1, & 0 \leq n < N/2 \\ 0, & \text{otherwise} \end{cases},$$

where N is even. Using the defining relation, it is easily determined that the discrete Fourier transform is

$$X[k] = \begin{cases} \frac{1-(-1)^k}{1-e^{-j(2\pi/N)k}}, & k = 0, 1, 2, \ldots, N-1 \\ 0, & \text{otherwise} \end{cases}$$

This is shown in Figure 4.10.

It is seen that there is no fundamental difference between the discrete Fourier series and the discrete Fourier transform (justifying the use of the same notation for both).

Whether one is dealing with the discrete Fourier series or the discrete Fourier transform is thus a matter of interpretation. If $x[n]$ and $X[k]$ are both assumed to be periodic, then the $X[k]$ are the discrete Fourier series coefficients of $x[n]$. On the other hand, if $x[n]$ and $X[k]$ are both assumed to be finite duration, then $X[k]$ is the discrete Fourier transform of $x[n]$.

Since from a formal viewpoint the quantities involved are periodic, the discrete Fourier series is perhaps the more fundamental interpretation. The properties of the discrete Fourier transform thus follow from those of the discrete Fourier series, and in a given situation are best interpreted by making all quantities periodic, invoking the corresponding property of the discrete Fourier series, and then making all quantities finite duration again.

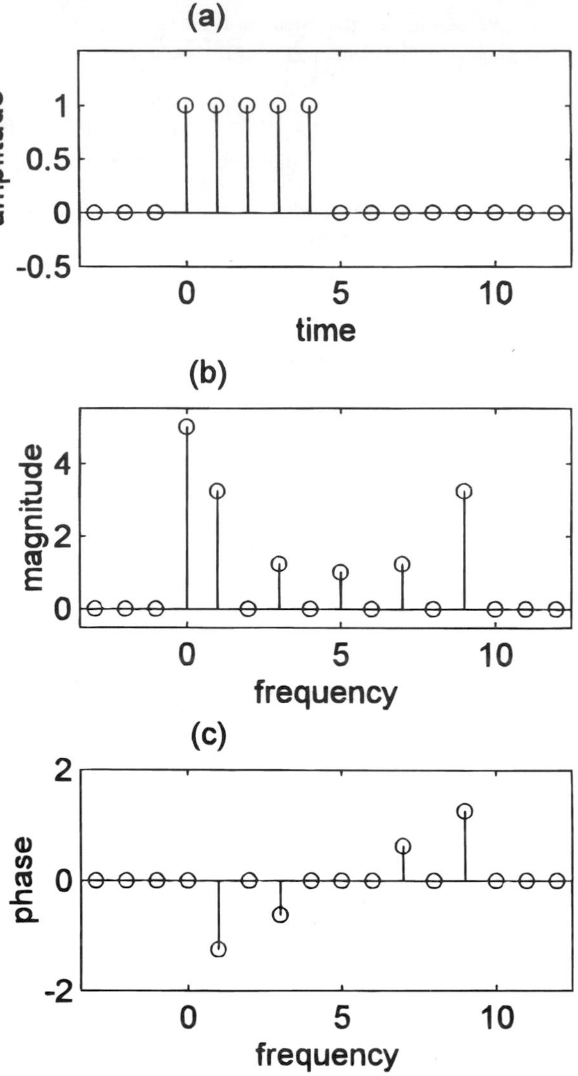

Figure 4.10 Discrete Fourier transform of a finite-duration signal: (a) signal, (b) magnitude spectrum, and (c) phase spectrum.

Fast Fourier Transforms

Calculation of the discrete Fourier transform would appear to be very straightforward and, indeed, it is. However, the number of arithmetic operations (additions and multiplications) necessary to compute even a moderately large discrete Fourier transform can be quite excessive. To overcome this difficulty, various clever algorithms have been developed for efficient calculation of the discrete Fourier transform. These algorithms are collectively known as fast Fourier transforms (FFTs).

The simplest FFT is probably the decimation-in-time algorithm. In this algorithm N is required to be a power of two. In deriving this FFT algorithm, the length-N transform is decomposed into two length-$N/2$ transforms. Each of these is decomposed into two length-$N/4$ transforms. This process is continued until one obtains all length-2 transforms. Because of their shape when represented in flow graph form, these simple length-2 calculations are known as butterflies. For the decimation-in-time FFT algorithm the number of arithmetic calculations is proportional to $N \log_2 N$. This compares with N^2 in the case of the direct evaluation of the discrete Fourier transform. For large N the savings in computation can be very significant. For instance, for $N = 2^{10} = 1024$, the number of computations is reduced by a factor of more than 100.

Discrete-Time Fourier Transforms

Periodic discrete-time signals may be represented by the discrete Fourier series and finite-duration discrete-time signals by the discrete Fourier transform. A larger class of discrete-time signals may be represented by the discrete-time Fourier transform.

A discrete-time signal $x[n]$ will have a discrete-time Fourier transform if it is absolutely summable, i.e., if

$$\sum_{n=-\infty}^{\infty} |x[n]| < \infty.$$

With proper interpretation, certain signals not satisfying this property will also have discrete-time Fourier transforms. These include steps, constants, and all periodic signals.

The discrete-time Fourier transform of $x[n]$ is defined as

$$X(e^{j\omega}) = DTFT\{x[n]\} = \sum_{n=-\infty}^{\infty} x[n]e^{-j\omega n}, \qquad -\infty < \omega < \infty.$$

It is noted that ω is a continuously varying real quantity, not an integer, despite the fact that $x[n]$ is a discrete-time signal. $x[n]$ may be recovered uniquely from $X(e^{j\omega})$ using the inverse discrete-time Fourier transform given by

$$x[n] = DTFT^{-1}\{X(e^{j\omega})\} = \frac{1}{2\pi}\int_{-\pi}^{\pi} X(e^{j\omega})e^{j\omega n}d\omega$$

Note that we have written $X(e^{j\omega})$ as a function of $e^{j\omega}$ for convenience, rather than as a function of ω. Some common discrete-time Fourier transforms are listed in Table 4.9.

Table 4.9 Common Discrete-Time Fourier Transforms

$x[n]$	$X(e^{j\omega})$
$\delta[n]$	1
1	$2\pi \sum_{m=-\infty}^{\infty} \delta(\omega - 2\pi m)$
$u[n]$	$\frac{1}{1-e^{-j\omega}} + \pi \sum_{m=-\infty}^{\infty} \delta(\omega - 2\pi m)$
$a^n u[n]$, $\lvert a\rvert < 1$	$\frac{1}{1-ae^{-j\omega}}$
$na^n u[n]$, $\lvert a\rvert < 1$	$\frac{ae^{-j\omega}}{(1-ae^{-j\omega})^2}$
$\sin(\omega_o n)$	$j\pi \sum_{m=-\infty}^{\infty} [\delta(\omega + \omega_o - 2\pi m) - \delta(\omega - \omega_o - 2\pi m)]$
$\cos(\omega_o n)$	$\pi \sum_{m=-\infty}^{\infty} [\delta(\omega + \omega_o - 2\pi m) + \delta(\omega - \omega_o - 2\pi m)]$

The quantity $X(e^{j\omega})$ is periodic in ω with a period of 2π. In fact, the expression for $X(e^{j\omega})$ is seen to be an exponential Fourier series with coefficients $x[-n]$ and with ω interpreted as the independent variable. While periodic and finite-duration signals have all their energy concentrated at N distinct frequencies, other signals allowed by the discrete-time Fourier transform have their energy spread out over, in general, all angular frequencies in the range $-\pi$ to π (or any range of width 2π).

The discrete-time Fourier transform is, in general, complex-valued, and may thus be written

$$X(e^{j\omega}) = |X(e^{j\omega})| \angle \phi(e^{j\omega}) = |X(e^{j\omega})| e^{j\phi(e^{j\omega})},$$

where $\phi(e^{j\omega}) = \arg\{X(e^{j\omega})\}$. For our assumed real-valued $x[n]$, $|X(e^{j\omega})|$ is an even function of ω and $\phi(e^{j\omega})$ is an odd function of ω, i.e., respectively,

$$|X(e^{-j\omega})| = |X(e^{j\omega})|, \qquad \phi(e^{-j\omega}) = -\phi(e^{j\omega}).$$

Thus the information contained in $X(e^{j\omega})$, $\omega < 0$, is redundant. In addition, since $X(e^{j\omega})$ is periodic with period 2π, it may be concluded that $X(e^{j\omega})$ is completely determined by $X(e^{j\omega})$, $0 \le \omega \le \pi$. When plotted versus ω, $|X(e^{j\omega})|$ is called the magnitude spectrum and $\phi(e^{j\omega})$ the phase spectrum. An example is shown in Figure 4.11.

Signals not satisfying the absolute summability condition stated above, will, if they are discrete-time Fourier transformable, normally contain frequency-domain impulses in their transforms. As a specific case, if $x[n]$ is a periodic signal with period N and with discrete Fourier series coefficients $X[k] = DFS\{x[n]\}$, then its discrete-time Fourier transform is given by

$$X(e^{j\omega}) = \frac{2\pi}{N}\sum_{m=-\infty}^{\infty}\sum_{k=0}^{N-1} X[k]\delta\left(\omega - \frac{2\pi}{N}k + 2\pi m\right).$$

For ω in the range $0 \le \omega < 2\pi$, this reduces to

$$X(e^{j\omega}) = \frac{2\pi}{N}\sum_{k=0}^{N-1} X[k]\delta\left(\omega - \frac{2\pi}{N}k\right),$$

with all energy clearly still concentrated (now, mathematically, by means of impulse functions) at the N frequencies $2\pi k/N$, $k = 0, 1, 2, \ldots, N-1$.

Some important properties of the discrete-time Fourier transform are summarized in Table 4.10.

z-Transforms

As seen above, the discrete-time Fourier transform is a generalization of the discrete Fourier series, in that it allows for the analysis of a broader range of discrete-time signals than does the discrete Fourier series. The z-transform is likewise a generalization of the discrete-time Fourier transform.

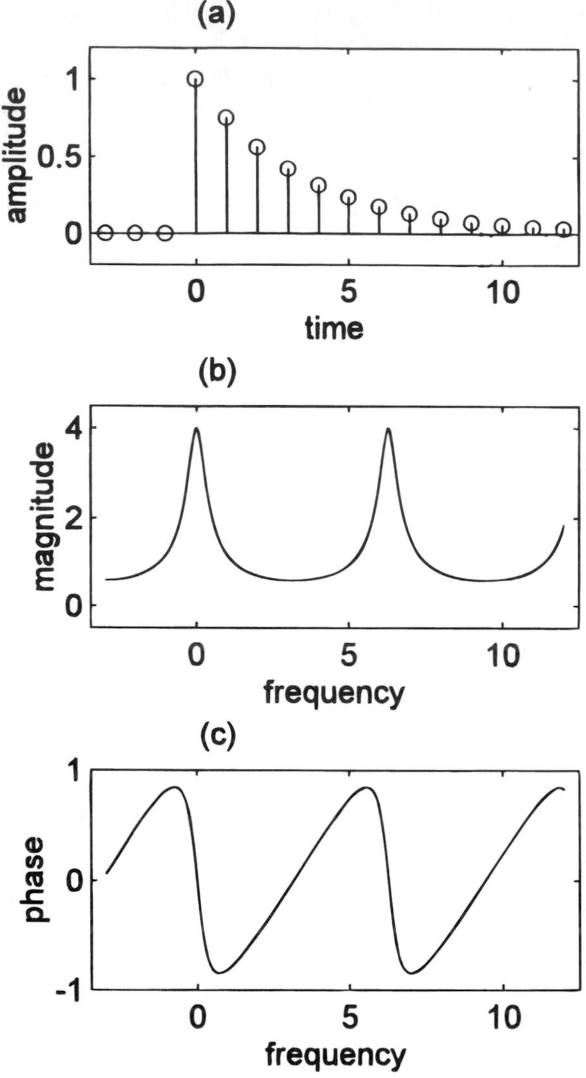

Figure 4.11 Discrete-time Fourier transform of $x[n] = (0.75)^n u[n]$: (a) signal, (b) magnitude spectrum, and (c) phase spectrum.

Table 4.10 Important Properties of Discrete-Time Fourier Transforms

Discrete-time signals	Discrete-time Fourier transforms
$Ax_1[n] + Bx_2[n]$	$AX_1(e^{j\omega}) + BX_2(e^{j\omega})$
$x[n - n_o]$	$X(e^{j\omega})e^{-j\omega n_o}$
$x[-n]$	$X(e^{-j\omega})$
$x_1[n]*x_2[n]$	$X_1(e^{j\omega})X_2(e^{j\omega})$

Note: $x[n], x_1[n], x_2[n]$ are arbitrary discrete-time signals with discrete Fourier transforms $X(e^{j\omega}), X_1(e^{j\omega}), X_2(e^{j\omega})$, respectively. A, B are arbitrary constants and n_o is an arbitrary integer constant.

A signal $x[n]$ will have a z-transform if

$$\sum_{n=-\infty}^{\infty} |x[n]r^{-n}| < \infty$$

Table 4.11 Common z-Transforms

$x[n]$	$X(z)$
$\delta[n]$	1
$u[n]$	$\frac{z}{z-1}, \lvert z\rvert > 1$
$a^n u[n]$	$\frac{z}{z-a}, \lvert z\rvert > \lvert a\rvert$
$na^n u[n]$	$\frac{az}{(z-a)^2}, \lvert z\rvert > \lvert a\rvert$
$\sin(\omega_o n)u[n]$	$\frac{z\sin(\omega_o)}{z^2 - 2z\cos(\omega_o) + 1}, \lvert z\rvert > 1$
$\cos(\omega_o n)u[n]$	$\frac{z^2 - z\cos(\omega_o)}{z^2 - 2z\cos(\omega_o) + 1}, \lvert z\rvert > 1$

for some real $r > 0$. Since $r = 1$ is allowed in this condition, if $x[n]$ has a discrete-time Fourier transform, it will also have a z-transform, but the converse is not true.

The z-transform of a signal $x[n]$ is defined as

$$X(z) = ZT\{x[n]\} = \sum_{n=-\infty}^{\infty} x[n]z^{-n},$$

where $z = re^{j\omega}$ is a complex variable. In general, this sum will converge only for certain choices of r. $x[n]$ may be recovered uniquely from $X(z)$ using the inverse z-transform given by

$$x[n] = ZT^{-1}\{X(z)\} = \frac{1}{2\pi j}\oint_C X(z)z^{n-1}dz,$$

where C is an appropriately chosen contour in the complex z plane. (This formula is rarely directly used in practice because of its complexity. Instead, tables of z-transform pairs are used, with complicated transforms first being decomposed into simpler ones using partial fraction expansion techniques.) Some common z-transforms are listed in Table 4.11.

For a signal $x[n]$ having a discrete-time Fourier transform $X(e^{j\omega})$, it is the case that

$$DTFT\{x[n]\} = ZT\{x[n]\}|_{r=1}$$

or

$$X(e^{j\omega}) = X(z)|_{z=e^{j\omega}},$$

where $X(z)$ is the z-transform of $x[n]$. (This explains our choice of notation for the discrete-time Fourier transform.) Thus, the discrete-time Fourier transform is simply the z-transform evaluated on the unit (radius one) circle in the z (complex) plane.

If the signal $x[n]$ is additionally assumed to be finite duration with length N, then

$$DFT\{x[n]\} = DTFT\{x[n]\}|_{\omega=(2\pi/N)k}$$

or

$$X[k] = X(e^{j\omega})|_{\omega=(2\pi/N)k}$$

where $X[k]$ is the discrete Fourier transform of $x[n]$. That is, the discrete Fourier transform is the discrete-time Fourier transform evaluated at equally spaced frequency values (or the z-transform evaluated at equally spaced points around the unit circle).

Some important properties of the z-transform are summarized in Table 4.12.

4.13 Discrete-Time Signal Processors

A discrete-time signal processor (system) is a device or algorithm (Figure 4.12) that acts on an input signal $x[n]$, modifying it in some manner to produce an output signal $y[n]$. This may be represented abstractly as

$$y[n] = \mathcal{H}\{x[n]\}.$$

(Here $x[n]$ and $y[n]$ should be thought of in their totality, rather than at specific instants n. Despite appearances, this notation does not imply that $y[n]$ is a function of $x[n]$ at only the same instant n. $y[n]$ could, for example, depend on all past values of $x[m]$, $m \leq n$.)

Only certain relatively simple classes of discrete time signal processors are usually considered in practice. In this treatment, we will discuss only those that can be described by linear constant-coefficient difference equations. Such signal processors are of necessity linear, time-invariant, and causal. They may be implemented, in hardware or software, using only summers, constant multipliers, and delay (or memory) elements. A system is linear if

$$\mathcal{H}\{Ax_1[n] + Bx_2[n]\} = A\mathcal{H}\{x_1[n]\} + B\mathcal{H}\{x_2[n]\}, \quad -\infty < n < \infty,$$

for all signals $x_1[n]$, $x_2[n]$, and all constants A,B. It is time-invariant if

Table 4.12 Important Properties of z-Transforms

Discrete-time signals	z-Transforms
$Ax_1[n] + Bx_2[n]$	$AX_1(z) + BX_2(z)$
$x[n - n_o]$	$X(z)z^{-n_o}$
$x[-n]$	$X(z^{-1})$
$x_1[n]*x_2[n]$	$X_1(z)X_2(z)$

Note: $x[n]$, $x_1[n]$, $x_2[n]$ are arbitrary discrete-time signals with z-transforms $X(z)$, $X_1(z)$, $X_2(z)$, respectively. A, B are arbitrary constants and n_o is an arbitrary integer constant.

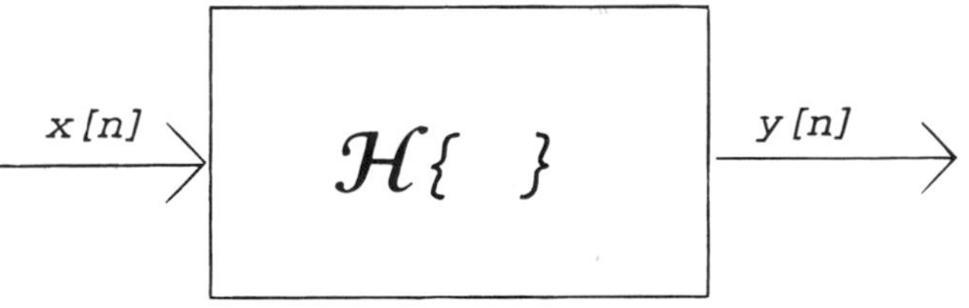

Figure 4.12 Discrete-time signal processing system.

$$y[n - n_o] = \mathcal{H}\{x[n - n_o]\}, \quad -\infty < n < \infty,$$

for all input signals $x[n]$ and corresponding output signals $y[n]$ and all integer constants n_o. A system is causal if $\mathcal{H}\{x[n]\}$ depends only on $x[m]$, $m \leq n$, for all signals $x[n]$.

4.14 Time-Domain Analysis of Discrete-Time Signal Processors

Under our assumptions, a discrete-time signal processor may be represented by a linear constant-coefficient difference equation. It may seem natural to write such an equation in terms of differences of the input and output, and this may be done. However, it is customary, and more convenient, to simply use delayed versions of the input and output. The equation then takes the form

$$\sum_{r=0}^{N} a_r y[n - r] = \sum_{r=0}^{M} b_r x[n - r].$$

Often this equation is normalized so that $a_o = 1$. The equation may thus be solved for $y[n]$ in terms of current and past values of $x[n]$ and past values of $y[n]$. As such, it may be implemented directly and solved for $y[n]$ recursively provided that $x[n]$ and a sufficient number of initial values of $y[n]$ are known. This equation may also be solved using z-transforms and other techniques.

An alternative means for describing our signal processor is based on the use of its impulse response. The impulse response is defined as

$$h[n] = \mathcal{H}\{\delta[n]\},$$

i.e., the output when the input is a unit impulse. Since we have assumed causality, it may be shown that $h[n]$ is a positive-time signal, i.e., that $h[n] = 0$, $n < 0$. Because the signal processor is linear and time-invariant, it may additionally be shown that for any input $x[n]$ the corresponding output $y[n]$ is given by the convolution of $h[n]$ and $x[n]$, namely

$$y[n] = \mathcal{H}\{x[n]\} = h[n] * x[n].$$

This is a very significant result, since it shows that all information regarding the signal processor (at least, from an input-output viewpoint) is contained in its impulse response.

A very desirable property of signal processors is stability. A system is said to be (bounded-input bounded-output) stable if whenever $x[n]$ is bounded (i.e., $|x[n]| \leq K_x < \infty$, $-\infty < n < \infty$, for some K_x) then the corresponding $y[n]$ is also bounded (i.e., $|y[n]| \leq K_y < \infty$, $-\infty < n < \infty$, for some K_y). (The alternative would normally be undesirable.) It is possible to easily determine stability directly from $h[n]$. Specifically, the signal processor is stable if and only if $h[n]$ is absolutely summable, i.e., if and only if

$$\sum_{n=-\infty}^{\infty} |h[n]| < \infty.$$

Unlike continuous-time systems, there are two distinctly different types of discrete-time systems, depending on the nature of $h[n]$. If $h[n]$ is a finite-duration signal, then the system is said to be FIR (i.e., to have a finite impulse response). If this is not the case, then the system is IIR (i.e., having an infinite impulse response). For an FIR system the difference equation reduces to

$$y[n] = \sum_{r=0}^{M} b_r x[n-r],$$

and $h[n]$ is a finite-duration signal of length $M+1$ with

$$h[n] = \begin{cases} b_n, & 0 \le n \le M \\ 0, & \text{otherwise} \end{cases}.$$

Since their impulse responses are clearly always absolutely summable, FIR signal processors are always stable.

4.15 Frequency-Domain Analysis of Discrete-Time Signal Processors

In the frequency (transform) domain, a signal processor is characterized by its transfer function $H(z)$. $H(z)$ may be determined by z-transforming the convolution relationship to obtain

$$Y(z) = H(z)X(z),$$

where

$$H(z) = ZT\{h[n]\}.$$

Thus the transfer function is the z-transform of the impulse response.

Alternatively, the input-output difference equation may be z-transformed to obtain

$$\sum_{r=0}^{N} a_r z^{-r} Y(z) = \sum_{r=0}^{M} b_r z^{-r} X(z),$$

from which

$$H(z) = \frac{\sum_{r=0}^{M} b_r z^{-r}}{\sum_{r=0}^{N} a_r z^{-r}}.$$

Thus $H(z)$ is a rational function in z^{-1}, which, if preferred, may also be written as a rational function in z. As such, the impulse response $h[n]$ will necessarily consist only of terms of the form

$$An^p a^n \cos(\omega n + \theta)u[n],$$

where $p \ge 0$.

When written as polynomials in z, the roots of the numerator are the zeros of $H(z)$ and those of the denominator are the poles of $H(z)$. It may be shown that a signal processor is stable if and only if all of its poles lie strictly inside the unit circle in the complex z plane (i.e., have magnitudes less than one). It is noted that all poles of an FIR signal processor are at the origin, which is thus seen to be consistent with our previous comment that such systems are always stable.

Discrete-time Fourier transforming the convolution relationship leads to

$$Y(e^{j\omega}) = H(e^{j\omega})X(e^{j\omega}),$$

where

$$H(e^{j\omega}) = DTFT\{h[n]\}.$$

$H(e^{j\omega})$ is called the frequency response of the signal processor, and is the discrete-time Fourier transform of the impulse response. The magnitude of $H(e^{j\omega})$ is the magnitude response and its angle is the phase response. It is seen that

$$H(e^{j\omega}) = H(z)|_{z=e^{j\omega}},$$

where $H(z)$ is the transfer function. That is, the frequency response is the transfer function evaluated on the unit circle in the complex z plane. Thus we may also write

$$H(e^{j\omega}) = \frac{\sum_{r=0}^{M} b_r e^{-j\omega r}}{\sum_{r=0}^{N} a_r e^{-j\omega r}}.$$

The frequency response of a signal processor is also useful for determining its output in the case of a periodic input. If $x[n]$ is a periodic input signal with period N (not the same N used in the difference equation) and with discrete Fourier series coefficients $X[k]$, then the output signal will also be periodic with period N. Its discrete Fourier series coefficients will be given by

$$Y[k] = H(e^{j(2\pi/N)k})X[k].$$

This relationship may also be used for the steady-state analysis of systems, i.e., systems that have been operating for a sufficiently long time that all transient terms may be neglected.

It is thus clear that the way a signal processor processes an input signal (either aperiodic or periodic) is determined by its frequency response. It is thus very informative to have this quantity available, particularly in visual form as embodied in its magnitude and phase spectra.

4.16 Discrete-Time (Digital) Filters

Just as in the case of continuous-time (analog) filters, a discrete-time (digital) filter is a signal processor designed to allow signal components of certain frequencies to pass through to the output while preventing input signal components of other frequencies from doing so. The frequency response $H(e^{j\omega})$ should ideally have magnitude one in the passbands and magnitude zero in the stopbands.

Common Filter Types

The common types of discrete-time filters are the same as those for continuous-time filters, the only difference being that the frequency response $H(e^{j\omega})$ is completely specified by its values in the frequency range $0 \le \omega \le \pi$ as opposed to $0 \le \omega < \infty$. The comments made regarding ideal and nonideal (practical) continuous-time filters apply directly to discrete-time filters as well.

Some significant differences with continuous-time filters exist regarding linear phase. Again, linear phase ensures that the shape of a signal is preserved in the filtering process. In discrete-time filters, however, true linear phase may in fact be achieved, but only if the filter is FIR. The FIR filter

$$H(z) = \sum_{r=0}^{M} b_r z^{-r}$$

will have linear phase if and only if

$$b_r = b_{M-r}, \qquad 0 \le r \le M,$$

i.e., if and only if the coefficients b_r are symmetric. It is important to point out that if M is odd then $H(z)$ will have a zero at $z = -1$, or equivalently $H(e^{j\omega})$ will be zero at $\omega = \pi$. This rules out the use of odd values of M in this situation if it is desired to construct a highpass or bandstop filter.

FIR Filter Design

Different techniques are employed for the design of FIR and IIR discrete-time filters, so these will be discussed separately. For simplicity, in the case of FIR filters, we will only consider linear phase filters with M even.

One method for designing FIR filters is known as windowing. This procedure begins by specifying the impulse response $h_i[n]$ of the desired ideal filter, assumed to have zero phase. Anticipating an even M, this quantity is given in Table 4.13 for the common filter types, with cutoff frequencies ω_c, ω_l, and ω_u.

A simple way to obtain an FIR filter from $h_i[n]$ is to truncate it at $n = \pm M$ and then shift the response to the right by $M/2$ to result in a causal linear phase FIR filter. However, since it is recalled that the $h_i[n]$ (actually $h_i[-n]$) are the Fourier series coefficients of $H(e^{j\omega})$ (with ω interpreted as the independent variable), this process amounts to Fourier series truncation, with its attendant Gibb's phenomenon oscillations.

Table 4.13 Common FIR Impulse Responses

Filter type	$h_i[0]$	$h_i[n],\ n \ne 0$
Lowpass	$\frac{\omega_c}{\pi}$	$\frac{\sin(n\omega_c)}{n\pi}$
Highpass	$1 - \frac{\omega_c}{\pi}$	$\frac{-\sin(n\omega_c)}{n\pi}$
Bandpass	$\frac{\omega_u - \omega_l}{\pi}$	$\frac{\sin(n\omega_u) - \sin(n\omega_l)}{n\pi}$
Bandstop	$1 - \frac{\omega_u - \omega_l}{\pi}$	$\frac{\sin(n\omega_l) - \sin(n\omega_u)}{n\pi}$

Table 4.14 Responses from Commonly Used Windows

Window	$w[n],\ 0 \le n \le M^{\dagger}$
Rectangular	1
Bartlett	$1 - \frac{2}{M}\left\lvert n - \frac{M}{2}\right\rvert$
Hanning	$\frac{1}{2} - \frac{1}{2}\cos\left(\frac{2\pi n}{M}\right)$
Hamming	$0.54 - 0.46\cos\left(\frac{2\pi n}{M}\right)$

† $w[n] = 0$ outside of this range.

Since these oscillations are normally unacceptable, a better approach is to reduce the effects of Gibb's phenomenon by multiplying the ideal impulse response by a tapered function known as a window, after first shifting the response to the right to produce causality. The impulse response $h[n]$ of the resulting filter is given by

$$h[n] = h_i[n - M/2]w[n], \qquad -\infty < n < \infty,$$

where $w[n]$ is a symmetric window satisfying

$$w[n] = \begin{cases} w[M-n], & 0 \le n \le M \\ 0, & \text{otherwise} \end{cases}.$$

The resulting filter is then

$$H(z) = \sum_{r=0}^{M} b_r z^{-r},$$

where

$$b_r = h[r], \qquad 0 \le r \le M.$$

Numerous choices for windows exist. Some commonly used windows are listed in Table 4.14 and shown in Figure 4.13.

It is noted that use of a rectangular window is equivalent to simply truncating the Fourier series. As we move down in the table, the windows are increasingly better at reducing the oscillations caused by Gibb's phenomenon. Generally, windows that

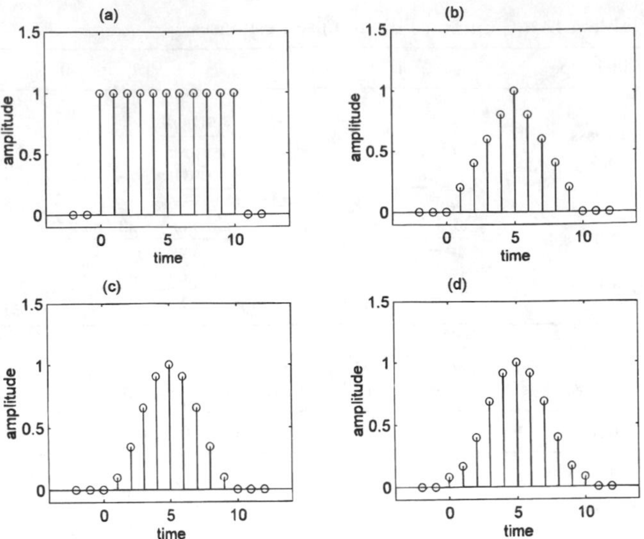

Figure 4.13 Four popular windows (shown for $M = 10$): (a) rectangular, (b) Bartlett, (c) Hanning, and (d) Hamming.

are good in reducing these oscillations do so at the expense of increased transition bandwidths.

As an example of using windowing to design a linear phase FIR filter, suppose we wish to construct a highpass filter with a cutoff frequency $\omega_c = \pi/4$ and a length $M = 30$. Choosing a Hanning window and following the procedure just discussed results in the filter

$$H(z) = \sum_{r=0}^{30} b_r z^{-r},$$

where

$$b_r = \frac{\sin\left((r-15)\dfrac{\pi}{4}\right)}{2(r-15)\pi}\left[\cos\left(\frac{\pi r}{15}\right) - 1\right],$$

$$r = 0, 1, \ldots, 30,\ r \neq 15,$$

and $b_{15} = 3/4$. Figure 4.14 illustrates the result of this design procedure.

While windowing has the advantage of being very simple, it has some significant drawbacks. These are the fact that there is no direct link with the tolerances specified by the designer and the fact that the tolerances in the various filter bands are not independently controllable.

A very popular method for FIR filter design, which does not suffer from these drawbacks, is the Parks-McClellan algorithm. Readily available and easy-to-use computer implementations of this algorithm exist. The Parks-McClellan algorithm results in FIR filters that are equiripple in all passbands and stopbands. The designer simply specifies the passband and stopband edges, the desired magnitude response in each band, the relative tolerances, and the filter order. The algorithm produces the resulting filter coefficients. Since the magnitude response in the transition bands is not the concern of the algorithm, the magnitude response in these regions must be separately checked to ensure acceptability.

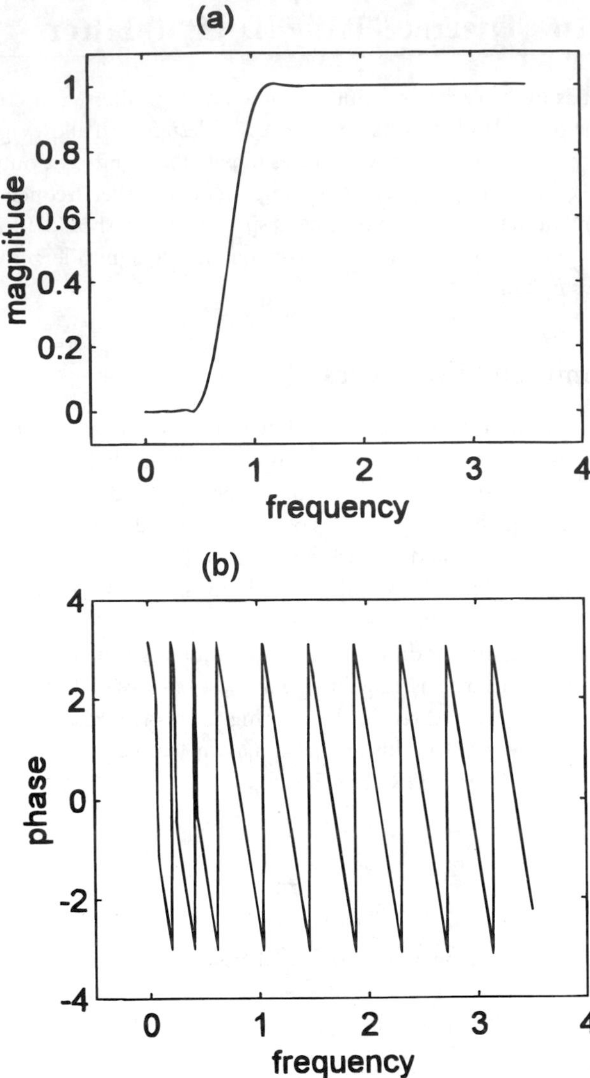

Figure 4.14 Highpass FIR linear phase filter (a) magnitude response and (b) phase response.

IIR Filter Design

The design of an IIR discrete-time filter usually involves the design of a continuous-time lowpass prototype filter, its transformation to a lowpass IIR discrete-time filter, and if necessary, a frequency transformation to produce the desired filter type. When simultaneously discussing continuous-time and discrete-time quantities, we will use Ω to represent continuous-time angular frequency while retaining ω for discrete-time angular frequency.

The design process begins by establishing the discrete-time filter specifications in precisely the same manner as in the case of continuous-time filters. These specifications are then translated to equivalent specifications on the continuous-time filter using the formula

$$\Omega = \tan\left(\frac{\omega}{2}\right)$$

to translate specific frequency values such as cutoff frequencies and band edges. Using techniques discussed earlier, a continuous-time lowpass prototype filter $H_p(s)$ is designed to meet the translated specifications. This filter is then transformed to a lowpass IIR discrete-time filter using the formula

$$H_{lp}(z) = H_p(s)\,|_{s=B(z)},$$

where

$$B(z) = \frac{z-1}{z+1}.$$

$B(z)$ is known as a bilinear transformation.

If a lowpass filter is desired, the design process is complete. If a highpass discrete-time filter $H_{hp}(z)$ is desired, it may easily be determined using the frequency transformation formula

$$H_{hp}(z) = H_{lp}(-z).$$

Assuming that the cutoff frequency of $H_{lp}(z)$ is ω_c, this results in $H_{hp}(z)$ having a cutoff frequency of $\pi - \omega_c$. Bandpass and bandstop filters may also be obtained in this manner, but these require more complicated frequency transformations which double the order of the filter.

4.17 Discrete-Time Analysis of Continuous-Time Signals

Because of their convenience and efficiency, discrete-time techniques are often employed in the analysis of continuous-time signals. In particular, an FFT is often used as a spectral analysis tool to determine the frequency content of a continuous-time signal. It is thus important to consider the relationship between continuous-time and discrete-time Fourier transforms. As in the previous section, Ω will be used for continuous-time angular frequency and ω for discrete-time angular frequency.

As discussed earlier, if the continuous-time signal $x(t)$, with Fourier transform $X(j\Omega)$, is impulse sampled with an angular sampling frequency Ω_s, a continuous-time signal $x_s(t)$ is produced with Fourier transform

$$X_s(j\Omega) = \frac{1}{T_s}\sum_{m=-\infty}^{\infty} X(j(\Omega - m\Omega_s)),$$

where $\Omega_s = 2\pi/T_s$ and T_s is the sampling period. Now, $x[n] = S\{x(t)\} = x(nT_s)$, i.e., the sampled version of $x(t)$, is a discrete-time signal. As such it will have a discrete-time Fourier transform $X(e^{j\omega})$, which may be shown to be given by

$$X(e^{j\omega}) = X_s\left(j\frac{\omega}{T_s}\right) = \frac{1}{T_s}\sum_{m=-\infty}^{\infty} X\left(j\left(\frac{\omega}{T_s} - m\Omega_s\right)\right).$$

As is true of all discrete-time Fourier transforms, $X(e^{j\omega})$ is periodic with period 2π, and is hence completely specified by its values in the range $-\pi \le \omega \le \pi$. Thus, if the original signal is bandlimited and if sampling is carried out at the Nyquist rate or greater, no aliasing will occur and this expression reduces to

$$X(e^{j\omega}) = X_s\left(j\frac{\omega}{T_s}\right) = \frac{1}{T_s} X\left(j\frac{\omega}{T_s}\right), \qquad -\pi \le \omega \le \pi.$$

This is justified since, as a result of the absence of overlap, only the $m = 0$ term in the infinite sum will be present in this frequency range. This shows that, under the circumstances stated (namely, no aliasing), the discrete-time Fourier transform $X(e^{j\omega})$ of the discrete-time signal $x[n] = x(nT_s)$ is simply an amplitude- and frequency-scaled version of the Fourier transform $X(j\Omega)$ of the continuous-time signal $x(t)$, with the frequencies related by $\Omega = \omega/T_s$.

Now, if the discrete-time signal $x[n] = x(nT_s)$ is additionally assumed to be of finite duration with length N, then, as stated earlier, the discrete Fourier transform $X[k]$ of $x[n]$ is given by

$$X[k] = X(e^{j(2\pi/N)k}), \qquad k = 0, 1, 2, \ldots, N-1,$$

and hence

$$X[k] = \frac{1}{T_s} X\left(j\frac{2\pi}{NT_s}k\right), \qquad k = 0, 1, 2, \ldots, N-1.$$

That is, the discrete Fourier transform values are amplitude-scaled samples of the continuous-time Fourier transform $X(j\Omega)$ at the frequency values $\Omega = (2\pi/NT_s)\,k$, $k = 0, 1, 2, \ldots, N-1$. (Because of redundancy, only the first half of these values, for $k = 0, 1, 2, \ldots, N/2 - 1$, are useful. Here N is assumed to be even.) This relationship clearly justifies the utility of using an FFT (to evaluate the discrete Fourier transform) in performing spectral analysis of a continuous-time signal.

Unfortunately, the assumption that $x[n]$ is of finite duration is at odds with the assumption that $x(t)$ is bandlimited. If $x(t)$ is bandlimited, it, and hence $x[n]$, must be of infinite duration. Thus, in reality, the expression for $X[k]$ is an approximation. However, this approximation may still be quite useful if applied carefully.

Several comments are in order regarding the use of an FFT for spectral analysis. Prior to sampling, it is often beneficial to pass the original signal through a continuous-time lowpass filter. Such an *anti-aliasing* filter helps to ensure that the signal to be sampled is bandlimited and free of high-frequency noise.

Additional zero-valued samples are sometimes appended to the end of the discrete-time signal obtained by the sampling process. This *zero padding* may be done to provide a signal length equal to a power of two for use with an FFT. It may also be done to increase the resolution of the Fourier transform. This may be seen by observing that, in the expression for $X[k]$, increasing N both provides more frequency samples and more closely spaced frequency samples.

Prior to taking an FFT, the discrete-time signal $x[n]$ may first be multiplied by a window $w[n]$, of the type discussed earlier. While this certainly alters the frequency content of the signal, this effect may be outweighed by the benefit gained due to the tapered nature of the window, which reduces unnatural discontinuities introduced by artificially time-limiting the signal.

Because of the frequency sampling that occurs when using an FFT, a phenomenon known as *leakage* may sometimes occur. This happens when a frequency component of the original signal falls between two frequencies of the FFT. The *energy* in this component is then distributed (*leaks*) to nearby frequencies, thereby somewhat obscuring the true frequency component.

Finally, it should be noted that other more sophisticated and generally better methods for spectral analysis exist, but these are beyond the scope of this discussion.

4.18 Discrete-Time Processing of Continuous-Time Signals

Because of the desirable properties of discrete-time systems, they are often used in circumstances in which the goal is to process (filter) a continuous-time input to produce a continuous-time output. This involves sampling to produce a discrete-time signal, discrete-time filtering, and finally, conversion of the resulting discrete-time signal to a continuous-time signal. Since the input and output signals are continuous-time, the process is, in fact, equivalent to continuous-time filtering (even though it is implemented in discrete-time).

Ordinarily, such a process begins by passing the original signal through a continuous-time anti-aliasing filter to ensure that the signal $x(t)$ to be processed is bandlimited (or, at least, nearly so). The signal $x(t)$ is then sampled at the Nyquist rate (or greater) to produce the discrete-time signal $x[n] = x(nT_s)$. This signal is then passed through the discrete-time filter $H(e^{j\omega})$ to produce an output $y[n]$. The signal $y[n]$ is then converted to a continuous-time signal $y(t)$ in such a manner that $y(nT_s) = y[n]$. This process (of digital-to-analog conversion) may be viewed as consisting of first generating $y_s(t)$, an impulse-sampled version of $y[n]$. The frequency response of $y_s(t)$ is given by

$$Y_s(j\Omega) = \frac{1}{T_s} \sum_{m=-\infty}^{\infty} Y(j(\Omega - m\Omega_s)).$$

To complete the process the signal $y_s(t)$ is passed through a "reconstruction" filter (a continuous-time lowpass filter with a gain of T_s in the passband) to remove all but the $m = 0$ term in the above expression.

Under these circumstances it may be shown that the equivalent continuous-time filter $H_e(j\omega)$ (with input $x(t)$ and output $y(t)$) has frequency response

$$H_e(j\Omega) = \begin{cases} H(e^{j\Omega T_s}), & -\frac{\pi}{T_s} < \Omega < \frac{\pi}{T_s}. \\ 0, & \text{otherwise} \end{cases}$$

That is, in the range $-\pi/T_s < \Omega < \pi/T_s$, $H_e(j\Omega)$ is simply a frequency-scaled version of the discrete-time frequency response $H(e^{j\omega})$ where the frequencies are related by $\omega = \Omega T_s$ or $\Omega = \omega/T_s$. It is noted that because of the requirement for bandlimiting, true highpass and bandstop filters cannot be constructed in this manner.

References

Ambardar, A. 1995. *Analog and Digital Signal Processing,* PWS, Boston, MA.

Antoniou, A. 1993. *Digital Filters: Analysis, Design, and Applications,* 2d ed., McGraw-Hill, New York, NY.

Baher, H. 1990. *Analog and Digital Signal Processing,* Wiley, New York, NY.

Gabel, R. A. and Roberts, R. A. 1987. *Signals and Linear Systems,* 3d ed. Wiley, New York, NY.

Ghausi, M. S., and Laker, K. R. 1981. *Modern Filter Design: Active RC and Switched Capacitor,* Prentice-Hall, Englewood Cliffs, NJ.

Jackson, L. B. 1989. *Digital Filters and Signal Processing,* 2d ed., Kluwer Academic, Boston, MA.

Jackson, L. B. 1991. *Signals, Systems and Transforms,* Addison-Wesley, Reading, MA.

Johnson, D. E. 1976. *Introduction to Filter Theory,* Prentice-Hall, Englewood Cliffs, NJ.

Lam, H. Y.-F. 1979. *Analog and Digital Filters: Design and Realization,* Prentice-Hall, Englewood Cliffs, NJ.

Lathi, B. P. 1992. *Linear Systems and Signals,* Berkeley-Cambridge, Carmichael, CA.

Oppenheim, A. V. and Schafer, R. W. 1989. *Discrete-Time Signal Processing,* Prentice-Hall, Englewood Cliffs, NJ.

Phillips, C. L. and Parr, J. M. 1995. *Signals, Systems, and Transforms,* Prentice-Hall, Englewood Cliffs, NJ.

Proakis, J. G. and Manolakis, D. G. 1988. *Introduction to Digital Signal Processing,* Macmillan, New York, NY.

Roberts, R. A. and Mullis, C. T. 1987. *Digital Signal Processing,* Addison-Wesley, Reading, MA.

Taylor, F. J. 1994. *Principles of Signals and Systems,* McGraw-Hill, New York, NY.

Ziemer, R. E., Tranter, W. H., and Fannin, D. R. 1993. *Signals and Systems: Continuous and Discrete,* 3d ed., Macmillan, New York, NY.

Data Acquisition and Measurement Systems

5 **Sensors** *Charles W. Einolf, Jr.* 97
Introduction • Passive Sensors • Active Sensors

6 **Measurement System Architecture** *Patrick L. Walter, David Ryerson, Otis Solomon, William Boyer, Richard Pettit, Belle Upadhyaya, Thaddeus Roppel, Ray P. Reed, Giridhar Mandyam, Neeraj Magotra, Samuel D. Stearns, Wes McCoy, Li-Zhe Tan, and Bernard C. Jiang* 103
Introduction • System Considerations • Signal Conditioning and Filtering • Signal/Data Transmission Components • Software Data Correction • Computers in Instrumentation Systems • Software for Instrumentation Systems • Calibration and Testing • Digital Signal Processing • Signal Pick-up and Interface Circuitry • Thermal Effects in Industrial Electronic Circuits • Lossless Waveform Compression • 3-D Measurement Techniques

5 Sensors

Charles W. Einolf, Jr.
Westinghouse Science and Technology Center

5.1 Introduction 97
5.2 Passive Sensors 98
5.3 Active Sensors 98
Analog Outputs • Digital Outputs • RS-232C • RS-422A • RS-485 • HART

5.1 Introduction

Sensors are devices which transform one physical quantity into another physical quantity. Sensors provide the link between the physical world and industrial electrical and electronic equipment. Sensors are found in nearly every electronic product available today. Sensors are also used extensively throughout industrial processes for monitoring, control, and protection. The purpose of this section is to provide an overview of available sensors for various industrial applications. Extensive reviews of sensors and their theory of operation can be found in the literature. Göpel et al. (1989a,b, 1992a–d, 1994) have prepared a seven-volume series on chemical, biochemical, thermal, magnetic, optical, and mechanical sensors.

The term transducer is often used interchangeably with sensor. The Instrument Society of America (ISA), which defines a sensor as synonymous with a transducer, first published Standard S37.1 in 1969 (ISA, 1969). This standard, *Electrical Transducer Nomenclature and Terminology,* defines a transducer (sensor) as "a device which provides a usable *output* in response to a specified *measurand.*" A *measurand* is defined as "a physical quantity, property, or condition which is measured." An *output* is defined as an "electrical quantity." This definition is specific to an electrical transducer. However, in the broader sense, a transducer could have an output which can be defined as a physical quantity, property, or condition. The ISA Standard also defines a nomenclature for describing a transducer (sensor) consisting of the following:

1. The noun "transducer".
2. A first modifier denoting the measurand.
3. When required, a second modifier restricting the measurand.
4. A third modifier denoting the electrical transduction principle in adjective form.
5. An optional fourth modifier denoting the mechanical link in the transducer or any noteworthy special feature.
6. When required, a modifier phrase restricting the modifier.

The construction of typical transducer nomenclature and examples of modifiers are shown in Table 5.1 (ISA, 1969).

Sensors, of course, are not limited to the measurement of physical quantities. They are also used to measure chemical and biological properties. Similarly, the range of useful outputs does not have to be restricted to electrical quantities. Lion (1961, 1962a–c) and Grandke (1989) have classified sensors in groups where the measurand into the sensor (input signal) and the output from the sensor (output signal) can be one of the following six signal types:

- Mechanical—e.g., length, area, volume, mass flow, force, torque, pressure, velocity, acceleration, position, acoustic wavelength, acoustic intensity.
- Thermal—e.g., temperature, heat, entropy, heat flow.
- Electrical—e.g., voltage, current, charge, resistance, inductance, capacitance, dielectric constant, polarization, electric field, frequency, dipole moment.
- Magnetic—e.g., field intensity, flux density, magnetic moment, permeability.
- Radiant—e.g., intensity, wavelength, polarization, phase, reflectance, transmittance, refractive index.
- Chemical—e.g., composition, concentration, oxidation/reduction potential, reaction rate, pH.

A sensor utilizes a physical or chemical transduction principle to convert from an input signal type to an output signal type as illustrated in Figure 5.1. A sensor may employ one or more of these principles for producing a practical output signal. Industrial electronic applications generally require an electrical output from a sensor. Table 5.2 shows examples of the physical and chemical transduction principles that can be utilized in sensors. The transduction principles have been organized in a matrix of the six signal types.

The selection of a sensor for a specific application requires the consideration of several factors. These factors include the static and dynamic characteristics of the sensor as well as the environment to which the sensor will be subjected. Examples of these characteristics are listed in Table 5.3. For a given application,

0-8493-8343-9/97/$0.00+$.50

Table 5.1 Construction of Typical Transducer Nomenclature and Examples of Modifiers

Main Noun	First Modifier Measurand (Examples)	Second Modifier (Restricts Measurand) (Examples)	Third Modifier (Electrical Transduction Principle) (Examples)	Fourth Modifier (Sensing Element, Special Features or Provisions) (Examples)	Range (Examples)	Units (Examples)
Transducer	Acceleration	Absolute	Capacitive	AC Output	0 to 1000	A
	Air Speed	Angular	Electromagnetic	Amplifying	±5	°C
	Attitude	Differential	Inductive	Bellows	−100 to +500	cm
	Attitude Rate	Gage	Ionizing	Bondable	−430 to −415	cm/s
	Current	Infrared	Photoconductive	Bonded		deg
	Displacement	Intensity	Photovoltaic	Bourdon-Tube		°F
	Flow Rate	Linear	Piezoelectric	Capsule		fps
	Force	Mass	Potentiometric	DC Output		Hz
	Heat Flux	Radiant	Reluctive	Diaphragm		ips
	Humidity	Relative	Resistive	Digital-Output		in
	Jerk	Surface	Strain Gage	Discrete Increment		K
	Light	Total	Thermoelectric	Dual-Output		kgf
	Liquid Level	Volumetric		Exposed Element		lb/min
	Mach Number			Frequency Output		m
	Nuclear Radiation			Gyro		mmHg
	Pressure			Integrating		N
	Speed			Self-Generating		%RH
	Sound Pressure			Semiconductor		psia
	Strain			Servo		psid
	Temperature			Switch		psig
	Torque			Toothed-Rotor		rad/s
	Velocity			Triaxial		
				Turbine		
				Ultrasonic		
				Unbonded		
				Vibrating-Element		
				Weldable		

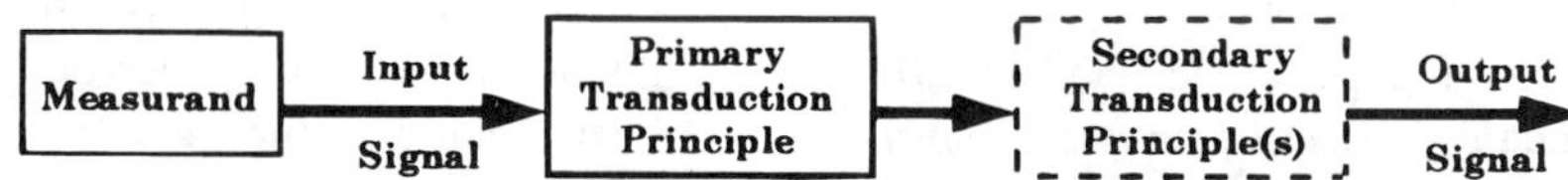

Figure 5.1 Sensor block diagram. A sensor may utilize one or more transduction principles in order to produce a useful output in response to a measurand.

any of these characteristics may need to be considered in the selection of a sensor.

5.2 Passive Sensors

A passive sensor is defined as a device which can convert physical/chemical properties into another property without an auxiliary energy source. The block diagram in Figure 5.1 illustrates a passive sensor since no external energy source is involved in the transduction from one physical quantity to another. An example of a passive sensor is the thermocouple. The thermocouple produces an output voltage proportional to the temperature at the thermocouple junction. Other examples of passive sensors are switches which transform mechanical movement into an electrical contact closure.

5.3 Active Sensors

An active sensor is a device which uses an auxiliary energy source to generate the desired output signal. The block diagram in Figure 5.2 illustrates the active sensor. The active sensor is typically chosen for measuring weak signals.

Analog Outputs

Active sensors which are used in industrial electronic systems typically provide standard electrical output signals in analog or digital format. Sensors providing analog outputs are either voltage or current loop types. The voltage output types vary greatly. A few examples are listed in Table 5.4. The current loop output types are more standard with the 4 to 20 mA type being more common today, where 4 mA corresponds to zero and 20 mA

Table 5.2 Physical and Chemical Transduction Principles

Output Signal / Input Signal	Mechanical	Thermal	Electrical	Magnetic	Radiant	Chemical
Mechanical	(Fluid) Mechanical and Acoustic effects (e.g., diaphragm, gravity balance, echo sounder)	Friction effects (e.g., friction calorimeter) Cooling effects (e.g., thermal flow meters)	Piezoelectricity Piezoresistivity Resistive, Capacitive, and Inductive effects Acousto-dielectric effects	Magneto-mechanical effects (e.g., piezomagnetic effect, magnetoelastic effect, Rowland disk)	Photoelastic systems (stress-induced birefringence) Interferometers Sagnac effect Doppler effect	
Thermal	Thermal expansion (e.g., bi-metallic strip, liquid-in-glass and gas thermometers, resonant frequency) Radiometer effect (light Mill)		Thermoelectric effects (e.g., thermoresistance, thermionic emission, superconductivity) Seebeck effect Pyroelectricity Thermal (Johnson) Noise		Thermooptical effects (e.g., in liquid crystals) Radiant emission	Reaction activation (e.g., thermal dissociation)
Electrical	Electrokinetic, Electrostrictive, and Electromechanical effects (e.g., piezoelectricity, electrometer, Ampere's law)	Joule (Resistive) heating Peltier effect	Charge collectors Langmuir probe Electrets	Biot-Savart's law, Electromagnetic meters and recorders	Electrooptical effects (e.g., Kerr effect) Pockels effect Electroluminescence	Electrolysis Electromigration
Magnetic	Magnetomechanical effects (e.g., Magnetostriction, Magnetometer) Joule effect Guillemin effect Wiedermann effect	Thermomagnetic effects (e.g., Righi-Leduc effect) Galvanomagnetic effects (e.g., Ettingshausen effect)	Thermomagnetic effects (e.g., Ettingshausen-Nernst effect) Galvanomagnetic effects (e.g., Hall effect, magnetoresistance)	Magnetic storage Barnett effect Einstein-de Haas effect de Hass-van Alphen effect	Magnetooptical effects (e.g., Faraday effect) Cotton-Mouton effect Kerr Magneto-Optic effect	
Radiant	Radiation pressure Crooke's light mill	Bolometer thermopile	Photoelectric effects (e.g., photovoltaic, photoconductive, photogalvanic, photodielectric effects)	Curie effect Radiation meter	Photorefractive effects Optical bistability	Photosynthesis-dissociation
Chemical	Hygrometer Electrodeposition cell Photoacoustic effect	Calorimeter Thermal conductivity cell	Potentiometry Conductimetry Polarography Amperometry Flame ionization Volta effect Gas sensitive field effect	Nuclear magnetic resonance	(Emission and absorption) spectroscopy Chemiluminescence	

Table 5.3 Sensor Characteristics to be Considered in the Selection of a Sensor

Static	Dynamic	Environmental
Accuracy	Transfer function	Operating temperature range
Conformance	Frequency response	Thermal cycling
Distortion	Impulse response	Storage temperature range
Grounding	Step response	Operating humidity range
Hysteresis		Operating altitude
Instability and drift		Shock
Isolation		Vibration
Linearity/ nonlinearity		Chemical protection
Minimum detectable signal		Explosion protection
Noise		EMI protection
Offset		RFI protection
Output impedance		
Range		
Repeatability		
Resolution		
Selectivity/specificity		
Sensitivity		
Span		
Threshold		

corresponds to full scale. Older sensors may provide 10 to 50 mA current loop outputs.

Sensors with 4 to 20 mA current loop outputs are most common in industrial environments. Current loops are particularly suited for industrial environments for the following reasons:

- The sensor can be located in a remote location where electrical power cannot be supplied.
- The sensor can be located in a hazardous environment since the voltage signals can be limited.
- The number of wires to the sensor can be reduced to two.
- Current loop signals are relatively immune to noise voltage pick up and are not degraded by long distance transmission.
- The sensor can be electrically isolated from the measurement instrumentation.

Digital Outputs

With the advent of the microprocessor and computer-based industrial control systems there has been an increasing demand

Table 5.4 Typical Output Ranges for Voltage and Current Sensor Types

Voltage	Current
0 to 10 volts	0 to 20 mA
0 to 5 volts	4 to 20 mA
±10 volts	10 to 50 mA
±5 volts	
±1 volt	

for sensors with digital output signals that can directly communicate with a computer. There are a number of choices available as illustrated by the examples in Table 5.5. These communication interfaces are based upon a variety of physical interface standards as well as proprietary protocols.

RS-232C

RS-232C is a communications interface standard initially used in telephone data communications (EIA, 1969). RS-232C has become the interface standard of choice for most personal computers and has found increasing popularity in sensor systems. RS-232C employs a single-ended voltage interface as illustrated in Figure 5.3a where a logic one (1) is a voltage between -15 to -3 volts and a logic zero(0) is a voltage between $+3$ to $+15$ volts. Sensor data is sent to the computer using a bit serial digital data transmission protocol. Since RS-232C is a single-ended interface, the distance between the transmitter and receiver must be short to minimize susceptibility to interference.

RS-422A

RS-422A is a differential voltage interface illustrated in Figure 5.3b, which extends both the maximum cable length between transmitter and receiver as well as the maximum data rate (EIA, 1978). The polarity of the transmission line determines the transmitted logic level and allows for reliable detection at the receiver at signal levels as low as 200 millivolts. The differential interface can tolerate significant attenuation of the signal. This allows for transmission over long distances at higher data rates. The differential interface is also suited to transmission over twisted-pair cabling which will reduce the susceptibility to interference.

RS-485

The EIA standard RS-485 extends the RS-422A standard by allowing up to 32 transceivers to be connected in a multidrop configuration on a single cable (EIA, 1983). Higher order protocols are used to provide organized utilization of the cable. The

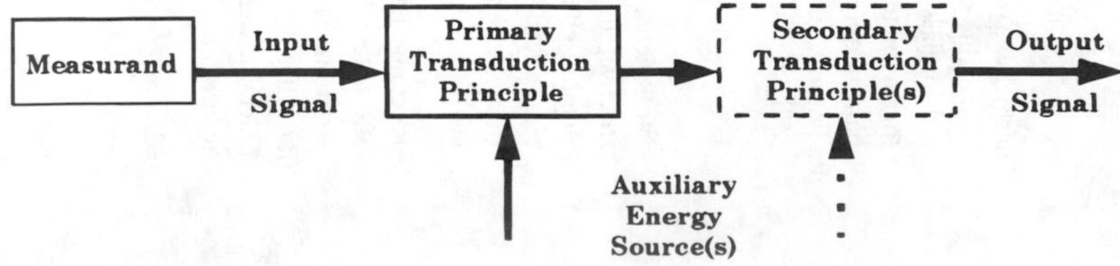

Figure 5.2 Active sensor block diagram. An active sensor requires one or more energy sources in order to utilize the transduction principles and produce a useful output in response to a measurand.

Table 5.5 Typical Digital Outputs available for Sensor Applications

Digital Output	Interface Type	Maximum Cable Length	Communications Type	Max. Data Rate
RS-232C	Single ended voltage	15 meters	Point-to-point	20 kbps
RS-422A	Differential voltage	1,200 meters	Point-to-point	10 Mbps
RS-485	Differential voltage	1,200 meters	Multidrop (up to 32 nodes)	10 Mbps
HART	Current loop	3,048 meters	Point-to-point or multidrop	1.2 kbps

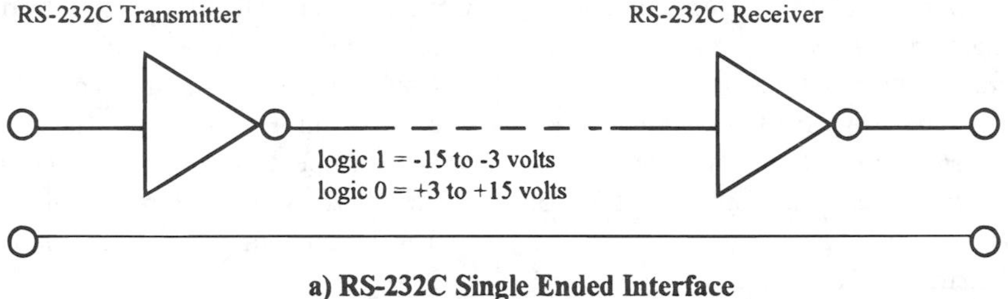

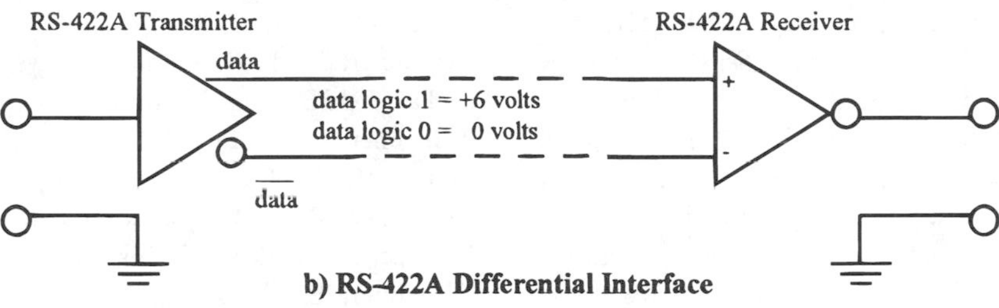

Figure 5.3 Electrical interface standards. Comparison between interface standards RS-232C and RS-422A.

RS-485 transceivers are designed to tolerate simultaneous contention on the cable.

HART

HART (highway addressable remote transducer) is one of many Fieldbus sensor communication protocols that have been proposed for remote data acquisition and control. The HART protocol retains the industry-standard 4 to 20 mA process control signal while simultaneously providing additional process information digitally. Digital communication occurs over the same analog loop with disrupting the process signal. The protocol uses the frequency shift keying (FSK) technique based on the Bell 202 communication standard. Digital communication is accomplished by superimposing a frequency signal over the 4 to 20 mA current.

The signal uses two individual frequencies of 1,200 and 2,200 Hz which represent logic 1 and 0, respectively. The two frequency levels form a sine wave that is superimposed over the 4 to 20 mA current loop. Since the average value of the sine wave is zero, no dc component is added to the 4 to 20 mA signal.

The HART protocol is an open protocol currently available from numerous sensor and process control manufacturers. Figure 5.4 illustrates an example of an active sensor system which provides both analog and digital outputs using the HART protocol. The sensor uses a chemical-to-electrical transduction to measure oxygen concentration in a combustion gas stream.

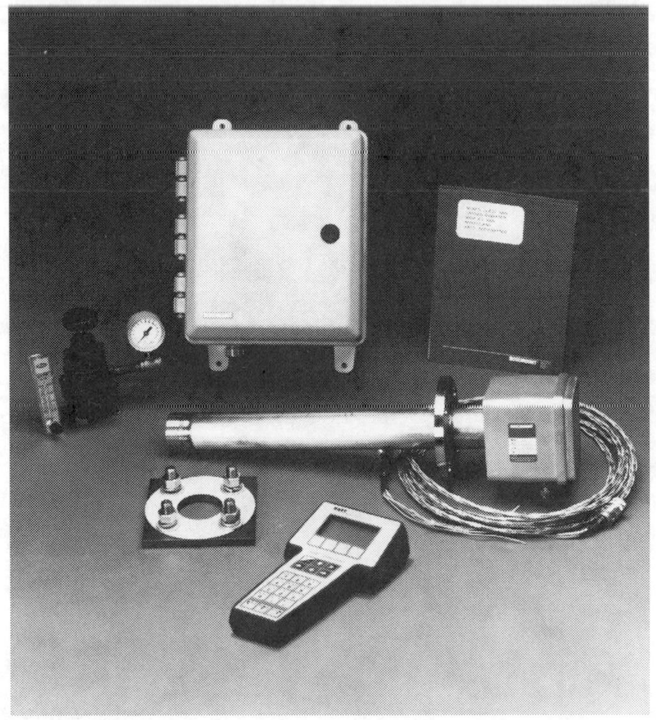

Figure 5.4 *In situ* Oxygen Analyzer. Transducer, oxygen concentration, absolute, zirconium oxide ionic conductive, HART output, 0 to 25, % O_2. (Photo courtesy of Rosemount Analytical, Inc.).

References

Electronic Industries Association, (EIA), 1969. *Interface Between Data Terminal Equipment and Data Communication Equipment Employing Serial Binary Data Interchange,* EIA Standard RS-232-C, Electronic Industries Association.

Electronic Industries Association, (EIA), 1978. *Electrical Characteristics of Balanced Voltage Digital Interface Circuits,* EIA Standard RS-422-A, Electronic Industries Association.

Electronic Industries Association, (EIA), 1983. *Standard for Electrical Characteristics of Generators and Receivers for Use in Balanced Digital Multipoint Systems,* EIA Standard RS-485, Electronic Industries Association.

Göpel, W., Hesse, J., and Zemel, J. H. eds., 1989a. *Sensors: A Comprehensive Survey: Volume 1: Fundamentals and General Aspects,* Verlagsgesellschaft mbH, Weinheim, Germany.

Göpel, W., Hesse, J., and Zemel, J. H. eds., 1989b. *Sensors: A Comprehensive Survey: Volume 5: Magnetic Sensors,* Verlagsgesellschaft mbH, Weinheim, Germany.

Göpel, W., Hesse, J., and Zemel, J. H. eds., 1992a. *Sensors: A Comprehensive Survey: Volume 2: Chemical and Biochemical Sensors, Part I,* Verlagsgesellschaft mbH, Weinheim, Germany.

Göpel, W., Hesse, J., and Zemel, J. H. eds., 1992b. *Sensors: A Comprehensive Survey: Volume 3: Chemical and Biochemical Sensors, Part II,* Verlagsgesellschaft mbH, Weinheim, Germany.

Göpel, W., Hesse, J., and Zemel, J. H. eds., 1992c. *Sensors: A Comprehensive Survey: Volume 4: Thermal Sensors,* Verlagsgesellschaft mbH, Weinheim, Germany.

Göpel, W., Hesse, J., and Zemel, J. H. eds., 1992d. *Sensors: A Comprehensive Survey: Volume 6: Optical Sensors,* Verlagsgesellschaft mbH, Weinheim, Germany.

Göpel, W., Hesse, J., and Zemel, J. H. eds., 1994. *Sensors: A Comprehensive Survey: Volume 7: Mechanical Sensors,* Verlagsgesellschaft mbH, Weinheim, Germany.

Grandke, T. and Hesse, J. 1989. Introduction, Vol.1: Fundamentals and General Aspects, in *Sensors: A Comprehensive Survey,* Göpel, W., Hesse, J., and Zemel, J. H., eds., 1–16, Verlagsgesellschaft mbH, Weinheim, Germany.

The Instrument Society of America, (ISA), 1969. *Electrical Transducer Nomenclature and Terminology,* ISA Standard S37.1, The Instrument Society of America, Reaffirmed, December 1982.

Lion, K. S. 1969. Transducers: problems and prospects, IEEE Trans. Ind. Electronics and Control Instrumentation, IECI–16 (1):2–5.

Scheibner, E. J. 1961. Solid-state physical phenomena and effects: Part I, IRE Trans. Component Parts, CP–8(4):133–151.

Scheibner, E. J. 1962a. Solid-state physical phenomena and effects: Part II, IRE Trans. Component Parts, CP–9(1):19–32.

Scheibner, E. J., 1962b, Solid-state physical phenomena and effects: Part III, IRE Trans. Component Parts, CP–9(2):61–74.

Scheibner, E. J., 1962c, Solid-state physical phenomena and effects: Part IV, IRE Trans. Component Parts, CP–9(3): 119–141.

6

Measurement System Architecture

Patrick L. Walter
Texas Christian University

David Ryerson
Sandia National Laboratories

Otis Solomon
Sandia National Laboratories

William Boyer
Sandia National Laboratories

Richard Pettit
Sandia National Laboratories

Belle Upadhyaya
University of Tennessee

Thaddeus Roppel
Auburn University

Ray P. Reed
Proteun Services

Giridhar Mandyam
University of New Mexico

Neeraj Magotra
University of New Mexico

Samuel D. Stearns
Sandia National Laboratories

Wes McCoy
Motorola Advanced Messaging Group

Li-Zhe Tan
Iterated Systems

Bernard C. Jiang
Yuan-Ze Institute of Technology

6.1 Introduction 103
6.2 System Considerations 104
6.3 Signal Conditioning and Filtering 105
Signal Conditioning Example • Signal Level Change • Differential Amplifiers and Common Mode Rejection • Filtering • Nonlinear Distortion • Analog Multiplexing • Minimizing Signal Conditioning Components • Smart Gauges • Signal Conditioning Summary
6.4 Signal/Data Transmission Components 119
Data Transmission via Copper Wire • Grounding and Shielding
6.5 Software Data Correction 122
Digital Filtering of Data During Data Reduction • Deconvolution Methods • Nonlinear Component Compensation
6.6 Computers in Instrumentation Systems 126
Data Aquisition Times • Instrument Control Interfaces • Operating System Software • Data Storage • Reliability and Maintenance
6.7 Software for Instrumentation Systems 129
Automated Data Acquisition • Performance Verification Testing • Troubleshooting Mode • Data Manipulation and Plotting Software • Data Files and Organization • Instrumentation Database • Operating System and Support Utility Software Requirements
6.8 Calibration and Testing 132
Modeling • Interpreting Specifications • Selection of Transducer • Uncertainty Analysis
6.9 Digital Signal Processing 138
Introduction • Data Acquisition from Sensor and Sampling Requirements • Time-Domain Analysis
6.10 Signal Pick-up and Interface Circuitry 146
Bridge Circuits • Operational Amplifier Circuits
6.11 Thermal Effects in Industrial Electronic Circuits 151
Introduction • Principles of Thermal-Electrical Effects • Analysis of Thermal-Electrical Circuit Effects • Summary
6.12 Lossless Waveform Compression 164
Introduction • Compressibility of a Data Sequence • Performance Criteria for Compression Schemes • The Decorrelation Compression Stage • The Coding Compression Stage • Conclusions
6.13 3-D Measurement Techniques 174
Contact Sensing Devices • Noncontact Sensing Devices

6.1 Introduction

Patrick L. Walter

Measurement systems are normally used to record and display signals from transducers that measure various physical phenomena, e.g., temperature, pressure, acceleration, velocity, displacement, flow rate, voltage, current, electromagnetic energy, and radiation. Some common uses for measurement systems are monitoring conditions in industrial facilities and collecting data from engineering tests and scientific experiments. In this section we will treat the instrumentation system as all the hardware and software components from the input of the transducers to the delivery of data to the customers or users in the format that they have requested. The output could be in the form of electronic displays, hardcopy printouts, or as data files on computer-readable media. In addition to handling the output signal, a measurement system may also be required to provide excitation or bias for some transducers.

0-8493-8343-9/97/$0.00+$.50

We discuss the information necessary for successful measurements. First presented is a general discussion on planning. Subsequent sections provide specific discussions on signal characteristics, transducer selection and types, transducer mounting, data multiplexing, cables, signal conditioning, data filtering, data recording, grounding and shielding, field calibration, and data validation.

6.2 System Considerations

Patrick L. Walter

In designing a new measurement system, the engineer must begin by determining the overall system level requirements. We divide considerations which should influence the design of overall measurement system into three categories: measurement objective, measurement environment, and resource management.

Planning should start with a clear written definition of the measurement objectives. The measurement system should have its design fully documented as well as have identification of system components performed. The measurement objectives include:

- Number of channels total and number for each type of transducer.
- Bandwidth of each type of transducer.
- Minimum and maximum full scale signal ranges.
- Signal duration/amount of memory required.
- Signal amplitude accuracy.
- Signal-to-noise ratio/dynamic range.
- Sampling rates.
- Time alignment among multiple signals.
- Gauge excitation requirements.
- Electrical power availability.
- Uncertainty.
- Reliability.
- System data throughput rates.
- Output data formats.
- Computer system compatibility requirements.

The measurement locations should clearly be identified and observations made as to any physical constraints which would influence transducer mounting. These observations should include such items as: will the transducer physically fit, will the transducer interfere with the operation of the facility or plant, and can cables be routed to and interfaced with the transducer.

The anticipated characteristics of the signal to be measured provide guidance for measurement system design. Estimates should be made as to peak amplitudes and frequencies to be recorded. Signal duration will dictate if a DC recording capability is mandated. If the signal is deterministic (i.e., its time history is of interest), both amplitude and phase characteristics of the frequency response for the measurement system must be considered. If the signal is stochastic, the statistics of the signal are of primary interest so that measurement system phase characteristics may not be important. The availability and quality of electrical power should be an early consideration in measurement planning. Gasoline or diesel generator power may be of lesser quality than commercial 60 Hz line power. Battery power may limit measurement system operating time but provide the most noise-free electrical environment due to ground isolation.

When large numbers of data channels are involved, and the data are to be transmitted long distances, economies can be gained by sharing a common data transmission path. This path can be a cable or radio-frequency transmitter. The path sharing (multiplexing) can occur in either time or frequency. Signal conditioning selection should first consider voltage gain requirements to assure compatibility with the data recorder. Adequate frequency response should be provided as well as high common mode rejection and fast overload recovery. Data filtering may be utilized to eliminate unwanted signal frequency content. The most common filters employed in measurement systems are flat amplitude (Butterworth) or linear phase (Bessel). Filter attenuation is described in terms of the number of filter poles where one pole produces 6 dB/octave attenuation. Data recording should consider ease of data retrieval and duration of retention. An oscilloscope may suffice for a single event test while a digitizer with a magnetic disk may be required for more sophisticated tests. After the measurement system is planned, shielding should be provided for radio-frequency and/or magnetic fields and single point grounding should be guaranteed. Before test, the transducers should be calibrated at the operating conditions of their intended use. The signal conditioning should have fidelity checks performed on it. Finally, techniques should be agreed upon for field calibration of the entire system through the data recorder. In any measurement system, it may be desired to design in "spy" channels to document measurement system noise. Special piezoelectric transducers with depolarized crystals and unpowered bridge transducers provide two examples of "spy" channels used for data validation. Considerations regarding final data reduction and presentation might include sampling times, interpolation techniques, computer program validation, final plot size, assignment of engineering units, auto- or fixed scaling of plot axes, coordinates definition, and so on.

The measurement environment imposes constraints on how or whether the measurement objectives may be obtained. Environmental operating conditions include:

- Temperature (steady-state and transient).
- Humidity.
- Magnetic and radio-frequency fields.
- Dust.
- Shock.
- Vibration.
- Acoustic pressure.
- Pressure (ambient).
- Radiation levels.
- Corrosive media.

The environmental operating conditions for the measurement system must be understood. Transducer and signal conditioning performance degrades with temperature, and at extreme temperatures system components can become inoperative. Temperature compensation designed into transducers for steady-state environments may not suffice in transient thermal environments. Humidity can degrade the performance of high impedance circuits. The circuits associated with piezoelectric transducers may require impedance to be maintained at hundreds or thousands of millions of ohms in the presence of humidity to successfully operate. Magnetic and radio-frequency fields can introduce unwanted signal content in recorded data. Depending on spectral content, it may or may not be possible to separate induced signals due to these fields from the measured response of the test item. Ambient and acoustic pressure can couple unwanted strain into transducer sensing elements. Nuclear radiation can influence transducer and signal conditioning performance since X rays and gamma rays provide additive noise while neutrons change gain and sensitivity by modifying the lattice structure of semiconductor and crystal elements. Corrosive media can limit the successful life of any experiment.

The resources available to achieve the management objective impose further restrictions on the system design. Resource management considerations include:

- Cost.
- Schedule.
- Manpower.

Programmatic constraints of cost, manpower, and schedule should also be understood and it must be determined if these constraints are compatible with the measurement objectives in conjunction with the constraints imposed by the measurement environment. For example, reliability and accuracy can both be improved by incorporating into the test additional measurement channels calibrated for different ranges; however, cost, manpower, and schedule may be influenced.

Meeting all of the system performance requirements in addition to those relating to cost and schedule will require careful attention to all system components and usually making tradeoffs to achieve optimum performance. The specific system components addressed in this section are

- Signal processing concepts.
- Signal conditioning.
- Data transmission.
- Data recorders.
- Computers.
- Software.
- Output data formats.

6.3 Signal Conditioning and Filtering

David Ryerson

Signal conditioning and filtering circuitry depends on the transducer and analog-to-digital converter. Horan (1993), Carstens (1991), Pallas-Areny and Webster (1991), Helfrick and Cooper (1990), Strock (1987), and Morrison (1984) contain further information on signal conditioning and filtering. Piezoresistive transducers usually operate from a DC power supply with signal gain provided by a DC differential amplifier. This type of amplifier is an electronic circuit whose input lines are isolated from its output lines, power ground, and chassis ground and whose output voltage is proportional to the differential input signal voltage. Inductive and capacitive transducers require an AC power supply and carrier amplifier/demodulator/lowpass filter for successful operation. This type of signal conditioning uses a reference signal from the power supply for synchronous demodulation. Electrical isolation between amplifier input and output is provided. Pallas-Areny and Webster (1991) and Morrison (1984) discuss how the signal conditioning requirements depend on the transducer.

Key performance parameters of transducer power supplies include:

1. Warm-up time—the time necessary for the power supply to deliver nominal output voltage at full-rated load.
2. Line regulation—the change in steady-state output voltage resulting from an input voltage change over the rated range.
3. Load regulation—the change in steady-state output voltage resulting from a full-range load change.
4. Load transient recovery—the time required for the output voltage to recover and stay within a specified band following a step change in load.
5. Stability—the long-term deviation in the output voltage.
6. Temperature coefficient—the change in output voltage per degree change in ambient temperature.

When using an AC power supply, fidelity of response to vibration and rapid mechanical motion requires an operating frequency at least ten times greater than the highest frequency to be measured. Most transducers, whether powered by AC or DC, require 1–10 volts excitation.

Key performance parameters of transducer amplifiers include:

1. Input impedance/output impedance—the minimum input impedance and maximum output impedance the amplifier will present when operated within its specification.
2. Source current—the bias current flowing through a circuit comprised of the amplifier input terminals closed through the source resistance.
3. Common mode rejection—the measure of the conversion of voltage common with both amplifier inputs to normal output signal.

4. Linearity—the maximum deviation from a specified straight line through the output versus input voltage characteristic.
5. Gain range—the maximum and minimum values of gain defined as the slope of a specified line through the output versus input voltage characteristic.
6. Gain stability with temperature—the change in gain as a function of ambient temperature.
7. Zero stability—the change in output voltage with temperature, pressure or humidity either referred to the amplifier input or its output.
8. Frequency response—the minimum frequency range over which the amplifier gain is within some deviation of a specified level for all specified gains.
9. Slew rate—the maximum rate at which the amplifier can change output voltage from the minimum to the maximum limit of linear output voltage range.
10. Settling time—the time following application of a step voltage input for the amplifier output to settle within a specified percentage of its final value.
11. Overload recovery—the time required for the amplifier to recover from a specified input signal overload.
12. Noise—the portion of amplifier noise referred to its input which varies with gain and the portion referred to its output which remains fixed with gain.

Analog instrumentation filters can be characterized as either lowpass, bandpass, band-reject, or highpass. Lowpass filters pass frequencies from DC to some high-frequency cutoff. Band-pass filters pass frequencies between a low- and high-frequency cutoff. Band-reject filters reject frequencies between a low- and high-frequency cutoff. Highpass filters pass frequencies above some low-frequency cutoff. Cutoff frequencies are specified in terms of the filter resonant frequency which is −3 dB for a Butterworth filter.

Modern filter theory involves the approximation of filter specifications by a frequency response function and the design of a network which realizes this frequency response function. The rate of attenuation of a filter is specified in decibels/octave. Each filter pole provides 6 dB/octave ultimate attenuation. The more common filters in current use are the Butterworth, Bessel, and Chebyshev.

Filters have many applications in measurement systems. Lowpass or band-reject filters may be used to eliminate transducer resonances from recorded signals. Highpass filters may be used to remove DC levels from resultant signals. Band-pass filters can eliminate harmonics and subharmonics from sinusoidal signals. Lowpass filters are required to constrain the spectral content of signals to be frequency multiplexed (premodulation filters) or time multiplexed (anti-aliasing filters).

Reproduction of the time history of a signal requires both flat amplitude response and linear phase response from a filter. Reproduction of stationary random vibration data requires only flat amplitude response so that filter phase characteristics can be ignored.

Analog filtering should occur as near to the front of the measurement system as possible. By placing the filtering in this location:

1. The system signal-to-noise ratio is enhanced since system front end gain can be increased.
2. System components are less likely to be overdriven.
3. The frequency bandwidth of measurement system components after the filter are smaller.

Signal Conditioning Example

For a digital data acquisition system, analog signal conditioning is required to match the sensor signal's amplitude, level, and frequency content to the needs of the analog-to-digital converter (ADC). Figure 6.1 is an example of a simple 4-channel memory-based data acquisition system used to gather shock data on a test object. We will use this system to explain the concepts to be discussed in this section. First, each block of this example will be discussed in summary defining the purpose of the block. Then we follow with a detailed discussion on each of the signal conditioning blocks of the system, including the amplifiers, filters, and multiplexer.

This system consists of sensors, amplifiers, anti-aliasing lowpass filters, an analog multiplexer, analog-to-digital converter, memory, and controller. The basic operation of the system is to take data from the accelerometers, digitize these data, and store the data in memory for later recovery. The analog signal conditioning part of this system starts at the differential amplifiers and goes through the analog multiplexer. The summary discussion of the system blocks follow.

This system is used to measure shock as a test object strikes or penetrates a media. The accelerometer gauges convert the acceleration measured at different locations on the test object to electrical signals. The accelerometer gauges used are Endevco Model 7270A (Performance Specification for Accelerometer Model 7270A, 1985) with full scale ranges of ±6,000 g (Earth's gravity) (Model 7270A-6K) or ±20,000 g (Model 7270A-20K). This is the maximum acceleration range for which the gauge response is linear. In addition, the larger the full scale range, the larger the acceleration that the gauge will take without breaking. These gauges consist of a full Wheatstone bridge of piezoresistors as shown in Figure 6.2.

The equation for this gauge is

$$V_o = \frac{dR}{R} V_S \tag{6.1}$$

where

V_o = Accelerometer output voltage
V_s = Accelerometer supply voltage
R = Bridge leg resistance
dR = Leg resistance change due to acceleration

Note that this bridge consists of active resistors on all four legs of the bridge. These resistors are semiconductor strain gauges

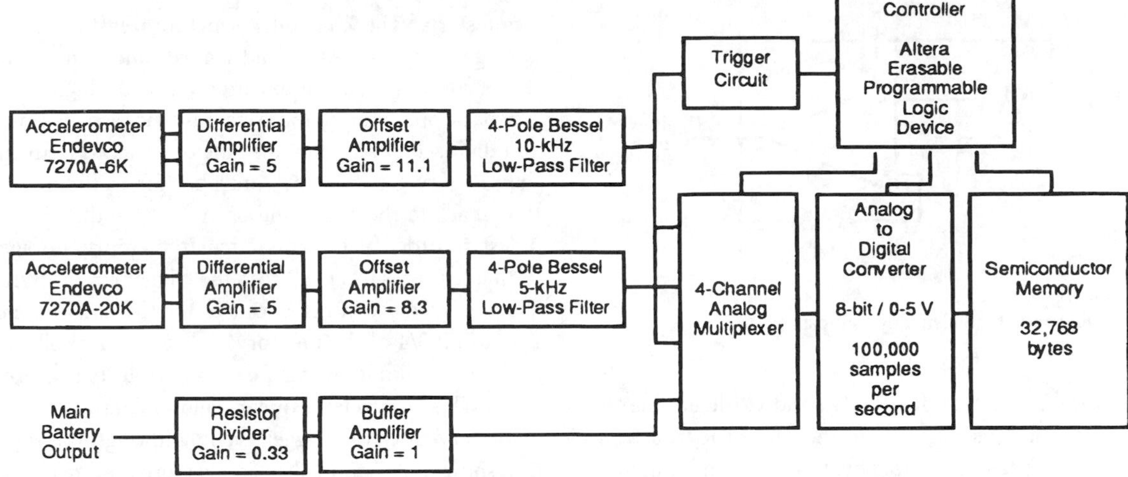

Figure 6.1 Memory-based data acquisition system block diagram.

built into a beam carrying a mass which responds to the acceleration applied to the gauge. Also note that the output voltage has a common mode component, i.e., an equal voltage appearing on both sides of the output voltage, which is equal to one half of the supply voltage.

The differential amplifier takes the output voltage generated by the accelerometer gauge, amplifies it by five and converts the signal to a single-ended voltage varying about zero volts (Ott, 1988; Van Valkenburg, 1982). Since this amplifier is differential, i.e., measures only the difference voltage between its two input terminals, it removes the common-mode voltage coming from the accelerometer. It will also remove any common mode noise that may be picked up by the cable connecting the gauge to the amplifier.

With 10-volt excitation, the 7270A-6K accelerometer puts out ±45 millivolts corresponding to an acceleration input of ±1,500 g and the 7270A-20K accelerometer puts out ±60 millivolts corresponding to an acceleration input of ±6,000 g. The ADC requires a 0- to- 5-volt input signal. Therefore, we must change the bipolar accelerometer signal to a unipolar signal for the ADC input. At the same time, we must apply gain to the signal to make it cover the ADC input range of 0 to 5 volts. The

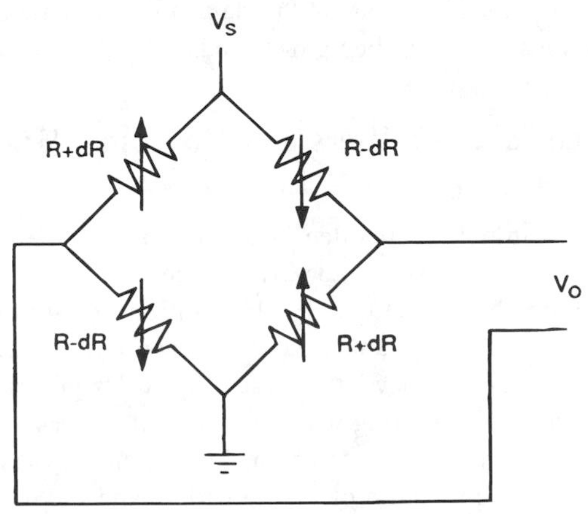

Figure 6.2 Accelerometer schematic.

offset amplifier applies this offset voltage as well as giving the remaining required gain. Note, that the differential amplifier applied some of the gain (a factor of 5).

The accelerometers are not used to the full dynamic range for which they were designed. The −6K accelerometer has a linear full-scale range of ±6,000 g of which we are only using ±1,500 g and the −20K accelerometer has a linear full-scale of ±20,000 g of which we are only using ±6,000 g. This was done to assure that the accelerometers would survive if larger than expected acceleration signals were applied. The disadvantage of not using the full available output of the accelerometer is that the signal may be noisier due to the higher analog gain required in the system.

A four-pole Bessel filter was used on each accelerometer channel to remove high frequencies in the acceleration data that could distort the signal by the sampling process (Horan, 1993). The aliasing effect that occurs when high frequency signals are not sampled fast enough is discussed later. The cutoff frequency of the first channel is set at 10 kHz and the cutoff frequency for the second channel is set to 5 kHz. Note, the first channel is sampled twice as fast as the second and third channels since it is connected to the multiplexer twice. This will be further discussed in the section on the multiplexer. Each channel is sampled at 1/4 of the analog-to-digital converter rate of 100,000 samples per second, which is 25,000 samples per second per channel. The cutoff frequency of the filter was set to 1/5 of the sampling rate to assure that the signal was well attenuated at the Nyquist frequency, 1/2 of the sampling frequency.

The system uses a Bessel filter. Bessel filters have a poor amplitude response around the cutoff frequency but a relatively linear phase response. The linear phase response keeps the distortion of the signal to a minimum in the time domain, which is the most desired response for this shock measuring system. Often systems use a Butterworth filter which gives a better amplitude response but a less linear phase response.

This filter was implemented with two stages of active Sallen-Key two-pole filters. The filter has unity gain and no signal offset. Figure 6.3 is a schematic of a Sallen-Key filter. As you can see, it is a fairly simple filter requiring an operational amplifier, two

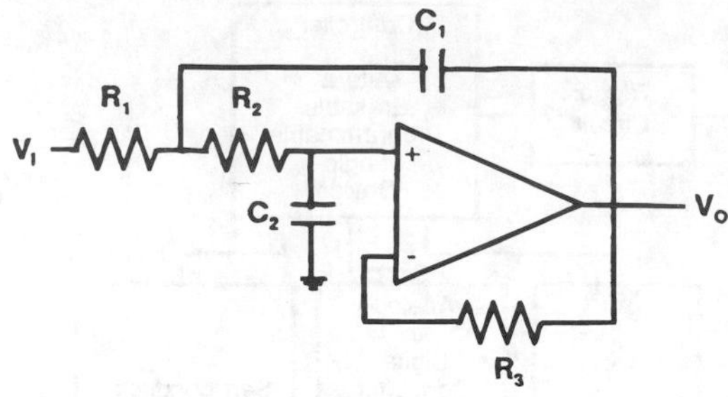

Figure 6.3 Sallen-Key low-pass filter.

capacitors and three resistors. R_1, R_2, C_1, and C_2 determine the filter parameters and R_3 is set equal to the sum of R_1 and R_2 to minimize the offset voltage generated by the operational amplifier bias currents. Equations for this circuit are described later.

The last channel consists of resistor divider and unity gain buffer amplifier. An anti-aliasing low-pass filter was not used since the DC level of the battery is the only information of interest and the frequency content of this signal is low because the battery voltage is nearly constant.

The analog multiplexer consists of a semiconductor switch which sequences between the four analog channels feeding one at a time into the analog-to-digital converter (Horan, 1993). The controller in this system causes each analog multiplexer input to be sampled in sequence at equal time intervals.

Note that the first accelerometer channel is fed into both the first and third analog multiplexer input. Measurements from this accelerometer are gathered at twice the rate of the other accelerometer to give it twice the frequency response. This channel is fed to two analog multiplexer inputs which are not adjacent to one another but are selected such that the signal's twice sample rate is still at equal time intervals between samples. If the sample intervals for this channel were not equal, increased frequency response could not be obtained.

The ADC takes a 0- to 5-volt signal and converts to an 8-bit digital data sample (Horan, 1993; Strock, 1987). An 8-bit ADC gives a resolution of 1 part in 2^8 or 256. Therefore, the analog signal will be resolved by this system to 0.4% of 0 to 5 volts or 20 millivolts. This ADC digitizes at a rate of 100,000 samples per second. Since we have four inputs to the analog multiplexer, each input will be sampled at 1/4 of the total sample rate, or 25,000 samples per second, except for the first analog channel which is fed into two multiplexer inputs giving 50,000 samples per second. The sample rate and the channel sampling order are determined by the controller.

The digital output of the ADC is written into a semiconductor memory for storage until after the test is over, the test unit is recovered, and the data is read out of the memory. This memory holds 32,768 bytes or, in this case since we have an 8-bit or 1 byte ADC, 32,768 samples of data. Since we are sampling at 100,000 samples per second, the memory holds 328 milliseconds of data.

The trigger circuit in Figure 6.1 detects when the first accelerometer channel signal exceeds a predefined absolute signal level. The output of this circuit is used to tell the controller when an event starts. The controller synchronizes the activities of the analog multiplexer, ADC, and semiconductor memory. The controller consists of an erasable programmable logic device (EPLD). The basic operation of the system is that each channel is sampled and digitized in sequence and the data stored in memory. When the memory gets full, the controller forces the memory pointer to go back to the beginning of memory and store data over the oldest recorded data. This circular memory storage continues until a trigger is received from the trigger circuit. When a trigger is received, the controller causes the system to collect enough data to fill 3/4 of the memory and cease data collection. So, 1/4 of the data samples in memory were collected before the trigger event. This controller sequence allows data to be recorded both before and after the trigger event. Although the signal lines are not shown, the controller is also used to read the data out of memory to a personal computer after the recovery of the data package.

Signal Level Change

The signal coming from the sensor must be of the correct level to match it to the ADC input amplitude range, for example 0 to 5 volts. The sensor often puts out a much different full-scale voltage, for example −20 millivolts to +20 millivolts. An amplifier is needed to scale the sensor signal level to the input range of the ADC. If the signal levels were not matched, the resolution of the data would be reduced by the ratio of the ADC full-scale input signal range to the sensor full-scale output range.

The amplifier furnishing the gain should have an amplitude response which is flat and a phase response which is linear versus frequency in the frequency band of interest. This will keep the amplifier from distorting the signal.

In addition to the sensor putting out the incorrect signal amplitude range to match the ADC input, it may also put out a DC bias different than that required by the ADC. Often the ADC will require only positive voltages and the sensor outputs are symmetrical plus and minus voltages. Therefore a voltage offset needs to be added to adjust the sensor bias to match the ADC required bias. Quite often, the gain and offset adjustments are combined in one amplifier or amplifier string. Again, the offset circuit must have flat amplitude and linear phase response to keep the signal from being distorted.

Differential Amplifiers and Common Mode Rejection

Differential amplifiers are often used to remove DC offsets which can be on the sensor signal, for example, the signal coming from a Wheatstone bridge (Ott, 1988). In addition, the sensor signal may be sent over a long cable which will pick up stray electromagnetic signals such as may be generated by 60-Hz power lines. A differential amplifier along with a good balanced cable, such as twisted pair, can reject much of this interference because it will be common to both sides of the signal line and the differential amplifier looks at only the difference between the signal line sides, giving good common mode rejection.

Today, typically operational amplifier circuits are used to generate the required signal conditioning amplifiers and filters. Figure 6.4 is a single-ended amplifier built with resistors and an operational amplifier. Single-ended means that the circuit takes the voltage V_I with respect to ground, or common, and amplifies it to generate the voltage V_o with respect to ground. The gain of this circuit is the ratio of R_F to R_I and its input impedance is equal to R_I. This circuit does invert the signal between the input and the output. R_B should be set equal to the parallel combination of R_F and R_I to minimize the offset voltage generated by the amplifier input bias currents.

Figure 6.5 is a modification of the previous circuit to generate an amplifier with a differential input. It works similarly to the previous circuit with the same gain except the input voltage is the difference between V_1 and V_2. Any signal common to V_1 and V_2 is not passed through the circuit. The input impedance is R_I plus R_F for the V_2 input and less than or equal to R_I plus R_F for the V_1 input depending upon the levels of the input signals. The output is a single-ended signal with respect to ground.

Because of the relatively low input impedance of the above two circuits, an instrumentation amplifier as shown in Figure 6.6 is often used. Its input impedance is a function of the operational amplifiers used and for a good amplifier is on the order of 10^{10} ohms.

This instrumentation amplifier has a gain of

$$Gain = \frac{V_O}{V_2 - V_1} = \left(1 + \frac{2R_2}{R_1}\right)\frac{R_4}{R_3} \tag{6.2}$$

This circuit has the advantage of having a differential input with high input impedance and generating a single-ended output. The high input impedance will not put an impedance load on

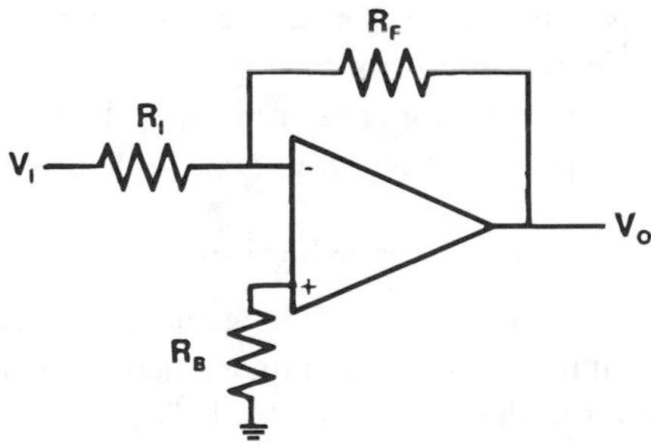

Figure 6.4 Single-ended amplifier.

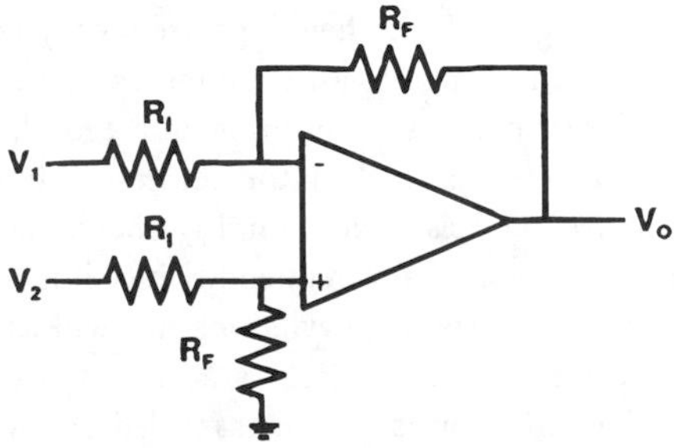

Figure 6.5 Differential amplifier.

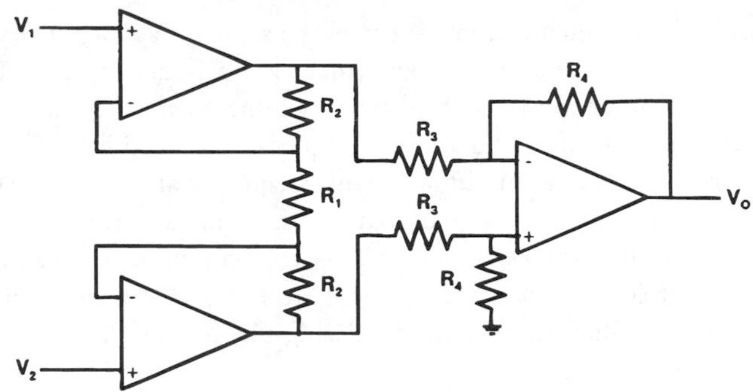

Figure 6.6 Instrumentation amplifier.

the sensor or signal being measured and therefore modify the measured signal. Instrumentation amplifiers are available as a single integrated circuit.

If a fixed input impedance, such as 50 ohms, is required at the input of the amplifier for impedance matching with a transmission line between the gauge and the amplifier, an instrumentation amplifier allows a well-characterized impedance be put across the inputs without affecting the amplifier circuit.

Filtering

Filtering is a tool to change the amplitude and phase of a signal as viewed in the frequency domain (Van Valkenburg, 1982; Williams, 1981). These changes are used to remove unwanted components in the signal and to adjust for signal distortion that may be added by the remainder of the system.

If the sensor signal is sent over long signal lines, whose frequency response is not flat, then a filter can be used to compensate for the frequency response of the signal lines to give a flat frequency response between the sensor and the ADC. An alternative to doing analog frequency compensation is to frequency compensate the data digitally after the data is recovered. The advantage of digital frequency compensation is that more stable digital filters can be built than analog filters. The disadvantage of digital filters in real-time systems is that digital filters are usually applied after data recovery rather than during the real-time process.

A high-pass filter can be used to remove DC components from the sensor signal. This can be very valuable if the DC component has a tendency to drift and is irrelevant to the measurement process. Some sensors do not have DC response, so bias removal can be used to take out drifting DC components of the signal conditioning circuitry.

Anti-Aliasing

The data acquisition systems that we are discussing are sampled data systems, that is, they sample the data signal at discrete time intervals. The sampling theorem by H. Nyquist of Bell Laboratories shows that to reconstruct a sine wave signal from uniform-rate discrete samples, it is necessary to sample the signal at least twice in each cycle of the signal (Horan, 1993). Therefore, to reconstruct a signal from a sampled-data set, the signal must have been originally sampled at a rate at least twice the highest frequency component in the signal. If any frequency components reside in the signal that have a frequency greater than the one-half the sample rate, called the Nyquist frequency, these components will have their frequency modified by folding their frequencies around the Nyquist frequency and will distort reconstruction of the original signal. This folding of the frequencies is called aliasing since a high-frequency component will alias itself as a low-frequency component. To eliminate this problem a low-pass filter, called an anti-aliasing filter, should be placed before the sampler.

Figure 6.7 shows what effect aliasing will make on a signal in the frequency domain. Notice that the signal centered around frequency $f_N + f_1$ is folded around the Nyquist frequency f_N to give a mirror image of the signal centered around the frequency $f_N - f_1$. This signal is added to the desired signal shown with a higher amplitude.

If we could build a perfect filter with unity response before the cutoff frequency and zero response after the cutoff frequency and also with no phase distortion, we could set its cutoff frequency equal to the Nyquist frequency and have our anti-aliasing filter. Unfortunately such a filter is unrealizable in the real world. Therefore, we must trade off filter specifications to best meet our needs. The specifications that we must trade off are cutoff frequency, pass-band amplitude response, transition bandwidth, stop-band amplitude response, and phase response.

In this section we will only discuss the number of poles in the lowpass filter and assume an idealized Bode plot amplitude response with 6 dB per frequency octave per pole amplitude roll-off and ignore the phase response. A Bode plot does not show the correct amplitude response of a real filter near the cutoff frequency, but it is a good approximation of a filter in the pass-band before the cutoff frequency and in the stopband after the cutoff frequency. Figure 6.8 shows this filter's Bode amplitude response and following the figure are definitions of terms used in this section.

The desired phase response is linear so that all frequency components of the signal are delayed by equal time amounts. This keeps the signal from being distorted. The time delay of a frequency component of a signal can be found by taking the negative ratio of the phase in radians to the frequency in radians

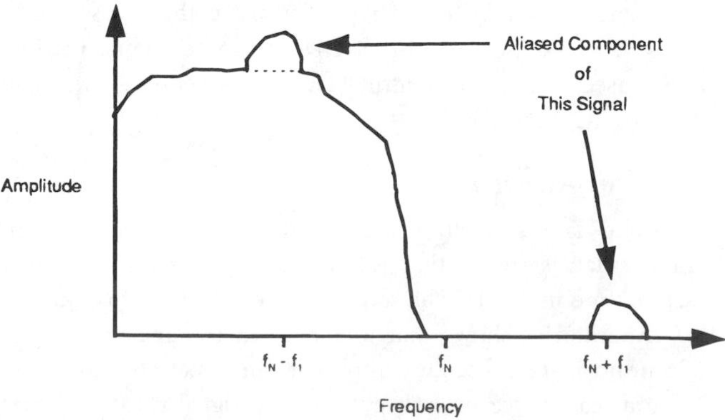

Figure 6.7 Aliased signal.

per second. This can be seen in Table 6.1, which gives phase shift and time delays for different sine wave signals.

Note that in Table 6.1 the time delay is held constant and the phase shift is adjusted to get this time delay. As the frequency increases the phase shift must so also increase proportionally to hold a constant time delay. This time delay versus frequency function is called phase delay.

An example of the distortion that can take place if all frequency components are not delayed by the same amount is to consider a square wave. It consists of a fundamental frequency and odd harmonics. Figure 6.9 shows a plot of only the fundamental and third harmonic of a one-hertz square wave with no time delay. Figure 6.10 shows a plot of the fundamental and third harmonic with equal time delays. Figure 6.11 shows a plot of the fundamental with 0.14 second delay and the third harmonic with 0.17 second delay. This is the typical delay one would get with four-pole Butterworth filter with a three Hz cutoff frequency. Note how the signal is distorted even though no modifications were made to amplitude levels of the two components.

Phase distortion is a function of the type of low-pass filter used with some types, i.e., Bessel, giving good (linear) phase response and poor (slowly changing) amplitude response while other filter types, i.e., Butterworth, giving poor (nonlinear) phase response and good (flat in the passband) amplitude response. In addition, as the number of filter poles is increased, the phase response typically becomes more nonlinear.

Anti-Aliasing Filter Definitions

f_S = Sampling frequency
f_N = Nyquist frequency = $f_S/2$
f_C = Anti-aliasing filter cutoff frequency
f_A = Aliasing frequency
R = Ratio of f_S to f_C
M = Number of poles in the anti-aliasing filter
N = Number of bits in the digitizer
A_F = Amplitude response of the anti-aliasing filter
A_N = Amplitude of one count of the digitizer

Anti-Aliasing Filter Selection

When selecting an anti-aliasing filter, we need to make several assumptions about the input analog signal frequency spectrum and the allowed aliasing level. The first assumption is that the frequency spectrum of the input analog signal is flat, that is, all of its frequency components have the same amplitude. Environmental signals typically differ from this assumption in two ways. First, signals have their high-frequency components attenuated more that their low-frequency components by the environmental signal transmission media. Second, measuring devices and structures upon which the devices are mounted may have resonant frequencies which amplify the signal frequency components near the resonant frequency. Since it is very difficult to account for these response deviations, we will assume a flat frequency spectrum.

The second assumption is that the sampled data will be digitized with a N-bit ADC and we will force the amplitude of the aliased signals to be below the amplitude of one count of the

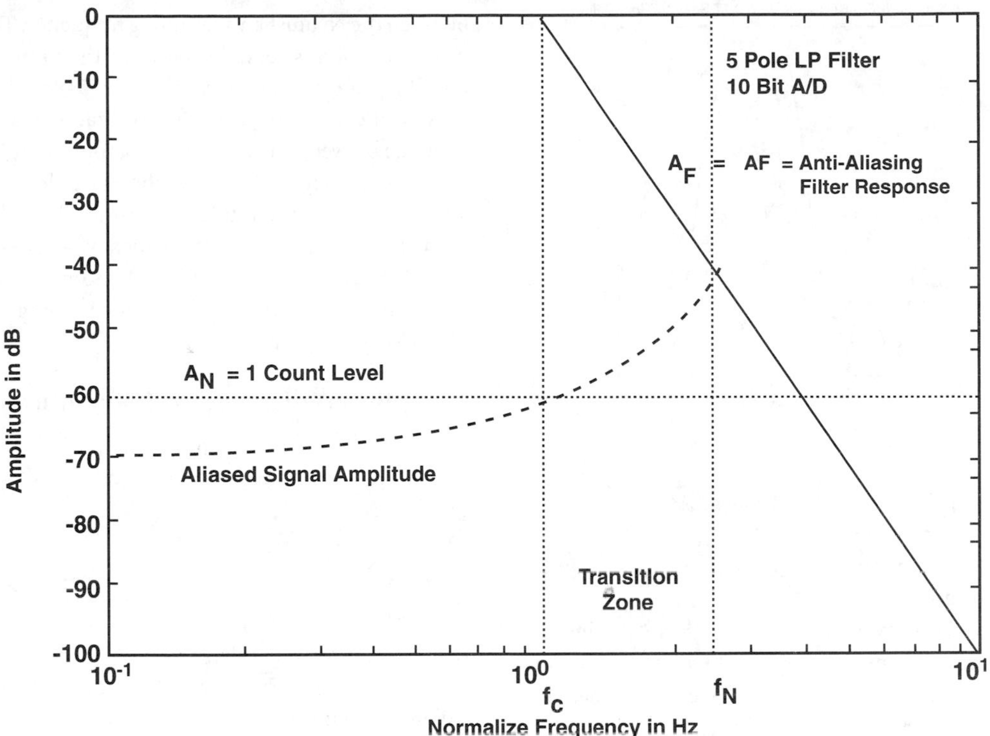

Figure 6.8 Bode plot of anti-aliasing filter amplitude response.

Table 6.1 Phase Shift and Time Delay vs. Frequency for Sine Waves

Frequency (Hz)	Period (milliseconds)	Time Delay (milliseconds)	Phase Shift (periods)	Phase Shift (degrees)
10	100	1	0.01	3.6
20	50	1	0.02	7.2
50	20	1	0.05	18
100	10	1	0.10	36
500	2	1	0.50	180
1000	1	1	1.00	360

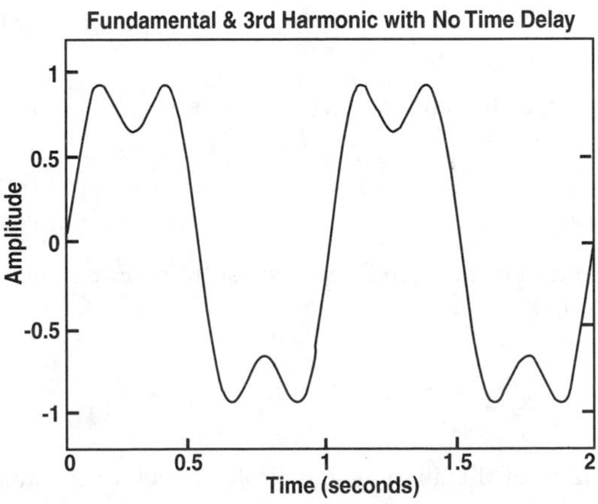

Figure 6.9 Signal with no delay.

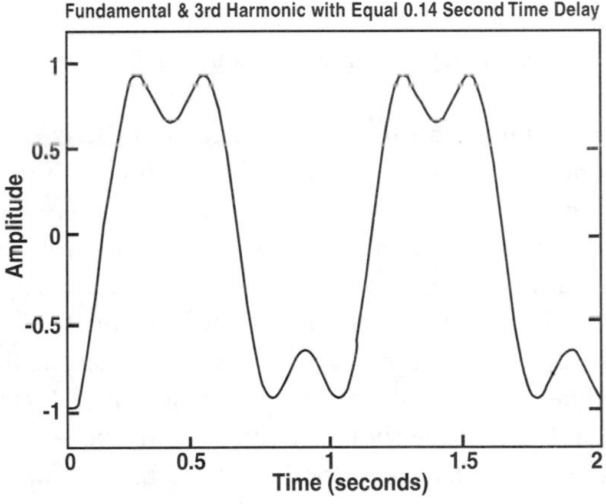

Figure 6.10 Signal with equal delay.

digitized signal at all frequencies either below the filter cutoff frequency or below the Nyquist frequency. Allowing aliasing in the transition zone between the Nyquist frequency and the filter cutoff frequency is less conservative and easier to implement. A digital filter could then be used on the digitized data to remove the signal components in the frequency transition zone between the cutoff and Nyquist frequencies. This digital filter must have a much steeper amplitude roll-off characteristic than the analog filter. The advantage of using digital filtering is that one can realize digital filters that will not distort the phase of the signal by doing such things as filtering the digital data both forward and backward in time. This is not possible with electrical analog

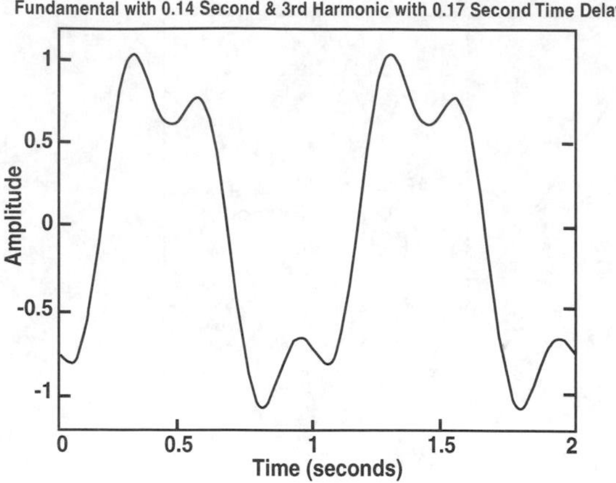

Figure 6.11 Signal with unequal delay.

filters. This digital filter method is called time reversal and is explained in Stearns and Rush (1990).

In the following sections we will discuss both no aliasing and aliasing in the frequency transition zone, but we will never allow aliasing below the filter cutoff frequency. Each bit of the ADC represents a factor of two in resolution and a factor of two in amplitude is a change of 6 dB. Therefore, an N-bit ADC would have a dynamic range of 6 × N dB. The anti-aliasing filter should attenuate the input signal by at least 6 times N dB for those frequencies which you do not want aliased.

The two parameters that we will be using in selecting the anti-aliasing filter will be the number of poles in the filter and the ratio of the sampling frequency to the filter cutoff frequency. These two parameters can traded off against each other to get the desired aliasing filtering. The advantage of more filter poles is that we can get data with frequencies closer to the Nyquist frequency. The disadvantage of more poles is the phase distortion will be increased and additional circuitry will be required to implement the filter. The second parameter used in selecting the anti-aliasing filter is the cutoff frequency of the filter. As the ratio of the sampling frequency to the filter cutoff frequency is increased, fewer poles will be required in the filter, but the sample frequency must be increased to get the desired measured frequency response. A rule of thumb often used for simple systems is to make this ratio equal to five.

An example of this process can be done for a 12-bit ADC in which we will allow no aliasing in the filter transition zone. A 12-bit ADC would required 6 * 12 = 72 dB of attenuation at the Nyquist frequency. Assume that we want to sample at four times the cutoff frequency of the anti-aliasing filter. This would have the Nyquist frequency at a factor of two, or one octave, above the cutoff frequency. At one octave, the filter will have an attenuation of 6 dB per filter pole. Therefore, the filter would need to have 72/6 = 12 poles. This is a high number of poles. Assume that want only a 6-pole filter. Then the attenuation per pole would be 72/6 = 12 dB. Since a filter gives an attenuation of 6 dB per octave per pole, the number of octaves between the filter cutoff frequency and the Nyquist frequency would equal 12/6 = 2 octaves. Another factor of two in frequency is needed between the Nyquist and sampling frequency. Therefore, 3 octaves or a factor of 8 is required between the sampling frequency and the filter cutoff frequency.

As you can see from the above example, a trade-off needs to be made between the ratio of the sampling frequency to the filter cutoff frequency and the number of poles. In addition, if one allows some aliasing into the transition band between the cutoff frequency and the Nyquist frequency, a fewer number of poles would be required for the filter. A more detailed analysis of this trade-off between filter poles and sampling frequency to filter cutoff frequency ratio follows.

Filter Design with Aliasing Allowed in the Transition Zone

The amplitude of one count of the signal for a N-bit digitizer is

$$A_N = -N * 6 \text{ dB} \tag{6.3}$$

with respect to full scale.

The amplitude response of an M-pole low-pass filter well beyond the cutoff frequency is

$$A_F = -20 * M * \log(f/f_C)\text{dB} \tag{6.4}$$

with respect to the pass-band response, f >> f_C, where log is the logarithm to base 10.

The aliasing frequency of a signal folded around the Nyquist frequency is

$$f_A = f_N - (f - f_N) = 2f_N - f = f_S - f, \qquad f_N < f < f_S \tag{6.5}$$

The frequency of the signal folded back to the anti-aliasing filter cutoff frequency is

$$f_A = f_C \tag{6.6}$$

which implies from the above equation that f = $f_S - f_C$. But

$$f_S = R * f_C \tag{6.7}$$

from the definition of R. Therefore, by substituting Equation 6.7 into Equation 6.6, we get

$$f = (R - 1) * f_C \tag{6.8}$$

The amplitude of the aliasing signal folded back to the anti-aliasing filter cutoff frequency from substituting Equation 6.8 into Equation 6.4 is

$$A_F = -20 * M * \log(R - 1) \tag{6.9}$$

Setting the amplitude of the aliasing signal less than or equal to one count of digitized signal we get

$$-20 * M * \log(R - 1) \leq -N * 6 \qquad (6.10)$$

This requires that the number of required poles for the anti-aliasing filter to be

$$M \geq 3 * N/(10 * \log(R - 1)) \qquad (6.11)$$

Table 6.2 lists the required number of poles for different N-bit digitizers and sampling frequency to filter cutoff frequency ratios.

One of the rules of thumb often used is that the ratio of the sampling frequency to filter cutoff frequency be five. For that special case, the number of poles required for the anti-aliasing filter is equal to one-half the number of bits in the digitizer.

Often one has a given number of ADC bits and a fixed number of poles in the anti-aliasing filter. Table 6.3 lists the required sampling frequency to filter cutoff frequency ratio to eliminate aliasing in the filter pass-band.

Filter Design with Aliasing Not Allowed in the Transition Zone

The previous analysis was based on the assumption that aliasing components will be accepted in the transition frequency zone between the filter cutoff frequency and the Nyquist frequency. If that assumption is not allowed in a system, then the number of poles will have to be increased. This increase can be calculated by using the previous equations except replacing

$$f = f_S - f_C \quad \text{with} \quad f = f_S/2 \text{ for Equation 6.6} \qquad (6.12)$$

This will give

$$f = (R/2) * f_C \text{ for Equation 6.8} \qquad (6.13)$$

Finally, Equation 6.11 will be replaced with

$$M \geq 3 * N/(10 * \log(R/2)) \qquad (6.14)$$

Therefore, the number of poles would be increased by the factor

$$\log(R - 1)/\log(R/2) \qquad (6.15)$$

Table 6.4 gives a listing of this pole increase factor as a function of the sampling frequency to cutoff frequency ratio.

Table 6.5 gives the required sampling frequency to filter cutoff frequency ratio to allow no aliasing in the frequency transition zone between the filter cutoff frequency and the Nyquist frequency.

The above has described a procedure for selecting the number of poles for an anti-aliasing lowpass filter in a digital data acquisition system. Both aliasing and no aliasing in the frequency transition zone between the filter cutoff frequency and the Nyquist frequency were considered. If aliasing components are not acceptable in this transition zone, calculations were made to show that the number of poles in the anti-aliasing filtering would be increased by a factor of approximately 1.5 as shown in Table 6.5.

Anti-Aliasing Filter Distortion

Filters distort the signal that is put through them (Van Valkenburg, 1982; Williams, 1981). Some distortion is desired and other distortion is not. Typically we like the filter to remove certain frequency components of the signal without modifying the phase response. This is not possible with real filters. This section discusses the different type of distortion we get with analog filters.

Commonly used anti-aliasing filters are Bessel and Butterworth types. A Bessel filter gives near linear phase-shift for frequencies

Table 6.2 Required Anti-Aliasing Filter Poles with Aliasing Allowed in Transition Zone

	No. of ADC Bits				
f_s/f_c	8	10	12	14	16
2.5	14	17	21	24	28
3	8	10	12	14	16
4	5	7	8	9	10
5	4	5	6	7	8
6	4	5	6	6	7
7	3	4	5	6	7
8	3	4	5	5	6
9	3	4	4	5	6
10	3	4	4	5	5

Table 6.3 Sampling Frequency/Filter Cutoff Frequency (f_s/f_c) with Transition Zone Aliasing Allowed

	No. of ADC Bits				
No. of Poles	8	10	12	14	16
1	252.19	1001.00	3982.07	15849.93	63096.73
2	16.85	32.62	64.10	126.89	252.19
3	7.31	11.00	16.85	26.12	40.81
4	4.98	6.62	8.94	12.22	16.85
5	4.02	4.98	6.25	7.92	10.12
6	3.51	4.16	4.98	6.01	7.31
7	3.20	3.68	4.27	4.98	5.85
8	3.00	3.37	3.82	4.35	4.98
9	2.85	3.15	3.51	3.93	4.41
10	2.74	3.00	3.29	3.63	4.02

Table 6.4 Pole Number Multiplier Factor for No Transition Zone Aliasing

f_s/f_c	No. Poles Increase Factor
2.5	1.82
3	1.71
4	1.58
5	1.51
6	1.46
7	1.43
8	1.40
9	1.38
10	1.37

Table 6.5 Required Sampling Frequency/Filter Cutoff Frequency (f_s/f_c) for No Transition Zone Aliasing

	No. of ADC Bits				
No. of Poles	8	10	12	14	16
1	502.38	2000.00	7962.14	31697.86	126191.47
2	31.70	63.25	126.19	251.79	502.38
3	12.62	20.00	31.70	50.24	79.62
4	7.96	11.24	15.89	22.44	31.70
5	6.04	7.96	10.50	13.84	18.24
6	5.02	6.32	7.96	10.02	12.62
7	4.40	5.37	6.54	7.96	9.70
8	4.00	4.74	5.64	6.70	7.96
9	3.70	4.31	5.02	5.86	6.83
10	3.48	3.99	4.58	5.26	6.04

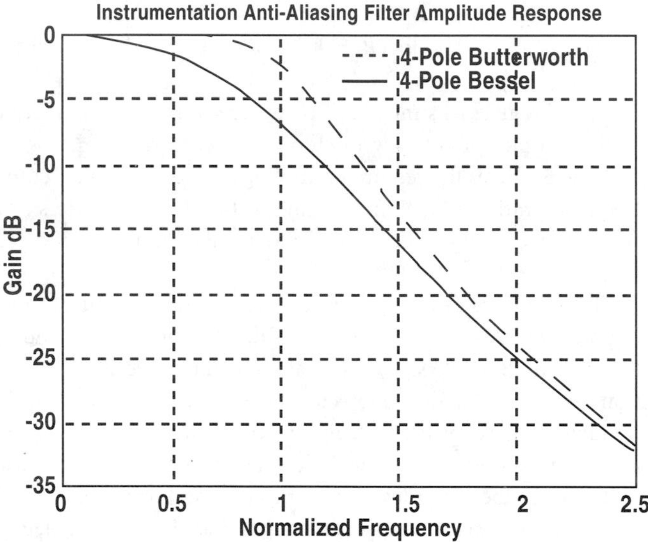

Figure 6.12 Anti-aliasing filter amplitude response.

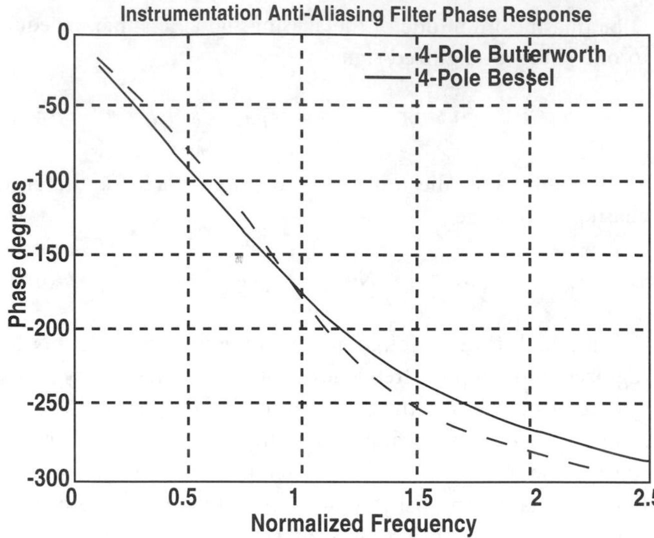

Figure 6.13 Anti-aliasing filter phase response.

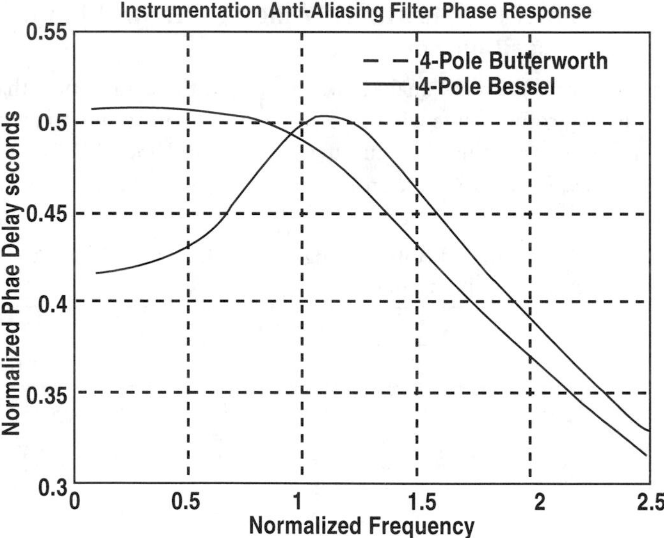

Figure 6.14 Anti-aliasing filter time delay.

below the cutoff frequency. Butterworth filters give a fairly flat amplitude response for frequencies below the cutoff frequency. Figure 6.12 through 6.14 show the amplitude, phase, and time delay responses, respectively, for anti-aliasing four-pole Bessel and Butterworth filters.

Note that the two different filters have different signal attenuation at the normalized frequency of one which is the cutoff frequency. This is due to the fact that they have different mathematical models. Notice that their responses are equal at 0 Hz and asymptotically equal at high frequencies which indicate that they have the same cutoff frequency.

Figure 6.15 shows the effect of 2-kHz Bessel and Butterworth four-pole filters on a square one-millisecond pulse of amplitude 1000 with a total record time of 5 milliseconds. Note that the Butterworth filter has a faster rise-time but more overshoot. The Bessel filter does not have this overshoot which is caused by the different time delay presented by the Butterworth filter to the third harmonic versus the fundamental frequency component of the square pulse. Comparision of this figure with Figure 6.11 shows the same overshoot effect due to varying time delay with frequency as seen in Figure 6.14. Because the Bessel filter has a flatter time delay response in the passband, it gives a better response for our applications which require as little distortion as possible in the time domain. The simulation was run at 100,000 samples per second to reduce aliasing problems in the simulation.

Nonlinear Distortion

Another form of distortion often encountered in data acquisition systems is amplifier clipping caused by input signals that are too large. One can always reduced the gain of the signal conditioning to help minimize clipping, but then the signal will not use the full range of the ADC and the signal can get lost in noise. We often do not know the maximum signal levels that the data acquisition will see in operation. So we have to make the best

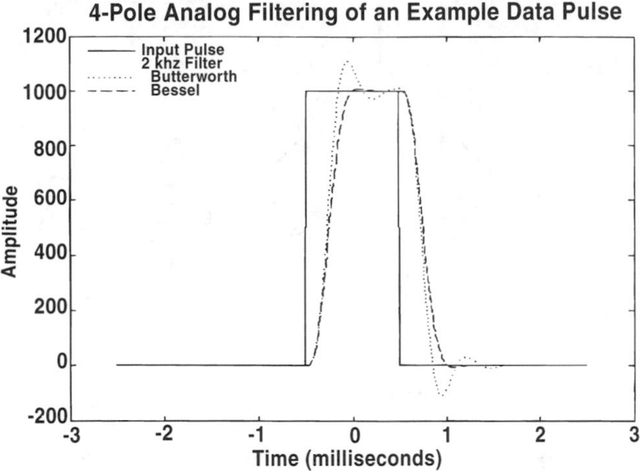

Figure 6.15 Analog filtering of sample pulse.

Table 6.6 Endevco 7270A Accelerometer Specifications

Model	Range (± kg)	Typical Sensitivity with 10-V Excitation (mV/g)	Typical Resonant Frequency (kHz)	Minimum Resonant Frequency (kHz)
−200K	200	1	1200	800
−60K	60	3	700	400
−20K	20	10	350	220
−6K	6	30	180	120
−2K	2	100	90	60

prediction that we can and then correct any clipping that occurs during data acquisition after recovering the data.

One of the worst problems to detect is clipping which is smoothed by the signal conditioning filters. This section will discuss an example of such a problem where we use an accelerometer gauge with a very low damped resonant response at high frequencies. An undamped accelerometer gauge can have a resonant frequency response which is 100 times greater than the low-frequency response where we are trying to measure our signal. The resonant frequency is on the order of five times the upper usable frequency. If the gauge is excited by a very fast rise time acceleration pulse, it can excite this resonant frequency. Since our anti-aliasing filter has a cutoff frequency much less than the resonant frequency, it removes most of the resonant frequency. But this is often after a previous amplifier has clipped the resonant frequency. We will used the Endevco 7270A accelerometer gauge as the input sensor in this example.

Table 6.6 gives specifications for different models of the 7270A accelerometer from the Performance Specification for Accelerometer Model 7270A (1985). Each of these accelerometers has a resonant frequency at which the output of the accelerometer can be very high if it is excited with a fast rise-time mechanical shock pulse. The resonant frequency of the 7270A-20K accelerometer is typically 350 kHz, which is considerably higher than our typical upper frequency of interest, typically 3.5 kHz. This section will give ways of removing the resonant frequency signal before it gets into the amplifiers. But first let us set what can

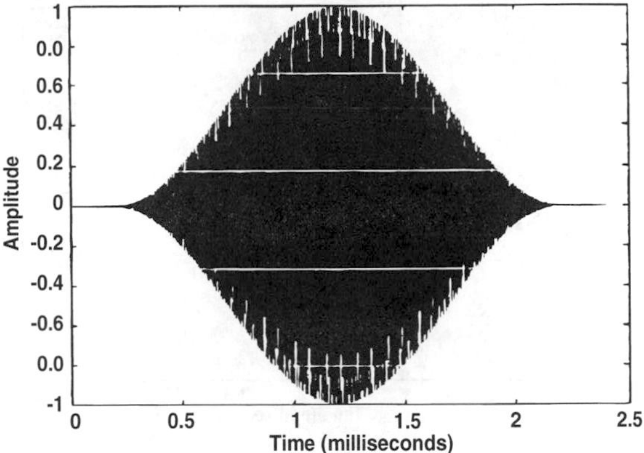

Figure 6.16 Modeled acceleration signal a(t).

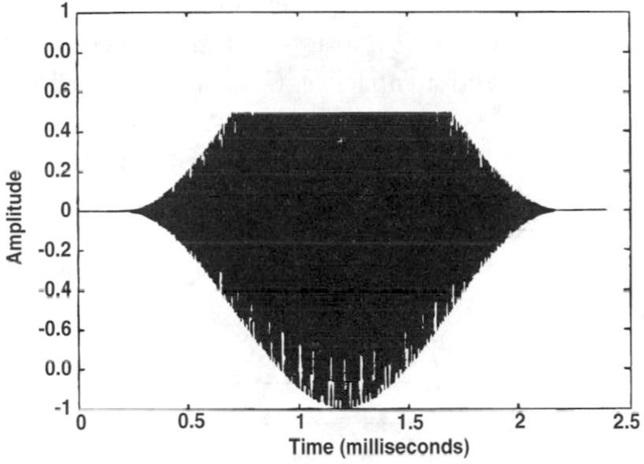

Figure 6.17 Clipped accelerometer signal.

happen if that resonant frequency gets into the amplifiers and causes clipping.

To see what would happen to the output of the Bessel filter if a 350-kHz signal is clipped and filtered, we simulated this condition using the software package PC Matlab as follows. First, we generated a pulse of 350 kHz sine wave signal using the following equation:

$$a(t) = 0.5 * [1 - \cos(2 * \pi * 500 * t)] * \sin(2 * \pi * 350 * 10^3 * t) \qquad (6.16)$$

A plot of this two-millisecond pulse with 200 microseconds of leading and trailing zeroes is given in Figure 6.16. The 350-kHz signal cannot be seen because of the plot resolution. This signal was then clipped on one side as shown in Figure 6.17. Finally, the clipped signal was passed through four-pole Bessel filter with a cutoff frequency of 3.5 kHz. The output of the Bessel filter is shown in Figure 6.18. As you can see, the filter removed the 350-kHz signal but left a distorted low-frequency pulse. Without clipping, the signal in Figure 6.18 should be a straight line. Passing the unclipped signal through the filter model verified that a straight line was generated.

If the signal was filtered before it was amplified, this clipping distortion would be reduced or eliminated. It is possible to add a single-pole lowpass filter between the accelerometer and first

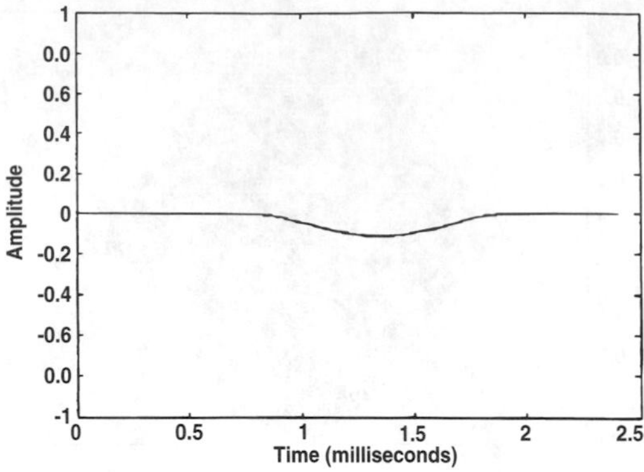

Figure 6.18 Filtered clipped signal.

amplifier. A schematic of the accelerometer with such a filter is shown in Figure 6.19. Notice that the accelerometer is a resistance bridge and a capacitor *C* has been placed across the bridge output.

The equation for the response of this circuit is

$$V_o(f) = \frac{\dfrac{dR}{C * (R + dR) * (R - dR)}}{j * 2 * \pi * f + \dfrac{R}{C * (R + dR) * (R - dR)}} * V_s \quad (6.17)$$

where
V_s = Supply voltage
V_o = Output voltage
f = Frequency
R = Bridge leg resistance
dR = Leg resistance change due to acceleration
C = Filter capacitor

This equation is of the form of a single-pole low-pass filter. Since *dR*, typically 10 ohms for a full-scale output, is much less than *R*, typically 550 ohms, this equation can be simplified as:

$$V_o(f) = \frac{\dfrac{1}{R * C}}{j * 2 * \pi * f + \dfrac{1}{R * C}} * \frac{dR}{R} * V_S$$

$$= \left(\frac{\omega_o}{j * \omega + \omega_o}\right) * \frac{dR}{R} * V_s \quad (6.18)$$

where ω_o is the resonant or cutoff frequency of the filter in radians per second. The time constant for the one-pole filter is *RC* and the filter's 3-dB cutoff frequency (f_o) in hertz is

$$f_o = \frac{\omega_o}{2 * \pi} = \frac{1}{2 * \pi * R * C} \quad (6.19)$$

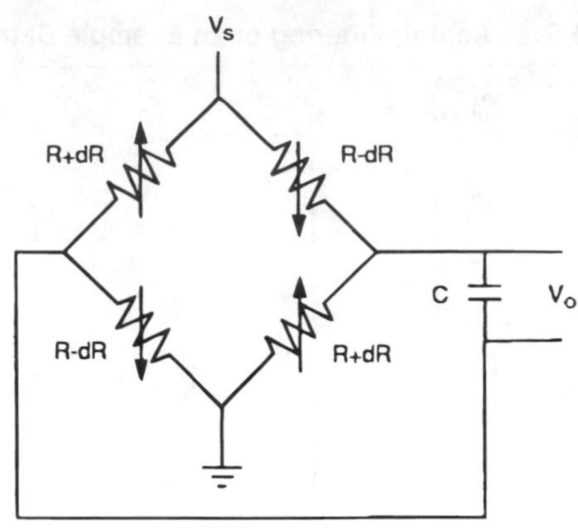

Figure 6.19 Accelerometer schematic with filter capacitor.

The gain in decibels of the one-pole filter can be written as:

$$G(f) = 20 * \log\left(\left|\frac{\dfrac{1}{R * C}}{j * 2 * \pi * f + \dfrac{1}{R * C}}\right|\right) \quad (6.20)$$

If we do not want the capacitor *C* to significantly affect the signal in the Bessel filter passband, a good rule of thumb is to make f_o equal to three times the Bessel low-pass filter cutoff frequency. An example of this is

$$\text{Bessel } f_o = 3.5 \text{ kHz}$$

$$R = 550\ \Omega$$

$$C = 0.027\ \mu\text{F}$$

Then:

$$f_c = 10.7 \text{ kHz}$$

$$G(3.5 \text{ kHz}) = -0.44 \text{ dB}$$

$$G(350 \text{ kHz}) = -30.3 \text{ dB} = \text{voltage factor of } 33$$

Therefore, the accelerometer's resonant frequency response is attenuated by 30.3 dB.

A better way to remove the undesired resonant frequency is to design a five-pole Bessel filter using the *RC* of the accelerometer and attached capacitor to get one pole and the four-pole filter Bessel filter to get the remaining four poles. This requires that the components in the present four-pole Bessel filter be changed to work with the preceding one-pole section to get an integrated five-pole Bessel filter. The advantage of this approach is that the one-pole *RC* filter section will have a lower cutoff frequency and give more attenuation of the accelerometer resonant frequency response. This approach will also minimize the phase errors applied to the signal.

The Laplace transfer function for a five-pole Bessel filter from Van Valkenburg's (1982; Table 10.3) work is:

$$H(s) = \frac{14.2725}{s^2 + 6.70391s + 14.2725} \frac{18.15631}{s^2 + 4.64934s + 18.15631} \times \frac{3.64674}{s + 3.64674} \tag{6.21}$$

The resonant frequency of this filter is the fifth root of the product of the numerators, which is

$$\omega_o = \sqrt[5]{945} = 3.93628 \text{ radians/second} \tag{6.22}$$

The gain at ω_o is −8.86 dB and the phase is −221.6 degrees.

To change $H(s)$ for any desired frequency ω_c, we will replace s with:

$$s = \frac{\omega_o}{\omega_c} s \tag{6.23}$$

where ω_c = desired cutoff frequency in radius/second.

The resulting Laplace transfer function for the 5-pole Bessel filter is

$$H(s) = \frac{0.921145\ \omega_c^2}{s^2 + 1.70311\ \omega_c s + 0.921145\ \omega_c^2} \times \frac{1.17181\ \omega_c^2}{s^2 + 1.18115\ \omega_c s + 1.17181\ \omega_c^2} \frac{0.926443\ \omega_c}{s + 0.926443\ \omega_c} \tag{6.24}$$

This equation consists of two second-order (two-pole) low-pass sections and one first-order (single-pole) low-pass section with the following parameters:

$$H(s) = \frac{\omega_{o1}^2}{s^2 + d_1 * \omega_{o1} s + \omega_{o1}^2} \frac{\omega_{o2}^2}{s^2 + d_2 * \omega_{o2} s + \omega_{o2}^2} \times \frac{\omega_{o3}}{s + \omega_{o3}} \tag{6.25}$$

where

$$\omega_o = 2 * \pi * f_o = \text{resonant frequency} \tag{6.26}$$

$$d = \text{damping coefficient} \tag{6.27}$$

In our systems we implement the above equation with two Sallen-Key active filters and a passive RC filter. The schematic of a two-pole Sallen-Key active filter is given in Figure 6.20.

The equations for this filter are

$$f_o = \frac{1}{2\pi\sqrt{R_1 C_1 R_2 C_2}} \tag{6.28}$$

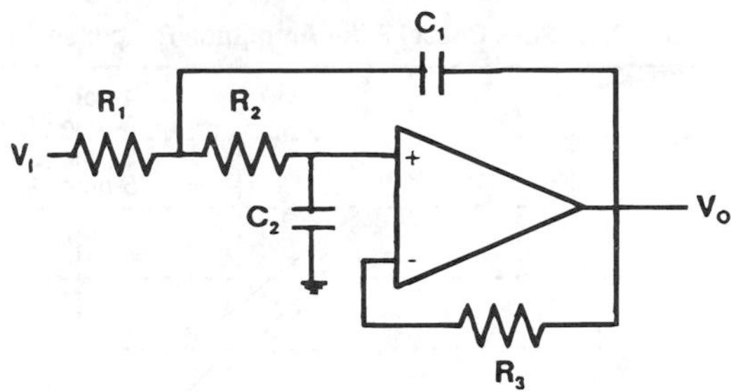

Figure 6.20 Sallen-Key two-pole low-pass filter.

Table 6.7 Five-Pole Bessel Filter Parameters

	f_{c1}/f_c	d_1	f_{c2}/f_c	d_2	f_{c3}/f_c
Desired	0.9598	1.7745	1.0825	1.0911	0.9264
Actual	0.9588	1.773	1.0783	1.091	0.9686

f_c = desired 5-pole Bessel filter cutoff frequency
f_{c1} = 1st 2-pole section cutoff frequency
d_1 = 1st 2-pole section damping
f_{c2} = 2nd 2-pole section cutoff frequency
d_2 = 2nd 2-pole section damping
f_{c3} = 1-pole section cutoff frequency

$$d = (R_1 + R_2)\sqrt{\frac{C_2}{R_1 R_2 C_1}} \tag{6.29}$$

The filter also has unity gain at DC.

By matching up the terms of Equations 6.24 and 6.25, substituting in Equations 6.28 and 6.29, and using commercially available components, we get the filter parameters as shown in Table 6.7 and the commercially available component values as shown in Table 6.8.

The capacitor selected to be put on the output of the accelerometer was calculated assuming that the accelerometer output resistance is 550 Ω. The specification for the 7270A accelerometer is that the output impedance is 550 Ω ± 200 Ω. We examined the data sheets on many of the accelerometers that we have and found them to deviate less than ± 100 Ω. The following curves in Figure 6.21 and Figure 6.22 show the amplitude and time delay response of the combined five-pole filter with three different resistor values from 450 Ω to 650 Ω. The response does not

Table 6.8 Five-Pole Bessel Filter Components
Section 1 C_1 = 2700 pF C_2 = 1500 pF
Section 2 C_1 = 3000 pF C_2 = 820 pF

Cutoff Frequency (kHz)	Section 1 R_1 (kΩ)	Section 1 R_2 (kΩ)	Section 2 R_1 (kΩ)	Section 2 R_2 (kΩ)	Accelerometer Capacitor (μF)
1.0	150	45.3	127	69.8	0.27
2.4	63.4	18.7	52.3	29.4	0.12
3.5	43.2	12.7	35.7	20.0	0.082
4.8	31.6	9.53	26.1	14.7	0.056
6.0	25.5	7.50	21.0	11.5	0.047
7.0	21.5	6.49	18.2	10.0	0.039
10.0	15.0	4.42	12.7	7.15	0.027

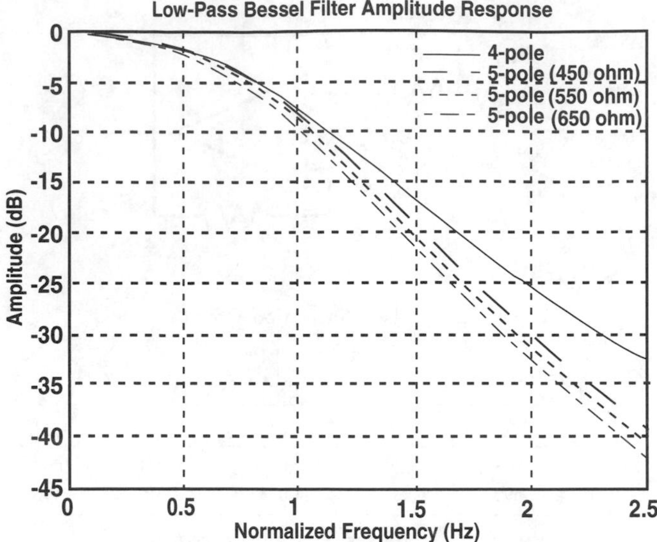

Figure 6.21 Five-pole Bessel filter amplitude response.

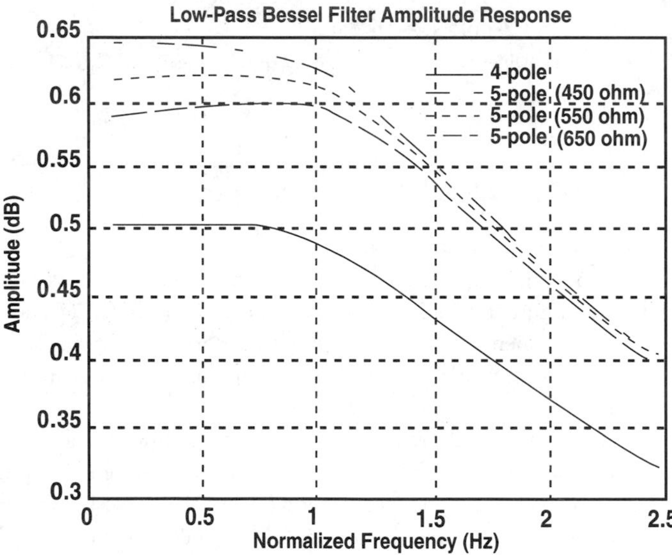

Figure 6.22 Five-pole Bessel filter delay response.

deviate enough for the different accelerometer resistances to make it necessary to match the filter to the accelerometer.

The attenuation of the resonant frequency signal response by the capacitor across the accelerometer is

$$\text{Bessel } f_c = 3.5 \text{ kHz}$$
$$R = 550\ \Omega$$
$$C = 0.082\ \mu\text{F}$$
$$f_c = 3.53 \text{ kHz}$$
$$G(3.5 \text{ kHz}) = -2.97 \text{ dB}$$
$$G(350 \text{ kHz}) = -39.9 \text{ dB} = \text{voltage factor of } 99$$

A few key points of the four- and five-pole Bessel filter responses are given in Table 6.9.

To convert the normalized time delay to actual time delay, divide the normalized time delay by the actual cutoff frequency in Hz.

A capacitor added across the output of the accelerometer in our data acquisition systems will significantly reduce the possibility of clipping distortion of the accelerometer resonant output. The selection of the capacitor along with the components in the remaining active filter sections must be done carefully to keep the signal from being distorted. The conversion of the data acquisition system from a four-pole Bessel filter to a five-pole Bessel filter is the best solution for two reasons. It will significantly reduce the accelerometer resonant frequency output and it will also give additional anti-aliasing filtering of the signal before it is digitized.

Analog Multiplexing

Analog multiplexing is taking the multiple analog input channel signals and switching them so that one signal at a time is fed into the ADC. The multiplexer is an electronic rotary switch that feeds one signal into the ADC for digitization and then switches to the next analog channel in sequence. Since the multiplexer is switched to only one channel at a time, the analog signals are not digitized at identically the same time, but are digitized in sequence with all channels digitized within one complete rotation time of the multiplexer. To digitize two signals at the same time, one inserts sample-and-hold circuits between the filters and the multiplexer. The design of the controller must be modified to send a hold signal to the sample-and-hold circuit at the appropriate time in the data acquisition cycle. If an analog channel is input to more than one multiplexer input to get a higher sampling rate, make sure that the sample-and-hold is not done simultaneously, but is done at the correct equal time intervals.

The multiplexer output and any circuitry between the output and the ADC input must have a much larger bandwidth than the bandwidth of each analog input to the multiplexer. The reason is that the multiplexer output keeps changing from one channel to the next and these channels do not necessarily have the same signal level at any given instance. But the input to the ADC must stabilize to the new channel level before the digitization takes place. Therefore, the multiplexer output and following circuitry must have a large bandwidth to allow this signal stabilization.

Multiplexers consist of a bank of electronic analog switches with their outputs tied together. Electronic analog switches have nonzero resistance and nonzero capacitance. The output load will also have a nonzero resistance and a nonzero capacitance. The combination of these resistances and capacitances will have a finite bandwidth which determines the response of the signal going into the ADC.

Analog multiplexers have a nonzero resistance between the input and the output. In addition, they are built on a common integrated circuit substrate. If one channel has an input level that exceeds the multiplexer input level specification, the signal will typically feed into other channels causing crosstalk between the channels. Some means should be used to guarantee that this maximum input level is not exceeded. This can consist of amplitude limiting circuitry or the amplifiers feeding the multiplexer input must be set up to not exceed this maximum voltage.

Table 6.9 Bessel Filter Response

Normalized Frequency (Hz)	4-Pole Attenuation (dB)	4-Pole Phase Shift (degrees)	4-Pole Normalized Time Delay (seconds)	5-Pole Attenuation (dB)	5-Pole Phase Shift (degrees)	5-Pole Normalized Time Delay (seconds)
0.5	1.7	91	0.506	1.8	112	0.622
1.0	7.5	177	0.492	8.7	221	0.614
1.5	16.5	235	0.435	20.2	296	0.548
2.0	25.1	267	0.371	31.2	337	0.468
2.5	32.3	286	0.318	40.3	360	0.400
3.0	38.4	299	0.277	48.0	376	0.348

Minimizing Signal Conditioning Components

One of the questions often asked is whether signal conditioning is required on each signal before it goes through the analog multiplexer or switch, or can the system be simplified by putting a single signal conditioning system after the multiplexer output. The normal answer is that conditioning is required on each channel before the multiplexer. The reason is that every time the multiplexer switches between the different channels, a step occurs in the signal coming from the multiplexer because the channels are not at the same level at any instant. If the signal conditioner were after the multiplexer, the signal conditioner would have to respond very fast to this step change so that the conditioned signal would have settled to its final value before being digitized by the ADC. But the signal conditioner contains amplifiers and filters that cannot respond this fast. In fact, the anti-aliasing filters are designed to not change very fast between samples. About the only items that could be put after the multiplexer are DC offsets assuming that the offset of each signal is the same, and also maybe an amplifier. But this amplifier must have a very wide bandwidth and fast slewing rate.

Smart Gauges

So called smart gauges are beginning to be introduced into the world of instrumentation. These gauges incorporate signal conditioning, analog-digital converter and have a serial digital output. The digital signal may be sent to either a multiplexer or a memory via either fiber optic or copper cable. The high-level digital signal is inherently much more immune to noise than low-level analog signals. The amplifier in such a system can be much simpler and cheaper than a remote amplifier because it can have greatly reduced common mode rejection and overvoltage requirements.

Signal Conditioning Summary

Signal conditioning is the process of converting the analog signals from the different sensors to signals that are electrically compatible with the ADC. One item that we have not considered yet is the different signal conditioning on each separate analog channel. If we need to have the different analog channels time correlated, then we must assure that the phase response of each channel is the same. Linear phase response on each channel is not sufficient if one wants equal time delay on all channels. Since time delay is proportional to the slope of the phase response, it is necessary that each channel have the same slope to the linear phase response. In many applications it is not necessary that all channels have the same time delay as long as the response is well characterized, then the channel difference effects can be removed in the processing of the data.

6.4 Signal/Data Transmission Components

Otis Solomon and William Boyer

As noted earlier instrumentation systems are used to record analog signals representing physical phenomena. The gauges are often physically separated from some or all of the components of the recording system by distances ranging from less than a meter to thousands of kilometers. When the gauges are all within the same facility the distances are more typically tens to thousands of meters. This is the situation we are most concerned with here. There are a number of different possibilities for transmitting signals in this situation. We will consider copper wire and fiber optic cabling systems. We will not consider radio frequency transmission. The method chosen will depend primarily on the bandwidth of the signal and on the combination of desired dynamic range and electromagnetic or radiation environment that the signal must pass through. Fiber optic cable systems have the following advantages and disadvantages compared to copper cable:

Advantages

- Immune to electromagnetic noise pickup.
- Noncorrosive fibers.
- Provides electrical isolation between gauge and recording system.
- Wider bandwidth achievable for a given length.
- Smaller, lighter, more flexible, and easier to install long runs.

Disadvantages

- Less dynamic range.
- Active transmitter and receiver required.
- More expensive to purchase.
- Expensive test and splicing equipment may be required.

Data Transmission via Copper Wire

The two types of data transmission via copper wire are twisted shielded pair and coaxial cable. Low-level, low-frequency signals are generally transmitted using a twisted shielded pair cable (TSP). This is because TSP cable provides for electromagnetic noise cancellation and shielding for both low and high frequencies. TSP also provides adequate bandwidth at a reasonable price. Finally the cable is flexible and relatively easy to install. The twisting in TSP causes cancellation of noise induced by magnetic fields. Because both conductors are very close to each other for the entire run, they will pick up the same noise signal as they pass through the facility environment. This common mode noise signal can then be canceled by using differential amplifiers. Differential amplifiers can have common mode rejection ratios (CMRR) as high as 80 dB. However it is very difficult to match the two conductor runs and the termination resistors to achieve this high cancellation level. It is particularly difficult when the cable run is terminated in its characteristics impedance. This is usually about 100 ohms. In this case small differences in conductor resistance will result in common mode signals being converted into normal mode signals. It is desirable to terminate cable runs in their characteristic impedance when the electromagnetic environment has impulsive noise sources. This is because the impulses are absorbed by the matched termination impedance. Otherwise the noise pulses will be reflected back and forth along the cable until they are absorbed by the cable's DC resistance.

The shield around the twisted pair provides for some high-frequency noise signal attenuation. TSP cable can be obtained either as a single pair within a single shield or as multiple individually shielded pairs inside an outer braid shield. Multipair cables are cheaper to install. However the close packing can cause crosstalk problems if high-level signals are mixed with low-level signals. Crosstalk occurs when a signal from one circuit couples or interferes with a different circuit. Also the internal shields tend to be very thin.

For signal bandwidths more than about 1 MHz and distances more than 10 meters coaxial cable must be used. The type of cable will be a function of the length of the run and the bandwidth of the signals to be transmitted. Table 6.10 shows the attenuation and capacitance per foot for a few commonly used coaxial cable types.

In general larger diameter cables are required to maintain a given bandwidth as distance is increased. A major reason for this is that the skin depth decreases and losses increase in the center conductor as signal frequencies increase. Also cables are dispersive; signals at different frequencies travel at different velocities. At very high frequencies, i.e., a few gigahertz, transverse propagation modes can exist in large diameter cables. This negates the gain of the larger diameter.

Table 6.10 Cable Characteristics

Cable Type	Attenuation at 400 MHz (dB/100 ft)	Capacitance (pF/ft)
RG-58C (0.195″ OD with polyethylene dielectric)	14.0	30.8
RG-214 (0.425″ OD with polyethylene dielectric)	5.5	30.8
RG-331 (0.600″ OD foam polyethylene dielectric)	2.2	25.4

Cable frequency response compensation may be used if the desired bandwidth cannot be achieved for any practical cable type. Compensation may be done either via hardware equalizers or by software post processing. Equalizers have been used to provide up to 20 dB of high end frequency response compensation. However these devices are normally passive and attenuate the signal by the compensation factor. For example a 10 dB equalizer attenuates the DC and low-frequency components of a signal by 10 dB. The attenuation decreases at higher frequencies, and the device produces a net flat response out to the frequency where the original cable response becomes greater than 10 dB.

Software processing can also be used to compensate for high-frequency losses in cables. This is also known as deconvolution. Software compensation is limited to about 8–10 dB. This limitation is due to the fact that all known high-frequency waveform recorders introduce high-frequency noise into the output as they digitize signals. Deconvolution is a mathematically unstable operation in the presence of noise. Deconvolution techniques must incorporate some mechanism for limiting amount of high-frequency noise they amplify in trying to recover the original signal.

Differential data recording may be used to help eliminate noise pickup in high-frequency systems just as it was using TSP for low-frequency systems. To do this two cables must be run for each signal. This method is particularly effective if the gauge is designed to output a balanced signal. Since the majority of high-frequency waveform recorders are single-ended, a balun transformer must be used to subtract the two input signals and produce a single-ended output. In order for this to work the cable transit times must be identical to the order of a few hundred picoseconds. Also the cables must be run next to each other so they will both pick up the same common mode noise signal as they pass through the environment.

Grounding and Shielding

A well-designed measurement system should contain precautions against electrical interference. Ott (1988) divides noise sources into three categories:

1. Intrinsic noise sources that arise from random fluctuations within physical systems such as thermal and shot noise.
2. Human noise sources such as motors, switches, digital electronics, and radio transmitters.
3. Noise due to natural disturbances such as lightning and sunspots.

Industrial facilities normally contain many Category 2 noise sources such as 60 Hz from lighting and heavy equipment, impulsive noise sources produced by switching transients, and other transient and/or continuous wave noise from radio frequency transmitters. When noise causes a circuit to malfunction or degrades the performance of a circuit, it is called interference. Electromagnetic compatibility (EMC) is the ability of electronic equipment to function properly in a specified electromagnetic environment without interfering with other electronic equipment.

More detailed discussions on shielding are contained in Ott (1988) and Meiksin and Thackray (1984). AC voltages have electric fields associated with them. Circuit theory assumes that electric fields are confined to capacitors. AC currents have magnetic fields associated with them. Circuit theory assumes that magnetic fields are confined to inductors. Stray electric and magnetic fields as well as electromagnetic fields can couple to circuits. This coupling can be minimized if electromagnetic field sources such as solenoids, motors, power transformers are physically separated as far as possible from the measurement system. Shields or metallic enclosures provide a means to reduce the effects of stray fields. Electric fields are much easier to deal with than magnetic fields especially at low frequencies. When the field source is near the circuit, electric and magnetic fields are analyzed separately. When the field source is far from the circuit, the electric and magnetic fields must be considered together as an electromagnetic field. Distance to the field source is measured in terms of the wavelength of the source radiation. Far means distances greater than one-sixth of a wavelength.

Shields reduce stray fields through two mechanisms: reflection, and absorption or penetration loss. A Faraday cage or electrostatic shield completely surrounds an electronic system with a conductive material. Normally the conductive material is copper or aluminum. Reflection is the primary loss mechanism for electrostatic shields. Low-frequency magnetic fields are not reflected very effectively by electrostatic shields. To shield against low-frequency magnetic fields, one must use a high-permeability magnetic material such as mumetal. At high frequencies above 10 MHz, shields of nonmagnetic material attenuate the electromagnetic fields. At high frequencies, effective shielding is guaranteed by completely sealing the circuit. The shield must be solid to avoid leakage through seams, joints and holes.

A susceptible circuit can be shielded to keep the enclosed components from exposure to external fields. Electrostatic coupling occurs due to a time varying potential difference between two conductors coupled electrically by stray capacitance. Electrostatic shielding provides a conducting surface for the termination of electrostatic flux lines, but need not be magnetic material. Stranded braid, mesh, and screens of good electrical conductors such as aluminum or copper are good electrostatic shields. Most shielded cables use copper braids as the outer conductor and shield.

Magnetic shielding has differing requirements than electrostatic shielding. The success of magnetic shielding is partially attributable to short circuiting of magnetic flux lines by low-reluctance paths and partially attributable to self-cancellation due to opposing fields set up by eddy currents. The shielding is comprised of high-permeability material such as mumetal and Permalloy and should be as thick and as free of holes as possible. A good magnetic shield is a good electrostatic shield, but the converse is not true.

The magnitude of all electrical interference is proportional to cable impedance. High-impedance lines are more susceptible to undesirable pickup than low-impedance lines. Cable impedance therefore should be kept as low as measurement system time constants and loading effects will permit.

Current coupled noise is associated with the situation where signal currents share a common path with other currents, either by intention or otherwise. Any impedance in this path causes the non-signal currents to develop extraneous signals which are interjected through the measurement system. Before formulating a correction for this situation, a definition of ground is required.

An ideal ground is a reference plane of zero impedance so that no potential difference exists between any two points on the plane. Since zero impedance is not physically obtainable, potentials do exist between points on a ground plane. Thus, a ground point is often chosen for a reference as opposed to a ground plane.

Grounding in instrumentation systems is a very important consideration. There are two types of grounds: safety grounds and signal grounds. For safety, in the United States, the chassis or enclosure for electric equipment is grounded. Ungrounded electrical equipment can create severe shock hazards through stray impedances and insulation breakdown. In the United States, the National Electrical Code (NEC) describes AC power distribution and wiring standards. A signal ground is an equipotential surface that serves as a reference potential for a circuit. Below we discuss signal but not safety grounds.

A ground loop is formed when a common connection in a measurement system is grounded at more than one point. If circulating ground currents cause a potential difference e_{gnd} to develop between the grounding points, the normal input signal will be modulated by hum and pickup from such a potential difference. The voltage seen by the input amplifier is the sum of the normal input signal e_1 in series with the hum voltage e_{gnd}. It is not uncommon for ground potential differences of several volts to exist between grounded electrical items only a few feet away. For physically small electronic systems operated at low frequencies, ground-loop interference can be eliminated by avoiding multiple grounds on the signal circuit. Ground the entire circuit at one point: the input to the data recorder. In multichannel systems, this is the only location which will not cause ground loops between channels.

At high frequencies, a single point ground system does not work. As the frequency increases, the inductance of a ground wire causes its impedance to climb to unacceptable levels. At high frequencies, multipoint ground systems are employed to provide multiple low-impedance paths to ground. The noise currents are shunted to ground before they can travel very far in the circuit. Each circuit of the system is connected to the ground plane. The connections must be short to minimize their impedances.

The major goal of a grounding scheme is to reduce the noise-induced currents flowing on the cable shields. These currents produce noise voltages on the inside of the shield and then onto the internal signal conductors via the shield transfer impedance (Ott, 1988):

$$Z_T = \frac{1}{I_S}\left(\frac{dV}{dl}\right)$$

where Z_T is transfer impedance in Ω/m, I_S is the shield current, V is the voltage induced between the inner conductors and the shield, and l is the length of the cable. A transfer impedance of 0.001 Ω/m means that 1 A of noise current will induce 0.001 V of noise for every meter of cable.

A simple-minded, but easy-to-understand, model for noise coupling is obtained by applying Faraday's law to a loop of wire in a time-varying magnetic field. The result relates the induced voltage V_N to the negative rate of change of the magnetic field normal to the loop:

$$V_N = \omega BA \cos(\theta)$$

where B is the RMS value of the flux density varying sinusoidally at frequency ω, A is the area of the loop, and θ is the angle of the loop with respect to the magnetic field. The amount of current produced is simply the voltage divided by the impedance of the loop. Thus by not having a conductive path around the loop, i.e., infinite impedance, the noise current can be eliminated. Noise pickup can also be reduced by minimizing the area of the loop. This is the fundamental reason to eliminate ground loops in electronic systems. For small systems this is achieved by implementing a single point ground system. This simple model is only valid when the size of the cable system is small compared to the wavelength of the interference. When this is not the case a traveling wave analysis must be used.

Many instrumentation engineers espouse the use of floating recording systems and grounded gauges. Floating a system means that the ground wire in the AC electrical feed is not attached to the metal surface surrounding the equipment, e.g., the racks and chassis. This violates standard electrical wiring codes. However this practice can be done legally if the system incorporates a warning light that flashes when the ground is disconnected. In a floating system, the hazard is that when a fault in the wiring causes the hot electrical lead to touch the metal frame, the circuit breaker will not trip. This creates a shock hazard for personnel. This is particularly serious for systems powered by 220 VAC or 440 VAC. Floating recording systems also requires that all other paths to ground except the cable shields to the gauges be eliminated. This requires using nonconductive conduit near the recording facility for utilities such as telephone, intercom, fire alarms, and computer networks. Also all structural conductive paths must be eliminated. This philosophy of grounding a system only at the gauge will not be effective when the gauges are spread over a large facility and feed into a common recording system. Thus one might try to float all the gauges and shields and ground the recording system. This is very difficult if the cable shields have to pass through metal bulkheads in the facility.

6.5 Software Data Correction

William Boyer and David Ryerson

Digital Filtering of Data During Data Reduction

Filters distort the signal that is put through them (Van Valkenburg, 1982; Williams, 1981). Some distortion is desired and other distortion is not. Typically we like the filter to remove certain frequency components of the signal without modifying the phase response. This is not possible with real filters. This section discusses the different types of distortion we get with digital filters.

Data are commonly digitally filtered at data reduction time. Stearns and Rush (1990) contains information on how to design digital filters. A filter often used for this purpose is a four-pole Butterworth. A data record can be run through the filter both forward and backward in time. This time reversal is discussed in detail in Stearns and Rush (1990). Such a filter gives a filter response equivalent to eight poles of amplitude and zero phase shift. This is a noncausal filter, i.e., a filter that can produce output before input. This happens because time is run backward. The advantage of this method is that no phase distortion is introduced into the signal. The disadvantage is that amplitude changes occur before the input pulse starts. Often the user of the data has trouble understanding how a system can generate an output before an input is received, which is true in causal real systems.

To see the effect of filtering, a one-millisecond pulse of amplitude 1000 was generated and filtered on a computer at 100,000 samples per second. The data record extends from −2.5 milliseconds to +2.5 milliseconds with the pulse center at time equal zero. This pulse was run through a computer simulation of a 2-kHz analog Bessel filter. Next it was run through one of five different digital filters as defined above with cutoff frequencies varying from 2 kHz to 200 Hz. The results are plotted in Figure 6.23.

Note that the 1-kHz digital filter causes the pulse to have a higher amplitude than the original pulse. This is caused by the

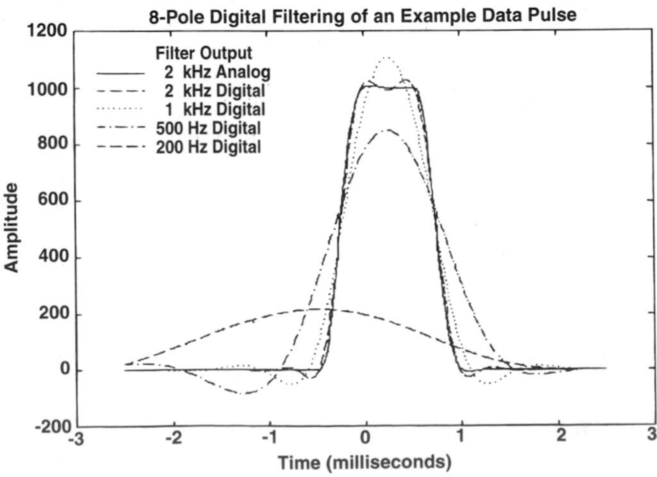

Figure 6.23 Digital filtering of 5-ms pulse record.

removal of the third harmonic in the pulse and leaving only the fundamental frequency. The filter at 500 Hz starts attenuating the pulse significantly. When studying a pulse of width w in seconds, it probably is not wise to filter this pulse with a cutoff frequency lower than 1/w Hz. Even with this cutoff frequency, the undershoot at the beginning and end of the pulse is approaching five percent of the amplitude of the pulse and the center of the pulse is magnified by approximately ten percent.

This plot can be used to evaluate the effect on any pulse width by varying the cutoff frequency inversely to the change in pulse width. For example, if the pulse width was 500 microseconds instead of one millisecond, the filter cutoff frequencies to give the same effect would be double the frequencies shown on the attached plots.

Figure 6.24 takes the input pulse from Figure 6.15 and extends it with leading and trailing zeroes to give a data record that extends from −5.0 milliseconds to +5.0 milliseconds instead of −2.5 milliseconds to +2.5 milliseconds. This shows that filtering errors can be reduced by adding zeroes to the beginning and end of the data record. What this does is reduce the filter startup effect. The most significant startup distortion in Figure 6.23 is in the 200-Hz filtered pulse although there is a little distortion in the 500-Hz pulse. If the cutoff frequency of the filter is f_c in Hz, then one should probably have at least $1/f_c$ second of data before and after the pulse. For a 200-Hz filter, the pulse should have five milliseconds of leading and trailing data. This is approximately what Figure 6.24 has.

Additional evidence of the start up problem without sufficient leading and trailing zero data is shown in Figure 6.25 and Figure 6.26. This two figures were generated by filtering the pulse forward then backward and by filtering the pulse backward then forward. The results should be the same except for any filter startup problems. Note that there is considerable difference for the 5-millisecond record but little difference for the 10-millisecond record. The 10-millisecond record has sufficient leading and trailing zeroes to limit most of the filter start up problem.

What these plots show is that false data can be produced by a digital filter. Ringing pulses can be produced around the main pulse that are not meaningful. This effect becomes worse as the filter cutoff frequency is lowered.

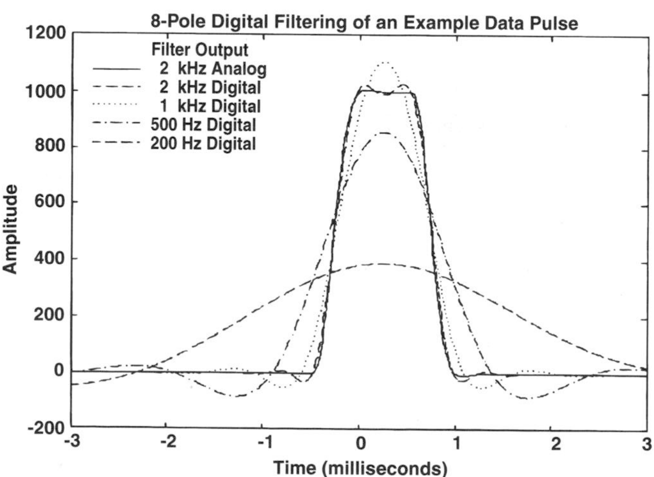

Figure 6.24 Digital filtering of 10-ms pulse record.

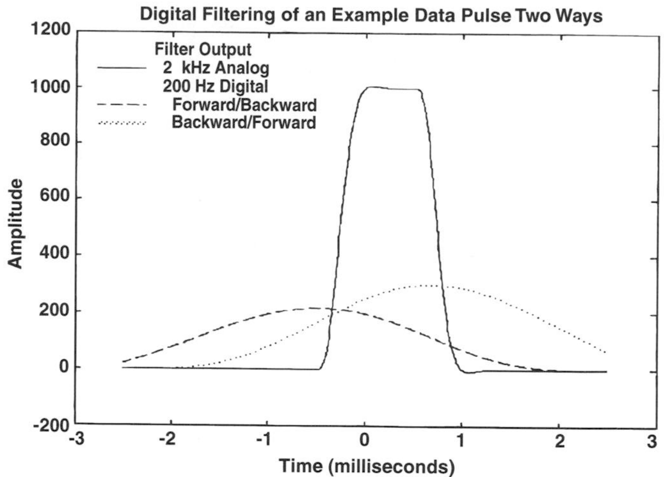

Figure 6.25 Digital filtering of 5-ms pulse record in 2 different directions.

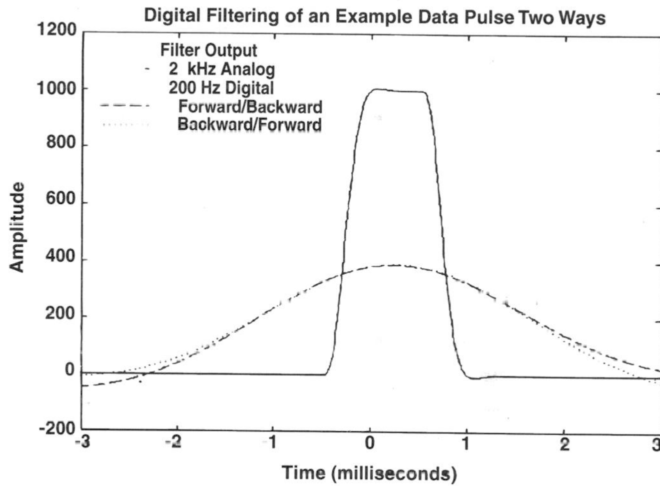

Figure 6.26 Digital filtering of 10-ms pulse record in 2 different directions.

Filtering of data will distort that data. If one knows what type of distortion occurs with a given filter, then intelligent decisions can be made as to when and what type of filter to use.

Deconvolution Methods

As noted above deconvolution may be required to compensate for high-frequency losses in an instrumentation system. The losses may occur in the transducer, the cable system, signal conditioning components or in the recorder. Ideally a data recording system should be designed so that the bandwidth is adequate for the signals to be recorded. When this cannot be achieved, then some other method must be used to restore the signal bandwidth. The bandwidth may be restored either by hardware equalizers or by software deconvolution. We will limit our discussion in this section to software deconvolution. We will also strongly orient the discussion to high-frequency system bandwidth limitations imposed by coaxial cables. However the method described will also be applicable to the more general case where calibration signals with sufficient bandwidth can be injected into an instrumentation system and recorded with enough high-frequency information to characterize the system.

Coaxial cables attenuate the high-frequency components of the signals they transmit. If the attenuation is not too severe, these high-frequency components can be recovered by the process of software deconvolution. A qualitative cable frequency response curve is shown in Figure 6.27. We will model the cable system as a time invariant linear system with a frequency response $H(\omega)$. If the system could be characterized adequately and if the data could be recorded without introducing any noise, then the original signal could be recovered by simply applying a filter with an inverse filter compensation function frequency response $H^{-1}(\omega)$. An example of a cable inverse function is shown in Figure 6.28.

The problem becomes much more difficult in the presence of digitizer induced noise. Figure 6.28 shows that the gain of $H^{-1}(\omega)$ steadily increases. Data signals typically have frequency content that is band limited below the bandwidth of the recorder. All waveform recorders introduce quantization error during the digitization process. For large signals, the spectrum of quantization error is often white. Waveforms recorders can introduce noise in addition to quantization error. The noise floor includes both quantization error and other noise sources. For a white noise floor, the spectrum of the net recorded signal looks like Figure 6.29. Note that at high frequencies the spectrum is dominated by the digitizer noise. If we were to apply the inverse filter compensation function to this signal, at some we point we would only be amplifying noise. Thus the high-frequency gain of the compensation function must be constrained in some manner. Designing such a constrained function is the heart of deconvolution theory.

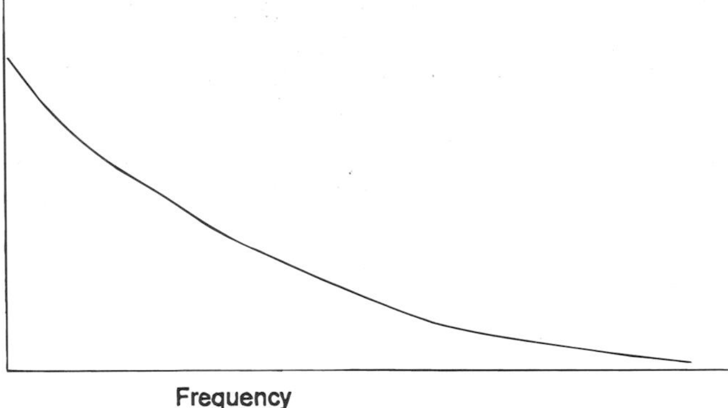

Figure 6.27 Typical cable frequency response.

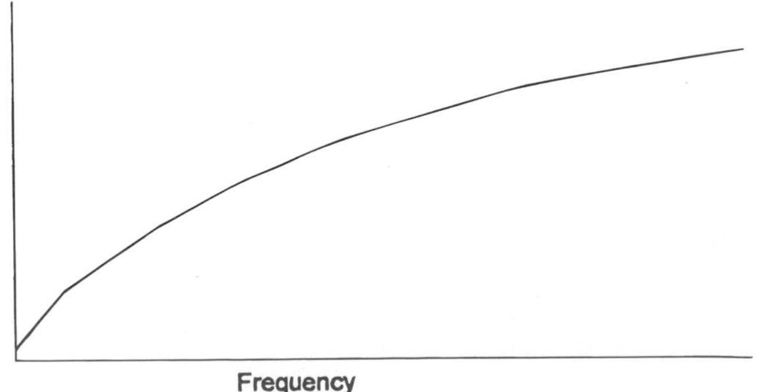

Figure 6.28 Inverse of cable frequency response.

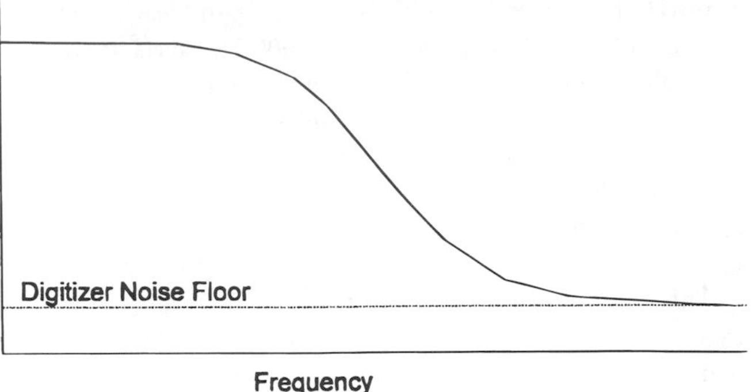

Figure 6.29 Typical recorded signal spectrum with white noise floor.

There has been and continues to be much research in this area. A few representative papers are given in the references. In the rest of this section we will briefly describe one practical method that has been used successfully in many instrumentation systems for many years (Boyer, 1982). The method is known as the Toe Truncation Method. The "recipe" for implementation is as follows:

1. Apply a "fast rising" step pulse to the input of the recording system. The risetime of the pulse must be fast enough to characterize the system, e.g., if the system has a 1-GHz bandwidth, then the pulse must have at least a 300-picosecond risetime. The input pulse will be called the undegraded signal.
2. Record the undegraded pulse as close to its output as possible, i.e., with minimal cable length. There are two options at this point. In option 1 the undegraded pulse is recorded using an instrument that has a bandwidth significantly greater than the recorders normally used in the system, e.g., a digital sampling oscilloscope. In option 2 the undegraded pulse is recorded by the same type of recorded by the same type of recorder used in the system. If option 1 is used, then the frequency response of both the cable and the recorder can be compensated. For option 2 only the cable can be compensated. If option 2 is used, then many instances of the pulse should be recorded and then averaged to reduce digitizer noise. Denote this signal by $f(t)$. An example of a good undegraded step pulse is shown in Figure 6.30.
3. Record the step pulse at the output of the system. This will be called the degraded signal and will be denoted by $g(t)$. The signal should be recorded by the same type of digitizer normally used in the system. In order to implement the Toe Truncation algorithm, the degraded

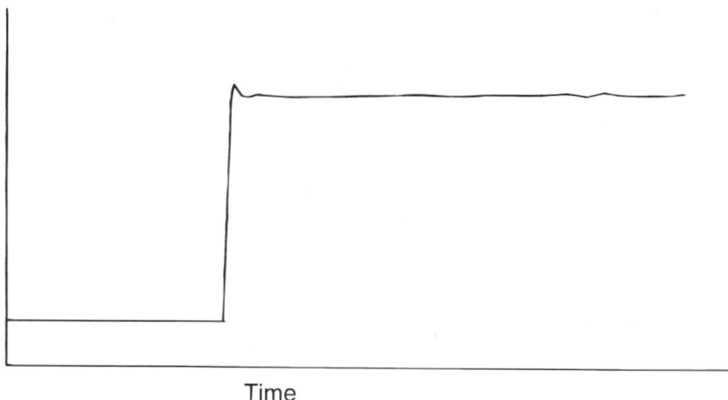

Figure 6.30 Undegraded step pulse signal.

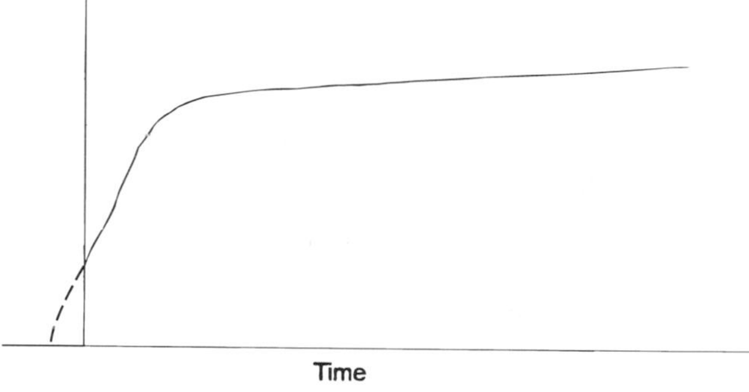

Figure 6.31 Toe truncated degraded step pulse signal.

pulse must have at least 10 points on the risetime. The signal may have to be resampled to produce a sufficient number of samples. The resampling process should be one that maintains the bandwidth of the signal. Also the degraded and undegraded signals must have the same sample interval.

4. The Toe Truncation algorithm consists of truncating the degraded signal by eliminating the leading samples of the record up to some fraction of the final value, e.g., 20%. An example of a truncated degraded response pulse is shown in Figure 6.31. In this case the samples represented by the dotted line before t = 0 are discarded and not used in the deconvolution calculation.
5. Apply the discrete time deconvolution equation to produce a candidate compensation function.

$$h_i^{-1} = \frac{1}{g_1}\left(f_i - \sum_{j=2}^{i} h_{j-i+1} g_j\right) \tag{6.30}$$

6. Apply the candidate compensation function to the original degraded signal to estimate the undegraded signal as:

$$\tilde{f}_i = \sum_{j=1}^{i} h_{j-1+1}^{-1} g_j \tag{6.31}$$

(This is also the equation for compensating real test data.)

7. Observe the result. If the truncation level was set at too low of a fraction of the final value, the resulting signal will be overcompensated. This may well make the estimated undegraded signal extremely noisy. In extreme caes the computation will produce values that exceed the maximum floating point value and "blow up." In less severe overcompensation the estimated signal will have excessive overshoot. If the truncation is set too low, the resulting estimated signal will have too slow of a risetime. The proper or best toe truncation level may be found by iterative trial-and-error. An alternative method is to try all possible truncation levels and compute the risetime and overshoot for each case. The user must then observe a few candidate cases of small risetime and small overshoot to determine the best one. An example of a successful deconvolution using this method is shown in Figure 6.32.
8. The resulting compensation function, h_i, is then convolved with recorded data signals using (6.31) to produce compensated data records. The compensation function should then be tested on a variety of test signals. The test signals should have varying risetimes, pulse widths, and amplitudes.

Potential new users of this deconvolution method are often critical of the subjective trial-and-error nature of the algorithm described above. Based on our research of deconvolution methods and experience, such subjectivity is an explicit or implicit property of all known effective and practical deconvolution processes. There are deconvolution methods that will automatically produce a result that is optimum in some sense, e.g., least squared error. The problems with these methods are that that they are either unacceptable from a human data interpretation standpoint and/or that the methods require invalid assumptions or some other information that is not available.

Nonlinear Component Compensation

A nonlinear amplitude response function is a possible aberration of one or more components in an instrumentation system. The nonlinearity may be caused by the transducer or by some instrumentation system component. For example, a common nonlinearity is gain compression at high signal levels in amplifiers and solid state switching or multiplexing circuits. In the frequency domain, such nonlinearities will generate harmonics. If the frequencies of interest are high enough, the harmonics may be filtered out by the recorder's input circuitry. When recording transient, pulse-like signals the nonlinear effects will appear as amplitude distortions. This type of nonlinear aberration can be compensated for somewhat by characterizing the nonlinear response function and correcting for it in software postprocessing.

One method of characterizing a nonlinear response function is to inject a number of known calibration signal levels and recording the output. Both positive and negative polarity calibration levels should be used if the system is intended to record bipolar signals. When possible, DC calibration signals should be

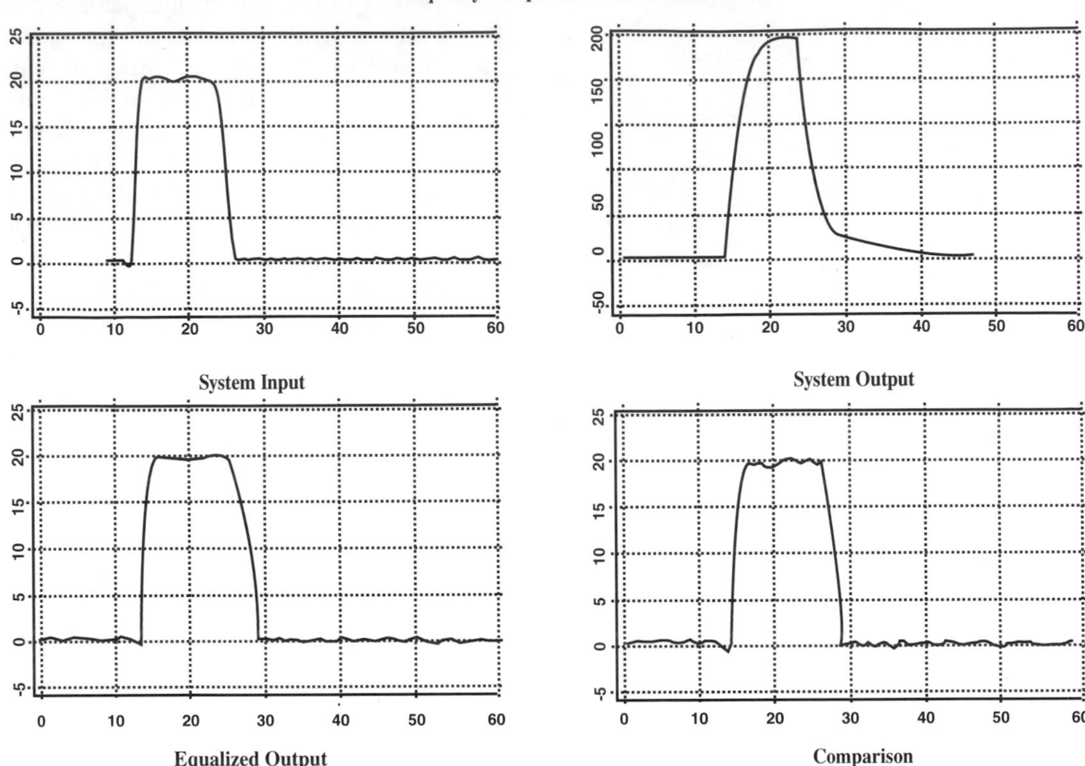

Figure 6.32 Example of deconvolution.

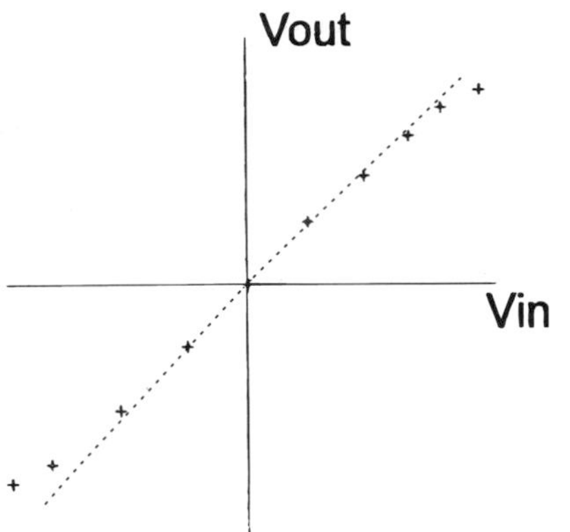

Figure 6.33 Example of nonlinear component response function.

used, because, their levels are generally known better than pulses. However, if the system is AC coupled or if large DC signals may cause thermal problems, then varying pulse levels can be used. If the nonlinearity of the response function is not too severe, then it can be adequately characterized, e.g., to within a few percent, by a relatively small number of calibration levels, e.g., 10. An example is shown in Figure 6.33. To use this coarse characterization to correct digitized data, the response curve must be significantly subsampled to produce output data points that match the resolution of the waveform digitizer. This may be done either by fitting the characterization to a known analytical model or by interpolating the response curve. One effective method for doing this is cubic spline interpolation (Stroud, 1974). Another more *ad hoc* method which has been used effectively is to fit two parabolas through each set of four adjacent points and compute the interpolated value as the average of the two parabola values at each desired point (Bavole, 1970).

6.6 Computers in Instrumentation Systems

William Boyer

This section describes some of the considerations in selecting and designing a computer or distributed network to support an instrumentation system. Modern large instrumentation systems are built using computer programmable hardware. Systems consist of some or all of the following types of programmable hardware: data recorders, instrumentation amplifiers, attenuators, switching matrices, multiplexers, calibration signal generators, trigger and timing signal generators, range timing units, etc. Choosing the proper computer system is critical to the success of an instrumentation system.

The computer system includes both hardware and the operating system software. The hardware includes CPUs, data storage devices, terminals, printers, and networking and communications hardware. The operating software includes the operating system, device drivers, and some or all of the following common utilities: window-manager, editors, compilers, linkers, library generators, networking support, disk maintenance, computer automated software engineering (CASE) tools, performance monitors, etc. The operating software does not include the custom or commercial software related to operating the data recording hardware. This software is covered in Section 6.7.

The two major features of an instrumentation computer system are: it must control and read data from the data recorders and other programmable hardware and provide the data to the facility users and it must provide for maximum efficiency of facility operation. The computer system must provide the some or all of the following specific capabilities:

- Meet specific data acquisition time requirements.
- Run software and provide the hardware interfaces for control and readout of data recording and other ancillary instruments.
- Output data to customers and users in various computer readable formats and media.
- Maintenance of the instrumentation setup databases.
- On-line storage of an appropriate amount of data.
- Off-line storage and retrieval of all relevant archived data.
- User data manipulation at dedicated terminals or workstations.
- Hardcopy plotting.
- Hardware and software interfaces to other facility subsystems.
- Software development and maintenance environment.
- Support other instrumentation system operations as required, e.g., calibration, report generation, etc.
- High reliability and efficient maintainability.

Small, simple instrumentation systems can meet requirements using a single computer. Modern personal computers (PCs) are very powerful and can support up to about 25 input digitized data channels if data throughput requirements are not too stringent. The next level computer used in instrumentation systems is the multitasking workstation class such as made by traditional minicomputer manufacturers, e.g., SUN, DEC, Hewlett Packard, Silicon Graphics, etc. These workstations can have two to three times the computing power of a PC. Even larger systems will require multiple computers configured in a client/server distributed local area network (LAN). Such a system will be different from a classical client/server network in which the server computer node is supporting users at various client nodes. In a distributed instrumentation system the server will act as a master and will control the operation of various slave nodes that in turn are controlling the programmable hardware in the system. The nodes on a LAN communicate among each other via an ethernet hardware link and some higher level message-passing protocol such as TCP/IP, LAT, DECnet, Novell, etc. It is not uncommon to have a network that uses multiple protocols simultaneously.

If the instrumentation system is spread over a geographical area of more than a few miles, it will likely have to be configured as a wide area network (WAN) consisting of a number of LANs. WANs require digital communications devices such as routers, bridges, hubs, concentrators, multiplexers, and modems to efficiently manage message traffic among the LANs.

Designing an efficient client/server instrument control system is a complex process. All of the requirements must be met. The following sections give more details on the key performance requirements.

Data Acquisition Times

Data acquisition time is a major factor in selecting the computer hardware and defining the network architecture. The data acquisition time is composed of the various operations performed by the system:

- Send programmable settings to recording and ancillary equipment.
- Automated programmable equipment checkout and calibration.
- Read raw data from recording instruments.
- Produce fully processed data.
- Display data per customer requirements.
- Transfer data to local mass storage and/or off-line archival storage media.

The definition of fully processed data depends on the customer's requirements. At a minimum it is typically raw recorder data, possibly in counts, converted either to volts output from a gauge or possibly a conversion of volts to physical phenomenon engineering units into the gauge. It may also include other digital signal processing steps such as filtering, integration, parameter estimation, or corrections for nonlinearities.

There will be many tradeoffs that must be considered in designing a network to meet acquisition time requirements. Some typical bottlenecks that must be addressed are

- Data transfer rates between the data recorders and the computer.
- Transferring data among the various nodes in a distributed system.
- CPU power as measured in floating point operations per second (FLOPS) in applying numerically intensive processing algorithms to raw data.
- Transferring data to on-line storage.

A thorough system trade-off analysis is required to make optimum processing allocation decisions. It may be better to do all processing on local controller nodes and then send both raw and processed data to the server for transfer to on-line storage. Or it may be more efficient to send only raw data to the server if the amount of processing is small or if the server has much more CPU power than the controller nodes. Local ethernet network speeds of 10 megabits/sec are common. By going to FDDI speeds of 100 megabits/sec may be realized. Some applications may require dual-ported disk drives where both the server and the clients have direct access to the disk. If a system is to incorporate an interactive data manipulation software package, this package must also operate at good ergonomic speeds.

Instrument Control Interfaces

The computer system must have the proper interfaces to control the data recording and other supporting programmable hardware. There are two fundamentally different kinds of programmable hardware: self-contained, standalone "rack and stack" devices and modular devices that plug into a crate and rely on the crate for power supplies and the digital data transfer bus.

Currently one of the most common interfaces for programmable equipment is IEEE 488.2 general-purpose instrumentation bus (GPIB). This is a bit-parallel, byte-serial interface. This means that data is transferred one 8-bit byte at a time. GPIB uses an asynchronous hand-shaking protocol to transfer data. Up to 15 devices including the controller board in the computer may be put on each bus. Thus, a large system will have many busses and controllers. Devices from different manufacturers may be put on the same bus. GPIB can support data transfer rates up to about 600 Kbytes/sec. However, the asynchronous hand-shaking protocol means that data transfer speed is determined by the slowest device on the bus. It is not uncommon to find that actual data transfer rates are well under 100 Kbytes/sec. GPIB is used both for standalone and crate/module programmable instrumentation devices. The main advantage of GPIB is that it is supported by a wide variety of equipment. There are some disadvantages. Data transfer rates can be very slow. Using the standard cables the total cable length should not exceed 50 ft. (The distance limitations may be overcome using fiber optic bus extenders.) The standard GPIB cable is relatively expensive, unwieldy to install, and prone to failure.

There is also a relatively new standard instrumentation interface known as VXI (VME extensions for instrumentation). VXI is a crate system. There are various size crates. The most common size is size C. A C-size crate can support up to 12 modules that measure approximately 1″ wide by 10″ highly by 14″ deep. Each crate must be interfaced to a computer via a controller module that resides in Slot 0. Common controllers used are GPIB, MXI bus (multisystem extension interface), and embedded PCs packaged in a module. In the case of a PC controller, an ethernet protocol can then be used to communicate with a server node. Although VXI is several years old, the introduction of VXI compatible instruments has been relatively slow.

A related crate-based instrumentation system is VME. There are not many instrumentation system modules available in the VME form factor. VME is mainly used for purely digital applications. The VXI standard is a superset of VME designed especially for the analog instrumentation world.

Another crate-based instrumentation standard is CAMAC. A CAMAC crate will hold up to 24 instrument modules plus a controller module. The crate slots are only about 0.6″ wide, and thus many modules require 2 or more slots. CAMAC has been used since the mid sixties. Despite its age there are many types of CAMAC instrumentation modules available. Common types of CAMAC controllers are custom parallel interfaces, GPIB, various serial modules, and a PC or workstation CPU packaged in a module.

Operating System Software

The operating system software is another critical feature of any computer or network. In many cases, the specific operating system, and thus its capabilities, will be determined when the computer hardware is selected. There are some options in the case of PCs, e.g., MS-DOS/Windows, Windows NT, OS-2, or Windows 95. MS DOS/Windows is most widely used and is best supported by third-party software vendors. However, the other three do provide multitasking capabilities. Most multitasking workstations today use some type of UNIX operating system. This is convenient for users since they only have to learn one operating system command set for use on several different brands of computers. There are also many common system management features among different UNIX implementations. However, the latest version and resulting support by third-party software differs greatly among the various manufacturers at any given time. The most important features of the computer operating software for a large distributed system are

- Multiple simultaneous tasks in a window environment.
- Strong file transfer and remote task execution support for a mature networking protocol.
- Minimal number of system parameters to tune.
- Strong set of utilities for file management, software development and maintenance, and office, automation word processor, database, spreadsheet, forms management, desktop publishing, etc.
- Mechanism for limiting access to files and system privileges.
- Multiuser and multiprogramming operating environment.

In addition to performance concerns, the computer system designer must take into account the long-term system management tasks. The designer should avoid mixing different operating systems in the same computer network. It is even undesirable to mix different versions of the same operating system such as UNIX. The reason for this is that different manufacturers will be at different versions of the operating system. The computer network will inevitably contain a significant amount of third-party software which will also be continuously upgraded with new versions. It is a major challenge to find a version of even one operating system that is compatible with a wide variety of third-party software. Upgrading commercial software is also very system manager intensive. One should expect that many hours, days or even weeks of debugging will be required any time any software version upgrade is made. Nor is it viable to adopt an "If it isn't broke, don't fix it" attitude and ignore the upgrades. Supplier support for old software decays rapidly after the introduction of new versions. Customer support personnel adopt an attitude of "Install the latest version, then call me back" whenever a user calls with a problem about old software. System managers need to exercise judgment in upgrading commercial software. If a new version, e.g., an upgrade from a Rev. 4.7 to 5.0 contains many errors, wait for Rev. 5.0.1 to avoid dissatisfied users and other problems associated with software deficient in design and testing.

In the era of single-user, single-tasking PCs running MS-DOS, there is a question of whether a multitasking operating system is really needed. The ability to perform many tasks simultaneously will greatly enhance the instrumentation system's operational efficiency. For example, recording system operators can run calibrations, update the database, generate reports, retrieve and archive data, etc., simultaneously. Customers can use this ability simultaneously to run batch data analysis macros, look at data interactively, and develop and debug new macros. Programmers can simultaneously edit a program while watching a debugger or program display and also be running compiles and links in the background.

Access to files and system privileges should be limited in such a way that eliminates or minimizes the chance that a user can inadvertently destroy a critical file. For example customers should not have delete privileges for master data files or for production software files. The protection system must be designed so that all users can perform their jobs as efficiently as possible.

Data Storage

The computer must have sufficient on-line storage for the following operations:

- Operating system including scratch, swap and spool files.
- Instrumentation database and control files.
- Instrument control and data analysis applications software elements: source, executables, run macros, libraries, CASE files, documentation, help files, etc.
- User specified amount of recorded data including: raw data, processed data, log files, calibration data.
- Utility software including adequate storage for user generated input and output files.
- Adequate storage to support software development and maintenance.

Access time for on-line data storage must be consistent with data acquisition time requirements. Magnetic disks remain the best choice for most applications. Despite predictions that other technologies such as optical storage will eventually overtake magnetic storage, magnetic storage packing density, effective access time, and cost/megabyte are all continuing to improve dramatically. Any new computer should be purchased with as much magnetic storage capacity as technology and cost will allow.

The computer system may be required to have the capability of storing and retrieving archived data off-line. The off-line data storage must utilize a media that minimizes storage volume and uses a proven long-term data retention technique. Some applications may require off-line storage for decades. Nine-track magnetic tape has historically provided the best interchangeability among different and upgraded computer systems. However the physical packing density is relatively poor and the drives are expensive. The best packing densities are currently available on optical disks. Archiving data to a CD-ROM is becoming a viable option.

Even though extremely long data retention times are possible on archival media, one must be concerned with the ability of newer computer systems to support the disk and tape drives required to read older media. No matter what medium is chosen, it is wise to allocate resources for periodic archive media conversions as technology advances.

Reliability and Maintenance

Reliability requirements will be different for different applications. Some applications may be so critical that it is unacceptable for the computer system to go down under any circumstances. Such systems will have to be designed using fully redundant auto-fail-over hardware components. At a lesser level of reliability, the instrumentation computer system may be required to have on-hand sufficient spare equipment such that no failure of any single component in the system will prevent the system from collecting data for a specified time period. Computer system performance may be critical enough that maintenance for all computing hardware must be available on-site with a maximum response time on the order of hours.

The computer system may have stringent cooling and humidity environment requirements. If the facility environment is not compatible with common commercially available computer hardware, it may be required to house the computer system as well as the rest of the instrumentation system in its own room with its own dedicated air conditioning and humidity control systems. In extremely high electromagnetic noise environment cases it may be necessary to house the instrumentation electronics in an RF-shielded enclosure.

When applicable, the computer system should be designed with an expandable architecture to support more data recorders with longer memories. Such expansion should not increase the data acquisition time.

6.7 Software for Instrumentation Systems

William Boyer

Modern instrumentation systems that utilize digitizing data recorders and control computers also require a significant amount of software to run the operation. This section describes some of the considerations and requirements in selecting or developing the control software for an instrumentation system. Requirements are given for the following types of system software:

- Automated data acquisition.
- Instrumentation database maintenance.
- Troubleshooting and calibration support.
- Data reduction and manipulation.
- Data file formats, storage and organization.

There are two major options for obtaining software for an instrumentation system. The system designer can either purchase a turnkey off-the-shelf package or develop a custom software system for the application. The first option is most viable when all of the programmable instrumentation equipment is purchased from the same vendor. Vendors often sell software packages that will provide most necessary features. These turnkey programs typically run on a personal computer (PC). They may not meet all of the performance requirements suggested in the following sections.

The second option is to develop custom software. Again there are two types of development efforts. First, for small systems, there are some high-level languages that make software development relatively easy. One popular method uses graphical languages, i.e., all programming is done by selecting and connecting icons that represent normal program steps. The developer builds up a hierarchy of screens in the course of the development process. Software can be developed very efficiently using these graphical languages for systems that have small numbers of many different types and brands of instruments. Various versions of enhanced Basic with graphical user interface enhancements are also good candidate high level languages for small systems.

The high-level languages described above are currently not well suited for large, complex systems. Graphical language programs become very difficult to track once more than four or five hierarchical layers have been created. They do not have the richness of programming constructs that would allow a software engineer to develop a large complicated system that would be feasible to maintain in the long run. Object oriented design (OOD) software engineering methods are particularly effective for developing large-complex applications. C++ is a commonly used programming language for implementing OOD software.

Typically there will be heavy usage of commercial software even where custom software is required for an instrumentation system. It is often more cost-effective to use commercial software for data bus controller drivers, plotting and data analysis, the instrumentation setup database, general-purpose utilities, etc.

Ease of long-term maintenance is a critical factor in designing a custom software system. A large instrumentation system normally will undergo many modifications and upgrades in the course of its lifetime. The software should be designed to provide for such modifications with minimal effort. In particular, the software should be designed to support the addition of new programmable signal conditioning equipment, new data recorders with more bits of resolution, longer memories, more channels per unit and other desirable features, new computer components, new signal processing algorithms, etc. Such expansion is often extremely difficult with turnkey software packages. In addition to a good design technique, it is also important for maintainability to use programming languages, operating systems, and supporting third party software with a long history of acceptance and popularity. There may be some short term advantages in using inexpensive but obscure, unproven products. In the long term, lack of support and lack of programmer knowledge of such products can make a software package unmaintainable.

Earlier it was noted that stringent data acquisition times for a large instrumentation system may require a client/server or master/slave computer system architecture. Such an architecture will greatly impact design and implementation of the software. This type of an architecture will almost certainly require that custom software be developed.

The key features of the major software elements in an instrumentation system are described in more detail in the following sections.

Automated Data Acquisition

The major software element in an instrumentation system will be the software to automate the acquisition of data. This software package should automate as much as possible the processes of setting up and checking out the recording system and then acquiring and processing data. Since human and hardware errors will inevitably occur, the software should also incorporate sophisticated error checking and recovery capabilities. This software should provide the following basic functions:

- Read instrument setup information from a file or database which we will call the instrumentation database. The desirable properties of the instrumentation database are described later in this section.
- Initialize all programmable instruments and download and verify by readback when possible all programmable settings.
- Where applicable verify system performance by routing test or calibration signals to the data recorders, acquiring the test data, reading this data, and checking as many performance features as possible for all of the data channels.
- Enable the system to begin collecting data.
- Read data from recorders. Store this "raw" data to the on-line and/or archive storage media.
- Process raw data as required by the customer to produce fully processed data. Store the processed data to the on-line and/or off-line archive storage media.
- If required, produce "Quick-Look" and/or final production plots of the raw data.
- Provide for extensive error checking and informative messages especially for hardware communications, file access, and network access problems. Write status and all error and diagnostic messages to a log file(s).

In addition to the automated data collection capabilities described above, the software system should also provide interactive troubleshooting capabilities to resolve any problems detected during the system operation. Some additional comments on the features outlined above are given below.

Performance Verification Testing

The performance verification test is an important step in recording high-quality data on large systems, especially when the test or process being monitored is expensive or provides a critical function.

Everything that can possibly be checked should be checked in the recorded test signal. If no test signal is applied the software should still acquire baseline records and check such features as the baseline level, baseline noise, and any time alignment information such as a time mark pulse. If a test or calibration signal is applied, the software should check for proper signal amplitude, time of occurrence, risetime, duration, period, etc.

The output data from the hardware is called raw data. This data is not ready for users to view and interpret. Software converts raw data to fully processed data. In some systems this can involve a significant amount of signal processing software. In order to provide a full-service capability to customers it is often desirable for the instrumentation software to convert raw recorded data into datasets that represent the time histories of the physical phenomena that excited the various transducers as accurately as possible. It is also desirable to have all of the data channels properly aligned in time.

To accomplish the above the data processing software should perform some or all of the following functions:

- Convert raw recorder data from binary counts to voltage input to the recorder using one of the following volts/count conversion factor methods: nominal based on selected recorder sensitivity, generated from internal recorder calibration data, generated by using an external calibration signal.
- Correct for any known systematic sampling time errors, e.g., inaccurate or varying sampling clock.
- Place data at correct time of occurrence using some combination of the following: IRIG time stamps, signal and trigger cable delays, trigger generator delay times, and time mark pulses.
- Apply any required filtering. This might be a band stop or low pass filter to help eliminate noise. In very wideband systems it is often necessary to perform a deconvolution process to compensate the recording system frequency response for high-frequency cable or amplifier losses. Deconvolution and filtering methods are described in Section 6.5.
- Apply corrections for other data transmission and signal conditioning gains and attenuations. These are generally linear, DC scalar operations. However it may be required to compensate for some nonlinear signal transformation. In order to compensate for a nonlinear transformation, the function must be characterized and software written to invert the process. It may also be desirable to compensate for transducer conversion factors at this stage. Methods for nonlinear corrections are discussed in Section 6.5.

The output from this stage should be in engineering units or volts in a floating point file. The data at this stage will be referred to as "fully processed data". It may be desirable to utilize data compression techniques to reduce storage requirements for processed data files.

Troubleshooting Mode

Large data recording systems, e.g., >100 channels, will often encounter errors in setting up and reading out instruments. This is especially true for systems using state-of-the-art data recorders. Robust, easy-to-use, intelligent troubleshooting software is very important for efficient facility operation. This troubleshooting mode will also be essential for initially setting up the recording system for each new activity. The major requirements of the troubleshooting mode are outlined below:

- Interactive control of all instruments including the capability of sending commands, interogating settings, and reading data. Interactive control should be implemented by a well-designed graphical user interface employing pop-up menus.
- Ability to overwrite settings in the instrumentation setup database.
- Ability to plot data in raw or processed format.
- Incorporate all interactive signal processing capabilities described below.

Data Manipulation and Plotting Software

The instrumentation system should include software that provides both interactive and batch mode data manipulation and plotting capabilities. The following data manipulation capabilities are required:

- Arithmetic operations on a single waveform, e.g., integrate, differentiate, Fourier transform, exponentiate, time shift, time and amplitude truncation, resample, filtering, logarithm, dB. Also provide add, subtract, multiply, or divide by a constant.
- Arithmetic operations on two waveforms, e.g., add, subtract, divide, multiply, convolve, deconvolve. The software should automatically place the two waveforms on a common time base with a common duration before performing the requested operation. The software should support these operations on complex number type waveforms, e.g., when calculating in the frequency domain.
- Measure pulse parameters such as start time, risetime, baseline, topline, amplitude, width, baseline noise, topline noise, etc.
- Perform statistical analyses (max., min., average, median, standard deviation) of measured pulse parameters over multiple waveforms from the same test and multiple waveforms from several tests.
- A macro capability should also be provided.

Plotting software is required for the following:

- Plot raw data up to the full digitizer record length in either counts, or volts.
- Include legend data such as test name and number, important data recorder settings, calibration information, and pulse parameter measurements.

- Provide for interactive panning and zooming in both the horizontal and vertical direction.
- Provide for multiple overlayed waveforms with different curves differentiated with colors and line styles.
- Proivde for multiple plots per page.
- Provide for user annotation of plots.
- Selective hardcopy of interactively generated plots.
- User selection of axis extent and size, graticule types, plot colors, fonts, and character sizes.
- To minimize plotting time, data arrays should be "thinned" by a appropriate min/max algorithm to a resolution compatible with the plotting device.

Data Files and Organization

The software system should have an intuitive naming mechanism for files and directories associated with experiments. The file and directory naming should support multiple types of experiments or tests. It should be easy to the point of being obvious how to locate a given file for a given experiment. The file system should be designed to make maximum use of operating system file manipulation commands, particularly wild cards. The raw and processed data files should either contain all instrumentation database information in local headers; or, perhaps more desirable, the files should contain pointers to the setup information in the database itself.

Instrumentation Database

A properly designed instrumentation database and support software are essential to efficient and flexible operation of a large complex data recording system. The instrumentation database should contain the following minimum information:

- All programmable settings for all programmable instruments to be used on a given test including waveform recorders, trigger generators, time mark generators, time interval digitizers, automatic calibration and test signal generators and coax switches.
- Full description of the control path for every programmable device *in the entire system* including GPIB bus number, device address, subchannel address, VXI/VME/CAMAC slot addresses, coax switch signal routing information, switch relay driver addresses, etc.
- Signal processing options for production waveform data which include signal conditioning attenuations, cable resistances, time/recorder memory interval of interest, software frequency response compensation filter, quick-look plotting priority, time alignment information, and information for converting from gage output volts to input engineering units.
- Cable lengths, trigger delays, time marker delays, time interval digitizer channel to be used for time alignment calculations.
- Test identification information including experimenter name, signal identification, plot labels, etc.

The instrumentation database maintenance software should have the following attributes:

- Specialized screens for text data entry.
- Check validity and consistency of as many data entries as possible.
- Ability to recall old test setup configurations from the data archive.
- Provide a variety of user-defined-format database reports.
- Graphical user interface easily interpretable symbols and pop-up menus for data entry.

Operating System and Support Utility Software Requirements

In addition to the applications software used for data acquisition and processing, the system designer must pay close attention to the software capabilities of the computer operating system chosen to run the facility. Some important features are outlined below:

- Multiuser and multiprogramming support.
- Support for as much *directly addressable* memory as feasible by using either virtual and/or physical memory. The system should not require that large programs be broken up into overlays or that large data arrays be treated specially to fit within directly addressable memory.
- Strong support for a mature networking protocol. Network support should include file transfer, remote task execution, mailbox, etc.
- Minimal amount of tuning system parameters to make the system work.
- Multiple simultaneous tasks in a window environment.
- Compilers, linkers, smart editors, and symbolic debuggers for the programming languages to be used. Programming support should also include a code management system and other computer aided software engineering (CASE) tools.

6.8 Calibration and Testing

Richard Pettit

Metrology is the study of the science of measurements. A measurable quantity is an attribute of a phenomenon, body or substance that may be distinguished qualitatively and determined quantitatively. Electrical examples of measurable quantities are current, voltage, resistance, capacitance, inductance, and power. A unit is a reference or standard quantity with which an unknown quantity is compared to determine its numerical value. Examples of units are the ampere, the volt, the ohm, the farad, the henry and the watt. The value of a quantity is expressed as a numerical value times a unit. The international system of units (SI) is now almost universally used in science and technology for expressing the values of quantities in experiments and manufacturing processes. The SI units are the second (s), meter (m), kilogram (kg), ampere (A), kelvin (K), mole (mol), and candela (cd) along with units derived from them. An example of a derived unit is the ampere second (A s) used to express the quantity electric charge. Measurement is a set of operations that determine the numerical value of an quantity with respect to a system of units. The measurand is the particular quantity whose value is determined by the measurement operations. The indicated value of the measurement process is the value attributed to the measurand by the measurement process. A calibration of a measuring instrument or system is a set of operation that establish, under specified conditions, the relationship between indicated values of measurands to the SI units. The most important task of national metrology institutes such as the National Institute of Standards and Technology (NIST) in the United States is realization, maintenance, and dissemination of the SI units.

Modeling

The initiation of a measurement system design usually starts from either one of two extremes. In one situation, the measurement system designer is given the job of instrumenting a well characterized process, which means that all the required parameters and their accuracy levels are well defined. Thus the system designer has explicit measurement system requirements that need to be translated into the characteristics of the transducers and data recording system. The uncertainty analysis of the measurement system must ensure that the accuracy meets the needs of the system requirements. In fact, the measurement system can be more accurate than the system requirements. If the system accuracy requirements are not state-of-the-art, the system designer may have a variety of different transducers and data recording systems to choose from, and will base his/her decisions on cost, implementation time, and use environment constraints. However, if the system requirements are state-of-the-art, then the judgment of the system designer becomes critical in the selection of the measurement system, since the overall operating characteristics of the transducers and recording system must be well matched to the experimental use environment, costs, time frames, etc. in order to meet requirements.

The second situation involves a system designer who must record a series of outputs for a given process by controlling and measuring a series of input parameters. While the range and control accuracy of the output parameters may be clearly specified, in many cases a model is developed which relates the input to output parameters (sensitivity coefficients) in order to define the required range and measurement accuracy of the input and output parameters. An example of this situation would be an electroplating operation with the output being an electroplated coating with a specific composition, thickness, density, and hardness, but the control parameters are the bath chemical composition, plating time, current density, and bath temperature. The system designers job of selecting the transducers which will measure the input parameters is more difficult since the sensitivity of the output to the controllable input parameters is not known. In some cases, the resulting accuracy requirements for the input parameters cannot be determined until the system is in place and sensitivity coefficients are measured experimentally. Many times the system designer develops the "best" measurement system that is currently available, based on his/her judgment. After the system is designed and modeled, a detailed uncertainty analysis can be performed in order to quantify the overall uncertainty in the output parameters. When the analysis is completed and all significant sources of uncertainty are identified, the resulting system control can be judged as being either adequate or in need of improvement. With the analysis and model completed, it is straightforward to determine the areas where the greatest improvement can be realized and the system redesigned in order to improve the final results.

Interpreting Specifications

Published specifications for transducers and instruments vary widely and can be difficult to interpret and understand. For example, a digital voltmeter specification usually has the following form: When used on the 11.0000 volt range (full scale), the 1 year accuracy of the meter is ±(0.025% of reading +0.005% of full scale) for operating temperatures from +18°C to +28°C. One must understand the uncertainty. Is the uncertainty worst case or statistical? If the uncertainty is statistical, it probably corresponds to some number of standard deviations of the measurements from some experiment. If so, is the data measured on every instrument or is it based on some set of nominal or typical instruments? Based on the above specification, the uncertainty at a voltage reading of 5.0000 volts is ±(5.000*0.00025 + 11.0000*0.00005) volts = ±(0.00125 + 0.00055) volts = ± 0.0018 volts (or 0.036%). (This is, of course, assuming that the voltmeter is within 1 year of its last calibration.) Furthermore, the temperature coefficient is stated as (0.001% of reading + 0.0015% of full scale)/°C for temperatures from 0 to 18°C and 28 to 50°C. Thus for a measurement of 5.0000 volts at 31°C (or 3°C above the normal specification limits), an additional uncertainty of (0.00001*5.0000 + 11.0000*0.000015)volts = (0.00005 + 0.000165) volts = 0.000215 volts must be added, to give a final uncertainty in the 5.0000 volt reading of ±(0.00180 + 0.000215) volts = ±0.002015 volts (or 0.0403%).

Transducers can have their own specification format, such as: typical accuracy of the pressure transducer is ±0.2% of full scale for 1 year for readings from 20% to 100% of full scale and do not require temperature compensation if maintained in the temperature range from 0 to 50°C. Thus, for a pressure reading of 15 psi from a gage with 0–30 psi range, the accuracy is ±(30*0.002) psi = ±0.06 psi over 1 year. If properly accounted for, the accuracy statement includes the uncertainty of the standards used to calibrate the pressure transducer, uncertainties in the resulting interpolation curve used to fit the resulting data (assuming that the transducer is not completely linear over the 20% to 100% range), and drift with time over a 1 year time period. In addition, it may be important to zero the transducer periodically by applying a zero pressure condition (through evacuation of the transducer, for example). In addition, while these specifications are stated as typical, individual calibration curves may be obtained from the calibration laboratory that would allow for a better accuracy.

Finally, it may be possible to calibrate the transducer and readout instrument as a matched pair so that when they are used together, the overall measurement accuracy can be improved. However, in this case, if one device should malfunction, a replacement transducer or readout instrument cannot be substituted and maintain the same accuracy. In many cases, the transducer and readout device are purchased as an integral unit and therefore must be both calibrated and used together. Sometimes the readout device may have a computer fit to the transducer output that corrects for nonlinearities in the transducer output with input quantity. By calibrating the device as a unit, the combined effects of the transducer/readout system are included in the calibration.

Selection of Transducer

The selection of a transducer is driven by many requirements, including the cost, range, stability (both over time and in the use environment), size, output, and required accuracy of the measurement. Typically, the transducer converts measurement units (pressure, acceleration, temperature, etc.) into an electrical signal. Therefore the relationship between the measured quantity and volts is critical to obtaining accurate data. While the manufacturer may provide transducer data that are typical of a type of transducer, individual transducers may show significant variations. Therefore the calibration of each transducer is critical to obtaining high-quality, meaningful data.

The accuracy specification of a transducer, unless defined in detail by the manufacturer, may have one of several meanings. For example, it may specify the uncertainty of the transducer that can only be achieved after the transducer is calibrated. In addition, it may relate to the best accuracy which can only be obtained by using the transducer in a well controlled, laboratory environment; the accuracy of the transducer may be degraded in the actual application environment. For example, the transducer may have a significant temperature dependence, so that the calibration value determined by the calibration laboratory at 23°C may change by 5% in the use environment of 18°C. This can be a real problem in actual testing environment where the transducer is cycled over a wide temperature, pressure, vibration, etc. range. Other factors affecting the accuracy of the transducer in the use environment must be understood, such as preconditioning of the transducer (some transducers provide different readings if exercised from zero to full scale at least once as opposed to never cycling before readings are obtained); effects of different liquids or gases; static versus dynamic conditions; calibration factors valid for only part of the range (e.g., 10% to full scale) etc.

Finally, the selection of the electronic readout instrumentation has to be considered in the overall accuracy requirements. Parameters such as grounding, thermal EMFs, frequency response, environmental sensitivity, drift and/or changes over the calibration interval, etc. must be considered.

Uncertainty Analysis

The purpose of this section is to discuss the process involved in an uncertainty analysis, to show the importance of these analyses, to give basic guidelines for their application and to present specific examples. There are four important prerequisites that must be fulfilled before the detailed uncertainties associated with a specific measurement can be determined:

- All instruments and transducers involved in the tests must be calibrated.
- The instrument and transducer calibrations must be valid during the period of time covering the tests.
- The calibrations must have traceability to national standards.
- The details of the testing techniques must be well understood (e.g., temperature effects, modeling assumptions and limitations, material properties, etc.).

Uncertainties have their basis in science and physics of the measurement and can be quantified. Uncertainty analyses do not assess the influence of human blunders, mistakes, or errors in the measurement process. Blunders, mistakes and errors by humans are nearly impossible to quantify. In analyzing the uncertainty of a measurement process for a particular physical phenomenon, one assumes that the process is followed without a blunder or mistake and that the process is appropriate for measuring the physical phenomenon. Metrologists prefer to use the term "uncertainty" to the term "error" since in general the error of a measurement is unknown because the true value of the standard used is unknown. However the uncertainty on the result of a measurement can be evaluated. On the other hand, if a measurement system is compared with a known reference standard using a procedure with negligible uncertainty, then the comparison may be viewed as determining the error or offset of the measurement system.

The quality of the data obtained from a test is directly related to the uncertainties involved in each part of the test. Without an uncertainty assigned to each measured quantity, the significance of the results are not known. With an increasing emphasis on obtaining better and more detailed results, increasing emphasis is placed on the quality of the data. Improvement in the uncertainty analysis is demanded by the use of more sophisticated transducers and measurement instruments, improvements in analytical models, the correlation of experimental results with real situations, and improvements in the basic measurement uncertainty of data systems, instruments, and transducers.

The starting point of an uncertainty analysis is the assigned uncertainty to the instruments and transducers through a calibration process. One basic aspect of this calibration is the traceability to accepted national measurement systems or standards, e.g., usually to those maintained by the National Institute of Standards and Technology (NIST), formerly the National Bureau of Standards (NBS). Thus the uncertainty of a calibration and its traceability are not independent or separate issues. As described in Belanger (1980), the following definition is given:

> "Measurements have traceability to designated standards if and only if scientifically rigorous evidence is produced on a continuing basis to show that the measurement process is producing measurement results (data) for which the total measurement uncertainty relative to national or other designated standards is quantified."

Thus measurements are traceable, not because of an unbroken chain of calibrations, but because the uncertainty or quality of the measurements at each step in the calibration process are scientifically defendable. The purpose of traceability is to ensure that measured quantities are "true" or they can be reproduced at different times, conditions, and places.

An uncertainty analysis cannot be described as a set of rules that are followed, since there are far too many conditions, instruments, and influence factors involved in each measurement. However, general guidelines and approaches can be discussed that allow a uniform framework for approaching an uncertainty analysis and for reporting the results of the analysis.

Traditionally, uncertainties have been divided into two groups: random and systematic. However, recently it has been found that the random/systematic formulation of uncertainties leads to confusion concerning classifying uncertainties and to situations where a source of uncertainty changes from random to systematic depending upon the viewpoint of the experimenter. For these and other reasons, recent international and national metrology experts have adopted the recommendation of dividing uncertainties into two new classifications called by the provisional names Type A and Type B. Type A represents uncertainties that can be evaluated using statistical techniques developed for random variables. Type B represents uncertainties that are evaluated using any other objective method. Starting in January 1994, the NIST has adopted this method of developing and expressing uncertainties in all of their calibration reports.

A Type A uncertainty is one evaluated using a statistical approach for a measurement system that is in statistical control. Type A uncertainty component should be reported as an estimated standard deviation (or variance) along with the number of degrees of freedom. For correlated uncertainties, an estimated covariance should also be given.

A Type B uncertainty is determined mostly by the judgment of the experimenter. Thus all factors that may affect the measured quantity are identified. Next the range or some other measure of the factors dispersion must be estimated. This dispersion can be estimated from theoretical arguments, from model calculations, from published results of others, or from direct experimental measurements. For each source of Type B uncertainty, the range must be converted into an estimated effective standard deviation for the factor. This requires the experimenter to decide on a probability distribution for the factor over the range in order to convert the range to an estimated standard deviation. If the range represents a 95% probability interval, for example, then the standard deviation is estimated an one-half the range. Another frequent choice is to assume that the expected values are uniformly distributed over the range, so that the standard deviation is given by the range divided by $\sqrt{3}$. This estimate is a common choice and is given special emphasis.

Once the Type A and Type B uncertainties are expressed as standard deviation, the final combined uncertainty is given by the square root of the sum of the standard deviations squared. Typically the reported uncertainty is multiplied by a factor K in order to convert the combined uncertainty into a probability interval which contains the true value with some probability. Thus, if the overall variance is well defined and the probability distribution is approximately Gaussian, then using $K = 2$ gives a 95% probability that the interval will contain the true value. Using a value $K = 3$ gives a 99% probability that the interval will contain the true value. For uniformity, NIST has chosen to report the overall uncertainty using a factor $K = 2$.

Sources of Uncertainty

The following list represents some sources to be considered when assigning uncertainties: instrument and transducers, measurement process, and statistical technique used.

The uncertainties associated with the measuring instrument and transducer are very important. An instrument's or transducer's uncertainty should be taken from its calibration certificate or report and should include corrections for the particular environment or use conditions for the equipment.

Factors to be considered as part of the measurement process include uncertainties or limitations in the analytical model used for data analysis; the use environment/condition that may be different from the calibration conditions and their effect on uncertainties; corrections and the uncertainty in the correction factor; limitations of the readout device; and uncertainties in interpolations or extrapolations of the calibration or measurement data.

Factors to be considered for statistical techniques include the validity of statistical techniques applied to analyze the data; anomalies and outliers should be eliminated only with careful considerations; resolution of measuring instruments; elimination of all controllable sources of variation in the measurement.

To ensure that a measurement process is in control, various comparison techniques can be used. These techniques identify any shift in the measurement equipment, indicate long-term drift, and quantify long-term data scatter. Useful comparison techniques include comparison with previous measurements; comparison of the results of two similar measurement systems; comparison with other measurement techniques on the same quantities; comparison with results from other laboratories; and comparison with data taken before and after the test to determine the effects of transportation, handling, or exposure to harmful environments.

There are many aspects of the use conditions that need to be understood, known, and/or estimated. These include an adequate description of the use conditions; a determination of changes in calibrations parameters to correct for use conditions; imposing restrictions on the use of instruments to limit uncertainties in the results; deciding to make corrections, combine uncertainties or imposing other analysis as needed.

A time period should be assigned after which the measurements and the associated uncertainty analyses are no longer valid. This can be determined from known drifts in the instruments or equipment, or from comparison measurements performed to detect changes in the measurement system.

The process of performing a measurement and evaluating the associated uncertainty involves considerable experience, judgment, and knowledge of the interaction of various parameters. Issues where judgment are required include the cost effectiveness of improving the measurement capabilities to reduce the associated uncertainties; estimates of the effect of wear, handling, or other conditions on the measurements; effects of the environment including temperature, humidity, vibration, corrosion, contamination, electromagnetic fields, etc.; variation in the test conditions or equipment over time.

At times results contain anomalies or data which lie outside the expected uncertainty bounds. These data need to be evaluated carefully in order to determine whether the results reflect problems in the measurement system, the material or system under study, or represent new, unexpected results. For example, a transducer calibrated in the vertical position may respond in a different way when mounted horizontally.

Value of an Uncertainty Analysis

An uncertainty analysis of a measurement system is essential for determining the validity of the data obtained. Often overlooked or postponed, the uncertainty analysis provides vital information on the operation, performance, and design of the measurement system. An uncertainty analysis points out the most critical parameters and instruments affecting the resulting data. Thus the analysis aids in optimizing or upgrading the measurement system in order to improve the results. Improvements in the system can be aimed at the equipment or transducers which contribute most to the final measurement uncertainty. When any changes are made to a measurement system or its environment, an updated uncertainty analysis can be easily determined by following the previous analysis. In documenting an uncertainty analysis, the governing equations and analytic approach used for determining the overall uncertainty must be included.

All derivations should be included and each parameter included in the final uncertainty value should be described. Factors considered insignificant should also be mentioned and the reasons for not including these factors discussed. All statistical techniques for calculation of Type A uncertainties should be documented and the combination procedures explained.

System Calibration

Manufacturers of components and test equipment publish data sheets to describe their products. The specification section of the data sheet lists the performance parameters. At least two types of performance parameters are listed: warranted or guaranteed, and typical or nominal. The service manual for test equipment contains the functional tests as well as performance tests. Functional tests verify that the internal hardware and circuits are operating as designed. Performances tests check the warranted performance parameters with calibrated standards and test equipment. The service manual does not necessarily contain a test procedure for each warranted parameter. For example, the input impedance on a Tektronix TDS 684A digitizing oscilloscope is warranted, but there is no procedure for verifying that this parameter meets its specifications in its service manual. Service manuals also contain descriptions of adjustments to perform on instruments that fail one or more performance test. The service manual suggests a calibration interval, which is how often the performance tests should be checked.

Calibration of a measurement system as opposed to a component or instrument requires some thought. The individual elements of the system can be calibrated separately and the calibration data can be combined into a statement about the calibration of the entire system. Calibration of an entire measurement system from a transducer to the data display is referred to as end-to-end calibration. This calibration does not replace the evaluation or calibration or individual components of the measurement system. A true end-to-end calibration occurs when a known artifact standard is measured with the system. All other procedures calibrate the measurement system without exercising the transducer. Such procedures cannot disclose problems with the transducer. Also, the transducer's impact on the total uncertainty of the measurement system must be assessed separately.

The minimum calibration for a measurement system should include a determination of its DC gain, offset and linearity. Its amplitude-frequency characteristic should be determined at two different input levels (e.g., 20 percent and 80 percent of full scale). To verify system linearity, the two resultant amplitude-frequency data sets, when normalized, should superpose one on top of the other. Additional points can be taken, either at 0 Hz or at some other frequency in a frequency interval where the system's amplitude-frequency response is unchanging, to further characterize linearity. The slope of the resultant input versus output line at this particular frequency is the system sensitivity. Linearity is a requirement of all measurement systems intended to record dynamic signals. Nonlinear measurement systems distort data by creating frequencies in their output signals not present in their input signals.

When a physical stimulus to the transducer cannot be applied immediately before test, it is still possible to simulate the transducer's output signal to the remainder of the measurement system. Following is one method to accomplish this simulation for potentiometric, AC and DC bridge, and piezoelectric transducers respectively.

A voltage insertion technique for potentiometric transducers requires physically switching a voltage level and source impedance into the measurement system equivalent to a known value of the system's stimulus. We consider a resistive transducer whose output resistance varies from 0 to 2,000 ohms. By switching the data acquisition system from the transducer to a precision resistor, one can calibrate this channel of the data acquisition system. By varying the value of the precision resistor, one can calibration at values which correspond to 10, 50, and 90 percent of the transducer's full scale output.

The insertion of a resistor of known value and low temperature coefficient of resistance (e.g., a metal film resistor) in parallel with one arm of an AC or DC bridge will result in a bridge unbalance equivalent to some value of the transducer's stimulus. This equivalency can be established when the transducer is calibrated prior to test. The arm being shunted should have no other parallel impedance (e.g., a balance circuit) across it. The insertion of the shunt resistance across the transducer electrically simulates the transducer's output in terms of signal level and characteristic impedance to the remainder of the measurement system.

On-line or *In Situ* Calibration

Manufacturers of instruments frequently include some type of self-calibration feature. Two common features are auto-zero and auto-calibration. These features do not replace calibration by a metrology laboratory. When properly used, these features enhance the accuracy of the instrument between calibration by a metrology laboratory.

Data acquisition systems are often set up to feed in a known reference zero level and measure the response of the system to this known level. This will tell how much offset is in the entire system. In some cases this offset is recorded for later use in data reduction. Alternatively, the offset amplifier circuit can be set to counteract this measured offset. This is known as auto-zero. Typically a system will perform this auto-zero on a timed basis or before some critical measurement.

In the example system in Figure 6.32, the shock signal is measured before, during, and after the shock event. Then the zero level is measured immediately before the event happens and is part of the data record. This zero level is used at data reduction time to adjust the signal level offset during the event. In addition, the zero level before and after the event can be compared to see if the accelerometer gauge shifted its zero during event. This can happen if the accelerometer gauge is mounted incorrectly and has its mounting loosened during the event.

Some data acquisition systems not only have an auto-zero, they also have a known calibration signal built in to check and possibly adjust the gain of the system. This calibration signal is often a stable resistor divider circuit connected across an accurate voltage reference. The system will normally have self-controlled switches which connect these reference signals into the front end of the system in place of the gauges. The system then measures these known reference levels and records or adjusts its gain to match the known response to these calibration levels.

References

Bauder, M. E. 1970. *Operations and Maintenance Manual for Program PLBRD,* Sandia National Laboratories Report TC-TM-70–114.

Belanger, B. C. 1980. Traceability: an evolving concept, *ASTM Standardization News,* 8(1).

Boyer, W. B. 1987. *Computer Compensation for Cable Signal Degradations,* Sandia National Laboratories Report SAND 87–3072.

Carstens, J. R. 1993. *Electrical Sensors and Transducers,* Regents/Prentice-Hall, Englewood Cliffs, NJ. A detailed look at sensors, devices to convert physical parameters to electrical signals.

Demler, M. J. 1991. *High-Speed Analog-to-Digital Conversion,* Academic Press, New York, NY.

Helfrick, A. D. and Cooper, W. D. 1990. *Modern Electronic Instrumentation and Measurement Techniques,* Prentice Hall, Englewood Cliffs, NJ. An introduction to data acquisition systems.

Horan, S. 1993. *PCM Telemetry Systems,* CRC Press, Boca Raton, FL. An introduction to data acquisition systems and the transmission of that data by pulse code modulation (PCM) telemetry systems.

Horowitz, P. and Hill, W. 1989. *The Art of Electronics,* Cambridge University Press, Cambridge, MA. General purpose reference on electronics technology.

International Standards Organization, *International Vocabulary of Basic and General Terms in Metrology,* ISO Standards Handbook, 2d ed., International Standards Organization,

The Institute of Electrical and Electronics Engineers, 1994. *IEEE Standard 1057 for Digitizing Waveform Recorders,* The Institute of Electrical and Electronics Engineers, New York, NY.

Meiksin, Z. H. and Thackray, P. C. 1984. *Electronic Design with Off-the-Shelf Components,* Prentice Hall, Englewood Cliffs, NJ. An introductory treatment of op amps, analog-to-digital converters, and grounding and shielding. The discussions are much more tutorial and explanatory than most treatments.

Morrison, R. 1984. *Instrumentation Fundamental and Applications,* John Wiley & Sons, New York, NY. Contains information on transducers and their connections, the plant environment, signal conditioning, and calibration of instruments.

Morrison, R. 1986. *Grounding and Shielding Techniques in Instrumentation,* John Wiley & Sons, New York, NY. Contains a nice chapter on differential amplifiers.

Ott, H. W. 1988. *Noise Reduction Techniques in Electronic Systems,* John Wiley & Sons, New York, NY. Explains the differences between safety and signal grounds.

Pallay-Areny, R. and Webster, J. G. 1991. *Sensors and Signal Conditioning,* John Wiley & Sons, New York, NY.

Performance Specification for Accelerometer Model 7270A, Endevco PS7270A, Rev. A, 12–04–85, San Juan Capistrano, CA.

Schnell, L., ed. 1993. *Technology of Electrical Measurements,* John Wiley & Sons, New York, NY. This book is a nice collection of tutorial papers on measurement theory and technology. The chapters on modeling and measurement, time and frequency measurement, and electrical units are especially good.

Sclater, N. 1991. *Wire and Cable for Electronics,* McGraw-Hill, New York, NY. Describes conductor material, insulation material, cable manufacture, twisted pairs, coaxial cable, triaxial cable, twin axial cable, fiber optic cable and connectors.

Stearns, S. D. and Rush, D. R. 1990. *Digital Signal Analysis,* Prentice Hall, Englewood Cliffs, NJ.

Strock, O. J. 1987. *Introduction to Telemetry,* Instrument Society of America, Triangle Park, NC. A self-teaching book on data acquisition and telemetry systems.

Stroud, A. H. 1974. *Numerical Quadrature and Solution of Ordinary Differential Equations,* ch. 2, Springer-Verlag, New York, NY.

Van Valkenburg, M. E. 1982. *Analog Filter Design,* Holt, Rinehart & Winston, New York, NY. A detailed look at analog filter design. This book can be used in the design of anti-aliasing filters.

Williams, A. B. 1981. *Electronic Filter Design Handbook,* McGraw-Hill, New York, NY.

Witte, R. A. 1993. *Electronic Test Instruments,* Prentice Hall, Englewood Cliffs, NJ. This book describes how to measure electrical quantities with instruments.

6.9 Digital Signal Processing

Belle Upadhyaya

Introduction

Signal processing deals with the analysis of stationary and non-stationary measurements acquired from sensors used for monitoring, control, and diagnostics of process systems. Digital signal processing (DSP) is the technology of analyzing signals that are digitized or discretized at a certain specified sampling rate. The primary goal of DSP is to extract information about the operation of a component or a system with applications to: monitoring, diagnostics, control, predictive maintenance, estimation of performance-related parameters, pattern recognition, sensor validation, and many others.

The area of signal processing grew out of the needs in aerospace, communications and defense applications beginning in the late 1930s. Because of the advent of fast and efficient digital computers and advances in data acquisition and instrumentation systems, the DSP technology has been changing continuously and has made tremendous advances. DSP is being used in all areas of science and engineering. Some examples are: aerospace industry, communications industry, underwater acoustics, vibration analysis, power plant applications, chemical, metals and other process industries, medical diagnostics, robotics, and image and pattern analysis.

This section presents the fundamentals of data acquisition and analysis of process signals, and system monitoring using the DSP techniques. The emphasis is on processing stationary random signals.

Data Acquisition from Sensor and Sampling Requirements

The steps in data acquisition consist of the following:

1. Sensor placement and verifying its output.
2. Signal preconditioning such as filtering and amplification.
3. Signal recording using analog or digital devices.
4. Data qualification and storage.

Figure 6.34 shows a schematic of a data acquisition system where the signals are sampled and stored in a PC. If a large number of signals has to be digitized, a solid state multiplexer may be used to reduce the number of digitizing channels and the cabling requirement.

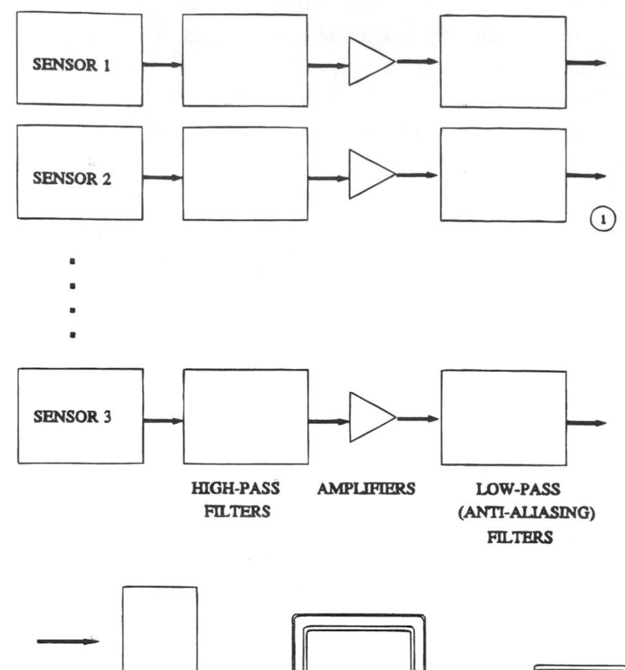

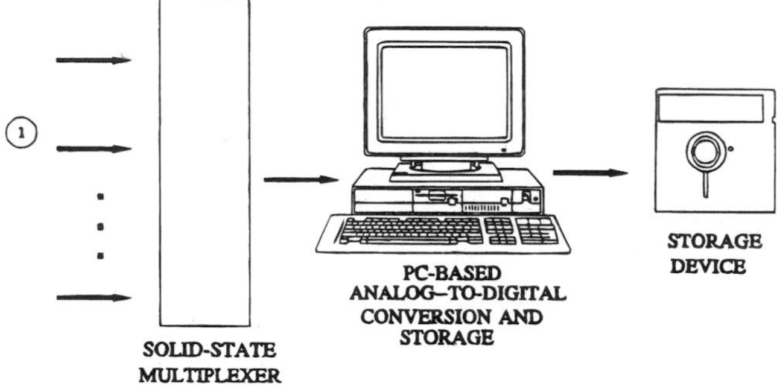

Figure 6.34 Data acquisition system showing signal conditioning devices, signal multiplexer and PC-based A/D conversion.

Data Collection/Measurement

Physical variables are converted by transducers into proper electrical outputs, generally voltage. Examples of sensors used in process industry are thermocouples, resistance temperature detectors, pressure transmitters, accelerometers (for mechanical vibration), neutron detectors, thickness gauges, strain gauges, current probes, and many others. Nonlinearities, phase shifts, and gain changes are sources of error in an instrument channel.

Signal Conditioning

Before signals from the sensor system are digitized, it is necessary to prepare the signal input to the analog-to-digital (A/D) converter. Noise filtering and signal amplification are two of the conditioning important to improve the quality of the recorded data.

A high-pass (HP) filter is used to remove frequency components below a certain value, f_H (Figure 6.35). The purpose of the HP filtering is to remove DC levels and very-low-frequency components that are not indicative of the dynamics of the system being monitored. An example of high-pass filter setting for a

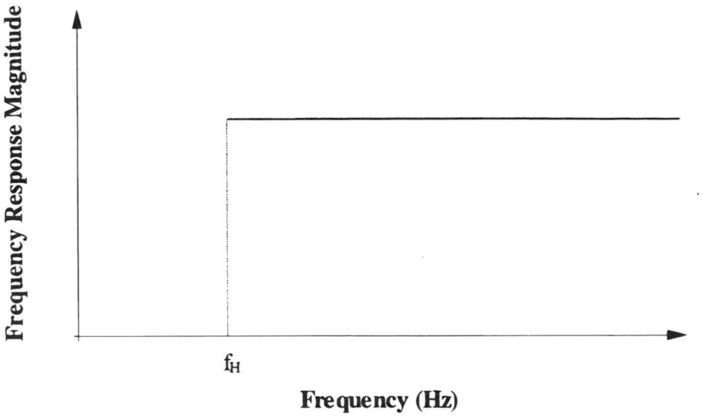

Figure 6.35 Illustration of high-pass (HP) filtering.

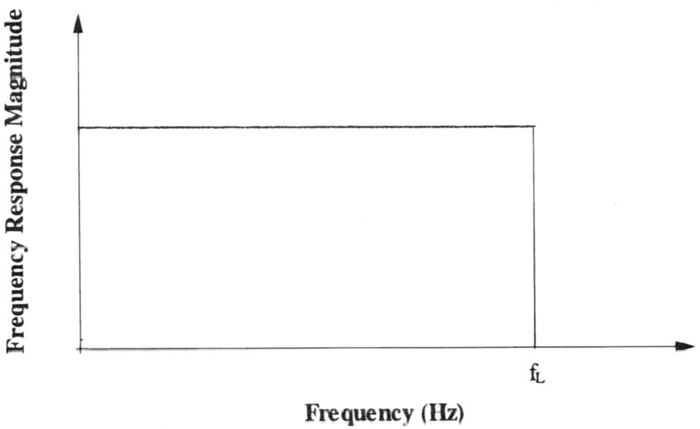

Figure 6.36 Illustration of low-pass filtering. The combination of HP and LP filters results in a band-pass filter. The pass band $= f_L - f_H$.

thermocouple or a pressure sensor is 0.01 Hz. A low-pass (LP) filter is used to remove frequency components above a certain value, f_L (Figure 6.36). A low-pass filter is also referred to as an anti-aliasing filter (see Data Sampling Criterion). The purpose of LP filtering is to remove frequency components of a signal above a specified frequency. The HP filtering enables us to amplify the dynamic component of the signal, so that the accuracy in A/D conversion is very high. The LP filtering removes unnecessary higher frequency components and enables us to choose a sampling rate that avoids any "distortions" in the digitized signal. An example of low-pass filter setting for a pressure transmitter signals is $f_L = 20$ Hz.

Before the filtered signal is digitized, the signal is amplified using drift-free amplifiers. The purpose of signal amplification is to increase the signal voltage range to match with the A/D converter range, so that the resolution of the digitized signal is high, and the quantization error is low. Figure 6.37 illustrates the effect of signal conditioning.

Filter and amplifier hardware devices are manufactured by several companies in the United States, Japan and Europe. Analog low-pass and high-pass filters with four to eight pole Butterworth filter design are very common. These filters have good frequency cut-off characteristics and also minimize the side-band effects

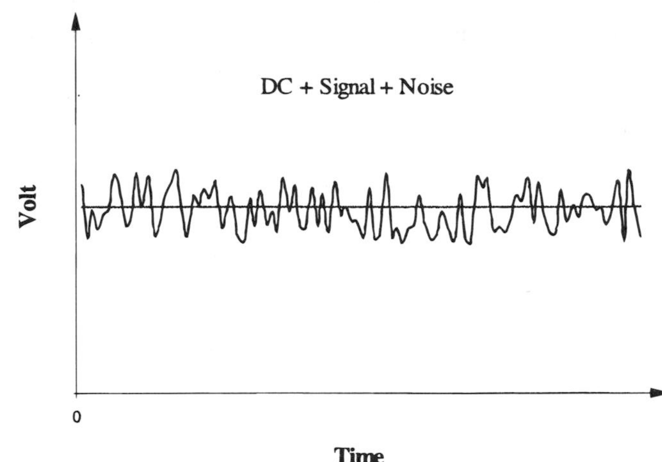

Figure 6.37 Illustration of signal conditioning using high-pass and low-pass filtering and amplification.

in the frequency-domain behavior. Vendors of predictive maintenance equipment and companies specializing in signal conditioning equipment, supply filters, amplifiers and other devices to meet required specifications.

Multiplexing of Signals

If the number of signals to be digitized is large compared to the number of A/D converters, it is often desirable to use a device to transmit signals using a single cable but interrupt the transmission of signals at certain preset time intervals. To be more effective in data digitization, the configuration shown in Figure 6.34 must be used. A multiplexer allows the flow of signals in time sequence, instead of simultaneously.

Signal mutliplexing after amplification (high-level multiplexing) and the use of high-speed solid state switches have the advantages of minimizing source impedance, switch resistance, frequency limitation and thermal-induced voltages. The capital cost of signal conditioning and high-performance multiplexers is high, but will pay off in the long term. Multiplexing also avoids the need for running individual cables from instrument channels to the end recording devices.

Analog-to-Digital Conversion

Analog-to-digital converters (ADCs) are used to transform an analog signal into a discrete or sample valued signal, with a specified number of samples per second. This sampling rate and the design of the ADC must be such as to minimize the error in the digitized data.

In an analog-to-digital conversion, the value of a signal is converted into a discrete level. The resolution of the discrete level is defined by ADC output code as n bits. The number of discrete levels (or bins) represented by this ADC is 2^n. For a 12-bit ADC, the number of discrete levels is $2^{12} = 4096$. If FS is the full scale voltage, then the minimum level change or least-significant bit (LSB) size is

$$LSB = \frac{FS}{2^n}. \text{ (Also called a quantum)}$$

If $FS = 10$ volts, and $n = 12$

$$LSB = 10/4096 = 0.00244 \text{ volt}$$

All analog values within a given *quantum* are represented by the same digital code (such as the binary code), which is generally used as the mid-range value or threshold (Tompkins and Webster, 1988). Since the ADC output can differ from the threshold by as much as $\pm 1/2$ LSB and still be represented by the same output code, there is an inherent quantization uncertainty of $\pm 1/2$ LSB in any digitizing process. The only way to reduce this is to increase the number of bits, n.

Figure 6.38 shows the conversion relationship for a 3-bit ADC. The LSB is 1/8 FS and the output is quantized into eight distinct levels from 0 to 7/8 FS (or FS-1 LSB).

Successive-approximation A/D converters are widely used for interfacing with computers. Some of the features include high-resolution capability (16 bit), high throughput rate (MHz), fixed conversion time independent of voltage level, and each conversion is independent of the previous because the internal logic is cleared at the start of a conversion.

Many data acquisition and analysis devices are available in the market. ADC are available for different throughput rates, number of channels, and other requirements. The sensor outputs are interfaced with a PC through a vendor supplied RS 232 serial interface. For details of the operation of an ADC see Tompkins and Webster (1988) and Sheingold (1986).

IEEE 488 is a general purpose instrument bus (GPIB) for interfacing a system controller (on a PC) with other devices, or for one device in the bus to communicate with other devices on the same bus. A GPIB may be used to trigger various instruments for data acquisition at specified time intervals.

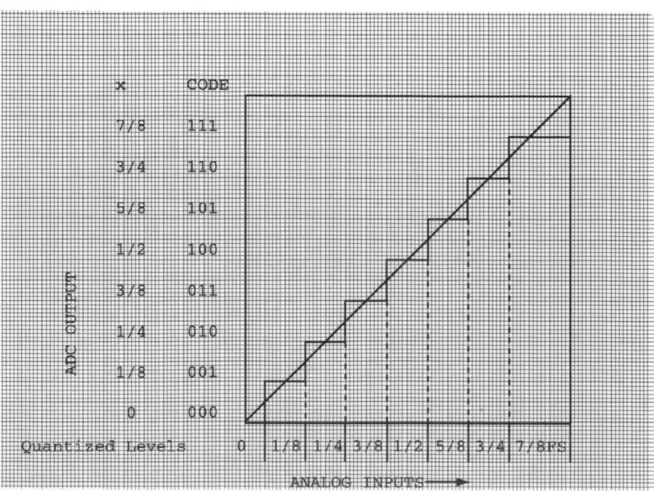

Figure 6.38 Analog-to-digital converter characteristics of a 3-bit ADC. (*Source*: Tompkins, W. J. and Webster, J. G., 1988. *Interfacing Sensors to the IBM® PC*, Prentice-Hall, Englewood Cliffs, NJ. With Permission.)

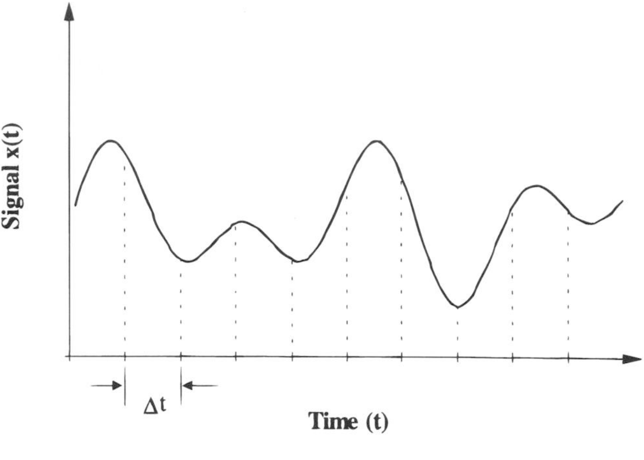

Figure 6.39 Sampling of continuous signal.

Data Sampling Criteria

The sampling criterion establishes the minimum rate at which a signal must be sampled in order to avoid errors in the estimation of signatures from sensor data. Figure 6.39 shows a sample of a signal. Δt is the uniform sampling interval.

Nyquist Criterion. Let f_{max} be the maximum frequency (Hz) contained in a signal. For a given signal to be sampled (or digitized) without loss of information, the sampling rate must be at least twice the maximum frequency of the signal. If f_s is the sampling frequency, then

$$f_s \geq 2f_{max}$$

Since $f_s = 1/\Delta t$, the maximum sampling interval

$$\Delta t \leq \frac{1}{2f_{max}}$$

EXAMPLE 6.1: If $f_{max} = 20$ Hz, then the sampling frequency (number of samples/second) $f_s > 40$ Hz. The corresponding sampling interval

$$\Delta t \leq \frac{1}{40} = 0.025 \text{ sec}$$

Aliasing. If the highest frequency in the original signal exceeds $f_s/2$, then the resulting frequency-domain (and time-domain) signatures will have errors due to overlapping (or folding) of adjacent frequencies. This effect is termed *aliasing*, because the components at higher frequencies will appear in the frequency band $(0, f_s/2)$.

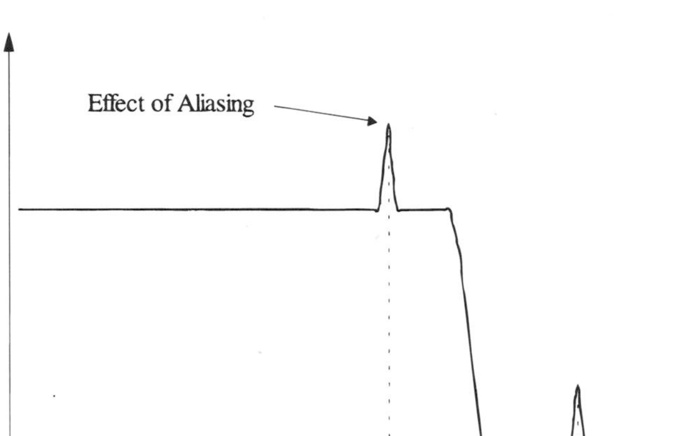

Figure 6.40 Demonstration of aliasing.

For any frequency f in the range

$$0 \leq f \leq \frac{f_s}{2}$$

the higher frequencies which are aliased with f are given by

$$(f_s \pm f), (2f_s \pm f), \ldots, (nf_s \pm f), \ldots$$

The maximum information frequency $f_s/2$ is called the Nyquist folding frequency. The Fourier transform of a discrete signal is periodic as a function of frequency. This is a fundamental property of discrete-time signals, and as a result the computed signatures repeat and may overlap if the sampling frequency f_s is not sufficiently large.

Example of Aliasing. Let us assume that a certain signal was sampled at 100 samples/second, that is, $f_s = 100$ Hz. The Nyquist folding frequency is then $f_s/2 = 50$ Hz. It so happened that the original signal had a frequency component at $f_h = 60$ Hz. Since $f_s/2 < 60$ Hz, the component at 60 Hz shows at frequency $f \leq 50$ Hz. This frequency is calculated from the relationship

$$f_s \pm f = 60$$

$$\rightarrow f = 100 - 60 = 40 \text{ Hz (see Figure 6.40)}$$

Remarks.

1. The sampling rate established by the Nyquist criterion is the minimum rate at which a signal must be digitized. For some problems it is necessary to digitize the signal at a rate greater than $2f_{max}$. This must be determined by the user. When anti-aliasing filters are used, the frequency at which the filter is set is not the same as the maximum signal frequency. In this case the sampling frequency must be

$$f_s = 2.5\, f_c \text{ to } 3\, f_c$$

where f_c is the filter cut-off frequency.

2. This section presented data acquisition directly in the digital form. In many applications analog tape recorders are used to acquire plant data. The analog signals may then be digitized at any desired sampling rate.
3. If a signal is modulated (amplitude or frequency modulation) it may be necessary to demodulate the signal. Both software and hardware devices are available for signal demodulation. This is often the case with accelerometers and electric current probes.

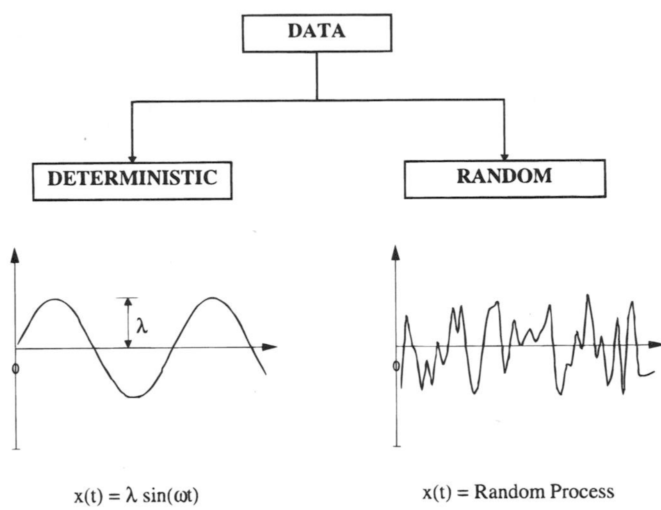

Figure 6.41 Types of signals.

Time-Domain Analysis

The computation of various signatures in the time domain is described in this section.

Stationary Random Signals

Random Signal. A random signal (process) cannot be described by an explicit mathematical relationship, because *each observation of the phenomenon is unique* (Figure 6.41).

EXAMPLE 6.2: Temperature fluctuations of flow through identical pipes with identical flow and temperature. The fluctuations of the temperature in each of the pipes has a unique realization at a given time.

Each of the time histories T_1, T_2, T_3 (see Figure 6.42) observed over a finite time interval is a *sample* record. The collection of all possible sample functions generated by a random phenomenon is called a *random process* or a *stochastic process.*

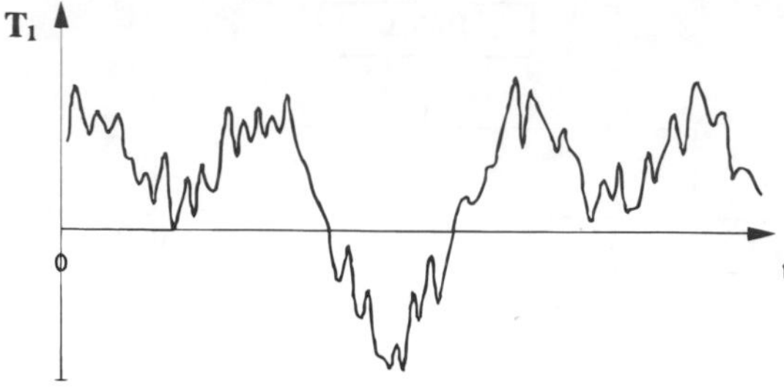

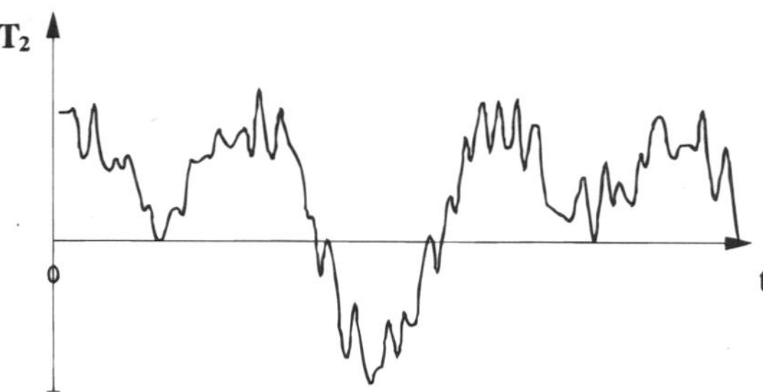

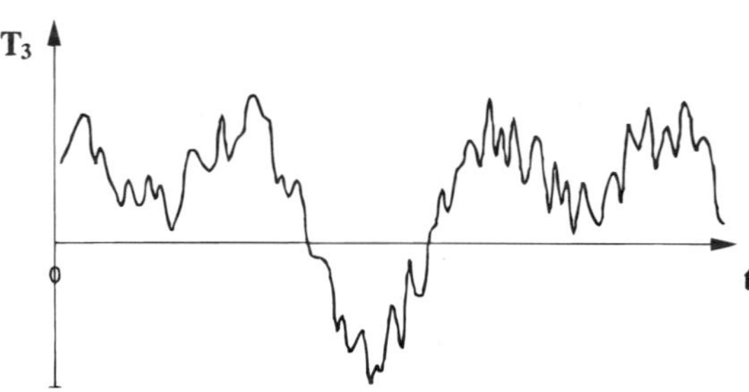

Figure 6.42 Sample records of RTD (temperature) outputs of similar flow systems, or the same temperature measurements in a given flow.

Stationary Random Signal A random signal is said to be stationary if all its statistics (such as mean value, correlations, RMS value, frequency spectrum) do not change as a function of time. A random process (signal) is said to be ergodic if ensemble averages are equal to time averages.

Figure 6.43 shows an ensemble (collection) of sample functions forming the velocity of individual dust particles in a given volume. Assuming each sample function is equally likely the ensemble averages are defined by

$$\text{Mean value } m_x(t_1) = \lim_{N\to\infty} \frac{1}{N} \sum_{k=1}^{n} x_k(t_1)$$

and the autocorrelation function

$$R_x(t_1, t_1 + \tau) = \lim_{N\to\infty} \frac{1}{N} \sum_{k=1}^{\alpha} x_k(t_1) x_k(t_1 + \tau)$$

x_k (t_1) corresponds to samples k and at time t_1.

- If $m_x(t_1)$ and R_x $(t_1, t_1 + \tau)$ are functions of time t_1, then the random process is said to be *nonstationary.*

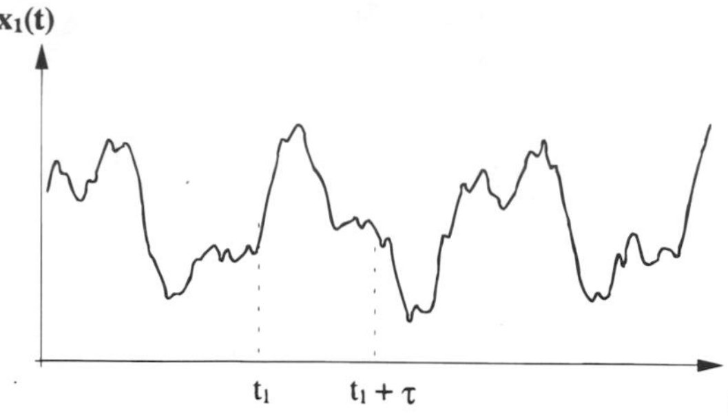

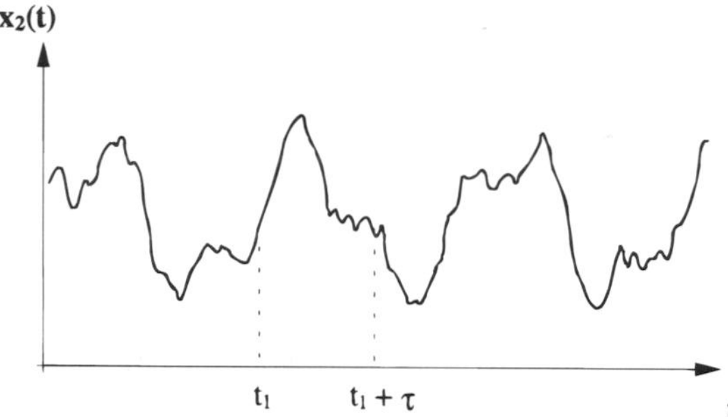

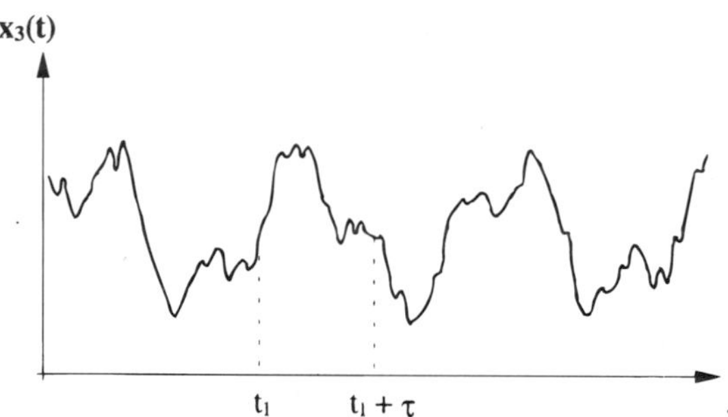

Figure 6.43 Ensemble of sample functions of a random process.

- For the case when m_x (t_1) and R_x $(t_1,t_1 + \tau)$ do not vary as time t_1 varies, the random process is said to be *weakly stationary* or *stationary in the wide sense.*
- For wide-sense stationarity

$$m_x(t_1) = m_x = \text{constant}$$

$$R_x(t_1, t_1 + \tau) = R_x(\tau);$$

$R_x(\tau)$ is only a function of the time difference τ.

- Thus for stationary random processes, ensemble (or probabilistic) averages can be estimated using time averages (over one time sample).

The time average over one sample function is given by

$$m_x = \frac{1}{T} \int_0^T x(t)\,dt$$

$$R_x(\tau) = \frac{1}{T} \int_0^T x(t)x(t + \tau)\,dt$$

Time-Domain Signatures

Several signatures or characteristics of a stationary signal are defined and illustrated. The purpose is to make an estimate

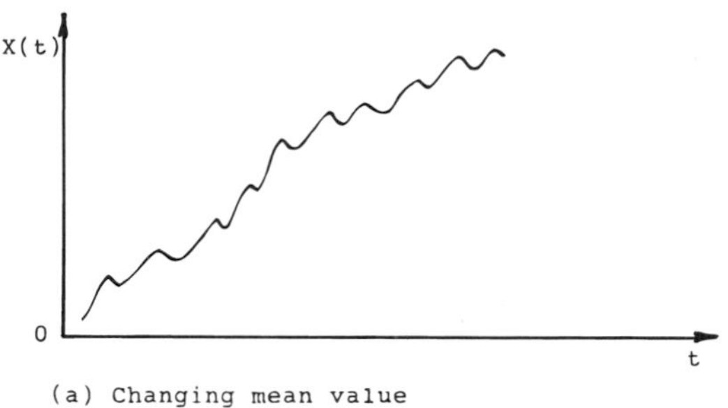

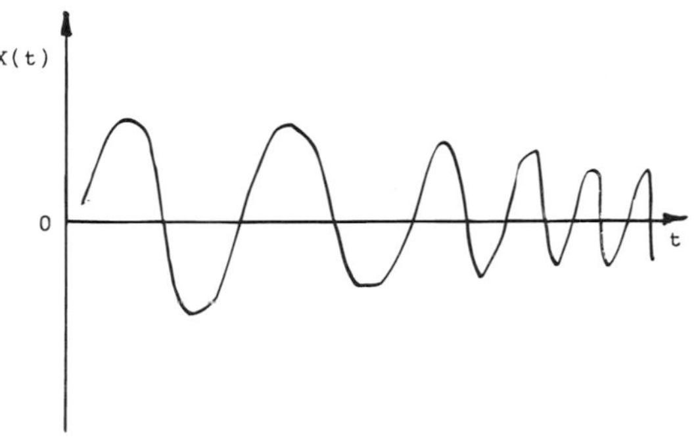

Figure 6.44 Examples of nonstationary processes.

of the desired signature from a set of samples of x(t) and is denoted by

$$S = (X_1, X_2, \ldots, X_n)$$

The number of data points, N, is referred to as the sample size.

Mean Value

$$m_x = \frac{1}{T}\int_0^T x(t)dt = \frac{1}{N}\sum_{i=1}^{N} x_i$$

Mean-Squared Value

$$\psi_x^2 = \frac{1}{T}\int_0^T x(t)^2\, dt = \frac{1}{N}\sum_{i=1}^{N} x_i^2$$

Root-Mean-Squared (RMS) Value
RMS value is the more commonly used signature and is given by

$$RMS = \Psi_x = \left\{\frac{1}{N}\sum_{i=1}^{N} x_i^2\right\}^{1/2}$$

Variance and Standard Deviation
Variance

$$\sigma_x^2 = \frac{1}{N}\sum_{i=1}^{N}(x_i - m_x)^2$$

σ_x is the *standard deviation* of x(t).

Remark.

1. If the mean value of a signal $m_x = 0$, then the RMS value and the standard deviation are the same.
2. The RMS value of a pure sinusoidal waveform of amplitude A is given by

$$RMS_{(\text{sine})} = \frac{A}{\sqrt{2}} = 0.707A$$

Probability Density Function

Any random process can be completely described by its probability distribution. One may consider the probability density function (PDF) as the histogram of a signal. The most common PDF is the Gaussian or normal density function. This density function is given by

$$p_G^{(x)} = \frac{1}{\sqrt{2\pi\sigma^2}}\exp\left\{-\frac{(x-m)^2}{2\sigma^2}\right\}$$

The Gaussian density is more common in power plant applications. p_G (x) has a bell-shaped characteristic with m = mean value of x, σ^2 = variance of x. Figure 6.45 shows a typical Gaussian (normal) density function.

In computing the PDF, the sample values of x(t) are used to estimate the histogram of x(t). The actual range of the signal is divided into bins of equal interval, and the samples of x(t) belonging to a given bin are counted. This is repeated for all the bins and the fractional count Nx/N is plotted as a function of x. Figure 6.46 shows the PDF of random temperature fluctuations measured by a resistance temperature detector (RTD). Any skewness in the computed PDF (from its symmetry) indicates possible anomaly in the signal.

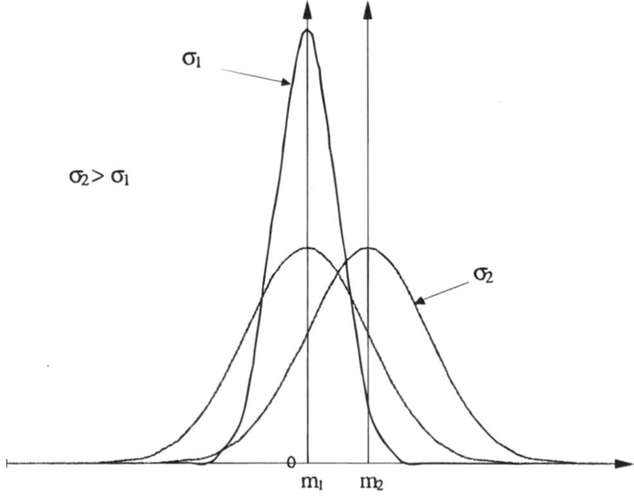

Figure 6.45 Illustration of Gaussian density function for the cases of different mean values and variances.

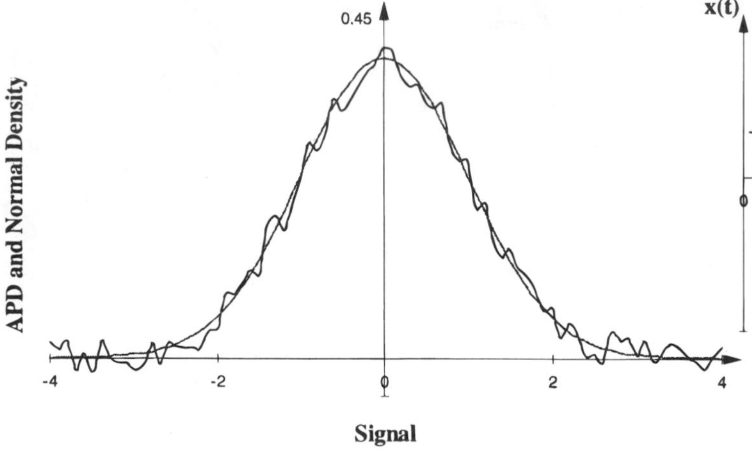

Figure 6.46 Probability density function of a random temperature signal from an RTD. This is compared with an ideal Gaussian density function.

The skewness can be determined by computing the third moment of the signal. The normalized third moment is given by

$$\text{Skewness} = \left\{\frac{1}{N}\sum_{i=1}^{N}(x_i - m_x)^3\right\}/\sigma_x^3$$

For a symmetric distribution skewness = 0.

Correlation Functions

Autocorrelation function of a signal x(t), and cross correlation between two signals x(t) and y(t) provide the relationship for one or two signals when they are separated by a time displacement.

Autocorrelation Function. Autocorrelation function of a signal x(t) describes the general time dependence of signal values separated by a time lag τ. This is illustrated in Figure 6.47. For a stationary process the autocorrelation function $R_{xx}(\tau)$ is defined as

$$R_{xx}(\tau) = \lim_{T\to\infty}\frac{1}{T}\int_0^T x(t)x(t+\tau)\,dt$$

$R_{xx}(\tau)$ provides insight into the existence of periodic signals, the frequency bandwidth of the signal, and the memory of a system. For less time-dependent signals the autocorrelation function decays to zero faster than for signals with large "memory."

For sample size N, the autocorrelation at lag $\tau = k\Delta t$ is calculated as

$$R_{xx}(k) = \frac{1}{N}\sum_{i=1}^{N-k}(x_i - m_x)(x_{i+k} - m_x)$$

Remarks. When the total number of samples is very large, it is often suggested to divide the total signal into several blocks of data, each of length N samples. Then, average the correlations over the number of blocks of data.

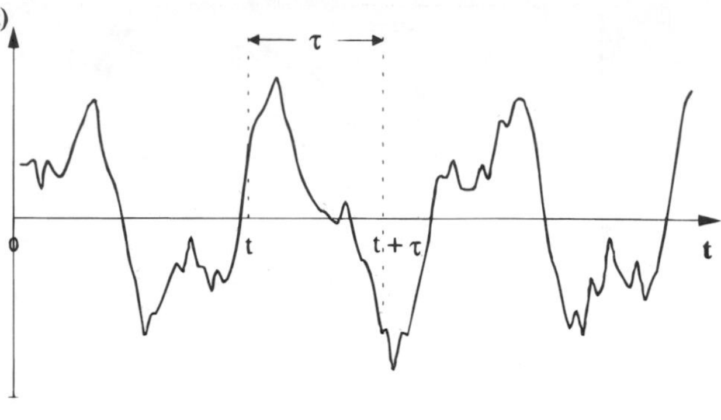

Figure 6.47a Illustration of correlation lag.

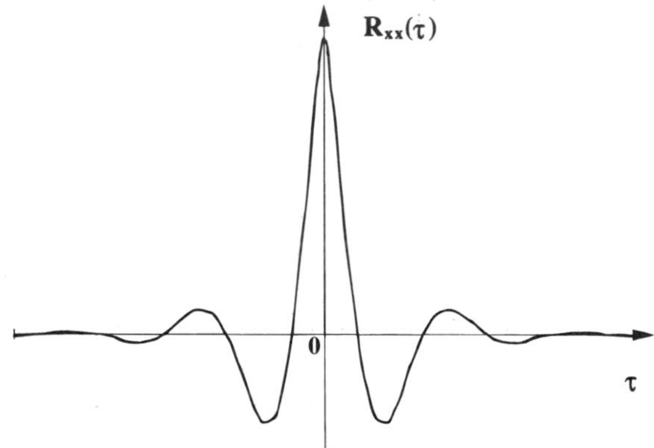

Figure 6.47b Autocorrelation of a random signal x(t).

$R_{xx}(\tau)$ has several important properties.

- $R_{xx}(\tau) = R_{xx}(-\tau)$, symmetric function of τ.
- $R_{xx}(0) > 0$, positive for zero lag.
- $|R_{xx}(\tau)| \leq R_{xx}(0)$, maximum at zero lag.

Figures 6.48a and 6.48b show the autocorrelation functions of signals recorded from a thermocouple at the core-exit of a pressurized water reactor (PWR), and an in-core neutron detector in a boiling water reactor (BWR). The autocorrelation function of the temperature signal is similar to an exponential function and the neutron detector signal has a small oscillatory characteristic.

Cross Correlation Function. Cross correlation function relates the dependency of one signal at time t, with another signal at time $t + \tau$. Figure 6.49 illustrates the time lag between x(t) and y(t). The cross correlation is defined by

$$R_{xy}(\tau) = \lim_{T\to\infty}\frac{1}{T}\int_0^T x(t)y(t+\tau)\,dt$$

Numerical computation at lag $k\Delta t$ is given as

$$R_{xy}(k) = \frac{1}{N}\sum_{i=1}^{N-k}(x_i - m_x)(y_{i+k} - m_y)$$

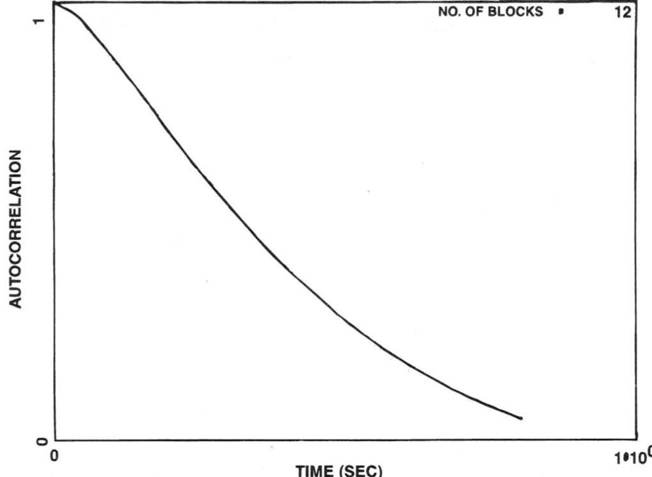

Figure 6.48a Autocorrelation function of a core-exit thermocouple signal from a PWR.

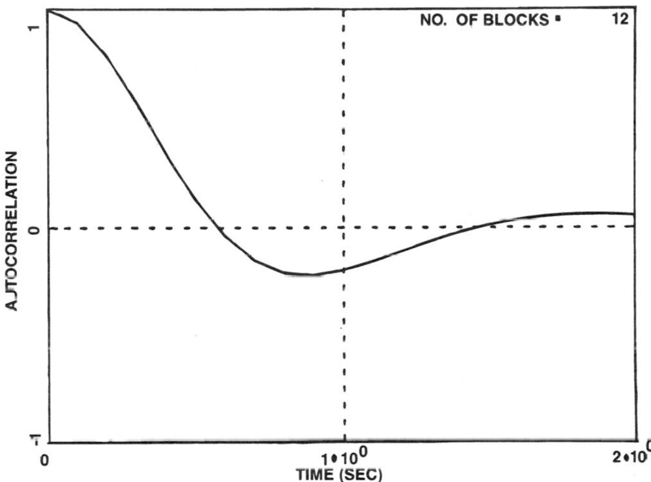

Figure 6.48b Autocorrelation function of an in-core neutron detector (APRM) signal from a BWR.

A normalized value of cross correlation is defined using the correlation coefficient as

$$\rho_{xy}(\tau) = \frac{R_{xy}(\tau)}{\{R_{xx}(0)R_{yy}(0)\}^{1/2}}, \qquad |\rho_{xy}| \leq 1$$

Because of using two signals x(t) and y(t), the information content in R_{xy} (τ) is greater than in R_{xx} (τ) or R_{yy} (τ) individually. The cross correlation may be used for estimating time delays and transmission path of an acoustic source from one point to another point.

Auto- and cross-correlation functions can also be treated as compression of information, thus reducing a large sample size to a few hundred correlation values.

In estimating the time delay between two signals, the cross correlation function attains a maximum value at a lag corresponding to the transport lag between the two detector locations.

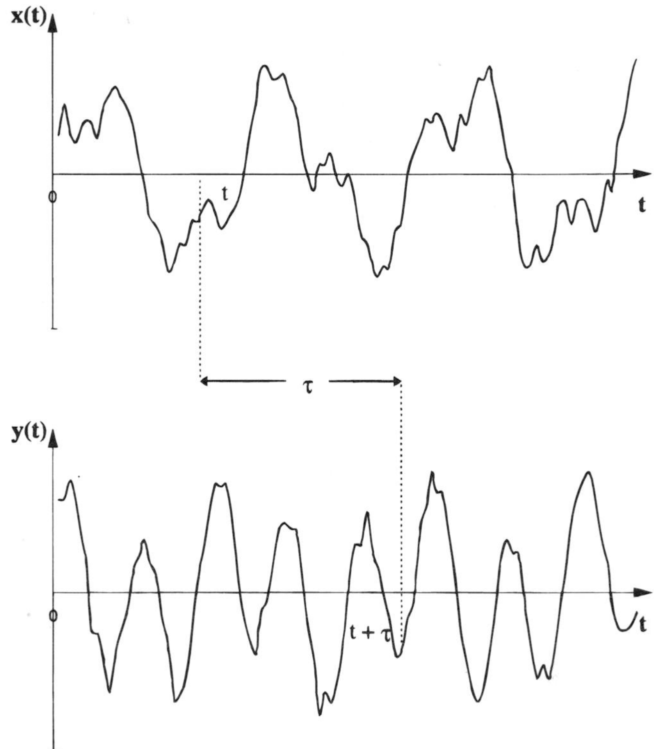

Figure 6.49 Illustration of time lag for cross correlation between two signals x(t), y(t).

This is illustrated in Figure 6.50. Note that the cross correlation is not a symmetric function of lag τ. Thus the order of multiplication, that is, $R_{xy}(\tau)$, $R_{yx}(\tau)$, must be distinguished as antisymmetric functions of τ.

White Noise

A random process x(t) is said to be a white noise process if the adjacent values of x(t) are uncorrelated with each other. Emission of neutrons from a source may behave like a white noise process. The correlation function of a white noise signal is a delta function with

$$R_{xx}(0) = \frac{1}{T}\int_0^T x^2(t)\, dt$$

being the only nonzero value. Any general random process may be generated by passing white noise *through* a given transfer function.

References and Suggested Additional Reading Material

Bendat, J. S. and Piersol, A. G. 1986. *Random Data: Analysis and Measurement Procedures,* Wiley-Interscience, New York, NY.

Hamming, R. W. 1989. *Digital Filters,* Prentice-Hall, Englewood Cliffs, NJ.

Randall, R. B. 1987. *Frequency Analysis,* Bruel & Kjaer, Larson & Sons A/S, Denmark.

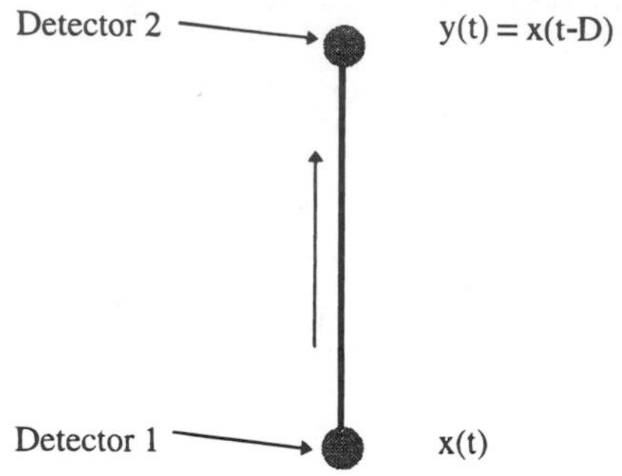

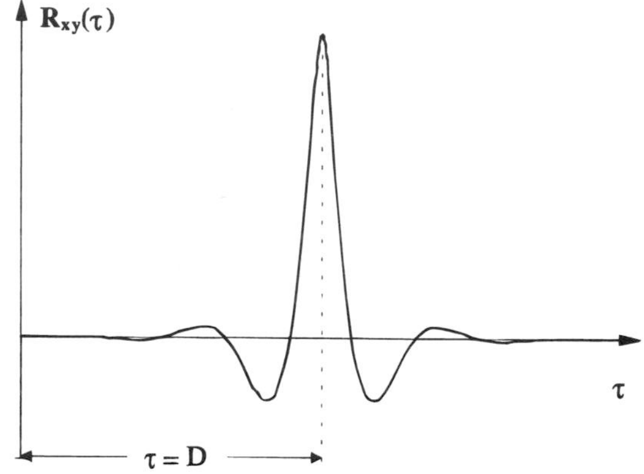

Figure 6.50 Illustration of cross correlation between two signals for the case of transport delay between x(t) and y(t).

Oppenheim, A. V. and Schafer, R. W. 1975. *Digital Signal Processing,* Prentice-Hall, Englewood Cliffs, NJ.

Sheingold, D. H. (Ed.) 1986. *Analog-to-Digital Conversion Handbook,* Prentice-Hall, Englewood Cliffs, NJ.

Stearns, S. D. and Hush, D. R. 1990. *Digital Signal Analysis,* Prentice-Hall, Englewood Cliffs, NJ.

Taylor, J. L. 1986. *Computer-Based Data Acquisition Systems,* Instrument Society of America.

Tompkins, W. J. and Webster, J. G. 1988. *Interfacing Sensors to the IBM ® PC,* Prentice-Hall, Englewood Cliffs, NJ.

Widrow, B. and Stearns, S. D. 1985. *Adaptive Signal Processing,* Prentice-Hall, Englewood Cliffs, NJ.

6.10 Signal Pick-up and Interface Circuitry

Thaddeus Roppel

In this section, we discuss the most commonly employed components and circuits for converting raw sensor signals to voltage levels suitable for computer data acquisition, conditioning, and recording. We discuss typical applications, sources of error and provide references for more detailed investigation.

The two major categories of circuits addressed here are bridge circuits and operational amplifier circuits. These are often used together in typical applications.

Bridge Circuits

Bridge circuits are widely used as sensor interface circuits. The principle employed is to measure the offset that results from the deviation of a sensor element (e.g., resistance) from its nominal value. Bridges can form part of a null-seeking circuit, in which feedback is used to force the sensor element back to its nominal value. In this case, the output is measured as the amount of feedback required to null the bridge. Such null-seeking circuits are employed, for example, in accelerometers, where the feedback is used to keep the moving element at zero deflection.

Resistive Bridges

The resistor *Wheatstone bridge* circuit consists of four legs, and requires an AC or DC voltage bias. Each leg contains one resistor. Figure 6.51 shows a resistive bridge powered by a DC bias source V_{DC}. The output of the bridge is the difference between the voltages V_A and V_B indicated in Figure 6.51.

Single-Element, Half-Active and Full-Active Configurations. The bridge circuit shown in Figure 6.51 has three fixed resistors R_f and a single active (varying) resistor $R_x = R_f + \Delta R$. Shortly we will look at two common configurations which use two or four active elements. The variation ΔR is usually the raw sensor response, e.g., the change in resistance of a strain gauge, an RTD (resistance temperature detector), or a photoconductor. We desire to relate the bridge voltage V_{OUT} to the measurand (temperature, strain, etc.). This is readily accomplished by employing the rule of *voltage division* from basic electric circuit theory. The total voltage dropped across a series of resistors divides proportionally among the individual resistors. Referring back to Figure 6.51, we observe that when $\Delta R = 0$, all four legs of the bridge contain the same resistance. This is called the *balanced* condition. Using voltage division, we see that when $\Delta R = 0$ in Figure 6.51 then $V_A = V_{DC}/2$ and $V_B = V_{DC}/2$. Thus

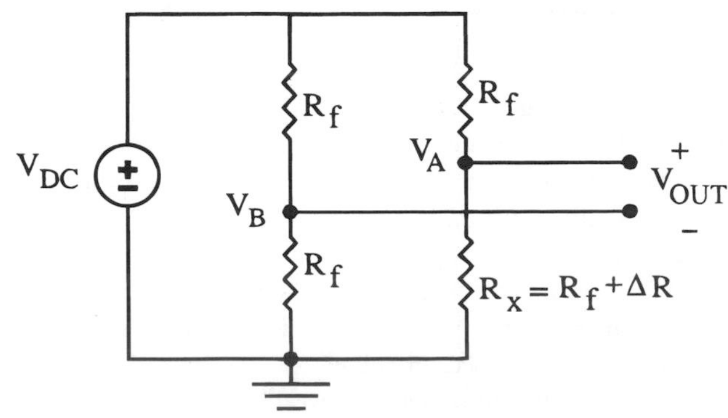

Figure 6.51 Resistive bridge with a single active element.

$V_{OUT} = (V_A - V_B) = 0$. Applying voltage division to the general case, we obtain V_{OUT} in terms of ΔR:

$$V_{OUT} = \frac{R_f + \Delta R}{(R_f + \Delta R) + R_f} V_{DC} - \frac{1}{2} V_{DC} \tag{6.32}$$

When ΔR is much smaller than R_f, this result can be approximated by the simpler *linear approximation:*

$$V_{OUT} \approx \frac{1}{4}\left(\frac{V_{DC}}{R_f}\right)\Delta R \tag{6.33}$$

This approximation shows that for small variations in ΔR, the bridge output is proportional to ΔR. Furthermore, the magnitude of the output is proportional to the supply voltage and inversely proportional to the fixed resistance.

Some sensors provide two matched active elements. In this case, a half-active bridge can be used, as shown in Figure 6.52. The linear approximation for the half-active bridge is

$$V_{OUT} \approx \frac{1}{2}\left(\frac{V_{DC}}{R_f}\right)\Delta R \tag{6.34}$$

Strain gauges used for pressure sensors and accelerometers are commonly provided with four active elements which are physically arranged on the sensing diaphragm so that, under measurement conditions, two elements experience tension and two elements experience compression. Consequently, two elements increase in resistance and two elements decrease in resistance. Using these four active elements, a full-active bridge can be constructed, as shown in Figure 6.53. In this case, the output is proportional to ΔR with no approximation required:

$$V_{OUT} = \frac{V_{DC}}{R_f}\Delta R \tag{6.35}$$

Linearization and Sources of Error

Traditionally, linear response has been a crucial requirement for sensors. However, the widespread use of computerized data acquisition and signal conditioning is gradually eliminating this requirement by making it possible to correct for nonlinearity by using calibration curves generated by table look-up or calculation. Nevertheless, in any system that does not use this approach, the sources of nonlinearity must be considered. These sources may be divided into those which are mathematically predictable and those which are process and environment dependent. Mathematically predictable sources include the bridge circuit equations, the effects of self-heating, and the nonlinear relationship between ΔR and the measurand. Common process and environment dependent sources include resistor mismatches, lead and contact resistance, magnetic and galvanic potentials, leakage currents, and temperature, humidity, and aging effects. Figure 6.54 shows a single-element bridge where the active element is connected by means of long leads having nonnegligible resistance. A typical application would be interfacing an RTD, where the RTD element must be placed at the temperature source, but the bridge is maintained at room temperature to avoid TCR (temperature coefficient of resistance) induced drifts in the fixed resistors. In this case, the lead wire resistance can be compensated by using a three-wire connection as shown in Figure 6.54 The lead-wire voltage drops I_wR_w cancel since they are in opposite legs of the bridge. The sense wire carries no current and thus $V_{OUT} = V_A - V_B$, as desired. This approach is effective in the usual case of equal lead lengths and wire types.

The most accurate conventional method of measuring resistance is the four-wire Kelvin probe, illustrated in Figure 6.55. Although not really a bridge technique, it is included in this section for completeness. A reference current is forced to flow through R_x, and the voltage is measured using two additional sense leads. If the sense leads are connected to a high-impedance

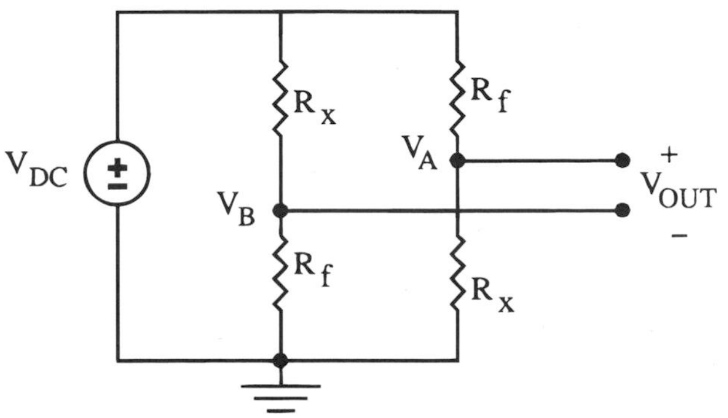

Figure 6.52 Half-active bridge.

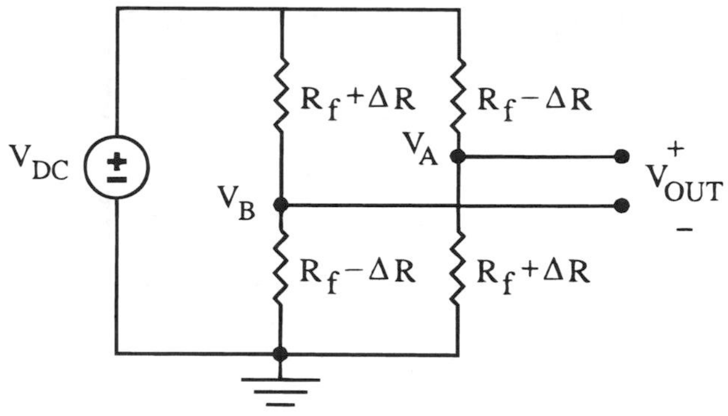

Figure 6.53 Full-active bridge.

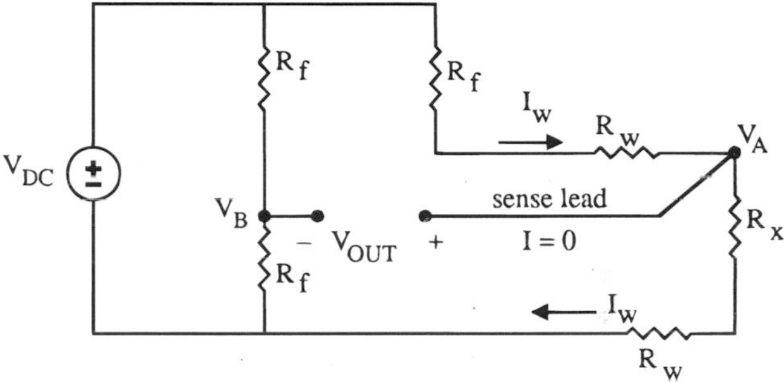

Figure 6.54 Three-wire bridge.

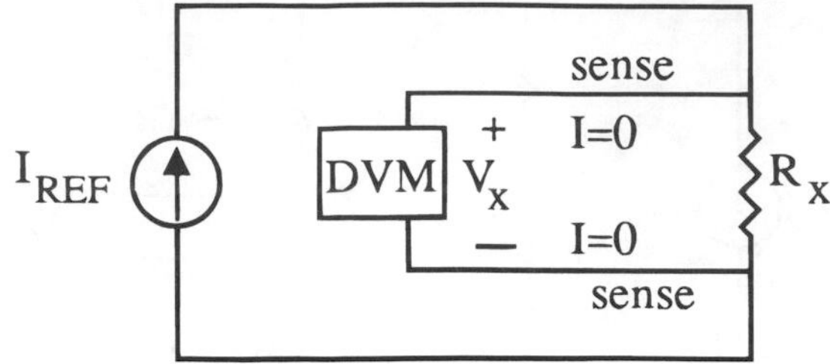

Figure 6.55 Four-wire Kelvin probe.

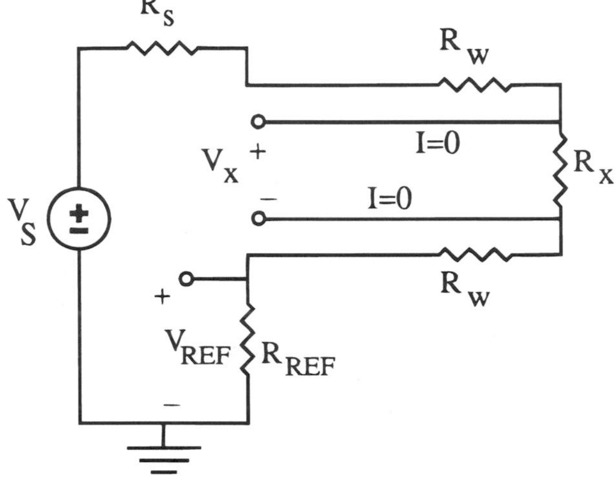

Figure 6.56 Four-wire resistance measurement using a voltage source instead of a current source.

input (e.g., digital voltmeter, instrumentation amplifier, or analog-to-digital converter), they carry essentially zero current so no voltage drop is induced. The voltage V_x is thus directly proportional to R_x. This is true regardless of whether the leads are matched in length and type.

A modification of this approach that is especially suitable for multichannel analog-to-digital converter (A/D) based systems is shown in Figure 6.56. Here, the need for a precision current source is eliminated. Instead a precision resistor R_{REF} is used to sample the current. The A/D conversion system must acquire both V_x and V_{REF}, one of which requires a differential measurement, for a total of three single-ended channels. The resistance R_x can be calculated from the measured voltages:

$$R_x = \left(\frac{R_{REF}}{V_{REF}}\right)V_x \tag{6.36}$$

The choice of the series resistor R_S involves a tradeoff. As R_S is made smaller, the power dissipation in R_x increases thus contributing to self-heating error. As R_S is made larger, V_{REF} and V_x are reduced in magnitude, which reduces the achievable resolution and the signal-to-noise ratio.

Impedance Bridges

The same principle employed in the resistance bridge can be used for sensor elements that are capacitive or inductive. Such elements are sometimes used in position, proximity, and accelerometer sensors. A capacitance bridge is shown in Figure 6.57. Employing AC voltage division, we find the output voltage is

$$V_{OUT} = -\frac{1}{4}\left(\frac{V_{AC}}{C_f}\right)\Delta C \tag{6.37}$$

The negative sign in Equation 6.37 indicates that, for the polarities indicated in Figure 6.57, the output sine wave is 180° out of phase with the supply voltage for positive ΔC.

Operational Amplifier Circuits

The integrated circuit operational amplifier (op-amp) is perhaps the most widely used interface component in sensor systems. Flexibility and low cost are key reasons. The op-amp circuit symbol is shown in Figure 6.58. To simplify circuit diagrams, the DC power supplies (V+) and (V−) are normally not shown. Models are also available that use a single supply voltage. Most of the selection process for an op-amp model involves tradeoffs among input impedance, bandwidth, slew rate, noise, and other parameters, *vs.* cost. Examples of several applications that demand high performance op-amp models will be discussed later in this section.

Voltage Amplifying Op-Amp Circuits

The operational amplifier is always used in a feedback configuration. All of the circuits of interest here use negative

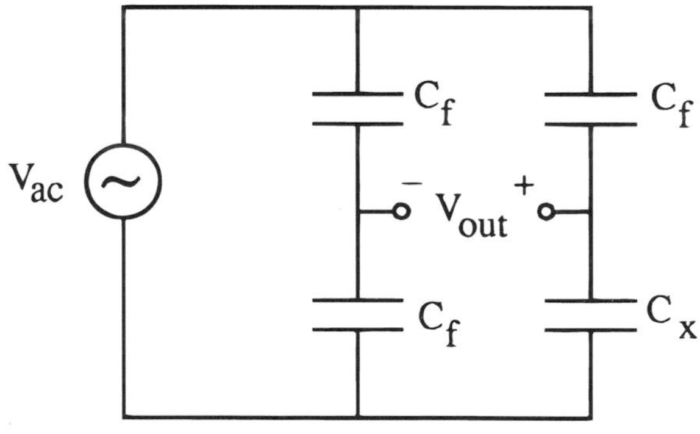

Figure 6.57 Capacitance bridge using an AC bias voltage.

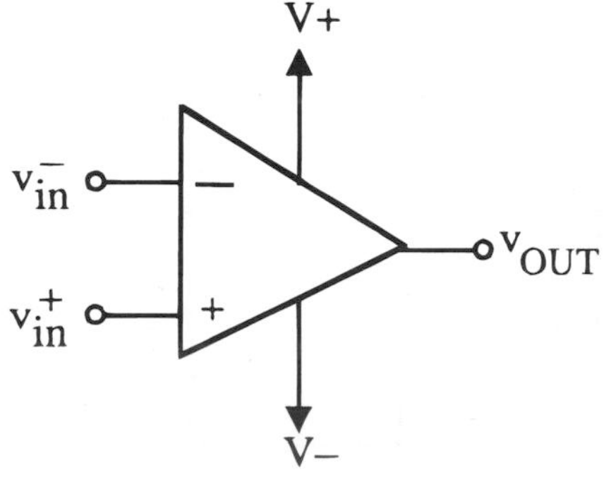

Figure 6.58 Operational amplifier circuit symbol. V+ and V− are the DC power supplies.

feedback. The two types of negative feedback op-amp circuits are *inverting* and *noninverting*. When negative feedback is employed, the input terminals are forced to have the same voltage, to a very close approximation, due to the high gain of the op-amp. This condition is called a virtual short-circuit. Analysis of op-amp circuits is simplified by assuming the input currents and dc offset voltages are negligible, and that the differential voltage gain (usually 10,000 to 100,000 volts/volt) is infinite. Caution must be used though, because high-precision measurements can be affected by these assumptions, and more careful analysis must then be employed.

The Unity-Gain Buffer

A very useful op-amp circuit is the unity-gain buffer shown in Figure 6.59.

Because of the virtual short at the input, the output voltage is equal to the input voltage. The key features of this circuit are

1. Negligible input current, resulting in no loading of the input source.
2. Gain of exactly one with no component trimming required.
3. Load current is supplied by the power supplies via the op-amp, thus isolating the load from the input source.

This circuit is often used as a "first line of defense" in protecting a sensor or bridge output from loading by the acquisition circuitry.

Noninverting Op-Amp Circuits

The resistive noninverting op-amp configuration is shown in Figure 6.60. The gain of this circuit is:

$$\frac{v_{OUT}}{v_{IN}} = 1 + \frac{R_f}{R_1} \tag{6.38}$$

Key features of this circuit are the gain can be adjusted by using a potentiometer for R_f or R_1, and the input resistance is very high. The effects of offsets generated by nonzero input currents can be reduced by approximately a factor of ten by using a resistor

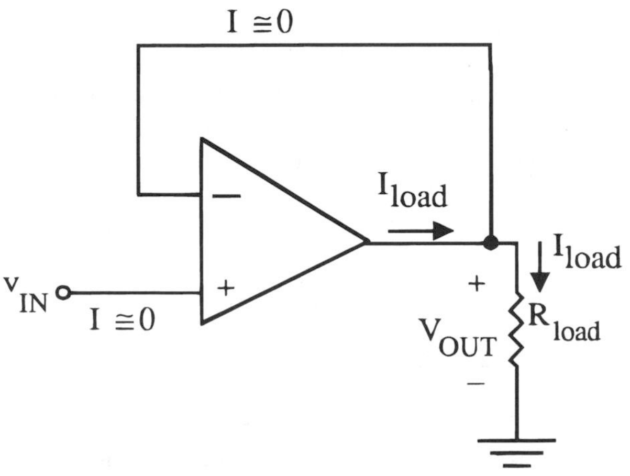

Figure 6.59 Unity-gain op-amp buffer circuit.

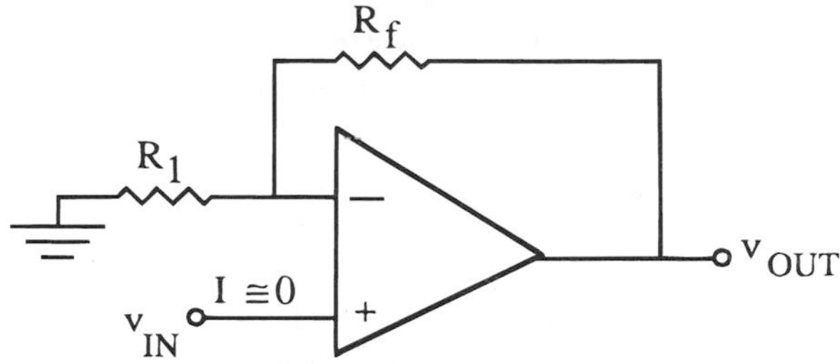

Figure 6.60 Noninverting op-amp circuit.

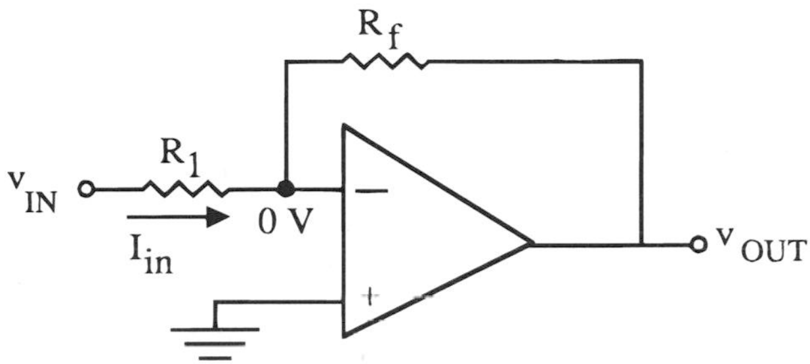

Figure 6.61 Inverting op-amp circuit.

R_B in series between the input and the (+) terminal, where R_B is equal to the parallel combination of R_f and R_1:

$$R_B = \frac{1}{\frac{1}{R_f} + \frac{1}{R_1}} \tag{6.39}$$

Inverting Op-Amp Circuits

The resistive inverting configuration is shown in Figure 6.61. The gain of this circuit is

$$\frac{v_{OUT}}{v_{IN}} = -\frac{R_f}{R_1} \tag{6.40}$$

The input resistance of this circuit is R_1. Figure 6.62 shows a modified circuit in which the (+) terminal of the op-amp is connected to a voltage V_{REF} rather than to ground. The output is

$$v_{OUT} = -\frac{R_f}{R_1} v_{IN} + \left(1 + \frac{R_f}{R_1}\right) V_{REF} \tag{6.41}$$

As for the noninverting case, in order to minimize the effects of input currents at the op-amp terminals, a resistor R_B equal to the parallel combination of R_f and R_1 should be inserted in series between the (+) terminal and the reference voltage, even if the reference used is ground.

The Instrumentation Amplifier

The instrumentation amplifier is a three op-amp circuit with the following useful features: very high input impedance, high voltage gain, differential input, low output resistance, and gain adjustment provided by a single resistor. The circuit is shown in Figure 6.63. The output voltage is given by

$$v_{OUT} = \frac{R_o}{R_2}\left(1 + \frac{2R_1}{R}\right)(v_1 - v_2) \tag{6.42}$$

The resistor R can be adjusted to provide control of the overall differential voltage gain. Three pairs of matched fixed resistors are required for R_o, R_1, and R_2.

Integrators

Signal integrators find wide use in sensor interfacing. They are primarily used for averaging and noise cancellation. They are also used in PID (proportional-integral-differential) process controllers. Electronic integrator circuits tend to suffer from offsets and drift, so software integration is preferred in installations where data is acquired by a computer, and where sufficient processing power and time are available.

The basic op-amp integrator circuit is shown in Figure 6.64. The (−) input terminal of the op-amp is forced to virtual ground (0 volts) by the feedback. Since $i_1 = i_2$, we can write the following equation for $v_{OUT}(t)$:

$$v_{OUT}(t) = v_{OUT}(t_o) - \frac{1}{RC}\int_{t_o}^{t} v_{IN}(\tau)\,d\tau \tag{6.43}$$

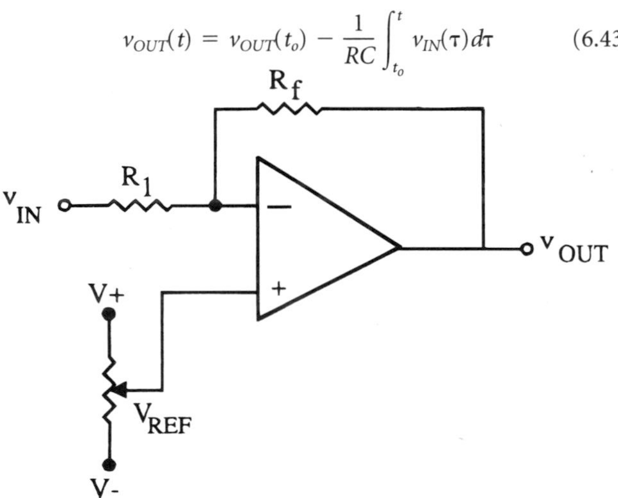

Figure 6.62 Inverting op-amp circuit with adjustable offset.

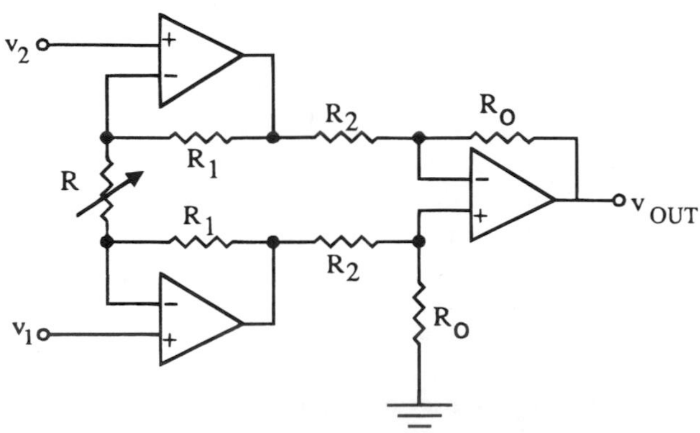

Figure 6.63 Instrumentation amplifier circuit.

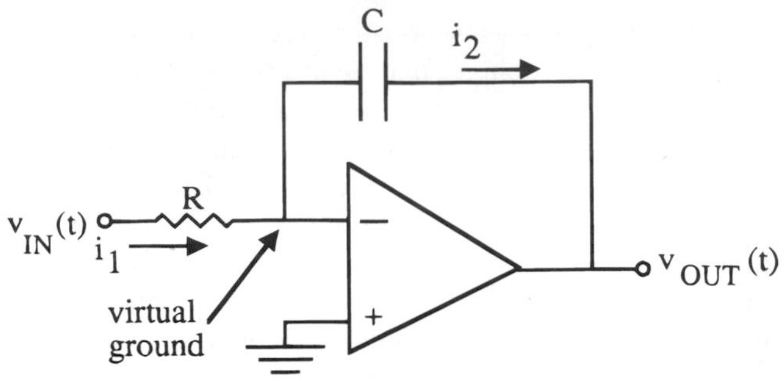

Figure 6.64 Op-amp integrator circuit.

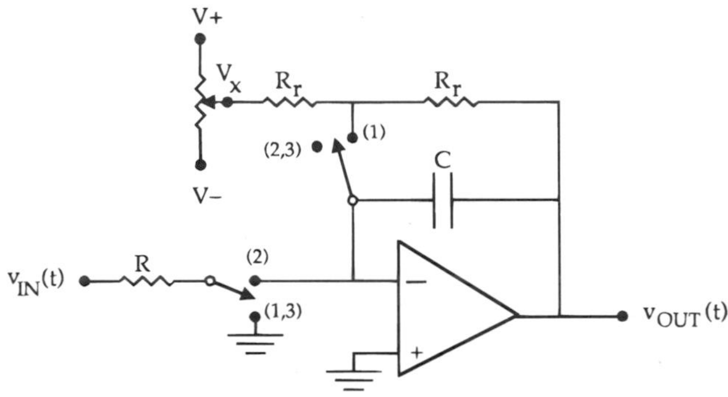

Figure 6.65 Op-amp integrator with reset (1), integrate (2), and hold (3) modes.

where $v_{OUT}(t_o)$ is the initial output voltage at time t_o. A significant practical limitation of the op-amp integrator is that the circuit integrates its own DC offsets, resulting in a nonzero *drift rate.* In order to minimize this undesirable effect, it is important to select a low-offset model op-amp, and the op-amp should be carefully nulled.

To prevent saturation of the integrator output, it is usually necessary to provide a means for periodically resetting the integrator. Figure 6.65 illustrates a technique for adding switch-controlled reset, integrate, and hold capability based on the inverting integrator of Figure 6.64. A potentiometer is used to provide V_x, which is the negative of the desired reset voltage. Alternatively, a fixed voltage could be used, or a voltage supplied by a computer-controlled digital-to-analog converter could be used. In most practical applications the switches are electronic switches or relays controlled by a timing circuit or computer.

In position 1 (*reset*), the output settles to the reset value $v_{OUT}(t_o) = -V_x$ with a time constant $\tau_r = R_rC$. In position 2 (*integrate*), the output follows Equation 6.43. In position 3 (*hold*), the output is constant at $v_{OUT}(t_f)$, where t_f is the time at which the switches are switched from *integrate* to *hold.* The ability of the circuit to hold the output constant depends upon the drift rate of the integrator and the capacitor leakage current.

High-Precision Op-Amp Circuits

Interfacing to high-impedance sensors and sensors with low output levels requires the use of high-precision op-amps. High-precision op-amps are manufactured for applications that require

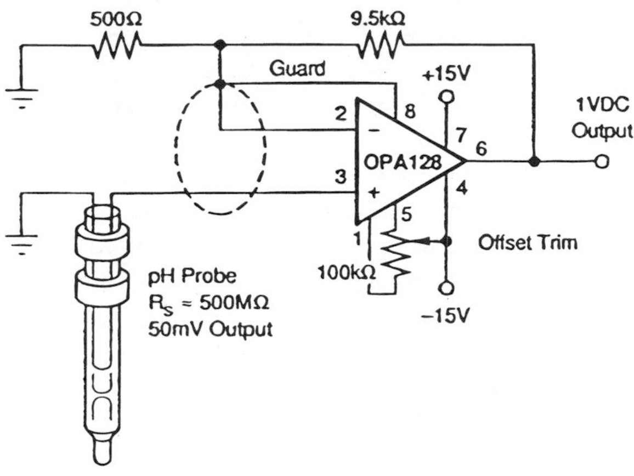

Figure 6.66 High-impedance ($10^{15}\Omega$) amplifier. © 1989–1994 Burr-Brown Corp. Reprinted with permission of Burr-Brown Corp., Tucson, AZ.

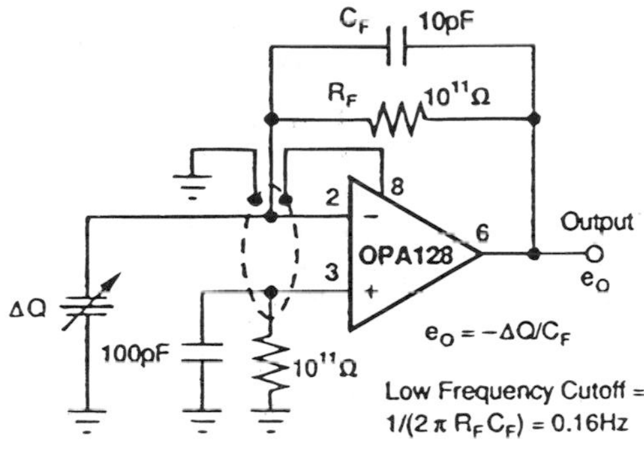

Figure 6.67 Piezoelectric transducer charge amplifier. © 1989–1994 Burr-Brown Corp. Reprinted with permission of Burr-Brown Corp., Tucson, AZ.

various combinations of low-noise, low input current, high bandwidth, and low drift (temperature insensitivity). An example of a high-precision op-amp is the Burr-Brown OPA128. The specifications for this model include subpicoampere input currents, as well as a maximum output drift of 5μV/°C. In order to take advantage

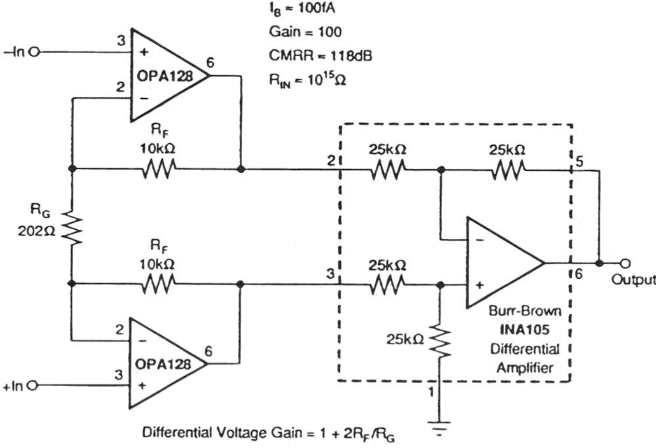

Figure 6.68 FET input instrumentation amplifier for biomedical applications. © 1989–1994 Burr-Brown Corp. Reprinted with permission of Burr-Brown Corp., Tucson, AZ.

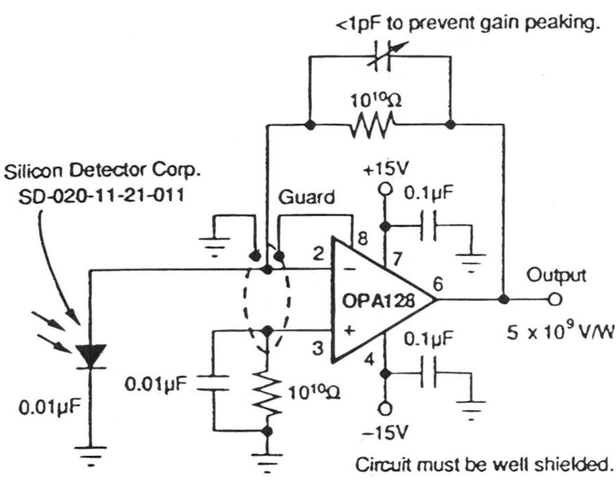

Figure 6.69 Sensitive photodiode amplifier. © 1989–1994 Burr-Brown Corp. Reprinted with permission of Burr-Brown Corp., Tucson, AZ.

of the extremely small input current characteristic of this op-amp, it is necessary to take precautions against leakage currents associated with board layout and interconnection. The manufacturer provides detailed information on board layout, guarding, and wiring types for use with this model. Figures 6.66 through 6.69 illustrate typical applications for this op-amp.

References

Irwin, J. D. 1993. *Basic Engineering Circuit Analysis,* 4th ed. Macmillan, New York, NY.

Omega Engineering, Inc. 1992. *The Temperature Handbook (Section Z),* Omega Engineering, Stamford, CT.

Omega Engineering, Inc. 1992. *The Pressure, Strain, and Force Handbook (Section Z),* Omega Engineering, Stamford, CT.

Traister, Robert J. 1989. *Operational Amplifier Circuit Manual,* Academic Press, San Diego, CA.

6.11 Thermal Effects in Industrial Electronic Circuits

Ray P. Reed

Introduction

Temperature is often incidental to electronic circuit application. Yet temperature, its magnitude, duration, distribution, and rate of change can have very important incidental effects in industrial circuits. Printed circuits, computer chips, Hall effect devices, measuring instruments, control systems, Wheatstone bridges, piezoelectric sensors, and semiconductor strain gauges are all examples of industrial electronic applications that can be adversely affected by electrothermal effects.

Within some range, temperature acceptably affects the electrical properties of all components. But, beyond characteristic temperature limits circuit behavior becomes anomalous. If the effect is reversible, components revert to normal behavior when returned to appropriate temperatures. At greater temperature extremes, components can be irreversibly changed in their electrical characteristics or even destroyed.

In addition to these familiar passive temperature effects, energy is converted between thermal and electrical forms by three quite distinct categories of physical effect: *thermoelectric, thermomagnetoelectric,* and *pyroelectric.* In industrial electronics these energy conversions can be applied usefully, but each of these also can occur as noise effects that must be deliberately controlled if significant.

Principles of Thermal-Electrical Effects

The Joule Effect

This most familiar of all thermal-electrical effects is only one of several mechanisms for conversion of electrical energy to heat (Pollock, 1990; Harman and Honig, 1986; deGroot, 1966; Seitz, 1940). Electrical current converts electrical to thermal energy. The familiar relation between electric current, I, a resistance, R, and the rate of heat evolution, dH/dt, is expressed by Joule's law:

$$dH = RI^2\ dt. \tag{6.44}$$

The Joule effect can be applied efficiently and directly for electrical heating, but it cannot be used for cooling. Indeliberately, in proportion to local circuit resistance, it elevates the temperature of circuit components and so affects their electrical characteristics through their temperature dependent parameters. Less obviously, it can also result in spurious emf indirectly through other thermal-electrical effects by affecting temperature distribution in a circuit or component. Also, Joule heating is superposed on other electrical to thermal effects wherever current exists and it can aggravate them.

Thermoelectric Effects

The next most familiar conversion between thermal and electrical energy is the set of three interrelated thermoelectric phenomena: *Seebeck effect, Peltier effect,* and *Thomson effect.* Despite their familiarity, widespread use, and universal occurrence in industrial electronics, the nature of these thermoelectric effects is very often misrepresented and their importance is understated, even in the current technical literature. The three thermoelectric effects are often confused with each other and with the quite different *contact potential* phenomenon (Reed, 1982, 1993; Pollock, 1990, 1991; Harman and Honig, 1986; deGroot, 1966; Seitz, 1940). They are often misunderstood by instrument designers and users in avoidable ways that can degrade performance in practical application. The effects are illustrated in Figure 6.70. Thermoelectric effects occur in all electrically conducting materials that are not at uniform temperature. They occur in solids, liquids, and gases. They are most familiar in metallic materials but they occur also in conducting polymers. The effects are strongest in semiconductors. They occur normally in superconductors at temperatures above critical superconducting temperature thresholds, T_c. Along with resistivity, they vanish in the immediate vicinity of 0 K in all materials, but the effects vanish over significant temperature spans only in superconductors at all temperatures below T_c. In elemental metals T_c is less than 10 K. However, in special "high-temperature" superconductor alloys T_c can exceed 100 K. (Pollock, 1990; Doss, 1989).

Absolute Thermoelectric Properties. For individual materials, the strengths of these thermoelectric effects are characterized respectively by the temperature dependent *Seebeck coefficient,* $\sigma(T)$, *Peltier coefficient,* $\pi(T)$, and *Thomson coefficient,* $\tau(T)$. To distinguish that these coefficients are intrinsic transport properties of individual materials they are called "*absolute*" rather than "*relative*" properties (Reed, 1982, 1993; Pollock, 1991, 1990; Harman and Honig, 1986; deGroot, 1966; Seitz, 1940). Absolute thermoelectric properties have been derived from Thomson effect measurements from near 0 K for reference materials such as lead, copper, platinum, and other elemental metals. Representative values of absolute thermoelectric coefficients are presented in Table 6.12.

Relative Thermoelectric Properties. Even though these physical transport properties are of individual materials, in application it is often possible and, where possible, usually is more convenient to employ relative coefficients between pairs of materials that are temperature related in a circuit. For example, the *relative Seebeck coefficient* for a segment of any material, m, relative to reference material, r, that shares the same endpoint temperatures is (Reed, 1982, 1993, 1992; Wang, 1992; Pollock, 1991, 1990; Harman and Honig, 1986; Kinzie, 1973; deGroot, 1966; Seitz, 1940):

$$\sigma_{mr} \equiv \sigma_m - \sigma_r, \text{ from which} \tag{6.45}$$

$$\sigma_{rm} = -\sigma_{mr}, \text{ and} \tag{6.46}$$

$$\sigma_m = \sigma_{mr} + \sigma_r \tag{6.47}$$

Corresponding relations hold for Seebeck emfs and the other thermoelectric properties as well. Absolute properties that are unknown for particular materials can be deduced from properties

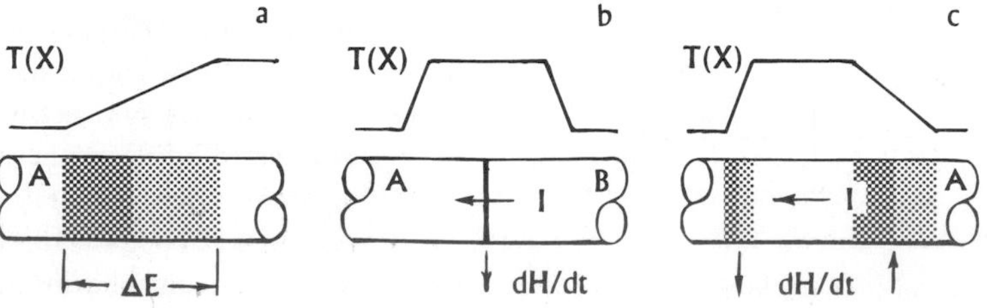

Figure 6.70 The three thermoelectric effects: (a) Seebeck, (b) Peltier, and (c) Thomson.

Table 6.12 Representative Values of Absolute Seebeck Thermoelectric Coefficients of Some Materials Used in Industrial Electronic Circuits (Reed, 1992, 1993; Wang, 1992; Kiazie, 1993)

	Seebeck Coefficient, μV/°C			
	0°C	20°C	100°C	400°C
Lead	0.03×10^{-3}	0.05×10^{-3}	0.08×10^{-3}	0.11×10^{-3}
Tin	0.03×10^{-3}	0.06×10^{-3}	0.09×10^{-3}	0.12×10^{-3}
Copper	1.72	1.82	2.23	3.85
Silver	1.42	1.50	1.84	4.07
Gold	2.3	2.1	2.0	2.3
Tungsten	1.9	4.1	6.7	12.1
Chromium	13.2	14.4	15.3	17.3
Nickel	−7.0	−9.7	−12.4	−15.0
Platinum	−4.2	−7.2	−9.7	−13.1
Brass	0.7	0.82	1.33	1.95
Kovar	0.20	0.20	0.19	0.02
Manganin	1.37	1.39	1.45	1.95
Nichrome	20.84	20.24	17.85	11.89
Silicon	−408	417	−455	−502
Germanium	−303			
CuO	−696			
Cu_2O	−474 − 1150			
Mn_2O_3	−385			

Note: Values reported in the literature are for nominal materials that may not be well documented as to composition and state. They are presented only to allow estimates of plausible Seebeck emf contributions. Specific values should be determined for critical applications.

relative to a reference material for which the absolute property is known (Equation 6.47) (Reed, 1993, 1992; Wang, 1992).

The Kelvin Relations. The three thermoelectric effects are thermodynamically connected by relations first noted by Lord Kelvin (William Thomson) (Reed, 1982, 1993; Pollock, 1991, 1990; Harman and Honig, 1986; deGroot, 1966; Seitz, 1940):

$$\tau_m = -T(d\sigma_m/dT) \tag{6.48}$$

$$\tau_m = T\sigma_m \tag{6.49}$$

where T, here, is the absolute temperature and the single subscripted coefficients are for the absolute properties of individual materials, m. The relations hold for relative coefficients as well. If the temperature-dependent value of any one of the coefficients is known then it is known indirectly also for the other two through the Kelvin relations.

The Seebeck Effect. The most familiar thermoelectric effect is the phenomenon recognized by T. J. Seebeck and first reported in 1823. The Seebeck effect is the occurrence of an electrical potential difference between any distinct pair of points of an electrically conducting material that are simultaneously at different temperatures (Pollock, 1991, 1990; Reed, 1982, 1993).

The Seebeck coefficient for a homogeneous material m is

$$\sigma_m(T) = \lim_{\Delta T \to 0} \Delta E/\Delta T \tag{6.50}$$

where ΔE is the Seebeck emf between the two points and ΔT is the temperature difference between them. The Seebeck effect is the only *thermoelectric* effect that converts heat to electrical form. Widely misunderstood, the Seebeck effect is not a junction effect as its occurrence does *not* depend on the association of dissimilar materials. Nevertheless, the Seebeck effect is most easily observed and is always applied as a differential effect in circuits of two or more dissimilar materials. Though the phenomenon is not a junction effect, it is true that the temperatures of interfaces (i.e., real junctions) between thermoelectrically dissimilar materials do indirectly control the *net* Seebeck emf in circuits (Pollock, 1991, 1990; Reed, 1982, 1993).

Most widely used in the application of thermocouples as self-generating sources of voltage in thermometry, the phenomenon is also used for the thermal generation of electricity as in miniature nuclear generators and flameout sensors for gas pilot lights. In the measurement of low-level signals of millivolt or smaller order the Seebeck effect is a most troublesome and insidious noise source that must be minimized by equipment or experiment design based on an authentic understanding of the effect.

The Peltier Effect. The Peltier effect (1834) is an absorption or expulsion of heat at any isothermal interface (a junction) between dissimilar conductive materials through which current flows (Reed, 1982, 1993; Pollock, 1990, 1991; Harman and Honig, 1986; deGroot, 1966; Seitz, 1940). Every electrically conductive material is characterized by an absolute Peltier coefficient that expresses the capacity of that material to convey heat transported by electric current. Nevertheless, the Peltier effect is a junction effect as its occurrence does depend on the transition between dissimilar materials that have different capacities to transport the heat energy of the electrical current. Unlike the Joule effect, the Peltier effect allows cooling as well as heating for application. Also, unlike the Joule effect, the net heat flow is localized to the immediate vicinity of any junction between dissimilar materials. In a circuit of several dissimilar materials, each junction is heated or cooled in proportion to the current through that junction, at the temperature of the individual junction (independent of other junction temperatures), and to the relative Peltier coefficients of joined materials. The absolute Peltier coefficient for each individual material m is defined by the relation:

$$\pi_{km}(T) = (dH/dt)/I \tag{6.51}$$

At an interface, the net heat conveying capacity of the joined materials is expressed by the relative Peltier coefficient $\pi_{ab}(T) = \pi_a - \pi_b$. The rate of heat flow is proportional to current. The sense of heat flow depends on the direction of current. Where some junctions in a circuit are cooled by the current, other junctions in that circuit may simultaneously be heated by the same current. In industrial electronics, the most common application is for cooling small heat sensitive components but sizeable thermoelectric refrigerators for cooling food and producing ice and temperature controllers for calibration are commercially available. The Joule effect is much more efficient for heating alone, but thermostatic temperature control can be simplified by using the Peltier effect as either heating or cooling of a single extended junction can be accomplished electrically by regulating the magnitude and direction of current through it.

The Thomson Effect. The Thomson effect is the absorption or expulsion of current transported heat along an electrically conducting material wherever the temperature varies with position (Reed, 1982, 1993; Pollock, 1990, 1991; Harman and Honig, 1986; deGroot, 1966; Seitz, 1940). The Thomson effect (1856) was first predicted, and then years later was demonstrated, by William Thomson. The rate and the sense of heat conversion depend on the magnitude and on the sense of both current and the local temperature gradient. The absolute Thomson coefficient for an individual material, m, depends on absolute temperature and is

$$\tau_m(T) = (dH_\tau/dt)/(I\ dT/dx). \tag{6.52}$$

Rarely employed in practice, the most notable use of the Thomson effect is in the indirect determination of the absolute Seebeck and Peltier coefficients of individual materials.

Similarities and Differences between Thermoelectric Effects. In common with the Seebeck effect, the Thomson effect occurs within conducting materials apart from junctions. Unlike the Seebeck effect, which exists even in open circuit, where there is no current there can be neither Peltier nor Thomson effect. To the contrary, the Seebeck effect is best used for sensing under series "open circuit" conditions to avoid resistive voltage drop and the slight effect of current-induced Peltier heating or cooling of junctions. The Peltier and Thomson effects relate only to the conversion from electrical to thermal energy form, not thermal conversion to electrical. There is no Peltier emf. There is no Thomson emf. Where there is no spatial temperature variation there can be neither Seebeck effect nor Thomson effect.

Thermomagnetoelectric Effects

By combining electric current, temperature gradient and magnetic field, considering both longitudinal and transverse behavior, and both isothermal and adiabatic boundary conditions, it is possible to define several hundred physical effects (Harman and Honig, 1986). Generally, these magnetic field related effects are classified as some nonstandardized combination and sequence of the forms thermo-, galvo-, galvano-, electro-, and magneto- effects. Four of these magneto- related effects that have been named and studied are the Nearnst, Ettinghausen, and Righi-Leduc effects that are related to the more familiar and widely applied Hall effect.

Three of these thermomagnetoelectric effects relate to a conversion between thermal and electrical energy forms. However, they are distinct from thermoelectric effects in that an essential magnetic field is operative in the conversion through the influence of the Lorentz force (Pollock, 1990; Harman and Honig, 1986; deGroot, 1966; Seitz, 1940). The values of thermoelectric coefficients of some materials are dependent on the strength of a magnetic field to which they are exposed. That incidental magnetic dependence, where the longitudinal thermoelectric coefficient is modulated by the magnetic Lorentz force, is classified as a thermomagnetoelectric effect. As with thermoelectric effects, the thermomagnetoelectric effects are sometimes misrepresented and misnamed in the technical literature. Little used in application, in special circumstances (as in broadly extended planar conductors that are exposed to a substantial normal magnetic field and in Hall effect applications) the effects can be troublesome in producing significant noise. The effects are illustrated in Figure 6.71.

The Hall Effect. In a strip of conducting material m of thickness l_y, length l_x, and width l_z exposed to an electric current I_x in the x direction and simultaneously to a magnetic field of strength B_y in the orthogonal y direction, a Hall emf ΔE_z is created in a direction perpendicular to each. The Hall coefficient is

$$\gamma_m = (\Delta E_z l_z)/(I_x B_y) \tag{6.53}$$

As the Hall voltage is proportional to both current and magnetic field, it can be used to sense either quantity. In such applications, the other three effects occur as potential noise sources.

The Nearnst Effect. The Nearnst effect is the occurrence of an electrical potential gradient, dE_z/dz, or net difference, ΔE_z, that is caused by and arises perpendicular to both temperature gradient and magnetic field that are imposed simultaneously (Pollock, 1990; Harman and Honig, 1986; deGroot, 1966; Seitz, 1940). The Nearnst coefficient is

$$\nu_m = -\Delta E_z/(dT/dx\ H_y l_y) \tag{6.54}$$

Unlike the Seebeck effect that also results from the temperature gradient, the Nearnst effect exists only with the magnetic field.

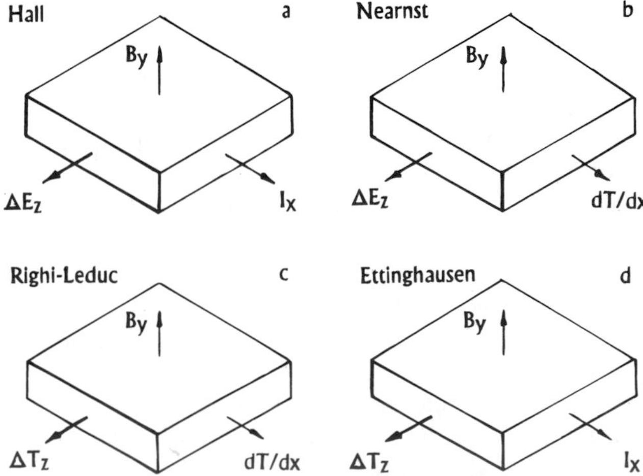

Figure 6.71 Thermomagnetoelectric effects: (a) Hall, $\Delta E_z = f(I_x,B_y)$, (b) Nearnst, $\Delta E_z = f(dT/dx,B_y)$, (c) Righi-Leduc, $\Delta T_z = f(dT/dx,B_y)$, and (d) Ettinghausen, $\Delta T_x = f(I_x,B_y)$.

The Righi-Leduc Effect. This is the occurrence of a temperature gradient, dT/dz, in response to an orthogonal applied temperature gradient, dT/dx, and magnetic field, B_y (Pollock, 1990; Harman and Honig, 1986; deGroot, 1966; Seitz, 1940). B_y. The Righi-Leduc coefficient is

$$\lambda_m = -\Delta T_z/(dT/dx\ B_y\ l_y) \tag{6.55}$$

The Ettinghausen Effect. The Ettinghausen effect, converse to the Nearnst effect, is the occurrence of a transverse temperature gradient that is caused by and arises perpendicular to a simultaneous electric current and magnetic field (Pollock, 1990; Harman and Honig, 1986; deGroot, 1966; Seitz, 1940). The Ettinghausen coefficient is

$$\epsilon_m = -(\Delta T_z l_z)/(I_x B_y) \tag{6.56}$$

where I_x is current, B_y is magnetic field, and ΔT_z is the resulting transverse temperature difference.

The Nearnst effect can directly produce an electrical noise signal while the Ettinghausen and Righi-Leduc effects can produce noise emf only indirectly by creating a temperature difference that results in a Seebeck emf in dissimilar material circuits. Representative thermomagnetoelectric coefficients are presented in Table 6.13.

Pyroelectric Effects

Some crystalline materials that lack structural symmetry exhibit electrical effects (pyroelectricity) in response to temperature and its distribution or change (Lang, 1974; Mattiat, 1971; Cady, 1964; Washburn, 1926). All pyroelectric materials also are piezoelectric (responding to strain with electrical charge—a source of microphonic noise). However, not all piezoelectric materials are pyroelectric. Of thirty two crystal symmetry classes, only ten are pyroelectric (twenty can be piezoelectric.) These materials are most noted for their piezo- or ferroelectric activity and for passive dielectric properties.

The pyroelectric response can be very large as in commonplace polycrystalline ceramics such as barium titanate and lead zirconate titanate. Various temperature states result in the occurrence of electric charge on certain crystallographic faces. In electrical components, the charge is collected on electrodes applied to particular crystallographic faces of oriented crystal cuts. The nature of the effects is influenced by mechanical constraint. Depending on the application circuit, the terminal voltage can relate to temperature value, gradient, or rate of change.

The pyroelectric property may be spontaneous, as in some single crystals such as tourmaline, or it may require poling of polycrystalline materials by prolonged application of an electric field to align crystal domains. The pyroelectric effects are of several distinct types (Figure 6.72). They can occur as significant incidental effects in materials that are used widely in industrial electronics as in capacitors, filters, oscillators, or sensors. Common materials that can have substantial pyroelectric character include single crystal lithium niobate, and tourmaline, polycrystalline ceramics such as barium titanate and lead zirconate titanate, and ferroelectric polymers such as polyvinylidene difluoride and its copolymers (Lang, 1974).

Tourmaline is a naturally occurring mineral that is inherently pyroelectric. Lithium niobate is grown as a synthetic single crystal but pyroelectric behavior is introduced by poling under an electric field. Barium titanate and lead zirconate titanate are uniformly polycrystalline ceramics while polyvinylidene difluoride

Table 6.13 Representative Thermomagnetoelectric Coefficients*

Material	Hall Coefficient [m-volts / amp-tesla]	Nearnst Coefficient [volts / °C-tesla]	Righi-Leduc Coefficient [1 / tesla]	Ettinghausen Coefficient [°C-m / amp-tesla]
Ag	-8×10^{-11}	-43×10^{-17}	-4×10^{-18}	-2×10^{-16}
Al	-3×10^{-11}	$+39 \times 10^{-18}$	—	$+1 \times 10^{-16}$
Au	-7×10^{-11}	-181×10^{-18}	-3×10^{-18}	-1×10^{-16}
Cd	$+6 \times 10^{-11}$	—	$+1 \times 10^{-18}$	-3×10^{-16}
Cu	-6×10^{-11}	-27×10^{-17}	-2×10^{-18}	-2×10^{-16}
Fe	$+100 \times 10^{-11}$	-86×10^{-17}	$+4 \times 10^{-18}$	-43×10^{-16}
Ni	-60×10^{-11}	$+53 \times 10^{-16}$	-2×10^{-11}	$+30 \times 10^{-16}$
Pb	-5×10^{-11}	-5×10^{-18}	—	—
Pt	—	—	-1×10^{-19}	—
Si	—	$+33 \times 10^{-15}$	3×10^{-18}	3×10^{-10}
Sn	—	-4×10^{-18}	—	—
Te	—	$+36 \times 10^{-14}$	$+40 \times 10^{-18}$	$+10 \times 10^{-12}$
W	—	-100×10^{-17}	$+2 \times 10^{-18}$	—
Zn	$+3 \times 10^{-11}$	-73×10^{-18}	$+1 \times 10^{-18}$	-3×10^{-16}
Source**	†	‡	‡	‡

*For illustration of relative values only.

**Adapted from † Seitz, F. 1940. *Modern Theory of Solids,* McGraw-Hill, New York, NY and ‡ Washburn, E. W., ed., 1926. *International Critical Tables of Numerical Data—Physics, Chemistry, and Technology,* Vol. VI, McGraw-Hill, New York NY. Source values are for temperatures and magnetic fluxes in the vicinity of 30°C and 1 tesla. Properties and test conditions were not well documented. For application, confirm appropriate properties. For the reader's convenience, physical units have been converted to SI conventions. They have not been corrected to 1990 revised voltage and temperature scales.

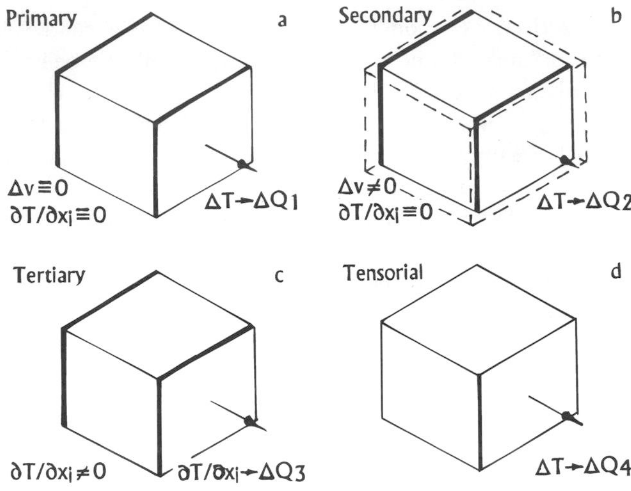

Figure 6.72 Pyroelectric effects: (a) primary vectorial pyroelectricity, (b) secondary vectorial pyroelectricity, (c) tertiary pyroelectricity, and (d) tensorial pyroelectricity.

contains crystallites within an amorphous matrix. Polycrystalline materials must be electrically poled to introduce active pyroelectricity of useful strength but they could also have some remanent polarization introduced incidentally in processing or use. Often used polarized as active charge sources, these materials also are widely used in passive applications as dielectric materials and as coatings. If incidentally active, they can introduce substantial electrical noise from temperature changes.

Vectorial Pyroelectric Effects. The ordinary pyroelectric effect is a linear reversible change in polarization due to a uniform change of temperature with the electrical field held constant. Primary and secondary forms of the effect are directional or vectorial effects due to temperature change. The vectorial effects are distinguished by the state of mechanical constraint of the crystal.

Primary Pyroelectricity. The primary pyroelectric effect is the occurrence of charge in response to a homogeneous change of temperature in a crystal that is mechanically *constrained* from changing dimension (Lang, 1974; Cady, 1964; Washburn, 1926). The vectorial pyroelectric coefficient relates to a vector component, P_i, of the change of polarization per unit of temperature change. The primary pyroelectric coefficient is

$$(p_i)_1 = d(P_i)_1/dT \qquad (6.57)$$

Secondary Pyroelectricity. The secondary pyroelectric effect is the occurrence of change in response to a homogeneous change of temperature in a crystal that is *free* to deform (Lang, 1974; Cady, 1964; Washburn, 1926).

$$(p_i)_2 = d(P_i)_2/dT \qquad (6.58)$$

Total Pyroelectricity. In the general state of deformation the crystal may be constrained in some directions and not in others. The sum of the primary and secondary effects is

$$(p_i)_t = (p_i)_1 + (p_i)_2 \qquad (6.59)$$

Despite the fact that there are other pyroelectric effects that are not included, the sum of primary and secondary vectorial effects is usually called the *total pyroelectric effect.* It is this property that usually is inferred by the less specific generic term "pyroelectric coefficient".

In particular application geometries the pyroelectric response is determined by the crystallographic cut, electrode geometry, and inherent mechanical constraint. In such instances, the relevant piezoelectric effect is assumed without distinction and simply the symbol, p_i, is used, absent a statement of type. Representative pyroelectric coefficients are compared in Table 6.14.

Tertiary Pyroelectricity. Charge generated in a piezoelectric crystal due to a temperature gradient, $\partial T/\partial x_i$, analogous to the Seebeck thermoelectric effect, is classified as the tertiary pyroelectric effect. This effect can occur in any piezoelectric crystal, even those that are not piezoelectric under homogeneous temperature change (Lang, 1974; Cady, 1964; Washburn, 1926). The strength is given by the coefficient:

$$(p_i)_3 = \Delta Q/(\partial T/\partial x_i) \qquad (6.60)$$

The tertiary pyroelectric effect is a substantial noise source in pyroelectric experiments where temperature uniformity is not maintained. Therefore, it can contribute to pyroelectrically induced noise in industrial circuits. Both secondary and tertiary pyroelectricity are indirectly of piezoelectric origin as induced by temperature or its gradient. They both can be of greater magnitude than the primary pyroelectric effect.

Tensorial Pyroelectricity. A real, but extremely small, tensorial pyroelectric effect occurs in all crystal classes except the cubic system (Lang, 1974; Cady, 1964; Washburn, 1926). In a

Table 6.14 Representative Values of Pyroelectric Coefficients of Some Crystalline and Polycrystalline Materials Used in Industrial Electronic Circuits Near Room Temperature (Lang, 1974; Cady, 1964; Washburn, 1926)

Material	p_{tot} μC/m²K
BeO	3.4
Tourmaline	4
PVF	10
PVDF	20–40
$VF_2 - VF_3$ copolymer	30–40
γ-Nylon 11	30–40
$BaTiO_3$	200
PZT-5	60–500

Note: For illustration of relative values only. Values are not well documented with regard to material properties or test conditions. For application, confirm appropriate properties.

uniformly heated crystal the tensorial pyroelectric effect produces charges of like sign at edges at the ends of particular crystal axes (Lang, 1974; Cady, 1964; Washburn, 1926). The tensorial effect is negligible as a noise source in industrial circuits.

The Electrocaloric Effect. Analogous to the distributed Joule heating effect and inverse to the pyroelectric effects, the electrocaloric effect in crystals transforms electrical energy into heat in response to a change in electric field (Cady, 1964). The effect is significant only at very low temperature and in the vicinity of a ferroelectric phase transition. Under ordinary circumstances it is insignificant in industrial electric circuits. It can be a significant source of error in cryogenic measurements.

Analysis of Thermal-Electrical Circuit Effects

Most thermal-electrical effects are experienced localized to discrete circuit components where the source of noise signal is readily recognized and controlled. The consequences of such effects is determined by conventional circuit analysis. Distinctively, the Seebeck thermoelectric effect can subtly be distributed over an entire electrical circuit and effectively can arise not from junctions but from widely separated parts of a circuit that have no obvious association. Its insidious nature makes it an often troublesome noise source. As all nonisothermal conductors are active sources of Seebeck emf, where low level signals are of interest all circuits must be recognized as thermoelectric circuits. A special method of circuit analysis is required to treat thermo electric circuits that, even if incidental, occur universally in modern industrial electronic application.

Thermoelectric Circuit Analysis

The analysis of thermoelectric circuits for application or for noise control is consistent with, but different from, the customary analysis of electric circuits. In ordinary circuit analysis, wiring is viewed merely as a passive conduit to convey signals between components. Thermoelectric circuits must be analyzed differently. This is because "conductors" can be unrecognized variable Seebeck emf sources that subtly change in their nature, sensitivity, and polarity in response to normally changing temperature distribution. Simplistic thermoelectric models, idealized to homogeneous two-material series circuit thermocouples, present the Seebeck effect in ways that may be physically unrealistic. There is a simple, practical, more general, and physically authentic method that is better adapted to the analysis of complex industrial thermoelectric circuits. (Reed, 1982)

The Functional Model of Thermoelectric Circuits

Behavioral rules for thermocouple circuits, which had long been recognized individually, were first formally codified as a set by Roeser (Reed, 1993, 1982). His narrow intent was to simplify the practice of thermoelectric thermometry by providing a concise set of only three rules sufficient for thermometry. He loosely termed the set the "Laws of Thermocouple Circuits" (Pollock, 1991, 1990; Reed, 1992, 1993). These "laws" are actually corollaries to a single Seebeck law that fully describes the Seebeck effect in a way well adapted to circuit analysis. Though very widely parroted, the laws, as stated by Roeser, conceal as much as they reveal about practical behavior of the more general circuit. They are simplistic in addressing only circuits effectively of only two ideal materials, inappropriate in focusing on current rather than voltage, and restrictive in only indirectly applying to more general practical situations.

A more general, yet simple, approach is available. It applies to the understanding and analysis of realistic compound circuits as are encountered in industrial electronics (Pollock, 1991, 1990; Reed, 1982, 1993). It is reviewed here. The important understanding of the Seebeck effect as it relates to industrial electronics is based on four features that focus on function:

1. A basic Seebeck source element.
2. A single Seebeck law.
3. Four practical corollaries from that law.
4. A graphic $T(X)$ depiction for visualization.

The Seebeck Source Element. The Seebeck phenomenon is a three dimensional effect of vector character. Nevertheless, thermoelectric effects as most often encountered in industrial electronics occur in conductors that can be treated as isotropic one-dimensional filaments. The effects usually occur at low frequency. Except for unusual circumstances, for example where thin film or very tiny thermocouples are exposed to very rapid pulsed heating, the effect of capacitive and inductive parameters is negligible. Therefore, the basic circuit element usually needs include only DC resistance (Figure 6.73). Often ignored, that resistance may be important in causing voltage drop in industrial circuits that carry significant current. Thermoelectric effects may be superposed in powered circuits and those with parallel branches.

With these restrictions, the basic thermoelectric circuit element is a nonideal voltage source (a segment within a filament) of material, m, such as a wire, circuit land, narrow foil, or chip of a given resistance with a voltage sensitivity characterized by a Seebeck coefficient. Note that segment endpoints a and b need not, but may, coincide with physical endpoints of the conductor. Each basic segment, of arbitrary length, is thermoelectrically homogeneous (i.e., it is momentarily uniform in its absolute Seebeck coefficient, $\sigma_m(T)$, over the full span of the segment, and the endpoint temperatures are both known but they may both vary with time. Individual conductors of industrial circuits that are of a single nominal material are frequently inhomogeneous

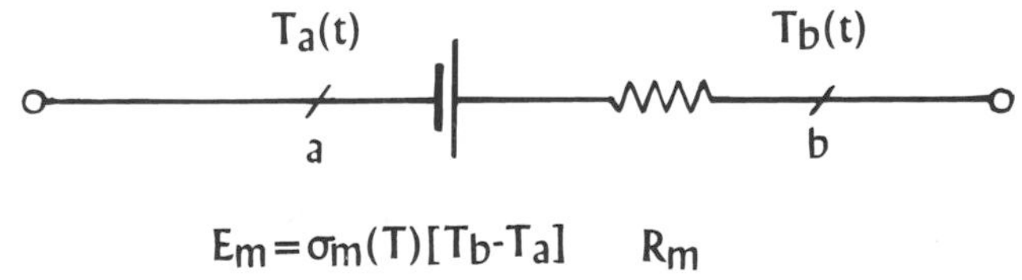

Figure 6.73 The basic Seebeck thermoelectric emf source element.

(i.e., different locations within a conductor are characterized by significantly different $\sigma(T)$) particularly where thermoelectric behavior is merely incidental. Chemically identical materials can have very different Seebeck coefficients due, for example, to mechanical strain (or to very small thickness as in thin films.) Any conductor (whether globally homogeneous or inhomogeneous) is viewed as composed of one or more such homogeneous segments. Note that, distinctively, the magnitude and also the polarity of each Seebeck emf source element depend momentarily on both temperatures, unlike an electrochemical cell.

The Seebeck Law. For such individual circuit segments, the sensitivity of the Seebeck effect is defined, as in Equation 6.50, by the simple differential Seebeck law:

$$\sigma(T) = dE(T)/dT \tag{6.61}$$

Recall that the Seebeck emf, $E(T)$, is the electrical potential difference between a pair of segment endpoints that are not necessarily the conductor endpoints. Note particularly that the temperature differential is the temperature increment between those same endpoints, *not* a homogeneous change of a local temperature. Alternately, this fundamental law can be expressed as:

$$dE(T) = \sigma(T)dT \tag{6.62}$$

or else as the definite integral,

$$E(T) = \int_{T_a}^{T_b} \sigma_x(T)dT \tag{6.63}$$

The temperature-dependent Seebeck coefficient may, itself, be dependent on other environmental variables, such as strain or magnetic field. However, any dependences of σ other than on temperature, though possibly significant, are not the Seebeck effect.

Corollaries to the Seebeck Law. The simplicity of the Seebeck law, Equation 6.62, conceals many practical implications. Four particularly revealing corollaries have been stated formally elsewhere (Pollock, 1991, 1990; Reed, 1982, 1993). These corollaries were first applied to thermoelectric thermometry. They are even more important in the analysis of industrial circuits. In conjunction with a simple plot of junction temperatures, T_i, versus junction relative positions in the circuit, X_i, these corollaries force the attention to the circuit locations that are critical in analysis and application. The corollaries are intended only to be understood to provide insight. They are not to be memorized. They apply fully to all series electrical circuits and, in most regards, to circuits containing closed loops. Their very real practical significance is best understood through application to a variety of practical circuit circumstances such as those that will be presented.

The Corollary of Functional Roles. Every electric circuit is composed of segments that function either as net Seebeck emf source elements, or as "conductors" that, effectively, contribute no net Seebeck emf, and of interfaces between thermoelectrically dissimilar materials that function as "real junctions".

(Note: This corollary, stated in more specific detail elsewhere and complemented informally below, identifies those several circumstances where conducting segments, individually or in pairs, function effectively as passive elements rather than as *net* Seebeck emf sources. Nonisothermal resistors, inductors, and all other electrically conductive circuit components (that may have other intended electrical functions) must also be recognized as incidental sources of Seebeck emf (Pollock, 1991, 1990; Reed, 1982, 1993).

The Corollary of Functional Determinacy. Thermoelectric roles of segments are variable as determined by the momentary distribution of the temperatures only of all real junctions of a circuit and cannot be independently determined by circuit design or by material choice alone.

The Corollary of Temperature Determinacy. The net Seebeck emf of any circuit depends on the temperatures of all real junctions and is independent of temperature distribution within its homogeneous segments. (The temperature of any junction can be deduced from the net Seebeck voltage only if the simultaneous temperatures of all other junctions are known.)

The Corollary of Seebeck emf. The net Seebeck emf of any Seebeck source element of material, *m*, is

$$E_m(T) = \int_{T_a(X_a)}^{T_b(X_b)} \sigma_m(T)dT, \text{ or} \tag{6.64}$$

$$= E_m[T_b(X_b)] - E_m[T_a(X_a)] \tag{6.65}$$

It is emphasized that the corollaries of temperature determinacy and of Seebeck emf apply only to thermoelectrically homogeneous segments for which, local to each, $\sigma(T)$ is a known momentarily fixed function that applies to the entire segment.

For many materials of general interest, standardized characteristics are defined in polynomial form (Reed, 1982, 1993). For convenience, the results, Equation 6.65, of the indicated integration for emf from the nonlinear Seebeck coefficient are tabulated relative to a fixed reference temperature, T_a, for table lookup (Reed, 1982, 1993). The limits of integration expressed in the corollary of Seebeck emf are shown jointly in explicit terms of both temperature and position at the segment endpoints to emphasize the important fact that the temperature integration is between spatial locations where the corresponding segment endpoint temperatures happen to exist.

From the corollary of Seebeck emf, Equation 6.64 or 6.65, and from Equation 6.47, it follows that over any temperature interval where the Seebeck coefficient, either absolute or relative, happens to be essentially constant,

$$E_m \approx \sigma_m \cdot (T_b - T_a), \text{ or} \tag{6.66}$$

$$E_{mr} \approx \sigma_{mr} \cdot (T_b - T_a). \tag{6.67}$$

The Thermoelectric Circuit T/X Sketch. The behavior of thermoelectric circuits has been depicted in a variety of planar presentations that relate different pairs of thermoelectric variables such as $E(T)$ or $E(X)$. However, the subtly powerful T/X plane representation, Figure 6.74, is not customary. Nevertheless, the corollaries reveal a strong practical motivation for visualizing thermoelectric circuits in the $T(X)$ plane (Pollock, 1991, 1990; Reed, 1982, 1993). This plot generally differs from the related $E(X)$ representation because thermoelectric $E(T)$ is significantly nonlinear.

The "spatial" variable, X, represents relative position or junction sequence around the circuit as the circuit is traversed progressively from one terminal to the other. As physical dimension (or temperature gradient) is not important in this analysis it is convenient to express X normalized to the range from 0 at one terminal to 1 at the other or else by numbers, $i = 1$ to n, that reflect the sequence in which junctions occur in progressively traversing the circuit. The key temperature/position pairs are plotted for all real junctions of the circuit. The informal sketch is not used for graphic calculation. It need not be to scale nor to proportion. It is drawn for visualization alone.

Thermoelectric Circuit Analysis

The procedure for analysis is conducted in simple steps that are suggested by the corollaries. The corollary of functional determinacy, in its more formal and detailed statement, identifies several circumstances that render certain segments or combinations of them thermoelectrically neutral, contributing no net Seebeck emf, so that such segments, individually or in pairs, can be eliminated from analysis by simple inspection. For example, as a circuit is traversed progressively from an initial to a final terminal, the Seebeck emfs of any pair of segments of like $\sigma(T)$ that cross a temperature zone in opposite directions cancel. Therefore, across that temperature span the pair functions effectively as conductors only. They contribute no *net* emf over that temperature interval so they can be eliminated from thermoelectric analysis by inspection. Also, any isothermal segment or any segment with its endpoints at the same temperature functions effectively only as a conductor.

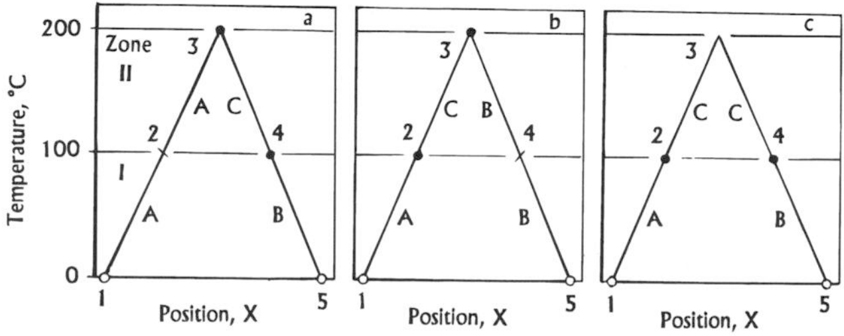

Figure 6.74 T/X sketches for three different temperature distributions around the same thermoelectric circuit. (Material C could have an insignificant Seebeck coefficient, $\sigma_C = 0$.)

Execution of the T/X Sketch. The simple process of creating the plot is illustrated by a series circuit of only three dissimilar materials, Figure 6.74a. Other examples will also illustrate the unique benefit of T/X representation.

1. Sketch temperature, T, and relative position, X, axes.
2. Beginning at one terminal, plot the temperature and relative positional sequence of each terminal and real junction around the circuit.
3. Connect adjacent pairs of points with line segments in the sequence they occur in progressing around the circuit.
4. Identify each line segment with the material that it represents.
5. Draw isothermal lines across the diagram through each terminal and each real junction to bound relevant temperature zones. Label each zone.
6. Indicate, with a tic mark, all incidental intersections of these isotherms with the line segments. (These intersections within segments are defined as "virtual junctions" that function, for this analysis only, effectively as junctions that define temperature zones.)
7. For each temperature zone, apply to all segments that occupy the zone the absolute or relative temperature/voltage relation in Equation 6.65 depending on whether absolute or relative Seebeck emf information is available for the materials.
8. Sum the relative voltages from all temperature zones noting the momentary polarities. This sum yields the net emf from all temperature zones from all segments between the terminals.

The External Circuit. The analysis has focused on the Seebeck emf from the described portion of the circuit (the "internal" circuit) between the circuit terminals. However, it must be recognized that the Seebeck law and its corollaries apply equally to the generally indefinite monitoring or recording circuit that must be used to observe the circuit terminal voltage. That "external" monitoring circuit is necessarily a portion of the entire thermoelectric circuit. Therefore, the external circuit can contribute irrelevant Seebeck and other emf to the terminal voltage unless its contribution is suppressed or compensated by appropriate hardware or numerical means. If the external circuit is unchanging in temperature distribution and in incidental emf, the constant external voltage can be compensated by offset.

Examples of Analysis

The T/X presentation forces the attention to the physical location and identification of the individual sources of Seebeck emf. The T/X presentation is valid, but of course it is trivial, for an idealized simplistic circuit of only two homogeneous elements for which Equation 6.64 with 6.47 directly yields the net Seebeck emf. For more complex circuits, the T/X sketch makes it clear that in simple *series* circuits segments within each temperature zone of a circuit occur only in one or more *pairs*. This always allows the use of relative Seebeck relations. Note that each segment pair is defined by the temperature span between the endpoints regardless of their proximity in the circuit. The thermally paired segments usually are not contiguous in the circuit and often they are widely separated.

An *odd* number of segments can occupy a particular temperature zone only in circuits with electrically *paralleled* segments (i.e., with loops in which a real junction node is shared by at least three dissimilar segments.) In such instances, ignored by the traditional "thermocouple laws", it is necessary to use individual leg Seebeck properties for at least one of the segments. Examples for three different realistic arrangements illustrate these very important points with compound electronic circuits. A brief set of absolute Seebeck emfs for these materials, to be used in the numerical examples, is given as Table 6.15.

A Series Circuit of Three Materials. Improperly viewing thermoelectric circuits as ordinary electrical circuits it has been suggested that "thermoelectric insertion error" introduced by a component in low level measuring instruments can be avoided by material choice merely by using a component with insignificant Seebeck coefficient. The profound error of this presumption is illustrated by the simple circuit of Figure 6.74. For illustration, this circuit is made of realistic electric circuit materials *A* (copper) and *B* (constantan, a resistor material). Representative Seebeck properties for these materials are given in Table 6.15. The inserted component is of hypothetical material *C*. Component *C* arbitrarily has a Seebeck coefficient that is negligible over the temperature range of application (lead, tin, and magnesium have very small absolute Seebeck coefficients near room temperature.) The intent is to avoid thermoelectric error due to unwanted emf contribution from the inserted "neutral" segment, *C*.

Figure 6.74 shows three different temperature distributions imposed on the same circuit that is viewed under open circuit conditions. With the temperature distribution of Figure 6.74a, note that only a portion of the material *A* segment is thermally paired with all of segment *B* across Zone *I* over the 0–100°C temperature span. Material *A* has a "virtual" junction, labeled 2, where the temperature along it happens to be at 100°C as defined by the real junction labeled 4. Null component *C* is paired with a segment of *A* from 100°C to 200°C. In Figure 6.74a the net Seebeck emf would be

$$\begin{aligned} E_{\text{net}} &= \left\{ E_A\Big|_{0°\text{C}}^{100°\text{C}} - E_B\Big|_{0°\text{C}}^{100°\text{C}} \right\}_I + \left\{ E_A\Big|_{100°\text{C}}^{200°\text{C}} - E_C\Big|_{100°\text{C}}^{200°\text{C}} \right\}_{II} \\ &= [(217.9 - 0.0) - (-4060.6 - 0.0)] \\ &\quad + [(480.3 - 217.9) - (0.0 - 0.0)] \\ &= 217.9 + 4060.6 + 262.4 - 0.0 \\ &= 4540.9\ \mu V \end{aligned} \tag{6.68}$$

Next, simply interchanging the temperatures of a pair of junctions while considering the same circuit and materials produces a very different net Seebeck emf. The modified temperature distribution causes different material combinations to span the two temperature zones (Figure 6.74b). In this instance, for brevity, the tabulated relative Seebeck emf can be used for Zone I. Now, the net Seebeck emf is

$$\begin{aligned} E_{\text{net}} &= \left\{ E_{AB}\Big|_{0°\text{C}}^{100°\text{C}} \right\}_I + \left\{ E_C\Big|_{100°\text{C}}^{200°\text{C}} - E_B\Big|_{100°\text{C}}^{200°\text{C}} \right\}_{II} \\ &= [4278.5 - 0.0]_I + [0.0 - (-4747.1)]_{II} \\ &= 9025.6\ \mu V \end{aligned} \tag{6.69}$$

In Figure 6.74c, terminals 2 and 3 are forced to the same temperature but the highest temperature of segment *C* is at the same 200°C maximum. The symmetric contribution from material *C*, Zone II, producing a null net Seebeck emf, can be eliminated by inspection regardless of material (the corollary of functional roles). The net Seebeck emf, directly from the table of relative Seebeck emf values is

$$E_{net} = \left\{ E_{AB}\Big|_{0°\text{C}}^{100°\text{C}} \right\} = 4278.5\ \mu V \tag{6.70}$$

Table 6.15 Brief Table of Absolute and Relative Seebeck Coefficients and Seebeck emfs for Copper and Constantan Materials Referenced to 0°C*(Reed, 1992; Wang, 1992; Kinzie, 1973; Burns, 1993).

	Absolute				Relative	
	Copper, A		Constantan, B		Copper/Constantan, AB	
Temperature* T, °C	σ_A (T) μV/°C	E_A (T) μV	σ_B (T) μV/°C	E_B (T) μV	σ_{AB} (T) μV/°C	E_{AB} (T) μV
0	5.895	0.0	−36.885	0.0	38.748	0.0
50	7.800	99.2	−40.618	−1936.6	42.820	2035.7
100	9.394	217.9	−47.557	−4060.6	46.785	4278.5
150	10.633	345.0	−50.308	−6359.0	50.158	6704.1
200	11.875	480.3	−50.308	−8807.7	53.150	9288.1
250	13.135	631.4	−52.609	−11381.9	55.800	12013.4
300	14.305	799.3	−54.573	−14062.6	58.088	14861.9

*ITS-90 temperature scale

The three net voltages (Equations 6.68–6.70) are very different only because of the temperature distribution of the real junctions, not changed materials, around the circuit. Only the situation of Figure 6.74c produces the intended Seebeck voltage. Clearly, material selection alone cannot control the output. Not only might special material C be costly or otherwise be unsuitable, the approach is ineffective. Moreover, it can *increase* rather than decrease error. Note, however, in Figure 6.74c, that merely by maintaining 2 and 3 at the same temperature, the thermoelectric contribution of homogeneous segment C is necessarily eliminated regardless of the temperature distribution between 2 and 3 and independent of the material of C. Because junction temperatures 2 and 3 are equal it is irrelevant that material C has a null Seebeck coefficient. Recall that this nulling of the effect of any Seebeck emf from C is the result intended by ineffectually employing the "neutral" material C. The important practical lesson is that elimination of spurious Seebeck emf can be uniformly effected only by deliberate temperature control rather than by material choice (corollary of temperature determinacy.)

An Electronic Circuit of Many Materials. Industrial electronic circuits typically involve many substantially dissimilar materials. Unintended thermoelectric circuit behavior often is not recognized. Many materials, particularly semiconductors, as well as silicon, germanium, and selenium, have very large Seebeck coefficients. Many conductor components are plated or multilayered or have undetermined Seebeck coefficients. The temperature distribution around such circuits usually is not controlled so it may vary significantly throughout circuit warm-up and during use. Substantial temperature differences can exist, even over small distances, along the circuit in small com ponents introducing very steep gradients. Maintaining such circuits continually isothermal, independent of their temperature level, is the only effective general approach to minimizing spurious Seebeck emf.

A realistic circuit (Figure 6.75a) that was adapted from a thermometric instrument manufacturer's literature involves many materials, of diverse Seebeck coefficient, that are typical of industrial electronic circuits. The circuit is intended to produce Seebeck emf only from the Type T (copper/constantan) thermocouple connected to input terminals, g and i. In thermoelectric temperature measurement the magnitude of net Seebeck voltage is critical to accuracy. Any inappropriate thermoelectric emf from portions of the circuit other than the thermocouple is unwanted and can result in large error. Because some components are of small dimension and junctions may not be closely spaced, the circuit is sensitively vulnerable to very small shifts of temperature distribution. Figure 6.75b shows the intended temperature distribution. Note that, except for the thermocouple, all pairs are of like circuit materials that are presumed to share the same pair of endpoint temperatures. From the corollary of functional roles it is evident that, in this idealized temperature controlled situation, Seebeck emf from all segments except the thermocouple is suppressed, as intended.

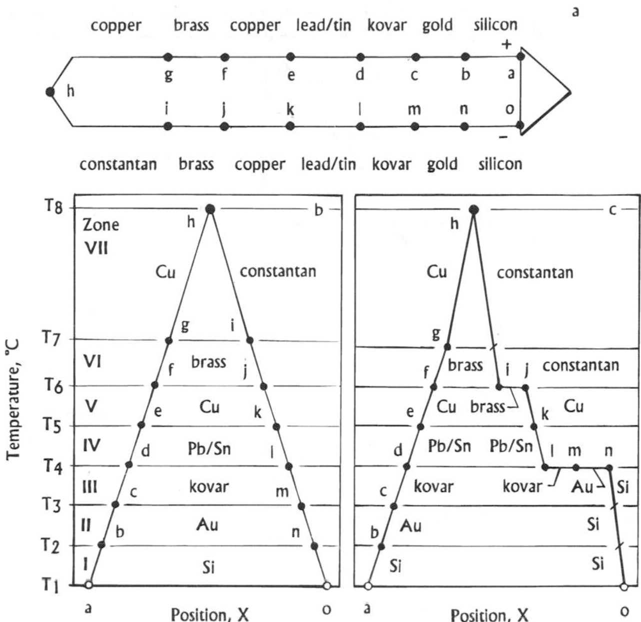

Figure 6.75 An electronic circuit of many materials: (a) electrical schematic, (b) T/X sketch of the intended temperature distribution, and (c) of an uncontrolled temperature distribution.

However, a small uncontrolled redistribution of junction temperatures can thermally pair unintended segments and seriously affect the net output rendering any thermocouple calibration irrelevant. For example, Figure 6.75c shows a plausible tempera ture redistribution that pairs portions of the silicon circuit element with silicon, gold, and kovar segments. Also, a portion of the constantan leg of the thermocouple is thermally paired with uncalibrated brass rather than with the intended calibrated copper leg. All isothermal segments produce no Seebeck emf. Had the silicon segment been a semiconductor, as in many modern electronic circuits, the error could have been substantially greater. For example, semiconductor strain gauges are particularly vulnerable to such errors. In this realistic example note that, due to temperature distribution, the errors arise from indeliberate pairing of unrelated and uncalibrated segments that are widely separated in the circuit schematic and have no intended emf source function in the instrument.

Electrically Paralleled Thermoelectric Circuit. A final example (Figure 6.76) extends the application of the series T/X plot to realistic electronic circuits that have paralleled branches. Such arrangements are commonplace where dissimilar conductors are laminated, plated (and with irregular plating thickness), indeliberately shunted due to degraded insulation, etc.

Under series open circuit conditions at the terminals, where there can be no current, there is no resistive voltage drop. Only with paralleled segments is it possible to have an *odd* number of segments sharing a given temperature interval. In such instances it is necessary to use Seebeck emfs of individual legs in solving the circuit. Where loops span a temperature zone, the paralleled Seebeck sources result in self-induced loop current, therefore voltage drop and a slight temperature shift at junctions

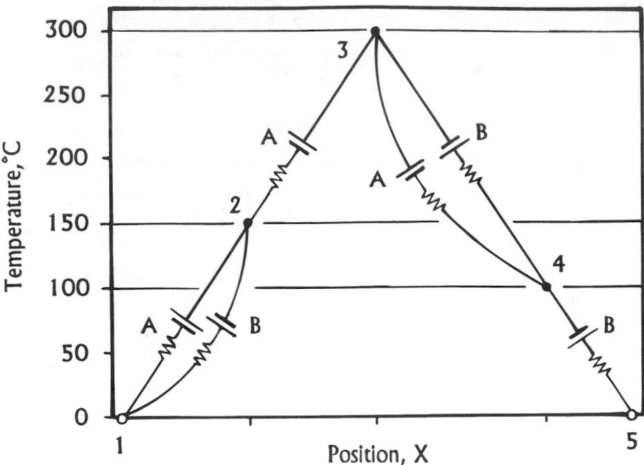

Figure 6.76 *T/X* sketch for a thermoelectric circuit with legs paralleled by shunting with dissimilar materials.

from the Peltier effect. (Usually, the Peltier junction self heating effect is significant only in powered circuits with significant current.)

Figure 6.76 represents an intended *AB* pair in which a portion of each leg, incidentally, has been shunted unsymmetrically by segments of the other material. Again, to simplify the example, the materials are chosen to be *A* (copper) and *B* (constantan). For illustration, only the two materials are involved. Without the shunting of the basic *AB* pair the expected open circuit terminal voltage would be:

$$E_{AB}\Big|_{0°C}^{300°C} = 14{,}861.9\ \mu V \tag{6.71}$$

If both legs were shunted across the full temperature span the net Seebeck open circuit terminal voltage would be nulled. However, in Figure 6.76, the *A* material is shunted only between 0°C and 150°C by the *B* material and between 100°C and 300°C the *B* leg is shunted by material *A*. These shuntings affect the net terminal voltage.

Even if the external circuit between 1 and 5 is open, so there is no current in the external circuit, the two loops each can have a significant local circulating Seebeck current. Therefore, in the two loops between 1 and 2 and between 3 and 4 there is the possibility of a slight Peltier shift of temperature of junctions 1, 2, 3, and 4. That Peltier temperature shift, as small, will be ignored in the numerical example. The net Seebeck voltage between 1 and 2 and between 3 and 4 may also be substantially affected by the resisitive voltage drop. Resistance has a substantial effect. In practice, the resistances of segments is a function of the temperature interval that they span. For simplicity, the resistances for Seebeck source materials *A* (copper) and *B* (constantan) are assumed to have constant realistic resistance values of 9.1 and 288.9 mΩ between each pair of junctions, respectively. These are room temperature resistances of 20 AWG wires about 30 cm long. Effects from self-generated Seebeck current could be significant in applications where the net Seebeck emf is to be used and often even where noise only is of concern.

Loops complicate the circuit analysis, yet the thermoelectric principles are the same for every segment. It is valid to consider Seebeck voltage contributions zone by zone. However, in instances such as this example it is simpler to focus on the complicating loops and to treat them separately over the full temperature range spanned by each. The individual absolute Seebeck emfs for each leg must be used. From Table 6.15, the following absolute Seebeck emf values are required:

$$E_A\Big|_{0°C}^{150°C} = 345.0\ \mu V, \qquad E_B\Big|_{0°C}^{150°C} = -6359.0\ \mu v,$$

$$E_A\Big|_{150°C}^{300°C} = 454.3\ \mu V, \qquad E_B\Big|_{0°C}^{100°C} = -4060.6\ \mu V,\ \text{and}$$

$$E_A\Big|_{100°C}^{300°C} = 581.4\ \mu V, \qquad E_B\Big|_{100°C}^{300°C} = -10{,}002.0\ \mu V$$

Because of the Seebeck currents and relative values of resistances, the loop-paired elements have effective relative Seebeck emfs that are different from open circuit paired values that apply in purely series circuits. Considering each loop separately and using the conventional Kirchhoff relations, for the loops with nodes at 1 and 2, and at 3 and 4 the loop currents are

$$I_{12} = \frac{\{E_A|_{0°C}^{150°C} - E_B|_{0°C}^{150°C}\}}{\{R_A|_{0°C}^{150°C} + R_B|_{0°C}^{150°C}\}} = \frac{6704.0\ \mu V}{0.298\ \Omega} = 22.497\ mA \tag{6.72}$$

and

$$I_{43} = \frac{\{E_A|_{100°C}^{300°C} - E_B|_{100°C}^{300°C}\}}{\{R_A|_{100°C}^{300°C} + R_B|_{100°C}^{300°C}\}} = \frac{10583.4}{0.298\ \Omega} = 35.515\ mA. \tag{6.73}$$

The corresponding effective relative Seebeck emfs are:

$$V_{12}|_{0°C}^{150°C} = 345.0 - (22497\cdot 0.0091) = 140.3\ \mu V \tag{6.74}$$

and

$$V_{43}|_{100°C}^{300°C} = -10{,}002.0 + (35515\cdot 0.2889) = 258.3\ \mu V \tag{6.75}$$

Therefore, the net open circuit terminal Seebeck voltage is:

$$\begin{aligned} V_{15} &= V_{12}|_{0°C}^{150°C} + E_A|_{150°C}^{300°C} - V_{43}|_{100°C}^{300°C} - E_B|_{0°C}^{100°C}, \\ &= (140.3) + (454.3) - (258.3) - (-4060.6) \\ &= 4396.9\ \mu V \end{aligned} \tag{6.76}$$

The difference between terminal voltages in the unshunted and shunted states is

$$(V_{15})_u - (V_{15})_s = 14{,}861.9 - 4396.9 = 10465.0\ \mu V. \qquad (6.77)$$

Note that, in the temperature zone between 100°C and 150°C, the four segments, two of material *A* and two of *B*, do not directly cancel as they would in a series circuit because they lie in paralleled loops. In this example (with realistic materials), the effect of thermoelectric shunting is large. Two factors aggravate the error for this material pair. Both the resistances and the absolute Seebeck emfs of the copper and constantan materials are very dissimilar. The effect of shunting was substantial in both loops. The shunting of material *A* by *B*, loop 1–2, had relative lesser effect on the intended emf from *A* but the shunting of *B* by *A*, loop 3–4, had very large effect on the intended emf because the magnitude of the intended contribution of *B* was greatly reduced and also because the intended sign of that contribution was reversed.

Most industrial circuits are of much greater complexity than this illustration. Manual solution of such networks is tedious but an analog DC network analysis program such as the public domain code, SPICE, can easily be programmed to include the Seebeck emf source model (Figure 6.73) so that circuits of any complexity and any temperature distribution can readily be addressed.

Significance of the Thermoelectric Analysis In addition to any other intended electrical functions, all nonisothermal electrical circuits are also thermoelectric circuits in which Seebeck emf occurs. The consequences of paralleling in thermoelectric circuits can be significant. The analysis method presented here applies to all industrial circuits whether intentionally or incidentally thermoelectric in function. In typical applications, the temperature ranges encountered may be much smaller but the Seebeck coefficients of some practical circuit materials can be very much greater than the simple materials used for these examples. The examples are realistic.

Complex thermoelectric circuits must be analyzed by a special method. In industrial circuits, the goal of thermoelectric analysis is to assure that only intended segments, but all intended segments, contribute to the net Seebeck emf. In the suppression of spurious emf from electronic applications, circuit analysis is to assure that the net Seebeck emfs are tolerable under all plausible temperature distributions, to guide identification of potentially troublesome elements, and to suggest practical solutions.

The principles of the Seebeck effect that are essential to circuit analysis are very simple. Complex circuits can be readily evaluated by the use of the practical tools described here. In industrial circuit applications that involve signals of millivolt or lower order where accuracy is important, the contribution of Seebeck emf can be significant. Active hardware or software compensation is only hypothetically possible. The value of spurious Seebeck emf usually is unknown and is variable, therefore, it must be avoided by temperature control rather than by electrical compensation.

Seebeck emf is universally nulled only if every segment of the circuit, including connections, is of the identical homogeneous material or else if they are isothermal, regardless of homogeneity or the temperature level. In practice, the elimination of spurious emf from single or paired segments is most easily accomplished merely by assuring homogeneity and that appropriate terminals are maintained at the same temperature by proximity or heat sinking. The *T*/*X* sketch presented here forcefully reveals segments where such an approach can be fruitful.

Summary

Most modern industrial circuits and their components include many different materials. Some circuits combine metals, semiconductors, conducting polymers, single crystals, and polycrystalline ceramics. Some such commonplace materials, individually or in combination, can introduce spurious emf or charge of significant magnitude from thermoelectric, thermomagnetoelectric, or pyroelectric thermal-electrical effects that can degrade or interfere with proper circuit operation. Many modern circuits operate at subvolt or much lower signal level. Unexpected voltages can arise from a variety of incidental temperature effects. To systematically overcome potential noise, the physical noise source must be recognized, properly understood, and correctly addressed. Some of the significant physical mechanisms are unfamiliar as they are not introduced in customary course work or discussed in the familiar literature. If recognized, they may be incorrectly understood. This article surveyed some thermally excited noise sources that have proved troublesome in practical circuits subjected to uncontrolled operational temperatures. For the most common thermal source of low-level DC electrical noise, the Seebeck effect, the special analysis presented is necessary to identify, evaluate, and overcome the unwanted noise source. For the several other effects, normal electrical circuit analysis can be applied.

References

Burns, G. W., Scroger, M. G., Strouse, G. F., Croarkin, M. C., and Guthrie, W. F. 1993. Temperature-electromotive force reference functions and tables for the letter-designated thermocouple types based on the ITS-90, NIST Monograph 175, U. S. Department of Commerce, Washington, DC.

Cady, W. G. 1964. *Piezoelectricity,* Dover, New York, NY.

deGroot, S. R. 1966. *Thermodynamics of Irreversible Processes,* 1st ed., North-Holland, Amsterdam.

Doss, J. D. 1989. *Engineer's Guide to High-Temperature Superconductivity,* John Wiley & Sons, New York, NY.

Harman, T. C. and Honig, J. M. 1986. *Thermoelectric and Thermomagnetic Effects and Applications,* McGraw-Hill, New York, NY.

Kinzie, P. A. 1973. *Thermocouple Temperature Measurement,* Wiley-Interscience.

Lang, S. B. 1974. *Sourcebook of Pyroelectricity,* Gordon & Breach, London.

Mattiat, O. E. 1971. *Ultrasonic Transducer Materials,* Plenum Press, New York, NY.

Pollock, D. D. 1990. *Physics of Engineering Materials,* Prentice-Hall, Englewood Cliffs, NJ.

Pollock, D. D. 1991. *Thermocouples, Theory, and Properties,* CRC Press, Boca Raton, FL.

Reed, R. P. 1982. Thermoelectric Thermometry—a Functional Model, *Temperature, Its Measurement and Control in Science and Industry,* American Institute of Physics, 5(2):915–922.

Reed, R. P. 1992. Absolute Seebeck thermoelectric characteristics—principles, significance, and applications, *Temperature, its Measurement and Control in Science and Industry,* American Institute of Physics, 6(2):503–508.

Reed R. P. 1993. *Manual on the Use of Thermocouples in Temperature Measurement, MNL-12, 4th edition* Ch. 2, Park, R. W., ed., American Society for Testing and Materials, Philadelphia, PA.

Seitz, F. 1940. *Modern Theory of Solids,* McGraw-Hill, New York, NY.

Wang, T. P. 1992. Absolute Seebeck coefficients of metallic elements, *Temperature, Its Measurement and Control in Science and Industry,* American Institute of Physics, 6(2):509–514.

Washburn, E. W. (ed.) 1926. *International Critical Tables of Numerical Data—Physics, Chemistry, and Technology,* Vol. VI, McGraw-Hill, New York, NY.

6.12 Lossless Waveform Compression

Giridhar Mandyam, Neeraj Magotra, Samuel D. Stearns, Li-Zhe Tan, and Wes McCoy

Introduction

Compression of waveforms is of great interest in applications where efficiency with respect to data storage or transmission bandwidth is sought. Traditional methods for waveform compression, while effective, are lossy, with inexact reconstruction. In certain applications, even slight compression losses are not acceptable. For instance, real-time telemetry in space applications requires exact recovery of the compressed signal. Furthermore, in the area of biomedical signal processing, exact recovery of the signal is necessary not only for diagnostic purposes but also to reduce potential liability for litigation. As a result, interest has increased recently in the area of lossless waveform compression.

There are many techniques for lossless compression. The most effective of these belong to a class of coders commonly called *entropy coders.* These methods have proven effective for text compression, but perform poorly on most kinds of waveform data, as they fail to exploit the high correlation that typically exists among the data samples. Therefore, preprocessing the data to achieve decorrelation is a desirable first step for data compression. This yields a two-stage approach to lossless waveform compression, as shown in the block diagram in Figure 6.77. The first stage is a "decorrelator," and the second stage is an entropy coder (Stearns, et al., 1993). This general framework covers most classes of lossless waveform compression.

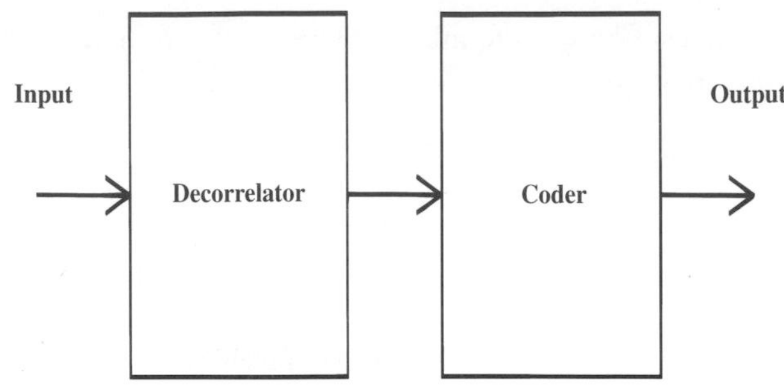

Figure 6.77 Two-stage compression.

Compressibility of a Data Sequence

The compressibility of a data sequence is dependent on two factors: the amplitude distribution of the data sequence, and the power spectrum of the data sequence. For instance, if a single value dominates the amplitude distribution, or a single frequency dominates the power spectrum, then the data sequence is highly compressible. Four sample waveforms are depicted in Figure 6.78 along with their respective amplitude distributions and power spectra. Their characteristics and compressibility are given in Table 6.16. Waveform 1 displays poor compressibility characteristics, as no one particular value dominates its amplitude distribution and no one frequency dominates its power spectrum. Waveform 4 displays high compressibility, as its amplitude distribution and its power spectrum are both nonuniform.

Performance Criteria for Compression Schemes

There are several criteria for quantifying the performance of a compression scheme. One such criterion is the reduction in *entropy* from the input data to the output of the decorrelation stage in Figure 6.77. The entropy of a set of K symbols $\{s_0, s_1, \ldots, s_{K-1}\}$, each with probability $p(s_i)$, is defined as (Blahut, 1990)

$$H_p = \sum_{i=0}^{K-1} p(s_i)\log_2[p(s_i)] \text{ bits/symbol} \qquad (6.78)$$

Entropy is a means of determining the minimum number of bits required to encode a sequence of data symbols, given the individual probabilities of symbol occurrence.

When the symbols are digitized waveform samples, another criterion is the *variance* (mean-squared value) of the zero-mean output of the decorrelation stage. Given an N-point zero-mean data sequence $x(n)$, where n is the discrete time index, the variance σ_x^2 is calculated by the following equation:

$$\sigma_x^2 = \frac{1}{N-1}\sum_{i=0}^{N-1} x^2(n) \qquad (6.79)$$

This is a much easier quantity to calculate than entropy; however, it is not as reliable as entropy. For instance, only two values

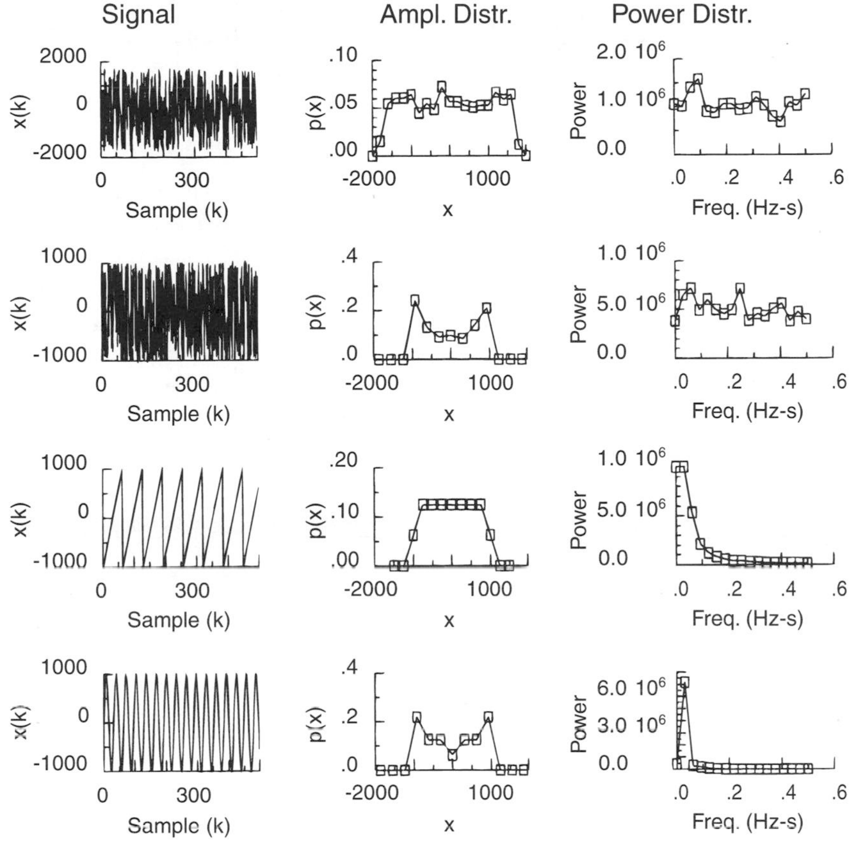

Figure 6.78 Example waveforms: waveform 1 on top; waveform 4 on bottom.

Table 6.16 Compressibility of Example Waveforms

Waveform	Amplitude	Spectrum	Compressibility
1	Uniform	White	Low
2	Nonuniform	White	Some
3	Uniform	Nonwhite	More
4	Nonuniform	Nonwhite	Waveform

might dominate a sample sequence, yet these two values may not be close to one another. If this is the case, the entropy of the data sequence is very low (implying good compressibility), while the variance may be high. Nevertheless, for most waveform data, using the variance of the output of the decorrelator stage to determine compressibility is acceptable, due to the approximately white Gaussian nature of the output of the decorrelation stage. This nature results from arguments based on the *central limit theorem* (Papoulis, 1984), which basically states that the distribution of the sum of independent, identically distributed random variables tends to a Gaussian distribution as the number of random variables added together approaches infinity.

The *compression ratio,* abbreviated as c.r., is the ratio of the length (usually measured in bits) of the input data sequence to the length of the output data sequence for a given compression method. This is the most important measure of performance for a lossless compression technique. When comparing compression ratios for different techniques, it is important to be consistent by noting information that is known globally and not included in the compressed data sequence.

The Decorrelation Compression Stage

Several predictive methods exist for exploiting correlation between neighboring samples in a given data sequence. These methods all follow the process shown in Figure 6.79: the same decorrelating function is used in compression and reconstruction, and this function must take as input a delayed version of the input and output sequences, as shown. Some of these techniques are described in the following sections.

Linear Prediction

A one-step linear predictor is a nonrecursive system that predicts the next value in a data sequence by using a weighted sum of a prespecified number of samples immediately preceding the sample to be predicted. The linear predictor does not contain the feedback path in Figure 6.79, and is thus a special case of Figure 6.79. Given a sample sequence of length K, $ix(n)$ $(0 \leq n < K)$, one can design a predictor of order M by using M predictor coefficients $\{b_i\}$ to find an estimate $i\hat{x}(n)$ for each sample in $ix(n)$:

$$i\hat{x}(n) = \sum_{i=0}^{M-1} b_i ix(n - i - 1) \tag{6.80}$$

Obviously, M should be much less than K to achieve compression, because $\{b_i\}$ must be included with the compressed data. The estimate $i\hat{x}(n)$ is not the same as the original value; therefore a residue sequence is formed to allow exact recovery of $ix(n)$:

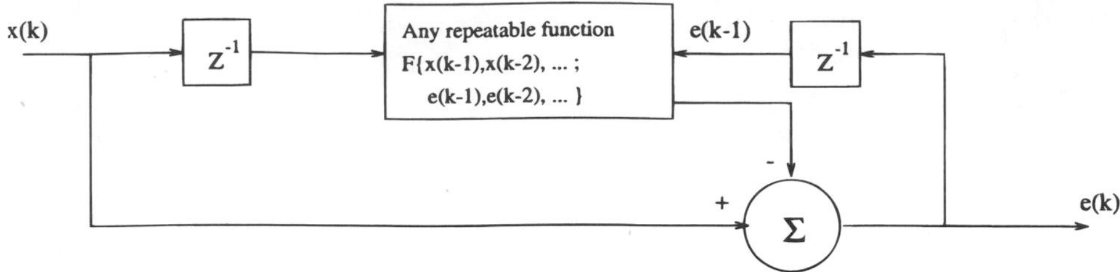

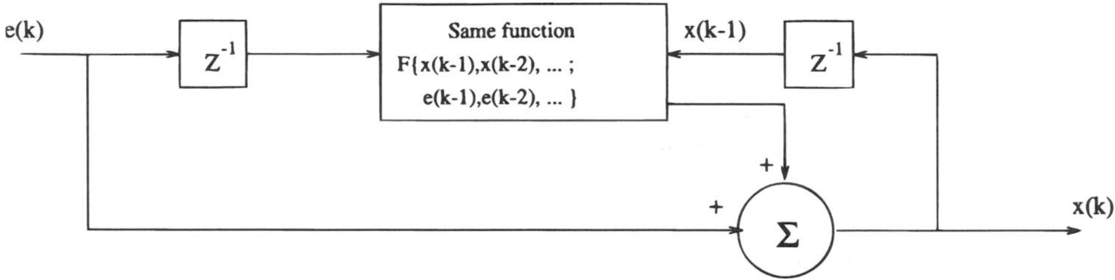

Figure 6.79 Prediction and recovery of waveform data.

$$ir(n) = ix(n) - i\hat{x}(n) \tag{6.81}$$

If the predictor coefficients are chosen properly, the entropy of $ir(n)$ should be less than the entropy of $ix(n)$. Choosing the coefficients $\{b_i\}$ involves solving the *Wiener-Hopf* equations:

$$\begin{aligned} R_{0,0}b_0 + R_{0,1}b_1 + \cdots + R_{0,M-1}b_{M-1} &= R_{M,0} \\ R_{1,0}b_0 + R_{1,1}b_1 + \cdots + R_{1,M-1}b_{M-1} &= R_{M,1} \\ &\vdots \\ R_{M-1,0}b_0 + R_{M-1,1}b_1 + \cdots + R_{M-1,M-1}b_{M-1} &= R_{M,M-1} \end{aligned} \tag{6.82}$$

where R_{ij} is the average over n of the product $ix(n)ix(n + (i - j))$. This can be represented as the matrix-vector product

$$Rb = p \tag{6.83}$$

where R is the M by M matrix defined in Equation 6.82, b is the M by 1 vector of predictor coefficients, and p is the M by 1 vector in Equation 6.83. Equation 6.84 can be solved by a variety of techniques involving the inversion of symmetric matrices (Widrow and Stearns, 1985).

The original data sequence $ix(n)$ can be exactly recovered using the predictor coefficients $\{b_i\}$, the residue sequence $ir(n)$, and the first M samples of $ix(n)$ (Stearns, et al., 1993). This is accomplished by the recursive relationship

$$ix(n) = ir(n) + \sum_{i=0}^{M-1} b_i ix(n - i - 1) \qquad M \leq n \leq K - 1 \tag{6.84}$$

If the original data sequence $ix(n)$ is an integer sequence, then the predictor output can be rounded to a nearest integer and still form a residue sequence of comparable size. In this case, $ir(n)$ is calculated as

$$ir(n) = ix(n) - NINT\left\{\sum_{i=0}^{M-1} b_i ix(n - i - 1)\right\} \tag{6.85}$$

where $NINT\{\}$ is the *nearest integer* function. Similarly, the $ix(n)$ data sequence is recovered from the residue sequence as

$$ix(n) = ir(n) + NINT\left\{\sum_{i=0}^{M-1} b_i ix(n - i - 1)\right\} \qquad M \leq n \leq K - 1 \tag{6.86}$$

where it is presumed that the $NINT\{\}$ operation is performed (at the bit level) exactly as in Equation 6.84.

Determination of Predictor Order. Formulating an optimal predictor order M is crucial to achieving optimal compression (Stearns et al., 1993), because there is a trade-off between the lower variance of the residual sequence and the increasing overhead due to larger predictor orders. There is no single approach to the problem of finding the optimal predictor order; in fact, several methods exist. Each of the methods entail finding the sample variance of the zero-mean residues $ir(n)$ as determined by Equation 6.80 for an order M:

$$\sigma_{ir}^2(M) = \frac{1}{K - M - 1} \sum_{i=M}^{K-1} ir^2(i) \tag{6.87}$$

One of the easiest methods for finding an optimal predictor order is to increment M starting from $M = 1$ until $\sigma_M^2(n)$ reaches a minimum, which may be termed as the *minimum variance criterion* (MVC). Another method, called the *Akaike information criteria* (AIC) (Marple, 1987), involves minimizing the following function:

$$AIC(M) = K\ln(\sigma_{ir}^2(M)) + 2M \tag{6.88}$$

The $2M$ term in the AIC serves to "penalize" for unnecessarily high predictor orders. The AIC, however, has been shown to be statistically inconsistent, so the *minimum description length* (MDL) criterion has been formed (Marple, 1987):

$$MDL(M) = K\ln(\sigma_{ir}^2(M)) + M\ln(K) \tag{6.89}$$

A method proposed by Tan (1992) involves determining the optimal $\alpha(M) = 0.5\log_2\sigma_{ir}^2(M)$ number of bits necessary to code each residual. Starting with $M = 1$, M is increased until the following criterion is no longer true:

$$(K - M)\Delta\alpha(M) > \Delta\beta(M) \tag{6.90}$$

where $\Delta\alpha(M) = -[\alpha(M) - \alpha(M - 1)$ and $\Delta\beta(M)]$ represents the increase in overhead bits for each successive M (due mainly to increased startup values and predictor coefficients). There are several other methods for predictor-order determination; none has proven to be the best in all situations.

Quantization of Predictor Coefficients. Excessive quantization of the coefficients $\{b_i\}$ can reduce the c.r. and affect exact recovery of the original data sequence $ix(n)$. On the other hand, unnecessary bits in the representation of $\{b_i\}$ will also decrease the c.r. The representation proposed by Stearns et al. (1993) is briefly outlined here. Given predictor coefficients $\{b_i\}$, an integer representation is

$$ib_i = NINT\{2^{N_{ib}-I-1}b_i\}; \qquad 0 \le i < M \tag{6.91}$$

where N_{ib} is the number of bits for coefficient quantization and I is the number of integer bits $INT\{b_i\}_{MAX}$, where $INT\{\}$ is the maximum integer less than or equal to the operand. For prediction, rather than using the calculated coefficients $\{b_i\}$, modified predictor coefficients are used:

$$b_i^* = \frac{PREC\{ib_i\}}{2^{N_{ib}-I-1}}; \qquad 0 \le i < M \tag{6.92}$$

where $PREC\{\}$ is a function which converts the operand to whatever maximum precision is available, depending upon the processor one is using. It is desirable to find N_{ib} such that $\sigma_{ir}^2(M)$ in Equation 6.87 remains roughly the same for either the set of coefficients $\{b_i\}$ and $\{b_i^*\}$. A coefficient ib_i can be represented as

$$ib_i = 2^{N_{ib}-I-1}b_i + \epsilon_i \tag{6.93}$$

where ϵ_i is an error function such that $|\epsilon_i| \le 0.5$. The coefficient quantization error is

$$\Delta b_i = b_i^* - b_i = \epsilon_i 2^{N_{ib}-I-1} \tag{6.94}$$

Using Equation 6.85 to form residues, the residue error can be represented as (Stearns, et al., 1993)

$$\begin{aligned} |\Delta ir(n)| &= |ir(n) - ir^*(n)| \\ &= \left| NINT\left\{ \sum_{i=0}^{M-1} \frac{\epsilon_i}{2^{N_{ib}-I-1}} ix(n-i-1) \right\} \right| \end{aligned} \tag{6.95}$$

It can also be seen that

$$\begin{aligned} \left| \sum_{i=0}^{M-1} \frac{\epsilon_i}{2^{N_{ib}-I-1}} ix(n-i-1) \right| &< \sum_{i=0}^{M-1} \frac{|\epsilon_i|}{2^{N_{ib}-I-1}} |ix(n-i-1)| \\ &< \frac{1}{2} \frac{M|ix(n)|_{\max}}{2^{N_{ib}-I-1}} \end{aligned} \tag{6.96}$$

If the constraint $|\Delta\{ir(n)\}_{\max}| \le 1$ is imposed, then the following inequality results:

$$N_{ib} \ge 1 + (I + 1) + INT\left\{ \log_2 \frac{M|ix(n)|_{\max}}{2} \right\} \tag{6.97}$$

As long as this minimum bound is met, residue error will be minimal.

Selection of Data Frame Size K. Selecting a proper frame size is very important in the process of linear prediction. Each frame contains fixed data, including $\{ib_i\}$ and other coding parameters, plus an encoded version of the residues, $ir(n)$. Hence, decreasing the frame size toward zero ultimately causes the c.r. to decrease toward zero due to the fixed data. On the other hand, a very large frame size may require an unreasonably high amount of computation with little gain in compression performance; moreover, if the data is nonstationary, large frame sizes may actually provide worse compression performance than smaller data frame sizes, due to highly localized statistical behavior of the data. As an example, a 10,000-point seismic waveform was compressed using linear prediction along with arithmetic coding (described in the section on the coding compression section) (Tan, 1992) for different frame sizes; the results are given in Table 6.17. In this table, two compression ratios are given, due to the fact that the original data was quantized to 32 bits, but the data samples occupied at the most 12 bits; the compression ratio corresponding to 32-bit quantization is *CR1* and for 12-bit quantization is *CR2*. As can be seen from the results, the compression ratio tends to increase with increasing frame size, but the gain becomes smaller with larger frame sizes. This indicates that one should look at the effects of different frame sizes when using linear prediction, but one should also weigh compression gains with the increasing computational complexity associated with larger frame sizes. We also note that the monotonically increasing behavior of the compression ratio with increasing frame size is relevant only to the seismic data sequence used in

Table 6.17 Effects of Frame Size on Compression Performance (Seismic Waveforms)

Frame Size	CR_1	CR_2
50	5.8824	2.2059
100	6.8966	2.5862
500	8.2645	3.0992
1000	8.5837	3.2189
2000	8.8594	3.3223
3000	9.2094	3.4535
4000	9.2007	3.4503
5000	9.2081	3.4530
6000	9.2307	3.4615
7000	9.2317	3.4619
8000	9.2219	3.4582
9000	9.2783	3.4794
10000	9.3002	3.4876

this example; other types of waveform data may display different results.

Adaptive Linear Filters. While the method of linear prediction presented previously in **Linear Prediction** is effective, it suffers from the problem of finding a solution to the Wiener-Hopf equations in Equation 6.83, which becomes increasingly computationally expensive with larger data block sizes. Adaptive FIR filters have been proposed and used successfully as a way of solving this problem (Widrow and Stearns, 1985). With a fixed filter size, *M*, there is a variety of stochastic gradient methods to adapt the filter. One common method is discussed here, the normalized least mean square (NLMS) algorithm (Haykin, 1991). Once again, the sample sequence index is denoted by *n*, and the set of predictor coefficients $\{b_i\}$ is now time-varying, and is represented by the column vector $b(n) = [b_0(n) \ldots b\_M\text{-}1(n)]^T$, where $[\ ]^T$ is the transpose operator. If the input to the filter is the vector $ix(n) = [ix(n - 1) \ldots ix(n - M)]^T$, then a time-varying residue (McCoy et al., 1994) is given by

$$ir(n) = ix(n) - NINT\{b^T(n)ix(n)\} \qquad (6.98)$$

If two fixed parameters, a smoothing parameter *b* and a convergence parameter *u*, are specified, then $b(n)$ is computed iteratively as follows:

$$b(n + 1) = b(n) + \mu(n)ir(n)ix(n) \qquad (6.99)$$

$$\mu(n) = \frac{u}{\sigma_M^2(n)} \qquad (6.100)$$

$$\sigma_M^2(n) = \beta\sigma_M^2(n - 1) + (1 - \beta)(ir^2(n - 1)) \qquad (6.101)$$

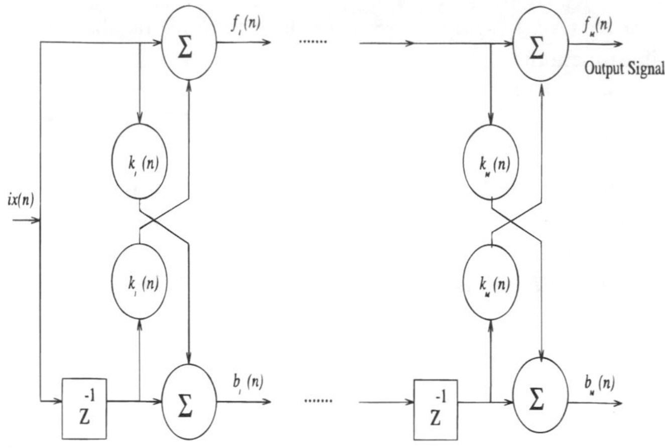

Figure 6.80 Gradient adaptive lattice (GAL) filter.

To reverse the algorithm without loss (McCoy et al., 1994), the following equation may be used along with Equations 6.99–6.101.

$$ix(n) = ir(n) + NINT\{b^T(n)ix(n)\} \qquad (6.102)$$

Therefore, one needs the initial predictor coefficients $b(0)$, the initial data vector $ix(0)$, and the residue sequence $ir(n)$ to reconstruct the original $ix(n)$ sequence exactly. Moreover, in this approach, the coefficients $b(n)$ or the data vectors $ix(n)$ do not have to be transmitted at all after start-up. Referring to Figure 6.79, we note that the functions used in encoding must be repeated exactly in decoding. Therefore all quantities used in Equations 6.99–6.101, that is, u, β, $b(0)$ and $ix(0)$, must be stored in the compressed data frame exactly as they are used in the encoding process, so that they may be used in the same way in the decoding process. We also note that here it is not necessary to break the data up into frames as was the case with the linear predictor method described in **Linear Prediction,** and that in general this method requires less overhead than the linear predictor method.

Lattice Filters

While the NLMS adaptive filter is effective, it sometimes displays slow convergence, resulting in a residue sequence of high-variance. This has lead to the use of a class of adaptive filters known as *lattice filters.* Although the implementation of lattice filters is more involved than that of the NLMS algorithm, faster convergence usually makes up for the increased complexity. A simple *M*-stage adaptive lattice filter is known in Figure 6.80; this filter is known as the *gradient adaptive symmetric lattice* (GAL) filter. The update equations for this filter are (McCoy et al., 1994)

$$b_i(n) = b_{i-1}(n - 1) + k_i(n)f_{i-1}(n) \qquad (6.103)$$

$$f_i(n) = f_{i-1}(n) + k_i(n)b_{i-1}(n - 1) \qquad (6.104)$$

$$k_i(n) = k_i(n-1) + \alpha \frac{f_i(n-1)b_{i-1}(n-2) + f_{i-1}(n-1)b_i(n-1)}{\sigma_{i-1}^2(n-1)} \quad (6.105)$$

$$\sigma_{i-1}^2(n-1) = \beta\sigma_{i-1}^2(n-2) + (1-\beta)(f_{i-1}^2(n-1) + b_{i-1}^2(n-2)) \quad (6.106)$$

$$f_M(n) = ix(n) + NINT\left\{\sum_{i=0}^{M-1} b_i(n-1)k_i(n-1)\right\} \quad (6.107)$$

where α is a convergence parameter, β is a smoothing parameter, $f_i(n)$ is the *forward prediction error,* and $b_i(n)$ is the *backward prediction error.* The recovery equation is (McCoy et al., 1994)

$$ix(n) = f_M(n) - NINT\left\{\sum_{i=0}^{M-1} b_i(n-1)k_i(n-1)\right\} \quad (6.108)$$

An improvement on the GAL is the recursive least-squares lattice (RLSL) filter (Haykin, 1991). The Mth-order forward and backward prediction error coefficients, $\eta_M(n)$ and $\psi_M(n)$ respectively, are given by

$$\eta_M(n) = \eta_{M-1}(n) + \Gamma_{f,M}(n-1)\psi_{M-1}(n-1) \quad (6.109)$$

$$\psi_M(n) = \psi_{M-1}(n-1) + \Gamma_{b,M}(n-1)\eta_{M-1}(n) \quad (6.110)$$

$$\Gamma_{f,M}(n) = -\frac{\Delta_{M-1}(n)}{B_{M-1}(n-1)} \quad (6.111)$$

$$\Gamma_{b,M}(n) = -\frac{\Delta_{M-1}(n)}{F_{M-1}(n)} \quad (6.112)$$

$$\Delta_{M-1}(n) = \lambda\Delta_{M-1}(n-1) + \gamma_{M-1}(n-1)\psi_{M-1}(n-1)\eta_{M-1}(n) \quad (6.113)$$

$$F_{M-1}(n) = \lambda F_{M-1}(n-1) + \gamma_{M-1}(n-1)\eta_{M-1}^2(n) \quad (6.114)$$

$$B_{M-1}(n) = \lambda B_{M-1}(n-1) + \gamma_{M-1}(n)\psi_{M-1}^2(n) \quad (6.115)$$

$$\gamma_M(n) = \gamma_{M-1}(n) - \frac{\gamma_{M-1}^2(n)\psi_{M-1}^2(n)}{B_{M-1}(n)} \quad (6.116)$$

The parameter λ is a fixed constant arbitrarily close to, but not equaling 1. The residues are computed as (McCoy et al., 1994)

$$ir(n) = ix(n) + NINT\left\{\sum_{i=1}^{M} \Gamma_{f,i}(n-1)\psi_{i-1}(n-1)\right\} \quad (6.117)$$

The RLSL usually will perform better than the GAL or the NLMS algorithms. As an example, the seismic waveforms given in Figures 6.80 and 6.81 were each subjected to all three methods; the error variances are given in Table 6.18. As can be seen, the RLSL outperformed the NLMS and GAL algorithms in nearly every case.

Transform-Based Methods

Although linear predictor and lattice methods are effective, both methods require significant computation and have fairly complex implementations, due in large part to the requirements for long frame sizes for linear predictors and long predictor sizes and startup sequences for lattice filters. This has led to the consideration of transform-based methods for lossless waveform compression. Such methods have been applied to lossless image compression (Mandyam et al., 1995), and provide performance comparable to linear-predictor methods with relatively small frame sizes.

Given the data vector $ix_i = [ix(Ni-1) \ldots ix(Ni-N)]^t$, where i refers to ith data frame of size N, an N-point transform of ix_i can be found from

$$z_i = T_{N\times N} ix_i \quad (6.118)$$

where $T_{N\times N}$ is an N by N unitary transform matrix. The term *unitary* refers to the fact that $T_{N\times N}^{-1} = T_{N\times N}^{T}$. Many unitary transforms have proven to be effective in decorrelating highly correlated data, and finding the inverse transform for a unitary transform is simple. Most transform-based waveform compression schemes achieve compression by quantization of the transform coefficients, i.e., the elements of z_i in Equation 6.118. While effective, such compression schemes are lossy. One would thus like to form a lossless method based on transforms to take advantage of existing hardware for lossy methods. To do this, one could retain a subset of transform values and reconstruct from these. For instance, if only M real values are retained ($M<N$), then a new vector results:

$$z_i' = [z_i(0) \cdots z_i(M-1) \quad 0 \cdots 0]^T \quad (6.119)$$

where z'_i is also an N by 1 vector. A residue vector can now be formed:

$$ir_i = ix_i - T_{N\times N}{}^T z_i' \quad (6.120)$$

This residue vector should be of lower entropy than the original data vector.

While precise quantization of the transform coefficients is desirable in lossy compression schemes, the problem is not as complex in lossless schemes. A method proposed in (Mandyam et al., 1995) involved quantizing the entries of $T_{N\times N}$ uniformly. A consequence of this type of quantization is that the transform is no longer unitary, and in order to perform the inverse transform operation, the inverse of the quantized transform matrix must be found explicitly. However, this method is computationally

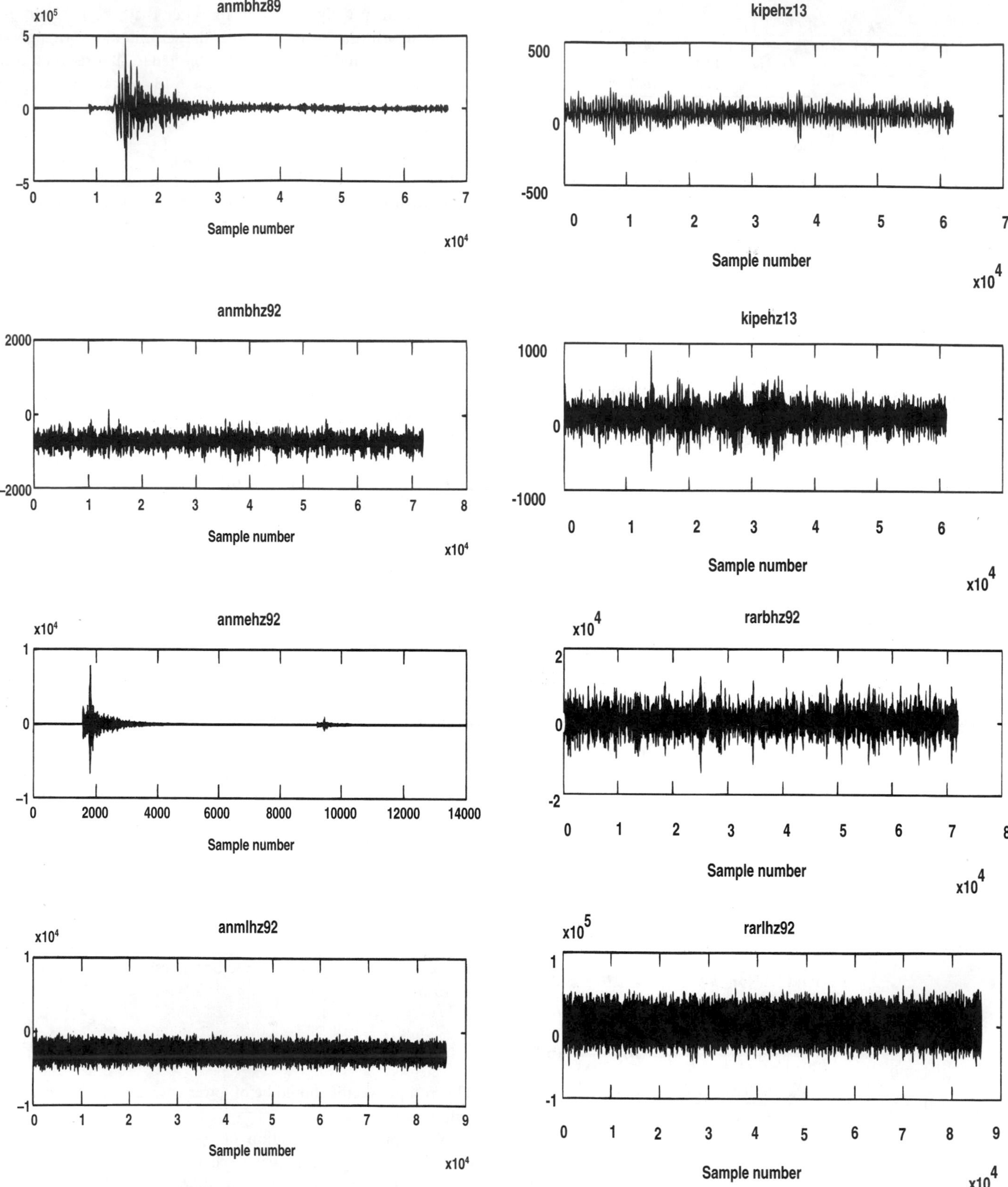

Figure 6.81 Seismic data base, part A.

Figure 6.82 Seismic data base, part B.

Table 6.18 Residue Variances

File	Input	NLMS	GAL	RLSL
anmbhz89	$1.87e^{9}$	$3.85e^{4}$	$6.18e^{1}$	$2.07e^{1}$
anmbhz92	$3.18e^{2}$	$2.26e^{1}$	$1.06e^{1}$	$9.24e^{0}$
anmehz92	$7.84e^{4}$	$2.17e^{4}$	$2.28e^{4}$	$1.82e^{4}$
anmlhz92	$4.44e^{5}$	$9.40e^{4}$	$7.61e^{4}$	$4.54e^{4}$
kipehz13	$2.45e^{3}$	$6.74e^{0}$	$8.22e^{0}$	$5.67e^{0}$
kipehz20	$1.52e^{4}$	$9.56e^{1}$	$7.17e^{1}$	$4.45e^{1}$
rarbhz92	$1.02e^{7}$	$2.05e^{3}$	$4.36e^{2}$	$2.71e^{2}$
rarlhz92	$1.76e^{8}$	$1.42e^{7}$	$2.21e^{7}$	$2.34e^{7}$

more efficient than linear-prediction, since the inverse transform matrix needs to be calculated only once.

As an example, the speech waveform (male speaker, sampled at 8 KHz and quantized to 8 bits) in Figure 6.83 was compressed using a transform method and also the linear-predictor method described in **Linear Prediction.** Both methods used the same frame size ($N = 8$) and same number of coefficients for reconstruction ($M = 2$). The linear predictor solved the Wiener-Hopf equation by the *Levinson-Durbin* method (Widrow and Stearns, 1985). The transform used was the *discrete cosine transform* (DCT), with the N by N DCT matrix defined as:

$$T_{N\times N_{ij}} = \frac{1}{\sqrt{N}} \quad i = 0, \qquad j = 0 \cdots N-1 \tag{6.121}$$

$$= \sqrt{\frac{2}{N}} \cos \frac{(2i+1)j\pi}{2N}$$

$$i = 1 \cdots N-1, \quad j = 0 \cdots N-1 \tag{6.122}$$

The entries of the DCT matrix were quantized to five digits past the decimal point. The resulting residue sequence for the linear predictor had a variance of 163.5, while for the DCT the residue variance 122.0 (the original data had a variance of 350.8). For further comparison, an RLSL filter with 2 stages was used, which yielded a residue variance of 160.5. When an 8-stage RLSL filter was employed, the residue variance reduced to 96.8. Moreover, when a frame size of 100 and a predictor length of 8 was used, the linear predictor residue variance fell to 94.5 (the DCT's performance under this scenario worsened).

These three methods were also compared using the speech waveform in Figure 6.84, a female speaker sampled at 20 KHz and quantized to 16 bits. For ($N = 8$) and ($M = 2$), the residue variance was 380890 for the DCT method, 523620 for the linear predictor, and 355950 for the RLSL. The speech waveform had a variance of 8323200. For ($N = 100$) and ($M = 8$), however, the linear predictor's residue variance fell to 141080; for 8 stages, the RLSL filter yielded a residue variance of 141550. The DCT's performance worsened under this scenario. One may conclude that transform-based methods can possibly be less efficient in compressing data quantized to a high number of levels.

Transform-based methods are worth examining for lossless applications; they are simple and can use existing hardware. Unlike linear predictors, they do not require large frame sizes for adequate performance, and they do not require complex implementations and startup values.

The Coding Compression Stage

The second stage of waveform compression involves coding the residue sequence using an entropy coder. The term *entropy coder* comes from the goal of entropy coding: the length in bits of the encoded sequence should approach the number of symbols times the overall entropy of the original sequence in bits per symbol. Several entropy coders exist; the two most widely used types, Huffman and arithmetic coders, will be discussed.

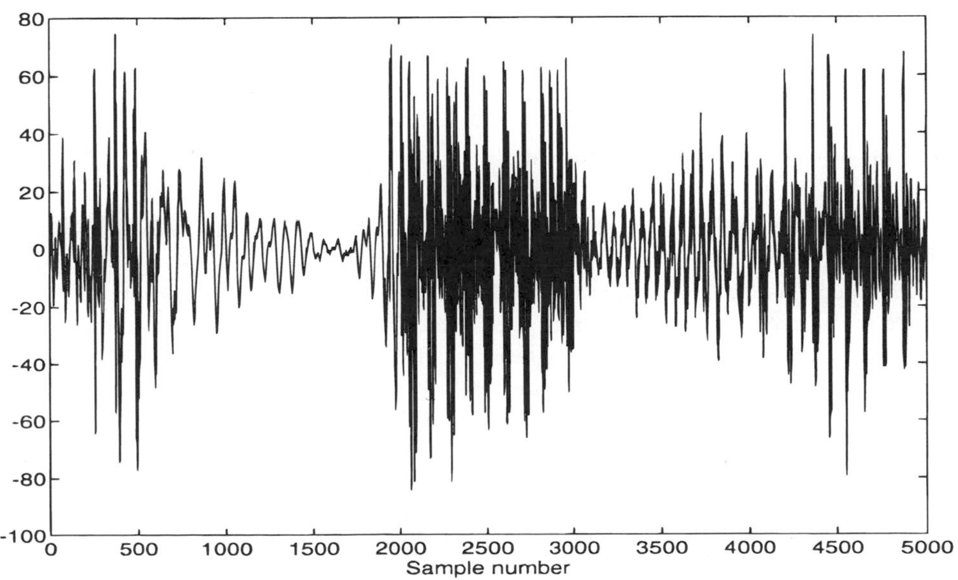

Figure 6.83 Speech waveform: 8 KHz, 8-bit sampling.

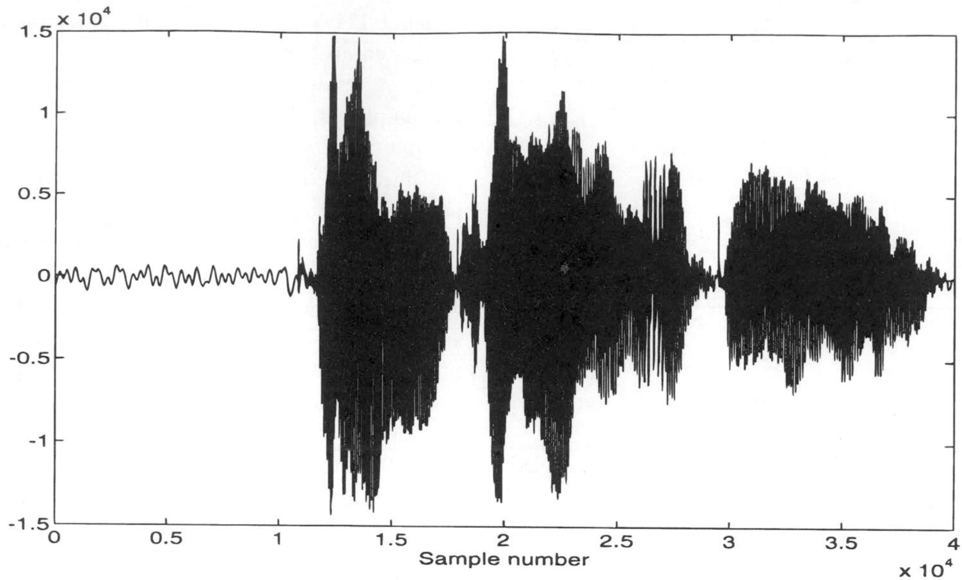

Figure 6.84 Speech waveform: 20 KHz, 16-bit sampling.

Table 6.19 Symbol Probability Table with Codes from Figure 6.85

Symbol	Probability	Code
a	0.1	000
b	0.1	001
c	0.2	01
d	0.3	10
e	0.3	11

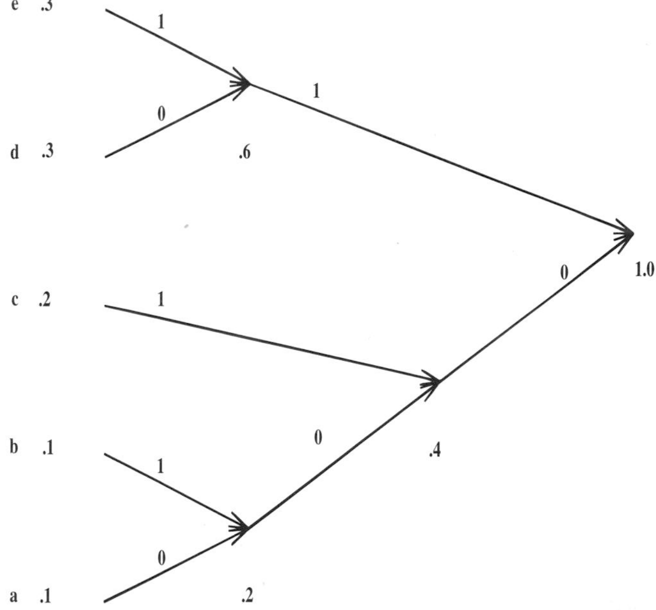

Figure 6.85 Huffman tree for Table 6.19.

Huffman Coding

Given the integer residue sequence $ir(n)$, $M<n<K-l$, one can determine "probabilities" (that is, *frequencies*) of a particular integer value occurring in the residue sequence. Each integer value appearing in $ir(n)$ is called a *symbol;* if the set of symbols $\{s_i\}$ occurs in $ir(n)$ with corresponding probabilities $\{p_i\}$, then a code can be formed as follows (Jain, 1989):

1. Order the symbols using their respective probabilities $\{p_i\}$ in decreasing order, assigning each symbol probability a node (the code will be constructed as a binary tree); these nodes are called *leaf nodes.*
2. Combine the two nodes with the smallest probabilities into a new node whose probability is the sum of the two previous node probabilities. Each of the two branches going into the new node will be assigned either a one or a zero. Repeat this process until there is only one node left, the *root node.*
3. The code for a particular symbol can be determined by reading all the branch values sequentially starting from the root node and ending at the symbol leaf node.

As an example, the Huffman code for the symbol table given in Table 6.19 is shown in Figure 6.85. Given a sequence of Huffman-coded data and an associated symbol probability table, each symbol can be decoded uniquely.

Huffman coders are divided into two classes: fixed Huffman coders and adaptive Huffman coders. Fixed Huffman coding involves using a static symbol table based either on the entire data sequence or on global information. Adaptive Huffman coding involves forming a new code table for each data sequence and encoding the table in addition to the data sequence. Alternatively, the adaptive Huffman coder may switch at intervals between previously selected code tables, indicating at each interval the selected table. Adaptive Huffman coders generally exhibit better performance in terms of the c.r. achieved, yet suffer from increased overhead. In real-time applications, fixed Huffman coders work more quickly and have simpler hardware implementations (Venbrux et al., 1992).

Arithmetic Coding

Although Huffman coding is attractive because of its simplicity and efficiency, it suffers from the fact that each symbol must have a representation of at least one bit. Normally, this is not a problem; however, in the case where the probability of a symbol approaches one, Huffman coding becomes inefficient. This leads to the concept of *arithmetic coding* (Rissanen and Langdon, 1979; Witten et al., 1987). The fundamental idea behind arithmetic coding is the mapping of a string of symbols to an interval in the interval [0.1] on the real line. The process begins by assigning each symbol s_i to a unique interval in (0,1) of length p_i. For example, assume an alphabet of symbols {a,b,c,} with respective probabilities {0.3,0.4,0.3} and corresponding intervals assigned as in Table 6.20. Then the string "ab" is coded as follows:

1. The symbol "a" puts the string in the interval [0,0.3].
2. The symbol "b" implies that the string occupies the middle 40% of the interval [0,0.3], i.e., [0.09,0.21].
3. Finally, any number is selected from this interval, e.g., 0.09.

This pedagogical example does not really demonstrate any of the compression abilities of arithmetic coding—a two-symbol string was simply mapped to a two-digit number—however, as the string length increases, arithmetic coding produces nearly optimal results.

Different Implementations of Arithmetic Coding. As pointed out in Witten et al. (1987), the above approach to arithmetic coding suffers from two problems: incremental transmission is not possible (i.e., the entire sequence must be coded before transmission), and the representation for the symbol mapping table is cumbersome and can produce significant overhead. The first problem can be alleviated by transmitting the top-most bit during coding when the top-most bits at each end of the interval are equal (i.e., the interval has become sufficiently narrowed) (Witten et al., 1987). The second problem may be solved by maintaining a running, reversible symbol mapping table which starts with equal initial probabilities for each possible symbol. For instance, if each symbol is 8 bits long, then the symbol mapping table normally has 256 entries, each with initial probabilities of 1/256.

As symbol sizes grow larger, it eventually becomes impractical to maintain a table with entries for every possible symbol (Stearns, 1995). For instance, if a particular waveform residue sequence $ir(n)$ spans the range (ir_l, ir_H), then the sequence requires a minimum of B bits per sample for accurate representation with $ir_H - ir_L < 2^B - 1$. If B is large, say greater than 10, then maintaining a symbol table with $2^B - 1$ entries would be highly inefficient. In Stearns (1995), a modified version of arithmetic coding is described to address this type of situation. In this version, the interval (ir_l, ir_H) is divided up into N_f successive intervals, each of length 2^{Nr} successive values. Rather than a symbol mapping table, an interval mapping table is developed from the number of symbols present in each interval, i.e., a frequency table $f(1{:}N_f)$. Each frequency in this table is represented by N_h bits, where $N_h = \log_2[\max f(n)]$. Each residue $ir(n)$ is assigned an interval number from 1 to N_f, denoted by $ir_I(n)$. Then each residue can be represented as

$$ir(n) = ir_L + 2^{N_r}(ir_I(n) - 1) + ir_0(n) \tag{6.123}$$

where $ir_0(n)$ is the offset of $ir(n)$ in interval $ir_I(n)$, a value between 0 and $2^{Nr} - 1$. Therefore, the compressed data is composed of three parts: (1) an overhead portion containing linear prediction parameters in addition to N_r, N_f, N_h, and $f(1{:}N_f)$, (2) an arithmetically coded sequence of interval numbers is $ir_I(n)$, and (3) an offset sequence $ir_o(n)$, which can be represented by a minimum of N_r bits per value.

For this modified version of arithmetic coding, an optimal value of N_r is required. It is argued (on the basis of the central limit theorem) that for most kinds of waveforms, the decorrelated residue sequence is approximately Gaussian. This assumption has been confirmed experimentally with data such as speech and seismic waveforms. With Gaussian residues, an empirically derived formula for the optimal N_r is (Stearns, 1995)

$$N_r = \max\left\{0, NINT\left\{\log_2\left[\frac{R}{K^{0.3}\sqrt{2}}\right]\right\}\right\} \tag{6.124}$$

where K is the data frame size and R is the range of the residue sequence, i.e., $ir_H - ir_l$.

This modified arithmetic coding method was tested with respect to the waveforms in Figures 6.81 and 6.82 using the linear prediction method given in **Linear Prediction.** The performance of the method was determined by comparing the average number of bits needed to store the residue sequence, bps_{ir}, with the average number of bits per sample of the modified arithmetic coder output, bps_{iy}; the results are given in Table 6.21.

Table 6.20 Symbol Mappings for Arithmetic Coding

Symbol	Probability	Interval
a	0.3	[0,0.3]
b	0.4	[0.3,0.7]
c	0.3	[0.7,1.0]

Table 6.21 Modified Arithmetic Coding Results

File	M_{min}	M_{max}	bps_{ix}	bps_{ir}	bps_{iy}
anmbhz89	2	8	15.75	5.42	4.26
anmbhz92	2	6	10.08	5.01	3.78
anmehz92	2	7	7.93	6.93	5.55
anmlhz92	3	3	12.70	11.97	10.56
kipehz13	1	4	8.15	4.90	3.48
kipehz20	2	8	10.03	6.03	4.87
rarbhz92	2	8	14.33	7.42	6.20
rarlhz92	2	5	17.00	15.98	14.71

Conclusions

The two-stage lossless compression scheme in Figure 6.77 has been presented and developed. Different linear filter implementations of the first stage were presented: linear predictors, adaptive filters, and lattice filters. The lattice filters, particularly the RLSL filter, displayed fast convergence and are desirable for fixed-order predictors. Transform-based decorrelation was also described, which displayed some advantages over filter methods; in particular, ease of implementation and superior performance for small data frame sizes. The second stage was discussed with respect to Huffman and arithmetic coding. Modifications to basic arithmetic coding, which are often necessary, were described.

While only a few decorrelation and symbol coding methods have been discussed here, many more exist. The particular implementation of two-stage compression depends on the type of data being compressed. Experimentation with different implementations is often advantageous.

References

Blahut, R. E. 1990. *Principles and Practice of Information Theory,* Addison-Wesley, Menlo Park, CA.

Haykin, S. 1991. *Adaptive Filter Theory,* Prentice-Hall, Englewood Cliffs, NJ.

Jain, A. K. 1989. *Fundamentals of Digital Image Processing,* Prentice-Hall, Englewood Cliffs, NJ.

Mandyam, G., Ahmed, N., and Magotra, N. 1995. A DCT-based scheme for lossless image compression, *SPIE/IS&T Electronic Imaging Conference,* February, San Jose, CA.

Marple, S. L. 1987. *Digital Spectral Analysis with Applications,* Prentice-Hall, Englewood Cliffs, NJ.

McCoy, J. W., Magotra, N., and Stearns, S. 1994. Lossless predictive coding, *37th IEEE Midwest Symp. Circuits and Systems,* August, Lafayette, LA.

Papoulis, Athanasios 1984. *Probability, Random Variables, and Stochastic Processes,* McGraw-Hill, New York, NY.

Rissanen, J. and Langdon, G. G. 1979. Arithmetic coding, *IBM Journal of Research and Development,* 23(2):149–162, March.

Stearns, S. D. 1995. Arithmetic coding in lossless waveform compression, *IEEE Trans. Signal Processing,* August, 1995.

Stearns, S. D., Tan, L.-Z., and Magotra, N. 1993. Lossless compression of waveform data for efficient storage and transmission, *IEEE Trans. Geoscience and Remote Sensing,* 31(3):645–654.

Stearns, S. D., Tan, L.-Z., and Magotra, N. 1992. A bi-level coding technique for compressing broadband residue sequences, *Digital Signal Processing,* 2(3):146–156, July.

Tan, L.-Z. 1992. *Theory and Techniques for Lossless Waveform Data Compression,* Ph.D. Thesis, The University of New Mexico.

Venbrux, J., Yeh, P.-S., and Liu, M. N. 1992. A VLSI chip set for high-speed lossless data compression, *IEEE Trans. Circuits and Systems for Video Technology,* 2(4):381–391, December.

Widrow, B. and Stearns, S. D. 1985. *Adaptive Signal Processing,* Prentice-Hall, Englewood Cliffs, NJ.

Witten, I. H., Neal, R. M., and Cleary, J. G. 1987. Arithmetic coding for data compression, *Communications of the ACM,* (6):520–540.

Woodward, P. M. 1964. *Probability and Information Theory,* 2d ed., p. 25, Pergamon Press, Elmsford, NY.

Further Information

Information on advances in entropy coding can be found in the *IEEE Transactions on Information Theory.* Information on waveform coding can also be found in the *IEEE Transactions on Geoscience and Remote Sensing,* the *IEEE Transactions on Speech and Audio Processing,* and the *Journal of the Acoustic Society of America.* Information on statistical signal analysis and stationarity issues can be found in the text *Random Signals: Detection, Estimation, and Data Analysis* by K. Sam Shanmugan and Arthur M. Breipohl (Wiley, 1988).

6.13 3-D Measurement Techniques

Bernard C. Jiang

This section provides a review of various 3-D measurement techniques. Three-dimensional measurement refers to the ability to measure an object's position, orientation, or tracking a moving object three dimensionally. The principles of operation, accuracy and repeatability of measurement systems, and any special features are summarized. A robot's position or movement may be used as an example, as it represents a typical object's 3-D information of its location.

Different devices for measuring spatial location have been developed (Jiang, 1988; Warnacke et al., 1985; Lau, 1984; RAACC, 1984). Basically, a spatial location device should be able to spot or trace an object, stationary or moving, with specified accuracy, under various operating conditions. In addition, the device's repeatability and measuring speed should optimally be 10 times higher (practically, at least 3 times higher) than that of the robot (called the Rule of Ten) (Warnacke and Schiele, 1984). The measuring device should also have a storage and output mechanism where the data can be stored and used for further analysis. Broadly speaking, the methods adapted for using these systems can be classified as contact sensing and non-contact sensing. Measurement techniques based on the type of instrument, such as the Theodolite System, SELSPOT, WATSMART, or the ExpertVision system, are also presented. The principles of operation are briefly explained, and the advantages and limitations of each technique are also discussed.

Contact Sensing Devices

The simplest way to measure the spatial location of an object is by using a contact sensor (e.g., touch-sensor, probe) with feedback capability. When the programmed robot makes contact with the sensor, a signal is generated (by displacement in the case of a probe, and by voltage in the case of a transducer). The displacement or voltage thus produced is a direct measure of the position of the object. Two examples of such contact sensing devices are dial indicators and linear variable differential transformers (LVDTs) (Degarmo et al., 1984; Kreamer, 1984; Lembke and Jones, 1983; Leu and Dukovski, 1987; Lipp and Maul, 1987; McEntire, 1976; Vira and Lau, 1986; Warnecke et al., 1985; Warnecke and Schiele, 1984). Brief descriptions of these methods are presented below.

Technique 1—Dial Indicator Method

Dial indicators are mechanical devices for measuring distance variations. Linear displacement of the measuring tip is converted into rotational movement, which is amplified by a gear train combination and displayed by a pointer rotating on a dial (Norton, 1982). Figure 6.86a shows a common dial indicator structure. The dial indicator method is a widely used, low-cost solution for displacement measurement. This device has been used to measure robot and sensor accuracy (Brown, 1988), as well as to determine repeatability by varying operating pressure and stroke length for a simple-structured robot (Warnecke and Schiele, 1984). By properly arranging three or more dial indicators, three-dimensional (3-D) data can be obtained (Figure 6.86b).

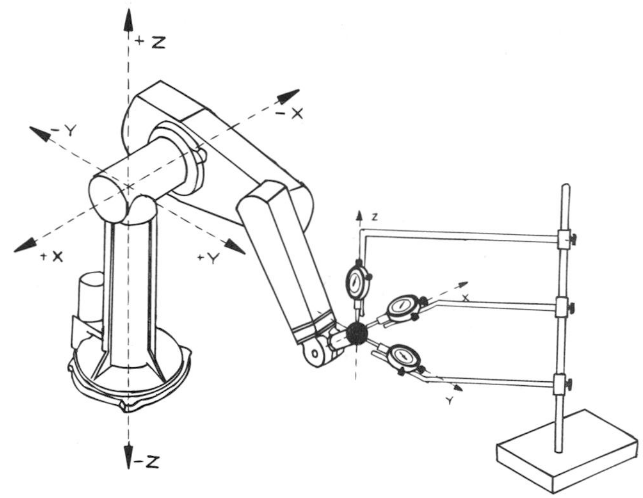

Figure 6.86b Dial indicator setup for 3-D measurement.

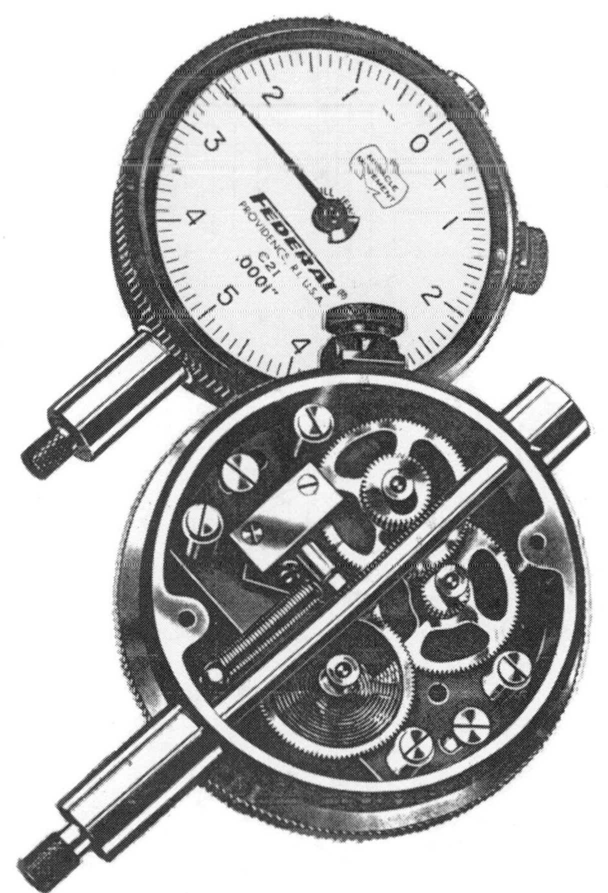

Figure 6.86a Dial indicator construction. (Courtesy of Federal Products Corp.)

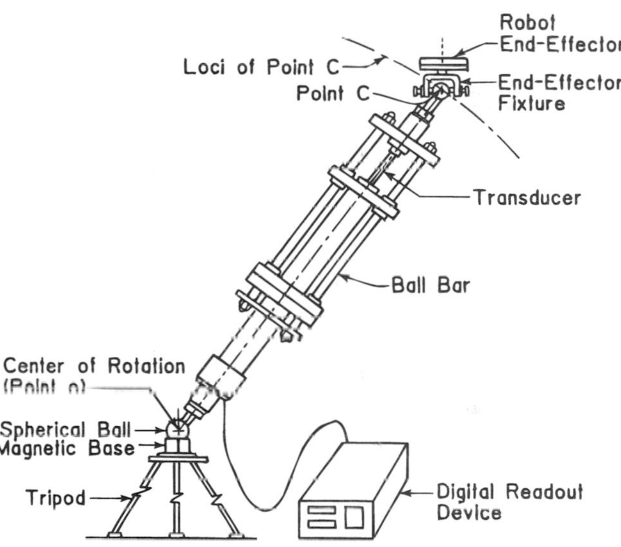

Figure 6.87 Extensible ball bar method. *Source:* Degarmo, E. P. et al. 1984. *Materials and Processes in Manufacturing,* 6th ed., Macmillan, New York, NY. With permission.

Technique 2—Extensible Ball Bar Method

The extensible ball bar technique was developed by Vira and Lau (1986) at the National Bureau of Standards (Figure 6.87). The operation of the extensible ball bar is similar to a dial gauge. A high-precision steel ball of 2.54 cm diameter is fixed to each end of a spring-loaded extensible bar. A probe is installed inside the bar to sense the linear displacement (a maximum of 5 cm) between the balls. One end of the ball bar is attached to a universal joint, which is mounted to the robot's end effector, while the other end is magnetically held in a socket, where it is free to rotate. During the measurement, the robot is programmed to move in an arc whose radius is the nominal length of the bar (Vira and Lau, 1986). The deviation between the actual and programmed paths of the robot is continuously sensed and recorded by the probe. Thus the instantaneous position of the robot's end effector is obtained.

Technique 3—Coordinate Measuring Machine (CMM)

A CMM usually consists of the following elements:

1. Probes—Can be hard, soft (electronic), or noncontact probes.
2. Motion mechanism—Can be as a rotating shaft, an air bearing, or a ball bearing.
3. Displacement measuring devices—Can be a linear encoder, moire fringe, or an induction-type encoder.
4. Motion transmission mechanism—Can be a ball screw, gear train, belt, or roller and shaft.

There are several types of CMMs (Figures 6.88a and 6.88b). Cantilever-type CMMs employ three movable components moving along mutually perpendicular guideways. They are usually the smallest in size and lowest in cost and occupy a minimum of floor space. Bridge-type CMMs are similar to cantilever-type but more stable and more accurate. Column-type CMM is similar in construction to accurate jig boring machines. The column moves in a vertical (Z) direction only, and a two-axis saddle permits movement in the horizontal (X and Y) directions. Gantry-type CMMs also employ three movable components moving along mutually perpendicular guideways. An extra support is provided to allow an open space for large parts (e.g., automobile bodies) inspection. Horizontal arm-type CMMs provide maximum flexibility of the measurement range. By incorporating a rotary table, four-axis capability is obtainable.

Technique 4—LVDT Method

The LVDT is an electromechanical contact gauge. A typical linear LVDT construction is shown in Figure 6.89 (Norton, 1982). A metal core resides in a sensor housing that consists of one primary winding and two secondary windings. The linear variation (e.g., displacement) of the contact head causes a proportional difference in the output voltage of the internal coils. Although the resolution of each individual sensor can be very high (e.g., 0.0025 mm), the range of the LVDTs is usually very limited (e.g., 2 mm in Deere & Company Technical Center, 1987). Three LVDTs can be arranged orthogonally to determine the 3-D coordinates of an object. Riley (1987) constructed a pose (position and orientation) measurement system using nine LVDTs. Leu and Dukovski (1987) employed a similar system to conduct a robot accuracy study.

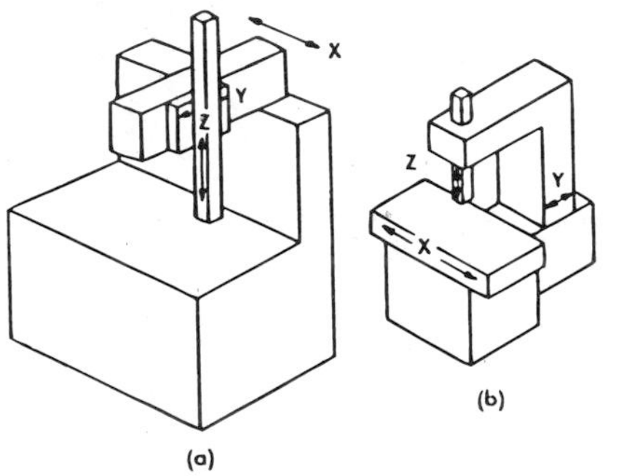

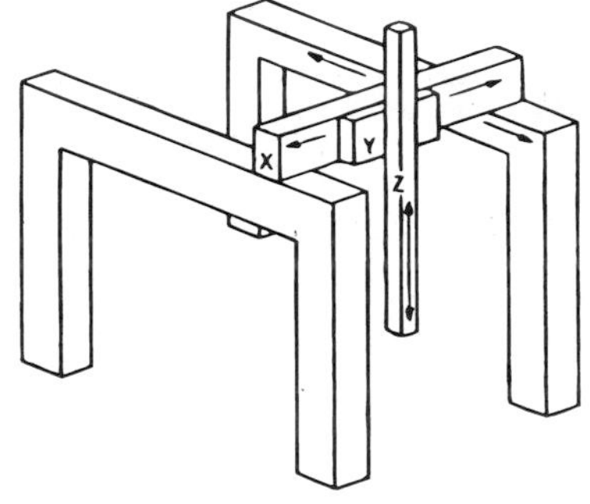

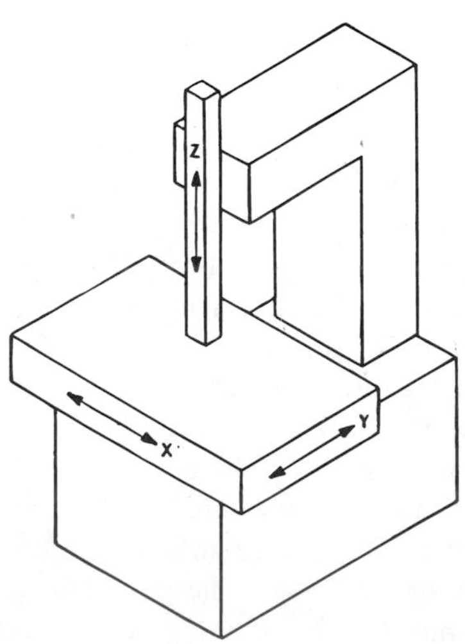

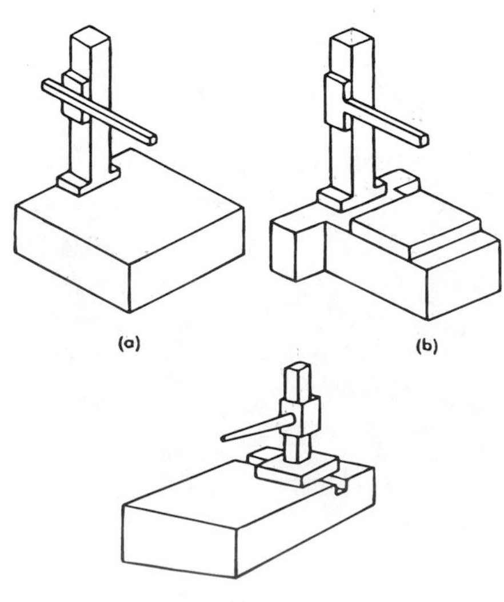

Figure 6.88a CMM types—I.

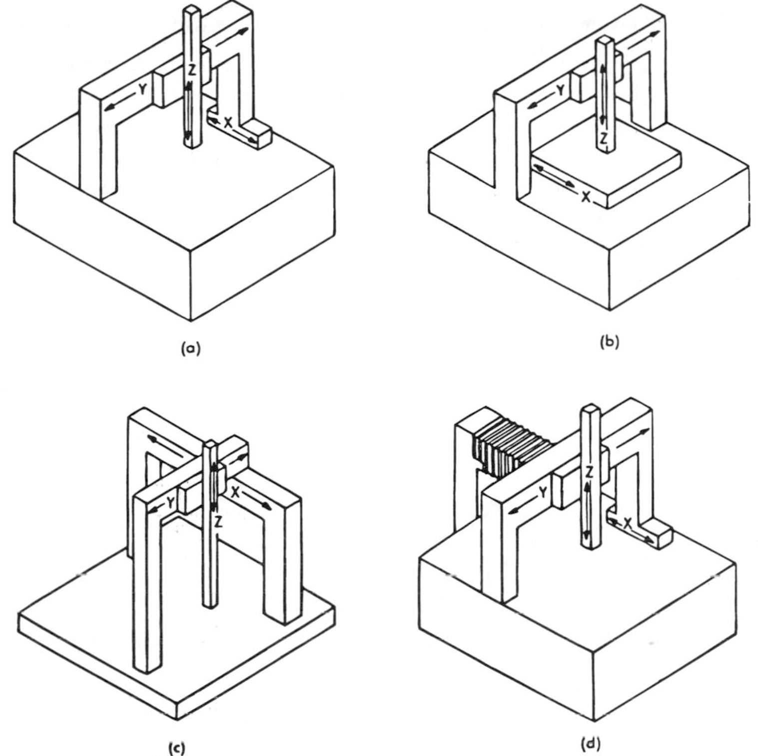

Figure 6.88b CMM types—II.

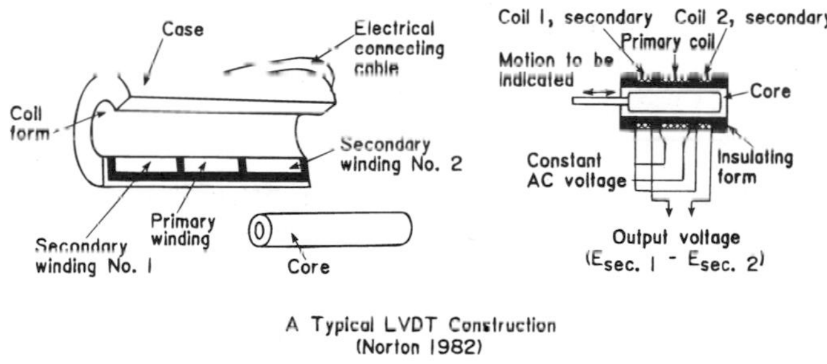

Figure 6.89 A typical LVDT construction. *Source:* Norton, H. N. 1982. *Sensor Analyzer Handbook,* Prentice-Hall, Englewood Cliffs, NJ. With permission.

Noncontact Sensing Devices

The noncontact sensors locate a 3-D position by detecting changes (i.e., electrical, optical, or acoustic) from a known reference location. Acoustic-based, optical-based, and electromagnetic-based systems are the types of noncontact systems discussed in this section. An advantage of using a noncontact sensing device is that the measuring devices do not touch the robot to be measured; therefore, measurement error is reduced by keeping the measuring device stationary.

Acoustic-Based System

Acoustic systems work on the principle of time domain or sonar. High-frequency sound is transmitted in pulses to the object whose position is to be measured. A receiver is positioned to receive the reflected pulses. When the velocity of the sound waves in air and the time taken to hit the object and reflect back are known, the distance can be computed. Special electronics are built into the transmitter and receiver to account for inertia. Two or more of the transmitters and receivers are used to measure the 3-D location.

Technique 5—Acoustic Range Sensor

Figure 6.90 shows a block diagram of the polyvinylidene floride (PVF_2) acoustic ranging system (Schoenwald and Martin, 1984). Two acoustic ranging transducers of PVF_2 are mounted on the gripper of a PUMA Unimate robot. The transmitter has a coil transformer for impedance matching. The receiver feeds its signal to an FET preamplifier. The signal is then further amplified, and a detector circuit produces a positive envelope

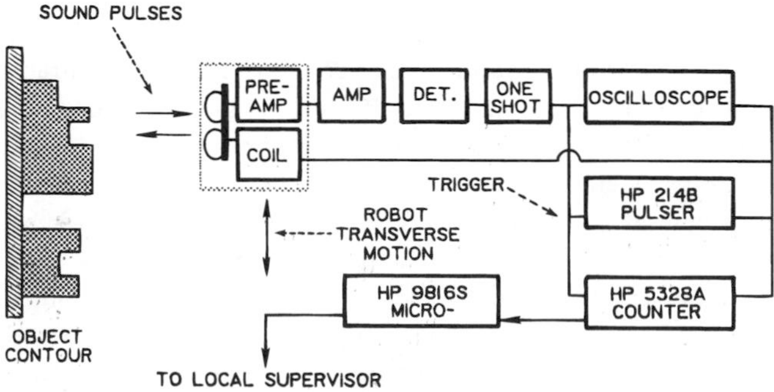

Figure 6.90 Acoustic range sensor system. *Source:* Shoenwald, J. S. and Martin, J. F., 1984. An acoustic ranging system for robot position control and tracking, *Proc. Robot West Conf.,* Anaheim, CA.

pulse. After measuring the time delay T between the transmitted pulse and the received echo, the distance L from the transducer pair to the target is computed by:

$$L = V(T - T_0)/2$$

The term T_0 is the delay arising from the electronics, and V is the velocity of sound. The received echo is then used to retrigger the pulse generator. As the distance to the target is reduced, the pulse frequency increases in inverse proportion (Schoenwald and Martin, 1984).

Optical-Based Systems

The operating principle of optical-based systems is as follows. The optical sensors are positioned to detect the light sources placed on the end effector and thereby determine its coordinates. The optical sources may be light-emitting diodes (LEDs), infrared rays, or ambient light. Optical-based systems can be categorized into active and passive sensing systems (Norton, 1982). Active systems are those that have active communication between the sensors and the generators. Passive systems usually work in ambient light.

Active-Video Systems

Active-video systems are those that send pulses or other forms of signals (infrared, laser, fiber optic light, etc.) to the sensing and measuring unit. SELSPOT and WATSMART belong to this category. Ishii et al. (1987) presented an application using a 3-D active sensing system for teaching robot paths. Tucker et al. (1987) reported a pose (X, Y, Z coordinates and pitch, roll, yaw angles) measurement sensor based on the detection of the reflective fiber optic light source. This sensor can simultaneously measure the position and orientation of a quasi-static object to within 0.13 mm and 0.1 mrad.

Technique 6—SELSPOT and WATSMART System

SELSPOT's robot check system (Selcom Corp., 1985) and Waterloo Spatial Motion Analysis and Recording Techniques (WATSMART, Northern Digital Inc., 1986) are the commercially available systems for 3-D noncontact digitizing active-video techniques.

The object to be monitored is fitted with a number of infrared markers (LEDs). Each marker is attached to a distributer (strober) through wires and is activated for a brief period of 65 μs by the controller (Figure 6.91a). For 3-D recording, at least two cameras must be used simultaneously. The camera lens focuses an image of the marker on a two-dimensional photosensitive sensor called the lateral effect photodiode. The resulting signals are amplified, converted to 12-bit digital values in the camera, and transmitted to the camera's controller board in the system cabinet. Here, the data is transformed into 3-D coordinates by a direct linear transformation (DLT) technique. A computer (e.g., an IBM/PC-AT) drives the camera and strobing controllers. A schematic of the camera setup is shown in Figure 6.91b.

Technique 7—Laser Systems

In this method, reflected laser beams are detected by a sensor, and the resulting data are used in computations to determine positional and orientational accuracy. Laser-based systems demonstrate higher accuracy and faster measurements than other noncontacting systems. As reported by Brown (1985), there are five techniques available for measuring distance or coordinates by using a laser system:

1. The time-of-flight technique involves the creation of a pulse, which is made to reflect from the target and return to the measuring device.
2. The phase modulation technique involves a collimated laser beam that is modulated to produce an amplitude-varying wave.

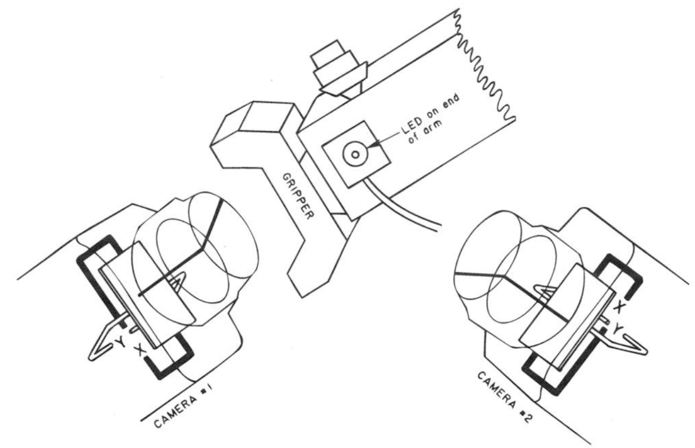

Figure 6.91a Positioning sensors (cameras) and LED.

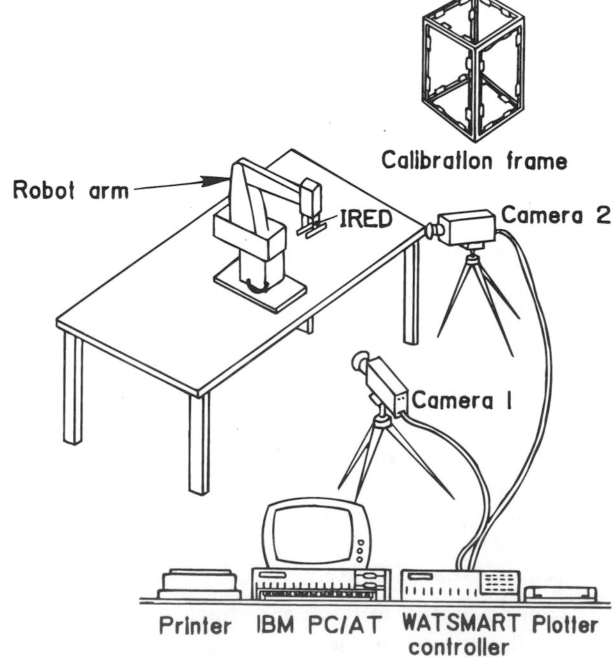

Figure 6.91b SELSPOT and WATSMART systems setup.

3. The triangulation technique involves the measurement of angles via a camera or a laser probe.
4. The optical encoder technique involves the measurement of an encoder's angular position by sensing the laser light passing through.
5. The interferometry technique is based on the interference of waves that are made to travel different paths.

Of these five techniques, optical encoders have the disadvantage that they must be mechanically coupled to the device being measured, causing motion constraint and intrinsic measurement error. The time-of-flight and phase modulation techniques yield low measuring resolution (Brown, 1985). Therefore, only interferometry and triangulation techniques are discussed below.

The laser interferometry technique is based on the phenomenon of phase change caused when light waves are made to travel in different paths. By recombining these waves, an interference pattern (in Doppler signals) is produced from which the number of quarter waves of path difference as the target is moved can be counted (Chandra et al., 1986). By analyzing this interference, the exact location of the object can be determined. A schematic diagram of a laser interferometer is shown in Figure 6.91.

Lau et al. (1988, 1985) at the National Institute of Standards and Technology developed a single-beam 3-D laser tracking system. This system is commercially available. It consists of a single-beam laser interferometer system for distance (length) measurement and an encoder system for angle measurement. By combining the distance and angle measurements, an object's pose data can be obtained. Another commercially available 3-D laser tracking system employs three laser beams (called "trilateration") to track a retroreflector that is attached to a moving target. With multiple beams, no angular information is required to determine the exact 3-D location (Brown, 1988; Chandra et al., 1986).

Measurement of spatial location with the aid of a laser interferometer was experimented with at IPA in Stuttgart where a tracking controller was developed (Warnecke and Schiele, 1984).

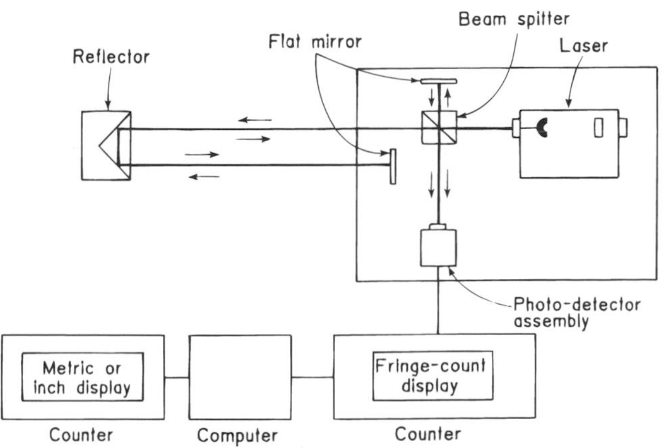

Figure 6.92 Laser interferometer. *Source:* Norton, H. N. 1982. *Sensor Analyzer Handbook,* p. 104, Prentice-Hall, Englewood Cliffs, NJ. With permission.

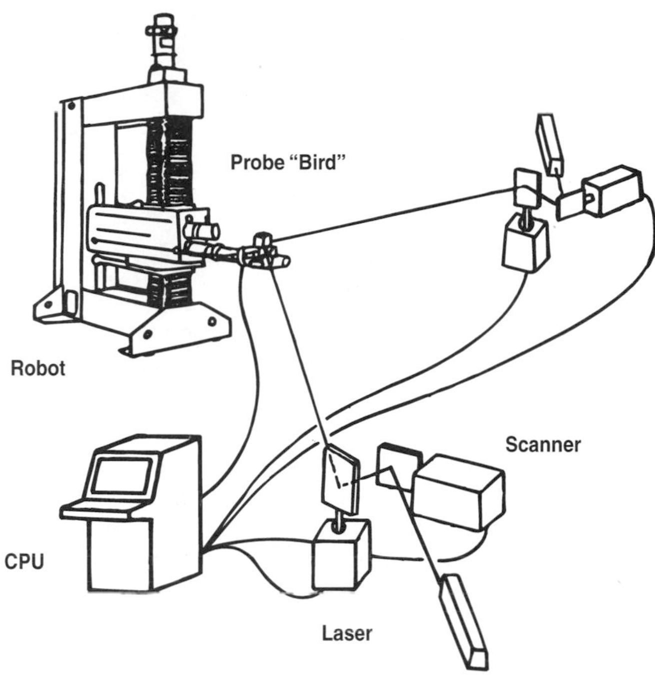

Figure 6.93 The Bird system. *Source:* Tamio et al. 1983. *Position and Orientation Measurement of a Moving Object by CCD Photo Array Sensors,* University of Tokyo, 7–3–1, Hongo, Bunkyo-ku, Tokyo, Japan. *Proc. 4th Int. Conf. Assembly Automation,* 1983. With permission.

Technique 8—The Bird System

The Bird system was developed by Dr. Tamio et al., University of Tokyo in Japan (Tamio et al., 1987). The aim of their research was to develop a measurement system to estimate the position and orientation of an object that moves in 3-D space.

Two laser beams in a cross shape are projected on the object and a cubic probe is fixed to the end effector. Two charged coupled device (CCD) ring sensors are installed on the adjacent planes of a probe that detects structured laser beams. Figure 6.93 shows the general framework of the system. A scanner is used as a tracking system.

The rotation is obtained on the basis of the angle between the two projections of the beams; the translation (i.e., the position) is determined with the help of triangulation. Various components that go into the Bird system are

1. CCD Sensors: Since translation and rotation can be computed from information obtained from adjacent planes, two 64-bit ring sensors are installed as a probe on two neighboring planes of a cube. To prevent noises, a read-out amplifier for the CCD is installed on two corresponding planes of the probe. The output from the CCD sensors are transformed into binary data.
2. Cross-Shaped Beams: The cross shape of the helium-neon (He-Ne) laser beams (0.1 mm wide by 4 mm long) helps in determining the orientation of the object. In order to produce cross-shaped beams, a laser with two cylindrical lenses is used.

3. Scanners: To keep the laser projected on the center of the rings, a tracking system called scanners is used. It has two mirrors that are actuated by galvanometers. The scanners have a 30° rotation angle, 3.7 gm-cm^2 inertia, and 130-Hz natural frequency.

Technique 9—Optical Scanner System

This system was developed by Gilby et al. at the University of Surrey in Guildford, Surrey, England (Gilby et al., 1984; Parker and Gilby, 1982/3), for testing the dynamic performance of a robot arm.

The principle of operation of this system is to follow the target with two separate beams of light and record their position with respect to their source location. The general layout of the system is shown in Figure 6.94. The instrument consists of a moving target that is rigidly fixed to the robot arm and two similar static measurement units. Each of these units (called subsystems), consists of optical and mechanical components and a control unit. Two separate beams of light (one from each subsystem) are made to follow the moving target, and the relative position of the target from the reference source is continuously recorded. To effect this tracking, the distances between the center of the target and the center lines of the beams are measured by light-sensitive detectors. Error signals from these detectors are used to correct the directions of the light beams so that they point directly at the optical center of the target. To enable higher tracking speed the instrument has been designed for low inertia.

The beam that is generated by the He-Ne laser is first expanded to about 8 mm in diameter and then passed through the beam-splitter and on to the center of a plane mirror attached to the optical scanner. This scanner is able to rapidly rotate the mirror through an angle of approximately 20° about an axis that is perpendicular to the beam's direction. The beam reflected by this mirror impinges upon the axis of a second mirror also rotated by an optical scanner. This axis of this second mirror is parallel to the initial direction of the laser beam. After reflection by both mirrors, the beam emerges from the subsystem in a direction that is determined by the rotations of the mirrors. By suitably rotating the shafts of the two optical scanners, the emergent beam is directed towards the target.

Passive-Video Systems

Passive systems are those that do not send pulses or other forms of signals to the measuring unit, but rather work on ambient light. The simplest form of a passive system uses television cameras to sense a robot's two-dimensional position and then transform it into 3-D coordinates by using triangulation (Figure 6.95a). Related technical background for these systems can be found in Gennery (1986), Fang and Huang (1984), Tsai et al. (1982), Tsai and Huang (1981), Yamimovsky and Cunningham (1978). Motion analysis systems are commercially available. An application using visual information to control a high-precision manipulator is described by Tamio et al. (1987). Application of a commercially available, gray-scale vision system in defining a forging part is discussed in Kelly (1982).

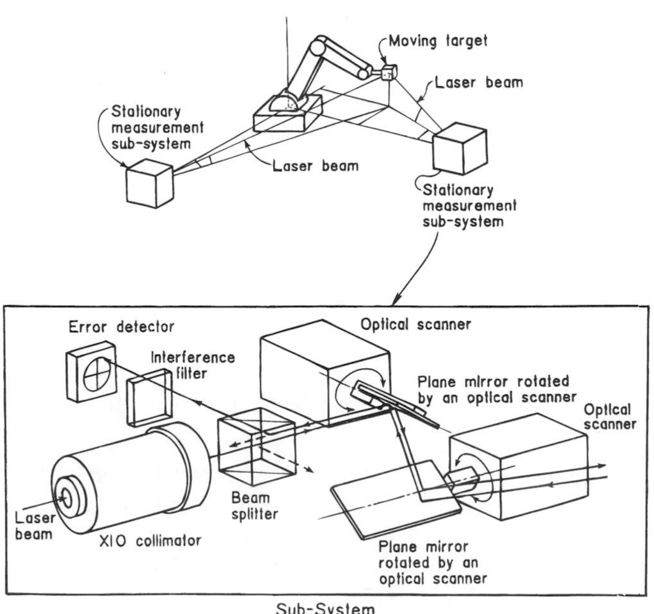

Figure 6.94 Optical scanner system. *Source:* Parker, G. A. and Gilby, J. 1982/3. *An Investigation of Robot Arm Position Measurement Using Laser Tracking Techniques,* University of Surrey, Dept. of Mech. Eng., Surrey, England. *Proc. 1st Robotics Europe Conf.,* Brussels, Belgium, 1984. With permission.

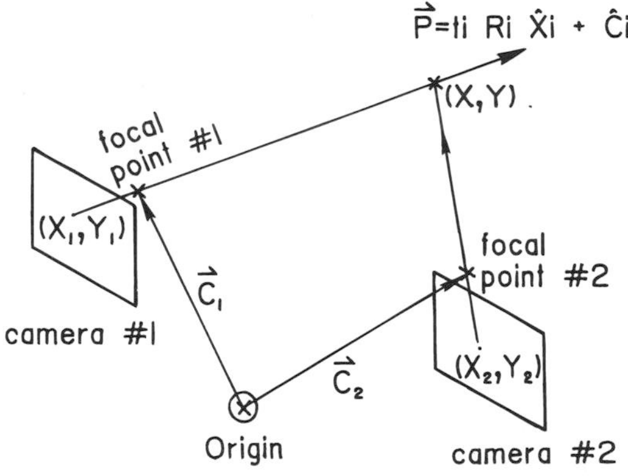

Figure 6.95a Principle of measuring a spatial location using an optical system.

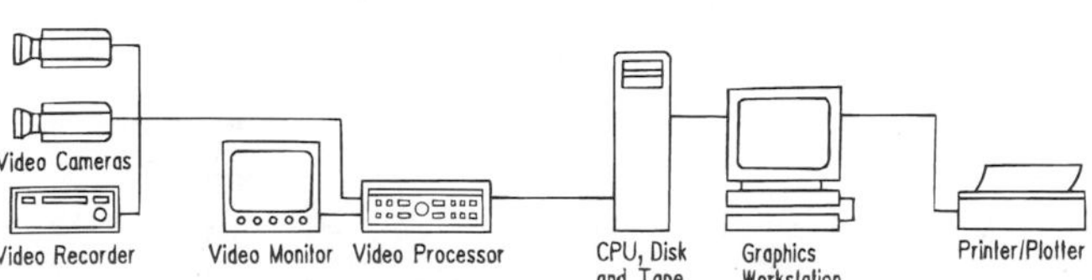

Figure 6.95b ExpertVision system. *Source:* Motion Analysis Corp., Santa Rosa, CA. With permission.

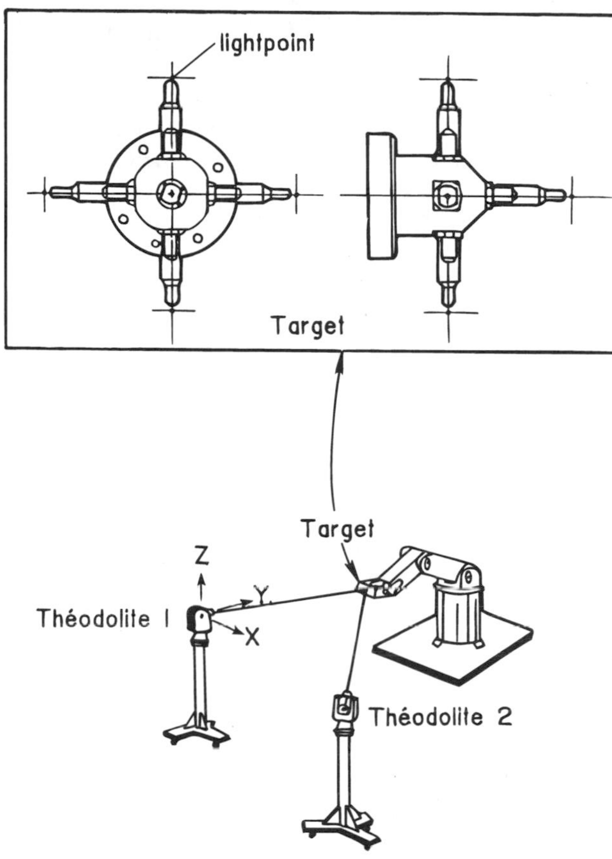

Figure 6.96 Theodolite system. *Source:* Warnecke, H. J. and Schiele, G. 1984. Performance characteristics and performance testing of industrial robots—state of the art, *Proc. 1st Robotics Europe Conf.*, Brussels, Belgium. With permission.

Technique 10—ExpertVision—Motion Analysis System

The Motion Analysis System (called ExpertVision) is marketed by Motion Analysis Corporation, Santa Rosa, California (Figure 6.95b). This is a video-based system that receives the input through a video recorder. One or more retroreflective markers can be attached to the robot's joints. The data from this unit is processed by a video processor. Further image analysis is done to determine the centroids of each marker, and the tracking results are displayed or plotted in graphic or tabular form.

The motion of an object can either be viewed directly through a video camera (RS-170/330) or be played back on a video recorder. This video data flows to the video processor (VP 100), where it is reduced to position-related centroids. Later, it is converted to time-space paths. The statistics software analyzes the position parameters and executes many mathematical operations. A mouse/menu editor provides easy interaction with the raw or processed data. The system comes with a SUN Workstation (SUN 2/20) (which supports a wide variety of visual formats including annotated graphics) or an IBM PC as a controller.

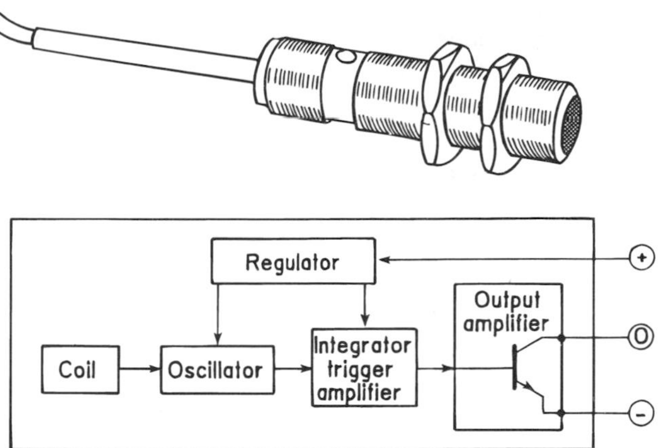

Figure 6.97a Eddy-current type proximeter. *Source:* Norton, H. N. 1982. *Sensor Analyzer Handbook*, Prentice-Hall, Englewood Cliffs, NJ. With permission. Courtesy of Micro Switch, A Honeywell Division.

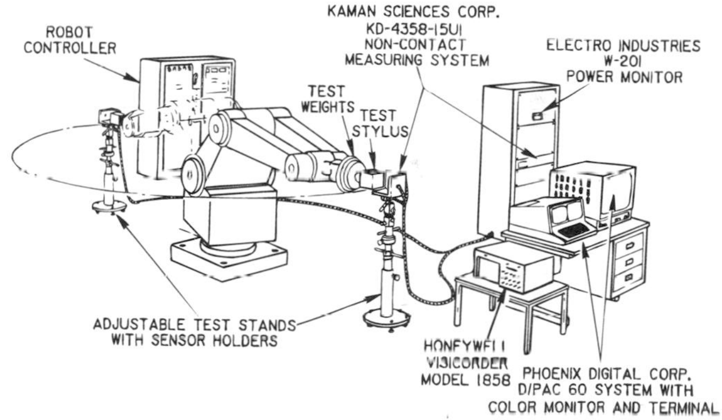

Figure 6.97b Proximeter system setup. *Source:* RAACC 1984. *Robot Assessment Program*, RAACC, Ford Motor Co., Dearborn, MI. With permission.

Technique 11—Theodolite System

The theodolite system is noncontact measuring system employing a triangulation technique (Ackerson, 1987; Ackerson et al., 1987; Whitney et al., 1986; Ackerson and Harry, 1985; Warnecke and Schiele, 1984). Two theodolites are used to determine the spatial location of an object, using a 3-D triangulation principle. The setup (Figure 6.96) consists of two theodolites, the respective positions of which are calibrated by means of landmarks obtained from the graduations of rulers or the vertices of a calibrated trihedron. To show the position of the robot, a special target made of five arms (or other forms) ending in lightpoints is fixed to the robot. To measure the position of the lightpoint in the space of the robot, both theodolites are aimed at the same light point. Then azimuth and elevation angles are recorded. This operation is repeated for three positions of the end effector being measured. By processing the collected data, the position and orientation of the object are determined.

Electromagnetic-Based System. The spatial location of the sensor is determined by the change in electricity (current or voltage) measured by the transducer. The analog output from this transducer is converted to digital output and stored for further processing and analysis.

Technique 12—Proximity Transducer System

A typical eddy-current type proximeter is shown in Figure 6.97a. The measured object (robot's end effector) absorbs the electromagnetic field generated in the sensing coil by the eddy currents. This changed signal is amplified, and the amplified signal triggers a switch (Norton, 1982). By properly arranging multiple proximeters, the pose of an object can be measured (RAACC, 1986; Warnecke and Schiele, 1984).

Chandra et al. (1986) developed a method for measuring a robot's spatial location with the help of a proximeter installed on the robot's end effector. The measuring head, which is fixed to the robot's end effector (see Figure 6.97b) has three mutually perpendicular plates. A probe (proximeter transducer system) which is attached to each plate of the measuring head can sense the relative distance, between the probe's end and the target surface of the reference stand. Three-dimensional coordinates are determined after a analog/digital conversion.

A similar device was developed by Tucker et al. (1987). Here, three fiber optic sensors were installed on a pose measurement device to sense the position and orientation of a robot's end effector.

References

Ackerson, D. S. and Harry, D. R. 1985. Theory, experimental results, and recommended standards regarding the static positioning and orienting precision of industrial robots, *Robotics and Computer-Integrated Manufacturing,* 2(3/4):247–259.

Ackerson, D. S. 1987. Uses of the WILD C.A.T. 2000 Electronic Theodolite System with Robot-Based Manufacturing Systems, Personal communication.

Ackerson & Associates Inc. 1987. *Robot Accuracy Study Conducted for NBS—Overview,* Vol. 1, West Chester, PA.

Brown L. B. 1985. A random-path laser interferometer system, *Int. Congress on Applications of Laser and Electro-Optics,* San Francisco, CA.

Brown, L. B. 1988. Laser trilateration system—a new coordinate measurement system improves accuracy of CMMs, *SME Clinic on Precision Metrology with Coordinate Measuring Systems,* June, Gaithersburg, MD.

Chandra, M. J., Rosenshine, M., and Soyster, A. L. 1986. Analysis of robot positioning error, *Int. J. Prod. Research,* 24(5):1159–1169.

Chesapeake Laser Systems, Inc., 1987. *3-D Laser System,* Product information.

Deere & Company Technical Center 1987. *John Deere Repeatability Analyzer,* Product description, Moline, IL.

Degarmo, E. P., Black, J. T., and Kohser, R. A. 1984. *Materials and Processes in Manufacturing,* 6th ed., MacMillan, New York, NY.

Expert Vision, Automated Motion Analysis System, Motion Analysis Corporation, Santa Rosa, CA.

Fang, J. and Huang, T. S. 1984. Some experiments on estimating the 3-D motion parameters of a rigid body from two consecutive image frames, *IEEE Trans. Pattern Analysis and Machine Intelligence,* PAMI–6(5):545–554.

Gennery, D. B. 1986. *Stereo Vision for the Acquisition and Tracking of Moving Three-Dimensional Objects. Techniques of 3-D Machine Perception,* 53–74. Rosenfeld, A., ed., Elsevier Science Publishers, New York, NY.

Gilby, J. H., Mayer, R., and Parker, C. A. 1984. Dynamic performance measurement of robot arms, *Proc. 1st Robotics Europe Conference,* 31–44. Brussels, Belgium.

Ishii, M., Sakane, S., and Mikami, Y. 1987. A 3-D sensor system for teaching robot paths and environments, *Int. J. Robotics Research,* 6(2):45–59.

Jiang, B. C., Black, J. T., and Duraisamy, R. 1988, A review of recent developments in robot metrology. *J. Man. Systems,* 7(4):339–357.

Kelley, R. B. 1982. Pose refinement vision, *Proc. Int. Conf. Robot Vision and Sensory Controls,* 379–388, Stuttgart, Germany.

Kreamer, W. C. 1984. *Measuring Robot and Sensor Accuracy,* Technical Paper MS84–1038, Society of Manufacturing Engineers.

Lau, K., Dagalakis, N., and Myers, D., 1988. Testing, in the *International Encyclopedia of Robotics Applications and Automation,* John Wiley & Sons, New York, NY.

Lau, K. and Hocken, R. J. 1984. A survey of current robot metrology methods, *Ann. CIRP,* 32(2):485–488.

Lau, K., Hocken, R., and Haynes, L. 1985. Robot performance measurements using automatic laser tracking techniques, *Robotics and Computer-Integrated Manufacturing,* 2(3): 227–236.

Lembke, J. R. and Jones, L. L. 1983. *Measurement of Robot Accuracy Using the Latin Square Three-dimensional Ball Plate,* final report, Bendix Corp., Kansas City, MO.

Leu, M. C. and Dukovski, V. 1987. Robot accuracy and its improvement—an experimental investigation, *Proc. 15th North Am. Manufacturing Research Conf.,* 676–681, Bethlehem, PA.

Lipp, A. R. and Maul, G. P. 1987. Performance testing on a prototype low technology robot, *Int. J. Prod. Research,* 25(4):549–559.

McEntire, R. H. 1976. Three-dimensional accuracy measurement methods for robots, *Ind. Robot,* 3(3):105–112.

Northern Digital Inc. 1986. WATSMART Product/Technical Description, Waterloo, Ontario, Canada.

Norton, H. N. 1982. *Sensor and Analyzer Handbook,* Prentice Hall, Englewood Cliffs, NJ.

Parker, G. A. and Gilby, J. 1982/3. *An Investigation of Robot Arm Position Measurement Using Laser Tracking Techniques,* University of Surrey, Dept. of Mech. Eng., Report.

Riley, D. L. 1987. Robot calibration and performance specification determination, *Proc. Robots 11/17th Int. Symp. Ind. Robots,* 10.1–10.17, Chicago, Il.

Robotics and Automation Applications Consulting Center

(RAACC) 1986. *Robot Assessment Program,* Ford Motor Company, Dearborn, MI.

Robotics and Automation Applications Consulting Center (RAACC) 1984. *Robot Assessment Program,* Dearborn, MI.

Schoenwald, J. S. and Martin, J. F. 1984. *An acoustic ranging system for robot position control and tracking,* paper presented at the Robot West Conference, Anaheim, CA.

Selcom Corp. 1985. *SELSPOT Hardware System,* Product information, Troy, MI.

Tamio, A., Tatsuya, E., and Shohji, M. 1983. *Position and Orientation Measurement of a Moving Object by CCD Photo Array Sensors,* University of Tokyo, 7-3-1, Hongo, Bunkyo-ku, Tokyo, Japan.

Tsai, R. Y. and Huang, T. S. 1981. Estimating three-dimensional motion parameters of a rigid planar path, *IEEE Trans. Acoustics, Speech, and Signal Processing,* ASSP-29(6):1147–1152.

Tsai, R. Y., Huang, T. S., and Zhu, W. 1982. Estimating three-dimensional motion parameters of a rigid planar path, II: Singular Value Decomposition, *IEEE Trans. Acoustics, Speech, and Signal Processing,* ASSP-30(4):525–534.

Tucker, M., Perreira, N. D., and Nyman, D. H. 1987. A pose measurement sensor, *Proc. Robots 11/17th Int. Symp. Ind. Robots,* 2.83–2.98, Chicago, IL.

Vira, N. and Lau, K. 1986. Design and testing of an extensible ball bar for measuring the positioning accuracy and repeatability of industrial robots, *Proc. 14th North Am. Manufacturing Research Conf.,* Minneapolis, MN.

Warnecke, H. J., Schraft, R. D., eds., and Wanner, M. C. 1985. "Performance Testing", in *Handbook of Industrial Robotics,* Nof, S. Y., 158–166, John Wiley & Sons, New York, NY.

Warnecke, H. J. and Schiele, G. 1984. Performance characteristics and performance testing of industrial robots—state of the art, *Proc. 1st Robotics Europe Conf.,* p. 17, Brussels, Belgium.

Whitney, D. E., Lozinski, C. A., and Rourke, J. M. 1986. Industrial robot forward calibration method and results, *J. Dynamic Systems, Measurement, and Control,* 108:1–8.

Yakimovsky, Y. and Cunningham, R. 1978. A system for extracting three-dimensional measurements from a stereo pair of TV cameras, *Computer Graphics and Image Processing,* 7:195–210.

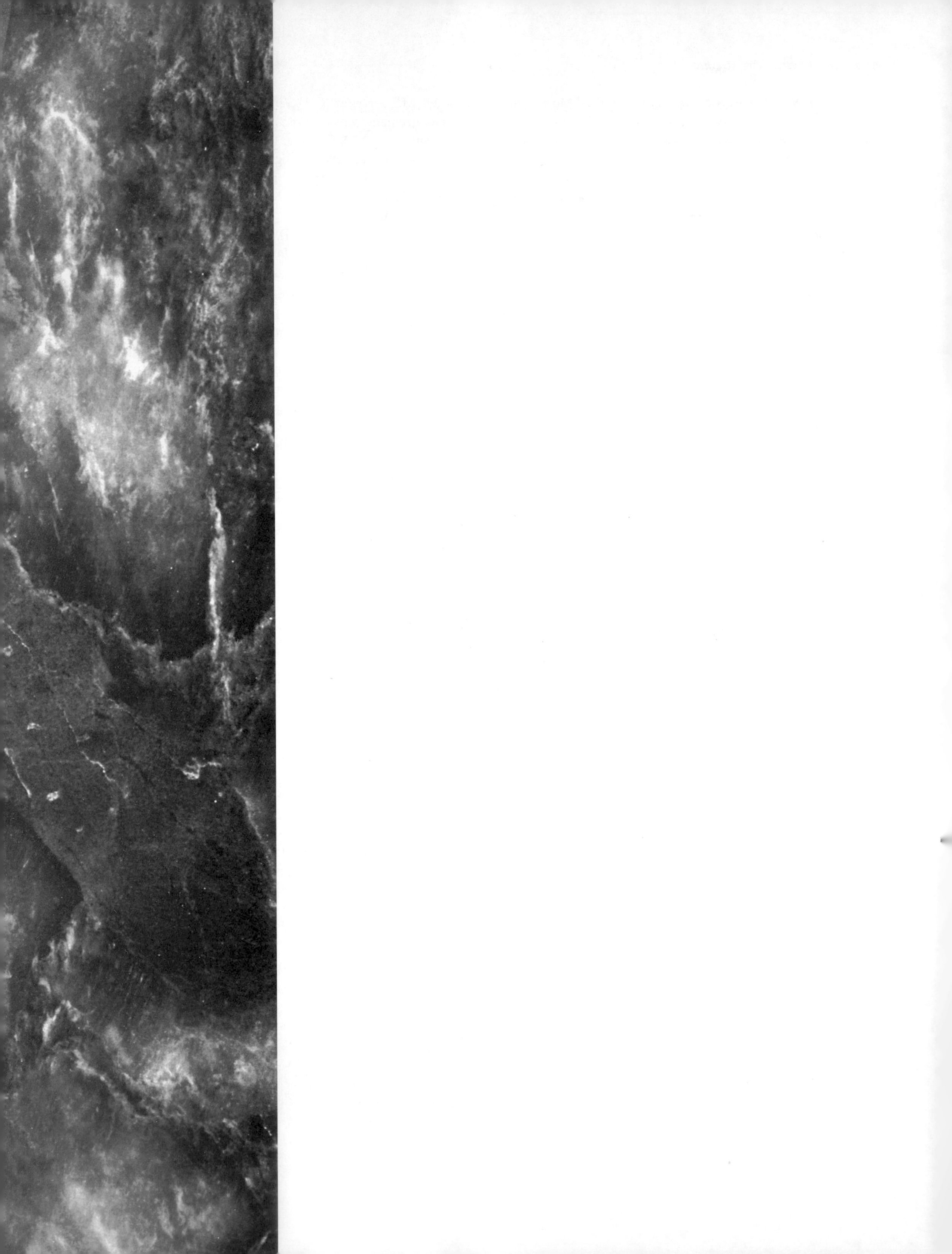

Power Electronics

7 **Introduction to Power Electronics** *Janos Bencze* 187
Introduction • Power Supplies • Electric Drives • Application Examples • Future Trends

8 **Overview: Devices and Components** *Malay Trivedi, Sameer Pendharkar, and Krishna Shenai* 195
Introduction • Diode • Thyristor • Transistors • New Devices

9 **Devices and Components** *Imre Ipsits and Karoly Kurutz* 203
Power Diodes • Power Bipolar Junction Transistors (BJTs) • Passive Networks

10 **Power MOSFETs** *Vrej Barkhordarian* 218
Introduction • Static Characteristics • Dynamic Characteristics • Applications

11 **Insulated Gate Bipolar Transistors** *Michael Robinson, Richard Francis, Ranadeep Dutta, and Al Diy* 229
Introduction • Basic Structure and Operation • Design Considerations • Requirement for Anti-parallel Diode • Comparison Between the Power MOSFET, IGBT, and MCT • IGBT Data Sheet Parameters • Appendix: Typical IGBT Data Sheet

12 **Conversion** *Attila Karpati, István Nagy, and Sándor Halász* 244
AC-DC Converters • DC-DC Converters • DC-AC Conversion • AC-AC Conversion • Resonant Converters

13 **Motor Drives** *Takamasa Hori, Hiroshi Nagase, Mitsuyuki Hombu, M. F. Rahman, Khiang-Wee Lim, Ronald H. Brown, Sándor Halász, and Jozsef Borka* 288
Control Systems and Applications • DC Motor Control Systems • Induction Motor Control Systems • Synchronous Motor Control Systems • PM Synchronous Motor Control • Step Motor Drives • Servo Drives • Switched Reluctance Motor Drives

14 **Main Disturbances** *James Stanislawski, Gerry Heydt, Prasad Enjeti, Laura Steffek, John Hecklesmiller, Dave Layden, and Brian Young* 349
Power Quality • Reactive Power and Harmonics Compensation • New Power Converters • Uninterruptible Power Supplies (UPS)

15 **Electromagnetic Compatibility for Drives** *Walt Maslowski* 377
Compatibility: Emissions and Immunity

7

Introduction to Power Electronics

Janos Bencze
EKA Factory/EI Appliance and Materials SC

7.1 Introduction 187
What Is Power Electronics? • Applications
7.2 Power Supplies 189
7.3 Electric Drives 190
7.4 Application Examples 191
7.5 Future Trends 194

7.1 Introduction

Power electronics is that field of electrical engineering which deals with the conversion and switching of electrical energy for power applications. The switching techniques take advantage of the inherent features of power semiconductor devices. Within this context, power is referenced in the widest sense and ranges from VA to MVA.

Power electronics finds broad applications within the entire field of electric energy systems. The great extent to which power electronics is applied is due to a number of advantages which electronics apparatus generally has over its electromechanical counterparts. Power electronics equipment is cheaper, lighter and smaller, has greater efficiency and generally is more available. It can be easily controlled by microprocessors and computers. Problems which earlier were completely intractable can now be solved by using power electronics.

The following material will guide us, step-by-step, through the technology of power electronics.

This section provides an overview of the area of power electronics and illustrates the complexity of this applied science.

Chapter 8 introduces the different basic "building blocks" of power electronics. These elements are the so-called active and passive devices. The active devices are semiconductors which range from simple diodes, thyristors and transistors, to the up-to-date power devices with very complicated structures and self turn-off capabilities, like IGBTs (isolated gate bipolar transistors), MCTs (MOS controlled thyristors), SITs (static induction transistors), and SITHs (static induction thyristors). The passive devices and components play an important role—typically that of protection.

The following section deals with the different types of converters. Active and passive devices and components are the basic parts of these items. The converters convert energy from some given form to that which is required. The features and behavior of these different type of converters as well as the most frequently used methods of energy conversion are presented.

An electric drive is one of the most important elements in power electronics; Chapter 13 introduces many of the drives that are used in practice.

It is important to note that interactions exist between power converters and AC supply lines. Power electronics equipment, because it is switching electrical energy, introduces disturbances on the AC lines such as reactive power and higher harmonics. These disturbances must be controlled within certain limits. Chapter 14 introduces the standards for allowed disturbances and methods of dealing with their effects. Finally, this section illustrates—especially in the case of sensitive loads—how to protect the load from the different active disturbances, by reactive power and harmonics compensation, and through the use of uninterruptible power supplies (UPS) which are another important area of power electronics.

Chapter 15 gives some details about system engineering.

What is Power Electronics?

Power electronics is an applied science. It deals with the conversion of electric energy using power semiconductor devices and the requisite switching. It also includes the required control and measurement techniques. The fraction of electric energy which is available after conversion is constantly increasing. Thus power electronics represents an important link between power generation and the load, as shown in Figure 7.1. Furthermore, this area is growing in significance as the demand to control and convert electric energy increases.

Power electronics today is one of a wide range of well defined technologies, which aids in adapting electric energy to the needs of the user. To meet technical demands, power electronic devices and systems should be highly efficient, small in size, light weight, controllable with good dynamic response, highly reliable and economical to manufacture. In addition, acceptable power quality

0-8493-8343-9/97/$0.00+$.50

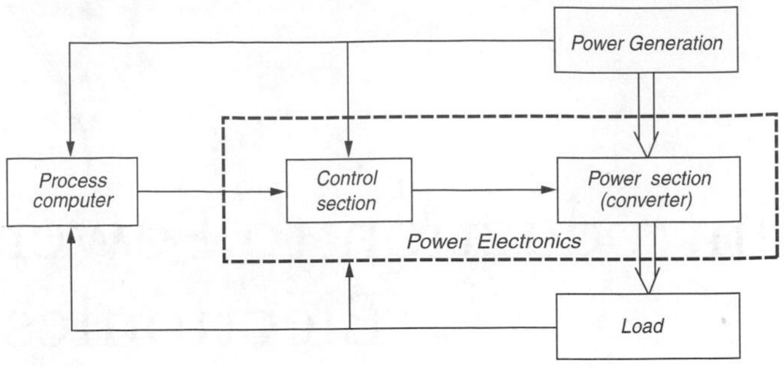

Figure 7.1 Power electronics.

and sufficiently low radiation interference are also characteristics of growing importance.

Power electronics applications span a wide range extending from household equipment to space systems, including such typical areas as the transmission and distribution of electric energy and the use of electric drives.

As illustrated in Figure 7.1, power electronic equipment can be subdivided into two distinct parts. The first part is the power section, composed of power semiconductor devices, which perform the energy conversion from AC to AC, AC to DC, DC to AC, or DC to DC. The second part (control section) employs such things as discrete elements, ICs (integrated circuits), ASICS (application-specific integrated circuits) or microcomputers that function in a very low powel level.

The different types of converters are illustrated graphically in Figure 7.2. These devices may also be classified according to the nature of the commutation, which defines the manner in which the power devices of the converter are turned on and off. Commutation is very important because it specifies the way in which current is transferred from one power device to another within the converter. There is, however, a short period of time called the overlapping period when two devices are simultaneously conducting. The turn-on and turn-off of a given power device depend upon how the trigger pulses are generated and the source of commutation voltage, respectively, as shown in Figure 7.3. In addition, the lower half of this figure illustrates the possible ways in which to affect turn-on and the upper portion of the figure shows the ways to affect turn-off.

The turn-on processes for the power devices are relatively simple. The devices are turned on when the instantaneous voltage on the device is positive and the clocked trigger pulses are present.

The turn-off processes are quite different. For "line commutation" the line voltage is the source of commutation. The commutation will occur when the instantaneous value of one of the line voltages will be higher than the previous one, and the next device is turned-on. For "self-commutation," an auxiliary power source, referred to as the forced commutation circuit, ensures the voltage for commutation. Self-commutation also occurs in converters which have power devices with turn-off capability. The state-of-the-art converter technology uses only these latter types of power semiconductor devices. For "load commutation" the load generates the voltage for the commutation, e.g., a synchronous machine, etc.

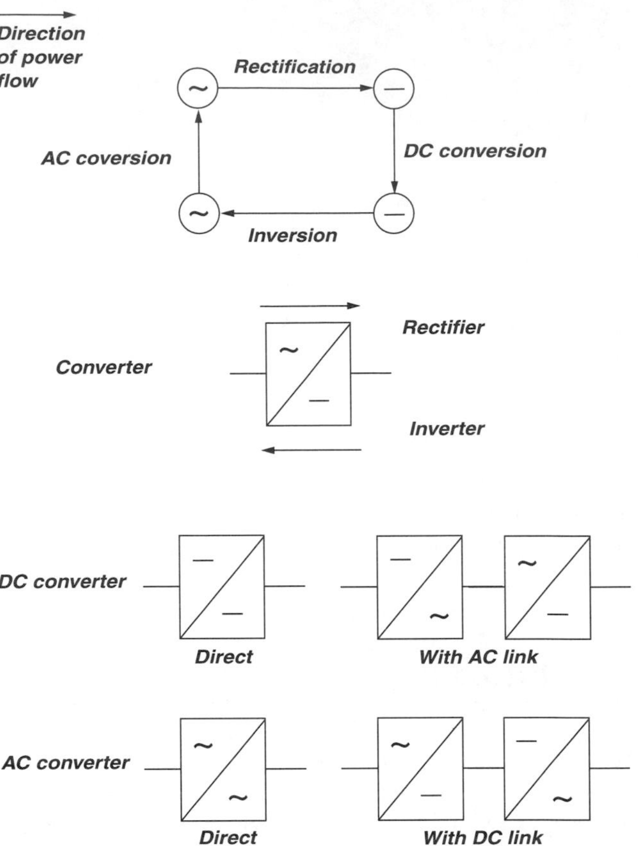

Figure 7.2 Conversion of electrical power.

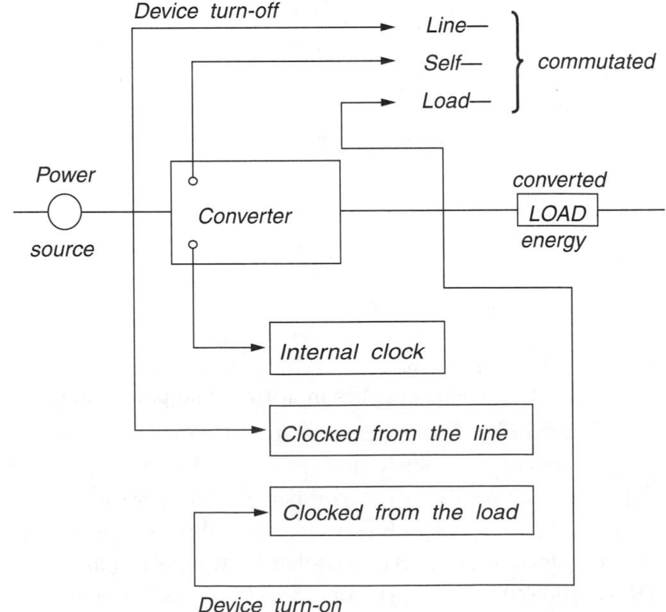

Figure 7.3 Classification of converters.

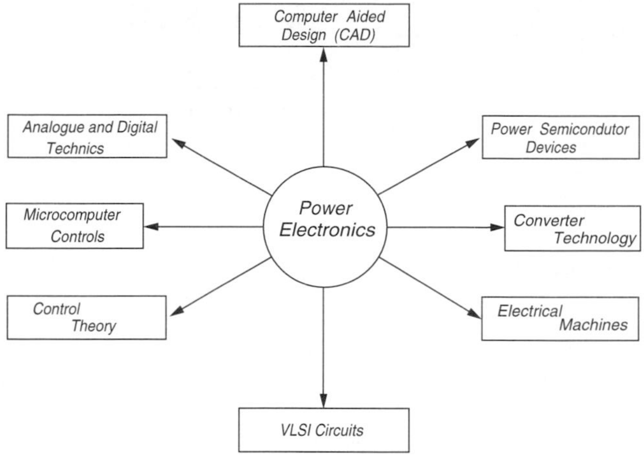

Figure 7.4 Interdisciplinarity of power electronics.

Power electronics is an interdisciplinary field of applied science with tentacles which extend into a number of diverse areas as shown in Figure 7.4. Thus, the design and development of power electronic equipment involves the use of power devices, converter circuits and electric machines as well as the remaining technologies exhibited in the figure.

Although the details of the various devices are discussed in some detail in the remaining sections of the chapter, one obtains a better physical view of the device's operation under certain simplifications. For example, if it is assumed that the switching device is ideal, then there is no leakage current in the off-state, no voltage drop between anode and cathode in the on-state and both switching time and overlapping are negligible, i.e., zero. However, when simulating power electronic systems, we must decide whether these simplifications are valid, since they will not be in every case.

Applications

The applications of power electronics are practically limitless. They cover essentially all aspects of electrical engineering since in most cases, the controlled conversion of electrical energy is included.

The primary areas are

- Standard power supplies in a very wide power range.
- Uninterruptible power supplies (UPS) in a range from hundreds of watts to hundreds of kilowatts.
- High-voltage direct current transmission (HVDC) in a range from ten to hundreds of megawatts.
- Reactive power compensation and active power filters in the kilowatt to megawatt range.
- Power conditioning.
- Induction heating.
- Industrial drives including traction, servo, stepper and switch mode reluctance motors (SMR).
- Special power supplies for
  - Atomic accelerators.
  - Electrochemical processes.
  - Welding.

The conservation of energy and the attendant issue of environmental pollution control are yet other important applications of power electronics. To improve our standard of living, the demand for energy, particularly in electrical form, continues to increase. Today, the major portion of this energy is generated from fossil fuel plants with their associated environmental impact problems such as acid rain, air pollution and global warming—the so-called greenhouse effect. The alternate pollution-free energy sources that demonstrate some promise are such things as wind power and photovoltaics. Power electronics play important roles in each of these technologies in that they are an effective catalyst in reducing power consumption

As the list above indicates, there are two primary categories of applications for power electronics: power supplies and electric drives. It is important to emphasize that the difference between these two groups lies primarily in the control section. The experts in this area are not unanimous in their thinking about this subject, and some of them treat power electronics and electric drives as two independent fields of technology.

7.2 Power Supplies

A power supply is a power electronics system, composed of a controlled converter, which converts the input voltage, in both quantity and quality, into that which is required by the load. From this point of view, the power supply can be an AC/DC, AC/AC, DC/AC, or DC/DC controlled converter. Within this context, power supplies cover an extremely wide power range which extends from fractions of watts to many megawatts.

One of the most important subgroups of power supplies is the switched mode power supply (SMPS). They are normally employed in low power applications, typically in the range from a few watts to several kilowatts. They are very important because there are literally millions of them in service being used in such things as personal computers and TV sets, as well as the control section of a power electronics system. This special class of power supplies uses a relatively high chopping frequency, in the 20 to 100kHz range, and as a result the equipment is light in weight and small in size and exhibits good controllability and little acoustic noise.

Another important subgroup is the uninterruptible power supply (UPS). Certain pieces of electronic equipment such as computers and telecommunication systems, and certain businesses, e.g., hospitals, cannot tolerate disturbances on the AC power lines. These facilities must have continuous, good-quality electric supply in order to operate in a fail-safe manner. In most cases, the UPS is connected to an AC power line. Then the use of a tandem connection of an AC/DC converter and a DC/AC converter can be used to supply the load. Under these conditions, a battery bank can be connected to the DC link to provide energy during those time intervals when the supply network is cutoff, as shown in Figure 7.5. During normal operation, a battery charger keeps the battery bank fully charged in case it is needed.

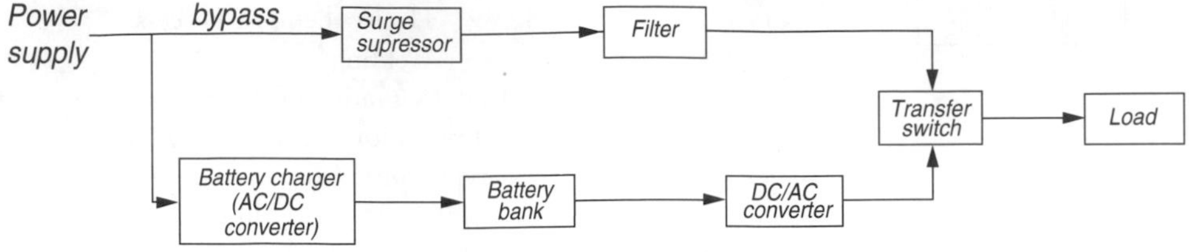

Figure 7.5 Block diagram of a general UPS.

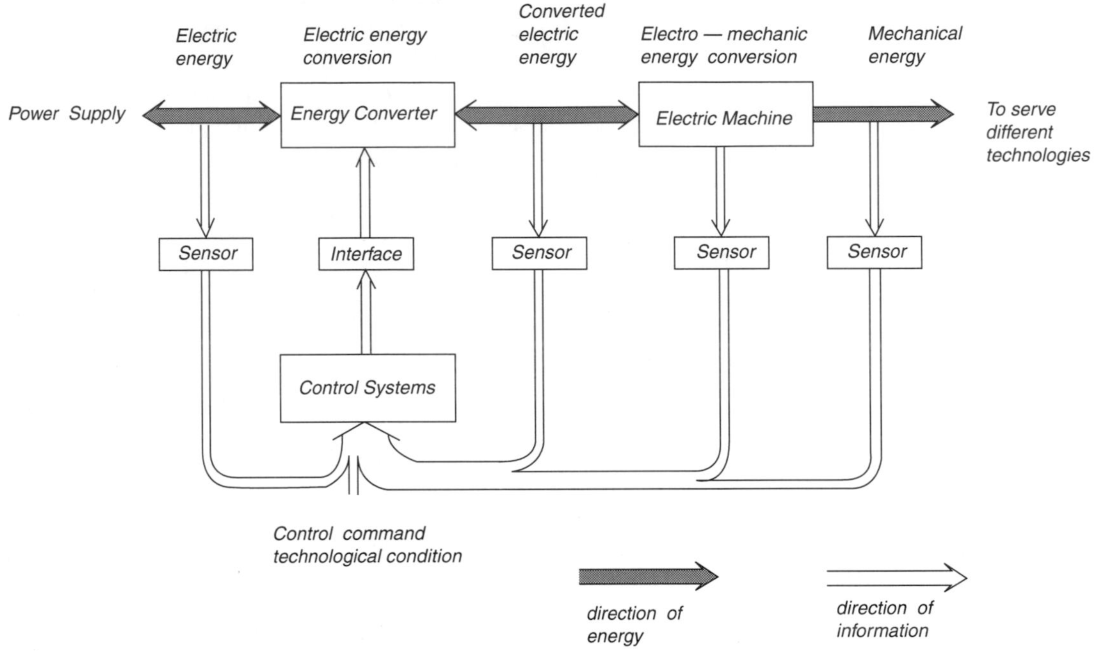

Figure 7.6 Schematic diagram of a generalized variable speed drive.

Bypass circuitry is also present so that the load can be continuously supplied during maintenance or in the case of some breakdown. Although there are a number of different UPS system configurations which are dependent upon load requirements, most of them are similar to the system described here.

In general, power supplies for such things as welding, induction heating, electrochemical processes, heat controls, lasers and the like, all differ in their input and output parameters. The one common feature that they possess is that of one or more steps of electric energy conversion to ensure the proper quantity and quality of power necessary to fulfill the output requirements.

7.3 Electric Drives

An electric drive is a system which consists of one or more electric motors and the control equipment necessary to govern their performance. The control equipment may or may not include various rotating electric machines.

Electric drives convert electric energy into some form of controlled mechanical energy in order to meet certain requirements. These requirements may be, for example, speed control for a rolling mill, position control for a robotic arm or torque control for a winder. Figure 7.6 illustrates the basic functional parts of a generalized electric drive system. Both energy and information flow are exhibited on this figure, as well as the different features involved in the various types of drives.

The DC drive has the advantage of a relatively simple power circuit and control system. The AC/DC, or in a few cases DC/DC, converter supplies the armature of the DC machine. The traditional types of DC drives commutate naturally using the AC supply lines. The disadvantage of this machine is a lower reliability factor derived from the brushes and the commutator.

The asynchronous, squirrel cage machine drive has the advantage of being a very simple and robust machine. The disadvantage of this drive is that the power and control circuits are more complicated and expensive. Most drives of this type have an input rectifier which produces the DC link and a DC/AC converter which produces a voltage that varies in magnitude and frequency. The DC/AC converter requires the use of power devices with a turn-off capability. This type of drive exhibits very good reliability.

The synchronous drive lies somewhere between the DC and AC drives. The machine itself is more reliable than its DC counterpart. It uses a supply-side converter, a machine-side converter and a DC link between them. The two converters are similar. In a normal situation, the supply-side converter commutates using the AC power source and the machine-side converter commutates from the load. The power circuit is much simpler than that employed in the asynchronous squirrel cage drive.

In addition to those configurations described above, there are a number of hybrid configurations. A simple and reliable one is a machine with a permanent magnet (PM) excitation. A relatively new configuration is the SMR drive, which employs a doubly-salient, single-excited motor. This is a simple and reliable system. Other special electric drives include stepper motor drives and servo and robotics drives.

7.4 Application Examples

It is a well-known fact that the efficiency of an extra-high-power electric power generator is optimum in the range of the rated output power. Since the rated power of such a generator is in the 100 to 1000 MW range, each 1% of the power represents 1 to 10 MW and is therefore very important. Power consumption is changing significantly during a 24-hour period. Therefore, power is supplied by a combination of basic power plants and peak power plants. The former provide the basic level of power and the latter are used only to fulfill peak power requirements. To satisfy the peak requirements, gas turbine-generator sets are used because they can be brought on-line in a very short interval of time.

Starting is a problem with gas turbine-generator sets since up to approximately two-thirds of rated speed, it has no torque. In the past, special help was provided by a special starting machine. Today, however, this is done with power electronics. Figure 7.7 illustrates a static frequency converter (a so-called machine commutated inverter drive described in the section on synchronous motor control) for starting a 160 MVA generator-gas turbine set. The generator itself is used as a start-up motor. The rated power of the equipment is 1.8 MW; the input voltage is 1300 V at 50 Hz; the output parameters are 0–1300 V and 0–36 Hz and the start-up time is 6 min from 0 to 20 Hz and another 6 min from

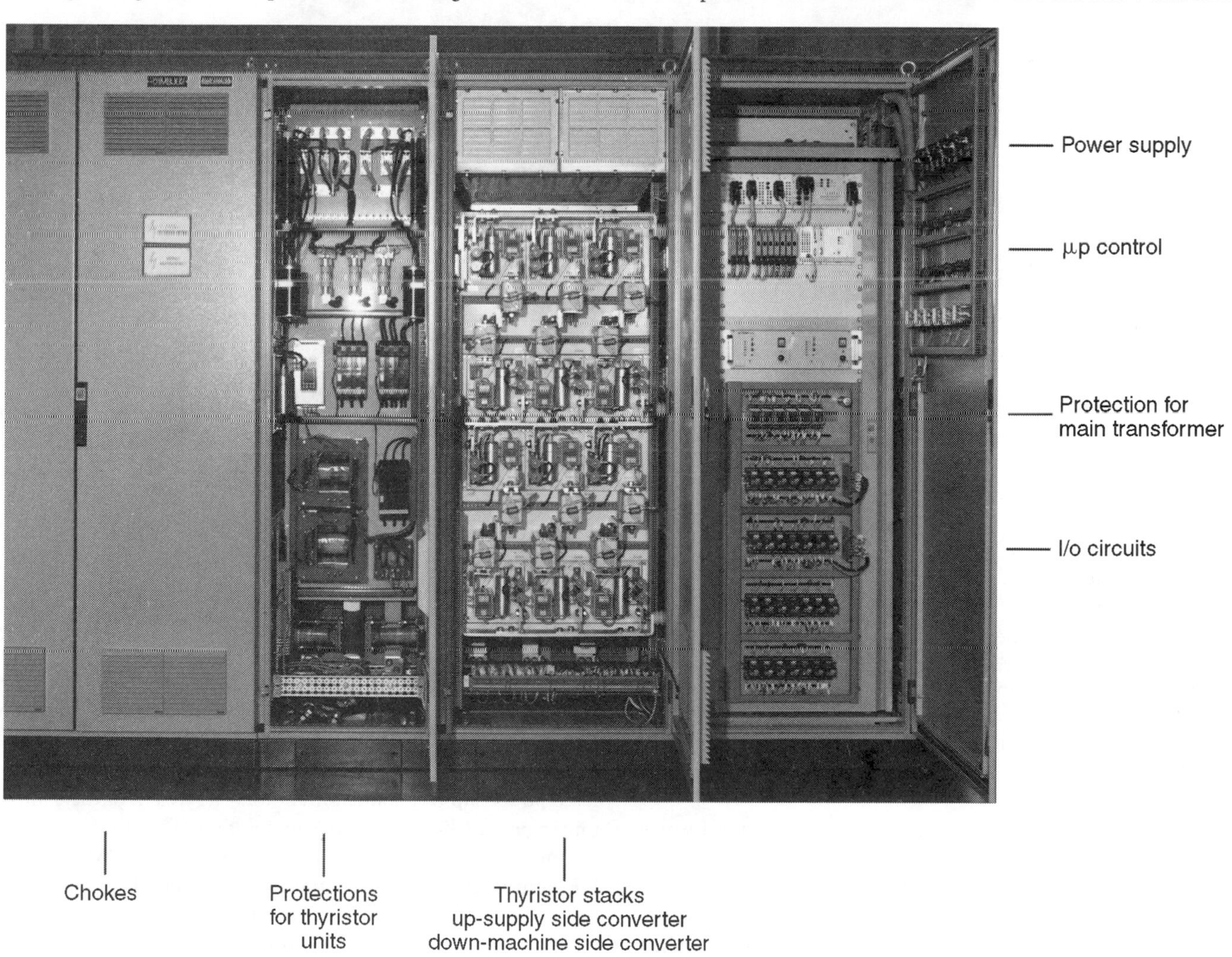

Figure 7.7 Gas-turbine starter.

Figure 7.8(a) SIMONTRANS HE Control unit for 225A starter phase angle control with electronic rotating field reversal for cranes.

35–40 Hz. From this speed the gas turbine accelerates the set. This particular drive was made by Ganz Ansaldo, Budapest, Hungary.

It is interesting, from a historical perspective, to mention the slip ring wound rotor induction motor as an application example. While the induction motor is a very practical, reliable, robust and cheap electric machine, it is difficult to change its speed. However, the slip ring version of this machine provides an opportunity to change speed easily by using different resistors in the rotor circuits. Although this was an effective technique from a practical standpoint, as the resistance of the rotor increased, the speed decreased and the energy loss of the system increased. Therefore, when the modern quasi-loss-free controlled electric drives appeared, the slip ring induction motor was quickly forgotten.

The first renaissance of this induction motor occurred when the slip recovery drive system started to use static converters instead of different circuit configurations with auxiliary machines. However, it was found that the static cascade drive contaminates the AC supply with reactive power and higher harmonics, and the slip rings and brushes, as moving parts, required frequent maintenance and therefore this induction motor was once again discarded.

The slip ring induction motor drive is currently in its second renaissance. The so-called stator phase angle control with electronic field reversal drive is a state-of-the-art drive which was specially developed for use with cranes. The development of this drive was made possible by technological advances which solved the maintenance problem associated with the slip rings and brushes. This system is a good compromise, both technically and economically, among the variable speed three-phase AC drives, the closed-loop controlled DC drives and the frequency-converter fed drives.

An example of the second renaissance of the slip ring induction motor drive system is the SIMOTRANS HE drive series developed by Siemens. Figure 7.8a,b illustrates the basic unit for a 115 kW rated power slip ring induction motor. The weight of the control unit is only 21 kg, and the height, width and depth dimensions in mm are 495, 260, and 250, respectively. The SIMOTRANS HE control unit performs closed-loop and open-loop control of three-phase AC crane motors with slip ring rotors, and consists of a power section, a control section and a hoisting module. The power section is a compact three-phase thyristor AC controller. Two additional thyristor modules present in the shunt arm of the power section allow reversal of the rotating field, and hence

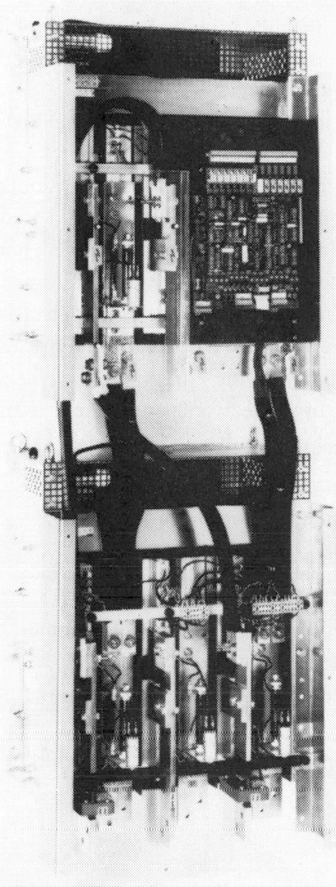

Figure 7.8(b) The big brother of Figure 7.8(a). Unit for 1150A.

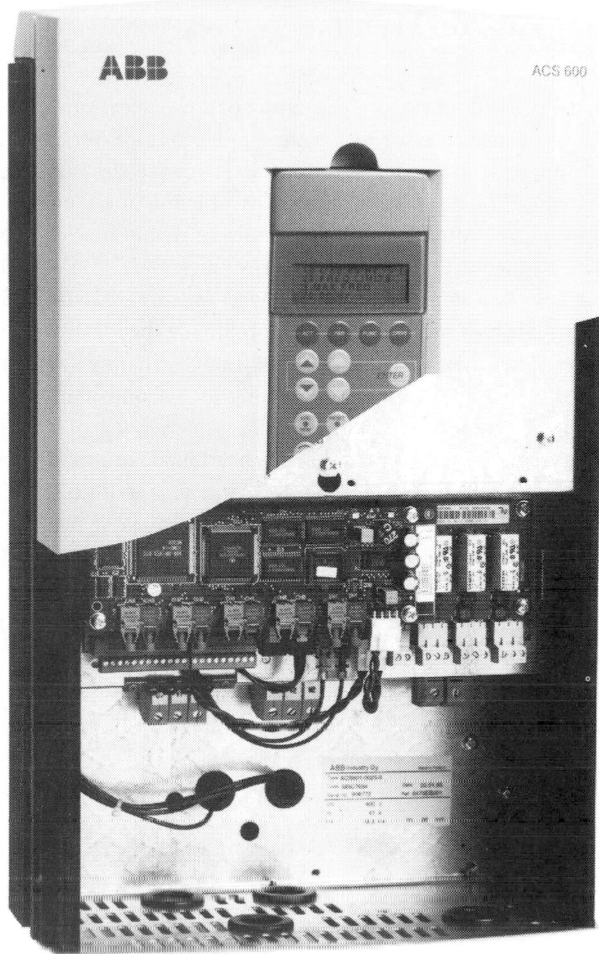

Figure 7.9 A state-of-the-art frequency converter with the remote controller for man-machine communication.

four-quadrant operation of the drive. The control module contains the trigger equipment and pulse distributor for the thyristors, as well as the power supply which generates the control voltage and the synchronizing voltage for the trigger equipment. All the open- and closed-loop controls are on the hoisting module, which is mounted on top and easily accessible. All parameters and setpoints can be adjusted by this module. Thus, through the use of state-of-the-art technology, complex high power equipment such as this can be made in small size with light weight.

The ultimate in controlled electric drives is the frequency converter fed squirrel cage induction motor drive (see the section on induction motor control). This machine is very attractive because its construction is simple, robust, and uses a simple rotor and is easily and economically produced in large scale production. The only problem with this machine is the complicated frequency converter required to change motor speed. However, in this age of modern power semiconductor devices with turn-off capability and state-of-the-art control circuits with microprocessors, it is relatively easy to implement the power and control section equipment necessary for frequency converters.

There are many applications for these drives, in the range from several hundred watts to several hundred kilowatts and beyond, where the only reliable solution is the frequency converter fed squirrel cage induction motor drive.

In general, the requirement for any state-of-the-art drive is that it be suitable for any drive application from pumps and fans to conveyors, cranes, extruders, winders and the like, without the added cost of tachometer feedback. They should be flexible and have easily programmable drive features, and also be able to receive both analog and digital signals and fulfill the requirements of Fieldbus, Profibus or Modbus and/or any other standards. Furthermore, the electromagnetically compatibility (EMC) requirements must be satisfied also.

A drive series which fulfills the stringent drive requirements stated above, is the ACS 600 from ABB. This drive is shown in Figure 7.9. The power range for this series is 2.2 to 315 kW for squirrel cage induction motors. The control system employs the so-called direct torque control (DTC) technology, with outstanding system integration and communication capabilities to meet virtually any drive application. Communication with the drive is made simple through a removable control panel which provides a great deal of information in English as well as nine other languages. The optional control panel can be remotely located or mounted on the drive enclosure. One panel can control up to 31 drives, and therefore this system was designed with the future in mind.

7.5 Future Trends

Because of the interdisciplinary nature of power electronics, many fields of science sponsor its advancements. As a result, each year great progress is made, and the last two decades have been significant. The most important fields which define the progress of power electronics are power devices. In the area of power devices, which are the tools of energy conversion, the advancement has been enormous. New devices, such as IGBTs, MCTs, SITs and SIThs have been able to approximate the ideal switching devices. They have features such as high switching frequency, turn-off capability, very low switching losses, and high-voltage and high-current capability.

The tremendous progress that has been made in microprocessor technology, which is the tool for control, is well known. The continuous increases in speed and capacity make it possible to implement complex systems which were unthinkable just a few years ago. They are not only able to satisfy all control requirements, but are capable of fault diagnosis and other functions like man-machine communication as well.

The supporting technologies such as mounting, heat control and isolation techniques provide for a more efficient use of the devices, as well as a more attractive and practical mechanical construction.

One of the major trends in the advancement of power electronics equipment is the rapidly increasing switching frequency. This increase in switching frequency results in smaller components, lower power requirements, less volume and weight, better controllability and faster response time. The higher switching frequency also makes it possible to realize true sinusoidal input and output voltages and currents in the power electronics equipment without any reactive power generation. Furthermore, the efficiency and reliability of the equipment will also increase.

In the future power electronics performance must be differentiated into the following three categories:

- Output parameter performances.
- Man—machine communication interface performance.
- Environmental performance, in a broad sense.

To meet the demands made by these three categories, the key word for the future will be integration—integration not only in the field of control devices, but the integration in one package of control and power devices together with the passive elements.

8

Overview: Devices and Components

8.1 Introduction 195
8.2 Diode 195
8.3 Thyristor 196
8.4 Transistors 197
Bipolar Junction Transistor (BJT) • Metal-Oxide-Semiconductor Field Effect Transistor (MOSFET)
8.5 New Devices 199
Insulated Gate Bipolar Transistor (IGBT) • MOS-Controlled Thyristor (MCT) • Emitter Switched Thyristor (EST) • Smart Power™ Technologies • Other Material Technologies • Devices for Resonant Power Conversion

Malay Trivedi
University of Wisconsin-Madison

Sameer Pendharkar
Texas Instruments, Dallas

Krishna Shenai
University of Illinois at Chicago

8.1 Introduction

Power semiconductor devices are used as switches in power electronics applications. Ideal switches block arbitrarily large forward and reverse voltages with zero current flow in the off-state, conduct arbitrarily large currents with zero voltage drop in the on-state, and have negligible switching time and power loss. Material and design limitations prevent semiconductor devices from behaving as ideal switches. It is important to understand the operation of these devices to determine how much the device characteristics can be idealized. Available semiconductor devices could be either controllable or uncontrollable. In an uncontrollable device such as the diode, on-and off-states are controlled by circuit conditions. Devices like BJT, MOSFET, and so forth, can be turned on and off by control signals, and hence, are controllable switches. Thyristors belong to an intermediate category where they can be latched on by a control signal, but their turn-off is governed by external circuit conditions. This section presents a brief summary of the terminal characteristics, and the circuit performance of contemporary power devices.

8.2 Diode

Figure 8.1 shows the cross section and circuit symbol of a typical power rectifier. A p-n junction is formed when an n-type region in a semiconductor crystal abuts a p-type region in the same crystal. Rectification properties of the device result from the presence of two types of carriers in a semiconductor—electrons and holes. When a junction is formed, majority carriers on each side of the junction diffuse across it to the other side resulting in an electric field that acts as a barrier to further diffusion.

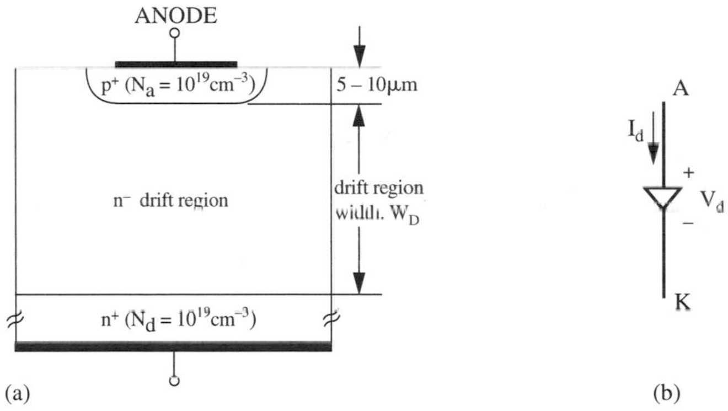

Figure 8.1 P-*i*-N diode (a) cross section and (b) circuit symbol.

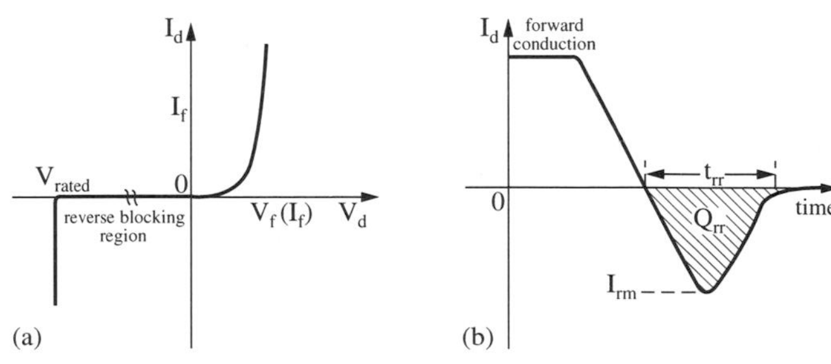

Figure 8.2 Typical P-*i*-N diode (a) static I-V characteristic and (b) reverse recovery waveform.

When an external potential is applied, the equilibrium condition is disturbed and carriers begin to flow across the junction. Typical I-V characteristic for the diode is shown in Figure 8.2a. When the diode is forward biased (p-region more positive with respect to n-region), charges get accumulated in the drift region, and

0-8493-8343-9/97/$0.00+$.50

the diode begins to conduct in high level injection with only a small forward voltage, of about one volt across it. Under reverse bias (p-region more negative with respect to n-region), the device conducts negligibly small leakage current until the reverse breakdown voltage is reached. The drift region supports the reverse voltage. Typical diodes used in power applications are required to have large breakdown voltages. This requires large drift regions implying high resistance. The drift region thickness can be reduced considerably by using a heavily doped n^+ substrate. The device is then a P-*i*-N diode; *i* denotes the lightly doped drift region. Figure 8.1a shows the typical doping density and dimensions of power diodes. Drift region doping density and thickness are controlled by the reverse breakdown voltage requirements.

Typically, a diode turns on much faster than the power circuit transients. It can then be considered an ideal switch during turn-on. During conduction, charges get accumulated in the drift region and need to be removed during turn-off. Charge removal takes place by a reversal in current direction for a reverse recovery time t_{rr} as indicated in Figure 8.2b. The reverse recovery characteristics of a diode can be quite significant in many circuit applications as they affect the maximum speed of operation, result in additional power loss and also introduce stress on other circuit components.

Because of charge accumulation in the drift region during conduction, the P-*i*-N diodes show significant reverse recovery current. Also, there is a built-in voltage at the junction between the p- and the n-doped regions. A majority carrier diode may be formed by making a rectifying junction between a metal and a semiconductor. These are the Schottky diodes, shown in Figure 8.3. Schottky diodes have low forward voltage drop and good reverse recovery performance. They have the disadvantage of large leakage in the off-state. Merged P-*i*-N-Schottky (MPS) structures have been proposed to utilize the advantages of both P-*i*-N and Schottky diodes. MPS diodes offer high switching speed and conduct high current densities.

Diode parameters can be modified for various applications. Fast-recovery diodes can be made to have improved reverse recovery characteristics at the cost of higher forward voltage drop. On the other hand, diodes with very low forward voltages can be fabricated by sacrificing reverse recovery performance.

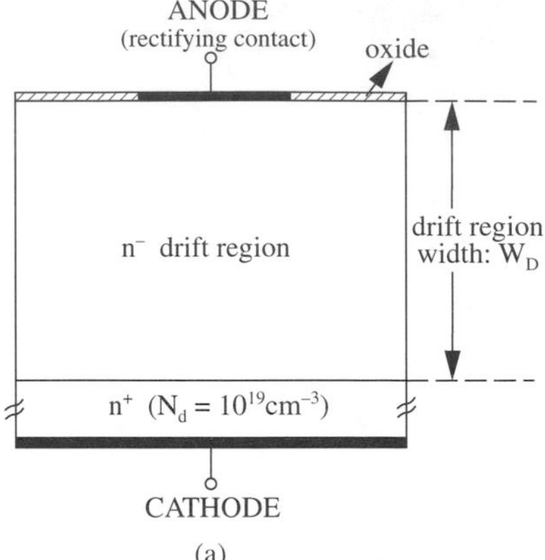

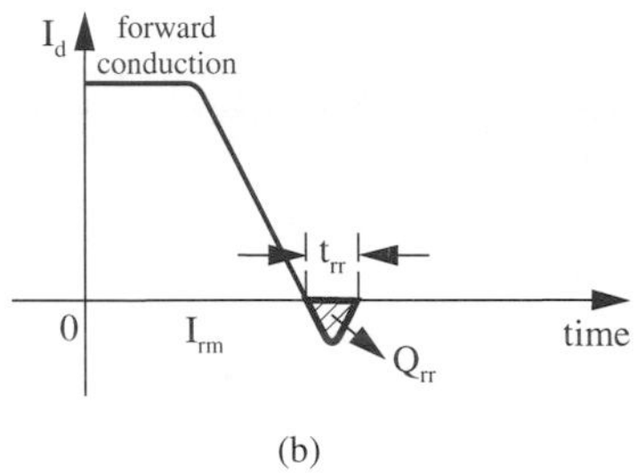

Figure 8.3 Schottky diode (a) cross section and (b) typical reverse recovery waveform.

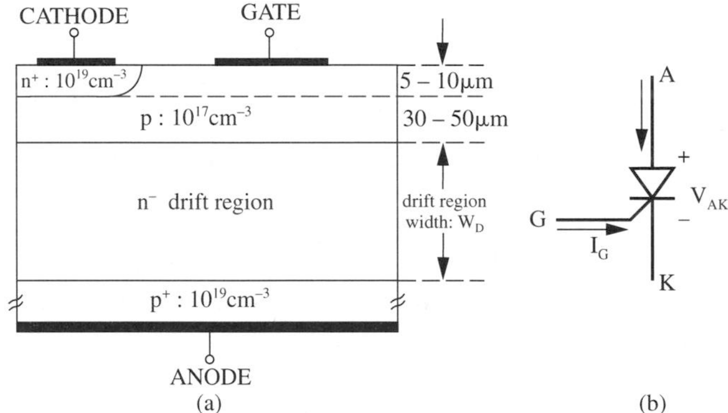

Figure 8.4 Thyristor (a) cross section and (b) circuit symbol.

8.3 Thyristor

The cross section and circuit symbol of a thyristor are shown in Figure 8.4. Forward current flows from the anode (A) to the cathode (K). The thyristor is a three-junction p-n-p-n device. The I-V characteristic of a thyristor is shown in Figure 8.5. When a forward polarity voltage is applied, junction J_2 (between the gate and drift region) is reverse biased and the thyristor is in the off-state as shown in Figure 8.5. The device can be triggered into latchup by injecting hole current from the gate electrode. The forward voltage drop of a thyristor is very small. Among the power semiconductor devices known, the thyristor shows the lowest forward voltage drop for large current densities. Once the device begins to conduct, it cannot be turned off by the gate. The thyristor turns off only when anode current is forced to reverse direction under the influence of the circuit. Under reverse bias, junctions J_1 (cathode and the gate) and J_3 (anode and drift region) block the voltage. Very small leakage current flows until the device breaks down.

A simple circuit application for the thyristor and the corresponding waveforms is shown in Figure 8.6. The thyristor can be turned on at any instant during the positive half-cycle by applying the gate pulse. When the source voltage goes negative, the thyristor current tries to reverse polarity causing the device to turn off. A thyristor exhibits reverse recovery characteristics just like a diode. However, the important parameter for a thyristor is the turn-off time interval t_q, called the circuit-commutated-recovery-time, defined in Figure 8.6c from the zero crossover of the current to the zero crossover of the voltage. Application of a forward

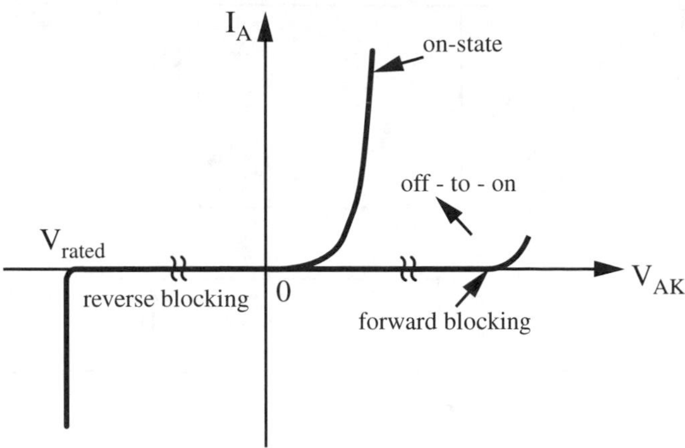

Figure 8.5 Thyristor static I-V characteristics.

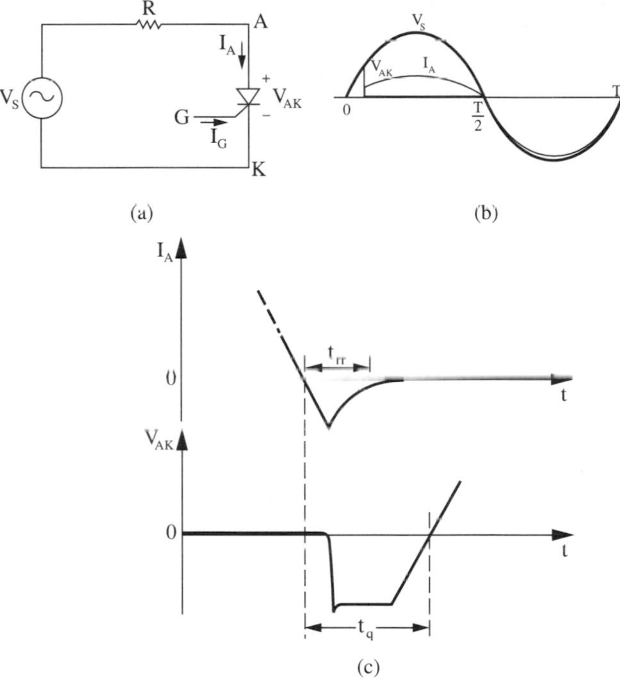

Figure 8.6 Thyristor (a) circuit, (b) waveforms, (c) turn-off time interval t_q.

voltage to the thyristor during the voltage recovery phase may lead to premature device turn-on resulting in possible damage to the device.

Thyristors come in various forms depending on the application requirements. They can be made to handle large voltages and currents with low on-state forward drop. Such thyristors typically find applications in circuits requiring rectification of line-frequency voltages and currents such as phase-controlled rectifiers. Faster recovery thyristors can be made at the expense of increased on-state voltage drop. Some thyristors, called the gate turn-off thyristors (GTOs) also have gate turn-off capability that is achieved by making some modifications in the basic thyristor structure. There are other variations such as the reverse-conducting thyristors (RCTs), asymmetrical silicon-controlled-rectifiers (ASCRs), and so forth, that are applied in many modern power electronics applications.

8.4 Transistors

Transistors fall in the category of controllable switches. They are three-terminal devices, one of the terminals being the control electrode. These devices include BJTs, MOSFETs, GTOs, and so forth. This subsection considers the conventional devices, the BJT and the MOSFET.

Bipolar Junction Transistor (BJT)

The cross section and circuit symbol for an *npn* bipolar junction transistor are shown in Figure 8.7, and its steady-state static I-V characteristics are shown in Figure 8.8. BJTs are current-controlled devices. The device is turned on by injecting a current through the base B. This causes electrons to flow from the emitter to the collector. A family of curves can be obtained by varying the base current, as shown in Figure 8.8.

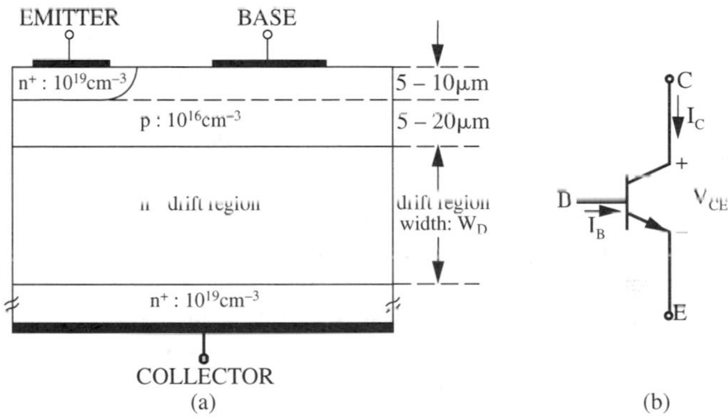

Figure 8.7 (a) Cross section and (b) circuit symbol for an npn BJT.

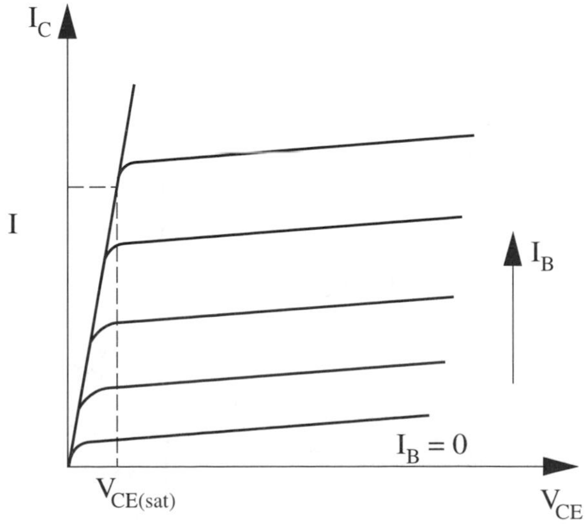

Figure 8.8 BJT static I-V characteristics.

For power circuit applications, BJTs are operated in saturation, with a small on-state voltage, usually 1–2 V, to keep the conduction power loss quite small. Since BJTs are current controlled, base current must be supplied continuously to keep them in the on-state, resulting in significant power loss. Power BJTs have a low current gain, usually not exceeding 10. The gain can be enhanced by connecting the transistors in a Darlington configuration, as shown in Figure 8.9, wherein the output of one transistor drives the other transistor. The resulting increase in current-handling capability is accompanied by a deterioration in forward voltage and switching speed.

The clamped inductive-load circuit in Figure 8.10a represents a typical switching application of BJTs. The external circuit determines the collector current that can flow in the on-state. A minimum amount of charge needs to be built-up in the drift region of the BJT to support the current. A minimum base current needs to be provided to establish and maintain this charge. The switching waveforms during BJT turn-on are shown in Figure 8.10b. After an initial delay due to B-E capacitance, charge in the base begins to build-up and the collector current rises to its on-state value with the voltage being clamped by the diode. The voltage begins to fall abruptly. The rate of fall slows down as charges are injected into the drift region reducing the gain. This phase increases the overall device turn-on time. All the stored charge needs to be removed during turn-off. Hence, BJTs have significant turn-off delay due to stored charge in the long drift region. Turn-off process can be made faster by reversing the direction of the base current. The current reversal is usually achieved gradually since abrupt reversal leads to a long current tail, thus increasing switching losses.

For higher current or voltage handling capabilities, devices can be connected in parallel or series. Connecting BJTs in parallel can lead to instabilities due to negative temperature coefficient of the on-resistance. During conduction, if the temperature of one device rises, its resistance reduces causing it to pass more current than other transistors which in turn leads to a further rise in current. This leads to a thermal-runaway condition that destroys the device.

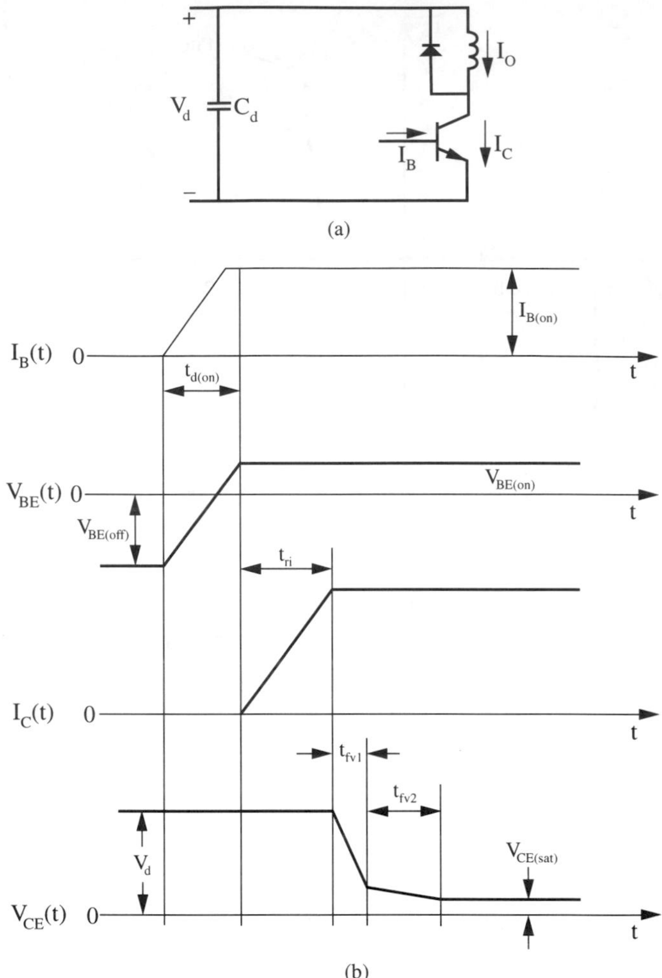

Figure 8.10 (a) Clamped inductive load circuit and (b) BJT turn-on waveforms.

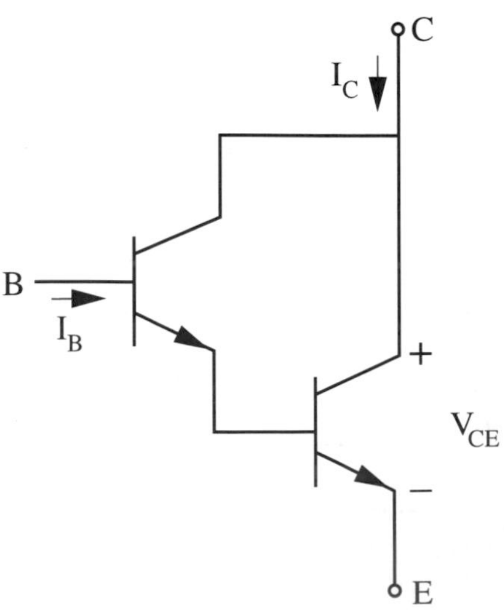

Figure 8.9 BJT Darlington pair.

Metal-Oxide-Semiconductor Field Effect Transistor (MOSFET)

The cross section and circuit symbol of a vertical diffused n-channel MOSFET are shown in Figure 8.11a and b, respectively. MOSFET is a voltage-controlled device as opposed to BJT which is a current-controlled device. Application of a gate voltage above a threshold value, V_{th}, creates a conduction channel from the drain to the source and the device approximates a closed switch in the fully on-state condition. Application of a drain voltage then forces electrons to flow through the channel into the drift region. The MOSFET is a majority carrier device with only one type of carrier (electrons for the n-channel MOSFET shown in Figure 8.11) responsible for current flow. I-V characteristics are shown in Figure 8.11c. Increasing gate voltage reduces the resistance of the channel, thus increasing the current level.

MOSFETs require continuous application of gate-source voltage above V_{th} to be in the on-state. The gate is isolated from the

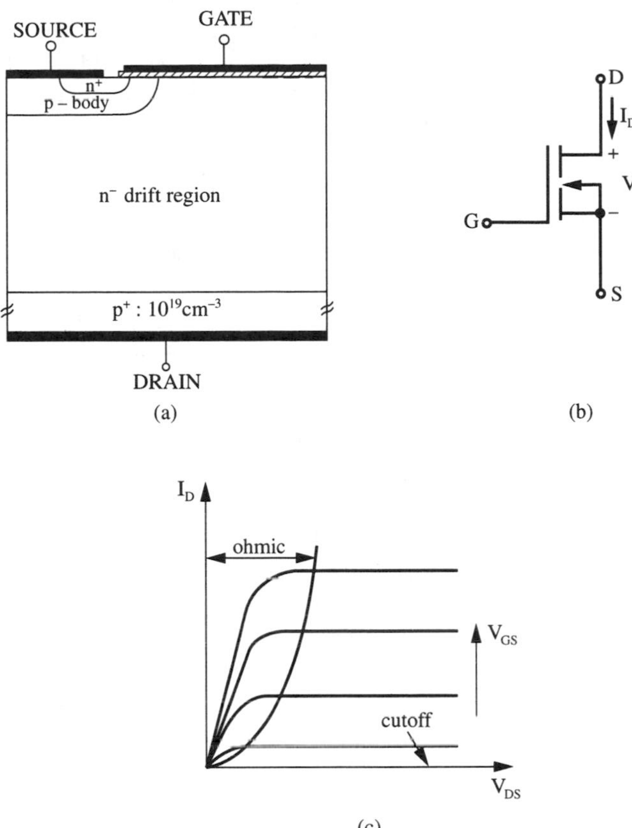

Figure 8.11 MOSFET (a) cross section, (b) circuit symbol and (c) typical static I-V characteristics.

semiconductor by an insulator. Gate current flows only during transition from on- to off-state and vice versa when the gate capacitance charges or discharges. MOSFETs have relatively short switching times and correspondingly small switching losses. Switching times in MOSFETs are governed by the time required for the device capacitances to charge and discharge.

The on-state resistance r_{ds} (on) of the MOSFET between the drain and source increases rapidly with the device blocking voltage rating, the dependence being nearly cubic. Because of this, only devices with small voltage ratings are available that have low on-state resistance, and hence, small conduction loss.

Positive temperature coefficient for the on-state resistance permits paralleling of MOSFETs without risk of instability. The device conducting higher current heats up, and is thus forced to share its current equitably with the MOSFETs in parallel with it.

8.5 New Devices

Although BJTs have low on-state resistance, they are current-controlled devices resulting in significant losses in the control circuitry. On the other hand, MOSFETs involve relatively simple gate drive circuits, but have significant losses during forward conduction at higher blocking voltages. Recently, efforts have been made to incorporate MOS-gated control with bipolar conduction characteristics. This has led to the invention of several hybrid structures. Among the most widely used new devices are the insulated gate bipolar transistor (IGBT) and MOS-controlled thyristor (MCT). Another type of device that is being developed is the emitter switched thyristor (EST). These devices are briefly described below.

Insulated Gate Bipolar Transistor (IGBT)

A novel way of combining the best qualities of BJT and MOSFET is to form a monolithic Darlington pair in which MOSFET is used to supply base current to the BJT. The device so formed is termed as the insulated gate bipolar transistor (IGBT). It is also known as COMFET (conductivity modulated FET), IGT (insulated gate transistor) and bipolar-mode MOSFET.

Figure 8.12a and b show a vertical cross section and the circuit symbol of a generic n-channel IGBT. The structure is similar to a vertical diffused MOSFET. The main difference is the p^+ layer at the back that forms the collector of the IGBT. The forward blocking of the device is essentially the same as in the MOSFET. When a positive gate voltage is applied, an inversion layer is formed under the gate which facilitates the injection of electrons from the emitter into the drift region. These electrons supply the base current for the *pnp* bipolar transistor. Positive voltage at the collector causes hole injection into the drift region. This bipolar conduction mode gives the device its excellent on-state properties. Just like a power diode, the drift region conductivity

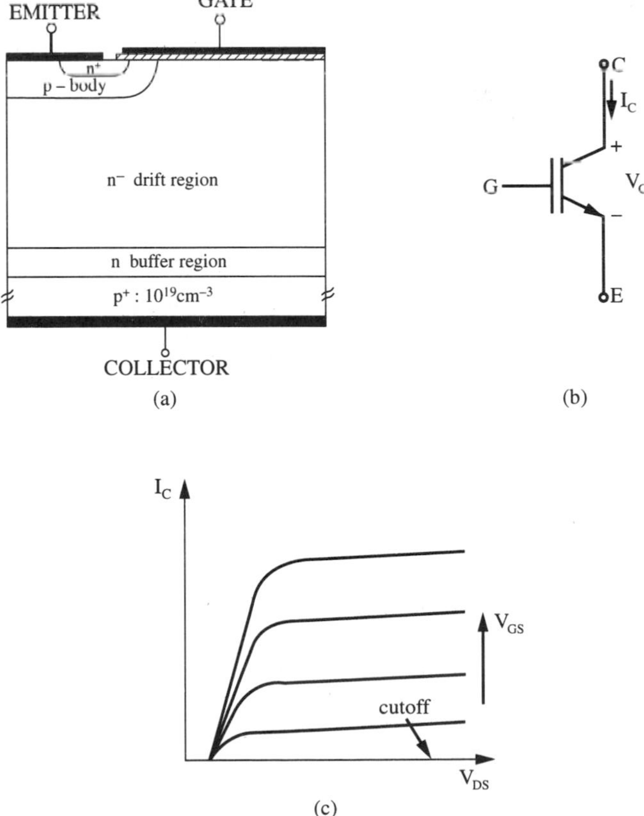

Figure 8.12 IGBT (a) cross section, (b) circuit symbol and (c) static I-V characteristics.

is modulated significantly during conduction which helps reduce the on-state resistance, unlike the power MOSFET. Figure 8.12c shows the typical I-V characteristics of the IGBT. Current saturation is governed by the saturation of the MOS channel through which the base current flows. A parasitic thyristor is formed in the device by junctions J_1, J_2 and J_3 in Figure 8.12 where J_1 is the junction between n^+ and p-body regions, J_2 is that between the p-body and n^- drift region and J_3 is the junction formed between the n-type buffer region and the p^+ substrate. If this thyristor gets latched, the device loses its gate control capability and cannot be turned off by controlling the gate voltage. The commercially available devices today have a very high latch-up immunity.

The turn-on switching characteristics of an IGBT, for the clamped inductive-load circuit shown in Figure 8.10a, are essentially similar to those of a MOSFET. However, the turn-off performance of the IGBT is significantly different from that of the MOSFET. Typical turn-off waveforms of the IGBT are shown in Figure 8.13, for the standard chopper circuit. The turn-off current shows two distinct regions. The first phase corresponds to an almost instantaneous fall in current due to the abrupt shutting off of the MOS channel. The second phase is the so-called "tail current" phase during which the excess carriers in the drift region are removed primarily by recombination. The second phase makes the device slower compared to a MOSFET and is primarily responsible for the turn-off losses. The n^+ layer (called buffer layer) near the collector in Figure 8.12a is generally used to get a better trade-off between conduction and switching characteristics. An optimized buffer layer and use of lifetime killers can be used to achieve turn-off times less than 1 μs.

MOS-Controlled Thyristor (MCT)

In the on-state, the IGBT operates as a bipolar transistor driven by a MOSFET. The on-state characteristics can be significantly improved by thyristor-like operation. This is achieved by the MOS-controlled thyristor (MCT) shown in Figure 8.14a. Application of a negative gate voltage creates a p-channel under the gate. This channel provides the gate current to the vertical n-p-n-p thyristor that latches due to regenerative feedback and conducts a large forward current with small forward voltage drop. The I-V characteristics of an MCT are shown in Figure 8.14b. The characteristics are similar to the on-state behavior shown in Figure 8.14. The p-channel loses control over MCT performance once the thyristor latches on, and hence, the I-V characteristics are independent of gate voltage beyond the threshold voltage V_{th} of the channel. MCTs have very high forward current densities and correspondingly large charge in the drift region. MCT turn-off is accomplished by applying a high voltage of reverse polarity at the gate. This creates an n-channel under the gate. If the resistance of the channel is small enough, it will divert the electron current from the thyristor. This effectively raises the holding current of the thyristor, thus forcing it to turn off.

Although MCTs have significantly improved current-handling capability, they do not match the switching performance of IGBTs. This is due to the increased resistance in the device during turn-on and higher current density. During turn-off, the thyristor may fail to turn off if resistance of the diverting channel is high. Also, the channel has to be formed uniformly and abruptly in order to effectively cutoff the feedback in the thyristor.

Emitter Switched Thyristor (EST)

The MCT has an uncontrolled on-state performance, and hence, its forward-biased safe operating area (FBSOA) cannot be defined. The emitter switched thyristor (EST) has been proposed to achieve control over the thyristor on-state conduction. The cross section of an EST is shown in Figure 8.15. The EST has a N^+ floating emitter region in addition to the basic IGBT structure. When the inversion channel is formed on application of a gate voltage, device behavior is identical to an IGBT. At higher current levels, hole flow in the p-base region under the floating region forward biases the p-n^+ junction causing the vertical thyristor to latch up. The thyristor current flows through the MOS channel before reaching the emitter terminal. This provides a control over the thyristor current as resistance of the channel can be

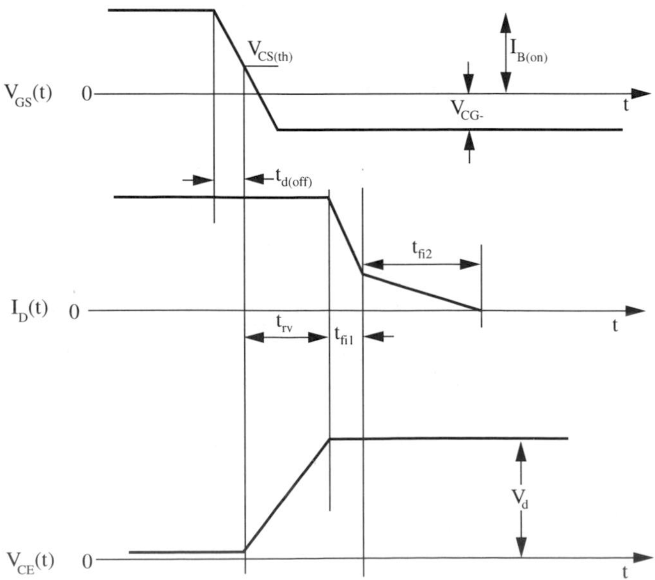

Figure 8.13 Turn-off waveforms of IGBT in clamped inductive load circuit.

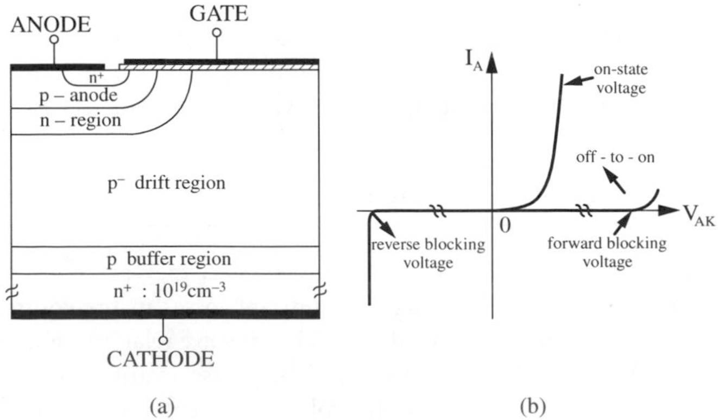

Figure 8.14 MCT (a) cross section and (b) typical I-V characteristics.

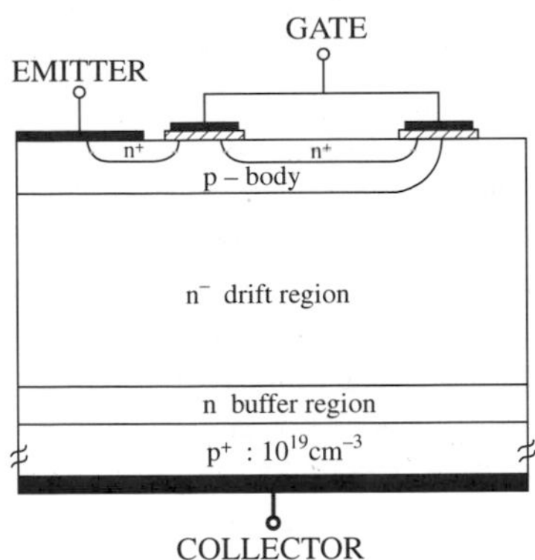

Figure 8.15 Cross section of emitter switched thyristor (EST).

varied by changing the gate voltage. On-state thyristor operation of EST results in much higher current density than IGBT. EST also shows current saturation like an IGBT.

The switching mechanism in an EST is identical to an IGBT. However, because of a higher level of drift region conductivity modulation, larger turn-off power loss results.

Smart Power™ Technologies

The development of MOS-gate-driven power devices has greatly simplified the gate drive circuits. This has made it possible to integrate the gate drive circuit into a monolithic chip. With the current technology, it is also possible to add other functions, such as protection against adverse operating conditions and logic circuits to interface with microprocessors. Figure 8.16 shows a typical Smart Power(TM) IC layout for application in the low-power range. In a Smart Power™ control chip, the sensing and protection circuits are usually implemented using analog circuits with high speed bipolar transistors. These circuits must sense adverse temperatures, currents and voltages. This circuitry helps detect situations like thermal-runaway, impact ionization, insufficient gate drive, and so forth.

Today's Smart Power™ chips are manufactured using a junction isolation technology. Efforts are in progress to make lateral structures with high breakdown voltages with thin epitaxial layers. The dielectric isolation technology is being perfected to replace the junction isolation technology in order to achieve fewer parasitics, compactness and higher degree of intergration.

Other Material Technologies

There are numerous power circuit applications in which the device temperature can rise significantly above room temperature. Further, there are several applications which require devices which can handle large blocking voltages exceeding 5 kV. Inherent material limitations make it impossible to use silicon-based devices beyond 200°C. This has led to the search for new materials that can handle large voltages. Wide bandgap materials with large carrier mobilities are ideally suited for such applications. Presently, SiC appears to be the most promising material to replace Si for high-voltage high-temperature applications. Power switches and rectifiers fabricated from silicon carbide offer tremendous promise for reduction of power losses in the device. Much work needs to be done before a manufacturable technology can be commercialized.

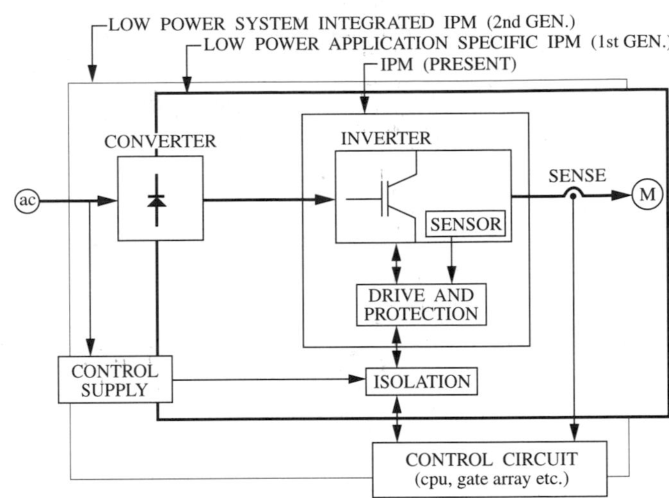

Figure 8.16 Schematic of an intelligent power module (IPM).

Devices for Resonant Power Conversion

Recent advances in circuit topologies promise tremendous advances in developing low-cost, high-efficiency power electronic circuits. For example, resonant power conversion using zero voltage switching (ZVS) and zero current switching (ZCS) are two such circuit topologies that are being investigated. Recent work has shown that the power semiconductor devices employed in ZVS and ZCS conditions behave significantly different compared to conventional hard-switching applications. New device designs and optimization techniques are in development.

References

Adler, M. S., Owyang, K. W., Baliga, B. J., and Kokosa, R. A. 1984. The evolution of power device technology, *IEEE Trans. Electron. Devices,* ED-31:1570–1591.

Azuma, M. and Kurata, M. 1988. GTO thyristors, *Proc. IEEE,* 76:419–427.

Baliga, B. J., Adler, M. S., Love, R. P., Gray, P. V., and Zommer, N. 1984. The insulated gate transistor: a new three-terminal MOS controlled bipolar power device, *IEEE Trans. Electron. Devices,* ED-31:821–828.

Baliga, B. J. 1990. The MOS-gated emitter switched thyristor, *IEEE Electron. Device Let.,* EDL-11:75–77.

Bauer, F., Hollenbeck, H., Stockmeier, T., and Fichtner, W. 1991. Current handling and switching performance of MOS controlled thyristor (MCT) structures, *IEEE Electron. Device Let.,* EDL-12:297–299.

Benda, H. and Spenke, E. 1967. Reverse recovery processes in silicon power rectifiers, *Proc. IEEE,* 55:1331–1354.

Blitcher, A. 1976. *Thyristor Physics,* Spring Verlag, New York, NY.

Blitcher, A. 1981. *Field Effect and Bipolar Power Transistor Physics,* Press, New York, NY.

Chudobiak, W. J. 1970. The saturation characteristics of NPνN power transistors, *IEEE Trans. Electron. Devices,* ED-17:843–852.

Clemente, S. and Pelly, B. R. 1981. Understanding the power MOSFET switching performance, *Proc. IEEE Ind. Applications Soc. Meeting,* Abs. 32.B, 763–776.

Darwish, M. N. and Board, K. 1984. Optimization of breakdown voltage and on-resistance of VDMOS transistors, *IEEE Trans. Electron. Devices,* ED-31:1769–1773.

Fischer, K. J. and Shenai, K. 1995. Effect of bipolar turn-on on the static current-voltage characteristics of scaled vertical power DMOSFETs, *IEEE Trans. Electron. Devices,* 42(3):555–563.

Ghandhi, S. K. 1977. *Semiconductor Power Devices,* Wiley, New York, NY.

Grover, R. J. 1982. Epi and Schottky diodes, in *Semiconductor Devices for Power Conditioning,* 331–56, Sting, R. and Roggwitter, P., eds., Plenum Press, New York, NY.

Herlet, A. and Raithel, K. 1966. Forward characteristics of thyristors in the fired state, *Solid State Electronics,* 9:1089–1105.

Hower, P. L. 1973. Optimum design of power transistor switches, *IEEE Trans. Electron. Devices,* ED-20:426–437.

Hu, C. 1979. A parametric study of power MOSFETs, *IEEE Power Electronics Specialists Conference Record,* 385–395.

Hu, C., Chi, M. H., and Patel, V. M. 1984. Optimum design of power MOSFETs, *IEEE Trans. Electron. Devices,* ED-31:1693–1700.

Longini, R. L. and Melngailis, J. 1963. Gate turn-on of four layer switch, *IEEE Trans. Electron. Devices,* ED-10:178–185.

McGrath, E. J. and Navon, D. H. 1977. Factors limiting current gain in power transistors, *IEEE Trans. Electron. Devices,* ED-24:1255–1259.

Pelly, B. R. 1983. Power MOSFETs—a status review, *Int. Power Electronics Conference Record,* 19–32.

Pendharkar, S., Winterhalter, C., and Shenai, K. 1995. A behavioral circuit simulation model for high-power GaAs Schottky diodes, *IEEE Trans. Electron. Devices,* 42(10):1847–1854.

Powerex IGBTMOD and Intellimod—*Intelligent Power Modules Applications and Technical Data Book,* 1994.

Russell, J. P., Goodman, A. M., Goodman, L. A., and Nielson, J. M. 1983. The COMFET: a new high conductance MOS gated device, *IEEE Electron. Device Let.,* EDL-4:63–65.

Shenai, K. 1990. Potential impact of emerging semiconductor technologies on advanced power electronic systems, *IEEE Electron. Dev. Let.,* 11(11):520–522.

Shenai, K. 1990. Optimally scaled low-voltage vertical power DMOSFET's for high-frequency power switching applications, *IEEE Trans. Electron. Devices,* 37:(4)1141–1153.

Shenai, K. 1990. Novel refractory contact and interconnect metallizations for high-voltage and smart-power applications, *IEEE Trans. Electron. Devices,* 37(10):2207–2221.

Shenai, K. 1992. Optimized trench MOSFET technologies for power devices, *IEEE Trans. Electron. Devices,* 39(6):1435–1443.

Shenai, K., Scott, R. S., and Baliga, B. J. 1989. Optimum semiconductors for high-power electronics, *IEEE Trans. Electron. Devices,* 36:1811–1823.

Temple, V. A. K. 1986. MOS controlled thyristors—a new class of power devices, *IEEE Trans. Electron. Devices,* ED-33: 1609–1618.

Trivedi, M., Pendharkar, S., and Shenai, K., 1996. Switching characteristics of IGBTs and MCTs in power converters, *IEEE Trans. Electron. Devices,* 43(11).

Tu, L. and Baliga, B. J. 1993. Controlling the characteristics of the MPS rectifier by variation of area of Schottky region, *IEEE Trans. Electron. Devices,* 40:1307–1315.

Wheatley, C. F. and Einthoven, W. G. 1976. On the proportioning of chip area for multistage Darlington power transistors, *IEEE Trans. Electron. Devices,* ED-23:870–878.

Widjaja, I., Kurnia, A., Shenai, K., and Divan, D. M. 1995. Switching dynamics of IGBTs in soft-switching inverters, *IEEE Trans. Electron. Devices,* 42(3):445–454.

Wolley, E. D. and Bevacqua S. F. 1981. High speed, soft recovery, epitaxial diodes for power inverter circuits, *IEEE Industrial Applications Society Meeting Digest,* 797–800.

9

Devices and Components

Imre Ipsits
Technical University of Budapest

Karoly Kurutz
Technical University of Budapest

9.1 Power Diodes 203
Static Characteristics • High-Voltage Diodes • Dynamic or Switching Characteristics • Significant Properties or Parameters of Diodes
9.2 Power Bipolar Junction Transistors (BJTs) 211
Static Characteristics and Ratings
9.3 Passive Networks 215
General Description • Resistors • Capacitors • Inductances

9.1 Power Diodes

Imre Ipsits

The majority of the power semiconductor diodes are junction diodes, therefore the discussion will be concentrated on these types. The junction power diodes are two-terminal pn junction devices. Figure 9.1 shows the simple basic structure of the diode and its symbol. The anode and cathode electrode are connected through ohmic metal contacts to the p and n type layer of the diode.

Static Characteristics

Diodes conduct in the forward direction and practically block all current flow in the reverse direction. The static characteristic of the diode is shown in Figure 9.2. In static forward biasing, the diode has a forward voltage drop that is practically independent of the current and is typically about 1V for a Si diode. The forward characteristic may be approximately represented by the slope resistance r_F and threshold voltage $V_{(TO)}$, so the voltage drop is $v_F = V_{(TO)} + r_F i_F$. The maximum allowable current depends on the cross section of the pn junction, and is limited by the power losses (dissipated power) produced in the crystal and the maximum allowable virtual junction temperature, T_{jmax}. Under reverse bias the diode conducts a small leakage current in the range of micro- or milliampere while in the blocking state. The reverse voltage is limited by the breakdown voltage V_{BD}. In the breakdown area, there is a rapid increase in reverse current; therefore, the operation of the diode generally must avoid this condition in order to prevent failure of the device. The avalanche diode is a special type of diode with so homogeneous a cross section that the avalanche breakdown occurs simultaneously over the main portion of the junction cross section. The avalanche diodes are able to operate in the breakdown area without failure during a short time. Therefore, these devices are transiently voltage proof within a specified limit.

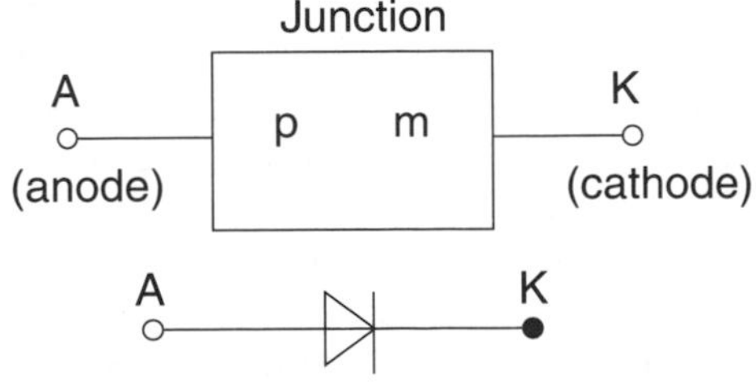

Figure 9.1 Basic structure and electrical symbol of junction diode.

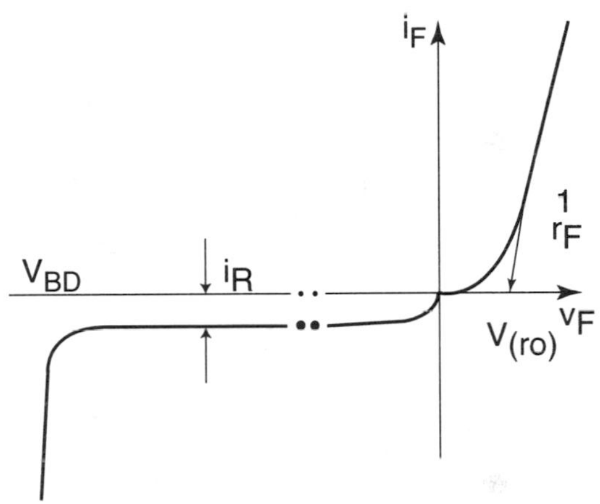

Figure 9.2 Static characteristic of the diode.

High-Voltage Diodes

The higher-voltage diodes have a larger forward voltage drop than their lower-voltage counterparts of equal current rating. For the sake of decreasing the trade-off between breakdown voltage and forward voltage drop, the three-layer diodes (or pin diodes) have been developed. Figure 9.3 shows a sketch of the arrangement of a three-layer diode. Between the highly doped p^+ and n^+ layer there is a very lightly doped n^- layer, the so-called drift layer. Forward biasing the p^+ and n^+ layers injects

0-8493-8343-9/97/$0.00+$.50

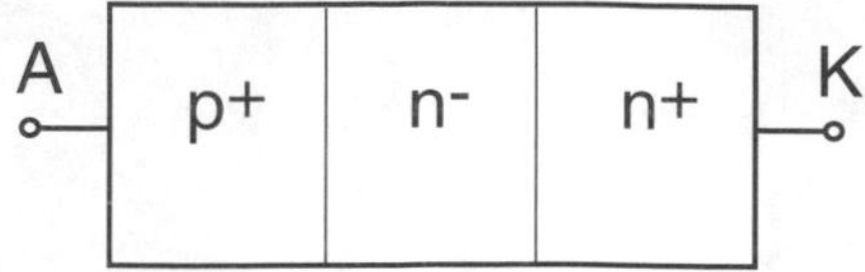

Figure 9.3 High-voltage three-layer diode structure.

enough charge carriers into the drift layer to modulate (increase) its conductivity, so the forward voltage drop will be lower. In the reverse direction, the depletion layer extends through the drift region and will be in contact with the n^+ layer. This is the so-called punch-through effect, which modifies the electric field profile, whereby the breakdown voltage increases. The maximum breakdown voltage range of three-layer diodes may reach 6–7kV.

The characteristic of the diode is temperature dependent, both in the forward and reverse directions. The forward voltage drop, caused by a constant current, decreases linearly with temperature. The temperature coefficient is approximately −2mV/°C. The reverse current changes almost exponentially with the temperature. The reverse current approximately doubles every 10°C. The temperature dependence of the breakdown voltage depends on the breakdown mechanism, namely avalanche or Zener. The temperature coefficient of breakdown voltage is approximately ±10mV/°C, which can be neglected in higher breakdown voltage values. The values of the temperature coefficients are valid for the Si devices at T = 300K.

Dynamic or Switching Characteristics

In normal operation, the diodes are generally used as a rectifier so the biasing from forward to reverse and vice versa is changing periodically. In a forward-biased diode, majority carriers from the p− and n− regions cross the junction to become minority carriers on the other side. The carriers, maintained in the steady-state condition, cannot be removed instantaneously when the forward bias is changed to a reverse bias. In other words, the diode cannot block reverse voltage until these carriers have been removed or have recombined. The temporary charge storage effect of minority carries is called the reverse recovery transient or phenomenon.

Typical current and voltage waveforms are shown in Figure 9.4. Initially the diode conducts a forward current, I_{FM}, and has a forward voltage drop, V_F, of about 1V. The decay rate of the current, $-di_F/dt$, is a function of the supply voltage v and the parameters of the circuit (R,L). When reverse voltage is applied to a diode, which has been conducting, the minority carriers near the pn junction are swept back across the junction, and there is a brief interval, the storage time, t_s, of high reverse current before the diode begins to block reverse voltage. t_s, which represents the time between the zero crossing and the peak of the reverse recovery current, I_{RRM}, is due to the charge storage in the depletion region of the junction. The fall time, t_f, is due to the remaining charge storage in the bulk of the semiconductor crystal. The total or just simply the reverse recovery time, t_{RR}, is composed of t_s and t_f, $t_{RR}=t_s+t_f$. The current-time area associated with reverse recovery current is the reverse recovery charge, Q_{RR}, shaded in Figure 9.4. As long as t_s is partly circuit parameter dependent, t_f is exclusively determined by device properties, first of all by the minority carrier lifetimes. If t_f is small, the reverse recovery current falls very rapidly and may generate a large transient overvoltage due to high values of L di/dt. The ratio $S = t_f/t_s$, as a "softness"-factor, is an indicator of the probability of excessive voltage transients when the diode recovers. The S values of a "soft" recovery diode are about unity, indicating lower overvoltages, whereas a "snappy" recovery diode has a smaller softness-factor, which produces a larger voltage overshoot. To restrict overvoltages to below the permissible voltage of the diode, it is necessary to use a protection circuit, i.e., so-called snubber circuit (in Figure 9.4 R_s-C_s with dashed line). The R_s-C_s values can be computed from a knowledge of the circuit parameters and device reverse recovery parameters.

In the catalogs, the manufacturers provide a figure of reverse recovery charge, Q_{RR}, as a function of $-di_F/dt$ and I_{FM} and the softness-factor, S (Figure 9.5.). On the basis of Figure 9.4, we can determine the other parameters of the reverse recovery phenomenon:

$$I_{RRM} = t_s di_F/dt; \qquad Q_{RR} = \frac{1}{2} I_{RRM} t_{RR}; \qquad S = t_f/t_s;$$
$$t_{RR} = t_s + t_f \tag{9.1}$$

From these equations:

$$t_{RR} = \sqrt{\frac{s^2+1}{s+1}\frac{2Q_{RR}}{di_F/dt}}; \qquad I_{RRM} = \sqrt{\frac{2Q_{RR}di_F/dt}{1+s}} \tag{9.2}$$

If the softness-factor is a small value, $S << 1$, i.e., a "snappy" diode, and $t_{RR} \approx t_s$, then approximately:

$$t_{RR} = \sqrt{\frac{2Q_{RR}}{di_F/dt}}; \qquad I_{RRM} = \sqrt{2Q_{RR} \cdot di_F/dt} \tag{9.3}$$

The parameters of the reverse recovery transient (I_{RRM}, t_{RR} and S) are also temperature dependent and all of them increase with temperature. In some detailed catalogs, we can find t_{RR} as a function of $-di_F/dt$, where the parameter is temperature, and I_{RRM} as a function of di_F/dt where the parameter is the temperature or the forward current, I_{FM}, prior to turn-off. The reverse recovery performance of the diode is very important to the design of converter circuits. It is more critical in the various types of high frequency DC-DC, DC-AC and resonant converter circuits.

The switching properties of the diode are also influenced by the forward recovery, but their importance is smaller, when compared to reverse recovery. In an ideal case, the waveshape of the voltage and current are shown in Figure 9.6. The diode voltage first rises to a value of V_{FR}, which is much higher than the static voltage drop, V_{FO}, and then decays to V_{FO}. V_{FR} is the forward

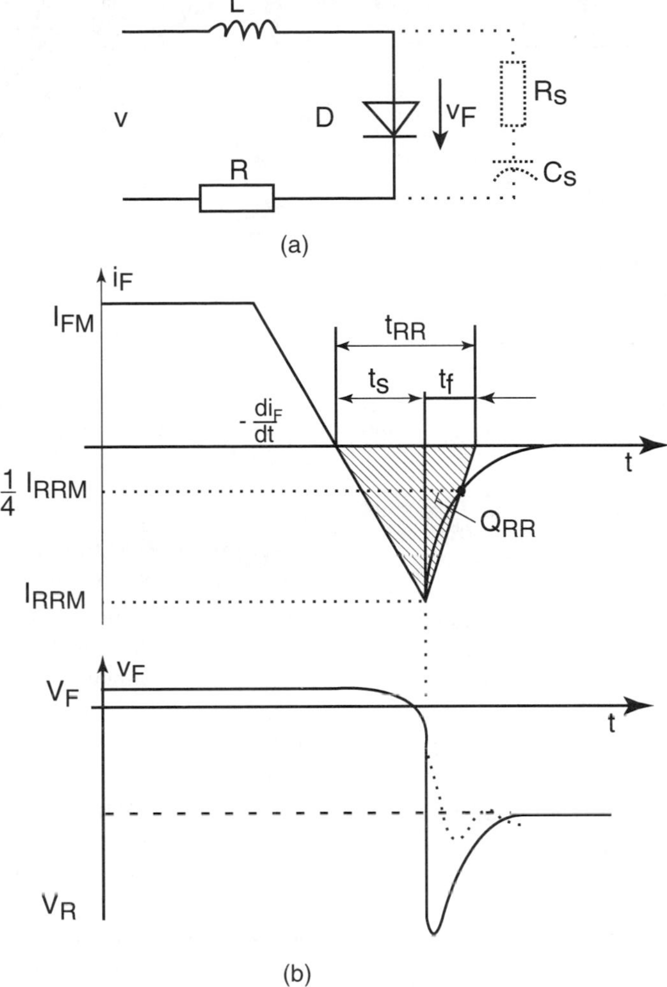

Figure 9.4 Reverse recovery phenomenon of junction diode: (a) circuit, (b) current and voltage waveforms.

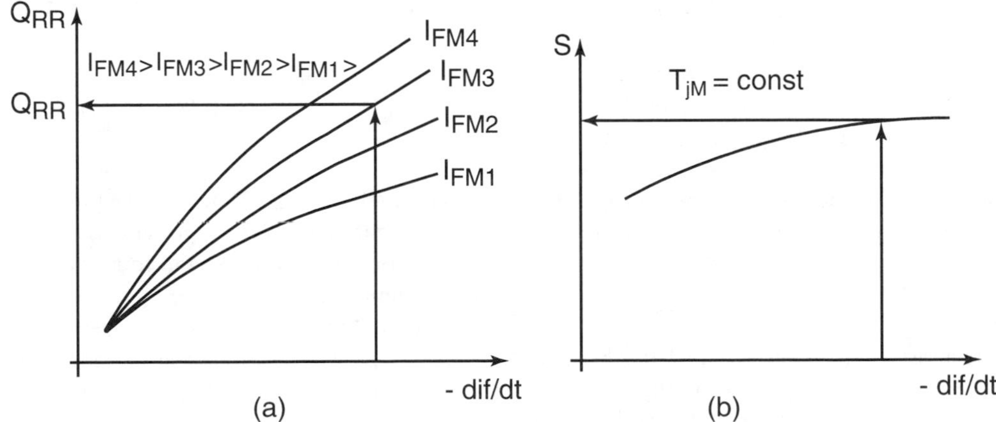

Figure 9.5 Reverse recovery parameters. (a) Typical reverse recovery charge vs. $-di_F/dt$. (b) Typical softness-factor: $S = t_f/t_s$ vs. $-di_F/dt$.

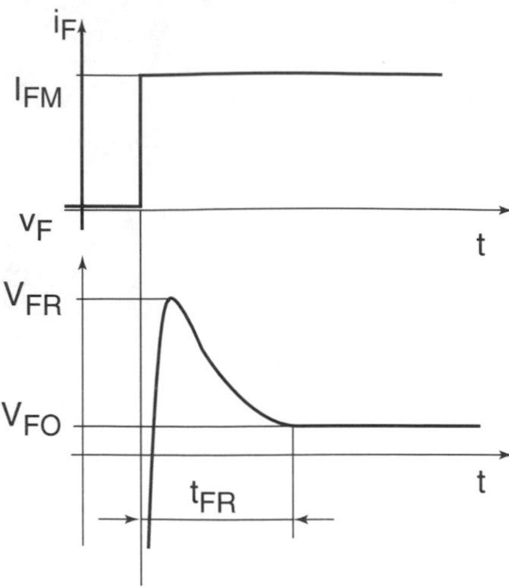

Figure 9.6 Forward recovery phenomenon of a junction diode. Current and voltage waveforms.

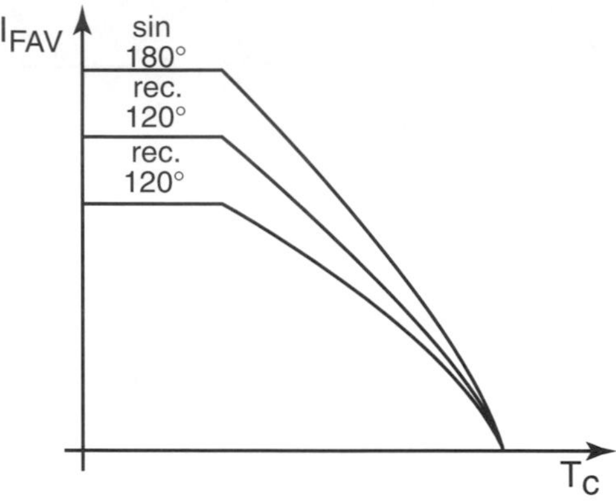

Figure 9.7 Typical maximum average forward current vs. case temperature.

recovery voltage and t_{FR} is the forward recovery time. A simplified explanation of the transient is that in the first instant there are no excess charge carriers in the depletion-and diffusion layers of the junction—and in the drift layer of the pin diode—and therefore the ohmic resistance of the layer is higher. The forward current "charges" the layers with excess carriers, so the resistance is reduced. Naturally the process is more complex, because the current could not change instantaneously and the induced voltage of the intrinsic inductances of the diode is not neglectable. The typical value of the forward recovery time, t_{FR}, is near the microsecond range.

Significant Properties or Parameters of Diodes

The semiconductor diodes are characterized by the manufacturers in terms of certain parameters. These parameters are nonlinear and dependent on a number of factors. The manufacturers normally provide characteristic curves for important parameters in the data sheets. Catalog data include rated values (that is, the operating values recomended by the manufacturers), limiting values (which must not be exceeded under any circumstance), and minimum, maximum, and average values for certain sets of conditions. The data describing the properties of diodes may be electrical within forward and reverse, temperature, and mechanical data.

Forward Data

From the characteristic of the diode we know the threshold voltage, $V_{(TO)}$, and the slope resistance r_F. Both of them are given at T_{jmax} temperature. The maximum average forward current, I_{FAVM}, is the maximum permissible value of average forward current at a specified temperature. These data are normally referred to half sine waves at a case temperature ($T_c = 60–120°C$). Figure 9.7 shows the typical variation of I_{FAVM}, with the maximum case temperature, for half sine waves and for square waves with 120° or 60° conduction angles during a period. The I_{FAVM} value of a diode is generally the first rating to be seen on a data sheet.

The maximum RMS forward current, I_{FRMSM}, is the maximum permissible RMS value of forward current, taking into account the electrical and thermal stresses on all the individual parts of diode. The electrical or thermal limited properties of the diode depend on the configurations of the diode. The lead-type configurations are mainly RMS limited, but the stud-type and the disk-types (press-pak or hockey-puck types) are, above all, thermal limited. In the thermal limited configuration, the diode current must be progressively reduced to keep the junction temperature within T_{jmax}, which ranges from 125°C to 180°C. T_{jmax}, which is a very important parameter of the diode, is the maximum average junction temperature in steady state operation that the diode can withstand without failing due to thermal runaway. T_{jmax} influences the ratings of most parameters, such as the forward current values. The actual operating current rating of a diode depends on the power losses. In normal rectifier operation, the average power losses (the dissipation power) of the diode—neglecting the reverse and switching losses—can be calculated from an approximate forward characteristic of the diode

$$P_D = P_{FAV} = V_{TO}I_{FAV} + r_T I^2_{FRMS} = V_{TO}I_{FAV} + r_T F^2 I^2_{FAV} \quad (9.4)$$

where $F = I_{FAV}/I_{FRMS}$ is the form factor.

The Figure 9.8 shows the power losses as a function of average current for half sine waves and square waves with 120° and 60° conduction angles.

Due to the power losses, heat is generated within the power device. This heat must be transferred from the device to the cooling medium (generally the ambient air) to maintain the operating junction temperature within the specified range. To improve the heat transfer, diodes are mounted on a heat sink. The heat passes through a number of component parts of different materials and geometries. Thermally, these parts can be modelled using their thermal resistances and capacitances as shown in Figure 9.9a. All of the parts have a thermal time constant: $\tau_{th} = R_{th} \cdot C_{th}$. In a steady state operating condition, the thermal capacitances can be neglected, so the electrical analog model of

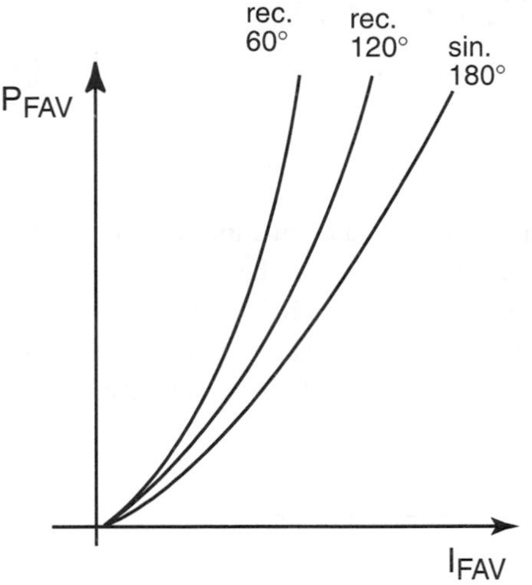

Figure 9.8 Typical forward loss characteristics.

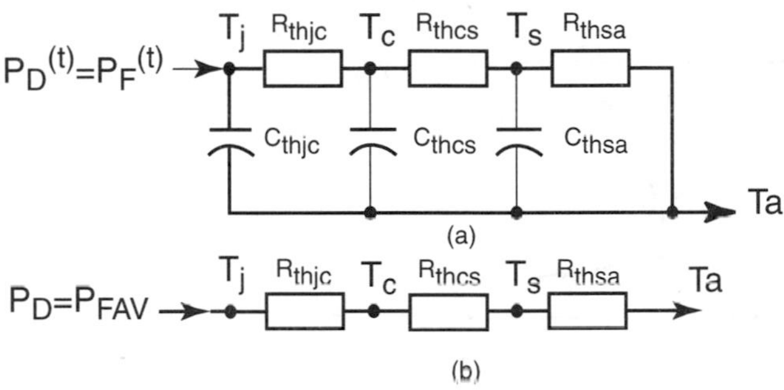

Figure 9.9 Simplified thermal impedance vs. time.

the heat transfer is shown in Figure 9.9b. The junction temperature of the device, T_j, can be calculated by

$$T_{j\max} \geq T_j = T_a + P_D(R_{thjc} + R_{thcs} + R_{thsa}) \qquad (9.5)$$

where

T_a = ambient temperature (°C)

R_{thjc} = thermal resistance from junction to case (°C/W)

R_{thcs} = thermal resistance from case to sink (°C/W)

R_{thsa} = thermal resistance from sink to ambient (°C/W)

P_D = dissipated power (power losses) (W)

R_{thjc} and R_{thcs} are normally specified in the catalog, P_D is known, so the required thermal resistance of the heat-sink can be calculated for the specified ambient temperature T_a. Then a heat sink must be chosen which will meet the R_{thsa} value requirement.

Overloadability of Diodes

The overload currents, which are permissible for short time overloads or for intermittent duty, may be calculated by means of a transient thermal impedance under pulse conditions, so that the virtual junction temperature at no instant exceeds the maximum value. A typical transient thermal impedance curve for a diode is shown in Figure 9.10, where Z_{thjc} is the transient thermal impedance from junction to case and Z_{thjs} is the thermal impedance from junction to heat sink. The curves arise because of thermal capacity effects. The thermal impedance of a power device is very small for a short time duration, compared to the steady-state value of $Z_{thjc} = R_{thjc}$ and $Z_{thjs} = R_{thjs}$, and as result the junction temperature of devices varies with the instantaneous power loss. When a diode experiences a square-wave pulse, P_{DO}, the rise in junction temperature begins immediately, but the maximum temperature depends on the duration of the power pulse, t_1. The junction temperature may be expressed quantitatively as

$$T_j(t_1) = T_{jo} + P_{DO}Z_{tht}(t_1) \qquad (9.6)$$

where

T_{jo} = the average junction temperature before the P_{DO} pulse

$Z_{tht}(t_1)$ = the transient thermal impedance junction to case at time t_1

To calculate the junction temperature after the power pulse is completed, at $t_2 > t_1$, the superposition technique is used, and the effect of the positive power pulse is canceled by applying a negative power pulse of the same amplitude, when the original

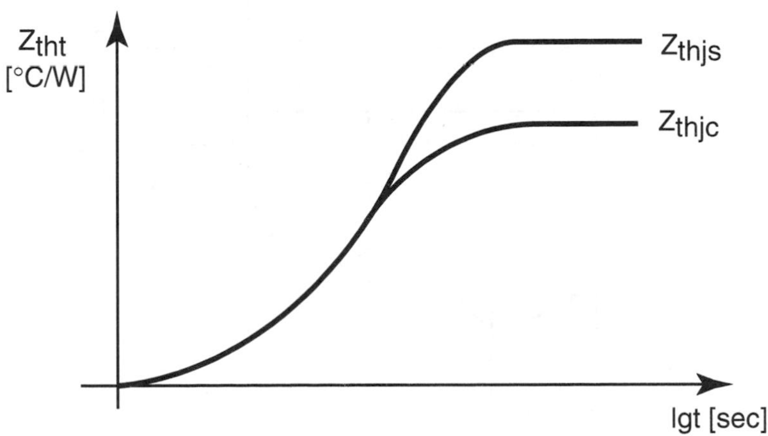

Figure 9.10 Transient thermal impedance vs. time.

power pulse terminates. The superposition technique can be seen in Figure 9.11, and the junction temperature at t_2 is given by the expression

$$T_j(t_2) = T_{jo} + P_{DO}[Z_{tht}(t_2) - Z_{tht}(t_2 - t_1)] \quad (9.7)$$

To compute the junction temperature after three different power pulses in a pulse train, the superposition technique can also be used. The pulse train, and the junction temperature time function are shown in Figure 9.12. The junction temperature at the instant t_5 is

$$T_j(t_5) = T_{jo} + P_{D1}[Z_{tht}(t_5) - Z_{tht}(t_5 - t_1)] + P_{D2}[Z_{tht}(t_5 - t_2) - Z_{tht}(t_5 - t_3)] + P_{D3}[Z_{tht}(t_5 - t_4)] \quad (9.8)$$

The previously reviewed calculation method can be extended to optional power waveforms. The waveform, of any shape, can be represented approximately by rectangular pulses of equal or different duration, with the amplitude of each pulse being equal to the average amplitude of the actual pulse duration as shown

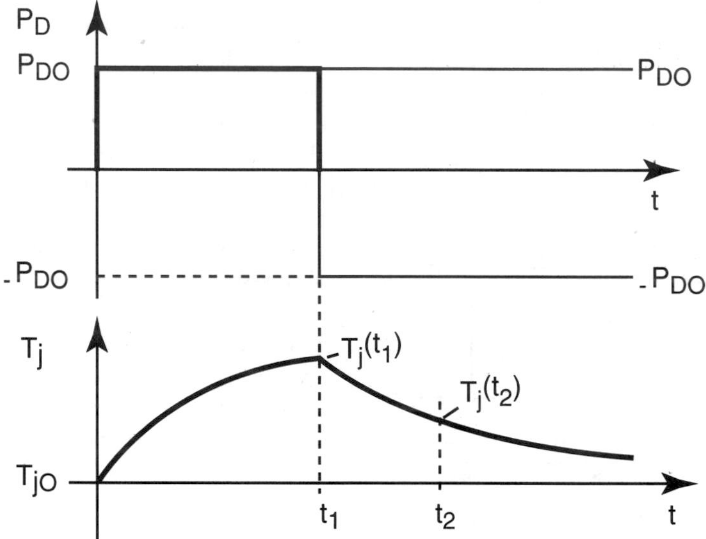

Figure 9.11 Superposition technique used to calculate the T_j for a power pulse.

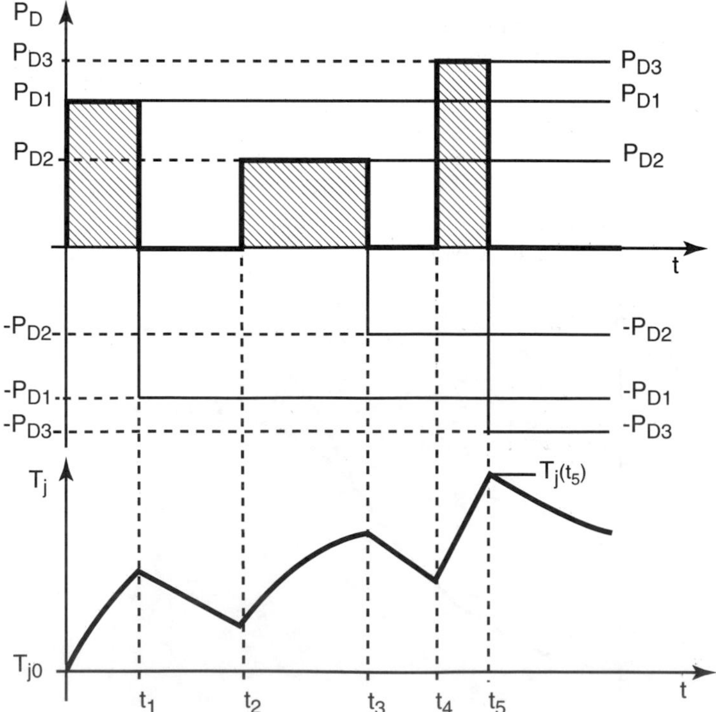

Figure 9.12 Superposition technique used to calculate the T_j for power pulse train.

in Figure 9.13. The junction temperature at the end of m-th pulse can be found from:

$$T_j(t_m) = T_{jo} + \sum_{n=1,2\cdots}^{m} P_{Dn}[Z_{tht}(t_m - t_{n-1}) - Z_{tht}(t_m - t_n)] \quad (9.9)$$

where

P_{Dn} = the n-th pulse amplitude
$(t_m - t_{n-1})$ = the duration between the end of the waveshape and the start of the n-th pulse
$(t_m - t_n)$ = the duration between the end of wave shape and the finishing of the n-th pulse

The accuracy of such approximations can be improved by increasing the number of pulses and thus reducing the duration of each pulse.

Transient thermal impedance calculations generally assume that the rectifier is mounted on a heat sink, and that no change in case temperature occurs during the time of the power pulses. This assumption is valid if the duration of the power pulse(s) is negligible compared to the thermal time constant of the heat sink. The boundary pulse duration can be seen in Figure 9.10, when the curve of Z_{tht} is separated into two curves. For longer power pulses, the change of case temperature should be considered.

When analyzing the overloadability of diodes, other characteristic current values can be found in the catalog:

Maximum rated surge forward current, I_{FSM}, is the maximum permissible nonrepetitive instantaneous value of a single current impulse in the form of a sinusoidal half-wave at the network frequency, without subsequent reverse voltage. Repetition is permissible only after a minimum interval, while the junction temperature is reduced to the rated maximum junction temperature, during which the diode can be loaded by rated average current and voltage. The surge current may be repeated up to 100 times in a similar fashion. After 50 to 100 such loadings, irreversible alterations in the characteristics and/or failures may occur.

Maximum overload forward current, $I_F(_{OV})_M$, is the maximum permissible value of the forward current with which the diode may by loaded in short time duty. It is given in the overload characteristic as the peak current value of a sinusoidal half-wave at the network frequency. The various initial conditions are the parameters for the group of curves. Typical overload characteristics are given in Figure 9.14, where curve a gives the overload ability after a no-loaded state, and curve b gives similar data after the nominal loaded state. Another form of the overload characteristic is shown in Figure 9.15. This data indicates the repetitive peak reverse voltage after the overload is terminated. The I_{FSM} values, as parameters, are given at various junction temperatures.

Maximum rated value of $\int i^2dt$. The $\int i^2dt$ value for the diode is given so that suitable fuses can be selected to protect the diode against damage and short circuit. The $\int i^2dt$ value of the fuse, over a specified time interval and for a specified supply voltage, must be less than the value for the diode.

Maximum peak repetitive forward current, I_{RFM}, is the maximum value of the peak forward current which may not be exceeded even with a nonsinusoidal waveform, e.g., for a capacitive load, or charging a battery in quasi-steady-state operation.

Reverse Data

Repetitive peak reverse voltage, V_{RRM}, is one of the most important properties of the diode, defined as the sustainable voltage level of repetitive transient reverse voltage, which may be blocked by the diode at rated temperature. The applied voltage peak value is essentially lower than V_{RRM} The safety factor is generally between 1.5 and 2.5. The diodes, within an identical type, are selected on the basis of the V_{RRM} level.

Nonrepetitive peak reverse voltage, V_{RSM}, can exceed the repetitive rating (max 25%) or be equal to V_{RRM}, depending on the

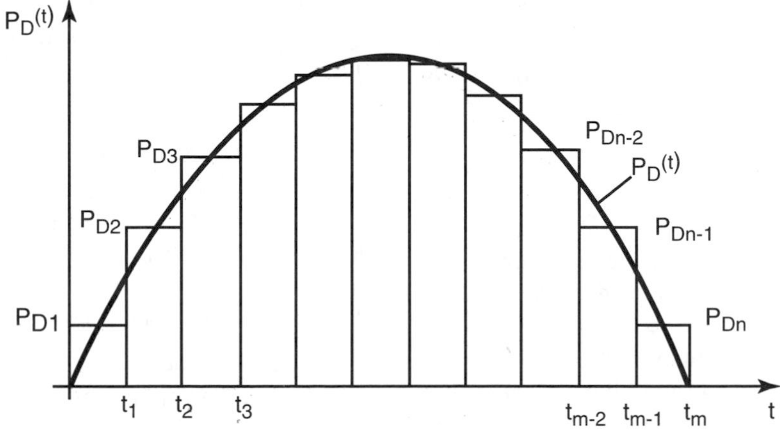

Figure 9.13 Approximation of a continuous power pulse by rectangular pulses.

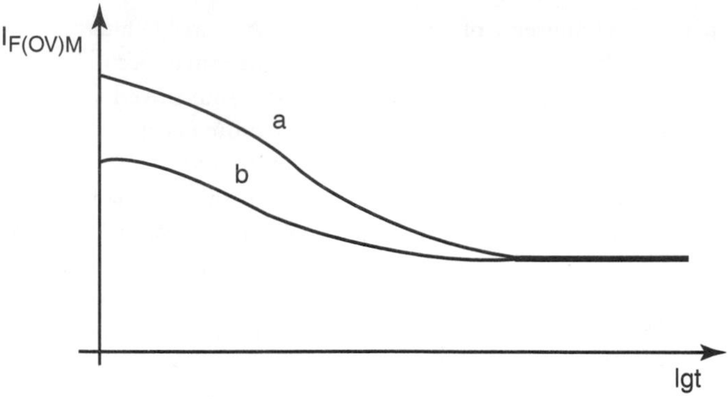

Figure 9.14 Short duration overload characteristics of power diode.

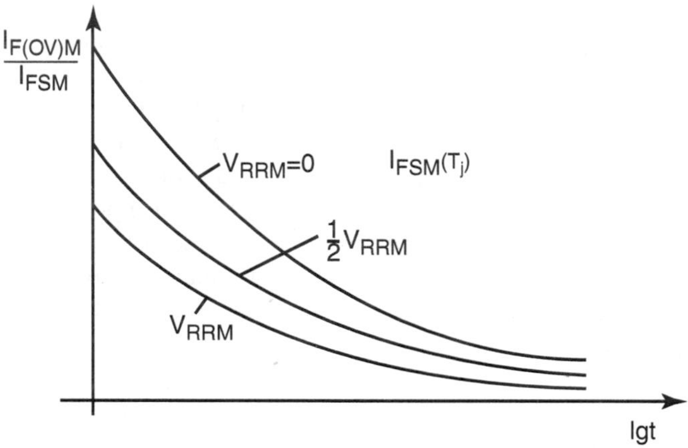

Figure 9.15 Limiting overload characteristics of power diode.

manufacturer. After a nonrepetitive surge, the diode has to block V_{RRM} on the next negative half-cycle of the voltage. V_{RSM} cannot exceed the leakage current of the diode significantly, because it would cause device destruction.

With the exception of an avalanche-type diode, the overvoltage transient cannot reach the diode breakdown voltage, V_{BD}. In a circuit, if it is expected that the transient voltage will exceed V_{BD}, we have to use protection devices, e.g., voltage suppressors, for protection against the transient voltage.

The reverse recovery properties of the diode also belong to the reverse data. These characteristics have been discussed previously within the scope of Dynamic or Switching Characteristics, and therefore will not be repeated here.

Thermal Data

Most of the thermal data for diodes has been discussed within the scope of Forward Data. Other typical data are: maximum junction temperature, T_{jmax}, the case temperature, T_c, the heat-sink temperature, T_s, and the ambient temperature, T_a. A new element of data, the storage temperature range, T_{stg}, is the range between two temperature limits within which the diode may be stored without any electrical loading. It is typically $-40°C$ to 150°C, and is limited by the mechanical tension, which results from the different temperature coefficients of materials inside the diode.

Other thermal data are the various types of thermal resistances R_{th}, and the transient thermal impedance, Z_{tht}. In high-power applications, heat sinks are more effectively cooled by liquids: water or oil.

Switching data for the diode result from forward and reverse recovery phenomenon. These have been discussed in the Dynamic or Switching Characteristics part of this section.

Mechanical Data

This group of properties includes the weight, dimensions, tightening torque (when mounting the diode to a heat-sink), and vibration resistance (subject to acceleration in all three directions at a frequency of 50Hz with a usual value of $5 \cdot 9.81\ m/s^2$).

Power Diode Types

Depending on the application requirements, various types of diodes are available. The basis of the classification is generally the recovery characteristics and manufacturing techniques. On the basis of the recovery behaviors, the power diode can be classified into three categories:

1. Line-frequency diodes.
2. Fast recovery diodes.
3. Schottky diodes.

The first two groups are bipolar junction diodes, and thus

they are minority carrier devices, while the Schottky diodes are majority carrier devices.

Line-Frequency Diodes. The line-frequency, general-purpose or normal diodes are generally manufactured by diffusion. They have relatively high reverse recovery time, typically higher than 25μs, and are used in low-frequency applications where recovery time is not critical. These diodes are used as rectifiers in line commutated converters and other low-frequency applications up to 1kHz. These diodes are available with current ratings up to several kiloamperes and voltage ratings up to several kilovolts. Morover, they can be connected in series and parallel, to increase the reverse blocking and current carrying capability. With these arrangements, one can meet the desired voltage and current requirements; however, it is necessary to use voltage- and current sharing protection circuits.

Fast Recovery Diodes. Fast recovery diodes have low recovery time, normally less than 5μs. They are used in DC-DC converter (DC-chopper) and DC-AC converter (inverter) circuits, where the speed of recovery is, in most cases, critically important, because the reverse recovery current is increasing the current load of the switching devices (transistors, thyristors).

The higher-voltage (above 600V) fast diodes are manufactured by diffusion. These diodes cover voltage ratings up to 3kV, current ratings up to 1.5kA and the reverse recovery times are between 0.5 and 5μs. For voltage ratings below 600V, the epitaxial technology provides faster switching speed than the diffused one. The FREDs (fast recovery epitaxial diodes) have a narrower base width, resulting in shorter recovery times in the range of 50 to 200ns, and smaller recovery charge. Besides the technology, the most important tool is the decrease in carrier lifetime. The short carrier lifetime increases the forward voltage drop to about 1.2 to 1.6V and causes an abrupt cutoff of the reverse current just after its maximum. This action also produces high overvoltage on the leakage inductances in the circuit. These latter difficulties have reduced the development of special soft recovery diodes. Figure 9.16 shows the recovery of a line-frequency, a fast and a soft recovery diode.

Schottky Diodes. The charge storage problem associated with a junction diode is eliminated in a Schottky diode. Figure 9.17 shows the basic structure and electrical symbol of a

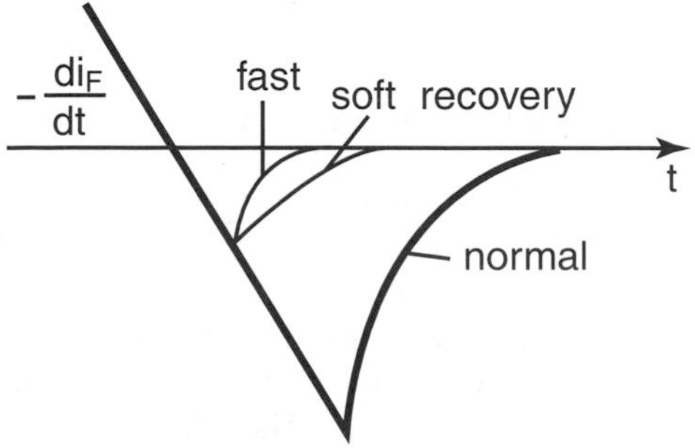

Figure 9.16 The reverse recovery current by normal, fast and soft recovery diodes.

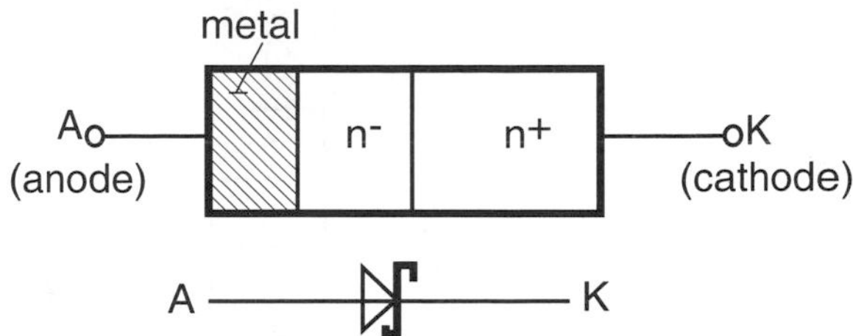

Figure 9.17 The basic structure and electrical symbol of a Schottky diode.

Schottky diode. In the Schottky diode, the nonlinear layer is produced by setting up a "barrier potential" with a contact between a metal and semiconductor layer. The "barrier potential" depends on the majority carriers only, and as a result there are no excess minority carriers to be swept out or recombined. From a practical standpoint there is no storage effect like a pn junction. The small recovery effect comes from the self-capacitance of this type of diode. Schottky diodes have a relatively low forward voltage drop in the range from 0.3 to 0.6V. The maximum forward current can be several hundered amperes. In the reverse direction, the Schottky diode has a reverse leakage current, which is larger than that of a comparable Si junction diode in the range of 1mA–10mA and exponentially dependent on the temperature. The maximum allowable voltage is limited to 100V.

Schottky diodes are used in high-frequency converter circuits, and they are ideal for high current, low voltage power supplies.

9.2 Power Bipolar Junction Transistors (BJTs)

Imre Ipsits

Bipolar junction transistors are continuously controllable, three-terminal active devices. When used in industrial electronic applications such as power converters, they are primarily used as switches. The BJTs have three main layers in a common crystal: emitter E, base B, and collector C, which are the names of the terminals connected to each layer. The E is a highly-, the B is a lightly- and the C is a very lightly-doped layer. In modern transistors, the collector layer is divided into two parts: a very lightly-doped drift layer, and a highly-doped substrate layer. This arrangement for the collector is unavoidable in high-voltage power transistors. Inside of the transistor there are two junctions, one between the B-E and the other between the B-C. Depending on the sequence of the layers, the BJTs can be either an npn or pnp type. Figure 9.18 shows the simplified one-dimensional structures and electrical symbols of the npn and pnp transistors. The operation of the two transistor types does not differ significantly, and so we will concentrate on the npn type, because they are the most commonly used for high-voltage and high-current applications.

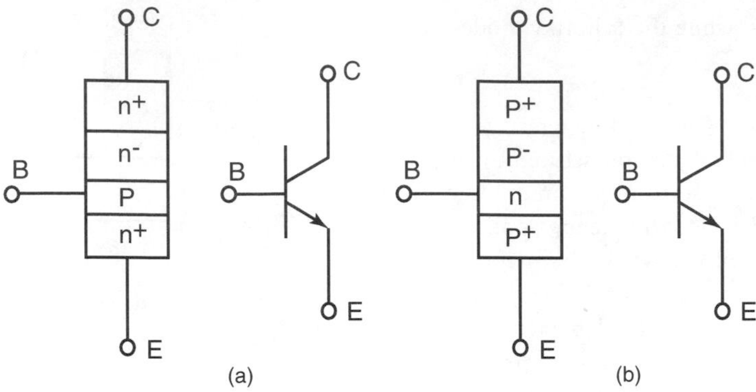

Figure 9.18 Simplified structure and electrical symbols for (a) npn and (b) pnp power transistors.

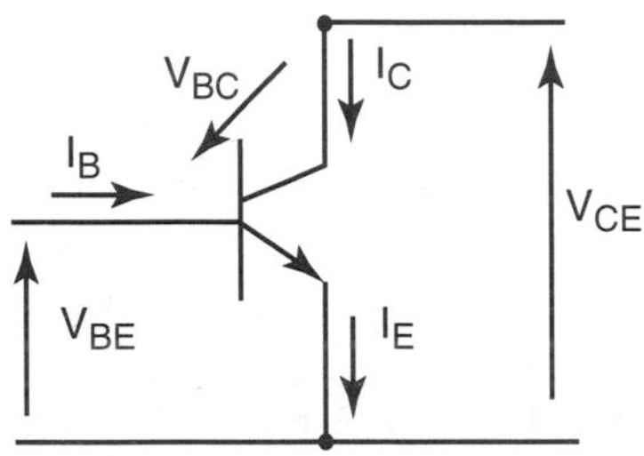

Figure 9.19 Common-emitter configuration of an npn transistor.

Although in various circuit applications we can use the transistors in common-emitter, common-base and common-collector configurations, the common-emitter configuration, which is shown in Figure 9.19, is generally used in converter circuits. In this figure the subscripts identify the electrode currents and the voltage terminals.

There are three operating regions for bipolar transistors, depending on the manner in which the B-E and B-C junctions are biased. In the cutoff region, both junctions are reverse biased, the transistor is in the off-state and a small leakage current can flow. The boundary state occurs when $V_{BE} = O$ or approximately $I_B = O$. In the active region, the B-E junction is forward biased and the B-C junction is reverse biased. The B-E junction acts as a source of mobile, in npn types, electron carriers, which enter the base region. Most of these electron minority carriers diffuse through the base region, which is very narrow, and arrive at the B-C junction. Consequently, a large current may flow in a reverse biased B-C junction resulting from carrier injection from a nearby E-B junction. This is the basis of the transistor effect, and it is realized only when the two junctions are close enough physically to interact in the manner described. These carriers are swept into the collector region by the electric field in the reverse biased C-B junction. Some electrons recombine in the base region and do not reach the collector, and they form the main part of the base current. If the B-E circuit is open, and the collector emitter terminal is supplied by a $V_{CE} > 0$ voltage, a small collector-to-emitter leakage current I_{CEO} flows ("O" means the base is open). In the common-emitter connection, the base current, I_B, is the input current, which controls the collector current, I_C, as an output current. The ratio of I_C to I_B is the current transfer ratio, or common-emitter DC current-gain, β *or* h_{FE}. So the collector current has two components:

$$I_C = \beta I_B + I_{CEO} \tag{9.10}$$

If the transistor is operated in some other configuration, e.g., common-base, where I_E is the input and I_C is the output current, the relationship between I_C and I_E can be derived from Kirchhoff's current law, $I_E = I_C + I_B$, and Equation 9.10:

$$I_C = \frac{\beta}{1+\beta} I_E + \frac{I_{CEO}}{1+\beta} = \alpha I_E + I_{CBO} \tag{9.11}$$

where α is the common-base DC current gain, and I_{CBO} is the collector leakage current with open emitter. The typical range for α is 0.95 to 0.99 and the value for β or h_{FE} typically lies in the range from 20 to 100. In the active region, the transistor can be used as a linear amplifier. In the saturation region both the B-E and B-C junctions are forward biased; therefore, there will be an excess of minority charge carriers in the base region. The base current is high and the collector-emitter voltage, $V_{CE(sat)}$, is low, depending on the I_C amplitude (see the later section on switching properties of BJTs.). At the boundary between the active and saturation regions, $V_{CB} = 0$. The saturation and cutoff regions are very important operational areas in the switching mode, because the transistor, when used as a switch, usually traverses the saturation (on-state) and the cutoff (off-state) regions.

Static Characteristics and Ratings

Figure 9.20 shows the common-emitter output characteristics: I_C versus V_{CE} for fixed values of I_B. Several features of these characteristics should be noted.

These characteristics illustrate the three operational regions (the saturation and cutoff regions are shaded). Figure 9.20b shows the saturation area in more detail. The $1/R_d$ line is the boundary between the active- and quasi-saturation areas. In the quasi-saturation area, the forward biased B-C (pn^-) junction injects holes into the drift layer and more and more parts of the drift

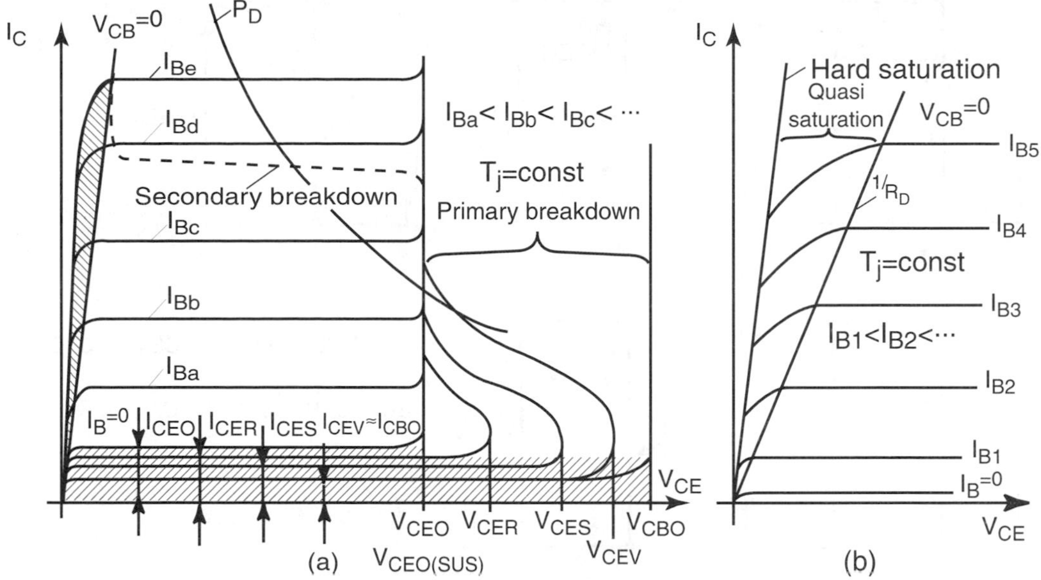

Figure 9.20 Common-emitter output characteristics of BJTs: (a) Showing primary and secondary breakdown. (b) Showing quasi-saturation.

layer are conductivity modulated, so the resistance of the drift layer and $V_{CE(sat)}$ will decrease. When the conductivity modulated area covers the whole area of the drift layer, hard saturation of the transistor is obtained. In the hard saturation area, the $V_{CE(sat)}$ voltage is minimum and neither I_C nor $V_{CE(sat)}$ changes essentially, but the excess charge carrier density—both in the base and drift layers—will be increased. The power dissipation in hard saturation is minimized in comparison to quasi-saturation. At constant base current, I_C is decreased and the h_{FE} current gain decreases.

Transistor Voltage Ratings

The permissible voltages of the transistor are limited by avalanche breakdown or primary breakdown. The highest voltage rating for the transistor is V_{CBO}, when the B-C junction is reversed biased with the emitter open circuited, and the small I_{CBO} leakage current is rapidly increasing. In the common-emitter configuration there are a number of collector-to-emitter voltage ratings, each of them specified by particular base-emitter conditions. (See Figure 9.21, where the individual leakage currents and the corresponding collector-to-emitter voltage ratings are marked.) If the base terminal is opened the leakage current flows across the B-E junction and, as a result of the transistor effect, it is amplified by the current gain. The total leakage current is $I_{CEO} = (1 + \beta)I_{CBO}$ (see Equation 9.11). For this condition, the collector-to-emitter breakdown voltage is $V_{CEO} < V_{CBO}$. The V_{CEO} is a maximum voltage that can be sustained across the transistor, *when $I_B > 0$ and the transistor is carrying substantial collector current.* This voltage is usually labelled $V_{CEO(sus)}$. Under other conditions, the collector leakage current is divided between the base and emitter terminals. The collector-to-emitter breakdown voltages are designated by V_{CER}, if between the B-E terminals there is a resistance R, V_{CES}, if between the B-E terminals there is a short circuit, and V_{CEV}, if the B-E junction is reverse biased by an external voltage, V (see Figure 9.21 c, d, e and Figure 9.20a).

The above-mentioned breakdown voltages limit the voltage blocking capabilities of transistors at low values of I_C. In the cases of V_{CEV}, V_{CES} and V_{CER}, when the breakdown process commences, and the leakage current rises, more and more of the leakage current flows through the emitter terminal, so the breakdown voltages more closely approach the $V_{CEO(sus)}$ value, which does not change with I_C. The $V_{CEO(sus)}$ is a characteristic value and a measure of the voltage capability of the transistor if the I_C and V_{CE} are simultaneously high, e.g., an inductive load when the transistor is in a switching mode. The voltage blocking capability of the device can be extended beyond $V_{CEO(sus)}$ with proper control of the operating points below the breakdown area and the proper base drive and protection conditions.

The emitter-base breakdown voltage V_{EBO} is a low value, typically 6V, because the emitter layer is highly doped and independent of the operating modes (regions) of the transistor. The emitter-base leakage current with open collector, I_{EBO}, is one or two orders of magnitude higher than I_{CBO}.

There is another limit of the voltage which is tied to the current. It is the secondary breakdown, which appears on the output characteristics as a sudden drop in the V_{CE} voltage at large collector currents (see the dotted curve in Figure 9.20a). We have tried to emphasize that the secondary breakdown does not originate from avalanche breakdown of a pn junction. The reason for the secondary breakdown will be explained in the section on current and power ratings.

The transistor current and power ratings depend on the cross section of the junctions and the rated junction temperature, which is typically 150 °C for Si transistors. The continuous or DC current rating, I_{CM}, depends on the cooling methods and the dissipation powers, and both of them must ensure that the junction temperature does not exceed the rated value. The peak

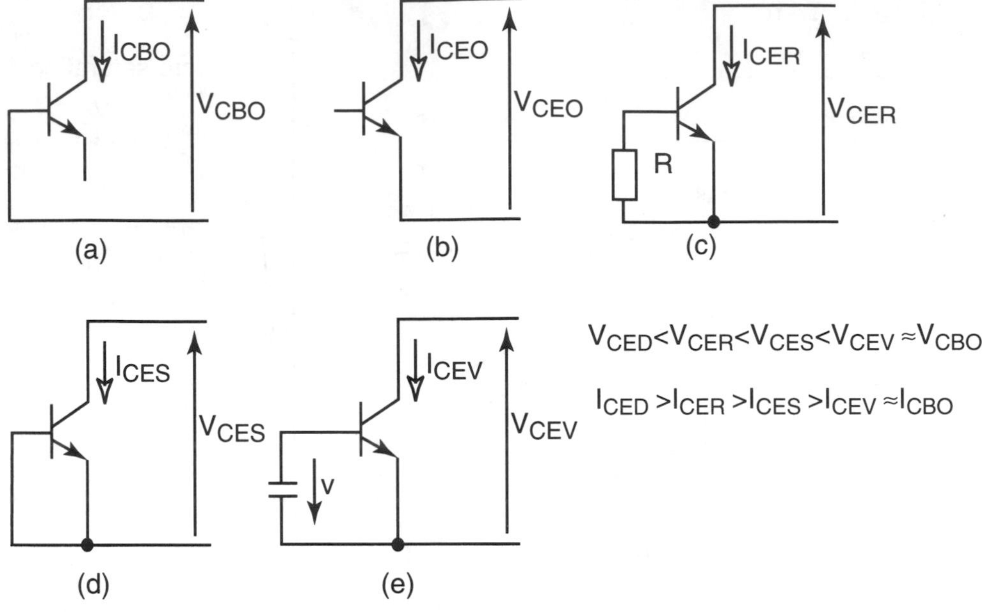

Figure 9.21 Breakdown voltages and leakage currents of bipolar transistors.

or pulsed current rating, I_{CPM}, is limited by the fusing current of the bonding wires within the device packages. The total maximum dissipation power, P_{Dtot}, is also limited and is specified for a case temperature of generally $T_c = 25$ °C. For most practical applications, this case temperature is too low, and therefore at higher values of T_c the parameters have to be derated. The manufacturers provide a derating curve to determine the permissible dissipation at a particular case temperature. (See the "safe operating area" discussion). In a steady-state condition or a DC working mode, the power dissipation of the transistor is equal to the B-E and B-C junction dissipations.

$$P_D = V_{BE}I_E + V_{BC}I_C \cong V_{CE}I_C \qquad (9.12)$$

The limiting value of P_D in the output characteristics is a hyperbola as shown in Figure 9.20a.

The secondary breakdown limit is provided as catalog data, and it is one of the limits in the "safe operating area". The boundary of the secondary breakdown area in the output characteristics of the transistor can be used to determine the secondary breakdown current value, $I_{CS/B}$, depending on the case temperature. The origin of the secondary breakdown effect can be explained on the basis of Figure 9.22. This figure shows a planar structure of a BJT, in which the I_B current flow is illustrated. The base current produces a lateral voltage drop in the high-resistance base region. In the passive region of the base, where there is no carrier injection, the voltage drop can be represented by an ohmic resistance. In the active region, the voltage drop reduces the forward bias voltage across the B-E junction, from the periphery near the base contact to the center of the active region. As a result, the carrier injection falls off from the edge of the base inward. This results in a nonuniform

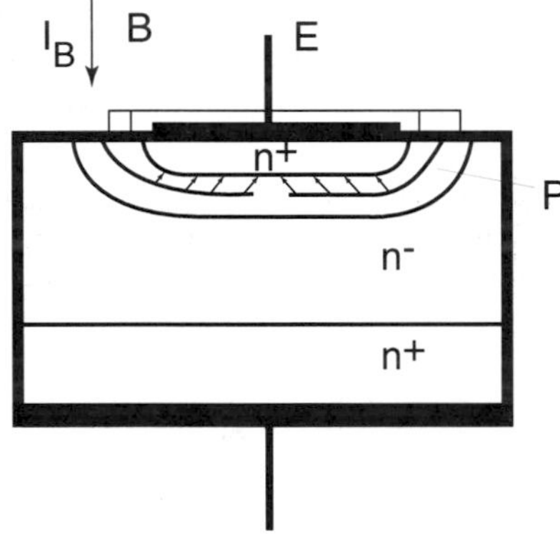

Figure 9.22 Planar structure of an npn BJT.

current distribution, higher current density near the edge, and is known as the current crowding effect. The probability of this effect becomes greater as the cross section of the junction becomes larger and is a function of the current loadability of the transistor. This effect can be produced by the electrical inhomogenieties in the cross section of the crystal in the steady-state condition, or the finite spreading and contraction time for the current conduction through the cross section at the turn-on and turn-off process in switching mode applications. The current crowding areas, where the current density is substantially larger than the surrounding areas and the voltage is common, may cause localized thermal runaway, because the current crowding effect is a regenerative process. In the current crowding areas, there is classic positive feedback in which the power dissipation leads to an increase in temperature, which leads to further increases in power dissipation, and so on, until the device is destroyed. The consequence of the positive feedback

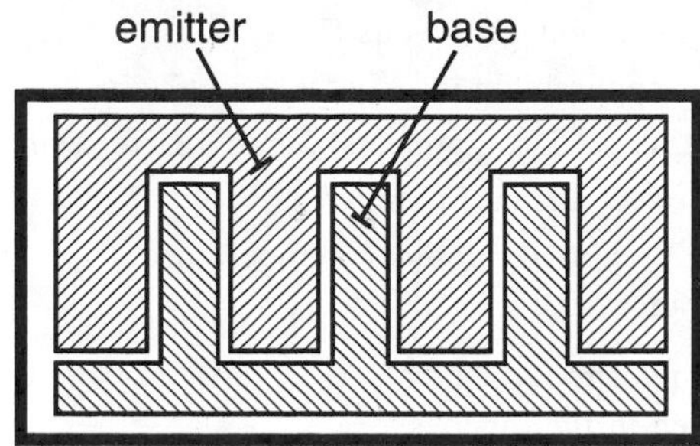

Figure 9.23 Top section view of interdigitated geometry between B and E.

is that local temperature may grow very quickly to unacceptably high values, and therefore the local area can sometimes be melted. Power BJTs have great susceptibility to secondary breakdown (thermal runaway), and therefore one must certainly avoid this effect. To minimize this effect, power transistors are manufactured with an interdigitated geometry between the base and emitter, as shown in Figure 9.23, so as to produce a high periphery-to-area ratio for the emitter.

It is important to note that all of the minority carrier devices, e.g., junction diodes, BJTs, and thyristors, are susceptible to thermal runaway because their resistivity has a negative temperature coefficient.

In addition to the junction temperature, the collector current is limited by the decrease in current gain β, at higher collector current values. Figure 9.24 shows a typical variation in DC current gain β as a function of collector current for different junction temperatures and constant voltage V_{CE}. The gain limit for the continuous collector current rating is that current which corresponds to a minimum acceptable value of β.

9.3 Passive Networks

Karoly Kurutz

General Description

In the area of power electronics, the nonlinear devices described in sections 9.1 and 9.2 have special importance; however, the linear elements, e.g., resistors, capacitors, and inductors also have special functions. Their necessary values can be calculated using the *Standard Handbook for Electrical Engineers* (12th ed., Section 2) published by McGraw-Hill, 1987. These discrete elements can be characterized primarily by their units, e.g., ohms, farads, and henrys, but to select the most preferable element it is necessary to evaluate, from any catalog, their additional properties as well. In other cases, such as fast switching industrial electronic devices, other considerations may influence the operation. In the following, a short summary of the various aspects involved will be provided.

Resistors

Resistors are determined by the following parameters:

- Resistance in ohms (Ω)
- Rated dissipation in watts (W)
- Tolerance (%)
- Covering and coating material
- Guaranteed protection
- Rated insulation in volts (V)
- Operating temperature range (°C) and temperature coefficient (ppm/°C)

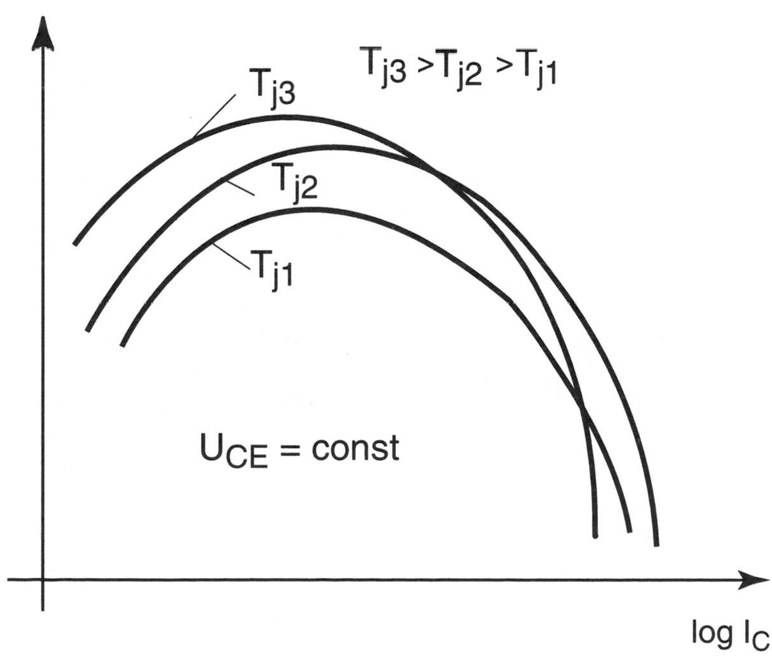

Figure 9.24 Common-emitter DC current gain as a function of collector current.

In the case of adjustable resistances such as potentiometers and trimmers with 3 terminals, linearity (%) is also an important issue. Additional data for higher-accuracy of resistors are the insulation resistance and in the case of wire wound resistors, the inductance.

For high-power (>2 W) resistors in small sizes, suitable for use under harsh environmental conditions, there are elements available with silicon or vitreous enamel coatings, which are wound onto a high-purity ceramic substrate. The wire terminals are welded to the caps on both sides to ensure a safe connection. Higher-power resistors use ceramic covers or aluminum housings. Their rated dissipation is up to 300 W, with a tolerance of 5%, temperature coefficient between 75 and 300 ppm/°C, an insulation resistance up to 10^9 Ω, and an operating temperature range from −55 to +300°C. In power electronic systems, resistors are employed in snubber networks connected in parallel with semiconductor devices in order to protect them.

Resistors used in lower-power (<2 W) applications, e.g., in control networks, are produced on ceramic bodies by carbon film or metal film methods. A wide range of these high-stability resistors are suitable for precision and semi-precision applications. An extremely low current-noise level with a low temperature coefficient and a close tolerance make them reliable in electronic circuits. A color code can be baked onto the body, and important data can be written on the coatings which protect the resistor against abrasion and chipping. These resistors are characterized in an ambient temperature range from −55 to 120°C by a 5% tolerance, and a temperature coefficient of 50 to 200 ppm/°C with 1000 Ω insulating resistance.

In the case of current controlled and protected power electronic devices, shunt resistances are also used. Their most important properties are the extremely low temperature coefficients and high accuracy. Their data are the rated current and the voltage in mV-s.

Capacitors

Capacitors are frequently used in power electronics. The following list outlines some of the ways in which they are employed:

- To store peak electrical energy in pulsing circuits to maintain the voltage essentially constant.
- To store electrical energy for forced commutation in some thyristor circuits.
- To delay the voltage rise on long turn-off-time semiconductors by connecting R-C components to their terminals.
- To transmit impulses between circuits on different voltage levels in control units.

The capacitance is essentially proportional to the area of the two conductor surfaces (A) facing one another and the dielectric factor (ϵ) of the insulator material, and is inversely proportional to the distance between them.

The important data are

- Capacitance in farads (F) (or rather in 10^{-6} F = μF, 10^{-9} F = nF, or 10^{-12} F = pF)
- The rated voltage in volts (V)
- Their applicability in only DC or AC circuits
- In case of AC components, the loss factor (tan δ)
- The rated current in amperes (A)
- The permitted ambient temperature (°C) and their tolerance

In some high-frequency cases, their resistance and inductance are also given in the catalogs.

Capacitors used only for DC represent nonlinear elements, which can be manufactured with a special polar aluminum electrolytic material between the plates. This material creates a very thin aluminum-oxide insulating sheet on the positive metal, enabling it to produce a high capacitance in a relatively small volume. Their maximum rated voltage is generally 400 V. They have a guaranteed lifetime of 10^5 hours and a minimum leakage current. In the presence of an AC ripple voltage whose amplitude is less than the maximum rated voltage for the capacitor, the maximum value of tan δ at 100 Hz and 20°C is 0.2. The maximum capacitance is dependent upon rated voltage, and if the rated voltage is small, the capacitance may be extremely high, e.g., in the mF range. Tantalum capacitors used only for DC have a maximum rated voltage of 35 V and a relatively high capacitance up to 100 μF in spite of their very small sizes. Both products can be used in an ambient temperature range between −50 to +85°C.

Because of their self-healing ability, when an overvoltage occurs during operation, the metallic surface evaporates where the break-through takes place. In case of higher overvoltages, some of these capacitors incorporate a nonresettable safety device to disconnect the internal electrical connections in the event that excessive pressure is built up inside of the metal housing. These capacitors are manufactured with a tolerance of 10% and a maximum capacitance of 20 μF at a rated AC voltage of 450 V. Their maximum voltage rise/fall time may be 20 V/μs, over a temperature range of −25 to +85°C.

Other polypropylene capacitors are manufactured with a very low-loss dielectric material, which makes them suitable for continuous use at high AC voltages. Because of their fast impulse-rise times, they exhibit excellent performances at high frequencies. These capacitors are normally coated with a hard epoxy resin which is water repellent and flame retardant. Polystyrene capacitors operate at a lower temperature than polyester and polycarbonate types but they are smaller in size and are manufactured with smaller tolerance.

For relatively low-capacitance applications (in the pF range) ceramic capacitors are used. They are useful at relatively high frequencies up to 1 MHz, rated voltages up to 15 kV and possess a low tan δ. Higher capacitances in the nF range can be reached with rated voltages down to 100 V. Some of them are manufactured as trimmer capacitors which enable them to be tuned.

Inductances

Inductances in power electronic circuits are normally used for the following purposes:

- To suppress harmonics in DC networks.
- To ensure the equal division of dynamic currents in parallel connected branches.
- To reduce the penetration of higher harmonics into the supply network, caused by power electronic devices.
- To store magnetic energy for any purpose, e.g., in case of resonant circuits.
- To decrease current rise.

Since the permeability (μ) of iron is at least four grades higher than air or most other materials, for concentrated inductance the coils are placed around iron cores on bobbins. For inductances operating in DC circuits, it is necessary to place an airgap perpendicular to the magnetic flux lines into the core to avoid DC biasing. AC components cause eddy current and hysteresis loss in the iron core, and therefore inductors used in this application have to be laminated. The sheet thickness of most silicon alloyed iron should be between 0.35 to 0.1 mm to reduce losses. Some sheets are grain-oriented to achieve more determined magnetic saturation, e.g., in magnetic amplifiers. Some of these grain-oriented iron cores are manufactured in wound form with cut and polished iron surfaces. If higher harmonics than the supply frequency exist, ferrites (powdered-cores) are used.

Inductive coils are determined by:

- Their inductance in henrys (H)
- Resistance in ohms (Ω)
- Rated voltage of the insulation in volts (V)
- Rated current in amperes (A) in accordance with the material and cross section of the coil conductor

It is also important to note that in special high-frequency cases, the capacitance between the terminals of the coil may affect the behavior of the circuit.

Inductances are seldom manufactured for commercial purposes because of the many special requirements; however, iron cores in the form of E and I shapes, or pot form ferrites with and without an airgap and coil bodies (bobbins) made from different insulator materials in sizes and forms fitting to the iron core, are available in many forms.

10

Power MOSFETs

10.1 Introduction 218
10.2 Static Characteristics 220
Breakdown Voltage • On-Resistance
10.3 Dynamic Characteristics 224
Switching and Transient Response • Gate Charge • dV/dt Capability
10.4 Applications 227
Portable Electronics and Wireless Communication • Automotive

Vrej Barkhordarian
International Rectifier

10.1 Introduction

The metal-oxide-semiconductor field-effect transistor (MOSFET) is the most commonly used active device in very large scale integrated (VLSI) circuits. Figure 10.1 shows the device schematic, current-voltage characteristics, transfer characteristics and device symbol for a MOSFET. It is a lateral device and though very suitable for integration into integrated circuits, it has severe limitations at high power levels. The power MOSFET design is based on the original field-effect transistor and, since its invention in the early 1970s, has gone through several evolutionary steps. The processing of power MOSFETs is very similar to that of today's VLSI circuits though the device geometry is significantly different from the design used in these circuits. Power MOSFETs are commonly used as switches in power electronic applications.

The invention of the power MOSFET was partly driven by the limitations of bipolar power transistors which, until recently, were the devices of choice in power electronics applications. Although it is not possible to define absolutely the operating boundaries of a power device, we will loosely refer to the power device as any device which is capable of switching at least 1A. The bipolar power transistor is a current-controlled device and a large base drive current as high as one fifth of the collector current is required to keep the device in the on state. Also, higher reverse base drive currents are required to obtain fast turn-off. Despite the very advanced state of manufacturability and lower costs of bipolar power transistors, these limitations have made the base drive circuit design more complicated and hence more expensive. There are two further limitations to the bipolar power transistor. First, both electrons and holes contribute to conduction in BJTs. Presence of holes with their higher carrier lifetime causes the switching speed to be several orders of magnitude slower than for a power MOSFET of similar size and voltage rating. Secondly, the BJTs suffer from thermal runaway. The forward voltage drop of a BJT decreases with increasing temperature causing diversion of current to a single device when several devices are paralleled. Power MOSFETs, on the other hand, are majority carrier devices with no minority carrier injection. They are superior to the BJTs in high-frequency applications where switching power losses are important and can withstand simultaneous application of high current and voltage without undergoing destructive failure due to second breakdown. Power MOSFETs can also be paralleled easily since the forward voltage drop increases with increasing temperature, ensuring an even distribution of current among all components. However, at high breakdown voltages ($>\sim$200V) the on-state voltage drop of the power MOSFET becomes higher than that of a similar size bipolar device with a similar voltage rating, making it more attractive to use the bipolar power transistor at the expense of worse high-frequency performance. Figure 10.2 shows the present current-voltage limitations of power MOSFETs and BJTs. New materials, structures and processing techniques are expected to push these limits out over time. A relatively new device which combines the high-frequency advantages of the MOSFET with the low on-state voltage drop of high voltage BJTs is the insulated-gate-bipolar-transistor (IGBT).

MOSFETs used in integrated circuits are lateral devices with gate, source and drain all on the top of the device and with current flow taking place in a path parallel to the surface. Although this design lends itself to integration, it is not suitable for discrete power device applications due to large distances required between source and drain in order to maintain isolation. Having all three terminals at the upper surface makes the metallization and isolation of terminals more complicated from the processing point of view. The vertical double diffused MOSFET solves this problem by using the substrate of the device as the drain terminal. Figure 10.3 shows the schematic diagram and the circuit symbol for an n-channel power MOSFET. When a positive bias greater than the threshold voltage is applied to the gate, the silicon surface in the channel region is inverted and a current starts to flow between the source and drain. For gate voltages of less than V_{th}, no surface inversion occurs in the channel and the device remains in the off-state. The current in this device flows horizontally

0-8493-8343-9/97/$0.00+$.50

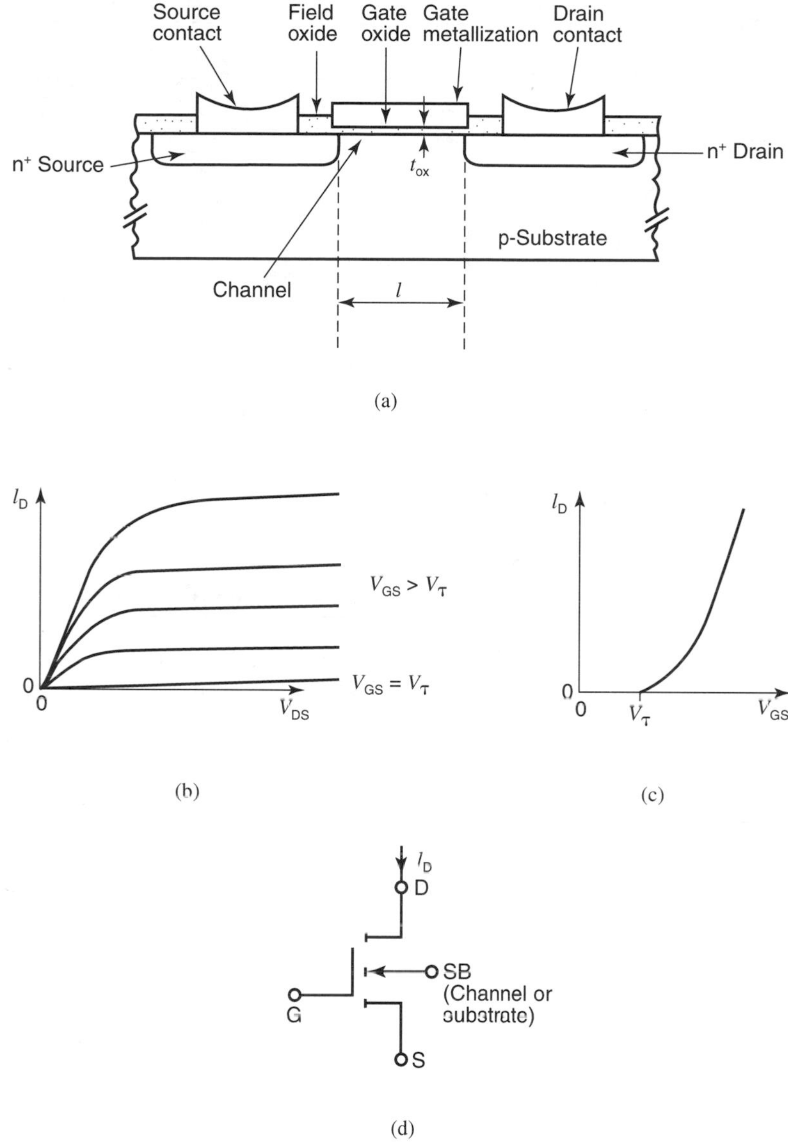

Figure 10.1 (a) Schematic diagram, (b) current-voltage characteristics, (c) transfer characteristics, and (d) device symbol for an n-channel enhancement mode MOSFET.

along the inverted channel first and then vertically between the drain and source. The term "double-diffused" refers to the two consecutive ion implantation steps using the poly as a mask. For an n-channel device, the regions formed by double implant and subsequent diffusion are first p-type to define the channel and then n-type to define the source. The p-body implant is performed in a separate step. The terms "body drift" and "body-drain" diodes are used interchangeably to denote the p-n junction formed by this p-body implant and the drift region.

Figure 10.4 shows the physical origin of the parasitic components in an n-channel power MOSFET. The parasitic JFET appearing between the two body implants restricts current flow when the depletion widths of the two adjacent body diodes extend into the drift region with increasing drain voltage. Poly linewidth and the epi layer resistivity under the poly are two important design parameters for minimizing the JFET effect. The parasitic BJT can make the device susceptible to unwanted device turn-on and premature breakdown. The base resistance R_B has to be minimized through careful design of the doping and distance under the source region. These two components and the parasitic resistances are discussed further in the next sections. There are several parasitic capacitances associated with the power MOSFET as shown in Figure 10.4. C_{GS} is the capacitance due to the overlap of the source and the channel regions by the polysilicon gate and is independent of applied voltage. G_{GD} is made up of two parts. The first part is the capacitance associated with the overlap of the polysilicon gate and the silicon underneath in the JFET region. The second part is the capacitance associated with the depletion region immediately under the gate. C_{GD} is a nonlinear function of voltage and is discussed further

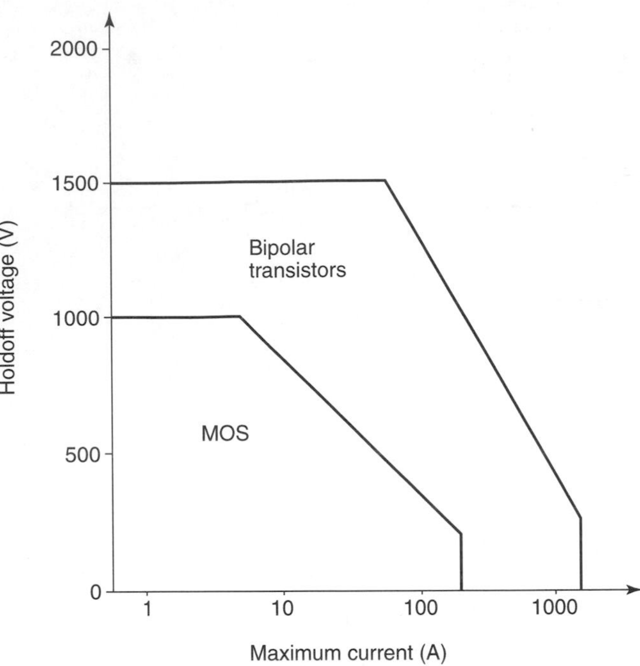

Figure 10.2 Current-voltage limitations of MOSFETs and BJTs.

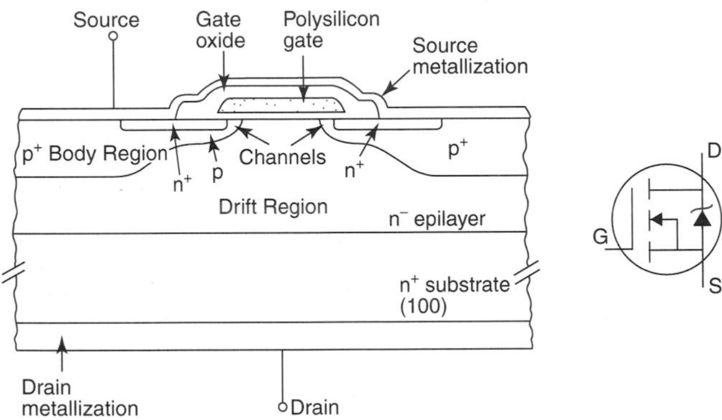

Figure 10.3 Schematic diagram for an n-channel power MOSFET and the device symbol.

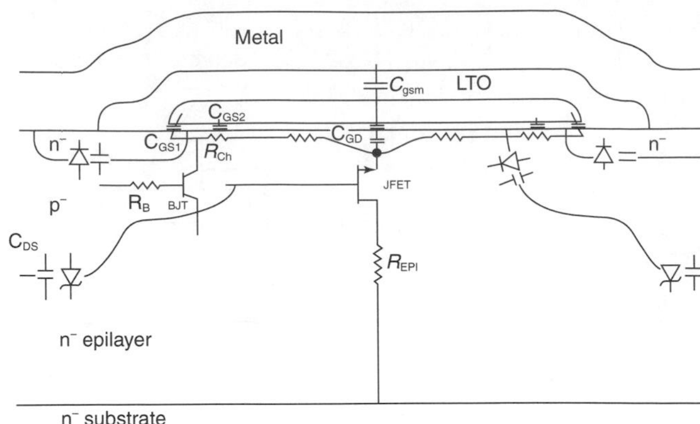

Figure 10.4 The origin of parasitic components for a power MOSFET.

in the "Dynamic Characteristics" section. Finally, C_{DS} is the capacitance associated with the body-drift diode and varies inversely with the square root of the drain-source bias.

There are currently two designs of power MOSFETs. These are usually referred to as the planar and the trench designs. The planar design has already been introduced in the schematics of Figures 10.3 and 10.4. Two variations of the trench power MOSFET are shown in Figure 10.5. The V-groove device is fabricated by etching a groove in the silicon after the double diffusion step. The use of an anisotropic etch results in the sides of the groove to be at an angle of 54.7° to the surface of the wafer. Etching stops when the groove sides, which are planes, reach each other. The gate oxide and gate poly or metallization are then grown in the groove followed by the source metallization. Current crowding at the apex of the V groove reduces current handling capability. In a truncated V-groove design, the anisotropic etch is stopped before this point is reached. The trench technology has the advantage of higher cell density but is more difficult to manufacture compared with the planar device.

10.2 Static Characteristics

One of the important features of the power MOSFET is the very high input impedance which simplifies the gate drive circuitry

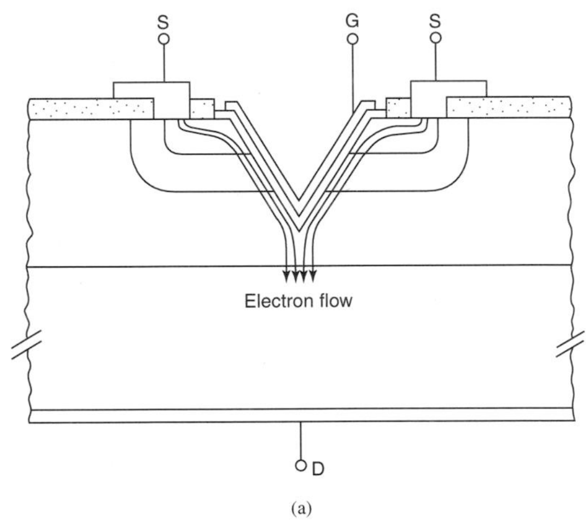

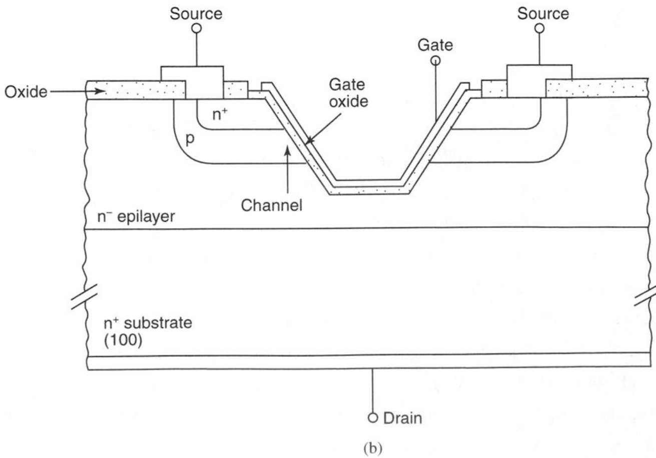

Figure 10.5 Schematic diagram of (a) V-groove trench MOSFET showing the current crowding at the apex and (b) truncated V-groove design.

and reduces the cost. It is a voltage-controlled device with no gate current flow during operation. Figure 10.6 shows I–V characteristics of an enhancement mode (normally off) power MOSFET. Data sheets contain typical graphs which can be used to determine if the device is in the fully on state or in the constant-current region for a given value of gate bias and drain current. Temperature effect on threshold voltage (about 6mV/°C reduction) and the difference between typical values of parameters and the maximums should be taken into account.

Breakdown Voltage

This is the drain voltage at which the reverse-biased body-drift diode breaks down and a significant current starts to flow between the source and drain by the avalanche multiplication process, while the gate and source are shorted together. Breakdown voltage, BV_{DSS}, is normally measured at a drain current of 250μA. For drain voltages below BV_{DSS} and with no bias on the gate, no channel is formed under the gate at the surface and the drain voltage is entirely supported by the reverse-biased body-drift p-n junction. There are two related phenomena which can occur in poorly designed and processed devices. These are punch-through and reach-through.

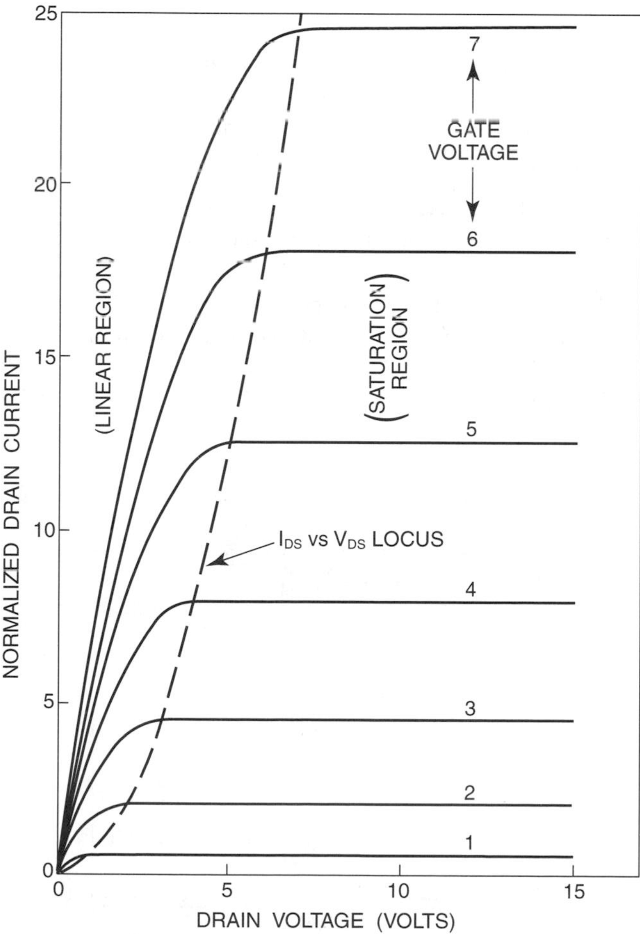

Figure 10.6 Current-voltage characteristics of a power MOSFET.

Punch-through is observed when the depletion region on the source side of the body-drift p-n junction reaches the source region at drain voltages below the rated avalanche voltage of the device. This provides a current path between source and drain and causes a soft breakdown characteristic as shown in Figure 10.7. The leakage current flowing between source and drain is denoted by I_{DSS}. Careful selection and optimization of the doping profile used in the fabrication of a power MOSFET is therefore very important. Figure 10.8 shows a typical diffusion profile for a power MOSFET. The surface concentration of the body diffusion and the channel length (distance between the two p-n junctions formed by the source diffusion and the channel diffusion) will determine whether punch-through will occur or not. There are trade-offs to be made between on-resistance R_{dson} which requires shorter channel lengths and punch-through avoidance which requires longer channel lengths. An approximate equation giving the depletion region width as a function of silicon background doping is given by:

$$W \approx \sqrt{\frac{4\epsilon_s KT}{q^2 N_A} \ln\left[\frac{N_A}{n_i}\right]} \tag{10.1}$$

where ϵ_s is semiconductor permittivity, K is Boltzmann's constant, T is temperature in K, q is electronic charge, N_A is background doping and n_i is the intrinsic carrier density.

Also, higher channel implant dose is beneficial from the punch through point of view since depletion width will be smaller, but the R_{dson} will suffer through reduced carrier mobility. The design of the doping profile involves choosing channel and source implant doses, diffusion times and temperatures that give a desired thresh old voltage while simultaneously minimizing R_{dson} and I_{DSS}. Optimizing these performance parameters with manufacturability in mind is one of the challenges of power MOSFET design.

The reach-through phenomenon, on the other hand, occurs when the depletion region on the drift side of the body-drift p-n junction reaches the epilayer-substrate interface before avalanching takes place in the epi. Once the depletion edge enters

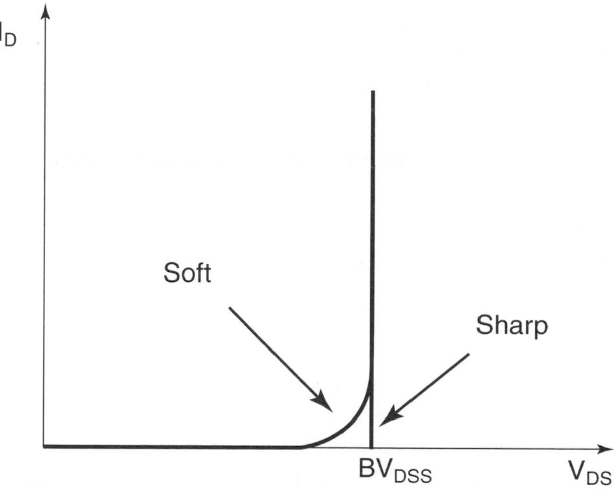

Figure 10.7 Breakdown characteristics of a power MOSFET showing the ideal (sharp) and non-ideal (soft) behaviors.

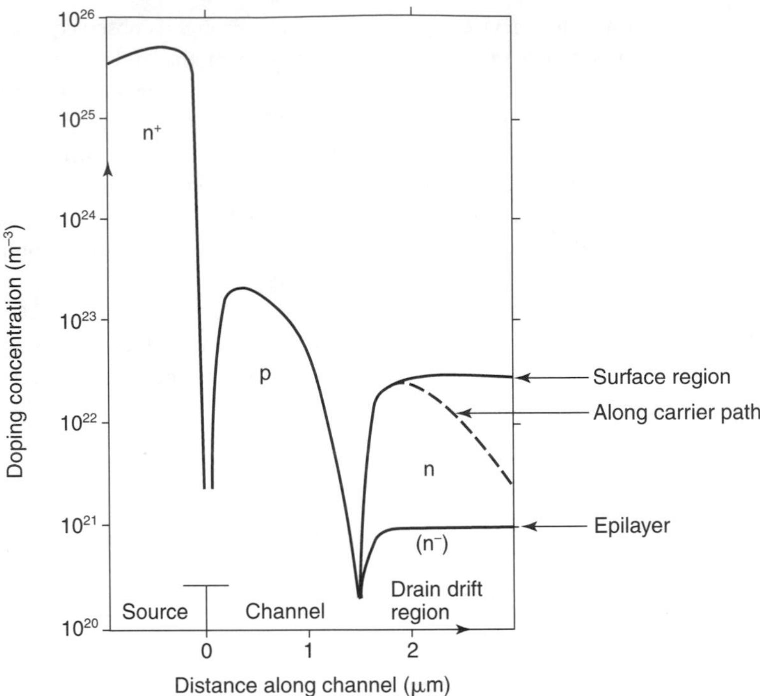

Figure 10.8 Typical doping profile of a power MOSFET, in a direction parallel to the device surface. Threshold voltage is determined by the peak carrier concentration in the channel region.

the high carrier concentration substrate, a further increase in drain voltage will cause the electric field to quickly reach the critical value of 2×10^5 V/cm at which avalanching begins.

Other factors that affect the breakdown voltage of power MOSFETs for a given epitaxial layer include termination design, cell spacing (poly line width) and curvature of the body diode depletion region in the epi which is a function of diffusion depth. Power MOSFETs are designed such that avalanche breakdown occurs in the active area first.

On-Resistance

The on-state resistance of a power MOSFET is made up of several components as shown in Figure 10.9.

$$R_{dson} = R_{source} + R_{ch} + R_A + R_J + R_D + R_{sub} + R_{wcml} \quad (10.2)$$

where

R_{source} = Source diffusion resistance
R_{ch} = Channel resistance
R_A = Accumulation resistance
R_J = The "JFET" component-resistance of the region between the two body regions
R_D = Drift region resistance
R_{sub} = The substrate resistance. Wafers with resistivities of up to 20 mΩ-cm are used for high-voltage devices and less than 5mΩ-cm for low-voltage devices.
R_{wcml} = Sum of Bond **W**ire resistance, **C**ontact resistance between the source and drain metallization and the silicon, **M**etallization resistance and **L**eadframe contributions. These are normally negligible in high-voltage devices but can become significant in low-voltage devices.

Figure 10.10 shows the relative importance of each of the components to R_{dson} over the voltage spectrum. As can be seen, at high voltages the R_{dson} is dominated by epi resistance and the JFET component. This component is higher in high-voltage devices due to the higher resistivity or lower background carrier concentration in the epi. At lower voltages, the R_{dson} is dominated by the channel resistance and the contributions from the metal to

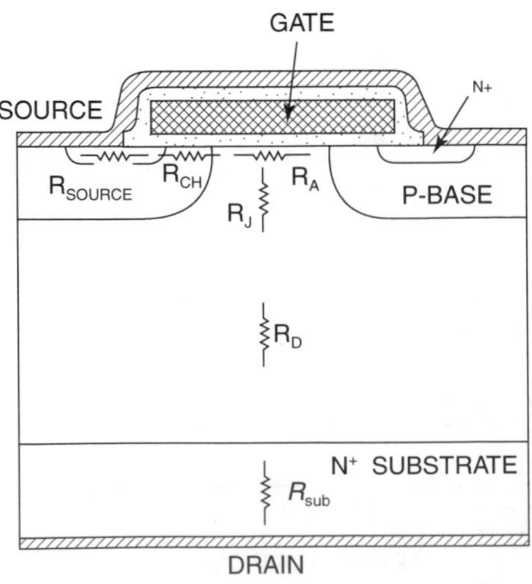

Figure 10.9 The origin of the internal resistances in a power MOSFET.

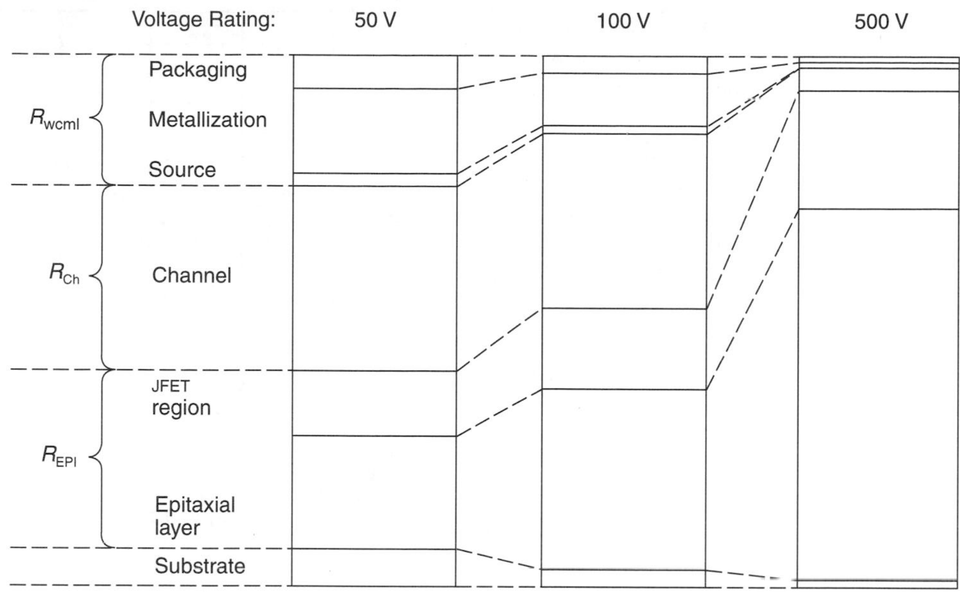

Figure 10.10 Relative contributions to R_{dson} in devices with different voltage ratings.

semiconductor contact, metallization, bond wires and leadframe. The substrate contribution becomes more significant for lower breakdown voltage devices.

Transconductance

This parameter is a measure of the sensitivity of drain current to changes in gate-source bias and is defined as:

$$g_{fs} = \left(\frac{\Delta I_D}{\Delta V_{gs}}\right) V_{ds} \text{ constant} \quad (10.3)$$

i.e., the gradient of the I_d vs. V_{gs} graph. In the saturation region, g_{fs} is given by:

$$g_{fS} = \mu C_{ox} \frac{W}{L} (V_{gs} - V_{th}) \quad (10.4)$$

This parameter is normally quoted for a V_{gs} that gives a drain current equal to about one half of the maximum current rating value and for a V_{DS} that ensures operation in the constant current region. With mobility μ fixed for a given semiconductor, the design parameters influencing transconductance of a MOSFET are gate width W, channel length L, and gate oxide thickness t_{ox} and hence C_{ox}. Gate width is the total polysilicon gate perimeter of the cellular structure and increases in proportion to the active area as the cell density increases. The cell density has increased over the years from around half a million per square inch in 1980 to around eight million for planar MOSFETs and around 12 million for the trench technology at the present time. The limiting factor for even higher cell densities is the photolithography process control and resolution which allows contacts to be made to the source metallization in the center of the cells.

Reduced channel length is beneficial to both g_{fs} and on-resistance, with punch-through as a trade-off. The lower limit of this length is set by the ability to control the double-diffusion process and is around 1–2μm today. Finally, reductions in gate oxide thickness give higher C_{ox} and higher g_{fs}. The reduction in oxide thickness will reduce V_{th} unless channel implant dose is increased which in turn will cause a higher R_{dson}. Ultimately, the lower limit of t_{ox} is set by the maximum gate-source voltage rating. This is ±30V for high-voltage devices and ±20V for lower-voltage logic-level devices used in portable electronic applications.

Threshold Voltage

This is defined as the minimum gate electrode bias required to strongly invert the surface under the poly and form a conducting channel between the source and the drain regions. V_{th} is usually measured at a drain-source current of 250μA. A value of 2–4V for high-voltage devices with thicker gate oxides and logic-compatible values of 1–2V for lower-voltage devices with thinner gate oxides are common. With power MOSFETs finding increasing use in portable electronics and wireless communications where battery power is at a premium, the trend is towards lower values of R_{dson} and V_{th}. Gate oxide quality and integrity become major issues as the gate oxide thickness is reduced to achieve lower V_{th}. An approximate expression for V_{th} is given by:

$$V_{th} \approx \frac{\sqrt{4\epsilon_s KTN_A \ln(N_A/n_i)}}{(\epsilon_{ox}/t_{ox})} + \frac{2KT}{q}\ln(N_A/n_i) \quad (10.5)$$

where ϵ_{ox} and t_{ox} are oxide permittivity and thickness and the other parameters are defined in Equation 10.1.

Processing methods used and their influence on the chemistry of the silicon surface have pronounced effects on V_{th}. Fixed and mobile surface and interface charges as well as charges in the gate oxide act to change the value of V_{th} from the intended value. Therefore, control of these charges in the process is necessary for obtaining consistent V_{th} values in production. Also, the presence of mobile charges away from the gate oxide and oxide/silicon interface may find their way to the device surface over the lifetime of the device and cause a gradual shift in V_{th}. For example, sodium ions in the low-temperature oxide (LTO) or in the metallization can cause a shift in V_{th} by changing the charge distribution at the interface. Accelerated life-tests are used by manufacturers to evaluate new processes and also to monitor V_{th} shift in production. Monitoring and control of contamination in the clean room equipment are routinely carried out by capacitance-voltage measurements of test diodes.

In real devices, V_{th} is altered by the unequal metal and semiconductor work functions. Denoting the barrier height between the metal and silicon oxide as ϕ_B, the work function difference is given by:

$$q\phi_{ms} = q\phi_B + q\chi_o - (q\chi + Eg/2 + q\psi_B) \tag{10.6}$$

where ψ_B is the potential difference between the intrinsic and Fermi levels in the semiconductor; χ and χ_o are the semiconductor and oxide electron affinities and Eg is the semiconductor band-gap energy.

Taking into account this effect and also the various fixed and mobile charges that may alter the value of V_{th} from that given above, the expression for V_{th} becomes:

$$V_{th} = \phi_{ms} + 2\psi_B - \left(\frac{Q_s + Q_{ss} + Q_I + Q_{FC}}{C_{ax}}\right) \tag{10.7}$$

where

Q_s = Surface charge, is a function of surface potential and determines channel conductivity

Q_{ss} = Interface state charge (typically $10^{10} - 10^{12}$ cm^{-2}), caused by dangling bonds at the semiconductor surface. These can charge and discharge with changes in the surface potential.

Q_I = Charge due to mobile ions in the oxide

Q_{FC} = Fixed surface charge at the silicon-oxide interface

It is worth mentioning that the success of silicon devices lies partly in the low density of these interface states which is due to the existence of native oxide in silicon as opposed to other semiconductors such as GaAs where such a native oxide does not exist and oxide layers have to be deposited with several orders of magnitude higher interface state densities.

Diode Forward Voltage (V_F or V_{SD})

This is the guaranteed maximum forward drop of the body-drain diode at a specified value of source current. Figure 10.11 shows a typical I–V characteristic for this diode at two temperatures. p-Channel devices have higher values of V_F due to the higher contact resistance between metal and p-silicon compared with n-type silicon. Maximum values of 1.6V for high-voltage devices (>100V) and values of 1.0V for low-voltage devices (<100V) are common.

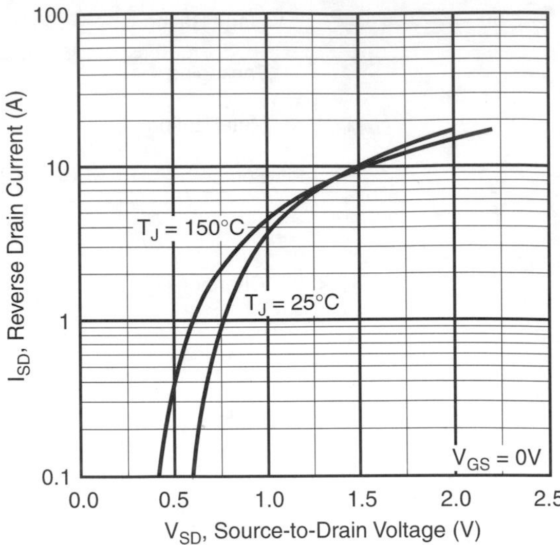

Figure 10.11 Typical source-drain (body) diode forward voltage characteristics.

Power Dissipation

The maximum allowable power dissipation which will raise the die temperature to the maximum allowable when the case temperature is held at 25°C is an important parameter and is given by

$$P_d = \left(\frac{T_{jmax} - 25}{R_{thJC}}\right) \tag{10.8}$$

where T_{jmax} is the maximum allowable temperature of the p-n junction in the device (normally 150°C or 175°C) and R_{thJC} is the junction to cause thermal impedance of the device.

10.3 Dynamic Characteristics

Switching and Transient Response

When the MOSFET is used as a switch, its basic function is to control the drain current by the gate voltage. Figure 10.12 shows the transfer characteristics and an equivalent circuit model often used for the analysis of MOSFET switching performance. For a detailed discussion of this topic see Chapter 4 in Grant and Gower (1989). The following is a summary of the important points.

The switching performance of a device is determined by the time required to establish voltage changes across capacitances and current changes in inductances. R_G is the distributed resistance of the gate and is approximately inversely proportional to active area. Values of around 20 Ω-mm^2 are common for the product of R_G and active area for polysilicon gates. L_S and L_D are source and drain lead inductances and are around a few tens of nH. The physical origin of the capacitances C_{GS}, C_{GD}, and C_{DS} were discussed in the introduction of this chapter regarding the device schematic shown in Figure 10.4. The typical values of input (C_{iss}), output (C_{oss}) and reverse transfer (C_{rss}) capacitances given in the data

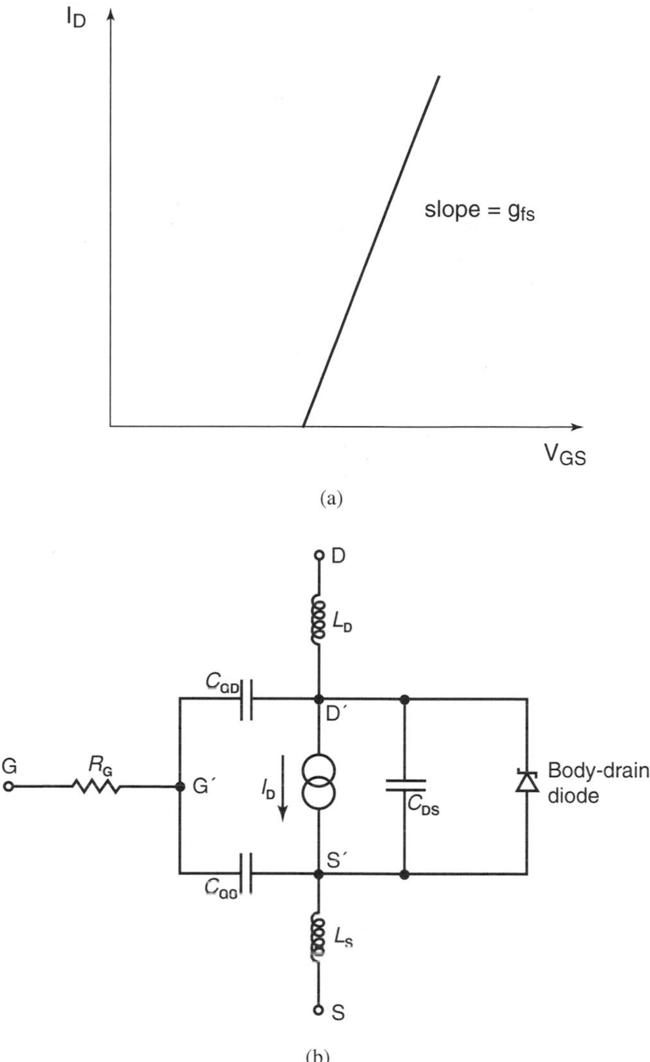

Figure 10.12 (a) Transfer characteristics and (b) an equivalent circuit diagram showing the MOSFET parasitic components that have the greatest effect on switching speed.

sheets are used by circuit designers as a starting point in determining circuit component values. The data sheet capacitances are defined in terms of the equivalent circuit capacitances as:

$$C_{iss} = C_{GS} + C_{GD}, \quad C_{DS} \text{ shorted}$$

$$C_{rss} = C_{GD}$$

$$C_{oss} = C_{DS} + C_{GD}$$

The gate-to-drain capacitance C_{GD} is a nonlinear function of voltage and is the most important parameter since it provides a feedback loop between the output and the input of the circuit. C_{GD} is also called the Miller capacitance since it causes the total dynamic input capacitance to become greater than the sum of the static capacitances.

Figure 10.13 shows a typical switching time test circuit. Also shown are the components of the rise and fall times with reference

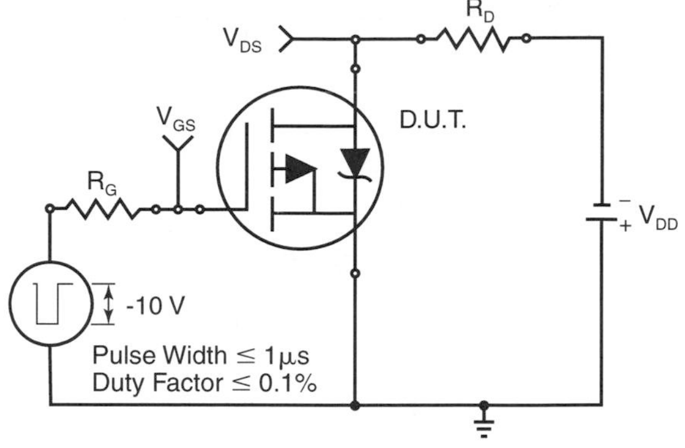

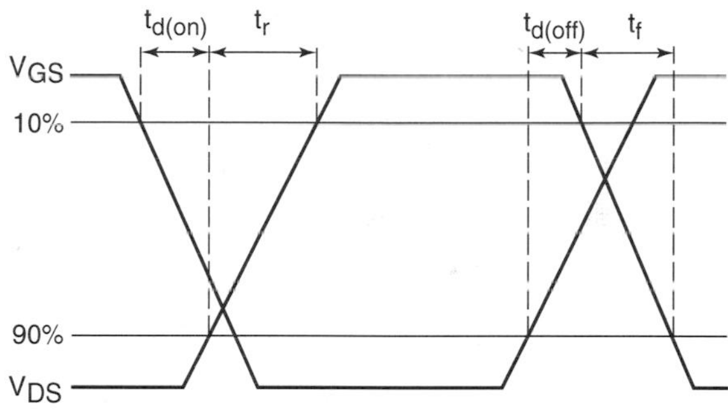

Figure 10.13 Switching time test circuit and resulting V_{GS} and V_{DS} waveforms.

to the V_{GS} and V_{DS} waveforms. Turn-on delay, $t_{d(on)}$, is the time taken to charge the input capacitance of the device before drain current conduction can start. Similarly, turn-off delay $t_{d(off)}$ is the time taken to discharge the capacitance after the gate is switched off.

Gate Charge

Although input capacitance values are useful, they do not provide accurate results when comparing the switching performances of two devices from different manufacturers. Effects of device size and transconductance make such comparisons more difficult. A more useful parameter from the circuit design point of view is the gate charge rather than capacitance. Most manufacturers include both parameters on their data sheets. Figure 10.14 shows a typical gate charge waveform and the test circuit. When the gate is connected to the supply voltage, V_{GS} starts to increase until it reaches V_{th}, at which point the drain current starts to flow and the C_{GS} starts to charge. During the period t_1 to t_2, C_{GS} continues to charge, the gate voltage continues to rise and the drain current rises proportionally. At time t_2, C_{GS} is completely charged and the drain current reaches the predetermined current I_D and stays constant while the drain voltage starts to fall. With reference to the equivalent circuit model of the MOSFET shown in Figure 10.14, it can be seen that with C_{GS} fully charged at t_2, V_{GS} becomes

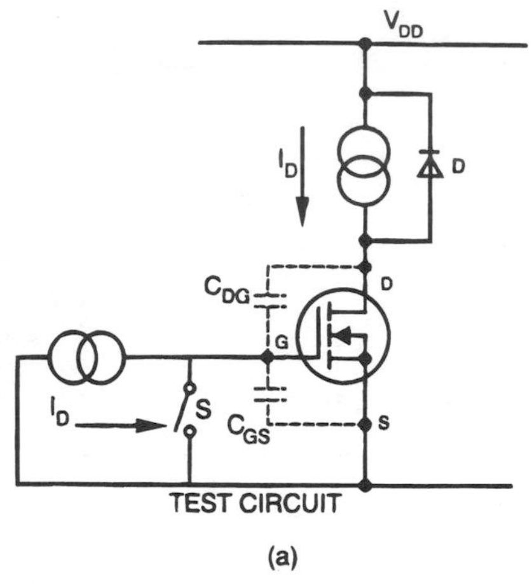

(a)

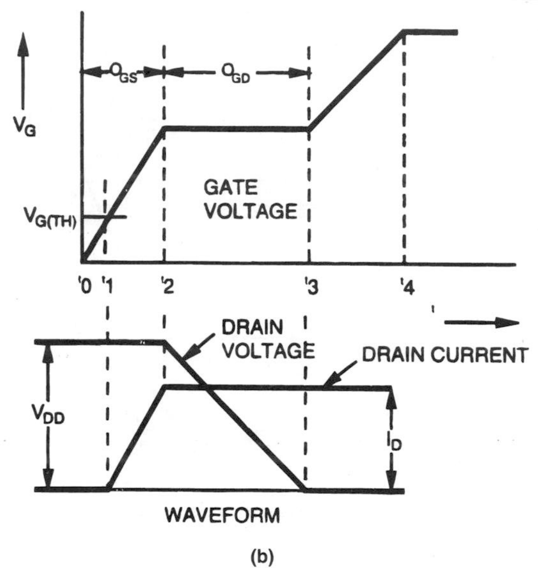

(b)

Figure 10.14 (a) Gate charge test circuit and (b) resulting gate and drain waveforms.

constant and the drive current starts to charge the Miller capacitance C_{GD}. This continues until time t_3. Note that the charge time for the Miller capacitance is larger than that for the gate to source capacitance C_{GS}, due to the rapidly changing drain voltage between t_2 and t_3 (current $= C\ dV/dt$). Once both of the capacitances C_{GS} and C_{GD} are fully charged, the gate voltage V_{GS} starts increasing again until it reaches the supply voltage at time t_4. The gate charge ($Q_{GS} + Q_{GD}$) corresponding to time t_3 is the bare minimum charge required to switch the device on. Good circuit design practice dictates the use of a higher gate voltage than the bare minimum required for switching and therefore the gate charge used in the calculations is Q_G corresponding to t_4.

The advantage of using gate charge is that the designer can easily calculate the amount of current required from the drive circuit to switch the device on in a desired length of time; since $Q = CV$ and $I = C\ dV/dt$ then Q = time × current. For example, a device with a gate charge of 20nC can be turned on in 20μs if a current of 1mA is supplied to the gate or it can turn on in 20ns if the gate current is increased to 1A. These simple calculations would not have been possible with input capacitance values.

dV/dt Capability

This is also called the peak diode recovery and is defined as the maximum rate of rise of drain-source voltage allowed. If this rate is exceeded then the voltage across the gate-source terminals may become higher than the threshold voltage of the device, forcing the device into the current conduction mode and under certain conditions a catastrophic failure may occur. There are two possible mechanisms by which a *dV/dt* induced turn-on may take place. Figure 10.15 shows the equivalent circuit model of a power MOSFET, including the parasitic BJT. The first mechanism of *dv/dt* induced turn-on becomes active through the feedback action of the gate-drain capacitance C_{GD}. When a voltage ramp appears across the drain and source terminals of the device, a current I_1 flows through the gate resistance R_G by means of the gate-drain capacitance C_{GD}. R_G is the total gate resistance in the circuit and the voltage drop across it is given by:

$$\begin{aligned} V_{GS} &= I_1 R_G \\ &= R_G C_{GD}\left(\frac{dV}{dt}\right) \end{aligned} \tag{10.9}$$

When the gate voltage V_{GS} exceeds the threshold voltage of the device V_{th}, the device is forced into conduction. The *dV/dt* capability for this mechanism is thus set by

$$\left(\frac{dV}{dt}\right) = \frac{V_{th}}{R_G C_{GD}} \tag{10.10}$$

It is clear that low V_{th} devices are more prone to *dV/dt* turn-on. The negative temperature coefficient of V_{th} is of special importance in applications where high-temperature environments are present. Also, gate circuit impedance has to be chosen carefully in order to avoid this effect. C_{GD} is an internal device parameter and is determined by the overlap area between poly gate and silicon and gate oxide thickness. Higher gate oxide thicknesses reduce C_{GD} and also increase V_{th}, both advantageous to *dV/dt* rating, as long as the higher V_{th} is acceptable in the application.

The second mechanism for the *dV/dt* turn-on in MOSFETs is through the parasitic BJT as shown in Figure 10.16. The capacitance associated with the depletion region of the body diode extending into the drift region is denoted as C_{DB} and appears between the base of the BJT and the drain of the MOSFET. This capacitance gives rise to a current I_2 which flows through the base resistance R_B when a voltage ramp appears across the drain-source terminals. With analogy to the first mechanism, the *dV/dt* capability of this mechanism is given by

$$\left(\frac{dV}{dt}\right) = \frac{V_{BE}}{R_B C_{DB}} \tag{10.11}$$

If the voltage that develops across R_B is greater than about 0.7V, then the base-emitter junction is forward-biased and the parasitic BJT is turned on. Under the conditions of high *dV/dt* and large

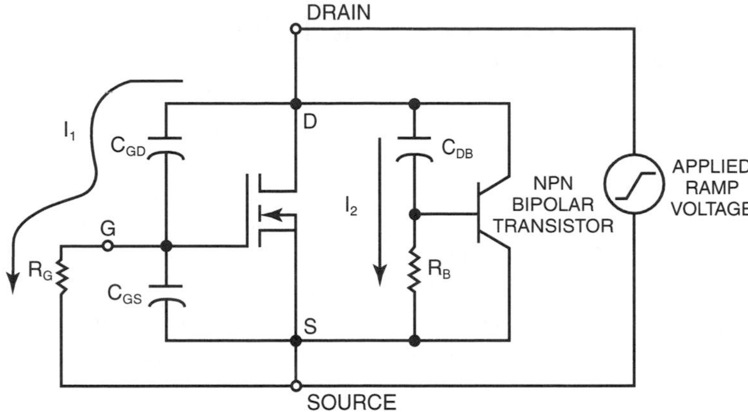

Figure 10.15 Equivalent circuit model of a power MOSFET showing the two possible mechanisms for *dV/dt*-induced turn-on. *Source*: Baliga, B. J. 1987. *Modern Power Devices.* © 1987 John Wiley & Sons, Inc., New York. Reprinted by permission of John Wiley & Sons, Inc.

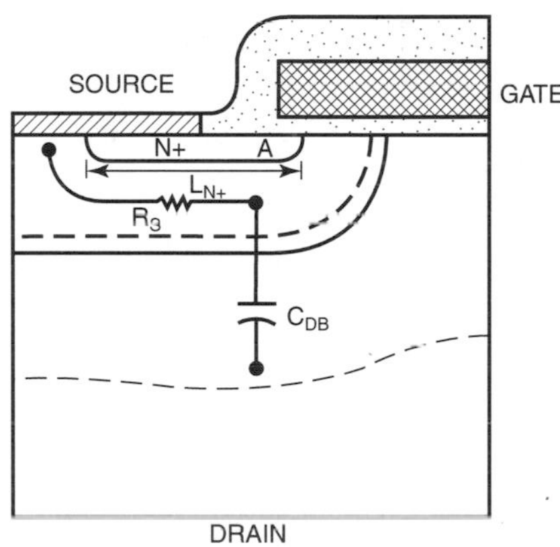

Figure 10.16 Physical origin of the parasitic BJT components which may cause *dV/dt*-induced turn-on in a power MOSFET. *Source*: Baliga, B. J. 1987. *Modern Power Devices.* © 1987 John Wiley & Sons, Inc., New York. Reprinted by permission of John Wiley & Sons, Inc.

values of R_B, the breakdown voltage of the MOSFET will be limited to that of the open-base breakdown voltage of the BJT. If the applied drain voltage is greater than the open-base breakdown voltage, then the MOSFET will enter avalanche and may be destroyed if the current is not limited externally.

Increasing *dV/dt* capability therefore requires reducing the base resistance R_B by increasing the body region doping and reducing the distance the current I_2 has to flow laterally before it is collected by the source metallization. As in the first mode, the BJT related *dV/dt* capability becomes worse at higher temperatures since R_B increases and V_{BE} decreases with increasing temperature.

10.4 Applications

The following are two of the major markets where power MOSFETs are finding increasing applications as either logic-controlled or analog switches.

Portable Electronics and Wireless Communication

With the recent advances in the portable electronic products, low R_{dson}, logic level surface mount power MOSFETs are experiencing explosive demand. A portable computer, for example, uses power MOSFETs in the AC-DC converters, the DC-DC converters and voltage regulators, load management switches, battery charger circuitry, and reverse battery protection. Required features of MOSFETs in these applications are small size, low power dissipation, and low on-resistance for extended battery life. Reduction of both conduction and switching losses are important considerations in the design of MOSFETs aimed at this market.

Automotive

Mechanical contact breakers have mostly been replaced by semiconductor devices in ignition circuits in modern cars. A suitable semiconductor device must be capable of blocking high voltages in a severe environment where line voltage surges are common due to the opening and closing of switches and the connection and disconnection of inductive loads during maintenance and loose connections. Bipolar transistors with their susceptibility to secondary breakdown are not suited whereas power MOSFETs with avalanche capability are ideally suited. Voltage transients are clamped by the avalanching of the MOSFET without the need to use any external protection circuits.

In 12V battery vehicles the most commonly used MOSFETs are rated at 50V or 60V breakdown voltages. The significant guard-banding is necessary in order to avoid device failure due to the alternator producing high voltages after shedding a heavy load.

The other features of power MOSFETs which make them suitable for the automotive applications are high *dV/dt* ratings, high-temperature performance, ruggedness and high reliability. Logic level, surface mount devices with low R_{dson} have recently found application in this field. The smaller footprint of surface mounts offers space savings and the lower R_{dson} does away with the need to parallel devices to reduce on-resistance. This in turn translates into fewer device counts and heat-sinks which lowers the overall cost.

In addition to ignition control, power MOSFETs are used in anti-lock brake (ABS) systems, electronic power steering (EPS) systems, air bags, electronic suspension, and numerous motor control applications such as power windows, power seats, radiator fan, wipers, fuel pump, etc.

References

Baliga, B.J. 1987. *Modern Power Devices,* John Wiley & Sons, NY.

Grant, D.A. and Gower, J 1989. *Power MOSFETs—Theory and Applications,* John Wiley & Sons, NY.

International Rectifier, 1995, *HEXFET Power MOSFET Designer's Manual—Application Notes and Reliability Data,* International Rectifier, El Segundo, CA.

Oxner, E.S. 1982. *Power FETs and Their Applications,* Prentice-Hall, Englewood Clliffs, NJ.

Sze, S.M. 1981. *Physics of Semiconductor Devices,* John Wiley & Sons, NY.

11

Insulated Gate Bipolar Transistors

Michael Robinson
International Rectifier Corporation

Richard Francis
International Rectifier Corporation

Ranadeep Dutta
International Rectifier Corporation

Al Diy
International Rectifier Corporation

11.1 Introduction ... 229
What are IGBTs • Typical Applications of IGBTs • Voltage, Current, Frequency Ratings • Commonly Available Packages

11.2 Basic Structure and Operation ... 230

11.3 Design Considerations ... 232
Cell Design • Design of the PNP Portion of the IGBT • Minority Carrier Lifetime Control • Device Performance Trade-off Considerations

11.4 Requirement for Anti-parallel Diode ... 236

11.5 Comparison Between the Power MOSFET, IGBT, and MCT..... 236

11.6 IGBT Data Sheet Parameters ... 237

11.7 Appendix: Typical IGBT Data Sheet ... 238

11.1 Introduction

This paper describes IGBTs (insulated gate bipolar transistors), where they are used, how they work, and expected advances in technology.

During the 1990s and well into the next century the interest in and use of devices such as IGBTs will continue to increase rapidly, driven by the need to convert, control, and conserve electrical energy. Driving the increased use of these power semiconductor switches are macroeconomic factors such as the increasing cost of electrical power as well as power generation shortages in developing countries. Furthermore, government regulations concerning noise generated by electronic equipment and then fed back into the power grid will increase demand for power switches which are used to minimize this disturbance.

IGBTs and their close cousin the power MOSFET represent one of the fastest growing segments of all semiconductors, including ICs and memory. This growth is fueled in part by applications which would not be economically feasible without IGBTs. For many new applications, IGBTs are the enabling technology. Additional growth stems from the fact that IGBTs are able to replace power MOSFETs, SCRs, and bipolar transistors in existing applications.

What are IGBTs

IGBTs are three-terminal, silicon-based, power semiconductor switches used primarily to control or convert electrical energy. Other competing technologies are power MOSFETs, bipolar junction transistors (BJTs), silicon-controlled rectifiers (SCRs), gated turn off thyristors (GTOs), and MOS-controlled thyristors (MCTs). Each technology is defined and limited by blocking voltage, current rating, and switching frequency.

IGBTs combine the best of MOSFET and bipolar transistor technology (Figure 11.1). Just as with a power MOSFET the IGBT is gated (turned on and off) by applying a voltage to the gate terminal. However in a BJT, the device is turned on and off by applying current to the base terminal. The IGBTcan be easily driven by an IC whereas a bipolar device would require substantial power dissipation at the input as well as considerable circuit complexity to turn it on and off. For applications above approximately 200V, the IGBT can offer lower on resistance than a power MOSFET because it makes use of the minority carrier injection that characterizes bipolar devices.

Typical Applications of IGBTs

In most power conversion and control applications, the IGBT performs the switch function (Figure 11.2).

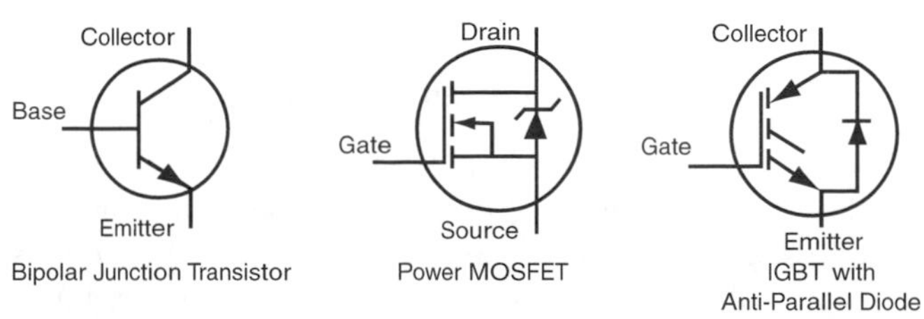

Figure 11.1 Circuit symbols for bipolar junction transistor, MOSFET, and IGBT.

0-8493-8343-9/97/$0.00+$.50

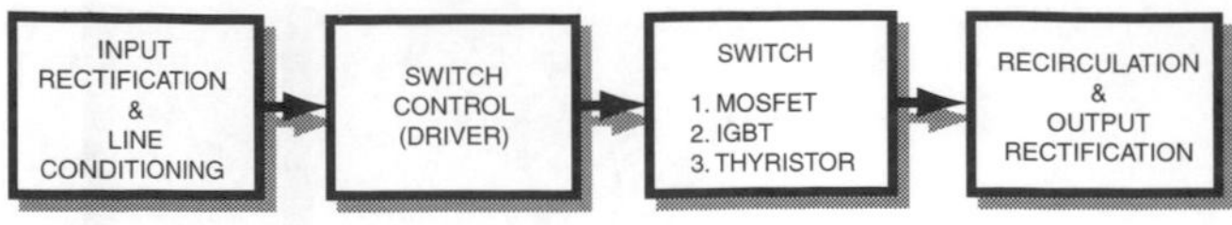

Figure 11.2 The four functions used in most power control and power conversion systems.

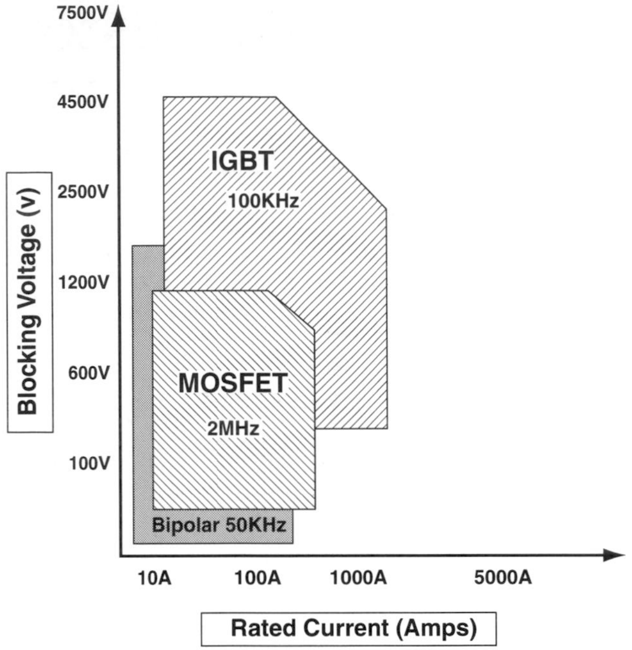

Figure 11.3 Switch technologies versus blocking voltage, current rating, and upper switching frequency. Note that upper frequency does not apply to full current/voltage range.

The most common application of IGBTs are motor speed control and power conversion. IGBTs can be found in AC and DC motor controls used in HVAC, refrigeration, factory conveyor systems, forklifts, elevators, robotics, and washing machines. Typical uses of IGBTs in power conversion are: uninterruptible power supplies (UPS), high-frequency welding, high-voltage/high-current power supplies, power factor correction (PFC), and induction heating. Automotive applications include solid state ignition, motor drives in electric vehicles, and battery chargers.

As blocking voltage capability increases with improvements in technology, IGBTs will find their way into the control and distribution of electrical power transmission and electric powered locomotives.

Voltage, Current, Frequency Ratings

Choosing one power switch technology over another usually involves a comparison of blocking voltage, current rating, on-state losses, and switching losses (see Figure 11.3). The most common IGBT blocking voltages are 250V, 500V, 600V, 1200V, 1700V with more recent developments at 2500V, 3500V, and 4500V.

Current ratings and on-state losses are usually a matter of the size (active area) of the semiconductor chip. The most common sizes run from 2mm × 2mm to 14mm × 14mm. Usable currents for a single chip run from 8A up to 200A. For higher currents, chips can be paralleled, although this presents problems at higher frequencies.

Each switch technology has its trade-offs, and operating frequency is one of them. Improving operating frequency within a technology will usually result in a reduced current rating or increased on-state losses. Conversely, increasing blocking voltage will negatively impact frequency and on-state losses. Device designers and product marketers strive to determine the optimum trade for each end market. Quite often, newly introduced products are nothing more than a different optimization rather than a breakthrough in technology.

Figure 11.3 shows a current rating versus blocking voltage graph showing the range covered by each technology and the corresponding switching frequency limit.

Commonly Available Packages

IGBTs can be purchased in wafer form or as individual chips, but more often they are housed and protected in a variety of packages (Figure 11.4).

The most simple package is called a "discrete" and holds only one IGBT chip. The next level of complexity is that of the "CoPak" which consists of a single package containing both the IGBT chip and an ultrafast diode (FRED) chip. Discretes and CoPaks are typically limited to 1200V and 70A. It is very common for designers to parallel discretes and CoPaks.

To achieve yet higher voltage and current ratings semiconductor manufacturers parallel chips in packages called modules. Typical ratings can run from 100A to over 1000A. Blocking voltages can reach 4500V. Modules provide an economical method for equipment designers to handle and interface to the sensitive semiconductor chips inside.

Modules can also be configured into simple power circuits such as half bridges, chopper, and three-phase switch (6 Paks), etc. Integrated modules include both the switch function as well as input rectification function. Intelligent power modules (IPMs) can include all the above plus gate driver ICs, thermal and current sense functions and much more.

11.2 Basic Structure and Operation

The IGBT, first introduced as the insulated gate rectifier (IGR) (Baliga et al., 1982), and also called the conductivity modulated field effect transistor (COMFET) (Russel et al., 1983), is a monolithic integration of a wide-base bipolar junction transistor and a MOSFET. The structure of an IGBT (Figure 11.5) is similar to that of a double diffused (DMOS) power MOSFET, the major difference being that the N+ drain region of the MOSFET is replaced by a P+ collector region. The DMOS gate structure of the IGBT is formed on an N− drift region connected to a P+ collector. The P+ collector is the key element that makes bipolar current flow possible in an IGBT.

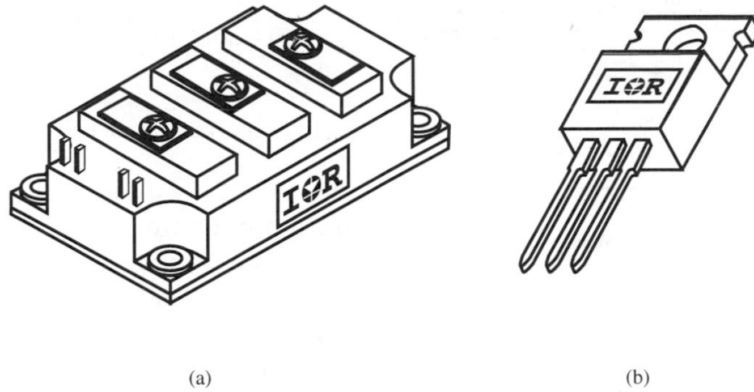

Figure 11.4 (a) Typical IGBT module (not to scale) containing both IGBTs and FREDs. Ratings for modules range from 100A to over 1000A. (b) Typical package used for discrete and CoPak IGBTs.

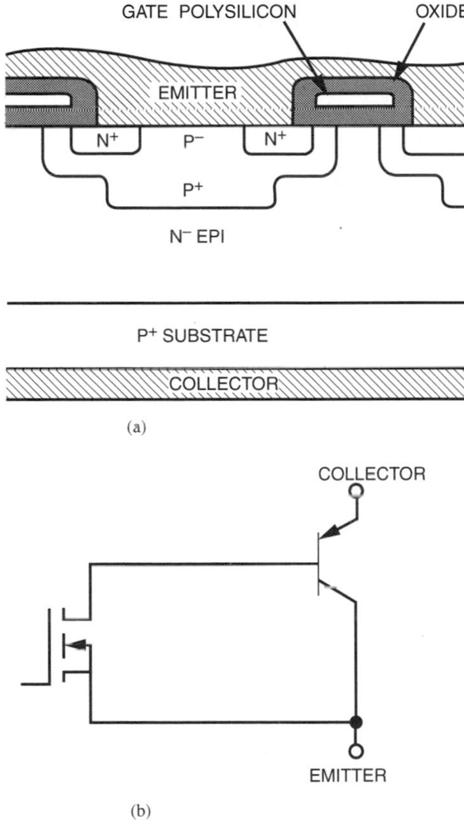

Figure 11.5 Cross section of a symmetrical IGBT and equivalent circuit, (a) device structure and (b) equivalent circuit.

The device operation of an IGBT, in its simplest terms, is that of a vertical bipolar pnp transistor, the current flow through which is controlled by the potential applied to the MOS gate. Consider the emitter terminal to be tied to the ground potential for all modes of operation. For forward current to flow between the emitter and collector, the gate and collector must both be biased positively. The forward biased collector junction can inject holes into the N^- region only as long as electrons are available to neutralize them. When the gate potential exceeds the threshold voltage required to invert the surface of the MOS region beneath the gate, the channel thus formed provides a path for electrons to flow into the N^- drift region. These electrons maintain space charge neutrality, when the positively biased P+ collector begins to inject holes into the N− drift region. Vertical current flow is thus initiated between the emitter and collector. To describe this on-state operation simply, the IGBT behaves like a vertical wide-base pnp transistor, the base current to which is supplied through a MOSFET whose source is shorted to the collector of the pnp transistor. It should be pointed out, to avoid any confusion, that the emitter of the equivalent pnp transistor is in fact the collector terminal of the IGBT. Unless otherwise mentioned, all references to the term "collector" in this text imply the collector terminal of the IGBT.

When the collector potential is small (~2–3V), and the gate bias sufficiently larger than the threshold voltage, the IGBT exhibits the on-state characteristics of a wide-based bipolar transistor in high-level injection (Hefner and Blackburn, 1988). However, when the potential drop across the inversion layer is comparable to the difference between the gate bias and the threshold voltage, pinchoff of the channel occurs (Sze, 1981). In this mode of operation, the base current supply through the inversion layer cannot increase with further increase in collector bias, and therefore, neither can the collector current. The reason for this is that the P+ layer injects holes only to the extent that can be neutralized by the electrons brought in as base current through the channel. Therefore, when channel pinchoff causes the base current to saturate, the collector current saturates as well. The collector saturation current is simply the channel saturation current at pinchoff (Sze, 1981) multiplied by a factor of $(1 - \alpha_{pnp})^{-1}$, where α_{pnp} is the common base gain of the pnp section. It is important to note from this qualitative development that the forward output characteristics of the IGBT are completely gate controlled and exhibit saturation. Figure 11.6. shows typical forward characteristics of an IGBT as a function of gate-to-emitter voltage.

During turn-off of the IGBT, the gate is shorted to the emitter, and the path for majority carrier base current is cutoff since there is no channel. No vertical current can flow as there is no base current supply to the pnp transistor. Any positive potential applied to the collector of the IGBT is supported by the junction between the P well and the N− drift region, until the open base, collector-emitter breakdown BVCEO (Sze, 1981) of the pnp transistor is reached. This is referred to as the forward blocking

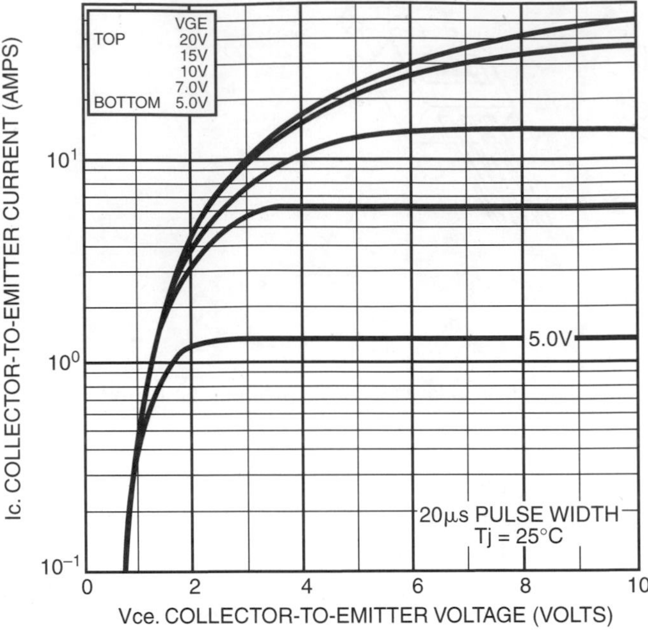

Figure 11.6 Forward characteristics (collector current versus collector voltage for different gate voltage) of a typical IGBT (IRGBC40F).

mode of operation. The collector junction provides reverse blocking capability, supporting negative potential applied to the collector terminal.

The gain, α_{pnp}, of the pnp section, is a key parameter in determining the conducting and switching properties of the IGBT. The common base pnp gain is given by Sze (1981)

$$\alpha_{pnp} = 1/\cosh\,(W/L_a)$$

where W is the undepleted base width of the pnp transistor and L_a is the ambipolar diffusion length (Sze, 1981). It is important to note that a higher value of the pnp gain decreases the conduction loss, but increases the switching loss of an IGBT.

To initiate turn-off of the IGBT, the gate is shorted to the emitter, which quickly removes the MOS channel, and hence, the base current supplied to the pnp transistor. Once the base drive is removed the device behaves like an open base pnp transistor and the excess carriers in the n base decay by electron-hole recombination.

11.3 Design Considerations

An IGBT device is made up of a multiplicity of DMOS cells which work in parallel (Figure 11.7). The performance of the IGBT depends not only on the design of the cells but also on the silicon starting material, the wafer fabrication process, and the means used to control the gain of the pnp transistor.

Cell Design

The base drive to the pnp transistor is provided by the channel current and reducing the channel resistance will increase the base drive and enable greater efficiency during forward conduction. The channel resistance is given by (Sze, 1981):

$$R_{ch} = L/Z\mu C_{ox}(V_G - V_T)$$

where L is the channel length, Z is the channel width, μ is the mobility of electrons in the channel, V_G the gate voltage and V_T the threshold voltage of the device. It can be decreased by increasing the channel width and reducing the channel length. Thus increasing the cell density and reducing the junction depth of the channel region will improve the forward conduction characteristics of the device. Improvement in performance of IGBTs in recent years is largely due to improved processing and design which exploit this simple relationship.

As cell densities increase and the polysilicon gate width is reduced, the area between the cells, referred to as the JFET region, becomes narrower, resulting in an increased resistance of this region, which degrades the device performance. By doping this region with an n type dopant to reduce the resistance, this problem is minimized, enabling the use of higher cell densities. Figure 11.8 illustrates the effect of reducing the channel resistance on the switching loss versus VCE(on) trade-off curve.

Design of the PNP Portion of the IGBT

Earlier in the section on basic structure and operation, we described how the gain of the pnp transistor portion of the IGBT had a major effect on the switching characteristics. A number of different designs have been used to control the switching losses

- Punch-through.
- Collector shorted.
- Symmetrical.

Figure 11.9 compares the three types of device construction for devices designed to have breakdown voltages 200V to 1500V. Punch-through devices are the preferred design at higher voltages (>2000V) where the growth of epitaxial silicon with high enough resistivity and thickness becomes difficult. Collector shorting becomes the favored means of controlling the pnp gain, and in some applications where switching losses are not important, a diffused P collector region is the simplest most cost-effective structure.

Punch-Through Devices

The punch-through IGBT structure shown in Figure 11.9b is the most widely used for devices with BVCES < 1700V. It is characterized by a relatively thin ($\sim$10 μm) heavily doped ($\sim 10^{17}$ atoms cm^{-3}) N buffer layer at the collector end of the N^- epi. The presence of the buffer layer speeds up the switching process for two reasons. The injection efficiency of the collector junction is reduced, and so is the effective minority carrier lifetime in the n base, both due to a much higher doping density in the buffer (typically 3–4 orders of magnitude higher than the epi concentration for a high-voltage IGBT). For the same two reasons, the conduction loss is adversely affected, however, and can be offset by using a thinner epi, a facility that the asymmetrical IGBT

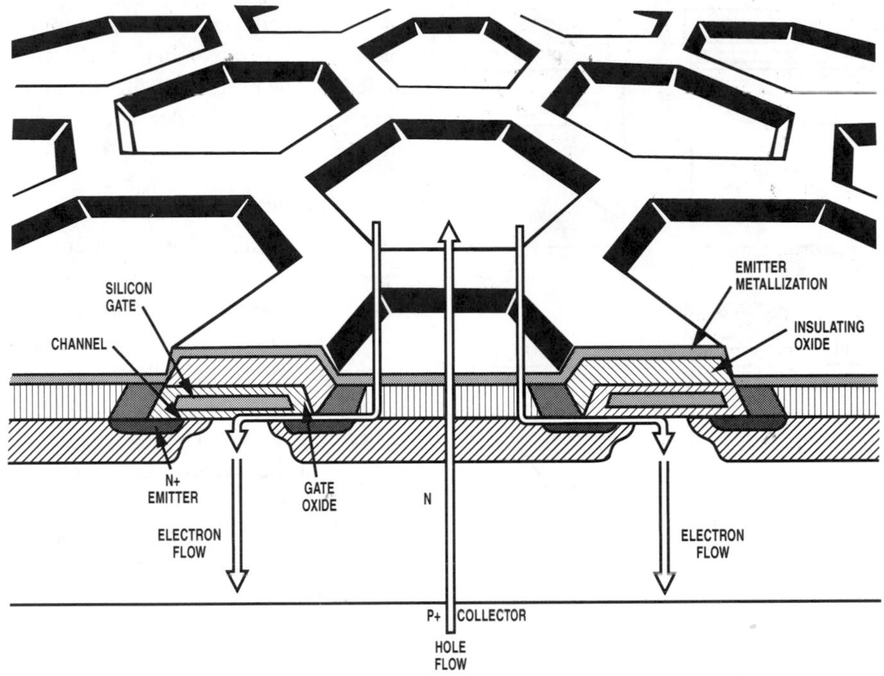

Figure 11.7 Top view and cross section showing the cellular structure of an IGBT.

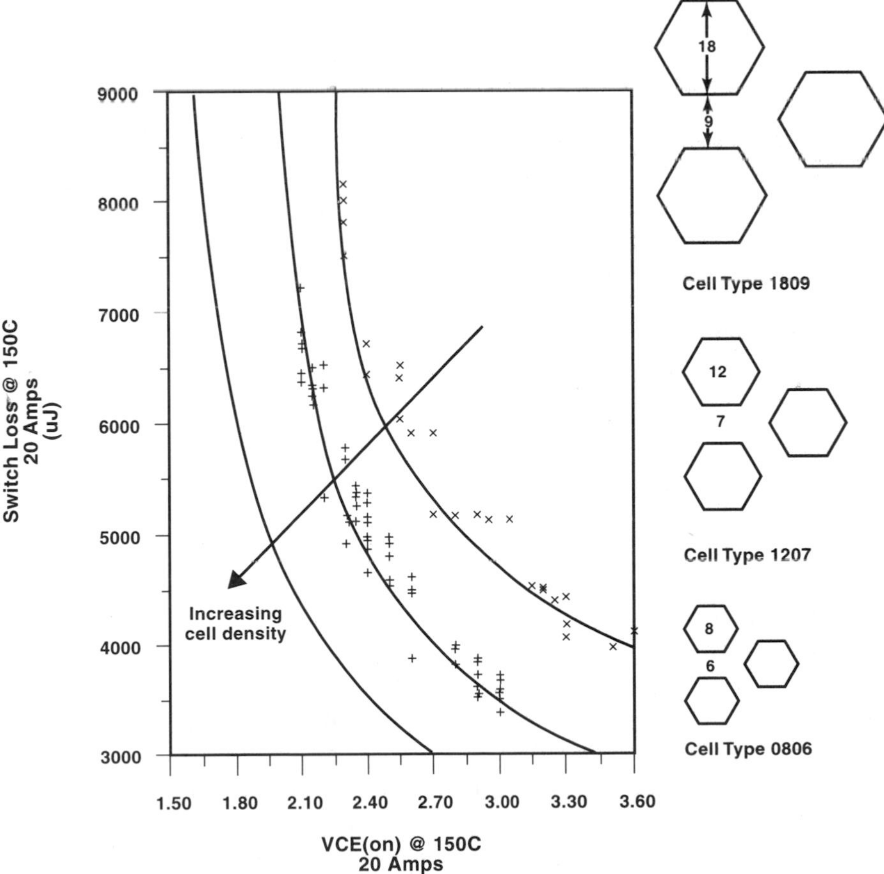

Figure 11.8 Dependence of the IGBT technology performance curve (VCE(on) versus switching energy loss) on cell density.

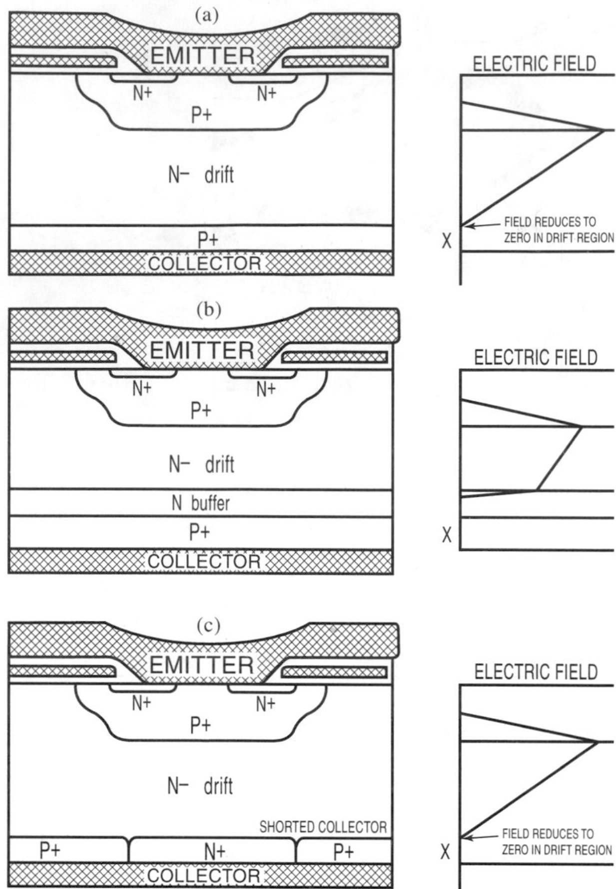

Figure 11.9 Cross section and electric field profile in forward blocking mode for (a) symmetrical IGBT, (b) punch-through IGBT, and (c) collector-shorted IGBT.

provides. Figure 11.9 shows the electric field distribution in the forward blocking mode for the different types of IGBTs, and illustrates how, by using a lower doped and thinner epi in the asymmetric structure, the same forward blocking voltage as in the symmetric can be attained. In essence, the presence of a buffer layer permits the net base width to be narrower than a corresponding symmetric structure with similar blocking, conducting, and switching properties.

The gain, α_{pnp}, of the pnp section, a key parameter in determining the conducting and switching properties of the IGBT, is dominated by the physical properties of the buffer region—its thickness, doping density, and minority carrier lifetime. The reason for that is the dopant density per unit area, or Gummel number (Sze, 1981), in the buffer layer is typically two orders of magnitude higher than that in the epi.

In addition to the effects of the buffer layer, reducing the minority carrier lifetime within the base of the pnp transistor will reduce the pnp gain of the IGBT and reduce the turn off switching loss. Devices designed for a particular application will have well defined VCE(on) and switching loss requirements, and both lifetime and buffer layer specification will be chosen with care.

The two epitaxial layers for the punch-through IGBT are grown on a P+ substrate wafer which is thick enough that the final wafers can be processed without difficulty thus giving this structure an additional edge over other methods for low-voltage devices.

Collector-Shorted IGBTs

The structure of a collector-shorted IGBT is shown in Figure 11.9c. The thickness of the n layer is greater and the resistivity lower than for the punch through device as there is no buffer to prevent the depletion layer from punching through to the collector. The collector-emitter efficiency and hence the gain of the pnp is controlled by providing a short bypassing the collector junction. The gain of the device is controlled by the fraction of shorted area. Unlike punch-through devices, the gain is not sensitive to junction temperature and the switching loss at elevated temperature is the same as that at room temperature. There is no need to reduce the minority carrier lifetime, and hence it is possible to obtain a good trade-off between switching losses and VCE(on).

Manufacture of such devices requires that the back side of the silicon wafers be patterned and, as wafers need to be at least 300 μm thick in order to be robust enough to survive processing, this makes such devices impractical for ratings of less than 2000V.

Symmetrical IGBTs

These devices offer an advantage for higher-voltage devices where switching losses are not important but high breakdown voltage, low on-state drop and cost are. In order to improve the switching losses of these devices the efficiency of the collector can be reduced by making it a very shallow, lightly-doped region and reducing the minority carrier lifetime. Other than in the area of cost, the symmetrical IGBT has little advantage and both the punch-through and collector-shorted devices have superior switching behavior.

Minority Carrier Lifetime Control

Reducing the minority carrier lifetime in the device can significantly improve its switching losses. The primary ways of reducing the minority carrier lifetime are

- Electron irradiation.
- Proton implantation.
- Platinum diffusion.
- Gold diffusion.

Electron Irradiation

Electron irradiation is carried out using 12-MeV electrons that penetrate the entire depth of the silicon. It is carried out when the wafer processing steps are complete. Doses of between 3 and 20MR are used depending on the speed of the device being produced. Considerable damage is produced in the process so as to severely degrade the IGBT characteristics unless they are recovered by annealing at 300°C. The damage is distributed uniformly throughout the device and hence the concentration of recombination centers is the same in all parts of the IGBT. After electron irradiation and annealing a deep level is generated at 0.35eV below the conduction band. This is the dominant recombination center which determines device performance. The relationship between the minority carrier lifetime and the electron irradiation dose is given by:

$$1/\tau = 1/\tau_0 + K\phi$$

where τ is the lifetime after irradiation, τ_0 is the initial carrier lifetime before irradiation, and K is a constant which depends on the energy of the electron irradiation, the silicon resistivity and temperature (Taylor, 1987). The dose used depends on the required device performance; obviously, high doses result in low switching losses and high on-state drop.

Annealing of the electron irradiation damage is a very important consideration when selecting an assembly process for soldering the IGBT die to its heat sink. The relationship for the rate of annealing is

$$N_t(t)/N_t(0) = \exp(-\{t \cdot 2.1 \cdot 10^{11}\} \exp[-1.9 \text{ eV}/kT])$$

where $N_t(t)$ is the deep level concentration after time t, $N_t(0)$ is the initial concentration, and T is the annealing temperature in K (Sun, 1977). Because of these annealing effects, it is important during the assembly of electron irradiated parts that temperatures be kept as close to 300°C for as short a time as possible.

Proton Implantation

Proton implantation, an alternative method of lifetime killing, is also carried out after the wafer fabrication steps are complete (Taylor, 1987; Mogro-Campero et al., 1986). Its advantage lies in the fact that a band of damage can be placed at a specific depth into the silicon rather than uniformly as in the case of electron irradiation. By placing the recombination centers in an appropriate location in the n base of the device the switching and VCE(on) of the devices can be optimized. This type of implantation is carried out at between 0.5 and 4MeV using proton doses of between 5×10^{11} and $5 \times 10^{12}/\text{cm}^2$, and just like electron irradiation a post implantation anneal at 300°C is necessary.

A typical proton implanted profile is illustrated in Figure 11.10 showing the damage profile using 3-MeV protons implanted into silicon. As far as annealing during assembly is concerned the behavior of proton implanted and electron irradiated devices are similar. A limitation of the proton implantation method is that it requires specialized ion implantation equipment which is not readily available for commercial use.

Platinum and Gold

The heavy metals platinum and gold have also been used extensively to reduce the minority carrier lifetime in silicon. Usually a thin layer (<200 Å) of platinum or gold is deposited on one side of the wafer and diffused into the silicon at a temperature between 850°C and 1000°C for 10 to 30 minutes. The diffusion of gold and platinum is very difficult; both are prone to considerable variation and the metals tend to precipitate at dislocations or defects; these precipitates can result in hot spot formation during device operation, which can be destructive.

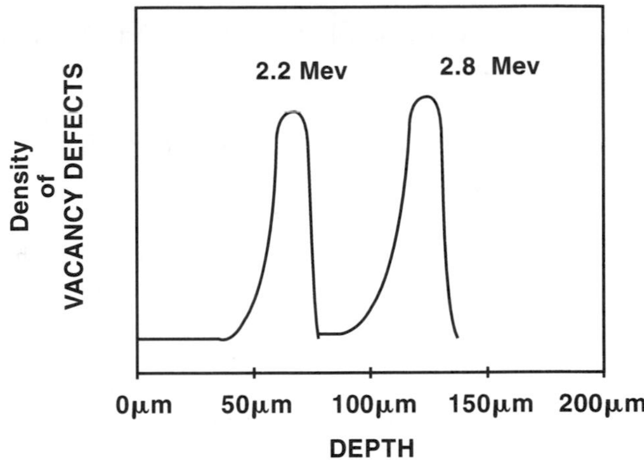

Figure 11.10 Example of proton implantation range in silicon.

Comparison of Lifetime Reduction Methods

Gold diffusion results in the formation of a dominant recombination center at 0.54eV below the conduction band at about mid band gap, whereas platinum results in a deep level 0.42eV above the valence band. As a consequence, gold-doped devices exhibit high elevated temperature leakage current whereas the leakage for platinum-doped devices is much lower. The dominant level for electron irradiation is 0.35 eV below the conduction band and results in much lower leakage current than gold but significantly larger than platinum. Because of the high leakage current gold diffusion is rarely used in IGBTs. A detailed comparison of the various lifetime reduction techniques can be found in Baliga (1987).

For devices up to 900V electron irradiation provides the best trade-off between switching losses and VCE(on). The only disadvantage lies in the leakage current. At 1200V and above, platinum provides the best trade-off. Electron irradiation provides a convenient means of reducing carrier lifetime with none of the difficulties associated with heavy metal diffusion.

Device Performance Trade-off Considerations

IGBTs are used in a number of different applications and the design of the IGBT must be appropriate. In applications where the IGBT is operated at low frequencies (<1000Hz) it is important that the VCE(on) of the device be very low and the switching losses are comparatively unimportant. For higher-frequency applications such as motor drives operating in the range between 5 and 8kHz the IGBT needs lower switching losses and low VCE(on). For high-frequency applications (>10kHz) low switching losses are critical and the VCE(on) may be high.

Three classes of IGBT switching speeds are usually available to accommodate these requirements. The slow switching speed device typically undergoes no lifetime killing process, the fast device having some lifetime killing and the ultrafast device being heavily lifetime killed.

11.4 Requirement for Anti-parallel Diode

Unlike power MOSFETs, IGBTs do not have a body diode which acts anti-parallel when the device is used in an inverter circuit. So an external fast recovery diode needs to be used. The characteristics of the anti-parallel diode used are very critical. During turn-on of the IGBT the reverse recovery current of the diode across the complementary IGBT device appears as an additional component and contributes significantly to the turn on losses in the IGBT. A typical reverse recovery waveform is shown in Figure 11.11. It is very important that, under very high dI/dt conditions during turn on (up to 2000A/μs), the diode exhibits low reverse recovery. In order to prevent oscillation in the circuit, it is desirable that the reverse recovery of the diode be soft, i.e., t_b be greater than t_a. In addition, during the reverse recovery phase a voltage spike equal to L dI(REC)/dt appears across the IGBT and diode, where L is the stray inductance. Thus, the reverse recovery needs to be soft to prevent a large voltage spike, VRM, that is potentially destructive to the devices.

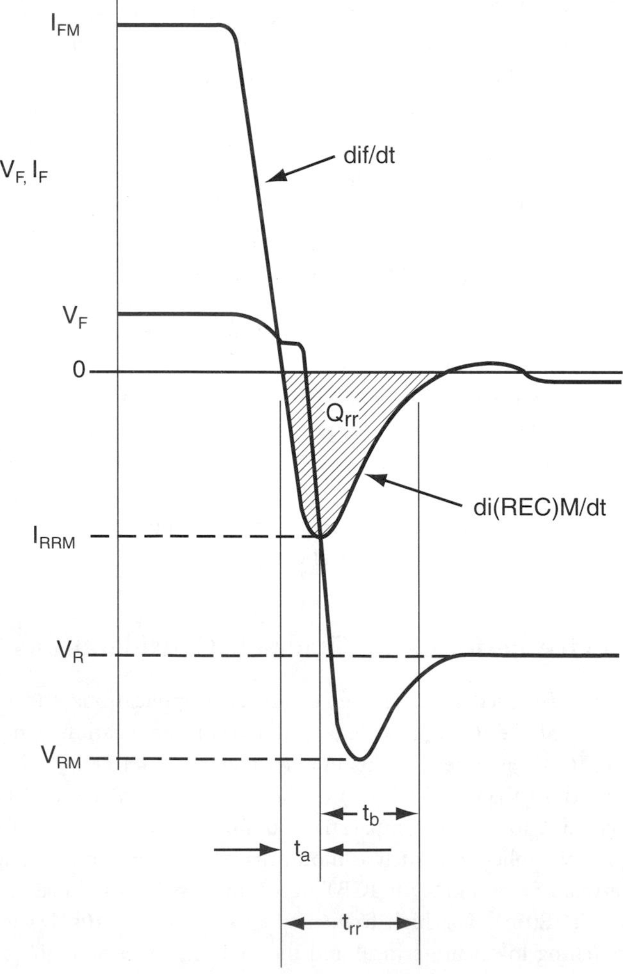

Figure 11.11 Reverse recovery waveform of a fast recovery epitaxial diode (FRED).

11.5 Comparison Between the Power MOSFET, IGBT, and MCT

The concept of the IGBT structure is an offshoot of that of a power DMOSFET (Figure 11.12.), a device that was first commercially available in 1979. The power MOSFET offers the same advantages of high input impedance and gate controlled output characteristics as does the IGBT. In applications requiring blocking voltages less than 200V, the power MOSFET is the chosen switch due to lower on-state power dissipation and much superior switching speed. The latter attribute arises from its unipolar operation. The switching mechanism is limited by the speed of gate signal propagation (typically a few nanoseconds), since unipolar charge carrier relaxation time is much shorter, typically picoseconds. Switching in bipolar devices, on the other hand, is limited by the much slower process of recombination, typically a few tenths to a few microseconds, depending on the level of lifetime killing. However, in a power MOSFET, lack of space charge neutralization due to the presence of only one carrier type is the reason why the on-state power dissipation in the bulk region is directly proportional to its thickness, like a linear resistor. For this reason, the power MOSFET drops out of favor in applications above 200V.

The on-state losses in a high-voltage IGBT, although significantly lower than in a corresponding MOSFET, are larger than those in a thyristor, a 4-layer pnpn structure. The reason is that a thyristor, due to its ability to inject oppositely charged

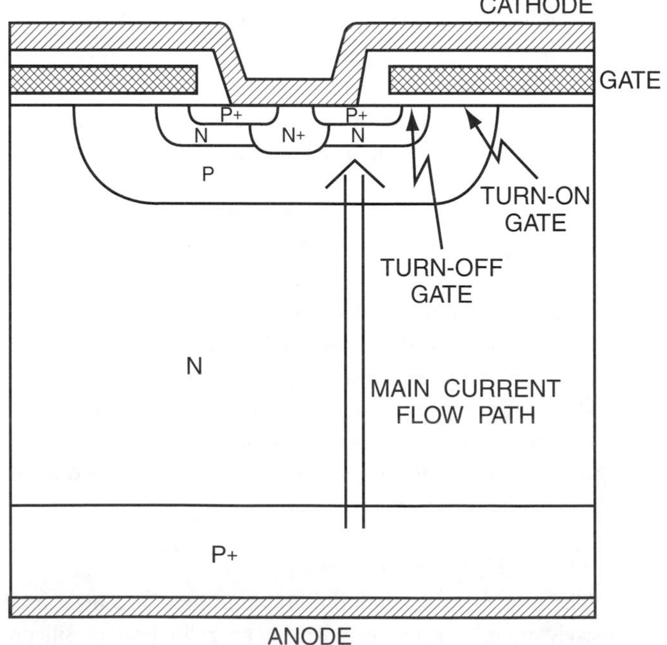

Figure 11.12 Cross section of a MOS-controlled thyristor. (*Source:* Temple, V.A.K. 1984. MOS-controlled thyristors, *IEDM,* abstract 10.7, 282–285.)

carriers from its two end contacts, provides a greater degree of conductivity modulation by excess carriers. The MOS-controlled thyristor (MCT) (Figure 11.12), which combines this feature with the advantage of MOS gated switching capability, was first proposed by Temple in 1984 (Taylor, 1987). To turn off the device, a negative potential is applied to the gate with respect to the cathode so that an inverted p channel is created. Holes entering the P base can then bypass the N + P junction and reach the cathode contact through the P base—p channel—P+ path. The turn-off process depends sensitively on the channel resistance. The primary reason the MCT has not enjoyed commercial success is that its turn-off current density drops off significantly with an increasing number of cells, with increasing anode voltage, increasing temperature and under inductive loads.

11.6 IGBT Data Sheet Parameters

A typical IGBT data sheet begins on the following page. Some of the important device parameters included in the data sheet are

- The Collector-to-Emitter Breakdown Voltage—BVCES, is the breakdown voltage of the IGBT and is defined at a specific leakage current.
- The Emitter-to-Collector Breakdown Voltage—BVECS, is the reverse breakdown of the device, and in the case of punch-through IGBTs is typically 15 to 30V.
- The Collector-to-Emitter Saturation Voltage—VCE(on), is the on state voltage drop of the device for particular value of forward current and V_{GE}, usually 15V. The VCE(on) is a very important parameter that determines the conduction losses in the IGBT. This parameter is dependent on temperature, V_{GE} and IC, the collector current.
- The Gate Threshold Voltage—$V_{GE(th)}$, is the gate voltage relative to the emitter at which collector current begins to flow in the IGBT. The data sheet value is given for a specific collector current, usually 250μA.
- The Total Switching Loss—E_{TS}, is the total energy associated with turning the device on and off, and is a very important parameter together with the VCE(on) for determining the amount of heat dissipation in the device.

For a more comprehensive review of an IGBT data sheet the reader is referred to Clemente (1992).

References

Baliga, B. J. et al., 1982. The insulated gate rectifier: a new power switching device, *IEDM Tech. Dig.*, 264–267.

Baliga, B. J. 1987. *Modern Power Devices,* John Wiley & Sons New York, NY.

Clemente, S. et al., 1992. *IGBT Characteristics*, IR Applications Note AN-983A.

Hefner, A. R. and Blackburn, D. L. 1988. An analytical model for steady-state and transient characteristics of the power IGBT, *Solid State Electron.*, 31(10):1513–1532.

Mogro-Campero, A. et al., 1986. *IEEE Trans. Electron Dev.*, ED-33(11).

Russel, J. P. et al., 1983. The COMFET: a new high conductance MOS gated device, *IEEE Electron Dev. Lett.*, EDL 4:63 65.

Sun, Y. E. 1977. IEEE—IAS Mtng, October.

Sze, S. M. 1981. *Physics of Semiconductor Devices*, 2d ed., John Wiley & Sons, New York, NY.

Taylor, P. D. 1987. *Thyristor Design and Realization,* John Wiley & Sons, New York, NY.

Temple, V. A. K. 1984. MOS controlled thyristors, *IEDM*, abstract 10.7, 282–285.

11.7 APPENDIX: Typical IGBT Data Sheet

INSULATED GATE BIPOLAR TRANSISTOR — UltraFast IGBT

Features

- Switching-loss rating includes all "tail" losses
- Optimized for high operating frequency (over 5kHz) See Fig. 1 for Current vs. Frequency curve

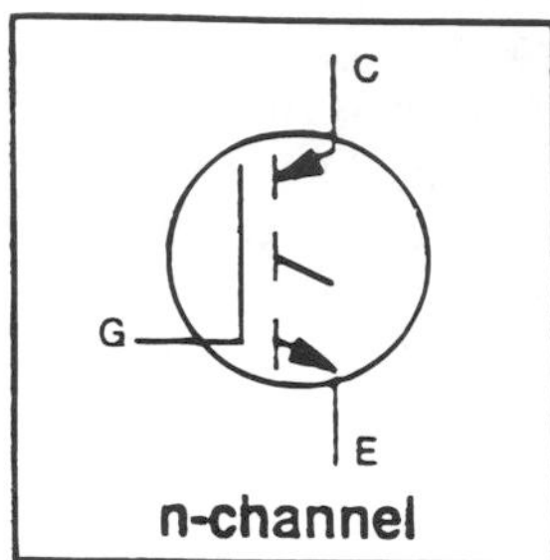

$V_{CES} = 600V$

$V_{CE(sat)} \leq 3.0V$

$@V_{GE} = 15V, I_C = 20A$

Description

Insulated Gate Bipolar Transistors (IGBTs) from International Rectifier have higher usable current densities than comparable bipolar transistors, while at the same time having simpler gate-drive requirements of the familiar power MOSFET. They provide substantial benefits to a host of high-voltage, high-current applications.

TO-220AB

Absolute Maximum Ratings

	Parameter	Max.	Units
V_{CES}	Collector-to-Emitter Voltage	600	V
I_C @ $T_C = 25°C$	Continuous Collector Current	40	A
I_C @ $T_C = 100°C$	Continuous Collector Current	20	
I_{CM}	Pulsed Collector Current ①	160	
I_{LM}	Clamped Inductive Load Current ②	160	
V_{GE}	Gate-to-Emitter Voltage	±20	V
E_{ARV}	Reverse Voltage Avalanche Energy ③	15	mJ
P_D @ $T_C = 25°C$	Maximum Power Dissipation	160	W
P_D @ $T_C = 100°C$	Maximum Power Dissipation	65	
T_J T_{STG}	Operating Junction and Storage Temperature Range	-55 to +150	°C
	Soldering Temperature, for 10 sec.	300 (0.063 in. (1.6mm) from case)	
	Mounting torque, 6-32 or M3 screw.	10 lbf•in (1.1N•m)	

Thermal Resistance

	Parameter	Min.	Typ.	Max.	Units
$R_{\theta JC}$	Junction-to-Case	—	—	0.77	°C/W
$R_{\theta CS}$	Case-to-Sink, flat, greased surface	—	0.50	—	
$R_{\theta JA}$	Junction-to-Ambient, typical socket mount	—	—	80	
Wt	Weight	—	2.0 (0.07)	—	g (oz)

Source: International Rectifier Corporation, El Segundo, CA.

11.7 APPENDIX: Typical IGBT Data Sheet (Continued)

Electrical Characteristics @ T_J = 25°C (unless otherwise specified)

	Parameter	Min.	Typ.	Max.	Units	Conditions	
$V_{(BR)CES}$	Collector-to-Emitter Breakdown Voltage	600	—	—	V	V_{GE} = 0V, I_C = 250μA	
$V_{(BR)ECS}$	Emitter-to-Collector Breakdown Voltage ④	20	—	—	V	V_{GE} = 0V, I_C = 1.0A	
$\Delta V_{(BR)CES}/\Delta T_J$	Temp. Coeff. of Breakdown Voltage	—	0.63	—	V/°C	V_{GE} = 0V, I_C = 1.0mA	
$V_{CE(on)}$	Collector-to-Emitter Saturation Voltage	—	2.2	3.0	V	I_C = 20A	V_{GE} = 15V
		—	2.7	—		I_C = 40A	See Fig. 2, 5
		—	2.3	—		I_C = 20A, T_J = 150°C	
$V_{GE(th)}$	Gate Threshold Voltage	3.0	—	5.5		$V_{CE} = V_{GE}$, I_C = 250μA	
$\Delta V_{GE(th)}/\Delta T_J$	Temperature Coeff. of Threshold Voltage	—	-13	—	mV/°C	$V_{CE} = V_{GE}$, I_C = 250μA	
g_{fe}	Forward Transconductance ⑤	11	18	—	S	V_{CE} = 100V, I_C = 20A	
I_{CES}	Zero Gate Voltage Collector Current	—	—	250	μA	V_{GE} = 0V, V_{CE} = 600V	
		—	—	1000		V_{GE} = 0V, V_{CE} = 600V, T_J = 150°C	
I_{GES}	Gate-to-Emitter Leakage Current	—	—	±100	nA	V_{GE} = ±20V	

Switching Characteristics @ T_J = 25°C (unless otherwise specified)

	Parameter	Min.	Typ.	Max.	Units	Conditions
Q_g	Total Gate Charge (turn-on)	—	51	67	nC	I_C = 20A
Q_{ge}	Gate - Emitter Charge (turn-on)	—	8.9	11		V_{CC} = 400V See Fig. 8
Q_{gc}	Gate - Collector Charge (turn-on)	—	20	33		V_{GE} = 15V
$t_{d(on)}$	Turn-On Delay Time	—	25	—	ns	T_J = 25°C
t_r	Rise Time	—	21	—		I_C = 20A, V_{CC} = 480V
$t_{d(off)}$	Turn Off Delay Time	—	90	190		V_{GE} = 15V, R_G = 10Ω
t_f	Fall Time	—	43	120		Energy losses include "tail"
E_{on}	Turn-On Switching Loss	—	0.34	—	mJ	
E_{off}	Turn-Off Switching Loss	—	0.41	—		See Fig. 9, 10, 11, 14
E_{ts}	Total Switching Loss	—	0.75	1.6		
$t_{d(on)}$	Turn-On Delay Time	—	25	—	ns	T_J = 150°C,
t_r	Rise Time	—	23	—		I_C = 20A, V_{CC} = 480V
$t_{d(off)}$	Turn-Off Delay Time	—	174	—		V_{GE} = 15V, R_G = 10Ω
t_f	Fall Time	—	140	—		Energy losses include "tail"
E_{ts}	Total Switching Loss	—	1.4	—	mJ	See Fig. 10, 14
L_E	Internal Emitter Inductance	—	7.5	—	nH	Measured 5mm from package
C_{ies}	Input Capacitance	—	1500	—	pF	V_{GE} = 0V
C_{oes}	Output Capacitance	—	190	—		V_{CC} = 30V See Fig. 7
C_{res}	Reverse Transfer Capacitance	—	17	—		f = 1.0MHz

Notes:

① Repetitive rating; V_{GE}=20V, pulse width limited by max. junction temperature. (See fig. 13b)

② V_{CC}=80%(V_{CES}), V_{GE}=20V, L=10μH, R_G= 10Ω, (See fig. 13a)

③ Repetitive rating; pulse width limited by maximum junction temperature.

④ Pulse width ≤ 80μs; duty factor ≤ 0.1%.

⑤ Pulse width 5.0μs, single shot.

11.7 APPENDIX: Typical IGBT Data Sheet (Continued)

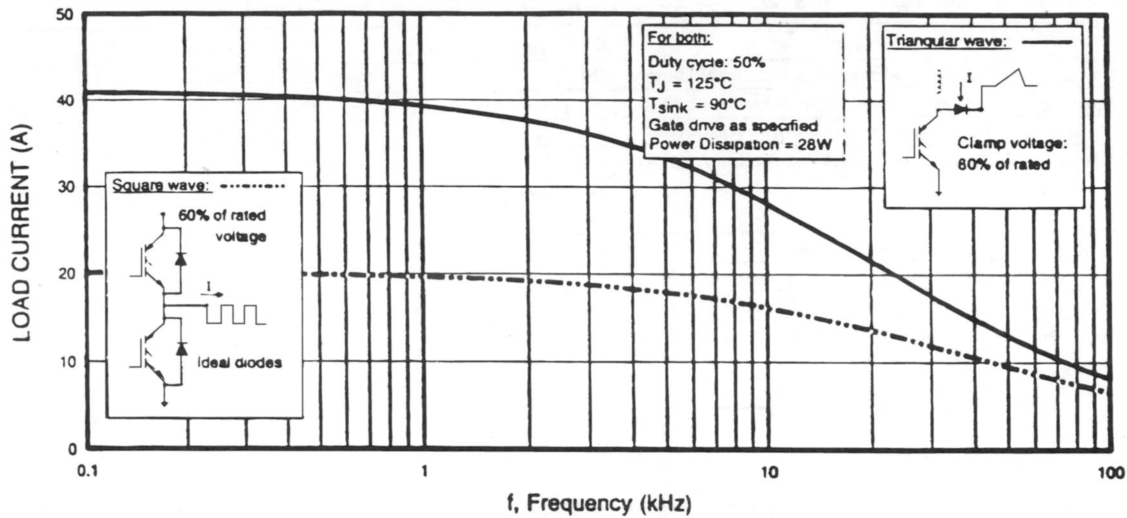

Fig. 1 - Typical Load Current vs. Frequency
(For square wave, $I = I_{RMS}$ of fundamental; for triangular wave, $I = I_{PK}$)

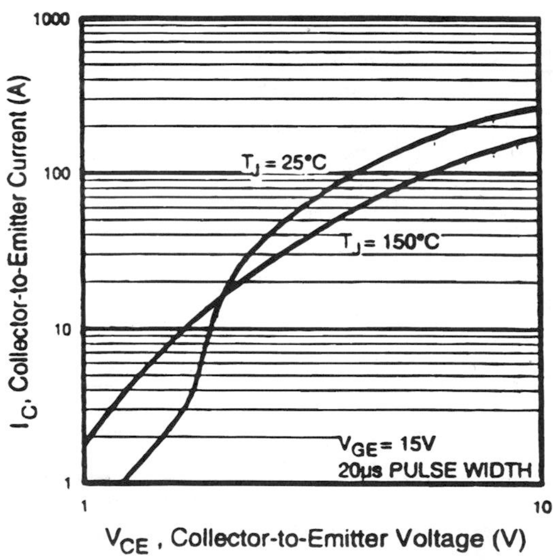

Fig. 2 - Typical Output Characteristics

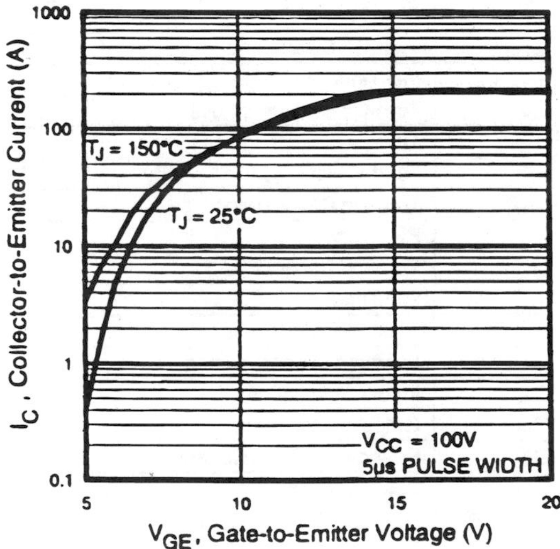

Fig. 3 - Typical Transfer Characteristics

11.7 APPENDIX: Typical IGBT Data Sheet (Continued)

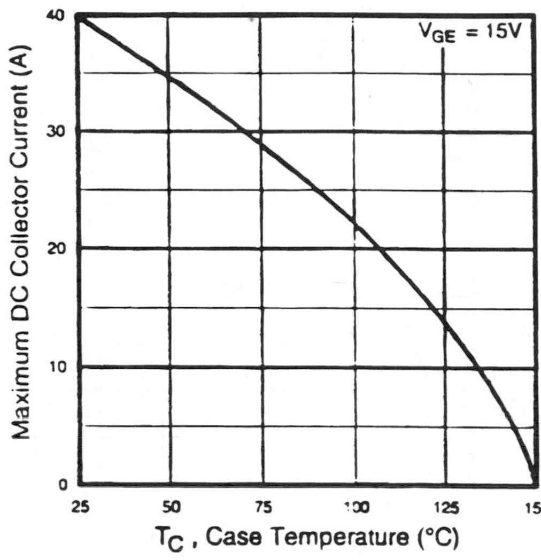

Fig. 4 - Maximum Collector Current vs. Case Temperature

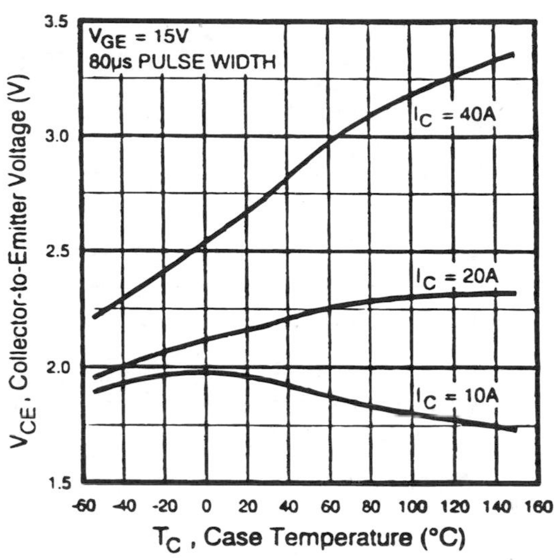

Fig. 5 - Collector-to-Emitter Voltage vs. Case Temperature

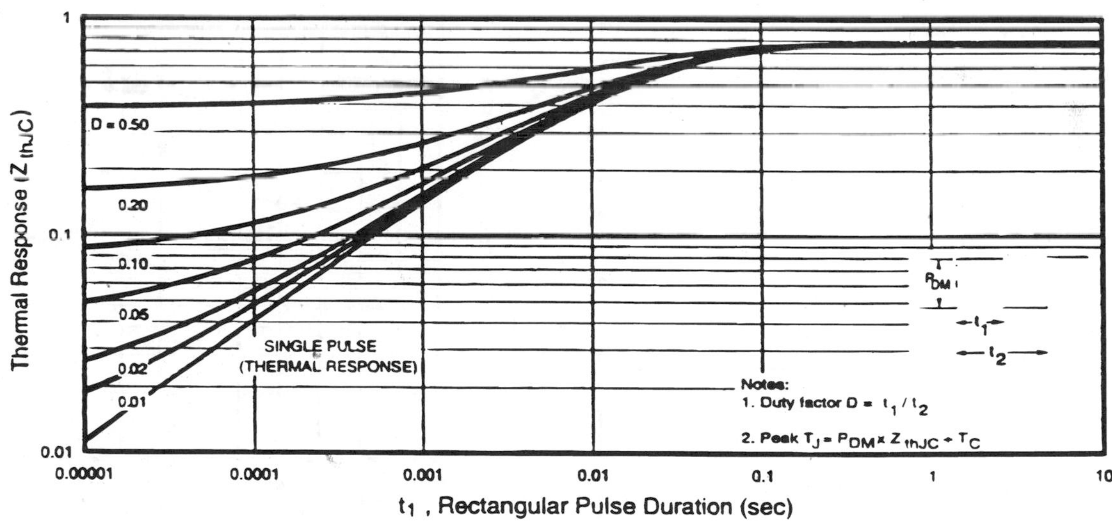

Fig. 6 - Maximum Effective Transient Thermal Impedance, Junction-to-Case

11.7 APPENDIX: Typical IGBT Data Sheet (Continued)

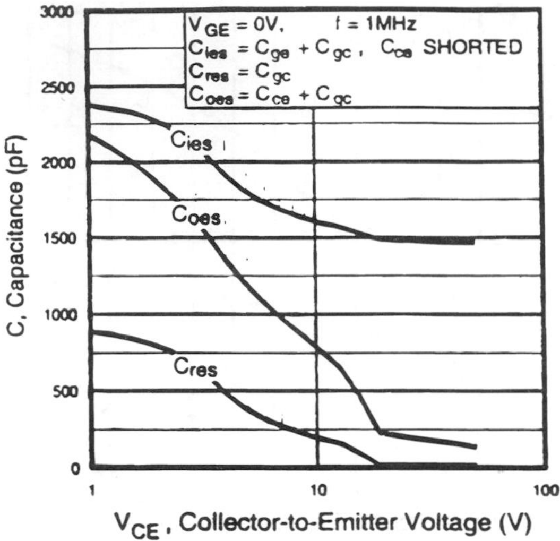

Fig. 7 - Typical Capacitance vs. Collector-to-Emitter Voltage

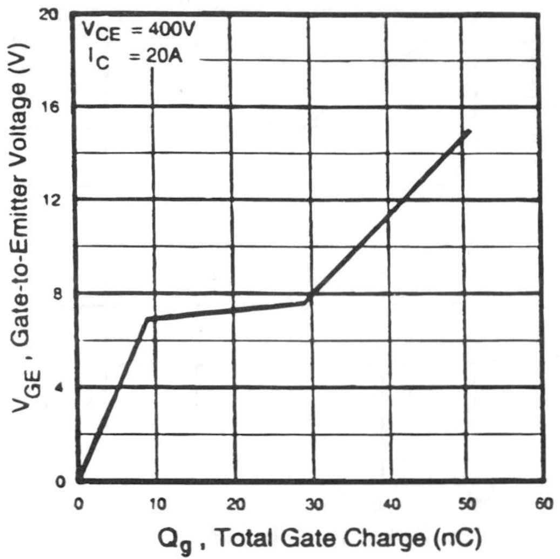

Fig. 8 - Typical Gate Charge vs. Gate-to-Emitter Voltage

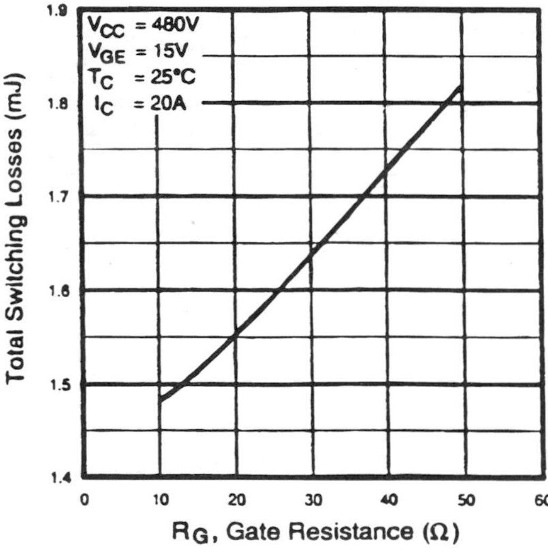

Fig. 9 - Typical Switching Losses vs. Gate Resistance

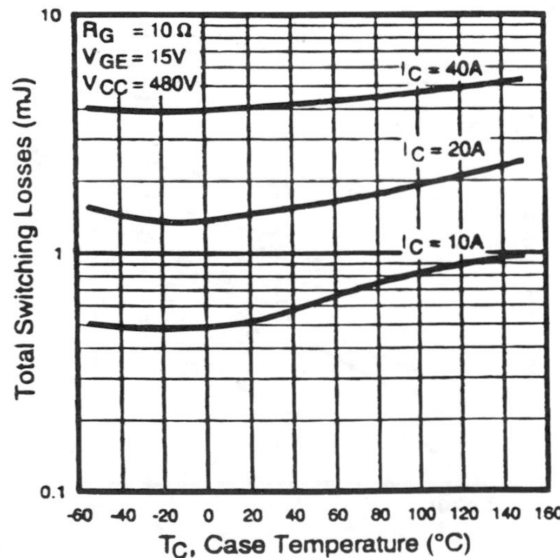

Fig. 10 - Typical Switching Losses vs. Case Temperature

11.7 APPENDIX: Typical IGBT Data Sheet (Continued)

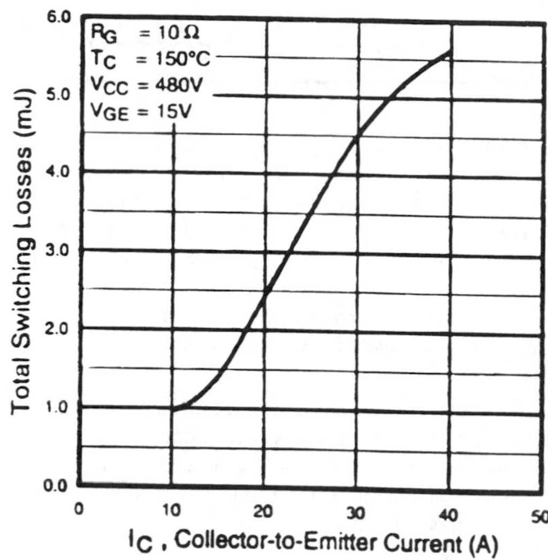

Fig. 11 - Typical Switching Losses vs. Collector-to-Emitter Current

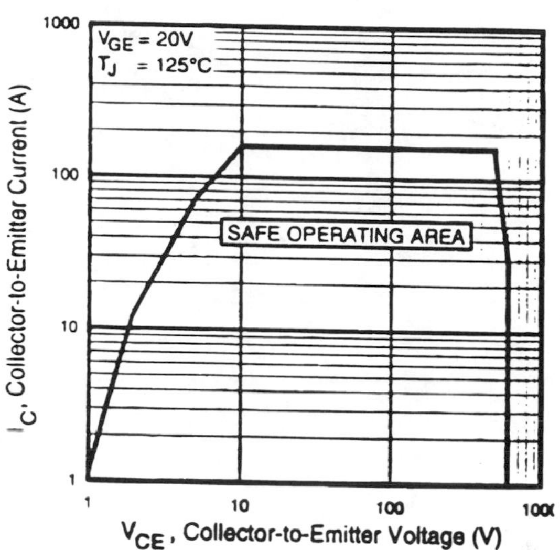

Fig. 12 - Turn-Off SOA

12

Conversion

12.1 AC-DC Converters 244
Letter Symbols Most Frequently Used in the Analysis of Converter Circuits • Basic Definitions and Calculation Methods • Rectifier and Inverter Working Mode of Line Commutated Converter Circuits • Connections of the Most Important Types of Line Commutated Converters • Calculation of Primary Current and Power Factor • Harmonics of AC Line Current and DC Terminal Voltage and Current • Waveforms and Data for Various Converter Circuits • Modern Converter Applications

12.2 DC-DC Converters 253
Introduction • Switch Mode Conversion Concept • Output Current Sourced Converters • Output Voltage Sourced Converters • Fundamental Topological Relationships • Bilateral Power Flow • Isolated DC-DC Converters • Control

12.3 DC-AC Conversion 263
Basic DC-AC Converter Connections (Square-Wave Operation) • Control of the Output Voltage • Harmonics in the Output Voltage • Filtering of Output Voltage • Practical Realization of Basic Connections • Special Realizations (Application of Resonant Converter Techniques)

12.4 AC-AC Conversion 273
Cycloconverters • Matrix Converters

12.5 Resonant Converters 276
Introduction • Survey of Second Order Resonant Circuits • Load Resonant Converters • Resonant Switch Converters • Resonant DC-Link Converters with ZVS

Attila Karpati
Technical University of Budapest

István Nagy
Technical University of Budapest

Sándor Halász
Technical University of Budapest

12.1 AC/DC Converters

Attila Karpati

The most common AC/DC converters are line commutated converter circuits. An alternating voltage generator or generator system is required on the AC terminal to realize the commutation, which primarily ensures the expedient operation of the circuit. It is possible to connect passive and/or active electrical networks to the DC terminal of the converter.

The switching devices of the modern converters are semiconductors. When only diodes are used as switching elements, the converters are uncontrolled rectifiers.

If the switching elements are controllable, they are in most cases thyristors (SCR), in which case we have controlled rectifiers. Half-controlled rectifiers, in which only a portion of the semiconductor elements are controllable, are often used.

A special case is comprised of converters with a free-wheeling diode on the DC terminal. The main direction of the power flow is from the AC terminal to the DC terminal (rectification), but in fully controlled converters without a free-wheeling diode the inverse power flow from the DC to the AC terminal is also possible (inverter operation). In the latter case it is necessary to have an energy producing generator on the DC side.

The classifications of the converter circuits are generally based on the phase number, way number and pulse number. The phase number of the converter is equal to the phase number of the AC power supply, usually one or three. The way number, depending on the directions of current flow in the secondary of the converter transformer, may be one or two. Figure 12.1 illustrates the so-called one-way and two-way converter connections. In most cases one-way connections are disadvantageous for the converter transformer, because of the unidirectional magnetizing on the secondary side. The pulse number is the ratio of the frequency of the lowest-order AC voltage harmonic across the DC terminals, to the frequency of the AC terminal voltage. At higher pulse number the order of the AC harmonics in the output voltage and in the primary current will be higher; the amplitude of the single harmonics and the resulting harmonic content (distortion) on both terminals will be smaller. The number of different converter circuits is large. Currently the one- and three-phase two-way connections, also known as bridge or 1Ph2W2P and 3Ph2W6P connections, and their combinations are used.

0-8493-8343-9/97/$0.00+$.50

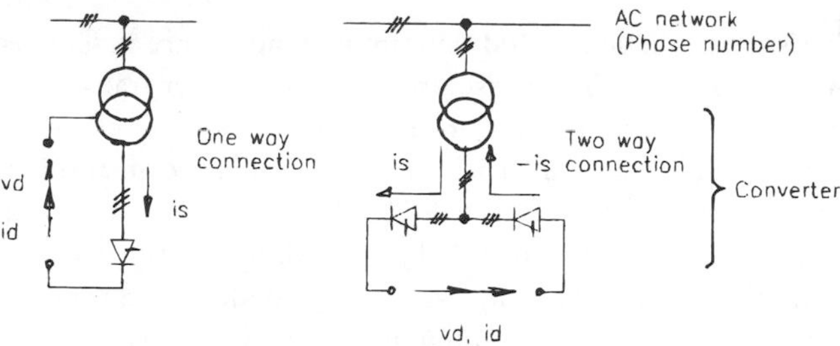

Figure 12.1 One-way and two-way converter connections.

The shorthand notation used here is as follows: Ph indicates the number of phases; W defines the number of ways and P specifies the number of pulses. In our latter case, we are talking about a three-phase, two-way six-pulse converter connection. At very low DC voltage, the application of one-way connections, also known as midpoint connections, may be advantageous. For one-way connections, the single-phase solution and the three-phase solutions with interphase transformer are recommended.

The internal design arrangements of the line commutated converters contain individual commutating groups. One individual commutating group includes all those switching elements that, under normal operating conditions, commutate among themselves, independently of other elements. The practical converter systems, depending on the required power, consists of one or more commutating groups. These may be connected in parallel or in series, as required. If the converter is either uncontrolled, half-controlled or full-controlled with free-wheeling diode (also known as bypass diode), only one polarity of the DC voltage is possible. In this case the power flows from the AC network to the DC terminal. These types of converters belong to the one-quadrant converter category.

The fully controlled converters, without a free wheeling diode are two-quadrant converters. They can work as rectifiers or as inverters. In the first case the power flows from the AC network to the DC terminal; in the second case the direction of the power flow is reversed. Because of the presence of semiconductor elements only one polarity of the DC current is possible; therefore, in the rectifier and inverter working modes the polarity of the DC voltage average value is reversed.

In some applications four-quadrant converters are necessary, and hence both polarities of the current and voltage must be permitted on the DC terminal. These types of converters consist of two two-quadrant converter sets, which are inverse parallel connected (see Figure 12.10). This configuration makes possible two directions of DC current. In a four-quadrant converter the firing angle of the thyristors changes periodically, and thus the average value of the DC terminal voltage also changes periodically. Special control of the firing angles ensures a variable frequency on the output terminal AC voltage, i.e., a Cycloconverter, direct frequency changer. The maximum possible output frequency is half that of the network frequency.

The DC terminal voltage also contains superimposed AC components (ripple), and for this reason filter circuits on the DC side are necessary in most cases. The elements of the filter circuits are inductors, and capacitors. At high output power the application of series inductors ensures the required smoothing of the DC current in most cases. At lower power, the application of capacitors for filtering the output voltage is more characteristic, but in practice mixed solutions are also used.

The AC network current of the line commutated converters is nonsinusoidal. When changing the DC terminal voltage by firing angle control the phase delay of the fundamental network current also changes and the converter requires reactive power from the AC network. The harmonic components depend on the pulse number of the converter. At increasing pulse number the harmonic content of the line current decreases. Due to high costs, greater than six pulse numbers will only be used in the MW range.

Because of the harmonic problems and the need for reactive power, especially in large converter systems, reactive power compensation and harmonic filtering are necessary on the AC terminal.

Letter Symbols Most Frequently Used in the Analysis of Converter Circuits

I_d Average value of the direct current at the DC terminal of the converter.

$I_{ThAV/RMS}$ Average/RMS value of thyristor current.

$I_{DAV/RMS}$ Average/RMS value of diode current.

I_p RMS value of the primary current of the converter transformer.

I_s RMS value of the secondary current of the converter transformer.

I_1 RMS value of the fundamental component of the network current.

I_{1P} RMS value of the active component of I_1.

I_{1Q} RMS value of the reactive component of I_1.

L_c Resulting commutating inductance.

L_d Inductance in the DC circuit.

p Pulse number of the converter; ratio of fundamental DC terminal ripple frequency to the frequency of the AC network.

P_d Average power at the DC terminal of the converter.

$V_{A,B,C}$ RMS values of the phase-to-neutral voltage system at the converter AC terminals.

V_d Generally the average value of the voltage at the DC terminal of the converter.

$V_{di\alpha}$ The average value of the voltage at the DC terminal of the converter in ideal case (with no commutation overlap) and at firing angle α.

$V_{d\alpha}$ The average value of the voltage at the DC terminal of the converter at firing angle α and with commutation overlap.

V_s RMS value of phase-to-neutral voltage at the converter AC terminals.

X_c Resulting commutating reactance, at the network frequency.

α Phase-control angle (firing angle) measured from the natural commutating points.

- δ Recovery angle of the thyristor at the network frequency.
- φ Displacement angle between the fundamental component of the converter line current and the associated phase voltage.
- $\cos(\varphi)$ Displacement factor of the fundamental harmonic of the converter line current.
- λ Network power factor of the converter; ratio of the active power to the apparent power.
- μ Commutation overlap angle.
- μ_0 Commutation overlap angle at $\alpha = 0$.
- v Distortion factor of the line current; ratio of the RMS value of the fundamental harmonic to the total RMS value.
- ω Angular frequency of the AC network.

Basic Definitions and Calculation Methods

To explain the basic definitions and calculation methods used in converter systems, the Three-Pulse Midpoint Circuit (also known as the 3Ph1W3P circuit) will be used. This, the most often used individual commutating group, is the basic circuit of various three-phase converter systems. The circuit is shown in Figure 12.2a. The following investigations assume the use of ideal thyristors or diodes. In addition, infinite inductance is assumed in the DC circuit. The resistances of the converter transformer and various current paths are neglected. The three-phase AC supply network is symmetrical, and in a first approximation the impedances are also neglected. The time functions for steady state are shown in Figures 12.2b, c. The line commutated converter circuits have generally two working modes, the continuous current mode and the discontinuous current mode. In the continuous current mode in steady state the current in the DC circuit, i_d, can never be equal to zero. In the discontinuous current mode i_d will be periodically, for a given time interval, zero. The periodicity of i_d is given by the pulse number.

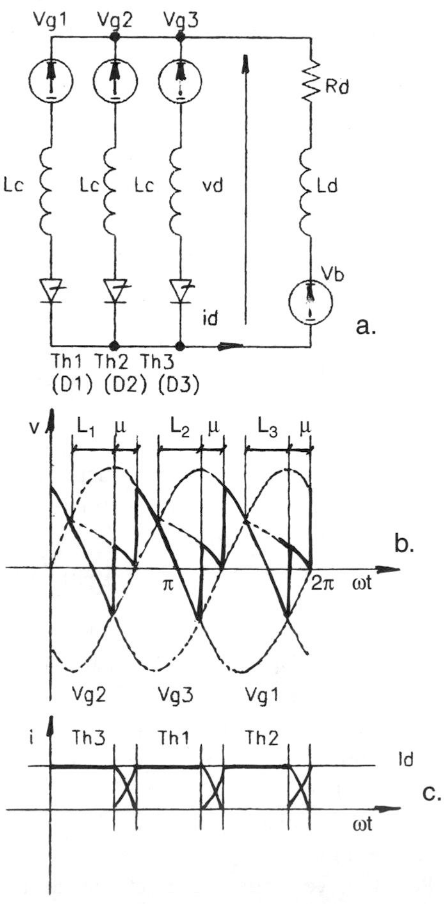

Figure 12.2 Three-pulse midpoint circuit.

In our case, due to the infinite inductance in the DC circuit the current i_d has no ripple, and the converter works in the continuous current mode. In line commutated converter circuits the commutation process, namely the change in current conduction from one thyristor to another, is very important. By turning on a new thyristor a short circuit, also known as a commutating circuit, comes into being. This short circuit contains the two thyristors and their series connected L_c inductances as well as the voltage generators.

Generally, the commutation can only begin if the polarity of the difference of the generator voltages in the commutating circuit ensures that the current of the newly turned-on thyristor increases. The velocity of the current change is, in practice, limited by the difference in the generator voltage and the inductance L_c of the commutating circuit. For this reason, the change in the current conduction requires a finite time. The process is known as commutation overlap, and its duration is given as the overlap angle, μ, in electrical degrees. Theoretically it is possible to simplify the investigation, if we assume that L_c is equal to zero. In this case the commutation is ideal. At the turn-on of the new thyristors their current increases immediately to I_d and the current in the others decreases to zero in a similar fashion.

In our investigations, it will be assumed that during commutation only two thyristors conduct current. That is why L_c is equal to the asymmetrical short-circuit inductance of the arrangement. The main part of this inductance is given by the short-circuit impedance of the converter transformer.

In most of the practical cases, we can assume that the L_c inductances in each phase have the same value. The firing angles of the thyristors are measured from the natural commutating points of the system. In our case, these points are the positive crossover points of the phase voltages as shown in Figure 12.2b. Because of the finite firing angles, the commutation is delayed. The time functions of the thyristor currents are shown in Figure 12.2c.

The relationship between the firing-delay angle α and the overlap angle μ for a one-pulse independent commutating group is given by the following equation:

$$\cos(\alpha) - \cos(\alpha + \mu) = \frac{I_d \cdot \omega \cdot L_c}{\sqrt{2} \cdot V_s \cdot \sin\left(\frac{\pi}{p}\right)}$$

In case of instantaneous commutation ($L_c = 0$), the determination of the DC terminal voltage is very simple. v_d is always equal to the generator voltage of the current conducting thyristors. After calculating the average value, we obtain the following formula for $V_{di\alpha}$:

$$V_{di\alpha} = V_{di0} \cdot \cos(\alpha)$$

where

$$V_{di0} = \sqrt{2} \cdot V_s \cdot \frac{p}{\pi} \cdot \sin\left(\frac{\pi}{p}\right)$$

In case of real commutation, the time function of V_d changes as shown in Figure 12.2b. During commutation, because of the symmetry of L_c in each phase, v_d follows the mathematical average value of the generator voltages. If only one thyristor is conducting current, the output voltage is equal to the generator voltage in the branch of the conducting thyristors, because there is no voltage drop on the L_c inductances since L_d is assumed infinite. Finally, because of the commutation, the output voltage decreases. After integration we obtain for a one independent p-pulse commutating group the following formula for $V_{d\alpha}$:

$$V_{d\alpha} = V_{di\alpha} - g_x$$

$$g_x = \frac{I_d \cdot \omega \cdot L_c}{\frac{2 \cdot \pi}{p}}$$

where g_x is the voltage drop caused by the commutation.

The average value of the thyristor current for a one-p-pulse commutating group can be calculated using the formula:

$$I_{\text{ThAV}} = \frac{I_d}{p}$$

The RMS value, in the case of ideal commutation, can be calculated by the formula:

$$I_{\text{ThRMS}} = \frac{I_d}{\sqrt{p}}$$

In the case of real commutation, the use of one correction factor is necessary and thus

$$I_{\text{ThRMS}} = \frac{I_d}{\sqrt{p}} \cdot c(\alpha, \mu)$$

Rectifier and Inverter Working Mode of Line Commutated Converter Circuits

The DC terminal voltage of the fully controlled individual commutating group in an ideal case depends upon the firing voltage α as follows:

$$V_{di\alpha} = V_{di0} \cdot \cos(\alpha)$$

Accordingly, at instantaneous commutation in the range $0 \leq \alpha \leq \pi/2$, V_d is positive, and in the range $\pi/2 \leq \alpha \leq \pi$, V_d is negative. Because of the presence of the semiconductor elements in the converter, the current I_d flows only in one direction. The DC terminal power of the converter is also positive or negative, depending on α. In the case of $P_d < 0$, that is, $0 \leq \alpha \leq \pi/2$, the converter is working as a rectifier and power flows from the AC terminal to the DC terminal. In the case of $P_d > 0$, that is, $\pi/2 \leq \alpha \leq \pi$, the converter is working as a inverter and power flows from the DC terminal to the AC terminal.

Under these considerations, we assume that the converter is working in a continuous current mode, and there is a finite I_d current in the DC circuit. This means, for example, that when operating as an inverter to produce energy, which is fed back to the DC terminal through the AC network, a DC voltage generator is also employed.

In practical cases, it is not possible to use the above-mentioned upper limit of $\alpha = \pi$ because of the overlap angle μ of the circuit and the recovery angle δ of the thyristors. Theoretically the upper limit for α is calculated by the formula $\alpha + \mu + \delta \leq \pi$. But in practice, in most cases it is sufficient if in inverter operation the maximum firing angle α is limited to $5 \cdot \pi/6$.

The above-mentioned considerations are valid for all types of fully controlled line commutated converter circuits.

Connections of the Most Important Types of Line Commutated Converters

The most frequently used one-way connections are the two-pulse midpoint connection (1Ph1W2P), the three-pulse midpoint connection (3Ph1W3P), and the six-pulse midpoint connection (3Ph1W6P) with interphase transformer.

The two pulse midpoint connection can be seen in Figure 12.3. The two opposite voltages necessary for its operation are produced by a center-tapped transformer.

The three-pulse midpoint connection is shown in Figure 12.4. The three-phase voltage system, which is necessary for the operation of the converter, is generally produced by a three-pulse transformer. At the primary winding of the transformer, both the star (Y) and the delta connections are permitted. Theoretically it would be possible to connect the semiconductors directly to

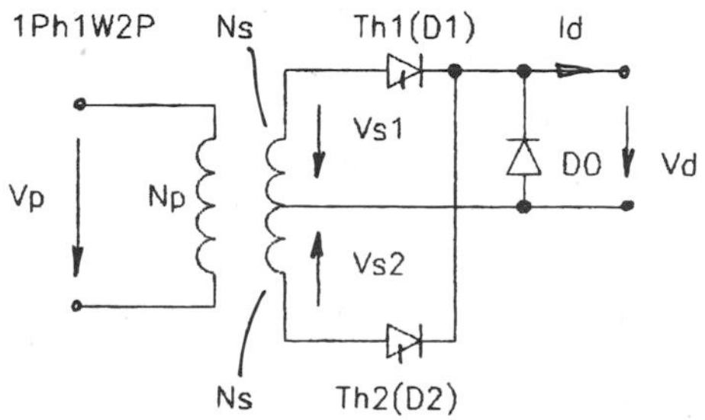

Figure 12.3 Two-pulse midpoint connection.

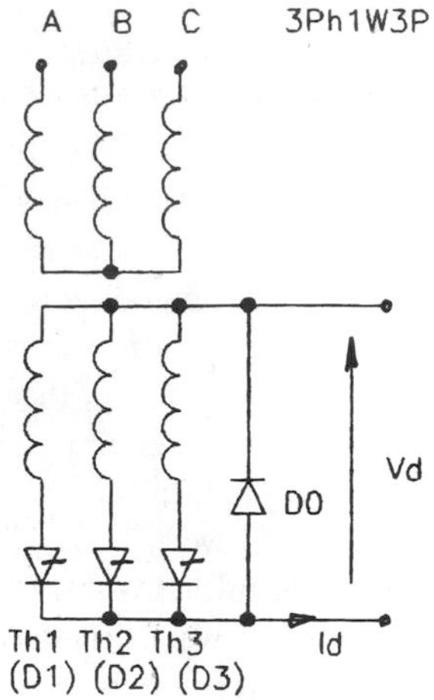

Figure 12.4 Three-pulse midpoint connection.

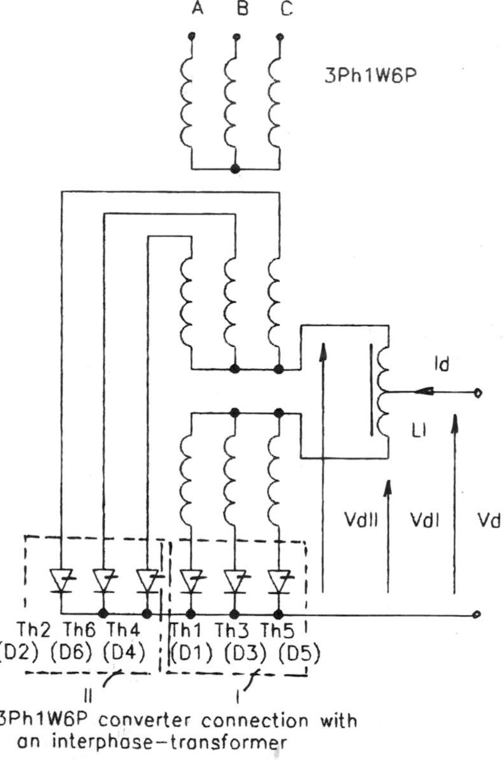

Figure 12.5 Six pulse midpoint connection.

one three-phase AC network, but because of the DC component in the AC supply network this solution is generally undesirable.

The six-pulse midpoint connection with interphase transformer is shown in Figure 12.5. The six-phase voltage system, which is necessary for the operation of the converter, is produced by one three-phase transformer with two star-connected winding systems on the secondary side. The center-tapped interphase transformer is connected to the star points of the secondary systems. The connection is one example of the application of two individual commutating groups. In this parallel connection, the interphase transformer is used because of the difference in AC voltage between the two commutating groups. The application of this connection is advantageous at smaller DC terminal voltages and higher DC currents.

All of these connections can be either controlled or uncontrolled. In the controlled versions, the application of free-wheeling diodes is possible. They are shown enclosed in dotted lines in the figures. If they are employed, inverter operation is not possible because $V_d \geq 0$; but with firing angle control, the reactive and the harmonic components of the AC network current will be reduced.

The most frequently used two-way connections are the two-pulse bridge connection (1Ph2W2P connection) and the six-pulse bridge connection (3Ph2W6P). At greater DC power the parallel and series combination of the six-pulse bridge connections are often used to get twelve-pulse converter systems.

The double-way two-pulse bridge connection is illustrated in Figure 12.6. The connection contains two commutating groups. Full-controlled, half-controlled and uncontrolled versions are possible. In the half-controlled version, two diodes and two thyristors are used. The two diodes ensure one free-wheeling branch for the DC circuit. In addition, in this half-controlled version, inverter operation is not possible because $V_d \geq 0$.

The six-pulse bridge connection is illustrated in Figure 12.7. The circuit is the series connection of two 3Ph1W3P commutating groups. Full-controlled, half-controlled and uncontrolled versions are possible. In the half-controlled version, one commutating group contains thyristors, the other diodes. In this case, inverter operation is not possible because $V_d \geq 0$.

The application of converter transformers is not absolutely necessary in bridge connections. Furthermore, when transformers are not used, the converter is connected to the AC network and series impedances are used to reduce the short circuit currents. This solution is typically used at small output power levels.

At high output power levels, above ~3 MW, it is necessary to increase the pulse number. This is possible by using more

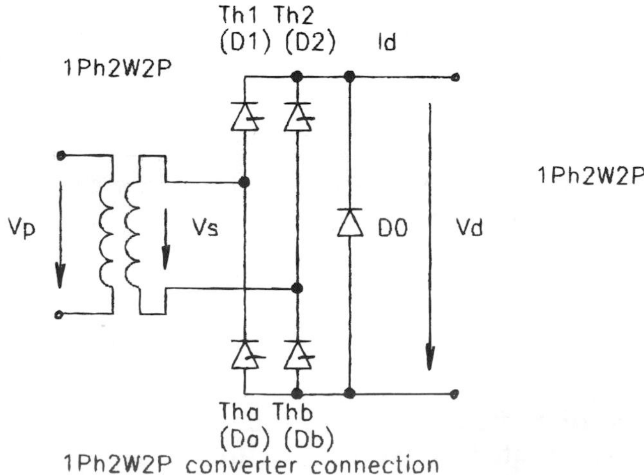

Figure 12.6 Double-way two-pulse bridge connection.

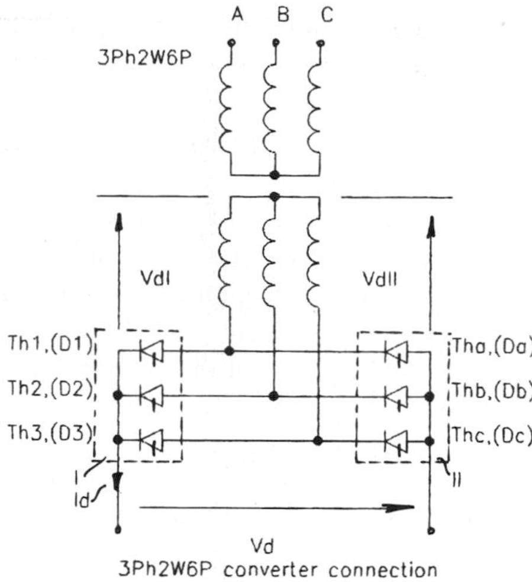

Figure 12.7 Six-pulse bridge connection.

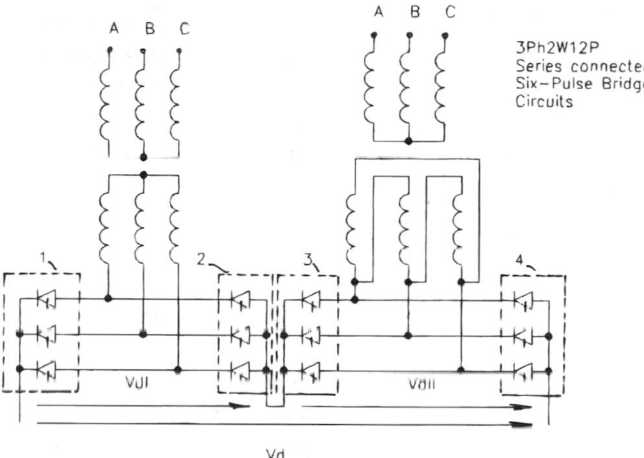

Figure 12.8 Two six-pulse bridge circuits are series connected.

than two series- or parallel-connected commutating groups in the converter.

A typical solution is illustrated in Figure 12.8, where two six-pulse bridge circuits are series connected (four series connected commutating groups). Symmetrical control of the switching elements and a suitable displacement of the voltages of the two secondary windings ($c \cdot \pi/6$, where c is an odd number) results in a twelve-pulse converter. We can also obtain a 12-pulse converter, when two six-pulse bridge circuits are connected in parallel. This resulting converter is the series-parallel connection of four commutating groups. The suitable displacement of the secondary voltages, as shown previously, is again necessary. The converter is illustrated in Figure 12.9.

A typical four quadrant converter is illustrated in Figure 12.10. The two six-pulse bridge circuits are inverse parallel connected. The positive current is supplied by converter I, and negative current by converter II. Both converters can operate as either a rectifier or an inverter; for this reason both directions of voltage and current are possible on the DC terminal.

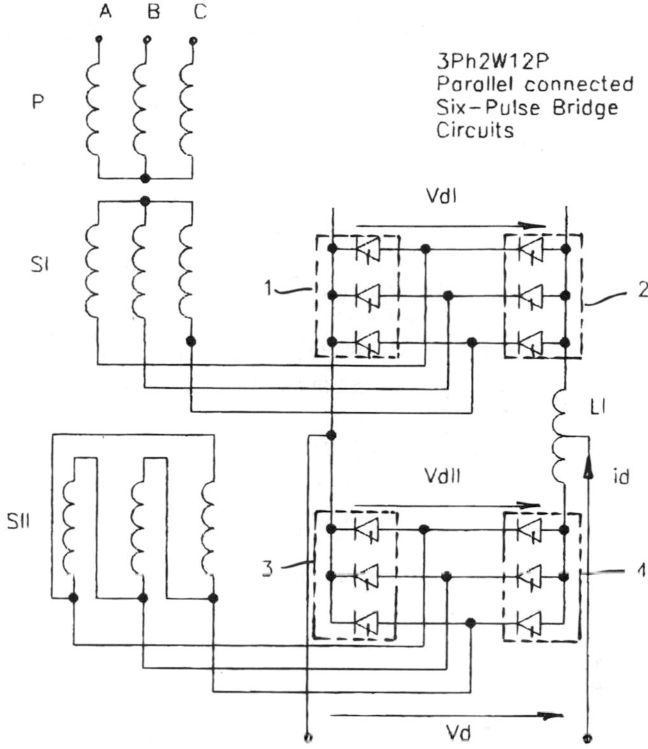

Figure 12.9 Two six-pulse bridge circuits are parallel connected.

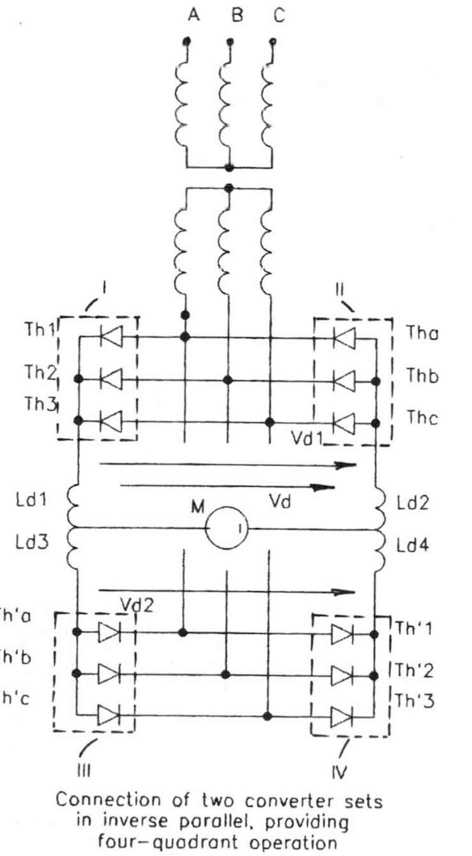

Figure 12.10 Two six-pulse bridge circuits are inverse parallel connected.

Calculation of Primary Current and Power Factor

The calculation of primary current is based upon the assumption of an ideal transformer, i.e., the magnetizing current and the iron-core losses are negligible.

In this calculation the excitation law is used; however, because of the zero-sequence and DC current excitations on the secondary side of the transformer, the primary and secondary excitations on one leg of the iron core are not always equal to zero.

In the two-way solutions there are no zero sequence or DC components in the secondary current, and therefore the excitation law can be used for each leg in the classical form, i.e., $i_p \cdot N_p + i_s \cdot N_s = 0$. Furthermore the calculation of the primary current by utilizing knowledge of the secondary currents is straightforward.

In one-way converters the main problem is that the secondary currents have DC components.

In the 1Ph1W1P solution the converter transformer will be periodically saturated because of the DC component, and that is why this type of converter should not be used in power electronics.

In other one-way converters, the classical form of the excitation law is not always valid and depends upon both the type of iron core and the secondary winding system. In this case the excitation law has the form $i_p \cdot N_p + i_s \cdot N_s = \Theta_0$ where Θ_0 is the residual excitation of the leg. There are converter circuits where $\Theta_0 = 0$. In other converter circuits Θ_0 is equal to a constant value, and there is an application where Θ_0 is alternating.

In general the case $\Theta_0 \neq 0$ should be avoided, because of the magnetic bias of the legs, which entails a considerable increase in the primary current of the transformer, especially if low saturation induction is used. For this reason one should use: the 1Ph1W2P circuit, if the windings are placed on two legs in a zigzag connection on the secondary side, and the 3Ph1W6P circuit if the primary side is connected in delta.

In both the 3Ph1W6P circuit with interphase transformer and the 1Ph1W2P circuit, when all of the windings are placed on the same leg, $\Theta_0 = 0$.

The network power factor of the converter is defined by the classical definition, i.e., the ratio of the active power to the apparent power. Because the network current is nonsinusoidal, the power factor has two parts and is of the form $\lambda = \nu \cdot \cos(\varphi)$. φ is the displacement angle between the fundamental components of the network current and the associated phase voltage. ν is the distortion factor of the network current and is defined as the ratio of the RMS value of the fundamental harmonic to the total RMS value. This definition assumes that the network voltage is sinusoidal and only currents and voltages with equivalent frequency produce active power. The size and cost of the converter transformer are proportional to its apparent power. Because at the converter transformer, the apparent power of the primary and secondary sides are not always equal, an average value will be used. This value is known as the rated apparent power of the transformer and is calculated by the formula

$$S_t = \frac{S_p + S_s}{2}$$

Harmonics of AC Line Current and DC Terminal Voltage and Current

The AC line current of the converters is non-sinusoidal and the instantaneous value of the output voltage of the converter is not constant (see Figure 12.11). In this chapter, the main features of the harmonics of the AC line current and the DC side voltage will be discussed.

Harmonics of AC Line Current

In practice, all harmonics with an integer number appear in the AC line current of the converter. The harmonics can be

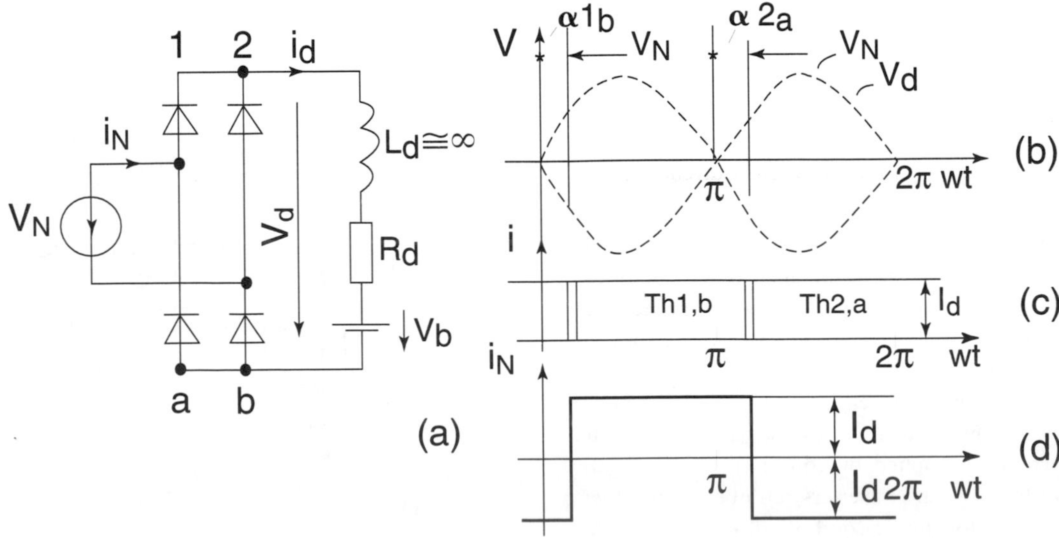

Figure 12.11 AC line current of converters is non-sinusoidal.

divided into two parts. The characteristic harmonics are in the first classified part, while the non-characteristic harmonics are classified in the second. The network currents of the symmetrical converters (s) supplied from a symmetrical AC network contain only characteristic harmonics. In practical cases, the r.m.s. values of the non-characteristic harmonics are rather small.

With regard to the line current of a converter circuit it is useful to note the following practical rules:

- The order of the current harmonics is given by the pulse number. In a p-pulse circuit, the AC line current contains harmonics of order $n = cp + 1$ (where p is the pulse number and $c = 0, 1, 2 \ldots$).
- The r.m.s. value of the characteristic harmonics depends upon the type of DC load and the average value of the DC current.
- Assuming that infinite inductance exists in the DC circuit, and that commutations are ideal, the r.m.s. value of the fundamental component of the AC line current and its harmonics are related as follows: $I_2/I_1 = 1/n$. (The Muller-Lubeck amplitude rule.) The amplitude of the AC harmonic currents is reduced by overlapping.
- By increasing the pulse number certain harmonics will be omitted, but the r.m.s. value of the remaining harmonics related to the fundamental current remain unchanged (see Table 12.1).
- If the voltage on the DC side is filtered by capacitance, the per unit amplitude of the AC line harmonics is strongly correlated to the AC side input impedance of the converter. If a small input impedance is used, the instantaneous maximum value of the AC line current can be very high.
- In practice the amplitudes of the non-characteristic harmonics of the AC line current are generally only some thousandth part of the fundamental current, except the third harmonic, which can increase to approx. 1 percent per unit value.

The non-characteristic harmonics are caused by the asymmetry of the converter system, such as the asymmetries of the firing angles: of the inductances in the commutating circuits, etc.

Table 12.1 The Per Unit Amplitude of the AC Current Harmonics

n	$p = 2$	$p = 3$	$p = 6$	$p = 12$
1	1.000	1.000	1.000	1.000
2	—	0.500	—	—
3	0.330	—	—	—
4	—	0.250	—	—
5	0.200	0.200	0.2007	—
6	—	—	—	—
7	0.143	0.143	0.143	—
8	—	0.125	—	—
9	0.111	—	—	—
10	—	0.100	—	—
11	0.091	0.091	0.091	0.091
12	—	—	—	—

Harmonics of DC Terminal Voltage and Current

With reference to the ripple of the converters DC side voltage, it is useful to note the following practical rules:

- With high power converters series inductance is generally used to reduce the ripple of the DC current. The harmonic content of the DC side voltage of the converter remains unchanged.
- At lower power levels, i.e., ≤ 100 kW, filtering with capacitance is often used. This filtering decreases the ripple of the DC side voltage.
- The order of the harmonics in the DC side voltage is given by the following equation:

$$n = cp$$

(where $c = 0, 1, 2, \ldots$ and p is the pulse number).

- The amplitudes of the harmonics in the DC side voltage will be calculated by Fourier analyses. Their amplitudes are approximately in direct proportion to the inverse of the harmonic order.
- By increasing the pulse number certain harmonics will be omitted, and the ripple of the DC side voltage strongly decreases.
- The r.m.s. values of the DC side current components can be calculated by the following formula:

$$I_{dn} = V_{dn}/Z_n$$

(where V_{dn} is the r.m.s. value of the nth harmonic voltage on the output and Z_n is the harmonic impedance of the load at the given frequency).

Waveforms and Data for Various Converter Circuits

Presented in Table 12.2 is a compilation of the most familiar uncontrolled converter connections, together with their characteristic current waveforms and the more important relationships needed for their dimensioning. These facts are only valid in case of infinite series inductance on the DC side.

Modern Converter Applications

The application of forced commutation and fully controllable semiconductor elements makes the solution of special problems possible. At present the most important problems are the reducing or the elimination of the harmonics in the AC line current. The most important types of these converters can be divided into two groups.

Group 1

Converters, which contain on their input an uncontrolled rectifier and thus supplies a special controlled DC-DC converters

Table 12.2

Converter connection		Single-phase center-tap connection (1Ph2W2P)	Single-phase bridge connection (1Ph2W2P)	Star-star connection	Three-phase delta/zigzag connection
Phase number		Line-side 1, Valve-side 2	1	Line-side 3, Valve-side 3	Line-side 3, Valve-side 3
Circuit symbol		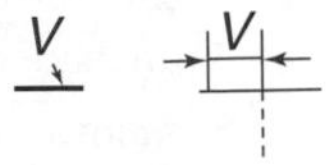	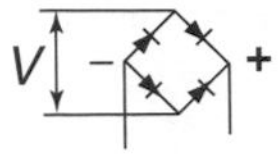		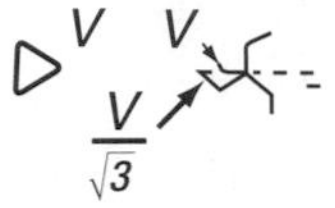
Valve current	Waveform	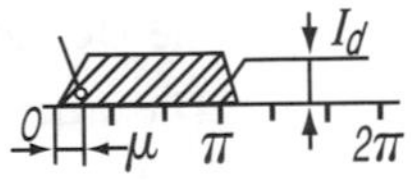	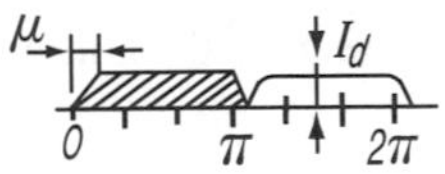	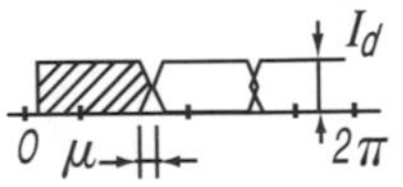	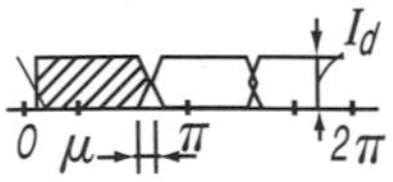
	Mean value	$1/2\ I_d = 0.500\ I_d$	$1/2\ I_d = 0.500\ I_d$	$1/3\ I_d = 0.333\ I_d$	$1/3\ I_d = 0.333\ I_d$
	R.m.s. value (without overlap)	$1/\sqrt{2}\ I_d = 0.707\ I_d$	$1/\sqrt{2}\ I_d = 0.707\ I_d$	$1/\sqrt{3}\ I_d = 0.577\ I_d$	$1/\sqrt{3}\ I_d = 0.577\ I_d$
	Overlap correction	$\sqrt{1 - 2\Psi(\alpha, \mu)}$	$\sqrt{1 - 2\Psi(\alpha, \mu)}$	$\sqrt{1 - 3\Psi(\alpha, \mu)}$	$\sqrt{1 - 3\Psi(\alpha, \mu)}$
Current in valve-side winding	Waveform	Conformable to valve current		Conformable to valve current	Conformable to valve current
	R.m.a. valve (without overlap)	Conformable to valve current		Conformable to valve current	Conformable to valve current
	Overlap correction	Conformable to valve current		Conformable to valve current	Conformable to valve current
	Waveform	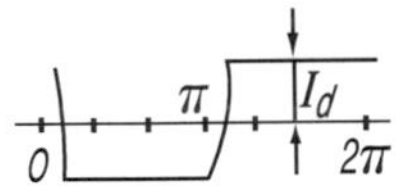		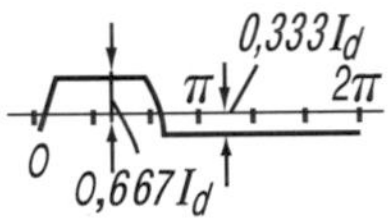	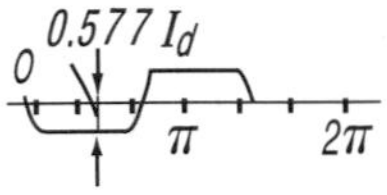
	R.m.s. value without overlap	I_d		$\frac{\sqrt{2}}{3} I_d = 0.471\ I_d$	$\frac{\sqrt{2}}{3} I_d = 0.471\ I_d$
	Overlap correction	$\sqrt{1 - 4\Psi(\alpha, \mu)}$		$\sqrt{1 - 9/2\Psi(\alpha, \mu)}$	$\sqrt{1 - 9/2\Psi(\alpha, \mu)}$
	Form factor ($\mu = 0$)	1		—	$\sqrt{\frac{3}{2}} = 1.23$
Line current	Waveform	Conformable to line-side current	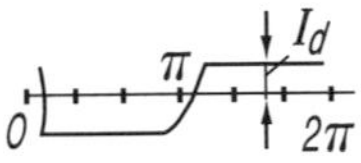	Conformable to line-side current	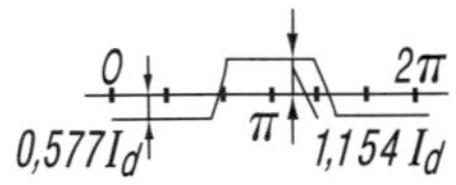
	R.m.s. value without overlap	Conformable to line-side current	I_d	Conformable to line-side current	$\sqrt{\frac{2}{3}}\ I_d = 0.815\ I_d$
	Overlap correction	Conformable to line-side current	$\sqrt{1 - 4\Psi(\alpha,\mu)}$	Conformable to line-side current	$\sqrt{1 - 9/2\Psi(\alpha, \mu)}$
	Form factor ($\mu = 0$)	Conformable to line-side current	1	—	—

(see Figure 12.12a). (The following types of DC-DC converters are primarily used: Buck, Boost, Buck-Boost, SEPIC, Full-Bridge, Flyback.) The DC-DC converter works in continuous (Figure 12.12a) or in discontinuous (Figures 12.12b and c) current mode. In the first case, its control is a cascade control, where the external loop controls the output DC voltage. The internal loop controls the input current of the DC-DC converter. The reference value for the current controller has an absolute sine curve. The amplitude of the sine function is given by the voltage controller. The internal control loop controls the input current of the DC-DC converter such that it approximates the absolute sine curve. The switch on and off times of the semiconductor element(s) are given in DC-DC converter by PWM or two-position control. In the second case, in discontinuous current mode, only one control loop and PWM are used. For a half period of the line voltage the $t_{(on)}$ time is approximately constant, therefore the amplitudes of the input current pulses of the DC-DC converter changes with the absolute sine function of the AC line voltage. The resulting harmonic content of the AC line current will be relatively small. In discontinuous current mode resonant converter connections are also used. In this case the AC line current consists of half sine pulses directly on the rectifier input. During the

Table 12.3

Converter connection		Single-phase center-tap connection (1Ph1W2P)	Single-phase bridge connection (1Ph1W2P)	Star-star connection	Three-phase delta/zigzag connection
Average rated apparent power of transformer	Line-side windings (without overlap)	$\frac{\pi}{2\sqrt{2}} = 1.11\ P_d$		$\frac{2\pi}{3\sqrt{3}} = 1.21\ P_d$	$\frac{2\pi}{3\sqrt{3}} = 1.21\ P_d$
	Overlap correction	$\frac{\sqrt{1 - 4\Psi.0°, \mu)}}{\cos^2 \mu/2}$		$\frac{\sqrt{1 - 9/2\Psi(0°, \mu)}}{\cos^2 \mu/2}$	$\frac{\sqrt{1 - 9/2\Psi(0°, \mu)}}{\cos^2 \mu/2}$
	Valve-side windings (without overlap)	$\frac{\pi}{2} = 1.67\ P_d$		$\frac{\sqrt{2}}{3}\pi = 1.48\ P_d$	$\pi\frac{2}{3}\sqrt{\frac{2}{3}} = 1.71\ P_d$
	Overlap correction	$\frac{\sqrt{1 - 2\Psi(0°, \mu)}}{\cos^2 \mu/2}$		$\frac{\sqrt{1 - 3\Psi(0°, \mu)}}{\cos^2 \mu/2}$	$\frac{\sqrt{1 - 3\Psi(0°, \mu)}}{\cos^2 \mu/2}$
	Average rated apparent power (without overlap)	$\frac{1.11 + 1.57}{2} = 1.34\ P_d$		$\frac{1.21 + 1.48}{2} = 1.35\ P_d$	$\frac{1.21 + 1.71}{2} = 1.46\ P_d$
Average rated apparent power of interphase transformer	At operating frequency	—		—	—
	As an equivalent transformer	—		—	—
Power factor $\mu = 0$;		$\frac{2 \cdot \sqrt{2}}{\pi} \cdot \cos \alpha$;	$\frac{2 \cdot \sqrt{2}}{\pi} \cdot \cos \alpha$;	$\frac{3\sqrt{3}}{2\pi} \cdot \cos \alpha$	$\frac{2\sqrt{3}}{2\pi} \cdot \cos \alpha$
$\mu \neq 0$		$\frac{0.9}{2} \frac{\cos(\alpha + \mu) + \cos(\alpha)}{\sqrt{1 - 4\Psi(\alpha\mu)}}$	$\frac{0.9}{2} \frac{\cos(\alpha + \mu) + \cos(\alpha)}{\sqrt{1 - 4\Psi(\alpha, \mu)}}$	$\frac{0.826}{2} \cdot \frac{\cos(\alpha + \mu) + \cos(\alpha)}{\sqrt{1 - (9/2)\Psi(\alpha, \mu)}}$	$\frac{0.826}{2} \cdot \frac{\cos(\alpha + \mu) + \cos(\alpha)}{\sqrt{1 - (9/2)\Psi(\alpha, \mu)}}$
DC voltage		$\left(\frac{2\sqrt{2}}{\pi} = 0.9\right) \cdot V \frac{\cos(\alpha + \mu) + \cos(\alpha)}{2}$	$\left(\frac{2\sqrt{2}}{\pi} = 0.9\right) \cdot V \frac{\cos(\alpha + \mu) + \cos(\alpha)}{2}$	$\left(\frac{3}{\pi}\sqrt{\frac{3}{2}} = 1.17\right) \cdot V \frac{\cos(\alpha + \mu) + \cos(\alpha)}{2}$	$\left(\frac{3}{\pi}\sqrt{\frac{3}{2}} = 1.17\right) \cdot V \frac{\cos(\alpha + \mu) + \cos(\alpha)}{2}$

network half period these pulses are dispersed such that the harmonic content of the AC line current will be optimal.

Group 2

Converters where the input rectifier is directly modified. Fully controllable semiconductor elements or thyristors with forced commutation are usually applied. Single-phase and three-phase (see Figure 12.13) realizations and on the DC side filtering with capacitance or inductance are also used. The switch on and off times of the semiconductor elements are given by PWM or two-position control. In all of these cases, in the AC line an additional filter circuit which filters the harmonic currents in the necessary degree must be used.

12.2 DC-DC Converters

István Nagy

Introduction

The DC-DC converters illustrated in Figure 12.14 are used to interface two DC systems and control the flow of power between them. Their basic function in a DC environment is similar to that of transformers in AC systems. Unlike transformers, the ratio of the input to the output, either voltage or current, can continuously be varied by the control signal and this ratio can be higher or lower than unity.

The DC-DC converter is called a chopper in high-power applications. They are used for DC motor control mainly in battery supplied vehicles and in the following applications: for electric cars, airplanes, and spaceships, where on-board regulated DC power supplies are required. In general, DC-DC converters are employed as power supplies in sensors, controllers, transducers, computers, commercial electronics, electronic instruments as well as a variety of technologies which include plasma, arc, electron beam, electrolytic, nuclear physics, solar energy conversion, and the like. The DC-DC converters are constructed of electronic switches and sometimes include inductive and capacitive components, all of which are normally followed by a low-pass filter. There are a number of classifications for these converters which are dependent upon the input impedance, $\overline{Z}_i$, of the low-pass filter as shown in Figure 12.15 (Rashid, 1993). The converter is either output current sourced in which case $\overline{Z}_i \simeq j\omega L$ or output voltage sourced such that $\overline{Z}_i = -j/\omega C$; in which case either the

Table 12.4

Converter connection		Delta six-phase star connection		Interphase-transformer connection		Three-phase bridge connection	Interphase-transformer connection	
		Line-side	Valve-side	Line-side	Valve-side		Line-side	Valve-side
Phase number		3	6	3	6	3	3	12
Circuit symbol				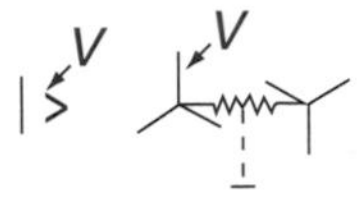		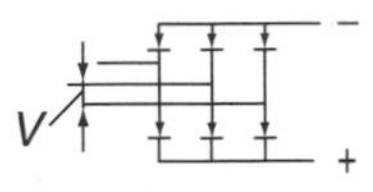	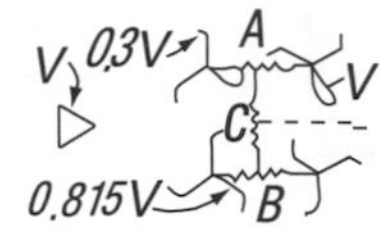	
Valve current	Waveform	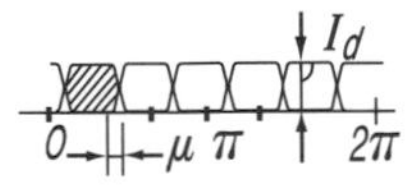		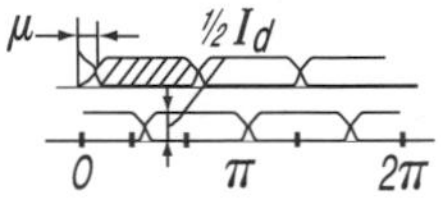		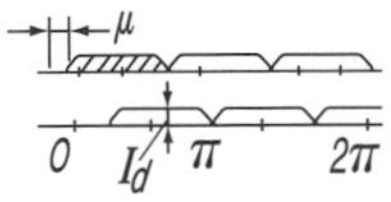	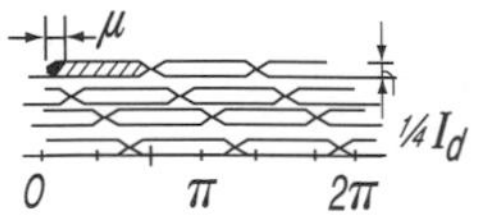	
	Mean value	$1/6\ I_d = 0.167\ I_d$		$1/6\ I_d = 0.167\ I_d$		$1/3\ I_d = 0.333\ I_d$	$1/12\ I_d = 0.083\ I_d$	
	R.m.s. value (without overlap)	$\frac{1}{\sqrt{6}} I_d = 0.408\ I_d$		$\frac{1}{2\sqrt{3}} I_d = 0.280\ I_d$		$\frac{1}{\sqrt{3}} I_d = 0.577\ I_d$	$\frac{1}{4\sqrt{3}} I_d = 0.144\ I_d$	
	Overlap correction	$\sqrt{1 - 6\Psi(\alpha, \mu)}$		$\sqrt{1 - 3\Psi(\alpha, \mu)}$		$\sqrt{1 - 3\Psi(\alpha, \mu)}$	$\sqrt{1 - 3\Psi(\alpha, \mu)}$	
Current in valve-side winding	Waveform	Conformable to valve current		Conformable to valve current			Conformable to valve current	
	R.m.s. value (without overlap)	Conformable to valve current		Conformable to valve current			Conformable to valve current	
	Overlap correction	Conformable to valve current		Conformable to valve current			Conformable to valve current	

output current or voltage is designed to be ripple free, i.e., constant in one switching cycle. Some DC-DC converters permit power flow in only one direction, others implement bidirectional power flow. Depending upon the direction of the output current and voltage, the converters can be classified into five classes as shown in Figure 12.16. One-quadrant (classes A and B), two-quadrant (classes C and D) and four-quadrant operation can be realized.

Hard switched and soft switched or resonant converters exhibit one other classification. In the first (second) group the power loss is high (low) in switching as a result of the non-zero voltage and current (zero voltage and/or current) on the switches at the initialization of the switching action as described in Section 12.5.

The step-down or buck converter can only reduce, while the step-up or boost converter can only increase the average output voltage in comparison with the input voltage. The step-up/down or buck-and-boost converter produces an output voltage that is either lower or higher than the input voltage. DC-DC converters are built with and without electrical isolation. The former usually incorporate both a DC-AC and an AC-DC converter in cascade as well as a transformer at the terminals of the AC signals for electrical isolation. The transformer turns ratio is also utilized for bridging a larger gap between the input and output voltage.

There is (not) a direct path between the input and output terminals in the direct (indirect) converter. Although these converters may operate in either a continuous or discontinuous current conduction mode, only the continuous current conduction mode will be discussed here.

Switch Mode Conversion Concept

The ripple-free DC voltage shown in Figure 12.17a or the ripple free current shown in Figure 12.17b is periodically chopped by the switch S. By changing the duty ratio $D = T_{on}/T$, the average value of either waveform can be varied continuously. The ratio of the switching frequency $f_s = 1/T$ to the frequency of the external signals is large enough to remove the switching frequency component from the signals.

Output Current Sourced Converters

The load circuit is given in Figure 12.15a. The input voltage v_1 and the load current i_2 are assumed to be ripple free in all cases. The circuit configurations and the time functions for the output voltage v_2 and the input current i_1 are illustrated in Figures 12.18 and 12.19. The voltage ratio in class A (class B) is $V_2/V_1 = D$ $(V_2/V_1 = 1 - D)$. If switch S_p (S_n) is turned on and off while the other switch remains off, the circuit configuration for class C operates like class A (B) in the first (second) quadrant. If the load is connected across the terminals of the positive switch $S_p - D_p$ as shown in the figure by the dotted line, the converter operates either in the first or third quadrants. Classes D and E can be operated with either bipolar or unipolar voltage switching. In the first case, two switches located diagonally in the circuit diagram are simultaneously turned on and off as a pair (see Figure 12.18e and Figure 12.19b). Operation with unipolar voltage switching is achieved by shifting the turn-on-off process in

Table 12.5

Converter connection		Delta six-phase star connection	Interphase-transformer connection	Three-phase bridge connection	Interphase-transformer connection
Current in line-side winding	Waveform	0, π, 2π, I_d	0, 0,500 I_d, π, 2π		0,483 I_d, 0, π, 2π, 0,558 I_d, 0,279 I_0
	R.m.s. value (without overlap)	$\frac{1}{\sqrt{3}} I_d = 0.577\ I_d$	$\frac{1}{\sqrt{6}} I_d = 0.408\ I_d$		$\frac{\sqrt{3}+1}{4\sqrt{3}} I_d = 0.395\ I_d$
	Overlap correction	$\sqrt{1 - 6\Psi(\alpha, \mu)}$	$\sqrt{1 - 3\Psi(\alpha, \mu)}$		$\sqrt{1 - 1.61\Psi(\alpha, \mu)}$
	Form factor ($\mu = 0$)	$\sqrt{3} = 1.73$	$\sqrt{\frac{3}{2}} = 1.23$		$\frac{3\sqrt{2}}{\sqrt{3}+2} = 1.14$
Line current	Waveform	0, I_d, π, 2π	0,500 I_d, 0, I_d, π, 2π	0, I_d, π, 2π	0,483 I_d, 0, π, 2π, 0,965 I_d, 0,835 I_d
	R.m.s. value (without overlap)	$\sqrt{\frac{2}{3}} I_d = 0.815\ I_d$	$\frac{1}{\sqrt{2}} I_d = 0.707\ I_d$	$\sqrt{\frac{2}{3}} I_d = 0.816\ I_d$	$\frac{\sqrt{3}+1}{4} I_d = 0.682\ I_d$
	Overlap correction	$\sqrt{1 - 3\Psi(\alpha, \mu)}$	$\sqrt{1 - 3\Psi(\alpha, \mu)}$	$\sqrt{1 - 3\Psi(\alpha, \mu)}$	$\sqrt{1 - 1.61\Psi(\alpha, \mu)}$
	Form factor $\mu = 0$	$\sqrt{\frac{3}{2}} = 1.23$	$\frac{3}{2\sqrt{2}} = 1.06$	$\sqrt{\frac{3}{2}} = 1.23$	$\frac{3\sqrt{2}}{\sqrt{3}+2} = 1.14$
Average rated apparent power of transformer	Line-side windings without overlap	$\frac{\pi}{\sqrt{6}} = 1.28\ P_d$	$\frac{\pi}{3} = 1.05\ P_d$		$\frac{\pi(\sqrt{3}+1)}{6\sqrt{2}} = 1.01\ P_d$
	Overlap correction	$\frac{\sqrt{1 - 6\Psi(0°, \mu)}}{\cos^2 \mu/2}$	$\frac{\sqrt{1 - 3\Psi(0°, \mu)}}{\cos^2 \mu/2}$		$\frac{\sqrt{1 - 1.61\Psi(0°, \mu)}}{\cos^2 \mu/2}$
	Valve-side windings	$\frac{\pi}{\sqrt{3}} = 1.81\ P_d$	$\frac{\sqrt{2}\pi}{3} = 1.48\ P_d$		$\frac{\pi(\sqrt{3}+1)}{3\sqrt{3}} = 1.65\ P_d$
	Overlap correction	$\frac{\sqrt{1 - 6\Psi(0°, \mu)}}{\cos^2 \mu/2}$	$\frac{\sqrt{1 - 3\Psi(0°, \mu)}}{\cos^2 \mu/2}$		$\frac{\sqrt{1 - 3\Psi(0°, \mu)}}{\cos^2(\mu)/2}$
	Average rated apparent power	$\frac{1.28 + 1.81}{2} = 1.55\ P_d$	$\frac{1.05 + 1.48}{2} = 1.26\ P_d$		$\frac{1.01 + 1.65}{2} = 1.33\ P_d$
Average rated apparent power of interphase transformer	At operating frequency	—	$0.214\ P_d \ldots (3f)$		(*A* and *B*) $0.107\ P_d \ldots (3f)$ (C) $0.0496\ P_d \ldots (6f)$
	As an equivalent transformer	—	$\frac{0.214 \times (3)^{1/1.6}}{6} = 0.071\ P_d$		$0.107 \times (3)^{(1/1.6)}/6 = 0.035\ P_d$ $0.050 \times (6)^{(1/1.6)}/12 = 0.013\ P_d$
Power factor $\mu = 0$; $\mu \neq 0$		$\frac{3}{\pi} \cdot \cos(\alpha)$; $\frac{0.955}{2} \cdot \frac{\cos(\alpha + \mu) + \cos(\alpha)}{\sqrt{1 - 3\Psi(\alpha, \mu)}}$	$\frac{3}{\pi} \cdot \cos(\alpha)$; $\frac{0.955}{2} \cdot \frac{\cos(\alpha + \mu) + \cos(\alpha)}{\sqrt{1 - 3\Psi(\alpha, \mu)}}$	$\frac{3}{\pi} \cdot \cos(\alpha)$; $\frac{0.955}{2} \cdot \frac{\cos(\alpha + \mu) + \cos(\alpha)}{\sqrt{1 - 3\Psi(\alpha, \mu)}}$	$\frac{6\sqrt{2}}{\pi(\sqrt{3}+1)} \cdot \cos \alpha$; $\frac{0.988}{2} \cdot \frac{\cos(\alpha + \mu) + \cos(\alpha)}{\sqrt{1 - 1.61\Psi(\alpha, \mu)}}$
DC voltage		$\left(\frac{3\sqrt{2}}{\pi} = 1.35\right) \cdot V \frac{\cos(\alpha + \mu) + \cos(\alpha)}{2}$	$\left(\frac{3}{\pi}\sqrt{\frac{3}{2}} = 1.17\right) \cdot V \frac{\cos(\alpha + \mu) + \cos(\alpha)}{2}$	$\left(\frac{3\sqrt{2}}{\pi} = 1.35\right) \cdot V \frac{\cos(\alpha + \mu) + \cos(\alpha)}{2}$	$\left(\frac{3}{\pi}\sqrt{\frac{3}{2}} = 1.17\right) \cdot V \frac{\cos(\alpha + \mu) + \cos(\alpha)}{2}$

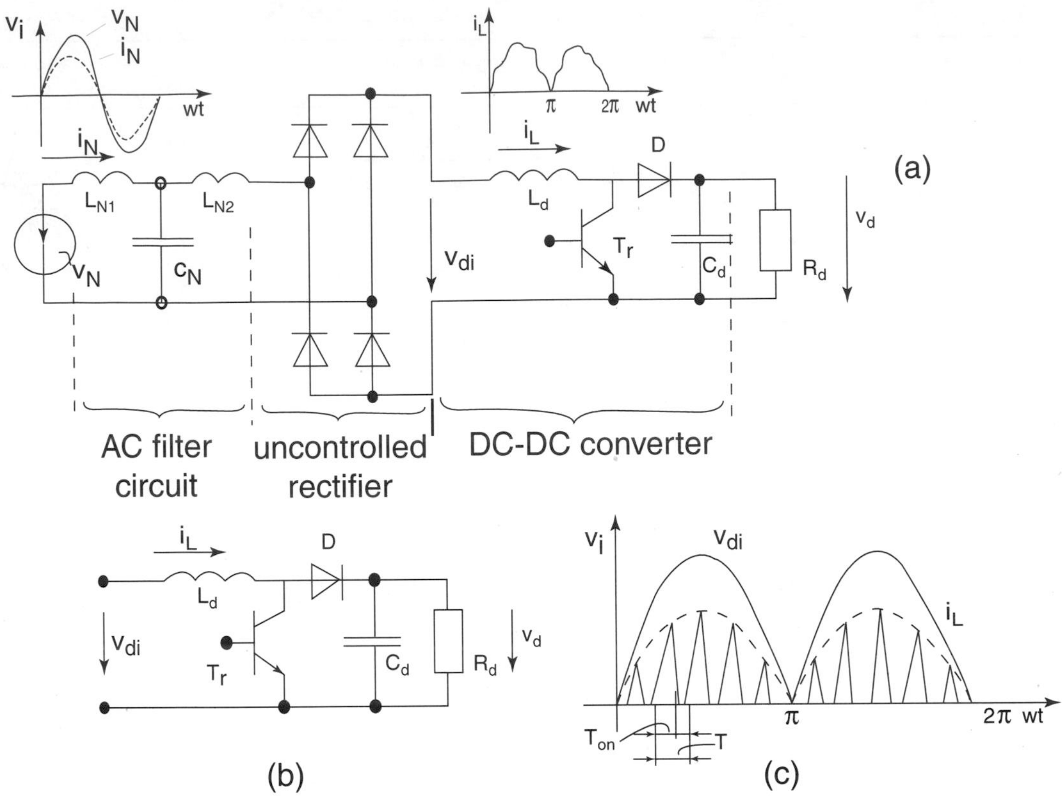

Figure 12.12 Uncontrolled rectifier with DC-DC converters.

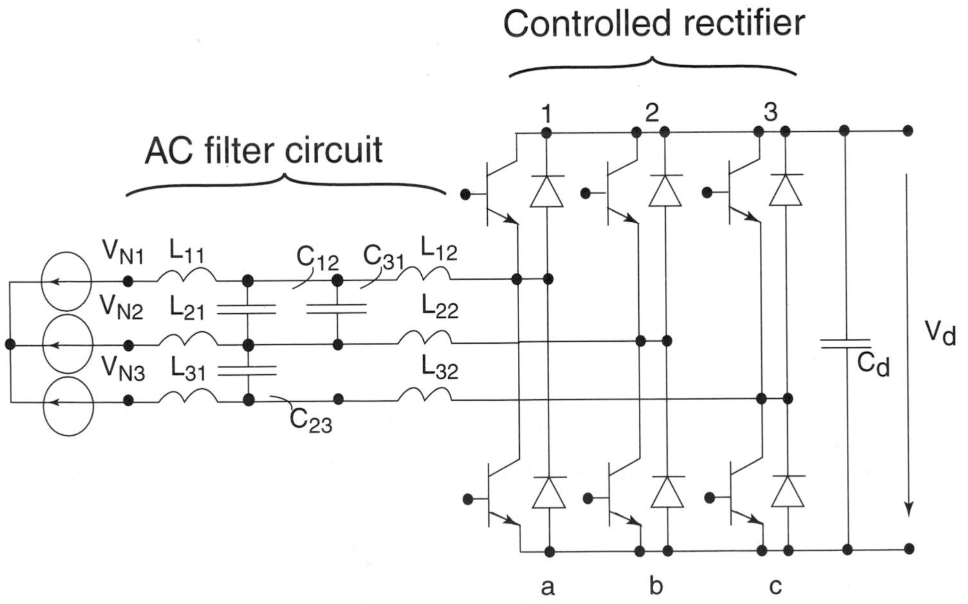

Figure 12.13 Three-phase directly modified rectifier.

these switches by half a cycle (see Figure 12.18f and Figure 12.19c). Figure 12.19 shows the time functions for both bipolar (Figure 12.19b) and unipolar (Figure 12.19c) voltage switching in all four quadrants. The conducting device is either the switch turned on or its antiparallel diode. The turned-on switches and the conducting diodes, in bracket, are shown in the figures. The conducting diode is always indicated along with the switch that is turned-on. Both for bipolar and unipolar voltage switching the average output voltage is $V_2 = (2D - 1)\ V_1$ where D is the duty ratio of switch S_{p1}, S_{n2}.

Assuming the switches have an identical switching frequency, the unipolar voltage switching produces better output voltage

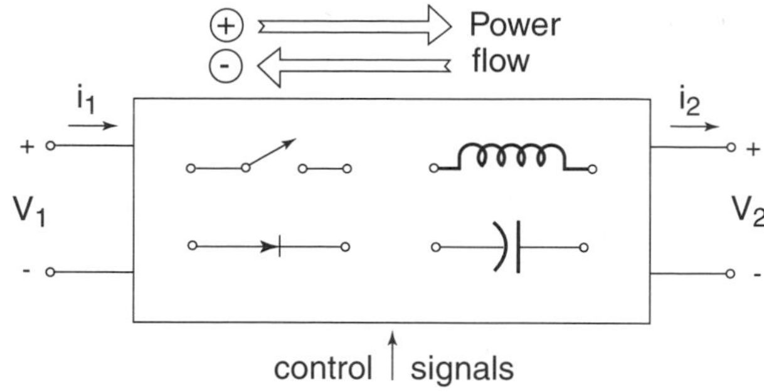

Figure 12.14 DC-DC converters.

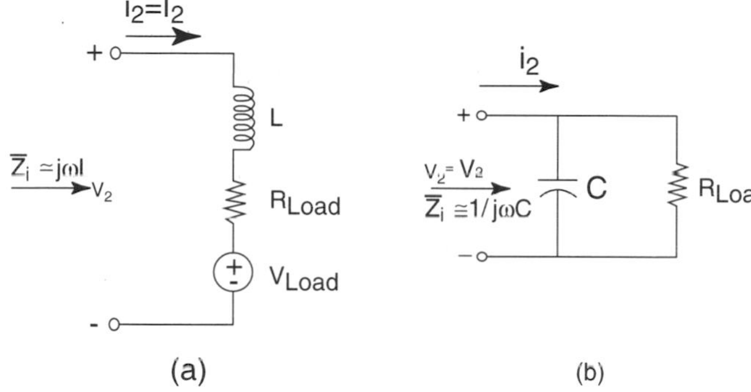

Figure 12.15 Basic low-pass filters.

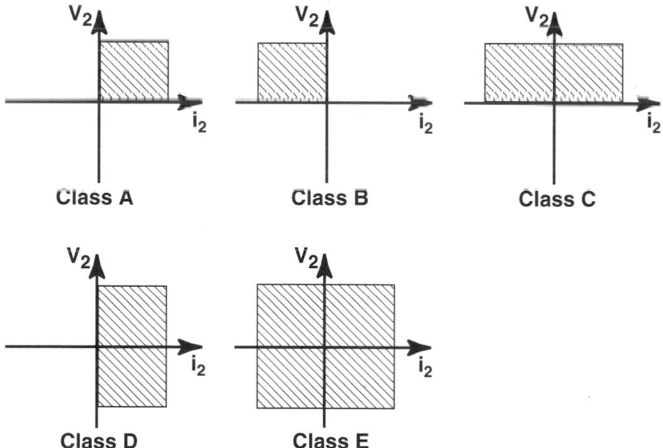

Figure 12.16 Unidirectional (class A and B) and bidirectional (class C, D and E) power flow.

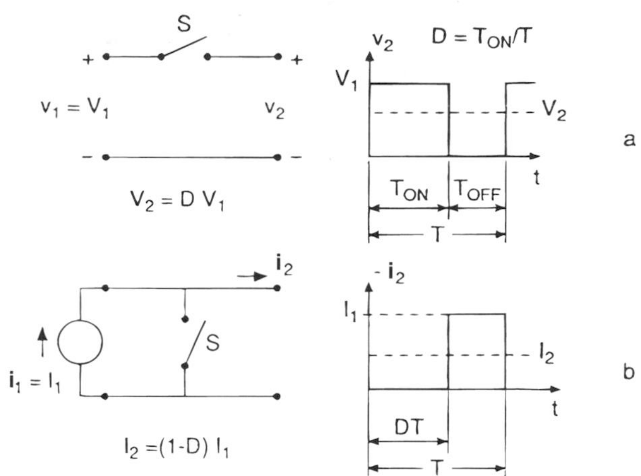

Figure 12.17 Switch mode conversion concept.

and input current waveforms as well as a better frequency response, since the "effective" switching frequency of the two waveforms is doubled and the ripple amplitude is halved.

Class E can be converted to a class C or class D configuration by appropriate control, e.g., continuously turning on S_{n2}, the antiparallel-connected S_{n2} and D_{n2} constitute a short circuit and S_{p2} and D_{p2} are equivalent to an open circuit. First- and second-quadrant operations are achieved with the waveforms shown in Figure 12.19a and 12.19b. On the other hand, by continuously turning on S_{n1}, in which case switch S_{n1} and D_{n1} constitute a short circuit and S_{p1} and D_{p1} are equivalent to an open circuit, third- and fourth-quadrant operations are accomplished.

Since the converters are ideally lossless, the current ratio for all configurations is $I_2/I_1 = V_1/V_2$.

Output Voltage Sourced Converters

In what follows, it will be assumed that L and C are large enough to eliminate switching frequency components from the terminal variables v_1, i_1 and v_2, i_2. Furthermore, the relation between the average input and output voltage can be derived by using the simple fact that the time integral of the inductor voltage v_2 over one period must be zero.

Direct Converters

The circuit configurations and the time functions for the buck (step-down) and boost (step-up) converters are shown in Figure 12.20. By turning on the switch *S* in interval *D* in the buck converter, the diode becomes reverse biased and the input supplies energy to both the load and the inductor *L*. If the same action is repeated in the boost converter, energy is supplied only to the inductor *L*. If switch *S* is turned off in interval $(1 - D)$, the inductor current flows through the diode in the buck converter transferring some of its stored energy to the load, while in the boost converter the energy is forced toward the output both from the inductor and the input through the diode as a result of the inductor current even though $V_2 > V_1$.

Indirect Converters

The circuits and time functions for the buck and boost (step-up/down) and Čuk converters are shown in Figure 12.21. Note that the polarity of the output voltage is negative. Turning on the switch *S* in interval *D* reverse biases the diode. In Figure 12.21a, energy is supplied from the input to the inductor *L* and from the capacitor C to the load. In Figure 12.21b, energy is supplied from the input to inductor *L*, and from capacitor C to the load as well as inductor L_2.

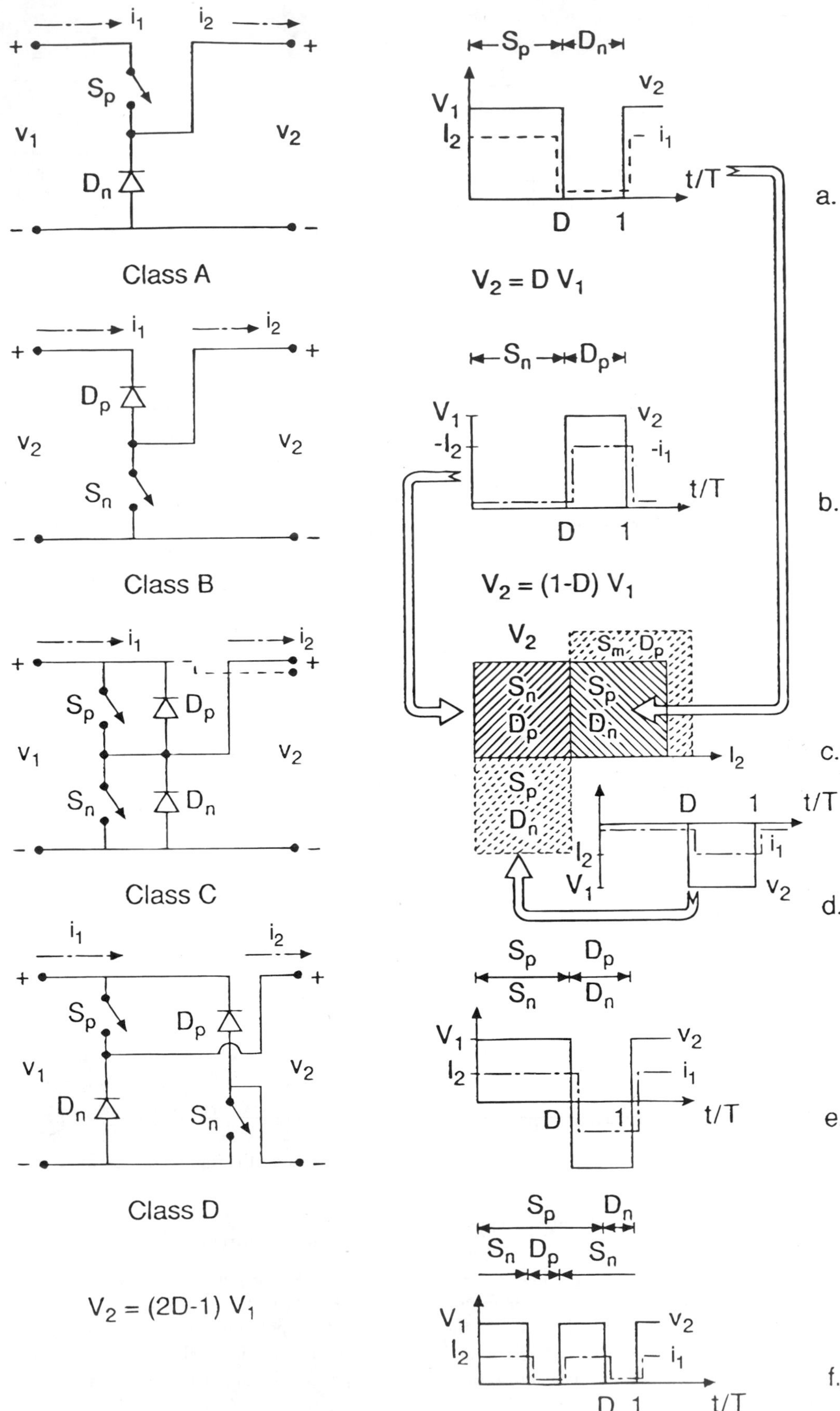

Figure 12.18 Configuration and time functions of class A, B, C and D converters.

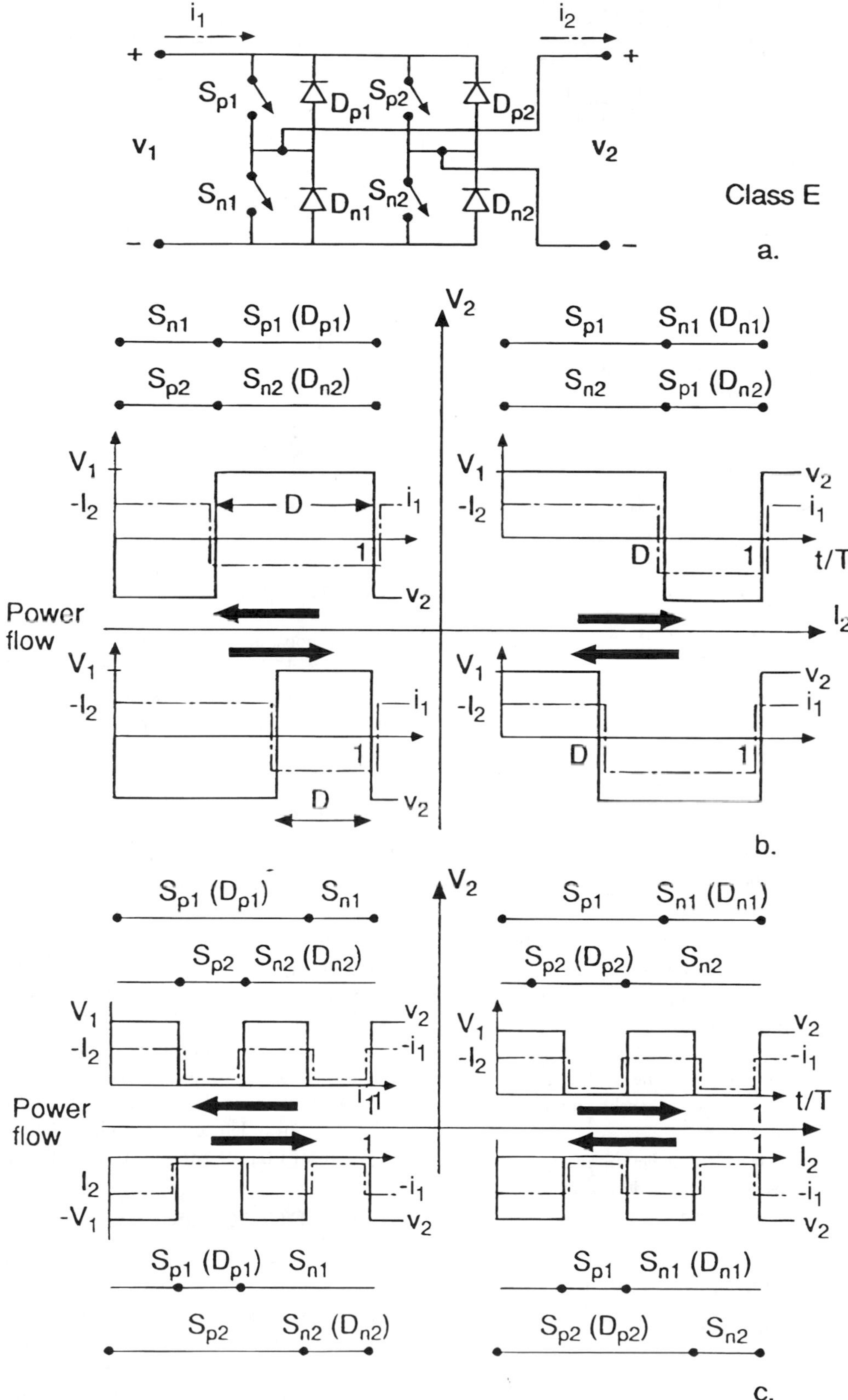

Figure 12.19 Configuration of class E converter (a). Time function of bipolar (b) and unipolar (c) voltage switching.

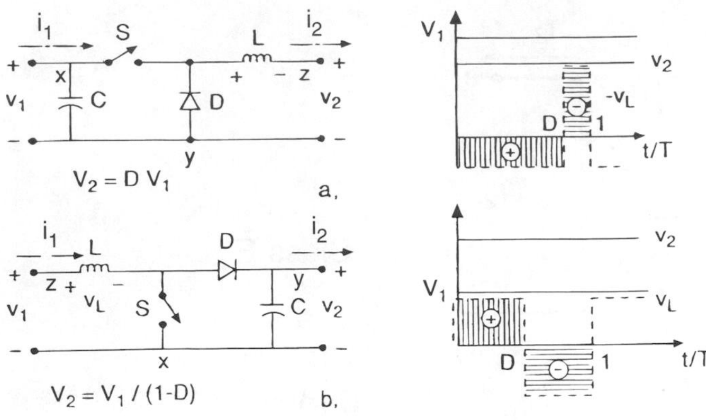

Figure 12.20 Configuration and time functions of buck (step-down) (a) and boost (step-up) (b) converters.

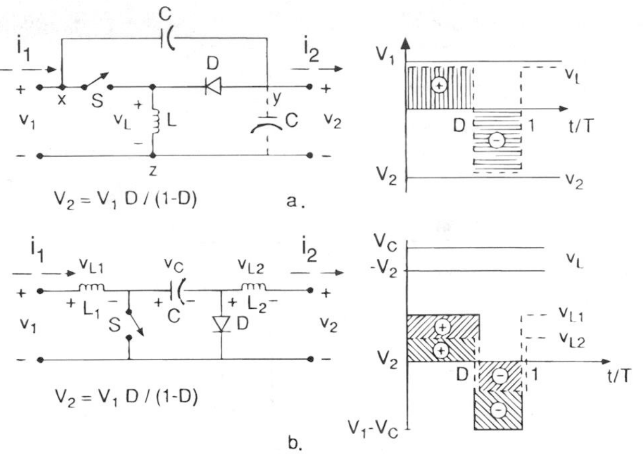

Figure 12.21 Configuration and time functions of buck-and-boost (step-up/down) (a) and Čuk (b) converters.

If the switch S is turned off in interval $(1 - D)$, the diode conducts current. On the one hand, the energy stored in inductor L_2 is transferred to the load. Also, in Figure 12.21a, energy is supplied from the input to the condenser C. In Figure 12.21b, energy is supplied to the capacitor C from the input and inductor L_1. The relation V_2/V_1 is the same for the buck and boost and Čuk converters. The output voltage V_2 can be either smaller or larger than V_1.

The capacitor C can be placed either between terminals x and y or y and z without changing the operation of the buck-boost converter. In both cases, the voltages v_1, v_2 and v_{xy} are ripple-free.

Fundamental Topological Relationships

The basic circuit, the so-called canonical switching cell (CSC), that is common to buck, boost and buck-and-boost converters is shown in Figure 12.22 (Kassakian et al., 1992). It uses a double-throw switch which satisfies the condition that the two switches—transistor and diode in the four converters—be neither on nor off simultaneously. The CSC is the basic building block for a large number of DC-DC converters in addition to those discussed in the previous section. The different converter configurations,

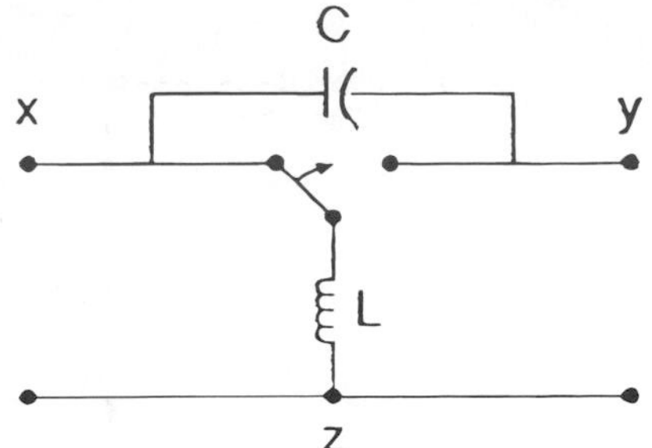

Figure 12.22 Canonical switching cell (CSC).

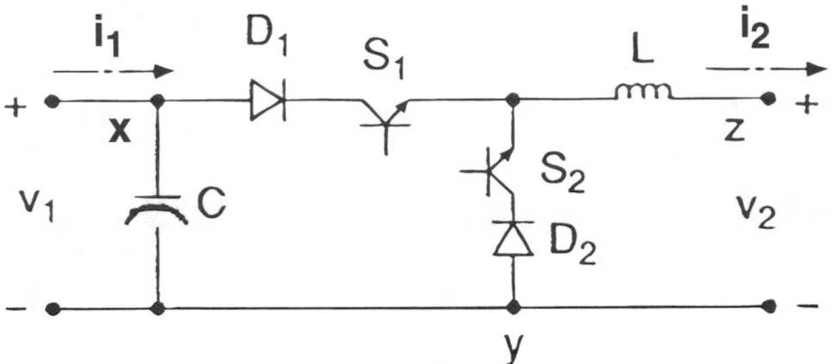

Figure 12.23 Configuration providing bidirectional power flow.

i.e., buck, boost and buck-and-boost, are dependent upon both the way in which the CSC is connected to the external system and the implementation of switches. It can be shown that the Čuk converter can easily be derived from CSC as well (Kassakian, et al., 1992).

Bilateral Power Flow

The power can flow only from left to right in the configurations discussed in **Output Voltage Sourced Converters.** However, bilateral power flow is required in some applications. Figure 12.23 shows the implementation of the switches within the CSC for bilateral power flow under the conditions that the polarity of the two external voltages (currents) can (cannot) change. Assuming $V_1 > 0$ and $V_2 < 0$, ($V_1 < 0$ and $V_2 < 0$), transistor S_2 (S_1) can be kept continuously on. Control is achieved by switching the other transistor. When S_2 (S_1) is continuously on the configuration works as a buck (boost) converter and the power flows from left to right (right to left). The converter can operate in quadrant I and IV like class D converters.

Isolated DC-DC Converters

Each of the basic converters can only accommodate one input and one output with input and output sharing a common reference line. To overcome these limitations, an isolation transformer is added to the DC-DC converters. An additional benefit achieved through the application of a transformer is the reduction of component stresses when the conversion ratio V_2/V_1 is far from unity. Isolated DC-DC converters can be classified according to the core excitation of their transformer:

- In unidirectional core excitation, the flux density B and the magnetic field strength H can be of only one polarity.

- For example: The forward converter which is derived from the buck converter and the flyback converter derived from the buck-and-boost converter belong to this group. They are called "single-ended" converters, also, because power is forwarded through the transformer in only one polarity of the primary voltage.
- In the bidirectional core excitation, B and H can have both positive and negative polarity. Push-pull, half-bridge and full-bridge inverter topologies belong to this group. They are called "double-ended" converters, as well, because power is forwarded through the transformer in both polarities of the primary voltage.

Single-Ended Forward Converter

The basic configuration for this converter and the associated time functions are shown in Figure 12.24. The losses and the leakage inductance of the transformer are neglected and it is modelled by an ideal transformer with the turns ratio N:1 and a parallel magnetizing inductance L_m.

By ignoring the magnetizing current i_m, i.e., $L_m = \infty$, and assuming $N = 1$, the operation of the configuration in Figure 12.14 is the same as that of the buck converter (Figure 12.20a). By changing N, the voltage ratio is simply altered.

The magnetizing current cannot be ignored in a practical forward converter. Assuming the magnetizing current $i_m(0) = 0$ at the beginning of the period (Figure 12.24b), the DC voltage $v_p = V_1$ in interval DT during the on time of switch S, causes the magnetizing current i_m and the flux density to increase in a linear fashion, reaching their peaks at $t = DT$. Power is delivered through the transformer and diode D_2 to the load and inductor L. By turning off switch S, current i_m is diverted from S to the clamping circuit consisting of D_1, R, and C_R. Assuming an approximately constant clamping voltage $v_c = V_c > V_1$, the primary voltage of the transformer is $-(V_c - V_1) < 0$ for time $t \geq DT$ and the current begins decreasing linearly. Diodes D_2 and D become reverse and forward biased, respectively. In steady-state the magnetizing current i_m must reach zero prior to, or at time, $t = T$ and the core is reset. The required maximum value of the duty ratio D_{max} determines the minimum value of the clamping voltage $V_{c\ min}$ since the relationship for the voltage-time area, i.e., $(V_c - V_1)(1 - D)T \geq V_1DT$, must be satisfied. The higher the value of D_{max}, the bigger $V_{c\ min}$ must be.

The energy stored in the magnetizing inductance by $i_m = I_{mp}$ is partially dissipated in the resistance R. At high power, the resistance R can be replaced by a DC-DC converter to recover the magnetizing energy. The clamping function can be implemented with a Zener diode or the addition of a tertiary winding on the transformer. In the latter case, the winding has to be connected in series with a diode either across the input or output terminals of the converter in such a way that the magnetizing energy is supplied back to the input or output circuit during the off interval of switch S.

Single-Ended Hybrid-Bridge Converter

In contrast to the single switch and the clamping circuit of the forward converter shown in Figure 12.24, the single-ended hybrid-bridge converter has two switches turned on and off simultaneously and two diodes performing the clamping function on the primary side of the transformer as shown in Figure 12.25. Otherwise this circuit is the same as that of the forward converter. The two converters operate in a similar manner as shown in Figures 12.24b and 12.25b, and the transformer core is excited unidirectionally. However, the magnetizing current i_m is flowing through diode D_1 and D_2 in the off interval and the primary voltage is clamped at $v_p = V_1$. i_m decays to zero at $t = 2DT$ and the maximum value of the duty ratio is $D_{max} = 0.5$.

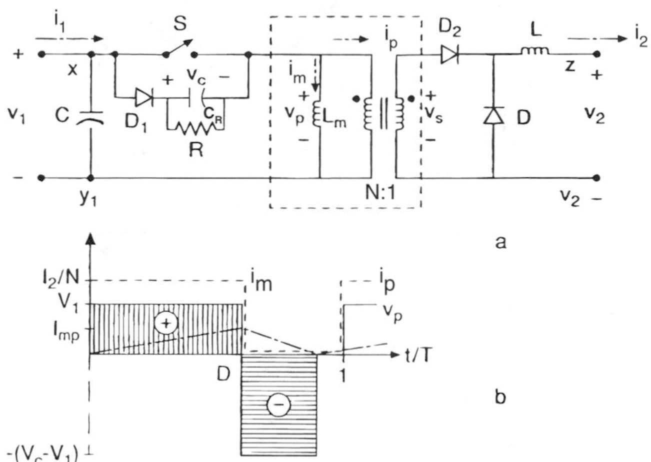

Figure 12.24 Configuration and time functions (b) of forward converter.

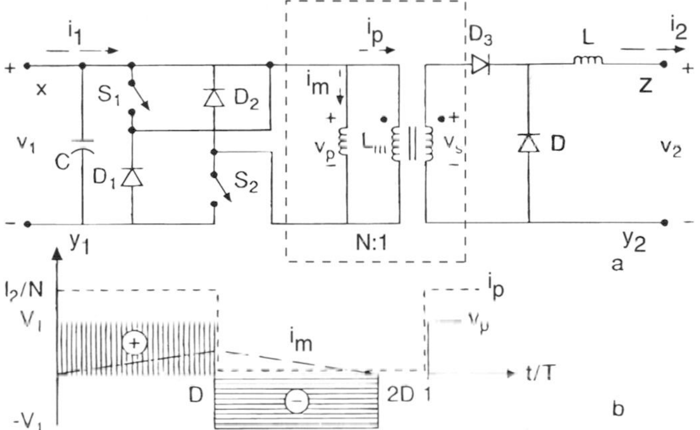

Figure 12.25 Configuration and time functions (b) of hybrid bridge converter.

Flyback Converter

The circuit, which employs the same transformer as that used in the forward converter, and the time functions shown in Figure 12.26 reveal the basic similarity between the flyback converter and the buck-and-boost converter shown in Figure 12.21a. Ignoring the leakage inductances of the transformer, its operation is identical to that of the non-isolated buck-and-boost converter except for the transformer effect. Unlike the forward converter the transformer magnetizing inductance stores energy

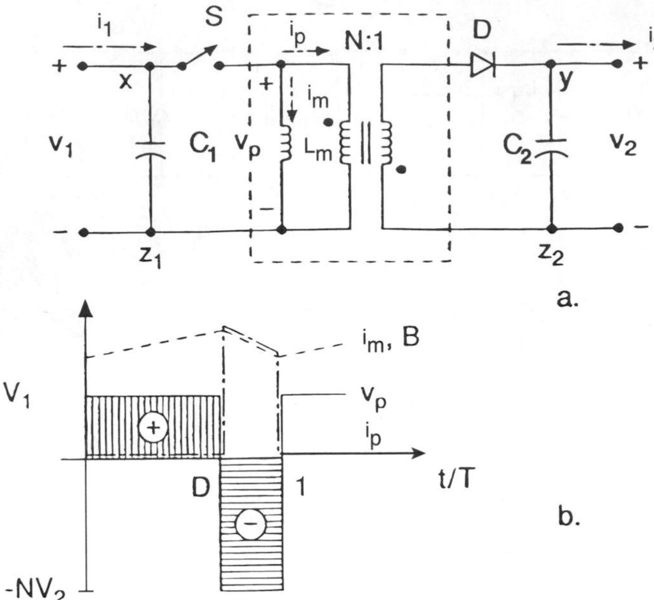

Figure 12.26 Configuration (a) and time functions (b) of flyback converter.

during the on interval of switch *S*. This energy is transferred during the off interval through the transformer and the diode *D* to the load.

Flyback converters are applied in television receivers with a very high turns ratio to produce a high voltage "to fly-back" the horizontal beam on the screen in order to start the next line. The flyback converter is single-ended and the transformer core is excited unidirectionally.

Double-Ended Isolated Converters

The transformer core for these types of converters is excited bidirectionally. This group of converters includes the push-pull shown in Figure 12.27a the half-bridge shown in Figure 12.27b and the full-bridge shown in Figure 12.27c. All three converters generate a high-frequency AC voltage without any DC component across the primary of the transformer by turning on-and-off the switches periodically according to the pattern shown in Figure 12.27d. The AC voltage v_s is rectified by a diode bridge in all three converters as shown in Figure 12.27e.

Control

There are basically three control methods as illustrated in Table 12.6.

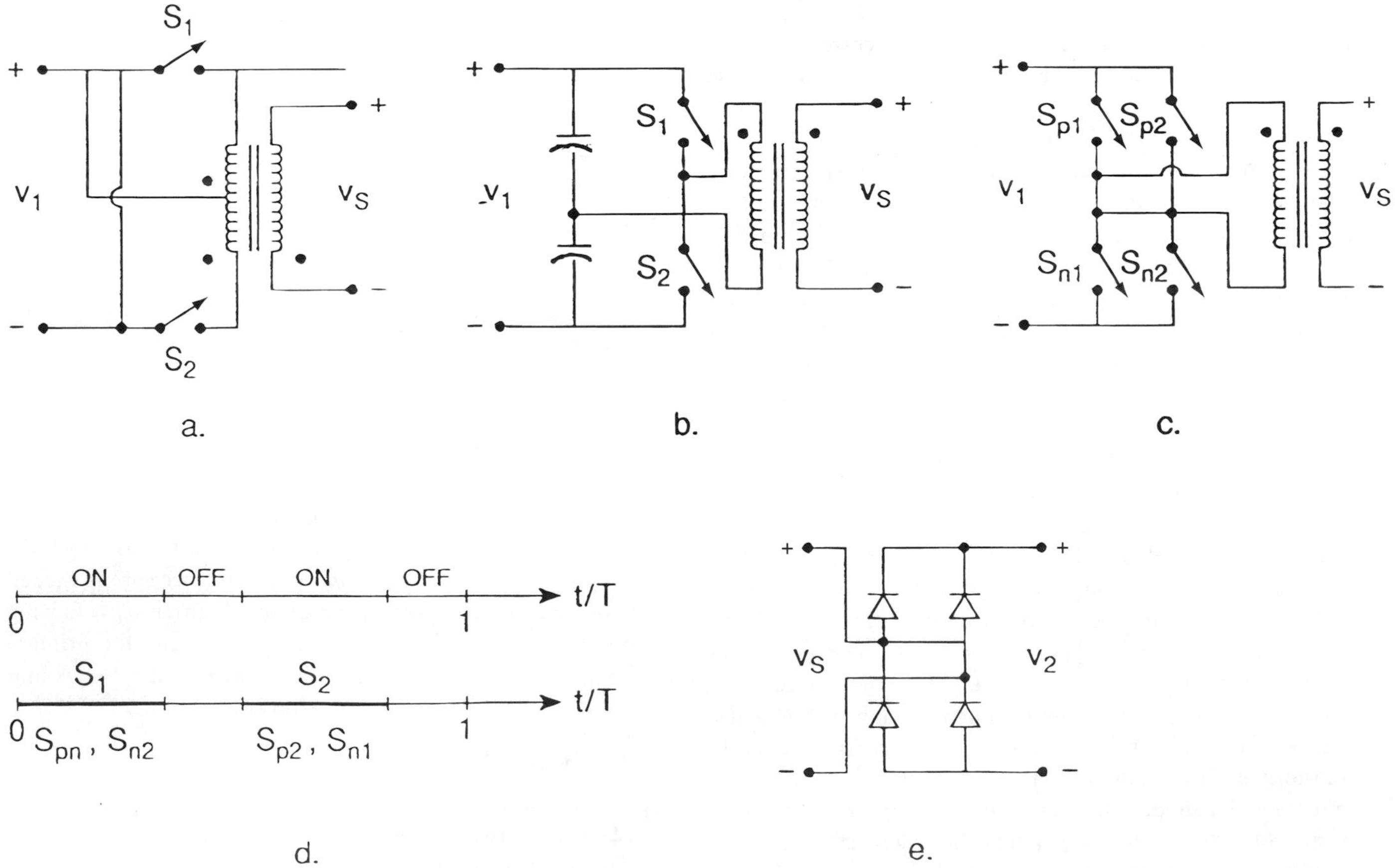

Figure 12.27 Push-pull (a), half-bridge (b), full-bridge (c) double-ended (d) converter and their control (e).

Table 12.6 Control Methods of Converters

Constant	Controlled
Period T	T_{on}, T_{off} or duty ratio T_{on}/T
Pulse width T_{on}	T, T_{off} or frequency f = $1/T$
Pulse pause T_{off}	T, T_{on} or frequency f = $1/T$

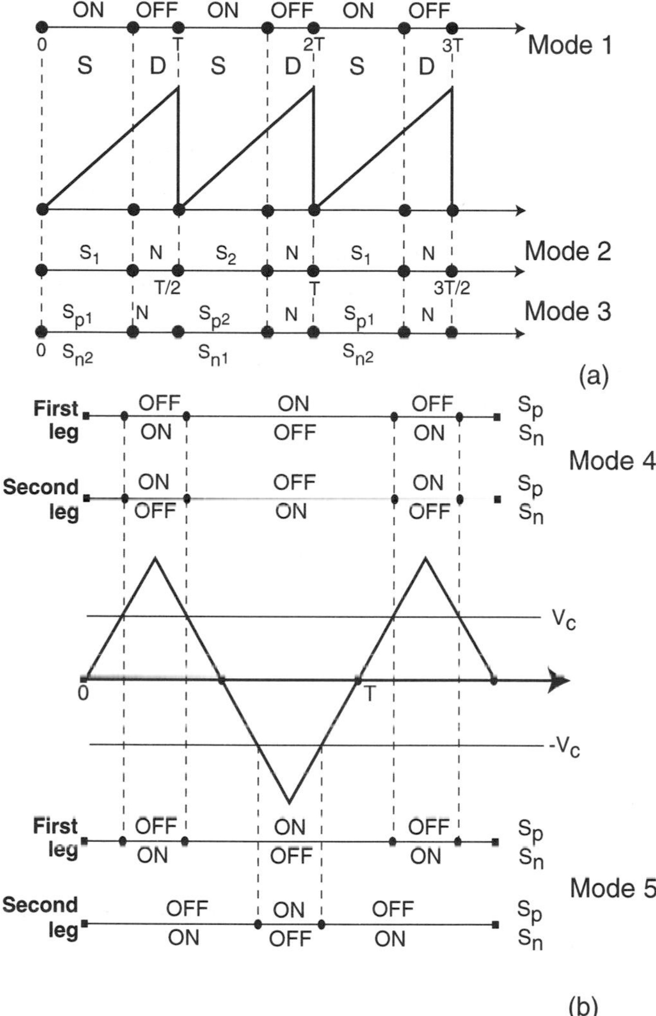

Figure 12.28 Control modes of converters.

The first control method is referred to as pulse-width modulation (PWM). Five PWM control modes of DC-DC converters are shown in Figure 12.28. There is only one controlled switch in mode 1, two switches in mode 2 and four switches in modes 3, 4 and 5. Control mode 1 is applied in class A, B and C converters as well as in the buck, boost, buck-and-boost, Čuk, single-ended forward and flyback converters. Control mode 2 is applied in isolated converters, single-ended hybrid bridge, push-pull and half-bridge converters. Mode 3 is used in isolated double-ended full-bridge converters. Modes 4 and 5 are applied in class D and E converters as well as the nonisolated full-bridge converter for bipolar and unipolar voltage switching, respectively.

Note the basic difference between control modes 2 and 3 and modes 4 and 5. In modes 2 and 3 there are intervals when none of the controlled switches are turned on. In modes 4 and 5, one controlled switch is always on in each leg. In other words, two switches are never off nor on simultaneously in one leg. Switch S_p (S_n) in the first leg is controlled together with S_n (S_p) in the second leg in mode 4. On the other hand, the control of switch S_p (S_n) in the first leg is shifted by half a cycle to the control of switch S_n (S_p) in the second leg in mode 5.

References

Kassakian, J. G., Schlecht, M. F., and Verghese, G. C. 1992. *Principles of Power Electronics,* Addison-Wesley, Reading, MA.

Mohan, N., Undeland, T. M., and Robinsons, W. P. 1989. *Power Electronics,* John Wiley & Sons, New York, NY.

Rashid, M. H. 1993. *Power Electronics,* Prentice-Hall International, London, UK.

Severns, R. P. and Blomm, G. E. 1985. *Modern DC-to-DC Switchmode Power Converter Circuits,* Van Nostrand Reinhold Electrical/Computer Science and Engineering Series, Van Nostrand Reinhold, New York, NY.

12.3 DC-AC Conversion

Attila Karpati

The DC-AC converters, also known as inverters and shown in Figure 12.29, produce an AC voltage from a DC input voltage. The frequency and amplitude produced are generally variable. In practice, inverters with both single phase and three phase outputs are used, but other phase numbers are also possible. Electric power usually flows from the DC to the AC terminal, but in some cases reverse power flow is possible. These types of inverters, where the input is a DC voltage source, are also known as voltage-source inverters (VSI). The other type of inverter is the current-source inverter (CSI), where the DC input is a DC current source. These converters are used primarily in high-power AC motor drives.

Basic DC-AC Converter Connections (Square-Wave Operation)

This section presents a short summary of the main types of voltage-source DC-AC converter connections and a brief description of their functions. At the end of this subsection is also given a current-source converter configuration with its short

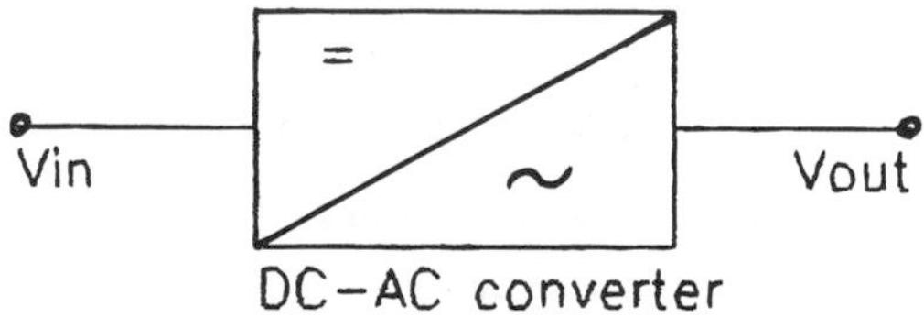

Figure 12.29 DC-AC converter.

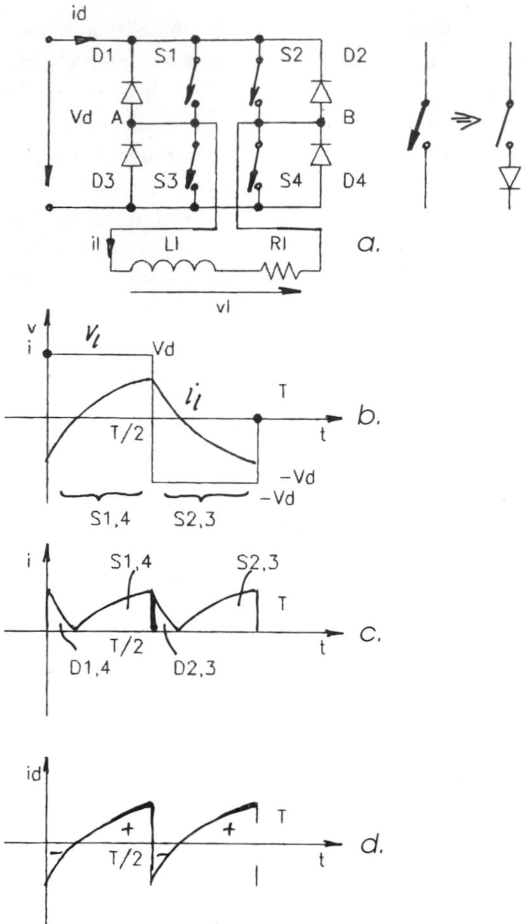

Figure 12.30 Voltage-source, single-phase, full-bridge inverter connection.

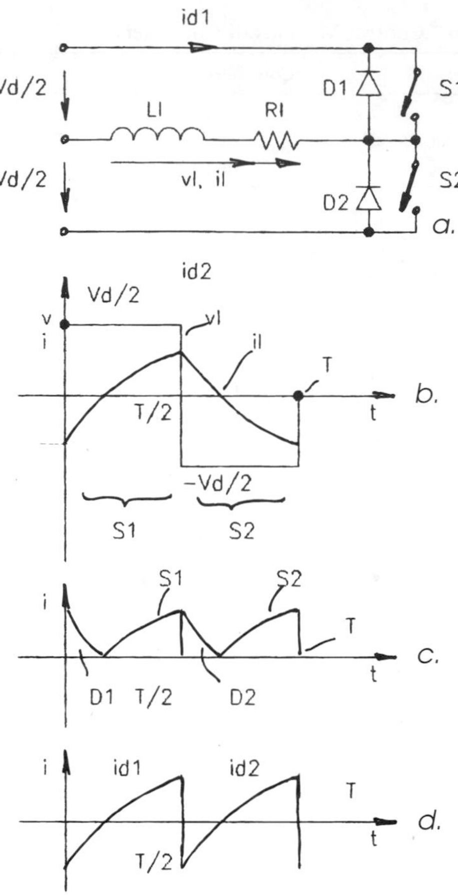

Figure 12.31 Voltage-source, single-phase, half-bridge inverter connection.

description. It is assumed that the circuits incorporate ideal semiconductor switches.

The most frequently used types of single-phase inverters are full-bridge inverters, as shown in Figure 12.30a, the half-bridge inverters, as shown in Figure 12.31a, and push-pull inverters, as shown in Figure 12.32a.

The switching sequences for the switches and the most important time functions for the full-bridge, half-bridge and push-pull inverters during square-wave operation, can be seen in Figures 12.30 through 12.32.

It is assumed that the load on the output consists of a series resistance and inductance. The three-phase basic inverter configuration is the full-bridge connection shown Figure 12.33a. The loads are assumed to be symmetrical inductances in the three phases. The switching-sequences of the switches and the most important time functions at square-wave operation are demonstrated in Figure 12.33b–g.

One can draw the following conclusions from these figures:

- The output voltage is non-sinusoidal.
- Due to the presence of the freewheeling diodes, the output voltage is independent of the direction of the load current, and is only dependent on the on and off state of the switches.
- The semiconductor switches and freewheeling diodes form two rectifiers. They are connected in inverse parallel. The semiconductor switches make the energy flow from the DC side to the AC side possible. The freewheeling diodes allow the reverse situation.
- Accordingly, the freewheeling diodes are necessary if the converter outputs are connected to loads, which require either reactive power or effective power feedback. In the case of reactive power, the direction of the power flow in the converter changes periodically (see the i_B currents in Figures 12.30–33).

A three-phase current-source inverter configuration is shown in Figure 12.34a. The switching sequences of the switches and the most important time functions are demonstrated in Figure 12.34b.

Because of jumps in the output current, capacitors must be used, which are connected in parallel to the load. In most cases, the current-source inverters use thyristors as switching devices, and the aforementioned capacitances are the energy storage elements of the quenching circuits.

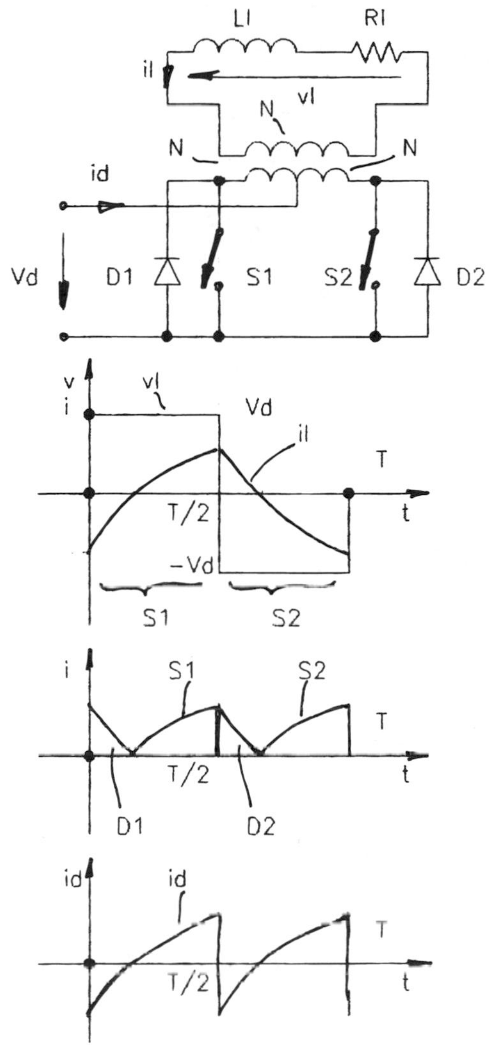

Figure 12.32 Voltage-source, single-phase, push-pull inverter connection.

Control of the Output Voltage

In voltage-source inverters, the output voltage is controlled by following methods:

- In inverters with square-wave operation, voltage changes on the DC side.
- Voltage cancellation, which is feasible in single-phase full-bridge inverters.
- Sinusoidal pulse-width modulation (sinusoidal PWM), with bipolar and unipolar voltage switching.
- Programmed harmonic elimination switching.
- Tolerance band control.
- Fixed-frequency control.

In current-source inverters, the output is controlled by changing the input DC voltage. In most cases the DC voltage is changed by a controlled rectifier or DC chopper. In voltage cancellation the T_{on} and T_{off} times of the switches in the two legs of the full-bridge connection are shifted to one another as shown in Figure 12.35. The RMS value of the AC voltage can be changed between 0 and a maximum value, as defined by the square-wave operation. This is a very simple method, in which the switching frequency of the semiconductor elements is equal to the output frequency, but the harmonic content of the AC side voltage is rather high. Therefore, it is the preferred method used in converters with high-frequency output. At lower output frequency, i.e., at 60 Hz, other methods are used, and the switching frequency of the semiconductors is much higher than the output frequency. This method allows for extensive reduction of the harmonic content in the output voltage or current. In inverter circuits the sinusoidal PWM is used to minimize the output harmonic content. The basic principle employed in a one-phase half-bridge converter with bipolar voltage switching is demonstrated in Figure 12.36. The switches S_+ and S_- work with an internal frequency, which is much higher than the output frequency. The on and off state of the switches is determined by the crossover points of the triangular comparison signal V_{tri} and the sinusoidal control signal V_{cont}. The sinusoidal control signal causes constant changes in the duty ratio of the switches S_+ and S_- during the half-period of the output so that the harmonic content of the output is minimized. The output voltage or current can be changed by varying V_{cont}.

The most important definitions are as follows:

The amplitude-modulation ratio:
$m_a = V_{contM}/V_{tri}$
The frequency-modulation ratio:
$m_f = f_s/f_1$
where f_s is the internal switching frequency and f_1 is the frequency of the fundamental of the output.

At small m_f ($m_f \leq 21$), synchronous PWM should be used, namely m_f should be an integer and V_{cont} and V_{tri} are synchronized to one another. (Asynchronous PWM in the $m_f \neq$ integer output produces subharmonics of the fundamental frequency, which are generally undesirable.)

At large values of m_f ($m_f > 21$), the amplitudes of subharmonics caused by asynchronous PWM are small. Therefore asynchronous PWM may be used, except in AC motor drives, if the frequency approaches zero. In this case, small subharmonic voltages can also occur as well as high and undesirable currents.

In the case of $m_a < 1.0$, the sinusoidal PWM operates in the linear range. The amplitude of the fundamental frequency component varies linearly with m_a. In this range, the maximum value of the fundamental is less than the allowable maximum, which is achieved by overmodulation, with $m_a > 1$. In this range, the relation is not linear between m_a and the fundamental. The allowable maximum value is given by square-wave operation. The relation between the fundamental and m_a is illustrated in Figure 12.37.

The operating principles for sinusoidal PWM with unipolar voltage switching for a full-bridge inverter can be seen in Figure

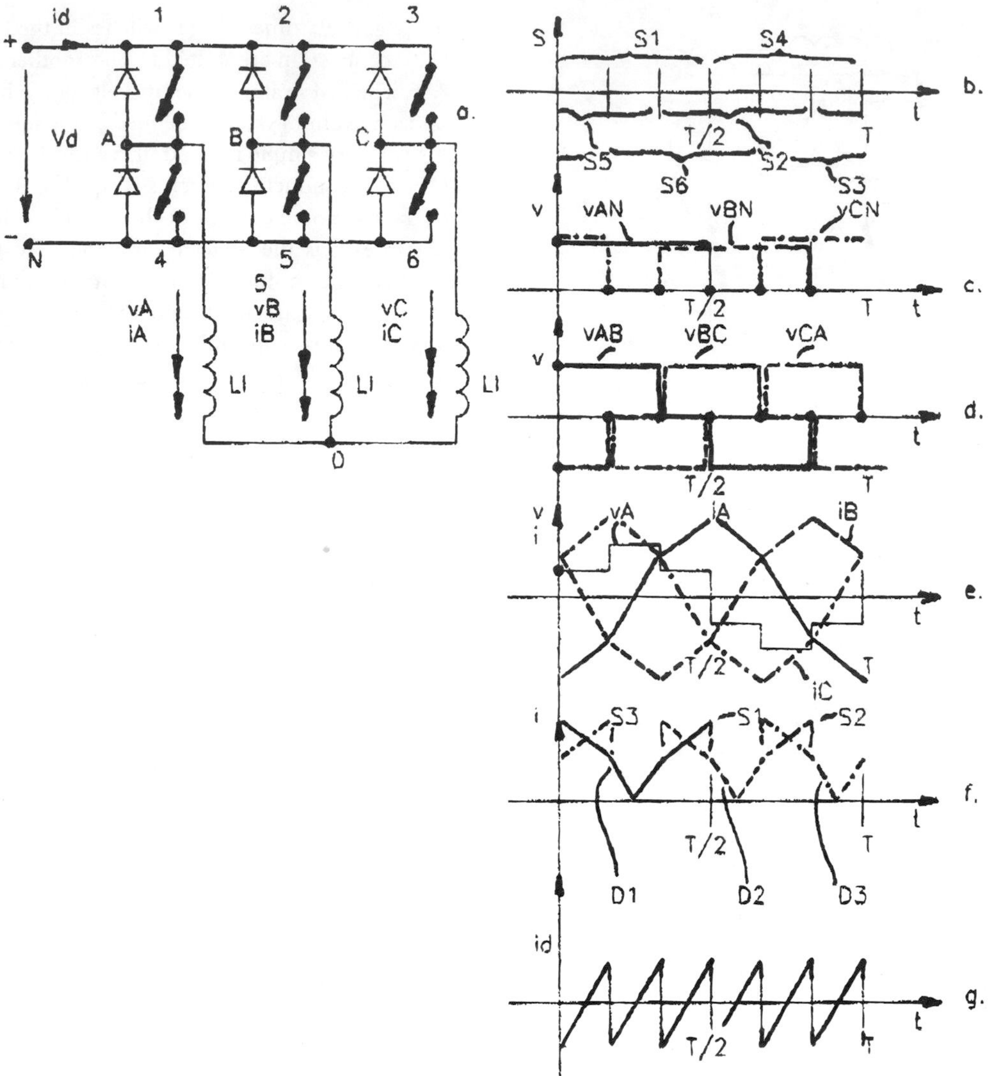

Figure 12.33 Voltage-source, three-phase, bridge inverter connection.

12.38. The two legs of the inverter are not switched simultaneously, and are controlled separately. For this reason, two control signals, V_{cont} and $-V_{cont}$ are used. The advantage of this method is that of "effectively" doubling the switching frequency, which results from the cancellation of certain harmonic components.

The operating principles for sinusoidal PWM with three-phase inverters are shown in Figure 12.39. To control the three legs of the bridge connection, three control signals are used, $V_{cont,A}$ $V_{cont,B}$ and $V_{cont,C}$. The fundamental of the output as a function of m_a is given in Figure 12.40.

In the case of overmodulation, low-order harmonics appear, and therefore the above-mentioned natural sampling method is used only until the output fundamental voltage becomes equal to 78.5% of its maximum possible value. For a three-phase system, this situation can be improved by using a reference wave with the addition of a third harmonic, as shown in Figure 12.41. In the output phase voltages, the third harmonics have in all of the phases the same time functions (zero sequence components), and, therefore, cannot produce current.

For programmed harmonic elimination switching, the moments of the semiconductor switching are calculated so that the lower harmonics will be eliminated. This method permits the elimination of undesirable lower harmonics, without a very high resulting switching frequency. Therefore, the power losses in the converter can be reduced.

The principles of tolerance band control (following control) can be seen in Figure 12.42. The difference between the reference value and the actual value will be directed to one comparator with a tolerance band. The output of the comparator controls the switches in the inverter so that the above-mentioned difference will not be greater than that required. At the sinusoidal output, the reference value has the required sinusoidal form, and the actual value fluctuates along the curve. The switching frequency varies in a large interval and depends on the AC side load and the input DC voltage. The controlled variable can be the output voltage or current.

The principles of fixed-frequency control are shown in Figure 12.43. The difference between the reference value and actual

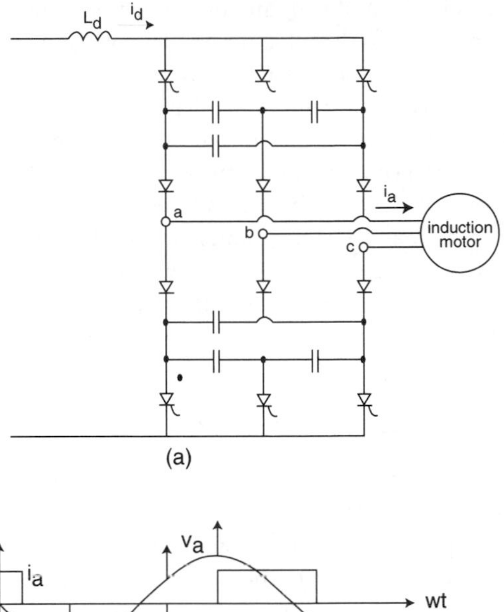

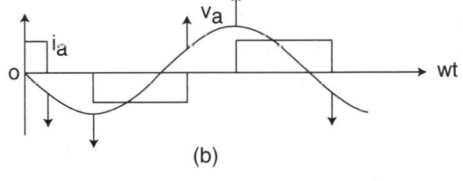

Figure 12.34 Three-phase current-source inverter circuit.

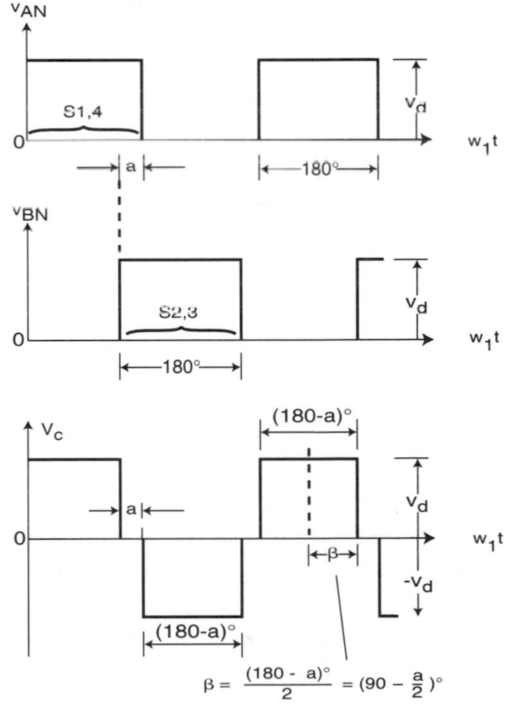

Figure 12.35 Voltage cancellation by full-bridge connection.

value will be directed to a regulator. The regulator output is the control signal, V_{contr}, which is compared to a triangular waveform, V_{tri}, with the switching frequency f_s. The switching moments are specified by the crossover points of the two signals. This type of control circuit is also used in following control. At the sinusoidal output, the reference value has the required sinusoidal form.

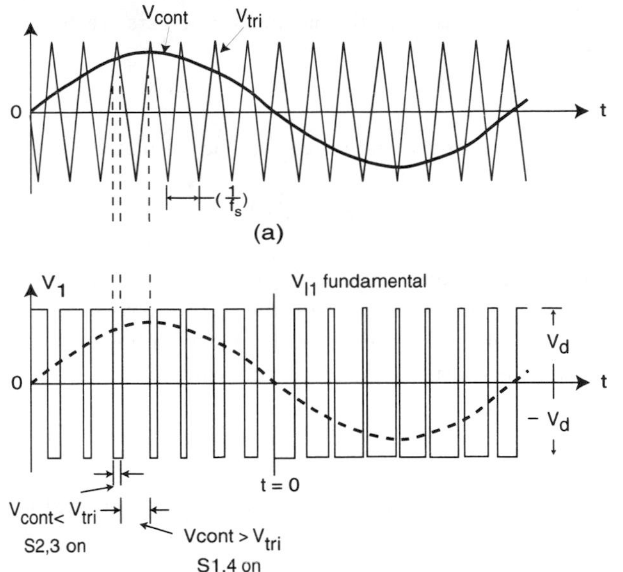

Figure 12.36 Pulse-width modulation with bipolar voltage switching.

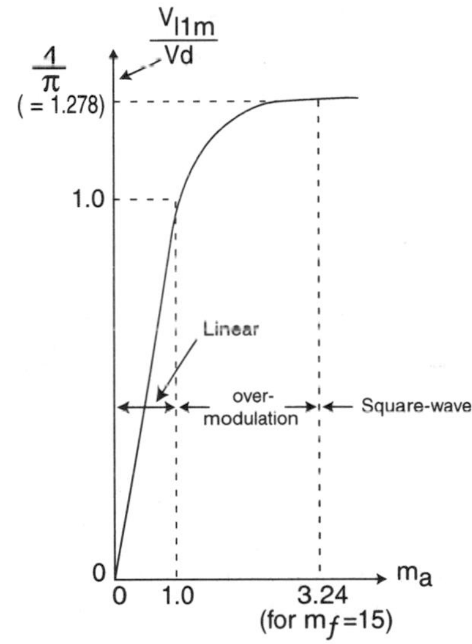

Figure 12.37 Voltage control by varying m_a.

Harmonics in the Output Voltage

The harmonics in the output voltage depend primarily on the control method for the output voltage. For inverters with square-wave operation, the harmonic content is constant. For single-phase inverters the harmonic numbers are

$$n = 1, 3, 5, 7, \ldots$$

The amplitude of the nth harmonic can be calculated for the full-bridge inverter by the following formula:

$$V_{onrmso} = 0.9V_d/n$$

For three-phase inverters the harmonic numbers are

$$n = 6c \pm 1, \qquad \text{where } c = 1, 2, 3, \ldots$$

The RMS value of the line voltage can be calculated as follows:

$$V_{onrms} = 0.78 V_d / n$$

For voltage cancellation in a single-phase full-bridge inverter, the harmonic numbers are the same as those for square wave operation, but the amplitude of the output voltage harmonics vary with the control angle in the following form:

$$V_{onrms} = V_{onrmso} \sin(n \cdot \beta)$$

For sinusoidal PWM with bipolar voltage switching and $m_a \leq 1.0$, the harmonic numbers are

$$n = jm_f \pm k$$

where the fundamental frequency is denoted by $n = 1$. For odd values of j, only even values of k are possible, and vice versa.

The harmonic spectrum is presented in Figure 12.44. In case of overmodulation, the harmonic content will be higher, as shown in Figure 12.45.

For sinusoidal PWM with unipolar voltage switching, the harmonic content is less than that for bipolar voltage switching, due to the cancellation of some harmonics, as shown in Figure 12.46.

For sinusoidal PWM with three-phase inverters, the harmonic spectrum of the output voltage is given in Figure 12.47.

For programmed harmonic elimination switching the harmonics of lower order are eliminated. In three-phase bridge inverters, the 5th, 7th, 11th, and 13th harmonics are usually eliminated.

For tolerance band control, the switching frequency varies in a large interval. Therefore, the frequencies of the harmonic spectrum and the harmonic amplitudes are not constant.

Filtering of Output Voltage

As was demonstrated in the previous section, the output voltage is not sinusoidal. If AC voltage with low distortion is necessary, and the output frequency is constant, (for example in uninterruptible power supplies), output voltage filter circuits are used to decrease distortion. Reducing the internal frequency of the inverters results in greater filtering problems. The solution of the filtering problems is the most difficult in line frequency inverters with voltage cancellation. In this case, the use of large and complicated output filters is necessary as shown in Figure 12.48. The basic principle is simple. The filter circuit is a frequency-dependent voltage divider. Under ideal conditions, the

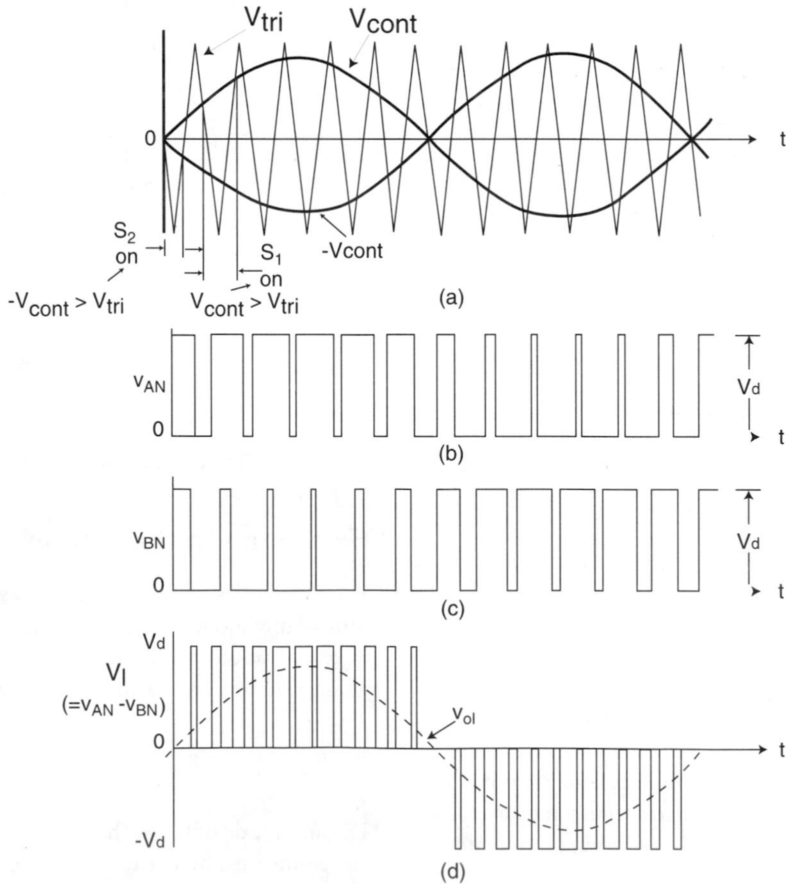

Figure 12.38 Pulse-width modulation with unipolar switching.

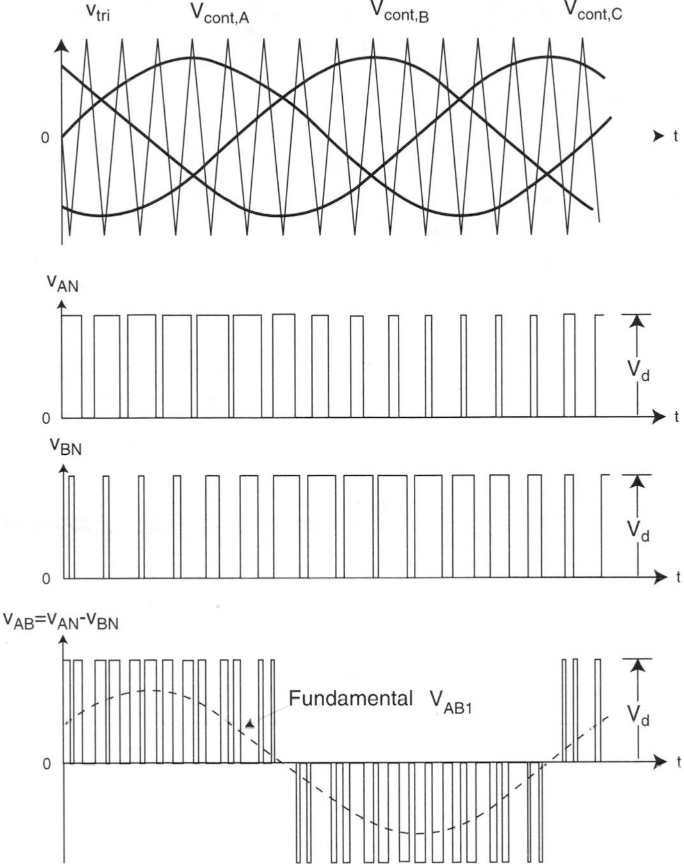

Figure 12.39 Three-phase PWM waveforms.

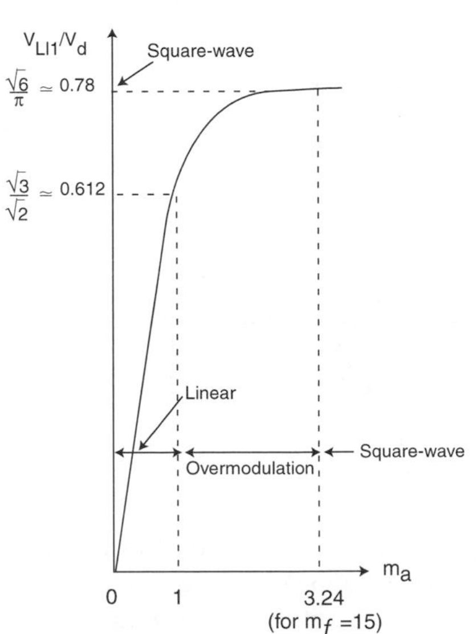

Figure 12.40 Three-phase inverter $V_{LI1} = V_{LI1}(m_a)$.

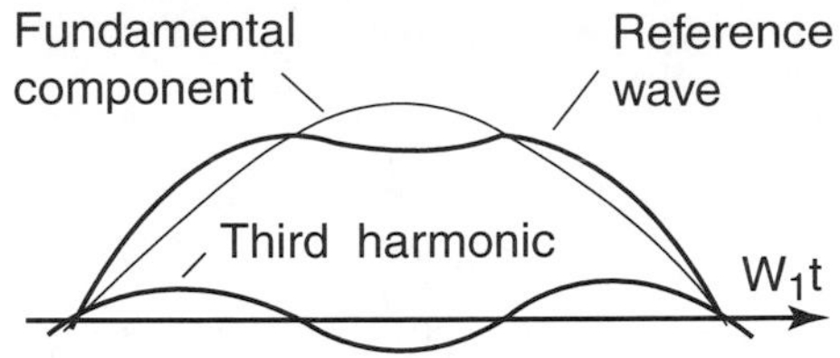

Figure 12.41

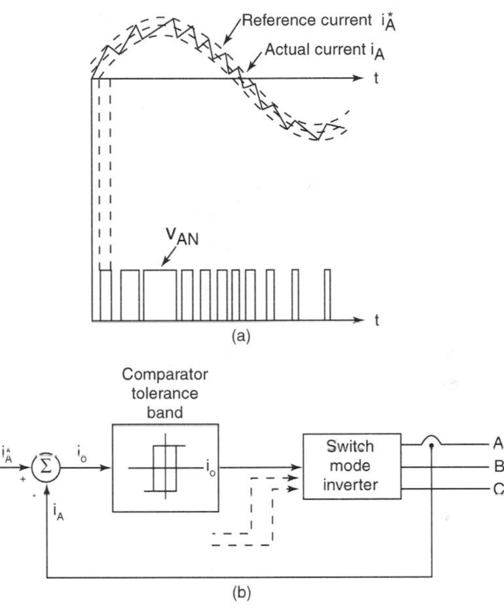

Figure 12.42 Tolerance-band current control.

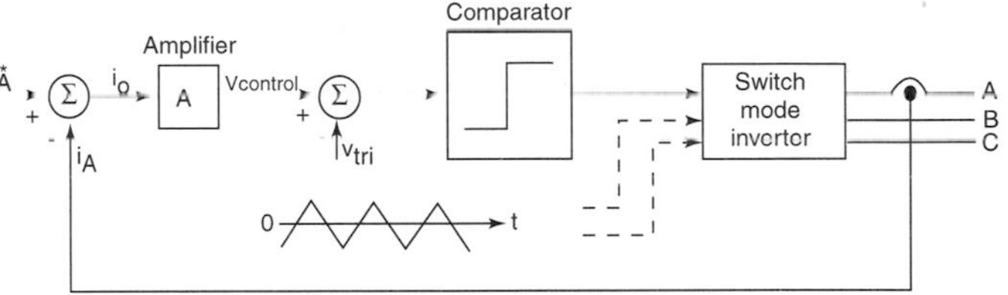

Figure 12.43 Fixed-frequency current control.

transfer ratio (V_{out}/V_{in}) for the fundamental is equal to one, and for the other harmonics is equal to zero. In the basic version of the filter circuit (Figure 12.48) the ideal behavior is approximated using a series resonant circuit in the input of the filter, and a parallel resonant circuit in the output. Both circuits are tuned to the fundamental frequency. Therefore, the transfer ratio for the fundamental is equal to one, and the inverter is not loaded with the reactive power of the parallel output capacitance. For the harmonics, the series impedance increases with frequency, and the parallel impedance decreases. This effect ensures a certain reduction in the harmonic voltages. If this reduction is not adequate, series resonant circuits will be connected in parallel with the output, which are tuned to various harmonic frequencies. The resulting output will be short-circuited at the chosen frequencies. The dynamic behavior of this filter circuit is not good at load jumps because of the large number of energy-storage

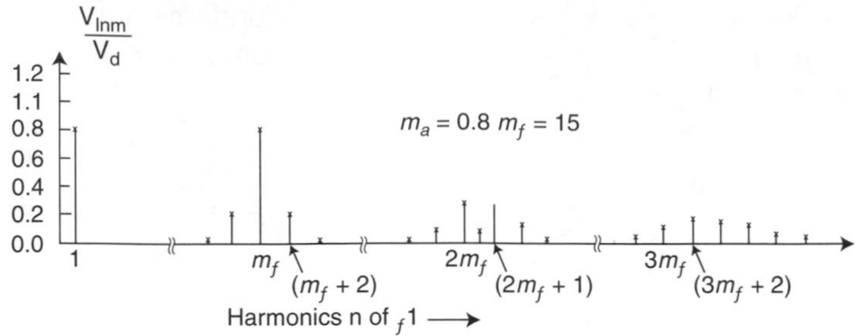

Figure 12.44 Single-phase full-bridge.

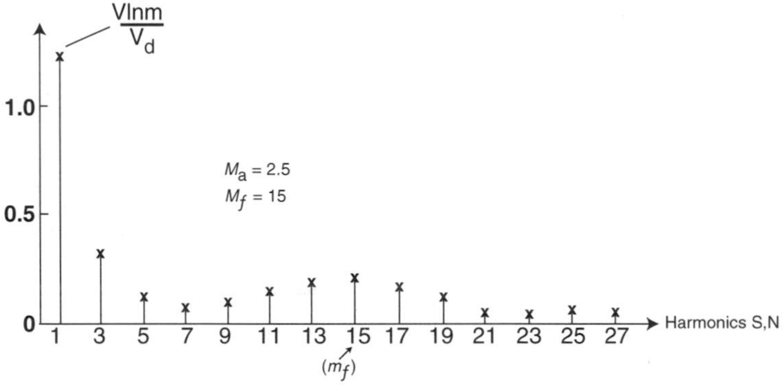

Figure 12.45 Single-phase full-bridge with bipolar switching, harmonic spectrum.

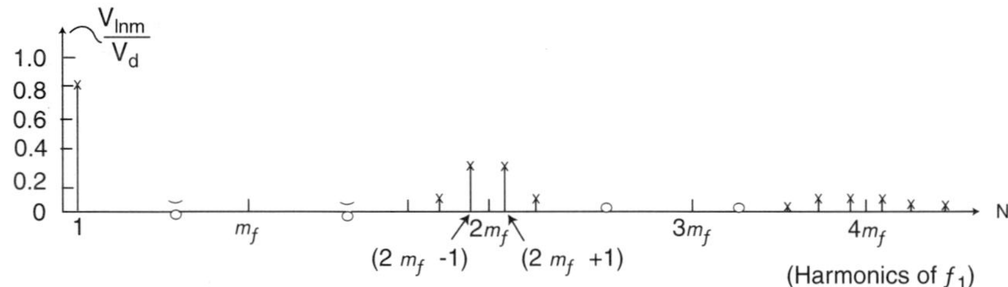

Figure 12.46 Single-phase full-bridge with unipolar switching, harmonic spectrum.

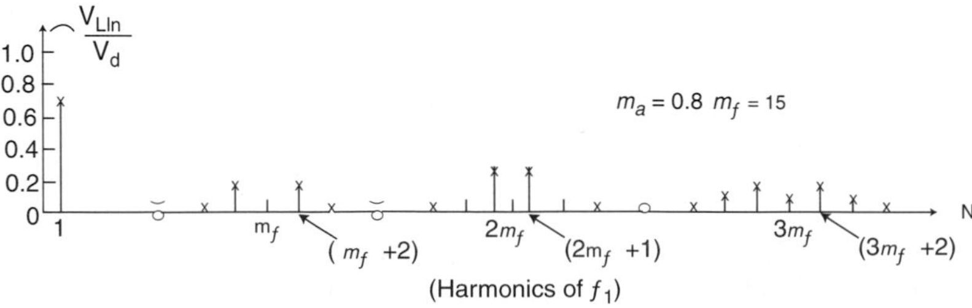

Figure 12.47 Three-phase PWM, harmonic spectrum.

elements. Since modern converter circuits are used with a high internal frequency (e.g., 20 kHz at PWM), the necessary filter circuit is simpler. The simplified filter circuit in Figure 12.49 is currently utilized. If an output transformer is also used, the transformer values are calculated such that the series inductance of the filter circuit is given by the transformer's leakage inductance and the parallel inductance is equal to the transformer's magnetizing inductance. To ensure the required magnetizing inductance, the application of an air gap in the iron core is necessary. Using modern converter techniques, low distortion levels (a few percent) and very good dynamic behavior (5–10% overshoot at load jumps) can be achieved.

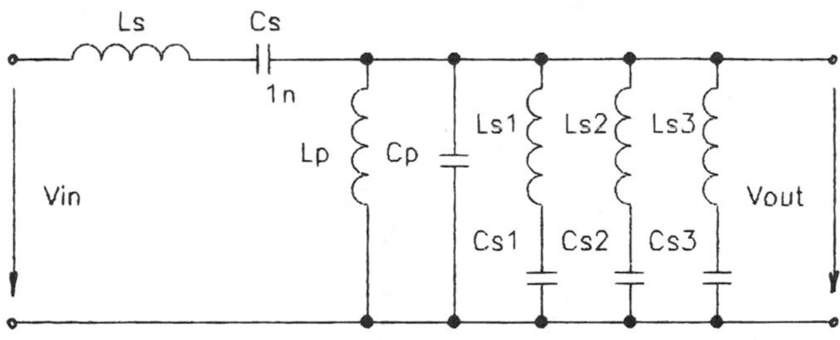

Figure 12.48 Basic filter circuit.

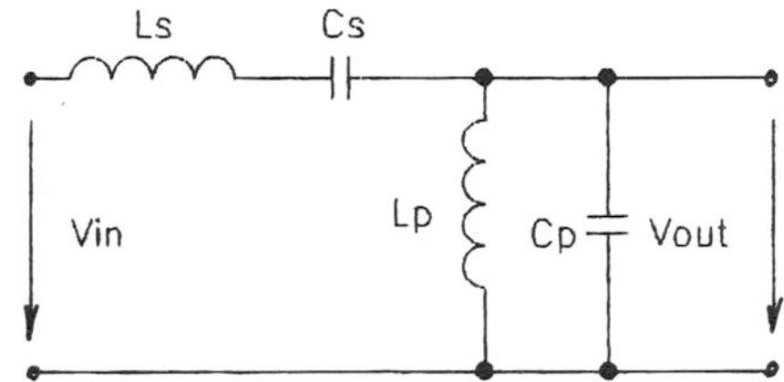

Figure 12.49 Simplified filter circuit.

Practical Realization of Basic Connections

Bipolar transistors, IGBTs, and FETs are generally used in modern converters with ≤ 100kW output power. At higher power, the application of GTOs and thyristors are common. If thyristors are used, the connection must be completed by quenching circuits to turn off the current conducting thyristor. The energy necessary to turn off the thyristor is stored in capacitors. The basic connections with thyristors are used for frequencies up to 1–2 kHz. With IGBTs a frequency of ~20 kHz is attainable. If FETs are used, 100 kHz frequency is normal, but equipment with 500 kHz frequency is also possible.

Special Realizations (Application of Resonant Converter Techniques)

Certain types of DC-AC converters use series or parallel resonant circuits. They are known as resonant converters, which can be subdivided into the following groups:

- Load resonant converters, i.e., current-source parallel resonant and voltage-source series resonant DC-to-AC inverters.
- Resonant switch converters. ZVS-CV DC-to-AC Inverters.
- Resonant converter connections, used in electrical drives
  Auxiliary resonant-commutated pole inverters.
  Parallel and series resonant DC-link converters.
  Active clamped parallel resonant DC-link inverters.
  Parallel and series resonant AC-link converters.

In load resonant converters the load is completed by capacitance to a resonant circuit. In the current-source inverters a capacitance is connected in parallel with the load. The circuit

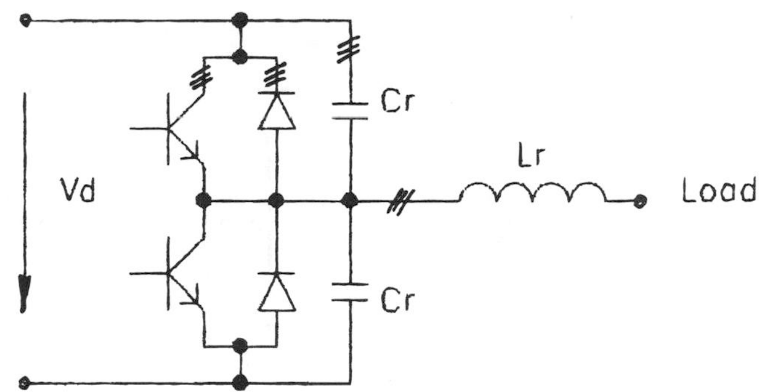

Figure 12.50 ZVS-CV DC-to-AC inverter.

and time functions are in steady-state as shown in Figure 12.50. The connection operates as a line commutated circuit in the inverter working mode; however, the voltage on the parallel resonant circuit ensures commutation. The power can be controlled by changing the value of V_d. A controlled rectifier is generally used for this purpose. This connection is typically applied in induction heating.

For voltage-source inverters the capacitance is connected in series with the load. If converter thyristors are used, the circuit and the time functions in steady-state are shown in Figure 12.51. The quenching of the thyristor is ensured by the voltage drop across the freewheeling diode, which is connected to the thyristor in inverse parallel. The output frequency is less than the series resonant frequency. The output power is usually controlled by changing the output frequency. If semiconductor elements, e.g., IGBT, FET etc., which can be turned off by a gate signal are used, the output frequency can be equal to or greater than the resonant frequency. In the latter case, the switching losses are smaller. The output power can be controlled by changing the output frequency or voltage, V_1. In the latter case, voltage cancellation can be utilized.

In resonant switch converters, resonant circuits are connected to the semiconductors to ensure soft switching and to reduce the switching losses. In practice, zero current switching (ZCS) and zero voltage switching (ZVS) are possible. Because the voltage on the semiconductors increases with simple ZVS, clamped voltage (CV) versions are used. A simplified version of a three-phase ZVS-CV DC-to-AC inverter is shown in Figure 12.52. The transistor's switching is done at zero voltage on the capacitances, which are connected in parallel to the transistors.

The most important types of inverters used with electrical drives are the three-phase bridge connections. Solutions for the realization of soft switching are briefly described below. The auxiliary resonant-commutated pole inverter is shown in Figure 12.53. It is a traditional voltage-source inverter, which contains switched resonant circuits, with components L_r, C_r, $T_{r,1,2}$, for each leg. The resonant circuits and the switch control ensure that the additional circuits operate only during switching in the main bridge, which guarantee soft switching for the semiconductor elements.

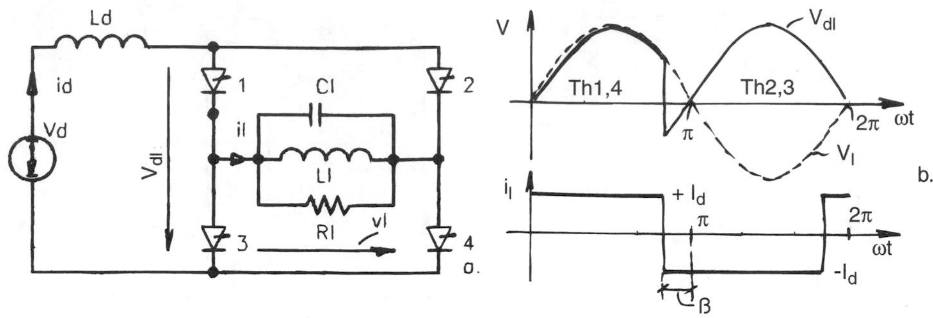

Figure 12.51 Current source parallel resonant inverter.

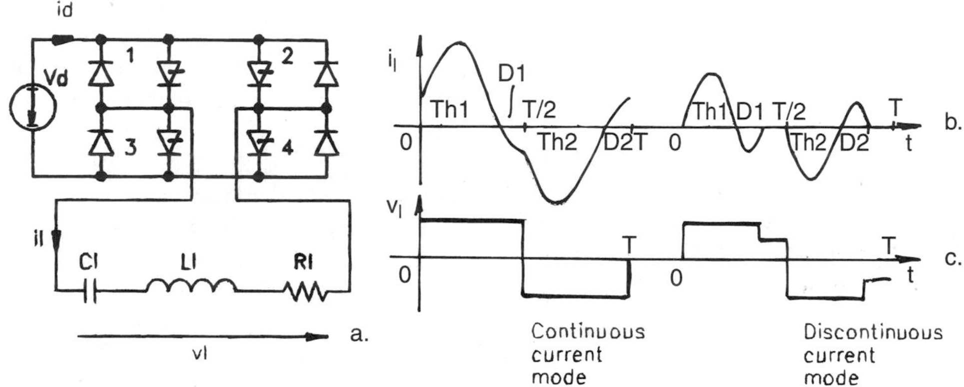

Figure 12.52 Voltage-source series resonant converter.

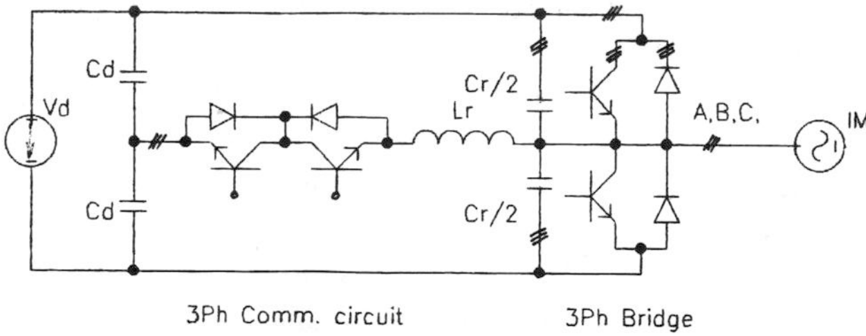

Figure 12.53 Auxiliary resonant commutated pole inverter.

The parallel resonant DC-link converter is shown in Figure 12.54. An AC voltage on the input DC voltage is superimposed using the resonant circuit L_r, C_r, so that V_r will be periodically zero. When V_r equals zero, the semiconductor elements in the output bridge are switched (ZVS) which results in soft switching. The resonant circuit is excited by the periodic common turn-on of all elements in the output bridge.

The series resonant DC-link converter is shown in Figure 12.55. It is a traditional current-source inverter which contains a series resonant circuit. Therefore, an AC component is superimposed on the DC current which ensures that the current in the bridges will be periodically zero. The semiconductor elements are switched when the current is equal to zero (ZCS). A suitable control strategy ensures that the network and output current are approximately sinusoidal. Thyristors or GTOs are used as semiconductor elements which can also operate in the reverse voltage direction.

The active clamped parallel resonant DC-link inverter is shown in Figure 12.56. (It is a parallel resonant DC-link inverter containing a clamping circuit, C_{cl}, T_{cl}, to limit the maximum voltage on the semiconductor elements.)

The AC-link resonant converters are a special type of converter. The parallel resonant AC-link converter is shown in Figure 12.57. Suitable operation of the switches and parallel resonant circuit ensure that there is a high-frequency AC voltage on the input of the output bridge. The output voltage with the required frequency and small harmonic content is defined by suitably linking the half-periods of the input pulses.

The series resonant AC-link converter is shown in Figure 12.58. The suitable operation of the switches and series resonant circuit ensures that a high-frequency AC current is present in the input of the output bridge. The output current with the required frequency and small harmonic content is defined by suitably linking the half-periods of the input pulses.

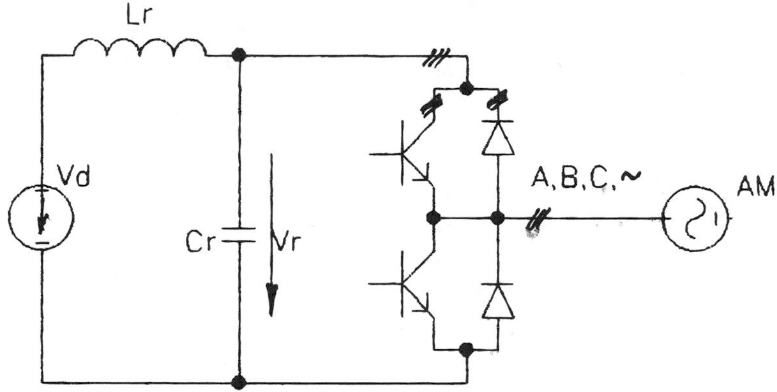

Figure 12.54 Parallel resonant DC-link converter.

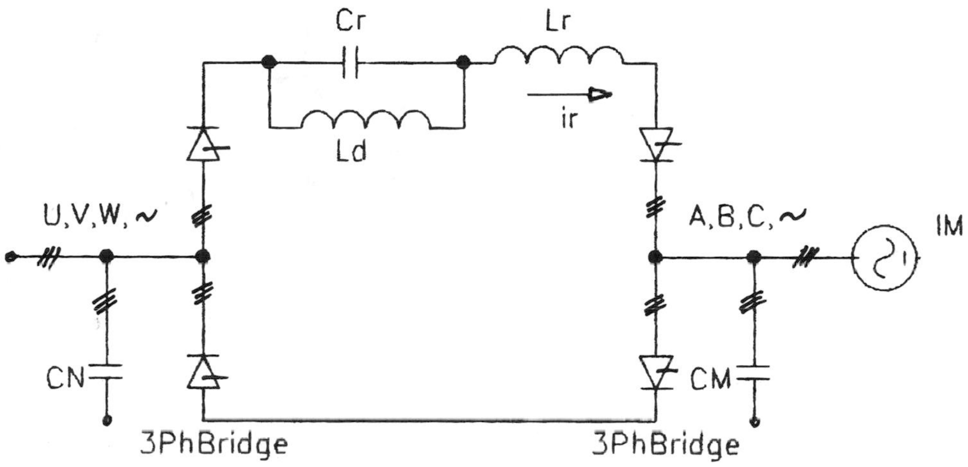

Figure 12.55 Series resonant DC-link converter.

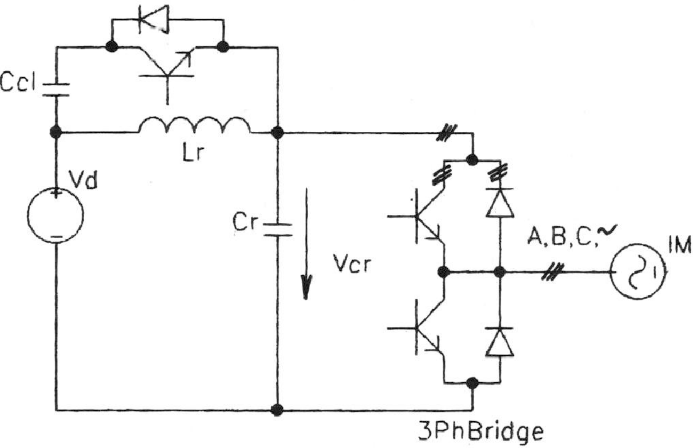

Figure 12.56 Active clamped parallel resonant DC-link converter.

12.4 AC-AC Conversion

Sándor Halász

AC-AC converters as shown in Figure 12.59 are frequency converters. They produce an AC voltage in which both the frequency and voltage can be varied directly from the AC line voltage, e.g., from a 60 Hz or 50 Hz source.

There are two major classes of AC-AC, or so-called direct static frequency converters, as shown in Figure 12.59.

1. Cycloconverters, which are constructed using naturally commutated thyristors. The commutation voltage is ensured by the supply voltage. These are so-called line commutated converters.
2. Matrix converters, which are constructed using full-controlled static devices, such as transistors or GTOs (gate turn-off thyristors).

Cycloconverters

In Figures 12.60 and 12.61, the two typical types of cycloconverters are presented. In the first case there are two three-phase midpoint controlled rectifiers connected back-to-back. The second case shows two three-phase bridge rectifier converters connected back-to-back. Both are used for three-phase to three-phase conversion. In Figure 12.62 the single-phase output voltage and current waves are presented for the bridge rectifier circuits. The output voltage V_a and current I_a have V_{a1} and I_{a1} fundamental components with ϕ_1 phase displacement and numerous harmonics. Because of the load inductance, the current harmonics will

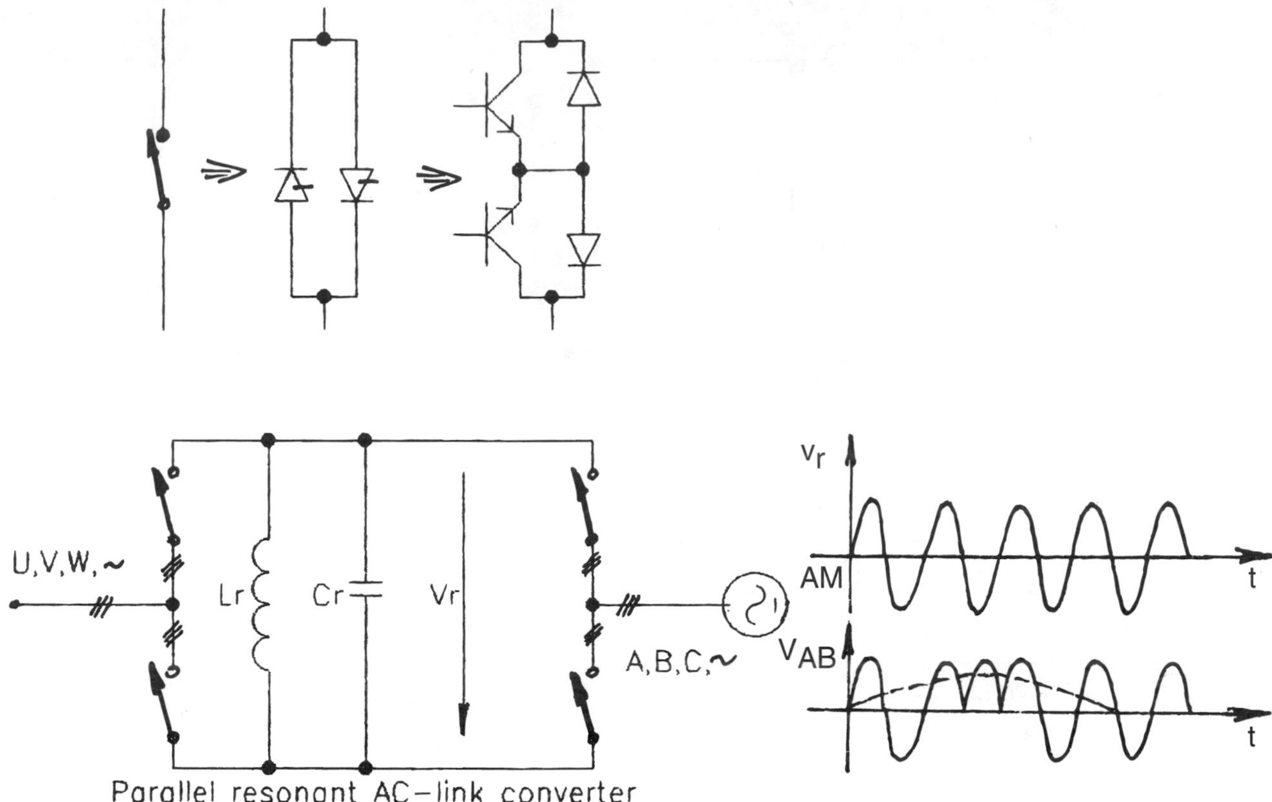

Figure 12.57 Parallel resonant AC-link converter.

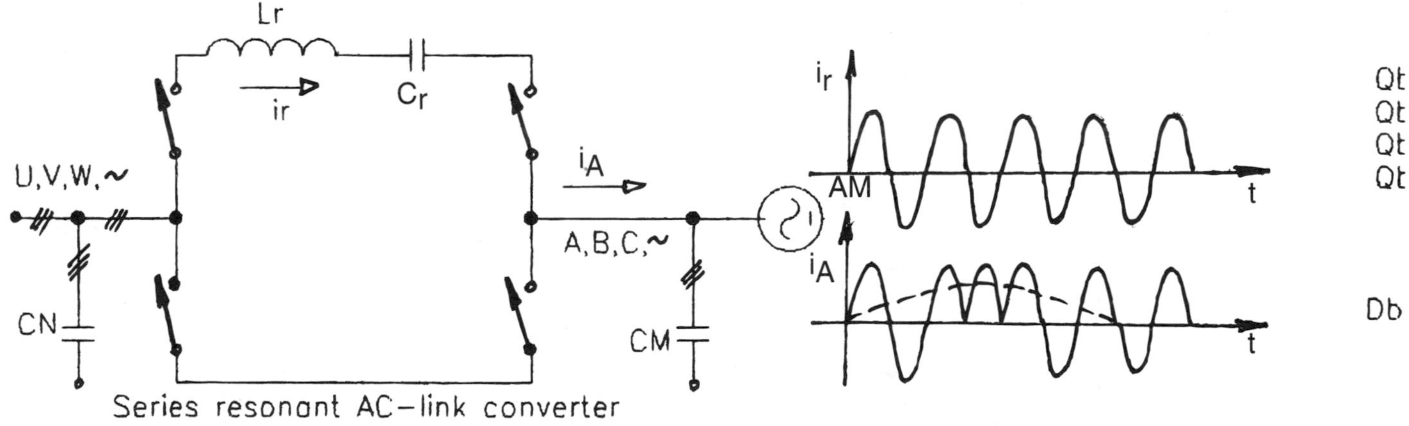

Figure 12.58 Series resonant AC-link converter.

be significantly lower than the voltage harmonics. The firing angles are α_P and α_N for the P and N converters, respectively. In general, the controls are designed so that only the thyristors of either the P or N converter is firing, which produces a current in the desired direction. During this period the other converter is blocked. When the current changes direction, both converters must be blocked for a short time as described in Section 12.1.

It is possible to operate without blocking the converters. In this case, their average voltage must be the same, and therefore the relation $\alpha_P = 180 - \alpha_N$ is valid. However, additional inductances are necessary to limit the circulating currents between two converters since the instantaneous voltages of the two converters differ from one another.

The phase control of the P and N converters is modulated by a sine or trapezoidal wave. The content of the harmonics for sine modulation is lower; however, the maximum value of the output voltage is lower than that for trapezoidal modulation. During every cycle of the output voltage both of the converters must work as rectifiers and inverters.

The shape of the output voltage goes from bad to worse with an increase in the output voltage and the output frequency. If the frequency reaches the well-defined value the current harmonics

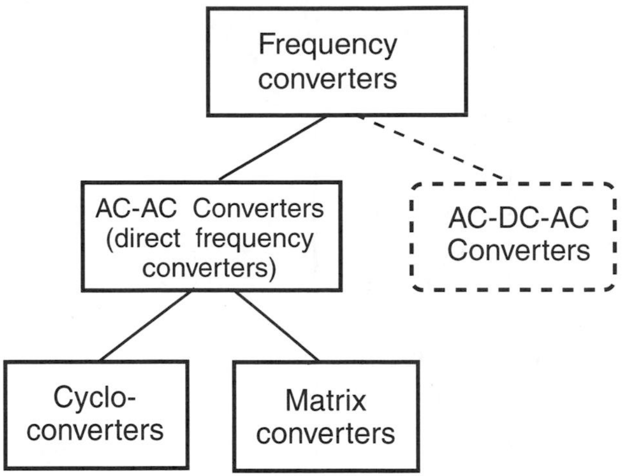

Figure 12.59 Classification of frequency converters.

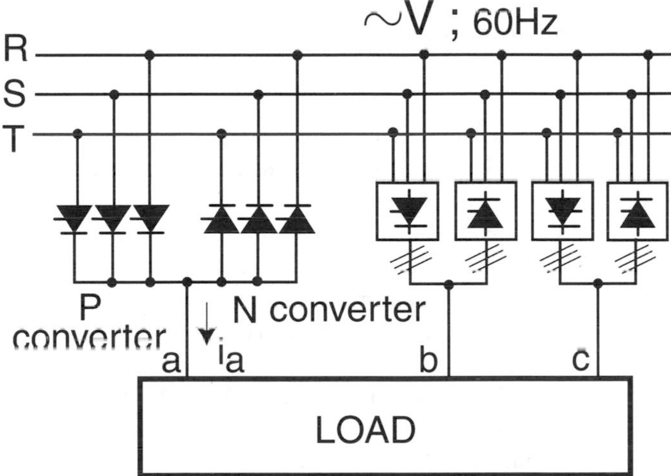

Figure 12.60 Cycloconverter scheme with three-phase midpoint controlled rectifier.

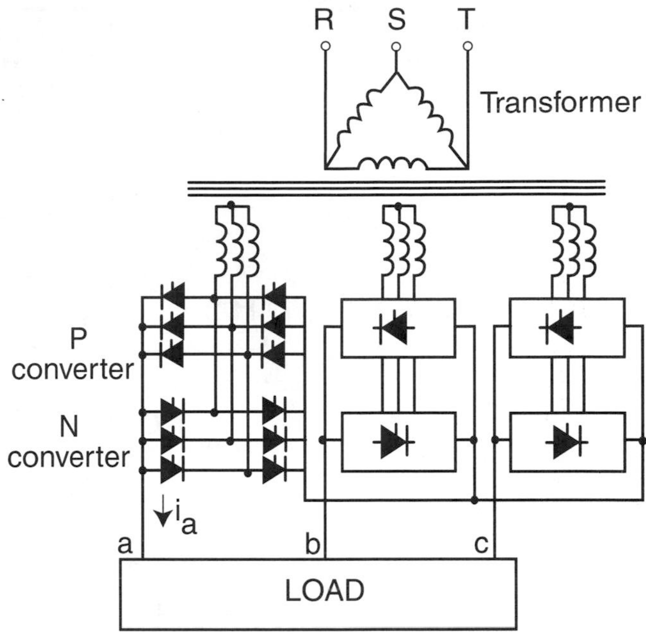

Figure 12.61 Cycloconverter scheme with three-phase bridge controlled rectifier.

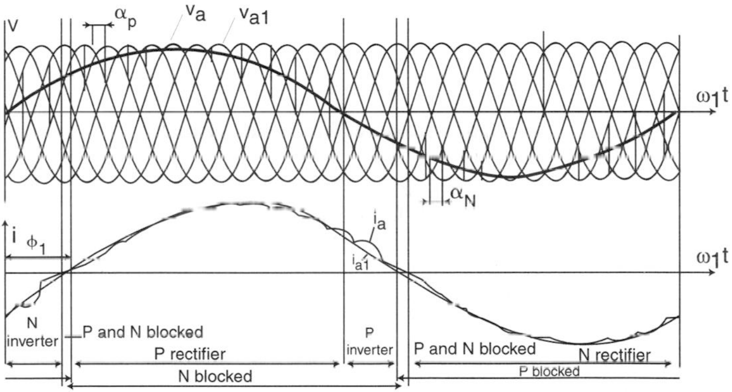

Figure 12.62 Voltage and current vs time for cycloconverter with three-phase bridge converters.

become unacceptable. This frequency is usually 33% of supply frequency for three-phase (Figure 12.60) and 50% for three-phase bridge (Figure 12.61) converters.

The cycloconverter is usually used for three-phase, high-power, low-speed synchronous motor drives and rarely employed for induction motor drives.

Matrix Converters

The three-phase to three-phase matrix converter is presented in Figure 12.63. Using the bidirectional switches, any phase of the load can be connected to any phase of the input voltage, e.g., the zero value of the load phase voltages is maintained by connecting all the load phases to the same input phase. Using pulse-width modulation techniques, the load voltage and the load frequency are controlled from zero to their maximum values. The maximum voltage is usually close to the input voltage, but the maximum frequency can be several times that of the input frequency and is only limited by practical considerations. The bidirectional switches must be capable of permitting current flow in either direction. In Figure 12.64 one possible configuration of the bidirectional switch is shown.

Matrix converters require the use of numerous switches and well-established control methods. Some additional elements are necessary for the safe commutation of the bidirectional switches. These disadvantages of matrix converters prevent their use in industrial applications.

References

Guggi, L. and Pelly, B. R. 1976. *Static Power Frequency Changes*, John Wiley & Sons, New York, NY.

Pelly, B. R. 1976. *Thyristor Phase-Controlled Converters*, John Wiley & Sons, New York, NY.

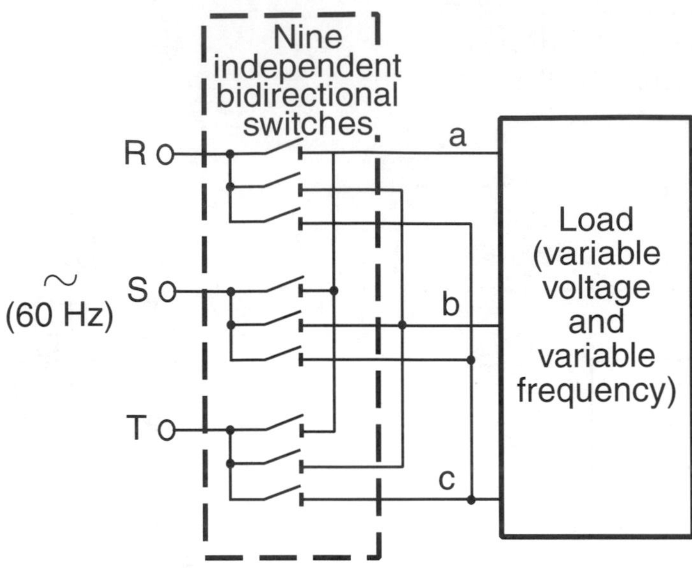

Figure 12.63 Three-phase to three-phase matrix converter.

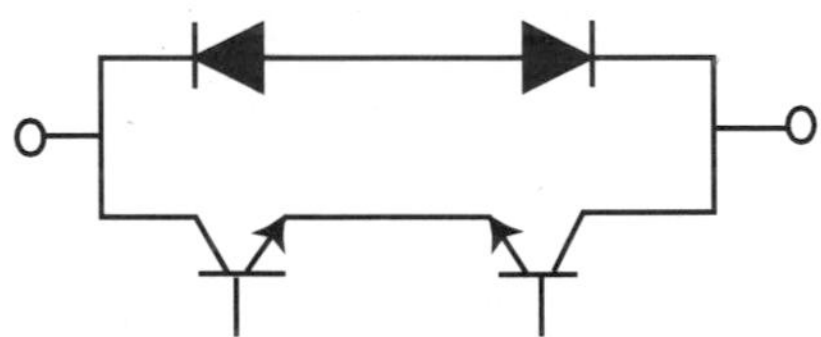

Figure 12.64 Bidirectional switch.

12.5 Resonant Converters

István Nagy

Introduction

Resonant converters connect a DC system to an AC system or another DC system and control both the power transfer between them and the output voltage or current. They are used in such applications as: induction heating, very high frequency DC-DC power supplies, sonar transmitters, ballasts for fluorescent lamps, power supplies for laser cutting machines, ultrasonic generators, etc.

There are some common features characterizing the behavior of most, or at least some, of these elements. DC-DC and DC-AC converters have two basic shortcomings when their switches are operating in the switch mode. During the turn-on and turn-off time, high current and voltage appear simultaneously in and across the switches producing high power losses in them, that is, high switching stresses. The power loss increases linearly with the switching frequency. To ensure reasonable efficiency of the power conversion, the switching frequency has to be kept under a certain maximum value. The second shortcoming in a switching mode operation is the electromagnetic interference (EMI) generated by the large dv/dt and di/dt values of the switching variables. The drawbacks have been accentuated by the trend which is pushing the switching frequency to higher and higher range in order to reduce the converter size and weight.

The resonant converters can minimize these shortcomings. The switches in resonant converters create a square-wave-like voltage or current pulse train with or without a DC component. A resonant L-C circuit is always incorporated. Its resonant frequency could be close to the switching frequency or could deviate substantially. If the resonant L-C circuit is tuned to approximately the switching frequency, the unwanted harmonics are removed by the circuit. In both cases the variation of the switching frequency is one of the means for controlling the output power and voltage.

The advantages of resonant converters are derived from their L-C circuit and they are as follows: sinusoidal-like wave shapes, inherent filter action, reduced dv/dt and di/dt and EMI, facilitation of the turn-off process by providing zero current crossing for the switches and output power and voltage control by changing the switching frequency. In addition, some resonant converters e.g., quasi-resonant converters, can accomplish zero current and/or zero voltage across the switches at the switching instant and reduce substantially the switching losses. The literature categorizes these converters as hard switched and soft switched converters. Unlike hard switched converters the switches in soft switched converters, quasi-resonant and some resonant converters are subjected to much lower switching stresses. Note that not all resonant converters offer zero current and/or zero voltage switchings, that is, reduced switching power losses. In return for these advantageous features, the switches are subjected to higher forward currents and reverse voltages than they would encounter in a nonresonant configuration of the same power. The variation in the operation frequency can be another drawback.

First, a short review of the two basic resonant circuits, series and parallel, are given. Then the following three types of resonant converters are discussed:

- Load resonant converters.
- Resonant switch converters.
- Resonant DC link converters.

Survey of Second Order Resonant Circuits

The parallel resonant circuit is the dual of the series resonant circuit (Figure 12.65). The series (parallel) circuit is driven by a voltage (current) source. The analog variables for the voltages and currents are the corresponding currents and voltages (Figure 12.65). Kirchhoff's voltage law for the series circuit

$$v_i = v_L + v_R + v_C = i_i\left(sL + R + \frac{1}{sC}\right) \tag{12.1}$$

and Kirchhoff's current law for the parallel circuit

$$i_i = i_L + i_R + i_C = v_i\left(\frac{1}{sL} + \frac{1}{R} + sC\right) \tag{12.2}$$

have to be used. The analog parameters for the impedances are the corresponding admittances (Figure 12.65). The input current for the series circuit is

$$i_i = Y_s(s)v_i = \frac{1}{Z_s(s)}v_i \tag{12.3}$$

and the input voltage for the parallel circuit is

$$v_i = Z_p(s)i_i \tag{12.4}$$

where the input admittance is

$$Y_s(s) = \frac{1}{R}\frac{2\xi_s Ts}{T^2s^2 + 2\xi_s Ts + 1} \tag{12.5}$$

and the input impedance is

$$Z_p(s) = R\frac{2\xi_p Ts}{T^2s^2 + 2\xi_p Ts + 1} \tag{12.6}$$

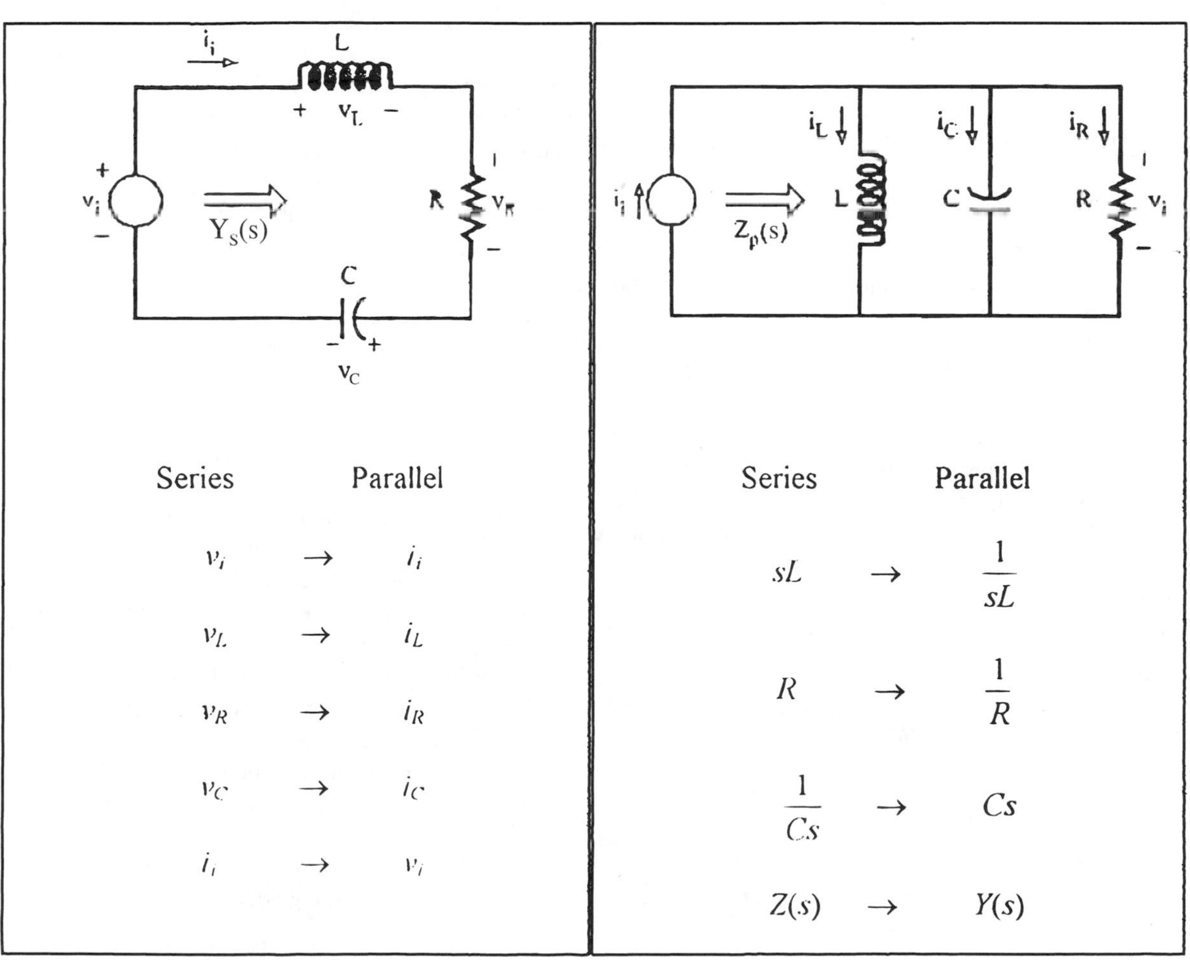

Figure 12.65 Dual circuits.

Table 12.7 Parameters

	Series	Parallel
Time constant	$T = \sqrt{LC}$	
Resonant angular frequency	$\omega_0 = 2\pi f_0 = \frac{1}{T}$	
Damping factor	$\xi_s = \frac{1}{2}\frac{R}{\omega_0 L} = \frac{1}{2}\omega_0 CR$	$\xi_p = \frac{1}{2}\frac{\omega_0 L}{R} = \frac{1}{2}\frac{1}{\omega_0 CR}$
Characteristic impedance	$Z_0 = \sqrt{L/C}$	
Damped resonant angular frequency	$\omega_d = \omega_0\sqrt{1-\xi_s^2}$	$\omega_d = \omega_0\sqrt{1-\xi_p^2}$
Quality factor	$Q_s = \frac{1}{2\xi_s}$	$Q_p = \frac{1}{2\xi_p}$

The time constant and the damping factor ξ together with some other parameters are given in Table 12.7. ξ must be smaller than unity in Equations (12.5) and (12.6) to have complex roots in the denominators, that is, to obtain an oscillatory response.

When v_i is a unit step function, $v_i(s) = 1/s$, the time response of the voltage across R in the series resonance circuit from Equations (12.3) and (12.5) is

$$Ri_i(t) = 2\xi_s T\left[\frac{1}{T\sqrt{1-\xi_s^2}}e^{-\xi_s t/T}\sin(\sqrt{1-\xi_s^2})t/T\right] \tag{12.7}$$

$$= 2\xi_s T f(t/T)$$

or for $\xi_s = 0$

$$i_i(t) = \frac{1}{\omega_0 L}\sin\omega_0 t \tag{12.7a}$$

that is, the response is a damped, or for $\xi_s = 0$ undamped, sinusoidal function.

When the current changes as a step function in the parallel circuit, $Ri_i(s) = 1/s$, the expression for the voltage response v_i is given by the right side of equation 12.7, as well, since $RY_s = Z_p/R$. Of course, now ξ_s has to be replaced by ξ_p. The time function $f(t/T)$ for various damping factors ξ is shown in Figure 12.66.

Assuming sinusoidal input variables, the frequency response for series circuit is

$$\frac{R\bar{i}_i}{\bar{v}_i} = R\bar{Y}_s(j\nu) = \frac{1}{1+jQ_s(\nu-1/\nu)} = \frac{1}{\bar{D}_s(\nu)} \tag{12.8}$$

and for parallel circuit is

$$\frac{\bar{v}_i}{R\bar{i}_i} = \frac{1}{R}\bar{Z}_p(j\nu) = \frac{1}{1+jQ_p(\nu-1/\nu)} = \frac{1}{\bar{D}_p(\nu)} \tag{12.9}$$

where

$$\nu = \omega/\omega_0.$$

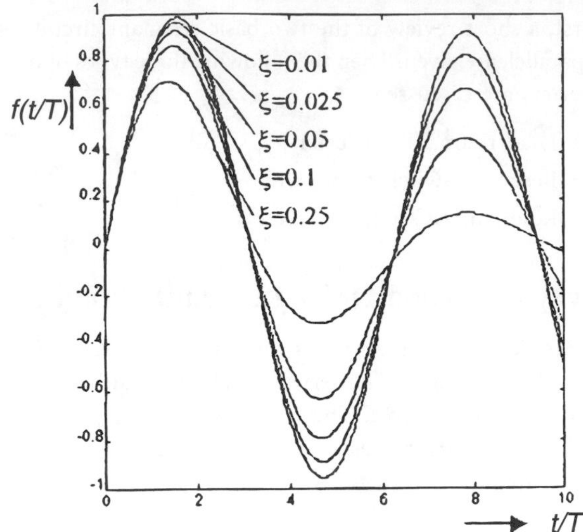

Figure 12.66 Time response of $f(t/T)$. ($T = 1$).

Both circuits are purely resistive at resonance: $\bar{v}_i = R\bar{i}_i$ when $\nu = 1$.

The plot of the amplitude and phase of the right side of equation 12.8 and equation 12.9 as a function of ν are shown in Figure 12.67. The voltage across R and its power can be changed by varying ν. When Q is high, a small change in ν can produce a large variation in the output.

The voltage across the energy storage components, for instance, across L in the series circuit, is

$$\frac{\bar{v}_L}{\bar{v}_i} = \frac{j\nu Q_s}{\bar{N}_s(\nu)} \tag{12.10}$$

and the currents in the energy storage components, for instance, in L in the parallel circuit, is

$$\frac{\bar{i}_L}{\bar{i}_i} = \frac{Q_p}{j\nu\bar{N}_p(\nu)} \tag{12.11}$$

The voltages (currents) of the energy storage components in series (parallel) resonant circuits at $\nu = 1$ is Q times as high as the input voltage (current) (Table 12.8). If $Q = 10$ the capacitor or inductor voltage (current) is 10 times the source voltage (current).

The value of L and C and their power rating is tied to the quality factor. The higher the value of Q, the better the filter action, that is, the attenuation of the harmonics is better and it is easier to control the output voltage and power by a small change in the switching frequency. The definition of Q is

$$Q = \frac{2\pi * \text{ Peak stored energy}}{\text{Energy dissipated per cycle}} \tag{12.12}$$

Using this definition, the expressions for Q are given in Table 12.8 where I_p and V_p are the peak current in the inductor and peak voltage across the capacitor, respectively. For a given output

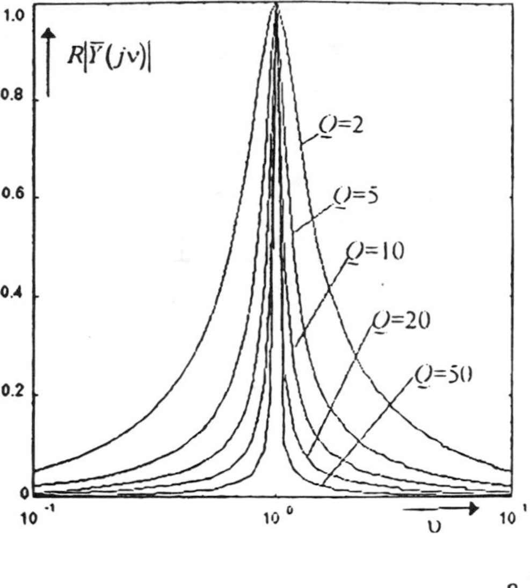

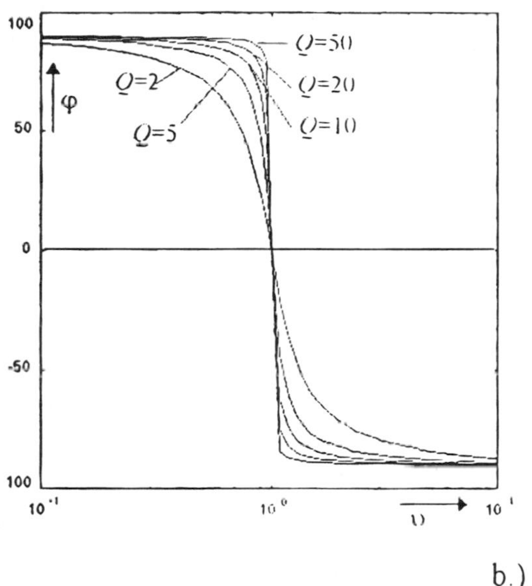

Figure 12.67 Frequency response of $[R\bar{Y}(jv)]$. Amplitude (a), phase (b).

Table 12.8 Resonance, $\omega = \omega_0$

Series	Parallel
$\dfrac{\bar{v}_c}{\bar{v}_i} = -jQ_s$	$\dfrac{\bar{i}_C}{\bar{i}_i} = jQ_p$
$\dfrac{\bar{v}_L}{\bar{v}_i} = jQ_s$	$\dfrac{\bar{i}_L}{\bar{i}_i} = -jQ_p$
$Q_s = \dfrac{2\pi\left(\frac{1}{2}LI_P^2\right)}{\left(\frac{1}{2}RI_P^2\right)\frac{1}{f_0}}$	$Q_p = \dfrac{2\pi\left(\frac{1}{2}CV_P^2\right)}{\left(\frac{1}{2}\frac{V_P^2}{R}\right)\frac{1}{f_0}}$

power, the energy dissipated per cycle is specified. The only way to obtain a higher Q is to increase the peak stored energy. The price paid for a high Q is the high peak energy storage requirements in both the inductor and capacitor.

Load Resonant Converters

In these converters the resonant L-C circuit is connected in the load. The currents in the switching semiconductors decay to zero due to the oscillation in the load circuit. Four typical converters are discussed:

1. Voltage source series resonant converters.
2. Current source parallel resonant converters.
3. Class E resonant converters.
4. Series and parallel loaded resonant DC-DC converters.

Input Time Functions

As a result of the on-off action of the switching devices, the frequently produced time functions of the input variable at the terminals of the ringing load circuit are shown in Figure 12.68. The input variable x_i can be either voltage in series resonant converters (SRC) or current in parallel resonant converters (PRC) and it can be unidirectional (Figure 12.68a) or bidirectional (Figure 12.68b and 12.68c). The ringing load is excited by a variable (Fig. 12.68a) which is constant in the interval $\alpha \leq \omega_s t \leq \pi - \alpha$ and short-circuited in the interval $\pi + \alpha \leq \omega_s t < 2\pi - \alpha$, where ω_s is the switching angular frequency. The circuit is interrupted during the rest of the period. The interruption interval shrinks to zero when $\omega_s \geq \omega_d$. The input variable is square-wave and a quasi-square-wave in Figures 12.68b and 12.68c, respectively. The RMS value of the fundamental component is

$$X_{irms} = \frac{4}{\pi\sqrt{2}} X_p \cos\alpha \tag{12.13}$$

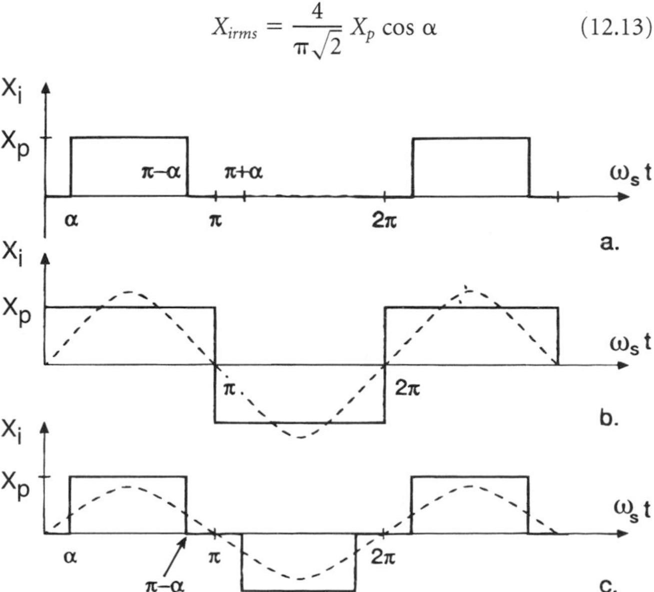

Figure 12.68 Frequently used input time functions.

The output variable changes in proportion to the input. Varying the angle α provides another means of controlling the output besides the switching frequency f_s.

Series Resonant Converters

Series resonant converters (SRC) can be implemented by employing either unidirectional (Figure 12.69) or bidirectional (Figure 12.70) switches. The unidirectional switch can be a thyristor, GTO, bipolar transistor, IGBT, etc., while these devices with an antiparallel diode or RCT (reverse conducting thyristor) can be used as a bidirectional switch.

Depending on the switching frequency f_s, the wave shape of the output voltage v_o can take any one of the forms shown in Figure 12.71 using the circuit in Figure 12.69. The damped resonant frequency f_d is greater than f_s in Figure 12.71a, $f_s < f_d$; equal to f_s in Figure 12.71b; $f_s = f_d$; and smaller than f_s in Figure 12.71c, $f_s > f_d$. *S1* and *S2* are alternately turned on. The terminals of the series resonant circuit are connected to the source voltage V_{dc} by *S1* or short-circuited by *S2*. When neither of the switches are on, the circuit is interrupted. The voltage across the terminals of the series resonant circuit follows the time function shown in Figure 12.68a for $f_d > f_s$, and in Figure 12.68b for $f_d \leq f_s$, respectively. By turning on one of the switches, the other one will be force commutated by the close coupling of the two inductances.

The configuration shown in Figure 12.70 can be operated below resonance, $f_s < f_d$ (Figure 12.72a); at resonance, $f_s = f_d$ (Figure 12.72b); and above resonance, $f_s > f_d$ (Figure 12.72c). The voltage, v_i, across the terminals of the series resonant circuit is square wave. The harmonics of the load current can be neglected for high Q value. The output voltage v_o equals its fundamental component v_{o1}. The *L-C* network can be replaced by an equivalent capacitor (inductor) below (above) resonance and by a short-circuit at resonance. The circuit is capacitive (inductive) below (above) resonance and purely resistive at resonance (Figure 12.72). The output voltage $v_o \cong v_{o1}$ is leading (lagging) the input voltage v_i below (above) resonance and in phase at resonance. Negative voltage develops across switches *S1* and *S2* during diode conduction and can be utilized to assist the turn-off process of switches *S1* and *S2*.

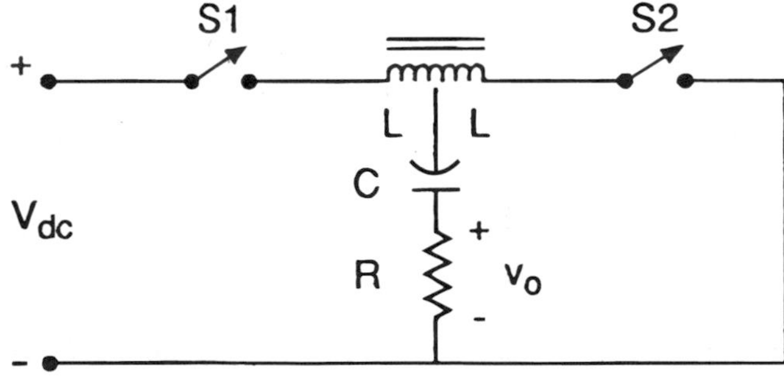

Figure 12.69 Series resonant converter (SRC) with unidirectional switches.

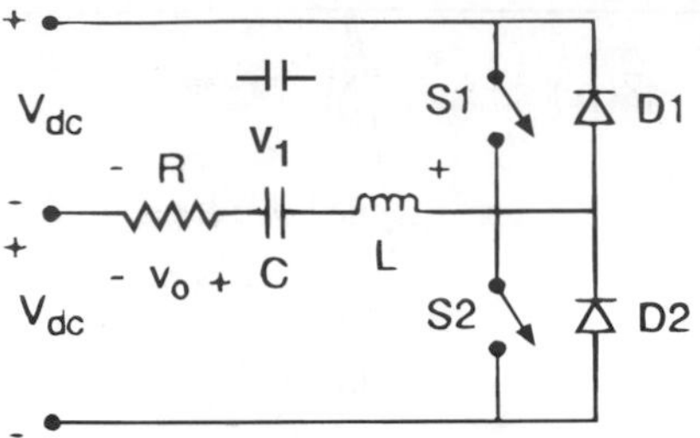

Figure 12.71 SRC with bidirectional switches.

No switching loss develops in the switches at $f_s = f_d$ (Figure 12.72b) since the load current will be passing through zero exactly at the time when the switches change state (zero current switching). However, when $f_s < f_d$ or $f_s > f_d$ the switches are subjected to lossy transitions. For instance, if $f_s < f_d$ the load current will flow through the switch at the beginning of each half-cycle and then commutate to the diode when the current changes polarity (Figure 12.72a). These transitions are lossless. However, when the switch turns on or when the diode turns off, they are subjected to simultaneous step changes in voltage and current. These transitions therefore are lossy ones. As a result, each of the four devices is subjected to only one lossy transition per cycle.

The bridge topology (Figure 12.73) extends the output power to a higher range and provides another control mode for changing the output power and voltage (Figure 12.74).

Discontinuous Mode

Converters with either unidirectional or bidirectional switches can be controlled in a discontinuous mode as well. In this mode, the resonant current is interrupted in every half-cycle when using unidirectional switches (Figure 12.71a) and in every cycle when using bidirectional switches (Figure 12.75). The power

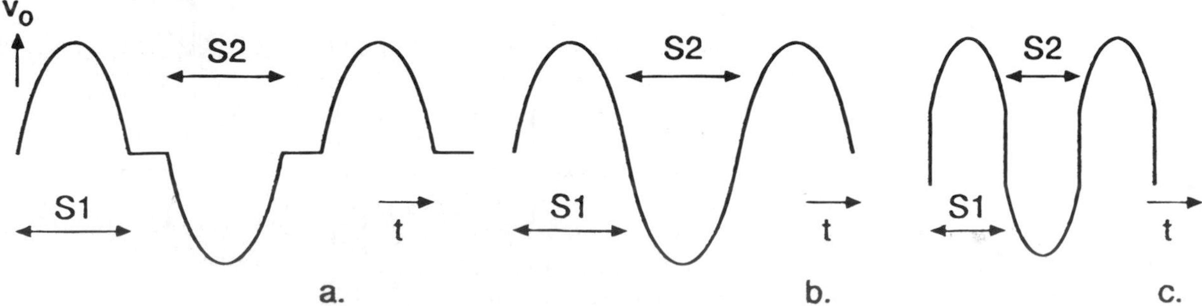

Figure 12.70 Output voltage waveforms for Figure 12.69, $f_s < f_d$ (a), $f_s = f_d$ (b), $f_s > f_d$ (c).

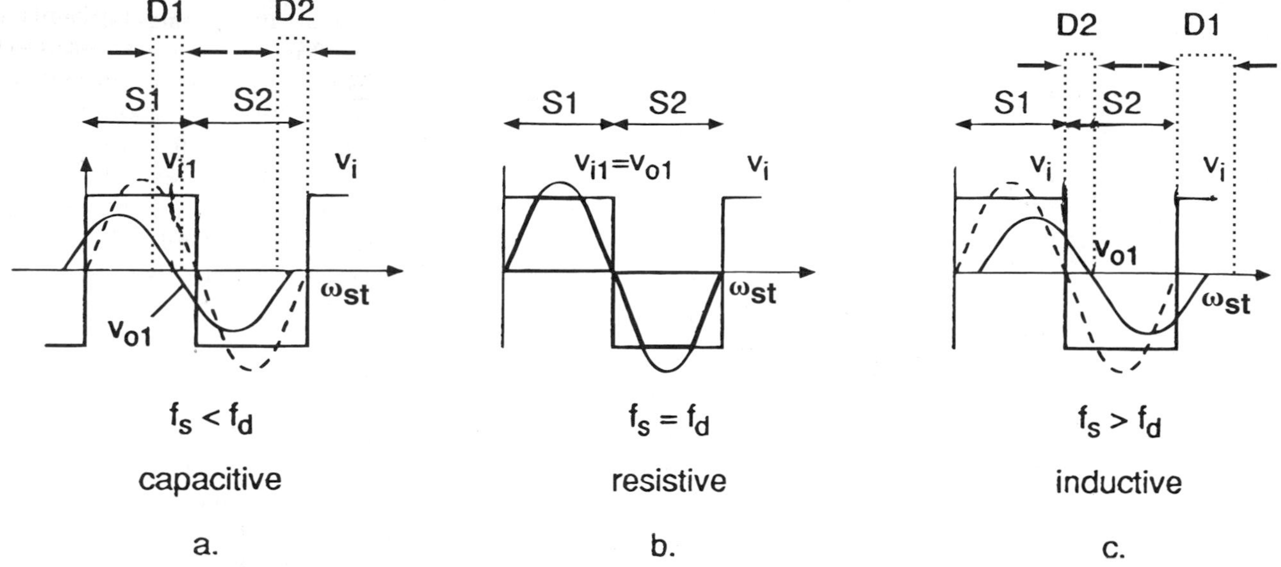

Figure 12.72 Output voltage waveforms for Figure 12.70.

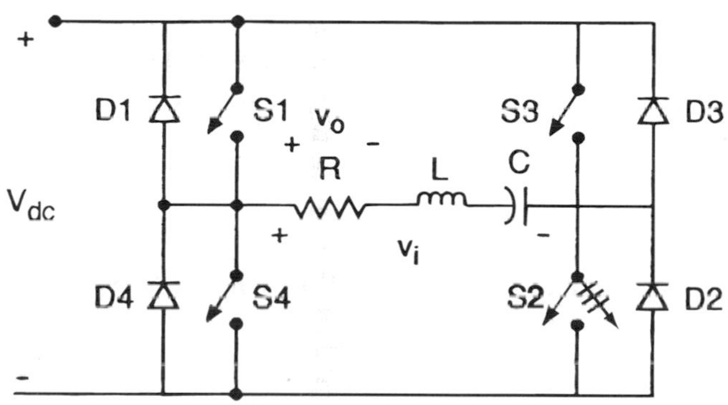

Figure 12.73 SRC in bridge topology.

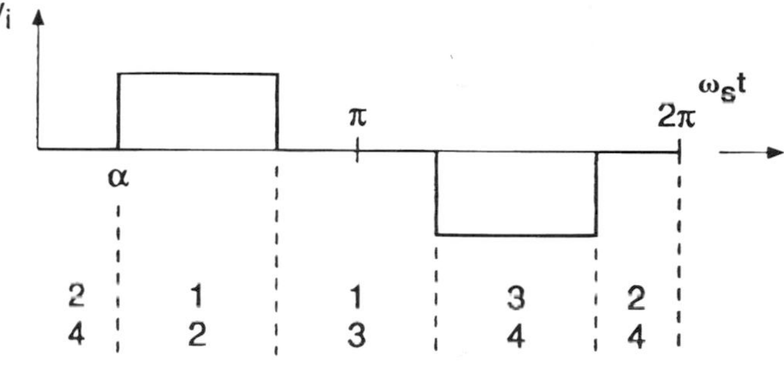

Figure 12.74 Quasi square-wave voltage for output control.

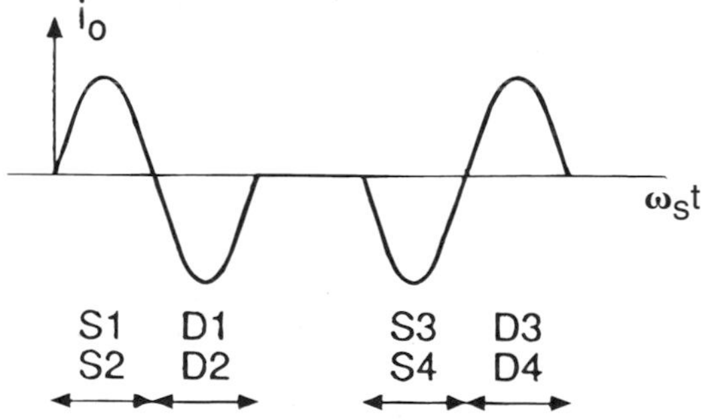

Figure 12.75 Discontinuous mode for bridge topology.

is controlled by varying the duration of the current break as it is done in duty ratio control of DC-DC converters. Note that this control mode theoretically avoids switching losses because whenever a switch turns on or off its current is zero and no step change can occur in its current as a result of the inductance L. The shortcoming of this control mode is the distorted current waveform. In some applications, such as induction heating and ballasts for fluorescent lamps, the sinusoidal waveform is not necessary.

Parallel Resonant Converters

The parallel resonant converters (PRC) are the dual of the series resonant converters (SRC) (Figure 12.76). The bidirectional switches must block both positive and negative voltages rather than conduct bidirectional current. They are supplied by a current source and the converters generate a square wave input current i_i that flows through the parallel resonant circuit (Figure 12.77). They offer better short-circuit protection under fault conditions than the SRCs with a voltage source.

When the quality factor Q is high and f_s is near resonance, the harmonics in the R-C-L circuit can be neglected. For $f_s < f_d$, the parallel L-C network is, in effect, inductive. The effective inductance shunts some of the fundamental components of the input current i_{i1} and a reduced leading current i_{i1} flows in the load resistance (Figure 12.77a). For $f_s = f_d$, the parallel L-C filter looks like an infinitely large impedance. The total current i_{i1} passes through R and the output voltage v_{o1} is in phase with i_{i1} (Figure 12.77b). Since $v_{o1}=0$ at switching instants, no switching loss develops in the switching devices. For $f_s > f_d$, the L-C network is an equivalent capacitor at the fundamental component of i_{i1}. A part of the input current flows through the equivalent capacitor and only the remaining portion passes through the resistor R developing the lagging voltage v_{o1} (Figure 12.77c). As

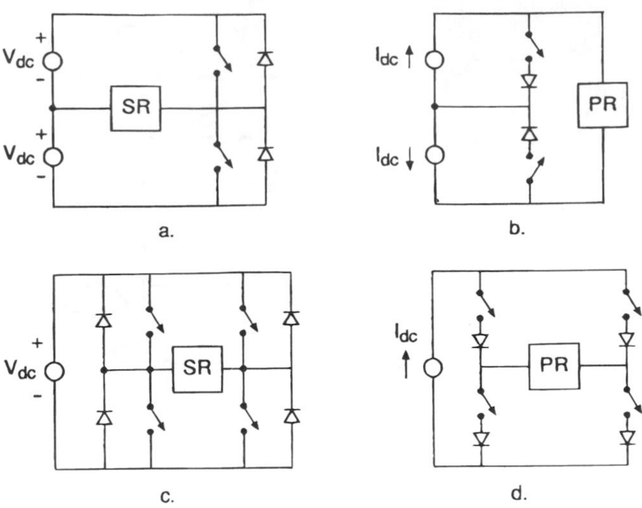

Figure 12.76 SRC and PRC are duals.

a result of the current shunting through the equivalent L_e and C_e, the voltage v_{o1} is smaller in Figure 12.77a and 12.77c than in Figure 12.77b, although i_{i1} is the same in all three cases. The current source is usually implemented by the series connection of a DC voltage source and a large inductor (Figure 12.78a). The bidirectional switch is implemented in practice for SRCs with the anti-parallel connection of a transistor-diode or thyristor-diode pair (Figure 12.78b) and for PRCs with the series connection of a transistor-diode pair or thyristor. The condition $f_s > f_d$ must be met for PRCs in order for the thyristor to be commutated. By turning on one of the thyristors, a negative voltage is imposed across the previously conducting one, forcing it to turn off (Figures 12.76b and 12.77c). If $f_s > f_d$ and a series transistor-diode pair is used, the diode will experience switching losses at turn-off and the transistor will experience losses at turn-on (Figure 12.77c).

Class E Converter

The class E converter is supplied by a DC current source (Figure 12.78a) and its load R is fed through a sharply tuned series resonant circuit ($Q \geq 7$) (Figure 12.79a). The output current i_o is practically sinusoidal. It uses a single switch (transistor) which is turned on and off at zero voltage. The converter has low—theoretically zero—switching losses and a high efficiency of more than 95% at an operating frequency of several ten kHz. Its output power is usually low, less than 100 W, and it is used mostly in high-frequency electronic lamp ballasts.

The converter can be operated in optimum and in suboptimum modes. The first mode is explained in Figure 12.79. When the switch is on (off) the equivalent circuit is shown in Figure 12.79b (12.79c). In the optimum mode of operation the switch (capacitor) voltage, $v_T = v_{C1}$, decays to zero with a zero slope; $I_{dc} + i_o = i_{C1} = 0$. Turning on the switch at t_0, a current pulse $i_T = I_{dc} + i_o$ will flow through the switch with a high peak value; $\hat{I}_T \cong 3I_{dc}$ (Figure 12.79d). Turning off the switch at $t = t_1$, the capacitor voltage builds up reaching a rather high value: $\hat{V}_C = 3.5V_{dc}$ and eventually falls back to zero at $t = t_0 + T$ (Figures 12.79e, d). The average value of v_T, and that of the capacitor voltage v_C, is V_{dc}. The average value of i_T is I_{DC} while there is no DC current component in i_o. In the non-optimum mode of operation, $i_{C1} < 0$ when v_T reaches zero value and the diode D is needed.

The advantage of the class E converter is the simple configuration, the sinusoidal output current, the high efficiency, the high output frequency and the low EMI. Its shortcomings are the high peak voltage and current of the switch and the large voltages across the resonant L-C components.

Series and Parallel Loaded Resonant DC-DC Converters

The load R can be connected in series with L-C or in parallel with C in series resonant converters. The first case is called a series loaded resonant (SLR) converter while the second one is called a parallel loaded resonant (PLR) converter. When the converter is used as a DC-DC converter, the load circuit is built up by a transformer followed by a diode rectifier, a low-pass filter and finally the actual load resistance. The resonant circuit makes possible the use of a high-frequency transformer reducing its size and the size of the filter components in the low-pass filter.

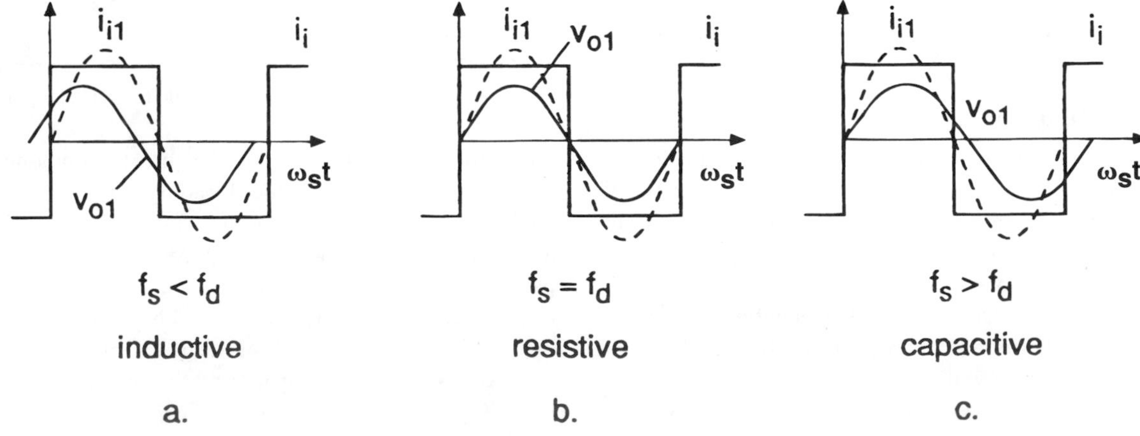

Figure 12.77 Waveforms for PRC.

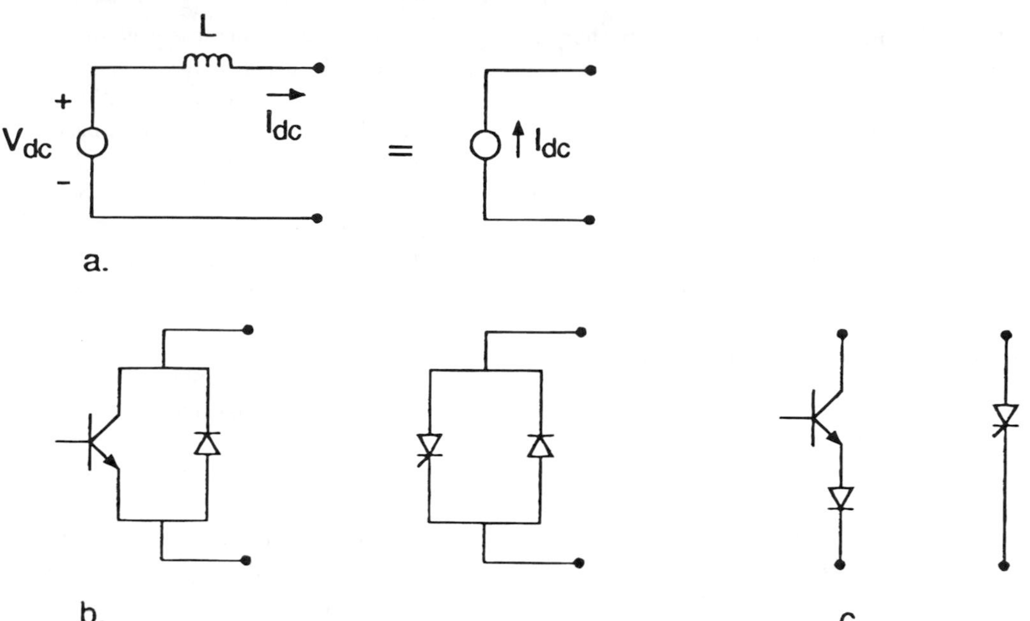

Figure 12.78 Implementation of current source (a). Implementation of bidirectional switch for SRC (b) and for PRC (c).

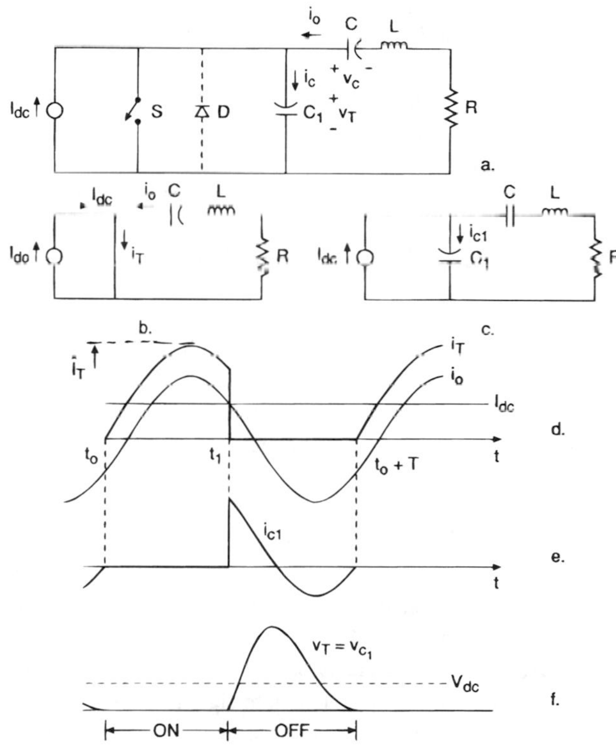

Figure 12.79 Class E resonant converter (optimum mode).

The properties of the SLR and PLR converters are quite different in some respects. Without the transformer action, the SLR converter can only step-down the voltage (Equation 12.8) while the PLR converter can both step-up and step-down (in discontinuous mode of operation) the voltage. The step-up action can be understood by noting that the voltage across the capacitor is Q times higher than that across R in the SRC. The PLR converter has an inherent short-circuit protection when the capacitor is shorted due to a fault in the load. The current is limited by the inductor L.

Resonant Switch Converters

The trend to push the switching frequency to higher values, to reduce size and weight and to suppress EMI led to the development of switch configurations providing zero-current-switching (ZCS) or zero-voltage-switching (ZVS). As a result of having zero current (voltage) during turn-on and turn-off in ZCS (ZVS), the switching power loss is greatly reduced. The L-C resonant circuit is built around the semiconductor switch to ensure ZCS or ZVS. Sometimes the undesirable parasitic components, such as the leakage inductance of the transformer and the capacitance of the semiconductor switch, are utilized as components of the resonant circuit. Two ZCS and one ZVS configurations are shown in Figure 12.80. The switch S can be implemented for unidirectional and bidirectional current (Figure 12.81). Converters using

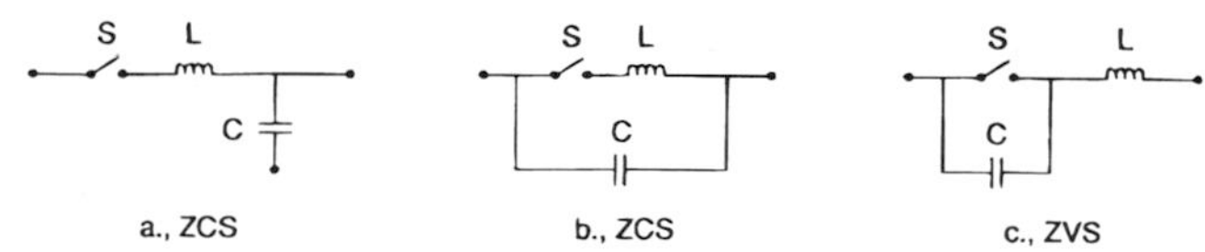

Figure 12.80 ZCS (Fig. a, and b) and ZVS (c) configurations.

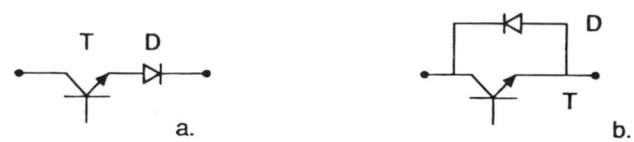

Figure 12.81 Switch for unidirectional (a) and for bidirectional (b) current.

ZCS or ZVS topology are termed resonant switch converters or quasi-resonant converters.

ZCS Resonant Converters

A step-down DC-DC converter using the ZCS configuration shown in Figure 12.80a is presented in Figure 12.82a. Switch S is implemented as shown in Figure 12.81a. The L_f–C_f are sufficiently large to filter the harmonic current components. Current I_o can be assumed to be constant in one switching cycle. Four equivalent circuits associated with the four intervals of each cycle of operation are shown in Figure 12.82b and 12.82c together with the waveforms.

Interval 1 ($0 \leq t \leq t_1$): Both the current i_L in L and the voltage v_C across C are zero prior to turning the switch on at $t = 0$. The output current flows through the freewheeling diode *D1*. After turning the switch on, the total input voltage develops across L and i_L rises linearly *ensuring ZCS* and soft current change. The interval 1 ends when i_s reaches I_o and the current conduction stops in *D1* at t_1.

Interval 2 ($t_1 \leq t \leq t_2$): The *L-C* resonant circuit starts resonating and the change in i_L and v_C will be sinusoidal (Figure b and c). Interval 2 has two subintervals. The capacitor current $i_C = i_L - I_o$ is positive in $t_1 \leq t \leq t_2'$ and v_C rises; while it is negative in $t_2' \leq t \leq t_2$, v_C falls. The peak current is $\hat{I}_L = I_o + V_{dc}/Z_o$ at $t = t_m$ and peak voltage is $\hat{V}_c = 2V_{dc}$ at $t = t_2'$. V_{dc}/Z_o must be larger then I_o otherwise i_L will not swing back to zero.

Interval 3 ($t_2 \leq t \leq t_3$): Current I_L reaches zero at t_2 and the switch is turned off *by ZCS*. The capacitor supplies the load current and its voltage falls linearly.

Interval 4 ($t_3 \leq t \leq t_4$): The output current freewheels through *D1*. The switch is turned on at t_4 again and the cycle is repeated.

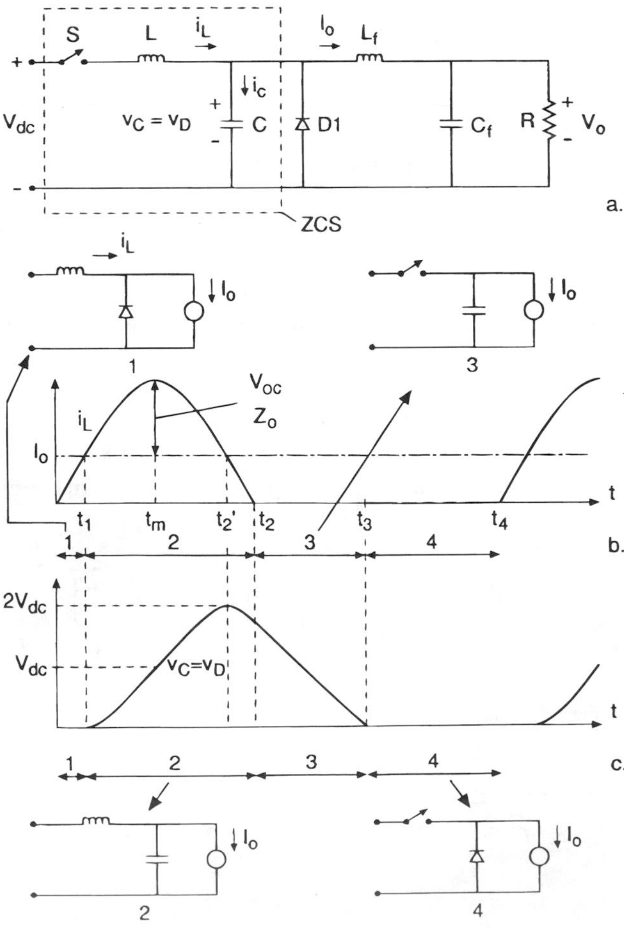

Figure 12.82 ZCS resonant converter.

The output voltage V_o will equal the average value of voltage v_C. V_o can be varied by changing the interval $t_4 - t_3$, that is, the switching frequency.

Applying the ZCS configuration shown in Figure 12.80b, rather than that shown in Figure 12.80a, the operation of the converter remains basically the same. The time function of the switch current and the *D1* diode voltage will be unchanged. The C capacitor voltage will be $v_C = V_{dc} - v_D$.

ZVS Resonant Converter

A ZVS resonant and step-down DC-DC converter is shown in Figure 12.83a and is obtained from Figure 12.82a by replacing the ZCS configuration with the ZVS configuration shown in Figure 12.80c. Note, that the bidirectional current switch is used. This converter's operation is very similar to that of the ZCS converter. The waveform of v_C is the same as the one for i_L in Figure 12.82b and the waveform of i_L is the same as the one for v_C when the ZCS configuration shown in Figure 12.80b is used. I_o = *const.* in one cycle can be assumed again.

Interval 1 ($0 \leq t \leq t_1$): S is turned off at $t = 0$. The constant $i_L = I_o$ current starts passing through the capacitor C. Its voltage v_C rises linearly from zero to V_{dc}. ZVS occurs.

Interval 2 ($t_1 \leq t \leq t_2$): Diode *D1* turns on at t_1. The *L-C* circuit starts resonating through *D1* and the source. Both v_C and i_L are changing sinusoidally. When i_L drops at zero v_C reaches its peak value: $\hat{V}_C = V_{dc} + Z_o I_o$. The voltage v_C reaches zero at t_2. The load current must be high enough so that $Z_o I_o > V_{dc}$; otherwise v_C will not reach zero and the switch will have to be turned on at non-zero voltage.

Interval 3 ($t_2 \leq t \leq t_3$): Diode D turns on. It clamps v_C to zero and conducts i_L. The gate signal is reapplied to the switch. V_{dc} develops across L and i_L increases linearly up to I_o, which is reached at t_3. Prior to that, the current i_L changes its polarity at t_3' and S begins to conduct it.

Interval 4 ($t_3 \leq t \leq t_4$): Freewheeling diode *D1* turns off at t_3. It is a soft transition because of the small negative slope of the current i_D. Current I_o flows through S at t_4 when S is turned off and the next cycle begins.

Diode voltage v_D develops across *D1* only in intervals 1 and 4 (Figure 12.83d). Its average value is equal to V_o which can be varied by interval 4, or in other words, by the switching frequency.

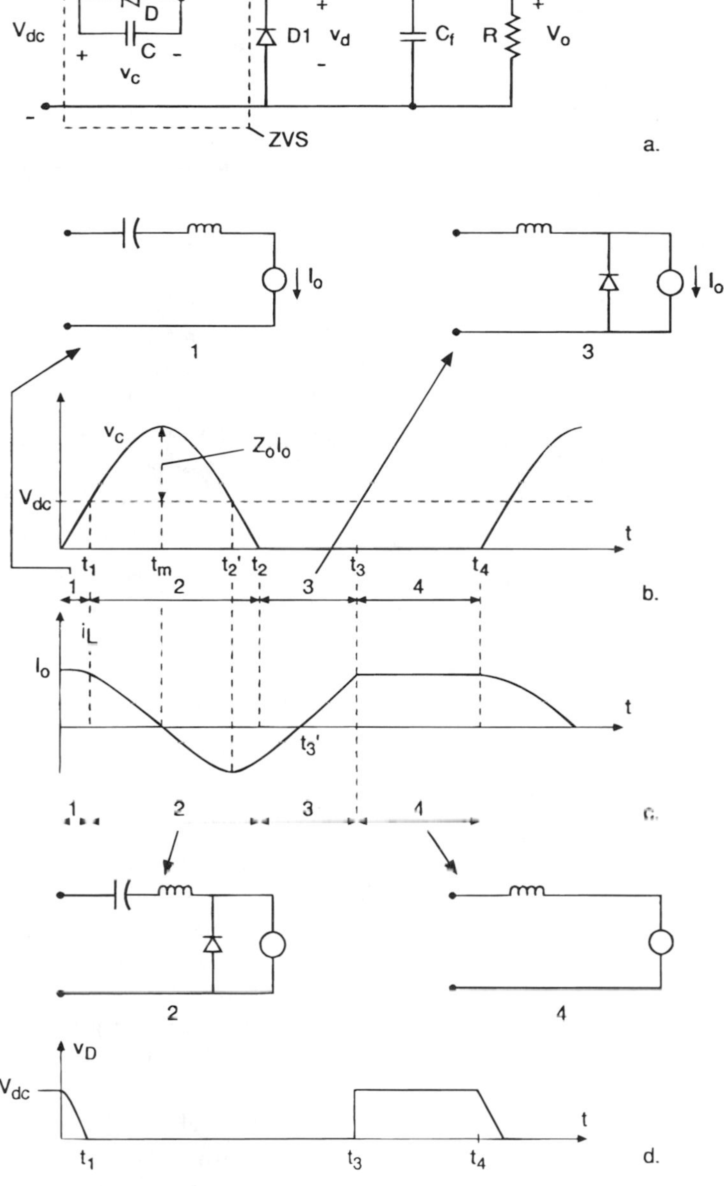

Figure 12.83 ZVS resonant converter.

Summary and Comparison of ZCS and ZVS Converters

The main properties of ZCS and ZVS are highlighted as follows:

- The switch turn-on and -off occurs at zero current or at zero voltage which significantly reduces the switching losses.
- Sudden current and voltage changes in the switch are avoided in ZCS and in ZVS, respectively. The *di*/*dt* and *dv*/*dt* values are rather small. EMI is reduced.
- In the ZCS, the peak current $I_o + V_{dc}/Z_o$ conducted by *S* must be more than twice as high as the maximum of the load current I_o.
- In the ZVS, the switch must withstand the forward voltage $V_{dc} + Z_oI_o$ and Z_oI_o must exceed V_{dc}.
- The output voltage can be varied by the switching frequency.
- The internal capacitances of the switch are discharged during turn-on in ZCS which can produce significant switching loss at high switching frequency. No such loss occurs in ZVS.

Two-Quadrant ZVS Resonant Converters

One drawback in the ZVS converter, shown in Figure 12.83, is that the switch peak forward voltage is significantly higher

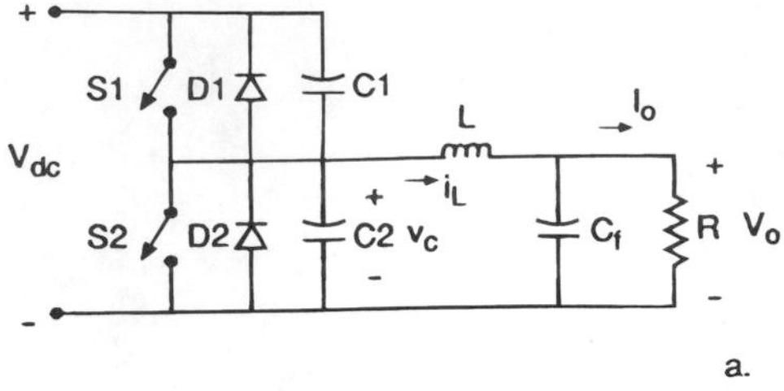

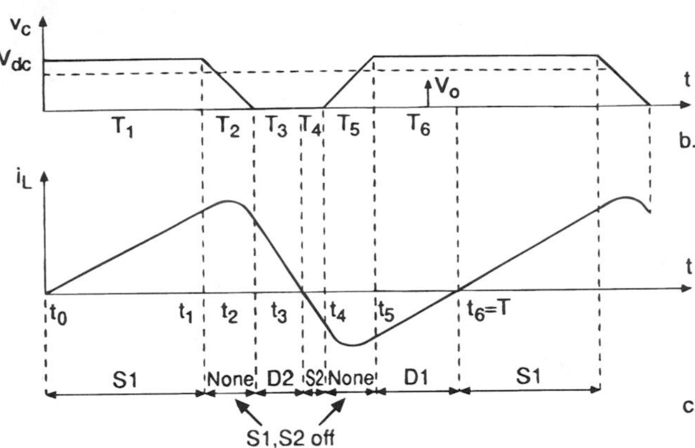

Figure 12.84 Two-quadrant ZVS resonant converter.

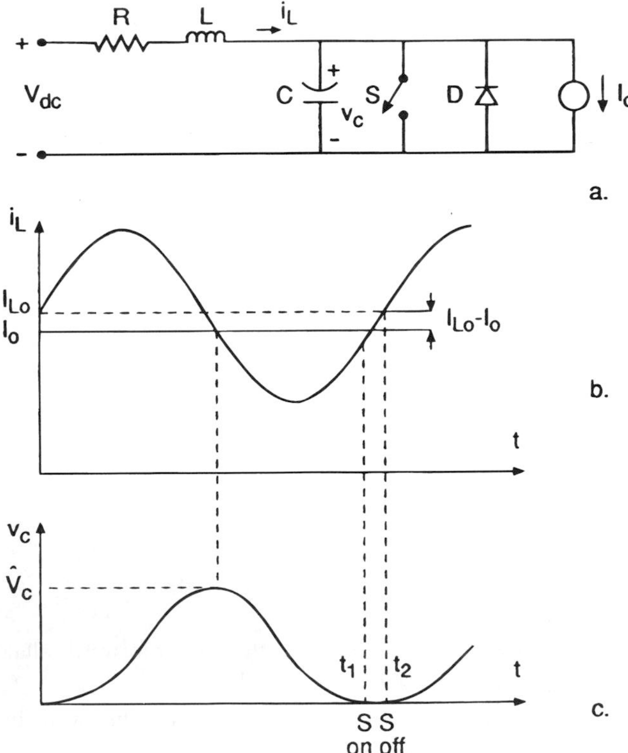

Figure 12.85 Resonant DC link converter.

than the supply voltage. This drawback does not appear in the two-quadrant ZVS resonant converter where the switch voltage is clamped at the input voltage. In addition, this technique can be extended to the single phase and the three-phase DC-to-AC converter to supply an inductive load.

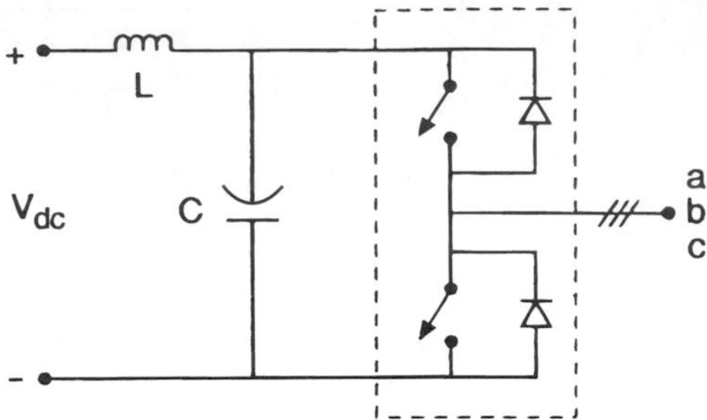

Figure 12.86 Resonant DC link converter for three-phase PWM-VSI.

The basic principle will be presented by means of the DC-DC step-down converter shown in Figure 12.84a. Two switches, two diodes and two resonant capacitors $C1 = C2 = C$ are used. The voltage V_o can be assumed to be constant in one switching period because C_f is large. The current i_L must fluctuate in large scale and must take both positive and negative values in one switching cycle. To achieve this operation L must be rather small. One cycle consists of six intervals.

Interval 1. S1 is on. The inductor voltage is $v_L = V_{dc} - V_o$. i_L rises linearly from zero.

Interval 2. S1 is turned off at t_1. None of the four semiconductors conducts. The resonant circuit consisting of L and the two capacitors connected in parallel is ringing through the source and the load. Now the impedance $Z_o = \sqrt{2L/C}$ is high (C is small) and the peak current will be small. The voltage across *C2* approximately changes linearly and reaches zero at t_2. As a result of *C1*, the voltage across *S1* changes slowly from zero.

Interval 3. D2 conducts i_L. The inductor voltage v_L is $-V_o$. i_L is reduced linearly to zero at t_3. *S2* is turned on in this interval when its voltage is zero.

Interval 4. S2 begins to conduct, v_L is still $-V_o$ and i_L increases linearly in a negative direction.

Interval 5. S2 is turned off at t_4. None of the four semiconductors conduct. A similar resonant process occurs as in interval 2. As a result of *C2*, the voltage across *S2* rises slowly from zero to V_{dc}.

Interval 6. v_C reaches V_{dc} at t_5. *D1* begins to conduct i_L. The inductor voltage $v_L = V_{dc} - V_o$ and i_L rises linearly with the same positive slope as in interval 1 and reaches zero at t_6. The cycle is completed.

The output voltage can be controlled by PWM at a constant switching frequency. Assuming that the intervals of the two resonant processes, that is, interval T_2 and T_5, are small compared to the period T, the wave shape of v_C is of a rectangular form. V_o is the average value of v_C and, therefore, $V_o = DV_{dc}$, where D the duty ratio: $D = (T_1 + T_6)/T$. Here T is the period: $T \cong T_1 + T_3 + T_4 + T_6$. During the time DT either *S1* or *D1* is on. Similarly, the output current is equal to the average value of i_L.

Resonant DC Link Converters with ZVS

To avoid the switching losses in the converter, a resonant circuit is connected between the DC source and the PWM inverter. The basic principle is illustrated by the simple circuit shown in Figure 12.85a. The resonant circuit consist of the L-C-R components. The load of the inverter is modelled by the I_o current source. I_o is assumed to be constant in one cycle of the resonant circuit.

Switch S is turned off at $t = 0$ when $i_L = I_{Lo} > I_o$. First, assuming a lossless circuit ($R = 0$), the equations for the resonant circuit are as follows:

$$i_L = I_o + \frac{V_{dc}}{Z_o}\sin\omega_0 t + (I_{L0} - I_0)\cos\omega_0 t \qquad (12.14)$$

$$v_C = V_{dc}\,(1 - \cos\omega_0 t) + Z_0(I_{L0} - I_0)\sin\omega_0 t \qquad (12.15)$$

where

$$\omega_o = 1/\sqrt{LC} \text{ and } Z_o = \sqrt{L/C}$$

To turn on and off the switch at zero voltage the capacitor voltage v_c must start from zero at the beginning and must return to zero at the end of each cycle (Figure 12.85c). Without losses and when $I_{Lo} = I_o$, the voltage swing must start off and return to zero peaking at $2V_{dc}$. However, when $R \neq 0$ which represents the losses, the voltage swing is damped and v_C would never return to zero under the condition $I_{Lo} = I_o$. To force v_C back to zero a value of $I_{Lo} > I_o$ must be chosen (Figure 12.85b). This condition adds the term $Z_o(I_{Lo} - I_o)\sin\omega_o t$ into the right side of Equation 12.15 and thus v_C can reach zero again. By controlling the time $t_2 - t_1$, in other words, the on-time of switch S, both $I_{Lo} - I_o$ and the peak voltage $\hat{V}_c$ are regulated (Figure 12.85c).

This principle can be extended to the three-phase PWM voltage source inverter(VSI) shown in Figure 12.86. The three cross lines indicate that the configuration has three legs. Any of the two switches and two diodes in one leg can perform the same function which is done by the antiparallel connected S-D circuit in Figure 12.85a. All of the six switches can be turned on and off at zero voltage in Figure 12.86.

References

Kassakaian, J. G., Schlecht, M. F., and Verghese, G. C. 1992. *Principles of Power Electronics,* Addison-Wesley, Reading, MA.

Mohan, N., Undeland, T. M., and Robinsons, W. P. 1989. *Power Electronics,* John Wiley & Sons, New York, NY.

Ohno, E. 1988. *Introduction to Power Electronics,* Clarendon Press, Oxford, UK.

Rashid, M. H. 1993. *Power Electronics,* Prentice-Hall International, London, UK.

13

Motor Drives

Takamasa Hori
Mie University

Hiroshi Nagase
Hitachi Ltd.

Mitsuyuki Hombu
Hitachi Ltd.

M.F. Rahman
University of New South Wales

Khiang-Wee Lim
University of New South Wales

Ronald H. Brown
Marquette University

Sándor Halász
Technical University of Budapest

Jozsef Borka
Hungarian Academy of Sciences

13.1 Control Systems and Applications 288
13.2 DC Motor Control Systems 289
Torque-Speed Characteristics of DC Motors (DCM) • Ward Leonard Control System and Thyristor Leonard Control System • Chopper Control System
13.3 Induction Motor Control Systems 294
Torque-Speed Characteristics of Induction Motor, Speed Control Methods • Vector Control • Speed Sensorless Vector Control • Slip-Power Recovery Control
13.4 Synchronous Motor Control Systems 315
Torque-Speed Characteristics • Speed Control of a Synchronous Motor
13.5 PM Synchronous Motor Control 319
Construction and Operation • PM Synchronous Motor Model • Field-Oriented Control
13.6 Step Motor Drives 331
Types and Operation of Step Motors • Step Motor Models • Control of Step Motors
13.7 Servo Drives 341
DC Drives • Induction Motor Drives
13.8 Switched Reluctance Motor Drives 344
Construction and Operation • Supply Units • Controls

13.1 Control Systems and Applications

Takamasa Hori

There are two types of motors. One is the DC motor (DCM) driven by a DC power source and the other is the AC motor (ACM) driven by an AC power source. Generally, the DC motor has been used as a speed controlled motor because of the ease of controlling DC voltage by power semiconductor circuits, while the AC motor has been used as a constant speed motor, because of the difficulty in controlling AC voltage and/or frequency. But the advent of high-speed microprocessors and power devices has resulted in the enhanced control performance of AC motors that are controlled by a frequency changer, e.g., transistor or GTO thyristor inverter, or thyristor cycloconverter. Table 13.1 shows some typical industrial applications of motor speed control systems.

Motor control systems have been widely used in industrial control systems, home electric appliances, information equipment, factory automation (FA)/office automation (OA) equipment, subway/urban train car drives, and so on. They are used for improving both the function and performance of machines in processing lines through high-speed and/or fast response, for energy savings through variable speed operation, and for reducing motor maintenance by using AC motors in place of DC motors. The use of motor control systems has been enhanced due to the progress in power electronics technology (power switching and control by power semiconductor devices), microelectronics technology (ICs such as microprocessors) and their application technologies.

Controlled motors, in practice, range from the small 1 W actuator used in information equipment to those that are very large and used in 400 MW variable speed pumped storage power plants. Control by thyristor is mainly employed in large capacity motors, while the control of middle or small capacity motors (less than several hundred kW) is carried out by transistors.

References

Bose, B. K. 1992. Recent advances in power electronics, *IEEE Trans.*, PEL-7(1):2–16.

Heuman, K. 1990. Power electronics-state of the art, *IPEC-Tokyo Record*, 11–20.

Holtz, J. 1993. Speed estimation and sensorless control of AC drives, *Proc. IEEE/IES IECON'93*, 649–654.

Hori, T. 1994. Industrial applications of motor control systems in Japan—past and future, *Proc. Int. Conf. Industrial Technology (ICIT'94)*, O15–O23, December 5.

0-8493-8343-9/97/$0.00+$.50

Table 13.1 Some Typical Industrial Applications of Motor Speed Control Systems

Motor	Control	Speed control system	Industrial applications
DC motor (DCM)	Voltage control	Thyristor Leonard	General purpose for industrial applications, iron rolling mill, winder, process line, locomotive, large crane, extruder, paper machine, elevator
		Chopper	Subway/urban train, electric vehicle, OA/FA equipment, industrial vehicle (forklift),
Induction motor (IM)		Primary voltage (for squirrel cage motor)	Crane, pump, fan, elevator (small capacity motor)
		Slip power recovery (for IM with secondary winding) (Scherbius control) (Kraemer control)	Pump, fan, cement kiln, paper machine (large capacity motor)
Induction motor (IM) Synchronous motor (SM)	Frequency control	Inverter Cycloconverter	General purpose for industrial applications, elevator, process line, table roller, subway/urban train, fan, blower, pump, compressor, extruder, paper machine, cement kiln, iron rolling mill, pumped storage power plant
Synchronous Motor (SM)		Thyristor motor (commutatorless motor)	Large pump, fan, blower, compressor, extruder, iron rolling mill, starter of large SM/SG
		Transistor motor	Home electric appliances, servo motor, OA/FA equipment

Nishihara, M. 1990. Power electronics diversity, *IPEC-Tokyo Record,* 21–28.

Tadakuma, S. et al. 1993. Historical and predicted trends of industrial AC drives, *Proc. IEEE/IES IECON'93,* 655–661.

13.2 DC Motor Control Systems

Takamasa Hori

Torque-Speed Characteristics of DC Motors (DCM)

Figure 13.1 shows the various types of DC motor (DCM), classified by the field winding excitation.

Figures 13.1a to 13.1c are used mainly for speed control. Figures 13.1a and 13.1c are for large capacity industrial applications. Figure 13.1b is for forklifts, electric train cars and locomotives which need the characteristics of a series motor, that is, high torque at low speed and low torque at high speed.

In particular, Figure 13.1a, motor with a permanent magnet for excitation, is widely used as a servo motor, robot motor or a small capacity OA/FA motor, controlled by a full-bridge chopper control system.

Figure 13.2 shows the circuit of a DC motor with separate excitation.

The following equations describe the circuit in this figure:

$$\left.\begin{aligned} v &= e + R_a i_a + L_a di_a/dt \\ \tau &= K i_f i_a \\ e &= K i_f \omega \\ \tau - \tau_1 &= J\, d\omega/dt + D\omega \end{aligned}\right\} \quad (13.1)$$

where J is the moment of inertia of the motor and load (machine), and D is the coefficient of viscosity.

Applying the Laplace transformation to Equation 13.1 yields

$$\left.\begin{aligned} V &= E + (R_a + L_a s) I_a \\ T &= K I_f I_a \\ E &= K I_f \Omega \\ T - T_1 &= (Js + D)\Omega \end{aligned}\right\} \quad (13.2)$$

From Equation 13.2, Figure 13.3 can be drawn, and the dynamic characteristics of the DC motor in the transient state can be analyzed.

In steady state, the DC motor equations are

$$\left.\begin{aligned} V &= E + R_a I_a \\ T &= K I_f I_a \\ E &= K I_f \Omega \end{aligned}\right\} \quad (13.3)$$

Assuming that the mechanical loss is negligible, the torque-speed characteristics are shown in Figure 13.4 as a function of terminal voltage.

Ward Leonard Control System and Thyristor Leonard Control System

The Ward Leonard control system, shown in Figure 13.5 has been used as a speed control system in industrial applications for many years. An ACM (synchronous motor or induction motor) is mechanically coupled with a DC generator (DCG) rotating at constant speed. The DCG terminals are electrically connected with the DCM terminals, and the DCG terminal voltage is regulated by the DCG field current I_g, which is supplied to the DCM

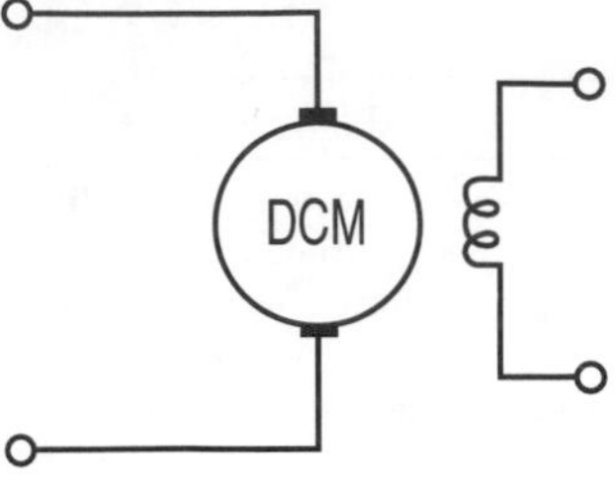

(a) Separate Excitation

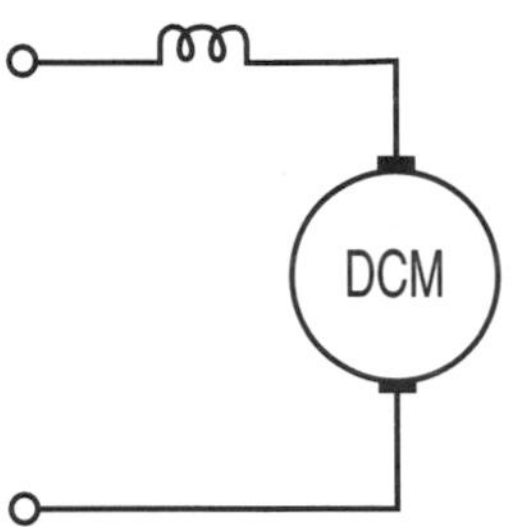

(b) Series Excitation

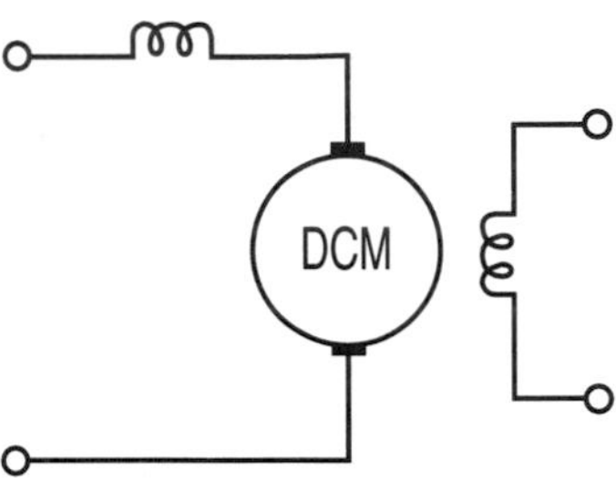

(c) Compound Excitation
(Separate plus Series)

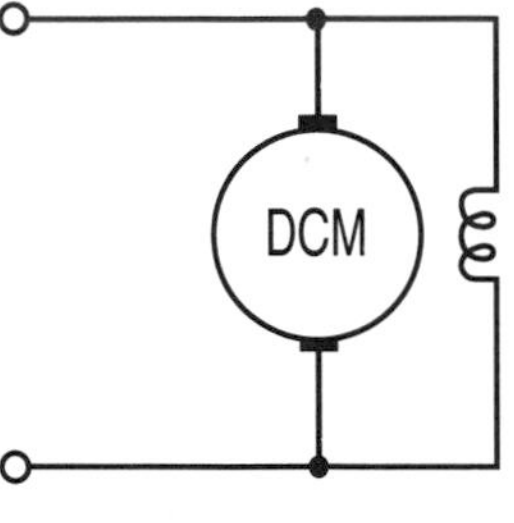

(d) Shunt Excitation

Figure 13.1 Classification of DC motors by field excitation method.

terminals for controlling the DCM's speed. The field current I_m of the DCM is constant at low speed, but controlled at high speed for constant output power control of the DCM. This Ward Leonard control system is still used for elevator control because of the high control reliability.

On the other hand, because of the progress made in thyristor control techniques during the 1960s, DC motor control had been modernized by using thyristor power converters instead of motor-generator sets (ACM and DCG) in order to improve performance, efficiency and maintenance.

Figure 13.6 shows the thyristor Leonard control system for an iron rolling mill drive which needs four-quadrant speed-torque operation. The parallel thyristor bridge converter circuits in back-to-back connection provide positive or negative voltage adjustment contactlessly and continuously for performing normal, reverse, accelerated, or decelerated motor operation. In the case of a large capacity motor, each arm of the three-phase bridge circuit has a number of parallel thyristor elements.

The logic circuit for changeover of the bridge circuit is indispensable since a gate pulse generator (GPG) is employed for distributing the firing pulse for thyristors among each bridge circuit under circulating current-free conditions. The general type of logic circuit employed for changeover is based on a combination of the polarity of a speed error and the absence or presence of current. By using a high-response automatic current controller it is possible to provide a high-performance automatic speed controller. The terminal voltage E of the DC motor is controlled by the firing angle (α) which is controlled through a three-phase thyristor bridge circuit ($E \propto \cos \alpha$).

To achieve a faster response of the speed controller, the fully digitalized thyristor Leonard control system, shown in Figure 13.7, is employed, with a compensation algorithm for the nonlinearity of the thyristor circuit caused by current continuity and discontinuity.

Figure 13.8 shows the four-quadrant operation of a thyristor Leonard control system. When the thyristor control angle α is smaller than 90° ($\alpha < 90°$), the thyristor converter is operated as a rectifier and the dc motor operates as a motor, but when $\alpha > 90°$, the thyristor converter is operated as an inverter, and the DC motor operates as a generator. The rectifier and motor operations transfer AC electric power to the motor, through the converter, as load or acceleration power. The inverter and generator operations transfer mechanical load or inverter power back to the AC power source line, through the converter, as regenerative or braking power.

Chopper Control System

Chopper control systems, called DC-DC converters, are widely used in DC motor drives for battery forklift and subway/urban train cars whose power supplies are derived from an AC power supply using a diode bridge rectifier. There are several DC-DC converter circuits for DC motor control, the most common of which are step-down (buck) chopper circuits, and step-up (boost) chopper circuits as shown in Figure 13.9. Transistors are used in the chopper circuit for control of a small capacity DC motor (DCM), while in the control of a large capacity DC motor, thyristors are used.

The chopper is repeatedly closed for a time T_1 and opened for a time T_2. To reduce the current ripple in the motor, the cycle time T is selected as $L/R >> T$.

1. In the Case of a Step-down Chopper
   During T_1, the DC voltage E is supplied to the load through the chopper, and during T_2, the DC motor current flows through the freewheeling diode D. The

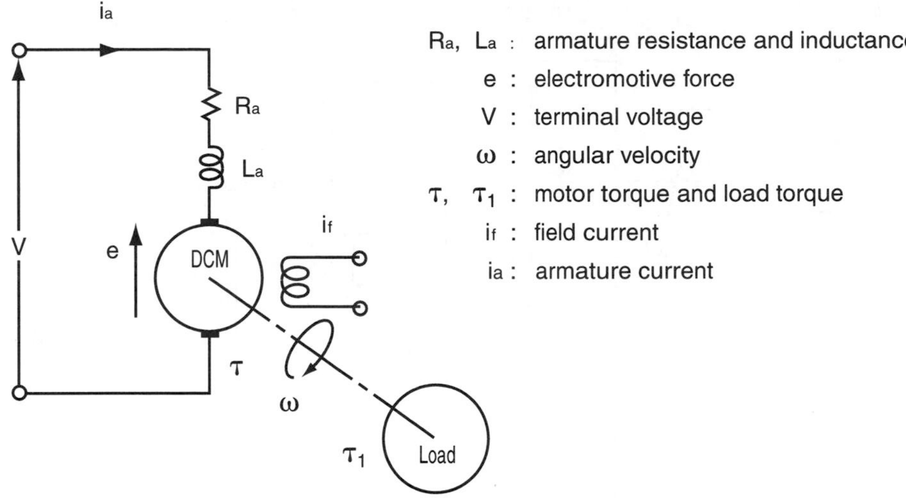

Figure 13.2 Equivalent circuit of a DCM with separate excitation.

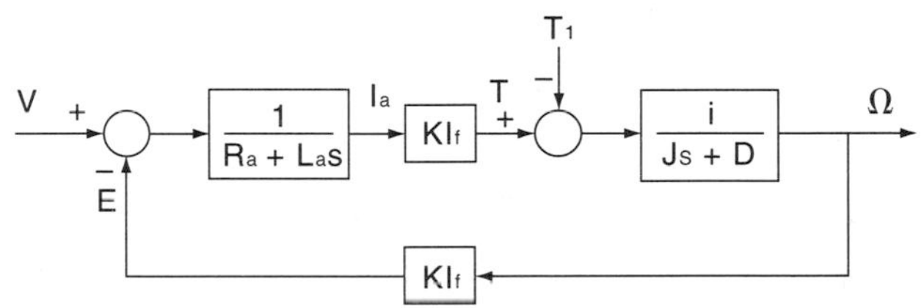

Figure 13.3 Block diagram of DC motor.

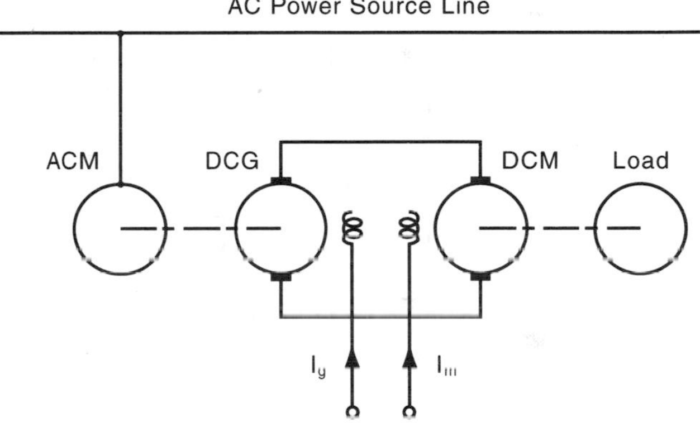

Figure 13.5 Ward Leonard control system.

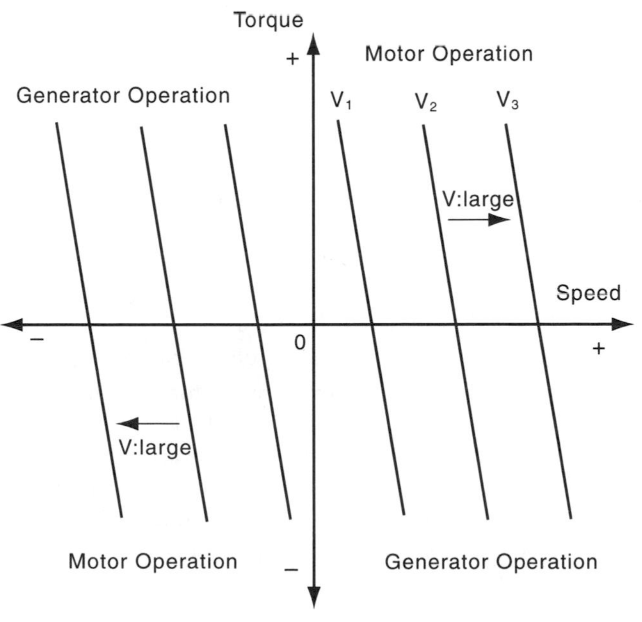

Figure 13.4 Torque-speed characteristics of DC motor.

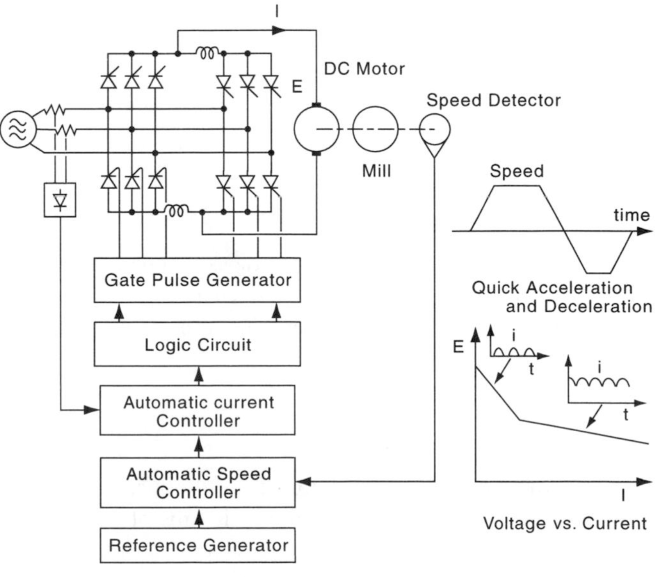

Figure 13.6 Thyristor Leonard control system.

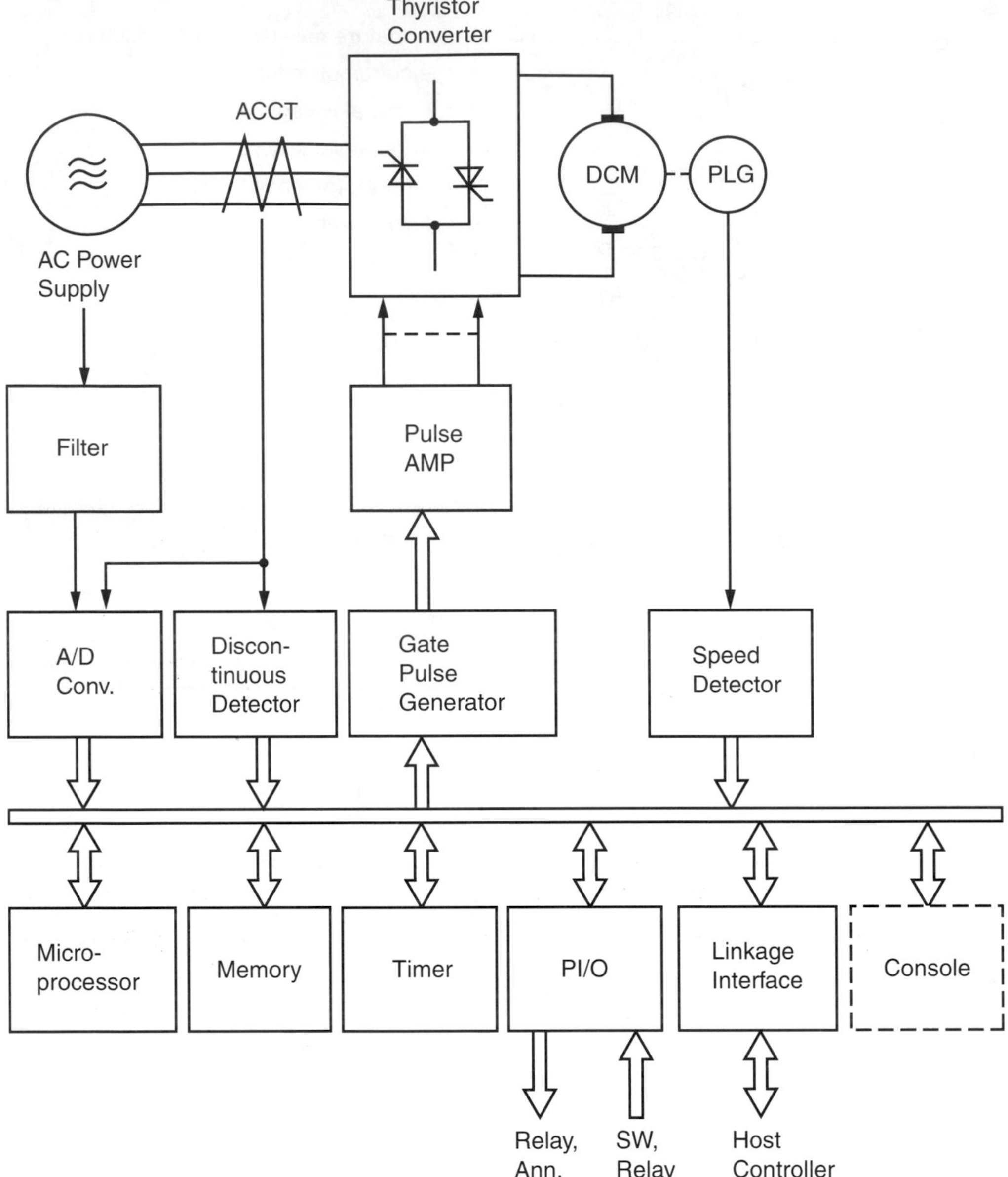

Figure 13.7 Fully digitalized thyristor Leonard control system (microprocessor-based digital speed regulator).

DC motor voltage E_m can be controlled downward from the source voltage E by changing the on-to-off time ratio of the chopper. The average voltage E_m is given by

$$E_m = \alpha E \tag{13.4}$$

where α is the duty factor of the chopper and is given by $\alpha = T_1/T$, where T is the cycle time of the chopper.

2. In the Case of a Step-up Chopper
When chopper is on during T_1, the energy from the DC generator (DCG) is stored in the inductor L which is connected to the DCG. When the chopper is off during T_2, the inductor current is forced to flow to the DC power source E through the diode. For steady-state operation, there must be zero average voltage across the inductor. Therefore, the positive volt-seconds on L during the on-interval T_1 must be equal to the negative volt-seconds when the chopper is open.
In general

$$E_m T_1 = (E - E_m) T_2 \tag{13.5}$$

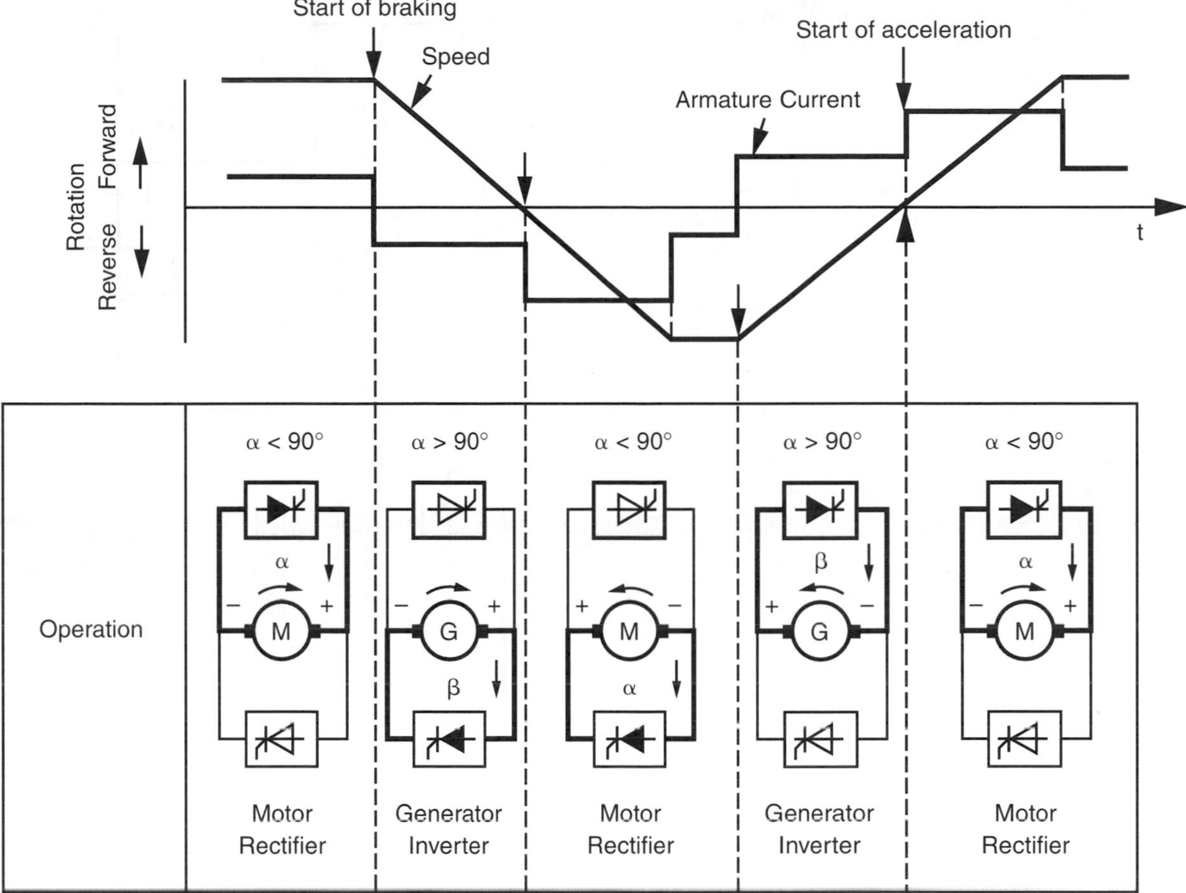

Figure 13.8 Four quadrant operation of a thyristor Leonard control system.

and thus

$$E_m = (1 - \alpha)E \quad \text{or} \quad E = E_m/(1 - \alpha) \qquad (13.6)$$

where

$$\alpha = T_1/T, \qquad T = T_1 + T_2$$

In this case, the maximum average value of E_m is approximately equal to the DC voltage E, i.e., $E_m \lesssim E$. Therefore, this step-up chopper is used to produce a voltage higher than the voltage E_m. When the motor slows down, the mechanical energy of the load returns to the power source through the chopper, this is the so-called regenerative braking in which the DC motor operates as a generator.

In the case of robotics and OA/FA equipment, four-quadrant speed control of a DC motor with permanent magnet excitation is necessary for quick acceleration and deceleration of the speed, forward and reverse rotation, and precise positioning. Figure 13.10 shows a full-bridge chopper control system for this four-quadrant operation.

The average DC motor voltage may be controlled by operating the chopper at a fixed frequency with a variable on-time, operating with a constant on-time using a variable chopping (switching) frequency, or employing a combination of these pulse-modulation techniques to control the duty cycle of the chopper, such as PWM (pulse width modulation).

References

Konishi, T. et. al. 1980. A performance analysis of microprocessor-based control systems applied to adjustable speed motor drives, *IEEE Trans.*, IA-16 (3):378–387, May/June.

Ohmae, T. et al. 1980. A microprocessor-controlled fast response speed regulator with dual mode current loop for DCM drives, *IEEE Trans.*, IA-16(3):388–394, May/June.

Ohmae, T. et al. 1982. A microprocessor-controlled high-accuracy wide-range speed regulator for motor drives, *IEEE Trans.*, IE-29(3):207–211, August.

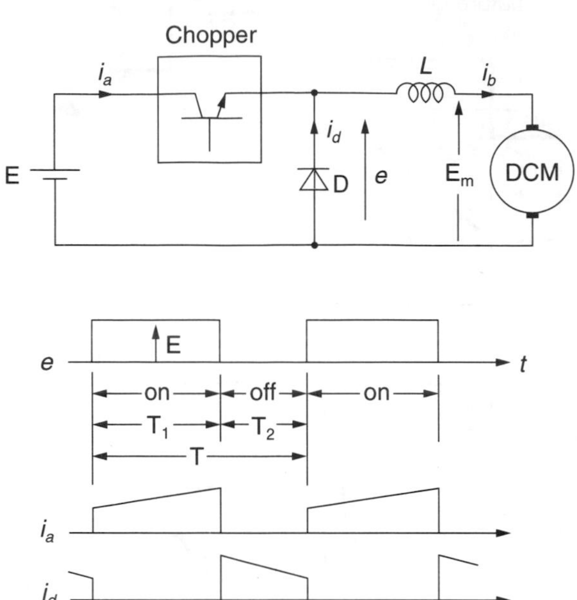

(a) Step-Down Chopper Circuit (Buck Converter)

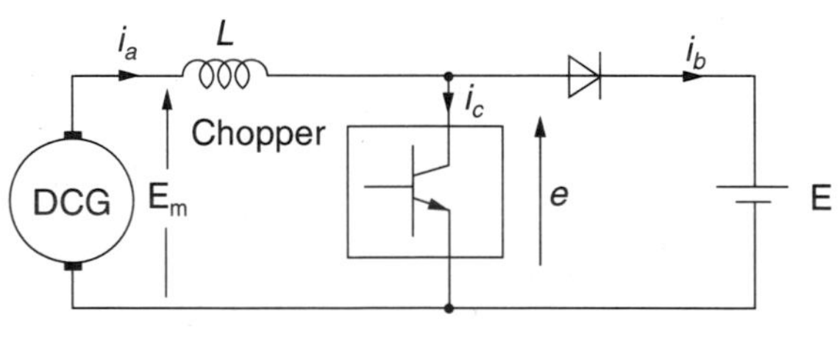

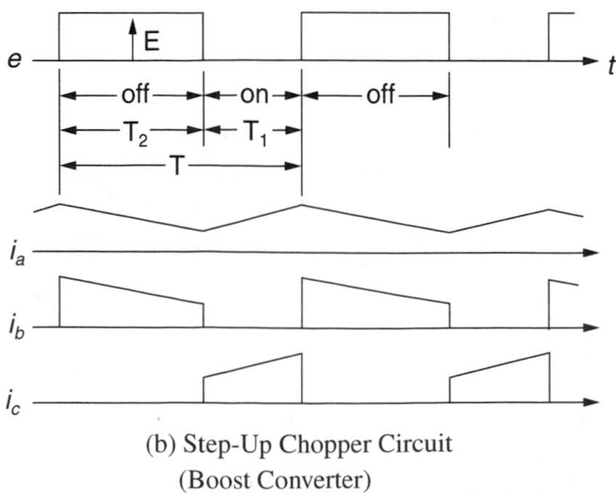

(b) Step-Up Chopper Circuit (Boost Converter)

T_1 : duration time of chopper on-state
T_2 : duration time of chopper off-state
T : one-cycle time of chopper

Figure 13.9 The principle of chopper control.

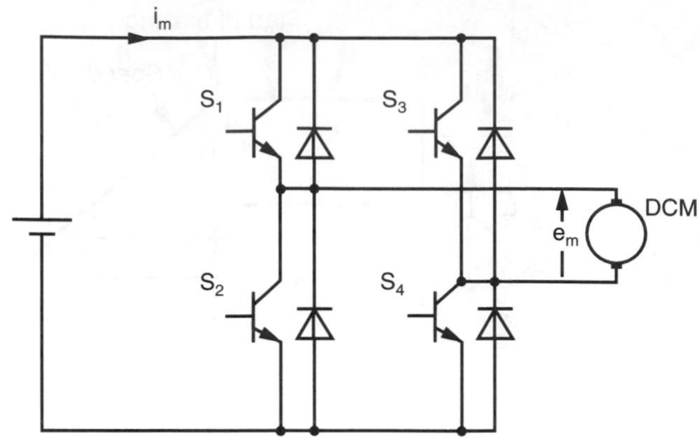

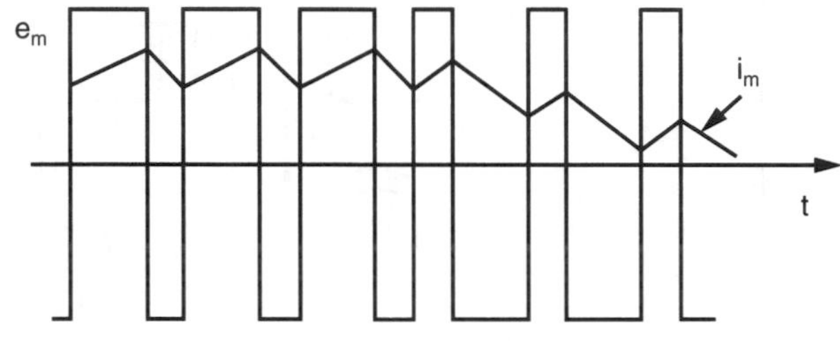

Figure 13.10 A full-bridge chopper control system.

13.3 Induction Motor Control Systems

Takamasa Hori, Hiroshi Nagase, Mitsuyuki Hombu

Torque-Speed Characteristics of Induction Motor, Speed Control Methods

Figure 13.11 illustrates an equivalent circuit for an induction motor (IM). The basic equations for the torque-speed characteristics are derived from this figure for three-phase induction motors.

a. The secondary input power P_2 is

$$P_2 = \frac{3sr_2'E_1^2}{(r_2')^2 + (s\omega_1 l_2')^2} = 3I_2'^2\frac{r_2'}{s} = 3V_2'I_2' \qquad (13.7)$$

b. The output power

$$P_0 = P_2 - 3I_2'^2 r_2' = (1 - s)P_2 \qquad (13.8)$$

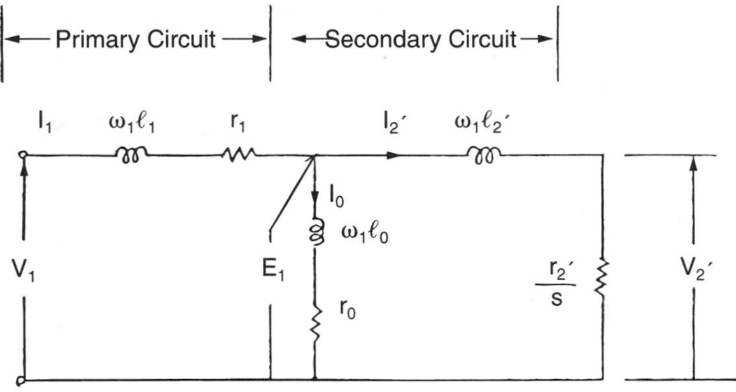

V_1 : terminal voltage / phase
I_1 I_2' I_0 : primary, secondary, exciting current
$\omega_1 = 2\pi f_1$: angular frequency of terminal voltage
ℓ_1 ℓ_2' : primary, secondary leakage inductance / phase
r_1 r_2' : primary, secondary resistance / phase
ℓ_0 : exciting inductance / phase
r_0 : equivalent resistance of exciting loss / phase
s : slip

Figure 13.11 Equivalent circuit of induction motor/phase.

c. The rotating angular frequency ω is

$$\left.\begin{aligned} \omega &= 2\pi(1-s)\frac{f_1}{p} = 2\pi\frac{N}{60} \\ \omega_1 &= 2\pi f_1 \\ \omega_0 &= \frac{2\pi f_1}{p} = 2\pi\frac{N_0}{60} \end{aligned}\right\} \tag{13.9}$$

where f_1 is the frequency in Hz of the terminal voltage, N is the rotating speed in rpm, N_0 is the synchronous speed in rpm, and p is the number of pairs of poles.

d. The relationship between torque T in newton meters and output power P_0 in Watts is

$$P_0 = \omega T \tag{13.10}$$

e. The torque T is proportional to the secondary input power P_2,

$$T = \frac{p}{2\pi f_1} P_2 \tag{13.11}$$

f. The voltage E_1 is proportional to main linkage flux Φ_1,

$$E_1 = \omega_1 \Phi_1 = 2\pi f_1 \Phi_1 \tag{13.12}$$

g. Neglecting the primary impedance voltage drop, V_1 is

$$\left.\begin{aligned} V_1 \approx E_1 &= 2\pi f_1 \Phi_1 \\ \Phi_1 &\propto V_1/f_1 \end{aligned}\right\} \tag{13.13}$$

If V_1/f_1 is controlled at a constant value, Φ_1 also is essentially constant.

h. The torque T is shown as a function of sf_1 and Φ_1 by the expression

$$T = 3p \cdot \frac{2\pi f_1 s r_2' \Phi_1^2}{(r_2')^2 + (2\pi f_1 s l_2')^2} \tag{13.14}$$

Figure 13.11 and the equations above indicate that IM torque is controlled by V_1, f_1, r_2 and V_2.
The various speed control systems of an induction motor (IM) are listed as follows:

- Primary voltage control of a cage rotor type IM.
- Frequency control of a cage rotor type IM.
- Secondary resistance control of a wound rotor type IM.
- Slip power recovery control of a wound rotor type IM.

The torque-speed characteristics for each of these control systems are shown in Figure 13.12.

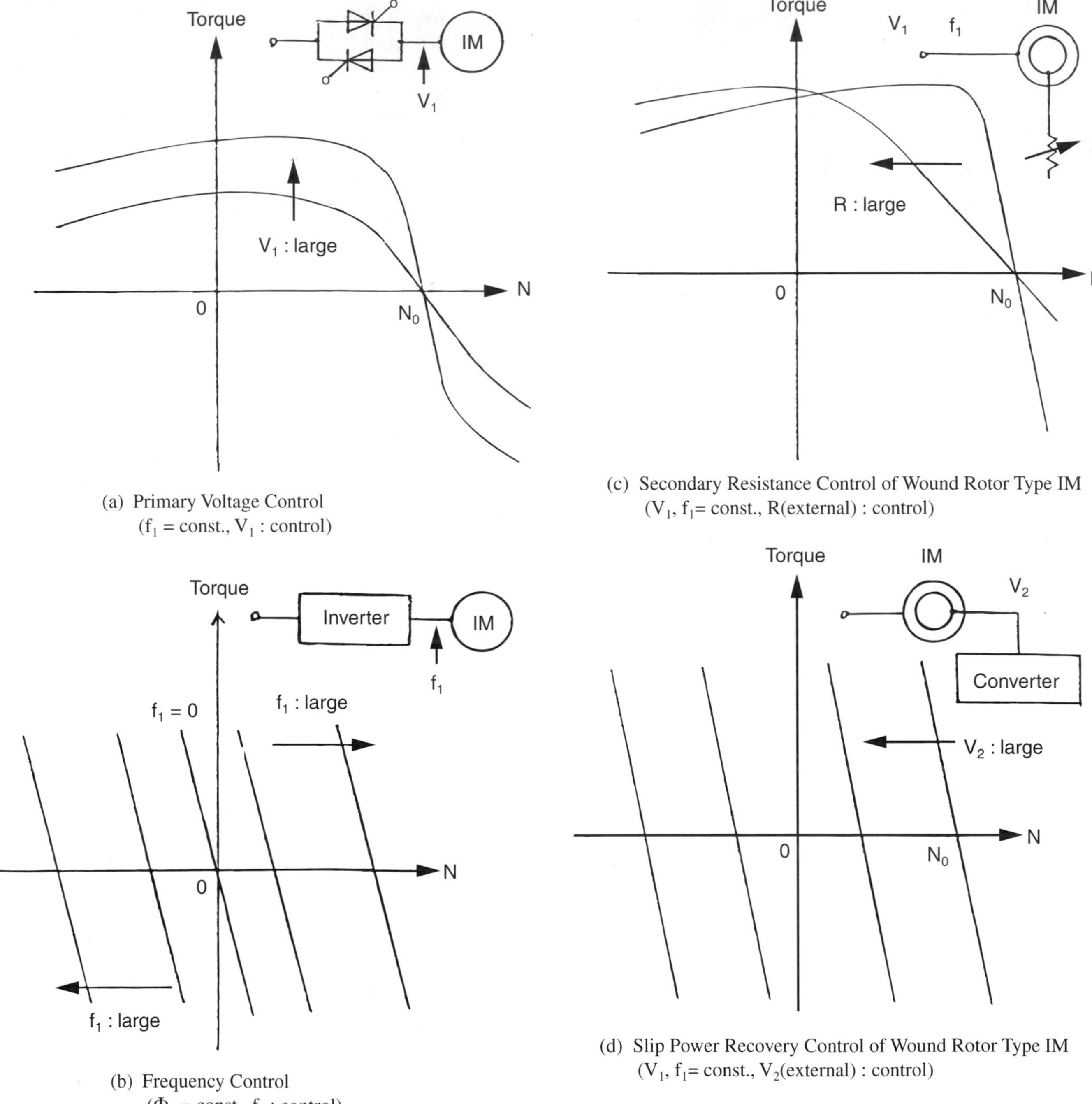

Figure 13.12 Torque-speed characteristics of IM control systems.

Primary Voltage Control System

The primary voltage control system consists of an unbalanced voltage control system (one-phase or two-phase control) and a balanced voltage control system (three-phase control).

Figure 13.13 shows the circuit configuration for each system. The primary voltage control system used for low-speed operation deteriorates in its power factor and diminishes in its ratio of torque to the square of the primary current. As a result, this control system is applicable for small capacity cage rotor type induction motors which have high-slip characteristics.

Although the unbalanced voltage control system which provides one-phase or two-phase control has the defect that the current flowing through the motor becomes unbalanced, its circuit construction may be simplified. In contrast, although the balanced system is able to balance the primary current, it requires a complicated circuit to do so. In practice, the primary voltage

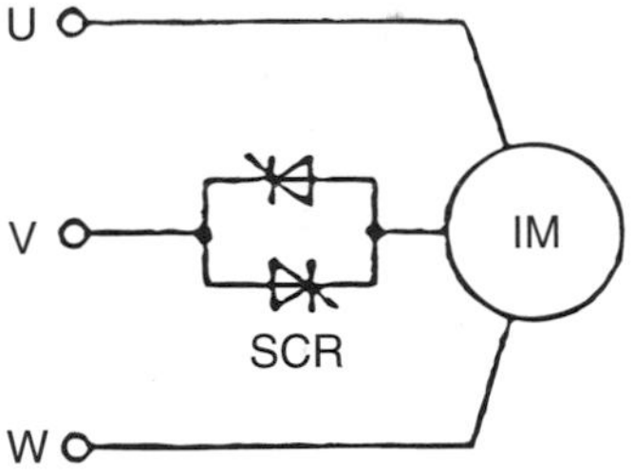

(a) One-phase control

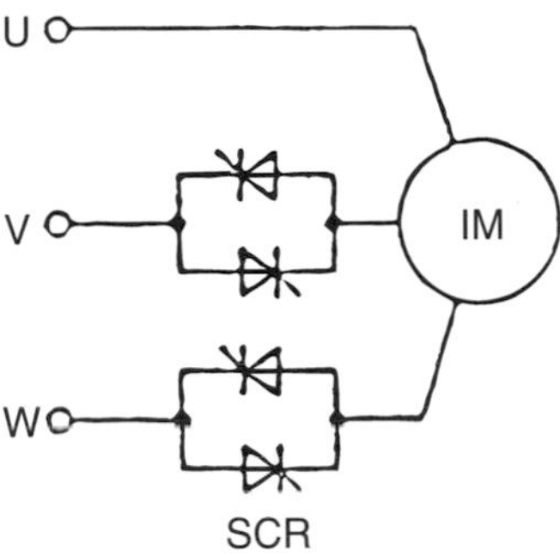

(b) Two-phase control

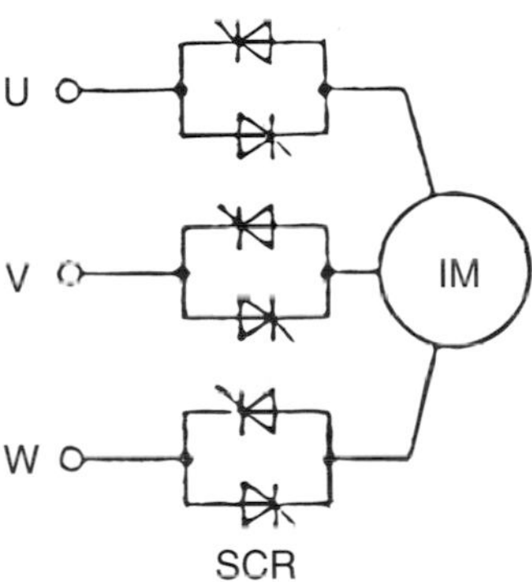

(b) Three-phase control

IM : Induction motor,
SCR : Thyristor

Figure 13.13 Circuit configuration for primary voltage control.

control system is used primarily for applications which permit the resultant effect on the power source to be ignored. The balanced control system is used for driving fans which require no braking torque. The unbalanced control (two-phase control for acceleration speed control, and deceleration speed control by a plugging operation or DC dynamic braking) is employed for elevator drives.

Frequency Control System

It is very easy to control the speed of an IM by changing the power source frequency f_1 continuously, since the synchronous speed of an IM is a function of f_1. Figure 13.14 is an example of a circuit configuration for frequency control of an IM. A rectifier circuit converts AC power to DC power and the amplitude of the DC voltage is controlled through a chopper circuit. The DC voltage is introduced to a voltage source type inverter (INV.) the output voltage of which is controlled either in a square-wave mode or in a pulse width modulation (PWM) mode.

By avoiding magnetic saturation of the linkage flux Φ_1, the ratio V_1/f_1, of inverter output voltage V_1 to frequency f_1, is kept almost constant. But at low speed, the impedance voltage drop in the primary circuit of an IM cannot be neglected. Therefore the practical ratio of V_1/f_1 at low speed is set at a higher value than that at high speed, as shown in Figure 13.15.

Figure 13.16 illustrates the configuration for both open loop and closed loop control. The former is used for controlling multiple IM groups connected in parallel and driven by one set of inverters (INV.), such as in a table roller drive, while the latter is used for realizing fast response by a single induction motor (IM) such as in an iron rolling mill drive. The latter control system is called a "vector control" system or "field oriented control" system (see **Vector Control**). Frequency control systems for an IM are widely used for industrial applications.

Figure 13.17 illustrates a system configuration for a cycloconverter driving a large capacity IM with a vector control loop, such as that used for an iron rolling mill drive. In the cycloconverter type control system, it is not feasible to produce a frequency which is higher than that of the power supply because the frequency produced results from a power supply composed of the combined waveforms of the original power supply.

Secondary Resistance Control System

The secondary resistance control system of wound rotor type IMs has been employed for many years. A liquid resistor is used for large capacity IMs such as those used with fans and pump drives, and step-by-step resistance control by mechanical contactor or thyristor switch is used for small capacity IMs such as those used with crane drives.

Figure 13.18 shows the circuit configuration and the torque-speed characteristics for secondary resistance control.

Slip Power Recovery Control System

The slip power recovery control system of wound rotor type IMs is one of the secondary voltage control systems. In a squirrel cage type IM, the corresponding slip power is dissipated in the rotor circuit, but in this control system, slip power is controlled by the voltage added to the slip rings of the IM through the converter (rectifier). High-efficiency operation at low speed can be achieved.

Figure 13.19 shows both the Kraemer and Scherbius types of slip power recovery control systems. The Kraemer system applies the secondary slip power of the IM in the form of mechanical power to the DC motor (DCM), while the Scherbius system applies this secondary slip power in the form of electric power to AC power source. The former, because of its constant output characteristics, has been used for driving pumps; although it is seldom used today because of its use of a DC motor, which is contradictory to recent tendencies toward maintenance-free

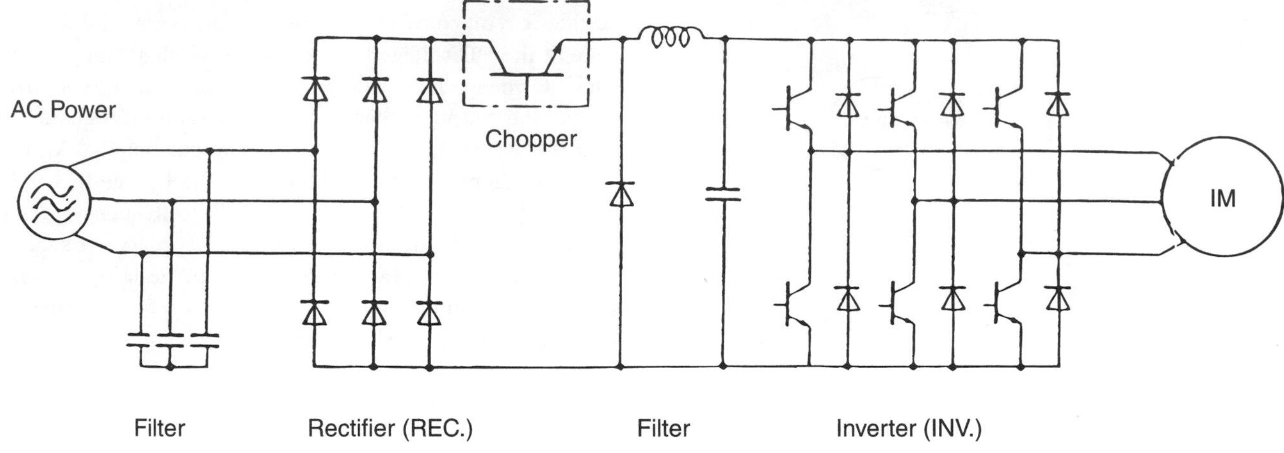

Figure 13.14 Circuit configuration for frequency control (non-regenerative).

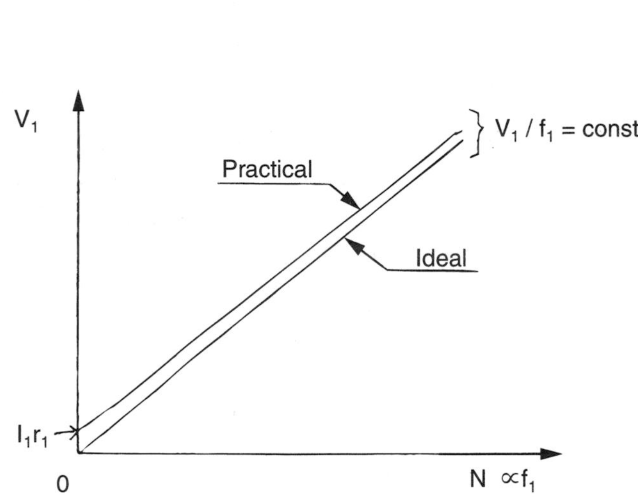

Figure 13.15 The relationship between V_1 and f_1 (controlled).

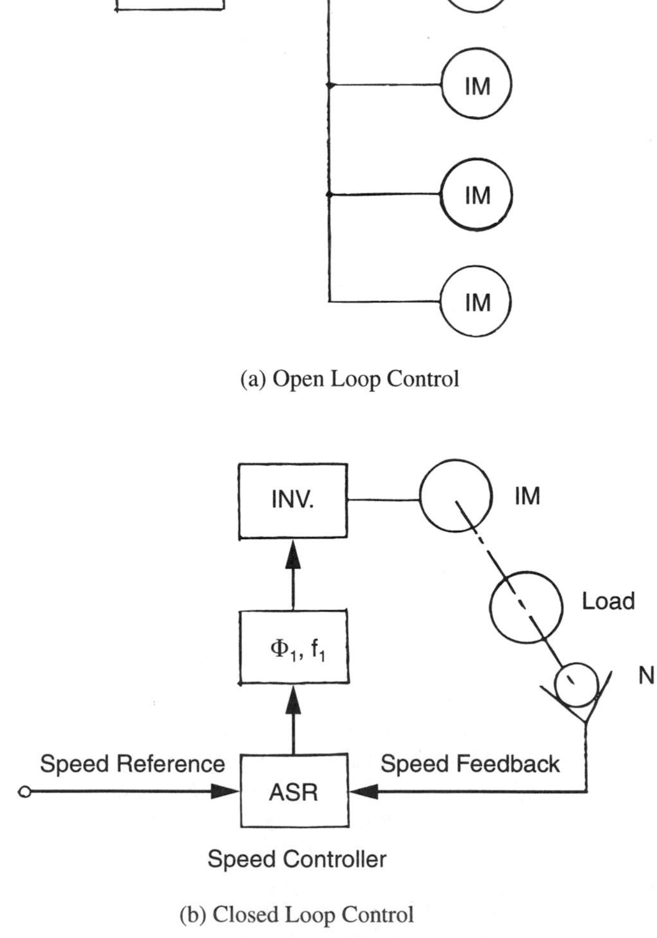

Figure 13.16 Open loop and closed loop control.

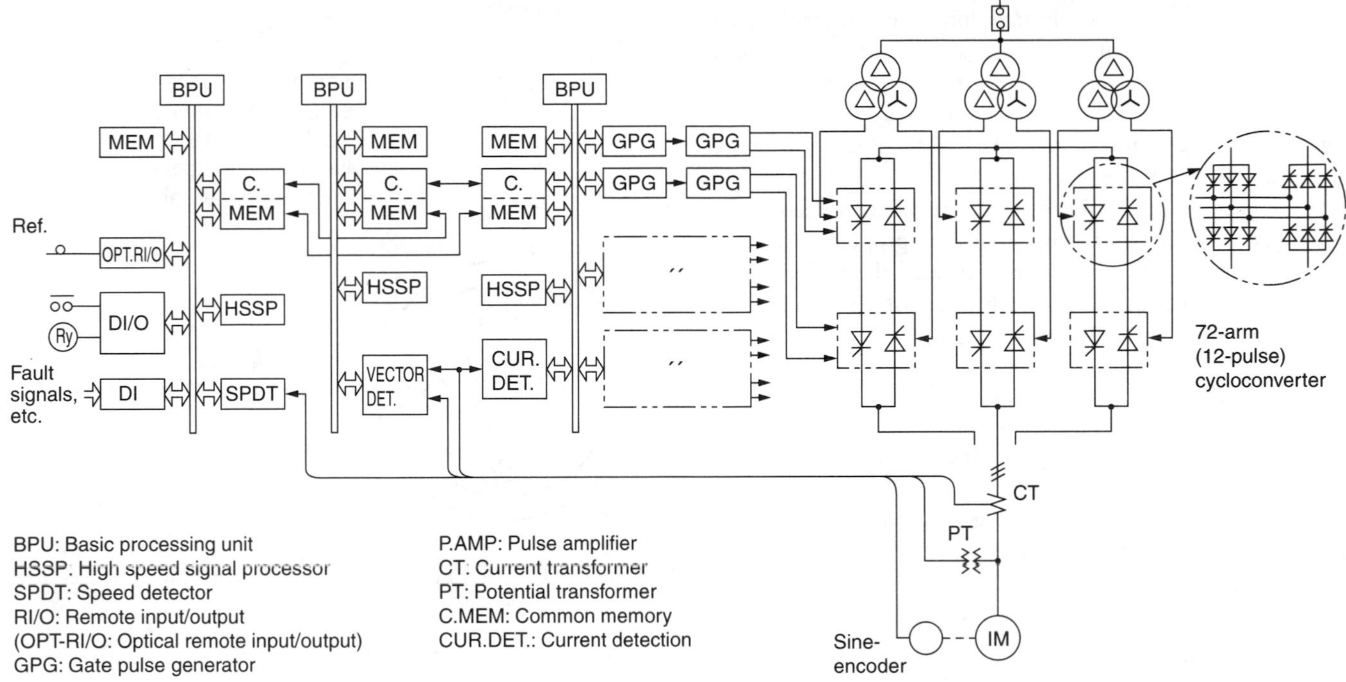

Figure 13.17 System configuration of cycloconverter.

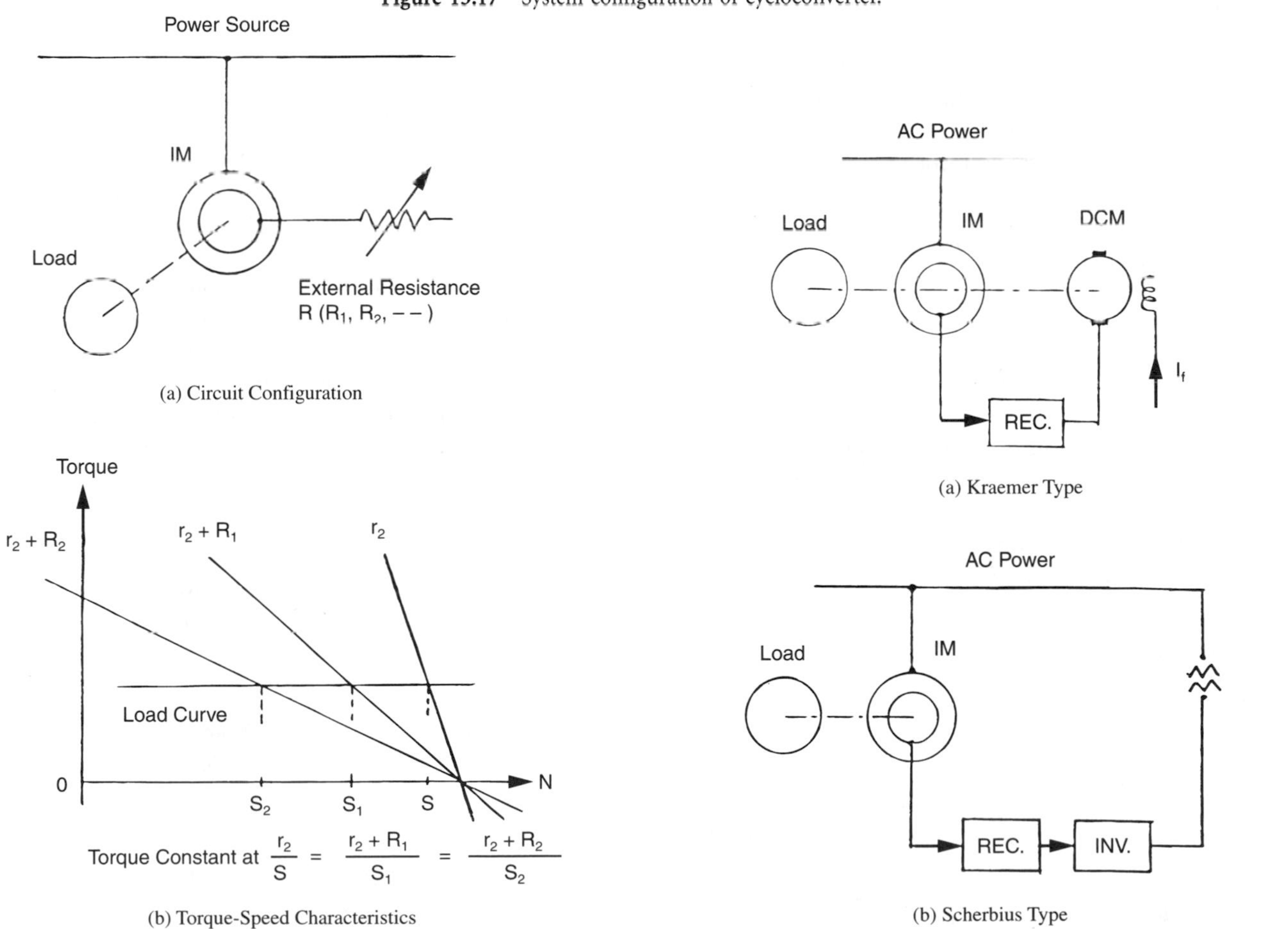

Figure 13.18 Secondary resistance control.

Figure 13.19 Slip power recovery control.

operation. The latter has constant torque characteristics, i.e., when the motor current is held constant, its torque becomes constant, and is extensively used for driving fans and pumps. A rectifier (REC) is used to convert slip power into direct current, which is returned to the DC motor (DCM) or ac power source by the inverter (INV.).

The higher the DC voltage on the rectifier, the slower the motor speed. This DC voltage, which is proportional to the secondary slip voltage, is varied by adjusting the field current I_f of the DCM or the firing angle of the INV, which permits speed control (see **Slip Power Recovery Control**).

Vector Control

A variable speed system offering both the ruggedness of an induction motor and the controllability of a DC motor is desired. Such a system is the vector control system for an induction motor using an inverter. Vector control makes it possible to respond quickly and control torque accurately (Bose, 1993; Abraham, 1986; Blaschke, 1971; Hasse, 1968).

Basic Concept of Vector Control

In the case of an induction motor, the principles of generating torque can be considered to be the same as those used in a DC motor (Blaschke, 1971). Figure 13.20 illustrates these principles. For a DC motor, the magnetic flux Φ is generated by the field current I_f as shown in Figure 13.20a. The armature current I_a, which generates torque, flows through the armature coil and intersects with the magnetic flux Φ. This interaction generates a torque τ_e which can be expressed as follows:

$$\tau_e = \Phi I_a \qquad (13.15)$$

By holding the magnetic flux Φ constant, torque can be controlled by simply controlling I_a.

The situation in an induction motor is quite different because its voltage and current are AC quantities. With reference to Figure 13.20b, by observing the machine on the rotating field side and decomposing the primary current into its magnetizing component I_0 ($\propto \Phi$) in the direction of magnetic flux and its component

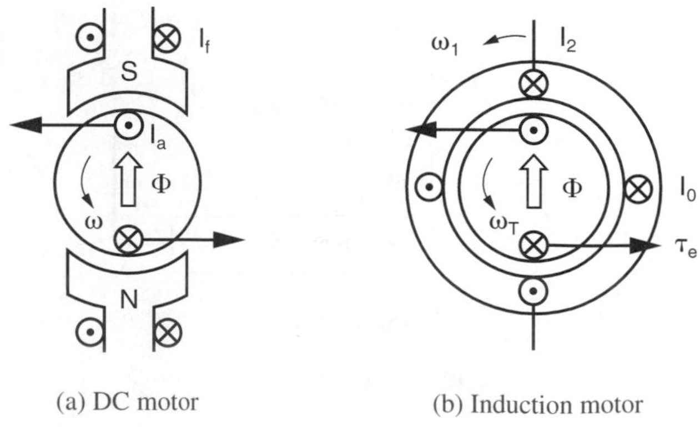

Figure 13.20 The Principles of generating torque in motors.

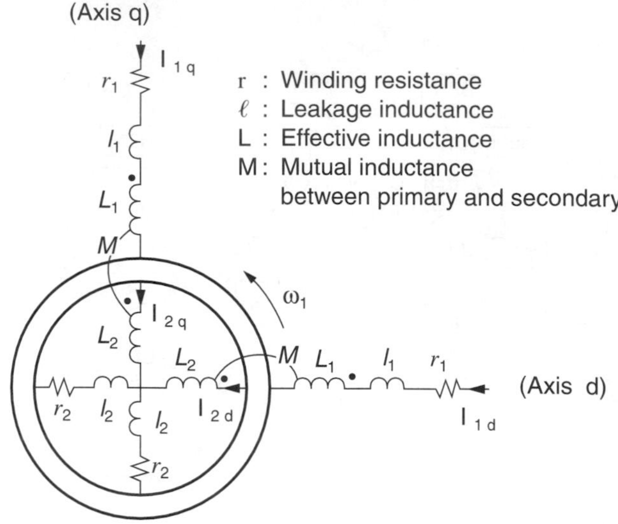

Figure 13.21 Model of induction motor by indication in axes *d, q*.

I_2 which is perpendicular to this flux, the generated torque τ_e can be expressed as

$$\tau_e = \Phi I_2 \qquad (13.16)$$

Vector control of torque is based on this concept, and by controlling I_0 and I_2 independently, this technique can control torque accurately with quick response.

Principles of Control

A theoretical analysis of this control technique can be performed using circuit equations (Miyairi, 1981).

Converting various induction motor quantities onto two intersecting axes (axes *d, q*) of a rotating field coordinate produces the model shown in Figure 13.21. In this model, magnetic flux Φ, voltage E and torque τ_e are expressed using the following general equations:

$$\begin{matrix} \Phi_{1d} \\ \Phi_{1q} \\ \Phi_{2d} \\ \Phi_{2q} \end{matrix} = \begin{matrix} l_1 + L_1 & 0 & M & 0 \\ 0 & l_1 + L_1 & 0 & M \\ M & 0 & l_2 + L_2 & 0 \\ 0 & M & 0 & l_2 + L_2 \end{matrix} \cdot \begin{matrix} I_{1d} \\ I_{1q} \\ I_{2d} \\ I_{2q} \end{matrix} \qquad (13.17)$$

$$\begin{matrix} E_{1d} \\ E_{1q} \\ E_{2d} \\ E_{2q} \end{matrix} = \begin{matrix} r_1 + P(l_1 + L_1) & -(l_1 + L_1)\omega_1 & PM & -M\omega_1 \\ (l_1 + L_1)\omega_1 & r_1 + P(l_1 + L_1) & M\omega_1 & PM \\ PM & -M\omega_s & r_2 + P(l_2 + L_2) & -(l_2 + L_2)\omega_s \\ M\omega_s & PM & (l_2 + L_2)\omega_s & r_2 + P(l_2 + L_2) \end{matrix} \cdot \begin{matrix} I_{1d} \\ I_{1q} \\ I_{2d} \\ I_{2q} \end{matrix} \qquad (13.18)$$

$$\tau_e = p(\Phi_{2q} I_{2d} - \Phi_{2d} I_{2q}) \qquad (13.19)$$

where

M = Mutual inductance,
L = Effective inductance,
l = Leakage inductance,
r = Resistance,
ω_1, ω_s = Primary and slip frequency,
p = Number of pole pairs,
P = Differential operator,
τ_e = Torque,
$\text{Affix}_{1,2}$ = Primary and secondary quantities,
$M = L_1 = L_2, E_{2d} = E_{2q} = 0$ (squirrel cage motor).

Vector control is a system that enables high-speed torque control by maintaining constant magnetic flux, even in a transient state. In other words, it controls torque by maintaining a constant secondary magnetic flux Φ_2. Examining the direction of the magnetic flux on axis d with Φ_2 constant illustrates that:

$$\Phi_{2d} \equiv \Phi_2$$

$$\Phi_{2q} = 0 \tag{13.20}$$

Therefore, substituting Equation 13.20 into Equations 13.17 and 13.18 yields an expression of the currents I_{1d}, I_{1q}, I_{2d}, I_{2q} as a function of Φ_2:

$$I_{1d} = ((1 + PT_2)/M)\Phi_2 \tag{13.21}$$

$$I_{1q} = (T_2\omega_s/M)\Phi_2 \tag{13.22}$$

$$I_{2d} = -P\Phi_2/r_2 \tag{13.23}$$

$$I_{2q} = -(\omega_s/r_2)\Phi_2 \tag{13.24}$$

where

$$T_2 = (l_2 + M)/r_2 = \text{the secondary time constant} \tag{12.25}$$

Substituting Equations 13.20 to 13.24 into Equation 13.19 yields the torque τ_e:

$$\tau_e = p\left(\frac{M}{1 + PT_2} I_{1d}\right)\left(\frac{M}{l_2 + M} I_{1q}\right)$$

$$= p\frac{M}{l_2 + M}\Phi_2 I_{1q} \tag{13.26}$$

This equation indicates that by holding I_{1d} constant (i.e., magnetic flux Φ_2 constant), the torque τ_e is proportional to the current component I_{1q}, without delay.

Therefore, holding I_{1d} constant with reference to this flux axis and controlling the component I_{1q} of the primary current that is perpendicular to it in proportion to the desired torque, this torque can be generated in proportion to I_{1q}.

Basic Control Method

In order to implement the foregoing control principles, all that is required is to maintain the secondary magnetic flux

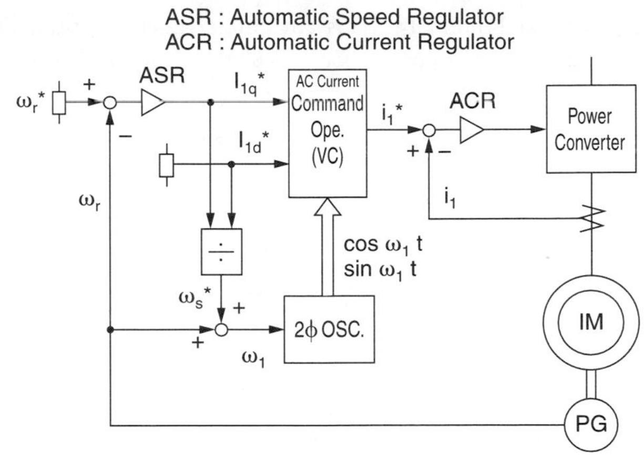

Figure 13.22 Basic circuit construction of a slip frequency type vector control.

Φ_2 constant (i.e., I_{1d} constant), and adjust I_{1q} to control torque. This control can be achieved using slip frequency type vector control (i.e., indirect vector control) or flux detection type vector control (i.e., direct vector control) as indicated in Nagase, et al. (1983).

Slip Frequency Type Vector Control. This method of control relies on the conditions specified in Equation 13.20, i.e., keeping the secondary magnetic flux Φ_2 constant (Hasse, 1968). Thus, this control maintains the relationship between the primary current component I_{1d} and the slip frequency ω_s as specified in Equation 13.22, and then controls the primary frequency f_1 using this slip frequency ω_s. The basic control circuit is shown in Figure 13.22, and illustrates that the current command for achieving a constant secondary magnetic flux is I_{1d}^* (magnetizing current command). The current command I_{1q}^* (torque current command) perpendicular to the secondary magnetic flux Φ_2 can be obtained from the automatic speed regulator (ASR) output. The flux Φ_2 is given by:

$$\Phi_2 = \frac{M}{1 + pT_2} I_{1d}^* \tag{13.27}$$

Conversely, the primary angular frequency ω_1 can be determined from Equation 13.22 as

$$\omega_s^* = \frac{M}{T_2\Phi_2} I_{1q}^* = \frac{1}{T_2} \cdot \frac{I_{1q}^*}{I_{1d}^*} \tag{13.28}$$

$$\omega_1 = \omega_s^* + \omega_r \tag{13.29}$$

where ω_r = the rotational angular frequency of the motor (electric angle, $\omega_r = p\,N$, where N is the actual rotational angular speed). Primary current command i_1^* is AC, and is obtained from the vector sum of I_{1d}^* and I_{1q}^*. This current is generated from the output signals sin $\omega_1 t$, cos $\omega_1 t$ of a two-phase oscillator (2ϕ OSC).

$$i_1^* = I_{1d}^* \sin \omega_1 t + I_{1q}^* \cos \omega_1 t$$
$$= I_1^* \sin(\omega_1 t + \theta) \tag{13.30}$$

$$I_1^* = \sqrt{[(I_{1d}^*)^2 + (I_{1q}^*)^2]}$$
$$\theta = \tan^{-1}(I_{1q}^*/I_{1d}^*) \tag{13.31}$$

When the primary current i_1^* is generated in this manner, the actual current i_1 is controlled to follow i_1^* through action of the automatic current regulator (ACR). A PWM inverter is usually used as the power converter.

In this way, slip frequency type vector control is used as a speed control system with a minor current control system such as a Leonard system. Because the induction motor is an AC motor, its operation differs from that of a DC motor in that the primary frequency is determined by Equations 13.28 and 13.29, and the current is obtained as an AC current of the form described in Equation 13.30. Because I_{1d}^* is a flux command and the primary current if obtained as the sum I_{1d}^*, and I_{1q}^*, the oscillator output signal becomes a signal with the same phase as the flux. For this reason, this system may be considered as one in which the phase of the magnetic flux is determined in a feed-forward manner. This system is characterized by its use as a general-purpose induction motor.

Flux Detection Type Vector Control. Flux detection type vector control is a system which controls torque through the separate use of magnetizing current and torque current in order to establish the conditions in Equation 13.26. For that reason, this system performs control based on the phase of the magnetic flux (Blaschke, 1971).

Since control is performed through command of the respective current components with reference to the phase of the magnetic flux, it is necessary to obtain the real phase of the motor's magnetic flux. Therefore, the phase of the magnetic flux must be detected because control is based on this quantity.

The basic diagram illustrating this type of control is shown in Figure 13.23. A flux detector is installed inside the motor to detect the phases of the magnetic flux (sin $\omega_1 t$, cos $\omega_1 t$). With regard to the results obtained, only the reference signal of the phases of the magnetic flux (sin $\omega_1 t$, cos $\omega_1 t$) is different from that used in slip frequency type vector control. The remaining parts of the circuit diagram are the same as shown in Figures 13.22 and 13.23.

Because this system uses phase detection of the magnetic flux, it is characterized by very accurate control.

System Comparisons

Table 13.2 can be used to compare both systems. The slip frequency type, for which a general-purpose motor is applicable,

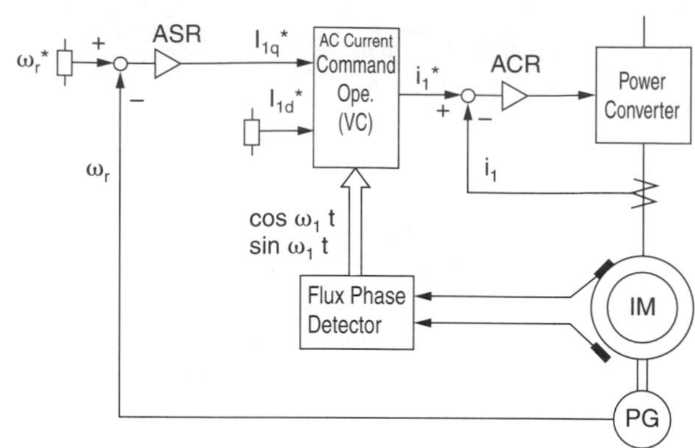

Figure 13.23 Basic circuit construction of a flux detection type vector control.

is constructed using a simple circuit because it has no flux detection circuit. However, the motor parameters are necessary for calculating the slip frequency. For that reason, this system may be affected by changes in parameters, which result in deteriorating characteristics (Nagase et al., 1984).

On the other hand, the problem with flux detection type control is that, although it is very accurate, it is difficult to use a general-purpose motor with this system due to the necessity of detecting the flux phase.

It is also possible to detect the phase of the magnetic flux from the terminal voltage of the motor. For example, systems for estimating the secondary flux phase from the primary voltage, primary current and rotating speed by using an observer are also being studied (IEEJ Technical Report No. 416, 1992; Leonhard, 1988). Such methods present a problem regarding operation in the low-speed range due to the difficulty of accurately detecting voltage in this range.

Both types of systems have their own advantages and disadvantages as previously discussed. The slip frequency type is advantageous because of its simple construction and possible use of a general-purpose motor. For these reasons, the slip frequency type control is preferred as the basic control method for vector control. The flux detection type is used to compensate for the disadvantages which occur in slip frequency type control.

Moreover, as indicated above, ideal current control is assumed as a pre-condition for these control systems. Therefore, the circuit construction should be designed for quick response of the current control system (Nagase et al., 1987).

Equivalent Circuit Relationships

Consider now the equivalent circuit for the control system described in the previous section. Figure 13.31a shows a general equivalent circuit of an induction motor (see Control Method for details). The equivalent circuit Figure 13.31b is one in which the voltage and current have been transformed using α (primary/secondary winding ratio) $= M/L_2$. Both circuits are essentially the same. Figure 13.31b is convenient for clearly expressing the phase relationship which exists between the current components. This diagram illustrates the relationship between the primary

Table 13.2 Comparison of Vector Control Systems

Item \ System	Flux detection type: Direct detection of the magnetic flux of the motor	Flux detection type: Detection of the magnetic flux from the primary voltage	Slip frequency type
Accuracy of the vector control	Accurate	Accurate	Subject to the effects of changes in motor parameters
Application to general-purpose motors	Difficult	Possible	Possible
Operation in a very low-speed range	Possible	Difficult	Possible
Simplicity of the circuit	Slightly complicated (Magnetic Flux detector required)	Slightly complicated (Flux detection method required)	Relatively easy

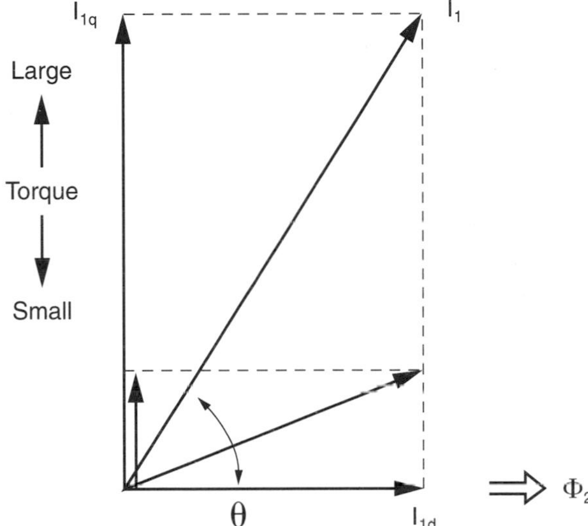

Figure 13.24 Vector diagram of current.

current I_1 and its components $I_{1d}\,(= I_0)$ and $I_{1q}\,(= I_2')$ as shown in Figure 13.24.

This vector diagram yields the following relationships which are the same as those in Equation 13.31.

$$I_1^* = \sqrt{[(I_{1d}^*)^2 + (I_{1q}^*)^2]}$$
$$\theta = \tan^{-1}(I_{1q}^*/I_{1d}^*) \qquad (13.32)$$

Moreover, the relationship between primary and secondary circuits in Figure 13.31b yields the same relationship as described in Equation 13.28, i.e.,

$$\omega_1 M' I_{1d} = (r_2'/s)\, I_{1q}, \quad \text{i.e.}$$
$$\frac{I_{1q}}{I_{1d}} = \omega_s T_2 \qquad (13.33)$$

where $I_0 = I_{1d}$ and $I_2 = I_{1q}$

Equations 13.32 and 13.33, indicate that vector control may be considered as a system which always satisfies the equivalent circuit in Figure 13.31b.

Industrial Applications

Vector control produces quick-response, high-accuracy control of induction motors. The following examples describe in the vector control of a servo system and a steel rolling mill drive system.

Application to a Servo System (Sugiura, et al., 1990) Figure 13.25 illustrates the application of vector control to a servo system. Since the servo system requires quick response, a control microprocessor (H8/532) and a DSP (μPD77C) are employed. Figure 13.26 shows the speed control characteristics of this system. The speed response frequency is approximately 100 Hz, which is comparable to that of a DC servo. Performance roughly equivalent to that a DC motor can also be obtained with this system, even with an induction motor.

Application to a Steel Rolling Mill Drive System (Sukegawa, et al., 1991) A steel rolling mill drive system requires quick-responses and high-accuracy control. Figure 13.27 illustrates the application of vector control to a 2000 kW induction motor. This system employs a multi-inverter construction because it is a large capacity drive. It uses an I_d-I_q current control loop to realize high-accuracy current control. This control loop makes it possible to equalize real current components and hence this loop produces high-accuracy torque control.

Figure 13.28 shows the speed response characteristics. Note that a good speed response is obtained for this rolling mill drive system application.

In summary, vector control can be applied in a variety of fields which require quick-response and high-accuracy control.

Speed Sensorless Vector Control

Speed sensorless vector control is a control system that incorporates both simplicity of V/f control for an ordinary inverter system with high-performance controllability. This scheme can also use a general-purpose induction motor because no speed sensor is used.

Background and Objectives

The general-purpose inverter is commonly used for variable speed control of general-purpose motors. This inverter application extends not only to wind-power and hydraulic machinery such as pumps, fans and air conditioners, but also to machine tools, hoists and cranes. V/f control is sufficient for applications such as pumps and fans, but the torque characteristics during low-speed operation may pose a problem when V/f control is applied to machine tools and cranes (Kin, 1992).

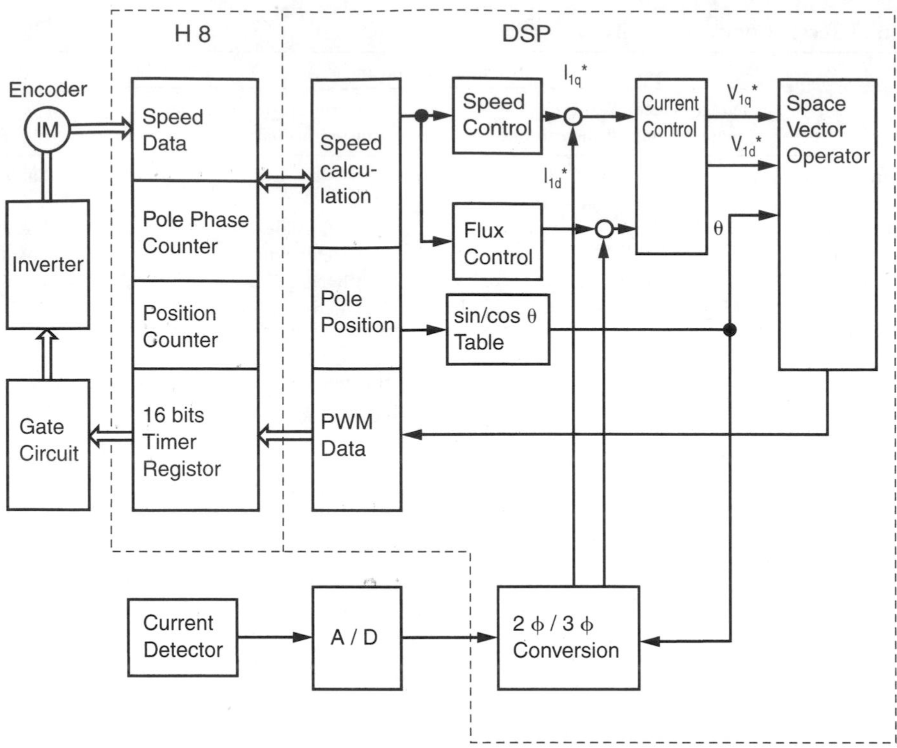

Figure 13.25 Block diagram of servo system.

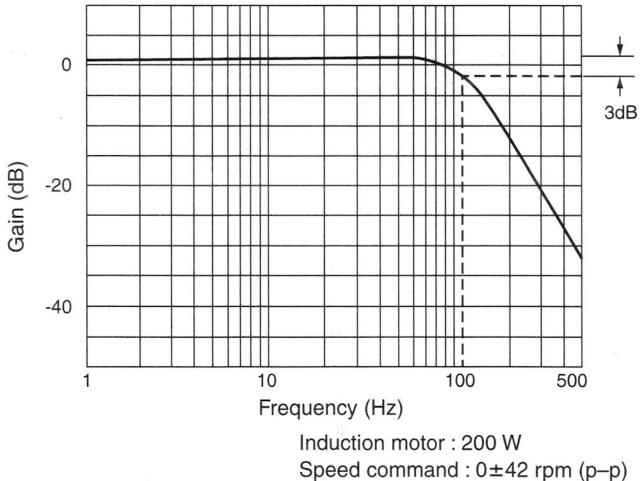

Figure 13.26 Frequency characteristics of speed control system.

In such cases, vector control should be applied. It is difficult to use a general-purpose motor due to the necessity of mounting a speed sensor on it. Moreover, a special circuit and cables are required for speed feedback, which makes a general-purpose motor less convenient to use.

This is why a control system offering the high performance of vector control and the convenience of a general-purpose inverter is needed. Speed sensorless vector control is a control system which meets this need. General-purpose motors can also be used. Figure 13.29 illustrates the positioning of speed sensorless control. From a technical viewpoint, speed sensorless vector control is positioned between V/f control and vector control.

Figure 13.30 shows the torque characteristics of both V/f control and speed sensorless vector control. Figure 13.30b shows the characteristics of a V/f control system. This chart indicates that this system generally exhibits large variations in speed and has difficulty operating in the low-speed range. Conversely, the speed sensorless system in Figure 13.30a offers excellent torque characteristics throughout the entire speed range, with little variation in speed. Table 13.3 compares the typical characteristics of V/f control and speed sensorless control. This comparison indicates an improvement in characteristics with speed sensorless vector control (Fujii et al., 1994).

Consequently, the proliferation of general-purpose inverter applications has advanced the sophistication of control technology. Table 13.4 summarizes this progress of control technology for the general-purpose inverter.

Control Method

Several control methods have been developed for speed sensorless vector control. They can be classified into two major categories: methods realized by using the Model Reference Adaptive System (MRAS), and methods developed from conventional classical control. The MRAS methods are still in research stage, while methods based upon the second approach are in practical use (IEEJ Technical Report No. 416, 1992; Kin, 1992). Thus the second method will be introduced here.

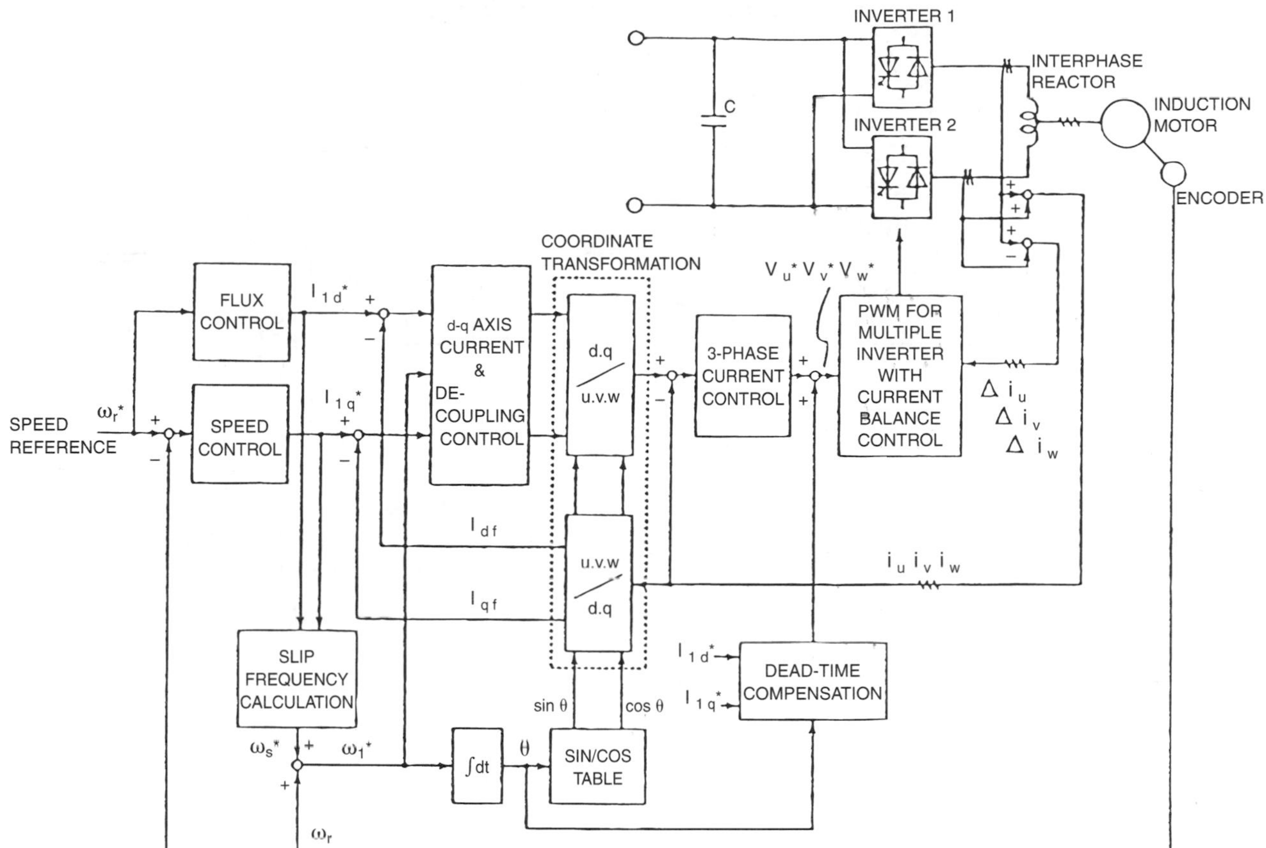

Figure 13.27 Control block diagram of a multiple inverter for an induction motor drive.

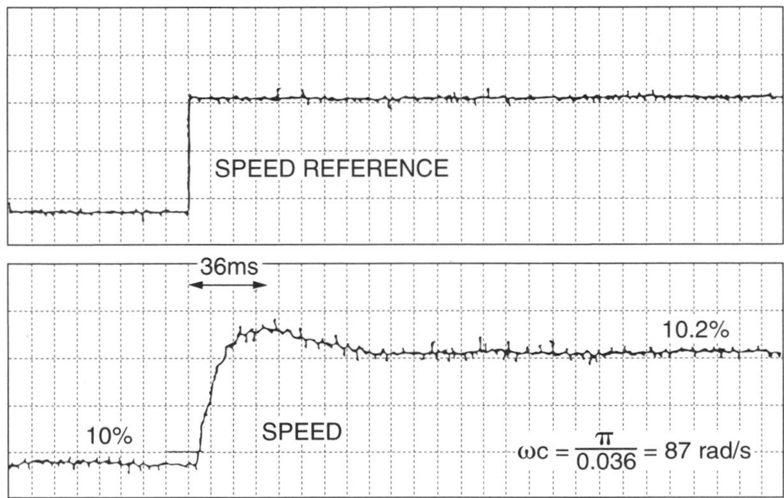

Figure 13.28 Step response of motor speed.

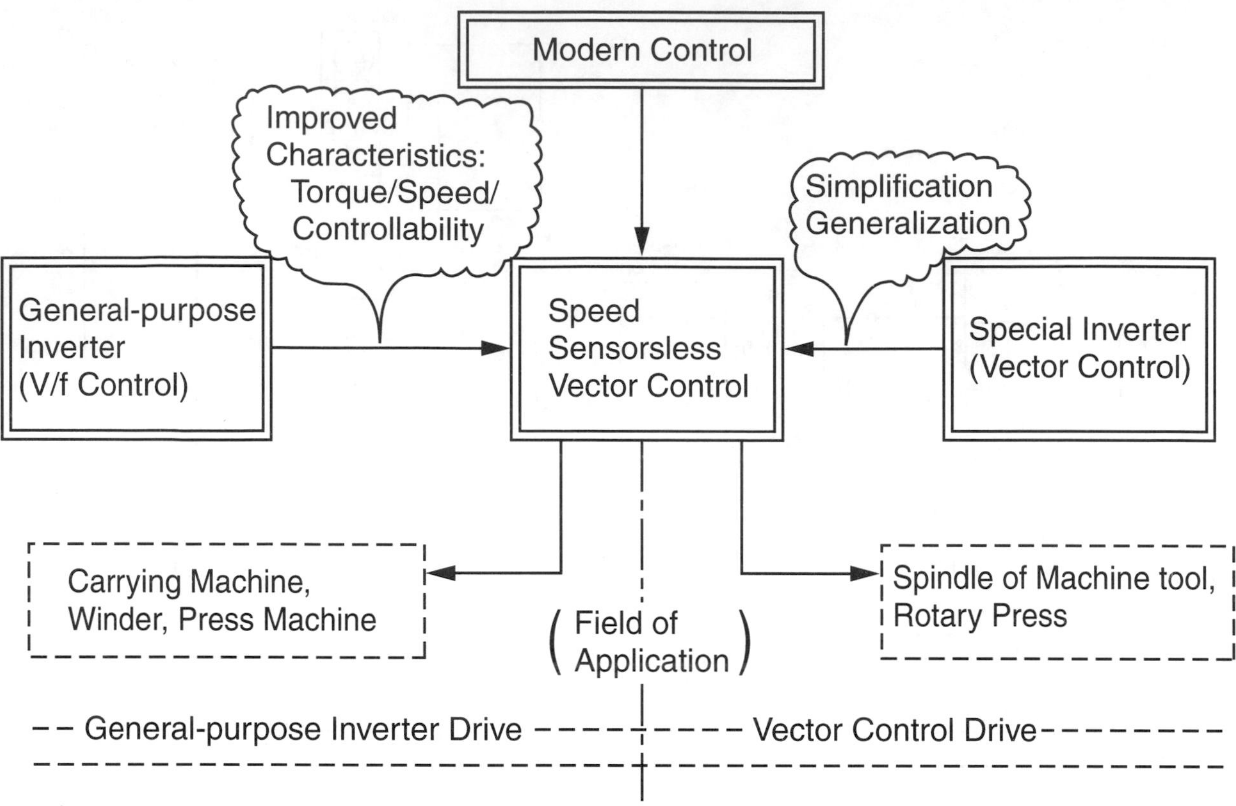

Figure 13.29 Conceptual explanation of an approach to sensorless vector control.

Figure 13.31a shows a general equivalent circuit of an induction motor. The value of α, i.e., primary/secondary winding ratio, can be selected as desired. There, the selection is made to eliminate secondary leakage inductance. Under these conditions, the equivalent circuit shown in Figure 13.31b has the following parameters.

$$\alpha = M/L_2 \tag{13.34}$$

$$M' = M^2/L_2 \tag{13.35}$$

$$l_2' = l_2 M/L_2 \tag{13.36}$$

$$r_2' = r_2\,(M/L_2)^2 \tag{13.37}$$

Voltage and frequency control are performed to establish the equivalent circuit of Figure 13.32b.

Figure 13.32 illustrates the block diagram of a control system based on this concept. Figure 13.33 shows a vector diagram of the motor voltage and current under these conditions. The control unit construction is based upon the following analysis.

Voltage components V_{1d}^*, V_{1q}^* on the axes d, q are calculated based upon the estimated value of the induced electromotive force $E = (\omega_1^*(M/L_2)\,\Phi_2^*)$ and the estimated value of the drop in leakage inductance $= (\omega_1^*(l_1 + l_2')\,I_1)$. Therefore,

$$\begin{aligned} V_{1d}^* &= r_1 I_{1d}^* - \omega_1^*\,(l_1 + l_2') I_{qf} + \Delta V \\ V_{1q}^* &= r_1 I_{qf} + \omega_1^*\,(l_1 + l_2') I_{1d}^* + \omega_1^*\,(M/L_2)\,\Phi_2^* \\ &\approx r_1 I_{qf} + \omega_1^*\,L_1\,I_{1d}^* \end{aligned} \tag{13.38}$$

where ΔV is the output of the automatic current regulator (ACR_2).

$$\Phi_2^* = M I_{1d}^* \tag{13.39}$$

And I_{qf} is the current component along the q axis for the detected primary current.

The voltage commands of each phase (V_u^*, V_v^*, V_w^*) are calculated by VC as shown in Figure 13.32 according to the following equation:

$$V_u^* = V_1^*\sin(\theta^* + \delta) \tag{13.40}$$

where

$$V_1^* = \sqrt{[(V_{1d}^*)^2 + (V_{1q}^*)^2]}$$

$$\theta^* = \int \omega_1^*\,dt \text{ is the phase reference} \tag{13.41}$$

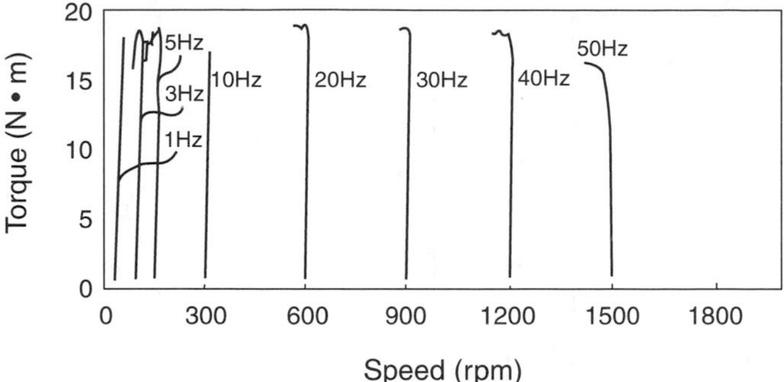

(a) SPEED SENSORLESS VECTOR CONTROL

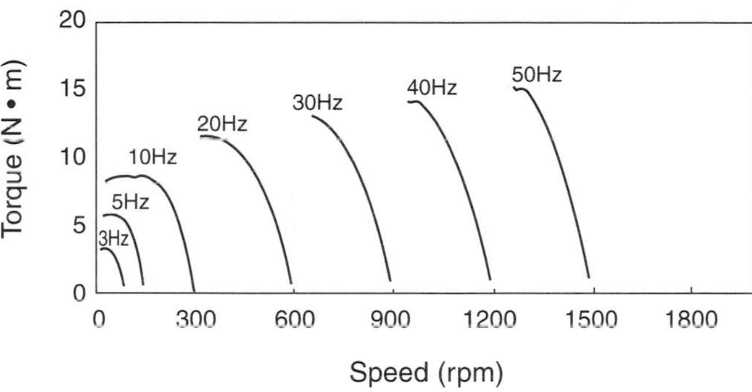

(b) V / F CONTROL

Figure 13.30 Torque characteristics (1.5 kW, 4 poles).

Table 13.3 Improved Characteristics of Sensorless Vector Control

Performance item	V/f control (an example)	Sensorless vector control
Range of speed control, torque 150% or higher	1:5	1:50 or higher
Speed accuracy	>10%	±1%
Speed response	—	20 rad/s or higher

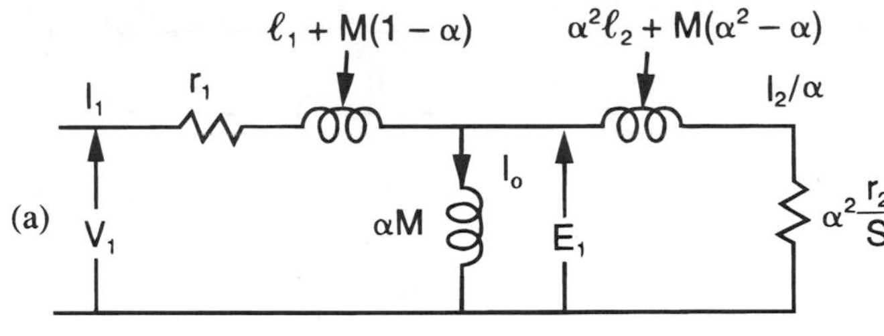

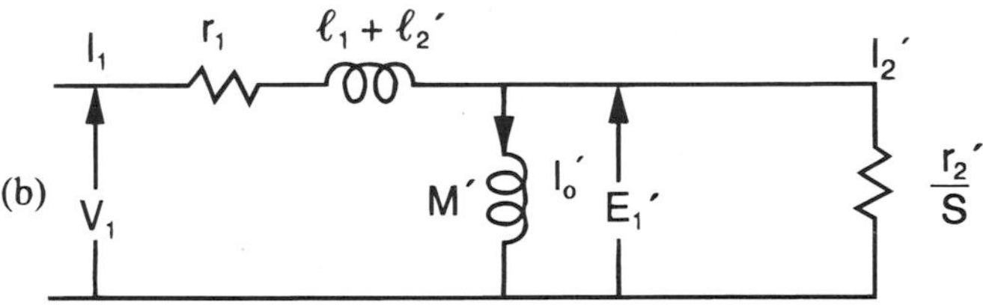

Figure 13.31 Equivalent circuits of an induction motor.

$$\delta = -\tan^{-1}(V_{1d}^*/V_{1q}^*)$$

In a voltage type PWM inverter, control is performed in proportion to the voltage of respective phases (V_u^*, V_v^*, V_w^*).

The speed control unit operation is based on the following principle:

The estimated speed value $\hat{\omega}_r$ is calculated from Equation 13.42 using the frequency $\omega_1{}^*$ and the estimated value of slip frequency $\hat{\omega}_s$, i.e.,

$$\hat{\omega}_r = \omega_1^* - \hat{\omega}_s \tag{13.42}$$

The automatic speed regulator (ASR) performs speed control by using PI control and produces as an output the current I_{1q}^*. Under these circumstances, constant magnetic flux is maintained, so that

$$\hat{\omega}_s = K_\omega I_{qf} \tag{13.43}$$

Table 13.4 Progress in Control Technology of General Purpose Inverter

Generation	Main control system	Main method	Main construction	Problems & characteristics	Period
First generation	V/f control	Subharmonic control	Mixed analog-digital circuit	Incomplete PWM waveform; Vulnerability to instantaneous power failure; Insufficient protective functions	Early 1980s
Second generation	V/f control	Magnetic flux control system; Voltage vector control system	Full digital	Improvement of above-mentioned disadvantage; Automatic restarting; Insufficient low-speed torque; Inaccurate speed	Late 1980s
Third generation	Sensorless control	Method unique to respective manufacturers or MRAS method	Full digital or DSP system	Improvement of above-mentioned disadvantage	Around the mid-1990s

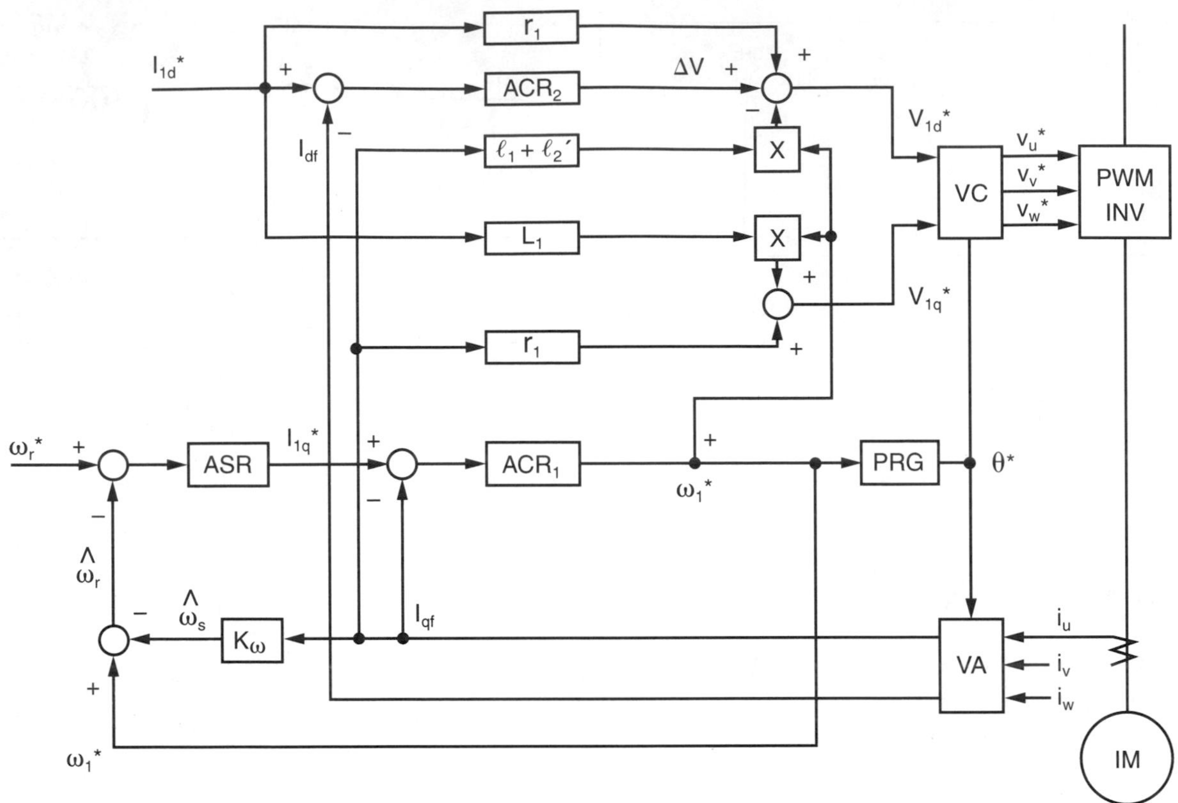

Figure 13.32 Configuration of the control system.

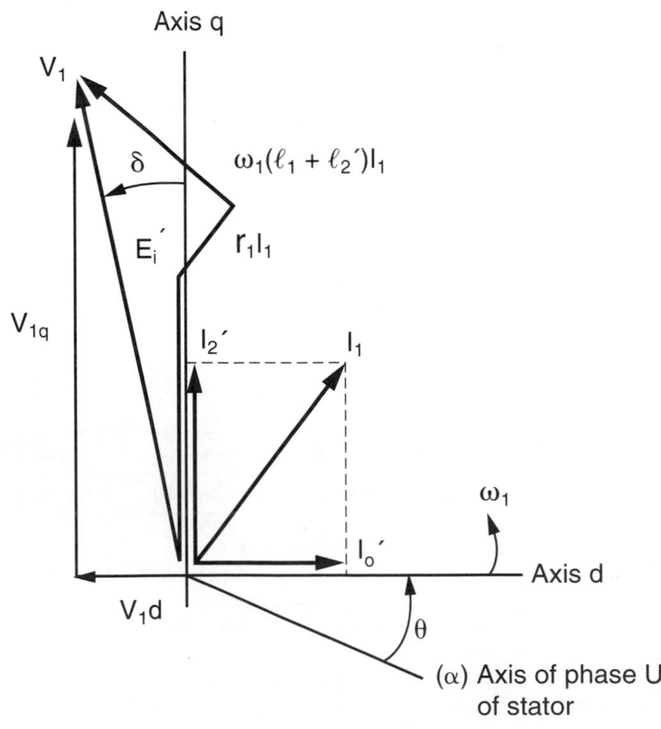

Figure 13.33 Vector diagram of current and voltage.

The current components (I_{df}, I_{qf}), needed for the control system, are calculated by the unit VA using the following equation:

$$\begin{matrix} I_{df} \\ I_{gf} \end{matrix} = \begin{matrix} \cos\theta^* & \sin\theta^* \\ -\sin\theta^* & \cos\theta^* \end{matrix} \cdot \begin{matrix} i_u \\ (i_v - i_w)/\sqrt{3} \end{matrix} \tag{13.44}$$

where i_u, i_v, i_w are the primary currents of the motor.

The results of this control scheme are shown in Figure 13.30. Sensorless control reduces speed variations and provides for stable operation even in the low-speed range.

Figure 13.34 displays the motor current-torque characteristics (Fujii et al., 1994). In V/f control, a relatively large V/f value is set in the low-frequency area (called boost control) to compensate for insufficient torque. While current in the rated torque range is small when the boost is large, excessively large current flows under reduced loads. On the other hand, large currents exist in the rated torque range, when the boost is small. Conversely, speed sensorless vector control enables optimal control because the current components are always properly controlled. In other words, the motor current diminishes due to the large torque/current value ratio. This improves the overall efficiency of the entire system.

As indicated by using speed sensorless vector control, speed variations diminish, torque characteristics improve, and the torque/current value ratio increases. All these characteristics serve to reduce inverter capacity.

Industrial Applications

General-purpose inverters, based on V/f control, are used in many fields due to advantages such as convenience and low

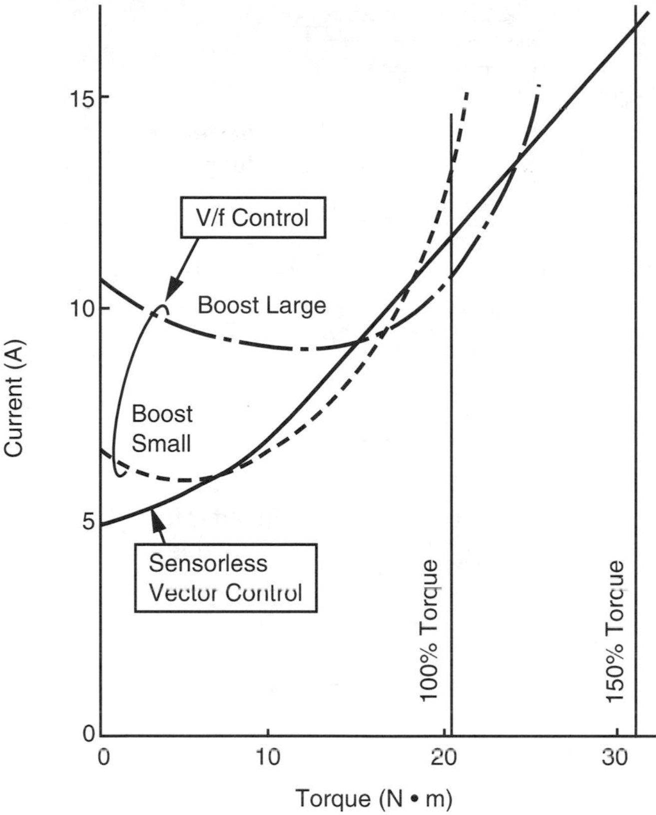

Figure 13.34 Motor current—torque characteristics (Motor : 3.7 kW, 4 P., f_1 = 10 Hz).

cost. Insufficient torque in low-speed operations may be a problem however, depending upon the purpose of the application.

By adopting a speed sensorless vector control system for general-purpose inverters, the number of applications can be further expanded. The following illustrates the effective use of the characteristics of speed sensorless vector control (Fujii et al., 1994; Okuyama, 1991).

Application to a Hoist. The hoist is a mechanical system that lifts cargo against gravity, and therefore must have sufficient torque characteristics to overcome the gravitational pull. DC machines and induction motors with vector control using sensors have been utilized extensively. In addition, the use of a general-purpose inverter with V/f control requires the use of an inverter with much greater capacity than is conventionally required.

Figure 13.35 illustrates a hoist system in which a general-purpose inverter is used. The hoist must satisfy the following specifications:

1. Prevent cargo from stalling: A hoist must generate large torque at low speed. In particular, it must generate sufficiently large starting torque to prevent the cargo from dropping momentarily when the mechanical brake is released at the start of operation, which requires large torque at the low speed.
2. Stop accurately: Recently, there has been a growing demand for smoother hoist positioning control. To improve the stopping accuracy, a hoist must secure the necessary torque during low-speed operation, and maintain a small speed variation ratio.

   Moreover, since the hoist operates in the motoring mode for lifting and in the regenerative mode for lowering, the speed variation ratio should be small in both directions (i.e., motoring and regeneration).

Speed sensorless vector control is an effective method to satisfy these demands.

Applications to a Three-Dimensional Parking Lot. The three-dimensional parking lot, employing an elevator carrier mechanism, is becoming more common in urban areas. Figure 13.36 shows the construction drawing of an elevator type three-dimensional parking lot. The drive system has a lifting motor for the vehicle carrier and an auxiliary motor for the turntable. The performance required for such a parking lot includes smooth operation, reduced loading/unloading time, and improved operability. For these reasons, the drive system must feature low-noise and small torque as well as small speed variation.

Speed sensorless vector control satisfies these specifications.

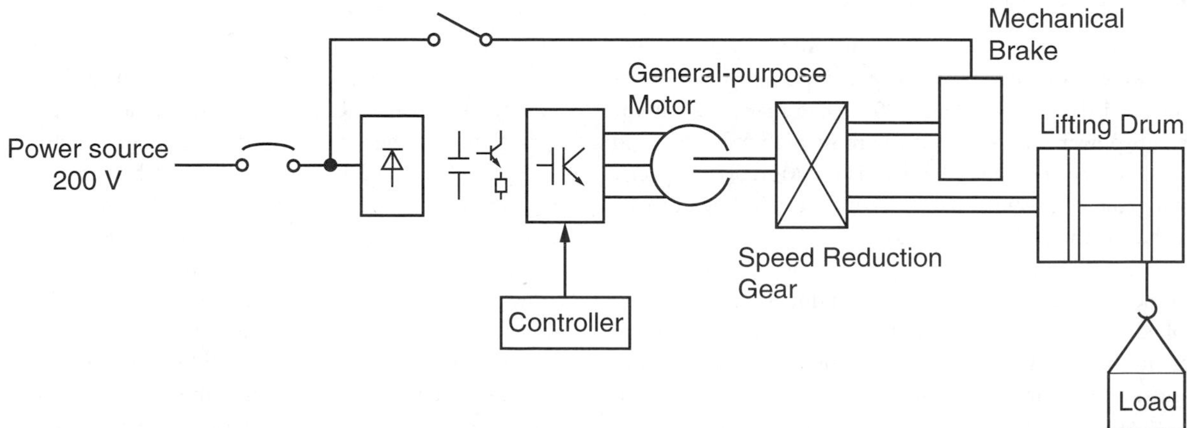

Figure 13.35 System construction of a hoist.

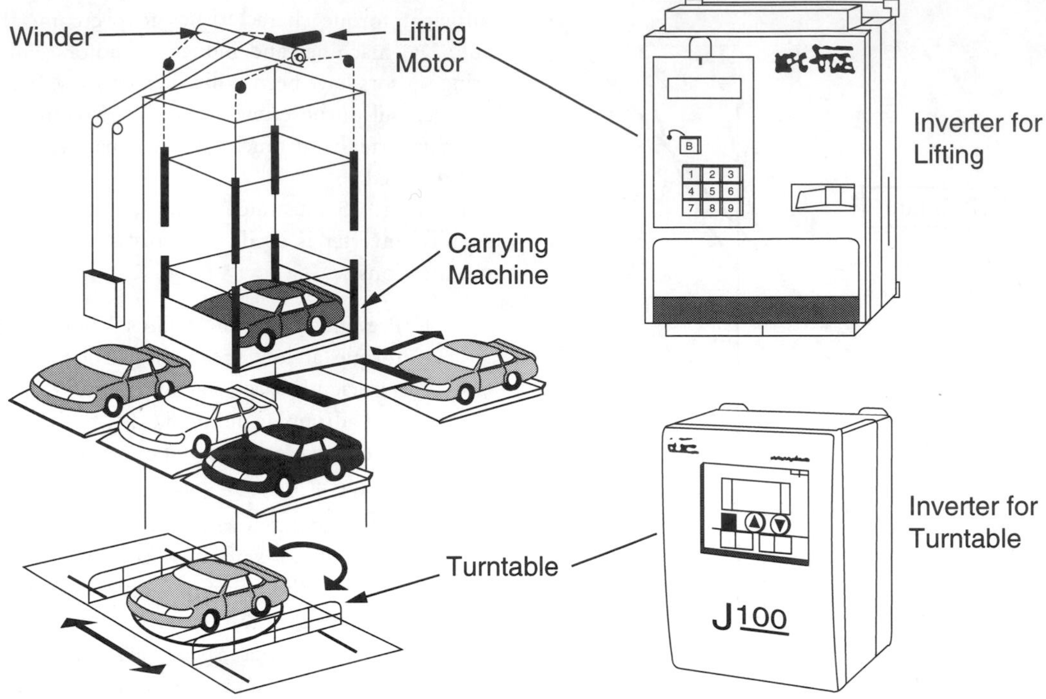

Figure 13.36 Construction drawing of 3-dimensional parking lot.

Applications to Conveyors. The inverter can be used in a conveyor to obtain advantages such as free-speed setting and easy positioning control. A conveyor, which requires large torque during low-speed operation like the hoist, must have small speed variations against changes in load torque.

A conveyor may often be used continuously, even in the low-speed range. It is, therefore, necessary to maintain large, continuous, allowable torque in the low-speed range. Speed sensorless vector control can generate the required torque with optimal current due to the large torque/current ratio described in Figure 13.34. It provides a sharp improvement in the continuous allowable torque compared with conventional V/f control, as shown in Figure 13.37.

Slip-Power Recovery Control

Speed control of wound-rotor induction motors by slip-power recovery is widely applied to large capacity pump or fan drives (Jifuku, et al., 1987; Noda, et al., 1981). The main reasons for choosing slip-power recovery control are its high efficiency and good adjustable speed control characteristics. Moreover, capacity of the converter connected to the secondary windings can be less than that of the induction motor.

Many kinds of slip-power recovery control schemes have been put into practical use. In all of them, the induction motor speed is controlled by regenerating the slip-power to an AC power source or mechanical shaft through the converter connected to the secondary windings of the motor. The total loss in each scheme is the sum of the induction motor loss and the converter loss, principally. Therefore, the total loss is very small compared with that in the rotor resistance control scheme. As a result, the efficiency of the slip-power recovery control schemes becomes very high. Since the handling power of the converter is proportional to the slip of the induction motor, the converter capacity is 40% of the motor capacity when the speed control range is selected from 60 to 100% of the synchronous speed. For practical uses, it is necessary to consider a margin of 10 to 20% for the converter capacity.

Figure 13.38 shows the static Scherbius control which is the most popular of the slip-power recovery control schemes (Hori, et al., 1972). The speed can be adjusted by regenerating the slip-power of the wound rotor induction motor, which is rectified by a diode bridge and then is inverted again to AC power by a thyristor bridge, to the AC power source. In the static Scherbius control, the slip is proportional to the DC link voltage which decreases with larger control angle of advance of the thyristor bridge. That is, the motor speed increases with larger control angle of advance. In the figure, power flow is in one direction from the secondary side of the motor to the AC power source. So, the regenerative braking torque cannot be generated and operation over the synchronous speed is impossible.

Figures 13.39 and 13.40 are schemes which remove the drawback of the first scheme in Figure 13.38. In Figure 13.39, the diode bridge in Figure 13.38 is changed to a thyristor bridge 1 (Zimmermann, 1977), and a cycloconverter is used as a power converter in Figure 13.40 (Chattopadhyay, 1978). These schemes are called supersynchronous static Scherbius control. Since the power flow from the secondary side of the motor to the AC power source is reversible, getting the regenerative brake torque and operating over the synchronous speed are possible. Since the induced voltage in the secondary windings of the motor is very low around the synchronous speed, the commutation failure

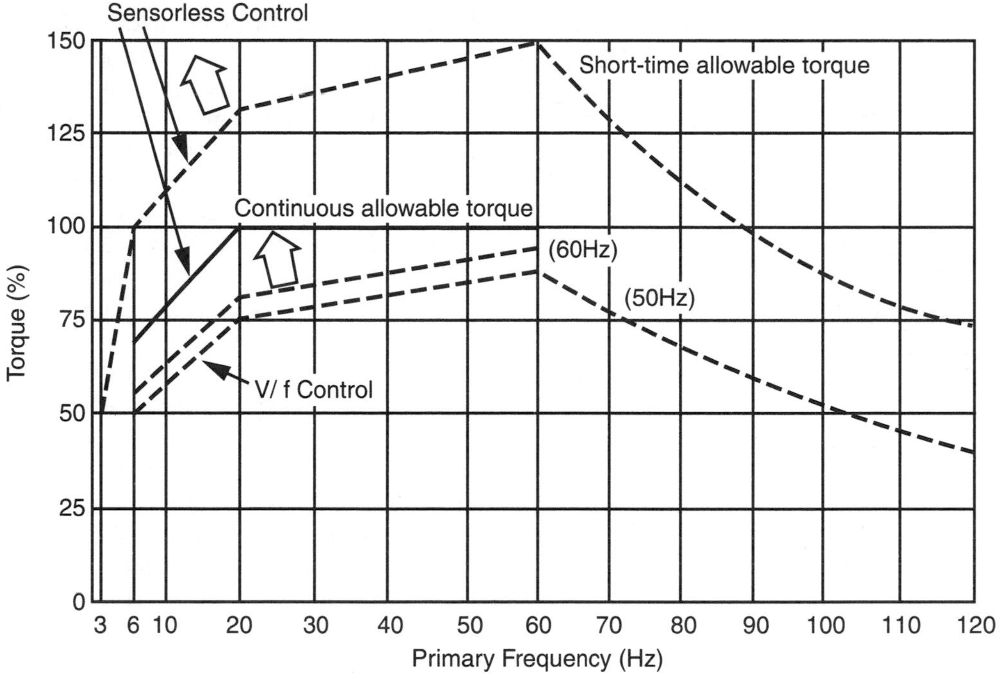

Figure 13.37 Continuous allowable torque characteristics.

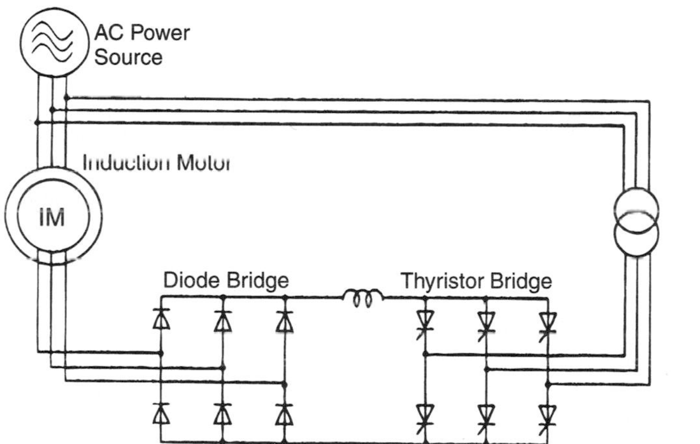

Figure 13.38 Static Scherbius control.

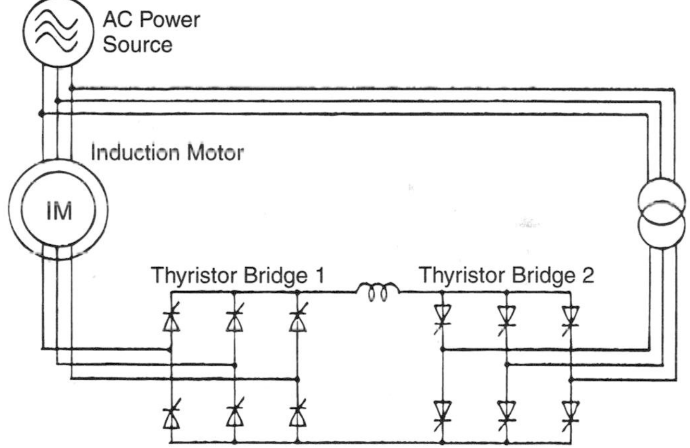

Figure 13.39 Supersynchronous static Scherbius control using DC link converter.

occurs in thyristor bridge 1 in Figure 13.39. Therefore, thyristor bridge 1 should be commutated by letting the current flow intermittently with phase control of thyristor bridge 2. Though an 18-arm cycloconverter is used in Figure 13.40, a 36- or 72-arm cycloconverter can also be used.

Since the power factor of the thyristor bridge or the cycloconverter to the AC power source is low in Figures 13.38 to 13.40, it is necessary to connect a power-factor-correcting capacitor to the AC power source lines.

Figures 13.41 and 13.42 show the static Scherbius controls which use self-extinction devices such as GTO thyristors (GTOs) or bipolar junction transistors (BJTs) or insulated gate bipolar transistors (IGBTs) instead of thyristors. In these cases, it is possible to operate for GTO bridge connected to the AC power source at unity power factor. Therefore, there is no need to connect the power-factor-correcting capacitor which is needed in the schemes of Figures 13.38 to 13.40. In the case of Figure 13.41, getting the regenerative braking torque and operating over the synchronous speed are possible in the same manner as for Figures 13.39 and 13.40, because the power flow from the secondary side of the motor to the AC power source can be reversibly controlled.

A buck-boost chopper is used in Figure 13.42. This chopper keeps the DC voltage in the GTO bridge constant by operating as a buck (step-down) regulator in the low-speed range in which the slip is large and acting as a boost (step-up) regulator in the high-speed range in which the slip is small. As a result, the voltage rating of the GTO bridge can be reduced to a proportional value

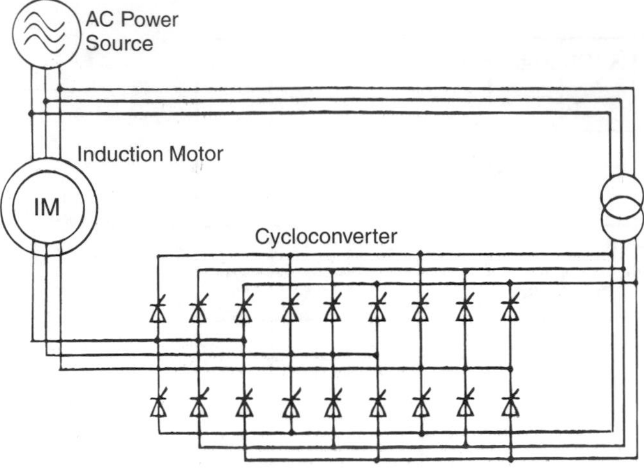

Figure 13.40 Supersynchronous static Scherbius control using cycloconverter.

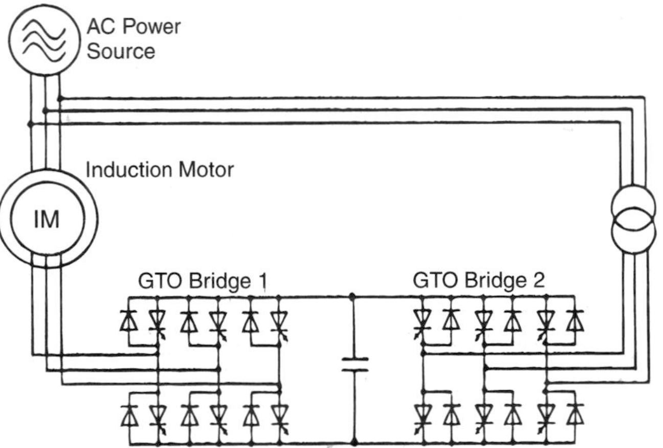

Figure 13.41 Supersynchronous static Scherbius control using DC link GTO converter.

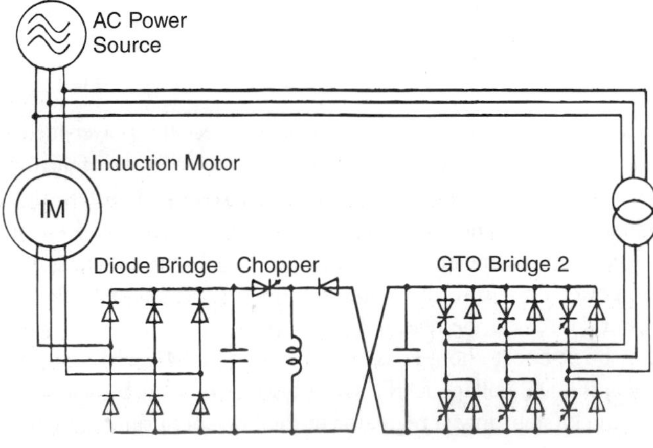

Figure 13.42 Static Scherbius control using chopper and GTO converter.

for the speed control range which is about 30 to 40% of the induction motor secondary voltage at speed = 0, or slip = 1.0. Since the starting resistor can be omitted and the capacity of the GTO bridge, which is a major part of the power converter, can be reduced to a low limit, the total size of the equipment can be decreased. In the case of Figure 13.42, getting the regenerative braking torque and operating over the synchronous speed are impossible, the same as in Figure 13.38.

Generally, the power converter capacity is selected as a value proportional to the speed control range, so it is necessary to start and accelerate the induction motor to the control range speed by using a separately installed resistor. In Figure 13.40 or 13.41, however, it is possible to start the induction motor without a starting resistor by the following method. At first, the primary windings of the induction motor are shorted, and the motor is accelerated to the speed control range by using the converter for excitation connected to the secondary windings of the motor. Next, the primary windings of the motor are opened and connected to the AC power source. Finally, AC power with the slip frequency is supplied to the secondary windings by the converter for excitation, and the starting operation is finished. Then, the induction motor speed is adjusted by commands.

Figure 13.43 shows the brushless Scherbius control which dispenses with maintenance and inspection for the slip-ring area of the induction motor (Noda et al., 1974). This brushless speed control is realized by using a cascade connection of two wound-rotor induction motors. The secondary converter is connected to the stator of IM2 as shown in the figure. The configuration of the converter is the same as that in Figure 13.38. Of course, converters in Figures 13.39 to 13.42 can also be used. In the cascade connection, both rotors are mechanically coupled and the phase sequences are electrically inversely connected. The cascade induction motor acts as one induction motor with a pole of (P1 + P2), where P1 and P2 are poles of IM1 and IM2, respectively.

Figures 13.44 and 13.45 show Kraemer control which controls the speed of the induction motor by regenerating the slip-power to the mechanical shaft. In Figure 13.44, the induction motor speed can be adjusted by regulating the induced voltage of the DC motor with the field current (Honda et al., 1966). In this

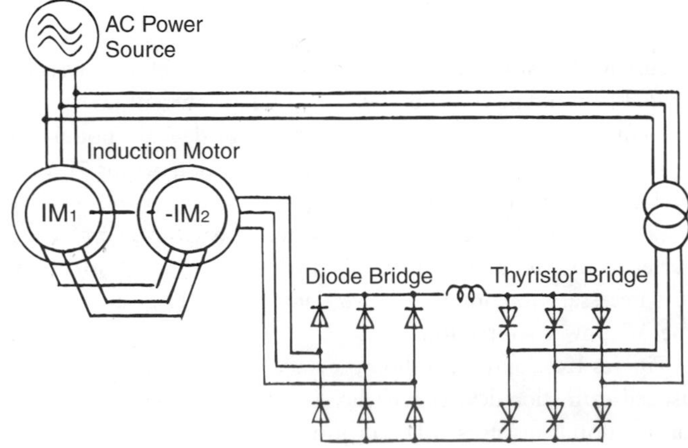

Figure 13.43 Brushless Scherbius control.

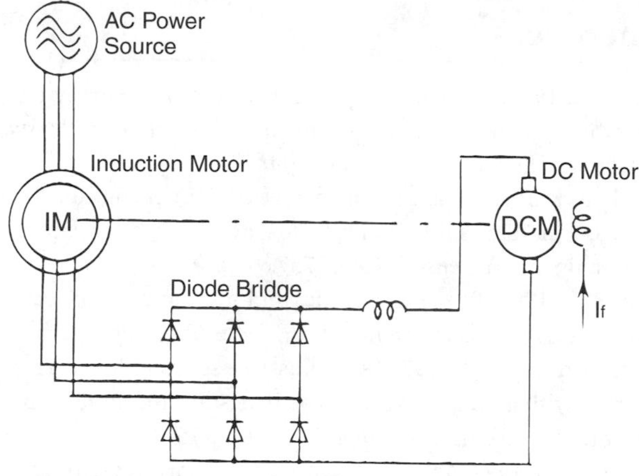

Figure 13.44 Static Kraemer control.

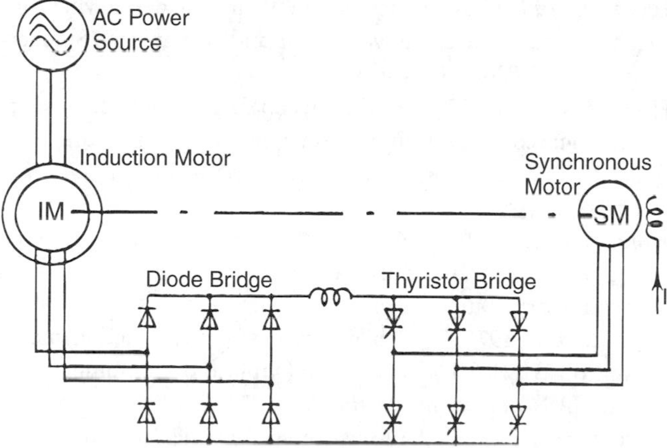

Figure 13.45 Commutatorless Kraemer control.

case, maintenance and inspection for the commutator of the DC motor is very complicated. Therefore, this is a weak point if labor is to be saved. Figure 13.45 shows a scheme in which the DC motor in Figure 13.44 is replaced by a commutatorless motor consisting of a synchronous motor and a thyristor bridge (Hori et al., 1976). Since a commutatorless motor is used, the scheme is called the "commutatorless" Kraemer control and it has better maintainability compared with the conventional DC motor Kraemer control shown in Figure 13.44. The induction motor speed can be adjusted by regulating the field current of the synchronous motor. But, the speed is generally controlled by regulating the firing angle of the thyristor bridge. Since the induced voltage frequency of the synchronous motor varies in proportion to the induction motor speed, a gate pulse generator which can regulate the firing angle following the frequency change is required. The commutatorless Kraemer control can continue normal operation without interruption even during instantaneous power failure, while the conventional Scherbius control cannot operate through instantaneous power failure. Even if the power supply to the synchronous motor field circuit is lost by power failure, the large field time constant (second order) allows the field current to continue flowing. So, the induced voltage of the synchronous motor can be kept (slightly decreased) even during power failure, and commutation failure of the thyristor bridge is prevented. Therefore, even during instantaneous power failure, stable continuous operation is possible. In this case, it is necessary to suppress the overvoltage generated in the secondary windings of the induction motor at instantaneous power failure.

In the conventional Scherbius control shown in Figure 13.38 or 13.39, commutation failure of the thyristor bridge occurs during the instantaneous power failure, because the AC power source voltage, or commutation voltage of the thyristor bridge, drops. Moreover, an extremely high overvoltage is induced in the secondary windings of the induction motor during the power failure and at recovery. Because of the overvoltage and commutation failure, an overcurrent flows in the secondary circuit and the operation must be stopped for protection. And there is a fear that the power converter consisting of the diode and thyristor bridges may be damaged by the overvoltage.

Figure 13.46 shows the power-failure-free static Scherbius control by which continuous operation is possible even during instantaneous power failure (Honbu et al., 1977). When the power failure occurs, it is rapidly detected. And then, the firing angle of the thyristor bridge is shifted to the maximum value (about 150 degrees) and the current in the thyristor bridge is reduced to zero by the power failure detection signal. After this operation, the gate firing signals of the thyristor bridge

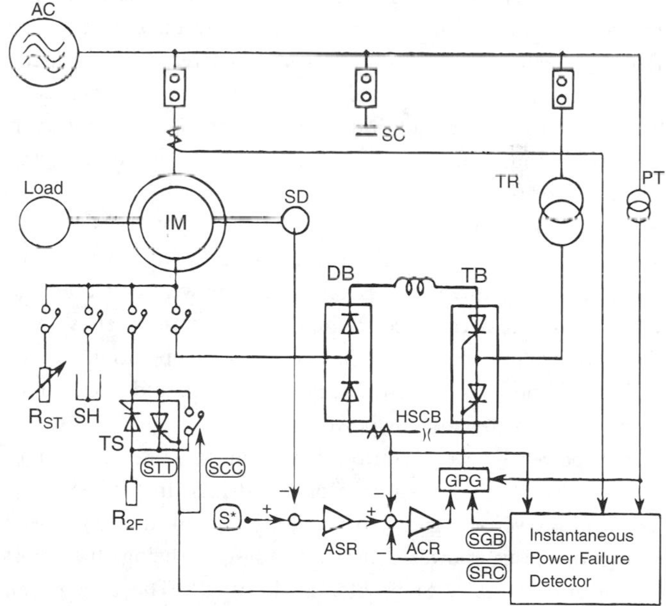

AC; AC Power Source
Load; Pump, Fan, Fly-Wheel etc.
SD; Speed Detector
PT; Potential Transformer
TB; Thyristor Bridge
SH; Short Circuit Equipment
TS; Thyristor Switch
R_{2F}; Resistor for Reducing Overvoltage
ASR; Automatic Speed Regulator
S*; Speed Command
SCC; Signal for Closing Contactor
SBG; Signal for Gate Block
SC: Power-Factor -Correcting Capacitor
IM; Induction Motor
TR; Secondary Exciting Transformer
DB; Diode Bridge
R_{ST}; Starting Resistor
HSCB; High-Speed Circuit Breaker
GPG; Gate Pulse Generator
ACR; Automatic Current Regulator
STT; Signal for Turning -On Thyristor
SRC; Signal for Reducing Current

Figure 13.46 Power-failure-free static Scherbius control.

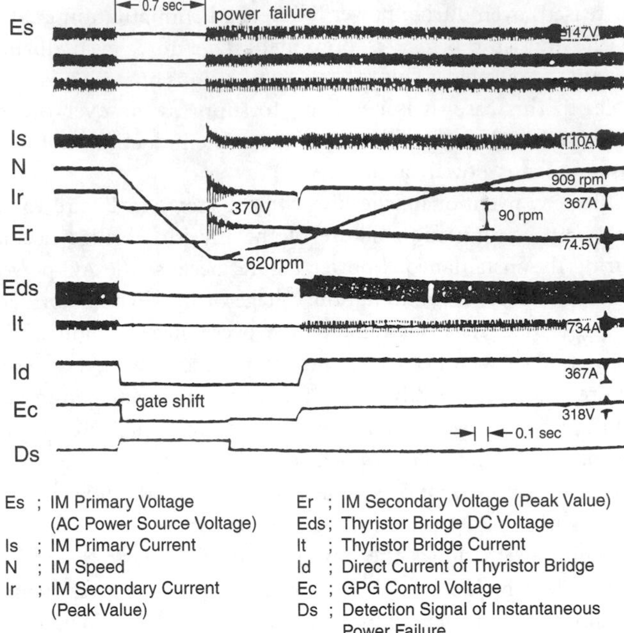

Figure 13.47 Operation oscillogram in power-failure-free static Scherbius control.

are blocked. The thyristor switch, connected to the induction motor secondary windings, is closed at the instant of power failure detection. With this operation, the overvoltage induced during the power failure and at recovery can be suppressed. To execute the above-mentioned operations in a stable manner, the power source for control circuits must be secured during power failure.

Figure 13.47 shows an oscillogram of the power-failure-free static Scherbius control. In this case, operational waveforms obtained when three phases of the AC power source are all opened and then recovered to the normal condition are shown. From the figure, it is seen that continuous operation can be realized without an extraordinary overvoltage and overcurrent even during instantaneous power failure.

Slip-power recovery control has been generally applied to industrial fields such as pump or fan drives. In the 1980s, its applications were expanded to power utilities, an adjustable speed pumped-storage power generation system being the most important example (Sugimoto et al., 1988). The system can improve the electric power network stability by adjusting the generator/motor speed and regulating the electric power consumption continuously in the pumping operation at night. Not only can the system be operated at the speed which gives the maximum efficiency in the generating mode, but it also easily regulates the active and reactive power to the electric power network. In the adjustable speed pumped-storage power generation system, the slip-power recovery control using a 72-arm cycloconverter, or DC link GTO converter as shown in Figure 13.40 or 13.41 is put into practice.

References

Abraham, 1986. Control of squirrel cage motor, a survey on the methods with regard to the history, *Evolution and Modern Aspects of Induction Machines Conf. Rec.*

Blaschke, F. 1971. Das Prinzip der Feldorientierung, die Grundlage für die TRANSVECTOR—Regelung von Drehfeldmaschinen, *Siemens-Z*, 45, p. 757.

Bose, B. K. 1993. Power electronics and motion control-technology status and recent trends, *IEEE Trans. Ind. Appl.*, 29(5):902.

Chattopadhyay, A. K. 1978. An adjustable speed induction motor drives with a cycloconverter type thyristor-commutator in the rotor, *IEEE Trans. Ind. Appl.*, IA-14(2):116–122.

Fujii, H. et al. 1993. Application of speed sensorless vector control to carrying machines: *Conf. Rec. IEEJ '93 Annual Meeting*, S10–25, in Japanese.

Hasse, K. 1968. Zum dynamischen Verhalten der Asynchronmaschine bei Betrieb mit variabler Standfrequenz und Standerspannug: *ETZ–A*, 89, H4 p. 77.

Hombu, M. et al. 1977. Continuous operation methods of static Scherbius control system at instantaneous power failure, *Proc. 2d IFAC Symp. Control in Power Electronics and Electrical Drives*, 547–558.

Honda, K. et al. 1966. 6200kW raw water transfer pumping equipment of Asaka purification plant, Tokyo Metropolitan Water Works Bureau, *Hitachi Review*, 15:60–65.

Hori, T. et al. 1972. The characteristics of an induction motor controlled by a Scherbius system (application to pump drive), *IEEE/IAS 7th Annual Meeting Rec.*, 775–782.

Hori, T. et al., 1976. Commutatorless Kraemer control system for large-capacity induction motors for driving water service pumps, *IEEE IAS 11th Annual Meeting Rec.*, 822–828.

Hosoda, H. et al. 1986. A new concept high performance large-scale AC drive system-cross current type cycloconverter fed induction motor with high-performance digital vector control, *IEEE/Annual Meeting Rec.*, 229–234.

IEEJ 1992. *High-performance Technology of Variable-speed Induction Motor Drive Systems*, Technical Report of IEEJ, Part II, No. 416, in Japanese.

Jifuku, Y. et al. 1987. Large capacity AC variable speed drive systems for electric power and general industrial use, *Hitachi Review*, 36(1):21–28.

Kin, T. 1992. Sensorless vector control of induction motor, *IEEJ Journal*, 112(3):167, in Japanese.

Leonhard, W. 1988. Field-orientation for controlling AC machine—principle and application, *IEE Conf. Rec.*, p. 277.

Miyairi, S. ed. 1981. *Variable Speed Drives of AC Motor by Power Electronics*, Publishing Dept. of Tokyo-denki-daigaku, in Japanese.

Mori, S. et al. 1995. Commissioning of 400 MW adjustable speed pumped storage system for Ohkawachi hydro power plant, *CIGRE Symp. Tokyo on Power Electronics in Electronic Power Systems*, May, pp. 520–04.

Morino, N. et al. 1979. Application of thyristor motor to steel rolling mill main drive system, *Hitachi Review*, 28(5):250–254.

Nagase, H. et al. 1983. Theory of vector control, *Conf. Rec. IEE J'83 Annual Meeting,* S8–2, in Japanese.
Nagase, H. et al. 1984. High-performance induction motor drive system using a PWM inverter, *IEEE Trans. Ind. Appl.,* IA–20(6):1482.
Nagase, H. et al. 1987. A design method of current control loop on vector control of induction motor, *IEEJ Trans.,* 107D(12):1491, in Japanese.
Noda, J. et al. 1974. Brushless Scherbius control of induction motors, *IEEE IAS 9th Annual Meeting Rec.,* 111–118.
Noda, J. et al. 1981. Speed control systems for wound-rotor induction motors, *Hitachi Review,* 30(5):263–268.
Ohmae, T. et al. 1993. Hitachi's role in the area of power electronics for transportation, *Proc. IEEE/IES IECON'93,* 714–718.
Okuyama, T. 1991. Current status and application example of speed and voltage sensorless vector control, *Conf. Rec. IEEJ '91 Annual Meeting,* S9–15, in Japanese.
Okuyama, T. et al. 1986. A high performance speed control scheme of induction motor without speed and voltage sensors, *IEEE IAS'86 Conf. Rec.,* p. 106.
Okuyama, T. et al. 1990. A simplified vector control system without speed and voltage sensors—effect of setting errors of control parameters and their compensation, *Scripta Technica,* p. 129.
Saito, K. et al. 1987. AC drive systems for rolling mills, *Hitachi Review,* 36(1):13–20.
Sugi, K. et al. 1983. A micro computer-based high capacity cycloconverter drive for main rolling mill, *Proc. 1983 IPEC-Tokyo,* 744–755.
Sugimoto, O. et al. 1988. Developments targeting 400 MW class adjustable speed pumped hydro plant and commissioning of 18.5MW adjustable speed hydro plant, *ASME 5th Int. Symp. Hydro Power, Fluid Machinery,* FED-68:71–79.
Sugiura, Y. et al. 1990. Fully digital servomotors common to SM/IM, *Conf. Rec. Semiconductor Power Conversion Committee,* SPC-90-48, in Japanese.
Sukegawa, T. et al. 1991. A multiple PWM GTO line-side converter for unity power factor and reduced harmonics, IEEE IAS'91 Conf. Rec. p. 279.
Ueda, A. et al. 1983. GTO inverter for AC traction drives, *IEEE Trans.,* IA-19(3):343–348.
Zimmermann, P. 1977. Super-synchronous static converter cascade, *Proc. 2d IFAC Symp. Control in Power Electronics and Electrical Drives,* 559–566.

13.4 Synchronous Motor Control Systems

Takamasa Hori

Torque-Speed Characteristics

A synchronous motor (SM) is defined as a machine in which the average speed of normal operation is exactly proportional to the frequency of the system to which it is connected. The synchronous rotating speed N_0 is

$$N_0 = \frac{60f}{p}\,[\text{rpm}]$$

where f is the frequency of the SM terminal voltage and p is the number of pole pairs.

Figure 13.48 shows the torque-speed characteristics of an SM when the frequency f is varied by a frequency changer such as an inverter or cycloconverter. If the frequency f is adjusted continuously, the motor speed can be controlled smoothly.

Figure 13.49 shows both the open loop and closed loop control configurations. The open loop control system is used for the precise speed control of a multi-parallel SM with a permanent

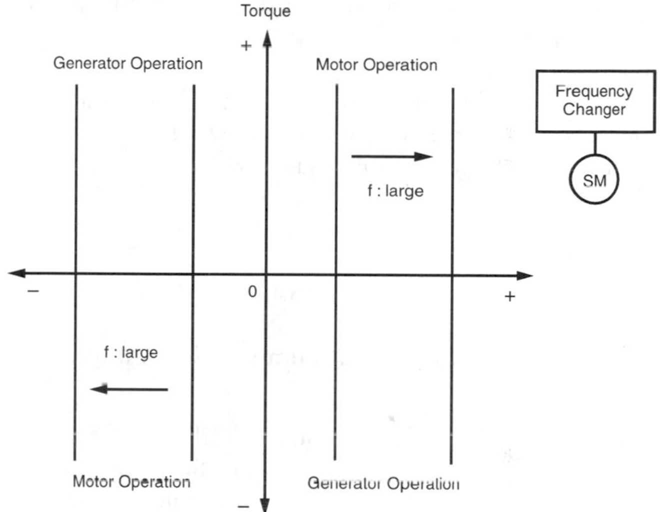

Figure 13.48 Torque-speed characteristics of a SM.

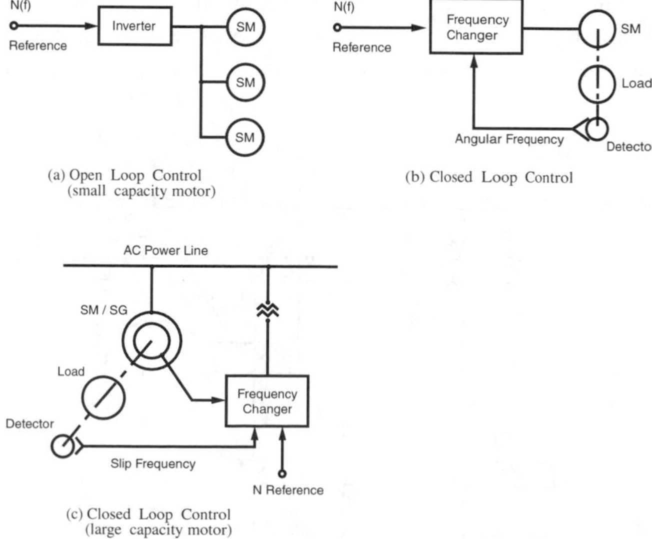

Figure 13.49 Open loop and closed loop control systems.

magnetic field, controlled by one set of inverters. The typical applications for this system are pot motor drives in the textile industry, and centrifugal drives. This system is effective in cases where the load characteristics are clearly known, and the motor normally rotates at a certain frequency. The closed loop control system is indispensable in cases where rapid acceleration and deceleration are required. The detector is coupled to the mechanical shaft and generates an angular frequency signal, corresponding to the field pole position, which is needed to acquire a timing signal for controlling the frequency changer of the inverter or cycloconverter type.

There are two types of the closed loop control systems: the primary frequency control system and the secondary frequency control system. The former system, when the frequency changer is composed of a thyristor current source type frequency changer, is called a thyristor motor. In the latter case, the SM/SG has an AC exciter composed of a variable frequency changer, instead of DC exciter, and is used for variable speed operation in a pumped storage power plant.

The transistor motor is a small capacity synchronous motor combined with a transistor inverter and widely used for home electric appliances and OA/FA equipment.

Speed Control of a Synchronous Motor

Thyristor Motor or Commutatorless Motor

The thyristor motor may be considered as a commutatorless motor in which the function provided by the brushes and commutator of a DC motor is replaced by the thyristor current source type frequency changer and the field pole position detector (distributor).

To provide this motor with characteristics equivalent to those of a DC motor, it is necessary that the motor has polyphase windings and each arm of the frequency changer connected to each phase of the motor performs on-off operations in accordance with the signals delivered by the detector.

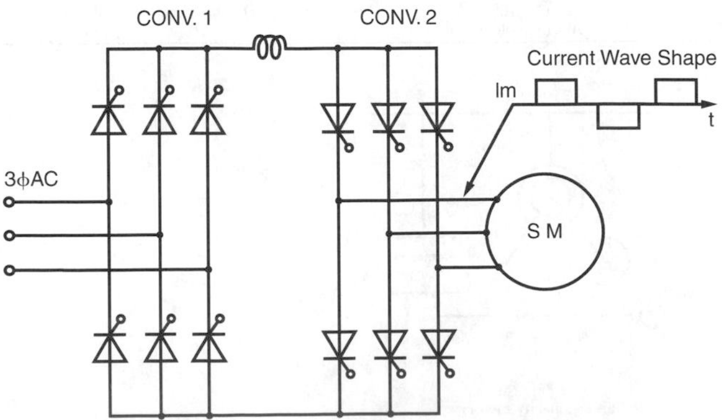

Figure 13.50 Main circuit of a thyristor motor (6-arm current source type inverter).

Figure 13.50 shows a typical main circuit for a thyristor motor with a 6-arm current source type inverter (CONV.2). In practice, a three-phase synchronous motor is used by permitting a certain amount of pulsating torque. The current flowing into each phase of the motor (SM) is a square-wave of a 120-degree electric angle. The counterelectromotive force of the SM is proportional to the rotating speed and therefore is small at motor start-up. During low-speed operation, its voltage cannot perform commutation for the thyristor in CONV.2. To perform stabilized commutation of CONV.2 at low speed, the following method, shown in Figure 13.51, is applied.

1. The thyristor to be commutated in CONV.2 is determined by the relative position between the field pole position and the armature winding; i.e., the signal delivered by the detector (distributor).
2. Commutation from one thyristor to another is performed after shifting the control angle α of CONV.1 toward the inverter side ($\alpha > 90°$) in order to reduce the DC current to zero, and after completion of commutation, the DC current is increased by CONV.1. That is, during every commutation of each thyristor in CONV.2, the DC current is forced to zero by CONV.1.
3. When the motor develops a rotating speed which is more than 10% of the rated speed, natural commutation by a counterelectromotive force of the SM becomes possible.

Figure 13.52 shows both the main circuit and control circuit configurations of a thyristor motor used for the control of the main exhaust fan in blast furnace sintering equipment. In this system, the detector is connected to the motor terminals and detects its field pole position by measuring the motor terminal voltage.

Large Capacity Synchronous Motor Drive

There are two types of frequency changer for variable speed drive in AC motors. One is the inverter, which is used for industrial applications of small and medium capacity AC motors, and the other is the cycloconverter, which is used for large capacity AC motor drives such as an iron rolling mill drive in the 5–10 MW class. In the cycloconverter type control system, it is impossible to produce a frequency higher than that of the power supply because the frequency produced results from a power supply composed of the combined waveforms of the original power supply. The maximum frequency, in practical situations, is determined by the distortion of the voltage wave shape and the torque ripple acceptable to the SM and load.

Approximately 20–30Hz in output frequency is produced in a 36-arm cycloconverter with circulating current control, and below 20 Hz, with circulating current-free control.

Figure 13.53 shows the main circuit of a 36-arm voltage source type cycloconverter with circulating current-free control which can produce three-phase sinusoidal armature current for a motor. Armature current is effectively converted to torque with a minimum of torque ripple.

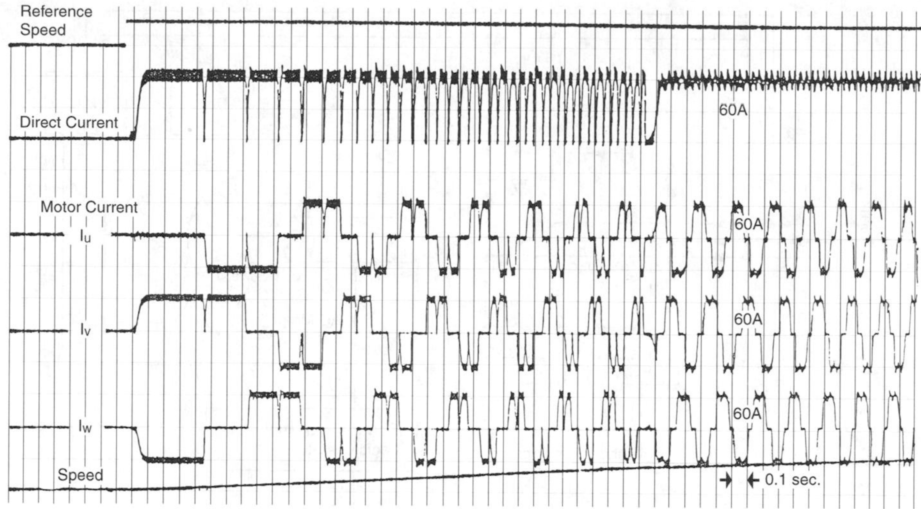

Figure 13.51 Thyristor motor current at low speed for starting.

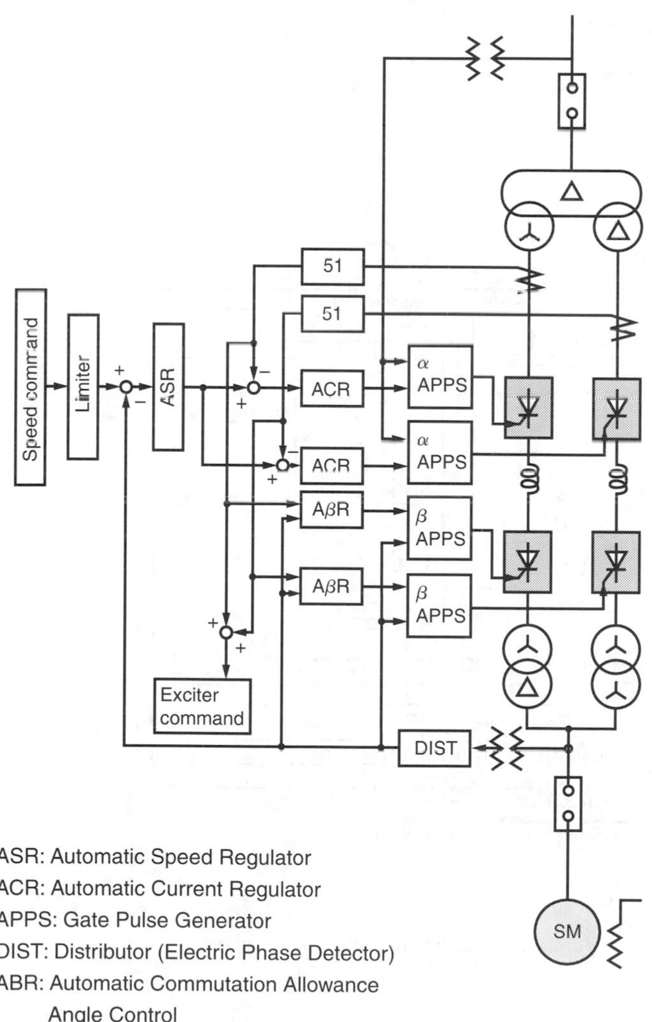

Figure 13.52 Main circuit and control circuit configuration of a thyristor motor.

Variable Speed Drive of a Pumped Storage Power Plant

In pumped storage hydroelectric power plants, the operating speed of the conventional SM/G system with a DC field exciter is fixed at synchronous speed and cannot be controlled. The thyristor motor starting system is used for easy starting and braking of a SM/G as shown in Figures 13.50 and 13.51.

The total operating efficiency is poor because of the difference in efficiency of the water turbines between the pumping and generating stages. If the operating speed can be controlled, the total efficiency can be improved.

Figure 13.54 illustrates the driving system for a 400 MW class SM/G with a cycloconverter type AC exciter in practical use.

The frequency changer operating as an AC field exciter instead of a DC field exciter is connected to the rotor circuit through collector rings. Inverter type frequency changers are also available. The smaller the variable speed range around the synchronous speed, the smaller the capacity of an AC field exciter. In the pumped storage power plant, the variable speed range around the synchronous speed may be 10–20% of synchronous speed.

References

Furuya, S. et al. 1995. Large capacity GTO inverter-converter for double-fed adjustable speed system, *CIGRE Symp. Tokyo on Power Electronics in Electronic Power Systems* May, 530–04.

Jifuku, Y. et al. 1987. Large capacity AC variable speed drive systems for electric power and general industrial use, *Hitachi Review*, 36(1):21–28.

Kita, E. et al. 1993. Development of the 395 MVA, 330–390 min^{-1} adjustable speed generator-motor, *Proc. JSME-ASME Ind. Conf. Power Engineerings -93 (ICOPE-93)*, September, p. 115.

Mori, S. et al. 1995. Commissioning of 400 MW adjustable speed pumped storage system for Ohkawachi hydro power plant, *CIGRE Symp. Tokyo on Power Electronics in Electronic Power Systems*, May, 520–04.

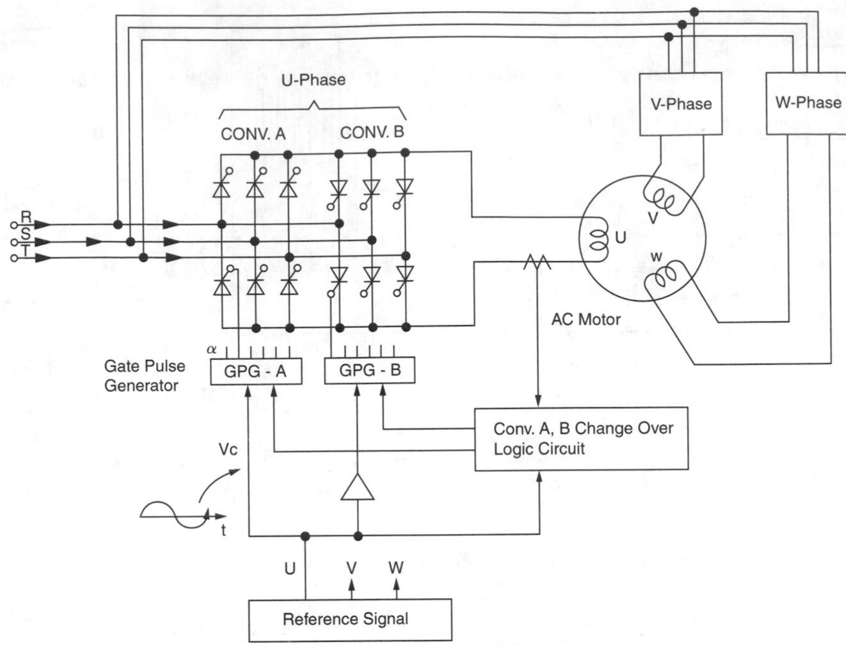

Figure 13.53 Main circuit of a thyristor motor (36-arm voltage source type cycloconverter).

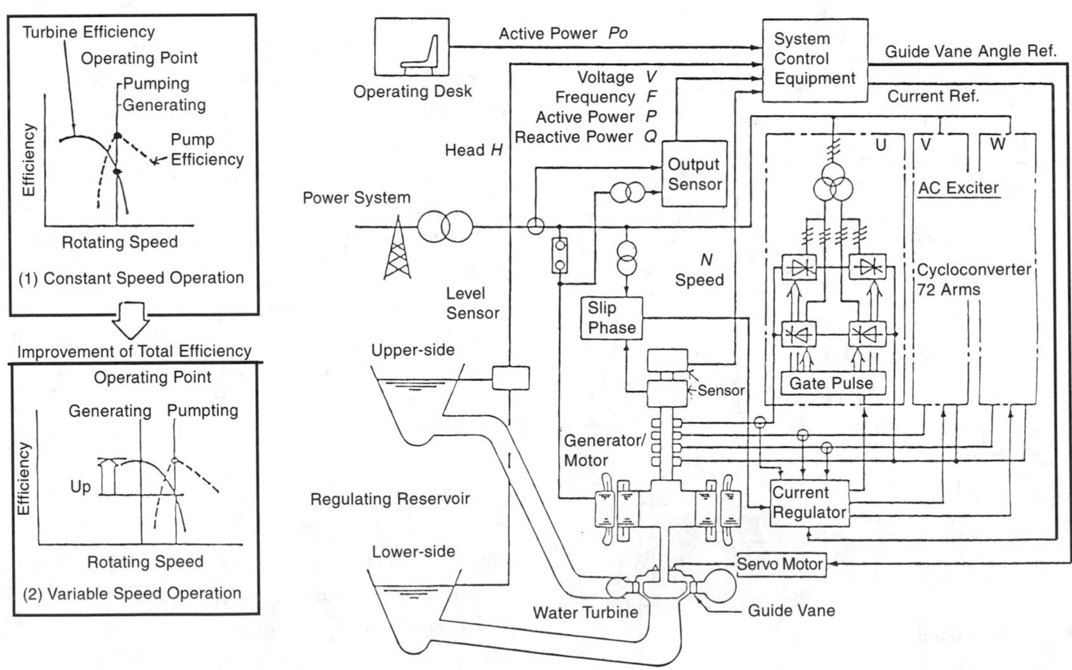

Figure 13.54 Driving system of a generator/motor for a pumped storage hydroelectric power plant.

Morino, N. et al. 1979. Application of thyristor motor to steel rolling mill main drive system, *Hitachi Review,* 28(5):250–254.

Saijo, T. et al. 1981. Characteristics of linear synchronous motor drive cycloconverter for maglev vehicle ML-500 at Miyazaki test track, *IEEE Trans.,* IA-17(5):533–543.

13.5 PM Synchronous Motor Control

M. F. Rahman and Khiang-Wee Lim

Historically, the control of the permanent magnet synchronous motor (PMSM) was based on the self-controlled synchronous motor drive (Harashima et al., 1979; Le-Huy et al., 1982). These were and are used in many applications with power levels beyond the rating of conventional DC motor drives and the controllability of induction motor drives. The situation changed with the advent of the rare-earth samarium cobalt magnets in late 1970s and the less rare, and hence less costly, neodymium-iron-boron magnets in the mid 1980s. With these magnets it became possible to develop a rich variety of permanent magnet motors utilizing the same principles of control (Binns, 1984; Chalmers, 1985; Jahns, 1986). For small motors, up to a few kilowatts, permanent magnet excitation is cheaper than the conventional electromagnetic excitation. At the same time, there is considerable reduction in motor size (volume) and inertia with attendant increases in specific power and torque (Miller, 1989). These gains come from the high-energy product of the new magnet materials. Another advantage is that no mechanical brushes are required either in the armature or in the field circuits. These advantages, coupled with the availability of a number of gate controlled transistor switches at the low power levels, resulted in widespread adoption of the PMSM for many applications in machine tools, robotics, high density disk drives and other drives used in automation applications.

The PMSM drive can easily be given the characteristics and ease of control of a conventional DC machine without the problems of the latter. The underlying principle behind this requires the rotor position of the motor to generate the switching signals for an inverter. This technique is called self-control or self-synchronization of an inverter-fed drive (Slemon, 1974). For simple spindle drives, the rotor position sensors can consist of just three discrete sensors of Hall or optocoupler type devices for a three-phase motor. For more accurate point-to-point positioning, the sensor needs to have a much higher accuracy. This sensor also serves as the position synchronizer for the inverter.

The synchronous motor can be driven from a voltage source or a current source. In either case, when controlled in the self-synchronized manner, it behaves as a DC motor, slowing down slightly with load, producing a torque proportional to the current and a speed proportional to the voltage supplied to the motor (Morimoto, 1990). The section on **Construction and Operation** begins with a description of various rotor structures of the permanent magnet synchronous motor. This is followed by a description of the mechanism of torque production and the operating principle for DC motor-like operation of the PMSM. The dynamic model of the motor will be described in **PM Synchronous Motor Model** while **Field-oriented Control** outlines a technique for the dynamic control of a PMSM.

Construction and Operation

The typical permanent magnet synchronous motor has a stator similar to that of an induction motor with three-phase windings (Slemon, 1994; Slemon, 1992). Unlike the induction motor, the field flux is provided by permanent magnets in the rotor. Figure 13.55 shows the cross section of a permanent magnet motor. Permanent magnet synchronous motors are generally classified according to how the magnets are embedded in the rotor. A

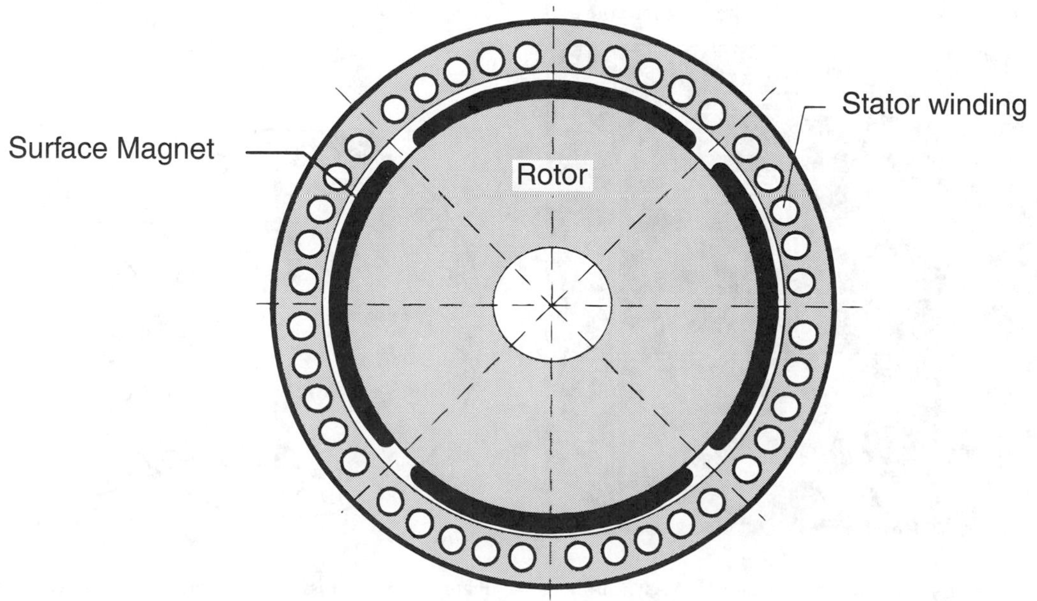

Figure 13.55 Cross-sectional view of surface magnet permanent magnet synchronous motor.

popular construction which lends itself easily to control is the surface magnet synchronous motor in which the magnets are attached to the surface of the laminated rotor by epoxy adhesives. This construction results in a large air gap and low stator inductances which are constant for all positions of the rotor.

As the rotor moves, the moving flux in the air gap induces a voltage (*the back emf*) in each of the stator windings. The large air gap of the motor effectively decouples the stator and the rotor fields so that the air gap flux, and hence the back emf, is not affected by currents in the stator windings (i.e., *low armature reaction*). As will be shown later, this severely limits the maximum speed of the surface magnet motor.

PMSMs are also classified by the shape of the back emf waveforms (Miller, 1989). By distributing the stator windings and by adjusting the span of the rotor magnets, largely sinusoidal back emf waveforms can be induced in the stator winding when the rotor runs at a constant speed. Figure 13.56 shows the measured back emf of a motor with surface mounted magnets and sinusoidal distribution of stator windings. It shows some deviation from the ideal sinusoid, with the back emf waveform for each stator phase consisting of a fundamental and some odd-order harmonics, the amplitudes of which depend on the speed of the rotor.

A variant from the sinusoidal permanent magnet motor is the trapezoidal emf motor in which the stator windings are distributed evenly rather than sinusoidally. The rotor magnet spans are nearly full pitch. Figure 13.57 shows the ideal back emf waveform of such a motor. It is a trapezoid with a flat top for two thirds of each half cycle. This construction simplifies some of the control electronics for constant speed applications such as spindle drives (Le-Huy, 1985).

In the interior permanent magnet (IPM) motor, also known as the buried magnet motor, the magnets are embedded inside the rotor iron. Figure 13.58 shows the cross sections of two popular interior magnet motors. With this construction, the air gap is small. As a result, the magnetic flux due to the stator current can significantly modify the net flux in the air gap. This armature reaction is more prominent along the magnet pole axis

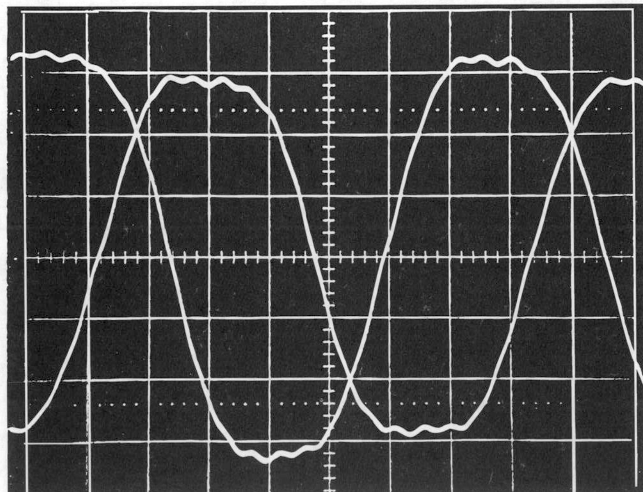

Figure 13.56 Back emf of typical surface magnet PM synchronous motor.

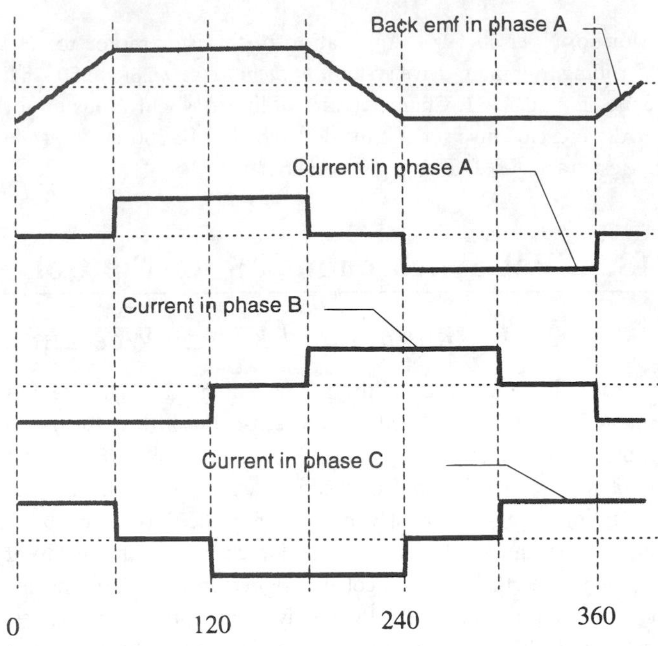

Figure 13.57 Back emf and currents in ideal trapezoidal permanent magnet synchronous motor.

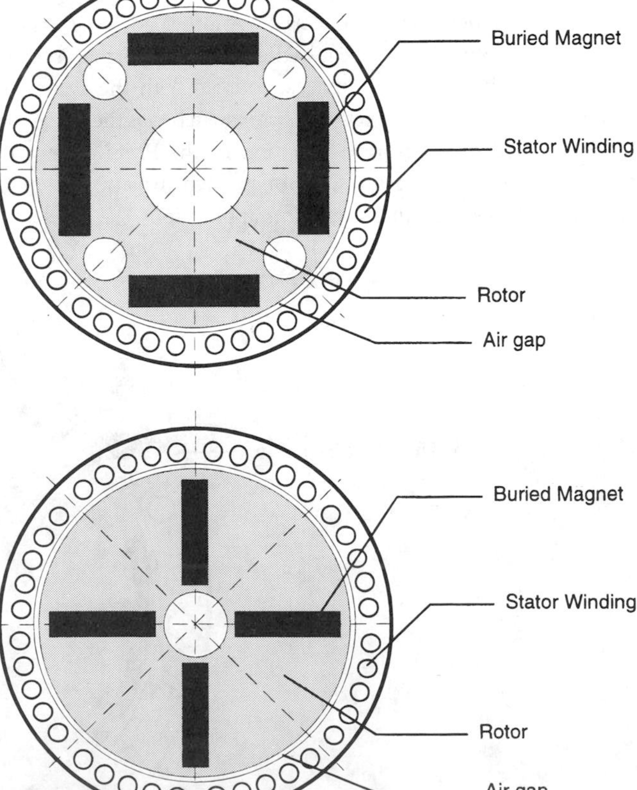

Figure 13.58 Cross-sectional view of two interior magnet permanent magnet synchronous motors.

(also called the *d-axis*) where the reluctance is small, and less along the quadrature axis (also called the *q-axis*) where the reluctance is large. This is known as the saliency of the poles. As a result of the small air gap, the air gap flux can be reduced by current control (Jahms, 1987; Sneyers, 1985). This allows an operating mode whereby the top speed of the motor can be increased by as much as five times the base speed and is one of the main features of this motor (Maeminn and Jahns, 1991). In most practical motors, this pole saliency also makes the back emf highly non-sinusoidal (see Figure 13.59), even for a motor with stator windings which has a sinusoidal distribution. In such machines, the tendency of magnetic paths to minimize reluctance provides another torque mechanism, called reluctance torque which may be exploited through control.

A variety of other specialist constructions exist, designed to optimize particular features. Examples are the axial flux disc motor (Rahman et al., 1994; Caricchi, 1992), shown in Figure 13.60, which minimizes stator inductance and rotor inertia and the imbricated rotor motor which concentrates the air gap flux so as to use less magnet material or low-cost ferrite magnets (Binns and Wong, 1984).

The most common magnetic materials in use today are those from the rare-earth family, such as samarium cobalt, the less expensive neodymium-iron-boron and the inexpensive ceramic magnets such as ferrites. Magnetic circuit design for the motor is strongly influenced by the demagnetization characteristics of the magnet material used (Zijlstra, 1984). Figure 13.61 shows the demagnetization characteristics at two temperatures for NdFeB and ceramics. Important factors to be considered in magnetic circuit design include the worst-case temperature, maximum demagnetizing mmf encountered while in operation, the operating point of the magnets and the saturation of the core material.

The upper temperature limit of the new high-energy magnets (NdFeB) is from 100–140°C. This is not a constraint on motor design provided the motor is not used in a region with a high ambient temperature. The present generation of magnetic materials, such as the sintered NdFeB, have remanent flux density ($B_r \approx 1.2$ T) which allows for the use of conventional core material without any need for flux concentration. In addition, the coercive force ($H_c \approx 970$ kA/m) is such that, for a well-designed magnetic circuit, there is no danger of demagnetization even when the peak current capacities of the appropriate switching devices are fully utilized. In modern machine design, the interplay of these design factors can be analyzed with the help of a two-or three-dimensional finite element analysis package (Cavicchi, 1993; Binns, 1993). Figure 13.62, for instance, shows the magnetic field within an interior magnet machine, computed from a finite element analysis.

The total losses in a PMSM is typically only about 50–60% that of a similarly sized induction motor (Slemon, 1994) and is largely confined to the outer stator from where it is easily dissipated to the environment. The resistivity of the magnetic material is about 85 times that of copper so that losses in the rotor magnets are small.

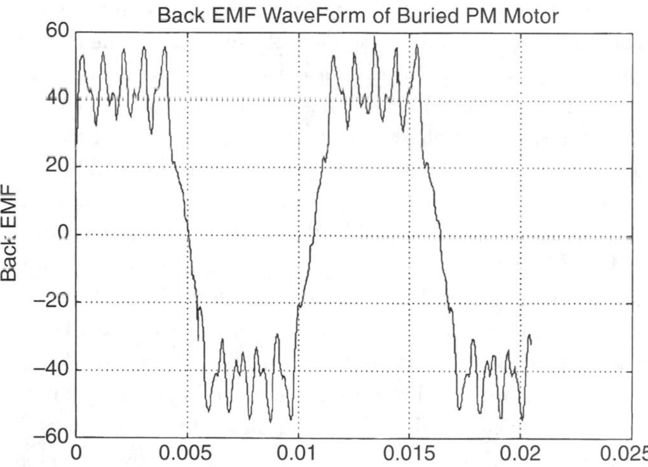

Figure 13.59 Back emf of typical buried magnet PM synchronous motor.

Figure 13.60 Components of a low-inertia axial flux motor.

Voltage Source Operation—Steady-State Model

Consider a conductor in one of the stator windings in a surface magnet motor, shown in Figure 13.55. As the rotor rotates at a steady speed, the stator conductors experience a magnetic

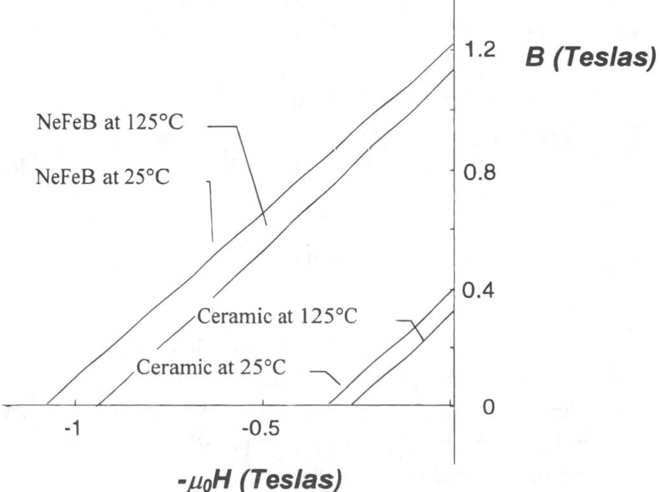

Figure 13.61 Demagnetization characteristics for permanent magnet materials.

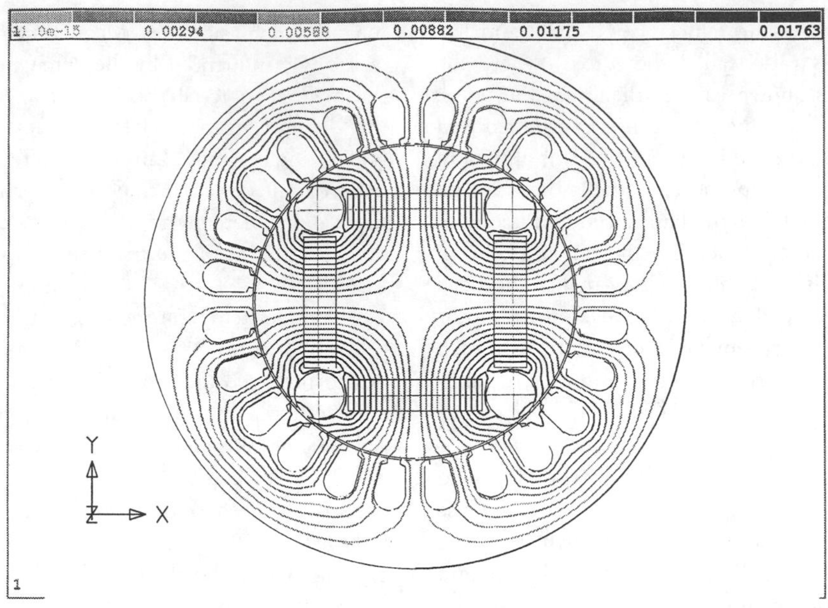

Figure 13.62 Magnetic field of an interior magnet machine from a finite element analysis.

field, B *(in teslas)*, moving past. A torque is thus generated in each of the current carrying conductors according to the familiar expression:

$$T = i(B \times l)r \tag{13.45}$$

where i is the current in the conductor, l is the length of the conductor and r is its radial distance from the center of the shaft. The net torque is the sum of contributions from all the conductors in the stator.

In an ideal 3-phase motor, the distribution of the stator windings and the magnet span in a surface magnet motor results in a sinusoidal back emf for each phase, which is displaced from that of each other phase by 120° electrical. Consider steady-state operation where the rotor runs at velocity ω_r. Then the instantaneous value of the back emf for each phase is

$$\begin{aligned} e_A &= E_m \sin\theta \\ e_B &= E_m \sin\left(\theta - \frac{2\pi}{3}\right) \\ e_C &= E_m \sin\left(\theta - \frac{4\pi}{3}\right) \end{aligned} \tag{13.46}$$

where $\theta = \omega t + \delta$ is the angle of the rotor flux axis with respect to a stationary stator reference frame, $\omega = p\omega_r$ is the frequency of the supply voltage, t is time and p is the number of pole pairs. The amplitude of the back emf for each phase, E_m, is proportional to the rotor flux and the velocity, i.e.,

$$E_m = K\phi\omega \tag{13.47}$$

where ϕ is the rotor flux (in Wb) and K is a constant reflecting stator parameters such as the length and number of conductors.

Figure 13.63 shows the equivalent circuit for one phase of such a motor with a sinusoidal voltage, V_{rms}, applied to each phase of the motor. Neglecting harmonics, sinusoidal currents of frequency ω flow in each phase winding. Figure 13.64 shows the corresponding phasor diagram. Here it is assumed that the air gap of the surface magnet motor is uniform. The phase inductance, L_s, is then constant and independent of rotor position. It is convenient to consider that the current has two orthogonal components with one component, I_q (along the *q-axis*), aligned with the back emf. The other component, I_d, is along the magnetic pole (d-) axis of the rotor. From Figure 13.64, the rms value of the current in phase with the back emf is $V/X_s \sin\delta$ where X_s is the synchronous reactance of the motor and the stator resistance is neglected. δ is the angle between the sinusoidal applied voltage and the sinusoidal back emf. The power developed in each phase of a motor is the product of the rms value of the back emf and the rms value of the current which is in phase

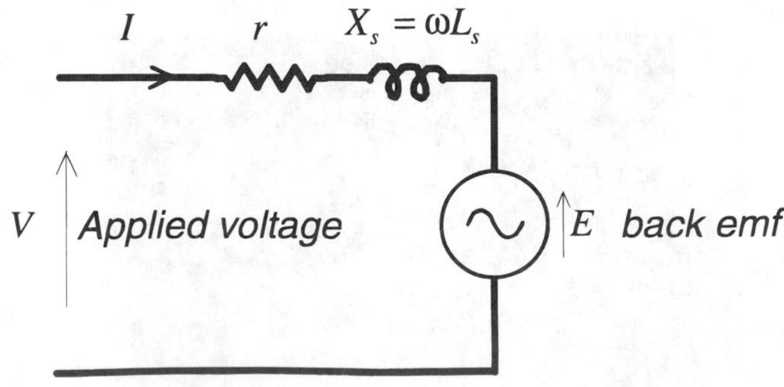

Figure 13.63 Equivalent circuit for one phase of PMSM.

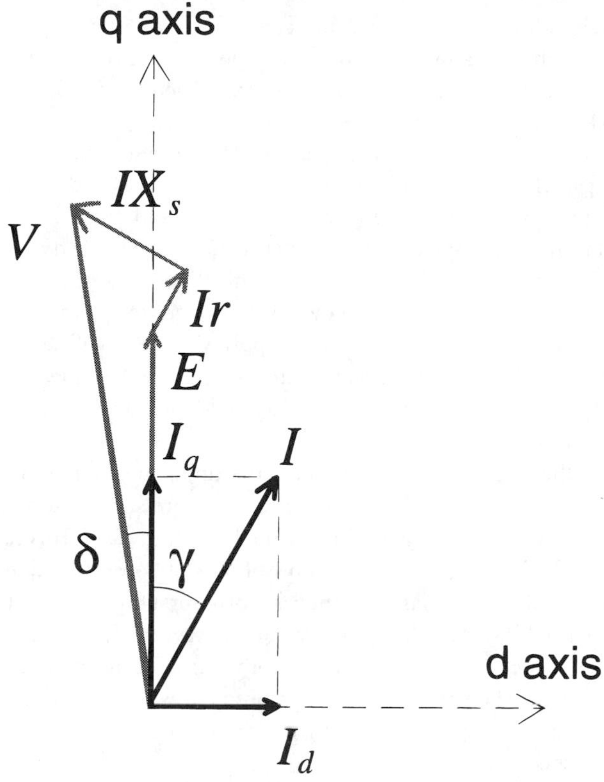

Figure 13.64 Phasor diagram for surface magnet motor.

with the back emf. Electrical power, EI_q, is the developed mechanical power, i.e., the product of rotor angular velocity and torque. Thus the total torque for three phases, in the case of the sinusoidal surface magnet motor, can be written as (Slemon, 1992).

$$T = \frac{3}{\omega_r} E \frac{V}{X_s} \sin\delta = \frac{3p}{\omega} E \frac{V}{\omega L_s} \sin\delta = \frac{3pK\phi}{\sqrt{2}L_s} \frac{V}{\omega} \sin\delta \; Nm \tag{13.48}$$

where

$E = E_m/\sqrt{2}$ = rms value of the back emf in each phase (volts)
V = rms supply voltage (volts)
ω_r = rotor angular velocity
$\omega = p\omega_r$ = supply frequency (rad/sec)
p = number of pole pairs
L_s = synchronous inductance (henry's)
δ = angle between the input voltage phasor and the back emf phasor, the power angle (radians)

In contrast, the pole saliency in an interior magnet motor produces significantly different inductances along the magnet pole (d-) and quadrature (q-) axes. Thus the reactance in the equivalent circuit of Figure 13.63 is a function of rotor position. Ignoring harmonics in the back emf waveforms and assuming that the inductance variations are sinusoidal with position, the phasor diagram of Figure 13.65 can be used to show the relationship between fundamental components of the currents and voltages.

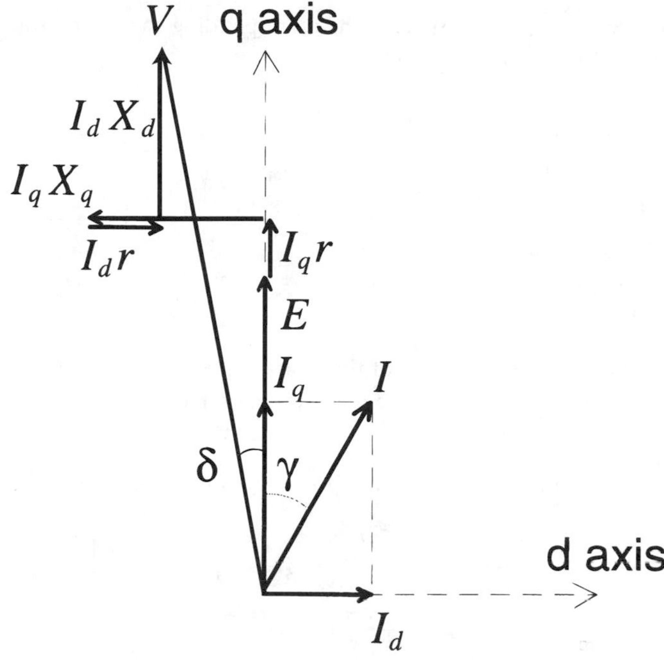

Figure 13.65 Phasor diagram for interior magnet motor.

In terms of the *d-axis* stator inductance, L_d, and the *q-axis* stator inductance, L_q, the torque of the interior magnet motor can be shown to be (Slemon, 1992).

$$\begin{aligned} T &= \frac{3}{\omega_r}\left[\frac{EV}{\omega L_d}\sin\delta + \frac{V^2}{2}\left(\frac{L_d - L_q}{\omega L_d L_q}\right)\sin 2\delta\right] \\ &= 3p\left[\frac{K\phi}{L_d\sqrt{2}}\frac{V}{\omega}\sin\delta + \frac{V^2}{\omega^2}\left(\frac{L_d - L_q}{2L_d L_q}\right)\sin 2\delta\right] Nm \end{aligned} \tag{13.49}$$

This expression for torque is also based on an assumption of negligible stator resistance. The first term is the electromagnetic torque similar to that for the surface magnet motor. The second term represents reluctance torque which is due to the saliency of the rotor.

The expressions for steady-state torque developed in this subsection (Equations 13.47–13.49) together indicate that if the voltage-to-supply frequency ratio V/ω is kept constant, the motor will develop the same maximum torque at all speeds. This is an operating characteristic which the PMSM shares with the induction motor (Murphy and Turnball, 1989).

Current Source Operation—Steady-State Model

This subsection describes the self-synchronous mode of operation that leads to DC motor-like characteristics. If the phase currents supplied to the motor are synchronized with the back emf waveforms of each phase (shown in Equation 13.46) at a phase angle γ with respect to the corresponding back emf waveform, then we have a *current source drive* with

$$\begin{aligned} i_A &= I_m \sin(\theta - \gamma) \\ i_B &= I_m \sin\left(\theta - \frac{2\pi}{3} - \gamma\right) \\ i_C &= I_m \sin\left(\theta - \frac{4\pi}{3} - \gamma\right) \end{aligned} \tag{13.50}$$

where I_m is the amplitude of the phase current. The total developed mechanical power produced by the three phases is

$$Power = \left\{ \begin{array}{c} E_m \sin\theta \cdot I_m \sin(\theta - \gamma) + E_m \sin\left(\theta - \frac{2\pi}{3}\right) \\ \cdot I_m \sin\left(\theta - \frac{2\pi}{3} - \gamma\right) \\ + E_m \sin\left(\theta - \frac{4\pi}{3}\right) \cdot I_m \sin\left(\theta - \frac{4\pi}{3} - \gamma\right) \end{array} \right\} \tag{13.51}$$

Assuming that electrical power output is equal to the mechanical power output which is the product of rotor angular velocity and torque, we get

$$Torque = \frac{electrical\ power}{rotor\ velocity} = \frac{\frac{3}{2} E_m I_m \cos\gamma}{\omega_r} = \frac{3}{2} Kp\phi I_m \cos\gamma \tag{13.52}$$

Thus, the electromagnetic torque produced by the motor for this mode of operation is proportional to the amplitude I_m of the current, and is independent of the rotor position. This is similar to the characteristics of a conventional DC machine. The developed torque only has a DC component and no torque pulsations exist. In practice, the back emf waveforms are not exactly sinusoidal and as a result the torque output is dependent on rotor position, leading to significant torque pulsations (Le-Huy et al., 1985; Ng et al., (1988).

If the phase current I delivered to the motor is synchronized to be in phase with the back emf E, i.e., $\gamma = 0°$, then the highest torque per ampere is obtained. This is equivalent to placing the stator mmf at right angle to the rotor mmf. The choice of $\gamma = 0°$ defines the constant torque region of operation of the drive.

Note that the preceding description is independent of velocity. So long as the rotor position is known and the currents of Equation 13.50 are applied, the torque is unidirectional and constant. Thus starting a PMSM is not an issue if θ is available. As the motor approaches its base speed, the amplitude of the back emf approaches that of its voltage supply. The current amplitude can only increase further with a corresponding decrease in back emf E. For the PMSM (Equation 13.47) this requires a decrease in speed or air gap flux.

For small air gap machines, it is possible to reduce the back emf by advancing the input stator current (i.e., $\gamma < 0°$ in Equation 13.50), thus creating a component of phase current that *reduces* the air gap flux. As a result the speed can increase for the same back emf. This is the so-called constant power region of operation where torque is traded off for speed. However, this reduction of air gap flux is only possible when the reluctance of the airgap is small so that armature reaction can take hold.

From the phasor diagram of Figure 13.65, it is worth noting that forcing current to be in phase with the back emf ($\gamma = 0°$) also leads to less than unity power factor operation of the motor at constant velocity, because the supply voltage V will lead the current I. Advancing the stator currents ($\gamma < 0°$) will lead to a better overall power factor and hence is adopted when this is an overriding concern.

In the trapezoidal emf machine, the uniform distribution of the windings and the span of the magnets are such as to make the emf waveforms trapezoidal. If the back emf waveforms have perfectly flat tops with a duration of 120°, the ideal (i.e., zero torque pulsation) phase current waveform for a three-phase star-connected PMSM will be a quasi-square waveform, as indicated in Figure 13.57. This results in some simplification in drive electronics (Finney, 1988). The trapezoidal waveform motor is well suited to spindle type drives where the drive operates at fixed speeds.

To realize current source operation with maximum torque per unit current, a mechanism is required for delivering the currents specified in Equation 13.50 to the stator with $\gamma = 0$. The magnitude of the torque generated is then proportional to the stator current. The following subsection describes a current feedback system commonly used.

Current Feedback Control System

A typical mechanism for delivering the required stator current to the motor is shown in the block diagram of Figure 13.66. The diagram shows a feedback system where the measured signals are the current in each phase and the rotor position. An inverter acts as a power amplifier which converts a DC voltage supply into a 3-phase variable voltage source to the motor. The inverter is driven by a set of three current controllers (or two controllers for a balanced system) where the current controller for one phase is indicated. The reference signal to the current controller shown is the product of the magnitude of the desired current, I_m, and the sine of the rotor position, θ. The input to this current controller is the current error, the difference between the measured phase current and the desired phase current. Under ideal conditions, the current controllers adjust the output voltages

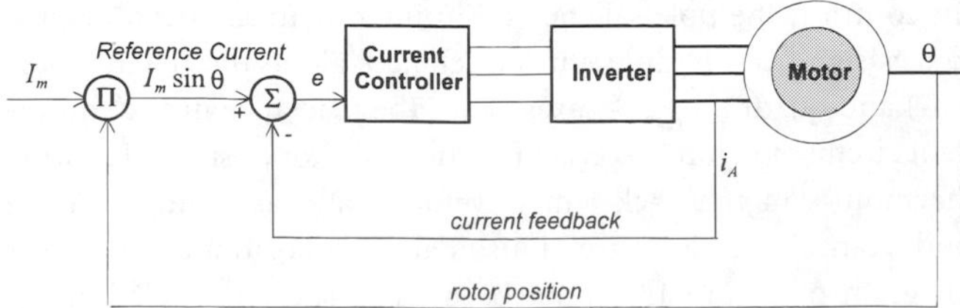

Figure 13.66 Block diagram of torque/current control scheme.

from the inverter to the motor so that there are no current errors and the stator currents become $I_m\sin\theta$, $I_m\sin(\theta - 2\pi/3)$ and $I_m\sin(\theta - 4\pi/3)$, respectively. This ensures that the stator currents meet the conditions for maximum electromagnetic torque (i.e., $\gamma = 0$).

The feedback system as a whole then serves as an ideal current source to the motor. The following paragraphs describe the function of each block in this feedback system.

Inverter and Current Controllers

Figure 13.67 is a block diagram of the major components of an inverter. The inverter consists of 3 pairs of complementary switches and associated drive circuits. Each pair of switches is connected to one phase of the motor at one end and to a DC voltage supply at the other end. The choice of switching device depends on the speed of switching and the power rating required. For low-power applications, power MOSFETs are commonly used. In addition, an inverter may have turn-on and turn-off snubber circuits and an overcurrent protection circuit. To prevent the two complementary switches from turning on at the same time, a hardware cross-over protection circuit is used to provide an interlock. Typically, the DC supply to the inverter is obtained from a three-phase AC source rectified by a six-pulse uncontrolled rectifier, followed by a simple LC filter circuit. Further detail on inverter topologies and design may be found in Section 12.3 and (Finney, 1988).

When the switching device in one leg of an inverter pair is turned on, the corresponding motor terminal is connected either to the positive or negative of the DC voltage supply. It is the task of the current controller to manage the switches of the inverter so that the desired current flow occurs in the motor. The simplest current controller consists of a comparator. Here each phase current is compared to the corresponding current reference defined in Equation 13.50. These current signals are shown in Figure 13.67. When the magnitude of the measured current is greater (less) than that of current reference by a preset amount (the hysteresis band), the corresponding inverter leg is switched to the negative (positive) terminal of the DC source. The hysteresis band is a lower bound on the current ripple magnitude. Conventional hysteresis controllers are usually implemented with analog circuits to achieve high bandwidth. This method of current control is popular because of its simplicity and good dynamic response. Drawbacks of this method include the varying switching frequency, hence an increased difficulty in controlling harmonic currents, and the difficulty in relating the hysterisis band to motor parameters, switching frequency and noise (Bose, 1990).

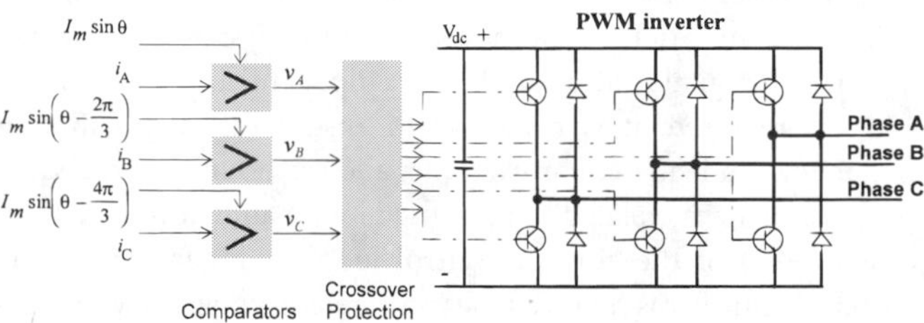

Figure 13.67 Block diagram of voltage source inverter and hysterisis controller.

In many low- and medium-power applications, it is now preferable to use a pulse width modulated (PWM) inverter (Finney, 1988), (Holtz, 1994). The operation of a simple PWM inverter is illustrated in Figures 13.68 and 13.69. The hysterisis controllers in Figure 13.67 are replaced by linear controllers. A typical linear controller is the proportional plus integral controller shown in Figure 13.70. Each controller output modulates the on time of

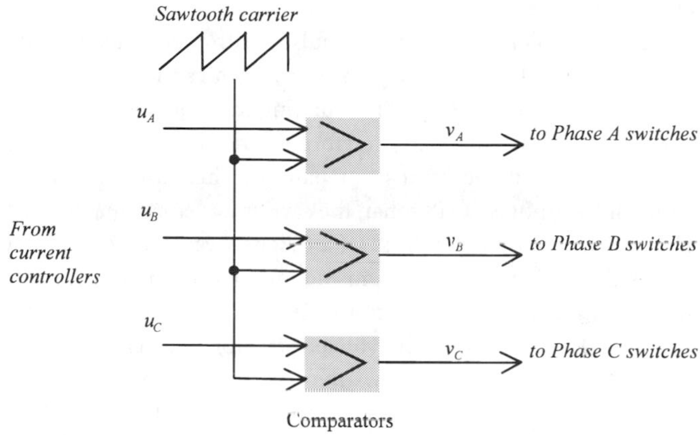

Figure 13.68 Pulse width modulation for inverter.

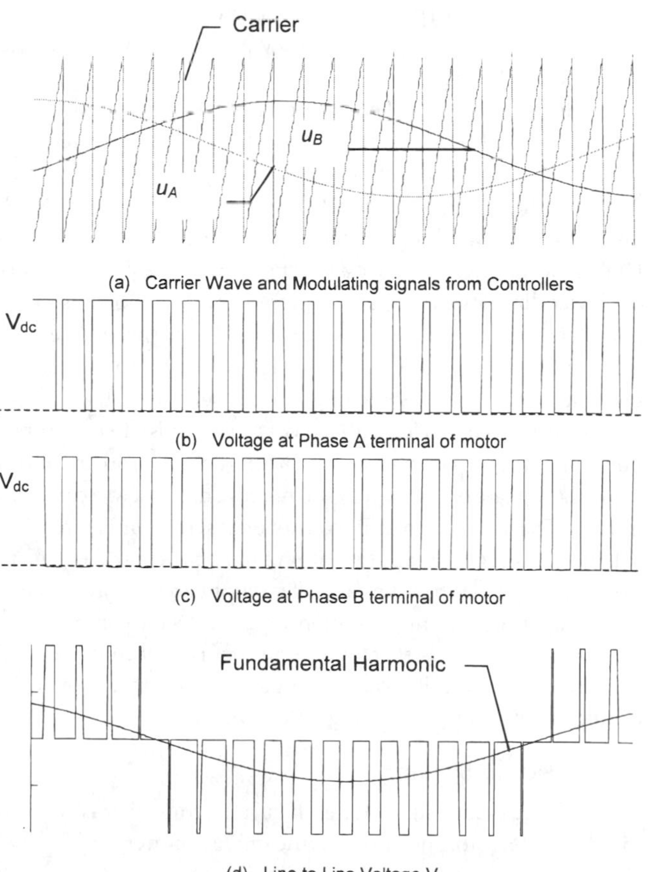

Figure 13.69 Inverter waveforms for PWM control.

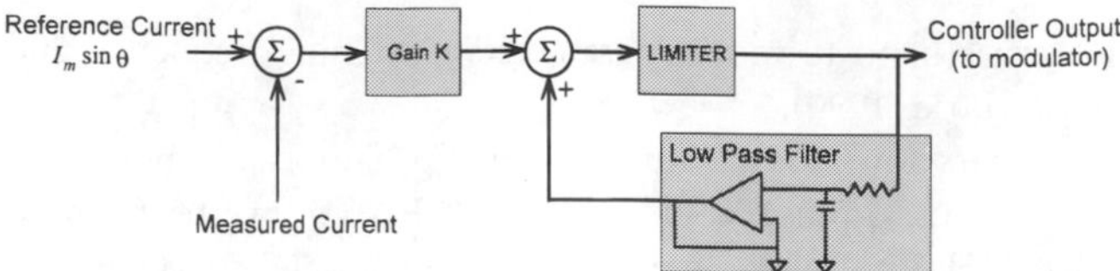

Figure 13.70 Block diagram of a simple proportional plus integral current controller.

the corresponding set of switches to generate the required voltage at the motor terminal.

In Figure 13.68 one method of pulse width modulated switching is illustrated. A high-frequency sawtooth carrier wave signal is compared with the desired modulating waveform to determine the switching instants, and therefore the resultant pulse widths. The upper leg of each transistor pair is switched on when the controller output is greater than the carrier waveform magnitude. Figure 13.69 shows a typical set of signal waveforms. Part (a) shows the carrier waveform and the modulating signals, u_A and u_B, which are the outputs of controllers corresponding to phases A and B. Part (b) shows the voltage waveform at terminal A of the motor while part (c) shows the voltage at terminal B of the motor. Figure 13.69(d) shows the resulting line to line voltage V_{ab}. If an adequate carrier frequency and sinusoidal modulation is chosen, the fundamental component of this voltage will be the effective line to line voltage seen by the motor. For example, at a speed of 3000 rpm, the phase current in a 4-pole motor would have a fundamental frequency of 100 Hz. With a carrier frequency set to 10 kHz, we obtain a carrier-to-modulating signal frequency ratio of 100. Thus the carrier frequency and its harmonics are well beyond the bandwidth of the current loop.

The PWM inverter described is relatively simple and has been used in many applications. Conventional low-power PWM inverters use switching frequencies in the range 5–20 kHz. Higher-power inverters have a lower switching frequency because of device limitations. In applications where audio noise is an issue, carrier frequencies close to 20 kHz are needed. Carrier frequencies up to 20 kHz in medium-power applications are possible with modern switching devices such as IGBTs.

The current controller with a PWM inverter is often a proportional plus integral controller (PI). There is an independent controller for each phase of the motor (two phases only for a wye connected motor). The structure of a PI controller is shown in Figure 13.70. When the controller output is within limits, the controller behaves as a PI controller. When the controller output exceeds preset values, it is limited to prevent saturation of the integral component of the controller and acts effectively as a proportional controller with an offset. This controller is easily implemented, either with analog circuits or in software on a digital controller.

Position Sensing

The reference signal for each current controller is synchronized to the rotor position. Some manufacturers equip their servo motors with a synchro or brushless resolver (Boyes, 1980) to provide high angular position resolution. These devices are also capable of producing a velocity feedback signal. Coupled with a synchro to digital converter, very high angular position resolution up to 16 bits is available. The synchro-resolver is thus useful not only for synchronization, but also as the sensing element in the outer motion control loops.

Where a resolver is not available an absolute optical position encoder can be used. This is a natural choice when the current controller is implemented on a microprocessor or digital signal processor because the encoder output is usually built to be microprocessor compatible. Six to ten bit absolute encoders are readily available. *Monostrophic* codes such as the Gray-code are often used to minimize errors. With a higher resolution encoder, it is also relatively simple to estimate the velocity from the position (Rahman, 1992; Brown et al., 1992). Incremental encoders of higher resolution driving a counter can also be used with external circuits which uses an index signal to reset the counter.

Current Sensors

The feedback control system shown in Figure 13.66 requires that each line current be measured. This contrasts with trapezoidal motors where at any one time only two motor phases have current flowing and it is sufficient to measure the DC link current. If the 3-phase PMSM has star connected windings, it suffices to measure 2 line currents as the sum of the three currents is zero.

To ensure simultaneous sampling of the two currents, sample-and-hold circuits or parallel conversion should be used in a digital implementation. The current sensors should have sufficient bandwidth for high-performance current control. A simple sensing solution is to use a resistor in series with a phase winding. The voltage across the resistor is amplified by isolation amplifiers. As the motor currents have considerable noise, a low-pass filter is usually necessary.

Digital Realization of Current Control

With the rapid developments in the computing capabilities of microprocessors and digital signal processors, it is now feasible to implement the controller described above as a fully digital system (Rahman et al., 1993; Low et al., 1994). Many low-cost microcontrollers have built-in PWM generators which allow direct digital interfacing to the inverter drive circuits. The optical absolute position encoder delivers rotor position in a digital format. The only other system components required for digital implementation are the analog to digital converters required for current sensing. Either two analog to digital converters are used or a single ADC with appropriate sample-and-hold arrangements.

This section briefly examines some of the design issues which arise with a digital realization. When a digital realization of Figure 13.70 is required, it is necessary to ensure that the sampling frequency is adequate. For example, a lower bound on the sampling frequency is related to the closed loop bandwidth (Franklin et al., 1990) of the current control loop. In Figure 13.70, the current controller is required to track a reference signal which has a frequency component up to the maximum speed of operation of the motor. Thus, a motor rated at 3000 rpm with 4 pole pairs would have a rated electrical speed of 100 Hz. Allowing for some operation above rated speed, consider a closed loop bandwidth of about 133 Hz. Following (Franklin et al., 1990). This suggests a minimum sampling frequency of 1.3 kHz to 2.6 kHz. The lower range is realizable using conventional 16 bit microcontrollers while the upper range would require digital signal processors (Le-Huy, 1994). Using the latter, sampling frequencies up to 12.5 kHz have been reported for high-performance applications.

Digital realization also requires some care with anti-aliasing filters for current. Ideally, if a sampling frequency of 1.3 kHz is used, the filtered current should have no significant frequency components above 650 Hz. A filter with this cutoff would still allow measurements of up to the 5th current harmonic at maximum speed for a 4-pole motor. In practice, it is difficult to achieve sufficient rolloff with analog low-order low-pass filters. Some non-linear filtering to remove noise spikes after sampling is often necessary.

Limitations on Control Performance

Several factors limit the tracking capability of the conventional current controller described above. A fundamental limit is the DC link voltage supplied to the inverter. A large DC voltage increases the *dynamic headroom* of the inverter. However, an upper limit on the DC link voltage is set by the voltage ratings of the power switching devices used in the inverter.

Another limit on current tracking is the influence of the back emf. Ideally, for a sinusoidal motor, this back emf is a sinusoidal function of the rotor position so that the torque (Equation 13.52) is independent of position. Then an increase in the proportional gain of the current controller would reduce the tracking error in the current. In practice, the stator winding distribution is not ideal and there are spatial harmonics in the back emf. This places a limit on the proportional gain of the current controller as excessive gain will also amplify the harmonics in the torque generated, leading to increased torque ripple.

At steady state, the current references and the back emf waveforms are all sinusoidal signals. Furthermore, this back emf acts as a relatively slow time-varying disturbance in the current loop (Holtz, 1994; Low, 1994). Thus, even with the integral action in a PI controller, this (non-constant) disturbance is only partially compensated for, and there will be a phase lag between the reference and actual phase currents. As a result, the stator field is not always in quadrature to the rotor field.

To compensate for this disturbance, it is possible to augment each current controller with an additional term in the current controller which uses measurements of velocity and rotor position. This is illustrated for the PI controller in Figure 13.71 (Low, 1994; Rahman et al., 1992). The effect of the back emf term is attenuated with an additional feedback signal which is proportional to the rotor velocity and is synchronized to the rotor position. This augmented feedback uses the rotor position and the velocity signal, required for the outer velocity loop. The back emf compensation gain factor, K_{emf}, is the ratio of motor voltage constant to the inverter static gain (Low, 1994). The scheme shown in Figure 13.71 can be implemented with analog circuits or in software.

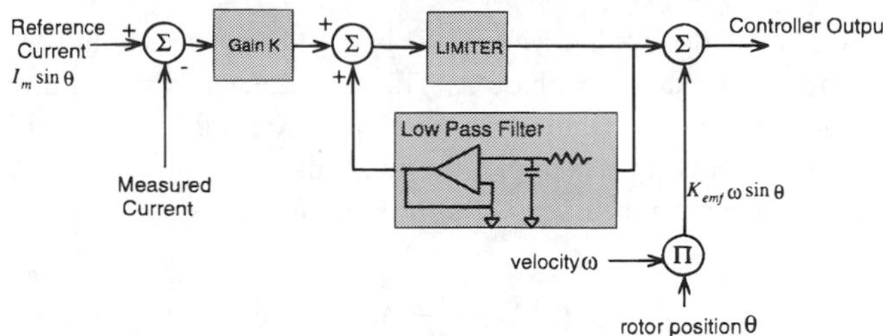

Figure 13.71 Emf feedback into the current loop.

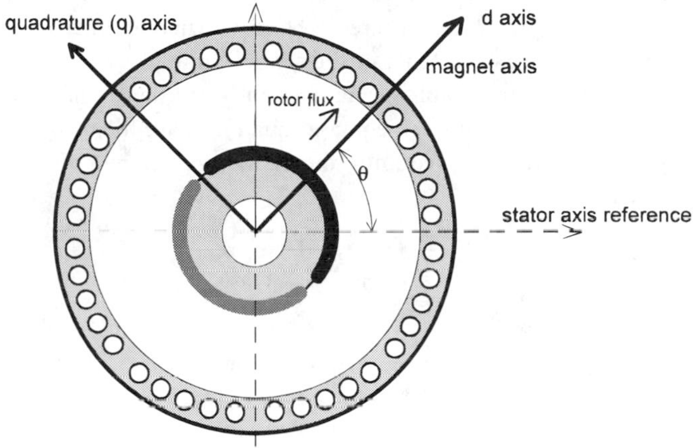

Figure 13.72 Rotating reference frame for 2-pole motor.

PM Synchronous Motor Model

The previous section has discussed the control of torque with the help of a steady-state model of the PMSM. To address the issue of torque and flux control necessary for good dynamic response, for example, in servo control applications, a dynamic model of the PMSM is required.

The control of a permanent magnet motor is more elegantly described when the dynamic equations of each phase are transformed to an orthogonal reference frame fixed to the rotor. A popular reference frame is the *d-q* reference frame, illustrated in Figure 13.72. The *d-axis* of the frame is oriented on the center of the rotor flux. For this frame of axes (Vas, 1992),

$$\begin{bmatrix} i_d \\ i_q \end{bmatrix} = \frac{2}{3}\begin{bmatrix} \cos\theta & \cos\left(\theta - \frac{2\pi}{3}\right) & \cos\left(\theta - \frac{4\pi}{3}\right) \\ -\sin\theta & -\sin\left(\theta - \frac{2\pi}{3}\right) & -\sin\left(\theta - \frac{4\pi}{3}\right) \end{bmatrix}\begin{bmatrix} i_A \\ i_B \\ i_C \end{bmatrix} \tag{13.53}$$

where $\theta = p\theta_r$, θ_r is the instantaneous rotor position, p is the number of pole pairs, and i_d and i_q are the components of the stator currents i_A, i_B and i_C in the *d-q* reference frame.

The voltage applied to each phase of the stator is the sum of the voltage drop across the winding resistance and the emf induced in the winding by flux linkages. Thus for phase A, we have

$$v_A = ri_A + \frac{d}{dt}(\Psi_A) \tag{13.54}$$

where v_A is the applied voltage, i_A is the phase current and Ψ_A is the flux linkage for phase A. This flux linkage is the sum of rotor permanent magnet flux and the flux resulting from the stator currents. For example, the flux linkage in phase A for a surface magnet machine would have the form:

$$\psi_A = L_{ss}i_A + M_{AB}i_B + M_{AC}i_C + \psi \cos \theta \quad (13.55)$$

where L_{ss} is the self-inductance, M_{ij} represents the mutual inductances and the last term is the flux linkage from the permanent magnets on the rotor. There is clearly dynamic interaction between the phase currents in the stator windings.

From the symmetry of the windings and assuming a sinusoidal distribution of windings, the phase equations have the following form when transformed into the rotor *d-q* reference frame (Vas, 1992).

$$v_d = ri_d + \frac{d}{dt}\psi_d - \omega\psi_q$$

$$v_q = ri_q + \frac{d}{dt}\psi_q + \omega\psi_d \quad (13.56)$$

$$\psi_d = L_d i_d + \psi \quad \text{: direct axis flux linkage}$$

$$\psi_q = L_q i_q \quad \text{: quadrature axis flux linkage} \quad (13.57)$$

where

i_d, i_q = *d-* and *q-axis* components of the armature current
v_d, v_q = *d-* and *q-axis* component of the terminal voltage
ψ = a flux-linkage constant due to permanent magnet
r = armature resistance
L_d, L_q = *d-* and *q-axis* components of armature inductance
ω = electrical angular velocity

This transformation also assumes that the magnetic circuit is linear and that the back emf and inductance variations are sinusoidal quantities.

Neglecting rotational losses, the instantaneous power output is the sum of the products of currents and flux linkages in quadrature, i.e., $\omega\Psi_d i_q - \omega\Psi_q i_d$ and

$$\textit{Torque per pole pair, } T = \psi i_q + (L_d - L_q)i_d i_q \quad (13.58)$$

The first term of Equation 13.58 represents the electromagnetic torque produced by the permanent magnet flux. The second term represents reluctance torque. For a surface magnet motor, the reluctance torque is zero since the uniform airgap renders L_d and L_q equal. Note that in this case, the developed mechanical torque is only from the product of the flux linkage in the *q-axis* and the *q-axis* current.

The model of Equations 13.54–13.57 shows the strong dynamic coupling between the current and flux quantities in the two axes. This interaction is represented in a block diagram form in Figure 13.73 for a surface magnet machine. Here the inductance is constant with position, $L_d = L_q = L_s$. The figure shows that the instantaneous direct axis current, while not contributing to developed torque, nevertheless affects the quadrature current.

Under the ideal steady-state conditions described in the previous subsection, where the stator currents are sinusoidal and in phase with the back emf ($\gamma = 0$), i_d, the *d-axis* component of the stator currents is zero and the *q-axis* component of the current is I_m. Control of torque is then equivalent to controlling i_q only. This characteristic was first developed from the steady-state model in **Current Source Operation—Steady-State Model** and illustrated the DC motor-like operation at steady state. The present development shows that if it is possible to control i_d instantaneously to zero, then the dynamic response will also be that of a full flux DC motor. Nevertheless, it is clear from Figure 13.73 that there is considerable dynamic interaction between the *d* and *q-axis* currents and fluxes. This can also be seen in Equation 13.56, where the *d-axis* current can be regarded as a dynamic disturbance in the second differential equation. This has to be accounted for in applications requiring full dynamic control.

From the block diagram of Figure 13.73, this system can be regarded as one where two inputs, v_d and v_q are manipulated to control two interacting currents, i_d and i_q. When the independent PI current controllers described previously are used to control the individual stator currents directly, this interaction is implicitly ignored, resulting in less than ideal control performance. This can be shown in experimental data when the controller of **Current Feedback Control System** is applied to a surface magnet machine. Figure 13.74 shows the resulting stator currents in the *d-q* reference frame, when under no load conditions, a step change is demanded from a velocity control system. This leads to a time-varying torque/current demand. Figure 13.74 shows that while the steady-state value of the *d-axis* current is zero, the transient value of the *d-axis* current is clearly non-zero, even when the back emf compensation described in **Limitations on Control Performance** is included. Thus the dynamic response is not necessarily that of the DC motor implied by Equation 13.56.

For an IPM motor, there is in addition a reluctance torque which is due to the saliency of the rotor. The large value of L_d in an IPM provides a means for modifying the *d-axis* flux linkage, ψ_d, by controlling i_d (Equation 13.57). To reduce flux linkage, for instance, i_d should be negative. This is equivalent to advancing the stator currents (setting γ negative in Equation 13.52). This in turn reduces the back emf of the motor which enables the

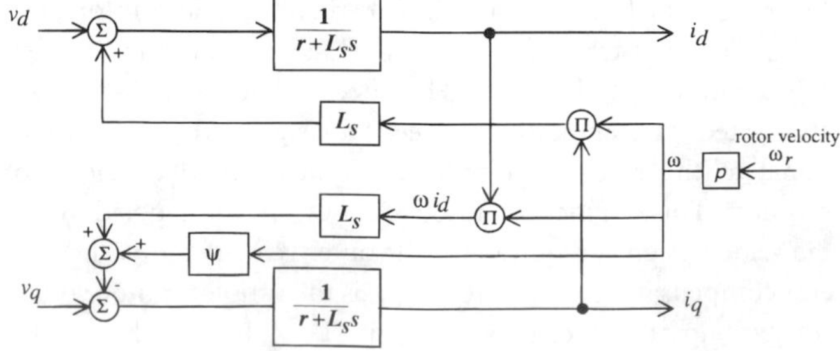

Figure 13.73 Block diagram of PMSM dynamics in *d-q* reference frame.

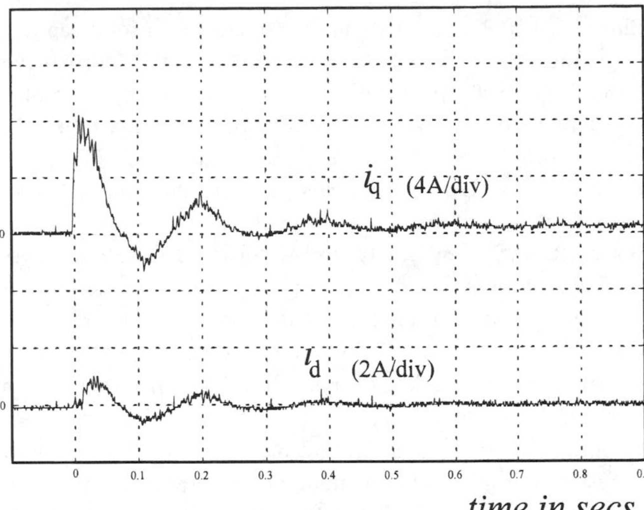

Figure 13.74 *d-* and *q-axis* current for independent phase current control with back emf compensation.

speed range of the motor to be increased. Unlike the surface magnet motor, the control law for torque of the IPM must take into account both electromagnetic and reluctance torque (Equation 13.58). In particular, for high-torque and high-efficiency operation of interior magnet motors, i_d and i_q should both be controlled according to the maximum torque per ampere characteristic.

Field-oriented Control

In **Construction and Operation,** the basic principle of torque production in the PMSM has been described. The instantaneous torque generated is related to the magnitude of the stator current flux and its angle with respect to the rotor flux. The control scheme described in **Current Feedback Control System** is capable of manipulating both the phase and magnitude of the stator currents so that the stator flux is at an optimum angle and magnitude for torque production at steady state. Field-oriented control can be regarded as extending this to manipulating the stator currents dynamically so as to optimize torque production at all times. This is most elegantly described in the *d-q* reference frame.

Given measurements of the rotor position, the stator currents can be transformed to i_d and i_q using dedicated hardware or with simple table lookup mechanisms in a microprocessor. An advantage of this reference frame is that at a steady speed, the currents and voltages are no longer sinusoidal functions of position.

Improved current control is possible by implementing the controller in this transformed reference frame (Low, 1994; Vas, 1992; Jahns, 1994). A simple scheme, illustrated in Figure 13.75 is to use two PI controllers, one for each of the quadrature currents. In Figure 13.73, the *d-axis* current dynamics has a cross-coupling disturbance signal due to i_q and the rotor speed. In the steady state, i_q and the speed ω are constant. It is thus possible for the integral action in the PI controller to reject the effects

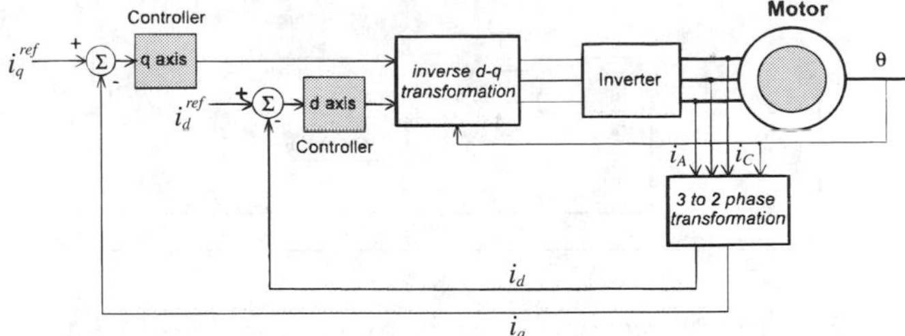

Figure 13.75 Control in the rotating reference frame.

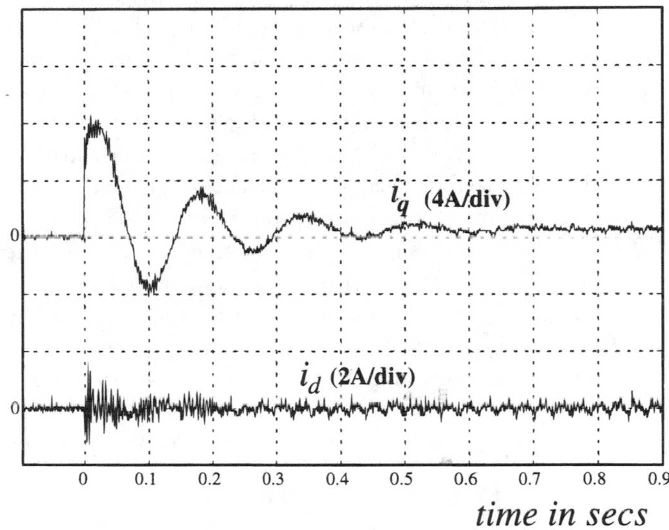

Figure 13.76 *d-* and *q-axis* current for control in the *d-q* reference frame.

of this constant disturbance to the *d-axis* current. Note that this is usually not possible with the set of independent current controllers because the disturbance in that context is a time-varying back emf. Consequently, the latter scheme usually leads to a phase lag between the desired current and the actual phase currents.

On the *q-axis,* there are two disturbance signals. One is due to the back emf of the machine and is speed dependent. At steady speed, the back emf is a constant value and can thus also be eliminated with a PI controller for the *q-axis* current. The other disturbance is due to the cross-coupling of i_d and the rotor speed. When operated up to rated speed, the desired *d-axis* current is zero for a surface magnet motor. With a good *d-axis* controller, this disturbance term is quickly attenuated. For an interior magnet motor, however, the desired *d-axis* current is not zero, even below rated speed. In that case, the PI controller may not be adequate and some form of decoupling control should be considered.

Figure 13.76 shows the resulting *d-* and *q-axis* currents when this simple field oriented controller is applied to the same motor and under the same conditions as the data of Figure 13.74. Here the reference *d-axis* current is zero. With a relatively simple set of PI controllers, the perturbation in *d-axis* current has been almost eliminated. This yields a set of stator currents much closer to the ideal case and at all instants of time (Low, 1994).

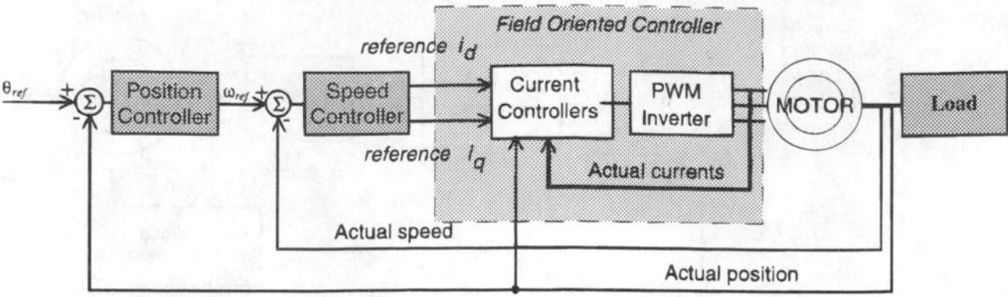

Figure 13.77 Position and speed control loops for a PMSM.

The field oriented control scheme described is equivalent to independent control of torque and flux. This allows a broader range of operation. Control of i_d is equivalent to modifying air gap flux if L_d is significant, Equation 13.57, as is the case in an interior magnet machine. Control of i_q is equivalent to control of torque. With flux control, the motor can easily be operated beyond rated speed. To operate above the rated speed while maintaining the terminal voltage of the machine, the air gap flux is reduced in inverse proportion to the speed. This is done by increasing the desired value of i_d to some non zero value. At the same time, it is necessary to ensure that the total stator current is kept at rated value. This it is also necessary to reduce the quadrature component of the stator current, i_q. As a result, the output torque falls inversely with speed and output power becomes constant. This mode of operation is variously known as field weakening, flux weakening, or constant power operation.

Motion Control with PMSM

The field oriented controller described above is concerned with the independent control of flux and torque in the motor. In a motion control application, the flux and torque commands would come from external velocity and position loops which are added in a typical cascaded loop configuration as shown in Figure 13.77. Here the output of the speed controller is the described torque or current of magnitude, I_m (i_q) and the desired flux (i_d). The reference signal to the speed controller in turn is the output of the outermost position control loop.

It is now common for manufacturers to provide digital microprocessor based systems for implementing the outer position and velocity loops (Leonhard, 1986). With field oriented control, these external loops are very similar for DC motors. Thus it is possible to provide a common design for both classes of machines. Using digital controllers allows many other functions to be implemented in software. These include self diagnostics and simple user interfaces. These factors, taken together with the increasing standardisation of hardware, should lead to a reduction in per unit drive material cost.

References

Alwmon, G. R. 1994. Electrical machines for variable frequency drives, *Proc. of IEEE,* 82(8):1123–1139.

Binns, K. J. 1984. Permanent magnet motors for inverter fed drive, *Drive/Motors/Control,* 101–5, 1984.

Binns, K. J. and Wong, T. M. 1984. Analysis and performance of a high field permanent magnet synchronous motor, *Proc. of IEE, Pt. B,* 131:252–258.

Binns, K. J. 1993. Major design parameters of a solid cannned pm motor with skewed magnets, *IEEE Proc. Pt. B,* 140:161–165.

Bose, B. K. 1990. An adaptive hysteresis-band current control technique of a voltage-fed PWM inverter for machine drive system, *IEEE Trans. Industrial Electronics,* 37:402–408.

Boyes, G. S. ed. 1980 *Synchro and Resolver Conversion,* Memory Devices Ltd, U.K., ISBN 0-91650-06-0.

Brown, R. H., Schneider, S. C., and Mulligan, M. G. 1992. Analysis of algorithms for velocity estimations from discrete position versus time data, *IEEE Trans. Industrial Electronics,* 39:11–19.

Cavicchi, F., Crescimbini, F., Honorati, and Santini, 1992. Performance and evaluation of axial flux PM generator, *Proc. of ICEM* 2:761–765.

Cavicchi, F., Crescimbini, F., and Santini, E., 1993. Optimum design of ironless stator winding for axial flux PM machines, *IEEE Conference on Electric Machines and Drives, EMD '93,* Oxford University, Sept. 1993.

Chalmers, B. J., Hamed, S. A., and Baines, G. 1985. Parameters and performance of high field PM synchronous motor for variable frequency operations, *Proc. of IEEE Pt. B,* 117–124, 1985.

Finney, D. 1988. *Variable Frequency AC Motor Drive Systems,* Peter Peregrinus Ltd., ISBN 0-86341-114-2.

Franklin, G. F., Powell, J. D., and Workman, M. L. 1990, Digital control of dynamic systems, ed. Addison-Wesley, New York.

Harashima, F., Naitoh, H., and Haneyoshi, T. 1979. Dynamic performance of self-controlled synchronous motors fed from current source inverters, *IEEE Tans. Ind.* Appl. IA-15(1): 36–46.

Holtz, J. 1994. Pulse width modulation for electronic power conversion, *Proc. of IEEE,* 82:1194–1214.

Jahns, T. M., Kliman, G. B., and Neumann, T. 1986. Interior permanent magnet synchronous motors for adjustable speed drives, *IEEE Trans. Ind. Appl.* IA-22, 738–747.

Jahns, T. M. 1987. Flux weakening regime opeation for an interior pm synchronous motor drive, *IEEE Trans. Ind. Appl.,* 23:681–689.

Jahns, T. M. 1994. Motion control with permanent magnet AC machines, *Proc. of IEEE,* 82(8):1241–1252.

Le-Huy, H. Jakubowicz, A., and Perret, R. 1982. A self-controlled synchronous motor drive using terminal voltage sensing, *IEEE Trans. Ind. Appl.,* IA-18(1):116–135.

Le-Huy, H., Perret, R., and Feuillet, R. 1985. Minimization of torque ripple in brushless DC motor drives, *IEEE Trans. Ind. Appl.,* IA-23:748–55.

Le-Huy, H. 1994. Microprocessors and digital IC's for motion control, *Proc. of IEEE,* 82(8):1140–1163.

Leonhard, W. 1986. Microcomputer control of high dynamic performance AC drives - A survey, *Automatica,* 22(2):1–19.

Low, K. S. 1994. Control Strategies for High Performance Permanent Magnet Synchronous Motor Brushless Drive, doctoral dissertation, University of New South Wales, Sydney 2052, Australia.

Low, K. S., Lim, K. W., and Rahman, M. F. 1994. A fully digital vector controlled permanent magnet synchronous motor servo drive, *Proc. of Australasia University Power Engineering Conference (AUPEC '93)*, 29 September–1 October, University of Wollongong, Australia, 1:27–32.

Maeminn, S. R. and Jahms, T. M. 1991. Control techniques for improved high speed performance of interior pm synchronous motor drives, *IEEE Trans. Ind. Appl.* IA-27, 997–1004.

Miller, T. J. E. 1989. *Brushless Permanent-Magnet and Reluctance Motor Drives*, Oxford University Press, New York, 1989.

Morimoto, S., Takeda, Y., and Hirasa, T. 1990. Current phase control methods for permanent magnet synchronous motors, *IEEE Trans. on Power Electronics*, 5(2):133–139.

Murphy, J. M. D., and Turnbull, F. G. 1989. *Power Electronic Control of AC Motors*, Pergamon Press, New York.

Ng, B. H., Rahmnan, M. F., Low, T. S., Lim, K. W. 1988. An investigation into the effects of machine parameters on torque pulsations in a brushless DC drive, *Proc. of IEEE Industrial Electronics, Control and Instrumentation Conference (IECON)*, pp. 749–754.

Rahman, M. F., Lim, K. W., and Low, K. S. 1992. Velocity sensing techniques using discrete position sensors—A comparative study, *Proc. of Second International Workshop on Advanced Motion Control (AMC'92)* 16–18 March, Japan, pp. 311–316.

Rahman, M. F., Lim, K. W., and Low, K. S. 1992. Evaluation of an Emf injection scheme for a permanent magnet brushless DC drive, *Proc. of IEEE International Symposium on Power Electronics (ISPE'92)*, 25–29 May, Xian, China, pp. 625–629.

Rahman, M. F., Low, K. S., and Lim, K. W. 1993. Software torque/current controllers for permanent magnet synchronous motor drive, *Proc. of Fifth European Conference on Power Electronics and Applications (EPE'03)*, 13–16 September, Brighton, U.K., 4, 249–254.

Rahman, M. E., Goris, M. and Lim, K. W. 1994. Analysis and design of a brushless DC controller for an axial flux disc rotor synchronous motor, *Proc. of ICEM '94 Conference*, Paris.

Slemon, G. R., Dewan, S., and Wilson, J. 1974. Synchronous motor drive with current source inverter, *IEEE Trans. on Ind. Appl.*, IA-10(3).

Slemon, G. R. 1992. *Electrical Machines and Drives*, Addison Wesley, New York, Chap. 11.

Sneyers, B. D. O., Novotny, W., and Lipo, T. A. 1985. Field weakening of buried permanent magnet synchronous motor drive, *IEEE Trans. Ind. Appl.*, IA-21, 398–407.

Vas, P. 1992. *Electrical Machines and Drives: A Space Vector Theory Approach*, Oxford University Press, New York.

Zijlstra, H. 1984. Nd-Fe Permanent Magnets—Their Present and Future, IV Mitchell, Elsivier Applied Science Publishers, London, pp 5–11, 1984.

13.6 Step Motor Drives

Ronald H. Brown

Step motors are used in many low-lost positioning applications due to their inherent ability to stop at discrete positions and follow position versus time profiles while being controlled open loop. A step motor is a synchronous machine, but historically has been used almost exclusively in positioning and position tracking applications. Recently, however, some types of step motors have been applied in variable speed drives applications.

Step motors can be driven without feedback to stop at discrete angular positions, known as detent positions. The number of detent positions can be as low as 12 steps or detent positions per revolution to 400 or 500 or more steps per revolution. The location accuracy of the detent positions vary typically within 5 percent of the step size. The repeatability of the motor is also high, in that the rotor can start on one position, move away to other positions, and then return to within typically 3 percent of step size of the original position.

The motors with relatively few steps are typically used in higher-speed applications, where the motors with many steps per revolution are often used in high-torque, low-speed, direct drive applications or applications where many repeatable discrete positions are required. Step motors have been successfully applied in many applications such as computer peripherals (e.g., disk drives, pen plotters, and printers), office machines (e.g., copiers, scanners), automotive (e.g., seat positioning, speed control), aerospace (e.g., flap control, starter-generators), industrial (e.g., robots, scanners, machine tools), to name a few.

Many step motor drives are driven with digital pulses, thus it is easy to interface and control step motors from computers or micro-controllers without digital to analog circuitry. For control purposes, the step motor and drive can be thought of as a digital to angular position converter.

It is perhaps easier to understand how a step motor works than any other rotating machine. However, the mathematical models of step motors are nonlinear, since the inductance and torque vary sinusoidally with position. The non-linear nature of the models requires that the engineer carefully design the controllers for the motors. In the following section the three types of step motors are discussed along with the operation and drive circuits of each. The mathematical models of each type is discussed in the second section. In the third section, the control of step motors is presented.

Types and Operation of Step Motors

The three common types of step motors are the variable reluctance, the permanent magnet, and the hybrid step motors. The variable reluctance and the hybrid step motors are double salient structures, i.e., teeth on both the stator and the rotor, with multiple windings or phases on the stator and no windings on the rotor (thus brushless machines). The variable reluctance step motors can have three, four, or even five phases, while the permanent magnet and the hybrid step motors usually have two phases.

Principle of Operation: A first understanding of the principle of operation can be easily seen from the variable reluctance step motor, but is common for all types. When a single winding or phase is energized, the motor generates a torque in the direction to align the rotor teeth with the teeth of the energized phase. The torque generated by current in a single phase, for example, phase A, is

$$T_e = -k_T i_A \sin(N_r\theta) \quad (13.59)$$

where T_e is the generated torque, k_T is the torque constant, i_A is the current in phase A, N_r is number of electrical cycles per mechanical revolution, and θ is the mechanical rotor position. The cross-sectional view of a variable reluctance step motor is shown in Figure 13.78, showing only the phase A winding. The rotor is shown in the aligned position, which is the detent position. The generated torque by the current in phase A versus rotor position for this motor is shown in Figure 13.79. If a load on the rotor were to displace the rotor in the positive θ or clockwise direction, the generated torque would act in the direction to realign the rotor. From Figure 13.79, we see that a displacement in the positive θ direction generates negative torque, which will push the rotor in the negative direction, thus trying to restore the detent position. If a load on the rotor were to displace the rotor in the negative θ or counterclockwise direction, the generated torque would also act in the direction to realign the rotor. From Figure 13.79, we see that a displacement in the negative θ direction generates positive torque, which in turn is in the direction to restore the rotor to the detent position.

Now, suppose the current in phase A is de-energized and phase B is energized. It is apparent from Figure 13.78 that the rotor will rotate 15° in the positive direction so that rotor teeth will align with the phase B stator teeth. Next, suppose that phase B is de-energized and phase C is energized. The rotor will rotate 15° additionally in the positive direction, aligning phase C stator teeth with rotor teeth. One more phase switching, de-energizing phase C and energizing phase A causes the rotor to rotate yet an additional 15° in the positive direction. Now the rotor has moved a total of 45° in the positive direction. This detent position is at 45° in Figure 13.79. Detent positions occur where the torque versus position curve crosses zero torque with negative slope.

The torque versus position for all three phases of the motor (energized one at a time) are shown in Figure 13.80. By exciting the motor phases in order during intervals of positive torque, as shown in Figure 13.81, the motor can be made to run in the positive direction. Conversely, by exciting the motor phases in reverse order during intervals of negative torque, the motor can be made to run in the opposite direction.

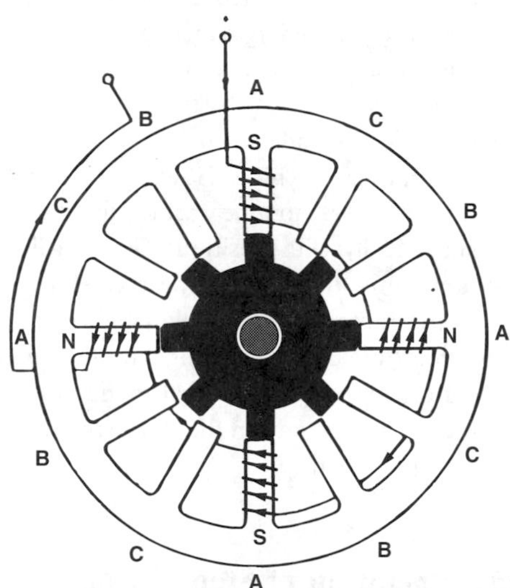

Figure 13.78 Cross-sectional view.

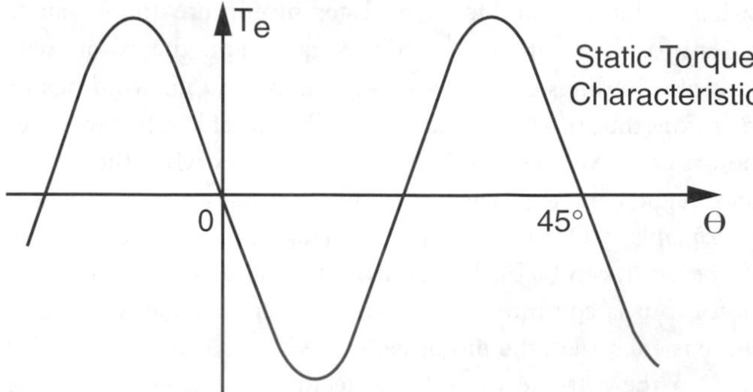

Figure 13.79 Static torque characteristic.

Variable Reluctance Step Motor

As mentioned above, the variable reluctance step motors can have three, four, or even five phases. The mode of operation discussed above is common to all variable reluctance step motors, and will not be repeated here. The motor discussed above is known as a 12/8 variable reluctance step motor, in that the stator has 12 teeth and the rotor has 8 teeth. This motor takes 15° steps and has 24 steps per revolution. The number of detent positions per revolution can be as low as 12 for a 6/4 motor and as high as 200 or 400 steps per revolution. The size of the motor can range from small fractional HP, with detent torque in the few ounce-inch range to 10 HP or more. The larger variable reluctance step motors are more commonly called switch reluctance motors and are usually used in variable speed applications.

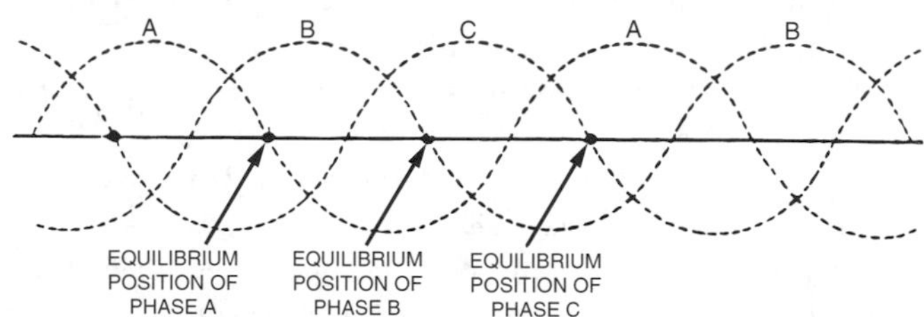

Figure 13.80 Static torque characteristics for all three phases.

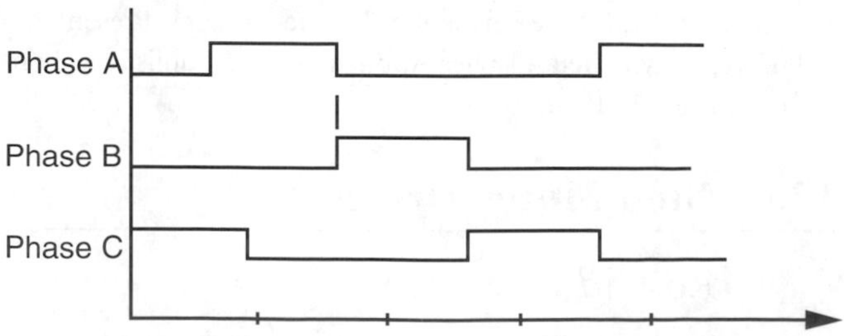

Figure 13.81 By exciting the motor windings during the positive portions of their torque curve, the motor can be made to produce nonzero average torque.

Drive Circuits for Variable Reluctance Step Motors

The drive circuits are quite simple for the variable reluctance step motor. Figure 13.82a shows the simplest drive circuit for one phase (each phase requires its own identical drive circuit). The transistor acts as a switch, either off or on. When the switch is on, the current flows from the supply, through the phase winding, through the switch, to ground. When the switch is off, the current in the winding cannot drop to zero instantaneously due to the winding inductance. A path for the decay current is provided through the diode. The voltage across the winding when the switch is on is the supply voltage, V_s (neglecting the voltage drop across the switch). The current in the phase takes time to reach the value of V_s/R, where R is the phase resistance due to the phase inductance, L. Neglecting the back EMF, the phase current is approximately:

$$i_A \approx \frac{V_s}{R}(1 - e^{-(R/L)t}) \tag{13.60}$$

as shown in Figure 13.82b. When the switch is off, the diode forms a short-circuit and the current flows through the diode, thus the voltage across the winding is zero (neglecting the voltage drop across the diode). Thus the current decays to zero with the L/R time constant. Once the current decays to zero, neglecting leakages, the winding is open-circuited and no current flows.

The drive circuit in Figure 13.83a can be used for higher-performance operation of the variable reluctance step motor. In this circuit, both transistors act as switches. When both switches are on, the current flows from the supply, through the top switch, through the phase winding, through the bottom switch, to ground. When both switches are off, the current in the winding cannot drop to zero instantaneously due to the winding inductance, so the current flows from ground through the lower diode, through the winding, through the top diode, and back into the supply. The voltage across the winding when both switches are on is the supply voltage (neglecting the voltage drop across the switches). When the switches are off, the voltage across the winding is the negative of the supply voltage (neglecting the voltage drop across the diodes). Thus the current decays towards $-V_s/R$ with the L/R time constant. Once the current decay reaches zero the diodes block and, neglecting leakages, the winding is open-circuited and no current flows. The current decay time is much less with this circuit than with the circuit in Figure 13.82. The current waveform of this operation of this circuit is shown in Figure 13.83b.

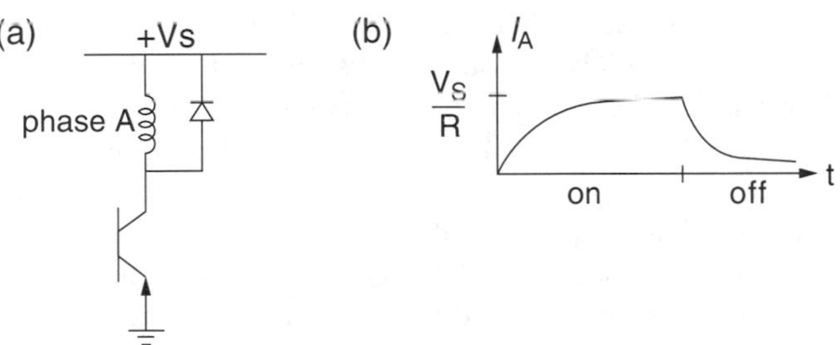

Figure 13.82 Drive circuit and phase current for one phase.

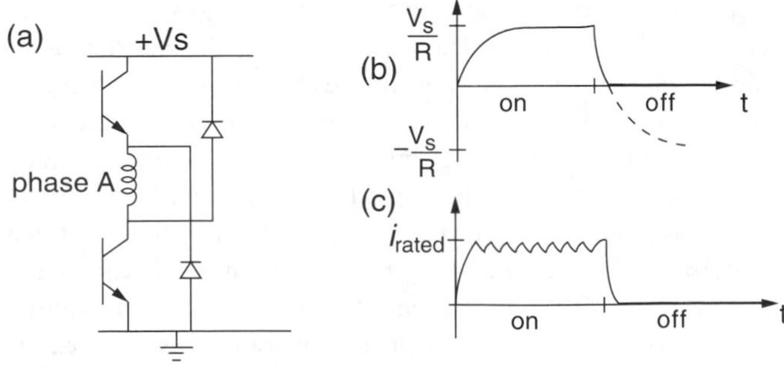

Figure 13.83 Drive circuit and current waveforms for higher-performance operation.

Even greater performance of the variable reluctance step motor can be achieved with the circuit in Figure 13.83a. V_s is set five to ten times larger than the motor rated voltage and the current is controlled by "chopping" one of the switches on and off. When the phase is energized, the current rises with the L/R time constant towards V_s/R as before, but now V_s/R is five to ten times larger than rated current, so the current reaches rated current much sooner. At this time, one of the two switches is then turned off, allowing the current to decay towards zero. A short time later, the switch is turned on again until current reaches rated current again. This process is repeated until the phase is to be de-energized. The current waveforms for this operation of the circuit is shown in Figure 13.83c.

Permanent Magnet (Can-Stack) Step Motor

The permanent magnet step motor has a smooth, permanent magnet rotor. The rotor is constructed to have many pairs of magnetic poles. The windings are not wrapped around poles as in the variable reluctance motor, but around the circumference of the air gap. The stator poles are wrapped around the windings to form north and south magnetic poles to attract and repel the magnetic poles on the rotor. Two sets of windings and stator poles are required, with each set of stator poles offset by half a tooth pitch. This motor usually has a low number of steps or detent positions per revolution, and detent positions are less accurate than the other types of motors. The motor does have an unenergized detent torque.

The permanent magnet step motor has two phases, but can be wound in two different ways. If the motor is wound unifilar, that is one winding per phase, bidirectional currents are required for proper operation. With bifilar windings, that is two windings or a center tapped winding per phase, unidirectional currents can be used to run the motor.

Hybrid Step Motor

The hybrid step motor can be described as two 2-phase (unidirectional) variable reluctance step motors put together with an axially mounted permanent magnet between the rotors. The magnetic flux paths are 3-dimensional, aligned axially between the rotor halves and radially in the air gaps. The possible winding configurations are similar to the permanent magnet type motor, either unifilar, requiring bidirectional currents, or bifilar, requiring only unidirectional currents. As with variable reluctance motors, the number of steps per revolution typically range from 24 to 400. As with permanent magnet motors, there is some un-energized detent torque.

Drive Circuitry for Permanent Magnet and Hybrid Step Motors

The drive circuits for the permanent magnet and hybrid step motors are different than the drive circuits for the variable reluctance step motor. The drive circuit also depends on if the motor is wound unifilar or bifilar. Bifilar wound motors require fewer drive circuit components than for the unifilar wound motors, but at most only half the phase winding is energized at one time.

Figures 13.84 and 13.85 show partial drive circuits for unifilar wound step motors. The circuits shown are for only half or one winding of the motor. A second identical circuit is needed for the other motor winding. The drive circuit in Figure 13.84, known as a half-H-bridge, requires both positive and negative voltage supplies. The transistors act as switches, connecting one end of the phase winding to either the positive supply voltage, $+V_s$, or the negative supply voltage, $-V_s$. When switch Q_1 is on, the current flows from $+V_s$ through the switch, through the phase winding to ground. When switch Q_1 is off, the current cannot instantaneously drop to zero due to the winding inductance, thus the decay current flows through diode D_2 from $-V_s$ to the winding to ground. Once the current decays to zero, the diode blocks the current and, neglecting leakage, the winding is open-circuited. When Q_1 is on, the phase voltage is $+V_s$ (neglecting the voltage drop across the switch). When Q_1 is off, the phase voltage is $-V_s$ (neglecting the voltage drop across the diode) until the current decays to zero, then the phase is open-circuited. Similarly, when switch Q_2 is on, the current flows from ground through the phase winding in the opposite direction as before, through the switch to $-V_s$. When switch Q_2 is off the decay current flows through diode D_1 to $+V_s$ from the winding from ground.

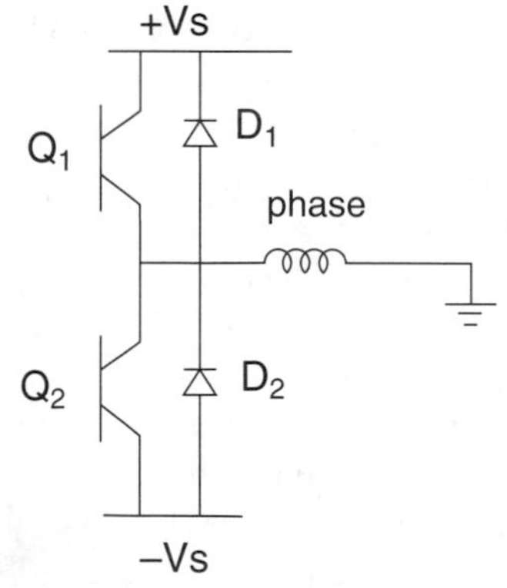

Figure 13.84 Half-H-bridge drive circuit.

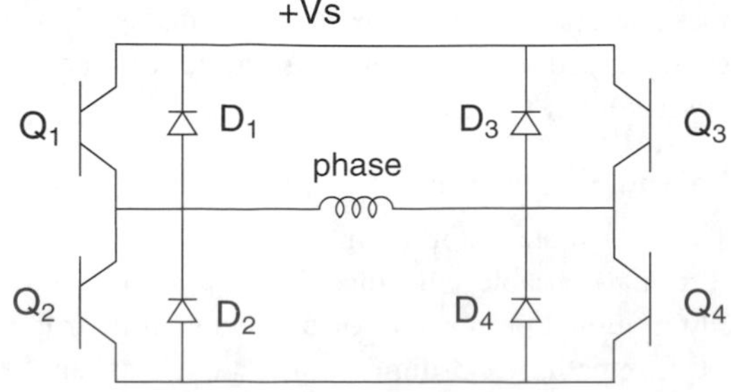

Figure 13.85 H-bridge drive circuit.

The drive circuit in Figure 13.85, known as an H-bridge, requires only a positive supply voltage. The four transistors act as switches, connecting each end of the phase winding to either the positive supply voltage, $+V_s$, or ground. When switches Q_1 and Q_4 are on, the current flows from $+V_s$ through Q_1, through the phase winding through Q_4 to ground, applying $+V_s$ across the winding. When switches Q_1 and Q_4 are off, the current in the winding cannot instantaneously drop to zero due to the winding inductance, thus the decay current flows through diodes D_2 and D_3, applying $-V_s$ across the winding until the current decays to zero, when the diode blocks the current and, neglecting leakage, the winding is open-circuited. Similarly, when switch Q_2 and Q_3 are on, the current flows from $+V_s$ through Q_3, through the phase winding in the opposite direction as before, through Q_2 to ground, applying $-V_s$ across the winding. When switches Q_2 and Q_3 are off, the decay current flows through diodes D_1 and D_4, applying $+V_s$ across the winding until the current decays to zero, when the diode blocks the current and, neglecting leakage, the winding is open-circuited.

Higher motor performance can be achieved from the circuit in Figure 13.85 if the supply voltage is set at five to ten times the motor rated voltage and the phase currents are regulated by chopping either switch Q_1 or Q_3. For example, when Q_1 and Q_4 are on, the phase current rises towards V_s/R with the L/R time constant, where R is the winding resistance and L is the winding inductance. When the phase current reaches rated current, Q_1 is off and the circuit path is through D_2, the phase winding, and Q_4, thus the applied voltage is zero (neglecting the diode and transistor voltage drops), and the current now starts to decay towards zero. A short time later, Q_1 is turned on again and the current builds towards rated current. Once the current reaches rated current, the cycle is repeated until the phase is to be de-energized.

Figures 13.86 and 13.87 show drive circuits for bifilar wound step motors. Both of these circuits are known as inverse diode clamped drive circuits. The circuits shown assume the center tapped winding configuration and are for only one bifilar winding of the motor. A second identical circuit is needed for the other motor winding. It is best to think of a bifilar wound step motor as having four phases, A, B, C, and D, but unlike with the variable reluctance step motor, phases A and C are inversely mutually coupled and phases B and D are inversely mutually coupled.

The drive circuit in Figure 13.86 assumes that the supply voltage is set to motor rated voltage. The transistors in the circuit act as switches. When Q_a is on, the current flows from the supply through the phase A, through Q_a to ground. When Q_a is off, and since the current in the bifilar winding cannot decay to

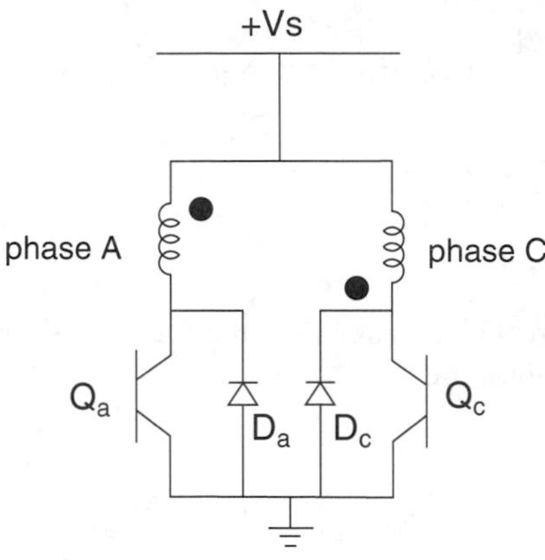

Figure 13.86 Inverse diode clamped drive circuit.

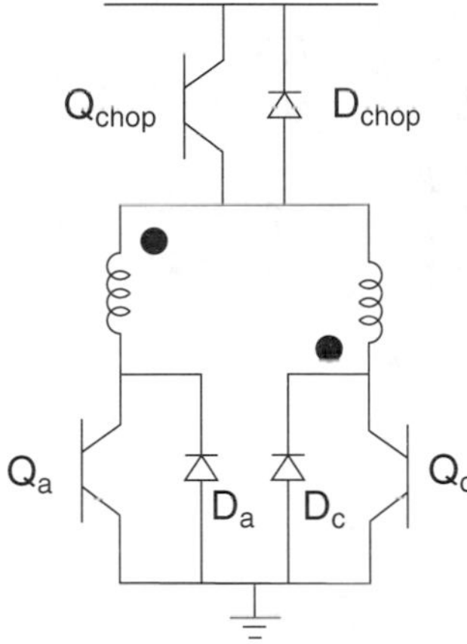

Figure 13.87 Drive circuit of Figure 13.86 as a chopper drive.

zero instantaneously, the mutual coupling in the bifilar winding couples the current to phase C, where the current flows from ground up through the diode, D_c, through phase C, into the supply. This applies $-V_s$ across the phase C, causing the current to decay quickly.

The drive circuit in Figure 13.87 works the same way as the drive circuit in Figure 13.86 when Q_{chop} is on. The addition of Q_{chop} and the additional diode allows the inverse diode clamp drive circuit to be a chopper drive. The supply voltage is set to five to ten times the motor rated voltage and when either leg of the circuit is on, the current is regulated using Q_{chop}. When Q_{chop} is off, the phase current drops to half of its original value, half of the conducting current couples to the opposite phase, and the current flows up through the clamp diode in the opposite phase, backwards through the opposite phase, through the on phase, and through the on phase transistor.

Step Motor Models

When a constant current is passed through one phase of a step motor, the motor generates a torque. This torque is typically a sinusoidal function of rotor displacement from the detent position that causes the rotor to minimize this displacement. When the phases of the motor are excited so that the motor "runs," the generated torque is still a function of position and current, but the current becomes a varying quantity, dependent on time, position, velocity, and of course, the drive circuit and drive scheme. Selection of a motor, drive circuit, and drive scheme depends on predicting the performance and the dynamic torque-speed characteristics of a particular motor with a drive circuit and drive scheme. These performances of step motors can be predicted to within reasonable accuracy using mathematical models for both the motor and drive circuit. Ways to model the motor and drive circuit are presented in this section. As with modeling most physical systems, more accurate models produce more accurate results. Tradeoffs between accuracy and simplicity is also discussed with each model.

The variable reluctance step motor model needs to be modeled separately from the permanent magnet and hybrid step motor models, and separate models are needed for unifilar and bifilar windings.

Variable Reluctance Step Motor Model

Precise mathematical modeling of variable reluctance step motors requires knowledge of both the geometry of the machine and of the ferromagnetic material characteristics. These requirements are often relaxed and assumptions are made to simplify the model to a set of nonlinear differential equations.

Assumption 1. The ferromagnetic material does not saturate. This is a poor assumption for variable reluctance step motors in that the motors are usually operated with a high degree of saturation. This assumption is replaced after the "non-saturated" model is presented.

Assumption 2. The inductance for each phase varies sinusoidally around the circumference of the air gap, for example the phase A inductance is $L_A(\theta) = L_0 + L_1\cos(N_r\theta)$. This assumption required Assumption 1, otherwise, L_A is a function of both θ and i_A.

The terminal voltage for phase A can be found using Faraday's law as:

$$V_A = R_A i_A + \frac{d\lambda_A}{dt} \tag{13.61}$$

where V_A is the terminal voltage, R_A is the winding resistance, i_A is the winding current, and λ_A is the phase flux linkages. Since $\lambda_A = L_A\, i_A$:

$$\frac{d\lambda_A}{dt} = L_A\frac{di_A}{dt} + i_A\frac{dL_A}{dt} = L_A\frac{di_A}{dt} - L_1 i_A \omega N_r \sin(N_r\theta) \tag{13.62}$$

where the first term is the magnitizing voltage and the second term is the speed voltage. Equation 13.64 can be rewritten as

$$\frac{di_A}{dt} = \frac{1}{L_A}V_A - \frac{R_A i_A}{L_A} + \frac{L_1 N_r}{L_A} i_A \omega \sin(N_r\theta) \tag{13.63}$$

The differential equations for the remaining phases are the same as the above equations, replacing the subscripts with the appropriate phase letter. The inductances for the other phases, however, need to be shifted in position. For a three phase motor, the inductances are

$$L_B(\theta) = L_0 + L_1 \cos\left(N_r\theta - \frac{2\pi}{3}\right)$$

$$L_C(\theta) = L_0 + L_1 \cos\left(N_r\theta - \frac{4\pi}{3}\right) \quad (13.64)$$

For a four-phase motor, the inductances are

$$L_B(\theta) = L_0 + L_1 \cos\left(N_r\theta - \frac{\pi}{2}\right)$$

$$L_C(\theta) = L_0 + L_1 \cos(N_r\theta - \pi) \quad (13.65)$$

$$L_D(\theta) = L_0 + L_1 \cos\left(N_r\theta - \frac{3\pi}{2}\right)$$

The mechanical equations can be found from Newton's law and conservation of energy. Newton's law states

$$J\frac{d\omega}{dt} = T_e - T_L - B\omega \quad (13.66)$$

where J is the rotor and load moment of inertia, ω is the rotor velocity in mechanical radians per second, T_e is the torque generated by the motor, T_L is the load torque, and B is the rotor and load viscous friction coefficient. Using conservation of energy, the torque generated by i_A for the variable reluctance step motor assuming no magnetic saturation is

$$T_A = -\frac{L_1N_r}{2}\sin(N_r\theta)i_A^2 \quad (13.67)$$

where T_A is the torque generated by the current in phase A.

Summarizing, the differential equations for the three-phase variable reluctance step motor are

$$\frac{di_A}{dt} = \frac{1}{L_A}V_A - \frac{R_A}{L_A}i_A + \frac{L_1N_r}{L_A}i_A\omega\sin(N_r\theta)$$

$$\frac{di_B}{dt} = \frac{1}{L_B}V_B - \frac{R_B}{L_B}i_B + \frac{L_1N_r}{L_B}i_B\omega\sin\left(N_r\theta - \frac{2\pi}{3}\right)$$

$$\frac{di_C}{dt} = \frac{1}{L_C}V_C - \frac{R_C}{L_C}i_C + \frac{L_1N_r}{L_C}i_C\omega\sin\left(N_r\theta - \frac{4\pi}{3}\right) \quad (13.68)$$

$$\frac{d\omega}{dt} = \frac{T_e}{J} - \frac{T_L}{J} - \frac{B}{J}\omega$$

$$\frac{d\theta}{dt} = \omega$$

where

$$T_e = -\frac{L_1N_r}{2}\left[\sin(N_r\theta)i_A^2 + \sin\left(N_r\theta - \frac{2\pi}{3}\right)i_B^2 + \sin\left(N_r\theta - \frac{4\pi}{3}\right)i_C^2\right] \quad (13.69)$$

The differential equations for the four-phase variable reluctance step motor are

$$\frac{di_A}{dt} = \frac{1}{L_A}V_A - \frac{R_A}{L_A}i_A + \frac{L_1N_r}{L_A}i_A\omega\sin(N_r\theta)$$

$$\frac{di_B}{dt} = \frac{1}{L_B}V_B - \frac{R_B}{L_B}i_B + \frac{L_1N_r}{L_B}i_B\omega\sin\left(N_r\theta - \frac{\pi}{2}\right)$$

$$\frac{di_C}{dt} = \frac{1}{L_C}V_C - \frac{R_C}{L_C}i_C + \frac{L_1N_r}{L_C}i_C\omega\sin(N_r\theta - \pi)$$

$$\frac{di_D}{dt} = \frac{1}{L_D}V_D - \frac{R_D}{L_D}i_D + \frac{L_1N_r}{L_D}i_D\omega\sin\left(N_r\theta - \frac{3\pi}{4}\right) \quad (13.70)$$

$$\frac{d\omega}{dt} = \frac{T_e}{J} - \frac{T_L}{J} - \frac{B}{J}\omega$$

$$\frac{d\theta}{dt} = \omega$$

where

$$T_e = -\frac{L_1N_r}{2}\left[\sin(N_r\theta)i_A^2 + \sin\left(N_r\theta - \frac{\pi}{2}\right)i_B^2 + \sin(N_r\theta - \pi)i_C^2 + \sin\left(N_r\theta - \frac{3\pi}{2}\right)i_D^2\right] \quad (13.71)$$

The above model does not account for magnetic saturation of the ferromagnetic material used to construct the variable reluctance step motor. A common and effective way to account for the magnetic saturation is to replace the torque expressions with an expression that is linear, instead of quadratic, in phase current. For example, the torque due to the current in phase A is modeled as:

$$T_A = -k_T\sin(N_r\theta)i_A \quad (13.72)$$

Equation 13.69 is replaced by

$$T_e = -k_T\left[\sin(N_r\theta)i_A + \sin\left(N_r\theta - \frac{2\pi}{3}\right)i_B + \sin\left(N_r\theta - \frac{4\pi}{3}\right)i_C\right] \quad (13.73)$$

and Equation 13.71 is replaced by

$$T_e = -k_T\left[\sin(N_r\theta)i_A + \sin\left(N_r\theta - \frac{\pi}{2}\right)i_B + \sin(N_r\theta - \pi)i_C + \sin\left(N_r\theta - \frac{3\pi}{2}\right)i_D\right] \quad (13.74)$$

where k_T is the torque constant, equal to the zero speed one phase on holding torque.

Bifilar-Wound Hybrid Step Motor Model

A two-phase bifilar-wound hybrid step motor is wound with two windings or one center-tapped winding per pole. A positive current in one center-tapped winding will cause the magnetic flux to align in one direction, while a positive current in the other half of the winding causes the flux to align in the reverse direction. In both cases, the current can be supplied through the center tap. Thus, this motor can be driven from a single, or unipolar, supply. For convenience, the bifilar-wound hybrid step motor is considered to have four phases, with each center-tapped winding consisting of two opposite phases. One center-tapped winding consists of phases A and C, the other consists of phases B and D. Nearly perfect flux coupling exists between phases A and C as well as between phases B and D, whereas practically no flux coupling exists between the two separate winding pairs. As a result, the flux linkage in the k-th phase is due to current in the k-th phase winding, the current in the other half of the winding pair, and the flux due to the permanent magnet. These relationships are

$$\begin{aligned}\lambda_A &= \lambda_{AA} + \lambda_{AC} + \lambda_{AF}\\ \lambda_B &= \lambda_{BB} + \lambda_{BD} + \lambda_{BF}\\ \lambda_C &= -\lambda_A\\ \lambda_D &= -\lambda_B\end{aligned} \quad (13.75)$$

where λ_k is the total flux in winding k: λ_{kj} is the flux in winding k due to the j current winding and λ_{kf} is the flux in winding k due to the permanent magnet.

If the per phase inductances, L_k, are assumed to be equal, i.e., in the k-th phase $L_k = L$ for all k, with the assumption of no saturation, and with eddy currents neglected, the flux linkage due to self-inductance is $\lambda_{kk} = L\ i_k$ and flux linkage in the opposite (j-th) phase due to mutual inductance is $\lambda_{jk} = -\lambda_{kk}$ where k and j are phases on the same pole. With these relationships between flux linkage and current, Equation 13.75 reduces to

$$\begin{aligned}\lambda_A &= L(i_A - i_C) + \lambda_{AF}\\ \lambda_B &= L(i_B - i_D) + \lambda_{BF}\end{aligned} \quad (13.76)$$

where λ_{AF}, λ_{BF} are the flux linkages due to the permanent magnet given by $\lambda_{AF} = k_o\cos(\theta)$ and $\lambda_{BF} = k_o\sin(\theta)$; where k_o is the flux constant due to the permanent magnet; θ is the rotor angular position in electrical radians; and $\theta = N_r\theta_m$, where θ_m is the mechanical position or displacement of the rotor with respect to the detent position of phase A which is at $\theta = 0$ radians. N_r is the number of rotor teeth (one mechanical period $= N_r$ electrical period).

The phase voltages, being the voltages measured on the motor terminals, are modeled as

$$V_k = R_k i_k + \dot{\lambda}_k \quad (13.77)$$

where k = A, B, C, D. R_A, R_B, R_C, and R_D represent the resistance of phases A, B, C, and D, respectively, which are not assumed equal, and the super-dot denotes derivative with respect to time.

From Equations 13.76 and 13.77, four differential equations corresponding to the four phases can be derived with the flux linkages as the state variables; however, two of these four state variables are dependent since the flux linkage in phase C is equal to the opposite of that in phase A and flux linkage in phase D is equal to the opposite of that in phase B, as stated in Equation 13.75. Only two differential equations are necessary to model the flux linkages in the motor, one for each half of the motor.

Let the flux linkages in phase A and phase B be the two state variables. Flux linkages are used as state variables instead of currents because step discontinuities can occur in the current when a phase is switched, while the flux linkages are continuous in time. For phase A, replacing for i_C and i_A (for $k = A,C$) in Equation 13.76, replacing for λ_C from Equation 13.75, and rearranging terms yields

$$\begin{aligned}\dot{\lambda}_A &= \frac{1}{2}V_A\left[1 - \frac{R_A - R_C}{R_A + R_C}\right] - \frac{1}{2}V_C\left[1 + \frac{R_A - R_C}{R_A + R_C}\right] - \frac{R_A R_C(\lambda_A - \lambda_{AF})}{L(R_A + R_C)}\\ \dot{\lambda}_B &= \frac{1}{2}V_B\left[1 - \frac{R_B - R_D}{R_B + R_D}\right] - \frac{1}{2}V_D\left[1 + \frac{R_B - R_D}{R_B + R_D}\right] - \frac{R_B R_D(\lambda_B - \lambda_{BF})}{L(R_B + R_D)}\end{aligned} \quad (13.78)$$

The torque generated by the motor can be modeled by:

$$T_E = \frac{k_T}{L}(-\lambda_A\sin(\theta) + \lambda_B\cos(\theta)) - k_m\sin(4\theta) \quad (13.79)$$

where

k_T is the torque constant
k_m is the maximum torque due to the permanent magnet

Drive Circuit Modeling

The easiest and most common way to model a motor and drive circuit is to model the model as presented above with all the resistances equal to the phase resistance, and the phase voltages as V_s (or $-V_s$ when appropriate) when a phase is on and either 0 or $-V_s$, whichever is appropriate when a phase is off. However, this models the unenergized phases as short-circuited, which, after the current has decayed, would be better modeled as an open circuit. In general, this has the effect of underpredicting the performance of the motor, especially at higher speeds.

When more accurate motor and drive system models are indicated, the phase voltages should be set to V_s, and the transistors and diodes modeled as variable resistors. Conducting transistor and diode resistances can be set to zero and non-conducting transistor and diode resistances can be set to a large value.

Control of Step Motors

Successful application of a step motor to a positioning or position tracking application requires careful attention to the control of the step motor. In this section, techniques are discussed that show how to increase the torque, double the number of detent positions, and open-loop control characteristics. The drive circuits for the various types of step motors and winding configurations was discussed in **Types and Operation of Step Motors.**

Excitation of Step Motors

Although energizing one phase at a time is the simplest way to control the step motor, greater performance from the motor is possible by exciting two phases at a time, or by switching between one phase on and two phases on at a time. This latter switching scheme is known as half-stepping.

One Phase On Excitation: By exciting the phases one at a time, the motor will move from detent position to detent position, for example, A-B-C-D-A-B-C ... for a four-phase motor.

Two Phase On Excitation: When two phases are energized at a time, the torque curves for the individual phases add. The stable detent position is halfway between the detent positions of the motor when the phases are energized one at a time. By exciting the phases two at a time, the motor will move from the new detent position to new detent position, for example, AB-BC-CD-DA-AB-BC-CD ... for a four-phase motor, where AB is the position halfway between detent position A and detent position B. Exciting the four-phase motor with two phases on at a time produces $\sqrt{2}$ more torque but consumes twice the power of exciting the motor one phase on at a time. Exciting a three-phase motor with two phases on at a time produces no additional torque but also consumes twice the power of exciting the motor one phase on at a time.

Half-Stepping Excitation: Switching the excitation alternately between one phase on and two phases on is called half-stepping excitation. This mode of operation doubles the number of detent positions of the motor, in that all of the one phase on detent positions and all of the two phase on detent positions are available. For example, by exciting a four-phase motor using half-stepping, the rotor can be made to step from detent position to detent position such as A, AB-B-BC-C-CD-D-DA-A ... .

Open-Loop Control

The most common way to control a step motor is open loop, that is, without position and/or velocity feedback. Once the motor and drive scheme have been chosen, the step command sequence must be chosen. The following examples illustrate some of the characteristics and hazards of open-loop controlled step motors.

The One-Step Move: Figure 13.88a illustrates the position versus time and Figure 13.88b illustrates the velocity versus time for a one-step move achieved by de-energizing *phase A* and energizing phase B. The characteristic position overshoot and ringing can be seen from these plots. Figure 13.88c shows the velocity versus position or phase-plane plot of this one-step move. Observe that the peak overshoot is almost half a step. The peak velocity reaches over 500 steps per second.

A Six-Step Move at a Rate of 50 Steps per Second: Figure 13.89a shows the position versus time and Figure 13.89b shows the velocity versus time for a six-step move at a rate of 50 steps per second. The stair step in Figure 13.89a is the commanded position, the straight line in Figure 13.89b is commanded velocity. We can see the motor is moving in discrete steps, in that the position and velocity profiles are almost six individual single-step responses with overshoots close to half a step and peak velocities over 500 steps per second and close to -400 steps per second.

A Six-Step Move at a Rate of 500 Steps per Second: Figure 13.90a shows the position versus time and Figure 13.90b shows the velocity versus position for a six-step move at a rate of 500 steps per second. The stair step in Figure 13.90a is the commanded position and the straight line in Figure 13.90b is the commanded velocity. In these figures, we no longer see the individual step responses in that the next phase is energized before the rotor has reached the peak overshoot for the previously energized phase. The velocity profile shows that the motor is still not running at a constant speed, but the wild velocity oscillations are gone. The characteristic ringing is seen at the end of the move.

A Six-Step Move at a Rate of 1000 Steps per Second: Figure 13.91a shows the position versus time and Figure 13.91b shows the velocity versus position for a six-step move at a rate of 1000 steps per second. The stair step in Figure 13.91a is the commanded position and the straight line in Figure 13.91b is the commanded velocity. In these figures, we see that the rotor is unable to keep up with the commanded position. A step motor is a synchronous machine, and can generate non-zero average torque only at synchronous speed. Typically, the motor will just vibrate when it looses synchronism. Here we see that the motor is attracted to the wrong detent position after the move is completed, which is four steps away from the desired position.

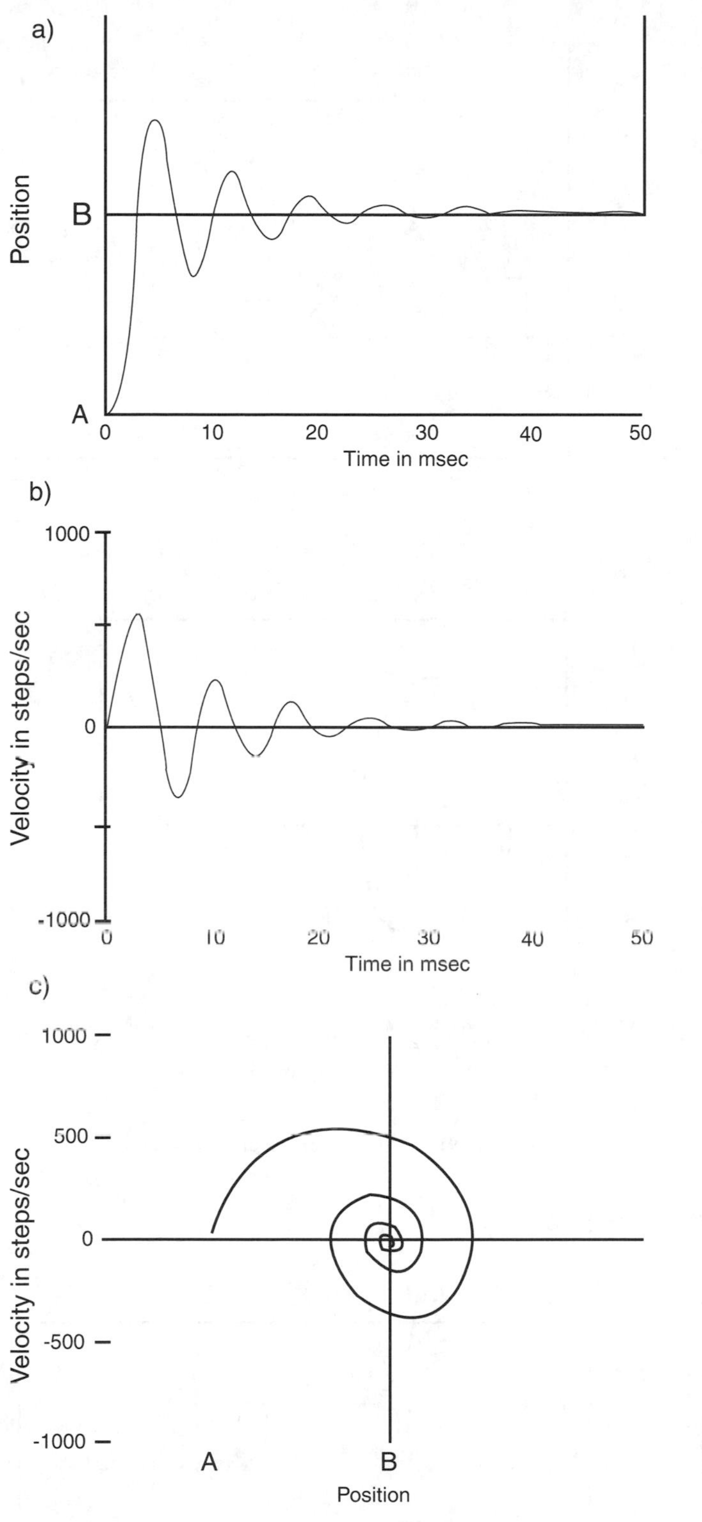

Figure 13.88 The one-step move. a) position vs. time. b) velocity vs. time.

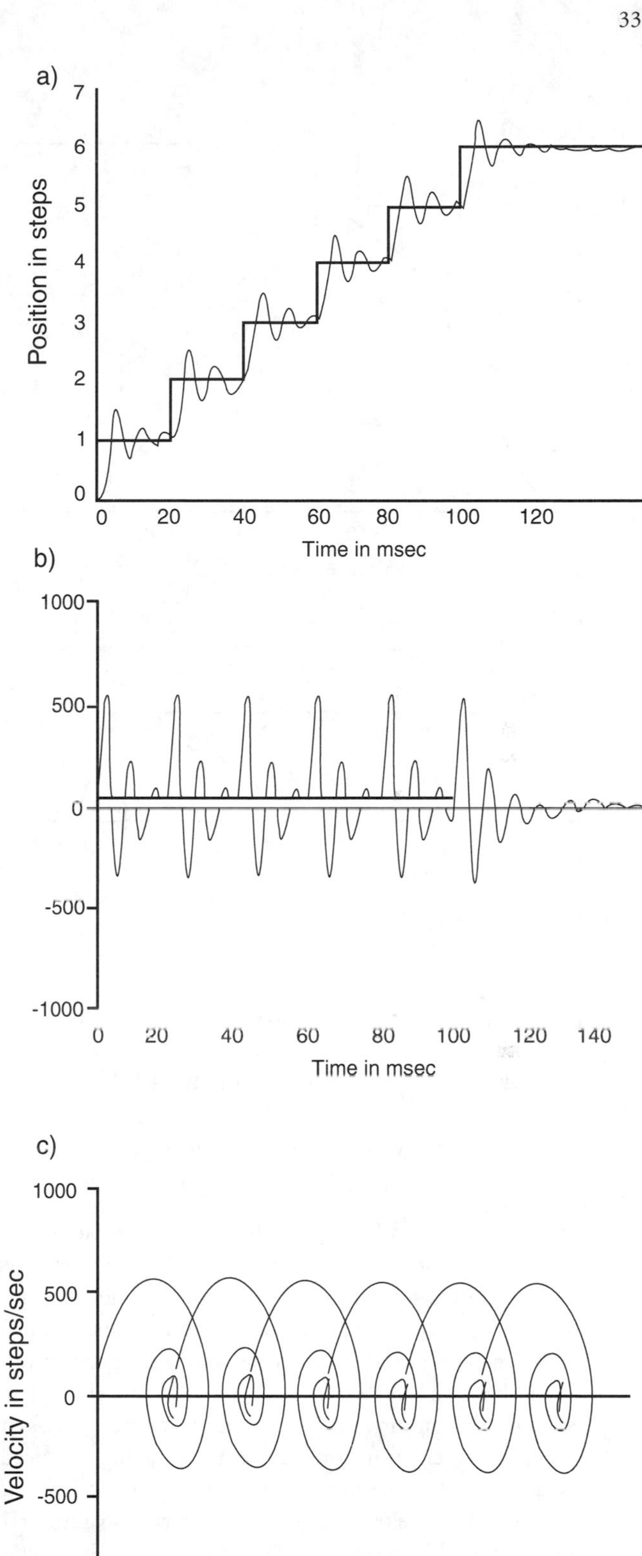

Figure 13.89 A six-step move at 50 steps/sec.

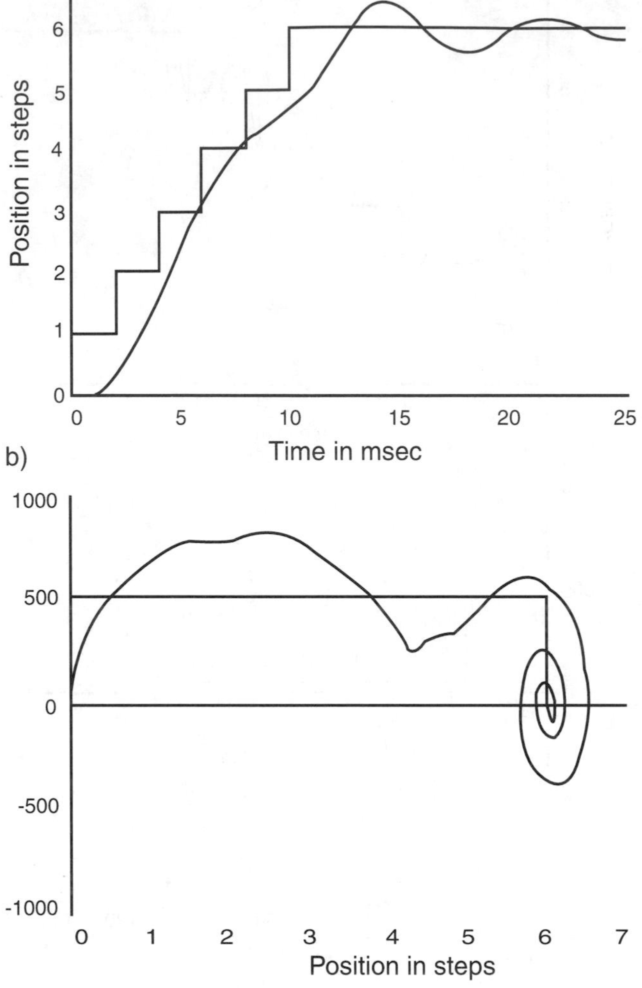

Figure 13.90 A six-step move at 500 steps/sec.

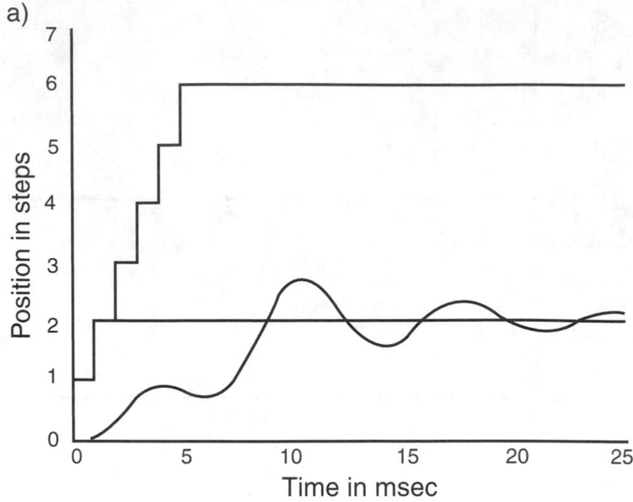

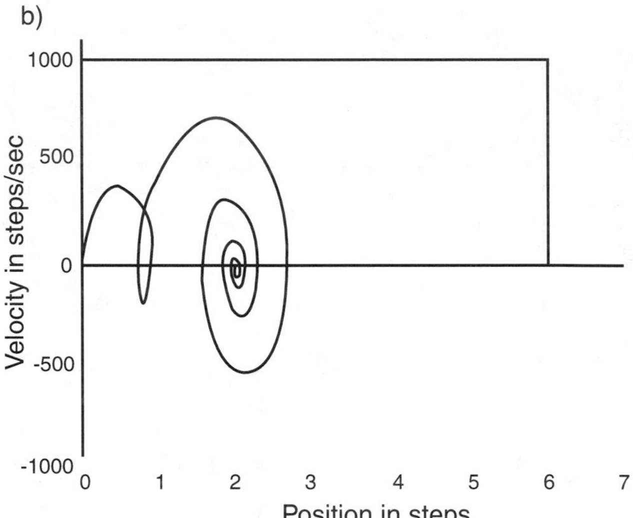

Figure 13.91 A six-step move at 1000 steps/sec.

Error-free Start-stop Rate: From the above examples we see that the motor can run at speeds up to 500 steps per second, but cannot start at a speed of 1000 steps per second. The error-free start-stop rate for this motor is between 500 and 1000 steps per second. Speed greater than the error-free start-stop rate can be achieved by starting the motor at or below the error-free start-stop rate and accelerating the motor up to a higher speed. Near the end of the move, the motor must decelerate to a speed at or below the error-free start-stop rate or it may not be able to stop at the desired position, but somewhere beyond.

Low-frequency Resonance: When a step motor is run at or near the natural frequency of its one-step response, synchronism with the commanded position can be lost due to low-frequency resonance.

Mid-frequency Resonance: Figure 13.92 illustrates another problem known as mid-frequency resonance. In this position versus time plot, the motor is started out at 400 steps per second, well below its error-free start-stop rate, and

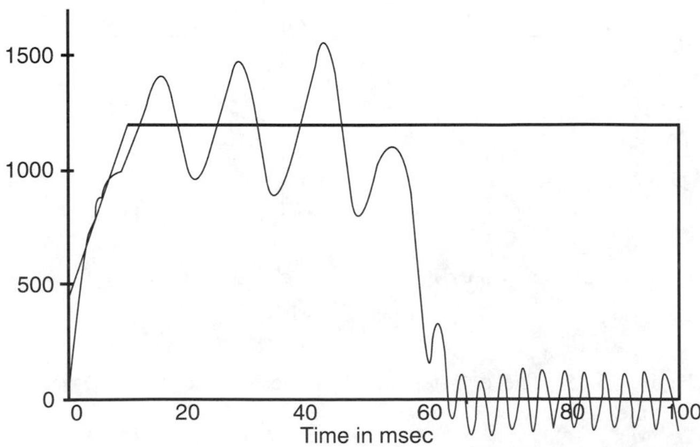

Figure 13.92 Mid-frequency resonance.

accelerated up to a speed of 1200 steps per second. At this speed, large oscillations occur in the velocity until, finally, the rotor loses synchronism with the commanded signal. This phenomenon can be avoided by accelerating through this speed range.

References

Kuo, B. C., ed. 1979. *Incremental Motion Control,* vol. 1 and 2, SRL Publishing Co.

Leenhouts, A. C. 1986. *The Art and Practice of Step Motor Control,* Intertec Communications.

Miller, T. J. E. 1993. *Switched Reluctance Motors and Their Control,* Oxford University Press, Oxford, UK.

13.7 Servo Drives

Sándor Halász

A significant and very special class of industrial drives are those that are used for position control. These drives are typically called servo drives and the intelligent control of these drives is often called motion control. Some of the application areas of servo drives are machine tool servos, robotic actuator drives, electric vehicles, computer disk drives and the like. The power level for these drives usually range below 20–30 kW; however, drives with slightly lower control quality usually have power levels below 50–60 kW.

Servo drives must meet several quality requirements, such as

1. High dynamic response which can be realized only with special control schemes and special motors with a high torque/inertia torque ratio.
2. Smooth torque production in order to achieve smooth rotation and the elimination of position angle oscillations.
3. High reliability with quick maintenance and repair.
4. Robust control, i.e., the ability of the drives to tolerate wide swings in load inertia or motor parameters.

As a result of these quality requirements, the price of servo drives can be several times that of common industrial drives of the same size.

The control scheme for servo drives usually consists of three subordinate loops as shown in Figure 13.93. The first and most inner one is the current control loop. The Y_I transfer function of the current controller is generally chosen in such a manner that the current closed loop must have a cutoff angular frequency of $\omega_{0I} \geq 1000$ r/s. The second control loop, referred to as the speed loop, usually has a closed loop control band width of $\omega_{0\omega} \geq 300$ r/s. The outer loop, or position control loop, must accurately follow the position reference. All the control loops, as a rule, are proportional integral (PI) controllers; the position controller is the only one that often employs proportional, sometimes proportional differential controller. But if a parameter of the system changes in some reasonable fashion, e.g., the inertia torque in a robotics application, this control system cannot achieve fast and accurate position control without overshoot. In this case, other types of control schemes have been used such as feed-forward, optimal, and sliding mode control, topics which are described in Chapter 7.

In some applications the servo drives require only torque control for positioning, e.g., in robotics applications. In this case, the torque control loop becomes the outer loop. For most drives a proportionality exists between torque and motor current; therefore, in this case torque control means current control. All the control loops, at present, usually employ digital control. Only the current loop, at high operating frequencies, is sometimes implemented with analog circuits. About 10% of all servo drives are used in single applications. In machine tools applications, where there are several axes of control, all the servo drives have a common DC supply, which is obtained from a standard AC supply through a common rectifier as shown in Figure 13.93. When electric motors are used, they are either permanent magnet (PM) DC motors, permanent magnet synchronous motors or, rarely, induction motors. For low-power applications stepping motors can be used, but this type of motor exhibits a considerable amount of torque pulsation. The switched reluctance motor (SRM) can also be used. The permanent magnet for DC and synchronous motors is manufactured from various types of ferrite (strontium ferrite, hard ferrite etc.), ceramic or samarium cobalt.

Servo motors normally come with different built-in sensors, e.g., encoder or resolver for position control and tachometer for velocity control. The motors generally are of a rugged design.

DC Drives

At present most servo drives are DC drives. The servo motor with permanent magnet excitation permits a 400–1000% torque overload. The torque limitation areas are shown in Figure 13.94. Area I is the continuous operating area; area II is the intermittent operating area and area III can be used only for accelerating and decelerating. These areas are limited by absolute maximum speed, an absolute commutation limit and the peak stall torque. The speed and the torque (current) control loops must take into account these limitations of the DC servo motors. DC servo motors have such low torque (size) ratings that normally their rated voltage must be less than 100–200 V. Therefore, DC servo drives when supplied from an AC source use a transformer for the creation of the supply with a reasonably rated value of voltage.

The motor supply circuits are shown in Figure 13.93. The 4-quadrant transistor chopper with a commutation frequency of 5–20 kHz ensures a very good dynamic control of the motor with very little, usually below 1–2 microseconds, dead time (the time between turn-off of one transistor and turn-on of the next). The transistors are either MOSFETs (metal oxide semiconductor field effect transistors) or IGBTs (insulated gate bipolar transistors).

The servo drives are normally unable to return the braking energy to the AC supply: this energy is lost in the DC circuit

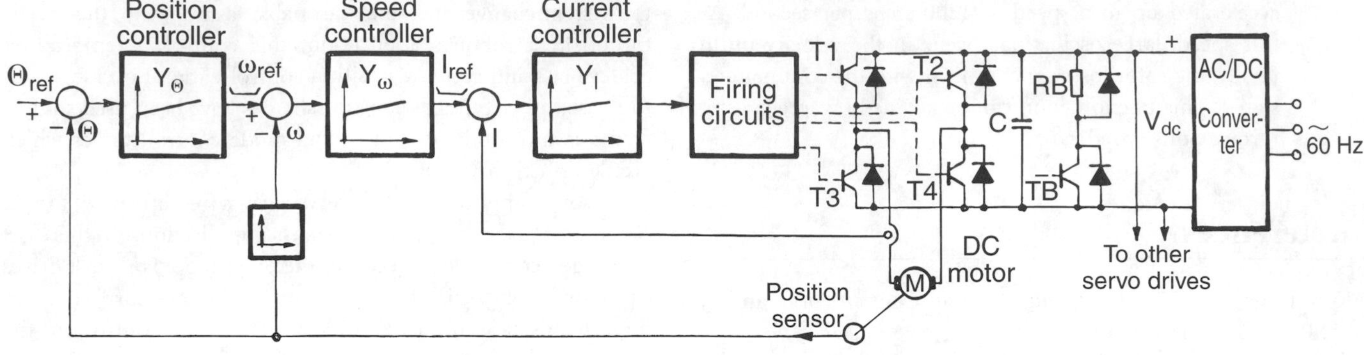

Figure 13.93 DC servo drive (with control scheme).

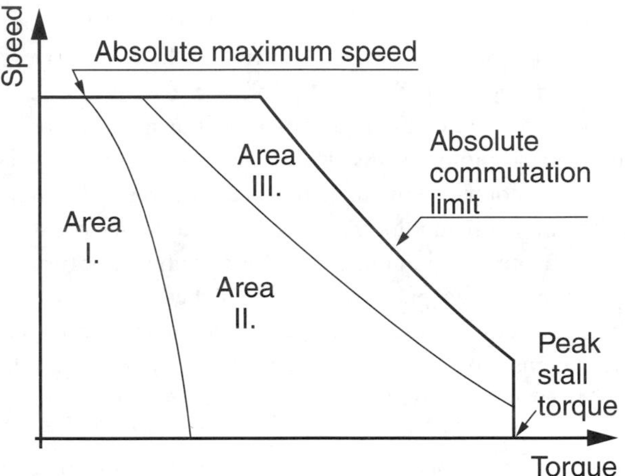

Figure 13.94 Limitations on the operating areas of the DC servo motors.

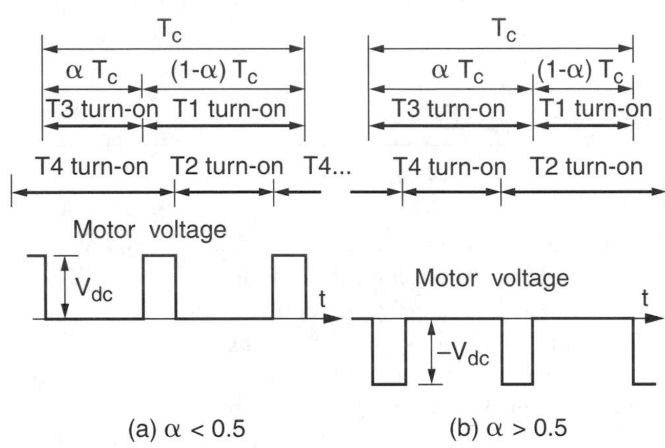

Figure 13.95 DC chopper transistor control.

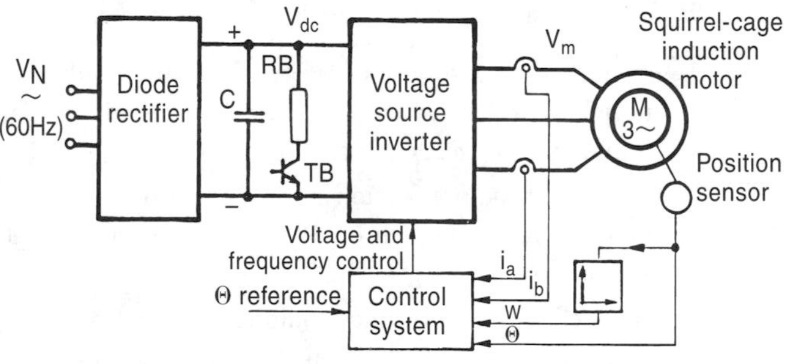

Figure 13.96 Servo drive with induction motor.

resistance, i.e., RB in Figure 13.93. The resistor current is controlled by the transitor TB. During braking, DC current flows through the capacitor C and the DC voltage increases. When this voltage achieves its maximum permitted value, the transitor TB turns on, and DC current flows through resistance RB and then the DC voltage decreases to the minimum value when transitor TB turns off. Thus, during braking, the DC line voltage is maintained between a maximum and a minimum value.

The control scheme for the chopper transistors is presented in Figure 13.95. A so-called overlapping control is commonly used. The T_C time-cycle is derived from the times αT_C and $(1 - \alpha)\ T_C$. The *T1* and *T4* transistors turn on during the time period $(1 - \alpha)\ T_C$, and the *T2* and *T3* transistors turn on during the time period αT_C; however, the turn-on and turn-off times of the odd and even transistors are shifted, i.e., overlapping, by $(1–2)\ T_C/2$ as shown in Figure 13.95. If α is between 0 and 0.5 the motor voltage is positive and if α is between 0.5 and 1.0 the motor voltage becomes negative as illustrated in Figure 13.90. A very important advantage of this control scheme is that the motor voltage and current waveforms repeat twice during one period (T_C) of the transistor control.

As a result of the high-frequency control, the motor current (and torque) consists only of high-frequency harmonics with very low amplitudes, usually under 1% of the motor's rated current. This means that in both the transient and steady states the motor current and torque consists of virtually only a DC component and therefore there are no speed (or position) oscillations.

Induction Motor Drives

The induction servo motor with a squirrel cage rotor has very small rotor inertia torque, high reliability and it is very economical. However, the control system for the induction motor is very complicated, expensive and the quality of the control is sensitive to motor parameter changes. Therefore this motor is not widely used.

The typical supply circuits for the induction servo motor are shown in Figure 13.96. The AC supply voltage feeds the diode rectifier which creates the DC link. The DC link consists of the capacitor C, braking resistor RB and transistor TB. The control of the DC voltage during the braking operation is performed in the same manner as that for DC drives. The voltage source inverter is usually constructed with IGBT transistors and very fast parallel diodes. In the last several years, the use of IGBT modules with six transistors and six diodes has been the preferred configuration.

The drive does not need a transformer since high-voltage motors are available. If the AC phase voltage is V_N (RMS value), then the DC link voltage will be $V_{dc} \cong 3/\pi\sqrt{6}V_N$ and the maximum possible motor phase voltage will be

$$V_m = \frac{2}{\pi}\frac{1}{\sqrt{2}} \quad V_{dc} = 3\frac{\sqrt{12}}{\pi^2} \quad V_N \cong 1.05V_N \tag{13.80}$$

Hence, if the rated voltage of the motor is equal to the AC supply voltage, then as a result of the voltage drop in both the rectifier and the inverter, the motor can operate with a rated flux between 0 Hz and the approximate frequency of the AC supply.

A position control system usually uses the indirect field oriented principle (see Section 13.3). The rotor flux is generated by the two phase currents as well as the speed, as shown in Figure 13.96. The calculation is a function of the rotor time constant, which is dependent upon both the rotor resistance and rotor inductance. Variations in these parameters must be taken into consideration; however, the identification of the parameter changes is very complicated.

PM Synchronous Motor Drive

The permanent magnet synchronous motor is much more expensive than the squirrel cage induction motor, but the control system of the PM synchronous motor drive is much simpler than that used for the induction motor. When compared to DC motors, PM synchronous motors normally have less inertia torque and require less maintenance. As a result of these features, PM synchronous servo drives have become one of the most popular types of servo drives. The converter circuits for PM synchronous motor drives are identical to those for induction motors as shown in Figure 13.96. PM synchronous motors, such as induction servo motors, are usually manufactured for high voltage and therefore transformers are not required in their use.

There are the two classes of PM synchronous motors:

1. Those with a square flux density distribution along the rotor air gap surface, as shown in Figure 13.97a, which produces a trapezoidal back-EMF (electromotive force) in the stator coil—the so-called trapezoid PM machines.
2. Those with a sinusoidal flux density distribution as shown in Figure 13.97b, which produces a sinusoidal back-EMF—the so-called sinusoidal PM machines.

In the trapezoidal machines the angle β illustrated in Figure 13.97a is the width of the magnet. In general, β≅ 180°. In Figure 13.98 the trapezoidal machine with β = 180° is presented and a two-pole machine is assumed. In steady-state the machine is rotated with constant synchronous speed which is a function of the number of pole pairs p and the frequency of the stator supply f_1

$$\omega_1 = \frac{2\pi f_1}{p} = \text{const} \tag{13.81}$$

Consider Figure 13.98a or Figure 13.98b where the machine is expanded along the stator air gap surface. In the range $-60° \leq \omega_1 t \leq 60°$, the a phase conductors are located under the maximum flux density B, i.e., $a+$ is under +B and $a-$ is under −B. Hence in this timeframe in the a phase the maximum value of the back-EMF E_a is induced as shown in Figure 13.98c. For $\omega_1 t \geq 60°$, E_a begins to decrease since the $a+$ conductors (or $a-$) are in the flux density of different directions. As shown in Figure 13.98b at $\omega_1 t = \pi/2$ half of the a phase coil will be under a positive and the other half under a negative value of the flux density; therefore, at this time $E_a = 0$. As a result this analysis indicates that the back-EMF-time function is a trapezoidal shape of the form shown in Figure 13.98c.

Suppose that the drive control only permits stator current to flow in two phases at any time. With reference to Figure 13.98a, positive current is supplied to phase a and negative current is supplied to phase b. The resulting stator phase currents are shown in Figure 13.98d. This current distribution is achieved by the appropriate phase current commutations through the use of a position sensor signal (once for every 60°). The motor torque will be

$$T = cBI \tag{13.82}$$

where c is a motor constant. The torque will not have ripples if the current is constant and this constant current is ensured by

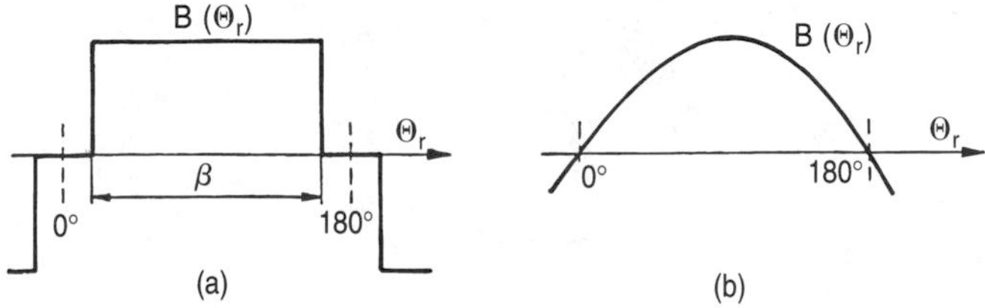

Figure 13.97 Two types of the PM synchronous servo motors: (a) with square flux density distribution (trapezoidal PM machines); (b) with sinusoidal flux density distribution (sinusoidal PM machines).

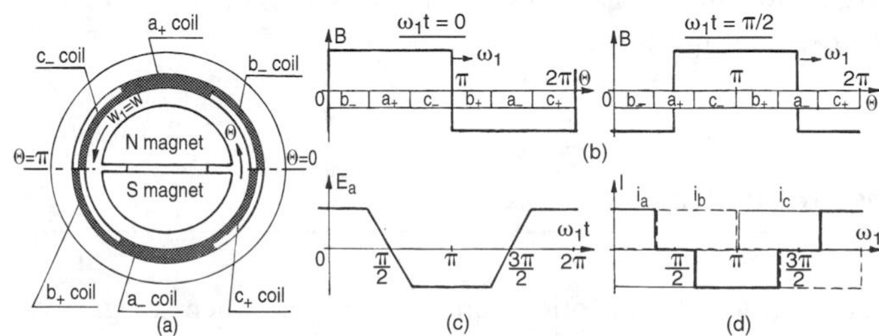

Figure 13.98 Trapezoidal PM machines with β = 180°: (a) motor construction; (b) flux density displacement at $\omega_1 t = 0$ and $\omega_1 t = 90$; (c) back-EMF vs time; (d) motor phase currents vs time.

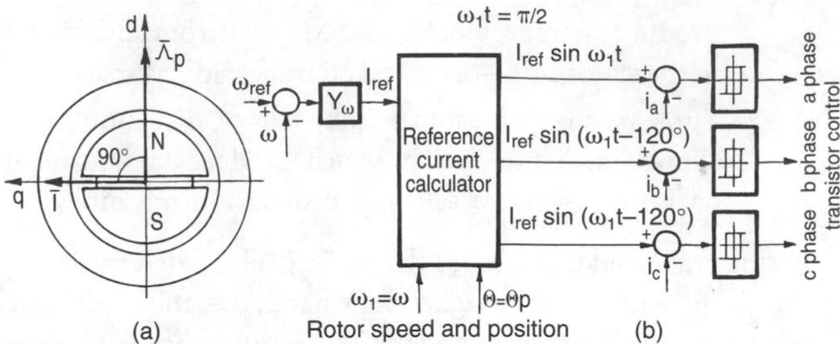

Figure 13.99 Sinusoidal PM machines: (a) pole flux and current vector orientation; (b) control schemes.

DC current control just as it is for DC servo drives. But now under control are only the transistors which belong to the two current conducting phases. As a result of the high-frequency current control, the torque is essentially constant; however, during the phase current commutations, i.e., every $\omega_1 t = 60°$, current control is not possible. Hence, torque oscillations occur at a frequency of $6f_1$, which is a considerable disadvantage of trapezoidal machines.

In the sinusoidal PM machines the sinusoidal flux density distribution will produce a constant torque only if the phase currents are also sinusoidal. The sinusoidal values can be characterized by vectors as shown in Figure 13.99. The $\overline{\Lambda}_p$ pole flux linkage vector and the $\bar{I}$ current vector will produce the torque:

$$T = c_1 \overline{\Lambda}_p x \bar{I} = c_1 \Psi_p I \sin(\Lambda_{\hat{p}} I) \qquad (13.83)$$

where c_1 is a constant. Therefore, if the angle between these two vectors is equal to 90°, as shown in Figure 13.99, the torque is maximized. Current control is normally achieved as shown in Figure 13.99b. The position sensor signal requires the creation of three sinusoidal phase current reference signals, which generate the current vector with a 90° displacement from the pole flux vector. The three Schmidt triggers ensure two-point phase current control with the desired hysteresis. Because the phase current hysteresis is very small the motor torque ripples are very high frequency and have very small values. The important advantage of sinusoidal machine drives is that there are no torque oscillations with $6f_1$ frequency, as is the case with trapezoidal machines.

References

Jahns, T. M. 1994. Motion control with permanent-magnet AC machines, *Proc. IEEE,* 82(8):1241–1255, August, special issue.

Kenjo, T. and Nagamori, S. 1985. *PM and Brushless DC Motors,* Clarendon, Oxford, UK.

Miller, T. J. E. 1993. *Brushless Permanent-Magnet and Reluctance Motor Drive,* Clarendon, Oxford, UK.

13.8 Switched Reluctance Motor Drives

Jozsef Borka

Construction and Operation

Construction

A switched reluctance motor (SRM) is a doubly salient, single-excited motor. This means that it has salient poles on both the rotor and the stator, but only the stator carries windings. The rotor has no windings, magnets, or cage windings, but it is built up from a stack of salient-pole laminations as shown in Figure 13.100. The simplified principle of operation is as follows: when current is passed through the phase (stator) windings, the rotor tends to align with the stator poles, i.e., it produces a torque that tends to move the rotor to a minimum-reluctance position. When a rotor pole is approaching the aligned position of the excited stator pole, positive (motoring) torque is produced regardless of the direction of the current.

The classical forms of SRM are those with stator/rotor pole numbers of 6/4 and 8/6, but others are also possible. The relationship between speed and fundamental frequency is obtained from the fact that if the poles on the stator are wound oppositely in pairs to form phases, then each phase produces a torque pulse on each passing rotor pole, thus the fundamental frequency in one phase will be

$$f_1 = n.N_r \quad \text{Hz} \qquad (13.84)$$

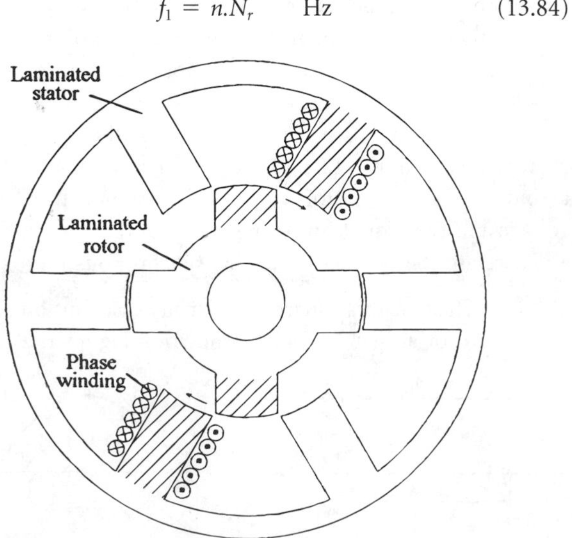

Figure 13.100 Cross section of SRM. *Source*: Miller T. J. E. 1993. *Brushless Permanent-Magnet and Reluctance Motor Drives*, Oxford University Press, U. K. Used by permission.

if n is the speed of rotation in r/s and N_r is the number of rotor poles.

If there are q phases, then the step angle is

$$\epsilon = 2.\pi/q.N_r \qquad \text{radians} \tag{13.85}$$

The number of stator poles usually exceeds the number of rotor poles.

Operation Principle

The SRM cannot start or run from an AC voltage source; it is normally necessary to use a shaft position sensor for commutation and speed feedback. The simplified diagram of an SRM drive is illustrated by Figure 13.101. The phase winding and the switching circuit of only one of the four stator phases are shown in this figure. For the motoring operational mode, a stator phase must be excited when a pair of opposite rotor poles is approaching its poles, and it must be turned off before rotor and stator poles actually come into alignment. The continuous rotation of the rotor is obtained by exciting the stator poles sequentially; the rotor rotates in a direction opposite to that of stator phase excitation. Thus, the direction of rotation is determined by the excitation sequence of the stator poles.

The stator winding inductance varies cyclically with the rotor position and has an angular periodicity

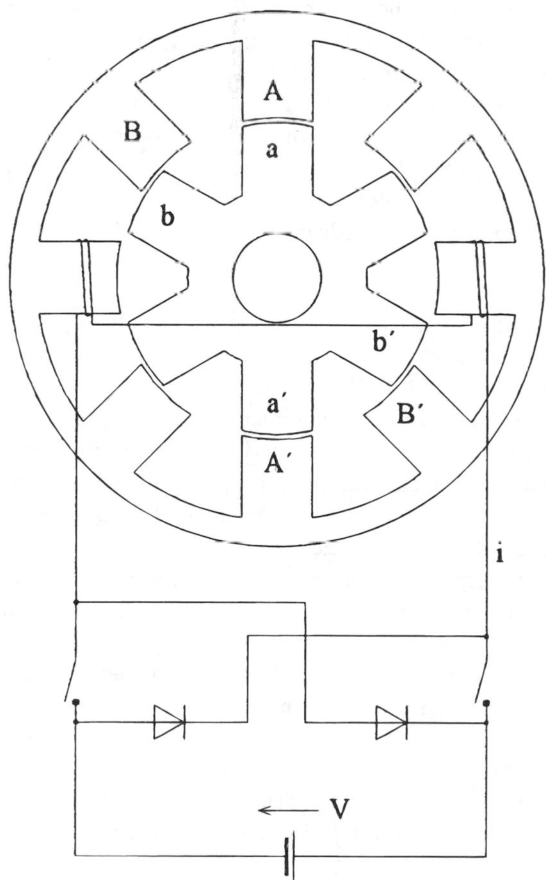

Figure 13.101 Elements of a 4-phase SRM showing one circuit.

$$\Theta_{cy} = 360/N_r \qquad \text{degrees} \tag{13.86}$$

An excitation current pulse is applied during each cyclic variation of inductance and the idealized current curve is shown in Figure 13.102. If magnetic saturation is neglected, then the torque is proportional to the product of i^2 and $\partial L/\partial\Theta$. In accordance with that, Figure 13.102 shows the active regions for motoring (positive) torque and generating (negative) torque for the case with constant current excitation.

The "real" current waveform is shown in Figure 13.103 for high-speed operation. This figure shows three important angles or rotor positions, Θ_{on} which is the turn-on angle, usually unaligned position; Θ_{off} which is the turn-off angle, usually a little before aligned position; and Θ_{ext} which is the current extinction angle. If $\Theta_{ext} - \Theta_{off}$ is too large, the current, when $\partial L/\partial\Theta$ is negative, produces a negative or braking torque. Thus, Θ_{off} should be carefully chosen so that the maximum torque can be obtained.

Fundamental Equations and Equivalent Circuit

As has been shown previously, the inductance of each motor phase is a function of the rotor position. It varies between a minimum and a maximum value. The variation depends on the motor construction. We assume that the variation of the phase inductance is linear; furthermore the mutual inductance between motor phases is negligible and consequently those can be considered independently. It is also assumed that the motor

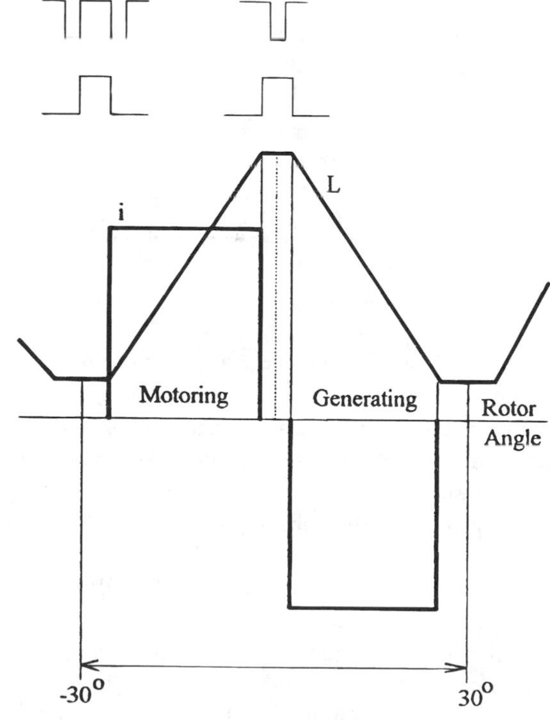

Figure 13.102 Motoring and generating regions of the phase inductance cycle. *Source*: Miller T. J. E. 1993. *Brushless Permanent-Magnet and Reluctance Motor Drives*, Oxford University Press, U. K. Used by permission.

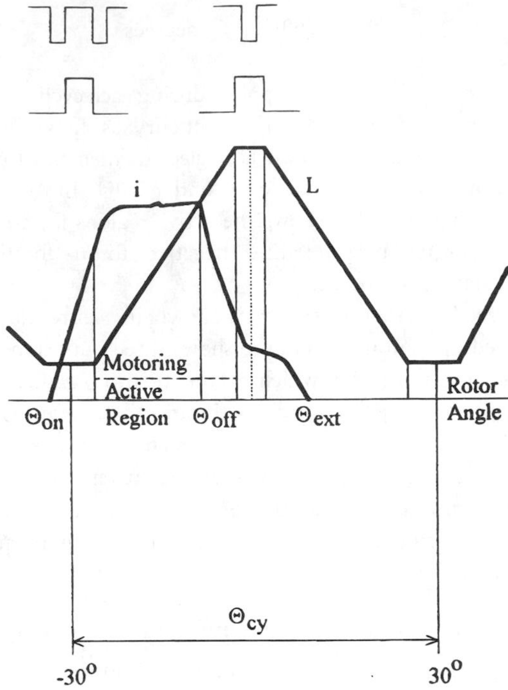

Figure 13.103 Current and inductance variation.

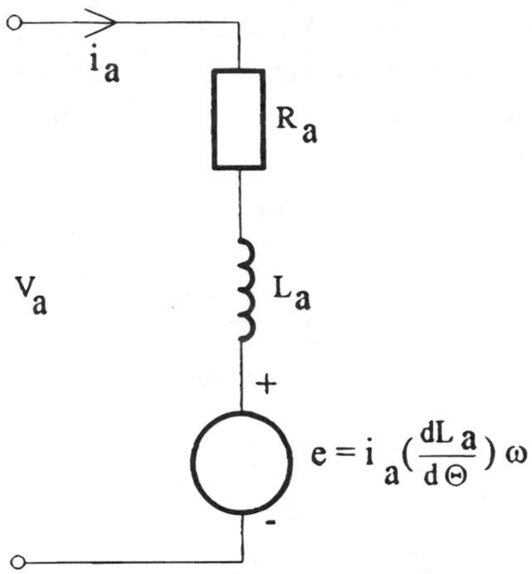

Figure 13.104 An equivalent circuit for one phase.

is not saturated, thus linear analysis can be used for simplicity. In practice, most SRMs operate in a saturated condition to provide high output torque.

The voltage equation for phase (a) can be written as:

$$v_a = i_a.r_a + L_a(\Theta).di_a/dt + i_a.\omega.dL_a(\Theta)/d\Theta \quad (13.87)$$

This equation gives an equivalent circuit containing a resistor, an inductance and a counter emf as shown in Figure 13.104. In general, this equation is similar to that of other motors. It can be seen, however, that the emf is a function of the current, the rotation speed, and the inductance variation. Depending on the rotor position and the current waveform, the equivalent circuit can change from being mainly inductance to mainly emf within one stroke in this simplified case.

In SRMs the torque is produced by the variation of the motor reluctance with the rotor position. The developed electromagnetic torque can be calculated, taking the variation of the system co-energy into consideration. If the motor is not saturated and the mutual inductance between the windings is negligible, then the instantaneous torque produced by one winding will be

$$T = 1/2.i^2.dL/d\Theta \quad (13.88)$$

where i is the motor instantaneous current.

Supply Units

Since the torque is independent of the current direction because of the square relation, the SRM can operate with unipolar currents or converters, but the converter drive signals must be synchronized with the rotor position. The use of unipolar converter circuits has a number of advantages over the corresponding circuits for AC motors or PM brushless motors, which require alternating currents.

Figure 13.105 shows some of the different, possible circuit arrangements for supplying SRMs. They are well suited for use with power transistors or GTOs. The phases are independent, and in this respect the SRM converter differs from an AC motor inverter in which the motor windings are connected between the midpoints of adjacent inverter phaselegs. With SRM converters the windings are in series with the switches, providing valuable protection against faults. If the phase number is signed by q, then the number of switches in the different arrangements for supplying the SR motor may be q, $q + 1$, $2q$. All arrangements have advantages and disadvantages.

Controls

Control Problems

In the motoring operational mode, the pulses of phase current must coincide with a period of increasing inductance, i.e., when a pair of rotor poles is approaching alignment with the excited stator poles. The timing and dwell of the current pulse determine the torque, efficiency and other motor parameters. In DC and permanent magnet (PM) brushless motor the torque/ampere relation is more or less constant, but in SRMs no such simple relationship occurs, naturally. With fixed firing angles, there is a monotonic relationship between the average torque and the rms phase current. This is a square relation neglecting the saturation; however, if we take saturation into account, the connection will be much more complicated. This may cause some problem in the feedback control system, but a "near-servo quality" dynamic performance can be achieved.

Angle Control, Shaft Position Sensing

The commutation requirement of an SRM is very similar to that of the PM brushless motor. The shaft position sensing

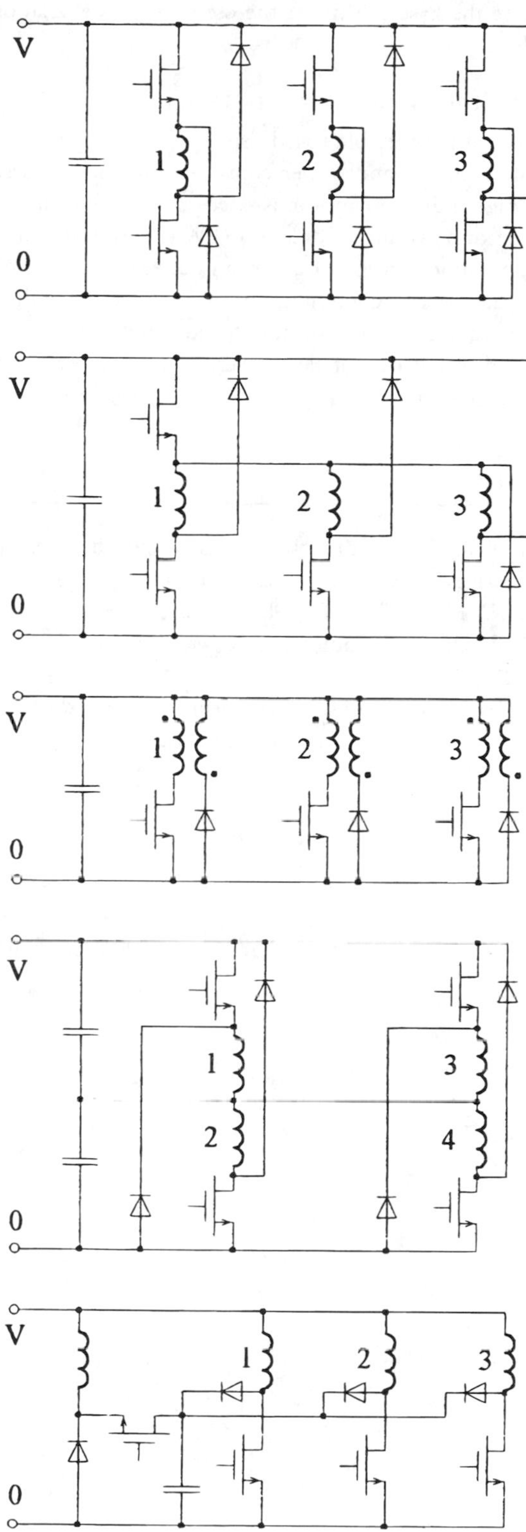

Figure 13.105 Converter circuits for SRM.

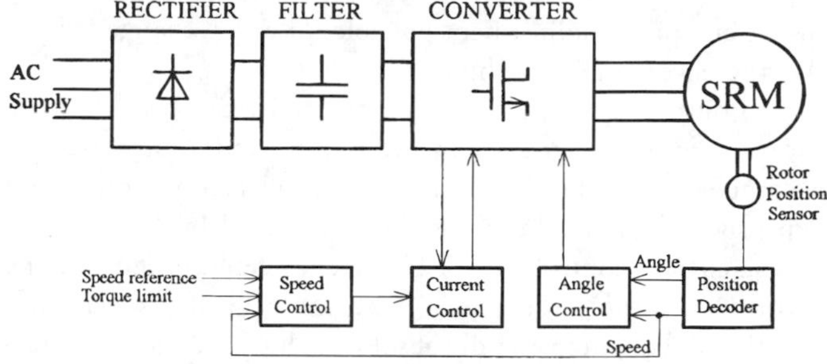

Figure 13.106 Block scheme of SRM control circuit.

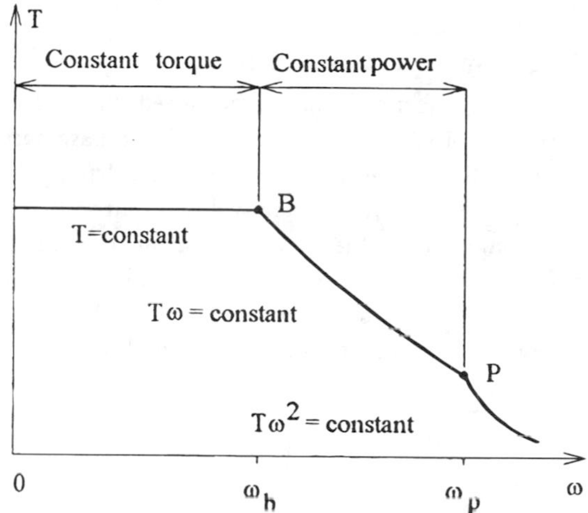

Figure 13.107 General torque speed characteristic of SRM.

and the decoding logic are also very similar, and similar arrangements can be used in some cases. In position or speed servos, optical encoders or resolvers may be used to perform all the functions of providing commutation signals, speed and position feedback. Much has been made of the undesirability of the shaft position sensor because of the associated cost, the space requirement, and the possible extra source of potential failure. Operation without a shaft sensor is possible, but achieving good performance requires a considerable amount of extra complexity in the controller.

Speed and Current/Torque Control

The general structure of a SRM control scheme is much the same as that of a PM brushless drive, as shown in Figure 13.101. Fundamentally, the SRM drive has two control loops. The outer loop is the speed loop and the inner loop is the current/torque loop. There is an additional loop for digital angle control. In case of a four-quadrant operational mode, a brake chopper is necessary to consume the recuperated energy. Hardware and software together implement the two-loop control. A speed feedback signal is calculated from the variation of position. In a general case, the controller is of PID (Proportional, Integral and Derivative elements) type, preferably PDF type (an integral element with Proportional and Derivative Feedback). This algorithm can be well applied to control the speed of drives, as it makes an optimal adjustment possible for a step change of both the reference signal and the load.

The output signal of the speed controller can be used for a torque command signal to ensure fast dynamics, but inserting a torque control loop makes the controller complicated and expensive. A good solution could be achieved by using only a simple current control loop. The current reference signal can be formed from the torque signal with the help of a square-root function after composing the absolute value of the torque reference signal. The motoring and generating operational modes of SRM drives can be determined from the sign of the torque reference signal and the momentary direction of rotation. The change in operational mode is performed by the digital angle controller.

Torque-speed Characteristics

The general form of the torque-speed characteristics is shown in Figure 13.102. For speeds below the base speed the torque is only limited by the motor current. Up to the base speed, it is possible to achieve any value of current up to the maximum value. The value of current at an operating point depends on the load characteristics, speed, and control strategy. Operating in this speed range provides considerable freedom in design to obtain smooth torque, which simplifies the control. Only above the base speed can the constant power relation be achieved.

Application Field

Most of the variable speed drives produced today consist of brushless motors and power converters. In many cases, the squirrel cage induction motor is used and is controlled by a voltage fed pulse width modulation (PWM) inverter. For most of the applications, the SRM drive can be seen as a competitive system to the PWM induction motor drive, especially in the field of high-speed and/or large starting torque applications, i.e., fan and pump drives, drives for electric vehicles, machine tool drives, conveyor drives, etc.

References

Borka, J. et al. 1993. Control a spects of switched reluctance motor drives, *ISIE '93, Budapest, Hungary*, pp. 286–300.

Bose, B. K., Miller, T. J. E. 1985. Microcomputer control of switched reluctance motors, *IEEE/IAS Annual Meeting*, pp. 542–547.

Miller, T. J. E. 1989. *Brushless Permanent-Magnet and Reluctance Motor Drives*, Oxford University Press, U.K.

14

Main Disturbances

James Stanislawski
National Power Laboratory

Gerry Heydt
Arizona State University

Prasad Enjeti
Texas A&M University

Laura Steffek
Best Power

John Hecklesmiller
Best Power

Dave Layden
Best Power

Brian Young
Best Power

14.1 Power Quality 349
What is Power Quality? • Sources of Power Quality Information • Power Problems • Varying Effects of Low Voltage Power Disturbances • Protecting Equipment from Power Problems

14.2 Reactive Power and Harmonics Compensation 352
Introduction • Reactive Power • Reactive Power Compensation • Reactive Power in Nonsinusoidal Circuits • True Power Factor Versus Displacement Power Factor • Reduction of Harmonic Signal Amplitudes • Harmonic Filters and Power Factor Compensation

14.3 New Power Converters 363
Single-Phase Active Power Factor Correction • Improved Three-Phase Utility Interface Converters

14.4 Uninterruptible Power Supplies (UPS) 367
UPS Functions • Static UPS Topologies • Rotary UPSs • Alternate AC and DC Sources

14.1 Power Quality

James Stanislawski

As sensitive electronic loads proliferate on commercial utility grids, the concern over power quality also increases.

What is Power Quality?

Power quality is the degree to which the utilization and delivery of electrical power affect the performance of electrical equipment. Any power line disturbance that affects the performance of sensitive electronic equipment is said to be related to power quality.

Some power line disturbances are created by day-to-day utility operations, but these account for only part of the problem. Often, power line disturbances are created by equipment within a building or by acts of nature.

To determine how many of the disturbances might affect computer hardware or other sensitive microprocessor-based devices, a susceptibility curve was created. The curve (Figure 14.1) can be found in the CBEMA/IEEE 446–1987 document. The vertical scale on the CBEMA curve depicts voltage levels from zero to +300% of nominal. The horizontal scale depicts duration in either "cycles" or seconds. The shaded area represents voltage and duration combinations that are considered "safe" for sensitive electronics. Events outside the shaded area may cause data errors, equipment malfunctions or equipment damage.

Each year the average location experiences 289 power line deviations which fall outside the limits of the CBEMA curve, according to data collected in a study by National Power Laboratory (NPL) (Dorr, 1994). Any of these power line deviations can corrupt data or damage computers or other sensitive electronic equipment.

Sources of Power Quality Information

The Institute of Electrical and Electronics Engineers (IEEE) recently released standard P1100-1992, *IEEE Recommended Practice for Power and Grounding Sensitive Electronic Equipment,* more commonly known as the Emerald book. In part, this book discusses power quality problems at the user or facility level and recommends devices that can be used to protect electronic equipment and data.

The appendix to the IEEE Emerald book, though not part of IEEE standard 1100–1992, combines the data from two comprehensive U.S. power quality studies—the Allen-Segall (IBM) study conducted from 1969–1972 (Allen and Segall, 1974) and the Goldstein-Speranza (AT&T) study conducted from 1977–1979 (Goldstein and Speranza, 1982). However, it should be noted that these studies were conducted before computer switch mode power supplies and other nonlinear loads became common. These types of loads, which have proliferated on the utility grids, can introduce severe power line disturbances in facility branch circuits.

More recent data is available from a five-year study begun in 1990 by National Power Laboratory (NPL). The purpose of the

0-8493-8343-9/97/$0.00+$.50

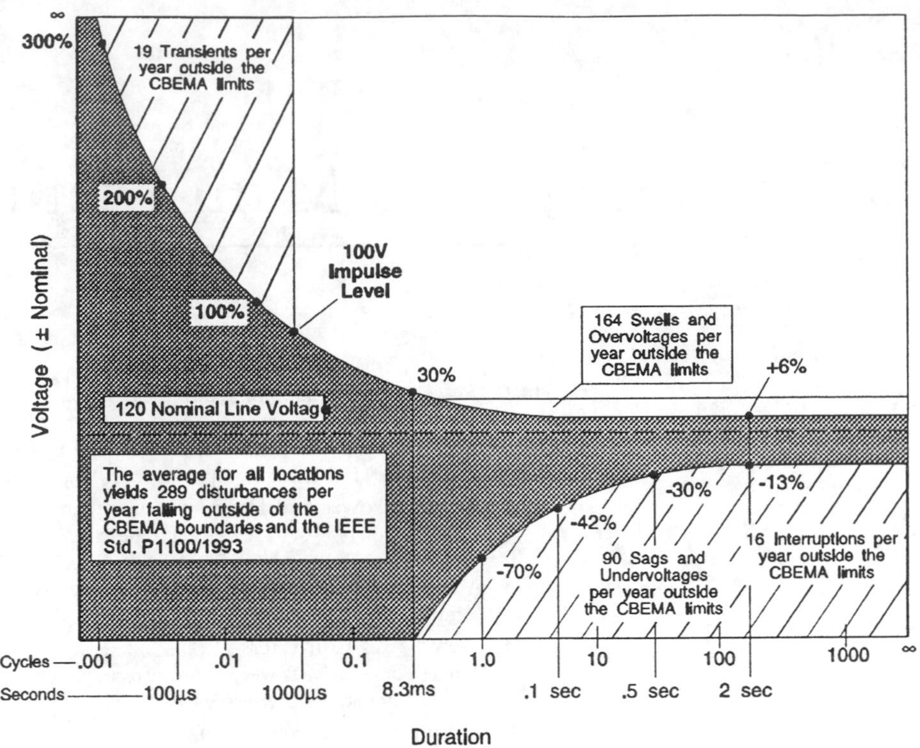

Figure 14.1 The CBEMA curve.

NPL study is to provide a large, well-defined database that profiles power quality at typical power usage points (Jurewicz, 1990). The NPL study collects single phase, line-to-neutral data at standard wall receptacles, where power disturbances can affect computers and other electronic equipment.

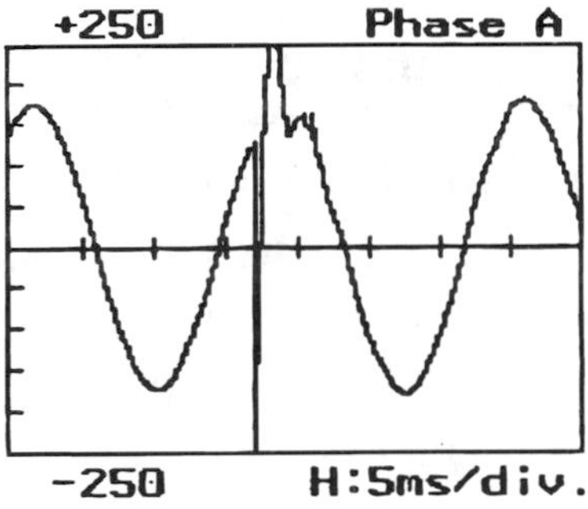

Figure 14.2 High-energy transients.

Power Problems

A number of different kinds of power problems occur on building power lines, including transients, high voltage events, low voltage events, outages and harmonic distortion. These power problems are usually random and unpredictable and can occur any time, anywhere.

The information cited in this section is based on 1057 site-months of NPL data collected from 112 randomly selected sites in the North American continent.

Transients

Transients, sometimes called spikes or impulses, are sharp, brief discontinuities of the AC waveform. Transients may be of either polarity and may be additive to or subtractive from the nominal waveform.

High-energy transients (Figure 14.2), most commonly caused by utility capacitor switching, can damage electronic components, microchips, and networked equipment thousands of feet away from the origin of the transient.

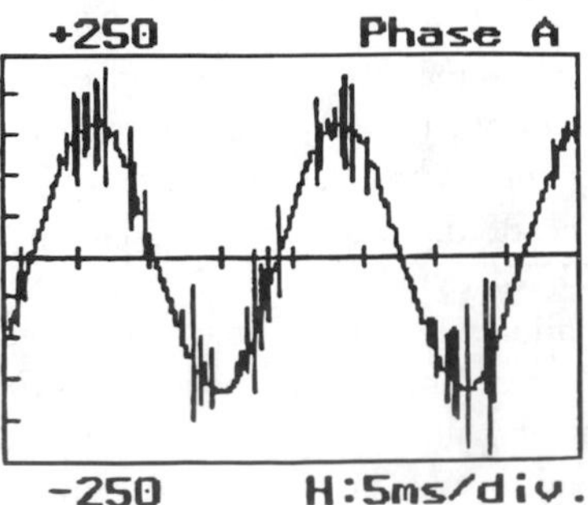

Figure 14.3 Data corrupting transients.

Data corrupting transients (Figure 14.3) typically do not have much energy, but because of their high frequency nature they can travel to the heart of delicate processors or couple to data

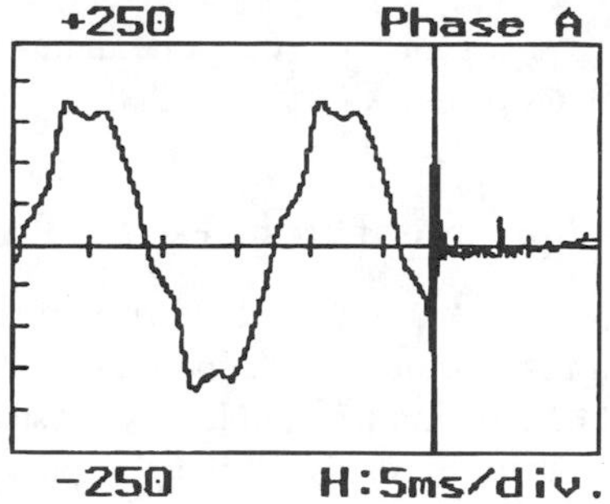

Figure 14.4 Transients caused by lightning strikes.

transmission cables and cause data corruption. Data corrupting transients are usually created by the switching on and off of electrical components or equipment inside the building.

Transients caused by lightning strikes (Figure 14.4) are rare but are usually the most damaging. Transients caused by lightning strikes are both high energy and high frequency in nature, and they therefore cause both equipment damage and data corruption.

According to the NPL study, the average site experiences over 1000 transients[1] per year, which constitute over 50 percent of all recorded power disturbances. Of these 1000+ transients, 19 go outside of the CBEMA limits (Dorr, 1994).

High-Voltage Events

High-voltage events are voltage increases above the recommended level for a piece of electronic equipment.

High-voltage events have a number of causes. Usually, high-voltage events are created when large loads like air conditioners and motors are disconnected—the sudden decrease in demand creates a high-voltage event. Sometimes, auxiliary generators supply unstable voltage during start-up and create high-voltage disturbances. Occasionally, high-voltage events within a building occur because the utility voltage tap at the building entrance is set too high.

High-voltage events can cause overheating of electronic components and premature failure of electronic components.

According to the NPL study, each year the average site experiences 164 high-voltage disturbances outside of the CBEMA limits (Dorr, 1994).

Low-Voltage Events

Low-voltage events (Figure 14.5) are voltage decreases below the recommended level for a piece of electronic equipment.

Low-voltage events often occur when large loads such as elevators, air conditioners or motors turn on—such equipment needs a lot of energy when first started and extracts energy out of the utility line, causing a low-voltage condition. Occasionally, low-voltage events are caused by short circuits on utility power lines.

Low-voltage events can cause unexplained system resets, process control shutdowns or errors, and data errors. These types of equipment problems occur because many microprocessor-based systems continue to operate even when voltages dip below the level at which logic chips begin to behave erratically.

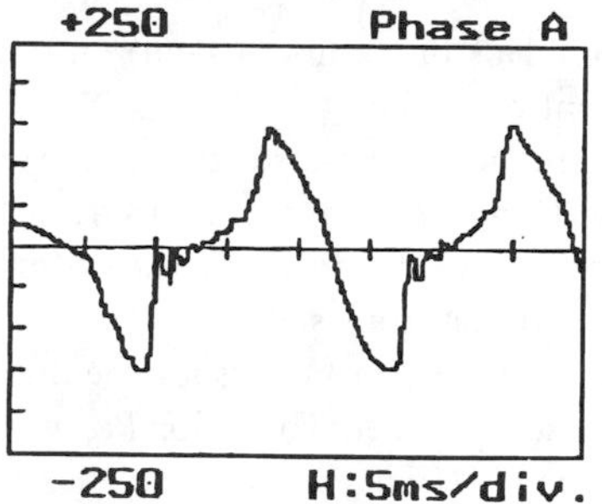

Figure 14.5 Low-voltage events.

According to the NPL study, each year the average site experiences 90 low-voltage events outside of the CBEMA limits (Dorr, 1994).

Outages

Outages are complete losses of voltage.

Outages have a number of causes, including tripped circuit breakers, downed power lines, and utility faults. Interestingly, NPL study data indicate that lightning strikes account for many short-term outages—such outages occur when utility protection devices open the line to prevent the high energy transient from reaching the building load equipment.

According to the NPL study, each year the average site experiences 16 outages outside the CBEMA limits, the least frequently recorded power disturbance. The number of long-term outages (>30 minutes) at a given site varies widely. In the NPL study, some locations did not experience any long-term outages, while other locations experienced as many as six in one year (Dorr, 1992).

Harmonic Distortion and IEEE Standard 519

Harmonic distortion (Figure 14.6) is discontinuity in a pure sine wave (Waller, 1994).

Harmonic distortion is often caused by equipment such as computers, peripherals, and adjustable-speed drives. Since most microprocessor-based electronic equipment does not draw current continuously, it distorts the AC input voltage when actually

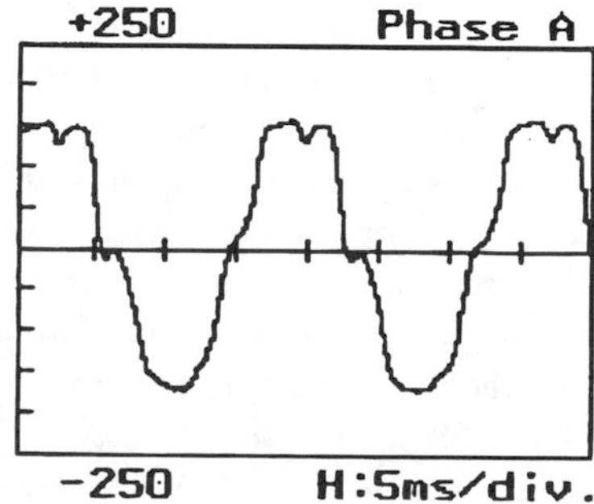

Figure 14.6 Harmonic distortion.

[1] The NPL study defines a transient as an event with 100 to 6000 volts peak amplitude which lasts 0.5 to 2048 microseconds.

drawing current. This discontinuous current draw is known as harmonic current distortion.

Excessive harmonic current distortion can overload the building service transformer and create an electrical fire hazard.

About one out of every ten sites monitored in the NPL study has a problem with harmonics.

To address harmonic distortion issues, the IEEE has developed standard P519–1992. Named The *IEEE Recommended Practices and Requirements for Harmonic Control in Electrical Power Systems,* it originally provided recommendations for applying capacitors for power factor correction for the presence of power converters and was first published in 1981.

A draft of the new standard published in 1992 tries to provide the first consensus standard which recommends dividing responsibility for harmonic control between individual customers (harmonic current limits) and the utility company (harmonic voltage limits) (IEEE P509, 1992).

IEEE-P519 limits a customer's current distortion based on the relative size of the load and the supply voltage based on the voltage level. IEEE-P519 is concerned with system-wide harmonics and how it effects the utility interface. The standard acknowledges that the utility does not have an unlimited capacity to absorb user-generated harmonic currents. The limit on harmonics attempt to prevent users from using up the utilities capacity and reducing voltage distortion problems (Waller, 1994).

Varying Effects of Low Voltage Power Disturbances

The effects of low voltage power disturbances vary widely for different types of sensitive electronic equipment. Some equipment may ride through several cycles of power interruption without experiencing noticeable problems. Other equipment can withstand only a few milliseconds of power interruption before problems occur. The capacity to withstand power interruptions partly depends upon the AC input voltage at the time of a power loss; if the power line voltage is 100 volts right before a power interruption occurs, a given piece of equipment may drop out faster than it would have if the power line voltage had been 120 volts. The capacity to withstand power interruptions also depends upon the quality of the equipment design. However, since ride-through capacity contributes to total product cost and does not influence the buying decisions of most consumers, many electronic equipment manufacturers reduce costs at the expense of ride-through capacity.

The CBEMA curve (Figure 14.1) defines limits for computer equipment, but some types of electronic equipment are sensitive to low voltage power disturbances which fall *within* the CBEMA boundaries. To obtain a sample of low voltage power disturbances that will affect most sensitive electronic equipment, the NPL data was filtered to include all events $\leq$90 Vrms with a duration $\geq$2 cycles. Each year, the average site experiences 38 such power disturbances (Dorr, 1994).

As shown in Figure 14.7, most $\leq$90 Vrms power disturbances have very short durations: 41% last for less than ten cycles, 63% last for less than one second, and 89% last for less than one minute. Only 11% last for more than one minute, and just 4% last for 30 minutes or more.

Protecting Equipment from Power Problems

The IEEE Standard P1100–1992 includes a summary of the performance features of various types of power conditioning equipment (Figure 14.8). An uninterruptible power system (UPS, IEEE P1100 1992) protects for almost all types of power problems.

Power protection requirements vary depending on the protected system's importance. The level of power protection or power conditioning for any specific application should be evaluated carefully. With the proper power protection, any location should be able to significantly limit equipment exposure to power problems and improve system uptime to acceptable levels.

References

Allen, G. W. and Segall, D. 1974. *Monitoring of Computer Installations for Power Line Disturbances,* C74 199-6, IEEE PES.

Dorr, D. 1992. Power interruptions: frequency and impact, *PCIM/Power Quality '92.*

Dorr, D. 1994. *Point of Utilization Power Quality Study Results,* 94CH34520 6–7803–1993–1/94, IEEE IAS.

Goldstein, M. and Speranza, P. D. 1982. *The Quality of U.S. Commericial Power,* INTELEC, CHI818–4/82–0000–002B.

Institute of Electrical and Electronics Engineers, Inc. 1992. *P509–1992 IEEE Recommended Practice for Harmonic Control in Electrical Power Systems,* Institute of Electrical and Electronics Engineers, Inc., New York, NY.

Institute of Electrical and Electronics Engineers, Inc. 1992. *P1100–1992 IEEE Recommended Practice for Power and Grounding Sensitive Electronic Equipment,* Institute of Electrical and Electronics Engineers, Inc., New York, NY.

Jurewicz, R. E. 1990. *National Power Laboratory Power Quality Study 1990–1995,* INTELEC CH2928–0/90/0000/0443.

Waller, M. 1994. *Harmonics,* Prompt Publications, Indianapolis, IN.

14.2 Reactive Power and Harmonics Compensation

Gerry Heydt

Introduction

Reactive power is a concept of alternating current (AC) circuits that results from the definition of power (as the product of voltage and current). Reactive power is related to instantaneous energy storage in inductors and capacitors in the circuit. For sinusoidal AC circuits of a fixed, single frequency (e.g., 60 Hertz), no reactive power can exist in a purely resistive circuit. This is the case since reactive power is a phenomenon of instantaneous energy storage and this can occur only for inductive and capacitive circuits. Reactive power in AC circuits is also related to bus

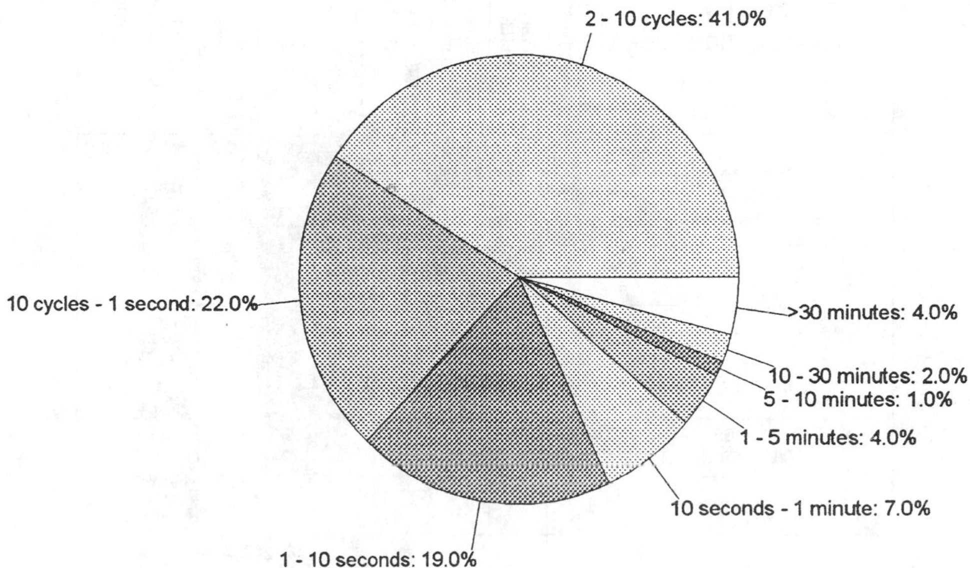

Figure 14.7 Duration: events 90 Vrms and lower.

voltage magnitude and losses. The term 'bus' is used by power engineers to refer to a power system node—often three phase in transmission and primary distribution circuits. A bus voltage exists phase to phase and phase to neutral in three phase cir cuits. It is a general principal that when reactive power is not transmitted (or distributed) over circuit conductors, better system response is obtained. This means that there are lower losses and lower voltage drops when reactive power is *not* carried by the circuit conductors. The most common power engineering symbol for reactive power is Q. However, the reader needs to be cautious that in the older literature and in the literature of power system revenue metering, the symbol Q may be used for some other AC circuit parameter related to reactive power: for example, the product of root mean square voltage and current amplitude and the sine of the phase angle plus 60 degrees may be referred to as Q. Also, in radio engineering, Q refers to the quality factor of inductors—the ratio of their impedance to resistance—and this has little relationship to reactive power.

This article also deals with distorted but periodic voltages and currents in power systems. Table 14.1 gives some commonly used indices which are used to describe these signals. The main application of these indices are:

- *Total harmonic distortion* is the most commonly used index of the distortion of a wave. The THD of load currents can be well above 100% for some nonlinear loads (e.g., a compact fluorescent electronic ballast lamp). Typical THDs for bus voltages are a few percent for subtransmission and transmission voltages, and 3–5% for distribution voltages.
- *Total demand distortion* is the THD of a load current where the denominator of the index is taken to be the circuit rating rather than the fundamental frequency current content. The TDD is used in standards and recommended practice specifications for giving limits on harmonic distortion of load current. Unlike THD of the load current, the TDD is usually in the 5–50% range.
- *Crest factor* is a measure of the amplitude of the voltage (for dielectric stress assessment) or current (for conductor heating assessment).
- *Root mean square* (RMS) is a commonly used measure of current or voltage amplitude. Direct reading RMS meters are used for measurement of this quantity. It is a measure of the overall amplitude of the fundamental component of a wave *and* all its harmonics.
- *Telephone influence factor* is an index much like the THD and TDD that gives a measure of the distortion of a wave; however, the TIF has weights used in the definition in such a way that the index is weighted for signals in the audio frequency range. Thus TIF is a measure of the potential of a voltage or current to interfere with audio signals.
- *C message weight index* is a measure of the interference of power conductors with communications circuits.
- $I \cdot T$ and $V \cdot T$ *product* are indices of the interference of power conductors with communications conductors. Unlike the THD and TDD, the $I \cdot T$ and $V \cdot T$ products are not unit-less ratios: they are fully measures of both the frequency content of the current and voltage, and the amplitude of the current and voltage. The terms $kI \cdot T$ (or kIT) and $kV \cdot T$ (kVT) refer to 1000 times the $I \cdot T$ and $V \cdot T$ respectively. The T in these designations refers to the telephone influence factor weights.

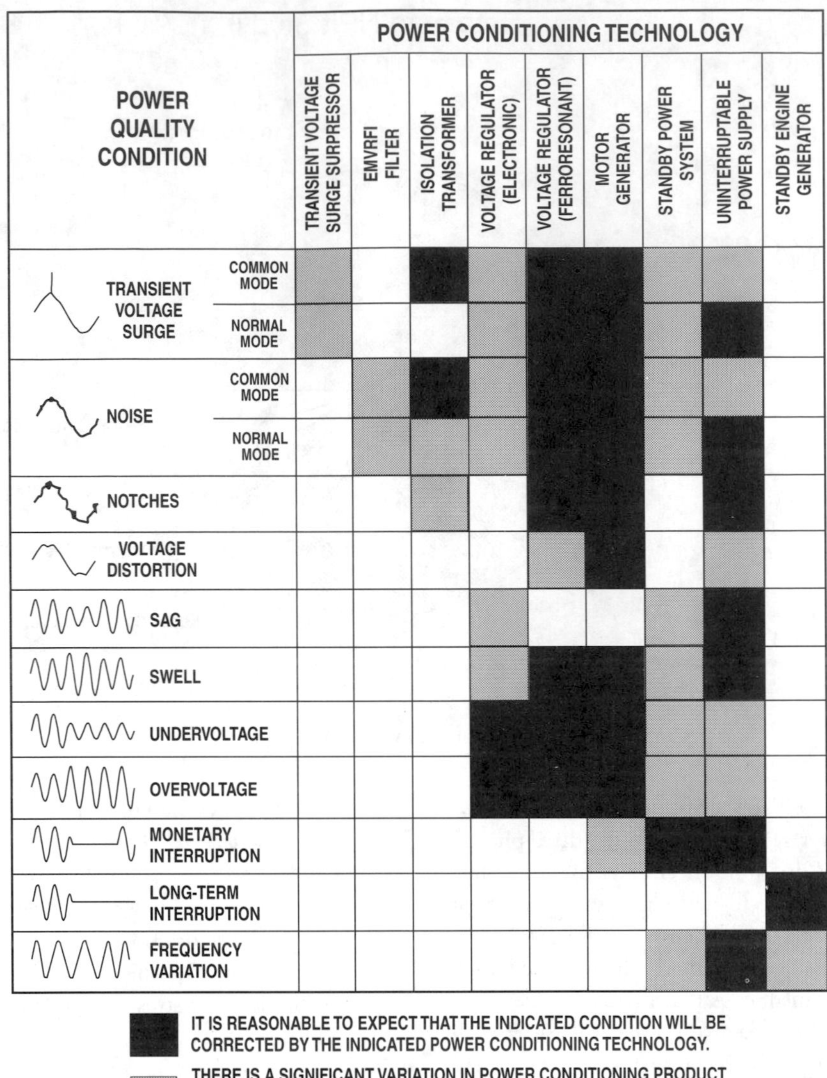

Figure 14.8 The IEEE standard P11D0—1992.

For three phase signals, these indices are occasionally broken into their symmetrical components. Table 14.2 shows some of these definitions.

Reactive Power

The concept of reactive power comes from the alternating current circuit theory developed at the early part of the twentieth century. For purely sinusoidal signals, that is no nonlinear components such as rectifiers, when voltages and currents in the power circuit are given by simple sinusoidal waves,

$$v(t) = \sqrt{2}\, V_{rms} \cos(\omega t)$$

$$i(t) = \sqrt{2}\, I_{rms} \cos(\omega t + \phi)$$

These signals are depicted in Figure 14.9. For this case, note that the current reaches a peak later than the voltage. The term *lagging* is used to refer to the current for this case. For the case shown, current lags voltage by ϕ. The angle ϕ is termed the *power factor angle.* The power factor angle lags when the current lags the voltage. The reverse situation, namely when voltage lags current, is said to be a case in which *current leads voltage.* Table 14.3 contains a summary of leading and lagging cases for simple sinusoidal AC circuits.

The instantaneous power $p(t)$ is related to the instantaneous voltage and current for a load by

$$p(t) = v(t)i(t)$$

For the sinusoidal case of current lagging by angle ϕ,

$$\begin{aligned} p(t) &= \sqrt{2}\, I_{rms} \cos(\omega t - \phi)\sqrt{2}\, V_{rms} \cos(\omega t) \\ &= V_{rms} I_{rms} \cos(\phi) + V_{rms} I_{rms}(\cos(\phi)\cos(2t) \\ &\quad + \sin(\phi)\sin(2t)) \end{aligned}$$

Table 14.1 Commonly Used Indices for Periodic, Nonsinusoidal Voltages and Currents in Power Systems

Quantity	Definition
Total harmonic distortion (THD)	$THD = \frac{\sqrt{\sum_{h=2}^{\infty} I_h^2}}{I_1}$
Total demand distortion (TDD)	$TDD = \frac{\sqrt{\sum_{h=2}^{\infty} I_h^2}}{I_{rated}}$
Crest factor (CF)	$CF = \frac{I_{peak}}{I_{rms}}$
Root mean square (RMS)	$I_{rms} = \sqrt{I_1^2 + I_2^2 + I_3^2 + \cdots}$
Telephone influence factor (TIF)	$\frac{\sqrt{\sum_{h=1}^{\infty} (w_h I_h)^2}}{I_{rms}}$
I · T product	$\sqrt{\sum_{h=1}^{\infty} (w_h I_h)^2}$
V · T product	$\sqrt{\sum_{h=1}^{\infty} (w_h V_h)^2}$
C-message weight index	$\frac{\sqrt{\sum_{h=1}^{\infty} (e_h I_h)^2}}{I_{rms}}$

In these formulas, the harmonic components of current are shown as subscripts. Most formulas shown applied to current may also be applied to voltage. The notation c_h and w_h refer to the C-message and telephone influence factor weights, respectively. For 60, 180, 300, and 420 Hz, the C-message weights are 0.0017, 0.0333, 0.1500, and 0.3100, respectively, and the TIF weights for these frequencies are 0.5, 30.0, 225, and 650.

This time function is depicted in Figure 14.10. Note in this figure that:

- The instantaneous power is a double frequency function as compared to the supply voltage and current (e.g., 120 Hz for a 60 Hz system).
- The average value of p(t) is called the active power or the real power, and its value is $V_{rms} I_{rms} \cos(\phi)$.
- It is possible for p(t) to be negative for some time. This means that the load is, in fact, sending power to the source. This can only happen if the load contains energy storage elements (inductors, capacitors).
- If the power factor angle is 90 degrees, there is no active power in the circuit, but there is still instantaneous power, and this alternates between storage and recovery of energy in the load. When the power factor angle is 90 degrees (purely inductive or capacitive case), the wave shown in the figure is centered on the time axis.
- If the power factor angle is zero, the wave depicted in the figure never goes negative. In this case there is no storage of energy and the load is all resistive.

The cosine of the power factor angle is key to many calculations, and for this reason it is called the *power factor.* The sine of the power factor angle is called the *reactive power factor.* Table 14.4 shows some cases power factor for AC circuits.

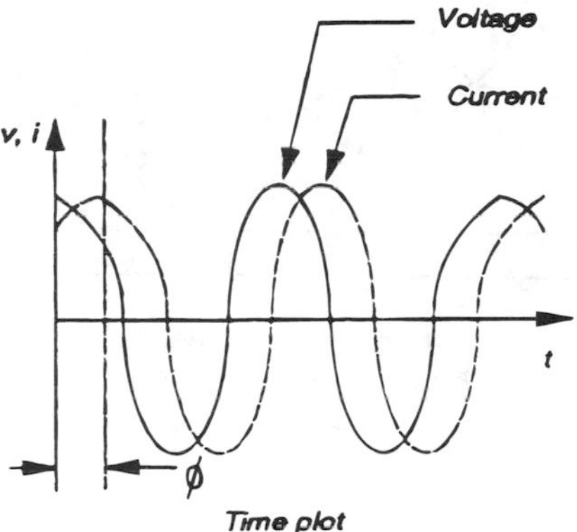

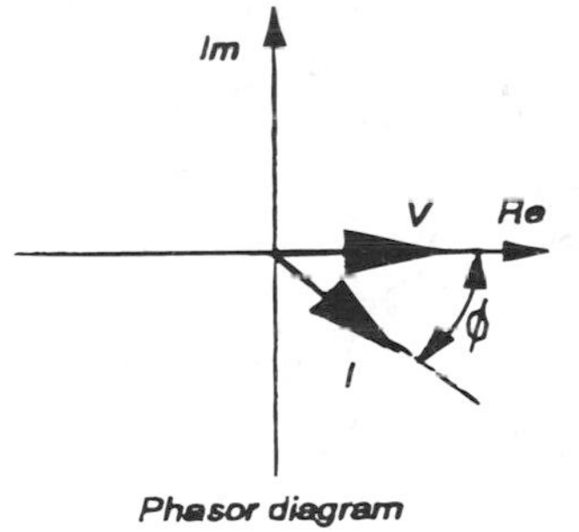

Figure 14.9 Voltage and current waves in a purely sinusoidal circuit. Current lags voltage by ϕ.

Table 14.2 Three-phase Applications of Indices for Nonsinusoidal Periodic Voltage and Currents

Index	Definition	Application
Balanced I · T product	The I · T product calculated only the positive and negative sequence currents at all frequencies	Communications circuit interference (see IEEE Standard 368)
Residual I · T product	The I · T product calculated on the zero sequence currents at all frequencies	Communications circuit interference (see IEEE Standard 368)
Residual V · T product	The V · T product calculated using zero sequence voltages	Assessment of harmonic impact on wye connected capacitors
Balanced TIF	The TIF calculated only the positive and negative sequence currents at all frequencies	Communications circuit interference
Residual TIF product	The TIF product calculated only the zero sequence currents at all frequencies	Communications circuit interference
THD	The THD can be calculated using only positive and negative sequences (balanced THD) or only zero sequence (residual THD)	Assessment of power quality

Table 14.3 Leading and Lagging Relationships for Simple AC Circuits

Load	Load current
Purely resistive	Current in phase with supply voltage
Pure inductive	Current lags supply voltage by 90 degrees
Purely capacitive	Current leads voltage by 90 degrees
Partially inductive	Current lags voltage
Partially capacitive	Current leads voltage

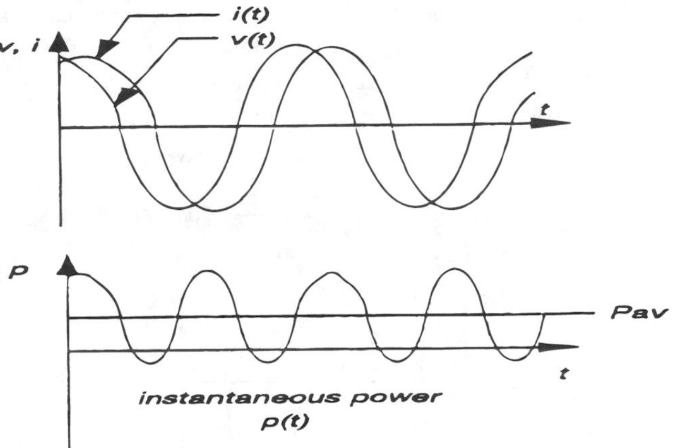

Figure 14.10 Instantaneous power in an AC circuit.

Reactive power is a measure of how energy is stored in the load. Because active power (or average power, or real power) is proportional to the cosine of the power factor angle, the reactive power was defined as somewhat of an opposite phenomenon, namely proportional to the sine of the power factor angle. Thus, for simple, sinusoidal AC circuits,

$$Q = V_{rms} I_{rms} \sin(\phi)$$
$$= V_{rms} I_{rms} (rpf)$$

In this definition, note that the voltage and current are the *amplitudes* of the bus voltage and load current measured in root mean square (rms) volts and amperes. Dimensionally, the units of Q are the same as those of P, but to distinguish between the two, P is expressed in watts and Q in volt-amperes-reactive or VArs. Table 14.5 summarizes these and other definitions for simple AC circuits. For three-phase circuits, the formulas in Table 14.6 are used. In this table, it is assumed that the three-phase circuit is balanced. In these tables, the notation (*) refers to *complex conjugation* of a phasor. For the unbalanced three-phase case, it is necessary to calculate P, Q, and S on a per phase basis for each of the phases, and addition of these quantities gives the three-phase value. In general, the power factor and reactive power factor will be different in each of the phases for the unbalanced case.

Table 14.4 Power Factor and Reactive Power Factor

Load	Power factor	Reactive power factor	Current
Purely resistive	1.0	0.0	In phase with $v(t)$
Purely inductive	0.0	1.0	90 degrees lagging $v(t)$
Purely capacitive	0.0	-1.0	90 degrees leading $v(t$
Partially inductive	0.0 − 1.0	0.0 − 1.0	Lags $v(t)$
Partially capacitive	0.0−1.0	0.0−(−1.0)	Leads $v(t)$

The measurement of power related quantities depends on the application: field measurement, revenue measurement, laboratory testing. Table 14.7 gives typical methods. The clear trend is toward digital measurement of all quantities. The most common instrumentation is the use of revenue induction disk meters. These are single- or three-phase meters that measure $p(t)$ integrated over time.

Reactive Power Compensation

The term *reactive power compensation* refers to the placement of sources of reactive power, usually capacitive reactive power, near load buses such that the circuit conductors serving those loads do not have to carry reactive power to the load. The practice of reactive power compensation results in lower line current; and this results in lower voltage drop in the line. Because of the phase of the line voltage drop, this voltage drop is large when produced by loads that are reactive. In other words, when loads have large reactive power demands, the load bus voltage will be low. The converse is also true: when loads generate reactive power, the load bus voltage will actually be boosted in amplitude over that of the sending end.

The main principal of reactive power compensation is that reactive power demands of the load should be made up locally. This means that when a load has lagging power factor, a capacitor should be placed in parallel with the load to *correct the power factor* (i.e., make the net demand zero) to keep the load bus voltage amplitude at rating. When the load has leading power factor, it is necessary to place an inductor in parallel with the load to correct the power factor. Resistive loads do not require power factor correction.

Because most loads are partially inductive, that is they have lagging power factor, it is usual to require capacitors for power factor correction. These capacitors should be located as close to the load as feasible in order to minimize the effect of line voltage drops. Also, power factor correction will reduce circuit active power losses. This is a consequence of lower line current. In many cases, it is infeasible to place the power factor correction capacitors at the load bus. For this reason capacitors in the distribution circuit are common—and capacitors in the transmission circuit may also be used. Table 14.8 shows several types of power factor correction measures.

The several types of power factor correction devices are:

- *Fixed shunt capacitors* may be located in the transmission system, the subtransmission system, or the distributions system. In the transmission system, these capacitors are three phase, located at substations, and generally made up of rack-like mounted individual units. The sizing of transmission shunt capacitors is generally done be system planning, and the design values are intended to provide

Table 14.5 Power Definitions (Single-Phase Circuits)

Quantity (and synonyms)	Symbol	Relationships	Units
Active power (real power, average power)	P	$P = \lvert V_{rms} \rvert \lvert I_{rms} \rvert \cos(\phi) = \lvert V_{rms} \rvert \lvert I_{rms} \rvert pf$ $= \sqrt{S^2 - Q^2}$	Watt (W)
Reactive power	Q	$Q = \lvert V_{rms} \rvert \lvert I_{rms} \rvert \sin(\phi) = \lvert V_{rms} \rvert \lvert I_{rms} \rvert rpf$ $= \sqrt{S^2 - P^2}$	VAr
Power factor	pf	$\cos(\phi)$	None, often represented as a percentage
Reactive power factor	rpf	$\sin(\phi)$	None
Complex power	S	$S = VI^*$	Voltamperes (VA)
Apparent power	| S |	$\lvert S \rvert = \lvert V_{rms} \rvert \lvert I_{rms} \rvert = \sqrt{P^2 + Q^2}$	Voltamperes (VA)

Table 14.6 Power Definitions (Three-Phase Circuits)

Quantity	Symbol	Relationships	Units
Active power (real power)	P	$P = 3\lvert V_{ln} \rvert \lvert I_{phase} \rvert \cos(\phi)$ $= 3\lvert V_{ln} \rvert \lvert I_{phase} \rvert pf$ $= \sqrt{3}\lvert V_{ll} \rvert \lvert I_{line} \rvert pf$ $= \sqrt{S^2 - Q^2}$	Watt (W)
Reactive power	Q	$Q = 3 \lvert V_{ln} \rvert \lvert I_{phase} \rvert \sin(\phi)$ $= 3 \lvert Vrubln \rvert \lvert I_{phase} \rvert rpf$ $= \sqrt{3} \lvert V_{ll} \rvert \lvert I_{line} \rvert rpf$ $- \sqrt{S^2 - P^2}$	VAr
Power factor	pf	$\cos(\phi)$	Often represented as a percentage
Reactive power factor	rpf	$\sin(\phi)$	None
Complex power	S	$S = 3\ V_{ln} I^*_{phase}$ $= \sqrt{3}\ -30^\circ V_{line} I^*_{line}$	Voltamperes (VA)
Apparent power	|S|	$\lvert S \rvert - 3 \lvert V_{ln} \rvert \lvert I_{phase} \rvert$ $= 3 \lvert V_{ll} \rvert \lvert I_{line} \rvert$ $= \sqrt{P^2 + Q^2}$	Voltamperes (VA)

Table 14.7 Typical Power Measurement Techniques

Quantity	Phases	Basis of measurement (typical)	Technique
Active power	1	D'Arsonval wattmeter method (analog) or voltage and current signals modulating a pulse train (digital)	Wattmeter, either analog or digital
	3	Two wattmeter method	Wattmeters, either analog or digital
Reactive power	1	Wattmeter with voltage coil input shifted by 90 degrees by R-L-C circuit	Wattmeter, either analog or digital
	3	"VAr transformer" which shifts wattmeter voltage coil input by 90 degrees	Wattmeters, either analog or digital
Power factor	1 or 3	Watt-hours divided by volt-ampere-hours	Induction disk watt-hour meter and software
Reactive power factor	1 or 3	Usually not instrumented	
Complex power	1 or 3	Usually not instrumented	
Apparent power	1 or 3	Volts times amps	Voltmeter, ammeter

Table 14.8 Power Factor Correction

Location	Reason	Hardware used
Generation system	To maintain generation bus voltage at set point	• Generator excitation is adjusted to produce the required reactive power—and thereby hold the generation bus at the set point • Exciter used to set Q
Transmission system	To maintain transmission buses at rating and to minimize line losses	• Fixed shunt capacitors • Switched shunt capacitors • Synchronous condensers • Static VAr compensators
Subtransmission system	To maintain subtransmission buses at rating and to minimize line losses	• Fixed shunt capacitors • Switched shunt capacitors • Synchronous condensers • Static VAr compensators
Distribution system	To maintain load busses at rated value and to minimize line losses	• Fixed shunt capacitors • Switched shunt capacitors • Capacitors on timers • Static VAr compensators
At load	Holding bus voltage at rating and avoid revenue charges due to low power factor	•Fixed shunt capacitors • Switched shunt capacitors • Synchronous motors • Static VAr compensators

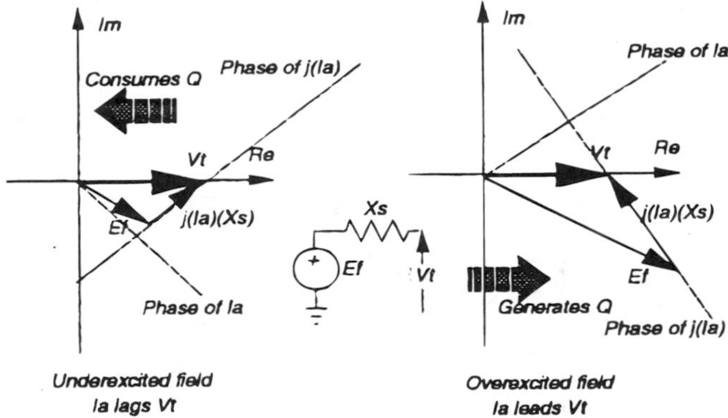

Figure 14.11 Synchronous condenser phasor diagrams.

adequate reactive power compensation over a wide range of load conditions. Fixed capacitors are usually *switched* either by remote operation, radio signals, or timers. The status of a fixed capacitor may be reported to a supervisory system. In the latter case, records are kept of the status of a capacitor. Typical values of capacitor, represented in terms of VArs injected, are: 40 MVA (transmission), 15 MVA (subtransmission); 500 kVA (distribution system).

- *Synchronous condensers* are rotating machines, they are similar to synchronous motors, but the shaft of the machine is not brought out of the case as in the case of a synchronous motor. The machine demands very little active power because there is very little shaft load. The machine field excitation may be varied to produce different armature circuit power factors. For example, when the machine is underexcited (i.e., its internal $|E_f|$ is smaller than the terminal voltage magnitude (Figure 14.11), the load current will lag the supply voltage and the device acts as nearly a perfect inductor. For the case of overexcitation, $|E_f|$ is greater than the supply bus voltage, and the phasor diagram shown in Figure 14.11 applies. In this case the supply current leads the terminal voltage, and the machine acts like a capacitor. Therefore, by control of the machine excitation, it is possible to vary the reactive power demand from capacitive to inductive. In this way, an excitation controller may be used to continuously adjust the injected reactive power. The main advantage of this machine is its continuous reactive power variation capability. Also, very large synchronous condensers have been constructed (e.g., in the 300 MVA range and higher) for cases in which the reactive power demand is very high. The disadvantages are: high cost, required maintenance, large size.
- *Generator excitation systems.* In the same way that synchronous condensers may be used to generate reactive power, so may large synchronous generators. This is not usually done because these generators are intended to generate active power. If the machine capacity is used to generate Q, the P must be reduced in order to stay within rated S. Many large system generators are used to generate some reactive power and the control is the voltage regulator. The generation bus voltage amplitude is compared to a set point value, and the excitation is adjusted to raise or lower the bus voltage. This corresponds to raising or lowering, respectively, the reactive power generated.
- *Static VAr compensators* are capacitor-inductor combinations which are electronically switched (see Figure 14.12). If additional VArs are needed, a control system is used to adjust the silicon controlled rectifiers shown in the figure so that the capacitors are switched in. If less reactive power is needed (or if inductive VArs must be consumed), the electronic switches are controlled so

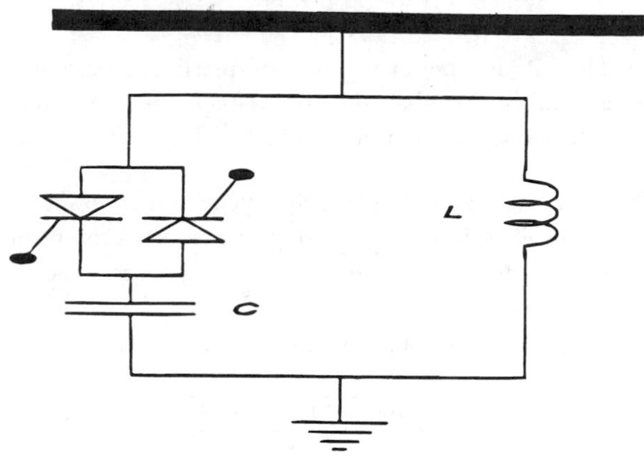

Figure 14.12 A single-phase static VAr compensator (SVC).

that the capacitor is out of the circuit, and the shunt inductors are in the circuit. The advantage of SVCs are their ability to continuously control the required reactive power. The disadvantages mainly stem from the electronic switching involved, the cost of the electronics and control circuits, the distorted waves produced by these switches, and the relatively low capacity of the SVC for the cost of the unit.

- *Synchronous motors* are often used in industrial applications because if these machines are operated overexcited, they generate VArs. The phenomenon is the same as that of a synchronous condenser as described earlier. The advantage of synchronous motors is that they do useful tasks at the same time as correcting the power factor. The disadvantages are cost, efficiency, and the fixed speed of operation.

In industrial power distribution, the essence of power factor correction is the addition of reactive power sources, usually shunt capacitors, in order to make the local load power factor closer to unity. Usually, power factor correction is not done to make the power factor unity *exactly* because the economics of most industrial settings is that correction to 86–92% power factor lagging is the economic optimum. This is studied by considering the advantages of correcting the power factor versus the cost of the shunt capacitors. Some power engineering consultants have the policy of correction to very near unity power factor; some use lower power factor as a goal. It is never expedient to correct power factor beyond unity power factor—i.e., to leading power factor.

Reactive Power in Nonsinusoidal Circuits

The foregoing remarks about reactive power are valid for circuits that have only sinusoidal voltages and currents of one frequency. When electronic switching circuits are used in power engineering applications, the following phenomena occur:

- For circuits in steady state, voltages and currents will often be periodic but nonsinusoidal.
- For these voltages and currents, Fourier series exist, and these Fourier series are made up of terms which are sinusoids of frequency $h\omega_o$ *where ω_o is the fundamental frequency related to the period T by*

$$\omega_o = \frac{2\pi}{T}$$

- The frequencies $h\omega_o$ are called *harmonics* and they are integer multiples of ω_o.
- The total active power associated with a voltage and current given by the Fourier series

$$v(t) = \sum_{h=1}^{\infty} a_h \cos(h\omega_o + \phi_h)$$

$$i(t) = \sum_{h=1}^{\infty} b_h \cos(h\omega_o + \theta_i)$$

is given by the sum of the powers of the several harmonics,

$$P_{total} = \sum_{h=1}^{\infty} a_h b_h \cos(\phi_h - \theta_h)$$

where a and b are expressed in rms volts and amps, respectively.

- If reactive power is similarly defined (i.e., sum of the reactive powers of the several harmonics), the total P and total Q does *not* result in the complex power S in the expression $S = \sqrt{P^2 + Q^2}$. This is because there is also a "power" associated with voltages and currents of different frequencies. This is called the *distortion power.* Table 14.9 summarizes these expressions. Other definitions of *reactive power* for the nonsinusoidal case have been proposed, but none have been fully accepted by the power engineering community.

Table 14.9 Formulas for Power Calculations for Nonsinusoidal Periodic Voltages and Currents

Quantity	Symbol	Calculations
Fourier coefficients	a_h, b_h	$a_h = \frac{2\pi}{T}\int_0^T v(t)\cos(h\omega_h t)dt$
		$b_h = \frac{2\pi}{T}\int_0^T i(t)\cos(h\omega_h t)dt$
Rms voltage and current	V_{rms}, I_{rms}	$V_{rms} = \sqrt{\sum_{h=1}^{\infty} a_h^2}$
		$I_{rms} = \sqrt{\sum_{h=1}^{\infty} b_h^2}$
Active power	P	$P_{total} = \sum_{h=1}^{\infty} a_h b_h \cos(\phi_h - \theta_h)$
Reactive power	Q	$Q_{total} = \sum_{h=1}^{\infty} a_h b_h \sin(\phi_h - \theta_h)$
Distortion power	D	$D = \sqrt{S^2 - P^2 - Q^2}$
Apparent power	$\lvert S \rvert$	$\lvert S \rvert = \lvert V \rvert \lvert I \rvert$

Voltages and currents assumed to be in rms volts and amps.

Because the reactive power in nonsinusoidal circuits depends strongly on the degree of the distortion of the current wave (the voltage wave is rarely distorted more than 5%), the harmonic distortion of the current wave, defined as

$$THD_I = \frac{\sqrt{\sum_{h=2}^{\infty} b_h^2}}{b_1}$$

(where THD_1 is the total harmonic distortion of the current wave, often expressed as a percent) usually determines the level of distortion power *D*.

True Power Factor vs. Displacement Power Factor

For sinusoidal currents and voltages,

$$pf = \frac{P}{S}$$
$$= \cos(\phi)$$

However, for distorted but periodic voltages and currents, this expression is ambiguous because it is unclear which phase angle should be used: the voltage-current phase angle in the fundamental, or that in one of the harmonics. For the case of nonsinusoidal voltages and currents, there are two widely used power factor definitions:

True power factor

$$tpf = \frac{P_{total\ all\ harmonics}}{|V_{rms}||I_{rms}|}$$

Displacement power factor

$$dpf = \frac{P_{fundamental}}{|V_{fundamental}||I_{fundamental}|}$$

The comparison of the two definitions is as follows:

- The true power factor is more a measure of the total power handling capability of the circuit, inclusive of all harmonic effects.
- The displacement power factor is easier to measure.
- The traditional measurement of power factor in the electric utility industry is usually a relatively narrow band measurement, and it is approximately the displacement power factor.
- Digital instrumentation is capable of measurement of the true power factor.
- For sinusoidal circuits, the TPF and DPF are the same.
- The TPF is always less than or equal to the DPF.
- Power factor multipliers used in revenue rate structures will be larger (i.e., greater dollar charge to the customer) if TPF is used to derive that multiplier.
- The TPF is a better measure of distribution circuit loss and utilization than the DPF. This is the case since the TPF takes into account the effect of all circuit harmonics.

The following is an example of the calculation of power factor in a nonsinusoidal circuit. Note that the harmonic decomposition of voltage and current are needed—including phase angles. This is a single phase case.

For this case, the voltage and current THDs are

$$THD_V = \frac{\sqrt{200^2 + 17^2 + 7^2 + 2^2}}{4161}$$
$$= 0.0483 = 4.83\%$$
$$THD_I = \frac{\sqrt{5.1^2 + 3.0^2 + 1.9^2 + 1.0^2}}{51}$$
$$= 0.1234 = 12.34\%$$

The rms voltage and current are

$$V_{rms} = \sqrt{4161^2 + 200^2 + 17^2 + 7^2 + 2^2} = 4165.8\ V$$
$$I_{rms} = \sqrt{51^2 + 5.1^2 + 3.0^2 + 1.9^2 + 1.0^2} = 51.4\ A$$

The displacement power factor is

$$DPF = \cos(\phi_1) = \cos(-21°) = 0.9336$$
$$= 93.36\%\ \text{lagging}$$

The total active power delivered is

$$P_{total} = (4161)(51)\cos(-21°) + (200)(5.1)\cos(-54°)$$
$$+ (17)(3.0)\cos(-145°)$$
$$+ (7)(1.9)\cos(-46°) + (2)(1.0)\cos(20°)$$
$$= 198.68\ \text{kW}$$

The true power factor is

$$TPF = \frac{P_{total}}{|V_{rms}||I_{rms}|}$$
$$= \frac{198.68\ \text{kW}}{(4165.8)(51.4)\ \text{VA}}$$
$$= 0.9279$$

Reduction of Harmonic Signal Amplitudes

An issue apart from power factor correction and bus voltage regulation is the reduction of harmonic signal amplitudes in distribution circuits that experience nonlinear loads. These cases are those in which loads are rectifiers, adjustable speed drives,

Table 14.10 Measures to Reduce Harmonics in Power Systems

Measure	Implementation	Main application
Tuned filter	Use of an L-C filter, usually tuned slightly below the desired rejection frequency	Adjustable speed drives, high-power converters
Bandpass filter	Usually an R-L filter with a cutoff frequency below the lowest frequency to be rejected. Often a highpass filter.	High-power converters, to attenuate high-frequency harmonics (e.g., above the seventeenth)
Shunt capacitor	A simple shunt capacitor	Inexpensive solution for low-harmonic penetration
Transformer leakage reactance	The leakage reactance of a distribution transformer will provide some isolation and attenuation of harmonic load currents	Inexpensive solution for low-harmonic penetration. Also for some adjustable speed drive applications
Series choke	Simple series inductor inserted in the line	Adjustable speed drive applications
Phase multiplication	For three-phase rectifiers and inverters, the use of alternating wye and delta connections to obtain higher pulse order. Also, possibly the use of other transformer connections to obtain pulse order greater than 12	Three-phase adjustable speed drives, three-phase rectifiers
Alternation of three phase connections	For three-phase, six-pulse loads. To effectively create a twelve-pulse load from several six-pulse loads	Three-phase adjustable speed drives, three-phase rectifiers
Active filters	Intentional injection of harmonic signals—out of phase with ambient signals—for cancellation	Not in common use

electronic ballast fluorescent lighting, magnetic devices in saturation, fluorescent lighting with magnetic ballasts, arc furnaces, and other nonlinear loads. It is desirable to reduce the harmonic content of the load current because of the following factors:

- The harmonic load currents cause low true power factor.
- The harmonic load currents cause increased losses in both primary and secondary conductors due to increased levels of I^2R.
- The harmonic load currents cause increased distribution transformer losses. These are due to increased eddy current losses as well as increased core losses. The IEEE Standard C57.110 describes a way to estimate these losses.
- The harmonic currents cause heating of the distribution transformer.
- The harmonic currents *may* cause malfunctions of protective relaying.
- Because harmonic currents may be in zero sequence, these currents flow in the neutral circuit. The result is higher losses in distribution circuits and the possibility of unacceptably high neutral-to-ground voltages.

Harmonic Filters and Power Factor Compensation

Because of the cited problems and losses caused by harmonic signals, there has been established limits on harmonic voltages and currents. The Institute of Electrical and Electronic Engineers (IEEE) has adopted a Recommended Practice, IEEE 519, which suggests limits for voltage and current harmonics in transmission and distribution systems. Also, the International Electrotechnical Commission (IEC) in Geneva, Switzerland has adopted IEC 555 which contains limits of harmonic components in load currents. In order to reduce losses, reduce the problems associated with harmonics, and comply with accepted recommendations and standards, several measures may be taken. Table 14.10 shows some of these measures.

The most important of these measures are described below.

Tuned R-L-C filters. The tuned L-C filter is a simple measure to attenuate harmonic voltages at power system busses (usually distribution busses). For a series inductor/capacitor placed from the bus to ground (to neutral in a three phase system if connected in wye, or line-to-line for the delta connection), the formula for the resonant frequency is simply

$$f_r = \frac{1}{2\pi\sqrt{LC}}$$

Any inductor has resistance, and the L-C combination is, in fact, an R-L-C circuit. The quality factor of the inductor (often denoted Q, not to be confused with reactive power) is the ratio of inductive reactance to resistance, $\omega L/R$. Thus, for a low resistance coil, the quality factor will be high. At resonance, the R-L-C filter becomes simply a resistance from the bus to ground. Therefore, the most effective filters are those with high quality factor. The quality factor may be expressed at either the power frequency or the resonant frequency. The useful parameter to determine the effectiveness of the tuned filter is the quality factor at the resonant frequency. The R-L-C filter will behave as a net capacitive tie to ground below the resonant frequency. Thus, a filter designed to attenuate harmonics of the power frequency will give capacitive VArs at the power frequency. The R-L-C filter attenuates harmonics and acts as a shunt capacitor as well. Many tuned filter designs have as a specification the capacitive VAr support at the power frequency. The R-L-C filter will also result in active power loss in the resistance of the inductor: at the power frequency this loss is effectively due to the current that charges the capacitor passing through the resistance of the coil. The important parameter in this regard is the quality factor of the coil at the power frequency. One way to avoid this power loss is to bypass the inductor with a second L-C series combination. The second L-C is tuned to the power frequency so that the current at this frequency simply does not pass through the main inductor. The main application areas of tuned L-C filters are, for distribution circuits, to attenuate harmonics caused by a variety of nonlinear loads (e.g., adjustable speed drives, power rectifiers); and for transmission circuits, high-voltage DC applications. For distribution circuits, the added VAr support often moves the financial considerations in the direction of a tuned filter because the filter acts not only to attenuate harmonics but also to support the bus voltage. Figure 14.13 shows several alternative filter designs.

Bandpass filters are devices of lower cost than tuned filters, but they usually do not give the same attenuation as the tuned counterpart. A bandpass filter might be simply a power resistor in series with a capacitor. This R-C combination is placed from the bus to ground. The effectiveness of the combination is best at high frequencies, and this is essentially a lowpass filter. The disadvantages are losses in the resistor (which can be reduced at an added cost by placing a tuned circuit in parallel with the resistor to bypass the power frequency), and the often ineffective level of attenuation. The advantages are the simplicity of design and the fact that the filter can be a wideband filter (unlike the tuned L-C filter which is designed for a single frequency). Another advantage is lower cost.

Shunt capacitors are often a natural way to attenuate harmonics. These units are used for voltage support and power factor correction, and they provide low-impedance paths to harmonic currents. The capacitive reactance drops linearly with frequency rise. Therefore a shunt capacitor will provide a path of reactance 1/5 of that at the power frequency. The main advantage of this solution to harmonic problems is the dual use of the unit—as power factor correction as well as harmonic filtering. Other advantages are simplicity of design and low cost. The main disadvantage is that the needed shunt capacitor to give the desired attenuation (especially at lower harmonics such as the third) is simply too large to be practical. The attenuation provided by a capacitor that is designed for power factor correction may be too low. In some cases, if the bus voltage is severely distorted, the capacitor current will exceed the rating of the unit; for such a case, a tuned filter is clearly favored.

Transformer leakage reactance is the series reactance of a transformer. At the power frequency, this reactance is usually in the range of 5–15% of the transformer reactance base. At harmonic h, this reactance is multiplied by a factor of *h*. Therefore, a transformer tends to isolate loads from the line at higher frequencies. This phenomenon comes at no added cost—it is a natural phenomenon of all transformers. The distribution transformer tends to isolate loads from the line at higher frequencies. The disadvantage or reliance on transformer leakage reactance for harmonic isolation is that some loads, such as certain adjustable speed drives and most rectifiers, act as harmonic current sources; these sources are not attenuated by leakage reactance of the supply transformer. On the other hand, loads that have high-frequency components of voltage at the transformer secondary will experience harmonic attenuation

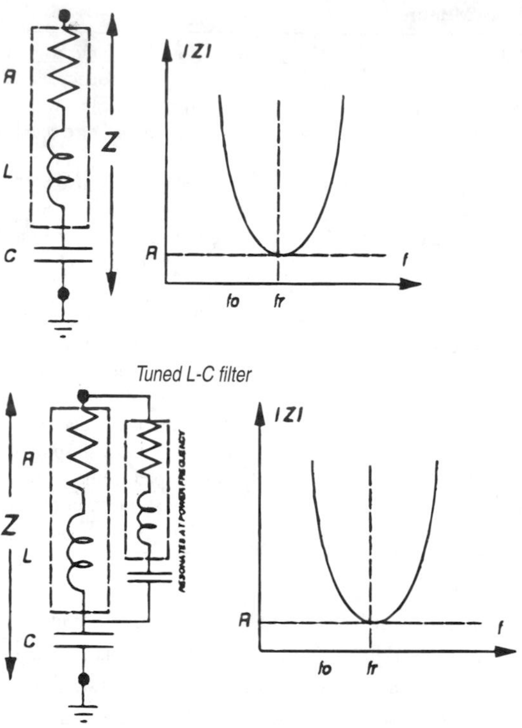

Tuned L-C filter with bypass

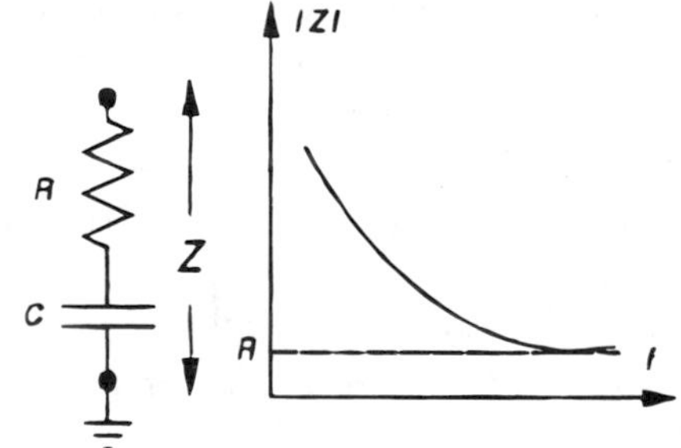

High pass (current) filter

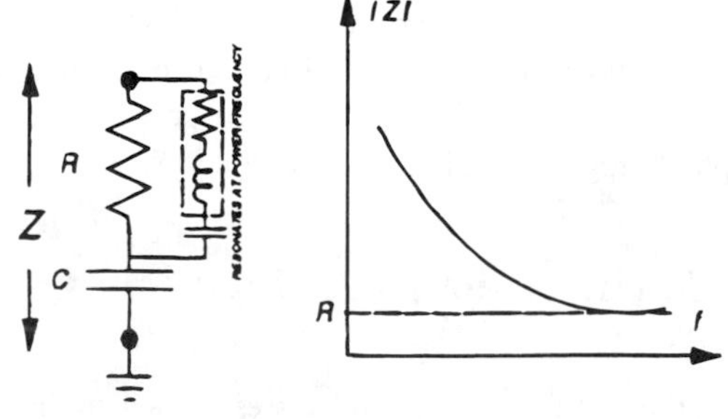

High pass filter with bypass

Figure 14.13 Alternative passive filter designs for harmonics in power systems.

Table 14.11 Predominant Harmonic Currents in Ideal Three-Phase Rectifiers of Various Pulse Order

	Harmonic order					
Pulse order	5	7	11	13	17	19
6	X	X	X	X	X	X
12			X	X		
18					X	X

by the transformer. The use of a *series choke* likewise is a means of isolation of the load from the supply bus. Especially for adjustable speed drives, often a few ohms inserted in the supply feeder (in the form of a series choke) will attenuate harmonics. The disadvantage of a series choke is the concomitant bus voltage regulation degradation.

Phase multiplication is a term that refers to the transition of lower pulse order devices to higher pulse order devices. For example, a six-pulse rectifier operating from a three-phase supply might use a wye/wye connection. The predominant harmonics are of order $6n \pm 1$. However, if a second identical six-pulse rectifier is served from the same bus, connected to the bus by a wye/delta connection, certain harmonic currents in the supply side cancel and the remaining harmonics are of order $12n \pm 1$. Table 14.11 shows the predominant harmonics present in the AC supply current of various pulse order rectifiers. It is evident that as the pulse order increases, the harmonic impact is less. This is due to the fact that the lowest-order harmonic increases with increasing pulse order, and the approximate magnitude of that harmonic is the inverse of the harmonic order. Thus the lowest order harmonic of an ideal six-pulse rectifier is about 1/5 or 0.2000 whereas the lowest harmonic current for a twelve-pulse bridge is 1/11 or (0.0999). It is possible to use alternating three-phase connections, that is, alternating between wye/wye and wye/delta, to effectively produce the effect of phase multiplication.

Active filters are devices that intentionally inject harmonic currents into a circuit in such a way as to cancel existing harmonics in the circuit. The method is very effective at low-power levels (i.e., for communications and other low power electronic circuits), but at high-power levels, the costs for the control circuitry may be high. The results, on the other hand, can be impressive: harmonic cancellation may result in very low levels of harmonic distortion. The main types of active filters are the series type in which a voltage is added in series with an existing bus voltage. The series voltage is out of phase with the bus voltage harmonics and therefore cancellation occurs. The other type of active filter is the parallel type in which a harmonic current is injected into the bus. The injected current cancels the line current harmonics. Active filters are not in widespread use because of their cost.

14.3 New Power Converters

Prasad Enjeti

In this section new power converters are discussed which are more immune to mains disturbances and draw sinusoidal currents. A typical diode rectifier interface to convert utility AC power to DC is the most common utility interface to several power electronic equipment such as: computers, office automation systems, AC/DC motor drives, induction heating, high current power supplies, etc. These loads are commonly termed as nonlinear loads and draw excessive current harmonics from the mains. Nonlinear loads contribute to poor power quality, cause excessive neutral currents and contribute to voltage distortions.

In addition to these the diode type utility interface also suffers from the following drawbacks:

- The power available from the wall outlet is reduced to about two thirds. This is mainly due to presence of current harmonics which increases the rms value.
- The DC output is susceptible to utility disturbances. Such as voltage sag/swell greatly effects the DC output and contributes to malfunction.
- If mains isolation becomes necessary this approach results in a large transformer.
- The EMI filter employed in the input must be designed for higher peak pulse currents.

In view of these drawbacks, some new power converters are discussed in this section which exhibit superior characteristics. Practical applications for these schemes are also shown.

Single-Phase Active Power Factor Correction

In this section, two approaches are discussed. The first approach, shown in Figure 14.14, employs one semiconductor switch (Sbst). Figure 14.15 shows the two modes of operation when the switch

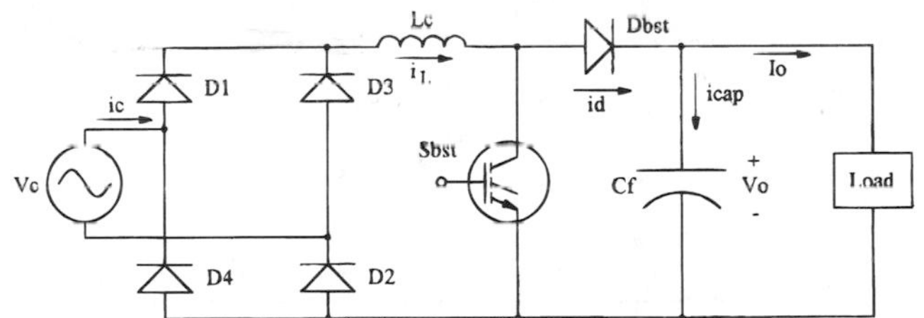

Figure 14.14 Bridge boost converter.

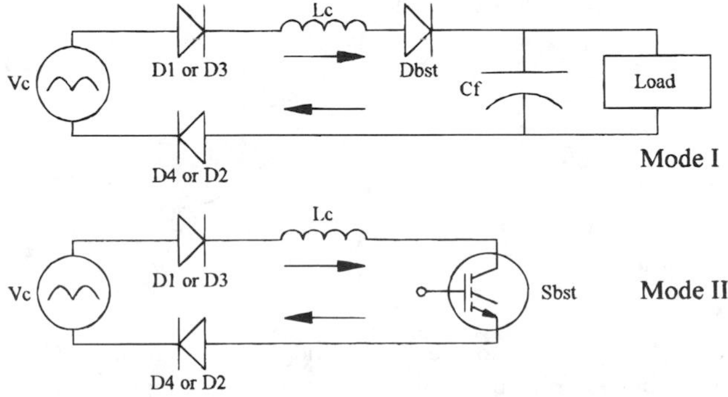

Figure 14.15 Bridge boost mode I and mode II equivalent circuits.

Sbst is off and on, respectively. In Mode I, the switch Sbst is off and the inductor current flows to the capacitor and the load. In Mode II, the switch Sbst is on and the current increases in the inductor. The repeated on/off operation of the switch Sbst in pulse width modulation (PWM) mode shapes the input current. Figure 14.16 shows the simulation results of this circuit. Note the near sinusoidal shape of the input current and unity input power factor. Several integrated circuits are currently available for practical implementation of this approach and are described in Strassberg (1991), Klein and Nalbant (1990) and Kit (1989). The Unitrode UC3842 and MicroLinear ML4812 are two such commercially available integrated circuits. *Unitrode Application Notes* contain detailed application notes and practical design examples for power factor corrected power supplies, electronic ballasts, etc.

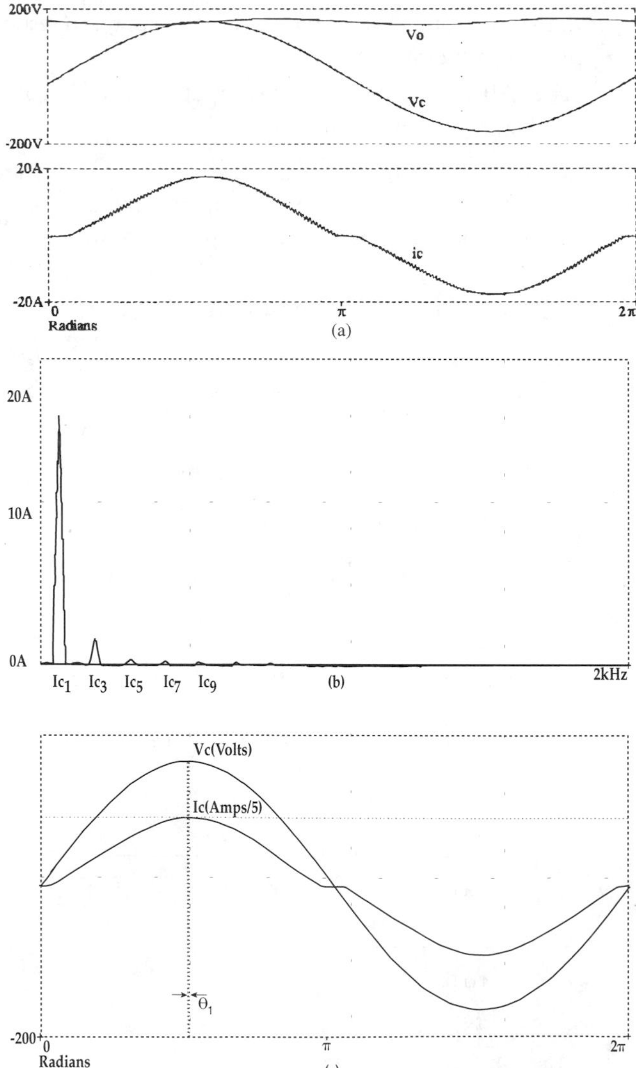

Figure 14.16 Simulation results of bridge boost design. (a) Output voltage and current; (b) harmonic spectrum of the input current; (c) displacement power factor measurement.

Figure 14.17 illustrates a modified circuit topology employing two semiconductor switches and is called a modified boost converter. Comparing Figure 14.14 and Figure 14.17, it can be noted that the modified boost converter has one less diode and hence one less semiconductor device in series with the power flow path. Also the inductor is located on the AC side. Further, the two switches can receive the same PWM gating signal. Figure 14.18 shows the simulation results of this scheme.

Figure 14.19 shows a full-bridge active power factor correction scheme. Four active semiconductor switches are employed in this scheme. This topology permits bidirectional power flow between the AC and DC sides. This approach is most suitable for motor drive applications in which regenerative braking is important. Figure 14.20 shows the PWM gating scheme for this approach. Figure 14.21 shows the experimental performance of this scheme. The input current is near sinusoidal in shape.

Tables 14.12 and 14.13 show the per-unit rating of all of the components employed in the three schemes. Table 14.14 illustrates the application aspects of the three single-phase active power factor correction approaches.

Improved Three-Phase Utility Interface Converters

Modern high-power electronic loads such as variable speed AC motor drives, uninterruptable power supplies in emergency and standby systems, magnet power supplies, etc. employ a three-phase diode bridge rectifier with filter capacitor on the DC side as an interface to the electric utility. The diodes conduct only for short periods of time and this causes significant harmonic currents in the utility lines. Figure 14.22 shows the line current waveform and harmonics drawn by a 250HP variable speed AC motor pump load. Notice the discontinuous nature of the current with 63% total harmonic distortion (THD). In this section several schemes are discussed to provide improved characteristics.

Figure 14.23 shows a passive scheme. A three-phase star/delta transformer is interconnected between the AC and DC side as shown. This interconnection, in combination with 120° conduction of each diode generates a circulating third harmonic current (I_f) in the loop. The transformer connected on the AC side is a short circuit for third harmonic currents and this scheme drastically improves the utility line current waveform. The transformer

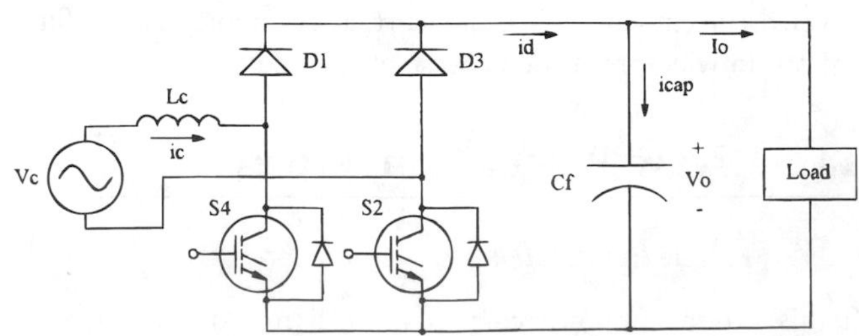

Figure 14.17 Modified boost converter.

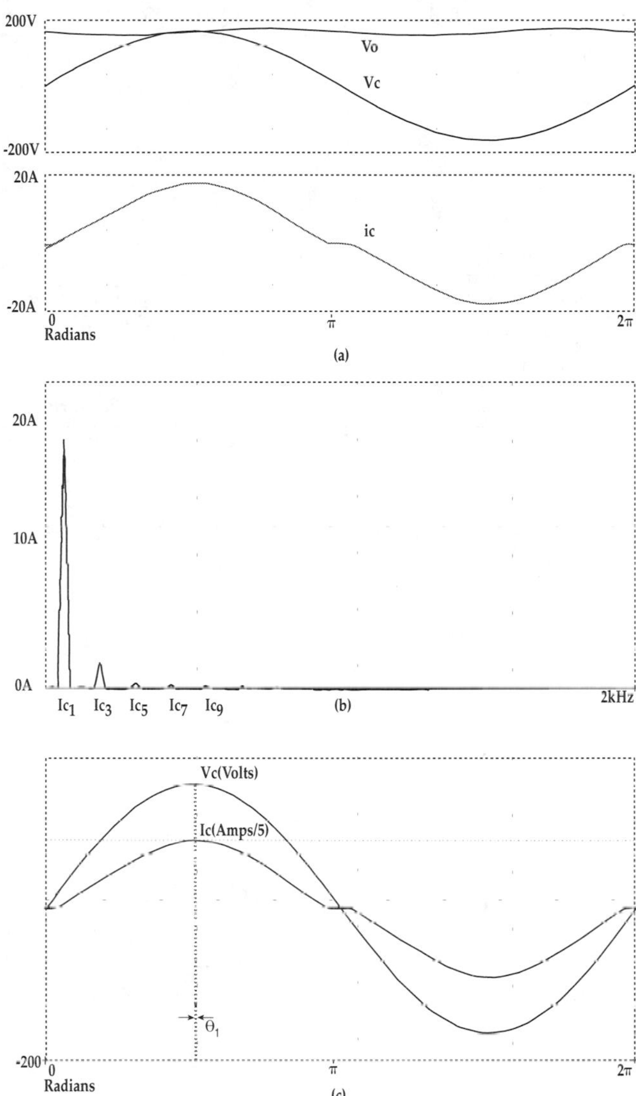

Figure 14.18 Simulation results of modified boost converter. (a) Voltage and current; (b) harmonic spectrum of input current; (c) displacement power factor measurement.

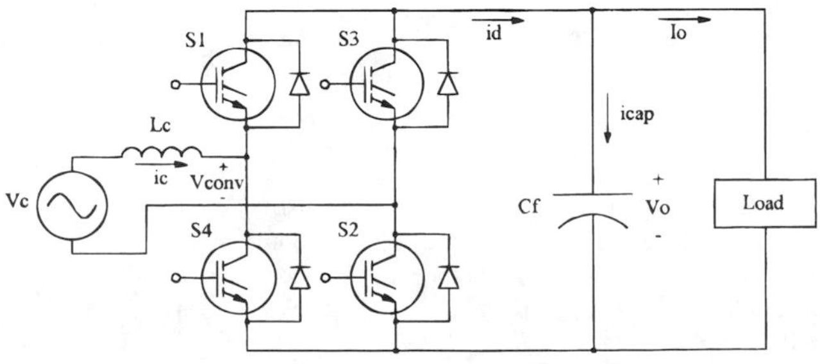

Figure 14.19 Full bridge converter.

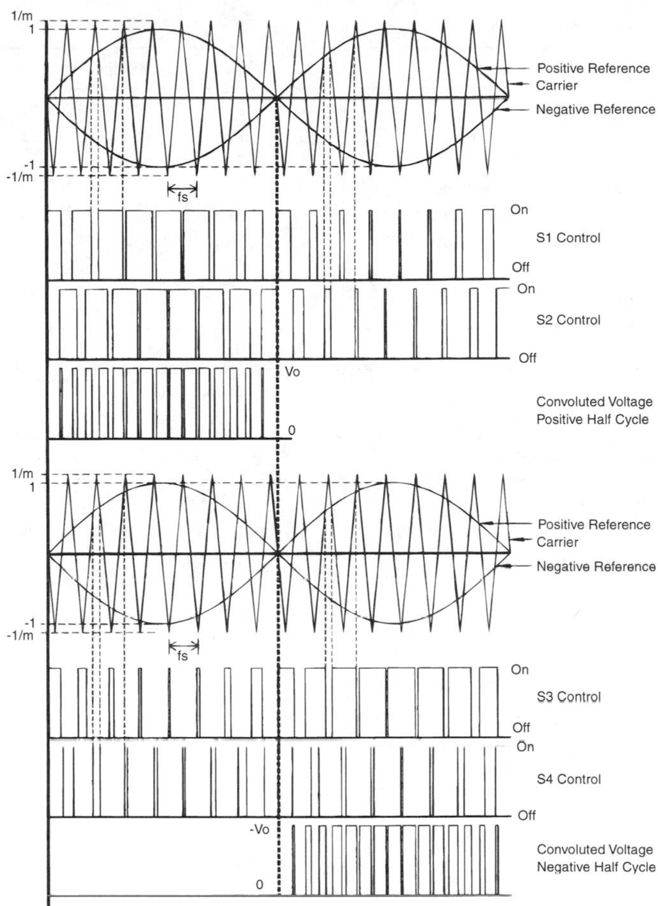

Figure 14.20 PWM gating signals for bidirectional full-bridge converter.

can be altered to a zigzag configuration and thus a smaller size and the advantages of this approach are

- The scheme is passive and does not interfere with the rectification process of the diodes.
- The resulting utility line current is near sinusoidal.
- The circulating third harmonic current is automatically generated.

Enjeti et al. (1994) gives all of the necessary design equations along with a design example. Figure 14.24a–d show the experimental performance of this scheme.

Figure 14.25 shows another high-performance diode rectifier type utility interface with polyphase auto-transformer arrangement (Choi et al., 1996). The proposed scheme is essentially a 12-pulse diode rectifier which guarantees the cancellation of 5,7 harmonic components in the utility line currents. An auto-transformer shown in Figure 14.25a provides the necessary phase shift (30 degrees) and the two interface reactors on the DC side ensure independent operation of the two rectifier bridges. Figures

Table 14.12

Item	Bridge-Boost	Modified Boost	Bi-directional Full-Bridge
Stages Required	2	1	1
Total Number of Switches	1	2	4
Total Number of Diodes	5	2	0
Total Number of Devices	6	4	4
Devices in Power Flow Path	3	2	2
Bi-directional Power Flow	No	No	Yes
Output Voltage Range	$\sqrt{2} - \frac{\sqrt{2}}{0.8}$ p.u.	$\sqrt{2} - \frac{\sqrt{2}}{0.8}$ p.u.	$\sqrt{2} - \frac{\sqrt{2}}{0.8}$ p.u.
Components	**Bridge-Boost**	**Modified Boost**	**Bidirectional Full-Bridge**
X_L	2666/(m·fs) p.u.	2666/(m·fs) p.u.	1333/(m·fs p.u.)
X_C	$0.2/m^2$ p.u.	$0.2/m^2$ p.u.	$0.2/m^2$ p.u.
Stage 1 Device Ratings	**Bridge-Boost**	**Modified Boost**	**Bidirectional Full-Bridge**
Diode Peak Current	$\sqrt{2}$ p.u.	$\sqrt{2}$ p.u.	N.A.
Diode Peak Voltage	$\sqrt{2}$ p.u.	$\frac{\sqrt{2}}{m}$ p.u.	N.A.
Diode RMS Current	$\frac{1}{\sqrt{2}}$ p.u.	$\sqrt{\frac{m^2 \cdot 3}{8}}$ p.u.	N.A.
Diode RMS Voltage	$\frac{1}{\sqrt{2}}$ p.u.	$\sqrt{\frac{4 - m^2}{2 \cdot m^2}}$ p.u.	N.A.
Switch Peak Current	N.A.	$\sqrt{2}$ p.u.	$\sqrt{2}$ p.u.
Switch Peak Voltage	N.A.	$\frac{\sqrt{2}}{m}$ p.u.	$\frac{\sqrt{2}}{m}$ p.u.
Switch RMS Current	N.A.	$\sqrt{\frac{4 - m^2 \cdot 3}{8}}$ p.u.	$\sqrt{\frac{4 - m^2 \cdot 3}{16}}$ p.u.
Switch RMS Voltage	N.A.	$\frac{1}{\sqrt{2}}$ p.u.	$\frac{1}{m}$ p.u.
Antiparallel Diode Peak Current	N.A.	$\sqrt{2}$ p.u.	$\sqrt{2}$ p.u.
Antiparallel Diode Peak Voltage	N.A.	$\frac{\sqrt{2}}{m}$ p.u.	$\frac{\sqrt{2}}{m}$ p.u.
Antiparallel Diode RMS Current	N.A.	$\frac{1}{\sqrt{2}}$ p.u.	$\sqrt{\frac{4 + m^2 \cdot 3}{16}}$ p.u.
Antiparallel Diode RMS Voltage	N.A.	$\frac{1}{\sqrt{2}}$ p.u.	$\frac{1}{m}$ p.u.
Switch Utilization Ratio	1/8	m/8	m/8

Table 14.13

Stage 2 Device Ratings	Bridge-Boost	Modified Boost	Bidirectional Full-Bridge
Diode Peak Current	$\sqrt{2}$ p.u.	N.A.	N.A.
Diode Peak Voltage	$\sqrt{2}/m$ p.u.	N.A.	N.A.
Diode RMS Current	$\sqrt{\frac{m^2 \cdot 3}{4}}$ p.u.	N.A.	N.A.
Diode RMS Voltage	$\sqrt{\frac{2}{m^2} - 1}$ p.u.	N.A.	N.A.
Switch Peak Current	$\sqrt{2}$ p.u.	N.A.	N.A.
Switch Peak Voltage	$\sqrt{2}/m$ p.u.	N.A.	N.A.
Switch RMS Current	$\sqrt{\frac{4 - m^2 \cdot 3}{4}}$ p.u.	N.A.	N.A.
Switch RMS Voltage	1 p.u.	N.A.	N.A.
Switch Utilization Ratio	m/4	N.A.	N.A.

Table 14.14

Application	Bridge-Boost	Modified Boost	Bidirectional Full-Bridge
Computer Power Supply	Yes	Yes	Yes
Uninterruptible Power Supply	No	No	Yes
Battery Charger	Yes	Yes	Yes
Appliances	Yes	Yes	Yes
Fluorescent Lamps	Yes	Yes	Yes
High-Intensity Discharge Lamps	Yes	Yes	Yes
Photovoltaic Energy System	No	No	Yes
Wind Energy System	No	No	Yes
AC Motor Drive	Some	Some	Yes
Inverters	Some	Some	Yes
Single-Phase Active Power Filter	No	No	Yes
Low-Switching Harmonic Converter	No	No	Yes
High-Power Application	No	No	Yes

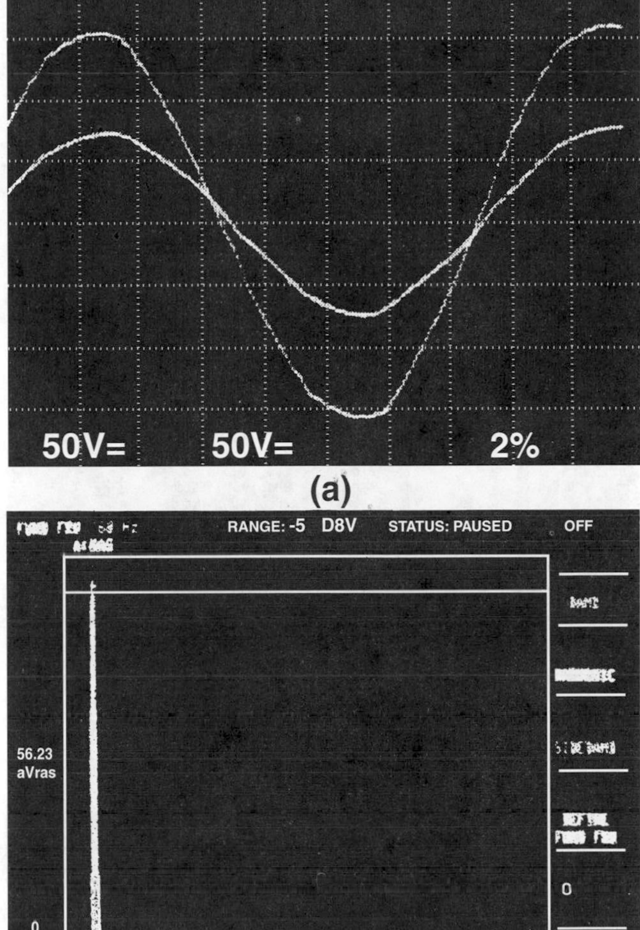

Figure 14.21 Experimental results of full-bridge converter shown in Fig. 14.19. (a) Input voltage and line current; (b) harmonic spectrum of input current.

14.25b and 14.25c show the vector diagram and the winding arrangement on a three limb core. Choi et al. (1994) details the necessary design equations. The resulting kVA rating of the autotransformer is only 0.18 per unit of the total output power. This results in 82% reduction in transformer size, weight and cost compared to a conventional 12-pulse isolation transformer approach. Figure 14.25d shows the experimental results from a 10kVA proto-type system.

Figure 14.26 shows a switched mode rectifier employing IGBT (insulated gate bipolar transistor) switches. The switches are operated in PWM mode at high frequency. This converter is capable of drawing near sinusoidal input currents at unity input power factor, and in addition, the power flow is reversible permitting regenerative braking of the AC motor load. Commercial AC motor drives with the PWM rectifier type utility interface are becoming available. Baldore series 22H line regenerative, 460V motor controller boasts 3% THD and compliance to IEEE 519 distortion limits (Baldore Electric Co., 1995).

References

Baldore Electric Company 1995. Baldore Series 22H line regenerative motor drive, Baldore Electronic Company, Fort Smith, AR.

Choi S., Enjeti, P., and Pitel, J. 1996. Polyphase transformer arrangements with reduced kva capacities for harmonic current reduction in rectifier type utility interface, *IEEE* '96, 680–690.

Enjeti et al. 1994. A new approach to improve power factor and reduce harmonics in a three phase diode rectifier type utility interface, *IEEE Trans. Ind. Appl.,* November/December, 1557–1564.

Kit, S. 1989. Power factor correction for single phase input power supplies, *PCIM,* December, 18–24.

Klein, and Nalbant, M. K. 1990. Power factor correction—incentives, standards and techniques, *PCIM,* June, 26–31.

Strassberg, 1991. Power factor corrected switching power supplies, *EDN,* April, 90–100.

Unitrode Corporation 1996. *Unitrode Application Notes,* Unitrode Corporation, Merrimack, NH.

14.4 Uninterruptible Power Supplies (UPS)

Laura Steffek, John Hecklesmiller, Dave Layden, and Brian Young

With the proliferation of electronic loads like computers, the incidence of power quality-related problems is growing. As a result, the uninterruptible power supply (UPS) market has grown significantly in the last few years. What follows is an overview of UPS functions and descriptions of common types of UPSs and backup power sources.

UPS Functions

The primary purpose of a UPS is to provide conditioned, continuous power to its load. Another UPS function that is of growing importance in today's market is system integration, or the ability to communicate over a network to facilitate the monitoring and orderly shutdown of loads.

Power Conditioning

A UPS provides continuous, regulated power to its load, under all conditions of the utility power line. Unlike other types of power conditioning equipment, a UPS provides power during outages. Typically, a UPS will provide back up power for 10 or 15 minutes, although longer times are possible with large battery strings or a DC generator.

A UPS will also correct for high- and low-voltage events, known as surges and sags. This regulation is provided either electronically or by a tapped transformer or a ferroresonant transformer.

Normal mode, or line to line, transients are prevented from

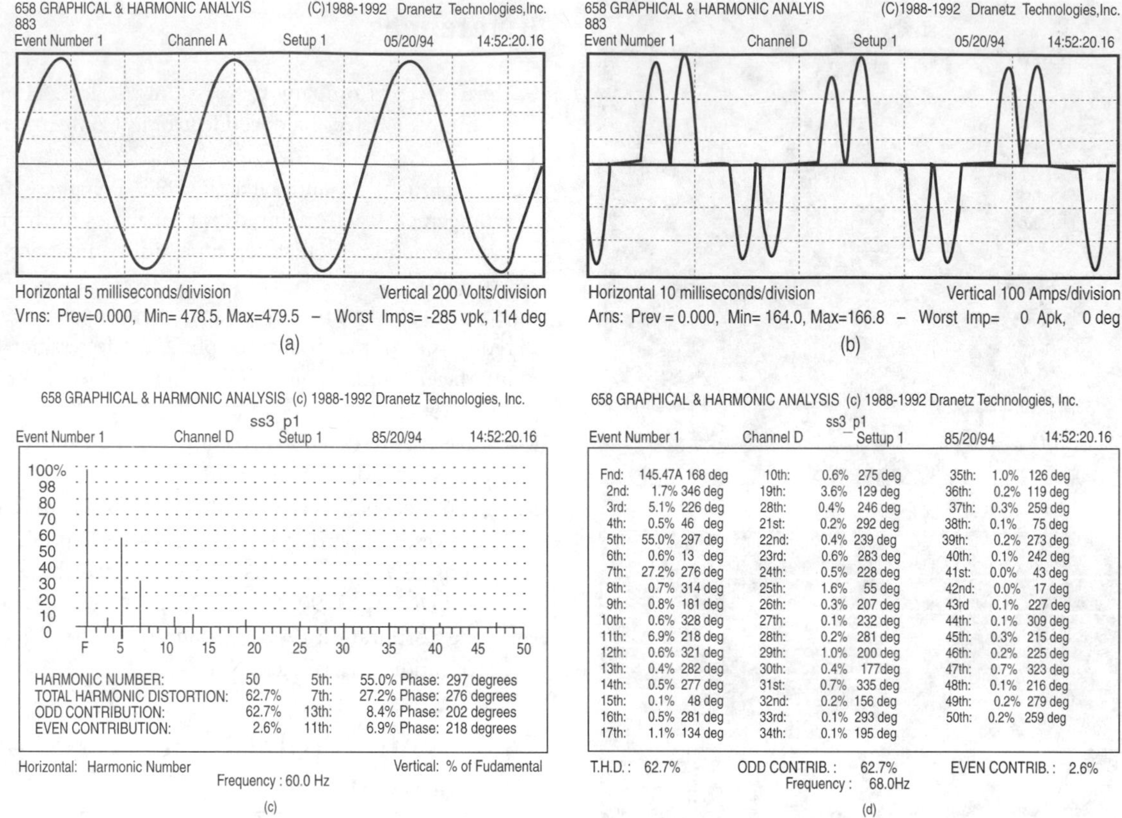

Fnd:	145.47A	168 deg	10th:	0.6%	275 deg	35th:	1.0%	126 deg
2nd:	1.7%	346 deg	19th:	3.6%	129 deg	36th:	0.2%	119 deg
3rd:	5.1%	226 deg	28th:	0.4%	246 deg	37th:	0.3%	259 deg
4th:	0.5%	46 deg	21st:	0.2%	292 deg	38th:	0.1%	75 deg
5th:	55.0%	297 deg	22nd:	0.4%	239 deg	39th:	0.2%	273 deg
6th:	0.6%	13 deg	23rd:	0.6%	283 deg	40th:	0.1%	242 deg
7th:	27.2%	276 deg	24th:	0.5%	228 deg	41st:	0.0%	43 deg
8th:	0.7%	314 deg	25th:	1.6%	55 deg	42nd:	0.0%	17 deg
9th:	0.8%	181 deg	26th:	0.3%	207 deg	43rd	0.1%	227 deg
10th:	0.6%	328 deg	27th:	0.1%	232 deg	44th:	0.1%	309 deg
11th:	6.9%	218 deg	28th:	0.2%	281 deg	45th:	0.3%	215 deg
12th:	0.6%	321 deg	29th:	1.0%	200 deg	46th:	0.2%	225 deg
13th:	0.4%	282 deg	30th:	0.4%	177deg	47th:	0.7%	323 deg
14th:	0.5%	277 deg	31st:	0.7%	335 deg	48th:	0.1%	216 deg
15th:	0.1%	48 deg	32nd:	0.2%	156 deg	49th:	0.2%	279 deg
16th:	0.5%	281 deg	33rd:	0.1%	293 deg	50th:	0.2%	259 deg
17th:	1.1%	134 deg	34th:	0.1%	195 deg			

Figure 14.22 Experimental voltage and current waveforms of a three-phase, 250HP variable speed pump. (a) Line to line voltage V_{ab}; (b) line current I_a; (c) and (d) harmonic spectrum of line current I_a.

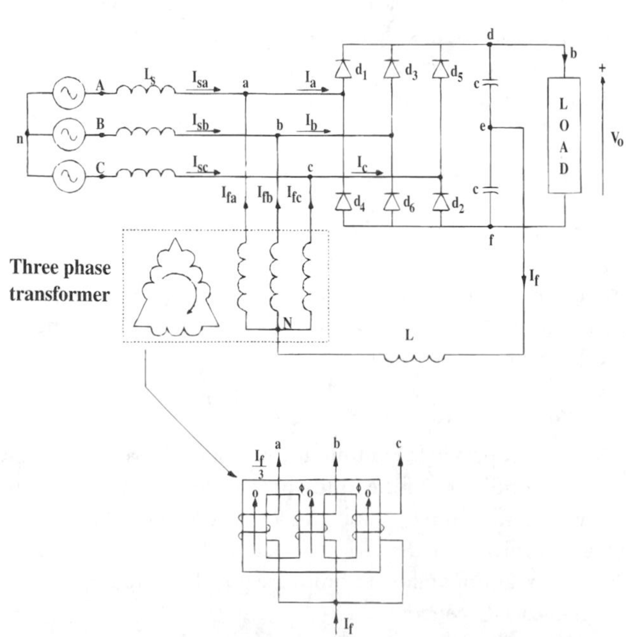

Figure 14.23 An approach to improve power factor and reduce input harmonic currents of three-phase diode rectifier type utility interface.

reaching the load. This is accomplished either with filter components, or in a double conversion UPS, by converting the AC to DC and then back to AC. There is quite some variation in the ability of UPS systems to protect the load from common mode, or line to ground, transients. Safety agency requirements preclude most forms of common mode transient protection. The best common mode transient suppression is achieved with an isolation transformer. Some UPSs have isolation transformers and some do not.

System Integration

The industrial electronics environment is very similar to the typical office LAN/WAN environment when it comes to using a UPS to provide power protection for industrial-grade PCs, PLCs and other equipment which make use of any form of microprocessor control.

The fact that a UPS only provides a finite amount of battery backup during an extended power outage, should encourage us to take certain precautions to prevent the corruption and loss of data once the UPS reaches a point where it can no longer support the load equipment.

Certain methods may be used to communicate to the load equipment when a power outage has occurred and in extreme cases, when a low battery condition exists. The load equipment should be configured to react to critical UPS conditions by saving data and preparing the system for a safe shutdown. Creating this

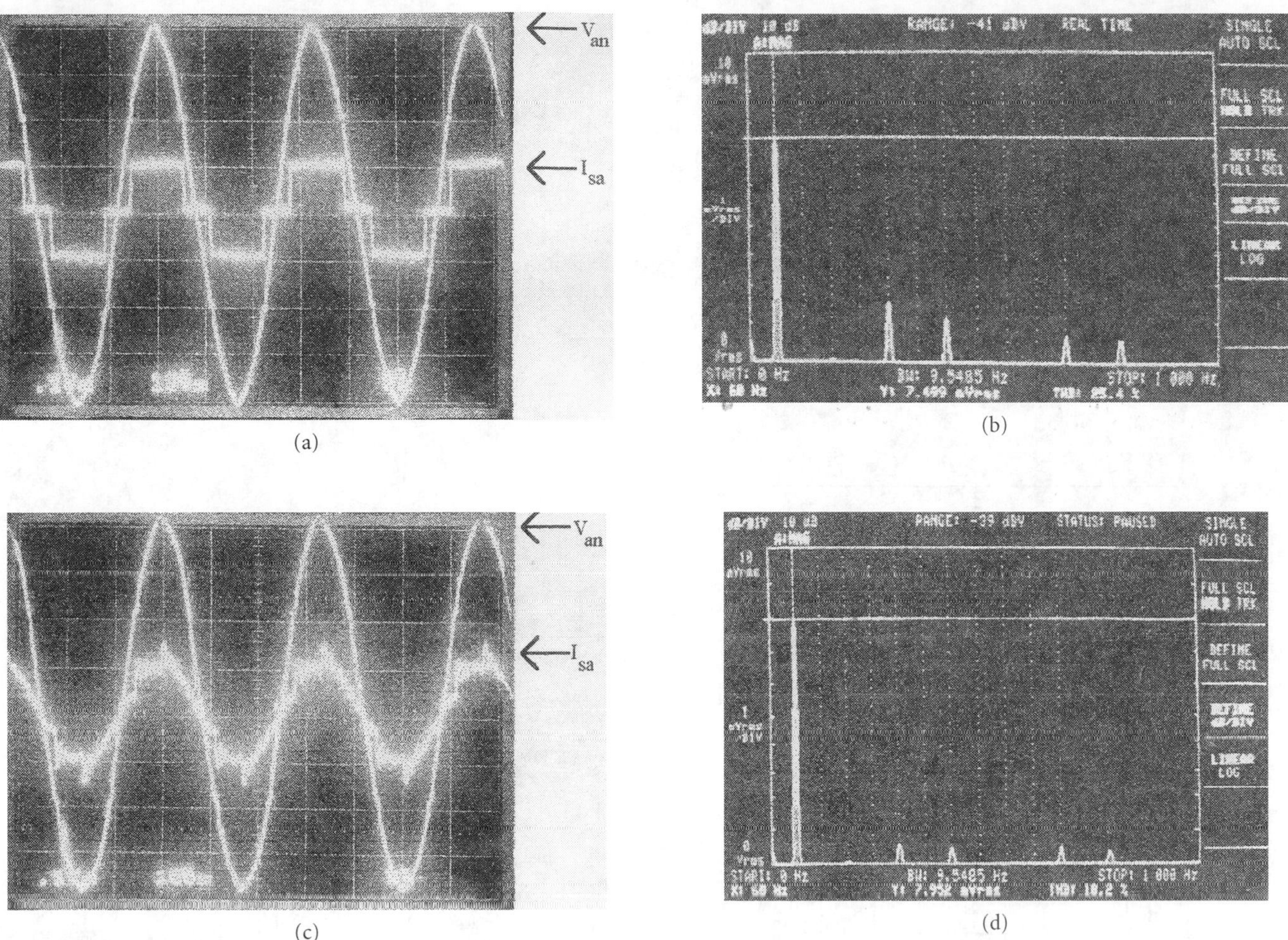

Figure 14.24 a) Line to neutral voltage V_{an}, and line current I_{sa} without the proposed scheme. Notice the square-wave nature of I_{sa}. Scale 50V/div, 20A/div, 5ms/div; b) Frequency spectrum of input current I_{sa}; c) Line to neutral voltage V_{an} and line current I_{sa} with the proposed approach (Figure 14.23). Notice the near sinusoidal wave shape of I_{sa}. Scale 50V/div, 20A/div, 5ms/div; d) Frequency spectrum of input current I_{sa} shown in Figure 14.24c.

"communication" between UPS and the load equipment is called UPS integration. There are several ways that the UPS can be integrated. These methods may be classified into three integration categories:

Basic
Enhanced
Network

No matter what integration methodology is utilized, four items are required to integrate the UPS. First, the UPS must have a communication port. Second, the equipment being protected must also have a communication port. Third, some medium (cabling) must be used to connect the two together. Finally, some form of software must be used to monitor the UPS and provide the appropriate actions relevant to specific UPS conditions.

Basic. The first and most common integration method communicates the status of the UPS via contact closures. Typically, normally open or normally closed relay contacts are used to signal two UPS conditions to the load equipment. These conditions are "AC Failure" and "Low Battery." An "AC Failure" should be signalled by the UPS whenever a power failure condition exists for more than 5 seconds. The "Low Battery" signal exists when a minimum of 2 minutes of battery runtime remains to support the load. However, most UPS manufacturers allow this setpoint to be programmed by the user to allow more time to shutdown the system.

In most cases, the software to monitor the UPS is provided as a part of the computer's operating system. The UPS manufacturer typically provides the cable and appropriate setup information required to connect the two together.

Note that UPS manufacturers often substitute open-collector type circuits in place of actual relays, to provide the UPS signals. The user should pay close attention to this detail if they choose to build their own interface cable, since current is only allowed to pass in one direction through an open-collector circuit.

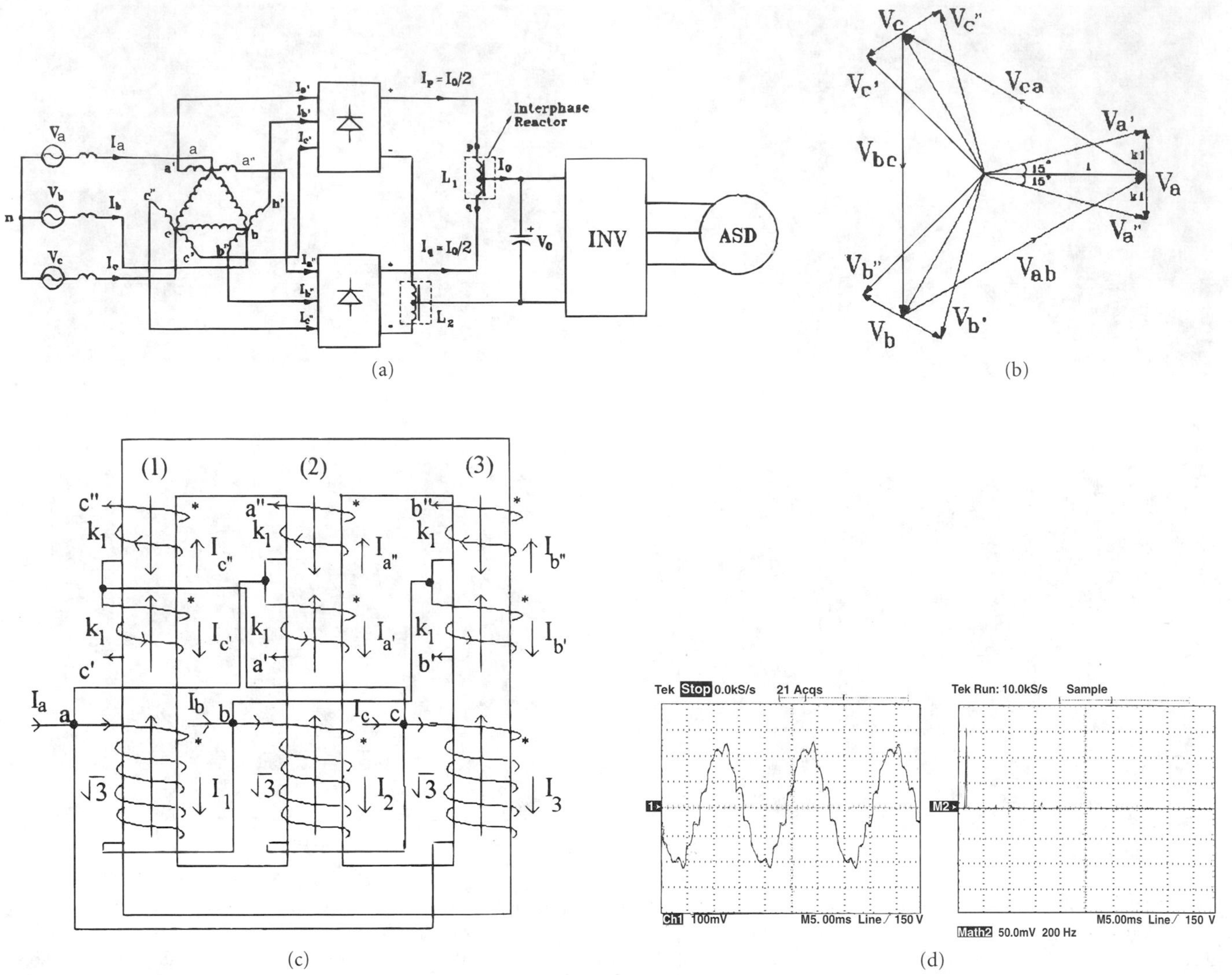

(a)

(b)

(c)

(d)

Figure 14.25 (a) Optimized 12-pulse diode rectifier with polyphase auto-transformer arrangement; (b) Vector diagram; (c) Auto-transformer winding configuration; (d) Experimental utility line current (10A/div) and its frequency spectrum.

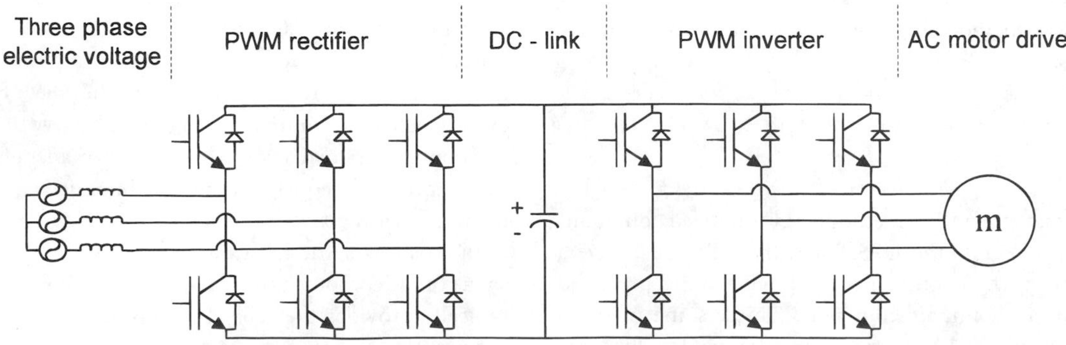

Figure 14.26 Three phase PWM rectifier/inverter motor drive system draws clean input power from electric utility.

Enhanced. To provide more than just the basic UPS status information, many UPS manufacturers have chosen to offer RS232 and other forms of serial communication which allow real-time UPS data to be monitored by software running on the load equipment. Instead of knowing only that a power failure has occurred or that a low battery condition exists, the user may now know how much calculated runtime is available and the measured battery voltage at any given time. Other data values are typically available which represent the input and output voltage, percent of full load, UPS temperature, as well as many others.

Since the way this UPS data is presented is usually proprietary, the UPS manufacturer most often supplies the software to run on the protected load. Because the software is capable of monitoring real time data from the UPS, a GUI (graphical user interface) is typically used to portray the data using easy-to-read digital displays and historical graphs.

Network. The size and complexity of today's local and wide area networks has led to an increase in the use of network management tools to monitor and control network devices. The Simple Network Management Protocol (SNMP) has become the defacto standard for network management and is backed by many network management software products including SunNet Manager, HP-OpenView, IBM's Netview/6000 and Novell's NMS.

Today, many UPS manufacturers offer software or a software/hardware combination that effectively makes the UPS a network peripheral. In some cases an internal or external network adapter is provided that through its own microprocessor and associated components effectively translates proprietary UPS data and commands into a format which is compatible with the SNMP standards set forth by a working group of the Internet Engineering Task Force (IETF). This group recently adopted a standard database of UPS-related information, called a Management Information Base (MIB) for all UPS products. The official IETF document that describes this MIB is RFC-1628 which is available on the Internet.

The SNMP-capable UPS provides three basic functions when communicating with a network management station. It responds to "get" requests by replying back to the management console with a value corresponding to the requested MIB variable. It responds to "set" requests by allowing the UPS configuration to be changed by the management console. And it broadcasts unsolicited alarm "traps" to the network management console alerting the network administrator to the existence of potential power problems.

The UPS in an industrial environment presents a new challenge to the integrator due to the existence of many different industrial network protocols. In some cases, many of the same protocols exist that are present in the office LAN environment, but they are often joined by such protocols as SP50 and PROFIBus, which are adaptations of Field Bus. Other industrial control protocols include FIP (Factory Instrumentation Protocol), MAP (Manufacturing Automation Protocol), and Echelon's LonTalk. UPS manufacturers have not yet built-in direct connections to these industrial networks. In some cases, protocol adapters are available that translate the RS232 information from the UPS into the required network protocol. Future developments from UPS vendors may enhance and simplify the UPS's connectivity in the industrial environment.

Static UPS Topologies

A static UPS is one which relies on power electronics, rather than a motor generator, to provide power to the load. Most UPSs today are of this type.

There are several basic UPS topologies, each of which has its advantages and disadvantages. The terms "on-line UPS" and "off-line UPS" have commonly been used to describe some UPS topologies. Unfortunately, UPS manufacturers have not been able to agree on the meaning of these terms, leading to confusion among users. Terms which are more descriptive of the differences between various topologies are double conversion UPS, line interactive UPS, and standby power supply.

Double Conversion UPS

A double conversion UPS (Figure 14.27) first rectifies incoming AC line to a DC voltage, then inverts that DC voltage to provide an AC output. During normal operation, the rectifier is providing current to charge the batteries and also to the inverter. The inverter supports the load and provides regulation of the output voltage and frequency. In the event that line is lost or deviates from the specified input voltage and frequency tolerances, the inverter uses the batteries as an energy source and operates until the batteries are depleted or line is restored (See Figure 14.28 for typical response.)

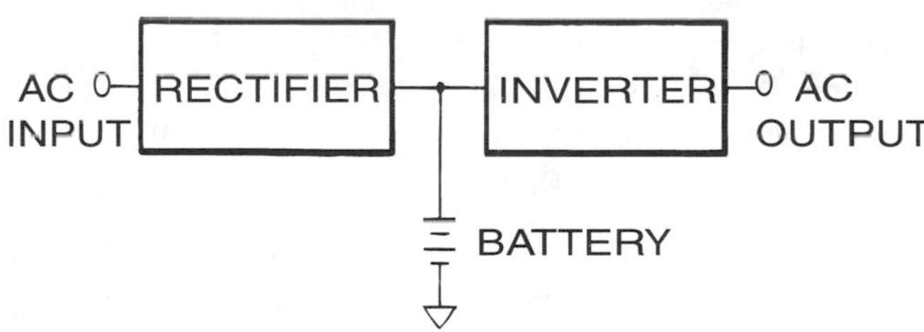

Figure 14.27 Double conversion UPS. There is no disruption in output power when the UPS transfers from its line source to battery power, because the inverter is always operating.

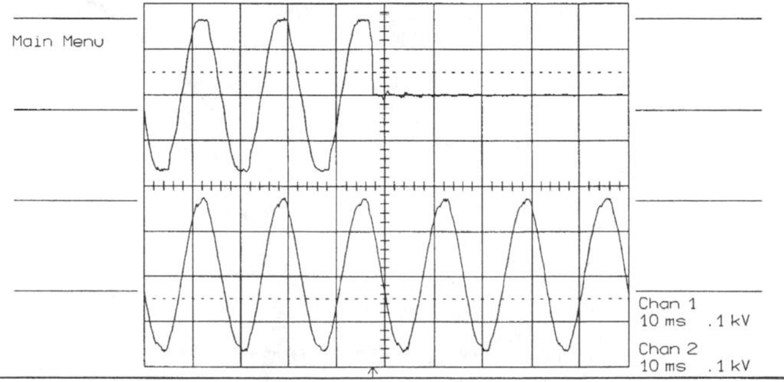

Figure 14.28 Typical double conversion UPS response to a power disturbance. Top trace: AC input; bottom trace: AC output. Courtesy National Power Laboratory of Best Power, a unit of General Signal.

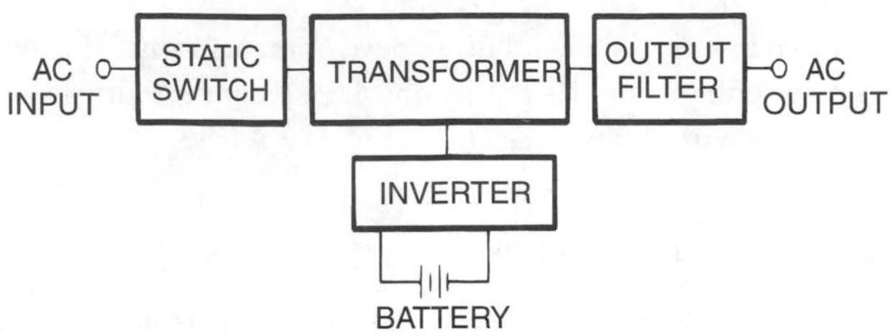

Figure 14.29 Typical line-interactive UPS.

Some double conversion UPS have an automatic bypass switch. This switch connects the load to the AC source in the event of a UPS failure. It may also be used to help support a temporary overload that the inverter cannot support alone.

Traditionally, phase-controlled thyristor rectifiers have been used in double conversion UPSs. These rectifiers cause distortion of the input current and voltage waveforms. Distorted current waveforms can cause excess neutral currents in the building wiring, and distorted voltage waveforms can cause problems in other equipment on the same circuit. Some newer rectifier designs use pulse width modulation (PWM) techniques to reduce waveform distortion. These techniques can result in harmonic distortion levels of 5% or less.

Line-Interactive UPS

In normal operation (Figure 14.29), the AC input passes through a filter or transformer to the load. The inverter is normally not supporting the entire load, but may be used to buck or boost the line voltage, or even fill in "notches" of the incoming line voltage waveform on a subcycle basis. It is this ability of the inverter to interact with line that gives the line-interactive UPS its name. The inverter does not support the load unless there is a power outage, or the AC input falls outside the specified voltage and frequency tolerances (See Figure 14.30 for typical response.)

The key to a line interactive unit is its ability to respond to line disturbances quickly. This is necessary to insure that power is supplied continuously to the load. Some energy is stored in the magnetics and output filter which can support the load for a short time. The static switch must open quickly and the inverter become active before that energy is lost to the load.

Voltage regulation during line operation may be achieved by phase-controlling the inverter, by using a tapped transformer, or by using a ferroresonant transformer.

Line-interactive UPSs do not themselves cause harmonic distortions on the utility line. However, they may or may not pass harmonic load currents to the input. Ferroresonant-based units correct the current harmonic distortion of the load and present a near sinusoidal current waveform to the utility line. Other line-interactive units provide little harmonic correction. This is of diminishing importance as computer power supplies are being redesigned to reduce the harmonic currents they cause, to meet the requirements of standards such as IEC 555-2.

Standby Power Supplies

Standby power supplies (SPS) (Figure 14.31) are not properly called UPS because they do not provide continuous power to the load. A standby power supply is similar to a line interactive UPS in that the inverter is not normally supporting the load. However, when the load is transferred from line to inverter, an interruption in power occurs due to the break time of the transfer switch. Typically this switching device is an electromechanical relay and takes several milliseconds to open or close. The minimum operation on inverter is usually several seconds, as compared to the subcycle control possible with a line interactive UPS (See Figure 14.32 for typical SPS response.)

Standby power supplies are typically low-cost products and provide minimum levels of voltage regulation and line conditioning. They usually provide square wave or stepped-square wave outputs on inverter, rather than the sinewave outputs provided by most double conversion and line interactive products. They

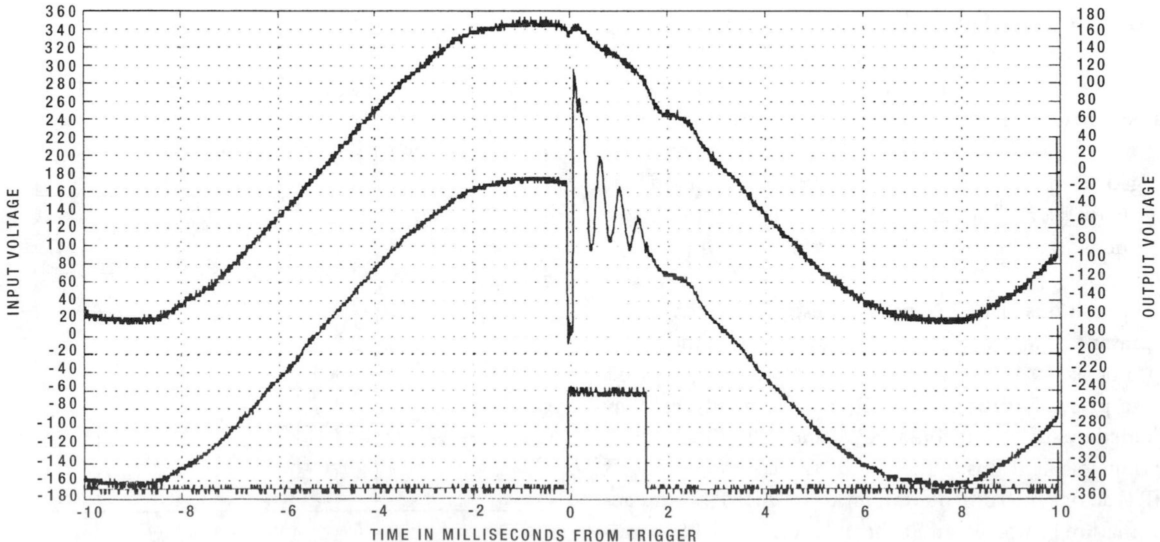

Figure 14.30 Typical response of a line interactive UPS to a power disturbance. Top: AC output (right scale); middle: AC input (left scale); bottom: inverter active signal (no scale). Courtesy Best Power, a unit of General Signal.

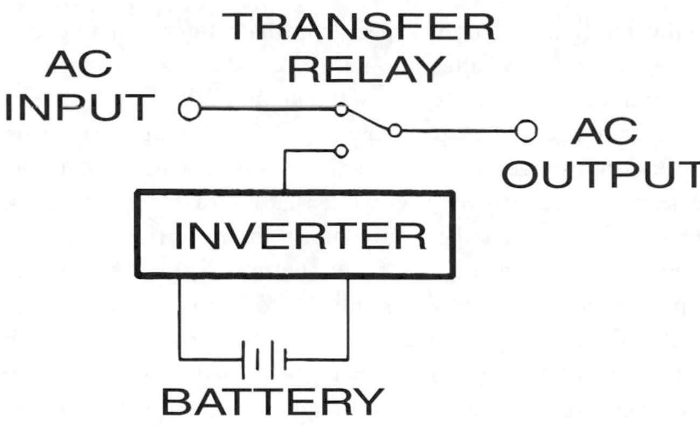

Figure 14.31 Standby power supply.

are most appropriately used in less critical applications, where power interruptions several milliseconds in duration and voltage fluctuations can be tolerated.

Rotary UPSs

The earliest form of UPS is the rotating, or rotary UPS. Motor and generator combinations have provided uninterruptible power since circa 1950. These early systems offered excellent isolation and fairly good overall performance. They consist of little more than a DC motor coupled to an AC generator. Rectified line normally powers the DC motor. Power switch-over to batteries occurs when the utility (line) fails. Due to the inertia of the rotating mass, switch-over times on the order of 0.3 seconds are typical. In practicality, however, the decay in frequency is usually more of a problem than the decay in voltage. To remedy this problem, a supplemental flywheel increases the inherent ride-through to one second or longer. Thus large mechanical contactors are acceptable to make the power transfer from line to battery. However, modern systems use power semiconductors to do the switching. The mechanical coupling between motor and generator can be either direct, where the components share a common shaft, or by belt. Belt drives, while less efficient, do allow for different speeds between the motor and the generator. Rotary UPSs are currently available in sizes from 35KVA up to 1000KVA.

As have other forms of UPSs, the rotary UPS has continued to evolve. Most rotary UPSs today use AC induction motors instead of the DC motor, as AC motors do not require brush maintenance. A typical modern system rectifies and controls incoming AC to charge batteries. The batteries then power a simple three-phase inverter. This inverter, which requires no commutation or voltage-regulation circuitry, drives the induction motor. An added benefit is that this system requires no flywheel for energy storage, as there is no transfer time from line to battery power. Figure 14.33 depicts the block diagram of a typical rotary UPS.

Some of the newer rotating UPSs combine the motor and generator on one stator, and apply a DC field to the rotor. This scheme makes a very compact and cost-effective system. Other advances include the introduction of a "pole-writing" generator. In this topology, there are no pole windings as such. The poles of the generator write on a ferrite stator with varying position and frequency, depending on the speed of the rotor. Pick-up coils read these poles and use them to produce the AC output, much as a tape recorder records a signal then plays it back. This design can give as much as fifteen seconds of ride-through. Frequency and voltage stability are excellent.

A recent entry is a line-interactive rotary UPS. The line-interactive rotary UPS uses a normally free spinning unloaded synchronous motor, with an additional motor or engine used for back-up power. The synchronous motor has an over-excited field connected to a tapped line inductor. The motor acts as a synchronous capacitor, thereby providing power factor correction. When power fails, the mass of the synchronous motor powers the load until the engine comes up to speed and can assume the load.

Advantages of rotating UPSs include unmatched isolation, and the ability to use many different sources of energy. Single- or three-phase AC power, or power from a turbine or diesel engine

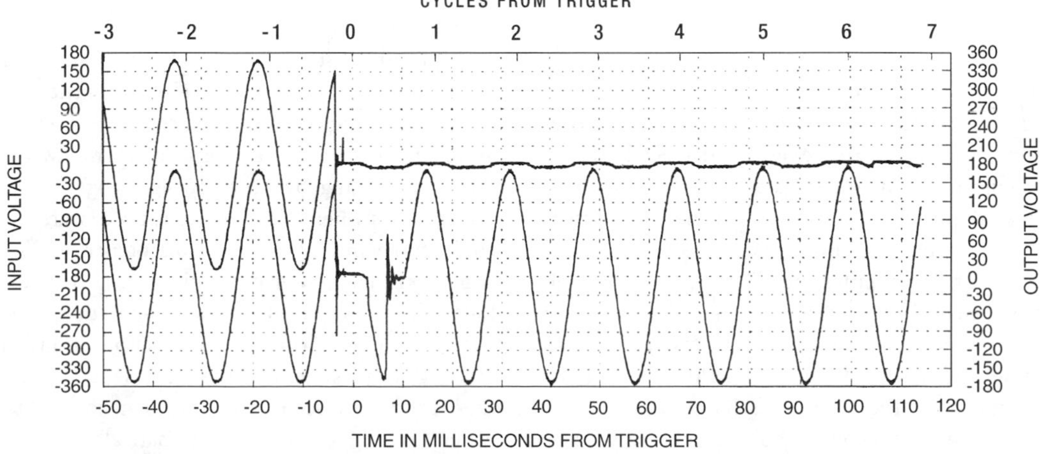

Figure 14.32 Standby power supply response to power disturbance. Top trace: AC input; bottom trace: AC output. Courtesy National Power Laboratory of Best Power, a unit of General Signal.

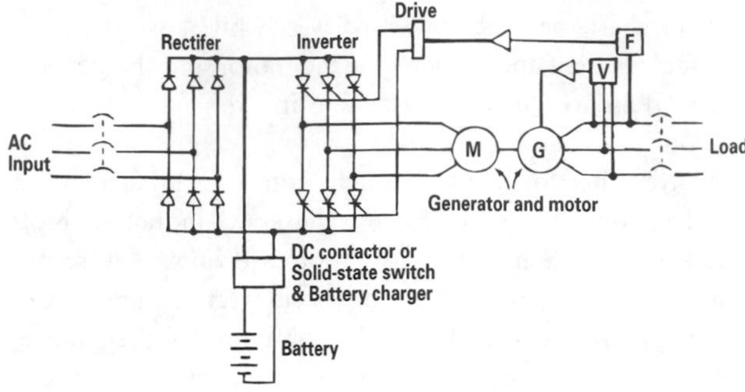

Figure 14.33 Typical rotary UPS.

can provide rotation. Reliability is excellent; with a demonstrable MTBF that exceeds 10^6 hours. High thermal inertia means that the UPS can sustain very heavy overloads for a short period of time. The units are efficient, with typical efficiencies running from 84–88%. Some newer designs can exceed 90% efficiency. For example, a 500KVA Uniblock demonstrates an efficiency of over 94% at load. As with all rotary UPSs, the rotating mass offers a degree of frequency stability as well as immunity to small load fluctuations.

Disadvantages include an inherent difficulty in starting the system into high-surge loads. It is difficult to make a completely redundant system, and some maintenance, such as bearing replacement, will require shutdown. Rotary UPSs usually cannot start from the inverter, requiring a secondary motor to start rotation.

The rotary UPS is a very practicable system for any application requiring a medium- to- high-powered premium UPS system that is reliable and cost-effective.

Alternate AC and DC Sources

Most UPSs use utility line for the AC source and batteries for the DC source. Batteries are a critical but often misunderstood component that bear further mention. Some installations use alternate power sources that are described below.

Batteries

Both flooded and valve regulated lead acid (VRLA) batteries are commonly used in UPS applications. Wet cell batteries require maintenance of the electrolyte level and special precautions to prevent build up of hydrogen gas. VRLA batteries have become increasingly popular in the last few years because of their relative ease of installation and maintenance.

All batteries require some maintenance. Battery terminals and connections should be checked for cleanliness and tightness. The batteries should be discharged periodically to test for battery capacity. End of life is usually defined as a 20% loss of the specified battery capacity at the desired discharge rate.

Battery life may be degraded by several factors, the greatest of which is battery temperature. Battery life is typically reduced by 50% for every 10°C increase in its temperature. Note that the battery temperature may be significantly higher than the ambient temperature of the room, especially if the battery is in an enclosure. Other factors affecting life include the charging method, the number of discharge cycles, the depth of discharge, the rate of discharge, and the ripple voltage across the battery terminals.

Battery storage life is also temperature dependent. Batteries experience a self discharge at a rate that increases with temperature. This self discharge is in addition to any current drain the UPS may have when it is off. UPS batteries should be charged upon receipt and every six months of storage after that, or more often if the storage temperature exceeds 25°C. If a battery is stored longer than this without being recharged, then a phenomenon called sulfation will occur. Sulfation is the formation of lead sulfate on the battery plates. This lead sulfate is an insulator and causes a loss of battery capacity. Many battery users have stored their batteries for long periods of time, only to find that at installation those batteries have no useful capacity at all. Most of the lost capacity can be recovered by exercising the batteries with repeated charge/discharge cycles, preferably at a high charge rate.

DC Generators

Direct current (DC) power generation has developed over the years to become a viable replacement for batteries in a variety of applications. These applications range from remote island power, such as railroad signal and switching, to uninterruptible power system backup and even lighting applications. Anywhere batteries are traditionally used, a DC generator can be installed to reduce the battery requirement or work in conjunction with alternate power sources such as solar.

To a great extent, the DC generator of today has changed from the days of maintenance intensive brushes and commutators to highly efficient rectified systems. Now, instead of relying on the brush and commutator to perform the rectification process, AC alternators and diodes are used to produce near “battery quality” DC power. Reduced maintenance requirements through elimination of brushes, commutators and slip rings are a few of the obvious advantages of a solid state rectified system. High frequency alternators and rectifier assemblies provide years of reliable service in less floor space than traditional AC generators or batteries.

Applications such as railroad switch and signal locations are examples of the versatility of DC power generation. Traditionally signal maintenance staff would replace a discharged battery with a recharged battery every few days. This was required to keep the trains rolling by properly signaling the track’s availability. Even at sites where solar power sources were utilized, dark days could force increased signal maintenance due to low battery conditions. DC power generation has successfully demonstrated long term battery backup and cooperation with other alternate energy sources. Some railroad applications have adopted a “cycling duty” system to maintain signal integrity. By allowing the battery to discharge or the solar charger to operate, a DC generator can be used to automatically start and recharge the battery when required. This reduces the site maintenance requirement to about twice a year.

UPS backup applications for DC generation has also proven

a viable alternative to large strings of batteries. By installing a "minimum" battery, most short term power outages can be supported. For the long duration power outages, DC power generation can be used. DC power generation generally requires only a connection to the two battery terminals for the system to operate. Through these two battery connections the battery condition is monitored and the DC power generator will start and run automatically to provide long term reliable DC power to the UPS inverter. Using the DC generator topology, oversized AC generators are not required. Generally the AC generator manufacturers recommend oversizing the generator to reduce the power factor induced by many UPS installations. With a DC generator, the induced power factor problem is not possible.

DC power generation has even grown into the lighting arena. By using a DC power source for floor lighting, HID bulb life is increased substantially. In some cases, this increase can be tenfold. Lighting manufacturers have doubled their lamp warranties due to DC power. In conjunction with these installations, DC power users have discovered the many utility rebates and rate credits available for peak-shaving with DC power generators. Generator run times as low as 50 hours per year can qualify for utility power reduction programs.

Telecommunications applications such as remote offices and cellular radio sites also enjoy the reliability of DC power generation. Even the information superhighway is powered by DC generators, bringing the benefits of the new technology to your door step.

Figure 14.34 shows two examples of DC generators available for operation with UPSs.

Superconducting Magnetic Energy Storage

Superconducting magnetic energy storage (SMES) systems are a relative newcomer on the field of back up power systems. SMES systems store DC energy in a superconducting magnetic coil. The niobium-titanium coil is cooled by liquid helium to 4.2K or by superfluid helium to 1.8K.

SMES units are used to provide large amounts of power for short durations. This is useful in industrial applications where even momentary power disturbances can cause expensive equipment downtime and production losses. Commercially available units store 0.3 to 1 kW-hour and are rated for 0.75 to 1.5 MW.

A complete SMES system is functionally the same as a traditional UPS with a more conventional DC storage element. AC line is fed to the load under normal operating conditions. A line fault detector monitors the AC line, and if line is unacceptable, disconnects line from the load. The magnetic storage element provides DC power to an inverter, which in turn supports the load. When acceptable line returns, the load is transferred back to the line source.

SMES has several advantages over batteries. The expected life of a SMES unit is claimed to be as long as 30 years, compared to 10 years or less for batteries. A SMES can be recharged completely in several minutes and the charge-discharge cycle can be repeated thousands of times without degrading the magnet.

AC Generators

What of extended autonomy, where utility may be out for hours at a time? For many applications where the AC line quality is unimportant, the AC generator is still a viable alternative for

Figure 14.34 DC generators available for operation with UPSs. (Courtesy Best Power Technology, Inc.)

extended-run applications. With the potentially unlimited run-times obtainable, the AC generator is certainly attractive. But with today's more sensitive loads, the AC generator may not be the best solution. It is well known in the industry that the AC generator suffers from poor regulation, and unless the unit is very large in comparison to the load, will also exhibit poor frequency stability. So, many users will run a hybrid combination of the AC generator and UPS to power critical loads for prolonged times.

A generator has its own set of maintenance issues. Aside from fuel, oil and water requirements, the engine must be run periodically to maintain a degree of readiness. Any engine, especially a gasoline engine, must be run occasionally to keep moving parts lubricated, and fuel must be treated against the formation of varnish or bacterial growth which can restrict fuel flow. Generators are usually kept outside, so shelters must be built and in many areas, cold-weather starting packages must also be used.

Sometimes, compatibility problems between AC generators and UPSs occur. Double-conversion is usually the most trouble-free of the different topologies of UPSs when used with an AC generator. As the name implies, the power for the double-conversion UPS is converted twice—once from AC to DC (for the batteries) and then from DC back to AC. This scheme assures that no matter what is happening on the input, the output can be controlled precisely. A line-interactive or single-conversion UPS which typically passes line through to the output must incorporate design features to accommodate AC generator operation. The primary trade-off is the desensitization of the UPS to the fluctuating inputs. Often, generator outputs are far from sinusoidal and rather unstable, so the voltage window in which the UPS stays on utility power must be widened. Also, the out-of-frequency window must also be widened, and the tracking capabilities of the phase-locked-loop must be increased. If these alterations are not taken, protracted inverter runs result, depleting the batteries and thus negating the purpose of the AC generator. Needless to say, the output reflects these widened windows. As most modern UPSs can be adjusted to function with an AC generator, the user must assess the impacts of the somewhat-diminished performance. The vast majority of loads will function acceptably.

References

DeWinkel, C., Losleben, J., and Billman, J. 1993. Recent applications of superconductivity magnet energy storage, *Proc. Power Quality Conf.*, 462–469.

Griffith, 1989. *Uninterruptible Power Supplies: Power Conditioners for Critical Equipment,* Marcel Dekker, New York, NY.

Platts and Aubyn, 1992. *Uninterruptible Power Supplies,* Peter Perigrinus, Stenenage, Herts, UK.

The Institute of Electrical and Electronic Engineers, 1992. *IEEE Recommended Practice for Powering and Grounding Sensitive Electronic Equipment* (IEEE Emerald Book), IEEE Std. 1100–1992, Institute of Electrical and Electronic Engineers, New York, NY.

15
Electromagnetic Compatibility for Drives

Walt Maslowski
Allen-Bradley Company

15.1 Compatibility: Emissions and Immunity 377
Standards • Emissions • Immunity

15.1 Compatibility: Emissions and Immunity

Electromagnetic compatibility (EMC) is defined as "the ability of an equipment or system to function satisfactorily in its electromagnetic environment without introducing intolerable electromagnetic disturbance to anything in that environment" (IEC 1000–1–1, 2.1).

This definition implies that electrical equipment will generate electrical disturbances or emissions, which will be called electromagnetic interference (EMI) if it interferes with other equipment. The definition also implies that the electrical equipment under consideration must also be immune to electromagnetic disturbances already existing in the environment. The compromise between allowable emissions from equipment and the necessary immunity of equipment defines the compatibility levels between all electrical equipment in the environment.

There are two levels of electromagnetic compatibility (EMC) in drives, the levels which will allow the drive and the system to operate correctly, and the levels which will allow the drive and system to meet agreed upon international standards. To meet the first level will require some ingenuity on the part of the designer and a minimum amount of changes in both hardware and software. To meet the second and more restrictive levels will require planning and design with respect to EMC. The more restrictive limits will almost certainly require additional hardware.

The International Electrotechnical Commission (IEC) is in the process of releasing sections of a comprehensive standard on EMC that is denoted as IEC 1000, *Electromagnetic Compatibility (EMC)*. IEC 1000 is developing into an excellent EMC guide for tutorial purposes, setting standards levels, and offering practical solutions to meet those standards levels. In this article, all definitions will be taken from IEC 1000–1–1, *Electromagnetic Compatibility (EMC)—Application and Interpretation of Fundamental Definitions and Terms* and indirectly from IEC 50, Chapter 161, *International Electrotechnical Vocabulary, Electromagnetic Compatibility.* EMC is a rule that equipment designers must follow to limit emissions generated into the surrounding environment and to limit equipment response to the same electromagnetic environment.

Standards

The European Union standards governing the recommended limits for emissions are in a state of flux as of this writing. At this time there are generic emissions and immunity standards available. These are EN50081, *Electromagnetic Compatibility—Generic Emission Standard,* Part 1, Residential, Commercial, and Light Industry and Part 2, Industrial Environment and EN50082, *Electromagnetic Compatibility—Generic Immunity Standard—* Part 1, Residential, Commercial, and Light Industry and Part 2, Industrial Environment. These are the standards to be used today if an electric drive were to meet the European Union Electromagnetic Compatibility (EMC) Directive. Within the IEC, a product specific standard for drives is being worked upon that is at this time denoted as SC22G/WG4 EMC Standards for Power Drive Systems. This standard, when approved, will recommend both emissions and immunity levels for drives. It is expected that when approved or if CENELEC should approve the recommendation of SC22G/WG4, this document will become the EMC standard for drives to meet in the European Union.

The International Special Committee on Radio Interference (CISPR) has been the traditional source of Radio Frequency Emissions standards for many years. CISPR has recommended that limits be imposed in the radio frequency ranges from 9 kHz to 400 GHz. Please note that in most CISPR standards only the frequency ranges from 150 kHz to 1.0 Ghz have agreed upon limits, while other frequency ranges are "under consideration". Standards such as CISPR 11 (1990), "Limits and methods of measurement of electromagnetic disturbance characteristics of industrial, scientific and medical (ISM) radio-frequency equipment," recommend levels of emissions for certain residential and industrial environments, while CISPR 16 (1987), "CISPR specification for radio interference measuring apparatus and measuring method," suggests specifications for instrumentation used in the measurement of conducted and radiated emissions.

IEC 1000 has as its scope the range of frequencies from DC (0 Hz) to about 400 GHz and this standard includes in it's

0-8493-8343-9/97/$0.00+$.50

purview, power line phenomena. Within IEC 1000 are several sections devoted to the power utility, commenting upon such phenomena as harmonics, voltage sags, swells, and flicker. Other sections address radio frequency emissions and EMC equipment immunity.

The FCC has several standards limiting the amount of conducted and radiated energy electrical equipment may generate. FCC Rule 15 is such a standard, and Subpart J is concerned with the electromagnetic interference (EMI) computers may emit. However, most industrial equipment is exempted from Rule 15, unless the equipment is interfering with communications or navigation, whereupon the equipment will be forced to comply with the concept of non-interference. FCC Rule 18 has as its scope industrial, scientific, and medical equipment (ISM), very similar to CISPR 11, but exempts most industrial equipment.

Immunity standards were relatively few to the point of being non existent until the publication of the IEC 801 series. This standard consists of several parts that deal with phenomena such as electrostatic discharge (ESD); radiated emissions from intentional and unintentional radiators; electrical fast transient/burst waveforms of the type that is found in power distribution systems when AC contractors or circuit breakers interrupt current creating an ionized air gap; surge voltages that simulate direct or induced lightning strikes or circuit interruptions; high-frequency voltage disturbances on the power distribution system created by conducted emissions from other equipment; low-frequency voltage disturbances such as harmonics, sags and swells; and power frequency magnetic fields. Today, IEC 1000-4-1 *Electromagnetic Compatibility (EMC)*, Part 4, Testing and Measurement Techniques, Section 1, Overview of Immunity Tests, Basic EMC publication gives an excellent tutorial on low-, medium-, and high-frequency immunity testing.

Emissions

Electromagnetic emission is defined as "the phenomenon by which electromagnetic energy emanates from a source" (IEC1000–1–1, 2.1). Emissions may be divided into two types, conducted and radiated emissions. Conducted emissions can be defined to be common mode (CM) disturbances with respect to ground or differential mode (DM) disturbances from line to line. The frequency range for common mode (CM) conducted emissions is from 9 kHz–30 MHz, but is typically defined between 150 kHz–30 MHz. Differential mode (DM) conducted emissions cause disturbances in lower frequency ranges between 0 Hz–9 kHz. These type of disturbances are sometimes referred to as harmonics, sags, swells, or flicker, among other names.

Radiated emissions are defined to be the electromagnetic far field electric field intensity emanating from the equipment. The frequency range for radiated emissions is generally from 30 MHz–400 GHz, but is typically defined from 30 MHz–1.0 GHz.

Conducted Emissions

As was mentioned previously, conducted emissions can be divided into two classifications, differential mode (DM) and common mode (CM). DM emissions includes the phenomenon known as "harmonics" when it is referred to the lower frequencies from 50 or 60 Hz up to 10 kHz (the 200th harmonic). Differential mode (DM) noise or disturbance is also generated along with common mode (CM) disturbance. DM disturbances are easier to filter because of RC snubbers within the power structure and within the power distribution system. CM disturbances have been more difficult to filter because of the difficulty in predicting the parasitic capacitance between physical electrical conductors and ground.

The reason for being concerned about CM electromagnetic disturbances is that the disturbing current and voltages will be conducted within the power distribution system. The power distribution system is a widespread geographic network where there will always be some physical segment tuned to a particular frequency as either a monpole, dipole, or loop radiating antenna. When a particular frequency excites one of the antenna systems, radiation can result as possible interference into susceptible equipment.

Parasitic capacitances always exist due to the presence of conductors at different potentials separated by a medium with a given electrical permittivity. The predominant parasitic capacitances are in the motor where the electrical windings are in close proximity to the iron stator frame, the motor supply cables in close proximity to ground, and components in close proximity to ground (semiconductor modules) within the power structure of the drive.

The parasitic capacitances include, but are not limited to:

1. Motor winding capacitance to frame.
2. Motor cable capacitance to conduit, raceway or ground.
3. Power semiconductor module transistor capacitance to heatsink.
4. Any switched mode power supply (SMPS) component capacitance to heatsink or ground.
5. Power system cabling or busduct capacitance to ground.

Electrical current will always flow due to the relationship:

$$i(t) = Cp\, dv(t)/dt \tag{15.1}$$

This means that the higher the $dv(t)/dt$, the higher the currents induced in the parasitic capacitances. The components that can generate such high $dv(t)/dt$ are BJT, IGBT, and MOSFET transistors, in the order of increasing switching speeds. As an example, depending upon the current rating, IGBTs at rated current can have voltage fall times in the order of 100 nanoseconds to 1000 nanoseconds. These fall times, when operating from a 650 V DC bus, ranges from 6500 V/μsec to 650 V/μsec. Assuming that there exists a parasitic capacitance of 10 nanofarads from a motor winding to ground, and ignoring any series inductance for the moment, one can calculate a peak current of 65 A during the period that the $dv(t)/dt$ is at 6500 V/μsec. These currents are the CM noise current that flow through the ground plane and will get back to the source of the $dv(t)/dt$ by any path that is available. One of these paths is the power distribution system. The equipment that these types of components are used in are switch mode power supplies (SMPS), uninterruptible power supplies (UPS), and drives. The amplitude spectra of the CM current depends on the switching frequency, the magnitude of voltage, and how the switching elements are configured.

Part of the ground currents will flow back directly to the drive by means of parasitic capacitances (which is the source), and the remainder into the power distribution system conductors by means of the distributed conductor parasitic capacitances. The currents that flow in the power distribution system will have the highest probability of resonating in a tuned monopole or dipole, which will radiate into the environment and possibly interfere with other electrical equipment.

Conducted Emissions Measurements

Rather than repeat information that the involved reader is almost required to have in his/her possession, it is best to refer to standards that give explanations and specifications for the measurement instrumentation and the measurement process. CISPR 16 (1987), "CISPR specification for radio interference measuring apparatus and measuring method," is the best international standard for the specification of the spectrum analyzer and the line impedance stabilization networks (LISN) that one must use in order to determine the levels of CM disturbances. The test setup and methodology is also discussed. Unfortunately, the LISNs that are specified in CISPR 16 go up to only 100 amps. A voltage probe is specified, but is relegated to be used for *in situ* measurements. At this time, there is no international agreement upon LISN specifications beyond 100 amps.

CISPR 11 (as an example) gives limits to conducted emissions in the frequency range of 150 kHz–30 MHz. The limits are given in dB(μV), in both quasi-peak (QP) and average numbers. The levels for conducted emissions are given in decibels referenced to 1.0 μV. Hence, conducted emissions is a measurement of the disturbance rms voltages. The term dB(μV) is defined as:

$$\text{dB}(\mu\text{V}) = 20\log(\text{Vd}/1\ \mu\text{V}) \tag{15.2}$$

Where Vd is the measured disturbance voltage at a given frequency in volts. This equation takes measured voltages in rms values, and forms a ratio with respect to 1 μV, and calcualting 20 times the logarithm to the base 10 of the ratio will give comparative numbers. The term quasi-peak is defined in CISPR 16 and is basically a circuit with asymmetrical charge and discharge time constants that takes into account higher magnitude disturbances at discrete frequencies.

Typical data taken on drives with SMPSs and IGBT power structures indicate that at 150 kHz the magnitudes of conducted emissions will be in the 60 dB(μV) to 150 dB(μV), depending upon power ratings of the drives. The lower dB numbers are for the lower power ratings. CISPR 11 limits for classes A and B are 79 dB(μV) QP and 66 dB(μV) QP, respectively. Clearly, the drive with 60 dB(μV) QP will pass both class A and B in CISPR 11, while the drive with 150 dB(μV) QP will need a lot of work.

Past experience has shown that conducted emissions generated by high power IGBT PWM AC drives with SMPS are difficult to economically reduce.

Conducted Emissions Mitigation

There are a number of procedures one may take to meet the limits for conducted emissions. Basically, one identifies the path by which the EMI current is flowing, and by inserting impedance into that path and/or by decreasing the impedance in a more desirable path, eliminate or contain EMI current to a path that will not generate EMI voltages in the power distribution system. Looking at these various paths from the section on conducted emissions we have:

1. Motor winding capacitance to frame. Possible solutions will include the use of common mode inductors and/or differential mode inductors at the output terminals of the drive. The inductors will insert impedance into the circuit so that the current from the dv(t)/dt will be reduced. It is important to be aware that the use of these CM or DM inductors may increase bearing voltages, with deleterious results. The designer must also be aware of the transmission line effects of voltage addition between a transmission line and a load whose impedance is other than the characteristic impedance of the motor windings. Such a discontinuity may puncture the motor winding insulation if the electric field intensity is high enough. Shielded cable or cable contained in steel conduit is another method to reduce currents in the ground plane. Where CM and DM inductors will reduce EMI currents, the use of shielded conductors or conductors contained in conduit will reduce the impedance of the EMI current return path, reducing the EMI current in the ground plane. The use of this method will actually increase the magnitude of these EMI currents slightly.
2. Motor cable capacitance to conduit, raceway or ground. Possible solutions for this path are the same as for motor winding capacitances.
3. Power semiconductor module transistor capacitance to heatsink. The best solution would be to add CM capacitors between the ±DC bus to ground. This will allow the EMI current to have a low impedance path from ground to the source (power semiconductor switches).
4. Any switched mode power supply (SMPS) component capacitance to heatsink or ground. The best possible solution would be CM capacitors between the ±DC SMPS bus to ground. This will allow the EMI current to have a low-impedance path from ground to the source (power semiconductor switches).
5. Power system cabling or busduct capacitance to ground. The solution for this capacitance is to insert a power line EMI filter with DM and CM series inductors and DM and CM parallel capacitors. The series inductors are to be towards the utility, while the capacitors are towards the drive. The intent of the power line EMI filter is to provide a low-impedance path from the ground plane to the source and to provide a high-impedance path from the ground plane through the utility conductors to the source.

Components for EMI Filters

A brief note on filter components: Capacitors and inductors are frequency sensitive. That is, if one measures capacitors and inductors over a wide frequency range whose upper limit extends into the MHz range, one will find that capacitors and inductors will have reduced capacitance and inductance while exhibiting ever higher resistance. LCR meters can provide data extending into the very-low-MHz range while network analyzers can provide higher frequency data.

Inductor cores tend to be ferrite while capacitor dielectrics tend to be ceramic or high-quality film. All capacitor leads are to be kept as short as possible in order to keep the ESL to a minimum.

Radiated Emissions

As mentioned earlier, radiated emissions are typically defined from 30 MHz–1000 MHz. The typical sources of radiated EMI are from digital circuitry with oscillators in the MHz range. It is not necessary for these clock frequencies to be above 30 MHz, because the harmonics of these frequencies will also effectively radiate.

The method of radiation is simplistically modeled using the differential monopole antenna and the loop antenna. These models find their analogs in the loops and dead end traces prevalent in printed circuit boards. Other radiators are found in long pigtails of shielded cables, slot antennas in the electronic equipment cabinetry formed by ventilation slots, wire and cable access openings, and door openings.

Radiated Emissions Measurements

Rather than repeat information that the readers are almost certain to have in their possession, it is best to refer to standards that give explanations and specifications for the measurement instrumentation and the measurement process. CISPR 16 (1987), "CISPR specification for radio interference measuring apparatus and measuring method," is the best international standard for the specification of the spectrum analyzer and the antenna systems that one must use in order to determine the levels of radiated disturbances. An additional criteria is the ambient electromagnetic. This ambient must be at least 6 dB(μV/m) QP below the limits that must be met. The acceptable environment for the radiated emissions test is the open field test site. Preliminary tests may be made in shielded chambers. The test setup and methodology is also discussed. The levels for radiated emissions are given in dB(μV/m), in quasi-peak (QP) or decibels referenced to 1 μV/m. The accepted dimensions for the electric field intensity is in volts/meter, hence the radiated emissions is a measurement of the radiated electric field intensity. The term dB(μV/m) is defined as:

$$\mathrm{dB}(\mu\mathrm{V/m}) = 20\log((\mathrm{Vd/m})/1\ \mu\mathrm{V/m}) \tag{15.3}$$

where Vd/m is the measured disturbance voltage at a given frequency in volts/meter. This equation takes the measured electric field intensity in rms values, and forms a ratio with respect to 1 μV/m, and calculating 20 times the logarithm base 10 of the ratio will give comparative numbers. The term *quasi-peak* is defined in CISPR 16 and is basically a circuit with asymmetrical charge and discharge time constants that takes into account higher-magnitude disturbances at discrete frequencies.

Radiated Emissions Mitigation

There are two methods to mitigating radiated emissions; reduction of the source voltage, current, or electric field intensity; and the shielding of the existing radiated electric field intensity.

1. Reduction of the source voltage, current or electric field intensity is generally not an option, since such mitigation methods will reduce the operational reliability of the source circuits or make the circuitry inoperative.
2. Shielding is generally the only option available. This can be accomplished using ground planes on printed circuit boards, surface mount components, and shielding some of the components such as oscillators and adjacent components. Other methods are the elimination of radiating monopole and loop antennas, such as shortening the pigtails of drain wires of shielded cable, use of all metal housings for connectors that are well grounded, keeping loops to be a minimum area, keeping slot geometry cutoff frequencies higher than the radiating frequencies and harmonics, and using metal shorting fingers in cabinet door slots.

Immunity

Immunity is defined as "the ability of a device, equipment or system to perform without degradation in the presence of an electromagnetic disturbance" (IEC 1000–1–1, 2.1). In this case, the frequency range is from DC to daylight, or from 0 Hz, to very high frequencies. In general, rather than employ some sort of continuous wave frequency generator to test for immunity, immunity tests are centered on naturally occurring phenomena and man-made entities. The immunity tests are found in IEC 1000, Part 4, where there are presently contemplated up to 24 separate sections. These 24 sections are a combination of guidelines and tests. There are 21 separate tests planned in this part. Since it will be impossible to review all 21 test specifications in this document, it will be necessary for the reader to invest in a copy of IEC 1000, Part 4 and to keep it current as the test specifications become available. Several of the most important immunity tests are relatively mature and will be reviewed.

Electrostatic Discharge (ESD)

This test is specified in IEC 1000–4–2. Some engineers view successful ESD testing as being second in difficulty only to successful high-altitude nuclear electromagnetic pulse (HEMP) testing. ESD phenomena is generally classified as high-magnitude conducted and radiated disturbances in the frequency range of 300 to 900 MHz. The source is generally triboelectrical in nature, the phenomena being generated by human activity on synthetic material or by moving machinery, in particular by rapidly moving plastic web. Some of the electrostatic potentials generated by machinery can reach as high as 100 kV unless properly grounded.

The test equipment is easily obtainable commercially, and the test procedure is relatively uncomplicated.

The objective of the test is to have the equipment under test (EUT) be able to withstand as high an ESD simulation, both air and contact discharge, as the standards mandate. The levels range from 2 kV to 8 kV for contact discharges, and from 2 kV to 15 kV for air discharges.

The ability to withstand the higher levels of ESD depend upon sheet metal conductivity and grounding, good grounding practices, filtering of circuitry, the use of delay registers in digital circuitry, and redundancy of software read and write commands.

Radiated EMI

This test is specified in IEC 1000–4–3. The rationale for this test is to withstand the emissions from intentional emitters such as nearby transmitters and personal communications devices, as well as unintentional emitters. With today's printed circuit board technology employing the use of surface mount components and multilayer board designs, this test is relatively benign.

The test consists of a radio frequency power amplifier with properly tuned antennas radiating electromagnetic energy at the equipment under test. The test equipment is easily obtainable, but the use of a shielded chamber, absorbing chamber, or open field test site is necessary. The test will consist of both a continuous wave (CW) test where the magnitude of the frequency is constant and an amplitude modulated test where the frequency is amplitude modulated at a given frequency much lower than the carrier. The frequency range is given as from 26 MHz to 1.0 GHz. The electric field intensity magnitudes are given as 1, 3, and 10 V/m. Modifications to this test may require the operation of a portable transmitter such as a walkie talkie operating at 900 MHz within a meter or so of the equipment under test.

If the EUT responds to radiated EMI, shielding and the reduction of monopole and loop antennas in printed circuit boards and discrete wiring are the possible mitigation means.

Electric Fast Transient/Burst (EFT/B)

This test is specified in IEC 1000–4–4. The source of this interference is the high-current interruption of circuits by contactors and circuit breakers causing ionized airgaps resulting in high frequency conducted interference.

The generator and coupling/decoupling circuitry for this test are commercially available from any one of several domestic or foreign sources. The test procedure is very easy to accomplish.

The objective of the test is to have the equipment under test withstand high magnitudes of EFT/B applications. The magnitudes range from 1.0 kV to 4.0 kV applied on power lines and control lines.

Proper shielding of control wires and/or the use of small signal EMI filters are recommended. Capacitive filtering is recommended for input and output power lines for successful testing.

Surge Voltage

This test is specified in IEC 1000–4–5. The source of this interference can be from direct or indirect lightning strikes on power transmission and distribution systems, short circuit interruptions in power distribution systems, or circuit interruptions using fast interruption techniques such as vacuum contactors.

The generator and coupling/decoupling circuitry for this test are commercially available from any one of several sources. The test procedure is easy to accomplish.

The objective of the test is to have the equipment under test withstand as high a magnitude of surge voltage as possible using a number of waveforms. The magnitudes range from 1.0 kV to 4.0 kV applied on the power distribution input lines. The source impedance of the generator and coupling circuitry can be varied to simulate high or low energy transients.

To pass this test, the use of transient voltage suppressors is recommended. These may be metal oxide varistors (MOV), capacitors, or gas discharge apparatus.

IV

Factory Communications

16 **Evolution of Factory Communication** *W. Timothy Strayer and Carmen M. Pancerella* 385
Point-to-Point Communications • Network Communications • Advantages of Network Interconnection • Communications Requirements for Distributed Systems

17 **Open Systems Interconnection Basic Reference Model** *Robert M. Hines* 389
Introduction • Physical Layer • Datalink Layer • Network Layer • Transport Layer • Session Layer • Presentation Layer • Application Layer

18 **Local Area Networks** *Alfred C. Weaver, John W. Sublett, Jean-Dominique Decotignie, Patrick Pleinevaux, Robert W. Christie, and Curtis L. Moffit* 394
Ethernet and IEEE 802.3 Contention Bus • IEEE 802.5 Token Ring • IEEE 802.4 Token Bus • Fieldbus • Fiber Distributed Data Interface (FDDI) • Asynchronous Transfer Mode

19 **Manufacturing Automation Protocol (MAP)** *Juan Pimentel* 417
History • Purpose • Description • Standards Used • Example of Use

20 **Essential Communications Protocols** *Bert J. Dempsey and Debapriya Sarkar* 427
Datalink Protocols • Network Protocols • Transport Layer Protocols

16

Evolution of Factory Communication

W. Timothy Strayer
Sandia National Laboratories

Carmen M. Pancerella
Sandia National Laboratories

16.1 Point-to-Point Communications 385
16.2 Network Communications 386
16.3 Advantages of Network Interconnection 387
16.4 Communications Requirements for Distributed Systems 388

In the early part of this century, manufacturers found that competitiveness could be gained from the amortization of production costs over large quantities of parts. Factories were built to produce long runs of the same part; retooling was expensive and infrequent. Today, competitiveness is driven by the demand for small-batch and custom parts without sacrificing the quality or cost-effectiveness of parts produced in large quantities. Recent publications specifically refer to the need for agile manufacturing systems (Agile, 1991), focusing on "improving flexibility and concurrence in all facets of the production process, and integrating differing units of production across a firm, or among firms, through integrated software and communication systems" (National Science Foundation, 1993). To do this, traditional barriers must be breached, allowing concurrency of part design, validation of manufacturability, increase in precision, and cooperation among various elements of the factory hierarchy. Communication at all levels is critical to achieve the goals of agile manufacturing.

Factory floor communication differs from conventional office communication in several ways. Special consideration must be given to the fact that factory floors, unlike clean office buildings, are harsh environments with many unfriendly features, like dust, water, electromagnetic interference (EMI), and other forms of radiation. These can cause noise or disruption in electronic transmission. Selection of physical media and topology, and choice of error control and fault tolerance procedures, must be made with respect to this environment.

The types of communication necessary for factories also differ from those found in more genteel settings. Traditionally, there is a strong relationship between a machining device and its controller, so early communication paradigms focused on command-control messages over point-to-point links. Eventually, long sets of wires, called busses, were used to connect devices with controllers over shared media, reducing cabling costs and complexity. As devices become more intelligent, and agents are used to effect distributed control, the simple command-control message exchange will no longer be appropriate. Indeed, as more elements in the factory become involved in the design and fabrication process, the requirements for the communication between and among these elements will continue to evolve.

16.1 Point-to-Point Communications

The simplest and most direct route between a controller and the device it controls is point-to-point wiring. Wiring begins at one point (a controller) and ends at another point (a device). Such direct connection does not share the medium with other devices, and hence does not have to deal with such issues as contention and priority.

Point-to-point wiring requires installing cables, possibly of different specifications, between each device and controller. Some connections between devices and controllers require very long individual cable runs. If controllers need to be relocated, all devices connected to those controllers may require new cable runs. Furthermore, wiring from old installations must either be removed or abandoned, both of which are potentially costly.

The traditional communication task is merely to send a short instruction to a machine from a controller, and possibly for the machine to send an acknowledgment back to the controller. The only requirements of this type of connection are ruggedness and reliability. Today, machines and controllers are more sophisticated. One new computer technology at the factory-floor level, open-architecture manufacturing (Wright, 1994), adds flexibility and quality assurance to the manufacturing process, enabling the transmission of sensor data from a machine back to a controller. Thus, communication between a controller and machine tool is no longer predictable, and large amounts of sensor data may need to be transmitted to other manufacturing components in the system.

A major drawback of point-to-point interconnection is its inability to scale. For example, computer integrated manufacturing (CIM) is occurring throughout all levels of the modern day factory. This requires communication paths between cell- and

0-8493-8343-9/97/$0.00+$.50

shop-level controllers. Point-to-point connectivity in this situation is ineffective and inefficient. Fortunately, network technology exists to solve these connectivity problems.

16.2 Network Communications

Simplistically, a network is an interconnection of hosts attached to a common transmission where any medium host in the network can transmit to any other host. The topology of the physical media, and what protocols are used to mediate access to the media, are among the factors that differentiate various approaches to networking.

A network's topology is a description of the layout of the wiring plant. Figure 16.1 shows some common network topologies. A bus is a length of wire with hosts either daisy-chained or otherwise attached. A star interconnection has a central switch with each host connected directly to the switch (the switch may in fact be a host with switching logic). A ring configuration is so-called because each host is connected to two neighbors in such a way that a circular path is created. While these three are the most common topologies, other hybrid and irregular configurations also exist.

Busses

Bus topologies have several properties, most notable of which is the simplicity of the wiring scheme. Since the starting and ending points are not related, as in a ring, the length of wire connecting a set of hosts can be minimized. If the hosts are passively attached, removing or adding hosts does not disturb the network operation. However, since a single wire connects any two hosts, a break in the wire either destroys or bifurcates the network. Further, the single wire is a shared resource, and contention for that resource must be resolved.

Examples of bus networks used in factory communication are the several Fieldbus standards, including MIL-STD-1553. These standards were developed specifically to replace point-to-point wiring for controller-to-device communication. Ethernet and IEEE 802.4 Token Bus are more modern examples of bus standards. Although not developed for the factory, Ethernet has become somewhat of a de facto standard because of its widespread use. Token bus is specified in the manufacturing automation protocol (MAP, see Chapter 19) as the network of choice for factories because of its noise reduction and deterministic delivery properties.

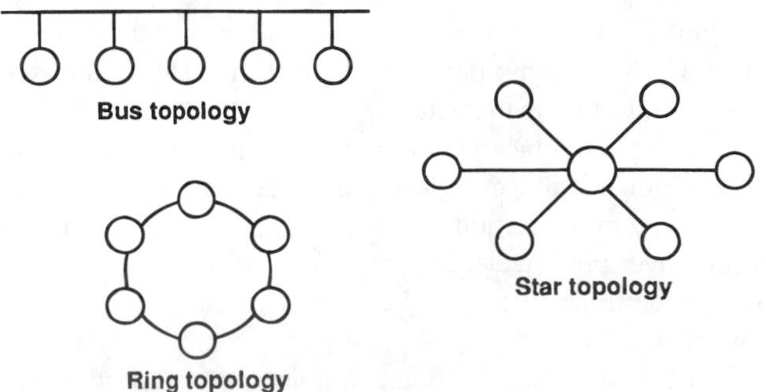

Figure 16.1 Example network topologies.

In MIL-STD-1553, contention for the bus is resolved by a central bus controller. In Ethernet, arbitration is decentralized, depending on a backoff algorithm when two or more hosts try to use the network at the same time. Token bus controls access by having the hosts pass around a token, or transmission permit, so that only the host with the token can use the network. Fieldbus, Ethernet, and token bus are described in more detail in Chapter 18.

Stars

Star topologies eliminate the problem of bifurcation since breaking one wire only separates one host from the network. Problems arise, however, if the central switch is disabled, since this switch is required for the network to operate. The central switch is, therefore, a single point of failure. The switch is also a single resource and point of contention. A well-designed switch will have capacity to handle several simultaneous demands; in this way, pairs of hosts can communicate as though each pair were connected by a dedicated point-to-point wire.

Star topologies are becoming increasingly popular with advances in high-speed switching fabrics. Asynchronous transfer mode (ATM) networks are examples of how telephony and computer networking technologies are becoming intertwined. Vendors are also finding markets for older protocols repackaged in switched, rather than shared, media. These include switched Ethernet and switched fiber distributed data interface (FDDI).

Rings

Ring topologies are like busses with the ends joined. Messages are transmitted in one direction around the ring. Either the transmitter or the receiver must remove the message, or the message would be propagated indefinitely around the ring. (On a bus the message is removed at the ends, possibly with special hardware.)

Rings have several interesting properties. Often contention for the ring is resolved by passing around a token, as with token bus; since the ring already physically exists, determining where to pass the token next is much less complex than when the ring is only logical. Rings also have fault tolerant properties. If a ring is broken, it can degenerate into a bus. Some standards specify that two rings are used, one flowing in the opposite direction of the other (Figure 16.2). When a single break occurs here (Figure 16.3), the second ring is used. When a break occurs in both rings, the two loose ends fold together to form a single ring (Figure 16.4).

IEEE 802.5 Token Ring and FDDI, discussed more fully in Chapter 18, are examples of standards using ring topologies.

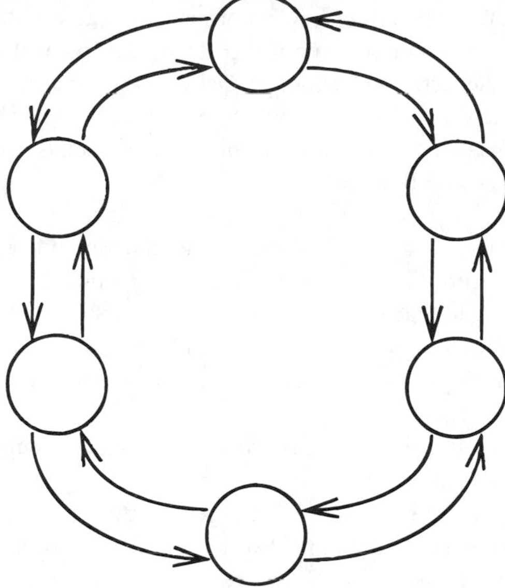

Figure 16.2 Two-ring topology.

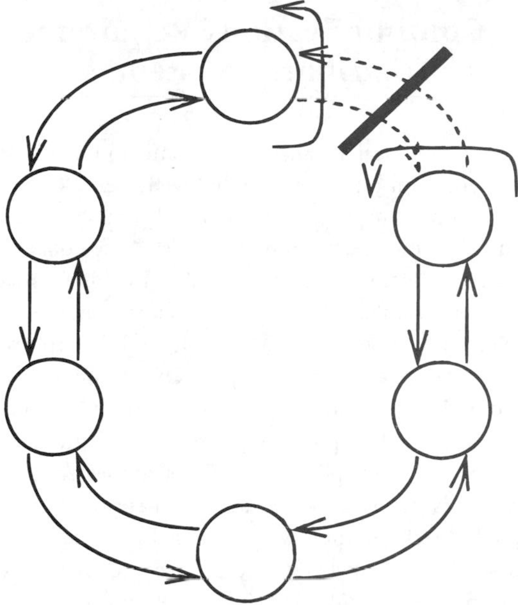

Figure 16.4 Two-ring topology with a break in both rings.

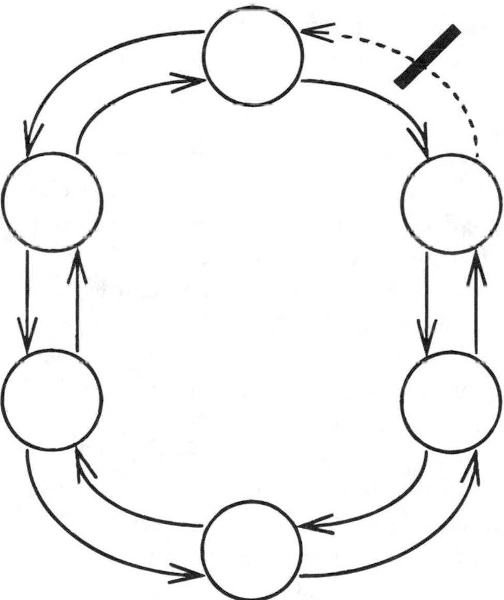

Figure 16.3 Two-ring topology with single break.

16.3 Advantages of Network Interconnection

Perhaps the most striking advantage of networks over point-to-point wiring is the ability to add hosts to the network without substantial wiring growth. In a fully connected point-to-point wiring plant, the addition of one new host to n existing ones requires n new wires. In the bus and ring topologies, no new wires are required, and only one wire is needed for the star topology. Even so, most networks have some upper limit to the number of hosts and the total length of the wires due to physical constraints such as propagation delay and signal degradation.

In addition, all of the devices that populate a network have the ability to communicate with one another. This opens the possibility of cooperation of devices. With feedback from manufacturing cells, more effective schedules can be devised, and the time to market reduced. Achieving such connectivity with point-to-point wiring is prohibitively complex and costly.

However, moving from dedicated communication lines to a shared medium can cause problems, especially with respect to performance. Point-to-point lines have well-known delay characteristics; when multiple devices are competing for a shared resource, delays may no longer be deterministic, or even bounded. Ethernet, for instance, uses a binary exponential backoff scheme when two or more hosts attempt to transmit at once. This essentially delays these hosts in hope that the next time they attempt to send, they will not collide. This delay, on successive collisions, may be indefinitely long.

The designs of modern networks address delay and throughput issues in many different ways. Token-based protocols use the token to regulate use of the medium, and hence bound the time until access. Switched networks rely on the speed of the switching fabric to avoid contention.

Since network connectivity enables cooperation between hosts, a concept called *distributed systems* has emerged. In a distributed system, work or data or both are divided and doled out to the members of the system, effectively creating a single system with aggregate computing or manufacturing capability. In order to achieve this level of cooperation, however, the network and the communication protocols used within the network must meet certain requirements.

16.4 Communications Requirements for Distributed Systems

With the technological advances in transmission media and fast, extremely powerful but affordable hardware (e.g., 486 and Pentium PCs), communication networks and protocols are playing important roles in the development of distributed data and control systems in factories of the future. Distributed systems in some form are useful at all levels of the factory hierarchy. Distributed object models, such as Microsoft's OLE (Brockschmidt, 1995), Object Management Group's CORBA (Object Management Group, 1991), and others, permit the encapsulation of both data and functionality, providing a convenient mechanism for the decomposition of applications into distributed environments. Current trends in industry and academia suggest coupling the distributed object paradigm with artificial intelligence, through expert systems and knowledge bases, to provide proactive objects, or intelligent agents (Culkosky, 1993). This approach is an essential foundation for agile manufacturing, where intelligent agents facilitate virtual design, process planning, manufacturability analysis, rapid prototyping, and process control.

Communication technology is at the heart of these advances. Distributed systems place certain requirements on the network and its communication protocols. Generally, networks link autonomous or semi-autonomous systems together in order to accomplish some common goal. Research in distributed systems has shown that communication patterns are mostly transactional in nature. A request to a server is made by a client; the server considers the request and eventually issues a response. For example, a database query is a client-server interaction.

Communication protocols, however, tend to fall into one of two categories: support for long-lived, completely reliable data transfers, or support for finite size best-effort message delivery. The former type is called a virtual circuit, or connection, and the latter type is called a datagram. Protocols which support only one or the other of these communication paradigms are not well-suited to support transactional communication patterns. As a consequence, protocols like Stanford's Versatile Message Transaction Protocol (Cheriton, 1989) were developed to provide explicit support for distributed systems.

In addition to the transactional paradigm, distributed systems in a factory must provide support for message priority, security, process migration, and delivery assurance. Message priorities are necessary to differentiate between essential and nonessential traffic. Security is required as factories are connected to one another and to other firms; authentication of request, validity of response, and espionage are vital issues for open factories and design houses. Process migration, or addressing stability, is required if servers (e.g., the intelligent agents) are able to move (possibly to a new cell or new set of machining devices). Migration of objects would occur if the old placement of the object becomes ineffective or inefficient. Delivery assurance is required to prevent situations where the communicants may be left in ambiguous states because one side of the transaction is not certain if the other side finished its work.

Another requirement for distributed systems, both on the factory floor during part fabrication and in design offices during part conception, is for real-time communication. Floor devices cooperating to machine a part must know exactly where moving arms and cutting tools are. Position information, if delayed, is useless or harmful, so networks must ensure timing constraints are met.

Timing constraints also exist when part designers, using collaborative tools, are working together to design a part while not being physically colocated. Distributed multimedia is a rapidly growing application area in networking, where the issues involve the real-time delivery of voice and video data streams to the collaborators' workstations.

Factory communication is evolving from simple point-to-point command-control relationships between floor devices and their controllers, to distributed systems where elements of the factory, from devices to design tools, are interconnected. Future factories will use this interconnectivity to reduce costs, and reduce the time to market, increasing competitiveness.

References

Agile Manufacturing Enterprise Forum 1991. *21st Century Manufacturing Enterprise Strategy: An Industry-Led View,* Two Volumes, Lehigh University, Bethlehem, PA.

Brockschmidt, Kraig, Inside OLE 2, Second Edition, Microsoft Press, Redmond, Washington, 1995.

Cheriton, David R. and Williamson, Carey L. 1989. VMTP as the transport layer for high-performance distributed systems, *IEEE Communications Magazine,* June.

Cutkosky, Mark R., Engelmore, Robert S., Fikes, Richard E., Genesereth, Michael R, Gruber, Thomas R., Mark, William S., Tenenbaum, Jay M., and Weber, Jay C. 1993. PACT: an experiment in integrating concurrent engineering systems, *IEEE Computer,* 26(1):28–37.

National Science Foundation, Directorate for Engineering 1993. *Agile Manufacturing Initiative* Program Solicitation, August.

Object Management Group, 1991. *The Common Object Request Broker: Architecture and Specification,* OMG Document #91.12.1, December, Object Management Group, Framingham, MA.

Wright, Paul K. 1994. Principles of open-architecture manufacturing, ESRC 94–26, October, Engineering Systems Research Center, University of California at Berkeley.

17

Open Systems Interconnection Basic Reference Model

Robert M. Hines
University of Virginia

17.1 Introduction 389
17.2 Physical Layer 389
17.3 Datalink Layer 390
17.4 Network Layer 390
17.5 Transport Layer 391
17.6 Session Layer 392
17.7 Presentation Layer 392
17.8 Application Layer 392

17.1 Introduction

One of the major problems that surfaced in the 1970s was the inability to share data among computers built by different manufacturers. As a result, the International Organization for Standardization (ISO) formed a group that developed a basic reference model (BRM) for Open Systems Interconnection (OSI) (International Organization for Standardization, 1994). This seven-layer model identified the operations and services that would be required to interconnect heterogeneous computers over an arbitrary network topology.

The BRM advocates a layered approach to help minimize the complexity of designing communication systems. The principles which were used by ISO to help guide the development of the model include:

1. Create a boundary between layers where interaction across the boundary can be minimized.
2. Create separate layers only to handle functions which are substantially different, and put similar functions into the same layer.
3. Use past experience to suggest where boundaries should be created.
4. Make the layers modular so that one can be redesigned without affecting the layers directly above and below it.
5. Create boundaries where it may be useful to standardize the interface.

It is important to remember that there is a distinction between real systems and the OSI reference model. The model is a specification of what a communication system should do, divided into manageable layers. It does not dictate or suggest specifically what protocols should be used or how the functions in the layer should be implemented. A real system may be missing one or more of the layers, or an actual protocol may contain functionality that spans two or more layers. In essence, it should be remembered that the OSI model is, indeed, a model designed to ease the task of the system designer and implementor.

The following sections will individually discuss each of the seven layers of the model (Figure 17.1.)

17.2 Physical Layer

The physical layer is the lowest layer in the OSI Reference Model. It is responsible for the transmission of data packets across the physical network medium which forms a connection between two or more datalink entities. Bits are passed down from the datalink layer to the physical layer and transmitted across the physical medium from one system to another. The physical layer must also be listening for any transmissions so that they may be passed up to the datalink layer when they are received.

Specifically, the physical layer is responsible for:

1. Activating and deactivating physical connections.
2. Transmitting bits over the physical connection.
3. Multiplexing two or more physical connections on the same link.
4. Dealing with physical management activities such as error control and link activation.

The type of signaling used is not defined by the standard and can range from electrical or optical signaling to broadcast radio signaling. Since the type of signaling is directly related to the medium over which it travels, different types of media are usable for the physical layer. There must be agreement at the physical layer in terms of signaling rate and transmission medium in order for two or more systems to communicate.

0-8493-8343-9/97/$0.00+$.50

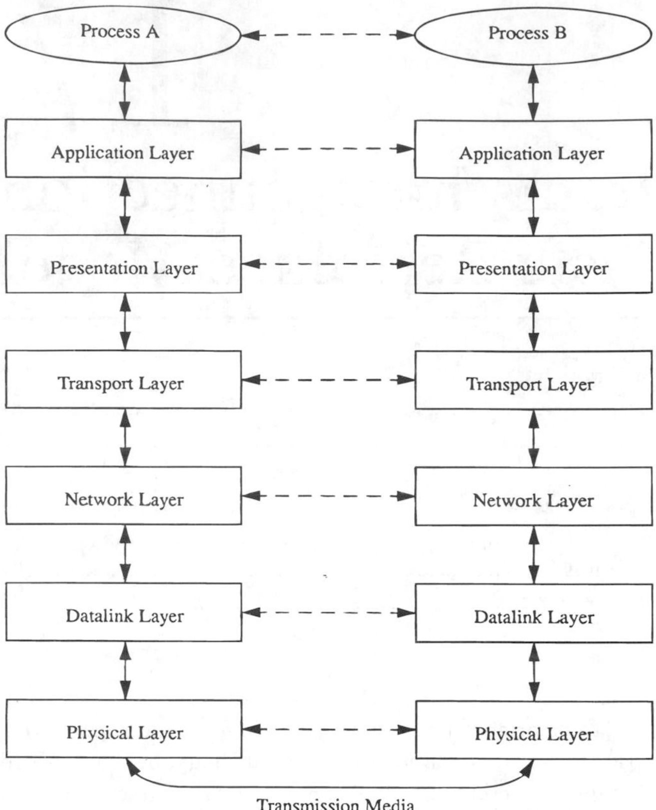

Figure 17.1 ISO OSI reference model.

The following are provided by the physical layer to the datalink layer:

Physical connections—The transparent transmission of bits between datalink entities across the physical network media.

Physical service data units—The unit of data which is manipulated by the physical layer.

Physical connection endpoint identifiers—The identifier used by datalink entities to refer to physical connection endpoints.

Data circuit identification—Unique identifiers which specify the physical connections between two datalink entities.

Sequencing—Bits are delivered in the order that they are transmitted.

Fault condition notification—Datalink layer entities are notified of any detected physical errors.

Quality of service parameters—Parameters which specify the type of service available at the physical layer including error rate, service availability, transmission rate, and transit delay.

17.3 Datalink Layer

The datalink layer is designed to provide a means for raw transmission from one physical node to another which is free from errors. The functions of the datalink layer are:

1. Establishing and releasing datalink connections, possibly over multiple physical links.
2. Delimiting and transmitting datalink service units (frames).
3. Maintaining the proper sequence of frames across a datalink connection.
4. Detecting and possibly correcting transmission, format, and operational errors.
5. Dynamically controlling the rate of frame receipt.

The datalink layer is passed data-link-service-data-units, also known as frames, by the network layer. These frames are converted into bit streams which are passed down to the physical layer. Since the physical layer only deals with bits, the datalink layer is responsible for delimiting the frame boundaries with special bit patterns so that they may be processed by the receiving datalink layer. Additional encoding is necessary to prevent these bit patterns from occurring in the data portion of the frame and causing confusion.

Communication at the datalink level can take place either in connectionless mode or connection mode. Functionality is provided for controlling all aspects of this communication between stations, including the establishment, maintenance, and release of connections, and the transfer of frames. The interface provided shields the network layer from dealing with details of transmitting on the physical link between the nodes.

Another issue at the datalink layer (and other layers above) is how to keep a faster transmitting station from swamping a slower receiver with data. Some sort of flow control mechanism must be implemented at this layer to prevent this from happening.

The datalink layer provides the following to the network layer:

Datalink addresses—Unique addresses provided by the datalink layer for communication between network entities at the datalink level.

Datalink connection—A dynamically established and released mechanism which provides a way for data transfer to occur between network entities.

Datalink service data units—The unit of data exchange at the datalink layer. These are commonly known as frames.

Datalink connection endpoint identifiers—The identifier used by the network layer to refer to datalink endpoints.

Error notification—The network layer is notified when an unrecoverable error is detected.

Quality of service parameters—Parameters which specify the quality of service available at the datalink layer, including mean time between errors, residual error rate, service availability, delay, throughput.

Reset—Forces the datalink entity into a known state.

17.4 Network Layer

The network layer is responsible for providing transparent information transfer between transport layer entities. This involves hiding underlying differences in the physical and datalink portions of subnets and creating a well-defined, consistent network service.

Messages are received from the transport level and divided into network service data units, commonly called packets. The packets are passed to the datalink layer and from there to the physical layer and onto the physical network. Symmetrically, when the data reaches its destination, it is passed up through the physical and datalink layers to the network layer. After performing its functions, the network layer passes the data to the transport layer above it.

Some of the key issues for the network layer are network addressing, packet routing, and the fragmentation and reassembly of packets. Problems can occur when transferring a packet from one type of network to a different type of network. The packet from the first network may be too large or of a different format than the packets carried by the second network. The network layer is responsible for solving these problems and allowing different types of networks to be connected.

The specific functions of the network layer include:

1. Determining a route from the sending network entity to the receiver and forwarding the packet between each hop along this route.
2. Connecting two transport level entities.
3. Optimizing the use of datalink connections by multiplexing multiple network level connections onto them.
4. Fragmentation and/or assembly of packets to facilitate their transfer.
5. Error detection and recovery.
6. Delivery of packets in sequence over a given connection when requested by transport level entities.
7. Maintaining the same level of service at each end of a network connection even though it may span subnets of differing quality.

The network layer provides the following to the transport layer:

Network addresses—Unique addresses provided for the communication between transport level entities.

Network connections—The means to transfer data between transport level entities, and the means to manage these point-to-point connections are provided.

Network connection endpoint identifiers—The identifier used by transport entities to refer to physical connection endpoints.

Network service data unit transfer—The data unit which is handled by the network layer is transparently transferred over a connection between transport level entities.

Quality of service parameters—Parameters which specify the type of service available at the network layer include residual error rate, service availability, reliability, throughput, transit delay, and delay for connection establishment.

Error notification—Errors are reported to the transport layer.

Expedited network service data unit transfer—An optional, faster, means of exchanging information over a network connection.

Reset—An optional service which allows all packets in transit to be discarded and the network entity on the other end to be notified.

Release—Release of a network level connection due to a request by a transport level entity.

17.5 Transport Layer

The purpose of the transport layer is to control end-to-end transport of data between two session level entities at a specified service quality. Data is accepted from the session level, divided into smaller units (segmented) if necessary, and delivered according to the user's specification (e.g., reliably or unacknowledged). The transport layer is not concerned with data routing and relay functions since they are handled by the network layer.

Specific responsibilities of the transport layer include:

1. Establishment and release of transport level connections including negotiation of quality of service parameters.
2. Mapping transport level addresses to network level addresses.
3. Multiplexing and splitting transport level connections to maximize the use of network level service.
4. Reliable, in-order delivery of messages when requested by a session level entity.
5. Blocking, segmenting and concatenating messages.
6. Error detection and recovery.
7. Providing a mechanism for expedited data transfer.

It is possible for the network level service to lose packets under some circumstances. Some applications, such as a file transfer program, require that all packets arrive intact. Consequently, some implementations of transport level protocols may have a mechanism to determine if a packet is missing and ask for its retransmission. In other applications losing a packet does little harm. To fill this need, implementations of transport level protocols which don't guarantee reliable delivery of packets are also available.

The transport layer provides the following services to the session layer:

Transport connection establishment—A connection can be established between two session level entities with a service quality negotiated at the time of establishment.

Transport connection release—The means to release a session level connection which notifies the receiver is provided.

Data transfer—A service to transfer transport data units is provided which operates at the negotiated service quality. If the service quality cannot be maintained, the connection is closed and all receivers are notified.

Expedited data—A service for faster information exchange is provided.

17.6 Session Layer

The session layer defines a standard used to establish and terminate sessions between user processes.

The specific functions of the datalink layer are:

1. Providing at one-to-one mapping between a session level connection and a transport level connection at a given time. Over the long run this mapping may change and a single transport connection may support multiple consecutive session connections.
2. Exertion of back pressure over the transport level flow control mechanism since there is no session level peer flow control.
3. Re-establishment of a transport level connection in the event of a failure to maintain a connection between session layer entities.

Other important functions of the session layer include synchronization and token management. These sessions are connections between cooperating presentation layer entities that exchange data and synchronization information.

Synchronization means that during the communication between the sessions, checkpoints are identified which signify some sort of completed operation. If the session fails, it can be restored to one of these checkpoints without data loss. A concrete example of this could be a file transfer program which is aborted due to an error half-way through the transfer of a large file. Since checkpoints are built into session layer, the transfer can be resumed at the last checkpoint, which minimizes the retransmission of data.

The session layer provides the following to the presentation layer:

Session connection establishment—Enables two presentation level entities to establish a session level connection.

Session connection release—A mechanism for orderly release of a session level connection between two presentation level entities without loss of data. The connection may also be aborted by one of the presentation entities, but data may be lost.

Normal data transfer—The normal transfer of session service data units between two presentation level entities.

Expedited data transfer—High-priority transfer of data between two presentation level entities.

Token management—Allows presentation level entities to control who is using specific critical functions.

Session connection synchronization—Permits presentation level entities to define synchronization points and the services to reset to them with a possible loss of some data.

Exception reporting—Presentation level entities are notified of any exceptional conditions.

17.7 Presentation Layer

The presentation layer is concerned with delivering data and information in a format or representation which can be understood by the receiver. If the two systems communicate using the same syntax, the presentation layer functions can be minimal or absent. Unfortunately, this is not always the case. The presentation layer functions may provide services such as conversion from one character representation to another, data compression, and data encryption. So, the presentation layer preserves the semantics of the message, while changing the syntax into something that the receiving system can understand.

An example of a case where the presentation layer may be useful is when one system which stores data in big-endian representation is communicating with a system which stores data in little-endian representation. There are a number of ways that this conversion could take place, but the best way to handle this is to convert the data into some sort of universal format. This is a good solution because it is very extensible; when a new type of system is added, it only has to be able to understand how to convert to and from the standard format, instead of understanding all of the different formats used by other systems on the network.

The functions which are specified for the presentation layer in the standard include:

1. Negotiation and re-negotiation of the syntax used for communication.
2. Conversion to and from the syntax of the application and the syntax chosen for transfer.
3. Ability to access session layer services.

The following services are provided to the application layer:

Session services—The application layer is able to access all session layer services in the form of presentation layer services.

Syntax transformation—Conversion of the end system syntax into a transfer syntax. This may include encryption or compression.

Selection of transfer syntax—Initially selecting the syntax to be used in the data transfer, and possibly modifying it later during the transmission.

17.8 Application Layer

The seventh (top) layer of the OSI reference model is not the user's application, but is instead a set of common functions and utilities that are believed to be generally useful for communication. Typical types of communication which may take place between application layers include virtual terminal protocol and file transfer protocols.

The services which are provided by the application layer include:

Identification—The identification of communication partners by name of address, or some other description.

Quality of service—Determination of the quality of service parameters required for the application including acceptable error rate, response time and cost.

Synchronization—The synchronization of participating applications.

Error recovery—Agreement on responsibility for error recovery.

Security—Agreement on issues of security including authentication, access control, and data integrity.

Transfer syntax—Selection of the syntax used for transfer of information between communicating systems.

References

International Organization for Standardization 1994. *Information Processing Systems—Basic Reference Model,* Draft International Standard 7498.

18

Local Area Networks

Alfred C. Weaver
University of Virginia

John W. Sublett
University of Virginia

Jean-Dominique Decotignie
EPFL-LIT

Patrick Pleinevaux
University of Virginia

Robert W. Christie
University of Virginia

Curtis L. Moffit
Newbridge Networks, Inc.

18.1 Ethernet and IEEE 802.3 Contention Bus 394
History • Goals and Non-Goals • Architecture • Protocol • Performance • Summary

18.2 IEEE 802.5 Token Ring 396
Frame Format • Tokens and Access Control • Normal Transmission • Priority Transmission and Stacking Stations • Error Correction • Hardware Issues • Summary

18.3 IEEE 802.4 Token Bus 400
History • Architecture • Topology • Network Access • Token and Message Frames • Dynamic Network Membership • Error Handling • Access Classes • Summary

18.4 Fieldbus 403
Examples of Use • Description • Solutions • Suitability and Performance

18.5 Fiber Distributed Data Interface (FDDI) 408
Overview • Description • Suitability • Performance • Applications

18.6 Asynchronous Transfer Mode 412
Introduction • Characteristics • Network Architecture • ATM Cells • ATM Protocols • Transmission Media • Benefits • New Applications • Integrated Networks • Suitability • Performance • Example of Use

18.1 Ethernet and IEEE 802.3 Contention Bus

Alfred C. Weaver

History

The University of Hawaii devised a collision-based network access protocol called Aloha for use in an inter-island packet radio network. Using this protocol, stations would transmit whenever they had a packet ready, which sometimes resulted in simultaneous transmissions called collisions. Packets involved in a collision were simply lost, and the transmitter eventually retransmitted unacknowledged packets. It is easy to show that this simple protocol cannot use more than 18% of the available network bandwidth.

Robert Metcalfe and Michael Boggs, working for Xerox Corporation, adapted this basic algorithm for use in a local area network. They added the concept of carrier sense, in which a station senses the state of the network and defers if the network is busy. They added collision detection, the ability to sense a collision and truncate a transmission as soon as the collision is detected. Finally, they introduced an adaptive retransmission scheme called binary exponential backoff for rescheduling collided transmissions. The result was Ethernet (Metcalfe, 1976).

In 1982, the Ethernet specification was submitted to the IEEE 802.3 committee for consideration as an international standard. After making some changes to the packet structure and field definitions, the committee adopted the new specification as the IEEE 802.3 (1982) contention bus.

Goals and Non-Goals

In the process of developing the idea into a product, the designers set the following goals:

- Simplicity—Features which would complicate the design without substantially contributing to the other goals were excluded.
- Low cost—From the beginning, the goal was to package the network access protocol as a VLSI chip set.
- Compatibility—To eliminate the possibility of incompatible variants of Ethernet, there are no options at the data-link layer.
- Addressing flexibility—Network addresses allow messages to be sent to any single station, a group of stations, or to all stations.
- Fairness—All stations have equal access to the network.
- Progress—No single station can block other stations indefinitely.

0-8493-8343-9/97/$0.00+$.50

High speed—The signaling rate was set at a constant 10 Mbits/s, which was high at the time.

Low delay—At low loads the network itself introduces negligible delay to frame transfer times (not true at high loads).

Stability—Throughput does not decrease as network load increases.

Layered architecture—Although Ethernet predates the OSI model, it does isolate the physical layer and the medium access control sublayer.

In addition to this set of goals, there was a specific set of non-goals, i.e., characteristics which Ethernet would *not* have:

Full duplex operation—Message transmission is instead accomplished by broadcast on a bidirectional bus.

Error control—Error detection is provided, but recovery is the responsibility of higher level software.

Security—The cable is easy to tap and data is not encrypted.

Speed flexibility—There is only one signaling rate.

Priority—There is no priority system for either messages or stations.

Hostile users—There is no mechanism to protect the network from a malicious user.

From this set of goals and non-goals, it is clear that the Ethernet product (and the IEEE 802.3 contention bus) as well was designed for use in a low load, non-hostile, non-real-time environment such as office automation.

Architecture

IEEE 802.3 is an international standard for baseband communication using a bus topology and Carrier Sense Multiple Access with Collision Detect (CSMA/CD) as the network access mechanism. Signaling on the medium uses Manchester encoding for the data; as a result, any one transmission produces constant energy on the bus regardless of the data stream. Multiple simultaneous transmissions produce "collisions" and are detected by sensing a rise in bus energy level.

IEEE 802.3 specifies a 50 ohm baseband coax cable (Fig. 18.1) as the acceptable network medium. It provides low impedance and allows a large number of stations to be connected to the medium through simple coax taps. Any one network segment is limited to a length of 500 meters, although segments can be interconnected by bidirectional repeaters (Fig. 18.2). Even so, the maximum end-to-end separation of any pair of devices is limited to 2,500 meters, and the whole network is limited to 1,000 attached devices. Transmission speed is fixed at 10 Mbits/s, although that is not a fair indication of the overall network throughput due to the operational characteristics of CSMA/CD access schemes; because of collisions, throughput is always less than the signaling speed.

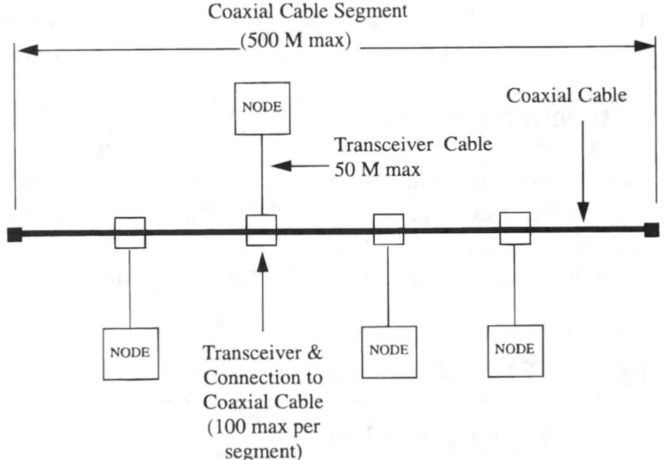

Figure 18.1 IEEE 802.3 contention bus—single segment.

Protocol

Figure 18.3 shows a possible transmission scenario for an arbitrary pair of nodes A and B. The x axis shows the passage of time from left to right; the two horizontal displays on the y axis are the transmissions of two arbitrary nodes called A and B. At time to, the bus is idle. Shortly thereafter node A senses the channel, finds it idle, and transmits message A1. Transmission finishes and the bus goes idle. Somewhat later node A has another message to send; it senses the channel, finds it idle, and begins transmitting message A2. At approximately the same time, node B has a message to transmit. Because node B is physically distant

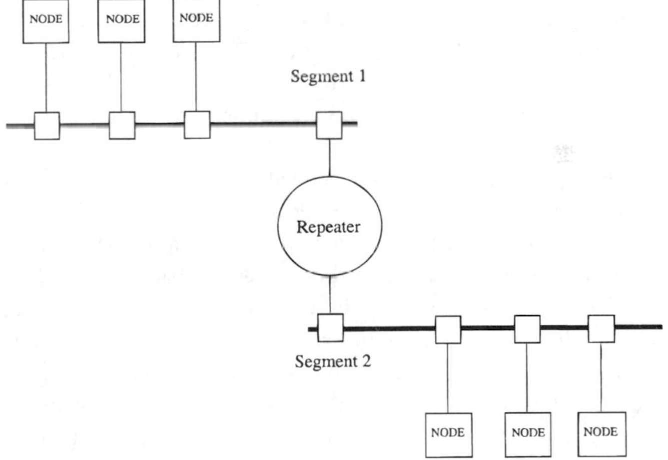

Figure 18.2 IEEE 802.3 contention bus—two segments.

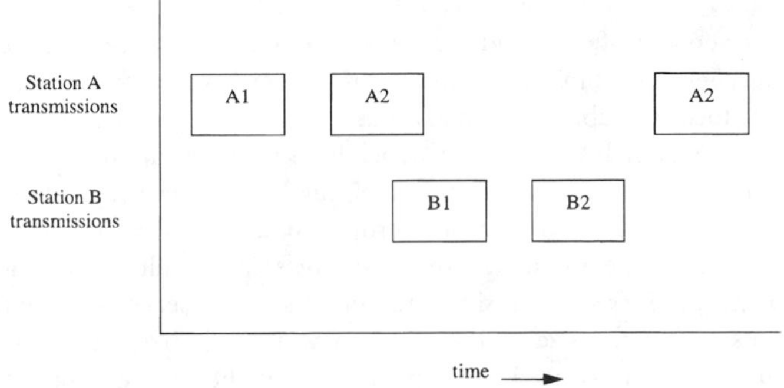

Figure 18.3 CSMA/CD transmission sequence.

from node A (by up to 2,500 meters), it listens to the channel and hears it idle because the message from A2 has not yet physically propagated from node A to node B. Thus node B legitimately begins transmission, but messages A2 and B1 overlap, or collide. Both nodes sense the collision, stop transmitting, transmit a short packet called a *jam packet* to reinforce the collision, and finally enter a rescheduling discipline called binary exponential backoff. It is this backoff mechanism (discussed shortly) which gives the contention bus its characteristic performance. After a delay mandated by the backoff algorithm, message B1 is retransmitted (this time without collision), and at a still later time message A2 is retransmitted.

The binary exponential backoff algorithm is unique among retransmission schemes. It works as follows: For the i-th collision suffered by this message, compute:

```
j := min (i,10)
k := random (0..(2^j-1))
wait k slot times (equals 512*k bit times)
sense channel
transmit when channel is next idle
```

For each of the first ten collisions, the delay time is a random number of *slot times* (multiples of 512 bit times which, at 10 Mbits/s, makes a slot time about 50 microseconds) where the upper bound of the random number field is increasing (binary) exponentially with each successive collision. Thus, on the first collision, the wait might be 0 or 50 microseconds; on the tenth it is between 0 and 150 milliseconds. Collisions 11 through 15 are treated the same as the tenth. If there is a sixteenth collision, then it causes the packet to be abandoned and an error to be reported to upper level software.

This algorithm is clearly intended to optimize the channel performance at low loads. When loading is light, collisions are rare, and those which do occur are subject to short backoffs (retransmission delays). As load increases, the backoff procedure artificially drives down the network offered load by forcing retransmissions to spread out over longer and longer time periods. This approach directly implements the classic tradeoff between throughput and delay—namely, that network throughput can be increased at the cost of increasing average message delay.

Performance

As network offered load increases, throughput increases until it reaches a maximum; at this point it remains steady (i.e., the protocol is stable). The maximum throughput is less than the bus capacity by an amount dependent primarily upon the bus length and the message size. As bus length increases, propagation delay on the medium increases proportionately and the "collision window" (the amount of time each message is vulnerable to a collision) increases, causing throughput to decrease. For any fixed bus length the size of the collision window is fixed; thus for increasing message size the percentage of the message subject to collision decreases. As a result, throughput increases with increasing message length.

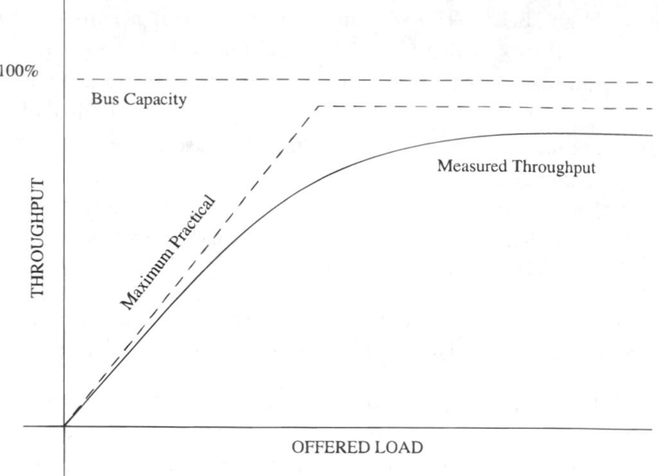

Figure 18.4 Throughput vs. offered load for the contention bus.

The general form of the throughput curve is shown in Figure 18.4. Throughput is defined to be the ratio of bits actually delivered per unit time to the maximum number of bits which could have been carried by the medium in the same unit time.

Summary

The IEEE 802.3 contention bus (Ethernet) was designed to support office automation environment. It was intended to operate with low average offered loads, where those loads are bursty and asynchronous. Throughput is linear with offered load at low loads, then drops off and becomes asymptotic at high loads. The protocol is stable (i.e., throughput is non-decreasing as offered load increases). Throughput is sensitive to offered load, propagation delay, and packet size; the latter two parameters effectively determine the maximum throughput achievable on a particular network. Despite its fixed signaling rate of 10 Mbits/s, throughput is always less than that amount due to the contention-based protocol.

The contention bus is of limited utility in the real-time control of the factory since its message delivery latency can have high variance and it lacks any data prioritization mechanism. However, it works well for the non-real-time operations and thus becomes the basis for TOP (Technical Office Protocol), discussed later.

References

Carrier Sense Multiple Access with Collision Detection. ANSI/IEEE Standard 802.3, 1982.

Metcalfe, R. M. and Boggs, D. R. 1976. Distributed packet switching for local computer networks, Communications of the ACM.

18.2 IEEE 802.5 Token Ring

John W. Sublett

The IEEE 802.5 standard (IEEE802.5; Werner, 1983) is the specification for a LAN which uses the token-passing access method

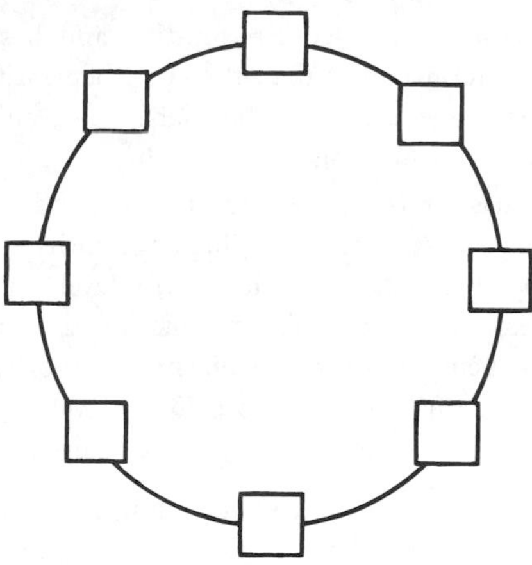

Figure 18.5 Token ring network topology.

to efficiently share the transmission medium. The specification includes both 4 Mbits/s and 16 Mbits/s varieties with a maximum of 250 stations per ring. A token ring is a series of stations attached by unidirectional links to form a ring (Fig. 18.5). Information travels from one station to the next around the ring, with each station forwarding the information to the next. Access to the transmission medium is controlled by the passing of a token (a special bit pattern).

Initially, the ring is idle. Each station is in repeat mode, forwarding each bit as it is received. A one bit delay is introduced for each station attached to the ring. When the ring is idle, the token circulates continuously around the ring waiting to be claimed. In order for a station to transmit, it must first obtain the token.

A station wishing to transmit monitors the bitstream awaiting the arrival of the token. When the free token is recognized, the station captures it and transfers its information onto the ring. The information includes the address of the destination station. The destination station will recognize its address in the frame header and copy the information from the ring and also forward it to the next station. When the transmitted frame returns to the transmitter, the transmitter removes the frame and generates a new free token. Thus, the ring is made available for other stations to access. The token passing access method assures that only one station on the ring will be transmitting at a time.

Frame Format

The frame format is shown in Figure 18.6 and is taken from the IEEE Standard 802.5:

Starting Delimiter (S)—A special sequence of symbols used to denote the start of a frame. It includes violations of the differential Manchester encoding scheme so that it will not be confused with any other field.

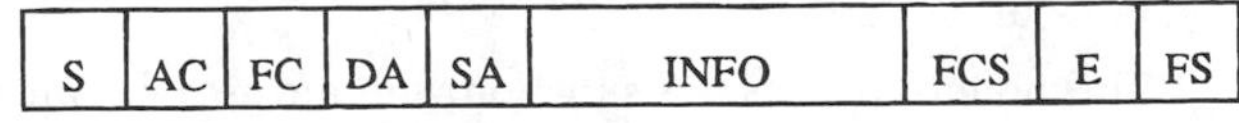

Figure 18.6 Token ring frame format.

Access Control Field (AC)—A one byte field containing information relevant to proper token passing operation. A more detailed discussion follows.

Frame Control Field (FC)—A one byte field used to indicate the type of the frame. A frame may contain either application data or control data.

Destination and Source Addresses (DA,SA)—An address field can be 2 or 6 bytes, but the address length must be uniform across an entire LAN. The destination address indicates which station(s) is to receive the transmitted frame. The source address serves a dual purpose: to indicate which station transmitted the frame, and to help ensure that only one station is transmitting at a time (discussed later).

Information field (INFO)—The information field may be zero or more bytes. It can contain application data or control data. The contents of the information field are determined by the frame control field.

Frame Check Sequence (FCS)—The frame check sequence is a 32-bit sequence based on a standard generator polynomial of degree 32. It is computed using the contents of the FC, SA, DA, and INFO fields and thus protects these fields against corruption.

Ending Delimiter (E)—A special sequence of symbols used to denote the end of a frame. It includes intentional violations of the differential Manchester encoding scheme so that it will not be confused with any other field. It also contains a special bit to indicate bit errors. The error-detected bit is transmitted as a 0. If a station detects an error in the frame based on the FCS, it sets the error-detected bit to 1 to indicate that the frame is in error.

Frame Status Field (FS)—a one byte field used as an acknowledgment mechanism indicating proper receipt of a frame at the destination. The frame status field contains two important values in duplicate (for robustness), since they are not covered by the FCS. The address-recognized bits (A) and the frame- copied bits (C) are used to indicate the actions of the destination station. The A and C bits are initially 0 when the frame leaves the transmitter. If a station recognizes its address in the DA field, it sets both of the A bits to 1. If the destination station copies the frame from the ring, it also sets both of the C bits to 1. In this way, the bits indicate to the transmitter whether the destination station is active and whether or not the frame was successfully received by the destination station.

Tokens and Access Control

The token consists of a starting delimiter (S), followed by the access control field (AC) and the ending delimiter (E). The AC field contains all information relevant to the token itself.

The Access Control Field is defined in the IEEE Standard 802.5 and is shown in Figure 18.7:

Priority bits—three bits indicate the current service priority level of the ring. The priority level is raised according to the priority of the frames waiting for transmission.

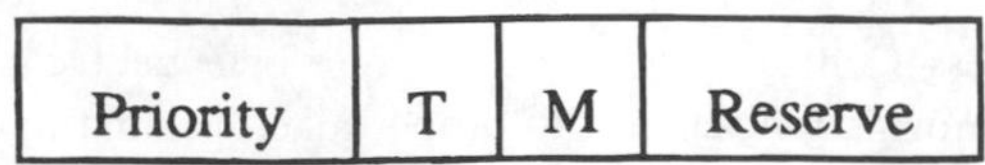

Figure 18.7 Access control field.

Token bit (T)—indicates the status of the token (1 = busy, 0 = free). A busy token is actually just a frame header in transit.

Monitor bit (M)—used by the active monitor to ensure that a frame does not make more than one round-trip around the ring before it is removed. If this occurs, then a busy token will be continuously circulating, preventing access to the ring by any station.

Reserve bits—three bits used to reserve the token for a given priority level. A reservation can only be made if it is higher than the current reservation level.

Normal Transmission

A transmitting station claims the free token by setting the token bit to 1. It then begins transmitting frames until either it has no more frames to transmit, or the token holding time (THT) expires. The station monitors the incoming bitstream for the return of its transmitted frame header. When the header returns, the SA field is checked for accuracy against the station's address. If the two addresses match then a new token is generated. The station does not reenter repeat mode until the entire frame has been stripped from the wire. This is done by transmitting fill while receiving the remainder of the transmitted frame.

Priority Transmission and Stacking Stations

The goal of the priority mechanism in the 802.5 standard is to provide fairness to all stations transmitting at the same priority level. Eight priority levels are supported using three bits, ranging from 000 (lowest) to 111 (highest). Initially, the token is generated with the lowest priority so any station may claim the token. If station A has a frame of priority Px waiting for transmission, it must request a token of that priority. To do this, the station sets the reservation field in the AC field of a passing frame to Px. This can only be done if the token has not already been reserved for a higher priority level. When station B (the current transmitter) receives its frame header with the reservation bits set, it releases a new token at the new priority level and stores both the old priority level and the new one. Having raised the service priority of the ring, station B becomes a stacking station. It is responsible for lowering the ring to the previous priority as soon as possible. Upon receiving the free token, station A now transmits its priority frames until either it has no more priority frames to transmit or the token holding time expires. After transmitting, station A has a few options. (Remember that station A is not currently a stacking station.)

1. If it has completed transmitting and has received no reservations, the token is generated at the current priority level with a reservation value of 000.
2. If it has completed transmitting and has received a reservation higher than the current priority level, the token is generated at the higher level and station A becomes a stacking station.
3. If it has not completed transmitting, and the token has not been reserved for a higher priority, the token is generated at the current priority level with Px in the reservation field. This prevents lower priority frames from being transmitted while higher priority frames are still waiting. (Note: The priority level of the ring is never lowered by a non-stacking station.)

Since station B is still a stacking station, it claims every free token with priority equal to the value it stored in order to try to lower the ring service priority. If a token is found with a reservation lower than the current highest stacked priority, the stacking station lowers the token priority to its highest stacked priority and removes that priority from its stack. Once the stacking station has emptied its priority stack, it no longer acts as a stacking station. In this way, the priority level of the ring is raised and lowered according to the priority of the waiting frames.

Error Correction

In a system where all access to the transmission medium is controlled by the passing of a single token, it is easy to see that errors involving token operation can cause the entire system to fail. Such errors must be detected and corrected in a timely manner for the entire system to run smoothly and reliably. A single station called the active monitor takes on the role of monitoring token operation. It is the responsibility of the active monitor to detect token errors and correct them if possible. Possible token errors are:

1. Circulating busy token—Due to some error, a busy token circulates around the ring without being removed by any station. The ring becomes unavailable to all stations.
2. Loss of token—The token has been somehow destroyed so that no station can gain access to the ring.
3. Multiple available tokens—Multiple free tokens on the ring allow multiple stations to transmit simultaneously.

Circulating Busy Token

This error is detected using the monitor bit (M) in the AC field of the frame header. Every transmitted frame initially has the monitor bit set to 0. When a frame passes by the monitor station, the monitor bit is set to 1. If the monitor station detects a frame with its monitor bit set to 1, that frame has made an entire trip around the ring without being removed. The monitor clears the ring by transmitting fill for a given amount of time. A free token is then generated.

Loss of Token

The monitor detects this situation with the use of a timer. The timer value must be greater than the maximum token rotation time in order to prevent false timeouts. If the number of stations on the ring is N, the maximum token rotation time (MTRT) is defined as follows:

$$MTRT = D * (N - 1)(THT),$$

i.e., the round-trip delay of the ring times one token holding time for every other station on the ring. This timer is reset every time a starting delimiter is detected. If a timeout occurs, it indicates that there is no longer a valid frame circulating on the ring. The ring is cleared by transmitting fill and a new token is generated.

Multiple Tokens Available

This situation is detected by the offending stations. When a station transmits, it waits for the frame header to return. If the returning frame header does not contain the transmitter's address in the SA field, it must be the case that more than one station is transmitting. Each transmitting station will recognize this and abort transmission. Further, no offending station will generate a free token. In this way, the problem becomes a lost token error. The active monitor will then recognize this and handle the error as previously specified.

Any station on the ring has the capability to be the active monitor station. At any given moment, there is one active monitor, and N-1 standby monitors. While the monitor station is active, it periodically transmits an *active monitor present* frame. This allows the remaining stations to recognize when the active monitor is no longer functioning. In the event that the active monitor fails, some other station must assume the role of active monitor. This process is carried out in a distributed fashion. Each station has a timer which is reset each time an *active monitor present* frame is received. If this timer expires, then each standby monitor begins transmitting Claim Token frames. If a standby monitor receives a Claim Token frame with its own SA, it then becomes the new active monitor.

Hardware Issues

Serious Failures/Beaconing

Occasionally, serious failures will occur which disrupt ring operation. These are known as hard failures and must be accurately detected to facilitate repair. Some examples of hard failure are a broken cable or an out-of-control or failed station. When a station or link fails, the station immediately downstream will detect the failure. The detecting station will transmit beacon frames. This indicates to all other stations that the token protocol is not functioning. The beacon frame contains the address of the beaconing station as well as the last received address of its upstream neighbor. All stations receive the beacon frame and therefore the location of the failure is determined. Once the failure is isolated, further action may be taken.

Neighbor Identification

For the beaconing frame to transmit the address of its upstream neighbor, a station must periodically notify its downstream neighbor of its address. This is done by taking advantage of the broadcast *active monitor present* (AMP) and *standby monitor present* (SMP) messages. Both messages are broadcast messages (received by all stations). As usual, when a station receives the frame, it sets the A and C bits in the FS field to ones. Since this will occur the first time the frame is received, only the first downstream station from the transmitter will receive the frame with zeros for the A and C bits. So, when a station receives an AMP or SMP frame with the A and C bits set to 0, it knows it came from its upstream neighbor. The downstream node copies the SA field and remembers it as its upstream neighbor to be used in any beacon frames.

Alternative Topology

To make the overall operation of the ring more robust, the ring may be constructed as a star. Logically the operation is the same, but a wiring concentrator lies in the middle with each station dually connected to it. The order of the ring is not altered. This topology has more flexibility, however, since failing stations may be bypassed easily at the concentrator without significant rewiring. Additional stations may also be easily added at any point in the ring since all stations connect to a common point.

Latency Buffer

One restriction on the ring is that an idle ring must be able to hold an entire 24-bit token. If not enough stations are attached to the ring to create this much latency, the monitor station is responsible for introducing enough delay to compensate for the shortfall. Each station is equipped with a 30-bit buffer to serve this purpose. In addition to providing the ring with the minimum latency, the buffer is also used to compensate for phase jitter accumulation. The overall ring timing is controlled by the master oscillator of the active monitor station. However, since the ring is actually made up of individual point-to-point links, the signal can get out of phase as it travels around the ring. This effect may be magnified at each station around the ring. The latency buffer compensates for this by receiving bits according to the receive clock, but always transmitting according to the active monitor's local clock.

Summary

The 802.5 token ring is well suited for office automation and some real-time control applications. The use of a wiring concentrator together with its error detection functions makes it easy to maintain and operate. The priority mechanism allows traffic to be divided into classes to ensure prompt service for important information. Most importantly, because of the efficiency of the access method, token ring networks achieve high utilization even under extremely heavy load.

References

Strole, N. C. 1983. A local communications network based on interconnected token access rings: a tutorial, *IBM Journal of Research and Development*, 27 (5) September.

Token Ring Access Method and Physical Layer Specifications, IEEE Standard 802.5.

Werner et al. 1983. Architecture and design of a reliable token-ring network, *IEEE J. Selected Areas in Communications,* SAC–1 (5) November.

18.3 IEEE 802.4 Token Bus

Alfred C. Weaver

History

The IEEE 802.4 subcommittee was charged with developing a local area network standard which addressed quite a different environment from that of the contention bus. From the beginning, the area of application for the token bus was factory communication. There were at least three specific requirements which the token bus had to meet and which the contention bus could never meet:

1. Allow a broadband signaling option to accommodate both voice and video in addition to data.
2. Provision for priority classes for data.
3. The capability to tune the network so as to guarantee delivery times for high priority messages.

The standard was adopted by IEEE and ANSI in 1984 and has since been adopted by ISO as Standard 8802.4. Compared with the contention bus and the token ring, this standard is the most complicated.

The token bus is of particular interest to factory communication because it provides the data link and physical layers of the Manufacturing Automation Protocol (MAP). While 802.4 exists in its own right outside of MAP, the two are often taken to be synonymous (they are not). The role of MAP is covered separately in Chapter 19.

Architecture

The 802.4 standard defines the

1. Electrical and physical characteristics of the transmission medium.
2. Electrical signalling method used.
3. Frame format required.
4. Actions of a station upon receipt of a data frame or a token.
5. Services provided by the Medium Access Control (MAC) sublayer of the OSI Data Link Layer.

Topology

The token bus is specified for three types of physical layer implementation. One type is a single-channel, phase-continuous-FSK (frequency shift keying) implementation as shown in Figure 18.8. This system encodes information by frequency modulating a carrier and impressing that signal onto a 75 ohm cable trunk. In this version, only one information signal can be carried at a time without disruption. The data signaling rate is 1 Mbps and the standard calls for the mean bit error rate at the MAC interface to be less than 10^{-8}, with a mean undetected bit error rate of less than 10^{-9}.

A second type is a single-channel, phase-coherent-FSK scheme. Phase-coherent FSK is a particular form of frequency shift keying in which the two signalling frequencies are integrally related to the data rate and transitions between the two signaling frequencies are made at zero crossings of the carrier waveform. The signaling rates are standardized at 5 Mbits/s and 10 Mbits/s for phase-coherent-FSK systems, with mean bit error rates of less than 10^{-8} at the MAC interface and with an undetected bit error rate of less than 10^{-9}.

The third type of implementation is the broadband bus. This type uses multilevel duobinary AM/PSK modulation, i.e., an RF carrier is both amplitude modulated and phase shift keyed, then pulses are shaped to reduce the frequency spectrum required for their transmission, and more than two distinct amplitude levels are utilized. The 802.4 standard calls for a three-level duobinary AM/PSK system capable of encoding one MAC symbol per baud; its appendix discusses extensions to two and three MAC symbols per baud. The required detected bit error rate is lower than 10^{-8} and the required undetected bit error rate is lower than 10^{-9}. The data signaling speeds are 1, 5, and 10 Mbits/s.

The broadband bus requires the use of a headend remodulator as shown in Figure 18.9. Each station is equipped with an RF modem which transmits in the "reverse" direction to the frequency translator at the head of the cable. There all transmissions are frequency shifted and retransmitted in a different frequency spectrum on the outbound or "forward" channel. The frequencies

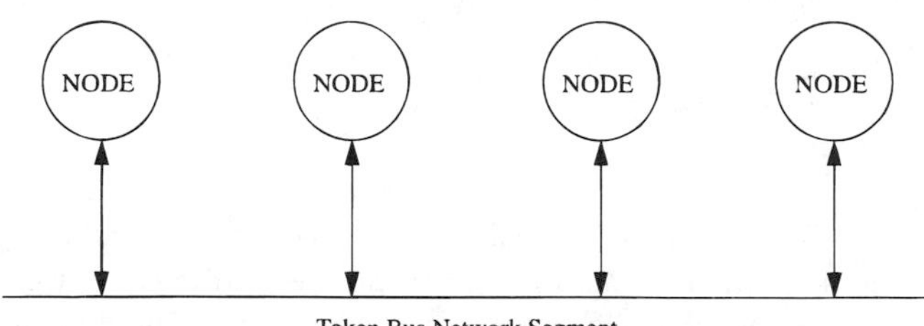

Figure 18.8 Single-channel phase-continuous-FSK network; Single-channel phase-coherent-FSK network.

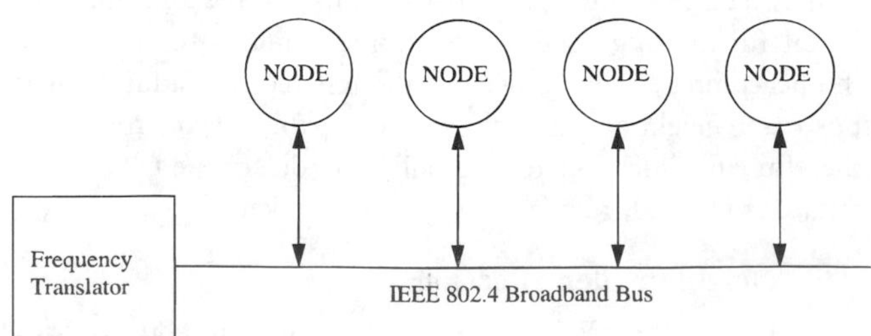

Figure 18.9 Broadband bus implementation.

to be used are chosen from the standard "North American 6 MHz Mid-split Channels" specification which pairs certain reverse channel frequencies with certain forward channel frequencies. All transmitters on a single 802.4 network share one frequency while all receivers share a different frequency.

Every station requires its own RF modem for its transmitter and receiver. This is a source of considerable cost in the station's network interface. Although there is only one headend remodulator in the normal network, it is fairly expensive. Since the frequency translator is a potential single point of failure, it is often available in a dual-redundant form, with either manual or automatic switchover. The duplicate headend plus switchover gear makes the dual-redundant system more than double the cost of a single-headed system.

Network Access

Network access is controlled by a special data element called the token. The station which holds the token is momentarily the network master and may transmit messages, or even engage in a subprotocol, for a limited period of time. When a station has transmitted all of its enqueued messages, or when certain timers expire, it must pass the token to a known successor. This orderly progression of the token from station to station thus forms a logical ring on a physical bus. The network supports four priorities of messages, called access classes.

The station's interface to the network is its MAC (Medium Access Controller), and the MAC is responsible for token recognition, passing, and regeneration after loss, message encapsulation and framing, service of the four priorities, and error control and recovery.

At network startup, each station is assigned a unique logical address. During the startup period each station adds itself to the logical ring using a contention process (see the section on dynamic membership). When ring membership has been established, token passing begins and the participating stations share the network capacity sequentially. Tokens are passed in order of descending station addresses, with the lowest addressed station then passing the token to the highest addressed station.

Token and Message Frames

The token is a crucial data element since it alone confers bus mastership. It is a data message with several fields, and even in its shortest form requires 96 bits. Each token frame carries the address of the token's destination as well as the address of the token's source (the sender). The addressing scheme used in all 802.x networks permits both short (16-bit) and long (48-bit) addresses. Thus, these two fields, source and destination, may each be either 16 or 48 bits in length, but they must be the same length. Token length is also dependent upon the network's transmission rate; at 5 Mbits/s the preamble field is 8 bits, at 10 Mbits/s it is 24 bits. The token consists of these fields:

preamble (one or more octets depending upon bus speed)
start delimiter (1 octet)
control information (1 octet)
destination address (2 or 6 octets)
source address (2 or 6 octets)
optional data (0 or more octets)
frame check sequence (4 octets)
end delimiter (1 octet)

Message frames have the same format as tokens, with the control information field denoting "data message." There are various types of protocol messages used for controlling the token passing sequence; they all have the same format and the control information field defines their type and function.

Dynamic Network Membership

The protocol automatically handles dynamic ring membership. Each station has a counter, called the inter_solicit_count, which is initialized to a network-wide variable called max inter_solicit_count. The inter_solicit_count is decremented once per token receipt, just prior to passing the token. If the network load is not too high and this counter counts downs to zero, then the counter is reset and the station sends to its normal successor a special protocol message called solicit_successor whose purpose is to determine if there is a currently passive station which wishes to become active. The solicit_successor message contains the address of the token holder as its source address and the address of its successor as its destination address. Any station whose address falls within this range may respond with another control message called set_successor. The source address of the set_successor frame provides the address of the station wishing to join the ring.

There are only three possible responses to solicit_successor:

1. No response. In this case there are no stations in the allowable address range waiting to join, so the station passes the token normally.
2. Exactly one response. In this case the current token holder passes the token to its new successor, who will later pass it on to the token holder's old successor. In this way the new station has been successfully "patched" into the logical ring.
3. Many responses. This case is active if the sender sends a solicit_successor message and then hears a collision (i.e., many responses). To resolve the collision, the token holder enters a loop in which it sends a resolve_contention frame and then opens four "response windows." Now only stations who participated in the original collision may respond. These contending stations reply in response window 0, 1, 2, or 3 depending upon the complement of the most significant two bits of their address. Whenever a contending station hears a valid set_successor frame, it drops out of the contention process. If this first loop does not produce a winner, then the loop is repeated using the second most significant pair of address bits, and so on. If there is still no clear winner after examining all pairs of address bits,

then the contending stations must have a duplicate address. In that case an error notification is made, the contending stations select a random response window in which to reply, and one more try is made.

While the handling of dynamic membership is indeed complicated, all that complexity is bundled into the protocol, and hence into the VLSI chip set which implements the MAC, and so is of little concern to the user.

Leaving the logical ring is much simpler. Each station maintains two Boolean variables called in_ring_desired and any_send_pending. The former is true whenever the station is or wishes to be a ring member; the latter is true whenever the station has a message enqueued for transmission. To leave the ring, the station sets in_ring_desired false. As soon as both flags are false (i.e., all queued data has been transmitted), the station waits for the token, then sends its predecessor a set_successor message which provides its predecessor with the address of its successor. Then it passes the token to its successor, thereby removing itself from the ring. To rejoin, a station goes through the dynamic membership procedure described above.

Error Handling

The protocol can recover from most common errors. There is a timer which records how long it has been since the station last saw a token. That timer is initialized as the worst case token rotation time, and if it expires the token must be lost. In that case the station is permitted to regenerate a token. Of course, two stations might time out simultaneously and both generate tokens; in that case, if a station holding a valid token hears another valid token on the network, the station destroys its own token.

Other common sense precautions are taken after each token pass. The only legal activity on the bus after a token pass is either another token pass or a message transmission. If a station passes the token and hears silence, it passes the token again. If the token pass fails a second time, the station issues a who_follows frame to ask who is its successor's successor. If there is a positive reply, then the token is passed and the normal successor is now out of the logical ring. If there is no reply to who_follows, the query is made once more. If that fails also, then a major ring error is assumed, so the station attempts to reform the ring by issuing a solicit_any frame to which any station can reply. If there is any valid response, the token passes to the responder. If there is no response, the station assumes that itself is in error (e.g., deaf receiver), rather than the whole network, and enters an internal "off-line" state.

Simple hardware provides a "jabber inhibit" function which resolves the error of "stuck transmitter." Every legal transmission is bounded by the maximum frame size of 8192 octets. Every station monitors its own transmissions. If a station detects its own transmission to exceed approximately one-half second, a hardware switch disconnects the transmitter from the bus. The transmitter can only be reconnected by a hardware reset or a protocol command to reset.

Access Classes

IEEE 802.4 supports four priorities, called access_classes. In descending order of priority they are called:

1. Synchronous.
2. Urgent asynchronous.
3. Normal asynchronous.
4. Time available.

Of these, only synchronous messages are guaranteed a particular level of service; the others are carried on a "best effort" basis. Any 802.4 station must either implement the synchronous class only or else it must implement all four priority classes. Stations of both types may be intermixed on a network. Note that the concept of priority applies to a message, not to a station.

When a station receives a token, it resets a timer to a network-wide value called the High_Priority_Token_Hold_Time (HPTHT), which defines how long a station may serve its synchronous class. This value is loaded into the token_holding_timer and service begins. The synchronous class is served until either this timer expires or the station has transmitted everything in its synchronous queue. If a station is transmitting when the timer expires, the station is allowed to finish the message currently in progress. Therefore, a station can overrun its service time by up to one message frame (a maximum of 64K bit times). If the station is not implementing the priority option, then the token is passed to this station's successor.

If the station is implementing the priority option, it attempts to serve each of its asynchronous classes in order of priority. Associated with each of the three asynchronous access classes is a variable called the Target_Rotation_Time and a timer called the token_rotation_timer. The timer records how long it has been since this station last received the token at this access class and the station compares this value to the Target_Rotation_Time. If the last token rotation, as measured by this access class at this station, took less time than the associated Target_Rotation_Time for this class, then the difference between those two values is loaded into the token_holding_timer and service of this access class proceeds. Service of this access class halts when its queue is empty, or when this timer expires (an overrun of up to one message frame is allowed as explained previously). The station cycles through its three asynchronous classes in order, applying this algorithm to its respective variables and timers. When the time_available class has been examined and possibly serviced, then the token is passed to this station's successor.

The result is that the network token rotation time (the time to complete one circuit of all stations and all access classes) varies with the offered load, and is responsive to the loadings at each priority class as well.

Summary

The advantage of the IEEE 802.4 token bus is its integration of audio, video, and real-time computer data over a single cable plant. The utility of the protocol comes from its having been selected as the datalink and physical layers of the Manufacturing

Automation Protocol. As a network protocol, it provides robust, real-time (i.e., latency bounded) service in its synchronous data class and non-real-time or "best effort" service in its three asynchronous data classes. When compared to a token ring (IEEE 802.5 or FDDI), its hardware components are expensive and would only be justified by the need to reuse an existing in-factory cable plant or by the need to upgrade to the MAP protocols.

References

Token Passing Bus Access Method. ANSI/IEEE Standard 802.4, 1984.

18.4 Fieldbus

Jean-Dominique Decotignie

Field busses are "true" communication networks intended to replace point-to-point, digital or analog, connections from control systems to the process. They link field devices such as sensors, actuators or operator interfaces to control devices. Sensors, whether intelligent or not, are used to measure quantities such as flow, level, pressure, speed, position, torque or temperature. In the process control domain, these devices are often called transmitters. Actuators might be as simple as valves or hydraulic jacks, or they might be more intelligent such as motor regulators and drives or small weld controllers. Operator interfaces usually consist of small operator panels or display units as well as data logging units. Control devices might be PLCs (Programmable Logic Controllers), CNCs (Computerized Numerical Controllers for machine tools), robots or process controllers. As there is some confusion in the use of the term "fieldbus," we will define it as an "industrial local area network that enables interconnection and interoperation of field devices, whether intelligent or not, and the first level of automation devices." In particular, there is sometimes a tendency to call fieldbus the networks that are used to interconnect automation devices in a cell. This is clearly the domain covered by networks such as Mini-MAP or FAIS (see Chapter 19) that are not fieldbusses but cell networks. This does not mean that fieldbusses may not be used for this purpose. It is, however, not their prime target. Advantages gained by replacing direct analog links by an industrial network are numerous (Pleinevaux, 1988). They are, however, not sufficient to push for such a radical change. The main incentive is economic. In process control, there is a need to go digital for remote calibration and maintenance to drastically reduce commissioning time and mean time to repair (MTTR). In manufacturing, reduction of cabling related costs (planning, installation, repair and change) is of paramount importance. We first show some examples of fieldbus use, then describe the functionalities that fieldbusses should offer. Finally, we sketch the solutions adopted by the different fieldbus proposals and give some information on the performance obtained with current solutions.

Examples of Use

Fieldbusses may be used in several ways:

To replace point to point cables for remote inputs and outputs.

To interconnect distributed intelligent sensors and actuators.

To implement a distributed real-time control system where all, or part of, the task of the lowest level automation devices such as CNCs, PLCs, and process controllers is distributed with the inputs and outputs.

Let us illustrate these three categories of use by three examples.

As a first example, let us take a PLC that controls a process with a number of sensors and actuators. Each sensor and actuator is connected to an I/O board of the PLC using a point-to-point link transporting an analog or a digital signal depending on the device. Cabling may be reduced and eased by putting a fieldbus interface and the necessary I/O interface electronics close to a device or a group of devices (Figure 18.10). The PLC is also connected to the fieldbus which is used to transfer the state of the sensors and actuators. Besides cabling, this solution exhibits several advantages. The signal-to-noise ratio for analog instruments is improved because conversion is performed close to the device. It is possible to add a monitoring or a maintenance computer without duplicating sensors (Figure 18.10) as the information is available on the bus. Cable integrity may be checked as part of the fieldbus protocol.

The second example illustrates the use of a fieldbus to interconnect intelligent devices. Figure 18.11 sketches the main functions of the CNC real-time part with their hierarchical relationships and the main flows of data between them. The axes coordinator generates the set point data (interpolation) that are sent periodically to each axis controller. In return, it receives status information that may be used to slow down all the axes when a temporary problem occurs or even stop them in case of emergency. Axis controllers perform all functions related to a single axis: position or velocity control, current loop, reference search and sometimes local interpolation. This means that all sensors and actuators belonging to a single axis are connected to the corresponding controller. These are position and/or velocity sensors, end of

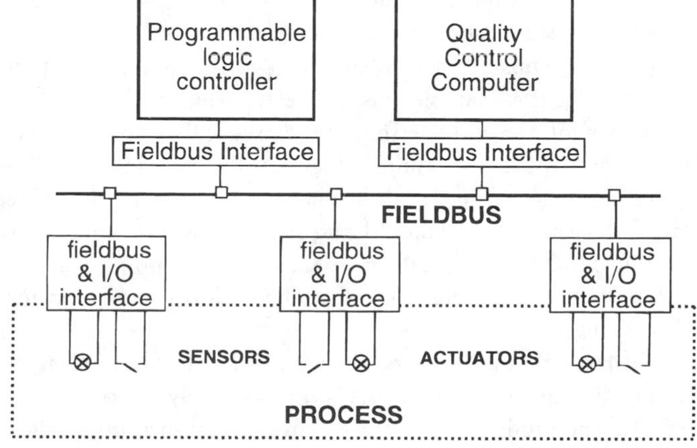

Figure 18.10 Fieldbus used for remote inputs and outputs.

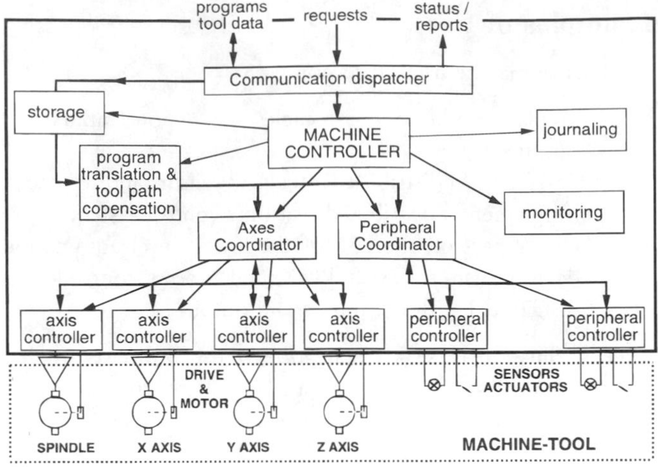

Figure 18.11 Architecture of the real-time part of a CNC with main data (thick lines) and control flows.

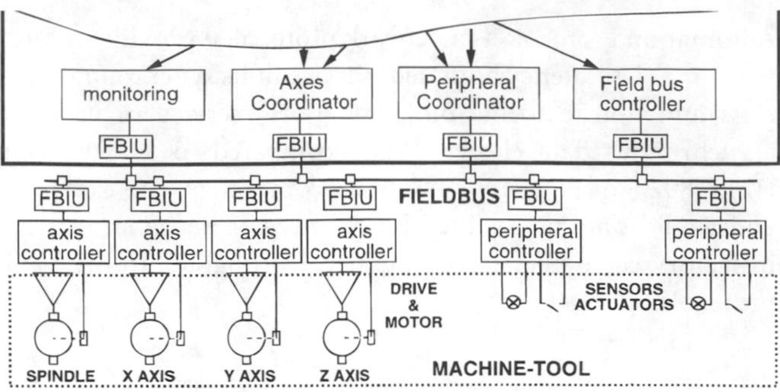

Figure 18.12 Lower part of the architecture of the real-time part of a CNC after introduction of the fieldbus (FBIU: Fieldbus Interface Unit).

track switches, reference switches, and motor drive. Sensors are polled periodically and set point data are sent to the drive with the same period. Periods range from less than a millisecond to a few milliseconds.

The peripheral coordinator handles coordinated actions on the peripherals (lubrication, tool change, spindle rotation, spindle orientation, etc.) and dispatches the requested actions to the appropriate peripheral controller. Each peripheral controller handles the sequences corresponding to a given peripheral according to the requests received from the peripheral coordinator and indications from sensors. They act on the actuators accordingly. This is normally done in a periodic manner with a period ranging from 20 to 50 milliseconds. The axis coordinator and the peripheral coordinator return abnormal conditions that may not be treated locally and completion reports to the machine controller. This one also handles operations that require a synchronization between axes and peripherals, i.e., tool change. When looking at Figure 18.11, a fieldbus may be introduced at three different levels: above the axes and peripheral coordinators, between the peripheral and axis controllers and the sensors and actuators, or between the coordinators and the peripheral and axis controllers. The best choice is to select the third case in which real-time constraints are important but can be met. Each controller (axis or peripheral) is embedded in the device it controls. The device may then be considered as intelligent. Besides cabling reduction as seen in the first example, this architecture (Figure 18.12) brings many benefits. Separate testing of the axes and peripherals may be performed before mounting them on the machine-tool. The same applies to options. Maintenance is eased because the entire faulty device may be removed and replaced by a new one very quickly. In addition to real-time data such as set point data, the fieldbus has to transport all the configuration information for the intelligent devices.

The last example illustrates the use of a fieldbus to implement a fully distributed system. It is a simple assembly station (Figure 18.13). The product comes on a conveyor clamped on a pallet. The positioning system then attaches the pallet to a known geometrical reference. A pick-and-place manipulator takes a single component from the separator and places it on the product. Finally, the positioning system releases the pallet. Components are taken by the feeder from a raw stock supply (vibrating bowl) and presented in a fixed orientation by the separator. The station is hence composed of a number of functions (positioner, manipulator, feeder, separator), each corresponding to a physical entity. Automation of such a station is usually done by a PLC to which all sensors and actuators are connected.

Using a fieldbus, automation may be approached in a very different way. First, all the functions encountered in assembly stations, as well as the flow of information between them, were identified and classified. Each function corresponds to a physical entity and the corresponding control sequence depends on the hardware technology used to implement the function. The idea is to put a small control system dedicated to each function and pre-programmed according to the technology, resulting in what we call a module. As the information exchanged between the functions is known and independent from the technology and the assembled product, a given assembly station may be built by putting together the necessary modules. Then a fieldbus may be used to transport the information exchanged between the modules. Each piece of information is assigned an identifier and a module needs to know only which information it produces and which it consumes. For example, in Figure 18.14 the operator module needs to know that it produces its status and uses the status of the positioner and of the separator. Each information produced is broadcast on the fieldbus and available to all modules.

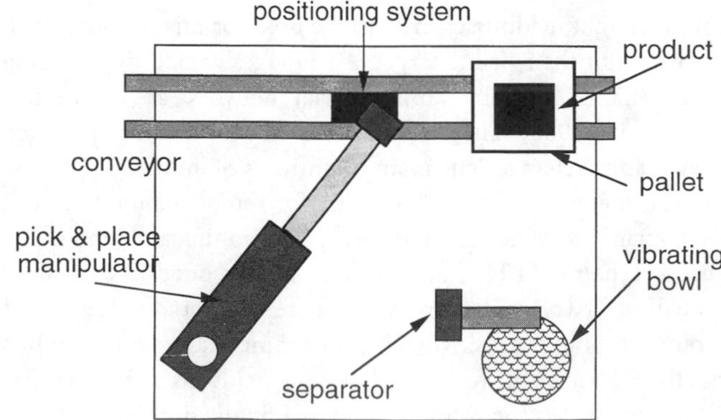

Figure 18.13 Simple assembly station.

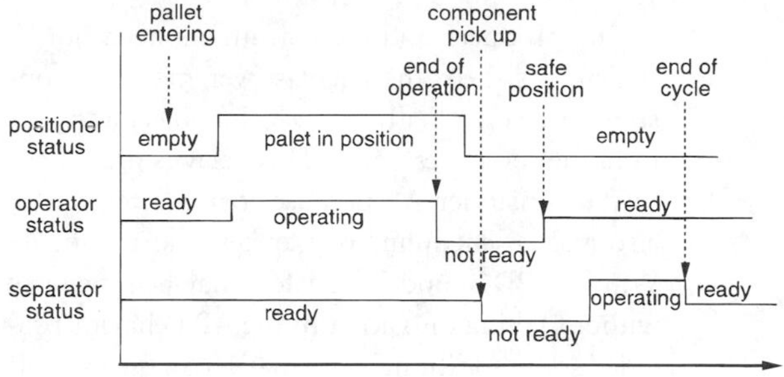

Figure 18.14 Sequence of operations and information exchange in a simple assembly station.

The assembly station is hence built from a set of cooperating applications using the fieldbus to exchange information. One of the main advantages of this architecture is that the manufacturer of a module may embed knowledge of his device in the module control system and need not furnish this knowledge to the designer of the assembly station. The latter does not need to worry about the technology of the module and may select the most appropriate module. The above examples were taken in the manufacturing field because the author is more familiar with it. Equivalent cases may be found in process control where fieldbusses have also to comply with intrinsic safety requirements.

Description

To be used in applications such as those described above, fieldbusses have to offer a number of functions that are not found in conventional industrial networks.

Event- vs. Time-Triggered Systems

Fieldbusses have been designed to ensure communication and interoperation at the lowest layers of the production hierarchy (Decotignie, 1993a). The primary function of these lower layers is to ensure a given production under constraints or objectives. For this purpose, they have to "diagnose" the process under control, that is to determine the state of this process and its evolution. According to this diagnosis, control systems act upon the process to satisfy the given objectives or constraints. Diagnosis requires knowledge of the evolution of the process; that is, for a computer that functions digitally, the acquisition of the sequence of process states. The notion of sequence implies that the time relation between states should be known. This can be achieved in two manners (Kopetz, 1991).

In the first approach, the control system watches the evolution of the process at predefined discrete instants. This perception of time allows a precise definition of the notion of simultaneity of events and of synchronization. Systems following this approach are often called sampled data or time-triggered systems. In the second approach, the control system tracks the evolution of the process "continuously." The definition of simultaneity must then rely upon timestamps provided by physical clocks. It may also be replaced by ordering or causality relations between events. Such systems are often referred to as "event-triggered systems." Both approaches are equivalent and can be found in the field.

From the transmission point of view, the first approach leads to cyclic or periodic state transfers. It also means that transmission occurs whether or not a signal has changed. Under normal conditions, the average bandwidth is hence more important than when using the event approach in which transfers only take place at event occurrence. On the other hand, event-driven systems require a more reliable transfer protocol because events must not be lost. State transfer need not be as reliable because the value will be transmitted again during the next cycle and, in case of transmission failure, the application needs to keep only the previous value and wait for the next cycle to get a new one. By taking this property into account, some fieldbus proposals suppress immediate retransmissions and acknowledgments. They may hence use a simple broadcast mechanism. The corresponding bandwidth savings is usually not sufficient to render state-based transfers competitive, in terms of bandwidth, with event-based transfers. The main advantage of state-based transfers is its predictability. All transfers are known in advance and may be easily scheduled in such a way that all timing requirements are fulfilled. Furthermore, in case of emergency, there is no additional traffic whereas in the event-driven approach a large number of transfer requests occur simultaneously. In the event-driven approach, to cope with the response time requirements even in emergency conditions, a bandwidth much higher than required under normal conditions should be reserved. This means that, in most of the cases, event-based transfers require a total bandwidth higher than the one necessary for state-based transfers. The former are often referred to as aperiodic transfers and the latter as periodic transfers.

Event Ordering, Time and Space Consistency

If control devices are sampled data systems, that is if they sample the states of the devices at predefined discrete instants, the communication network must be able to indicate whether values transmitted are time and space consistent. Time consistency means that the set of values available at a given control device corresponds to samples of the states of lower-level devices at the same sampling instant. Time consistency only applies to sampled data systems. It may be ensured by simultaneous sampling orders signaled by the indication of transfer of synchronization variables on the network. This is, however, not sufficient as shown in the following example. Let us assume that an application process on a sensor receives a sampling order. It samples the sensor state accordingly. In an on-request transfer, client-server mode, this application process requests the transfer of the sampled state. Due to traffic scheduling, this request may not be satisfied before the next sampling order is transferred on the fieldbus. The application process to which the sample has been sent shall hence receive the sample only after the next sampling order and may well consider that it corresponds to the next cycle which is false. Other examples may be found that also show the necessity to accompany the transferred value with some indication of the cycle in which the value was sampled (Decotignie, 1993). This indication may well be the value of the synchronization variable.

In event-triggered systems, the required information on temporal relationships (simultaneity and ordering) between events relies upon timestamps. Events are stamped by the device on which they are detected and the timestamp is transferred together with the event itself. To afford reliable relationship information, local clocks must be synchronized using proper algorithms (Cristian, 1989) which generate additional traffic.

Space consistency means that the different copies, transmitted to different controllers, of a value resulting from the sampling of a state are identical or correspond to the same sampling instant. It also applies to event-triggered systems. Space consistency is often ensured by special algorithms in distributed systems where a notion of consensus is defined. It may be implemented in a more effective way inside the communication network, especially if only an indication of the consistency status is required (Decotignie, 1993). Time consistency is mandatory for a good "diagnosis" of the process while space consistency is necessary when coordinated actions are required from a group of control devices. The latter is also necessary when redundant control systems are implemented.

Communication Models

Most existing networks, whether industrial or not, conform to the so-called "client-server" model. In this model, an application called the client requests another application, the server, to provide a given service. In return, the server replies with the results of execution of the service. A typical example is a write request in which the client indicates the object to be written along with the new value of the object. The server executes the request and, in its reply, indicates whether the write operation has succeeded or not. This is a process view of the interactions. If we take a data approach of the transactions, another model, the Producer-Distributor-Consumer (PDC) model, may be defined. With this model, each object has a single producer and one or more consumers of the transferred object values. The producer is an Application Process (AP) responsible for producing the value of an object. The consumers are the APs that need the values to perform their tasks. The distributor is an AP responsible to transfer the values from their producer to all their consumers (users). This PDC model has some similarities with the distributed shared memory model but differs from the Publisher/Subscriber model (Oki et al., 1993) in that the messages (object values) are not queued. This means that a newly produced value will overwrite the previous one, and a newly transmitted value will overwrite the previously received one. This solution is possible because object values have limited time validity. In a control loop or a PLC application, when a new value is produced, the previous one is no longer useful, for the new one reflects more accurately the process state. In other applications such as data acquisition or data logging processes, consumer APs are assumed to react quickly enough to capture all received values.

For object state transfer, the PDC model offers several advantages over the conventional client-server model (Thomesse, 1993):

The producer and the consumers need not be synchronized. In the client-server model, a consumer must explicitly request the object and wait until the server responds. This delay is prejudicial to the consumer AP because it needs to wait. It also makes scheduling of transfers more difficult. With the PDC model, transfers may be scheduled without taking into account the AP behavior provided some additional information on the variable value is given to the consumers. Due to this property, worst case performance may be assessed more easily.

Flow control is not necessary because of the overwrite property.

The distributor can handle simultaneous transfers from a producer to several consumers, whereas in the client-server model, each consumer would have to invoke separate requests to the producer.

In some cases, transfers are related, for example when several values correspond to the same sampling instant. If the different variable values are produced on different nodes and potentially need to be transferred to more than one user node, transfers need to be scheduled globally. Such an operation is much more easily handled by the distributor in the PDC model.

Finally, the PDC model reverses somewhat the responsibilities. The responsibility to check the transfer success is given to the consumers, while it is given to the client in the client-server model. For example, an actuator acting as a server will never receive any indication as long as the transmission fails. The same would apply in the PDC model, where the actuator is a consumer, except if some information is given to the consumer in order to check the temporal validity of the value it receives. On the basis of this information, it may take appropriate countermeasures.

Time Constraints

Fieldbusses are often exposed to severe time constraints. As seen above, cycle durations as low as one millisecond are requested. This means that as many as 20,000 transfers per second may be requested. With the client-server model, it is usual to define the time constraints as the response time, that is the time elapsed between a request from a client and the arrival of the corresponding response to the same client. This imposes severe constraints on the response time of the server and translates into an end-to-end transfer time, from client node to server node or vice versa, on the network. As the server response time cannot be assessed, predictable response times are impossible to guarantee. The Producer-Distributor-Consumer model suppresses the constraint on the information provider response time because the information has been stored in the Producer network stack before transfer takes place. Transfers may then be scheduled in a predictable way with a high degree of confidence. Due to the possibility of errors, a 100% guarantee may never be given.

In industrial communication, information that has not arrived at its destination before a given time is often useless to the control application. In fieldbusses, the idea of response time is thus replaced by the concepts of validity time or freshness. Validity time can be defined as the delay after production, within which the given piece of information means something to its users. Freshness is the status of a piece of information that is still usable. The same piece of information may have different validity times depending on the user. It means that the network should provide a way for a user to know the age of the piece of information. This issue applies to sampled data control systems as well as to systems whose behavior is based upon events. For event- triggered systems the timestamp may be used to know the age. Sampled data, or time-triggered, systems may take advantage of the synchronization variable value if this has some relation with the cycle number (Decotignie, 1993).

Device, Information and Traffic Characteristics

In most applications, the number of sensing and actuating device ranges from 20 to 100 with up to 4,000 devices in large plants. However, in such an extreme case, most of these devices will be connected through concentrators which reduces the number of network nodes. Apart from configuration and maintenance data, field devices receive or produce very small pieces of data (1 bit to 5 bytes). Traffic is essentially cyclic with up to 10,000 devices polled every second. Acyclic traffic is usually lower with up to 200 messages per second under normal operating conditions and bursts up to 100 events in 50 ms in case of emergency. In most applications, different cycle durations may coexist with a lower bound around 1 millisecond. Time and space consistency on lists of variables as well as validity time for single variables have to be provided in most cases. For event transmission (acyclic traffic), timestamps with resolution better than 1 ms are mentioned. It should be emphasized that, contrary to common thoughts, time and traffic constraints for manufacturing and for process control are very similar.

Solutions

Proposals for fieldbus implementations are numerous (Pleinevaux and Decotignie, 1988; Jones 1992). Most of them have been proposed by vendors to fulfill partial requirements or provide a quick solution in a given field. Some of these suffer from unacceptable restrictions that hinder their use outside their initial market. Two general purpose proposals, PROFIBUS and FIP, are national standards, respectively, in Germany (PROFIBUS standard DIN 19245, 1991) and France (NORME FRANCAISE, 1990). They are the most serious contenders for international recognition as open, general-purpose fieldbusses. They are very different in their approach to the problem. The former is more suited to event-triggered systems while FIP is more adequate for sampled data systems. A good description of FIP is available in Leterrier (1992). Bender et al. (1993) gives a good survey of PROFIBUS (but unfortunately based on an old version of the standard). Most of the ideas developed in those two proposals have been kept in the international IEC 1158 standard proposal. Most of the proposals conform to a collapsed OSI model that only keeps layers 1, 2 and 7 (physical, datalink, and application). The necessary functionalities of suppressed layers have been redistributed in those three layers.

Physical Layer

Most of the fieldbusses offer a general topology in which several bus segments may be interconnected using repeaters. To connect devices on a segment, spurs of up to a few meters are often allowed. Bus lengths range from a few meters to 2000 m. The preferred medium is shielded twisted pair but fiber optic and radio are also available. In the future, the possibility to mix wired and wireless devices will be offered. On copper and optical fibers, transmission bit rates of 31.25 Kbit/s, 1 Mbit/s and 2.5 Mbit/s have been standardized. Manchester encoding is used. Compliance to intrinsic safety requirements is also defined.

Data Link Layer

To ensure deterministic transfer times, either token bus or centralized medium access control is used. The second solution has the advantage of being simple and more predictable. It is also more reliable if active backup access controllers are provided. It is, however, less efficient when most transfers are on-demand which is the case for event driven systems. In such a case, the first solution is more adequate but unable to comply with periodicity requirements. For all these reasons, the international standard implements a combination of both mechanisms in which a central access controller may delegate a token for a given duration. The Logical Link Control sublayer offers services to transfer data with or without acknowledgement in a periodic or acyclic way. In the latter case, multicast and broadcast capabilities are provided. It also implements the functions required to indicate temporal and spatial consistency. The international standard data link layer even implements a clock synchronization scheme that offers the capability to timestamp the transferred information. Finally, some fieldbusses offer the internetworking functionality at the data link layer level. Internetworking is then restricted to non-time-critical transfers.

Application Layer

MMS (Manufacturing Message Specification) has influenced most of the application layers for fieldbus. Usually, only a subset of MMS is used. For fieldbusses which adopt the client-server model, this is the only application layer defined. This model is perfectly suited for non-real-time transfers such as configuration, calibration and testing. As seen above, it is less suited for real-time transfers. That is why some fieldbusses such as FIP define a separate application layer for real-time transfers according to the Producer-Distributor-Consumer model. Defined services allow one to read periodic (cyclic) variables with indication of validity time plus time and sometimes space consistency, write periodic variables that have to be transmitted to multiple users, send events or alarms with timestamping, or synchronize actions between devices.

Suitability and Performance

Fieldbusses find their application in many fields such as manufacturing, process control, building automation or embedded networks in cars, trains and planes. They may replace point-to-point links to field devices in a cost effective way. They also provide a number of new possibilities. Performance of up to 20,000 periodic transfers and 5,000 acyclic transfers per second may be obtained with currently existing fieldbusses.

Due to their particular place, fieldbusses need to offer a number of functionalities absent in other industrial networks. As flavored in the third example, the full benefits of this technology require a radical change in the control architecture.

References

Bender, K. et al. 1993. *PROFIBUS: The Fieldbus for Industrial Automation,* Prentice Hall, Englewood Cliffs, NJ.

Cristian, F., 1989. Probabilistic clock synchronization, *Distributed Computing,* 3:146–158.

Decotignie, J.-D., 1993. Fulfilling temporal constraints in fieldbus, *Proc. 19th Annual Con. IEEE Industrial Electronics Society, IECON'93,* pp. 519–524, Nov. 15–19; Maui, Hawaii.

Decotignie, J.-D. and Pleinevaux, P., 1993. A Survey on industrial communication networks, *Annales des Telecommunications,* invited paper, 48(9–10):435–448.

Jones, J., 1992. How do you get cheap distributed control? *Control and Instrumentation,* 24(4):57–61.

Kopetz, H., 1991. Event-triggered versus time-triggered real-time systems, *Operating Systems in the 90s and Beyond,* LNCS 563, 87–101, Springer-Verlag, New York, NY.

Leterrier, Ph., 1992. *FIP Tutorial,* WorldFIP, P.O. Box 13867, Research Triangle Park, NC 27709, USA.

NORME FRANCAISE NF C 46-601 to C 46-607, 1990. Bus FIP pour echange d'information entre transmetteurs, actionneurs et automates, AFNOR, Paris, France. An English version of these documents is available from WorldFIP, P.O. Box 13867, Research Triangle Park, NC 27709, USA.

Oki, B. et al., 1993. The information bus(r)—an architecture for extensible distributed systems, *Proc. 14th ACM Symp. Operating Systems Principles,* 58–68, Dec. 5–8, Asheville, NC.

Pleinevaux, P. and Decotignie J.-D., 1988. Time critical communications networks: field busses, *IEEE Network Magazine,* 2(3):55–63, May.

Powell, D., 1986. Dependable architectures for real-time local area networks, *Proc. Advanced Seminar on Real-Time Local Area Networks,* 53–84, Bandol, France.

PROFIBUS standard DIN 19245 part I, II & III, translated from German, PROFIBUS Nutzerorganisation e.V., Haid-und-Neu Strasse 7, D-76131 Karlsruhe, Germany, 1991.

Thomesse, J.-P., 1993. Time and industrial local area networks, *Proc. COMPEURO'93,* 365–374, May 24–27, Paris, France.

18.5 Fiber Distributed Data Interface (FDDI)

Robert W. Christie

Overview

The Fiber Distributed Data Interface (FDDI) is an ANSI standard for a Local Area Network (LAN) using an optical medium and employing a token ring protocol. Operating at 100 Mbits/s, an FDDI ring is faster than Ethernet (10 Mbit/s) by an order of magnitude, and twenty-five times faster than an IEEE 802.5 token ring operating at 4 Mbit/s. Originally, FDDI was developed for use as a high speed link between mainframes and large storage devices (a back-end interface), as a LAN backbone, and as a front-end for LANs which needed more performance than was available via Ethernet. FDDI-II extends the services of the original standard and offers the ability to have isosynchronous data traffic in conjunction with the synchronous and asynchronous data of the original standard. This isosynchronous capability lends itself to the transfer of real time data such as digital voice, video and control information (Ross, 1989).

A discussion of the general design characteristics of FDDI will follow. Then the interface will be described in terms of the Open Systems Interconnection (OSI) Reference Model. The last three sections discuss the performance, suitability, and applications of FDDI.

Description

The rationale for using a ring topology is due to the nature of fiber optic communication. The use of a bus would require a receiver to detect several sources at the same time. Although this is possible, fiber optic communication is better suited to point-to-point communication. However, in a single ring network the failure of one host would mean the whole LAN would fail. FDDI avoids this pitfall by using dual-attached counter-rotating rings, concentrators, and station by-pass switches to dramatically improve the reliability of the ring.

A station directly attached to an FDDI ring must have two physical connections, one for each counter-rotating ring. The two rings transmit data in opposite directions. One ring is considered the primary ring and is used to transmit data under normal situations, while the other secondary ring typically remains idle. This is known as a Dual Attached Station (DAS), or a station with two peer ports. In the event that either a single host or part of the cabling fails, the FDDI adapters on either side of the failure loop back by using the secondary ring to transmit data. This loop-back technique effectively isolates the fault in the system. See Figure 8.15 from (Ross, 1989).

A station may also be connected to a ring via a concentrator. A concentrator acts as a virtual ring for all stations connected to it. The advantage of using a concentrator is that multiple hosts connected to it may fail without causing segmentation of the

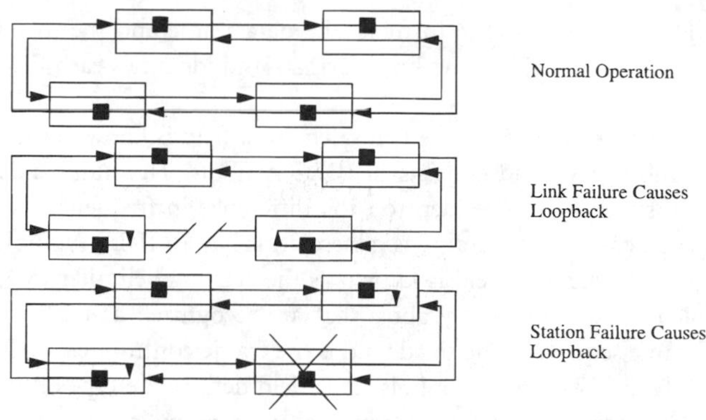

Figure 18.15 Reconfiguration of counterrotating rings.

main ring. The disadvantage is that if the concentrator fails the connectivity among all attached stations is lost. Stations may be either singly-attached or dually-attached to the concentrator. A Single Attached Station (SAS) has only one peer port on one ring which it uses to connect to the concentrator. A DAS connects both ports to the concentrator with the secondary ring remaining idle unless the first ring is damaged.

A third way to increase reliability is to use a station bypass switch. A station bypass switch keeps the ring up even if a host is turned off or damaged. Using these three techniques allows for a variety of ring configurations which will tolerate a certain level of failure (Ross, 1989).

FDDI uses a token ring algorithm which is similar to the IEEE 802.5 standard. However, the FDDI algorithm differs in some ways so that it may maximize its efficiency (Stallings, 1994).

Token ring control as implemented in IEEE 802.5 is based on the idea of there being a small "token" packet that circulates around the ring. When there is no contention for the ring, then the token packet is labeled as free. If a host wishes to transmit then it waits until it sees the token packet. Once the host receives the token packet it changes the label of the token packet to busy, and this packet then becomes header information for its frame. The host then retransmits the (now busy) token packet and frame. The transmitting station does not place a new free token on the ring until it has finished sending its data and the leading edge of its frame has returned to the sending host. Once this occurs the next host downstream from the sender will receive a free token giving it the opportunity to transmit data. By using this notion of a free/busy token only one host will ever have control of the ring. FDDI alters this algorithm to achieve higher efficiency and allow for multiple traffic priorities through the use of restricted and unrestricted control tokens (Stallings, 1994).

There are two main differences between the FDDI token protocol and the 802.5 standard. FDDI does not use an altered token packet as the header for its frames. Instead, a host is allowed to begin transmission of its frame as soon as it has completely received the token packet. Also, an FDDI host posts a new free token as soon as it has finished sending data. Otherwise, at high data rates the efficiency of the ring would be lowered because the host would have to wait until it received its own frame before releasing a token. FDDI also has a capacity allocation mechanism that supports both synchronous and asynchronous traffic (Stallings, 1994).

FDDI Protocol Stack

The FDDI standard can be divided into five parts which fall into the physical and datalink layers of the OSI Reference Model. These five parts consist of the logical link control (LLC), the medium access control (MAC), the physical (PHY) layer, the physical medium dependent (PMD) layer, and station management (SMT). Figure 18.16 shows the FDDI protocol stack.

Physical Layer

The physical layer is divided into two sublayers, Physical Medium Dependent and Physical Layer Protocol. The PMD provides the digital point-to-point communication between hosts on an FDDI network. The PMD defines and characterizes the fiber optic transmitter and receivers, cables, connectors, and physical hardware used to make a point-to-point connection between FDDI hosts. The PHY provides services that allow a connection between the PMD and datalink layer. Such services include clock synchronization between transmitters and receivers, and the encoding and decoding of symbol streams to bit streams or vice versa.

Data encoding in the physical layer is done with a 4B/5B scheme, that is, each 4 bits of data are encoded as a 5-bit sequence called a symbol. Therefore, in order to transmit at a data rate of 100 Mbits/s it is necessary to have a signal rate of 125 Mbits/s. This scheme was chosen over light intensity modulation encoding because the latter scheme does not provide a means for synchronization. Before transmission, an FDDI signal is still further encoded using non-return-to-zero-inverted (NRZI) to provide more reliability in signal recovery. By using the five-bit pattern for encoding four bits of data the signal is assured of never having more than three zeros transmitted in a row. This feature provides an acceptable level of signal synchronization. Because only 16 of the 32 values are used in the 5-bit encoding, the rest of the values are either said to be invalid or are given a special meaning (Stallings, 1993).

There are six different types of symbols represented in the 5-bit encoding scheme used by FDDI. Sixteen of the symbols are

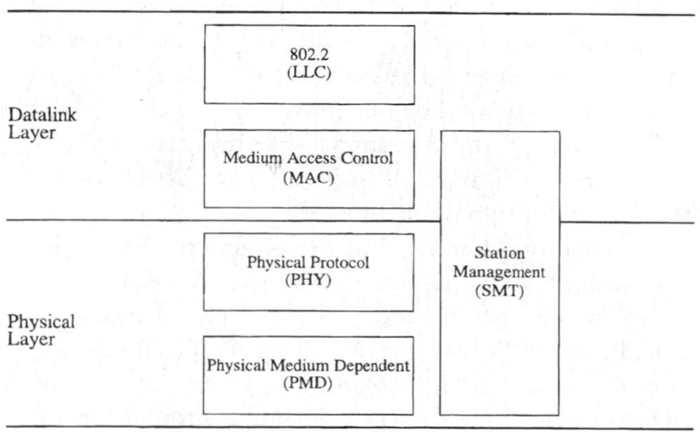

Figure 18.16 FDDI protocol stack.

data symbols representing 4-bit data sequences. Three symbols are line state symbols known as quiet, idle, and halt. The quiet symbol indicates that there are no transitions occurring on the line. The halt symbol indicates a logical break in the activity on the line. The idle symbol indicates the normal condition of the line between transmissions. Four symbols are control symbols used to represent the starting and ending delimiters of a frame. Two symbols are control indicators which indicate a logical one (set) and a logical zero (reset). The last eight symbols are invalid assignments, which should only be received because of an error condition or during clock synchronization (ANSI X3T9.5, 1987).

Synchronization of clocks around a ring is another concern. The clocking allows the receiver to sample the incoming signal at the correct time in order to recover the bit stream. Hosts can resynchronize their clocks through the use of the transition sequences in signals that have been encoded with the 4B/5B encoding scheme. This deviation in clocks at the different hosts is also known as timing jitter. The main problem with timing jitter is that the bit length of the ring will start to vary while frames are repeated from host to host. In FDDI each host has its own clock, and timing jitter is dealt with by using what is known as an elastic buffer. Each FDDI host has a fixed frequency clock which it uses when transmitting data. Each host also has a variable frequency clock which is used by the receiver in order to track an incoming frame. The elasticity buffer is placed at the receiver of an FDDI host to compensate for differing frequencies between the incoming and outgoing clocks. If the clock frequency for the outgoing frame is less than the clock frequency as it comes in then bits need to be dropped from the frame, and if the outgoing clock's frequency is greater then bits need to be added. The elasticity buffer is used in each host to do this compensation. When a frame is created at the MAC, sixteen idle symbols are prepended to the frame. The elasticity buffer in the hosts around the ring either add or subtract idle symbols depending on the difference in the clock frequencies. Due to FDDI specifications of clock stability, an elasticity buffer of 10 bits will ensure the proper transmission of frames up to 4500 bytes in length (Ross, 1986).

Datalink Layer

The datalink layer is also divided into two sublayers, the Medium Access Control and the Logical Link Control. The MAC layer provides fair access to the medium, address recognition, and the generation and verification of the frame check sequence. Its main function is to deliver frames. This includes inserting the frames on the ring, repeating frames that are already on the ring, and removing frames that have circled the ring once. The LLC provides three possible services to the next protocol above. These services are an unacknowledged connectionless service, an acknowledged connectionless service, or a connection-oriented service. This common functionality is then provided via a Service Access Point (SAP) to higher layer protocols. FDDI supports the use of the 802.2 logical link control. A general overview of the FDDI token ring protocol was given in the introduction, so this section will cover the frame format and then some of the specific characteristics of the MAC protocol.

FDDI has two frame formats, the data frame and the token frame. Figure 18.17 shows both formats and identifies each field in the frame.

The data frame has nine fields. The first field is known as the preamble (PA), and consists of IDLE symbols. The number of IDLE symbols is dependent on the difference in frequencies of host clocks. When a frame is generated it has 16 IDLE symbols (64 bits). The next field is known as the starting delimiter (SD) which consists of two symbols that are recognized as the start of frame sequence. The third field is the frame control field (FC) which consists of two symbols. This field defines the type of the frame, such as whether the frame is synchronous or asynchronous. It also specifies the length of the address, and the content of the frame. The fourth and fifth fields of the frame are the destination address (DA) and source address (SA). These fields may be either 48 bits or 16 bits depending on the addressing mode specified in the frame control field. The info (INFO) field follows, and contains data whose meaning is determined by the contents of the frame control field. The INFO field is of variable length but its length is limited by the maximum data unit size of 4500 bytes, which includes all fields of the frame plus two bytes of the preamble. The frame check sequence (FCS) follows the INFO field. The FCS is a 32-bit cyclic redundancy code for the frame. The error detection code covers the last 4 bits of the starting delimiter up to the end of the FCS field. The frame ending delimiter (ED) follows the FCS, and contains two symbols (1 byte) so that the frame retains byte boundaries. The frame status field (FS) contains an arbitrary number of control indicator symbols. The first three symbols must be in the field. The first symbol is the error detected indicator. It is reset (logical zero) by the sender of the frame. If an error in the frame is detected by another host then this symbol is set (logical one). The next symbol is the address recognized indicator. This symbol is reset (logical zero) on transmission. If another host recognizes the destination address of the frame as one of its own, then it sets the symbol (logical one). The third symbol is the frame copied indicator which is transmitted by the originator of the frame as the reset symbol. If another host recognizes this destination address as its own and copies the packet, then it sets this symbol to logical one. The token frame format contains the preamble, the starting delimiter, the frame control, and the ending delimiter.

There are two types of traffic in FDDI, synchronous and asynchronous. During an initialization phase of the ring a bidding process occurs to decide the value of the target token rotation time (TTRT), which is the lowest value bid from all stations.

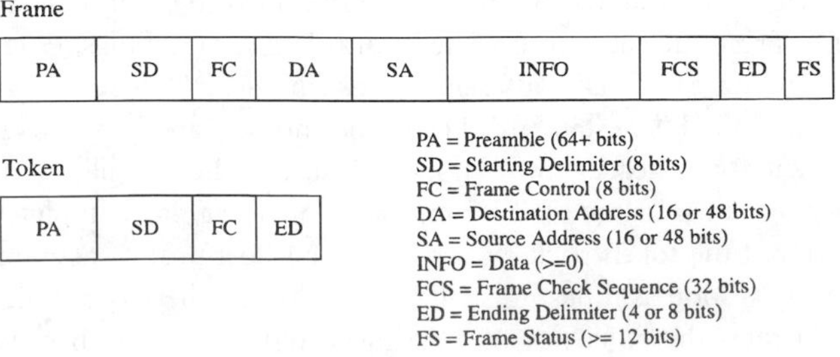

Figure 18.17 Frame and token field formats.

Each host must request a synchronous allocation block via the station management protocol (SMT). If a host does not support synchronous allocation (it is optional), then the host may only transmit in the asynchronous mode. In synchronous mode, if a host has synchronous frames that are ready to be transmitted, then it may capture the next available token. In asynchronous mode the host may only capture a token if the time that has passed since the last time it saw a token is less than the TTRT. Also, multiple priorities are available for asynchronous traffic, because more restrictive timing boundaries may be placed on the TTRT at each host, and a host may use what is known as a restricted token. When a station captures a regular token (non-restricted), it may send its frames, and then release a restricted token. All other hosts may send synchronous traffic when they receive this token, but only the host who was the destination of the previous frame may send asynchronous traffic. These two stations can continue to be the only two stations sending asynchronous traffic by holding the restricted token. Also, this new token doesn't violate the timing protocol because all other hosts are still able to send frames synchronously.

The benefit of using the timed token rotation (TTR) protocol is that a host can negotiate bandwidth and latency guarantees for its synchronous traffic. The initialization process of the ring accepts the lowest bid as the TTRT, thus ensuring that there is a low guaranteed response time for synchronous traffic. The worst case will have the token arriving no more than two times the TTRT after the last token arrived. The use of the TTR protocol also affects ring performance. Depending on the choice of the TTRT value, one can either have a small latency between response times, or one can get a very high ring utilization under heavy loads (Ross, 1986).

The FDDI Station Management specification is the fifth component of the FDDI interface. It spans both the physical and datalink layer of the OSI reference model, and provides coordinating services so that the different processes in an FDDI adapter will work together effectively. Some of the services provided by SMT are ring management, connection management including the initialization and configuration of stations, plus the insertion and removal of stations from the ring, and fault isolation and recovery (ANSI X3T9.5, 1987).

FDDI-II

FDDI-II is an upward-compatible extension to FDDI with the key difference being the ability to support circuit-switched traffic as well as the previously supported packet-switched data. FDDI-II supports a mechanism that allows a constant data rate connection to be set up between two hosts which is well suited for the transmission of constant data streams such as digitized voice. This service is provided by having an 8 kHz rate imposed on a certain frame type in the ring. Hosts may then request to acquire spots in this frame for their use. Each of these slots in the frame has a 125 microsecond latency between reappearances. Thus a circuit switched connection would be composed of a few of these regularly repeating sections from this frame. This mode of transmission is also called isochronous, and can be used in conjunction with the regular modes of FDDI service so long as all hosts are using FDDI-II (Ross, 1989).

Suitability

The synchronous and asynchronous modes of FDDI allow the interface to handle a wide variety of traffic types. Traffic such as file transfers and e-mail are well suited for the asynchronous mode of FDDI, while control information for systems is more suited to the synchronous mode because this service gives an upper bound on the time before the arrival of the next free token. The high throughput (100 Mbits/s) of FDDI allows it to be used as a backbone for slower LANs. Because of their high throughput requirements, FDDI is also a good choice for multimedia data streams. In settings where electromagnetic interference is strong, FDDI is a good choice since fiber optic cable is insensitive to EMI.

Performance

The calculations for FDDI LANs were based on there being 1000 physical connections with a total path of 100 km in fiber. This translates into 100 km for each counter-rotating ring of 500 dual-attached stations. The light emitting diodes used with FDDI will emit light with wavelengths in the 1300 nm range. The transmission mode for FDDI is multimode, meaning the light will typically reflect off the walls of the cable before being received by the destination. With the constraints above, FDDI will have an upper bound of 100 Mbits/s for throughput. Assuming the hardware is of high quality, there should be less than a 5% degradation in sustainable throughput at the upper level interface to FDDI (Ross, 1986).

Applications

A typical use for FDDI is as a backbone for multiple Ethernet LAN segments. FDDI's high throughput compared to Ethernet (10 Mbits/s) keeps packets from being dropped when there is high utilization on a number to the Ethernet segments.

Another use for FDDI is for control signaling in distributed systems. Factory control systems that require guaranteed bandwidth and an upper bound on delivery latency can get both by using FDDI's synchronous mode.

References

ANSI X3T9.5, 1986. *FDDI Token Ring Media Access Control (MAC),* Draft Proposed American National Standard.

ANSI X3T9.5, 1987. *FDDI Physical Layer Medium Protocol (PHY),* Draft Proposed American National Standard.

ANSI X3T9.5, 1987. *FDDI Station Management (SMT),* Draft Proposed American National Standard.

ANSI X3T9.5, 1989. *FDDI Physical Layer Medium Dependent,* Draft Proposed American National Standard.

Ross, F. 1986. FDDI—a tutorial, *IEEE Communications Magazine,* 24(5).

Ross, F. 1989. An overview of FDDI: the fiber distributed data interface, *IEEE J. Selected Areas in Communications,* 7(7).

Stallings, W. 1993. *Networking Standards: A Guide to OSI, ISDN, LAN and MAN Standards,* Addison-Wesley, Reading, MA.

Stallings, W. 1994. *Data and Computer Communications,* Macmillan, New York, NY.

18.6 Asynchronous Transfer Mode

Curtis L. Moffit

Introduction

Asynchronous Transfer Mode (ATM) is a high performance networking technology currently being developed under the direction of the ATM Forum, a consortium of networking software, hardware, and workstation vendors. Suited for Local and Wide Area Networks (LANs and WANs), ATM has evolved from the vision set forth in the Integrated Services Digital Network (ISDN). The end goal of ISDN is to create a single networking infrastructure to carry all types of digital traffic (the so-called "Information Superhighway"). These services include traditional computer data (electronic mail, file transfers) and voice traffic (telephone) as well as future applications (multimedia, visualization). This section serves to give a quick overview of ATM concepts and explains how it may be used in a factory setting; for a more detailed overview, refer to Alles (1993).

ISDN's initial development was done on Narrowband ISDN (N-ISDN) which carried digital traffic at 144 Kilobits per second (Kbits/s) and 1.5 Megabits/s (Mbits/s). The creation of more advanced applications quickly demonstrated the need for more network bandwidth and Broadband ISDN (B-ISDN) was developed to handle this need. ATM was chosen as the Broadband transmission mechanism because it has certain characteristics that are very important for the types of services ISDN must provide. These characteristics include both low latency and scalability. That is, ATM is designed to carry gigabits of aggregate bandwidth at very fast speeds over both large and small areas.

As a networking technology, the design of ATM is significant because it will impact both LANs and WANs. Currently, WANs are composed of many hardware devices (bridges, routers, hubs) using different media (Ethernet, FDDI) that are interconnected across long distances. However, ATM WANs are based on a single networking fabric and public communication companies need only deploy one type of digital infrastructure to provide all of the ISDN services; there is no need for many different networking devices. ATM will also have a great impact on LANs due to its scalability and the large amount of bandwidth it can deliver to a single user. Currently, LANs are very limited in the amount of bandwidth they can provide to an end user and in the number of users they can support before the network becomes saturated. In this way, ATM provides solutions to local and wide area networking problems and, thus, will gain widespread acceptance in these areas as well.

Characteristics

ATM uses small, fixed size cells (53 bytes) for carrying network traffic. This is unlike traditional computer networks which use variable length packets for transmitting data and causes a shift in the networking paradigm. Packets can be very large (many kilobytes) and parsing a packet header is often difficult and complex. ATM cell headers, however, are simple enough to be parsed in hardware. The small size helps reduce the amount of delay uncertainty in cell delivery. This allows ATM to carry many different types of traffic. It also allows the network to do traffic shaping so that nodes can be guaranteed a required quality of service (delay and bandwidth guarantees).

ATM is also fundamentally different from traditional networks because it is inherently connection-oriented. That is, a connection between both end nodes is established before any data is actually sent. Packet switching networks (and most LAN technologies) are connectionless; they inject data into the network without establishing an explicit connection. Packet switching networks use routers containing special hardware and header parsing software that determines where to forward each packet. However, once an ATM connection has been established, cells are self-routed to the destination via connection identifier.

Each ATM cell contains a VPI (Virtual Path Identifier) and VCI (Virtual Circuit Identifier) that identifies the connection it belongs to. The VPI/VCI mapping is used across the network and allows virtual connections to flow through any number of intermediate switches. A desirable characteristic of the virtual channel mechanism is that it prevents misordering of cells. In ATM networks, multiple paths to the same destination on the same circuit do not exist. Thus, all cells sent on a connection are received at the destination in the same order they are sent.

Virtual circuits allow flexible and dynamic bandwidth reservation while virtual paths serve as a higher level aggregation mechanism (they bundle together a large number of virtual circuits). This is unlike Time Division Multiplexing (TDM) which uses static bandwidth reservation by allocating a slot in the data stream by position. ATM virtual circuits can reserve a specific amount of bandwidth over a period of time but do not dictate the position of cells in the data stream like TDM. Thus, bandwidth is only reserved when a node has data to send and this gives ATM the efficiency of packet switching. In addition, since cells can be switched in hardware, communication can be performed at much higher speeds (in the hundreds of Mbits/s) than current LAN technologies. ATM may also prove to be more cost effective since the necessary hardware interfaces will be simpler and cost less. Furthermore, since ATM is not tied to a particular medium, it is highly versatile. It is designed to operate over links of many different speeds, unlike FDDI, Ethernet, and other technologies.

Network Architecture

The general architecture of an ATM network is shown in Figure 18.18. Note that no particular organization is assumed. That is,

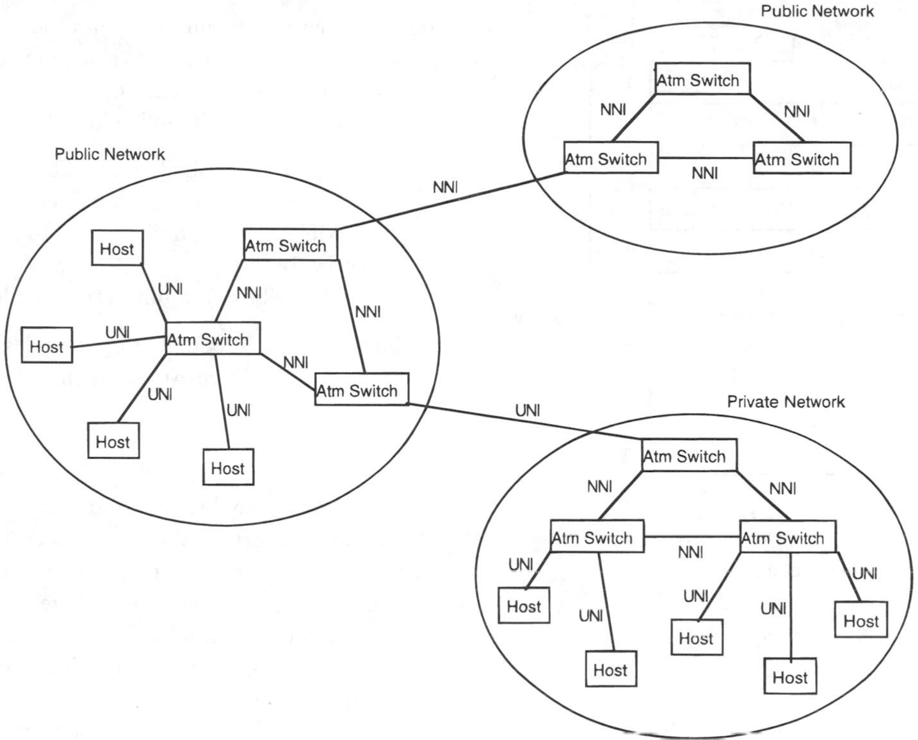

Figure 18.18 ATM network architecture.

there is no more or less important piece in the network and no hierarchy of components is imposed on the structure. It simply consists of end nodes, switches connected to end nodes, and switches connected to switches over public and private domains. As shown in the figure, the User-to-Network Interface (UNI) is used to connect nodes to switches and private switches to public switches. The Network-to-Network Interface (NNI) is used to interconnect switches within a domain (either public or private but not between both). These interfaces provide the connectivity that is needed for ISDN services.

The overall architecture of ATM networks is fundamentally different from packet switching networks. As noted previously, connectionless networks do not suffer from connection setup time and can be faster in cases where traffic volume is low and infrequent. In addition, end stations require less intelligence since all they have to do is put packets on the medium. This causes the greatest amount of complexity and expense to be relegated to the network itself, creating the need for many special hardware devices. The packets must also be complex because each one has to contain enough information to fully route it to its destination. ATM, on the other hand, does away with many of these problems. It trades greater complexity in the network for greater complexity in end stations. Network switches do not care what type of data is being sent; they only relay cells on the correct circuits. The simplicity of switch hardware and cells makes ATM networks scalable to very high speeds. It also allows nodes to dictate to the network the desired characteristics of the circuit instead of relying on the network for best effort service. In this way, quality of service requirements can be placed on peak bandwidth usage, the rate and burst characteristics of cells flowing through a circuit, and the admission of end stations into the network.

ATM Cells

Cells are the only way to transmit data in an ATM network. Each cell is 53 bytes and consists of a 5 byte header and 48 byte payload. Figure 18.19 shows an ATM cell and the corresponding fields in a UNI cell header (the NNI header is not shown but is similar). The following fields are defined:

CLP—Cell Loss Priority. Indicates whether or not a cell can be dropped when the network is congested.

GFC—Generic Flow Control. Used to control UNI traffic flow, it is unused in the NNI.

HEC—Header Error Control. A checksum computed over the header fields to detect possible errors.

PTI—Payload Type Indicator. Tells what type of data is carried in the payload.

VCI—Virtual Channel Identifier. Identifies the channel with which the cell is associated.

VPI—Virtual path identifier. Identifies the path with which the cell is associated.

ATM Protocols

ATM also supports protocols for delivering cells called ATM Adaptation Layers (AALs). These AALs are designed to carry the four current classes of traffic:

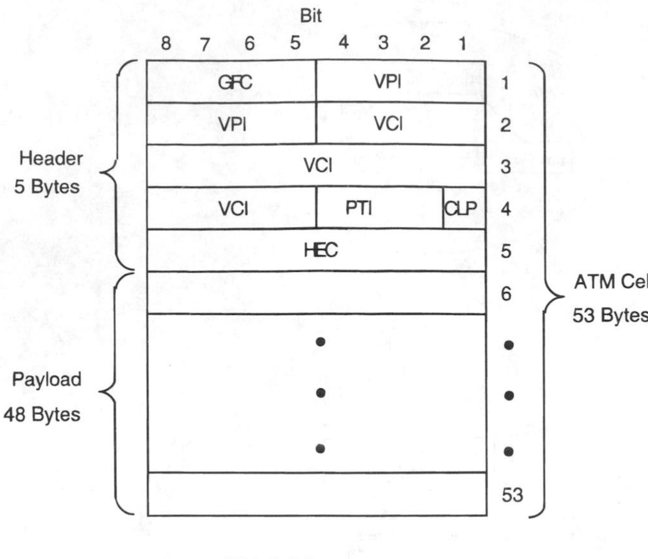

Figure 18.19 ATM cell structure.

Class A—Constant Bit Rate (CBR), connection oriented, synchronous traffic (AAL 1).
Class B—Variable Bit Rate (VBR), connection oriented, synchronous traffic (AAL 2).
Class C—VBR, connection oriented, asynchronous traffic (AAL 3/4 and 5).
Class D—Connectionless packet data (AAL 3/4).

Each AAL requires its own header information within each cell (4 bytes) and this reduces the payload per cell to 44 bytes. The header information is needed to provide segmentation and reassembly of buffers in a node's memory. When large chunks of data need to be transmitted, they must be divided into 44 byte pieces so that they can be transmitted as cells. The header information allows the receiving node to reassemble the cell payloads into the original buffer. These protocols provide the necessary networking functionality for applications that require the different classes of service. Other protocols such as TCP/IP are being supported for interoperability with existing internetworks. Network management services will be provided by the Interim Layer Management Interface (ILMI) which is based on the Simple Network Management Protocol (SNMP) and is used to manage a single UNI. Switches will support one management entity for each UNI in order to control all of its interfaces.

Transmission Media

ATM has many physical transmission medias that it can use. In particular, the Synchronous Optical Network (SONET) is a set of physical layers that ATM networks can use. SONET is a standard for fiber optic transmission lines that specifies how framing is done and how bytes are transmitted. The base rate is set at 51.84 Mbits/s for STS-1 on OC-1 fiber cables. STS-x and OC-x are fiber lines which run at x times the base transmission rate. For example, STS-3 on OC-3 cables runs at approximately 155 Mbits/s. Among the different physical layer standards, ATM networks can use any of the following:

STS-3 over OC-3 single mode fiber (155 Mbits/s).
STS-3 over OC-3 multimode fiber (155 Mbits/s).
100 Mbits/s multimode fiber from the FDDI physical layer (TAXI).
DS-3 (45 Mbits/s) public carrier media.

Single and multimode fiber optic cables are becoming widely available and less expensive as the need for them increases.

Benefits

ATM provides many benefits over conventional networks. One of the most important advantages is that it can support high speed interfaces for new applications. Increases in computing and networking power can be exploited to provide multimedia and visualization applications. Another important benefit is that ATM allows a single networking fabric for all classes of traffic. Previously, different networks were needed to provide specific types of services and the cost of network duplication was very high. However, ATM was designed with this end goal in mind and can handle applications that require the attributes of both packet and circuit switched networks.

New Applications

Because they are based on shared media technology, traditional LANs are always bandwidth limited. Applications on a particular end station can only receive a fraction of the total bandwidth available on the network. Scalability is also a problem in LANs because they become saturated very quickly with hosts and their network traffic. A large amount of sustainable aggregate bandwidth is necessary to provide end users with new multimedia based applications. Video transmissions contain rapid sequences of very large images which require low end-to-end and interframe delay. With ATM, these new real time voice and video communication services are made possible. More importantly, ATM is scalable and these services can be deployed across WANs, especially across interconnections between public and private networks.

Integrated Networks

Many networks exist for carrying various types of traffic. The public telephone network uses Time Division Multiplexing (TDM) for transporting voice traffic. TDM is best for traffic that is time sensitive and generated at regular intervals. It is unsuited for bursty data traffic since TDM allocates the bandwidth whether it is used or not. On the other hand, packet switched networks are best for applications that generate variable amounts of data at irregular intervals where some delay is tolerable. In such a situation, packet switching is more efficient than TDM since bandwidth is only used when it is needed. However, packet networks are not suited for time sensitive traffic since only one node has access to a shared media network at once. That is, when one node is transmitting, all other nodes are blocked from accessing the network even though their traffic may be more time critical.

The disparity in these technologies have necessitated the creation and deployment of multiple physical networks. Both TDM and packet networks require different types of specialized hardware devices. Packet switching requires particularly complex switching hardware and software due to the difficulty involved in parsing the many types of protocol headers. ATM provides one physical network that can efficiently handle all kinds of traffic in a cost-effective way; there is no need for multiple networks and complex hardware devices.

Suitability

ATM is a technology that is suitable for combining all factory communications and placing them on a single networking infrastructure. Because it is expandable, one ATM network is capable of carrying all of the voice, video, and computer data traffic that a factory would need for almost any reason. This is especially true in situations where multimedia based applications can increase factory productivity. Real time voice and video communications can vastly change the way in which work is done in such a setting.

Further advances in visualization applications will also allow real time modeling of complex processes. These applications also have the potential to change the way factories operate on the most basic level. All factory devices can feed data to computers that monitor their operation on the same physical network that all communication and visualization processes use. Traditional packet switching networks can handle simple data transfer services and are adequate for information that is not time critical. However, if a factory needs computer data along with video and visualization services, conventional packet networks will not be able to provide them. Furthermore, if a factory runs processes that are dependent upon time critical delivery of data, only ATM networks will be able to provide the necessary quality of service guarantees. Also, with an ATM network in place, communication between distant factories will also be possible through connections into the public networking infrastructure. In this way, both factory personnel and processes will be able to communicate over long distances whenever necessary.

Performance

The performance of ATM networks far surpasses that offered by traditional LANs and WANs. In addition to being able to provide performance guarantees, ATM brings large amount of bandwidth to a single user. Note that since ATM links from an end station to a switch are not shared, the full bandwidth of the link can be offered to the node. Thus, fiber optic links like OC-3 will bring a full 155 Mbits/s to the desktop. For raw ATM cells, this gives 140.8 Mbits/s (155.52 Mbits/s $\times$ 48/53) of usable throughput. For applications that use AALs, 129.1 Mbits/s (155.52 Mbits/s $\times$ 44/53) of throughput is available. At these speeds, a new cell can be transmitted or received every 2.7 microseconds (Traw and Smith, 1993) and cell delivery delay times are extremely low. Even over very long distances, such as across the continental United States, maximum delay times only reach 15 milliseconds (Kleinrock, 1992) (30 ms round-trip). Very fast network access and low end-to-end delays allow fine-grained control of factory processes. That is, the high performance nature of ATM allows control and automation of processes to be done in real time. Furthermore, since ATM is not tied to a particular medium, faster interfaces and fiber optic links can be used as they become available. Thus, if a factory needs more bandwidth to operate efficiently, an upgrade to faster fiber optic cables and switch hardware will be much easier than replacing an entire packet switched network.

Examples of Use

As an example, consider a plastics manufacturing facility. Such a factory could reap all of the benefits of a single ATM network. Fiber optic cables are placed in protective conduit and run throughout the building. This includes the manufacturing areas (that hold the actual equipment) and all of the administrative offices that run the business. Within the manufacturing areas, the computers that control and monitor the complex chemical processes can do so over the network in real time. The communication between sensors in the equipment and the computers that control them is nearly instantaneous. This allows for real-time modeling of the processes and quick reaction capabilities when a production error occurs. For example, consider the machine that actually mixes the chemicals to produce the raw plastic material. Assume that it must constantly and reliably monitor the chemical reactions in very fine-grained time intervals (on the order of every millisecond). No traditional LAN can supply performance guarantees needed for such a process.

Previously, a completely separate and dedicated high speed link was needed. However, with ATM, this time critical data can be placed on the same set of cables with all other communications. In addition, terminals can be placed throughout the manufacturing areas that allow visualization of the processes going on inside of the equipment itself. Such visualization would allow a worker to verify that the equipment is functioning correctly and help to figure out the cause of problems when it is not. In the event of a malfunction, the worker can call the manager and forward the visualization imagery to him or her for evaluation. The terminal also provides real-time voice and video communications as they discuss the problem. Note that all of these services are provided by only one network; all of the traffic for each service is placed on the same set of fiber optic cables. Furthermore, the single network provides all communication services; there is no need to run computer network cables, equipment control cables, and phone lines. This also includes both internal communication (factory wide and adjoining offices) and external communication (to public branches and other factories). Having a single integrated networking structure is very cost efficient, not only in equipment costs but also in maintenance costs. Also note that the kinds of personal interaction and communication are more robust and varied; it can be accomplished via traditional voice phones or by real-time voice and video transmissions over arbitrarily long distances. All of these features provide an overwhelming case for bringing ATM networks into the factory of the future.

References

A. Alles 1993. Tutorial: ATM in private networking, *Hughes Lan Systems,* Interop.

Kleinrock, L., 1992. The latency/bandwidth tradeoff in gigabit networks, *IEEE Communications Magazine,* April.

Traw, C. and Smith, J. 1993. Hardware/software organization of a high-performance ATM host interface, *IEEE J. Selected Areas in Communications,* II(2).

19

Manufacturing Automation Protocol (MAP)

Juan R. Pimentel
GMI Engineering and Management Institute

19.1 History 417
19.2 Purpose 417
19.3 Description 418
MAP Architecture • MAP 3.0 Architecture • MAP/EPA Architecture • Mini-MAP Architecture • Network Architecture
19.4 Standards Used 420
IEEE 802.4 Physical Layer • MAP Physical Layer • MAP Data Link Layer • Logical Link Control (LLC) Sublayer • Medium Access Control (MAC) Sublayer • MAP Network Layer • MAP Transport Layer • MAP Session Layer • MAP Presentation Layer • MAP Application Layer
19.5 Example of Use 426

19.1 History

The Manufacturing Automation Protocol had its origin at the General Motors (GM) Corporation in the late 1970s and early 1980s. MAP moved significantly towards becoming an important set of protocols when its version 3.0 was demonstrated at the Enterprise Networking Event at Baltimore, Maryland, in June 1988. These data communication specifications, based on international standards, have been an important element for computer and intelligent device based automation. Standards have impacted many areas over the years from automobiles to electricity to telephony. But what made the MAP specification unique was the influence of users to establish the direction for development of standards based products that they would use.

During 1980, General Motors, among others, recognized the significant need for computer communication standards and decided to do something about it. A small group of engineers from seven GM divisions formed the MAP Task Force to pursue the ISO in the development of the OSI reference model. Shortly after, during 1982, a small GM staff was formed to act as the nucleus for a worldwide effort which was to emerge. The Task Force quickly recognized the potential of users to help make standardized computer communication occur. During these initial stages, the MAP Task Force also worked closely with the IEEE Project 802 in the development of the Token Bus Protocol (IEEE 802.4). Later on, all the IEEE standards developed by the so called Project 802 were adopted by the ISO as international standards.

Soon, in early 1984, the MAP Users Group (with the Society of Manufacturing Engineers as Secretariat) was formed. Then the first successful demonstration of MAP occurred at the 1984 National Computer Conference (with the help of the National Bureau of Standards, now NIST, and the Industrial Technology Institute of Ann Arbor, Michigan). Seven vendors participated at the GM sponsored booth to demonstrate some of the standards implemented at the time. At Autofact'85, 21 vendors helped demonstrate MAP version 2.1. Shortly after, many companies began implementing MAP products. In 1986, the Technical and Office Protocol (TOP) Users Group was formed under the sponsorship of Boeing, and the effort became known as MAP/TOP. TOP is a companion specification to MAP intended for technical (e.g., CAD design) and office applications. At about the same time, MAP/TOP became an international Users Group. Later on, a World Federation of MAP/TOP users groups was formed representing many users in several countries. Other large organizations such as the Corporation for Open Systems and the European ESPRIT-SPAG-CCT joined the movement. All of these efforts have helped many countries toward the implementation of MAP/TOP version 3.0, the first practical OSI network.

The initial MAP document was issued in October of 1982, with major additions incorporated in 1984. MAP 2.0 was issued in February 1985, MAP 2.1 in March 1985, MAP 2.2 in August 1986, and MAP 3.0 in April 1987.

19.2 Purpose

Recently, there has been much interest in interconnecting computers and intelligent devices of various kinds. By intelligent devices we mean any device with some form of intelligence capable of sending or receiving information to or from other computers or intelligent devices. Examples of intelligent devices include personal computers, laser printers, file servers, computer

0-8493-8343-9/97/$0.00+$.50

workstations, and digital telephones. The emphasis of MAP, however, is to interconnect devices used as components of modern manufacturing systems such as vision systems, robots, graphic stations, programmable logic controllers, numerically controlled machines, and others.

The reasons for the interest in multidevice interconnection are twofold: first, all intelligent devices of interest have some kind of embedded processor or computer whose prices are plummeting at unbelievable rates; second, applications such as manufacturing can benefit immensely from an interconnected system of its main elements. The latter reason forms the foundation of modern manufacturing concepts such as computer integrated manufacturing (CIM).

There are several types of networks in existence. Probably the two best known are telecommunication networks and office automation networks. MAP however, belongs to the class of manufacturing networks. In the beginning, the interconnection of computers and intelligent devices was done in an adhoc manner. Vendors had their own methods for interconnecting their devices. The problems that this situation posed were numerous with, perhaps, the major one being the incompatibility of multivendor equipment.

There are two key elements to the solution of this problem. The first one is to view a network as a collection of several layers, and the second one is to have a few well recognized standards applicable to each layer. MAP solved this problem by adopting the OSI reference model, a collection of seven well known layers, and to use international standards at each layer.

Perhaps the two main features of the MAP specification is its use in international standards and interoperability of devices. All MAP protocols are in fact ISO international standards. In some cases, the ISO developed the standard in question, and in others, it adopted standards developed by another organization.

Interoperability means that devices implementing the same protocols as specified by the corresponding standards are able to communicate with one another regardless of the hardware used or the implementation details. Interoperability allows network hardware from different manufacturers to communicate with one another.

19.3 Description

Map Architecture

In this section, we provide an overview of the different architectures that have been developed for MAP. We only describe the latest architectures since earlier ones have been mostly used as stepping stones to produce more stable architectures. Furthermore, the architectures described in this section have been used for actual implementations. The architectures described are MAP 3.0, MAP-EPA, and mini-MAP. Earlier architectures include MAP 2.0, MAP 2.1, and MAP 2.2.

The term architecture used in the context of this section is used to indicate station architecture. A station architecture specifies the number, name, configuration, functions, protocols, and protocol options of all layers which taken together allow network stations to communicate with one another. The layer architecture should also specify how the functions of each layer relate to one another, and also the relationship to the OSI reference model. Other terms which are sometimes used to denote station architecture are node architecture, layer profile, or protocol stack.

MAP 3.0 Architecture

Whereas MAP 2.2 and MAP 2.1 were very similar, MAP 3.0 provides significant enhancements and it is not upward compatible with MAP 2.2. In addition to enhancements in virtually all layers, perhaps the major enhancements over MAP 2.2 are the inclusion of the Manufacturing Message Specification (MMS), the use of use of the Association Control Service Element (ACSE), the incorporation of the ISO presentation layer, and the specification of an application layer interface with application processes. With the exception of directory services, network management, and the application layer interface, all the standards are treated in detail in Pimentel, 1990. The MAP 3.0 architecture is depicted in Figure 19.1.

In addition to the network management, directory service, and application layer interface specifications, the MAP 3.0 specification also contains a detail specification for broadband transmission systems including topology, coaxial cables, system design, performance, installation, test, and maintenance procedures.

MAP/EPA Architecture

One approach to have a simpler architecture (for performance reasons) and still maintain ISO compatibility is to use a dual architecture. As depicted in Figure 19.2, the MAP/EPA architecture follows this approach. In addition to the dual architecture, perhaps the most significant feature of MAP/EPA is the use of an LLC class 3 sublayer. Class 3 operation allows a responder to issue an acknowledgment immediately, without the need to wait for the token (immediate acknowledgment). Crucial to the operation of a MAP/EPA station is a version of MMS called mini-MAP MMS and its direct interface to the data link layer which is known as LSAP bindings. Optionally, a user may use a mini-MAP object dictionary or a user defined application layer protocol (instead of mini-MAP MMS).

Normally, a MAP/EPA station communicates with similar stations using the three layer path (i.e., application, data link, and physical). When the station communicates with an ISO node, the seven layer path is used. As noted, the dual architecture allows the MAP/EPA station to behave as an ISO station though only when needed.

The carrierband modulator accepts a data rate of 5 Mbps and delivers a signal into a 75-ohm coaxial cable.

Mini-Map Architecture

If it is not important to be ISO compatible but still have a simple architecture, then a mini-MAP architecture could be used. As

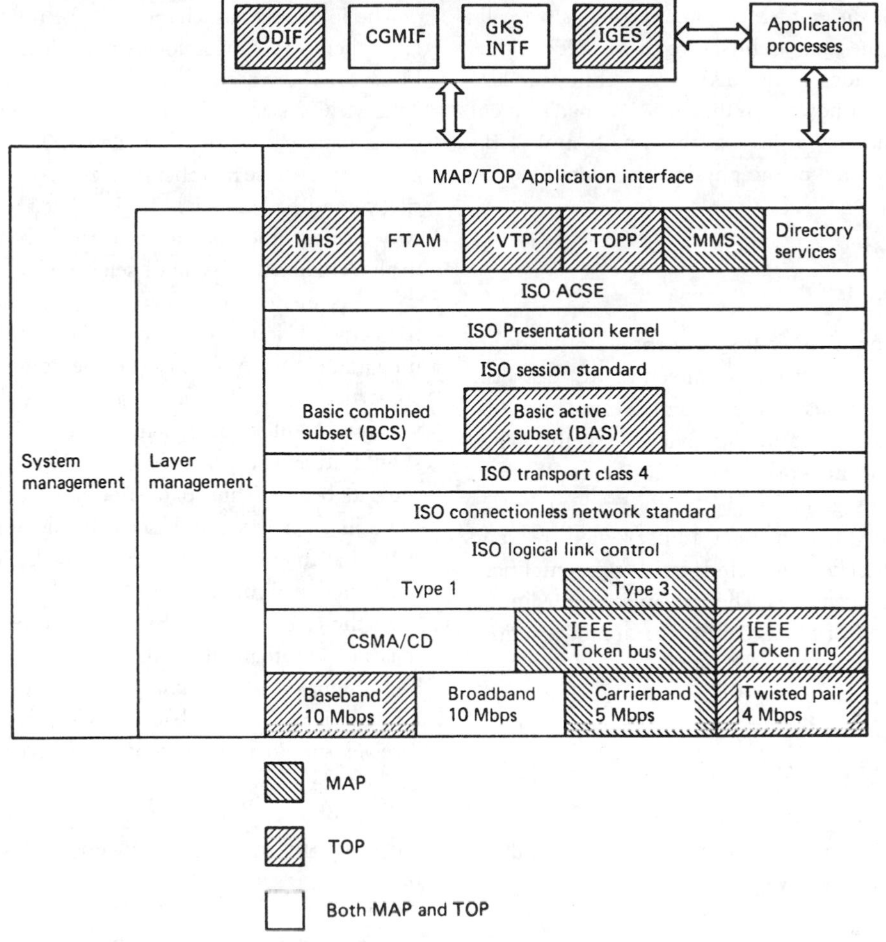

Figure 19.1 MAP/TOP 3.0 station architecture.

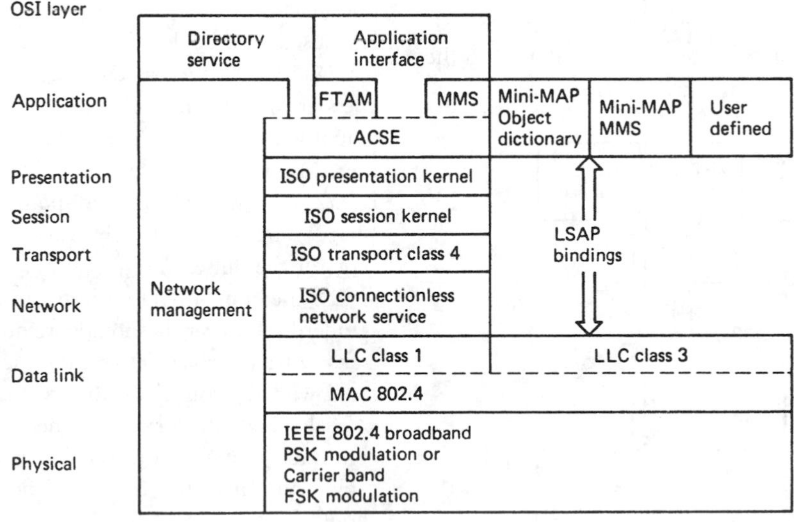

Figure 19.2 MAP/EPA station architecture.

depicted in Figure 19.3, the mini-MAP architecture is basically one of the portions of the dual architecture of MAP/EPA, thus the same observations made on the MAP/EPA architecture also applies here. One difference however is that network management in a mini-MAP environment (at least at the time being) works in an environment consisting of the physical and data link layers only.

Network Architecture

In addition to the station architecture, one should also specify a network architecture. By network architecture we mean the topological configuration, transmission system devices, transmission media, and station configuration which taken together, allow end user applications to interwork even when the applications are on different physical segments.

Possible topologies for MAP includes a bus, rooted tree, and a star. Transmission system devices include head ends, amplifiers, taps, bridges, routers, and gateways. The transmission medium is broadband and baseband (for carrierband) coaxial cable. Stations can be configured using the MAP 2.2, MAP 3.0, MAP/EPA, or mini-MAP architectures.

19.4 Standards Used

Because of its unique role in MAP, we describe the IEEE 802.4 standard for the physical layer in some detail.

IEEE 802.4 Physical Layer

The IEEE 802.4 standard provides three options for encoding and modulation based on AM/PSK, phase coherent FSK, and phase continuous FSK modulation schemes. Only the AM/PSK option is described in detail here. The reader is referred to the IEEE 802.4 standard for details on the FSK schemes.

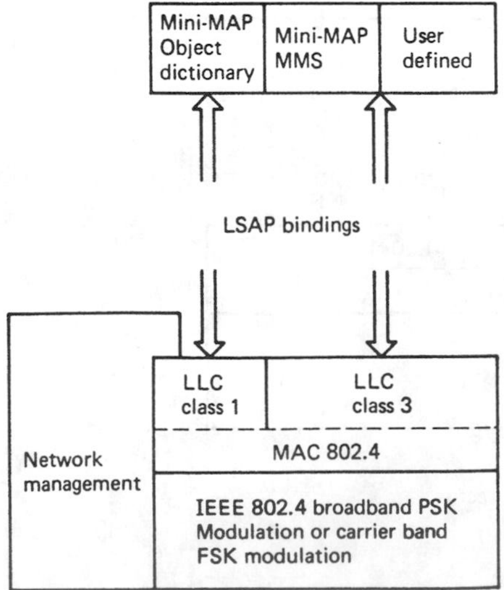

Figure 19.3 Mini-MAP station architecture.

The following is exchanged at the interface with the data link layer: clock (Phy_clock and Tx_clock) and symbols (Tx_MAC_symbol and Rx_MAC_symbol). The Tx_MAC_symbol is a set whose members are: zero, one, non_data, pad_idle, and silence. The Rx_MAC_symbol is another set whose members are: zero, one, non_data, pad_idle, silence and bad_signal. Notice that the only difference in the two sets of symbols is that the element bad_signal is not present in the Tx_MAC_symbol set.

The elements zero and one correspond to data bits 0 and 1 respectively. Non_zero symbols are used to indicate the beginning and ending of frames (i.e., delimiters). Pad_idle is used to construct a special sequence sent before the data (i.e., preamble) which is useful for bit synchronization. Data (either zero or one) should always be preceded by delimiters which indicate where the data begins. Thus, data should not be preceded by pad_idle symbols. A new symbol is used called silence, which indicates that what follows after preamble (pad_idle symbols) is neither data nor a delimiter.

At the receiver side, when the decoder decodes a symbol and it is not any that the transmitter sent, then it is considered a bad_signal. A bad_signal is also useful for indicating hardware malfunctions. When a bad_signal is received by a receiver, the receiver should resynchronize the decoding process as rapidly as possible.

Symbols received at the data link-physical layer interface are encoded using a duobinary encoder. Data symbols (i.e., zero and one) receive a different treatment as compared to all other symbols in that they are applied to a scrambler resulting in a scrambled zero or one MAC_symbol. The duobinary encoder allows the representation of incoming MAC_symbols as another set of PHY_symbols whose elements are denoted as {0}, {2}, and {4}, thus it is a ternary encoder. The main advantage of the duobinary encoder is that in cooperation with the specific modulator used for broadband modulation (AM/PSK), it allows better bandwidth efficiency as compared with other combinations of encoder-modulator. The AM/PSK modulator is a combined amplitude modulation/phase shift keying modulation in which the AM component uses three different amplitude levels of 0, max/2, and max which correspond to the encoder outputs of {0}, {2}, and {4} respectively, where max is the maximum amplitude for the pulse.

The medium interface configuration is not specified in the standard. However, it should provide ac coupling between the transmitter (or receiver) and the medium. The station is interfaced with the communication medium by means of a flexible coaxial cable with a small diameter known as drop cable. The communication medium is a larger diameter, semi-rigid coaxial cable with characteristic impedance $Zo = 75$, thus all loads should be terminated in 75 ohms to avoid reflections on the cable. The cable can be used to carry both ac power and rf signals because the frequency components of these two signals do not overlap. Bidirectional and unidirectional couplers constitute the medium attachment.

At the receiver end, the bandpass filter rejects signals from channels other than the one used by the network. Since the AM/

PSK modulation scheme does not carry any information in the PSK component, it is not necessary to recover phase information. Thus, only AM demodulation is needed. The AM demodulator specified is of the full-wave rectifier type. The decoder performs functions which are the inverse of that of the duobinary encoder (with the exception of the bad_signal symbol).

The IEEE 802.4 physical layer provides the following four primitives:

- PHY_MODE.request(mode).
- PHY_DATA.request(Tx_MAC_symbol).
- PHY_DATA.indication(Rx_MAC_symbol).
- PHY_NOTIFY.request.

Any station in the network can function as a repeater station. Some means are required to indicate whether a station is originating the messages or is simply relaying messages from other segments. Such means are provided by the PHY_MODE.request(mode) primitive with mode = {originating, repeating}.

The PHY_DATA.request primitive is passed from the data link layer to the physical layer to request the transmission of a symbol on the communication medium. The physical layer, upon receipt of a PHY_DATA.request will encode and transmit the symbol using the appropriate signaling technique (i.e FSK or AM/PSK modulation). At the destination station, the recently arrived symbol is received and decoded by the corresponding physical layer and presented to its corresponding data link layer by means of the PHY_DATA.indication(Rx_MAC_symbol) primitive.

When the data link layer of a station is receiving data from another station and detects the end of frame, it is convenient to notify the corresponding physical layer. The notification is done using the primitive PHY_NOTIFY.request. The notify primitive is useful because it allows the physical layer to be prepared for detecting silence, preamble, or entering a high speed acquisition mode for synchronization purposes.

MAP Physical Layer

The physical layer of the MAP architecture consists of a backbone network with gateways, routers, and bridges connecting to other MAP and possibly some non-MAP subnetworks. Although an important issue, the network topology is left to the individual plant requirements. The two best known topologies applicable to MAP networks are the star and the tree (Pimentel, 1990). The physical medium recommended as the network backbone in MAP 2.2 is a coaxial cable capable of supporting many channels (i.e., broadband). The modulation technique is AM-PSK with duobinary encoding at a data rate of 10 Mbps. Channels on the broadband cable are chosen according to the mid-split channel allocation scheme although a high-split scheme is also possible. To improve reliability, redundant media is recommended.

For applications not requiring broadband communications, a single channel network based on a phase coherent, Frequency Shift Keying (FSK) modulation scheme at a data rate of 5 Mbps is recommended. The corresponding network is referred to as a carrierband network. Installations using carrierband networks require that disturbance noise be carefully controlled. Specifically, it is recommended that the highest root mean square (RMS) noise level be −10 dBmV. Since carrierband subnetworks and broadband backbones use different modulation, encoding and data rates, their interconnection require appropriate routers, or bridges to perform the required speed, modulation, and encoding conversions.

Broadband technology offer the following advantages:

- Broadband allows multiple networks to coexist simultaneously on the same medium with each network occupying a different channel. Frequency division multiplexing allows networks to use any desired channel. The channels can be dedicated or switched.
- In addition to supporting data transmission, broadband also supports voice and video transmission using the same or separate channels. Some applications benefiting from this option include security surveillance, closed circuit television (CCTV), teleconferencing, and education.
- Broadband can also be used to support other network types (e.g., CSMA/CD).

Carrierband technology has the following advantages:

- Since a single channel is used, the network interfaces are simpler and more inexpensive when compared to broadband interfaces. For example, channel cross talk noise is not a problem because there is no interference with other channels. In addition, modulation schemes can be simpler since there is no attempt to have high bandwidth efficiency.
- Unlike broadband, Carrierband technology is passive in nature because there is no need for a Head End and amplifiers. Thus a higher degree of reliability is anticipated when compared with broadband technology.
- Because there are no Head End or amplifiers, signals travel directly from source station to destination station thus incurring in lower propagation delays when compared with the broadband case. The savings in delays are attributed to Head End and amplifier delays. Although the savings is in the order of tens of microseconds, it can make a difference for some time critical applications.
- Because of its passive nature, carrierband is limited to shorter distances and fewer stations than broadband. The limitations are 32 stations per segment, and 1 Km between most distant communicating stations.

MAP Data Link Layer

Since the MAP network architecture is based on the IEEE 802 protocols, its data link layer is divided into two sublayers: logical link control (LLC) and medium access control (MAC). The LLC sublayer corresponds to the IEEE 802.2 standard under the class I or class III operation whereas the MAC sublayer corresponds to the IEEE 802.4 standard. The IEEE 802.2, 802.3, 802.4, and 802.5 standards are also ISO standards with identifiers 8802/2, 8802/3, 8802/4, and 8802/5 respectively.

Recall that the main function of the MAC sublayer is to manage the access to a shared channel and that the main functions of the LLC sublayer is to perform error control, addressing, and flow control in order to ensure reliable data transmission between nodes. For the MAP 2.2 network architecture, the MAC sublayer is the token bus protocol specified in the ISO 8802/4 standard whereas the LLC sublayer is a subset of the ISO 8802/2 protocol known as class I (connectionless oriented) services. Alternatively, for applications requiring improved performance class (acknowledged, connectionless oriented) is recommended. The MAP 3.0 network architecture specifies ISO 8802/2 class III (acknowledged connectionless) services.

Logical Link Control (LLC) Sublayer

Class I operation of the IEEE 802.2 protocol operation specifies a connectionless type of service. Initially, the MAP network was designed with a connectionless type of service for layers 1, 2, and 3 and with a connection oriented type of service for layers 4, 5, 6, and 7. As pointed out earlier, the IEEE 802.2 with class I operation is a very simple protocol. One reason for its simplicity is that it could be readily incorporated into VLSI designs. The disadvantages of a simple data link layer protocol are: delivery of data cannot be guaranteed, data could arrive at the destination in a sequence different to the one it was sent (because of transmission errors and subsequent retransmissions), no flow control is exercised, and only a limited number of error recovery procedures is available. For the MAP network however, the physical layer alone provides low error rates; thus a complex data link layer is not required. Accordingly, MAP uses the simplest data link layer protocols offered by the IEEE 802.X standards, namely IEEE 802.2 class I and class III.

As noted earlier, the LLC sublayer of the MAP network architecture is the ISO 8802/2 standard protocol using class I services to provide connectionless communications, or when required by certain applications, class III. Since the ISO 8802/2 class I protocol does not provide message sequencing, acknowledgments, flow control, or error recovery, other layers in the MAP network architecture must provide these important functions. As noted earlier, the IEEE 802.2 standard does provide other options known as class II and class IV operation based on a connection oriented service known as data link connection. However, classes II and IV are not part of MAP.

The IEEE 802.2 class I operation basically performs data exchange without establishing a data link connection. The IEEE 802.2 class III operation provides unacknowledged connectionless and/or acknowledged connectionless services. In the acknowledged connectionless type (i.e., immediate response mechanism) data units are exchanged without the need for the establishment of a data link connection. The LLC protocol is responsible for providing acknowledgments for individual protocol data units (PDUs) regardless of whether they carry user information.

The following reasons for the selection of the ISO 8802/2 LLC protocol are cited by MAP designers (MAP, 1986):

- It supports data transfers at very high rates.
- It can be used on multiple media (twisted pair, coaxial cable, optical fibers, etc).
- It provides connectionless services.
- It provides an acknowledged single frame service.
- If widely accepted, VLSI chips will emerge and substantially reduce the cost of network interfaces.

Medium Access Control (MAC) Sublayer

The MAC sublayer of the MAP architecture specifies the token bus protocol. At the time the token bus protocol was selected as the medium access method for the MAP architecture, MAP designers offered the following reasons for the selection: (MAP, 1986)

- Token bus was the only protocol presently supported on broadband by the IEEE 802 project.
- Many protocols used by programmable device vendors were already somewhat token bus based.
- The token bus protocol supported a message priority scheme.
- In the token bus protocol, high priority messages are delivered within a specified time limit. The time limit can be easily determined knowing details of a network configuration (e.g., number of stations, etc.).

In the MAP architecture, the MAC frame address field length is 48 bits to allow for a sufficient number of stations on the network. Although a particular MAP network segment is not expected to have as many stations as allowed by the 48 bit field, it is expected that the 48 bits is sufficient to identify stations not only in a particular segment but also between segments, local area networks, and wide area networks (i.e., a global address). For this purpose, the 48 bit address field can be segmented into a network address and a station address.

MAP Network Layer

The main function of the network layer is to provide message routing between end nodes, either in the same subnetwork or any other subnetwork regardless of their location. Additional functions include addressing, congestion control, and message segmentation. The MAP network layer performs the routing function by converting global address information into routing information, maintaining message routing tables and/or algorithms, and switching each incoming message to its proper outgoing path. Associated with the above operations are special procedures and/or databases for translating global addresses into the required routing information. The databases are part of the MAP directory services.

End-to-end routing is performed in the MAP network by means of a global routing function as specified in the ISO DIS 8473 Connectionless Mode Network Service (CLNS) also referred to as the CLNS protocol. The CLNS protocol is an internetworking protocol for allowing packets to traverse multiple LANs regardless of the routing algorithm used in each LAN. Specifically, the protocol provides exchange of data and control information (CI) via connectionless transmission, IPDU (Internet Protocol Data Unit) information encoding, procedures for interpreting the information, and a formal specification to be met to be in compliance with the standard.

The major functions of the CLNS protocol as used in the MAP architecture are PDU segmentation/reassembly, routing, and congestion control. Segmentation is performed when the PDU size is greater than a predefined maximum size. At the receiver, the initial PDU is reconstructed from the derived PDUs. Routing is performed using a hierarchical routing algorithm with tables and routing control centers (RCCs). The routing algorithm has provisions for complete source routing and partial source routing. Congestion control is performed by simply throwing packets away (i.e., packet discarding). The protocol machine discards packets when it encounters a violation of the protocol, a PDU with incorrect checksum, not enough buffers, or a PDU header which cannot be analyzed. The PDU format at the network layer is shown in Figure 19.4.

Depending on whether the communication nodes belong to a single subnetwork (i.e., segment) two subsets of the CLNS protocol are defined: Inactive Network Layer Protocol, and Full Conformance Protocol. The Inactive Network Layer Protocol is used when the following three conditions are met.

1. The source and destination stations are on the same subnetwork.
2. The size of transport protocol data unit (TPDU) is less than the size of maximum link protocol data units (LPDU) at both ends of the communication.
3. The destination network service access point (NSAP) is accessible via an inactive network layer header.

When any of the above conditions are not true, the full Conformance Network Protocol is used.

MAP supports the following type 1 functions of the CLNS protocol.

- PDU composition and decomposition.
- Header format analysis.
- PDU lifetime control.
- PDU routing and forwarding.
- PDU segmenting, reassembly, and discard.

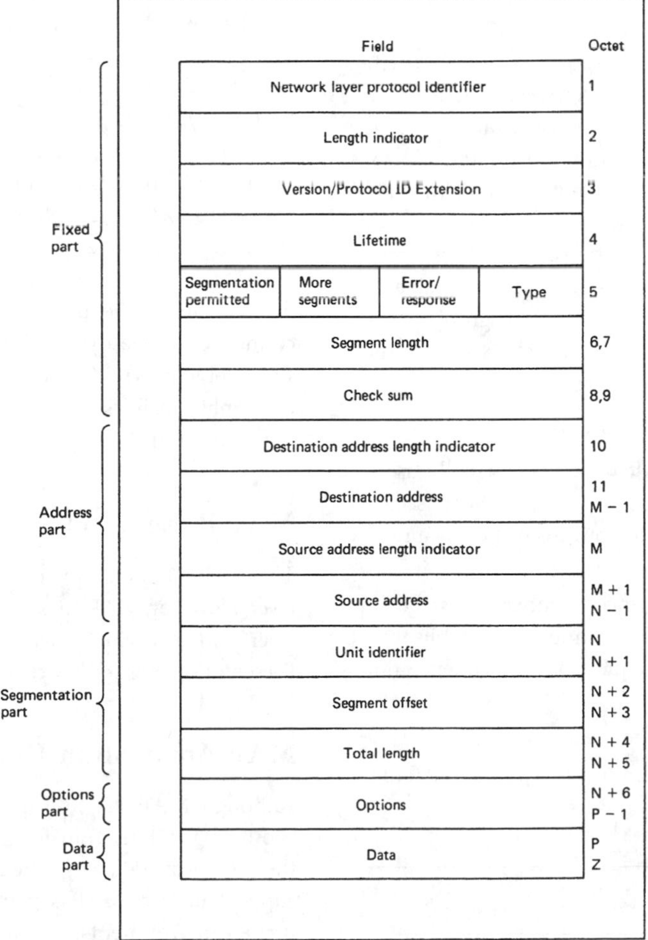

Figure 19.4 ISO Network Layer Protocol data unit.

- Error reporting.
- PDU header error detection.

The implementation of the CLNS protocol type 2 functions is not required. Type 2 functions include security and complete source routing. If a network implementation does not support a type 2 function which is requested by an incoming packet, the packet is discarded and an error report PDU is sent back to the source station.

Although CLNS protocol type 3 functions are also optional, unlike type 2 functions, when type 3 functions are requested by an incoming packet the packet is not discarded. Rather, the function parameters are recognized as valid but ignored. Thus, if type 3 functions are not implemented, a type 3 request should be processed as if the request were a type 1 and the type 3 parameters ignored. Type 3 functions include padding, partial source routing, priority, record route, and quality of service maintenance.

MAP Transport Layer

The MAP architecture uses a subset of the ISO 8073 Class 4 transport specification. Since the MAP data link layer does not provide flow and error control functions it is up to the MAP transport layer to provide these functions. This is one of the main reasons why the MAP transport layer specifies Class 4, the largest and most complex class of transport services as specified by ISO standard. As explained in the previous section, in addition to providing flow and error control functions, the ISO Class 4 transport provides the ability to multiplex user transmissions, recover from network layer errors and failures, and the ability to support datagrams.

Since the ISO standard classifies all transport services into two groups, connection management and data transfer, Table 19.1 shows the MAP service requirements from each group.

In addition, the MAP subset of the class 4 ISO transport protocol has the following characteristics.

- Ability to negotiate a Class 2 service from Class 4.
- The following parameter list which is normally used as acceptance criteria are ignored if received: security, acknowledgment time, throughput, transit delay, priority, and residual error rate.
- Both 7-bit and 31-bit sequence numbers must be supported and negotiated during connection establishment. All implementations shall request 31-bit sequence numbers in the CR TPDU.
- Expedited data support is required.
- Checksums are required in connect requests (CR TPDU) per ISO rules but are optional on data transfer (DT PDU). Using checksums for data transfer is application dependent and negotiated during connection establishment.
- With the exception of a connection request service, a parameter which is not defined in any other transport service or not implemented at the receiver, is treated as a protocol error. In the connection request case the parameter is ignored.
- A specific implementation is required to support the concurrent use of multiple transport connections via one or more TSAPs.
- Whereas all transport machines must receive concatenated TPDUs, sending concatenated TPDUs is optional.
- Although the transmission of flow control information (i.e., AK TPDUs) by the transmitter is optional, the receiver must recognize flow control information and act in accordance with the ISO specification.

Table 19.1

CONNECTION MANAGEMENT	DATA TRANSFER
T_CONNECT	T_DATA
T_DISCONNECT	T_EXPEDITED_DATA

MAP Session Layer

The MAP session layer is a subset of the ISO IS 8327 Session protocol standard. Specifically, the MAP architecture requires that its session layer support the minimum subset of the ISO standard which is composed of the Kernel functional unit and the Duplex functional unit where the available options are not required. The following are additional characteristics of the MAP Session layer.

In order to support the communication of messages with priority, a MAP implementation of the Session layer must utilize the Transport expedited service. Sending urgent data using T_DATA is considered a protocol error. Transport connections are not reused (i.e., multiplexed). To avoid incompatibilities with default values, the user requirements parameter in a Session connection request is required. Likewise, to avoid potential collision problems with the Session release service, it is recommended that only the initiating application may request the release of the connection.

MAP Presentation Layer

The MAP 2.1 and 2.2 specifications do not require a presentation layer. However, MAP 3.0 specifies a subset of the ISO presentation layer standard which involves basically the functionality of the Presentation Kernel Functional Unit.

MAP Application Layer

Although MAP 2.2 and TOP 1.0 both use the ISO standard protocol for File Transfer, Access and Management (FTAM), their application layers differ significantly in the other protocols supported. Since MAP is primarily intended to support manufacturing environments, its application layer must have all the required functionality required by manufacturing applications. Because the ISO did not have an appropriate application layer protocol for manufacturing at the time the MAP architecture was under development, the MAP task force at General Motors developed one called the manufacturing message format standard (MMFS). The MAP 2.1 and 2.2 architectures specify MMFS. Later on, MMFS evolved into a more sophisticated protocol called Manufacturing Message Service (MMS) which is part of MAP 3.0.

The Manufacturing Message Service (MMS)

MMS is a specific application layer protocol which enables communication among intelligent devices in manufacturing applications. The services provided by MMS are classified into the following functional groups:

Environment and general management.
Virtual Manufacturing Device (VMD) support.
Domain management.
Program invocation management.
Variable access.
Semaphore management.
Operator communication.
Event management.
Journal management.

The specific services provided under each functional group are listed below.

Environment and general management

Initiate
Conclude
Cancel

Virtual Manufacturing Device (VMD) support

Status
Get Name List
Identify
Unsolicited Status
Get Capability List
Rename

Domain management

Initiate Download Sequence
Download Segment
Terminate Download Sequence
Initiate Upload Sequence
Upload Segment
Terminate Upload Sequence
Request Domain Download
Request Domain Upload
Load Domain Content
Store Domain Content
Delete Domain
Get Domain Attributes

Program invocation management

Create Program Invocation
Delete Program Invocation
Start
Stop
Resume
Reset
Kill
Get Program Invocation Attributes
Obtain File

Variable access

Read
Write
Information Report
Get Variable Access Attributes
Define Named Variable
Define Scattered Access
Get Scattered Access Attributes
Delete Variable Access
Define Named Variable List
Get Named Variable List Attributes
Delete Named Variable List
Define Named Type
Get Named Type Attributes
Delete Named Type

Semaphore management

Take Control
Relinquish Control
Define Semaphore
Delete Semaphore
Report Semaphore Status
Attach to Semaphore
Report Pool Semaphore Status
Report Semaphore Entry Status

Operator communication

Input
Output

Event management

Define Event Condition
Delete Event Condition
Get Event Condition Attributes
Report Event Condition Status
Alter Event Condition Monitoring
Trigger Event
Define Event Action
Delete Event Action
Get Event Action Attributes
Report Event Action Status
Define Event Enrollment
Delete Event Enrollment
Alter Event Enrollment
Report Event Enrollment Status
Get Event Enrollment Attributes
Acknowledge Event Notification
Event Notification
Get Alarm Summary

Attach to Event Condition
Get Alarm Enrollment Summary

Journal management

Read Journal
Write Journal
Initialize Journal
Create Journal
Delete Journal
Report Journal Status

FTAM

The incorporation of FTAM services into the MAP 2.2 architecture is done in two phases. Phase 1 includes the file transfer kernel, read, write, and limited file management functional units and the kernel group of the virtual filestore attributes. In addition, phase 1 supports the transfer of both binary and ASCII text file formats and the remote creation and deletion of files. Not supported by phase 1 is the concatenation function and the F_BEGIN_GROUP and the F_END_GROUP services. Phase 2 supports file access capabilities and an enriched set of file formats.

The following are additional details of FTAM and its implementation for MAP 2.2. The access structure type is of the unstructured type only (i.e., no flat and hierarchical types are used). The FTAM implementation is mapped directly into Session services without incorporating the use of the Common Application Service Element (CASE) where the F_INITIALIZE service corresponds to the S_CONNECT service and the F_TERMINATE service corresponds to the S_RELEASE service. No concurrency on file operations are provided since the file attributes governing them are not supported. Thus, it is up to each implementation to support the degree of concurrency desired for local files. However, in order to avoid potential problems, two rules are recommended: first, a file may be involved in several transfers simultaneously only if accessed in read mode; and second, for a file involved in a transfer in a write mode, any request for access (i.e., a read or write) to the file should be rejected regardless of whether the request is local or remote.

The application layer of MAP also includes directory services for providing a name to address mapping.

19.5 Example of Use

General Motors has installed many networks based on the MAP specification. One of the latest installations is in the Pontiac assembly plant in Michigan. The MAP network is used in conjunction with gateways for supporting applications involving facilities monitoring systems, and production control.

References

Pimentel, J. R. 1990. *Communication Networks for Manufacturing*, Prentice Hall, Englewood Cliffs, NJ.

20

Essential Communications Protocols

Bert J. Dempsey
University of North Carolina

Debapriya Sarkar
University of Virginia

20.1 Datalink Protocols ... 427
High-Level Data Link Control • Logical Link Control
20.2 Network Protocols ... 429
Purpose • Internet Protocol (IP) • Connectionless Network Protocol (CLNP) • Summary
20.3 Transport Layer Protocols ... 434
User Datagram Protocol • Transmission Control Protocol • ISO Transport Protocol • Xpress Transport Protocol

20.1 Datalink Protocols

Bert J. Dempsey

Datalink layer protocols address the need to manage and control signaling over a physical medium to create data communication services. In particular, two directly connected stations capable of transmitting and receiving signals require the following types of functions for effective data communication.

Framing: Data are grouped into blocks called "frames" during data communication. The link layer manages framing on the physical link.

Error control: Bit errors introduced during transmission over the link are detected at the link layer. Reliable datalink protocols perform retransmissions to recover frames lost or corrupted in transmission.

Flow control: A reliable link layer protocol regulates the sending of frames such that the receiving station does not suffer buffer overflows.

Miscellaneous control: Various forms of control information concerning the state of communicating stations can be exchanged using the link layer protocol.

Addressing: On a multipoint line, the identity of the communicating stations must be communicated.

Link management: Messages to establish and terminate data exchanges are necessary for connection oriented services at the datalink layer.

The specific functions required of the datalink protocol vary with the characteristics of the network. The most important characteristic is the physical topology of the network. The network may be either a direct physical link between two stations (point-to-point) or a physical link with multiple attached stations (multipoint). Multipoint communication has two subcases: a multidrop link has a master station with a set of secondary stations whereas a multiaccess link has a set of stations that share a single physical link in a distributed fashion. (The IEEE 802 broadcast local area networks are multiaccess links in this terminology.) Other important characteristics include the signaling speed and directional capability (unidirectional (half-duplex) or bidirectional (full-duplex)) of the physical medium and the services required by the datalink layer users.

In this section, we will discuss key aspects of two datalink layer protocols, the high-level data link control (HDLC) protocol developed by the International Organization for Standardization (ISO 3309) and the logical link control (LLC) found in all IEEE 802 LANs (IEEE 802.2). HDLC is a complex protocol since it is designed to accommodate a wide range of physical networks and to provide different services to the datalink user. We will discuss it here only to give a flavor for its functionality and protocol mechanisms. LLC is, by contrast, a relatively simple protocol (it was modeled after a subset of HDLC) designed specifically for multiaccess link networks. Actually, LLC is only one-half of the datalink layer in IEEE 802 LANs. The other half is the medium access control protocol. We have examined these protocols in detail in Section 20.3, e.g., the CSMA/CD protocol defined for IEEE 802.3 (Ethernet), the token ring protocol defined for IEEE 802.5, etc.

High-Level Data Link Control

The ISO standard high-level data link control (HDLC) protocol is very similar to a number of other widely used datalink standards. It is virtually identical to the advanced data communication control procedures (ADCCP) developed by the American National Standard Institute (ANSI X3.66). Two other datalink standards are subsets of HDLC: the link access procedure, balanced (LAP-B) in the CCTTT X.25 packet-switched network standard and the synchronous data link control (SDLC) used by IBM. Finally, the IEEE 802 logical link control (LLC) used in

0-8493-8343-9/97/$0.00+$.50

IEEE 802 broadcast local area networks is similar in form and function to a subset of HDLC.

HDLC defines three types of stations, two link configurations, and three data transfer modes of operation (Stallings, 1991). The station types include primary, secondary, and combined stations. Primary stations control the operation of the link. Frames issued by primaries are called commands. Secondary stations operate under the control of a primary station. They issue frames known as responses. A primary station maintains a separate logical link to each secondary station. Combined stations have both the features of the primary and the secondary stations. The link configurations are unbalanced and balanced. Unbalanced mode is used in point-to-point as well as multipoint operation, and unbalanced mode supports one primary with one or more secondary stations using full-duplex or half-duplex transmission. Balanced mode is used only in point-to-point communication. It consists of two combined stations using full-duplex or half-duplex transmission.

The three data transfer modes of HDLC are normal response mode (NRM), asynchronous response mode (ARM), and asynchronous balanced mode (ABM). NRM and ARM are for unbalanced configurations. NRM represents a strict master-slave relationship between the primary and the secondaries. That is, a primary may initiate data transfer to a secondary, but a secondary can send data to a primary only in response to a poll from the primary. ARM, by contrast, allows the secondary to initiate data transfers to the primary. The ARM mode is rarely used. Finally, ABM is for balanced configurations in which either combined station may initiate data transfer without requesting permission from the other station. While NRM is often used in point-to-point links, ABM makes more efficient use of a full-duplex point-to-point link.

To give the flavor of the protocol mechanisms in HDLC, we discuss below the way in which the HDLC protocol provides each of the datalink layer functions listed above.

Framing

HDLC delimits frames on the physical medium by sending a special bit pattern, 01111110, at the beginning and end of each frame. Receiving stations must recognize this pattern as the frame delimiter. This technique leads to a problem, namely that the special pattern may appear in the user data carried in an HDLC frame. The HDLC protocol has special mechanisms for fixing this problem, known as bit stuffing.

Error Control

Each HDLC frame has a header field to carry a data integrity check, specifically a 16-bit cyclic redundancy check (CRC). The CRC is placed in the frame at the sender, and the receiver recalculates it to determine if any bits in the frame were corrupted during the transmission. Corrupted frames are discarded. An HDLC option allows for using a 32-bit CRC to improve reliability.

In its reliable modes, HDLC uses retransmission mechanisms. HDLC frames carry sequence numbers that allow a receiving station to communicate to the sender which frames have been correctly received. This mechanism forms the basis for acknowledgment-retransmission handshaking schemes, so-called automatic repeat request (ARQ) techniques, used to detect frame loss and perform retransmissions.

Flow Control

The HDLC sender and receiver use a sliding window protocol for flow control. That is, using sequence numbers carried in each frame, the HDLC receiver informs the sender of the current number of available frame buffers at the receiver. (More detail on sliding window protocols is given in the section in this chapter on transport layer protocols.)

Miscellaneous Control

Miscellaneous control messages in HDLC include, for example, a TEST frame that is echoed by the receiver in order for the sender to verify physical connectivity. Several control frames related to recovery from various protocol error conditions are defined.

Addressing

Each HDLC frame carries the address of the secondary station that transmitted or is to receive the frame. Addressing is unnecessary for point-to-point communication, though address fields are still present in the header for consistency.

Link Management

HDLC defines explicit commands and corresponding responses that allow, for example, a primary station to manage the establishment and termination of links with secondary stations.

Logical Link Control

In the IEEE 802 committee, the work on broadcast local area networks divided the datalink layer protocol into two parts: the medium access control protocol and the logical link control (LLC) protocol. In an 802 LAN, the physical topology of the network is a multiaccess link with distributed medium access control. The LLC protocol (IEEE 802.2) provides supplemental link layer functions over this structure. LLC is closely modeled on HDLC, with the format of the control section of the LLC header identical to that in HDLC.

The LLC standard specifies three types of service to LLC users: unacknowledged connectionless service, acknowledged connectionless service, and connection-oriented service. The unacknowledged connectionless service simply provides for sending and receiving frames. The acknowledged connectionless allows a user to send a frame and receive an acknowledgment without the overhead of setting up a (link layer) connection. This service is supported by defining two frames that are not found in the HDLC protocol. The connection-oriented service provides a reliable connection, providing flow control, error control, and sequencing. The connection-oriented service is closely modeled after HDLC ABM mode.

In practice the unacknowledged connectionless mode of LLC is most often used. It provides a simple framing service for LLC users and handles the addressing appropriate for LAN communication. Addressing consists of LLC Service Access Points (SAPs). Each LLC frame carries the destination and source SAPs for this frame. These 8-bit numbers, along with the MAC addresses of the sending and receiving stations, identify the communicating endpoints. As with the 802 MAC protocols, one-to-many (multicast) and one-to-all (broadcast) addressing is supported. By using LLC SAPs, an LLC user (e.g., a network layer protocol) is provided an interface to the network through which it may send and receive frames. The LLC layer handles the multiplexing of outgoing frames from concurrent senders on a single station and the demultiplexing of incoming frames for concurrent receivers at a station.

References

Stallings, W. 1991. *Data and Computer Communications,* Macmillan, New York, NY.

20.2 Network Protocols

Debapriya Sarkar

Purpose

Network protocols, as the name suggests, correspond to the network layer of the ISO OSI architecture (see Chapter 17), and are responsible for implementing the functionalities of the network layer. A network protocol is the lowest level of the protocol stack which is concerned with the end-to-end transmission of packets. It provides a uniform network-wide service to the transport layer protocols. The service provided may be either connection-oriented (e.g., X.25) or connectionless (e.g., CLNP or IP).

In order to achieve the goal of providing a uniform network-wide end-to-end service to the transport layer protocol, the network layer protocol must maintain information about the topology of the network and perform the necessary routing functions to deliver packets to the appropriate destinations. It should also maintain information about the traffic level on the routes and perform flow and congestion control. In addition to the above functions, the network protocol has to perform another set of functions if the underlying network is not homogeneous but comprises interconnections of a set of heterogeneous subnetworks. In particular, the network protocol must be capable of providing either connectionless or connection-oriented service over a set of subnetworks providing both connectionless and connection-oriented services. Also, the network protocol has to carry packets across subnetworks with possibly conflicting address formats and packet sizes.

In the following subsections we shall study in detail two important network protocols that are in current use: the connectionless network protocol (CLNP) which has been defined by ISO as the standard network layer protocol conforming to the OSI model, and the Internet protocol (IP), which is the network layer protocol developed for the Internet.

Internet Protocol (IP)

The present version of the DOD Standard Internet Protocol was developed in 1981 and was based on the earlier versions of the ARPA Internet Protocol (Defense Advanced Research Projects Agency, 1981). IP was designed for use in interconnected systems of (possibly heterogeneous) computer networks. IP is a network layer protocol which provides the functionalities needed for transmitting blocks of data, called datagrams, transparently across an internet-work from a source to a destination. It provides no other functionality to the transport layer, and in particular, does not provide end-to-end error control, acknowledgments, retransmission, flow or congestion control, or the other functionalities specified in the OSI Network Layer Specification.

Overview

The Internet protocol exists locally on every host machine connected to a local network running IP as its network protocol. The individual networks may be linked together by gateways also running IP. When a source application wishes to transmit a message, it prepares the data and submits it to the local IP module along with the Internet address of its destination. The IP module creates a datagram, generates the header and fills in the data. It determines the local network address that corresponds to the destination Internet address supplied by the application, and sends it to the data link layer to be transmitted across the network. The data link layer creates a local network header, appends the IP datagram and sends the packet via the local network to the local network address specified by the IP module. The data link layer of the local destination strips off the local network header and delivers the datagram to the local IP module. The IP module checks the destination field of the IP header to see if the datagram has to be forwarded to another destination or if it is addressed to itself. In the former case, the IP module determines the forwarding address, fragments the datagram into several smaller pieces if required, and passes the datagram(s) down to the data link layer to be processed as before. In the latter case, the IP module checks to see if the datagram requires reassembly, performs reassembly functions if needed, removes the IP header from the datagram and provides the data to the overlaying transport layer for further processing. At this point (unless errors have occurred), the data delivered to the transport layer of the receiver is identical to that received by the IP module of the sender. Note that the IP module itself does not do any error-checking of the data; it is the responsibility of the transport layer to perform the requisite error detection/correction.

IP Datagram Format

The IP datagram format is depicted in Figure 20.1. A short description of each of the fields follows.

Version (4 bits): The version field indicates the version of the IP protocol being followed.

IHL (4 bits): The Internet header length (IHL) field is the length of the IP header in 32-byte words. Because the IP header has a minimum length of 20 bytes, the minimum

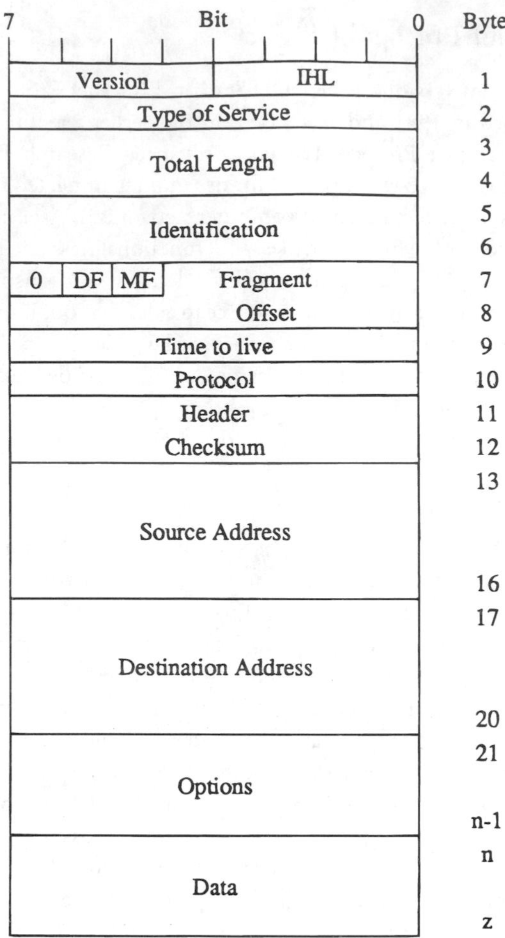

Figure 20.1 IP data PDU.

Precedence	D	T	R	0	0

Figure 20.2 Type of service flags.

IHL value of a correct header is 5. The maximum IP header length is 15 words (60 bytes).

Type of service (8 bits): The type of service flags define abstract parameters of quality of service. These values may be used to guide the choice of actual parameters of the underlying subnetwork services. The structure of the type of service flags is shown in Figure 20.2.

Bits 0–2: used for encoding the precedence as shown in Table 20.1.
Bit 3: if set, indicates low delay desired.
Bit 4: if set, indicates high throughput desired.
Bit 5: if set, indicates high reliability desired.
Bits 6–7: reserved for future use.
The type of service flags are used only if the underlying subnetwork supports the different classes of service.

Total length (16 bits): the length of the datagram (including header) in bytes. The field size limits the maximum IP datagram length to be 65,535 bytes. The actual maximum size is dependent on the type of hosts and networks. However, the protocol specification makes it mandatory for all hosts and networks to support a datagram size of at least 576 bytes.

Identification (16 bits): an identifier used by the destination IP module to reassemble fragmented datagrams. During fragmentation, the IP module must ensure that all fragments of a particular datagram carry the same Identification value to aid reassembly at the destination.

Flags (3 bit): This field contains three control flags.
Bit 0: reserved, must be set to zero.
Bit 1: (DF)—If set to 1, then don't fragment the datagram.
Bit 2: (MF)—Set to 0 to indicate the last fragment of a datagram. All other fragments must have this bit set to 1. The MF flag serves as an additional check against the total length field to ensure correct reassembly.

Fragment offset (13 bits): This field, measured in units of 8 bytes, denotes the position of the fragment in the original datagram. The initial fragment has an offset of 0. The definition of the fragment offset field ensures that the minimum fragment size is 64 bytes.

Time to live (8 bits): This field is actually a counter used to limit the amount of time a packet is allowed to remain in the network. The counter is decremented by each IP module as it is processed, and the datagram must be destroyed when the counter reaches 0. The unit of time is one second, allowing a maximum packet lifetime of 255 seconds.

Protocol (8 bits): This field indicates the protocol that should be used to interpret the data part of the datagram at the transport layer. Examples of such protocols include TCP, UDP, XTP, etc. (see section 6.5.3).

Header checksum (16 bits): This is a checksum on the header only, assuming a value of 0 for the checksum field itself. This field is recomputed every time the header is modified (for time to live changes or fragmentation) and is verified at every point where the IP header is processed.

Source address and destination address (32 bits): These fields are covered in detail in the discussion on addressing functions.

Options (variable length): The options field may or may not appear in an individual datagram, although each IP module is required to implement the options. They may be used for security, source routing, error reporting or other purposes. They afford

Table 20.1 Precedence Codes

111	Network Control
110	Internetwork Control
101	CRITIC/ECP
100	Flash Override
011	Flash
010	Immediate
001	Priority
000	Routine

a convenient way to extend newer versions of the protocol without needing to allocate header fields for rarely used information.

Functions of IP

The IP module has to perform three major functions. Firstly, since it has to transport datagrams through an internetwork of individual subnetworks, it has to deal with a single, uniform, internetwork-wide addressing scheme. Secondly, the protocol has to route the individual datagrams through a number of intermediate systems to its destination. Thirdly, the protocol may have to route the datagram over networks with maximum packet sizes that are smaller than the current size of the datagram. In such cases, IP has to fragment the datagram into smaller pieces, and reassemble them at the destination. We now discuss these three functions in greater detail.

Addressing

In IP, source and destination addresses are of fixed length (4 bytes each). Commonly, the IP address comprises a network field and a host field, which indicate the specific network and the specific host respectively. Since IP has to run over a large variety of networks with widely different numbers of hosts in each, the address field is encoded so as to allow the specification of this diversity. There are four different formats of addresses, as shown in Figure 20.3. The first format allows up to 128 large networks with 2^{24} (about 16 million) hosts each. The second allows 16,384 medium-sized networks with up to 65,536 hosts and the third allows 2^{21} (about 2 million) small networks with 256 hosts each. The fourth format allows multicast transmission to a group of hosts. A fifth format, in which addresses begin with 1111, is reserved for future use.

Routing

In order to perform routing functions, each IP module maintains a representation of the entire Internet. This database also keeps track of the delays on each line. The delay metric is used to calculate the shortest path to every other IP module. Every IP module measures the average delay on each of its lines over a 10-second period. The results of these measurements are then broadcast to all other routers using a flooding algorithm. When an IP module receives a packet that it has to transmit or forward, it checks the routing table to determine the next node to send the packet, and transmits the packet.

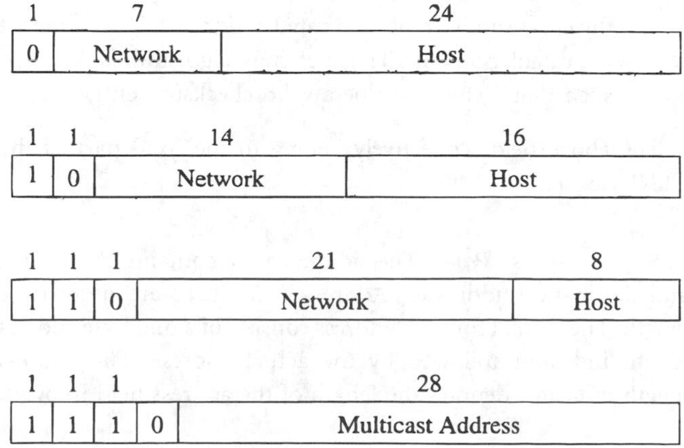

Figure 20.3 IP address formats.

Fragmentation and Reassembly

These functionalities are needed in the Internet Protocol because different networks allow different sizes of datagrams. If a datagram has to be routed through a network whose maximum permitted packet size is smaller than the datagram, then the original datagram must be fragmented (divided) into pieces small enough to be handled by the network.

The fragmentation and reassembly functions impose certain limits on the size of an IP datagram. IP specifies that the smallest fragment size is 8 bytes. It also specifies that every IP module should be able to transport a datagram of size 68 bytes without fragmentation; this is because the maximum IP header is 60 bytes and the minimum fragment is 8 bytes. Every IP module must be able to receive a datagram of 576 bytes, either in one piece or as fragments requiring reassembly.

An IP module initiates a fragmentation procedure if a datagram exceeds the maximum transmission size of the next network on which it must be transmitted. If the DF field is set, then the datagram is split into multiple fragments, each smaller than the maximum transmission size of the next network. The identification field and the source and destination addresses of the original datagram are copied into the corresponding fields of each fragment. The MF flag is set to one for all fragments except the last. The fragment offset is calculated for each fragment.

An IP module that receives a datagram sets up a reassembly buffer. The buffer identifier is a concatenation of the source and destination addresses, the protocol field and the identification field. If the datagram has not been fragmented (indicated by zero in both the fragment offset and the MF fields), the reassembly resources are released and further processing takes place. If a datagram is a fragment of a larger datagram, then the data part is copied into the appropriate location of the reassembly buffer. When all the fragments have been received the original datagram is reassembled and processing continues. The reassembly buffer has a timer associated with it, which keeps track of the maximum time to live of the fragments received. If this timer expires before the entire datagram is reassembled, then the entire buffer is discarded.

Internet Message Control Protocol (ICMP)

The Internet message control protocol is often considered to be a part of IP but is actually a distinct protocol. ICMP is used to monitor the Internet and report unusual occurrences in the network. There are about a dozen message types defined in ICMP. Each message is encapsulated in an IP datagram. A brief discussion of the various message types follows:

Destination unreachable: used when an IP module cannot locate the destination or the packet has the DF flag set and is of a larger size than the maximum transmission size of the next network.

Time exceeded: sent when a packet is discarded because its time to live expires.
Parameter problem: there is an illegal entry in one of the header fields.
Source quench: used to throttle sources which are sending packets too fast.
Redirect: used when a packet appears to have been routed incorrectly.
Echo request: used to determine if a given destination is reachable and alive.
Echo reply: used to reply to a echo request message.

Connectionless Network Protocol (CLNP)

The connectionless network protocol (CLNP) [ISO94] was specified to provide connectionless- mode network services conforming to the OSI network layer specifications. It was based on the IP protocol and provides quite similar services. The protocol performs a suite of functions which enable it to present the appearance of a uniform connectionless-mode network services over either identical or dissimilar interconnected subnetworks. Some of the underlying subnetworks may not support all the services needed by the network layer to provide connectionless services to the transport layer. In such a case, it is the task of CLNP to provide subnetwork-specific functions that would provide the appearance of a uniform internetwork.

Overview

The mechanism for transporting a datagram, referred to as a protocol data unit or PDU in CLNP nomenclature, from a source to a destination is exactly the same as for IP. The CLNP module at the source receives a packet from the transport layer, creates a PDU of the requisite size, segmenting or padding if required, locates the correct address to send to and hands the PDU to the data link layer for transmission. The intermediate CLNP modules direct the PDU closer to its destination, performing segmentation whenever required. The destination CLNP module reassembles the PDU and passes it up to the transport layer.

A key difference from IP is that CLNP defines subsets of the full protocol which may be used with higher efficiency in certain predetermined network configurations. The Inactive Network Layer Protocol Subset is used when it is known that the source and destination reside on the same subnetwork and none of the functions provided by the full protocol is needed to provide connectionless service to the end systems. The nonsegmenting protocol allows the CLNP header to be reduced by eliminating the segmentation part of the header. This protocol is used when the underlying subnetworks are capable of transporting PDUs without requiring segmentation. (Segmentation in CLNP performs the same function as fragmentation in IP.)

CLNP PDU Format

There are actually four different PDU types. However, they are sufficiently similar so that we shall discuss only the data PDU format. The data PDU format, which is very similar to the IP datagram format is illustrated in Figure 20.4. The fields may be divided into five different categories as shown. The summary of the various fields follows.

Fixed Part:

Network layer protocol identifier (8 bits): This field identifies the particular network protocol in use.
Length indicator (8 bits): This field indicates the header length in bytes. The maximum header size is limited to 254 bytes, with the value 255 being reserved for future expansion. This is a much larger header size than in IP.
Version/protocol identifier extension (8 bits): This field is set to 1 to identify the standard version 1 of the CLNP.
Lifetime (8 bits): This field gives the time to live of the datagram in units of 500ms, half that of an IP packet.
Flags (3 bits)

Bit 7: SP—If set to 1, then the datagram may be segmented by an intermediate CLNP module. A 0 value indicates that the datagram may not be segmented. In this case the CLNP header does not contain the segmentation part.
Bit 6: MS—this flag is identical to the IP MF flag.
Bit 5: E/R—If this flag is set to 0, then an Error Report PDU is not generated.

Type code (5 bits): This field identifies the type of the datagram. The types of PDUs are listed below:

11100—Data PDU
00001—Error PDU
11110—Echo Request PDU
11111—Echo Response PDU

Segment length (16 bits): This field specifies the length of the entire PDU (both header and data) in bytes.
PDU checksum (16 bits): The checksum covers the entire PDU header. The value 0 is reserved to indicate that the checksum is to be ignored. A non-zero entry indicates that the checksum is to be computed and the PDU discarded if the calculation fails. The checksum algorithm is designed such that 0 can never be a valid checksum entry.

The above fields collectively belong to the fixed part of the CLNP header.

Address Part The address part contains the source and destination addresses. Addresses in CLNP are of variable length. The format for each address consists of a one-byte address length indicator followed by the actual address. The address length indicator denotes the length of the address field in bytes.

Segmentation Part This part of the PDU header is present only if the SP flag is not set to 0.

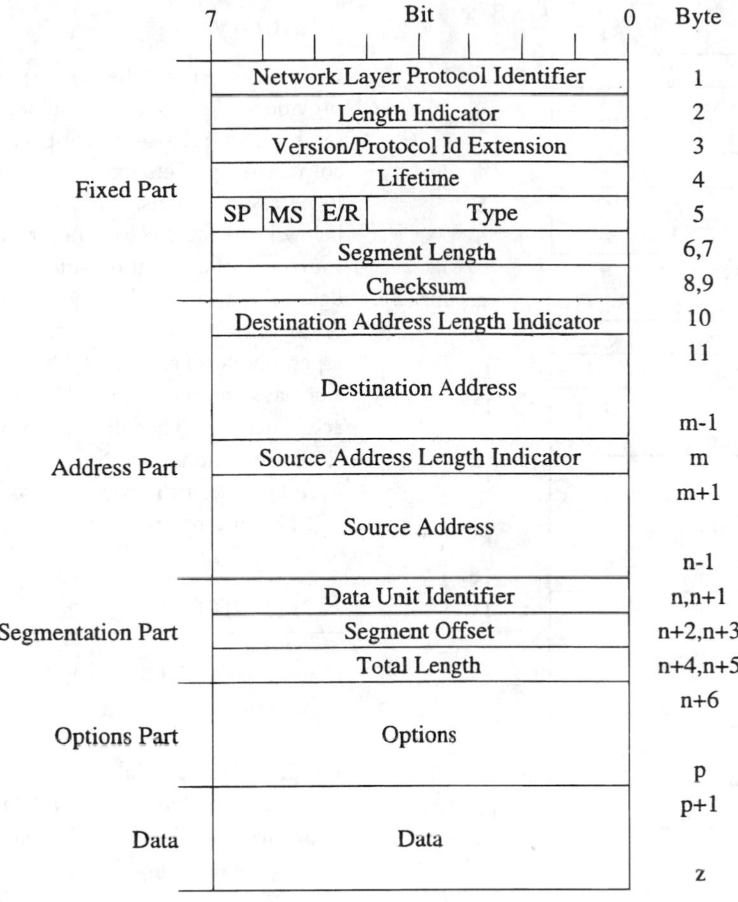

Figure 20.4 CLNP data PDU.

Data unit identifier (16 bits): This field gives a unique identifier for the original PDU so that it can be correctly reassembled following segmentation.

Segment offset (16 bits): This field indicates the relative position of a segmented PDU in the original PDU. For the first segment and the original PDU, the offset is set to 0. The unit of the offset is 8 bytes as in the IP protocol.

PDU total length (16 bits): This field indicates the total length of the original PDU in bytes, including both header and data.

The segmentation part is omitted in the nonsegmenting subset of CLNP for obvious reasons.

Options Part This is a variable length field which may or may not be present in a PDU. If present it may contain one or more parameters which specify such functions as header padding, security, source routing, quality of service and priority.

The echo request PDU and the echo response PDU have the same format as the data PDU. The error report PDU differs from the data PDU in that the former has an extra field—the reason for discard field (Figure 20.5). This field contains a binary encoding of the error that caused the error report PDU to be generated. Also the destination address field contains the address of the CLNP module that originated the error message.

Functions of CLNP Unlike IP, which has a limited set of functions, CLNP has a large set of functions. However, implementations of CLNP are not required to support the complete set of functions. The CLNP functions may be divided into three broad types:

Type 1: Functions in this category must be supported by all implementations. The functions include PDU composition and decomposition, segmentation, reassembly, forward PDU, discard PDU, header error detection, header format analysis, and error detection and reporting. Of these, all but the last two functions are similar to the functions of IP. The header format analysis function is required by CLNP to detect whether a received packet is using the full CLNP protocol or a subset of it such as the inactive network layer protocol subset or the nonsegmenting subset. Unlike IP, CLNP has a comprehensive error reporting function. An error report PDU is generated when a Data PDU with its E/R flag is dropped by an IP module. The report details the reason for which the packet was dropped. While IP does not support any error reporting functions, the functionalities provided by the CLNP error detection function are provided to IP by the ICMP.

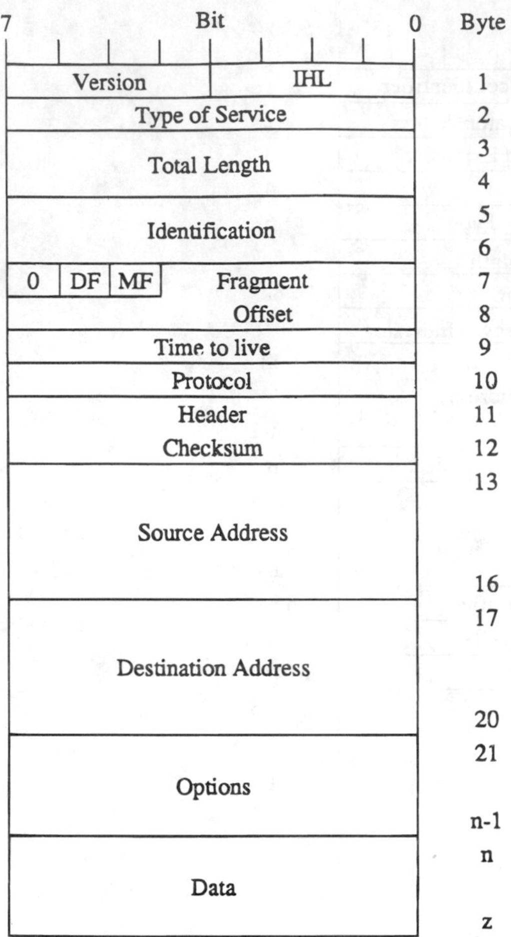

Figure 20.5 CLNP error report PDU.

Type 2: These functions may or may not be supported by a CLNP module. However, if a module receives a PDU which selected an unsupported Type 2 function, then the PDU will have to be discarded and an error report PDU generated. Examples of such functions include security, complete source routing, echo request and echo response. Again, the echo request and echo response functions are similar to those in ICMP. CLNP provides for the use of protection functions such as data origin authentication, confidentiality and integrity. However, it does not specify any standard for these functions. The options field in the header may be used to indicate the security mechanism being employed. The complete source routing function allows the originator of the PDU to specify the exact path that the PDU may take. Analogous to this function is the complete mute recording function which is used to record the path taken by the PDU.

Type 3: These functions also need not be supported. However, a CLNP module cannot discard a PDU because it selected an unsupported Type 3 function. In this case, the function is not performed and the processing of the PDU ignores the selection of those option. Functions such as priority, QoS maintenance and congestion notification fall under this category.

Summary

In summary, both the two network protocols, IP and CLNP, provide packet routing, fragmentation/segmentation and reassembly, and time-to-live control. On the sending side, the protocol receives packets from the transport layer above it, performs its necessary actions, and delivers datagrams to the datalink layer below it. At the receiver the process is reversed. In intermediate routers, the destination address of each datagram is used to determine which (of many) outgoing networks will be used for retransmission of the packet. IP has very limited functionalities, especially compared to CLNP, and requires the use of other associated protocols (e.g., ICMP) for smooth functioning in an actual network. The full CLNP protocol is base on IP and provides similar functionalities. Subsets of CLNP are also defined for use in certain specific network scenarios, where all the functionalities of CLNP are not required.

References

Defense Advanced Research Projects Agency 1981. *Internet Protocol.* DARPA Network Working Group Report RFC-791, USC Information Sciences Institute, September.

ISO/IEC International Standard 1994. *Information Technology—Protocol for Providing The OSI Connectionless-Mode Network Service.* ISO/IEC 8473–1:1994 (E).

20.3 Transport Layer Protocols

Bert J. Dempsey

In a typical local area communication protocol stack, the transport layer has special importance for two reasons. First, for most applications, the natural interface to the communications subsystem is at the transport layer. Thus, the transport layer defines the network services available. Second, in many protocol stacks, the transport layer protocol provides the mechanisms to ensure end-to-end reliable communication since the network layer and datalink layer protocols used in local area communication are most often of an unreliable connectionless flavor (e.g., IP and the unacknowledged connectionless mode of LLC).

In this section we will discuss the services, typical uses, and caveats of four widely used transports. From the Internet protocol suite, we discuss the connectionless transport user datagram protocol (UDP) and the connection-oriented transport transmission control protocol (TCP). We also include a section on the connection-oriented transport protocol class 4 (TP4) from the ISO protocols, and finally we discuss the next-generation transport, the Xpress transport protocol (XTP).

User Datagram Protocol

The Internet model includes two transport protocols, UDP and TCP. Both protocols assume the connectionless network layer protocol, the Internet Protocol (IP), as the underlying network service. UDP provides a very simple service, namely unreliable connectionless data delivery. Each UDP datagram sent by an application is treated as an independent message by the network. One implication is that messages may not arrive at the receiver in the order in which they were sent by the sending application. TCP provides a very different service, namely a reliable, bi-directional connection. TCP users are assured of reliable, sequenced delivery of their data.

Addressing in the Internet suite consists of service access points called ports. A port in the Internet protocol terminology is a 16-bit number that identifies an access point for an application to either UDP or TCP. The UDP port space and the TCP port space are logically independent, though system applications using both TCP and UDP are often assigned the same port. Certain low-numbered ports, e.g., 1 to 1023, are reserved for system usage while higher-numbered ports are assigned on-demand by the network interface to applications. Certain services in the Internet have been assigned well-known port numbers in the reserved range. On most Unix systems, the well-known ports and their services are identified in a file named/etc/services.

In the UDP communication model, an application submits a message to the protocol that produces a UDP datagram, which causes one IP datagram to be sent over the network. The size of the largest UDP datagram that a user can send is theoretically 65,507 bytes. This number comes from observing that the maximum size of an IP datagram is 65,535 bytes due to the 16-bit length field in IP and then substracting away bytes needed for the IP and UDP headers. However, in practice, other system limitations bound the size of UDP datagrams. First, a popular programming interface for UDP is known as sockets. There are a number of operating system-dependent variations of this interface, but all contain a call to set the size of the receive and send buffers. The size of these buffers limits the maximum UDP datagram. The default value for most systems is 8192 bytes. Additionally, implementations of the Internet protocols in many operating systems often limit the maximum UDP datagram to significantly less than the theoretical maximum.

While UDP is an unreliable protocol, the UDP user may enable data checksumming to ensure that messages are not corrupted during transmission. For large amounts of data, checksumming can be a significant performance issue. Good implementations hide the cost of checksumming in data copy operations. This implementation trick allows the checksumming operation to be done as the data is being copied, thus avoiding a separate pass through the data for checksumming alone.

For performance reasons, UDP allows the user to disable the checksumming feature. In local area communications with no IP routers in the end-to-end communication path, the reliability mechanisms at the datalink layer (e.g., CRCs) provide protection against transmission errors. In this case checksumming is only valuable to detect data corruption that takes place inside the endsystem computer itself, which is very rare. By contrast with UDP, the reliable transport TCP performs a mandatory checksum on all the data sent in a TCP connection.

It is important for the reader to understand the implications of UDP being a datagram protocol, namely that any number of reasons can result in the loss or reordering of data in the network. In some environments, for example, the UDP sockets programming interface allows the receiver to specify the maximum number of bytes to return in each network read call. If a UDP datagram arrives that is larger than this maximum, some systems truncate the excess data without notification to the user. Other systems truncate and notify the receiver; other systems do not truncate the data, but return the excess data in subsequent reads. The point is that the user of UDP must be prepared to deal with data loss and must understand the system-dependent behavior of the UDP interface (Stevens, 1994).

Transmission Control Protocol

Unlike the message-oriented, unidirectional data delivery offered by UDP, TCP provides a reliable byte stream service over a full-duplex virtual circuit connection. Users view the reliable virtual circuit provided by TCP as a pipe; after a connection is established, bytes are pushed into the pipe and flow out the remote end in the same order. User data is not structured by the transport service; instead a remote user must understand how to interpret the arriving byte stream.

The reliable byte stream paradigm in TCP requires a number of protocol functions. Note that the protocol function list here bears a strong resemblance to that for the datalink layer protocol HDLC. The mechanisms used in HDLC (and other connection-oriented protocols) are indeed quite similar. In this section, to provide the reader with a deeper understanding of TCP and its performance, we describe these mechanisms in some detail and focus specifically on the TCP algorithms.

Segmentation and Reassembly

Since the user data may be arbitrarily large, TCP must break the user data into segments suitable for transport across the network. A TCP protocol header is attached to each segment and it is this information that is passed to IP for transport across the network as an IP datagram. At the receiving side, the TCP headers are stripped off, and the segments are reassembled for delivery to the user. Segmentation and reassembly are transparent to the protocol user. (Note that TCP sends segments containing no user data in order to exchange control messages between the transmitter and receiver.)

Connection Establishment and Termination

TCP uses a three-way handshake for connection establishment. That is, the TCP endpoint requesting the connection sends a segment with a special flag set to the remote TCP endpoint. Like a UDP endpoint, a TCP endpoint is identified by a port and each segment carries the source and destination port of that segment. The remote endpoint responds with a segment acknowledging the opening segment and opening the reverse channel. Finally, the requesting endpoint sends a third segment that acknowledges the second one. To terminate a connection, four segments are required. A close message and its acknowledgment are used for each direction in the full-duplex connection.

Even in a LAN environment, these handshaking procedures may represent a significant latency overhead for delay-sensitive applications. Such applications must rely on connectionless transport such as UDP or move to next-generation transports such as the Xpress transport protocol, which uses implicit connection set-up as well as handshaking techniques.

Flow Control

Mismatches between a transmitter's ability to deliver data to the network service and the receiver's ability to process and buffer that data can severely degrade communication efficiency. In order to match the capabilities of the sender and the receiver, TCP uses a sliding window flow control mechanism.

Under this scheme feedback from the receiver dictates the maximum amount of unacknowledged data that can be outstanding at the transmitter. Each segment sent from the TCP receiver contains the offered window, i.e., the portion of the sequence space in which the transmitter may have unacknowledged segments outstanding. After the transmitter sends all the segments in the offered window, it must wait for an acknowledgment to arrive from the receiver that increases the offered window. The size of the offered window is typically fixed, representing the amount of buffer space available at the receiver for this connection. The receiver normally increases the offered window by acknowledging some number of segments and thus "sliding" the fixed-size window forward in the sequence space. That is, if the fixed-size offered window were [100,600], then an acknowledgment from the receiver of all data up to byte 300 would result in an offered window of [300,800].

A typical value for the sliding window in current (1994) workstations is 8192 bytes. The sockets interface to TCP allows the TCP user to set the size of the receive buffer, which translates into the size of the offered window. For bulk data transfers, increasing the window will likely improve performance.

Error Control

Error control for TCP includes data corruption in transmission as well as reordering, loss, or duplication of segments in the network. TCP segments are in general transported as IP datagrams over internetworks in which loss and reordering are not infrequent. While these hazards are rare in local area communications, they do occur, making a reliable transport protocol valuable.

For data integrity, in the TCP header attached to each segment, there is a mandatory 16-bit checksum over that segment. The TCP receiver must verify data integrity by recalculating the checksum and matching the result to that in the header of the segment. In order to avoid incorrect reassembly due to segments being reordered in the network, a sequence number is assigned to each byte of user data, and the sequence number of the first byte of each segment is carried in the TCP header. This sequence number also provides the mechanism for detecting and discarding duplicate segments arriving at the receiver.

For segment loss, TCP implements a retransmission algorithm. Specifically, whenever a segment is transmitted, a timer is set. If the timer expires before that segment is acknowledged, then the segment is retransmitted. TCP has algorithms to set the value of the retransmission timer based on the measured roundtrip time in the network. However, the value of the retransmission timer is generally relatively long. (A typical default starting value is 6 seconds.) Thus the loss of a segment can result in a considerable performance hit as the transmission stalls waiting for the retransmission timer to expire. As a result, an algorithm called "fast retransmit and fast recovery" has been developed. Using this algorithm, whenever the TCP transmitter receives a small number of consecutive acknowledgments indicating out-of-order segments arriving at the receiver, the transmitter assumes a segment loss. The presumed missing segment is retransmitted without waiting for the retransmission timer to expire.

Urgent Data

TCP has an option whereby the TCP user can indicate that some "urgent data" is in the data stream. TCP notifies the remote (receiving) application that urgent mode has been entered. Whenever a segment is received with urgent data in it, the receiving TCP side sets a pointer to this urgent data, and the receiving application can use this urgent pointer to find its urgent data. The urgent data feature is most often used in interactive applications when an abort or reset command needs to be passed to the remote side.

Since TCP was designed for use in a wide range of network environments and applications, some tuning of the protocol through the network interface is often appropriate. One example that we have noted above is changing the size of the offered window to improve the performance of bulk data transfers. Another example has to do with interactive traffic. As an optimization to reduce the number of small messages in an internetwork, an algorithm known as Nagle's algorithm is often implemented to regulate the behavior of a TCP connection for applications that exchange small, interactive messages (i.e., keystrokes in a remote login session). Nagle's algorithm specifies that a TCP endpoint never have more than one "small" segment unacknowledged at any time, where "small" is typically defined as any packet that is less than the maximum packet size. While the TCP sender awaits an acknowledgment of a small segment, any new user data is buffered for transmission in the next packet. This algorithm improves the efficiency of network transfers since many fewer packets are used to transmit the data stream than if packets were issued whenever user data became available. Also, the algorithm is ingeniously self-clocking in that new packets are introduced into the network as quickly as acknowledgments arrive. On fast LANs, the packet rate will be high while on slow WANs the packet rate will be low. Nagle's algorithm is inappropriate, however, in cases where an interactive application does not follow the expected traffic pattern. The sockets interface to TCP allows the user to turn off Nagle's algorithm using the TCP_NODELAY option.

ISO Transport Protocol

The open systems interconnect architecture has defined five classes of connection-oriented transport protocol numbered 0 through 4. The differences among these five service classes are related to the type and quality of service required by the Session Layer and, more directly, the quality of service provided by the network. ISO transport protocol class 4 (TP4) assumes that the network layer provides an unreliable datagram delivery service. That is, it operates under the same assumptions about the network as TCP does. The TP4 mechanisms are a superset of the other transport classes.

TCP was well-established at the time of the development of the ISO transport protocol standards, and TCP influenced TP4 heavily. Consequently, TP4 resembles TCP in most of its protocol mechanisms: it is a connection-oriented sliding window protocol that uses positive acknowledgments and timer-based retransmission. One important difference between the TCP and TP4 is the network model for the service interface. That is, TCP assumes its user is an application and the service interface is ultimately system dependent. TP4 assumes a session layer entity as its user and the service interface is explicitly defined in terms of a small set of parametrized service calls.

Since much of the discussion on TCP mechanisms applies to TP4, we do not discuss TP4 in detail.

Xpress Transport Protocol

The Xpress transport protocol (XTP) is a next-generation transport protocol that runs over IP. Beginning in 1987, its development was shaped by the lessons learned from the first decade of TCP deployment and experimental protocol efforts. XTP was designed to take full advantage of the high bandwidth and low latency of emerging networks while providing new functionality for emerging applications. A primary feature of the protocol is configurability. Using the XTP protocol options, an XTP user can construct a number of different network services, including the unreliable connectionless service offered by UDP and the reliable connection-oriented offered by TCP. In order to achieve this flexibility, the XTP designers developed new protocol mechanisms and reengineered conventional mechanisms, both individually and in their interrelationships. In this section we will describe the important novel protocol functions in XTP and their application in a factory communication environment.

Connection Establishment and Termination

XTP uses both TCP-like packet exchanges (handshaking) and timer-based connection establishment and termination techniques. These techniques are especially important for efficient support of transaction-based communications, that is, reliable, bi-directional exchange of small amounts of data.

Orthogonal Control Algorithms

As noted above, the configurability of XTP for different communication needs is a major feature of the protocol. An important design philosophy in XTP was to provide a set of orthogonal options from which the user may select those necessary for its communication needs. This allows for an explicit trade-off of performance and functionality.

XTP has options for disabling its flow and error control (including checksumming) algorithms in a connection. Thus, for example, a user can create a connection-oriented service in which lost packets are not retransmitted (no error control). Such a service is useful to emerging applications involving real-time traffic such as voice and video. For these applications, the delay due to a retransmission is often more disruptive than a "hole" in the data stream. Note that this unreliable connection service is much more powerful than, for example, using UDP to send each message in the stream. The XTP unreliable connection provides sequencing, segmentation/reassembly, and all the other selectable options of an XTP connection (e.g., an urgent data mechanism).

Multicast

Perhaps the most important functionality in XTP that is not available in conventional transports is a reliable one-to-many (multicast) connection-oriented service. An XTP multicast connection has the same protocol functionality as the XTP point-to-point connection with the major exception of bi-directional communication: a multicast connection is a unidirectional data flow from a source to multiple receivers.

A transport layer multicast offers efficient use of processing cycles and memory in a station and bandwidth on the network. In IEEE broadcast LANs, datalink layer multicast addressing enables a sender to transmit a single copy of the data for reception by multiple receives. In applications with large data streams (e.g., video) or with large receiver sets, using a set of point-to-point connections becomes infeasible.

Reliability in the case of multiple receivers implies management of the set of receivers involved in a multicast connection. XTP provides different algorithms for managing the receiver set since the semantics of reliable delivery vary between applications. One feature in XTP that has proven very valuable is its in-progress join procedure, which allows a new receiver to join an ongoing multicast distribution without disturbing the progress of the other receivers. This feature supports the ability of new receivers to become active at any time and tap into an existing multicast data flow in a fashion transparent to other receivers.

Rate Control

Flow control concerns the problem of a transmitter sending more data than the receiver can process at one time. Rate control addresses a similar, but independent, problem, namely restraint of the transmitter from sending data faster than the destination station (or some routing node in the network between the transmitter and its receiver) can receive the data. In a LAN this problem is most often seen in the form of slow network hardware interface cards being overrun by fast transmitters.

XTP implements a sliding window protocol for flow control. It also provides the user with two rate control parameters. The "rate" parameter specifies in bytes per second the maximum rate at which data can be emitted while the "burst" parameter specifies the maximum number of bytes in any one burst of data, i.e., the maximum data transmitted all at once. With these parameters, the XTP user can control with fine granularity communication between stations of widely different capabilities.

Priority

Each message submitted to XTP can be labeled with a priority value; this priority value is carried in each XTP packet that transports the message. XTP end-systems (transmitters and receivers) process packets in priority order. The XTP priority scheme must be mapped onto priority mechanisms in lower layer protocols; for example, a research project at the University of Virginia (USA) has developed an "XTP-aware" IP router that processes incoming packets according to their XTP priority.

References

Stevens, W. R., 1994. *TCP/IP Illustrated, Volume 1: The Protocols,* Addison-Wesley, Reading, MA.

Xpress Transport Protocol 1994. *XTP Forum, Protocol Specification Version 4.0* available at http://www.ca.sandia.gov/xtp/xtp.html.

V

System Control

21 **Control System Fundamentals** *A. S. Hodel* 443
Modeling • Controller Design • Intelligent Control • Other Control Approaches

22 **Modeling for System Control** *A. John Boye and William L. Brogan* 447
Introduction • Analytical Modeling • Defining the Problem • Determining the System Components • Writing the System Equations • Verifying the Model • Empirical or Experimental Modeling

23 **Basic Feedback Concept** *T. H. Lee, C. C. Hang, and K. K. Tan* 453
Beneficial Effects of Feedback • Analysis and Design of Feedback Control Systems • Implementation of Feedback Control Systems

24 **Stability Analysis** *N. K. Sinha* 456
Stability Analysis for Linear Systems • Stability of Linear Time-Invariant Continuous-Time Systems • Stability of Linear Time-Invariant Discrete-Time Systems • Nonlinear Systems

25 **PID Control** *James C. Hung* 470
Introduction • Classical PID Control (Ziegler-Nichols Tuning) • Remarks

26 **Bode Diagram Method** *John Parr* 474
Bode Diagram Analysis • Mathematical Model Determination • Correlation of Frequency Response and Time Response • Shaping the Cutoff Response • Compensator Design • Design for Digital Systems

27 **The Root Locus Method** *Robert J. Veillette and J. Alexis De Abreu-Garcia* 490
Motivation and Background • Root Locus Analysis • Compensator Design by Root Locus Method • Examples

28 **Pole Placement Design** *Michael Greene and Victor Trent* 504
Pole Placement • State Observation • Discrete Implementation

29 **The Smith Predictor Technique** *John Y. Hung* 511
Background—Control of Processes Having Time Delay • Basic Principle of the Smith Predictor • A Smith Predictor Design Example

30 **Internal Model Control** *James C. Hung* 513
Basic IMC Structures • IMC Design • Discussion

31 **Model Predictive Control** *Jay H. Lee* 515
Overview • Applications

32 **Dynamic Matrix Control** *James C. Hung* 522
The Dynamic Matrix • Output Projection • Control Computation • Remarks

33 **Disturbance Observation-Cancellation Technique** *Kouhei Ohnishi* 524
Why Estimate Disturbance? • Plant and Disturbance • Higher-Order Disturbance Approximation • Disturbance Observation • Disturbance Cancellation • Examples of Application • Conclusions

34 **Phase-Locked Loop-Based Control** *Guan-Chyun Hsieh* 529
Introduction • Configurations of PLL Applications • Analog, Digital, and Hybrid PLLs • Popular PLL Integrated Circuits (ICs)

35 **Variable Structure Control Technique** *Vadim Utkin* 535
Introduction • Mathematical Aspects • Sliding Mode Control Design • Chattering Problem • Control of Manipulators • Control of Mobile Robots • Control of Railway Wheelset • Control of Torsion Oscillations of a Flexible Shaft • DC Motors • Control of DC Motors Based on a Reduced-Order Model • Conclusion

36 Digital Computation *James R. Rowland* 545
System Response • Numerical Integration Formulas • Exact Difference Equations for Linear Systems • Summary

37 Digital Control *John Y. Hung and Victor Trent* 553
Introduction • Discretization of Continuous-Time Systems • Discretization of the Servomotor System • Frequency Domain Design through the *w*-Transform • Root Locus Design on the Unit Circle • Simulation Comparisons

38 Estimation and Identification *Thomas S. Denney, Jr.* 559
Kalman Filters • Other Types of Kalman Filters • Identification

39 Fuzzy Logic-Based Control *Mo-yuen Chow* 564
Introduction to Intelligent Control • DC Motor Dynamics • Fuzzy Control • Conclusion and Future Direction

40 Neural Network-Based Control *Dian-cheng Zhang* 572
Control Configuration • Design Procedure

41 Programmable Logic Control (PLC) *Ernst Dummermuth* 587
Basic Concepts • Hardware Components • PLC Real-Time Operating Systems • Software Components • PLC Communications • Selecting the Right PLC

42 Adaptive Control *Stephen T. Hung* 593
Introduction • Update Strategies • Direct Adaptive Control • Indirect Adaptive Control • Adaptive/Self-Tuning Behavior • Summary

43 Hardware Compensating Networks *Royce D. Harbor and Charles L. Phillips* 609
Continuous Compensation • Other Compensation Procedures

44 μ-Synthesis and Analysis *Dan Bugajski, Dale Enns, Mike Jackson, Blaise Morton, and Gunter Stein* 613
Defining the Interconnection Structure • H_∞-Synthesis • μ-Analysis and D Scales • D-K Iteration • Changing Weights • Compensator Model Reduction • Summary

21
Control System Fundamentals

A. S. Hodel
Auburn University

21.1 Modeling 443
21.2 Controller Design 444
21.3 Intelligent Control 445
21.4 Other Control Approaches 445

Automatic control systems have played an important role in many of the technological developments of this century. From commonplace items such as computer disk drives and automobile cruise control units to high-performance military aircraft and satellite systems, automatic control systems impact many important areas in current engineering practice. The use of automatic control systems dates back to antiquity (Mayr, 1970), although much of the related theory and practice were developed during the past century. Some commonly used textbooks on automatic control systems are *Feedback Control of Dynamic Systems* (Franklin et al., 1991), *Feedback Control Systems* (Phillips and Harbor, 1991), *Automatic Control Systems* (Kuo, 1995), *Digital Control System Analysis and Design* (Phillips and Nagle, 1995).

Fundamental features of any control system are shown in Figure 21.1. The central focus of the control system is the system to be controlled, or the plant. The plant could be an electric motor in a washing machine, a chemical distillation column, or a high-performance aircraft. Associated with the plant are actuators, input devices that respond to command signals u in order to modify plant behavior or characteristics, and sensors, output devices that describe plant behavior through output signals y. The plant has several key features.

1. Left to its own natural (open-loop) behavior, the plant does not perform as desired. For example, an unregulated electric motor does not hold a steady speed, or perhaps it maintains an undesired speed. Similarly, a distillation column does not naturally maintain the desired temperature and pressure necessary for optimal performance. High-performance fighter aircraft are designed to be inherently unstable when operating in an uncontrolled open-loop manner.
2. The plant is acted upon by external disturbances w_1, not under user control. Examples of external disturbances are load torque on an electric motor or wind gusts on an aircraft.
3. Plant sensors provide imperfect measurements due to noise signals w_2. These noise signals could be deterministic (bias) or nondeterministic.

For the plant to behave in a desired fashion, it is necessary to design an automatic control system. Part of this task may be to require that a set of regulated variables z tracks a command reference signal w_3, e.g., a distillation column set point. For most purposes, w_3 may be treated as a disturbance signal, since it is not under direct control of the control system shown. Notice that the regulated variables z need not be the same as the measured variables y.

21.1 Modeling

To design an automatic control system, it is necessary to develop some form of mathematical model that attempts to predict plant behavior. The model may be a set of differential equations, difference equations, an artificial neural network that has been trained to mimic plant behavior based on recorded observations, or "fuzzy" rules taken from human intuition. Each of these concepts is treated in Chapters 22, 37, 39, and 40, respectively.

System identification and model validation techniques are an important part of the development of any plant model. Even

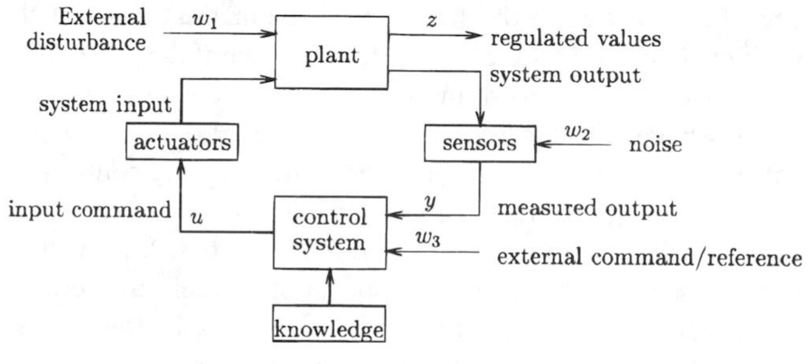

Figure 21.1 Typical control system block diagram.

0-8493-8343-9/97/$0.00+$.50

with the best system identification and model validation techniques, plant models are inherently inaccurate. Thus, any properly designed control system must be designed such that it is *robust*, i.e., it fulfills design objectives in the presence of modeling inaccuracy. To perform this task, it is highly desirable to have some measure of its accuracy, so that a control system may be designed to accommodate uncertainties in the plant model. Modeling errors or inaccuracies are typically dealt with in one of two fashions:

1. Treat unmodeled features as a part of the disturbance signal w_1, e.g., load torque on a motor, wind gusts on an aircraft, gear backlash, amplifier saturation, etc.
2. Use the given plant model P as a nominal plant model P_0 associated with an uncertainty bound Δ. That is, it is assumed that, for example, $P = P_0 + \Delta$ where the magnitude of Δ does not exceed some limit. Uncertainties of this type occur when certain constant parameters of a plant are known to lie in some given range, or perhaps when certain dynamics are neglected to simplify a model description. Given a reasonable description of both the nominal plant P_0 and the associated anticipated errors Δ, the control engineer attempts to design a controller that achieves closed loop desired performance levels plants with dynamics "nearby" the nominal plant model (robust control).

External disturbances w_1 may be incorporated into the plant model as stochastic (random) processes as in linear-quadratic Gaussian (LQG) control design, or as deterministic (but unknown) signals as in H_2 and H_∞ optimal control. Roughly speaking, H_2 optimal controller design minimizes the energy in the closed-loop system impulse response from the external signals $w = [w_1\ w_2\ w_3]^T$ to a set of regulated variables z (see Figure 21.1), while H_∞ optimal control minimizes the maximum "gain" from w to z over all sinusoidal disturbances w. When disturbances are properly modeled in the given framework, a good control system design will achieve the desired system performance even in the presence of external disturbances. Further discussion of this topic is given in Chapter 24, Chapter 26, Chapter 27, and Chapter 38.

Treatment of plant uncertainties Δ can be dealt with in either standard robust control techniques (Cusumano et al., 1988) such as μ synthesis (Chapter 44), or by adaptive control (Chapter 42). The issue involved in the selection of one or the other is highly problem dependent. One rule of thumb is that if the plant uncertainty can be described in terms of a given structure with unknown coefficients (parametric uncertainty), then adaptive control (or, more generally, nonlinear time varying control) can provide strong advantages over a fixed controller (e.g., linear time invariant controllers provided by μ synthesis). μ Synthesis is primarily used in aerospace applications, where the concerns are dynamic uncertainty (neglected dynamics in the nominal plant model), rapid response times and robustness over a wide range of rapidly changing conditions. Adaptive control has been used in some process control settings and in slower varying systems where adaptation algorithms can track system parameters reliably.

21.2 Controller Design

Once a plant model including nominal plant dynamics, plant uncertainty, sensors, actuators, disturbances, and noise has been developed, a control engineer may design a control system based on the plant and any other knowledge he/she may have about the system. The control system is designed with any number of objectives in mind, e.g., stability over some set of nominal plant perturbations (gain and phase margins), closed loop settling time of 10 seconds, minimization of energy expended, etc. During the past 50 years there has been an explosion of practical approaches and theoretical results on how and why to go about control system design in various situations. Classical control techniques (Franklin et al., 1991, Chapter 25, and Chapter 26; see also Kuo, 1995) make use of frequency domain analysis, i.e., the response of a plant at some operating point to sinusoidal inputs. Frequency domain techniques assume that a plant may be (locally) modeled as a linear, time-invariant differential equation

$$\frac{d^n y}{dt^n} + a_{n-1}\frac{d^{n-1}y}{dt^{n-1}} + \cdots + a_1\frac{dy}{dt} + a_0 y = b_m\frac{d^m u}{dt^n} + b_{m-1}\frac{d^{m-1}u}{dt^{n-1}} + \cdots + b_1\frac{du}{dt} + b_0 u \quad m \leq n$$

subject to some initial conditions $y(0)$, $y'(0)$, etc. Such a model is called a continuous time system and is studied in the frequency domain through the use of the Laplace transform. Frequency domain techniques can also be applied to plants controlled by a digital computer; in this case, the plant is modeled as a linear, time-invariant difference equation

$$y_{k+n} + a_{n-1}y_{k+n-1} + \cdots + a_1 y_{k+1} + a_0 y_k = b_m u_{k+m} + \cdots b_1 u_1 + b_0 u_0 \quad m \leq n$$

Such a plant model is called a discrete time system and is studied in the frequency domain through the use of the Z transform (Chapter 37, Digital Control; see also Franklin et al., 1990; Kuo, 1992; Phillips et al., 1995.)

The digital computer has also made it possible to make use of alternative design strategies based on state-space control (Brogan, 1991.) State-space representations of a continuous time system take the form

$$\frac{dx}{dt} = f(x, u, w, t) \qquad y = g(x, u, v, t) \tag{21.1}$$

where x is a vector of system states, u is a vector of system inputs, w and v are system disturbances and noise, respectively, and t is time. Equation 21.1 is a nonlinear, time-varying ordinary differential equation. A control system for the plant (Equation 21.1) may be obtained by computing a linear approximation

$$\frac{d\Delta x}{dt} = A\Delta x + B\Delta u \qquad \Delta y = C\Delta x + D\Delta u$$

of the plant about a desired operating point x_0, u_0, where $\Delta_x = x - x_0$, etc. The design may employ classical techniques, optimal control (Athans and Falb, 1966; Bryson and Ho, 1969; Anderson and Moore, 1971, 1990; Sage and White, 1977, etc.), nonlinear control (Isidori, 1989), adaptive Control (Goodwin and Sin, 1984), or any number of other options. State-space controller design, also called multivariable controller design, allows the designer to treat a system in a unified framework rather than attempting to tune each input-output channel individually. Many of these techniques are outlined in this chapter.

Typical state space approaches result in controllers with an unacceptably large number of states. In this case, it is desirable to design reduced order controllers, or, when applicable, with decentralized control. A reduced order controller may be designed by one of three methods:

1. Approximating a plant model by a lower order model, then designing a robust controller for the reduced plant model.
2. Direct approximation of a predesigned controller.
3. Direct design.

Common state-space controller and plant reduction techniques are given in (Glover, 1984; Moore, 1981). Direct controller design techniques are being developed based on the work in (Hyland and Bernstein, 1984). Further discussion of this topic is given in the survey article Anderson and Liu 1989.

Decentralized control is used in large, distributed systems, such as power networks and large, flexible structures (Aoki, 1972; Sandell et al., 1978; Davison, 1984; West-Yukovich et al., 1984; Collins et al., 1991). While a decentralized controller may yield a preferable controller implementation, its design is more complex since it is more difficult to treat the issues of controllability and observability (see Chapter 28 for more discussion of controllability and observability).

21.3 Intelligent Control

As computing power has increased over the past 10–20 years, new control systems have been developed based on artificial neural networks (Chapter 40; see also Werbos, 1992) and/or fuzzy logic (Chapter 39; see also Zadeh, 1965). Since these control system designs are inspired by models of biological learning and intelligence, they are often referred to as intelligent actuators, or simply intelligent control. Many techniques exist for the design of intelligent controllers, but only recently have results been made available that provide measures of system stability and performance. A neuro-fuzzy controller can be regarded as a static, nonlinear mapping from the plant sensor outputs y to the actuator inputs u. Adaptive fuzzy control has also been proposed and used (see Kwong et al., 1994; Wang, 1993), and associated stability analysis is being developed (Wang, 1994).

Intelligent control has been applied in numerous applications and commercial products, from video cameras to high-performance aircraft (Dai, 1992), (Kwong et al., 1994). Intelligent control provides a framework in which to combine quantitative performance measures with human heuristics to achieve acceptable (although not necessarily optimal) control performance. Further, since intelligent control inherently makes use of nonlinear interpolation, it is potentially capable of providing robustness not available through other control design techniques.

21.4 Other Control Approaches

There are many other approaches to control system design that may be used fruitfully. These include "receding horizon," or model predictive control (Chapter 31, see also Morari and Zafiriou, 1989), variable structure control (Chapter 35), etc. Model predictive control solves a dynamic optimization on-line, and is typically associated with process control, where the high-speed control loops associated with aerospace applications are not required. Variable structure controllers, or "sliding mode" controllers, allow the use of "on-off" actuators to cause a plant being controlled to track a desired state-space trajectory.

Many industrial control systems require sequencing control such as that provided by programmable logic controllers (Chapter 41). Prior to controller implementation, controller verification typically is done through simulation on a digital computer, including hardware in-the-loop simulation. Controllers of this nature often appear in manufacturing assembly lines or in process control settings. While one would expect that programmable logic controllers would be application specific, general-purpose PLCs are available that can be adapted easily to a given application.

References

Anderson, B. D. O. and Liu, Y. 1989. Controller reduction: Concepts and approaches, *IEEE Trans. Autom. Control*, 34(9)802–812.

Anderson, B. D. O. and Moore, J. B. 1971. *Linear Optimal Control*, Prentice Hall, Englewood Cliffs, NJ.

Anderson, B. D. O. and Moore, J. B. 1990. *Optimal Control: Linear Quadratic Methods*, Prentice Hall, Englewood Cliffs, NJ.

Athans, M. and Falb, P. L. 1966. *Optimal Control: An Introduction to the Theory and Its Applications*, McGraw-Hill, New York, NY.

Aoki, M. 1972. On feedback stabilizability of decentralized dynamic systems, *Automatica*, 8:163–173.

Brogan, W. L. 1991. Modern Control Theory, 3d ed., Prentice Hall, Englewood Cliffs, NJ.

Bryson, Jr., A. E. and Ho, Y. C. 1969. *Applied Optimal Control*, Blaisdell Publishing, New York, NY.

Collins, Jr., E. G., Phillips, D. J., and Hyland, D. C. 1991. Robust decentralized control laws for the ACES structure, *IEEE Control Syst. Magazine*, 11(4):62–70.

Cusumano, S. J. and Poolla, K. 1988. *Adaptive* robust control: A new approach, Technical report, manuscript.

Dai, Han-Sen 1992. Systems Identification Using Neural Networks, Ph.D. thesis, UCLA.

Davison, E. J. 1984. The decentralized control of large scale systems, *Adv. in Large Scale Systems,* 1:61–91.

Franklin, G. F. Powell, J. D., and Emami-Naeini, A., 1991. *Feedback Control of Dynamic Systems,* 2d ed., Addison-Wesley, Reading, MA.

Franklin, G. F., Powell, J. D., and Workman, M. L. 1990. Digital *Control of Dynamic Systems,* 2d ed., Addison-Wesley, Reading, MA.

Glover, K. 1984. All optimal Hankle-norm approximations of linear multivariable systems and their L_∞-error bounds, *Int. J. Control,* 39(6):1115–1193.

Goodwin, C. G. and Sin, K. S. 1984. *Adaptive Filtering Prediction and Control,* Prentice Hall, Englewood Cliffs, NJ.

Hyland, D. C. and Bernstein, D. S. 1984. The optimal projection equations for fixed-order dynamic compensation, *IEEE Trans. Autom. Control,* AC-29(11):1034–1037.

Isidori, A. 1989. *Nonlinear Control Systems,* 2d ed., Springer-Verlag, New York, NY.

Kuo, B. C. 1992. *Digital Control Systems,* 2d ed., Harcourt, Brace, Jovanovich, Orlando, FL.

Kuo, B. C. 1995. *Automatic Control Systems,* 7th ed., Prentice Hall, Englewood Cliffs, NJ.

Kwong, W. A., Passino, Kevin M., Laukonen, E. G., and Yurkovich, S. 1994. Expert supervision of fuzzy learning systems with applications to reconfigurable control for aircraft, Proc. 33rd *IEEE Conf. Decision and Control,* 4116–4121, Lake Buena Vista, FL, December 14–16.

Mayr, O. 1970. *The Origins of Feedback Control,* MIT Press, Cambridge, MA.

Moore, B. C. 1981. Principal component analysis in linear systems: Controllability, observability and model reduction, *IEEE Trans. Autom. Control,* AC–26:17–32.

Morari, M. and Zifiriou, E. 1989. *Robust Process Control,* Prentice Hall, Englewood Cliffs, NJ.

Phillips, C. L. and Harbor, R. D. 1991. *Feedback Control,* 2d ed., Prentice Hall, Englewood Cliffs, NJ.

Phillips, C. L. and Nagle, H. T. 1995. *Digital Control System Analysis and Design,* 3d ed., Prentice Hall, Englewood Cliffs, NJ.

Sage, A. P. and White, C. C. 1977. *Optimum Systems Control,* 2d ed., Prentice Hall, Englewood Cliffs, NJ.

Sandell, N. R., Varaiya, P. P., Athans, M., and Safonov, M. G. 1978. Survey of decentralized control methods for large scale systems, *IEEE Trans. Autom. Control,* AC–23:108–128.

Wang, L. 1993. Stable adaptive control of nonlinear systems, *IEEE Trans. Fuzzy Syst.,* I(2):146–155.

Wang, L. 1994. *Stable Adaptive Control of Nonlinear Systems,* Prentice Hall, Englewood Cliffs, NJ.

Werbos, P. 1992. Neurocontrol and supervised learning: An overview and evaluation, *Handbook of Intelligent Control: Neural, Fuzzy, and Adaptive Approaches,* 65–90, Van Nostrand Reinhold, New York, NY.

West-Vukovich, G. S., Davison, E. J., and Hughes, P. C. 1994. The decentralized control of large flexible space structures, *IEEE Trans. Autom. Control,* AC–29(10):866–879.

Zadeh, L. A. 1965. *Fuzzy sets, Information and Control,* 8:338–353.

22
Modeling for System Control

22.1	Introduction	447
22.2	Analytical Modeling	447
22.3	Defining the Problem Dynamic or Static • Distributed Parameter or Lumped Parameter • Determinist or Stochastic • Continuous Time or Discrete Time • Linear or Nonlinear • Timing the Variant or Timing the Invariant	448
22.4	Determining the System Components	448
22.5	Writing the System Equations	449
22.6	Verifying the Model	450
22.7	Empirical or Experimental Modeling	451

A. John Boye
University of Nebraska-Lincoln

William L. Brogan
University of Nevada, Las Vegas

22.1 Introduction

A mathematical model of a physical system is a representation of the system by a set of mathematical equations. As stated by others, modeling is a "technique of system analysis and design using mathematical or physical idealizations of all or a portion of the system. Completeness and reality of the model are dependent on the questions to be answered, the state of knowledge of the system, and its environment" (Jay, 1984).

Mathematical models are constructed for a wide range of systems. They are used for physical engineering systems such as electrical, mechanical, hydraulic, pneumatic, and thermal, as well as for systems in the areas of applied sciences, life sciences and medicine, social sciences, and business and management (Sandquist, 1985; Fowkes and Mahony, 1994; Ljung and Glad, 1994). Models are used when it is not possible or practical—due to complexity, expense, or safety—to build the actual system, then test and modify it until desired performance is achieved. A model is a substitute for the actual system when designing, testing, and modifying various components of the system. The model is used to examine the system without having to do any physical experiments. Obviously, it would be easier (and safer) to study and modify the operation of a new controller for the space shuttle using a mathematical model than it would be to actually use the space shuttle itself.

Obtaining a good mathematical model of a system is perhaps the most important task in mathematical modeling. No mathematical model is exact. Certain basic assumptions are made. Because of this, it is crucial that the model includes all of the necessary critical factors for the particular problem. While the model must include all critical factors in order to be adequate for the particular needs, it should not be overly complex. Completely describing the performance and operation of a system will require a large number of equations that need to be simplified. At the same time however, they must not be overly simplified so that important properties are omitted. A good model is one that is both accurate and simple.

The knowledge of the particular problem, the skills, and the practical experience of the engineer play an important role in creating a good model. Some have said that "the development of the models of the physical systems involved is from 80 to 90 percent of the effort required in control system analysis and design" (Phillips and Harbor, 1991). The importance of obtaining the correct system model cannot be stressed enough. Many problems that arise in applying control theory to real applications are directly related to the use of an incorrect or inappropriate plant model. Developing an accurate model is perhaps the most important and difficult aspect of control engineering (Houpis and Lamont, 1992). Ultimately, how well the model meets the two possibly conflicting criteria of simplicity and accuracy will be decided by the performance of the actual physical system.

There are two basic approaches to modeling: analytical modeling and empirical or experimental modeling. Both will be discussed below, with the emphasis on analytical modeling.

22.2 Analytical Modeling

There are four basic steps in the process of analytical modeling.

1. Define and structure the problem.
   - What is the purpose and intended use of the model?
   - What are its limitations and boundaries?
   - What assumptions can be made (linear, time-invariance, etc.)?
2. Determine the system components.
   - What are the inputs and outputs of the system?
   - What are the variables?
   - What are the constants?

0-8493-8343-9/97/$0.00+$.50

- What are the important system parameters?
- What are the fundamental physical laws affecting the system?
- Divide the system into subsystems whose behaviors are known. How do the various subsystems interact?
- Draw a block diagram or signal flow graph of the system.

3. Write the necessary equations to describe the system using:
   - The fundamental physical relations.
   - The conservation and compatibility laws.
4. Verify the model. The model needs to be evaluated and its performance validated. This can be done analytically or by computer simulation using a more complete (or more realistic) model. Such verification by simulation will not catch errors common to both models, and the ultimate verification must be based on experiment. If the model is not accurate enough for the particular application, then it needs to be corrected by returning to one of the above steps and modifying accordingly.

These four steps will be discussed in the following sections.

22.3 Defining the Problem

The first step in constructing a mathematical model is to determine how the system can be described given its intended purposes. As mentioned earlier, a number of assumptions or simplifications should be made about the system and the signals for the purposes of creating the model.

A system can be defined in a number of different ways.

Dynamic or Static

A *dynamic* system is one that is described by differential or difference equations, while a *static* system is one that is described only by algebraic equations.

Distributed Parameter or Lumped Parameter

A *distributed parameter* system is one that is described by partial differential equations. Systems described this way are a function of both time and space. Examples include such things as the voltage on a transmission line or a displacement on a flexible structure. *A lumped parameter* system is one described by ordinary differential or difference equations. Often a distributed parameter system is approximated as a lumped parameter system. This is usually done in control system applications.

Deterministic or Stochastic

A *deterministic* system is a system with no random parameters or inputs. In other words, the system is known exactly. On the other hand, a *stochastic* system is one in which at least one parameter or input is affected by a random disturbance or noise. The external signals that influence a system also have to be modeled. They too will either be deterministic or stochastic. These random disturbances affecting the system parameters or inputs could be known and measurable, known and unmeasurable, or unknown. Smoothing, filtering or estimation techniques are used to get an accurate response for a stochastic system.

Continuous Time or Discrete Time

A *continuous time* system is one that is represented by differential equations, while a *discrete time,* or *sampled data* system is one that is represented by difference equations. While most systems are actually continuous, they are often modeled as discrete so that a digital computer may be used for design and/or control of the system. Also, systems may be considered affected by continuous changing events, sometimes called *change oriented,* or by discrete changing events, called *discrete event oriented.*

Linear or Nonlinear

A *linear* system is one that is represented by linear equations, while *a nonlinear* system is one represented by nonlinear equations. In reality every real-world system is nonlinear, however, these are often approximated by a linear system in the operating range of interest. For many purposes this approximate linear model is sufficient for control system design and gives acceptable performance.

Time-Variant or Time-Invariant

Finally, a *time-variant* system is one in which the system parameters vary with time, while a *time-invariant* system is one in which the system parameters do not vary with time. Quite often systems are approximated as time invariant. This is especially true if the parameters only change slowly with time.

22.4 Determining the System Components

Systems can be described in terms of *across* and *through* variables, inductive and capacitive storage elements, and energy dissipaters. Similarities between diverse systems allow systems to be described by analogous circuits (Cannon, 1967; Dorf and Bishop, 1995). A comparison between the various types of engineering systems is shown in Table 22.1 (Dorf and Bishop, 1995). Of course, the actual values of the system parameters need to be identified. This may be simply an educated estimate of the parameter values or it may entail the implementation of some kind of system identification technique using experimental data.

Block diagrams and signal flow graphs are often used here to better represent and simplify the system. Such diagrams help show the interactions within the system itself. Bond graphs are also occasionally used for this purpose (Ljung and Glad, 1994).

Table 22.1 Summary of Describing Differential Equations for Ideal Elements

Type of Element	Physical Element	Describing Equation	Energy E or Power P	Symbol
	Electrical inductance	$v_{21} = L\dfrac{di}{dt}$	$E = \dfrac{1}{2}Li^2$	v_2 L i v_1
Inductive storage	Translational spring	$v_{21} = \dfrac{1}{K}\dfrac{dF}{dt}$	$E = \dfrac{1}{2}\dfrac{F^2}{K}$	v_2 K v_1 F
	Rotational spring	$\omega_{21} = \dfrac{1}{K}\dfrac{dT}{dt}$	$E = \dfrac{1}{2}\dfrac{T^2}{K}$	ω_2 K ω_1 T
	Fluid inertia	$P_{21} = I\dfrac{dQ}{dt}$	$E = \dfrac{1}{2}IQ^2$	P_2 I Q P_1
	Electrical capacitance	$i = C\dfrac{dv_{21}}{dt}$	$E = \dfrac{1}{2}Cv_{21}^2$	v_2 i C v_1
	Translational mass	$F = M\dfrac{dv_2}{dt}$	$E = \dfrac{1}{2}Mv_2^2$	F v_2 M v_1 = constant
Capacitive storage	Rotational mass	$T = J\dfrac{d\omega_2}{dt}$	$E = \dfrac{1}{2}J\omega_2^2$	T ω_2 J ω_1 = constant
	Fluid capacitance	$Q = C_f\dfrac{dP_{21}}{dt}$	$E = \dfrac{1}{2}C_fP_{21}^2$	Q P_2 C_f P_1
	Thermal capacitance	$q = C_t\dfrac{d\tau_2}{dt}$	$E = C_t\tau_2$	q $\mathcal{T}_2$ C_t $\mathcal{T}_1$ = constant
	Electrical resistance	$i = \dfrac{1}{R}v_{21}$	$P = \dfrac{1}{R}v_{21}^2$	v_2 R i v_1
	Translational damper	$F = fv_{21}$	$P = fv_{21}^2$	F v_2 f v_1
Energy dissipators	Rotational damper	$T = f\omega_{21}$	$P = f\omega_{21}^2$	T ω_2 f ω_1
	Fluid resistance	$Q = \dfrac{1}{R_f}P_{21}$	$P = \dfrac{1}{R_f}P_{21}^2$	P_2 R_f Q P_1
	Thermal resistance	$q = \dfrac{1}{R_t}T_{21}$	$P = \dfrac{1}{R_t}T_{21}$	$\mathcal{T}_2$ R_t q $\mathcal{T}$

Nomenclature

- *Through-variable:* F = force, T = torque, i = current, Q = fluid volumetric flow rate, q = heat flow rate.
- *Across-variable:* v = translational velocity, ω = angular velocity, v = voltage, P = pressure, T = temperature.
- *Inductive storage:* L = inductance, $1/k$ = reciprocal translational or rotational stiffness, I = fluid inertance.
- *Capacitive storage:* C = capacitance, M = mass, J = moment of inertia, C_f = fluid capacitance, C_t = thermal capacitance.
- *Energy dissipators:* R = resistance, f = viscous friction, R_f = fluid resistance, R_t = thermal resistance.

Source: Dorf, R. and Bishop, R. 1995. Modern Control Systems, 7th ed. © 1995 by Addison-Wesley Publishing Company. Reprinted by permission.

22.5 Writing the System Equations

Once the proper form of the elements is selected, these analogs, along with the law of conservation or continuity (relating the *through* variables) and the law of compatibility (relating the *across* variables), are used to write the differential or difference equations. Linear equations are sufficient for many system problems. Symbolic algebra computer programs such as MACSYMA Maple, Matlab, and Mathematica are becoming increasingly popular in manipulating these equations.

There are two approaches to representing these equations. The first approach represents the system as one nth order differential or difference equation. As an example, a continuous time system with a single input, u, and a single output, y, would be represented by:

$$\frac{d^n y}{dt^n} + a_{n-1}\frac{d^{n-1}y}{dt^{n-1}} + \cdots + a_1\frac{dy}{dt} + a_0 y = b_0 u + b_1\frac{du}{dt} + \cdots + b_m\frac{d^m u}{dt^m} \quad (22.1)$$

Taking the Laplace transform of Equation 22.1, and rearranging, the transfer function can be obtained as:

$$\frac{Y(s)}{U(s)} = H(s) = \frac{b_m s^m + b_{m-1}s^{m-1} + \cdots + b_1 s + b_0}{s^n + a_{n-1}s^{n-1} + \cdots + a_1 s + a_0} \quad (22.2)$$

The other approach to these equations is to use the state-space model. Here state variables and matrices are used to represent the system by n first order differential or difference equations. Using this approach, Equation 22.1, with $m = n$, can be written as the following state equation:

$$\dot{\mathbf{x}}(t) = \begin{bmatrix} -a_{n-1} & 1 & 0 & 0 & \cdots & 0 \\ -a_{n-2} & 0 & 1 & 0 & \cdots & 0 \\ \vdots & \vdots & \vdots & \vdots & \vdots & \vdots \\ -a_1 & 0 & 0 & 0 & \cdots & 1 \\ -a_0 & 0 & 0 & 0 & \cdots & 0 \end{bmatrix} \mathbf{x}(t) + \begin{bmatrix} b_{n-1} - a_{n-1}b_n \\ b_{n-2} - a_{n-2}b_n \\ \vdots \\ b_1 - a_1 b_n \\ b_0 - a_0 b_n \end{bmatrix} u(t) \quad (22.3)$$

and output equation:

$$y(t) = [1 \quad 0 \quad 0 \quad \cdots \quad 0]\mathbf{x}(t) + \mathbf{b}_n u(t) \quad (22.4)$$

Equations 22.3 and 22.4 represent just one of many valid forms of the state and output equations, but (for linear systems) they all can be represented by the compact matrix form:

$$\dot{\mathbf{x}} = \mathbf{Ax} + \mathbf{Bu}$$

and

$$y = \mathbf{Cx} + \mathbf{Du}$$

Use of direct digital control and digital computers necessitates that the system be described in a digital form by use of difference equations. These discrete system difference equations can be obtained in two different ways. The system difference equations can be written directly in discrete form from the original system, or the continuous equations can be written first, then approximated using any of a number of methods, such as forward or backward differences. The conversion to the transfer function uses the z-transform instead of the Laplace transform. As an example, a discrete time system with a single input, u, and a single output, y, would be represented by:

$$\begin{aligned} y(k+1) + a_0 y(k) + a_1 y(k-1) + \cdots + a_n y(k-n) \\ = b_0 u(k+1) + b_1 u(k) + b_2 u(k-1) \\ + \cdots + b_m u(k+1-m) \end{aligned} \quad (22.5)$$

or, after taking the z-transform and rearranging, the transfer function can be obtained as:

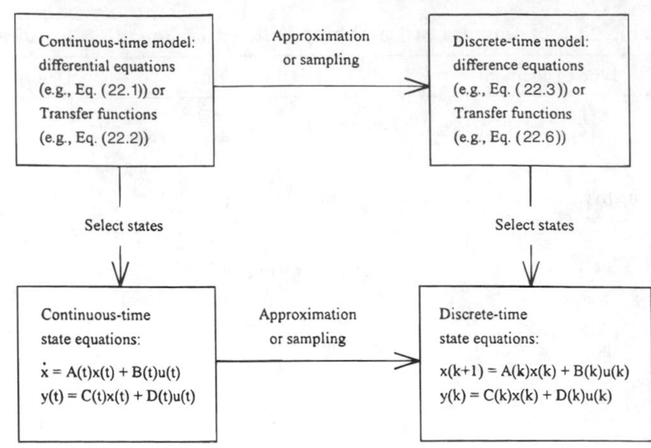

Figure 22.1 State variable modeling paradigm (*Source*: Brogan, W. L. 1991. *Modern Control Theory*, 3rd ed. Prentice-Hall, Englewood Cliffs, NJ. Used with permission.)

$$\frac{Y(z)}{U(z)} = H(z) = \frac{b_0 + b_1 z^{-1} + \cdots + b_m z^{-m}}{1 + a_0 z^{-1} + \cdots + a_n z^{-(n+1)}} \quad (22.6)$$

These equations also can be written in state variable form similar to what was done for the continuous case. As illustrated in Figure 22.1 (Brogan, 1991), these discrete time state equations can be obtained by either approximating the original differential equations and then selecting the states, or by approximating the continuous time state equations.

For a nonlinear system, the state and output equations will be given by:

$$\dot{\mathbf{x}} = \mathbf{f}(\mathbf{x}, \mathbf{u}, t)$$

and

$$\mathbf{y} = \mathbf{h}(\mathbf{x}, \mathbf{u}, t)$$

The transfer function is of no value in the nonlinear case.

22.6 Verifying the Model

The final and perhaps the most important step in the process of analytical modeling is verifying the model's accuracy. There is always a limited, valid range for any model and it is necessary to be certain that the model is realistic for the specific problem and purpose. If the model is not acceptable, then a return to one or more of the previous steps is necessary to redefine or reformulate the problem.

While this verification can be done analytically or by computer simulation, it is most easily done by simulating the system using a computer. One of the main advantages in using computer simulation is that equations do not necessarily need to be in closed form. However, when using a computer for model simulation, it is necessary that the model be in digital form. A continuous system problem must be converted to a discrete one. This conversion will either have to be done by the engineer or, as is usually the case, by the particular software used for the simulation. In either case, numerical methods will need to be used which introduce their own unique approximations. Consequently, it is important to realize that even if the simulation shows that the particular model is sufficient, the model still may have differences with the actual physical system.

Computer simulations may be done using a higher level language such as FORTRAN or C, using spreadsheets, or using any one of the many software packages such as ACSL, CONTROL-C, CSAP, MATLAB, MATRIX-X, and PROGRAM CC. Besides simulating the model of the system, these software packages are very useful for designing the control system and verifying its performance as well. Most modern software packages can also handle a variety of system types and have good graphic capabilities.

22.7 Empirical or Experimental Modeling

In addition to analytical modeling, the second basic approach to modeling is empirical or experimental modeling. Empirical or experimental modeling assumes a model structure and then uses experimental measurements to estimate the parameter values so that the model best fits the data. Such modeling can be done either on or off line. Adaptive systems, for example, carry out the estimations on line.

Usually time series models are used for empirical modeling. These models are classified as autoregressive (AR), moving average (MA) and the combination of the two known as ARMA. For these time series models, the difference equations that relate the inputs to the outputs are of the form given in Equation 22.5. The z-transform transfer function is given in Equation 22.6. For the MA (all zero) model, all $a_i = 0$, for the AR (all pole) model all $b_i = 0$, except b_0. The ARMA model has both poles and zeros.

EXAMPLE 22.1:

A model of an electromechanical system consisting of a separately excited direct current (dc) motor is developed here. For this example, assumptions are made to model the dc motor as a linear, lumped parameter, deterministic, continuous time, time invariant system (Brogan, 1991, Dorf and Bishop, 1995). A schematic of one is shown in Figure 22.2. Separately excited dc motors may be connected either as field-controlled or armature-controlled. For field-controlled, the armature current, i_a, is held constant and the motor is controlled by varying the field current, i_f. For armature-controlled, the field current, i_f, (and therefore the flux, ϕ_f) is held constant and the motor is controlled by varying the armature current, i_a. Armature-controlled motors are more stable and therefore used more often. Referring to Figure 22.2, the torque, $T(t)$, produced by the armature is proportional to the armature current, or

$$T = K_t i_a \tag{22.7}$$

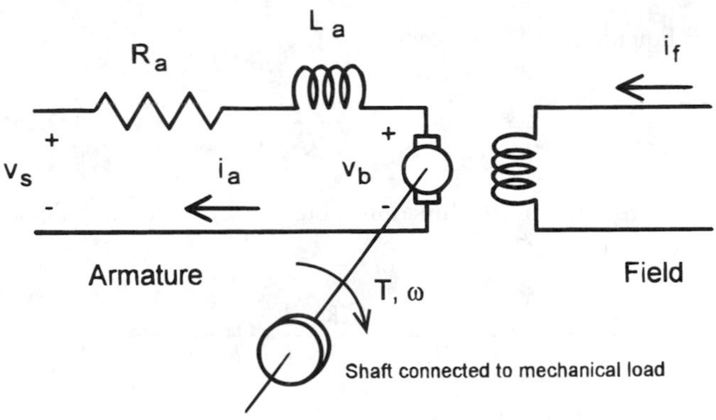

Figure 22.2 Separately excited dc motor.

where K_t is the torque constant. Writing Kirchhoff's voltage law at the armature,

$$v_s = R_a i_a + L_a \frac{di_a}{dt} + v_b \tag{22.8}$$

where v_b is the back emf, and

$$v_b = K_b \omega \tag{22.9}$$

where K_b is the back emf constant and ω is the motor's speed. Since the position of the armature, θ, is related to the speed by $\omega = d\theta/dt$, Equation 22.9 can be written as

$$v_b - K_b \frac{d\theta}{dt} \tag{22.10}$$

Substituting Equation 22.10 into Equation 22.8

$$v_s = R_a i_a + L_a \frac{di_a}{dt} + K_b \frac{d\theta}{dt} \tag{22.11}$$

With a mechanical load on the motor, the motor torque is used to accelerate the total inertia of the motor and load, J, and in overcoming the friction, which is assumed proportional to the angular velocity, $F\omega$, or

$$T = J\frac{d\omega}{dt} + F\omega = J\frac{d^2\theta}{dt^2} + F\frac{d\theta}{dt} \tag{22.12}$$

Combining Equation 22.7 with Equation 22.12

$$i_a = \frac{J}{K_t}\frac{d^2\theta}{dt^2} + \frac{F}{K_t}\frac{d\theta}{dt} \tag{22.13}$$

Taking the Laplace transform, Equation 22.11 becomes

$$V_s = R_a I_a + L_a s I_a + K_b s\Theta \tag{22.14}$$

and Equation 22.13 becomes

$$I_a = \frac{J}{K_t} s^2\Theta + \frac{F}{K_t} s\Theta \tag{22.15}$$

Therefore, the transfer function, obtained by combining Equations 22.14 and 22.15, is

$$\frac{\Theta}{V_s} = \frac{K_t}{JL_a s^3 + (JR_a + FL_a)s^2 + (FR_a + K_t K_b)s}$$

The state-space model of this system can also be written. One such configuration assigns the states as $x_1 = \theta$, $x_2 = d\theta/dt$, and $x_3 = d^2\theta/dt^2$. This leads to the state equation

$$\dot{\mathbf{x}} = \begin{bmatrix} 0 & 1 & 0 \\ 0 & 0 & 1 \\ 0 & a_{32} & a_{33} \end{bmatrix} \mathbf{x} + \begin{bmatrix} 0 \\ 0 \\ b_3 \end{bmatrix} v_s$$

where

$$a_{32} = -\frac{FR_a + K_t K_b}{JL_a}, \qquad a_{33} = -\frac{JR_a + FL_a}{JL_a}$$

and

$$b_3 = \frac{K_t}{JL_a}.$$

If the output is the armature velocity, $\omega = x_2$, then the output equation is

$$y = [0 \quad 1 \quad 0]\mathbf{x} + [0]v_s$$

It should be emphasized that this, as in all models, is only an approximation to the true system transfer function and state-space representation, since a number of assumptions and approximations are used.

References

Brogan, W. L. (1991) *Modern Control Theory,* 3d ed., Prentice-Hall, Englewood Cliffs, NJ.

Cannon, R. H. (1967) *Dynamics of Physical Systems,* McGraw-Hill, New York, NY.

Dorf, R. C. and Bishop, R. H. (1995) *Modern Control Systems,* 7th ed. Addison-Wesley, Reading, MA.

Fowkes, N. D. and Mahony, J. J. (1994) *An Introduction to Mathematical Modelling,* John Wiley & Sons, New York, NY.

Houpis, C. H. and Lamont, G. B. (1992) *Digital Control Systems: Theory, Hardware, Software,* 2d ed., McGraw-Hill, New York, NY.

Jay, F., ed. (1984) *IEEE Standard Dictionary of Electrical and Electronics Terms,* 3d ed. The Institute of Electrical and Electronics Engineers, New York, NY.

Ljung, L. and Glad, T. (1994) *Modeling of Dynamic Systems,* P. T. R., Prentice Hall, Englewood Cliffs, NJ.

Phillips, C. L. and Harbor, R. D. (1996) *Feedback Control Systems,* 3rd ed., Prentice Hall, Englewood Cliffs, NJ.

Sandquist, G. M. (1985) *Introduction to System Science,* Prentice Hall, Englewood Cliffs, NJ.

23

Basic Feedback Concept

T. H. Lee
National University of Singapore

C. C. Hang
National University of Singapore

K. K. Tan
National University of Singapore

23.1 Beneficial Effects of Feedback 454
Reduction of the System Sensitivity • Improvement of the Transient Response • Reduction of the Effects of External Disturbance and Noise • Improvement of the System Stability
23.2 Analysis and Design of Feedback Control Systems 455
23.3 Implementation of Feedback Control Systems 455

Control systems can be broadly classified into two basic categories, open-loop control systems and closed-loop control systems. While both systems are configured to provide a desired system response, they differ in their physical configurations as well as system performance. An open-loop control system, typically shown in Figure 23.1, utilizes a controller or control actuator in order to obtain the desired response of the controlled variable y. The controlled variable is often a physical variable such as speed, temperature, position, voltage, or pressure associated with a process or a servomechanism device. An input signal or command r is applied to the controller $C(s)$, whose output acts as the actuating signal, e which then actuates the controlled plant $G(s)$ to drive the controlled variable y towards the desired value. The external disturbance or noise affecting the plant is represented in the figure by a signal n, appearing at the input of the plant. An example of an open-loop control system is the speed control of some types of variable-speed electric drills. The user will move the trigger which activates an electronic circuit to regulate the voltage to the motor. The shaft speed is strongly load-dependent, so the operator must observe and counter any speed deviation through the trigger without any help from the system.

In contrast to an open-loop control system, a closed-loop control system utilizes an additional measurement and feedback of the actual output to compare the actual output with the desired output response. A simple closed-loop feedback control system is shown in Figure 23.2. The difference between the output y of the plant $G(s)$ under control and the reference input r, is amplified by the controller $C(s)$, and used to control the plant so that

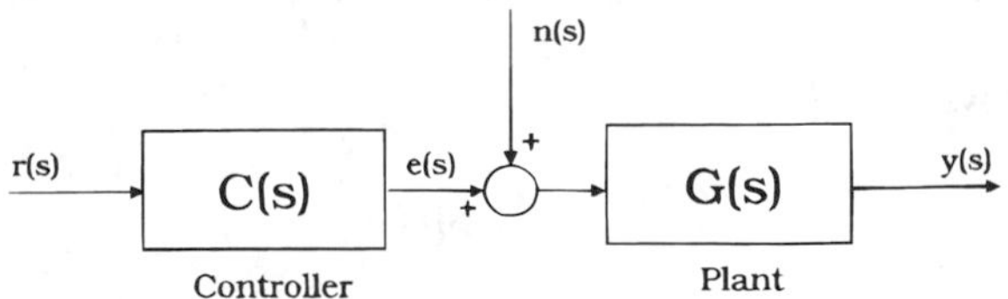

Figure 23.1 Block diagram of an open-loop control system.

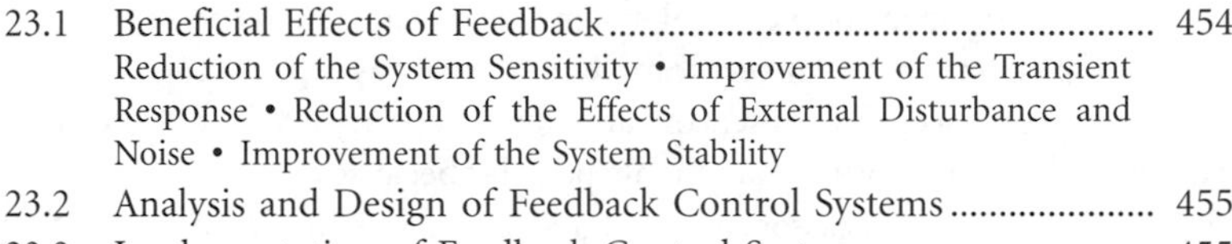

Figure 23.2 Block diagram of a closed-loop control system.

the difference is continually reduced. Figure 23.2 represents the simplest type of feedback systems and it forms the backbone of many other more complex feedback systems such as those used in cascade control, Smith predictor control, Internal Model Control, etc. It is also pertinent to note that human beings often inherently employ a feedback system in many of our every-day activities. For example, while driving the car, the driver gets visual feedback on the speed and position of the car by looking at a speedometer and checking through the mirrors and windows.

A simple position control system for a dc motor is shown in Figure 23.3. In the figure, solid lines are drawn to indicate electrical connections, and dashed-lines shows the mechanical connections. A position command Θ_r is dialed into the system and it is converted into an electrical signal V_r representative of the command via an input potentiometer. A gear connection on the output shaft transfers the output angle Θ_o to a shaft feeding back the sensed output Θ_f into an electric signal V_f which thus represents the output variable Θ_o. The power amplifier amplifies

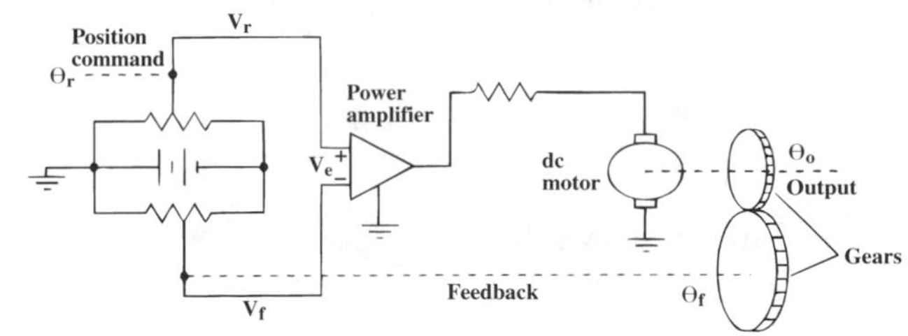

Figure 23.3 A simple position control system for a dc motor.

0-8493-8343-9/97/$0.00+$.50

the difference V_e between V_r and V_f and drives the dc motor. The motor turns until the output shaft is at a position such that V_f is equal to V_r. When this situation is reached, the error signal V_e is zero. The output of the amplifier is thus zero, and the motor stops turning.

23.1 Beneficial Effects of Feedback

Compared to an open-loop control system, feedback systems or closed-loop systems are relatively complex and costly as expensive measurement devices such as sensors are often necessary. Despite the cost and complexity, they are highly popular because of many beneficial characteristics associated with feedback. The continual self-reduction of system error is merely one of the many effects that feedback brings upon a system. We shall now show that feedback also has other effects on the system performance characteristics such as reduction of the system sensitivity, improvement of the transient response, reduction of the effects of external disturbance and noise, and improvement of the system stability.

Reduction of the System Sensitivity

To construct a suitable open-loop control system, all the components of the open-loop transfer function $G(s)C(s)$ must be very carefully selected so that they respond accurately to the input signal. In practice, all physical elements will have properties that change with environment and age. For instance, the winding resistance of an electric motor changes as the temperature of the motor rises during operation. The performance of an open-loop control system will thus be extremely sensitive to such changes in the system parameters. In the case of a closed-loop system, the sensitivity to parameter variations is reduced and the components can be less accurate without affecting the performance too much. To illustrate this, consider the open-loop system and the closed-loop system shown in Figures 23.1 and 23.2, respectively. Suppose that, due to parameter variations, $G(s)$ is changed to $G(s) + \Delta G(s)$, where $|G(s)| \gg |\Delta G(s)|$. Then, in the open-loop system shown in Figure 28.1 the output is given by

$$y(s) + \Delta y(s) = [G(s) + \Delta G(s)]C(s)r(s)$$

Hence, the change in the output is given by

$$\Delta y(s) = \Delta G(s)C(s)r(s)$$

In the closed-loop system shown in Figure 23.2,

$$y(s) + \Delta y(s) = \frac{[G(s) + \Delta G(s)]}{1 + [G(s) + \Delta G(s)]C(s)} C(s)r(s)$$

or

$$\Delta y(s) \approx \frac{\Delta G(s)}{1 + G(s)C(s)} C(s)r(s)$$

The change in the output of the closed-loop system, due to the parameter variations in $G(s)$, is reduced by a factor of $1 + G(s)C(s)$. Thus, feedback systems have the desirable advantage of being less sensitive to inevitable changes in the parameters of the components in the control system.

Improvement of the Transient Response

One of the most important characteristics of control systems is their transient response. Since the purpose of control systems is to provide a desired response, the transient response of control systems often must be adjusted until it is satisfactory. Feedback provides such a mean to adjust the transient response of a control system and thus a flexibility to improve on the system performance. Consider the open-loop control system shown in Figure 23.1 with $G(s) = K/Ts + 1$ and $C(s) = a$. Clearly, the time constant of the system is T. Now consider the closed-loop control system shown in Figure 23.2 with the same feedforward transfer function as that for Figure 23.1. It is straightforward to show that the time constant of this system has been reduced to $T/1 + Ka$. The reduction in the time constant implies a gain in the system bandwidth and a corresponding increase in the system response speed.

Reduction of the Effects of External Disturbance and Noise

All physical control systems are subject to some types of extraneous signals and noise during operation. These signals may cause the system to provide an inaccurate output. Examples of these signals are thermal noise voltage in electronic amplifiers and brush or commutator noise in electric motors. In process control, for instance considering a room temperature control problem, external disturbance often occurs when the doors and windows in the room are opened or when the outside temperature changes. A good control system should be reasonably resilient under these circumstances. In many cases, feedback can reduce the effect of disturbance/noise on the system performance. To elaborate, consider again the open-loop and closed-loop systems of Figures 23.1 and 23.2, respectively, with a nonzero disturbance/noise component n, at the input of the plant $G(s)$. Assume that both systems has been designed to yield a desired response in the absence of the disturbance/noise component, n. In the presence of the disturbance/noise, the change in the output for the open-loop system is given by

$$\Delta y(s) = G(s)n(s)$$

For the closed-loop control system, the change in the output is given by

$$\Delta y(s) = \frac{G(s)}{1 + G(s)C(s)} n(s)$$

Clearly, the effects of the disturbance/noise has been reduced by a factor of $1 + G(s)C(s)$ with feedback.

Improvement of the System Stability

Stability is a notion that describes whether the system will be able to follow the input command in a reliable manner, and feedback can be used to enhance system stability. In a non-rigorous manner, a system is said to be unstable if its output is out of control or increases without bound. A more thorough treatment on stability will be given in a later section. Here, it suffices to remark that feedback is a powerful methodology that can be used to substantially improve overall system stability. In fact, in many situations, feedback constitutes the most efficient means to provide an originally unstable system with nice and substantially improved stability properties.

23.2 Analysis and Design of Feedback Control Systems

There are numerous analysis and design techniques for feedback control systems. Depending on the complexity of the plant and the control objectives, one might want to employ a linear design method such as the classical root-locus or frequency response method for the controller, or resort to nonlinear design methods involving adaptive control, optimal control, variable structure systems, etc. These techniques are fields in their own rights and entire books can be dedicated to each one of them. In later sections, more will be discussed on these topics. There are also many tools available currently for computer-based simulation and design of feedback control systems such as MATLAB, MATRIX$_x$, SIMNON, PROGRAM CC, VISSIM, WORKBENCH, LABVIEW etc.

23.3 Implementation of Feedback Control Systems

Feedback control systems may be implemented via analog circuits or with digital computers. An electronic analog controller typically consists of operational amplifiers, resistors and capacitors. The operational amplifier is often used as the function generator, and the resistors and capacitors are arranged to implement the transfer functions of the desired control combinations. An analog implementation of control system is more suited to fairly simple systems such as those using a three-term PID controller as $C(s)$ in the feedback system of Figure 23.2. As the target control system design becomes more complex, the analog implementation will become increasingly bulky and tedious to handle. In these cases, a digital controller will be more useful. Microprocessor-based digital controllers have now become very popular in industrial control systems. There are many reasons for the popularity of such controllers. The power of the microprocessor provides advanced features such as adaptive self-tuning, multivariable control, and expert systems. The ability of the microprocessor to communicate over a field bus or local area network is another reason for the wide acceptance of the digital controller. A digital controller measures the controlled variable at specific times which are separated by a time interval called the sampling time, Δt. Each sample (or measurement) of the controlled variable is converted to a binary number for input to a digital computer. Based on the sampled information, the digital computer will execute an algorithm to calculate the controller output. More will be dealt with on the subject of digital control in a later section.

References

Dorf, R. C. 1974. *Modern Control Systems,* 2nd ed., Addison-Wesley, Reading, MA.

D' Souza, A. F. 1988. *Design of Control Systems,* Prentice Hall Inc., Englewood Cliffs, NJ.

Miron, D. B. 1989. *Design of Feedback Control Systems,* Technology Publication.

Ogata, K. 1990. *Modern Control Engineering,* Prentice Hall Inc., Englewood Cliffs, NJ.

24

Stability Analysis

24.1 Stability Analysis for Linear Systems 456
24.2 Stability of Linear Time-Invariant Continuous-Time Systems ... 456
The Routh-Hurwitz Criterion • Relative Stability • Stability under Parameter Uncertainty • Stability Analysis from Frequency Response
24.3 Stability of Linear Time-Invariant Discrete-Time Systems 463
The Routh-Hurwitz Criterion • The Jury Stability Test • Stability Analysis from Frequency Response
24.4 Nonlinear Systems 466
Linearization • Lyapunov's Method

N. K. Sinha
McMaster University

To be useful, a control system must be stable. We shall start with a definition of stability, and proceed with methods for determining whether a given system is stable or not. In practice, it is not sufficient to answer the question by "yes" or "no", since it is also desirable to know how far a stable system is from the verge of instability, and what may be done to stabilize an unstable system.

24.1 Stability Analysis for Linear Systems

Before getting into a formal definition of stability that applies to all systems, we shall first give a general definition that is easy to understand and applies to any *linear* system.

DEFINITION: 24.1 A linear system is said to be stable if its output is bounded for all possible bounded inputs.

24.2 Stability of Linear Time-Invariant Continuous-Time Systems

We shall first apply this definition to a single-input single-output linear time-invariant continuous-time system. The input-output relationship for such a system can be expressed either by a linear differential equation with constant coefficients, or by its *transfer function.* Alternatively, the differential equation can be decomposed into a set of first-order differential equations (also called *state equations*). From the requirement that the output be bounded for all possible bound inputs, we can make the following statements which are all equivalent.

1. A single-input single-output linear time-invariant con-tin-uous-time system will be stable if all the poles of its transfer function have negative real parts.
   This follows from the fact that poles in the right half of the s-plane will cause the natural response of the system to be unbounded for any input, whereas poles on the $j\omega$ axis of the s-plane will cause the output to be unbounded for certain bounded inputs. For example, if the transfer function has a pole at the origin of the s-plane, then application of a step input (which is bounded) will cause the forced response of the system to be a ramp function, which is unbounded. Similarly, if the transfer function has poles at $\pm j\alpha$, then application of the input $\sin \alpha t$, will cause the forced response to have components of the form $t \sin(\alpha t + \phi)$, where ϕ is a constant. Clearly, this is unbounded.
2. A single-input single-output linear time-invariant continuous-time system will be stable if its impulse response, $w(t)$ satisfies any of the conditions given below

$$\text{(i)} \quad \lim_{t\to\infty} w(t) \to 0$$

$$\text{(ii)} \quad \int_0^\infty w^2(t) < \infty \tag{24.1}$$

   These are seen to be the direct consequence of the fact that $w(t)$ is the inverse Laplace transform of the transfer function, all the poles of which have negative real parts.
3. A single-input single-output linear time-invariant continuous-time system described by the state equations

$$\dot{x} = Ax + Bu \tag{24.2}$$

0-8493-8343-9/97/$0.00+$.50

will be stable if and only if all the eigenvalues of A have negative real parts.

This is obvious if we recall that the poles of the transfer function are the eigenvalues of the matrix A.

These results are easily extended to multi-input multi-output linear time-invariant continuous-time (MIMOLTICT) systems. Since such a system is described by a transfer function matrix, statement 1 must apply to every element of the transfer function matrix, and statement 2 must apply to each impulse response (that is the inverse Laplace transform of each element of the transfer function matrix). No modifications is needed if the system is described by state equations, and statement 3 can be used directly.

From statements 1 and 3, it would appear that one would have to find all the roots of the denominator of the transfer function (this is also the characteristic polynomial of the matrix A, that is, the determinant of sI—A). However, this is not necessary. To determine stability, we need only find out if this polynomial has any root in the right half of the s-plane. The Routh-Hurwitz criterion enables us to obtain this information with much less effort. A brief description is given below.

The Routh-Hurwitz Criterion

The Routh-Hurwitz criterion was developed independently by A. Hurwitz (1895) in Germany and E.J. Routh (1892) in the United States. Let the characteristic polynomial (the denominator of the transfer function after cancellation of common factors with the numerator) be given by

$$\Delta(s) = a_0 s^n + a_1 s^{n-1} + \cdots + a_{n-1}s + a_n \tag{24.3}$$

Then the Routh table is obtained as follows:

$$\begin{array}{c|cccc} s^n & a_0 & a_2 & a_4 & \cdots \\ s^{n-1} & a_1 & a_3 & a_5 & \cdots \\ s^{n-2} & b_1 & b_3 & b_5 & \cdots \\ s^{n-3} & c_1 & c_3 & c_5 & \cdots \\ \vdots & & & & \\ s^0 & h_1 & & & \end{array}$$

where the first two rows are obtained from the coefficients of $\Delta(s)$. The elements of the following rows are obtained as shown below.

$$b_1 = \frac{a_1 a_2 - a_0 a_3}{a_1} \tag{24.4}$$

$$b_3 = \frac{a_1 a_4 - a_0 a_5}{a_1} \tag{24.5}$$

$$\vdots$$

$$c_1 = \frac{b_1 a_3 - a_1 b_3}{b_1} \tag{24.6}$$

$$c_3 = \frac{b_1 a_5 - a_1 b_5}{b_1} \tag{24.7}$$

and so on.

In preparing the Routh table for a given polynomial as suggested above, some of the elements may not exist. In calculating the entries in the line that follows, these elements are considered to be zero. The procedure will be clearer from the examples that follow.

The Routh-Hurwitz criterion states that the number of roots with positive real parts is equal to the number of changes in sign in the first column of the Routh table. Hence, the system is stable if and only if there are no sign changes in the first column of the table.

EXAMPLE 24.1:

Consider

$$\Delta(s) = s^4 + 5s^3 + 20s^2 + 30s + 40 \tag{24.8}$$

The Routh table is as follows:

$$\begin{array}{c|ccc} s^4 & 1 & 20 & 40 \\ s^3 & 5 & 30 & \\ s^2 & 14 & 40 & \\ s^1 & 220/14 & & \\ s^0 & 40 & & \end{array}$$

No sign changes in the first column indicate no root in the right half of the s-plane, and hence a stable system.

EXAMPLE 24.2:

Consider

$$\Delta(s) = s^3 + 2s^2 + 4s + 30 \tag{24.9}$$

The Routh table is as follows:

$$\begin{array}{c|cc} s^3 & 1 & 4 \\ s^2 & 2 & 30 \\ s^1 & -11 & \\ s^0 & 30 & \end{array}$$

Two sign changes in the first column indicate two roots in the right half of the s-plane. Hence, this is the characteristic polynomial of an unstable system.

Two special difficulties may arise while obtaining the Routh table for a given characteristic polynomial.

Case I. If an element in the first column turns out to be zero, it should be replaced by a small positive number, ε, in order to complete the table. The sign of the elements of the first column is then examined as ε approaches zero.

EXAMPLE 24.3:

Consider

$$\Delta(s) = s^5 + 2s^4 + 3s^3 + 6s^2 + 12s + 18 \quad (24.10)$$

The Routh table is shown below, where zero in the third row is replaced by ε.

s^5	1	3	12
s^4	2	6	18
s^3	ε	3	
s^2	$6 - 6/\varepsilon$	18	
s^1	$\dfrac{18\varepsilon - 18 - 18\varepsilon^2}{6\varepsilon - 6}$		
s^0	18		

As ε approaches zero the first column of the table may be simplified to obtain

s^5	1
s^4	2
s^3	ε
s^2	$-6/\varepsilon$
s^1	3
s^0	18

and two sign changes indicate two roots in the right half of the s-plane.

Case II. Sometimes an entire row of the Routh table may be zero. This indicates the presence of some roots that are negative of each other. In such cases, we should form an auxiliary polynomial from the row preceding the zero row. This auxiliary polynomial contains only alternate powers of s, starting with the highest power indicated by the leftmost column of the row, and is a factor of the characteristic polynomial. The number of roots of the characteristic polynomial in the right half of the s-plane will be the sum of the number of right-half-plane roots of the auxiliary polynomial and the number determined from the Routh table of the lower-order polynomial obtained by dividing the characteristic polynomial by the auxiliary polynomial.

EXAMPLE 24.4:

Consider

$$\Delta(s) = s^5 + 6s^4 + 10s^3 + 35s^2 + 24s + 44 \quad (24.11)$$

The Routh table is as follows:

s^5	1	10	24
s^4	6	35	44
s^3	25/6	50/3	
s^2	11	44	
s^1	0		← zero row

The auxiliary row gives the equation

$$11s^2 + 44 = 0 \quad (24.12)$$

Hence, $s^2 + 4$ is a factor of the characteristic polynomial, and we get

$$q(s) = \frac{\Delta(s)}{s^2 + 4} = s^3 + 6s^2 + 11s + 6 \quad (24.13)$$

The Routh table for $q(s)$ is as follows:

s^3	1	11
s^2	6	6
s^1	10	
s^0	6	

Since this shows no sign changes in the first column, it follows that all roots of $\Delta(s)$ are in the left half of the s-plane, with one pair of roots $s = \pm j2$ on the imaginary axis. Consequently, $\Delta(s)$ is the characteristic polynomial of an oscillatory system. According to our definition, this is an unstable system.

EXAMPLE 24.5:

Consider the case when

$$\Delta(s) = s^6 + 2s^5 + 9s^4 + 12s^3 + 43s^2 + 50s + 75 \quad (24.14)$$

The Routh table is as follows:

s^6	1	9	43	75
s^5	2	12	50	
s^4	3	18	75	
s^3	0	0		

From the third row, the auxiliary polynomial is

$$p(s) = 3(s^4 + 6s^2 + 25) = 3(s^2 + 2s + 5)(s^2 - 2s + 5) \quad (24.15)$$

which has two roots in the right half of the s-plane. Also,

$$q(s) = \frac{\Delta(s)}{p(s)} = s^2 + 2s + 3 \quad (24.16)$$

The Routh table for $q(s)$ is as follows:

s^2	1	3
s^1	2	
s^0	3	

Hence, $\Delta(s)$ has two roots in the right half of the s-plane, and is the characteristic polynomial of an unstable system.

Relative Stability

Application of the Routh criterion, as discussed above, only tells us whether the system is stable or not. In many cases, we need more information. For example, if the Routh test shows that a system is stable, we often like to know how close it is to instability, i.e., how far from the $j\omega$-axis is the pole closest to it. This information can be obtained from the Routh criterion by shifting the vertical axis in the s-plane to obtain the p-plane, as shown in Figure 24.1.

Hence, in the polynomial $\Delta(s)$, if we replace s by $p - \alpha$, we get a new polynomial $\Delta(p)$. Applying the Routh test to this polynomial will tell us how many roots $\Delta(p)$ has in the right half of the p-plane. This is also the number of roots of $\Delta(s)$ located in the region to the right of the line $s = -\alpha$ in the s-plane.

EXAMPLE 24.6:

Consider the polynomial

$$\Delta(s) = s^3 + 10.2s^2 + 21s + 2 \tag{24.17}$$

The Routh table shown below tells us that the system is stable.

s^3	1	21
s^2	10.2	2
s^1	20.804	
s^0	2	

Now we shall shift the axis to the left by 0.2, that is,

$$p = s + 0.2 \tag{24.18}$$

or

$$s = p - 0.2 \tag{24.19}$$

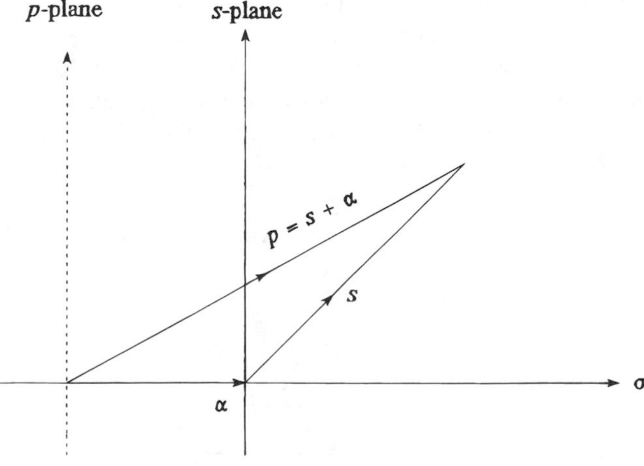

Figure 24.1 Shift of the axis to the left by α.

Hence,

$$\begin{aligned}\Delta(p) &= (p - 0.2)^3 + 10.2(p - 0.2)^2 + 21(p - 0.2) + 2\\ &= p^3 + 9.6p^2 + 17.04p - 1.8\end{aligned} \tag{24.20}$$

The following Routh table shows that $\Delta(p$ has one root in the right half of the p-plane. This implies one root between 0 and -0.2 in the s-plane. It follows that shifting by a smaller amount would enable us to get a better idea of the location of this root.

p^3	1	17.04
p^2	9.6	−1.8
p^1	17.227	
p^0	−1.8	

Hence, although $\Delta(s)$ is the characteristic polynomial of a stable system, it has a root between 0 and -0.2 in the s-plane. This indicates a large time constant, greater than 5 seconds.

Stability under Parameter Uncertainty

In our discussion so far, we have assumed that the characteristic polynomial $\Delta(s)$ for the system is known precisely. In practice, for a product to be economical, some tolerance must be allowed in the values of the components. As a result, for most systems, the coefficients a_i of the characteristic polynomial

$$\Delta(s) = a_0 s^n + a_1 s^{n-1} + \cdots + a_{n-1}s + a_n \tag{24.21}$$

are unknown except for the bounds

$$\alpha_i \le a_i \le \beta_i, \qquad i = 0, 1, \ldots, n \tag{24.22}$$

Consequently, the system will be stable if the roots of $\Delta(s)$ will be in the left half of the s-plane for each set of values of the coefficients within the ranges defined by Equation 24.22. This is called *robust parametric stability.*

According to an interesting result proved by Kharitonov (1978), it is necessary to investigate the stability for only four polynomials formed from Equations 24.21 and 24.22. We shall now discuss how one can obtain these four *Kharitonov polynomials*. These are defined in the form of the polynomial

$$\Delta(s) = c_0 s^n + c_1 s^{n-1} + \cdots + c_{n-1}s + c_n \tag{24.23}$$

where the coefficients, c_i are defined in a pairwise fashion, c_{2k}, c_{2k+1}, $k = 0, 1, \ldots, m$, where

$$m = \begin{cases} \dfrac{n}{2} & \text{if } n \text{ is even} \\ \dfrac{n-1}{2} & \text{if } n \text{ is odd} \end{cases} \tag{24.24}$$

Based on the pairwise assignment of the coefficients, the four polynomials are assigned a mnemonic $\{k = \text{even}, k = \text{odd}\}$ name as shown below.

$$\Delta_1(s)[\max, \max; \min, \min] \quad c_{2k} = \begin{cases} \beta_{2k} & \text{if } k \text{ is even} \\ \alpha_{2k} & \text{if } k \text{ is odd} \end{cases}$$

$$c_{2k+1} = \begin{cases} \beta_{2k+1} & \text{if } k \text{ is even} \\ \alpha_{2k+1} & \text{if } k \text{ is odd} \end{cases}$$

$$\Delta_2(s)[\min, \min; \max, \max] \quad c_{2k} = \begin{cases} \alpha_{2k} & \text{if } k \text{ is even} \\ \beta_{2k} & \text{if } k \text{ is odd} \end{cases}$$

$$c_{2k+1} = \begin{cases} \alpha_{2k+1} & \text{if } k \text{ is even} \\ \beta_{2k+1} & \text{if } k \text{ is odd} \end{cases}$$

$$\Delta_3(s)[\min, \max; \max, \min] \quad c_{2k} = \begin{cases} \alpha_{2k} & \text{if } k \text{ is even} \\ \beta_{2k} & \text{if } k \text{ is odd} \end{cases}$$

$$c_{2k+1} = \begin{cases} \beta_{2k+1} & \text{if } k \text{ is even} \\ \alpha_{2k+1} & \text{if } k \text{ is odd} \end{cases}$$

$$\Delta_4(s)[\max, \min; \min, \max] \quad c_{2k} = \begin{cases} \beta_{2k} & \text{if } k \text{ is even} \\ \alpha_{2k} & \text{if } k \text{ is odd} \end{cases}$$

$$c_{2k+1} = \begin{cases} \alpha_{2k+1} & \text{if } k \text{ is even} \\ \beta_{2k+1} & \text{if } k \text{ is odd} \end{cases}$$

Proofs of this remarkable result can be found in Kharitonov (1978), Bose (1985) and Minnichelli et al. (1989). We shall illustrate its usefulness by a simple example.

EXAMPLE 24.7:

Consider the feedback control system shown in Figure 24.2 for the case of unity feedback, that is $H(s) = 1$. We shall assume that the transfer function of the forward path is given by

$$G(s) = \frac{K}{s(s+a)(s+b)} \tag{24.25}$$

where, the uncertainty ranges of the parameters K, a and b are as given below:

$$K = 10 \pm 2, \ a = 3 \pm 0.5 \text{ and } b = 4 \pm 0.2 \tag{24.26}$$

We shall determine the stability of the closed-loop system using Kharitonov's four polynomials. First we determine the characteristic polynomial of the closed-loop system,

$$\begin{aligned} \Delta(s) &= a_0 s^3 + a_1 s^2 + a_2 s + a_3 \\ &= s^3 + (a+b)s^2 + abs + K \end{aligned} \tag{24.27}$$

From the given ranges on the parameters, the coefficients of the characteristic polynomial are

$$\begin{aligned} a_0 &= 1 \\ 6.3 \le a_1 &\le 7.7 \\ 9.5 \le a_2 &\le 14.7 \\ 8 \le a_3 &\le 12 \end{aligned} \tag{24.28}$$

We now obtain the four polynomials as

$$\begin{aligned} \Delta_1(s) &= s^3 + 7.7s^2 + 9.5s + 8 \\ \Delta_2(s) &= s^3 + 6.3s^2 + 14.7s + 12 \\ \Delta_3(s) &= s^3 + 7.7s^2 + 14.7s + 8 \\ \Delta_4(s) &= s^3 + 6.3s^2 + 9.5s + 12 \end{aligned}$$

Application of the Routh-Hurwitz criterion to each of these four polynomials shows that there is no sign change in the first column of Routh table for all of them. This shows that the system in this example will be stable for all possible sets of parameters within the ranges given by Equation 24.28.

Stability Analysis from Frequency Response

The method described above assumes that the transfer function of the closed-loop system is known, and is in the form of a rational function of the complex frequency variable s, that is the ratio of two polynomials of finite degree. In many practical situations, the transfer function may not be known, or it is not a rational function, as will be the case when the forward path of a closed-loop system contains an ideal delay. In such cases, one can use the Nyquist criterion for stabliity to investigate the stability of the closed-loop system from the frequency response of the open-loop transfer function. The main advantage here is that the frequency response can be obtained experimentally if the transfer function is not known.

Consider the closed-loop system shown in Figure 24.2. The main reason for possible instability of this system is that although it is designed as negative feedback, it may turn out to be positive feedback if at some frequency a phase lag of 180° occurs in the loop. If the feedback at this frequency is sufficient to sustain oscillations, the system will act as an oscillator. The frequency response of the open-loop system can therefore be utilized for investigating the stability of the closed-loop system. From superficial considerations, it will appear that the closed-loop system

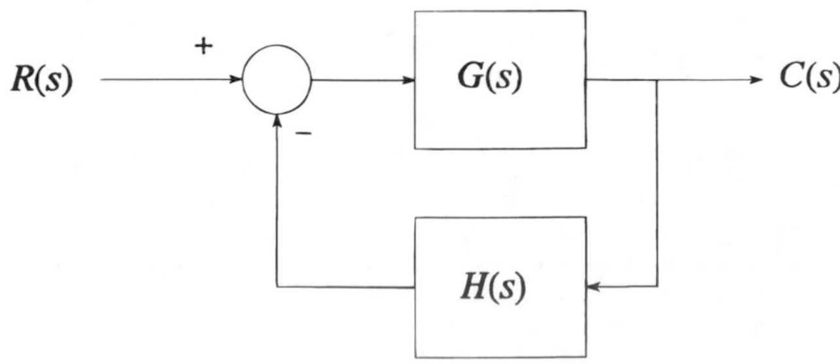

Figure 24.2 A closed-loop system.

shown in Figure 24.2 will be stable, provided that the open-loop frequency response $GH(j\omega)$ does not have gain of 1 or more at the frequency where the phase shift is 180°. However, this is not always true.

For example, consider the polar plot of the frequency response shown in Figure 24.3.

For two values of ω, the phase shift is 180° and the gain is more than one (at points A and B), yet this system will be stable when the loop is closed. A thorough understanding of the problem is possible by applying the criterion of stability developed by H. Nyquist in 1932. It is based on a theorem in complex variable theory, due to Cauchy, called the principle of the argument, which is related to mapping of a closed path (or contour) in the s-plane for a function $F(s)$.

The overall transfer function of the system shown in Figure 24.2 is given by

$$T(s) = \frac{G(s)}{1 + G(s)H(s)} \tag{24.29}$$

To find out if the closed-loop system will be stable, we must determine whether $T(s)$ has any pole in the right half of the s-plane (including the $j\omega$-axis), that is whether $F(s) = 1 + G(s)H(s)$ has any root in this part of the s-plane. For this purpose, we must take a contour in s-plane that encloses the entire right half plane as shown in Figure 24.4(a).

If $F(s)$ has a pole or zero at the origin of the s-plane or at some points on the $j\omega$-axis, we must make a detour along an infinitesimal semicircle, as shown in Figure 24.4(b). This is called the *Nyquist contour,* and our object is to determine the number of zeros of $F(s)$ inside this contour.

Let $F(s)$ have P poles and Z zeros within the Nyquist contour. Note that the poles of $F(s)$ are also the poles of $GH(s)$, but the

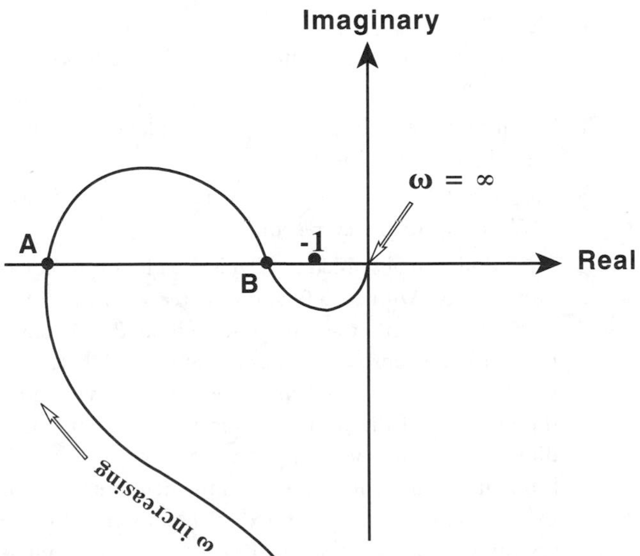

Figure 24.3 Polar plot of the frequency response of a conditionally stable system.

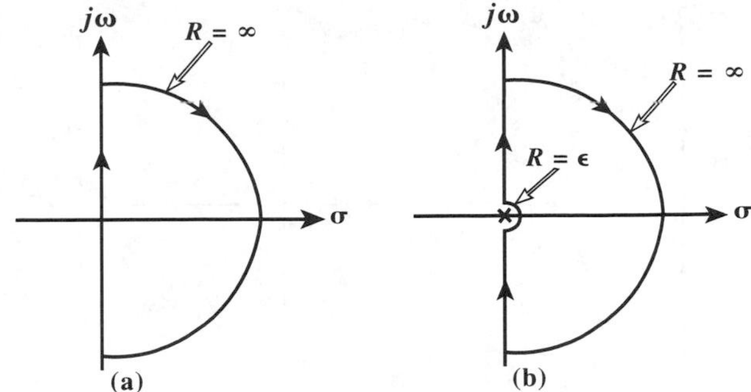

Figure 24.4 The Nyquist contour.

Table 24.1 Frequency Response of the Transfer Function given by Equation 24.31

ω	Gain	Phase shift	ω	Gain	Phase shift
0	0	0°	3.0	0.90	−158.8°
0.25	5.77	−24.0°	4.123	0.476	−180°
0.6	5.18	−46.3°	5.0	0.309	−191.9°
1.0	3.72	−82.9°	10.0	0.052	−226.4°
2.0	1.76	−130.2°	20.0	0.007	−247.4°

zeros of $F(s)$ are different from those of $GH(s)$, and not known. For the system to be stable, we must have $Z = 0$, that is the characteristic polynomial must not have any root within the Nyquist contour.

From the principle of the argument, a map of the Nyquist contour in the F-plane will encircle the origin of the F-plane N times in the clockwise direction, where

$$N = Z - P \tag{24.30}$$

Thus, the system is stable if and only if $N = -P$, so that Z will be zero. Further, note that the origin of the F-plane is the point $-1 + j0$ in the GH-plane. Hence, we get the following criterion in terms of the loop transfer function $GH(s)$.

A feedback system will be stable if the number of counter-clockwise encirclements of the point $-1 + j0$ by the map of the Nyquist contour in the GH-plane is equal to the number of poles of GH(s) inside the Nyquist contour in the s-plane.

The map of the Nyquist contour in the GH-plane is called the *Nyquist plot* of $GH(s)$. The polar plot of the frequency response of $GH(s)$, which is the map of the positive part of the $j\omega$-axis, is an important part of the Nyquist plot and can be obtained experimentally or by computation if the transfer function is known, The procedure of completing the rest of the plot will be illustrated by a number of examples.

EXAMPLE 24.8:

Consider the transfer function

$$GH(s) = \frac{60}{(s + 1)(s + 2)(s + 5)} \tag{24.31}$$

The frequency response is readily calculated, and is given in Table 24.1. Sketches of the Nyquist contour and the Nyquist plot are

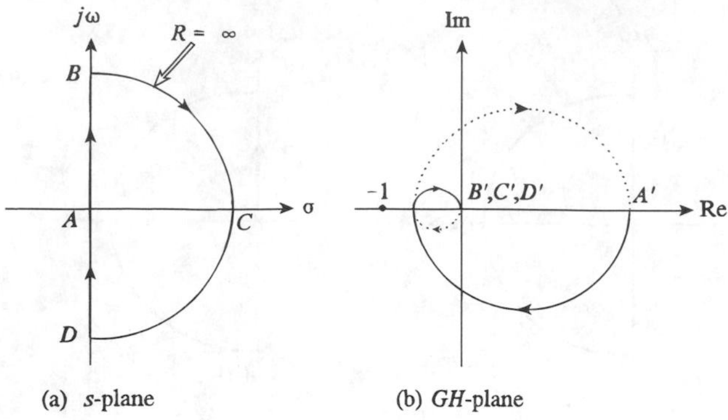

Figure 24.5 Nyquist contour and plot for the transfer function given by Equation 24.31.

shown in Figure 24.5. The various steps of the procedure for completing the Nyquist plot are given below.

1. Part *AB* in the *s*-plane is the $j\omega$-axis from $\omega = 0$ to $\omega = \infty$, and maps into the polar plot $A'B'$ in the *GH*-plane.
2. The entire infinite semicircle *BCD* maps into the origin of the *GH*-plane, since as $s \to \infty$, $GH(s) \to 0$.
3. The part *DA*, which is the negative part of the $j\omega$-axis maps into the curve $D'A'$. It may be noted that $D'A'$ is the mirror image of $A'B'$ about the real axis. This follows from the fact that $GH(-j\omega)$ is the complex conjugate for $GH(j\omega)$. Consequently, it is easily sketched from the polar plot of the frequency response.

The Nyquist plot does not encircle the point $-1 + j0$ in the *GH*-plane. Hence we have $N = 0$. Also, the function $GH(s)$ does not have any pole inside the Nyquist contour. This makes the system stable since $Z = N + P = 0$.

If we increase the open-loop gain of this system by a factor of 2.15 (which is 1/0.476 and will make the gain 1 with phase-shift 180°), or more, the Nyquist plot will encircle the point $-1 + j0$ twice in the *GH*-plane, as shown in Figure 24.6. Consequently, for this case, $N = 2$. Since *P* is still zero, the closed-loop system is unstable, with two poles in the right half of the *s*-plane.

EXAMPLE 24.9:

Consider the open-loop transfer function

$$G(s) = \frac{K}{s(s+2)(s+6)} \tag{24.32}$$

Since this time we have a pole at the origin of the *s*-plane, the Nyquist contour must take a small detour around this pole while still attempting to enclose the entire right half of the *s*-plane, including the $j\omega$-axis. Hence, we get the semicircle *EFA* of infinitesimal radius, ϵ. The resulting Nyquist contour and plot are shown in Figure 24.7. The various steps in obtaining the Nyquist plot are described below.

1. Part *AB* of in the *s*-plane is the positive part of the $j\omega$-axis from $j\epsilon$ to $j\infty$ and maps into the polar plot of the

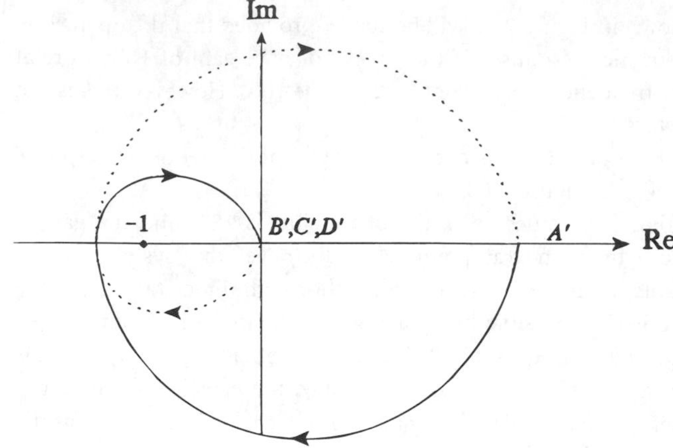

Figure 24.6 Nyquist plot for the transfer function given by Equation 24.31 if the open-loop gain is increased by a factor of 2.15 or more.

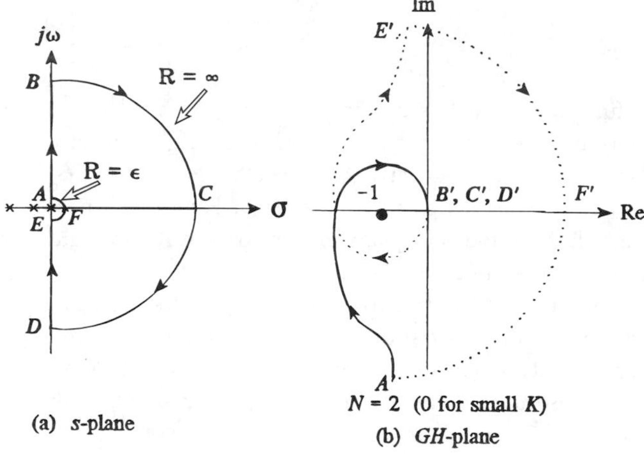

Figure 24.7 Nyquist plot for the transfer function given by Equation 24.32.

frequency response $A'B'$ in the *GH*-plane (shown as a solid curve).

2. The infinite semicircle *BCD* in the *s*-plane maps into the origin of the *GH*-plane.
3. The part *DE* (negative $j\omega$-axis) maps into the image $D'E'$ of the frequency response $A'B'$.
4. The infinitesimal semicircle *EFA* is the plot of the equation $s = \epsilon e^{j\theta}$, with $\epsilon \to 0$, and θ increasing from $-90°$ to $90°$. It maps into the infinite semicircle, $E'F'A'$, since $GH(s)$ can be approximated as $K/12\epsilon e^{j\theta} = Ke^{-j\theta}/12\epsilon$ for small *s*. Note that $s = \epsilon e^{j\theta}$ in the denominator will cause the traversal of the infinite semicircle in the opposite direction due to the change in the sign of θ as it is brought into the numerator from the denominator. For the plot shown, $N = 2$ and $P = 0$; consequently, the system will be unstable, with $Z = 2$. If the gain is sufficiently reduced, the point $-1 + j0$ will not be encircled, with the result that the system stable, since $N = 0$ and $Z = 0$.

EXAMPLE 24.10:

We shall reconsider the transfer function of the preceding example but this time we shall take a different Nyquist contour. The detour around the pole at the origin will be taken from its left, as shown in Figure 24.8(a). The resulting Nyquist plot is shown in Figure 24.8(b). The essential difference with Figure 24.7 is in the map of the infinitesimal semicircle in the Nyquist contour, which is an infinite semicircle in the counterclockwise direction. Consequently, we have $P = 1$. Also, the Nyquist plot encircles the point $-1 + j0$ once in the clockwise direction, giving us $N = 1$, so that $Z = N + P = 2$, and the system will be unstable. Also, reducing the gain sufficiently will cause the Nyquist plot to encircle the point $-1 + j0$ once in the counterclockwise direction, making $N = -1$, with the result that now we shall have $Z = N + P = 0$. Thus, the system will be stable for small values of K. Both of these results agree with those obtained in the previous example, as expected.

24.3 Stability of Linear Time-Invariant Discrete-Time Systems

The necessary and sufficient condition for the stability of a discrete-time system is that all the poles of its z transfer function lie inside the unit circle of the z-plane. This fact follows from the transformation $z = e^{sT}$, which maps the left half of the s-plane inside the unit circle.

Keeping this in mind, the two basic techniques studied for continuous-time systems—(a) the Routh-Hurwitz criterion, and (b) the Nyquist criterion—can be used with slight modifications for discrete-time systems.

The Routh-Hurwitz Criterion

The Routh-Hurwitz criterion, discussed earlier, is an attractive method for investigating system stability without requiring evaluation of the roots of the characteristic polynomial. Since it enables us to determine the number of roots with positive real parts, it cannot be used directly for discrete-time systems, where we need to find the number of roots outside the unit circle. It is possible to use the Routh-Hurwitz criterion to determine if a polynomial $Q(z)$ has roots outside the unit circle by using the *bilinear transformation*

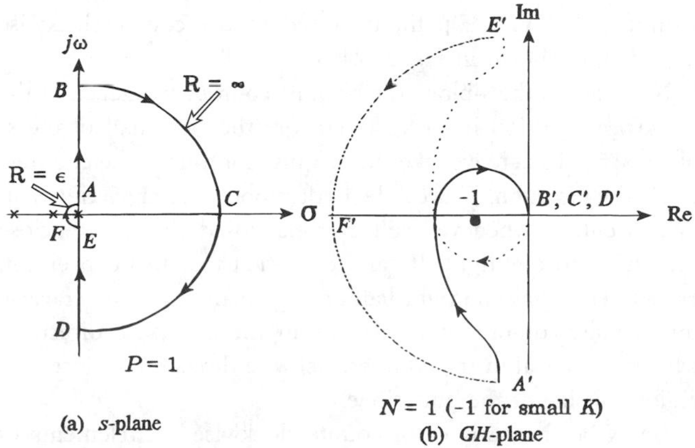

Figure 24.8 Nyquist plot for the transfer function given by Equation 24.32 with the Nyquist contour enclosing the pole at the origin.

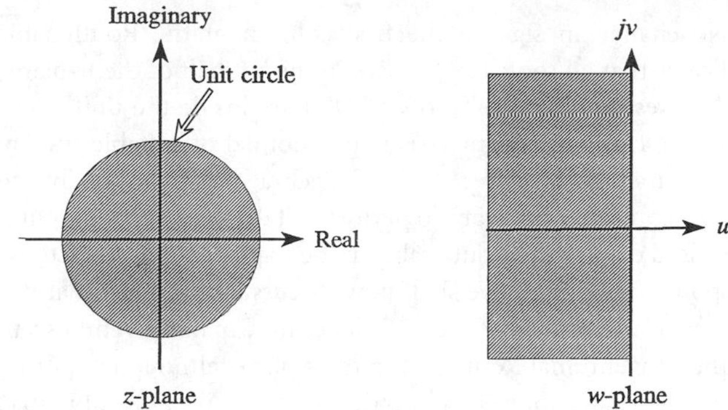

Figure 24.9 Mapping in bilinear transformation.

$$w = u + jv = \frac{z+1}{z-1}, \text{ or } z = \frac{w+1}{w-1} = \frac{u+1+jv}{u-1+jv} \tag{24.33}$$

(also called the Möbius transformation) which maps the unit circle of the z-plane into the imaginary axis of the w-plane and the interior of the unit circle into the left half of the w-plane. This is shown in Figure 24.9. We can now apply the Routh criterion to $Q(w)$ as in the s-plane.

EXAMPLE 24.11:

The characteristic polynomial of a discrete-time system is given by

$$Q(z) = z^3 - 2z^2 + 1.5z - 0.4 = 0 \tag{24.34}$$

Then

$$Q(w) = \left(\frac{w+1}{w-1}\right)^3 - 2\left(\frac{w+1}{w-1}\right)^2 + 1.5\left(\frac{w+1}{w-1}\right) - 0.4 = 0$$

or

$$(w+1)^3 - 2(w+1)^2(w-1) + 1.5(w+1)(w-1)^2 \\ - 0.4(w-1)^3 = 0$$

This can be further simplified to obtain the polynomial

$$0.1w^3 + 0.7w^2 + 2.3w + 4.9 = 0 \tag{24.35}$$

The Routh table is shown below.

w^3	0.1	2.3
w^2	0.7	4.9
w^1	1.6	
w^0	4.9	

No change in sign in the first column of the Routh table indicates that all roots of $Q(w)$ are in the left half of the w-plane, and correspondingly, all roots of $Q(z)$ are inside the unit circle. Hence, $Q(z)$ is the characteristic polynomial of a stable system.

This method requires a lot of algebra, but it is possible to write a computer program to perform all the necessary computations and display the Routh table in the w-plane for a characteristic polynomial $Q(z)$. We shall now discuss another test that can be performed directly in the z-plane, in a manner reminiscent of the conventional Routh test in the s-plane, although requiring more terms and more computation than the usual Routh table. Still, it requires less effort than the test based on the bilinear transformation described above.

The Jury Stability Test

Let the characteristic polynomial be given by

$$Q(z) = a_0 z^n + a_1 z^{n-1} + \cdots + a_{n-1} z + a_n, \quad a_0 > 0 \quad (24.36)$$

Then we form the array shown in the following table.

a_0	a_1	a_2	$\ldots\ a_k$	$\ldots\ a_{n-1}$	a_n	
a_n	a_{n-1}	a_{n-2}	$\ldots\ a_{n-k}$	$\ldots\ a_1$	a_0	$\alpha_n = a_n/a_0$
b_0	b_1	b_2	$\ldots\ b_{n-k}$	$\ldots\ b_{n-1}$		
b_{n-1}	b_{n-2}	b_{n-3}	$\ldots\ b_{k-1}$	$\ldots\ b_0$		$\alpha_{n-1} = b_{n-1}/b_0$
c_0	c_1	c_2	$\ldots\ c_{n-k}$	$\ldots$		
c_{n-2}	c_{n-3}	c_{n-4}	$\ldots\ c_{k-2}$	$\ldots$		$\alpha_{n-2} = c_{n-2}/c_0$
.	.		$\ldots$.	$\ldots$		
.	.		$\ldots$.	$\ldots$		
.	.		$\ldots$.	$\ldots$		

The elements of the first and second rows are the coefficients of $Q(z)$, in the forward and the reverse order, respectively. The third row is obtained by multiplying the second row by $\alpha_n = a_n/a_0$ and subtracting this from the first row. The fourth row is the third row written in the reverse order. The fifth row is obtained in a similar manner, that is by multiplying the fourth row by b_{n-1}/b_0 and subtracting this from the third row. The sixth row is the fifth row written in the reverse order. The process is continued until there are $2n + 1$ rows. The last row has only one element.

Jury's stability test states that if $a_0 > 0$, then all roots of $Q(z)$ are inside the unit circle if and only if b_0, c_0, d_0, etc. (that is, the first element in each odd-numbered row) are all positive. Furthermore, the number of negative elements in this set is equal to the number of roots of $Q(z)$ outside the unit circle and a zero element implies a root on the unit circle.

EXAMPLE 24.12:

The Jury table for the characteristic polynomial, $Q(z)$, given in example 24.11 is shown below.

1	−2	1.5	−0.4	
−0.4	1.5	−2	1	$\alpha_3 = -0.4$
0.84	−1.4	0.7		
0.7	−1.4	0.84		$\alpha_2 = 5/6$
0.2567	−0.2333			
−0.2333	0.2567			$\alpha_1 = -0.909$
0.0445				

Since the coefficients a_0, b_0, c_0 and d_0 are all positive, it follows that all roots of $Q(z)$ are inside the unit circle and it is the characteristic polynomial of a stable system.

It is possible to reduce the number of rows required in the Jury table from $2n + 1$ to $2n - 3$ by including the following conditions

$$Q(1) > 0 \quad (24.37)$$

$$(-1)^n Q(-1) > 0$$

This is especially useful if the degree of $Q(z)$ is two, since in that case, we need only apply the tests in Equation 24.37, along with the requirement that $|a_2| < a_0$. The results of Example 24.12 could have been obtained with much less effort using these tests.

Stability Analysis from Frequency Response

Just as in the case of continuous-time systems, we can use the frequency response of an open-loop discrete time system to investigate stability after the loop is closed. Again, the main advantage is that one does not have to know the transfer function of the system, since the frequency response can be obtained experimentally. Only a minor modification makes it possible to extend the Nyquist criterion to discrete-time systems.

To apply the Nyquist criterion to determine stability of discrete-time systems, we need to determine the number of roots of $1 + GH(z)$ in the region *outside the unit circle* of the z-plane, which is the map of the right-half of the s-plane. The Nyquist contour in this case is the unit circle of the z-plane, with detours around poles on the path, traversed in the counterclockwise direction, as shown in Figure 24.10.

Note that in the s-plane the Nyquist contour was taken in the clockwise direction in order to enclose the right half of the s-plane, whereas here we take the Nyquist contour in the counterclockwise direction. This can be understood by appreciating that, by convention, whenever we traverse a closed-curve, we enclose the region to our right. It can be verified that this convention leads to enclosing the right half of the s-plane when we traverse the Nyquist contour in that plane in the clockwise direction, whereas traversal in the counterclockwise direction encloses the region outside of the unit circle.

Let N be the number of counterclockwise encirclements of the Nyquist plot of the point $-1 + j0$ in the $GH(z)$-plane. The Nyquist criterion states that for a stable system, we must have $N = -P$, where P is the number of poles of $GH(z)$ inside the

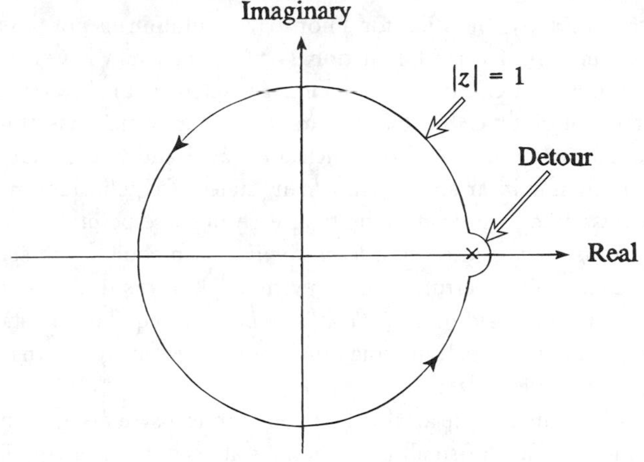

Figure 24.10 The Nyquist contour in the z-plane.

Nyquist contour (that is, in the region to our right while we traverse the contour in the z-plane).

EXAMPLE 24.13:

The Nyquist criterion will be illustrated using the unity-feedback system, the transfer function for which, is given below.

$$G(z) = \frac{0.6(z + 0.4)}{(z - 1)(z - 0.4)} \tag{24.38}$$

Since the transfer function has a pole at $z = 1$, the Nyquist contour will be as shown in Figure 24.10, with a detour around that point. On this detour

$$z = 1 + \epsilon e^{j\theta}, \qquad \epsilon << 1 \tag{24.39}$$

and

$$G(z)\,|_{z=1+\epsilon e^{j\theta}} \approx \frac{(0.6)(1.4)}{\epsilon e^{j\theta}(0.6)} = \frac{1.4}{\epsilon} e^{-j\theta} \tag{24.40}$$

This leads to an arc of infinite radius on the Nyquist plot. For z on the unit circle, we have

$$G(z)\,|_{z=e^{j\omega T}} = \frac{0.6(e^{j\omega T} + 0.4)}{(e^{j\omega T} - 1)(e^{j\omega T} - 0.4)} \tag{24.41}$$

In Equation 24.41, $0 < \omega T < 2\pi$. Since $G(e^{j\omega T})$ is the complex conjugate of $G(e^{-j\omega T})$, it is necessary to calculate the frequency response only for $0 < \omega T < \pi$. Some values of the frequency response are shown in Table 24.2.

The polar plot of the frequency response is the map of the upper half of the unit circle of the z-plane. The lower half maps as the mirror image of this response about the real axis. The plot is completed from the map of the detour, which is a semicircle of infinite radius, obtained from Equation 24.39. Figure 24.11 shows the Nyquist plot for this system, with the Nyquist contour as in Figure 24.10. From the Nyquist plot, it is evident that the system is stable. It can, however, be forced into instability if the gain is increased by a factor of 2.5 so that the point $-1 + j0$ is enclosed within the Nyquist plot.

Table 24.2 Frequency Response of the Transfer Function given by Equation 24.38.

ωT	M	ϕ	ωT	M	ϕ
0.1	13.91	−98.3°	1.2	0.69	−166.0°
0.2	6.83	−106.4°	1.4	0.52	−173.3°
0.3	4.43	−114.3°	1.6	0.41	−179.4°
0.4	3.20	−121.8°	1.8	0.33	−184.4°
0.5	2.45	−128.9°	2.0	0.26	−188.2°
0.6	1.94	−135.5°	2.2	0.22	−190.6°
0.7	1.57	−141.6°	2.4	0.18	−191.5°
0.8	1.30	−147.3°	2.6	0.16	−190.6°
0.9	1.09	−152.5°	2.8	0.14	−187.9°
1.0	0.93	−157.4	π	0.13	−180.0°

The concepts of gain margin and phase margin, can be utilized with discrete-time systems without any change. Moreover, they may be used either in the z-plane, or in the w-plane (using the bilinear transformation due to Tustin). For instance, in the problem of Example 24.13, the gain margin is 2.5 and the phase margin is about 25°.

Alternatively, one may transform $G(s)$ to the w plane using the bilinear transformation due to Tustin, i.e.,

$$w = \frac{2}{T}\frac{z - 1}{z + 1}, \qquad z = \frac{2 + wt}{2 - wT} \tag{24.42}$$

The resulting transfer function in the w-plane is obtained as

$$G(w) = \frac{0.6(28 + 0.6w)(20 - w)}{2w(12 + 1.4w)} = \frac{-0.36w^2 - 9.6w + 336}{2.8w^2 + 24w} \tag{24.43}$$

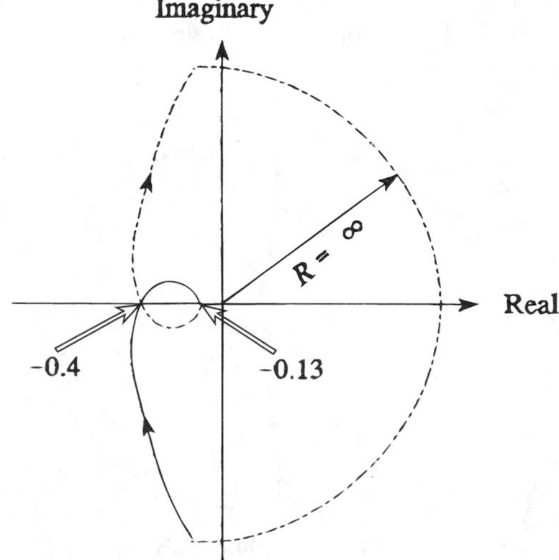

Figure 24.11 Nyquist plot for Example 24.13.

The Nyquist contour in the w-plane and the corresponding Nyquist plot are shown in Figure 24.12. As in Figure 24.11, the system is stable with gain margin 2.5 and phase margin about 25°.

24.4 Nonlinear Systems

So far we have only considered linear systems. Since most physical systems are nonlinear, it is important to study their stability properties as well. This is complicated due to the fact that superposition does not apply to nonlinear systems. Consequently, whereas stability in a linear system does not depend upon the nature or the magnitude of the input, such is not the case with nonlinear systems. As the general problem is rather complicated, we shall only study the case of *autonomous* systems, that is the case with the forcing function $u = 0$.

For proper understanding of stability, we must first recognize that all systems have one or more states of equilibrium. A state of equilibrium may be defined as a point in the state space where, in the absence of any input (forcing function) or disturbance, the time derivative of the state vector is zero. In other words, given the state equation of a system

$$\dot{x} = f(x, u) \tag{24.44}$$

the equilibrium states will be all vectors x for which

$$f(x) = 0 \tag{24.45}$$

For a linear system, Equation 24.45 reduces to the form

$$Ax = 0 \tag{24.46}$$

One solution of Equation 24.46 is $x = 0$, or the origin of the state-space. This is the only possible solution for a stable autonomous linear system, since A is only allowed to have eigenvalues with negative real parts for such a system, and a zero eigenvalue

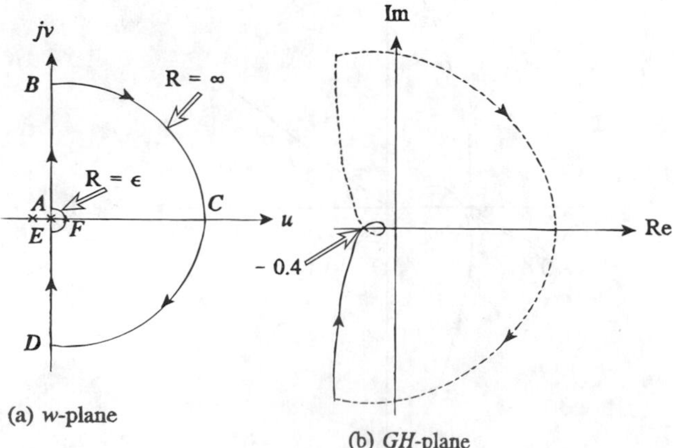

Figure 24.12 Nyquist plot in the w-plane.

(the necessary condition for a non-trivial solution) is not possible. On the other hand, Equation 24.44 will generally have many solutions. An example of a nonlinear system with three states of equilibrium is the bistable multivibrator, which has three equilibrium states, two of which are stable and one unstable. For any nonlinear system with many states of equilibrium, it is necessary to investigate stability at each of these points.

A system is said to have *local stability* at an equilibrium state if, after a small perturbation, it eventually returns to that state.

A system is said to have *global stability* at an equilibrium state if, for any perturbation (small or large), it eventually returns to the that state.

We can investigate local stability of nonlinear systems by examining the effect of small perturbations at each point of equilibrium. This can be done by obtaining an approximate linear model at each of these points and testing it for stability. This is described below.

Linearization

Linearization is based on the Taylor series expansion of a nonlinear function about an operating point. For example, consider a nonlinear function, $f(x)$. It can be written as

$$f(x) = f(x_0) + \left.\frac{df}{dx}\right|_{x=x_0}(x - x_0) + \left.\frac{d^2f}{dx^2}\right|_{x=x_0}\frac{(x - x_0)^2}{2!} + \cdots \tag{24.47}$$

We get a linear approximation of Equation 24.47 if we ignore all terms except the first two. Clearly, this will be a good approximation if either $(x - x_0)$ is very small, or the higher order derivatives of f are very small. This is the main idea behind the incremental linear models used for the analysis of electronic circuits.

We shall now generalize this to the case of the state equations for nonlinear systems. Assuming that the dimension of x is n, and that of u is m, we can write Equation 24.47 as

$$\dot{x} = f(x, u) = \begin{bmatrix} f_1(x, u) \\ f_2(x, u) \\ \vdots \\ f_n(x, u) \end{bmatrix} \tag{24.48}$$

Ignoring the higher-order terms in the Taylor series expansion of this vector differential equation leads to the linearized model (assuming $x_0 = 0$, that is, the coordinates have been transformed to make the origin the point of equilibrium, and $u_0 = 0$)

$$\dot{x} = Ax + Bu \tag{24.49}$$

where

$$A = \begin{bmatrix} \frac{\partial f_1}{\partial x_1} & \frac{\partial f_1}{\partial x_2} & \cdots & \frac{\partial f_1}{\partial x_n} \\ \frac{\partial f_2}{\partial x_1} & \frac{\partial f_2}{\partial x_2} & \cdots & \frac{\partial f_2}{\partial x_n} \\ \vdots & \vdots & \cdots & \vdots \\ \frac{\partial f_n}{\partial x_1} & \frac{\partial f_n}{\partial x_2} & \cdots & \frac{\partial f_n}{\partial x_n} \end{bmatrix} \qquad (24.50)$$

and

$$B = \begin{bmatrix} \frac{\partial f_1}{\partial u_1} & \frac{\partial f_1}{\partial u_2} & \cdots & \frac{\partial f_1}{\partial u_m} \\ \frac{\partial f_2}{\partial u_1} & \frac{\partial f_2}{\partial u_2} & \cdots & \frac{\partial f_2}{\partial u_m} \\ \vdots & \vdots & \cdots & \vdots \\ \frac{\partial f_n}{\partial u_1} & \frac{\partial f_n}{\partial u_2} & \vdots & \frac{\partial f_n}{\partial u_m} \end{bmatrix} \qquad (24.51)$$

A and B are said to be Jacobian matrices. Again, this linear model is valid only for small deviations around the point of equilibrium. Nevertheless, it can be utilized for the investigation of local stability around the point of equilibrium (for the autonomous) case by simply applying the Routh criterion to the characteristic polynomial of A. The following example will illustrate the main idea behind this approach.

Note that in Equation 24.49, we have used x instead of the deviation $x - x_0$. This is a common practice, even if x_0 is not the origin of the state space. It is to be understood that x represents the variation of the state from the point of equilibrium, or "set-point" in the terminology of process control.

EXAMPLE 24.14:

Consider the following second-order nonlinear differential equation:

$$\frac{d^2x}{dt^2} + 2x\frac{dx}{dt} + 2x^2 - 4x = 0 \qquad (24.52)$$

a. Determine the points of equilibrium.
b. Investigate the stability of the system near each point of equilibrium.

Solution We shall first derive state equations for the given differential equation. Let $x_1 = x$ and $x_2 = \dot{x}$. Then we obtain

$$\left.\begin{aligned} f_1 &= \dot{x}_1 = x_2 \\ f_2 &= \dot{x}_2 = -2x_1x_2 - 2x_1^2 + 4x_1 \end{aligned}\right\} \qquad (24.53)$$

The points of equilibrium are obtained by setting the two derivatives in Equation 24.53 to zero, and are readily found as (0,0) and (2,0). The Jacobian matrix is obtained as

$$\begin{bmatrix} \frac{\partial f_1}{\partial x_1} & \frac{\partial f_1}{\partial x_2} \\ \frac{\partial f_2}{\partial x_1} & \frac{\partial f_2}{\partial x_2} \end{bmatrix} = \begin{bmatrix} 0 & 1 \\ -4x_1 - 2x_2 + 4 & -2x_1 \end{bmatrix} \qquad (24.54)$$

Hence, for the equilibrium point given by $x_1 = 0$, $x_2 = 0$, we have

$$A = \begin{bmatrix} 0 & 1 \\ 4 & 0 \end{bmatrix} \qquad (24.55)$$

The characteristic polynomial for this case is

$$\det(sI - A) = \begin{vmatrix} s & -1 \\ -4 & s \end{vmatrix} = s^2 - 4 \qquad (24.56)$$

indicating an unstable system. For the equilibrium point given by $x_1 = 2$, $x_2 = 0$, we have

$$A = \begin{bmatrix} 0 & 1 \\ -4 & -4 \end{bmatrix} \qquad (24.57)$$

The characteristic polynomial for this case is

$$\det(sI - A) = \begin{vmatrix} s & -1 \\ 4 & s + 4 \end{vmatrix} = s^2 + 4s + 4 \qquad (24.58)$$

which shows that the system is stable in the neighborhood of this point.

Lyapunov's Method

Note that the approach described above enables us to examine local stability of an autonomous nonlinear system at its points of equilibrium. Investigation of global stability is more difficult. As stated earlier, many different classes of stability have been defined for nonlinear systems. Here, we shall discuss stability in the sense of Lyapunov.

Consider a region ε in the state space enclosing an equilibrium point x_0. Then this is a point of stable equilibrium provided that there is a region $\delta(\varepsilon)$ contained within ε such that any trajectory starting in the region δ does not leave the region ε. This is shown if Figure 24.13 for a system with two states.

Note that with this definition it is not necessary for the trajectory to approach the point of equilibrium. It is only required that the trajectory be within the region ε. This permits the existence of oscillations of limited amplitude, like limit cycles.

Lyapunov's direct method provides a means for determining the stability of a system without actually solving for the trajectories in the state space. It is based on the simple concept that the

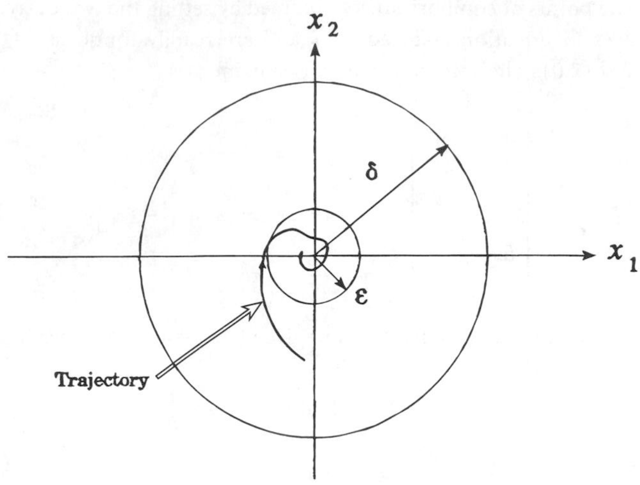

Figure 24.13 Stability in the sense of Lyapunov.

energy stored in a stable system cannot increase with time. Given a set of nonlinear state equations, one first defines a scalar function $V(x)$ that has properties similar to energy and then examines its derivative with respect to time.

THEOREM 24.1 A system described by $\dot{x} = f(x)$ is asymptotically stable in the vicinity of the point of equilibrium at the origin of the state space if there exists a scalar function V such that

1. $V(x)$ is continuous and has continuous first partial derivatives at the origin.
2. $V(x) > 0$ for $x \neq 0$ and $V(0) = 0$.
3. $\dot{V}(x) < 0$ for all $x \neq 0$.

Note that these conditions are *sufficient but not necessary* for stability. $V(x)$ is often called a Lyapunov function.

THEOREM 24.2 A system described by $\dot{x} = f(x)$ is unstable in a region Ω about the equilibrium at the origin of the state space if there exists a scalar function V such that

1. $V(x)$ is continuous and has continuous first partial derivatives in Ω.
2. $V(x) \geq 0$ for $x \neq 0$ and $V(0) = 0$.
3. $\dot{V}(x) > 0$ for all $x \neq 0$.

Again it should be noted that these conditions are *sufficient but not necessary.*

EXAMPLE 24.15:

Consider the system described by the equations

$$\begin{aligned}\dot{x}_1 &= x_1 - x_2 \\ \dot{x}_2 &= -x_1 - 2x_2^3 \end{aligned} \tag{24.59}$$

If we make

$$V = x_1^2 + x_2^2 \tag{24.60}$$

which satisfies conditions 1 and 2, then we get

$$\begin{aligned}\dot{V} &= 2x_1\dot{x}_1 + 2x_2\dot{x}_2 = 2x_1(x_1 + x_2) + 2x_2(-x_1 - 2x_2^3) \\ &= -2x_1^2 - 4x_2^4 \end{aligned} \tag{24.61}$$

It will be seen that $\dot{V} < 0$ for all nonzero values of x, and hence the system is asymptotically and globally stable.

EXAMPLE 24.16:

Consider the system described by

$$\begin{aligned}\dot{x}_1 &= -x_1(1 - 2x_1x_2) \\ \dot{x}_2 &= -x_2 \end{aligned} \tag{24.62}$$

Let

$$V = \frac{1}{2}x_1^2 + x_2^2 \tag{24.63}$$

which satisfies conditions 1 and 2. Then

$$\dot{V} = x_1\dot{x}_1 + 2x_2\dot{x}_2 = -x_1^2(1 - 2x_1x_2) - 2x_2^2 \tag{24.64}$$

Although it is not possible to make a general statement regarding global stability for this case, it is clear that $\dot{V}$ is negative if $1 - 2x_1x_2 > 0$. This defines a region of stability in the state-space, bounded by all points for which $x_1x_2 < 0.5$.

Note that our inability to find a function which satisfies all the conditions of Theorem 24.1 does not establish that the system is not globally stable.

Evidently, the main problem with this approach is the selection of a suitable function $V(x)$ such that its derivative is either positive definite or negative definite. Unfortunately, there is no general method that will work for every nonlinear system. In his referenced work, Gibson (1963) has described the variable gradient method for generating Lyapunov functions. It will not be described here but the interested reader may refer to section 8.1 of the Gibson text. Atherton (1982) and Sinha (1994) provide further information on Lyapunov's direct method.

References

Atherton, D. P. 1982. *Nonlinear Control Engineering,* Van Nostrand-Reinhold, London.

Bose, N. K. 1985. A systematic approach to stability of sets of polynomials, *Contemporary Mathematics,* 47:25–34.

Gibson, J. E. 1963. *Nonlinear Automatic Control,* McGraw-Hill, New York, NY.

Kharitonov, V. L. 1978. Asymptotic stability of an equilibrium position of a family of systems of linear differential equations, *Differential'nye Uraveniya,* 14:1483–1485.

Minnichelli, R. J., Anagnost, J. J., and Desoer, C. A. 1989. An elementary proof of Kharitonov's stability theorem with extensions, *IEEE Trans. on Automatic Control,* AC–34:995–998.

Sinha, N. K. 1994. *Control Systems,* 2nd ed., Wiley Eastern, New Delhi.

25

PID Control

25.1 Introduction 470
25.2 Classical PID Control (Ziegler-Nichols Tuning) 470
First Technique • Second Technique
25.3 Remarks 472

James C. Hung
University of Tennessee, Knoxville

25.1 Introduction

PID control was developed in the early 1940s for controlling processes of the first-order-lag-plus-delay (FOLPD) type. The type is modeled by the transfer function

$$G_P(s) = \frac{Ke^{-\tau s}}{1 + Ts} \tag{25.1}$$

where K is the d-c gain of the process, T is the time constant of the first order lag, and τ is the transport-lag or delay. The controller generates a control signal which is proportional to the system error, its time integral, and its time derivative.

A PID controller is often modeled in one of the following two forms.

$$G_C(s) = K_P + \frac{K_I}{s} + K_D s \tag{25.2}$$

$$G_C(s) = K_P\left(1 + \frac{1}{T_I s} + T_D s\right) \tag{25.3}$$

The effect of each term on the closed-loop system is intuitively clear. The proportional term is to affect system error and system stiffness; the integral term is mainly for eliminating the steady-state error; the derivative term is for damping oscillatory response. Note that having the pure integration term makes the PID controller an active network element. Dropping appropriate terms in these equations gives P, PD, or PI controllers. Determination of controller parameters for a given process is the controller design.

Two commonly used PID control configurations are shown in Figure 25.1. In Figure 25.1 (a), the entire PID forms a cascade compensator, while in Figure 25.1 (b), the derivative part is in the feedback. The latter is sometimes more convenient for practical reasons. For example, in motor control, the motor shaft may already have a tachometer, so the derivative output is already available.

25.2 Classical PID Control (Ziegler-Nichols Tuning)

Two celebrated Ziegler-Nichols techniques were developed in the early 1940s for tuning PID control of FOLPD systems (Ziegler and Nichols, 1942). Both techniques are very easy to do, making them very popular.

First Technique

The first technique can be performed on-line and does not require knowledge of the model parameters. It involves an adjustment procedure which can be carried out by a technician. The result is often satisfactory but can be crude. Equation 25.2 is the controller model for this technique. The tuning procedure is given as follows.

To begin, set K_P, K_I, and K_D all to zero.

Increase the value of K_P until oscillation occurs. At this setting, record

$$K_{max} = K_P \quad \text{and} \quad T = \text{oscillation period.}$$

For PID controller, set:

$$K_p = 0.6\ K_{max}$$
$$K_I \le 2K_p/T$$
$$K_D \ge .125\ K_p T$$

For PI controller, set:

$$K_p = 0.45 K_{max}$$
$$K_I \le 1.2\ K_p/T$$

For PD controller, set:

$$K_p = 0.65\ K_{max}$$
$$K_D \ge .125\ K_p T$$

For P controller, set:

$$K_p = 0.5\ K_{max}$$

0-8493-8343-9/97/$0.00+$.50

Second Technique

The second technique consists of two steps:

1. The determination of the model parameters *K*, τ, and *T*, from a step response test;
2. Setting controller parameters.

The steps can be computerized which have been incorporated into some of the commercial PID controllers (Rovira et al., 1969).

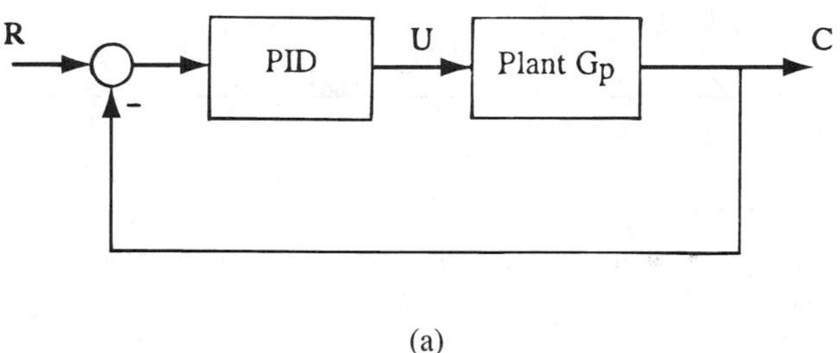

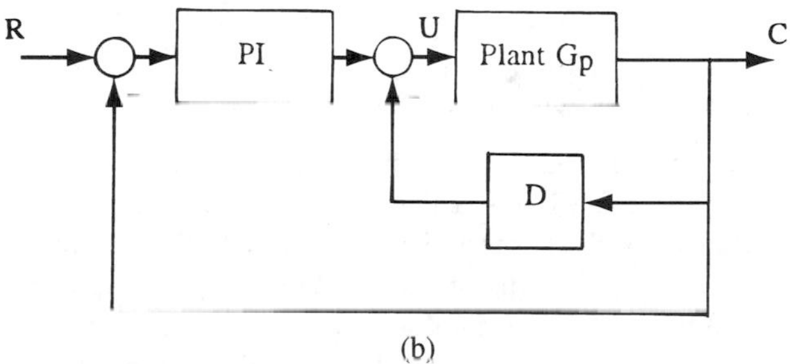

Figure 25.1 Two configurations for PID controlled systems: (a) cascade compensation; (b) cascade and feedback compensation.

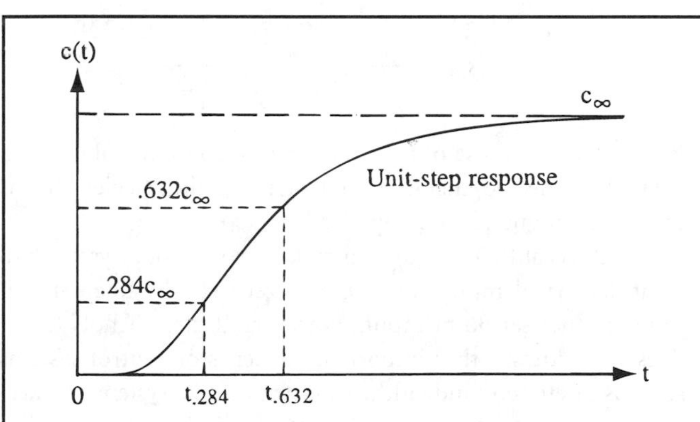

1. $K = c_\infty - c(0)$
2. Locate $t_{.284}$ and $t_{.632}$, the time instants when $c(t) = .284c_\infty$ and $c(t) = .632c_\infty$, respectively.
3. Solve T and τ from the two equations

$$t_{.284} = \tau + \frac{T}{3} \quad \text{and} \quad t_{.632} = \tau + T .$$

Figure 25.2 A graphical FOLPD parameter identification method.

Unit-step response equation

$$c(t_i) = K[1 - e^{-(t_i - \tau)/T}]$$

$$K = c_\infty - c(0)$$

Let $\quad x(t_i) = \ln[1 - \frac{c(t_i)}{K}] = (\tau - t_i) \qquad i = 1, 2, ---$

Choose $\tau = \tau_1$, and determine T by least-square using n measurement $x(t_i)$, i = 1 to n.

$$\mathbf{x} = \begin{bmatrix} x(t_1) \\ \bullet \\ \bullet \\ x(t_n) \end{bmatrix} = \begin{bmatrix} (\tau_1 - t_1) \\ \bullet \\ \bullet \\ (\tau_1 - t_n) \end{bmatrix} \frac{1}{T} = \mathbf{h}\frac{1}{T}$$

$$\frac{1}{T} = (\mathbf{h}^T\mathbf{h})^{-1}\mathbf{h}^T\mathbf{x}$$

Compute $\quad y(t) = K\,[1 - e^{-(t - \tau_1)/T}]$

$$J = \sum_{i=1}^{n} [c(t_i) - y(t_i)]^2$$

Repeat the above process using different τ's until a minimum J is reached. The values of τ and T at the minimum J is the desired values.

Figure 25.3 A least-square based FOLPD parameter indentification algorithm.

$$K_P = \frac{A_P}{K}\left(\frac{\tau}{T}\right)^{B_P}$$

$$T_I = \frac{T}{A_I\left(\frac{\tau}{T}\right)^{B_I}} \text{ for regulator,} \quad T_I = \frac{T}{A_I + B_I\left(\frac{\tau}{T}\right)} \text{ fr tracker.}$$

$$T_D = T\,A_D\left(\frac{\tau}{T}\right)^{B_D}$$

R/T	A_P	B_P	A_I	B_I	A_D	B_D
P	.902/ -	-.985/ -	- / -	- / -	- / -	- / -
PI	.984/.586	-.986/-.916	.608/1.03	-.707/-.165	- / -	- / -
PID	1.435/.965	-.921/-.855	.878/.796	-.749/-.147	.482/.308	1.137/.929

R/T : (value for regulator) / (value for tracker)

Figure 25.4 Formulas and data for tuning PID family of controllers.

The result obtained from the second technique is more refined. Equation 25.3 is the controller model for this technique. The process parameters are determined from its step response. Graphical techniques as well as computer techniques are available for such determination. Figure 25.2 shows a typical graphical techniques and Figure 25.3 gives a least-square based computer alogorithm. The parameter setting is done by using the formula

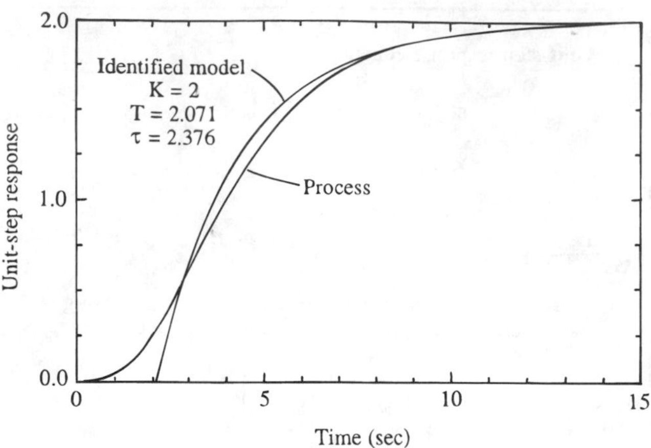

Figure 25.5 Unit-step responses of a process and its identified FOLPD model.

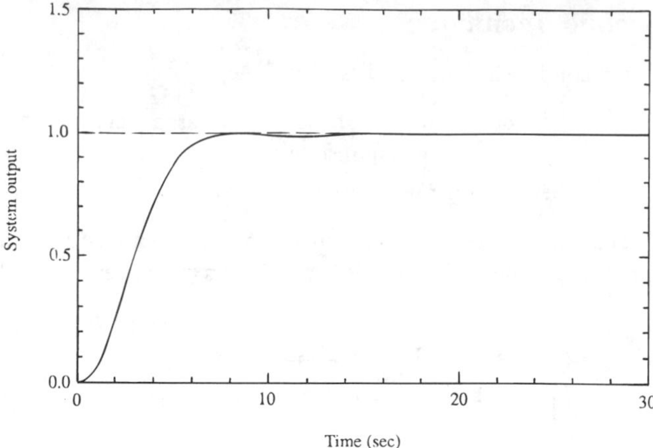

Figure 25.7 Unit-step input response of a PID controlled tracking system.

and the table contained in Figure 25.4 where values in the table were established empirically (Cheng and Hung, 1985).

Both techniques have become classics. Details of PID tuning and some further development can be found in Cheng and Hung, 1985 and Franklin et al. 1994.

EXAMPLE 25.1:

The unit-step response of a process is shown in Figure 25.5. The computer algorithm of Figure 25.3 is used to identify the parameters of the FOLPD model. The unit-step response of the identified model is also shown in Figure 25.5 where values of the identified parameters are also shown. Computer formulas of Figure 25.4 are used to tune the PID controller. For regulator control, $K_P = .814$, $T_I = 2.442$, and $T_D = .980$; for tracking control, $K_P = .543$, $T_I = 3.558$, and $T_D = .644$. Comparing the two sets of parameter values, one sees that the tracking control system is more damped than the regulator system. The closed-loop regulator sysem responding to a disturbance is shown in Figure 25.6; the closed-loop tracking system responding to a unit-step input is shown in Figure 25.7.

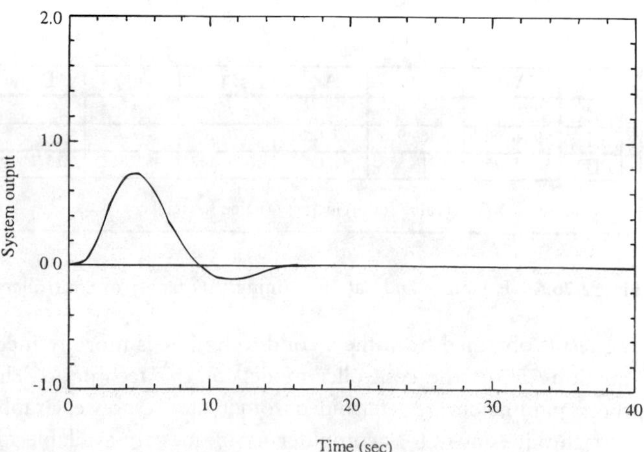

Figure 25.6 Unit-step disturbance response of a PID controlled regulator.

25.3 Remarks

The classical PID techniques have been very useful for process control, since many industrial processes can be well approximated by FOLPD models. The techniques are also effective for processes modeled by other forms of transfer functions, as long as their input-output characteristics resemble that of FOLPD. In general, these techniques can be applied to any linear process whose step response does not have overshoot. In fact, the true transfer function of the process in the above example is not a FOLPD one. It is given by

$$G_P(s) = \frac{2}{(s+1)^2(s+2)(s+.5)}$$

But the step response of this transfer function resembles that of a FOLPD one. So the PID control using the Ziegler-Nichols tuning technique can be applied.

The classical PID control can make a closed-loop system have a satisfactory damping ratio ξ and zero steady-state error in response to a set-point input. However, Ziegler-Nichols tuning does not address other important concerns of control systems, such as risetime, bandwidth, robustness, and system stiffness. Therefore, the approach is for very limited objectives.

It should be pointed out that a PID controller can be designed using a Bode plot, root-locus or other technique to achieve more sophisticated objectives.

References

Cheng, G. S. and Hung, J. C. (1985) A least-square based self-tuning of PID controller, *Proc. IEEE Southeastcon*, IEEE Catalog no. 85CH2161-8, p. 325.

Franklin, G. F., Powell, J. D., and Emami-Naeini, A. (1994) *Feedback Control of Dynamic Systems*, 3d ed. Addison-Wesley, Reading, PA.

Rovira, A. A., Murrill, P. W., and Smith, C. L. (1969) Tuning controllers for set-point changes, *Instruments and Control Systems*, 42(12):67.

Ziegler, J. G. and Nichols, N. B. (1942) Optimum setting for automatic controllers, *Trans. ASME*, 64(11):759.

26

Bode Diagram Method

26.1 Bode Diagram Analysis 474
Mathematical Development • The Asymptotic Bode Plot • Asymptotic Magnitude Plots • Asymptotic Phase Plots • Complex Poles and Zeros

26.2 Mathematical Model Determination 478

26.3 Correlation of Frequency Response and Time Response 480
Stability Determination • Transient Response • Steady-State Performance

26.4 Shaping the Cutoff Response 481

26.5 Compensator Design 482
Phase-Lead and Phase-Lag Compensators • Phase-Lag Compensator Design • Analytical Method of Phase-Lead and Phase-Lag Compensator Design • Analytical Method for Phase-Lag Compensator Design • Lag-Lead Compensators • PID Compensators

26.6 Design for Digital Systems 486
Emulation Design Methods • Bilinear Approximation (Tustin's Method) • Matched Pole-Zero Method • Direct Discrete Design via W-Plane • Digital Controller Implementation • Summary

John Parr
University of Evansville

Control systems used in industrial applications are probably most commonly designed using frequency response techniques. There are some practical reasons for this. Situations in which frequency response design methods are particularly useful occur when the system's mathematical model is unknown or when there is a considerable amount of uncertainty about the model. If the system to be controlled is stable in open-loop operation and if its frequency response can be measured experimentally, then a controller can be designed using frequency response techniques without the development of a mathematical model.

The Bode diagram method of compensator design uses the Bode plot, developed by H. W. Bode, to display the frequency response of the system. The frequency response data is analyzed and compensators are designed based on the phase margin and gain margin derived from the Nyquist stability criterion. This methodology is one of the classical feedback control system design techniques. Both Bode and Nyquist developed their respective frequency response analysis techniques while working for Bell Laboratories in the 1930s.

26.1 Bode Diagram Analysis

Mathematical Development

The Bode diagram method of control system design can be applied without knowledge of the system's transfer function. However, to discuss the mathematical background for the technique, a transfer function model is used. Consider the system shown in Figure 26.1, which has open-loop transfer function $G(s)H(s)$. In general form

$$G(s)H(s) = \frac{a_m s^m + a_{m-1}s^{m-1} + \cdots + a_1 s + a_0}{b_n s^n + b_{n-1}s^{n-1} + \cdots + b_1 s + b_0}$$

In product-of-sums form the general form, open-loop transfer function is

$$G(s)H(s) = \frac{K(s - z_1)(s - z_2)\cdots(s - z_m)}{(s - p_1)(s - p_2)\cdots(s - p_n)}$$

where the z_i and p_i are, respectively, the zeros and poles of the open-loop system. The open-loop frequency response is found by evaluating the open-loop transfer function for values which lie on the $j\omega$ axis in the s-plane.

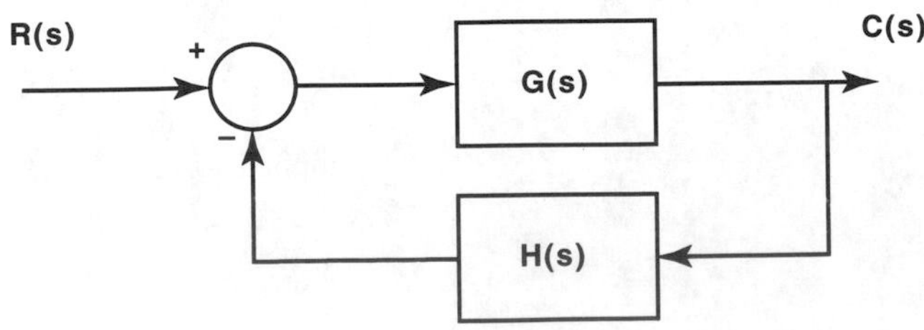

Figure 26.1 A system with feedback.

0-8493-8343-9/97/$0.00+$.50

$$G(j\omega)H(j\omega) = \frac{K(j\omega - z_1)(j\omega - z_2) \cdots (j\omega - z_m)}{(j\omega - p_1)(j\omega - p_2) \cdots (j\omega - p_n)}$$

It is apparent that frequency responses are generally complex valued.

To make the following explanation less complicated, consider all poles and zeros to be real-valued. Complex-valued poles and zeros will be considered separately.

For real-valued poles and zeros, each factor of the numerator enumerator and denominator factor can be expressed in polar form

$$j\omega - z_i = Z_i\angle\phi_i \quad \text{and} \quad j\omega - p_i = P_i\angle\theta_i$$

Where

$$Z_i = \sqrt{\omega^2 + z_i^2}, \qquad P_i = \sqrt{\omega^2 + p_i^2}$$

$$\phi_i = \tan^{-1}\left(\frac{\omega}{-z_i}\right) \quad \text{and} \quad \theta_i = \tan^{-1}\left(\frac{\omega}{-p_i}\right) \tag{26.1}$$

The frequency response function in terms of the magnitude and phase angle of each factor in the numerator and denominator is

$$G(j\omega)H(j\omega) = \frac{K(Z_1\angle\phi_1)(Z_2\angle\phi_2) \cdots (Z_m\angle\phi_m)}{(P_1\angle\theta_1)(P_2\angle\theta_2) \cdots (P_n\angle\theta_n)}$$

The magnitudes and phase angles of numerator and denominator factors are combined to express the frequency response in terms of total magnitude and phase shift.

$$|G(j\omega)H(j\omega)| = \frac{KZ_1Z_2 \cdots Z_m}{P_1P_2 \cdots P_n} = \frac{K\prod_{i=1}^{m} Z_i}{\prod_{i=1}^{n} P_i} \tag{26.2}$$

and

$$\angle G(j\omega)H(j\omega) = (\phi_1 + \phi_2 + \cdots + \phi_m) - (\theta_1 + \theta_2 + \cdots + \theta_m) = \sum_{i=1}^{m} \phi_i - \sum_{i=1}^{n} \theta_i \tag{26.3}$$

To evaluate the magnitude of the system response at a particular frequency, ω_0, calculate the value of P_i corresponding to each pole and the value of Z_i corresponding to each zero from Equations 26.1. Equation 26.2 is then applied to calculate the magnitude of the system's response to excitation at the frequency ω_0, $|G(j\omega_0)H(j\omega_0)|$. This value is the magnitude of the output of the system when the input is $1\cos(\omega_0 t)$.

The Bode magnitude plot displays magnitude in decibels (dB) vs. frequency in radians per second on a logarithmic scale. The decibel is defined as a power ratio power gain in dB $= 10\log_{10}(P_{out}/P_{in})$. Since power is proportional to the square of voltage, the normalized power gain in dB $= 10\log_{10}(V^2_{out}/V^2_{in}) = 20\log_{10}(V_{out}/V_{in})$ or

$$|G(j\omega)H(j\omega)|_{dB} = 20\log_{10}|G(j\omega)H(j\omega)|$$

The Bode phase plot displays the phase shift between the input and the output signal in degrees. This phase shift is a result of the time delay in processing the signal through the system. The frequency response phase angle is plotted vs. frequency on a $\log_{10}$ scale.

Figure 26.2 shows both the magnitude and phase Bode plots for a system with the transfer function

$$G(s)H(s) = \frac{100(s+1)}{s(s+10)} \tag{26.4}$$

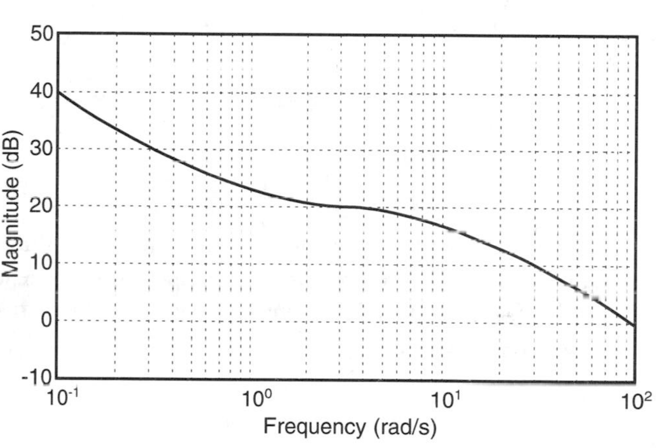

(a)

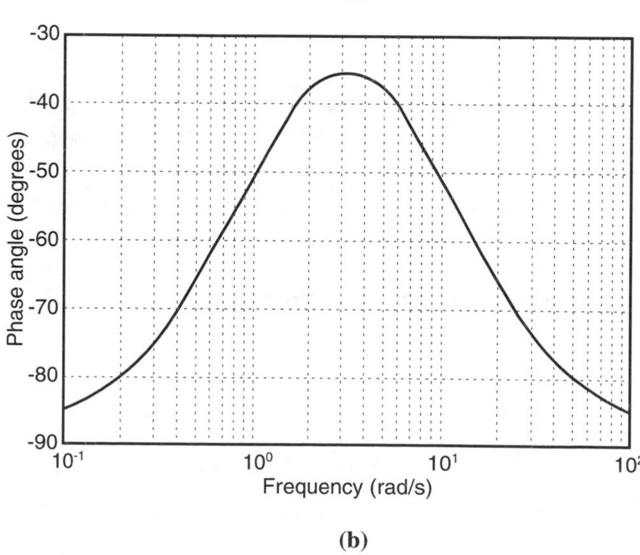

(b)

Figure 26.2 Bode plot of magnitude and phase frequency spectra.

The Bode plot can be plotted either by mathematically evaluating the frequency response from the transfer function or by calculating and plotting the output-input relationship from experimentally obtained measurements.

The Asymptotic Bode Plot

The asymptotic Bode plot is a useful tool for control system design and analysis. The asymptotic Bode plot is a straight-line approximation of the Bode diagram as described above. Again, consider all poles and zeros of the transfer function to be real-valued in order to simplify the explanation of the method. The treatment of complex poles and zeros is discussed separately. To prepare to sketch the asymptotic Bode plot first manipulate the transfer function into the form

$$G(j\omega)H(j\omega) = \frac{K_B(1 - j\omega/z_1)(1 - j\omega/z_2)\cdots(1 - j\omega/z_m)}{(1 - j\omega/p_1)(1 - j\omega/p_2)\cdots(1 - j\omega/p_n)} \quad (26.5)$$

where

$$K_B = \frac{K\prod_i z_i}{\prod_i p_i}$$

In (7.5.x) the z_i and p_i determine frequency response of their respective factors of the transfer function. To make this frequency relationship more explicit, define *break frequencies* of the respective zeros and poles as

$$\omega_{zi} = -z_i \quad \text{and} \quad \omega_{pi} = -p_i$$

After substituting these variables Equation 26.5 becomes

$$G(j\omega)H(j\omega) = \frac{K_B(1 + j\omega/\omega_{z1})(1 + \omega/\omega_{z2})\cdots(1 + j\omega/\omega_{zm})}{(1 + j\omega/\omega_{p1})(1 + j\omega/\omega_{p2})\cdots(1 + j\omega/\omega_{pn})}$$

Asymptotic Magnitude Plots

The magnitude in decibels is

$$\begin{aligned}|G(j\omega)H(j\omega)|_{\text{dB}} = {} & 20\log_{10}|K_B| + 20\log_{10}|1 + j\omega/\omega_{z1}| \\ & + 20\log_{10}|1 + j\omega/\omega_{z2}| \\ & + \cdots + 20\log_{10}|1 + j\omega/\omega_{zm}| \\ & - 20\log_{10}|1 + j\omega/\omega_{p1}| - 20\log_{10}|1 \\ & + j\omega/\omega_{p2}| - \cdots - 20\log_{10}|1 + j\omega/\omega_{pn}|\end{aligned} \quad (26.6)$$

To create the asymptotic, straight-line Bode plot, each pole or zero term in the summation is approximated as

$$20\log_{10}|1 + j\omega/\omega_1| \approx \begin{cases} 20\log_{10}(1) = 0, & \omega < \omega_1 \\ 20\log_{10}(\sqrt{2}), & \omega = \omega \\ 20\log_{10}(\omega) - 20\log_{10}(\omega_1), & \omega > \omega_1 \end{cases}$$

The asymptotic magnitude plot is produced by sketching this straight-line approximation of each term and summing them graphically. The asymptotic Bode magnitude plot for a real-valued zero of the transfer function is shown in Figure 26.3. Figure 26.4 shows the asymptotic magnitude plot for a real-valued pole. The asymptotic magnitude plot for a pole shows a negative slope for frequencies above the break frequency because of the minus sign that precedes each denominator term in Equation 26.6.

In the case of zeros at the origin in the *s*-plane, we will have terms of

$$20\log_{10}|j(\omega)^N| = 20N\log_{10}|\omega|,$$

where N is the number of zeros at the origin. This will result in the asymptotic plot shown in Figure 26.5. Poles of the transfer function located at the origin in the *s*-plane will cause a term of

$$20\log_{10}\left|\frac{1}{(j\omega)^N}\right| = -20N\log_{10}|\omega|$$

The asymptotic magnitude plot for a pole at the origin is shown in Figure 26.6.

It is noteworthy that each zero of the transfer function produces a straight-line asymptote with a slope of +20 dB/decade in the magnitude Bode plot. Similarly it is noted that each pole produces a straight-line asymptote with a slope of −20 dB/decade. It is also significant that each cumulative slope in the

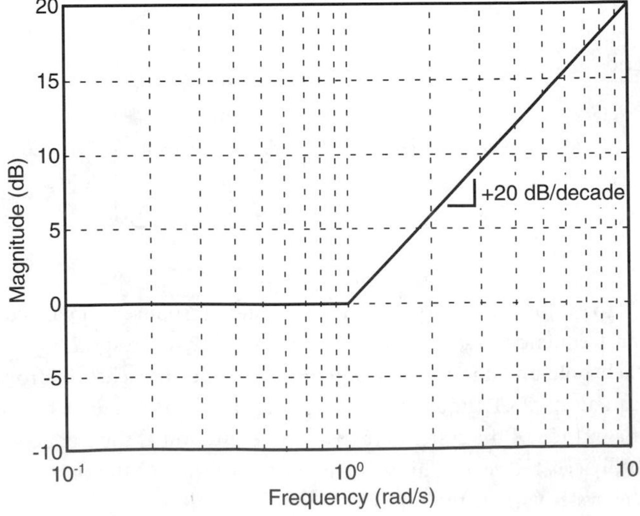

Figure 26.3 Asymptotic magnitude Bode plot for a zero.

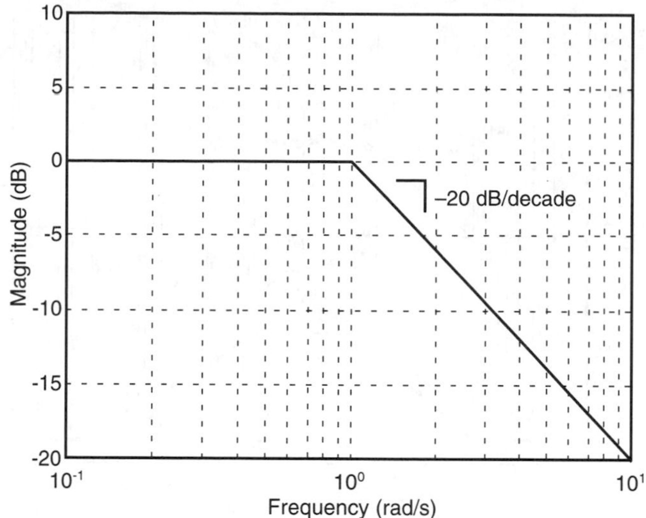

Figure 26.4 Asymptotic magnitude Bode plot for a pole.

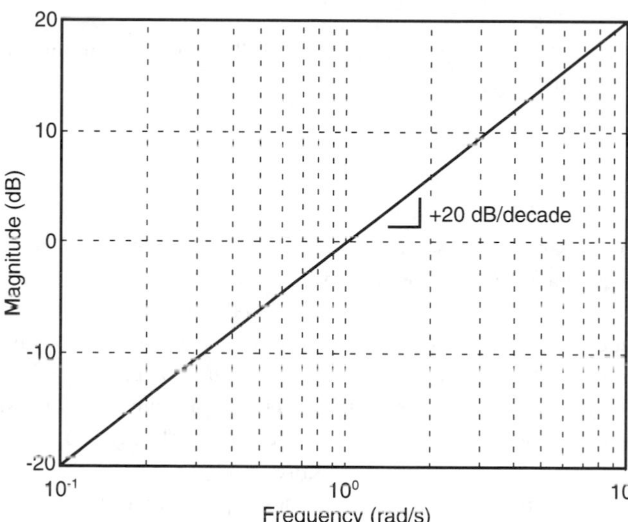

Figure 26.5 Asymptotic magnitude Bode plot for a zero at the origin.

asymptotic magnitude Bode plot is an integer multiple of ± 20 dB/decade.

For example, the transfer function given in Equation 26.4 is rewritten as

$$G(j\omega)H(j\omega) = \frac{10(1 + j\omega/1)}{j\omega(1 + j\omega/10)}$$

The magnitude in decibels is calculated from

$$|G(j\omega)H(j\omega)|_{dB} = 20\log_{10}|10| + 20\log_{10}|1 + j\omega/1| - 20\log_{10}|j\omega| - 20\log_{10}|1 + j\omega/10|$$

The asymptotic Bode diagram is shown in Figure 26.7. The dashed lines in the figure represent the asymptotic approximations of the individual factors of the transfer function. The total approximation of the system's magnitude frequency response is shown by the solid lines.

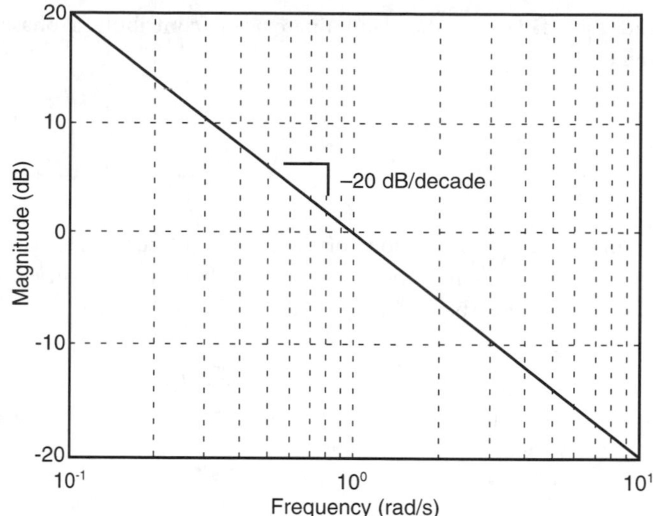

Figure 26.6 Asymptotic magnitude Bode plot for a pole at the origin.

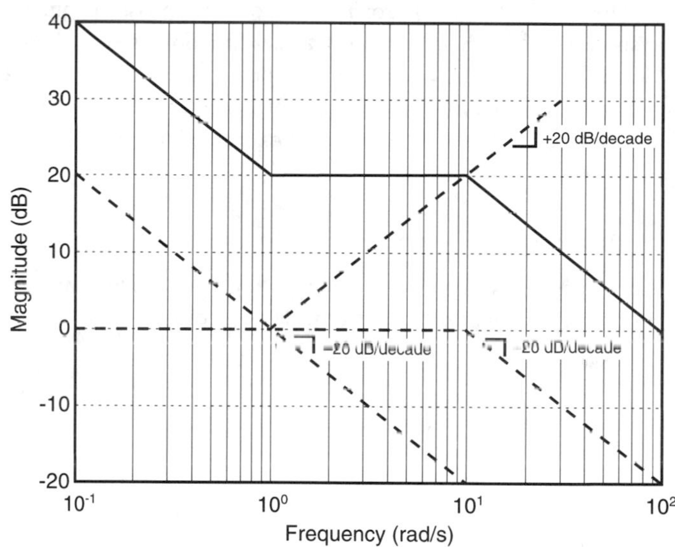

Figure 26.7 Asymptotic magnitude Bode plot.

Asymptotic Phase Plots

Asymptotic phase plots are based on the phase angle of the frequency response

$$\angle G(j\omega)H(j\omega) = (\phi_1 + \phi_2 + \cdots + \phi_m) - (\theta_1 + \theta_2 + \cdots + \theta_m) = \sum_{i=1}^{m}\phi_i - \sum_{i=1}^{n}\theta_i$$

As implied in the equation, the total phase response is the algebraic sum of the phase responses of the individual numerator and denominator factors of the transfer function. Each factor of the numerator will contribute a phase angle of

$$\phi_i = \tan^{-1}\left(\frac{\omega}{\omega_{z_i}}\right)$$

and each factor of the denominator will contribute a phase angle of

$$-\theta_i = -\tan^{-1}\left(\frac{\omega}{\omega_{Pi}}\right)$$

Straight-line approximations of the arctangent curves are used to produce the asymptotic phase plot. Each arctangent curve is approximated as shown in Figure 26.8, where

$$\tan^{-1}\left(\frac{\omega}{\omega_i}\right) \approx \begin{cases} 0^\circ, & \omega \le \omega_i/10 \\ 45^\circ, & \omega = \omega_i \\ 90^\circ, & \omega > 10\omega_i \end{cases}$$

In the case of zeros or poles of the transfer function at the origin in the s-plane the phase asymptote is plotted as $+90^\circ$ for each zero at $s = 0$, and -90° for each pole at $s = 0$, at all frequencies.

The straight-line approximation of the phase frequency response of the system with the transfer function given by Equation 26.4 is shown in Figure 26.9, where the asymptotic phase plot for the individual factors are shown as dashed lines and the total phase frequency response is approximated by the solid lines.

Complex Poles and Zeros

Complex zeros and poles result in quadratic factors in the numerator or denominator respectively. The quadratic factors are of the general form

$$s^2 + 2\zeta\omega_n s + \omega_n^2 = (s + \zeta\omega_n - j\omega_n\sqrt{1 - \zeta^2})$$
$$(s + \zeta\omega_n + j\omega_n\sqrt{1 - \zeta^2})$$

where $0 \le \zeta \le 1$.

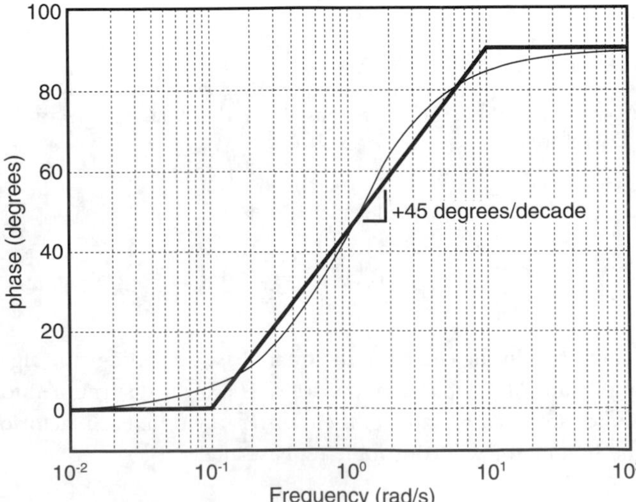

Figure 26.8 Asymptotic phase Bode plot for a zero.

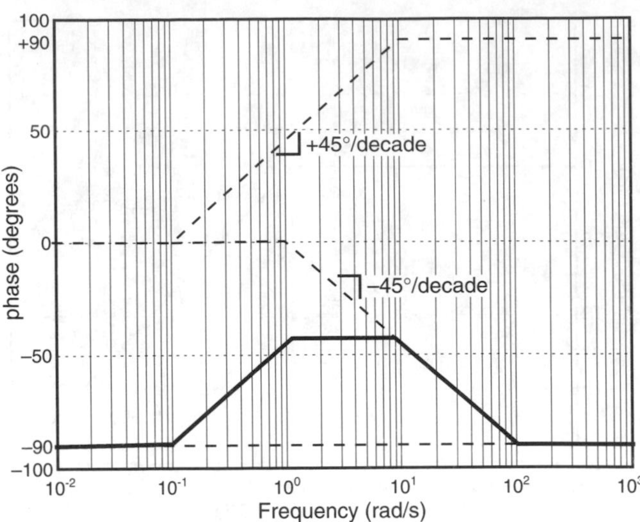

Figure 26.9 Asymptotic phase Bode plot.

We substitute $s = j\omega$ and algebraically rearrange the form of the quadratic factor to

$$\omega_n^2(1 - (\omega/\omega_n)^2 + j2\zeta(\omega/\omega_n))$$

Figure 26.10 (a) shows the magnitude frequency response for a complex-conjugate set of poles. Figure 26.10 (b) shows the phase frequency response. The figures would be inverted for complex-conjugate zeros. Figure 26.10 shows that the shape of both magnitude and phase curves is a function of ζ. A straight-line asymptotic approximation of either the magnitude or phase response of quadratic roots can lead to large errors. Straight-line approximations are sometimes found to be acceptable when $\zeta > 0.3$; however, it can be seen from Figure 26.11 that at $\zeta = 0.3$, there will be significant error in the phase approximation. For $\zeta < 0.3$, either exact curves or approximations of the curves shown in Figure 26.10 should be used.

$$20\log_{10}|1 - (\omega/\omega_n)^2 + j2\zeta(\omega/\omega_n)|$$
$$\cong \begin{cases} 20\log_{10}(1) = 0, & \omega << \omega_n \\ 20\log_{10}(2\zeta), & \omega = \omega_n \\ 20\log_{10}(\omega) - 20\log_{10}(\omega_1), & \omega >> \omega_n \end{cases}$$

26.2 Mathematical Model Determination

In situations where frequency response data can be experimentally collected, but no mathematical model of the system is available, an approximate mathematical model can be developed by plotting the experimental frequency response data in a Bode plot and sketching in straight-line asymptotes as shown in Figure 26.11. The straight-line asymptotes in the magnitude plot are always at integer multiples of ±20 dB/decade. Straight-line approximations in the phase plot are always at integer multiples

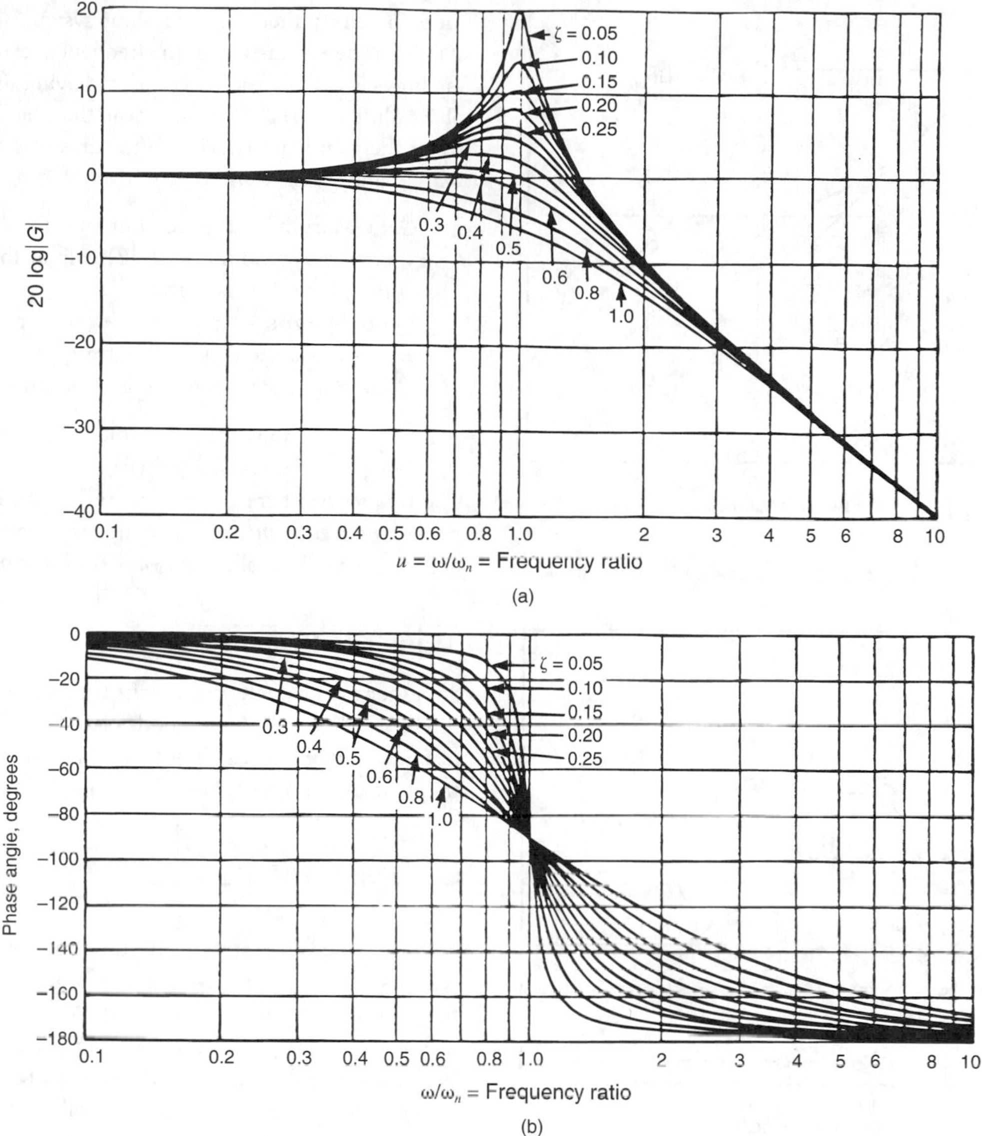

Figure 26.10 Bode diagram of $G(j\omega) = [1 + (2\zeta/\omega_n)j\omega + (j\omega/\omega_n)^z]^{-1}$. Copied from *Modern Control Systems,* 4th ed., Addison-Wesley, 1986.

of ±45°/decade. Each intersection of straight-line asymptotes in the magnitude plot is interpreted as occurring at the break frequency of a pole or zero of the transfer function. The phase plot is used to confirm that all poles and zeros of the system have negative real parts. If a pole or zero of the system is located in the right half of the s-plane the phase plot will show non-minimum phase angles.

For example, the smooth curves in Figure 26.11 represent experimentally measured frequency response data for a system. Straight-line asymptotes are sketched along the frequency-response curves. By observing the intersections of the magnitude asymptotes, it is seen that the system has a break frequency from a zero at $\omega = 2$ rad/s and break frequencies from poles at $\omega = 0.1$ rad/s and $\omega = 50$ rad/s.

The low-frequency magnitude of 30 dB is considered to be produced by a constant multiplying factor in the transfer function such that

$$30 \text{ dB} = 20 \log_{10}(K_B)$$

Therefore

$$K_B = 10^{1.5} = 31.6.$$

The estimated transfer function is

$$G(j\omega)H(j\omega) = \frac{31.6(1 + j\omega/2)}{(1 + j\omega/0.1)(1 + j\omega/50)}$$

or

$$G(s)H(s) = \frac{79.1(s + 2)}{(s + 0.1)(s + 50)}$$

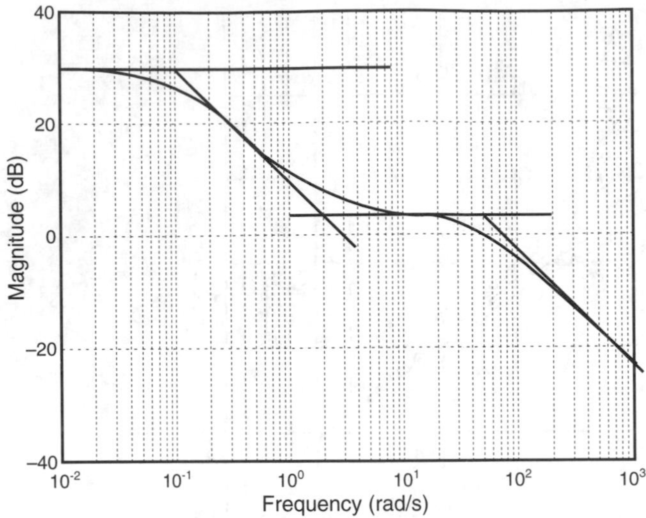

Figure 26.11 Mathematical model determination from Bode plot.

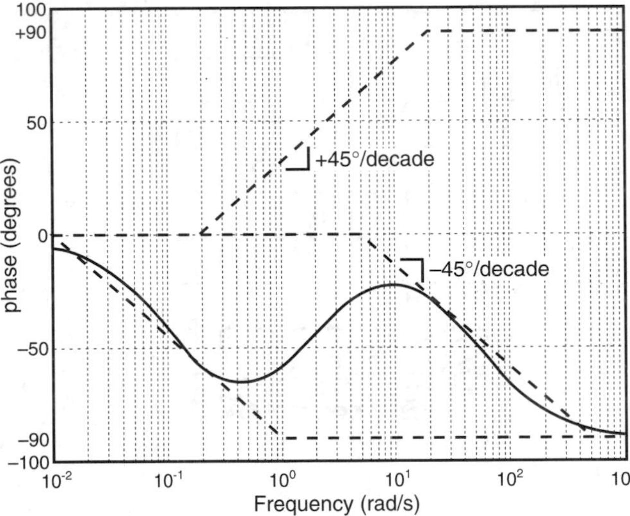

Figure 26.12 Phase plot for the example.

Figure 26.12, which shows the measured phase response and the asymptotic phase plot of the estimated transfer function, discloses no contradiction. This confirms that the zero and the poles are in the left half of the *s*-plane.

26.3 Correlation of Frequency Response and Time Response

Stability Determination

The stability of a feedback system can be determined from the Bode plot of the frequency response of the open-loop system. This stability determination is based on the Nyquist criterion for stability. If the open-loop frequency response shows gain greater than or equal to unity when the phase shift is $-180°$ then the closed-loop system is not stable. Or if the phase shift of the open-loop response is $-180°$ or more negative than $-180°$ when the gain is unity, then the closed-loop system is not stable. Relative stability is determined from the frequency response by determining how close the response comes to having unity gain when the phase shift is $-180°$ and how near the phase shift is to $-180°$ when the gain is unity. These measures of stability are the *gain margin* and *phase margin* respectively.

Gain Margin = the amount of gain that will cause the system to have unity (0 dB) gain at the frequency where the phase shift is $-180°$.

Phase Margin = the amount of negative phase-shift which will cause the system to have a phase shift of $-180°$ at the frequency where the gain is unity (0 dB).

Figure 26.13 illustrates the measurement of gain margin (labeled g_m) and phase margin (labeled ϕ_m) from the Bode plot. The frequency where the gain margin is measured is called the *phase-crossover frequency.* The frequency at which the phase margin is measured is called the *gain-crossover frequency.*

Transient Response

Some general characteristics of the transient response of a system can be determined from the Bode plot of its *closed-loop* frequency response. Figure 26.14 shows the magnitude Bode plot of the frequency response of an underdamped closed-loop system.

$$T(s) = \frac{G(s)}{1 + G(s)H(s)} = \frac{100}{s^2 + 6s + 100}$$

As observed in Figure 26.10, the Bode plot of the frequency response of a complex-conjugate set of poles shows a peak at the natural frequency, ω_n, when the damping ratio, ζ, is 0.707 or less. The Bode plot of the closed-loop frequency response can be compared with Figure 26.10 to obtain an estimate of ζ. A larger resonant peak indicates a smaller ζ and therefore more overshoot in the transient step response. The peak value in the frequency response of an underdamped second-order system

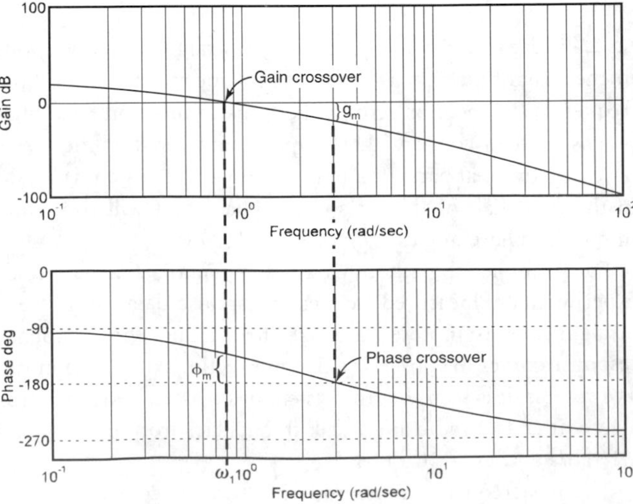

Figure 26.13 Gain margin and phase margin.

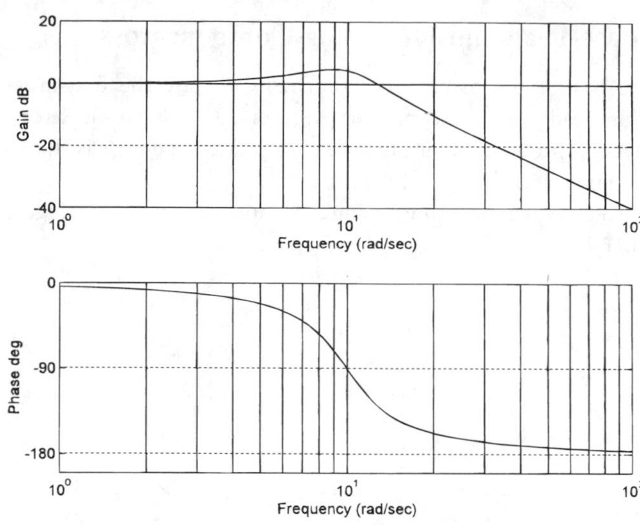

Figure 26.14 Frequency response of an underdamped system.

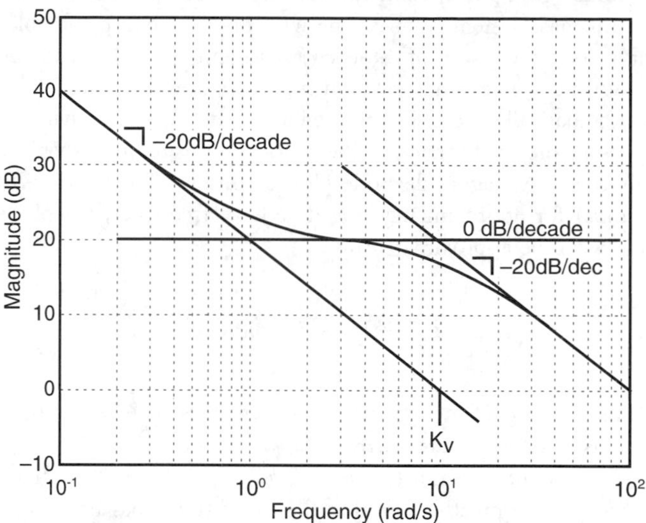

Figure 26.15 Frequency response of a type 1 system.

occurs near the natural frequency, ω_n. For a second-order system, or a system that has dominant second-order characteristic roots, the natural frequency is within a factor or two of the gain-crossover frequency. Therefore a larger value of ω_n indicates a higher bandwidth. A higher bandwidth corresponds to a shorter time constant and therefore shorter rise time and settling time.

Note that here we are referring to the Bode plot of the frequency response of the closed-loop system rather than the open-loop system as we have discussed elsewhere. The closed-loop frequency response can be determined from open-loop frequency response data (and vice versa) by use of a *Nichols chart* (D'Azzo and Houpis, 1988).

Steady-State Performance

The *system type* for steady-state error analysis can be determined from the Bode plot of the open-loop frequency response. The slope of the magnitude plot at low frequencies indicates the system type. A low-frequency slope of 0 dB/decade indicates no poles of $G(s)H(s)$ at $s = 0$, and therefore a type 0 system. The magnitude of the Bode plot at low frequencies in dB is 20 $\log_{10} K_p$ where K_p is the position-error constant. The steady-state error for a unit-step input, $r(t) = u(t)$, is

$$e_{ss} = \frac{1}{1 + K_p}$$

A low-frequency slope of the magnitude plot of -20 dB/decade indicates a single pole of $G(s)H(s)$ at $s = 0$ and therefore a type 1 system. A straight-line projection of this slope to intersect the 0 dB axis as shown in Figure 26.15 yields the value of the velocity error constant, K_v. The steady-state error for a unit-ramp input, $r(t) = t\,u(t)$, is

$$e_{ss} = \frac{1}{K_v}$$

A low-frequency slope of the magnitude plot of -40 dB/decade indicates two poles of $G(s)H(s)$ at $s = 0$ and therefore a type 2 system. A straight line projection of this slope to intersect the 0 dB axis provides the datum, ω_a, needed to calculate the acceleration-error constant

$$K_a = \frac{1}{\omega_a^2}$$

The steady-state error for a parabolic input, $r(t) = \frac{1}{2}t^2u(t)$, is

$$e_{ss} = \frac{1}{K_a}$$

26.4 Shaping the Cutoff Response

Analysis of the magnitude and phase plots at frequencies near cutoff provides valuable information about the performance of the system. Some important rules for evaluating a systems performance are based on the shape of the frequency response plots near cutoff.

Although it is impossible to give values for the gain margin and phase margin that will provide satisfactory performance for every system, one rule of thumb that applies to many systems states that a gain margin $\geq$ 8 dB and a phase margin $\geq$ 45 degrees are usually required (Phillips and Harbor, 1991).

The gain and phase margins are not useful design parameters for some systems. For example, the phase plot for first- or second-order systems never cross the -180 degree line. Therefore the gain margin is infinite and cannot be used for design consideration. For high-order systems there may exist more than one frequency at which the gain is unity or at which the phase is -180 degrees. Therefore, the definition of the gain margin or phase margin will require clarification in these cases.

Bode's *gain-phase relationship* states that for any stable, minimum-phase system (that is, one with no right half-plane poles or zeros), the phase of $G(j\omega)H(j\omega)$ is uniquely related to the magnitude. The exact relationship is somewhat complicated mathematically and is seldom used in practice. However, in many cases a simplified version of the relationship is useful. When the slope of the magnitude Bode plot is constant (*at n × 20 dB/decade*) for approximately one decade of frequency, the phase angle can be approximated by

$$\angle G(j\omega)H(j\omega) \cong n(90^\circ)$$

until the slope of the magnitude plot changes. This relationship is illustrated by Figure 26.16

For stability we must have $\angle G(j\omega)H(j\omega) > -180$ degrees. Therefore we must ensure that the magnitude does not have a slope such that $n < -1$ for a decade or more before the gain crossover. The phase margin will usually be sufficient if the slope of the magnitude plot is -20 dB/decade for one decade centered at the gain crossover frequency. If the slope of the magnitude plot is -20 dB/decade for one decade below and one decade above the gain crossover frequency the system has a phase margin of approximately 90 degrees (Franklin, et al., 1994).

26.5 Compensator Design

Compensators are used to shape the frequency response of the system so that it displays desirable characteristics as described in the previous section. Compensators can be designed to change the slope of the magnitude plot in the vicinity of the gain crossover, to add positive phase shift, to allow for increased low frequency gain, to increase the system type, or to achieve a combination of these effects. Methods for designing compensation networks to achieve these effects are discussed in this section.

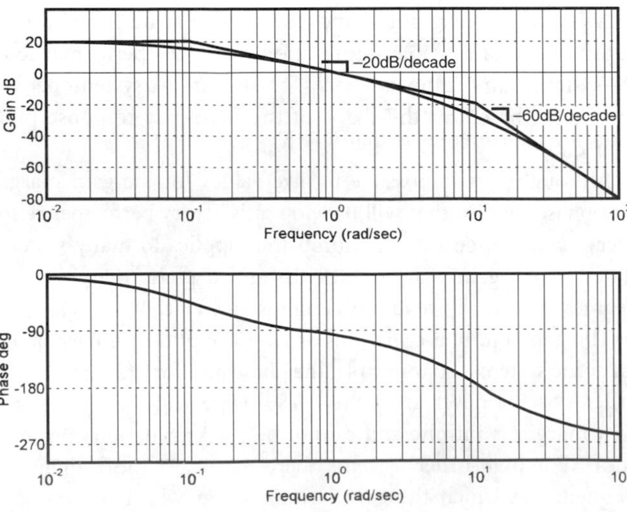

Figure 26.16 Bode gain-phase relationship.

Phase-Lead and Phase-Lag Compensators

The basic types of compensation networks are the first-order phase-lead and phase-lag compensators. An electronic circuit that will produce either phase-lead or phase-lag effects is shown in Figure 26.17.

The frequency response function for the circuit of Figure 26.17 is

$$G_c(j\omega) = K_c \frac{1 + j\omega/\omega_z}{1 + j\omega/\omega_p}$$

where

$$\omega_z = \frac{1}{R_1 C_1}, \quad \omega_p = \frac{1}{R_2 C_2} \quad \text{and} \quad K_c = \frac{R_2 R_4}{R_1 R_3}$$

The choice of electronic component values determines whether the circuit produces a phase-lead or phase-lag effect. If $R_1C_1 > R_2C_2$ then $\omega_z < \omega_p$ and the circuit is a phase-lead compensator. If $R_1C_1 < R_2C_2$ then $\omega_z > \omega_p$ and the circuit is a phase-lag compensator. Bode plots of the frequency response of phase-lead and phase-lag compensators are shown in Figures 26.18 and 26.19, respectively.

Phase-Lag Compensator Design

Figure 26.20 shows the effect of a phase-lag compensator on the open-loop frequency response of a system. The dashed line

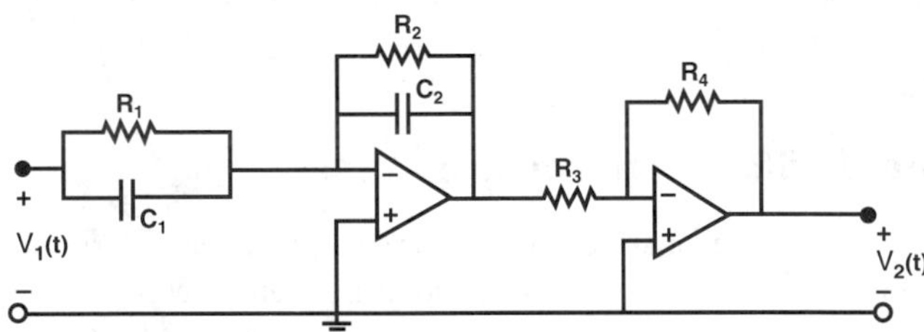

Figure 26.17 A phase-lag/phase-lead circuit.

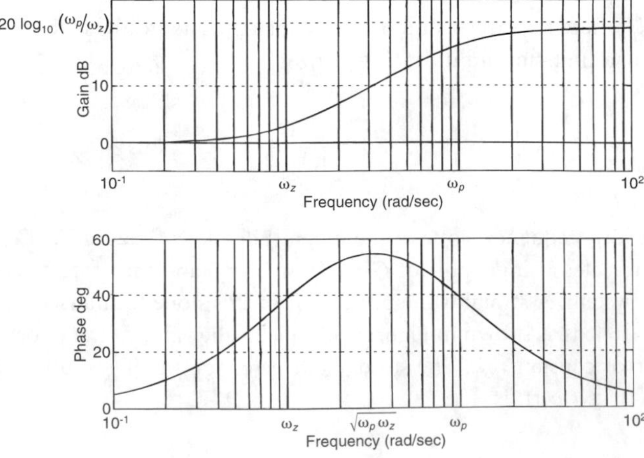

Figure 26.18 Frequency response of a phase-lead compensator.

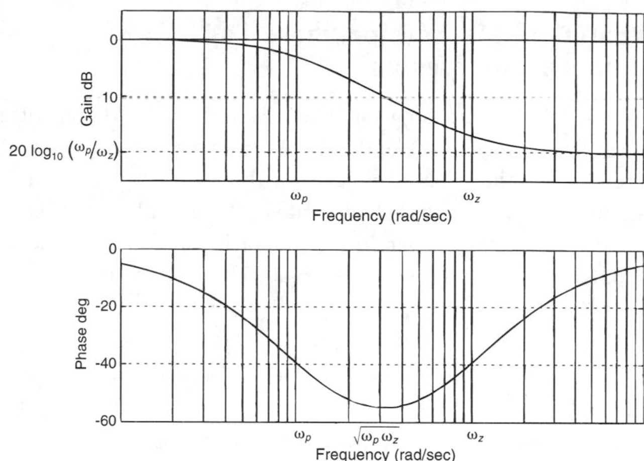

Figure 26.19 Frequency response of a phase-lag compensator.

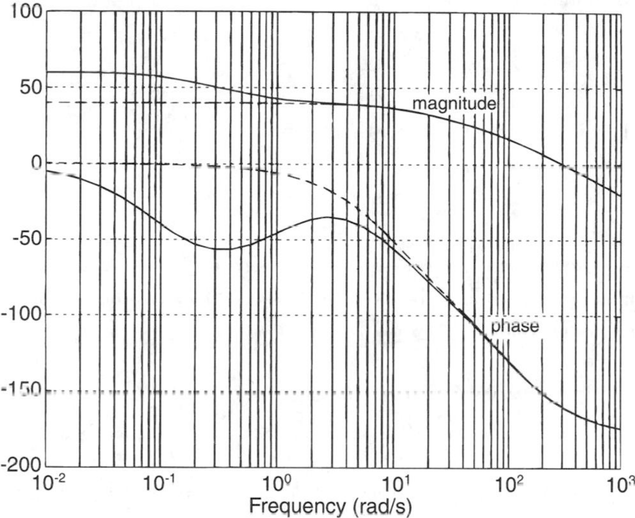

Figure 26.20 Frequency response with phase-lag compensation.

represents the frequency response of the uncompensated system. The solid line represents the frequency response of the system with phase-lag compensation. Notice that the negative phase shift caused by the compensator takes effect at relatively low frequencies and is almost insignificant at the gain-crossover frequency. The low-frequency gain of the compensated system is greater than that of the uncompensated system by 20 $\log_{10}(K_c)$. The high-frequency gain of the compensated system is lower than that of the uncompensated system by 20 $\log_{10}(K_c\omega_p/\omega_z)$. These observations lead to a relatively straightforward design method for phase-lag compensators.

1. Determine the dc gain needed from the compensator to provide satisfactory steady-state error performance. Set K_c to provide the necessary gain.
2. Determine the phase margin that will ensure satisfactory transient performance.
3. From the uncompensated system frequency response:
   a. Find the frequency, ω_1, at which the uncompensated system has a phase shift of $-180° + \phi_m + 5°$. (The added 5° corrects for the residual negative phase-shift contributed by the compensator at the new gain-crossover frequency.)
   b. Record the magnitude of the uncompensated system frequency response at that frequency, $|G(j\omega_1)H(j\omega_1)|$.
4. Choose $\omega_z = \omega_1/10$.
5. Calculate $\omega_p = \dfrac{\omega_z}{|G(j\omega_1)H(j\omega_1)|}$.

Analytical Method of Phase-Lead and Phase-Lag Compensator Design

C. L. Phillips (Phillips and Harbor, 1991) provides an analytical method for the design of both phase-lead and phase-lag compensators. Application of this method yields values for ω_z and ω_p such that the compensated system will have a specified phase margin at a specified gain-crossover frequency. It must be noted, however, that this method does not ensure stability. Therefore after the design is completed, the frequency response of the compensated system must be analyzed to determine stability.

For either phase-lead or phase-lag compensators, the pole and zero break frequencies are given by

$$\omega_z = \frac{\omega_1 \sin\theta}{M^{-1} - \cos\theta} \quad \text{and} \quad \omega_p = \frac{\omega_1 \sin\theta}{\cos\theta - M}$$

where

ω_1 is the chosen gain-crossover frequency
$\theta = -180° + \phi_m - \angle G(j\omega_1)H(j\omega_1)$
ϕ_m is the desired phase margin
$M = K_c\,|G(j\omega_1)H(j\omega_1)|$
K_c is chosen to provide adequate low frequency gain

The key to this design method is in the choice of the gain-crossover frequency, ω_1. The following criteria provide guidance in the selection ω_1.

When a phase-lead type compensator is desired, the compensator must be designed to provide positive phase shift (phase lead) at the gain crossover frequency. Therefore part of the consideration for choosing ω_1 is that $\theta > 0$.

$$\theta = -180° + \phi_m - \angle G(j\omega_1)H(j\omega_1) > 0$$

$$\angle G(j\omega_1)H(j\omega_1) < -180° + \phi_m$$

Recall that the frequency response of a phase-lead compensator also has gain greater than unity at the gain-crossover frequency of the system.

$$|G_c(j\omega_1)| > 1$$

Therefore, $M < 1$, or in other terms,

$$|G(j\omega_1)H(j\omega_1)| < 1/K_c$$

A further consideration is the desire to avoid designing an unstable controller. It is normally unsatisfactory to have either a pole or the zero of the controller in the right half of the s-plane. Therefore

$$\omega_p = \frac{\omega_1 \sin\theta}{\cos\theta - M} > 0$$

Since both $\sin\theta$ and $\cos\theta$ are positive when $\theta > 0$,

$$\cos\theta - M > 0$$

or

$$|G(j\omega_1)H(j\omega_1)| < \frac{\cos\theta}{K_c}$$

This supersedes the previous constraint since $\cos\theta \leq 1$.

We also desire that

$$\omega_z = \frac{\omega_1 \sin\theta}{M^{-1} - \cos\theta} > 0$$

This requires that

$$M^{-1} - \cos\theta > 0$$

or in other terms

$$|G(j\omega_1)H(j\omega_1)| < 1/K_c \cos\theta$$

However, since $\cos\theta \leq 1$, this is superseded by the previous constraint that

$$|G(j\omega_1)H(j\omega_1)| < \frac{\cos\theta}{K_c}$$

In summary, the criteria for selection of ω_1 for a phase-lead controller are

$$\angle G(j\omega_1)H(j\omega_1) < -180° + \phi_m$$

and

$$|G(j\omega_1)H(j\omega_1)| < \frac{\cos\theta}{K_c}$$

where ϕ_m is the chosen value of phase margin, $|G(j\omega_1)H(j\omega_1)|$ and $\angle G(j\omega_1)H(j\omega_1)$ are magnitude and phase angle, respectively, of the uncompensated system frequency response at ω_1 and $\theta = -180° + \phi_m - \angle G(j\omega_1)H(j\omega_1)$.

Analytical Method for Phase-Lag Compensator Design

The criteria for choosing ω_1 that will lead to the design of a phase-lag compensator are developed in much the same way as those for the phase-lead compensator discussed above. A phase-lag compensator contributes negative phase shift at the gain-crossover frequency. Therefore ω_1 must be chosen such that $\theta < 0$. For phase-lag design ω_1 is usually chosen such that $-5° \leq \theta < 0°$.

$$\theta = -180° + \phi_m - \angle G(j\omega_1)H(j\omega_1) < 0$$

$$\angle G(j\omega_1)H(j\omega_1) > -180° + \phi_m$$

Recall that the frequency response of a phase-lead compensator also has gain less than unity at the gain-crossover frequency of the system.

$$|G_c(j\omega_1)| < 1$$

Therefore, $M > 1$, or in other terms,

$$|G(j\omega_1)H(j\omega_1)| > 1/K_c$$

A further consideration is the desire to avoid designing an unstable controller. The pole of the controller should be placed in the left half of the s-plane. Therefore

$$\omega_p = \frac{\omega_1 \sin\theta}{\cos\theta - M} > 0$$

Since both $\sin\theta$ is negative and $\cos\theta$ is positive when $\theta < 0$,

$$\cos\theta - M < 0$$

or

$$|G(j\omega_1)H(j\omega_1)| > \frac{\cos\theta}{K_c}$$

However, this is superseded by the previous constraint since $\cos\theta \leq 1$. To avoid placing the zero of the compensator in the right half of the s-plane it is required that

$$\omega_z = \frac{\omega_1 \sin\theta}{M^{-1} - \cos\theta} > 0$$

This requires that

$$M^{-1} - \cos\theta < 0$$

or in other terms,

$$|G(j\omega_1)H(j\omega_1)| > 1/K_c \cos\theta$$

Since $\cos\theta \le 1$, this supersedes the constraint that $|G(j\omega_1)H(j\omega_1)| > 1/K_c$.

In summary, the criteria for selection of ω_1 for a phase-lag controller are

$$\angle G(j\omega_1)H(j\omega_1) > -180^\circ + \phi_m$$

and

$$|G(j\omega_1)H(j\omega_1)| > \frac{1}{K_c \cos\theta}$$

where ϕ_m is the chosen value of phase margin, $|G(j\omega_1)H(j\omega_1)|$ and $\angle G(j\omega_1)H(j\omega_1)$ are magnitude and phase angle, respectively, of the uncompensated system frequency response at ω_1 and $\theta = -180^\circ + \phi_m - \angle G(j\omega_1)H(j\omega_1)$.

Lag-Lead Compensators

When compensation by neither phase-lag nor phase-lead produces the desired system performance, a phase-lag controller and a phase-lead controller are sometimes connected in cascade. This is known as a *lag-lead compensator.*

A design procedure for lag-lead controllers is followed. The phase-lag section of the compensator is designed to maintain the low-frequency gain required for satisfactory steady-state error performance and to help realize part of the gain margin. The phase-lead section is then designed to achieve the specified phase-margin and increase the bandwidth to provide satisfactory transient performance.

PID Compensators

Proportional plus integral plus derivative (PID) controllers are probably the most commonly used type of compensators in feedback control systems today. PID compensators are used almost exclusively in industrial control systems. A classical method of tuning PID controllers is described in Section 5. Here we consider the design of PID controllers using frequency response data from the open-loop system.

A block diagram of a PID compensator is shown in Figure 26.21. The transfer function of the controller is

$$G_c(s) = K_P + \frac{K_I}{s} + K_D s = \frac{K_D s^2 + K_P s + K_I}{s}$$

The transfer function shows that the PID controller has two zeros and one pole.

The Bode plot of the frequency response of a PID controller is shown in Figure 26.22. In the frequency response we can observe an effect similar to the phase-lag compensator at low frequencies. The PID compensator has negative phase shift at low frequencies and the low-frequency gain is high. At high frequencies the PID frequency response shows a phase lead effect. The high-frequency gain is increased and the phase shift is positive. The PID compensator provides a combination of phase-lag and phase-lead effects and can be considered to be a lag-lead controller.

PI and PD compensators are simplifications of the PID. The PI is derived from the PID with $K_D = 0$. The PD is a PID with $K_I - 0$.

$$G_{PI}(s) = K_P + \frac{K_I}{s} = K_I\left(\frac{1 + s/\omega_i}{s}\right), \qquad \omega_i = \frac{K_I}{K_P}$$

$$G_{PD}(s) = K_P + K_D s = K_P(1 + s/\omega_d), \qquad \omega_d = \frac{K_P}{K_D}$$

Frequency spectra for PI and PD compensators are shown in Figures 26.23 and 26.24, respectively.

The PI is considered a special case of phase-lag compensation with negative phase shift and decreased gain at high frequencies. The PI is especially useful for eliminating steady-state error because the pole at the origin increases the system type.

The PD is a special case of phase-lead compensation. It provides positive phase shift and increased gain at high frequency. In practice the PD compensator usually is implemented as a phase lead by adding a pole to limit the high frequency gain. This is discussed below.

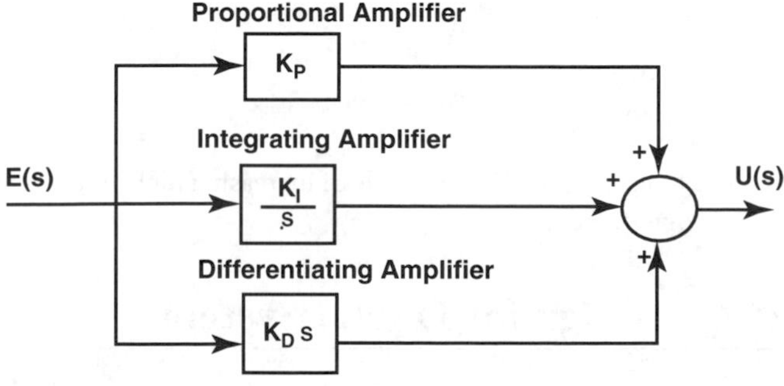

Figure 26.21 Block diagram of a PID controller.

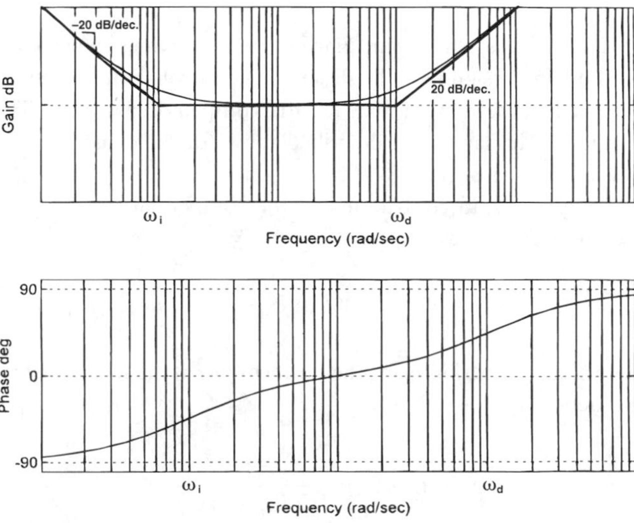

Figure 26.22 Frequency response of a PID compensator.

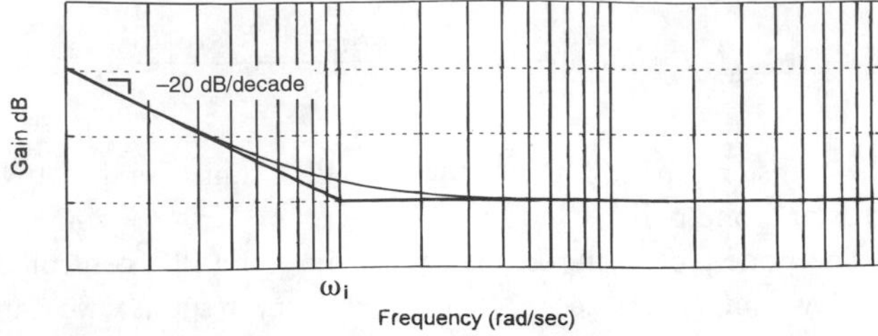

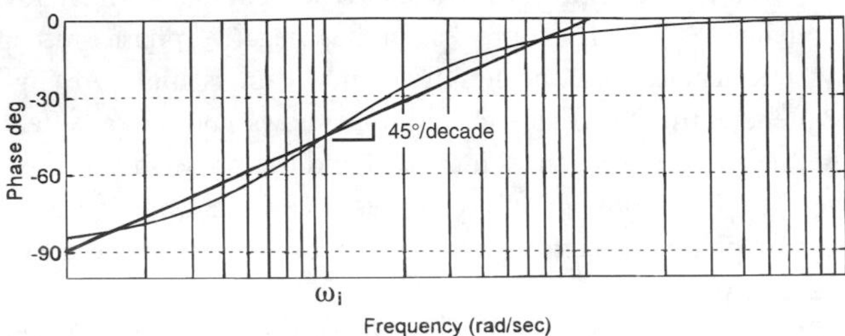

Figure 26.23 Frequency spectra of a PI compensator.

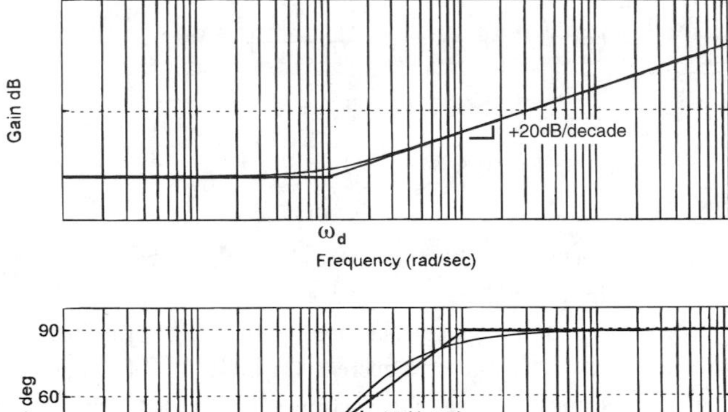

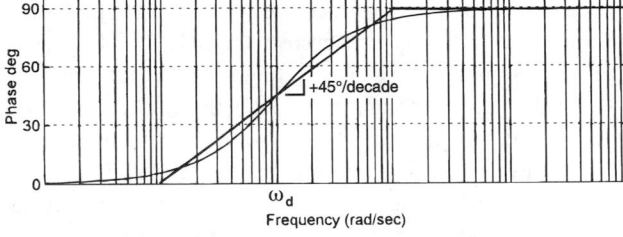

Figure 26.24 Frequency spectra of a PD compensator.

C. L. Phillips (Philips and Harbor, 1991) provides an analytical method for designing PID, PI and PD controllers to produce a specified phase margin, ϕ_m, for the compensated system. This method is similar to that described above for the phase-lead and phase-lag controllers.

The phase shift contributed by the compensator at the gain-crossover frequency, ω_1, is

$$\angle G_c(j\omega_1) = \theta = -180^\circ + \phi_m - \angle G(j\omega_1)H(j\omega_1)$$

and the gain contributed by the compensator at the gain-crossover frequency is

$$|G_c(j\omega_1)| = \frac{1}{|G(j\omega_1)H(j\omega_1)|}$$

The design of a PID type controller amounts to establishing the gains of the three terms. Phillips provides two equations for the three gains:

$$K_P = \frac{\cos\theta}{|G(j\omega_1)H(j\omega_1)|}$$

and

$$K_D - \frac{K_I}{\omega_1} = \frac{\sin\theta}{|G(j\omega)H(j\omega_1)|}$$

This method requires that we determine the desired phase margin, ϕ_m, and select a desired gain crossover frequency, ω_1, such that $\theta \le 90^\circ$. In choosing ω_1 consideration should be given also to the effect on the system bandwidth. Larger values of ω_1 will provide wider bandwidth and therefore faster response.

These equations can be used for PID, PI and PD compensator design as follows:

For a PD compensator the integrator gain K_I is set to zero. Then K_P and K_D are calculated from the equations.

For a PI compensator the derivative gain, K_D is set to zero. Then K_P and K_I are calculated from the equations.

For full PID design the value of K_I is usually calculated to achieve the required steady-state error performance. K_P and K_D can then be calculated from the equations.

In practice, the high-frequency gain of the PID or PD compensators can cause problems because of high-frequency noise in the signal. To eliminate this potential problem, it is common to design the PID compensator so that it has a second pole.

$$G_{PID}(s) = \frac{K_Ds^2 + K_Ps + K_I}{s(1 + s/\omega_{pd})}$$

$$G_{PD}(s) = \frac{K_Ds + K_P}{(1 + s/\omega_{pd})}$$

The location of this second pole is adjusted so that its break frequency is well outside the bandwidth of the system, but so that it adequately limits the high-frequency gain. The PID design procedure is completed as described above without regard to the location of the second pole. The second pole location is then determined. A typical choice is to make the break frequency of the second pole

$$\omega_{pd} \cong 10\omega_d$$

in order to limit the +20 dB/decade of increasing high-frequency gain to one decade.

26.6 Design for Digital Systems

Frequency response techniques for the direct design of digital controllers accomplished by transforming the system frequency response function to the w-plane using the bilinear transformation. However, in many cases digital controllers are designed by emulating a continuous controller design. In this section the emulation methods for the design of digital controllers are considered first.

Emulation Design Methods

Emulation methods of digital controller design consist of designing a continuous-time controller and then approximating that design with a digital filter. Emulation methods usually produce good results if the sampling frequency is at least 20 times the bandwidth of the system. These methods can be used with confidence if the sampling frequency is 30 or more times the bandwidth of the system.

Bilinear Approximation (Tustin's Method)

The bilinear approximation, or Tustin's method, is based on replacing each s in the continuous-time controller transfer function with the z-transform equivalent of trapezoidal integration.

$$s = \frac{2}{T}\left(\frac{1 - z^{-1}}{1 + z^{-1}}\right)$$

where T is the sampling period.

For example, the continuous-time controller with transfer function

$$G_c(s) = \frac{K_c(s + a)}{s + b}$$

is approximated by the digital controller with transfer function

$$D(z) = K_c \frac{\frac{2}{T}\left(\frac{1 - z^{-1}}{1 + z^{-1}}\right) + a}{\frac{2}{T}\left(\frac{1 - z^{-1}}{1 + z^{-1}}\right) + b} = K_c \frac{aT + 2 + (aT - 2)z^{-1}}{bT + 2 + (bT - 2)z^{-1}}$$

Matched Pole-Zero Method

The matched pole-zero method for emulating a continuous-time controller with a digital filter is based on mapping the s-plane poles and zeros of the controller into the z-plane using

$$z = e^{sT}$$

For example, the continuous-time controller with transfer function

$$G_c(s) = \frac{K_c(s + a)}{s + b}$$

is approximated by the digital controller with transfer function

$$D(z) = \frac{K_d(z - e^{-aT})}{z - e^{-bT}}$$

The gain K_d is calculated to make the dc gain of the digital controller equal to that of the continuous-time controller.

$$\frac{K_d(1 - e^{-aT})}{1 - e^{-bT}} = K_c \frac{a}{b}$$

therefore

$$K_d = K_c \frac{a(1 - e^{-bT})}{b(1 - e^{-aT})}$$

Direct Discrete Design via W-Plane

When the sampling frequency is less than 10 times the bandwidth of the system, the emulation methods of digital compensator design often provide unsatisfactory results. When the sampling frequency is less than 20 times the bandwidth of the system, it is sometimes necessary to use the direct discrete design method in order to achieve the desired control.

In the direct discrete design the system is modeled based on its input and output at the sampling instants using the z-transform.

The z-transform transfer function for a system like the one shown in Figure 26.25 can be found from the Laplace transform transfer functions using the zero-order-hold equivalent. The z-transform transfer function of the combined plant and D/A converter is

$$G(z) = (1 - z^{-1})z\left\{\frac{G_P(s)}{s}\right\}$$

The discrete-time system is represented by the block diagram shown in Figure 26.26. The z-transform transfer function is

$$\frac{C(z)}{R(z)} = \frac{D(z)G(z)}{1 + D(z)G(z)}$$

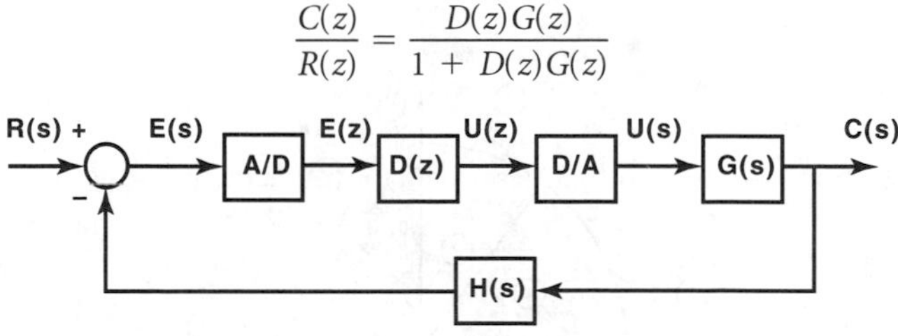

Figure 26.25 A digital control system.

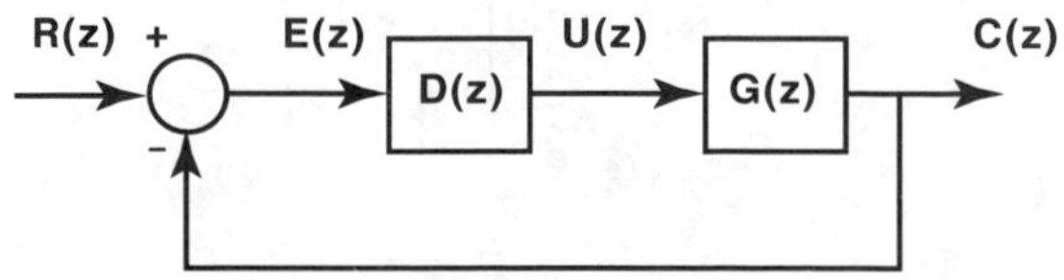

Figure 26.26 A digital control system.

The continuous-time system frequency response techniques and equations presented above can be used in the design of discrete-time controllers. This is accomplished by transforming the discretized system into the w-plane using the bilinear transformation.

$$z = \frac{1 + \frac{T}{2}w}{1 - \frac{T}{2}w}$$

The w-plane is a complex plane similar to the s-plane. The design and analysis tools that are used in the s-plane for continuous-time systems can be used in the w-plane for discrete-time systems. The w-plane variable, like the Laplace transform variable, s, is complex.

$$w = \sigma_w + j\omega_w$$

The w-plane frequency variable, ω_w, is not the frequency of excitation in radians per second. This frequency variable is mathematically defined so that the interior of the unit circle in the z-plane maps into the left half of the w-plane. Therefore the unit circle in the z-plane becomes the $j\omega_w$ axis in the w-plane as shown in Figure 26.27. The relationship between ω_w and the actual radian frequency is

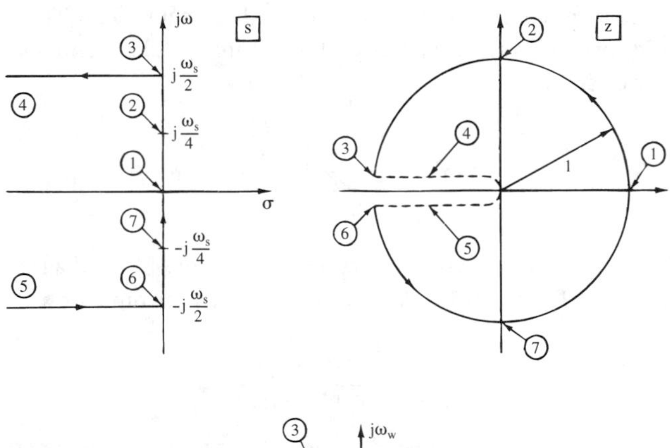

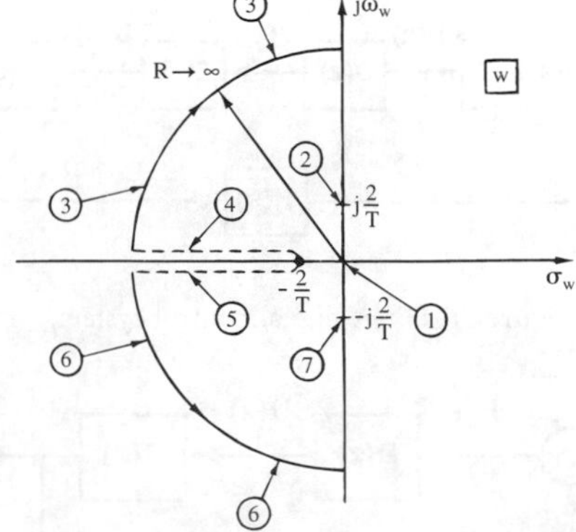

Figure 26.27 Mapping from s-plane to z-plane to w-plane. Copied from Phillips, C.L. and Nagle, H.T., *Digital Control System Analysis and Design*. 3rd ed., Prentice Hall, Englewood Cliffs, NJ, 1995.

$$\omega_w = \frac{2}{T}\tan\left(\frac{\omega T}{2}\right)$$

By evaluating the equation we can show that

$$\omega_w \cong \omega \text{ for } \omega \leq \frac{2\pi}{10T} = \frac{\omega_s}{10}$$

However, as ω approaches $\omega_s/2 = \pi/T$, ω_w approaches infinity.

For Bode plot design techniques we deal with the w-plane frequency response, $G(j\omega_w)$. The Bode plot of the w-plane frequency response is sketched from $G(w)$ using the same techniques as we described for the s-plane frequency response. All of the same rules apply. The only difference is that the frequency variable is ω_w instead of ω.

An alternative method of calculating the frequency response of the discretized system is by evaluating

$$G(z)\,|_{z=e^{j\omega T}} = G(e^{j\omega T}), \qquad 0 \leq \omega < \omega_s/2$$

The Bode plot can then be constructed by plotting 20 $\log_{10}$ $|\,G(e^{j\omega T})\,|$ (dB) and $\angle G(e^{j\omega T})$ in semilogarithmic plot with the frequency variable scaled to

$$\omega_w = \frac{2}{T}\tan\left(\frac{\omega T}{2}\right)$$

on the logarithmic frequency axis.

Either method of constructing the Bode plot of the w-plane frequency response should yield the same result. Some computer programs, MATLAB and PROGRAM CC, for example, provide easy methods of computing the w-plane frequency response using one or both of the methods described above.

The techniques discussed above for the design of phase-lead, phase-lag and PID compensators in continuous-time systems are also used to design controllers in the w-plane. After completing the controller design in the w-plane to determine the compensator transfer function, $G_c(w)$, the z-transform compensator is found by applying the inverse bilinear transformation

$$w = \frac{2}{T}\left(\frac{z-1}{z+1}\right)$$

$$D(z) = G_c(w)\,|_{w=\frac{2}{T}\left(\frac{z-1}{z+1}\right)}$$

Digital Controller Implementation

After completing the design of the discrete-time compensator using either the emulation method or the direct discrete design method, the z-transform transfer function, $D(z)$ must be converted into a discrete-time difference equation for implementation.

The difference equation that must be solved to implement the controller in a digital processor is found from

$$D(z) = \frac{U(z)}{E(z)}$$

as shown in Figure 26.26.

For example, let

$$D(z) = \frac{U(z)}{E(z)} = \frac{K_d(z - a)}{z - b} = \frac{K_d - aK_d z^{-1}}{1 - bz^{-1}}$$

then

$$U(z) - bU(z)z^{-1} = K_d E(z) - aK_d E(z)z^{-1}$$

The inverse z-transform yields the difference equation

$$u[k] = bu[k - 1] + K_d e[k] - aK_d e[k - 1]$$

to be implemented in the digital controller.

Summary

This section discusses linear system compensator design using the Bode plot of the frequency response. Both continuous-time and discrete-time control systems are considered.

Design methods are presented for several types of compensators including phase-lag, phase-lead, PI, PD, and PID.

Both emulation methods and direct discrete-time design methods for the design of digital controllers are presented.

Defining Terms

Bode diagram: A graph of the gain magnitude and phase frequency response of a linear system. Plotted with frequency on a logarithmic scale. Gain magnitude commonly plotted in decibels (dB). Phase usually plotted in degrees.

Compensator: A compensator is an electrical network or other components added to a system to allow the system to achieve specified performance characteristics

Gain-crossover frequency: At the gain-crossover frequency the open-loop gain frequency response of a system is unity (0 dB) and decreasing as frequency increases.

Gain margin: The gain margin is a measure of relative stability. The amount of gain that will cause the system to have unity (0 dB) gain at the frequency where the phase shift is $-180°$ is called the gain margin.

Lag-lead compensator: A phase-lag compensator cascaded with a phase-lead compensator exhibits negative phase shift at low frequencies and positive phase shift at high frequencies.

Phase-crossover frequency: At the phase crossover frequency the open-loop phase frequency response of a system is -180 degrees.

Phase-lag compensator: In a phase-lag network the phase angle of the frequency response is always negative.

Phase-lead compensator: In a phase-lead network the phase angle of the frequency response is always positive.

Phase margin: The phase margin is a measure of relative stability. The amount of negative phase shift which will cause the system to have a phase shift of $-180°$ at the frequency where the gain is unity (0 dB).

Proportional + integral + derivative (PID) compensator: The PID, or three-term compensator for continuous-time systems consists of a proportional gain amplifier, an integrating amplifier and a differentiating amplifier connected in parallel.

References

D'Azzo, J. J. and Houpis, C. H. (1988) *Linear Control System Analysis & Design Conventional and Modern,* 3d ed., McGraw-Hill, New York, NY.

Franklin, G. F., Powell, J. D. and Emami-Naeini, A. (1994) *Feedback Control of Dynamic Systems,* 3d ed., Addison-Wesley, Reading, PA.

Phillips, C. L. and Harbor, R. D. (1991) *Feedback Control Systems,* 2d ed., Prentice Hall, Englewood Cliffs, NJ.

Phillips, C. L. and Nagle, H. T. (1995) *Digital Control System Analysis and Design,* 3d ed., Prentice Hall, Englewood Cliffs, NJ.

27

The Root Locus Method

Robert J. Veillette
University of Akron

J. Alexis De Abreu-Garcia
University of Akron

27.1 Motivation and Background 490
27.2 Root Locus Analysis 490
27.3 Compensator Design by Root Locus Method 495
27.4 Examples 497

27.1 Motivation and Background

The form of the natural response of any linear dynamic system is determined by the complex-plane locations of the poles and zeros of the system transfer function. Any negative real pole of the s-domain transfer function of a continuous-time system corresponds to an exponential mode of response; the rate of decay of the exponential depends on the magnitude of the pole. Any pair of complex-conjugate poles in the left half-plane corresponds to an exponentially decaying oscillatory mode of response; the real part of the poles determines the rate of decay, whereas the imaginary part determines the frequency of oscillation. Any pole or poles in the right half of the complex plane correspond to an unbounded (exponentially growing) mode of response; the system is then said to be unstable.

The pole locations of a feedback control system are a crucial consideration in its design. Loosely speaking, if a system is to respond to inputs quickly and without excessive oscillation, it should be designed such that its poles lie far enough into the left-half plane. A more detailed discussion relating the pole and zero locations of a system transfer function to the time-domain system response and control design specifications is presented in Franklin et al. (1994) (118–138).

Root locus analysis was first developed by Evans (1948) as a means of determining the set of positions (loci) of the poles of a closed-loop system transfer function as a scalar parameter of the system varies over the interval from $-\infty$ to ∞. It constitutes a powerful tool in the design of a compensator for a feedback control system. The compensator poles and zeros are chosen to shape the branches of the root locus; then the compensator gain is chosen to place the closed-loop poles at the desired positions along the branches. By displaying the closed-loop poles, the root locus analysis complements the frequency-domain methods, which provide information on the magnitude and phase of the system frequency response. It is especially useful for systems that are unstable or marginally stable in open loop, since the frequency-domain methods are more cumbersome for such systems.

27.2 Root Locus Analysis

Problem definition

Consider the closed-loop feedback control system shown in Figure 27.1. The quantity $KG(s)H(s)$ is referred to as the open-loop transfer function of the system. Usually, the poles and zeros of the open-loop transfer function are known, or easily determined. Moreover, these poles and zeros do not depend on the gain K. If $G(s)$ and $H(s)$ are each expressed as the ratio of polynomials in s as

$$G(s) = \frac{n_G(s)}{d_G(s)}, \qquad H(s) = \frac{n_H(s)}{d_H(s)}, \tag{27.1}$$

then the poles of the open-loop transfer function are the solutions of

$$d_G(s)d_H(s) = 0, \tag{27.2}$$

and the zeros of the open-loop transfer function are the solutions of

$$n_G(s)n_H(s) = 0. \tag{27.3}$$

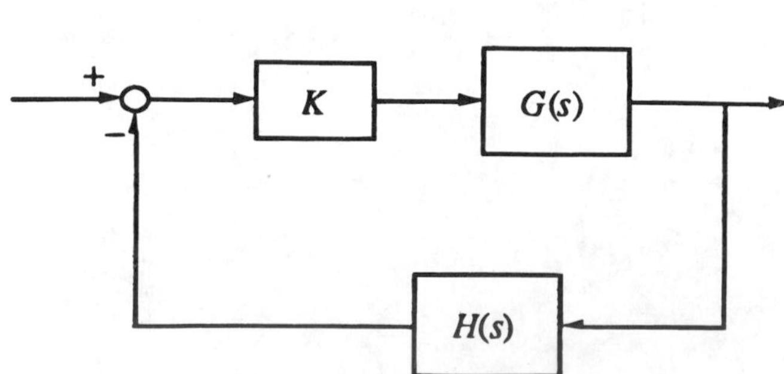

Figure 27.1 Closed-loop system.

0-8493-8343-9/97/$0.00+$.50

The transfer function of the closed-loop system is

$$T(s) = \frac{KG(s)}{1 + KG(s)H(s)} = \frac{Kn_G(s)d_H(s)}{d_G(s)d_H(s) + Kn_G(s)n_H(s)}. \quad (27.4)$$

The poles of $T(s)$ are the solutions of the characteristic equation

$$1 + KG(s)H(s) = 0, \quad (27.5)$$

or of the equivalent characteristic equation

$$d_G(s)d_H(s) + Kn_G(s)n_H(s) = 0, \quad (27.6)$$

which clearly depend on the gain K. It is of interest to determine how the closed-loop poles move in the s plane as K varies. The root locus method is a means of plotting these poles as K ranges from $-\infty$ to ∞.

DEFINITION: 27.1 The *root locus* (sometimes called the *complete root locus*) of the system in Figure 27.1 is the set of loci of the poles of the closed-loop system as the gain K ranges from $-\infty$ to ∞.

In order to determine and plot the root locus of the system, write Equation 27.5 as

$$\begin{aligned} KG(s)H(s) &= -1 \\ &= 1\angle(2k+1)\pi, \qquad k = 0, \pm 1, \pm 2, \ldots \end{aligned} \quad (27.7)$$

which is equivalent to the two simultaneous conditions

$$|KG(s)H(s)| = 1 \quad (27.8a)$$

$$\angle G(s)H(s) = \begin{cases} (2k+1)\pi, & K > 0 \\ 2k\pi, & K < 0 \end{cases}, \qquad k = 0, \pm 1, \pm 2, \ldots \quad (27.8b)$$

The conditions (Equation 27.8) are sometimes given as an alternative definition of the root locus. The condition (Equation 27.8a) is known as the *magnitude condition,* and the condition (Equation 27.8b) is known as the *angle condition.*

Since K can be any real number, any point s automatically satisfies the magnitude condition; therefore, the magnitude condition is mostly useful for determining the value of K that corresponds to a point s already known to lie on the root locus.

It is the angle condition that determines the locus. According to the angle condition, a point s is on the root locus if the angle of $G(s)H(s)$ is any multiple of π. If the angle is an odd multiple of π, then s is said to be on the *standard root locus* (sometimes called simply the *root locus*), which is the set of solutions of Equation 27.5 for $K > 0$; if the angle is an even multiple of π, then s is said to be on the *complementary root locus,* which is the set of solutions of Equation 27.5 for $K < 0$. For the standard root locus, the angle condition reduces to

$$\angle G(s)H(s) = (2k+1)\pi, \qquad k = 1, \pm 1, \pm 2, \ldots \quad (27.9)$$

We concentrate here on the construction of the standard root locus. The development of the complementary root locus is similar to that of the standard root locus, and is not discussed. A summary of the guidelines for sketching the complementary root locus is given in Franklin et al. (1994) (286–287). For a complete treatment of the complementary root locus, interested readers are referred to Kuo, 1995 (472–509).

Development of rules for constructing root locus

Considering the availability of special-purpose software for control systems analysis, it can be argued that the most convenient approach to plotting a root locus is to use a computer. Nevertheless, the ability to sketch a root locus by hand is essential. It affords confidence that a computer-generated solution is free of errors. In addition, it improves the intuition necessary for designing a feedback controller—a process of altering a system to obtain the desired root locus.

The intuition and understanding that attend the ability to sketch the root locus by hand and the use of the computer for precise root locus calculations are complementary tools in feedback control design. It is possible, with hardly any calculations, to get a clear idea about the general form of the root locus of even a complicated system. However, except for simple systems, precise calculations of all the detailed features of the locus are best done by computer.

The rules for plotting the root locus are now developed. The development is intended to be intuitively convincing and complete, but not too mathematical. Other developments, with varying degrees of mathematical rigor, may be found in Franklin et al. (1994) (249–260), Kuo (1995) (477–505), Nise (1995) (364–378), D'Azzo and Houpis (1988) (225–235), and Miron (1989) (119–132). This presentation starts with the rules that are the simplest, and therefore the most useful for sketching the root locus by hand, and proceeds to the rules that are more complex and tedious to apply. Generally, the rules that are more tedious to apply are also less important for making an approximate sketch.

RULE 27.1 The number of branches of the root locus is equal to the number of poles of the open-loop transfer function $G(s)H(s)$.

As K varies, each of the closed-loop poles moves along a continuous path in the complex plane, called a "branch" of the root locus. Thus the number of branches is equal to the number of closed-loop poles for any value of K. But the number of closed-loop poles is equal to the number of open-loop poles, since the characteristic polynomials in Equations 27.2 and 27.6 have the same degree. (It is assumed that both $G(s)$ and $H(s)$ are proper—i.e., that neither $G(s)$ nor $H(s)$ has more zeros than poles.)

RULE 27.2 The root locus is symmetric about the real axis in the s plane.

For any physical system, all the coefficients of the open-loop transfer function, and hence of the characteristic Equation 27.6, are real. As a result, for each K, any complex closed-loop poles occur in conjugate pairs; therefore, so do the root locus branches.

RULE 27.3 The root locus starts (for $K = 0$) at the poles of $G(s)H(s)$, and ends (for $K = \infty$) at the zeros of $G(s)H(s)$, including the zeros at infinity.

It seems reasonable from an intuitive point of view that the branches of the root locus should originate from the open-loop poles. When $K = 0$, there is no feedback to alter the open-loop system dynamics. This is also easy to establish mathematically. Simply observe that setting $K = 0$ makes the closed-loop characteristic Equation 27.6 identical with the open-loop characteristic Equation 27.2.

Let the open-loop transfer function have n poles and m zeros. Then Rule 27.3 claims that m of the branches of the root locus terminate at the finite zeros of $G(s)H(s)$, while the other $n - m$ branches go off to infinity as $K \to \infty$. This claim also seems reasonable. According to the magnitude condition, as K approaches infinity, $|G(s)H(s)|$ must approach zero. To establish the claim mathematically, write Equation 27.6 as

$$\frac{d_G(s)d_H(s)}{K} + n_G(s)n_H(s) = 0. \tag{27.10}$$

It is clear that each of the m open-loop zeros will satisfy Equation 27.10 as $K \to \infty$. Both terms on the left-hand side of Equation 27.10 vanish, the first one since $d_G(s)d_H(s)$ is bounded and $1/K \to 0$, and the second one since the open-loop zeros are defined by Equation 27.3. Therefore, m of the branches of the root locus must terminate at the open-loop zeros. To see that the other $n - m$ branches go to infinity, note that the left-hand side of Equation 27.10 is a polynomial of degree n, whose $n - m$ highest-order coefficients decrease as K increases. As these coefficients go to zero, the polynomial approximates $n_G(s)n_H(s)$ in a larger and larger region in the s plane. The additional $n - m$ roots must lie outside this growing region, where the highest-order terms of the polynomial still dominate.

RULE 27.4 A given point on the real axis is included in the root locus if the number of real poles and zeros to the right of that point is odd.

This rule is a direct consequence of the angle condition. Consider, for example, the open-loop system pole-zero configuration shown in Figure 27.2. The intervals on the real axis included in the root locus, as given by Rule 27.4, are indicated by dark line segments. To derive this result, write the open-loop transfer function as

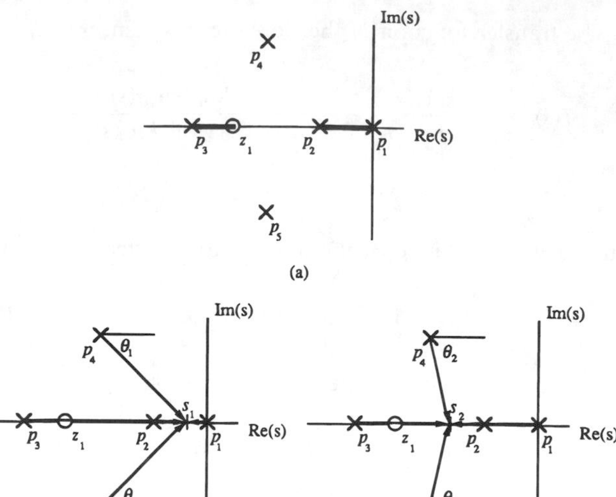

Figure 27.2 The angle condition for points on the real axis: (a) the real-axis locus; (b) test point on the locus; (c) test point not on the locus.

$$G(s)H(s) = \frac{(s - z_1)}{(s - p_1)(s - p_2)(s - p_3)(s - p_4)(s - p_5)}. \tag{27.11}$$

Then the angle of the open-loop transfer function is given by

$$\angle G(s)H(s) = \angle(s - z_1) - \angle(s - p_1) - \angle(s - p_2) - \angle(s - p_3) - \angle(s - p_4) - \angle(s - p_5). \tag{27.12}$$

Now consider the test point s_1 on the real axis, and the vectors $s_1 - z_1$, $s_1 - p_1$, etc., shown in Figure 27.2(b). The angle of the open-loop transfer function at s_1 is computed according to Equation 27.12 from the angles of these vectors as

$$\angle G(s_1)H(s_1) = 0 - 180° - 0 - 0 - \theta_1 - (-\theta_1) = -180°. \tag{27.13}$$

Therefore, s_1 satisfies the angle condition Equation 27.9, and lies on the root locus. By contrast, consider the test point s_2. As shown in Figure 27.2(c), the angle of the open-loop transfer function at s_2 is

$$\angle G(s_2)H(s_2) = 0 - 180° - 180° - 0 - \theta_2 - (-\theta_2) = -360°. \tag{27.14}$$

Therefore, s_2 does not satisfy the angle condition Equation 27.9, and does not lie on the root locus.

In general, for any test point on the real axis, complex-conjugate pairs of poles or zeros contribute zero net angle; real poles and zeros to the left of the test point also contribute zero angle;

but real poles and zeros to the right of the test point contribute angles of $\pm 180°$ each. Therefore, the angle of $G(s)H(s)$ at a test point on the real axis depends only on the real poles and zeros to the right of the test point; the test point satisfies the angle condition if the number of those poles and zeros is odd.

RULE 27.5 The $n - m$ branches of the root locus that go to infinity for large K asymptotically approach straight lines at angles

$$\theta_k = \frac{(2k+1)\pi}{n-m}, \qquad k = 0, 1, 2, \ldots, n-m-1, \quad (27.15)$$

where n is the number of open-loop poles, and m is the number of open-loop zeros, with $n > m$.

This rule is illustrated in Figure 27.3, where the asymptotes of the root locus branches are illustrated for several values of $n - m$. It is derived by applying the angle condition for test points s of large magnitude—i.e., test points that are far from the origin, and far from all the finite open-loop poles p_i and zeros z_i, as shown in Figure 27.4. For such a test point, each angle $\angle(s - z_i)$ and $\angle(s - p_i)$ is approximately equal to $\angle s$. The angle condition Equation 27.9 may be written as

$$\sum_{i=1}^{m} \angle(s - z_i) - \sum_{i=1}^{n} \angle(s - p_i) = -(2k+1)\pi \quad (27.16)$$

for some integer k. For large s, Equation 27.16 is approximated by

$$\sum_{i=1}^{m} \angle(s) - \sum_{i=1}^{n} \angle(s) = -(2k+1)\pi, \quad (27.17)$$

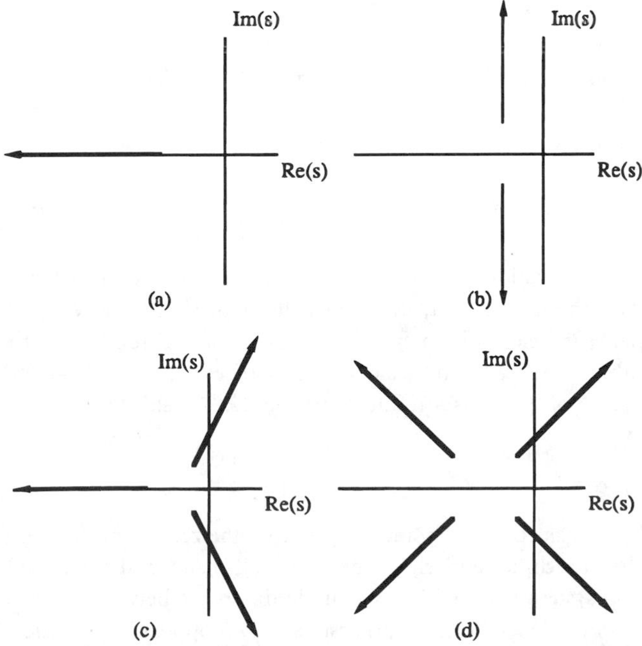

Figure 27.3 Asymptotes of the root locus branches: (a) $n - m = 1$; (b) $n - m = 2$; (c) $n - m = 3$; (d) $n - m = 4$.

or

$$m\angle(s) - n\angle(s) = -(2k+1)\pi. \quad (27.18)$$

Solving for $\angle s$ yields the condition Equation 27.15.

RULE 27.6 The intersection of the asymptotes lies on the real axis, and is given by

$$\sigma = \frac{\sum_{i=1}^{n} p_i - \sum_{i=1}^{m} z_i}{n-m}, \quad (27.19)$$

where p_i and z_i denote the open-loop poles and zeros, respectively.

Note that if $n - m = 0$, then there are no asymptotes, and that if $n - m = 1$, then there is a single asymptote at an angle of 180°. In either of these cases, the "intersection of the asymptotes" is irrelevant; therefore, only the situation where $n - m \geq 2$ needs to be considered. The illustrations in Figure 27.3 indicate that, for $n - m \geq 2$, the roots that go to infinity are balanced about some point on the real axis. This point, the "center of gravity" or average of those roots, is the intersection of the asymptotes.

The center of gravity of the roots that go to infinity may be found by taking advantage of a simple property of polynomials: The sum of the roots of any polynomial

$$q(s) = s^n + a_1 s^{n-1} + \cdots + a_n \quad (27.20)$$

is simply the coefficient a_1. (This is easily verified by writing $q(s)$ in factored form.) In the case where $n - m \geq 2$, the second coefficient of the closed-loop characteristic polynomial

$$q_{cl}(s) = d_G(s)d_H(s) + Kn_G(s)n_H(s) \quad (27.21)$$

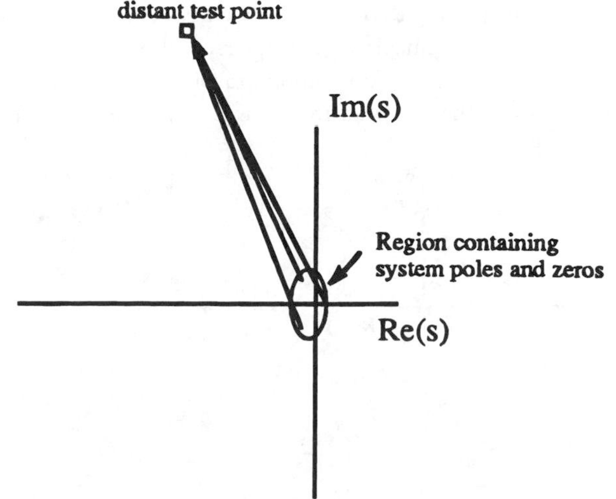

Figure 27.4 The angle condition for test points of large magnitude.

is the same as that of the open-loop characteristic polynomial $d_G(s)d_H(s)$; therefore, the sum of all the closed-loop poles does not depend on K. For $K = 0$, the closed-loop poles r_i are at the open-loop poles; therefore, the sum of the closed-loop poles is established as

$$\sum_{i=1}^{n} r_i = \sum_{i=1}^{n} p_i. \tag{27.22}$$

for all K. As $K \to \infty$, m of the roots are found at the open-loop zeros z_i, and the other $n - m$ roots are on the asymptotes. Therefore the sum of the roots is also given by

$$\sum_{i=1}^{n} r_i = \sum_{i=1}^{n} z_i + (n - m)\sigma, \tag{27.23}$$

where σ is the average of the roots on the asymptotes. Comparison of Equation 27.22 and Equation 27.23 yields

$$\sum_{i=1}^{n} z_i + (n - m)\sigma = \sum_{i=1}^{n} p_i, \tag{27.24}$$

which in turn yields Equation 27.19.

RULE 27.7 The points at which the root locus intersects the $j\omega$ axis, and the corresponding values of K, may be determined using the Routh-Hurwitz stability test, or the magnitude and phase plots of the open-loop system.

The first method consists simply of multiplying out the closed-loop characteristic polynomial (Equation 27.21), writing its coefficients as functions of K, and computing the Routh-Hurwitz table to determine those values of K for which roots cross the $j\omega$ axis. The second method is based on the angle and magnitude conditions for the root locus. An accurate (computer-generated) phase plot of the open-loop transfer function will show the frequency or frequencies at which the phase equals an odd multiple of 180°. By the angle condition, these frequencies correspond to the values on the imaginary axis intersected by the root locus. The magnitudes of the open-loop transfer function at these frequencies can then be read from an accurate magnitude plot.

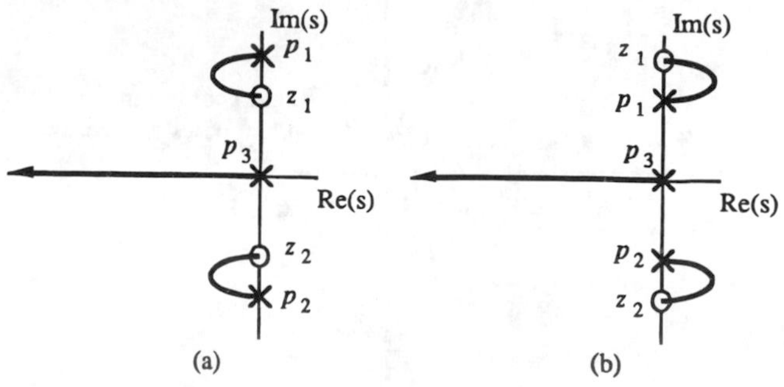

Figure 27.5 Root locus plots demonstrating angles of departure and arrival: (a) $\theta_{z_1} = -180$; (b) $\theta_{z_1} = 0$.

By the magnitude condition, these magnitudes are the reciprocals of the values of K that correspond to the $j\omega$-axis crossings.

RULE 27.8 The angle of departure of the root locus from a complex pole p_1 is given by

$$\theta_{p1} = \sum_{i=1}^{m} \angle(p_1 - z_i) - \sum_{i=2}^{n} \angle(p_1 - p_i) + 180°, \tag{27.25}$$

where $p_1, \ldots, p_n$ are the open-loop poles and $z_1, \ldots, z_m$ are the open-loop zeros. The angle of arrival of the root locus at a complex zero is given by

$$\theta_{z1} = \sum_{i=1}^{n} \angle(z_1 - p_i) - \sum_{i=2}^{m} \angle(z_1 - z_i) - 180°. \tag{27.26}$$

Equation 27.25 and Equation 27.26 are found by applying the angle condition to test points near the complex pole or zero. This calculation is useful for determining the position of a short branch of the root locus between a complex pole and a complex zero lying close together.

As an example, Figure 27.5 shows two pole-zero configurations for which the position of the locus between a pole and a zero is critical to determining the closed-loop system stability. The approximate position is easily determined by computing the angle of departure from the pole or the angle of arrival at the zero. For Figure 27.5(a), the angle condition near the zero z_1 is

$$\theta_{z1} + \angle(z_1 - z_2) - \angle(z_1 - p_1) - \angle(z_1 - p_2) - \angle(z_1 - p_3) = -180°, \tag{27.27}$$

or

$$\theta_{z1} + 90° - (-90°) - (90°) - (90°) = -180°, \tag{27.28}$$

which gives

$$\theta_{z1} = -180°. \tag{27.29}$$

This calculation implies that the branch approaches the zero from the left. In fact, the branch lies entirely in the left half-plane; the closed-loop system is stable. For Figure 27.5(b), the only difference in the calculation is that $\angle(z_1 - p_1) = 90°$, instead of $-90°$; as a result, Equation 27.27 yields

$$\theta_{z1} = 0. \tag{27.30}$$

This means that the branch approaches the zero from the right. The branch lies entirely in the right half-plane, and the closed-loop system is unstable. The subtle difference between the two systems in Figures 27.5(a,b) results in two opposite conclusions concerning the closed-loop system stability. The computation of angle of departure or angle of arrival is helpful in this case for accurately reaching these conclusions.

RULE 27.9 Points where the root locus breaks away from the real axis (breakaway points) or breaks in to the real axis (break-in points) are the real values of s on the root locus that satisfy

$$\frac{d[G(s)H(s)]}{ds} = 0. \tag{27.31}$$

At a breakaway point, two branches of the root locus on the real axis meet for a certain value of K, and then leave the axis as complex-conjugate pairs as K increases. Considering only the real axis, the value of K corresponding to the roots reaches a relative maximum at the breakaway point. By the magnitude condition, this implies that $G(s)H(s)$ reaches a relative minimum on the real axis at a breakaway point. Similarly, $G(s)H(s)$ reaches a relative maximum at a break-in point. Hence, a breakaway or break-in point is a point on the real locus for which the derivative of $G(s)H(s)$ is zero. The condition (Equation 27.31) may be checked directly by hand; or the function $G(s)H(s)$ may simply be plotted for real s to find its relative minima and maxima on the real axis.

The same derivative rule presented for finding real breakaway and break-in points may also be used for finding complex breakpoints. A complex breakpoint is a point off the real axis at which two or more branches of the root locus meet, resulting in a repeated root for the corresponding value of K, and split up again.

Steps for Sketching the Root Locus

As a summary, the steps for sketching the root locus of a system are now presented. These steps do not always need to be taken sequentially, but may be used as general guidelines for the root locus method of analysis. It is assumed that the open-loop transfer function is proper.

1. Write the open-loop transfer function with numerator and denominator polynomials in factored form.
2. Plot the open-loop poles and zeros in the s plane.
3. Fill in all intervals of the real axis that lie to the left of an odd number of real poles and zeros (Rule 27.4).
4. Sketch the asymptotes of the $n - m$ branches that go to infinity (Rule 27.5 and Rule 27.6).
5. Start a root locus branch at each open-loop pole (Rule 27.1). Keeping in mind that the locus is symmetric with respect to the real axis (Rule 27.2), try to visualize how each branch will arrive at a zero, or approach an asymptote as it goes to infinity (Rule 27.3).
6. If uncertain about the position of certain branches of the locus, it may be helpful to compute approximate angles of departure from or angles of arrival at complex poles or zeros (Rule 27.8).
7. If the exact frequency or gain corresponding to a $j\omega$-axis crossing is critical, or if uncertain whether a $j\omega$-axis crossing occurs, apply the Routh-Hurwitz test to the closed-loop characteristic polynomial, or plot accurate frequency-response graphs (magnitude and phase) to determine it (Rule 27.7).
8. If the exact location (or the existence) of a breakaway or break-in point is critical, use the derivative rule (Rule 27.9) or a computer plot of the locus to determine it.

Remember that the angle condition may be checked at any point in the complex plane to determine whether that point lies on the root locus. If a given point in the vicinity of the open-loop poles and zeros satisfies the angle condition only *approximately* (not exactly), then it is safe to assume that it lies *near* the root locus, at least in the sense that a modest perturbation in the plant could result in a closed-loop pole there.

27.3 Compensator Design by Root Locus Method

An essential aspect of feedback control design is the introduction of a compensator to alter the characteristics of the open-loop transfer function. In a few cases, a satisfactory design may result simply from introducing a gain into the loop. Then, a straightforward root locus analysis is sufficient to determine the value of the gain required to achieve the desired transient response. In most cases, however, a satisfactory design requires the introduction of a dynamic compensator. Then, the overall system transient response depends upon the selection of the pole and zero locations, as well as the gain, of the compensator.

A root locus design method consists of selecting the compensator poles and zeros to position the locus branches, and then selecting the compensator gain to place the roots at desired positions along the branches. The first part of the design is the more complicated, and requires first of all an understanding of how adding poles and zeros affects the root locus.

Effect of Adding a Pole to the Open-loop Transfer Function

The addition of poles to the open-loop transfer function increases the number of branches in the root locus and pushes the closed-loop poles to the right in the s plane. This is a consequence of the angle condition. It is clear from Equation 27.12 that a pole of the open-loop transfer function contributes a negative angle to $G(s)H(s)$ for points s in the upper half-plane. For a test point in the upper half-plane to the right of all the open-loop poles and zeros, an additional open-loop pole will cause the angle of $G(s)H(s)$ to become more negative; therefore, the angle of $G(s)H(s)$ will reach $-180°$ for test points at smaller angles in the s plane. In other words, at least some branches of the root locus will lie farther to the right.

The effect on the root locus of adding poles to two different open-loop transfer functions is illustrated in Figures 27.6(a) and 27.6(b). For each system, the original root locus is shown, followed by the root locus with first one, and then two additional poles. For both systems, the root locus lies farther to the right as more open-loop poles are added.

Adding poles to a system has an undesirable effect on the root

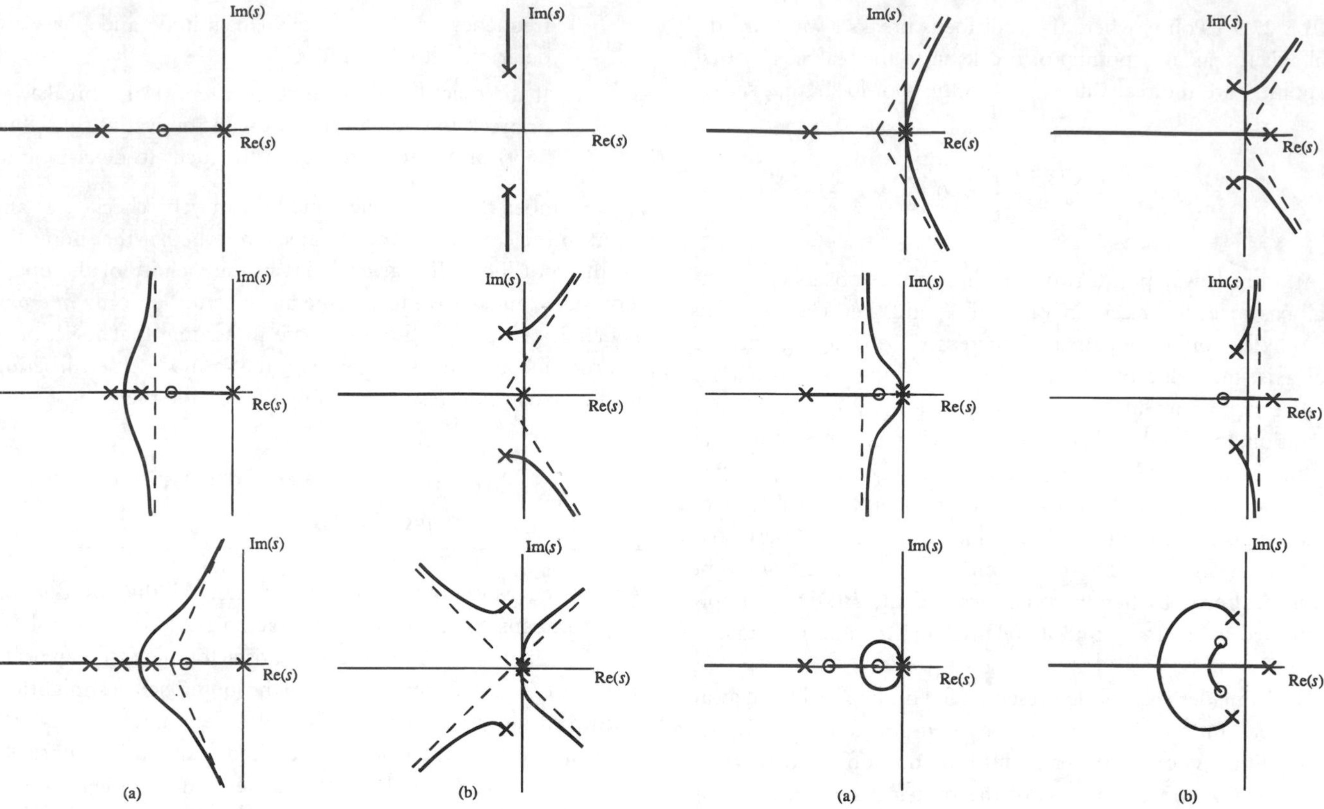

Figure 27.6 The effect of additional poles on the root locus.

Figure 27.7 The effect of additional zeros on the root locus.

locus. In particular, for systems of higher relative degree $n - m$, the feedback tends to be more destabilizing. On the other hand, frequency-domain analysis shows that the addition of a pole or poles near the origin can improve the low-frequency performance of the closed-loop system. In any case, for a system with relative degree $n - m \geq 2$, care must be taken to keep the feedback gain sufficiently low that all the closed-loop poles lie in the left half-plane.

Effect of Adding a Zero to the Open-Loop Transfer Function

The addition of zeros to the open-loop transfer function pulls the closed-loop poles to the left in the s plane. Also, each zero attracts a branch of the root locus to terminate there for $K = \infty$. In short, the effect of additional zeros is roughly the opposite of that of additional poles.

The effect on the root locus of adding zeros to two different open-loop transfer functions is illustrated in Figures 27.7(a) and 27.7(b). For each system, the original root locus is shown, followed by the root locus with first one, and then two additional zeros. For both systems, the root locus lies farther to the left as more open-loop zeros are added.

Adding zeros to a system has a desirable effect on the root locus. Systems with low relative degree (i.e., $n - m \leq 1$) can be stable for all values of the feedback gain. Even a system with some open-loop poles in the right half-plane can be stabilized using feedback, if there are some zeros in the left-half plane. On the other hand, a zero that has a stabilizing effect on the system may cause a deterioration of the low-frequency performance of the system. Such an undesirable effect can be discerned by use of frequency-domain analysis techniques.

The Effect of a Lead Compensator

If the branches of the root locus of a given system lie too far to the right, they could be moved to the left if it were possible to introduce an additional zero to the open-loop transfer function. Of course, a proper compensator has at least as many poles as zeros; so, adding zeros requires also adding poles. When the effect of a zero is desired, the design calls for a lead compensator. A lead compensator has a transfer function of the form

$$G_c(s) = \frac{K(s+a)}{(s+b)}, \tag{27.32}$$

with $0 < a < b$. It is called a lead compensator because its frequency response $G(j\omega)$ has a positive angle (phase lead) for all frequencies $\omega > 0$.

The effect of a lead compensator on the root locus of two different open-loop transfer functions is illustrated in Figures 27.8(a) and 27.8(b). (These are the same two systems used to illustrate the effect of additional zeros in Figure 27.7.) For each system, the original root locus is shown, followed by the root locus with the lead compensator. The angles of the asymptotes

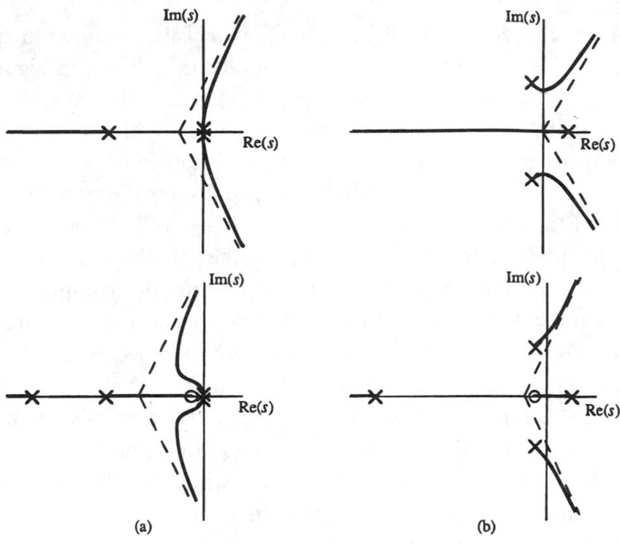

Figure 27.8 The effect of a lead compensator on the root locus.

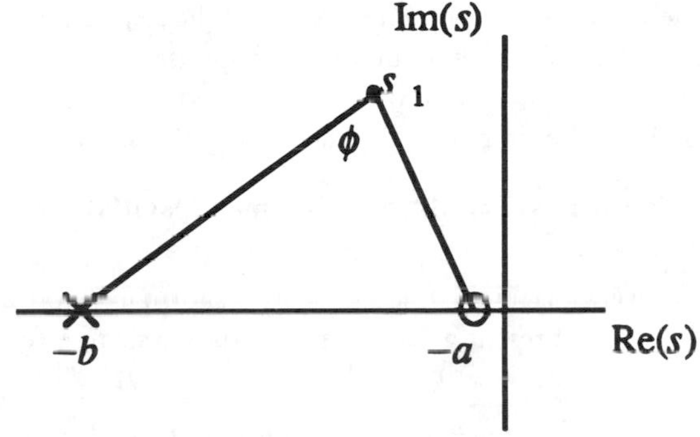

Figure 27.9 Graphical determination of the angle of a lead compensator at a point s_1.

of the root locus branches remain unchanged; however, the intersection of the asymptotes is moved to the left. In effect, the zero of the compensator, being locally more influential than the pole, initially draws the root locus to the left.

Lead Compensator Design

The goal of the lead compensator design is to improve the transient response of the system over that of the uncompensated system, in order to obtain adequate speed of response and damping. In many cases, the speed and damping of the system response can be correlated to the locations of the dominant closed-loop poles, which are the poles nearest the origin of the s plane, but not too close to a zero of the forward transfer function $G(s)$. Generally speaking, the larger the magnitude of the dominant poles, the faster the system response; and the larger the angle of the dominant poles, measured counterclockwise from the positive $j\omega$ axis, the more heavily damped (i.e., the less oscillatory) the system response. Therefore, the goal of the lead compensator design may be restated as the placement of the dominant poles at a desired magnitude and angle.

Given an open-loop transfer function $G_p(s)$, and assuming the transient response of the uncompensated closed-loop system is not satisfactory, one method for the design of a lead compensator may proceed as follows:

1. Choose the desired locations (magnitude and angle) of the dominant poles in the s plane. Assume the dominant poles are a complex-conjugate pair, and call the desired upper half-plane location s_1.
2. Find $\angle G_p(s_1)$.
3. Choose the parameters a and b (i.e., the pole and zero) of the compensator such that

$$\angle \frac{(s_1 + a)}{(s_1 + b)} + \angle G_p(s_1) = \pm 180°. \tag{27.33}$$

   Then, according to the angle condition, the point s_1 will lie on the compensated root locus.
4. Choose the parameter K (i.e., the gain parameter) of the compensator such that

$$\left| \frac{K(s_1 + a)}{(s_1 + b)} G_p(s_1) \right| = 1. \tag{27.34}$$

   Then, according to the magnitude condition, s_1 will be a pole of the closed-loop system.

There is some design freedom inherent in Step 3 above. The angle of the compensator at s_1 is found graphically as the angle between the line segments connecting s_1 with the compensator pole and zero; see Figure 27.9, where this angle is labelled ϕ. It is clear that the choice of the compensator pole and zero that yield a particular angle is not unique. An additional criterion may be imposed to fix the two parameters a and b. For example, a may be chosen to cancel a real pole of the plant near (not on) the $j\omega$ axis, and then b determined to satisfy the condition (Equation 27.33). Or a and b may be chosen together to minimize the required value of K in Step 4, while still achieving the desired angle. A graphical method for this approach is given in D'Azzo and Houpis (1988) (370–371).

Note that the steady-state performance is not considered in the design of the lead compensator as presented. The steady-state performance of the system depends not on the closed-loop pole locations, but on the open-loop dc gain of the system, which cannot easily be visualized using the root locus technique. If necessary, the steady-state performance of the system may be improved by use of a lag compensator; however, the lag compensator design is best done by frequency-domain techniques, where the open-loop gains can be clearly seen.

27.4 Examples

The use of the root locus method as a design technique is now illustrated by use of two examples. The details of the construction of the loci are omitted for the most part, since they are routine and can be filled in by the reader, either by hand or by computer.

Compensation of an Inertial System

Consider the feedback system shown in Figure 27.10. By comparison with Figure 27.1, $G(s) = G_p(s)G_c(s)$ and $H(s) = 1$. The plant is represented by

$$G_p(s) = \frac{1}{s^2}, \tag{27.35}$$

which describes an inertial system, such as a mass in linear motion with a force input along the direction of motion. The compensator is to have the form

$$G_c(s) = \frac{K(s+1)}{(s+p)}, \tag{27.36}$$

where $K > 0$ and $p > 0$ are design parameters. The design procedure consists of choosing p so that the branches of the root locus assume a desired shape, and then choosing K to place the closed-loop poles at the desired positions along the branches. The root locus is given in Figure 27.11 for several values of p. Some of the corresponding closed-loop pole locations and step responses are shown in Figures 27.12 and 27.13.

For $p = 1$, the pole and zero of the $G_c(s)$ cancel, and the compensator becomes just a gain K. This choice corresponds to the usual first step for a compensator design. The open-loop transfer function is

$$G(s) = G_p(s)G_c(s) = \frac{K}{s^2}, \tag{27.37}$$

which has no zeros, and two poles at the origin. The relative degree of $G(s)$ is two, so the asymptotes of the loci are at angles of $\pm 90°$ in the complex plane. As the gain K is increased, the closed-loop poles simply move along the $j\omega$ axis away from their starting points at the origin, as shown in Figure 27.11(a). Hence, the system is not effectively stabilized for any value of K. The system response will be characterized by undamped oscillations. Increasing gain results in increasing frequency of oscillation.

For $p = 3$, $G_c(s)$ takes the form of a lead compensator, and the open-loop transfer function is

$$G(s) = \frac{K(s+1)}{s^2(s+3)}. \tag{27.38}$$

The relative degree of $G(s)$ is still two, so the root locus asymptotes are still at angles of $\pm 90°$; however, the zero of the compensator attracts the locus somewhat, so that the vertical asymptotes intersect the real axis at $\sigma = -1$. The root locus is shown in Figure 27.11(b). All the closed-loop poles are in the left half-plane for all values of $K > 0$. (Frequency-domain analysis would show that the compensator has added stabilizing phase lead to the open-loop transfer function.) The pole locations and the step responses of the closed-loop system with $p = 3$ are shown in Figure 27.12 for $K = 10$.

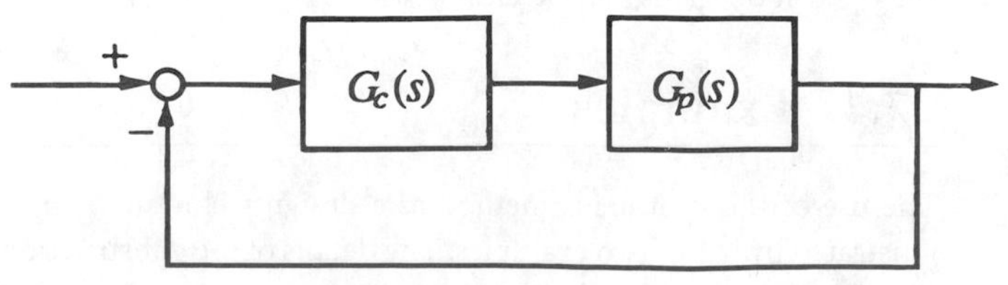

Figure 27.10 Form of closed-loop system for examples.

To obtain better speed and damping, the pole of the compensator can be moved farther to the left. With the compensator pole more distant, the compensator zero attracts the locus more strongly to the left. (Frequency-domain analysis would show that the stabilizing phase lead of the compensator is increased.) Figures 27.11(c–e) show the root locus plots for $p = 6$, 9, and 12, respectively. For $p > 9$, the complex-conjugate branches of the locus are attracted to the left and downward sufficiently that they meet the real axis, resulting in break-in and breakaway points on the root locus.

The pole locations and the step responses of the closed-loop system with $p = 12$ are shown in Figure 27.13 for $K = 40$, 45, and 60. The pole locations near the break-in and breakaway points are extremely sensitive to the gain chosen, but the step responses are not. Note that the overshoot in the step responses results from the presence of the zero at $s = 1$, which dominates the system response since it lies to the right of all the closed-loop poles.

Compensation of an Undamped Oscillatory System

Consider again the feedback system shown in Figure 27.10. Let the plant represent a forced spring-mass system, given by

$$G_p(s) = \frac{K}{(s^2+1)}, \qquad K = 1. \tag{27.39}$$

Since the natural response of the system is oscillatory without damping, it is desired to design a compensator to satisfy certain performance specifications. The form of the compensator is

$$G_c(s) = K_p + K_d s + \frac{K_i}{s}, \tag{27.40}$$

where K_p, K_d, and K_i are design parameters. In the typical design procedure K_p and K_d are chosen to meet transient response specifications and K_i is chosen to satisfy steady-state requirements. In this particular situation K_i may also be used to position the third (real) pole to control the percent overshoot somewhat. The open-loop transfer function $G(s)$ of Figure 27.10 is

$$G(s) = G_p(s)G_c(s) = \frac{K_d s^2 + K_p s + K_i}{s(s^2+1)}. \tag{27.41}$$

Note that when only the proportional mode of the compensator is used—i.e., when K_d and K_i are set to zero—the open-loop transfer function is

$$G(s) = G_p(s)G_c(s) = \frac{K_p}{(s^2+1)}. \tag{27.42}$$

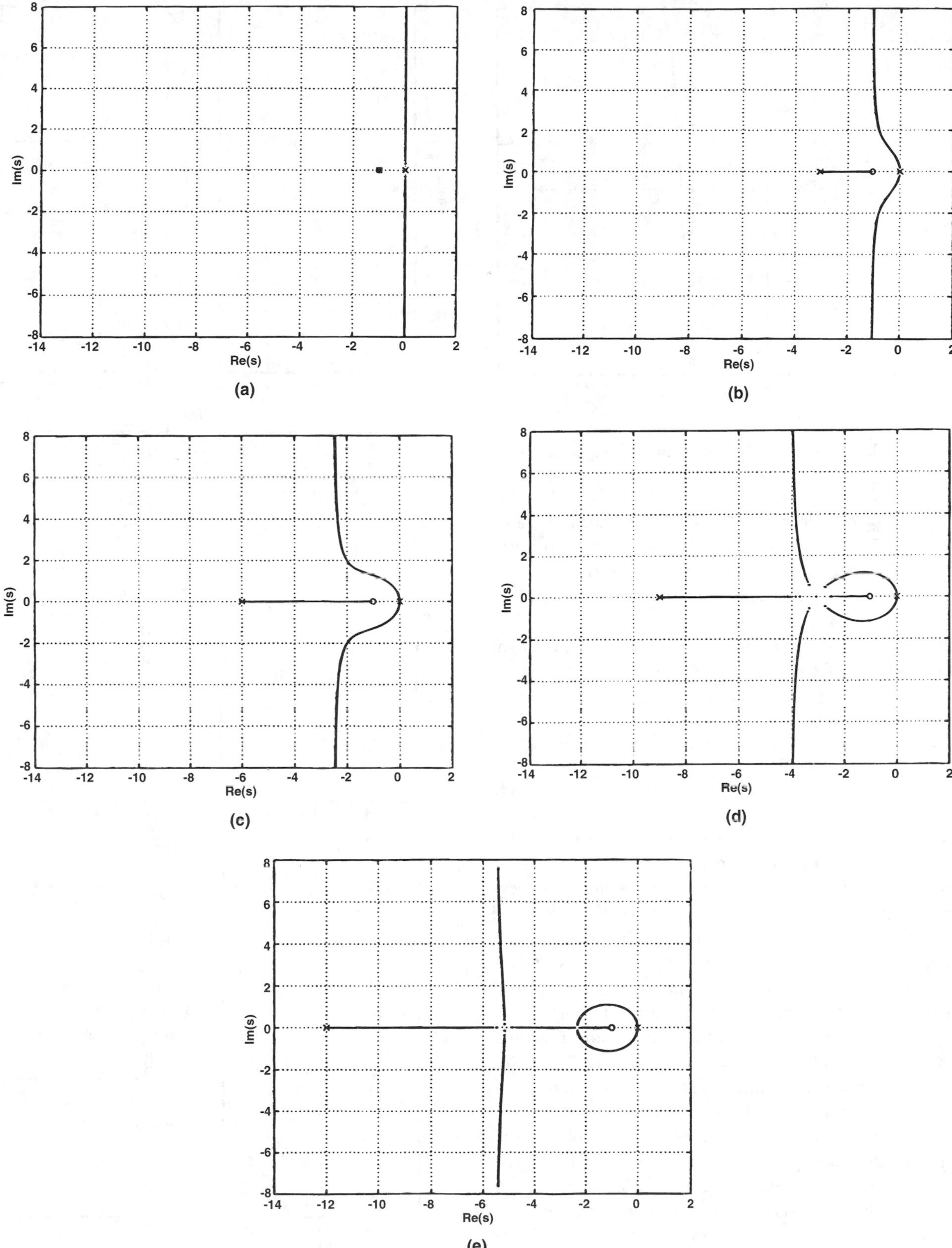

Figure 27.11 Root locus plots for inertial plant with lead controller: (a) $p = 1$; (b) $p = 2$; (c) $p = 6$; (d) $p = 9$; (e) $p = 12$.

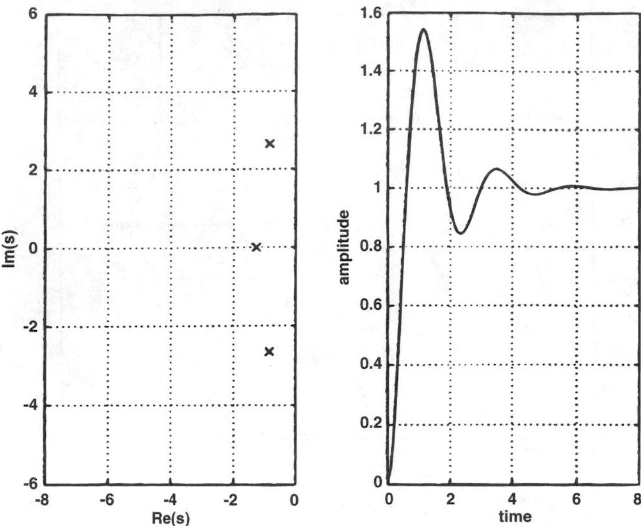

Figure 27.12 Closed-loop pole locations and step response, $p = 3$, $K = 10$.

In this case the closed-loop system behaves in much the same way as the inertial system just considered, with $p = 1$. Moreover, as K_p increases the frequency of oscillations of the system response also increases. Hence, the system response always results in an undamped oscillation for all values of K_p.

If all three compensator parameters are included in the design, the closed-loop transfer function $T(s)$ is given by

$$T(s) = \frac{G(s)}{1 + G(s)} = \frac{K_d s^2 + K_p s + K_i}{s^3 + K_d s^2 + (1 + K_p)s + K_i}. \tag{27.43}$$

The system is third order; by choice of the three compensator parameters, the three closed-loop poles may be placed arbitrarily. Two degrees of freedom may be taken up by placing a complex-conjugate pair of poles at $s_{1,2} = -3 \pm j3$, to obtain a damping ratio of 0.707 and a damped natural frequency of 3 rad/sec. Assuming this pole-pair dominates the system response, this is equivalent to requiring about 5% overshoot and an undamped natural frequency of 4.2420 rad/sec. If the third (real) system pole is positioned at $s = -\alpha$, then the compensator gains may be written as $K_p = 17 + 6\alpha$, $K_d = 6 + \alpha$, and $K_i = 18\alpha$. Table 27.1 gives these gains, along with the closed-loop transfer-function coefficients, for several values of α.

For small nonzero values of α, the closed-loop pole close to the origin results in a sluggish system response with a long rise time. Still, the overshoot is almost 20%. As α increases, the rise time decreases, but the overshoot increases to a maximum of about 25% for $\alpha \approx 3$. For larger values of α, the complex-conjugate pair of closed-loop poles dominates the response, and the overshoot decreases to near 4%. This is clearly illustrated in Figure 27.14 which shows the step response of the closed-loop system for various values of α. Figure 27.14 also shows the root locus plot for the system as the plant gain K in Equation 27.39 varies from 0 to ∞. The s-plane locations $-3 \pm j3$ and $-\alpha$ are

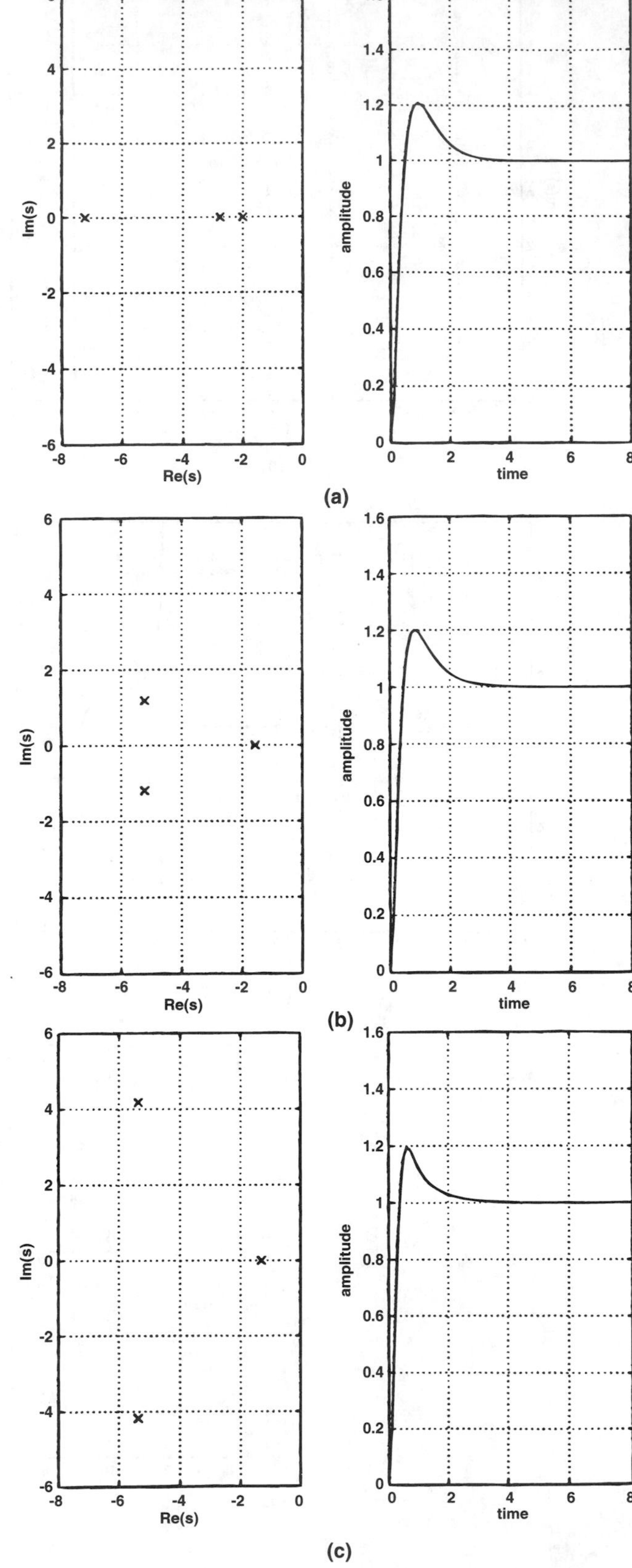

Figure 27.13 Closed-loop pole locations and step responses for $p = 12$: (a) $K = 40$; (b) $K = 45$; (c) $K = 60$.

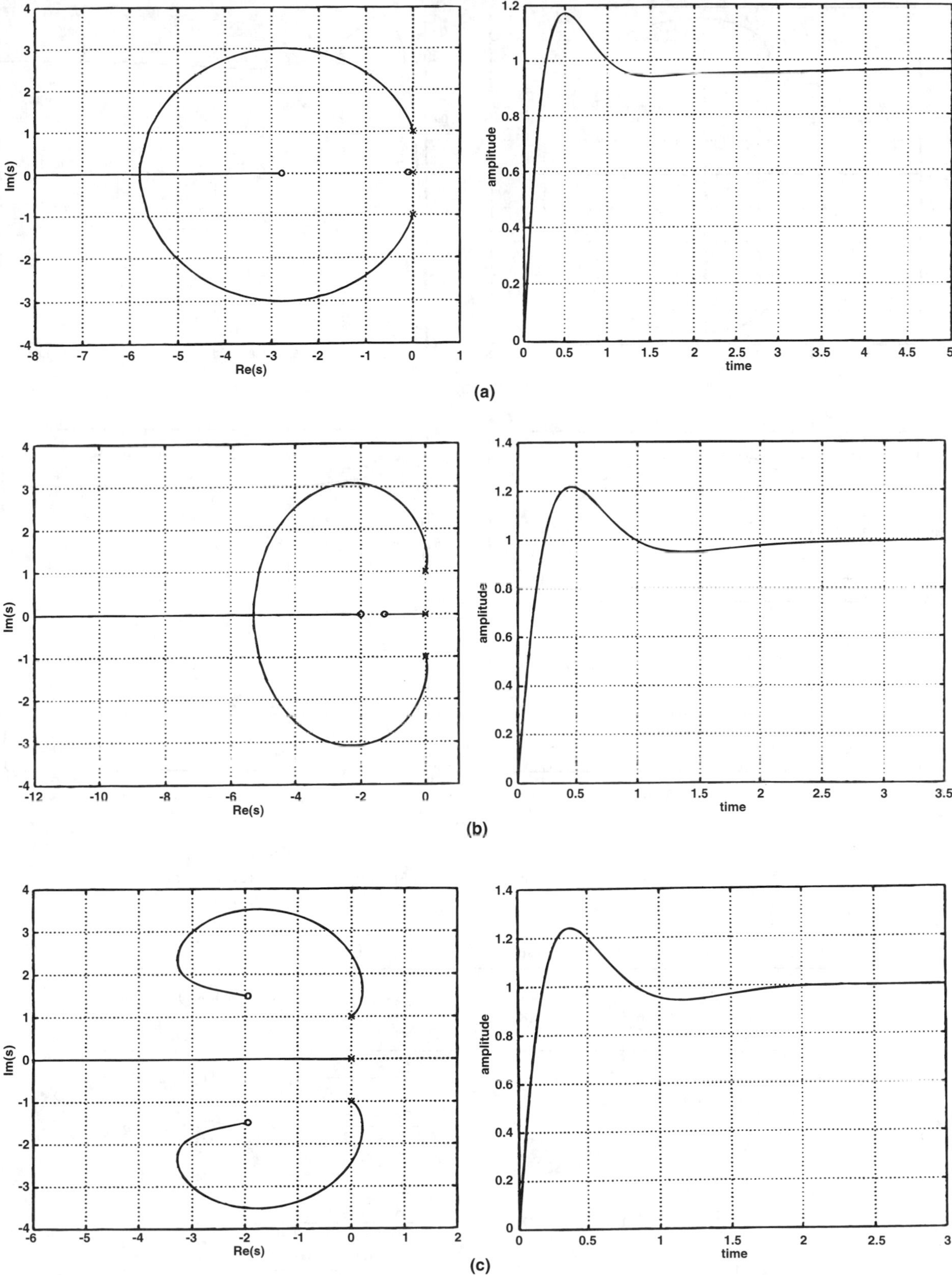

Figure 27.14 Root locus and step response: (a) $\alpha = 0.1$; (b) $\alpha = 1$; (c) $\alpha = 3$.

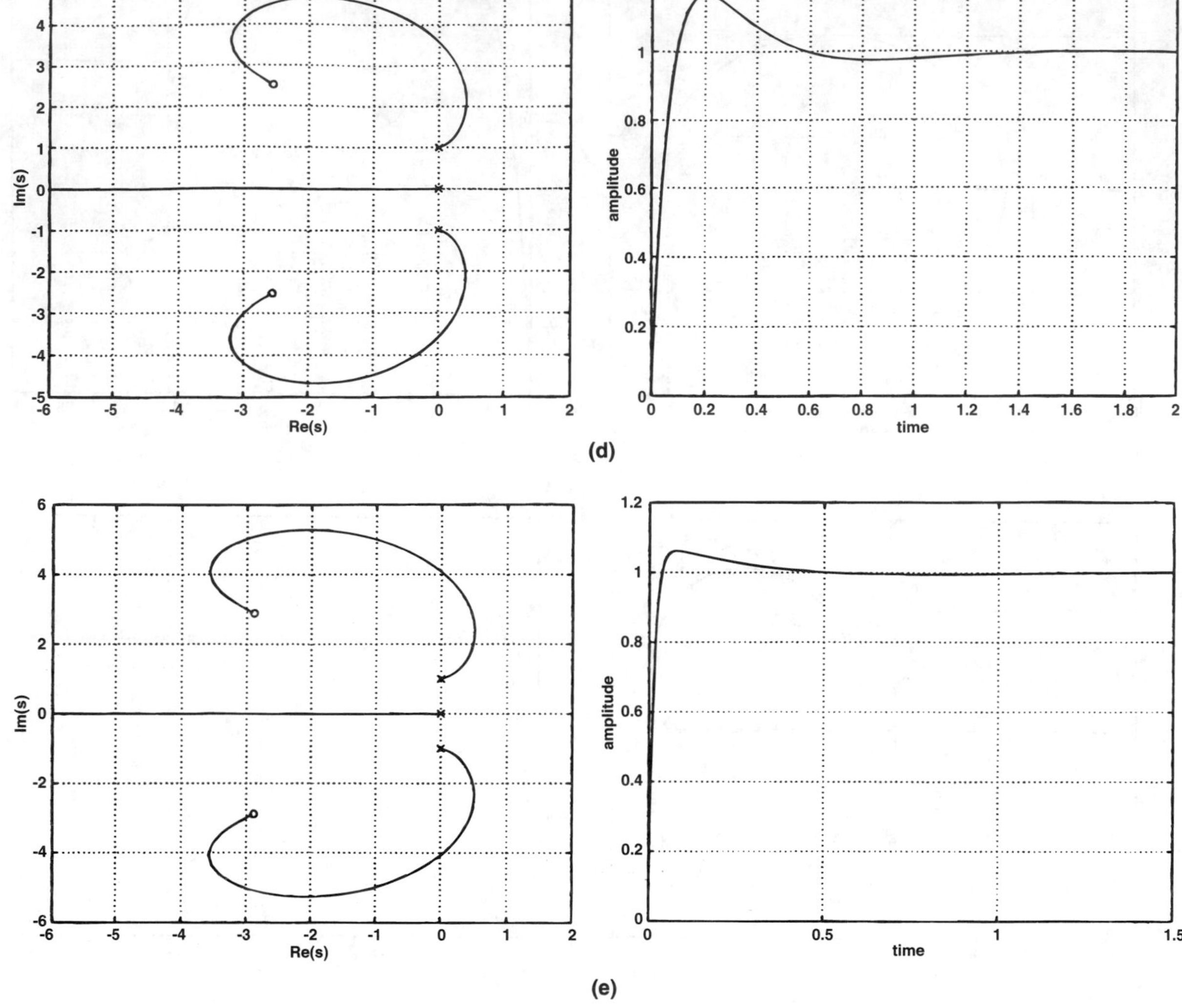

Figure 27.14 (Continued) Root locus and step response: (d) $\alpha = 15$; (e) $\alpha = 75$.

Table 27.1 Summary of PID Design Results

α	K_p $17 + 6\alpha$	K_d $6 + \alpha$	K_i 18α	$T(s)$ $\dfrac{K_d s^2 + K_p s + K_i}{s^3 + K_d s^2 + (1 + K_p)s + K_i}$	Root locus and step response
0.1	17.6	6.1	1.8	$\dfrac{6.1s^2 + 17.6s + 1.8}{s^3 + 6.1s^2 + 18.6s + 1.8}$	Figure 27.14a
1	23	7	18	$\dfrac{7s^2 + 23s + 18}{s^3 + 7s^2 + 24s + 18}$	Figure 27.14b
3	35	9	54	$\dfrac{9s^2 + 35s + 54}{s^3 + 9s^2 + 36s + 54}$	Figure 27.14c
15	107	21	270	$\dfrac{21s^2 + 107s + 270}{s^3 + 21s^2 + 108s + 270}$	Figure 27.14d
75	467	81	1350	$\dfrac{81s^2 + 467s + 1350}{s^3 + 81s^2 + 468s + 1350}$	Figure 27.14e

included in each root locus plot, corresponding to the nominal value $K = 1$. Clearly, the gains may be chosen to position the closed-loop system real pole such that the system response overshoot is suppressed. However, as α increases the steady-state error coefficient to a ramp input also increases. Therefore, there is a trade-off between steady-state performance and transient response behavior.

References

D'Azzo, J. J. and Houpis, C. H. 1988. *Linear Control System Analysis and Design: Conventional and Modern,* 3d ed., McGraw-Hill, New York, NY.

Evans, W. R. 1948. Graphical Analysis of Control Systems, *Trans. Am. Inst. Electrical Engineers,* 67:547–551.

Franklin, G. F., Powell, J. D., and Emami-Naeini, A. 1994. *Feedback Control of Dynamic Systems,* 3d ed., Addison-Wesley, Reading, MA.

Kuo, B. C. 1995. *Automatic Control Systems,* 7th ed., Prentice Hall, Englewood Cliffs, NJ.

Miron, D. B. 1989. *Design of Feedback Control Systems,* Harcourt Brace Jovanovich, Orlando, FL.

Nise, N. S. 1995. *Control System Engineering,* 2nd ed., Benjamin Cummings, Redwood City, CA.

28

Pole Placement Design

Michael Greene
Auburn University

Victor Trent
Auburn University

28.1 Pole Placement 504
28.2 State Observation 506
28.3 Discrete Implementation 509

The basic idea of pole placement design is really quite simple. If a system is composed of N state variables and every state variable is available as a measurable output, the closed-loop roots in a system to be controlled by feedback can be arbitrarily placed using N independent feedback gain terms. A state space description of the open loop system may be employed which can be found from the transfer function, $G(s)$. Remember,

$$G(s) = \frac{P(s)}{D(s)} = \frac{s^w + b_{w-1}s^{w-1} + \cdots + b_1 s + b_0}{s^n + a_{n-1}s^{n-1} + \cdots + a_1 s + a_0} = \frac{Y(s)}{U(s)} \tag{28.1}$$

where

$$W < N$$

By substituting

$$Y(s) = P(s)X_1(s) \tag{28.2}$$

we obtain

$$D(s)X_1(s) = U(s) \tag{28.3}$$

Likewise, we define

$$\begin{aligned} sX_1 &= X_2(s) \\ sX_2 &= X_3(s) \\ &\vdots \\ sX_{n-1} &= X_n(s) \end{aligned} \tag{28.4}$$

which implies the time domain description:

$$\begin{aligned} \dot{x}_1 &= x_2 \\ \dot{x}_2 &= x_3 \\ &\vdots \\ \dot{x}_{n-1} &= x_n \end{aligned} \tag{28.5}$$

substituting and solving for x_N we obtain:

$$\dot{x}_n = -a_0 x_1 - a_1 x_2 - \cdots - a_{n-1} x_n + u(t) \tag{28.6}$$

Likewise, we can obtain the output equation from Equation 28.2 as

$$y(t) = b_0 x_1 + b_1 x_2 + \cdots + b_{w-1} x_{w-1} + b_w x_w \tag{28.7}$$

These equations can be expressed in matrix notation by defining

$$\boldsymbol{x} = \begin{bmatrix} x_1 \\ x_2 \\ \vdots \\ x_n \end{bmatrix} \tag{28.8}$$

The state space equations are therefore:

$$\begin{aligned} \dot{\boldsymbol{x}} &= A\boldsymbol{x} + Bu \\ y &= C\boldsymbol{x} \end{aligned} \tag{28.9}$$

where

$$\mathbf{A} = \begin{bmatrix} 0 & 1 & 0 & \cdots & 0 \\ 0 & 0 & 1 & \cdots & 0 \\ \vdots & \vdots & \vdots & & \vdots \\ -a_0 & -a_1 & -a_2 & \cdots & -a_{n-1} \end{bmatrix}$$

$$\boldsymbol{B} = \begin{bmatrix} 0 \\ 0 \\ \vdots \\ 1 \end{bmatrix} \qquad C = [b_0 \quad b_1 \quad \cdots \quad b_w] \tag{28.10}$$

28.1 Pole Placement

Suppose we desire to place the closed loop roots such that:

$$\begin{aligned} \alpha_c(s) &= (s - s_1)(s - s_2) \cdots (s - s_n) \\ &= s^n + \alpha^{n-1}s^{n-1} + \cdots + \alpha_0 = 0 \end{aligned} \tag{28.11}$$

0-8493-8343-9/97/$0.00+$.50

and we further define a control law using all the state variables (Phillips and Harbor, 1991):

$$u = -\boldsymbol{K}\boldsymbol{x}(t) \tag{28.12}$$

Combining Equation 28.12 with Equation 28.9 we obtain the closed loop description of the regulator system:

$$\dot{\boldsymbol{x}} = [\boldsymbol{A} - \boldsymbol{B}\boldsymbol{K}]\boldsymbol{x} \tag{28.13}$$

When the description of the plant is in controllable canonical form Equation 28.13 becomes:

$$\dot{\boldsymbol{x}} = \begin{bmatrix} 0 & 1 & 0 & \cdots & \cdot & \cdot \\ 0 & 0 & 1 & \cdots & \cdot & \cdot \\ \vdots & \vdots & & & & \\ -a_0 - k_1 & -a_1 - k_2 & \cdot & \cdots & -a_{n-1} & -k_N \end{bmatrix} \boldsymbol{x} \tag{28.14}$$

whose characteristic equation is

$$s^N + (a_{N-1} + k_N)s^{N-1} + \cdots + (a_1 + k_2)s + a_0 + k_1 = 0 \tag{28.15}$$

Equating coefficients of Equations 28.11 and 28.15, we obtain the desired control gains to place the closed loop roots as desired

$$\begin{aligned} k_1 &= -a_0 - \alpha_0 \\ k_2 &= -a_1 - \alpha_1 \\ k_3 &= -a_2 - \alpha_2 \\ &\vdots \\ k_N &= -a_{N-1} - \alpha_{N-1} \end{aligned} \tag{28.16}$$

or in vector form:

$$\boldsymbol{K}_{i+1} = \boldsymbol{\alpha}_i - \boldsymbol{a}_i \tag{28.17}$$

EXAMPLE 28.1: A second-order system

Consider the second-order system in controllable canonical form shown below:

$$\boldsymbol{A} = \begin{bmatrix} 0 & 1.0 \\ -1.0 & -0.8 \end{bmatrix} \qquad \boldsymbol{B} = \begin{bmatrix} 0 \\ 1 \end{bmatrix}$$

$$\boldsymbol{C} = [1.0 \quad 0] \qquad \boldsymbol{D} = [0] \tag{28.1e}$$

Pole-placement controller design will be used to place the closed-loop system poles so that the natural frequency of the system is doubled ($\omega_{nd} = 2$ rad/sec) and the integral of the time multiplied by absolute error (ITAE) is minimized. The desired characteristic equation for the compensated closed-loop system is given by (Dorf, 1983):

$$\alpha_c(s) = s^2 + 2.828s + 4 = 0 \tag{28.2e}$$

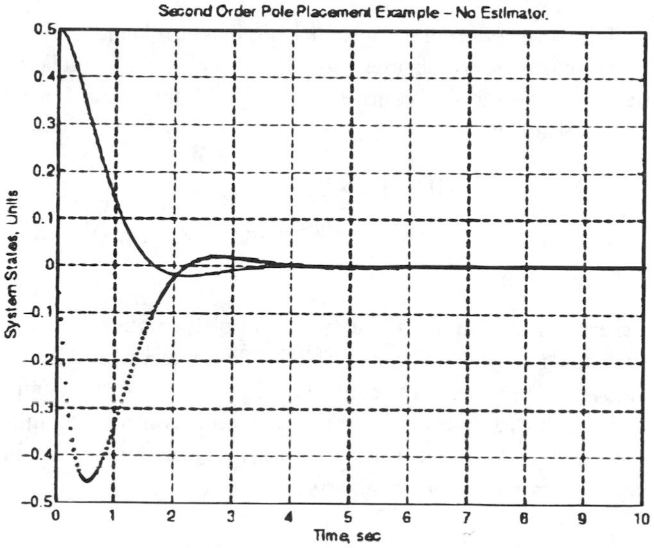

Figure 28.1 System states vs. time for a second-order system with full state feedback and no state estimation: – position state variable, ... velocity state variable.

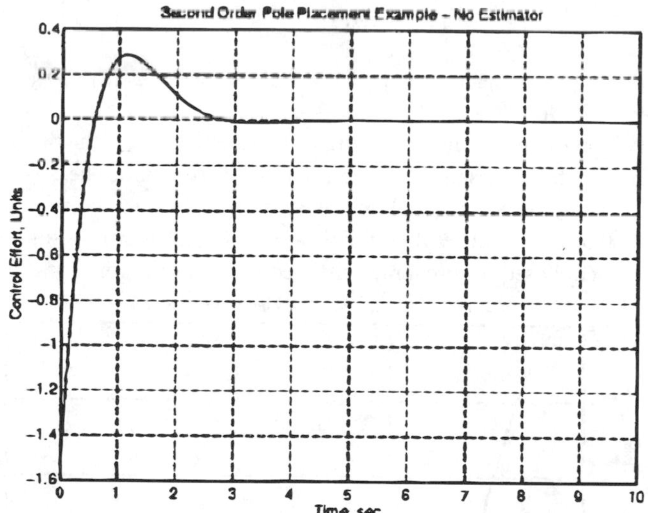

Figure 28.2 Control effort vs. time for a second-order system with full state feedback and no estimation.

Additionally, since $N = 2$:

$$\boldsymbol{K} = [k_1 \quad k_2] \tag{28.3e}$$

The objective is to design a controller of the form of Equation 28.12 which yields a closed-loop characteristic equation equivalent to Equation 28.15 or:

$$s^2 + (.8 + k_2)s + (1 + k_1) = 0 \tag{28.4e}$$

Equating the coefficients of Equation 28.3e to the coefficients of the desired closed-loop system characteristic equation of Equation 28.2e yields:

$$K = [3.0 \quad 2.028] \qquad (28.5e)$$

If the description of the system is not in controllable canonical form or the order becomes too large, Ackerman's formula can be used to find the control gains, K (Phillips and Harbor, 1991). Namely:

$$K = [0, 0, \ldots, 0, 1][\boldsymbol{B}, \boldsymbol{AB}, \boldsymbol{A}^2\boldsymbol{B}, \ldots, \boldsymbol{A}^{N-1}\boldsymbol{B}]^{-1}\alpha_c(\boldsymbol{A}) \qquad (28.18)$$

Ackermann's formula is the typical implementation found in such tools as MATLAB or MATRIX_x. The matrix $[B, AB, A^2B, \ldots, A^{N-1}B]$ is called the controllability matrix and clearly must be invertible for pole placement (or any other control technique) to be viable. A system whose controllability matrix is invertible is called an input controllable system.

EXAMPLE 28.2: A fourth-order system

Consider the fourth-order system where:

$$A = \begin{bmatrix} 0 & 1.0 & 0 & 0 \\ 0 & 0 & 1.0 & 0 \\ 0 & 0 & 0 & 1.0 \\ 0 & -4.01 & -4.21 & -1.2 \end{bmatrix} \qquad B = \begin{bmatrix} 0 \\ 0 \\ 0 \\ 1.0 \end{bmatrix}$$

$$C = [1.0 \quad 0 \quad 0 \quad 0] \qquad D = [0] \qquad (28.6e)$$

Pole-placement controller design will be used to place the closed-loop system poles using the dominant-poles method. The closed-loop system poles are placed at $s1 = -0.5 + j0.5$, $s2 = -0.5 - j0.5$, $s3 = -1.0 + j1.7$ and $s4 = -1.0 - j1.7$ and the resulting desired closed-loop characteristic equation in the form of is

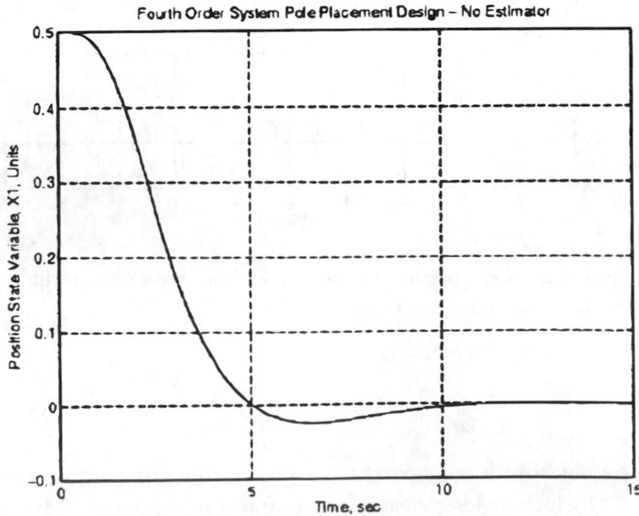

Figure 28.3 Position state variable vs. time for a fourth-order system with full state feedback and no state estimation.

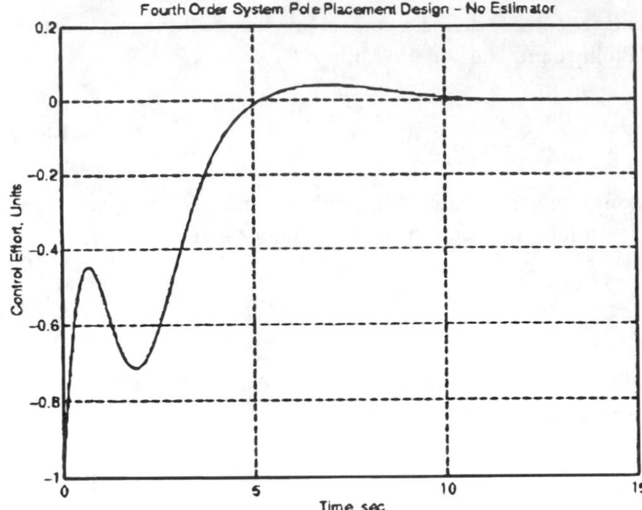

Figure 28.4 Control effort vs. time for a fourth-order system with full state feedback and no state estimation.

$$\alpha_c(s) = s^4 + 3.0s^3 + 6.39s^2 + 4.89s + 1.9450 \qquad (28.7e)$$

The form of the control law is given by Equation 28.12 and Ackermann's formula (Equation 28.18) may be used to solve for the control gains:

$$K = [1.945 \quad 0.880 \quad 2.180 \quad 1.800] \qquad (28.8e)$$

28.2 State Observation

While pole placement is extremely powerful, the major drawback is that every state variable is required for implementation. Since each state variable may not even exist in the system, feedback comprised of the state vector must be replaced by output feedback for pole placement to be a useful tool. The concept of state observation does just this (Luenberger, 1964).

In state observation we build a model of the plant with a state space description:

$$\dot{\boldsymbol{q}} = \boldsymbol{Aq} + \boldsymbol{B}u + \mathbf{G}(y - \hat{y}) \qquad (28.19)$$

where q is our estimate of the state vector, $x(t)$. Since $\hat{y} = Cq$,

$$\dot{\boldsymbol{q}} = (\boldsymbol{A} - \boldsymbol{GC})\boldsymbol{q} + \boldsymbol{B}u + \boldsymbol{G}y \qquad (28.20)$$

we see that the dynamics of the estimator are set by matrix $A - GC$. We further define the error between state vector and state estimate as:

$$\boldsymbol{e} = \boldsymbol{x} - \boldsymbol{q} \qquad (28.21)$$

obtaining,

$$\dot{\boldsymbol{e}} = \dot{\boldsymbol{x}} - \dot{\boldsymbol{q}} = \boldsymbol{Ae} - \boldsymbol{G}(y - \hat{y}) = (\boldsymbol{A} - \boldsymbol{GC})\boldsymbol{e} \qquad (28.22)$$

which demonstrates that the error dynamics between state vector and state estimate are set by the roots or eigenvalues of $A - GC$.

Now, in lieu of using the state vector for feedback, if we use the estimate, $q(t)$, we obtain:

$$\dot{x} = (A - BK)x + BKe \tag{28.23}$$

demonstrating that the closed loop system dynamics are set by $A - BK$ as originally designed by the pole-placement controller.

Finally, the state observer implemented is

$$\dot{q} = (A - GC - BK)q + Gy \tag{28.24}$$

which is driven only by the system output, $y(t)$.

If the system dynamics are expressed in observable canonical form

$$A = \begin{bmatrix} -a_{n-1} & 1 & 0 & \cdots & 0 \\ -a_{n-2} & 0 & 1 & \cdots & 0 \\ \vdots & \vdots & \vdots & & \vdots \\ -a_0 & & \cdots & & 1 \end{bmatrix} \text{ and}$$

$$C = [1, 0, \ldots, 0] \tag{28.25}$$

the dynamics are set by $A - GC$ and are usually set such that they are 2 to 5 times faster than the system dynamics set by $A - BK$ and will be designated $\alpha_e(s)$.

If

$$A - GC = \begin{bmatrix} -a_{n-1} - g_n & 1 & 0 & \cdots & 0 \\ -a_{n-2} - g_{n-1} & 0 & 1 & \cdots & 0 \\ \vdots & & & & \\ -a_0 - g_1 & 0 & 0 & & 1 \end{bmatrix}$$

where (28.26)

$$G = \begin{bmatrix} g_n \\ g_{n-1} \\ \vdots \\ g_1 \end{bmatrix}$$

the characteristic equation of $A - GC$ is

$$s^n + (a_{n-1} + g_n)s^{n-1} + \cdots + (a_o + g_1) = 0 \tag{28.27}$$

If the desired characteristic equation, $\alpha_e(s)$, is defined as:

$$\alpha_e(s) = s^n + \alpha_{n-1}s^{n-1} + \cdots + \alpha_o \tag{28.28}$$

then equating coefficients we obtain:

$$a_{i-1} + g_i = \alpha_{i-1} \tag{28.29}$$

EXAMPLE 28.3: A second-order example

Consider the second-order system presented in Example 28.1. The second-order system may be expressed in observable canonical form Equation 28.25 as:

$$A_o = \begin{bmatrix} -0.8 & 1.0 \\ -1.0 & 0 \end{bmatrix} \qquad B_o = \begin{bmatrix} 0 \\ 1 \end{bmatrix}$$

$$C_o = [1 \quad 0] \qquad D_o = [0]. \tag{28.9e}$$

The observable canonical form of the system state equations shown in Equation 28.9e was obtained by using the similarity transform:

$$T = \begin{bmatrix} 1 & 0 \\ -0.8 & 1 \end{bmatrix} \tag{28.10e}$$

to convert the state space description of the system from controllable canonical form to observable canonical form.

The pole-placement controller designed in Example 28.1 is to be implemented on the second order system, however, it is assumed that the state variables are not available for measurement and must be estimated for system control.

The error dynamics for this estimator (Equation 28.20) are set *two times faster* than the system dynamics (note: $w_{nd} = 2.0$ rad/sec); hence, the desired estimator characteristic equation is

$$\alpha_e(s) = s^2 + 8s + 16 \tag{28.11e}$$

and using Equation 28.27 the eigenvalues of the estimator dynamics may be calculated and the resulting characteristic equation is given by Equation 28.28:

$$s^2 + (0.8 + G_1)s + (1.0 + G_2) = 0 \tag{28.12e}$$

Equating coefficients of Equations 28.11e and 28.12e, the estimator gains are

$$G = \begin{bmatrix} 7.2 \\ 15.0 \end{bmatrix} \tag{28.13e}$$

Finally, the control gains given by Equation 28.5e must be transformed into the observable canonical space. This is accomplished using the similarity transform of Equation 28.10e:

$$Ko = KT \tag{28.14e}$$

where Ko is the controller gain matrix (to be used with the observable canonical state description), K is the controller gain matrix of Equation 28.5e and T is the similarity transform of Equation 28.10e.

Comparing Figures 28.5 and 28.6 with Figures 28.1 and 28.2 one notes an increase in the oscillation amplitude for the position and velocity state variables, as well as for the control effort for the system employing the state estimates in the control. This is due to the convergence properties of the error dynamics of the system (see Equation 28.20).

If the description of the system is not in observable canonical form or the order becomes too large, Ackermann's formula can be used to find the observer gains, G (Phillips and Harbor, 1991). Namely:

$$G = \alpha_e(A)[C, CA, CA^2, \ldots, CA^{N-1}]^{-1}[0, 0, \ldots, 0, 1]^T \quad (28.30)$$

Ackermann's formula is the typical implementation found in such tools as MATLAB or MATRIX$_X$. The matrix $[C, CA, CA^2, \ldots, CA^{N-1}]$ is called the observability matrix and clearly must be invertible for state observation (or any other control technique) to be viable. A system whose observability matrix is invertible is called an input observable system.

EXAMPLE 28.4: A fourth-order system

Consider the fourth order system presented in Example 28.2. The pole-placement controller designed in Example 28.2 is to be implemented on the fourth-order system; however, it is assumed that the state variables are not available for measurement and must be estimated for system control.

The error dynamics for this estimator Equation 28.20 are set *five times faster* than the system dynamics (note: $w_{nd} = 1.9723$ rad/sec) hence the desired estimator characteristic equation is:

$$\alpha_e(s) = (s + 9.86)^4 = 0 \quad (28.15e)$$

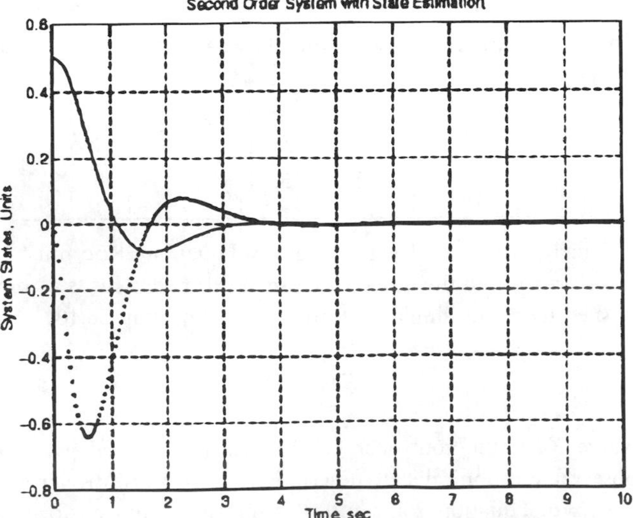

Figure 28.5 System states vs. time for a second-order system with full state feedback and full state estimation.: — position state variable; . . . velocity state variable.

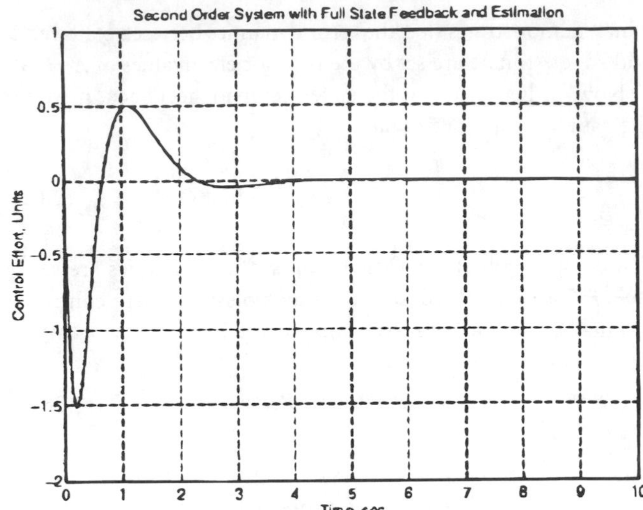

Figure 28.6 Control effort vs. time for a second-order system with full state feedback and full state estimation.

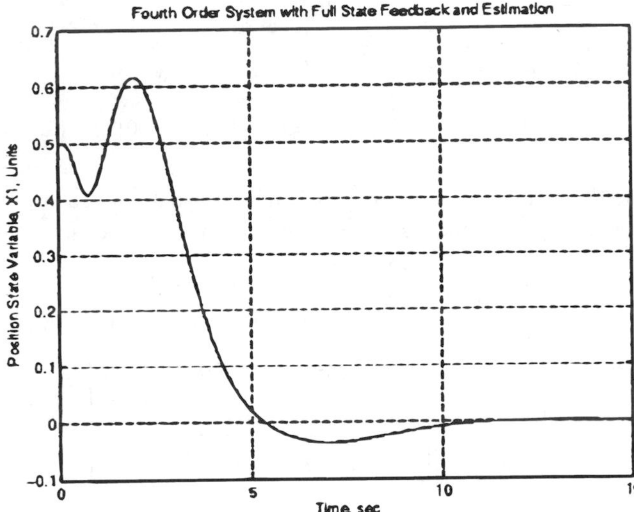

Figure 28.7 Position state variable vs. time for a fourth-order system with full state feedback and full state estimation.

The estimator gains may be calculated using Ackermann's formula (Equation 28.30):

$$G = \begin{bmatrix} 38.2 \\ 533.4 \\ 3031.0 \\ 3421.4 \end{bmatrix} \quad (28.16e)$$

Once again, the convergence properties of the error dynamics (Equation 28.22) and the time for convergence of the state estimates to the state vector values affect the transient response of the system. Comparing Figures 28.7 and 28.8 with Figures 28.3 and 28.4 one notes a marked difference in the transient behavior of both the position state variable and the control effort. Once convergence of the state estimates to the state vector values has occurred, system regulation is achieved.

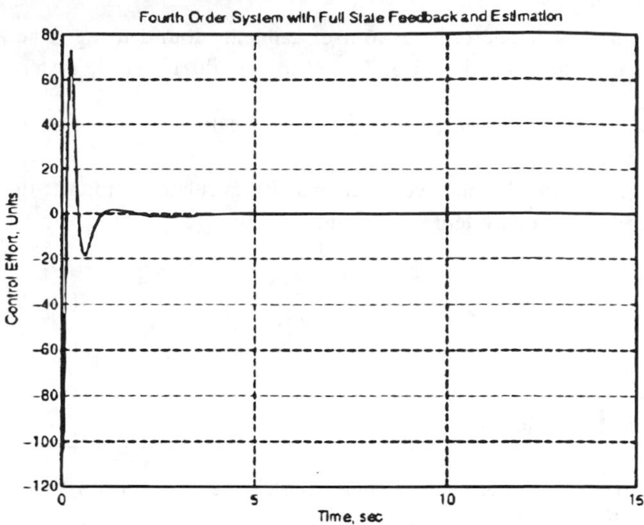

Figure 28.8 Control effort vs. time for a fourth-order system with full state feedback and full state estimation.

28.3 Discrete Implementation

It should be noted that to implement a state observer, the dynamics of the plant need to be replicated in hardware. This hardware would essentially comprise an analog computer the same order as the plant to be controlled. Since such systems are difficult to build and tune, a discrete implementation is preferred. If the sampling frequency is extremely fast compared to the closed-loop system roots, then the analog controller/observer can be digitized by using the step-invariant method (see Chapter 32). A general (but not always successful) rule of thumb is that the sampling frequency be at least 50 times the highest natural frequency found in the closed-loop system. As this rule of thumb is neither a guarantee or always possible, we will consider designing the controller and observer in discrete time.

The solution to the state equations can be found as (Phillips and Nagle, 1995):

$$\mathbf{x}(t) = \boldsymbol{\phi}(t - t_0)\mathbf{x}(t_0) + \int_{t_0}^{t} \boldsymbol{\phi}(t - \tau)\mathbf{B}_0 u(\tau)d\tau \quad (28.31)$$

where

$$\boldsymbol{\phi}(t) \triangleq \text{state transition matrix} = \zeta^{-1}[(sI - A)^{-1}] \quad (28.32)$$

Note that by letting $t = kT + T$ and $t_0 = kT$ this solution is over one sampling period, T. A zero order hold (ZOH) placed between the controller and plant holds the input to the plant constant from kT to $kT + T$ and allows the $u(\quad)$ term to be taken outside the integral of Equation 28.31 implying a vector difference equation

$$\mathbf{x}(k + 1) = \mathbf{A}_D\mathbf{x}(k) + \mathbf{B}_D\mathbf{u}(k) \quad (28.33)$$

where

$$\mathbf{A}_D = \boldsymbol{\phi}(T), \qquad \mathbf{B}_D = \int_{kT}^{kI+T} \boldsymbol{\phi}(kT + T - \tau)\mathbf{B}d\tau \quad (28.34)$$

Thus, these equations are the discrete-time equivalents of Equation 28.9 and Equation 28.10. Design of the controller and observer proceed exactly according to Equation 28.11 to Equation 28.18 and Equation 28.19 to Equation 28.30 except that the characteristic values of the discrete system matrix A_D are transformed from continuous time by

$$z = e^{sT} \quad (28.35)$$

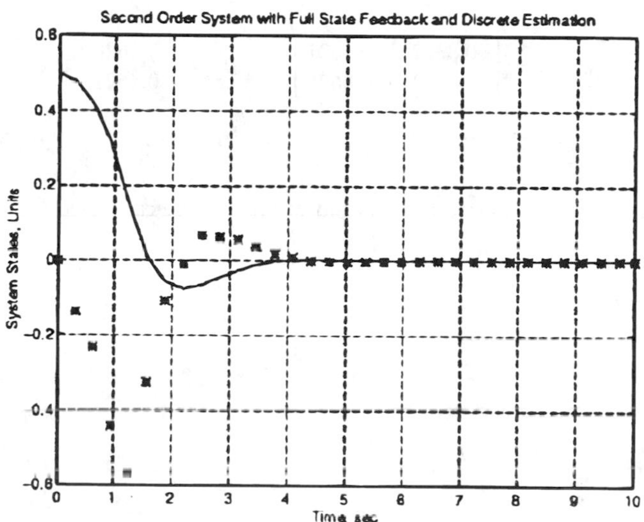

Figure 28.9 System states vs. time for a second-order system with full state feedback and discrete estimation.: — position state variable; *velocity state variable.

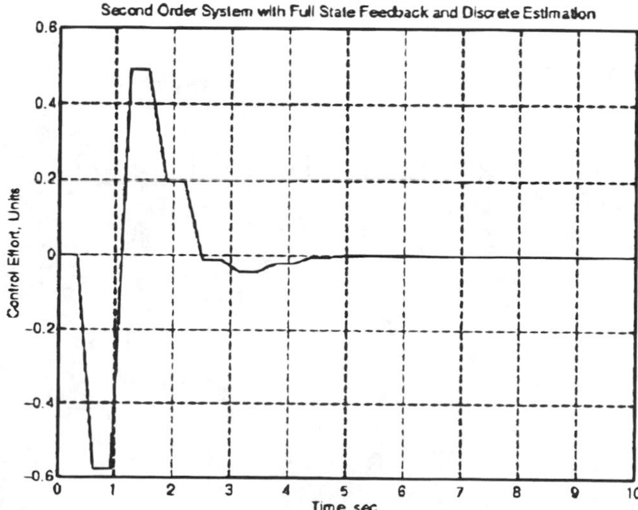

Figure 28.10 Control effort vs. time for a second-order system with full state feedback and discrete estimation.

The control gains, K, are then found using Ackermann's control formula (Equation 28.18) and the observer gains found using Equation 28.30. Again, we set observer roots 2 to 5 times faster than the fastest the closed-loop roots. The observer equation, however, are discrete difference equations which can be implemented in code in a computer with an A/D and D/A converters.

EXAMPLE 33.5: A second-order system with discrete implementation

Consider the second-order system presented in Figure 28.1. The second-order system is converted to a discrete-time representation using the *ZOH equivalence method* and a sample frequency of *ten times* the natural frequency of the system ($T = 0.6283$ s), hence:

$$\mathbf{A}_D = \begin{bmatrix} 0.8372 & 0.4621 \\ -0.4621 & 0.4675 \end{bmatrix} \qquad \mathbf{B}_D = \begin{bmatrix} 0.1628 \\ 0.4621 \end{bmatrix}$$

$$\mathbf{C}_D = [1.0 \quad 0] \qquad \mathbf{D}_D = [0] \tag{28.17e}$$

The desired discrete-time root locations may be calculated using Equation 28.35, and the control gains are found using Ackermann's formula (Equation 28.18) for the discrete-time system:

$$\mathbf{K}_D = [1.1666 \quad 1.2900] \tag{28.18e}$$

Similarly, the observer gains may be calculated using the discrete-time equivalent of Equation 28.30:

$$\mathbf{G}_D = \begin{bmatrix} 1.1427 \\ -0.1389 \end{bmatrix} \tag{28.19e}$$

References

Dorf, R. C. 1983. *Modern Control System*, 3d ed., Addison-Wesley, Reading, Mass.

Luenberger, D. G. 1964. Observing the state of a linear system, *IEEE Trans. Military Electr.*, MIL–8:74–80.

Phillips, C. L. and Harbor, R. D. 1991. *Feedback Control Systems*, 2d ed., Prentice Hall, Englewood Cliffs, NJ.

Phillips, C. L. and Nagle, H. T. 1995. *Digital Control System Analysis and Design*, 3d ed., Prentice Hall, Englewood Cliffs, NJ.

29

The Smith Predictor Technique

John Y. Hung
Auburn University

29.1 Background—Control of Processes Having Time Delay............ 511
29.2 Basic Principle of the Smith Predictor.. 511
29.3 A Smith Predictor Design Example.. 512

29.1 Background—Control of Processes Having Time Delay

Many industrial processes are characterized by some dead time and a dominant time constant. An example is the control of temperature in a heat exchanger process, where the sensing element may be separated from the actuator by some physical distance (Figure 29.1). A change in the steam valve position results in a fluid temperature change; the change is detected by the sensor as the fluid flows by. Time is required to transport the fluid along the system. Hence, the delay is sometimes referred to as transport lag or delay. Another example is a pneumatic-based control system in which the controller and actuator are separated by a long air pipe. A change in the controller output pressure must travel the length of the pipe to reach the actuator. In both of these illustrations some time delay or dead time exists in the control loop. Time delay introduces additional phase lag in a feedback loop, thus reducing the phase margin and relative stability. Feedback control of processes having time delay may be characterized by a bounded oscillation or "hunting" of signals. In extreme cases, the time delay may make stabilization very difficult to achieve.

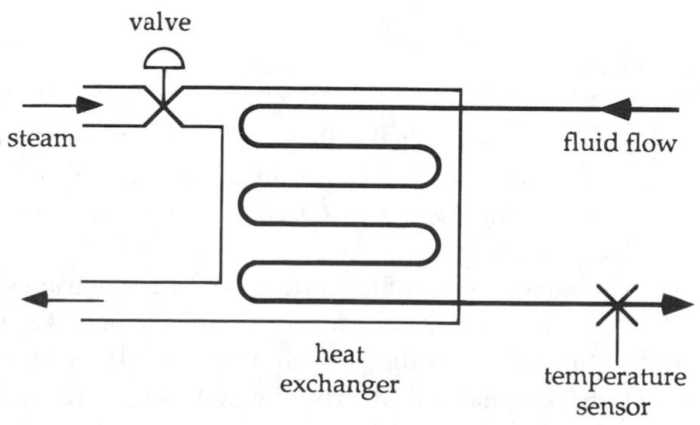

Figure 29.1 A heat exchanger process.

The Ziegler-Nichols tuning rules (Chapter 24) form one of the most well-known procedures for tuning the proportional-integral-derivative (PID) controller in many single-loop industrial processes. However, it has also been found that Ziegler-Nichols tuning procedures are best suited for processes where the ratio of apparent dead time and the dominant time constant is small. Dead time compensation is recommended for systems having significant ratios of dead time to dominant time-constant. In particular, some method of prediction is desired to counteract the destabilizing effects of delay. The idea of prediction is that information about future changes in the measured signal is derived, so that control is based on this predicted value. A simple way to predict the future value of a measured signal is to extrapolate along the derivative or slope of the feedback signal. Hence, the derivative term of the PID controller offers some predictive effect. Unfortunately, the derivative action of a PID controller alone is often insufficient for large time delays. In addition, the differentiation of feedback signals is often not meaningful in systems with large time delay.

An alternative to prediction based on feedback information is to simulate the process within the controller. A model of the process is required, which typically requires a minimum of three parameters. A common transfer function model for an industrial process having time delay is

$$\frac{Y(s)}{U(s)} = \frac{K}{\tau s + 1} e^{-sT}. \tag{29.1}$$

Here, $Y(s)$ and $U(s)$ are the process output and input, respectively. The process gain is denoted by K, the dominant time constant is τ, and the time delay is T (seconds). Among several methods for predictive control, the most well-known is the Smith Predictor (Smith, 1958).

29.2 Basic Principle of the Smith Predictor

The basic principle behind the Smith Predictor is quite easy to explain. Consider the feedback system shown in Figure 29.2.

0-8493-8343-9/97/$0.00+$.50

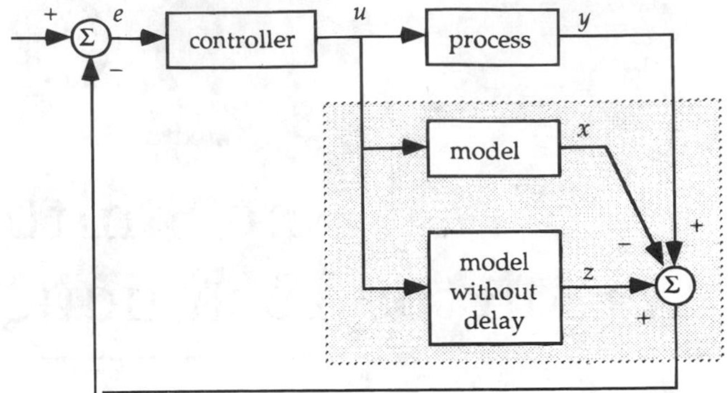

Figure 29.2 The Smith Predictor.

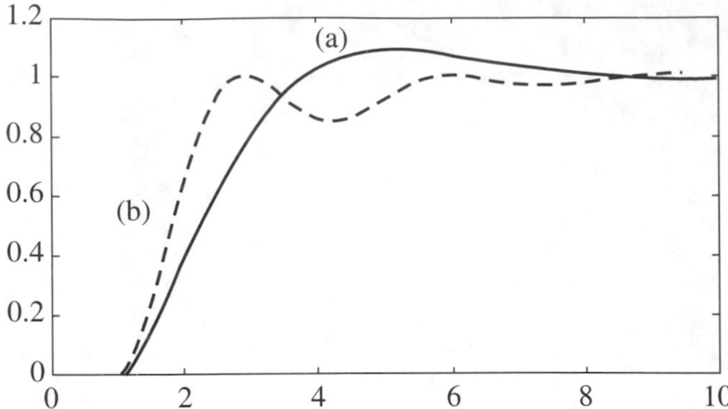

Figure 29.3 Comparison of output responses: (a) Smith Predictor with PI controller; (b) Ziegler-Nichols tuned PID controller.

The original system consists of the process with time delay, the feedback path, and the controller that must be designed. Selection and tuning of a suitable controller would be greatly simplified if the effects of process delay are canceled. Stated another way, it is desired to design a controller based only on that portion of the process that does not include the time delay. To achieve this, two blocks are added to the system as indicated by the shaded area in Figure 29.2. The block named "model" describes the process. Under ideal circumstances, with perfect modeling of the process, the model output x exactly cancels the process output y. Hence, the feedback signal consists only of the output z, which is from the block named "model without delay". In practice, the actual process is seldom identical to the model. Care must be exercised in estimating the time delay and dominant time constant.

29.3 A Smith Predictor Design Example

Referring to Figure 29.2, let the process be described by the transfer function:

$$\frac{Y(s)}{U(s)} = \frac{10}{0.25s^2 + s + 1} e^{-s} \tag{29.2}$$

which has time delay of 1 s, and dominant time constant of 0.5 s. Two controllers are compared for this example. In the first case a PID controller:

$$\frac{U(s)}{E(s)} = P + \frac{I}{s} + Ds \tag{29.3}$$

is tuned using the Ziegler-Nichols method (the shaded portion in Figure 29.2 is not present). Ziegler-Nichols tuning gains are $P = 0.0877$, $I = 0.0468$, $D = 0.0411$. In the second case, the Smith Predictor method is used. The Smith Predictor method introduces the shaded portion of Figure 29.2. The predictor model is chosen to have the form described in Equation 29.1.

Predictor model parameters can be obtained in practice by performing some system identification technique on the process, such as a step response test. For this example, the parameters are chosen to give the predictor model:

$$\frac{X(s)}{U(s)} = \frac{10}{0.5s + 1} e^{-1.1s}. \tag{29.4}$$

Note the differences between the process and the predictor model with regard to dynamic order and time delay. These differences are introduced in the example to illustrate the effect of modeling errors. The design model without delay is simply:

$$\frac{Z(s)}{U(s)} = \frac{10}{0.5s + 1}. \tag{29.5}$$

Since prediction is used, derivative action is not necessary for time delay compensation. Therefore, a simpler proportional-integral (PI) controller is tuned for the design model without delay Equation 29.5. PI controller gains are tuned using the root locus method, yielding $P = 0.05$ and $I = 0.1$. Shown in Figure 29.3 are the output responses using the Smith Predictor (solid curve) and the Ziegler-Nichols tuning (dashed curve). The ratio of process time delay to dominant time constant equals two, so it is challenging to tune the PID controller. For this particular example, the Smith Predictor method performs well despite slight mismatches between the model and process time delay.

References

Hagglund, T. 1992. A predictive PI controller for processes with long dead times, *IEEE Cont. Syst. Magazine*, 12–1:57–60.

Smith, O. J. M. 1958. *Feedback Control Systems*, McGraw-Hill, New York, NY.

Other techniques for controlling processing with significant dead times include the Internal Model Control method (Chapter 30), and prediction methods based on linear extrapolation of the feedback signal. All of these methods are related, and a good explanation is available in Hagglund (1992).

30
Internal Model Control

James C. Hung
University of Tennessee, Knoxville

30.1 Basic IMC Structures 513
30.2 IMC Design 514
30.3 Discussion 514

Internal Model Control (IMC) is an attractive method for designing a control system if the plant is inherently stable. The method was first formally reported in Garcia and Morari (1982). Since then, additional research results on this subject have been reported in journals and conferences. The method has been formulated using Laplace transform, and therefore is a frequency-domain technique. The concept is summarized here for a single-input-single-output (SISO) system.

30.1 Basic IMC Structures

A single-degree-of-freedom (SDF) IMC structure is shown in Figure 30.1. Notice that the plant model, G_{pm}, is explicitly included in the overall controller. When the actual plant G_p matches the plant model G_{pm}, there is no feedback signal and the control is open-loop.

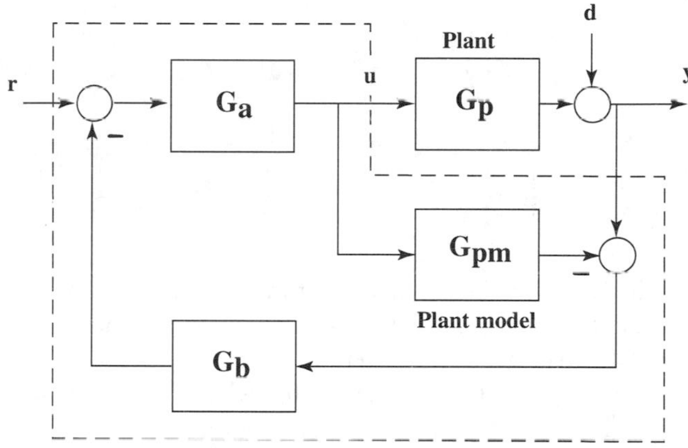

Figure 30.1 SDF internal model control.

$$Y(s) = G_aG_pR(s) + (1 - G_aG_p)D(s)$$

The design of a cascade compensator G_a for a specified response to input r is simple. Clearly, the stability of both the plant and the compensator is necessary and sufficient for the system to be stable. The system is a feedback system for disturbance d, thus attenuating its effect.

It should be pointed out that, corresponding to each IMC controller G_a of Figure 30.1 an equivalent controller G_c of a conventional control structure, shown in Figure 30.2 exists. In fact,

$$G_c = \frac{G_a}{1 - G_aG_{pm}}$$

Independent reference input response and disturbance rejection can be achieved by adopting a two-degree-of-freedom (TDF) IMC structure as shown in Figure 30.3. In this structure, the controller consists of the plant model G_{pm} and two compensators G_a and G_b. When $G_p = G_{pm}$, the system responses to reference input r and to disturbance d are given, respectively, by

$$Y_r(s) = G_aG_pR(s)$$

and

$$Y_d(s) = (1 - G_aG_bG_p)D(s)$$

where compensators G_a and G_b can be adjusted independently to achieve the desired individual responses. Notice that these responses are linear in G_a and G_b, making compensator design easy.

When there are differences between the plant and the model, a feedback signal exists and can be used for achieving system robustness. Under this condition,

$$Y_r(s) = G_aG_pR(s)/Q$$

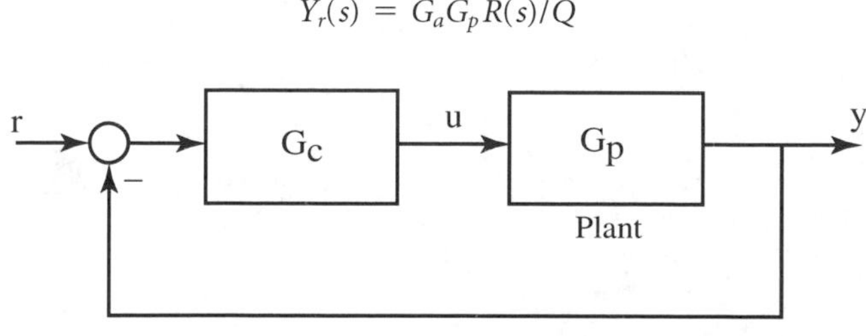

Figure 30.2 Conventional control.

0-8493-8343-9/97/$0.00+$.50

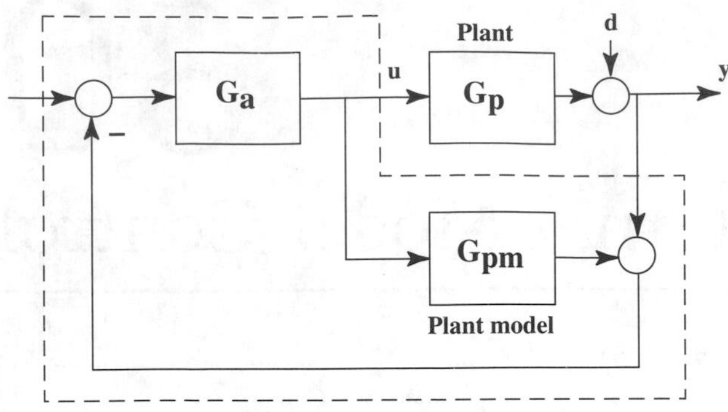

Figure 30.3 TDF internal model control.

and

$$Y_d(s) = (1 - G_a G_b G_p) D(s)/Q$$

where

$$Q = 1 + G_a G_b (G_p - G_{pm})$$

Q is the characteristic polynomial of the system. The condition for stability include: all roots of the characteristic equation are on the left-half-plane; and there is no pole zero cancellation among G_a, G_b, and $(G_p - G_{pm})$ on the right-half-plane. Robust stability is achieved by an appropriate choice of G_b.

30.2 IMC Design

Two-degree-of-freedom IMC design usually begins with the design of the forward compensator G_a for a desired input-output response. Then the feedback compensator is designed to achieve a specified system stiffness with respect to disturbances and a specified robustness with respect to model error.

30.3 Discussion

In the IMC approach, the controller is easy to design and the IMC structure provides an easy way to achieve system robustness.

For the MIMO case and for an in-depth treatment of IMC, readers are referred to Morari and Zafiriou, (1989) and the reference list therein.

One limitation of the IMC method is that it cannot handle plants which are open-loop unstable. As mentioned before, a stable plant using a stable compensator results in a stable IMC system. An equivalent controller can be determined for the conventional control configuration giving the same response characteristics. On the other hand, an unstable plant may be stabilized by a controller using the conventional control configuration; however, the corresponding IMC control system is not stable. As an example, consider a system in the conventional control configuration. The system consists of an unstable plant

$$G_p = G_{pm} = \frac{1}{(s-1)(s+5)}$$

and a cascade controller

$$G_c = 18$$

The closed-loop system is stable having a transfer function

$$G_{CL} = \frac{18}{s^2 + 4s + 13}$$

The equivalent single-degree-of-freedom compensator G_a of the IMC is given by

$$G_a = \frac{G_c}{1 + G_c G_{pm}} = \frac{18(s-1)(s+5)}{s^2 + 4s + 13}$$

but it does not help to stabilize the system. Any noise entering the system at the plant input will drive the plant output to infinity. Here one sees that the stability of a system is, in general, implementation dependent.

References

Garcia, C. E. and Morari, M. 1982. Internal model control. 1. A unifying review and some new results, *Ind. Eng. Chem. Process Des. Dev.*, 21–2:308–323.

Morari, M. and Zafiriou, E. 1989. *Robust Process Control,* Prentice-Hall, Englewood Cliffs, NJ.

31

Model Predictive Control

31.1 Overview 515
Introduction • Basic Concepts
31.2 Applications 516
A Prototypical MPC Algorithm for State-Space Linear Systems • Other Applications

Jay H. Lee
Auburn University

31.1 Overview

Introduction

Model predictive control (MPC) is an attractive tool for dynamic optimization and control of multi-variable systems with time-varying performance requirements and constraints. Recent proliferation of MPC research and application is generally credited to the seminal paper by Cutler and Ramaker (1980) that demonstrated the merits and potential of such a tool for the process industry. Independently from the research done in the process control community, a second branch of MPC emerged the objective of which is adaptive control; it has also found wide-spread use (Clarke 1987a b). Even though some differences exist in the details of the formulation, the underlying concept of various MPC techniques is the same and can be stated as follows:

> Perform an open-loop optimization for a fixed time horizon at each time step on the basis of online, updated estimates for plant states and parameters.

Because the open-loop optimization is repeated at every sample time after a feedback update, the methodology is also referred to as "receding horizon control" or "open-loop optimal feedback control." The purpose of this paper is to present the basic concept of MPC and to provide a short tutorial. Clearly, there can be variations and extensions of the particular version of MPC presented here. Readers interested in more general results are referred to a recent survey paper by Garcia et al. (1989) that provides a fairly comprehensive list of references on the subject.

Basic Concepts

There are several commercial versions of MPC offered by different vendors. They are similar in their main structure, but differ in details. The main structure is shown in Figure 31.1. Information about the process at the k_{th} sample time instant is contained in the state vector x_k, which is either directly measured or estimated using available measurements. Starting from state estimate $x_{k|k}$, one can develop prediction of process outputs over some time

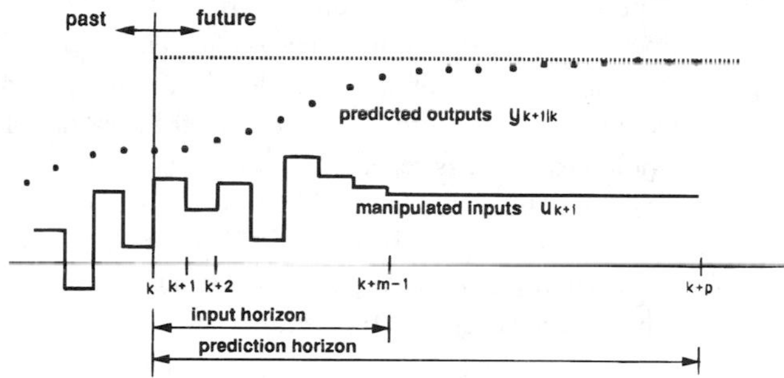

Figure 31.1 The schematic representation of basic concept of MPC.

horizon. Then process outputs are expressed as functions of manipulated variables that are to be changed at discrete time instances over a chosen input horizon (the inputs are assumed to return to zero or more typically assumed to remain constant beyond the input horizon). The prediction equation hence takes the following form:

$$\begin{aligned} y_{k+1|k} &= f_1(x_{k|k}, u_k) \\ y_{k+2|k} &= f_2(x_{k|k}, u_k, u_{k+1}) \\ &\vdots \quad \vdots \quad \vdots \\ &\vdots \quad \vdots \quad \vdots \\ y_{k+p|k} &= f_p(x_{k|k}, u_k, u_{k+1}, \ldots, u_{k+m-1}) \end{aligned} \tag{31.1}$$

where u and y are vectors containing the manipulated inputs and process outputs. p and m represent the number of time steps in the prediction horizon and the control horizon. The future manipulated variables are decided so that the predicted outputs follow the reference in a desirable manner, while satisfying given operating constraints. This is done by defining a loss function $\Psi(\mathcal{E}_k, \mathcal{U}_k)$ and solving the following optimization:

$$\min_{\mathcal{U}_k} \Psi(\mathcal{E}_k, \mathcal{U}_k)$$

such that (31.2)

$$\mathcal{C}^u \mathcal{U}_k \geq \mathcal{C}_k$$

0-8493-8343-9/97/$0.00+$.50

In the above, $\mathcal{E}_k = \mathcal{R}_{k+1|k} - \mathcal{Y}_{k+1|k}$ is the future error vector and

$$\mathcal{Y}_{k+1|k} = \begin{bmatrix} y_{k+1|k} \\ y_{k+2|k} \\ \vdots \\ \vdots \\ y_{k+p|k} \end{bmatrix}; \quad \mathcal{U}_k = \begin{bmatrix} u_k \\ u_{k+1} \\ \vdots \\ \vdots \\ u_{k+m-1} \end{bmatrix};$$

$$\mathcal{R}_{k+1|k} = \begin{bmatrix} r_{k+1|k} \\ r_{k+2|k} \\ \vdots \\ \vdots \\ r_{k+p|k} \end{bmatrix} \tag{31.3}$$

$r_{k+l|k}$ denotes the reference for the output at time $k + \ell$ projected at time k. The inequality in Equation 31.2 represents the constraints imposed on the manipulated variables and predicted outputs. Only u_k, the first of the open-loop optimal control sequence, is implemented on the real plant. At the next sample time, the whole procedure is repeated, i.e., a new state estimate $x_{k+\ell|k+1}$ is obtained, the horizons are shifted forward by one step and another optimization is carried out. This strategy is called *receding horizon control* or *open-loop optimal feedback control* and represents the core idea for all MPC techniques.

31.2 Applications

A Prototypical MPC Algorithm for State-Space Linear Systems

In this section, we present a prototypical MPC algorithm based on a linear state space model. The algorithm is a generalization of the technique presented by Lee et al. (1993) that uses the more restrictive step response model. The particular algorithm is chosen here for tutorial, since it displays the essence of MPC in a general and transparent manner and can be viewed as the current state-of-the-art.

Model

In this section we will assume that the model is given as the following linear state-space difference equation:

Process:

$$x_k^p = A_p x_{k-1}^p + B_p u_{k-1} + F_p d_{k-1} \tag{31.4}$$

$$y_k = C_p x_k^p + D_p d_k \tag{31.5}$$

Measurements:

$$\hat{y}_k = C_p x_k^p + D_p d_k + v_k \tag{31.6}$$

The external inputs u, d and v denote manipulated inputs, load disturbances and measurement noise respectively. Such a linear model may be obtained after linearization of a first principles model around a chosen operating point or realization of an input/output model constructed from an identification experiment.

When the model is obtained by linearizing and discretizing a first principles model, d contains disturbance variables that are not measured, but changes substantially during operation. To develop optimal estimation and prediction, their statistical properties need to be defined. Most commonly, these variables are characterized through appropriate stochastic differential or difference equations. In this tutorial, we will adopt the following form of the disturbance model:

$$x_k^w = A_w x_{k-1}^w + B_w w_{k-1} \tag{31.7}$$

$$w_k = w_{k-1} + \Delta w_k \tag{31.8}$$

$$d_k = C_w x_k^w \tag{31.9}$$

In the above, Δw_k is an independent, identically distributed (i.i.d.) sequence (i.e., white noise sequence). Hence, w_k is an integrated white noise sequence that can be viewed as a series of random steps whose amplitude changes are i.i.d. The reason for including the integrators in the disturbance model is that, in chemical processes (to which MPC has been most successfully applied), most disturbances are persistent in nature. Including integrators in the disturbance model is necessary to obtain controllers with integral action that reject constant disturbances. It is further assumed that A_w has all the eigenvalues inside unit disk.

It is worthwhile to discuss how such a model may also be obtained from an identification experiment. In general, the following structure is used to fit the input/output data:

$$y_k = G(q^{-1}, \theta)u_k + H(q^{-1}, \theta)w_k \tag{31.10}$$

where $G(q^{-1}, \theta)$ and $H(q^{-1}, \theta)$ are stable transfer functions written in terms of backwardshift operator q^{-1} and θ is the unknown model parameter vector. The exogenous signal w_k is assumed to have "persistent" characteristics (as in the CARIMA model) and therefore modelled as an integrated white noise sequence. Standard identification algorithms such as the Prediction Error Method may be applied after rewriting the model such that the exogenous signal is white noise:

$$\Delta y_k = G(q^{-1}, \theta)\Delta u_k + H(q^{-1}, \theta)\Delta w_k \tag{31.11}$$

Then, resulting $G(q^{-1}, \theta)$ and $H(q^{-1}, \theta)$ are realized as (A_p, B_p, C_p) and (A_w, B_w, C_w). Also, $F_p = 0$, $D_p = I$ in this context since $d_k = H(q^{-1}, \theta)w_k$.

Augmenting Equations 31.4 through 31.6 with Equations 31.7 through 31.9 gives

$$\begin{bmatrix} x_k^p \\ x_k^w \end{bmatrix} = \begin{bmatrix} A_p & F_p C_w \\ 0 & A_w \end{bmatrix} \begin{bmatrix} x_{k-1}^p \\ x_{k-1}^w \end{bmatrix} + \begin{bmatrix} B_p \\ 0 \end{bmatrix} u_{k-1} + \begin{bmatrix} 0 \\ B_w \end{bmatrix} w_{k-1} \tag{31.12}$$

$$\hat{y}_k = [C_p \quad D_p C_w] \begin{bmatrix} x_k^p \\ x_k^w \end{bmatrix} + v_k \tag{31.13}$$

The state-space model Equations 31.12 and 31.13 may be non-minimal if the model was obtained by fitting an input/output data set to a model structure containing auto-regressive terms (which introduces common poles among $G(q^{-1}, \theta)$ and $H(q, \theta)$). In this case, a minimal dimension model may be found via model reduction.

Equations 31.12 and 31.13 can be rewritten in terms of differenced inputs as follows:

$$X_k = \Phi X_{k-1} + \Gamma_u \Delta u_{k-1} + \Gamma_w \Delta w_{k-1} \tag{31.14}$$

$$\hat{y}_k = \Xi X_k + v_k \tag{31.15}$$

where

$$X_k = \begin{bmatrix} \Delta x_k^p \\ x_k^w \\ y_k \end{bmatrix}$$

$$\Phi = \begin{bmatrix} A_p & F_p C_w & 0 \\ 0 & A_w & 0 \\ C_p A_p & C_p F_p C_w + D_p C_w A_w & I \end{bmatrix} \tag{31.16}$$

$$\Gamma_u = \begin{bmatrix} B_p \\ 0 \\ C_p B_p \end{bmatrix}; \qquad \Gamma_w = \begin{bmatrix} 0 \\ B_w \\ D_p C_w B_w \end{bmatrix}$$

$$\Xi = [0 \quad 0 \quad I] \tag{31.17}$$

Δ variable represents the change in the variable from the previous sampling time (e.g., $\Delta x_k^p \equiv x_k^p - x_{k-1}^p$). The above differencing of the model together with the inclusion of integrators in the disturbance model is a standard way to ensure that the resulting controller has integral action. When integral action is not desired, neither steps would be necessary.

State Estimation

General Case: State/Output Disturbance Model. When Δw and v are independent, identically distributed (i.i.d.) Gaussian sequences, the optimal state estimator for model (Equations 31.14–31.15) is the Kalman filter of the following form (see Åström and Wittenmark, 1984 for details):

$$X_{k|k-1} = \Phi X_{k-1|k-1} + \Gamma_u \Delta u_{k-1} \tag{31.18}$$

$$X_{k|k} = X_{k|k-1} + K_f\{\hat{y}_k - \Xi X_{k|k-1}\} \tag{31.19}$$

In the above, notation $\{\cdot\}_{k|\ell}$ stands for an optimal estimate at time k using measurements up to time ℓ (more specifically, it represents the conditional mean of the variable at time k with conditions given by measurements collected up to time ℓ). K_f is the optimal filter gain computed according to the formula

$$K_f = \Sigma \Xi^T (\Xi \Sigma \Xi^T + R)^{-1} \tag{31.20}$$

where Σ is the solution to the following algebraic Riccati equation (ARE):

$$\Sigma = \Phi \Sigma \Phi^T + \Gamma^w Q (\Gamma^w)^T - \Phi \Sigma \Xi^T (\Xi \Sigma \Xi^T + R)^{-1} \Xi \Sigma \Phi^T \tag{31.21}$$

In the above, Q and R are the covariance matrices for Δw and v respectively. A practical drawback of using the Kalman filter is the necessity to specify the covariance matrices for Δw and v. Since this information is rarely available in practice, these are *de facto* tuning parameters that are adjusted to obtain desired closed-loop responses. The difficulty, of course, is that this results in many parameters to adjust. An alternative to using the Kalman filter is to use the observer gain obtained through pole placement. Pole placement is rarely used these days because it is difficult to determine the proper location of all observer poles *a priori*. In Lee et al. (1994), the two approaches are merged by analyzing how the Kalman filter places the observer poles according to the given signal-to-noise ratio and using the pole location as a tuning parameter.

Special Case: Output Disturbance Model. In many cases, modelling state disturbances is very difficult. Since MPC needs prediction for the future behavior of outputs and not of all states, it is convenient to lump the effect of all disturbances and express it directly at the output (i.e., set $F_p = 0$, $D_p = I$). For most chemical processes, the effect of load disturbances at each output is well described through integrated white noise passed through a first-order lag (see Figure 31.2). If we further assume that no information is available on the correlation among

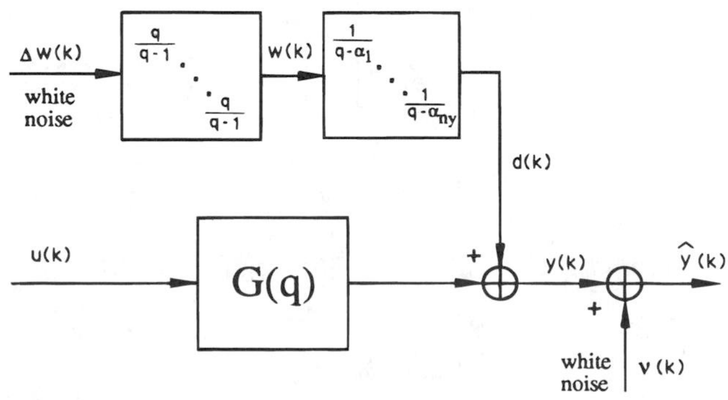

Figure 31.2 Block diagram representation for disturbances in the output channels described by integrated white noise passed through first-order lags.

disturbance effects for different outputs, we can set $B_w = I$, $C_w = I$. Then Equation 31.14 has the structure

$$\begin{bmatrix} \Delta x_k^p \\ \Delta x_k^w \\ y_k \end{bmatrix} = \begin{bmatrix} A_p & 0 & 0 \\ 0 & A_w & 0 \\ C_pA_p & A_w & I \end{bmatrix} \begin{bmatrix} \Delta x_{k-1}^p \\ \Delta x_{k-1}^w \\ y_{k-1} \end{bmatrix} + \begin{bmatrix} B_p \\ 0 \\ C_pB_p \end{bmatrix} \Delta u_{k-1} + \begin{bmatrix} 0 \\ I \\ I \end{bmatrix} \Delta w_{k-1} \quad (31.22)$$

where

$$A_w = \text{diag}\{\alpha_1, \ldots, \alpha_{n_y}\}; \qquad 0 \le \alpha_i < 1 \quad (31.23)$$

Δw is a Gaussian i.i.d. sequence with a diagonal covariance matrix, i.e.,

$$E\{\Delta w \Delta w^T\} = \text{diag}\{q_1, \ldots, q_{n_y}\} \quad (31.24)$$

If the model is developed via an identification experiment, α_i may be obtained along with other model parameters by fitting the identification data to the following structure (for the ith output):

$$(\Delta y_k)_i = G_i(q^{-1}, \theta_i)\Delta u_k + \frac{1}{1 - \alpha_i q^{-1}} (\Delta w_k)_i \quad (31.25)$$

Then, (A_p, B_p, C_p) are found by realization of $G(q^{-1}, \theta)$ as explained before.

Let us assume that the measurement noise at each output is also an i.i.d. Gaussian vector sequence of a diagonal covariance matrix, i.e.,

$$E\{vv^T\} = \text{diag}\{r_1, \ldots, r_{n_y}\} \quad (31.26)$$

Then, for open-loop stable systems (i.e., A_p has all eigenvalues strictly inside unit disk), it can be shown that the optimal filter gain K_f is parameterized in terms of an n_y-dimension real vector whose elements lie in [0, 1]. More specifically,

$$K_f = \begin{bmatrix} 0 \\ F_b \\ F_a \end{bmatrix} \quad (31.27)$$

where

$$F_b = \text{diag}\{(f_b)_1, \ldots, (f_b)_{n_y}\}; \qquad F_a = \text{diag}\{(f_a)_1, \ldots, (f_a)_{n_y}\} \quad (31.28)$$

$$(f_b)_i = \frac{(f_a)_i^2}{1 + \alpha_i - \alpha_i (f_a)_i} \quad \text{for} \quad 1 \le i \le n_y \quad (31.29)$$

and

$$(f_a)_i \to 0 \quad \text{as} \quad q_i/r_i \to 0 \quad (31.30)$$

$$(f_a)_i \to 1 \quad \text{as} \quad q_i/r_i \to \infty \quad (31.31)$$

We emphasize that the filter gain expression (Equation 31.27) is valid only for *stable* systems. For unstable systems, the gain is not stabilizing and a stabilizing solution of the ARE must be found and used to calculate the filter gain according to Equation 31.20. If A_p contains any eigenvalue on the unit circle, the corresponding modes of Δx must be excited by Δw by including additional external disturbances so that a stabilizing solution to ARE (Equation 31.21) exists (see Appendix E of Goodwin and Sin, 1984 for conditions for a strong solution of ARE to be stabilizing). In addition, all the controlled variables must be measured (or estimated from an independent output estimator) for such an estimator gain to be effective; no inference can be made on unmeasured output since any information on how the measured and unmeasured outputs are correlated is absent.

Although we made a number of assumptions in arriving at the parameterization (Equation 31.27) for the optimal filter gain, they are not too restrictive for most process control problems and should be useful in many practical situations. We further note that, even where a fundamental disturbance model is available, one may still include the output disturbance model in Figure 31.2 in addition to account for the effects of model errors and other external signals not represented in the fundamental model. Including the output disturbances of Equation 31.22 puts integral action on all the output channels and ensures offset-free control even when there is a significant mismatch between the model and the actual process.

Prediction

The following optimal multistep prediction equation can be easily developed using the state estimate:

$$\mathcal{Y}_{k+1|k} = \mathcal{S}^x X_{k|k} + \mathcal{S}^{\mathcal{U}} \Delta \mathcal{U}_k \quad (31.32)$$

where

$$\mathcal{Y}_{k+1|k} = \begin{bmatrix} y_{k+1|k} \\ y_{k+2|k} \\ \vdots \\ \vdots \\ y_{k+p|k} \end{bmatrix} \qquad \Delta\mathcal{U}_k = \begin{bmatrix} \Delta u_k \\ \Delta u_{k+1} \\ \vdots \\ \vdots \\ \Delta u_{k+m-1} \end{bmatrix}$$

$$\mathcal{S}^x = \begin{bmatrix} \Xi\Phi \\ \Xi\Phi^2 \\ \vdots \\ \vdots \\ \Xi\Phi^p \end{bmatrix} \quad (31.33)$$

$$\mathcal{S}^{\mathcal{U}} = \begin{bmatrix} \Xi\Gamma_u & 0 & \cdots & 0 \\ \Xi\Phi\Gamma_u & \Xi\Gamma_u & \cdots & 0 \\ \vdots & \vdots & \ddots & \vdots \\ \vdots & \vdots & \ddots & \ddots \\ \Xi\Phi^{p-1}\Gamma_u & \Xi\Phi^{p-2}\Gamma_u & \cdots & \Xi\Phi^{p-m}\Gamma_u \end{bmatrix}$$

$y_{k+\ell|k}$ represents the optimal prediction of $y_{k+\ell}$ based on the measurements at $t = k$. We also allowed the flexibility of suppressing the last $p - m$ input moves (i.e., we assumed that $\Delta u_{k-m} = \ldots = \Delta u_{k+p-1} = 0$).

Control Move Calculation

Unconstrained Case. A popular criterion used for control move calculation is the following quadratic performance objective (as in QDMC by Garcia and Morshedi, 1984):

$$\min_{\Delta u_{k+i}, i=1,\ldots,m} \left\{ \sum_{i=1}^{p} \|\tilde{\Lambda}_y(r_{k+i|k} - y_{k+i|k})\|_2^2 + \sum_{i=0}^{m-1} \|\tilde{\Lambda}_u \Delta u_{k+i}\|_2^2 \right\} \tag{31.34}$$

$r_{k+i|k}$ is the output reference vector for time $k + i$ projected at time k. $\tilde{\Lambda}_y$ and $\tilde{\Lambda}_u$ are weighting matrices. They are chosen to be same for every time step in most cases. The optimization objective can be rewritten as

$$\min_{\Delta\mathcal{U}_k} \{\|\Lambda_y(\mathcal{Y}_{k+1|k} - \mathcal{R}_{k+1|k})\|_2^2 + \|\Lambda_u \Delta\mathcal{U}_k\|_2^2\} \tag{31.35}$$

where

$$\mathcal{R}_{k+1|k} = [r_{k+1|k}^T, \ldots, r_{k+p|k}^T]^T \tag{31.36}$$

$$\Lambda_y = \mathrm{diag}(\overbrace{\tilde{\Lambda}_y, \ldots, \tilde{\Lambda}_y}^{p}); \qquad \Lambda_u = \mathrm{diag}(\overbrace{\tilde{\Lambda}_u, \ldots, \tilde{\Lambda}_u}^{m}) \tag{31.37}$$

In the absence of constraints, the above least squares problem can be solved analytically and the receding horizon control law based on objective (Equation 31.35) and prediction equation (Equation 31.32) can be shown to be the following state feedback control law:

$$\Delta u_k = -L_{MPC} X_{k|k} + K_{MPC} \mathcal{R}_{k+1|k} \tag{31.38}$$

where

$$L_{MPC} = -K_{MPC} \mathcal{S}^x \tag{31.39}$$

$$K_{MPC} = [I \;\; 0 \;\; \cdots \;\; 0]((\mathcal{S}^{\mathcal{U}})^T \Lambda_y^T \Lambda_y \mathcal{S}^{\mathcal{U}} + \Lambda_u^T \Lambda_u)^{-1} (\mathcal{S}^{\mathcal{U}})^T \Lambda_y^T \Lambda_y \tag{31.40}$$

Constrained Case. Constraints are generally imposed on the process inputs and outputs and can be expressed mathematically as inequality constraints:

- Manipulated Variable Constraints

$$u_{k+i}^{\mathrm{low}} \le u_{k+i} \le u_{k+i}^{\mathrm{high}} \qquad \text{for} \quad i = 0, \ldots, m-1 \tag{31.41}$$

- Manipulated Variable Rate Constraints

$$|\Delta u_{k+i}| \le \Delta u^{\max} \qquad \text{for} \quad i = 0, \ldots, m-1 \tag{31.42}$$

- Output Variable Constraints

$$y_{k+\ell}^{\mathrm{low}} \le y_{k+\ell|k} \le y_{k+\ell}^{\mathrm{high}} \qquad \text{for} \quad \ell = 1, \ldots, p \tag{31.43}$$

After some straightforward algebraic manipulation, the above constraints can be expressed together as

$$\mathcal{C}^u \Delta\mathcal{U}_k \ge \mathcal{C}_k \tag{31.44}$$

$$\mathcal{C}^u = \begin{bmatrix} I_L \\ I_L \\ -I \\ I \\ -\mathcal{S}^{\mathcal{U}} \\ \mathcal{S}^{\mathcal{U}} \end{bmatrix} \qquad I_L = \underbrace{\begin{bmatrix} I & 0 & \cdots & 0 \\ I & I & \cdots & 0 \\ \vdots & \vdots & \ddots & \vdots \\ I & I & \cdots & I \end{bmatrix}}_{m} \tag{31.45}$$

$$\mathcal{C}_k = \begin{bmatrix} u_{k-1} - u_k^{\mathrm{high}} \\ \vdots \\ u_{k-1} - u_{k+m-1}^{\mathrm{high}} \\ u_k^{\mathrm{low}} - u_{k-1} \\ \vdots \\ u_{k+m-1}^{\mathrm{low}} - u_{k-1} \\ -\Delta u_k^{\max} \\ \vdots \\ -\Delta u_{k+m-1}^{\max} \\ -\Delta u_k^{\max} \\ \vdots \\ -\Delta u_{k+m-1}^{\max} \\ \mathcal{S}^x \Delta X_{k|k} - \mathcal{Y}_{k+1}^{\mathrm{high}} \\ -\mathcal{S}^x \Delta X_{k|k} + \mathcal{Y}_{k+1}^{\mathrm{low}} \end{bmatrix} \tag{31.46}$$

$\mathcal{Y}_{k+1}^{\mathrm{high}}$ and $\mathcal{Y}_{k+1}^{\mathrm{low}}$ represent vectors containing the upper and lower bounds on $\mathcal{Y}_{k+1|k}$, i.e.,

$$\mathcal{Y}_{k+1}^{\mathrm{high}} = [(y_{k+1}^{\mathrm{high}})^T \;\; (y_{k+2}^{\mathrm{high}})^T \;\; \cdots \;\; (y_{k+p}^{\mathrm{high}})^T]^T \tag{31.47}$$

$$\mathcal{Y}_{k+1}^{\mathrm{low}} = [(y_{k+1}^{\mathrm{low}})^T \;\; (y_{k+2}^{\mathrm{low}})^T \;\; \cdots \;\; (y_{k+p}^{\mathrm{low}})^T]^T \tag{31.48}$$

Minimization of objective (Equation 36.35) with constraint (Equation 31.44) is a quadratic programming (QP). At every sampling time, an optimizer solves the QP on-line to compute the optimal input move sequence $\Delta\mathcal{U}_k$ and the first move Δu_k is implemented.

Infinite Horizon MPC. One drawback of using finite horizon objective (Equation 31.35), as elucidated by Bitmead et al. (1990), is that the stability of the resulting state feedback control law depends on the choice of tuning parameters such as the prediction/control horizons (p and m) and input/output weights ($\tilde{\Lambda}_y$ and $\tilde{\Lambda}_u$). This presents much inconvenience and difficulty in designing and tuning MPC controllers, since the stability must be checked for each different set of the parameters.

Recently, the following infinite horizon objective has been proposed as an alternative (Rawlings and Muske, 1993):

$$\min_{\Delta u_{k+i}, i=1,\ldots,m} \left\{ \sum_{i=1}^{\infty} \|\tilde{\Lambda}_y (r_{k+i|k} - y_{k+i|k})\|_2^2 + \sum_{i=0}^{m-1} \|\tilde{\Lambda}_u \Delta u_{k+i}\|_2^2 \right\} \tag{31.49}$$

The advantage is that, under the above objective, the resulting feedback law is guaranteed to be stable. This is true even when the constraints are imposed in the algorithm. If m is chosen large enough, the unconstrained control law closely approximates the classical LQ solution. (Equation 36.49) can be reformulated as a finite horizon objective:

$$\min_{\Delta u_{k+i}, i=0,\ldots,m-1} \left\{ \sum_{i=1}^{m-1} \|\tilde{\Lambda}_y (r_{k+i|k} - y_{k+i|k})\|_2^2 + \sum_{i=0}^{m-1} \|\tilde{\Lambda}_u \Delta u_{k+i}\|_2^2 + [X_{k+m|k}^T \quad r_{k+m|k}^T] \tilde{\Lambda}_X \begin{bmatrix} X_{k+m|k} \\ r_{k+m|k} \end{bmatrix} \right\} \tag{31.50}$$

where

$$\tilde{\Lambda}_X = \sum_{i=0}^{\infty} \{(\Phi_e^i)^T \Xi_e^T \tilde{\Lambda}_y^T \tilde{\Lambda}_y \Xi_e \Phi_e^i\}$$

and

$$\Phi_e = \begin{bmatrix} \Phi & 0 \\ 0 & I_{n_y} \end{bmatrix}; \qquad \Xi_e = [-\Xi \quad I_{n_y}] \tag{31.51}$$

$\tilde{\Lambda}_X$ is in the form of an observability grammian and can be calculated by solving a Lyapunov equation. If Φ_e contains unstable modes that do not belong to the null space of $\tilde{\Lambda}_y \Xi_e$, the resulting infinite weights should be interpreted as constraints to "zero in" these modes (defined by the Jordan decomposition) at time $k + m$ and should be reexpressed as state constraints. For stable processes, under the finite impulse response assumption (i.e., the impulse response settles after n time steps), Equation 31.49 is equivalent to finite horizon MPC with the following particular setting of the parameters:

$$\min_{\Delta u_{k+i}, i=0,\ldots,m-1} \left\{ \sum_{i=1}^{m-1} \|\tilde{\Lambda}_y (r_{k+i|k} - y_{k+i|k})\|_2^2 + \sum_{i=0}^{m-1} \|\tilde{\Lambda}_u \Delta u_{k+i}\|_2^2 \right\} \tag{31.52}$$

with constraint $\tilde{\Lambda}_y (r_{k+m+n|k} - y_{k+m+n|k}) = 0$. The constraint can be replaced by choosing the terminal output weight to be large.

Other Applications

Certainly some variations are possible on the prototypical MPC algorithm just presented. For instance, commercial MPC software that several vendors offer are based on a truncated step response model or a finite impulse response model. In addition, feedback updates are done in an open-loop fashion instead of using a rigorous closed-loop state estimator like the Kalman filter. The step response model or the finite impulse response model can be viewed as a special case of the state-space model that we assumed in this paper. The open-loop feedback update suffers from many drawbacks such as not being applicable to unstable systems and inferential control problems. Furthermore, generalization to the case involving complex stochastic disturbances and measurement noise is not as straightforward. By noting the similarity between MPC and classical optimal control techniques like linear quadratic Gaussian (LQG), significant enhancements have been made to the original MPC algorithm that is still being used by the industry.

MPC techniques using nonlinear models have also been developed and applied (see Biegler and Rawlings, 1991 for a survey). We are met with two main difficulties when we attempt to use a nonlinear model for MPC. First, state estimation becomes more involved since the state vector no longer stays Guassian preventing construction of an optimal recursive estimator. Standard techniques like the Kalman filter are no longer applicable. Nonlinear estimation techniques are available, but are computationally expensive and lack robustness (Jazwinski, 1970). Second, optimization for control move computation is computationally demanding since it is no longer a quadratic optimization with linear constraints. It is commonly recommended that nonlinear models are discretized via orthogonal collocation and expressed as algebraic constraints. This introduces nonlinear algebraic constraints and leads to a nonlinear programming (NLP) problem for which off-the-shelf software can be utilized. However, the computational requirement is often prohibitively large preventing any significant practical application of this type of tool thus far. A more practical approach based on successive linearization for state estimation and control computation is proposed and applied in a recent paper by Lee and Ricker (1994).

References

Åström, K. J. and Wittenmark, B. 1984. *Computer Controlled Systems: Theory and Design,* Prentice Hall, Englewood Cliffs, NJ.

Biegler, L. T. and Rawlings, J. B. 1991. Optimization Approaches to Nonlinear Model Predictive Control, *Proc. of CPC-IV,* 543–571, San Padre Island, TX.

Bitmead, R. R., Gevers, M., and Wertz, V. 1990. *Adpative Optimal Control: The Thinking Man's GPC,* Prentice Hall, Englewood Cliffs, NJ.

Clarke, D. W., Mohtadi, C., and Tuffs, P. S. 1987a. Generalized predictive control—Part I. The basic algorithm, *Automatica,* 25:137–148.

Clarke, D. W., Mohtadi, C., and Tuffs, P. S. 1987b. Generalized predictive control—Part II. Extensions and interpretations, *Automatica,* 23:149–160.

Cutler, C. R. and Ramaker, B. L. 1980. Dynamic matrix control—a computer control algorithm, *Proc. Automatic Control Conf.*, San Francisco, CA, Paper WP5-B.

Garcia, C. E. and Morshedi, A. M. 1984. Quadratic programming solution of dynamic matrix control (QDMC), *Proc. Am. Control Conf.*, San Diego, CA.

Garcia, C. E., Prett, D. M., and Morari, M. 1989. Model predictive control: theory and practice—a survey, *Automatica*, 25:335–348.

Goodwin, G. C. and Sin, K. S. 1984. *Adaptive Filtering, Prediction and Control*, Prentice Hall, Englewood Cliffs, NJ.

Jazwinski, A. H. 1970. *Stochastic Processes and Filtering Theory*, Academic Press, San Diego, CA.

Lee, J. H., Morari, M., and Garcia, C. E. 1990. State-space interpretation of model predictive control, *Automatica*, 30:707–717.

Lee, J. H. and Ricker, N. L. 1994. Extended Kalman filter based nonlinear model predictive control, *Ind. Eng. Chem. Res.*, 33:1530–1541.

Rawlings, J. B. and Muske, K. R. 1993. The stability of constrained receding horizon control, *IEEE Trans. Autom. Cntrl.*, 38:1512–1516.

32
Dynamic Matrix Control

32.1 The Dynamic Matrix 522
32.2 Output Projection 522
32.3 Control Computation 523
32.4 Remarks 523

James C. Hung
University of Tennessee, Knoxville

Dynamic matrix control (DMC) is a method suitable for the control of processes. The method was developed by control professionals in the oil industry in the 1970s. It is suitable for control by a computer and, since the concept is intuitively transparent, is attractive to technical people with a limited background in control theory. It is intended for linear processes, including both single-input-single-output (SISO) and multi-input-multi-output (MIMO) cases. The method provides a continuous projection of a system's future output for the time horizon required for the system to reach a steady state. The projected outputs are based on all past changes in the measured input variables. The control effort is then computed to alter the projected output to satisfy a chosen performance specification. The method is intended for the control of a linear process which is open-loop stable and does not have pure integration. Cutler (1982) gives a very lucid description of the concept. The method is based on the step-response characteristics of a process, and therefore, is a time-domain technique. The concept is summarized here for a SISO system.

32.1 The Dynamic Matrix

A key tool of the DMC is the *dynamic matrix* which can be constructed from a process' unit-step response data. Consider a SISO discrete-data process. Its output at the end of the kth time interval is given by the following convolution summation

$$y_k = \sum_{i=1}^{k} h_{k-i+1}\delta u_i$$

where h_i is the discrete-data step response and δu_i is the incremental step input at the beginning of the ith time interval. For $k =$ 1 to n, the following *dynamic matrix equation* can be formed.

$$\begin{bmatrix} y_1 \\ y_2 \\ \vdots \\ \vdots \\ y_n \end{bmatrix} = \begin{bmatrix} h_1 & 0 & 0 & 0 \\ h_2 & h_1 & h_1 & 0 \\ h_3 & h_2 & h_2 & 0 \\ \vdots & \vdots & \vdots & \vdots \\ h_n & \underbrace{h_{n-1}}_{h} & \cdots & \cdots \end{bmatrix} \begin{bmatrix} \delta n_1 \\ \delta n_2 \\ \cdots \\ \delta n_m \end{bmatrix}$$

or

$$\mathbf{y} = H\delta\mathbf{u}$$

The nxm H matrix shown is the so-called *dynamic matrix* of the SISO process and its elements are called dynamic coefficients. This matrix is used for projecting the future output. The vector at the left-hand side of the equation is called the *output projection vector.*

32.2 Output Projection

Assume that the control computation is repeated in every time interval. Under this condition, only the most current input change δu_1 is involved in the output projection computation and the H matrix becomes a column vector **h.** The iterative projection computation steps proceed as follows:

1. Initialize the output projection vector **y** by setting all its elements to the currently observed value of the output.
2. At the beginning of a subsequent time interval, shift the projection forward one time interval and change its designation to $\mathbf{y}^*$.
3. Compute the change in the input from the last time interval to the present.
4. Compute the changes in the output vector by using the input change and the dynamic matrix equation.
5. Compute the updated output projection vector **y** using the *updating equation.*

$$\mathbf{y} = \mathbf{y}^* + \mathbf{h}\delta u_1$$

6. Loop back to step 2 the next time interval.

Since the process is assumed to be open-loop stable, errors due to erroneous initialization will diminish with time. In general, the projected output values do not match the observed values due to disturbances. The discrepancy provides the feedback data for output vector adjustment.

0-8493-8343-9/97/$0.00+$.50

32.3 Control Computation

Let the process setpoint (reference input) be r, the output error at time instant i is given by

$$e_i = r - y_i$$

For $i = 1$ to n, e_i form an error vector **e.** The incremental control input vector $\delta\mathbf{u}$ needed to eliminate the error **e** can be obtained by one of a variety of methods depending on the performance criterion chosen. For example, it can be obtained by solving the equation

$$\mathbf{e} = H\delta\mathbf{u}$$

Since the output response from the last input must have time to reach steady state, the H matrix will have more rows than columns. A solution for $\delta\mathbf{u}$ can be obtained by using the classical least-square formula

$$\delta\mathbf{u} = (H^T H)^{-1} H^T \mathbf{e}$$

Figure 32.1 shows the block diagram of a DMC system. The DMC block performs output projection and control computation.

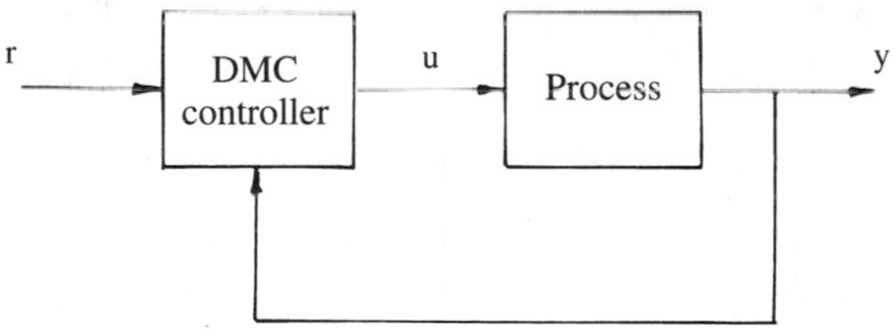

Figure 32.1 Dynamic matrix control (DMC).

32.4 Remarks

Extension to MIMO processes is conceptually straight-forward. For a 2-input-2-output process, the dynamic matrix equation has the form

$$\begin{bmatrix} \mathbf{y}_1 \\ \mathbf{y}_2 \end{bmatrix} = H \begin{bmatrix} \delta\mathbf{u}_1 \\ \delta\mathbf{u}_1 \end{bmatrix}$$

where $\mathbf{y}_1$ and $\mathbf{y}_2$ are the two output projection vectors, $\delta\mathbf{u}1$ and $\delta\mathbf{u}2$ are the two incremental step-input vectors, and H is the dynamic matrix of the process.

Merits of DMC are as follows:

1. It is easy to understand and easy to apply without involving in sophisticated mathematics.
2. Coefficients of the dynamic matrix can be obtained by testing, without relying solely on the mathematical model of the process.
3. Since the transport lag characteristics can easily be included in the dynamic matrix, the design of control is independent of whether there is transport lag or not.
4. When disturbances are observable, feed-forward control can conveniently be implemented via the DMC structure.

A demerit of DMC is that it is limited to open-loop bounded-input-bounded-output (BIBO) stable type of processes.

Theoretical work has been done on DMC by control researchers since the early 1980s. The additional results obtained have further revealed the capability of the method and its comparison to other control methods. Prett and Garcia (1988) is a good source for additional information.

References

Cutler, C. R. 1982. Dynamic matrix control of imbalanced systems, *ISA Trans.* 21–1:1–6.

Prett, D. M. and Garcia, C. E. 1988. Chapters 5 and 6, *Fundamental Process Control,* Butterworths.

33

Disturbance Observation-Cancellation Technique

Kouhei Ohnishi
Keio University

33.1 Why Estimate Disturbance? 524
33.2 Plant and Disturbance 524
33.3 Higher-Order Disturbance Approximation 526
33.4 Disturbance Observation 526
33.5 Disturbance Cancellation 526
33.6 Examples of Application 527
33.7 Conclusions 528

33.1 Why Estimate Disturbance?

One simple and effective robust control technique is disturbance observation-cancellation. As explained in the previous Section, it is necessary to have modes of disturbance in the controller for proper regulation. The internal model principle assures only the steady-state convergence of error. It is effective to run parallel with feed-forward compensation for faster response. Feed-forward compensation needs future disturbance signal beforehand. However, from a practical control viewpoint, the signal of only one or two steps in future are sufficient. The low order disturbance observer estimates an equivalent disturbance (or a modified disturbance) only several steps ahead. The estimated disturbance added to input cancels out the disturbance. This function is the same as feed-forward compensation and improves the transient performance to the disturbance, as well as, the steady-state operation of the plant. Since the equivalent disturbance includes parameters variation, the entire controller is expected to be robust against not only the disturbance, but also the parameters variation. As a result, the controlled plant seems as if it had nominal parameters and no disturbances. That is why it is worthwhile to estimate the equivalent disturbance. Also, the controller is simple and applicable to practical controllers. Let us consider the system as simply as possible.

33.2 Plant and Disturbance

The first goal is to estimate the additive disturbance in a linear system. For this purpose, we assume that the linear system has a single-input and single-output (SISO). Without losing generality, such a linear system is represented in the companion form as in Equation 33.1. Here d is an additive disturbance as a scalar function in the companion form.

$$\dot{\mathbf{x}} = \mathbf{A}\mathbf{x} + \mathbf{b}u + \mathbf{e}d \qquad (33.1)$$

$$y = \mathbf{c}\mathbf{x}$$

Here, $\mathbf{x}$ is a state vector of plant in the form of the following.

$$\mathbf{x} = \begin{bmatrix} x_1 \\ x_2 \\ \vdots \\ x_n \end{bmatrix}$$

y is an output and u is an input of the system, respectively. $\mathbf{A}$ is a system matrix in the form of the following.

$$\mathbf{A} = \begin{bmatrix} 0 & 1 & 0 & \cdots & 0 \\ 0 & 0 & 1 & \cdots & 0 \\ \vdots & \vdots & \vdots & \ddots & \vdots \\ 0 & 0 & 0 & \cdots & 1 \\ -a_1 & -a_2 & -a_3 & \cdots & -a_n \end{bmatrix}$$

$\mathbf{b}$ is a distribution vector in the form of the following.

$$\mathbf{b} = \begin{bmatrix} 0 \\ 0 \\ \vdots \\ 0 \\ K \end{bmatrix}$$

0-8493-8343-9/97/$0.00+$.50

$\mathbf{e}$ is a distribution vector of disturbance in the form of the following.

$$\mathbf{e} = \begin{bmatrix} 0 \\ 0 \\ \vdots \\ 0 \\ 1 \end{bmatrix}$$

$\mathbf{c}$ is an observation vector in the form of the following.

$$\mathbf{c} = [c_1 \quad c_2 \quad \cdots \quad c_m \quad 0 \quad \cdots \quad 0]$$

If d is zero, then the system is controllable and observable and is represented in the following transfer function.

$$\frac{Y(s)}{U(s)} = K\frac{c_m s^m + c_{m-1}s^{m-1} + \cdots + c_2 s + c_1}{s_n + a_n s^{n-1} + a_{n-1}s^{n-2} + \cdots + a_2 s + a_1} \tag{33.2}$$

Generally the system dynamics matrix $\mathbf{A}$ and the distribution vector $\mathbf{b}$ include the variable parameters in their elements. They are the sum of the nominal values and their variations, respectively.

$$\begin{aligned} \mathbf{A} &= \mathbf{A}_0 + \Delta\mathbf{A} \\ \mathbf{b} &= \mathbf{b}_0 + \Delta\mathbf{b} \end{aligned} \tag{33.3}$$

Here, $\mathbf{A}_0$ is a nominal system matrix in the form of the following.

$$\mathbf{A}_0 = \begin{bmatrix} 0 & 1 & 0 & \cdots & 0 \\ 0 & 0 & 1 & \cdots & 0 \\ \vdots & \vdots & \vdots & \ddots & \vdots \\ 0 & 0 & 0 & \cdots & 1 \\ -a_{01} & -a_{02} & -a_{03} & \cdots & -a_{0n} \end{bmatrix}$$

$\Delta\mathbf{A}$ is a variation of $\mathbf{A}$ in the form of the following.

$$\Delta\mathbf{A} = \begin{bmatrix} 0 & 0 & 0 & \cdots & 0 \\ 0 & 0 & 0 & \cdots & 0 \\ \vdots & \vdots & \vdots & \ddots & \vdots \\ 0 & 0 & 0 & \cdots & 0 \\ -\Delta a_{01} & -\Delta a_{02} & -\Delta a_{03} & \cdots & -\Delta a_{0n} \end{bmatrix}$$

$\mathbf{b}_0$ is a nominal distribution vector in the form of the following.

$$\mathbf{b}_0 = \begin{bmatrix} 0 \\ 0 \\ \vdots \\ 0 \\ K_0 \end{bmatrix}$$

$\Delta\mathbf{b}$ is a variation of $\mathbf{b}$ in the form of the following.

$$\Delta\mathbf{b} = \begin{bmatrix} 0 \\ 0 \\ \vdots \\ 0 \\ \Delta K \end{bmatrix}$$

It is noted that the variation of dynamic matrix $\mathbf{A}$ is equal to the variations of the coefficients in the lowest column of $\mathbf{A}$. Also the variation of $\mathbf{b}$ is substantially equal to the variation of forward gain K. The system equation is transformed into Equation 33.4.

$$\begin{aligned} \dot{\mathbf{x}} &= (\mathbf{A}_0 + \Delta\mathbf{A})\mathbf{x} + (\mathbf{b}_0 + \Delta\mathbf{b})u + \mathbf{e}d \\ &= \mathbf{A}_0\mathbf{x} + \mathbf{b}_0 u + (\Delta\mathbf{A}\mathbf{x} + \Delta\mathbf{b}u + \mathbf{e}d) \end{aligned} \tag{33.4}$$

The third term is the sum of the disturbance and the parameter variation effect. It is possible to define a scalar function as an equivalent disturbance.

$$\tilde{d} = d + \mathbf{e}^t(\Delta\mathbf{A}\mathbf{x} + \Delta\mathbf{b}u) \tag{33.5}$$

$\tilde{d}$, termed equivalent disturbance, includes not only the unknown disturbance but the unknown parameter variations. By substituting Equation 33.5 into Equation 33.1, Equation 33.6 holds.

$$\begin{aligned} \dot{\mathbf{x}} &= \mathbf{A}_0\mathbf{x} + \mathbf{b}_0 u + \mathbf{e}\tilde{d} \\ y &= \mathbf{c}\mathbf{x} \end{aligned} \tag{33.6}$$

Since $\tilde{d}$ is a function of time, it is expanded into power series of time. If $\tilde{d}$ is slower compared to system dynamics, it is approximated in the following form.

$$\frac{\mathrm{d}^{(p)}\tilde{d}}{\mathrm{d}t^p} = 0 \tag{33.7}$$

Equation 33.7 is easily combined to Equation 33.6. The results are

$$\begin{aligned} \dot{\tilde{\mathbf{x}}} &= \tilde{\mathbf{A}}_0\tilde{\mathbf{x}} + \tilde{\mathbf{b}}_0 u \\ y &= \tilde{\mathbf{c}}_0\tilde{\mathbf{x}} \end{aligned} \tag{33.8}$$

Here $\tilde{\mathbf{x}}$ is an augmented state vector in the form of the following.

$$\tilde{\mathbf{x}} = \begin{bmatrix} x_1 \\ x_2 \\ \vdots \\ x_u \\ \tilde{d} \\ \dot{\tilde{d}} \\ \ddot{\tilde{d}} \\ \vdots \\ \tilde{d}^{(p-1)} \end{bmatrix}$$

$\tilde{\mathbf{A}}_0$ is an augmented system matrix in the form of the following.

$$\tilde{\mathbf{A}}_0 = \overbrace{\begin{bmatrix} 0 & 1 & 0 & \cdots & 0 & 0 & 0 & \cdots & 0 \\ 0 & 0 & 1 & \cdots & 0 & 0 & 0 & \vdots & 0 \\ \vdots & \vdots & \vdots & \ddots & \vdots & \vdots & 0 & \vdots & 0 \\ 0 & 0 & 0 & \cdots & 1 & 0 & 0 & \vdots & 0 \\ -a_{01} & -a_{02} & -a_{03} & \cdots & -a_{0n} & 1 & 0 & \vdots & 0 \\ 0 & 0 & 0 & \cdots & 0 & 0 & 0 & \vdots & 0 \\ \vdots & \vdots & \vdots & \vdots & \vdots & \vdots & \vdots & \vdots & \vdots \\ 0 & 0 & 0 & \cdots & 0 & 0 & 0 & \cdots & 0 \end{bmatrix}}^{n+p}$$

$\tilde{\mathbf{b}}_0$ is an augmented distribution vector in the form of the following.

$$\tilde{\mathbf{b}}_0 = \left.\begin{bmatrix} 0 \\ 0 \\ \vdots \\ 0 \\ K_0 \\ 0 \\ \vdots \\ 0 \end{bmatrix}\right\} n+p$$

$\tilde{\mathbf{c}}_0$ is an augmented observation vector in the form of the following.

$$\tilde{\mathbf{c}}_0 = \overbrace{[c_1 \quad c_2 \quad \cdots \quad c_m \quad 0 \quad \cdots \quad 0]}^{n+p}$$

In Equation 33.8, an equivalent disturbance is treated as if it were a state variable. This is the key point in the design process. Equation 33.8 is the same to Equation 33.1. However, Equation 33.8 does not seem to have any disturbance, nor any parameter variations. The difference is the size of dimension. Clearly the controllability is lost; fortunately, however, the observability is preserved. It is possible to construct an observer which estimates an equivalent disturbance $\tilde{d}$. Such an observer is called a disturbance observer. Once the equivalent disturbance is estimated or identified, it is possible to synthesize an input u to include a signal to cancel the equivalent disturbance. This is the principle of the disturbance observation-cancellation technique. Sometimes this technique is called "zeroing" or "cancellation." The details of the above explanation are developed one by one in the following.

33.3 Higher-Order Disturbance Approximation

To regard disturbance as a state variable, the disturbance should have certain dynamics. From an analysis in the previous chapter, an equivalent disturbance is a function of time.

Since we do not know $\tilde{d}$ *a priori*, we should estimate it as closely as possible. From the point of digital control, we need $\tilde{d}$ only a few steps ahead of every control sampling time. This means that we will apply an approximation by a low-order polynomial of time to $\tilde{d}$ in every sampling time. For example, if $\tilde{d}$ is approximated by piecewise rectangular lines, the derivatives are zero almost everywhere. This is the case of $p = 1$ in Equation 33.7. Similarly, if the function is approximated by piecewise straight lines, the second derivatives are zero. By increasing the fitness of the function by a $(p - 1)$-order polynomial, we will get Equation 33.7, which means the equivalent disturbance of $(p - 1)$ steps ahead is estimated. From a practical standpoint, p less than 3 gives a good enough approximation. In the case of Equation 33.7, the augmented states and assorted matrix are corresponding to Equation 33.8.

33.4 Disturbance Observation

Since the linear system with additive disturbance is represented in the form of Equation 33.8, it is possible to construct the (reduced-order) observer which estimates $\tilde{d}$ whose p-th derivative is zero. Using Equation 33.8, various observers can be constructed. A reduced-order observer is designed by Gopinath's method whose order is $n + p - m$. A full-order observer whose order is $n + p$ is also applicable. There are several design procedures for the design of the above observer. Most of them, including Gopinath's method, are found in other sections of the book or in the references. An example applied to motion system is shown later. Please note that the dynamics of any observer are specified arbitrarily; good results are obtained by careful thought of pole allocation of the observer.

33.5 Disturbance Cancellation

Once an equivalent disturbance is estimated, the input will be designed to have two parts.

$$u = u^{ref} + u^{dis} \tag{33.9}$$

The first term is a driving input for a nominal plant which has only nominal parameters without disturbance. The second term is a compensation input to regard the original plant as a nominal plant without disturbance. The second term is synthesized so that the equivalent disturbance is cancelled by the feedback of the estimated equivalent disturbance $\tilde{d}$. Clearly u^{dis} is determined by the following equation.

$$\begin{aligned} u^{dis} &= -(\mathbf{b}_0^t\mathbf{b}_0)^{-1}\mathbf{b}_0^t\mathbf{e}\hat{\tilde{d}} \\ &= -\frac{1}{K_0}\hat{\tilde{d}} \end{aligned} \tag{33.10}$$

Equation 33.10 cancels out the real unknown equivalent disturbance. Since $\hat{\tilde{d}}$ is estimated with lag elements inside, the difference between the real value and the estimated value of the equivalent disturbance will converge to *zero* in steady state. The convergence velocity depends on the identification process, i.e., the poles of the disturbance observer. Equation 33.10 is the direct result of this section.

33.6 Examples of Application

Disturbance observation and cancellation techniques were realized originally in motion control systems. The mechanical system driven by the dc motor has the following dynamic equation.

$$J\frac{d\omega}{dt} = K_t I_a - T_l \tag{33.11}$$

Here,

J : inertia about motor shaft

K_t : torque constant

T_l : load torque

Suppose only the position of the motor shaft is detected by the rotary encoder. Then the output is written in the following form.

$$y = \theta = \int \omega dt \tag{33.12}$$

The companion form with the equivalent disturbance, shown in Equation 33.8, has the following elements.

$$\tilde{d} = -\frac{T_l}{J} + \left(\frac{K_t}{J} - \frac{K_{t0}}{J_0}\right) I_a \tag{33.13}$$

$$\mathbf{A} = \begin{bmatrix} 0 & 1 \\ 0 & 0 \end{bmatrix}$$

$$\mathbf{b} = \begin{bmatrix} 0 \\ \frac{K_t}{J} \end{bmatrix}$$

$$\mathbf{c} = [1 \quad 0]$$

Suppose the disturbance will be sufficiently slow in one sampling time. It is possible to assume the following equation according to the previous consideration.

$$\frac{d\tilde{d}}{dt} = 0 \tag{33.14}$$

The augmented equation is

$$\frac{d}{dt}\begin{bmatrix} \theta \\ \omega \\ \tilde{d} \end{bmatrix} = \begin{bmatrix} 0 & 1 & 0 \\ 0 & 0 & 1 \\ 0 & 0 & 0 \end{bmatrix}\begin{bmatrix} \theta \\ \omega \\ \tilde{d} \end{bmatrix} + \begin{bmatrix} 0 \\ \frac{K_{t0}}{J_0} \\ 0 \end{bmatrix} I_a \tag{33.15}$$

A disturbance observer which estimates $\tilde{d}$ is derived by Gopinath's algorithm as follows.

$$\hat{\tilde{d}} = k_1\theta + z_1$$

here, z_1 satisfies:

$$\frac{d}{dt}\begin{bmatrix} z_1 \\ z_2 \end{bmatrix} - \begin{bmatrix} 0 & -k_1 \\ 1 & -k_2 \end{bmatrix}\begin{bmatrix} z_1 \\ z_2 \end{bmatrix} + \begin{bmatrix} -k_1 k_2 \theta \\ (k_1 - k_2^2)\theta + \frac{K_{t0}}{J_0} I_a \end{bmatrix} \tag{33.16}$$

Here the two poles of the observer α and β, which are arbitrarily allocated in the complex plane, satisfy the following equation.

$$\alpha + \beta = -k_2 \tag{33.17}$$
$$\alpha\beta = k_1$$

Modifying Equation 33.16, we get

$$\hat{\tilde{d}} = \frac{k_1}{s^2 + k_2 s + k_1}\left(s^2\theta - \frac{K_{t0}}{J_0} I_a\right) \tag{33.18}$$
$$= \frac{k_1}{s^2 + k_2 s + k_1}\tilde{d}$$

$\hat{\tilde{d}}$ is estimated through a second-order lag system whose poles are α and β. The physically extended load disturbance T_{dis} is defined as

$$T_{dis} = -T_l - \Delta J\frac{d\omega}{dt} + \Delta K_t I_a \tag{33.19}$$

After some manipulations, we get

$$T_{dis} = J_0\tilde{d} \tag{33.20}$$

Equation 33.19 means that the equivalent disturbance substantially has the terms of the following three terms respectively.

- mechanical load ($-T_l$)
- varied self-inertia torque ($-\Delta J d\omega/dt$)
- torque pulsation generated by the motor ($\Delta K_t I_a$)

Instead of $\tilde{d}$, it is possible to use T_{dis} in the motion system. One example of the disturbance cancellation technique realized in the motion system is shown in Figure 33.1. Here the tachometer

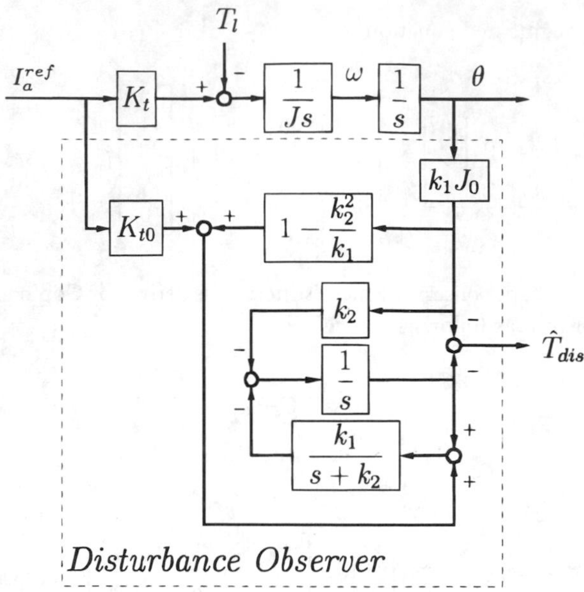

Figure 33.1 A disturbance observer realization in motion system.

is used instead of encoder. The direct solution of disturbance cancellation in the case of using a rotary encoder is left to be solved by readers. By comparing Fig. 33.1 with the direct solution, the readers will appreciate the physical meaning of the disturbance cancellation in the motion system.

33.7 Conclusions

The disturbance observation and cancellation technique is now widely used in industrial controllers including position servo controller, video head controller, UPS, industrial robots, pneumatic servo controller, automobile active controller and so on. The main reason is that higher performance is achieved by a simple and inexpensive controller. For example, the inverse dynamics in the controller of a robotic manipulator will be saved by using Figure 33.1 at every joint. It is possible to extend this technique to other industrial plants. In most cases, the performance of robustness and linearization are expectedly superior to other controllers; however, it is necessary to use a fast CPU as a controller to realize a short sampling time.

References

K. Ohnishi et al. 1982. Torque Control of DC Motor Observer, *Proc. of IEEJ Technical Meeting on Rotating Machinery*, vol. RM-82–83.

K. Ohnishi, 1987. New Development of Servo Technology in Mechatronics, *Trans. IEEJ*, vol. 107-D, No. 1, 83–86.

K. Ohnishi et al. 1996. Motion Control for Advanced Mechatronics, *IEEE/ASME Trans on Mechatronics*, vol. 1-1, 56–67.

34

Phase-Locked Loop-Based Control

Guan-Chyun Hsieh
National Taiwan Institute of Technology

34.1 Introduction ... 529
Basic Concept of PLL • Phase Detector • Voltage-Controlled Oscillator • Loop Filter and Other Subsystems
34.2 Configurations of PLL Applications ... 532
34.3 Analog, Digital, and Hybrid PLLs ... 533
34.4 Popular PLL Integrated Circuits (ICs) ... 533

34.1 Introduction

An early description of phase-locked loop (PLL) appeared in papers by Appleton (1923) and de Bellescize (1932). The advent of PLL has contributed to coherent communication systems without the Doppler shift effect. In the late 1970s, the theoretical description of PLL was well established (Blanchard, 1976; Garder, 1979), but PLL did not achieve widespread use until much later, because of the difficulty in realization. With the rapid development of integrated circuits in the 1970s, applications of PLL were widely used in modern communication systems. Since then, PLL has progressed significantly and has turned its earlier professional use in high-precision apparatuses to its current use in consumer electronics products. It has enabled modern electronic systems to improve performance and reliability, especially in common electronic appliances used daily.

In the 1970s, researchers in the control field first turned their attention to PLL for synchronous motors (Volpe, 1970). Since then, a phase-locked servo system (PLS) was rapidly developed for AC and DC motors servomechanisms, with the analog PLL ICs (More, 1973; Tal, 1977) Over the past ten years, rapidly developed high-performance digital ICs and microprocessors have resulted with strong motivation for PLSs implementation in the digital domain. New types of controllers for increasing PLS features were then developed to accomplish an easy-use and easy-control strategy for AC and DC servo drives (Margaris and Petridis 1985; Hsieh et al, 1987; Hsieh, 1989; Li and Hsieh, 1992).

Basic Concept of PLL

A PLL is a device which causes a system to track with another one. It keeps an output signal synchronized in frequency and phase by using a reference input signal. More precisely, the PLL is simply a servo system, which controls the phase of its output signal in such a way that the phase error between the output phase and the reference phase is reduced to a minimum. The functional block diagram of a PLL, shown in Figure 34.1, consists of a phase detector (PD), a loop filter (LF), and a voltage-controlled oscillator (VCO). We presume that x_i and x_o are respectively the input and the VCO signals, which can be expressed as (Blanchard, 1976; Garder, 1979)

$$x_i(t) = A\cos(\omega_i t + \theta_i) \tag{34.1}$$

$$x_o(t) = B\cos(\omega_o t + \varphi_o) \tag{34.2}$$

The angular frequency of the input signal is ω_i and ω_o is the VCO central angular frequency. θ_i and φ_o are phase constants. If the loop is initially unlocked and the phase detector has a sinusoidal characteristic, the significant output signal $v_e(t)$ at the PD is given by

$$v_e(t) = K_d \cos[(\omega_i - \omega_o)t + \theta_i - \varphi_o] \tag{34.3}$$

where K_d is the gain of the PD and the higher-frequency item $\omega_i + \omega_o$ is negligible here due to the rejection of the LF. After a period of time sufficiently long for transient phenomena, it will be observed that the VCO output signal x_o has become synchronous with the input signal x_i. Signal x_o can then be expressed as

$$x_o(t) = B\cos(\omega_i + \phi_o) \tag{34.4}$$

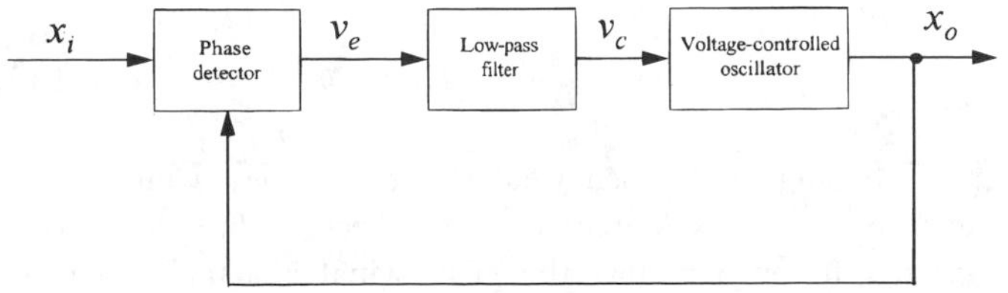

Figure 34.1 Basic topology of the phase-locked loop.

0-8493-8343-9/97/$0.00+$.50

From Equations 34.1 and 34.4, the quantity φ_o in Equation 34.2 becomes a linear function of time expressed as

$$\varphi_o = (\omega_i - \omega_o)t + \phi_o \quad (34.5)$$

and the PD output signal $v_e(t)$ in Equation 34.3 becomes a DC signal given by

$$v_e(t) = K_d \cos(\theta_i - \phi_o) \quad (34.6)$$

The LF is of the low-pass type so that the controlled signal $v_c(t)$ is given by

$$v_c(t) = v_e(t) = K_d \cos(\theta_i - \phi_o) \quad (34.7)$$

The VCO is a frequency-modulated oscillator, whose instantaneous angular frequency ω_{inst} is a linear function of the controlled signal $v_c(t)$, around the central angular frequency ω_o, i.e.,

$$\begin{aligned}\omega_{inst} &= \frac{d}{dt}(\omega_o t + \varphi_o)\\ &= \omega_o + K_v v_c(t) \quad (34.8)\end{aligned}$$

and

$$\frac{d\varphi_o}{dt} = K_v v_c(t) \quad (34.9)$$

where K_v is the VCO sensitivity. From Equations 34.5, 34.6, and 34.9, we have

$$\omega_i - \omega_o = K_d K_v \cos(\theta_i - \phi_o) \quad (34.10)$$

from which we have

$$\phi_o = \theta_i - \cos^{-1}\frac{\omega_i - \omega_o}{K_d k_v} \quad (34.11)$$

Substituting Equation 34.11 into Equation 34.6, we obtain

$$v_e = \frac{\omega_i - \omega_o}{K_v} \quad (34.12)$$

Equation 34.12 clearly shows that it is the DC signal $v_c = v_e$ that changes the VCO frequency from its central value ω_o to the input signal angular frequency ω_i, i.e.,

$$\omega_{inst} = \omega_o + K_v v_c = \omega_i \quad (34.13)$$

If the angular frequency difference $\omega_i - \omega_o$ is much lower than the product $K_d K_v$, Equation 34.11 becomes $\theta_i - \phi_o \approx \cos^{-1}0 = \pi/2$. It indicates that the VCO signal is actually in phase quadrature with the input signal while the loop is in lock. Strictly speaking, the phase quadrature actually corresponds to $\omega_i = \omega_o$. For this reason, we substitute the phase constant ϕ_o for the constant θ_o, so that $\theta_o = \phi_o - \pi/2$. Then

$$\begin{aligned}v_e &= K_d \cos(\theta_i - \phi_o)\\ &= K_d \sin(\theta_i - \theta_o) \quad (34.14)\end{aligned}$$

The difference $\theta_i - \theta_o$ is the so called phase error between the two signals, which is null when the initial frequency offset is null. When the difference $\theta_i - \theta_d$ is sufficiently small, the following approximation is used:

$$v_e \approx K_d(\theta_i - \theta_o) \quad (34.15)$$

Another interpretation for the signals in Equations 34.1 and 34.2 can also be represented by

$$x_i(t) = A \sin(\omega_i t + \theta_i) \quad (34.16)$$

$$x_o(t) = B \cos(\omega_o t + \varphi_o) \quad (34.17)$$

The phase detector output can be represented by $v_e(t) = K_d \sin[(\omega_i - \omega_o)t + \theta_i - \varphi_o]$ when the loop is out of lock and the $v_e = K_d \sin(\theta_i - \theta_o)$ when it is working, with

$$\theta_i - \theta_o = \sin^{-1}\frac{\omega_i - \omega_o}{K_d K_v} \quad (34.18)$$

The product $K = K_d K_v$ is referred to as the loop gain. When the difference $|\omega_i - \omega_o|$ exceeds the loop gain K in a sinusoidal-characteristic PD, a proper θ_o for lock can no longer be found by means of Equation 34.18. The synchronization no longer maintains and the loop falls out of lock.

Phase Detector

The phase detector (PD) in PLL can be described by two categories, sinusoidal and square signal phase detectors. The sinusoidal PD inherently has phase-detected interval $(-\pi/2, +\pi/2)$. It operates as a multiplier, which is a zero memory device. The square signal PDs are implemented by sequential logic circuits. Sequential PDs contain memory of past crossing events. They can generate PD characteristics that are difficult or impossible to obtain with multiplier circuits. Sequential PDs are usually built up from digital circuits and operate with binary, rectangular input waveforms. Accordingly, they are often called digital phase detectors. The characteristics of the square signal PDs are of the linear type over the phase-detected interval $(-\pi/2, +\pi/2)$ for triangular PD, $(-\pi, +\pi)$ for sawtooth PD, and $(-2\pi, +2\pi)$ for sequential phase/frequency detector (PFD). Their characteristics are depicted in Figure 34.2 (Garder, 1979). All curves of

Figure 34.2 are shown with the same slope at phase error $\theta_e = \theta_i - \theta_o = 0$, which means that the different PDs all have the same factor K_d. Increased PD output capability provides a larger tracking range, larger lock limit, than those are obtainable from a sinusoidal PD.

Voltage-Controlled Oscillator

The voltage-controlled oscillators (VCO) used in the PLL is similar to those used for other applications, such as modulation and automatic frequency control. The main requirements for the VCO are phase stability, large frequency deviation, high modulation sensitivity K_v, linearity of frequency versus control voltage, and capability to accept wideband modulation. The phase stability is in direct opposition to all other four requirements. Four types of VCO commonly used are given in order of decreasing stability: crystal oscillators (VCXO), resonator oscillators, RC multivibrator, and YIG-tuned oscillators (Blanchard, 1970; Garder, 1979). The phase stability can be enhanced by a number of factors: high Q in the crystal and circuit, low noise in the amplifier portion, temperature stability, and mechanical stability. Remarkably, much of the phase jitter of an oscillator arises from noise in the associated amplifier. If a wider frequency range is required, an LC oscillator must be used. In this application, the standard Hartley, Colpitts, and Clapp circuits appear. Tuning is accomplished by a varactor. At microwave frequencies, YIG-tuned Gunn oscillators have become popular. Tuning of the YIG-tuned oscillator is accomplished by altering a magnetic field.

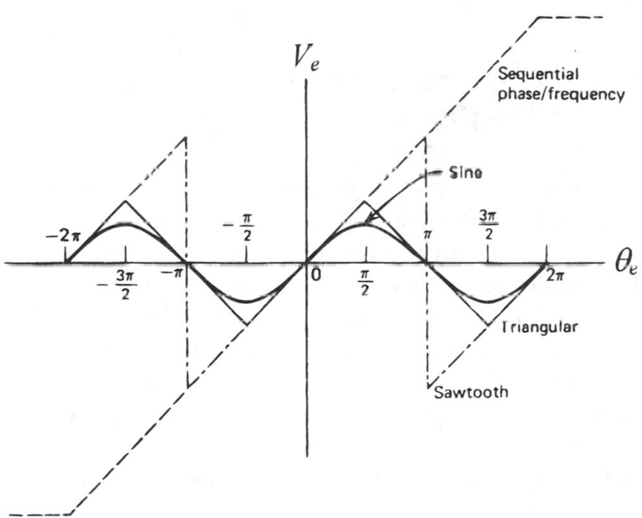

Figure 34.2 Characteristics of the phase detector (taken from Gardner, F. M. 1979 *Phaselock Technique*, 2nd ed. Wiley, New York). *Source:* Moore, A. W. 1973. Phase-locked loops for motor-speed control, *IEEE Spectrum,* IECI–24(2):118–125.

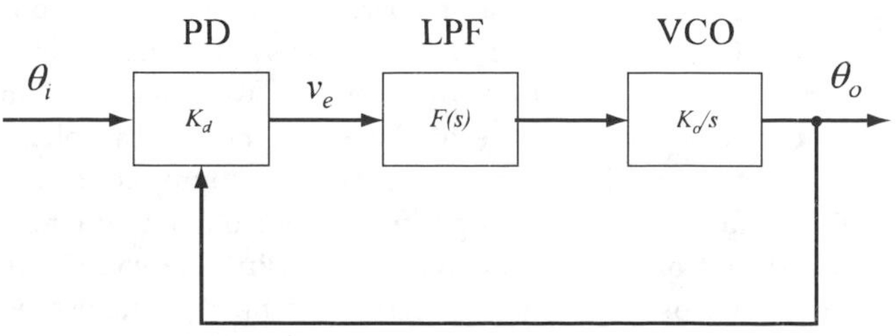

Figure 34.3 Linear model of the phase-locked loop.

Loop Filter and Other Subsystems

The loop filter (LF) in PLL is of a low-pass type. It is used to suppress noise and high-frequency signal components from phase error θ_e and provide a DC-controlled signal for the VCO. We assume that the loop is in lock, that the PD is linear and the PD output voltage is proportional to phase error

$$v_d \approx K_d(\theta_i - \theta_o)$$

The phase error voltage v_e is filtered by the loop filter. The servo scheme of the PLL in linear locking state is shown in Figure 34.3, where $F(s)$ is the loop transfer function. According to servo terminology, the type of the loop is determined by the number of perfect integrators within the loop. Any PLL is at least a type one loop because of the perfect integrator inherently in the VCO. If the loop filter contains one perfect integrator, then the loop is type two. A second-order PLL with a high-gain active filter can be approximately a type two loop, whereas a PLL with a passive filter is type one. The widely used passive and active filters for the PLL are shown in Figure 34.4. For the passive filter, the closed-loop transfer function $H'(s)$ of the PLL is

$$H'(s) = \frac{K_o K_d (s\tau_2 + 1)/\tau_1}{s^2 + s(1 + K_o K_d \tau_2/\tau_1) + K_o K_d/\tau_1} \tag{34.19}$$

For the active filter, the closed-loop transfer function $H''(s)$ is found to be

$$H''(s) = \frac{K_o K_d (s\tau_2 + 1)/\tau_1}{s^2 + s(K_o K_d \tau_2/\tau_1) + K_o K_d/\tau_1} \tag{34.20}$$

For convenience in description, these transfer functions Equations 34.19 and 34.20 may be represented as

$$H'(s) = \frac{s(2\zeta\omega_n - \omega_n^2/K_oK_d) + \omega_n^2}{s^2 + 2\zeta\omega_n s + \omega_n^2} \tag{34.21}$$

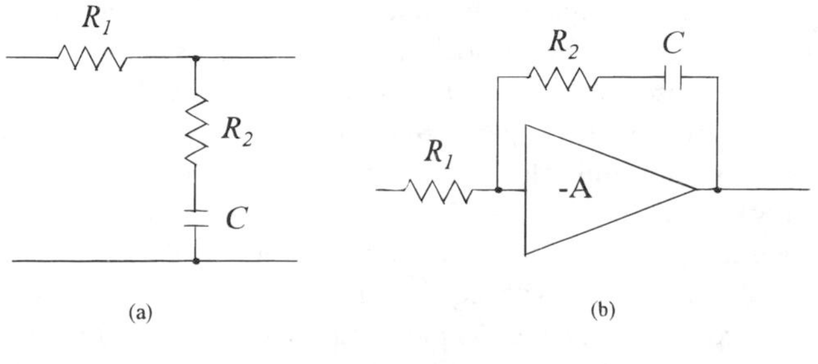

Figure 34.4 (a) Passive filter and (b) active filter.

and

$$H'(s) = \frac{2\zeta\omega_n s + \omega_n^2}{s^2 + 2\zeta\omega_n s + \omega_n^2} \tag{34.22}$$

in which ω_n is the natural frequency of the loop and ζ is the damping ratio. The relevant parameters for a passive filter are

$$\omega_n = \left(\frac{K_v K_d}{\tau_1}\right)^{1/2}, \zeta = \frac{1}{2}\left(\frac{K_v K_d}{\tau_1}\right)^{1/2}\left(\tau_2 + \frac{1}{K_v K_d}\right),$$

$$\tau_1 = (R_1 + R_2)C, \text{ and } \tau_2 = R_2 C$$

and for an active filter are

$$\omega_n = \left(\frac{K_v K_d}{\tau_1}\right)^{1/2}, \zeta = \frac{\tau_2}{2}\left(\frac{K_v K_d}{\tau_1}\right)^{1/2} = \frac{\tau_2 \omega_n}{2},$$

$$\tau_1 = R_1 C, \text{ and } \tau_2 = R_2 C$$

The two transfer functions are nearly the same if $1/K_v K_d << \tau_2$ in the passive filter. The open-loop transfer function of any PLL is given by

$$G(s) = \frac{K_v K_d F(s)}{s} \tag{34.23}$$

the closed-loop transfer function can be given by

$$H(s) = \frac{G(s)}{1 + G(s)} \tag{34.24}$$

We define the DC gain of the loop as

$$K_D = K_v K_d F(0) \tag{34.25}$$

A large value of K_D is usually required for achieving a good performance of the loop (Garder, 1979). We define a hold-in range of a loop as $\Delta\omega_H = K_D$. If the input frequency closes sufficiently to VCO frequency, a PLL locks up with just a phase transient; there is no cycle slipping prior to lock. The frequency range over which the loop acquires phase to lock without slips is called the lock-in range of the PLL. In a first-order loop, the lock-in range is equal to the hold-in range; but for the second- or higher-order loops, the lock-in range is still less than the hold-in range. Besides, there is a frequency interval, smaller than the hold-in interval and larger than the lock-in interval, over which the loop will acquire lock after slipping cycles for a while. This interval is called the pull-in range. Their relations are indicated in Figure 34.5. To ensure stable tracking, it is common practice to build loop filters with equal numbers of poles and zeros. At high frequencies the loop is indistinguishable from a first-order loop with gain $K = K_o K_d F(\infty)$. As a fair approximation, we can say that the higher-order loop has the same lock-in range as the

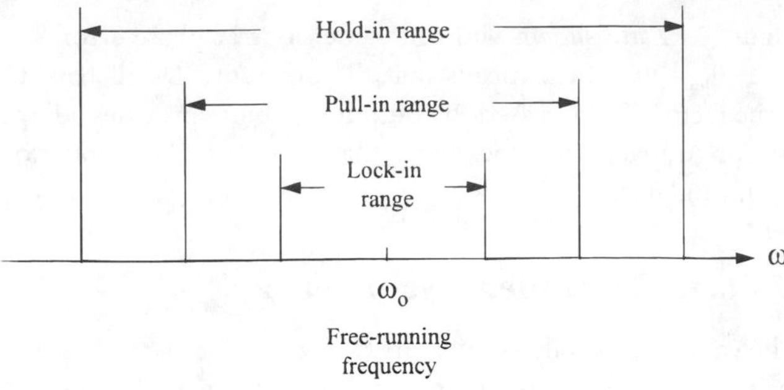

Figure 34.5 Scope of the dynamic limits of a PLL.

equivalent-gain, first-order loop. The lock-in limit of a first-order loop is equal to the loop gain. We argue here that a higher-order loop has nearly the same lock limit. The lock-in range $\Delta\omega_L$ can be approximately estimated as

$$\Delta\omega_L \approx \pm K_d K_v F(\infty) \tag{34.26}$$

Acquisition of frequency in PLL is more difficult, slower, and requires more design attention than does phase acquisition. Remarkably, the self-acquisition of frequency is known as frequency pull-in, or simply pull-in; the self-acquisition of phase as phase lock-in, or lock-in.

34.2 Configurations of PLL Applications

Integrated phase-locked loops developed since 1970s are versatile systems, which are suitable for use in a variety of frequency selective demodulation, signal conditioning or frequency synthesis applications. Standard configurations of PLL in communications are well developed and widely used for FM, AM, video, signal processing, commercial apparatus, control systems, telecommunication systems, etc. The use of PLL with analog PLL IC (NE565) developed by Signetics for a synchronous motor and for a DC motor began in the 1970s (Volpe, 1970; Moore, 1973). A phase-locked servo system (PLS) is certainly a frequency feedback control configuration that continually maintains motor speed or motor position by tracking the phase and frequency of the incoming reference signal which corresponds to the input command, such as speed or position. Basically, the PLS configuration is a combination of a phase detector, a loop filter, and a servo motor, as shown in Figure 34.6, in which the servo motor operates like the voltage-controlled oscillator (VCO) in PLL as shown in Figure 34.1. Between 1985–1994, a variety of controllers were developed to raise the servo performance of the PLS. They are a voltage pump controller (VPC) and an adaptive digital pump controller (ADPC) for DC motor speed servo control, a frequency-pumped controller (FPC) for DC motor position servo control, a phase-controlled oscillator (PCO) for stepping motor speed servo control, and a microcomputer-based variable slope pulse pump controller (VSPPC) for stepping motor position control

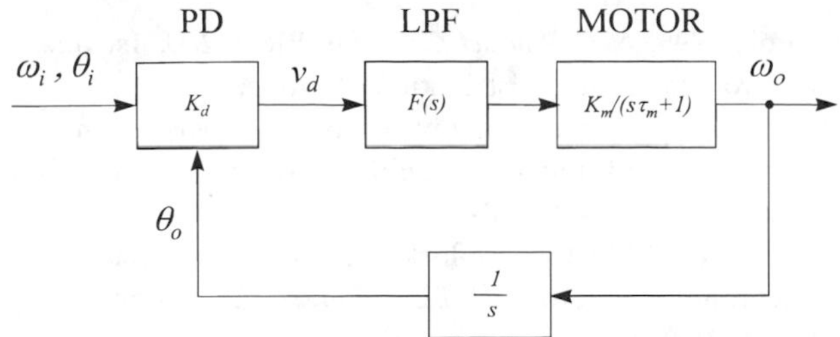

Figure 34.6 Linear model of the phase-locked servo loop.

(Margaris and Petridis 1985; Hsieh et al, 1987; Hsieh, 1989; Li and Hsieh, 1992). The future trend is to develop an integrated PLL as a modulus device including the phase detector, prescaler, and programmable counter; even including the VCO (Motorola, 1993).

34.3 Analog, Digital, and Hybrid PLLs

Due to rapid progress in digital ICs in the 1970s, the integrated PLL has been developed for filtering, frequency synthesis, motor speed control, frequency modulation, demodulation, signal detection, and a variety of other applications. PLL ICs have become important communication circuits. PLLs can be analog or digital, but most of them are composed of both analog and digital components. The first analog PLL ICs (NE565 and CD4046) were developed by Signetics and RCA in the 1970s. They are integrated on a chip including phase detector, loop filter and VCO. The phase detector is a multiplier, whose phase-lock range is from $-\pi/2$ to $+\pi/2$. To extend the lock-in range of PLL, the Motorola digital phase/frequency comparator (PFD), MC4044, capable of phase-detection range from -2π to $+2\pi$ was proposed in 1972 (Moore, 1973). A hybrid PLL combining discrete analog and digital components was then developed. The versatile PLL's components for raising the locking performance were developed in succession. Now, it is a future-exploited trend to combine phase detector, prescaler, programmable counter, and VCO into a modular device to be a digital PLL. The difficulty in realization is how to constitute a VCO on a chip suitable for operating at higher frequencies, above 2.5 GHz, by semiconductor technology (Motorola, 1993).

34.4 Popular PLL Integrated Circuits (ICs)

A variety of ICs for PLL are available from semiconductor manufacturers. The techniques widely used to implement the PLL are TTL, CMOS, and ECL. Today, fully integrated PLL on a single chip can operate at frequencies of up to 35 MHz (e.g., Exar XR-215 PLL). Other higher-frequency PLLs are easily achieved by combining sub-PLL ICs (including only phase detector, prescaler, and programmable counter) and discrete higher-frequency VCO. It is an important trend to realize fully integrated higher-frequency PLL including VCO into a modulus device in the near future. Motorola and Plessey are developing versatile PLL ICs operating at frequencies of over 2.5 GHz. The Motorola MC 12210 is a 2.5-GHz bipolar monolithic series input phase-locked loop synthesizer with phase-swallow function. It is designed as a high-frequency local oscillator of an RF transceiver in handheld communication applications. This PLL can operate at a minimum supply voltage of 2.7 V for input frequencies up to 2.5 GHz with a typical current drain of 9.5 mA. A dual modulus prescaler is integrated to provide either a 32/33 or 64/65 divide ratio (Motorola, 1994). New Motorola VCOs, MC12147 and MC12149, operating frequency up to 1.1 GHz were also defined in 1995 (Motorola, 1995). The GEC Plessey SP5070 and SP5655 are, respectively, 2.4-GHz and 2.7-GHz single-modulus frequency synthesizers for use in satellite TV receivers, high-IF cable tuning systems, and C-band with frequency doubling mixer. Both PLL synthesizers operate at a low power consumption (5 V and 30 mA). The SP5655, capable of standard I^2C BUS control format, contains two addressable current-limited outputs and four addressable bidirectional open collector ports one of which is a 3-bit ADC. The information on these ports can be read by the I^2C BUS. The device has a sub-device programmed by applying a specific input voltage to one of the current-limited outputs. This enables two or more synthesizers to be used in a system (Plessey, 1994).

References

Appleton, E. V. 1922–1923. Automatic synchronization of triode oscillators, *Proc. Cambridge Phil. Soc.*, 21 III:231.

de Bellescize, H. 1932. La Reception Synchrone, *Onde Electr.*, 11:230–240.

Blanchard, A. 1976. *Phase-locked loops: application to coherent receiver design*, John Wiley & Sons, New York, NY.

Garder, F. M. 1979. *Phaselock Techniques*, 2d ed., Wiley, New York, NY.

GEC Plessey, *Consumer IC Handbook*, GEC Plessey Semiconductors.

Geiger, D. F. 1981. *Phase Lock Loops for DC Motor Speed Control*, John Wiley & Sons, New York, NY.

Hsieh, G. C. 1989. A study on position servo control systems by frequency-locked technique, *IEEE Trans. Ind. Electron.*, vol IE-36, pp. 365–373, Aug.

Hsieh, G. C., Wu, Y. P., Lee, C. H., and Liu C. H. 1987. an adaptive digital pump controller for phase locked servo systems, *IEEE Trans. Ind. Electron.*, vol. IE-34, pp. 379–386, Aug.

Lai, M. F., Hsieh, G. C., and Wu, Y. P. 1995. Variable slope pulse pump controller for stepping position servo control using frequency-locked technique, *IEEE Trans. Ind. Electron.*, June.

Li, J. C. and Hsieh, G. C. 1992. A phase/frequency-locked controller for stepping servo systems, *IEEE Trans. Ind. Electron.*, 39(2) 379–386, April.

Lindsey, W. C. and Chie, C. M. 1981. A survey of digital phase-locked loops, *Proc. IEEE*, 69(4):410–431.

Margaris, N. and Petridis, V. 1985. Voltage pump phase-locked loops, *IEEE Trans. Ind. Electron.*, vol. IE-32, pp. 41–49, Feb.

Moore, A. W. 1973. Phase-locked loops for motor-speed control, *IEEE Spectrum,* (4):61–67.

Motorola 1993. *Communications Device Data,* REV 1.1, Motorola Literature Distribution, Arizona.

Motorola 1994. *Communications Device Data,* REV 1.1, Motorola Literature Distribution, Arizona.

Motorola 1995. *New Product Calendar,* BR1332/D, 1st quarter, Motorola Literature Distribution, Arizona.

Tal, J. 1977. Speed control by phase locked servo system—new possibilities and limitations, *IEEE Trans. Ind. Electron. Contr. Instrum.*, IECI–24(2):118–125.

Volpe, G. T. 1970. A phase-locked loop control system for a synchronous motor, *IEEE Trans. Automatic Contr.*, AC–15(2):88–95.

35

Variable Structure Control Technique

Vadim Utkin
Ohio State University

35.1 Introduction 535
35.2 Mathematical Aspects 536
Sliding Modes in Control Systems • Regularization • Equivalent Control Method • Sliding Mode Existence Conditions
35.3 Sliding Mode Control Design 538
Design Procedure • Regular Form • Control in Linear Systems
35.4 Chattering Problem 540
35.5 Control of Manipulators 540
35.6 Control of Mobile Robots 541
35.7 Control of Railway Wheelset 541
35.8 Control of Torsion Oscillations of a Flexible Shaft 542
35.9 DC Motors 542
35.10 Control of DC Motors Based on a Reduced-Order Model 543
35.11 Conclusion 544

35.1 Introduction

Variable structure control systems consist of a set of subsystems or structures with continuous controls and are supplied with an appropriate switching logic. Due to varying the system structure in the course of control process, the control actions are discontinuous functions of the system state and inputs (disturbances and reference inputs). Mathematical, design, and application aspects of the systems with discontinuous control actions constitute the scope of this paper. First of all let us present the major arguments showing why the class of discontinuous control systems provides such an efficient tool for solving the entire class of control problems for complex dynamic plants.

If discontinuities in control are deliberately introduced on some surfaces in the system state-space then motions in a sliding mode may occur. This motion features several attractive properties and will be the principle operation mode in the systems under study.

The state trajectories in the sliding mode belong to the intersections of the surfaces where control components undergo discontinuities. Since the trajectories are in the manifold of lower dimension than that of the original system, the sliding mode equation will be of a reduced order as well. In most practical systems, sliding motion does not depend on control and is determined merely by the plant dynamics and the position (or equations) of the discontinuity surfaces. This enables decoupling of the initial control design problem into independent lower dimension subproblems, wherein the control is designed to enforce sliding mode while the required dynamics of the motion in the sliding manifold is designed by an appropriate choice of its equation. As a result, the control design may be simplified significantly when controlling plant described by high-order differential equations which may be serious obstacles for applications of efficient analytical techniques and computational methods.

A specific feature of a sliding mode is that, under certain conditions, it may become insensitive to variations of control plant dynamics and disturbances. It is essential that, unlike continuous systems with non-measured disturbances in which the insensitivity condition implies use of infinite gains, the same effect is attained by finite control actions in discontinuous systems.

Finally, the technological aspect of using discontinuous control systems should be mentioned. To improve performance, electric inertialess actuators are widely used now. They are built around power electronic elements which may operate in a switching mode only. Therefore, even as we employ continuous control algorithms, the control itself is shaped as a high-frequency discontinuous signal with a mean value equal to the desired continuous control. A more natural way is to employ control algorithms oriented towards the use of discontinuous controls, which is the case for sliding mode control.

0-8493-8343-9/97/$0.00+$.50

35.2 Mathematical Aspects

A number of processes in mechanics, electrical engineering, and other areas are characterized by the fact that the right-hand sides of the differential equations of their dynamics feature discontinuities with respect to the current process state. A typical example of such a system is a dry (Coulomb) friction mechanical system whose resistance force may take either of two sign-opposite values depending on the direction of motion. This situation is often the case in control systems where the wish to improve the system performance, minimize power consumed for the control process, restrict the range of possible variations of control parameters, etc., leads to control as discontinuous functions of the system state vector and the system inputs (reference signals and disturbances).

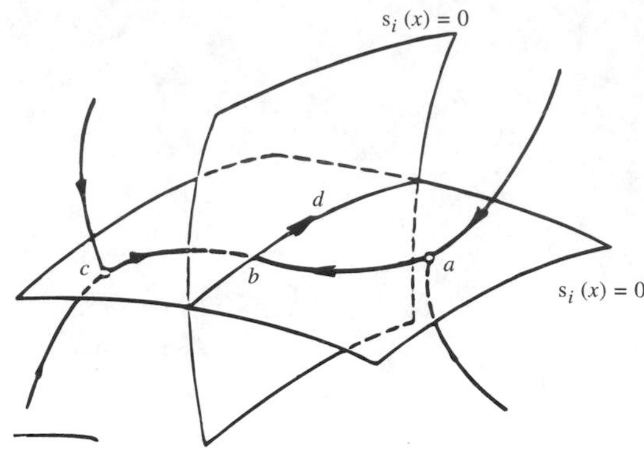

Figure 35.2 Sliding mode in the intersection of discontinuity surfaces. *Source:* Utkin, V. I. 1992. *Sliding Modes in Control and Optimization*, Springer-Verlag, New York. Used by permission.

Sliding Modes in Control Systems

The motion in control systems with discontinuous control actions may be described by differential equations

$$\begin{aligned} &\dot{x} = f(x, t, u), \qquad x \in R^n, \ u \in R^m \\ &u^T = (u_1, \ldots, u_m) \\ &u = \begin{cases} u^+(x, t), & \text{if } \ s(x) > 0 \\ u^-(x, t), & \text{if } \ s(x) < 0, \\ & \text{componentwise} \end{cases} \\ &s^T(x) = (s_1(x), \ldots, s_m(x)), \end{aligned} \tag{35.1}$$

where f, u^+, u^-, and s_i are continuous functions of the state.

The state vector of such systems may stay on one of the discontinuity surfaces $s_i = 0$ or their intersections. For example, the system-state vector trajectories belong to the discontinuity surface $s_i = 0$ if the state velocity vectors $f(x, t, u)$ are oriented towards the surface in its vicinity (Figure 35.1).

Another example illustrates both the motion in one of two discontinuity surfaces (arcs *ab* and *cd*) and their intersection (arc *bd*) (Figure 35.2).

As evidenced by these examples, the motion trajectories which belong to the set of discontinuity surfaces are singular since they differ from the system trajectories for any combination of continuous controls $u_i^+(x, t)$ and $u_i^-(x, t)$.

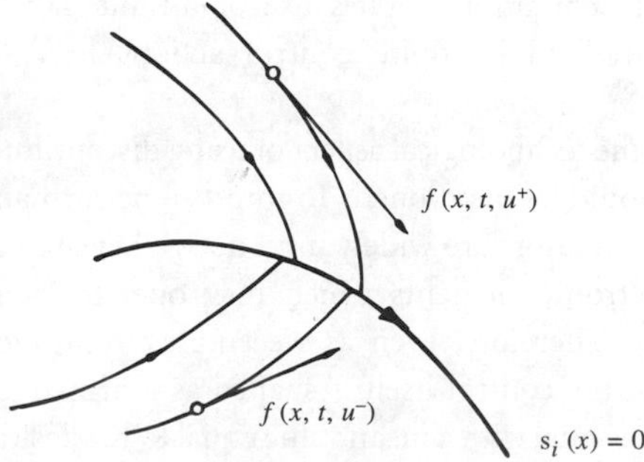

Figure 35.1 Sliding mode in discontinuity surface. *Source:* Utkin, V. I. 1992. *Sliding Modes in Control and Optimization*, Springer-Verlag, New York. Used by permission.

The motion in discontinuity surfaces is refered to as *sliding mode*. The problem of the sliding mode existence will be treated later; here we note that sliding mode does exist on a discontinuity surface whenever the distance to the surface and its time derivative are of opposite signs:

$$\lim_{t \to -0} \dot{s} > 0 \quad \text{and} \quad \lim_{t \to +0} \dot{s} > 0 \tag{35.2}$$

The mathematical description of sliding modes requires development of special techniques. The solution to the differential equation

$$\dot{x} = f(x, t) \tag{35.3}$$

is known to exist and be unique if a Lipschitz constant L may be found such that for any two vectors x_1 and x_2

$$\|f(x_1, t) - f(x_2, t)\| \leq L\|x_1 - x_2\|. \tag{35.4}$$

It is evident that the condition of Equation 35.4 does not hold for the dynamic system with discontinuous control (Equation 35.1). Indeed, if points x_1 and x_2 are on different sides of the discontinuity surface and $\|x_1 - x_2\| \to 0$, then Equation 35.4 is not true for any fixed value of L. Therefore, at least formally, some additional efforts are needed to find the equation governing sliding motions in a discontinuity manifold.

Regularization

The behavior of the system (Equation 35.1) in sliding mode with state trajectories in the discontinuity surfaces cannot be adequately described in terms of the classical theory of differential equations. To solve this problem, various special ways are usually suggested to reduce the original problem to the form which yields a solution close, in a sense, to that of the original problem, which allows the use of classical analysis techniques. Such substitution of the problem is usually called *regularization*.

Consider a regularization scheme which makes use of the boundary layer approach. Assume that a sliding mode occurs in the intersection of all discontinuity surfaces

$$s(x) = 0, \qquad s^T(x) = [s_1(x), \ldots, s_m(x)] \tag{35.5}$$

Substitute the ideal model (Equation 35.1) with a more accurate one

$$\dot{x} = f(x, t, \tilde{u}) \tag{35.6}$$

which takes into account all possible imperfections in the new control, ũ, e.g., delay, hysteresis, switching device inertiality, imprecision of measuring devices, dynamic mismatches, etc. As a result of the account of such imperfections, the solution to Equation 36.6 exists in the conventional sense (for example, right-hand side discontinuity points are isolated or a Lipschitz constant exists for the function $f(x,t,\tilde{u})$). However the state trajectories are not confined to the manifold $s = 0$ (Equation 35.5) but run in some neighborhood of this manifold, which is the "cost" of such regularization

$$\|s(x)\| < \Delta, \qquad \|s(x)\| = (s^T s)^{1/2} \tag{35.7}$$

where Δ is a small number depending on the introduced imperfections.

We do not specify the types of imperfections and consider the entire class of functions $\tilde{u}$ leading to the motion in the vicinity (Equation 35.7). If the value of Δ tends to zero then the motion in the boundary layer (Equation 35.7) tends to the ideal sliding mode. The motion equation obtained in such a limiting procedure will be regarded as the equation of the ideal sliding mode along the intersections of all discontinuity surfaces (Equation 35.5).

The solution on the discontinuity surfaces does exist and is unique if the above limiting procedure yields an unambiguous result regardless of the type of function $\tilde{u}$ and the way the limiting procedure is performed. Otherwise the equations describing the system motion outside the discontinuity surfaces (Equation 35.1) give an ambiguous description of this motion along their intersection.

Equivalent Control Method

A formal procedure will be described below to obtain equations of sliding mode along the intersection of a set of discontinuity surfaces (Equation 35.5) for the system (Equation 35.1).

Assume that a sliding mode exists on the manifold (Equation 35.5). Let us find a continuous control such that under the initial position of the state vector on this manifold, it yields identical equality of the time derivative of the vector $s(x)$ on the system (Equation 35.1) trajectories to zero

$$\dot{s} = Gf(x, t, u) = 0 \tag{35.8}$$

where the rows of the $(m \times m)$ matrix $G = (\partial s/\partial x)$ are the gradients of the functions $s_i(x)$.

Assume that the solution (or a number of solutions) of the algebraic system (Equation 35.8) with respect to m-dimensional control does (or do) exist. Substitute this solution, hereinafter referred to as *equivalent control* $u_{eq}(x, t)$, in the system (Equation 35.1) for control u:

$$\dot{x} = f(x, t, u_{eq}) \tag{35.9}$$

It is quite obvious, by virtue of Equation 35.8, that a motion starting in $s[x(t_0)] = 0$ will proceed along the trajectories which lie in the manifold $s(x) = 0$.

The above procedure will be called the *equivalent control method* and Equation 35.9 obtained as a result of this method will be regarded as the sliding mode equation describing the motion in the intersection of surfaces $s_i(x) = 0$, $i = 1, \ldots m$.

Consider now the equivalent control method procedure for an important particular case of a nonlinear discontinuous system such that the right-hand side of its differential equation is a linear function of control

$$\dot{x} = f(x, t) + B(x, t)u \tag{35.10}$$

The equivalent control (Equation 35.8) for the system (Equation 35.10) may be written as

$$\dot{s} = Gf + GBu = 0 \tag{35.11}$$

Assuming that matrix GB is nonsingular for all x and t, we find the equivalent control from Equation 35.11

$$u_{eq}(x, t) = -[G(x)B(x, t)]^{-1}G(x)f(x, t) \tag{35.12}$$

Substituting this control into Equation 35.10 yields the equation

$$x = f - (GB)^{-1}Gf \tag{35.13}$$

which describes the sliding mode motion in the manifold $s = 0$.

The above procedures for obtaining the sliding equations are just postulated. The validity of the equivalent control method for the the system linear with respect to control Equation 35.10 may be substantiated via the boundary layer regularization method (Utkin, 1992).

Sliding Mode Existence Conditions

Major attention in the previous sections was paid to methods of describing the motion along discontinuity surfaces and their intersection. The equations of sliding mode, found in the previous sections, only indicate the possibility for this type of motion to exist. Let us consider now the conditions for further motion to be a sliding mode, should an initial state be on the intersection of discontinuity surfaces.

From the viewpoint of geometrical representation of motion, we are interested in the trajectories in the neighborhood of the intersection of discontinuity surfaces such that in a small deviation from this intersection state vector would always come back to the intersection. This is the exact statement of the problem of asymptotic stability of motion in a nonlinear dynamic system. Hence, the existence conditions may be formulated in terms of the stability theory:

For the manifold $s = 0$ to be a sliding manifold it is sufficient that a continuously differentiable positive-definite with respect to s function $v(s, x, t)$ exists such that its time derivative on the system trajectories is negative.

To obtain the existence conditions the equation of the motion projection on subspace s should be regarded

$$\dot{s} = Gf + GBu \tag{35.14}$$

It can be shown that sliding mode exists if the matrix $[GB + (GB)^T]$ is positive definite and the control

$$u = -M(x, t)\, sign\, s \tag{35.15}$$

with

$$(sign s)^T = (sign s_1, \ldots, sign s_m) \tag{35.16}$$

$$M(x, t) > \alpha \| u_{eq} \| \tag{35.17}$$

$M(x, t)$ and α are scalar function and parameter.

35.3 Sliding Mode Control Design

Design Procedure

The basic idea behind the design of systems with discontinuous control based on deliberate introduction of a sliding mode, regardless of the criterion of functioning, is in the following:

- A sliding mode with desired dynamics in a certain sense is obtained by an appropriate choice of discontinuity surface.
- A control is chosen so that sliding mode in the intersection of those discontinuity surfaces is enforced, i.e., the trajectories having reached this manifold never leave it.

At the first step of the design procedure let us use the equivalent control method to write the equation of sliding mode in the manifold $s = 0$

$$\dot{x} = f - B(GB)^{-1}Gf \tag{35.18}$$

$$s = 0 \tag{35.19}$$

The last equation enables one to define m components of the state vector, x, as functions of the remaining $(n - m)$ components. Substituting these functions into the system (Equation 35.18) and dropping m equations in this system we obtain sliding mode equations in the form the $(n - m)$-th order system:

$$\dot{x}_1 = F(x_1), \tag{35.20}$$

where $x_1 \in R^{n-m}$ and function F depends on the sliding manifold equation. The desired dynamics of the sliding mode may be provided by a proper choice of the vector $s(x)$.

The second step of the suggested design procedure is in finding an m-dimensional control such that the origin in the m-dimensional subspace $s_1, \ldots, s_m$ is a stable equilibrium point.

Thus deliberate introduction of sliding modes allows the control design problem to be decoupled into two subproblems of lower dimensions, $(n - m)$ and m, which may be treated independently.

Control of the dynamic plants operating under uncertain conditions is one of the main problems of control theory. Let vector $h(x, t)$ in the motion equation

$$\dot{x} = f + Bu + h(x, t) \tag{35.21}$$

characterize all factors whose influence on the control process should be eliminated. As follows from the equivalent control method, the sliding mode equation in the manifold $s(x) = 0$ is of form

$$u_{eq} = -(GB)^{-1}(Gf + Gh) \tag{35.22}$$

$$\dot{x} = f - B(GB)^{-1}Gf + (I_n - B(GB)^{-1}G)h \tag{35.23}$$

Let $\mathcal{B}(x, t)$ be a subspace formed for each point x, t by the base vectors of matrix $B(x, t)$. For the sliding mode motion (Equation 35.23) to be invariant with respect to the vector $h(x, t)$ it is sufficient that

$$h(x, t) \in \mathcal{B}(x, t) \tag{35.24}$$

The condition (Equation 35.24) means that there exists m-dimensional vector $\lambda(x, t)$ such that

$$h(x, t) = B(x, t)\lambda(x, t) \tag{35.25}$$

The direct substitution of vector $h(x, t)$ in the form of Equation 35.25 into the sliding mode Equation 35.23 demonstrates that, if the condition (Equation 35.24) is obeyed, invariant motions occur in the sliding mode. The enforce sliding mode in the manifold $s = 0$ no need to measure the disturbance vector $h(x, t)$; according to the condition (Equation 35.17) only a range of variations or an upper estimate function is needed.

Regular Form

The sliding mode control design principle may be explained easily for the motion equations, linear with respect to control

$$\dot{x} = f(x) + B(x)u, \; x \in \mathcal{R}^n, \; u \in \mathcal{R}^m$$
$$\text{rank } B = m < n \tag{35.26}$$

represented in the so-called regular form

$$\dot{x}_1 = f_1(x_1, x_2), \qquad x_1 \in \mathcal{R}^{n-m}$$
$$\dot{x}_2 = f_2(x_1, x_2) + B_2(x_1, x_2)u, \quad x_2 \in \mathcal{R}^m$$
$$\det B_2 \neq 0 \tag{35.27}$$

The first equation does not depend on control while the second one has the same order as the control and the matrix B_2 is nonsingular. At the first stage the state subvector x_2 is handled as a control and designed as a function of the subsystem state vector in accordance with some performance criterion

$$x_2 = g(x_1) \tag{35.28}$$

At the second stage the discontinuous control u is designed to enforce sliding mode in the manifold

$$s = x_2 - g(x_1) = 0 \tag{35.29}$$

The condition for sliding mode to exist is equivalent to the stability condition of the projection of the overall motion on the subspace $s = 0$ governed by

$$\dot{s} = f_2 - Gf_1 + B_2 u, \qquad G^T := \frac{\partial g}{\partial x} \tag{35.30}$$

The state vector of the system (Equation 35.27) will reach the manifold $s = 0$ after a finite time interval if the matrix $B_2 + B_2^T$ is positive-definite and control u is of form

$$u = M_0(x_1, x_2)\text{sign}(s)(\text{componentwise}) \tag{35.31}$$

with

$$M_0(x_1, x_2) > \|B_2^{-1}(f_2 - Gf_1)\| \tag{35.32}$$

After the sliding mode starts the system motion will be governed by the equation

$$\dot{x}_1 = f_1(x_1, g(x_1)) \tag{35.33}$$

with the desired dynamics.

The design procedure has been decoupled into two subproblems of lower dimension: design of the desired dynamics in the system of the $(n-m)$-th order and enforcing sliding mode using Equation 35.30 of the m-th order.

It should be underlined that the sliding mode (Equation 35.33) does not depend on the functions in the right-hand side of the second equation in Equation 35.27, and an upper estimate of the functions in the motion equations rather than their exact values is needed to enforce sliding mode with the desired dynamics.

Control in Linear Systems

The design procedure is demonstrated in this section for linear time-invariant multidimensional systems

$$\dot{x} = Ax + Bu \tag{35.34}$$

where A and B are constant matrices and rank $B = m$.

For any controllable system there exists a linear feedback $u = Fx$, (F is a constant matrix) such that the eigenvalues of the feedback systems take any desired values and as a result the system exhibits the desired dynamic properties. The eigenvalue placement task may be solved in the framework of sliding mode control technique dealing with a reduced order system.

Since rank $B = m$, the matrix B may be partitioned (after reordering the state vector components) as

$$B = \begin{bmatrix} B_1 \\ B_2 \end{bmatrix} \tag{35.35}$$

where $B_1 \in R^{(n-m)\times m}$, $B_2 \in R^{m\times m}$, and $\det B_2 \neq 0$. Then the nonsingular coordinate transformation

$$\begin{bmatrix} x_1 \\ x_2 \end{bmatrix} = \begin{bmatrix} I_{n-m} & -B_1 B_2^{-1} \\ 0 & B_2^{-1} \end{bmatrix} x \tag{35.36}$$

reduces the system equation to the regular form (Equation 35.25)

$$\dot{x}_1 = A_{11}x_1 + A_{12}x_2$$
$$\dot{x}_2 = A_{21}x_1 + A_{22}x_2 + u, \tag{35.37}$$

where $x_1 \in R^{n-m}$, $x_2 \in R^m$, and A_{ij} are constant matrices $(i, j = 1, 2)$.

It follows from controllability of A, B that the pair (A_{11}, A_{12}) is controllable as well (Utkin, 1992). Handling x_2 as an m-dimensional intermediate control in the $(n-m)$-dimensional first subsystem in Equation 35.37, all $(n-m)$ eigenvalues may be assigned arbitrarily by a proper choice of matrix C in

$$x_2 = -Cx_1 \tag{35.38}$$

To provide the desired dependence between the components x_1 and x_2 of the state vector, the sliding mode should be enforced in the manifold

$$s = x_2 + Cx_1 = 0 \tag{35.39}$$

where s is the difference between the desired and real values of x_2.

For a piecewise linear discontinuous control

$$u = -\alpha|x|\,\text{sign}\, s$$
$$|x| = |x_1| + \cdots + |x_n| \tag{35.40}$$
$$(\text{sign}s)^T = (\text{sign}s_1, \ldots, \text{sign}s_m)$$
$$\alpha \text{ is constant positive value}$$

calculate time derivative of a positive-definite function $V = {}^1/_2 s^T s$:

$$\dot{V} = s^T[(CA_{11} + A_{12})x_1 + (CA_{12} + A_{22})x_2] - \alpha|x||s| \tag{35.41}$$

It is evident that there exists such a positive value of α that the time derivative $\dot{V}$ is negative for any $x \neq 0$ which validates convergence of the state vector to the manifold $s = 0$ and existence of sliding mode with the desired dynamics.

Again the design task has been decoupled into two independent subproblems of lower dimensions. The quadratic optimization problem also may be decoupled within the framework of the above design procedure (Utkin, 1992).

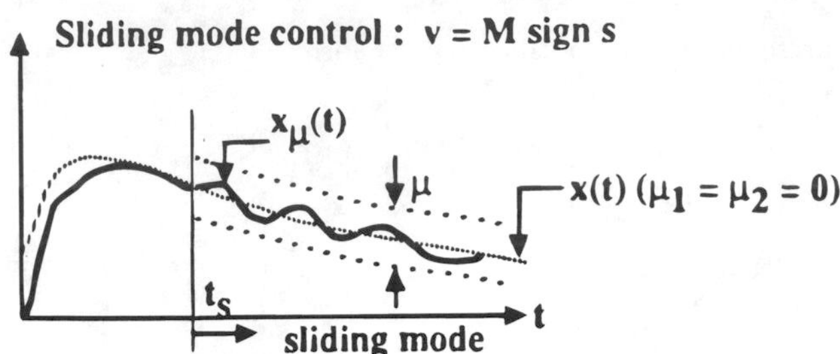

Figure 35.5 Chattering in system with discontinuous control. *Source:* Utkin, V. 1993. Variable structures systems and sliding modes—a state-of-the-art assessment. In K. K. D. Young, ed., *Variable Structure Control for Robotics and Aerospace Applications*, Elsevier Science Publishers, Amsterdam. Used by permission.

35.4 Chattering Problem

The subject of this section is important whenever we intend to establish the bridge between the recommendations of the theory and applications. Bearing in mind that the control has a high-frequency component, we should analyze the robustness or the problem of correspondence between an ideal sliding mode and real-life processes at the presence of unmodeled dynamics. Neglected small time constants (μ_1 and μ_2 in Figure 35.3) in plant models, sensors, and actuators lead to dynamics discrepancy (z_1 and z_2 are the unmodeled-dynamics state vectors).

In accordance with singular perturbation theory (Kokotovic, et al., 1976) in systems with continuous control, a fast component of the motion decays rapidly and a slow one depends on the small time constants continuously (Figure 35.4). In discontinuous control systems the solution depends on the small parameters continuously as well. But unlike continuous systems, the switchings in control excite the unmodeled dynamics, which leads to oscillations in the state vector at a high frequency (Figure 35.5). The oscillations, usually called chattering, are known to result in low control accuracy, high heat losses in electrical power circuits, and high wear of moving mechanical parts. These phenomena have been considered as serious obstacles for applications of sliding mode control in many papers and discussions. A recent study (Bondarev et al., 1985) and practical experience showed that the chattering caused by unmodeled dynamics may be eliminated in systems with asymptotic observers (Figure 35.6). In spite of the presence of unmodeled dynamics, ideal sliding arises. It is described by a singularly perturbed differential equation with solutions free from a high-frequency component and close to those of the ideal system (Figure 35.7). As shown in Figure 35.6 an asymptotic observer serves as a bypass for the high-frequency component, therefore the unmodeled dynamics are not excited. Preservation of sliding modes in systems with asymptotic observers enabled successful application of the of sliding mode control.

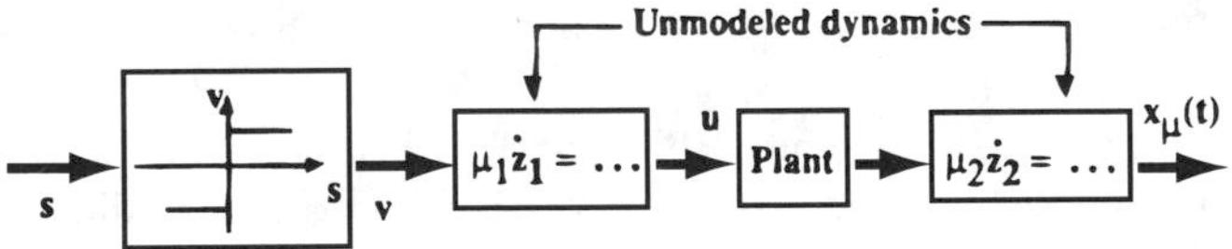

Figure 35.3 Unmodeled dynamics of actuator and sensor. *Source:* Utkin, V. 1993. Variable structures systems and sliding modes—a state-of-the-art assessment. In K. K. D. Young, ed., *Variable Structure Control for Robotics and Aerospace Applications*, Elsevier Science Publishers, Amsterdam. Used by permission.

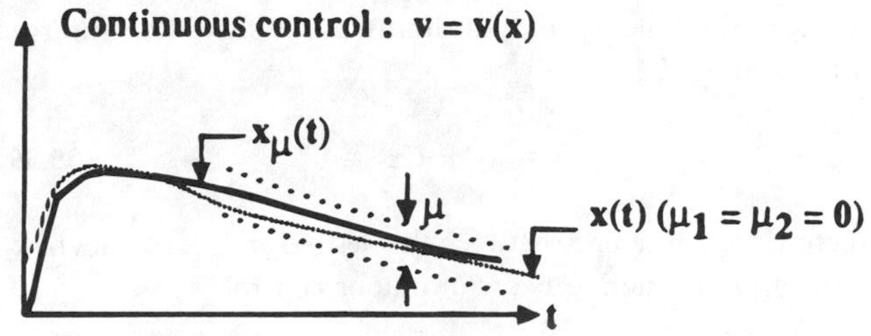

Figure 35.4 System with continuous control. *Source:* Utkin, V. 1993. Variable structures systems and sliding modes—a state-of-the-art assessment. In K. K. D. Young, ed., *Variable Structure Control for Robotics and Aerospace Applications*, Elsevier Science Publishers, Amsterdam. Used by permission.

35.5 Control of Manipulators

As an example of sliding mode control design, a multilink manipulator may be considered:

$$M(q)\ddot{q} + f(q, \dot{q}, t) = u \tag{35.42}$$

where q is the state vector, u is control torque vector, and $M(q)$ is the positive-definite inertia matrix. Let $q = p$, $\dot{q} = v$; then

$$\dot{p} = v, \qquad M(p)\dot{v} = -f(p, v, t) + u \tag{35.43}$$

The motion equations are in the regular form with p and v as the vectors x_1 and x_2 in Equation 35.27. First of all, the vector v handled as a control is designed as a function (for example linear) of p to provide the desired dynamics

$$v = -Cp \tag{35.44}$$

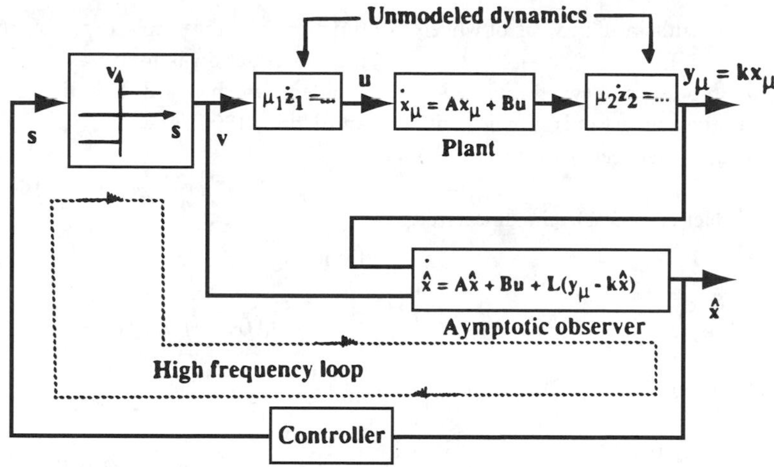

Figure 35.6 Control system with asymptotic observer. *Source:* Utkin, V. 1993. Variable structures systems and sliding modes—a state-of-the-art assessment. In K. K. D. Young, ed., *Variable Structure Control for Robotics and Aerospace Applications*, Elsevier Science Publishers, Amsterdam. Used by permission.

where C is a scalar positive value or diagonal matrix with positive elements. Secondly, since the matrix M is positive-definite, the discontinuous control

$$u = M_0 \operatorname{sign}(s), \; s = v + Cp \tag{35.45}$$

with high enough M_0 provides existence of sliding mode in the manifold $s = 0$ governed by

$$\dot{p} = -Cp \tag{35.46}$$

Apparently, after the sliding mode occurs, the system motion is described by a homogenous linear time invariant differential equation with solutions independent of the manipulator parameters and external forces.

35.6 Control of Mobile Robots

A sliding mode control strategy for tracking the gradient of an artificial potential field is introduced. The invariance property of sliding mode control for tracking the gradient of an artificial potential field is exploited. The key idea is to regard the velocity vector rather than the acceleration vector as the variable under control (Utkin, et al., 1991). An n-dimensional sliding manifold (Guldner and Utkin, 1993) is defined as

$$s = \dot{x} - v_d(t)\frac{\mathscr{E}}{\|\mathscr{E}\|} = 0 \tag{35.47}$$

where $\mathscr{E}(x)$ is the gradient to be tracked, and $v_d(t)$ is the scalar desired velocity along the gradient. For $v_d(t) = (2a_0 d(t))^{1/2}$ the motion of the system is restricted to the manifold Equation 35.47 and can be reduced to a description of the scalar distance $d(t)$ to the goal point (Utkin, et al., 1991):

$$\dot{d}(t) = -(2a_0 d(t))^{1/2} \tag{35.48}$$

The solution of Equation 35.48

$$d(t) = \left(-\left(\frac{a_0}{2}\right)^{1/2} t + d(0)^{1/2}\right)^2 \tag{35.49}$$

vanishes after a finite time interval:

$$d(t) \equiv 0 \quad \text{for} \quad t \geq \left(\frac{2}{a_0}\, d(0)\right)^{1/2} \tag{35.50}$$

Detailed analytical studies were performed with mobile robots operating in a planar workspace obstructed by obstacles.

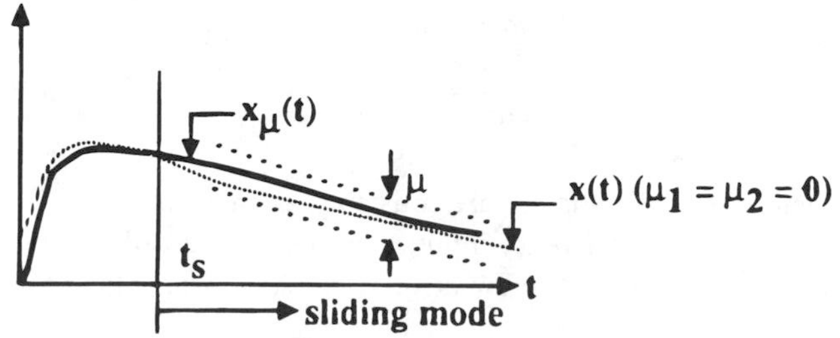

Figure 35.7 Sliding mode in system with observer. *Source:* Utkin, V. 1993. Variable structures systems and sliding modes—a state-of-the-art assessment. In K. K. D. Young, ed., *Variable Structure Control for Robotics and Aerospace Applications*, Elsevier Science Publishers, Amsterdam. Used by permission.

35.7 Control of Railway Wheelset

The control of a wheelset as the fundamental unit of a railway vehicle illustrates the application of the sliding mode control methodology for mechanical systems.

A rigid wheelset is governed by nonlinear differential algebraic equations. The equilibrium point becomes unstable for motions at high velocities and self-oscillations are generated in the system.

The self-oscillations are known to result in high wear of wheels and railways.

To suppress the oscillation modes a new technological approach has been proposed recently: a wheelset is supplied with a clutch and the interaction force u between two semiaxes is handled as a control action.

According to Jaschinski and Netter (1991), the linearized model in the subspace (y, ψ, ω),

$$\omega = \dot{\vartheta}_r - \dot{\vartheta}_l \tag{35.51}$$

may be described by the fifth-order system:

$$\begin{aligned} \ddot{y} &= L_{1,y}(y, \dot{y}) + L_{2,y}(\psi, \dot{\psi}) \\ \ddot{\psi} &= L_{\psi}(y, \dot{y}, \psi, \dot{\psi}) + \kappa \cdot \omega \\ \dot{\omega} &= L_{\omega}(y, \dot{y}, \psi, \dot{\psi}) + u \end{aligned} \tag{35.52}$$

with linear functions $L_{1,y}$, $L_{2,y}$, L_{ψ}, L_{ω}, where y, ψ, θ_r, θ_l are longitudinal deviation, yaw, and right and left wheel angle positions.

The system parameters depend on V and it becomes unstable for motions at high velocities.

The system (Equation 35.52) is represented in the regular form (Equation 35.27): the coordinate ω takes part of x_2 in Equation 35.27 may be handled as a control for the first two equations in Equation 35.52. Let "the control" ω be of form:

$$\kappa \cdot \omega = -\alpha_0 \psi - \alpha_1 \dot{\psi} - L_{\psi}(\cdot) \tag{35.53}$$

Then $\psi(t)$ as the solution to

$$\ddot{\psi} + \alpha_1 \dot{\psi} + \alpha_0 \psi = 0 \tag{35.54}$$

decays exponentially at the desired rate which can be guaranteed by a proper choice of α_0 and α_1. Since $L_{2,y}$ tends to zero and the coefficients in the linear form $L_{1,y}$ are negative, y tends to zero exponentially as well.

The control u should be designed such that sliding mode is enforced in the surface (Netter, et al., 1994)

$$s = \alpha_0 \psi + \alpha_1 \dot{\psi} + L_{\psi}(\cdot) + \kappa \cdot \omega \tag{35.55}$$

Then the condition (Equation 35.53) holds and the feedback system is asymptotically stable for any values of velocities.

35.8 Control of Torsion Oscillations of a Flexible Shaft

This section deals with the torsion oscillations in the one-dimensional flexible shaft with a load at the right end and a control torque applied to the left end.

The control torque u and deformation y of the shaft right end are considered as the system input and output and the transfer function via the Laplace transformation is found. Then the motion equation is found as a solution of the boundary value problem which is a differential-difference equation (Drakunov and Utkin, 1990).

$$m\ddot{y}(t) + m\ddot{y}(t - 2\tau) + a\dot{y}(t) - a\dot{y}(t - 2\tau) = 2u(t - \tau)$$

Denoting

$$s_1(t) = y(t),\ s_2(t) = \dot{y}(t),\ s_3(t) = m\dot{y}(t) + a\dot{y}(t)$$

and

$$s_4(t) = 2as_2(t - \tau) - s_3(t - \tau)$$

we obtain the motion equations

$$\dot{s}_1(t) = s_2(t), \qquad \dot{s}_2(t) = -\frac{a}{m} s_2(t) + \frac{1}{m} s_3(t) \tag{35.56}$$

$$s_3(t) = s_4(t - \tau) + 2u(t - \tau) \tag{35.37}$$

$$s_4(t) = -s_3(t - \tau) + 2as_2(t - \tau) \tag{35.58}$$

The motion equation may be represented in the regular form with s_3 as an intermediate control and the system is stabilized within the framework of the design procedure (Equations 35.27–35.33) (Drakunov and Utkin, 1990).

35.9 DC Motors

Control of electric drives is a challenging area of sliding mode applications (Utkin, 1993) From the point of controllability, a DC motor with constant excitation is the simplest. Its motion is governed by the second-order equation with respect to shaft angle speed n and current i with voltage u and load torque M_L as a control and a disturbance, respectively:

$$L\frac{di}{dt} = -Ri - K_e n + u \tag{35.59}$$

$$J\frac{dn}{dt} = K_m i - M_L \tag{35.60}$$

where L, J, R, K_e, and K_m are constant parameters.

Let $n_0(t)$ be a reference input; then the second-order motion equation with respect to the error $x_1 = n_0(t) - n$ is of the form

$$\begin{cases} \dot{x}_1 = x_2 \\ \dot{x}_2 = -a_1 x_1 - a_2 x_2 + f(t) - bu \end{cases} \tag{35.61}$$

where a_1, a_2, and b are positive parameters; $f(t)$ is a time function depending on $M_L(t)$, $n_0(t)$ and their time derivatives.

For discontinuous control

$$u = u_0 \text{ sign } s, \; s = cx_1 + x_2u_0, \; c - \text{const} \tag{35.62}$$

the error x_1 decays exponentially should sliding mode occur on the line $s = 0$ since the equation

$$cx_1 + \dot{x}_2 = 0$$

is linear and does not depend on $f(t)$.

It follows from

$$\dot{s} = cx_2 - a_1x_1 - a_2x_2 + f(t) - bu_0 \text{ sign } s$$

in the system (Equations 35.61 and 35.62) with

$$bu_0 > |cx_2 - a_1x_1 - a_2x_2 + f(t)| \tag{35.63}$$

that the s and $\dot{s}$ have opposite signs and the state reaches the sliding line $s = 0$ after a finite time interval. Equation 35.63 determines the voltage needed for enforcing sliding mode. As a result, the control error is steered to zero.

For implementation of control (Equation 35.62), angular acceleration is needed ($x_2 = \dot{n}_0 - \dot{n}$). Under the assumption that the angular speed n and the current i can be measured directly and the load torque varies slowly, i.e.,

$$\frac{dM_L}{dt} \approx 0 \tag{35.64}$$

a conventional Luenberger reduced-order observer may be designed:

$$\frac{d\hat{M}}{dt} = \frac{\ell}{J}(-\ell\hat{M} + \ell^2 n + \ell K_m i) \tag{35.65}$$

with ℓ as a constant and $\hat{M}$ as an estimate of $M = M_L + \ell n$. According to Equations 35.60, 35.64, and 35.65 the equation for the mismatch $\bar{M} = M - \hat{M}$ is of the form

$$\frac{d\bar{M}}{dt} = -\frac{\ell}{J}\bar{M}$$

By a proper choice of the gain ℓ, the desired convergence rate of $\bar{M}$ to zero (or $(\hat{M} - \ell n)$ to M_L) may be provided. It means that the load torque is known and the time derivative dn/dt may be found from Equation 35.60.

Similarly, the sliding mode control may be designed for position and torque control with or without measurement of the motor current. In addition to control of mechanical coordinates, optimization in accordance with a power consumption criterion may be provided for DC motors with a controlled excitation current.

35.10 Control of DC Motors Based on a Reduced-Order Model

The section on chattering was dedicated to sliding mode control in systems with unmodeled dynamics in which the partial motion components may be separated by rates and then the fast one is neglected. A similar situation may happen when controlling a DC motor with the mechanical motion being much slower than the electrical one. Formally it means that $L \ll J$ in Equations 35.59 and 35.60, which may be presented as

$$L\frac{di}{dt} = -Ri - K_e(n_0 - \Delta n) + u \tag{35.66}$$

$$J\frac{d\Delta n}{dt} = -K_m i + M_L + J\dot{n}_0 \tag{35.67}$$

with a control error $\Delta n = n_0 - n$. Let us write down formally the equation for Δn making L equal to zero. Then substitution of the solution of Equation 35.66 with $L = 0$,

$$i = -\frac{K_e}{R}(n_o - \Delta n) + \frac{1}{R}u \tag{35.68}$$

into Equation 35.67 yields

$$J\frac{d\Delta n}{dt} = \frac{K_m K_e}{R}(n_0 - \Delta n) - \frac{K_m}{R}u + M_L + J\dot{n}_0 \tag{35.69}$$

Equation 35.69 is taken as a reduced-order model of a DC motor. Within the framework of the model (Equation 35.69), discontinuous control $u = u_0\text{sign}\Delta n$, depending only on the control error (in contrast to Equation 35.62, depending also on its time derivative) for high enough u_0, provides the sliding mode in the "manifold" $\Delta n = 0$; this results in ideal tracking of the reference input $n_0(t)$ by the shaft rotation speed. However, as it was discussed in the section on chattering, the unmodeled dynamics (Equation 35.66) may excite nonadmissible chattering. Following the recommendations of the chattering problem section, chattering may be eliminated by using asymptotic observers.

Bearing in mind that $\dot{M}_L \approx 0$, let us design an asymptotic observer to estimate Δn (Equation 35.69) and M_L:

$$J\frac{d\hat{\Delta n}}{dt} = \frac{K_m K_e}{R}(n_0 - \hat{\Delta n}) - \frac{K_m}{R}u + \hat{M}_L + J\dot{N}_0 + \ell_1(\hat{\Delta n} - \Delta n) \tag{35.70}$$

$$\frac{dM}{dt} = \ell_2(\hat{\Delta n} - \Delta n) \tag{35.71}$$

where $\hat{\Delta n}$ and $\hat{M}_L$ are estimates for Δn and M_L, and the gains ℓ_1 and ℓ_2 are constant. Control is a discontinuous function of the error estimate:

$$u = u_0 \operatorname{sign} \hat{\Delta n} \tag{35.72}$$

The value of $\hat{\Delta n}$ and its time derivative (Equation 35.70) have different signs if

$$u_0 > \left| K_e(n_0 - \hat{\Delta n}) + \frac{R}{K_m}\hat{M}_L + \frac{JR}{K_m}\dot{n}_0 + \frac{\ell_1 R}{K_m}(\hat{\Delta n} - \Delta n) \right| \tag{35.73}$$

Existence of sliding mode means that the discontinuous control (Equations 35.72 and 35.73) reduces the error estimate $\hat{\Delta n}$ to zero. To derive a sliding mode equation in accordance with the equivalent control method (see above), the solution to $\frac{d\hat{\Delta n}}{dt} = 0$ with respect to control

$$u_{eq} = K_e n_0 + \frac{R}{K_m}\hat{M}_L + \frac{JR}{K_m}\dot{n}_z - \frac{\ell_1 R}{K_m}\Delta n$$

should be substituted into the system (Equations 35.66, 35.67 and 35.71) for control u:

$$L\frac{di}{dt} = -Ri + K_e\Delta n + \frac{R}{K_m}\hat{M}_L + \frac{JR}{K_m}\dot{n}_0 - \frac{\ell_1 R}{K_m}\Delta n \tag{35.74}$$

$$J\frac{d\Delta n}{dt} = -K_m i + M_L + J\dot{n}_0 \tag{35.75}$$

$$\frac{d\hat{M}_L}{dt} = -\ell_2 \Delta n \tag{35.76}$$

Equations 35.74, 35.75, 35.76, and $\dot{M}_L = 0$ describe the sliding in the manifold $\hat{\Delta n} = 0$. According to the theory of singular perturbed systems (Kokotovic et al., 1976) for $L \ll J$, the fast motion of the linear system may be neglected by zeroing the parameter L. Substitution of the solution to the algebraic Equation 35.74 ($L = 0$) with respect to i into Equation 35.75 results in a motion equation for Δn and $\bar{M}_L = \hat{M}_L - M_L$

$$J\Delta\dot{n} = -\frac{K_e K_m}{R}\Delta n - \bar{M} - \ell_1 \Delta n$$

$$\dot{\bar{M}}_L = -\ell_2 \Delta n.$$

Apparently the eigenvalues of the homogeneous system may be assigned at our will by a proper choice of ℓ_1 and ℓ_2 and the desired rate of the control error decay may be guaranteed.

The principle advantage of the reduced-order-based method is that the angle acceleration is not needed for designing sliding mode control.

35.11 Conclusion

This chapter has outlined the mathematical background and sliding mode control design philosophy oriented towards high-dimensional nonlinear systems operating under uncertainty conditions and has demonstrated its applicability to control of mechanical systems and electric motors.

An assessment of the scientific arsenal accumulated in the sliding mode control theory is beyond the scope of this chapter, therefore we confine ourselves to mentioning new research areas: discrete-time systems, infinite-dimensional (including distributed and time-delay) plants, adaptive control, state observers, Lyapunov technique-based design methods with applications for robotics, metal-cutting machine tools, aircraft, and combustion engines.

References

Bondarev, A. et al. 1985. Sliding modes in systems with asymptotic state observer, *Automation and Remote Control,* 46(6)–1:679–684.

Drakunov, S. V. and Utkin, V. I. 1990. Sliding modes in dynamic systems, *Int. J. Control,* 55:1029–1037.

Guldner, J. and Utkin, V. 1993. Sliding mode control for an obstacle avoidance strategy based on electric potential field, *Proc. IEEE Conf. Decision and Control,* 424–429, San Antonio, TX.

Jaschinski, A. and Netter, H. 1991. Non-linear dynamical investigations by using simplified wheelset models, *Vehicle System Dynamics,* 20:284–298.

Kokotovic, P., O'Malley, R., and Sannuti, P. 1976. Singular perturbation and order reduction in control theory, *Automatica,* 12:123–132.

Netter, H., Kortum, W., and Utkin, V. 1994. Sliding mode control of railway wheel set, *Proc. First ASCC,* 397–400, August 27–30, Tokyo.

Utkin, V. I. 1992. *Sliding Modes in Control and Optimization,* Springer-Verlag, New York, NY.

Utkin, V. 1993. Sliding mode control design principles and applications to electric drives, *IEEE Trans. Industrial Electronics,* 40(1):23–36.

Utkin, V., Drakunov, S., Hashimoto, H., and Harashima, F. 1991. Robot path obstacle control via sliding mode approach, *Proc. IEEE/RSJ International Workshop on Intelligent Robots and Systems,* 1287–1290, Osaka, Japan.

36

Digital Computation

James R. Rowland
University of Kansas

36.1 System Response 545
Unit-Sample Response • Unit-Step Response • Discrete Convolution
36.2 Numerical Integration Formulas 548
Euler's Formula • Runge-Kutta Formulas
36.3 Exact Difference Equations for Linear Systems 551
36.4 Summary 552

A digital control system is a discrete-time system for which the dynamics are described by a difference equation. Such systems occur when either the system to be controlled is itself a discrete-time system governed by a difference equation or the system is a continuous-time system which is to be controlled by discrete-level inputs from a microprocessor. Typical examples of these two cases are described in this section.

The focus of the section is twofold: to determine the time response of discrete-time systems by solving the difference equations which describe them and to determine difference equations which result from the digital simulation of continuous-time systems. First, methods are presented to solve for the unit-sample response and the unit step response of discrete time systems. Relationships between these two kinds of responses are shown, and convolution summation is presented to show how to determine the time response for an arbitrary input by using the unit-sample response. Second, numerical integration formulas are presented for the digital simulation of continuous-time systems. These formulas are applicable both to linear and nonlinear systems.

36.1 System Response

The response of a discrete-time system can be expressed as the convolution of its input $r(k)$ and its unit-sample response $h(k)$ as

$$y(k) = \sum_{n=-\infty}^{k} h(k-n)r(n) \tag{36.1}$$

where $y(k)$ is the system output. We next show how to determine $h(k)$ and illustrate the method with two examples.

Unit-Sample Response

Let the unit-sample input be denoted by $d(k)$, as shown in Figure 36.1. The input is one for $k = 0$ and zero for all other k, i.e,

$$d(k) = \begin{cases} 1 & \text{for} \quad k = 0 \\ 0 & \text{for} \quad k \neq 0 \end{cases} \tag{36.2}$$

When $d(k)$ is applied to a discrete-time system, the system response is referred to as the unit-sample response $h(k)$. The importance of obtaining $h(k)$ is that the system response for any input $r(k)$ can be expressed in terms of $h(k)$ and $r(k)$ as shown in Equation 36.1.

EXAMPLE 36.1:

Discrete-time systems are described by difference equations. Consider the difference equation given by

$$y(k+1) = (1/2)y(k) + r(k) \tag{36.3}$$

where $y(k)$ is the value of y at time k.

The unit-sample response is the solution of Equation 36.1 with $r(k) = d(k)$ from Equation 36.2. The resulting $y(k)$ is denoted by $h(k)$. To solve for $h(k)$, we note that

$$h(0) = y(0) = (1/2)y(-1) + d(-1) = (1/2)(0) + (0) = 0$$

$$h(1) = y(1) = (1/2)y(0) + d(0) = (1/2)(0) + (1) = 1$$

We use $h(1) = 1$ as the initial condition and seek to obtain the solution of $h(k+1) = (1/2)h(k)$. The solution has the form

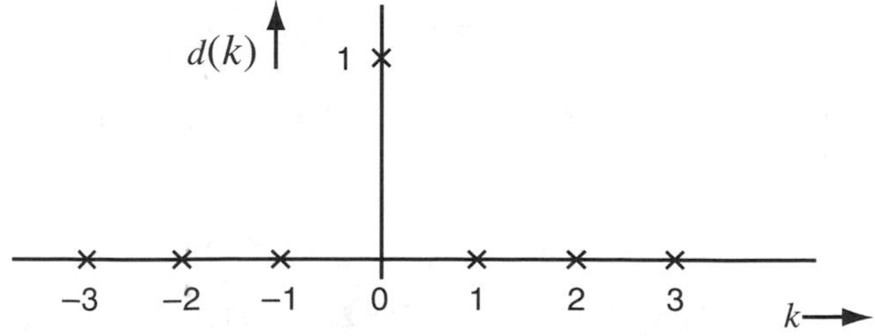

Figure 36.1 The unit-sample input $d(k)$.

0-8493-8343-9/97/$0.00+$.50

$C\alpha^k$, from which $C\alpha^{k+1} = (1/2)C\alpha^k$. Thus, $\alpha = 1/2$ and $h(k) = C(1/2)^k$. Using $h(1) = 1$, we have $1 = C(1/2)$ or $C = 2$. Therefore,

$$h(k) = \begin{cases} 2(1/2)^k = (1/2)^{k-1} & \text{for} \quad k \geq 1 \\ 0 & \text{for} \quad k < 1 \end{cases}$$

EXAMPLE 36.2:

As another example, consider the discrete-time system described by

$$y(k+2) = (3/4)y(k+1) - (1/8)y(k) + r(k) \quad (36.4)$$

Again, we seek to determine the unit-sample response $h(k)$.

The first step is to find the appropriate initial conditions to solve this second-order system. We note that

$$\begin{aligned} y(0) &= (3/4)y(-1) - (1/8)y(-2) + d(-2) \\ &= (3/4)(0) - (1/8)(0) + (0) = 0 \\ y(1) &= (3/4)y(0) - (1/8)y(-1) + d(-1) \\ &= (3/4)(0) - (1/8)(0) + (0) = 0 \\ y(2) &= (3/4)y(1) - (1/8)y(0) + d(0) \\ &= (3/4)(0) - (1/8)y(0) + (1) = 1 \\ y(3) &= (3/4)y(2) - (1/8)y(1) + d(1) \\ &= (3/4)(1) - (1/8)(0) + (0) = 3/4 \end{aligned}$$

Hence, we use $y(2) = 1$ and $y(3) = 3/4$ to solve

$$y(k+2) = (3/4)y(k+1) - (1/8)y(k)$$

for $k \geq 2$, since $r(k) = d(k) = 0$ for $k \geq 2$.

The form of the solution is $y(k) = C\alpha^k$. Therefore, we have

$$\begin{aligned} C\alpha^{k+2} &= (3/4)C\alpha^{k+1} - (1/8)C\alpha^k \\ \alpha^2 &= (3/4)\alpha - 1/8 \\ \alpha^2 - (3/4)\alpha - 1/8 &= 0 \\ (\alpha - 1/2)(\alpha - 1/4) &= 0 \\ \therefore \alpha &= 1/2,\ 1/4 \end{aligned}$$

and $h(k) = y(k) = C_1(1/2)^k + C_2(1/4)^k$. Using initial conditions, we obtain

$$h(2) = y(2) = 1 = C_1(1/2)^2 + C_2(1/4)^2$$

$$h(3) = y(3) = 3/4 = C_1(1/2)^3 + C_2(1/4)^3$$

from which $C_1 = 8$ and $C_2 = -16$. Therefore, we have $8(1/2)^k - 16(1/4)^k = 2(1/2)^{k-2} - (1/4)^{k-2}$, i.e.,

$$h(k) = \begin{cases} 2(1/2)^{k-2} - (1/4)^{k-2} & \text{for} \quad k \geq 2 \\ 0 & \text{for} \quad k < 2 \end{cases}$$

which completes the example. See Figure 36.2 for plots of $h(k)$ for these first two examples.

Unit-Step Response

Let $w(k)$ be the system response obtained when a unit-step input (Figure 36.3) is applied. Using superposition, we can express the following relationship between $h(k)$ and $w(k)$.

$$w(k) = \sum_{m=0}^{k} h(m) \quad (36.5)$$

$$h(k) = w(k) - w(k-1) \quad (36.6)$$

EXAMPLE 36.3:

Determine $w(k)$ both directly and by using Equation 36.5 for the discrete-time systems of Examples 36.1 and 36.2.

For the first-order system in Example 36.1, we again have $y(1) = 1$, but the form of the solution is

$$w(k) = y(k) = C(1/2)^k + y(\infty)$$

where $y(\infty)$ is the final (or steady-state) value which $y(k)$ approaches as $k \to \infty$. This value can be determined by setting $y(k) = y(k+1) = y(\infty)$ to yield

$$y(\infty) = (1/2)y(\infty) + 1$$

from which $y(\infty) = 2$. Using $y(1) = 1$ gives

$$w(k) = y(1) = 1 = C(1/2)^1 + 2$$

Therefore, $C = -2$ and

$$w(k) = \begin{cases} -2(1/2)^k + 2 & \text{for} \quad k \geq 1 \\ 0 & \text{for} \quad k < 1 \end{cases}$$

(See Figure 36.2 for a plot of $w(k)$ versus k). Using Equation 36.5 with $h(k) = (1/2)^{k-1}$ for $k \geq 1$ yields

$$w(k) = \sum_{m=1}^{k} (1/2)^{m-1} = \frac{1 - (1/2)^k}{1 - 1/2} = 2 - 2(1/2)^k \text{ for } k \geq 1$$

which agrees with the $w(k)$ just found. Moreover, we also see from Equation 36.6 that

$$\begin{aligned} h(k) &= [2 - 2(1/2)^k] - [2 - 2(1/2)^{k-1}] \\ &= 2(1/2)^{k-1} - 2(1/2)^k = 2(1/2)^k = (1/2)^{k-1} \text{ for } k \geq 1 \end{aligned}$$

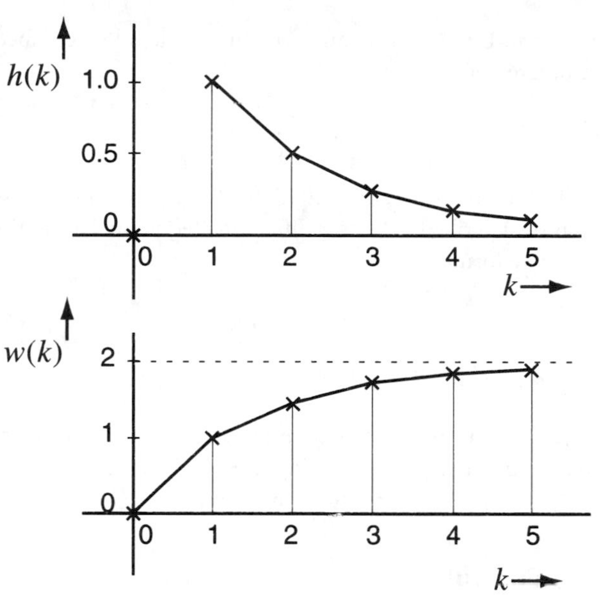

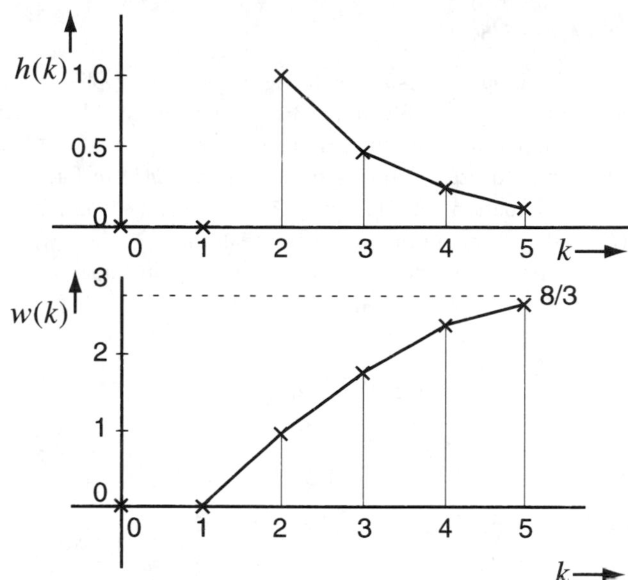

Figure 36.2 Plots of $h(k)$ and $w(k)$ versus k for Example 36.2.

For the second-order system of Example 36.2, we have

$$y(2) = (3/4)y(1) - (1/8)y(0) + d(0) = 1$$
$$y(3) = (3/4)y(2) - (1/8)y(1) + d(1) = 7/4$$

We solve for $y(\infty)$ from

$$y(\infty) = (3/4)y(\infty) - (1/8)y(\infty) + 1$$

to yield $y(\infty) = 8/3$. Thus,

$$w(k) = y(k) = C_1(1/2)^k + C_2(1/4)^k + 8/3$$

and

$$w(2) = y(2) = 1 = C_1(1/2)^2 + C_2(1/4)^2 + 8/3$$
$$w(3) = y(3) = 3/4 = C_1(1/2)^3 + C_2(1/4)^3 + 8/3$$

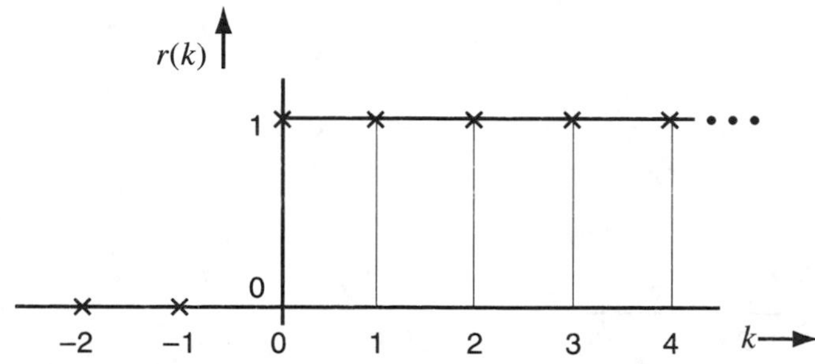

Figure 36.3 The unit-step input to obtain $w(k)$.

from which $C_1 = -8$ and $C_2 = 16/3$. Therefore,

$$w(k) = \begin{cases} -2(1/2)^{k-2} + (1/3)(1/4)^{k-2} + 8/3 & \text{for } k \geq 2 \\ 0 & \text{for } k < 2 \end{cases}$$

A plot is shown in Figure 36.2. Using Equation 36.5 with $h(k) = 8\,(1/2)^k - 16\,(1/4)^k$ for $k \geq 2$ we obtain

$$w(k) = \sum_{m=1}^{k} [2(1/2)^{m-2} - (1/4)^{m-2}]$$
$$= \frac{2 - 2(1/2)^{k-1}}{1 - (1/2)} - \frac{1 - (1/4)^{k-1}}{1 - (1/4)}$$
$$= -8(1/2)^k + (16/3)(1/4)^k + 8/3$$

which agrees with the $w(k)$ above. We also may use Equation 36.6 to obtain

$$h(k) = w(k) - w(k-1)$$
$$= [-2(1/2)^{k-2} + (1/3)(1/4)^{k-2} + 8/3]$$
$$- [-2(1/2)^{k-3} + (1/3)(1/4)^{k-3} + 8/3]$$
$$= 2(1/2)^{k-2} - (1/4)^{k-2} \text{ for } k \geq 2$$

which confirms previous results.

Discrete Convolution

We may use Equation 36.1 to solve for $y(k)$ when a given input $u(k)$ is applied.

EXAMPLE 36.4:

Let the input $u(k)$ shown in Figure 36.4 be applied to the discrete-time system of Example 36.1. Determine $y(k)$ by discrete convolution and sketch the result versus k.

We want to use the unit-sample response $h(k)$ to form $y(k)$ by using Equation 36.1. Figure 36.5 shows the steps in forming $h(k - n)$ and the limits on the summation.

Therefore, using the results of Figure 36.5 in Equation 36.1, for $k \leq 0$ we have

$$y(k) = \sum_{m=-\infty}^{k-1} (1/2)^{(k-n)-1}(2/3)^{-n}$$

$$= (1/2)^{k-1} \sum_{n=k-1}^{-\infty} (1/3)^{-n}$$

$$= (1/2)^{k-1} \sum_{\substack{m=1-k \\ (m=-n)}}^{\infty} (1/3)^{m} = (1/2)^{k-1} \frac{(1/3)^{1-k}}{1 - (1/3)}$$

$$y(k) = (2/3)^{-k} \quad \text{for} \quad k \leq 0$$

Note that $y(0) = 1$. For $k > 0$, we use Equation 36.1 to obtain

$$y(k) = \sum_{n=-\infty}^{-1} (1/2)^{k-n-1}(1)$$

$$= (1/2)^{k}\left[\sum_{n=-\infty}^{-1} (1/2)^{-n-1}(2/3)^{-n}\right] + \sum_{\substack{m=k-1 \\ (m=k-1-n)}}^{0} (1/2)^{m}$$

$$= (1/2)^{k}y(0) + \frac{1 - (1/2)^{k}}{1 - (1/2)}$$

$$y(k) = 2 - (1/2)^{k} \quad \text{for} \quad k > 0$$

A sketch of $y(k)$ is shown in Figure 36.6.

36.2 Numerical Integration Formulas

Difference equations can be formed for linear and nonlinear continuous-time systems when the input $u(t)$ is piecewise constant over the sampling interval of duration T. We use the notation $u(k)$ to denote the input over the interval $t_k \leq t < t_{k+1}$, where $t_{k+1} - t_k = T$. The output $y(k)$ is the value of y at the sampling instant $t = t_k$. Let a linear continuous-time system be described by the differential equation

$$\dot{y}(t) = ay(t) + u(t) \tag{36.7}$$

where the dot represents differentiation. Higher-order linear systems can be described by vector differential equations using the state variable format

$$\dot{\mathbf{x}}(t) = A\dot{\mathbf{x}}(t) + \mathbf{b}u(t) \tag{36.8}$$

$$y(t) = \mathbf{c}^{T}\mathbf{x}(t) + du(t)$$

The objective is to use numerical integration formulas to form differences equations corresponding to Equations 36.7 and 36.8 when $u(t)$ is a piecewise constant input $u(k)$.

Euler's Formula

The first-order formula referred to as Euler's formula uses the slope $\dot{y}(t)$ at time t_k to form

$$y(t_{k+1}) = y(t_k) + T\dot{y}(t_k) \tag{36.9}$$

which, from Equation 36.7, may be written simply as

$$y(k + 1) = y(k) + T[-ay(k) + u(k)]$$

$$= (1 - aT)y(k) + Tu(k)$$

From Equation 36.8, the vector form of Equation 36.9 is

$$\mathbf{x}(k + 1) = \mathbf{x}(k) + T\dot{\mathbf{x}}(k)$$

$$y(k) = \mathbf{c}^{T}\mathbf{x}(k) + du(k)$$

EXAMPLE 36.5:

Form the difference equations corresponding to

$$\dot{y}(t) = -y(t) + u(t)$$

and

$$\ddot{y}(t) + 3\dot{y}(t) + 2y(t) = u(t)$$

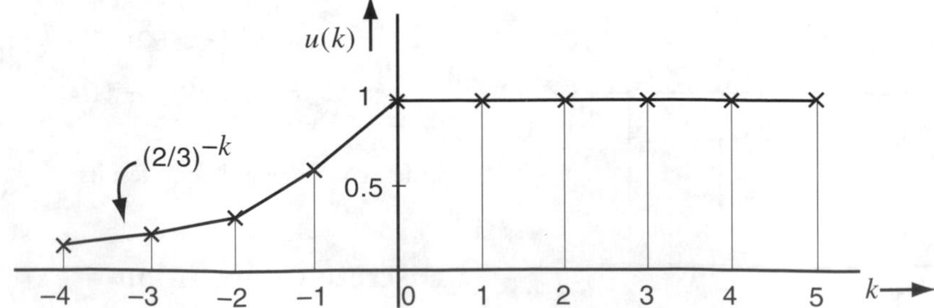

Figure 36.4 The input for Example 36.4.

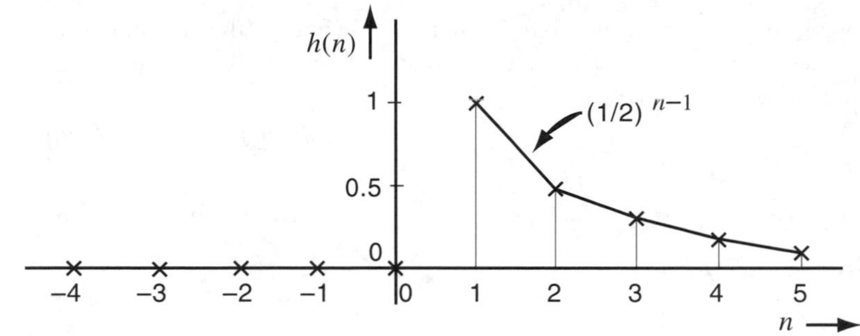

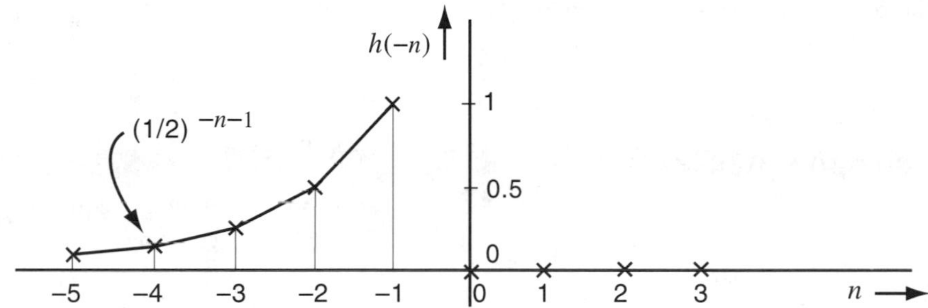

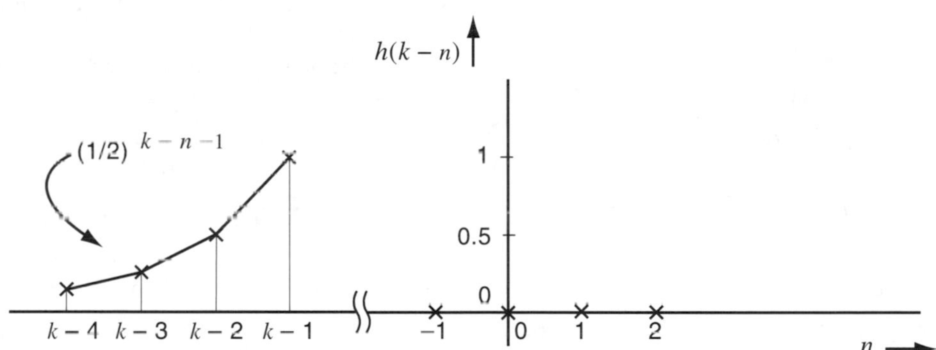

Figure 36.5 Forming $h(k - n)$ for Example 36.4.

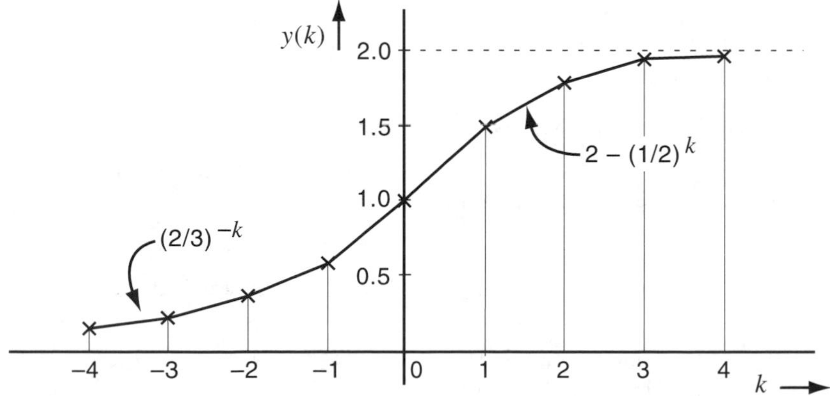

Figure 36.6 The output $y(k)$ versus k for Example 36.4.

when $u(t)$ is a piecewise constant scalar input $u(t_k)$ over the intervals $t_k \le t < t_{k+1}$. Use a sampling interval T of 0.1 seconds.

For the first equation, Euler's formula yields

$$\begin{aligned} y(k+1) &= y(k) + T\dot{y}(t_k) \\ &= y(k) + T[-y(t_k) + u(k)] \\ &= (1 - T)y(k) + Tu(k) \\ y(k+1) &= 0.9y(k) + 0.1u(k) \end{aligned}$$

where $T = 0.1$ seconds has been inserted.

For the second equation, we set $x_1(t) = y(t)$ and $x_2(t) = \dot{y}(t)$ to yield

$$\begin{aligned} \dot{x}_1(t) &= x_2(t) \\ \dot{x}_2(t) &= -2x_1(t) - 3x_2(t) - u(t) \end{aligned}$$

Using Euler's formula gives

$$\begin{aligned} x_1(k+1) &= x_1(k) + T[x_2(k)] \\ x_2(k+1) &= x_2(k) + T[-2x_1(k) - 3x_2(k) + u(k)] \end{aligned}$$

which can be written (for $T = 0.1$) as

$$\begin{aligned} x_1(k+1) &= x_1(k) + 0.1x_2(k) \\ x_2(k+1) &= -0.2x_1(k) + 0.7x_2(k) + 0.1u(k) \end{aligned}$$

The system output $y(k)$ is simply $x_1(k)$ for this system.

Runge-Kutta Formulas

A class of single-step formulas for numerical integration known as Runge-Kutta formulas are based on approximations using slopes at various points within the sampling interval. Euler's formula itself can be regarded as a first-order Runge-Kutta formula which forms an approximation for $y(k + 1)$ using simple rectangular integration.

A second-order Runge-Kutta formula (RK2) which uses trapezoidal integration requiring two slopes is given by

$$y(k+1) = y(k) + (T/2)[\dot{y}(t_{k+1}) + \dot{y}^*(t_{k+1})] \quad (36.10)$$

where $\dot{y}^*(t_{k+1})$ is obtained by evaluating the slope at t_{k+1} using $y^*(t_{k+1})$ from

$$y^*(t_{k+1}) = y(k) + T\dot{y}(t_k)$$

In other words, Equation 36.10 for RK2 combines the slope at time t_k and the approximate slope obtained by using a projection to time t_{k+1}. As an aside, the RK2 formula is also identified as Heun's method in the literature.

Computer libraries for numerical integration contain several versions of the fourth-order Runge-Kutta formula (RK4), which is based on Simpson's rule as a parabolic approximation for finding the area under a curve. Shown in Figure 36.7, together with Euler's formula and RK2, one RK4 formula is defined by

$$y(k+1) = y(k) + (T/6)[m_0 + 2m_1 + 2m_2 + m_3] \quad (36.11)$$

where m_0, m_1, m_2, and m_3 are slopes obtained from

$$\begin{aligned} m_0 &= \dot{y}[u(t_k), t_k] \\ m_1 &= \dot{y}[y(t_k) + (T/2)m_0, t_k + (T/2)] \\ m_2 &= \dot{y}[y(t_k) + (T/2)m_1, t_k + (T/2)] \\ m_3 &= \dot{y}[y(t_k) + Tm_2, t_{k+1}] \end{aligned}$$

EXAMPLE 36.6:

Form a vector difference equation for the second-order system of Example 36.5 given by

$$\ddot{y}(t) + 3\dot{y}(t) + 2y(t) = u(t)$$

when $u(t)$ is a piecewise constant scalar input $u(k)$ over $t_k \le t < t_{k+1}$. Let $T = 0.1$ seconds.

As in the previous example, we use a state variable form with $x_1(k) = y(k)$ and $x_2(k) = y(k+1)$. Thus, we have

$$\begin{aligned} \dot{x}_1(t) &= x_2(t) \\ \dot{x}_2(t) &= -2x_1(t) - 3x_2(t) + u(t) \end{aligned}$$

Using the RK2 formula in Equation 36.10, we first form

$$\begin{aligned} x_1^*(t_{k+1}) &= x_1(k) + T[x_2(k)] \\ x_2^*(t_{k+1}) &= x_2(k) + T[-2x_1(k) - 3x_2(k) + u(k)] \\ &= 2Tx_1(k) + (1 - 3T)x_2(k) + Tu(k) \end{aligned}$$

and then obtain

$$\begin{aligned} x_1(k+1) &= x_1(k) + (T/2)[\{x_2(k)\} + \{x_1(k) + Tx_2(k)\}] \\ x_2(k+1) &= x_2(k) + (T/2)[\{-2x_1(k) - 3x_2(k) + u(k)\} \\ &\quad + \{x_2(k) + T[-2x_1(k) - 3x_2(k) + u(k)]\}] \end{aligned}$$

Note that this result can be simplified to yield

$$\begin{aligned} x_1(k+1) &= (1 + T/2)x_1(k) + (T/2 + T^2/2)x_2(k) \\ &= 1.05x_1(k) + 0.055x_2(k) \\ x_2(k+1) &= (-T - T^2)x_1(k) + (1 - T - 3T^2/2)x_2(k) \\ &\quad + (T/2 + T^2/2)u(k) \\ &= -0.11x_1(k) + 0.885x_2(k) + 0.055u(k) \end{aligned}$$

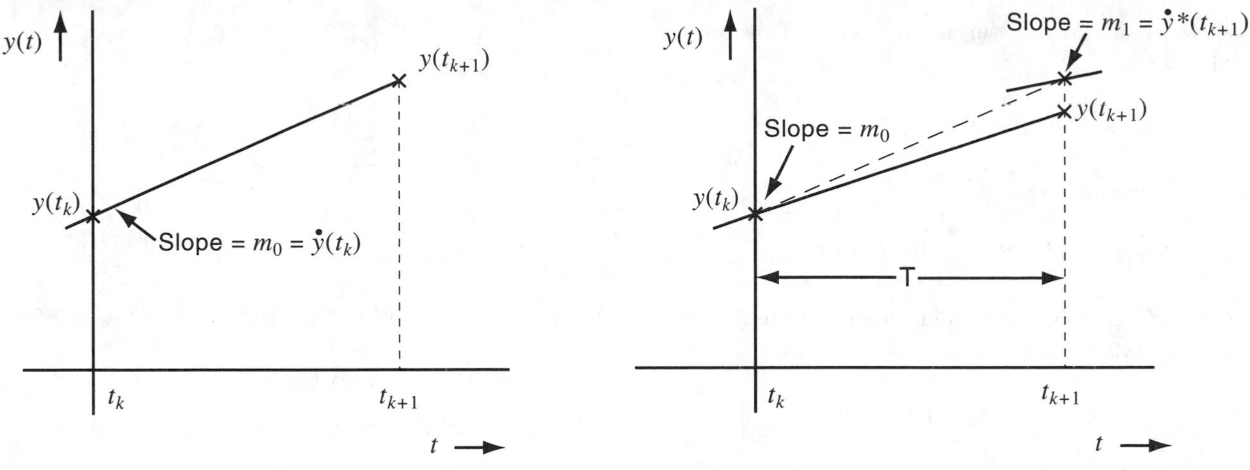

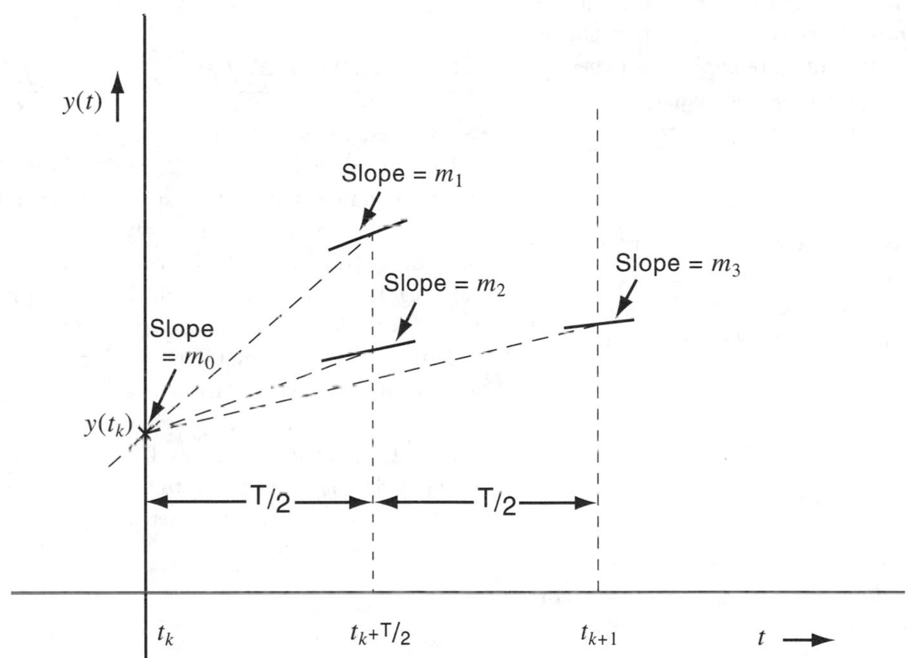

Figure 36.7 Geometrical descriptions of Euler's, RK2, and RK4 formulas.

Using the RK4 formula, we determine the slopes (vectors) as

$$m_0 = \begin{pmatrix} x_2(k) \\ -2x_1(k) - 3x_2(k) + u(k) \end{pmatrix}$$

$$m_1 = \begin{pmatrix} x_1(k) + (T/2)[x_2(k)] \\ x_2(k) + (T/2)[-2x_1(k) - 3x_2(k) + u(k)] \end{pmatrix}$$

$$m_2 = \begin{pmatrix} x_1(k) + (T/2)[x_1(k) + Tx_2(k)] \\ x_2(k) + T[x_2(k) + (T/2)\{-2x_1(k) - 3x_2(k) + u(k)\}] \end{pmatrix}$$

$$m_3 = \begin{pmatrix} x_1(k) + T[x_1(k) + (T/2) \\ \times \{x_1(k) + (T/2)[x_1(k) + Tx_2(k)]\}] \\ x_2(k) + T[x_2(k) + (T/2)\{x_2(k) + (T/2) \\ \times [x_2(k) + T\{-2x_1(k) - 3x_2(k) + u(k)\}]\}] \end{pmatrix}$$

We then use Equation 36.11 to obtain $x_1(k+1)$ and $x_2(k+1)$ and note that $y(k+1) = x_1(k+1)$.

We emphasize again that the Runge-Kutta formulas are applicable both for linear and nonlinear systems. We assume that the inputs are discrete levels, such as those provided as the output of a microprocessor.

36.3 Exact Difference Equations for Linear Systems

An exact solution exists in terms of a state transition matrix when the continuous-time system in Equations 36.7 and 36.8

are linear. Moreover, for time-invariant linear systems, we can express $x(t)$ as

$$\mathbf{x}(t) = e^{A(t-t_k)}\mathbf{x}(t_k) + A^{-1}[e^{A(t-t_k)} - I]\mathbf{b}u(t_k)$$

Thus, for $t = t_{k+1}$, we have

$$\mathbf{x}(k+1) = e^{AT}\mathbf{x}(k) + A^{-1}[e^{AT} - I]\mathbf{b}u(k) \quad (36.12)$$

Observe that e^{AT} contains an infinite number of terms when expressed in a Taylor series, i.e.,

$$e^{AT} = I + AT + A^2T^2/2 + A^3T^3/6 + \cdots$$
$$A^{-1}[e^{AT} - I] = T[I + AT/2 + A^2T^2/6 + \cdots] \quad (36.13)$$

For RK2 only terms in T through second-order are present, and for RK4 only terms in T through fourth-order appear. On the other hand, the Runge-Kutta formulas are applicable to general nonlinear systems for which exact difference equations cannot be determined.

EXAMPLE 36.7:

Form the exact difference equation for the second-order system of Examples 36.5 and 36.6 using Equation 36.12.

The continuous-time system can be expressed more conveniently as an uncoupled set of state variables to yield

$$\dot{x}_1(t) = -x_1(t) + u(t)$$
$$\dot{x}_2(t) = -2x_2(t) + u(t)$$
$$y(t) = x_1(t) - x_2(t)$$

where $x_1(t) = 2y(t) + \dot{y}(t)$ and $x_2(t) = y(t) + \dot{y}(t)$. The reason for this choice of state variables is that e^{AT} can be formed more easily because A is a diagonal matrix. We make this choice by observing that

$$Y(s)/U(s) = 1/(s^2 + 3s + 2)$$
$$= 1/(s+1)(s+2)$$
$$= 1/(s+1) - 1/(s+2),$$

where s is the Laplace-transformed variable. Using this uncoupled set of state variable yields

$$A = \begin{pmatrix} -1 & 0 \\ 0 & -2 \end{pmatrix}$$

from which

$$e^{AT} = \begin{pmatrix} e^{-T} & 0 \\ 0 & e^{-2T} \end{pmatrix}$$

and

$$A^{-1}(e^{AT} - I) = \begin{pmatrix} 1 - e^{-T} & 0 \\ 0 & (1 - e^{-2T})/2 \end{pmatrix}$$

Therefore, the exact (vector) difference equation we seek is

$$x_1(k+1) = e^{-T}x_1(k) + (1 - e^{-T})u(k)$$
$$x_2(k+1) = e^{-2T}x_2(k) + [(1 - e^{-2T})/2]u(k)$$
$$y(k) = x_1(k) - x_2(k)$$

36.4 Summary

We have examined difference equations to describe the dynamic behavior of digital control systems. We formed the unit-sample response and showed how to express the general system response as a convolution summation involving the system's input and its unit-sample response. We also considered Runge-Kutta numerical integration formulas to obtain difference equations as approximations for linear and nonlinear continuous-time systems.

Finally, we observe that while the incorporation of a feedback loop obviously changes the continuous-time or discrete-time equations, the procedures presented here are clearly applicable. The system dynamics will have been altered, i.e., improved by a properly designed controller, but the need to form and solve the resulting difference equation remains.

References

Brogan, William L. 1991. *Modern Control Theory,* 3rd ed., Prentice Hall, Englewood Cliffs, NJ.

Franklin, Gene F., Powell, J. David, and Workman, Michael L. 1990. *Digital Control of Dynamic Systems,* 2d ed., Addison-Wesley, Reading, MA.

Kuo, Benjamin, C. 1991. *Automatic Control Systems,* 6th ed., Prentice Hall, Englewood Cliffs, NJ.

Mayhan, Robert J. 1984. *Discrete-Time and Continuous-Time Linear Systems,* Addison-Wesley, Reading, MA.

Phillips, Charles L. and Nagle, H. Troy. 1990. *Digital Control Systems Analysis And Design,* 2d ed., Prentice Hall, Englewood Cliffs, NJ.

Rowland, James R. 1986. *Linear Control Systems,* John Wiley and Sons, New York, NY.

37

Digital Control

John Y. Hung
Auburn University

Victor Trent
Auburn University

37.1 Introduction 553
37.2 Discretization of Continuous-Time Systems 553
37.3 Discretization of the Servomotor System 554
37.4 Frequency Domain Design through the *w*-Transform 555
37.5 Root Locus Design on the Unit Circle 556
37.6 Simulation Comparisons 557

37.1 Introduction

Techniques to design a digital control system (discrete-time transfer function C(z)) can be grouped into two basic approaches. The first approach is to convert a continuous-time controller transfer function into a discrete-time controller. The approach is also known as *digital simulation of a controller* that has been designed using classical continuous-time models and techniques. The second group of approaches is to perform the design using discrete-time models. In other words, the conversion of a continuous-time *plant* is first performed. Then, the design is accomplished in discrete-time domain. Another interpretation of this class of approaches is that the plant is being simulated by the discrete-time model. Popular design methods include root locus analysis on the unit circle and frequency domain analysis. All classical approaches involve conversion of continuous-time functions to discrete-time, but they differ in where the conversion is executed. In the first case (controller simulation), the conversion is performed *after* a classical continuous-time design. In the second case (plant simulation), the conversion is performed on the plant model, *before* the design process. The difference is illustrated in Figure 37.1. These two classes of design are not equivalent, however, since different digital controllers may result from each approach.

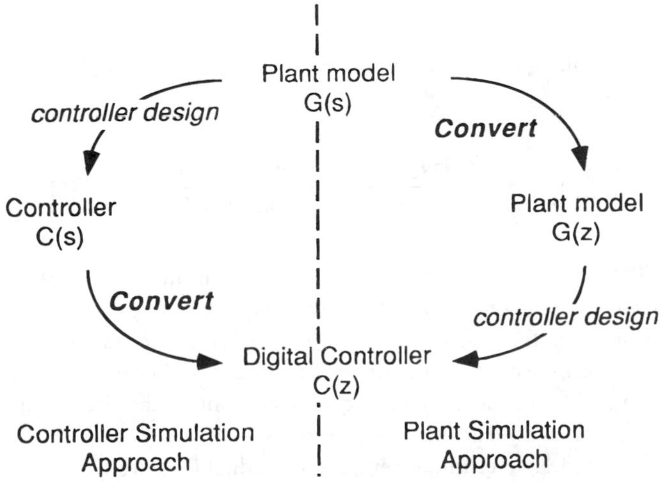

Some systems to be controlled are inherently discrete-time. In this case, the design model is already in discrete-time, so the model conversion is not required and the design process proceeds exactly as for the plant simulation approach.

In this chapter, the design of a digital controller using both approaches is reviewed. The example control system is a DC motor position control system, and the example demonstrates the design and implementation of a digital PID controller (see Figure 37.2). The first approach discussed is the conversion of an *s*-domain controller to a *z*-domain approximation. The second approach is to discretize the motor model, and then design the *z*-domain controller using frequency domain or root locus techniques.

Figure 37.1 Approaches for designing a digital controller, $C(z)$.

37.2 Discretization of Continuous-Time Systems

One approach used for the development of digital controllers is the conversion of continuous-time controller transfer functions into discrete-time representations. The techniques employed for discretizing a continuous-time system are based on digital approximation of continuous-time integrators. Some of the techniques include Euler's formula, Runge-Kutta equations (Chapter 36.2) and development of a zero-order hold (ZOH) equivalent for the continuous time system (Phillips and Nagle, 1995). Another digital simulation technique is the bilinear transformation or Tustin's transformation (Phillips and Nagle, 1995). The bilinear transformation is one of the simplest and most commonly used methods for converting a continuous-time transfer function into a discrete-time transfer function.

The bilinear transformation is the trapezoidal integration rule applied to a continuous time system:

0-8493-8343-9/97/$0.00+$.50

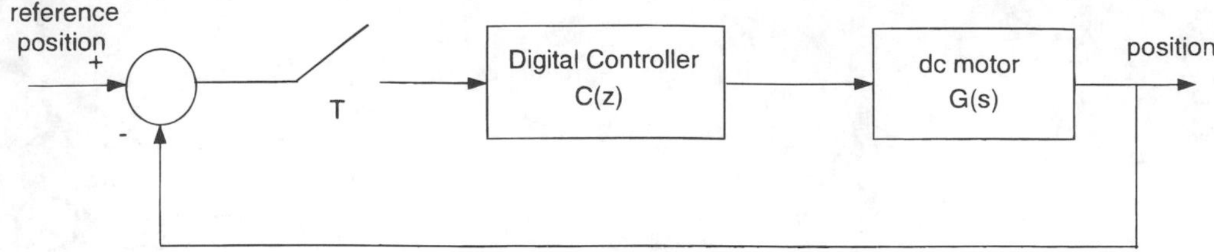

Figure 37.2 The example digital control system.

$$s = \frac{2}{T} \cdot \frac{(z-1)}{(z+1)} \quad (37.1)$$

where T is the sampling period. The bilinear transformation maps a portion of s-plane into the unit circle of the z-plane. The bilinear transformation introduces distortion in the frequency response near the sampling frequency. The frequency response distortion can be alleviated by prewarping the s-domain filter (Phillips and Nagle, 1995). A typical rule of thumb used for the bilinear transformation as a discretizing option is to maintain frequencies of interest in the transformation less than one tenth of the sampling frequency.

EXAMPLE 37.1

Consider the block diagram shown in Figure 37.2. The DC servomotor model is given by:

$$G(s) = \frac{36}{s(s+3)} \quad (37.2)$$

An analog PID controller has been designed for this system to yield a 70 degree phase margin at a frequency of 3.5 rad/sec (Chapter 25). Additionally, in order to make the controller transfer function proper, an instrumentation pole at $s = -300$ has been included with the derivative term of the controller. The continuous-time PID controller transfer function is

$$C(s) = 0.390459 + \frac{0.1}{s} + \frac{0.0710z1s}{(0.003333s + 1)} \quad (37.3)$$

The bilinear transformation is used to discretize the continuous-time PID controller transfer function. The sampling period selected for this conversion is $T = 0.01$ sec. Substituting Equation 37.1 into Equation 37.3 yields the digital controller transfer function:

$$C(z) = \frac{8.913479(z^2 - 1.947254z + 0.9473886)}{z^2 - 0.8z - 0.2} \quad (37.4)$$

37.3 Discretization of the Servomotor System

Another design approach used for the development of digital control systems involves the conversion of the continuous-time plant into a discrete-time model (plant simulation) and the design of a digital controller based on the discrete-time plant model. As was the case with the discretizing of the continuous-time controller in the previous section, many methods for converting the continuous-time plant into a discrete-time representation exist and are mentioned. The method employed in this example involves the use of a ideal sampler and zero-order hold (ZOH) model to represent the analog to digital (A/D) and digital to analog (D/A) conversion process (Phillips and Nagle, 1995).

EXAMPLE 37.2

As an example of the conversion of the continuous-time plant into a discrete-time model, consider the continuous-time DC servomotor model of Equation 37.2 in the system shown in Figure 37.2. The continuous-time plant is discretized using the ZOH equivalence technique and a sample period of $T = 0.01$ sec (Phillips and Nagle, 1995). The open-loop discrete-time plant transfer function is

$$G(z) = \mathcal{Z}\left[\frac{1 - \epsilon^{-sT}}{s} \cdot \frac{36}{s(s+3)}\right] \quad (37.5)$$

Employing the theory of residues, Equation 37.5 may be rewritten as:

$$G(z) = \frac{z-1}{z} \cdot \mathcal{Z}\left[\frac{-4}{s} + \frac{12}{s^2} + \frac{4}{s+3}\right] \quad (37.6)$$

and recalling the linearity property of z-transforms (Phillips and Nagle, 1995):

$$G(z) = \frac{1.782134 \times 10^{-3}(z + 0.9900499)}{(z-1)(z - 0.9704455)} \quad (37.7)$$

Once a discrete-time model of the continuous-time plant transfer function is available, controller designs may be performed in the discrete-time domain. Classical techniques for discrete-time controller design exist and are modifications of the continuous-time design techniques to account for the discretizing of the system. The next two sections will discuss two of the classical design techniques employed for digital controller design: frequency domain design through the w-transform and root locus design on the unit circle.

37.4 Frequency Domain Design through the w-Transform

Two approaches may be employed for the discrete-time design of digital controllers using the frequency domain. The first involves the generation of the frequency response of the discrete-time plant open-loop transfer function by using the relationship:

$$z = \epsilon^{sT} = \epsilon^{j\omega T} \tag{37.8}$$

The generation of the frequency response using this relationship is typically performed using a digital computer. Another approach which may be employed is to use the inverse bilinear transformation:

$$z = \frac{1 + (T/2)w}{1 - (T/2)w} \tag{37.9}$$

to convert the discrete-time open loop plant transfer function from the z-domain to the w-domain. Then the s-domain rules for construction of the frequency response apply. Since the technique involving the transformation of the discrete-time transfer function representation into the w-domain typically does not require a digital computer for generation of the frequency response, it is the method that will be presented in this section.

Consider the discrete-time open-loop transfer function representing the plant (Equation 37.7) developed using the ZOH equivalence method. If the bilinear transformation (Equation 37.9) is applied to this transfer function, the resulting w-domain transfer function representation is

$$G(w) - \frac{-4.499595 \times 10^{-6}(w + 40{,}000.6)(w - 200)}{w(w + 2.999775)} \tag{37.10}$$

The frequency response for the w-domain system of Equation 37.10 can be easily generated using the asymptotic approximations employed for generating continuous-time frequency responses (Chapter 21). If the design problem specifies a desired phase margin for the compensated system, then the magnitude and the angle of the compensated system at the 0 db crossing must satisfy the relationship:

$$C(j\omega_w)G(j\omega_w) = 1\angle 180 + \varphi_m \tag{37.11}$$

where φ_m is the desired phase margin, $C(j\omega_w)$ is the digital controller frequency response evaluated at ω_w (the compensated system 0 db crossing), and $G(j\omega_w)$ is the uncompensated plant frequency response evaluated at ω_w.

The PID w-domain controller transfer function may be expressed as:

$$C(w) = K_p + \frac{K_I}{w} + K_D w \tag{37.12}$$

where K_P is the proportional gain, K_I is the integral gain and K_D is the derivative gain term of the PID controller. At the design frequency, ω_w, the controller function is evaluated as:

$$K_P + j\left(K_D\omega_w - \frac{K_I}{\omega_w}\right) = |C(j\omega_w)|(\cos\theta + j\sin\theta) \tag{37.13}$$

where θ is $180 + \varphi m - \text{ang}(G(j\omega_w))$ (note: $\text{ang}(G(j\omega_w))$ is the angle of the uncompensated plant at $j\omega_w$). Equating the real and the imaginary parts of Equation 37.11 and 37.13, the resulting design equations for a w-domain PID controller are

$$K_P = \frac{\cos\theta}{|G(j\omega_w)|}$$

$$K_D\omega_w - \frac{K_I}{\omega_w} = \frac{\sin\theta}{|G(j\omega_w)|} \tag{37.14}$$

Once the controller is specified in the w-plane, the z-plane representation for the controller may be developed by noting that the bilinear transformation (Equation 37.9) may be solved for w as a function of z to yield:

$$w = \frac{2}{T}\cdot\frac{z - 1}{z + 1} \tag{37.15}$$

Note that the design constraints results in a set of two design equations with three unknowns. In the design of the PID controller one of the parameters is chosen and the remaining two parameters are solved for using (Equation 37.14).

EXAMPLE 37.3

Once again the DC servomotor model of Equation 37.2 will be used for the design of the digital controller. As was the case with Example 37.1, the PID compensated DC servomotor system should have a phase margin of 70 degrees at a frequency of 3.5 rad/sec. The frequency response for the w-plane representation of the plant (Equation 37.10) was generated and at $\omega_w = 3.5$ rad/sec, the magnitude of $G(j\omega_w)$ is 6.96 db and the angle of $G(j\omega_w)$ is -140.42 degrees. If $K_I = 0.1$, then using Equation 37.14:

$$K_P = 0.38696196$$

$$K_D = 0.0730806 \tag{37.16}$$

and incorporating the instrumentation pole into the controller after the design the w-plane representation for the controller is

$$C(w) = \frac{0.07437052w^2 + 0.3872953w + 0.1}{w(0.003333w + 1)} \tag{37.17}$$

If the inverse bilinear transformation (Equation 37.15) is substituted into Equation 37.17 then the digital controller transfer function $C(z)$ is

$$C(z) = \frac{9.15714(z^2 - 1.949116z + 0.9492468)}{z^2 - 0.8z - 0.2} \tag{37.18}$$

37.5 Root Locus Design on the Unit Circle

In addition to the frequency domain design techniques presented in previous sections, the design of a digital controller may be performed in the z-domain using root locus techniques. The concepts for digital controller design using root locus techniques are analogous to root locus compensator design for continuous-time systems (Sections 5.6.1–5.6.2; Phillips and Harbor, 1996; Dorf, 1992; and Franklin et al., 1991) with the exception that the design is performed in the z-plane. Since the boundary of stability in the z-plane is a unit circle (compared to the imaginary axis stability bound in the s-plane), the interpretation of the root locus in the z-plane differs from that of the s-plane even though the rules for constructing the root locus are identical in both domains.

Consider the DC servomotor system shown in Figure 37.2. The characteristic equation for this system is given by:

$$1 + C(z)G(z) = 0 \tag{37.19}$$

Equation 37.19 may also be expressed in terms of a magnitude and an angle as:

$$|C(z)G(z)| = 1 \tag{37.20}$$

and

$$\text{ang}(C(z)G(z)) = 180^\circ \tag{37.21}$$

where Equation 37.20 is called the magnitude criterion and Equation 37.21 is called the angle criterion (Phillips and Nagle, 1995). Any root of the system characteristic equation must satisfy Equations 37.20 and 37.21, hence any root location that is desired for the closed-loop system (possibly through compensation) must also satisfy the magnitude and angle criteria.

Typically, controller design in the z-plane using root locus techniques involves: conversion of a set of desired root locations specified in the continuous-time domain into their equivalent location in the discrete-time domain (Equation 37.8) and placement of the closed-loop system poles at these desired z-plane root locations through the development of controller constants. The design is satisfied if the compensated system root locus can be made to pass through the desired root locations, hence Equations 37.20 and 37.21 represent a set of design equations that must to satisfied to successfully realize the controller.

EXAMPLE 37.4

Consider the DC servomotor system shown in Figure 37.2 and modeled in continuous-time by the transfer function shown in Equation 37.2 and in discrete-time (using the ZOH equivalence method) by Equation (37.7). Additionally, consider the frequency domain design of Example 37.3. The characteristic equation of the closed-loop compensated system of Example 37.3 has a pair of complex conjugate roots located in the z-plane at $z_1 = 0.9731209 + j0.0225782$ and $z_2 = 0.9731209 - j0.0225782$, as well as two poles located on the real axis. For this example, a digital controller using root locus techniques in the z-plane will be designed to place a pair of the compensated system closed-loop characteristic equation at the locations specified by z_1 and z_2. Recall that the ZOH equivalent of the open-loop plant transfer function is given as (Equation 37.7):

$$G(z) = \frac{1.782134 \times 10^{-3}(z + 0.9900499)}{(z-1)(z-0.9704455)}$$

Substituting a desired root location (in this case z_1) into Equation 37.7 and expressing the result as a magnitude and an angle yields:

$$G(z)|_{z=z_1} = 4.38388 \cdot -222.5534^\circ \tag{37.22}$$

(Note that substituting z_2 into Equation 37.7 would yield the complex conjugate of Equation 37.21). Furthermore, the form of the digital controller, $C(z)$, used for this example is

$$C(z) = K_P + K_I \cdot \frac{T}{2} \cdot \frac{z+1}{z-1} + K_D \cdot \frac{z-1}{T_z} \tag{37.23}$$

where the integral term is the trapezoidal rule for numerical integration and the derivative term is the slope between sampling

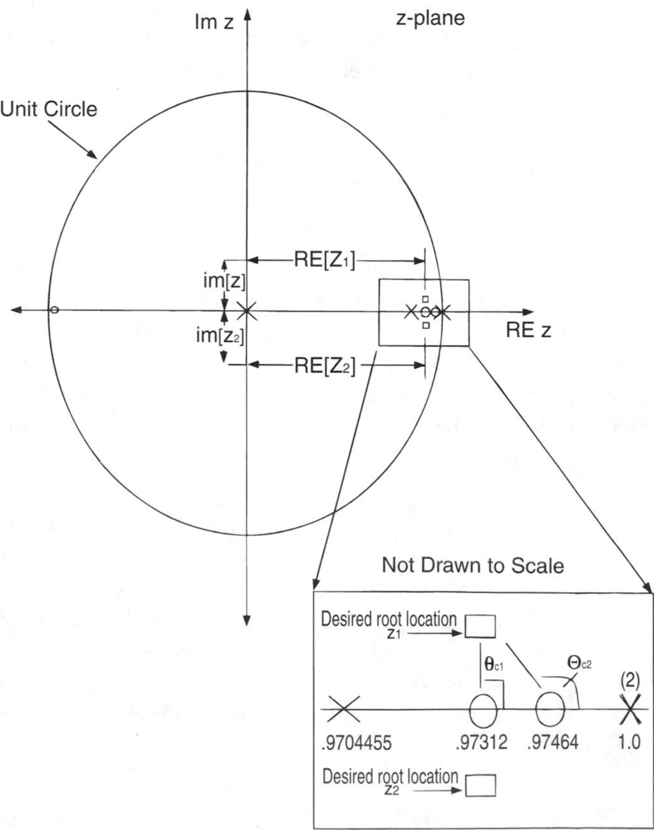

Figure 37.3 Open-loop pole-zero z-plane plot for the compensated system used in the root locus design example.

instances (a numeric approximation for the derivative). The controller will introduce a pole at the origin of the z-plane, a pole a 1 in the z plane, and two zeros whose locations will be defined by the constants K_I, K_P and K_D, and may be expressed as:

$$C(z) = \frac{K(Z - z_{C_1})(z - z_{C_2})}{z(z - 1)} \quad (37.24)$$

where K is a gain, and Z_{C_1} and Z_{C_2} are the controller zeros.

If the desired root location, z_1 is substituted into Equation 37.24, the result may be expressed as:

$$C(z)|_{z=z_1} = K(Z_1 - Z_{C_1})(Z_1 - Z_{C_2}) \cdot 0.341692/-141.2992° \quad (37.25)$$

and using Equations 37.22 and 37.25 the system characteristic Equation 37.19 may be rewritten as:

$$1 + K(Z_1 - Z_{C_1})(Z_1 - Z_{C_2})(1.49794/-3.853°) = 0 \quad (37.26)$$

Recall that in order for a root to lie on the root-locus, the root must satisfy both the magnitude criterion (Equation 37.20) and the angle criterion (Equation 37.21). From the angle criteria, it is apparent that the numerator of digital controller must provide a net angle of 183.853 degrees so that Equation 37.21 is satisfied with respect to angle. In order to simplify the calculations, one of the zeros of the controller, Z_{C_1}, will be placed on the real-axis of the z-plane and directly under the desired root location (Figure 37.3), hence, the angle from this zero to the desired root location is 90 degrees ($Z_{C_1} = 0.9731209$). The location of the second zero of the digital controller may be found using Equation 37.21.

$$\theta_{Z_{C_2}} = 180° + 3.853° - 90° = 93.853° \quad (37.27)$$

Equation 37.27 represents the angle from the zero placed in the z-plane to the desired root location z_1. Since the desired root location is known in the z-plane, trigonometry may be used to calculate the zero location in the z-plane.:

$$Z_{c_2} = \mathrm{RE}(z_1) + \frac{\mathrm{IM}(z_1)}{Y} \quad (37.28)$$

where RE(z_1) is the real part of the desired root location, IM(z_1) is the imaginary part of the desired root location, and $y = \tan(\theta_{C_2})$ if $\theta_{C_2} < 90$ degrees or $y = \tan(180 - \theta_{C_2})$ if $\theta_{C_2} > 90$ degrees. Substituting and solving the resulting zero location is $Z_{C_2} = 0.97464135$.

With the zero locations now defined, Equation 37.26 evaluated at the desired root location (z_1) may be written as:

$$K(0.065555/180°) = 1/180° \quad (37.29)$$

where K represents a variable gain. In order to solve for K, recall the magnitude criterion of Equation 37.20, hence:

$$K = \frac{1}{0.065555} = 15.2551 \quad (37.30)$$

and the controller transfer function, $C(z)$, is

$$C(z) = \frac{15.2551(z - 0.97464133)(z - 0.9731209)}{z(z - 1)} \quad (37.31)$$

If the individual constants, K_I, K_P, and K_D are desired, equate Equations 37.23 and 37.31 and solve.

37.6 Simulation Comparisons

Simulations of the compensated DC servomotor system were performed using MATLAB. As a basis for comparison, the analog system compensated with the analog controller (Equation 37.3) was also simulated. Recall that the design employing the bilinear transformation of the continuous-time controller and the w-plane design were both performed using a sample period $T = 0.01$ sec. Since this sample period is fast compared to the dynamics of the plant, the simulation results comparing the digital controllers to the analog controlled system essentially overlay one another.

To demonstrate the effect the sampling period has on the design of digital controllers, the designs of Examples 37.1 and 37.2 were repeated using a sampling period of 0.1 sec. The resulting digital controller transfer function using the bilinear transformation is

$$C(z) = \frac{1.727103(z^2 - 1.564887z + 0.5757429)}{z^2 - 0.125z - 0.875} \quad (37.32)$$

and the digital controller transfer function using the w-plane design method is

$$C(z) = \frac{2.00171(z^2 - 1.667466z + 0.6768327)}{z^2 - 0.125z - 0.875} \quad (37.33)$$

The system shown in Figure 37.2 was simulated with the analog controller of Equation 37.3 and the two digital controllers of Equations 37.32 and 37.33. The simulations were performed assuming a unit step input applied to the compensated DC servomotor system. The integration algorithm employed was a fixed-step Runge-Kutta routine and the step size used was 0.01 sec for the analog PID controlled system and 0.1 sec for the digitally controlled system. The results of the simulation are shown in Figure 37.4.

The simulation results indicate a change in the system response

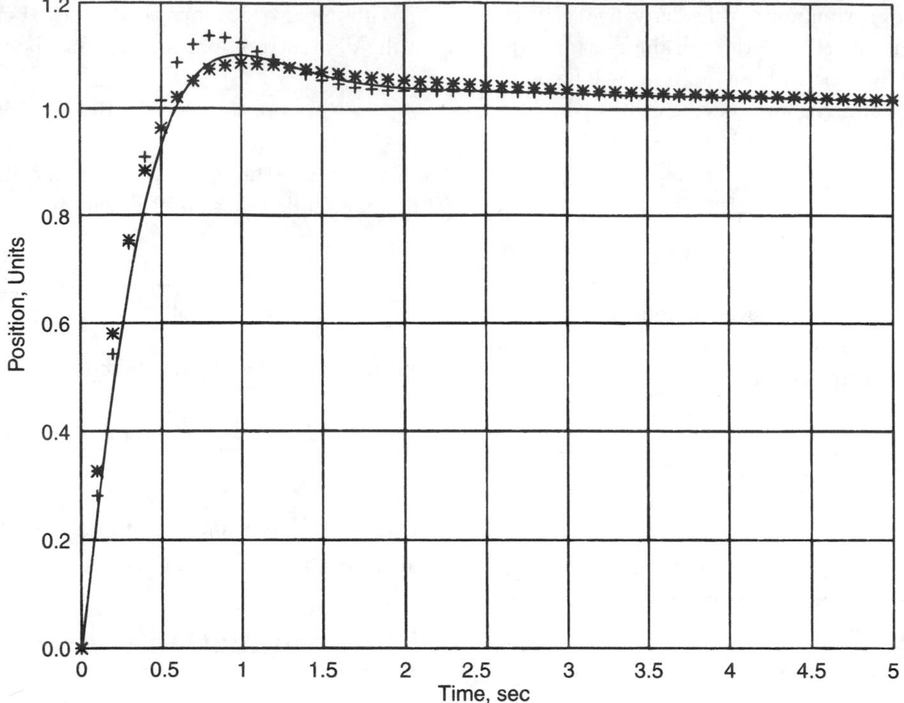

Figure 37.4 Step response of the compensated DC servomotor system: solid line, analog PID controller design; +, bilinear transformed controller design; *, *w*-plane controller design.

(as compared to the analog system) that is dependent on the sampling rate. The system compensated with the bilinear transformed PID controller has more overshoot then the analog PID controlled system. Although the *w*-plane designed filtered has decreased the overshoot (as compared to the analog PID controlled system), the settling time has increased. A further increase in the sampling period will tend to destabilize the system. In order to maintain satisfactory transient response characteristics in a digitally controlled system, it is usually recommended to sample as quickly as possible.

References

Dorf, Richard C. 1992. *Modern Control Systems,* 6th ed., Addison-Wesley, Reading, MA.

Franklin, Gene F., et al. 1991. *Feedback Control of Dynamic Systems,* 2d ed., Addison-Wesley, Reading, MA.

Phillips, C. L. and Nagle, H. T. 1995. *Digital Control System Analysis and Design,* 3d ed., Prentice Hall, Englewood Cliffs, NJ.

Phillips, C. L. and Harbor, R. D. 1996. *Feedback Control Systems,* 3d ed., Prentice Hall, Englewood Cliffs, NJ.

38
Estimation and Identification

38.1 Kalman Filters 559
System Equations and Assumptions • Discrete Kalman Filter Equations • Discrete Kalman Filter Examples • Discrete Kalman Filter with Sparse Measurements
38.2 Other Types of Kalman Filters 561
38.3 Identification 561
Least-Squares System Identification Example • Least-Squares System Identification: General Case

Thomas S. Denney, Jr.
Auburn University

Many control strategies presented in this chapter assume that the plant dynamics are known and the state vector can be measured at any given time. In most practical applications, however, certain parameters of the plant dynamics may be unknown, and there may be uncertainties (noise) in the measurement of the state vector and in the plant inputs. The problem of determining the state vector in the presence of noisy plant inputs and noisy measurements is called *state estimation.* The problem of determining unknown plant parameters is called *system identification.*

38.1 Kalman Filters

If the plant input is known and the state measurements are perfect (or nearly perfect), a deterministic Luenberger-type observer (Section 5.7) can be used to estimate the state vector at a given point in time. In this section we are concerned with the case where there are uncertainties (noise) in both the plant input and the state measurements. A commonly used approach to this problem is to use a *Kalman filter* to estimate the state vector. Kalman filters are a broad class of filters that can be applied in both continuous-time and discrete-time problems. Kalman filters are particularly useful in practical applications because they are recursive and can be used for both stationary and non-stationary noise processes.

System Equations and Assumptions

Consider the following discrete-time linear system

$$x_{k+1} = A_k x_k + B_k u_k + G_k w_k \tag{38.1a}$$

$$y_k = C_k x_k + v_k, \tag{38.1b}$$

where $k = 1, 2, \ldots$. x_0, w_k, and v_k are random variables with the following statistics

$$x_0 \sim (\bar{x}_0, P_0), \qquad w_k \sim (0, Q_k), \qquad \text{and} \qquad v_k \sim (0, R_k),$$

where the notation $\sim (\mu, \Sigma)$ means that the random variable has mean μ and covariance Σ. We assume that sequences $\{w_k\}$ and $\{v_k\}$ are white random processes uncorrelated with x_0 and each other. w_k, v_k, and x_0 are often assumed to have a Gaussian distribution, but this assumption is not necessary. The random vector x_0 models the uncertainty in the initial conditions of the system. The term $B_k u_k$ is the known part of the system input and $\{w_k\}$ models the uncertainty in the system input. The random process $\{v_k\}$ models the uncertainty in the system measurements. We assume that the system parameters A_k, B_k, G_k, C_k, $\bar{x}_0$, P_0, Q_k, and R_k are known. For the case when these parameters are not known, see section 38.3 and Gelb (1974). Note that the system parameters can change with time.

Discrete Kalman Filter Equations

Given the system equations and assumptions above, the objective of the Kalman filter is to compute a state vector estimate $\hat{x}_k$ based on the measurements y_k such that the estimation error covariance $P_k = \mathcal{E}\{\tilde{x}_k \tilde{x}_k^T\}$ is minimized, where $\mathcal{E}\{\cdot\}$ represents expected value and $\tilde{x}_k = x_k - \hat{x}_k$ is the estimation error. Let the Kalman filter have the following form

$$\hat{x}_{k+1}^- = A_k \hat{x}_k + B_k u_k \tag{38.2a}$$

$$\hat{x}_{k+1} = \hat{x}_{k+1}^- + K_{k+1}(y_{k+1} - C_{k+1}\hat{x}_{k+1}^-), \tag{38.2b}$$

where $\hat{x}_0 = \bar{x}_0$. It can be shown (Equations 38.1, 38.2) that with this initial condition, the estimation error has zero mean. The

0-8493-8343-9/97/$0.00+$.50

motivation for Equations 38.2 is as follows. Equation 38.2b *predicts* the state estimate $\hat{x}_{k+1}^-$ based on the plant dynamics. The predicted estimate is then *corrected* by Equation 38.2b based on the difference between actual measurement y_{k+1} and the predicted measurement $C_{k+1}\hat{x}_{k+1}$.

The amount of correction is determined by the *Kalman gain* matrix K_{k+1}, which is chosen such that the estimation error covariance is minimized. It can be shown (Gelb, 1974; Lewis, 1986) that the optimal K_{k+1} is given by

$$\begin{aligned} K_{k+1} &= P_{k+1}C_{k+1}^T R_{k+1}^{-1} \\ &= P_{k+1}^- C_{k+1}^T (C_{k+1}P_{k+1}^- C_{k+1}^T + R_{k+1})^{-1}, \end{aligned} \quad (38.3)$$

where P_{k+1}^- is the covariance of predicted estimate x_{k+1}^-. We see from Equation 38.3 that the Kalman gain matrix K_{k+1} is roughly proportional to the estimation error covariance P_{k+1} and inversely proportional to the measurement noise variance R_{k+1}. Therefore if the error covariance is increased, more weight is given to the measurement correction term $(y_{k+1} - C_{k+1}\hat{x}_{k+1})$ in Equation 38.2b. Conversely if the measurement noise is increased, more weight is given to the predicted estimate term $\hat{x}_{k+1}^-$.

P_{k+1} is the error covariance of the corrected state estimate $\hat{x}_{k+1}$ and is predicted and corrected according to the equations

$$P_{k+1}^- = A_k P_k A_k^T + G_k Q_k G_k^T \quad (38.4a)$$

$$P_{k+1} = (I - K_{k+1}C_{k+1})P_{k+1}^-. \quad (38.4b)$$

Note that the error covariance is only a function of the system parameters and not a function of the measurements y_k. As a result, the error covariance can be computed *before* any measurements are obtained.

Equations 38.2, 38.3, and 38.4 form the discrete Kalman filter. These equations are typically implemented in the following steps:

Initial Conditions:

$$\hat{x}_0 = \bar{x}_0, \qquad P_0 = P_0 \quad (38.5a)$$

Predict:

$$\hat{x}_{k+1}^- = A_k\hat{x}_k + B_k u_k \quad (38.5b)$$

$$P_{k+1}^- = A_k P_k A_k^T + G_k Q_k G_k^T \quad (38.5c)$$

Compute Kalman Gain:

$$K_{k+1} = P_{k+1}^- C_{k+1}^T (C_{k+1}P_{k+1}^- C_{k+1}^T + R_{k+1})^{-1} \quad (38.5d)$$

Correct:

$$\hat{x}_{k+1} = \hat{x}_{k+1}^- + K_{k+1}(y_{k+1} - C_{k+1}\hat{x}_{k+1}^-) \quad (38.5e)$$

$$P_{k+1} = (I - K_{k+1}C_{k+1})P_{k+1}^-. \quad (38.5f)$$

Discrete Kalman Filter Examples

To demonstrate the discrete Kalman filter, consider the following discretized model of a servomotor:

$$x_{k+1} = \begin{bmatrix} 1 & 0.0838 \\ 0 & 0.6942 \end{bmatrix} x_k + \begin{bmatrix} 0.5705 \\ 10.76 \end{bmatrix} u_k + \begin{bmatrix} 0 \\ 1 \end{bmatrix} w_k \quad (38.6a)$$

$$y_k = [1 \quad 0]x_k + v_k, \quad (38.6b)$$

where

$$x_0 = \begin{bmatrix} 0 \\ 0 \end{bmatrix}, \quad P_0 = \begin{bmatrix} 1 & 0 \\ 0 & 1 \end{bmatrix}, \quad Q_k = 1, \quad \text{and} \quad R_k = (1.5)^2.$$

The states x_1 and x_2 represent the angle (in degrees) and angular velocity (in degrees/second) of the motor shaft. The input to the servomotor consists of a known armature voltage u_k and a random disturbance w_k. Only the shaft angle is measured. We assume that the angle measurement errors have zero mean and a Gaussian distribution. This assumption means that 68% of the measurements are accurate to within ± 1.5 degrees of the true angle and 95% of the measurements are accurate to within ± 3.0 degrees.

The error covariance P_k and the Kalman gain matrix K_k for this example are plotted in Figure 38.1. Note that even for a constant coefficient system, both the error covariance and the Kalman gain matrix vary with time. In constant coefficient systems, however, P_k and K_k eventually reach a steady state. The initial values for both P_k and K_k are determined by the initial uncertainty in the state vector specified by P_0. The steady-state value is roughly proportional to the amount of process noise (Q_k) and measurement noise (R_k). An increase in either one will increase the steady-state error covariance. For this example the steady-state error covariance is

$$P = \begin{bmatrix} 0.32 & 0.23 \\ 0.23 & 1.88 \end{bmatrix}.$$

This means that in the steady state ($k > 15$ in this example), 68% of the Kalman filter estimates of shaft position are accurate to within ± 0.57 degrees and 68% of the shaft velocity estimates are accurate to within ± 1.37 degrees/second.

Discrete Kalman Filter with Sparse Measurements

In some applications state vector measurements are not available at each sample period. In these cases, the state estimate and error covariance are propagated by Equations 38.2a and 38.4a until a measurement becomes available and Equations 38.2b and 38.4b are applied. For example, consider the servomotor system in Equation 38.6 where shaft angle measurements are only available for $k = 5, 10, 15, \ldots$. A plot of the shaft angle error covariance $P(1, 1)$ versus sample period is shown in Figure 38.2. Between measurement times, the estimation error increases because the

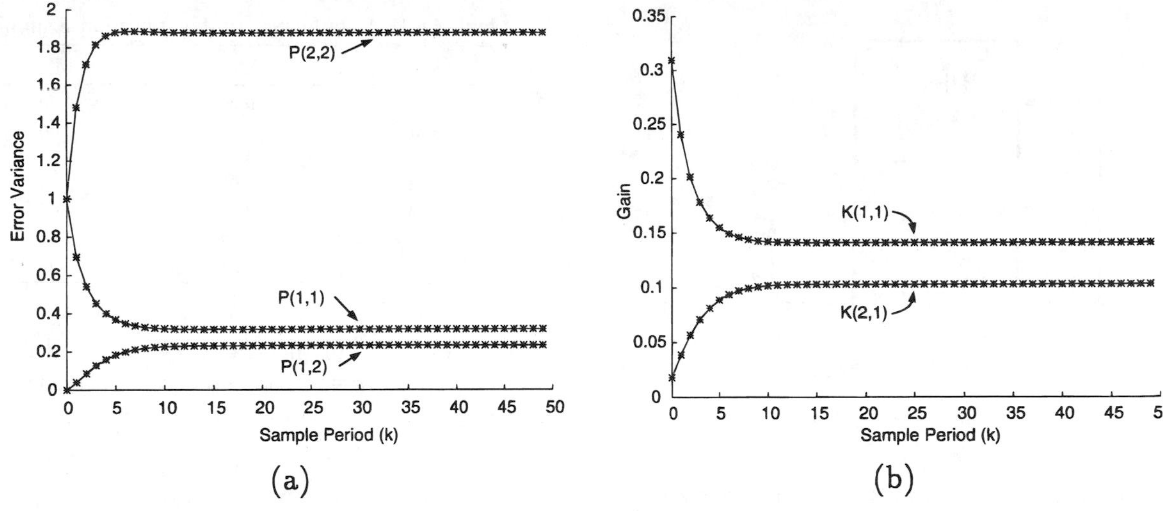

Figure 38.1 Servomotor example: (a) error covariance P_k versus sample period, (b) Kalman gain K_k versus sample period.

estimate is only based on the system model. When a measurement becomes available, the estimate and error covariances are corrected and the process is repeated.

38.2 Other Types of Kalman Filters

There are different formulations of the Kalman filter for different types of plant dynamics and measurements. Examples include (Gelb, 1974; Lewis, 1986) the *continuous Kalman filter* (continuous-time plant dynamics, continuous-time measurements), the *continuous-discrete Kalman filter* (continuous-time plant dynamics, discrete-time measurements), and the *extended Kalman filter* (nonlinear plant dynamics and measurements).

38.3 Identification

In this section we consider the case where the inputs and outputs of a system are known or can be measured, but certain parameters of the plant dynamics are unknown. The problem of determining these unknown parameters is called the *system identification* problem. There are many approaches to the system identification problem, and a comprehensive treatment is beyond the scope of this section. In this section we will develop a relatively simple approach called the *least-squares system identification method.* The reader is referred to Sage and Melsa (1971) and Ljung and Sonderstrom (1983) for more advanced methods.

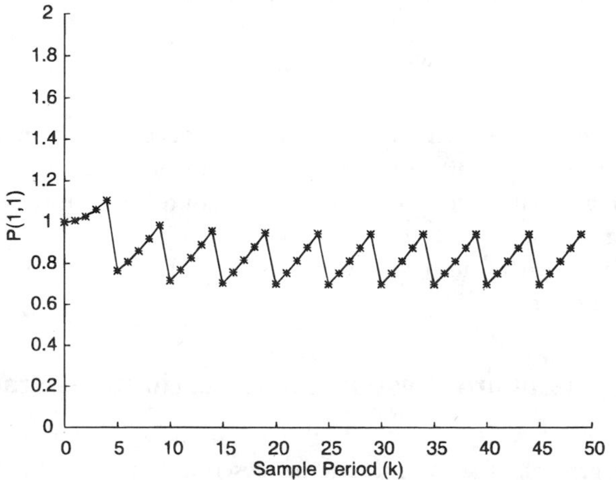

Figure 38.2 Shaft angle error covariance $P(1, 1)$ versus sample period for sparse measurement example.

Least-Squares System Identification Example

We develop the least-squares system identification method by way of an example. The servomotor model dynamics in Equation 38.6 can be expressed as the autoregressive moving-average (ARMA) model (Phillips and Nagle, 1995)

$$y_k = a_1 y_{k-1} + a_2 y_{k-2} + b_1 u_{k-1} + b_2 u_{k-2}, \qquad (38.7)$$

where u_k and y_k are the armature voltage and shaft angle at sample period k. Suppose that the servomotor is known to have the form of Equation 38.7, but the parameters a_1, a_2, b_1, and b_2 are unknown. We assume that the armature voltage u_k and the shaft angle y_k are either known (or can be measured), and we wish to identify the unknown parameters.

The least-squares system identification method is summarized in Figure 38.3. A known (or measured) input is applied to both the actual plant and to the plant model (Equation 38.7). The actual plant output y_k is measured. The estimated plant output $\hat{y}_k$ is computed from Equation 38.7 using the previous plant inputs u_{k-1} and u_{k-2} and the previous *actual* plant outputs y_{k-1} and y_{k-2}. The error is defined as

$$e_k = y_k - \hat{y}_k. \qquad (38.8)$$

A sequence of N such measurements are taken, and the unknown parameters are chosen such that the sum of the squared errors is minimized.

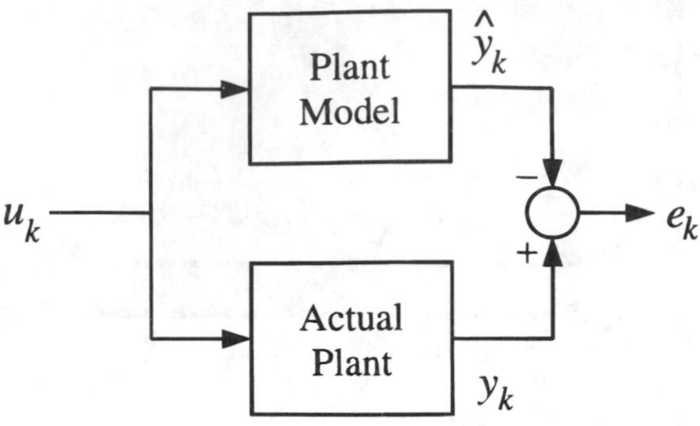

Figure 38.3 Least-squares system identification.

For example, if seven measurements ($y_0 - y_6$) are taken, the following estimated plant outputs are computed from Equation 38.7:

$$\hat{y}_2 = a_1 y_1 + a_2 y_0 + b_1 u_1 + b_2 u_0$$
$$\hat{y}_3 = a_1 y_2 + a_2 y_1 + b_1 u_2 + b_2 u_1$$
$$\hat{y}_4 = a_1 y_3 + a_2 y_2 + b_1 u_3 + b_2 u_2$$
$$\hat{y}_5 = a_1 y_4 + a_2 y_3 + b_1 u_4 + b_2 u_3$$
$$\hat{y}_6 = a_1 y_5 + a_2 y_4 + b_1 u_5 + b_2 u_4. \quad (38.9)$$

These outputs begin with $k = 2$ because the plant model output depends on the two previous outputs. Substituting Equation 38.9 into the error Equation 38.8 yields

$$e_2 = y_2 - a_1 y_1 - a_2 y_0 - b_1 u_1 - b_2 u_0$$
$$e_3 = y_3 - a_1 y_2 - a_2 y_1 - b_1 u_2 - b_2 u_1$$
$$e_4 = y_4 - a_1 y_3 - a_2 y_2 - b_1 u_3 - b_2 u_2$$
$$e_5 = y_5 - a_1 y_4 - a_2 y_3 - b_1 u_4 - b_2 u_3$$
$$e_6 = y_6 - a_1 y_5 - a_2 y_4 - b_1 u_5 - b_2 u_4. \quad (38.10)$$

Equation 38.10 can be written in matrix form

$$E(6) = Y(6) - F(6)\Theta, \quad (38.11)$$

where

$$E(6) = \begin{bmatrix} e_2 \\ e_3 \\ e_4 \\ e_5 \\ e_6 \end{bmatrix} \quad Y(6) = \begin{bmatrix} y_2 \\ y_3 \\ y_4 \\ y_5 \\ y_6 \end{bmatrix}$$

$$F(6) = \begin{bmatrix} y_1 & y_0 & u_1 & u_0 \\ y_2 & y_1 & u_2 & u_1 \\ y_3 & y_2 & u_3 & u_2 \\ y_4 & y_3 & u_4 & u_3 \\ y_5 & y_4 & u_5 & u_4 \end{bmatrix} \quad \Theta = \begin{bmatrix} a_1 \\ a_2 \\ b_1 \\ b_2 \end{bmatrix}.$$

Table 38.1 Data for System Identification Example

k	u_k	y_k
0	0.00	0.00
1	2.40	0.00
2	4.21	1.37
3	5.00	5.93
4	4.55	14.07
5	2.99	24.83
6	0.71	36.30

The (6) means that the matrices are constructed from inputs and outputs up to $k = 6$. The sum of the squared errors can be written as

$$J(6) = \sum_{k=2}^{6} e_k^2 = E(6)^T E(6). \quad (38.12)$$

The parameter vector Θ_{LS} that minimizes $J(6)$ is called the *least-squares estimate* and is given by

$$\hat{\Theta}_{LS} = [F(6)^T F(6)]^{-1} F(6)^T Y(6). \quad (38.13)$$

The inverse in Equation 38.13 will exist if the input u_k is *persistently exciting* (the input excites all modes of the system) and Θ is *identifiable* (all unknown parameters can uniquely determined from the measurements). See Sage and Melsa (1971) and Ljung and Sonderstrom (1983) for details.

For example, Table 38.1 shows the input applied to a servomotor system and the measured output. The matrices $Y(6)$ and $F(6)$ are

$$Y(6) = \begin{bmatrix} 1.37 \\ 5.93 \\ 14.07 \\ 24.83 \\ 36.30 \end{bmatrix} \quad F(6) = \begin{bmatrix} 0.00 & 0.00 & 2.40 & 0.00 \\ 1.37 & 0.00 & 4.21 & 2.40 \\ 5.93 & 1.37 & 5.00 & 4.21 \\ 14.07 & 5.93 & 4.55 & 5.00 \\ 24.83 & 14.07 & 2.99 & 4.55 \end{bmatrix}.$$

The least-squares estimate of the parameters is

$$\hat{\Theta}_{LS} = [1.6942 \quad -0.6942 \quad 0.5705 \quad 0.5052]^T.$$

These parameters in Equation 38.7 reproduce the data in Table 38.1 exactly. In practice, however, the estimated parameters will not match the data exactly because of measurement noise, modeling errors, etc. In some cases, the estimation accuracy can be improved by using more measurements to estimate the parameters.

Least-Squares System Identification: General Case

In general, if a system can be described by an n-th order ARMA model

$$y_k = a_1 y_{k-1} + a_2 y_{k-2} + \cdots + a_n y_{k-n}$$

$$+ b_1 u_{k-1} + b_2 u_{k-2} + \cdots + b_n u_{k-n}, \quad (38.14)$$

and all the a_i and b_i are unknown, the parameter vector is

$$\Theta = [a_1 \quad a_2 \quad \cdots \quad a_n \quad b_1 \quad b_2 \quad \cdots \quad b_n]^T. \quad (38.15)$$

If N measurements are taken, the least-squares estimate of the unknown parameters is given by

$$\hat{\Theta}_{LS} = [F(N)^T F(N)]^{-1} F(N)^T Y(N), \quad (38.16)$$

where

$$Y(N) = [y_n \quad y_{n+1} \quad \cdots \quad y_N]^T \quad (38.17)$$

and

$$F(N) = \begin{bmatrix} y_{n-1} & y_{n-2} & \cdots & y_0 & u_{n-1} & u_{n-2} & \cdots & u_0 \\ y_n & y_{n-1} & \cdots & y_1 & u_n & u_{n-1} & \cdots & u_1 \\ & \vdots & & & & \vdots & & \\ y_{N-1} & y_{N-2} & \cdots & y_{N-n} & u_{N-1} & u_{N-2} & \cdots & u_{N-n} \end{bmatrix}. \quad (38.18)$$

References

Gelb, A., ed. 1974. *Applied Optimal Estimation,* MIT Press, Cambridge, MA.

Lewis, F. L. 1986. *Optimal Estimation*, John Wiley and Sons, New York, NY.

Ljung, L. and Sonderstrom, T. 1983. *Theory and Practice of Recursive Identification,* MIT Press, Cambridge, MA.

Phillips, C. L. and Nagle, T. 1995. *Digital Control System Analysis and Design,* Prentice Hall, Englewood Cliffs, NJ.

Sage, A. P. and Melsa, J. L. 1971. *System Identification,* Academic Press, New York, NY.

39

Fuzzy Logic-Based Control

Mo-yuen Chow
North Carolina State University

39.1 Introduction to Intelligent Control ... 564
Fuzzy Control
39.2 DC Motor Dynamics ... 565
39.3 Fuzzy Control ... 566
Initial Fuzzy Rules and Membership Function Design • PI Controller • Borrowing PI Knowledge
39.4 Conclusion and Future Direction ... 570

39.1 Introduction to Intelligent Control

For the purposes of system control, much valuable knowledge and many techniques, such as feedback control, transfer functions (frequency or discrete-time domain), state-space time-domain, optimal control, adaptive control, robust control, gain scheduling, model-reference adaptive control, etc. have been investigated and developed during the past few decades. Different important concepts such as root locus, Bode plot, phase-margin, gain-margin, eigenvalues, eigenvectors, pole placement, etc. have been imported from different areas or developed in the control field.

However, most of these control techniques rely on system mathematical models in their design process. Control designers spend more time obtaining an accurate system model (through techniques such as system identification, parameter estimation, componentwise modeling, etc.) than in the design of the corresponding control law. Furthermore, many control techniques, such as transfer function approach, require the system to be linear and time invariant; otherwise, linearization techniques at different operating points are required to arrive at an acceptable control law/gain. With the use of system mathematical models, especially a linear time-invariant model, one can certainly enhance the theoretical support of the developed control techniques. However, this requirement creates another fundamental problem. How accurately does the mathematical model represent the system dynamics? In many cases, the mathematical model is only an approximated, rather than an exact, model of the system dynamics being investigated. This approximation may lead to a reasonable, but not necessarily good, control law for the system of interest.

For example, PI control, which is simple, well known and well suited for the control of linear time-invariant systems, has been used extensively for industrial motor control. The design process to obtain the PI gains is tied tightly to the mathematical model of the motor. Engineers usually first design a PI control based on a reasonably accurate mathematical model of the motor, then use the root locus/Bode plot technique to obtain suitable gains for the controller to achieve desirable motor performance. Then they need to tune the control gain on-line at the beginning of the use of the motor controllers to give acceptable motor performance for the real world. The requirement of gain tuning is mostly due to the unavoidable modeling error embedded in the mathematical models used in the design process. The motor controller may further require gain adjustments during on-line operations to compensate for the change in system parameters due to factors such as system degradation, change of operating conditions, etc. Adaptive control has been studied to address the changes in system parameters and has achieved a certain level of success. Gain scheduling has been studied and used in control loop (Teeter et al., 1994; Shamma and Athans, 1990) so that the motor can give satisfactory performance over a wide operating range. The requirement of mathematical models imposes artificial mathematical constraints on the control design freedom. Along with the unavoidable modeling error, the resulting control laws in many cases give an over-conservative motor performance.

There are also other control techniques, such as set-point control, sliding mode control, fuzzy control, neural control, that rely less on mathematical model of the system, but more on the designer's knowledge of the actual system. Especially, intelligent control has been attracting significant attention in the last few years. Different articles and experts' opinions have been reported in different technical articles. A control system which incorporates human qualities, such as heuristic knowledge and the ability to learn, can be considered to possess a certain degree of intelligence. Such an intelligent control system has an advantage over purely analytical methods because, besides incorporating human knowledge, it is less dependent on the overall mathematical model. In fact, human beings routinely perform very complicated tasks without the aid of any mathematical representations. A

0-8493-8343-9/97/$0.00+$.50

simple knowledge base and the ability to learn by training seem to guide humans through even the most difficult problems. Although conventional control techniques are considered to have intelligence in a low level, we want to further develop the control algorithms from the low-level to a high-level intelligent control, through the incorporation of heuristic knowledge and learning ability via the *fuzzy* and *neural network* technologies, among others.

Fuzzy Control

Fuzzy control is considered an intelligent control technique and has been shown to yield promising results for many applications that are difficult to handle by conventional techniques (Sugenon, 1985; Lee, 1990; Berenji and Khedkar, 1992; Lin and Lee, 1991; Gupta et al., 1986; Esogbue and Murrel, 1993; Tang and Mulholland, 1987; Sharpe et al., 1994). Implementations of fuzzy control in areas such as water quality control (Yagishita et al., 1985), automatic train operation systems (Yasunobu et al., 1987), traffic control (Pappis and Mamdani, 1977), among others, have indicated that fuzzy logic is a powerful tool in the control of mathematically ill defined systems which are controlled satisfactorily by human operators without the knowledge of the underlying mathematical model of the system. While conventional control methods are based on the quantitative analysis of the mathematical model of a system, fuzzy controllers focus on a linguistic description of the control action which can be drawn, for example, from the behavior of a human operator. This can be viewed as a shift from the conventional precise mathematical control to human-like decision making (Bezdek, 1993; Bellman and Zadeh, 1970; Zadeh, 1973 and 1979; Zimmerman, 1991), which drastically changes the approach to automate control actions.

39.2 DC Motor Dynamics—Assume a linear time-invariant system

This section will use a motor fuzzy controller design process to illustrate the use of fuzzy logic technology on industrial electronics applications.

Due to the popularity of DC motors for control applications, a DC motor velocity control will be used to illustrate the fuzzy control design approach. Readers are assumed to have a basic background on DC motor operations; otherwise, please refer to D'Azzo and Houpis, 1988 or Brogan, 1991. The fuzzy control will be applied to an actual DC motor system to demonstrate the effectiveness of the control techniques. We will describe briefly the actual control system set-up below.

The fuzzy controller is implemented on a 486 PC using the LabVIEW graphical programming package. The complete system setup is shown in Figure 39.1 and the actual motor control system is given in Figure 39.2. The rotation of the motor shaft generates a tachometer voltage which is then scaled by interfacing electronic circuitry. A National Instruments data acquisition board receives the data via an Analog Devices isolating backplane. After a control

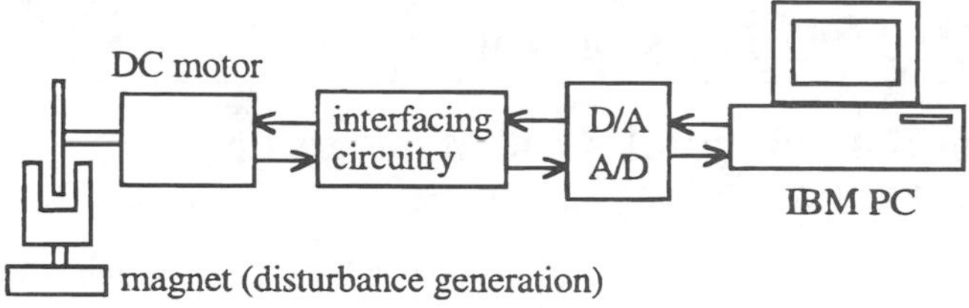

Figure 39.1 Schematic diagram of the experimental DC motor system.

Table 39.1 DC Motor Parameters

R_a	4.67 W
L_a	170e-3 H
J	42.6e-6 kg-m2
f	47.3e-6 N-m/rad/sec
K	14.7e-3 N-m/A
K_b	14.7e-3 V-sec/rad

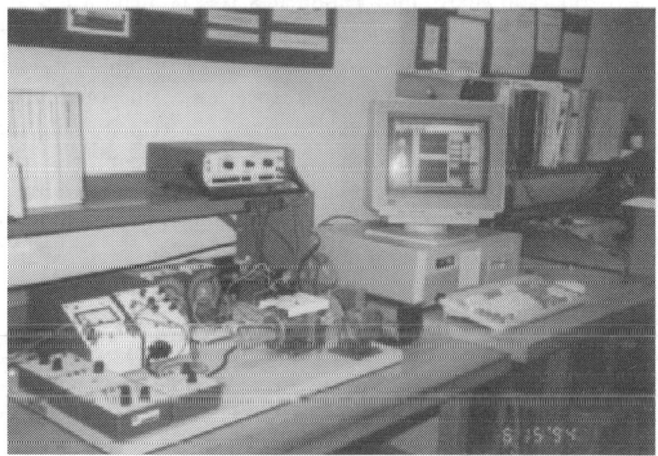

Figure 39.2 Motor control system setup.

value is computed, an output current is generated by the data acquisition board. The current signal passes through the backplane and is then converted to a voltage signal and scaled by the interfacing circuitry before being applied to the armature of the motor. Load disturbances are generated by subjecting a disc on the motor shaft to a magnetic field.

For illustration purposes, the control objective concentrates on achieving zero steady-state error and smooth, fast response to step inputs. These are popular, desired motor performance characteristics for many industrial applications. The parameters and their numerical values of the DC servomotor used for our simulation studies are listed in Table 39.1, obtained by conventional system identification techniques.

The parameters R_a and L_a are the resistance and the inductance of the motor armature circuit, respectively; J and f are the moment of inertia and the viscous-friction coefficient of the motor and load (referred to the motor shaft), respectively; K is the constant relating the armature current to the motor torque, and K_b is the constant relating the motor speed to the DC motor's back-EMF. The DC motor has an input operating range of $(-15,15)$ volts.

39.3 Fuzzy Control

Initial Fuzzy Rules and Membership Function Design

To design a fuzzy controller, first we will need to determine the inputs, outputs, universe of discourse, membership functions, and fuzzy rules. In this section, the input variables of the fuzzy controller are the error ($E = e(k)$), which is the difference between the DC motor speed and the reference speed, and the change in error ($CE = e(k) - e(k-1)$), where k is the time index. The output variable of the controller is the change in the control effort ($CU = u(k) - u(k-1)$).

The determination of the universe of discourse of the velocity error change and the control effort change is based on experience and knowledge of the DC motor. For example, the simulation results of the DC motor performance for different control laws based on estimated motor parameters is very helpful for the design of the fuzzy controller. The simulation results will give a rough idea of the response of the system, even though it does not give the exact system performance. Since the open loop simulations of the system result in a possible velocity range of -500 rad/sec to 500 rad/sec, the minimum and maximum possible values that the error can assume are -1000 rad/sec and 1000 rad/sec, respectively. Hence, the universe of discourse (operating range) of the velocity error spans between -1000 rad/sec and 1000 rad/sec. Based on these requirements, the maximum value of error change is then set to 5.5 rad/sec. Also, the maximum value for the control effort change is determined to be 1.5 volts. The universes of discourse of the fuzzy variables is then partitioned into seven quantization levels (fuzzy sets), each being described by a linguistic statement such as "big", "small", etc., as listed in Table 39.2. The number of partition levels chosen is a trade-off between the resolution of the quantization and the complexity of the design problem, and is often dependent on the designer's preference.

A fuzzy membership function requires assigning a real number in the interval (0,1) to every element in the universe of discourse. This number indicates the degree to which the element belongs to a fuzzy set, such as *big* or *small*. Fuzzy membership functions can have different shapes depending on the designer's preference and/or experience. Triangular and trapezoidal shapes are popular because of simple computations and the capture of the designers' fuzzy numbers sense. Again, the choice of membership functions is a subjective matter, but prior experience can provide some useful guidelines. For example, if the measurable data is disturbed by noise, then the membership functions should be sufficiently wide to reduce noise sensitivity (Kosko, 1992). Kosko (1992) also suggests that adjacent fuzzy-set values should overlap approximately 25%, and fine-tuning can be achieved by altering this overlap percentage.

Table 39.2 Fuzzy Set Definitions

PB	Positive Big
PM	Positive Medium
PS	Positive Small
ZE	Zero
NS	Negative Small
NM	Negative Medium
NB	Negative Big

Figure 39.3 shows the initial membership functions, which assign a real number in the interval (0,1) to every element in the universe of discourse, used for the motor control problem. This number indicates the degree to which the element belongs to a fuzzy set, such as big or small, used in the fuzzy velocity controller.

Notice that there is a rule for every possible combination of E and CE that may arise. Since E and CE both are partitioned into 7 fuzzy sets, the fuzzy rulebase table thus has a total of 49 entries, each one corresponding to a different combination of input fuzzy set values. These rules have the form:

$$\text{Rule } i\text{: if } E = A_{E,i} \text{ and } CE = A_{CE,i} \text{ then } CU = C_i, \tag{39.1}$$

where $A_{E,i}$ and $A_{CE,i}$ are the fuzzy set values of the antecedent part of rule i for E and CE, respectively. Likewise, C_i is the fuzzy set value of the consequent part of rule i for CU.

The fuzzification process, which is the transformation of crisp inputs to fuzzy set outputs, is accomplished by using the popular correlation-product inference method (Kosko, 1992). By the same token, these fuzzy set outputs were defuzzified with a centroid computation to generate an exact numerical output. With this method, the motor's current operating point numerical values E^0 and CE^0 are required. The inference method can be described mathematically as:

$$l_i = \min\{A_{e,i}(E^0), A_{ce,i}(CE^0)\}, \tag{39.2}$$

which gives the influencing factor of rule i on the decision-making process, and

$$I_i = \int C_i(CU)dCU, \tag{39.3}$$

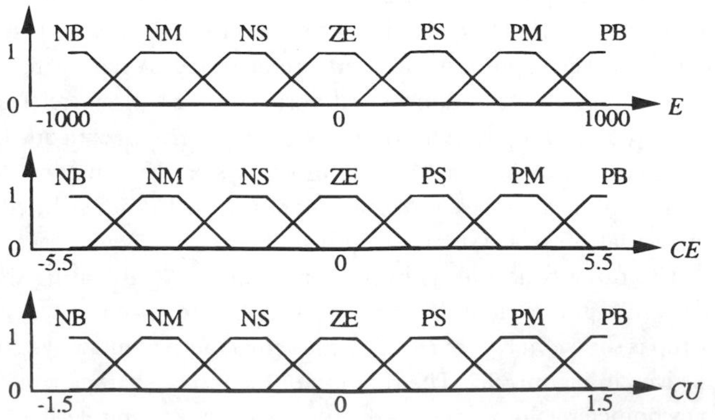

Figure 39.3 Initial membership functions used in the fuzzy velocity controller.

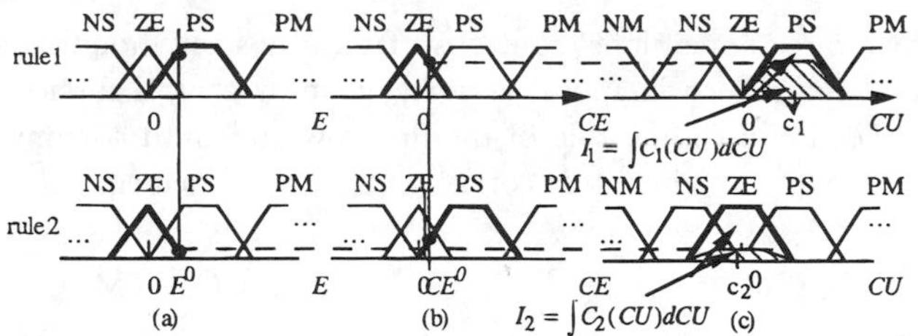

Figure 39.4 Graphical representation of the correlation-product inference method.

gives the area bounded by the membership function $C_i(CU)$; thus $l_i I_i$ gives the area bounded by the membership function $C_i(CU)$ scaled by l_i computed in Equation 39.2.

The centroid of the area bounded by $C_i(CU)$ is computed as:

$$c_i = \frac{\int CU * C_i(CU)dCU}{\int C_i(CU)dCU}, \tag{39.4}$$

thus $l_i I_i \times c_i$ gives the control value contributed by Rule *i*. The control value CU^0, which combine the control efforts from all *N* rules, is then computed as:

$$CU^0 = \frac{\sum_{i=1}^{N} l_i I_i \times c_i}{\sum_{i=1}^{N} l_i I_i}. \tag{39.5}$$

In Equations 39.2–39.5, the subscript *i* indicates the *i*-th rule of a set of *N* rules.

For illustration purposes, assume the motor currently has the current operating point E^0 and CE^0 and assume only two rules are used (thus $N = 2$):

$$\begin{aligned} &\text{if } E \text{ is PS and } CE \text{ is ZE then } CU = \text{PS, and} \\ &\text{if } E \text{ is ZE and } CE \text{ is PS then } CU = \text{ZE}. \end{aligned} \tag{39.6}$$

The correlation-product inference method described in Equations 39.2 and 39.3 and the area bounded by the inferred membership function $C_i(CU)$ are conceptually depicted in Figure 39.4. By looking at Rule 1, the membership value of *E* for PS, PS(E^0) is larger than the membership value of *CE* for ZE, ZE(CE^0), therefore:

$$l_1 = \text{ZE}(CE^0).$$

I_1 is the area bounded by the membership function PS on *CU* (the hatched and the shaded areas) in Figure 39.4(c), and $l_1 I_1$ is only the hatched area. c_1 is computed to give the centroid of I_1. The same arguments also apply to Rule 2. The defuzzification process is graphically depicted in Figure 39.5.

The scaled control membership functions (the hatched areas) from different rules, Figure 39.5(a), are combined together from all rules, which form the hatched area shown in Figure 39.5(b). The centroid value of the combined hatched area is then computed, to give the final crisp control value.

PI Controller

If only experience and control engineering knowledge are used to derive the fuzzy rules, the designers will probably be overwhelmed by the degrees of freedoms (number of rules) of the design, and many of the rule table entries may be left empty due to insufficient detailed knowledge to be extracted from the expert. To make the design process more effective, it will be helpful to have a structured way for the designer to follow in order to eventually develop a proper fuzzy controller.

In this section, we will illustrate how to take advantage of conventional control knowledge to arrive at a fuzzy control design more effectively. The PI controller is one of the most popular conventional controllers and will be used in this section as the technique to incorporate *a-priori* knowledge which will eventually lead to the fuzzy controller.

The velocity transfer functions of the DC motor control can be easily obtained from many textbooks (D'Azzo and Houpis, 1988; Brogan, 1991). The velocity transfer function can be derived as:

$$G_p(s) = \frac{\omega(s)}{e_a(s)} = \frac{K}{JL_a s^2 + (fL_a + JR_a)s + (fR_a + KK_b)}. \tag{39.7}$$

The general equation for a PI controller is

$$u(k) = u(k-1) + \left(K_p + \frac{K_i T_s}{2}\right)e(k) + \left(\frac{K_i T_s}{2} - K_p\right)e(k-1) \tag{39.8}$$

where K_p and K_i can be determined by the root-locus method (D'Azzo and Houpis, 1988). For velocity control $K_i = 0.264$ and $K_p = 0.12$ are chosen to yield desirable response characteristics which give an adequate trade-off between the speed of the response and the percentage of overshoot. The PI control surface over the universes of discourse of error (E) and error change (CE) is shown in Figure 39.6.

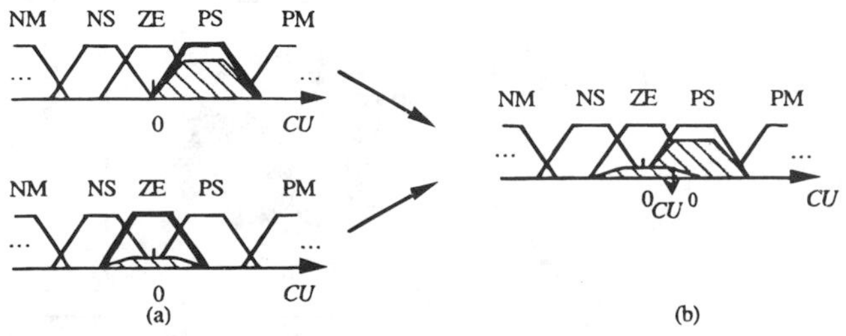

Figure 39.5 Graphical representation of the center-of-gravity defuzzification method.

Borrowing PI Knowledge

The PI control surface is then taken as a starting point for the fuzzy control surface. More specifically, a "fuzzification" of the PI control surface yielded the first fuzzy control surface to be further tuned. This starting point corresponds to the upper left surface of Figure 39.7 (the membership functions are identical in shape and size, and symmetric about ZE).

The initial fuzzy rulebase table (Figure 39.8) was specified by "borrowing" values from the PI control surface. Since the controller use the information (E, CE) in order to produce a control signal (CU), the control action can be completely defined by a three-dimensional control surface. Any small modification to a controller will appear as a change in its control surface.

For example, the rule of the first row and third column of Figure 39.8 (highlighted) corresponds to the statement:

$$\text{if } CE = \text{PB} \quad \text{and} \quad E = \text{NS} \quad \text{then} \quad CU = \text{PM},$$

which indicates that if the error is large and is gradually decreasing, then the controller should produce a positive medium compensating signal.

The initial fuzzy controller obtained will give a performance

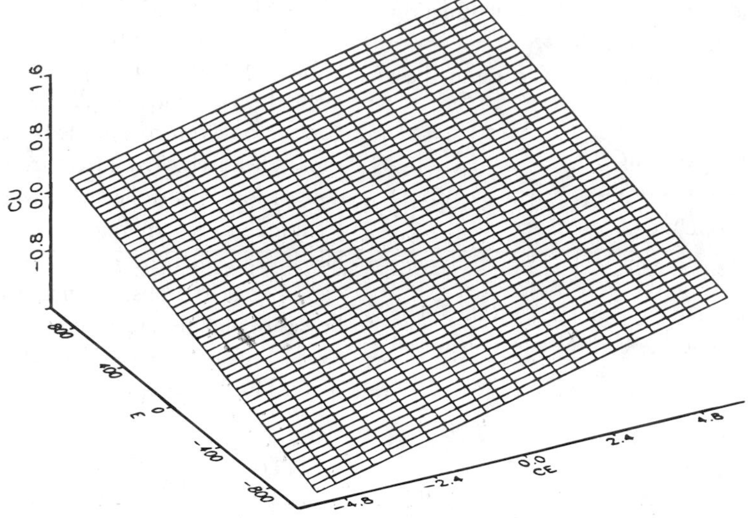

Figure 39.6 PI control surface.

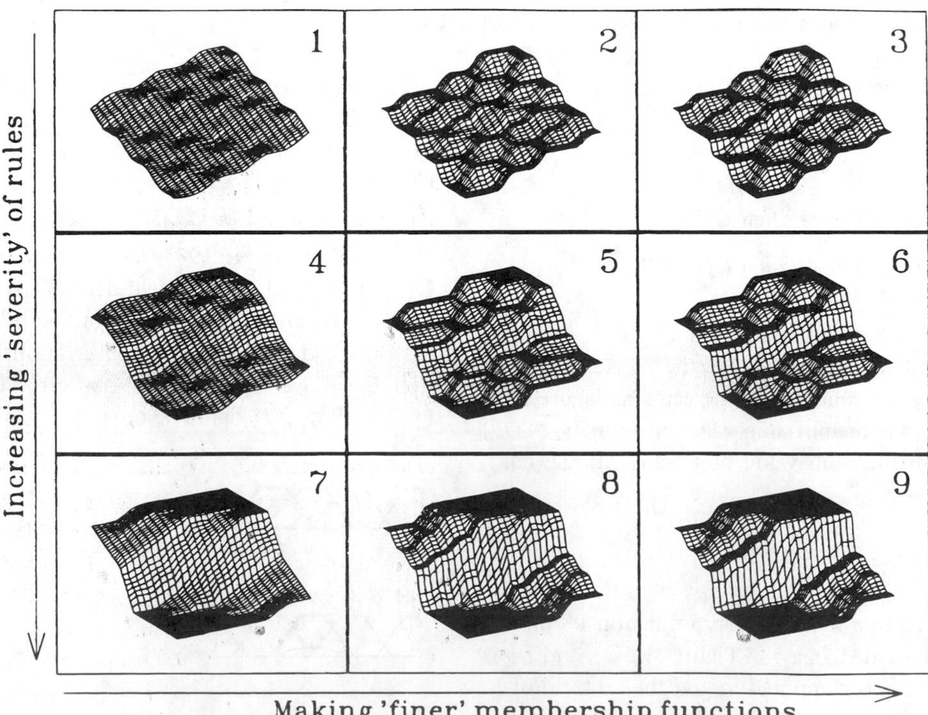

Figure 39.7 Effects of fine tuning on the fuzzy control surface.

CE \ E	NB	NM	NS	ZE	PS	PM	PB
PB	ZE	PS	PM	PB	PB	PB	PB
PM	NS	ZE	PS	PM	PB	PB	PB
PS	NM	NS	ZE	PS	PM	PB	PB
ZE	NB	NM	NS	ZE	PS	PM	PB
NS	NB	NB	NM	NS	ZE	PS	PM
NM	NB	NB	NB	NM	NS	ZE	PS
NB	NB	NB	NB	NB	NM	NS	ZE

Figure 39.8 Initial fuzzy rulebase table.

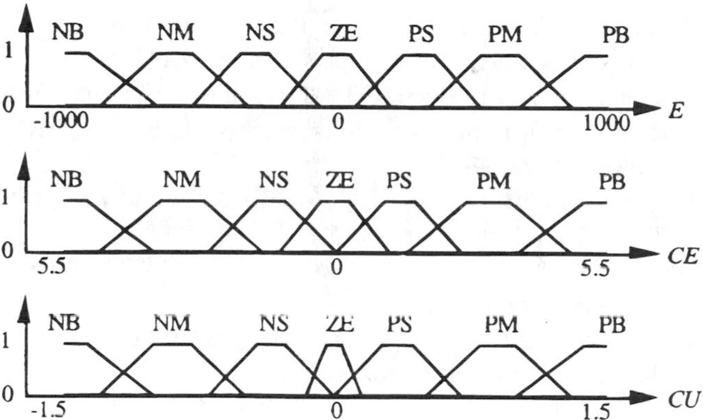

Figure 39.9 Intermediate membership functions used in fuzzy velocity controller.

similar to the designed PI controller. The controller performance can be improved by fine tuning the fuzzy controller while control signals are being applied to the actual motor. In order to fine tune the fuzzy controller, two parameters can be adjusted: the membership functions and the fuzzy rules. Both the shape of membership functions and the severity of fuzzy rules can affect the motor performance. In general, making the membership functions "narrow" near the ZE region and "wider" far from the ZE region can improve the controller's resolution in the proximity of the desired response when the system output is close to the reference values, thus improving the tracking performance. Also, performance can be improved by changing the "severity" of the rules, which amounts to modifying their consequent part. Figure 39.7 shows the changes in the fuzzy control surface brought upon by varying the membership functions and the fuzzy rules. The fine-tuning process begin with panel 1 of Figure 39.7. The initial control surface is similar to the PI control, which was used as a starting guideline for the fuzzy controller. The changes in control surfaces from the left hand side to right hand size signify the change of the shape of membership functions. The changes in control surfaces from top to bottom signify the change in rules.

In order to demonstrate the effect of fine-tuning the membership functions, we show a set of intermediate membership functions (relative to the initial one shown in Figure 39.7 (1)) Figure 39.9 and the final membership functions in Figure 39.10. Figure 39.9 and 39.10 show that some fuzzy sets, such as the ZE in *CU*,

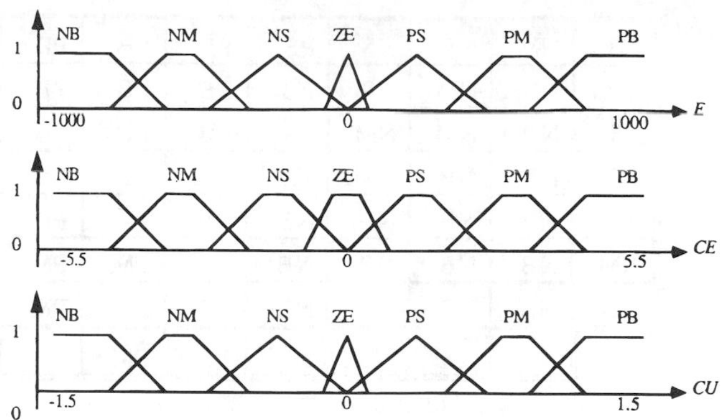

Figure 39.10 Final membership functions used in the fuzzy velocity controller.

CE \ E	NB	NM	NS	ZE	PS	PM	PB
PB	NS	NS	ZE	PB	PB	PB	PB
PM	NM	NS	ZE	PB	PB	PB	PB
PS	NM	NS	NS	PS	PM	PB	PB
ZE	NB	NM	NS	ZE	PS	PM	PB
NS	NB	NB	NM	NS	PS	PS	PM
NM	NB	NB	NB	NB	ZE	PS	PM
NB	NB	NB	NB	NB	ZE	PS	PS

Figure 39.11 Intermediate fuzzy rulebase table.

are getting narrower, which allows *finer* control in the proximity of the desired response, while the wider fuzzy-sets, such as the PB in *E*, permit *coarse* but *fast* control far from the desired response.

We also show an intermediate rule table and the final rule table during the fine-tuning process. The rule in the first row and third column (highlighted cell) corresponds to:

$$\text{if } CE = \text{PB} \quad \text{and} \quad E = \text{NS}, \quad \text{then} \quad CU = \underline{\quad\quad}.$$

In the initial rule table (Figure 39.8), CU = PM. However, during the fine-tuning process, we found that the rule should have different action in order to give *better* performance. In Figure 39.11, the CU becomes ZE and the final fuzzy rule table, CU = NS.

From (1) and (9) of Figure 39.7 it can be seen that gradually increasing the "fineness" of the membership functions and the "severity" of the rules can bring the fuzzy controller to its best performance level. The fine-tuning process is not difficult at all. The fuzzy control designer can easily get a "feel" of how to perform the correct fine-tuning after a few trials. Figure 39.7(9) exhibits the fuzzy control surface which yielded the best results. The membership functions and fuzzy rules which generated it are the ones of Figure 39.10 and Figure 39.12, respectively.

The performance of the controllers for the DC motor velocity control is shown in Figure 39.13 for two different references. The fuzzy controller exhibits better performance than the PI controller

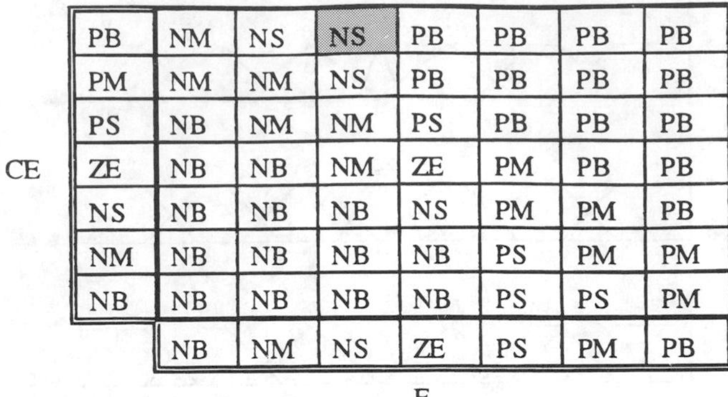

CE							
PB	NM	NS	NS	PB	PB	PB	PB
PM	NM	NM	NS	PB	PB	PB	PB
PS	NB	NM	NM	PS	PB	PB	PB
ZE	NB	NB	NM	ZE	PM	PB	PB
NS	NB	NB	NB	NS	PM	PM	PB
NM	NB	NB	NB	NB	PS	PM	PM
NB	NB	NB	NB	NB	PS	PS	PM
	NB	NM	NS	ZE	PS	PM	PB
				E			

Figure 39.12 Final fuzzy rulebase table.

because of shorter rise time and settling time. The fuzzy controller was thoroughly fine-tuned to yield the best performance.

39.4 Conclusion and Future Direction

This paper outlines the design procedures used in the design of fuzzy controllers for the velocity control of a DC motor. A comparison of these controllers with a classical PI controller was discussed in terms of the characteristics of the respective control surfaces. The PI control was shown to behave as a special case of the fuzzy controller, and for this reason it is more constrained and less flexible than the fuzzy controller. A methodology which exploits the advantages of each technique in order to achieve a successful design is seen as the most sensible approach to follow. A drawback of the fuzzy controller is that it cannot adapt. Recently, different researchers, including the author of this paper, have been investigating adaptive fuzzy controllers, many of them result in implementing the fuzzy controller in a neural network structure for adaptation, which is convenient and has fast computation. In fact, a very interesting area of exploration is the possibility of combining the advantages of fuzzy logic with those of artificial neural networks (ANN). The possibility of using the well-known learning capabilities of an ANN coupled with the ability of a fuzzy logic system to translate heuristic knowledge and fuzzy concepts into real numerical values may represent a very powerful way of coming closer to intelligent, adaptive control systems.

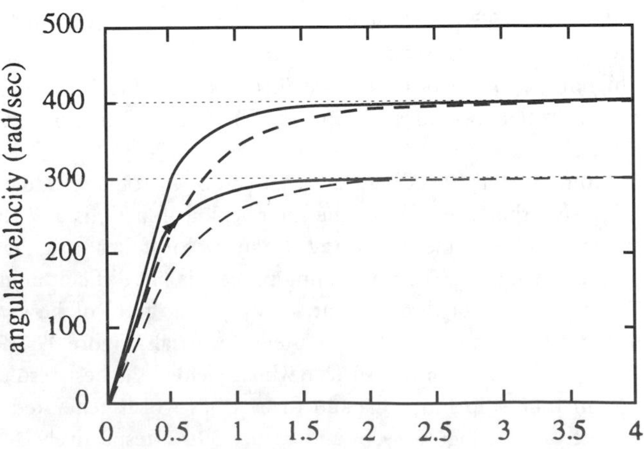

Figure 39.13 Velocity control (two references are shown).

Acknowledgment

The author of this paper would like to thank Mr. Alberico Menozzi and Mr. Jason Teeter for their contributions to this article.

References

Bellman, R. E. and Zadeh, L. A. 1970. Decision-making in a fuzzy environment, *Management Science*, 17:141–164.

Berenji, H. R. and Khedkar, P. 1992. Learning and tuning fuzzy logic controllers through reinforcements, *IEEE Trans. Neural Networks*, 3:724–740.

Bezdek, J. C. 1993. Fuzzy models—what are they and why?, *IEEE Trans. Fuzzy Systems*, 1:1–6.

Brogan, W. L. 1991. *Modern Control Theory*, 3d ed., Prentice Hall, Englewood Cliffs, NJ.

D'Azzo, J. J. and Houpis, C. H. 1988. *Linear Control System Analysis and Design*, 3d ed., McGraw-Hill, New York, NY.

Esogbue, A. O., and Murrel, J. A. 1993. Fuzzy adaptive controller using reinforcement learning neural networks, *IEEE Int. Conf. Fuzzy Systems.*

Gupta, M. M., Kiszka, J. B., and Trojan, G. M. 1986. Multivariable structure of fuzzy control systems, *IEEE Trans. Systems, Man, and Cybernetics*, 16:638–656.

Kosko, B. 1992. *Neural Networks and Fuzzy Systems: A Dynamical Systems Approach to Machine Intelligence*, Prentice Hall, Englewood Cliffs, NJ.

Lee, C. C. 1990. Fuzzy logic in control systems: Fuzzy logic controller, *IEEE Trans. Systems, Man, and Cybernetics*, 20:404–435.

Lin, C. and Lee, C. S. G. 1991. Neural-network-based fuzzy logic control and decision system, *IEEE Trans. Computers*, 40:1320–1336.

Pappis, C. P. and Mamdani, E. H. 1977. A fuzzy logic controller for a traffic junction, *IEEE Trans. Systems, Man, and Cybernetics*, 7:707–717.

Shamma, J. S. and Athans, M. 1990. Analysis of gain scheduled control for nonlinear plants, *IEEE Trans. Auto. Control*, 35:898–907.

Sharpe, R. N., Chow, M.-y., Briggs, S., and Windingland, L. 1994. A methodology using fuzzy logic to optimize feedforward artificial neural network configurations, *IEEE Trans. Systems, Man, and Cybernetics*, 24:760–768.

Sugenon, M. 1985. An introductory survey of fuzzy control, *Information Sciences*, 36:59–79.

Tang, K. L. and Mulholland, R. J. 1987. Comparing fuzzy logic with classical controller designs, *IEEE Trans. Systems, Man, and Cybernetics*, 17:1085–1087.

Teeter, J., Chow, M.-y., and Brickley, J. J. 1994. Use of a fuzzy gain tuner for improved control of a DC motor system with nonlinearities, *IEEE Int. Conf. Ind. Tech., Guangzhou, China.*

Yagishita, O., Itoh, O., and Sugeno, M. 1985. Application of fuzzy reasoning to the water purification process, *Industrial Applications of Fuzzy Control,* M. Sugeno, ed., North Holland, Amsterdam, pp. 19–40.

Yasunobu, S., Sekino, S., and Hasegawa, T. 1987. Automatic train operation and automatic crane operation systems based on predictive fuzzy control, *Second IFSA Congress,* Tokyo, Japan.

Zadeh, L. A. 1973. Outline of a new approach to the analysis of complex systems and decision processes, *IEEE Trans. Systems, Man, and Cybernetics,* vol. 3.

Zadeh, L. A. 1979. A theory of approximate reasoning, *Machine Intelligence,* J. Hayes, D. Michie, and L. I. Mikulich, eds., Halstead Press, New York, NY. pp. 149–194.

Zimmermann, H.-J. 1991. *Fuzzy Set Theory—and Its Applications,* Kluwer Academic Publishers, Norwell, MA.

40

Neural Network-Based Control

Dian-cheng Zhang
Hefei University of Technology

40.1 Control Configuration 572
Overview • Modes and Configurations of Control Systems Using ANNs
40.2 Design Procedure 580
Overview • Example of Design Procedure

40.1 Control Configuration

Overview

Over the past six decades great progress has been made from traditional control theory to modern control theory. The traditional control theory can only deal with the single-loop, single-variable and time-invariable coefficiency linear system. The emergence of modern control theory has pushed the control theory forward in depth and width. There are many features of modern control theory.

The treated problems have extended from single-loop mode to generalized mode. Many new theoritical bases have been developed, such as geometrical theory of linear system based on state space description, optimal control theory based on differential description and nonlinear control theory based on the differential manifold method. Many changes have been made in the modelling approaches: the generalized parametry estimation and system identification has replaced the direct modelling method based on physical features of a system. Along with the advance of the stochastic process and mathematical statistical theory, self-tuning and adaptive control have become a major direction of research in the control field. With the progress of mathematical tools, matrix theory and geometrical theory have been employed instead of the integral transformation method.

In the control field, automatic control theory and techniques have played important roles. With the progress of control theories, there are many new applications for automatic control with increased performance. However, modern technology is leading to increasingly complex systems with ever more demanding performance goals. These systems have often been under significant uncertainities and have nonlinear features that are difficult to describe exactly. The demands on control requirements have also increased. Systems must have the capabilities of processing intelligent information such as adaptive learning, and self-organization. Parallel discrete and distributed processing capabilities are also required. Perhaps systems must also have the capability for fault accomodation to operate successfully over long periods. It is therefore expected that control theory and techniques will undergo further progress and new approaches will be found to satisfy the above-mentioned demands. The difficulties that arise in the control of complex systems are mainly due to system complexity, and the presence of nonlinearities and uncertainties. Such systems are characterized by poor models, high dimensionality of the decision space, distributed sensors and decision makers, high noise levels, multiple subsystems, levels, time-scales and/or performance criteria, complex information patterns, overwhelming amounts of data, and stringent performance requirements. Even modern control theory is not in a position to cope with these situations. To address these problems in a systematic way, a number of methods have been proposed that are known as intelligent control theories or methodologies. Among these methods, perhaps the most attractive approach is that of artificial neural networks (ANNs) which experienced a resurgence in the late 1980s.

ANNs are large-scale, parallel, distributed processing, nonlinear dynamic systems. They have the common features of a general nonlinear system. ANNs also exhibit a surprising number of the human brain's characteristics. For example, they learn from experience, generalize from previous examples to new ones and abstract essential characteristics from input containing irrelevent data. There are other features due to the special structure of ANNs, such as high dimensionality, adaptability, self-organization, and variety of connections among neurals. Therefore, ANNs can use nonlinearity, learning, and generalization capabilities for application to advanced control. There are other features of ANNs that attract attention in the control field:

1. ANNs can generate input/output maps which can approximate, under mild assumptions, any function with any desired accuracy.
2. ANNs exploit parallel distributed information processing, hence have strong fault tolerance.
3. ANNs are well fitted for multi-information fusion and multimedia technology, may simultaneously combine

0-8493-8343-9/97/$0.00+$.50

quantative and qualitative information, and may be used conveniently in multi-input/output systems.

4. ANN computing may be implemented off-line or on-line to satisfy some control demands.

Modes and Configurations of Control Systems Using ANNs

There are several modes using ANNs in control:

1. Modes based on traditional control theory.
2. Neuromarphic control mode.
3. Mode combining conventional AI.
4. Intelligent control using ANNs.

The objectives of using ANNs to control may be generally classified into the following directions: using ANNs for learning functions to represent behavior control rules and systems, exploiting neural, computing, and learning strategy for robot manipulator position control, sensor-video camera-based position estimation, and robot trajectory planning, and to solve optimization problems in the control field and parametrics identification of control systems.

Models Based on Traditional Control Theory

This model uses an ANN to implement or replace complex algorithms of control, for example, the basic algebraic calculus of matrix and equations, optimization calculus, and parametric estimation algorithms. Using an ANN to solve these problems has the following advantages: higher speed of computing, simple controller structure, and suitability to hybrid systems.

Neuromorphic Control Mode

In this mode an ANN is used directly as a control tool. There are three classes of control systems according to the roles the ANN plays in the system.

1. Model-based control system in which the ANN plays model of the controlled object in the system, such as inner-model control, model reference adaptive control system, predict control, etc.
2. The ANN is used as controller of various types in the system.
3. The ANN is used as a tool for optimization computing.

Supervised Control. In many control cases human intervention is necessary; it is difficult to design a controller by conventional control techniques, since the controlled object, containing unknown factors such as nonlinearities, has complex features. Where operator skill plays the role of controller, the ANN may be used instead of the real controller to perform the control operations, because the ANN can approximately map from human perception to decision-making. Figure 40.1(a) shows the principle of supervised control; the ANN learns the mapping from sensor inputs to desired actions by adapting to a training set of examples of what it should do. This type of ANN control system has found use in the ANN-based fault diagnostic system in process control and the decision-making system in AI control. Figure 40.1(b) shows the learning scheme of a direct model of controller. The outputs of controller are taken as target for ANN learning; the learned ANN may take the place of a human operator as in Figure 40.1(a).

Inverse Control. In inverse control, an ANN learns the inverse dynamics of the controlled object; this learned ANN is placed before the controlled object, whose output equals the input of the ANN. Obviously, the necessary condition for inverse control is that the controlled object should be dynamically inversible, hence the inversibility of nonlinear system is still a difficult research problem. Figure 40.2 shows the principle of inverse control. The ANN is used in the loop or the neural controller is identified directly by other means. Figure 40.3(a) shows a basic architecture of a robot controller, in which an ANN forward direct controller (Figure 40.3(b)) is used to take advantage of the ANN capability. But the system may lose robustness at the beginning of the control, so an ANN observer (Figure 40.3(c)) performs the system identification to maintain the initial robustness of the system. Inverse ANN control has been used indirectly with optimal control (Figure 40.4). It has been used

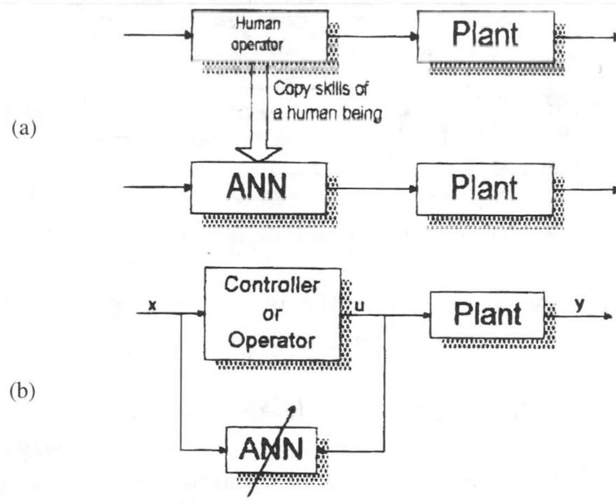

Figure 40.1 (a) Supervised control. (b) Direct learning mode of controller.

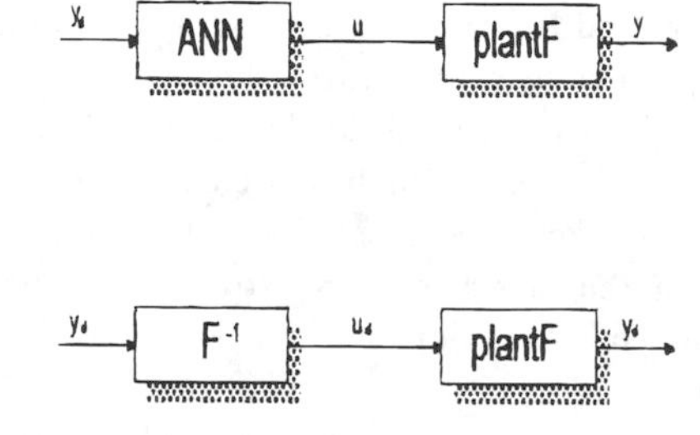

Figure 40.2 Inverse control.

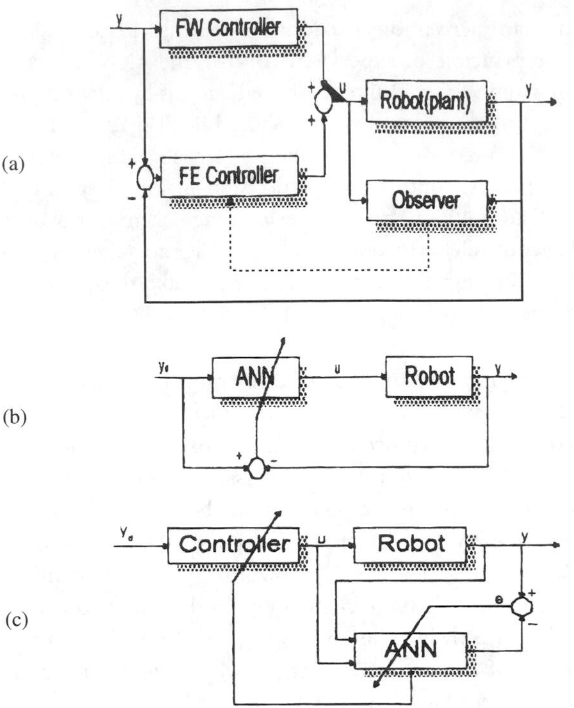

Figure 40.3 (a) Basic architecture of robot controller. (b) Forward direct control. (c) System observer.

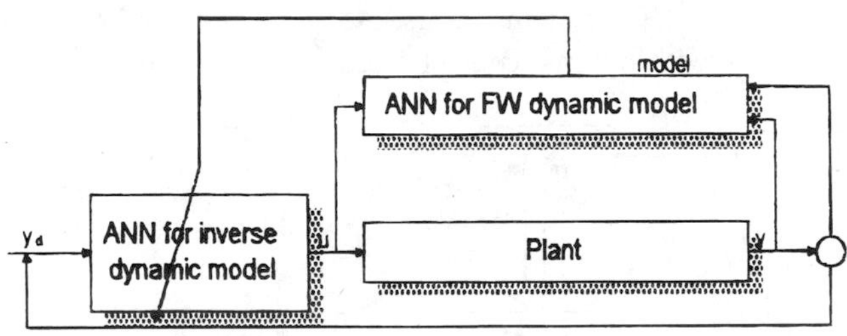

Figure 40.4 Indirect inverse controller.

indirectly for hybrid position/force control of the robot manipulator considering uncertainty of environment.

ANN Adaptive Control. In ANN adaptive control, an ANN is used in place of more classic mappings within the classic designs of adaptive control theory, the purpose of which is to realize a flexible controller capable of unknown parametric adaptation and self-tuning gain, etc. The ANN controller is expected to have the following capabilities: flexible structure to express a nonlinear system (this characteristic enhances the robustness of the controller) and learning capability due to flexible structure, which can give rise to new control scheme.

Control problems can be divided into two classes: regulation and tracking problem, in which the objective is to follow a reference trajectory, and the key issue is stability, and optimal control problem, in which the objective is to extremize a function of the controlled system's behavior, and the key issue is constrained optimization.

Adaptive methods can deal with these problems. When implementing adaptive control, ANNs are combined to both identify and control the controlled object. It is also possible to adaptively change the ANN controller based on an additional training signal, which indicates how well the system is performing, i.e., using a critic to help adjust the ANN controller parameters. Figure 40.5 shows a general ANN adaptive controller architecture, which exploits the ANN feature in the parameter-identification process only. Figure 40.6 shows a general control block diagram for model reference adaptive control (MRAC). Here the difference between the output of the actual system and the output of the reference model is the error signal e, which is used to adjust the coefficients of the feedback gain. Figure 40.7 shows a comparison between MRAC and ANN adaptive controls. In this scheme, the ANN uses an adaptive control method in a nonlinear system.

Figure 40.8 shows another ANN control architecture called a inner-model control system, in which a forward ANN model of a controlled object connects in parallel with the object; an inverse model of a controlled object is placed before the object. A linear filter is connected in the loop and the difference between the object and forward ANN model is applied to the input of the

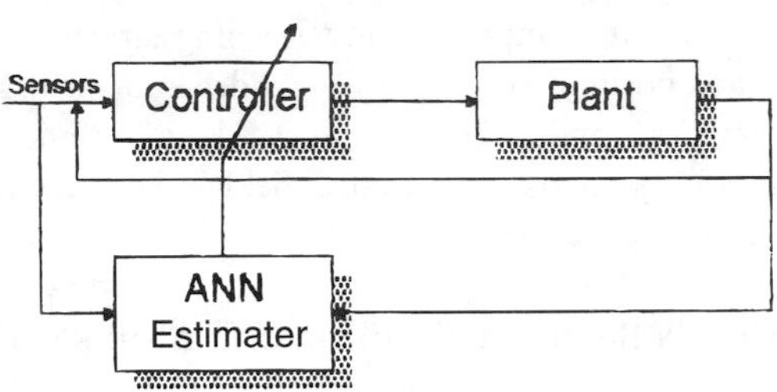

Figure 40.5 General ANN adaptive controller architecture.

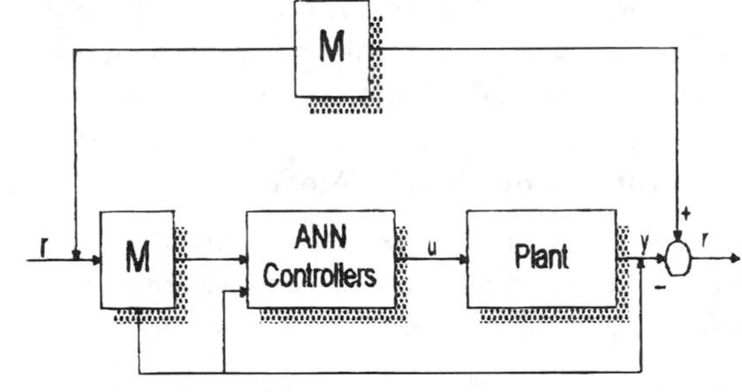

Figure 40.6 General control block diagram for MRAC.

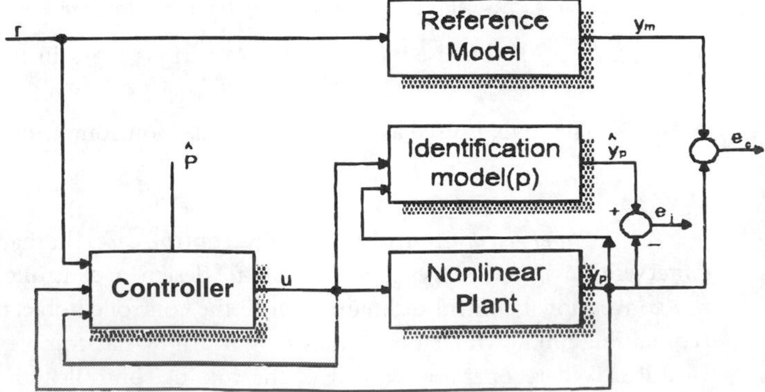

Figure 40.7 Comparison between MRAC and neural adaptive control.

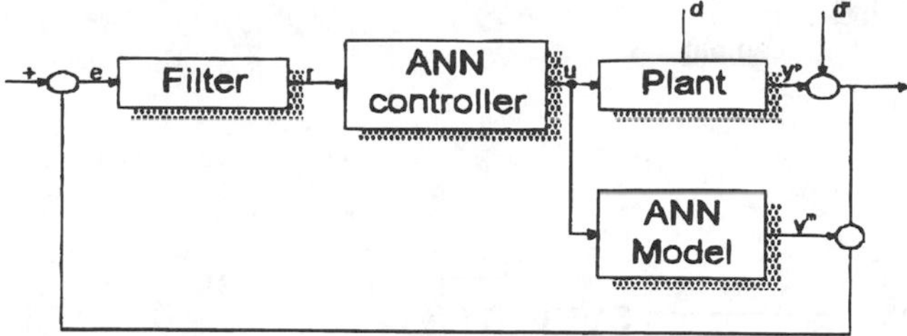

Figure 40.8 Inner-model control.

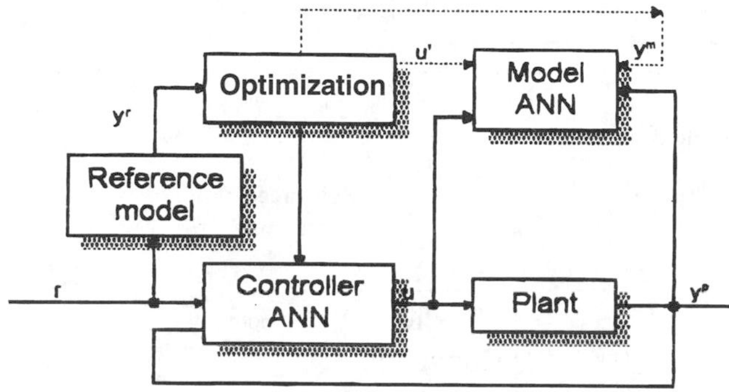

Figure 40.9 ANN predict control.

system as a feedback signal. This architecture is characterized by greater robustness and is easy for stability analyses.

Figure 40.9 shows an ANN predict control scheme, in which the ANN, as the object identified model, creates a predict signal and the control vectors are derived using an optimization algorithm. Therefore, the predict control is accomplished. Furthermore, after optimal control trajectory has been obtained, another ANN may be trained as controller that can approximate the control function. When the training process is completed, the trained controller may directly control the object in open-loop mode.

The ANN model may be used as a dynamic linear controller. The general structure of a controller using a recurrent ANN (Hopfield network) is shown in Figure 40.10. This is an example of adaptive control for a time-variable linear system using the optimization computing ability of the Hopfield network. The associate memory ability of the Hopfield network may be used to regulate PID parameters according to the change of system states.

Reinforcement learning (RL) is based on the common sense idea that if an action is followed by a satisfactory state of affairs, or by an improvement in the state of affairs, then the tendency to produce that action is strengthened. Hence, the RL method, as opposed to BP, is more attractive in that it replaces the teacher by performance measure from the environment to grade the goodness of the current actions. Measurement of performance

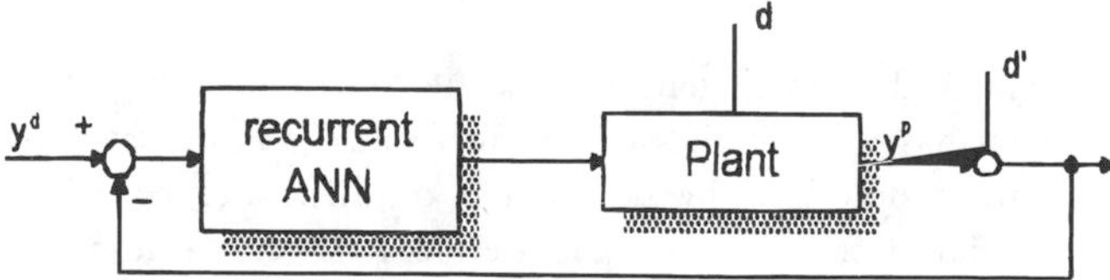

Figure 40.10 General structure of controler using recurrent ANN.

of a controller is feasible, hence on-line performance measurements can form the basis for adaptive ANN-based control by using an appropriate RL technique. Figure 40.11 shows a block diagram of an adaptive learning control, which consists of two neuronlike elements: an associative search element (ASE) and an adaptive critic element (ACE). The ASE generates an output pattern by receiving an evaluation from its environment in the form of scalar or reinforcement, updating the contents of its memory, and repeating this generate-and-test procedure. The state vector of the system is sampled and fed into the decoder, which transforms each state vector into an ANN-component binary vector, whose components are all zero except for a single one in the position corresponding to the state of the system at that instant. This vector is provided as input to the ASE, the adaptive element receives the signal through the reinforcement pathway and this information is used by the ASE. In cases in which only punishment is available the learning action needs to be more distinctive to ensure the least punishment to convergence. An ACE is introduced to overcome this problem. The central idea behind the ACE algorithm is that predictions are formed that predict not only present reinforcement, but also future reinforcement. Figure 40.12(a) shows the structure of the system and Figure 40.12(b) shows the structure of the controller.

Several neurocontroller schemes have been proposed to satisfy different requirements of a control system. Typically they may be divided into three types:

1. Serious connective type, in which an ANN replaces the FFC or FBC (Figure 40.13(a)).
2. Parallel connective type, in which an ANN is connected parallel with a conventional controller (Figure 40.13(b)).
3. Self-tuning type, in which an ANN is used to directly regulate the parameters of a conventional controller (Figure 40.13(c)). Figure 40.14 shows three types of neural controllers for a temperature control system.

Control Mode Using FAN (Fuzzy, AI, NN) Technology

Recently, traditional AI technologies, after going through more than thirty years of research and development, have been widely applied to various domains. The expert system technology, the name usually applied to AI, has found wide use in various fields with reasoning mechanisms, knowledge bases, and intelligent information processing. Meanwhile, the emergence of fuzzy rules made fuzzy set theory part of the AI family. Recently, much research has been tried to synthesize AI, fuzzy and ANN into fusion technology, namely FAN, whose concept is shown in Figure 40.15. Figure 40.16 shows the relations among the AI, fuzzy, ANN, and control in intelligent information processing. As we know, AI is good at logical reasoning, fuzzy can be used to process fuzzy information and make decision, and ANN has its features. The weak points in one aspect may be compensated by the advantages of the another aspect. Hence, using FAN technology in the control field has created new modes of control figurations.

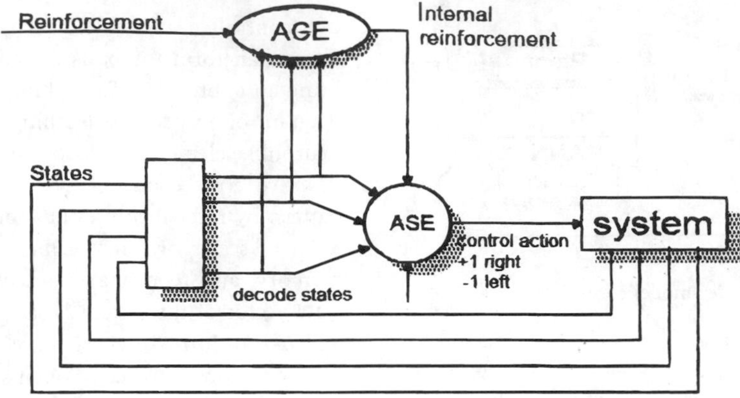

Figure 40.11 Adaptive learning control (ALC).

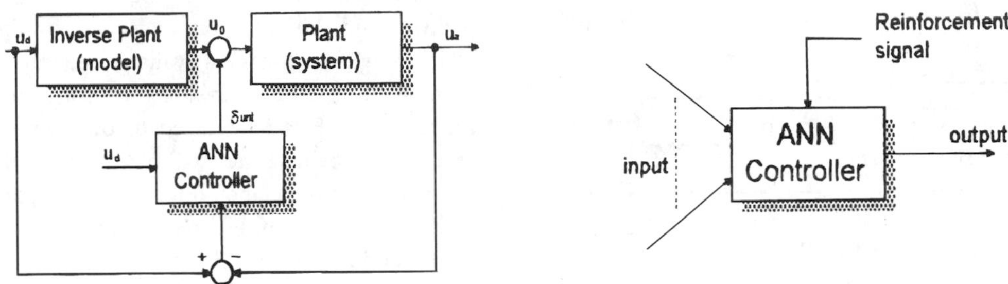

Figure 40.12 (a) Neuroadaptive controller. (b) Structure of the controller.

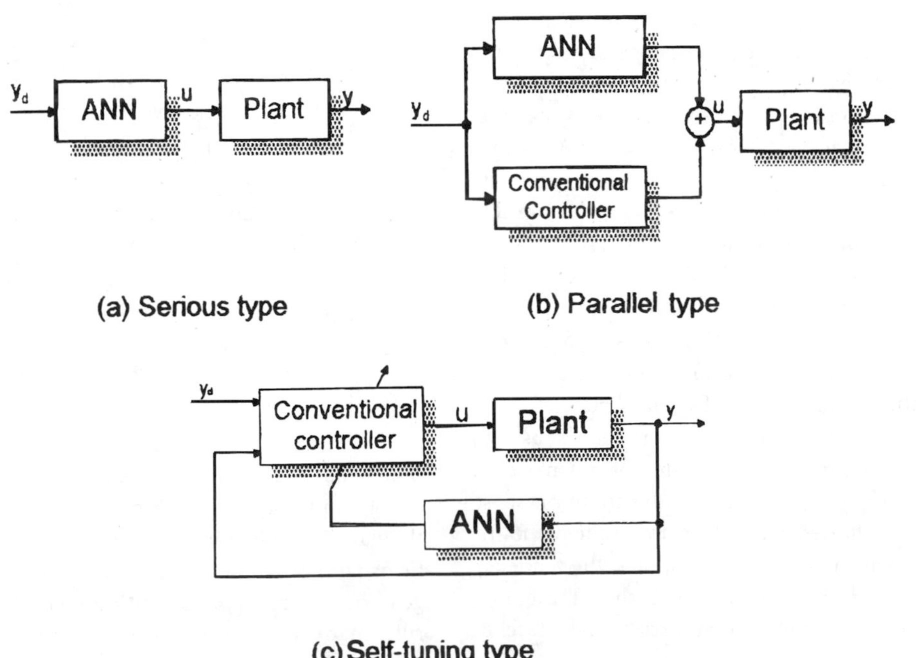

Figure 40.13 Basic types of ANN controller.

There are various configurations of this hybrid control mode. In general three approaches exist: conventional AI-based system, fuzzy set-based system, and ANN theory-based system.

ANN as Numerical/Symbolic Transformation Interface. In a hybrid system, an autonomous discrete control system, and an intelligent system, the controller in high layer usually has a symbolic or discrete form. In lower layer the system is described by differential or difference equations, hence, it is necessary to have an interface between the numerical layer and the symbolic layer to transfer the numerical information to symbolic or vice versa. This transformation process is

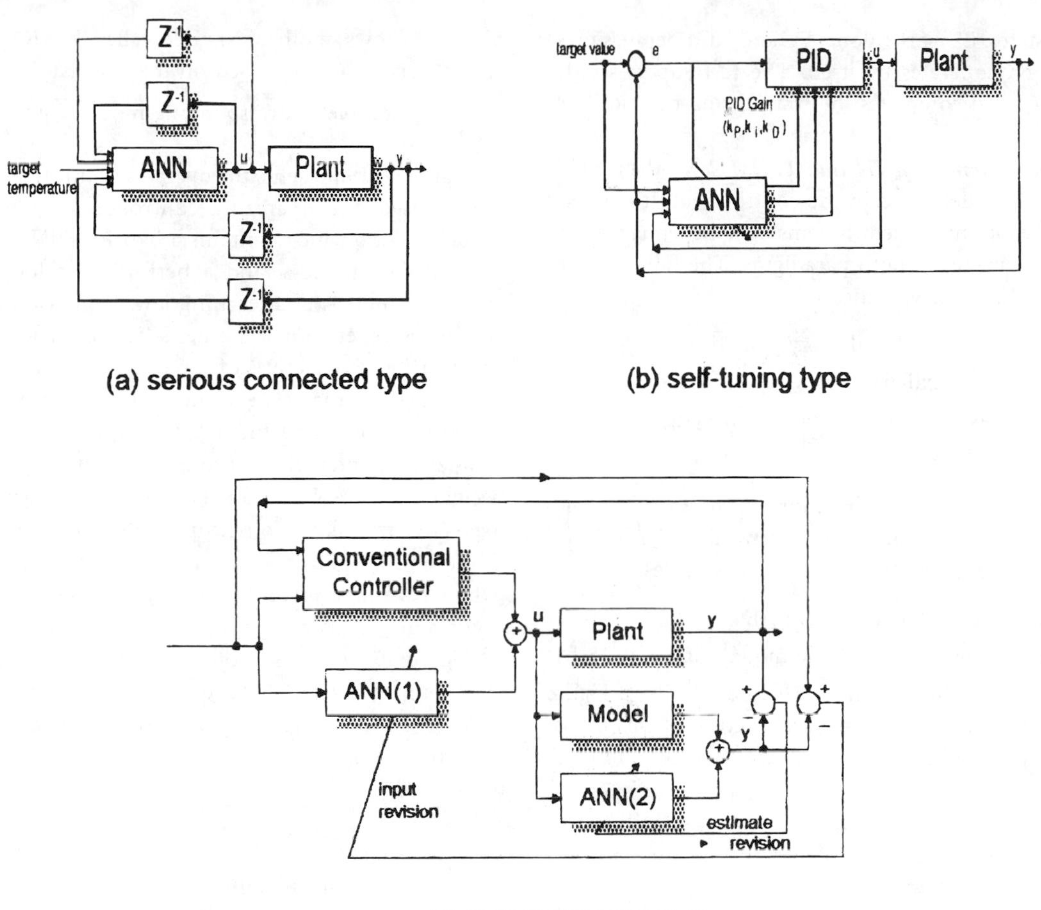

Figure 40.14 Neural controller structure of temperature control system.

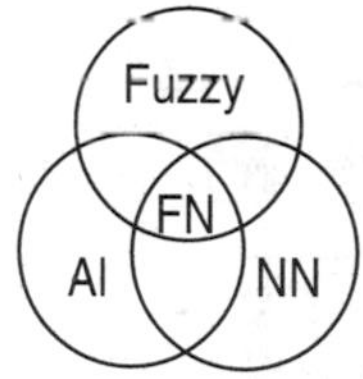

Figure 40.15 Concept of FAN technology.

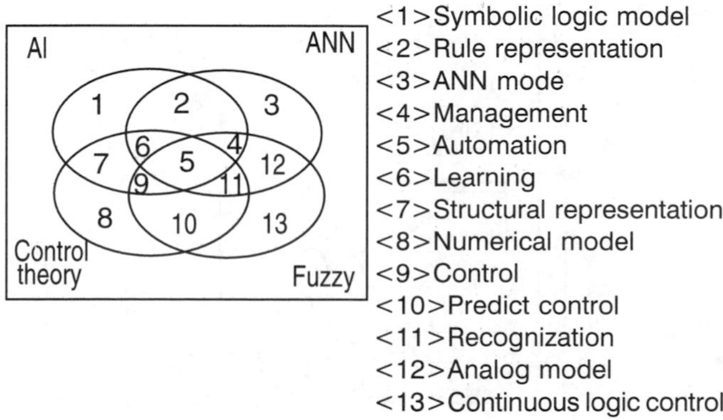

Figure 40.16 The relation of intelligent information processing.

similar to pattern recognition; ANN is well suited to this function.

Hybrid System of ES and ANN. ANN is good at rejective reasoning and an ES, or knowledge-based system, is skilled in explanative reasoning. Figure 40.17 shows the combined structure of ES and ANN in which a rule-based ES controller EC is first established to control the dynamic system P. Then the ANN controller is taught in real-time by EC the control functions more suitable for NC to implement. The operating monitor, or EM, supervises the operation of the system. The system can operate in

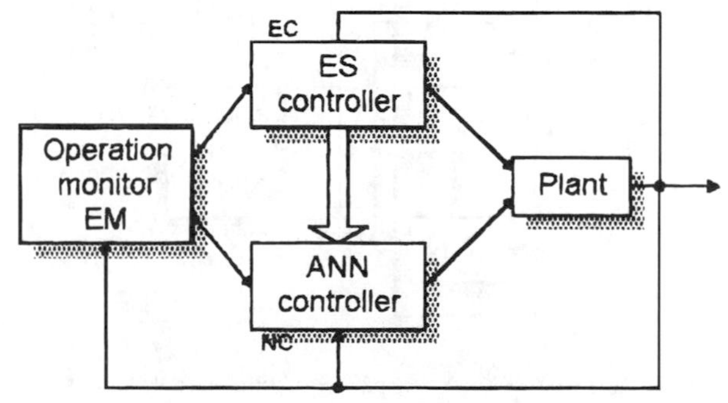

Figure 40.17 Hybrid system of ES and ANN.

following three states: ES works alone, EC and NC operate simultanteously, and NC operates by itself. The EM supervises the system conditions and completes the transferring functions.

Hybrid System of ANN and Fuzzy Controller. This hybrid approach is widely used in the control field. It is hoped that the capacity of the mixed systems will be enhanced by incorporating advantages of both paradigms. The following are the forms of ANN/fuzzy connections.

1. Connection of ANN and fuzzy system
2. Fuzzy used as auxiliary part for ANN system
3. ANN used as auxiliary part for fuzzy system
4. Fusion of ANN and fuzzy.

Among these forms, the fourth has more important advantages than the others. There are four modes of fusion of ANN and fuzzy for implementing reasoning and inference rules:

1. One-body mode (Figure 40.18(a))
2. Left part + right part + logic mode (Figure 40.18(b))
3. Left part + right part mode (Figure 40.18(c)). Figure 14.18(d) shows an example architecture of this mode, where fuzzy rules come from ANNs. Figure 40.18(e) shows another example called FAMOUS where fuzzy rules are expressed on the associate reasoning system
4. Complete corresponding mode (Figure 40.18(f))

Figure 40.19 presents the overall system structure of a hybrid ANN-based self-organizing controller. It consists of a basic feedback loop which contains a hybrid ANN-based controller, a controlled process, and a performance loop composed of reference models and learning law. It is assumed that neither a control expert nor a mathematical model of the process is available. The objectives of the overall system are to minimize the tracking error between the desired output specified by the reference model and the actual output of the process in the whole time interval of interest by adjusting the connection weights of the ANN, and meanwhile to construct control rule-bases dynamically by observing, recording, and processing the input and output data associated with the net controller. The whole system performs the two functions of control and learning simultaneously.

Figure 40.20 shows another hybrid system structure. It is a back-propagation, ANN-based fuzzy controller with self-learning teacher. Its structure is similar to that of a traditional fuzzy control system. However, this system works in two distinct modes:

Figure 40.18(a) One-body mode.

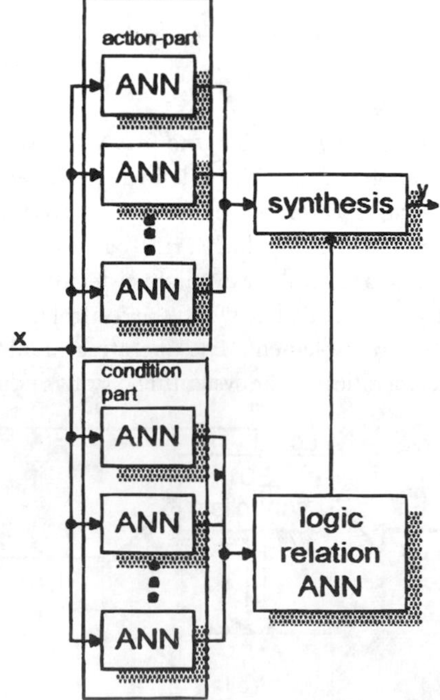

Figure 40.18(b) Condition-part + action-part + logic type.

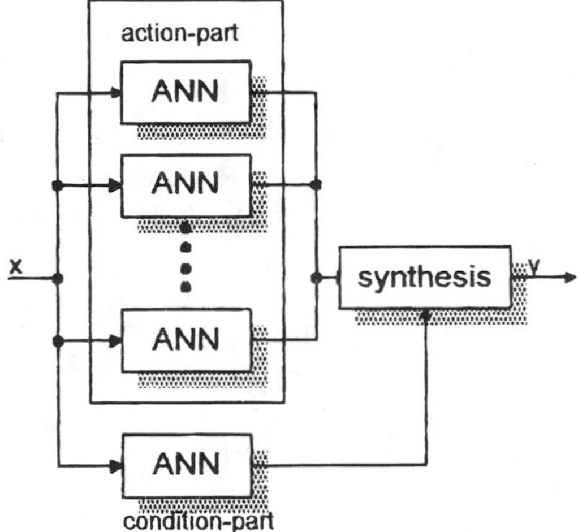

Figure 40.18(c) Condition-part+action-part type.

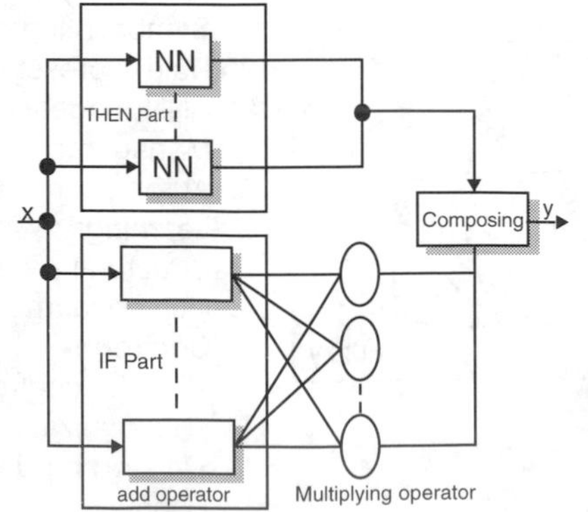

Figure 40.18(d) Example architecture.

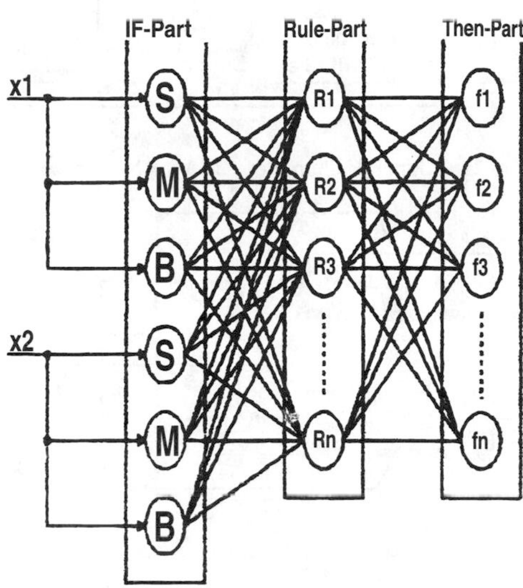

Figure 40.18(e) FAMOUS system.

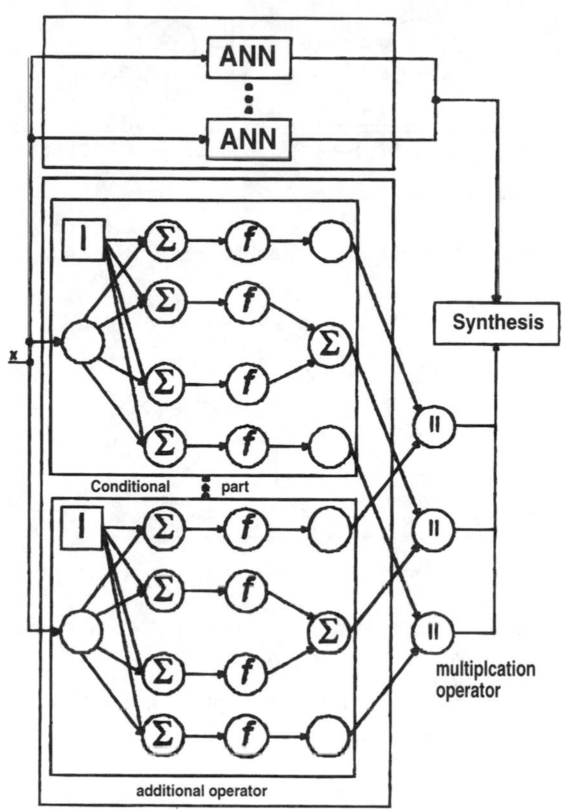

Figure 40.18(f) Complete corresponding type.

training and application. Assuming that the rule-base is given, the ANN network is trained off line by presenting all rules sequentially to the network. Then, the successfully trained network can be inserted in the control loop for on-line operation. Depending on methods for converting qualitative/linguistic labels into quantative/numerical values, the structure of the resulting controller is sufficiently different. There are at least two possibilities for doing this, namely, fuzzy set interpretation characterized by grade numbership functions and fuzzy number translation typically featured by central values and spread widths.

ANN-Based Intelligent Control

What is intelligent control? There is no exact definition, but in some authors' views we may have a working definition as follows:

An intelligent control system must be highly adaptable to significant unanticipated changes, therefore learning is essential. It must exhibit a high degree of autonomy in dealing with changes. It must be able to deal with significant complexity, and this leads to certain sparse types of functional architectures such as hierarchies. There are four approaches that have the potential for intelligent control:

1. Expert system as adaptive elements in the system
2. Fuzzy calculation as decision-producing element in system
3. ANN as compensation element in system
4. Synergetic system involving ES, fuzzy, and ANN

The attractive features of ANN in intelligent control are:

1. Self-adaptive and self-learning abilities for complex undetermined problems
2. Expressive abilities of any nonlinear functions
3. Rapid optimal computing ability for nonlinear dynamic of network
4. Distributed store ability, parallel processing, and composite ability for a large number of qualitative or quantative informations
5. Error tolerant ability

In general, ANN-based control may be considerated as a two-stage control problem: the first stage, pattern recognition, is used for system classification and selecting corresponding control parameters; the second stage, the learning controller, is used to implement self-adaptive algorithms.

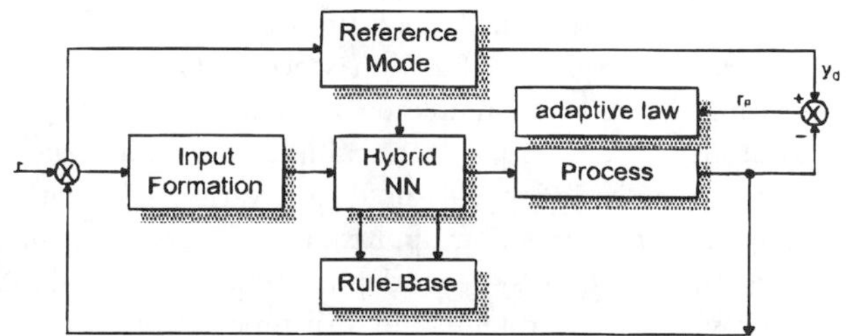

Figure 40.19 Overall system structure of a hybrid system.

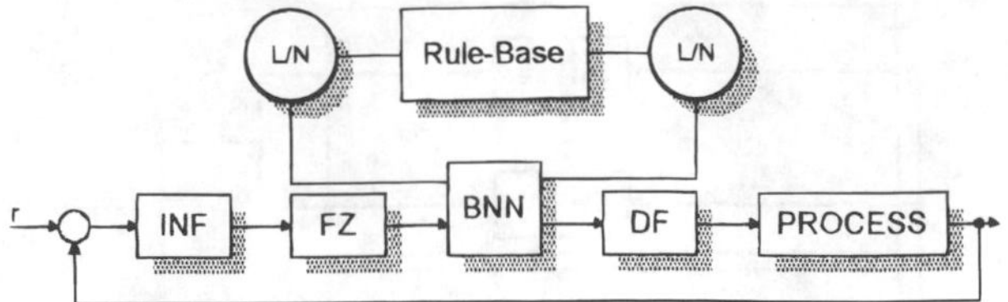

Figure 40.20 Back-propagation ANN-based fuzzy controller.

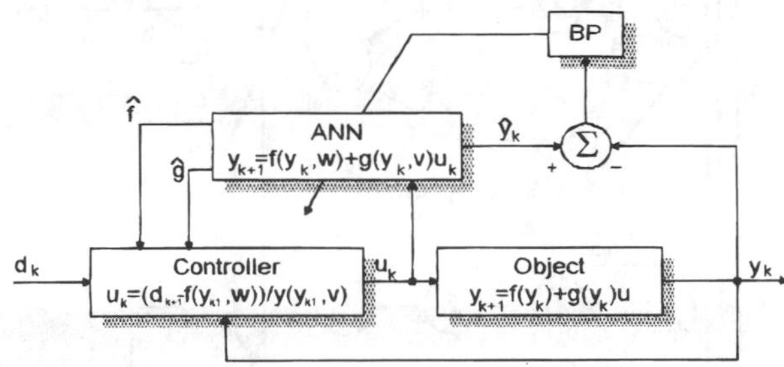

Figure 40.21 Self-tuning and adaptive ANN-based control.

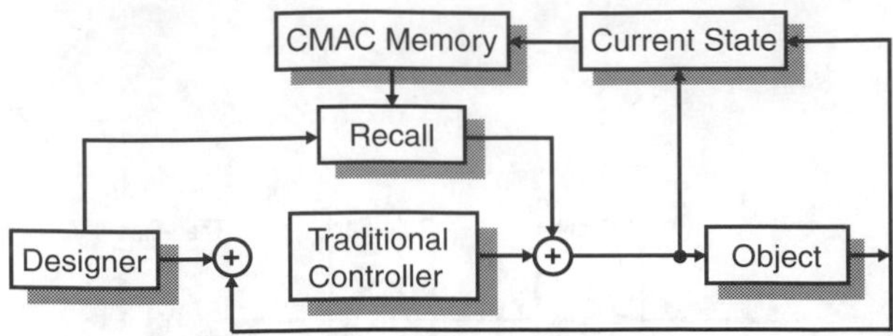

Figure 40.22 General scheme of CMAC ANN-based control.

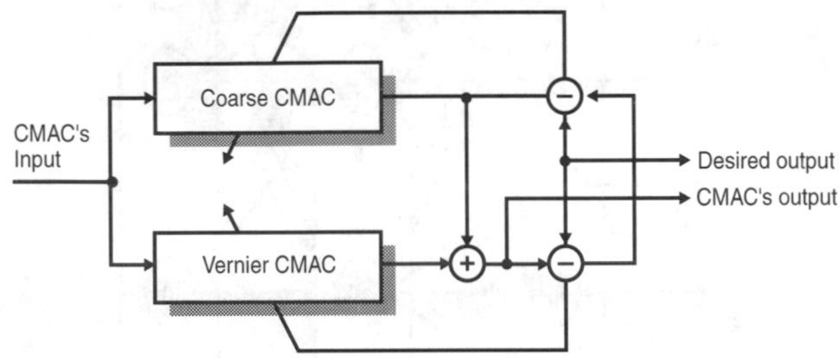

Figure 40.23 Multiple-CMAC ANN structure.

In terms of system functions accomplished, an intelligent control system may be divided into:

Intelligent adaptive control
Intelligent self-learning control
Intelligent self-organization control
Intelligent self-repair control

Among these types, the intelligent adaptive control has the common features of an intelligent control system.

The most acceptable schemes of adaptive control include:

Gain scheduling control
Model reference adaptive control (MRAC)
Self-tuning regulator (STR)
Self-learning system

The MRAC and STR are mainstays of the adaptive control algorithm. The structure schemes of MRAC and STR are shown in Figure 40.6 and Figure 40.13(c), respectively. As ANN may approximate any nonlinear functions or functionals, it allows the possibility of constructing nonlinear, self-tuning, and adaptive ANN-based controls. Figure 40.21 shows a structure scheme of a self-tuning and adaptive ANN-based control.

Two decades ago, a unique control approach based on a mathematical module called Cerebellar Model Arithmetic Computer (CMAC) was proposed by Albus. It is a perceptron-like associative memory that performs a table look-up of a nonlinear function over a particular region of function space. A CMAC network has the capability to learn an unknown nonlinear mapping given input/output sets, and to produce multiple outputs in response to multiple inputs. Today, it is used in a variety of situations including nonlinear function approximation, robot kinematic mapping, and real-time system adaptive control. CMAC-based ANNs provide great potential in real-time adaptive control applications. Figure 40.22 shows a general scheme of CMAC ANN-based control.

A multiple-CMAC network structure has been proposed as shown in Figure 40.23. This network consists of two CMAC nets, called coarse and vernier networks. This cascaded network may achieve faster learning than conventional ANNs and capture general trends and fine details of an unknown nonlinear mapping.

40.2 Design Procedure

Overview

In general, the procedure of application of ANN may be divided into following stages:

1. Analysis of controlled object, i.e., dynamically analyzing control system, determining its means of input information, making sure of constraint conditions, and clarifying criteria of evaluation.
2. Model design, i.e., collecting data, extracting learning data, choosing the structure of ANN, and selecting rules for learning.
3. Tuning stage, in this stage the learning parametry and structure of the ANN might be modified in the iterative process to get convergence of output error.

For designing an ANN-based control system, the first step is formulation of the desired control problem. It is necessary to describe the system behavior using mathematical equations. According to different system structures and control functions, these equations may be of differential/difference form, discrete event form, state space form, transfer function form, dynamic equation form, etc. However, all control problems involve manipulating a dynamic system input so its behavior meets a collection of specifications constituting the control objective.

In terms of objectives, the control problems may be divided into two major classes. The objective of some problems is defined in terms of a reference trajectory that the system's output should match or track as closely as possible. In this case, stability is the key issue. The objective of the other problem is to extremize a functional of the system's behavior. Here, the key issue is constrained optimization. Many control methods have been proposed for both problems among which the adaptive control methods are overwelmingly used. According to different features, objectives of the system and the proposed control method appropriate formulation of the problem may be used.

The next step after problem statement is to determine the ANN structure. More than thirty ANN modes have been proposed. In general, organization of these modes may be classified into two types: the feed-forward net which has a hierarchical structure that consists of some layers without interconnections between neurons in each layer and signals flow from input layer to output layer in one direction; the other type is the recurrent net in which multiple neurons are interconnected to organize the network. One of the main features of ANNs used in control is the ability to approximate the system input/output mapping with any function with any accuracy. Many factors may influence the approximating ability of ANNs. These factors include, besides the type of ANN, number of layers, number of neurons in each layer, individual neuron activation functions, and adopted learning algorithm.

The sigmoidal functions are most often used for the individual neuron activation function, but signum and Gaussian functions may be also used. As for the number of layers in an ANN, theory shows that any desired approximation can be accomplished with a multilayer network with only one hidden layer of neurons. The number of neurons in input and output layers mainly depends on system number of input signal and feedback signal, and the output of the system. The number of hidden layers is typically chosen based on empirical criteria and one may iterate over a number of networks to the ANN that has a reasonable number of neurons and accomplishes the degree of approximation. Some proposals have been made to find appropriate network structures and select acceptable numbers of hidden nodes. Adaptive learning method self-configuring network and variable-structure compitive network are examples of these proposals.

Selection of a learning algorithm is the key issue in ANN design for control application. Most of today's learning algorithms have evolved from the concepts of D. O. Hebb. An ANN using Hebbian learning will increase its weights according to the product of the excitation levels of the source and destination neurons. In symbols:

$$w_{ij}(n+1) = w_{ij} + \alpha \text{out}_i \text{out}_j$$

where

$w_{ij}(n)$ = the value of a weight from neuron i to neuron j prior to adjustment

$w_{ij}(n+1)$ = the value of a weight from neuron i to neuron j after adjustment

α = the learning-rate coefficient

out_i = the output of neuron i and input to neuron j

out_j = the output of neuron j

Another learning rule called delta rule or LMS (least mean square) method changes weights of net following presentation of an input/output pair p by

$$\Delta_p w_{ji} = \eta(t_{pj} - o_{pj}) i_{pj} = \eta \delta_{pj} j_{pi}$$

where t is the target input for the j-th component of the output pattern for pattern p, o is the j-th element of the actual output pattern produced by the presentation of input pattern p, i is the value of the i-th element of the input pattern, $= t_{pj} - o_{pj}$, and w_{ji} is the change to be made to the weight from the i-th to j-th unit following the presentation of pattern p. The delta rule essentially implements gradient descent in a sum-squared error for a linear function, so that it is guaranteed to find out the best set of weights in the case without hidden layers. Extending the standard delta rule to a layered feed-forward network with hidden layers, we may get the generalized delta rule as following. Let

$$E_p = \frac{1}{2} \sum_j (t_{pj} - o_{pj})^2$$

be the measure of the error on input/output pattern p and let $E = \Sigma \text{Ep}$ be the overall measure of the error. Here the weight sum of the output of previous layer is defined as

$$s_{pj} = \sum_i w_{ji} o_{pi}$$

The output,

$$o_{pj} = f_j(s_{pj})$$

uses the sigmoid function. Through derivations we may make the weight changes according to

$$\Delta_p w_{ji} = \eta \delta_{pj} o_{pj}$$

where $\delta_{pj} = (\text{t}_{pj} - \text{o}_{pj}) f'_j(s_{pj})$ for any output unit u_j, and $\sigma_{pj} = f'_j(s_{pj}) \Sigma\, \delta_{pk} w_{kj}$ whenever u_j is not an output unit.

Based on the above generalized delta rule, a flowchart of the back-propagation training algorithm is shown in Figure 40.24.

The approximation method with feed-forward network has the inherent difficulty that requires the dimension of the input/output signal space to be high when this state network is used to approximate dynamic behavior of the system. Consequently, the size of the network which will be able to provide acceptable approximation will be large; therefore, the training process will be problematic. Recurrent networks have a large potential of the approximation of nonlinear, dynamic behavior because the recurrent connections allow networks to produce complex time-varying outputs as a response to simple inputs. But it is difficult to find a common learning algorithm. Several learning methods, such as back-propagation through time, real-time recurrent learning, sensitivity equation method, and adjoint equation method, have been proposed.

Although the BP algorithm used in multilayer neural network has been widely accepted, there are some inherent limits, such as slow speed of learning, poor error tolerance, and incompleteness of algorithm. Some improved algorithms have been proposed, such as increased rate of convergence through learning

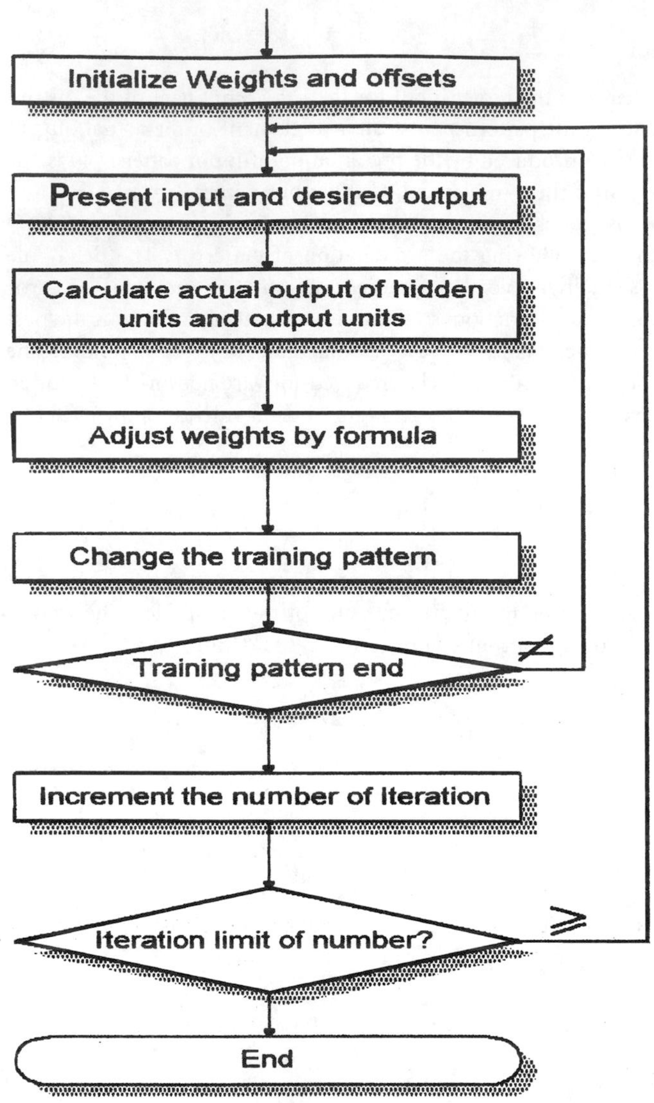

Figure 40.24 Flowchart of the back-propagation training algorithm.

rate adaptation, training feed-forward networks with the extended Kalman algorithm, etc. In recent years, some unsupervised learning paradigms, such as competitive learning, interactive activation, and adaptive resonance paradigms, have attracted more attention in the ANN field. These paradigms are expected to be exploited in the control field. A new stochastic learning algorithm based on simulated annealing in weight space has been proposed. It has been verified that it has convergence properties and can guarantee finding the optimal weights.

Example of Design Procedure

ANN-Based PID Learning Control

Formulation of Problem. The incremental form of conventional PID regulator is as follows:

$$\Delta u_n = Kp(e_n - e_{n-1}) + Kd(e_n - 2e_{n-1} + e_{n-2})$$

If the parameters Kp, Ki, Kd are properly chosen, the object can be effectively controlled.

ANN-Based PID Control Model. We can use a single ANN to form the PID control system as shown in Figure 40.25. Here we can choose $f_0\,(u) = u$ as the activate function and let $n = 3$, then the output of the ANN-based regulator may be expressed as follows:

$$\Delta u_n = w_1 x_1 + w_2 x_2 + w_3 x_3$$

here $x_i (i = 1, 2, 3)$ are the state variables, the values chosen which create great influence on control performance of the system. If we let

$$x_1 = e_n,\ x_2 = e_n - e_{n-1},\ x_3 = e_n - 2e_{n-1} + e_{n-2}$$

then we may obtain

$$\Delta u_n = w_1 e_n + w_2(e_n - e_{n-1}) + w_3(e_n - 2e_{n-1} + e_{n-2})$$

It is obvious that the values of w_i $(i = 1, 2, 3)$ can be adapted by regulating weights of the ANN, thus greatly improving the robustness of the controller.

Learning Algorithm. The essential function of an ANN controller is its learning algorithm, i.e., the rule to regulate w. Given the difference of target output and practical output e, we use an iterative gradient algorithm to minimize the mean square error; we have

$$E(k) = \frac{1}{2}\,[r(k) - y(k)]^2$$

$$w_i(k+1) = w_i(k) - \eta_i e(k)\,\frac{\partial e(k)}{\partial w_i(k)}, \qquad i = 1, 2, 3$$

where

$$\frac{\partial e(k)}{\partial w_i(k)} = -\frac{\partial r(k)}{\partial w_i(k)} = -\frac{\partial r(k)}{\partial u}\,\frac{\partial u}{\partial w_i(k)}$$

$$\frac{\partial e(k)}{\partial w_i(k)} = \frac{\partial y(k)}{\partial u}\,e(k)$$

$$\frac{\partial e(k)}{\partial w_2(k)} = -\frac{\partial y(k)}{\partial u}\,[e(k) - e(k-1)]$$

$$\frac{\partial e(k)}{\partial w_2(k)} = -\frac{\partial y(k)}{\partial u}\,[e(k) - 2e(k-1) + e(k-2)]$$

η_i is the rate of learning, $0 < \eta_i < 1$. If η_i is small enough then

$$\frac{1}{2}\,e^2(k) < \frac{1}{2}\,e^2(k-1)$$

Figure 40.25 ANN-based PID control system.

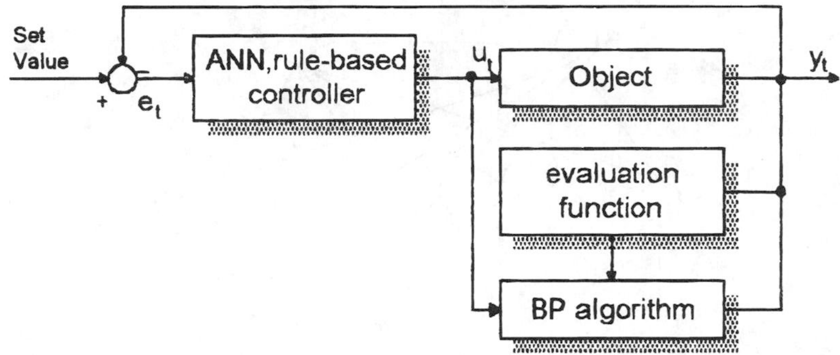

Figure 40.26 Structure of ANN rule-based controller.

which means that $e(k)$ tends to zero along with increasing of value of k. The learning algorithm is convergent.

ANN and Rule-Based Control

Description of Problem. Suppose we have the following dynamic system:

$$y(t+1) = f(y(t), u(t))$$

where $y(i)$ and $u(i)$ are output and input, respectively; $f(*,*)$ is a linear or weak nonlinear function.

Given a model of system, we may design the control rule that can regulate the deviation between target value and output as follows:

$$\text{if } e(t) \in E(i) \text{ and } \delta e(t) \in \delta E(j), \text{ then } \delta u(t) \in \delta U(k)$$

where $\delta e(t) = e(t) - e(t-1)$ is the deviation variation; $\delta u(t) = u(t) - u(t-1)$ is incremental input; and $E(i)$, $E(j)$, $U(k)$ are defined domains of $e(t)$, $e(t)$, and $u(t)$.

System Structure. There are some design methods to implement the above control rule. Figure 40.26 shows one of the ANN-based control structures. The ANN controller is shown in Figure 40.27. The input and output of the network are expressed as follows:

$$E_i(t) = \begin{cases} 1, & e(t) \in E_i \\ 0, & \text{otherwise} \end{cases} \qquad i = 1, 2, \ldots, p$$

$$E_i(t) = \begin{cases} 1, & \delta e(t) \in \delta E_i \\ 0, & \text{otherwise} \end{cases} \qquad i = 1, 2, \ldots, r$$

$$i(t) = \begin{cases} 1, & \delta u(t) \in \delta U_i \\ 0, & \text{otherwise} \end{cases} \qquad i = 1, 2, \ldots, q$$

Computing Formulas of Output and Change of Weights.

$$Z_i^K(t+1) = \varphi\left(\sum_{j=1}^{n_k-1} w_{ij}^k(t) z_j^{k-1}(t+1) + \Theta_i^k\right)$$

$$i = 1, 2, \ldots, n,\ k = 1, 2$$

$$\delta w_{ij}^k(t+1) = -\alpha(t+1) z_j^{k-1}(t+1)$$

$$k = 2, 1;\ i = 1, 2, \ldots, n;\ j = 1, 2, \ldots, n$$

where $\varphi(x) = 1/(1+e^{-x})$; w_j^k and θ_i^k are former weight and threshold;

$$d_i^2(t) = [(z_i^2 - o_i) z_i^2 (1 - z_i^2)](t), \qquad i = 1, 2, \ldots, n$$

$$d_i^1(t) = [\sum_{j=1}^{n_2} w_{ji}^2 d_j^2 z_j^1 (1 - z_j^1)](t), \qquad i = 1, 2, \ldots, n$$

$O(t)$ is the expecting mode of control action; $\alpha(t+1)$ is the learning factor computed by the following formula:

$$\alpha(t+1) = \begin{cases} K_1 c(t-1)/[c(t) + c(t-1)], & c(t) < c(t-1) \\ 0, & c(t) = c(t-1) \\ K_2 c(t)/[c(t) + c(t-1)], & c(t) > c(t-1) \end{cases}$$

where $0 < K_1 < 1$, $-1 < K_2 < 0$. $c(t) = g(|e_t|, |\delta e_t|) >= 0$, where $g(0, 0) = 0$, $|e_t|$ and $|\delta e_t|$ are rigorous incremental functions.

Control Algorithm.

Step 1. Establish ANN controller as shown in Figure 40.27.

Step 2. Measure output of object y and compute the deviation e and its change.

Step 3. Update the learning factor $\alpha(t+1)$, and train the

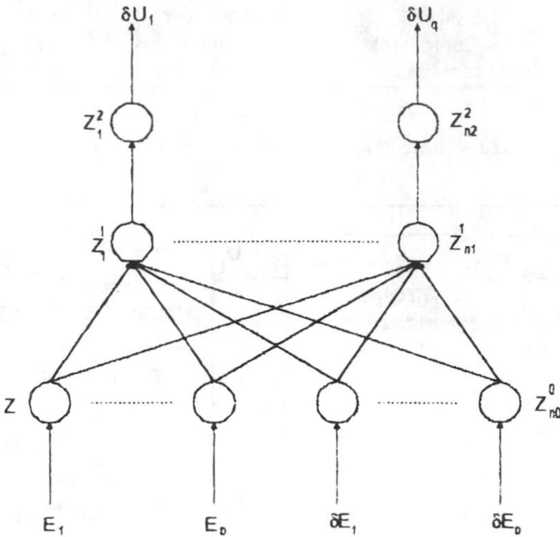

Figure 40.27 Structure of ANN used.

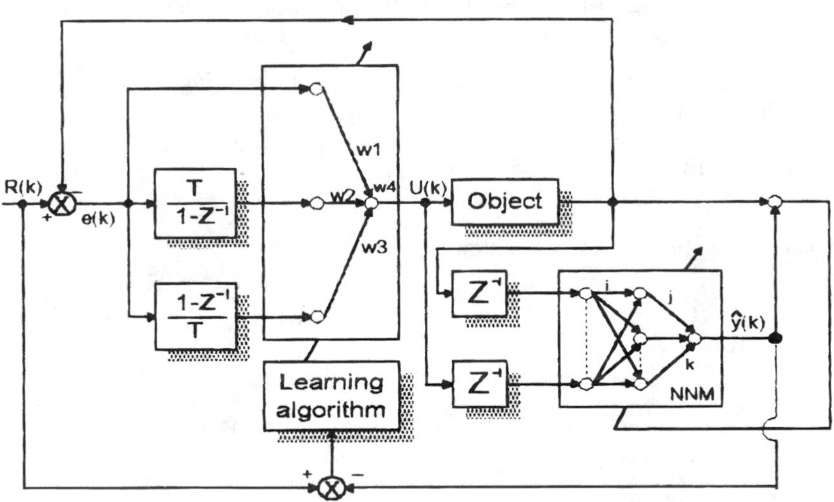

Figure 40.28 Structure of ANN intelligent PID control system.

controller by BP method, then compute the new weights of the network.

Step 4. Based on e_t and δe_t, compute the next mode of control change value $\delta U_i(t)$, then determine the practical incremental input δu_t.

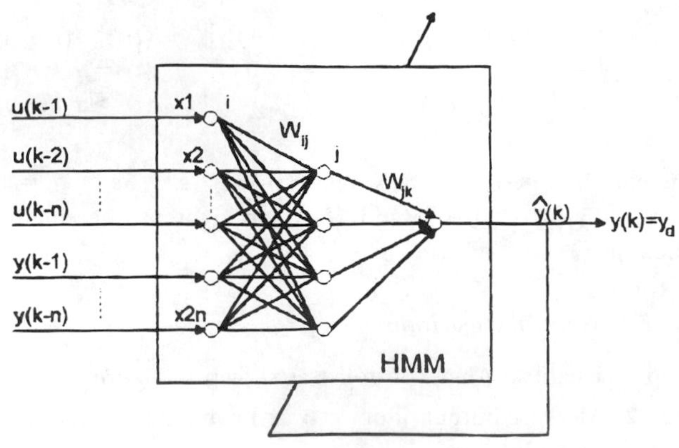

Figure 40.29 Structure of NNM.

Step 5. Return to step 2.

ANN-Based Intelligent PID Control

1. System Structure

System Identification Based on ANN. We suppose the discrete transfer function model of parameter estimation of the system to be as follows:

$$G(Z^{-1}) = \frac{a_1 Z^{-1} + a_2 Z^{-2} + \cdots + a_n Z^{-n}}{b_0 + b_1 Z^{-1} + \cdots + b_m Z^{-m}}$$

The corresponding difference equation is

$$y(k) = a_1 u(k-1) + \cdots + a_n u(k-n) - b_1 y(k-1) - \cdots - b_m y(k-m)$$

where $b_0 = 1$ and $u(k)$ and $y(k)$ are the respective k-th input and output signals. If the coefficients $[a_1, a_2, \ldots, a_n, b_1, b_2, \ldots,$

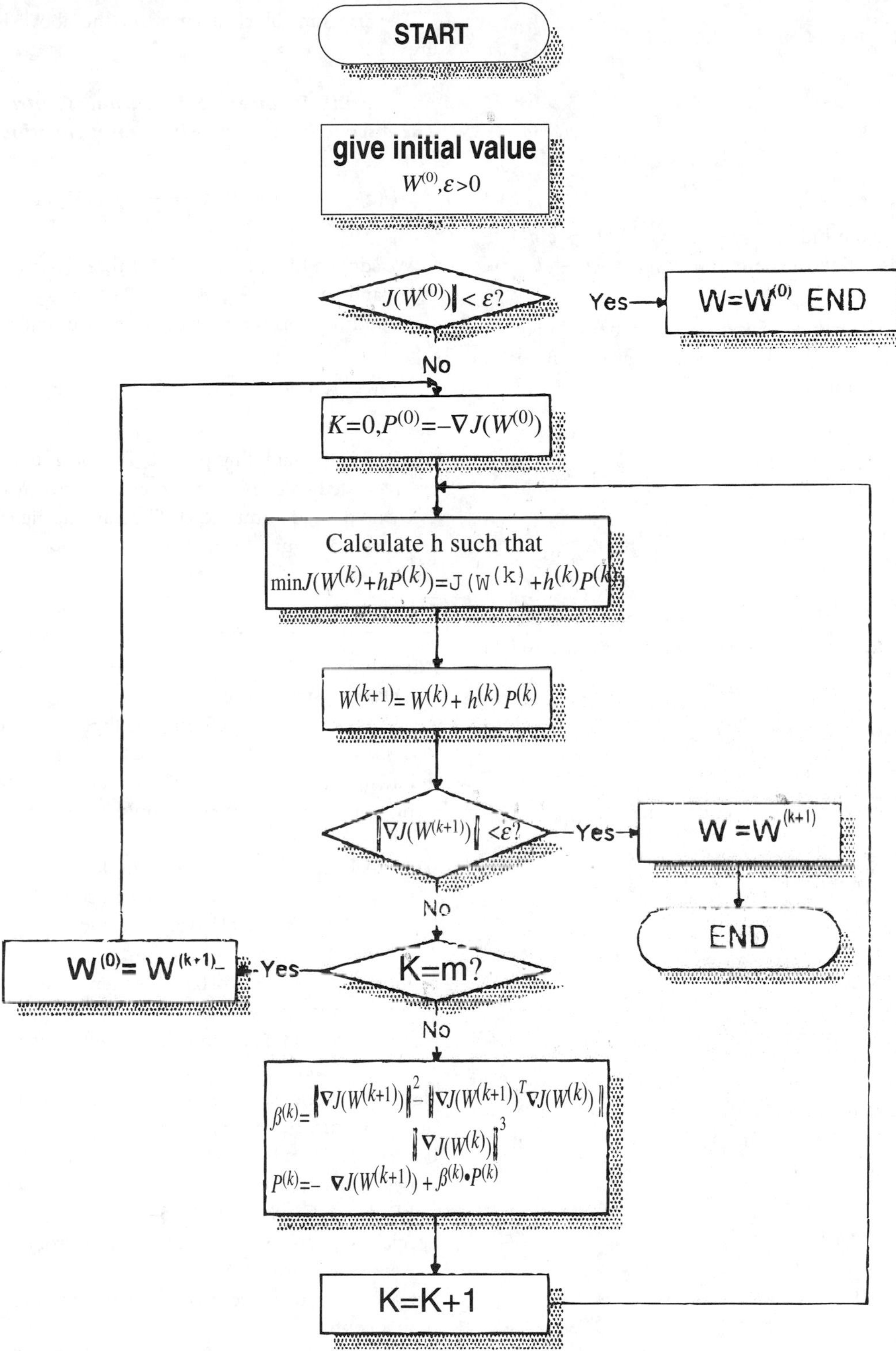

Figure 40.30. Program block diagram of the learning algorithm.

$b_m]$ are unknown, but there are a set of measured $u(k)$ and $y(k)$, then we may obtain the dynamic characteristic of the system by mapping input to output in an ANN with nonlinear. Figure 40.29 shows the structure of such NNM.

Suppose there are N nodes ($N = 2n, n$ is the order of system) in the input layer, the input vector of which is $x = [x_1, x_2, \ldots, x_{2n}]^T$; and H nodes in the hidden layer, the input and output of the j-th node in which are I_j and O_j, respectively; the output layer has m nodes ($m = 1$), the output of the k-th node in which is Y_k, W_{ij}, and W_{jk} are the weights from i-th node of the hidden layer to k-th node of the output layer. Then we may have the following network equation:

$$I_j = \sum_{i=1}^{N} w_{ij} \times x_j, \qquad i = 1, 2, \ldots, N$$

$$O_j = f(I_j)$$

$$I_k = \sum_{j=1}^{H} W_{jk} \times O_j, \qquad j = 1, 2, \ldots, H;\ k = 1$$

where the output of the hidden layer is $f(x) = (1 - e^x)/(1 + e^x)$, the activation function of the output layer is $f(I) = I$.

Learning Algorithm. Suppose under the action of the input pattern vector the error between the real output vector and desired output vector is

$$J(W) = \frac{1}{2} \mid Y_k - \hat{Y}_k \mid^2$$

$$J_k(W) = \frac{1}{2} \sum_{k=1}^{m} (Y_k - \hat{Y}_k)^2$$

where W is the weight value of the network and m is the set of input pattern. Then the formula for modifying the weight vector is

$$W^{(k+1)} = W^{(k)} + h^{(k)} \times P^{(k)}$$

where

$$P = \nabla J(W) = \sum_{k=1}^{m} (Y_k - \hat{Y}_k) \frac{\partial \hat{Y}_k}{\partial W}$$

and h can be found from the following formula:

$$\frac{dJ(W^{(k)} + hP^{(k)})}{dh} = 0$$

where $W^{(k)}$ is the weight vector of the k-th iteration and $P^{(k)}$ is the direction vector for search of the k-th step and can be calculated by the following formula:

$$P^{(k)} = \nabla J(W^{(k)} + \beta^{(k-1)} \times P^{(k-1)}$$

the conjugate coefficiency is

$$\beta^{(k-1)} = \frac{[\nabla J(W^{(k)}]^T \times [\nabla J(W^{(k)}] - [J(W^{(k)}]^T \times [J(W^{(k-1)})]}{[\nabla J(W^{(k-1)})]^T \times [\nabla J(W^{(k-1)})]}$$

The program block diagram of the above is shown in Figure 40.30

PID Parametry Self-Learning Control Based on ANN. The discrete form of the PID control equation is

$$U(z) = K_p e(t) + K \times \frac{T}{1 - Z^{-1}} + K_d((1 - Z^{-1})/T)E(z)$$

We adopt a BP ANN with 3-1-1 three layer as shown in Figure 40.30 and let $W_1 = K_p$, $W_2 = K_i$, $W_3 = K_D$. W^4 is a gain-amp stage and the corresponding PID control equation is

$$U(z) = W_4(W_1 + W_2 \times \frac{T}{1 - Z^{-1}} + W_3((1 - Z^{-1})/T)E(z)$$

where T is the sampling period. The weights (W_1, W_2, W_3, W_4) are regulated to minimize the error between output of object $Y(k)$ and desired output $R(k)$. The learning algorithm is the same as shown in Figure 40.30.

References

Autsaklis, P. J. 1992. Neural network in control system. *IEEE Control System*, April.

Autsaklis, P. J. 1994. Defining intelligent control. *IEEE Control System.* June.

Fukuda, T. and Shibat, T. Theory and applications of neural network for industrial control systems. *IEEE on IE*, 39(6)472–489.

Jiao-Li-Cheng. Application and implementation of Neural Network (in Chinese), *neural network control theory and application.* University of Electronic Science and Technology Press, Xi An. Chapter 3.

Motion and robot control by NN system control and information. *Int. J. Control.* 38(10)653–668.

Nareacira, K. et al. 1992. Intelligent control using neural network. *IEEE Control System.* April.

Neural network based learning adaptive control (in Japanese). *Measurement and Control.* 30(4)302–308.

Pon Hush, Chaouki AB Dallonh and Horue. 1993. The recursive neural net and its application in control theory computers in EE. *Int. J. Control.* 19(4)333–341.

Special section on NN for system and control *IEEE Control System Magazine.* April 1988.

Sutton, R. S. Reinforcement learning in direct adaptive optimal control.

Yamaguchi, T. et al. 1992. Self organizing control using fuzzy NN. *Int. J. Control.* 56(2)415–439.

41

Programmable Logic Control (PLC)

41.1 Basic Concepts ... 587
41.2 Hardware Components ... 588
Types of PLCs • Families of CPUs
41.3 PLC Real-Time Operating Systems ... 588
41.4 Software Components ... 590
41.5 PLC Communications ... 590
41.6 Selecting the Right PLC ... 591

Ernst Dummermuth
Rockwell Automation

41.1 Basic Concepts

Programmable logic controllers (PLCs) are generally employed in an automated factory to control actuators such as motors, solenoids, valves, indicators. Basically, an actuator is activated if a logic interconnect or equation that describes the condition is fulfilled. A sample of an equation that may control an output such as MOTOR1 is shown in Figure 41.1. The two vertical lines represent the power rail, e.g., a 115 V AC supply. If the contacts on the left side are in such a condition that power flows through them to the left terminal of MOTOR1 then the MOTOR1 will be turned on. In this case, the equation, also called rung, is true. A second rung is shown controlling Solenoid5 (SOL5). The names for the variables may also be chosen in accordance with the application, e.g., limit switch 3 may be called simply LS3.

Clearly the diagram of Figure 41.1 resembles a ladder and thus, these diagrams were commonly called ladder diagrams. The first successful PLCs retained this ladder diagram approach to programming, which made them relatively straightforward for plant operators and maintenance personnel to program. This programming language is called *Ladder Language* and is one of the standard features which distinguishes a PLC from a computer. This and three other standard features of a PLC are

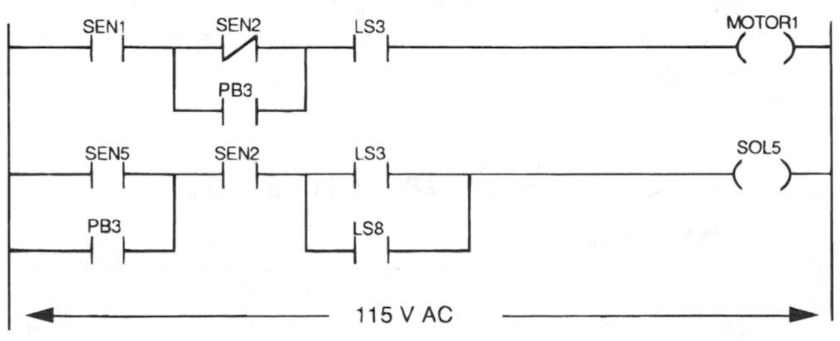

Figure 41.1 Basic programming language for PLCs.

1. Ladder Language for ease of programming.
2. Designed to operate in an industrial environment rather than the relatively benign office/home environment.
3. Standard discrete output modules which can supply several amperes to medium power AC and DC devices.
4. I/O modules and power supplies designed to operate in the presence of high EMI noise.

In an automated factory where conveyers, presses, handlers, shuttles, etc., are activated in a manner to produce or work on products, many of these rungs are implemented and are activated according to their describing equations. Several hundred rungs may be needed for a medium size machine. If a system has 100 actuators (outputs) then basically 100 rungs are needed, one for each output.

In the past, a large relay panel was built near the machine, and the required interconnects were hard-wired inside the panel. In many cases, this resulted in a terrible mess of wires. In addition, as can be seen from Figure 41.1, some conditions such as SEN2, LS3, PB3 may appear in multiple rungs. Intermediate relays with multiple isolated normally open and/or normally closed contacts had to be used to provide for repeated use of a condition. Additions or corrections to an already wired control were almost impossible to perform, and documentation was very hard to generate, maintain or even update for a modified requirement.

The idea of a programmable logic controller is to collect all inputs into a central processing unit (CPU), perform the required logic equations in accordance with a stored program, and then send the computed outputs to the actuators. The time needed by the CPU to evaluate all rungs is called the program scan time. Likewise, time needed to collect inputs and to update outputs is called the I/O scan time. While in local rack applications the I/O scan and the program scan are usually performed in sequence, in remote I/O applications program scan and I/O scan normally are done asynchronously and overlapping in time.

0-8493-8343-9/97/$0.00+$.50

Each sensor is wired to a specific input terminal of an input card, and each output terminal of an output card is wired to an actuator. Input cards may have room for 16, 32 or even 64 input connections while output cards are usually limited to 16 terminals per card. With this arrangement a very organized and control logic-independent wiring is achieved, no rewiring is ever needed. During expansion, additional connections are made to spare input/output terminals.

The stored program is usually prepared on a programming terminal that, with the help of graphic editors, allows the user to create and view the program in a form as shown in Figure 41.1. Corrections are easily made by replacing a variable name in a rung by the correct variable name or restructuring of a rung for a modified logic. Documentation is on-line since the terminal always displays the latest version. A completed program is then down loaded into the PLC for execution. A hard copy of all rungs may also be printed.

41.2 Hardware Components

A PLC usually comprises a card rack with a backplane and several slots for the insertion of different modules. Figure 41.2 shows a basic system including a power supply (PS), a CPU for the storage and execution of the logic program and various I/O cards. A programming terminal, usually a generic personal computer (PC), may be used in conjunction with the PLC. The terminal is connected to the PLC via a communication link during download or upload of a program, and can also be used for on-line monitoring or on-line editing of programs. Users can also do simulations on the programming terminal and thus test programs before actually running them on the application.

The programming terminal is not needed to control the machine and may be disconnected at any time. To extend the number of I/O points, additional I/O racks may be connected to the main rack. Depending upon the location of the additional I/O racks, a fast parallel cable may be used for distances up to 1 meter, a high-speed serial link for the 300 meter range or a slower-speed serial channel for the 3 km range. For applications demanding total ground separation, a serial fiber-optic link may be used.

Types of PLCs

Suppliers of PLCs generally offer several types as shown in Figure 41.3. One such type is the "shoe box" controller or micro-PLC, at very low cost, with a very limited number of input/output points. A second type may comprise a small size form factor, and may also allow for analog input/output cards, and an increased networking capability. A third type may use a medium size controller and additional racks with slots for optional coprocessors and for various peripheral expansions, a large controller may even include a hard disk, large on-board memory, including redundancy or hot back-up. As the number of I/O slots in a rack is limited, a typical system may use additional I/O racks connected to the main rack by a serial communication link. I/O scan times for these remote I/O racks are larger than for local scans because of time needed for communication.

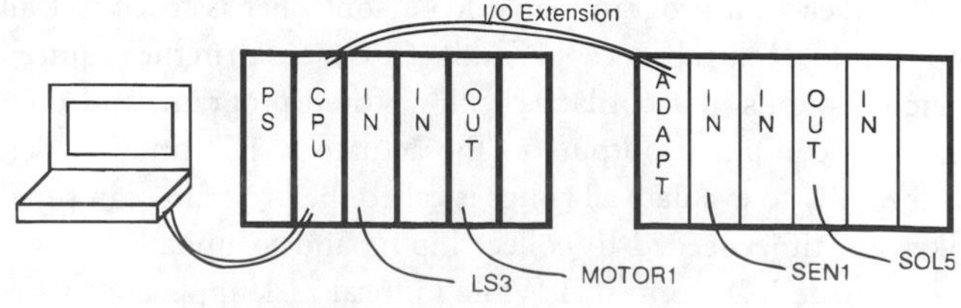

Figure 41.2 Hardware components.

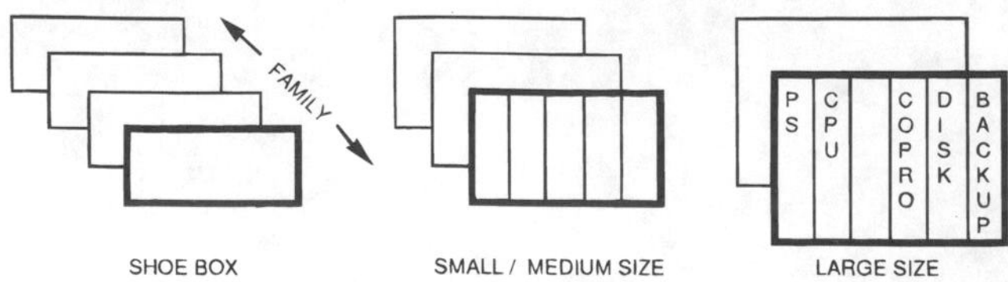

Figure 41.3 Different types of PLCs.

Families of CPUs

For every type of PLC, suppliers offer a family of CPUs that may be inserted into the specific card rack. So a medium-size PLC may have several different CPUs to choose from. Table 41.1 is a typical listing of specifications associated with various CPUs of one family and their I/Os. Important parameters are user memory size, maximum digital and analog I/Os, data table size, number of timers/counters, and scan times for program and I/Os. Generally, the data table size or the number of timer/counters are only limited by available memory. The columns in Table 41.1 with * are typical for a given family member. If, for example, CPU-A has a ladder program consuming 4K then only 2K remains for data table and/or timer/counters. Every counter or timer needs 4 words, so 64 counters require 256 words of memory; also for every 16 internal "coils" one additional word of memory is used.

In operation, the live inputs are collected by the input scan and placed into memory locations, called the input data table. The CPU during program scan examines these memory locations to resolve the state of each rung. The results of each rung are then written into other memory locations, called the output data table. The output data table is then moved to the live outputs by the output scan. When remote I/O is used, program scan and I/O scan are executed independently and overlapping in time, asynchronously to each other.

41.3 PLC Real-Time Operating Systems

The PLCs and their I/Os as a controlling system generally do not require any executive software. The operating system that

Table 41.1 Families of PLCs

Medium Processor Family	Maximum User Memory Words	Digital I/O Maximum (any mix)	Analog I/O Max.	Data Table Words*	Timers/ Counters*	Program Scan Time/Kword	I/O Scan Time/Rack
CPU-A	6K	512 32/slot 256 16/slot 128 8/slot	256 256 256	1K	64	2msec min 8msec typ	N/A
CPU-B	13K expand. to 21K	1024 (any mix) 1024 in + 1024 out (complem.)	1024	4K	128	2msec min 8msec typ	10msec@ 57.6Kb
CPU-F	32K	1024 (any mix) 1024 in + 1024 out (complem.)	1024	8K	512	0.5msec min 2msec typ	10msec@ 57.6Kb 7msec@ 115.2Kb 3msec@ 230Kb
CPU-H	64K	3072 (any mix) 3072 in + 3072 out (complem.)	3072	32K	2048	0.5msec min 2msec typ	10msec@ 57.6Kb 7msec@ 115.2Kb 3msec@ 230Kb

Table 41.2 Additional PLC Features

Medium Processor Family	Number of Remote/Ext Local I/O /DH + ports	Maximum Number of I/O Racks	Maximum Number of I/O Chassis			Number of RS-232/422/ 423 ports	EEPROM Backup	Battery-Backed RAM
			Total	Ext. Local	Remote			
CPU-A	1 DH+	4	1	0	0	0	Option	Yes
CPU-B	1 DH+ 1 Remote I/O Adapt or Scan	16	17	0	16	0	Option	Yes
CPU-F	2 DH+/Remote I/O Adapt or Scan	8	29	0	29	1	Option	Yes
CPU-H	4 DH+/Remote I/O Adapt or Scan	24	93	0	92	1	Option	Yes

invokes the program scan, the interpretation of the instructions, and the I/O scan are already in permanent memory. The user basically applies power to the controller, or in some cases, turns a switch to the "run" mode.

The operating system recognizes several interrupts, some of them are user controlled as shown in Figure 41.4 and explained below:

Main control program: runs as lowest priority if nothing else needs to be done. This mode is automatic, and all active tasks execute simultaneously (time-shared schedule).

Selectable timed interrupt: the user specifies a section of code to execute at programmed or configured time interval.

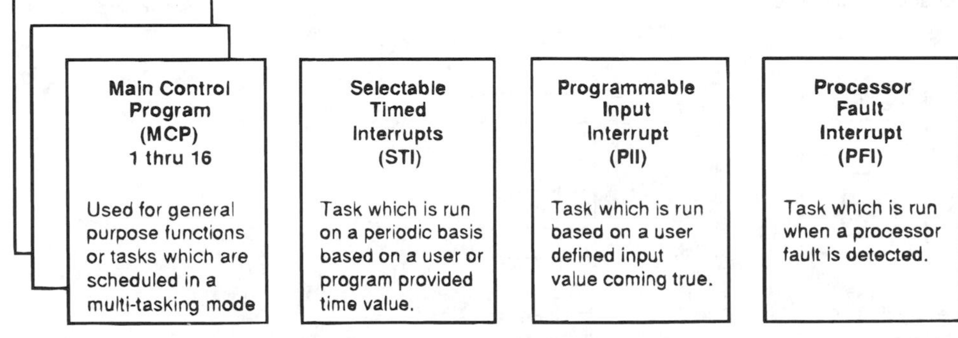

Figure 41.4 Real-time operating system.

Programmable input interrupt: the user may specify that if a certain input changes (e.g., from 0 to 1) that a specific program shall execute.

Processor fault interrupt: user task that runs when a processor fault is detected.

41.4 Software Components

The programming software is installed (for many PLCs under MS Windows™) into a PC programming terminal. This software enables the user to generate PLC programs and to down-load the resulting program into the PLC via a standard RS-232 PC port or via a supplier-specific data highway (DH) communication link. With the terminal connected during "run" the user can observe the current state of variables, may capture the time history of contacts, or modify the program on the fly.

The programming software is proprietary and is either provided by the PLC builder, or from some third party software houses that can compile to the machine code of certain PLCs. The various PLC programming languages are described in the IEC Standard 1131-3, and more and more suppliers now claim to be IEC 1131 compatible. In reality, many suppliers implement a subset of the standard, and, in their graphic representation as well as functionality, they are at least "close" to the standard.

Newer programming software allows the user to partition larger control tasks into smaller and more manageable tasks. The tasks may also be written in a variety of languages as needed by the application. In addition, scheduling of tasks may be keyed to features of the above mentioned operating system and/or under user software control. Figure 41.5 shows a variety of tasks using different language representations and execution control means.

MCP1 (main control program 1) shows a program in SFC (sequential function chart).

MCP2 uses SDS (smart diagnostic sequencer) instructions,

MCP3 uses MSG (message) instruction

MCP4 ladder diagram programming,

MCP11 and MCP12 use again SFC as language.

One MCP may, e.g., schedule or call another MCP.

The STI (selectable time interrupt) program is here shown with a process language instruction PID (proportional/integral/differential).

The PII (programmable input interrupt) routine shows again a ladder diagram as user language.

41.5 PLC Communications

Suppliers of PLCs provide communication links to program and/or interconnect several of their PLCs. This allows having local control centers, and still be able to link the controllers for a larger common task. Also, supervisory controllers may collect production data such as parts made, rejects, utilization. They also may provide production schedules and/or recipes.

Basically, all of these communication links are supplier proprietary designs with names such as data highway (DH) or xxx-net (e.g., ControlNet), or xxx-nec (NEtwork/Communication). In

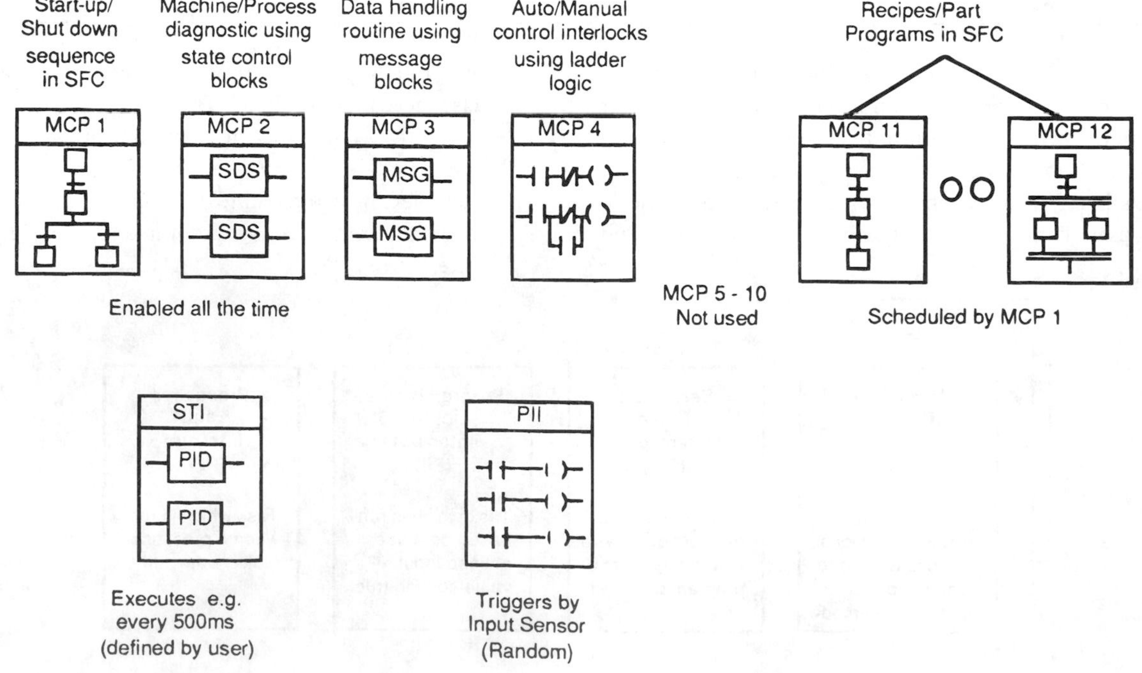

Figure 41.5 Different language representations.

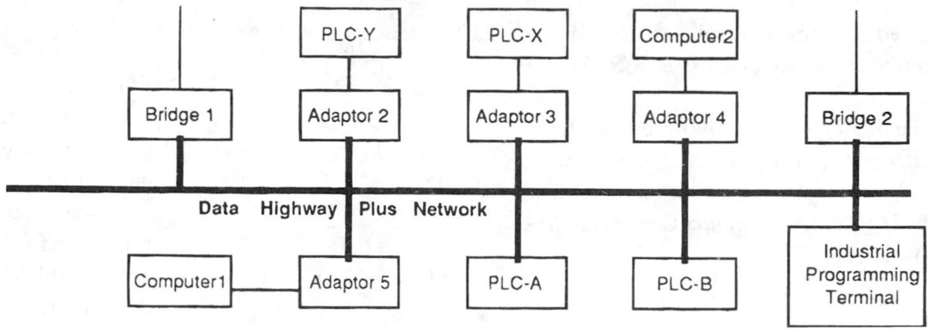

Figure 41.6 PLC intercommunication link (DH+).

some cases a standard physical layer such as RS-485 or one of the proposed IEC fieldbus physical layers is used. Another often used standard is the ISO IEEE 802.3 (Ethernet). Link layers, except for Ethernet (ISO IEEE 802.2), as well as higher level protocol layers are supplier specific. In cases were fiberoptics is used, the commonly used multidrop bus topology is replaced by a ring network.

The primary purpose of this PLC communication link is to exchange real-time control data between the various PLCs and computers and/or other networks (bridge). Figure 41.6 shows a typical network. The real-time requirement is the primary reason that suppliers want to maintain their own control over the access method, access mechanisms, package lengths, insertion and deletion of nodes on the fly, network start-up and recovery mechanisms.

The Data Highway Plus Network can be linked with Information Networks making plant-wide information management and control a reality. Computer hosts residing in the user's facility and automation products on the Data Highway Plus can share information.

Consideration for the various implementations are based on desired network length, number of drop-offs needed, device-specific commands, etc. Even though the cry for "open" communication is there, e.g., use of a standard network, no PLC supplier actually wants third party devices to "hitch a ride" on their real-time link, because of possible liability issues. The devices labeled Adaptor and/or Bridge are nodes that serve as message buffers. They basically perform a store/forward with protocol translation and adaptation to other links such as Ethernet, RS-232 with DF-1 protocol, or other PLC link layers and protocol.

41.6 Selecting the Right PLC

In general, the selection begins from the hardware side by counting the number of digital and/or analog I/Os. This gives an indication of the size of the I/O data table. Additional variables, as well as recipe data, plus timers/counters will require further data table memory. Each rung may control an output, and with approximately ten instructions per output, a first estimate of the program size is found. A tabulation of these features is given in Table 41.1.

The response time required by certain I/Os and the distance to these I/Os determines how many local or remote I/O racks are needed. Certain CPUs are designed to support many local I/O racks, others primarily support remote I/O racks or a combination thereof. Table 41.2 tabulates those additional features. The various CPUs differ also in the number of serial ports supported for remote I/O and the data highway (DH).

Another aspect of selecting the right PLC is based on the latest software features. A very clear, simple and efficient program environment including task management and object oriented programming as shown in Figure 41.5 may be the overwhelming deciding factor. Features that allow programs to be created "correctly" according to the machine's properties, diagnostics that capture and pinpoint machine or part problems, or actually predict and/or signal degradation may determine the choice of PLC. In many cases, these newer features can be compiled to run on established PLCs; in some cases, they may only be available with "new generation" CPUs.

Finally, the deciding factor in some applications is the ability to network several PLC to a programming terminal, to a supervisory computer, to a number of man/machine interfaces such as text or graphic displays and/or PC supported process displays.

References

Dummermuth, E. 1976 Multiprocessor based programmable controller. ISA/76 *International Conference and Exhibits* Oct. 11–14, Astrohall, Houston, Texas.

Dummermuth, E. 1977. A study of microprocessors' bit manipulating capabilities in programmable controls, *ISA/77 International Convention*, Oct 16–20, Niagara Falls, New York.

Dummermuth, E. 1981 Distributed processing with programmable controllers via data highway and remote input/output systems, *ISA/81 Conference and Exhibits*, March 23–26, St. Louis, Missouri.

Dummermuth, E. 1981. Ladder diagram programming, the basis for the acceptance of the programmable controller, 7th *Annual Advanced Control Conference*, Sept. 21–23, Purdue University, West Lafayette, Indiana.

Dummermuth, E. Distributed real time control, *6th IFAC Workshop on Distributed Computer Control Systems-DCCS 85*, May 20–22, Monterey, California.

Dummermuth, E. 1991 Fuzzy logic and programmable logic controllers, *ISA/91 Conference*, Oct. 28–31, Anaheim, CA.

Dummermuth, E. Distributed control using a serial communication link, *IFAC DCCS '94 Distributed Computer Control Conference*, Toledo, Spain Sept. 28–30.

US Patents	Issued	
3,942,158	Mar 02, 1976	Programmable logic controller
3,997,879	Dec 14, 1976	Fault processor for programmable controller with remote I/O interface racks.
4,118,792	Oct 03, 1978	Malfunction detection system for a micro-processor based programmable controller.
4,165,534	Aug 21, 1979	Digital control system with boolean processor (hardware co-processor for bit instruction)
4,291,388	Sep 22, 1981	Programmable controller with data archive
4,510,565	Apr 09, 1985	Programmable controller with intelligent position I/O modules
5,285,376	Feb 08, 1994	Fuzzy Logic Ladder Diagram Program for machine or process controller

42
Adaptive Control

42.1 Introduction 593
Controller Parameter Adjustment
42.2 Update Strategies 595
On-Line and Off-Line Parameter Adjustment Implementation
42.3 Direct Adaptive Control 599
Low-Order Controllers • Generation of Sensitivities • Full-Order Controllers • Adaptive Reference Model Adjustment
42.4 Indirect Adaptive Control 604
Self-Tuning PID Controllers • Åström and Wittenmark's Self-Tuning Regulator
42.5 Adaptive/Self-Tuning Behavior 606
Slow Drift Instability, Persistency of Excitation, and Fixes
42.6 Summary 607

Stephen T. Hung
Ford Motor Company

42.1 Introduction

Adaptive control schemes, in general, fit the mold of the scenario diagrammed in Figure 42.1. The control system itself includes a controller (including its supporting hardware and instrumentation), a performance index, and an adjustment algorithm. The overall system receives a command input $r(t)$, and the controller sends a control input $u(t)$ to the uncertain plant or process in order to effect an output signal $y_p(t)$. The process output signal feeds back to both the controller, for control input adjustment purposes, and the performance index, which measures how well the controller is performing its task. The adjustment algorithm, using the control input, process output, and plant performance measure, then substitutes or adjusts controller parameter values in order to improve system performance.

Some application scenarios also allow the measurement of parameters that characterize the system operational environment. Adjustment rules in such cases may incorporate such information if a need arises for a controller to discriminate between the effects of plant dynamics and the effects of environmental changes on plant output measurements. An alternative approach is for the designer to use the information to incorporate a model of the environment as part of an uncertain, augmented plant during the system design process.

Controller Parameter Adjustment

Realizations of the adaptive control scenario fall into two adjustment rule-based categories: indirect and direct adaptive control. The salient features of indirect adaptation are an explicit plant identification and a controller parameter set update that is based upon the identified plant parameter values. These two steps

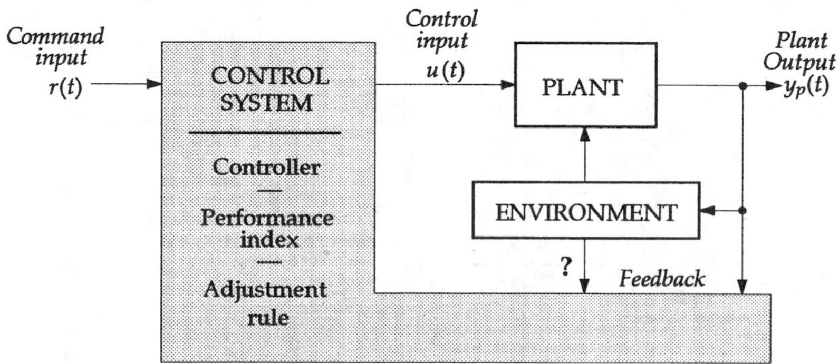

Figure 42.1 General topology for adaptive control.

0-8493-8343-9/97/$0.00+$.50

together form the adjustment rule. A topological diagram of an indirect adaptive control system appears in Figure 42.2. In this topology, an estimator first identifies the values of characterizing plant parameters through use of either physical sensing or recursive parameter identification routines that analyze input/output data and incrementally search for best-fit parameter values. The use of plant input/output data in the controller parameter determination process is *indirect*, since it results in only the intermediate step of plant identification. An update mechanism then takes into consideration the identified values of the plant parameters, instead of the system input/output measurements, in periodically determining the defining parameter values of what should be a new and better controller. A look-up table may be a sufficient update mechanism for determination of optimal controller parameter sets in scenarios that involve only a small number of operating regimes. Controller design formula calculations, alternatively, may be better, more-flexible options in application scenarios of greater complexity. The use of the results of the identification process, regardless of the update mechanism, arises from the logic of the certainty equivalence principle, the gist of which is that the identified plant parameter estimates are, for the purposes of design, equivalent to the true plant parameters.

A diagram of a direct adaptation topology appears in Figure 42.3. The primary differences between this topology and that of Figure 42.2 is the absence of an estimator for explicit process identification and the incremental update of controller parameters. The update law, in this case, is an error-driven recursive algorithm that calculates incremental changes in controller parameter values *directly* from command input, control input, plant output, and desired model output data. These stepwise changes are made so that the controller will minimize in gradual fashion the magnitude of an output error regardless of whether the causes for error are plant dynamics or environmentally based disturbances.

The salient features of direct adaptation are, thus, the *lack* of a plant parameter estimator and the error-driven incremental update. Elimination of the estimator correspondingly eliminates a set of computations, but necessitates the incremental update, which is a search for the right controller parameter values. Indirect adaptation does not involve a search for *controller* parameter values, but may include an incremental, recursive search for *plant* parameter values. Both types of adjustment involve some form of parameter identification. Direct adaptive control systems, in this sense, *implicitly* identify the plant.

The error-driven direct incremental update can be a mixed blessing. Adherents to indirect adaptation techniques argue that applications may require that the controller regulate the output error to zero, regardless of plant variation, and an absence of error results in little or no controller parameter adjustment in an error-driven update scheme. A directly-adapted closed-loop control system consequently may drift dangerously close to instability before the output error increases to a size sufficient to effect any corrective adjustments. This is not necessarily the case with indirect adaptation, since a change in identified parameters may or may not be accompanied by corresponding output error. Proponents of direct adaptation quite rightly argue, in return, that if uncontrollable environmental effects on the output are not extracted from the output, the plant parameter identifier can be fooled. Improper updates then will result from erroneous plant parameter estimates. The control design practitioner, in deciding whether direct or indirect adaptation is more appropriate, must choose between finding a reliable estimator and developing an acceptably sensitive update law. The best solution is context-specific to the application.

Perhaps the most commonly asked question about adaptive and self-tuning controls is, "Why bother with the complication of designing an adaptive or self-tuning control?" Most control systems are designed for operation across a set of many operating points during the execution of a task. It may be difficult to guarantee with certainty the sequence in which the system passes through these operating points. A fixed control law is, by nature, a compromise solution for control of any system that displays such uncertainty in its characterizing model. Such a solution may provide perfectly adequate results in many applications, but if the set of operating points is large, performance of such a compromise may not be acceptable in every circumstance.

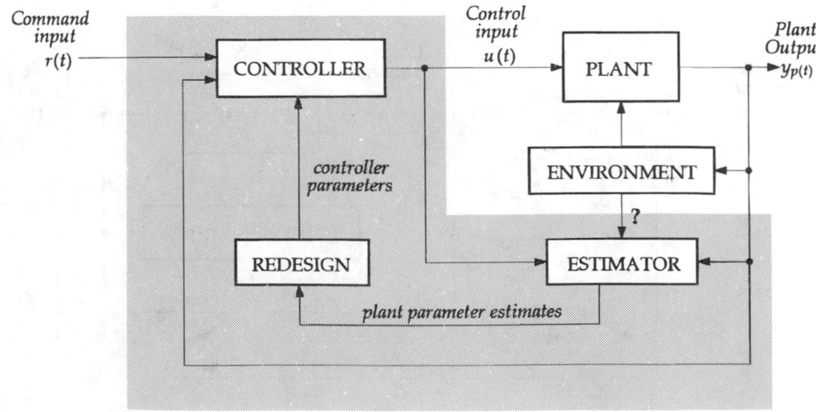

Figure 42.2 General topology for indirect adaptive control.

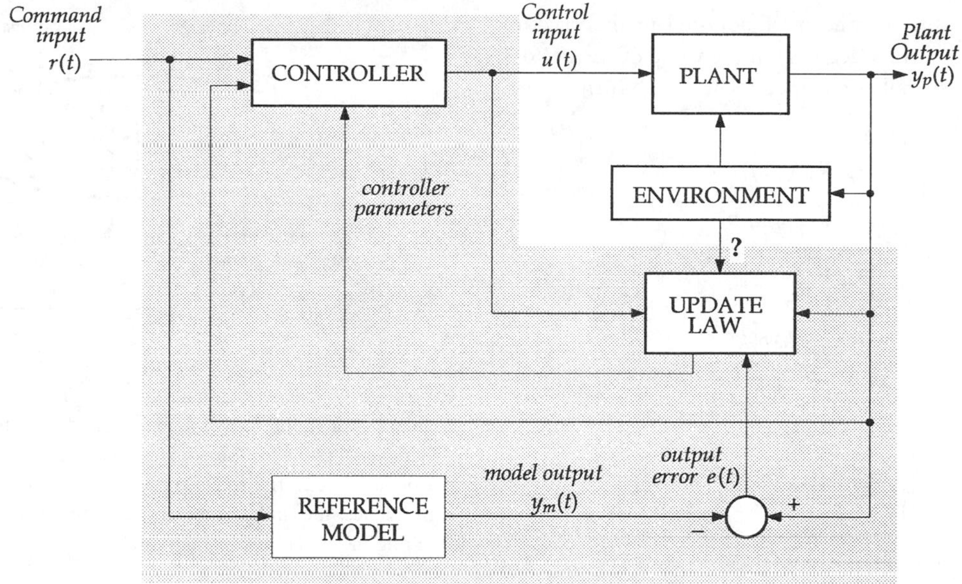

Figure 42.3 General topology for direct adaptive control.

The need for a system to automatically optimize its performance at any operating point is the motivation for considering an adaptive and/or self-tuning control system. The controller for such a system must simultaneously accomplish two tasks: it must control, or compensate, and it must adapt, or self-tune. The control objective may lean toward enhancement of command-following or regulation, or it may be limited to stability enforcement, or it may be a compromise that addresses multiple needs. The adaptation objective is, invariably, to counteract process variation by tuning the controller so that the control objective can be met.

Accomplishment of both control and adaptation hinges largely upon the proper consideration of both the source of process uncertainty or variation as well as the nature of desired performance optimality. The source of process uncertainty in its simplest form may be merely a parametric uncertainty such as that arising from process drift or continuous nonlinearities. A more drastic form of uncertainty is that of topological uncertainty. This may arise from discontinuous behavior resulting from a process fault or from sudden changes in operating environment. Knowledge of the nature of system uncertainty is essential for sound selection of a controller topology that will not only achieve the control objectives, but also have the adjustment mechanisms necessary for counteracting process variations.

The next question that control design practitioners often ask when first delving into adaptive and/or self-tuning control options is "What's the difference between adaptation and self-tuning?" Very little difference exists: practitioners of control design should find nothing wrong with using the words "adaptive" and "self-tuning" interchangeably except in those cases where one or the other is part of the name of a particular algorithm (e.g., Åström and Wittenmark's Self-Tuning Regulator).

42.2 Update Strategies

The updating action of an adaptive/self-tuning control algorithm is dependent upon the dynamics of the update law. Two types of update laws predominate: gradient and least-squares. The conceptual reason for using the gradient update law can be seen from the graph shown in Figure 42.4, which illustrates the dependency of the magnitude of a system output cost function $J(p, t)$ with respect to an adjustable parameter p. The objective of tuning is to find a setting of p that yields a minimal cost function magnitude.

The system is assumed to have a current parameter setting at $p = p_1$ and the plot of the cost function shows that the optimal parameter setting is of the value $p = p^*$. The gradient of the error function at $p = p_1$ is positive. The standard gradient update

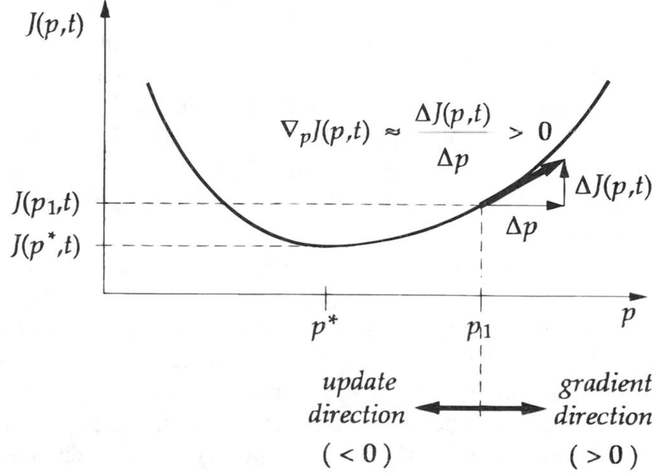

Figure 42.4 Gradient update concept. One-dimensional search: the parameter adjustment direction is in the *negative* gradient direction.

law commands that the parameter be adjusted in the *negative* gradient direction to yield a decrease in the value of the performance measure. This implies an update law of the form

$$\dot{p} = -g \cdot \nabla_p J(p, t), \tag{42.1}$$

in continuous time, or

$$\begin{aligned} p(k+1) &= p(k) - \Delta p(k) \\ &= p(k) - g \cdot \nabla_p J(p, k), \end{aligned} \tag{42.2}$$

in discrete time, where

$g \equiv$ update gain > 0, and

$k \equiv$ discrete time sampling index.

(Note: the gradient is the same as the partial derivative in the case of a scalar parameter.) The intuitive simplicity of the gradient update algorithm makes it quite popular, and it is quite reliable if the designer chooses the update gain g judiciously. This simplicity does lead, however, to a couple of drawbacks.

The first drawback is that the gradient following could be fooled into settling at a locally (as opposed to globally) optimizing setting if a small hump in $J(p, t)$ were to exist between the current setting $p = p_1$ and the optimal setting $p = p^*$. This is not that great a problem if the cost function can be rechosen or modified to eliminate the hump.

The second drawback arises in discrete-time application scenarios where the slopes of the performance measure near $p = p^*$ are too steep. Such a condition implies large values of the gradient function, $\nabla_p J(p, t)$, in a neighborhood close to the optimal setting. The possibility arises that the update will proceed in large update steps when small increments are needed. This problem is sometimes called the "football scenario" due to the elliptical shape of the performance index surface in the two-dimensional vector parameter case illustrated in Figure 42.5. The incremental gradient search consequently will repeatedly overshoot, bouncing back and forth, and take a potentially very long time to settle. Remedy for this ailment include the use of fixed update increments and the setting of the update gain, g, to small values.

Choosing the right magnitude of g for the gradient update law is almost an art form and is one reason some designers favor the faster converging least-squares update law. This update law has what could be described as an automatic scaling characteristic. Examination of a Taylor series expansion of a system output error reveals the logic behind its formulation. The simple system diagrammed in Figure 42.6 has an output $y(p, \text{k})$ whose discrete-time value is dependent upon an adjustable parameter p. An output value of $y^* = y(p^*, \text{k})$ is the optimal value of $y(p, \text{k})$ at any sampling instant k. A current parameter value p_c that differs

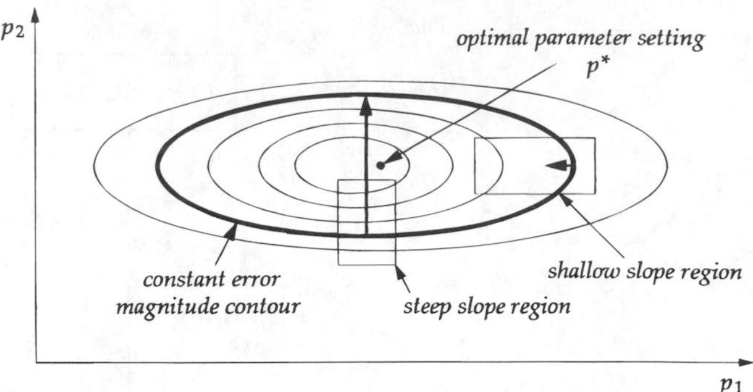

Figure 42.5 Gradient update concept. Two-dimensional search: the problem of a numerically ill-conditioned cost function. Performance cost increases as the contours move outward. For a given error magnitude, the update rate in the shallow-gradient region is low, resulting in slow convergence; the update rate in the steep-gradient region is too high, resulting in oscillations in the search.

from p^* by the amount Δp yields a suboptimal output value $y(p_c, \text{k})$. For an output error e defined as

$$e(p_c, \text{k}) = y(p_c, \text{k}) - y(p^*, \text{k}), \tag{42.3}$$

the current system output has a Taylor series representation

$$y(p_c, \text{k}) = y(p^*, \text{k}) + \nabla_p y(p^*, \text{k}) \cdot \nabla p + \text{higher order terms.}$$

A first-order approximation of the error can be made as

$$e(p_c, \text{k}) \approx \nabla_p y(p^*, \text{k}) \cdot \Delta p$$

and, in cases where Δp is considered small, as

$$e(p_c, \text{k}) \approx \nabla_p y(p_c, \text{k}) \cdot \Delta p. \tag{42.4}$$

The error is usually measurable, and the system output sensitivity $\nabla_p y(p, \text{k})$ is usually estimable. The error minimization problem, thus, becomes that of finding a parameter adjustment estimate, Δp. In the case of a squared error performance measure, where

$$J(p_c, \text{k}) \equiv \frac{1}{2} e^2(p_c, \text{k}) = \|\nabla y(p_c, \text{k}) \cdot \Delta p - e\|^2, \tag{42.5}$$

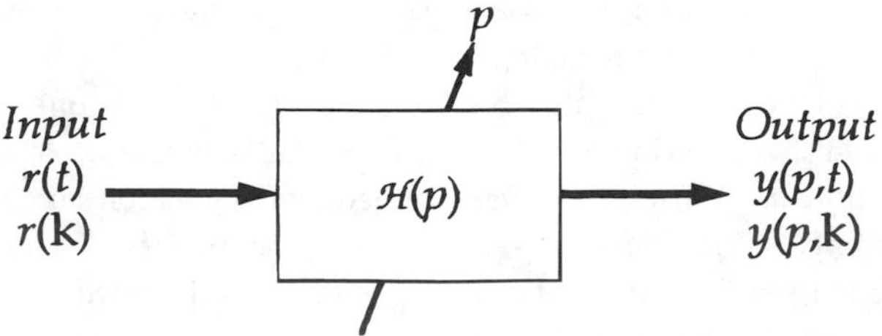

Figure 42.6 Simple single-input/single-output system with adjustable parameter p.

this task is that of finding the optimal update increment, Δp^*, which is the value of Δp that minimizes $J(p_c, k)$, the solution for which is given by

$$\Delta p^* \equiv \frac{e \cdot \nabla_p y(p_c, k)}{[\nabla_p y(p_c, k)]^2},$$

in the scalar parameter case, or

$$\Delta p^* \equiv [\nabla_p y(p_c, k)^T \nabla_p y(p_c, k)]^{-1} \cdot e \cdot \nabla_p y(p_c, k),$$

in the vector parameter case. Implementations of this update law in discrete time have the form

$$\begin{aligned} p(k+1) &= p(k) - \Delta p(k), \\ &= p(k) - g \cdot [I + \nabla_p y(p_c, k)^T \nabla_p y(p_c, k)]^{-1} \\ &\quad \cdot e \cdot \nabla_p y(p, k), \end{aligned} \tag{42.6}$$

where inclusion of the constant in the inverse prevents numerical overflows that might arise when the sensitivity $\nabla_p y(p, k)$ tends to very small values. (Note: the gradient is interpreted to be a row vector.) Recursive routines such as those described in Sastry and Bodson (1989) and Goodwin and Sin (1984) have been developed to avoid the computational burden of the inversions (or divisions) that appear in the update law. An analogous continuous-time update takes the form

$$\begin{aligned} \dot{p} = -g \cdot [I + \nabla_p y(p_c, t)^T \nabla_p y(p_c, t)]^{-1} \cdot e \\ \cdot \nabla_p y(p, t). \end{aligned} \tag{42.7}$$

This form is quite convenient for analytical purposes, but continuous multiplications and inversions can be difficult to accurately implement.

A relationship between the least-squares update of Equation 42.6 and the gradient update of Equation 42.2 can be found from examining the sensitivities of the error, as defined in Equation 42.3, and the cost function. These sensitivities have the form

$$\nabla_p e(p, k) = \nabla_p y(p, k),$$

$$\text{and } \nabla_p J(p, k) = \nabla_p \left[\frac{1}{2} e^2(p, k) \right] = e \cdot \nabla_p e.$$

The update in Equation 42.6, with these relationships in mind, may be expressed as

$$\begin{aligned} p(k+1) = p(k) - g \cdot [I + \nabla_p y(p_c, k)^T \nabla_p y(p_c, k)]^{-1} \\ \cdot \nabla_p J(p, k), \end{aligned}$$

which is identical to that in Equation 42.2 with the exception of a variable scaling factor. This scaling factor has a pleasant characteristic of speeding up the update when the magnitude of the gradient (slope) is small and slowing down the update when the magnitude of the gradient becomes large. Such a characteristic has the topological effect, depicted in Figure 42.7, of reducing the eccentricity of the elliptical cost function surface of Figure 42.5. This greatly reduces the overshoot problem that may arise with the gradient update of Equation 42.2. The update gain, g, consequently may be chosen larger and with less concern for precision, since a form of automatic gain correction will take effect when needed.

Limitations of the least-squares update do exist and result in restrictions on its use that are more stringent than those for the gradient scheme. Aside from increased demands on control system numerical and data-handling capabilities is a topological issue; namely, the convexity of the performance measure. The performance cost function contour that appears in Figure 42.8 illustrates this problem. This surface has a trough with a bend. Mere scaling cannot eliminate the bend in the surface. Change of scale through use of the least-squares update, in fact, exacerbates the steepness problem in one arm or the other of the performance surface, and, more often than not, results in numerical instabilities. The topological restriction on the least-squares update, therefore, is that the performance measure surface must be *convex*. In practical terms, this means that the surface should not possess any bends or arms whose ends are hidden from any other point on the surface.

Two remedies exist for dealing with the convexity restriction. The first remedy is to reparameterize the controller. This option requires more thought in the design of the controller, but usually

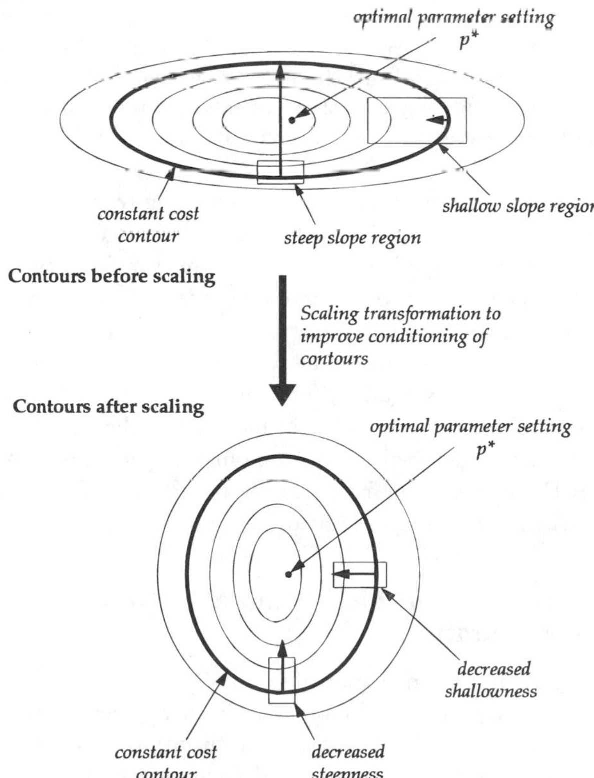

Figure 42.7 Normalizing effect of the variable scaling factor in the least-squares update. Performance cost increases as the contours move outward.

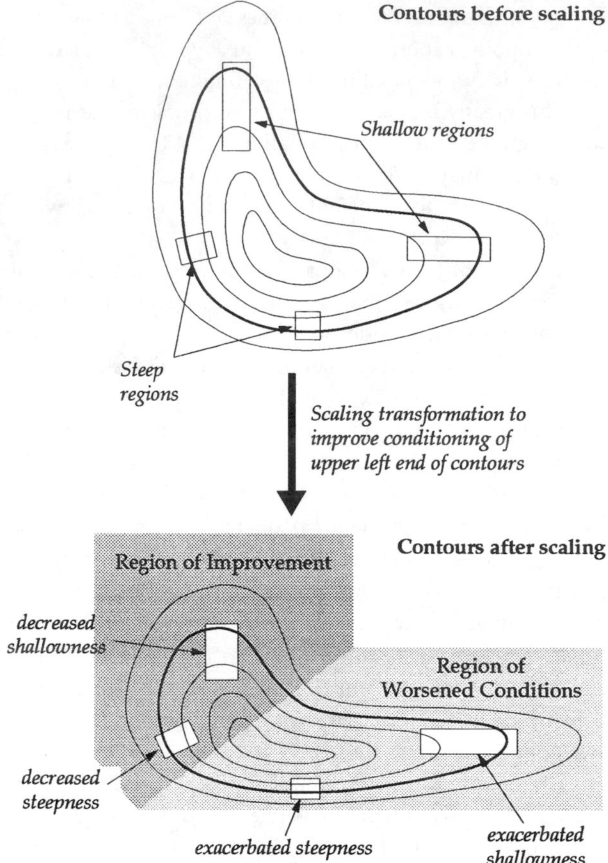

Figure 42.8 The nonconvexity problem: the bend in the constant-cost contours of the two-dimensional cost function surface makes it impossible to uniformly normalize gradients across the entire surface. Performance cost increases as the contours move outward.

leads to much simpler and smoother-running realizations. The second remedy is to employ a two-stage update. The first stage involves use of a gradient update with small update gain until the parameter is known to have descended to what might be called the trough of the performance surface. A second-stage search, employing a least-squares update, proceeds after this initial descent. If the second stage results in a rise out of the trough of the surface, as might occur in the bend of the trough, such a rise will yield an increase in the monitored performance measure. The two-stage update then returns to the first stage and the search procedure is repeated.

On-Line and Off-Line Parameter Adjustment Implementation

Strategies for implementation of self-tuning mechanisms fall into two broad categories: on-line and off-line parameter adjustment. The choice of an adjustment mechanism from one category or the other depends largely upon the factors of on-line cost and on-line flexibility needs.

Off-line self-tuning is well-suited to optimal, automated system scenarios such as factory tune-up procedures or maintenance calibration. The self-tuning mechanism is often a unit separate from the adjustable system controller. This facilitates the use of a common calibration unit by a collection of controllers. Such separability affords the possible use of rather sophisticated parameter adjustment laws and tuning safeguards, since design attention and implementation cost need to be lavished on only one auto-tuner. The resultant lack of need for an on-board tuning mechanism simplifies controller design and usually yields a controller of lower on-board cost. This form of tuning is sufficient for situations where system variation is constrained to a finite number of uncertain operating points, for each of which controller parameter settings can be precalculated and stored in memory. The flexibility of a controller tuned by off-line means is thus a function of on-board memory limitations. As an operational starting point, system initialization or commissioning calibration by off-line automated techniques should be considered a matter of natural course of development.

One form of off-line tuning uses the same form of gradient and least-squares recursive update laws as those used in continual on-line adaptation. The only difference is that the update mechanism is enabled only during tuning sessions where test input sequences are employed. These test sequences, which are not necessarily quiet or smooth, "exercise" the adaptive controller to confirm the robustness of self-tuning. The gradient and least-squares update laws also are applicable to another form of off-line tuning: namely, that based upon a window of sampled information instead of a recursive update at each individual sampling instant. Such an update mechanism is realized for an N-sample window by issuing an N-sample test input sequence and then creating a matrix $\mathbf{S}$ of the sensitivity samples, and a vector $\mathbf{E}$ of the error signal samples used in determining the measure of performance. These new variables are defined as

$$\mathbf{S} = \begin{bmatrix} \nabla_p y(p, 1) \\ \nabla_p y(p, 2) \\ \vdots \\ \nabla_p y(p, N) \end{bmatrix} \quad \text{and} \quad \mathbf{E} = \begin{bmatrix} e(p, 1) \\ e(p, 2) \\ \vdots \\ e(p, N) \end{bmatrix}.$$

With these definitions, a gradient update may be realized as

$$\Delta p^* = \mathbf{S}^T \mathbf{E}, \tag{42.8}$$

and a least-squares update may be realized as

$$\Delta p^* = (\mathbf{S}^T \mathbf{S})^{-1} \cdot \mathbf{S}^T \mathbf{E}. \tag{42.9}$$

These update laws provide adjustments that are based upon the averaging of directional information across the entire test sequence window. They are, thus, often more appropriate for tuning against quiet test input sequences where an accumulation of sampled data would be necessary to provide sufficient spectral richness in the tuning information.

On-line self-tuning implies use of a standard, near-optimal factory tune-up followed by continual tune-up during operation. The continual adaptation involves the ongoing use of an on-board tuning mechanism for every controller. The operational costs for on-board computation and memory are higher than those for off-line tuning, but these increased costs may be offset by an ability of the controller to adapt at any time and work across a wide envelope of uncertainty. This ability offers the important advantage of inherent compensation for system drift.

The choice of on-line versus off-line tuning ultimately will depend upon designer expectations of the range of operational uncertainty, the rate and frequency of operating point variation, and general maintenance accessibility. Greater range, rate, and frequency of uncertainty imply needs for greater flexibility that might favor on-line parameter adjustment. Greater maintenance accessibility affords increased maintenance frequencies that can greatly reduce the amount of time and uncertainty range for which a controller would need to adjust. Maintenance accessibility may thus make off-line tuning a perfectly viable option. Suitability is application-dependent.

42.3 Direct Adaptive Control

Low-Order Controllers

Various direct adaptive controllers can, and have been, built around the two basic update laws. The simplest of these are the low-order controllers. Most of these are derived as adjustable-parameter versions of the classical, fixed-parameter linear compensators that comprise combinations of elements such as the lead, lag, and lead-lag compensators, and PID controllers.

The majority of low-order controllers have a parallel comparison configuration, a first-order example of which is shown in Figure 42.9. This topology traces its roots back to the MIT-rule of Whitaker, Yamron, and Kezer (1958), who started the practice of calling it Model Reference Adaptive Control (MRAC). The topology, in general, involves sending a command input to both a control system and a reference model, which are connected in parallel. The control system in this particular example possesses a single adjustable parameter; namely, an adjustable gain θ. The plant output $y_p(t)$ may serve as an approximation at low frequencies of the control system's sensitivity to controller parameter variations. The outputs of the control system plant and reference model are compared to generate an output error that is used in conjunction with the sensitivity approximation to realize a gradient update law for the adjustable gain θ.

Generation of Sensitivities

The sensitivity approximation for the simple feedback loop in Figure 42.9 is easy to justify (albeit conditionally: low frequencies only) due to the simplicity of the system's topology. Most realistic systems, however, have greater complexity. The task of finding good sensitivity approximations for complex systems can present some difficulty. Kokotović developed a methodology in 1964 to simplify this task. Known as the sensitivity points technique, it is best illustrated through derivation of the controller parameter sensitivities of a commonly found negative feedback control system topology such as that shown in Figure 42.10. This system has a reference input, $R(s)$ and plant output, $Y(\alpha, \beta, s)$, uncertain plant blocks $W_1(s)$ and $W_2(s)$, and controller elements $K_1(\alpha, s)$ and $K_2(\beta, s)$ that are dependent upon the adjustable scalar parameters α and β. The Laplace transform representation for the plant output $Y(\alpha, \beta, s)$ may be expressed as

$$Y(\alpha, \beta, s) = \frac{K_1(\alpha, s)W_1(s)}{1 + K_1(\alpha, s)W_1(s)W_2(s)K_2(\beta, s)} \cdot R(s). \quad (42.10)$$

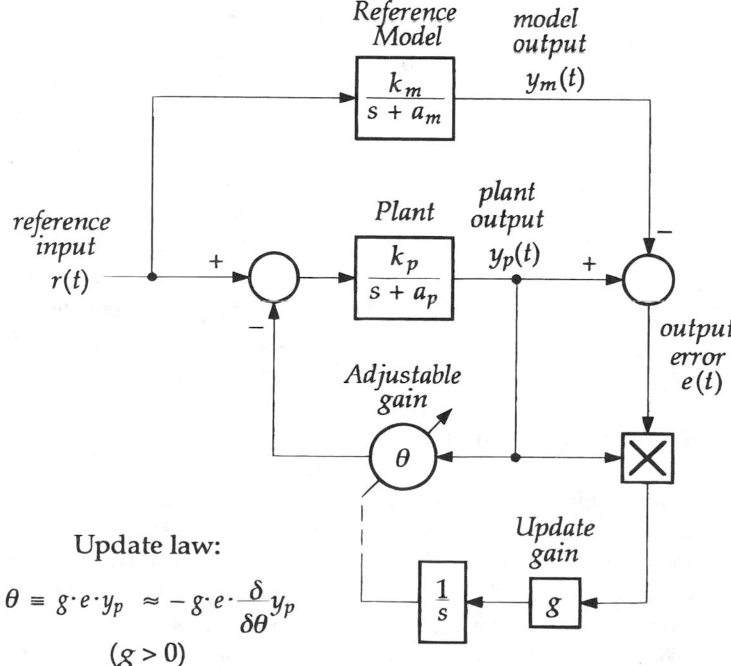

Figure 42.9 Parallel comparison topology for model-matching: Whitaker's model reference adaptive control (MRAC).

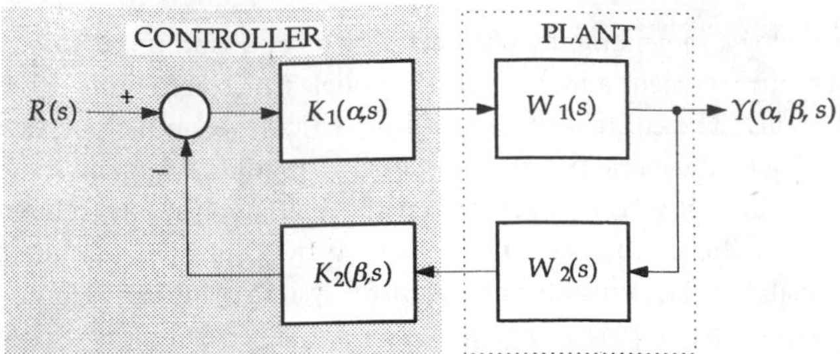

Figure 42.10 Single-input/single-output feedback control system with uncertain plant blocks $W_1(s)$ and $W_2(s)$, and controller elements $K_1(\alpha,s)$ and $K_2(\beta,s)$.

The corresponding controller parameter sensitivities are $(\delta/\delta\alpha)\, Y(\alpha_0, s) \equiv$ sensitivity of $Y(\alpha, \beta, s)$ with respect to the parameter α at $\alpha = \alpha_0$,

$$= \frac{1}{1 + K_1(\alpha_0, s)W_1(s)W_2(s)K_2(\beta, s)} \cdot \frac{1}{K_1(\alpha_0, s)} \cdot \frac{\delta}{\delta\alpha} K_1(\alpha_0, s) \cdot \frac{K_1(\alpha_0, s)W_1(s)}{1 + K_1(\alpha_0, s)W_1(s)W_2(s)K_2(\beta, s)} \cdot R(s), \tag{42.11}$$

$$= \frac{1}{1 + K_1(\alpha_0, s)W_1(s)W_2(s)K_2(\beta, s)} \cdot \frac{\delta}{\delta\alpha} \ln K_1(\alpha_0, s) \cdot Y(\alpha_0, \beta, s) \tag{42.12}$$

$(\delta/\delta\beta)\, Y(\beta_0, s) \equiv$ sensitivity of $Y(\alpha, \beta, s)$ with respect to the parameter β at $\beta = \beta_0$

$$= -\frac{K_1(\alpha, s)W_1(s)}{1 + K_1(\alpha, s)W_1(s)W_2(s)K_2(\beta_0, s)} \cdot \frac{1}{K_2(\beta_0, s)} \cdot \frac{\delta}{\delta\beta} K_2(\beta_0, s) \cdot \frac{K_1(\alpha, s)W_1(s)}{1 + K_1(\alpha, s)W_1(s)W_2(s)K_2(\beta_0, s)} \cdot R(s), \tag{42.13}$$

$$= -\frac{K_1(\alpha, s)W_1(s)}{1 + K_1(\alpha, s)W_1(s)W_2(s)K_2(\beta_0, s)} \cdot \frac{\delta}{\delta\beta} \ln K_2(\beta_0, s) \cdot Y(\alpha, \beta_0, s). \tag{42.14}$$

The expressions in Equations 42.12 and 42.14 imply that the respective sensitivities may be obtained by using the plant output $Y(\alpha, \beta, s)$ as the input to a second copy or model of the system. This second copy or model is referred to as the *sensitivity model*, $M(s)$. A topology for this process appears in Figure 42.11, where S_1 and S_2 are the *sensitivity points* (Kokotović, 1964) at which appropriate signals are picked off from the sensitivity model for use in calculating the sensitivities. The outputs of the sensitivity model, known as the sensitivity points signals, are filtered through *auxiliary filters* $A_i(s)$ that yield the actual sensitivity signals. Two adjustable parameters are present in this system; consequently, two auxiliary filters are used to recover the sensitivity signals:

$$A_1(s) = \frac{\delta}{\delta\alpha} \ln K_1(\alpha_0, s), \qquad \text{and} \qquad A_1(s) = \frac{\delta}{\delta\alpha} \ln K_1(\alpha_0, s).$$

The preference for realizing Equations 42.12 and 42.14, as opposed to Equations 42.11 and 42.13, arises from the fact that

$$\frac{\delta}{\delta p} \ln K(p, s) = \frac{1}{K(p, s)} \cdot \frac{\delta}{\delta p} K(p, s).$$

The implicit inversion of the compensator in the logarithmic filter makes it, in general, a lower order filter than that obtained by an ordinary partial derivative calculation.

An important implementation-oriented detail is that knowledge of the combined input/output characteristics of the sensitivity model and auxiliary filter are sufficient for obtaining sensitivity information from the system output signal $Y(\alpha, \beta_0, s)$. Consequently, although the scheme as depicted in Figure 42.11 involves the use of a copy or full-order model of the original control system, it is sufficient to use, instead, a sensitivity filter $F_i(s)$, possibly of reduced order, which closely approximates the gain and phase characteristics of the combined sensitivity model $M(s)$ and auxiliary filter $A_i(s)$ for each parameter subject to adjustment.

The development of sensitivity points as just described assumes that the control system may be modelled as locally linear and time-invariant (LTI). The updates in Equations 42.1, 42.6, and/or Equation 42.7, however, imply that the system is time-varying due to the change of the controller parameters. The signals in Equations 42.12 and 42.14 are, thus, *pseudosensitivities* that are close approximations of the true time-*varying* sensitivities. The practical significance of this detail is that, if g is chosen sufficiently small to yield parameter updates that are much slower than the process dynamics, the overall system may be analyzed as if the adjustable parameters were fixed, and, for design purposes, the pseudosensitivities may be substituted for the true sensitivities in an update law.

The ideal sensitivity filter comprises an ideally updated sensitivity model and an ideally updated auxiliary filter. If, however, some of the plant parameter variations are unmeasurable, this arrangement cannot be assumed to yield exactly accurate results. Luckily, such exactness is not an absolute necessity. Define the *tuned* sensitivity filter $F^*(s)$ as an ideal sensitivity filter whose controller parameters are set to *optimal* (not *actual!*) values. The self-tuning mechanism will converge to near-optimality if the phase of the actual sensitivity filter $F(s)$ is within $\pm 90°$ of the phase of the tuned sensitivity filter $F^*(s)$ at all frequencies in the bandwidth(s) of interest. If a tuning matchability condition is satisfied (i.e., tuning can totally eliminate the output error), the adaptive process will converge to exact optimality. If the phase conditions are met throughout the set of all possible operating points, a fixed filter approximate realization of $F^*(s)$ may thus be used not only to provide sufficient sensitivity information

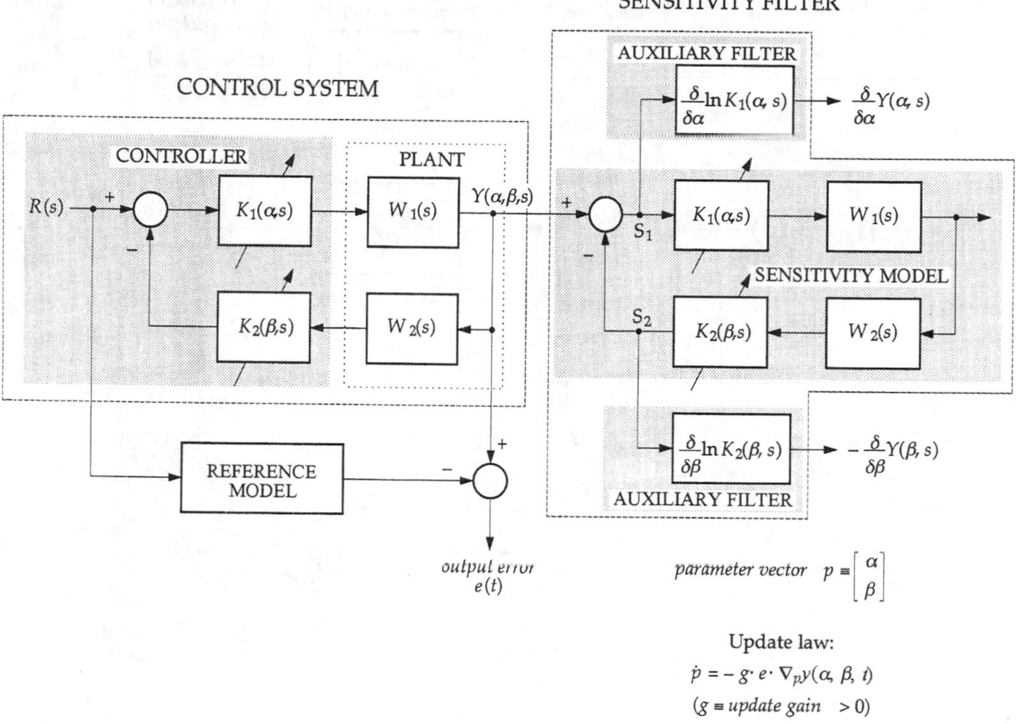

Figure 42.11 Topology for generation of the sensitivities $(\delta/\delta\alpha)\ Y(\alpha, s)$ and $(\delta/\delta\beta)\ Y(\beta, s)$ of the output $Y(\alpha, \beta, s)$ with respect to the adjustable controller parameters α and β. S_1 and S_2 are known as the sensitivity points corresponding to α and β, respectively.

but also to enhance tuning stability by suppressing signals of frequencies that fall outside of the bandwidth(s) of interest.

Full-Order Controllers

An inherent constraint on the capabilities of a low-order controller, regardless of whether it has fixed or self-tuning defining parameters, is a limited pole and/or zero placement capability. This is not a critical concern in the vast majority of applications where self-tuning may be helpful: control over a few dominant poles and/or zeros through tuning may be sufficient to attain adequate performance. Some situations may arise, however, where the entire closed-loop transfer function (not just the dominant dynamics) needs to emulate that of a reference model. This scenario implies the need to completely cancel plant dynamics while substituting desirable reference model dynamics and is the motivation behind the full-order direct adaptive controller.

A block diagram of a full-order control system appears in Figure 42.12. This topology provides sufficient flexibility so that the transfer function of the closed-loop control system can perfectly match the transfer function of the reference model output $W_m(s)$. The matchability objective, however, immediately creates a condition on the reference model: namely, that the model have the same number of poles n and number of zeros m as the plant transfer function

$$W_p(s) \equiv \frac{k_p N_p(s)}{D_p(s)}, \tag{42.15}$$

where $N_p(s)$ and $D_p(s)$ are monic (coefficient of the highest order term is unity), and k_p is a scalar gain.

The flexibility of the controller is a two-edged sword, since it implies that, if $N_p(s)$ has roots in the right half of the complex plane, the controller inherently will try to cancel the so-called "unstable zero" dynamics through the use of unstable controller poles. An introduction of instability of this form is not particularly enticing and leads to the constraining condition that the plant have no right half plane zeros. The reference model is parameterized in a fashion similar to that of the plant and, thus, must also have this characteristic. A constraint on the reference model arises as a result of this design assumption and the practical assumption that the reference model should be stable: namely, that the reference model must be minimum phase.

Three elements make up the controller: a scaling gain $k(t)$ (also known as a high-frequency gain), an interior loop in the feedforward path before the plant, and an outer feedback loop. The interior loop dynamics are defined by polynomials

$$C(s) = c_{n-2}s^{n-2} + c_{n-3}s^{n-3} + \cdots + c_1 s + c_0,$$

of order $n - 2$, and

$$\lambda(s) = s^{n-1} + \lambda_{n-2}s^{n-2} + \cdots + \lambda_1 s + \lambda_0,$$

of order $n - 1$. The outer loop feedback dynamics are defined by polynomials

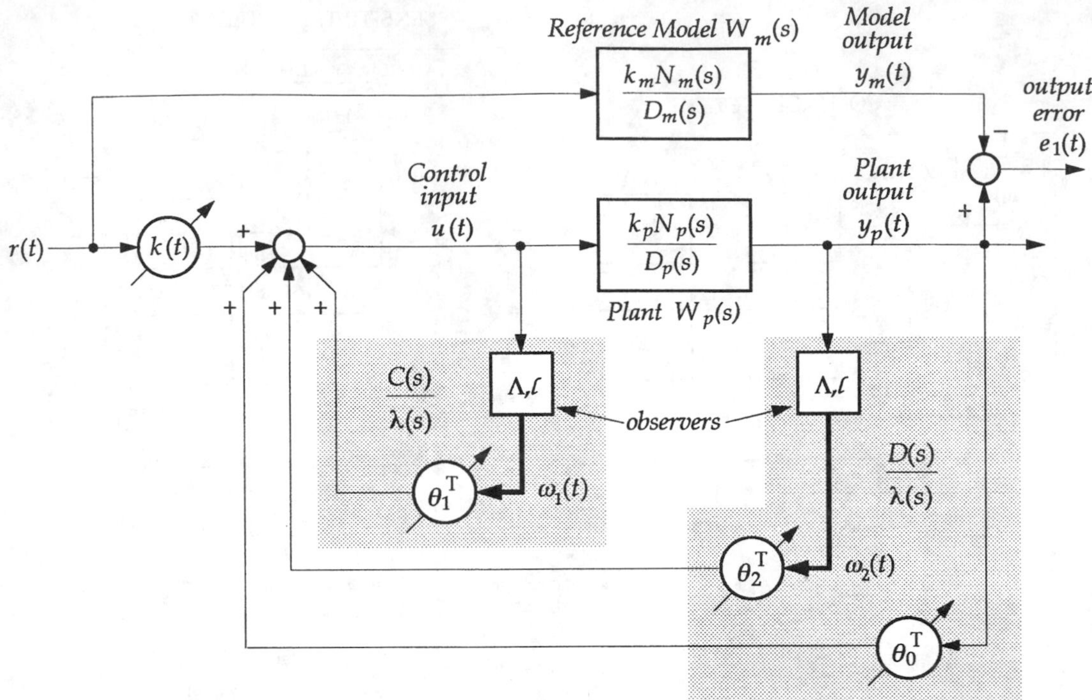

Figure 42.12 Topology for full-order, model-matching control.

$$
\begin{aligned}
D(s) &= d_{n-1}s^{n-2} + d_{n-2}s^{n-3} + \cdots + d_2 s + d_1 + d_0\lambda(s), \\
&= d_0 s^{n-1} + (d_0\lambda_{n-2} + d_{n-1})s^{n-2} \\
&\quad + \cdots + (d_0\lambda_1 + d_2)s + (d_0\lambda_0 + d_1)
\end{aligned}
$$

of order $n - 1$, and $\lambda(s)$.

The closed-loop transfer function (not including the parallel reference model path) can be written as

$$\frac{Y_p(s)}{R(s)} = \frac{kk_p N_p(s)\lambda(s)}{[\lambda(s) - C(s)]D_p(s) - k_p N_p(s)D(s)}. \tag{42.16}$$

The polynomial $\lambda(s)$ is chosen so that its roots include the zeros of the reference model (i.e., $\lambda(s) = N_m(s)$). The objective of the direct adaptation is to find and maintain optimal controller polynomials

$$C^*(s) = \lambda(s) - N_p(s) \quad \text{and} \quad D^*(s) = \frac{1}{k_p}[D_p(s) - D_m(s)] \tag{42.17}$$

and an optimal scaling gain $k(t) = k_m/k_p$ in order to attain a cancellation of plant dynamics $W_p(s)$ and the imposition of model dynamics $W_m(s)$. The variation of polynomial coefficients is achieved through use of the parameterization

$$
\begin{aligned}
\theta_1^T &= [c_0 \quad c_1 \quad \cdots \quad c_{n-1} \quad 0], \\
\theta_0 &= d_0, \quad \text{and} \\
\theta_2^T &= [d_1 \quad d_2 \quad \cdots \quad d_n \quad 0].
\end{aligned} \tag{42.18}
$$

The polynomial $\lambda(s)$ is then used to create a phase-variable canonical state-space model of a form such as

$$
\begin{aligned}
\dot{\omega}_1(t) &= \left[\begin{array}{c|c} 0 & I_{(n-2)\times(n-2)} \\ \hline \lambda_0 & \lambda_0 \cdots \lambda_{n-2} \end{array}\right]\omega_1(t) + \left[\begin{array}{c} 0 \\ \hline 1 \end{array}\right]u(t) \\
&\equiv \Lambda \cdot \omega_1(t) + l \cdot u(t).
\end{aligned} \tag{42.19a}
$$

Likewise,

$$\dot{\omega}_2(t) \equiv \Lambda \cdot \omega_2(t) + l \cdot y_p(t). \tag{42.19b}$$

The parameterizations in Equations 42.17 and 42.18 are combined to form the controller transfer functions

$$\frac{C(s)}{\lambda(s)} = \theta_1^T(sI - \Lambda)^{-1}l, \tag{42.20}$$

and

$$\frac{D(s)}{\lambda(s)}\theta_0 + \theta_2^T(sI - \Lambda)^{-1}l. \tag{42.21}$$

For those cases where the relative degree $\alpha = n - m = 1$, the two canonical blocks from Equation 42.19 serve as observers whose outputs are approximations of the controller parameter sensitivities corresponding to the controller parameter vectors θ_1 and θ_2. The reference input, $r(t)$, serves as the sensitivity of the scaling gain $k(t)$, and the plant output approximates the sensitivity of the controller parameter θ_0. A collection of these sensitivity approximations forms a *regressor vector*

$$\omega(t) = [r(t) \quad \omega_1(t) \quad y_p(t) \quad \omega_2(t)]^T \tag{42.22}$$

that corresponds to a *parameter vector*

$$\theta(t) = [k(t) \quad \theta_1 \quad \theta_0 \quad \theta_2]^T, \tag{42.23}$$

which together determine the control input to the plant

$$u(t) = \theta(t)^T \omega(t)$$

The sensitivity approximation holds due to the combination of assumptions that the plant is of relative degree 1, that the reference model is of minimum phase, and that the closed-loop control system dynamics are somewhat close to those of the reference model. These assumptions together imply that the closed-loop control system transfer function is strictly positive real (SPR), with phase within the range $\pm 90°$, and may be approximated as a constant. In the context of the sensitivity generation shown in Figure 42.11, one may picture the "control system" block as a constant approximation of the actual transfer function. *The control system of Figure 42.11 then becomes its own ideally updated sensitivity model.* The auxiliary filters are all constant relative to the constructions in Equations 42.20 and 42.21, which place all of the controller parameters in gain elements, as shown in Figure 42.13. Together with an output error defined as

$$e_1(t) = y_p(t) - y_p(t), \tag{42.24}$$

the sensitivity vector in Equation 42.19 can be employed to construct a gradient controller parameter vector update

$$\dot{\theta}(t) = -g \cdot \text{sgn}(k_p) \cdot e_1(t) \cdot \omega(t), \tag{42.25}$$

or least-squares update

$$\dot{\theta}(t) = -g \cdot \text{sgn}(k_p) \cdot [I + \omega(t)^T \omega(t)]^{-1} \cdot e_1(t) \cdot \omega(t), \tag{42.26}$$

where, as before, g is a small, positive update gain. Note that an assumption has been made that the sign of the scaling gain is either known or already identified.

This formulation of a full-order direct adaptive controller very elegantly molds the controller polynomial coefficient parameterization around the adaptive update, but it assumes that the plant is of relative degree $\alpha = 1$. The relative degree assumption allows

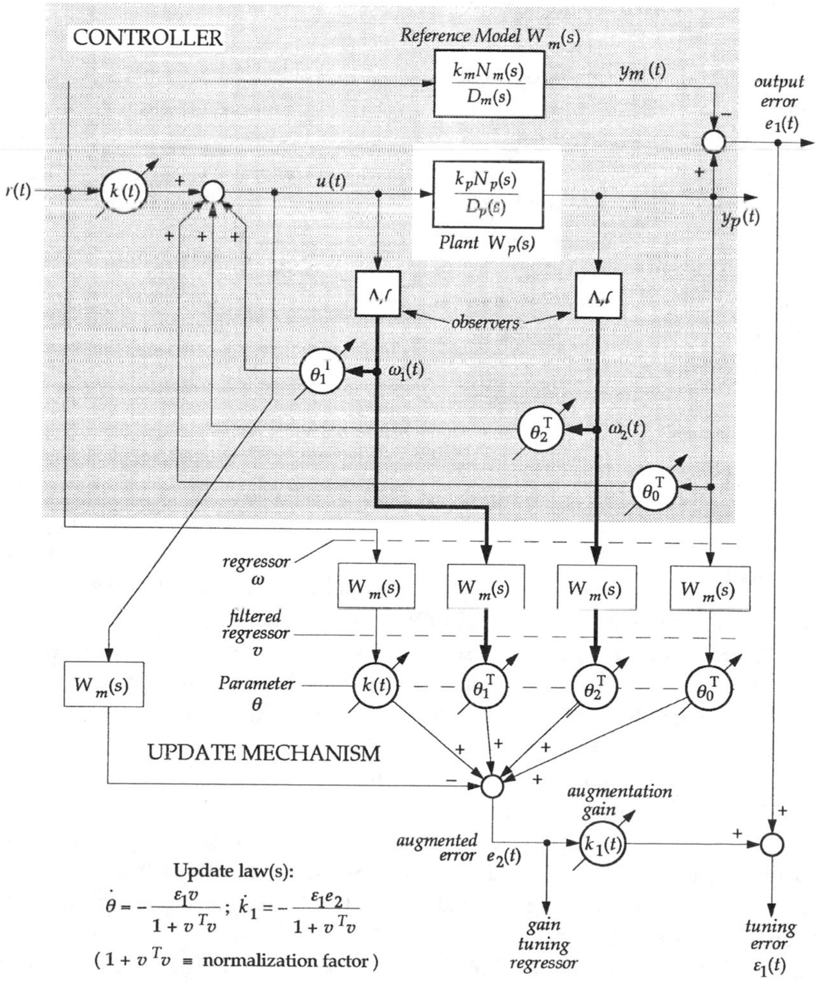

Figure 42.13 Output error direct adaptive control of a plant with relative degree $\alpha \geq 2$.

for the SPR assumption, which leads to the constant approximation of the closed-loop transfer function. Many real-life plants, however, do not obey this constraint. The modifications necessary for stable convergence of parameters in the most general case of relative degree $\alpha \geq 2$ are diagrammed in Figure 42.13. These changes contribute phase changes in the tuning error signal as well as the regressor of sensitivities. The update law in the relative degree 1 case used the raw regressor vector information. In the modified setting, the regressor signals are filtered through copies of the reference model (which happen to be optimally tuned approximations of the closed-loop control system transfer function). The new update law relies on this new, filtered regressor, v. Narendra and Lin arrived at the final topology of Figure 42.13 by constructing a new, *augmented error* signal that allows the generation of a Lyapunov function that assures boundedness of signals within the system. Note that this construction results in the need for an extra *augmentation gain* $k_1(t)$, which also requires tuning.

Adaptive Reference Model Adjustment

All of the direct adaptation topologies presented so far suffer from a common weakness: they strive to match the closed-loop system transfer function, $W_p(s)$, with that of a linear reference model. Too often, however, the best reference model will be nonlinear or only piecewise linear. Adaptive reference model adjustment marries the flexibility of the traditional model reference adaptive control topology to a plant identifier whose output provides information for updating the reference model, $W_m(s)$. Due to the use of process identification, this technique can no longer claim strict adherence to direct adaptation; conversely, it also does not use identified process parameter values in any explicit controller parameter design computation. Regardless of classification, this strategy allows the extension of MRAC model-matching flexibility to the realm of piecewise constantly linear reference models. Butler's 1992 text provides an overview of implementation aspects of MRAC, as well as refinements such as adaptive reference model adjustment.

42.4 Indirect Adaptive Control

Self-Tuning PID Controllers

The most prevalent adaptive or self-tuning controllers in active use today are probably those derived from the ubiquitous fixed-parameter PID controller commonly found in process control applications. PID controllers, along with their relatives, the PI and PD controllers, are classic examples of very simple low-order controllers that have long had the capability of providing satisfactory performance when tuned for any given operating point. This is well tailored for process control, where the local operating point of the controlled subprocess varies only a small amount as part of a much larger, multifaceted process that runs for long periods of time. The need for self-tuning is purely economic in nature and is usually performed on a periodic basis (as opposed to the continual searching of the direct adaptive controllers described above). Automated calibration of process control greatly reduces the manpower requirements necessary to maintain all of the local controllers at their respective optimum settings. Automation also allows for faster calibration to optimal performance after any set-point variations that may arise from commanded changes in the process. Åström's 1987 survey contains more detailed descriptions of all of the controllers discussed here.

Many of the currently available PID self-tuning mechanisms base their operations on a heuristic adjustment procedure known as the Ziegler-Nichols PID tuning rule that dates back to 1943 (Ziegler and Nichols, 1943) (many process control engineers actually keep a small card in their pockets with this simple rule imprinted on the card as a reference). The controller is assumed to be a single-input/single-output controller used in the location of the $K_1(\alpha, s)$ block of Figure 42.10. The adjustable parameter α of the $K_1(\alpha, s)$ block would be a vector of three gains, K_P, K_I, and K_D, whose relationship to the transfer function of the controller, renamed $K_{PID}(s)$, may be summarized as

$$K_{PID}(s) = K_P \cdot \left[1 + K_I \cdot \frac{1}{s} + K_D \cdot \frac{s}{1 + \tau_c s} \right], \quad (42.27)$$

where τ_c = a small positive instrumentation time constant.

The tune-up takes three steps:

1. The integral and derivative gains, K_I and K_D, respectively, are set to zero.
2. The proportional gain, K_P, is slowly increased until the onset of sustained oscillation in response to a test step or pulse input. The value of the gain is recorded as the critical gain, k_c, and the period of oscillation is recorded as the critical period, T_c.
3. From the critical gain k_c and the critical period T_c, the three controller parameters are calculated to be

$$K_P = \frac{k_c}{2}, \qquad K_I = \frac{2}{T_c}, \qquad K_D = \frac{8}{T_c}. \quad (42.28)$$

The settings dictated by the formulations in Equation 42.28 sometimes yield insufficient damping for the process at hand. It is not uncommon, therefore, to find the basic Ziegler-Nichols rule modified to suit the purposes of some local process. The Swedish-made SattControl auto-tuner operates using this tuning rule, as does the Foxboro EXACT adaptive regulator. The SattControl auto-tuner utilizes a relay feedback controller to effectively inject test steps in the control-loop system for the purposes of literal measurement of the critical gain and frequency. The Foxboro system, on the other hand, does not require the inducement of oscillation, but backcalculates the critical gain and frequency from the overshoot, damping ratio, and frequency of oscillation in response to an isolated test or disturbance step or pulse input.

The Leeds and Northrup adaptive regulator, the Electromax V, is a self-tuning PID that uses a pole-placement procedure, instead of the Ziegler-Nichols rule, to calculate the PID gains. A programmable test signal train, consisting of alternating positive and negative square pulses, is used create a sequence of pulse responses. A parameter estimator then performs a best fit of a second-order pulse response to the measured responses. The PID controller is then designed around the resultant second-order plant model using a pole placement formula. All three self-tuners can be quite effective for processes where the primary objective is to obtain a relatively constant output from a system with slight non-linearities.

Åström and Wittenmark's Self-Tuning Regulator

A most flexible form of the indirect adaptive controllers is the self-tuning regulator (STR), which was presented by Åström and Wittenmark in 1973. The principle of operation of this controller is to first recursively identify the plant from input/output data and then perform a polynomial-based pole-placement design. The plant and controller are described using the so-called autoregressive moving average (ARMA) models. For a plant with input $u(t)$, output $y(t)$, and random disturbance $\omega(t)$, a model may be constructed as

$$A(q^{-1})y(t) = q^{-d}B(q^{-1})u(t), \qquad (42.29)$$

where

q^{-1} = the unit delay operator (similar to z^{-1}, this notation allows for the mixing of delay operator and the *discrete* time index t in the same expression);

d = known finite integer time delay;

$A(q^{-1}) = 1 + a_1q^{-1} + \cdots + a_nq^{-n}$;

$B(q^{-1}) = b_0 + b_1q^{-1} + \cdots + b_mq^{-m}$; $b_0 \neq 0$.

The polynomial operator $A(q^{-1})$ contains the pole information of the model and, therefore, controls the rate of decay of output transients. This is the autoregressive portion of the model. The polynomial operator $B'(q^{-1})$ contains the zero information of the transfer function. It performs a moving average function on the input signal due to its structure as a sum of weighted delayed operand values. The values of the polynomial coefficients are attained through a recursive identification process, which might be described as follows.

The dynamics in Equation 42.29 can be reconfigured to express the current output $y(t)$ as the output of a regression model of the form

$$y(t) = [y(t-1)\ y(t-2) \cdots y(t-n)\ u(t-d)\ u(t-1-d) \cdots u(t-m-d)] \times \begin{bmatrix} -a_1 \\ -a_2 \\ \vdots \\ -a_n \\ b_0 \\ b_1 \\ \vdots \\ b_m \end{bmatrix} \qquad (42.30)$$

$$= \varphi(t)^T\theta,$$

where $\varphi(t)$ is known as the *regressor* vector and $\theta(t)$ is called the *parameter* vector. The identification objective is to find a best estimate $\hat{\theta}(t)$ of the true parameter vector. For off-line tuning or calibration this is accomplished over a fixed time window by first accumulating respective time histories of the output and the regressor:

$$Y(t) = \begin{bmatrix} y(1) \\ \vdots \\ y(t) \end{bmatrix}, \quad \text{and} \quad \Phi(t) = \begin{bmatrix} \varphi^T(1) \\ \vdots \\ \varphi^T(t) \end{bmatrix}.$$

A residual estimation error is then defined as the t-vector

$$E(t) = Y(t) - \Phi(t)\hat{\theta}.$$

The best estimate of $\hat{\theta}(t)$ is attained through minimization of the Euclidean norm of $E(t)$. The estimation task thus becomes that of finding the values of the optimal parameter vector, $\hat{\theta}^*$, the use of which would minimize the value of the cost function $\|Y(t) - \Phi(t)\hat{\theta}\|^2$. The solution for $\hat{\theta}^*$ is

$$\hat{\theta}^* = [\Phi(t)^T\Phi(t)]^{-1}\Phi(t)^TY(t). \qquad (42.31)$$

In on-line adaptive control, the time window is not fixed, but moves with time, so a recursive form of Equation 42.31 is needed. A matrix $P(t)$ is introduced for clarity of notation as

$$P(t) = [\Phi(t)^T\Phi(t)]^{-1}.$$

The least-squares estimate at time t can then be written as

$$\hat{\theta}(t) = \hat{\theta}(t-1) + P(t)\varphi(t)[y(t) - \varphi(t)\hat{\theta}(t-1)]. \qquad (42.32a)$$

The recursive update for $P(t)$ is given by

$$P(t) = [1 - K(t)\varphi^T(t)]P(t-1), \qquad (43.32b)$$

where

$$K(t) = P(t-1)\varphi(t)[I + \varphi^T(t)P(t-1)\varphi(t)]^{-1}. \qquad (42.32c)$$

(Note: For scalar $y(t)$, Equation 42.32c involves only a scalar inversion.) Detailed derivations of Equation 42.32 are found in the books by Åström and Wittenmark (1995), Goodwin and Sin (1984), and Sastry and Bodson (1989).

With the plant coefficients in hand, a pole placement design is carried out to attain a desired closed-loop transfer function

$$A_m(q^{-1})y(t) = q^{-d}B_m(q^{-1})r(t), \qquad (42.33)$$

through the use of a linear controller defined by the relationship

$$R(q^{-1})u(t) = T(q^{-1})r(t) - S(q^{-1})y(t). \qquad (42.34)$$

A topological diagram of the closed-loop control system appears in Figure 42.14.

The controller update process achieves pole placement through solution for the polynomials $R(q^{-1})$, $S(q^{-1})$, and $T(q^{-1})$ so that closed-loop transfer functions equate

$$\frac{q^{-d}B(q^{-1})T(q^{-1})}{A(q^{-1})R(q^{-1}) + q^{-d}B(q^{-1})S(q^{-1})} = \frac{q^{-d}B_m(q^{-1})}{A_m(q^{-1})}. \tag{42.35}$$

In order to avoid the unattractive introduction of unstable controller poles to cancel unstable or marginally stable zeros of the plant, the controller design process assumes a factorization of the plant numerator $B(q^{-1})$ of the form

$$B(q^{-1}) = B^{+}(q^{-1})B^{-}(q^{-1}), \tag{42.36a}$$

where $B^{-}(q^{-1})$contains all of the unstable zeros of the plant model. The reference model numerator $B_m(q^{-1})$ must preserve $B^{-}(q^{-1})$ as a factor, so

$$B_m(q^{-1}) = B_m^{+}(q^{-1})B^{-}(q^{-1}), \tag{42.36b}$$

where all of the roots of $B_m^{+}(q^{-1})$ are stable. The controller design process then achieves pole placement without cancellation of unstable zeros through setting

$$R(q^{-1}) = R_1(q^{-1})B^{+}(q^{-1}), \tag{42.37}$$

and algebraically solving the Diophantine equation

$$A(q^{-1})R_1(q^{-1}) + q^{-d}B^{-}(q^{-1})S(q^{-1}) = A_o(q^{-1})A_m(q^{-1}) \tag{42.38}$$

for the polynomial elements $R_1(q^{-1})$ *and* $S(q^{-1})$, where the polynomial $A_o(q^{-1})$ describes a set of ($\deg A(q^{-1}) - \deg B^{+}(q^{-1}) - 1$) fast instrumentation poles. The remaining polynomials needed for (34) are computed by reconstructing $R(q^{-1})$ through use of (37) and forming

$$T(q^{-1}) = A_o(q^{-1})B_m^{+}(q^{-1}). \tag{42.39}$$

The control design thus proceeds along the following steps:

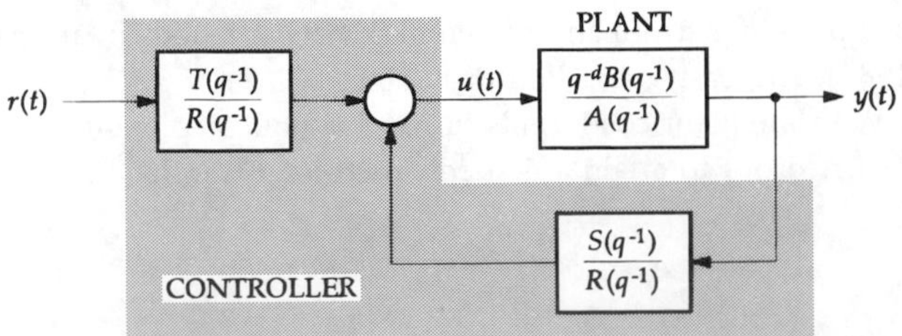

Figure 42.14 Control system topology of Åström and Wittenmark's Self Tuning Regulator (STR).

1. The plant is identified at each sampling instant using (32);
2. The plant and controller numerators are factored according to (36);
3. The Diophantine equation in (38) is solved to yield polynomials $R_1(q^{-1})$ and $S(q^{-1})$; and
4. The polynomials $R(q^{-1})$ and $T(q^{-1})$ are computed according to (37) and (39), and $R(q^{-1})$, $S(q^{-1})$, and $T(q^{-1})$, are substituted into (34) for the desired control law.

The order of the controller in (34) is the order of the polynomial $R(q^{-1})$, which from the condition following (38) is clearly no more than 2 deg $A(q^{-1})$. This scheme thus accomplishes pole placement with a controller of substantially lower order than the full-order direct adaptive controller. It does require, however, rather extensive computations for identification, an automated polynomial factorization, and algebraic solution of a polynomial Diophantine equation.

The STR in its original form appears as a one-step-ahead controller; i.e., the controller considers what it should output to the plant for just one sample instance. Generalized Predictive Control (GPC) is a generalization of this form to the multiple-step-ahead case. The consideration of multiple control samples into the future gives this algorithm a predictive characteristic which can be very robust with respect to variations in plant or process input/output delay. This characteristic may be particularly attractive for some process control applications. A good survey of applications and examples of STR and related algorithms, such as GPC, appears in Åström and Wittenmark's 1995 textbook on adaptive control.

42.5 Adaptive/Self-Tuning Behavior

The preceding discussions described the topological features of a few adaptive control topologies. It is equally important for the designer to understand the behavioral features of adaptive systems. Topological capabilities of a particular solution option must be balanced against self-optimizing and stability characteristics. The self-tuning behavior can be rigidly confined and controlled; such are the cases of the low-order adaptive controllers. This behavior also can be quite unrestricted, as in the case of the full-order direct adaptive controller. The designer should keep a few notes in mind.

Slow Drift Instability, Persistency of Excitation, and Fixes

A large number of adaptive control concepts had been developed by the early 1980s when an eye-opening paper by Rohrs et al. appeared at the Conference of Decision and Control in 1984 (CDC '84). This paper appeared again later, in archival form, as the *IEEE Transactions on Automatic Control* paper listed in the References. Rohrs et al.'s paper showed, through a collection of examples, that some of the most popular theoretical adaptive control algorithms of the day would tune themselves into instability in response to certain input conditions. Rohrs employed a third-order system with two directly adapted parameters and a first-order reference model, and pointed out two conditions for instability: inappropriate excitation and unmodelled dynamics.

Inappropriate excitation can be explained in very intuitive fashion. If a system is doing nothing, then it is yielding little or no input/output information and it is impossible to determine system stability or lack of stability. In one example, a step input was applied to the system. After the initial step response transient, the output error initially decayed, then suddenly diverged to infinity. The adjustable parameters had slowly drifted in response to the initially well-damped error; eventually they drifted to values leading to system instability. With too little information, it was impossible for the update mechanism to uniquely determine final values for the parameters.

The effects of unmodelled dynamics are that the adaptation mechanism may try to tune to match the wrong dynamics. Rohrs showed that application of a very high-frequency input to his system would yield an immediately divergent output error and parameter set. Information was available for tuning, but it was in the area of the spectrum where the third-order system included modes not present in the reference model. A phase mismatch occurred in this frequency range, and the adaptation immediately went awry.

The Rohrs examples spurred a period of investigation in the areas of adaptive behavior and adaptive system stability enhancement, or robustness. Anderson (1985), shortly after CDC '84, published an example of what is now referred to as "bursting." Anderson's simulations demonstrated that slow drift instability could occur in a second-order adaptive system in the *absence* of unmodelled dynamics. His system input was a constant (not much worse than Rohrs' step), in response to which the output error would cycle through alternating long periods of output error decay followed by short, large-magnitude bursts of output error. The adaptive system parameters would cycle through corresponding long periods of drift to values yielding instability followed by quick jumps back into the set of values yielding stability. The only information available to the adaptation mechanism during the slow drifts into instability was residual numerical noise that accompanied the decaying output error signal.

Anderson's and Rohrs' examples both demonstrated the impossibility of reliable adaptation in the absence of reliable tuning information. Realization of the need for information has led to a nearly universal current understanding that reliable adaptation demands *persistent excitation* of the system, and that this excitation must be of *sufficient richness* in its spectrum: i.e., the excitation spectrum must contain a sufficient number of distinct frequencies for the tuning mechanism to *uniquely* discriminate optimal parameter values. A common rule of thumb is that, for each significant frequency component in the system excitation, output magnitude and phase constitute two independent sources of tuning information. The tuning of n parameters, thus, requires an excitation with at least $\text{int}[(n+1)/2]$ distinct frequencies, where $\text{int}[\cdot]$ is the operator for truncation of real numbers to integers.

The stability problems of adaptive controllers prompted the development of a number of modifications to ensure stability. One such fix was the inclusion of the extra filters in the previously described full-order direct adaptive controller. Another was the imposition of polynomial-based constraints, as in the Diophantine equation-based indirect adaptive pole-placement controller.

About the same time that Rohrs was running his examples at MIT, Ioannou and Kokotović devised a very simple safety mechanism. They introduced the use of an integrator leakage term σ to the gradient parameter update equation so that the update in Equation 42.1 was modified to take the form

$$\dot{p} = -g \cdot e \cdot \nabla_p e - \sigma p. \tag{42.40}$$

The inclusion of the integrator leakage, now known as the σ-modification, is reminiscent of the common practice in digital controller coding of placing integrator poles at a value $z = 1 - \epsilon$, where ϵ = a very small positive number, in order to ensure integrator robustness in the presence of numerical noise. The effect of the leakage term is to draw the parameter toward a null value (or predetermined "safe" value) when the error and sensitivity data become sparse. The inherent, bounded drift counters any unplanned adaptive drift.

42.6 Summary

This description of adaptive and/or self-tuning control has touched upon only a few of many algorithms. Each algorithm has its own strengths and weaknesses: only the control system designer will know which strengths will outweigh which weaknesses. The discussion here has only briefly touched upon characteristics of algorithm topology and self-tuning behavior. Those interested in pursuing foundational background reference material probably should begin with some of the textbooks listed among the References. Narendra and Annaswamy comprehensively cover MRAC stability and convergence theory. Butler's text gives a good practical overview of MRAC application. Åström and Wittenmark's textbook is generally accepted as the standard reference for STR and related techniques. The other references listed offer good theoretical explanations of behavioral aspects of adaptation and self-tuning.

References

Anderson, B. D. O. 1985. Adaptive systems, lack of persistency of excitation, and bursting phenomena, *Automatica,* 21:271–276.

Åström, K. J., and Wittenmark, B. 1995. *Adaptive Control,* 2d ed., Addison-Wesley, Reading, MA.

Butler, H. 1992. *Model Reference Adaptive Control: from Theory to Practice,* Prentice-Hall, New York, NY.

Goodwin, G. C., and Sin, K. S. 1984. *Adaptive Filtering, Prediction and Control,* Prentice-Hall, Englewood Cliffs, NJ.

Ioannou, P. A., and Kokotović, P. V. 1984. Robust redesign of adaptive control, *IEEE Trans. Auto. Control,* AC–29:202–211.

Kokotović, P. V. 1964. Method of sensitivity points in the investigation and optimization of linear control systems, *Automation and Remote Control,* 25:1512–1518.

Narendra, K. S., and Annaswamy, A. 1989. *Stable Adaptive Systems,* Prentice-Hall, Englewood Cliffs, NJ.

Rhode, D. S., and Kokotović, P. V. 1991. Parameter Convergence Conditions Independent of Plant Order, *Advances in Adaptive Control,* K. S. Narendra, R. Ortega, and P. Dorato, eds., IEEE Press, New York, NY.

Rohrs, C. E., Valavani, L., Athans, M., and Stein, G. 1985. Robustness of continuous-time adaptive control algorithms in the presence of unmodeled dynamics, *IEEE Trans. Auto. Control,* AC–30(8):881–889.

Whitaker, H. P., Yamron, J., and Kezer, A. 1958. Design of model-reference adaptive control systems for aircraft, Rep. R–164, Instrumentation Laboratory, The Massachusetts Institute of Technology, Cambridge, MA.

Ziegler, J. G., and Nichols, N. B. 1943. Optimum settings for automatic controllers, *Trans. ASME,* 65:433–444.

43

Hardware Compensating Networks

Royce D. Harbor
University of West Florida

Charles L. Phillips
Auburn University

43.1 Continuous Compensation 609
43.2 Other Compensation Procedures 611

Compensation can be defined very generally as a modification made to a system for the purpose of improving the system's performance with respect to one or more specified characteristics. In analog electrical feedback control applications, compensation is generally achieved by adding to the system some form of network that is designed in such a way that it accomplishes the desired system performance. Such a network is called a *compensator.*

43.1 Continuous Compensation

In this chapter, we consider the physical realization of some of the more commonly-employed compensators. Generally, the most satisfactory compensator circuits are based upon operational amplifiers.

The basic operational amplifier circuit used to realize compensators, or any type of analog filters, is shown in Figure 43.1. In this circuit, $V_i(s)$ and $V_o(s)$ are the input and output voltages as functions of the Laplace-transform variable s, and $Z_i(s)$ and $Z_f(s)$ are the Laplace-domain input and feedback impedances, respectively. The transfer function of this circuit is

$$\frac{V_o(s)}{V_i(s)} = -\frac{Z_f(s)}{Z_i(s)} \tag{43.1}$$

Through proper choice of the impedances $Z_i(s)$ and $Z_f(s)$, we can realize phase-lead compensators, phase-lag compensators, and P, PI, PD, and PID compensators. It will be noted from the negative sign in Equation 43.1 that the circuit of Figure 43.1 causes a sign inversion. If this sign inversion is not acceptable in the physical system, a sign-inverting amplifier can be constructed from the circuit of Figure 43.1 by letting the input and feedback impedances be resistances. Such a configuration is pictured in Figure 43.2. In this circuit,

$$Z_i(s) = R_i \tag{43.2}$$

and

$$Z_f(s) = R_f \tag{43.3}$$

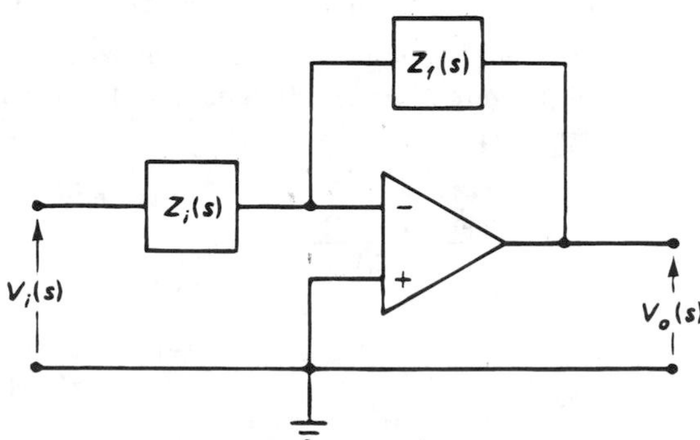

Figure 43.1 Operational amplifier circuit for compensators. (*Source:* C. L. Phillips and R. D. Harbor, *Feedback Control Systems,* 2d ed., Prentice-Hall, Englewood Cliffs, NJ, 1991, p. 260).

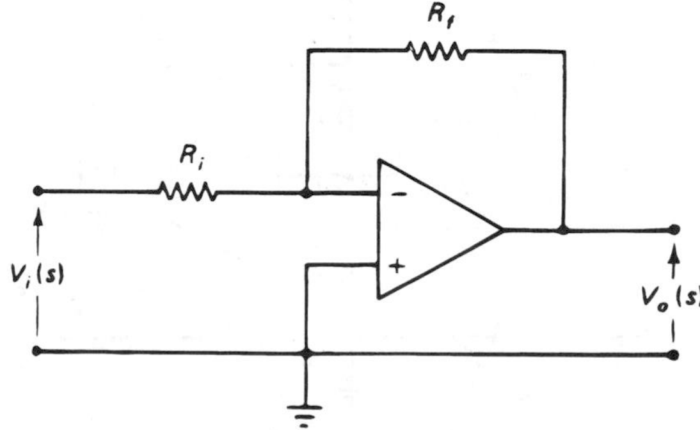

Figure 43.2 Circuit for constant-gain, sign-inverting amplifier. (*Source:* C. L. Phillips and R. D. Harbor, *Feedback Control Systems,* 2d ed., Prentice-Hall, Englewood Cliffs, NJ, 1991, p. 260).

0-8493-8343-9/97/$0.00+$.50

Then, Equation 43.1 becomes

$$\frac{V_o(s)}{V_i(s)} = -\frac{R_f}{R_i} \tag{43.4}$$

The transfer function is simply a negative constant. This configuration can be placed in cascade with either the input or the output of the circuit of Figure 43.1 to yield the overall transfer function

$$\frac{V_o(s)}{V_i(s)} = \frac{R_f Z_f(s)}{R_i Z_i(s)} \tag{43.5}$$

Clearly, if $R_i = R_f$, the circuit of Figure 43.2 functions only as a sign inverter.

We will now consider circuits through which we can realize various types of compensators. The circuit of Figure 43.3 can be used to realize first-order or zero-order transfer functions—that is, transfer functions having a denominator polynomial of degree one or zero. With reference to Figure 43.1, we note that both Z_i and Z_f consist of a resistor and capacitor in parallel. The impedance of a parallel RC circuit is given by

$$Z(s) = \frac{R(1/sC)}{R + 1/sC} = \frac{R}{RCs + 1} \tag{43.6}$$

By means of Equations 43.1 and 43.6, the transfer function of the circuit of Figure 43.3 is seen to be

$$\frac{V_o(s)}{V_i(s)} = -\frac{Z_f(s)}{Z_i(s)} = -\frac{R_f/(R_fC_fs + 1)}{R_i/(R_iC_is + 1)} = -\frac{R_f(R_iC_is + 1)}{R_i(R_fC_fs + 1)} \tag{43.7}$$

or,

$$\frac{V_o(s)}{V_i(s)} = -\frac{C_i(s + 1/R_iC_i)}{C_f(s + 1/R_fC_f)} = -\frac{K_c(s + \omega_o)}{(s + \omega_p)} \tag{43.8}$$

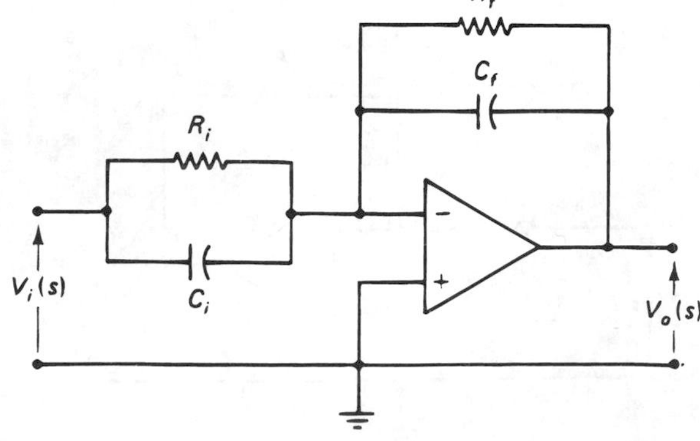

Figure 43.3 General compensator circuit. (*Source:* C. L. Phillips and R. D. Harbor, *Feedback Control Systems,* 2d ed., Prentice-Hall, Englewood Cliffs, NJ, 1991, p. 261).

By appropriate choice of resistor and capacitor values, the following types of compensators can be realized from the transfer function of Equation 43.8:

1. *Phase-lead.* The compensator will be phase lead if $\omega_o < \omega_p$, that is, if

$$\frac{1}{R_iC_i} < \frac{1}{R_fC_f}$$

or, equivalently,

$$R_iC_i > R_fC_f \tag{43.9}$$

2. *Phase-lag.* The compensator will be phase lag if $\omega_p < \omega_o$, that is, if

$$R_iC_i < R_fC_f \tag{43.10}$$

3. *Proportional.* If we remove both capacitors from the circuit, ($C_i = C_f = 0$), Equation 43.7 yields the constant gain of a proportional compensator.

$$\frac{V_o(s)}{V_i(s)} = -\frac{R_f}{R_i} = -K_P \tag{43.11}$$

It will be noted that removing the capacitors from the circuit of Figure 43.3 yields the circuit of Figure 43.2. Hence, it is not surprising that Equation 43.11 yields the same result as Equation 43.4.

4. *Proportional-plus-integral.* If we let R_f approach infinity (remove R_f from the circuit), Equation 43.8 yields the transfer function

$$\frac{V_o(s)}{V_i(s)} = -\frac{C_i(s + 1/R_iC_i)}{C_fs} = -\frac{C_i}{C_f} - \frac{1/R_iC_i}{s} = -\left(K_P + \frac{K_I}{s}\right) \tag{43.12}$$

which is the transfer function of a PI compensator.

5. *Proportional-plus-derivative.* If C_f is removed from the circuit of Figure 43.3 ($C_f = 0$), Equation 43.7 yields the transfer function

$$\frac{V_o(s)}{V_i(s)} = -\frac{R_f(R_iC_is + 1)}{R_i} = -\frac{R_f}{R_i} - R_fC_is = -(K_P + K_Ds) \tag{43.13}$$

which is the transfer function of a PD compensator.

6. *Proportional-plus-integral-plus-derivative.* The PID compensator cannot be realized with the circuit of Figure 43.3. Instead, consider the circuit of Figure 43.4.

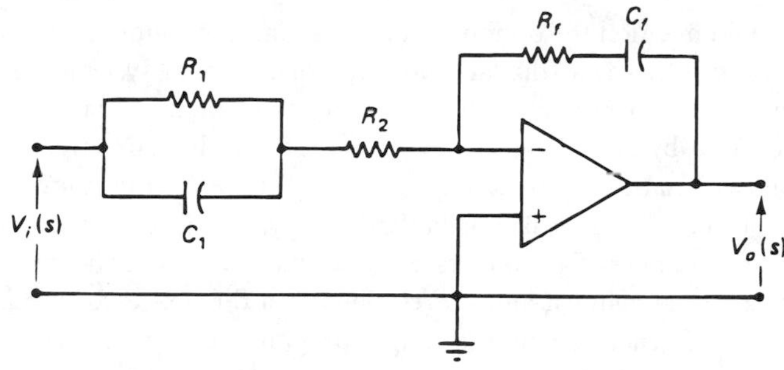

Figure 43.4 PID compensator circuit. (*Source:* C. L. Phillips and R. D. Harbor, *Feedback Control Systems,* 2d ed., Prentice-Hall, Englewood Cliffs, NJ, 1991, p. 262).

This circuit will realize a PID compensator that has an additional pole for the purpose of limiting high-frequency gain. With reference to Figure 43.1, the Laplace-domain input and feedback impedances are

$$Z_i(s) = \frac{R_1/sC_1}{R_1 + 1/sC_1} + R_2 = \frac{R_1 R_2 C_1 s + R_1 + R_2}{R_1 C_1 s + 1} \tag{43.14}$$

and

$$Z_f(s) = R_f + 1/sC_f \tag{43.15}$$

Substituting Equations 43.14 and 43.15 into Equation 43.1 yields

$$\frac{V_o(s)}{V_i(s)} = -\frac{(R_f C_f s + 1)(R_1 C_1 s + 1)}{C_f s(R_1 R_2 C_1 s + R_1 + R_2)}$$

$$= -\left[\frac{R_1 R_f C_1}{R_1 + R_2} s + \frac{R_1 C_1 + R_f C_f}{(R_1 + R_2)C_f} + \frac{1}{(R_1 + R_2)C_f}\frac{1}{s}\right] \frac{1}{\frac{R_1 R_2 C_1}{R_1 + R_2} s + 1}$$

$$= -\left(K_P + \frac{K_I}{s} + K_D s\right)\frac{1}{\tau s + 1} \tag{43.16}$$

where

$$K_P = \frac{R_1 C_1 + R_f C_f}{(R_1 + R_2)C_f} \qquad K_I = \frac{1}{(R_1 + R_2)C_f}$$

$$K_D = \frac{R_1 R_f C_1}{R_1 + R_2} \qquad \tau = \frac{R_1 R_2 C_1}{R_1 + R_2} \tag{43.17}$$

An example is now given to illustrate the realization of a compensator.

EXAMPLE 43.1

Suppose that a design procedure for compensating a feedback control system yields the transfer function of a phase-lag compensator to be

$$\frac{V_o(s)}{V_i(s)} = \frac{0.685(s + 0.190)}{s + 0.026} \tag{43.18}$$

This transfer function, exclusive of the negative sign, has the form of Equation 43.8; hence, it can be realized with the circuit of Figure 43.3 in cascade with a sign inverter based upon the circuit of Figure 43.2. Equating the parameters of Equation 43.18 to those of Equation 43.8, it is seen that

$$K_c = \frac{C_i}{C_f} = 0.685 \tag{43.19}$$

$$\omega_o = \frac{1}{R_i C_i} = 0.190 \tag{43.20}$$

$$\omega_p = \frac{1}{R_f C_f} = 0.026 \tag{43.21}$$

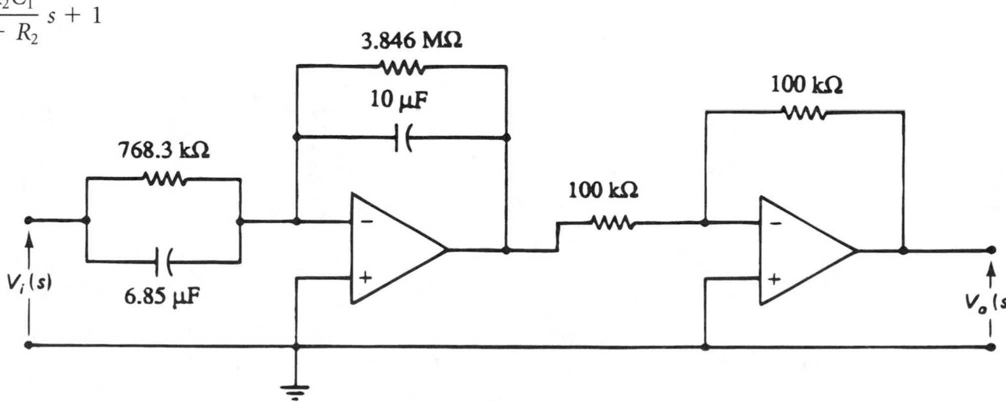

Figure 43.5 Realization of the transfer function of Equation 43.18. (*Source:* C. L. Phillips and R. D. Harbor, *Feedback Control Systems,* 2d ed., Prentice-Hall, Englewood Cliffs, NJ, 1991, pp. 260–261).

Equations 43.19, 43.20, and 43.21 represent three equations in four unknowns; hence one component can be chosen arbitrarily. If we select

$$C_f = 10\ \mu\text{F}, \tag{43.22}$$

then, from Equations 43.19, 43.20, and 43.21, respectively,

$$C_i = 0.685C_f = 6.85\ \mu\text{F} \tag{43.23}$$

$$R_i = \frac{1}{0.190C_i} = \frac{1}{(0.190)(6.85 \times 10^{-6})} = 768.3\ \text{k}\Omega \tag{43.24}$$

$$R_f = \frac{1}{0.026C_f} = \frac{1}{(0.026)(10 \times 10^{-6})} = 3.846\ \text{M}\Omega \tag{43.25}$$

If we select the input and feedback resistors for a sign-inverting amplifier each to be 100 kΩ, the transfer function of the compensator of Equation 43.18 is seen to be realized by the circuit of Figure 43.5.

43.2 Other Compensation Procedures

While this chapter has focused upon the realization of continuous (analog) compensators for continuous (analog) systems, we should mention the option of compensating continuous systems with discrete-data (digital) compensators. In such systems, the analog compensator of the type developed in this chapter is replaced by an analog-to-digital converter, a digital computer, and a digital-to-analog converter. During a time in which the cost of such components has fallen significantly, digital compensators offer a number of advantages. For example, the transfer function of a digital compensator corresponds to a difference equation. This difference equation is simply programmed into the computer; hence to change the transfer function of the compensator, it is necessary only to reprogram the computer. The reader is invited to explore the references for more detailed information on digital compensators.

References

Dorf, R. C. and Bishop, R. H. 1995. *Modern Control Systems,* 7th ed., Addison-Wesley, Reading, MA.

Kuo, B. C. 1991. *Automatic Control Systems,* 6th ed., Prentice Hall, Englewood Cliffs, NJ.

Phillips, C. L. and Harbor, R. D. 1996. *Feedback Control Systems,* 3rd ed., Prentice Hall, Englewood Cliffs, NJ.

Phillips, C. L. and Nagle, H. T. 1990. *Digital Control System Analysis and Design,* 2d ed., Prentice Hall, Englewood Cliffs, NJ.

Van de Vegte, J. 1986. *Feedback Control Systems,* Prentice Hall, Englewood Cliffs, NJ.

44

μ-Synthesis and Analysis[1,2]

Dan Bugajski
Honeywell Technology Center

Dale Enns
Honeywell Technology Center

Mike Jackson
Honeywell Technology Center

Blaise Morton
Honeywell Technology Center

Gunter Stein
Honeywell Technology Center

44.1 Defining the Interconnection Structure 614
Objective of μ-Synthesis
44.2 H_∞-Synthesis 615
44.3 μ-Analysis and D Scales 617
μ-Stability Criterion • Robust Performance Criterion
44.4 D-K Iteration 618
Deciding when To Stop the Iteration
44.5 Changing Weights 619
Perturbation Structure Analysis
44.6 Compensator Model Reduction 620
44.7 Summary 620

μ-Synthesis is a control-design technique used to optimize robust performance. Robustness of performance is a very stringent requirement on control system design. A control law can be called robust only if it provides acceptable performance for every system model within a specified, bounded uncertainty set. The uncertainty set includes, in addition to the nominal model, a multidimensional set of off-nominal or perturbed system models. The robust control objective is to achieve acceptable performance for all models in the uncertainty set with a single, fixed, linear time invariant (LTI) control law.

Compared with other control-design methodologies, μ-synthesis is one of the most demanding on the abilities of the designer, but it is also one of the most powerful. To achieve good results, the designer must understand the system and its performance objectives well enough to identify the sources of model uncertainty that have the greatest impact on the design objectives. The designer must also know how to represent the effects of these uncertainties within the μ-framework (see the various μ-references in the references). Finally, the designer must understand the techniques of optimal H_∞-design well enough to select weighting functions that drive the design to a reasonable solution.

The control designer who wishes to master μ-synthesis should expect to invest considerable time and effort during the learning stage. In this short article, we offer a brief introduction to the mechanics of the μ-synthesis procedure. The description given here is extracted from the design guidelines manual being prepared by Honeywell for the Air Force (*Multivariable Control Design Guidelines*, Bugajski, D. et al., in progress). The interested reader can consult that reference and others in the references for a more comprehensive introduction to the method (Balas, 1991, for example). Because the subject of the design guidelines report is aircraft flight control, some of the discussion below refers to aircraft systems. Most applications of μ-synthesis have been in the area of aerospace vehicle control, but the methodology is in no way limited to this class of problems.

In the analytical discussion that follows we assume the reader is familiar with the basics of linear systems theory and transfer functions of multi-input multi-output (MIMO) linear time invariant (LTI) systems. For a stable transfer function (matrix) T, we let $\|T\|$ denote its norm (in the H_∞-sense). The norm of the stable transfer function T is equal to the maximum over points jw on the imaginary axis of the maximum singular value $\overline{\sigma}(T(jw))$.

For clarity of presentation we have chosen to organize this article according to the steps taken in performing a μ-synthesis design. The steps in the μ-synthesis design procedure can be grouped as follows:

1. Defining the interconnection structure
2. H_∞-synthesis
3. μ-Analysis and rational approximation of D-scales
4. D-K iteration
5. Changing weights
6. Compensator model reduction

[1] Parts of this article have been excerpted from a draft of the design guidelines manual being prepared by Honeywell for Wright-Patterson Air Force Base under contract F33615-92-C-3607.

[2] AFOSR Contract F49620-92-C-0007 provided partial support for the preparation of this article.

0-8493-8343-9/97/$0.00+$.50

The basic ideas behind these steps are described in the subsections below.

44.1 Defining the Interconnection Structure

The first step in μ-synthesis is to construct the interconnection structure. A practical example of an interconnection structure is shown in Figure 44.2. The interconnection structure incorporates a set of subsystems, called design elements, into one large system that contains all the factors that the designer wishes to consider during the control design. Examples of design elements are models of the open-loop plant dynamics, models of desirable closed-loop dynamics, and models of frequency-dependent weighting functions. All the conventional control designer "knobs" are embodied in the weights and the input-output relations of the interconnection structure. The weights and other design elements in the interconnection structure are normalized, so that the uncertainty set is parametrized by the condition $\|\Delta\| \leq 1$, where Δ represents an unknown transfer function. Let P denote the transfer function associated with the interconnection structure.

The transfer function P has vector inputs and outputs arranged as follows

$$\begin{bmatrix} z \\ e \\ y \end{bmatrix} = P \begin{bmatrix} v \\ d \\ u \end{bmatrix} \tag{44.1}$$

where the z and v signals correspond to model uncertainty or perturbations, e is the generalized tracking error vector, d is the vector of external commands, disturbances and sensor noise, y is the measurement vector available to the controller K, and u is the vector of control actuation commands. (See Figure 44.1). The transfer function P can be written in terms of state space matrices A, B, C, D as follows:

$$P = C(sI - A)^{-1}B + D \tag{44.2}$$

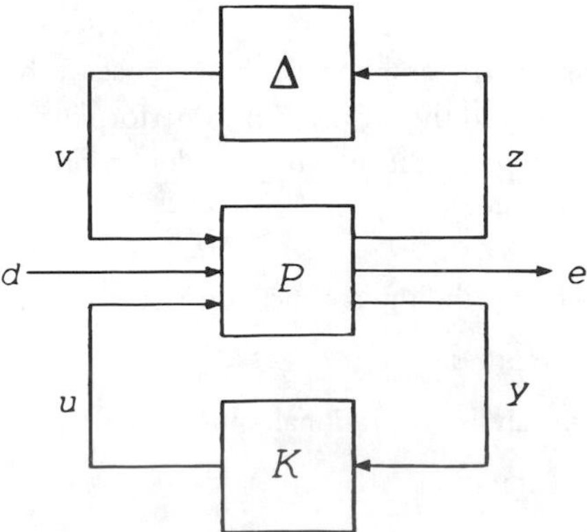

Figure 44.1 General feedback loop block diagram.

The control law (to be synthesized) imposes the following relation between u and y:

$$u = Ky \tag{44.3}$$

where K is also a transfer function. The goal of μ-synthesis is to produce a state-space realization of the transfer function K with the following property:

Objective of μ-Synthesis

Find K such that, for all Δ in a structured set with $\|\Delta\| \leq 1$, the transfer function from d to e in Figure 44.1 has norm less than 1.

The reader might not see immediately how to transform problems of practical interest into the form of Figure 44.1, in such a way that μ-synthesis will solve them. In fact, the methodology is broadly applicable once some special techniques for the construction of P are learned. For example, in constructing P, both feedback and precompensation portions of the control law are contained in K. Precompensation is that portion of the control law which acts on measured exogenous inputs (like pilot commands to an aircraft), and feedback is that portion of the control law which acts on the sensed state-measurements of the plant. In other words, the measurement output vector y of P can include exogenous commands as well as sensor feedback.

Generating the interconnection structure by hand from models of the design elements (Figure 44.2) is a tedious task, so software packages have been developed to automate production of the state space matrices for the interconnection structure. These tools take as inputs either algebraic or graphical descriptions of the design elements. In either case, the only mathematical operations required are simple matrix algebra.

As an illustration, a full interconnection structure used in aircraft control design is shown in Figure 44.2. Let us consider the design elements of that figure. Note the three boxes labeled Actuator Model, Aircraft Dynamics Model and $C_{Perf.model}$ (performance model). These represent LTI state-space models of the actuators, the aircraft dynamics and the desired response of the aircraft to the pilot commands. There are also nine boxes containing subscripted W symbols. These represent LTI state-space models of weighting functions used in three ways:

1. Represent frequency-dependent design objectives and constraints
2. Provide spectral coloration for noise and disturbance inputs
3. Normalize so that the uncertainty set is given by the condition $\|\Delta\| \leq 1$

The state-space realizations of these subscripted W systems and the three previously mentioned models are the deterministic design elements of the problem. The designer must specify the state-space realizations of these elements before the interconnection structure can be constructed.

The selection of the weighting functions is a challenging task that we do not address here. The interested reader is referred to

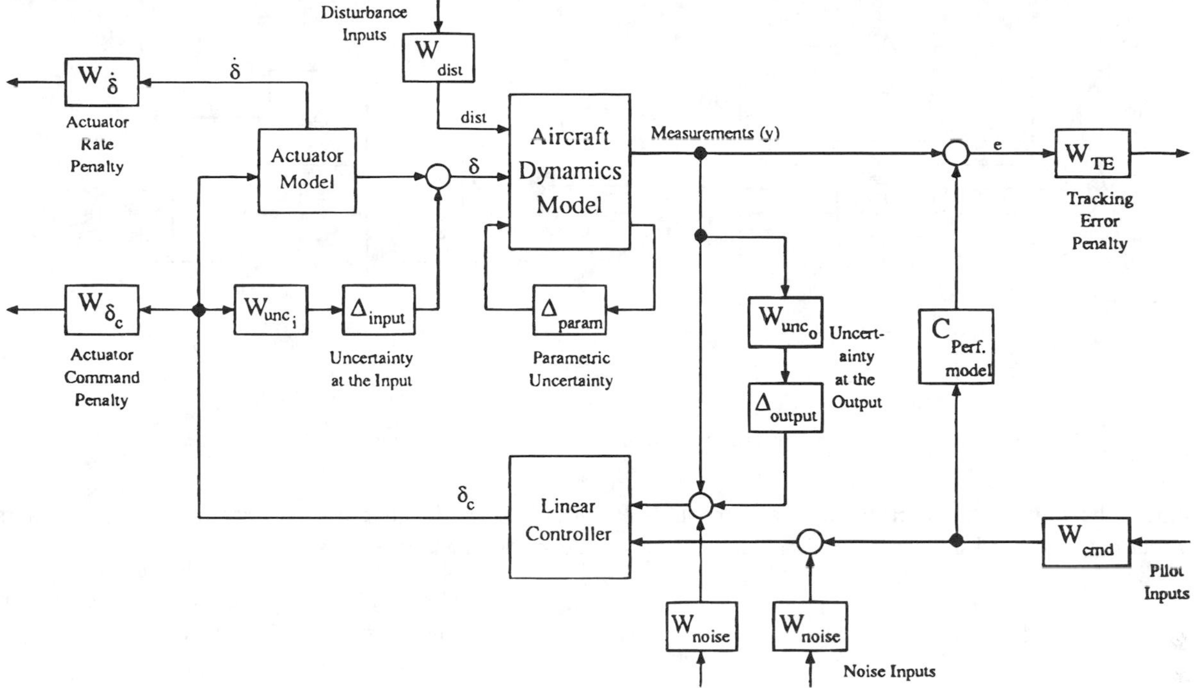

Figure 44.2 Example full interconnection structure.

the design guidelines (*Multivariable Control Design Guidelines*, Bugajski, D. et al., in progress) and other references in the reference section for information about weight selection.

There is a nondeterministic set of design elements in Figure 44.2 as well: the boxes containing subscripted Δ-symbols. These boxes represent the uncertainty in the system, and no state-space realizations are available for them. To build the interconnection structure, all we need to know about the Δ-blocks is the number of their inputs and outputs.

To transform the interconnection structure in Figure 44.2 into the form of Figure 44.1, the diagram is redrawn so that:

1. The box Linear Controller becomes K in Figure 44.1.
2. The subscripted Δ-blocks appear along the diagonal of Δ in Figure 44.1.

In this way the Δ-matrix inherits a block-diagonal structure relative to which the structured singular value will later be evaluated. The reader should note that each of the subscripted Δ-blocks might have its own block-diagonal structure, further refining the structure of the big Δ-block in Figure 44.1.

Each subscripted Δ-block plays a specific role in the problem. The block labeled Δ_{input} represents actuator uncertainty, Δ_{param} represents parametric uncertainty in the aircraft dynamics (e.g., uncertain aerodynamic coefficients), Δ_{output} represents uncertainty at the control system input (e.g., off-nominal sensor response). The only assumption made about the Δ-blocks is that they are stable transfer functions, of norm less than or equal to 1, having some specified block-diagonal structure.

44.2 H_∞-Synthesis

Once the interconnection structure has been realized, our next goal is to design an H_∞-optimal controller K. When K has been found we will be able to construct the closed-loop system M, called the perturbation structure as shown in Figure 44.3.

Recall that the goal of μ-synthesis is to find a compensator, K, that makes the transfer function from d to e smaller than 1 for all structured Δ of norm less than or equal to 1. At this stage of the process we temporarily ignore that objective and consider instead the problem to finding a controller K that makes the norm $\|M\|$ as small as possible. To understand this change in focus, the reader should be aware that if $\|M\| < 1$ then the closed-loop transfer function from d to e in Figure 44.3 is smaller than 1 whenever $\|\Delta\| \leq 1$ (verifying this fact is a simple exercise for those who are familiar with the small-gain theorem and the theory of L_2 signal spaces).

Let us concentrate on M. As shown in Figure 44.3, M is the transfer function consisting of the K feedback loop closed around P. In this way M depends on P and K. The external inputs to M are v and d and the external outputs are z and e. In block-matrix form:

$$\begin{bmatrix} z \\ e \end{bmatrix} = M \begin{bmatrix} v \\ d \end{bmatrix} = \begin{bmatrix} M_{11} & M_{12} \\ M_{21} & M_{22} \end{bmatrix} \begin{bmatrix} v \\ d \end{bmatrix} \tag{44.4}$$

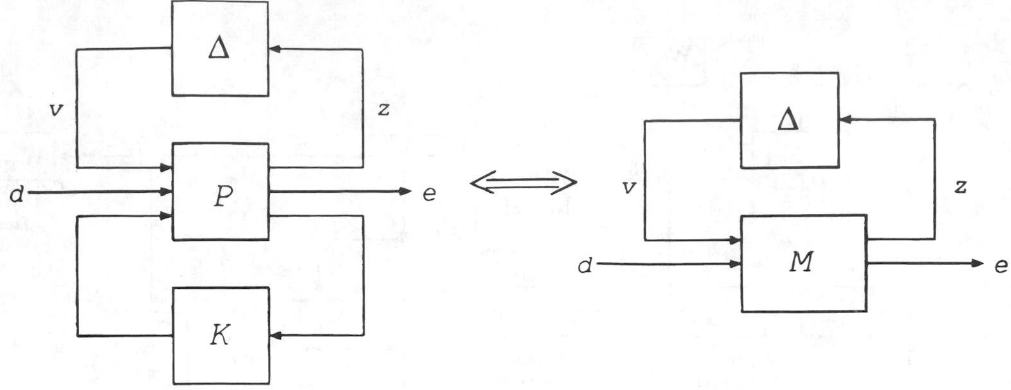

Figure 44.3 General feedback loop block diagram.

To compute the transfer function of M, as a function of P and K, partition P into blocks:

$$P = \begin{bmatrix} P_{11} & P_{12} \\ P_{21} & P_{22} \end{bmatrix} \tag{44.5}$$

where the partitions are such that

$$\begin{bmatrix} z \\ e \end{bmatrix} = P_{11}\begin{bmatrix} v \\ d \end{bmatrix} + P_{12}u \tag{44.6}$$

$$y = P_{21}\begin{bmatrix} v \\ d \end{bmatrix} + P_{22}u \tag{44.7}$$

The closed-loop response is given by

$$M = P_{11} + P_{12}K(I - P_{22}K)^{-1}P_{21} \tag{44.8}$$

as can be verified by substituting Equation 44.3 into Equations 44.6 and 44.7.

A stabilizing compensator that minimizes the H_∞-norm of the perturbation structure M is found with H_∞-synthesis. The technique used to perform H_∞-optimization is a search over a single variable, γ, called γ-iteration (Balas, 1991). This search can be automated and may be transparent to the user.

The input data required for H_∞-synthesis is the state space for P in a partitioned form:

$$\begin{bmatrix} A & B \\ C & D \end{bmatrix} = \begin{bmatrix} A & B_1 & B_2 \\ C_1 & D_{11} & D_{12} \\ C_2 & D_{21} & D_{22} \end{bmatrix} \tag{44.9}$$

where the partitions distinguish between the external inputs, (v, d) and control actuation (u), and between the external outputs (z, e) and sensors (y). Specifically, the columns of B_1, D_{11}, and D_{21} correspond to v and d, the columns of B_2, D_{12}, and D_{22} correspond to u, the rows of C_1, D_{11}, and D_{12} correspond to z and e, and the rows of C_2, D_{21}, and D_{22} correspond to y.

There are special rank conditions for D that must be satisfied for the H_∞-design process to give meaningful answers. These conditions are as follows:

1. Full rank $D_{12}^T D_{12}$, which means that control actuation at high frequencies is penalized.
2. Full rank $D_{21} D_{21}^T$, which means that the sensors are "noisy" at high frequencies.

To initialize the H_∞-design process, the user begins by selecting a sufficiently large value of γ. The technique described in Doyle (1989) is then used to find a controller K such that M has H_∞-norm less than or equal to γ, provided that such a controller exists.

The H_∞-synthesis procedure in Doyle (1989) involves the solution of two Riccati equations. From the solution of these Riccati equations two matrices are derived. There is an eigenvalue condition for these matrices that must be satisfied if a compensator can be found to make $\|M\| < \gamma$. If the Riccati equations have solutions for which the eigenvalue test is satisfied, we say that the necessary conditions are satisfied (n.b., the necessary conditions are also sufficient).

Denote by γ_{opt} the minimum achievable norm of M over all compensators K. If the necessary conditions are not satisfied then we conclude that γ was chosen too small, $\gamma < \gamma_{opt}$, and the value of γ is increased for the next H_∞-synthesis in the γ-iteration. If the necessary conditions are satisfied, the controller K is realized and we conclude that $\gamma \geq \gamma_{opt}$. If the stopping criterion is satisfied, (i.e., if the value of γ is known to be sufficiently close to γ_{opt}) then the γ-iteration is terminated, otherwise γ is decreased and the γ-iteration continues.

The state space matrices for K are obtained by elementary matrix operations involving the two Riccati solutions.

After the γ-iteration is finished and the compensator has been realized, the closed-loop transfer function, M, is computed for the H_∞-design. The theory guarantees that the closed-loop system will be stable, but numerical difficulties arise sometimes, so it is a good idea to check stability by computing and examining the poles of M at this point.

The next step of the procedure, μ-analysis, requires that the frequency response of M be computed for the frequency range of interest. Again, for a check of numerical consistency with theory, it is a good idea to examine the maximum singular value of the frequency response of M and verify that it is indeed less than or equal to γ.

44.3 μ-Analysis and D Scales

Having found K that minimizes $\|M\|$ we redirect our attention to the original objective of making the transfer function from d to e smaller than 1 for all $\|\Delta\| \leq 1$. From Figure 44.3 we observe that the transfer function from d to e has the form $H(s)$

$$H(s) = M_{22}(s) + M_{21}(s)\Delta(s)(I - M_{11}(s)\Delta(s))^{-1}M_{12}(s) \quad (44.10)$$

with the transfer functions M_{ij} as in Equation 44.4. We consider some conditions under which the transfer function $H(s)$ is stable for all relevant Δ.

Let $\mathcal{M}(N)$ denote the set of $N \times N$ complex matrices, and for all s suppose $\mathcal{M}_{11}(s) \in M(N)$ ($\mathcal{M}_{11}$ is an $N \times N$ transfer function). Recall that the matrix Δ has a block-diagonal structure. The size of Δ is $N \times N$, and the block structure is defined by a partition $J = \{j_1, \ldots, j_m\}$ of N, that is,

$$\sum_{k=1}^{m} j_k = N \quad (44.11)$$

Let D_J denote the set of block diagonal matrices:

$$\mathcal{D}_J = \left\{ \begin{bmatrix} \Delta(i_1) & 0 & \cdots & 0 \\ 0 & \Delta(i_2) & \cdots & 0 \\ \vdots & \vdots & \ddots & \vdots \\ 0 & 0 & \cdots & \Delta(i_m) \end{bmatrix} \Bigg| \Delta(i_j) \in \mathcal{M}(i_j) \right\} \quad (44.12)$$

Let $D_J(\delta)$ denote the set of all transfer functions Δ that take values in $\mathcal{D}_J$ and satisfy $\|\Delta\| < \delta$. Suppose for some value of δ there is a $\Delta \in \mathcal{D}_J(\delta)$ such that, for some s on the imaginary axis, $\mathrm{Det}(I - M_{11}(s)\Delta(s)) = 0$, where I is the $N \times N$ identity matrix. In that case, denote by δ_0 the minimum such δ; note that $\delta_0 > 0$. The function $\mu_J(M_{11})$ is defined:

$$\mu_J(M_{11}) = 1/\delta_0 \quad (44.13)$$

If $\mathrm{Det}(I - M_{11}(s)\Delta(s)) \neq 0$ for all s on the imaginary axis and all Δ with values in $\mathcal{D}_J$ then $\mu_J(M_{11}) = 0$.

The construction above defines a function μ_J for each positive integer N and each partition J of N. When N and J are fixed one writes $\mu(M_{11})$ to denote $\mu_J(M_{11})$. The function $\mu(M_{11})$ is called the structured singular value of M_{11}.

We can now state the μ-stability criterion:

μ-Stability Criterion

The transfer function $H(s)$ is table for all Δ in $\mathcal{D}_J(1)$ if and only if $\mu_J(M_{11}) < 1$.

The μ-stability criterion motivates us to consider the problem of computing $\mu_J(M_{11})$ for stable transfer functions M. When discussing the structured singular value for a particular partition J we often refer to the problem of computing $\mu_J(M_{11})$ by the number of blocks on the diagonal of Δ, i.e., the cardinality of J. For example, if $N = 8$ and $J = \{1, 3, 4\}$ we would refer to the computation of $\mu(M_{11})$ as a three-block problem.

A more general definition of μ (allowing repeated blocks in Δ) and some of its fundamental properties were first presented in Doyle (1982a). We have restricted attention to the case above to simplify the discussion.

It is easy to check that for the single-block partition, $J = \{N\}$, the function μ is simply the operator norm.

The μ-stability criterion tells us when H is guaranteed stable for all relevant Δ, but we want more—we want the size of $\|H\|$ small also. Specifically, the μ-synthesis objective will only be satisfied if $\|H\| < 1$ for all $\Delta \in \mathcal{D}_J(1)$. The theoretical solution to this more difficult problem turns out to be another μ-problem. Add one more block to Δ of size N_1 equal to Max(*dimension*(d), *dimension*(e)), and denote by J' the associated partition of $N + N_1$. Augment the input or output space of M with a block of zeros to make M_{22} square of size N_1. Then we have the robust performance criterion:

Robust Performance Criterion

K satisfies the μ-synthesis objective if and only if $\mu_{J'}(M) < 1$.

The μ-synthesis procedure would end here if we knew how to minimize $\mu(M)$ as a function of K. Unfortunately, the function $\mu(M)$ is difficult to compute as a function of anything, so more constructions are necessary to allow progress.

Consider $D(s)$, a stable, minimum phase transfer function that commutes with every transfer function Δ with values in $\mathcal{D}_J$. For the examples we are considering here, $D(s)$ has the form:

$$D(s) = \begin{bmatrix} d_1(s)I_{i_1} & 0 & \cdots & 0 \\ 0 & d_2(s)I_{i_2} & \cdots & 0 \\ \vdots & \vdots & \ddots & \vdots \\ 0 & 0 & \cdots & d_m(s)I_{i_m} \end{bmatrix} \quad (44.14)$$

where the scalar transfer functions $d_j(s)$ are stable, minimum phase and invertible.

Given such a $D(s)$ we can construct the transfer function $M_{scaled}(s)$ defined by:

$$M_{scaled}(s) = \begin{bmatrix} D(s) & 0 \\ 0 & I \end{bmatrix} \begin{bmatrix} M_{11}(s) & M_{12}(s) \\ M_{21}(s) & M_{22}(s) \end{bmatrix} \begin{bmatrix} D^{-1}(s) & 0 \\ 0 & I \end{bmatrix} \quad (44.15)$$

Now it is easy to check that if we replace M with M_{scaled} in Figure 44.3, the transfer function from d to e does not change. It might be the case, however, that $\|M_{scaled}\| < \|M\|$, and this is of interest for the following reason: if $\|M_{scaled}\| < 1$ then for all transfer functions in the structured set $\mathcal{D}_J(1)$ the transfer function from d to e has norm less than 1—that is, the μ-synthesis design objective is satisfied.

By the discussion just given we are led to consider the following problem: find a stable, minimum phase transfer function D that commutes with $\mathcal{D}_J$ and minimizes $\|M_{scaled}\|$. The solution to this problem can be approximated by a computational algorithm, which we describe next.

We use the fact that a stable transfer function realizes its norm at some point $j\omega$ on the imaginary axis. Pick a grid $\{s_k = j\omega_k\}$ on the imaginary axis and evaluate the transfer functions $M(s_k)$ at each gridpoint numerically. As noted in Doyle (1982a) we can compute constant matrices $D(s_k)$ that commute with $\mathcal{D}_J$ and minimize the norm $\overline{\sigma}\ (M_{scaled}(s_k))$ by a convex optimization. A stable, minimum phase transfer function D can then be found in the form of Equation 44.14 whose real-rational scalar diagonal components $d_i(s)$ have magnitudes at the gridpoints (approximately) equal to the corresponding values on the diagonals of the matrices $D(s_k)$. Automatic code for performing this D-interpolation is available in commercial software packages.

The D-interpolation leads to a computable upper bound for $\mu(M)$ that is useful for control synthesis. Let us define $\overline{\mu}(M)$ by the formula:

$$\overline{\mu}(M) = \inf_{D}\{\|M_{scaled}\|\} \tag{44.16}$$

where M_{scaled} is defined in Equation 44.15 for D as in Equation 44.14. It is not difficult to verify the inequalities:

$$\mu(M) \leq \overline{\mu}(M) \leq \|M\| \tag{44.17}$$

Ideally, μ-synthesis should find a controller K that minimizes the value $\mu(M)$. Because we are unable to solve that problem, as a compromise μ-synthesis finds a controller K that (locally) minimizes the value of $\overline{\mu}(M)$.

By the above inequalities, if the condition $\overline{\mu}(M) < 1$ is satisfied then the robust performance criterion is met and the controller K solves the μ-synthesis problem. In fact, if $\overline{\sigma}(M_{scaled}(s_k)) < 1$ for all gridpoints s_k, there was no need to perform the D interpolation at all. On the other hand, if $\overline{\sigma}(M_{scaled}(s_k)) > 1$ at some grid point, the D interpolation is necessary and we initiate the D-K iteration.

Some software tools that perform the D interpolation take as input, in addition to the values $D(s_k)$, a maximum order for the rational approximation and information about how accurate the fit should be with respect to frequency. If the maximum singular value of $M_{scaled}(s)$ is large and very sensitive to the D scales in a certain frequency range, then it makes sense to fit the D scales accurately in that region at the expense of errors where the maximum singular value is small and not very sensitive.

44.4 *D-K* Iteration

The first step of the D-K iteration is to replace the original interconnection structure P with a new one P_{scaled}, using the scaling function D obtained at the last stage of the μ-analysis step. Specifically,

$$P_{scaled} = \begin{bmatrix} D & 0 & 0 \\ 0 & I & 0 \\ 0 & 0 & I \end{bmatrix} \begin{bmatrix} P_{11} & P_{12} & P_{13} \\ P_{21} & P_{22} & P_{23} \\ P_{31} & P_{32} & P_{33} \end{bmatrix} \begin{bmatrix} D^{-1} & 0 & 0 \\ 0 & I & 0 \\ 0 & 0 & I \end{bmatrix} \tag{44.18}$$

where the partitioning of P is compatible with Figure 44.1.

The next step is to perform the H_∞-synthesis step with this new interconnection structure P. Note that the scaling function D typically contains some states, so the realization of the new P usually has a higher state-space dimension than the original. For this reason one does not want to approximate the D function too accurately in the non-critical frequency ranges.

The new H_∞-optimal K is found and the μ-analysis step is repeated. Either the new K satisfies the μ-synthesis objective or a new D is found. This alternating process of H_∞-synthesis and D-interpolation is the D-K iteration. With each iteration the value $\overline{\mu}(M)$ decreases, eventually approaching some asymptotic limit. The designer hopes that the value eventually becomes smaller than 1.

Design experience with aircraft and other control problems has shown that this D-K iteration is effective in synthesizing K to achieve high levels of robust performance. Unfortunately the iteration is not guaranteed to find the global minimum. In fact, an example in Doyle (1985) illustrates how the D-K iteration can fail to converge to the global minimum. The difficulties in this example arise because, despite the fact that both optimizations (over D and K) are convex, the combined problem is not necessarily convex.

Many of the steps of D-K iteration are highly automated in currently available software products.

Deciding when To Stop the Iteration

The steps of H_∞-synthesis of K and μ-analysis to find the D scales are repeated until a satisfactory result is obtained. The interconnection structure contains all of the scalings and weightings such that $\mu(M) < 1$ implies satisfactory performance, but smaller μ is even better. Plots are useful in determining whether a satisfactory result has been obtained. Typically we examine Bode plots (functions of s along the $j\omega$ axis) of the magnitudes of $\overline{\sigma}(M(s))$, $\overline{\mu}(M(s))$, $\overline{\mu}(M_{11}(s))$, $\overline{\sigma}(M_{22}(s))$ where the M_{ij} are as in Equation 44.15. Other tests will be described later in the section on changing weights. It is convenient to use the traditional log-log, as well as log-linear, and linear-linear plotting scales for these Bode plots.

Assuming that the numerical computations are well behaved as the γ-iteration progresses, the plot of $\overline{\sigma}(M(s))$ will become more and more 'flat' with respect to frequency. Since H_∞-synthesis minimizes the $\sup_\omega \overline{\sigma}(M(j\omega))$, the optimal solution (in theory) could always make any "peak" lower by raising a "valley" somewhere else. This "flatness" is actually achieved only over a limited range of frequencies because K 'rolls-off' (i.e., it has no D term in its state space realization). A consequence of this flattening process is that one or more compensator poles (depending on the dimensions of K) will go to infinity as the γ-iteration progresses.

As the D scales are fit tighter and tighter (and incorporated into P to get K) the upper bound $\|M\|$ for $\overline{\mu}(M)$ becomes tighter. Therefore, a comparison of the plots of $\overline{\sigma}(M(s))$ and $\overline{\mu}(M(s))$ is useful in determining whether a satisfactory result has been obtained. Here again it is reasonable to think in terms of a frequency ranges of interest. To reduce the upper bound one should concentrate on fitting D well in those frequency ranges where the upper bound is largest.

Experienced designers also rely on physical understanding of the design problem to evaluate their progress during the D-K iteration. Specifically, they decompose $\overline{\mu}(M(s))$ into its constituents and determine why it is large in different frequency ranges. This can often be accomplished by comparing the plots of $\overline{\mu}(M_{11}(s))$ and $\overline{\sigma}(M_{22}(s))$ with the plot of $\overline{\mu}(M(s))$. Note that $\overline{\mu}(M)$ is an upper bound for both $\overline{\mu}(M_{11})$ and $\overline{\sigma}(M_{22})$. In the frequency range where $\overline{\mu}(M_{11}(s)) \cong \overline{\mu}(M(s))$ robust stability is the main design driver. In the frequency range where $\overline{\sigma}(M_{22}(s)) \cong \overline{\mu}(M(s))$ nominal performance is the main design driver. At some frequencies (usually near crossover) the two design drivers might make equal contributions to $\mu(M)$. An understanding of the design drivers can help the designer in adjusting the weights, as described in the next subsection.

The upper-bound $\overline{\mu}$ is the primary estimate for μ used in the synthesis step, but there is also software available for computing lower bounds for $\mu(M)$. Usually these lower bounds are reasonably close to the upper bound $\overline{\mu}$, so it is possible to estimate the true value of $\mu(M)$ during the design process. To monitor the (approximate) value of $\mu(M)$ during the design process, the designer looks at both upper and lower-bound curves on the same Bode plot.

44.5 Changing Weights

If the D and K have converged and the μ-synthesis design does not meet the objectives ($\mu(M)$ is greater than 1) then the weights must be changed and the D-K iteration repeated. Analysis of the perturbation structure M is used to determine which input/output paths are driving the problem.

Perturbation Structure Analysis

The significant contributors to $\mu(M)$ can be assessed with Bode plots. For example, the contribution of a particular disturbance, or a group of disturbances can be assessed by selecting out the appropriate columns of M and plotting the maximum singular value of these columns regarded as a matrix-valued function of frequency. Likewise, the contribution of a particular tracking error, or a group of errors can be assessed by selecting out the appropriate rows of M and plotting the maximum singular value of these rows regarded as a matrix-funtion of frequency.

One of the rank conditions on D discussed in the H_∞-synthesis section is sometimes satisfied by introducing sensor noise into d. In some cases, it is desirable for this sensor noise to be regarded as negligible. This condition can be checked by plotting the maximum singular value of the columns of M corresponding to the sensor noise elements of d, and comparing with $\overline{\mu}(M)$. If $\overline{\sigma}(M_{sensor\ noise}) \ll \overline{\mu}(M)$, then the sensor noise makes no significant contribution to the final answer. In these cases, to simplify the analysis, we can eliminate the sensor-noise input (i.e., remove those columns) and perform the μ-analysis on an M-matrix of smaller size.

This same type of comparison can be done for other columns and rows of M. For example, the penalty associated with control actuation can be assessed by plotting the maximum singular value of the rows of M corresponding to the control penalty elements of e, and comparing with $\overline{\mu}(M)$. If $\overline{\sigma}(M_{control\ penalty}) \ll \overline{\mu}(M)$, then the control penalty makes no significant contribution to the final answer. In this case, we can remove the corresponding column of M. In many cases, the contribution of the control penalty to the final answer will be frequency dependent, e.g., actuation rate-limits in response to large pilot commands will be reflected by penalties (weights) on actuator use at high frequencies. In this case, the row (output) of M associated with the control penalty can be eliminated only when examining low-frequency problems.

Another useful technique is to zero-out subsets of the model perturbations, in groups of interest, as a diagnostic aid. For example, the effects of changes in stability derivatives can be studied independently of the high-frequency unmodeled dynamics (represented by a multiplicative perturbation). Again the idea is to find a smaller problem where $\bar{\mu}$ for the smaller problem agrees closely with $\overline{\mu}(M)$.

The methods described in the last few paragraphs are special cases of a general technique for identifying design limitations: μ-analysis of submatrices of M. For each submatrix we can plot singular values or $\overline{\mu}$ (as appropriate) as functions of frequency along the imaginary axis. The objective is to find, for each frequency range of interest, the smallest submatrix-problem that drives the design in that frequency range. Examination and comparison of submatrix Bode plots often clarifies the source of the problem when a satisfactory design cannot be achieved.

A diagnosis of what is causing $\mu(M)$ to be large in a particular frequency range is very useful in deciding how the frequency-dependent weightings are to be modified.

Sometimes the desired performance is not achievable for the modeled system. This situation arises when $\mu(M)$ remains greater than one after many D-K iterations. When this happens, if the result is not satisfactory it is necessary to change the problem. This means change the interconnection structure, or change the weighting transfer functions, or change the desired dynamics in such a way that a compensator can be found that produces $\bar{\mu}(M)$ less than unity.

Take an example of a two-block μ-synthesis (one for robust stability and one for nominal performance). If $\overline{\mu}(M)$ is greater than unity, and if the maximum singular value for nominal performance is close to $\overline{\mu}(M)$ but the maximum singular value for robust stability is much smaller, the weightings associated with nominal performance have to be changed.

An examination of Figure 44.2 shows that weighting functions usually appear at inputs or outputs of the interconnection structure. Input weightings scale entries in specific columns of the transfer-function matrix of the interconnection-structure while output weightings scale entries in specific rows. When faced with a problem for which μ-synthesis does not provide a satisfactory answer, the designer might be able to identify a particular entry or set of entries that must be reduced in magnitude before a solution can be found. To achieve this reduction the designer may introduce or modify selected input or output weightings. For each row and column the product of the associated weightings (input and output) scales the norm of the corresponding matrix entry. To relax a design requirement, either a row or column (output or input) weighting can be reduced. The designer must keep in mind that changing a row or column weighting to reduce emphasis on a particular entry will have an effect on other entries as well. A poor weight-adjustment strategy could relax the spec in a way that is not acceptable, so some care is required in adjusting row and column weightings.

44.6 Compensator Model Reduction

Once the μ-synthesis process is complete, the controller order is reduced by model reduction. The μ-analysis is then repeated to ensure that the model-reduced controller still meets the objectives.

If the γ iteration is repeated many times during the design, the maximum singular value of $M(j\omega)$ will be "flat" with respect to frequency. This implies that the compensator will have high frequency poles. In the limit as $\gamma \rightarrow \gamma_{opt}$ these poles go to infinity. These poles are often eliminated by a model reduction step. Indeed, it often happens that these poles have magnitudes far too large for implementation in practical hardware. A residualization procedure applied to a block diagonal realization can be used to eliminate high-frequency poles from the controller model. After compensator order reduction, the robust performance μ is recomputed to check if the model reduction was successful.

Further controller order reduction is usually necessary because, even after removing the high frequency poles, the order of K may be excessive. Truncated frequency-weighted balanced realizations or frequency-weighted Hankel model reduction are useful for these steps. The frequency-weighted balanced realization is discussed in Enns (1984), Hankel model reduction in Glover, (1984) and frequency-weighted Hankel model reduction in Latham (1986), Anderson (1986), Khou (1993).

44.7 Summary

We have presented the μ-synthesis/analysis approach to robust control design, concentrating on the steps in the process from a control-designer's viewpoint. We divided the design process into six steps and discussed the major issues associated with each one.

The two key mathematical steps in μ-synthesis are the solution of the optimal H_∞-control synthesis problem and the computation of the D scales in μ-analysis. Fortunately, the algorithms for solving these two problems have been automated to a great extent in commercially available software packages, so the control designer need not understand the details of the mathematical theory to perform μ-synthesis. It is enough for the designer to understand the basic results of these theories well enough to appreciate what it means to have a scaled perturbation structure with H_∞-norm smaller than 1. Intuitively, the basic concept is a generalization of the classical small gain theorem (Desoer, 1975).

The art of μ-synthesis lies in defining the interconnection structure. The problems of weight selection and translation of design requirements into the μ-framework are too complicated for this introductory article, so we focused our discussion on the mechanics of the synthesis process after the interconnection structure has been defined. There are many μ-synthesis designs presented in the literature (see References), from which techniques for defining the interconnection structure can be learned. We hope this introduction makes those examples and the general μ-methodology more accessible to the reader.

References

Ackerman, J. E. 1992. Does it suffice to check a subset of multilinear parameters in robustness analysis? *IEEE TAC*-37(4).

Anderson, B. D. O. 1986. Weighted Hankel-norm approximation: calculation of bounds, *Systems and Control Letters,* 7.

Balas, G., Doyle, J., Glover, K., Packard, A., and Smith R. 1991. *μ-Analysis and Synthesis Toolbox, User's Guide,* The MathWorks.

Banda, S. S., Yeh, H. H., and Heise, S. A. 1991. A surrogate system approach to robust control design, *Int. J. System Science,* 22(1).

Bernstein, D. S. and Haddad, W. M. 1989. LQG control with H_∞ performance bound: a Riccati equation approach, *IEEE TAC*-34(3).

Bode, H. W. 1945. *Network Analysis and Feedback Amplifier Design,* D. Van Nostrand.

Boyd, S. P. and Barratt, C. H. 1991. *Linear Controller Design, Limits of Performance,* Prentice Hall, Englewood Cliffs, NJ.

Boyd, S. P. and Yang, Q. 1989. Structured and simultaneous Lyapunov functions for system stability problems, *Int. J. Control,* 49.

Boyd, S. P., Balakrishnan, V., and El Ghaoui, L. 1992. Computing bounds for the structured singular value via interior point algorithms, *American Control Conference,* Chicago, IL.

Bugajski, D., Enns, D., Jackson, M., and Stein, G. *Multivariable Control Design Guidelines,* First Draft (in progress). Being prepared for the Air Force under contract number F33615–92–C–3607.

Chang, B. C., Ekdal, O., Yeh, H. H. and Banda, S. S. 1991. Computation of the real structured singular value via polytopic polynomials, *J. Guid. Control and Dynamics,* 14(1), January–February.

deGaston, R. R. E. and Safonov, M. G. 1988. Exact calculation of the multiloop stability margin, *IEEE TAC*–33, February.

DeMarco, C. L., Balakrishnan, V., and Boyd, S. 1990. Branch and bound methodology for matrix polytope stability problems, *IEEE CDC,* Honolulu, HI.

Desoer, C. A. and Vidyasagar, M. 1975. *Feedback Systems: Input-Output Properties,* Academic Press, New York, NY.

Doyle, J. C. 1978. Guaranteed margins for LQG regulators, *IEEE TAC*–23(4), August.

Doyle, J. C. and Stein, G. 1981. Multivariable feedback design: concepts for a classical/modern synthesis, *IEEE TAC*–26(1), February.

Doyle, J. C. 1982a. Analysis of feedback systems with unstructured uncertainties, *IEEE Proceedings,* 129–D(6) November.

Doyle, J. C., Wall, J. E. and Stein, G. 1982b. Performance and robustness analysis for structured uncertainty, *Proc. IEEE 20th Conf. on Decision and Control,* December.

Doyle, J. C. 1983. Synthesis of robust controllers and filters, *IEEE Conf. on Decision and Control,* San Antonio, TX.

Doyle, J. C. and Chu, Cheng-Chih, 1985. Matrix interpolation and H_∞ performance bounds, *1985 Am. Control Conf. Proc.,* Boston, MA, 19–21 June.

Doyle, J. C. and Packard, A. 1987. Uncertain multivariable systems from a state space perspective, *1987 Am. Control Conf. Proc.*, Minneapolis, MN, 10–12 June.

Doyle, J. C., Glover, K., Khargonekar, P. P., and Francis, B. A. 1989. State space solutions to standard H_2 and H_∞ control problems, *IEEE TAC*-34(8).

Doyle, J. C., Francis, B., A., and Tannenbaum, A. R. 1992. *Feedback Control Theory,* Macmillan, New York, NY.

Elgersma, M., Stein, G., Jackson, M., and Yeichner, J. 1991. Robust controllers for space station momentum management, *IEEE CDC,* Brighton.

Englehart, M. and Enns, D. F. 1991. A design comparison between dynamic inversion and μ-synthesis, Honeywell Internal Memo, August.

Enns, D. F. 1984. Model reduction for control design, Ph.D. dissertation, Dept. of Aeronautics and Astronautics Engineering, Stanford University, June.

Enns, D. F. 1990. Robustness of dynamic inversion vs. μ-synthesis: lateral-directional flight control example, *1990 AIAA Guidance, Navigation, and Control,* August.

Enns, D. F. 1991. Rocket stabilization as a structured singular value synthesis design example, *IEEE Control Systems Magazine,* 11(4), June.

Glover, K. 1984. All optimal Hankel-Norm approximations of linear multivariable systems, and their L_∞ error bounds, *Int. J. Control,* 39.

Horowitz, I. M. 1963. *Synthesis of Feedback Systems,* Academic,

Jackson, M. R. and Enns, D. F. 1990. Lateral-directional control of an aircraft using μ-synthesis, *1990 AIAA Guidance, Navigation, and Control Conf.*, Portland, OR.

Kailath, T. 1980. *Linear Systems,* Prentice Hall, Englewood Cliffs, NJ.

Khou, K. 1993. Frequency weighted L_∞ norm and optimal Hankel-Norm model reduction, preprint.

Krause, J., Morton, B., Enns, D. F., Stein, G., Doyle, J. C., and Packard, A. 1988. A general statement of structured singular value concepts, *1988 American Control Conf.,* Atlanta, GA.

Latham, G. A. and Anderson, B. D. O. 1986. Frequency weighted Hankel-Norm approximation of stable transfer function, *Systems and Control Letters,* 5.

McFarlane, D. and Glover, K. 1992. A loop shaping design procedure using H_∞ synthesis, *IEEE TAC*–37(6), June.

Moore, B. C. 1981. Principal component analysis in linear systems: controllability, observability and model reduction, *IEEE Trans. Auto. Control,* AC–26, February.

Morari, M. and Zafirou, E. 1989. *Robust Process Control,* Prentice Hall, Englewood Cliffs, NJ.

Morton, B. and McAfoos, R. 1985. A μ-test for robustness analysis of a real-parameter variation problem, *Proc. Am. Control Conf.,* Boston, MA.

Morton, B. 1985. New applications of μ to real-parameter variation problems, *CDC Proceedings,* Ft. Lauderdale, FL.

Moynes, J. and Stein, G. 1992. The approach for B-2 flight control algorithms, *Am. Control Conf.*, Chicago, IL.

Packard, A., Fan, M. K. H., and Doyle, J. 1988. A power method for the structured singular value, *IEEE CDC,* Austin, TX.

Packard, A., Doyle, J., and Balas, G. 1993. Linear multivariable robust control with a μ perspective, *ASME J. Dynamics, Measurements and Control,* June.

Ridgely, D. B., Valavani L. S., Dahleh, M. A., and Stein, G. 1992. Solution to the general mixed H_2/H_∞ problem—necessary conditions for optimality, *Am. Control Conf.*, Chigago, IL.

Rotea, M. A. and Khargonekar, P. P. 1991. Mixed H_2/H_∞ control: a convex optimization approach, *IEEE TAC*–36(7), July.

Sparks, A., Banda, S. S., and Yeh, H. H. 1990. A mixed H_2 and H_∞ approach to the Boeing 737 autopilot design problem, *AIAA Guid. and Control Conf.*, Portland, OR, August.

Stein, G. and Athans, M. 1987. The LQG/LTR procedure for multivariable feedback design, *IEEE TAC*-32. February.

Stein, G. and Doyle, J. C. 1991. Beyond singular values and loop shapes, *J. Guid. Control and Dynamics,* January–February.

Wonham, W. M. 1979. *Linear Multivariable Control: A Geometric Approach,* Springer Verlag, New York, NY.

Yeh, H. H., Banda, S. S., and Chang, B. C. 1992. Necessary and sufficient conditions for mixed H_2/H_∞ optimal control, *IEEE TAC*-37(3), March.

Yeh, H. H., Banda, S. S., and Sparks, A. G. 1991. Loop shaping in mixed H_2 and H_∞ optimal control, *Am. Control Conf.*, Boston, MA.

Yeh, H. H., Banda, S. S., Heise, S. A., and Bartlett, A. C. 1990. Robust control design with real parameter uncertainty and unmodelled dynamics, *J. Guid. Control and Dynamics,* 13(6), November–December.

Young, P., Newlin, M. P., and Doyle, J. C. 1992. Practical computation of the mixed μ problem, *Am. Control Conf.*, Chicago, IL.

Young, P., Newlin, M. P., and Doyle, J. C. 1992. Let's get real, *Proceedings of the Institute of Math and its Applications (of the University of Minnesota) on Robust Control Theory,* Minneapolis, MN.

Zames, G. and Francis, B. A. 1983. Feedback, minimax sensitivity and optimal robustness, *IEEE TAC*-28(5), May.

Zhou, K., Doyle, J. C., and Bodenheimer, B. 1989. Optimal control with mixed H_2 and H_∞ performance objectives, *Am. Control Conf.*, Pittsburgh, PA.

VI

Factory Automation

45 An Overview of Factory Automation *Richard Zurawski* 625
Introduction • New Technologies for Factory Automation

46 Types of Automated Manufacturing Systems *Ljubisa Vlacic, Walter Wong, and Theodore J. Williams* 629
The Hierarchial Model Presentation of Manufacturing Activities • Enterprise/Factory Integration • The Methodology for CIE/CIM • Architecture of Automated Manufacturing Systems • Implementations of Factory Automation Systems • Flexible Manufacturing Systems (FMS)

47 Production Management Architecture *Rakesh Nagi and Jean-Marie Proth* 653
Introduction • Production Management in the Sixties and Beyond • Components of the Hierarchical Production Management System • Long-Term Production Plan (LTPP) • Master Production Scheduling (MPS) • Capacity Requirement Planning (CRP) • MRP Philosophy • Application of the MRP • Conclusion

48 Production Management Techniques *Upendra Belhe and Andrew Kusiak* 663
Material Requirements Planning (MRP) • Manufacturing Resource Planning (MRPII) • Optimized Production Technology (OPT) • Toyota System and Just-in-Time • The Kanban Concept

49 Automated Manufacturing System Development Methodology *Richard Zurawski, MengChu Zhou, Sunderesh S. Heragu, Anthony D. Robbi, and Christopher M. Lucarelli* 669
Analysis of Functional Properties of Specification and Design Models of Industrial Automated Systems • Automated Manufacturing System Design Using Analytical Techniques • Discrete Event Simulation

50 Hybrid Systems and Control *Tarek M. Sobh* 706
Introduction • Discrete Event and Hybrid Observation Under Uncertainty • Conclusions

51 Virtual Manufacturing Environment *Robert G. Wilhelm* 718
Introduction • Scope for Virtual Manufacturing • Typical Applications • Emerging Technology

52 Signal Processing for Factory Production Lines *Rokuya Ishii* 723
Introduction • Examples of Signal Processing Systems

53 Robots *Ray Jarvis, Lindsay Kleeman, R. Andrew Russell, Marcelo H. Ang, Jr., Choon-seng Yee, Fathi Ghorbel, Miguel A. Salichs, Luis Moreno, Diego Gachet, Arthuro de la Escalera, Juan R. Pimentel, and Antal K. Bejczy* 730
Robots: Qualities and Capabilities • Robot Vision • Ultrasonic Sensors • Robot Tactile Sensing • A Robotic Sense of Smell • Actuators in Robotics and Automation Systems • Control • Mobile Robots • Teleoperators

45

An Overview of Factory Automation

Richard Zurawski
Swinburne University of Technology

45.1 Introduction 625
45.2 New Technologies for Factory Automation 626
Concurrent Engineering • Reverse Engineering and Rapid Prototyping • Virtual Reality • Virtual Prototyping • Multimedia

45.1 Introduction

The availability of considerable computer processing power, at a low cost, as well as substantial advancement in software development techniques, had a large impact on factory automation. Factory automation is concerned with automation of processes of the kind found in manufacturing, chemical process, power generation, and other industries. Automation aims at replacing, or complementing, or assisting human beings in activities requiring physical and/or intellectual effort. The manufacturing industry is the one which felt the most impact of the technological changes which took place in recent years. This is not surprising, as this industry is largely responsible for production of consumer goods. The growing international competition, in the area of consumer goods, forced companies to adopt many new technologies to retain competitiveness or gain new markets, mostly through low prices offered and increased customer satisfaction. As a result, the introduction of new technologies made it feasible to manufacture small quantities of a product, even on a one-off basis, without compromising quality or cost, and meeting customer's individual requirements. In addition, the time was reduced between the customer placing an order and the product being delivered. Nowadays, some companies offer telephone hotlines for placing orders, which are automatically translated into the relevant data, theoretically validated for manufacture feasibility, and then sent to the factory floor for fabrication (Strobel and Johnson, 1993).

Factory automation involves technologies which are rooted in mechanical engineering, electrical and electronic engineering, and computer science and engineering. Technologies used in manufacturing include automated manufacturing equipment and systems, control and monitoring systems, factory communication systems, and computer systems for automating procedures for design, planning, and decision making.

Examples of automated manufacturing equipment and systems are automatic machine tools and machining centers to process parts, automatic assembly machines, industrial robots, automated inspection systems for quality control, automatic material handling and storage systems, to mention a few.

Control and monitoring systems include machine control and monitoring, and shop floor control and monitoring. Machine control and monitoring is concerned with direct control of various operational parameters of the equipment such as numerically controlled machining tools and centers, robotic units for assembly and transportation, etc. The control at this level is done using embedded microprocessor based systems, programmable logic controllers (PLCs), and industrial computers. For example, the activities involved in control and monitoring of a machine tool include, among others, loading and execution of a control program, adaptive or intelligent control, control of the machine tool peripherals, collection of operating data on parts, diagnosis of machine tool, etc. Shop floor control and monitoring is concerned with planning and execution of a manufacturing job. Planning involves tool and fixture selection, machine tool selection, workpiece route assignment, machine programs selection, etc. Execution involves machine program loading, machine control, machine and tool monitoring, material and tool flow control, etc.

Communication plays a central role in factory automation by integrating activities performed at different levels of manufacturing system planning and control. Two types of networks can be found in use in manufacturing systems. They are data and control networks. Data networks are characterised by transmitting large data packets, with high data rates, relatively infrequently. This type of network is used to support activities involved in strategic and operations planning. Control networks, in contrast, are required to transmit small data packets, with relatively high transmission rates. They are used at machine level and shop floor control.

Computer systems for automating procedures for design, planning, and decision making include computer-aided design (CAD), computer-aided engineering (CAE), computer-aided planning (CAPP), computer-aided manufacturing (CAM), and computer-aided quality control (CAQC). CAD involves automatic engineering drawing, solid modeling, and product data

0-8493-8343-9/97/$0.00+$.50

management. CAE is concerned with product design analysis for automated manufacturing, with the objective to meet the required functionality as well as quality, while minimizing cost of manufacturing. A product may require redesigning, as result of CAE analysis, to meet specification and/or manufacturing constraints. CAPP involves computer-aided generation of a technological plan for a product manufacture. Product design data, generated by CAD, are used to derive manufacturing processes, used by CAM, according to appropriate algorithms or planning rules. CAM involves using manufacturing process plans for execution on the shop floor level. This includes control of production equipment, cutting tools and fixtures, material management, and maintenance. The activities supported by these systems play a central role in operations planning, which was mentioned before. CAE and CAPP employ computational techniques such as knowledge acquisition, rule-based and expert systems. Advance graphics, and in future virtual reality, will play an important role in CAE supporting the virtual prototyping. CAM uses, at different levels of the control hierarchy: logic and sequence control, adaptive control, intelligent control involving computational intelligence techniques such as fuzzy logic and neural networks based control. Discrete-event and hybrid control techniques play an important role on the factory floor. Hybrid systems involve both discrete and continuous state variables. Examples of hybrid systems are robotic assembly systems, manufacturing cells, flexible manufacturing systems, etc.

The objective of this sketchy overview of technologies involved in automated manufacturing systems was to give an appreciation for the extent of the technologies involved, and to provide a background for further reading by bringing into focus technologies involved in factory automation, some of which are extensively covered in this book. Another objective of this section is to introduce some of the technologies which have had or will have a considerable impact on factory automation, and which are not covered elsewhere in this handbook, or are covered in a different context.

45.2 New Technologies for Factory Automation

Concurrent Engineering

Concurrent engineering is a design philosophy and practice in which design and manufacturing engineering stages of a product life-cycle are handled simultaneously, instead of sequentially. This approach eliminates the major drawback of the traditional (sequential) design, which is the separation of the design and manufacturing engineering stages in a product life-cycle. The sequential approach, also termed the "over the wall" approach, inevitably leads to rework, as the design is done, in most cases, without taking into account constraints imposed by a manufacturing system. By adopting a concurrent engineering approach, the number of iterations involved in a product development, and with this, time for the product introduction cycle, are reduced considerably. Concurrent engineering requires a team effort, involving, at least, engineering design and manufacturing engineering. It may also involve other areas such as research and development, quality control, etc.

Concurrent engineering is essentially a framework for using various schemes, many of which are supported by software tools, facilitating the engineering design and manufacturing engineering. For example, design for manufacture rules capture characteristics of manufacturing processes to provide restrictions for the engineering design process. Design for assembly, design for disassembly, design for maintenance, and design for testing rules are examples of other schemes of assistance to a designer. On the other hand, computer-aided process planning schemes and tools aim at facilitating manufacturing engineering. Quality is also an important objective of concurrent engineering. The focus on quality is from the very beginning of the product development cycle. In addition to inspection and statistical process control, the quality, defined as a multifaceted concept, can be enforced by using approaches such as quality function deployment, robust design, and quality loss function, to mention a few. Quality function deployment relates customer's requirements to the product features. Robust design aims at reducing product sensitivity to various manufacturing and operational factors, which could reduce the product lifetime, instead of opting for a conservative design. Quality loss function is a measure of the loss of quality as a function of the deviation from a product characteristic's optimum value.

Reverse Engineering and Rapid Prototyping

The design of a new product must meet functional, performance, and aesthetical requirements. Although, the computer-aided designs can be generated, for simple parts, at the early stages of a product development, it is often not feasible for products with complex geometry. A part prototype has to be built using materials such as wood, plaster, etc. Once the design is approved, a computer model is created, using reverse engineering techniques, for redesign or manufacture. A number of techniques are available for mapping of the early prototype into a computer-aided design file. These techniques can be broadly classified into contact and noncontact methods. The contact methods are based on measurements performed either manually or using coordinate measuring machines. Newer techniques are based on electromagnetic or sonic digitizing. In the former technique, a magnetic field sensor, located on a hand-held stylus, senses the changes in the field intensity as the stylus is moved on the surface of a prototype, which is surrounded by magnetic field. In the sonic digitizing, ultrasonic impulses are emitted by the hand-held stylus as it is moved on the surface of a prototype. These impulses are then sensed by microphones. The noncontact methods are based on the projection of light or ultrasound onto the object and analyzing the reflected beam. These techniques can be divided into active and passive depending on the type of the (light or sound) source used.

The generated design file can then be used to obtain a prototype part, with production-like quality, using modern rapid prototyping techniques. This production-like quality allows the designer to check if prototype parts mate properly or function as expected. Several methods allowing for rapid prototyping have been developed. These methods are stereolithography, selective laser sintering, three-dimensional printing, laminated object manufacturing, and fused deposition modeling. In stereolithography, an *x-y* scannable ultraviolet laser beam sketches a cross section of the product on a photocurable polymer liquid, simultaneously curing it. By piling up successive layers, the part is built layer by layer. The process starts with the layer on the part support structure, and proceeds in a bottom-up fashion. The selective laser sintering method can be used to make a prototype out of the same material as the final production version, by replicating an investment-casting production process. In this method, the sintering creates a wax model by fusing wax powder with an infrared laser beam. The wax model is then fitted with flow gates and vents. The resulting model is dipped in the binder and a ceramic powder results in a ceramic shell. The wax is melted out, leaving a mold. Other materials used in this technique include metal-coated ceramics, PVC, and some metals. The three-dimensional printing also uses a powder deposition process for creating a cross section of the part. In this technique, the powder is treated with a binder, which is solidified by heating in a furnace once the part, built layer by layer, is complete. The laminated object manufacturing, as the name indicates, is based on gluing together successive layers of foils cut to the required shape. The fused deposition modelling uses melted material which is deposited on the existing layer, fuses with the layer, and then solidifies. Rapid prototyping techniques offer a considerable reduction of the lead time for the prototype fabrication compared with traditional approaches.

Virtual Reality

Virtual reality is expected to have a considerable impact on factory automation, affecting design of both product and process, as well as manufacturing. Virtual reality (or virtual environments) is an environment which allows its user, or viewer, to be immersed in a virtual space, with which he can intuitively interact, as well as manipulate virtual objects by using hand gestures. For this environment to be realized, a number of technologies are necessary to allow for varying the way things look depending on the viewer's gaze direction, and feeling the force interaction as a result of operations performed by the system's user in the virtual space.

Numerous areas of design and manufacturing are expected to benefit from virtual reality. Virtual prototyping is one of them. The virtual prototypes of parts will be checked for problems with mating. Force feedback will play an important role in those investigations carried out in the virtual space. The virtual part prototypes will be assembled into a system in order to investigate its functioning. In addition, the design of a system will be evaluated for the ease of assembly and disassembly, as well as maintenance. The prospective users of a system will be able to assess the ergonomic and performance aspects of the system operation. The virtual prototypes of factory systems, such as manufacturing and assembly cells, or even complete flexible manufacturing systems, will be validated for the correctness of the behavioural properties for various operational policies. Virtual reality will also provide support for in-job training, involving assembly and disassembly of complex products, or systems.

Virtual Prototyping

The design of new industrial automated systems, or planning for new tasks, requires that functional and performance requirements are considered. Before functionality validation and performance measurements are conducted on a real system, virtual prototypes, implemented in software, are used for this purpose. Experimenting with virtual prototypes allows the designers to check the system for the presence or absence of certain behavioral properties, and evaluate performance for different operational strategies and conditions. The construction of design models of industrial automated systems is not a trivial task, and requires a great deal of experience. It is time consuming, as well.

An approach, which is advocated to reduce the prototype development time, is based on object-oriented software development. The availability of objects allows for rapid prototype construction and evaluation. An example of this approach adopted for the development of real-time systems is ControlShell (1994), developed by RealTime Innovations Inc., Sunnyvale, CA. ControlShell provides facilities for object-oriented based construction and running real-time application. Applications are built from objects under a graphical editor. Applications may be run in simulation, real-hardware, or combinations of both. This allows for an incremental development of complex systems. The object-oriented approach supports reusability and reverse engineering, prerequisites for rapid design and evaluation. However, in order to take full advantage of the reusability of objects, the classification and definition of objects for different application domains will be required.

The combination of simulation of software prototype with advanced graphics will allow for visual validation of system operation (Freund and Rossmann, 1995). The above extended to virtual reality will allow the designer to walk around the factory floor visually inspecting operation of the prototype, representing a robotic assembly cell or a machining cell, for instance, and listening to sound generated by the virtual system.

Multimedia

The integration of various media such as text, graphics, animation, digital audio, and digital video into computer multimedia environments opened new prospects for supervisory control of highly automated industrial plants. The supervision and control of chemical process, manufacturing, and power generation plants involve monitoring and actuating a large number of control points, such as sensors, switches, actuators, etc. Before the electronic displays were introduced to the control rooms of these plants, monitoring the plant state was based on scanning large control boards displaying readings of sensors, states of switches, and positions of actuators, as well as warning and alarm light indicators. Any deviations from the required values or states required the operator to correct the state of the process involved by taking an appropriate action. This involved identifying the nature of the problem, and then actuating appropriate field devices, either remotely, by pressing buttons located on the control console, or informing the field staff to do so. In modern plants, monitoring is done automatically. Large control boards have been replaced by large cathode-ray tubes. The operator is automatically informed about any abnormal conditions. Integrated multimedia environments will play an important role here. They will allow the operator to visually inspect the affected plant area and listen to the sounds acquired from this area. By using hypermedia, the operator will be able to call a process or instrumentation logic diagram for trouble shooting, or contact an expert using video conferencing facilities integrated in the environment. Switching on and off process components can be done by pressing graphical buttons overlaid on the video picture. These buttons can be located adjacent to the devices manipulated, thus providing visual association between actual devices and virtual buttons. Additional information about the device state and pressure level, for instance, can also be overlaid on the video picture. This facility will be useful for preventive process monitoring during which the operator visually inspects various areas of the plant.

References

ControlShell; Guided Tour, Real-Time Innovations, Inc., 1994.

Freund, E. and Rossmann, J. 1995. Systems approach to robotics and automation, *Proc. IEEE Inter. Conf. on Robotics and Automation,* 3–13, Nagoya, Japan, May 21–27.

Strobel, R. and Johnson, A. 1993. Pocket pagers in lots of one, *IEEE Spectrum,* Sept., pp. 29–32.

46

Types of Automated Manufacturing Systems

46.1 The Hierarchical Model Presentation of Manufacturing Activities 629
46.2 Enterprise/Factory Integration 632
46.3 The Methodology for CIE/CIM 634
46.4 Architectures of Automated Manufacturing Systems 638
46.5 Implementations of Factory Automation Systems 641
Customer-Driven Manufacturing by Allen-Bradley • Integrated Factory Automation Solutions by Modicon • Integrated Factory Automation Solution by Siemens
46.6 Flexible Manufacturing Systems (FMS) 642
Definition of an FMS • FMS Equipment • Types of FMS • Production Modes • Implemention

Ljubisa Vlacic
Griffith University, Australia

Walter Wong
Queensland University of Technology, Australia

Theodore J. Williams
Purdue University

46.1 The Hierarchical Model Presentation of Manufacturing Activities

To meet its customer needs, the factory must be organized in such a way that it interfaces with its customers, suppliers and neighborhood. Therefore, the factory is presented here as a collection of the functional activities (the tasks) needed to carry out its primary mission, whether it is automated or not.

To present the real manufacturing decision making environment, Table 46.1 is reproduced here as a six-level factory automation model (FAM) for manufacturing activities (Williams, 1989).

The following is a detailed functional description of the manufacturing tasks provided in the form of the hierarchical model presentation. An example of the implementation of the hierarchical model from Table 46.1 is presented in Figure 46.2.

Level 0: Typically, this so-called equipment level covers the process, manufacturing and material handling equipments (i.e., reactors and distillation columns, machine tools and human operators, and supporting areas such as utilities and packaging).

Level 1: This level is aimed at directing and coordinating the activities of the shop floor level equipment in order to maintain direct control of the plant unit. Accordingly, this level is occupied by machine controllers, CNC controllers, process controllers and programmable logical controllers. This includes the determination of the values from sensors and detectors (position, weight, temperature, level, pressure, flow) as well as generation of the values for relays, valves, solenoids, hydraulic drivers, CRTs, printers, etc. In addition to its direct control function, this level is equipped with a logic to detect and respond to any emergency condition covering also an update of standby systems from this and lower levels. This shop-floor level transmits to higher levels information on unit production, raw material and energy use, and also provides an interface to the operator. A self-performed diagnostic test and diagnostics on lower level machines are further important characteristics of this manufacturing level.

Level 2: This so-called cell-level of the manufacturing hierarchy is aimed at optimizing operations of the lower level units. The optimization is in accordance with previously established production schedule(s) and operational schemes. It also collects and maintains data related to production line, raw material and inventory. Maintaining communication with higher and lower hierarchical levels and providing a capable operator interface are further significant tasks of this level of manufacturing hierarchy. It performs diagnostics on itself and on shop floor level equipment, as well as an update of standby system(s).

Level 3: A direct control function disappears from this level of manufacturing hierarchy. Instead, operational management activities are in place and aimed at optimizing working performance of the assigned section of the plant. It is assumed here that any plant, regardless of its type, can be divided into a number of sections, each of which is divided into further smaller production units. Following this, it can be said that levels 0, 1 and 2 deal with production units assigned to them while level 3 deals with the plant section, Figure 46.1.

0-8493-8343-9/97/$0.00+$.50

Table 46.1 Factory Automation Model

Hierarchy		Control	Responsibility	Basic functions
Level 5	Enterprise	Corporate Management	Achieving the mission of the enterprise and managing the corporate.	–Corporate management –Finance –Marketing & sales –Research & development
Level 4	Facility/ plant	Planning production	Implementation of enterprise functions, and planning and scheduling the production.	–Product design & Production –Production management (upper level) –Procurement (upper level) –Resources management (upper level) –Maintenance management (upper level)
Level 3	Section/area	Allocating and supervising materials and resources	Coordinating the production and supporting the jobs and obtaining and allocating resources to the jobs.	–Production management (lower level) –Procurement (lower level) –Resources management (lower level) –Shipping –Waste material treatment
Level 2	Cell	Coordinate multiple machines and operations	Sequencing and supervising the jobs at the shop floor, and supervising various supporting services.	–Shop floor production (cell level)
Level 1	Station	Command machine sequences and motion	Directing and coordinating the activity of the shop floor equipment.	–Shop floor production (station level)
Level 0	Equipment	Activate sequences and motion	Realization of commands to the shop floor equipment.	–Shop floor production (equipment level)

Source: Williams, T. J. (ed.) 1989. *A Reference Model for Computer Integrated Manufacturing (CIM). A Description from the Viewpoint of Industrial Automation,* Instrument Society of America, Triangle Park, NC. With permission.

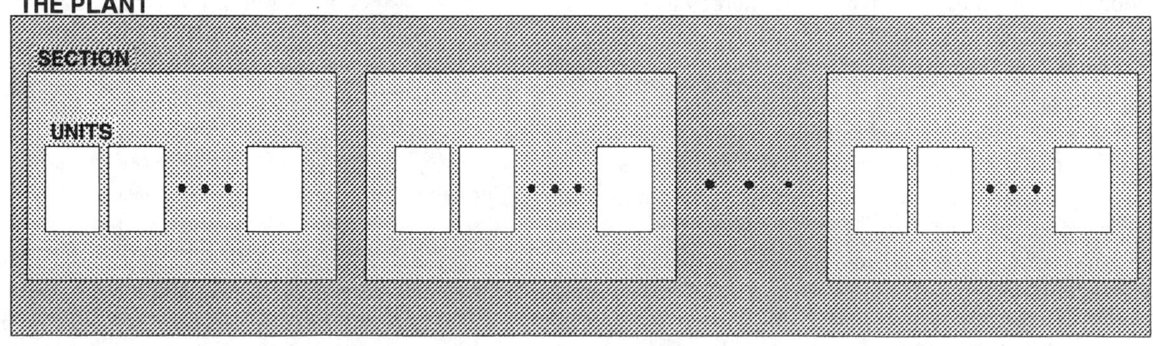

Figure 46.1 Decomposition of the manufacturing plant into sections and units.

A responsibility of level 3 is to carry out the schedule for all production related needs established by both the higher level consumers and locally (the so-called immediate production schedule). This section-level production schedule is also equipped with a diagnostic of self and lower level functions. In addition, a communication with higher and lower hierarchical levels is carried out by means of maintaining a range of data (raw material, energy, inventory, spare parts, manpower, personnel, etc.), and providing off-line analysis of data. A capable man-machine interface is a further duty of this level.

Level 4: As at level 3, no control actions are required between this level and level 5 of the manufacturing hierarchy. This level is responsible for the overall plant production scheduling, its set-up and real-time execution (what should be done in accordance with order stream received,

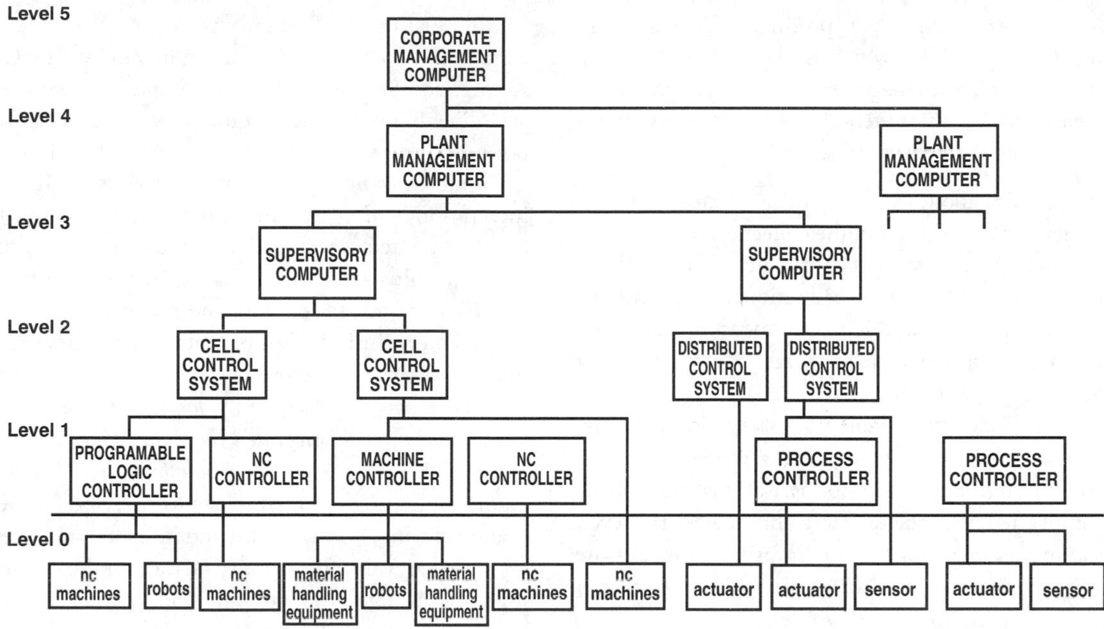

Figure 46.2 An example of partial implementation of the hierarchical model from Table 46.1. (*Source*: Williams, T. J. (ed.) 1989. *A Reference Model for Computer Integrated Manufacturing (CIM): A Description from the Viewpoint of Industrial Automation,* Instrument Society of America, Triangle Park, NC. With permission.)

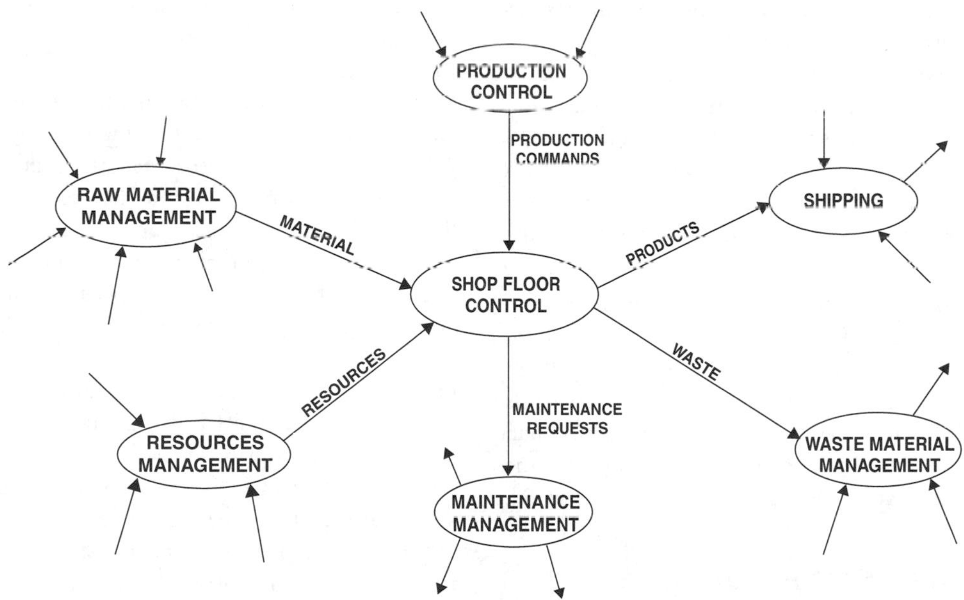

Figure 46.3 An example of the data flow type model.

power demand requirements, raw materials demands, spare part and energy sources constraints, etc.). A capability to cope with production interruptions at downstream units (plant sections and plant units) and dynamic coordination and synchronisation with production schedules for downstream units are the most important performance measures. Plant-wide data should be collected and maintained to assist in this function. This data covers raw material, spare parts, energy sources, machinery and their life history files, manpower, etc. In addition, self-diagnostics routines, and diagnostics of lower level machines and maintenance management are also provided.

Level 5: No control actions or production schedule are required from this level which is devoted to the tasks related to corporate management, marketing and sales, and research and development. Human innovative functions are essential to these tasks.

It remains to be pointed out that the Purdue's factory automation model is a generic description of the collection of functional entities which make up a particular factory and of their interaction through their assigned tasks and functional specifications. Manufacturing and material handling equipments are therefore not considered parts of the above decision and control hierarchy because of their non-generic nature.

The hierarchical model from Table 46.1 is applicable to any kind of manufacturing plant, independent of any given predetermined realizations in terms of system configurations and is open-ended in its ability to encompass new technologies. However, the hierarchical model does not provide information about the interrelationship of the tasks. For this, a data flow type model could be used (see Figure 46.3 for example).

Automatability requires that the operation of the task and its related physical equipment can be expressed in mathematical or computer program terms. If this is not possible, then the action should be carried out by a human being. In accordance with this definition, the hierarchical model from Table 46.1 can be interpreted as follows:

- Levels 0, 1 and 2 are characterized by repetitiveness in decision making, short response time and high reliability in decision making; the main objective of automation of all functions from these levels is to automate the decision making processes and to substitute the decision maker(s) with an appropriate computer-based automated system in such a way that the computer can operate relatively independently of human intervention in carrying out the assigned tasks;
- Levels 3, 4 and 5 are mostly oriented to human decision making; the objective of automation at these levels is not to replace human decision makers, but to support them in their everyday decision making; tasks from these levels are a type of ill-structured decision, and require human intervention for their implementation; consequently, the tasks from levels 3 and 5 can only be partially automated.

46.2 Enterprise/Factory Integration

Current trends in electronics and control systems technologies are providing the technical capability to facilitate computerization of the factory, its automation and integration. This has been described by the terms computer integrated manufacturing, (CIM) and computer integrated enterprise, (CIE). We will follow here a distinction between these two terms derived from the work of Theodore J. Williams (1989, 1994a).

The term *computer integrated manufacturing* (CIM) describes automation of all of the manufacturing tasks of the enterprise, i.e., automation of all tasks from levels 0, 1, 2 and 3 and some tasks from level 4 of the manufacturing hierarchy. The Computerised Automation System (CAS), i.e., computer-based control system, acts directly on the plant equipment to accomplish the needed tasks (which are related here to levels 0, 1 and 2 mainly).

The term *computer integrated enterprise* (CIE) describes the integration of *all* aspects of company operations including management related aspects split across all six levels of the hierarchical model illustrated in Table 46.1. By this definition all human activities are integrated with the manufacturing activities.

The enterprise is the company itself and its associated organisational entities. The factory is the manufacturing section (plant) of an enterprise. Consequently, *factory integration* is replaced by the term *computer integrated manufacturing*. Within the framework of this chapter, the terms a *factory* and *a manufacturing plant* will be used synonymously to describe a collection of the manufacturing, human-based and information system tasks necessary to carry out the primary mission of the factory in producing a marketable product. Figure 46.4 shows, in an abbreviated manner, the overall enterprise versus the manufacturing plant which is the main subject of this text. Figure 46.5 shows the same with more detail.

Control systems of the past were stand-alone items with no or very little interaction with "the rest" of the enterprise. Today many manufacturing plants must compete world-wide for their business and be tuned to the needs of their customers. The plant's direction as well as the people's role within the factory should therefore be justified before making major investments in CIE/CIM technology.

Piecemeal factory automation often results in a non-integratable island of automation rather than the integrated whole originally conceived and desired. Attempts at such factory automation to date have shown mixed, and often disappointing, results from the overall company goals point of view.

It is obvious that an integration of all manufacturing activities is unavoidable, and a factory automation issue must be observed as an integral part of the total enterprise integration strategy. Enterprise/factory integration is an extremely difficult problem, primarily because of the changing nature of automation and the extent of automatability across all hierarchical levels. Even so, it has long been a dream of practising industrial and/or control engineers to integrate the plant to be able to meet customer needs at low cost, and at maximum profit for the company involved.

A master plan for the total CIE system envisioned is absolutely necessary. The CIE master plan is essential to the development of long term integration, and includes a prioritized list of projects which will result in the implementation of the CIE/CIM system. These projects are implemented in phases as company resources permit and will result in the integrated system producing the best benefits. "Do it once, but do it right" is an idée fixe of today's manufacturing industry world-wide.

A decision on to what extent to automate the factory is, in fact, a part of a management decision which is not based on technical issues only but on social, political and economic issues as well. Where to draw the line on automation and how to set

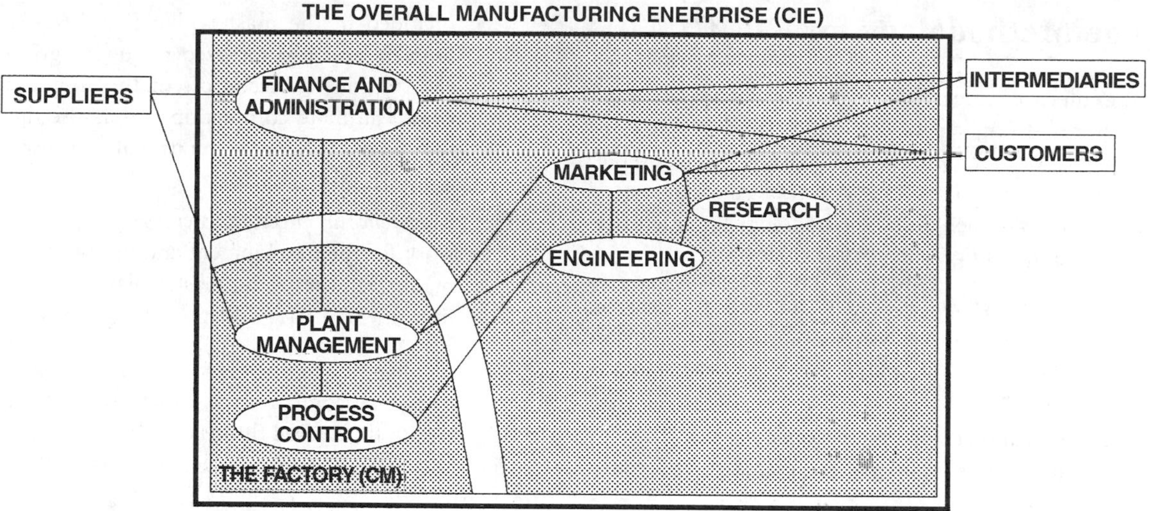

Figure 46.4 Relation of the computer integrated enterprise (CIE) to computer integrated manufacturing (CIM). *Source*: Williams, T. J. (ed.) 1989. *A Reference Model for Computer Integrated Manufacturing (CIM): A Description from the Viewpoint of Industrial Automation*, Instrument Society of America, Triangle Park, NC. With permission.

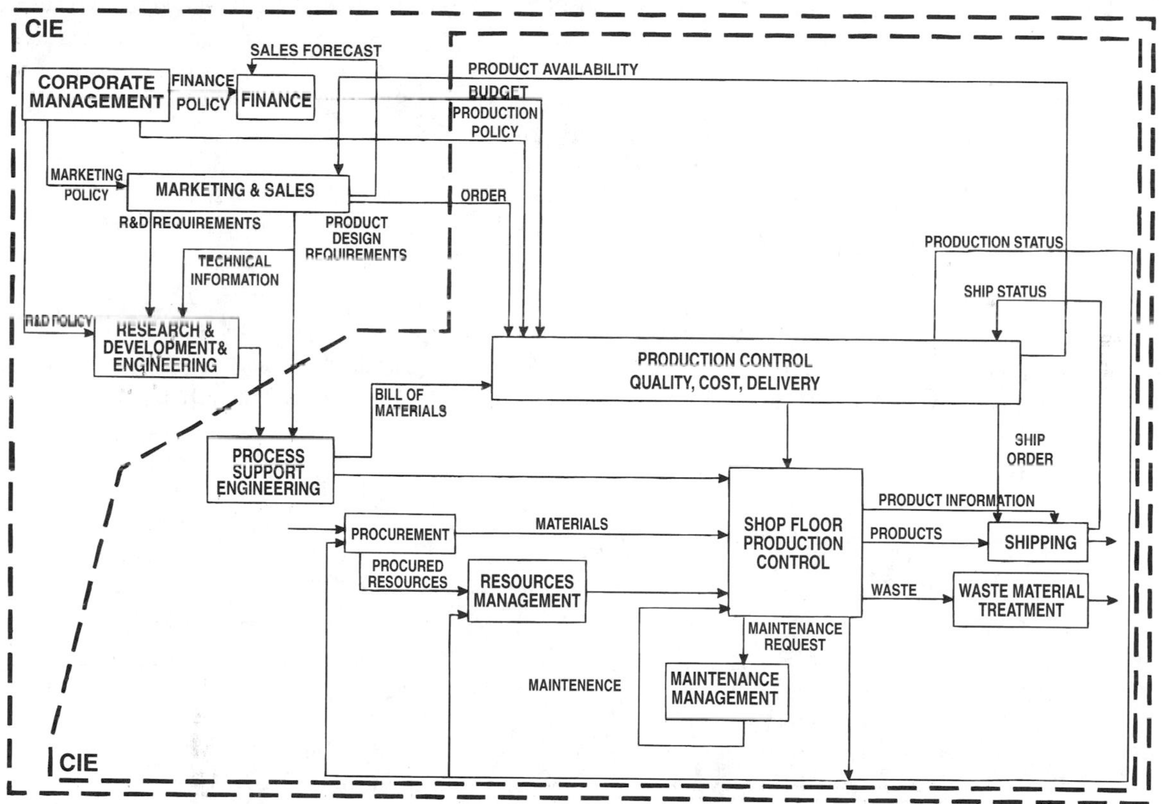

Figure 46.5 Scope for CIE and CIM. *Source*: Williams, T. J. (ed.) 1989. *A Reference Model for Computer Integrated Manufacturing (CIM): A Description from the Viewpoint of Industrial Automation*, Instrument Society of America, Triangle Park, NC. With permission.

up the role of people within the automated factory are just a few of the many questions which should be answered in solving factory automation problems.

The Purdue methodology for CIE/CIM is the only public domain CIE/CIM methodology available at the present time, which provides a basis for the treatment of human-implemented tasks together with manufacturing and information systems tasks in the automated factory. This is why the Purdue methodology, which has been developed by Purdue University Industry Consortium, will be presented shortly to provide the reader with the framework, the glue and explanations for factory automation (Williams, 1991, 1994a,b).

46.3 The Methodology for CIE/CIM

Implementing enterprise integration, even at the factory automation level, requires strong, forward-looking management support, a new pattern of organization, and a change in the overall company culture.

The basic principles of the Purdue methodology for enterprise integration are as follows (Williams, 1994a):

1. An overall detailed *master plan* for the desired transition is absolutely necessary before attempting to implement any CIE/CIM program, even at level 1 of the manufacturing hierarchy;
2. An extensive and detailed *instructional manual* is necessary to guide and simplify the preparation of the Master Plan;
3. Included in the master plan is a *CIE/CIM program proposal* and a prioritized set of integrated projects, each within the resources of the company involved, whose ultimate completion will assure the success of the finally desired operational integration of the enterprise;
4. A *reference architecture* is necessary to provide the framework for the development and use of the Instructional manual, the resulting master plan, and the ultimately implemented CIE/CIM program proposal;

The master plan compares the present state of the enterprise (the As-Is) with the desired future state (the To-Be) to characterize the modification path between them (the Transition; Figure 46.6).

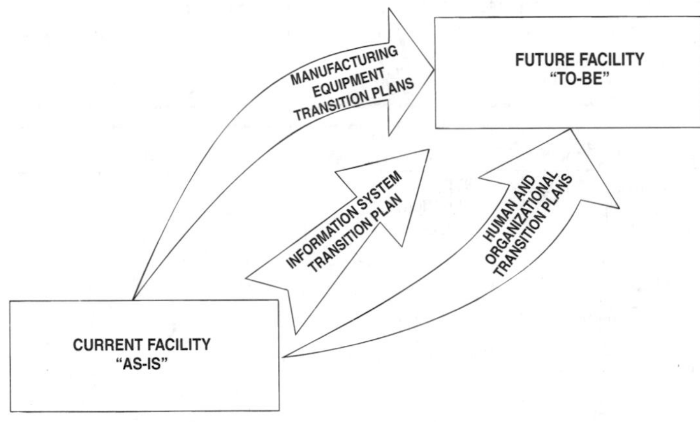

Figure 46.6 Transition Plan. *Source*: Williams, T. J. (ed.) 1994, *A Guide to Master Planning and Implementation for Enterprise Integration Programs*, Report No. 157, Purdue Laboratory for Applied Industrial Control, Purdue University, West Lafayette, IN. With permission.

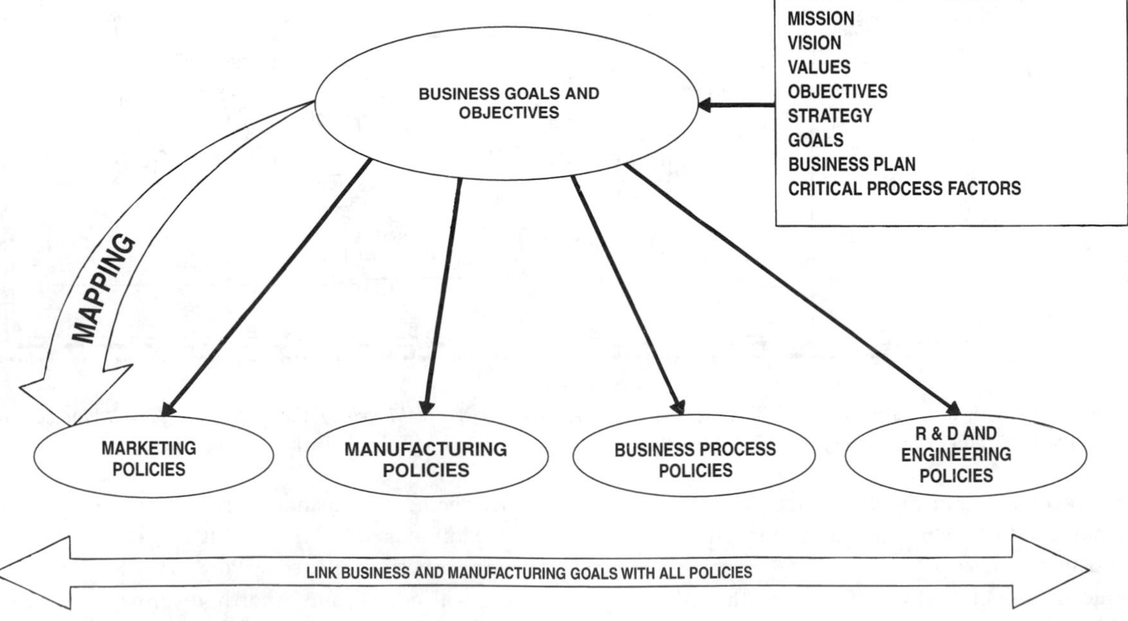

Figure 46.7 Link objectives with policies. *Source*: Williams, T. J. (ed.) 1994, *A Guide to Master Planning and Implementation for Enterprise Integration Programs*, Report No. 157, Purdue Laboratory for Applied Industrial Control, Purdue University, West Lafayette, IN. With permission.

Development of a formal master plan, (Figure 46.7) involves (Williams, 1994b):

- Affirming the critical success factors, goals and objectives of the enterprise.
- Identifying and defining all major projects and fast track opportunities.
- Investigating and recommending alternative solutions for key problems.
- Defining performance measurements.
- Developing resource requirements, costs, and an investment analysis.
- Defining intangible benefits.
- Prioritizing projects and opportunities based upon agreed to guidelines.
- Defining the organizational, procedural, and management impact on the facility.

The master plan includes:

- Definition of business functions and scope definition to the degree necessary to estimate the cost of the project(s).
- Definition of future systems architecture and standards.
- Definition of current state of manufacturing information and control systems.
- Allowance for modification to accommodate changes in business requirements, project results, and technology.
- Transition plan for each project.
- Cost benefit, justification, priority, risk assessment, and timing of the project(s).
- Performance tracking.
- Education and training plan.

At the top of the organization, strategies/policies are broad and deal with the mission, purpose, thrust, long term objectives,

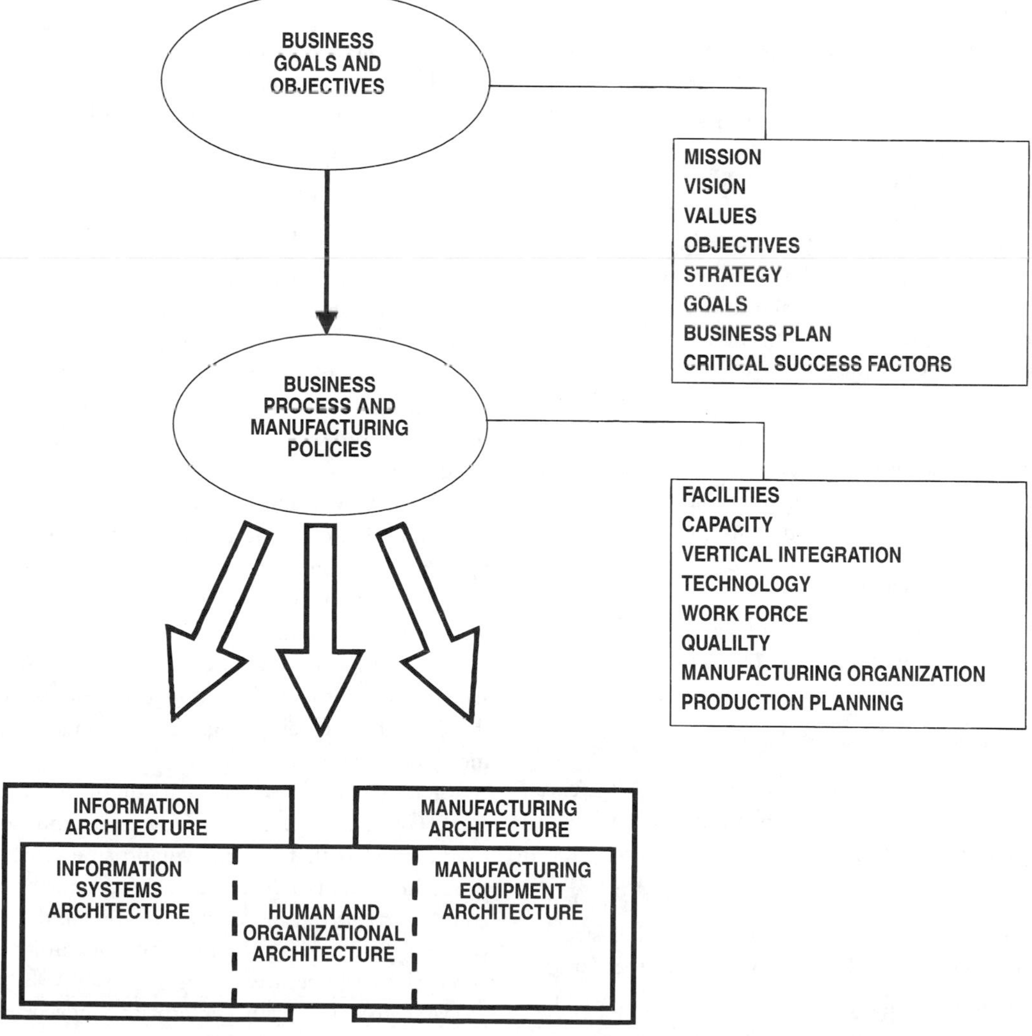

Figure 46.8 Mapping business and manufacturing policies. *Source*: Williams, T. J. (ed.) 1994, *A Guide to Master Planning and Implementation for Enterprise Integration Programs,* Report No. 157, Purdue Laboratory for Applied Industrial Control, Purdue University, West Lafayette, IN. With permission.

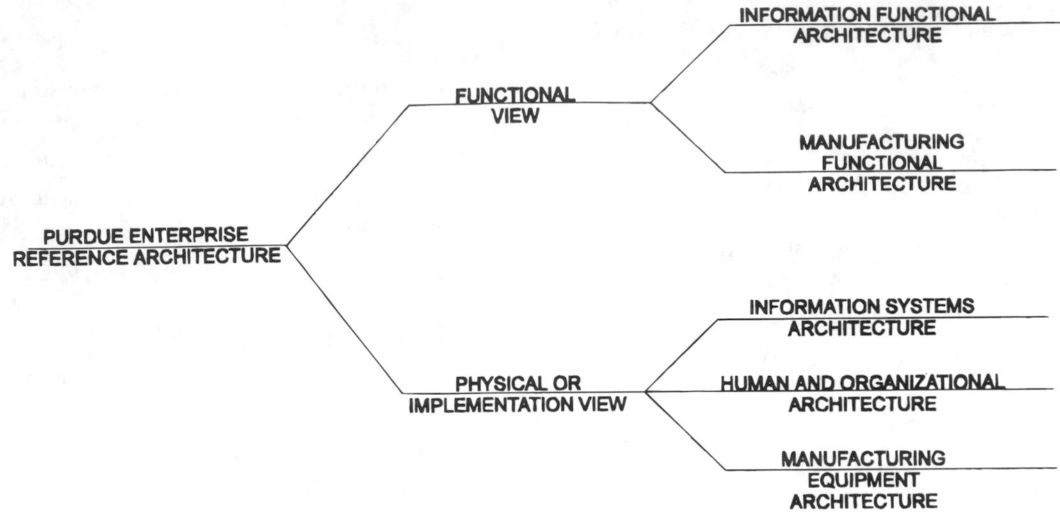

Figure 46.9 Breakdown of the Purdue enterprise reference architecture into views and functional and implementation sub-architectures. *Source*: Williams, T. J. 1991. *The Purdue Enterprise Reference Architecture,* Report No. 154, Purdue Laboratory for Applied Industrial Control, Purdue University, West Lafayette, IN. With permission.

and program strategies. Figure 46.8 illustrates how strategies at the top level of the enterprise percolate down to the lower, manufacturing level of the enterprise where policies deal with rules, procedures, and operating plans to carry out the actions.

Figure 46.8 also shows that all tasks in the manufacturing enterprise can be classified either (and only) as informational or as physical manufacturing, i.e., either involving the moving and transformation (use) of information or, conversely, the moving and transformation (use) of material and energy. Functionally, all tasks can be initially defined without reference to their method of implementation, i.e., whether they are conducted by humans or machines (or in what type of equipment or where). All of these latter considerations are implementation details.

Both the functional and implementation views are described by an enterprise reference architecture which should illustrate clearly all of the following aspects of the enterprise:

- Enterprise decision making.
- Enterprise activities.
- Enterprise business processes.
- Enterprise information exchange.
- Enterprise material and energy flows.

Following the principles of the *Purdue Enterprise Reference Architecture,* (Williams, 1994b) there are two important views of the enterprise reference architecture (Figure 46.9).

- *Functional view*—that collection of task modules (including their interconnectivity) which describes and illustrates the functions and their relationship to each other; and
- *Physical or implementation view*—a collection of the human organisations and of the physical hardware and software used to carry out all or part of the functions described and illustrated by the functional view of the enterprise.

Just as there are two categories of tasks, based upon the type of task transformation involved (information or material and energy), there are also two categories of functional architectures based on the types of tasks or functions involved (Figure 46.10):

1. Information functional architecture.
2. Manufacturing functional architecture.

The terms *information functional architecture* and *reference model* are synonymous in their current usage. In a totally automated factory (no humans involved) the two functional architectures would translate directly into two implementation architectures: information architecture and manufacturing architecture. These together would comprise the implementation or physical view of the enterprise.

Note that in Figure 46.10, the views of the architecture, the functional vs. the physical (implementation), are represented by the top vs. bottom part of the diagram. On the other hand, the question of category of architecture, information vs. manufacturing (or material and energy transformation) is represented by the left and right hand sides of the figure. Note that planning, scheduling, control and data management as used here includes the usual functional listings: engineering, marketing, management, research, etc.

A major point of the comparison of the enterprise architecture (Figure 46.10) and the decision and control hierarchy (Table 46.1) is that the right-hand side (customer service) of the architecture is at level 0 (manufacturing or customer service equipment) of the hierarchy. The remainder of the hierarchy (levels 1–5) describes the left-hand or information side of the enterprise architecture.

Every manufacturing enterprise, indeed every enterprise, which handles material and/or energy is composed of three parts:

- A physical part which actually accomplishes the service to the customer.

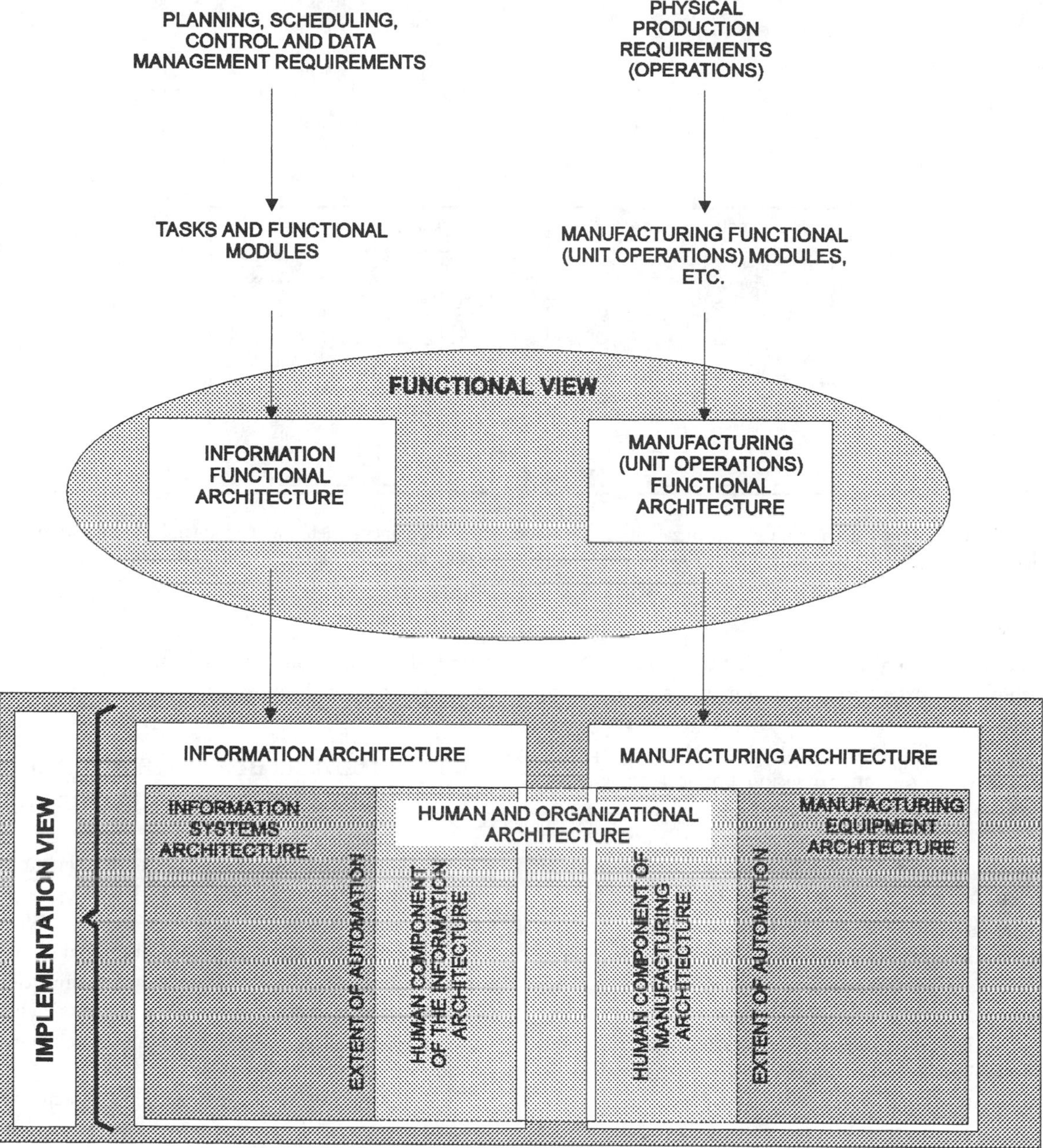

Figure 46.10 Functional and implementation views of the Purdue enterprise reference architecture. *Source*: Williams, T. J. (1994). The Purdue Enterprise Reference Architecture, *Computers in Industry*, 24, 141–158. With permission.

- An information part which monitors and controls the physical part in its accomplishment of the mission.
- A human part whose members may contribute to the tasks of either the physical part or the information part, or both these tasks.

Because of the necessary inclusion of humans in any implementation (Figure 46.10), there must be three separate physical or implementation architectures considered:

- The information systems architecture (the computers, communication equipment, interfaces, database facilities, etc.).
- The human and organizational architecture.
- The manufacturing equipment architecture (the processing and material handling equipment, etc.).

These then form an implementation view of the overall CIE/CIM system (Figure 46.10).

The lines separating the three implementation architectures in Figure 46.10 must now be defined (Figure 46.11).

There is a line which can be called the automatability line which shows the absolute extent of technology in its capability of actually automating the tasks and functions of the integrated enterprise. It is limited by the fact that many tasks and functions require human intervention and cannot be automated with presently available technology.

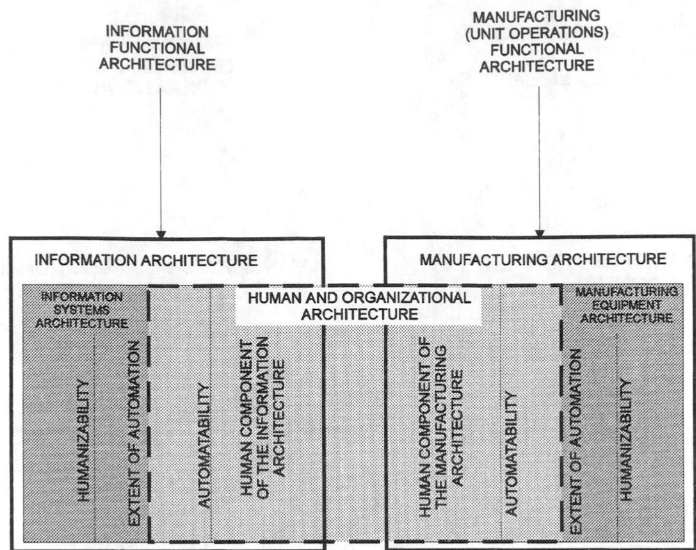

Figure 46.11 Relations of the automatability, humanizability and extent of automation lines in defining the human and organizational architecture. (*Source*: Williams, T. J. (1994a). The Purdue Enterprise Reference Architecture, *Computers in Industry*, 24, 141–158. With permission.)

There is another line which can be called the humanizability line which shows the extent to which humans can be used to actually implement the tasks and functions of the integrated enterprise. It is limited by human abilities in speed of response, breadth of comprehension, range of vision, physical strength, etc.

There is still a third line which can be called the extent of automation line which actually defines the boundary between the human and organizational architecture and the information systems architecture on the one hand, and between the human and organizational architecture and the manufacturing equipment architecture on the other. The extent of automation line shows the actual degree of automation carried out or planned in the CIE/CIM Program. The location of the extent of automation line has economic, social and technological factors in its determination.

An automatability line showing the limits of technology in achieving automation will always be outside the extent of automation line with respect to the automation actually installed. That is, for various reasons, not all of the technological capability for automation is ever utilized in any installation.

Note that for a totally automated plant both the automatability line and the extent of automation line will coalesce together and move to the right edge of the information architecture block and correspondingly to the left edge of the manufacturing architecture block. Therefore, the human and organizational architecture disappears and the information systems architecture and the manufacturing equipment architecture coincide with the information architecture and the manufacturing architecture, respectively.

Figure 46.12 presents a simple block diagram form of the Purdue enterprise reference architecture. The above cited references, listed at the end of this section, can be used for further reading about the enterprise integration.

When the scope of the overall enterprise integration is known, some particular issues of relevance to factory automation can be described, analyzed and discussed in detail in the light of the above presentation.

46.4 Architectures of Automated Manufacturing Systems

The distribution of information is an immanent property of manufacturing plants. Functional and topological distribution are specifically pronounced in process industry. With the development of effective data communication links, the picture of control systems for factory automation has drastically changed. Contemporary computer technology permits data processing within the control system to be run almost parallel to the natural flow of the manufacturing plant.

A specification of any control system for factory automation covers the aspects of its functions, its topology and its performance measures.

System associated functions comprise all functional tasks expected to be fulfilled by the control system toward the manufacturing plant (its production line(s) and operators). All functions expected to be carried out by the factory automation systems are already listed in Table 46.1. Depending on the applied level of automation, the automated manufacturing systems comprise some or all of the functions as follows: measurement, data acquisition and plant monitoring, open-loop control, closed-loop control, logic and sequence control, machining process control, motion control, plant supervisory control, production scheduling and control, acquisition of customer information, order statistics, order acceptance and terms checking, price calculation, capacity and order balancing, order dispatching, production orders and contracting reports, productivity analysis, turnover and profit/loss reporting, etc.

All these and other system functions are underpinned by databases of different sorts: real-time database which contains data corresponding to the current status of the system; rollback database which contains data concerning the system at a certain instant of time; and historical database which in fact snapshots databases in valid time (Popovic and Bhatkar, 1990). It can be seen that the databases always reflect the reality of the system to which they are related.

A capable man-machine interface (recently, the term *human interface* became preferred as the more adequate one) is a further important function of a control system for factory automation. The computer-oriented human activities discussed above imply that at least the following interfaces must be available for proper systems applications (Popovic and Bhatkar, 1990): computer-operator interface for generation, testing, documentation and maintenance of the system; plant-operator interface for monitoring and operating the plant at different hierarchical levels; production control and monitoring interface for plant management personnel; the enterprise monitoring and control interface for the enterprise management personnel.

System performance is the property of the system defined as the measure of its effectiveness in performing the functions associated with the system. The system performance can be measured against a number of criteria by applying the methods and techniques as for multicriteria-based decision analysis. System performance analysis is beyond the scope of this section, and cited literature can be used for further detailed reading (Vlacic and Matic, 1984; Vlacic, 1988, and 1989).

System topology (i.e., an architecture of the system) means the property of the system which defines its organization in the sense of defining the type of system components interconnection.

Local area networks are used to interconnect several components of the system placed at different locations in the automated factory. Either two nodes may be connected together directly or by a communication path and may have to pass via another unit, which relays the data to its destination. Three basic types of network topologies (star, ring and bus) are shown in Figure 46.13 (Olsson and Piani, 1992).

Other topologies have been designed following the above three basic structures in an attempt to meet the whole set of requirements. Although these requirements are versatile, the chosen topology should provide a guarantee that the system will perform well and in accordance with the specification given by the customer. System topology design is therefore a very complex task which should address a number of questions. For example, if bus topology (or its upgrade) is under analysis, then the following should be addressed: transmission medium, bus length, number of bus participants, bus arbitration, interoperability and interchangeability of bus participants, data transmission technique, medium access control procedure(s), access time, scan rate, error detection, signal isolation, final elements powering, redundancy, etc. (Vlacic, 1988).

There is a range of control solutions designed for automation of manufacturing plants on the market. The following section describes some of them.

46.5 Implementations of Factory Automation Systems

Customer-Driven Manufacturing by Allen-Bradley

In the company's publication on the role of control systems in customer-driven manufacturing, Allen-Bradley states: "As we move into a new era of manufacturing, automation components are becoming less and less critical, while the automation architecture that supports them is taking on dramatic importance." Allen-Bradley has identified the growing importance of the control system architecture as "a cohesive framework of highly configurable elements, interfaces and services that allows the creation of highly integrated solutions" (Allen-Bradley, March 1993).

However, no single company can offer the whole enterprise integration solution based on its products only. "If industry is to achieve greater customer-driven manufacturing, the multi-vendor solutions must become a part of the business culture." This has been a motto of Allen-Bradley's commitment to interoperability and the free exchange of data across the manufacturing hierarchy from Table 46.1. Allen-Bradley has established the Pyramid Solution Program by identifying complementary third party products and developing systems which can function on a variety of communication networks. The Pyramid Solution Program identifies products that are available throughout the world (robot controllers, welding controllers, intelligent sensors, control logics, process and operator interface, supervisory control, communication links, process operations management, and a range of other Windows, VME and UNIX related products) that can be used to complement Allen-Bradley's solutions and be tightly linked to the architecture of its factory automation system.

Figures 46.14 through 46.16 show the application of Allen-Bradley's Pyramid Integrator in the automation of a pharmaceutical plant, the metal industry, and the food industry respectively Figure 46.17 presents the architecture of the Allen-Bradley factory automation system.

Integrated Factory Automation Solutions by Modicon

Modicon is another manufacturer attempting to meet factory integration requirements. Figure 46.18 demonstrates the powerful link between Modicon 984 programmable controllers and host computers based on Modbus Plus Networking Strategy. This particular architecture has been developed in an attempt to meet one of the major requirements—to provide redundant communication and networking among all networked nodes, and to eliminate the risk of faulty software or accidental changes that could otherwise result in an unsafe condition. Redundancy is one of the most important requirements in the automation of chemical and nuclear power plants.

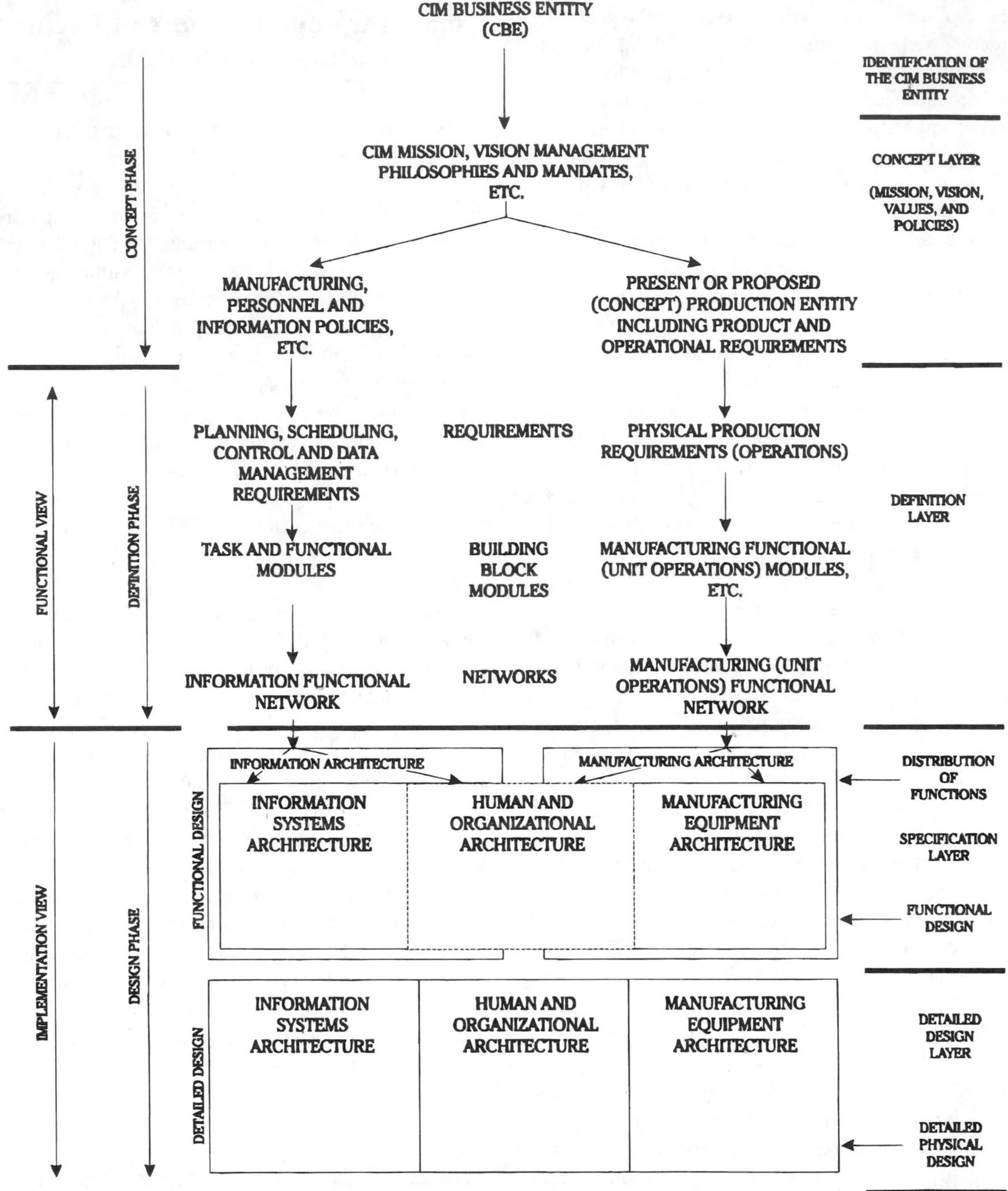

Figure 46.12 A simple block diagram form of the Purdue enterprise reference architecture. *Source*: Williams, T. J. (ed.) 1994. *A Guide to Master Planning and Implementation for Enterprise Integration Programs*, Report No. 157, Purdue Laboratory for Applied Industrial Control, Purdue University, West Lafayette, IN. With permission.

The system from Figure 46.18 corresponds to cell-level control of Table 46.1. Low-level nodes are occupied by Modicon 984 programmable controllers, and upper-level nodes by host computer(s). The same systems components can be interconnected in a slightly different way in order to design dual redundant architecture for a factory automation system, presented here in Figure 46.19.

Integrated Factory Automation Solution by Siemens

Figure 46.20 presents the simplified layout of the coating preparation paper mill plant, KNP, in the Netherlands (Miebs and Sonst, 1991). There are four main divisions: delivery and storage of raw materials, mixing, intermediate storage of end products, and machine vats.

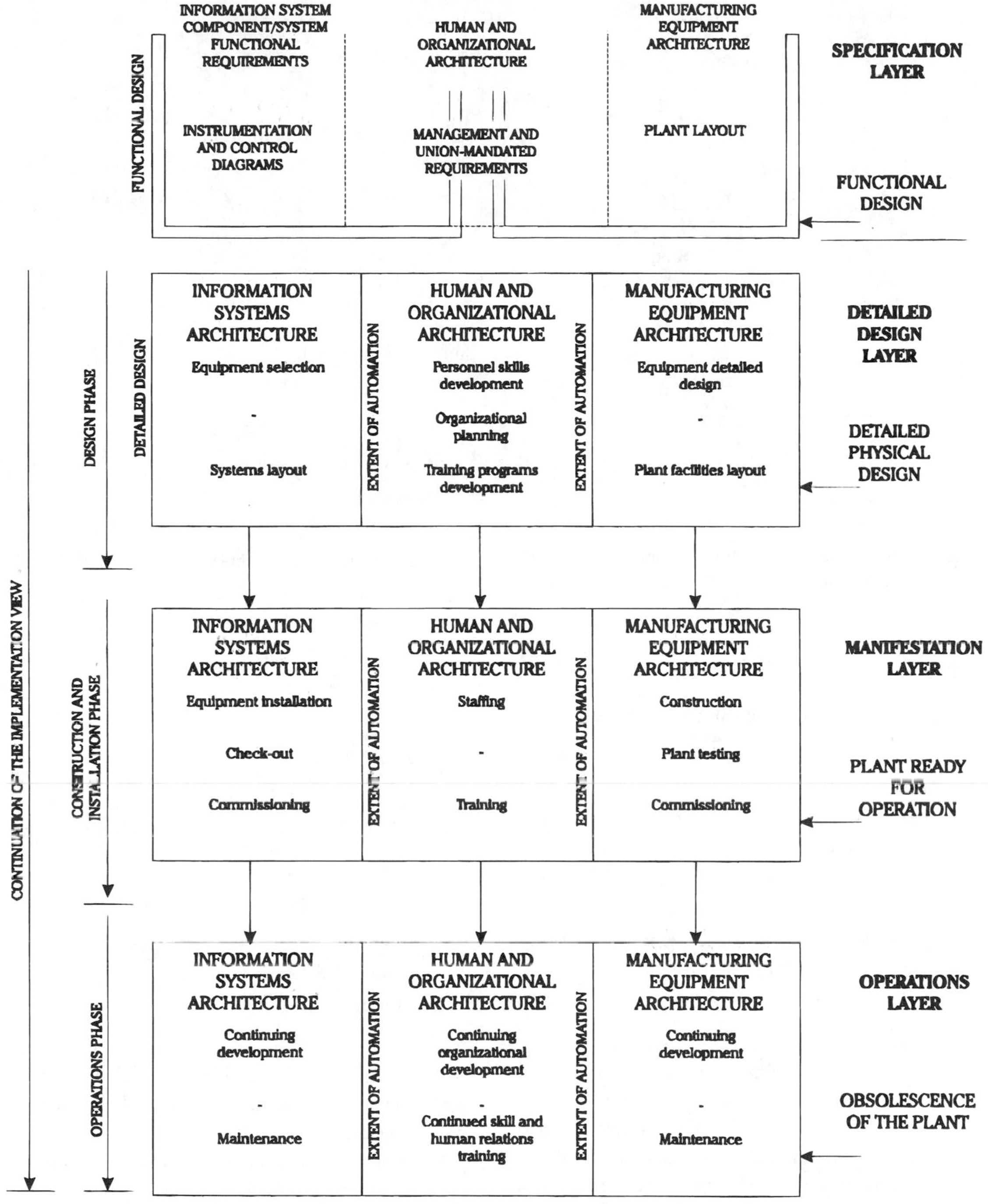

Figure 46.12 Continued

An overview of the installed TELEPERM M automation system is presented in Figure 46.21. Operation of the system is based on a software concept that employs a hierarchical structure described in Figure 46.22.

46.6 Flexible Manufacturing Systems (FMS)

Definition of an FMS

T. Yamazaki, the Japanese FMS Machine Tool builder, gives the following definition of a flexible manufacturing system: "FMS consists of three or more machining centers turning machines, fabricating centers, or the like, equipped with flexible automatic loading/unloading/transfer devices, and a method of monitoring tool conditions and replacement. The entire production scheduling and machining process is automatically supervised by a computer system."

The schematic diagram of a typical FMS is shown in Figure 46.23. The system provides great flexibility in terms of the types of parts and the process sequencing that can be used—a consequence of its high degree of automation. Material and information flow throughout the system is totally automated, and very little human intervention is required.

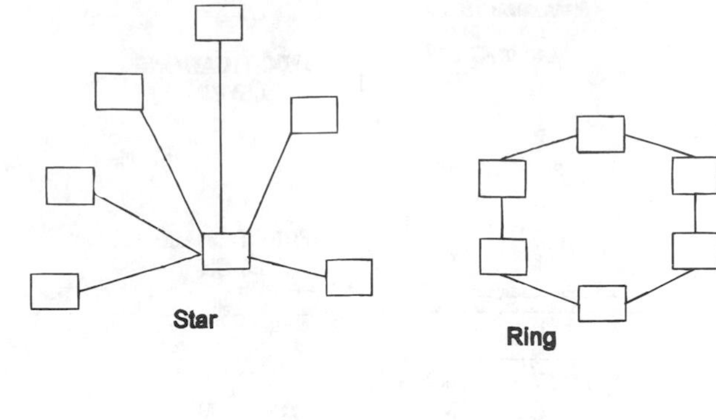

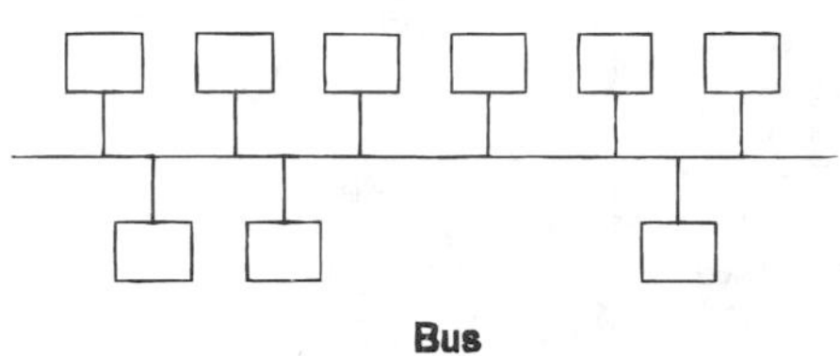

Figure 46.13 Network topologies. *Source*: Olsson, G. and Piani G. 1992. *Computer Systems for Automation and Control*, Prentice Hall, Englewood Cliffs, NJ. With permission.

FMS Equipment

A flexible manufacturing system would normally consist of:

- Two or more work stations with computer-controlled machine tools.
- An automated materials handling system for transporting the work-in-process inventory.
- Mechanisms for transferring the work-in-process inventory between the transportation systems and the machine tools.
- Storage by an automated storage and retrieval system for the work-in-process inventory and tooling.
- Central computer control of the entire process.

The manufacturing plant in an FMS will therefore consist of two types of capital equipment—primary and secondary (Figure 46.24). The primary equipment adds value to the raw material being manufactured. Secondary equipment is used to support the primary equipment in accomplishing its goal. The primary equipment consists of work centers, which physically machine a piecepart, and process centers, which assemble, inspect, test, clean, etc., the machine part.

Primary capital equipment
Work centers:

- CNC turning machines;
- machining centres;
- grinding machines;
- etc.

Process centers:

- robotic workstations;
- coordinate measuring machines;
- wash machines;
- communication system;
- etc.

Secondary capital equipment
Support systems:

- pallet, flexible fixture, load/unload stations;
- machine tool magazines/setting area;
- local metrology area;
- etc.

Support equipment:

- robots
- stores for pallet/fixture;
- pallet buffer stations;
- tool stores;
- raw material stores;
- transport system (AGVs, robots)
  –tooling;
  –pieceparts;
  –transport units;
- etc.

Types of FMS

There are a variety of architectures used in the construction of types of FMS. Parish (1993) has classified five types of FMS:

- Sequential FMS.
- Random FMS.
- Dedicated FMS.
- Engineered FMS.
- Modular FMS.

A *sequential FMS* manufactures one component batch type; then, planning and preparation is carried out for the next component batch type to be manufactured. It operates similarly to a small batch flexible transfer line.

A *random FMS* manufactures any random mix of component types at any one time.

A *dedicated FMS* continually manufactures, for extended periods, the same but a limited mix of component batch types.

The *engineered FMS* is common for the very early FMSs. This FMS type manufactures the same mix of components throughout its lifetime. It is a bespoke solution for an FMS user.

A *modular FMS,* with a sophisticated FMS host, enables an FMS user to expand FMS capabilities in a stepwise manner into any of the above four types of FMS.

The following are a few examples of modern and old FMS installations successfully developed in the past.

- The Hattersley Newman Hender FMS, installed to manufacture high- and low-pressure bodies and caps for water, gas and oil valves.

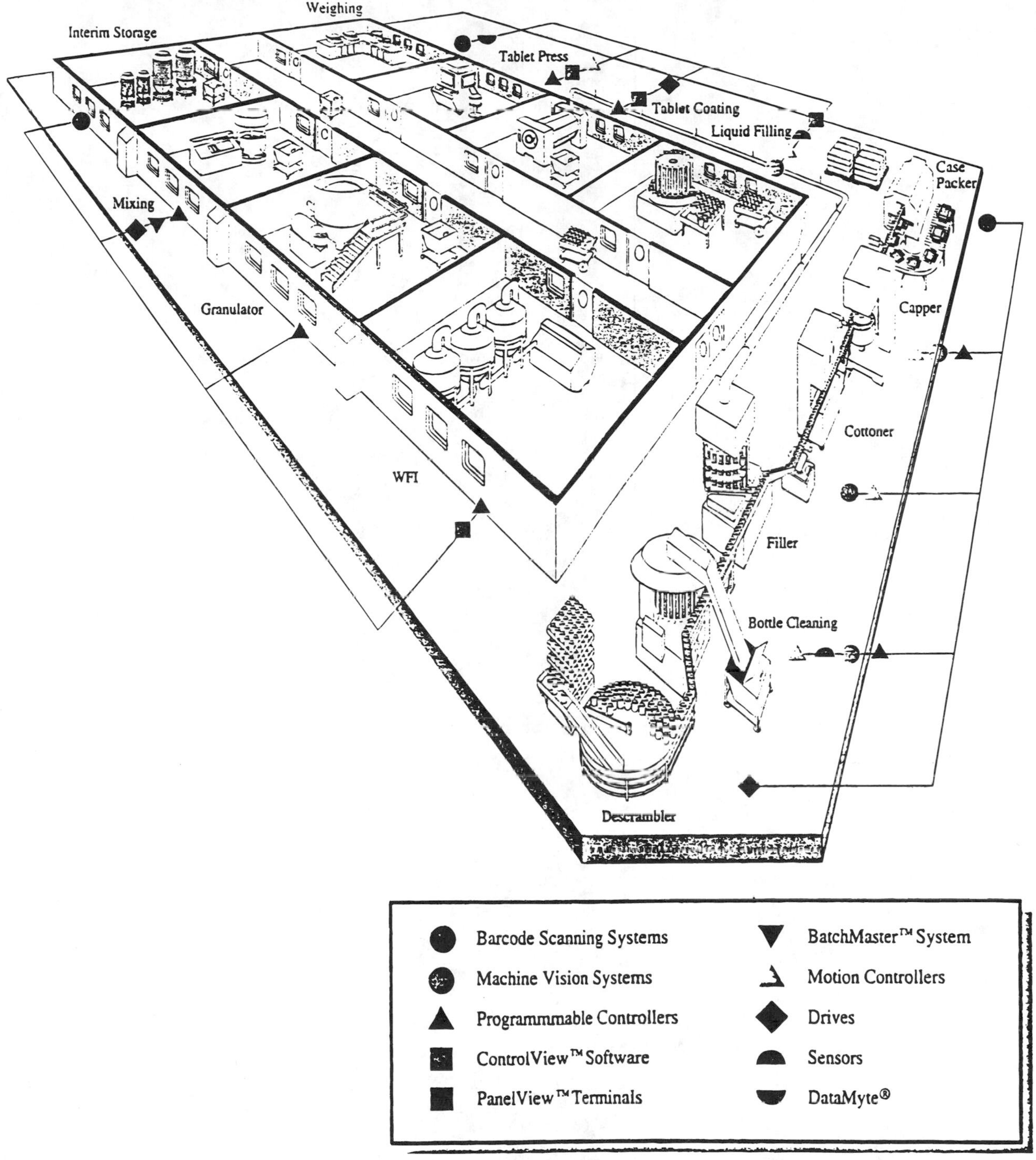

Figure 46.14 Allen-Bradley automation solution for a pharmaceutical plant. *Source*: Allen-Bradley, 1993. *Proven Ingredients for World-Class Manufacturing*, Report 6413, August, Allen-Bradley Company, Milwaukee, WI. With permission.

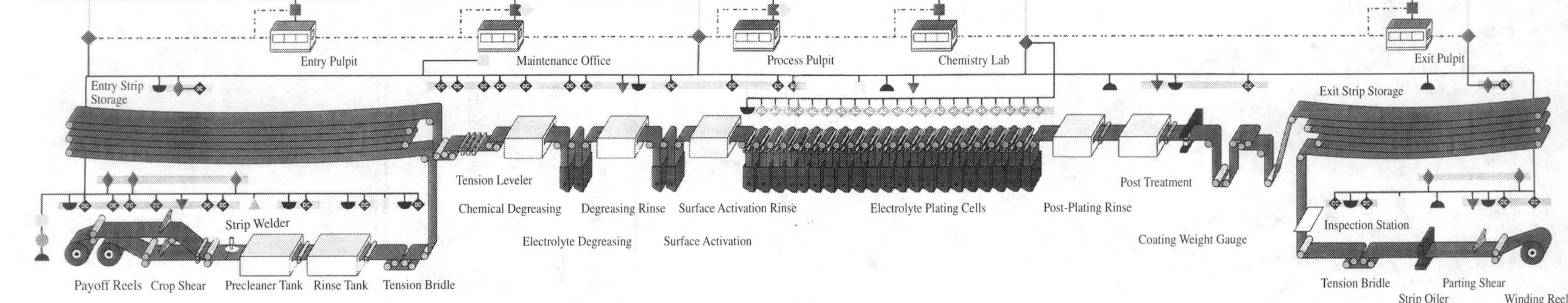

Figure 46.15 Allen-Bradley automation solution for the metal industry. *Source*: Allen-Bradley, 1993. *Put Us Through the Mill,* Publication 6414, October, Allen-Bradley Company, Milwaukee, WI. With permission.

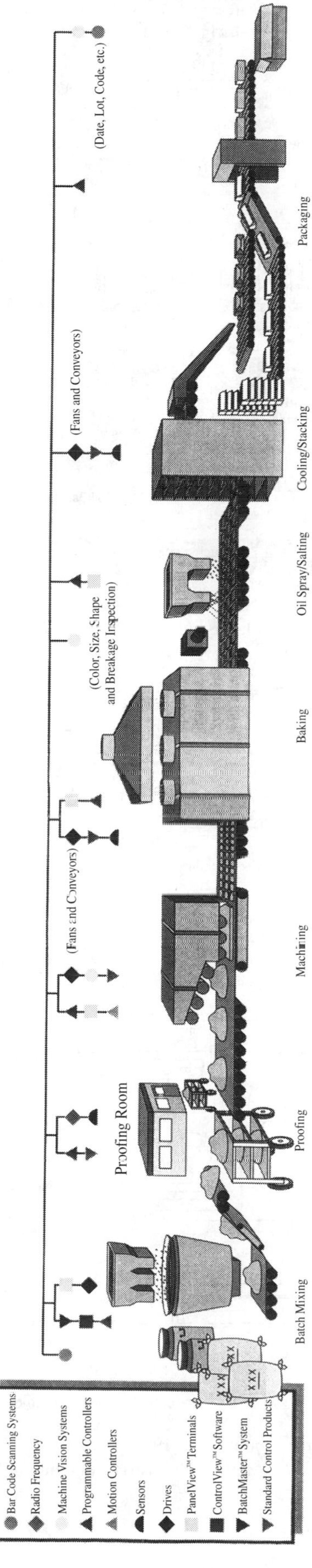

Figure 46.16 Allen-Bradley automation solution for the food industry. Allen-Bradley, 1992. *Appetizing Opportunities to Increase Productivity and Reduce Time to Market*, Publication 6461, November, Allen-Bradley Company, Milwaukee, WI. With permission.

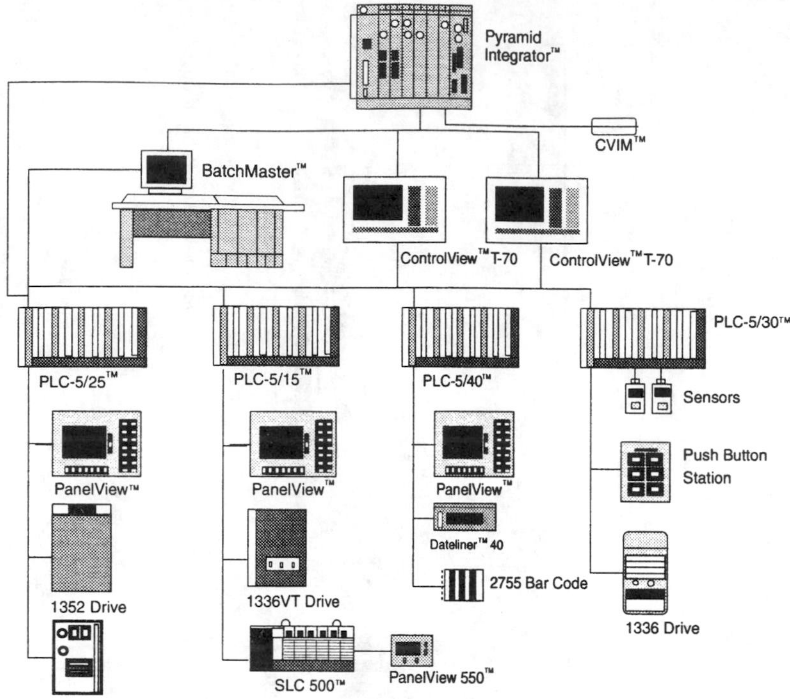

Figure 46.17 Allen-Bradley factory automation system. *Source*: Allen-Bradley, 1993. *Proven Ingredients for World-Class Manufacturing,* Report 6413, Allen-Bradley Company, Milwaukee, WI. With permission.

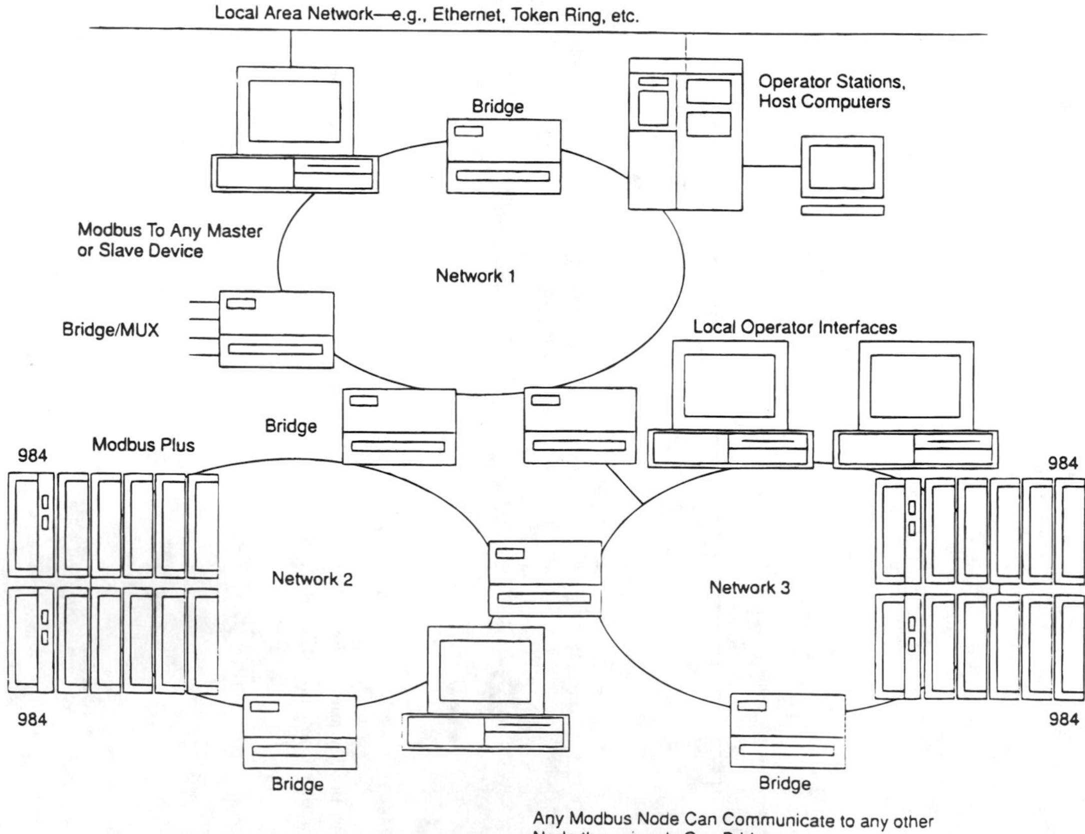

Figure 46.18 Modicon factory automation solution. *Source*: Edmonds, C. 1992. *Modbus Plus Networking: Dual Cables, Redundancy, and Multiple Networks,* Modicon Application Notes GM APPL 001, 159–168, Modicon Industrial Automation Systems, North Andover, MA. With permission.

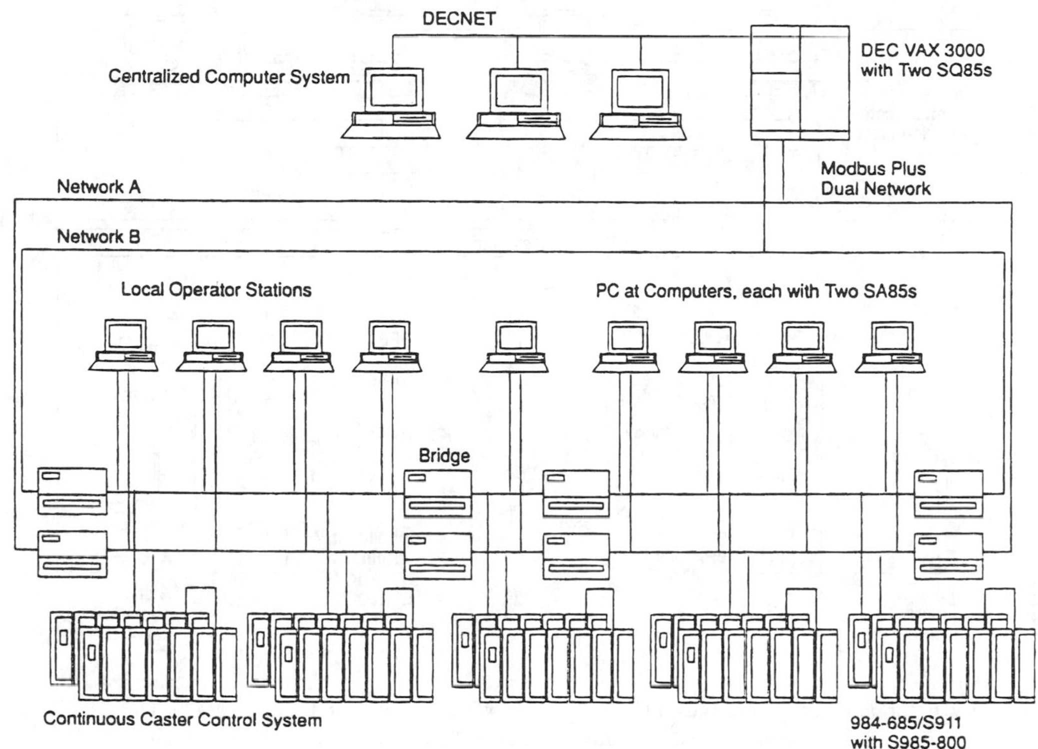

Figure 46.19 Dual redundant Modbus Plus networking. *Source*: Edmonds, C. 1992. *Modbus Plus Networking: Dual Cables, Redundancy, and Multiple Networks,* Modicon Application Notes GM APPL 001, 159–168, Modicon Industrial Automation Systems, North Andover, MA. With permission.

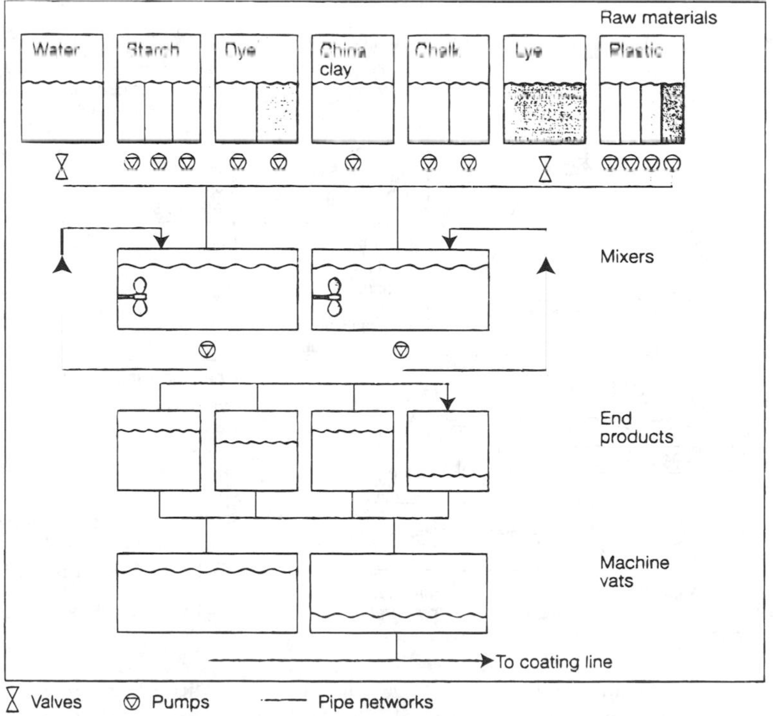

Figure 46.20 Plant layout at Maastricht. *Source*: Miebs, J. and Sonst, H. 1991. Flexible recipe handling and batch control at KNP Maastricht, *Engineering and Automation*, Vol. xiii, 6/9, 26–29, Siemens, Fürth, Germany. With permission.

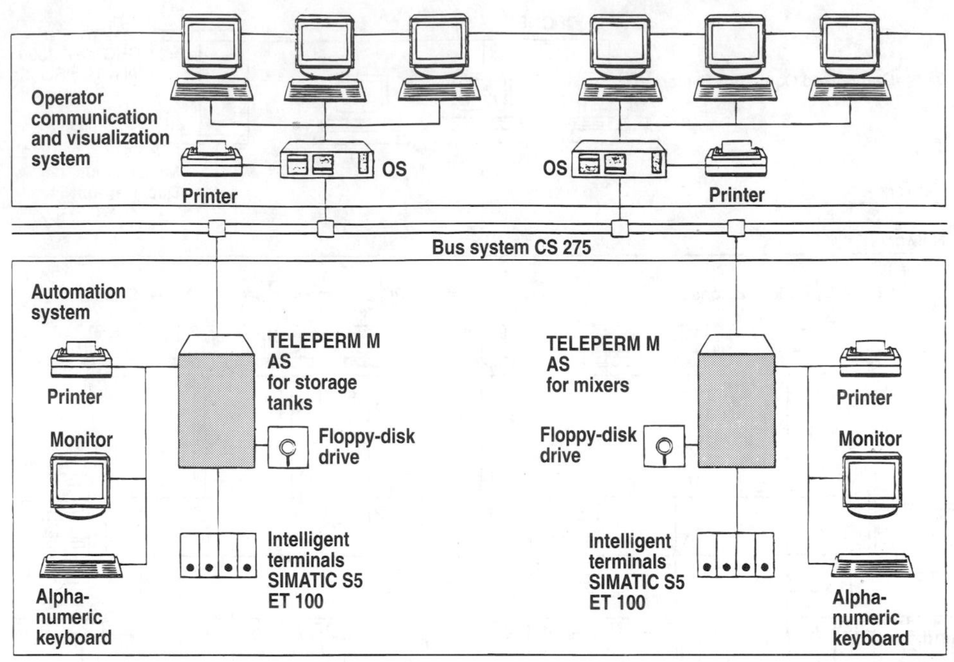

Figure 46.21 An overview of the automation system used by KNP. *Source*: Miebs, J. and Sonst, H. 1991. Flexible recipe handling and batch control at KNP Maastricht, *Engineering and Automation*, Vol. xiii, 6/9, 26–29, Siemens, Fürth, Germany. With permission.

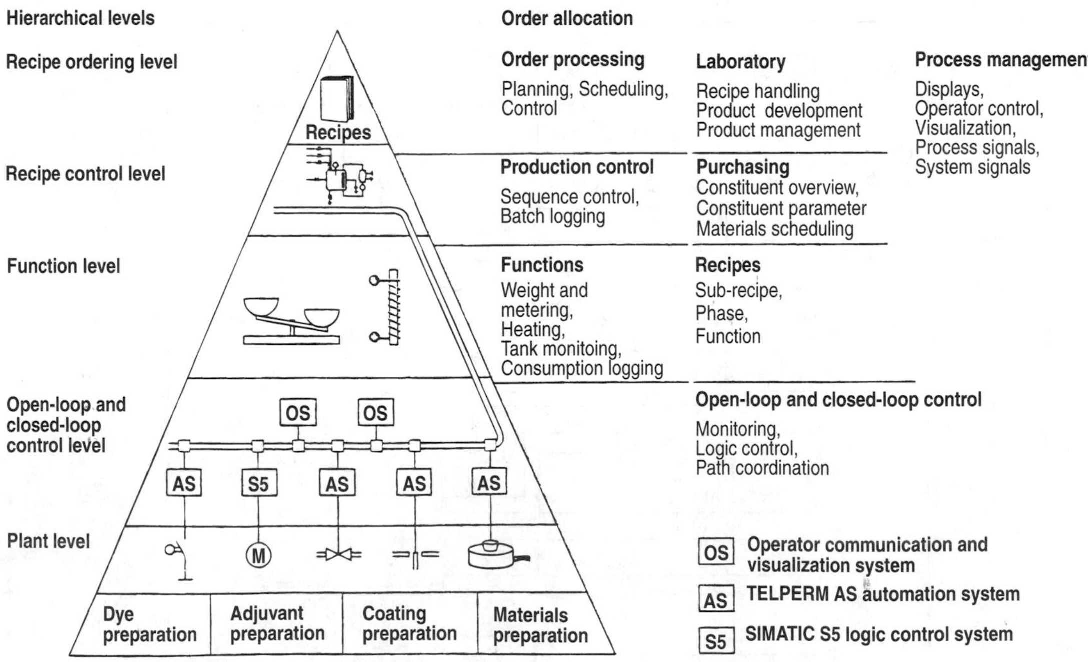

Figure 46.22 A hierarchical structure of the SIDRAS DCS 5000 SM 90 software. *Source:* Miebs, J. and Sonst, H. 1991. *Flexible Recipe Handling and Batch Control at KNP Maastricht,* Engineering and Automation, vol. xiii, 6/9, 26–29, Siemens, Fürth, Germany. With permission.

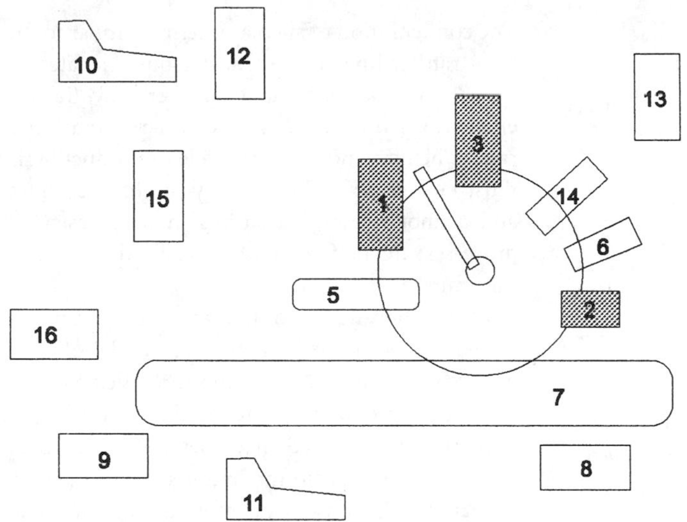

Figure 46.23 The layout of a typical FMS.

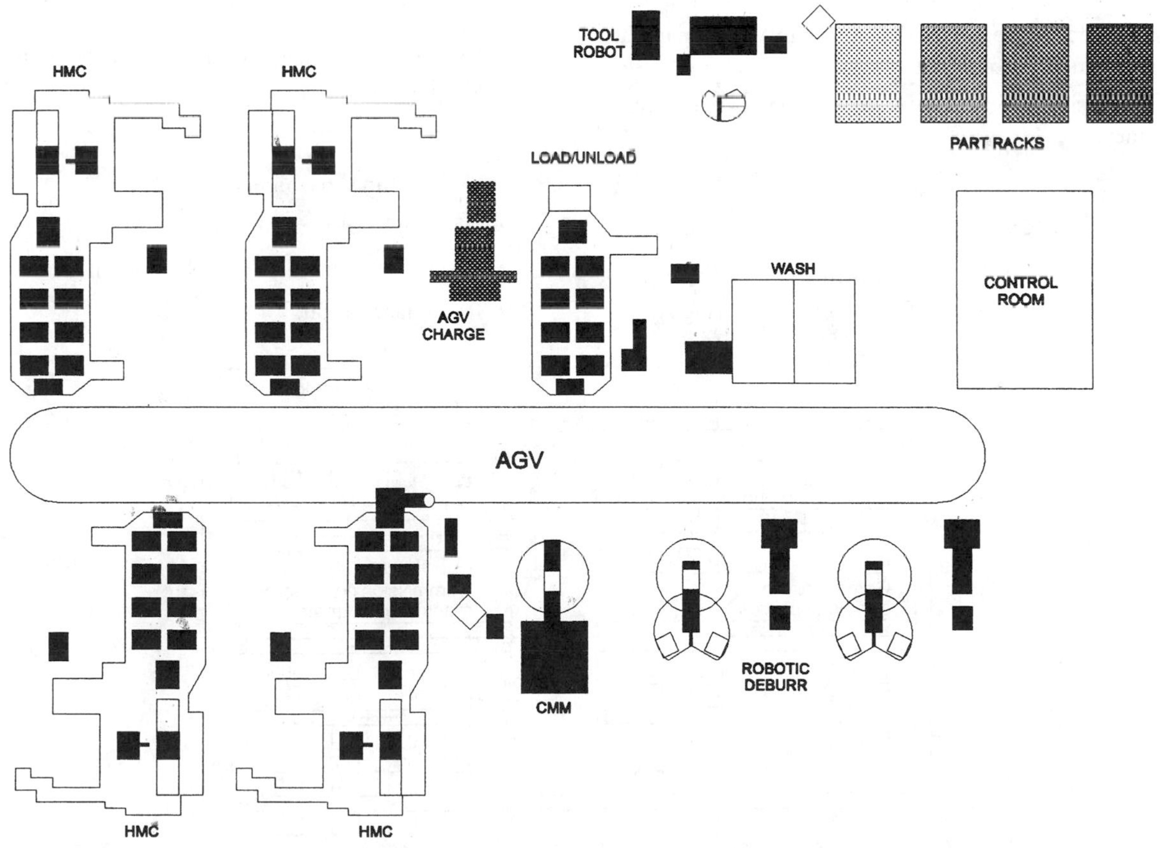

Figure 46.24 The equipment in a typical FMS.

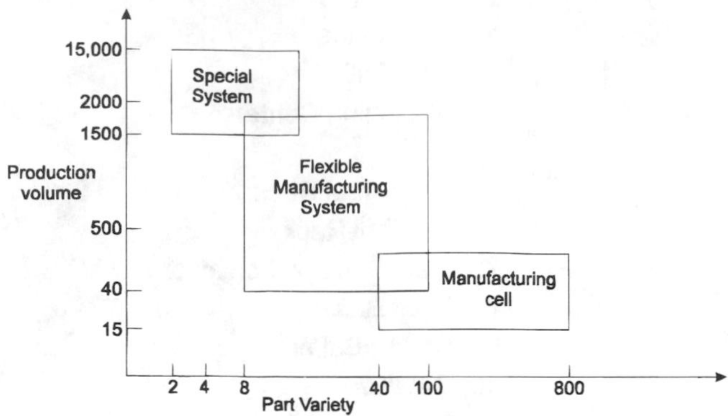

Figure 46.25 FMS production modes. *Source*: Groover, M. P. and Zimmens, E. W. 1984. *CAD/CAM: Computer-aided Design and Manufacturing*, Prentice-Hall, Englewood Cliffs, NJ. With permission.

- The Rover LM-500 FMS, for the manufacturing of 16-valve cylinder heads.
- The Vickers FMS, to manufacture a range of automobile power-steering servo-pumps.
- The Kraftwerk Union FMS, to manufacture turbine blades.
- The Makino Atsugi Plant—a monolithic FMS to manufacture parts for machine tools.
- The TI Machine Tools Cellular Turning FMS, to manufacture automobile components.
- The Hitachi Hybrid FMS, to manufacture electronic products.

Production Modes

The three modes of operations are shown in Figure 46.25.

The special system is the least flexible manufacturing system. The configuration of such a system is similar to the high production transfer line. These transfer lines (located mainly at levels 0 and 1 of the manufacturing hierarchy, Table 46.1) are very efficient when producing parts in large volumes at high output rates. The limitation of this mode of production is that the parts must be identical. These highly mechanized lines are inflexible and cannot tolerate variations in part design. The variety of processes are limited, and specialized machine tools are not uncommon.

At the opposite end of the mid-volume range is the manufacturing cell, positioned on levels 0, 1 and 2 of the manufacturing hierarchy, Table 46.1. It is the most flexible but generally has the lowest production rate of these three types. Stand-alone numerically controlled machines (NC machines, level 0, Table 46.1) are commonly found in this set-up as they are ideally suited for variation in workpart configuration. These machine tools are most appropriate for job shop and small batch manufacturing, because they can be conveniently reprogrammed to deal with product change-overs and part design changes.

The flexible manufacturing system covers the medium part variety and medium production volume range. A typical FMS will be used to process several part families, with 4 to 100 different part numbers being the usual case. Production rates per part would vary between 40 and 2000 per year. FMS offers the most challenging application of computer control to automate batch production manufacturing.

The FMS philosophy has been applied chiefly in the following manufacturing areas:

- Metal cutting machinery.
- Metal and sheet metal forming.
- Assembly (mechanical and electronic).
- Textile fabrication.

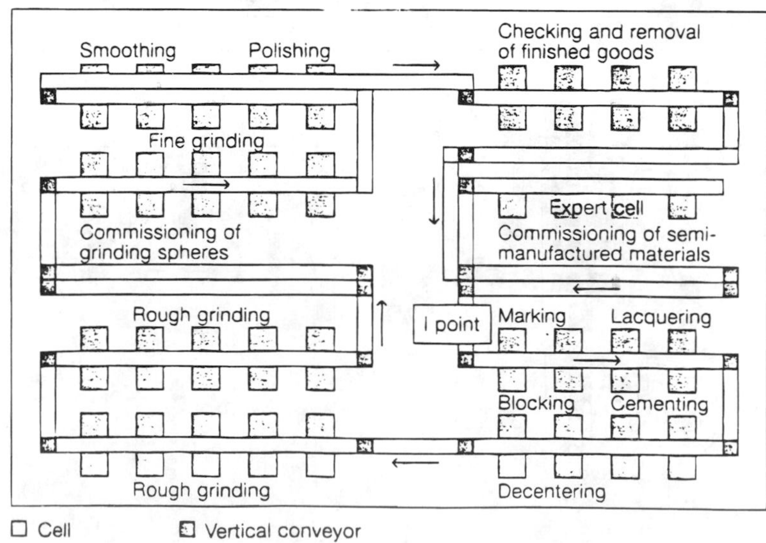

Figure 46.26 Schematic of materials flow between the individual operations. *Source*: Femppel, P., Rieder, G., and Stocker, A. 1994. Flexible manufacture of spectacle lenses at Zeiss in Aalon, *Engineering and Automation*, Vol. xvi, 1/94, 24–27, Siemens, Fürth, Germany. With permission.

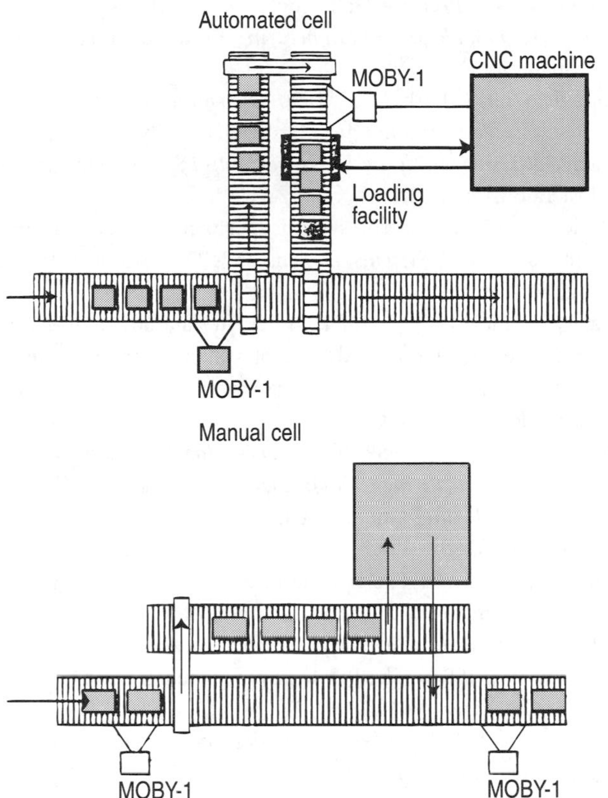

Figure 46.27 The design schematic of automatic and manual cells. *Source*: Femppel, P., Rieder, G. and Stocker, A. 1994. Flexible manufacture of spectacle lenses at Zeiss in Aalon, *Engineering and Automation*, Vol. xvi, 1/94, 24–27, Siemens, Fürth, Germany. With permission.

Its concept has also been applied to:

- Joining (including welding, glueing).
- Surface treatment.
- Inspection.
- Testing.

By far, the most widespread use of FMS technology has been in the machining areas of production.

Implementation

The following is a short description of the flexible manufacture of spectacle lenses at Zeiss in Aalen, Germany (Femppel, Rieder and Stocker, 1994).

The manufacture of spectacle lenses to prescription requires great flexibility in planning and executing customer orders. Optics specialists at Zeiss tackle this problem using computer-coordinated materials flow with distributed control logic—implemented with components and know-how from Siemens, Germany.

Figure 46.26 shows the schematic of materials flow between the individual operations. Figure 46.27 presents the design schematic of automatic and manual cells. MOBY-1 from Figure 46.27 is a mobile data carrier which contains all the processing and directional logic information.

Figure 46.28 presents an automation solution based on 6 programmable controllers and MOBY-1, linked to each other and to the production control computer.

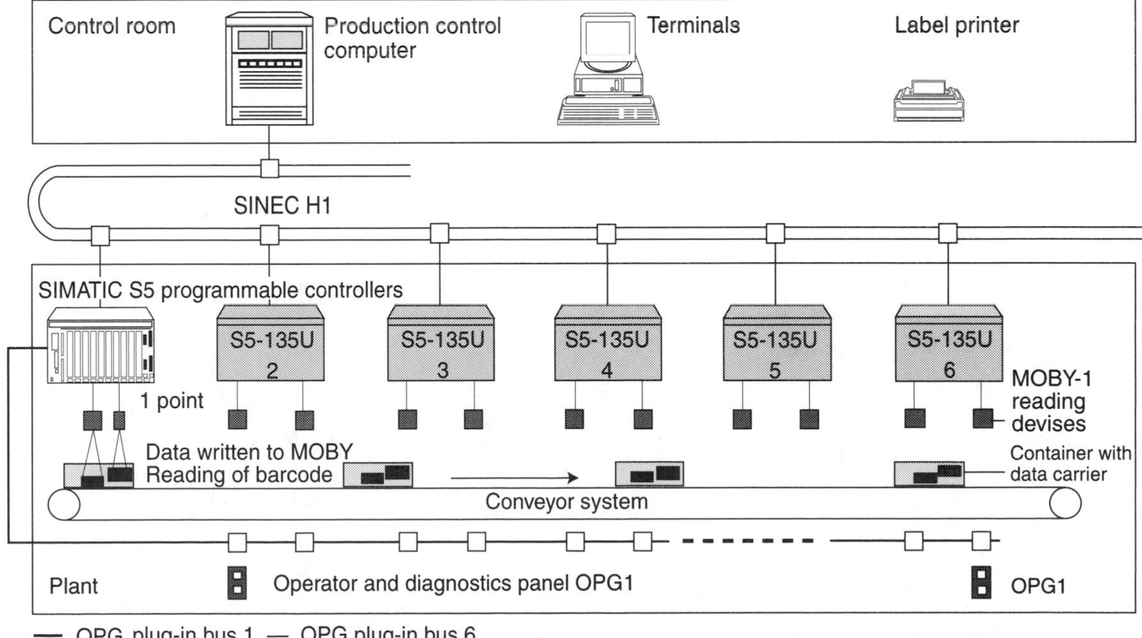

Figure 46.28 The overall plant automation solution. *Source*: Femppel, P., Rieder, G. and Stocker, A. 1994. Flexible manufacture of spectacle lenses at Zeiss in Aalon, *Engineering and Automation*, Vol. xvi, 1/94, 24–27, Siemens, Fürth, Germany. With permission.

References

Allen-Bradley, 1992. *Appetizing Opportunities to Increase Productivity and Reduce Time to Market,* Publication 6461, November, Allen-Bradley Company, Milwaukee, WI.

Allen-Bradley, 1993. *The Role of Control Systems in Customer-Driven Manufacturing,* Publication 5648, March, Allen-Bradley Company, Milwaukee, WI.

Allen-Bradley, 1993. *Proven Ingredients for World-Class Manufacturing,* Publication 6413, August, Allen-Bradley Company, Milwaukee, WI.

Allen-Bradley, 1993. *Put Us Through the Mill,* Publication 6414, October, Allen-Bradley Company, Milwaukee, WI.

Edmonds, C. 1992. *Modbus Plus Networking: Dual Cables, Redundancy, and Multiple Networks,* Modicon Application Notes GM APPL 001, 159–68, Modicon Industrial Automation Systems, North Andover, MA.

Femppel, P., Rieder, G., and Stocker, A. 1994. Flexible manufacture of spectacle lenses at Zeiss in Aalen, *Engineering and Automation,* vol. xvi, 1/94, 24–27, Siemens, Fürth, Germany.

Groover, M. P. and Zimmens, E. W. *CAD/CAM: Computer-aided Design and Manufacturing,* Prentice-Hall, Englewood Cliffs, NJ.

Miebs, J. and Sonst, H. 1991. Flexible recipe handling and batch control at KNP Maastricht, *Engineering and Automation,* vol. xiii, 6/91, 26–29, Siemens, Fürth, Germany.

Olsson, G. and Piani, G. 1992. *Computer Systems for Automation and Control,* Prentice Hall, Englewood Cliffs, NJ.

Parish, D. 1993. *Flexible Manufacturing,* Butterworth Heinemann, Stoneham, MA.

Popovic, D. and Bhatkar, V. P. 1990. *Distributed Computer Control For Industrial Automation,* Marcel Dekker, New York, NY.

Vlacic, Lj. 1988. *Multicriteria-based Control Systems Design,* Svjetlost Publishing, Sarajevo.

Vlacic, Lj. 1989. Control systems performance evaluation, in *Process Control Systems Design,* 257–272, Svjetlost Publishing, Sarajevo.

Vlacic, Lj. and Matic, B. 1984. Multicriteria analysis of dynamic properties of hierarchical control system topologies, in *Real Time Control of Large Scale System,* 124–130, University of Patras Press, Patras, Greece.

Williams, T. J. (ed.), 1989. *A Reference Model for Computer Integrated Manufacturing (CIM): A Description from the Viewpoint of Industrial Automation,* Instrument Society of America, Triangle Park, NC.

Williams, T. J. 1991. The Purdue Enterprise Reference Architecture, Report No. 154, Purdue Laboratory for Applied Industrial Control, Purdue University, West Lafayette, IN.

Williams, T. J. 1994a. The Purdue Enterprise Reference Architecture, *Computers in Industry,* 24, 141–158.

Williams, T. J. (ed.), 1994b. *A Guide to Master Planning and Implementation for Enterprise Integration Programs,* Report No. 157, Purdue Laboratory for Applied Industrial Control, Purdue University, West Lafayette, IN.

47

Production Management Architecture

47.1 Introduction ... 653
47.2 Production Management in the Sixties and Beyond ... 654
47.3 Components of the Hierarchical Production Management System ... 654
47.4 Long-Term Production Plan (LTPP) ... 655
General Management • Finance Department • Marketing Department • Production Department • Purchasing Department • Sales Department • Human Resource Department • Decisions under the Responsibility of the General Management • Decisions under the Responsibility of the Finance Department • Decisions under the Responsibility of the Purchasing Department • Decisions under the Responsibility of the Sales Department • Decisions under the Responsibility of the Human Resources Department
47.5 Master Production Scheduling (MPS) ... 656
Choice of the Horizon • Updating Period • Choice of the Types of Products to be Considered
47.6 Capacity Requirement Planning (CRP) ... 658
47.7 MRP Philosophy ... 658
Bill of Material
47.8 Application of the MRP ... 662
47.9 Conclusion ... 662

Rakesh Nagi
State University of New York at Buffalo

Jean-Marie Proth
Inria

47.1 Introduction

Production management is a difficult subject, mainly because of the following aspects of the production environment:

- Forecasting customers' requirements leads to results which are usually unreliable or, at least, not precise. This tendency is becoming more prevalent day after day, due to the uncertainty of the market.
- Delivery times of finished products are decreasing, due to worldwide competition. As a consequence, these times often become smaller than the sum of the manufacturing and the restocking times, which increases the difficulty of managing raw material inventories.
- The quantity of information to be handled has grown drastically during the past twenty years, due to the increasing complexity of the products. During the same period and for the same reason, the number of specialists involved in the production process has increased in the same proportions.

The above reasons do not constitute an exhaustive list of the difficulties encountered in production management, but they are probably among the most important ones.

To face these difficulties, three types of resources are required:

1. One powerful computer (or more) to handle the huge volume of information concerned.
2. A well-designed system to capture the right information in the right form, at the right time.
3. Well-trained specialists to take part in the management process. They are expected to have a solid background in computer sciences and applied mathematics, and a strong ability to understand and analyze production systems.

Above all, a stable and well designed structure to support the various components of the production management system is needed. This structure is usually a hierarchical structure. The goal of this section is to present the most commonly used of these structures.

Hereinafter, a gradual approach is adopted to establish the advantages of this hierarchical structure. It begins with providing some insight into the history of production management.

0-8493-8343-9/97/$0.00+$.50

47.2 Production Management in the Sixties and Beyond

The production management methods used until the beginning of the 1960s were based on the management of the inventories of raw materials, semi-finished products and finished products. These inventories were managed independently of each other, based on forecast requirements associated with each inventory. They were derived from the past sequence of requirements of this inventory, while taking into account the inventory and the backlogging levels at the instant the decision-making (DM) process started. Several inventory management approaches were used at that time, for instance the computation of the quantity to order at given points in time, or the definition of a minimal inventory level which triggered the ordering process.

It should be noted that these approaches, which consisted of a set of local management decisions, were the only way to manage a production system at that time, since the computers available where weak and unable to store and handle a huge volume of data.

The drawback with this approach is that it assumes a steady consumption and ignores links between the components of the same part. This may result in huge inventory levels for some components while others are backlogged, and lost efficiency of the safety stock associated to other components.

To illustrate this last aspect, assume that a given product is obtained by assembling five components, and that each safety stock is computed to satisfy the demand with a service ratio of 97%. In that case, the probability to be able to assemble a unit product at any time is $(0.97)^5 \approx 0.86$. Thus, the probability to be able to assemble a unit of product is close to 86%, which is far from adequate to guarantee the effective operation of the production system.

In the mid-1960s, a new philosophy was introduced to take advantage of the new possibilities of computers. This philosophy is still known under the acronym MRP which stands for material requirement planning. Since the MRP philosophy improved during the years, we now use MRP1 instead of MRP, and refer to the most recent issue of the MRP philosophy as MRP2, which stands for manufacturing resources planning. MRP2, unlike MRP1, attempts to take into account the capacity of the system.

The logic of the MRP approach is to decompose a product into a product structure bill-of-material (BOM) and to define, using explosion calculus, the requirements (quantities and timing) of components and raw materials. A complete description of MRP can be found in Orlicky (1975), and we will provide the basis of MRP in the next sections. Hereinafter, MRP will also be used to refer to MRP2.

Recently, some successful trials have been made to combine MRP2 and Japanese production management techniques. They are refered to as short cycle MRP.

The MRP approach and its successive improvements require increasingly powerful computers, but they are still limited. The limitation of the most common MRPs will be analyzed below.

47.3 Components of the Hierarchical Production Management System

At the highest level of the hierarchy, that is the strategic level, is the business plan (BP). It concerns the very long term (between 1 to 5 years, depending on the type of production performed). The goal of the BP is to orient long-term global production. Decisions are made on the families of products. The BP usually works on a rolling horizon. For instance, if the BP concerns a five year horizon, and the decisions are made for three-month periods, it may be reconsidered every three months for the next five years, to account for the random events which may have occured since the last review of the BP.

The BP includes the long-term production planning (LTPP) and the master production scheduling (MPS).

The LTPP elaborates a production plan whose goal is to define the quantities of each product family to manufacture at the horizon of the BP on each period of three to six months, depending on the horizon of the BP. This allows a rough evaluation of the cost of the resources required to meet the production decided before. This new activity is known as the long range resource planning (LRRP). The global production forecast and the financial evaluation of the required resources are gathered in the production plan, which concerns the strategic level.

The LTPP is the basis of the master production scheduling (MPS). The MPS can be considered as the translation of the LTPP into specific programs on a horizon which is smaller than the horizon of the BP. The periods on which decisions are made is also shorter than the periods used for the LTPP. The two main functions of the MPS are the medium range production and assembly planning (MRPAP) and the rough cut capacity planning (RCCP). The MPS in the interface between the strategic level and the tactical level. It is the most important input to the capacity requirement planning (CRP).

In turn, MPS and CRP, as well as the bill-of-material (BOM) and the state of the system, that is the levels of raw materials and work-in-process (WIP), are the basis of the MRP. The goal of the MRP is to decompose the BOMs of the parts to be manufactured and to decide how many of these components should be manufactured on each of the next elementary periods (days or hours), taking into account the customers' requirements, the WIP, and the (rough cut) capacity of the system. Often lot-sizing of components is also performed along with the explosion calculus. The result of the MRP approach is a list of operations to be performed on each resource (or on each group of resources) during each elementary period. This result is not a schedule, since it does not provide the order in which these operations must be performed.

The MRP belongs to the tactical level.

The general structure of the hierarchical production management system is given in Figure 47.1. As we can see in Figure 47.1, when the MRP approach fails for lack of capacity, we may have to return to the MPS level to reconsider the RCCP or the MRPAP.

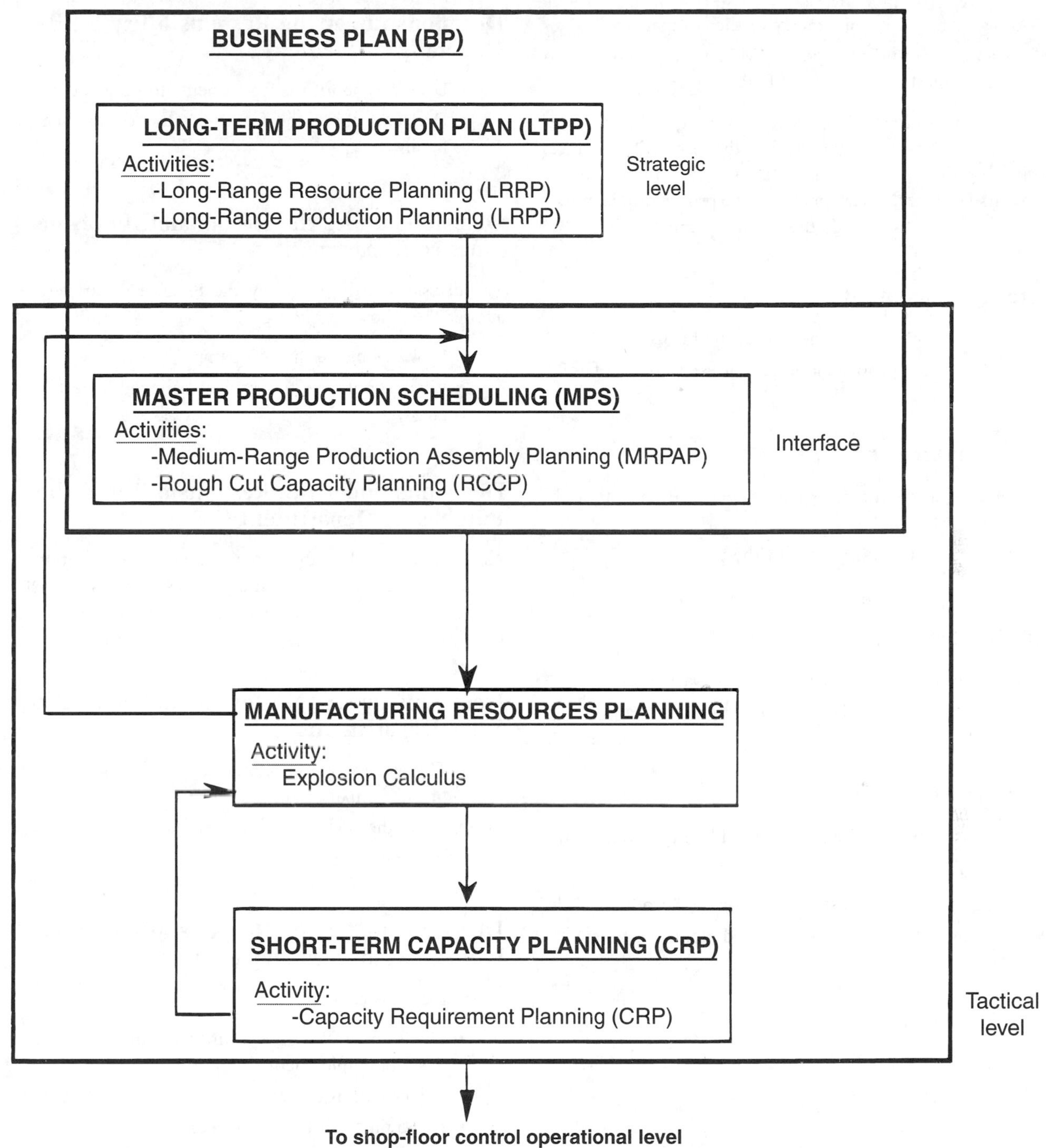

Figure 47.1 General structure of the hierarchical production management system.

We will now consider the different levels of the hierarchy and present in detail what is performed at each of these levels.

47.4 Long-Term Production Plan (LTPP)

The goal of the long-term production planning is to define the global production level, i.e., the quantity of each product family to manufacture on each period of the long term. The long term is usually one to five years, and a period is one to six months, depending on the type of company considered.

The long-term production planning takes into account the decisions made by the management of the company. Among the set of common decisions are long-term capacity planning (procurement of new resources and hiring) and long-term production smoothing (by using inventories to avoid frequent hiring/firing), to quote only two.

The LTPP obtained as output at this level is

- An agreement on the strategy of the company.
- An authorization given to the MPS to act according to the constraints fixed by the LTPP.

The acceptance of the LTPP is the responsibility of the management of the company. Its design involves the responsible finance, marketing, production, purchasing, sales and human resource departments. Each of the participants in the production planning activity is supposed to provide the following information.

General Management

- Forecast of the capacity needed at the horizon of the PP.
- Forecast of the profits and losses for each product family.

Finance Department

- Schedule of the future expenses decided by the production department.
- Evolution of the WIP levels in the past.

Marketing Department

- Evolution of the market needs and capacities.
- Evolution of the market plan based on competition.

Production Department

The following statistical information should be provided by the production department:

- Average production cost per unit for each product family.
- Percentage of global capacities used along the time.

Purchasing Department

- Evolution of the different components and raw materials purchased from outside.

Sales Department

- Evolution of historical sales volumes.

Human Resource Department

This department provides statistical information about:

- The evolution of the use and cost of man power.
- The evolution of the overtime.

Starting from the above information provided by each of the departments involved in the production planning, the activity consists of reaching a common decision on the following points:

Decisions under the Responsibility of the General Management

- Decisions about the investments to be made.
- Decisions about the future production of each product family.

Decisions under the Responsibility of the Finance Department

The decisions to be made by the Finance department should answer the following questions:

- How to finance the company?
- What should be the budget assigned to each product family?

Decisions under the Responsibility of the Purchasing Department

As a result of the LTPP, this department should have at its disposal a clear strategy concerning the purchasing costs as well as the acceptable delivery times. This will be the basis to negotiate with the suppliers.

Decisions under the Responsibility of the Sales Department

- The way the prices should evolve to balance the budget of the company.
- How the delivery times should evolve to face the competitors.

Decisions under the Responsibility of the Human Resources Department

These decisions concern:

- The future evolution of human resources in the company (level of employment, laying off).
- The evolution of overtimes.
- The training of the employees.
- The acceptable limits for a possible negotiation (in case of strike, for instance).

All the previous decisions are strategic decisions. This means that they are made in terms of general requirements and/or constraints, and each department is required to make its own decisions according to these requirements and/or constraints.

47.5 Master Production Scheduling (MPS)

The MPS is elaborated in the framework defined by the LTPP. We can consider the MPS as the translation of the LTPP in terms of production programs on quite short periods of time (a week, for instance).

The main characteristics of the MPS are

1. The horizon.
2. The length of the decision periods.
3. The updating period.
4. The types of products to be considered.

We investigate these characteristics in detail below.

Choice of the Horizon

The MPS is, in terms of the type of product or product family, that is, BOM or planning BOM, are considered. The usual rule to define the horizon of the MPS is the following:

"The horizon of the MPS is equal to twice the longest production cycle of the types of products considered."

The production cycle is the sum of three components:

The time needed to acquire the raw material and the components (purchasing times).

The time required to manufacture (set up, processing, queue and more time).

The time needed to prepare the products for shipping.

There are several reasons for this choice. This horizon is the minimum time required to develop some cooperation with the suppliers. It is also the minimum time required to perform the RCCP which consists of evaluating the feasibility of the MPS for, at least, the most important types of products to be manufactured, and adjusting the resources or the MPS if it turns out that the MPS is not feasible.

Three important remarks have to be made concerning the rule used to define the horizon:

- The production cycle is computed using purchasing, manufacturing and preparation times, which are those observed in the past: they include all the idle time during the production. As a consequence, the production cycle may be equal to several months.
- Even if the production cycle is equal to several months, it may be too short to adjust the long-term capacity of the system, since ordering a new machine, for instance, may need far more than one year. Thus, the rule proposed above should be considered as a proposal which can be modified according to the type of production and its environment.
- Within the horizon adjustment of worker levels and planning overtime should be possible to accomplish capacity adjustment in the medium term.

Choice of the Length of the Period

This value depends on the unit chosen to express the average manufacturing times. The week is often a good choice, but the period can be reduced to a day if the production cycle, and thus the horizon, is short. Note that, for a fixed horizon, the computation burden increases quickly as the length of the period decreases (i.e., number of periods increases). Note also that, in any case, the length of the period should be larger than the time needed to perform one operation on any of the products or components.

Updating Period

Two main reasons explain why it is necessary to update the MPSs. First, it is necessary to adjust the MPSs since endogenous, random events like machine breakdowns, quality problems, may perturb the execution of the former MPSs; second, we have to introduce the changes which occured in the customers' requirements, or in the delivery of raw materials or components (exogenous random events.)

Choice of the Types of Products to be Considered

The following rules are proposed by the American Production and Inventory Control Society (APICS):

- The BOM, that is, the detailed description of the raw materials and the components required, the operations to be performed, and the times associated to the previous tasks, should be known precisely.
- The number of BOMs selected must be quite small, since a large number of BOMs would make the system intractable.
- The demands (customers' requirements or forecasted requirements) should be known precisely for the selected BOMs.
- The selected BOMs should fully utilize the test resources selected, i.e., the resources which are considered as bottleneck resources.

The objective of these rules is to lead to a tractable set of data. The second rule is the most important, since it allows a quick and flexible computation of the MPS. In some companies, the model selected is known as the level of fewest configuration variables.

The goal of the MPS is, starting from the models designed according to the rules presented above, to define the products to be completed in each period and to evaluate the global capacities required to manufacture these products. Defining the quantities of products to be completed on each period is called medium range production and assembly planning (MRPAP).

This computation is based on the demands (forecast demands and customers' requirements) and on global capacities derived from statistics. The result of the MRPAP are the MPSs which contain, at least, for each product and each period:

- The sales forecasts.
- The customers' requirements.
- The planning of each product and each component; this planning consists of providing the period each product should be completed and the period each component should be delivered.

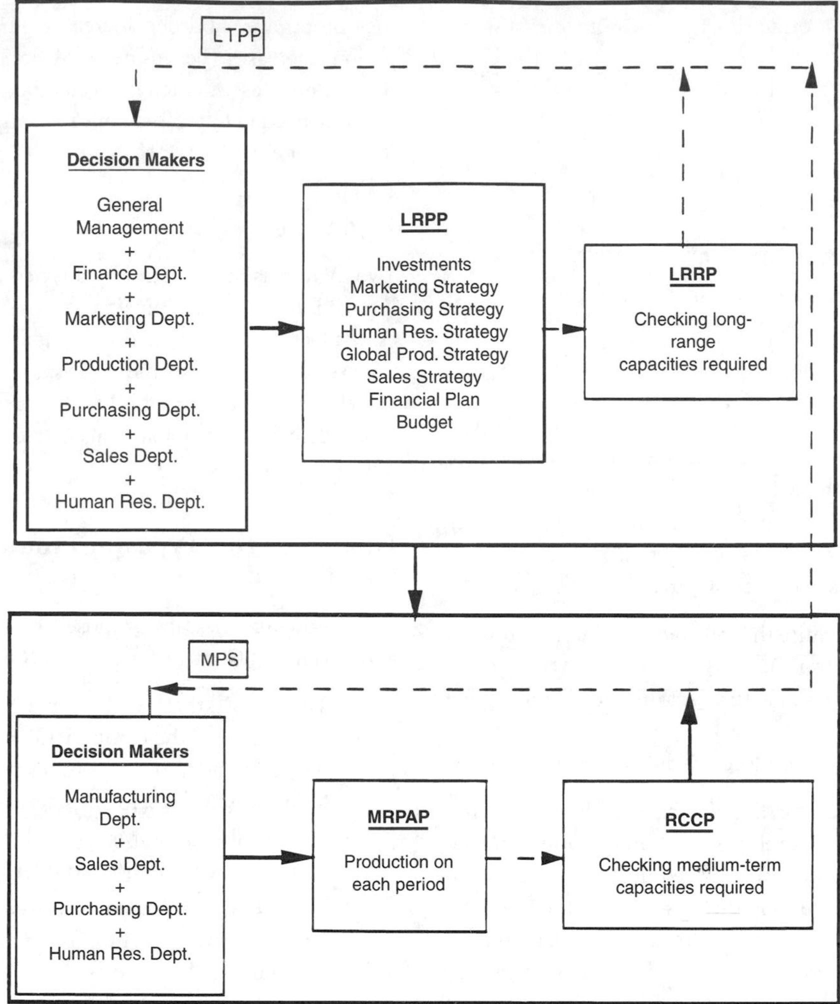

Figure 47.2 Long-term production plan and master production scheduling.

- The forecast availabilities, that is, the result of the following computation: Inventory level + Production decided − Customers' requirements.
- The production decided and which do not correspond to customers' requirements.
- The lot sizes.
- The safety stocks.

The capacities required to meet the production whose level has been defined by the MRPAP is derived from statistics. Using this information, each set of products to be manufactured is translated into resource loading on the horizon, or, on periods of the horizon.

The evolution of the capacities is the duty of the RCCP function. If some of the selected resources are overloaded, an adjustment is necessary. This adjustment is made by either adjusting resource levels or by modifying the production objectives.

The dynamics of the LTPP and the MPS are schematized in Figure 47.2. We notice that the results of the RCCP are transferred to the LTPP level in order to eventually adjust the information concerning the long-range capacities.

47.6 Capacity Requirement Planning (CRP)

This function is similar to the RCCP. The only difference is that the CRP applies on each detailed resource opposed to RCCP which may be performed over the bottleneck resources. The CRP can be considered as a part of the MRP.

47.7 MRP Philosophy

We first present the BOM which is a key input of MRP.

Bill-of-Material

The basic entity of the BOM is the *item*. An item is a component which may be purchased from suppliers or manufactured in the company. Semifinished products and finished products are items. A finished product is often called *end-item*. A BOM is associated with each end-item and to each item which are themselves assembled or manufactured semi-finished products. No BOM is associated with components purchased from suppliers.

Let A be an item obtained by assembling two items B, one item C and three items D. The BOM of A provides the whole information needed to know how to produce one entity A from items B, C and D, as shown in Figure 47.3.

The last column contains the name of the item considered. In turn, item C is obtained by assembling two components E and one component F, while D is obtained by performing operations on item G. The BOM of C and D are given in Figure 47.4.

Assuming that B, E, F, G are components purchased from the suppliers (or manufactured in a shop-floor which is managed independent of the shop floor considered), the structure of item A which is either an intermediate item (i.e., a semifinished product) or an end-item (i.e., a finished product) can be represented as in Figure 47.5.

End-item A is called the parent of items B, C and D. Similarly, C is the parent of E and F, and D is the parent of G. A parent is not necessarily an end-item. The representation given in Figure 47.5 is called a product structure tree. The drawback with this representation is its size when complex products are concerned.

Another representation of the BOM is the so-called indented bill-of-material. This representation is more convenient for computerizing and it progressively enumerates the branches of the product structure in a hierarchical manner, i.e., it identifies the level of the item and quantity required. Figure 47.6 presents the Indented Bill-Of-Material for the end item represented in Figure 47.5.

ITEM A		
Item	**Quantity**	**Name**
B	2	X
C	1	Y
D	3	Z

Figure 47.3 Bill-of-material of item A.

ITEM C		
Item	**Quantity**	**Name**
E	2	YX
F	1	YY

ITEM D		
Item	**Quality**	**Name**
G	1	ZX

Figure 47.4 Bill-of-material of items C and D.

Time-Phased Gross Requirements of Items

The end-items to be provided at the end of each period, as well as the lot or batch sizes, are given by the master production scheduling. The MRP method explodes each end-item quantity using the BOM and the item lead times defined hereafter. The result is the number of items of each type which should be available at the end of each period. An item can be made available by being manufactured in the company, thus being called a "make-item" or manufactured item. In this case, the lead time is the time needed to manufacture a reasonable number of such items. An item can also be delivered by a supplier, thus being called purchased item. In this case, the lead time is the time

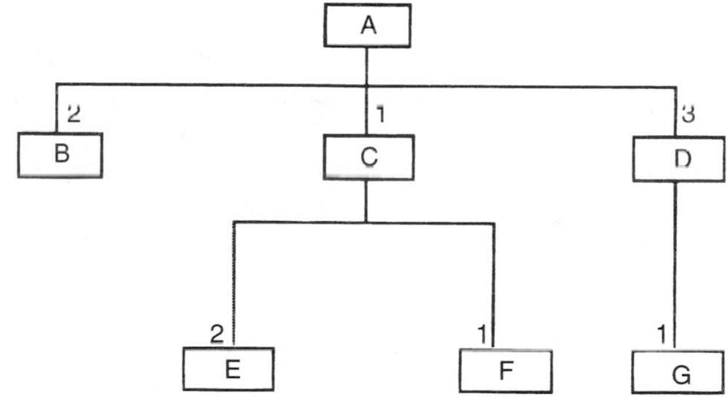

Figure 47.5 Product structure tree.

ITEM A			
Item level	**Item**	**Quantity**	**Name**
1	A	1	W
.2	B	2	X
.2	C	1	Y
..3	E	2	YX
..3	F	1	YY
.2	D	3	Z
..3	G	1	ZX

Figure 47.6 An indented bill-of-material.

Table 47.1 Lead Times of the Items Corresponding to End-Item A

ITEMS	A	B	C	D	E	F	G
LEAD TIMES	2	3	2	1	4	2	3

needed by the supplier to deliver a reasonable number of such items. This activity is known as the time-phasing activity.

An important comment should be made concerning manufacturing item lead times: they are extracted from the history of the production system, and this could be very far from the real manufacturing times; in fact, they include times spent waiting for a busy resource, times spent in queues and time spent waiting for or during material handling. They should be also large enough to allow providing not only one unit, but a number of products defined based on the number usually required. In other words, the lead times used to perform the time-phasing process are the results of the previous planning results. It should be noticed that this perpetuates the mistakes which may have been made in the past. For instance, an excessive lead time will be used indefinitely if, for some reasons, it has been introduced in the process. No warning mechanism exists to correct such a mistake.

The lead times are given in terms of number of periods. Let us consider, for instance, the end item A presented in Figures 47.4–6.

Assume that the lead times of the items are those shown in Table 47.1.

More recent MRP systems could employ dynamically computed lead-times, that are batch-size and possibly workload dependent. Assume also that some end-items A are due at the end of period 20. Then the result of the time-phasing activity is given in Figure 47.7. It is obtained by setting the completion time of A at the end of period 20, and by setting the components backward, according to the product structure tree given in Figure 47.5.

Most systems also provide offsetting capabilities, where all components of an assembly need not be available at the same time the assembly process begins. This allows some components to be further delayed and produced only when they are absolutely needed in the assembly steps.

Note that the horizon of the master production scheduling (MPS) should be greater than or at least equal to the longest cumulative lead time of the end items considered at the MPS level. In the example presented in Figure 47.7 for instance, the cumulative lead time is 8: it is the period which starts when items E starts to be manufactured (or is ordered), and ends when end-items A are completed.

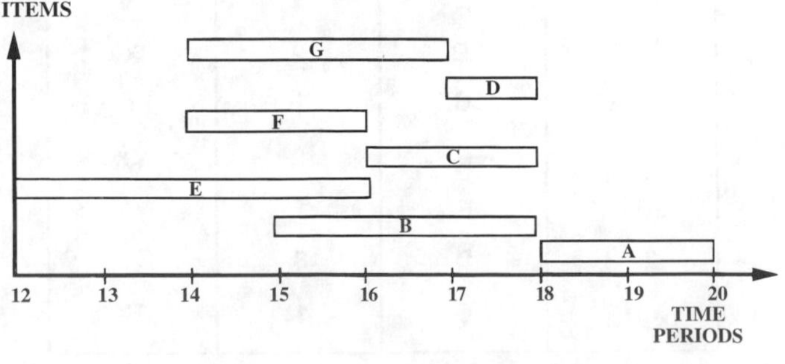

Figure 47.7 Time-phasing of the BOM.

As we can see, neither the number of end-items A to be made available, nor the number of items of various types required to complete end item A are taken into account in the time-phasing process. This means that the lead times considered are large enough to absorb any production or, in other words, to guarantee a sufficient flexibility of the management system to make it possible to face any customers' requirements. Note that the more flexible the management system, the more underutilized the resources. Although, as noted earlier, some recent systems have overcome this problem in different ways.

Assume that end-items A have been planned at the ends of periods 10 and 15 in quantities 60 and 100, respectively. These data are given by the MPS. The time-phasing corresponding to these requirements is given in Table 47.2, which is derived from the end-item quantities planned and the time-phasing of the BOM presented in Figure 47.7. Table 47.2 gives the number of items of each type required at the end of each period without considering existing inventories and scheduled receipts (i.e., without netting). A special column is provided for the backlogs of end-items, if any.

Time-Phased Net Requirement of Items

Gross requirements of items are the quantities which should be available by the end of the corresponding periods according to the MPS. These quantities may exceed the quantities to be produced, since items may be already available (i.e., in stock), or since some of them may have been ordered in the past (i.e., scheduled receipts). Thus, we have to derive the net requirements of items from their gross requirements. The time-phased net requirements of items corresponding to the gross requirements presented in Table 47.2 is given in Table 47.3.

The following hypothesis have been made to compute the data of Table 47.3:

1. The inventory levels of items A, B, C, D, E, F, G at the beginning of period 3 are respectively 20, 50, 0, 0, 0, 25, 100.
2. The scheduled receipts of all items is denoted as "On order" in the Table 47.3
3. The lead times are those given in Table 47.1

Table 47.2 Gross Requirements of Items for End-Item A without Level-by-Level Netting

Part and components	Lead times	Backlog	3	4	5	6	7	8	9	10	11	12	13	14	15
End item A	2									60					100
Items															
B	3							120					200		
C	2							60					100		
D	1							180					300		
E	4					120					200				
F	2					60					100				
G	3						180					300			

Table 47.3 The MRP Mechanism

Item			3	4	5	6	7	8	9	10	11	12	13	14	15
			Periods												
A	Lead time: 2	Required								60					100
		On order													
	On hand: 20	Available	20	20	20	20	20	20	20	0	0	0	0	0	0
	Lot size: 20	Net requ.								40					100
		Planned								40					100
		Release						40					100		
B	Lead time: 3	Required						80					200		
		On order				75									
	On hand: 50	Available	50	50	50	125	125	45	45	45	45	45	20	20	20
	Lot size: 25	Net requ.											155		
		Planned											175		
		Release								175					
C	Lead Time: 2	Required						40					100		
		On order													
	On hand: 0	Available	0	0	0	0	0	10	10	10	10	10	10	10	10
	Lot size: 50	Net requ.						40					90		
		Planned						50					100		
		Release				50					100				
D	Lead time: 1	Required						120							
		On order													
	On hand: 0	Available	0	0	0	0	0	0	0	0	0	0	0	0	0
	Lot size: 10	Net requ.						120					300		
		Planned						120					300		
		Release					120					300			
E	Lead time: 4	Required				100					200				
		On order		25											
	On hand: 0	Available	0	25	25	5	5	5	5	5	5	5	5	5	5
	Lot size: 10	Net requ.				75					195				
		Planned				80					200				
		Release	80				200								
F	Lead time: 2	Required			50					100					
	On order		25												
	On hand: 25	Available	25	50	50	0	0	0	0	0	0	0	0	0	0
	Lot size: 20	Net requ.									100				
		Planned									100				
		Release							100						
G	Lead time: 3	Required					120					300			
		On order													
	On hand 100	Available	100	100	100	100	5	5	5	5	5	5	5	5	5
	Lot size: 25	Net requ.					20					295			
		Planned					25					300			
		Release		25					300						

4. The lot sizes of items A, B, C, D, E, F, G, are, respectively, 20, 25, 50, 10, 10, 20, 25. The lot size of an item is the basic ordering unit or average batch-size. In other words, the system is only allowed to order a multiple of the lot. Most MRP systems also permit a lot-for-lot strategy in which exact requirements are produced; this is more prevalent in practice today.

The MRP process starts by the end-item, and more precisely by considering the gross requirement of the end-item. It uses the "On hand" data, which is the inventory level of the end item at the beginning of the first period (Table 47.3), the lot size and the lead time. The row denoted "Available" contains the inventory level. The "net requirement" row contains the quantities which remain to be produced to reach the gross requirement, taking into account the available quantities delivered to satisfy the net requirement. These quantities are computed taking into account the lot size. For instance, if the net requirement is 30 and the lot size 25, then the planned quantity will be 50 since we have to order at least twice the lot size to cover the net requirement.

The row "Release" contains the quantities ordered at each period. These quantities are the "Planned" quantities, translated according to the lead times.

Note that:

1. The quantity "on hand" can be negative in the case of backlog,
2. The quantities "on order" represent quantities ordered from outside, or quantities ordered from the shop-floor to replace product or components of bad quality. These quantities are provided to the MRP system, and are also referred to a scheduled receipts.

The MRP process continues to the next lower level, using the "Release" quantities computed at the end-item level to compute the "Required" quantities, taking into account the number of items required to produce one unit of end-item. For instance, for the example at hand, since the "Release" quantities of A are 40 at period 8 and 100 at period 13, the "required" quantities will be

- For item B, 80 at period 8 and 200 at period 13 since two units of item B are required to assemble one unit of item A.
- For item D, 120 at period 8 and 300 at period 13, since three units of item D are required to assemble one unit of item A.

Starting from these required values, we derive the values of the "Release" raw-materials in the same way as for the end-item. This process is respected until we reach the lowest level of the product structure.

If an item is used at more than one level of the final assembly, the values of its "Required" raw are obtained by cumulating the values derived from the values of the "Release" row of these items, and the "Release" is not computed until we have reached the lowest level.

47.8 Application of the MRP

The previous section was devoted to the explanation of the MRP mechanism. As we outlined, this mechanism applies to all the end-items to be produced, and the computations are linked since the same items may be the components of several other items. Nevertheless, the complexity of the computation remains very low. The reason is that items do not compete for the resources or, in other words, no in any capacity constraint is considered in the computation. As a consequence, the solution obtained may be infeasible, since some of the resources may be overloaded during some periods. The MRP systems usually provide the information concerning the load of the resources during each of the periods. If some resources are over loaded, several solutions may be applicable:

1. Reduce and/or increase the due dates of some end items.
2. Reduce and/or increase the due dates of some quantities of end items.
3. Cancel totally or partially some end-item requirements, and compute explosion calculus again with these new data.

The computation of the resource loads is a part of the capacity requirement planning (CRP).

47.9 Conclusion

Usually, the application of MRP in a company leads to the following advantages:

1. Reduction of finished product inventory and WIP levels.
2. Reduction of the production cycle if the lead times are well controlled.
3. Improvement of the reactivity of the system in case of unexpected demand or changes in the MPS.
4. Better use of the resources.
5. Better productivity.

An important drawback, which has been outlined before, is the tendency of MRP to perpetuate the use of excessive lead times defined in the past, since no mechanism exists to correct an overevaluated lead time assigned to a new item, for instance. On the other hand, underestimated lead-times assigned to items may cause excessive workload or violation of capacity limits. This also depends on the product mix and the state of the shop. It must be noted that lead times are a result of a schedule and therefore cannot be estimated in a static manner with even reasonable accuracy. We should also notice that the result of the MRP is not scheduling, since it does not assign tasks to resources and it does not define the starting time of the tasks. The output of MRP is more a production plan which provides in the number of each item to be manufactured during each period. This output does not guarantee the best utilization of the resources.

It is possible to enriched the MRP system by introducing a scheduling system which will perform a schedule inside each period, but such a system needs all the information concerning the resources as well as the manufacturing times, setup times, transportation times, etc, which is usually not provided to MRP or possible to incorporate if the simple backward scheduling algorithms are used. An interested reader is referred to the work of Agrawal (1996) for the discussion of such algorithms.

References

Agrawal, A., Harhalatkis, G., Minis, I., and Nagi, R. 1996. Just in-time production of large assemblies, *IIE Transactions, preprint.*

Bertrand, J. W. M., Wortmann, J. C., and Wijngaard, J. 1990. *Production Control, a Structural and Design Oriented Approach*, Elsevier, Amsterdam 1990.

David, G. B. 1974. *Management Information Systems,* McGraw-Hill, New-York, NY.

Orlicky, J. 1975. *Material Requirements Planning,* McGraw-Hill, New York, NY.

Proth, J. M. 1992. *Conception et Gestion des Systèmes de Production,* Presses Universitaires de France.

48

Production Management Techniques

Upendra Belhe
The University of Iowa

Andrew Kusiak
The University of Iowa

48.1 Material Requirements Planning (MRP) 663
Processing Frequency • Lot Sizing • Safety Stock and Safety Lead Time
48.2 Manufacturing Resource Planning (MRPII) 664
48.3 Optimized Production Technology (OPT) 665
48.4 Toyota System and Just-in-Time 666
48.5 The Kanban Concept 667
Six Rules for the Kanban • Materials Planning Using Kanban

Manufacturing management impacts the flow of materials and the set of process activities concerned with the procurement of raw materials, production of subassemblies, creation of products, and the distribution of them to consumers. Numerous interrelated management problems arise in this flow and they are handled by the manufacturing planning and control system. An accurate data base is required to use this system for routine decision making. The requirements on the design of the manufacturing and planning control system vary with the nature of the production process, customer expectations, and the management needs.

48.1 Material Requirements Planning (MRP)

Material requirements planning (MRP) is the central set of activities in material planning and control. The primary function of MRP is to take a period-by-period set of master production schedule requirements and produce a time-phased set of component/raw material requirements. In addition to the master production schedule, a bill of material and inventory status are the two basic inputs to MRP. A bill of material shows, for each higher level component numbers, all other component numbers required as direct elements. For example, for an electronic equipment, it shows that a power supply module is required. For each power supply module, the bill of material could be a printed circuit board, power transistor, heat sink, etc. The inventory status indicates how many power supply modules are on hand, how many of those are already allocated to existing needs, and how many have already been ordered.

The MRP data can be used to construct a time-phased requirement record for any component or subassembly. MRP translates overall production plans into the detailed individual steps required to accomplish those plans (Vollman et al., 1988).

The MRP time-phased record is the representation of the status and plans for any single part number in the manufacturing planning and control system. For example, the MRP record for 2.5 Ω resistors is shown in Figure 48.1.

This record provides the following information:

1. The anticipated demand for the resistors during the period.
2. Existing replenishment orders for the resistors due in at the beginning of the period.
3. The current and projected status for the resistors at the end of the period.
4. Planned replenishment orders for the resistors at the beginning of the period.

The first row in the MRP record shown in Figure 48.1 indicates time period. The convention used for developing MRP records is that the current time is the beginning of the first period. The initial available balance of 22 resistors is shown prior to period 1. The number of periods in the MRP record is called as the planning horizon. It indicates the number of future periods for which plans are made. In Figure 48.1, the planning horizon is 5 periods.

The second row, "gross requirements" indicates demand for resistors. The gross requirements are time-phased. It means that the requirements are stated on a period-to-period basis and not as aggregate or average. A gross requirement in a particular period means that the demand will be unsatisfied unless the quantity required is available during that period. This demand could be satisfied either from inventory or from a planned replenishment order.

The "scheduled receipt" row describes the status of any open orders for resistors. This row indicates quantities that have been ordered and the time they are expected to be received. The

0-8493-8343-9/97/$0.00+$.50

Period		1	2	3	4	5
Gross requirements			35		55	76
Scheduled receipts		100				
Projected available balance	22	122	87	87	32	56
Planned order release					100	
Lead time = 1 period Lot size = 100						

Figure 48.1 The basic MRP record for 2.5 Ω resistors.

scheduled receipts represent commitment. The timing convention for showing scheduled receipts is the beginning of the period.

The fourth row in the MRP record indicates "projected available balance." The timing convention for the projected available balance is the end of the period. For example in Figure 48.1, the balance of 22 is at the end of the period previous to period 1 and the balance of 122 is at the end of period 1. This means that the projected available balance shown at the end of a period is available to meet gross requirements in the next (and succeeding) periods.

The "Planned order release" row is determined from the projected available balance. A planned order release is created whenever projected available balance is not sufficient to satisfy the gross requirements. In other words, a planned order release is created to keep the projected available balance from becoming negative. For example, the projected available balance at the end of period 4 is 32 and gross requirement in period 5 is 76. Since the lead time is 1 period, a planned order is created at the beginning of period 4 to make 100 units available at the beginning of period 5.

Some of the technical issues to be considered in the design of MRP systems are discussed next.

Processing Frequency

As the new information becomes available the MRP records must be updated so plans can be adjusted to reflect these changes. The issue here is how frequently the records should be processed and whether all records should be processed at the same time.

One of the options is to process all records in one computer run, which is called a regeneration run. An alternative is called "net change" processing, where only those records that are affected by the new information are reconstructed. The appropriate frequency for processing the MRP time-phased records depends on the company, its products, and its operations.

The motivation for less frequent processing of new information is the computational cost, which can be especially high with regeneration. The problem with less frequent processing is that the component status expressed in the MRP record becomes increasingly inaccurate. More frequent processing of records results in fewer unpleasant surprises.

The net change approach can reduce computer time enough to make more frequent processing possible. On the other hand, daily processing of part of the MRP records could be computationally more expensive than weekly regeneration.

Lot Sizing

The time-phased information can be used in combination with economic, physical, vendor and other data to develop lot sizes for components or assemblies. The basic trade off usually involves elimination of one or more setups at the expense of carrying inventory longer. Many times the discrete lot sizes that are possible with MRP are more appealing than the fixed lot sizes that could be used.

Safety Stock and Safety Lead Time

Safety stock is a buffer of stock beyond the need to satisfy the gross requirements. This is illustrated in Figure 48.2 by incorporating safety stock for the 100 μF capacitors.

Safety lead time is a procedure where orders are released and scheduled to arrive one or more periods before necessary to satisfy the gross requirements. Figure 48.3 shows the MRP records for 100 μF capacitors with safety lead time of one period.

Both safety stock and safety lead time are used in practice. Safety stock tends to be used in an MRP system where uncertainty about quantities is the problem. On the other hand, safety time is used where the uncertainty is in the timing rather than the quantity.

48.2 Manufacturing Resource Planning (MRPII)

Manufacturing resource planning (MRPII) is an extension of MRP. Material requirements planning plans activities performed by functions such as production control, purchasing, and inventory control. In MRPII, these functions are linked together around a shared database so that the status information can be easily passed to the planning functions, and recommendations can be electronically linked to the release and execution functions. In addition to the functions considered by MRP, customer service and accounting applications are also considered in MRPII.

The characteristics of MRPII described in Wight (1981) are as follows:

Period		1	2	3	4	5
Gross requirements			20		35	16
Scheduled receipts		30				
Projected available balance	15	45	25	25	20	20
Planned order release			30	16		
Lead time = 2 periods Safety stock = 20						

Figure 48.2 The MRP record for 100 μF capacitors with safety stock of 20.

Period		1	2	3	4	5
Gross requirements			20		35	16
Scheduled receipts		30				
Projected available balance	15	45	25	45	20	4
Planned order release		20	10			
Lead time = 2 periods Safety stock = 20						

Figure 48.3 The MRP record for 100 μF capacitors with safety lead time of one period.

1. It has a "what if" capability and hence it can be used to simulate what would happen if various policy decisions were implemented. It simulates material requirements far enough in advance so that shortages can be predicted and prevented. It also simulates capacity requirements far enough in advance so that capacity problems can be prevented. MRPII gives management the information so that they have the time to make decisions.
2. It involves every facet of the company. MRPII is a whole company system that includes manufacturing, finance, marketing, engineering, purchasing, and distribution.

The best documented gains of MRPII are reduced inventory and improved customer service. Another advantage is that good scheduling improves productivity due to fewer material shortages. The broad implications of MRPII in the areas such as reduction of obsolescence, the productivity of engineering, and improved market penetration have not been studied in detail.

48.3 Optimized Production Technology (OPT)

The fundamental principle of OPT is that only the bottleneck operations (resources) are of critical scheduling concern (Vollman et al., 1988). The production output is limited by the bottleneck operations and throughput can only be increased by improving the utilization of the bottleneck facilities.

OPT first combines the data in the bill of material file with those in the routing file. As a result a network or extended tree diagram is constructed. The operational data attached to each part in the product structure is then combined with the master production schedule to form what is called a product network.

Each operation is defined in terms of the resources used, setup and processing times. The OPT files also include the data on the capacity, maximum inventory, minimum batch quantities, order quantities, due dates, alternative routings, and resource constraints. Product network and resource descriptions are then fed into a set of routines called BUILDNET and SERVE that identify the bottleneck resources. The BUILDNET routine combines the product network and resource information to form an engineering network. The SERVE routine determines backward schedule from the order due dates assuming infinite capacity for the resources.

Average expected loads on machine centers are determined using a rough cut capacity planning routine. These average loads are arranged in descending order and the most heavily loaded workcenters are studied. Then, a routine called SPLIT is used to split the OPT product network into two portions. The lower section is called the SERVE Network network, which includes all operations preceding the bottleneck resources. The upper portion is the OPT network, which incorporates all of the bottleneck resources and all succeeding operations.

One of the advantages of this split is that one can see where attention should be focused. Bottleneck capacity is used more extensively by finite loading of this small subset of workcenters. When this finite loading through bottleneck resources is completed, the result is a doable master production schedule. In short, OPT conceivably can take any master production schedule as input and determine the extent to what is doable.

For nonbottleneck resources OPT schedules are based on MRP logic. In such a case OPT reduces batch sizes to the point where some nonbottleneck resources become bottlenecks. This results in less work in process inventory and lead time is also reduced. This is achieved by overlapping schedules using unequal batch sizes for transferring and processing.

To maximize output from bottleneck operations, larger lot sizes are run. As a result, the percentage of nonproductive time devoted to setups in these workcenters is reduced. On the other hand, smaller batches are made at nonbottleneck workcenters. Calculation of the batch sizes is a part of the OPT procedure.

A transfer batch refers to the lot size that moves from operation to operation. A process batch is the total lot size released to the shop. The OPT distinguishes between a transfer batch and a process batch. Any differences are held in work in process inventories. Also, any operation cannot proceed until a transfer batch is built behind it.

OPT provides buffers for schedules for critical operations by using both safety stocks and safety lead time. When a sequence of jobs is scheduled on the same machine, safety timing is introduced between subsequent batches. In this way, a cushion is provided against variations adversely affecting the flow of jobs through the same operation.

48.4 Toyota System and Just-in-Time

An MRP system involves some guesswork. Customer demands need to be predicted in order to prepare the schedule and also it is required to guess the amount of time required for production to make the needed parts. Even though this system allows corrections to be made frequently, bad predictions result in excess inventories of some parts.

Manufacturing companies face the difficulty of reducing the production cost and improving the product quality. These products have three cost variables: materials, labor, and overhead. It is most important to use the correct manufacturing strategy to reduce these costs. Just-in-Time is a management philosophy that is continuously focused on integrating and streamlining the manufacturing system into the simplest possible process. It strives to minimize the elements in a manufacturing system that restrain productivity of the system.

The beginning of Just-in-Time can be traced back to the Toyota system. This system was implemented the first time in Toyota, Japan, and became successful in reducing inventory levels and improving quality. In this system material movement between workcenters follows three main rules:

1. Material is moved in a continuous flow rather than in a batch mode.
2. Material is moved in the smallest possible quantities.
3. Material is moved only when it is required by the next stage.

The Toyota system evolved into the Just-in-Time system designed to improve the efficiency of manufacturing organizations with minimum resources. It also improves quality, reduces inventory levels, and provides maximum motivation to solve problems as soon as they occur.

Just-in-Time is defined as a production system designed to eliminate waste in the manufacturing environment. Here, waste is described as anything that is not necessary for the manufacturing of the product or is in excess.

Most companies use master schedules and material requirement planning (MRP) to decide their production schedules and the movement of material in the factory. This system is referred to as a push system. Another system, called a pull system, uses bottom-up demand, which is driven by the consumption rates of parts in the production process. The goal of the pull system is to pull material required with minimum advance notice of production requirements from the customer. One of the most fundamental changes that Just-in-Time introduces in a manufacturing organization is the institution of a pull system instead of a push system. The pull system is favored because it eliminates unnecessary elements from the production system. These elements are primarily the cost of material labor diverted into inventory.

EXAMPLE 48.1 A company is building 80 units a day of power supply modules. Assume that a work order for a set of 400 transformers is released to the manufacturing floor on Monday to meet the weekly demand. According to the production schedule, we only need a set of 80 transformers on Monday, but we have released 400, producing an excess of 320 transformers that day, an excess of 240 transformers on Tuesday and so on. In a Just-in-Time system, these excess sets of transformers would be considered as waste, because they are not needed to produce the daily quota. In such a system, we would only release 80 transformers each day. For high-volume applications, the frequency of release could be increased. If the production rate increases tenfold, then we would be better off by releasing 400 transformers every four hours or 100 transformers every hour.

There are some major misconceptions about Just-in-Time manufacturing system (Lubben, 1988). One of the major misconceptions is that Just-in-Time is an inventory control system. Although inventory reduction is one of the key goals of a Just-in-Time system, it is much broader and affects the operation of many departments in the company. It is important not to reduce buffer inventories until the quality of parts reaches an acceptable level. The company should start solving quality problems with suppliers and the production process long before they start reducing inventory.

The second misconception is that Just-in-Time is a method used by the materials function to push inventory back into the supplier's shop, thereby forcing the supplier to carry the customer's inventory. Where the material is stored is irrelevant. The materials and resources required to make the part have been committed. The resources that could have been used to produce a needed product have been diverted into non-productive inventory.

Manufacturing organizations must have a clear procedure to issue materials to the production floor. There are three critical aspects involved in this procedure. First, the procedure must ensure that the materials issued are sufficient for the production build schedule. Second, the procedure must allow the company to track the materials moving through the production process. Third, the procedure must also allow analysis of the physical movement of the materials in the factory so as to be able to increase productivity and to reduce overhead.

48.5 The Kanban Concept

The word *kanban* means visual record and refers to a manufacturing control system developed and used in Japan. Toyota used the kanban system for many years as a means to communicate materials needs between two workcenters. The kanban is a mechanism by which a workstation signals the need for more parts from the preceding station. For example, consider the manufacturing process shown in Figure 48.4. There are three workcenters A, B and C cascaded together. The requirement for a certain number of finished goods is translated as a requirement of the corresponding number of units from workcenter B. This requirement is transferred backwards up to the material stock location. Then only the required number of raw material units are released to workcenter A, which finally are transformed into an exact requirement of the number of finished product units.

In the Toyota kanban system every part number has its own special container design to hold a precise quantity of that part. Usually this quantity is small. There are two cards or "kanban" for each container. Each kanban indicates the part number, container capacity and certain other information. One kanban, called a production kanban, serves the workcenter producing the part and the other kanban, known as the withdrawal kanban, serves the workcenter using it. Each container cycles from the producing workcenter to the using workcenter and back.

The withdrawal kanban travels between workcenters and is used to authorize the movement of parts from one workcenter to another. A withdrawal kanban must accompany the flow of material from one process to another. Once a withdrawal kanban fetches parts, it will stay with them all the time. Then, after the subsequent process consumes the last part of the lot, the kanban will travel again to the preceding process to fetch a new set of parts.

The production kanban's job is to release an order to the preceding process to build more parts. The production kanban goes into a queue with other production kanbans at the workcenter. After the workcenter builds the new parts, the production kanban travels back to the wait area until a new withdrawal kanban starts the cycle over again.

As described by the Japan Management Association (1986), the major functions of the kanban system are

1. Engage in standard operations at any time.
2. Give directions based on the actual conditions existing in the workplace.
3. Prevent addition of any unnecessary work for those engaged in start-up operations.

Six Rules for the Kanban

Kanban is a tool created to manage the workplace effectively. However, this tool should be used appropriately. The rules for operating kanbans are as follows:

1. *Do not send defective products to the subsequent process.* If a defective product is discovered, the highest priority must be given to the measures to prevent its recurrence. This is to make sure that such defects will not be produced again. If the defective products get mixed up with good products, they must be exchanged promptly.
2. *The subsequent process comes to withdraw only what is needed.* The subsequent process must withdraw materials from the preceding process at the time needed and in the quantity needed. A number of concrete steps are required to ensure that these withdrawals are not arbitrary, such as:
   a. No withdrawal without a kanban.
   b. Items withdrawn cannot exceed the number of kanbans submitted.
   c. A kanban must always accompany each item.
3. *Produce only the exact quantity withdrawn by the subsequent process.* This rule is a logical extension of the second rule. With this rule the process is able to restrict its inventory to the minimum. Not producing more than the number of kanbans and producing in the sequence in which the kanbans are received makes this rule operational. By observing the second and third rules, the production process can function in unison.
4. *Equalize production.* In order to produce the exact quantity, adequate facilities and personnel are required. As a result, the processes that cannot deal with the requirements resort to producing material ahead of time, which is a violation of the third rule. The fourth rule demands load smoothing.
5. *Kanban as a means of fine-tuning.* In using a kanban system, it is important to abide by the principle of load smoothing in production. Sudden changes in production demands cannot be handled by the kanban system. Kanban can only respond to the need for fine-tuning.
6. *Stabilize and rationalize the process.* The defective parts should not be sent to the subsequent processes and defective work is the result of not having sufficient

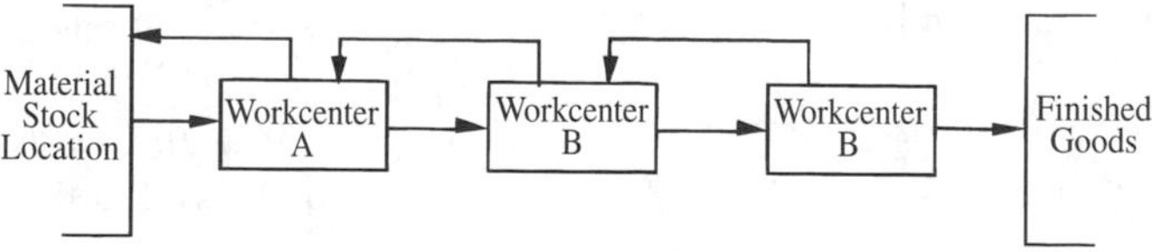

Figure 48.4 Communication needs between workcenters.

standardization and stabilization. Unless the process is rationalized and stabilized, adequate supply and quality cannot be maintained. This in turn supports the load-smoothing system of production.

Materials Planning Using Kanban

Kanbans are issued by the materials planner, who determines a lot size for each kanban. Additional kanbans are issued to increase production rate for a particular part or subassembly. The number of kanbans issued for a particular part is calculated using the following formula:

$$\text{Number of kanbans} = \frac{\text{Units daily demand} \times \text{Order cycle time} \times \text{Safety factor}}{\text{Lot size}}$$

The units daily demand refers to the daily production rate of the part. The order cycle time is the time it takes to process the part or to procure a purchased item. The safety factor is a percentage increase in the number of kanbans used as a precautionary measure for buffer inventories. The lot size is the number of parts to be fetched by the kanban if it is a withdrawal kanban or to be manufactured if it is a production kanban.

For example, assume that the production requirement for 100 Ω resistors is 6000 units a month. The cycle time for them is 14 days and the lot size used is 1000.

$$\text{Unit daily demand} = \frac{6000}{20} = 300 \text{ units/day}$$

Since the process is not stable, a safety factor of 1.25 is used. Using the formula for number of kanbans, we obtain:

$$\text{Number of kanbans} = \frac{300 \times 14 \times 1.25}{1000} = 5$$

This means that we need 4 kanbans to run the process and 1 extra kanban is used as a buffer until the process is steady and predictable.

As shown in the above example, the planner has global information about production levels. This information is used to calculate the number of kanbans required to support the schedule. Once the kanban system starts to operate, the planner will lose control of the status of the kanbans and therefore tracking kanban status is one of the main problems that a material planner faces in such a system. As the number of kanbans increases and the traffic becomes intense, the planner would spend considerable time in tracking kanban status.

One of the ways to reduce the additional work load in tracking kanbans is not to keep a kanban uniquely associated with a particular part lot. The only information required is the number of kanbans issued for a part and the size of the lot they represent. It should not matter which kanban numbers are at a particular location. For example, there are two kanbans for 250 Ω resistors in a particular location and these kanbans represent a lot of 100 resistors each. There is no need to know whether these are kanbans 2 and 4 or kanbans 3 and 6. The only important information is that there are two kanbans in that location with a total of two hundred 250 Ω resistors.

It is important that kanbans meet their estimated lead times, because this affects supply of parts to subsequent processes. One of the ways to track kanban lead times is to provide every workcenter with an estimate of how long it would take to process a kanban. Any deviations from that estimate are reported by the operators. Then, tracking would be based on the default cases. A simple kanban reporting system would require the operators to report the number of kanbans at hand and the number of kanbans that are past due. This reporting should be done at a particular time of the day. Such a report gives the material planner an overview of the status of the materials and provides the opportunity to detect supply problems in advance if the number of kanbans expected in a workcenter are not adequate to meet the production rate.

One of the important questions faced by some companies is whether the kanban system would work with their MRPII system or rather whether the kanban system could replace their MRPII system. MRPII gives manufacturing, sales, and finance a global picture of materials, capacity, and finance needs to meet the company's sales forecast. On the contrary, the kanban system is a bottom-up process that has a limited ability to generate an overall picture. In a kanban system, the MRPII system can provide top-down planning visibility to the different processes in the system. MRPII can be used to forecast the monthly build for the factory on a process-by-process basis.

Acknowledgment

This research has been partially supported by research grant DDM-9215259 from the National Science Foundation.

References

Hernandez, A. 1989. *Just-In-Time Manufacturing: A Practical Approach,* 52–72, Prentice Hall, Englewood Cliffs, NJ.

Japan Management Association (Ed.) 1986. *Kanban: Just-In-Time at Toyota,* 87–92, Productivity Press, Cambridge, MA.

Lubben, R. 1988. *Just-in-Time Manufacturing: An Aggressive Manufacturing Strategy,* 13–14, McGraw-Hill, New York, NY.

Vollman, T., Berry, W., and Whybark, D. 1988. *Manufacturing Planning and Control Systems,* Dow Jones-Irwin, Homewood, IL.

Wight, O. 1981. *MRP II: Unlocking America's Productivity Potential,* The Book Press, Brattleboro, VT.

49

Automated Manufacturing System Development Methodology

Richard Zurawski
Swinburne University of Technology

MengChu Zhou
New Jersey Institute of Technology

Sunderesh S. Heragu
Rensselaer Polytechnic Institute

Anthony D. Robbi
New Jersey Institute of Technology

Christopher M. Lucarelli
Rensselaer Polytechnic Institute

49.1 Analysis of Functional Properties of Specification and Design Models of Industrial Automated Systems 669
Introduction • Issues Involved in Behavioral Analysis of Models of Industrial Automated Systems • Description of Petri Nets • Properties of Petri Nets • Analysis Methods • Multirobot System: An Example

49.2 Automated Manufacturing System Design Using Analytical Techniques 677
Introduction • Models for System Design • Conclusions

49.3 Discrete Event Simulation 694
Introduction • Simulation Procedure • Simulation Models • Simulation Schemes and Tools • Simulation Examples • Concluding Remarks

49.1 Analysis of Functional Properties of Specification and Design Models of Industrial Automated Systems

Richard Zurawski and MengChu Zhou

Introduction

The growth in the complexity of modern industrial systems, such as production, process control, communication systems, etc., creates numerous problems for their developers. In the planning stage, one is confronted with the increased capabilities of these systems due to the unique combination of hardware and software, which operate under a large number of constraints arising from the limited system resources. In view of the capital-intensive and complex nature of modern industrial systems, the design and operation of these systems require modeling and analysis to select the optimal design alternative and operational policy. It is well known that flaws in the modeling process can contribute substantially to the development time and cost. The operational efficiency may be affected as well. Therefore, special attention should be paid to the correctness of the models that are used at all planning levels.

The development of industrial automated systems requires that both functional and performance requirements be met. Depending on the development stage of a system, the knowledge of either an approximate or exact (or both) performance may be required. For example, at the design stage, the approximate performance of the alternative design models is required to eliminate these proposals which are highly unlikely to meet the performance requirements when fully developed and implemented. Analytical techniques play an important role at this stage. They allow the designer to obtain the required performance measures, involving a relatively small time investment needed for the model construction and its solution. The selected design alternatives are then refined by increasing the level of the details to include in the model the actual operational policies and time characteristics. The model complexity, or the presence of heuristic algorithms, may prohibit the use of analytical techniques. The discrete-event simulation is, then, the only viable alternative for the performance evaluation, although it is an expensive and time-consuming technique.

In principle, there are two types of models which are used to represent industrial automated systems. These are models which allow for the system description only using, in most cases, graphical tools and associated modeling methodologies; models which allow for the system description and some form of analysis. The second group can be further divided into analytical and simulation-based modeling techniques. Examples of techniques and associated environments which allow for the representation only are SADT (Ross, 1977; Ross and Shoeman, 1977), IDEF-0 and its variants (Doumeints and Poumeyrol, 1987), and GRAI

0-8493-8343-9/97/$0.00+$.50

(Pun et al., 1985). The analytical techniques involve Markov processes, queuing networks, mathematical programming, perturbation analysis, etc. The simulation-based models are constructed by using application domain specific simulators, general-purpose simulation languages, and general-purpose programming languages.

The first type of models, typically, does not support any form of analysis. The models are constructed using purely and simply descriptive languages. This type of description has to be translated, in most cases manually, into executable models for performance evaluation. As a result, the executable model may differ in its functionality from the original model. In addition, the construction of the executable model may be as difficult and time consuming as the process of forming the descriptive model. This type of model is mostly used to help the system designer structure the knowledge of the system functionality. Top-down, bottom-up, and hybrid techniques are frequently used as the underlying methodologies in those approaches.

Due to the complexity of modern industrial automated systems, analytical models, to be tractable, must involve unrealistic simplifying assumptions which allow for the steady state analysis only. However, the steady state operation may never be reached in practice. This is due to changes in the production profile, or system failures. Also, since the analytical methods are incapable of handling synchronization, their applicability is restricted to systems which can be decomposed into independent subsystems.

The use of simulation is limited by the number of design alternatives that can be simulated in a reasonable time. Statistical properties of the simulation outputs require a relatively high number of simulation runs of each system configuration, preferably under every credibly conceivable scenario. This limits the designer's freedom to generate more alternative designs.

Perhaps the most severe limitation of all those approaches and tools is that they cannot be used for the formal verification of the model. Models which are not completely verified may result in implementations which are likely to exhibit some erroneous behavior in their operational phase. The simulation based verification of models of complex systems cannot be regarded as a substitute for the formal verification process. There are two major reasons for this. Simulation allows one to study only a limited number of states the simulation model can take on. The reasons were discussed above. The cost and time involved in generating even a limited set of states of a complex system can be excessive.

The focus of this section is on the analysis of the functional properties of models of industrial automated systems. The temporal aspects of the analysis are the subject of the next two sections which provide an overview of selected analytical methods and discrete-event simulation techniques.

Issues Involved in Behavioral Analysis of Models of Industrial Automated Systems

An important issue to be considered during analysis is whether there exists one-to-one functional correspondence between the informal requirements specification; typically expressed using a natural language, and its representation in the form of a model. The model can be informal, semiformal, and formal. The construction of models from informal requirements specifications is not a trivial task. It requires a great deal of modeling experience, as well as knowledge of the techniques assisting in the model construction. As a result, a model may differ considerably from its original specification. This is especially true when models of complex systems are involved. The existence of the one-to-one functional correspondence between an original requirements specification and a model allows one to project the analysis results, obtained for the model, onto the original description. This provides a feedback to the customers which can, in many instances, help the customers clarify their perception of the system. As a result, any changes required will be also reflected in the model. Another important issue to be addressed during the analysis stage is the completeness of the requirements specification. In most cases, the requirements specification defines the external functional behavior of a system. This is typically expressed in terms of the system input-output relationship. Inputs are generated by the environment of the system. Outputs are responses of the system to these inputs. If some inputs, generated by the environment of the system, are not included in the requirements specification, then the system will be unable to respond to these inputs when they occur during the system's normal operation. The completeness of the requirements is especially important in the case of safety-critical systems. In these systems, the incompleteness of the requirements specification may lead to catastrophic events occurring in the environment of the system. For instance, the occurrence of unanticipated states in the operation of a nuclear reactor may result in the failure of the control system to respond to them properly, or at all, thus potentially leading to the reactor failure. The consistency of the requirement specification is another issue to be considered during analysis. The inconsistency occurs when for a given permissible, temporal combination of inputs, a requirements specification allows for two or more different permissible temporal combinations of outputs. It is mainly due to a vague, incomplete, and frequently incorrect perception of the system functionality. The analysis of a model may in many cases help the system designer identify incompleteness and inconsistency of the requirements specification. This can manifest itself in various ways depending on the descriptive formalism used. The incompleteness and inconsistency of the requirements specification, in most cases, will have an impact on the logical correctness of the specification and the model. The lack of logical correctness may also arise from a vague perception of the functionality of a system resulting in incorrect assumptions regarding functional behavior of the system. In the system operational stage, this lack of correctness may result in deadlocks, buffer overflow and underflow, etc.

A tool that is useful for modeling, formal analysis, and design of discrete-event systems, examples of which are automated industrial systems, is Petri nets (Murata, 1989; Zurawski and Zhou, 1994). Petri nets were named after Carl A. Petri who created in 1962 a netlike mathematical tool for the study of communication with automata. Petri nets, as a graphical tool, provide a powerful communication medium between the user, typically a requirements engineer, and the customer. Complex requirements specifications, instead of using ambiguous textual descriptions or mathematical notations difficult to understand by the customer, can be represented graphically using Petri nets. This, combined with the existence of computer tools allowing for interactive graphical simulation of Petri nets, puts in the hands of the development engineers a powerful tool assisting in the development process of complex systems. As a mathematical tool, a Petri net model can be described by a set of linear algebraic equations, or other mathematical models reflecting the behavior of the system. This opens a possibility for the formal analysis of the model. This allows one to perform a formal check of the properties related to the behavior of the underlying system, e.g., precedence relations among events, concurrent operations, appropriate synchronization, freedom from deadlock, repetitive activities, and mutual exclusion of shared resources, to mention some.

In the remaining part of this section, Petri nets are used to introduce a number of concepts useful in studying selected aspects of logical correctness, or behavioral properties, of industrial automated systems.

Description of Petri Nets

A Petri net may be identified as a particular kind of a bipartite directed graph populated by three types of objects. These objects are places, transitions, and directed arcs connecting places to transitions and transitions to places. Pictorially, places are depicted by circles, transitions as bars or boxes. A place is an input place to a transition if there exists a directed arc connecting this place to the transition. A place is an output place of a transition if there exists a directed arc connecting the transition to the place. In its simplest form, a Petri net may be represented by a transition together with its input and output places. This elementary net may be used to represent various aspects of the modelled systems. For instance, input (output) places may represent preconditions (postconditions), the transition an event. Input places may represent the availability of resources, the transition their utilization, output places the release of the resources.

An example of a Petri net is shown in Figure 49.1. This net consists of five places, represented by circles, four transitions, depicted by bars, and directed arcs connecting places to transitions and transitions to places. In this net, place p_1 is an input place of transition t_1. Places p_2 and p_3 are output places of transition t_1.

To study dynamic behavior of the modeled system, in terms of its states and their changes, each place may potentially hold either none or a positive number of tokens, pictured by small solid dots, as shown in Figure 49.1. The presence or absence of a token in a place can indicate whether a condition associated with this place is true or false, for instance. For a place representing the availability of resources, the number of tokens in this place indicates the number of available resources. At any given time instance, the distribution of tokens on places, called Petri net marking, defines the current state of the modelled system. A marking of a Petri net with m places is represented by an $(m \times 1)$ vector M, elements of which, denoted as $M(p)$, are nonnegative integers representing the number of tokens in the corresponding places. A Petri net containing tokens is called marked Petri net. For example, in the Petri net model shown in Figure 49.1, $M = [1, 0, 0, 0, 0]$.

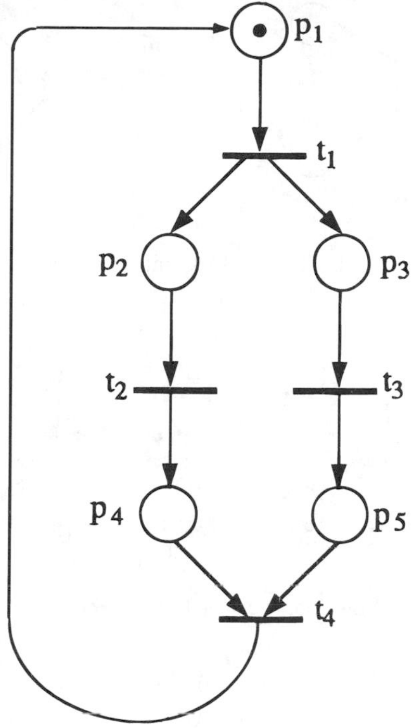

Figure 49.1 Example of graphical representation of a Petri net.

Formally, a Petri net can be defined as follows:

$$PN = (P, T, I, O, M_0)$$

where

$P = \{p_1, p_2, \ldots, p_m\}$ is a finite set of places,
$T = \{t_1, t_2, \ldots, t_n\}$ is a finite set of transitions, $P \cup T \neq \emptyset$ and $P \cap T = \emptyset$
$I: (P \times T) \rightarrow N$ is an input function that defines directed arcs from places to transitions, where N is a set of nonnegative integers
$O: (P \times T) \rightarrow N$ is an output function which defines directed arcs from transitions to places, where N is a set of nonnegative integers
$M_0: P \rightarrow N$ is the initial marking

If $I(p, t) = k$ ($O(p, t) = k$), then there exist k directed (parallel) arcs connecting place p to transition t (transition t to place p).

If $I(p, t) = 0$ ($O(p, t) = 0$), then there exist no directed arcs connecting p to t (t to p).

Frequently, in the graphical representation, parallel arcs connecting a place (transition) to a transition (place) are represented by a single directed arc labelled with its multiplicity, or weight k. This compact representation of multiple arcs is shown in Figure 49.2.

By changing distribution of tokens on places, which may reflect the occurrence of events or execution of operations, for instance,

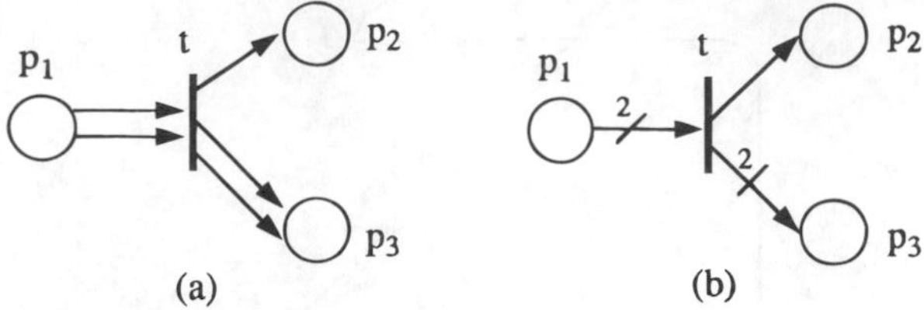

Figure 49.2 (a) Multiple arcs. (b) Compact representation of multiple arcs.

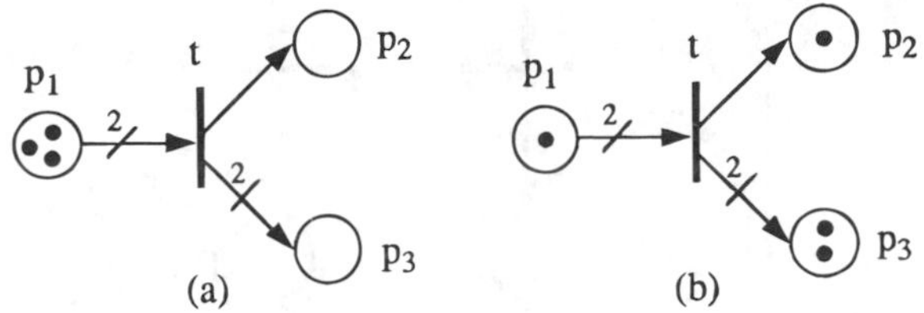

Figure 49.3 (a) Transition *t* enabled. (b) Enabled transition *t* fires.

one can study dynamic behavior of the modeled system. The following rules are used to govern the flow of tokens:

Enabling Rule A transition *t* is said to be enabled if each input place *p* of *t* contains at least the number of tokens equal to the weight of the directed arc connecting *p* to *t*.

Firing Rule (a) An enabled transition *t* may or may not fire depending on the additional interpretation, and (b) a firing of an enabled transition *t* removes from each input place *p* the number of tokens equal to the weight of the directed arc connecting *p* to *t*. It also deposits in each output place *p* the number of tokens equal to the weight of the directed arc connecting *t* to *p*.

The enabling and firing rules are illustrated in Figure 49.3. In Figure 49.3(a), transition t_1 is enabled as the input place p_1 of transition *t* contains three tokens, and $I(p_1, t) = 2$. The firing of the enabled transition *t* removes from the input place p_1 two tokens as $I(p_1, t) = 2$, and deposits one token in the output place p_2, $O(p_2, t) = 1$, and two tokens in the output place p_3, $O(p_3, t) = 2$. This is shown in Figure 49.3(b).

To illustrate how Petri nets can be used to model industrial automated systems, as well as selected properties of discrete-event systems such as concurrent activities, synchronization, mutual exclusion, etc., we consider an example of a multirobot system. In this example, robot arm R1 transfers parts from machining tool M1 to a buffer located in the common work area. Robot arm R2 transfers parts from the buffer to the machining tool M2. In order to avoid collision, only one robot can access the workspace at a time. The buffer has a space for a limited number of products. This system is represented by a Petri net model shown in Figure 49.4. The interpretation of places and transitions of Figure 49.4 is in Table 49.1.

In this model, places p_1, p_2, p_3 and transitions t_1, t_2, t_3 model activities of robot arm R1. Places p_4, p_5, p_6 and transitions t_4, t_5, t_6 model activities of robot arm R2. Transitions t_1 and t_4 represent concurrent activities of R1 and R2. Either of these transitions can fire before or after, or in parallel with the other one. The access to the common workspace requires synchronization of the activities of the arms in order to avoid collision. Only one robot arm should be allowed access to the common workspace at a time. This synchronization is accomplished by the mutual exclusion mechanism implemented by a subnet involving places p_7, p_3, p_6 and transitions t_2, t_3, t_5, t_6. Firing transition t_2 disables t_5, assuming t_5 is enabled, and vice versa. Thus only one robot arm can access the common workspace at a time. In addition, it is assumed that the buffer space is *b*. Thus, for instance, if p_8 is empty, then t_2 cannot be enabled. This prevents R1 from attempting to transfer to the buffer a part when there is no space in the buffer. Also, R2 cannot access the buffer if there is no part in the buffer; no token is present in place p_9.

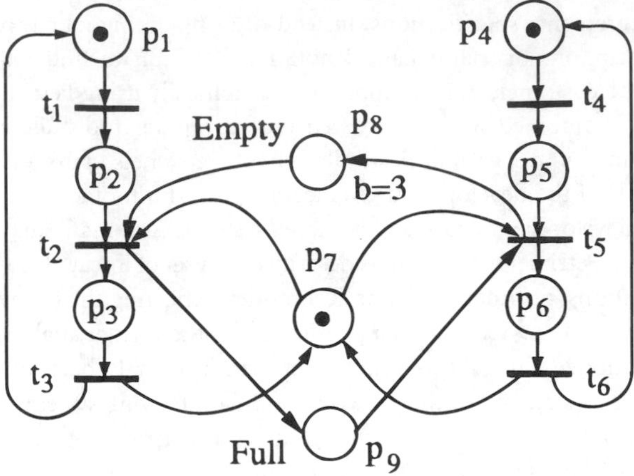

Figure 49.4 Petri net model of a multirobot system.

Table 49.1 Interpretation of Places and Transitions of the Petri Net Model of the Multirobot Assembly System

Place (with tokens on it)	Interpretation
p1(p4)	Robot R1 (R2) performs tasks outside the common workspace
p2(p5)	Robot R1(R2) waits for the access to the common workspace
p3(p6)	Robot R1 (R2) performs in the common workspace
p7	Mutual exclusion
p8(p9)	Number of empty (full) positions in buffer
Transition	**Interpretation**
t1(t4)	Robot R1 (R2) requests access to the common workspace
t2(t5)	Robot R1(R2) enters the common workspace
t3(t6)	Robot R1 (R2) leaves the common workspace

Properties of Petri Nets

Petri nets as a mathematical tool possess a number of properties. These properties, when interpreted in the context of the modelled system, allow the system designer to identify the presence or absence of the application domain specific functional properties of the system under design.

Reachability

An important issue in designing distributed systems is whether a system can reach a specific state or exhibit a particular functional behavior. For instance, using the example of a multi-robot system, the developer of a software controlling the operation of the system would need to know whether the control software allows for both robot arms to have access to the buffer at the same time. This undesirable state of the system operation can be reach through some sequences of elementary operations executed by robot arms. The existence of the execution paths in the control software allowing for this scenario may result in a collision during the system operation. Thus, it is essential to be able to identify this scenario in the both specification and design models prior to the control software implementation.

In general, when using Petri nets to represent specification and design models, the question is whether the system modeled with Petri nets exhibits all desirable properties, as specified in the requirements specification, and no undesirable ones. To find out whether the modeled system can reach a specific state as a result of a required functional behavior, it is necessary to find such a sequence of firings of transitions which would result in transforming a marking M_0 (a given marking) to M_i, where M_i represents the specific state, and the sequence of firings represents the required functional behavior. It should be noted that real systems may reach a given state as a result of exhibiting different permissible patterns of functional behavior. In a Petri net model, this should be reflected in the existence of specific sequences of transitions firings, representing the required functional behavior, which would transform a marking M_0 to the required marking M_i. The existence in the Petri net model of additional sequences of transitions firings which transform M_0 to M_i indicates that the Petri net model may not be exactly reflecting the structure and dynamics of the underlying system. This may also indicate the presence of unanticipated facets of the functional behavior of the real system, provided that the Petri net model accurately reflects the underlying requirements specification of the system. A marking M_i is said to be *reachable* from a marking M_0 if there exists a sequence of transitions firings which transforms a marking M_0 to M_i. For instance, in the Petri net model of the multi-robot assembly system shown in Figure 49.4, with $b = 3$, the state in which robot arm R1 performs tasks in the common workspace, with robot arm R2 waiting outside, is represented by the marking vector $M_i = (0, 0, 1, 0, 1, 0, 0, 2, 1)^T$. M_i can be reached from the initial marking M_0, where $M_0 = (1, 0, 0, 1, 0, 0, 1, 3, 0)^T$, by the following sequence of transitions firings—$t_1 t_2 t_4$. Also, there is no sequence of transitions firings which would result in marking $M_j = (0, 0, 1, 0, 0, 1, 0, 2, 1)^T$ which represents scenario in which both robots access the buffer at the same time. This implies that the model correctly implements the mutual exclusion mechanism allowing only one robot to have access to the buffer at a time. The set of all possible markings reachable from some initial marking is called the reachability set.

Boundedness and Safeness

The information storage areas can hold, without corruption, only a restricted number of pieces of data. In manufacturing systems, attempts to store more tools in the tool storage area than it is allowed may result in tool and equipment damage. Thus, it is important to be able to determine whether proposed control strategies prevent the overflows of these storage areas.

In Petri net models, places are frequently used to represent (data, equipment, tools) storage areas. The Petri net property which helps to identify in the modelled system the existence of overflows is the concept of *boundedness.* A Petri net is said to be *k-bounded* if the number of tokens in any place of the net is always less or equal to k (k is a nonnegative integer number) for every possible distribution of tokens on places (for every marking M reachable from some initial marking M_0). A Petri net is *safe* if it is *1-bounded.* The Petri net shown in Figure 49.5 is safe. In this net, no place can contain more then one token. An example of a Petri net which is unbounded is shown in Figure 49.6. This net is unbounded because place p_4 can hold an arbitrarily large number of tokens.

Conservativeness

In man-made systems, the number of resources in use is typically restricted by financial as well as other constraints. If

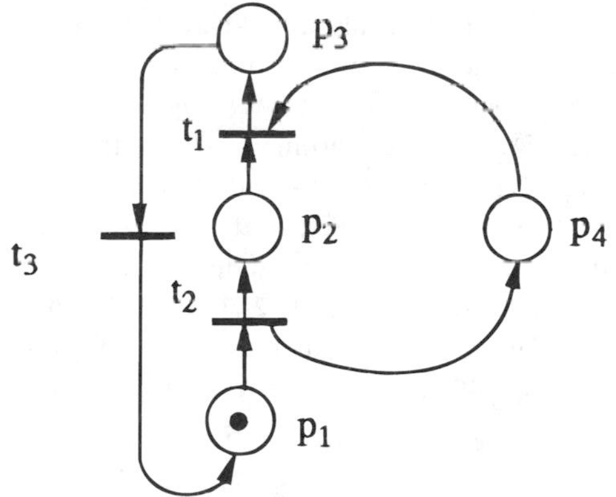

Figure 49.5 Petri net which is safe.

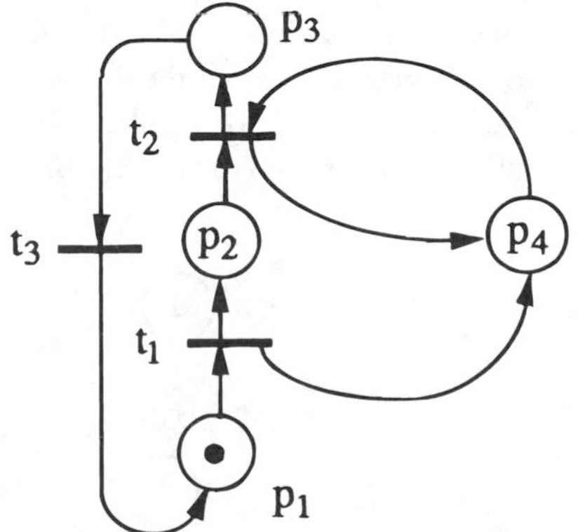

Figure 49.6 Petri net that is unbounded.

tokens are used to represent resources, the number of which in a system is typically fixed, then the number of tokens in a Petri net model of this system should remain unchanged irrespective of the marking the net takes on. This follows from the fact that resources are neither created nor destroyed, unless there is a provision for this to happen. For instance, a broken tool may be removed from the manufacturing cell, thus reducing the number of tools available by one. A Petri net is *conservative* if the number of tokens is conserved. From the net structural point of view, this can only happen if the number of input arcs to each transition is equal to the number of output arcs. However, in man-made systems, resources are frequently combined together so that a certain task can be executed. Then the resources are separated after the task is completed. For instance, in a flexible manufacturing system an automatic guided vehicle collects from a machining cell a pallet carrying products, and subsequently delivers it to the unload station where the vehicle and pallet are separated. This scenario is illustrated in Figure 49.7.

Transition t_1 models loading a pallet onto a vehicle; transition t_2 represents the pallet being delivered to the unload station and subsequently removed from the vehicle. Although the number of tokens in the net changes from two to one when t_1 fires, and then back to two tokens when t_2 fires, the number of resources in the system does not change. To overcome this problem, weights may be associated with places; thus, allowing for the weighted sum of tokens in a net to be constant. A Petri net is said to be *conservative* if there exists a vector w, $w = [w_1, w_2, \ldots, w_m]$, where m is the number of places, and $w(p) > 0$ (the weight of each place in the net is given by a nonnegative integer number different from zero), such that the weighted sum of tokens remains the same for each marking reachable from some initial marking. A Petri net is said to be strictly conservative if all entries of vector w are unity. The Petri net shown in Figure 49.7 is conservative since, for each marking, the weighted sum of tokens with respect to vector $w = [1, 1, 2, 1, 1]$ is costant and equal to two. An example of a Petri net which is not conservative is shown in Figure 49.6; place p_4 can hold an arbitrarily large number of tokens.

Liveness

In a flexible manufacturing system, a deadlock may occur for instance, when the input/output buffer of a machining tool holds a pallet with already machined products, and another pallet with products to be machined has been delivered to the buffer. Assuming that the buffer can hold one pallet only at a time, and an automated guided vehicle (AGV), for instance, has a space for one-pallet only, a deadlock occurs. The pallet with machined parts cannot be moved from the buffer to the AGV. The pallet with parts to be machined cannot be moved from the AGV to the buffer. Unless there is a provision in the control software for deadlock detection and recovery, a deadlock situation, initially confined to a small subsystem, may propagate to affect a large portion of a system. This frequently results in a complete standstill of a system.

The concept of liveness is closely related to the deadlock situation, which has been studied extensively in the context of operating systems. It can be shown that four conditions must hold for a deadlock to occur. These four conditions are

1. Mutual exclusion: a resource is either available or allocated to a process which has an exclusive access to this resource.
2. Hold and wait: a process is allowed to hold a resource(s) while requesting more resources.
3. No preemption: a resource(s) allocated to a process cannot be removed from the process, until it is released by the process itself.
4. Circular wait: two or more processes are arranged in a chain in which each process waits for resources held by the process next in the chain.

In the above example, all four conditions hold, with the buffer and AGV space for pallets regarded as resources. A Petri net modelling a deadlock-free system must be live. This implies that for all markings (representing states the actual system can take on), which are reachable from some initial marking, it is ultimately possible to fire any transition in the net by progressing through some firing sequence. The Petri net shown in Figure 49.5 is live. This requirement, however, might be too strict to represent some real systems or scenarios which exhibit deadlock-free behavior. For instance, the initialization of a system can be modelled by a transition (or transitions) which fires a finite number of times. After initialization, the system may exhibit a deadlock-free behavior, although the Petri net representing this system is no longer live as specified above. For this reason, different levels of liveness were introduced. For details see (Murata, 1989).

Analysis Methods

There are two fundamental methods of analysis of Petri net models. One is based on the coverability tree and the other one

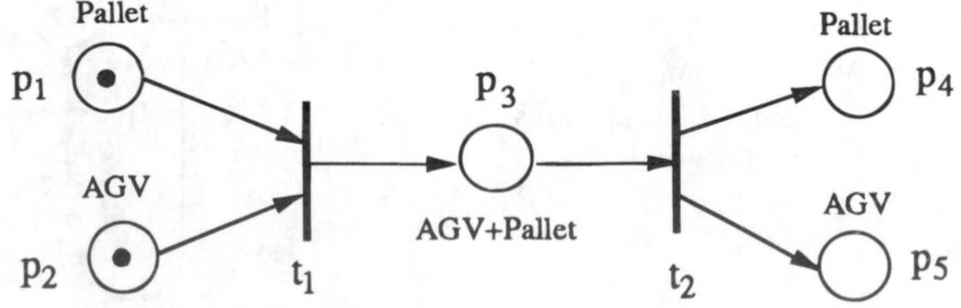

Figure 49.7 Petri net which is conservative with respect to w = [1, 1, 2, 1, 1].

on the matrix equation representation of a net. In addition to the two methods, a number of techniques were proposed to assist in the analysis of Petri net models. However, the discussion of these methods is beyond the scope of this presentation. In this subsection, we provide an overview of the approach based on the coverability tree, as it is the most widely used technique.

This approach is based on the enumeration of all possible markings reachable from some initial marking M_0. Starting with an initial marking M_0, one can construct the reachability set by firing all possible transitions enabled in all possible markings reachable from the initial marking M_0. In the coverability tree, each node is labelled with a marking; arcs are labelled with transitions. The root node of the tree is labelled with an initial marking M_0. The reachability set becomes unbounded for either of two reasons: the existence of duplicate markings, and a net is unbounded. In order to prevent the coverability tree from growing indefinitely large, two steps need to be taken when a tree is constructed. The first step involves eliminating duplicate markings. If on the path from the initial marking M_0 to a current marking M there is a marking M', which is identical to the marking M, then the marking M, as a duplicate marking, becomes a terminal node. The occurrence of a duplicate marking implies that all possible markings reachable from M have been already added to the tree. For unbounded nets, in order to keep the tree finite, the symbol ω is introduced. The symbol ω can be thought of as the infinity. Thus, for any integer n, $\omega + n = \omega$, $\omega - n = \omega$, $n < \omega$. In this case, if on the path from the initial marking M_0 to a current marking M there is a marking M', with its entries less or equal to the corresponding entries in the marking M, then the entries of M, which are strictly greater than the corresponding entries of M', should be replaced by the symbol ω. In some paths the existence of markings with the corresponding entries equal or increasing (as we move away from the root node) indicates that the firing sequence which transforms M' to M can be repeated indefinitely. Each time this sequence is repeated, the number of tokens on places labelled by the symbol ω increases. The coverability tree is constructed according to the following algorithm:

1.0. Let the initial marking M_0 be the root of the tree and tag it "new"
2.0. While "new" markings exist do the following:
3.0. Select a "new" marking M
3.1. If M is identical to another marking in the tree, then tag M "old", and go to another "new" marking
3.2. If no transitions are enabled in M, tag M "terminal"
4.0. For every transition t enabled in marking M do the following:
4.1. Obtain the marking M' which results from firing t in M
4.2. If on the path from the root to M, there exists a marking M'' such that $M'(p) \geq M''(p)$ for each place p, and $M' \neq M''$, then replace $M'(p)$ by ω for each p wherever $M'(p) > M''(p)$
4.3. Introduce M' as a node, draw an arc from M to M' labelled t, and tag M' "new"

The following example will illustrate the approach: Consider the net shown in Figure 49.8 and its coverability tree in Figure 49.9. For the given initial marking, the root node is $M_0 = (1,0,1,0)^T$. In this marking, transition t_3 is enabled. When t_3 fires a new marking is obtained: $M_1 = (1, 0, 0, 1)^T$. This is a "new" marking in which transition t_2 is enabled. Firing t_2 in M_1 results in $M_2 = (1, 1, 1, 0)^T$. Since $M_2 = (1, 1, 1, 0)^T, \geq M_0 = (1, 0, 1, 0)^T$, the second component should be replaced by the symbol ω. This reflects the fact that the firing sequence t_3t_2 may be repeated arbitrarily large number of times. In marking $M_2 = (1, \omega, 1, 0)^T$ two transitions are enabled; transition t_1 and transition t_3. Firing t_1 results in marking $M_3 = (1, \omega, 0, 0)^T$, which is a "terminal" node; there is no transition enabled in M_3. Firing t_3 results in a "new" marking $M_4 = (1, \omega, 0, 1)^T$, which enables

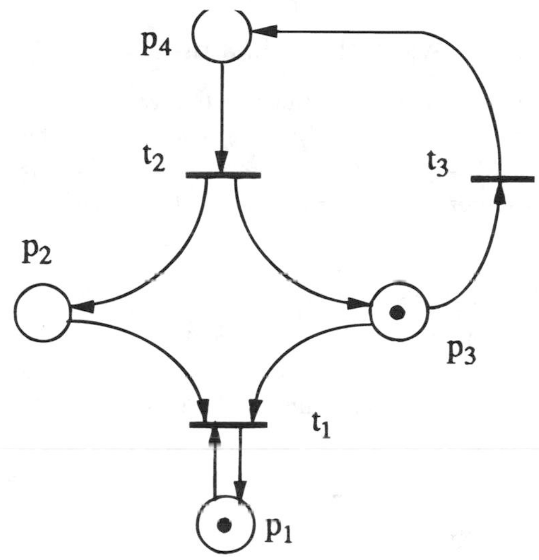

Figure 49.8 A Petri net model.

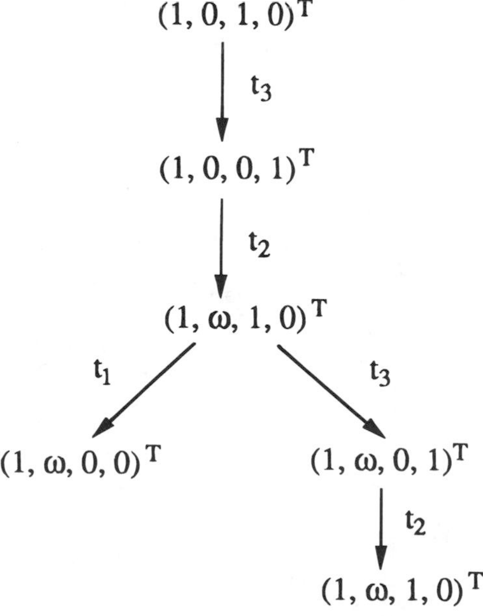

Figure 49.9 The coverability tree of the Petri net model shown in Figure 49.8.

transition t_2. Firing t_2 in M_4 results in an "old" node $M_5 = (1, \omega, 1, 0)^T$ which is identical to M_2.

A number of properties can be studied by using the coverability tree. For instance, if any node in the tree contains the symbol ω, then the net is unbounded since the symbol ω can become arbitrarily large. Otherwise, the net is bounded. If each node of the tree contains only zeros and ones, then the net is safe. A transition is dead if it does not appear as an arc label in the tree. However, since the symbol ω can become arbitrarily large, certain problems cannot be solved by studying the coverability tree only. For a bounded Petri net, the coverability tree contains, as nodes, all possible markings reachable from the initial marking M_0. In this case, the coverability tree is called the reachability tree. For a reachability tree any analysis question can be solved by inspection.

Multirobot System: An Example

In this section, we demonstrate how the reachability tree-based technique can be used to analyze a Petri net model of the multirobot system which is shown in Figure 49.4. Without losing generality, we assume $b = 1$. The reachability tree is shown in Figure 49.10.

Boundedness and Safeness

The Petri net shown in Figure 49.4 is bounded. This is evident from the reachability tree; no marking reachable from the initial marking M_0 contains the ω symbol. In addition, since, for each marking, no entry is greater than one, the net is safe. Two properties related to the operation of the actual system can be deduced from the boundedness property of the Petri net model. There is no buffer overflow; no provision for R1 to access the buffer area when it is full. Also, there is no buffer underflow; no provision for R2 to access the buffer area when it is empty. These properties follow from the net safeness. The entries in each marking, which represent the number of tokens in places p_8 and p_9, are either zero or one. In addition, if there is a token in one place, say p_8, then place p_9 is empty—and vice versa. If there is a token on place p_8, then transition t_5 cannot be enabled. This implies that robot R2 will not be allowed to access the buffer, which is empty at that time. If there is a token on place p_9, on the other hand, then transition t_2 cannot be enabled. This implies that robot R1 will not be allowed to transfer a part to the buffer, which is full at that time. Therefore, there is neither buffer overflow nor underflow.

Conservativeness

The Petri net shown in Figure 49.4 is conservative. From the reachability tree, the net is conservative with respect to vector $w = [1, 1, 2, 1, 1, 2, 1, 1, 1]$. The weighted sum of tokens remains the same for each marking reachable from the initial marking, and equals four. Assuming that tokens in the net of Figure 49.4 represent robots and space resources in the buffer, the implication of this property is that the number of resources in the system is constant, and does not depend on the state of the system.

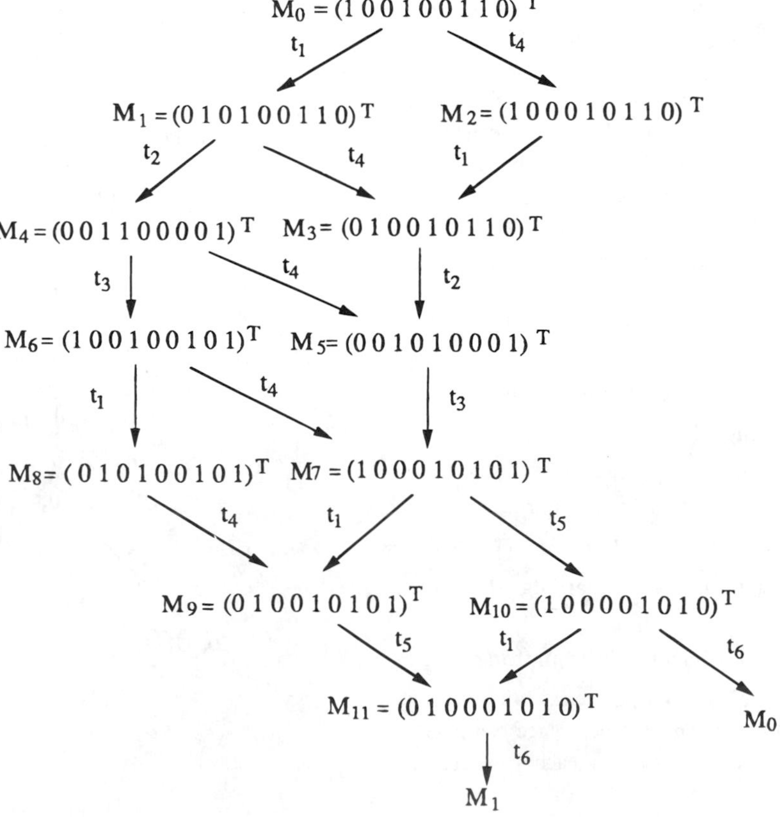

Figure 49.10 Reachability tree of the Petri net model shown in Figure 49.4.

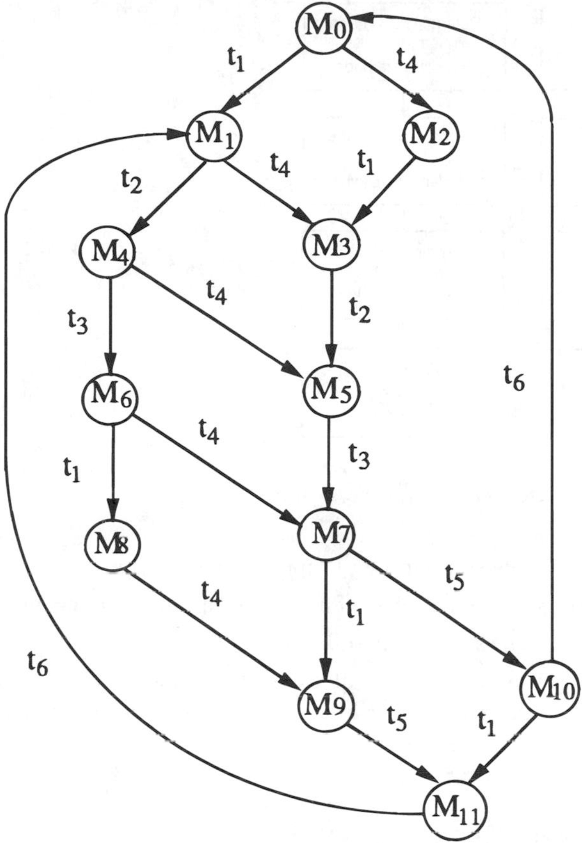

Figure 49.11 Reachability graph of the Petri net shown in Figure 49.4.

Liveness

The Petri net shown in Figure 49.4 is live; all transitions are live. Figure 49.11 shows a reachability graph of the Petri net of Figure 49.4. The reachability graph shown in Figure 49.11 is a directed graph consisting of a set of nodes and a set of directed arcs. The set of nodes represents all distinct labelled nodes in the reachability tree. The set of directed arcs, where each arc is labelled with a transition, represents all possible transitions between all distinct labelled nodes in the reachability tree. By inspection, the net is live since for any marking reachable from making M_0, it is possible to ultimately fire any transition by executing some firing sequence. As the net is live, the system cannot develop a deadlock state which might result in a standstill where no operation is possible.

References

Doumeints, G. and Poumeyrol, E. 1987. Computer aided design for advanced manufacturing systems, *New Technologies for Production Management Systems,* Yoshikawa H. (ed.), Elsevier Science Publishers, 131–160, New York, NY.

Murata, T., 1989. Petri nets, properties, analysis, and applications, Proc. *IEEE,* 1989(4):541–580.

Pun L., Doumeingts G., and Bourley, A. 1995. The GRAI approach to the structural design of flexible manufacturing systems, *J. Prod. Re.,* 23(6):1197–1215.

Ross, D. 1977. Structured analysis: a language for communicating ideas, *IEEE Trans. Software Eng.,* SE–3(1):6–14.

Ross, D. and Shoeman, K. 1977. Structured analysis for requirements definition, *IEEE Trans. Software Eng.,* SE–3(1):6–15.

Zurawski, R. and Zhou, M. 1994. Petri nets and industrial applications; a tutorial, *IEEE Trans. Industrial Electronics,* 41(6): 567–583.

49.2 Automated Manufacturing System Design Using Analytical Techniques

Sunderesh S. Heragu
Christopher M. Lucarelli

Introduction

The key to success of automated manufacturing systems is proper selection and effective use of manufacturing, service, and auxiliary equipment. Proper use of these resources has increased productivity significantly (Heragu and Kusiak, 1987). In recent years, the manufacturing industry has witnessed significant developments. The increase in the number and types of automated systems in the industry is a testament to the developments taking place. However, these developments have taken place at the expense of system design problems. Design problems have become even more complex and designers and users of automated manufacturing systems have developed new tools to cope with these problems. Here, we present some of these tools.

Manufacturing system design, which is a hierarchical combination of several problems is a complex activity. It involves solving a number of problems arranged in a hierarchy. Products to be manufactured, manufacturing processes to be used, number and types of manufacturing equipment (that are capable of performing the required processes) required, preliminary process plan development, determining the tooling and fixture requirements, layout of manufacturing cells and machines, material handling methods to be used, and number and types of specific material handling devices capable of performing the required material handling moves are some of the more important design questions that need to be addressed (Figure 49.12). Solving the manufacturing cell determination and cell layout problems is generally required only for manufacturing systems that produce a large number of components and for which manufacturing activities can be decomposed into almost mutually independent cells. For mass production or continuous production systems, we may bypass these two problems. It should be emphasized that the problems listed in Figure 49.12 are by no means exhaustive and the problems need not necessarily be solved in the order shown. In addition, it may be necessary at times to backtrack or iterate between two or more problems.

Before layout and material handling decisions are undertaken, the required *types* and *number* of manufacturing and support equipment must be known. This problem is sometimes referred to as the machine requirements problem or equipment

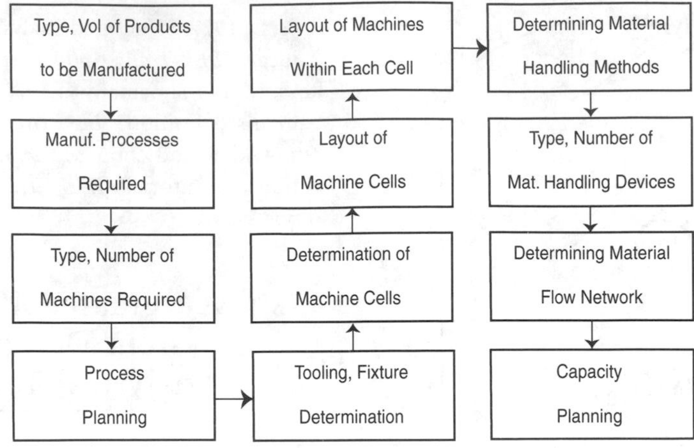

Figure 49.12 Typical design problems in automated manufacturing systems.

selection problem. In order to determine the required types of equipment, we must first know what types of basic manufacturing processes (e.g., forging, drilling holes, planing surfaces, finishing surfaces, and so on) are required and then match available equipment with these processes (e.g., forge is used for forging operations, radial drill presses or boring machines are suitable for drilling and enlarging holes, a planer is required for planing surfaces, a centerless grinder or rotary surface grinder may be used for surface finishing, etc.). Manufacturing equipment is typically classified into types based on their functional capabilities. The types of equipment may be lathes, horizontal milling machines, vertical milling machines, planers, shaping machines, vertical turret lathes, and so on (Miller and Schmidt, 1984).

By number of manufacturing equipment, we mean the number of pieces of each type of equipment available for manufacturing purposes. A company may have 3 lathes, 4 horizontal milling machines, 2 vertical milling machines, 1 planer, 1 shaping machine, 2 vertical turret lathes, and so on. Each type of equipment may require some support facilities as well. For instance, forging equipment will require heat treatment stations, painting stations will require drying equipment, etc. The number and type of support and auxiliary equipment depend upon the type of manufacturing equipment. Generally, if the type and number of manufacturing equipment is known, the required type and number of auxiliary equipment is also known. In fact, much manufacturing equipment is sold with the auxiliary equipment. While we need not know the exact number of each type of manufacturing, support, and auxiliary equipment, we must at least have a rough idea of what types of machines are capable of meeting our processing needs, what support facilities are required, and approximately how many of these are needed.

Automated manufacturing equipment, while capable of performing a variety of operations, is typically expensive. Thus, the manufacturing equipment selection problem is a critical one in the design of a system. By determining the right number and type of equipment, we can achieve the following benefits (Heragu and Kusiak, 1987):

- Make efficient use of capital equipment purchase budget.
- Make efficient use of maintenance and operating budgets.
- Increase machine utilization.
- Make efficient use of available space, as fewer equipment sufficient to meet manufacturing equipment now and in the future are purchased.

Models for System Design

In this section, for the most part, we assume that the type and volume of products to be manufactured has already been determined and that the candidate equipment capable of performing the required processes is known. The specific design problem we focus on is the manufacturing equipment selection problem. There are a number of models for determining the number and type of required manufacturing equipment and evaluating system designs. Examples are linear programming (LP), queuing and simulation models, and perturbation analysis. LP models are typically used to do a "rough-cut" determination of the number of required manufacturing equipment. Using queuing models we can improve this rough estimate by generating information concerning key performance measures for the initial design, examining them and suggesting changes to the initial design, so as to improve the performance measures. For example, a suitable queuing model may be used to examine the initial solution (system design) generated by the LP model and analyze its performance with respect to a number of criteria including work-in-process inventory buildup, machine utilization, production throughput rate, job flowtime, etc. Based on the performance analysis, the system designer may suggest changes to the design such as addition (purchase) of certain machines, deletion of certain others, and so on. Using the modified design, the system designer may then develop a detailed simulation model or stochastic Petri net model (among other tools) and analyze the system with respect to design and operational characteristics. Perturbation analysis is yet another tool that can be used in conjunction with the simulation tool to make the latter more efficient in analyzing the system performance. In Chapter 49.2, we cover the LP, queuing, and perturbation analysis based approaches to system design evaluation. Simulation and Petri net models are discussed in 49.3. Before discussing LP approaches, however, it is instructive to discuss traditional approaches and let the reader contrast them with modern LP and queuing based approaches to system design and planning.

Traditional Approach

Traditional approaches are very simple. Based on the number of products, desired production rate, production efficiency of the equipment required to process the products, standard processing times for the operations required on the products, and time for which machines are available, we can develop a simple formula to determine the number of manufacturing equipment required. One such formula originally presented in Shubin and Madeheim (1951) is presented below. Using the following notation:

- P desired production rate in units per day
- η efficiency of the machine
- τ time for which machine is available per day, in hours
- t time required to process one unit of product at the machine, in hours
- NM number of units of the machine required

the following formula (Equation 49.1) determines the number of units of machine required. Of course, we are assuming that only one product is processed on this machine.

$$NM = \frac{tP}{\tau\eta} \tag{49.1}$$

The above analysis can be easily extended to the case where we have more than one stage of production. To determine the desired production rate at each stage, a backward analysis is suggested (Miller and Schmidt, 1984). Based on the number of units required at the output of the last stage of production, we determine the number of units that must enter this last stage. This depends upon the percentage of scrap at the last stage of operation. For example, if:

- S_l scrap rate at stage l, expressed as a fraction
- N_{ol} Number of units required at the output of stage l

the number of units required at the input of stage l is given by

$$N_{il} = \frac{N_{ol}}{1 - S_l} \tag{49.2}$$

Performing a backward analysis for each stage of operation, we can determine the number of units of raw material required to produce the desired number of finished units. While the traditional approaches are simple, they have certain drawbacks. For example, they assume that each machine or workstation processes only one product. They do not take into account budget, overtime, floor space, and other constraints. When such constraints are imposed, mathematical programming approaches discussed in the next section are useful.

Linear Programming Approach

A number of linear programs have been used to model the manufacturing equipment selection problem. A detailed survey, classification, and comparison is provided in Miller and Davis (1977). These models cover a number of aspects of the equipment selection problem. Examples of the models are

- Dynamic resource allocation model developed by Miller and Davis (1978).
- Machine procurement model in Murty (1983).
- Aggregate production planning and machine requirements planning model in Behnezhad and Khoshnevis (1988).

In this section, an integer programming formulation of the problem is presented. Model M1 presented below minimizes the operating and purchase cost of the machines and material handling systems subject to budget and other constraints. It uses the following notation.

- O_i operation type i, $i = 1, 2, \ldots, o$
- M_i manufacturing equipment type i, $i = 1, 2, \ldots, m$
- P_i part type i, $i = 1, 2, \ldots, p$
- MH_i material handling system type i, $i = 1, 2, \ldots, n$
- c_{ij} cost of performing operation O_i on manufacturing equipment type M_j
- h_{ij} cost of handling part type P_i using material handling system type MH_j
- t_{ij} time required to perform operation O_i on manufacturing equipment type M_j
- s_{ij} time required to transport part type P_i using material handling carrier type MH_j
- τ_j time available on manufacturing equipment type M_j
- σ_j time available on material handling carrier type MH_j
- NO_i number of operations O_i to be performed
- NP_i number of units of part type P_i to be manufactured
- C_j cost of manufacturing equipment type M_j
- H_j cost of material handling system MH_j
- B total budget available
- x_{ij} number of operations O_i to be performed on manufacturing equipment type M_j
- y_{ij} number of units of part type P_i to be transported on material handling system type MH_j
- NM_j number of units of manufacturing equipment type M_j selected
- NMH_j number of units of material handling system type MH_j selected

The four parts of the objective function of model M1 minimize the operating and handling cost of the parts manufactured and procurement cost of manufacturing and material handling equipment. Note that h_{ij} can only be crudely estimated at this stage because the material handling cost depends upon the layout of manufacturing equipment, and the layout is not known yet. The layout is not known because we do not know the manufacturing equipment types and what quantities will be used on the factory floor. Hence, only a rough estimation is required at this stage. Also, note that NO_i depends upon the number of units of each part type being manufactured as well as the number of types of operation O_j required on each part type. The model does not account for set-up times, time that manufacturing and material handling equipment may have to wait for parts, etc. However, we are not making a final decision on the required number and type of equipment, and this decision will be revised in later stages using a queuing and/or simulation model. The model enables us to find a basic solution on which we can build a more concrete one later in the decision-making process.

Model M1. *Minimize*

$$\sum_{i=1}^{o}\sum_{j=1}^{m} c_{ij}x_{ij} + \sum_{i=1}^{p}\sum_{j=1}^{n} h_{ij}y_{ij} + \sum_{i=1}^{m} C_i NM_i + \sum_{i=1}^{n} H_i NMH_i \tag{49.3}$$

Subject to

$$\sum_{j=1}^{m} x_{ij} \geq NO_i, \qquad i = 1, 2, \ldots, o \tag{49.4}$$

$$\sum_{i=1}^{o} t_{ij}x_{ij} \leq \tau_j NM_j, \qquad j = 1, 2, \ldots, m \tag{49.5}$$

$$\sum_{j=1}^{n} y_{ij} \geq NP_i, \qquad i = 1, 2, \ldots, p \tag{49.6}$$

$$\sum_{i=1}^{p} s_{ij}y_{ij} \leq \sigma_j NMH_j, \qquad j = 1, 2, \ldots, n \tag{49.7}$$

$$\sum_{i=1}^{m} C_i NM_i + \sum_{i=1}^{n} H_i NMH_i \leq B \tag{49.8}$$

$$x_{ij} \geq 0 \text{ and integer}, \qquad i = 1, 2, \ldots, o, \quad j = 1, 2, \ldots, m \tag{49.9}$$

$$y_{ij} \geq 0 \text{ and integer}, \qquad i = 1, 2, \ldots, p, \quad j = 1, 2, \ldots, n \tag{49.10}$$

$$NM_j \geq 0 \text{ and integer}, \qquad j = 1, 2, \ldots, m \tag{49.11}$$

$$NMH_j \geq 0 \text{ and integer}, \qquad j = 1, 2, \ldots, n \tag{49.12}$$

Expression 49.4 ensures that the required number of operations are performed, whereas expression 49.6 ensures the required number of parts are transported. Expressions 49.5 and 49.7 require that the time available on each machine and material handling system not be exceeded. Of course, these two constraints ignore set-up and waiting times as mentioned earlier. Expression 49.8 ensures that the available budget is not exceeded. Expressions 49.9, 49.10, 49.11, and 49.12 ensure integer results. To obtain a near-optimal solution to the above model, we can relax the integer restrictions on x_{ij} and y_{ij}, solve the model, and round the resulting values of these variables up or down to the nearest integer. Doing so may cause infeasibility in the binding constraints, but can be overcome by increasing the resource constraints, for example, time available by a few units.

If the available budget is not known, then model M1 without expression 49.8 can be used for solving the manufacturing and material handling equipment selection problem. Other modifications to model M1 to suit various problem scenarios are possible. For example, if the manufacturing equipment and material handling equipment selection is split into two separate problems, model M1 can be modified to solve each. The corresponding two models and another model in which the operations requirements for each part is considered are discussed in Heragu (1997).

EXAMPLE 49.1

In this section, we solve a small example problem to demonstrate how model M1 may be used for the equipment selection problem. Consider the following:

An automobile engine cylinder manufacturing company which supplies high precision engines to a multinational car manufacturer plans to manufacture several models of the automobile engine cylinder. For planning purposes, it uses the following pseudoproducts—a "basic" engine cylinder, a high-technology model, an engine cylinder for sports cars, and a luxury car cylinder. The marketing department has demanded forecast figures which have been aggregated for the four pseudomodels. Thus, it has been determined that 2000, 1500, 1800, and 1000 units of the basic, high-tech, sports, and luxury models will be demanded during the next six months. The models require one or more of three operations, referred to as O_1, O_2, and O_3. There are three machine types and two material handling systems available for performing the three operations and transporting the models. These are denoted as M_1, M_2, M_3, and MH_1, MH_2, respectively. Each machine and handling system may be assumed to be available for 90 percent of the time. The cost of machines M_1, M_2, and M_3 are \$230,000; \$250,000; \$310,000 and that of the material handling carriers MH_1, MH_2 are \$90,000 and \$130,000. The available budget is \$10,000,000. Two matrices showing c_{ij} and h_{ij}, i.e., cost of performing operation O_i on manufacturing equipment type M_j and cost of handling part type P_i using material handling system type MH_j, respectively, are provided below.

$$c_{ij} = \begin{bmatrix} 6 & 12 & 8 \\ 4 & 20 & 4 \\ 12 & 10 & 5 \end{bmatrix}, \qquad h_{ij} = \begin{bmatrix} 10 & 5 \\ 12 & 6 \\ 18 & 9 \\ 6 & 3 \end{bmatrix}$$

The time required to perform operation O_i on manufacturing equipment type M_j is inversely proportional to the corresponding cost and can be determined using the left matrix above and the formula $t_{ij} = 10/c_{ij}$. Similarly, time required to transport part P_i on material handling system MH_j is inversely proportional to the corresponding cost and is determined using the right matrix above and the formula $s_{ij} = 1/h_{ij}$. The required number of each type of operation are: 6,000 for operation O_1, 6,000 for operation O_2, and 4,500 for operation O_3. The total time for which the machines and material handling systems are available is 333 units. Set up a model similar to model M1 and solve it using an available mixed-integer programming software. Determine the number and type of each manufacturing and handling equipment required.

Solution The model input file and solution output file obtained via LINDO (Schrage, 1986) are provided below.

```
Min 90000 NMH1 + 130000 NMH2 + 230000 NM1 + 250000 NM2
  + 310000 NM3 + 6 X11 + 12 X12 + 8 X13 + 4 X21 + 20 X22
  + 4 X23 + 12 X31 + 10 X32 + 5 X33 + 10 Y11 + 10 Y12
   + 12 Y21
  + 6 Y22 + 18 Y31 + 9 Y32 + 6 Y41 + 3 Y42
SUBJECT TO
  C1)    X11 + X12 + X13 >= 6000
  C2)    X21 + X22 + X23 >= 6000
  C3)    X31 + C32 + X33 >= 4500
  C4) - 300 NM1 + 1.67 X11 + 2.5 X21 + 0.83 X31 <= 0
  C5) - 300 NM2 + 0.833 X12 + 0.5 X22 + 2 X32 <= 0
  C6) - 300 NM3 + 1.25 X13 + 2.5 X23 + 2 X33 <= 0
  C7)    Y11 + Y12 >= 2000
  C8)    Y21 + Y22 >= 1500
  C9)    Y31 + Y32 >= 1800
 C10)    Y41 + Y42 >= 1000
 C11) - 300 NMH1 + 0.1 Y11 + 0.0833 Y21 + 0.056 Y31
   + 0.167 Y41 <= 0
 C12) - 300 NMH2 + 0.2 Y12 + 0.167 Y22 + 0.11 Y32
   + 0.33 Y42 <= 0
 C13) 90000 NMH1 + 130000 NMH2 + 230000 NM1 + 250000
   NM2 + 310000 NM3 <= 10000000
END
GIN    5
 OBJECTIVE FUNCTION VALUE
 1)      .1022041 + 08
VARIABLE          VALUE           REDUCED COST
   NMH1          2.000000         90000.000000
   NMH2           .000000        105454.546875
   NM1          14.000000        227516.312500
   NM2          26.000000        247181.562500
   NM3            .000000        307161.781250
   X11         210.420853             0.000000
   X12        5789.579102              .000000
   X13            .000000             0.860030
   X21          45.438900              .000000
   X22        5954.561035              .000000
   X23            .000000             2.954324
   X31        4500.000000              .000000
   X32            .000000              .523254
   X33            .000000             5.049860
   Y11        2000.000000              .000000
   Y12            .000000            11.363637
   Y21        1500.000000              .000000
   Y22           0.000000             7.663636
   Y31        1800.000000              .000000
   Y32           0.000000              .000000
   Y41        1000.000000              .000000
   Y42           0.000000            24.000002
```

As mentioned before, the values of X11, X12, X21, and X22 can be rounded up or down so that none of the constraints are violated and the rounded solution may be used as a good heuristic solution.

Queuing Theory Approach

As the name implies, queuing theory involves the study of queues or waiting lines. Unlike many other modeling methods, it is not process specific and can be used to model any dynamic system in which discrete events alter the state of the system. Queuing theory can be used to study any manufacturing or service system where a queue buildup occurs over time. Thus, the system can be an airport, a walk-in medical clinic, fast-food restaurants, machine shop, etc. In an airport, we see queue buildup occurring when departing airplanes wait for permission to take-off. Similarly, in a walk-in medical clinic, customers wait their turn for consultation with medical staff; in a fast-food restaurant customers wait to place an order and also to pickup their order; in a machine shop, jobs wait to be machined on an automated lathe, and so on. In this section, we will focus on the application of queuing theory to manufacturing environments and see how it can be used to model the automated manufacturing system design problem. Although it does not directly provide answers to questions such as how many pieces of a given machine type must we buy, it can be used to answer the following questions with respect to a given system design.

What is the expected number of parts waiting in a queue?
What is the expected time a part spends waiting in a queue?
What is the probability that a machine will be idle?
What is the probability of a queue being filled to capacity?

Based on the performance analysis of the current design, we can suggest modifications to it so that the modified design performs reasonably well.

In a queuing system, customers arrive by some arrival process and wait in a queue for the next available server. In the manufacturing context, customers could be parts, and servers could be machines or material handling carriers, or pick and place robots. When a server becomes available, a customer is selected by some queue discipline for service. Service is completed by an output process and the customer leaves the system. Figure 49.13 illustrates a simple queuing system.

The arrival process tells us how customers arrive at the queue. Generally, it is assumed that no more than one customer can arrive at any given instant of time. However, in some situations, especially in manufacturing, parts typically arrive in batches.

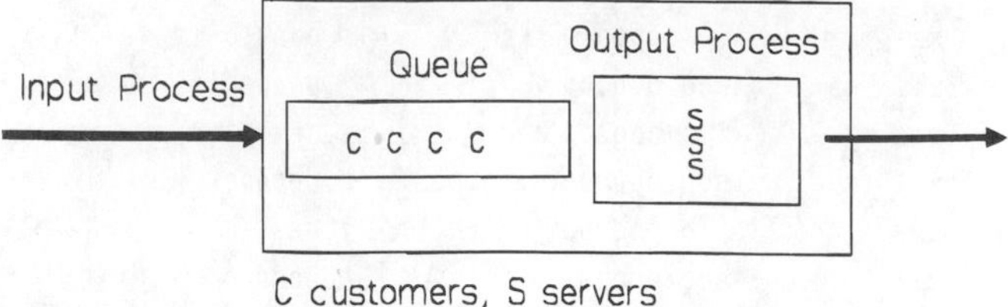

Figure 49.13 A queuing system.

Such arrivals are called bulk arrivals. Unless otherwise stated, the arrival process is assumed to be independent of the number of customers already in the system. The most common exception to this assumption occurs when the arrival rate decreases due to overcrowding, i.e., customers arrive at the system but do not enter it because the queue is too long. For example, a customer may not wait at a restaurant if there are many people already waiting. This phenomenon is referred to as balking. Another situation in which the arrival process depends upon the number of customers already in queue or service occurs when there is a finite calling population. Assume a machine repairman is assigned to the repair of five machines. If all of them have broken down and arrived for repair or are being repaired, no more arrivals can take place. On the other hand, if only one machine is currently being repaired, there is greater probability that another will break down and enter the queue soon. The calling source or calling population from which arrivals take place is said to be finite for such situations. In many models discussed in this chapter, we assume that the calling population is infinite.

An arrival process is usually described by the probability distribution of the number of arrivals in any interval of time. The most commonly used distribution is the Poisson distribution. As will be discussed later, if the number of arrivals in any interval of time follows a Poisson distribution, then the time between consecutive arrivals (known as the interarrival time) follows an exponential distribution.

The queue discipline is the method by which a part or customer is selected from a queue for service. When customers are served in the order in which they arrive, the queue discipline is first come, first served (FCFS). Other queue disciplines include last come, first served (LCFS), service in random order (SIRO), priority-based selection, etc.

The service process is typically described by a probability distribution. The most commonly used distribution is the exponential distribution. As with the arrival process, the service process is assumed to be independent of the number of parts already in the system. In other words, the service rate does not increase to accommodate a backlog. The service process may have one or more servers in series (in which case each customer has to go through a sequence of servers before service is completed) or in parallel (each customer visits one of many servers depending upon who is available at the time the customer is ready to depart the queue for service). The service rate is the number of customers served per unit time. Similarly, the arrival rate of a queuing system is usually given in terms of number of customers arriving per unit time.

Modeling of the Arrivals and Service Process. Suppose that the time to serve a customer is typically small, but an occasional customer requires extensive service. For such a situation, clearly the service time distribution can be approximated as an exponential distribution (Hillier and Lieberman, 1995). For example, jobs arriving at a machining center may require essentially the same machining process (and hence the setup and machining time), but every now and then a job requiring a large setup or machining time or both may arrive at the queue for processing. The reader should be cautioned that not all systems exhibit such a characteristic. In many systems, the type of service may be such that it is almost identical for every customer. In such a case, the actual service time will be close to the expected service time and hence, the exponential distribution cannot be used. However, we will assume that the interarrival times and service times follow an exponential distribution for most models discussed in this chapter. Exceptions are discussed in a separate section.

The exponential distribution has several properties that enable us to analyze queuing models rather easily. We discuss below two fundamental ones.

Property 1. The exponential distribution is "memoryless." If the interarrival time distribution is exponential, it means that the time until the next arrival is independent of how long it has been since the last arrival. Thus, it is completely random. This property of the exponential distribution is also referred to as the no-memory, Markovian, or forgetfulness property.

The no-memory property implies that to predict future arrivals, we do not have to remember how long it has been since the last arrival. If a customer's service in an S-server system has just begun, probability that his/her service will be completed is equal to the probability of the remaining customers' service being completed, although they may have begun to receive service much earlier. Thus, in a machine shop with two lathes (servers), if a job has just been assigned to lathe #2 while lathe #1 has been processing a job for the past ten or more minutes, the no-memory property implies that the service time of the job on lathe #2 has the same distribution as the remaining service time of the job on lathe #1. It is perhaps obvious to the reader now why the no-memory makes queuing models relatively easy to solve. If it were not for this property, it would be very difficult to compute the remaining service time distribution of the job on lathe #1. It has been shown that the exponential distribution is the only continuous probability density function (pdf) having the no-memory property (Feller, 1957). If the interarrival times are not exponential, we would have to use some other distribution, e.g., Erlang.

Property 2. If the interarrival time has an exponential distribution with parameter λ, then the number of arrivals occurring over a time period t, denoted by N_t, has a Poisson distribution with parameter λt. In such a case, the arrival process is said to be Poisson. The reverse is also true. See Gross and Harris (1985) for a complete derivation of this property.

Although the two properties above were discussed in the context of an arrival process, they hold for a service process also. To see this, the reader has to replace the terms interarrival time and arrival with service time and service completion, respectively. Thus, property 2 in the context of a service process says that the service time follows an exponential distribution if the service completion process is Poisson with parameter μ. Notice that the notation used for the arrival process parameter is λ, whereas for the service process it is μ.

Terminology and Notation. Depending upon the assumptions made regarding arrival/service rates, number of servers, queue capacity, discipline, and calling population size, we can have several models. It is therefore convenient to use a simple notation which will enable us to refer to each model concisely. The notation known as Kendall-Lee notation is the most widely used in the queuing literature. It was first devised by Kendall (1953) and modified later by Lee (1966). It clearly and concisely specifies the assumptions made about the system under study. The notation takes the following general form:

a/b/c/d/e/f, where

- a denotes the nature of the arrival process (An M in the a position, denotes that the arrival process is Poisson (or Markovian), or alternately, that the interarrival time follows an exponential distribution. Similarly, a D, G specify that the interarrival time is constant, follows General distributions, respectively. A general distribution refers to any arbitrary, but specified distribution representing independent and identically distributed (iid) random variables.)
- b denotes the service time distribution (Again, an M, D, G in the b position are used to denote the nature of the service time distribution.)
- c denotes the number of parallel servers ($S \geq 1$)
- d denotes the queue discipline (FCFS, LCFS, SIRO, GD, etc.)
- e denotes the maximum number of customers allowed in the system (system capacity) ($C \geq 1$)
- f denotes the size of the calling population (finite or infinite).

For example, a queuing model with exponentially distributed interarrival and service times, 4 servers, a first come first served queue discipline, and no limits on the queue capacity or calling population would have the notation M/M/4/FCFS/∞/∞. Since D, E, and F are typically assumed to be FCFS, infinite, and infinite, respectively, we sometimes use the notation a/b/c instead of a/b/c/d/e/f. The assumption here is that the queue discipline is FCFS, with no limits on queue and calling population sizes.

In an automated manufacturing system, parts are transported to a machine center for processing on a material handling carrier. If the material handling carrier is busy, it will wait in a material handling carrier queue until the latter becomes free. Then, the part is transported to the machine center. Once again, if the machine center is busy, the part will wait in a buffer until the machine is available. After processing, the part enters an output queue where it waits to be transported to the next operation. This process repeats until the last operation, after which the completed part is placed in a shipping area. The part processing time at each center includes any setup or dismantle time involved in the operation. A machine center may have one or more servers that can take the form of a worker, a machine, a combination of machine and worker, or any other resource. Assuming that there are S servers and S–1 or fewer jobs in the system, an incoming job will not have to wait for service. Parts correspond to customers and machines and material handling carriers correspond to servers, respectively. Thus, a queuing model can be readily used to formulate a manufacturing system design problem. In the sections to follow, we illustrate with numerical examples how queuing models can be used to provide specific information about the performance of a given system design. Using such information, the designer can optimize the design of a manufacturing system.

M/M/1/GD/∞/∞ Queuing Systems. The first queuing model discussed is one with a single server. Assume a process with constant part arrival and service completion rates. Specifically, let λ and μ be the arrival and service rate of parts per unit time. If the arrival rate is greater than the service rate, the queue will grow infinitely. The ratio of λ to μ is called the traffic intensity or utilization factor of the server and is denoted as $\rho = \lambda/\mu$. It is obvious that for the system to be in steady state, the traffic intensity must be less than 1.

The rate diagram for the above model is provided in Figure 49.14. It shows the possible transitions into and out of each state. Analysis of most queuing models is computationally feasible only when the system has reached steady state. Until steady state is reached, the system is said to be in a transient state and its analysis is extremely difficult. Therefore, this section deals only with steady-state analysis.

To analyze a queuing model which has reached steady state we make use of the following simple, but important, result. The mean rate of entering any state n is equal to the mean rate of leaving that state. This result (equation) is also called a balance equation or flow conservation equation. The derivation of this result is presented in Hillier and Lieberman (1995).

Let P_j be the steady-state probability of being in state j. For the M/M/1/GD/∞/∞ system, the steady-state conditions for the various states can be written by examining Figure 49.14. For state 0 in the rate diagram, the flow conservation equation requires $\mu P_1 = \lambda P_0$. Therefore, $P_1 = (\lambda/\mu)P_0$. Similarly, for state 1, $\mu P_2 + \lambda P_0 = \lambda P_1 + \mu P_1$. Hence, $P_2 = (\lambda/\mu)P_1 + (1/\mu)(\mu P_1 - \lambda P_0) = (\lambda/\mu)^2 P_0$. For state 2, $\mu P_3 + \lambda P_1 = \lambda P_2 + \mu P_2$ Therefore, $P_3 = (\lambda/\mu)P_2 + (1/\mu)(\mu P_2 - \lambda P_1) = (\lambda/\mu)^3 P_0$.

Thus, $P_n = (\lambda/\mu)^n P_0$, for $n = 1, 2, \ldots$ Substituting $\rho = (\lambda/\mu)$ and assuming $0 \leq \rho < 1$, we get $P_1 = \rho P_0$, $P_2 = \rho^2 P_0$, $\ldots$, $P_n = \rho^n P_0$, $\ldots$

We know that $P_0 + P_1 + \cdots + P_n + \cdots = 1$. Hence, $P_0(1 + \rho + \rho^2 + \cdots + \rho^n + \cdots) = 1$. Let $S = 1 + \rho + \rho^2 + \cdots + \rho^n + \cdots$. Then $\rho S = \rho + \rho^2 + \cdots + \rho^n + \cdots$ and $S - \rho S = 1$. Therefore, $S = 1/(1 - \rho)$ and $P_0 = 1 - \rho$. Since the value of P_0 is known, we can find all the remaining P_i's using $P_i = \rho^i P_0$, $i = 1, 2, \ldots$ Once the state probabilities are known,

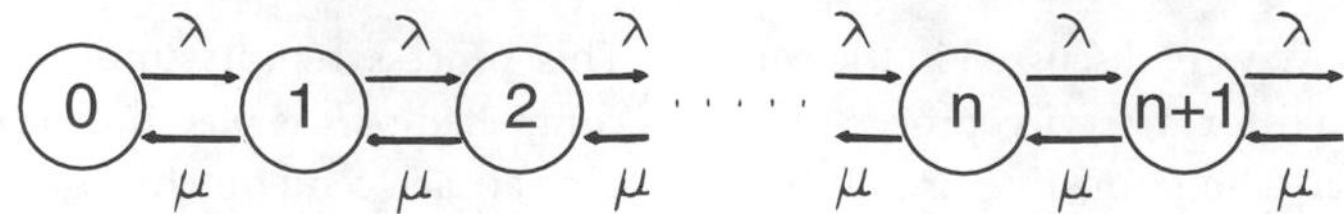

Figure 49.14 Rate diagram for an M/M/1/GD/∞/∞ queuing model.

the other operating characteristics of the queuing system can be calculated. To do so, we make use of one of the most fundamental results in queuing theory called Little's formula. Before we discuss this result, the following additional notation is necessary.

- L average number of customers in system
- L_q average number of customers in queue
- L_s average number of customers in service
- W average time a customers spends in system
- W_q average time a customer spends in queue
- W_s average time a customer spends in service
- λ mean arrival rate
- μ mean service rate

Little's formula (or Little's theorem or Little's law) is a simple one defining the relationship between the operational characteristics of a queuing system. It basically states that under steady-state conditions, the average number of customers in any queuing system is equal to the mean arrival rate times the average time spent in the system by a customer. This powerful result holds for any interarrival and service time distributions, any service discipline, and for any number of servers. The only requirement is that the queuing system must be in steady state.

Mathematically, Little's formula is given by $L = \lambda W$. It has also been proved that $L_q = \lambda W_q$ and $L_s = \lambda W_s$. For a rigorous proof of this result, see Little (1961), Stidham (1974) and Ross (1970). We "prove" it below in an intuitive manner. Assume a queuing system has Poisson arrivals with a mean arrival rate of λ. Then, there will be one arrival every $1/\lambda$ time units. Also assume that the current queue length is L customers and a new customer has just arrived. Then, average time spent by the customer in the system is L/λ which is equal to W, assuming W is the average waiting time in the system (i.e., waiting time in queue plus service time) for any customer, including the customer in question. Hence, $L = \lambda W$.

If we know the mean service time, then it is easy to see that $W = W_q + 1/\mu$. If any of the six key variables (L, L_q, L_s, W, W_q, W_s) is known, the remaining can be immediately determined using the above three equations. For a queuing model with S servers, we can analytically find the value of L, L_q and L_s using the following formulae.

$$L = \sum_{n=0}^{\infty} nP_n, \quad L_q = \sum_{n=S}^{\infty} (n - S)P_n \quad \text{and} \quad L_s = L - L_q.$$

Going back to the derivation of operating characteristics for the M/M/1 model, it is clear that the expected number of parts

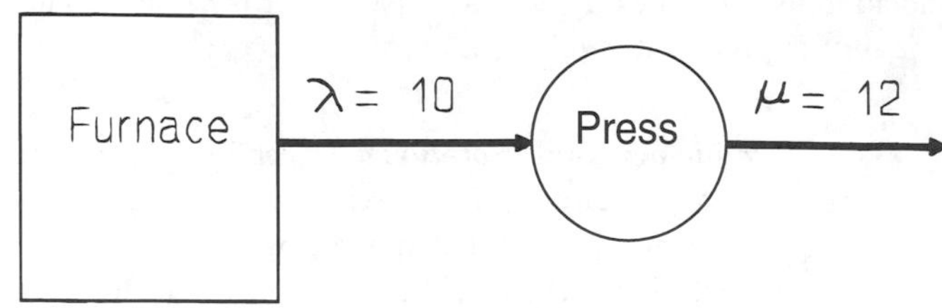

Figure 49.15 Furnace and press system.

at a workstation is simply the probability of a given state n multiplied by the probability of its occurrence. Thus,

$$L = \sum_{n=0}^{\infty} nP_n = \sum_{n=0}^{\infty} n\rho^n(1 - \rho)$$

$$= (1 - \rho) \sum_{n=0}^{\infty} n\rho^n$$

$$L_q = \sum_{n=1}^{\infty} (n - 1)P_n = \sum_{n=1}^{\infty} nP_n - \sum_{n=1}^{\infty} P_n$$

$$= L - (1 - P_0) = L - 1 + 1 - \rho = L - \rho$$

$$L_s = L - L_q = \lambda/(\mu - \lambda) - \lambda^2/(\mu(\mu - \lambda))$$

$$= \lambda[1/(\mu - \lambda) - \lambda/(\mu(\mu - \lambda))] = \lambda/\mu = \rho$$

Using Little's formula, we can then determine, W, W_q and W_s as $W = L/\lambda = 1/(\mu(1 - \rho))$, $W_q = \rho/(\mu(1 - \rho))$ and $W_s = 1/\mu$. We now have all the essential characteristics of the system under consideration. An example illustrating how the M/M/1 model can be applied to evaluate a system design is provided below.

EXAMPLE 49.2

Consider the forge press operation shown in Figure 49.15. After heating, parts arrive at the press with an exponentially distributed interarrival rate of one-tenth of an hour. The pressing process takes an average of 5 minutes.

Find the following:

1. The percentage of the time that the press is idle.
2. The average number of parts in the queuing system.
3. The average queue length.
4. The throughput time of the system.
5. The amount of time spent in the queue.

Solution The forge press operation can be modeled as an M/M/1 queuing system. The rate diagram for this problem is shown in Figure 49.16.

The arrival and departure rates are given as $\lambda = 10$ per hour and $\mu = 12$ per hour, respectively. The utilization factor is $\rho = \lambda/\mu = 0.833$.

1. The probability that the press is idle is the probability that there are no jobs being served or waiting. $P_0 = 1 - \rho = 0.167$ or the press is idle 16.7% of the time.
2. The average number of parts in the queuing system is $L = \rho/(1 - \rho) = 5$ parts. This is also the average work-in-process (WIP).
3. The average queue length is $L_q = \rho^2/(1 - \rho) = 4.167$ parts.
4. The throughput time (or time spent in the system) is $W = 1/(\mu(1 - \rho)) = 0.5$ hours. Since λ and μ were given in parts per hour, the throughput time is also in hours.
5. The amount of time spent in the queue is $W_q = \rho/(\mu(1 - \rho)) = 0.4167$ hours $= 25$ minutes.

M/M/S/GD/∞/∞ Queuing Systems. Consider a queuing system in which there are S servers and the ith server, $i = 1, 2, \ldots, S$, has an exponentially distributed service time with parameter μ_i. Then, it can be shown that the multiple server system performs just like a single-server system where the service time has an exponential distribution with parameter $\Sigma_{i=1}^{S} \mu_i$. Assume λ is the arrival rate and μ is the service rate for each of the S servers. Clearly, if $n \leq S$ customers are present in the system, all of them are in service. If $n > S$ customers are present, all S servers are busy and $n - S$ customers are waiting. The service rate for the entire queuing system is $\Sigma_{i=1}^{S} \mu_i$. Thus for an M/M/2 system with each server having a service rate of 5 parts per hour, the departure rate is 5 parts per hour if only one server is working. If both servers are working, the departure rate is 10 parts per hour.

The rate diagram for this system is shown in Figure 49.17.

The utilization factor for an M/M/S system is given by $\rho = \lambda/S\mu$. We can derive the operating characteristics of a multiple server system from the state probabilities in a similar manner as done for a single server system (see, for example, Hillier and Lieberman, 1995; Taha, 1992; Winston, 1994; Kleinrock, 1975; Gross and Harris, 1985).

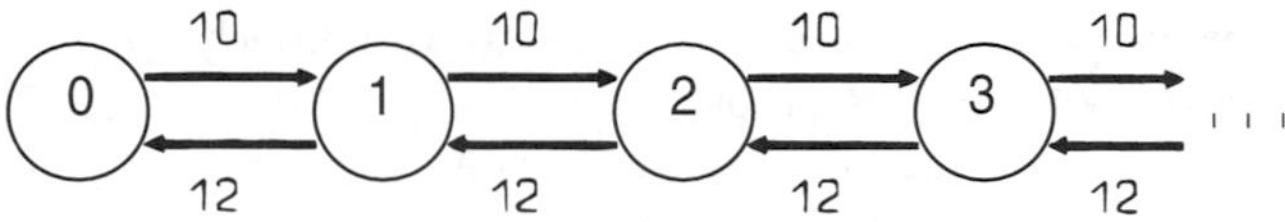

Figure 49.16 Rate diagram for example 49.2

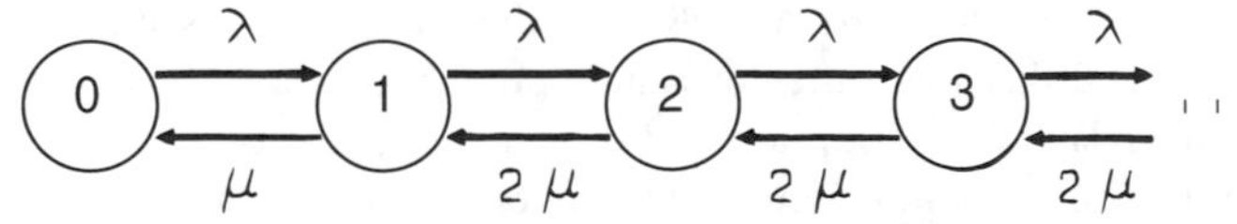

Figure 49.17 M/M/2 rate diagram.

Table 49.2 M/M/1/GD/∞/∞ and M/M/S/GD/∞/∞ Operating Characteristics

Model	M/M/1/GD/∞/∞	M/M/S/GD/∞/∞
P_0	$1 - \rho$	$1/[(\rho S)^S/(S!(1 - \rho)) + \Sigma_{n=0}^{S-1}((\rho S)^n/n!)]$
L	$\lambda/(\mu - \lambda)$	$L_q + \lambda/\mu$
L_q	$L - \rho$	$[(\rho S)^S P_0 \rho]/[S!(1 - \rho)^2]$
W	$1/(\mu(1 - \rho))$	$W_q + 1/\mu$
W_q	$\lambda/(\mu(\mu - \lambda))$	L_q/λ

Results for the basic M/M/1 and M/M/S models are summarized in Table 49.2. It should be noted that $\rho = \lambda/\mu$ for the single-server model, but $\rho = \lambda/(\mu S)$ for the multiple-server case. In both models, $P_n = c_n P_0$ and formulae for c_n are summarized below.

Single-server model:

$$c_n = \rho^n, \text{ for } n = 1, 2, \ldots$$

$$P_0 = 1 - \rho$$

Multiple-server model:

$$c_n = \begin{cases} \dfrac{1}{n!}\left(\dfrac{\lambda}{\mu}\right)^n & \text{for } n = 1, 2, \ldots, S \\ \dfrac{1}{S!S^{n-S}}\left(\dfrac{\lambda}{\mu}\right)^n & \text{for } n = S + 1, S + 2, \ldots \end{cases}$$

$$P_0 = \frac{1}{\displaystyle\sum_{n=0}^{S-1} \frac{(\lambda/\mu)^n}{n!} + \frac{(\lambda/\mu)^S}{S!}\left(\frac{1}{1 - \lambda(S\mu)}\right)}$$

EXAMPLE 49.3

In an effort to reduce work-in-process, the press in Example 49.2 is modified to have two servers. Find the following:

1. The average queue length.
2. The average number of parts in the queuing system (WIP).
3. The expected amount of time spent in the queue.
4. The throughput of the system.

Solution This new queuing system is an M/M/2 model. The rate diagram for this problem is shown in Figure 49.18. The arrival and departure rates are the same as in Example 2; $\lambda = 10$ per hour and $\mu = 12$ per hour. The 2 servers result in a utilization factor of $\rho = \lambda/S\mu = 0.417$.

The new idle time of the press is $P_0 = 1/[(\rho S)^S/S!(1 - \rho) + \Sigma((\rho S)^n/n!)] = 7/17 = 0.412$ or 41.2% of the time.

1. The expected queue length in the two server system is $L_q = [\rho(\rho S)^S P_0]/[S!(1 - \rho)^2] = 7/40 = 0.175$ parts.

2. The expected WIP is $L = L_q + \lambda/\mu = 121/120 = 1.0083$ parts.
3. The expected amount of time spent in the queue is $W_q = [(\rho S)^S P_0]/[S!S\mu(1 - \rho)^2] = 7/400 = 0.0175$ hours $= 1.05$ minutes.
4. The throughput time of the system is $W = W_q + 1/\mu = 0.10083$ hours $= 6.05$ minutes.

By comparing results from Examples 49.2 and 49.3, a system designer can make informed decisions as to which system design is better.

EXAMPLE 49.4

A workstation in a company receives parts from two sources—a subcontractor and a machining center within the company. The arrival rate of each is Poisson with parameter $\lambda/2$. The company has purchased another identical workstation in order to increase throughput. Each has a service rate of μ. The company has the following three ways of designing the system.

a. Have all the parts from the subcontractor visit the first workstation, and all the parts from the internal machining center visit the recently purchased workstation for processing.
b. Since the parts coming from the subcontractor are identical to those from the internal machining center, and the two workstations can be combined into one so as to have a service rate of 2μ, the company can have parts from both sources join a single queue and visit the combined workstation.
c. Combine the two arrivals into a single queue but keep the workstations separate.

Making the usual assumptions that services times are random variables following an exponential distribution, how would the queuing model representation of these designs differ?

Solution System (a) is an M/M/1 queue with parameters $\lambda/2$ and μ. System (b) is an M/M/1 queue with parameters λ and 2μ. System (c) is an M/M/2 queue with parameters λ and μ. The three systems are depicted in Figure 49.19.

It is known that the sum of two Poisson variables yields a Poisson variable. Since two Poisson arrivals are superimposed for systems (b) and (c), the arrival rate is $\lambda/2 + \lambda/2 = \lambda$. The effective arrival and service rates are λ and 2μ for all three systems. Using the formulae in Table 49.2, it can be shown that the average waiting times for system (a) is greater than that of system (c) which in turn is greater than that of system (b).

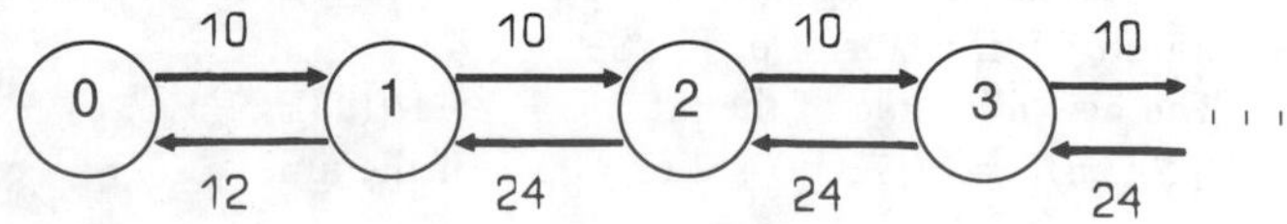

Figure 49.18 Rate Diagram for Example 49.3.

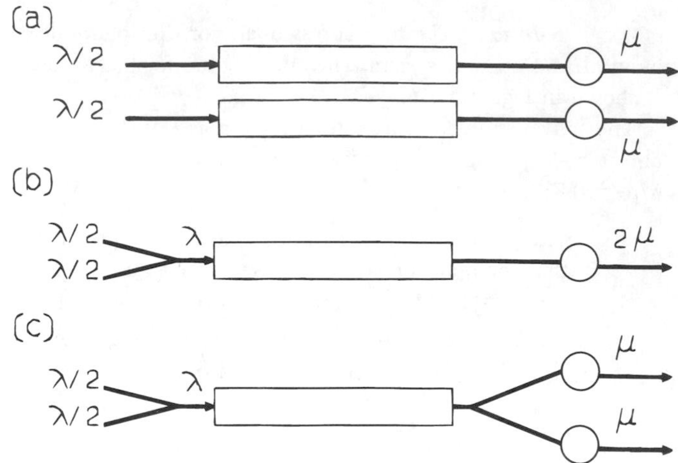

Figure 49.19 Three system designs.

However, when the number of parts waiting in queue (work-in-process inventory) is considered, it can be shown that system (c) is preferred to system (b) (Heragu, 1997).

Variations of the Basic Queuing Models. Excluding the birth-death model, we studied two basic queuing models in the previous two sections. Analytical results for the two models were easily obtained. These models serve as building blocks for other models in queuing theory. In this section, we will relax many of the assumptions made in the last two, one at a time, and where available, provide analytical results for each. Although derivations of the results are not provided, it is rather easy for the reader to prove the results for many of the models on his/her own. If necessary, we suggest that any standard queuing text such as Gross and Harris (1985) or Kleinrock (1975) be consulted.

Finite Queue Size Models. Thus far, we have assumed that the capacity of the queue is infinite. However, in many applications, this may not be true. For example, each machining center in a manufacturing system has a finite input buffer (waiting area for parts that need processing). Suppose that the queue can accommodate a maximum of C parts (customers). Then, the maximum number of servers to be used is also C, i.e., $S \leq C$. Also, arrival rate λ_n for states $n \geq C$ is equal to 0.

For all the previous models, we made the assumption that the ratio of the arrival rate to the service rate must be less than 1; otherwise the queue size would have grown infinitely. We need not worry about this problem here because due to the finite queue capacity, the queue size can never "blow up". Operating characteristics for this model can be calculated as shown in Taha (1992), amongst others.

Finite Source Models. All the models discussed so far assumed that the source or calling population was infinite. We now relax this assumption and present some results. Of course, if the calling population is finite, it does not make sense to have an infinite queue. So, we assume that the queue size is equal to the size of the calling population. The finite source model is also known as the machine repairman problem, as a machine repairman is usually assigned to a finite number of machines. Thus, the calling population is finite for such types of models. The results of such models are discussed in Bunday and Scraton (1980).

Nonexponential Models. Thus far, we have assumed that the interarrival and service times have an exponential distribution. While it enables us to find simple analytical results, it cannot be used in some real world situations, especially in many automated manufacturing systems. For example in a manufacturing system the arrival of parts may not be random, but scheduled to occur at regular intervals. The arrival process is no longer Poisson and the appropriate model is therefore D/M/S, assuming we have exponential service time distribution. Results for this and the E/M/S models are available in tabular form in Hillier, et al. (1981). Exact results for the more general G/M/1 and G/M/S models are available in Gross and Harris (1985). Unfortunately, although we can get exact results for the exponential service time (general interarrival time distribution models), we cannot for the general service time (exponential interarrival time distribution models). The latter are encountered more often in practice.

Priority-Based and Other Queuing Models. Consider a manufacturing system in which parts are processed not on FCFS basis, but based on the priority attached to them. Suppose that we have m types of jobs ranked so that job type 1 has the highest priority and job type m the lowest. The criteria for assigning priorities may be the importance of the job which itself may be a function of "value" to the company. Assume that jobs are selected for service in order of their priority. Thus, job type 1 is selected first; if the queue does not have any job of type 1, then a job of type 2 is selected, and so on. Within a priority level, however, jobs are selected for service on a FCFS basis. In other words, if there are no type 1 or 2 jobs in the queue, and we have 5 jobs of type 3, one of these is selected in the order in which they entered the queue. Further, once a job is selected for service, it will not exit the server (and the system) until its service is completed. In other words, its service will not be interrupted if a job with a higher priority happens to arrive after the service has begun. Such a priority rule is called the nonpreemptive rule. In contrast, the preemptive rule allows for a job to be "bumped" from service if a higher priority job arrives at the queue, even if the arrival occurred after the lower priority job began receiving its service.

For the single-server nonpreemptive case, we have results for models with Poisson arrivals and general service time distributions. For the multiple server case, we have results for models with Poisson arrivals and exponential service time distributions (see, for example, Taha, 1992). Variations of these models are treated in Jaiswal (1968).

In concluding this section, we wish to caution the reader that in addition to the models discussed so far, there are others for which results have been obtained. For many of these, results are available in the form of tables for various values of key variables and parameters. For example, results for models with certain state-dependent arrival and departure rates are available in table format in Conway and Maxwell (1961) and Hillier et al. (1964); results for the M/M/S/GD/N/N model and M/D/S/GD/∞/∞ are available in Peck and Hazelwood (1958) and Hillier et al. (1981).

Queuing Networks. The queuing models we studied in the previous sections assumed that the product undergoes one stage of service and then departs the system. While single-stage service models demonstrate the value of queuing theory in the design of manufacturing systems, multiple stage service models represent more realistic scenarios. These multiple stages can be modeled as networks of queues. Throughout the remainder of this section, we therefore consider networks of queues and derive analytical results for two types of networks. Before we do so, however, we define a queue network. Consider a network of i machine centers in which each center has S_i identical parallel servers (machines). In the most general case, parts arrive at each center i from an external source (e.g., a subcontractor) or another internal machine center, visit a series of other centers (perhaps including center i) and depart from some machine center. Different parts may enter (and exit) the system from different machine centers and may visit different numbers and sets of machine centers in different sequences to complete processing. Let us refer to machine centers as nodes. A network with the following properties is referred to as Jackson network (Jackson, 1957 and 1963).

(i) The arrival process from the external node to node i is Poisson with mean rate of γ_i.
(ii) The service time at node i follows an exponential distribution with parameter μ_i.
(iii) p_{i0} is the probability that a part will exit the network after completion of processing at node i and p_{ij} is the probability that a part will visit node j after completion of processing at node i. p_{i0} and p_{ij} are assumed to be known and independent of the state of the system.

Two types of Jackson networks are possible—open networks and closed networks. They are discussed in the following two subsections.

Open Queuing Network. In the open network model, jobs can arrive from an external source at one or more of the machine centers. As will be seen later, an open queuing network is simply an expansion of the single-machine system previously discussed. Although results are provided for the general m node open queuing network later, it is instructive to develop results for the two node network first. Consequently, we will consider an open network system with two single server machines in tandem (or series) as shown in Figure 49.20. Assume there is infinite capacity in front of each node (machine). Parts arrive at the network according to a Poisson process with parameter λ, i.e., the interarrival time follows an exponential distribution with mean $1/\lambda$ and each is served by machine 1 first and then machine 2, which have independent exponential service time distributions with parameters μ_1 and μ_2, respectively.

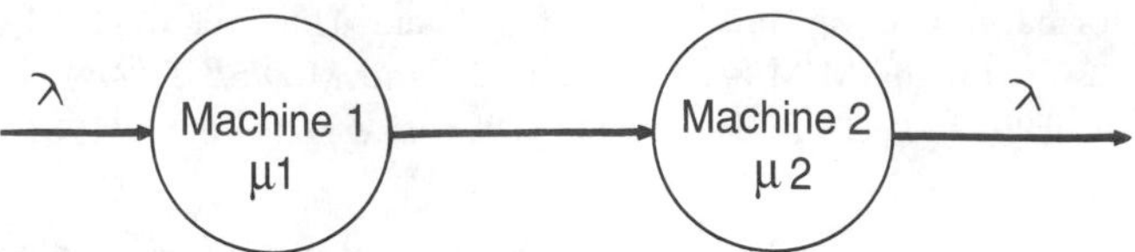

Figure 49.20 Two-machine system.

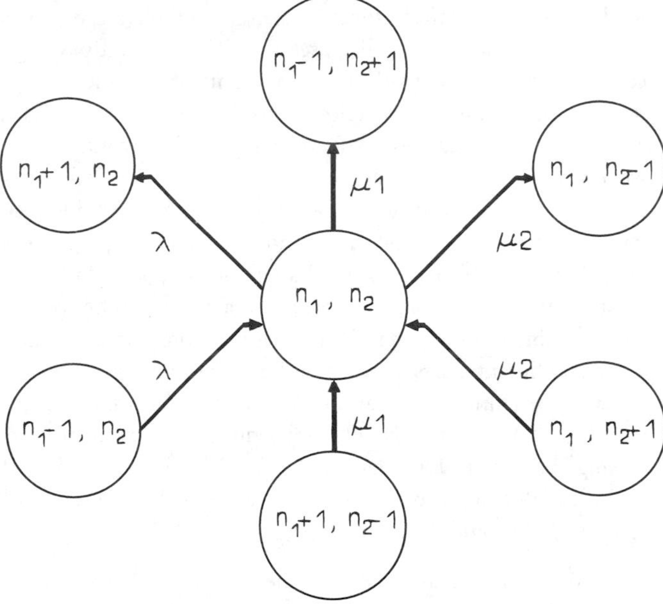

Figure 49.21 State diagram of 2-machine network. (*Source:* Adapted from Viswanadham, N. and Narahari, Y. 1992. In *Performance Modeling of Automated Manufacturing Systems,* p. 380. Prentice-Hall, Englewood Cliffs, NJ. With permission.)

Let N_{t1}, N_{t2} be the number of customers at nodes 1, 2, respectively, at time t. The stochastic process defining the number of parts at each of the nodes at time t, $t \geq 0$, is a Markov chain. Assuming we have reached steady state, let n_1 and n_2 represent the average number of parts at nodes 1 and 2, respectively. Since steady-state has been reached, we will denote $P(n_1,n_2)$ as the steady-state joint probability of having n_1, n_2 jobs at machines 1, 2, respectively. Whenever a service completion or arrival occurs, the state changes. In Figure 49.21, a partial steady-state diagram is shown for the general state in which we have n_1 and n_2 number of parts at nodes 1 and 2, respectively. Recall, we constructed rate diagram, developed state equations and determined steady-state probabilities for a single-machine system previously. Using the same principle, the steady-state probabilities can be calculated from the multiple-machine state diagram. As mentioned previously, steady-state implies that the rate out of state n is equal to the rate into state n. Of course, when defining the state of the system for the multiple machine system, both machines must be considered. The state of the system is the number of jobs at machine 1 and machine 2.

The steady-state probabilities of the network are generally referred to as product form solution, as it is a product of the individual steady-state probabilities derived earlier for the M/M/1 model. This remarkable result says that, in steady state, the joint probability of having n_1 parts at node 1 and n_2 in node 2 is simply the product of the individual probabilities of having n_1 parts at node 1 and n_2 parts at node 2. Hence, n_1, n_2 are independent of each other. Thus, the product form solution implies that the network behaves as if each node is an independent M/M/1 queue with parameters λ, μ_i for the two nodes 1 and 2. It should be cautioned that the network does not break up into two independent M/M/1 queues with the arrival process being a true Poisson process with parameters λ. In fact, it has been shown that this is not necessarily true (Disney, 1981). However, what the above result states is that despite the fact that the arrival process into each node is not a true Poisson process, the network can be treated as if it is made up of 2 independent M/M/1 nodes. Therefore, each machine in the network can be treated independently. This allows for a divide and conquer strategy of analyzing each machine in a 2-machine network as 2 single-machine queues. From the above product form solution, it can be shown that the mean number of parts in the system L is equal to $\rho_1/(1 - \rho_1) + \rho_2/(1 - \rho_2)$ and average time spent in the system W is equal to L/λ. (More details are provided in Gross and Harris, 1985; White et al., 1975; among others).

Although we demonstrated Jackson's result using a two machine center network, it can be expanded for any number of machine centers. Moreover, the above results hold even when the number of servers in each of these centers is greater than 1. Of course, all other conditions including infinite buffers in front of each machine, independent service time distributions at each node, identical machines at each machine center (node), and Poisson arrival process, must hold. In fact, even when there is a feedback loop, i.e., parts may reenter the system for rework. Jackson (1963) showed that the above product form solution will hold. However, when the infinite buffer assumption is violated, the product form result does not.

We conclude this section by providing steady-state results for the m-node Jackson network. Under the assumptions stated above, the analysis of an m node Jackson network is a three-step process. The first step is to calculate the effective arrival rate. Recall that each machine center in a network has two sources of arrivals. The first arrival source is from the external environment. The second arrival source is from other machines in the system. As before, let (i) the arrival process from the external node to internal node i ($i = 1,2, \ldots m$) be Poisson with mean rate of γ_i, (ii) the service time at (internal) node i follow an exponential distribution with parameter μ_i, (iii) each node have S_i identical machines, and (iv) p_{i0} be the probability that a part will exit the network after completion of processing at node i and p_{ij} be the probability that a part will visit node j after completion of processing at node i. Since the arrival at a node is made up of external as well as internal arrivals (the latter occurring as a result of service completions at other internal nodes) the arrival rate at node i ($i = 1,2, \ldots, m$) can be written as:

$$\lambda_i = \gamma_i + \sum_{j=1}^{m} p_{ji}\lambda_j$$

If p_{i0} and γ_j are greater than 0 for some i, j, and ρ_i given by $\lambda_i/(S_i\mu_i)$, for $i = 1,2, \ldots m$, is less than 1 (so that the queue does not grow infinitely at any node), then it can be shown that the product form solution will hold. Specifically, it can be shown that the joint probability of having $n_1, n_2, \ldots, n_m$ parts at nodes 1, 2, …, m is simply the product of the following probabilities—having n_1 parts at node 1, n_2 parts at node 2, … n_m parts at node m. Of course, the latter probabilities are obtained by treating each node as an independent M/M/S_i queue and $n_1, n_2, \ldots, n_m$ are independent of each other. Since, we are able to treat the above m-node Jackson network as m independent M/M/S_i queuing systems (for the same reasons mentioned in the two-node case), we can obtain the performance measures for the entire network using equations from Table 49.2.

Thus, the second step is to analyze each machine in the m-machine system independently. The number of servers in each center will determine whether the M/M/1 or M/M/S results are to be used. The final step is to combine the results from each machine center to analyze performance of the entire system.

In order to keep the presentation simple, we only present results for the M/M/1 case. The reader can get results similarly for the M/M/S_i case using Table 49.2. The average number of parts in the network and average waiting time for each part in the network, i.e., flowtime, are given by

$$L = \sum_{i=1}^{m} L_i = \sum_{i=1}^{m} \lambda_i/(\mu_i - \lambda_i), \qquad W = L/\lambda_{eff},$$

where λ_{eff}, the effective arrival rate into the network from the outside, is given by $\sum_{i=1}^{m} \gamma_i$. In addition to the above overall performance measures, we can obtain some specific performance measures at the individual nodes, e.g., the sojourn times, i.e., time spent by the part in traversing each node of the network (Viswanadham and Narahari, 1992).

We now present an example illustrating some variations of the open Jackson network model. The example shows the effect of a "rework" loop in a manufacturing system. It utilizes Jackson's result to calculate effective arrival rates from multiple arrival sources in a 4-machine network.

EXAMPLE 49.5

Figure 49.22 shows a 4-machine open network queuing system consisting of a furnace (F), a press (P), a rolling mill (M), and an inspection area (I). Each machine has a single server and the rejection rate is 5%. The part routing matrix is shown in Table 49.3.

Table 49.3

From/To	F	P	M	I	Exit
F	—	0.3	0.7	—	—
P	—	—	—	1.0	—
M	—	0.4	—	0.6	—
I	0.05	—	—	—	0.95

Given that the external arrival rate is $\lambda = 12$, and $\mu_F = 15$, $\mu_P = 12$, $\mu_M = 18$, and $\mu_I = 21$, determine the WIP and processing time of the system.

Solution Step 1. Calculate the effective arrival rate. From Table 49.3, we develop the following equations

$$\lambda_F = \lambda + 0.05\lambda_I = 12 + 0.05\lambda_I$$

$$\lambda_P = 0.3\lambda_F + 0.4\lambda_M$$

$$\lambda_M = 0.7\lambda_F$$

$$\lambda_I = 1.0\lambda_P + 0.6\lambda_M$$

From Figure 49.22, we can see that jobs are conserved (all jobs eventually leave the system). Therefore, the arrival rate for I is equal to the effective arrival rate at F. This can also be verified by solving the above four equations which have four unknowns. Solving them yields the following effective arrival rates. Service rates and utilizations at each center are also shown.

$$\lambda_F = 12.63, \qquad \mu_F = 15, \qquad \rho_F = 0.842$$

$$\lambda_P = 7.33, \qquad \mu_P = 12, \qquad \rho_P = 0.611$$

$$\lambda_M = 8.84, \qquad \mu_M = 18, \qquad \rho_M = 0.491$$

$$\lambda_I = 12.63, \qquad \mu_I = 21, \qquad \rho_I = 0.602$$

Step 2. Analyze each machine in the system independently. All the machines have only one server, therefore, we will use Table 49.2 for the operating characteristics of an M/M/1 queue.

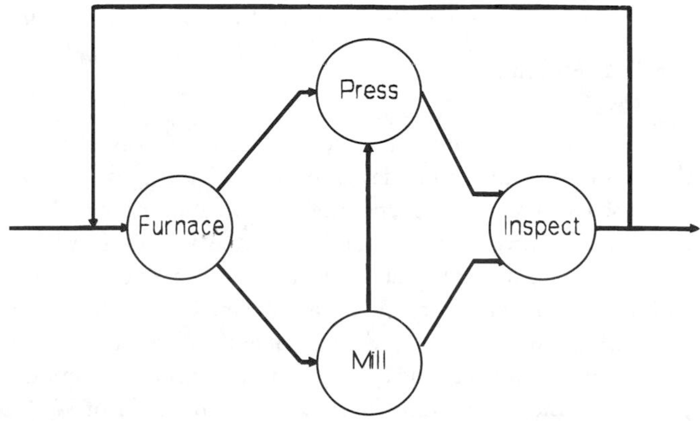

Figure 49.22 4-Machine system.

$$W_F = 1/(\mu_F(1 - \rho_F)) = 1/(15(1 - 0.842))$$
$$= 0.422 \text{ hours} = 25.33 \text{ minutes}$$
$$L_F = \rho_F/(1 - \rho_F) = 0.842/(1 - 0.842)$$
$$= 5.33 \text{ ingots}$$
$$W_P = 1/(\mu_P(1 - \rho_P)) = 1/(15(1 - 0.842))$$
$$= 0.422 \text{ hours} = 25.33 \text{ minutes}$$
$$L_P = \rho_P/(1 - \rho_P) = 0.611/(1 - 0.611)$$
$$= 1.57 \text{ ingots}$$
$$W_M = 1/(\mu_M(1 - \rho_M)) = 1/(18(1 - 0.491))$$
$$= 0.109 \text{ hours} = 6.6 \text{ minutes}$$
$$L_M = \rho_M/(1 - \rho_M) = 0.491(1 - 0.491)$$
$$= 0.97 \text{ ingots}$$
$$W_I = 1/(\mu_I(1 - \rho_I)) = 1/(21(1 - 0.602))$$
$$= 0.119 \text{ hours} = 7.2 \text{ minutes}$$
$$L_I = \rho_I/(1 - \rho_I) = 0.602(1 - 0.602)$$
$$= 1.51 \text{ ingots}$$

Step 3. Combine the results from each machine to analyze the performance of the entire system. Recall, that due to the "rework" loop we must find the expected number of visits (v_j).

$$v_j = \lambda_j/\lambda, \qquad v_F = \lambda_F/\lambda = 12.63/12 = 1.05$$
$$v_P = \lambda_P/\lambda = 7.33/12 = 0.61$$
$$v_M = \lambda_M/\lambda = 8.84/12 = 0.74$$
$$v_I = \lambda_I/\lambda = 12.63/12 = 1.05$$
$$L = L_F + L_P + L_M + L_I$$
$$= 5.33 + 1.57 + 0.97 + 1.51 = 9.38 \text{ ingots}$$
$$W = v_F W_F + v_P W_P + v_M W_M + v_I W_I$$
$$= 1.05(0.422) + 0.61(0.214) + 0.74(0.109) + 1.05(0.119)$$
$$= 46.8 \text{ minutes}$$

Closed Queuing Network. So far in our discussion of the uses of queuing in the design of manufacturing models, we have used WIP and processing time as evaluation criteria. In all the examples, there has been no limit on the amount of WIP in the system. A part arrived at a machine from an external source with rate γ_i and was processed in the system. This is what makes an open network queuing system "open." However, it is not typical for a manufacturing system to have an unlimited amount of WIP. In most manufacturing systems, a low level of WIP is maintained. The desired level of WIP may be a function of storage space, cash investment or various other factors. In fact, in a JIT system, which focuses on the elimination or at least minimization of all forms of waste, WIP is seen as a major source of waste and managers strive hard to keep the WIP level to a minimum. Kanbans are used to control WIP. For modeling such problems, the closed queuing network is used. A queuing network in which the WIP or level of jobs is fixed at some level n is called a closed network queuing network. A new part enters the network only when another leaves the network. Whereas WIP was an output statistic in an open network queuing model, it is a control parameter in the closed network. Another major difference between an open and closed network is that external arrival rates γ_i (i = 1,2, ...,m) are equal to 0.

Consider a simple manufacturing system in which we have two machines connected in series and each job first goes through machine 1 and then machine 2 as depicted in Figure 49.23. If we assume that there are n pallets in the system each containing one part (job) and that a new job is introduced into the system (loaded onto a pallet) only as another leaves the system (unloaded from a pallet), then we have a two-node (machine) closed queuing network model which always has a fixed number of jobs (n) in it. As long as the queue in front of each machine can hold at least $n - 1$ jobs, there will be no blocking. Deriving results for the no-blocking situation is much easier than doing so for the blocking case.

We now extend the above discussion to the multiple-node closed network. Assume that we have m-nodes each having a single server. Let n_i represent the number of jobs at node i after steady state has been reached. Since the total number of pallets is n, the number of jobs in the system at anytime is also n. Obviously, $\sum_{i=1}^{m} n_i = n$. For the two-node case, the number of possible combinations of jobs at the two nodes or the number of states $(0,n)$, $(1,n - 1)$, ... $(n,0)$ is equal to $n + 1$. In the m-node case, the number of states is given by $(n + m - 1)!/(n!(m - 1)!)$. Clearly, the computation of performance measures is likely to be tedious, because of the explosion in the number of states. Before we discuss this issue further, we provide two examples of how manufacturing systems may be modeled as closed m-node networks.

Consider a flexible manufacturing system in which a part is processed first on one of k machines and then on one of $m - k$ machines for the final processing step. After the first operation, the parts join a queue in front of a conveyor-robot material handling system and are then transported to one of $m - k$ machines for the final processing step. After the last operation is completed, another conveyor-robot system is used to unload completed parts and load new ones onto the pallet. The probability of the parts visiting each of the m machine is known. This problem may be modeled as a m + 2-node closed network as shown in Figure 49.24.

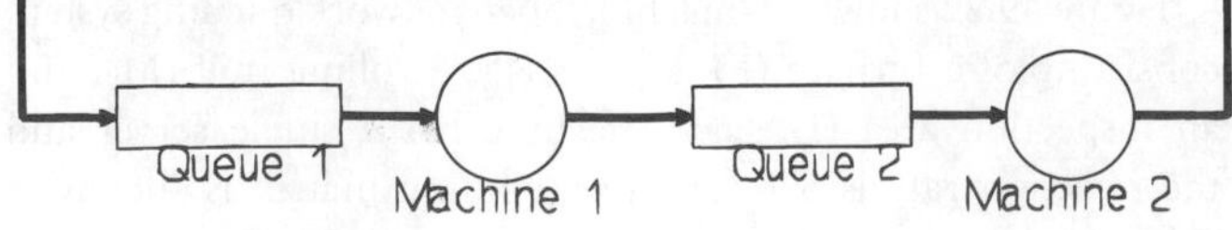

Figure 49.23 Closed network queuing system.

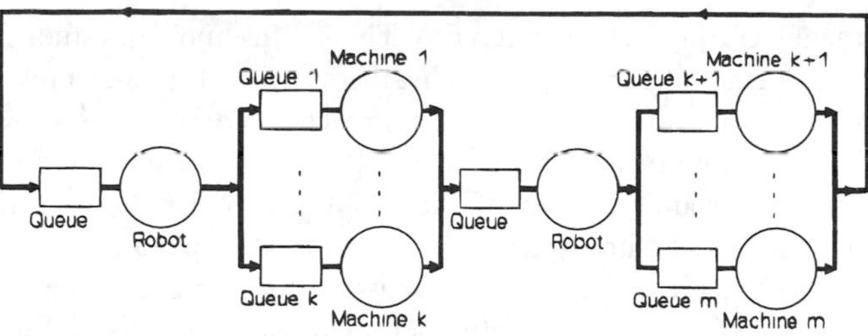

Figure 49.24 Closed network model of a simple automated manufacturing system.

Now consider an extension of the above example. A flexible manufacturing cell processes r different types of jobs and each unit of each type undergoes N_r processing steps. There are max $\{N_i: i = 1,2, \ldots r\}$ automated guided vehicles (AGVs). After the first operation, all the part types are sent to the first AGV queue for transportation to the second machine (operation); similarly, after the second operation, all the part types are sent to the second AGV queue for transportation to the third machine (operation), and so on. After the last operation, all the completed part types are removed from the pallets and the empty pallets are sent to the last AGV queue for transportation to their respective first operation, where new parts are loaded. Thus, the first transportation for each part type is done by the first AGV, the second transportation by the second AGV, and so on. There are several machines capable of doing the intermediate operations for part type r and the probability that part type r will visit a specific machine for the N_r^{th} operation is assumed to be known. Further, if the service rate at each node and each server is exponential with a known mean service rate, this problem can again be appropriately modeled as a multiclass, multinode closed network.

As discussed earlier, deriving results for the general m-node network is quite involved. Hence, we only present some basic results for a general m-node network, and refer the reader to Viswanadham and Narahari (1992) for derivation of the results and further details. The system considered here assumes that there are m nodes in the network and r types of jobs are processed in it. The processing of job type j at node i (which has a single server) is exponential with parameter μ_i^j, $i = 1,2, \ldots, m$, $j = 1,2, \ldots r$ and the routing probability p_{ij}^r that part type r will visit node j after operation at node i is known. The number of parts of type r at node i is n_i^r and N^r is the total number of parts of type r in the network. In other words, N^r is the number of pallets allotted to part type r. As before, $\sum_{i=1}^{m} n_i^r = N_i$. Defining T_i as the vector $(n_i^1, n_i^2, \ldots, n_i^r)$ corresponding to node i and T as $(T_1, T_2, \ldots, T_m)$, we can get the steady-state joint probability of having all possible combinations of parts and part-types at each node using the following product form solution.

$$P(T_1, T_2, \ldots, T_m) = (1/\text{NC}) \prod_{i=1}^{m} f(T_i),$$

where NC, $f(T_i)$ are a normalization constant, function of T_i, respectively.

Calculating the normalization constant on a computer can pose serious numerical difficulties, especially as n, m become large. Hence, the above method of optimally determining the steady-state joint probability (from which other performance measures of the closed network can be obtained) is not desirable. Instead, we have a heuristic method called the approximate MVA method, amongst others. MVA and other heuristic methods are discussed in several sources such as Viswanadham and Narahari (1992), Suri et al. (1991) and Seidmann et al. (1987).

Discrete Event Simulation

Discrete event simulation is one of the most popular performance evaluation and modeling tools that is widely used in manufacturing and nonmanufacturing systems. Its main attractive feature is that it can capture (model) all the complexities that one can imagine in a system. In fact, the level to which a simulation model can be made to reflect reality is limited only by the amount of programming and computer (hardware and software) costs and time. As Ho (1987) puts it, it is a brute force trial and error computer experimentation of the real system being studied. Thus, unlike the analytical methods studied so far, the output provided by a simulation model is a statistical estimate and not exact value of the performance measure being sought. However, it can be tailored to any specific system, can be made to generate estimates of a wide variety of performance measures and more important, evaluate time-variant behavior (Askin and Standridge, 1993).

A simulation model basically consists of three entity classes—resources, transactions and queues. In the manufacturing context, resources could be machines or material handling systems whose service is required by parts for the processing or transportation transaction. If a part has to wait for processing or transportation, it must wait in a queue. The attributes of a resource, transaction or queue define the state of the system. This state changes whenever an event occurs. Since these events occur at discrete, i.e., specific points in time, simulation models are often called discrete simulation models. There are several simulation languages available for developing simulation models. Examples are GPSS/H (Banks et al. 1989), ARENA (Systems Modeling Corporation, 1994), SLAM II (Pritsker, 1986), SIMNET II (Taha, 1992), etc. Simulation languages allow the user to develop models by tracing the sequence of steps taken by a part in the system. Another way is to make the modeling event driven, i.e., specification of the events that change the system state. Since simulation modeling is discussed in detail in a later chapter, we will not provide any further details here and conclude this section by mentioning that it is a versatile and powerful tool whose only disadvantage is the enormous amount of effort required to develop, validate and test the model. As mentioned in the beginning of this section, simulation is a tool that must be used to perform detailed analysis only after a rough cut or preliminary analysis has been done using linear programming and queuing approaches, respectively. This will drastically cut down the modeling effort.

Perturbation Analysis. The tools available for design and planning of automated manufacturing systems have been classified as being generative or evaluative (Suri, 1985). Generative tools such as the LP models discussed in this chapter and corresponding algorithms are those that find optimal or suboptimal decisions given a set of objectives and constraints. Evaluative tools evaluate decisions previously made (using generative tools or using other means) with respect to prespecified performance measures. It should be obvious to the reader at this point that the queuing and simulation models discussed previously fall into this category. The disadvantage with queuing models is that they provide average, steady-state values for the performance measures in an aggregate manner for relatively simple models. When we begin to increase complexity of the model, for example, by imposing finite queue capacity, state-dependent routing of parts within the system, incorporating specific scheduling policies, simultaneous resource sharing, nonstandard service mechanisms (queue disciplines), etc., queuing models are rather inadequate (Ho, 1987). The only tool that can effectively capture various complexities is simulation and we have to rely on this tool to perform detailed analysis. However, as pointed out in the previous section, simulation requires an enormous amount of programming, data input, model validation, testing, implementation and training effort.

Now suppose we have an analytical tool that will help our simulation experiment by suggesting at the end of a certain run in what direction a key decision parameter has to be changed. In other words, assume we have a tool that will tell us the sensitivity of a performance measure to small changes in the values of a decision parameter. For example, let us suppose that we can predict accurately what change a small decrease in the service time of a particular machine brings about in the part throughput rate? Based on this feedback, we can then decide how to improve the decision parameter and thereby use the simulation model more effectively. A costly way of estimating such sensitivity is of course to setup two simulation models one with the original service time and another with the revised time, run both models, obtain throughput data and use them to predict the effects of processing rate changes on production throughput rate. A more efficient way of estimating sensitivity is to use a relatively new tool called perturbation analysis. It can relatively accurately predict the sensitivity of performance measures to small changes in the values of a decision parameter. It has characteristics of evaluative and generative tools. One of the earliest papers on this topic was by Ho et al. (1979). The focus of perturbation analysis is on determining sensitivity of a given system with respect to small changes in one set of decision parameters by observing a single experiment. It does not require experimentation, excessive memory or computation time. The computations required are often very simple and information is required only from a single experiment of the system. For further details on this topic, we refer the reader to two papers, one by Ho (1987) and another by Suri (1989) which provide an excellent introduction and explore several research issues.

Conclusions

The purpose of this section has been to introduce the reader to relatively recent technologies that have been applied to solve the manufacturing system design problem. The technologies studied were linear programming, queuing theory, discrete event simulation and perturbation analysis approaches. Our emphasis has been on the queuing theory approach because of its ability to capture dynamic aspects of the problem. So that the reader may understand how queuing models may be applied to system design, we presented detailed discussions on the basic types of queuing models. There are numerous queuing models and we presented or derived results only for the ones for which analytical solution is possible. (Where it is not, the user must rely on simulation models which is discussed in detail in section 49.3.) In all the models studied, we were concerned primarily with the determination of waiting time (length) in (of) the system and queue.

Discussion of the basic queuing models allowed us to understand development of results for queuing networks which find extensive use in manufacturing systems analysis and design. Since our objective was to introduce the reader to developments in queuing and queuing network theory, we covered only some basic open and closed networks. Only one solution technique was discussed for the open networks. As mentioned previously, there are other techniques available and these are discussed in Viswanadham and Narahari (1992), Seidmann et al. (1987) and Suri et al. (1993), amongst others. For the reader interested in exploring queuing network applications to manufacturing systems, we suggest extensive survey articles such as those by Buzacott and Yao (1986, 1986a) as well as a number of research articles including those by Solberg (1977), Shantikumar and Buzacott (1981), Stecke and Solberg (1981), to name a few. There is abundant literature on the queuing network subject. In this section, we have provided a partial list. For an extensive bibliography on this subject, we refer the reader to Viswanadham and Narahari (1992) and Suri et al. (1993).

For the sake of continuity, we provided a brief description of discrete event simulation and perturbation analysis. The former is discussed in more detail in the following section. A thorough description of the latter is beyond the scope of this handbook and we refer the reader to Ho (1987) and Suri (1989) for an excellent introduction to this topic.

Acknowledgment

The books by Askin and Standridge (1993), Gross and Harris (1985), Hillier and Lieberman (1995), and Viswanadham and Narahari (1992) and the papers by Ho (1987), Suri (1989) and Suri, Sanders and Kamath (1993) have greatly helped us in putting the queuing and perturbation analysis sections of this chapter together. Some examples have been adapted from these sources to suit the design flavor of this chapter. We gratefully acknowledge the help we have received from the above sources.

References

Askin, R. G. and Standridge, C. R. 1993. *Modeling and Analysis of Manufacturing Systems*, Wiley, New York, NY.

Banks, J. Carson, J. S., and Sy, J. 1989. *Getting Started with GPSS/H*, Wolverine Software Corporation, Annandale, VA.

Behnezhad, A. R. and Khoshnevis, B. 1988. The effects of manufacturing progress function on machine requirements and aggregate planning, *Inter. J. Production Research*, 26:309–326.

Bunday, B. D. and Scraton, R. E. 1980. The G/M/r machine interference model, *European J. Operational Research*, 4:399–402.

Buzacott, J. A. and Yao, D. D. 1986. On queuing Network models of flexible manufacturing systems, *Queueing Systems*, 1:5–27.

Buzacott, J. A. and Yao, D. D. 1986a. Flexible manufacturing systems: a review of analytical models, *Management Science*, 32(7):890–905.

Conway, R. W. and Maxwell, W. L. 1961. A queuing model with state dependent service rate. *Ind. Eng.* 12:132–126.

Disney R. L. 1981. Queuing networks, *Proc. Symposia in Applied Mathematics*, 25:53–83.

Feller, W. 1957. *An Introduction to Probability Theory and its Applications*, Wiley, New York, NY.

Gross, D. and Harris, C. M. 1985. *Fundamentals of Queuing Theory*, Wiley, New York, NY.

Heragu, S. S. 1997. *Facilities Design*, PWS Kent West Publishing Company, Boston, MA.

Heragu, S. S. and Kusiak, A. 1987. Expert systems in manufacturing design, *IEEE Trans. Systems, Man and Cybernetics*. SMC–17 (6):898–912.

Hillier, F. S., Conway, R. W., and Maxwell, W. L. 1964. A multiple server queuing model with state dependent service rate, *J. Ind. Eng.* 15:153–157.

Hillier, F. S. and Lieberman, G. J. 1995. *Introduction to Operations Research*, McGraw-Hill, New York, NY.

Hillier, F. S. D., Avis, O. S., Yu, F., Fossett, L. Lo, F., and Reiman, M. 1981. *Queuing Tables and Graphs*, Elsevier, New York, NY.

Ho, Y. C. 1987. Performance evaluation and perturbation analysis of discrete event dynamic systems, *IEEE Trans. Auto. Control*, AC–32(7):563–572.

Ho, Y. C., Suri, R., Cao, X. R., Diehl, G. W., Dille, J. W., and Zazanis, M. 1984. Optimization of large multiclass (non-product-form) queuing networks using perturbation analysis, *Large Scale Systems Journal*, 7:165–195.

Ho, Y. C., Eyler, M. A., and Chien, T. T., 1979. A gradient technique for general buffer storage design in a serial production line. *Inter. J. Prod. Research*. 17(6):557–580.

Jackson, J. R. 1957. Networks of waiting lines, *Operations Research*, 518–527.

Jackson, J. R. 1963. Jobshop-like queuing systems, *Management Science*, 10:131–142.

Jaiswal, N. K. 1968. *Priority Queues*, Academic Press, New York, NY.

Kendall, D. G. 1953. Stochastic processes occurring in the theory of queues and their analysis by the Method of markov chains, *Annals of Mathematical Statistics*, 24:338–354.

Kleinrock, L. 1975, *Queuing Systems, Vol. I: Theory*, Wiley, New York, NY.

Lee, A. 1966. *Applied Queuing Theory*, MacMillan, New York, NY.

Leung, Y. T. and Suri, R. 1990. Performance evaluation of discrete manufacturing systems, *IEEE Control Systems Magazine*, June, 77–86.

Little, J. D. C. 1961. A proof for the queuing formula L=λW, *Operations Research*. 9:383–385.

Miller, D. M. and Davis, R. P. 1978. A dynamic resource allocation model for a machine requirements problem, *IIE Trans.* 10(3):237–243.

Miller, D. M. and Davis, R. P. 1977. The machine requirements problem, *Int. J. Prod. Research*, 15(2):219–231.

Miller, D. M. and Schmidt, J. W. 1984. *Industrial Engineering and Operations Research*, Wiley, New York, NY.

Murty, K. G. 1983. *Linear Programming*, Wiley, New York, NY.

Parzen, E. 1962. *Stochastic Processes*, Holden-Day, San Francisco, CA.

Peck, L. G. and Hazelwood, R. N. 1958. *Finite Queuing Tables*, Wiley, New York, NY.

Pritsker, A. A. B. 1986. *Introduction to Simulation and SLAM II*, Wiley, New York, NY.

Ross, S. 1970. *Applied Probability Models with Optimization Applications*, Holden-Day, San Francisco, CA.

Schrage, L. E. 1986. *Linear, Integer and Quadratic Programming with LINDO*, The Scientific Press, Palo Alto, CA.

Seidmann, A., Schweitzer, P. J., and Shalev-Oren, S. 1987. Computerized closed queuing network models of flexible manufacturing systems: a comparative evaluation, *Large Scale Systems*, 12:91–107.

Shantikumar, J. G. and Buzacott, J. A. 1981. Open queuing network models of dynamic job shops, *Int. J. Prod. Research*, 19:255–256.

Shubin, J. A. and Madeheim, H. 1951. *Plant Layout*, Prentice-Hall, New York, NY.

Solberg, J. J. 1977. A mathematical model of computerized manufacturing systems, *Proc. 4th Int. Research Conference on Production Research*, 22–30. Tokyo, Japan.

Stecke, K. E. and Solberg, J. J. (1981), Loading and control policies for a flexible manufacturing systems, *Inter. J. Prod. research*, Vol. 19, No. 19(5):481–490.

Stidham, S. 1974. A Last word on L=λW, *Operations Research*, 22(2):417–421.

Suri, R. 1985. An overview of evaluative models for flexible manufacturing systems, *Annals of Operations Research*, 3:13–21.

Suri, R. 1989. Perturbation analysis: the State of the art and research issues explained via the GI/GI/1 queue, *Proc. IEEE*, 77(1):114–137.

Suri, R. and Dille, J. W. 1985. A technique for on-line sensitivity analysis of flexible manufacturing systems, *Annals of Operations Research*, 3:381–391.

Suri, R. and Hildebrand, R. R. 1984. Modeling Flexible Manufacturing systems using mean value analysis, *J. Manufacturing Systems*, 3:27–38.

Suri, R., Sanders, J. L., and Kamath, M. 1991. Performance evaluation of production networks, in S. C. Graves, A. H. G. Rinooy Kan and P. Zipkin (Eds.), *Logistics of Production and Inventory: Handbook in Operations Research and Management Science*, Vol. 4, 1993.

Systems Modeling Corporation 1994. ARENA Reference Guide, Sewickley, PA.

Taha, H. A. 1992. *Operations Research: An Introduction,* MacMillan, New York, NY.

Viswanadham, N. and Narahari, Y. 1992. *Performance Modeling of Automated Manufacturing Systems,* Prentice-Hall, Englewood Cliffs, NJ.

Winston, W. L. 1994. *Operations Research: Applications and Algorithms,* Duxbury Press, Belmont, CA.

White, J. A., Schmidt, J. W., and Bennett, G. K. 1975. *Analysis of Queuing Systems,* Academic Press, New York, NY.

49.3 Discrete Event Simulation[1]

MengChu Zhou, Anthony D. Robbi, and Richard Zurawski

Introduction

A discrete event manufacturing system is one whose actions over time are driven by the occurrence of events. Examples of such events are the start or end of a machining operation, the arrival or departure of a part, and a robot picking up or loading a part. A discrete event model comprises system building blocks and a schedule or rules which govern the flow of work. Discrete event simulation is a process through which a discrete event model mimics the behavior of the system, event by event. Qualitative and quantitative data from this process are obtained to predict the behavior of the system and its level of performance.

Simulation has two basic motives, rapid prototyping to determine the correctness of system behavior and performance prediction. Some performance measures of interest are throughput, resource utilization, buffer capacity, yield, and effects of failures. Simulation studies are conducted using computers. The software model can capture all the dynamics and interactions of a real system. Since real manufacturing systems are expensive to build, simulation is an important means to predict performance accurately, investigate effects of parameter changes, identify bottlenecks, and choose the best design among alternatives.

Simulation Procedure

In discrete event simulation, the model contains entities or objects, attributes, events, activities, and the interrelationships among them. The collection of entities and their statuses define the system state. A system state may change only at discrete points in time. These changes are driven by the event occurrences, generally asynchronous. In manufacturing systems, an entity can be a machine, a robot, material (work in progress), and a controller with its schedule. The attributes of a machine include its operation rate, the nature of its operation, and its reliability. Event examples are the arrival of raw material, loading, unloading, the change of a tool, and the start of an operation.

The following steps are involved in discrete event system simulation as shown in Figure 49.25 (Banks, 1984; Rembold, 1993):

Goal definition and requirement specification. Determine the requirements specification and simulation goal of the manufacturing system under study or design. The goals are to determine the best system among several alternatives and to investigate the system behavior and performance. An appropriate level of detail is selected to match the modeling goal. For example, it could be to optimize

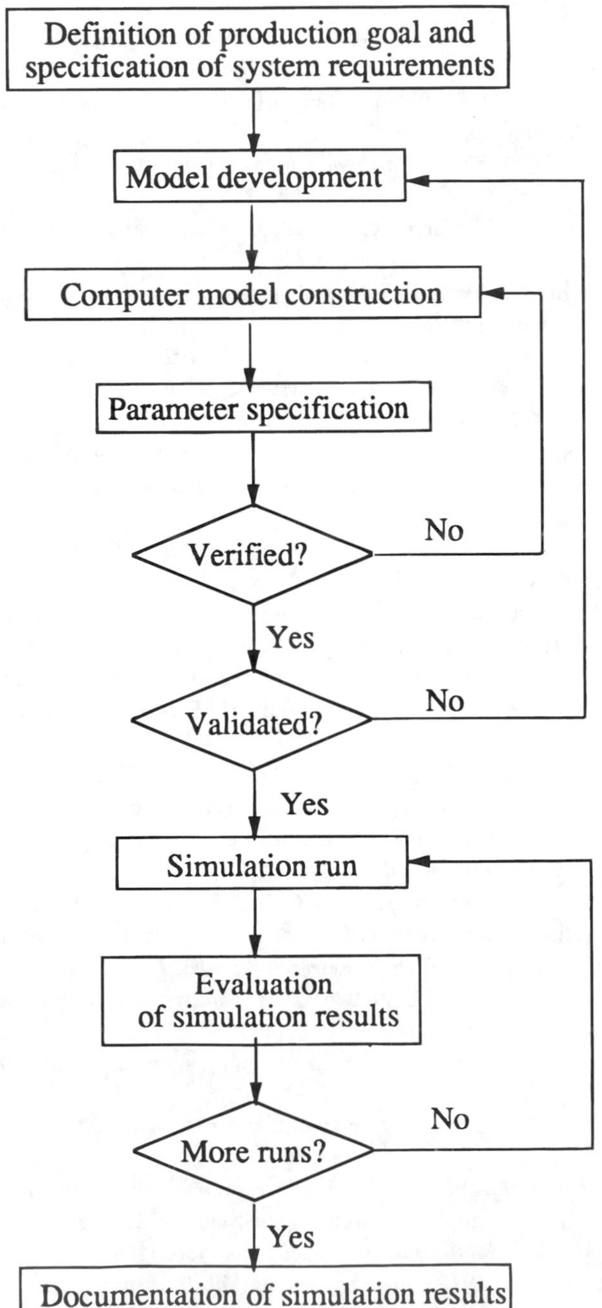

Figure 49.25 Steps involved in discrete event simulation.

[1] This work is supported by NSF Grant DMI-9410386 and the Center for Manufacturing Systems at New Jersey Institute of Technology.

the performance of an individual flexible manufacturing system (FMS) cell, or to optimize the performance of a shop floor containing many cells. Informal requirements have to be converted into formal requirements specifications which are understandable to both managers and designers or analysts. Simulation environments and related languages may be used to formulate requirements although they may differ from the language used to construct the simulation model.

Model development. Formulate a simulation model which could be a queueing model, a state-transition model such as a state machine and a Petri net, or object-oriented model. Construction of a simulation model can be a difficult task, requiring a modeler's understanding of the problem and modeling experience. Fortunately, there are software tools and frameworks available to facilitate the task. The model has to capture the essence of the real system without excessive detail. Hierarchical models which could represent different levels of detail are required for large scale factory automation systems. Simulation results can be presented to planners and managers for their decision making and serve the basis for design and real-time implementation of factory systems. Assumptions and simplifications have to be justified. Bottom-up, top-down, and hybrid methods are used in modeling large manufacturing systems to formulate a hierarchical model.

Computer model construction. Construct a computer representation of the model, e.g., program a mathematical formulation using a simulation language or general-purpose programming language, or build up a graphical model in a computer using a graphics editor and simulator in which simulation algorithms are embedded and not apparent to a user.

Data collection. Analyze all parameters involved in the model and specifications set, and collect data based on accounting data, experience, or from lower-level simulation runs. The quality of data has direct impact on the results obtained from the simulation. Thus care has to be taken during this process.

Simulation run. Run the computer model to verify its correctness. The model verification process ensures that the computer simulation models the system properly. Various animation techniques, discussed later, can facilitate this process. If the logic structure, inputs, and outputs are correctly represented in the computer model, verification is completed. Simple cases and common sense are used during this process. Validation ensures that the model is an accurate representation of a real system. Thus the model can be used as a substitute for the actual system for the prediction of performance with a high level of confidence.

Evaluation of simulation results. Obtain and evaluate simulation results. In this step statistical data analysis techniques may be needed to analyze the system simulation results and to validate them. Generally, the performances of two or more alternative system designs are compared.

Documentation. Document the input data, methods, simulation tools, computational time, and results. The results should be presented in graphs for patterns and trends over the parameters of interest. Histograms and barcharts are often used to present the simulation data pictorially.

The use of formal description techniques in the construction of simulation models allows for the verification of these models, with respect to their behavioral properties, using strict mathematical techniques. For example, Petri nets allow for the construction of simulation models, as well as their formal verification (Zhou, 1995; Zurawski, 1994). The benefit of this approach is that the correctness, if present, or absence of behavioral properties of the simulation model can be established at the early stages of the simulation system development (for details see Chapter 49.1). This results in increasing confidence in the validity of the simulation and computer models. Another advantage of this approach is in the reduction of the cost involved in redevelopment of the simulation and computer models, as a result of incorrect or missing behavioral properties identified during the test runs. However, the ultimate test for the validity of the simulation and computer models is the actual simulation run. The use of formal techniques can help only to increase the confidence in the validity of the computer model, with respect to behavioral properties. However, the computer model may differ from the simulation model, in terms of functionality. This may be due to considerable lack of compatibility between the simulation and computer model development environments. Or, simply, as a result of errors made during the computer model construction. The verification of the computer model for its functionality may be a time-consuming activity, especially for computer models of complex systems. For this reason, a more practical approach may be to develop computer models reflecting specific facets of the system functionality, rather then a model aiming at giving answers to questions reflecting complete system functionality. Another reason for using this approach is dictated by the need to verify the correctness of numerical results obtained during simulation. The simulation results can be compared with real-world data, if available, for existing systems, or compared with results produced by theoretical models. For complex systems, obtaining analytical solutions for models involving all facets of the system functionality may not be computationally trackable or practical. Queuing theory and models are used, in most cases, for obtaining the reference results. For details of queuing theory and models see Chapter 49.2. The use of queuing theory, as any other technique which yields results representing steady-state operation of the modeled systems, poses an additional problem. The effects of the initial bias have to be eliminated. These effects are due to the transient period, which follows the start of the simulation run, and influenced by the nature of the system simulated as well as the simulation environment. Testing for the closeness to the theoretical model is quite a complicated process. The discussion of this is, however, beyond the scope of this article.

A successful simulation of a factory design needs close cooperation with the factory personnel to ensure the model's correctness and to make the results acceptable to them (Rembold, 1993). With the increasing use of powerful PCs and workstations, graphic simulation and animation of manufacturing systems is possible. Its operation can then be viewed in real time and interactive simulation can be conducted. This type of simulation is most useful for debugging, fine manipulation, and material flow observance (Rembold, 1993). An example of such a graphic simulation is shown in Figure 49.26 where a robotic pick-place cell is animated and simulated. Different tasks can be simulated to check if they can be efficiently accomplished. The simulation is done with Silicon Graphic's IGRIP (Zhou, et al., 1994). Other tools for highly accurate animation include SLAMSYSTEM, Cinema SIMAN, and SIMFACTORY (Globle, 1990; Law, 1991; Mile, 1988; Pritsker, 1986). The following discussions focus on discrete event simulation models, schemes, and tools.

Simulation Models

There are three basic types of discrete event system models: queueing models, state-transition models whose representatives are state machines and Petri nets, and object-oriented models.

Queueing Models

A queueing model of a manufacturing system can be obtained if one treats resources such as machines and robots as servers, storage areas and conveyor systems as buffers (queues), and jobs or parts as customers. When strict, perhaps unrealistic and simplifying, assumptions are made for the model, analytic results can be derived for performance evaluation, as discussed

Figure 49.26 Simulation and animation of a Panasonic robotic assembly cell. *Source:* Zhou, et al., 1994. *Proc. 1994 IEEE Regional (NY/NJ) Control Conference,* 86–89. With permission.

in Chapter 49.2 of this handbook. When these assumptions cannot hold, simulation or approximation methods[2] (Kamath, 1994) can be used to derive the desired results.

State Machines

State-transition models include state machines, Petri nets, and high-level Petri nets. A system state is defined as the values of a set of variables necessary to describe the system at any time, relative to the objectives of the study (Banks, 1984). For example, a machine may be in one of the following states: busy, down, and idle. An event is an instantaneous occurrence that changes the system state. The relationship between states and events is described by a state machine, or state automaton. A new system state is a function of the previous state(s) and nature of the events which drive it to the new state.

A state machine can be represented by a state transition diagram which is a directed graph consisting of circles and arcs. Circles denote states and arcs denote events. An arc marked α connecting two circles marked s_i and s_j represents a transition from the current state s_i to s_j as a result of event α. A state machine which models a single machine is illustrated in Figure 49.27. At state *idle,* the occurrence of event **Start** brings the machine to *busy.* When the machine is busy, two events **Breakdown** or **Complete** may take place, but not simultaneously. Their occurrence leads the machine to either state *idle* or *down,* respectively. When the machine is down, event **Finish repair** brings the machine back to the *busy* state. A state machine model of a system clearly describes its evolution.

When a system is complex (contains a multiplicity of machines, AGVs, etc.), the graphical state machine representation may not be practical or possible due to the exponential state-space explosion problem[3] for many real industrial systems. For example, addition of a flexible AGV to a system can lead to a drastic increase in the number of its states since it can move to different locations, each of which represents a state. When multiple such AGVs are introduced, one can quickly see the exponential growth in the number of system states.

When time delays are introduced in the simulation, the event concept can be extended to include time. There are two ways to associate time delays with state machine models, as discussed below using the single-machine example.

Associate Time with the Events. For example, suppose it takes random time to trigger event **Start** at state *idle,* **Complete** and **Breakdown** at *busy,* and **Repair** at state *down.* At states *idle* and *down,* there are unique events which will occur, eventually. At state *busy,* depending upon the underlying physical processes, a policy has to be established to select an event and calculate its time of occurrence. An example is given below to illustrate this.

Suppose that at time k, the system enters state *busy* from *idle* or *down.* Based on the sampling of the random variables, **Complete** and **Breakdown** should happen after time intervals x and y (called event lifetime), respectively. If $x < y$, then **Complete** happens at time $k + x$ and the system enters state *idle.* If $y < x$, **Breakdown** takes place at time $k + y$, implying that the part is processed for y time units and the machine breaks down and needs repair. Once it is repaired after taking time, z, it enters state *busy* again. If the process is interruptible, then the simulator does not need to resample the random variable associated with event **Complete.** Its lifetime becomes $x - y$. Then $x - y$ and y' are compared where y' is a newly generated value from the random variable associated with event **Breakdown.** The event with the smaller lifetime is selected to take place. If the process is not interruptible, the random variable associated with event **Complete** is resampled. When only exponential random variables are involved, these two policies produce the same results due to the memoryless property of exponential random variables, i.e., knowledge of event age is irrelevant in determining the event residual lifetime. Otherwise, the policies produce different results.

Associate Time and Probability with the States. A random or fixed time delay is assigned to each state, implying that the delay starting at the moment the system enters the state has to pass before triggering any further events. Then if there are multiple events possible at the state, probabilities are used to select an event and determine a new state. In the example shown in Figure 49.27, it may take exponential random time to stay in state *idle,* a uniform random time to stay in state *busy,* and a uniform random time in state *down.* At state *busy,* probabilities Pr and $1 - Pr$ ($Pr > 0$) are assigned to events **Complete** and **Breakdown,** respectively.

Both timing techniques lead to Markov chains if all the time delays are exponentially distributed random variables. Thus analytical solutions can readily be obtained by solving a set of algebraic equations associated with Markov chains.

Petri Nets

A timed Petri net is a couple (PN, τ) where PN is an ordinary Petri net (see Chapter 49.1 of this book) and τ is a function assigning a real nonnegative number, τ_i, to each transition of the net. This number is the firing time of a transition.[4] Transitions are enabled according to the same rules that apply to classical Petri nets. Firing an enabled transition t_i causes a

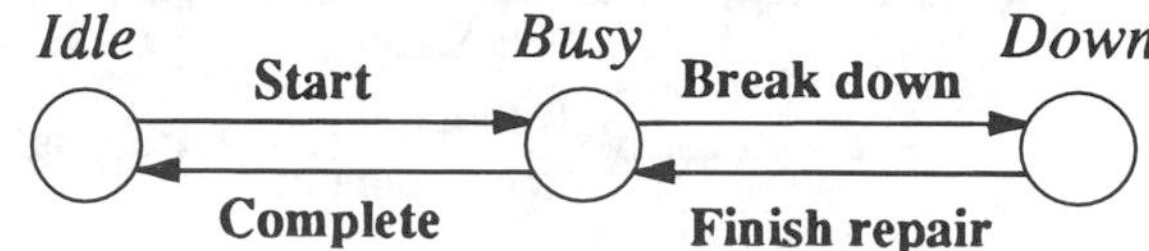

Figure 49.27 An example of state transition diagram for a machine.

[2] A heuristic can be applied to complex queueing models

[3] The number of system states is related exponentially to the number of system components.

[4] Time may also be associated with places leading to "timed place Petri nets," as opposed to "timed transition Petri nets" discussed in this section.

two-step marking change. Instantaneous with the enabling of t_i, the marking of its input places is decremented by one token per input arc. After τ_i time units, the marking of its output places is incremented by one token per output arc. Depending upon the underlying process that the transitions represent, a policy must be associated with the selection of one from competing events (conflict resolution), as discussed for state machine models before. A simulation model of a timed net requires a clock reference. In a timed net, the transition firing times may be either deterministic or stochastic. A generalized timed net may also contain instantaneous transitions. The instantaneous transitions are drawn as solid rectangles, and timed transitions are usually depicted as open rectangles.

To convert an ordinary net to a timed net is a straightforward process. One decides which activities require finite time and designates the corresponding transitions as timed. The system must also provide a means to enter the type and magnitude of the time delay. Certain activities may need to be separated in order to simulate correctly the system dynamics. If so, the affected net elements may need reinterpretation.

Simulation of a timed Petri net can help establish the behavioral properties and evaluate the temporal performance for a system under realistic assumptions. The operational policies and time characteristics for operations/activities have to be given based on the specifications of a system under design. Note that if all involved time delays are exponentially distributed, then the resulting models called stochastic Petri nets are, equivalent to Markov chains. Then performance data can be obtained analytically using software packages, e.g., SPNP (Ciardo, 1989).

Consider the operation of a pick-place robot[5] for component insertion (Zhou et al., 1994). A robot picks up a component and then places it in a desired position. The two events are **Picking-up** and **Placing.** The first can take place if a component is available and the robot is ready. As shown in Figure 49.28, two places, labeled p_1 and p_2, are used to represent these two conditions. Putting at least one token into each place represents that both conditions are true. Transition t_1 represents event **Picking-up.** The two arcs linking p_1 and p_2 to t_1 mean that the event **Picking–up** requires both a component (p_1) and the robot (p_2) to be available (logical AND). After the robot picks up a component, a new condition results, i.e., the robot holds the component, represented by place p_3. Then the event **Placing,** depicted by transition t_2, can occur. Arcs from t_1 to p_3 and from p_3 to t_2 show these relationships. Once the robot has placed the component, it is ready for the next pick-up operation. Thus the arc from t_2 to p_2 is created. Since the component is already inserted, no arc is formed from t_2 to p_1. Time delay functions could be associated with the events **Picking-up** and **Placing.** Note that a hierarchical model results if a transition or place represents lower level activities. For example, transition **Picking-up** can be expanded by a subnet modeling activities **Start, Move, Open gripper, Close gripper,** etc. This process can continue to the desired level.

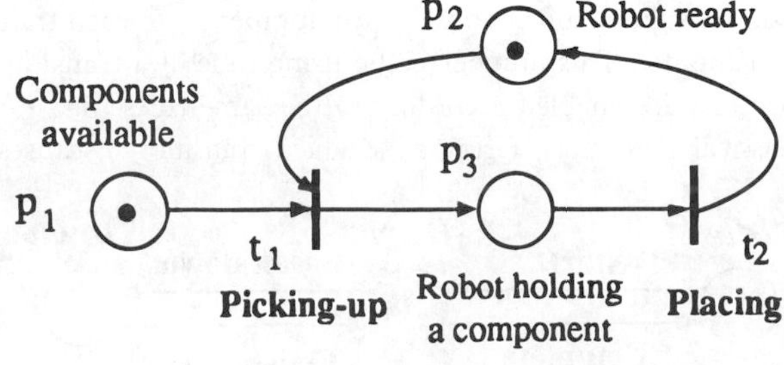

Figure 49.28 A simple PN example: a pick-place robot. *Source:* Zhou, and Robbi, 1994. In *Computer Control of Flexible Manufacturing Systems,* 207–230. Chapman and Hall, London. With permission.

High-level Petri Nets

By distinguishing among tokens in an ordinary Petri net and modifying Petri net execution rules, one creates a "colored" Petri net model.[6] Such a model is useful in describing a large factory with many similar subsystems consisting of machines, robots, AGVs, and parts. It is a more compact model than its ordinary Petri net counterpart; the complexity is hidden in the token attributes. CPN/Design is a commercially available tool for modeling and simulation using colored Petri nets. Other high-level Petri nets include Predicate-Transition and object-oriented nets associated with their development environments (Kordon, 1994).

Object-oriented Models

A manufacturing system can be viewed as a collection of objects with rules that govern their dynamics and interactions to generate desired objects (product). The objects can be represented graphically as simplified images, icons, and stored in a database as members of a class of similar objects sharing common properties. Example objects include robots, machines, machine tools, conveyors, etc. A simulation model can be built up by retrieving such objects from a database to represent a system model. Such models can lead to high reusage of simulation components. It should be noted that each object's dynamic behavior in such a model could be represented by other techniques, e.g., state machines and Petri nets.[7]

Comparisons

Queueing models offer mathematically concise models allowing development of analytic solutions for first-cut quick decision making under certain restrictions. It can be difficult to map a manufacturing system in terms of only queues, servers, and customers. The development of hierarchical models with different levels of detail is not straightforward.

State-transition models are easier to relate to a manufacturing system, allowing facile validation of a simulation model. When

[5] These robots are widely applied to assemble components to printed circuit boards.

[6] Color is a euphemism for a set of attributes which may be assigned to a token, and which may change as the token moves through the net.

[7] In fact, an object-oriented Petri net tool has been developed at NJIT by Liu, Juneja, and Robbi.

a system is complex, their visualization capability is diminished. Fortunately, they can be used for hierarchical modeling to facilitate system understanding and perform efficient simulation. Within state-transition models, state machines tend to be too complex for realistic industrial systems. They also have limited modeling capability in representing explicitly concurrent manufacturing operations. Petri nets offer a more powerful tool to handle discrete event dynamics and are more compact, in general. Colored Petri nets offer a more compact model than ordinary Petri nets at the sacrifice of clarity in some cases. Strict assumptions about the nature of their time delays and operation rules render all these models solvable analytically, i.e., no simulation is needed if computationally feasible.

Object-oriented models arose from the application of object-oriented technology to modeling and simulation of discrete event systems. Their basic elements are objects whose models, as well as their interactions, can be the ones discussed above. By using inheritance, data abstraction, dynamic binding and encapsulation, the objects can be reused for construction of new models.

Simulation Schemes and Tools

System simulation can be performed using either an event scheduling or a process-oriented scheme (Banks, 1984; Cassandras, 1993; Law, 1991). In the former, events are listed and scheduled to take place in temporal order, and the system state advances accordingly. The latter scheme better fits resource-contention environments, where each entity (e.g., part) undergoes a process as it flows through a discrete event system. A process can be treated as a sequence of functions or events triggered by an entity. Events are a fundamental concept for both schemes.

Event Scheduling Simulation

In the event scheduling scheme, the simulation process works as follows (Cassandras, 1993):

1. Set the initial values of the related variables including the initial feasible events and their scheduled times of occurrence in a so-called SCHEDULED EVENT LIST. The events are ordered by the system on a smallest-scheduled-time-first basis.
2. Remove the first entry from the SCHEDULED EVENT LIST.
3. Advance the clock time to the removed event's scheduled time.
4. Update the system state due to the occurrence of the event.
5. Delete any entries corresponding to infeasible events in the new state from SCHEDULED EVENT LIST.
6. Add new feasible events and their scheduled occurrence times (current time plus the time obtained from their corresponding random variable generators or set to the fixed delay for deterministic cases).
7. Reorder the updated SCHEDULED EVENT LIST based on smallest-scheduled-time-first and go to Step 2.

The event scheduling scheme is suitable for all the discrete event simulation models. When it is applied to a specific model, e.g., Petri nets, the algorithm can be easily modified to fit to it. The algorithm for Petri net simulation is used as an example, as follows:[8]

1. Form the set of all newly enabled transitions in the current state.[9] For each newly enabled transition with a random time delay, generate a delay based on its distribution and associate that delay with the transition. Associate appropriate fixed delays with the other newly enabled transitions.
2. Remove the required number of tokens from the input places of the above set. Add the newly enabled transitions to the to-be-fired list, in temporal order. Transitions on this list are not candidates for Step 1 above.
3. If the to-be-fired list is empty, stop (deadlock).
4. Determine the set of to-be-fired transitions with the minimum delay and advance the clock by that amount. Fire them by depositing the required number of tokens in output places, and removing them from the list. For each transition, log the information regarding its enabled time and increment the number of its firings.
5. Increment the simulation step counter. If the number of preset simulation steps is reached or if a state-dependent stopping criterion is met, stop. Otherwise, repeat Step 1.

The above algorithm, or equivalent, may be built into timed Petri net based graphical simulation environments. With a "Step Mode" the execution of a Petri net can be visualized on a work station screen.

Almost all the simulators using state-transition models are built based on an event scheduling scheme. General-purpose programming languages such as FORTRAN, C, and C++, and many simulation languages such as GASP, SIMAN, SIMSCRIPT, SLAM, and SIMFACTORY can be used to develop event scheduling simulation. While simulation code needs more skills and time to develop using a general-purpose programming language, the advantage is more efficient code and a saving of computer running time. The resultant code can be easily recompiled and run and reused by other computer platforms. Most simulation languages are supported by a commercially available software development toolkit including a graphical model editor, statistical analyzer, animation display, and report and graph generator. These systems greatly facilitate simulation modeling and execution.

Process-Oriented Simulation

In a discrete manufacturing system, a product is generated after it undergoes many steps. The sequence of these steps is called a manufacturing process. A process contains conditional

[8] There are other logically equivalent procedures, including ones with time delays associated with places rather than transitions.

[9] The initial state must contain tokens enabling at least one transition to get started.

logic so that the steps taken by a part may depend on its own attributes and the current state of the system (Askin, 1993). A model may contain multiple processes with a variety of types of entities. Since manufacturing resources may be shared among processes, an entity may have to compete and wait for service. Viewing discrete event systems as a number of entities undergoing processes results in the process-oriented simulation scheme. It contains the following components (Cassandras, 1993):

1. Entities: objects requesting service. Each entity type is characterized by a particular process it undergoes in the system it enters.
2. Attributes: information characterizing a particular individual entity of any type.
3. Process functions: the instantaneous actions taken by an entity that triggers this function in its process.
4. Time delay functions: the fixed or random delays taken by the entities.
5. Resources: objects providing service to entities.
6. Queues: sets of entities which are waiting for the use of a particular resource.

Every entity, usually a job, undergoes one or more following steps:

1. It arrives at the manufacturing system and enters a queue.
2. It requests service from a server; if the server is idle, the entity "seizes" that resource, otherwise it remains in the queue until the server becomes idle.
3. Once it seizes the server, it is served for some period of time.
4. When service is complete, it releases the server.
5. It leaves the system.

If an inspection station server is involved in the above model, a completed job or part is inspected to decide whether the part leaves a system marked "good" or "bad," or re-enters the system for rework in the case of some bad parts. A model containing such components can be used to perform yield analysis.

The process-oriented simulation scheme has gained popularity in factory automation community since the process viewpoint corresponds to the steps in a manufacturing process. Relevant simulation languages and tools include GPSS, SIMAN, SIMSCRIPT, and SLAM (Cassandras, 1993; Law, 1991; Pritsker, 1986; Russell, 1983).

Simulation Examples

To illustrate how to perform event scheduling simulation, C language is used to simulate a straightforward machining center and a state machine. SIMAN is used to show process-oriented simulation.

Event Scheduling Simulation of a Machine Center

The machining center is shown in Figure 49.29. It is modeled as a single-server queueing system to illustrate the above concepts. There is an input buffer for incoming parts, a queue. Assume that the raw part arrival process follows Poisson distribution characterized by part arrival rate λ and that the machining (or service) time follows a uniform distribution over $[a, b]$ where a is the minimum time and b is the maximum time to finish machining process. The goal is to find out the system throughput and machine center utilization under different parameters, i.e., λ, a, and b. The raw parts in the queue are machined on a first-come-first-served basis.

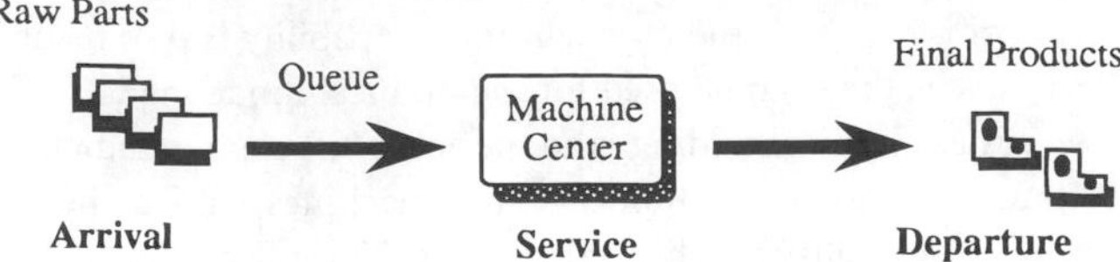

Figure 49.29 A single-server system example: A machining center.

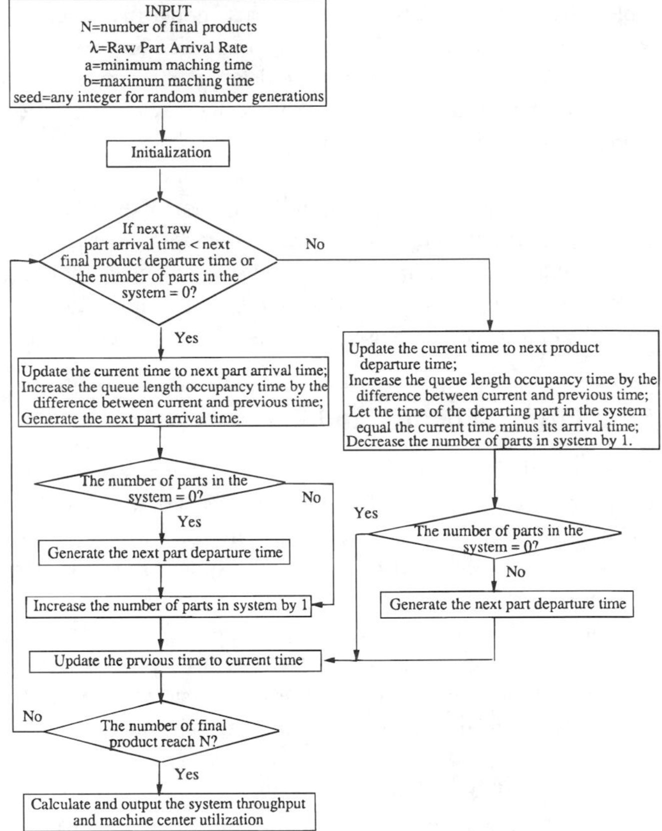

Figure 49.30 Flowchart for simulating a single-server system.

Two events in this system are raw part arrival and finished product departure. The number of parts in the queue and the status of the machining center define the state of the system. The occurrence of an event changes the system state. Events occur randomly according to the distributions given above. The flow diagram for the simulation program is given in Figure 49.30. The program works as follows:

1. The input module accepts the user's input data including the number of final products produced by the system

(used as a stopping criterion), the raw part arrival rate, minimum and maximum machining times, and an integer as seed for random number generation.

2. The initialization module sets the initial values of all the related variables used in the program.
3. If the next raw part arrival time is less than the next finished part departure time or the number of parts in the queue is zero, then the program handles the case that the arrival event occurs before the next departure event. Thus it updates the clock to the next part arrival time, increases the queue length occupancy time by the difference between current and previous times, and generates the next part arrival time. If the number of parts in the queue is zero, the finished part departure time for the newly arrived part can immediately be generated. Otherwise, the raw part enters the queue to wait for the machining center, and the number of parts in the queue is increased by one.
4. If the next raw part arrival time is greater than the next finished part departure time, the next event to take place is the departure of the finished part. The program updates the clock to next part departure time, increases the queue length occupancy time by the difference between the current and previous times, sets the time of the departing part in the system equal to the clock minus its arrival time, and decreases the number of parts in system by one. If there are any raw parts in the system, the next finished part departure time is scheduled for the first arriving part in the queue (FCFS policy).
5. Update the previous time with the current clock time.
6. Check if the stopping criterion is satisfied, i.e., the number of finished parts reaches the target number. If so, calculate the system throughput and machining center utilization. Otherwise, go to Step 3.

Note that this design is very elementary. There is no mechanism designed to handle the possible memory overflow (queue length overflow), and the design takes advantages of the existence of only two possible events in the system and the lack of concurrency. In general, as discussed before, a SCHEDULED EVENT LIST has to be set up, continuously reordered, and the event with the smallest scheduled time selected to take place next.

The program is coded as shown in Figure 49.31. Simulations are run for two cases with the following common data: $N = 1000$, $a = 1.0$, $b = 2.0$, and seed = 123. The results with $\lambda = 0.5$ are System Throughput = 0.500, and Machine Center Utilization = 0.752. Clearly the throughput is limited by the part arrival rate. When $\lambda = 1.0$, the System Throughput = 0.666, and Machine Center Utilization = 0.997. Now the machining rate limits production. Additional results can be obtained from the above program. Plots of system performance measures can be drawn with respect to different parameter values as illustrated in Figure 49.32. Real systems are generally much more complex than this example, and the benefits of simulation are more apparent.

Process-Oriented Simulation of a Machining Center

The above single-server queue system includes the process functions: part arrival, part machining, and part departure. It includes the following time delay functions: the interarrival time between successive raw parts and the machining time of a raw part. In this model, there is only one type of entity, a raw part. There is an input buffer for incoming parts, a queue. It can be programmed facilely using a dedicated simulation-language, such as SIMAN. A SIMAN program for the same system, consists of a Model file and an Experiment file (Cassandra, 1993; Pegden, 1990). In the Model file, the six steps a raw part undergoes in this example are described by six SIMAN blocks as shown in Table 49.4.

A Model file can also contain data collection blocks. Block COUNT operates as a counter. When an entity encounters a COUNT function, the discrete value of a counter is incremented. Its counterpart TALLY function records the time that elapses between two points in the process visited by an entity.

The Model file has to be accompanied by an Experiment file which provides the former with the parameters needed in a simulation run and specifies the desired outputs. Both the Model and Experiment files for the machining system example are given in Figure 49.33. In addition to the blocks discussed above, EXPO(Mean) function represents an exponential random variable with the expected value "Mean." MARK(attributeID) is used to record the time when an entity reaches a particular block function. In the example, this function is CREATE and attributeID is ArrivalTime. In other words, once a raw part enters the system, its arrival time is recorded in an attribute "ArrivalTime." UNIF(Min, Max) represents a uniform random variable over an interval starting at "Min" and ending at "Max."

In the Experiment file,

PROJECT is a header required to identify simulation experiments.

DISCRETE is used to establish upper bounds for a number of entities simultaneously present in the system. Reaching the bounds indicates possible queue overflow problems.

QUEUES specifies information related to all queues in the Model file. By default, FCFS policy is used.

RESOURCES specifies information related to all resources used in the Model file. By default, there is one and only one resource called "Machine" in this example.

ATTRIBUTE is used if attributes with symbolic names are used in the Model file. Since ArriveTime is used in the Model file, it is specified in the Experiment file in this example.

VARIABLES is used to specify all the symbolic variables appearing in the Model file. In the example, LAMDA, MIN, and MAX are specified with this function.

COUNTERS specifies information related to all counters in the Model file. Once a counter reaches its upper bound the simulation run is immediately ended.

TALLIES specifies information related to all tallies in the Model file. *TALLIES System Time* in the example generates

```
/*********************************************/
/* ar -- arrival time                        */
/* ar_index -- # of the arrival raw parts    */
/* de -- deparature times                    */
/* de_index -- # of the deparature products  */
/* qe -- queue length occupancy times        */
/* length -- queue length                    */
/* sy -- system times                        */
/* sys_index -- # of the deparature products */
/* c_t -- current times, p_t -- previous times */
/* getexp function is to get the exponential */
/* random number                             */
/* getuniform function is to get the         */
/* uniform random number                     */
/*********************************************/
#include "stdio.h"
#include "math.h"

double getexp(float lamda);
double getuniform(float a,float b);

main()
{
 float ar[1000],de[1000],qe[1000],sy[1000];
 float c_t,p_t;
 int  ar_index,de_index,sys_index,length;
 double u;
 double interval;
 int    n,seed,i;
 float  lamda;
 float  a,b;
 float  throughput,utilization;
 float  temp;

 printf("Please input N=");
 scanf("%d",&n);
 printf("Please input seed=");
 scanf("%d",&seed);
 printf("Please input lamda=");
 scanf("%f",&lamda);
 printf("Please input a=");
 scanf("%f",&a);
 printf("Please input b=");
 scanf("%f",&b);

/**initial**/
 for(i=0;i<1000;i++)
 {
  ar[i]=0;de[i]=0;sy[i]=0;qe[i]=0;
 }
 de_index=0;sys_index=0;
 srandom(seed);
 ar[0]=getexp(lamda);
 qe[0]=ar[0];
 ar_index=1;length=1;
 c_t=ar[0];p_t=c_t;

 /* ar[ar_index]=c_t+getexp(lamda); */
 de[de_index]=c_t+getuniform(a,b);
 de[de_index]=c_t+getexp(a);
```

```
while(sys_index!=n&&sys_index<1000)
  {
  if(ar[ar_index]<de[de_index]||length==0)
  {
    c_t=ar[ar_index];
    qe[length]=qe[length]+c_t-p_t;
    ar[++ar_index]=c_t+getexp(lamda);
     if(length==0)
/*          de[++de_index]=c_t+getexp(a); */
            de[++de_index]=c_t+getuniform(a,b);
length++;
   }
  else if(ar[ar_index]>de[de_index])
   {
    c_t=de[de_index];
    qe[length]=qe[length]+c_t-p_t;
    sy[sys_index++]=c_t-ar[de_index];
     length--;
     if(length!=0)
 /*     de[++de_index]=c_t+getexp(a); */
      de[++de_index]=c_t+getuniform(a,b);
}
   p_t=c_t;
  }

 for(i=0;i<ar_index;i++)
   printf("ar[%d]=%6.3f\n",i,ar[i]);
 for(i=0;i<de_index;i++)
   printf("de[%d]=%6.3f\n",i,de[i]);
 for(i=0;i<sys_index;i++)
   printf("sy[%d]=%6.3f\n",i,sy[i]);

 i=0;temp=0;
 while(qe[i]!=0)
 {
  temp=temp+qe[i];
   i++;
 }
 utilization=1.0-qe[0]/temp;
 throughput=(1.0*sys_index)/de[sys_index-1];

 printf("throughput=%10.3f\n",throughput);
 printf("utilization=%10.3f\n",utilization);
}

double getexp(float lamda)
{
 double u;
/*generate random number from 0~1*/
 u=(1.0*random())/2147483647.0;
 return( (-1.0/((float) lamda))*log(u));
}

double getuniform(float a,float b)
{
 double u;
/*generate random number from 0~1*/
 u=(1.0*random())/2147483647.0;
 return(a+(b-a)*u);
}
```

Figure 49.31 C program for simulating a single-server system.

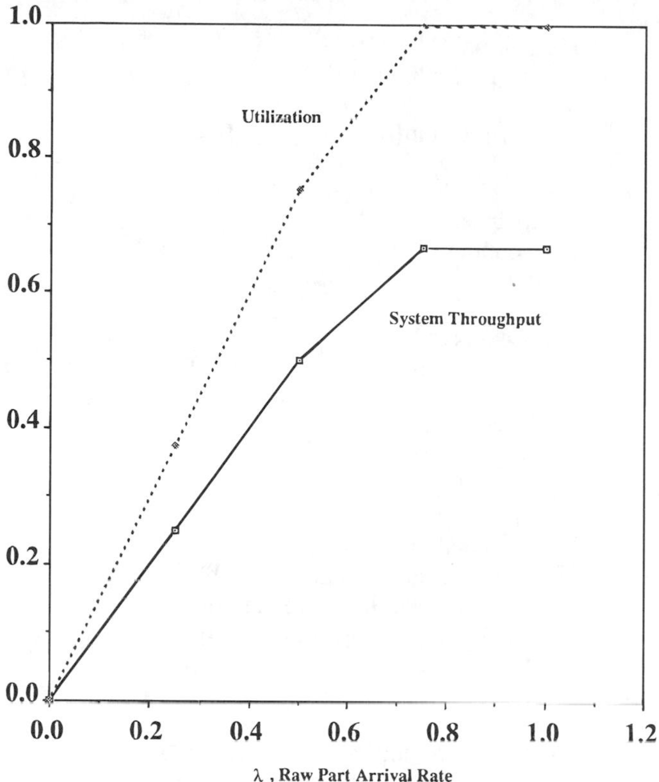

Figure 49.32 Simulation results for a single-server system.

Table 49.4 Process Steps and Their SIMAN Blocks

Process Step	SIMAN Block
1. A raw part arrives.	CREATE
2. The raw part enters queue.	QUEUE
3. The part requests machining from Machine; if idle, the part gets processed else it remains in the queue.	SEIZE
4. Once the part starts to be processed, it is in process for some period of time.	DELAY
5. When the machining is complete, it is released (idle).	RELEASE
6. The finished part leaves the system.	DISPOSE

statistics for the system time of raw parts, which includes the mean and deviation.

DSTATS collects data on certain time-dependent variables. NR(.) is a special SIMAN variable representing the number of idle (NR(0)) or busy (NR(1)) units of a resource. In the example, the machine utilization is recorded with *DSTATS: NR(1), Utilization.*

REPLICATE is used to automatically repeat simulation runs. *REPLICATE, 1, 0* means one simulation run and starting at time 0.

The execution of the above SIMAN code produces the clock TNOW at which the program is terminated. The system throughput is obtained by dividing 1000 by TNOW. It also produces its mean machine utilization and standard deviation.

Event Scheduling Simulation of A State Machine

Event scheduling simulation of a state-transition model is more suitable and convenient than process-oriented simulation

```
Simulation Model File of a M/U/1 system
BEGIN;
  CREAT: EXPO(1.0/LAMDA): MARK(ArrivalTime);
;     Raw part arrives and its arrival time is recorded in attribute "ArrivalTime"
   QUEUE, Queue; Wait in "Queue"
   SEIZE: Machine; Seize "Machine center" when idle
   DELAY: UNIF(MIN, MAX); Machining time delay
   RELEASE: Machine; Release "Machine center"
;     Tally part system time and store in "SystemTime"
   COUNT: NumProduct: DISPOSE;
;    Count final products and store in counter "NumProduct"
;    Then, leave the system
END

The accompanying Experiment file
BEGIN;
PROJECT, M/U/1 system;
DISCRETE, 1000; No more than 1000 concurrent entities
QUEUES: Queue;
RESOURCES: Machine;
ATTRIBUTES: ArrivalTime;
VARIABLES: LAMDAM, 0.5; Raw part arrival rate
  MIN, 1.0: ! Lower bound of uniform distribution
  MAX, 2.0: ! Upper bound of uniform distribution
COUNTERS: NumProduct, 1000; Stop after 1000 products
TALLIES: SystemTime;
DSTATS: NR(1), Utilization
REPLICATE, 1, 0; One simulation run
END;
```

Figure 49.33 SIMAN program for simulating a single-server system.

since in such a model, all possible events at a state are known. Take the state machine example shown in Figure 49.29. Assume that:

1. Triggering event **Start** at state *idle* takes exponential random time with rate λ_1.
2. Triggering event **Complete** at state *busy* takes a uniform random time with a and b as their lower and upper bounds.
3. Triggering event **Breakdown** takes exponential random time with rate λ_2.
4. Triggering event **Repair** at state *down* takes a uniform random time with c and d as their lower and upper bounds, and
5. When entering state *busy*, the random variables associated with events **Complete** and **Breakdown** are sampled, regardless of the history.

The C program for the example of Figure 49.29 is given in Figure 49.34. Given the following parameters: $N = 1000$, $\lambda_1 = 4.0$, $\lambda_2 = 0.5$, $a = 1$, $b = 2$, $c = 2$, $d = 3$, and seed $= 123$, the throughput is 0.183 (number of parts/time unit). The throughput is calculated by the number of occurrences of event **Complete** divided by the total time to finish them. When the breakdown rate becomes smaller, e.g., $\lambda_2 = 0.2$, the throughput is 0.339 (number of parts/time unit).

Examples of using Petri net-based simulation can be seen in (Zhou, 1994b). To create a simulation model based on this type of tool (Kordon, 1994; Meta, 1990; Zhou 1994b), one first draws the net by selecting and placing the net components, places, transitions, and arcs. These may be labeled to assist in understanding. Time-delay specifications are associated with all timed transitions, **Picking-up and Placing** in the Petri net

```
/*******************************************/
/* st -- start time array                     */
/* com -- completion time array               */
/* bk -- break-down time array                */
/* rp -- repair time array                    */
/* lam_st -- start rate                       */
/* lam_bk -- machine break-down rate          */
/* a -- lower bound of uniform random time of */
/*       completion                           */
/* b -- upper bound of uniform random time of */
/*       completion                           */
/* c -- lower bound of uniform random time of repair */
/* d -- upper bound of uniform random time of repair */
/* n -- given number of completions or final parts */
/* length -- current number of completions or final */
/*       parts                                */
/* seed -- any integer to initiate random number */
/*       generation                           */
/* bk_index -- current number break-downs     */
/* bk_index -- current number of repairs      */
/* c_t -- current times                       */
/* getexp function is to get the exponential  */
/* random number                              */
/* getuniform function is to get the          */
/* uniform random number                      */
/*******************************************/

#include "stdio.h"
#include "math.h"
double getexp(float lamda);
double getuniform(float a,float b);
main()
{
 float st[2000],com[2000],bk[2000],rp[2000];
 float lam_st,lam_bk;
 float a,b,c,d;
 float throughput;
 int   n,length;
 int   seed, i;
 int   bk_index,rp_index;
 float c_t;

 printf("Please input N=");
 scanf("%d",&n);
 printf("Please input seed=");
 scanf("%d",&seed);
 printf("Please input lamda value for start=");
 scanf("%f",&lam_st);
 printf("Please input lamda value for break down=");
 scanf("%f",&lam_bk);
 printf("Please input a=");
 scanf("%f",&a);
 printf("Please input b=");
 scanf("%f",&b);
 printf("Please input c=");
 scanf("%f",&c);
 printf("Please input d=");
 scanf("%f",&d);

/**initial**/
 for(i=0;i<2000;i++)
 {
  st[i]=0,com[i]=0,bk[i]=0,rp[i]=0;
 }
 length=0;
 rp_index=0;
 bk_index=0;
 c_t=0;

srandom(seed);
 while(length!=n&&length<2000)
 {
  st[length]=getexp(lam_st);
  c_t=c_t+st[length];
  while(1)
   {
/*   com[length]=getexp(a);  */
    com[length]=getuniform(a,b);
    bk[bk_index]=getexp(lam_bk);
    if(bk[bk_index]>com[length])
        break;
    else
     {
 /*    rp[rp_index++]=getexp(c); */
    rp[rp_index++]=getuniform(c,d);
        bk_index++;
      c_t=c_t+bk[bk_index-1];
      c_t=c_t+rp[rp_index-1];
     }
   }
  c_t=c_t+com[length];
   length++;
 }

 if(length==2000)
   throughput=(1.0*length)/c_t;
 else
   throughput=(1.0*n)/c_t;
 printf("throughput=%10.3f\n",throughput);
}

double getexp(float lamda)
{
 double u;
/*generate random number from 0~1*/
 u=(1.0*random())/2147483647.0;
 return( (-1.0/((float) lamda))*log(u));
}

double getuniform(float a,float b)
{
 double u;
/*generate random number from 0~1*/
 u=(1.0*random())/2147483647.0;
 return(a+(b-a)*u);
}
```

Figure 49.34 C program for state transition diagram in Figure 49.29.

example in Figure 49.28. The net is initialized with tokens in appropriate places, and stopping criteria are specified. In a complex situation, single stepping the net ensures a least minimal valid behavior. Then a run command is issued to execute the Petri net, and obtain an output file with performance results.

Concluding Remarks

The modeler should start with simple analytic models to get estimates, then progress to commercial simulation systems aimed at factory automation. When the modeled system is too difficult to fit in such a framework, then the flexibility of simulation with state-transition tools should be considered. When even these models become too hard to construct, then the utility of object-oriented approaches can be exploited.

Discrete event simulation permits computers to perform experiments with models of discrete manufacturing systems. Simulation models can conform closely to the discrete event dynamics of the real system, thus confirming the correctness of system behavior and providing accurate performance measures. Due to the complexity and many restrictions of the model systems, these models often have no analytical solution or require excessive simplifying assumptions to permit one. Simulation allows more realistic data and operational policies to be incorporated in a model. On the other hand, simulation suffers two major drawbacks compared with analytical methods (Kamath, 1994):

1. It is time-consuming to construct new simulation models and modify existing ones. Validation of simulation models is a difficult problem in general.
2. Execution may be computationally expensive, because a number of long runs are often needed to obtain statistically accurate performance results.

The research and development in object-oriented modeling, design and simulation addresses the first issue (Kamath, 1994). Simulation models developed using object-oriented technologies can gain high reusage. The effort to build up the object-oriented modeling and simulation tools and environments is surveyed in Kamath (1994). Examples include a library of objects called BLOCS/M in Objective-C (Adiga, 1990) and an icon-based simulation program generator called SmartSim (Ulgen, 1990).

With the emergence of parallel computers and high-speed networks, as well as supercomputers, parallel or distributed simulation effectively address the computation problem. Therefore, discrete event simulation promises to become more useful in factory automation design.

References

Adiga, S. and Gadre, M. 1990. Object-oriented modeling of a flexible manufacturing system, *J. Intelligent and Robotic Systems,* 3:147–165.

Askin, R. G. and Standridge, C. R. 1993. *Modeling and Analysis of Manufacturing Systems,* 409–428, Prentice-Hall, Englewood Cliffs, NJ.

Banks, J. and Carson II, J. S., 1984. *Discrete-Event System Simulation,* Prentice Hall, Englewood Cliffs, NJ.

Cassandras, C. G. 1993. *Discrete Event Systems: Modeling and Performance Analysis,* IRWIN, Boston, MA.

Ciardo, G. 1989. *Manual for the SPNP Package,* Duke University, February.

Globle, J. 1990. Introduction to SIMFACTORYII.5, *Proc. 1990 Winter Simulation Conf.,* 136–139, Piscatway, NJ.

Kamath, M. 1994. Recent developments in modeling and performance analysis tools for manufacturing systems, in *Computer Control of Flexible Manufacturing Systems,* 231–263, S. Joshi and G. Smith (Eds.), Chapman and Hall, London.

Kordon, F. and Kaim, W. E. 1993. *CPN/TAGADA Version 1.2 User Guide,* MASI Lab, Blaise Pascal Institute, University P. & M. Curie, Paris, France.

Law, A. M. and Kelton, W. D. 1991. *Simulation Modeling and Analysis,* McGraw-Hill, New York, NY.

Meta 1990. *Design/CPN,* Meta Software Corp., Cambridge, MA.

Miles, T., Sadowski, R. P., and Werner, B. M. 1988. Animation with CINEMA, *Proc. 1988 Winter Simulation Conf.,* 180–187, Piscatway, NJ.

Pegden, C. D., Shanon, R. E., and Sadowski, R. P. 1990. *Introduction to Simulation Using SIMAN,* McGraw-Hill, New York, NY.

Pritsker, A. A. B. 1986. *Introduction to Simulation and SLAM II,* 3d ed., Halster Press, New York, NY.

Rembold, U., Nnaji, B. O., and Storr, A. 1993. *Computer Integrated Manufacturing and Engineering,* Addison-Wesley, Wokingham, England.

Russell, E. C. 1983. *Building Simulation Models with SIMSCRIPT II.5,* CACI International, Los Angeles, CA.

Ulgen, O. M. and Thomasma, T. 1990. SmartSim: an object oriented simulation program generator for manufacturing systems, *Int. J. Production Research,* 28(9):1713–1730.

Zhou, D., Lubliner, C,. Cano, H., and Ma, X. 1994. Flashlight assembly through Panasonic robot, in *Proc. 1994 IEEE Regional NY/NJ Control Conference,* 86–89, Piscataway, NJ.

Zhou, M. C. and Robbi, A. D. 1994. Applications of Petri net methodology to manufacturing systems, in *Computer Control of Flexible Manufacturing Systems,* 207–230, S. Joshi and G. Smith (eds.), Chapman and Hall, London.

Zhou, M. C. (ed.) 1995. *Petri Nets in Flexible and Agile Automation,* Kluwer Academic Publishers, Boston, MA.

Zurawski, R. and Zhou, M. C. 1994. "Petri nets and industrial applications: A tutorial," *IEEE Trans. Industrial Electronics,* 41(6):567–583.

50

Hybrid Systems and Control

50.1 Introduction 706
50.2 Discrete Event and Hybrid Observation under Uncertainty 707
Hybrid and Discrete Event Dynamic Systems for Robotic Observation • DEDS for Modeling Observers • State Modeling and Observer Construction • Identifying Motion Events • Modeling and Recovering 3-D Uncertainties • Utilizing the Discrete Event Observer • Experiments
50.3 Conclusions 714

Tarek M. Sobh
University of Bridgeport

In this section we present an overview of selected issues involved in the development of complex discrete event and hybrid systems within the robotics, automation, and intelligent systems domains. We start by presenting an overview of discrete event and hybrid systems, and then illustrate the concept by using an example from the robotics and automation domain. The application discussed is for formulating an observer for manipulating agents.

50.1 Introduction

The area of hybrid systems, in which digital and analog devices and sensors interact over time, is attracting attention of researchers. The representation of states and the physical system condition includes continuous and discrete numerics, in addition to symbols and logical parameters. Most of the problems involving robotics, automation, and intelligent systems, as well as problems in other domains, fall within the description of hybrid systems. There are many issues that need to be resolved. Among them definitions for observability, controllability in general, stability and stabilizability, etc.

The underlying mathematical representation of complex computer-controlled systems is still insufficient to offer models which accurately capture dynamics of the systems over the entire range of their operation. We remain in a situation where one must trade-off the accuracy with the manageability of the models. Closed-form solutions of mathematical models are almost exclusively limited to linear system models. Computer simulation of nonlinear and discrete-event models provide a means for off-line design of control systems. The required system performance can only be guaranteed for the regions where the robustness conditions apply. These conditions may not apply during startup and shutdown or during periods of anomalous operation.

Recently, attempts have been made to model low and high-level system changes in automated and semiautomatic systems as discrete event dynamic systems (DEDS). Several attempts to improve the modeling capabilities focused on mapping the continuous world into a discrete one. However, results, obtained independently, indicate that large interactive systems evolve into states where minor events can lead to a catastrophe. Discrete event and hybrid system formulations have been used in many domains to model and control system state changes within a process. Some of the domains include: manufacturing, robotics, autonomous agent modeling, control theory, assembly and planning, concurrency control, distributed systems, hierarchical control, highway traffic control, autonomous observation under uncertainty, operating systems, communication protocols, real-time systems, scheduling, and simulation.

A number of tools and modeling techniques are being used to model and control discrete event systems in the above domains. Some of the modeling strategies include: Petri nets and state automata, state machines, stochastic (including markovian) and perturbation models, probabilistic modeling (uncertainty recovery and representation), queuing theory, and recursive functions.

A number of tools and graphical environments for simulating, analyzing, synthesizing, monitoring, and controlling complex discrete event and hybrid systems have been developed. A snapshot of one such environment is shown in Figure 50.1.

This software environment aids in the design, analysis and simulation of discrete event and hybrid systems. The environment allows the user to build a system using either finite state machines or Petri nets. The environment runs under X/Motif and supports a graphical DES (discrete event system) hybrid controller, simulator, and analysis framework. The framework allows for the control, simulation and monitoring of dynamic systems that exhibits a combination of symbolic, continuous, discrete, and chaotic behavior. The environment allows not only the graphical construction and mathematical analysis of various timing paths and control structures, but also produces C code to be used as a controller for the system under consideration.

0-8493-8343-9/97/$0.00+$.50

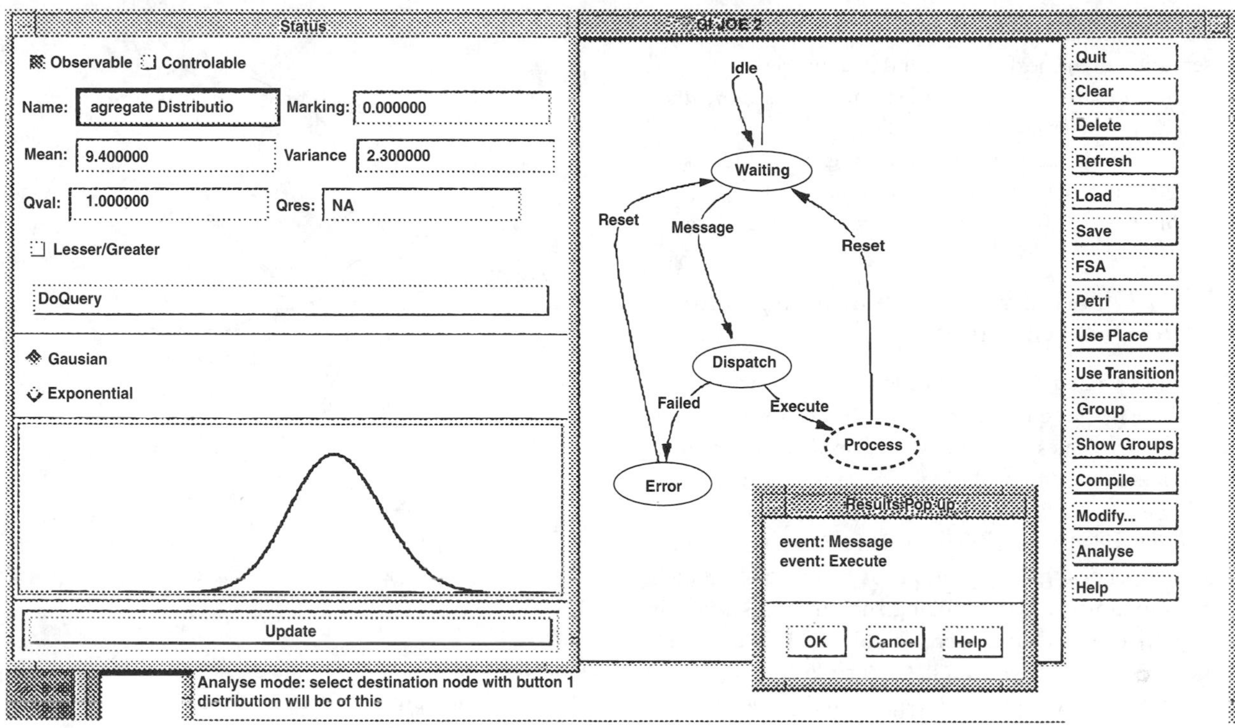

Figure 50.1 A stochastically timed FSM window during analysis.

50.2 Discrete Event and Hybrid Observation under Uncertainty

We discuss a representation for the general problem of observation through a discrete event and hybrid system framework. The system being studied can be considered as a "hybrid" one. This is due to the fact that we need to report on *distinct* and *discrete* visual states that occur in the *continuous, asynchronous,* and three-dimensional world, from two-dimensional observations that are sampled periodically. In other words, the system being observed and reported on consists of a number of continuous, discrete, and symbolic parameters that vary over time in a manner that might not be "smooth" enough for the observer, due to visual obscurities and other perceptual uncertainties.

The problem of observing a moving agent was addressed in the literature extensively. It was discussed in the work addressing tracking of targets and determination of the optic flow (Burt et al., 1989; Anandan, 1987; Horn and Schareck, 1981; Ullman, 1981); recovering 3-D parameters for different kinds of surfaces (Barron et al., 1990; Sobh and Wohn, 1989; Subbarao and Waxman, 1985; Longuet-Higgins and Prazday, 1981); and also in the context of other problems (Chaumette and Rives, 1990; Cucka and Sharma, 1990; Bajcsy, 1987; Aloimonos and Bandyopadhyay, 1987). However, the need to *recognize, understand,* and *report* on different visual steps within a dynamic task was not sufficiently addressed. In particular, there is a need for high-level symbolic interpretations of the actions of an agent. Those interpretations should attach meaning to the 3-D world events, as opposed to the simple recovery of 3-D parameters and the consequent tracking movements to compensate their variation over time.

In this work, we establish a hybrid system framework for the general problem of observation, recognition, and understanding of dynamic visual systems. The framework may be applied to different kinds of visual tasks. We concentrate on the problem of observing a manipulation process in order to illustrate the ideas and motive behind our framework.

We use a discrete event dynamic system as a high-level structuring technique to model the visual manipulation system. Our formulation utilizes all existing knowledge about the system and the anticipated actions in order to solve the observer problem. The resulting observer is efficient, stable, and practical. The model incorporates different hand/object relationships and the possible errors in the manipulation actions. It also uses different tracking mechanisms so that the observer can keep track of the workspace of the manipulating robot. A framework is developed for the hand/object interaction over time and a stabilizing observer is constructed. Low-level modules are developed for recognizing the "events" that cause state transitions within the dynamic manipulation system. The process uses a coarse quantization of the manipulation actions in order to attain an active, adaptive, and goal-directed sensing mechanism.

The work examines closely the possibilities for errors, mistakes, and uncertainties in the visual manipulation system, observer construction process, and event identification mechanisms. The work leads to a DEDS formulation with uncertainties in which state transitions and event identification are asserted according to a computed set of 3-D uncertainty models.

We motivate and describe a DEDS automation model for visual observation in the next section, and then proceed to formulate our framework for the manipulation process. We then develop efficient, low-level event-identification mechanisms for determining different manipulation movements in the system and for moving the observer. Next, the uncertainty levels are discussed. Some results from testing the system are enclosed.

Hybrid and Discrete Event Dynamic Systems for Robotic Observation

The general observation problem falls within the hybrid system domain, as there is a need to report, observe, and control *distinct* and *discrete* system states. There is also a need for recognizing the *continuous* 2-D and 3-D evolution of parameters. In addition, there should be a *symbolic* description of the current state of the system, especially in the manipulation domain.

We do not intend to give a solution for the general problem of defining, monitoring, or controlling such hybrid systems in general. What we intend to present in this work is a suitable framework for the class of hybrid systems encountered within the robotic observation paradigm. The representation we advocate allows for the symbolic, numeric, continuous, and discrete aspects of the observation task. We conjecture that the framework could be explored further as a possible basis for providing solutions for general hybrid systems representation and analysis problems.

We suggest employing a representation of discrete event dynamic systems, which is augmented by the use of a concrete definition for events. We also implement uncertainty modeling to achieve robustness and smoothness in asserting state and continuous event variations over time.

Dynamic systems are sometimes modeled by finite state automata with partially observable events together with a mechanism for enabling and disabling a subset of state transitions (Ozveren, 1989; Li and Wonham, 1988; Ramadge and Wonham, 1987). The reader is referred to those references for more information about this class of DEDS representation. We propose that such a DEDS skeleton is a suitable high-level framework for many vision and robotics tasks. In particular, we use a DEDS model as a high-level structuring technique for a system to observe a robot hand manipulating objects.

Discrete Event Dynamic Systems For Active Visual Sensing

An example of a high-level DEDS controller for part inspection can be seen in Figure 50.2. This finite state machine has some observable events that can be used to control the sequencing of the process. The machine remains in state A until a part is loaded. When the part is loaded, the machine transitions to state B where it remains until the part is inspected. If another part is available for inspection, the machine transitions to state A to load it. Otherwise, state C, the ending state, is reached. If an interruption occurs, such as a misloaded part or inspection error, the machine goes to state D, the error state.

Our approach uses DEDS to drive a semiautonomous visual sensing module. The module is capable of making decisions about the *visual state* of the manipulation process taking place. This module also provides both symbolic and parametric descriptions which can be used to observe the process *intelligently* and *actively.*

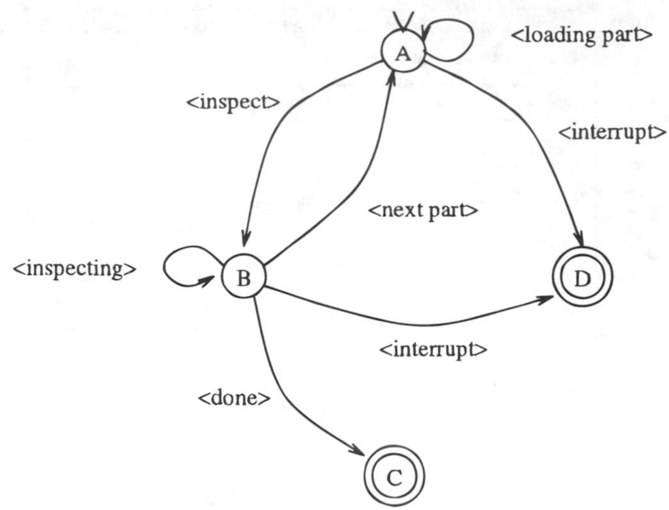

Figure 50.2 A simple FSM.

A DEDS framework is used to model the tasks that the autonomous observer system executes. This model is used as a high level structuring technique to preserve and make use of the known information about the manipulation process. The state and event description is associated with different visual cues; for example: appearance of objects, specific 3-D movements and structures, interaction between the robot and objects, and occlusions. A DEDS observer serves as an intelligent sensing module. It utilizes existing information about the tasks and the environment to make informed tracking movements and autonomous decisions regarding the state of the system.

For determining the current state of the system, the sequence of events should be observed. A decision should also be made regarding the state of the automaton. State ambiguities are allowed to occur, however, they are required to be resolvable after a bounded interval of events. In a *strongly output stabilizable* system, the state of the system is known at bounded intervals. Allowable events can also be controlled (enabled or disabled) in a way that ensures the return—in a bounded time interval—to one of a desired and known set of states (visual states in our case).

One of the objectives is to make the system strongly output stabilizable, and/or construct an observer to satisfy specific task-oriented visual requirements. Many 2-D visual cues for estimating 3-D world behavior can be used. Examples include: image motion, shadows, color, and boundary information. The uncertainty in the sensor acquisition procedure, and in the image processing mechanisms should be taken into consideration while computing the world uncertainty.

The observer framework can be utilized for recognizing error states and sequences. This recognition task will be used to report on *visually incorrect* sequences. In particular, if there is a predetermined observer model of a particular manipulation task under observation, then it would be useful to determine if something goes wrong with the exploration actions. The goal of this reporting procedure is to alert the operator or autonomously supply feedback to the manipulating robot so that it can correct its actions.

DEDS for Modeling Observers

DEDS can be considered as very suitable tools for modeling observers. In particular, in the manipulation observer domain, there is a need to recognize and report on distinct and discrete visual states. Those states might represent manipulation tasks and/or subtasks. The observer should have the ability to state a symbolic description of the current manipulation agent action. The coarse definition of DEDS states provide a means for such symbolic state descriptions.

The definitions for observers and the observer construction process for discrete event systems are very coherent with the requirements for an autonomous robotic observer. The purpose of DEDS observers is to be able to reconstruct the system state, which is exactly the requirement for a visual observer. The observer needs to recognize, report, and possibly act, depending on the visual manipulation state. The notions of controllable actions is easily mapped to some tracking and repositioning procedures that the robotic observer will have to undertake in order to "see" the scene from the "best" viewing position. The actions which the observer robot needs to perform depends on the sequence of "observable" events and the reconstructed state path.

Event descriptions in a visual observer are possibly a combination of different 2-D and 3-D visual cues. The visual primitives used in an observer domain could be motion primitives, matching measures, object identification processes, structure and shape parameters, and/or a number of other visual cues. The problem with a classical DEDS skeleton is that it does not allow for smooth state changes under uncertainty in recovering the events. We describe in the next sections techniques that facilitate the transition from a DEDS skeleton to a working hybrid observer for a moving manipulation agent. Stability and stabilizability issues are resolved in the visual observer domain by supplying suitable control sequences to the observer robot. Those control sequences are activated at intermittent points in time in order to "guide" the observer to the "desirable" set of visual states.

State Modeling and Observer Construction

Manipulation actions can be modeled efficiently within a discrete event dynamic system framework. It should be noted that we do not intend to *discretize* the workspace of the manipulating robot hand or the movement of the hand. We are merely using the DEDS model as a high-level structuring technique to preserve and make use of the known information about manipulation tasks. Furthermore, we also use all existing knowledge about the physical limitations of both the observer and manipulating robots. The high-level state definition permits the observer to recognize and report on symbolic descriptions of the task, and the physical relationships under observation. We avoid the excessive use of decision structures and exhaustive searches when observing the 3-D world motion and structure.

A bare-bones approach to solving the observation problem would have been to try and visually reconstruct the full 3-D motion parameters of the robot hand. The motion, shape, and structure of the different objects should also be recovered in 3-D. This process should be done in real time while the task is being performed. However, this formulation is inefficient, unnecessary, and for all practical purposes infeasible to compute in real time. In addition, the formulation does not provide any kind of interpretation for the *meaning* of the scene evolution, nor does it allow for any symbolic recognition for the task under observation. The limitation of the observer reachability, and the extensive computations required to perform the visual processing are motives behind attempting a different formulation. We view the problem as a hierarchy of task-oriented observation modules that exploits the higher-level knowledge about the existing system, in order to achieve a feasible mechanism of keeping the visual process under supervision.

State Space Modeling

We do a coarse quantization of the *visual manipulation actions,* which allows modeling both continuous and discrete aspects of the manipulation dynamics. State transitions within the manipulation domain are asserted according to probabilistic models. Those models determine at different instances of time whether the visual scene under inspection has changed its state within the discrete event dynamic system state space. Mapping the desired visual states to a DEDS skeleton is a straightforward procedure. We attach a DEDS automaton state to each meaningful visual state within a manipulation action. The quantization threshold depends on the application requirement. In other words, the state space can be expanded or contracted depending on the level of accuracy required in reporting and observing. A surgical operation step, performed by a robotic end effector, will obviously require an observer that reports (and possibly control the effector within a closed-loop visual system) with extreme precision. The observer for a robotic manipulator whose task is to pile up heaps of waste would, most likely, report in a crude fashion, thus needing a small number of states. The quantization threshold depends heavily on the nature of the task and the application requirements. The DEDS formulation is flexible, in the sense that it allows different precisions and/or state space models depending on the requirements.

The task of building DEDS automaton skeletons for observer agents can be performed either *manually* or *automatically.* In the manual formation case, the designer would have to draw the automaton model that best suits the task(s) under observation depending on the application requirements. The code for the state machine then needs to be implemented. Automatic construction of the state machine could be done by having a *learning* stage (Kuniyoshi et al. 1990, 1989) in which a mapping module would form the automaton. The building phase is performed before the actual observation process is invoked. The idea is to supply the module with sets of possible sequences in the form of *strings* of a certain language that the DEDS automaton should minimally accept. The language could be either supplied by an operator, in which case, the resulting automaton performance will depend on the relative skill of the operator, or through showing the module a sequence of visual actions and labeling those actions appropriately. The language strings should also be accompanied by a set of transitional conditions as event descriptions. The module would then produce the minimal DEDS automaton code complete with event and state descriptions that accepts the language.

We next discuss building the manipulation model for some simple tasks, then we proceed to develop the observer for these tasks. Formulating the models for the state transitions, the interstate continuous dynamics, and recovering uncertainty will be left for the sections that deal with the different uncertainty levels and event identification mechanisms.

Building the Model

The ultimate goal of the observation mechanism is to know at all (or most) of the time the current manipulation process. The fact that the observer will have to move makes one think of the stabilizability principle for general DEDS as a model for the tracking technique to be performed by the observer's camera.

In real-world applications, many manipulation tasks are performed by robots including, but not limited to, lifting, pushing, pulling, grasping, squeezing, screwing, and unscrewing of machine parts. Modeling all tasks, and also the possible order in which they are to be performed, is possible within a DEDS state model. The different hand/object visual relationships for different tasks can be modeled as the set of states X. Movements of the hand and object, either as 2-D or 3-D motion vectors, and the positions of the hand within the image frame of the observer's camera can be thought of as the events set Γ. The events cause state transitions within the manipulation process. Assuming, for the time being, that we do not have direct control over the manipulation process itself, we can define the set of admissible control inputs U as the possible tracking actions that can be performed by the hand holding the camera. Those actions can alter the visual configuration of the manipulation process (with respect to the observer's camera). Furthermore, we can define a set of "good" states where the visual configuration of the manipulation process enables the camera to keep track of and to know the movements in the system. Thus, it can be seen that the problem of observing the robot reduces to the problem of forming an output stabilizing observer (an observer that can always return to a set of "good" visual states) for the system under consideration.

It should be noted that a DEDS representation for a manipulation task is by no means unique. In fact, the degree of efficiency depends on the designer who builds the model for the task. Testing the optimality of a visual manipulation models is an issue that remains to be addressed. Automating the process of building a model was discussed in the previous section. As the observer identifies the current state of a manipulation task in a non ambiguous manner, it can then start using a practical and efficient way to determine the next state within a predefined set. Consequently, it can perform the necessary tracking actions to stabilize the observation process with respect to the set of good states. That is, the current state of the system tells the observer what to *look for* in the next step.

A Grasping Task

We present a simple model for a grasping task. The model is that of a gripper approaching an object and grasping it. The task domain was chosen for simplifying the idea of building a model for a manipulation task. It is obvious that more complicated models for grasping or other tasks can be built. The example shown here is for illustrative purposes.

As shown in Figure 50.3, the model represents a view of the hand at state 1 with no object in sight. At state 2, the object starts to appear. At state 3, the object is in the claws of the gripper, and at state 4, the claws of the gripper close on the object. The view as presented in the figure is a frontal view with respect to the camera image plane. However, the hand can assume any 3-D orientation so long as the claws of the gripper are within sight of the observer. An example of this would be the case of grasping an object resting on a tilted planar surface. This demonstrates the continuous dynamic aspects of the system. In other words, different orientations for the approaching hand are allowable and observable. State changes occur only when the object appears in sight, or when the hand encloses it. The frontal upright view is used to facilitate drawing the automaton only. It should be noted that these states can be considered as the set of good states E, since these states are the expected different visual configurations of a hand and object within a grasping task.

States 5 and 6 represent instability in the system, as they describe the situation where the hand is not centered with respect to the camera imaging plane. These states would occur when the hand and/or object are not in a good visual position with respect to the observer, as they tend to escape the camera view. These states are considered as "bad" states because the system will go into a nonvisual state unless we correct the viewing position.

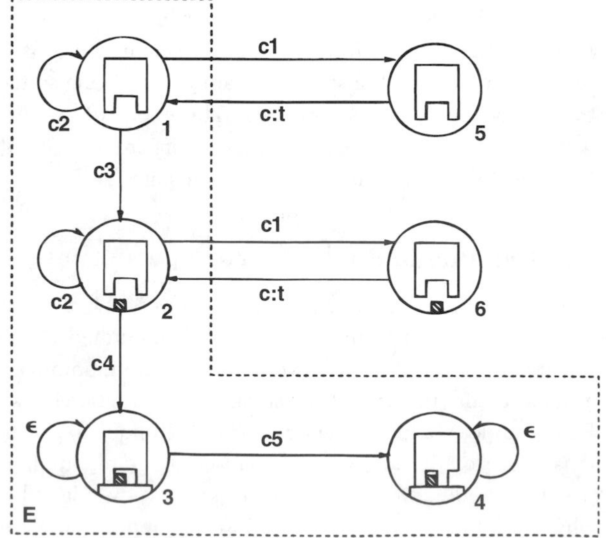

Figure 50.3 A model for a grasping task.

The set $X = \{1,2,3,4,5,6\}$ is the finite set of states. The set $E = \{1,2,3,4\}$ is the set of "good" states.

Some of the events are defined as motion vectors or motion vector probability distributions (as will be described later) that causes state transitions. Other events represent the appearance of the objects into the viewed scene. The transition from state 1 to state 2 is caused by the appearance of an object. The transition from state 2 to state 3 is caused by the hand enclosing an object. The transition from state 3 to state 4 is caused by the inward movement of the gripper claws. The transition from the set $\{1,2\}$ to the set $\{5,6\}$ is caused by movement of the hand as it escapes the camera view, or by the increase in depth between the camera and the viewed scene (the hand moving away from the camera). The self loops are caused by either the stationarity of the scene with respect to the viewer, or by the continuous movement of the hand as it changes orientations. In the next section, we discuss different techniques to identify the events. The controllable events denoted by "$: t$" are the tracking actions required by the hand holding the camera to compensate for the observed motion. Tracking techniques will later be addressed in detail. All the events in this automaton are observable and thus the system can be represented by the triple $G = (X, \Sigma, T)$, where X is the finite set of states, Σ is the finite set of possible events and T is the set of admissible tracking actions or controllable events.

It should be mentioned that this model of a grasping task could be extended to allow for error detection and recovery. Search states could be added in order to "look" for the hand if it is nowhere in sight. The purpose of constructing the system is to develop an observer for the automaton which will enable the determination of the current state of the system. Furthermore, the model will enable using the sequence of events and control to "guide" the observer into the set of good states E, and thus stabilizing the observation process. Disabling the tracking events will obviously make the system unstable with respect to the set $E = \{1,2,3,4\}$ (cannot get back to it). However, it should be noted that the subset $\{3,4\}$ is already stable with respect to E regardless of the tracking actions, that is, once the system is in state 3 or 4, it will remain in E. The whole system is stabilizable with respect to E. Enabling the tracking events will cause all the paths from any state to go through E in a finite number of transitions, and then will visit E infinitely often.

Developing the Observer

To know the current state of the manipulation process, we need to observe the sequence of events occurring in the system. Decisions must be made regarding the state of the automaton. State ambiguities are allowed to occur, however, they are required to be resolvable after a *bounded* interval of events. An observer has to be constructed according to the visual system for which we developed a DEDS model. The goal will be to make the system a stabilizable one, and/or construct an observer to satisfy specific task-oriented visual requirements. It should be noticed that events can be asserted with a specific probability (as will be described in the sections to come). Therefore, state transitions can be made according to prespecified thresholds that compliment each state definition. In the case of developing ambiguities in determining current and future states, the history of evolution of past event probabilities can be used to navigate backwards in the observer automaton till a strong match is perceived, a fail state is reached, or the initial ambiguity is asserted.

As an example, for the model of the grasping task, an observer can be formed for the system as shown in Figure 50.4. It can be easily seen that the system can be made stable with respect to the set E_O (the system always returns to that set). At the start, the state of the system is totally ambiguous. The observer can be "guided" to the set E_O consisting of all the subsets of the good states E as defined on the visual system model. It can be seen that by enabling the tracking event from the state (5, 6) to the state (1, 2), all the system can be made stable with respect to E_O. The singleton states represent the instances in time where the observer will be able to determine—without ambiguity—the current state of the system.

In the next section we shall elaborate on defining the different events in the visual manipulation system. We also discuss different techniques for event and state identification. A framework for computing the event uncertainty will be introduced.

Examples

Experiments were performed to observe the robot hand. The Lord experimental gripper is used as the manipulating hand. Different views of the gripper are shown in Figure 50.5. Tracking is performed for some features on the gripper in real time. The visual tracking system works in real time and a position control vector is supplied to the observer manipulator.

Some visual states for a grasping task using the Lord gripper, as seen by the observer camera, are shown in Figure 50.6. The sequence is defined by our model, and the visual states correspond to the gripper movements as it approaches an object and then grasps it.

The full system is implemented and tested for some simple visual action sequences. One such example is shown in Figure 50.7. The automaton encodes an observer which tracks the hand by keeping a fixed geometric relationship between the observer's camera and the hand, as long as the hand does not approach

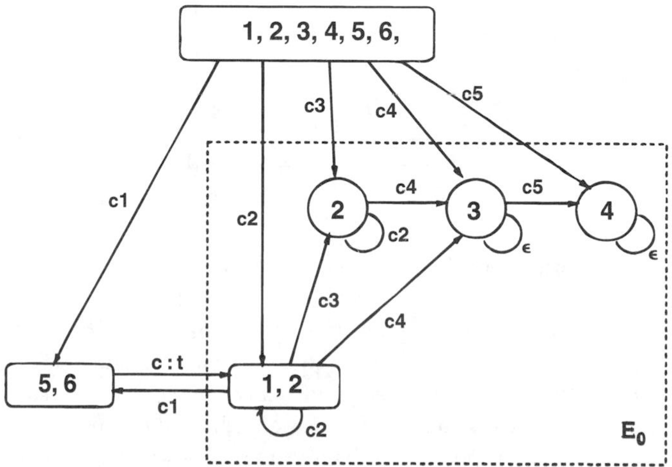

Figure 50.4 An observer for the grasping system.

Figure 50.5 Different views of the Lord gripper.

Figure 50.6 A grasping task as seen by the observer's camera.

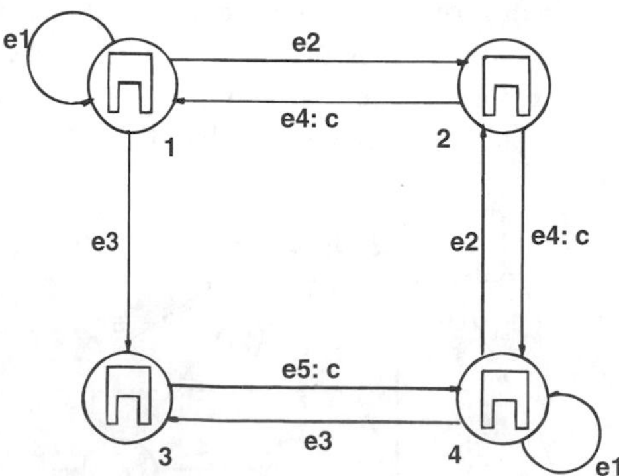

Figure 50.7 A model for a simple visual sequence.

the observer's camera rapidly. In that case, the observer tends to move sideways, that is, dodge and start viewing and tracking from sideways. It can be thought of as an action to avoid collision. State 1 represents the visual situation where the hand is in a centered viewing position with respect to the observer and viewed from a frontal position. State 2 represents the hand in a non-centered position and tending to escape the visual view, but not approaching the observer rapidly. State 3 represents a "dangerous" situation as the hand has approached the observer rapidly. State 4 represents the hand being viewed from sideways, and the hand is centered within the imaging plane.

Having defined the states, the events causing state transitions can be easily described. Event e_1 represents no hand movements. Event e_2 represents all hand movements in which the hand does not approach the camera rapidly. Event e_3 represents a large movement towards the observer. Events e_4 and e_5 are controllable tracking events. Event e_4 always compensates for e_2 in order to keep a fixed 3-D relationship, and e_5 is the "dodging" action where the observer moves to start viewing from sideways, while keeping the hand in a centered position. The events can thus be defined precisely as ranges on the recovered world motion parameters. For example, e_3 can be defined as any motion $V_Z \geq d_z$. Event e_1 is defined as any motion such that:

$$-\epsilon_x \leq V_X \leq \epsilon_x \wedge -\epsilon_y \leq V_Y \leq \epsilon_y \wedge -\epsilon_z \leq V_Z \leq \epsilon_z$$

It should be noted that defining e_1 in this manner helps a lot in suppressing noise. Having defined the events, the task reduces to computing the relevant areas under the probability distribution curves for the various 3-D motion parameters. State transitions are asserted and reported when the probability value exceeds a preset threshold. States 1 and 4 are considered to be the set of stable states. By enabling the tracking events e_4 and e_5, the system can be made stable with respect to that set.

The low level visual feature acquisition is performed on the MaxVideo pipelined video processor at frame rate. The state machine resides on a Sun SparcStation 1. The Lord gripper is mounted on a PUMA 560 arm and the observer's camera is mounted on a second PUMA 560.

Identifying Motion Events

We use the image motion to estimate the hand movement. This task can be accomplished by either feature tracking or by computing the full optic flow. The image flow detection technique we use is based on the sum of squared differences optic flow. The sensor acquisition procedure (grabbing images) and the uncertainty in image processing mechanisms for determining features are factors that should be taken into consideration when we compute the uncertainty in the optic flow.

One can model an arbitrary 3-D motion in terms of stationary-scene/moving-viewer as shown in Figure 50.8. The optical flow at the image plane can be related to the 3-D world as indicated by the following pair of equations for each point (x, y) in the image plane (Longuet-Higgins and Prazdny):

$$v_x = \left\{ x\frac{V_Z}{Z} - \frac{V_X}{Z} \right\} + [xy\Omega_X - (1 + x^2)\Omega_Y + y\Omega_Z]$$

$$v_y = \left\{ y\frac{V_Z}{Z} - \frac{V_Y}{Z} \right\} + [(1 + y^2)\Omega_X - xy\Omega_Y - x\Omega_Z]$$

where v_x and v_y are the image velocity at image location (x, y), (V_X, V_Y, V_Z) and $(\Omega_X, \Omega_Y, \Omega_Z)$ are the translational and rotational velocity vectors of the observer, and Z is the unknown distance from the camera to the object. In this system of equations, the only knowns are the 2-D vectors v_x and v_y. If we use the formulation with uncertainty, then the 2-D vectors are random variables with a known probability distribution. A number of techniques can be used to linearize the system of equations and to solve for the motion and structure parameters as random variables (Bajcsy and Sobh, 1991, 1990; Sobh and Wohn, 1989).

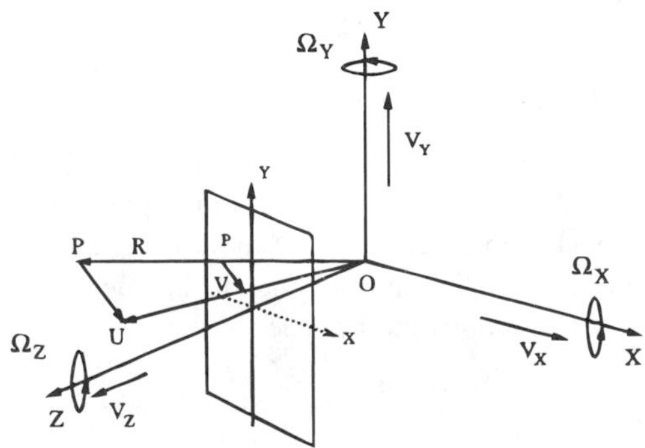

Figure 50.8 3-D formulation for stationary scene moving viewer.

Modeling and Recovering 3-D Uncertainties

The uncertainty in the recovered image flow values results from sensor uncertainties, noise, and the image processing techniques used to extract and track features. We use a static camera calibration technique to model the uncertainty in 3-D to 2-D feature locations. The strategy used to find the 2-D uncertainty in the features 2-D representation utilizes the recovered camera parameters and the 3-D world coordinates (x_w, y_w, z_w) of a known set of points. The corresponding pixel coordinates are then computed for points distributed throughout the image plane a number of times. We then find the actual feature pixel coordinates and construct 2-D histograms for the displacements from the recovered coordinates. The number of the experiments giving a certain displacement error would be the z axis of this histogram, while the x and y axis are the displacement error. The three-dimensional histogram functions are then normalized such that the volume under the histogram is equal to 1 unit volume and the resulting normalized function is used as the distribution of pixel displacement error.

The spatial uncertainty in the image processing technique is modeled by using synthesized images and corrupting them. We then apply the feature extraction mechanism to both kinds of images and compute the resulting spatial histogram for the error in finding features. The probability density function for the error in finding the flow vectors can thus be computed as a spatial convolution of the sensor and strategy uncertainties. We then eliminate the unrealistic motion estimates by using the physical (geometric and mechanical) limitations of the manipulating hand. Assuming that feature points lie on a planar surface on the hand, then we can develop bounds on the coefficients of the motion equations. These are second-degree functions in x and y in three dimensions, $v_x = f_1(x, y)$ and $v_y = f_2(x, y)$.

The 2-D uncertainties are then used to recover the 3-D uncertainties in the motion and structure parameters. The system is linearized by either dividing the parameter space into three subspaces for the translational, rotational, and structure parameters and solving iteratively; or using other linearization techniques, and/or assumptions to solve a linear system of random variables (Bajcsy and Sobh, 1991, 1990; Barron et al. 1990; Sobh and Wohn, 1989; Subbarao and Waxman, 1985; Ullman, 1983). As an example, the recovered 3-D translational velocity cumulative density functions for an actual world motion, $V_X = 0\ cm$, $V_Y = 0\ cm$ and $V_Z = 13\ cm$, is shown in Figure 50.9. It should be noted that the recovered distributions represent a fairly accurate estimation of the actual 3-D motion.

Utilizing the Discrete Event Observer

State transitions are asserted within the DEDS observer model according to the probability value of the occurrence of an event. Events are thus defined as ranges for the different parameters. The problem then reduces to computing the corresponding areas

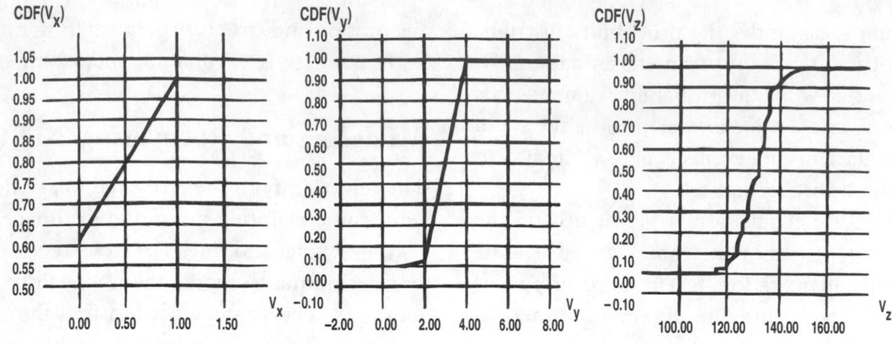

Figure 50.9 Cumulative density functions of the translational velocity.

under the refined distribution curves. An obvious way of using those probability values is to establish some threshold values, and assert transitions according to those thresholds. It might be the case that none of the obtained probability values exceeds the set threshold value, and/or all values are very low. In that case, there is a good chance that we are at the wrong automata state. The remedy to such problems can be implemented through time proximity, that is, wait for a while (which is to be preset) till a strong probability value is registered. Another technique is to *backtrack* in the observer automaton model till a high enough probability value is asserted, a fail state is reached, or the initial ambiguity is asserted. The backtracking strategy can be implemented using a stack-like structure associated with each state that has already been traversed. The stack includes a sorted list of the computed event probabilities, and a father-state variable.

Experiments

Experiments were performed to observe the robot hand. The low level visual feature acquisition is performed on the Datacube MaxVideo pipelined video processor at frame rate. The observer and manipulating robots are both PUMA 560's and the Lord experimental gripper is used as the manipulating hand.

The experiments were shot with three video camera. The right hand side of the images shows the actual observer and manipulation workspaces, and the different configurations as the experiment proceeds. The upper left corner shows the observer view, which is the set of images grabbed by the observer camera for processing. The lower left corner shows the observer state, that is, what the observer "thinks." A graphical representation of the different states and their change is used. Fail states are represented by an empty box. Figures 50.10 and 50.11 illustrate a manipulation experiment. In this sequence the hand tries to insert a peg in a hole. The observer approaches and focuses on the peg and hole when the peg gets nearer to the hole. State changes occur when the hole appears and when insertion is asserted.

Conclusions

We described a system for observing a manipulation process. The proposed approach can be generalized for other hybrid systems involving different kinds of quantization requirements. The use of discrete event dynamic systems with uncertainty modeling enables the observer to recognize tasks robustly. The proposed system also utilizes the *a-priori* knowledge about the task domain in order to achieve efficiency and practicality. The high level formulation allows for recognizing and reporting on the visual system state as a symbolic description of the observed tasks.

Thus, we have proposed a new approach to solving the problem of observing a moving agent. Our approach uses the formulation of DEDS as a high-level model for the evolution of the visual relationship over time. The proposed formulation can be extended to accommodate for more manipulation processes. Increasing the number of states and expanding the events set would allow for a variety of manipulating actions.

50.3 Conclusions

The control, analysis, modeling, synthesis, simulation, and monitoring of hybrid and discrete event systems are becoming more and more crucial in the current complex factory floor environments. We have discussed and presented hybrid systems through a problem related to robotics and automation for which discrete event and hybrid systems formulations play a significant role in the solution.

Acknowledgments

This work was supported in part by NSF grant CDA 9024721, and a University of Utah Research Committee grant. All opinions, findings, conclusions or recommendations expressed in this document are those of the authors and do not necessarily reflect the views of the sponsoring agencies.

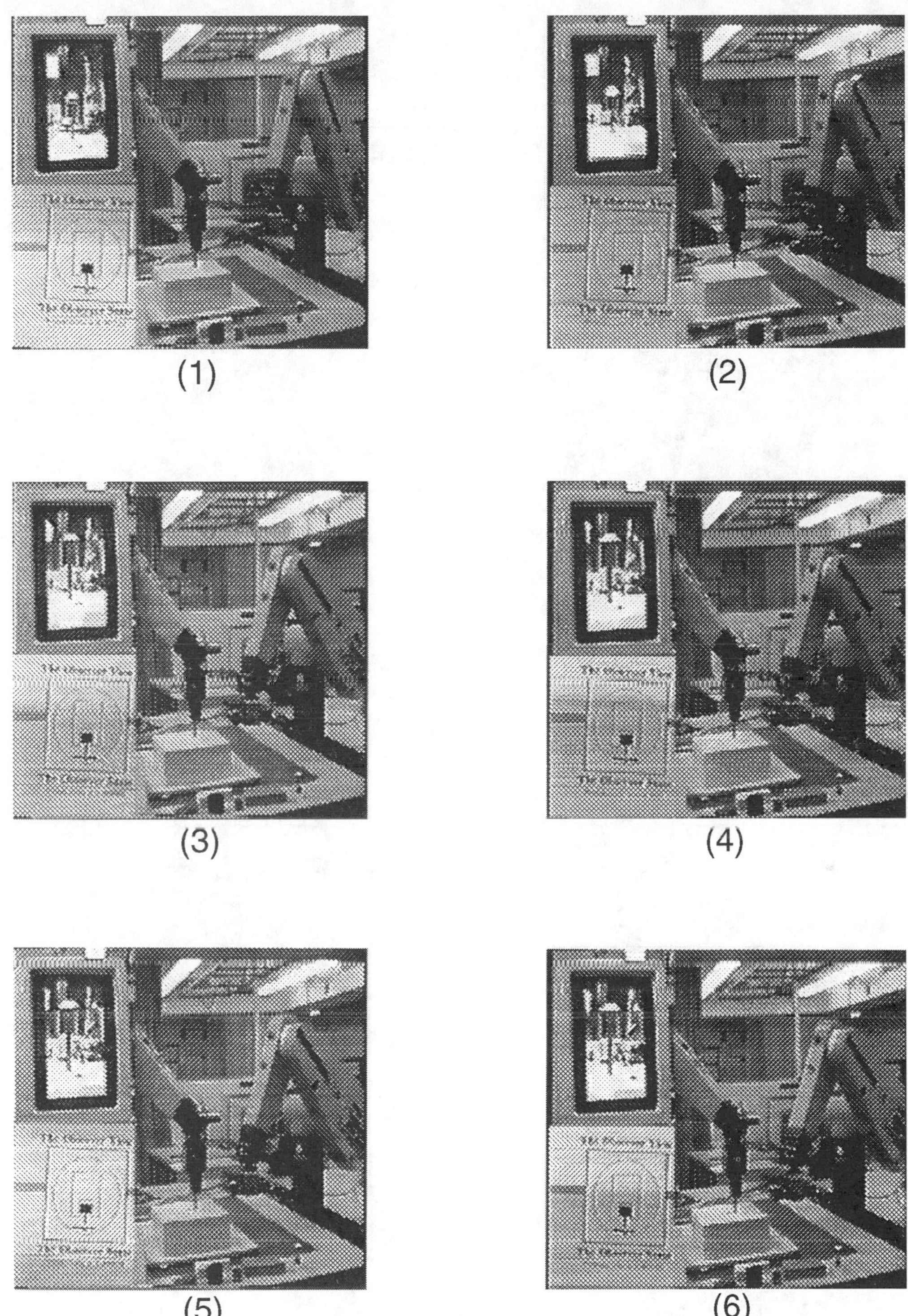

Figure 50.10 Observer state and view.

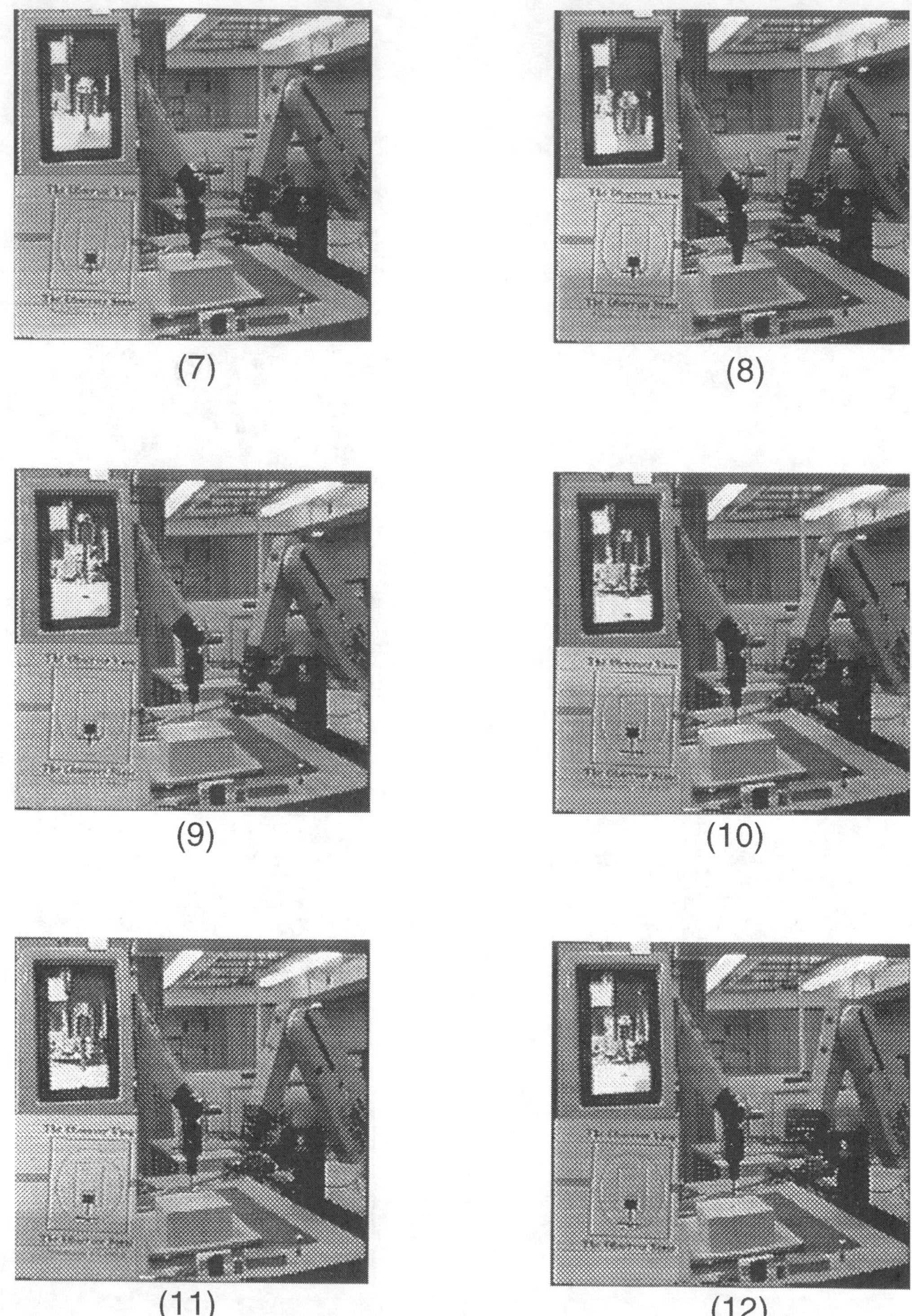

Figure 50.11 Continued.

References

Aloimonos, J. and Banyopadhyay, A. 1987. Active vision, *Proc. 1st Int. Conf. Computer Vision.*

Anandan, P. 1987. A unified perspective on computational techniques for the Measurement of visual motion, *Proc. 1st Int. Conf. Computer Vision.*

Bajcsy, R. 1988. Active perception, *Proc. IEEE,* 76(8).

Bajcsy, R. and Sobh, T. M. 1990. *A Framework for Observing a Manipulation Process,* Technical Report MS-CIS-90-34 and GRASP Lab. TR 216, University of Pennsylvania, June.

Bajcsy, R. and Sobh, T. M. 1991. *Observing a Moving Agent,* Technical Report MS-CIS-91-01 and GRASP Lab. TR 247, Computer Science Dept., School of Engineering and Applied Science, University of Pennsylvania, January.

Barron, J. L., Jepson, A. D., and Tsotsos, J. K. 1990. The feasibility of motion and structure from noisy time-varying image velocity information, *Int. J. Computer Vision.*

Burt, P. J. et al. 1990. Object tracking with a moving camera, *IEEE Workshop on Visual Motion,* March.

Chaumette, F. and Rives, P. 1990. Vision-based-control for robotic tasks, *Proc IEEE Int. Workshop on Intelligent Motion Control,* 2:395–400, August.

Hervé, J., Cucka, P., and Sharma, R. 1990. Qualitative visual control of a robot manipulator. *Proc. DARPA Image Understanding Workshop,* September.

Horn, B. K. P. and Schunk, B. G. 1981. Determining optical flow, *Artificial Intelligence,* 17:185–203.

Kuniyoshi, Y., Inaba, M., and Inoue, H. 1989. Teaching by showing: generating robot programs by visual observation of human performance, *20th ISIR.*

Kuniyoshi, Y., Inaba, M. and Inoue, H. 1990. Design and implementation of a system that generates assembly programs from visual recognition of human action sequences, IROS.

Li, Y. and Wonham, W. M. 1988. Controllability and observability in the state-feedback control of discrete-Event Systems, *Proc. 27th Conf. Decision and Control.*

Longuet-Higgins, H. C. and Prazdny, K. 1981. The interpretation of a moving retinal image, *Proc. Royal Society of London B,* 208:385–397.

Özveren, C. M. 1989. *Analysis and Control of Discrete Event Dynamic Systems: A State Space Approach,* Ph.D. Thesis, Massachusetts Institute of Technology. August.

Ramadge, P. J. and Wonham, W. M. 1987. Modular feedback logic for discrete event systems; *SIAM J. Control and Optimization,* September.

Sobh, T. M. and Wohn, K. 1989. Recovery of 3-D motion and structure by temporal fusion. *Proc. 2nd SPIE Conf. Sensor Fusion,* November.

Subbarao, M. and Waxman, A. M. 1985. *On The Uniqueness of Image Flow Solutions for Planar Surfaces in Motion,* CAR-TR-113, Center for Automation Research, University of Maryland, April.

Ullman, S. 1981. Analysis of visual motion by biological and computer systems, *IEEE Computer,* August.

Ullman, S. 1983. *Maximizing Rigidity: The incremental recovery of 3-D structure from rigid and rubbery motion,* AI Memo 721, MIT AI Lab.

51

Virtual Manufacturing Environment

Robert G. Wilhelm
University of North Carolina at Charlotte

51.1 Introduction 718
51.2 Scope for Virtual Manufacturing 718
51.3 Typical Applications 718
Unit Process • System Integration • Performance Modeling • Design
51.4 Emerging Technology 720

51.1 Introduction

Virtual manufacturing uses computer simulation in place of actual machines. The computer simulation predicts how a machine or system of machines will perform. The simulation can range from purely numeric calculations to very realistic animations where engineers and operators drive the simulation exactly as they would operate the actual machine or manufacturing process.

These computer simulations avoid many of the operating costs of actual machines and often run much faster. Considerable time savings can result when virtual manufacturing is used to design and test new machines, process steps, or systems. Safety, reliability, and measurement issues are often handled more easily.

Virtual manufacturing is rooted in the well established technologies of computer simulation, process modeling, virtual reality, and software engineering. Commercial software tools are now available for many types of virtual manufacturing applications. The scope of these applications ranges from simulation of tool-material interactions to integration of design and manufacturing facilities spread throughout the world.

51.2 Scope for Virtual Manufacturing

The simplest virtual manufacturing (VM) addresses single-unit processes such as metal cutting, etching, or welding. The process is simulated to find good operating parameters, estimate capacity, or troubleshoot.

Larger benefits come when VM integrates two or more components of a manufacturing system. To begin, simulated processes and controllers are used for each component. One by one, actual hardware replaces the simulated machines while configuration, programming, and testing are done. The simulated processes allow for cheap and varied test conditions with little danger to integration engineers. Later, the same simulators may be used to troubleshoot or optimize the manufacturing system.

VM is also used to predict how well a manufacturing system will run under different operating conditions. Simulations and actual data from the plant floor are used to predict throughput, yield, capacity, and other performance measures. These predictions are used to make production schedules, identify bottlenecks, and plan for capital improvements. In some cases, such as the power industry, complete simulators are also maintained for training and certification.

Design tools are now being developed so that VM can be used to design new manufacturing processes. From catalogs of materials, machines, processes, and control algorithms, a factory designer assembles a computer model of the manufacturing system. A computer simulation is automatically generated and computer animation is used to show the operation of the system. Control software for the system is tested against the simulation and performance is evaluated. Once the design is complete, a bill of materials and construction plan are generated automatically from the computer model. Maintenance and other life-cycle activities are also planned using the computer model.

VM is also being extended to allow dynamic configuration of manufacturing components. In a single plant, this might entail new routings between machines to switch from a process-oriented flow to a product-oriented flow. Additionally, machine breakdowns could be handled automatically by routing parts to different machines. Similarly, dynamic configuration can link manufacturing sites around the world. VM can also be used for benchmarking, outsourcing, and quality management.

Figure 51.1 illustrates development history for different virtual manufacturing applications and suggests the relative maturity of each type of application. Typical applications, as described below, show how VM is used.

51.3 Typical Applications

Four typical applications of virtual manufacturing (VM) are described below for technology available in 1995. The future of

0-8493-8343-9/97/$0.00+$.50

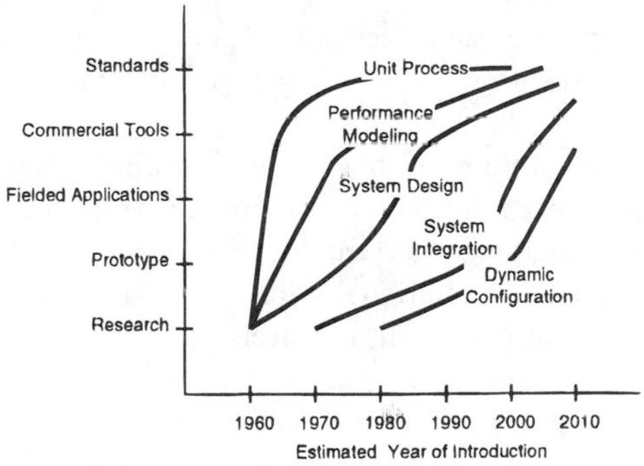

Figure 51.1 Scope for virtual manufacturing applications.

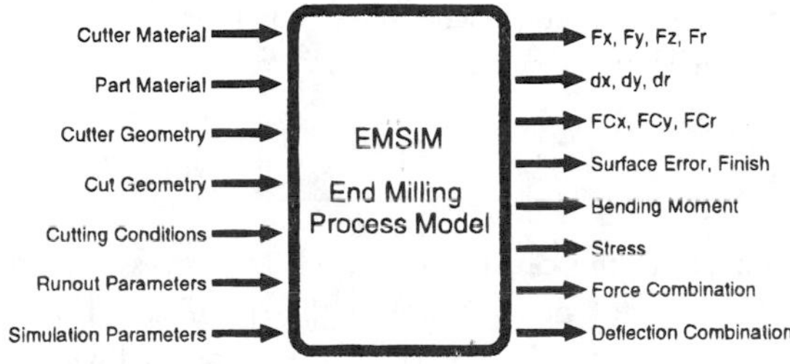

Figure 51.2 EMSIM machining process simulation model.

virtual manufacturing will be based on simulation and communication technologies that are only now under development.

Unit Process

One example of a VM unit process simulation is the EMSIM machining process simulation model which simulates the end milling process. Using EMSIM, process engineers find optimal cutting parameters for particular products requirements and costs. Machining simulations are absolutely necessary for parts that have complex cut geometries, flexible materials, and precisely located features.

EMSIM simulates the forces of an end milling operation. Working from a time sequence of force systems, accurate estimates can be made for the milling process parameters. Input and output data are depicted in Figure 51.2. Input data includes a description of the workpiece, the tool, and the cutting conditions. Output includes estimates of cutting forces, cutter deflection, and surface error of the part.

The milling simulator results can be obtained in seconds at a very low operating cost. By using the surface error estimates and force results, a manufacturing engineer can choose and test machining plans before the plans are sent to the shop floor. This practice replaces runoff procedures that require test parts to be cut on an actual machine tool. Costs due to machine time and tooling are avoided and dangers due to uncertain machining conditions are eliminated.

System Integration

Current applications of VM in systems integration use considerable amounts of virtual reality (VR) technology. Displays and sound generators provide visual or sound cues that can easily be recognized by operators. In some cases, computer simulations very closely mimic the operation of one machine in order that another may be rigorously tested.

At the cell level, a typical example is a multiaxis machining center with part loaders, complex tools, and many part programs. Initial input is part geometry and machine selection. Numerical control (NC) programs are generated for the parts to be machined in the cell and tested to insure that all machining paths can be accomplished by the machine. Each NC program is also tested for collisions and the machining time is measured.

It is common practice to complete the NC programming off-line and then do all checking on the machining center. With VM tools such as CIMSTATION or Virtual NC, much of the checking can be done with simulated machining centers. Figure 51.3 shows the graphical display for such a simulator. NC programs, tool geometry, and cutting stock dimensions are input for the simulation. As material is removed, the simulation display is updated to show the current shape of the part and the position of each part of the machine. As machine part interferences can be a problem with multiaxis machining, the simulator can also be programmed to check for interference that occurs away from the toolpoint.

At the system level, entire machines are often simulated so that communication between machines and the ensuing operations can be tested with minimal system downtime and danger. Figure 51.4 shows a sequential function chart for a system of machines and transporters. As a machine cycle is completed, a transporter is requested to move material to the next process step. The machine is commanded to a loading state until the part is moved to the transporter. The machine then signals that it is available for a different part and waits for delivery via a transporter.

The communication of command steps can be implemented with proprietary control and wiring designs but standard communication schemes are now available for these applications.

Figure 51.3 Virtual machining cell for system integration.

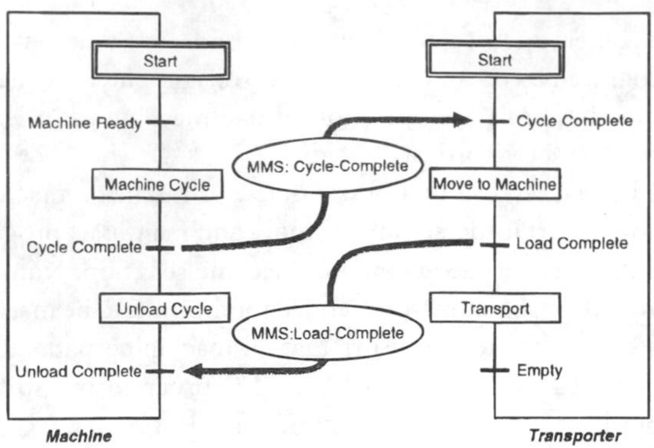

Figure 51.4 Simulated MMS services for system integration.

The manufacturing message specification (MMS) defines a standard set of messages that may be used on computer networks for remote control and monitoring of manufacturing machines and controllers. These messages are defined to operate with real and virtual manufacturing devices (VMDs). VMDs provide a standard way to implement virtual machines on a manufacturing network. During system integration, VMDs can be implemented on general purpose computers to simulate machine behavior. As real devices are added to the network, system behavior is checked using the VMDs or simulated machines.

For the machine and transporter example, the real transporters are tested while using simulated machines. At system initialization, the real transporter controller sends an MMS request to each machine asking for an update each time the state variable for cycle-complete changes. As a part is completed in the simulated machine, a MMS message with the new cycle-complete state is sent to the transporter controller. Transporter motion then continues until all loading and moving steps are complete. Note that testing time can be reduced since no time is actually required to complete the machine cycles.

Performance Modeling

VM techniques have long been used to model the overall performance of manufacturing plants. Typically, resources such as machines, transporters, and work areas are modeled as nodes in a network. Manufactured parts, repair orders, and other activities of the plant are modeled as events that pass between network nodes. Each resource and each event may have a set of rules that govern the time and resources needed for particular manufacturing activities. Processing times and the arrival of events is driven by random numbers generated in the simulation. The simulation is allowed to run for a fairly long time and various statistics are recorded for each type of part, job, repair, machine, etc.

The earliest simulations generated long reports of these statistics. Current simulation tools often display results by animating graphical models of the simulated system. Figure 51.5 shows the graphical display for an inventory system modeled with the EXTEND language. Similar results could be obtained with other common simulation tools such as SLAM and SIMON. During each simulation run, the icons of the display change shape as inventory increases and decreases. Figure 51.6 also shows the type of numerical output that is generated from the inventory simulation.

A recent application in the metals processing industry shows how performance modeling can be used to avoid cost. Plans for increased throughput suggested that in-process inventory would increase and require additional storage space. By using a simulation of the plant operation, engineers were able to find better operating procedures and reduce the in-process inventory and scale down capital improvement costs.

In another application in the electronics industry, large scale performance modeling simulations were used to dramatically increase the speed of production. The simulation results showed that balanced production scheduling was more important than increased machine capacity.

Design

To demonstrate the capabilities of a VM process design tool, we consider prototype software for the specification of control software. As shown in Figure 51.7, icons associated with various processes are copied to the design model. In this case, the icon for the stripper process is associated with a continuous simulation model that predicts the rate at which two products result from the continuous processing of four reactants. The simulation is based on a standard Eastman Kodak test problem.

To design a control algorithm, icons for program steps are also copied to the design model. The connections among icons indicate the sequence of program steps. Figure 51.8 shows the details of the control algorithm that can be seen by opening the *stripperControl* icon of Figure 51.7.

As different control algorithms are developed, their performance can be tested by executing the stripper process simulation. Figure 51.9 shows some of the output from a simulation run.

This control software design tool replaces the hardware simulators and plant testing cycles that are commonly used to develop process control software. Because many different algorithms can be tested without distributing the actual process, much more efficient control schemes are possible.

51.4 Emerging Technology

Even broader applications of virtual manufacturing (VM) are now under development. More realistic simulations with walk-through capability, testing toolkits for diagnosis, and dynamic configuration will each bring new efficiencies to manufacturing.

Manufacturing processes and systems are more complicated than the "virtual-worlds" that can currently be explored with virtual reality (VR) equipment. It will not be many years, however, before advances in computing technology bring VR to real applications in manufacturing. Heads-up and immersion displays will coherently present large amounts of process data to engineers and operators as they operate manufacturing processes. Virtual factories will be built to test new processes, tune manufacturing practices, and provide interactive training tools.

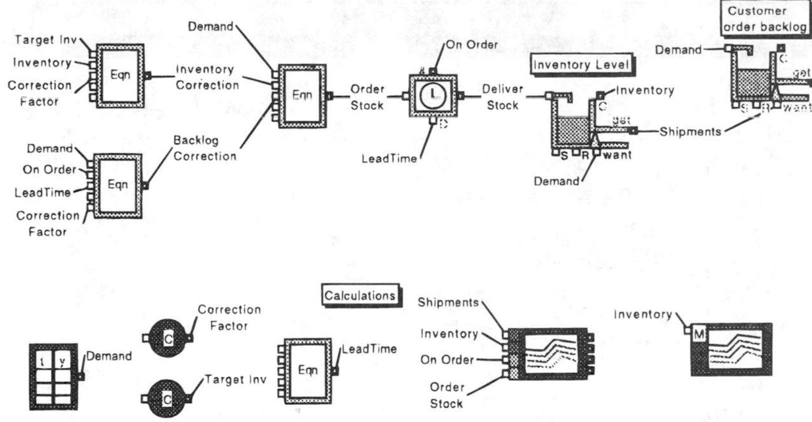

Figure 51.5 Simulating performance of an inventory simulation.

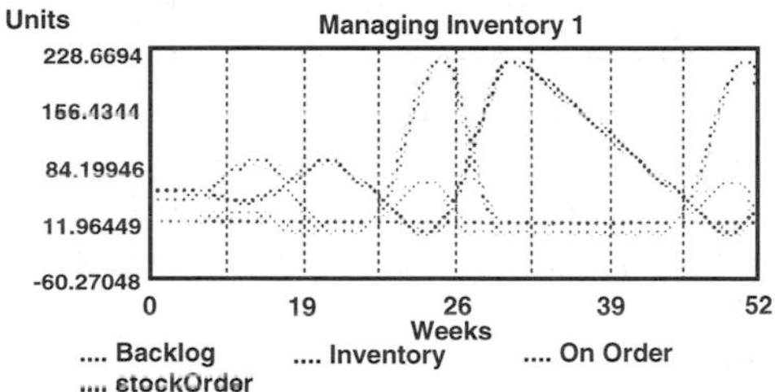

Figure 51.6 Output from inventory simulation.

Errors and breakdowns in current manufacturing systems are difficult to diagnose because it is costly, time consuming, and often dangerous to collect data that quantifies the symptoms of a problem. VM tools for process diagnosis will allow engineers to instrument "virtual-test-harnesses" for broad and inexpensive surveys of process symptoms. With simulators in hand, engineers will hypothesize possible trouble sources and run simulated experiments to test their guesses before going on with further operation of the real manufacturing system.

Factory network technology makes it possible to dynamically configure manufacturing systems according to production requirements. The same approach will soon be possible among factories distributed around the world. Within a plant, dynamic configuration allows machines to be grouped according to process or product flow patterns. For example, in a circuit board factory, each reflow oven may be assigned to a manufacturing cell or allowed to process parts from many different cells. A computer model of transportation protocols rather than a hard-tooled transportation system determines how parts are routed in the factory. Aside from benefits of load balancing this also provides for replacement machines during breakdowns or maintenance.

Between factories, the technology behind dynamic configuration can support testing and qualification for outsourcing. When a supplier bids to supply a part, the parent company can send requirements to the supplier plant electronically and ask for simulated results to estimate the time, cost, and quality performance for the plant. In other applications, such as load balancing, the presence of idle resources, such as finishing machines, in one facility will initiate precursor tasks, such as rough machining, in

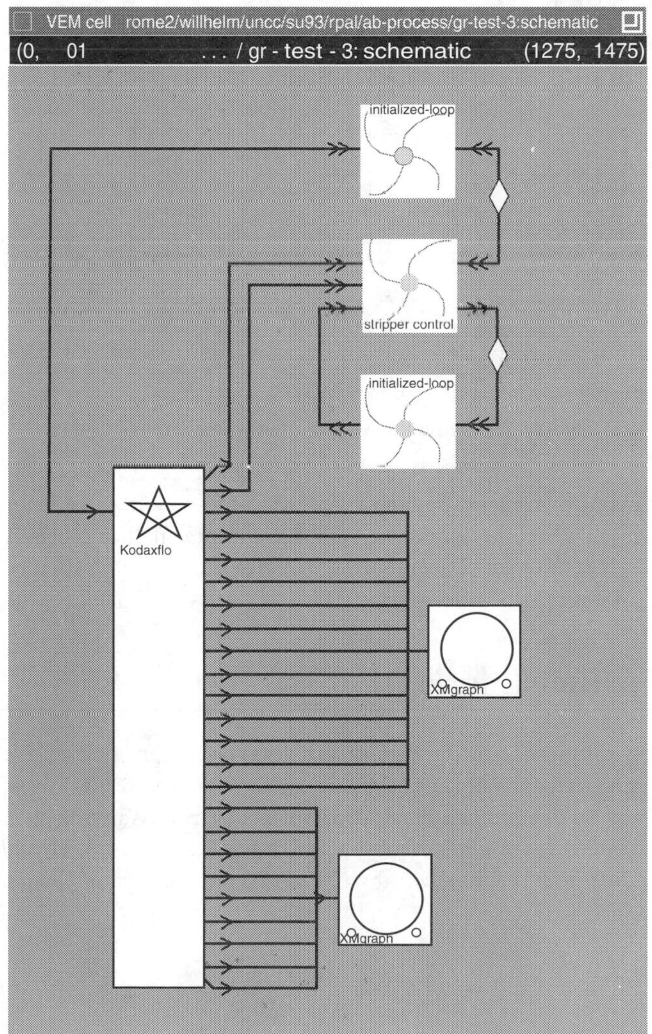

Figure 51.7 Design of process control for stripper.

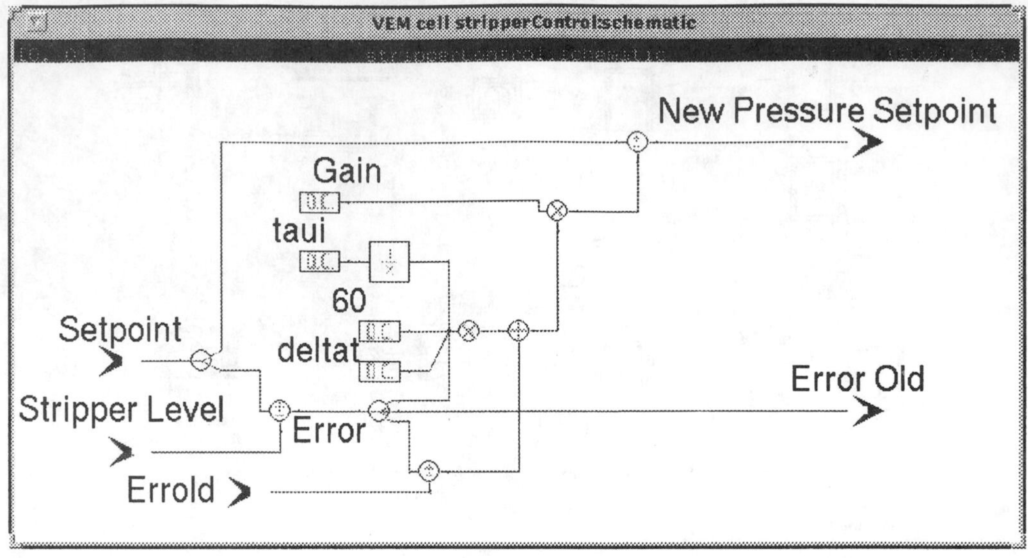

Figure 51.8 Control algorithm function block.

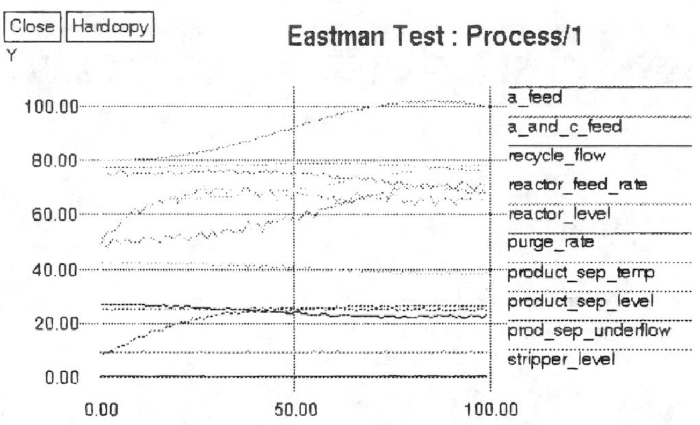

Figure 51.9 Simulation output for an algorithm design.

a neighboring plant. Factory to factory coordination, in real time, will also become the norm with VR equipment that concisely displays the state of neighboring factories and allows for detailed exploration when necessary.

Further Information

For discussions of the display and sensory equipment available for virtual reality applications, see *How Virtual Reality Works,* by Joshua Eddings, illustrated by Pamela Drury Watternmaker, and edited by Linda Jacobson, Ziff-Davis Press, 1994, or *Virtual Interface Environments,* by S. S. Fisher in *The Art of Human Computer Interface Design,* edited by B. Laurel, Addison-Wesley, 1990.

For a description of the EMSIM machining process simulation model and other process modeling techniques, see the World-Wide-Web (WWW) pages for the Machine-Tool Agile Manufacturing Research Institute at http://misled2.me.uiuc.edu/master.html.

For information on the manufacturing message specification (MMS), see ISO/IEC 9506, *Industrial Automation Systems—Manufacturing Message Specification,* Parts 1–6, 1990–1993, or *MAP and TOP Communications,* by A. Valenzano, C. Demartini, and L. Ciminiera, Addison Wesley, 1992 or the WWW page http://litwww.epfl.ch/ mms/mms.html.

For indepth information about simulation of system performance, see *Introduction to Simulation using SIMAN,* by C. Dennis Pegden, Robert E. Shannon, and Randall P. Sadowski, McGraw-Hill, 1995 or *Introduction to Simulation and SLAM II,* by A. Alan B. Pritsker, Wiley, 1994.

For discussion of VM software for design of manufacturing processes, See *Process Simulation for Computer-aided Factory Engineering,* by R. G. Wilhelm and A. J. Wilhelm in *Advances in Manufacturing Systems: Design, Modeling, and Analysis,* edited by R. S. Sodhi, Elsevier, 1994.

To monitor the progress of emerging technologies in VM, see the latest issues of *Manufacturing Engineering,* the *Proceedings of the World Congress of IFAC* and the *Proceedings of the ASME International Computers in Engineering Conference and Exhibition.*

52

Signal Processing for Factory Production Lines

52.1 Introduction 723
Processing Techniques
52.2 Examples of Signal Processing Systems 724
Transistor Wire-Bonding System • Wafer Inspection System • Automated Inspection of Thin Film Disk Heads • Automatic Visual Inspection of Soldered Parts • Assembled Printed Circuit Board Visual Inspection Machine • Color Vision Used in Industrial Applications • Automated Instrumentation of Fish Feature Recognition • 2-D Bar Code Reading System • Portable Image Signal Processing System

Rokuya Ishii
Yokohama National University

52.1 Introduction

In general, there are many different kinds of signals employed in factory production lines. These signals may be classified into 3 distinct categories, one-dimensional signals, two-dimensional (2-D) signals, and three-dimensional (3-D) signals. Typical examples of a one-dimensional signal are sound and vibration. A two-dimensional signal might be, for example, an image; and an example of a three-dimensional signal is a stereo-image. For efficient production, these signals must be processed in real time. The system that performs the real time processing of these signals should have the following specifications:

- High-speed signal processing: hardware and software for real-time response.
- A signal processing system that guarantees stable responses.
- High performance at reasonable cost.

Recently, a digital signal processor has been developed for processing one-dimensional signals (Chassaing and Horning, 1989). Using this processor, many different kinds of signals can be processed in real time. Therefore, one-dimensional signals are easily processed with digital signal processors.

Two-dimensional signal processing is in essence image processing. If each pixel in a 512 × 512 digital image is expressed by one byte, 256 k bytes are needed to store one image in memory. If it takes 1 microsecond to process each pixel, $^1/_4$ second is needed to process all pixels of an image. If more than one pixel is used for generating a processed pixel, more time is required. Therefore, many ideas for high-speed processing of digital images have been proposed. For example, parallel processing and pipeline processing with serial arranged processors have been employed to implement a high-speed processing system. On the other hand, many types of VLSI circuits have been developed for processing digital images in real time. Although these VLSI circuits are specifically designed for processing a digital image, it is very difficult to execute a complicated processing scheme in real time because the processing speed may be very slow.

The stability of a signal processing system is very important, because without it the system output will yield incorrect results. In addition, the system must be cost effective and a reasonable trade-off obtained between cost and performance.

Analog signals must be converted to digital signals prior to processing. Some typical equipment used in this conversion process are microphones, CCD area sensors, CCD line sensors, X-ray image intensifiers, X-ray line sensors, video-cameras, laser scanners, infrared cameras, and so on.

Processing Techniques

One-Dimensional Signal Processing

Sound and vibration are considered to be one-dimensional signals. These signals are typically used to diagnose a system in that they are used to detect abnormality and deterioration. The detection process is performed by analyzing the observed signal. The analysis calculation methods employed for the observed signal are classified into two categories.

1. Extraction of statistical characteristics.
2. Construction of a mathematical model for the observed signal.

The extraction of statistical characteristics involves calculating (a) the mean value for checking the variation of the bias, (b) the variance for calculating variations of the vibration level, (c) the power spectral density or the correlation for analyzing

0-8493-8343-9/97/$0.00+$.50

variations in the periodic portion of the signal, (d) the power spectrum for separating periodic parts from echo, and (e) the bi-spectrum for detecting variations in phase. These evaluation measures are very effective in detecting system performance.

Mathematical models for a system are obtained *a priori.* The output signal of the observed system is entered into the mathematical model and the output signal from the mathematical model is computed. If the system operates in a normal fashion, then white noise is observed as an output signal in the mathematical model. When certain types of abnormality occur, the output signal from the mathematical model is changed. Thus, by checking the output signal of the mathematical model, the abnormality can be detected from either the lack of white noise or the change in variance. This method is a suitable technique for on-line monitoring and is a very easy method for measuring abnormality quantitatively.

Two-Dimensional Signal Processing

An image signal is normally digitized using an A/D converter and entered into an image processing system. Prior to processing a digital image, an attempt is made to reduce distortion and to eliminate noise in order to improve the signal-to-noise ratio. Generally, an original image is distorted in the digitizing process.

These distortions may result from a blurring of the image and camera movement. Furthermore, color distortion and arithmetic distortion may occur. Additional signal processing can be used to remove the background of an image, emphasize contour lines in an image, and remove low-frequency portions of the image. Once this preprocessing is complete, the processing for extracting characteristics is executed.

Typical characteristics of a processed image can be expressed as arithmetical characteristics—for example, area, length, shape (circle, square . .); density characteristics—for example, color, thickness, etc.; and distance measurement—for example, the distance from the system to an object. Using these data, the signal processing system can recognize the object to control, e.g., a robot, by employing one of the following recognition algorithms:

1. Tree distinction discrimination.
2. Statistical distinction.
3. Fuzzy logic.
4. Neural network.

Some typical applications of an image processing system in a factory environment are

- Precise measurement of the area or length of an object in a given image.
- Image restoration for visualizing an object made invisible by noise and distortion.
- Visualization for the purpose of extracting physical information.
- Pattern recognition for inspection and fault detection.

Three-Dimensional Signal Processing

Computer vision is representative of three-dimensional signal processing. Using computer vision, objects are recognized using information obtained with a visual sensor. Data obtained with a visual sensor are two-dimensional images. Therefore, the three-dimensional object must be reconstructed from two-dimensional image information in a signal processing system. In the process of generating two-dimensional image information, some information about the three-dimensional object is obtained. Therefore, we need to examine a three-dimensional object by reducing the number of degrees of freedom in order to recognize it, and to observe the object from a number of different positions.

There are two methods of obtaining three-dimensional image information without coming in contact with an object: the passive method and the active method. The passive method generates three-dimensional information of an object by reconstructing a three-dimensional image from many two-dimensional images obtained under certain conditions. This method is very flexible and yields a considerable amount of information about an object without affecting the object's condition. However, the calculations are time-consuming and the measurement accuracy is very low.

On the other hand, the active method obtains information about an object by measuring the distance to the object using, e.g., an electromagnetic wave. Most useful methods use a laser and are based on the light projection method. Using this method, it is possible to measure the distance to an object, precisely and quickly. Very good characteristics can be obtained by employing this method, however, the measurement environment and objects that can be measured are limited. These methods are useful with robot vision systems in the automation industry.

52.2 Examples of Signal Processing Systems

Transistor Wire-Bonding System (Kashioka et al., 1989)

A wire-bonding system is illustrated in Figure 52.1. The image processor accepts the video signal from the TV camera and makes a comparison with a standard pattern by performing a high-speed correlation calculation. A bonding mechanism, positioned according to this calculation, uses a capillary tube of the bonding material to feed out and stretch gold wires between the terminals of the device and the corresponding outer leads. This assembly process is completed in less than 2 seconds.

The recognition algorithm employed in this process is realized using a combination of image processors and a computer as shown in Figure 52.2. The computer evaluates the distances and direction angles from the detected local pattern position by calculating correlations between standard patterns and two-dimensional image signals.

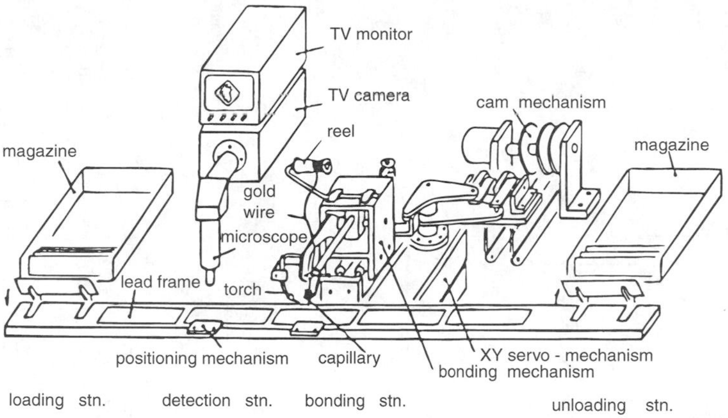

Figure 52.1 Configuration of bonding machine.

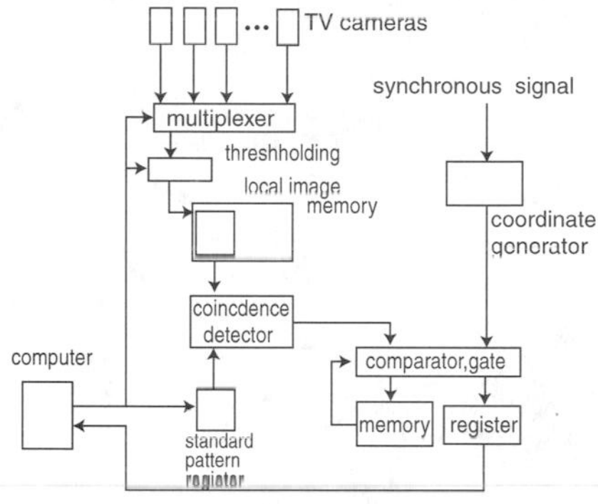

Figure 52.2 Visual image processor.

Wafer Inspection System (Yoda et al., 1988)

An automatic wafer inspection system is shown in Figure 52.3. An image signal of the wafer pattern is obtained through the linear array sensor, and is converted into 8-bit gray levels. The digitized image signal is continuously fed into a real-time correction circuit (RTCC) as well as a delay circuit (OPDC). Figure 52.4 shows the details of the image input mechanism. A wafer placed on a moving stage is illuminated by parallel light directed from a xenon arc lamp, and the surface pattern of the wafer is viewed by a CCD linear sensor. Automatic focusing is accomplished by using an air-micrometer, which blows dry air from outside an object lens onto the wafer surface. The output pressure signal from the air-micrometer controls a Z-stage which makes the gap between the lens and the wafer constant at all times. In very simple terms the image analysis method uses image subtraction (pixel-to-pixel in the adjacent periodic array) followed by thresholding to create a binary image. It then uses binary morphological operations (erosions, dilations, etc.) to filter out false defects and, in the process, converts the defects into binary blobs that are representative of their two-dimensional (2-D) shape. CAD design data are used to vary difference thresholds, numbers of erosions and dilations, and the size discrimination thresholds by level and region. In other words, it uses position dependent algorithms, driven by CAD data.

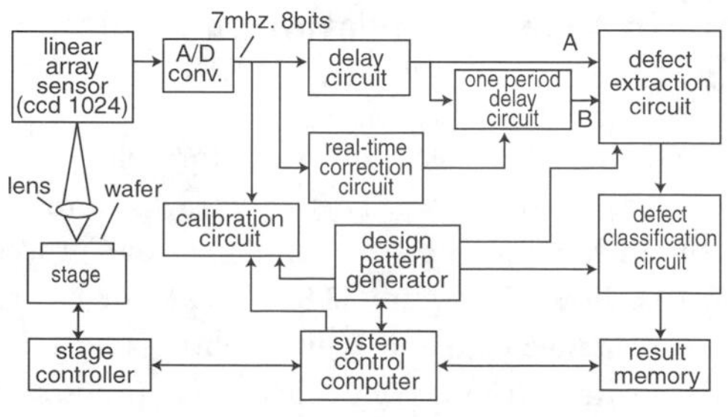

Figure 52.3 Configuration of the inspection system.

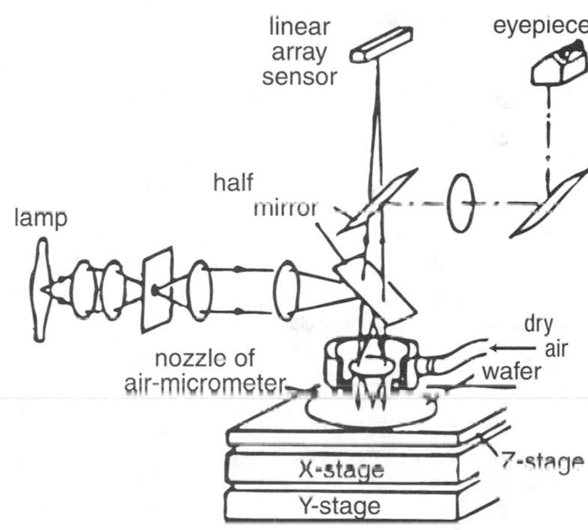

Figure 52.4 Large input mechanism.

Automated Inspection of Thin-Film Disk Heads (Sanz and Petkovic, 1988)

A digital visual inspection of thin-film disk heads is performed in a fully automated prototype system. The disk head, as shown in Figure 52.5, is 3 × 4 mm in size and the defect sizes are on

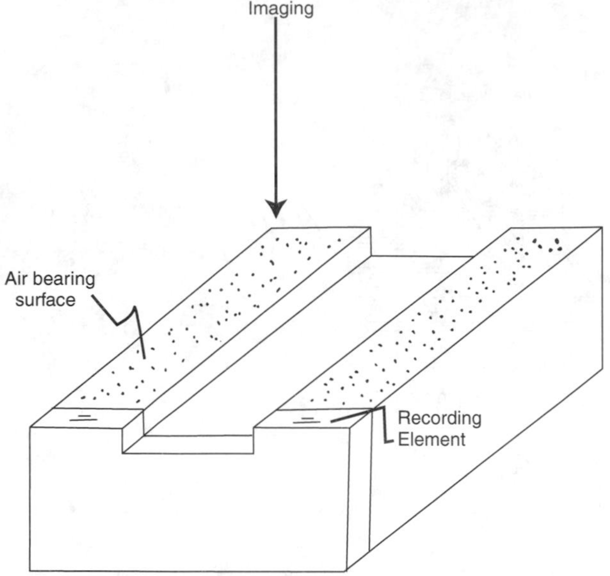

Figure 52.5 Disk head.

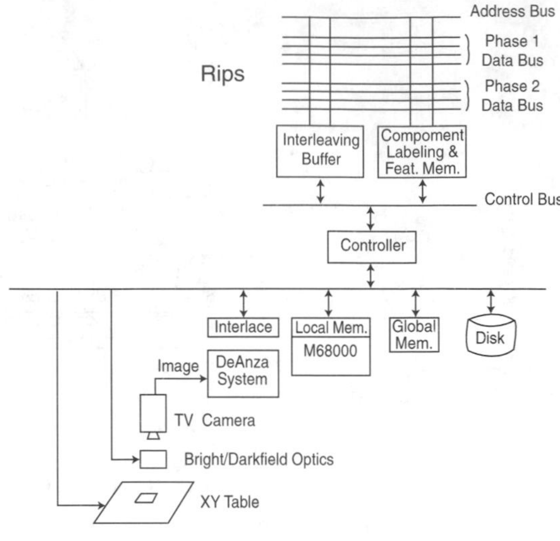

Figure 52.7 System architecture.

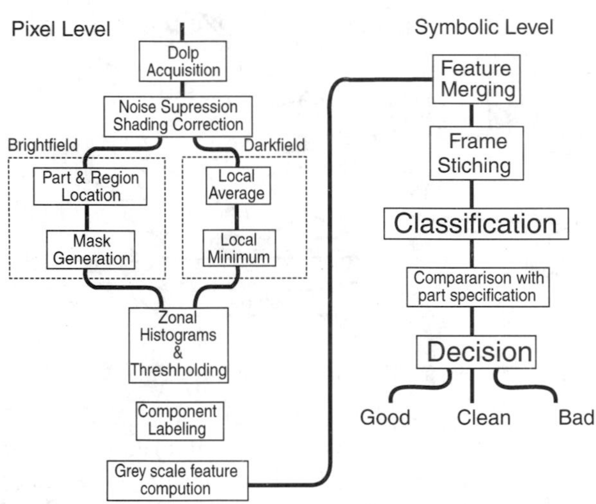

Figure 52.6 Life of a vista 1 image.

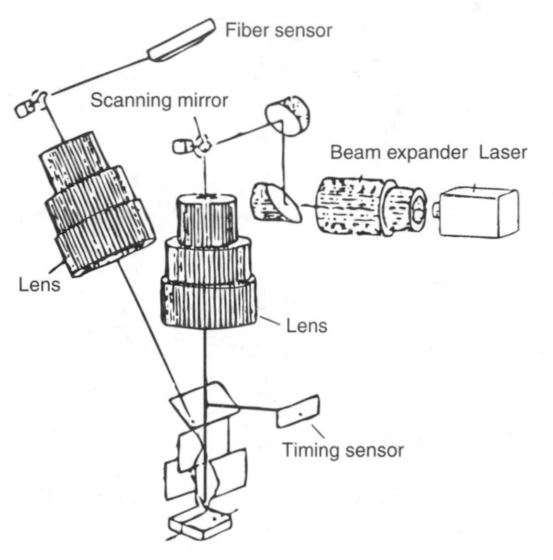

Figure 52.8 Schematic structure of 3-D sensor.

the order of tens of micrometers. The decision that a disk head is good or bad is made by applying a set of context-dependent rules to the detected and measured defects. In other words, the disk head inspection system must compute the exact location of the defects and determine the zones that they belong to in order to establish the relevance of the flaws. The overall image processing and analysis flow pattern is shown in Figure 52.6 and the system architecture is given in Figure 52.7. The system for visual inspection of thin-film heads consists of modules divided into three categories: image-to-image operations, image-to-symbolic transformation, and classification of symbolic objects. The first is devoted to image preprocessing operations, such as noise removal by multiple-frame averaging, shading correction to compensate for nonuniformities of the image setting, defect segmentation, etc. Image-to-symbol computations involve component labeling and feature extraction. The third step consists of defect classification, measurement, and comparison with inspection specifications.

Automatic Visual Inspection of Soldered Parts (Yoshimura and Okamoto, 1989)

An automatic visual inspection system for soldered parts was developed using a 3-D sensor. This system was composed of the 3-D sensor, an X-Y moving stage, and an image processor. The 3-D sensor is shown in Figure 52.8. This sensor has the ability to obtain 9-bit range data and 7-bit brightness data for 1024 × 1024 points within 0.85 seconds. The data processing unit is shown in Figure 52.9. This unit processes data for inspection at an average rate of 1.2×10^6 points per second.

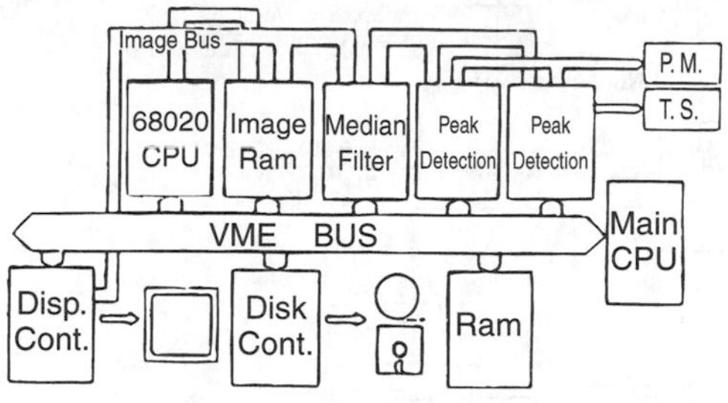

Figure 52.9 Block diagram.

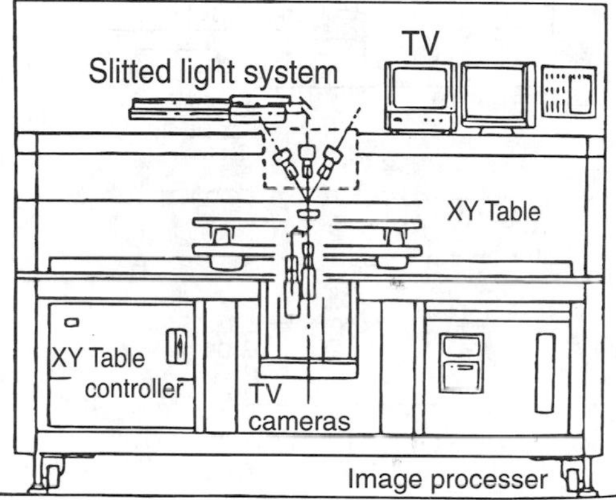

Figure 52.10 Assembled PCB visual inspection machine.

Assembled Printed Circuit Board Visual Inspection Machine (Hata et al., 1989)

A machine to automate the visual inspection of assembled printed circuit boards (PCBs) was developed and is shown in Figure 52.10. The machine detects a variety of defects in assembled PCBs, such as missing parts, polarity errors, improper names on ICs, and improper insertion of leads. A character recognition algorithm, called the multistep similarity vector comparison method, was implemented on the machine with an image signal processor. This system is shown in Figure 52.11. Figure 52.12 shows the methods used to light the object and to obtain an image. A slitted light beam and two TV cameras produced reliable results. Three types of algorithms were proposed to detect defects. One of them is shown in Figure 52.13.

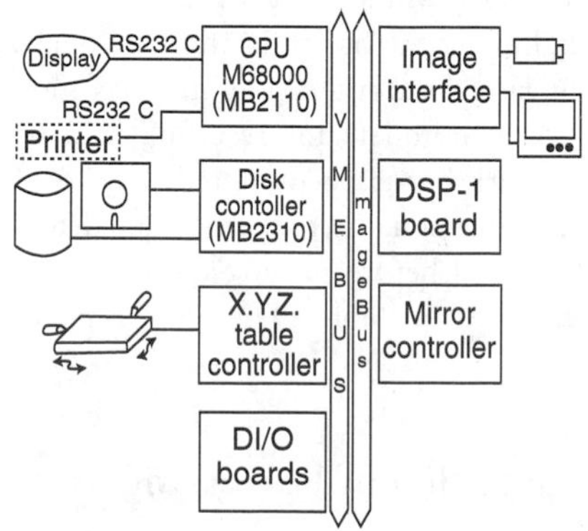

Figure 52.11 Structure of an image processor.

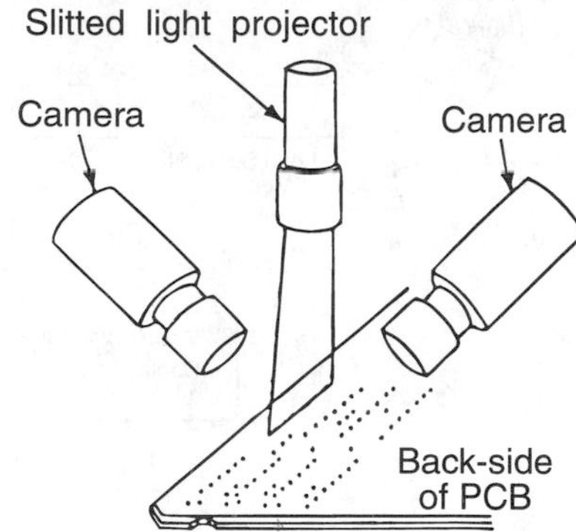

Figure 52.12 Optical systems for lead inspection.

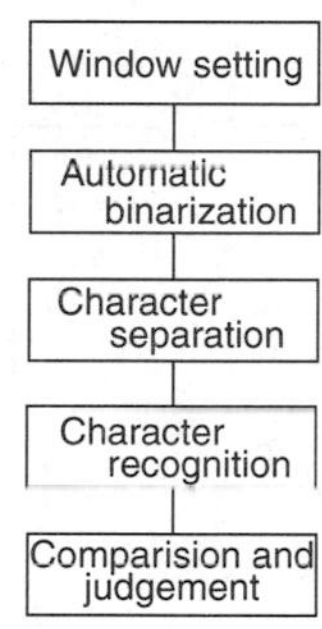

Figure 52.13 Flowchart of character recognition.

Color Vision Used in Industrial Applications (Mital et al., 1990)

One sample of a color vision system is the one shown in Figure 52.14. This vision system, for color object recognition, was implemented for sending the recognition results to a SIR-III robot controller which generates the appropriate signals to enable the robot to place objects in the proper bins.

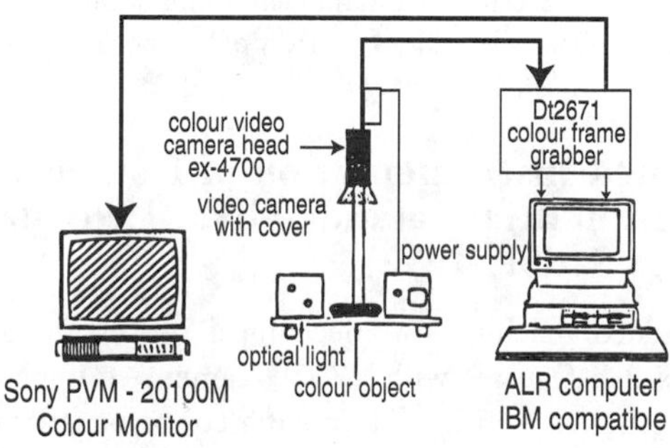

Figure 52.14 Sample of color vision system with DT-2871.

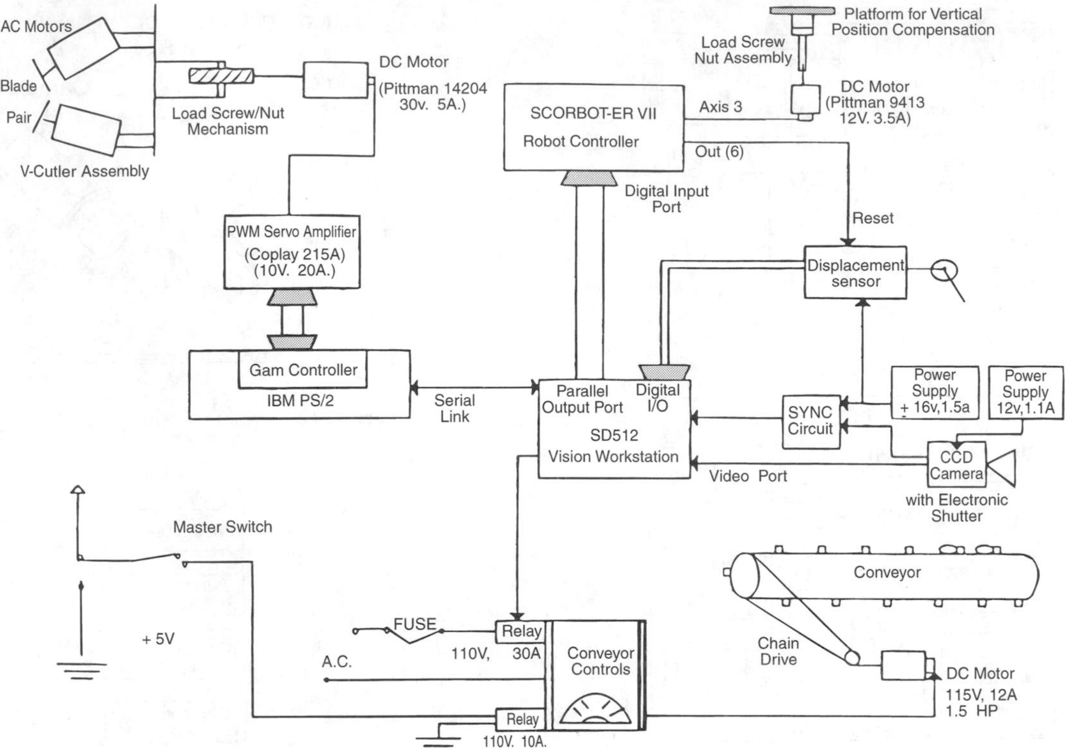

Figure 52.15 The automated system for fish processing.

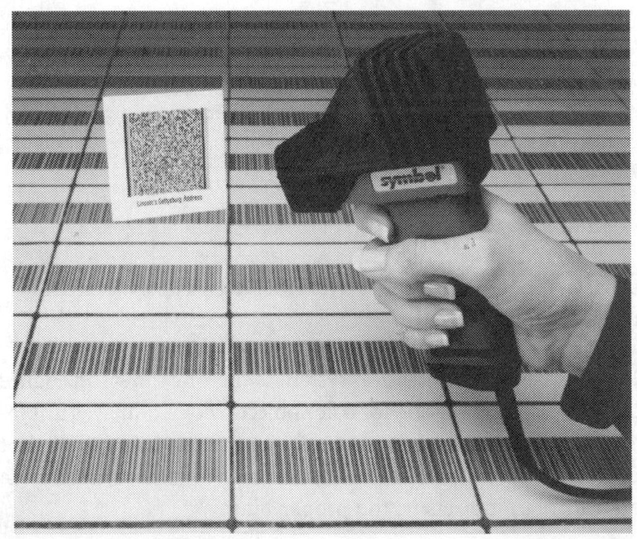

Figure 52.16 The 2-D bar code of the Gettysburg Address would require a 20-ft-long 1-D bar code. (Photo courtesy of Symbol Technologies, Inc., Holtsville, New York).

Automated Instrumentation of Fish Feature Recognition (Riahi and de Silva, 1990; de Silva and Saliba, 1992)

An automated machine for mechanical processing of salmon recognizes fish features with a CCD camera equipped with an electronic shutter. A fish on a moving conveyor is imaged by a CCD camera, the image is processed for feature recognition and a v-cut is performed for the head-removal operation. A dedicated vision system analyzes the image and determines the best lateral portion for the cutter blades. This information serves as the drive command for the motor controller of the cutter positioning system. This operation is performed at a speed consistent with the required throughput (more than two fish/s). The system is illustrated in Figure 52.15.

2-D Bar Code Reading System

The portable laser scanning system for reading the new two-dimensional bar code was developed with a digital signal processor. It has a high-density, high-capacity symbology which provides low-cost access to large amounts of information without referencing and external databases. This data file can store, in a machine-readable form, over a kilobyte of data. This system is shown in Figure 52.16. Wands and laser (moving beam) scanners pass focused light over the bar code and then analyze the reflection. When the laser encounters a light space, most of the light reflects back to a photodetector resulting in a large electrical signal. When a dark bar is encountered, a minimal amount of light returns and the signal is reduced. Edge detection circuitry squares up the signal before passing the signal to the decoder. This machine, called Portable PDF 1000, is made by Symbol Technologies, Inc.

Portable Image Signal Processing System

A visual image processing system with a digital signal processor was developed for the analysis of a binary image and a 256 color

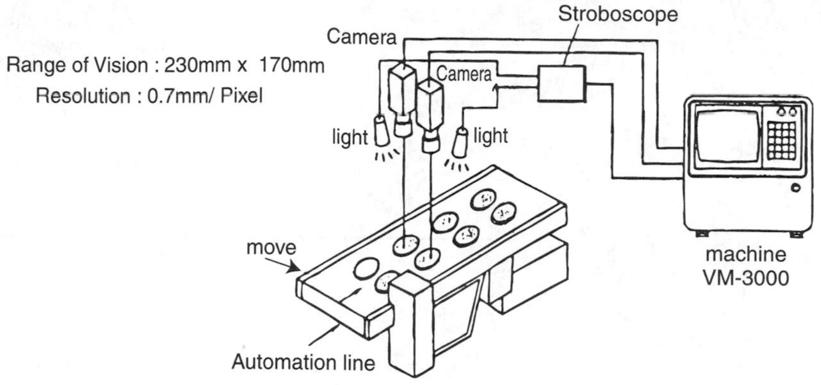

Figure 52.17 Vision Master VM-3000.

image. This system can be used for checking moving objects and inspecting them. This visual image signal processing system, called Vision Master VM 3000, is made by Shinko Electronic, Co., Ltd. and is shown in Figure 52.17.

References

Agin, G. J. 1980. Computer vision system for industrial inspection and assembly, *IEEE Computer,* 11–20, May.

Bartlett, S. L. et al. 1988. Automatic solder joint inspection, *IEEE Trans. Pattern Anal. and Machine Intell.,* 10(1):31–43.

Chassaing, R. and Horning, D. W. 1989. *Digital Signal Processing with the TM 320C25,* John Wiley & Sons, New York, NY.

de Silva, C. W. and Saliba, M. 1992. Instrumentation issues in the handling of fish for automated processing, *Proc. IECON'92,* 789–794.

Fujino, A. et al. 1992. An on-line tuning method for multi-objective control of elevator group, *Proc. IECON'92,* 795–880.

Gorden, S. J. et al. 1988. Real-time part position sensing, *IEEE Trans. Pattern Anal. and Machine Intell.,* 10(3):374–386.

Hata, S., Hagimae, K., Hibi, S., and Gunji, T. 1989. Assembled PCB visual inspection machine using image processor with DSP, *Proc. IECON'89,* 572–577.

Hirose, T. 1991 Measurement of location and orientation of vehicle by image processing, *SICE Trans.,* 27,(5):524–531, May. (in Japanese)

Image Processing Handbook, Shoko-do, 1988 (in Japanese).

Iwami, T. et al. 1989. Large area electron beam direct imaging technology for printed wiring boards, *Proc. IECON'89,* 550–561.

Kashioka, S., Ejiri, M., and Sakamoto, Y. 1976. A transistor wire-bonding system utilizing multiple local pattern matching techniques, *IEEE Trans. System, Man and Cybemetics,* SMC-6(8):562–57.

Mital, D. P., Leng, G. W., and Khwnag, T. E. Colour vision for industrial applications, *Proc. IECON'90,* 548–551.

Myers, W. 1980. Industry begins to use visual pattern recognition, *IEEE Computer,* 21–31, May.

Okamoto, H. 1993. Application example of color image processing in factory automation, *J. SICE,* 32(11):891–894. (in Japanese)

Riahi, N. and de Silva, C. W. 1990. Fast image processing techniques for the gill position measurement in fish, *Proc. IECON'90,* 476–781.

Sanz, J. L. and Petkovic, D. 1988. Machine vision algorithms for automated inspection of thin-film disk heads, *IEEE Trans. Pattern Anal. and Machine Intell.,* 10(6).

Watanabe, Y. 1989 Automated optical inspection of surface mount components using 2D machine vision, *Proc. IECON'89,* 584–588.

Yoda, H., Ohuchi, Y., Taniguchi, Y., and Ejiri, M. 1988. An automatic wafer inspection system using pipeline image processing techniques, *IEEE Trans. Pattern Anal. and Machine Intell.* (1):4–15.

Yoshimura, K. and Okamoto, S. 1989. A three-dimensional sensor for automatic visual inspection of soldered parts, *Proc. IECON'89,* 562–567.

Yokoya, N. 1994. Recent trends of computer vision, *J. ISICE,* 38(8):436–44. (in Japanese)

53 Robots

Ray Jarvis
Monash University

Lindsay Kleeman
Monash University

R. Andrew Russell
Monash University

Marcelo H. Ang, Jr.
National University of Singapore

Choon-seng Yee
National University of Singapore

Fathi Ghorbel
Rice University

Miguel A. Salichs
Universidad Carlos III

Luis Moreno
Universidad Carlos III

Diego Gachet
Universidad Carlos III

Arthuro de la Escalera
Universidad Carlos III

Juan R. Pimentel
GMI Engineering and Management Institute

Antal K. Bejczy
California Institute of Technology

53.1 Robots: Qualities and Capabilities 730
53.2 Robot Vision 732
53.3 Ultrasonic Sensors 738
Overview • Speed of Sound • Attenuation of Ultrasound due to Propagation • Target Scattering and Reflection • Beamwidth—The Round Piston Model • Transducers • Polaroid Ranging Module • Estimating the Echo Arrival Time • Estimating the Bearing to Targets • Specular Target Classification • Treatment of Rough Surfaces • Ultrasonic Beam-Forming Arrays
53.4 Robot Tactile Sensing 745
Whisker Sensors • Force/Torque Sensors • Skinlike Tactile Sensors
53.5 A Robotic Sense of Smell 749
Odor Discrimination • Odor Detection Systems • Marking and Detecting Odor Trails • Olfactory Sensors
53.6 Actuators in Robotics and Automation Systems 750
Overview • Direct Current Motors • Stepper Motors • Transmissions
53.7 Control 760
Introduction • Equations of Motion • Motion Control • Conclusions
53.8 Mobile Robots 773
Introduction • Robot Platforms • Kinematics • Control Architectures • Perception • Control of Mobile Robots • Navigation • Planning • Applications
53.9 Teleoperators 784
Introduction • Advanced Teleoperation • Anthropomorphic Telemanipulation • New Development Trends

53.1 Robots: Qualities and Capabilities

Ray Jarvis

The fascination many people have for robots is often associated with their supposedly humanlike qualities; a comparison with human capabilities is a good starting point in describing what robots are generally about.

There are essentially two kinds of robotic devices, those that are somewhat like the human arm (robotic manipulator) and are generally used in industry at a fixed location, and those that are mobile and are required to navigate through their defined workplaces efficiently and without collision.

In industrial settings, robotic manipulators are currently used for sorting, welding, deburring, painting, and manufacture; mobile robots, often referred to as automated guided vehicles (AGVs), are used to carry goods and components between workstations in a factory or warehouse.

Mechanically speaking, robots can be powerful and tireless. They are capable of repeating actions accurately and can survive hostile environments which are either uncomfortable or hazardous for humans. Many robot manufacturers are now able to design and produce units which are able to work for thousands of hours at high precision between maintenance sessions.

Robotic manipulators have a number of configurations according to how their links operate and the types of space they sweep out. Polar robotic manipulators have a rotation joint at the waist and an arm which can extend or contract in length and tilt up and down in a vertical plane. Cartesian robots are able to move their end-effectors (hands) independently in three

0-8493-8343-9/97/$0.00+$.50

orthogonal directions. Anthropomorphic robots (human arm like) have three main revolute joints, the waist, shoulder, and elbow, each with one degree of freedom. Scara robotic manipulators are very rigid in one plane to suit particular operations. In addition to being able to place the end of their arms in various positions in their work spaces, robotic manipulators also have degrees of freedom related to the orientation of the end-effector attached to the wrist. For example, a standard six degree of freedom anthropornorphic revolute jointed robotic manipulator would, in addition to the waist, shoulder, and elbow rotations, also have an end-effector attached to a wrist with another three degrees of freedom associated with yaw, pitch, and twist rotations. Also, the opening and closing of the end-effector would add other degrees of freedom (at least one).

The energy required to operate robotic manipulators is usually one of three kinds, electrical, hydraulic, or pneumatic. Since the efficiency, accuracy, and speed of a robotic manipulator is directly related to how a rigid set of links can be moved without carrying unnecessary loads, considerable thought at the design stage goes into how to keep the mass associated with the energy sources away from moving links, particularly the ones furthest away from the waist, and at the same time not introduce sources of imprecision such as gear-trains, wires, and long drive shafts. Hydraulic energy systems are often preferred over electric ones where very powerful but fast manipulators are required. However, through the use of direct drive motors (no gears), very high-precision and fast electric energy driven robotic manipulators have become very attractive, since such a system is highly controllable and has few moving parts. For smaller robotic manipulators, the use of shape memory alloy for tendonlike actuation is being carefully researched. For very small robotic manipulator devices (microrobots), processes similar to those used for very large scale integration (VLSI) circuitry in the microelectronics industry are being used to manufacture electrical motors and linear actuators. Friction wear is one of the difficulties to be overcome for such devices and magnetic bearing technology promises one solution to this problem.

The high accuracy and repeatability (not the same) enjoyed by many robotic manipulators derives from the precision of manufacture of their components, the linear and shaft encoder resolution of their joint/link position monitoring elements, and the quality of the control system used. Many very sophisticated control theoretic systems are now being applied in robotics; these are capable of better control than the classical proportional, integrative, and differential (PID) controllers that have been used for decades. Control quality is assessed in terms of speed, accuracy, and energy efficiency. The complexity of the dynamic distribution of load forces among the joints, links, and payload while a robotic manipulator is moving along an optimal trajectory at high speed represents a considerable challenge to the control system.

Automated guided vehicles (AGVs) have their distinct attributes while sharing some (particularly those relating to motion control) with robot manipulators. An important distinction derives from the different ways in which, on the one hand, the configuration of the links of a robotic manipulator and, on the other, the position and orientation of a AGV, are determined. Encoders on the joints of a robotic manipulator are used to monitor its configuration at all times; converting outputs from encoders to end-effector positions and orientations is just a matter of calculation. Unless an AGV is just a cart moving on toothed wheels along a rack, its exact position and orientation cannot be determined using the type of encoders used for robotic manipulators. The use of encoders on plain wheels (odometry) does not achieve the same result, since noncircularity under load, slippage, and ground irregularity all contribute to accumulative error so that eventually the position and orientation of the robotic vehicle becomes uncertain. Constraining the vehicle to a track specified by marks on the ground or signal wires buried beneath it does limit localization (position and orientation) error, but at the cost of severe path limitation and/or considerable site preparation. Realizing a free ranging AGV requires the use of localization instruments such as beacons or natural landmarks.

Outside uses in traditional factory environments mobile robots can be used to operate in rugged terrain, on and under water, in mines, space, and eventually, perhaps, inside the human body. Various means of locomotion include wheels, tracks, legs, hover, jets, and propellers. Power sources include combustion engine, electrical, and combinations of these and hydraulics.

The survivability of robotic manipulators and vehicles in hostile environments is a matter concerning the variety of materials which can be used in their construction, their rigidity and power, the choice of energy sources available, and the many ways by which the manipulator can be sealed from dangerous atmospheres and liquids in which they may be immersed. For example, a water-tight manipulator with a noncorrosive outer layer could work at a great depth in the ocean with electrical energy being provided from the surface, as part of a sea bed exploration mission on a tele-operated vehicle which could stay submerged for long periods. The use of robotic manipulators in handling radio-active materials or very high temperature components as well in the vacuum of deep space are other examples of hostile environment application.

An interesting complimentary activity to surviving hostile environments is the use of robots in clean rooms where human-centered contaminants cannot be tolerated, the robot being sterilized by methods not applicable to humans before being introduced to that environment. Also of interest is the potential use of microrobots for exploratory and surgical tasks within the human body, where small size and sterilization are critical so as to minimize the invasiveness of such operations.

Thus, overall, from the mechanical perspective, robots have certain advantages over humans. These include speed, power, endurance, accuracy, smallness, and capability of operating in hostile environments and not contaminating others. Cost-effectiveness must be gauged with respect to specific activities, but, in general, tasks which are repetitive and require precision and speed are those where the cost-effectiveness of using robots is often very great.

As working environments become more complex, less structured, and less controllable, more intelligence in humans, as well as machines, is needed to cope with the situation. For the type of intelligence associated with vast amounts of arithmetic calculations, computers are clearly superior to humans. There are other types of reasoning processes which, through developments in artificial intelligence, have been shown to be computationally feasible to a degree which is now, or will soon be, superior to human capabilities in the same area. However, in the area of perception, including speech recognition, vision, and tactile, force, and olfactory sensing, computational feasibility and capability have been slow in development. It is in these domains where nearly every human is expert by comparison.

Perception governs the way in which we deal with our environment in the sense of manipulating objects in it or navigating through it. To the extent that perceptive mechanisms must be particularly active during learning about an initially unknown environment and continuously when coping with an ever-changing one, robots must acquire the mechanical/electronic analogues of these human skills to accommodate to the types of unstructured environments that humans have mastered, and then apply these skills in a domain humans find hostile. This is the realm of *intelligent robotics*, which is the sensory and reasoning skill enhancement of robotic manipulators and mobile robots to enable them to carry out useful tasks in unstructured and variable environments.

Intelligent robotics can be defined as the melding of perception, reasoning, and action into an integrated mechanical/electronic device which is capable of operating on and/or in its environment to carry out useful tasks. The sensory aspects of this domain are described in sections to follow.

The reasoning aspects have, to a large extent, been covered in earlier sections. The most specific and generally low-level reasoning essential to both robotic manipulators and mobile robotics is path planning. Numerous examples of how to generate optimal trajectories abound in the literature. This problem is usually defined in terms of determining an optimal collision-free path from a given start point to a given goal point and a number of variations on this requirement to include search and exploratory modes. Optimality can be in terms of minimal length, minimal time, minimal energy, or combinations of these, perhaps with a modifying reliability factor based on safety tolerances. In initially unknown and/or time-varying environments sensor data acquisition and analysis are essential to allow piecewise optimal trajectories to be determined.

Communication between the subsystems of an intelligent robotic system and between separate systems, perhaps required to operate in cooperation, is an important requirement. A communication system, whether wired or wireless, ought to provide adequate bandwidth, high reliability, and consistency of protocol to maximize the efficiency and reliability of the whole system. For mobile robots, the appropriate distribution of sensors and computational support between on-board use and attached to stationary control stations is critically dependent on the communications requirements in relation to flow of command and sensory data in both directions. When multiple robots need to operate cooperatively, the means by which they communicate with each other, directly or via a central command station, is crucial to the success of such operations. In well-established industrial environments the need for using internationally standardized communication protocols throughout a manufacturing enterprise is now almost universally recognized as essential.

Thus, from the sensory/intelligence perspective, humans are still far in the lead by comparison with robots but the inspirations that has come from the study of intelligent and perceptive biological systems has led to a growing research effort in improving artificial perception and intelligence for use on robots to enable them to efficiently carry out a range of tasks which until now have been the exclusive domain of humans. The recent rapid improvement in affordable sensory and computational technology promises to deliver a variety of intelligent robot systems for use in unstructured environments in the manufacturing, mining, health-care, catering, and service industries and in the home to the marketplace within the next decade.

53.2 Robot Vision

Ray Jarvis

Seeing, in sighted humans, accounts for in excess of 85% of all sensory input data, the analysis of which supports our physical interactions with our everyday environment and our safe navigation through it as well as our appreciation of its richness as a quality of life component.

Not surprisingly, therefore, robot vision in its full realization would expand the scope of applicability of robotic systems, both manipulators and mobile vehicles, from the relative structure of classical industrial environments to the unstructuredness of natural and human-centered environments to embrace applications in mining, space exploration, undersea, service industries, healthcare, warehousing, mineral exploration, search and rescue, fire fighting, agriculture, forestry, recreation, catering, and domestic operations. Considerable research and development effort has gone into the theoretical as well as practical issues of robot vision (as a component of artificial intelligence). This effort continues today with the powerful support of rapidly improving and affordable sensor and computational technologies. This support promises an explosion of commercially viable vision-based robotic systems within the next decade.

Computer vision is usually distinguished from computer image processing by the explicit consideration of the three dimensionality of the world of objects in the former case, even when the extraction of the depth dimension may be by the analysis of one or more two-dimensional images. However, this distinction is not always acknowledged and many image processing procedures are claimed to be exercises in computer vision analysis.

Digital images are essentially two-dimensional arrays of values, each representing intensity. The dimensionality of the array specifies the spatial resolution of the image, whilst the number of data bits per cell specifies the intensity resolution. Each cell is referred to as a pixel (picture element). Three arrays of cells, one representing the red intensity, the second green, and the third blue, can collectively be regarded as a digital color image. Other color coding schemes such as hue, saturation, and intensity are also used. A typical digital color image may consist of three

512×512 element arrays with 8 bits per cell per array representing red, green, and blue values, respectively. The whole data structure is represented as 512×512×3 bytes (8 bits) of information. A color pixel is the set of three values at a particular spatial location in the arrays.

When range data is used or derived with respect to a single viewpoint, it is usually also represented in a rectangular array of cells, each being referred to as a rangel (range element). This type of representation is not strictly three dimensional, as it represents distances from the viewpoint to surfaces visible from that location. The surfaces of objects not visible because of self-obscuration or obstruction by other objects cannot be ranged from that viewpoint and are thus not represented in the range array. Since, for a typical 3D scene, approximately half the surface points cannot be ranged to from a single specific viewpoint, this representation is often referred to as $2^1/_2$D.

A full three-dimensional representation might take the form of a three-dimensional volumetric array with solid or surface occupancy indicated. Such a representation is made up of volume elements (voxels). Simple occupancy can be supplemented by values indicating color components or surface normal directionality or other property values of interest. A full 3D volumetric representation including surface properties such as color, normal directionality, and perhaps other attributes can take up a considerable number of memory locations in a computer. For example, a 512×5212×512 volume array with eight 8 bit property elements per cell would occupy 2^{30} bytes ($\approx$ 8 billion bits of memory).

Robot vision can be interpreted as that part of computer vision that is ultimately intended to be used to guide the actions of robots. The quality of a robot vision system is judged in terms of it providing timely, reliable, and accurate information, extracted from visual or range data (or both), which can be used to direct correct action of a robot in carrying out a specified task and the extent to which its structure permits its use over a wide range of possible tasks. A superior system should also be affordable, physically robust, safe, and energy efficient.

While vision strictly refers to 3D perception derived from images, the analysis of directly acquired range data is usually accepted as part of this domain.

In the context of intelligent robotics (a cooperative interplay of perception, reasoning, and action, Figure 53.1). Robot vision can be thought of as a powerful perceptual component. Since a robotic task can only be completed by also including reasoning (perhaps path planning and event sequencing) and action (perhaps following a planned path, avoiding obstacles, griping an object etc.), it is best to see robot vision as part of a whole system. In this way its quality can be judged in relation to its contribution to the whole task.

Two robotic task scenarios are used to illustrate the functions of useful vision systems. Whilst these examples are somewhat simplistic and traditional, they place vision unambiguously within a robot perception/reasoning/action loop to which humans can easily relate, thus improving the tutorial value of this exercise.

The first example is a robot/camera 'hand/eye' coordination task in which a robotic manipulator is used to pick up identified objects from a pile and sort them into groups of like objects (Figure 53.2).

A color camera and a rangefinder collect color intensity and range data in registered two-dimensional arrays which collectively results in a $2^1/_2$D representation of the 3D scene of the pile of objects to be dealt with. A database describing the types of objects likely to be encountered in terms of structure, geometry, and surface appearance is available.

Since, for reasonably high-resolution color intensity/range, large amounts of data have to be processed, the usual first task of the vision system is to segment the scene into its major components without referring to the database of object descriptions. Some type of "semantic-free" clustering procedure can be used to group pixels/ranges together in clumps which are homogeneous or continuous in some way. These initial clumps can be further refined, perhaps with some reference to broad distinguishing characteristics derived from the database. Once individual objects are more or less isolated, a search for a match with an object represented in the database can take place for each object in the scene. This is essentially a $2^1/_2$D to 3D matching process and can be very difficult to carry out reliably, particularly for free-form objects. Some objects may be oriented so that they cannot be unambiguously matched, even if they are not severely obscured by other objects. Severely obscured objects cannot be identified, in general, until the objects obscuring them are removed. At the end of the first pass of this process a number of objects, mostly among those on top of the pile will be identified and their poses (placement geometry) determined. Even objects not yet identified have to be taken account of in a volumetric occupancy sense when determining possible grip sites with respect to identified objects and planning approach, grip, and withdraw trajectory components for the manipulator, all of which should preferably be collision-free and optimal in some sense (e.g., minimum path length, time, or energy).

Each identified object that can be reached for and withdrawn from the pile is sorted into groups by the robot once it is safely removed from the pile. This cycle of perception, reasoning (planning), and action is repeated until the entire pile has been sorted.

An interesting variation of the above cycle is to discover grip sites for object removal and to carry out trajectory planning and actuation prior to identifying the object. In this way, the object might be more easily identified once it is physically separated from the pile, as not only has segmentation been carried out physically but the robot can present the object in various orientations to the vision system so that unambiguous identification can be carried out before completing the sorting action. If all objects must be handled (as is the case for this example) this is a good approach; however, in a situation where only a particular object in a pile must be retrieved, this post handling recognition approach will be cumbersome.

The type of vision system described above (both the pre- and post-handling recognition modes) is referred to as a model-reference system since an explicit database describing the various objects to be encountered is used for identifying scene objects by using matching search procedures. Most industrial robot vision

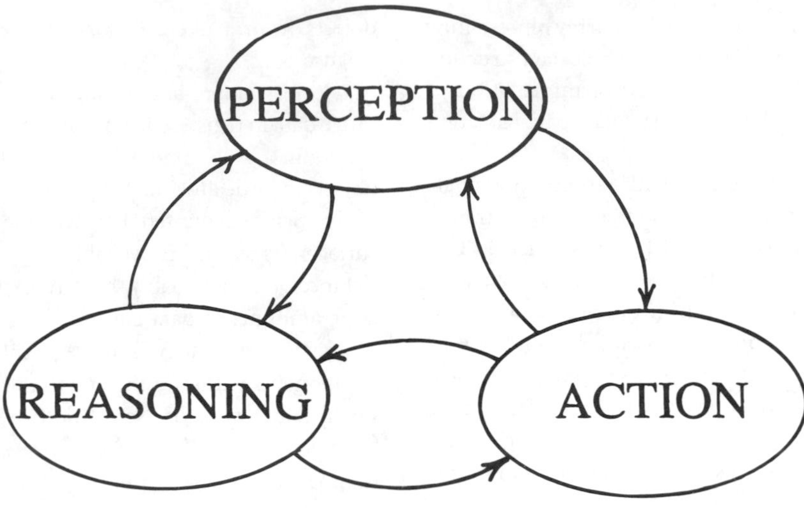

Figure 53.1 Intelligent robotics.

systems are of this type since model parameters can usually be expected to be available from computer aided design (CAD) data.

The second example concerns an autonomous mobile robot (Figure 53.3) equipped with beacon localization and range sensors, which is required to navigate through an initially unknown and slowly time-varying obstacle strewn space from a specified start position to a specified goal position via a quasi-minimal length collision-free path. Since a complete map of the environment is not initially available and any partial map generated incrementally using sensor data is subject to change, absolute global optimally of the path actually taken cannot be guaranteed; all that can be hoped is that what is known at any particular moment is taken fully into consideration at each stage of forward path planning.

At the start, the position and orientation of the robot is known; at any stage of its path towards its goal position and orientation

SCENE DESCRIPTION
TASK PLANNING
TASK SPECIFICATION
OBJECT RECOGNITION (SUPERVISED)
COLLISION FREE OBSTACLE AVOIDANCE
PARAMETER EXTRACTION
ROBOT CONTROL
SCENE SEGMENTATION (CLUSTERING)
SENSORY DATA ACQUISITION (INTENSITY, RANGE, TOUCH)
T.V. CAMERA
ROBOTIC MANIPULATOR
SCENE

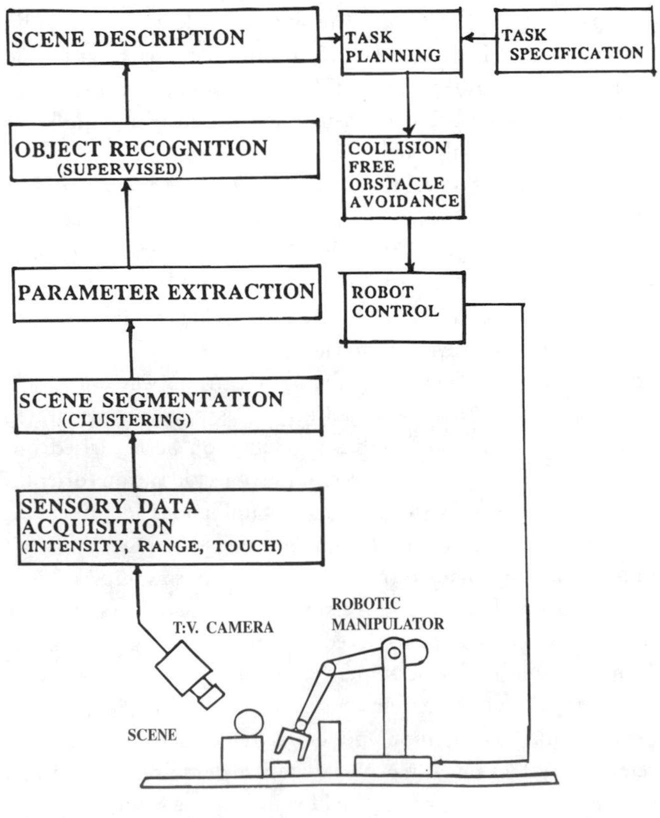

Figure 53.2 Robotic hand-eye coordination.

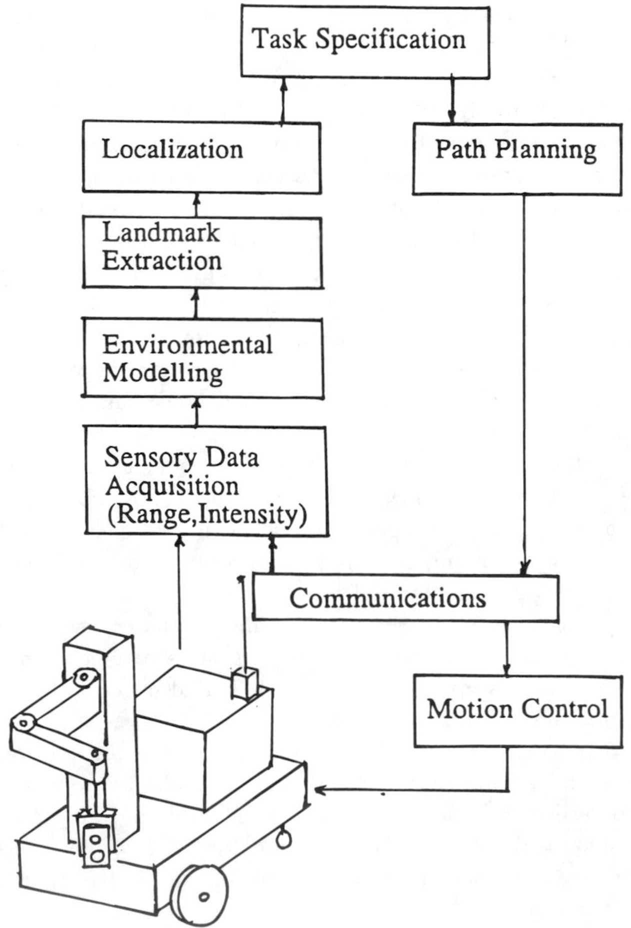

Figure 53.3 Autonomous mobile robot navigation.

data (localization) can be determined using a bar code localizer which determines the identity and angular placement of bar codes at known positions around the room (beacons).

The on-board rangefinder gathers position data from all surfaces visible all around the robot up to some maximum range measurable by the rangefinder. Obstacle identification is not required for the avoidance strategy. The floor projection of this range data is expanded in all directions by a distance equal to the radius of the robot (assumed circularly symmetric) plus a small collision tolerance amount which must accommodate the maximum range error. A high-resolution floor grid map of occupied and free space is built using the data so far available. Space whose occupancy status has not yet been determined is presumed empty until shown otherwise (the optimistic strategy).

A minimal length collision-free path is determined and the robot follows this path for a specified distance. Localization data collected during the move can be used to adjust the actual path to closely follow the planned path.

Range collection and incremental environmental modeling followed by path planning and partial actuation continues until the goal is reached. Change of occupancy status due to movement of obstacles must also be accommodated in the incremental map building process.

A more ambitious task would be to carry out the navigation task without the use of beacon localization, relying only on natural landmarks for localization. One way of doing this is to match range data against the map so far derived without actually identifying particular landmarks. The continuity of the robot's movement can simplify this process by constraining the search space of the match task. A more elegant approach would be to identify appropriate landmarks and to recognize these same landmarks as the robot moves; this clearly requires more of the vision system. Note that the recognition of a landmark for localization proposed does not mean that the landmark must be recognized in the model reference sense as was the situation for the first example.

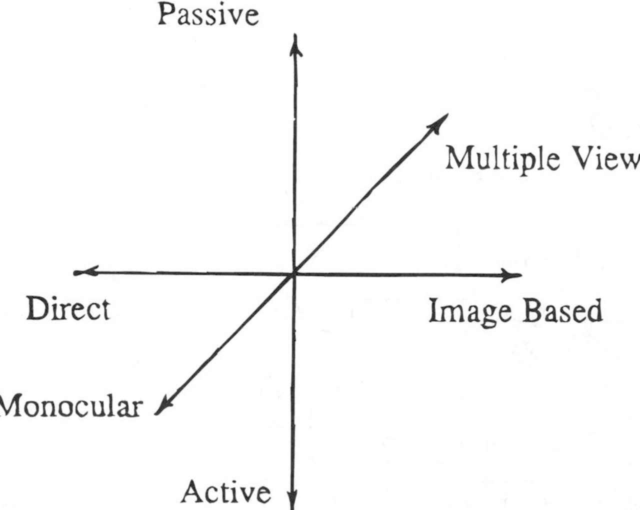

Figure 53.4 Rangefinding classes. *Source:* Jarvis, R. A. 1983. A perspective on range finding techniques for computer vision, *IEEE Trans. Pattern Anal. Mach. Intell.* PAMI-5 (2):122. Mar.© 1983 IEEE. With permission.

From the above two examples it is easy to see the important role vision can play in supporting robotic hand/eye coordination and autonomous robot navigation. In both cases, it was the 3D nature of the problem domain which made vision a particularly appropriate perceptive mechanism to apply, especially since non-contact analysis of the scenes involved was an important aspect of the tasks involved. Note also that vision may be applied both for volumetric modeling as well as shape recognition or both according to what is needed to solve the problem.

While instrumenting a system to collect or derive range data is only one part of robot vision, clearly the quality of such range data in terms of spatial density, accuracy, reliability, and timeliness is critical to the whole analysis process if it is to lead to robust plans for robots to follow. It is of interest to consider the variety of ways by which range might be extracted from a 3D scene.

Humans use a wide variety of depth queues, usually in selective combinations, to disambiguate many competing hypotheses about how images acquired by the retinas of two eyes might be explained in terms of 3D structures in the scene. Human depth cues include texture gradient, out-of-focus blur, aerial perspective, stereo disparity, binocular vergence, size constancy, and relative obscurance. Many robot vision ranging schemes have been inspired by human (and other animal) range extraction mechanisms. Some of these are better suited to robot vision than others and combinations can often be used to produce better results than can be obtained by using one approach alone. However, there are many ranging systems which are based on visual mechanisms not enjoyed by humans; these are also worth considering.

Three dichotomies define eight range-finding domains, not all of which are populated by feasible realizations. These dichotomies are (see Figure 53.4):

1. Active or passive.
2. Image based or direct.
3. Single (monocular) or multiple viewpoint.

A passive method uses only ambient lighting sources while an active system probes the scene with an energy beam or a structured light pattern. Image-based methods analyze pixel arrays to derive range while direct methods measure time-of-flight of known velocity light or ultrasonic energy beams to calculate range. Single-viewpoint methods use one fixed camera or one time-of-flight system while multiple-viewpoint methods are essentially based on triangulation. The more popular ranging modes within this categorizing structure will be briefly described below.

In indoor industrial settings, where the intrusion of contrived energy sources is not a severe disadvantage and where the contrived light source can be discriminated from the ambient light, the two most popular ranging methods are laser time-of-flight and striped lighting systems. Ultrasonic time-of-flight systems are also of interest and are covered in another part of the handbook.

Laser time-of-flight range instruments either measure the time it takes for a short pulse of laser light to beam to and be reflected back from a point on the surface of an object (Figure 53.5) or the phase shift encountered by a modulated continuous laser beam during the round trip (Figure 53.6). Since light travels at approximately 30 cm per nanosecond (10^{-9} seconds), subcentimeter accuracy range measurement requires time to be measured with approximately ± 10 picosecond (10^{-12} seconds) accuracy using the first approach. With both approaches, averaging over a number of cycles can be used to reduce range error but at the cost of increased measurement time.

Striped lighting ranging methods are triangulation based and rely on analyzing the distortion in the image of a stripe of light projected on the scene obtained from a camera displaced from the contrived light source in a direction perpendicular to the stripe light plane (Figure 53.7). Using many stripes simultaneously can lead to ambiguity in identifying the individual lines in the image, whilst using only one line at a time and sweeping this across the scene can be slow for high-density ranging, particularly if a standard frame rate video camera is used. An elegant compromise consists of collecting a set of images for a stripe lighting pattern set which allows each stripe to be binary coded amongst the images. For n stripes only $(\log_2 n) + 1$ images need be collected; doubling the number of stripes requires only one extra image.

In terms of the categorization scheme introduced earlier, the laser time-of-flight ranging method would be regarded as in the direct/active/single-view class while the striped lighting ranging method would be regarded as being in the image-based/active/multiple-view class. Other members of the active class are range from brightness (with contrived light sources) and range from attenuation (again with contrived light sources). Both time-of-flight and striped lighting systems have been used in industrial robotics for some years now.

Passive range methods have been actively researched over the last two decades, partly because their biological source of inspira-

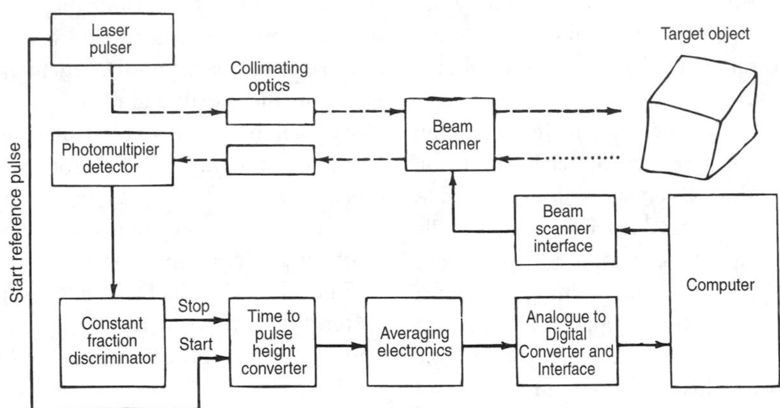

Figure 53.5 Direct pulse time-of-flight laser rangefinder. *Source:* Jarvis, R. A. 1983. A perspective on range-finding techniques for computer vision, *IEEE Trans. Pattern Anal. Mach. Intell.* PAMI-5 (2):122. Mar.© 1983 IEEE. With permission.

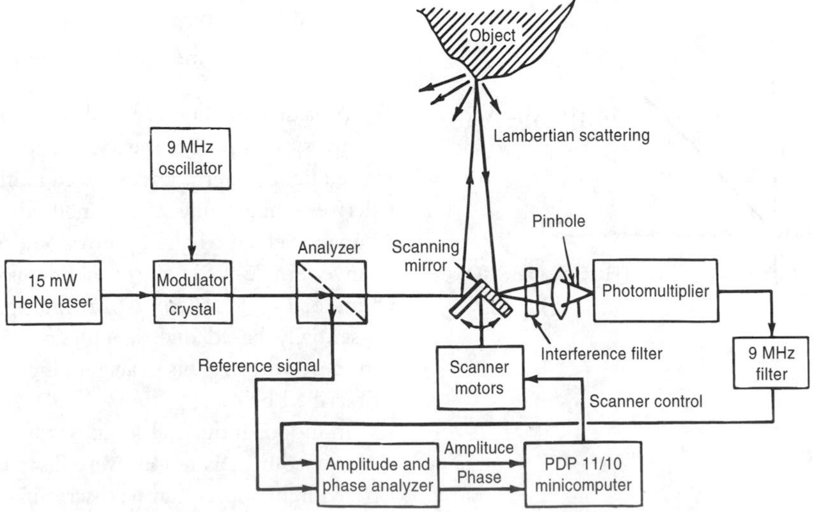

Figure 53.6 Continuous wave phase shift detection laser rangefinder. *Source:* Jarvis, R. A. 1983. A perspective on range-finding techniques for computer vision, *IEEE Trans. Pattern Anal. Mach. Intell.* PAMI-5 (2):122. Mar.© 1983 IEEE. With permission.

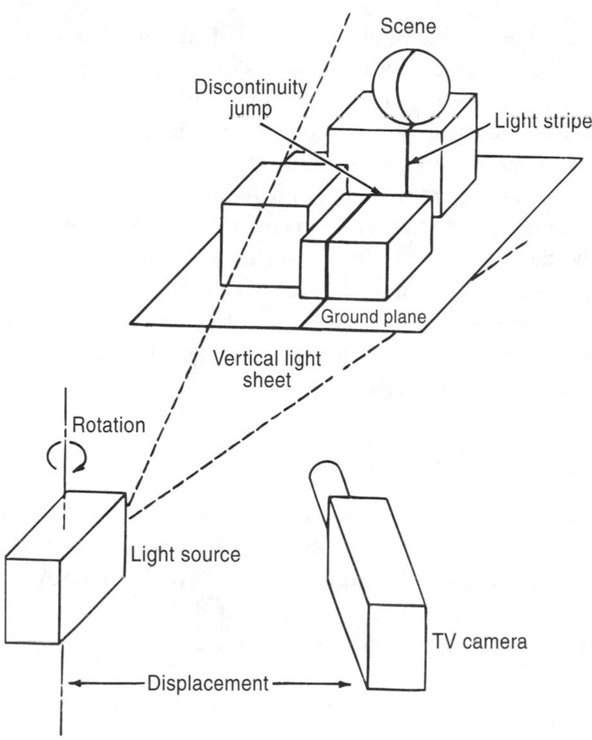

Figure 53.7 Striped light range sensing. *Source:* Jarvis, R. A. 1983. A perspective on range finding techniques for computer vision, *IEEE Trans. Pattern Anal. Mach. Intell.* PAMI-5 (2):122. Mar. © 1983 IEEE. With permission.

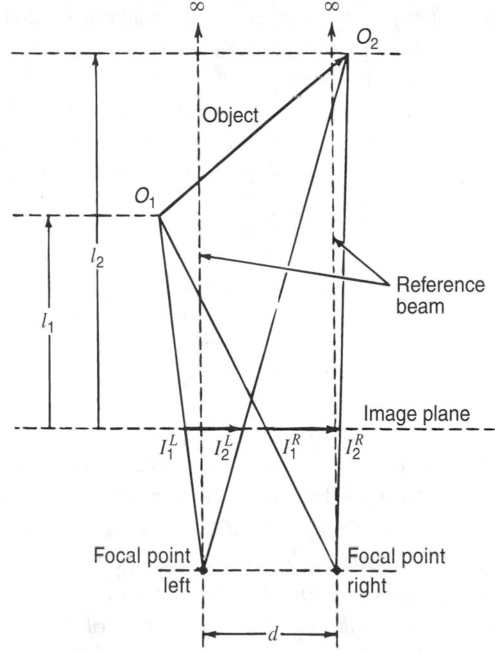

Figure 53.8 Lateral stereo matching geometry.

tion brings with it the confidence that these approaches have been tested through Darwinian natural selection over countless generations and partly because they promise very wide application scope and are intrinsically safe. Computational and microelectronic advances in recent times have brought the promise of fast and affordable general-purpose passive rangefinding closer to the marketplace for practical use in many industries and in the home with or without robotic implications.

Passive stereopsis is clearly the most popular approach to passive rangefinding. There are two stereopsis modes. Lateral stereopsis (Figure 53.8) refers to methods which use at least two cameras in a known baseline configuration in a plane perpendicular to the depth coordinate to derive range, while temporal stereopsis refers to methods where time sequences of images from a camera moving through the 3D scene are used to derive range. In either case there are two essential process components which must be dealt with since both methods rely on matching elements among images, using the shifts of location (disparity) as inverse range measures (scaled appropriately):

1. The extraction or definition of the elements to be matched among images.
2. The search for correspondence between those selected elements among the images.

The image elements chosen for matching can be edge points, lines, corners, area patches, regions, object segments, or entire objects. More preparatory processing is required by choosing more complex elements, and, in the case of entire objects, the segmentation of these objects may be simplified by the use of range data which is to be the result of the analysis, thus creating a potential deadlock situation. Thus, there is some advantage to extraction range matching the most primitive elements that can be discriminated clearly and matched unambiguously.

For lateral stereopsis, the use of more than the minimum number of two cameras can improve both accuracy and reliability. Special hardware for very fast extraction of multiple camera stereopsis range data is under active research and development.

Once dense, accurate, and reliable range data has been extracted, the resulting $2^1/_2$D representation of the 3D scene must be further processed. Full 3D scene representation can be in the form of an ordered or unordered set of 3D Cartesian coordinates with each point having a location in space as well as attributes such as surface color and normal vector data. Alternatively, a voxel based representation may be preferred. Other data structures have also been proposed. The appropriateness of a data structure used to represent 3D scenes is measured in terms of compactness, lack of ambiguity, easy access, and modifiability and its ability to support the application in mind.

The results of a complete scene analysis carried out by a robot vision system may take various forms but would generally provide information on the identity, pose, and placement of individual items and their functional interrelationships (e.g., support, adjacency, occlusion, linkage, containment, etc.). In particular applications perhaps a subset of this information would suffice; building vision systems to provide more than is required is hardly a sensible thing to do for a specific task, but stretching vision methodologies towards generalisation of applicability is a worthwhile endeavor also.

In summary, Robot Vision attempts to emulate human vision

to the extent of correctly interpreting the make-up of 3D scenes, with the purpose of guiding robotic manipulators and autonomous mobile robots in fulfilment of useful tasks; many methods of range extraction support these ambitious goals and rapidly improving sensory device technologies and computational systems are likely to lead to an explosion of commercially available systems within the next decade.

53.3 Ultrasonic Sensors

Lindsay Kleeman

Overview

Ultrasound is sound at frequencies above 20 kHz, the upper limit of audible human hearing. Ultrasonic sensing in robotics is popular due to the ability to directly achieve range sensing cheaply, simply, unobtrusively, and with low power consumption. Ultrasonic sensing is sometimes called sonar derived from sound navigation and ranging. The basic principle of ranging is the measurement of elapsed time between transmission and reception of ultrasonic wavefronts. From the speed of sound, c, to the elapsed time, t, the range is $ct/2$. The speed of sound varies with atmospheric conditions of temperature, humidity, and pressure as discussed below. The elapsed time can be measured by transmitting a short chirp and processing the echo to find the arrival time or using a continuous swept frequency transmission and examining the frequency of the echo. The former pulse echo technique is more common due to its simpler processing and hardware. The frequency, f, of transmission can range from 25 kHz to 500 kHz with corresponding wavelengths of ($\lambda = c/f$) approximately 14 mm to 1 mm. Finer discrimination of targets is obtained using smaller wavelengths with the disadvantage of much greater absorption losses during propagation in air as described below. Range measurement is limited to a maximum of around 10 m due primarily to absorption losses of air.

There are several limitations to ultrasonic sensing that must be well understood for effective and accurate use of the sensors. Due to the long wavelength in comparison with deviations in many surfaces, reflectors often behave in a specular manner to ultrasound—that is, the surface acts like a mirror with the angle of incidence and reflection being equal. This means, for example, that the return from a smooth wall will occur from one point where a normal to the wall intersects the sensor. As a sensor is scanned in pointing angle across the wall, returns will be obtained from this same point with reducing amplitude as the sensor points away from the wall. The range reported by the sensor at a particular angle can be from a reflection at a bearing equal to the angle plus the effective beamwidth of the sensor. Thus, a range scan obtained from an ultrasonic sensor can be misleading since there are typically large errors in the bearing of the plotted targets.

Using two receivers allows the bearing and ranges to specular targets to be estimated with accuracies of the order of 0.2° and 0.1 mm when matched filters are used to estimate pulse arrival times, described below. Using two transmitters and two receivers allows targets to be localized and classified into planes corners and edges. Three-dimensional targets can be similarly classified. The use of arrays of transducers allows narrow focussed beams, improved signal to noise ratio, and greater discrimination of targets at similar range and bearing.

This section starts with the underlying physical properties of ultrasound speed, attenuation in air, and scattering from targets. Properties of the transducers are then discussed. Estimating range, bearing, and target type are then examined. An overview of the treatment of rough surfaces is then presented and finally arrays are briefly discussed. References to more detailed treatment of these topics are provided.

Speed of Sound

The speed of sound varies significantly with atmospheric conditions, such as temperature, humidity, pressure, and altitude above sea level. The following formulae are derived from Equations 53.1 and 53.2. The speed of sound in dry air at sea level air density and one atmosphere pressure is given by

$$c_T = 20.05\sqrt{T_C + 273.16}\ ms^{-1} \tag{53.1}$$

where T_C is the temperature in degrees Celsius. Equation 53.1 is accurate to 1% for most conditions. A more accurate estimate, c_H, can be given if the relative humidity of air, h_r, in percent is known:

$$c_H = c_T + h_r[1.0059 \times 10^{-3} + 1.7776 \times 10^{-7}(T_C + 17.78)^3]\ ms^{-1} \tag{53.2}$$

Equation 53.2 is accurate to around 0.1% for temperatures of −30°C to 43°C and most atmospheric pressure conditions at sea level. When atmospheric pressure p_s is known, Equation 53.3 is more accurate:

$$c_P = 20.05\sqrt{\frac{T_C + 273.16}{1 - 3.79 \times 10^{-3}(h_r p_{sat}/p_s)}}\ ms^{-1} \tag{53.3}$$

where the saturation pressure of air, p_{sat}, is a function of temperature defined by Equation 53.8 below.

Attenuation of Ultrasound due to Propagation

Due to power spreading of a point source, the power of an ultrasonic pulse reduces with the square of the distance it travels in a lossless open medium. This translates to a *linear* reduction with distance of pressure, and hence voltage on a receiver, since power is proportional to the square of pressure or voltage. This power spreading applies in the far field (i.e., beyond $a^2/4\lambda$ where a is the diameter of the transduce and λ is the wavelength) which is beyond 20 mm for the Polaroid 7000 electrostatic transducer.

In reality, air absorbs energy from a propagating wave in the form of heat. Losses are affected by many factors, including air viscosity, heat conduction, molecular vibration modes, and composition of air in terms of nitrogen, oxygen, carbon dioxide, and water vapour. The attenuation, α, is a function primarily of frequency, temperature, and relative humidity and is expressed in nepers per meter. Thus the pressure after propagating a distance d (in meters) is multiplied by $e^{-\alpha d}$ (ignoring power spreading). The attenuation is most accurately determined by calibration, although there are empirical formulae available from the American National Standards (Piercy 1978) as given below in Equations 53.4–53.8 which are accurate to $\pm 10\%$ for temperatures 0 to 40°C, 10–100% relative humidity, pressure less than 2 atm, and frequency-to-pressure ratio 40–10^6 Hz atm^{-1}. The empirical formulae predict the attenuation coefficient, α (in N_p m^{-1}), in air for signal frequency f (in Hz), atmospheric pressure p_s (in Pa), temperature T (in K), and relative humidity h_r (in %) is

$$\begin{aligned}\alpha = {} & f^2[1.84 \times 10^{-11}(p_s/p_{s0})^{-1}(T/T_0)^{1/2} \\ & + (T/T_0)^{-5/2}\{1.278 \times 10^{-2} \\ & \times [\exp(-2239.1/T)]/(f_{r,O} + (f^2/f_{r,O})) \\ & + 1.068 \times 10^{-1}[\exp(-3352/T)]/ \\ & (f_{r,N} + (f^2/f_{r,N}))\}], \text{ in N/m} \qquad (53.4)\end{aligned}$$

where

$$f_{r,O} = (p_s/p_{s0})\{24 + 4.41 \times 10^4 h \times [(0.05 + h)/(0.391 + h)]\}, \text{ in Hz} \qquad (53.5)$$

$$f_{r,N} = (p_s/p_{s0})(T/T_0)^{-1/2}(9 + 350h \times \exp\{-6.142[(T/T_0)^{-1/3} - 1]\}), \text{ in Hz} \qquad (53.6)$$

$$h = h_r(p_{sat}/p_{s0})/(p_s/p_{s0}), \text{ in \%} \qquad (53.7)$$

$$\begin{aligned}\log_{10}(p_{sat}/p_{s0}) = {} & 10.795861[1 - (T_{01}/T)] \\ & - 5.02808\log_{10}(T/T_{01}) \\ & + 1.50474 \times 10^{-4} \\ & \times \{1 - 10^{-8.29692[(T/T_{01})-1]}\} \\ & + 0.42873 \times 10^{-3} \\ & \times \{-1 + 10^{4.76955[1-(T_{01}/T)]}\} \\ & - 2.2195983 \qquad (53.8)\end{aligned}$$

p_{s0} is the reference atmospheric pressure (101.325 kPa) and T_0 is the reference ambient atmospheric temperature of 293.15 K (20°C) and T_{01} is the triple-point isotherm temperature with the exact value of 273.16 K.

Figure 53.9 shows the dependency of absorption loss of air at 20°C on relative humidity and frequency. Note the steeply increasing losses as frequency increases and also the peak losses occurring at intermediate relative humidities.

Target Scattering and Reflection

Acoustic wave propagation is disturbed by changes in the acoustic impedance of the medium which is defined as the product of density and speed of sound of the medium.

Discontinuities of the acoustic impedance occur when the low impedance of air meets solid objects with high impedance and scattering of the acoustic wave results. When scattered waves are returned to a receiver, an echo is produced. There are two basic types of scattering: *reflection* and *diffraction*. Reflection occurs from smooth surfaces with the angle of incidence to the normal of the surface equalling the angle of reflection. *Smooth* is defined by the size of rough features of the surface being much smaller than the acoustic wavelength. Reflection from smooth surfaces is often called *specularity*. Diffraction occurs due to discontinuities of the surface, such as edges (i.e., where a smooth surface ends). Reflection from smooth planes and corners, and diffraction from edges will be considered here. More complex target profiles can be analyzed using approximations developed by Freedman (1962).

Reflections from a Plane

Smooth planes are considered here—rough planes are discussed below. It is assumed that the plane has no losses due to reflection. Solid materials such as glass, timber, perspex, and plaster are practically lossless reflectors. Softer materials such as cardboard, cloth, felt, and foam absorb ultrasonic energy.

In a lossless reflector, the plane reflector can be replaced by a virtual image of the transmitter or receiver. Figure 53.10 shows a transducer that acts as both transmitter and receiver. The transmitter is considered separately as a virtual image reflected in the plane as with a mirror in optics. The effect of the plane reflector is then identical to the separate receiver and a transmitter in the position of the virtual transmitter with no plane in between. It is equally valid to treat the receiver as the virtual image and the transmitter in the original position.

Reflections from a Concave Right Angled Corner

The corner is made up of two planes at right angles to each other. The virtual image of the transmitter in Figure 53.11 is thus made up of two reflections about each plane in turn. This amounts to reversing the image with respect to a plane reflection by reflecting through the line of intersection of the two perpendicular planes. The image reversal property can be exploited to allow an ultrasonic system to discriminate corners from planes as described below.

Diffraction from Edges

The edge, as shown in Figure 53.12, scatters the acoustic wave in the form of cylindrical wavefronts with center line along the edge. This results in more rapid power spreading than that of a plane at the same range. That is the amplitude of the received signal reduces as $\sqrt{1/range}$ times that of the plane reflector. Moreover, the edge presents a much reduced area of scattering the wave front than a plane which reduces the reflected amplitude further in comparison to the plane.

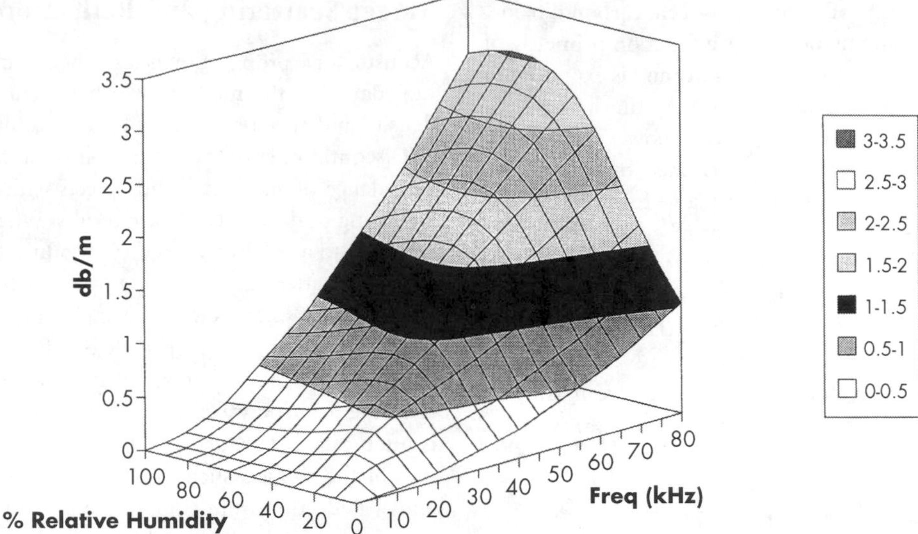

Figure 53.9 Absorption loss (dB/m) as a function of frequency and humidity in still air at 20°C. Data derived from Weast and Astle, 1978.

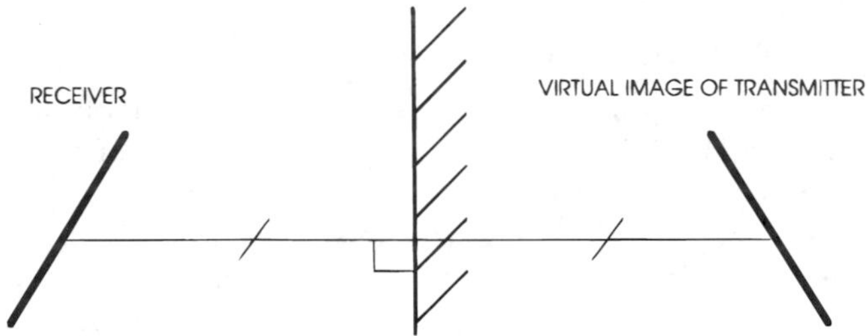

Figure 53.10 Virtual image formed by a plane.

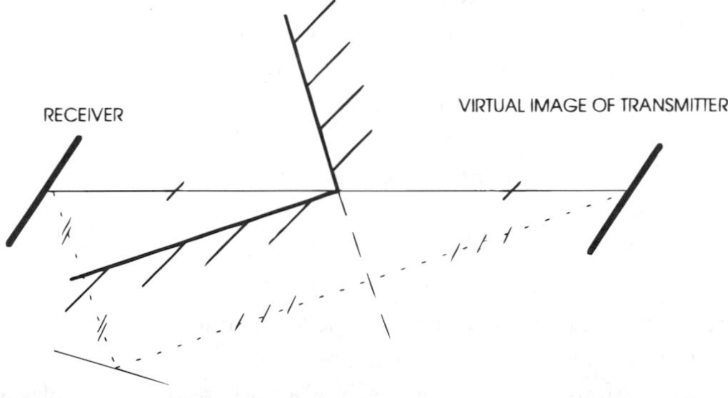

Figure 53.11 Virtual image formed by a corner.

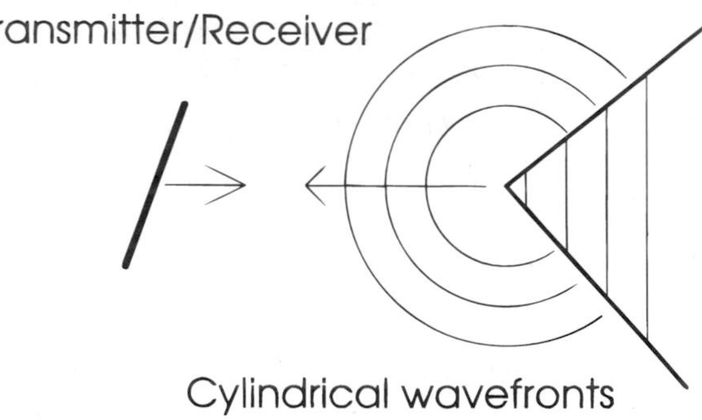

Figure 53.12 Scattering from an edge.

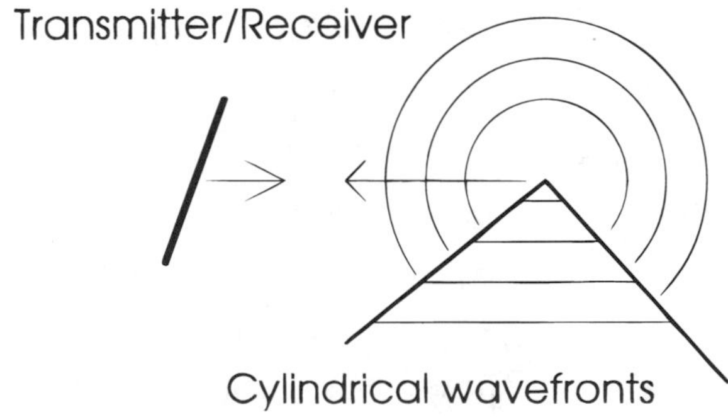

Figure 53.13 Scattering from a reversed edge.

Reverse edges, as shown in Figure 53.13, produce a sign reversed echo over the front on edge in Figure 53.12. This is due to the transition from higher acoustic impedance to lower. Moreover, Kuc and Viard (1991) report an approximately half-amplitude echo from the reverse edge compared to the front on edge.

Beamwidth—The Round Piston Model

Most ultrasonic transducers are round and can be modelled by a vibrating circular disk in an infinite baffle. The acoustic pressure, p, in the far field can then be derived (Morse and Ingard, 1991) in terms of the Bessel function J_1 as a function of the angle to the transducer normal, θ, wavelength λ, and the radius of the disk, a

$$p(\theta) = \frac{2J_1(ka\sin(\theta))}{ka\sin(\theta)} \qquad \text{where } k = 2\pi/\lambda \qquad (53.9)$$

An example of the radiation pattern is seen in the Polaroid data sheet in Figure 53.16 where side-lobes can be observed at 20° and 40° and nulls at 15° and 30°. Note that the physical principle

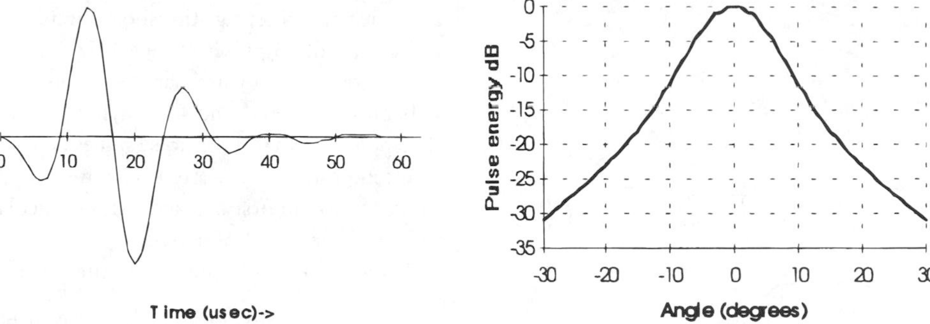

Figure 53.14 Received pulse shape from a plane reflector at 1 m range using Polaroid 7000 transducer and corresponding beam pattern for combined transmitter and receiver.

of reciprocity between transmitter and receiver implies that the same beam pattern applies to each. The half-width beam angle, θ_0 is the angle to the first off-axis zero in the Bessel function and is given by

$$\theta_0 = \sin^{-1}(0.61\ \lambda/a) \qquad (53.10)$$

This equation shows the relationship between transducer radius in wavelengths and the effective beamwidth. The larger the number of wavelengths across the transducer the narrower the beamwidth is.

Note that the radiation pattern in Equations 53.9 and 53.10 above correspond to exciting the transducer with a *continuous* sine wave at 50 kHz. In practice, *short* pulses are transmitted and thus contain a wider spectrum which blurs the radiation pattern. In the extreme case where the pulse is as narrow as practical the side lobes and nulls disappear altogether. This is illustrated in Figure 53.14 which shows the pulse, obtained by driving the transmitter with a 10 usec 300-0-300V pulse, and the *corresponding* combined transmitter receiver radiation patterns for a Polaroid 7000 electrostatic transducer. A Gaussian approximation can be employed as an effective approximation for the beam characteristics (Koc and Viard, 1991). The beamwidth of a transducer is defined by the angle difference for −3 dB attenuation relative to that at zero angle. Values other than −3 dB are also used in the literature. Targets outside the beamwidth may still return a weak echo, which may or may not be discernible above the background acoustic and amplifier noise.

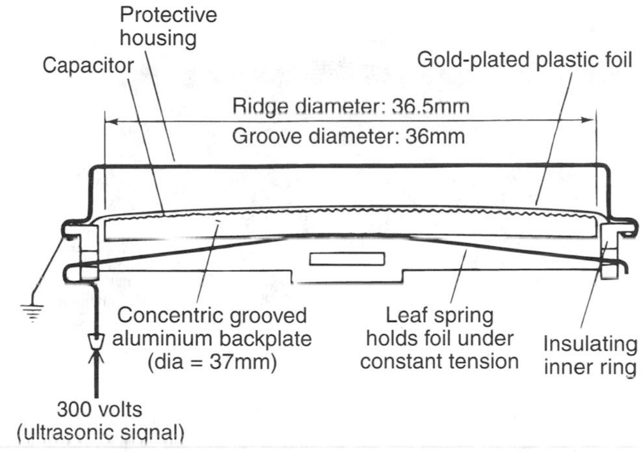

Figure 53.15 Construction of a Polaroid transducer. Source: McKerrow, P. J. 1991. *Introduction to Robotics,* Fig. 10.42, Addison-Wesley, Reading, MA.

Transducers

Two types of transducers are commonly employed: electrostatic and the piezoelectric. An example of an eletrostatic transducer is the Polaroid transducer constructed from a gold coated plastic foil membrane stretched across a round grooved aluminium backplate as shown in Figure 53.15. The membrane and the backplate form a varying capacitor. As a transmitter, the transducer membrane is vibrated by applying 0 to 300 V pulses across this capacitor. The charge induced by the 300 V on the capacitor causes an electrostatic attraction force between the membrane and the backplate. The same transducer also acts as a receiver when a 150 V DC bias voltage is applied to store opposing charges on the membrane and backplate. Incoming sound vibrates the membrane, changing the effective separation between the plates of the capacitor and hence the capacitance. The charge on the plates of the varying capacitance induce a voltage related to the incoming acoustic wave. The grooves on the backplate enhance the sensitivity of the transducer. Mounted on the front of the transducer is a protective grill—the author's experience suggests removing this, should the environmental conditions allow, to reduce losses and reverberation between the grill and the membrane. The −3 dB beamwidth of the transducer is of the order of 8°. The frequency bandwidth of a transmitter combined with receiver is of the order of 20 kHz for −3dB, centred at 50 kHz. The sensitivities of the transmitter and receiver allow range measurement up to approximately 10 meters for plane reflectors. Piezoelectric transducers can also act as both transmitters and receivers, although often manufacturers optimize the performance by offering separate devices. A piezoelectric resonant crystal mechanically vibrates when a voltage is applied across the crystal, and in reverse generates a voltage when mechanically vibrated. A conical concave horn is mounted on the crystal to acoustically match the crystal acoustic impedance to that of air.

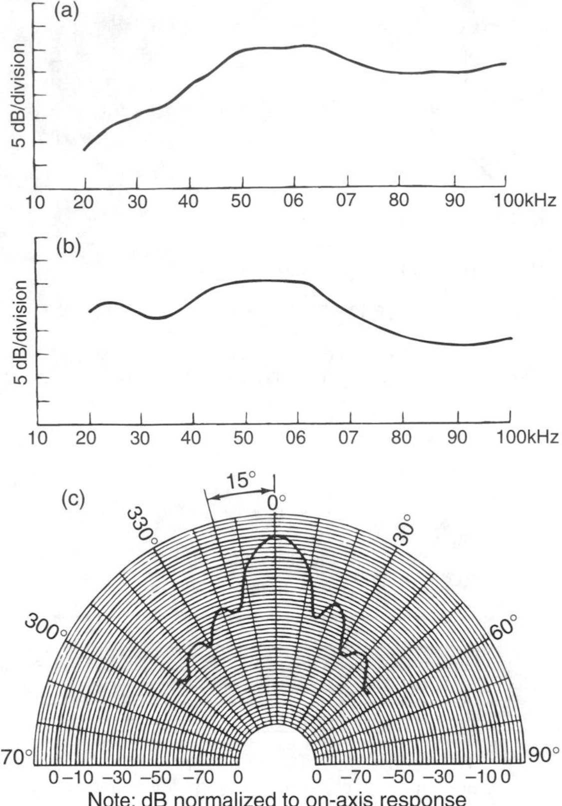

Figure 53.16 Characteristics of a Polaroid transducer—(a) transmitter frequency response; (b) receiver frequency response; (c) beam pattern at 50 kHz. *Source*: Mc Kerrow, P. J. 1991. *Introduction to Robotics*, Addison-Wesly, Reading, MA.

An example is the Murata MA40A5R/S receiver and sender transducers which operate at 40 kHz. This device has a diameter of 16 mm and a 60° beam angle for transmitter combined with receiver for −20 dB loss compared to the maximum sensitivity. The effective bandwidth of transmitter and receiver is only a few kHz due to the resonant nature of the crystals. This limits the envelope rise time of pulses to around 0.5 msec. An advantage is the ability to drive piezoelectric devices with low voltages, for example by connecting each terminal to complementary CMOS logic outputs. Piezoelectric transducers are also available at resonant frequencies from 25 kHz to 300 kHz.

In terms of performance, the electrostatic transducers offer the better sensitivity and bandwidth. Piezoelectric transducers, on the other hand, are simpler to drive due to the lower voltages necessary.

Polaroid Ranging Module

The Polaroid ranging module was developed originally as a range finder for autofocus cameras. It has found extensive application in robotics due to its simplicity, low cost, and availability.

The 6500 series sonar ranging module consists of a single Polaroid electrostatic transducer which acts as transmitter and receiver, driving electronics, receiver time-gain compensated amplifier and filtering, threshold circuitry, and echo output. A pulse train of 16 pulses at 49.4 kHz is transmitted and the echo is amplified with a gain that is stepped up with time. The signal is bandpass filtered and a decaying integrator applied before a threshold—so that short noise spikes do not trigger the threshold. The echo output is asserted once the threshold is triggered and the range to the first obstacle can be calculated from the elapsed time and the speed of sound.

There are some important limitations:

1. The beamwidth of the transducer must be taken into account when estimating the bearing of the obstacle.
2. Only the nearest returns are usually logged, thus masking of further obstacles occurs even though echoes are returned (it is however possible to log multiple echoes if they are spaced far enough apart and the received is reenabled after each echo).
3. Weak echoes can be produce errors in range due to the delay in charging the integration circuit.
4. The time-gain compensation can only be approximate due to the variability of absorption of ultrasound with temperature and humidity and also variability of scattering efficiency of different target types.
5. Multipath echoes can cause obstacles to appear in completely the wrong direction as can be seen in Figure 53.17.

Multiple (typically 16 to 24) ranging modules are often employed in a ring around a robot and simultaneously fired fire in noninterfering clusters. Work by Kuc (1990) suggests that most rings do not contain sufficiently many transducers to guarantee obstacle avoidance in the presence of rightangle edges. For example, to navigate in an environment with planes, corners and edges, and moving 0.28 m without colliding with obstacles, the

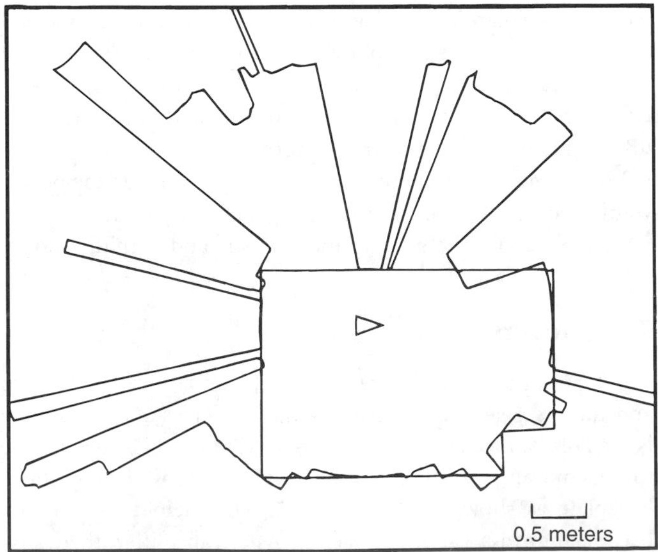

Figure 53.17 Polaroid range module scan of an indoor scene. *Source*: Leonard, J. J. and Durrant-Whyte, H. F. 1992. *Directed Sonar Sensing for Mobile Robot Navigation*, Kluwer Academic Publishers, New York, NY.

sonar sensor needed to be scanned at 2° steps. For a sonar ring this requires 180 ranging modules! Therefore sonar rings of 24 ranging modules are likely to miss obstacles in practice.

Estimating the Echo Arrival Time

Various techniques to estimate the arrival time of an echo in an ultrasonic sensor are considered in this section in order of processing complexity.

Thresholding

The arrival time is estimated as the time the echo exceeds a threshold. Pulses of low amplitude will give delayed estimates, due to the finite envelope rise time. Attempts to eliminate this amplitude dependency include the use of time-gain compensation as in the Polaroid ranging module. Since edges and planes, for example, return disparate echo amplitudes, time-gain compensation cannot help. Instead computer sampling of the echo envelope can then be used to threshold in software in proportion to the amplitude. Variability will still occur due to the pulse shape dependence on bearing angle. Since the arrival time is essentially based on one sample, noise on that sample can unduly corrupt the arrival time. Other estimation techniques based on many samples improve noise performance and are discussed next.

Curve fitting to the envelope

Discrete time samples of the echo envelope can be extracted either using Nyquist rate sampling and then processing or analogue full-wave rectification, low-pass filtering, and then lower-rate digital sampling. The rising edge of the envelope is then used to fit curves to, such as an increasing parabola, or a polynomial multiplied by a decaying exponential or maximum slope straight line of a constant number of consecutive samples. The rising edge is used since it offers the maximum slope and is least likely to be corrupted by overlapping pulses.

Matched Filtering

A matched filter is obtained by examining the cross correlation of the echo (containing noise) with the predicted pulse shape (no added noise). The arrival time corresponds to the time shifted position of the predicted pulse that gives a maximum in the cross correlation. Matched filtering can be applied if the sample rate of the echo exceeds twice the bandwidth of the echo signal. Theory of Radar (Woodward 1964) shows that the maximum likelihood estimate of the arrival time of the echo corrupted by additive white Gaussian noise is the matched filter. This means that it is usually the "best" estimator in practice. The problem with the matched filter is predicting the echo pulse shape, since it depends on the bearing angle to the target, dispersion in air of ultrasound, scattering properties of targets and transmitter and receiver characteristics. Linear models exist that accurately predict pulse shape and matched filtering has been implemented successfully (Kleeman and Kuc, 1994).

Estimating the Bearing to Targets

The range of ultrasonic reflectors can be determined by examining the echo waveform from one receiver. However, accurate *bearing* estimation requires two or more receivers. Figure 53.18 show two receivers observing the range to P which may be an edge reflector or a virtual image of a transmitter in a plane or corner. The bearing to P is given by

$$\theta = \sin^{-1}\left(\frac{d^2 + r_1^2 - r_2^2}{2dr_1}\right) \tag{53.11}$$

where d is the receiver separation, and r_1 and r_2 are the ranges to P from each receiver. In a cluttered environment with many closely spaced echoes, it may be difficult to determine corresponding pairs of arrival times on each receiver. This correspondence problem, as it is known, is minimized by employing closing spaced receivers, provided a sufficiently accurate arrival time estimator (e.g., matched filter) has been employed to prevent bearing inaccuracy.

Specular Target Classification

It is possible to classify targets into planes, concave right angle corners and edges using at least two transmitters and two receivers. Such an arrangement is shown in Figure 53.19, where T represents a transmitter and R a receiver. The algorithm for classification is based on the virtual image of a plane being reversed to that of a corner. An edge produces fixed bearing independent of the transmitter position. As illustrated in Figure 53.20, classification can be performed based on the difference, β, in bearing angles to the reflector from the two different transmitters fired in succession. The value of β is positive for a plane, negative for a corner and zero for an edge. An implementation of this approach (Kleeman and Kuc, 1994) can successfully classify reflectors to ranges of 8 m to accuracies of 0.2 mm range and 0.2° bearing. Three-dimensional targets can similarly be classified (Hong and Kleeman, 1994).

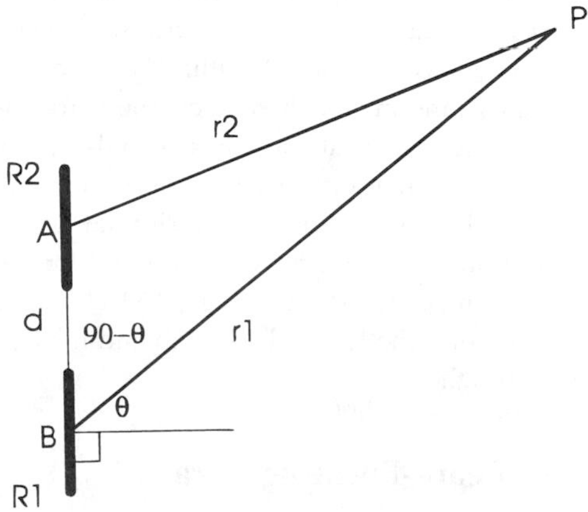

Figure 53.18 Two receivers used to estimate bearing to P.

Figure 53.19 Sensor arrangement for identification of planes, corners and edges.

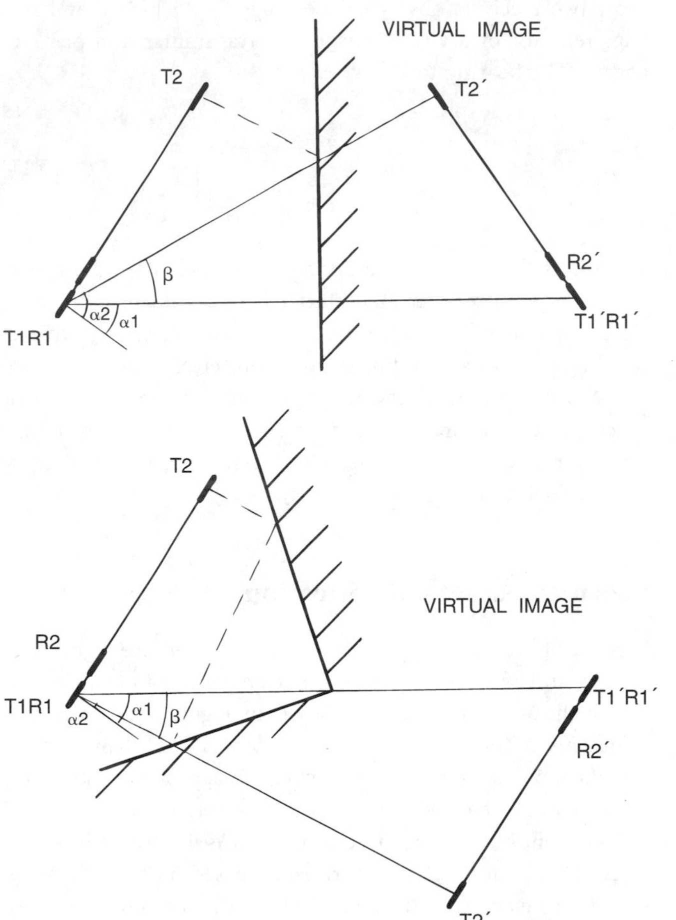

Figure 53.20 Virtual image configurations of a plane and a corner.

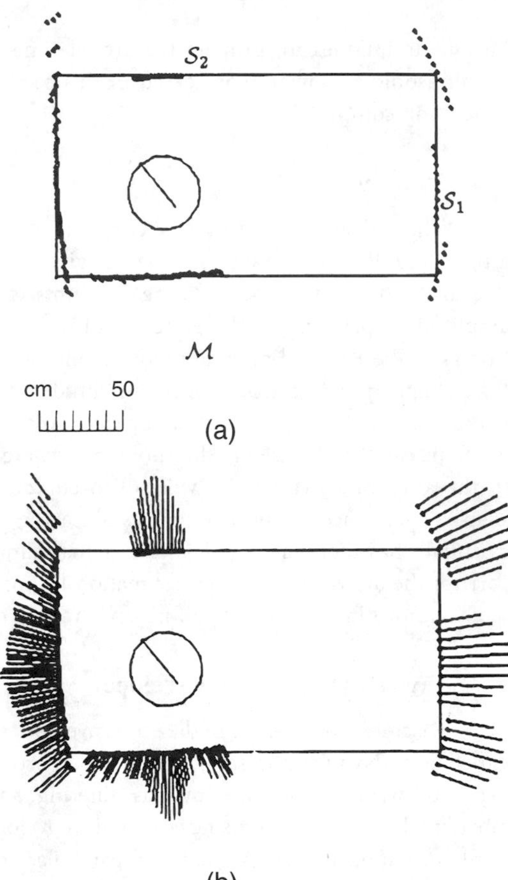

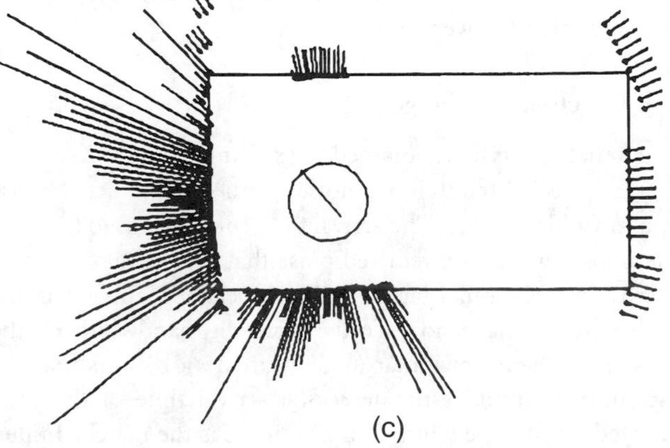

Figure 53.21 (a) Time-of-flight map; (b) echo-energy map; (c) echo-duration map. *Source*: Bozma, O. and Kuc, R. 1992. *IEEE/RSJ Int. Conf. Intelligent Robots and Systems,* 815–820.

Treatment of Rough Surfaces

Work has been done on the characterization of surfaces which are not specular. These rough surfaces produce echoes from a large number of points on the surface since surface features are comparable to the acoustic wavelength. The echo from rough surfaces contains random incoherent components and can be characterized using statistical measures of pulse duration and energy. A map containing energy, duration, and range (ENDURA) can be constructed to classify surface properties and then determine their positions (Bozma and Kuc, 1992). An example of these maps is shown in Figure 53.21. The right and top walls are smooth, the left wall is rough and the bottom wall is moderately rough.

Ultrasonic Beam-Forming Arrays

The sensor array shown in Figure 53.19 uses the full beam pattern of both the transmitter and receivers to collect echoes. Using

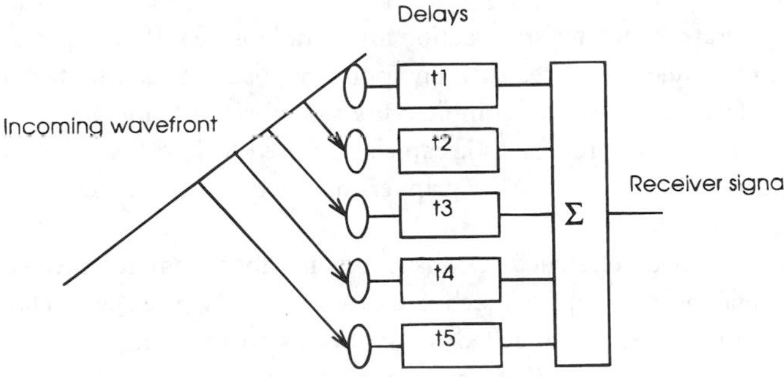

Figure 53.22 Phased receiving array.

beam-forming techniques, it is possible to restrict the effective beamwidth of either the transmitter or the receiver, so that targets at the same range but with different bearings can be discriminated. Targets falling within the beamwidth and at the same range cannot be discriminated with the sensor of Figure 53.19 since pulse overlap occurs. By using phase delays on an array of transmitters or phase delays and summer on an array of receivers as in Figure 53.22, the beam can be narrowed or even focussed to a point. The delay and summing can be performed electronically or with software processing of digitized echo waveforms. The advantage of phasing the transmitters is that more energy can be focussed into a narrow beam, thus increasing the range of the sensor. However, covering a wide bearing range requires many measurements, slowing the sensor. The advantage of receiver arrays is that one measurement can cover a large angle and processing can then be done on the same data for different bearing angles. For narrow band systems, such as is the case when piezoelectric transducers are employed, the separation of the elements of the array must be less than half a wavelength to avoid ambiguity in bearing estimation of receivers or grating lobes in transmitters. Wide-band systems can exploit envelope shape to overcome this limitation. An example of an ultrasonic array implementation can be found in Webb (1994).

References

Bozma, O. and Kuc, R. 1992. Characterizing the environment using echo energy, duration, and range: the ENDURA method, *IEEE/RSJ Int. Conf. Intelligent Robots and Systems,* 813–820 Raleigh, NC.

Freedman, A. 1962. A mechanism of acoustic echo formation, *Acustica,* 12:10–21.

Hong, M. I. and Kleeman, I. 1994. A low sample rate ultrasonic sensor system for the differentiation and location of common three-dimensional room features, MECSE–94–2, Monash University, Clayton, Australia.

Kleeman, I. and Kuc, R. 1994. An optimal sonar array for target localization and classification, *IEEE Int. Conf. Robotics and Automation,* 3130–3135, San Diego, CA.

Kuc, R. 1990. A spatial sampling criterion for sonar obstacle detection, *IEEE Trans. of Pattern Analysis and Machine Intelligence,* 12(7):686–690

Kuc, R. and Viard, V. R. 1991. A physically based navigation strategy for sonar-guided vehicles, *Int. J. Robotics Research,* 10(2):75–85.

Leonard, J. J. and Durrant-Whyte, H. R. 1992. *Directed Sonar Sensing for Mobile Robot Navigation,* Kluwer Academic Publishers, New York, NY.

McKerrow, P. J. 1991. *Introduction to Robotics,* Addison-Wesley, Reading, MA.

Morse, P. M. and Ingard, K. J. 1968. *Theoretical Acoustics,* McGraw-Hill, New York, NY.

Piercy, J. E. et al. 1978. American National Standard: Method for calculation of the absorption of sound by the atmosphere, ANSI SI-26-1978, American National Standards, Acoustical Society of America.

Poole, H. H. 1989. *Fundamentals of Robotics Engineering,* Van Nostrand, New York, NY.

Weast, R. C. and Astle, M. J. 1978. *CRC Handbook of Chemistry and Physics,* 59th ed. CRC Press, Boca Raton, FL.

Webb, P. F. 1994. *An Ultrasonics Based System for the Extraction of Range and Bearing Data for Multiple Targets,* Ph.D. thesis, University of Nottingham.

Woodward, P. M. 1964. *Probability and Information Theory with Applications to Radar,* 2d ed., Pergamon Press, Oxford, England.

53.4 Robot Tactile Sensing

R. Andrew Russell

Tactile sensing covers any method of sensing involving physical contact between the sensor and sensed object. For this reason it is also sometimes known as contact sensing. By its nature, tactile sensing is limited to contact and short-range situations. However, measurements gathered by tactile sensing usually require little processing to extract useful, unambiguous information. Mobile robots can use tactile sensing to warn of imminent collisions. The short-range but reliable information provided by touch is ideal as a last line of defense after obstacles have evaded longer-range sensors. Another aspect of tactile sensing is the measurement of contact forces. Many robotic tasks involve contact between the tool carried by the robot and external objects. By monitoring the resulting forces the quality and safety of these operations can be improved. In robotics there is a lot of interest in producing touch sensor arrays which mimic the sensing abilities of the human skin. Such sensors can determine the regions of contact between robotic fingers and the objects they grip. They can also measure temperature and thermal properties of the materials making up objects, their texture, and any tendency to slip. This wealth of information will prove useful during object recognition and manipulation tasks. Skinlike sensor arrays have also been developed to measure shear forces and the thermal properties of gripped objects. Individual, special purpose, sensors have also been designed to measure texture, slip, and incipient slip of a grasped object.

Whisker Sensors

A number of mobile research robots employ whiskers as a short-range form of obstacle detection. The Stanford Research Institute's "Shakey" robot was one of the first whisker equipped mobile robots. Whiskers have also been incorporated into legged robots, such as the four legged robot, Titan III, built at the Tokyo Institute of Technology, to detect proximity between their feet and the ground. This allows the feet to decelerate before ground contact. Several researchers have mounted whisker sensors on robot grippers and used them to locate small objects close to the gripper. Whisker sensors are simple and provide direct information about the close proximity of other objects. These attributes are useful in mobile robotics. However, their low information bandwidth (providing information about a single contact point) makes them unsuitable for all but very slow robotic manipulation tasks.

A simple whisker sensor could be no more complicated than a short length of piano wire passing through a small hole at the end of a metal tube. Figure 53.23 shows a cross-sectional view of such a sensor. When the wire is deflected it touches the metal tube thus completing an electrical circuit to signal the contact. More advanced designs can determine the position of contact along the length of the whisker.

Force/Torque Sensors

When two objects touch, the forces and torques generated by contact can be resolved into forces along three orthogonal axes and torques about these axes. In robotics it is often important to measure these six force/torque components and special purpose loadcells have been designed for this purpose. In practice, force/torque sensing load cells are commonly located between a robot manipulator arm and its tool or gripper. Here they can monitor and help control contact forces between the tool or gripper and external objects. In recent years, miniature force/torque sensors have been made for mounting in the fingertips of a robot hand. In this location, they can measure forces and torques generated during grasping and manipulation tasks. Applications for force/torque sensors are usually centered around manipulation tasks. By measuring applied force, a force/torque sensor can help maintain optimum contact force during robotic fettling, grinding and burnishing operations. During assembly operations jamming

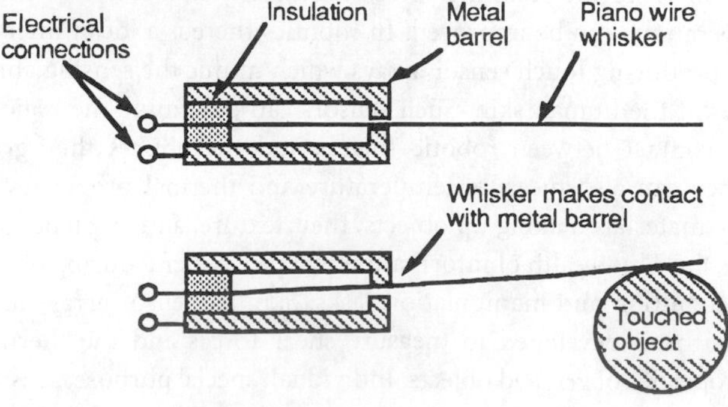

Figure 53.23 A simple whisker sensor.

generates characteristic reaction forces and torques. If these forces and torques are detected then appropriate actions can be taken to free the parts and complete the assembly. In principle, data from a force/torque sensor can also be used to find the point of contact between a tool or gripper mounted on the loadcell and the outside world.

The load-cell structure consists of a number of interconnected metal beams usually machined from a single piece of metal. Strain gauges are attached to the beams so that each gauge is sensitive to a different component of the applied forces and torques. Figure 53.24 shows a force/torque sensor machined from a single block of aluminium in the form of a spoked wheel. The outer rim of the wheel A is connected to the robot arm and the inner hub B to the gripper. There are four semiconductor strain gauges attached to each spoke of the wheel and they are wired in pairs as half-bridge circuits. The load-cell produces eight output signals and these can be related to the three components of force and three components of torque acting on the load cell.

Skinlike Tactile Sensors

Artificial sensory skins have been made containing arrays of contact sensors. These sensors can give an image of the area of contact between a robotic finger and a grasped object and many produce a graduated response depending upon the degree of indentation into the artificial sensory skin. Essentially, skinlike contact sensors measure deflection of the active sensor surface. Although numerous different designs of skinlike sensors have been developed over many years, they have seen little or no use in practical robotic applications. However, similar sensing technology has several biomedical applications including sensing the distribution of foot pressure on the ground and measuring the bite; the way in which the upper and lower teeth meet.

It is perhaps an indication of the lack of maturity in this area of robotic sensing that many different transducer mechanisms have been tried and still more are being continually proposed. Skinlike contact sensors have been designed which are based on simple switches, piezoresistivity, piezoelectricity, optical, opto-mechanical, fiber-optic, photoelasticity, magnetic, magnetoelastic, ultrasonic, capacitive, and electrochemical sensing methods. In such a brief outline it is impossible to cover the full field. However, four sensors will be mentioned. They are examples of devices which have shown sufficient promise to be developed to the stage of a commercial product.

Optomechanical

One commercially manufactured optomechanical sensor array employs a rubber skin with an array of mushroom-shaped projections molded into its surface. The head of the mushroom concentrates the applied force and the stalk acts as an optical shutter to reduce light transmission between a light-emitting diode and a photodetector as normal force increases. Construction of this sensor is quite labor intensive. All sensor sites contain a photoemitter and photodetector and they must be individually matched and trimmed with an associated resistor to equalize their responses. This sensing technique was used in the first

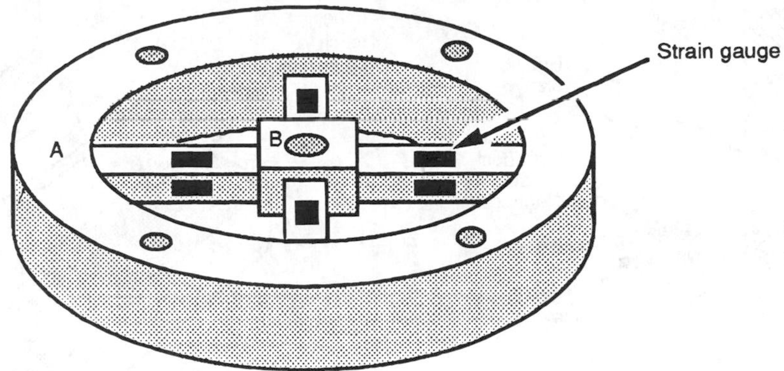

Figure 53.24 A 6-axis force/torque sensor.

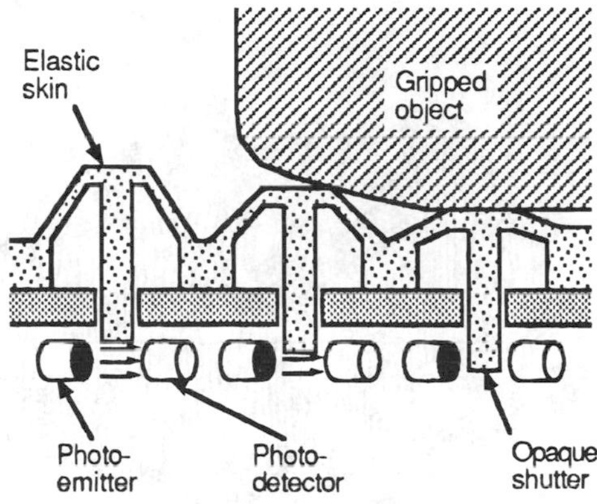

Figure 53.25 Optomechanical tactile sensor.

commercially available tactile sensor. Although each sensor site is quite complicated, sensor arrays were constructed with sensor spacings as small as 1.8 mm.

Frustrated Internal Reflection

This optical sensing technique is capable of producing very-high-resolution tactile images. A major component of the sensor is a light guide which may be a sheet of clear glass or plastic. Light introduced at one edge of the light guide propagates through the sheet by total internal reflection and emerges at the opposite edge (Figure 53.26). For total internal reflection to occur, light must strike the surface of the light guide at less than the critical angle. A sheet of reflective rubber material is suspended close to, but not touching, the light guide. This is the active area of the sensor. When an object presses against the rubber sheet the sheet deforms making contact with the light guide. At the area of contact, light leaves the light guide effectively illuminating the region of contact. Diffusely reflected light from the area of contact passes back through the guide and is received by a video camera. Using a flat rubber sheet gives a binary image which can only distinguish between contact and no contact. However, by embossing a pattern into the rubber surface, a certain amount of force information can also be distinguished.

This transduction technique produces very-high-resolution tactile images of the order of 256 by 256 taxels over a 2 cm by 2 cm area. However, the optical components including light source and television camera make this kind of sensor relatively bulky.

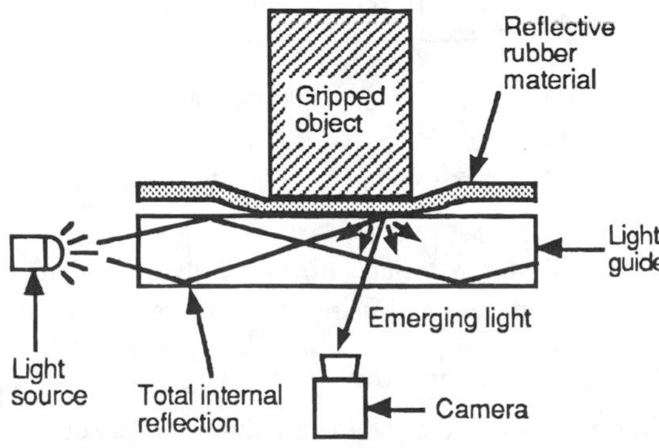

Figure 53.26 A tactile-sensor using the principle of frustrated internal reflection.

Piezoresistive

Piezoresistive sensors are relatively simple to make and provide a large electrical output which requires little or no amplification. Figure 53.27 shows a cross-sectional view of a single-sensor element. The sensor consists of two sheets of polyimide plastic film. On the inner surface of the lower sheet are deposited two metallic electrodes separated by a small gap. The upper plastic sheet is coated with conductive ink. Such inks consist of semiconducting particles like molybdenum disulphide held

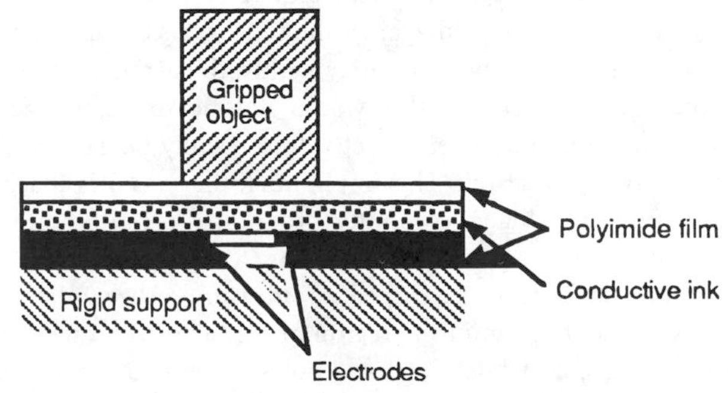

Figure 53.27 Cross-sectional through a single piezoresistive tactile sensor.

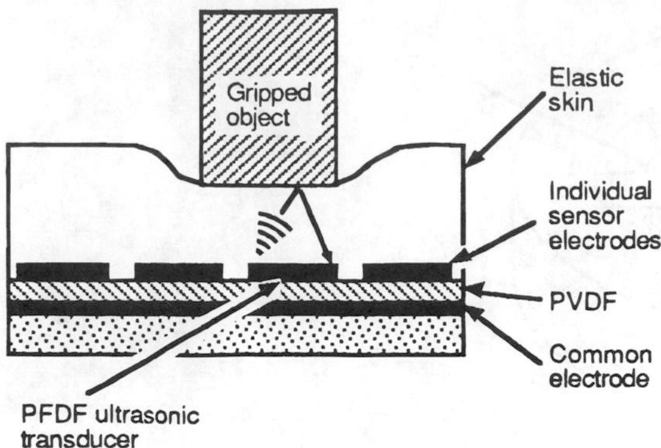

Figure 53.28 Cross-sectional view of a tactile sensor array using ultrasonic transduction.

together with a binder material such as acrylic resin. In this sensor the piezoresistivity is a surface effect. Pressure reduces the contact resistance between the electrodes and the ink. Resistance between the electrodes depends upon compressive force applied to the sensor. This technology can be used to make relatively simple and low cost touch sensor arrays.

Ultrasonic

Ultrasonic thickness gauges have been in use for many years to measure the thickness of paint layers, metal sheets, etc. An ultrasonic pulse is introduced into the sheet at one face, propagates through the thickness of the material and is reflected from the opposite face. The returning echo has traveled twice through the thickness of the material. Knowing the speed of propagation of the ultrasound pulse and measuring the propagation delay allows the thickness to be determined. This principle has been used to construct a tactile sensor by using ultrasonic transducers to measure the thickness of a flexible elastomer layer at many closely spaced points. The sensor is relatively simple consisting of a sheet of polyvinylidene fluoride (PVDF) piezoelectric plastic patterned with an array of electrodes. Overlying the sheet of PVDF is a layer of compliant elastomer which changes shape as objects indent into it. The electronics associated with the sensor generate, detect and time the ultrasonic pulses. Using this technology a 16 by 16 element ultrasonic sensor array has been manufactured with a 1.8 mm spacing between taxels.

Figure 53.29 shows a wire-frame plot of a typical touch sensor image. The object pressed against the sensor, a slotted electrical lug, is small enough the fit entirely within the active area of the sensor. For most objects this would not be the case. Tactile information about a larger object would usually be gathered by a process of active search. This implies that the sensor is attached to a gripper and that a computer program directs the gripper to search for the required information.

Sensors have been developed to measure other quantities besides skin deflection in the region of contact. When we touch metal it feels cold while wood and polystyrene foam feel warm, even though these materials are all at room temperature. The feelings of warm or cold are caused by different materials cooling our fingers at different rates. This gives us additional information about the kinds of materials things are made of and robotic sensors have been designed using the same principle.

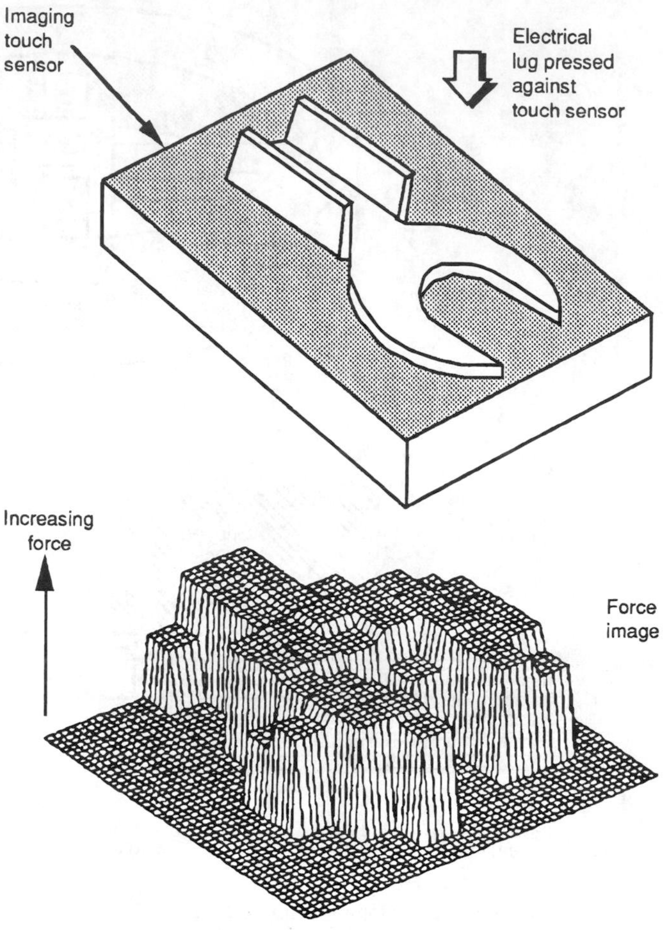

Figure 53.29 An example of a force image produced by pressing a slotted electrical lug against a 10×10 piezoresistive array sensor.

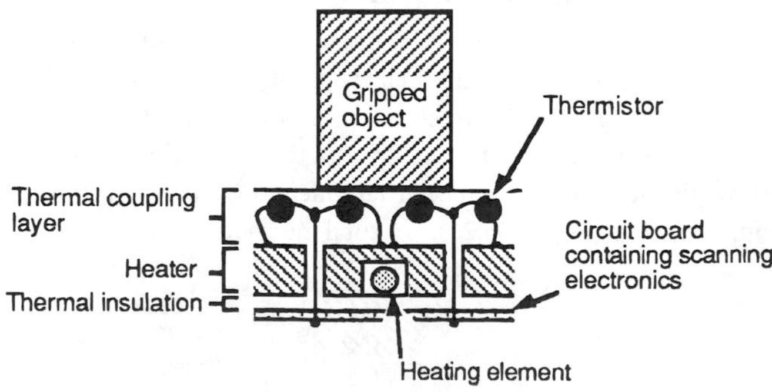

Figure 53.30 Cross-sectional view of a thermal sensor array.

Skinlike Thermal Sensor

In this sensor, a temperature stabilized heat source warms the sensor in the same manner that the blood supply warms the skin. As shown in Figure 53.30, a layer of material of known thermal conductivity couples the heat source to the touched

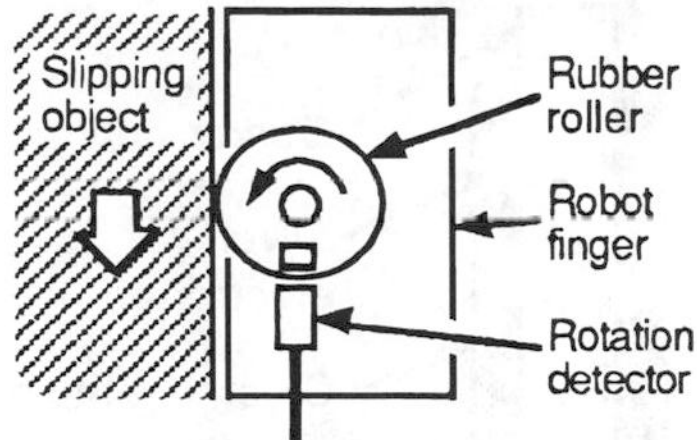

Figure 53.31 A sensor to determine slip.

object. An array of thermistors measure the contact point temperature over the sensor surface. These thermistors replace the thermally sensitive nerve endings in the human skin. When this sensor touches an object the temperature change in the thermistors gives information about the material properties of the object structure and the spatial extent of the temperature disturbance indicates the outline of the touched object.

Slip Sensing

The human skin can also determine surface texture and the onset of slip. Technological sensors have also been devised to give a robotic system similar capabilities. As an example Figure 53.31 shows a diagram of a special-purpose slip sensor. A free-wheeling rubber coated wheel presses against the gripped object. As this object starts to slip the wheel turns and a sensor detects the movement.

Many other tactile sensor techniques and systems are possible, e.g., sensors sensing multiple quantities at the same time. Also the design of robotic grippers to carry these sensors is an equally important consideration. Without appropriate grippers to position the sensors and associated control programs, tactile sensors have very few applications in robotics.

References

Beni, G. and Hackwood, S., eds. 1985. *Recent Advances in Robotics, Research,* Wiley, New York, NY.

Hirose, S., et. al. 1985. Titan III: A quadruped walking vehicle, *Robotics Research The Second International Symposium,* Hideo Hanafusa and Hirochika Inoue, eds. MIT Press, Cambridge, MA.

Nicholls, R. H., ed. 1992. *Advanced Tactile Sensing for Robotics,* World Scientific, Singapore.

Pugh, A., ed. 1986. *Robot Sensors Volume 2—Tactile and Non-Vision,* IFS Publications, UK.

Russell, R. A. 1990. *Robot Tactile Sensing,* Prentice Hall Australia, Sydney, Australia.

Webster, J. G., ed. 1988. *Tactile Sensors for Robotics and Medicine,* John Wiley and Sons, New York, NY.

53.5 A Robotic Sense of Smell

R. Andrew Russell

Compared to the other human senses the sense of smell has received relatively little attention from roboticists. However, there are many potential applications for an artificial sense of smell. These applications fall into the two broad categories of odor detection and odor discrimination.

Odor Discrimination

In a number of commercial situations, assessment of the quality of an odor is important. The aroma of foodstuffs such as bread, biscuits, coffee, beer, and wine are all important to the consumer. In the nonfood area, the success of perfume and scented products such as soap, washing powder, and air freshener all depend on their smell. Currently, quality control of these products is performed by humans. The human sense of smell is affected by age, health, and eating habits. For this reason, there would be many commercial applications for an artificial odor discrimination system.

Odor discrimination systems are being developed and have the structure shown in Figure 53.32.

The system comprises a number of odor sensors which are exposed to the odorant. Each sensor has a peak sensitivity to a different chemical species. However, their sensitivity is broad and overlaps between sensors. One specific odor will give a unique pattern of response from the odor sensors. This pattern is interpreted, in this case by a trained neural network, to identify the odor. The neural network may be trained using the back propagation algorithm by repeated exposures to the range of odors which the system will be required to discriminate. The individual sensors would typically be tin oxide or quartz crystal microbalance olfactory sensors.

Odor Detection Systems

Tracing leaks of poisonous or flammable materials is a potentially dangerous task which could be performed by a robotic system. Such a system would consist of a mobile robot platform equipped with appropriate odor sensors, a wind vane to measure wind direction, and a control program to search for and identify the source of the leak. An olfactory sense and the ability to mark odor trails on the ground could also form the basis for a robotic navigation system.

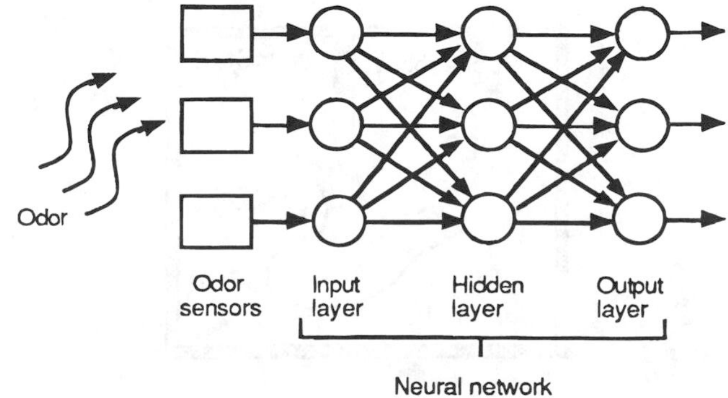

Figure 53.32 An odor discrimination system.

Marking and Detecting Odor Trails

Insects survive, feed, and reproduce very effectively—especially considering their relatively small nervous system. To operate effectively in the variable and unstructured real world environment, insects employ many strategies. Laying down and detecting odors is the basis of several of these strategies. Similar techniques can be implemented to improve the competence of robotic systems. As an example, a mobile robot may be required to explore a partially known or unknown environment. This could be the site of an accident where no accurate maps are available or an explosion has damaged the building. It is essential that the robot finds its way back to the starting point at the end of its mission. If the robot payed out an umbilical cable on the outward journey it could follow the cable to return to its starting point. However, umbilical cables tend to snag and the equipment to payout and retrieve an umbilical is bulky and cumbersome. The navigational effect of the cable can be gained by laying a trail of volatile chemical as illustrated in Figure 53.33. The robot can then return to its starting point simply by following the trail.

Solutions of odor chemicals can be layed on the ground using an applicator similar to a felt-tip pen. By sampling air from close to the floor surface the position of the odor trail can be found and used to guide a mobile robot.

Olfactory Sensors

Many techniques are available for detecting the concentration of volatile chemicals. For robotic applications the sensor must respond quickly to the target odor (responding in less than a second), be compact, and consume little power. The quartz crystal microbalance is one sensor which has proved suitable for robotic applications. It has a rapid response to changing chemical concentrations (responding in under one second), draws little power, is robust and inexpensive. The sensor uses a quartz crystal as a sensitive balance to weigh the odor molecules. A chemical coating on the crystal is chosen to have a specific affinity for the target odorant molecules. When air containing molecules of the target odor is drawn over the crystal, some of the molecules become temporarily attached to the coating. This increases the effective crystal mass and lowers its resonant frequency. A sensitivity of 1 Hz change in frequency for each part per million of odorant

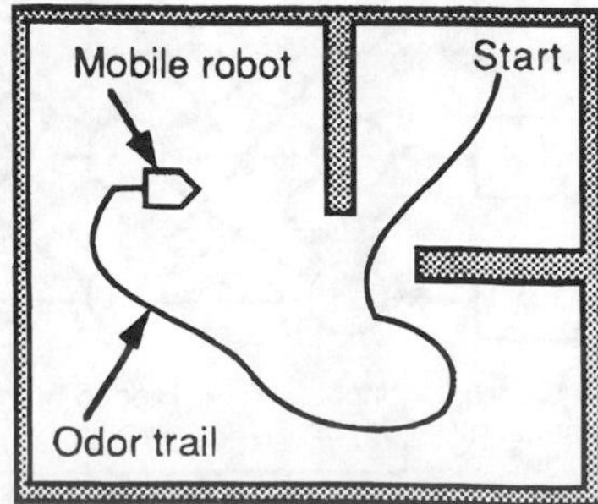

Figure 53.33 Finding the way back to the starting point—the virtual umbilical.

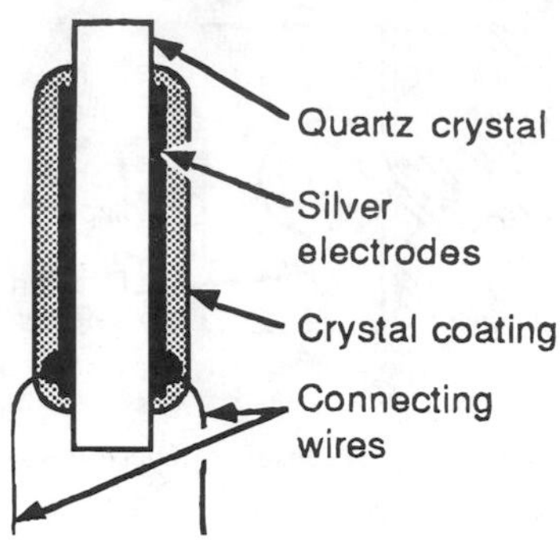

Figure 53.34 The quartz crystal microbalance olfactory sensor.

has been measured for this type of sensor. A cross-sectional view of the sensor crystal is shown in Figure 53.34.

The application of odor sensing to robotic systems is a relatively new development. However, the ability to detect and classify odors will be useful for service robots operating in hospitals, offices, and homes as well as special purpose units for quality control in food, perfume, and toiletries manufacture. The ability to mark and detect odor trails on the floor can also be used as an aid to robot navigation.

References

Deveza, R., Russell, R. A., Thiel, D., and Mackay-Sim, A. 1994. Odor sensing for robot guidance, *Inter. J. Robotics Research,* 13(3):232–239.

Gardner, J. W. and Bartlett, P. N., eds. 1992. *Sensors and Sensory Systems for an Electronic Nose,* Kluwer Academic Publishers, New York, NY.

Genovese, V., et. al., 1992. Self organizing behavior and swarm intelligence in a pack of mobile miniature robots in search of pollutants, *Proc. IEEE/RSJ Int. Conf. Intelligent Robots and Systems,* 1575–1582, Raleigh, NC, July 7–10.

King, W. H. 1964. Piezoelectric sorption detector, *Anal. Chem.,* 36(9):1735–1739.

Rozas, R., Morales, J., and Vega, D. 1991. Artificial smell detection for robotic navigation, *Proc. Fifth Int. Conf. Advanced Robotics,* 1730–1733.

53.6 Actuators in Robotics and Automation Systems

Marcelo H. Ang, Jr. and Choon-seng Yee

Overview

Actuators are devices that produce actions which typically are forms of motion or forces/torques exerted. The input to the actuator is some form of energy (usually electrical) and the output is mechanical motion or force exertion. Actuators are therefore key components in an electromechanical system such as a robotic manipulator. The input to the manipulator is a controlled amount of electrical energy and the result is the manipulator motion corresponding to the robotic task.

A robotic manipulator is a multi-degree-of-freedom mechanism consisting of a series of linkages that are connected in some fashion to provide dexterity in the manipulator end-effector (or "hand") which effects the manipulator task. Each link moves relative to each other through joints. Some joints may be passive, like pin joints, while other joints are coupled to actuators which cause relative motion between the links connected by these joints. The actuators, therefore, serve as the muscles of the manipulators.

There is a wide range of actuators used in industry. They vary depending on the load they actuate. In robotic manipulators, actuators can be in the form of motors that move the robot links. Actuators, such as solenoids, are also used to actuate small devices such as on/off switches. Actuators can be classified electric, hydraulic, and pneumatic, depending on the type of input energy they accept.

Motors are electric actuators that accept electrical energy and the output is mechanical rotation of the motor shaft (rotor). The resulting motion is effected electromagnetically where the electrical input and the magnetic fields inside the electric actuator cause the rotor to rotate. Solenoids are also electric actuators because the current flowing through the coil causes motion of a ferromagnetic cylindrical rod longitudinally located at the center of the coil. The ferromagnetic material inside electric actuators saturates at certain levels of magnetic flux density, thus limiting their torque or force capabilities (de Silva, 1989). The load to weight ratios of the electric actuators are therefore limited compared to hydraulic actuators.

Hydraulic actuators use very highly pressurized liquid to effect motion or force. The liquids are usually oils, because they are noncompressible, and the hydraulic pressures are in the order of a few thousand PSIs. Their load capabilities are at least an order of magnitude larger than electric actuators. Hydraulic actuators, therefore, have the best torque to weight ratios. Another important feature of hydraulic actuators is the high stiffness (viewed from the load side) provided due to the noncompressibility of oil as compared to a lower stiffness provided by an electromagnetic medium.

Pneumatic actuators use pressurized air as the actuating mechanism. Their advantage is that they offer a simple, low cost method for linear motion. Because air is compressible, they are not stiff and their responses are slow. Pneumatic actuators are therefore wellsuited for lighter load applications and lower performance servo-control applications.

Electric, pneumatic, and hydraulic actuators all find their use in robots. Hydraulic actuators are used in large, high performance robots with large payloads, while pneumatic actuators are used more in smaller, lower performance (and cost) robots for motions. Pneumatic actuators are ideal for applications requiring the robot end-effector (or hand) to move to fixed positions. They usually require no feedback of position, and mechanical stops are typically employed to define the fixed positions. If precise positioning is required, hydraulic or electric actuators with servo control are used. The advantage of hydraulic and pneumatic actuators are their ruggedness and safety. Because they are sealed, they can operate in harsh (dirty, wet, etc.) environments. They can also be operated in explosive environments, because unlike electric motors, there is no danger of brush arcing causing sparks. Their disadvantages compared to electric motors are their higher cost and maintenance requirements, higher noise levels, lower efficiencies, the need for fluid transport systems, and flammability of oils in hydraulic systems (Andeen, 1988). For applications that require medium loads and accurate positioning control, electric actuators provide the bet price/performance solution. Most robots today employ electric actuators to achieve accurate positioning control in light to medium load applications.

A special category of actuators are those that rely on special materials and their properties. These "exotic" actuators are not commonly used in robotic manipulators because of inadequate loading capabilities and slow response times (Andeenn, 1988). They are also very costly. Piezoelectric materials can be used as actuators because they change shape when subjected to an electric field. This change in shape can therefore effect motion. Magnetostrictive materials are similar to piezoelectric ones except that an applied magnetic field causes a change in shape. Thermal actuators rely upon expansion or contraction when subjected to temperature changes in a resistance type heating element inside the actuator. Another interesting actuator is the shape memory alloy which is a metal which adopts a "memorized" shaped when it achieves a certain transition temperature. Other actuators in this category include electrostatic actuators consisting of two conducting plates that are either electrostatically attracted to or repelled against each other depending on the DC voltages applied to the plates (Andeen, 1988).

Transmissions are also important in robots. Transmissions serve two purposes. First is to transmit the power output of the actuator to another location that is remote from the actuator. In a robot, for example, the actuators can be located at the base of the robot thus freeing the need for the robot links to carry the actuators. Second is to transform (i.e., amplify the load capability of the actuator at the expense of speed. Transmissions are gear systems, belts and pulleys, and sprockets and chains.

Actuators need peripheral circuitry to "drive" them. These circuities are called drivers or amplifiers. The drivers serve as an intermediate stage between the actuator input and the energy source. They also serve as interfacing circuits to allow input from a computer or other input device. In this section, we also explain the different drivers typically used with the actuators. We concentrate upon electric actuators (i.e., motors) because they are the most commonly used actuators in robotic manipulators. For completeness, we include discussions on the drives associated with the electric motors, and also transmission devices.

Direct Current Motors

Principle of Operation

DC motors in general operate based on Ampere's law of magnetic force, which explains and quantifies the force acting on a current carrying body (motor coil) in the presence of a magnetic field. According to Ampere's law of magnetic force, the force F acting on the body is the cross product of the current I flowing across the body and the magnetic flux density B of which the body is in, as shown in Figure 53.35. It is expressed in Equation 53.12 and the vector representation is shown on the right of Figure 53.35.

$$F = I \times B \tag{53.12}$$

Current I flowing through a conductor induces a force F that is normal to the current and the magnetic flux density B. The current is the electrical input to the DC motor, the induced force is coupled to the output shaft of the motor to produce the output torque. The clever arrangement of the flow of current, flux and induced force results in a smooth torque delivery of the motor.

DC Brushed Motor. A simple model of the DC motor can be seen in Figure 53.36. It consists of a pair of permanent magnets as the *stator* (which serves as the fixed housing of the motor) and a motor coil as the *rotor* (or rotating shaft) connected to a commutator. The commutator, which forms the heart of DC brushed motors, basically consists of 2 (or more) commutator bars (the half-cylinders in Figure 53.36) and 2 brushes, both made of conducting materials. The brushes are spring loaded to maintain contact with the cylinder formed by the commutator bars. Each of the two ends of the motor coil is connected electrically to an independent commutator bar. When the rotor coil rotates, the commutator cylinder (formed by all the commutator bars, insulated from one another) rotates together and the current flow changes according to the contact made by the brushes on the commutator bars, thus creating an alternating current flow in the coil, depending on the position of the coil.

In the example shown in Figure 53.36, there will be a time

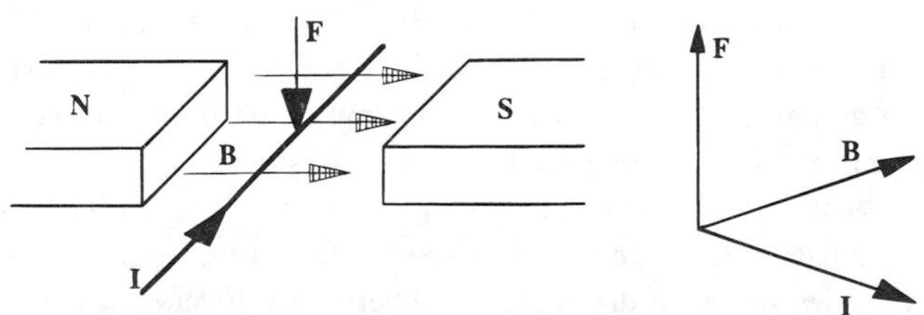

Figure 53.35 Ampere's law of magnetic force.

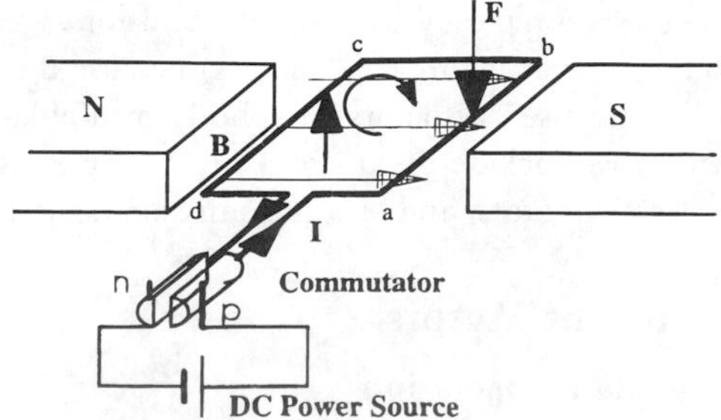

Figure 53.36 DC motor with commutator.

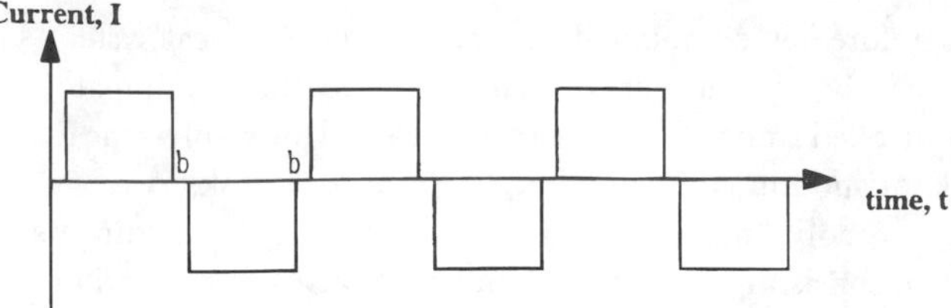

Figure 53.37 Current flow in motor coil.

when the brushes are in contact with both commutator bars, i.e., when the motor coil is perpendicular to the magnetic field lines. This will cause the power to short circuit, but the angular momentum of the rotor will bring the coil to move through this position, minimizing the short-circuit time. Alternatively, more than 1 set of coils can be used and the number of commutator bars increased accordingly to eliminate the short-circuit problem, which most motors nowadays implement.

With the commutator in place, the effective voltage across the motor coil, which is also equivalent to the current flowing across the motor coil with fixed resistance, will have an alternating effect as shown in Figure 53.37 for the segment ab of the motor coil. The small sections denoted by b shows the time when short circuit occurred.

DC Brushless Motors. DC brushless motors do not make use of commutators to regulate the power or current flowing into the coils (Dote, 1990). Instead, brushless motors regulate the current flow through semiconductor switches with position feedback of the motor shaft. An inherent characteristic of brushless motors is the requirement of sensors to sense the absolute angular position of the motor shaft. The construction of the brushless motor also differs from that of brushed motors. In brushed motors, permanent magnets are used as stators while the motor coils are attached to the rotor. In brushless motors, the stator normally consists of the motor coil windings while a permanent magnet takes over as the rotor.

Brushless motors have the following advantages over brushed motors (Dote, 1990):

1. Brushless motors have higher maximum speed and greater capacity because of the construction of the motor. Rotor shaft friction is reduced, because there is no need for commutator brushes. Furthermore, heat dissipation of the stator coils is more effective through the stationary motor housing or casing.
2. They work in less favorable surroundings. The size of brushless motors are comparatively smaller than brushed motors, making them suitable for compact applications, such as in robotic arms. Brushless motors provide much better torque to weight ratios, i.e., for the same weight of the motors, brushless motors provide much higher torque. The absence of commutator also eliminates the possibilities of sparks or arcing, making it safe for applications in locations with flammable gases such as in the petrochemical industries.

3. Since no commutator is used, brushless motors practically requires no maintenance. They also produce less noise compared to brushed motors.

Brushless motors are also similar in construction to induction motors or AC motors, but the characteristics of the two are different. Brushless motors have the characteristics of a DC brushed motor, namely having a linear speed-torque relationship when the power fed into the motor system is fixed or constant.

Brushed Motor Drives

In general, there are two popular types of driver circuits being used for the brushed DC motors: linear drive and the widely used pulse width modulation (PWM) drive (Electro-Craft Corp., 1980).

Linear Drive. As the name implies, a linear drive provides a continuous flow of current to the motor that is linearly proportional to the torque or speed required of the motor. Linear drive controls the amplitude of the voltage or current being sent to the motor. The simplest way of implementing linear drive in a closed loop system is with a single power transistor, as shown in the sample in Figure 53.38.

The current to the base of the transistor is controlled, thus controlling the current flowing through the motor. The input to the base of the transistor may come from the digital-to-analog (D/A) card or chip output of a microcontroller (or computer), or as shown in Figure 53.38, from the output of the tachometer in a speed control system. Linear drives are ideal for high-performance speed control systems. By adjusting the speed dial, the speed of the motor can be controlled by limiting the current to the base of the transistor, which is generated from the tachogenerator feedback of the system. The example above applies only for unidirectional motor speed. To perform reversible speed control, an additional reversible switch can be added to the circuit to switch the connections between the motor and the driver. The power rating of the transistor is appropriately chosen to match the current capacities or requirements of the motor.

To interface the linear drive to a digital controller such as a personal computer, the controller must have some form of analog input and output facilities such as the analog-to-digital and digital-to-analog converter (ADDA) card for the PC. The tacho feedback is read through the A/D converter into the PC for processing before an output through the D/A converter is generated. The

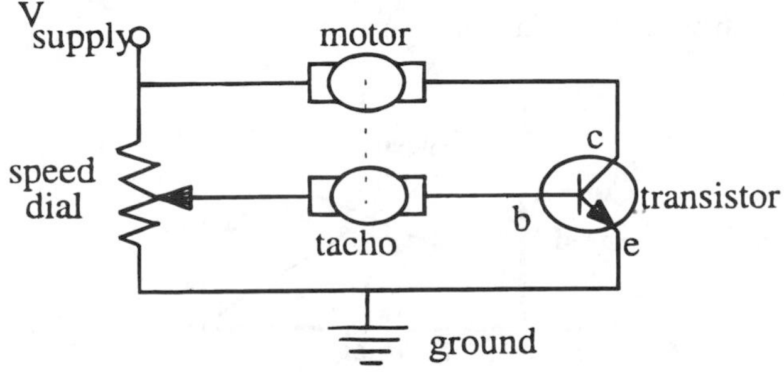

Figure 53.38 Simple linear drive for DC motor.

D/A output is then fed to the base pin of the transistor to trigger the motor motion.

Motor manufacturers also provide standard drivers for the different makes of motors. The input signals required are typically the bipolar analog command voltage and a TTL enable signal. The linear driver could also provide current feedback, tachogenerator feedback signals at the controller's disposal. Known motor manufacturers who provide motors with drivers are Maxon Motors from Switzerland and Baldor Motion Products of USA.

PWM Drive. The pulsed width modulated (PWM) drive is more popular among the motor driver circuits used. A PWM driver is driven by a single DC source, with an internal amplifier switching the power on and off at a fixed frequency and at a variable "firing angle" so that the average power (in terms of voltage and current) is controlled. The frequency of the output from the PWM system is determined by an external RC circuit as expressed in Equation 53.13 and the "firing angle" controlled by an analog input typically between the range of 0 to 3 V. A simple PWM driver circuit is shown in Figure 53.39 using the Motorola/TI TL.494 PWM driver chip.

$$f_{OSC} = 1.1 + (R_T \cdot C_T) \qquad (53.13)$$

The generation of output pulse from the PWM is shown in the timing diagram in Figure 53.40 with 3 different stages of input signal, two being the minimum and the maximum duty cycle and another at varying dead-time control signal.

The frequency of the PWM is normally set to the nonaudible range to keep the system quiet. A higher operating frequency would also ensure a more even distribution of power and smoother motion. In the example shown in Figure 53.39, the frequency used is about 1.1 kHz.

From Figure 53.40, the general operation of the PWM can be seen clearly. As the dead-time control input varies between 0 to 3 V, in the case of the TL494 being used here, the output PWM pulses changes from 0% duty cycle to 100% duty cycle linearly. The output PWM pulses generated then go through a dual full-bridge driver, the SGS L298N, which will translate the TTL pulse train into a DC pulse train with an amplitude of the motor rated voltage.

With the implementing of this PWM driver, the controller need only to provide the analog voltage for the dead-time control signal and a TTL direction signal. From the circuit shown, the controller could also read in the analog signal of the current drawn by the motor during the operation. Other forms of feedback can also be added, such as tachometer or digital encoder, which will only affect the controller but not the PWM driver directly. Motor manufacturers also produce PWM drivers for the different makes of motors. Typical input signals are the same as for the described PWM driver circuit.

Brushless DC Motor Drive

The driver for a DC brushless motor normally consists of a sine wave generator, a PWM driver array followed by a transistor

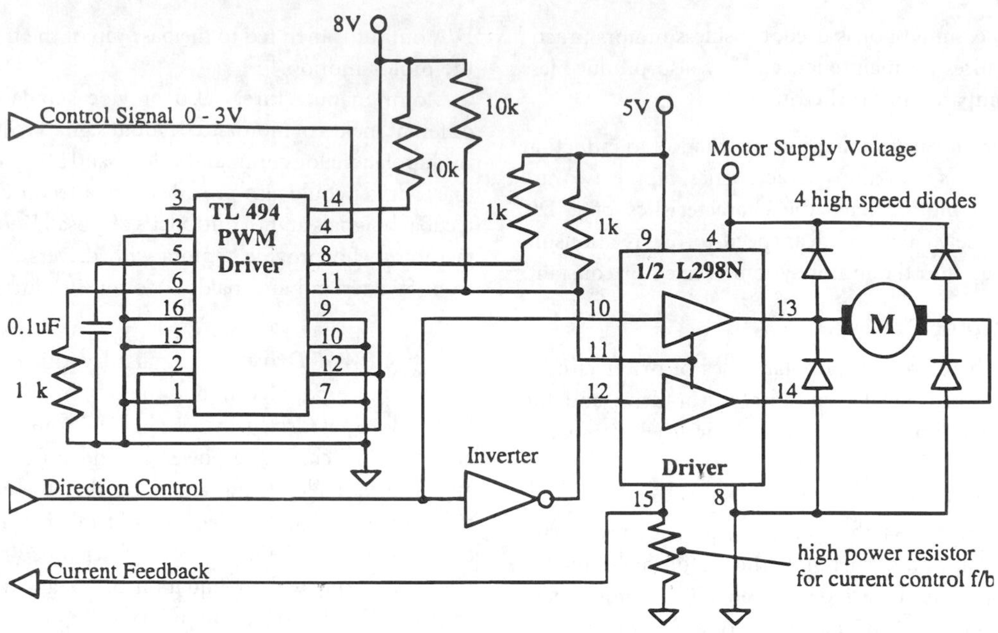

Figure 53.39 PWM circuit with driver and motor.

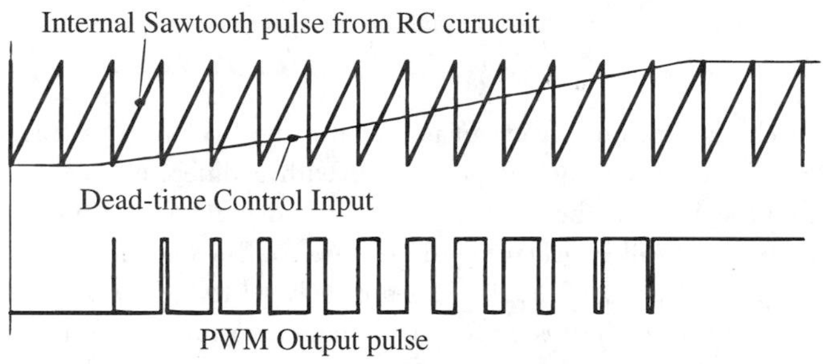

Figure 53.40 PWM output signal.

array and feedback devices attached to the brushless motor. The circuit has to read the position and (or) velocity feedback from the brushless motor, and interpret the signals in order to produce the output to the few (normally 3 or 4) stator coils. The overall control block diagram for a 3-phase brushless motor driver is shown in Figure 53.41.

Unlike an AC motor or induction motor, where the input to the actuator is an AC supply, the brushless motor uses a DC power source to drive the motor. A sine wave generator generates the alternating (or sinusoidal) waveform for each phase of the motor, after processing the feedback information on the position of the rotor shaft. In view of the processing power needed, either a microcontroller with EPROM or a DSP is used for this task. To maintain the DC motor characteristics, the sinusoidal waveform for each phase coil is fed through a PWM before sending the power into the motor.

There are also some other chips available in the market to drive brushless DC motors, such as the SGS L6230, which is a bidirectional 3-phase brushless DC motor driver, (SGS 1987) which can be implemented on its own or integrated with a microprocessor/controller.

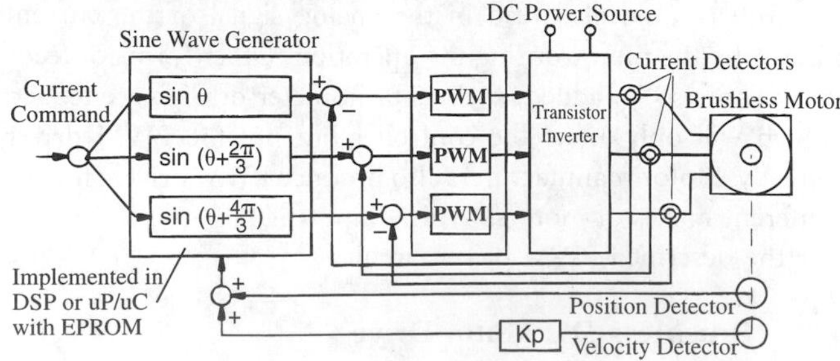

Figure 53.41 Control block diagram for brushless DC motor.

DC Motor Performance and Characteristics

As the magnetic flux of the permanent magnets for a typical DC permanent magnet motors, be it brushless or brushed, does not vary much, the speed-torque and the current-torque relationships are linear over the extended range of operation.

The maximum speed shown in Figure 53.42 is the no-load speed of the motor at a fixed voltage. The maximum torque on the *y* axis where the speed is zero is the stall torque. The current flowing through the motor coil is proportional to the torque or load applied onto the motor rotor. This simplifies the control system significantly in many applications.

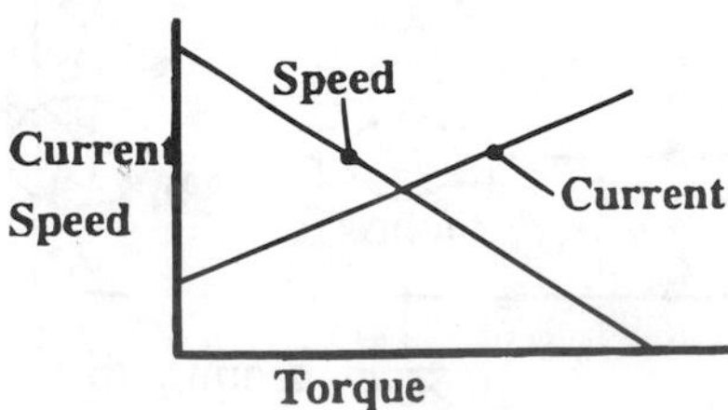

Figure 53.42 Typical DC motor characteristics.

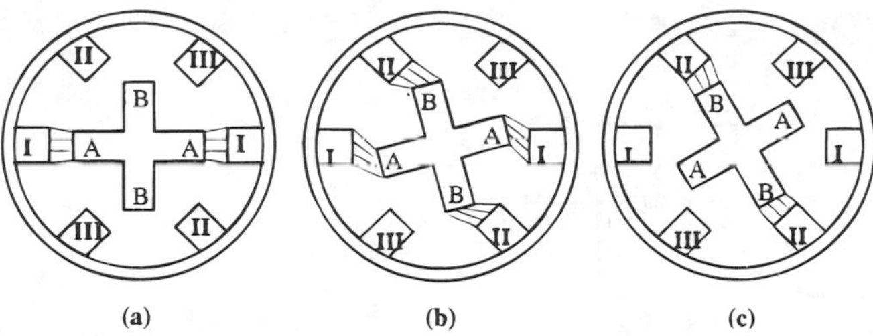

Figure 53.43 Simple stepper motor construction and operations.

Stepper Motors

Principle of Operation

Stepper motors are slightly different from DC motors in construction and application, although both categories make use of electromagnetic flux in coils for their operations. In a stepper motor, the construction is similar to that of a DC brushless motor, the coils are wound around mild steel cores with teeth like surfaces facing the rotor, attached to the motor housing as stator. The rotor is made up of either a mild steel multi-toothed core, sometimes with a permanent magnet at the center of the core. The total number of teeth for the stator is normally 1 pair more than the number of teeth at the rotor. A simple example can be seen in Figure 53.43.

From Figure 53.43, when coil number I is energized, the magnetic flux generated will pull the rotor teeth pair A towards teeth I, as shown in Figure 53.43(a). This is the first step in the stepper motor sequence. With coil I still activated, coil II is now energized as well, creating another flow of magnetic flux, which finds the shortest route through to teeth B of the rotor. As the path for the magnetic flux of coil II is much larger, teeth B is pulled towards teeth II, causing the rotor to rotate in the counterclockwise direction until both A-I and B-II strike a balance, moving the rotor 15 degrees to the position as shown in Figure 53.43(b). After this position is reached, coil I is de energized so that only the magnetic path B-II is present, moving the rotor another 15 degrees to the position shown in Figure 53.43(c), so that B is aligned to teeth II. This is a full step of the stepper motor and the steps repeat itself with the third (or even fourth) coil.

Stepper motors are normally used in open loop control systems. As shown in the example above, the stepper motor rotates at fixed steps for every change in the driver signal, i.e., from step (a) to step (b) in Figure 53.43. Unless the load applied to the stepper motor exceeds the rating of the motor, which causes it to slip, the performance of the motor should be very accurate. The typical step size of stepper motors is in the range of 0.9 and 1.8 degrees. By using a half-step driving, the step size could be halved, making the resolution of the stepper motor even finer.

There are a few types of stepper motors available in the market. A bipolar permanent magnet stepper motor has only 4 wires coming out from the motor, the AB pair and the CD pair as shown in Figure 53.44(a). The unipolar permanent magnet stepper motor, on the other hand, has at least 6 wires coming out of the motor, of which two of them are ground wires normally in black and white. Coils A and B shares one ground line and coils C and D shares the other ground line as shown in Figure 53.44(b). The color scheme for the wires varies widely for different manufacturers. One should always consult the manufacturer or the distributor for the specifications before connecting the wires to driver circuit. The driver circuit provides the current output to each of the coils of the stepper motor.

Stepper Motor Driver

The stepper motor in general requires 4 logical signals to activate the motion. The 4 signals are identified as A, B, C, and D channels. The signals sent to each of the channels are logical high (for ON state at the motor rated voltage) and logical low (for OFF state at 0 V) signals. These signals have to be sent in the particular sequence as described in the previous section.

One popular decoder chip being used for the stepper motor driver is the SGS L297, coupled with the dual full-bridge driver chip L298N, also from SGS. The standard connection for the stepper motor driver circuit is shown in Figure 53.45. (SGS 1987).

The SGS L297 is a stepper control that generates the required sequence of current flow through the coils. The sequence output of the L297 needs to be amplified at levels required by the motor, and this is accomplished by the L298 N driver chip.

The implementation of this driver circuit allows easy interface with other digital controllers such as a personal computer or a microcontroller. The driver requires essentially only 3 signals from the controller:

1. The CLOCK signal, which progresses the output stage by one step on the rising edge of the pulse received. This clock signal input to the driver is a pulse train whose frequency is linearly related to the angular velocity of the motor. For each pulse received, the motor rotates one step.
2. The CW/CCW signal, which decides whether to change the state of the output in the forward or backward direction, i.e., A–B–C–D for forward and D–C–B–A backward.
3. The ENABLE signal, which allows the driver chip to accept input signals and produce output pulses. A TTL high (+5 V) signal means get to work! (This signal could be permanently tied to high if it is a continuous system which does not require much setting up.)

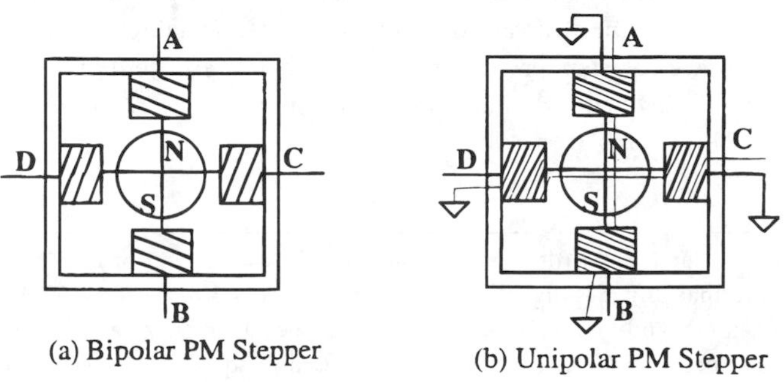

Figure 53.44 Bipolar and unipolar permanent magnet stepper motors.

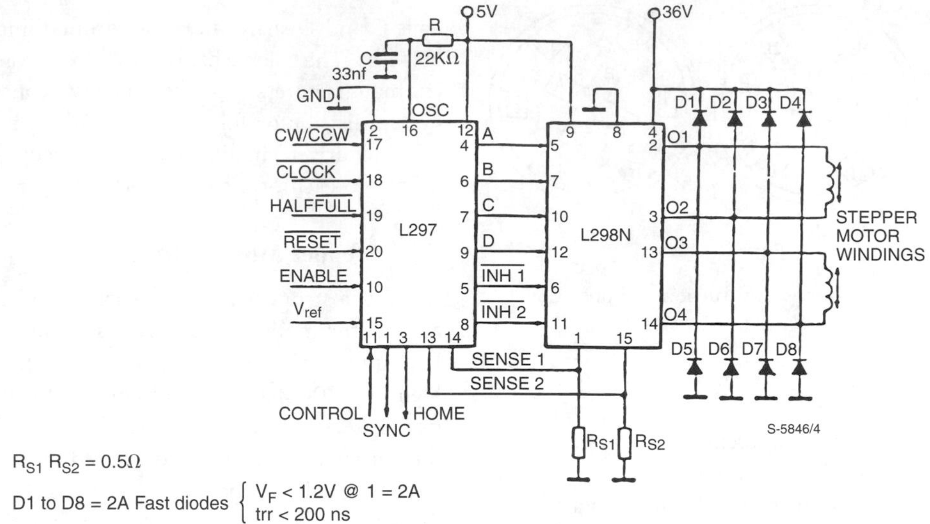

Figure 53.45 Two-phase bipolar stepper motor control circuit. *Source*: SGS 1987. *SGS Motion Control Application Manual.*

Another optional signal is the HALF/FULL signal, which determines whether to carry out full-step driving or half-step driving, depending on the resolution requirement of the system.

For interfacing with a personal computer, a digital I/O card such as the 8255 card needs to be installed to provide the digital signals to the driver chip. For position control, one of the 8255 ports can be used to provide the clock pulses at fixed interrupts with a counter keeping track on how much the motor have moved. Another 8255 port can be used separately to provide fixed signals to both the ENABLE and direction lines.

For speed control, the 8253 counters of the 8255 card can be configured to give a pulse train at fixed frequency until the configuration is changed.[1] By using it, the host PC can be free to perform other tasks. On top of that, one 8255 port is required to provide the direction and enable signals to the driver.

The same signals can also be used for stepper driver circuits supplied by the motor manufacturers and the prices are pretty reasonable. One added feature might be the option for two different pulse trains for the clockwise and counterclockwise directions.

Stepper Motor Performance and Characteristics

Stepper motors are excellent open loop control actuators. They are widely used in the PC industries such as printers, floppy disk drives. On the heavy industry side, most XY tables are driven by powerful stepper motors. They provide excellent interface capabilities with digital controllers such as microcontrollers and PCs.

On the other hand, there are some limitations to the implementation of stepper motors in certain systems. Stepper motors use fixed holding torque at every step and excessive load causes slippage in the motor positioning. In the same manner, the acceleration rate and the maximum velocity of the motor shaft is also limited by the load of the system.

One other characteristic is the resonance frequency of the stepper motor. For every stepper motor, there is a particular range of frequency where the motor would not function properly with full load. A typical speed-torque curve for the stepper motor is shown in Figure 53.46.

Transmissions

As mentioned in the introduction, transmissions amplify the load capabilities of actuators and/or allow the actuators to be remotely located to the devices being actuated. The following are transmissions used in machinery: gear systems, belts and pulleys, and sprockets and chains, with gear systems and cables and pulleys being more commonly used in robotic manipulators.

Belts, Cables, and Chains

Flat belts are common for large crowned pulleys to allow remote location of actuators, but are not appropriate for power

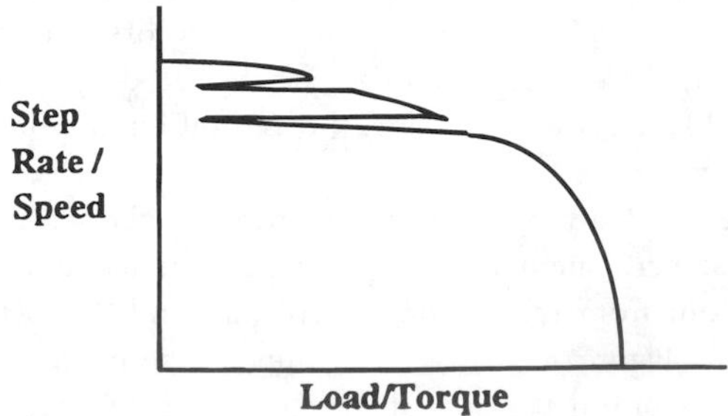

Figure 53.46 Stepper motor characteristics.

[1] Digital I/O cards typically contain at least one 8255 or equivalent chip that provides digital I/O (one 8255 provides 3 8-bit ports or 24 I/O lines), and at least one 8253 (or equivalent) timer (one 8253 chip provides 3 16 bit counters that can be used as square-wave generators and other.). One card with two 8255s and one 8253 costs less than US $40.00.

transmission. They are not applicable for robots. V belts have V-shaped cross sections that provide wedging action in the pulley, thus giving better power transmitting capabilities. However, belts can slip and this can be an advantage in robots because of the resulting overload protection (Andean, 1990). Timing belts have ridges that run in grooves on the pulleys so that no slip is ensured. Cables are similar to belts but have the advantage of being used in more than one plane. Some belts have studs on them to allow for timing and/or no-slip operations. Figure 53.47 shows a possible arrangement of a belt or cable to drive a robot link.

Although rarely used for power transmission, straps consisting of flat metal cables can be used as tendons to hold joints together as shown in Figure 53.48.

Gear Trains

Gear trains can be classified according to the relative geometric configurations of the gear shafts.

Parallel Shafts. The spur, helical, and herringbone gear arrangements have the gear shafts in parallel as shown in Figure 53.49. The spur gear is most common with teeth parallel to the centerline. The mating teeth makes instantaneous full width contact when meshing, thus resulting in possible vibrations and shakings. The helical gear, on the other hand allows gradual contact of the engaging teeth, and results in smoother load transmission. The helical gear, however, provides an axial force as a side effect, thus requiring thrust bearings to withstand this load. The herringbone gear have left- and right-hand helices cut into a single gear thus balancing the axial load in the gear itself.

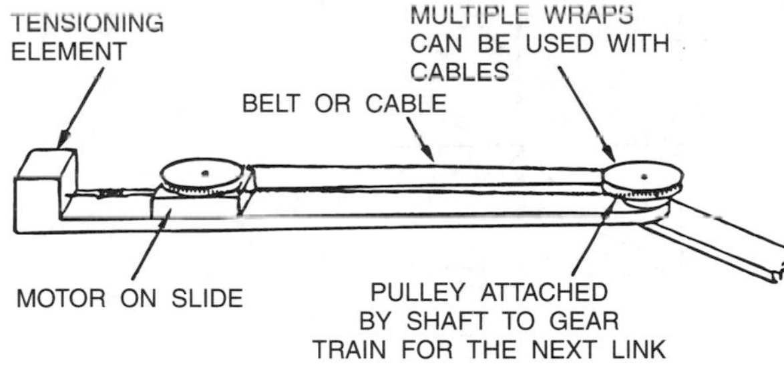

Figure 53.47 Belt or cable drive for a robot link. *Source*: Andeen, G. B. 1988. *Robot Design Handbook,* McGraw-Hill, New York, NY.

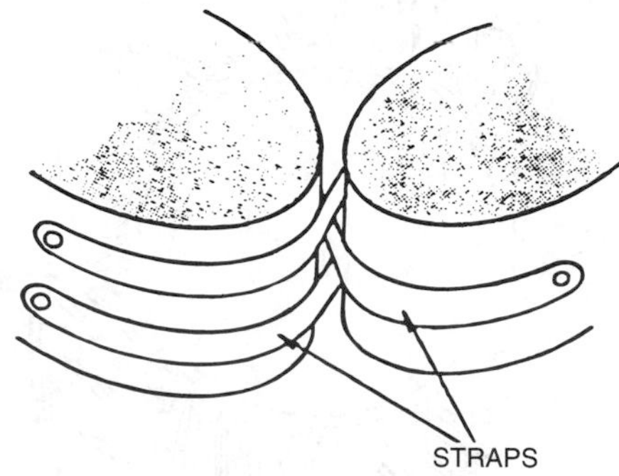

Figure 53.48 Flat metal cables used as tendons holding the joints together. *Source*: Andeen, G. B. 1988. *Robot Design Handbook,* McGraw-Hill, New York, NY.

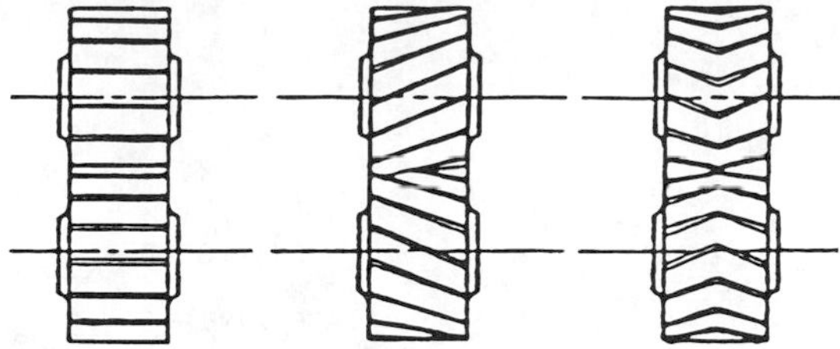

Figure 53.49 Spur (a), helical (b), and herringbone (c) gear couplings. *Source*: Andeen, G. B. 1988. *Robot Design Handbook,* McGraw-Hill, New York, NY.

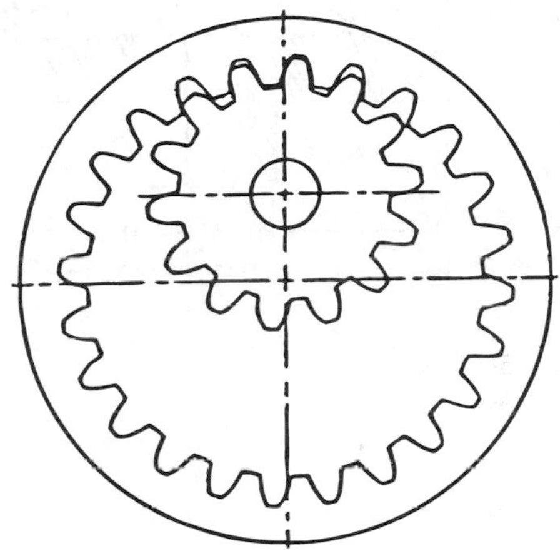

Figure 53.50 Internal gear arrangement. *Source*: Andeen, G. B. 1988. *Robot Design Handbook,* McGraw-Hill, New York, NY.

Another possible configuration for parallel gear shafts is the internal gear arrangement as shown in Figure 53.50. The advantage of this design is the compactness which makes it attractive for robots. This arrangement is commonly used for reduction gear and planetary gear arrangements.

Nonparallel Shafts. For gear shafts that are intersecting (i.e., shafts lie in a plane), bevel gears are used as shown in Figure 53.51. The teeth of bevel gears can be spiral too to acquire similar advantages of helical gears over spur gears.

Crossed-axis helical gears (Figure 53.52) are used when the gear shafts are nonparallel and nonintersecting. They are used solely for power transmissions in light load applications.

When the nonintersecting shafts are at right angles to each other, worm gears are used as shown in Figure 53.53, where the spur gear drives a worm gear.

Hypoid gears are similar to spiral bevel gears except the gear shafts are nonintersecting as shown in Figure 53.54.

Others. The rack and pinion type gearing arrangement is useful for transforming rotational motion to linear one. Teeth on a gear mates with teeth on a plane or rack as shown in Figure 53.55.

Another interesting transmission mechanism is the ball screw. Ball screws employ ball bearings incorporated between the mating surfaces resulting in friction-free transmissions.

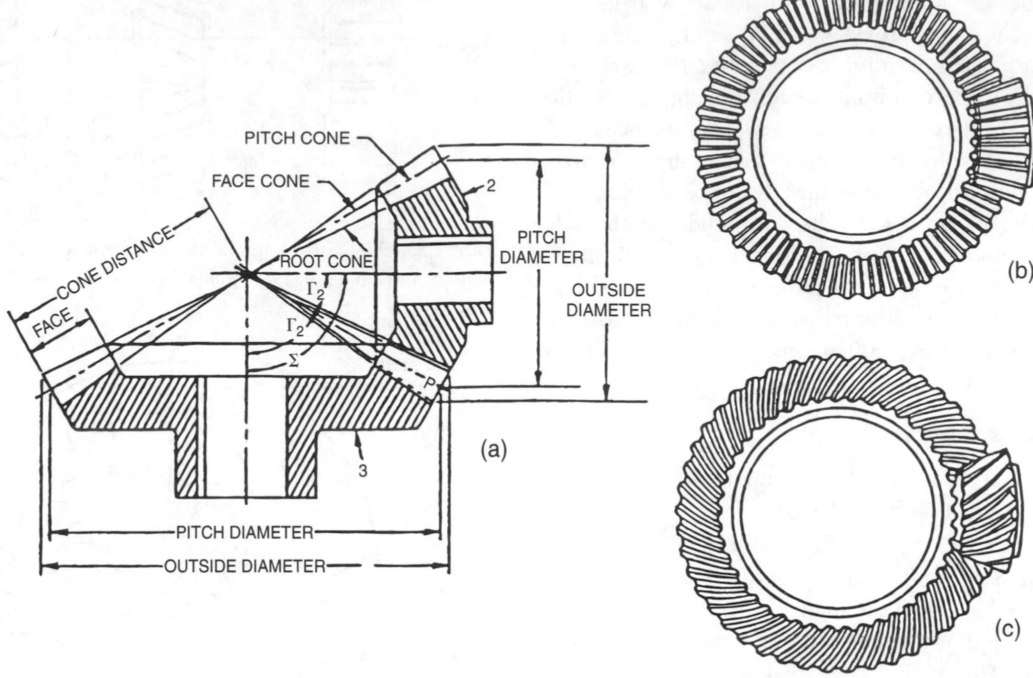

Figure 53.51 Bevel gear arrangement: (a) geometry; (b) straight and (c) spiral bevel gears. *Source*: Andeen, G. B. 1988. *Robot Design Handbook*, McGraw-Hill, New York, NY.

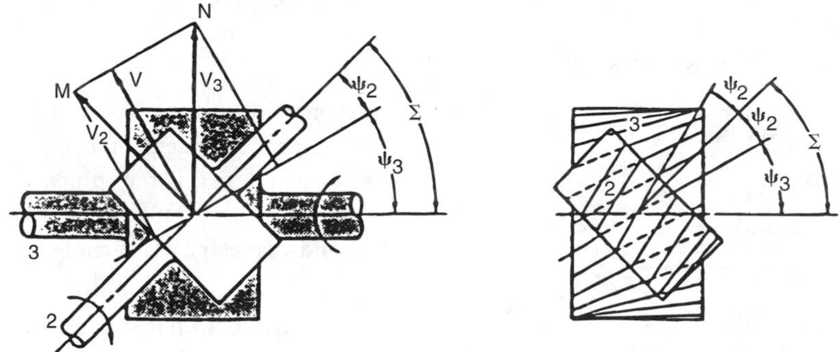

Figure 53.52 Crossed-axis helical gears. *Source*: Andeen, G. B. 1988. *Robot Design Handbook*, McGraw-Hill, New York, NY.

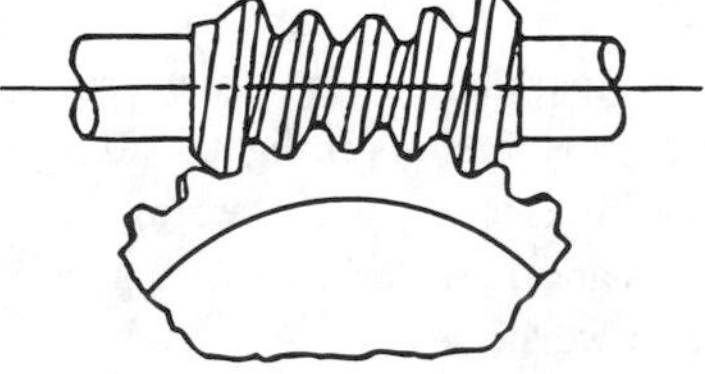

Figure 53.53 Worm gears used in nonintersecting shafts at right angles. *Source*: Andeen, G. B. 1988. *Robot Design Handbook*, McGraw-Hill, New York, NY.

There are other means of effecting motion direction transformations such as the one shown in Figure 53.56, wherein a linear motion from a hydraulic cylinder or ball screw causes rotation of the upper link about the hinged joint.

For large load capacities in a compact space, harmonic drives

Figure 53.54 Hypoid gears. *Source*: Andeen, G. B. 1988. *Robot Design Handbook*, McGraw-Hill, New York, NY.

Figure 53.55 Gear and rack arrangement. *Source*: Andeen, G. B. 1988. *Robot Design Handbook,* McGraw-Hill, New York, NY.

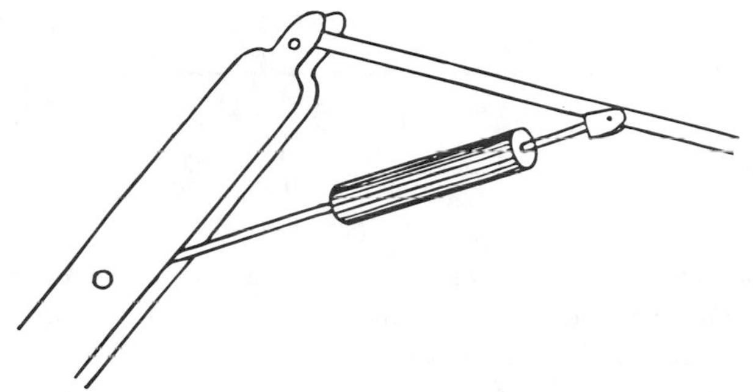

Figure 53.56 Linear to rotational motion using a hydraulic cylinder. *Source*: Andeen, G. B. 1988. *Robot Design Handbook,* McGraw-Hill, New York, NY.

are very useful in robots (de Silva, 1989; Andeen, 1988). Harmonic drives provide very high-speed reductions, e.g., off-the-shelf units available for 64:1 to 320:1 (Andeen 1988), and can therefore provide very high-torque. The principle of operation of a harmonic drives is shown in Figure 53.57. The harmonic drive consists of a rigid spline (circular spline) with internal teeth that forms the housing of the drive, an annular spline ("flexspline") that has external teeth that meshes with the internal teeth of the housing, and a wave generator that is driven by the actuator. The external radius of the annular spline is slightly smaller than the internal radius of the circular spline (housing). The annular spline is also called a flexspline, because it undergoes elastic deformation during the meshing process. The wave generator is the input side of the drive and is coupled

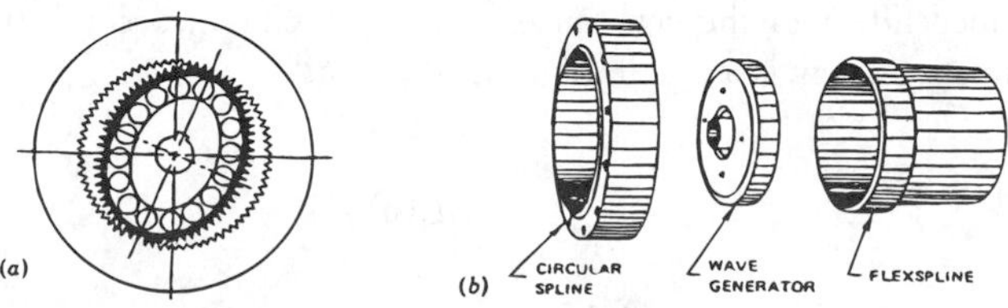

Figure 53.57 Harmonic drive assembled system (a) and its components (b). *Source*: Andeen, G. B. 1988. *Robot Design Handbook,* McGraw-Hill, New York, NY.

to the actuator; the flexspline or the circular spline are the output side of the drive and is coupled to the load. When the wave generator is rotated by an actuator, the flexspline is deformed thus engaging teeth in diametrically opposite points coincident with the major axis of the elliptical wave generator and disengaging points at the minor axis. The wave generator has a double ended cam in place that achieves the rotation of the major and minor axes of the ellipse. If the circular spline is fixed, and the load is coupled to the flexspline, rotation of the wave generator would produce a reverse rotation of the flexspline; and if the rigid spline has 202 teeth and flexspline 200 teeth, then a single revolution of the wave generator will precess the flexspline backward two teeth resulting in a velocity ratio of 100:1. If the flexspline is fixed and the load is coupled to the circular spline, then the input and output shafts rotate in the same direction. The relationship between input and output angular velocities is (Andeen, 1988):

$$\frac{\omega_{in}}{\omega_{out}} = \frac{N_o}{N_c - N_f}$$

where

N_o = number of teeth on output member (flexspline or circular spline)

N_c = number of teeth on circular spline

N_f = number of teeth on flexspline

Backlash. An inherent problem in gear transmissions is *gear backlash* which arises from the clearance between the tooth and tooth space. Backlash is the amount by which the width of a tooth space is wider than the thickness of the engaging tooth as references to the pitch circles (Andeen, 1988). Gear backlash has limited the positioning accuracy achievable by robots. With gear backlash, fine motion may not be achieved because small motions of the input side may not cause motion on the output side of the drive. Furthermore, when the teeth finally mesh, extreme stress due to impact occur and the inertia of the mechanisms may cause positioning errors. There are many ways to alleviate this problem. One way is to have gear preloading, i.e., to use an auxiliary torque motor so that the gear tends to favor one side of the gear tooth; or one can have spring loaded gears for light-duty applications.

References

Andeen, G. B., ed. 1988. *Robot Design Handbook,* McGraw-Hill, New York, NY.

de Silva, C. W. 1989. *Control Sensors and Actuators,* Prentice Hall, Englewood Cliffs, NJ.

Dote, Yauhiko and Kinoshita, Sakan. 1990. *Brushless Servomotors—Fundamentals and Applications,* Oxford Science Publications, Oxford, UK.

Electro-Craft Corp. 1980. *DC Motors, Speed Controls, Servo Systems—An Engineering Handbook,* Electro-Craft Corp., July.

Kenjo, Takashi, 1984. *Stepping Motors and Their Microprocessor Controls,* Oxford Science Publications, Oxford, UK.
SGS 1987. *SGS Motion Control Application Manual.*

53.7 Control

Fathi Ghorbel

Introduction

The motion control problem of robot manipulators consists of devising an appropriate control law that causes the joint variables to follow a prescribed trajectory making the end effector execute a specific task. Over the past decade, there has been a lot of progress in this area. Serial link (open kinematic chain) robot manipulators in particular have attracted the attention of many control/dynamics researchers which resulted in a remarkable rich and rigorous body of trajectory control results that range from the well known and simple independent-joint control laws, to more advanced robust and adaptive nonlinear techniques.

Several aspects of the motion control problem have been addressed in the literature including robustness of trajectory control strategies with respect to joint flexibility, link flexibility, and environment stiffness. Recent publication covering recent advances in this area include Spong and Vidyasagar (1989), Spong et al. (1993), and Lewis et al. (1993). In particular, the recent surveys of Ortega and Spong (1988) and Abdallah et al. (1991) review several adaptive and robust control strategies.

The purpose of this section is to describe and illustrate the use of some of the recent control strategies. An emphasis has been made to show how the structural properties of the equations of motion of robot manipulators play an important role in devising appropriate control laws. Examples have been used to illustrate different aspects of the topic and to show application of the results. Experimental results have been included to evaluate the performance of one particular control law.

Equations of Motion

Degrees of Freedom and Generalized Coordinates

Two important characteristics of a robot manipulator are the number of degrees of freedom (DOF) and the generalized coordinates. The number of DOF is defined as the number of coordinates which are used to specify the configuration of the manipulator minus the number of independent equations of constraints. It is a characteristic of the robot itself and does not depend upon the particular set of coordinates used to describe the robot's configuration. The configuration of the planar one-link manipulator shown in Figure 53.58 is fully specified by the coordinates x_A and y_A of point A in the x-y coordinate system, and the angle θ measured from the x axis to the axis passing through point A and the center of mass of the link. Let the link be constrained to rotate about an axis normal to the x-y plane passing through point A. The two equations describing the constraints are $x_A = x^*$ and $y_A = y^*$ where x^* and y^* are fixed coordinates. The planar one-link manipulator has three configuration coordinates and two constraint equations. Consequently, it has one degree of freedom.

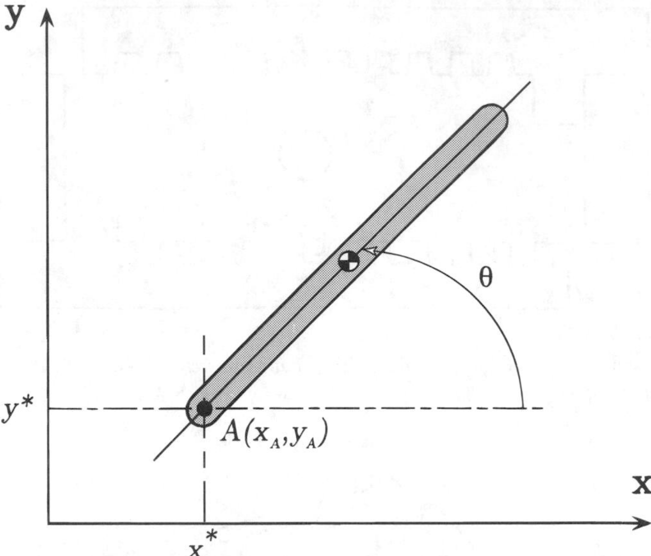

Figure 53.58 A one-link manipulator.

Any set of parameters which unambiguously represent the configuration of the manipulator can serve as a system of coordinates in a more general sense. Such parameters are known as generalized coordinates. The angle θ in Figure 53.58 serves as a generalized coordinate for the one-link manipulator.

Euler-Lagrange Equations of Robot Manipulators

A well known method to derive the equations of motion of mechanical systems is the use of the Euler-Lagrange equations (Greenwood, 1977)

$$\frac{d}{dt}\frac{\partial L}{\partial \dot{\mathbf{q}}} - \frac{\partial L}{\partial \mathbf{q}} = \mathbf{u}$$

where for an n-DOF manipulator, $\mathbf{q} = [q_1, q_2, \cdots, q_n]^T$ is the vector of generalized coordinates of the system, the Lagrangian $L = K - P$ is the sum of the kinetic energy K and the potential energy P of the manipulator, and $\mathbf{u} = [u_1, u_2, \cdots, u_n]^T$ is the vector of generalized forces driving the system.

An n-DOF robot manipulator has two special features (Spong and Vidyasagar, 1989): First, the potential energy $P = P(\mathbf{q})$ is independent of the generalized velocity vector $\dot{\mathbf{q}}$, Second, the kinetic energy is a quadratic function of $\dot{\mathbf{q}}$,

$$\begin{aligned} K &= \sum_{i=1}^{n}\sum_{j=1}^{n}\frac{1}{2}d_{ij}(\mathbf{q})\dot{q}_i\dot{q}_j \\ &\overset{\Delta}{=} \frac{1}{2}\dot{\mathbf{q}}^T D(\mathbf{q})\dot{\mathbf{q}}, \end{aligned}$$

where $D(\mathbf{q})$ is an $n \times n$ matrix called the inertia matrix. Application of the Euler-Lagrange equations to the case of n-DOF robot manipulator results in the following equation (Spong and Vidyasagar, 1989)

$$\sum_{j=1}^{n} d_{kj}(\mathbf{q})\ddot{q}_j + \sum_{i=1}^{n}\sum_{j=1}^{n} c_{ijk}(\mathbf{q})\dot{q}_i\dot{q}_j + \phi_k(\mathbf{q}) = u_k, \quad (53.14)$$

$$k = 1, 2, \ldots, n.$$

where

$$c_{ijk} = \frac{1}{2}\left\{\frac{\partial d_{kj}}{\partial q_i} + \frac{\partial d_{ki}}{\partial q_j} - \frac{\partial d_{ij}}{\partial q_k}\right\}$$

are known as the Christoffel symbols, and

$$\phi_k = \frac{\partial P}{\partial q_k}.$$

We commonly write the equations of motion (Equation 53.14) in matrix form in the following manner

$$D(\mathbf{q})\ddot{\mathbf{q}} + C(\mathbf{q}, \dot{\mathbf{q}})\dot{\mathbf{q}} + \mathbf{g}(\mathbf{q}) = \mathbf{u}, \quad (53.15)$$

where the vector $\mathbf{g}(\mathbf{q}) = [\phi_1, \phi_2, \cdots, \phi_n]^T$, and the elements c_{kj} of the matrix $C(\mathbf{q}, \dot{\mathbf{q}})$ are defined here as

$$c_{kj} = \sum_{i=1}^{n} c_{ijk}(\mathbf{q})\dot{q}_i$$

$$= \sum_{i=1}^{n} \frac{1}{2}\left\{\frac{\partial d_{kj}}{\partial q_i} + \frac{\partial d_{ki}}{\partial q_j} - \frac{\partial d_{ij}}{\partial q_k}\right\}\dot{q}_i. \quad (53.16)$$

Note that other choices of the $C(\mathbf{q},\dot{\mathbf{q}})$ matrix are possible.

EXAMPLE 53.1 (Spong and Vidyasagar, 1989):

Consider the planar two-DOF two-link manipulator depicted in Figure 53.59 where for $i = 1, 2$, m_i is the mass of link i, l_i denotes the length of link i, l_{ci} is the distance from the previous joint to the center of mass of link i, and I_i denotes the moment of inertia of link i about an axis perpendicular to the page and passing through the center of mass of link i. Let the generalized coordinates be $\mathbf{q} = [q_1, q_2]^T$. The computation of the kinetic energy is usually facilitated by setting an appropriate coordinate system to describe the kinematic relationships. For robot manipulators, the Denavit-Hartenberg convention is commonly used for this purpose (Spong and Vidyasagar, 1989). In fact, the kinetic energy for this example is given by

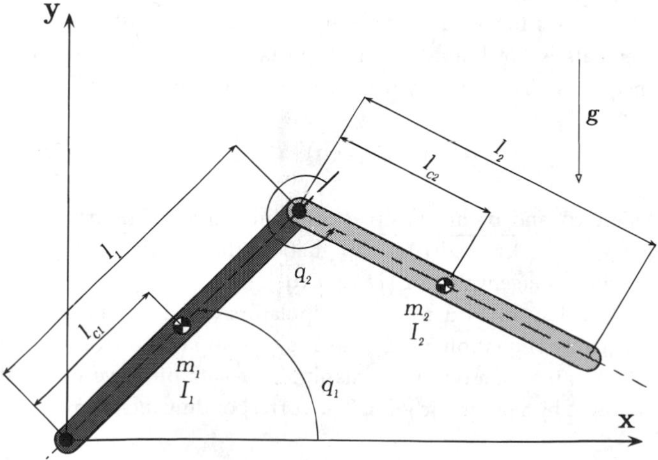

Figure 53.59 A planar two-link manipulator.

$$K = \frac{1}{2}\dot{\mathbf{q}}^T D(\mathbf{q})\dot{\mathbf{q}}$$

where the elements of $D(\mathbf{q})$, namely d_{ij}, $i, j = 1, 2$, satisfy

$$d_{11} = m_1 l_{c1}^2 + m_2(l_1^2 + l_{c2}^2 + 2l_1 l_{c2}\cos q_2) + I_1 + I_2$$

$$d_{12} = d_{21} = m_2(l_{c2}^2 + l_1 l_{c2}\cos q_2) + I_2$$

$$d_{22} = m_2 l_{c2}^2 + I_2.$$

The Christoffel symbols are therefore computed as follows:

$$c_{111} = \frac{1}{2}\frac{\partial d_{11}}{\partial q_1} = 0$$

$$c_{121} = c_{211} = \frac{1}{2}\frac{\partial d_{11}}{\partial q_2} = -m_2 l_1 l_{c2}\sin q_2$$

$$c_{221} = \frac{\partial d_{12}}{\partial q_2} - \frac{1}{2}\frac{\partial d_{22}}{\partial q_1} = -m_2 l_1 l_{c2}\sin q_2$$

$$c_{112} = \frac{\partial d_{21}}{\partial q_1} - \frac{1}{2}\frac{\partial d_{11}}{\partial q_2} = m_2 l_1 l_{c2}\sin q_2$$

$$c_{122} = c_{212} = \frac{1}{2}\frac{\partial d_{22}}{\partial q_1} = 0$$

$$c_{222} = \frac{1}{2}\frac{\partial d_{22}}{\partial q_2} = 0.$$

The potential energy of the two-link manipulator is the sum of the potential energy of each individual link so that

$$P(\mathbf{q}) = m_1 g l_{c1}\sin q_1 + m_2 g(l_1\sin q_1 + l_{c2}\sin(q_1 + q_2))$$
$$= (m_1 l_{c1} + m_2 l_1)g\sin q_1 + m_2 g l_{c2}\sin(q_1 + q_2),$$

where g is the gravitational acceleration. It follows that

$$\phi_1 = \frac{\partial P}{\partial q_1} = (m_1 l_{c1} + m_2 l_1)g\cos q_1 + m_2 g l_{c2}\cos(q_1 + q_2)$$

$$\phi_2 = \frac{\partial P}{\partial q_2} = m_2 g l_{c2}\cos(q_1 + q_2).$$

We now write the equations of motion in vector form (see Equation 53.15) using (Equation 53.16) for the computation of the matrix $C(\mathbf{q}, \dot{\mathbf{q}})$:

$$D(\mathbf{q}) = \begin{bmatrix} m_1 l_{c1}^2 + m_2(l_1^2 + l_{c2}^2 + 2l_1 l_{c2} \cos q_2) + I_1 + I_2 & m_2(l_{c2}^2 + l_1 l_{c2} \cos q_2) + I_2 \\ m_2(l_{c2}^2 + l_1 l_{c2} \cos q_2) + I_2 & m_2 l_{c2}^2 + I_2 \end{bmatrix},$$

$$C(\mathbf{q}, \dot{\mathbf{q}}) = \begin{bmatrix} -m_2 l_1 l_{c2} \sin q_2 \dot{q}_2 & -m_2 l_1 l_{c2} \sin q_2(\dot{q}_1 + \dot{q}_2) \\ m_2 l_1 l_{c2} \sin q_2 \dot{q}_1 & 0 \end{bmatrix},$$

$$\mathbf{g}(\mathbf{q}) = \begin{bmatrix} (m_1 l_{c1} + m_2 l_1) g \cos q_1 + m_2 g l_{c2} \cos(q_1 + q_2) \\ m_2 g l_{c2} \cos(q_1 + q_2) \end{bmatrix}.$$

Structural Properties of the Equations of Motion

Even though the equations of motion (Equation 53.15) of an n-DOF robot manipulator are generally very complex nonlinear differential equations, they have interesting structural properties which are exploited to facilitate control law design. We discuss four properties. First, the inertia matrix $D(\mathbf{q})$ is symmetric so that for each $\mathbf{q}$, $d_{ij}(\mathbf{q}) = d_{ji}(\mathbf{q})$, $i,j = 1,2, \cdots, n$. It is also positive definite meaning that all of the eigenvalues of $D(\mathbf{q})$ are strictly positive for any given $\mathbf{q}$, or equivalently, for each non-zero vector $\mathbf{x}$ of the same dimension as $\mathbf{q}$, the scalar $\mathbf{x}^T D(\mathbf{q})\mathbf{x}$ is strictly positive for any given $\mathbf{q}$. Consequently, the inertia matrix $D(\mathbf{q})$ and its inverse $D^{-1}(\mathbf{q})$ are bounded for any given $\mathbf{q}$. In Example 53.1, it is clear that the inertia matrix $D(q_2)$ is symmetric. It is also positive definite which can be easily verified by plotting the eigenvalues of $D(q_2)$ as q_2 varies. Second, there is an independent control input for each DOF. Third, all of the constant parameters of interest such as link masses, moments of inertia, link lengths, etc., appear in the equations of motion as coefficients of functions of the generalized coordinates. These coefficients consist of algebraic combinations of the parameters. By defining each coefficient as a new parameter θ_i, it is possible to write the equations of motion linear with respect to these parameters as follows:

$$D(\mathbf{q})\dot{\mathbf{q}} + C(\mathbf{q}, \dot{\mathbf{q}})\dot{\mathbf{q}} + \mathbf{g}(\mathbf{q}) = Y(\mathbf{q}, \dot{\mathbf{q}}, \ddot{\mathbf{q}})\boldsymbol{\Theta} = \mathbf{u} \qquad (53.17)$$

where $\mathbf{Y}(\mathbf{q}, \dot{\mathbf{q}}, \ddot{\mathbf{q}})$ is an $n \times r$ matrix of known functions, usually referred to as the regressor, and Θ is a vector of parameters of dimension r. In reference to Example 53.1, we define

$$\Theta^T = [\theta_1\ \theta_2\ \theta_3\ \theta_4\ \theta_5\ \theta_6\ \theta_7\ \theta_8\ \theta_9]$$
$$= [m_1 l_{c1}^2\ m_2 l_1^2\ m_2 l_{c2}^2\ m_2 l_1 l_{c_2}\ I_1\ I_2\ m_1 l_{c_1} g\ m_2 l_1 g\ m_2 l_{c_2} g],$$

$$\mathbf{Y}(\mathbf{q}, \dot{\mathbf{q}}, \ddot{\mathbf{q}}) = [Y_1\ Y_2],$$

$$\mathbf{Y}_1 = \begin{bmatrix} \ddot{q}_1 & \dot{q}_2 & \dot{q}_1 + \dot{q}_2 & 2\mathscr{C}_2\ddot{q}_1 + \mathscr{C}_2\ddot{q}_2 - 2\mathscr{S}_2\dot{q}_1\dot{q}_2 - \mathscr{S}_2\dot{q}_2^2 \\ 0 & 0 & \ddot{q}_1 + \ddot{q}_2 & \mathscr{S}_2\ddot{q}_1 + \mathscr{S}\mathscr{S}_2 \end{bmatrix},$$

$$\mathbf{Y}_2 = \begin{bmatrix} \ddot{q}_1 & \dot{q}_1 + \dot{q}_2 & \mathscr{C}_1 & \mathscr{C}_1 & \mathscr{C}_{12} \\ \ddot{q}_2 & \ddot{q}_2 & 0 & 0 & \mathscr{C}_{12} \end{bmatrix},$$

where $\mathscr{C}_i = \cos q_i$, $i = 1, 2$, and $\mathscr{C}_{12} = \cos(q_1 + q_2)$ and $\mathscr{S}_2 = \sin \mathbf{q}$. Note that the choice of the parameter vector Θ is not unique and the dimension of the parameter space depends on the particular choice of parameters. Fourth, define the matrix $\mathscr{N}(\mathbf{q}, \dot{\mathbf{q}}) = \dot{D}(\mathbf{q}, \dot{\mathbf{q}}) - 2\mathscr{C}(\mathbf{q}, \dot{\mathbf{q}})$. Then $\mathscr{N}$ is skew symmetric meaning that the components n_{jk} of $\mathscr{N}(\mathbf{q}, \dot{\mathbf{q}},)$ satisfy $n_{jk} = -n_{kj}$. Consequently, for any vector $\mathbf{x}$, the scalar $\mathbf{x}^T\mathscr{N}\mathbf{x} = 0$. In Example 53.1,

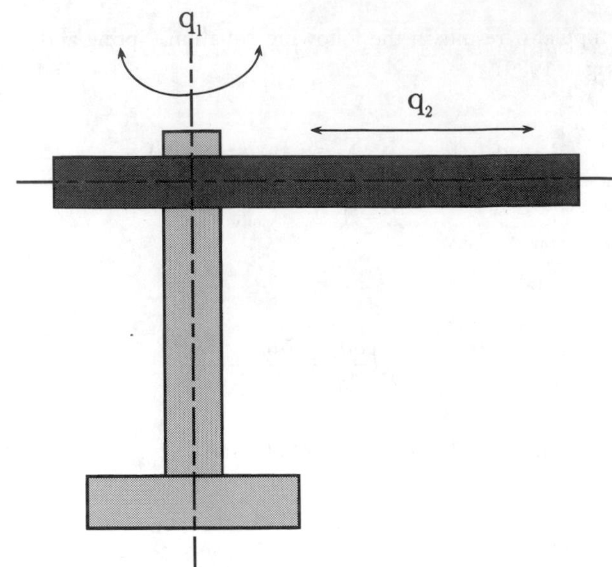

Figure 53.60 A Two DOF manipulator with unbounded inertia matrix.

$$\mathscr{N} = \dot{D} - 2\mathscr{C} = \begin{bmatrix} 0 & m_2 l_1 l_{c_2}\mathscr{S}_2(\dot{q}_2 + 2\dot{q}_1) \\ -m_2 l_1 l_{c_2}\mathscr{S}_2(\dot{q}_2 + 2\dot{q}_1) & 0 \end{bmatrix}$$

which is clearly skew symmetric.

Uniform Boundedness of the Inertia Matrix

The uniform boundedness of the inertia matrix is a typical assumption used in the design and analysis of several modern control laws for robot manipulators as will be demonstrated later. The inertia matrix $D(\mathbf{q})$ is said to be bounded if for each element $d_{ij}(\mathbf{q})$ of $D(\mathbf{q})$, there exists a constant $c < \infty$ such that $|d_{ij}(\mathbf{q})| \leq c$ for all $\mathbf{q}$. $D(\mathbf{q})$ is unbounded if at least one of its elements is not bounded. An important uniform boundedness inequality that is widely used in the robot control literature is

$$0 < \sigma_1 \leq \|D(\mathbf{q})\| \leq \sigma_2 < \infty, \qquad (53.18)$$

where σ_1 and σ_2 are positive constants, and the matrix norm $\|D(\mathbf{q})\| = \sqrt{\lambda_{\max}[D^T(\mathbf{q})D(\mathbf{q})]}$, that is, the square root of the maximum eigenvalue of $[D^T(\mathbf{q})D(\mathbf{q})]$.

It is easy to find robot manipulators which do not satisfy inequality (Equation 53.18). A simple example, shown in Figure 53.60, is that of a two-DOF manipulator with one revolute joint followed by a prismatic joint. The corresponding inertia matrix is

$$D(q_2) = \begin{bmatrix} m_2 q_2^2 + I_1 + I_2 & 0 \\ 0 & m_2 \end{bmatrix},$$

where m_2, I_1, and I_2 are the mass of link 2, inertia of link 1, and inertia of link 2, respectively. Note that $d_{11} = (m_2 q_2^2 + I_1 + I_2) \to \infty$ as $q_2 \to \infty$. Hence, there is no constant σ_2 such that (Equation 53.18) is satisfied for all possible values of q_2.

The class of robot manipulators for which the inertia matrix is uniformly bounded, hence satisfying Equation 53.18, are fully characterized in (Ghorbel et al. 1993) and is summarized here. The joint configurations of a robot manipulator which lead to a uniformly bounded inertia matrix satisfying Equation 53.18 are

1. All joints are prismatic ($\mathcal{PP} \ldots \mathcal{PP}$).
2. All joints are revolute ($\mathcal{RR} \ldots \mathcal{RR}$).
3. A series of prismatic joints followed by a series of revolute joints ($\mathcal{PP} \ldots \mathcal{PR} \ldots \mathcal{RR}$).
4. Configurations where the axis of translation of each prismatic joint j is parallel to all preceding revolute joints k.

It was shown in Ghorbel et al. (1993) that for the above class of robot manipulators, the constants σ_1 and σ_2 can be explicitly determined in terms of the Denavit–Hartenberg kinematic parameters as well as the inertia link parameters.

EXAMPLE 53.2:

The two-link manipulator shown in Figure 53.61 consists of a prismatic joint followed by a revolute joint ($\mathcal{PR}$). The kinematic (Denavit-Hartenberg) and dynamic link parameters are given in

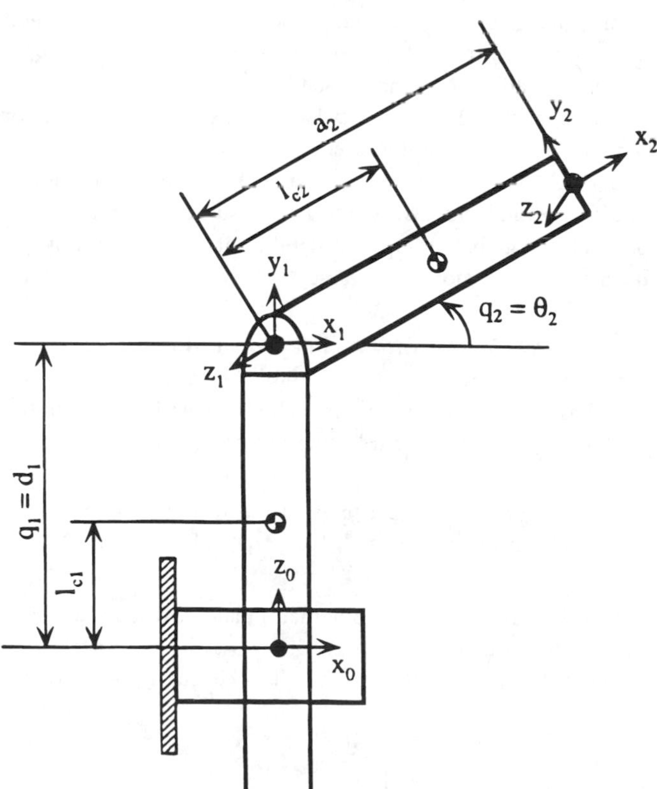

Figure 53.61 A two-link *PR* manipulator.

Table 53.1 Kinematic and Dynamic Link Parameters

Link i	Type	a_i	α_i	d_i	θ_i	q_i	m_i	l_{ci}	I_i
1	Prismatic	0	$\pi/2$	d_1	0	d_1	m_1	$d_1 - d_0$	diag $[I_{1x}\ I_{1y}\ I_{1z}]$
2	Revolute	a_2	0	0	θ_2	θ_2	m_2	l_{c2}	diag $[I_{2x}\ I_{2y}\ I_{2z}]$

Table 53.1 (see Spong and Vidyasagar (1989) for the definition of the Denavit–Hartenberg parameters a_i, α_i, d_i, and θ_i).

The inertia matrix is given by (Ghorbel et al. (1993))

$$D(\mathbf{q}) = \begin{bmatrix} m_1 + m_2 & m_2 l_{c_2} C_2 \\ m_2 l_{c_2} C_2 & m_2 l_{c_2}^2 + I_{2z} \end{bmatrix}.$$

Let $\bar{I}_2 = \max\,[I_{2x}, I_{2y}, I_{2z}]$. Then, the uniform bounds σ_1 and σ_2 of $D(\mathbf{q})$ are given by Ghorbel et al. (1993).

$$\sigma_1 = \frac{m_1(m_2 l_{c_2}^2 + I_{2z})}{m_1 + m_2(1 + l_{c_2}^2) + \bar{I}_2}$$

$$\sigma_2 = \frac{(m_1 + m_2(1 + l_{c_2}^2) + \bar{I}_2)^2 - m_1(m_2 l_{c_2}^2 + I_{2z})}{m_1 + m_2(1 + l_{c_2}^2) + \bar{I}_2}$$

For $m_1 = m_2 = 1$, $\bar{I}_2 = I_{2z} = 0.75$, $l_{c2} = 1$, we obtain $\sigma_1 = 0.47$, and $\sigma_2 = 3.28$.

Boundedness of the Derivative of the Gravity Vector

Another important boundedness inequality used in some of the control laws that are described in a later section concerns the derivative of the gravity vector. Specifically, it is typically assumed that a positive constant β exists such that

$$\left\| \frac{\partial \mathbf{g}(\mathbf{q})}{\partial \mathbf{q}} \right\| \leq \beta, \tag{53.19}$$

where the matrix norm

$$\left\| \frac{\partial \mathbf{g}(\mathbf{q})}{\partial \mathbf{q}} \right\| = \sqrt{\lambda_{\max}\left[\left[\frac{\partial \mathbf{g}(\mathbf{q})}{\partial \mathbf{q}} \right]^T \left[\frac{\partial \mathbf{g}(\mathbf{q})}{\partial \mathbf{q}} \right] \right]}.$$

It turns out that not all joint configurations insure the existence of a positive constant β such that inequality (Equation 53.19) is satisfied for all $\mathbf{q}$. A characterization of joint configurations of serial link robot manipulators for which inequality (Equation 53.19) is satisfied, and an explicit expression for β were recently proposed in Ghorbel and Gunawardana (1996) and Gunawardana and Ghorbel (1996).

Motion Control

Even though robot manipulators are very complex nonlinear systems, we show in this section how the structural properties just discussed can be exploited in a fundamental way to design control law strategies. First, we describe the design of PD-based control laws for the regulation problem, that is, the following of a constant trajectory. We consider different cases including full and simple gravity compensation, and adaptive PD control for the class of robot manipulators with bounded inertia matrix. Second, we present an adaptive passivity based control law for the tracking of time varying desired trajectories. Third, we introduce the problem of joint flexibility of the manipulator and discuss an adaptive composite control law that compensates for the joint flexibility. An experimental case study is presented.

PD Control

The results presented in this section deal with the design of PD-based control laws for regulation in which the control objective is stated as follows: Given a constant desired joint trajectory, namely $\mathbf{q}_d(t) = \bar{\mathbf{q}}_d$ and $\dot{\mathbf{q}}_d = 0$, design a PD-based control law such that $\mathbf{q}(t) \to \bar{\mathbf{q}}_d$ and $\dot{\mathbf{q}}(t) \to 0$ as $t \to \infty$.

Absence of Gravity. In the absence of gravity, i.e., when $\mathbf{g}(\mathbf{q}) = 0$, the equations of motion (Equation 53.15) become

$$\text{Plant: } D(\mathbf{q})\ddot{\mathbf{q}} + C(\mathbf{q}, \dot{\mathbf{q}})\dot{\mathbf{q}} = \mathbf{u}. \tag{53.20}$$

The two-DOF manipulator in Figure 53.59 satisfies Equation 53.20 when the x-y plane is horizontal. Consider an independent joint PD control scheme

$$\text{Control Law: } \mathbf{u} = K_p(\bar{\mathbf{q}}_d - \mathbf{q}) - K_d\dot{\mathbf{q}}, \tag{53.21}$$

where K_p and K_d are diagonal matrices with positive entries. When the control law (Equation 53.21) is used for the plant (Equation 53.20), the equilibrium $\mathbf{q} = \bar{\mathbf{q}}_d$, $\dot{\mathbf{q}} = 0$ is globally asymptotically stable, that is, for any initial conditions $\mathbf{q}_0$, $\dot{\mathbf{q}}_0$, the trajectories of the plant satisfy $\mathbf{q}(t) \to \bar{\mathbf{q}}_d$ and $\dot{\mathbf{q}}(t) \to 0$ as $t \to \infty$.

This result is classic (see for example Spong and Vidyasagas, 1989) and an outline of the proof is as follows: First, Equations 53.20 and 53.21 are combined and evaluated at steady state, i.e., at $\ddot{\mathbf{q}} = 0$, $\dot{\mathbf{q}} = 0$, giving an equilibrium point $\mathbf{q} = \bar{\mathbf{q}}_d$ and $\dot{\mathbf{q}} = 0$. To show asymptotic stability of the equilibrium point, we choose the following Lyapunov function candidate:

$$V = \frac{1}{2}\dot{\mathbf{q}}^T D\dot{\mathbf{q}} + \frac{1}{2}(\bar{\mathbf{q}}_d - \mathbf{q})^T K_p(\bar{\mathbf{q}}_d - \mathbf{q}).$$

The time derivative of V along the solution trajectories of the closed loop system (53.20)–(53.21) is (after some algebra)

$$\dot{V} = -\dot{\mathbf{q}}^T K_d\dot{\mathbf{q}} + \frac{1}{2}\dot{\mathbf{q}}^T[\dot{D} - 2C]\dot{\mathbf{q}} = -\dot{\mathbf{q}}^T K_d\dot{\mathbf{q}} \leq 0,$$

where the skew symmetry property of $\dot{D} - 2C$ was used. It follows that the equilibrium $\mathbf{q} = \bar{\mathbf{q}}_d, \dot{\mathbf{q}} = 0$ is globally stable. Invoking LaSalle's theorem (Vidyasagar, 1993), global asymptotic stability is established.

Presence of Gravity but No-Gravity Compensation. Reconsider now the general case when gravity is present so that the equations of motion are given by Equation 53.15 and choose the (PD with no-gravity compensation) control law (Equation 53.21), that is,

$$\begin{cases} \text{Plant:} & D(\mathbf{q})\ddot{\mathbf{q}} + C(\mathbf{q}, \dot{\mathbf{q}})\dot{\mathbf{q}} + \mathbf{g}(\mathbf{q}) = \mathbf{u} \\ \text{Control law:} & \mathbf{u} = K_p(\bar{\mathbf{q}}_d - \mathbf{q}) - K_d\dot{\mathbf{q}}, \end{cases}$$

where K_p and K_d are diagonal matrices of positive entries. Assuming that the resulting closed loop system is stable, it follows that at steady state, i.e., at $\ddot{\mathbf{q}} = 0$, $\dot{\mathbf{q}} = 0$, we have

$$\bar{\mathbf{q}}_d - \mathbf{q} = K_p^{-1}\mathbf{g}(\mathbf{q}). \tag{53.22}$$

Consequently, we conclude that if no gravity compensation is added to the PD controller, there is always a steady state error given by Equation 53.22 which could be made smaller by choosing high gain K_p.

PD Control with Full Gravity Compensation. Adding full gravity compensation to the previous case, we obtain

$$\begin{cases} \text{Plant:} & D(\mathbf{q})\ddot{\mathbf{q}} + C(\mathbf{q}, \dot{\mathbf{q}})\dot{\mathbf{q}} + \mathbf{g}(\mathbf{q}) = \mathbf{u} \\ \text{Control law:} & \mathbf{u} = K_p(\bar{\mathbf{q}}_d - \mathbf{q}) - K_d\dot{\mathbf{q}} + \mathbf{g}(\mathbf{q}). \end{cases}$$

The control law actually cancels the effect of the gravitational terms. This gives the same closed loop system as the no-gravity case. Consequently we conclude that the equilibrium $\mathbf{q} = \bar{\mathbf{q}}_d$, $\dot{\mathbf{q}} = 0$ is globally asymptotically stable.

Note that the PD control law with full gravity compensation requires at each instant the computation of the gravitational terms in $\mathbf{g}(\mathbf{q})$. If the system parameters in $\mathbf{g}(\mathbf{q})$ are unknown, due for example to inaccuracy in measurement or change in payload, the global asymptotic stability of the equilibrium $\mathbf{q} = \bar{\mathbf{q}}_d$, $\dot{\mathbf{q}} = 0$, is no more achievable.

PD Control with Simple Gravity Compensation. It turns out that under further assumptions on the gravity vector, global asymptotic stability is still achievable without computing $\mathbf{g}(\mathbf{q})$ at each instant. Consider the following controller

$$\begin{cases} \text{Plant:} & D(\mathbf{q})\ddot{\mathbf{q}} + C(\mathbf{q}, \dot{\mathbf{q}})\dot{\mathbf{q}} + \mathbf{g}(\mathbf{q}) = \mathbf{u} \\ \text{Control law:} & \mathbf{u} = K_p(\bar{\mathbf{q}}_d - \mathbf{q}) - K_d\dot{\mathbf{q}} + \mathbf{g}(\bar{\mathbf{q}}_d). \end{cases}$$

Consider the class of robot manipulators, discussed earlier, for which inequality (Equation 53.19) is satisfied (Ghorbel and Gunawardana, 1996), that is, there is a constant β such that

$$\left\|\frac{\partial \mathbf{g}(\mathbf{q})}{\partial \mathbf{q}}\right\| \leq \beta.$$

If the elements k_{pi} of the diagonal matrix K_p are chosen such that $k_{pi} > \beta$, $1 \leq i \leq n$, and K_d is a diagonal matrix with positive entries, then the equilibrium $\mathbf{q} = \bar{\mathbf{q}}_d, \dot{\mathbf{q}} = D$, is globally asymptotically stable, that is, $\mathbf{q}(t) \to \bar{\mathbf{q}}_d$ and $\dot{\mathbf{q}}(t) \to 0$ as $t \to \infty$ (Arimoto and Miyazeki, 1984; Korrami and Özgüner, 1988; Tomei, 1991).

Note that the difference between the full gravity compensation and the simple gravity compensation is that in the latter case the gravity term in the control law is computed at the desired position value $\bar{\mathbf{q}}_d$ only, hence it is constant and can be computed off-line. Therefore, the PD control law with simple gravity compensation doesn't require on-line model computation. Note that the result is valid only for the class of robot manipulators for which a constant β exists. Explicit expressions for β are discussed in Ghorbel and Gunawardana (1996). The proof of global asymptotic stability exploits in a fundamental way the skew symmetry of $\dot{D} - 2C$ and invokes LaSalle's Theorem (Vidyasagar, 1993).

Adaptive PD Control. For the class of robot manipulators with bounded inertia matrix for which there exist positive constants σ_1 and σ_2 satisfying Equation 53.18 (see Ghorbel et al., 1993, for full characterization of this class of robot manipulators and the computation of explicit bounds σ_1 and σ_2), an adaptive PD control law was proposed in Tomei (1991). Since the equations of motion are linear in the parameters as discussed earlier, the gravity vector can be expressed as $\mathbf{g}(\mathbf{q}) = Y_g(\mathbf{q})\Theta_g$ where the regressor matrix $Y_g(\mathbf{q})$ consists of known functions and the parameter vector Θ_g groups all the unknown parameters in $\mathbf{g}(\mathbf{q})$. Consider the following control law, and parameter update law

$$\begin{cases} \text{Plant:} & D(\mathbf{q})\ddot{\mathbf{q}} + C(\mathbf{q}, \dot{\mathbf{q}})\dot{\mathbf{q}} + \mathbf{g}(\mathbf{q}) = \mathbf{u} \\ \text{Control law:} & \mathbf{u} = K_p(\bar{\mathbf{q}}_d - \mathbf{q}) - K_d\dot{\mathbf{q}} + Y_g(\mathbf{q})\hat{\Theta}_g \\ \text{Parameter update law:} & \dot{\hat{\Theta}}_g = -\alpha Y_g^T(\mathbf{q})\left[\gamma\dot{\mathbf{q}} + \dfrac{2(\mathbf{q} - \bar{\mathbf{q}}_d)}{1 + 2(\mathbf{q} - \bar{\mathbf{q}}_d)^T(\mathbf{q} - \bar{\mathbf{q}}_d)}\right] \end{cases}$$

where α is a positive constant, K_p and K_d are constant diagonal matrices with positive entries, and γ satisfies

$$\gamma > \max\left\{\left[\frac{2\sigma_2}{\sqrt{\sigma_1\lambda_{\min[K_p]}}}\right], \left[\frac{1}{\lambda_{\min[K_d]}}\left(\frac{(\lambda_{\max[K_d]})^2}{2\lambda_{\min[K_p]}} + 4\sigma_2 + \frac{k_c}{\sqrt{2}}\right)\right]\right\}$$

where k_c is a constant that satisfies $\|C(\mathbf{q},\dot{\mathbf{q}})\| \leq k_c \|\dot{\mathbf{q}}\|$, $\lambda_{\min[K_p]}$ ($\lambda_{\min[K_d]}$) is the minimum eigenvalue of K_p (K_d) and $\lambda_{\max[K_p]}$ ($\lambda_{\max[K_d]}$) is the maximum eigenvalue of K_p (K_d). Then ($\mathbf{q}(t) - \bar{\mathbf{q}}_d$), $\dot{\mathbf{q}}(t)$, and $\hat{\Theta}_g(t)$ are bounded for $t \geq 0$. Moreover, $\mathbf{q}(t) \to \bar{\mathbf{q}}_d$ and $\dot{\mathbf{q}}(t) \to 0$ as $t \to \infty$ (Tomei 1991).

A Passivity-Based Control Law

In this section we present an adaptive passivity-based control law for the tracking problem in which the control objective is stated as follows: Given a desired joint trajectory, namely $\mathbf{q}_d(t)$, that is sufficiently many times continuously differentiable, design an adaptive control law such that $\mathbf{q}(t) \to \mathbf{q}_d(t)$ and $\dot{\mathbf{q}}(t) \to \dot{\mathbf{q}}_d(t)$ as $t \to \infty$ with all internal signals remaining bounded.

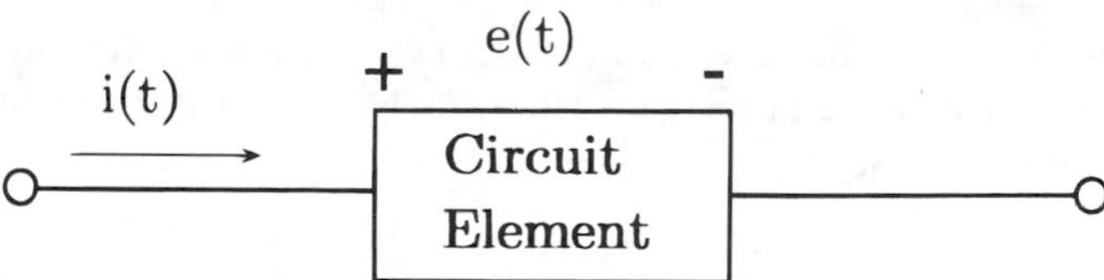

Figure 53.62 A circuit element.

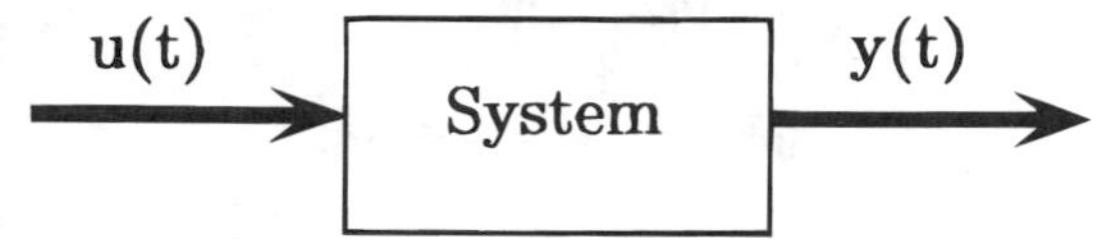

Figure 53.63 An input/output model.

Passivity. To introduce passivity, consider the circuit element shown in Figure 53.62 with positive current $i(t)$ and positive voltage $e(t)$. The power supplied to the element is $p(t) = e(t)i(t) \geq 0$. If at some instant $p(t)$ is negative, then the circuit element is supplying power to the rest of the circuit at that instance. Power is the time derivative of energy, hence the energy supplied to the circuit element over the interval time t_0 to t_1 is $\int_{t_0}^{t_1} p(t)dt = \mathscr{E}(t) - \mathscr{E}(t_0)$. Hence, the energy supplied to the circuit element at time t is $\mathscr{E}(t) = \mathscr{E}(t_0) + \int_{t_0}^{t_1} v(t)i(t)dt$. If $\mathscr{E}(t) \geq 0$ then energy is supplied to the system and we call the system *passive*. If, on the other hand, $\mathscr{E}(t) < 0$, the system supplies energy. For a mechanical system $p(t) = f(t)v(t)$ where $f(t)$ is force and $v(t)$ is velocity. It follows that $\mathscr{E}(t) = \mathscr{E}(t_0) + \int_{t_0}^{t_1} f(t)v(t)dt$. A mechanical system is passive if $\mathscr{E}(t) \geq 0$.

In a more general sense, consider the system shown in Figure 53.63 where the vector $\mathbf{u}(t)$ is the input to the system and the vector $\mathbf{y}(t)$ is the output. The system is said to be passive if $\int_0^\tau \mathbf{y}^T(t)\mathbf{u}(t)dt \geq -\gamma$ for all finite $\tau > 0$ and some finite $\gamma > 0$.

Passivity of Robot Manipulators. Consider the equations of motion of robot manipulators (Equation 53.15) which can be thought of as representing a system whose input is $\mathbf{u}$ and whose output is $\dot{\mathbf{q}}$. Then the equations of motion (Equation 53.15) define a passive mapping from the input $\mathbf{u}$ to the output $\dot{\mathbf{q}}$, that is $\int_0^\tau \dot{\mathbf{q}}^T(t)\mathbf{u}(t)dt \geq -\nu$ for some ν, and for all τ. This property is easily proved by considering the Hamiltonian of the system $\mathscr{H} = K + P$ which is the sum of the kinetic energy and the potential energy. Using the skew symmetry property of $\dot{D} - 2C$, it can easily be shown that $dH/dt = \dot{\mathbf{q}}^T\mathbf{u}$. Consequently, $\int_0^\tau \dot{\mathbf{q}}^T(t)\mathbf{u}(t)dt = H(T) - H(0) \geq -H(0)$ which proves the passivity property by setting $\nu = H(0)$.

A Passivity-Based Adaptive Control Law. Given a twice continuously differentiable reference trajectory $\mathbf{q}_d(t)$, consider the following control law

$$\mathbf{u} = \hat{D}(\mathbf{q})\mathbf{a} + \hat{C}(\mathbf{q}, \dot{\mathbf{q}})\mathbf{v} + \hat{\mathbf{g}}(\mathbf{q}) - K_D\mathbf{r}, \tag{53.23}$$

where $\hat{D}$, $\hat{J}$, $\hat{C}$ and $\hat{\mathbf{g}}$ represent the terms in Equation 53.15 with estimated values of the parameters, K_D is a diagonal matrix of positive gains,

$$\tilde{\mathbf{q}} = \mathbf{q} - \mathbf{q}_d,$$

$$\mathbf{v} = \dot{\mathbf{q}}_d - \Lambda\tilde{\mathbf{q}},$$

$$\mathbf{r} = \dot{\mathbf{q}}_d - \Lambda\tilde{\mathbf{q}},$$

$$\mathbf{r} = \dot{\mathbf{q}} - \mathbf{v} = \dot{\tilde{\mathbf{q}}} + \Lambda\tilde{\mathbf{q}},$$

$$\mathbf{a} = \dot{\mathbf{v}},$$

and Λ is a constant diagonal matrix. Substituting Equation 53.23 into Equation 53.15, and since $\dot{\mathbf{q}} = \dot{\mathbf{r}} + \mathbf{a}$ and $\dot{\mathbf{q}} = \mathbf{r} + \mathbf{v}$, we can write the combined system as

$$\begin{aligned} D\dot{\mathbf{r}} + C\mathbf{r} + K_D\mathbf{r} &= \tilde{M}\mathbf{a} + \tilde{C}\mathbf{v} + \tilde{\mathbf{g}} \\ &= Y(\mathbf{q}, \dot{\mathbf{q}}, \mathbf{v}, \mathbf{a})\tilde{\boldsymbol{\Theta}}, \end{aligned} \tag{53.24}$$

where $\tilde{\Theta} = \hat{\Theta} - \Theta$ is the parameter error. Note that the regressor function Y in Equation 53.24 does not depend on the manipulator acceleration, but only on $\mathbf{v}$ and $\mathbf{a}$, which depend on the velocity and acceleration of the reference trajectory. The parameter update law is given by

$$\dot{\hat{\boldsymbol{\Theta}}} = -\Gamma Y^T\mathbf{r} \tag{53.25}$$

where Γ is some symmetric, positive definite matrix. The control law (Equation 53.23) and the parameter update law (Equation 53.25), namely,

$$\begin{cases} \text{Plant:} & D(\mathbf{q})\ddot{\mathbf{q}} + C(\mathbf{q}, \dot{\mathbf{q}})\dot{\mathbf{q}} + \mathbf{g}(\mathbf{q}) = \mathbf{u} \\ \text{Control Law:} & \mathbf{u} = \hat{D}(\mathbf{q})\mathbf{a} + \hat{C}(\mathbf{q}, \dot{\mathbf{q}})\mathbf{v} + \hat{\mathbf{g}}(\mathbf{q}) - K_D\mathbf{r} \\ \text{Parameter Update Law:} & \dot{\hat{\Theta}} = -\Gamma Y^T\mathbf{r} \end{cases}$$

ensure the boundedness of all signals, preservation of passivity of the adaptive closed loop system, and the convergence of $\tilde{\mathbf{q}}$ and $\dot{\tilde{\mathbf{q}}}$ to zero, that is, $\mathbf{q}(t) \to \mathbf{q}_d(t)$ and $\dot{\mathbf{q}}(t) \to \dot{\mathbf{q}}_d(t)$.

This result can be proved by first showing that the mapping from input $-\mathbf{r}$ to output $Y\tilde{\Theta}$ is passive. Using expression (Equation 53.25), $\mathbf{r}^TY\tilde{\Theta} = -\dot{\tilde{\boldsymbol{\Theta}}}^{-1}\tilde{\boldsymbol{\Theta}}$, and hence,

$$\begin{aligned} -\int_0^\tau \mathbf{r}^TY\tilde{\boldsymbol{\Theta}}dt &= \int_0^\tau \dot{\tilde{\boldsymbol{\Theta}}}^T\Gamma^{-1}\tilde{\boldsymbol{\Theta}} \\ &= \frac{1}{2}\int_0^\tau \frac{d}{dt}[\tilde{\boldsymbol{\Theta}}^T\Gamma^{-1}\tilde{\boldsymbol{\Theta}}]dt \\ &= \frac{1}{2}\tilde{\boldsymbol{\Theta}}^T(\tau)\Gamma^{-1}\tilde{\boldsymbol{\Theta}}(\tau) - \frac{1}{2}\tilde{\boldsymbol{\Theta}}^T(0)\Gamma^{-1}\tilde{\boldsymbol{\Theta}}(0) \\ &\geq -\frac{1}{2}\tilde{\boldsymbol{\Theta}}^T(0)\Gamma^{-1}\tilde{\boldsymbol{\Theta}}(0) =: -\nu. \end{aligned}$$

Next we choose the Lyapunov function candidate V_1 defined by

$$\begin{aligned} V_1 &= \frac{1}{2}\mathbf{r}^TD\mathbf{r} + \tilde{\mathbf{q}}^T\Lambda^TK_D\tilde{\mathbf{q}} + \beta - \int_0^\tau \mathbf{r}^TY\tilde{\boldsymbol{\Theta}}dt \\ &= \frac{1}{2}\mathbf{r}^TD\mathbf{r} + \tilde{\mathbf{q}}^T\Lambda^TK_D\tilde{\mathbf{q}} + \frac{1}{2}\tilde{\boldsymbol{\Theta}}^T\Gamma^{-1}\tilde{\boldsymbol{\Theta}}. \end{aligned} \tag{53.26}$$

Using the skew symmetry property of $\dot{D} - 2C$, the time derivative of V_1 along the solution trajectories of Equation 53.24 can be shown to satisfy (after some algebra)

$$\dot{V}_1 = -\dot{\tilde{\mathbf{q}}}K_D\dot{\tilde{\mathbf{q}}} - \tilde{\mathbf{q}}\Lambda^TK_D\tilde{\mathbf{q}} \leq 0.$$

This proves stability, but additional arguments insure asymptotic stability (Ghorbel, 1990).

The above adaptive control law is due to Slotine and Li (1987). Note that it does not require acceleration measurement. Furthermore, the passivity property of the adaptive closed loop system is preserved which enabled us, using the Lyapunov function (Equation 53.26) along with the passivity property of the rigid robot, to conclude asymptotic stability.

Control of Flexible Joint Robot Manipulators

Joint Flexibility. A flexible joint robot model assumes that the actuators are elastically coupled to the links. The problem of joint flexibility, which arises from flexible couplings devices (harmonic drives, gear and bearing, fluid compressibility, etc.) between the rotor and the link, has been recognized as the major source of compliance in most present days manipulator designs. Experimental evidence (Sweet and Good, 1984) indicates that joint flexibility should be taken into account in both the modeling and control of manipulators if high performance is to be achieved. A model for flexible joint robot manipulators has been derived in Spong (1987) using the following assumptions:

- The kinetic energy of the rotor is due mainly to its own rotation. Equivalently, the motion of the rotor is a pure rotation with respect to an inertial frame.
- The rotor/gear inertia is symmetric about the rotor axis of rotation so that the gravitational potential of the system and also the velocity of the rotor center of mass are both independent of the rotor position.

Under the above assumptions, an n-link manipulator with revolute joints actuated by DC-electric motors and with elasticity of the joints modeled as linear torsional springs is suitably described by the following $2n$-dimensional set of differential equations (Spong, 1987):

$$D(\mathbf{q}_1)\ddot{\mathbf{q}}_1 + C(\mathbf{q}_1, \dot{\mathbf{q}}_1)\dot{\mathbf{q}}_1 + \mathbf{q}(\mathbf{q}_1) + K(\mathbf{q}_1 - \mathbf{q}_2) = 0 \tag{53.27}$$

$$J\ddot{\mathbf{q}}_2 - K(\mathbf{q}_1 - \mathbf{q}_2) = \mathbf{u}, \quad (53.28)$$

where the n-dimensional vectors $\mathbf{q}_1$ and $\mathbf{q}_2$ represent the link angles and rotor angles, respectively, $D(\mathbf{q}_1)$, $C(\mathbf{q}_1, \dot{\mathbf{q}}_1)\dot{\mathbf{q}}_1$, and $\mathbf{g}(\mathbf{q}_1)$ are as defined for the rigid model (Equation 53.15), J is a constant diagonal matrix of actuator inertias reflected to the link side of the gears, and K is a diagonal matrix representing the joint stiffness. For notational simplicity we will assume that all joint stiffness constants are the same in which case K may be taken as a scalar. In the limit as the joint stiffness K tends to infinity, we recover the n-dimensional rigid model (Spong, 1987).

$$[D(\mathbf{q}_1) + J]\ddot{\mathbf{q}}_1 + C(\mathbf{q}_1, \dot{\mathbf{q}}_1)\dot{\mathbf{q}}_1 + \mathbf{q}(\mathbf{q}_1) = \mathbf{u} \quad (53.29)$$

which is similar to the rigid model (Equation 53.15) in terms of $\mathbf{q}_1$.

It is easy to verify that of the four properties of rigid robot dynamics discussed earlier, positive definiteness $\begin{bmatrix} D(\mathbf{q}_1) & 0 \\ 0 & J \end{bmatrix}$ and uniform boundedness of its inverse, linearity in parameters, and skew symmetry of $d/dt\begin{bmatrix} D(\mathbf{q}_1) & 0 \\ 0 & J \end{bmatrix} - 2\begin{bmatrix} C & 0 \\ 0 & 0 \end{bmatrix}$ hold also in the case of flexible joint robots. The second property on the other hand fails to hold because the number of control inputs is n while there are $2n$ degrees of freedom.

It is easy to verify (even for a simple linear single-link arm) that the mapping $\mathbf{u} \rightarrow \dot{\mathbf{q}}_1$ (input $\rightarrow$ link velocity) is *not* passive. Consequently, the passivity based adaptive control law for rigid robots discussed earlier cannot be directly applied to flexible joint robots without further assumptions.

On the other hand, the dynamic Equations 53.27 and 53.28 of the flexible joint robot define a passive mapping $\mathbf{u} \rightarrow \dot{\mathbf{q}}_2$ (input $\rightarrow$ rotor velocity), i.e., $\int_0^\tau \dot{\mathbf{q}}_2^T(t)\mathbf{u}(t)dt \geq -\nu$ for some ν, and for all τ. This can be proved by defining the Hamiltonian $\mathcal{H}$ of Equations 53.27 and 53.28 which is the sum of the kinetic energy $K(\mathbf{q}, \dot{\mathbf{q}})$ and the potential energy $P(\mathbf{q})$

$$\begin{aligned} \mathcal{H} &= K + P \\ &= \frac{1}{2}\dot{\mathbf{q}}_1^T D(\mathbf{q}_1)\dot{\mathbf{q}}_1 + \frac{1}{2}\dot{\mathbf{q}}_2^T J\dot{\mathbf{q}}_2 + P_1(\mathbf{q}_1) \\ &\quad + \frac{1}{2}(\mathbf{q}_1 - \mathbf{q}_2)^T K(\mathbf{q}_1 - \mathbf{q}_2). \end{aligned}$$

It can be shown (Ghorbel, 1990) that $\int_0^\tau \dot{\mathbf{q}}_2^T \mathbf{u}dt = \int_0^\tau d\mathcal{H}/dt\, dt = \mathcal{H}(\tau) - \mathcal{H}(0) \geq -\mathcal{H}(0) =: -\nu$.

An implication of this fact is that we can directly apply the passivity-based adaptive control laws to control the rotor motion, and consider joint flexibility as unmodeled dynamics. But this would result in an accuracy problem since the control of link motion would become open loop, which is not acceptable for precision motion control.

It is clear that when the four properties discussed earlier are satisfied for rigid robots, they lead to a strong adaptive control result. On the other hand, the fact that the second property does not hold for the flexible joint case, and that the mapping $\mathbf{u} \rightarrow \mathbf{q}_1$ is not passive, greatly complicates the control problem of flexible joint robot manipulators.

Numerous approaches have been suggested in the literature for the control of flexible joint robot manipulators (Ghorbel, 1990). In this section, we only present one simple result based on singular perturbation method.

A Singular Perturbation Model For Flexible Joint Robots. We rewrite for convenience the dynamic Equations 53.27–53.28

$$D(\mathbf{q}_1)\ddot{\mathbf{q}}_1 + C(\mathbf{q}_1, \dot{\mathbf{q}}_1)\dot{\mathbf{q}}_1 + \mathbf{g}(\mathbf{q}_1) + K(\mathbf{q}_1 - \mathbf{q}_2) = 0 \quad (53.30)$$

$$J\ddot{\mathbf{q}}_2 - K(\mathbf{q}_1 - \mathbf{q}_2) = \mathbf{u}, \quad (53.31)$$

and define $\mathbf{z} := K(\mathbf{q}_2 - \mathbf{q}_1)$. The variable $\mathbf{z}$ therefore represents the torque transmitted through the joint. It is realistic to assume that the joint stiffness is large relative to other parameters in the system. We idealize the assumption of large joint stiffness by assuming that K is $O(1/\epsilon^2)$ where ϵ is a small parameter, so that we may write $K = 1/\epsilon^2 K_1$, where K_1 is $O(1)$.

The parameter ϵ can be interpreted as follows. First, it is obvious that ϵ is inversely proportional to the square root of the joint stiffness. The choice of the proportionality constant K_1 is dictated by design considerations. Roughly speaking, ϵ should be so that this proportionality constant is in the same range as other parameters (inertia, etc.) in the system. At the same time, ϵ should be small enough to ensure that the transient response of the boundary-layer system, defined in the following analysis, is sufficiently rapid.

Under the preceding assumptions, the dynamic Equations 53.30 and 53.31 are modified as follows:

$$D(\mathbf{q}_1)\ddot{\mathbf{q}}_1 + C(\mathbf{q}_1, \dot{\mathbf{q}}_1)\dot{\mathbf{q}}_1 + \mathbf{g}(\mathbf{q}_1) = \mathbf{z} \quad (53.32)$$

$$\epsilon^2 J\ddot{\mathbf{z}} + K_1\mathbf{z} = K_1(\mathbf{u} - J\ddot{\mathbf{q}}_1). \quad (53.33)$$

In this form, we clearly see how the joint force drives the rigid links, and how the link motion can excite the joint resonance. The system of Equations 53.32 and 53.33 is a singular perturbation of the rigid robot model (Equation 53.29). The link positions and velocities are the "slow" variables while the joint torques and torque rates are the "fast" variables. Equation 53.33 represents the *fast system.* When $\epsilon = 0$, which corresponds to infinite joint stiffness, Equations 53.32 and 53.33 become

$$D(\bar{\mathbf{q}}_1)\ddot{\bar{\mathbf{q}}}_1 + C(\bar{q}_1, \ddot{\bar{\mathbf{q}}}_1)\dot{\bar{\mathbf{q}}}_1 + \mathbf{g}(\bar{\mathbf{q}}_1) = \bar{\mathbf{z}} \quad (53.34)$$

$$K_1\bar{\mathbf{z}} = K_1(\mathbf{u} - J\ddot{\bar{\mathbf{q}}}_1) \quad (53.35)$$

where the overbar denotes that all the variables are computed at $\epsilon = 0$. From Equation 53.35 we obtain

$$\bar{\mathbf{z}} = \mathbf{u} - J\ddot{\bar{\mathbf{q}}}_1 \quad (53.36)$$

which, when substituted in Equation 53.34, yields the *slow reduced order system*

$$[D(\bar{\mathbf{q}}_1) + J]\ddot{\bar{\mathbf{q}}}_1 + C(\bar{\mathbf{q}}_1, \dot{\bar{\mathbf{q}}}_1)\dot{\bar{\mathbf{q}}}_1 + \mathbf{g}(\bar{\mathbf{q}}_1) = \mathbf{u} \tag{53.37}$$

which is just the rigid model (Equation 53.29) in terms of $\bar{\mathbf{q}}_1$.

Composite Control From Equation 53.33, we observe that the joint resonant modes are purely oscillatory and this, in fact, is largely the source of the problem associated with joint flexibility in robot control. The composite control approach can be explained intuitively then as follows: a fast feedback control law is first designed to damp the oscillations of the fast variables. Once the fast transients have decayed, the slow part of the system should appear nearly like the dynamics of a rigid robot, which can then be controlled using any number of techniques. The idea of *composite control,* therefore, is to set

$$\mathbf{u} = \mathbf{u}_s(\mathbf{q}_1, \dot{\mathbf{q}}_1, t) + \mathbf{u}_f(\mathbf{z}, \dot{\mathbf{z}}). \tag{53.38}$$

The term $\mathbf{u}_s$ is the *slow control,* and $\mathbf{u}_f$ is the *fast* control. Based on the previous decomposition of the flexible joint model, a reasonable choice for the fast control is

$$\mathbf{u}_f = K_v(\dot{\mathbf{q}}_1, \dot{\mathbf{q}}_2). \tag{53.39}$$

We choose K_v as a constant diagonal matrix such that $K_v = O(1/\epsilon)$, that is,

$$K_v = \frac{1}{\epsilon} K_2, \tag{53.40}$$

where K_2 is $O(1)$. Substituting the composite control (Equation 53.38) and the expression (Equation 53.40) into (Equations 53.32 and 53.33), we obtain

$$D(\mathbf{q}_1)\ddot{\mathbf{q}}_1 + C(\mathbf{q}_1, \dot{\mathbf{q}}_1)\dot{\mathbf{q}}_1 + \mathbf{g}(\mathbf{q}_1) = \mathbf{z} \tag{53.41}$$

$$\epsilon^2 J\ddot{\mathbf{z}} + \epsilon K_2\dot{\mathbf{z}} + K_1\mathbf{z} = K_1(\mathbf{u}_s - J\ddot{\mathbf{q}}_1). \tag{53.42}$$

It is important to note that the addition of the fast control (Equation 53.39) does not alter the slow system since at $\epsilon = 0$ the singularly perturbed system (Equations 53.41 and 53.42) reduces to (Equation 53.37). Thus the design of the slow control $\mathbf{u}_s$ is independent of the fast control. To control the slow reduced order system we now have two choices:

- We can use any number of techniques for the control of rigid robots to design the slow reduced order system. This will be presented briefly next. Details are found in Ghorbel (1990), Ghorbel and Spong (1990), and Ghorbel et al. (1989).
- We can use the integral manifold approach. In the integral manifold method, the slow control consists of the same rigid based component as above together with additional correction terms. This approach is not presented here. Details are found in Ghorbel and Spong (1992a,b; 1991) and Ghorbel (1990).

Using the adaptive passivity based algorithm of Slotine and Li (1987) discussed earlier for the design of the reduced order (rigid) system, the slow control and the parameter update law are therefore given by Equations 53.23 and 53.25 with $\mathbf{q} = \mathbf{q}_1$, that is,

$$\mathbf{u}_s = (\hat{D}(\mathbf{q}_1) + \hat{J})\mathbf{a} + \hat{C}(\mathbf{q}_1, \dot{\mathbf{q}}_1)\mathbf{v} + \hat{\mathbf{g}}(\mathbf{q}_1) - K_D\mathbf{r}, \tag{53.43}$$

where now

$$\tilde{\mathbf{q}}_1 = \mathbf{q}_1 - \mathbf{q}_d,$$
$$\mathbf{v} = \dot{\mathbf{q}}_d - \Lambda\tilde{\mathbf{q}}_1,$$
$$\mathbf{r} = \dot{\mathbf{q}}_1 - \mathbf{v} = \dot{\tilde{\mathbf{q}}}_1 + \Lambda\tilde{\mathbf{q}}_1,$$
$$\mathbf{a} = \dot{\mathbf{v}},$$

and

$$\dot{\hat{\mathbf{\Theta}}} = -\Gamma^{-1}Y^T\mathbf{r}. \tag{53.44}$$

Using standard results from singular perturbation theory (Kokotović et al., 1986), we may approximate the system of Equations 53.41 and 53.42 by using a quasi-steady-state system and a boundary layer system as follows. Note that at $\epsilon = 0$, Equation 53.42 reduces to 53.36 which, when substituted into 53.41 yields Equation 53.37. The latter, which represents the rigid model Equation 53.29 in terms of $\bar{\mathbf{q}}_1$, is called the quasi-steady-state system. From Tichonov's theorem (Kokotević et al., 1986), the joint force $\mathbf{z}(t)$ and the link angle $\mathbf{q}_1(t)$ satisfy

$$\mathbf{z}(t) = \bar{\mathbf{z}}(t) + \eta(\tau) + O(\epsilon)$$
$$\mathbf{q}_1(t) = \bar{\mathbf{q}}_1(t) + O(\epsilon)$$

for $t > 0$, where $\tau = t/\epsilon$ is the fast time scale, $O(\epsilon)$ denotes terms of order ϵ and higher, and η satisfies the boundary layer equation

$$J\frac{d^2\eta}{d\tau^2} + K_2\frac{d\eta}{d\tau} + K_1(I + JD(\mathbf{q}_1)^{-1})\eta = 0.$$

It follows that the flexible joint system (Equations 53.41 and 53.42) can be written up to $O(\epsilon)$ as

$$D(\mathbf{q}_1)\ddot{\mathbf{q}}_1 + C(\mathbf{q}_1, \dot{\mathbf{q}}_1)\dot{\mathbf{q}}_1 + \mathbf{g}(\mathbf{q}_1) = \mathbf{u}_s + \eta(t/\epsilon) \tag{53.45}$$

$$J\frac{d^2\eta}{d\tau^2} + K_2\frac{d\eta}{d\tau} + K_1(I + JD(\mathbf{q}_1)^{-1})\eta = 0. \tag{53.46}$$

Substituting the control law (Equation 53.43) into Equation 53.45) and using the update law (Equation 53.44), the system

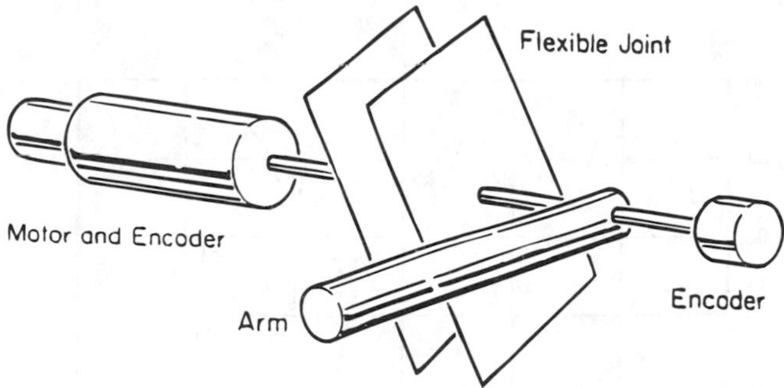

Figure 53.64 Sketch of experimental single-link flexible joint arm.

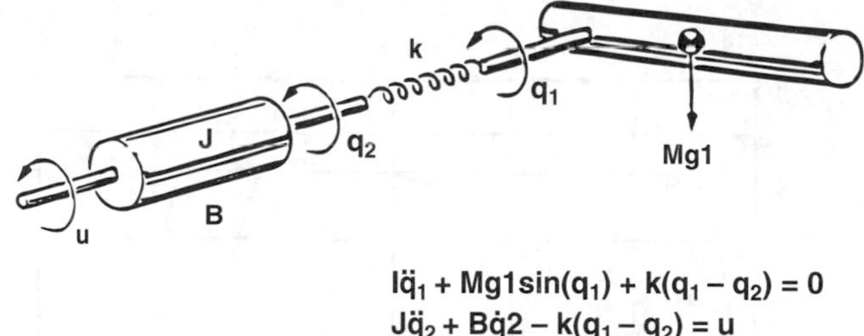

Figure 53.65 Model of experimental single-link flexible joint arm.

of Equations 53.45 and 53.46 can be written, after a little algebra, up to $O(\epsilon)$ as

$$M\dot{\mathbf{r}} + C\mathbf{r} + K_D\mathbf{r} = Y\tilde{\mathbf{\Theta}} + \eta(t/\epsilon) \quad (53.47)$$

$$J\frac{d^2\eta}{d\tau^2} + K_2\frac{d\eta}{d\tau} + K_1(I + JD(\mathbf{q}_1)^{-1})\eta = 0 \quad (53.48)$$

$$\dot{\hat{\mathbf{\Theta}}} = -\Gamma^{-1}Y^T\mathbf{r}. \quad (53.49)$$

We note that only the parameters of the rigid model are updated in this scheme. The joint stiffness and motor inertia need only be known with sufficient precision to determine K_2 to stabilize the boundary-layer system in the fast time scale. Typically these parameters can be identified with sufficient accuracy off-line and will not change with varying payloads.

The adaptive control scheme presented above has several attractive features with respect to its design and implementation. First, the overall complexity of the scheme is roughly the same as the rigid adaptive control scheme discussed earlier. Second, this control scheme exploits the two-time scale behavior in the system due to the relatively large joint stiffness, while, at the same time, it exploits the fundamental passivity properties of rigid robot dynamics. This is significant since the flexible joint robot dynamics do not possess the required passivity properties themselves as seen earlier. Third, the implementation of the full controller requires only joint position and velocity information. A detailed rigorous stability analysis of this scheme can be found in Ghorbel (1996) and Ghorbel and Spong (1990; 1989).

Example: An Experimental Single-Link Flexible Joint Arm. A single-link flexible joint arm is displayed in Figure 53.64 (Hung, 1989; Ghorbel et al., 1989). The flexible joint consists of two aluminum plates joined by extension springs. The actuator is a large DC motor connected directly to one plate. A hollow aluminum tube (1.5 inch diameter) about 18 inches long is connected to the second plate. Two incremental encoders provide feedback of the motor and link positions while velocity information is obtained by filtering the position feedback data. Parameter uncertainty is introduced by clamping payloads to the end of the arm. A payload that is approximately 40% of the nominal gravitational load of the arm is used in all experiments.

Table 53.2 Nominal Values of the Arm Parameters

Parameter	Value
Link inertia, I	0.031 $(kg - m^2)$
Rotor intertia, J	0.004 $(kg - m^2)$
Rotor friction, B_2	0.007 $(N - m - sec/rad)$
Nominal load, Mgl	0.8 $(N{-}m)$
Joint stiffness, k	7.13 $(N{-}m/rad)$

Table 53.3 Adaptive Composite Control System

Plant	$I\ddot{q}_1 + Mgl\sin(q_1) + k(q_1 - q_2) = 0$ $J\ddot{q}_2 + B_2\dot{q}_2 - k(q_1 - q_2) = u$
Control law	$u = u_s + u_f$ $u_f = K_v(\dot{q}_1 - \dot{q}_2)$ $u_s = \hat{\theta}_1 a + \hat{\theta}_2 sin(q_1) + B_2 v - K_d\tau$
Parameter update law	$\dot{\hat{\theta}}_1 = -\gamma_1 a\tau$ $\dot{\hat{\theta}}_2 = -\gamma_2 sin(q_1)\tau$

The dynamics of this system are modeled as (Figure 53.65)

$$I\ddot{q}_1 + Mgl\sin(q_1) + k(q_1 - q_2) = 0$$

$$J\ddot{q}_2 + B_2\dot{q}_2 - k(q_1 - q) = u$$

Nominal values for the arm parameters without a payload are shown in Table 53.2. This system is of the form of Equations 53.30 and 53.31, except for the nonzero damping at the joint. However, as we will see, this damping is not sufficient to stabilize the elastic oscillation of the joint. The related rigid model, obtained in the limit as $k \to \infty$, is

$$(I + J)\ddot{q}_1 + B_2\dot{q}_1 + Mgl\sin(q_1) = u_s. \quad (53.50)$$

The coefficient B_2 is known with sufficient precision and hence we can simply cancel in the control law. Only the inertia parameters are affected by varying payloads. We invoke the third property of the rigid model (Equation 53.50) (see early section) and write

Table 53.4 Desired Trajectory

$q_d(t)$	A(rad)	α (rad/sec)
$A - Ae^{-\alpha t}(1 + \alpha t)$	$\pi/2$	5

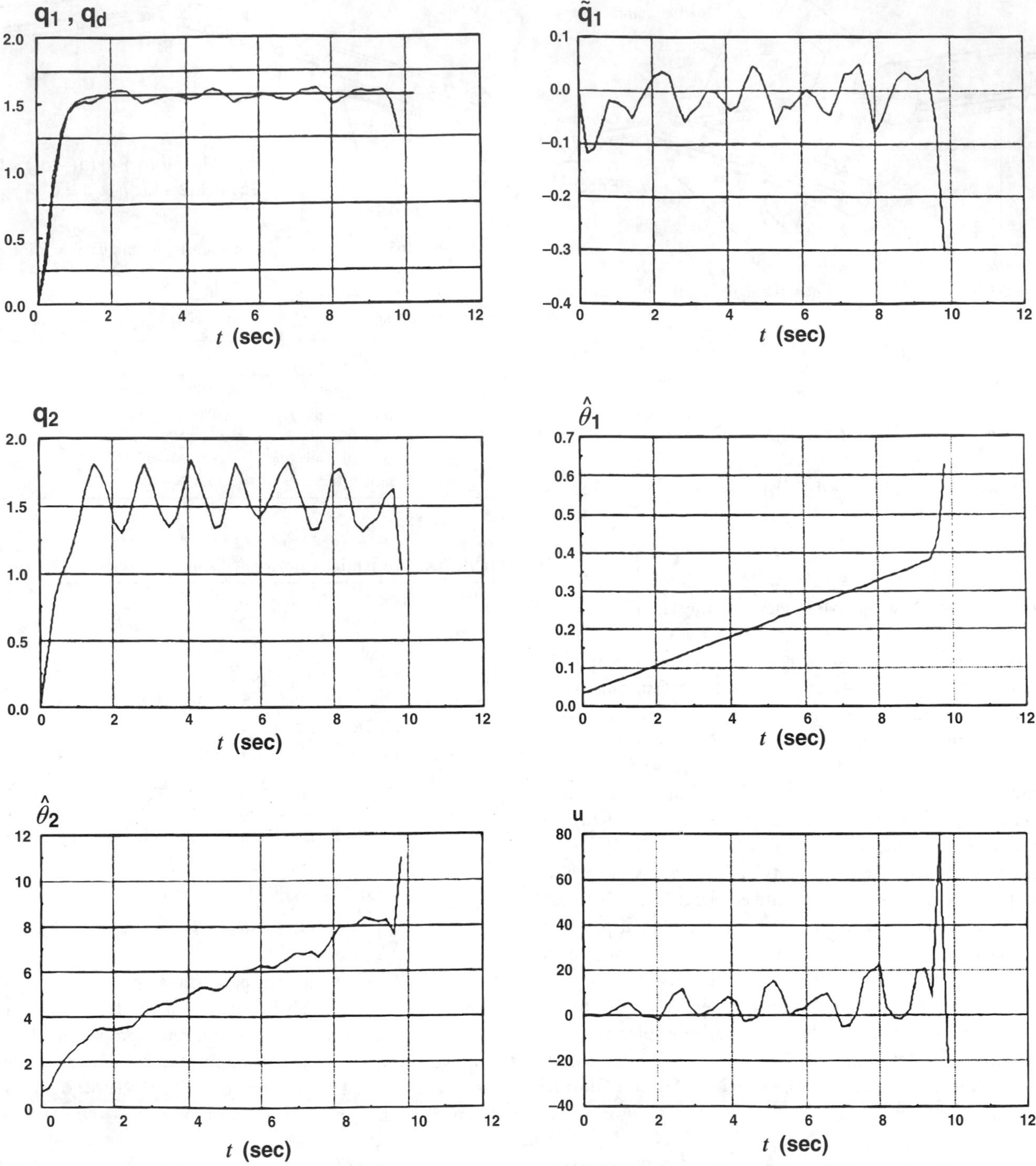

Figure 53.66 Response of flexible joint system with only adaptive rigid control.

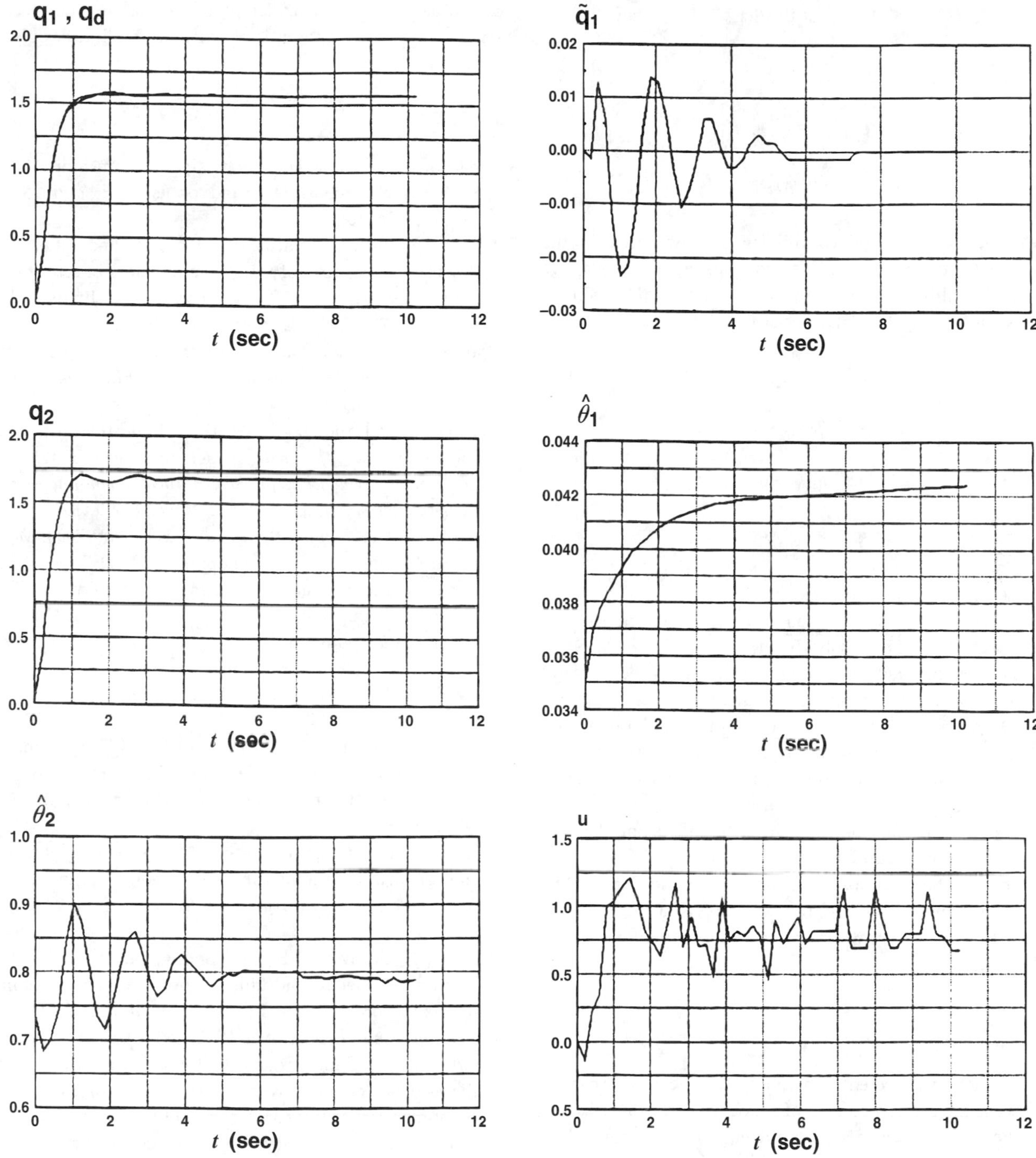

Figure 53.67 Response of flexible joint system with adaptive composite control.

$$(I + J)\ddot{q}_1 + B_2\dot{q}_1 + Mgl\sin(q_1) = Y\Theta + B_2\dot{q}_1$$
$$= u$$

$$\Theta = \begin{bmatrix}\theta_1 \\ \theta_2\end{bmatrix} = \begin{bmatrix} I + J \\ Mgl \end{bmatrix},$$
$$Y = [\ddot{q}_1 \ \sin(q_1)].$$

The design of the rigid control law is now based on the rigid model (Equation 53.50). Using the passivity based adaptive control law for this term, the complete description of the control system is shown in Table 53.3. Recall that

$$\tilde{q}_1 = q_1 - q_d,$$
$$v = \dot{q}_d - \lambda\tilde{q}_1,$$
$$r = \dot{\tilde{q}}_1 + \lambda\tilde{q}_1,$$
$$a = \ddot{q}_d - \lambda\dot{\tilde{q}}_1,$$

where $q_d(t)$ is the desired trajectory.

An experiment was run to demonstrate the effectiveness of the composite control idea (Ghorbel, 1990; Ghorbel et al., 1989). A desired trajectory $q_d(t)$ consisting of a smooth 90-degree rotation with the arm initially pointing straight down is considered (Table 53.4). First, we neglect joint flexibility and use a rigid joint based adaptive control law (i.e., $u = u_s$, $u_f = 0$) with link variables for feedback. The result is an unstable system. Figure 53.66 displays the unsatisfactory response of the link. The response of the flexible joint system is bounded only because the joint deflection is limited by mechanical stops. The effectiveness of the adaptive composite control is shown in Figure 53.67 where the response is stable and the tracking is satisfactory. The gains used in the above two runs of the experiment are shown in Table 53.5.

Conclusions

In this section, we presented a sample of the recent control methods for the motion control of robot manipulators. We discussed the structural properties of the equations of motion and showed how they are exploited to facilitate the design of control laws. PD-based control laws with full and simple gravity compensation as well as adaptive PD controllers were discussed. An adaptive control law exploiting the passivity property of robot manipulators was also presented and its robustness with respect to joint flexibility was discussed. Finally, an adaptive composite control law based on singular perturbations methods was presented and tested using an experimental one link flexible joint manipulator.

Table 53.5 Gain Values for Experiment

Gain	λ	K_d	K_v	γ_1	γ_2
Adaptive rigid control only	10	1	0	0.001	5
Adaptive composite control	10	1	0.34	0.001	5

References

Abdallah, C., Dawson, D., Dorato, P., and Jamshidi, M. 1991. Survey of robust control for rigid robots, *IEEE Control Systems,* February, 24–30.

Arimoto, S. and Miyazaki, F. 1984. Stability and robustness of PID feedback control of robot manipulators of sensory capability, *1st Int. Symp. on Robotics Research,* M. Brady and R. P. Paul, eds., 783–799, MIT Press, Boston, MA.

Ghorbel, F., Hung, J. H., and Spong, M. W. 1989. Adaptive control of flexible joint manipulators, *IEEE Control System Magazine,* 9(7):9–13.

Ghorbel, F. and Spong, M. W. 1992. Robustness of adaptive control of robots, *J. Intelligent Robotic Systems,* 6:3–15. (Also published in *Symposium on the Control of Robots and Manufacturing Systems,* Arlington, Texas, November 9, 1990.)

Ghorbel, F. 1990. *Adaptive Control of Flexible Joint Robot Manipulators: A Singular Perturbation Approach,* Ph.D. Thesis, Department of Mechanical Engineering, University of Illinois, Urbana—Champaign, IL.

Ghorbel, F. and Spong, M. W. 1991. Integral manifold corrective control of flexible joint robot manipulators: the known parameter case, *Proc. ASME Winter Annual Meeting,* December 1–6, Atlanta, GA.

Ghorbel, F. and Spong, M. W. 1992a. Adaptive integral manifold corrective control of flexible joint robot manipulators, *Proc. 1992 IEEE Int. Conf. on Robotics and Automation,* May 10–15, Nice, France.

Ghorbel, F. and Spong, M. W. 1992b. Adaptive integral manifold control of flexible joint robots with configuration invariant inertia, *Proc. 1992 American Control Conf.* June 24–26, Chicago, IL.

Ghorbel, F., Srinivasan, B., and Spong, M. W. 1993. On the positive definiteness and uniform boundedness of the inertia matrix for robot manipulators, *Proc. 32nd IEEE Conf. on Decision and Control,* December 15–17, San Antonio, TX.

Ghorbel, F. and Gunawardana, R. 1996. A Uniform bound for the jacobian of the gravitational force vector for a class of robot manipulators, *ASME Journal of Dynamic Systems, Measurement, and Control,* (to appear, accepted April 1996).

Greenwood, D. T. 1977. *Classical Dynamics,* Prentice Hall, Englewood Cliffs, NJ.

Gunawardana, R. and Ghorbel, F. 1996. The class of robot manipulators with bounded jacobian of the gravity vector, *Proc. of the IEEE International Conference on Robotics and Automation,* Minneapolis, Minnesota, April 22–28, 1996.

Hung, J. Y. 1989. *Robust Control of Flexible Joint Robot Manipulators,* Ph.D. Thesis, Department of Electrical and Computer Engineering, University of Illinois, Urbana-Champaign, IL.

Kokotović, P. V., Khalil, H. K., and O'Reilly, J. 1986. *Singular Perturbation Methods in Control: Analysis and Design,* Academic Press, London, UK.

Korrami, F. and Özgüner, U. 1988. Decentralized control of robot manipulators via state and proportional-integral feedback, *IEEE Conf. Robotics and Automation,* 1198–1203, Philadelphia, PA.

Lewis, F. L., Abdallah, C. T., and Dawson, D. M. 1993. *Control of Robot Manipulators,* Macmillan, New York, NY.

Ortega, R. and Spong, M. W. 1988. Adaptive motion control of rigid robots: a tutorial, *Proc. 27th Conf. Decision and Control,* 1575–1584, Austin, TX.

Slotine, J.-J. E., and Li, W. 1987. On the adaptive control of robot manipulators, *Int. J. Robotics Research,* 6(3):49–59.

Spong, M. W., Lewis, F. L., and Abdallah, C. T. 1993. *Robot Control, Dynamics, Motion Planning, and Analysis,* IEEE Press, New York, NY.

Spong, M. W. and Vidyasagar, M. 1989. *Robot Dynamics and Control,* John Wiley and Sons, New York, NY.

Spong, M. W. 1987. Modeling and control of elastic joint manipulators, *J. Dyn. Sys., Meas. and Control,* 109:310–319.

Sweet, L. M. and Good, M. C. 1984. Redefinition of the robot motion control problem: effects of plant dynamics, drive system constraints, and user requirements, *Proc. 23rd IEEE CDC,* Las Vegas, NV.

Tomei, P. 1991. Adaptive PD controller for robot manipulators, *IEEE Trans. Robotics and Automation,* 7(4).

Vidyasagar, M. 1993. *Nonlinear Systems Analysis,* 2nd ed., Prentice Hall, Englewood Cliffs, NJ.

53.8 Mobile Robots

Miguel A. Salichs, Luis Moreno, Diego Gachet, Arthuro de la Escalera, Juan R. Pimentel

Introduction

The objective of this section is to provide a brief overview of industrial mobile robots. Currently, the primary types of mobile robots are legged and wheeled. However, the industrial applications of the former type are limited, thus, we only address wheeled robots. The style of the presentation consists of presenting the main issues and important results corresponding to a major mobile robot topic. No attempt is made in deriving formulas or presenting related theory or algorithms. The choice of topics was dictated by the practical nature of the section. Some topics (e.g., manipulators) were not treated extensively as they are covered in traditional robots as opposed to mobile robots. Because of space considerations, some topics that some readers might expect to find in this section were not considered.

Robot Platforms

Frame

The frame is usually constructed of welded steel members with aluminum or plastic cover plates. Mobile robots designed to operate in industrial areas or indoors do not have suspension systems.

Base

Many mobile robots designed for handling industrial loads have rectangular bases. This kind of base presents more restrictions to movements than circular bases because of the different longitudinal and transversal dimensions.

Vehicles with circular bases are appropriate for cluttered environments because the circular base minimizes the size of the vehicle in a given working area. The main advantage is that space is better utilized when compared to mobile robots with rectangular bases. Because of the nature of its base, mobile robots with circular bases have maximum maneuverability.

Steering Configuration

The types of wheels used in mobile robots can be classified as follows: driving, steering, driving and steering, and passive. Depending on the physical arrangement of the wheels and the types of wheels used, it is possible to have different kinematic and dynamic characteristics, as well as different motion behaviors.

Vehicles are designed to maneuver in different ways; depending on the requirements, it is possible to select from several drive configurations (Figure 53.68). Of course the drive configurations must be compatible with the vehicle base.

Tricycle. The driving action can be through the front wheel and/or the rear wheels with the steering action being done by the front wheel. The minimum radius of curvature (defined as the ratio of linear speed over angular speed) depends on the distance between the front and rear axles. The tricycle configurations is normally used for forward motion. From the motion point of view, mobile robots with carlike configuration have similar capabilities.

Differential. Differentially steered vehicles have two

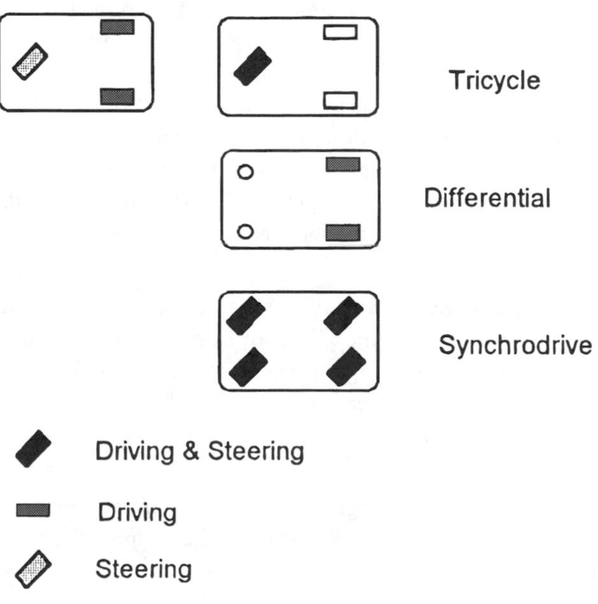

Figure 53.68 Steering configuration for mobile robots.

drive wheels which are responsible for driving and steering. The steering action is accomplished by having each wheel to rotate at different speeds. This type of configuration provides some additional advantages like forward and backward movements which can be performed at the same speed. In addition, the vehicle requires a smaller area to maneuver.

Synchrodrive. This technique, also known as all-wheel steering, has four steering wheels (two or four are also driving wheels). This configuration allows the vehicle to move transversally and a diagonal movement is also possible.

Manipulators

Depending on the mobile robot applications, there is a wide range of possible manipulation equipment on board the robot. If the application deals with material handling, the required tasks basically involve loading and unloading. For these types of applications, mobile robots have telescopic forks, conveyors or simply a platform.

Other applications require some kind of manipulation and these mobile robots include robotic manipulators on board. Such configuration needs additional control hardware for controlling the robot arm. Robot arms can be controlled in various ways: some difficult tasks need to be teleoperated and others can be done autonomously. In any case, it is necessary to collect appropriate sensory information form the environment.

Power

Mobile robots are typically powered by 24 or 48 V DC industrial batteries. Amp-hour requirements vary according to the mobile robot characteristics and the application.

There are two big groups of techniques for battery charging: opportunistic and full-cycle charging.

Opportunistic Charging. The opportunistic-charging technique involves the batteries being charged while the vehicle is waiting to perform (or is performing) a task. Time intervals during which the vehicle is stopped become possible charging intervals. The charge is done by means of a charging collector mounted on the vehicle that lowers the collector brushes to contact the floor-charger bus-plate. This charging method admits the use of maintenance free batteries.

Full-Cycle Charging. The full-cycle charging technique requires that the mobile robot be out of service and to go into a special battery-charging area. This is done when the battery is nearly discharged. There exists several charging methods for full-cycle charging.

Probe-type charging. In this technique the vehicle drives into a probe unit connected to an appropriate battery charger.

Bus-bar charging. The vehicle inserts its charging mast into a charging track connected to ac main supply and the charger is located on-board the vehicle.

Manual/Plug-in charging. The mobile robot is plugged in by an operator.

Change Out. If the mobile robot requires continuous service a battery exchange is the best solution, because it is much less expensive to have spare battery packs than to have spare vehicles.

Communications

Radio frequency communications is widely used for continuous communication. This technique allows different possibilities—the traditional one uses radio technology to establish a serial link between the ground station and the mobile robot (RS-232). Another possibility is to use the radio to create a Local Area Network. If the mobile robot is going to operate in indoor environments (hospitals, factories, etc.), it is also possible to use infrared links to communicate the mobile robot with the ground station or other vehicles.

In some situations, it can be desirable to send images to the ground station this could arise for instance if some operation requires teleoperation or operator decisions. These images can be sent through the conventional communication channels, which requires some kind of image compression, or can be communicated through a specific video communication channel.

Kinematics

To control a mobile robot, it is important to know the relationships between the actions on the actuators (e.g., linear and angular speed commands) and the movements of the robot. These relationships are used for two purposes. First to calculate the actions necessary to move the robot from one position to another. Second, to evaluate the displacements of the robot from the movements of the wheels (i.e., odometry).

In the following, we provide the kinematics equations that correspond to each of the steering configurations described in the section on robot platforms.

The driving signals are the steering wheel angle α and the linear speed of the driving wheel v, usually the front one (Figure 53.69). The kinematics equations of a point placed in the front wheel are

$$\dot{x}_F = -v\sin(\theta + \alpha)$$

$$\dot{y}_F = v\cos(\theta + \alpha)$$

$$\dot{\theta} = \frac{v}{L}\sin(\alpha)$$

The kinematics equations of a point placed in the middle of the rear axis are

$$\dot{x}_R = -v\cos(\alpha)\sin(\theta)$$

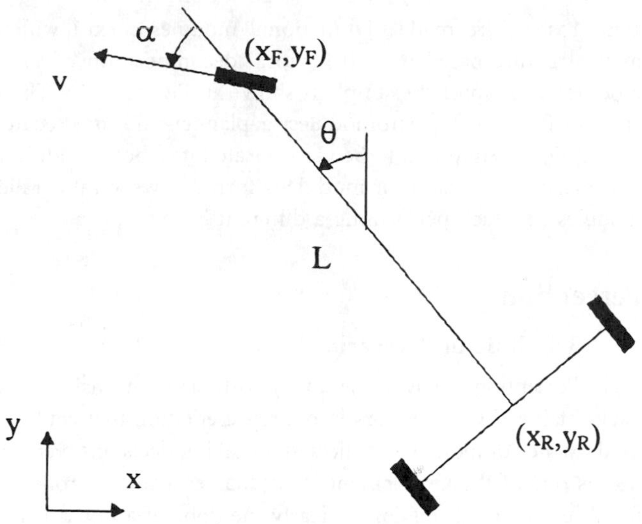

Figure 53.69 Tricycle steering configuration.

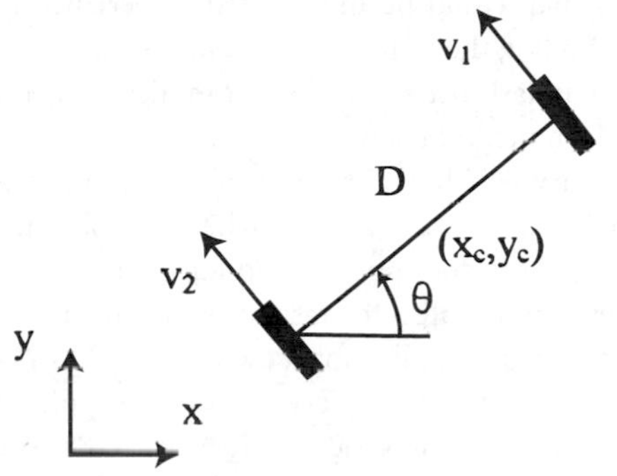

Figure 53.70 Differential steering configuration.

$$\dot{y}_R = v\cos(\alpha)\cos(\theta)$$

$$\dot{\theta} = \frac{v}{L}\sin(\alpha)$$

The driving signals are the linear speed of the two driving wheels: v_1, v_2 (Figure 53.70). The kinematics equations of a point placed in the middle of the driving wheels are:

$$\dot{x}_C = -\frac{v_1 + v_2}{2}\sin(\theta)$$

$$\dot{y}_C = \frac{v_1 + v_2}{2}\cos(\theta)$$

$$\dot{\theta} = \frac{v_1 - v_2}{D}$$

The driving signals are the linear speed v and the steering angle of the wheels α (Figure 53.71). The kinematics equations of any point in the driving structure are

$$\dot{x}_P = -v\sin(\alpha)$$

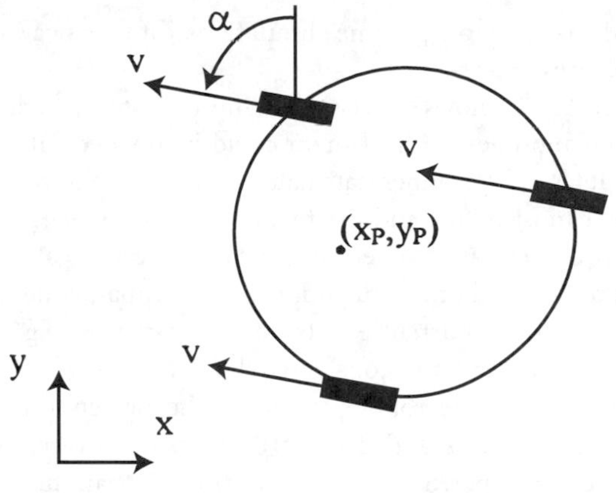

Figure 53.71 Synchrodrive steering configuration.

$$\dot{y}_P = v\cos(\alpha)$$

With this configuration, the robot orientation is independent of the actions on the driving wheels.

Control Architectures

The control system of a mobile robot must perform very complex tasks. Structuring the control system as a compact block would be impractical, because it would produce a system which is hard to design and develop. In addition, the capabilities of the final system would be very rigid, being quite difficult to modify. To reduce the complexity of the problem, the control system is decomposed following a modular approach. Two different criteria are frequently used to establish the role of the modules: functional and behavioral. In the functional approach each module perform a different function control system function (e.g., path planning). In the behavioral approach each module perform a different behavior of the robot capabilities (e.g., obstacle avoidance). An intermediate approach attempts to combine both philosophies. There is still an open discussion regarding the advantages and disadvantages of each philosophy.

There is another important consideration, related to the control architecture involving how and when the decisions about robot actions must be taken. There are two extreme alternatives: off-line decisions based mainly on the robot and environment models (planned control), and on-line decisions based mainly on real-time sensory information (reactive control). These considerations involve not only the control architecture, but also the design of the internal modules. It should be possible to design a pure, planned control system, independent of what kind of approach had been used to create the internal modules (behavioral or functional) and likewise for a pure reactive control system. But most of the control systems with functional modules work primarily in a planned way, and most of the control systems with behavioral modules work primarily in a reactive way. This is due not only to technical reasons, but also to historical reasons. Classical control architectures used functional modules and mainly planned control strategies. During the mid-80s, it was proposed to follow a behavioral and reactive approach, as a method to overcome some limitations of classical control architectures.

Figure 53.72 shows a classic functional control architecture with four modules; each of them could be divided into smaller submodules. The planner calculates in advance the routes that the robot must follow and the main actions to be performed in order to proceed with a certain mission. The navigator locates the robot and elaborates medium term commands taking into account local and current events or circumstances (e.g., mobile obstacles) that are not considered by the planner. The pilot actually controls the robot actuators. The perception module controls the sensors and process the sensory information. In many occasions, perception is not considered as an independent module, and its functions are distributed among the other modules. This architecture uses a hierarchical structure. Lower layers work on data with small levels of abstraction, and the time spent from the reception of a command to its execution is small. On the other hand, upper layers work on data with higher levels of abstraction, and the time spent from the reception of a command to its execution is longer.

Brooks was a pioneer showing how some of the problems of functional architectures could be solved by a behavioral decomposition of modules. He proposed an architecture called subsumption architecture (Brooks, 1986), in which behavioral modules work in parallel. In this way the control loop (i.e., the link from sensor information to actions) is shorter than that of classical functional architectures, where the information flows first from sensors to upper modules and then back to lower modules.

Each architecture, whether functional based or behavioral based, have advantages and disadvantages. There are some mobile robots (Salichs et al., 1993) that use a hybrid approach, where a macrostructure made of functional modules coexist with a microstructure based on a behavioral decomposition of some modules. A simplified example is shown in Figure 53.73. There are two functional macromodules: a planner and an executor (perception is supposed to be incorporated into both modules), but internally the executor module is formed by several parallel modules (B_i) each performing a different behavior.

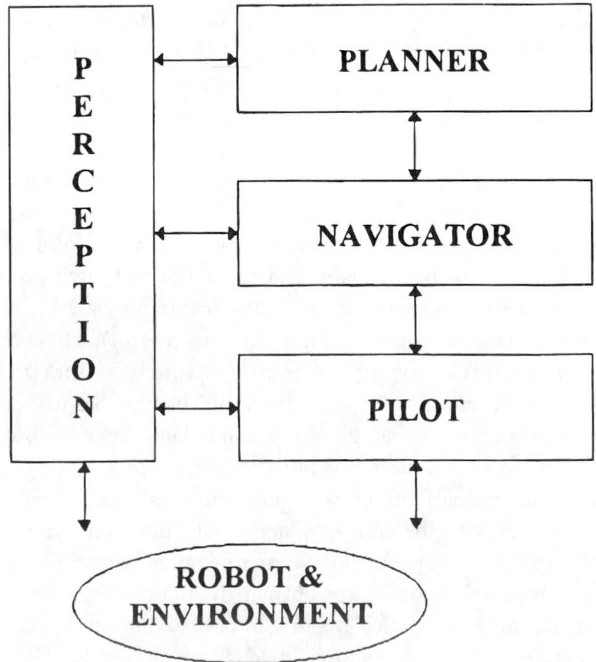

Figure 53.72 Control architecture with functional modules.

Perception

The Role of Perception

Perception involves a set of algorithms that transform the data obtained by the sensors into a representation that could be used by the components of the system taking decisions. Perception is one of the key elements that enables a mobile robot to fulfill its entrusted missions. Clearly the door of a house is not the most important part of a building, but what good is a house without doors? Think of being in a dark room in an unknown location or known with an approximation of three meters. Although a map could be memorized, experience tells us that the task of finding the switch to turn on the light could be quite laborious. The task becomes more complicated if the obstacles in the environment (tables, chairs, open windows) are not in their customary positions. The example continues being valid if it is referred to mobile robots even when the robot had a perfect control system if it does not have an adequate sensorial system, it would never accomplish the task for which the robot was built.

Regardless of the application task or how it is performed, a mobile robot has to be displaced from one point to another within an environment to accomplish the task. Regarding its environment, the mobile robot asks itself the following questions: where I am? and what surrounds me? Questions that any perceptual system must be able to answer with information on its location and information regarding possible obstacles (static or dynamic).

Perception in robotics has been perceived traditionally as a passive task, but recently researchers have arrived at the conclusion that it is necessary for it to be an active task; therefore, the

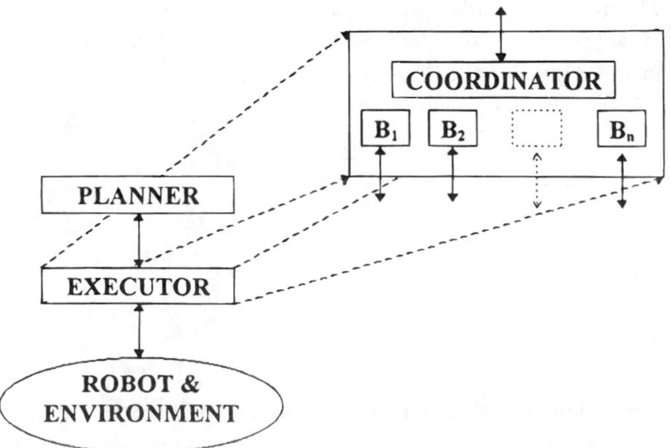

Figure 53.73 Control architecture with functional and behavioral modules.

perception tasks are decided by a planning system with explicit control of the robot's sensorial system. This has been done to prove that although the sensorial interpretation and the model of the environment are something fundamental for a robot that operates in the real world, the perception in robotics is, still, one of the weakest components in the current systems.

The characteristics that a designer should analyze when designing the sensorial system are size, power-consumption, simplicity, redundancy, capacity of operating in real time, capacity to detect all sorts of objects in the environment, resolution, accuracy, maximum and minimal effective distance, and angle of view. As one can imagine, the ideal sensor with optimum parameters for these characteristics does not exist. Therefore, a compromise between virtues and defects has to be made to choose the sensor depending upon the specific application for which the mobile robot is designed.

As far as the sort of sensors that can be used, the most commons are

- Proximity sensors: ultrasound, laser.
- Systems based on artificial vision: monocular, stereo, structured light.

And as far as the model of the sensorial information we have the following:

- Modeling according to occupancy maps.
- Modeling according to geometric characteristics.

Sensors

Ultrasound Sensors. Ultrasound sensors were initially developed for submarines. Their first use in land environments was automatic focus for cameras. The techniques were quickly applied to robot perception to obtain the distance from the robot to the various obstacles in the environment. The operation consists of emitting a pulse of a given frequency transmitted at the speed of sound and measuring of the time (designated flight time) elapsed until the possible echo is captured. For practical reasons, it is considered that the echo has been produced if a given percent of the emitted energy has been reflected by the object.

The advantages of ultrasound sensors are operational simplicity and price. Therefore, one can assemble a set of sensors to cover a large surface at a relatively low price. The usual physical layout of the sensors is to form a ring about the robot. The factors to consider when designing a perceptual system based on sensors of this type are the maximum and minimal distance that covers the sensor and the dispersion angle of the wave.

Ultrasound sensors have the following disadvantages:

- The spread speed of the sound is not constant, depending on various factors as the temperature and the dampness of the air.
- The echo amplitudes decrease inversely proportional to the fourth power of the distance to the object producing the echo. This causes the gain for the received wave to grow according to the elapsed time.
- The resolution due to the dispersion angle. For example, for an angle of 30° and an obstacle within 5 meters of the robot, the resolution of the object is about 2.7 meters.
- The problems associated in a multirobot system. If there are several robots working in the same environment, there would be interference among the sensors of one robot and the sensors from the other robots.
- Synchronization problems. If the robot has a ring of ultrasound sensors, necessary to cover the whole environment, they cannot all be fired at the same time, since the robot could not decide which sensor corresponds to which echo. Because of this, the sensors at opposed positions in the ring are fired simultaneously which results in a smaller sampling frequency.
- The dependency on the geometric form of the surface of objects producing the echo. Reflected energy is not the same on flat walls as it is on waved walls or corners.
- The dependency of the acoustic characteristics and size of the reflected object. There are several materials that absorb the ultrasound waves to produce a very attenuated response, or even worse, no response.
- The incidence angle. Depending on the orientation of the object with respect to the beam, the amount of reflected wave can be very small. Usually objects with a relative angle less than 15° are not detected.

The above objections are not intended to dismiss the use of ultrasound sensors in a perceptual system. They are the most economic of all, but they cannot be used as the only perceptual element. The ideal case is to have several ultrasound sensors to obtain approximate information about the environment and other sensors that obtain, with greater precision, more detailed information of the region of interest at that moment: the direction in which the robot is going to be moved, the goal zone, the path, etc. There is currently a great deal of research on combining the information coming from sensors of various types, an area called *sensor fusion.*

Lasers Sensors. Laser sensors provide distances to objects in front of the robot. These systems are much more accurate than ultrasound systems and they are primarily used in outdoor environments. The first laser used for the perception of a mobile robot was in 1972 for the Mars Rover developed at the JPL. It took 40 seconds to obtain an image of 64×64 points. Currently, the ALIS (Advanced Laser Imaging System) developed by the LETI (France) provides four images per second with a resolution of 120×150 pixels.

Depending upon the way in which the wave is transmitted, laser sensors are classified as pulse detection or amplitude modulation. In the former, the distance is obtained from the time elapsed between the emission and the receipt of a pulse. In the latter, the lasers emit an amplitude modulated wave, and the distance depends on the difference of the phase between the emitted and the received wave. Regarding the placing of sensors, there are two conventional structures. In the first case, the laser is on the ceiling of the vehicle on a revolving platform. In the second case, the laser is put on the front of the vehicle and, in conjunction with a mirror system, vertical and horizontal is performed obtaining the three-dimensional information of the environment.

The greatest drawback for using laser sensors is economical. However, they are beginning to be used by different groups for the perception in outdoor environments.

Vision Sensors. Another sort of perception is one based on computer vision. The proposed solutions are

- Stereo vision, which as in animals, uses the information provided by two cameras to obtain a three-dimensional information. The equations that derive the distance with respect to the points from each camera are simple. However, to find what point of an image corresponds to one of the second is not trivial. Several restrictions are used to limit the search space. For example, in the previous case, the coordinate of the points must be the same, but the computation time is high for applications in real time as in the case of the mobile robots.
- Analogous to the case of stereo vision is using just one camera where the movement of objects is actually used to obtain the information from the environment. This is done by obtaining the correspondence among some characteristics (corners, edges) that correspond to the projection of the same point in several images, or obtaining the "optical flow" or variation of the gray levels. The correspondence method gives results with low resolution and accuracy while the optical flow method has a higher resolution and accuracy. Because the optical flow method considers all of the image and not simply some characteristics, the computational cost is also high.
- Structured light. A laser line is projected and by triangulation with the points perceived by the camera the three-dimensional or two-dimensional information of the environment is obtained. It has the great advantage of fast speed and precision but the drawback of giving just distances to objects which are found on the same plane formed by the laser tip and the laser beam.

Modeling

As noted previously, the information provided by the sensors has to be interpreted in a way that a sufficiently structured knowledge of the space that surrounds the robot is obtained by the robot. Structuring this external information is accomplished through the construction of a model of the environment. This model is used to make decisions and to fuse the new information that the robot receives from its sensors, either from the same sensor in several time instants or from several sensors of the same or different type.

The two main approaches to the modeling problem are

Occupancy grids where the world is represented as a grid that represents the spatial information. This method is based on the use of an occupation map of the environment.

Geometric modeling which uses the geometric characteristics observed to compare and integrate observations, updating the model of the world with the location of the objects.

The choice of method is determined by the specific environment in which the mobile robot will work. Thus, in structured environments, such as those designed by humans, where it is easy to obtain the parameters of the objects in a geometric way the geometric modeling is preferred; in partial or unstructured environments, as a forest, the occupancy grid is preferred.

Range Sensor Model. For occupation maps, the sensor model is based on probability theory. If the sensor is ideal, the occupation probability could be interpreted in the following manner: in the previous cells to the distance r, the occupation probability is zero, in the cell located at a distance r the occupation probability is 1, and in the cells behind r the probability is 0.5, because the sensor does not give information. However, as sensors are never ideal, it is assumed that the model has a Gaussian distribution for the observations defined as a function of probability density p(z|r), which relates the reading z obtained by the sensor with the value of the actual distance r of an object to the sensor (Figure 53.74). This density function will be used to find the probability of the state of the cell in the occupancy map p[s (x) = occ|r](x). The probability p[s (x) = occ|r](x) indicates the probability that a cell is occupied by an object given that we have a sensor reading with a value r.

Occupancy Grids. In this model, the exact shape of the environment does not need to be known, which is the approach of the geometric modeling, but rather what parts of the environment are free and what parts are occupied by obstacles. There are two approaches for modeling the environment: cells or octrees.

In the cell model the world is a square grid, and the algorithm has to find which cells are occupied and which are free. The

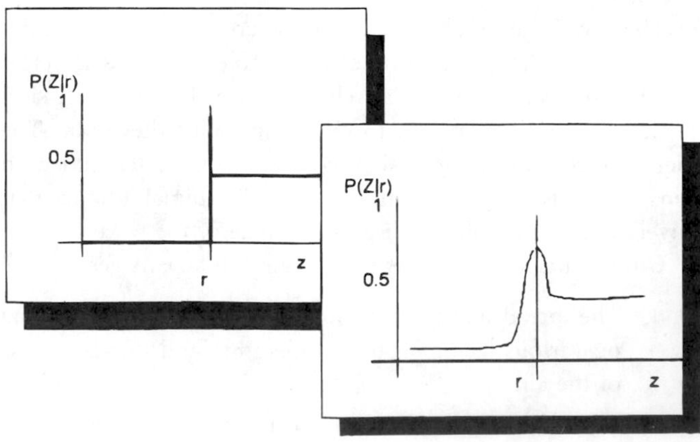

Figure 53.74 Sensor model.

advantage is its simplicity and its disadvantage is its low resolution. Since all the cells have the same dimensions, in some zones the resolution will be insufficient and redundant in others.

The quadtree (in 2D environments) or octree model (in 3D environments) avoids this inconvenience. Initially, the environment is considered as only one cell. If the surface covered by the cell is free or occupied, it is left as it is. But if part of that surface is free and part of it is occupied, then the surface is split into four cells. This process is continued by checking the previous condition again and splitting the surface into four other cells if necessary. The end condition is that all the cells represent occupied or free space, or that the desired resolution has been reached. A fine resolution is obtained at the edges of the obstacles. Its inconvenience is the need of good knowledge of the environment beforehand, which is not possible most of the time in mobile robot applications.

Every cell keeps some information of its state and the spatial information. This method developed by Moravec and Elfes combines the occupation probability with the spatial uncertainty. Each cell stores an estimate of the occupied probability. The uncertainty regarding the presence of objects surrounding a robot is represented by a distribution of probabilities within an occupancy map (Elfes, 1987).

This process requires three steps: the determination of the cells that should be updated, calculation of the current probability, and probability update.

Determination of the cells that should be updated. This is done based on the sensor reading and on the position of the sensor. When deciding what cells should be updated one must take into account a previous decision regarding what will be the reference center. There are two possibilities for the reference center: a set of fix cells in space or the center of the robot. In the former, we will have a map of cells fixed in space with the robot moving through it. In the latter, we will have a map of cells connected to the robot that is always in the center. The election between one and the other is determined by the application: if the environment in which the mobile robot is going to move is small, the fixed map will be chosen; however, if the environment is wide a connected map will be chosen since the interest is in the environment that surrounds the robot. Obviously, there can be information loss when the robot enters new places, but memory and computational cost is reduced. Therefore, the procedure for determining which cells should be updated consists of two steps. First, the sensor measurements and the position of the robot are captured through the sensory system. Second, if the displacement of the robot in some of the robot's two axes has gone beyond the dimensions of a cell, the map is displaced accordingly, initializing the new cells (if the robot is the center of reference). All connected cells between the robot and the obstacles which are detected by the sensors will be marked so that their probabilities can be updated.

Probability estimation. A given cell could be free (of obstacles), occupied, or unknown (i.e., partially occupied). Accordingly, one could assign cell occupancy probabilities of minimum (close to 0), maximum (close to 1), and intermediate (close to 0.5) values, respectively. The intermediate value is referred to as an indeterminate or uncertain value which does not mean that there is no information on the state of the cell. Obviously at the beginning, the occupancy probability of all the cells takes the intermediate value, but if a cell that had a high value of occupancy probability (because we have detected an obstacle in that cell) the probability value begins to reduce, because the obstacle moves; the decrease of the probability will be gradual, and at one instant its value will be uncertain. To calculate the current probability of the cell we use the model of the sensor described above that gives for every cell a probability value.

Updating the cells. The last step is the integration of the sensorial information, which is commonly accomplished through Bayes' theorem. In this recursive formulation, the previous estimate of the state of the cell is obtained directly from the occupancy map, and the value in the new period of time from the sensor model. One of the consequences of this method is that it behaves as a low-pass filter. Thus, for example, if an obstacle is detected n times, we should detect its absence n more times before returning to the indeterminate state. Since probability values are bounded, by using this method we can detect rapid changes in the environment. (Figure 53.75)

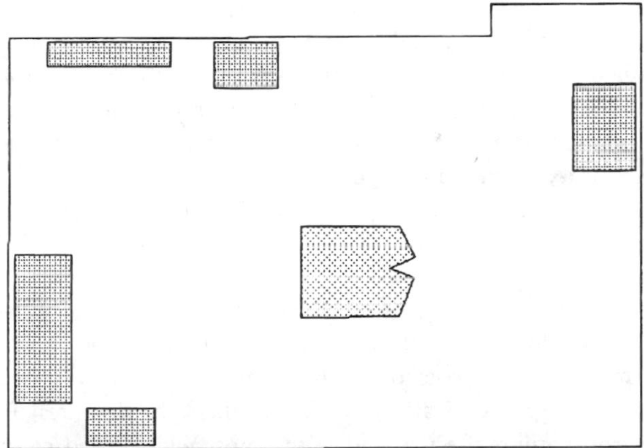

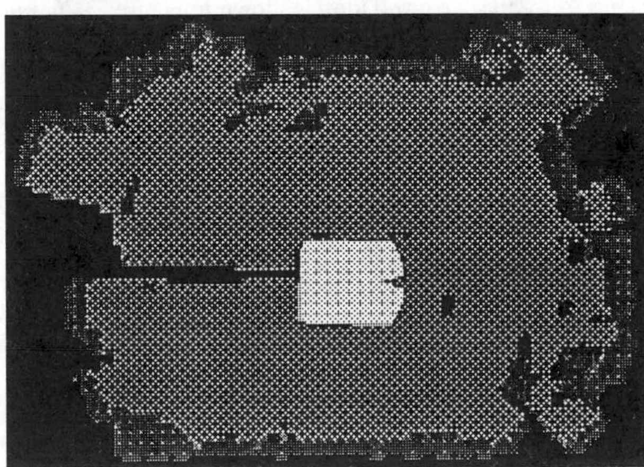

Figure 53.75 Occupancy map.

Geometric Modeling. Geometric modeling assumes *a priori* the geometric models of the objects that the sensors will perceive (lines, curves, circumferences, etc.). The goal is to obtain the parameters that define those models. The use of geometric modeling results from the introduction of artificial landmarks, such as circumferences, bar codes, etc., that have to be perceived by the robot. The use of geometric modeling also results from environments which are man made and there are several geometric forms present in a natural form (e.g., circular obstacles). Thus, for example, there are well known results for extracting lines from the information coming from revolving sensors (laser or ultrasounds) and from several ultrasonic sensors simultaneously. Similar results are available for vision sensors. Since an image is a two-dimensional projection of three-dimensional objects, two or more cameras, or the movement of one camera, are used to obtain several images corresponding to objects in different places. Once the lines are obtained, they are adapted to the previous models: walls, doors, windows, closets, cases, etc.

Control of Mobile Robots

Normally, the control of the movement of a mobile robot involves generation of commands for speed and orientation of the wheels depending on the task to be performed. In the classical control architectures, these commands are used to follow the path generated by the planner and to avoid unexpected obstacles. In the case of reactive navigation, there are no explicit commands to actuators. Instead, several small control modules are activated sequentially or in parallel. These modules provide speed and orientation commands for the actuators. In both cases, the most relevant techniques used to control the movements of the robot are the classical digital controllers (P, PI, PID), heuristic rules, fuzzy control, neural networks, and genetic algorithms.

Traditional Control

For the movement of a mobile robot, the classical control of the actuators is based on the implementation of PID (proportional, integral, derivative) type controllers. In this area, PID control is quite effective and is the most widely used control strategy. A feedback control loop is shown in Figure 53.76. where y denotes controlled output, u is the manipulating input, e is the error between the setpoint and the controlled output.

A standard continuous-time PID controller in Laplace form is given as:

$$U(s) = K\left(1 + \frac{1}{T_I s} + \frac{T_D s}{1 + T_D s/N}\right)E(s)$$

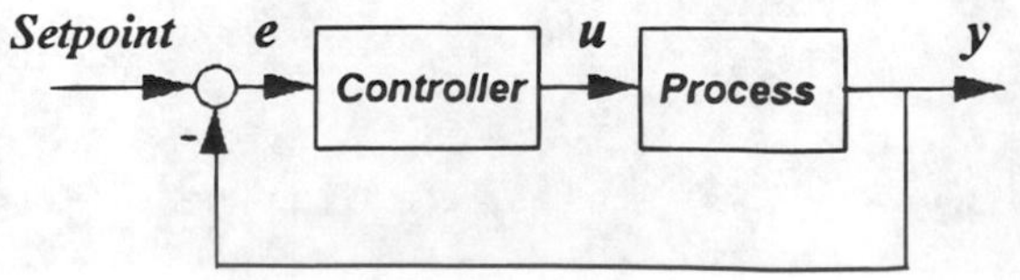

Figure 53.76 Feedback control loop.

where $U(s)$ is the process manipulating input, $E(s)$ is the error signal, K is the proportional gain, T_I is the integral time, and T_D is the derivative time.

When we wish to move the mobile robot from one point to another, the setpoint may be, for example, the *desired orientation*. The control system input may be the *curvature* (angular speed divided by linear speed, w/v) and the output is the *current orientation* (assuming for example constant linear speed) of the robot with respect to the goal point. The current output is provided by internal sensors of the robot, usually optical encoders (i.e., odometry). This approach is shown in Figure 53.77.

Figure 53.78 shows the result of this kind of control moving the robot from one point to another. The robot starts to move with a predefined orientation and location, in addition the coordinates of the goal point are known.

Depending on the task performed by the mobile robot, this control schema may be more complex and may include feedback by different kind of sensors (ultrasonic, laser, vision, etc).

Fuzzy Control

Another class of control paradigm is also used to control mobile robots is fuzzy control. In this case, the most common approach is to replace the classical PID controller with a set of linguistic rules. Suppose we need to follow a given path (sequence of pairs x,y). One implementation of the control for this task may use the angle error e (the angle between the vehicle axis and the line formed by the robot's rear axis middle point and the next path point), and the derivative of this angle ce as input variables. The output variables are the linear speed v and the curvature u of the mobile robot.

The fuzzy controller in this case attempts to minimize the angle error. When this angle error is zero, the mobile robot is going directly towards the destination point. Once the vehicle reaches this point, the next point is considered as a destination point. The process is continued until the last point of the path is reached.

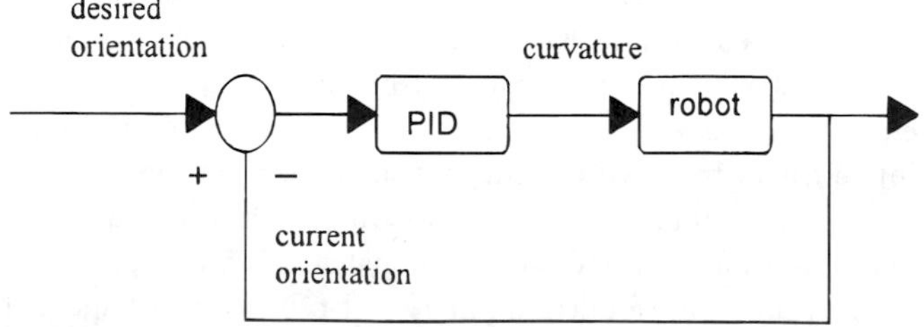

Figure 53.77 Classical control for the movement of a mobile robot.

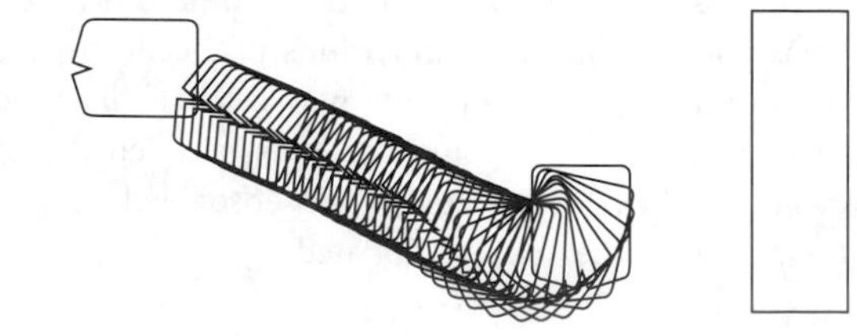

Figure 53.78 Control of the movement between two points with a PID controller.

The path is reasonably followed with linguistic rules of the form:

if (*e* small & *ce* small) then (*v* high & *u* small)
if (*e* small & *ce* medium) then (*v* medium & *u* small)
if (*e* small & *ce* big) then (*v* low & *u* small)
if (*e* medium & *ce* small) then (*v* medium & *u* medium)
if (*e* medium & *ce* medium) then (*v* medium & *u* small), etc.

Neural Network Control

The most recent approach to control is the use of artificial neural networks (ANN). The architecture of neural networks is defined by basic processing elements and the way they are interconnected. The basic element of a neural network is called a neuron, a node or a processing element. A basic model of a neuron consists of a weighted summer, a linear dynamic transfer function (in some cases), and a static nonlinear function. The backpropagation neural network based on the generalized delta learning rule has been widely used but with some disadvantages because the network first must learn the model of the plant. Recently some on-line learning algorithms for neural networks have been used to control mobile robots (Gachet et al., 1993).

Navigation

Control and navigation are important components of the overall architecture of intelligent, autonomous, mobile robots. Control and navigation involves the process of driving the robot from one point to another in a safe (obstacle avoiding) way using information provided by the sensory system. In this section, we review the two major approaches to this problem, classical and reactive/reflexive navigation. Perhaps one of the most important aspects of the navigational problem is the detection and avoidance of unexpected obstacles. This is essential for safe navigation and it is independent of the technique used to control the mobile robot.

Obstacle Detection and Avoidance

Suppose that we have a mobile robot moving along a predefined path or simply moving in a certain direction, and there is an obstacle blocking the movement of the robot. It is important to detect and avoid the obstacle using the robot's perception system. Ultrasonic sensing, computer vision, and laser rangefinders have been applied to this problem with varying degrees of success.

Ultrasonic data, explained in the section on perception are well suited for providing sensory information for obstacle avoidance purposes. In the simplest of obstacle avoidance systems, sonar echoes are used to detect the presence of an object within a giving range in front of the robot. This in turn causes the robot to stop. More sophisticated systems use the information in one of two ways: path replanning or reactive/reflexive navigation.

The path replanning for obstacle avoidance yields a path for the robot based on the environmentally sensed conditions. The scope of this path will typically extend for several meters. This approach is limited if there are moving obstacles present, because the path generated at an earlier time, fails to take into account the motion of the obstacles. Reactive/reflexive navigation, on the other hand, does not compute a path. Instead, it reacts immediately to perceived sensor information.

Actual robots have been constructed to demonstrate both methods. The grid-approach, described in Elfes, (1987), has been used to produce paths for a mobile robot after the ultrasonic sensors have filled a working grid model with information regarding the volumetric occupancy of the world. Potential fields have also been used for reactive/reflexive navigation using a repulsive force (or directly activating high-level "behaviors") generated by ultrasonically detected obstacles (Arkin, 1989; Brooks, 1986).

Computer vision and laser rangefinders have been used for obstacle detection (Wittaker et al., 1987). In this case, visual segmentation algorithms are applied to the range data resident in the incoming scanner image. The algorithms yield regions (obstacles) that can be shown to violate the ground-plane assumption and thus can be declared obstacles.

One problem with most of the obstacle detection techniques is their limited ability to detect pits and crevices (inverse obstacles) in the robot's path. Depressions in the path are not as readily discernible as upright obstacles due to the occlusion formed by the ground surface itself. Additional research is required to find more robust techniques to detect these more complex barriers to safe motion.

Classical and Behavioral Based Navigation

In the context of Figure 53.72 (described in Control Architectures), the navigator is the decision-making level. This implies that all information needed to make the decisions is concentrated here. The navigator has two main functions: planning and control, and learning.

The planning and control function is needed to calculate an output to the actuation system, corresponding to a specified task. In this case the planning is local and limited to a few meters based on sensor data inputs. The learning function is important if a robot to really behave intelligently. In this case the robot must learn from its actions (e.g., map building).

The planning stage uses processed sensor data, estimated robot position (normally provided by odometry), and the local map. By using the map stored in the world model (either an a *priori* given map or one built from local maps) a free path is planned (or replanned) with respect to the given task. Once the path, or at least a part of it, has been established, it has to be transformed into a trajectory. Here, boundary values of the robot parameters have to be taken into account. The knowledge about the present position, maximum accelerations, speeds and steering angles is provided by the robot model. The trajectory may be specified as a set of velocity profiles and desired positions. The final stage is the control the robot with respect to this trajectory.

At the other side of the behavioral architectures, depicted in Figure 53.73, the function of the navigator is performed by the *coordinator* module. In this case, there is not a local path to be followed. Instead, a sequential, switching, or parallel activation of the necessary behaviors (control modules) is performed. Each behavior generate robot control commands independent of one another which can be combined appropriately so that the robot can perform a given task. Examples of some implementations of this coordinator module can be found in Gachet et al. (1992), Arkin (1989), and Brooks (1986). As an example, the coordinator may be implemented as a set of heuristic rules for navigational purposes. The objectives of these rules is to calculate contributions of the primitive behaviors (simple control modules) to produce a free-collision navigation between two points as shown in Figure 53.79.

In the context of navigation, it is important to take into account the *relocalization* process of the mobile robot. The primary information about the localization of the robot is obtained by the evaluation of the movement of the wheels (odometry) (see Kinematics). However, this information is often innaccurate, because of the slippage of the wheels on the floor. To correct inaccuracies, we need additional information from other sensors (e.g., vision cameras and external marks) to correct the actual robot's coordinates from time to time automatically and when there is uncertainty about the robot position with respect to a global coordinated system.

Planning

This level receives from an operator the mission objectives and the initial description of the environment where the mission is going to be executed. The planning level generates the overall path which contains the rough description of the motion trajectory, speed trajectory, etc., within a relatively long time scale (lasting even the entire mission). Given the mission objectives and description, the planning system will have to plan a set of motion steps to achieve the objectives. This will require the analysis of the goal location, analysis of terrain map, and analysis of constraints. The world information will be interpreted to define singular points. This singular points or subgoals, or milestones correspond to the strategic regions that should be traversed to achieve the goal.

The strategic areas may include any descriptive entity of interest such as "door", "passage entrance", "rock", "hollow", "stairs", "elevator cabin", etc. If most of the space is traversable, the milestones of the path should be determined at the level of "planning". Thus, the process of planning consists of finding subgoals and milestones which should be achieved sequentially during the mobile robot operation.

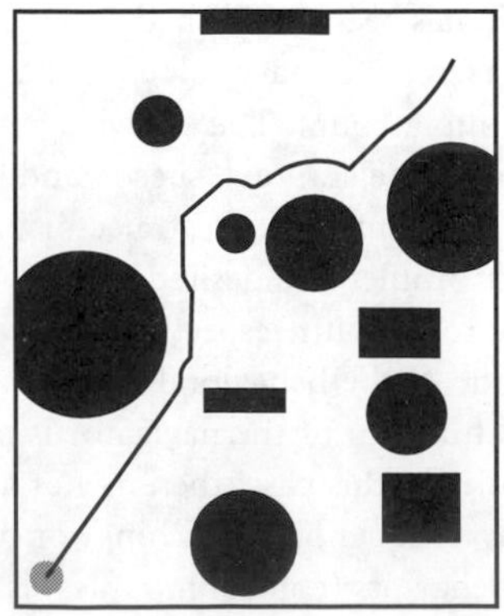

Figure 53.79 Navigation toward a goal.

Due to the multiplicity of possible path alternatives present in most real situations, some kind of cost function should be taken into consideration. This cost function should consider different parameters: the total time of motion, the energy consumed, the reliability and viability. The time will depend on distances and the average speed that is achievable within the region to be traversed. The energy consumed, and the values of reliability and viability are estimated according to the characteristic of the terrain to be traversed.

The typical approach towards the problem of path planning assumes a graph representation of the terrain. The problem of minimum-path planning on a graph is well known, and a variety of algorithms are applicable under different input specification.

There are several methods for solving the basic motion planning problem. Despite the external differences, the methods are based on a few general approaches: road map, cell decomposition, and potential field.

The road map approach to path planning consists of generating connecting cells within the mobile robot's free space into a network of one-dimensional curves called a road map. Once a road map has been constructed, it is used as a set of standardized paths. Path planning is thus reduced to connecting the initial and goal positions to points in the road map and searching in the road map for a path between these points. The constructed path, if one exists, is the concatenation of three subpaths: a subpath connecting the initial point to the road map, a subpath contained in the road map, and a subpath connecting the road map to the goal position. There are various methods based on this idea which compute different types of road maps called "visibility graph", "Voronoi diagram", "freeway net" and "silhouette".

The cell decomposition methods are perhaps the most extensively studied motion planning methods. They involve decomposing the mobile robot's free space into simple regions, called "cells", such that path between any two points in a cell can be easily generated. A nondirected graph representing the adjacency relation between the cells is then constructed and searched. This graph is called the "connectivity graph". Its nodes are the cells extracted from the free space. Two nodes are connected by a link only if the two corresponding cells are adjacent. A continuous free path can be computed from this sequence.

The cell decomposition method can be divided into exact and

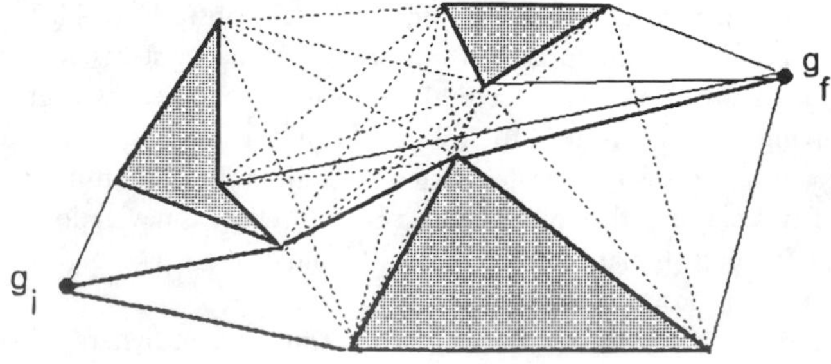

Figure 53.80 Visibility graph.

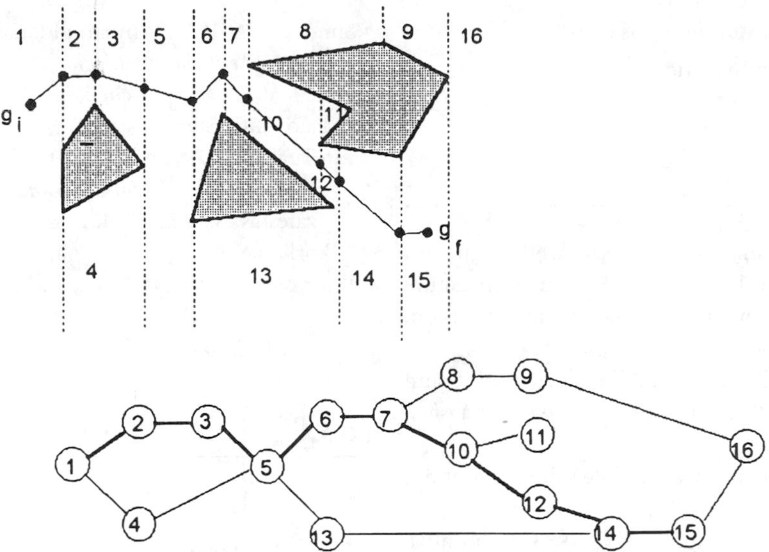

Figure 53.81 Exact cell decomposition method.

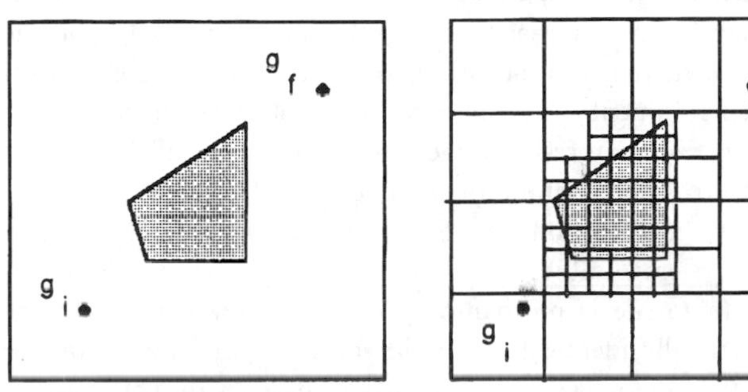

Figure 53.82 Approximate decomposition method.

approximate. The exact cell decomposition method partition the free space into cells whose union is exactly the free space. The boundary of a cell corresponds to a criticality of some sort, i.e., a sudden change in the constraints applying to the motion of the robot. The approximate cell decomposition methods produce cells of predefined shape (e.g., rectangloids) whose union is strictly included in the free space. The boundary of a cell does not characterize a discontinuity of some sort and has no physical meaning.

An approximate cell decomposition method is often used in a hierarchical fashion by using different resolution. For instance a 5 × 5 meters is used to find a path at the planning level, and a 0.5 × 0.5 meters resolution is used to find a path at the navigation level. Approximate methods may not be complete; but the precision of the approximation can be tuned. On the other hand, exact methods are more complicated than approximate ones. Because of this reason, the approximate methods are more used frequently.

Another approach involves a discretization of the space into a fine regular grid and search of this grid for a free path. This approach requires powerful heuristics to guide the search, since the grid is, in general, very large. There are several solutions to find a free path. One of the most successful ones is the "potential field" method.

The potential field methods consider the mobile robot as a point in space moving under the influence of an artificial potential produced by the goal destination and the obstacles. Typically the goal generates an "attractive potential" which pulls the robot toward the goal. The obstacles produce a "repulsive potential" which pushes the robot away from them. The negated gradient of the total potential is treated as an artificial force applied to the robot. The direction of this force is considered the better direction of motion. Since the potential field methods are descent optimization methods, they can get trapped into local minima of the potential function other than the goal destination. To cope with this problem, the basic potential field method is complemented with a mechanism allowing the method to escape from local minima.

Applications

Mobile robots are being used in some industrial applications with constrained environments (environments containing walls, halls, obstacles, etc.) or specific application domains (e.g., forests or farms). However, more research and development work needs to be done before mobile robots can move freely in a factory, home, farm, or military environment in a cost-effective manner. In particular, the area involving the *intelligent* connection between *perception* on the one hand and *planning, navigation,* and *control* on the other hand needs more development work before mobile robots can become truly practical. Current applications of mobile robots include both indoor and outdoor environments such as:

- Open quarries and construction.
- Forest and agriculture.
- Materials transport in manufacturing.

- Surveillance of indoor and outdoor sites.
- Disaster prevention and reaction.

References

Albus, J., McCain, J., and Lumia, R. 1987. NASA/NBS Standard Reference Model for Telerobot Control System Architecture (NASREM), NBS Tech. Note 1235, Robot System Division, National Bureau of Standards, Washington, D.C.

Ayache, N. and Faugeras, O. 1987. Building, registrating and fusing noisy visual maps, *Proc. Int. Conf. on Computer Vision*, London, UK.

Arkin, R. C. 1989. Motor schema-based mobile robot navigation, *Int. J. Robotics Research*, 8(4):92–112.

Barto, A. G., Sutton, R. S., and Anderson, C. 1983. Neuronlike elements that can solve difficult learning control problems, *IEEE Trans. Systems, Man, and Cybernetics* 13:835–846.

Brooks, R. A. 1986. A robust layered control system for a mobile robot, *IEEE J. Robotics and Automation*, RA (2) 14–24.

Davis, W., Jones, A., and Saleh, A. 1992. Generic architecture for intelligent control systems, *Computer-Integrated Manufacturing Systems*, 5(2):105–113.

Elfes, A. 1987. Sonar-based and real-world mapping and navigation, *IEEE J. Robotics Autom.*, RA-3:249–265.

Fritz, W., Martinez, R. G., Banque, J., Rama, A., Adobbati, R. E., and Sarno, M. 1989. The autonomous intelligent systems, *Robotics and Autonomous Systems*, 5:109–125.

Gachet, D., Salichs, M. A., Pimentel, J. R., Moreno, L., and la Escalera, de A. 1992. A software architecture for behavioral control strategies of autonomous systems, *Proc. IECON'92*, 1002–1007, San Diego CA.

Gachet, D., Pimentel, J. R., Moreno, L., Salichs, M. A., and Fernandez, V. 1993. Neural network control approaches for behavioral control of autonomous systems, *1st IFAC Int. Workshop on Intelligent Autonomous Vehicles*, 330–334, Southampton, UK.

Meystel, A. 1986. Theoretical Foundations of Decision Making Processes for Design and Control of Intelligent Mobile Autonomous Robots, Technical Report, Drexel University, Philadelphia PA.

Moreno, L., Moraleda, E., and Pimentel, J. R. 1993. Fuzzy supervision of behavioural primitives of autonomous systems, *Proc. IECON 1993*, Hawaii.

Nillson, N. 1984. Shakey the Robot, SRI International Tech. Note 323.

Pang, G. K. H. 1991. A framework for intelligent control, *Intelligent and Robotics Systems*, 4:109–127.

Pimentel, J. R., Puente, E. A., Gachet, D., and Pelaez, J. M. 1992. OPMOR: optimization of motion control algorithms for mobile robots, *Proc. IECON'92*, 853–861, San Diego CA.

Salichs, M. A., Puente, E. A., Moreno, L., and Pimentel, J. R. 1993. A Software Development Environment for Autonomous Mobile Robots, Ch. 8, *Recent Trends in Mobile Robots*, Y. F. Zheng, ed. World Scientific, River Edge, NJ.

Simon, H. A. 1983. Why should machines learn?, *Machine Learning: An Artificial Intelligence Approach*, Michalski, R. S., Carbonell, V. G., and Mitchell, T. M., Ed. 25–38, Tioga Publishing, Palo Alto, CA.

van Turennout, P., and Honderd, G. 1992. Navigation of a mobile robot, *Robotic Systems, Advanced Techniques and Applications*, Tzafestas, S., ed., 15–422, Kluwer Academic Publishing, New York, NY.

Wittaker, W., Turkiyyah, G., and Herbert, M. 1987. An architecture and two cases in range based modeling and planning, *Proc. IEEE Int. Conf. on Robotics and Automation*, 1991–1997.

53.9 Teleoperators

Antal K. Bejczy

Introduction

Historically, teleoperator systems were developed in the U.S.A. in the mid-1940s to create capabilities for handling highly radioactive material. Teleoperators allowed a human operator to handle radioactive material from a workroom separated by a one meter thick, radiation-absorbing concrete wall from the radioactive environment. The operator could observe the task scene through radiation resistant viewing ports in the wall. The development of teleoperators for the nuclear industry culminated in the introduction of bilateral force-reflecting master-slave manipulator systems. In these very successful systems, the slave arm at the remote site is mechanically or electrically coupled to the geometrically identical or similar master arm handled by the operator and follows the motion of the master arm. The coupling between the master and slave arms is a two-way coupling: inertia or work forces exerted on the slave arm can back-drive the master arm, enabling the operator to feel the forces that are acting on the slave arm. Force information available to the operator is an essential requirement for dexterous control of remote manipulators, since general-purpose manipulation consists of a series of well-controlled contacts between handling device and objects and also implies the transfer of forces and torques from the handling device to objects.

In a general sense, teleoperator devices enable human operators to remotely perform mechanical actions that usually are performed by the human arm and hand. Thus, teleoperators, or the act of teleoperation, extends the manipulative capabilities of the human arm and hand to remote, physically hostile, or dangerous environments. In this sense, teleoperation conquers space barriers in performing manipulative mechanical actions at remote sites, like telecommunication conquers space barriers in transmitting information to distant places.

In a more modern point of view, teleoperators are specialized robots, called telerobots, performing manipulative mechanical work remotely where humans cannot go or do not want to go. Following this viewpoint, current practice in advanced robotics is divided into two main areas: industrial robotics and telerobotics or robotic teleoperation. Industrial robots are used as an integral part of the manufacturing processes and within the frame of production engineering techniques to perform repetitive work in a structured factory environment. The characteristic control of industrial robots is a programmable sequence controller that functions autonomously with only occasional human intervention, either to reprogram or retool for a new task or to correct for an interruption in the work flow. Teleoperator robots, on the other hand, serve to extend, through mechanical, sensing, and computational techniques, the human manipulative, perceptive, and cognitive abilities into an environment that is either hostile to or remote from the human operator. Teleoperator robots or, in today's terminology, telerobots typically perform non-repetitive or singular, servicing, maintenance or repair work under a variety of environmental conditions ranging from structured to unstructured conditions. Telerobot control is characterized by a direct involvement of the human operator in the control since, by definition of task requirements, teleoperator systems extend human manipulative, perceptual and cognitive skills to remote places.

Continuous human operator control in teleoperation has both advantages and disadvantages. The main advantage is that overall task control can rely on human perception, judgement, decision, dexterity, and training. The main disadvantage is that the human operator must cope with a sense of remoteness, be alert to and integrate many information and control variables, and coordinate the control of one or two mechanical arms each having many (typically six) degrees of freedom—and doing all these with limited human resources. Furthermore, in many cases like space and deep sea applications, communication time delay interferes with continuous human operator control.

Modern development trends in teleoperator control technology are aimed at amplifying the advantages and alleviating the disadvantages of the human element in teleoperator control by the development and the use of advanced sensing and graphics displays, intelligent computer controls, and new computer-based human-machine interface devices and techniques in the information and control channels. The use of model and sensor data driven automation in teleoperation offers significant new possibilities to enhance overall task performance by providing efficient means for task-level controls and displays.

Automation in teleoperation is distinguished from other forms of automated systems by the explicit and active inclusion of the human operator in system control and information management. Such active participation by the human, interacting with automated system elements in teleoperation is characterized by several levels of control and communication, and can be conceptualized under the notion of "supervisory control" as discussed in Sheridan (1992). The human-machine interaction levels in teleoperation can be considered in a hierarchic arrangement: planning or high level algorithmic functions, motor or actuator control functions, and environmental interaction sensing functions. These functions take place in a task context in which the level of system automation is determined by the mechanical and sensing capabilities of the telerobot system, real time constraints on computational capabilities to deal with control, communication and sensing, the amount, format, content, and mode of operator interaction with the telerobot system, environmental constraints, like task complexity, and overall system constraints, like operator's skill or maturity of machine intelligence techniques.

Some advances have been made in teleoperator technology through the introduction of various sensors, computers, automation, and new human-machine interface devices and techniques for remote manipulator control. The development of advanced teleoperator technology is a challenging multidisciplinary effort. But, like the creation of a new tool, it is not a simple sum of other technologies. It represents a field of applied science and engineering on its own right, and requires its own experimental base.

The subsequent part of this chapter is focused on the description and some practical evaluation of an experimental advanced teleoperation (ATOP) system as an illustrative example of this evolving technology. This ATOP system was developed at the Jet Propulsion Laboratory (JPL) in the 1980s through 1993.

Advanced Teleoperation

The JPL ATOP system setting was conceived to provide a dual arm robot system together with the necessary operator interfaces to extend the two-handed manipulation capabilities of a human operator to remote places. The system setting intends to include all perceptive components that are necessary to perform sensitive remote manipulation efficiently, including nonrepetitive and unexpected tasks. The overall system is divided into two major parts: the remote robot work site and the local control station site, with electronic data and TV communication links between the two sites.

Remote Work Site

The remote site is a workcell. It comprises:

1. Two redundant eight-DOF arms (produced by AAI Company, Inc.) in a fixed base setting, each covering a hemispheric work volume, and each equipped with the latest JPL-developed model C smart hands that contain three-dimensional force-moment sensors at the hands' base and grasp force sensing at the base of the hand claws.
2. A JPL-developed control electronics and distributed computing system for the two arms and smart hands.
3. A computer controllable multi-TV gantry robot system with controllable illumination.

This gantry robot accommodates three color TV cameras, one on the ceiling plane, one on the rear plane, and one on the right side plane of the workcell. Each camera can be position controlled in two translational DOF in the respective plane, and in two orientation directions (pan and tilt) relative to the respective moving base. Zoom, focus, and iris of each TV camera can also be computer controlled. A stereo TV camera system is also available which can be mounted on any of the two side camera bases. The total size of the rectangular remote work site is about 5 m in width, about 4 m in depth, and about 2.5 m in height. (See Figure 53.83 for the ATOP remote workcell.)

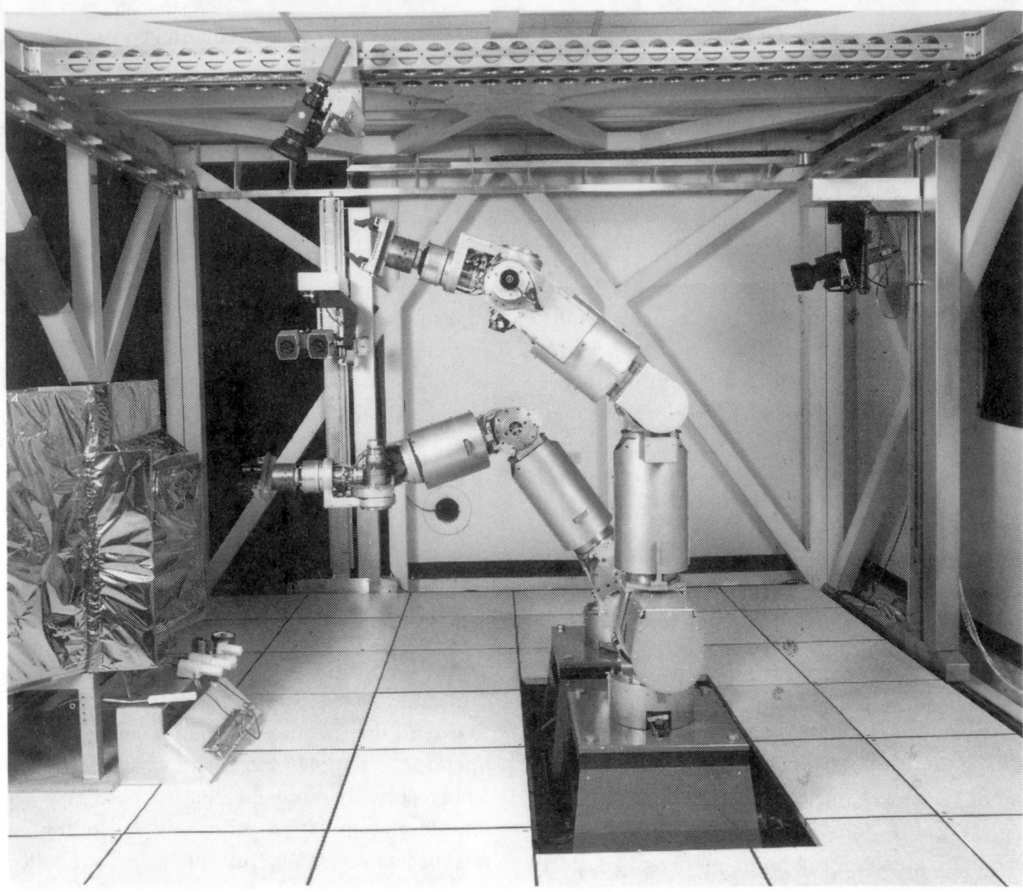

Figure 53.83 JPL ATOP dual-arm workcell with gantry TV frame.

Control Station

The control station site organization follows the idea of accommodating the human operator in all levels of human-machine interaction, and in all forms of human machine interfaces. Presently, it comprises:

1. Two general purpose force-reflecting hand controllers (FRHC).
2. Three TV monitors.
3. TV camera/monitor switchboards.
4. A manual input device for TV control.
5. Three graphics displays.

One of these graphics displays is connected to the primary graphics workstation (IRIS 4D/310 VGX) which is used for preview/predictive displays and for various graphical user interfaces (GUIs) in four-quadrant format. The second is connected to an IRIS 4D/70 GT workstation and is solely used for sensor data display. The third one is connected to a SUN workstation (SparcStation 10) and is used as a control configuration editor (CCE), which is an operator interface to the manipulators' control software based on an X-window environment. (See Figure 53.84 for the ATOP local control station.)

Hand Controllers

The human arm-hand system (hereinafter simply called hand) is a key communication medium in teleoperator control. With hand actions, complex position, rate, or force commands can be formulated and very physically written to the controller of a remote robot arm system in all workspace directions. At the same time, the human hand also can receive force, torque, and touch information from the remote robot arm-hand system. Furthermore, the human fingers offer additional capabilities to convey new commands to a remote robot controller from a suitable hand controller. Hand controller technology is, therefore, an important technology in the development of advanced teleoperation. Its importance is particularly underlined when one considers computer control which connects the hand controller to the remote arm system. The direct and continuous (scaled or unscaled) relation of operator hand motion to the remote robot arm's motion behavior in real time through a hand controller is in sharp contrast to the computer keyboard type commands which, by their very nature, are symbolic, abstract, and discrete (noncontinuous), and require the specification of some set of parameters within the context of a desired motion.

In contrast to the standard force-reflecting, replica master-slave systems, a new form of bilateral, force-reflecting manual control of robot arms has been implemented at the JPL ATOP project. The hand controller is a backdrivable six-DOF isotonic joystick. It is dissimilar to the controlled robot arm both kinematically and dynamically. But, through computer transformations, it can control the motion of any robot arm in six task space coordinates (in three position and three orientation coordinates). Forces and

Figure 53.84 JPL ATOP control station.

moments sensed at the base of the robot hand can backdrive the hand controller through proper computer transformations so that the operator feels the forces and moments acting at the robot hand while he controls the position and orientation of it. This hand controller can read the position and orientation of the hand grip within a 30-cm cube in all orientations, and can apply arbitrary force and moment vectors up to 20 N and 1.0 Nm, respectively, at the handgrip (two FRHCs are visible in Figure 53.84.) More details of the mechanical design of this hand controller and on hand controller technology in general can be found in Bejczy and Salisbury (1983) and Bejczy (1992). A computer-based control system establishes the appropriate kinematic and dynamic control relations between the FRHC and the robot arm. The FRHC can control any robot arm and can receive force/torque feedback from any robot arm equipped with a three-dimensional force-moment sensor at the base of the robot hand.

The computer-based control system supports four modes of manual control: position, rate, force-reflecting, and compliant control in task space (Cartesian space) coordinates. The operator, through an on-screen menu, can designate the control mode for each task space axis independently. The *position control mode* servos the slave position and orientation to match the master's. The *indexing function* allows slave excursions larger or smaller than the 30-cm cube hand controller work volume. In the *force-reflecting mode,* the hand controller is backdriven based on force-moment data generated by the robot and sensed during the robot hand's interaction with objects and environment. The *rate control mode* sets the slave endpoint velocity in task space based on the displacement of the hand controller. This is implemented through a software spring in the control computer of the hand controller. Through this software spring, the operator has a sensation of the commanded rate, and the software spring also provides a zero-referenced restoring force. The rate mode is useful for tasks requiring large translations. The *compliant control mode* is implemented through a low-pass software filter acting on the robot hand's force-torque sensor data in the hybrid position-force loop. This permits the operator to control a springy or less stiff robot. Active compliance with damping can be varied by changing the filter parameters in the software menu. Setting the spring parameter to zero in the low-pass filter will reduce it to a pure damper which results in a high stiffness hybrid position-force control loop.

Control System

The overall ATOP control organization permits a spectrum of operations between full manual, shared manual and automatic, and full automatic (call traded) control, and the control can be operated with variable active compliance referenced to force-moment sensor data. More on the overall ATOP control system can be found in Bejczy and Szakaly (1991a,b, 1990, 1989a,b, and 1987). The overall control/information data flow diagram (for a single arm) is shown in Figure 53.85. It is noted that the

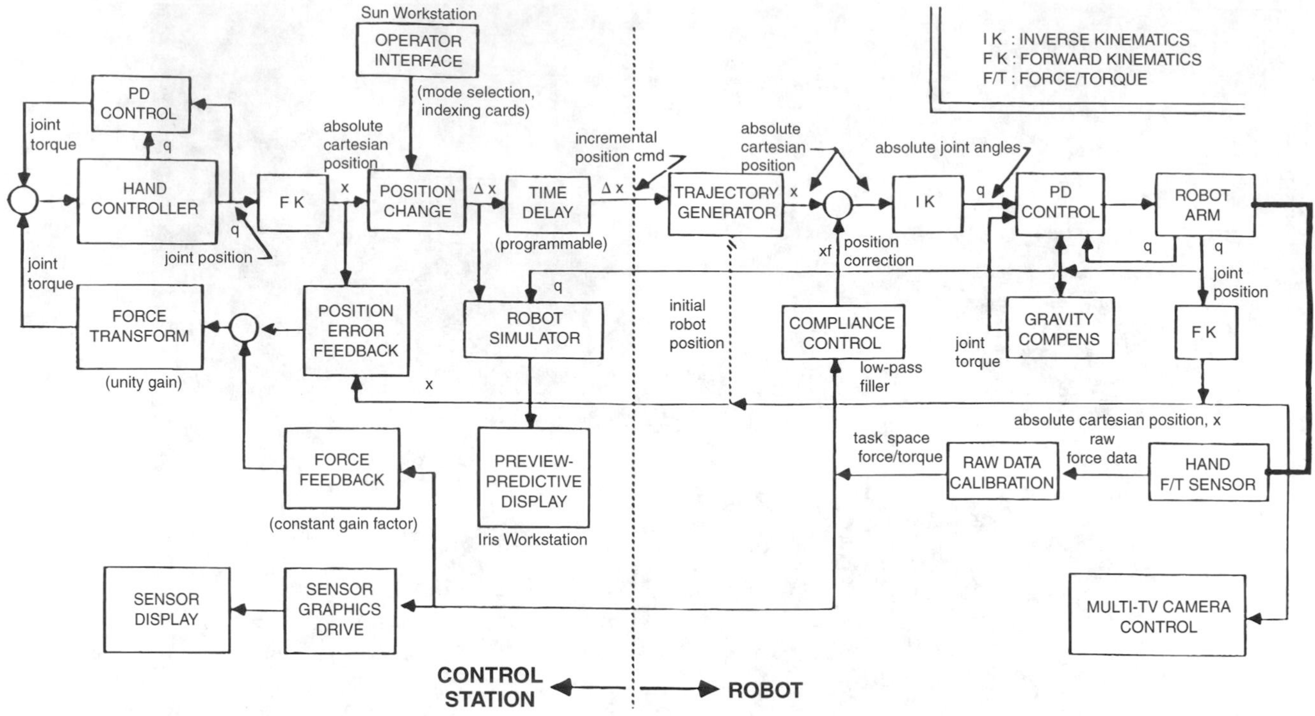

Figure 53.85 Control system flow diagram.

computing architecture of this original ATOP system is a fully synchronized pipeline, where the local servo loops at both the control station and the remote manipulator nodes can operate at a 1000-Hz rate. The end-to-end bilateral (i.e., force-reflecting) control loop can operate at a 200-Hz rate.

The data flow diagram shown in Figure 53.85 illustrates the organization of several servo loops in the system. The innermost loop is the position control servo at the robot site. This servo uses a PD control algorithm, where the damping is purely a function of the robot joint velocities. The incoming data to this servo is the desired robot trajectory described as a sequence of points at 1 ms intervals. This joint servo is augmented by a gravity compensation routine to prevent the weight of the robot from causing a joint positioning error. Because this servo is a first-order servo, there will be a constant position error that is proportional to the joint velocity.

In the basic Cartesian control mode the data from the hand controller are added to the previous desired Cartesian position. Form this the inverse kinematics generate the desired joint positions. The joint servo moves the robot to this position. From the actual joint position the forward kinematics compute the actual Cartesian positions. The force-torque sensor data and the actual positions are fed back to the hand controller side to provide force feedback.

This basic mode can be augmented by the addition of compliance control, Cartesian servo, and sticktion/friction compensation. Figure 53.86 shows the compliance control and the Cartesian servo augmentations. There are two forms of compliance, an integrating and a spring type (Figure 53.87). In integrating compliance the velocity of the robot end effector is proportional to the force felt in the corresponding direction. To eliminate drift a deadband is used. The zero velocity band does not have to be a zero force, a force offset may be used. Such a force offset is used if, for example, we want to push against the task board at some given force while moving along other axes. Any form of compliance can be selected along any axis independently. In the case of the spring-type compliance the robot position is proportional to the sensed force. This is similar to a spring centering action. The velocity of the robot motion is limited in both the integrating and spring cases.

There is a wide discrepancy between the robot response bandwidth and the force readings. The forces are read at a 1000-Hz sampling rate. The robot motion command has an output response at a 5-Hz bandwidth. To generate smooth compliance response, the force readings go through two subsequent filters. The first one is a simple averaging of ten force readings. This average is called 100-Hz force and is computed at a 100-Hz rate. From this 100-Hz force a 5-Hz force reading is computed by a first-order low-pass filter. This 5-Hz force reading is also computed at a 100-Hz rate. The 5-Hz force is used for compliance computations.

As shown in Figure 53.86, the Cartesian servo acts on task space (*X*, *Y*, *Z*, pitch, yaw, roll) errors directly. These errors are the difference between desired and actual task space values. The actual task space values are computed from the forward kinematic transformation of the actual joint positions. This error is then added to the new desired task space values before the inverse kinematic transformation determines the new joint position commands from the new task space commands.

A trajectory generator algorithm was formulated based on observations of profiles of task space trajectories generated by the operators manually through the FRHC. Three important

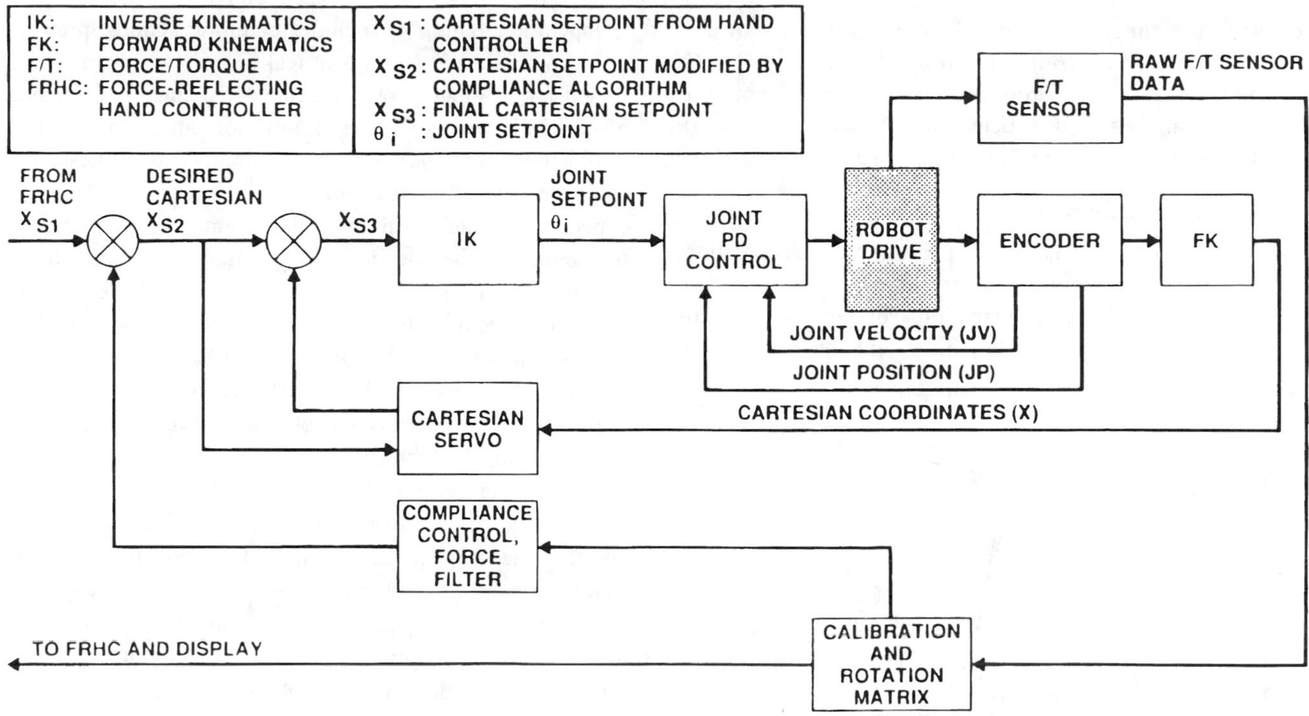

Figure 53.86 Control schemes: joint servo, Cartesian servo, compliance control.

features were observed in hand-generated task space trajectory profiles:

1. The operators always generated trajectories as a function of the relative distance between start point and goal point in the task space or, in general, as a function of the present position state relative to the desired position state of the end effector in the task space. In other words, the operators did not manually generate trajectories based on time (on clock signals).
2. The velocity-position phase diagrams of motion typically resembled a harmonic (sine) function.
3. Between the start and completion phases, the operator-generated trajectories typically attained a constant velocity profile.

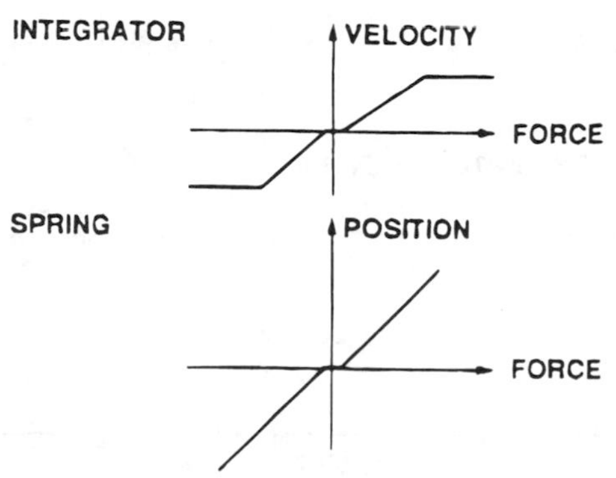

Figure 53.87 Compliance components and interpretations.

Based on these observations, we formulated a harmonic motion generator (HMG) with a sinusoidal velocity-position phase function profile as shown in Figure 53.88. The motion is parameterized by the total distance traveled, the maximum velocity, and the distance used for acceleration and deceleration. Both the accelerating and decelerating segments are quarter sine waves, with a constant velocity segment connecting them. This scheme still has a problem, the velocity being 0 before the motion starts. This problem is corrected by adding a small constant to the velocity function.

It is noted that the HMG discussed here is quite different from the typical trajectory generator algorithms employed in robotics which use a polynomial position-time function. Our algorithm generates the motion as a trigonometric (harmonic) velocity vs position function. The position vs. time and the corresponding

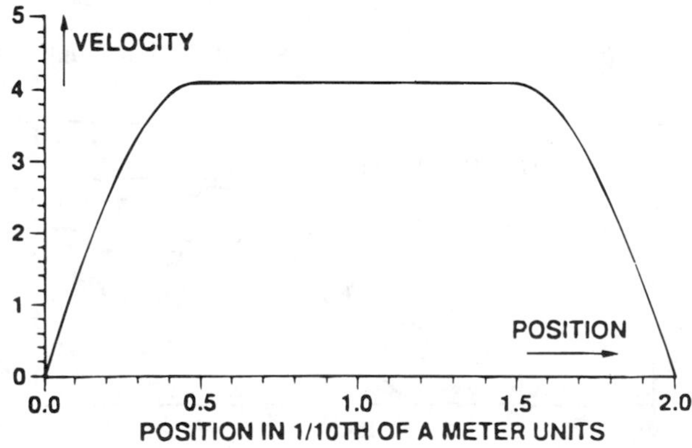

Figure 53.88 Harmonic motion generator velocity-position function.

velocity vs. time functions generated by the HMG are shown in Figure 53.89. More on performance results generated by HMG, Cartesian servo, and force-torque sensor data filtering in compliance control can be found in Bejczy and Szakaly (1991; 1990). Illustrative examples are shown in Figure 53.90 and Figure 53.91.

Computer Graphics

Task visualization is a key problem in teleoperation, because most of the operator's control decisions are based on visual or visually conveyed information. For this reason, computer graphics play an increasingly important role in advanced teleoperation. This role includes planning actions, previewing motions, predicting motions in real time under communication time delay, helping operator training, enabling visual perception of nonvisible events like forces and moments, and serving as a flexible operator interface to the computerized control system.

The actual utility of computer graphics in teleoperation depends to a high degree on the fidelity of graphics models that represent the teleoperated system, the task, and the task environment. In the past few years the JPL ATOP project developed high-fidelity calibration of graphics images to actual TV images of task scenes. This development has four major ingredients: first, the creation of high-fidelity three-dimensional graphics models of robot arms and objects of interest for robot arm tasks; second, the high-fidelity calibration of the three-dimensional graphics models relative to given TV camera two-dimensional image frames which cover the sight of both the robot arm and the objects of interest; third, the high-fidelity overlay of the calibrated graphics models over the actual robot arm and object images in a given TV camera image frame on a monitor screen; fourth, the high-fidelity motion control of robot arm graphics image by using the same control software that drives the real robot.

The high-fidelity fused virtual and actual reality image displays became very useful tools for planning, previewing, and predicting robot arm motions without commanding and moving the robot hardware. The operator can generate visual effects

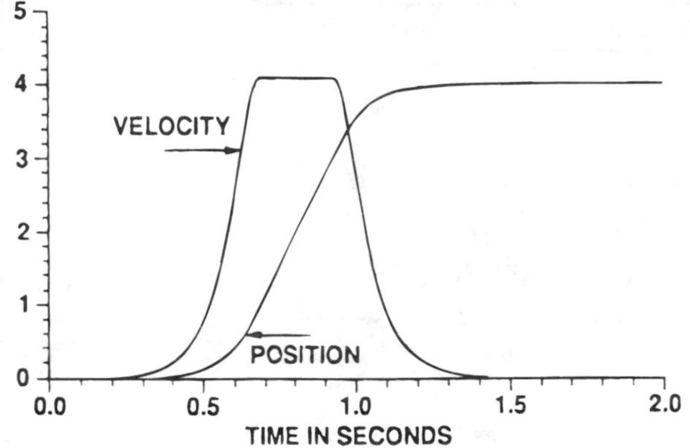

Figure 53.89 Harmonic motion generator position and velocity time functions.

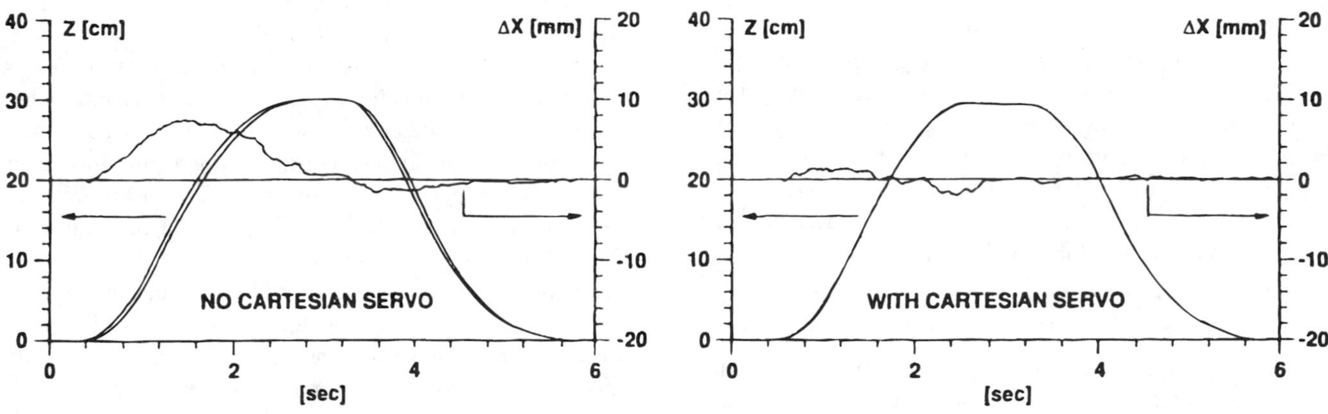

Figure 53.90 Vertical (Z) straight line trajectory from manual control and ΔX error.

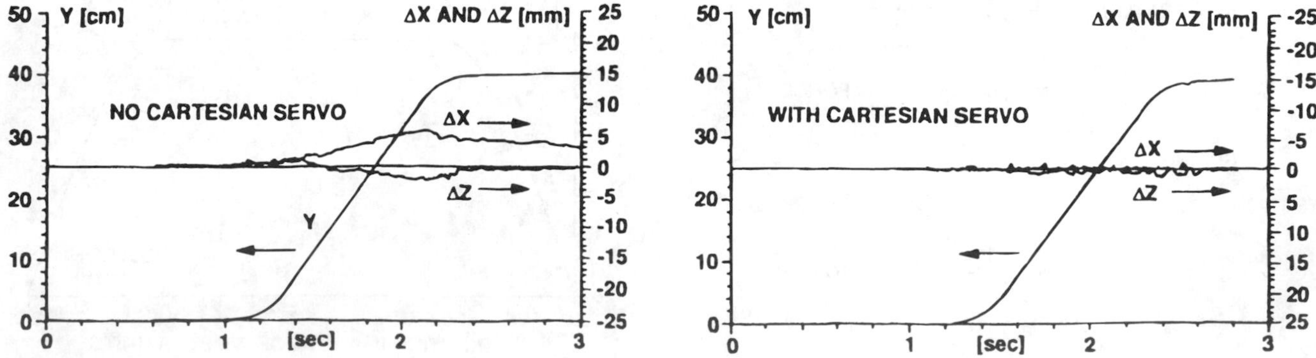

Figure 53.91 Horizontal (Y) straight line trajectory from harmonic motion generator and ΔX and ΔY errors.

of robot motion by commanding and controlling the motion of the robot's graphics image superimposed over TV pictures of the live scene. Thus, the operator can see the consequences of motion commands in real time, before sending the commands to the remotely located robot. The calibrated virtual reality display system can also provide high-fidelity synthetic or artificial TV camera views to the operator. These synthetic views can make critical motion events visible that are otherwise hidden from the operator in a given TV camera view or for which no TV camera view is available. More on the graphics system in the ATOP control station can be found in Fiorini et al. (1993), Kim (1993), Kim and Bejczy (1991), Bejczy and Kim (1990), and Bejczy et al. (1990).

High-Fidelity Graphics Calibration

A high-fidelity overlay of graphics and TV images of work scenes requires a high-fidelity TV camera calibration and object localization relative to the displayed TV camera view. Theoretically, this can be accomplished in several ways. For the purpose of simplicity and operator-controllable reliability, an operator-interactive camera calibration and object localization technique has been developed, using the robot arm itself as a calibration fixture, and using a nonlinear least-squares algorithm combined with a linear algorithm as a new approach to compute accurate calibration and localization parameters.

The current method uses a point-to-point mapping procedure, and the computation of camera parameters is based on the ideal pinhole model of image formation by the camera. In the camera calibration procedure, the operator first enters the correspondence information between the three-dimensional graphics model points and the two-dimensional camera image points of the robot arm to the computer. This is performed by repeatedly clicking with a mouse a graphics model point and its corresponding TV image point for each corresponding pair of points on a monitor screen which, in a four-quadrant window arrangement, shows both the graphics model and the actual TV camera image (see Figure 53.92). To improve calibration accuracy, several poses of the manipulator within the same TV camera view can be used to enter corresponding graphics model and TV image points to the computer. Then the computer computes the camera calibration parameters. Because of the ideal pinhole model assumption, the computed output is a single linear 4×3 calibration matrix for a linear perspective projection.

Object localization is performed after camera calibration by entering corresponding object model and TV image points to the computer for different TV camera views of the object. Again, the computational output is a single linear 4×3 calibration matrix for a linear perspective projection.

The actual camera calibration and object localization computations are carried out by a combination of linear and nonlinear least-squares algorithms. The linear algorithm, in general, does not guarantee the orthonormality of the rotation matrix, providing only an approximate solution. The nonlinear algorithm provides the least-squares solution that satisfies the orthonormality of the rotation matrix, but requires a good initial guess for a convergent solution without entering into a very time-consuming random search. When a reasonable approximate solution is known, one can start with the nonlinear algorithm directly. When an approximate solution is not known, the linear algorithm can be used to find one, and then one can proceed with the nonlinear algorithm. More on the calibration and object localization technique can be found in Kim (1994) and Kim and Bejczy (1993).

After completion of camera calibration and object localization, the graphics models of both the robot arm and the object of interest can be overlaid with high fidelity on the corresponding actual images of a given TV camera view. The overlays can be in wire-frame or solid-shaded polygonal rendering with varying levels of transparency, providing different task details. In the wire-frame format, the hidden lines can be removed or retained by the operator, depending on the information needs in a given task.

Graphics Operator Interface

The first graphic system as an advanced operator interface was aimed at parameter acquisition, an was handled as a teleoperation configuration editor (TCE) described in Lee et al. (1990). This interface used the concepts of windows, icons, menus, and a pointing device to allow the operator to interact, select, and update single parameters as well as a group of parameters. TCE utilizes the direct manipulation concept, with the central idea of having visible objects such as buttons, sliders, and icons that can be manipulated directly, i.e., moved and selected using the mouse, to perform any operation. A graphic interface of this type has several advantages over a traditional panel of physical buttons, switches, and knobs: the layout can be easily modified and its implementation cycle, i.e., design and validation, is significantly shorter than hardware changes.

The TCE, Figure 53.93, was developed to incorporate all the configuration parameters of an early single-arm version of the ATOP system. It was organized in a single menu divided into several areas dedicated to the parameters of a specific function. Dependencies among different graphical objects are embedded in the interface so that, when an object is activated, the TCE checks for parameter congruency. A significant feature of this implementation is its capability to store and retrieve sets of parameters via macro buttons. When a macro command is invoked, it saves the current system configuration and stores it in a function button which can later restore it. The peg-in-hole task, for instance, requires mostly translational motions but when holes have a tight clearance, a compliance is necessary. An appropriate macro configuration is one that enables x, y, and z axes, with position control in the approach direction and automatic compliance on the other two axes. This configuration can be assigned to a macro button and then recalled during a task containing a peg-in-hole segment.

The continuing work on a graphic system as an advanced operator interface was aimed at the data presentation structure of the interface problem, and, for that purpose, used a hierarchical architecture (Fiorini et al., 1993). This hierarchical data interface looks like a menu tree with only the last menu of the chain (the

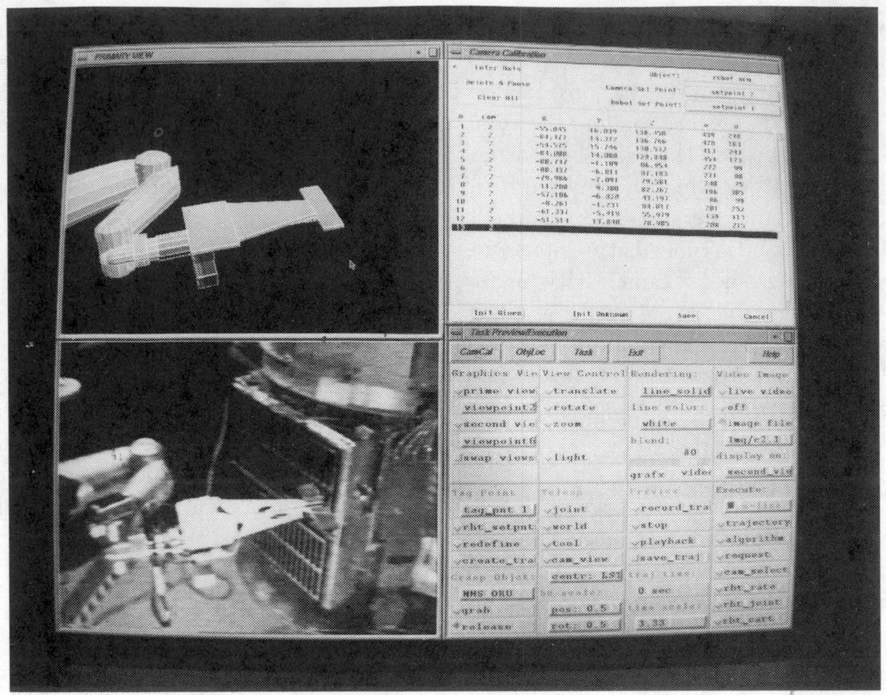

Figure 53.92 Graphics user interface for calibrating virtual (graphics) images to TV images.

Menu for file, diagnostic, help

LEFT ROBOT
On/Off line, Freeze, Neutral

System Access Levels
Operator, Monitor, Expert

LEFT ROBOT
On/Off line, Freeze Neutral

Save Config

Command Tranforms
Transform
No transform

Hand Controller Mounting
Vertical
Horizontal

Frame of Reference
World
Tool
Joint

Save Config

CONTROL MODES

All	Pos.	Rate	Spring	FFbk	Compl
X	□	□	□	□	□
Y	□	□			
Z	□	□			
Roll	□	□			
Pitch	□	□			
Yaw	□	□			

CONTROL GAINS

All, Motion Scaling, Force Scaling
X
Y
Z
Roll
Pitch
Yaw

DUAL ROBOT: full, partial save

Time Delay: __ sec. System Feedaback Messages Servo Rate: __ Hz

SYSTEM ON LINE !!

Figure 53.93 Schematic layout of the TCE interface.

leaf) displaying data. All the ancestors of the leaf are visible to clearly indicate the nature of the data displayed. The content of the leaf includes data or pictures and quickly conveys the various choices available to the operator. A schematic figure of this layout is shown in Figure 53.94. Parameters have been organized in four large groups that follow the sequence of steps in a teleoperation protocol. These groups are layout, configuration, tools, and execution. Each group is further subdivided into specific functions. The layout menu tree contains the parameters defining the physical task structure, such as the relative position of the robots and of the FRHC, servo rates, etc. The configuration menu tree contains the parameters necessary to define task phases, such as control mode and control gains. The tools tree contains the parameters and commands for the off-line support to the operator, such as planning, redundancy resolution, and software development. Finally, the execution tree contains commands and parameters necessary while teleoperating the manipulators, such as data acquisition, monitoring of robots, hand controllers and smart hands, retrieval of stored configurations, and camera commands.

Generic Task Experiments

Generic tasks are idealized, simplified tasks and serve the purpose of evaluating some specific ATOP features. In these experiments, described in detail in Hannaford et al. (1989), four tasks were used: attach and detach velcro, peg insertion and extraction, manipulation of three electrical connectors, and manipulation of a bayonet connector. Each task was broken down into subtasks. The test operators were chosen from a population with some technical background but not with an in-depth knowledge of robotics and teleoperation. Each test subject received 2–4 h of training on the control station equipment. The practice of individuals consisted of four to eight 30-min sessions.

As pointed out in Hannaford et al. (1989), performance variation among the nine subjects was surprisingly slight. Their backgrounds were similar (engineering students or recent graduates) except for one who was a physical education major with training in gymnastics and coaching. This subject showed the best overall performance by each of the measures. This apparent correlation between performance and prior background might suggest that potential operators be grouped into classes based on interest and aptitudes.

The generic task experiments were focused at the evaluation of kinesthetic force feedback vs. no force feedback, using the specific force feedback implementation techniques of the JPL ATOP project. The evaluation of the experimental data supports the idea that multiple measures of performance must be used to characterize human performance in sensing and computer-aided teleoperation. For instance, in most cases kinesthetic force feedback significantly reduced task completion time. In some specific cases, however, it did not, but it did sharply reduce extraneous forces. More information on the results can be found in Hannaford et al. (1991, 1989).

Application Task Experiments

Application tasks in a laboratory setting simulate some real-world use of ATOP. Two major application task experiments were performed: one without communication time delay and one with communication time delay.

The experiments without communication time delay were grouped around a simulated satellite repair task. The particular

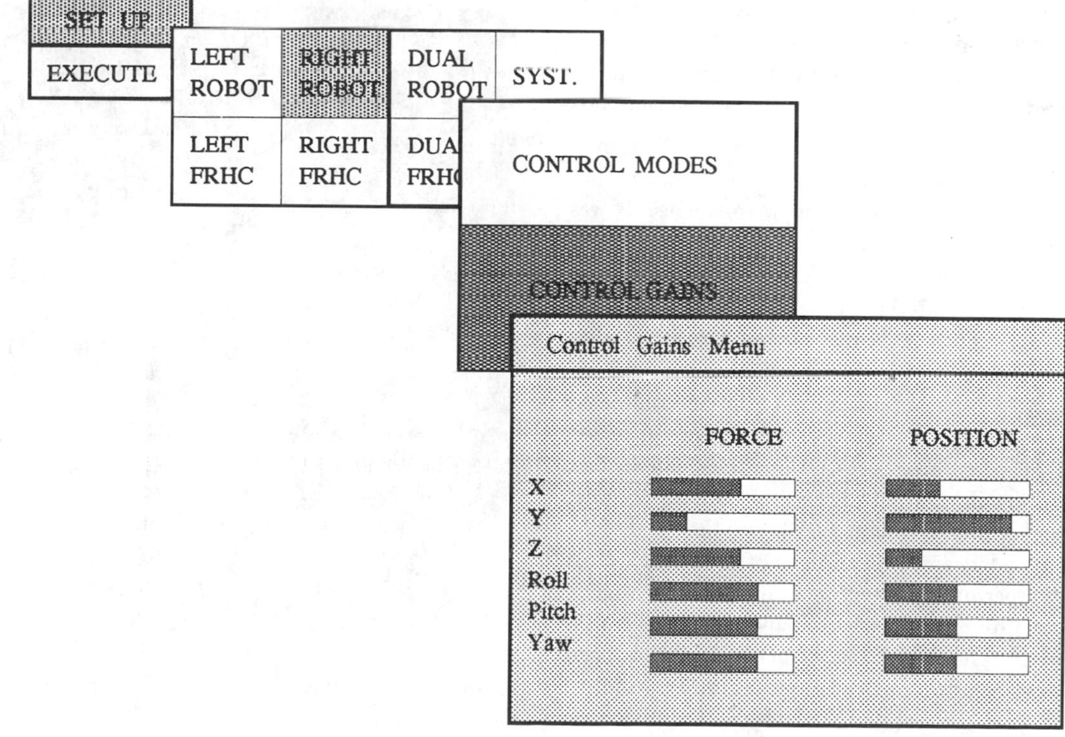

Figure 53.94 Schematic layout of the hierarchical data interface.

repair task was the duplication of the Solar Maximum Satellite Repair (SMSR) mission, which was performed by two astronauts in Earth orbit in the Space Shuttle Bay in 1983. Thus, it offered a realistic performance reference data base. This repair is a very challenging task, because this satellite was not designed for repair. Very specific auxiliary subtasks must be performed (e.g., a hinge attachment) to accomplish the basic repair which, in our simulation, is the replacement of the main electric box (MEB) of the satellite. The total repair, as performed by two astronauts in Earth orbit, lasted for about 3 h, and comprised the following set of substasks: thermal blanket removal, hinge attachment for MEB opening, opening of the MEB, removal of electrical connectors, replacement of MEB, securing parts and cables, replug of electrical connectors, closing of MEB, and reinstating thermal blanket. It is noted that the two astronauts were trained for this repair on the ground for about a year, including many underwater trainings in a buoyancy tank.

The SMSR simulation by ATOP capabilities was organized so that each repair scenario had its own technical justification and performance evaluation objective. For instance, in the first subtask-scenario performance experiments, alternative control modes, alternative visual settings, operator skills vs training, and evaluation measures themselves were evaluated. See details in Das et al. (1993, 1991). The first subtask-scenario performance experiments involved thermal blanket cutting and reinstating, and unscrewing MEB bolts. That is, both subtasks implied the use of tools. Figure 53.95 illustrates these experiments.

Several important observations were made during the aforementioned subtask-scenario performance experiments. The two most important observations are that:

1. The remote control problem in any teleoperation mode and using any advanced component or technique is at least 50% a visual perception problem to the operator, influenced greatly by view angle, illumination, and contrasts in color or in shading.
2. The training or, more specifically, the training cycle has a dramatic effect upon operator performance. It was found that the first cycle should be regarded as a familiarization with the system and with the task. For a novice operator, this familiarization cycle should be repeated at least twice. The real training for performance evaluation can only start after completion of a familiarization cycle. The familiarization can be considered complete when the trainee understands the system I/O details, the system response to commands, and the task sequence details. During the second cycle of training, performance measurements should be made so that the operator understands the content of measures against which the performance will be evaluated. Note, that it is necessary to separate each cycle and repetitions within cycles by at least one day. Once a personal skill has been formed by the operator as a consequence of the second training cycle, the real performance evaluation experiments can start. A useful criterion for determining the sufficient level of training can be, for instance, that of computing the ratio of standard deviation of completion time to mean completion time (that is computing the coefficient of variation). If the coefficient of variation of the last five trials of a subtask performance is less than 20%, then a sufficient level of training can be declared. In the subtask-scenario experiments quoted here, the real training, on the average, required one week per subject.

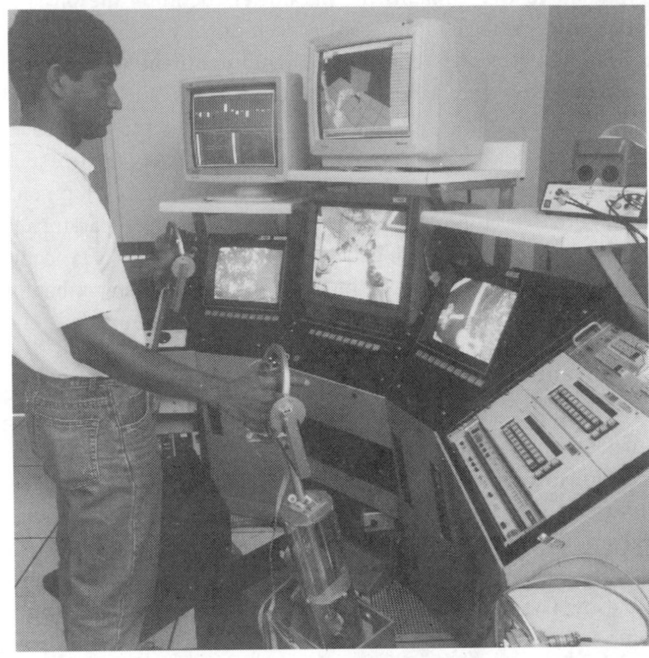

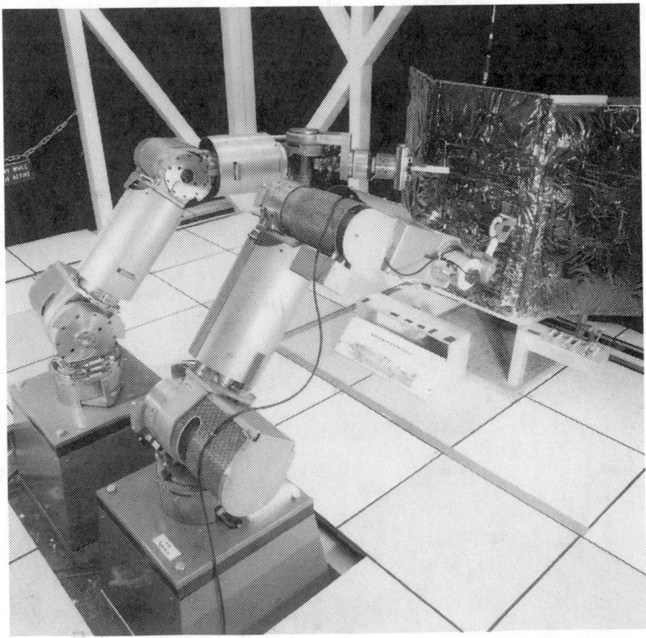

Figure 53.95 SMSR repair subtask simulation, reinstating the satellite's thermal blanket.

The practical purpose of training is, in essence, to help the operator develop a mental model of the system and of the task. During task execution, the operator acts through the aid of this mental model. It is, therefore, critical that the operator understands very well the response characteristics of the sensing and computer-aided ATOP system which has a variety of selectable control modes, adjustable control gains, and scale factors.

The procedure of operator training and the expected behavior of a skilled operator following an activity protocol offers the possibility of providing the operator with performance feedback messages on the operator interface graphics, derived from a stored model of the task execution.

A key element for such an advanced performance feedback tool to the operator is a program that can follow the evolution of a teleoperated task by segmenting the sensory data stream into appropriate phases.

A task segmentation program of this type has been implemented by means of a neural network architecture and it is able to identify the segments of a peg-in-hole task. (See details in Fiorini et al., 1993.) With this architecture, the temporal sequence of sensory data generated by the wrist sensor on the manipulators are turned into spatial patterns and a window of sensor observations which is related to the current task phase. A partially recurrent network algorithm was employed in the computation. Partially recurrent networks represent well the temporal evolution of a task, as they include in the input layer a set of nodes connected to the output units to create a context memory. These units represent the task phase already executed—the previous state. Several experiments of the peg-in-hole task have been carried out and the results have been encouraging, with a percentage of correct segmentations approximately equal to 65%. More on these experiments can be found in Fiorini et al. (1993) and Hannaford and Lee (1991).

The experiments with communication time delay, conducted on a large laboratory scale in early 1993, utilized a simulated life-size satellite servicing task which was set up at the Goddard Space Flight Center (GSFC) and controlled 4000 km away from the JPL ATOP control station. Three fixed TV camera settings were used at the GSFC worksite, and TV images were sent to the JPL control station over the NASA-Select Satellite TV channel at video rate. Command and control data from JPL to GSFC and status and sensor data from GSFC to JPL were sent through the Internet computer communication network. The roundtrip command/information time delay varied between 4–8 s between the GSFC worksite an the JPL control station, dependent on the data communication protocol.

The task involved the exchange of a satellite module. This required inserting a 45-cm-long power screwdriver, attached to the robot arm, through a 45-cm-long hole to reach the module's latching mechanism at the module's backplane, unlatching the module from the satellite, connecting the module rigidly to the robot arm, and removing the module from the satellite. The placement of a new module back to the satellite's frame followed the reverse sequence of actions.

Four camera views were calibrated for this experiment, entering 15–20 correspondence points in total from three to four arm poses for each view. The calibration and object localization errors at the critical tool insertion task amounted to about 0.2 cm each, well within the allowed insertion error tolerance. This 0.2 cm error is referenced to the zoom-in view (fovy = 8 deg) from the overhead (front view) camera which was about 1 m away from the tool tip. For this zoom-in view, the average error on the image plane was typically 1.2–1.6% (3.2–3.4% maximum error); a 1.4% average error is equivalent to a 0.2-cm displacement error on the plane 1 m in front of the camera.

The idea behind placing the high-fidelity graphics image over a real TV image is that the operator can interact with it visually in real time on a monitor within one perceptive frame when generating motion commands manually or by a computer algorithm. Thus, this method compensates in real time for the operator's visual absence from reality due to the time-delayed image. Typically, the geometric dimensions of a monitor and the geometric dimensions of the real work scene shown on the monitor are quite different. For instance, an 8-in.-long trajectory on a monitor can correspond to a 24-in.-long trajectory in the actual work space, that is, three times longer than the apparent trajectory on the monitor screen. Therefore, to preserve fidelity between a previewed graphics arm image and actual arm motions, all previewed actions on the monitor were scaled down very closely to the expected real motion rate of the arm hardware. The manually generated trajectories were also previewed before sending the motion commands to the GSFC control system to verify that all motion data were properly recorded. Preview displays contribute to operational safety. To eliminate the problem associated with the varying time delay in data transfer, the robot motion trajectory command is not executed at the GSFC control system until all the data blocks for the trajectory are received. An element of fidelity between the graphics arm image and actual arm motion was given by the requirement that the motion of the graphics image of the arm on the monitor screen be controlled by the same software that controls the motion of the actual arm hardware. This required the implementation of the GSFC control software in the JPL graphics computer.

A few seconds after the motion commands were transmitted to GSFC from JPL, the JPL operator could view the motion of the real arm on the same screen where the graphics arm image motion was previewed. If everything went well, the image of the real arm followed the same trajectory on the screen that the previewed graphics arm image motion previously described, and the real arm image motion on the screen stopped at the same position where the graphics arm image motion stopped earlier. After completion of robot arm motion, the graphics images on the screen were updated with the actual final robot joint angle values. This update eliminates accumulation of motion execution errors from the graphics image of the robot arm, and retains the robot arm graphics image position fidelity on the screen even after the completion of a force sensor referenced compliance control action.

The actual contact events (moving the tool within the hole and moving the module out from or into the satellite's frame) were automatically controlled by an appropriate compliance control algorithm referenced to data from a force-moment sensor at the end of the robot arm, implemented by the cooperating GSFC team and invoked by the JPL operator when needed.

The experiments have been performed successfully, showing the practical utility of high-fidelity predictive-preview display techniques, combined with sensor-referenced automatic compliance control, for a demanding telerobotic servicing task under communication time delay. More on these experiments and on the related error analysis can be found in Kim (1994) and Kim and Bejczy (1993). Figures 53.96a and 53.96b illustrate a few typical overlay views.

A few notes are in order here regarding the use of calibrated graphics overlays for time-delayed remote control.

1. There is a wealth of computation activities that the operator has to exercise. This requires very careful design considerations for an easy and user friendly operator interface to this computation activity.
2. The selection of the matching graphics and TV image points by the operator has an impact on the calibration results. First, the operator has to select significant points. This requires some rule-based knowledge about what is a significant point in a given view. Second, the operator has to use good visual acuity to click the selected significant points by the mouse.

Lessons Learned

The following general conclusions emerged so far from the development and experimental evaluation of the JPL ATOP.:

1. The sensing, computer- and graphics-aided advanced teleoperation system truly provides new and improved technical features. To transform these features into new and improved task performance capabilities, the operators of the system have to be transformed from naive to *skilled operators*. This transformation is primarily an undertaking of education and training.
2. To carry out an actual task requires that the operator follow a clear procedure or protocol which has to be worked out off line, tested, modified, and finalized. It is this procedure or protocol following habit that finally will help develop the experience and skill of an operator.
3. The final skill of an operator can be tested and graded by the ability to successfully recover from unexpected errors and complete a task.
4. The variety of I/O activities in the ATOP control station requires workload distribution between two operators. The primary operator controls the sensing and computer-aided robot arm system, while the secondary operator controls the TV camera an monitor system and assures protocol following. Thus, the *coordinated training of two cooperating operators* is essential to successful use of the ATOP system for performing realistic tasks. It is not yet known what a single operator could do and how. To configure and integrate the current ATOP control station for successful use by a *single operator* is challenging research and development work.
5. The problem of ATOP system development is not only to find ways to improve technical components and to create new subsystems. The final challenge is to integrate the improved or new technical features with the natural capabilities of the operator through appropriate human-machine interface devices and techniques to produce an improved overall system performance capability in which *the operator is part of the system* in some new way.

Anthropomorphic Telemanipulation

The robot arms employed in the JPL ATOP project are of the industrial type with industrial type parallel claw end effectors. This sets definite limits for the arms' task performance capabilities as dexterity in manipulation resides in the mechanical and sensing capabilities of the hands (or end effectors). We noted that, existing space manipulation tasks (except the handling of large space cargos) are designed for astronauts, including the tools used by astronauts. There are well over two hundred tools that are available today and certified for use by extra vehicular activity (EVA) astronauts in space. Motivated by these facts, an effort parallel to the ATOP project was initiated at JPL to develop and evaluate human-equivalent or human-rated dexterous telemanipulation capabilities for potential applications in space because all manipulation related tools used by EVA astronauts are human rated.

The general technical approach adopted in this project implies the following:

1. The master arm is a replica of the slave arm, and each arm has seven DOF.
2. The master arm is solidly attached to the operator's arm.
3. Forces acting on the slave arm can backdrive the master arm so that the operator can feel the forces/moments acting on the slave arm.
4. The slave hand is a humanlike fingered hand with a replica glovelike master controller attached solidly to the operator's hand.
5. Forces acting on the slave fingers can backdrive the fingers of the master glove so that the operator can feel the forces acting on the slave fingers.

The ability of the operator to feel forces acting at the remote slave site provides kinesthetic telepresence to the operator. This enables the operator to perform sensitive, force-compliant manipulation tasks with or without tools.

The actual design and laboratory prototype development included the following specific technical features:

1. The system is fully electrically driven.
2. The hand and glove have four fingers (little finger is omitted) and each finger has four DOF.
3. The base of the slave fingers follow the curvature of the human fingers' base on the hand.

a

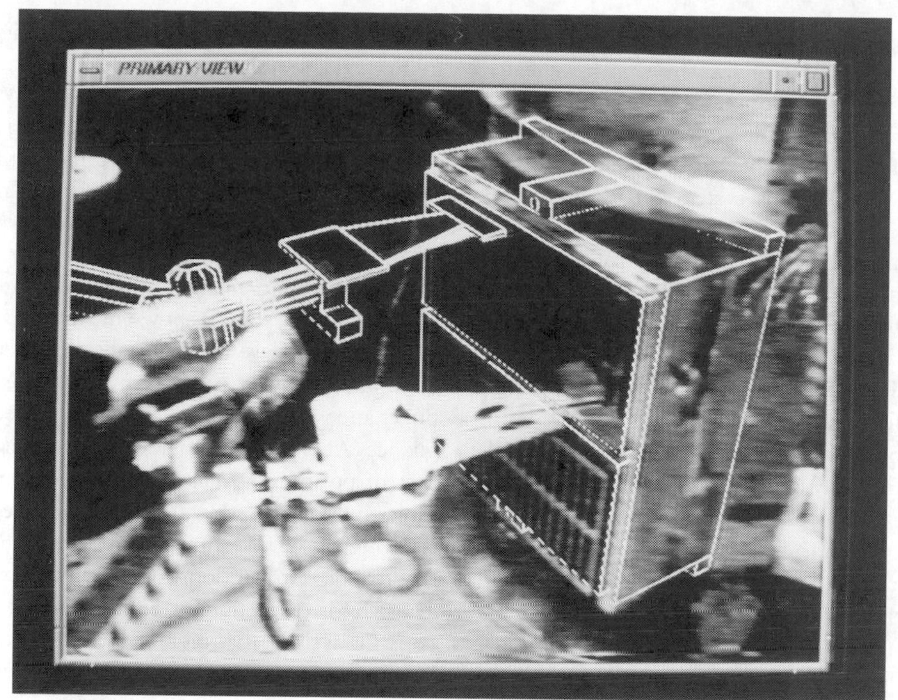

Figure 53.96a Predictive/preview display of end point motion.

b

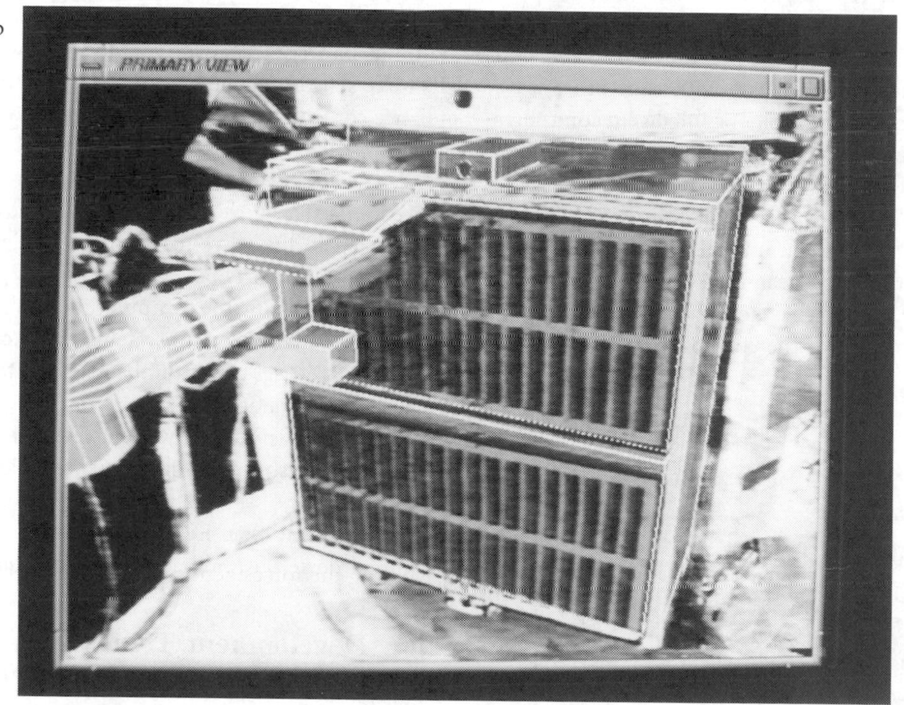

Figure 53.96b Status of predicted end point after motion execution, from a different camera view for the same motion shown in Figure 53.96.

4. The slave hand and wrist form a mechanically integrated closed subsystem, that is, the hand cannot be used without its wrist.
5. The lower slave arm which connects to the wrist houses the full electromechanical drive system for the hand and wrist (altogether 19 DOF), including control electronics and microprocessors.
6. The slave drive system electromechanically emulates the dual function of human muscles, namely, position and force control.

This implies a novel and unique implementation of active

compliance. All of the specific technical features taken together make this exoskeleton unique among the few similar systems. No other previous or ongoing developments have all the aforementioned technical features in one integrated system, and some of the specific technical features are not represented in any other similar systems at all. More on this system can be found in Jau (1992).

Currently, the JPL anthropomorphic telemanipulation system is assembled and tested in a terminus control configuration. In this configuration the master glove is integrated with our previously developed nonanthropomorphic six-DOF force-reflecting hand controller (FRHC), and the mechanical hand and forearm are mounted to an industrial robot (PUMA 560), replacing its standard forearm. The notion of terminus control mode refers to the fact that only the terminus devices (glove and robot hand) are of anthropomorphic nature, and the master and slave arms are nonanthropomorphic. The system is controlled by a high-performance distributed computer controller. Control electronics and computing architecture were custom developed for this telemanipulation system.

The present control electronics architecture for the master glove and the anthropomorphic hand/wrist is shown in Figure 53.97. It is comprised of PC-based computational engines, using TMS320C40 (C40) processors and two custom designed intelligent controllers. The interface to the FRHC and the PUMA upper arm joints is provided by two separate universal motor controllers (UMC). The UMC has been described previously in Bejczy and Szakaly (1987). The C40s communicate with each other via a single duplex communication channel. The intelligent controllers are based on the Texas Instrument TMS320C30 (C30). The C30 was selected for this task because of its low cost and high performance (33 MFLOPS). The C30 is very similar to the C40 except that it lacks the six high-speed communication ports. The two intelligent controllers are placed near the system's sensors, one is near the master glove, the other is near the anthropomorphic hand and wrist. The function of the controllers is to provide a sampling of analog signals, filter these signals, provide digital calibration of strain gages, model the actuator voltage-velocity curve, generate pulse-width modulated (PWM) signals, and communicate with the PC-based computational engine. All programs are written in the C language, using the SPOX Real-Time Operating System (Spectrum Microsystems) to facilitate the development of multipurpose programs. More on this system can be found in Jau et al. (1994).

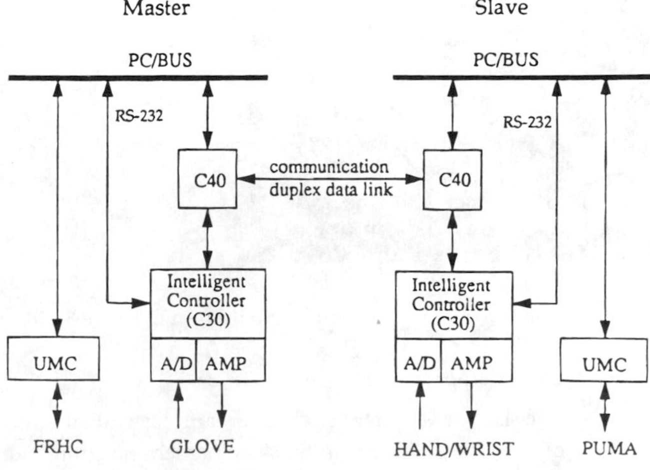

Figure 53.97 Control architecture overview.

The anthropomorphic telemanipulation system in terminus control configuration is shown in Figure 53.98. The master arm/glove and the slave arm/hand have 22 active joints each. The manipulator lower arm has five additional drives to control finger and wrist compliance. This active electromechanical compliance (AEC) system provides the muscle equivalent dual function of position as well as stiffness control. A cable links the forearm to an overhead gravity balance suspension system, relieving the PUMA upper arm of this additional weight. The forearm has two sections, one rectangular and one cylindrical. The cylindrical section, extending beyond the elbow joint, contains the wrist actuation system. The rectangular cross section houses the finger drive actuators, all sensors, and the local control and computational electronics. The wrist has three DOF with natural displacements similar to the human wrist. The wrist is linked to an AEC system that controls the wrist's stiffness. It is noted again that the slave hand, wrist, and forearm form a mechanically closed system, that is, the hand cannot be used without its wrist. A glove-type device is worn by the operator. Its force sensors enable hybrid position/force control and compliance control of the mechanical hand. Four fingers are instrumented, each having four DOF. Position feedback from the mechanical hand provides position control for each of the 16 glove joints. The glove's feedback actuators are remotely located and linked to the glove through flex cables. One-to-one kinematic mapping exists between the master glove and slave hand joints, thus reducing the computational efforts and control complexity of the terminus subsystem. The exceptions to the direct mapping are the two thumb base joints which need kinematic transformations.

The system is currently being evaluated, focusing on tool handling and astronaut equivalent task executions. The evaluation revealed the system's potential for tool handling but it also became evident that EVA tool handling operations in space require a dexterous, human-equivalent *dual* arm robot.

New Development Trends

Applications of teleoperators or telerobots are numerous, in particular in the nuclear and munitions industries, maintenance and reclaiming industries operating in hostile environments, and in industries that support space and underwater operations and explorations. Lately, robotics and teleoperation technology started breaking ground also in the *medical field*. Diagnostic and actual operative surgeries, including microsurgery and telesurgery within the general frame of *telemedicine*, seem to be receptive fields for potential use of robotic and teleoperator tools and

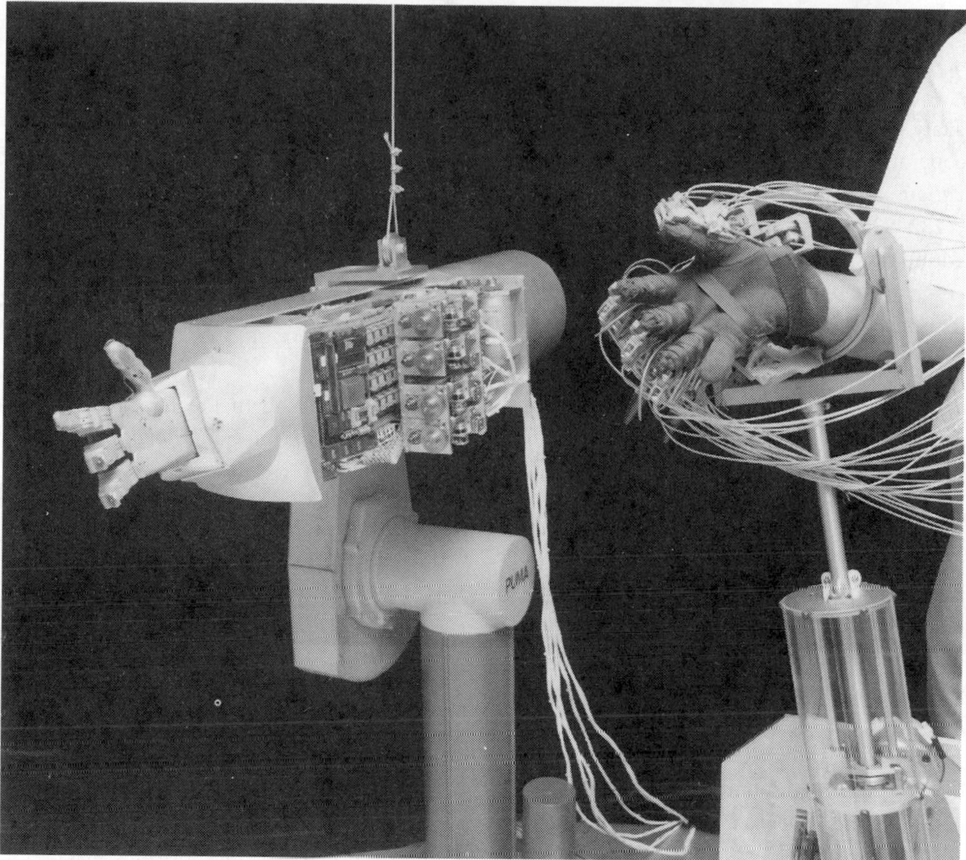

Figure 53.98 Master glove controller and anthropomorphic hand.

techniques. More details on this new medical application trend can be found in *Proceedings of the 1st Int. Symp. on Medical Robotics and Computer Assisted Surgery* (1994) and *Interactive Technology and the New Paradigm for Healthcare; Medicine Meets Virtual Reality Proceedings* (1995).

The application of robotic and teleoperator tools and techniques in the field of medicine is quite real. To demonstrate this, a surgeon in Milan, Italy very recently (in August, 1995) performed a live prostate telebiopsy on a real patient located at about 25 km from the surgeon. This was shown to about 350 attendees of a technical Congress in real time. The real technical point in this act was the application of telerobotic measurement and tool handling techniques. First, an automated calibration system determines the coordinates of a visually identified point of interest in an echographic image of a prostate, then, upon command from the surgeon, the robot arm inserts the biopsy tool to that point with high precision. (See the technical details in Sala (1995) and Rovetta and Sala (1995).) The claim, announced by the surgeon, was that this "telerobotic procedure" can provide better accuracy and can be performed in shorter time than a purely manual procedure. An additional technical point in this claim was that a low-cost PC-based telecommunication system in a limited bandwidth ISDN network setting can provide a satisfactory system capability for this type of "telerobotic procedure" in telemedicine.

Acknowledgment

This work was performed at the Jet Propulsion Laboratory, California Institute of Technology, under contract with the National Aeronautics and Space Administration.

References

Bejczy, A. K. 1992. Teleoperation: the language of the human hand, *Proc. IEEE Workshop on Robot and Human Communication*, Sept. 1–3, Tokyo, Japan.

Bejczy, A. K. and Salisbury, J. K. 1983. Controlling remote manipulators through kinesthetic coupling, *Computers in Mechanical Engineering*, 1(1); also, Kinesthetic coupling between operator and remote manipulator, *Proc. ASME Int. Computer Technology Conf.* Vol.1, 192–211, San Francisco, CA.

Bejczy, A. K. and Kim, W. S. 1990. Predictive displays and shared compliance control for time delayed telemanipulation, *Proc. IEEE Int. Workshop on Intelligent Robots and Systems* (IROS '90). 407–412, Tsuchiura, Japan.

Bejczy, A. K. and Szakaly, Z. F. 1987. Universal computer control system (UCCS) for space telerobots, *Proc. IEEE Int. Conf. Robotics and Automation*, March 30–April 3, Raleigh, NC.

Bejczy, A. K. and Szakaly, Z. 1990. Performance capabilities of a JPL dual-arm advanced teleoperation system, *Space Operations, Applications, and Research Symposium (SOAR '90) Proceedings,* 168–179, Albuquerque, NM.

Bejczy, A. K. and Szakaly, Z. 1991a. An 8-D.O.F. dual arm system for advanced teleoperation performance experiments, *Space Operations, Applications, and Research Symposium (SOAR '91),* NASA No. 3127, 282–293 Houston, TX; also Lee, S., and Bejczy, A. K., Redundant arm kinematic control based on parametrization, *Proc. IEEE Int. Conf. Robotics and Automation,* 458–465, Sacramento, CA. 1991.

Bejczy, A. K. and Szakaly, Z. 1991b. A harmonic motion generator for telerobotic applications, *Proc. IEEE Int. Conf. Robotics and Automation,* 2032–2039, Sacramento, CA.

Bejczy, A. K., Kim, W. S., and Venema, S. 1990. The phantom robot: predictive display for teleoperation with time delay, *Proc. IEEE Int. Conf. Robotics and Automation,* 546–550, Cincinnati, OH.

Bejczy, A. K., Szakaly, Z., and Kim, W. S. 1989a. A laboratory breaadboard system for dual arm teleoperation, *3d Annual Workshop on Space Operations, Automation and Robotics,* NASA Conf. Pub. 3059, 649–660, Johnson Space Center, Houston, TX.

Bejczy, A. K., Szakaly, Z., and Ohm, T. 1989b. Impact of end effector technology on telemanipulation performance, *3rd Annual Workshop on Space Operations, Automation and Robotics,* NASA Conf. Pub. 3059, 429–440, Johnson Space Center, Houston, TX.

Das, H., Zak, H., Kim, W. S., Bejczy, A. K., and Schenker, P. S. 1991. Performance experiments with alternative advanced teleoperator control modes for a simulated solar max satellite repair, *Proc. Space Operations, Automation and Robotics Symposium (SOAR '91) NASA No. 3127,* 294–301, July 9–11, Johnson Space Center, Houston, TX.

Das, H., Zak, H., Kim, W. S., Bejczy, A. K., and Schenker, P. S. 1993. Performance with alternative control modes in teleoperation, *PRESENCE: Teleoperators and Virtual Environtments,* 1(2):219–228.

Fiorini, P., Giancaspro, A., Losito, S., and Pasquariello, G. 1993. Neural networks for segmentation of teleoperation tasks, *PRESENCE: Teleoperators and Virtual Environments,* 2(1):66–81.

Fiorini, P., Bejczy, A. K., and Schenker, P. 1993. Integrated interface for advanced teleoperation, *IEEE Control Systems Magazine,* 13(5):15–20.

Hannaford, B. and Lee, P. 1991. Hidden markov model analysis of force-torque information in telemanipulation, *Int. J. Robotics Research,* 10(5).

Hannaford, B., Wood, L., Guggisberg, B., McAffee, D., and Zak, H. 1989. Performance evaluation of a six-axis generalized force-reflecting teleoperator, *Jet Propulsion Lab., JPL Pub. 89–18,* June 15, Pasadena, CA.

Hannaford, B., Wood, L., Guggisberg, B., McAffee, D., and Zak, H. 1991. Performance evaluation of a six-axis force-reflecting teleoperation, *IEEE Trans. Systems, Man and Cybernetics,* 21(3). *Interactive Technology and the New Paradigm for Healthcare; Medicine Meets Virtual Reality Proceedings,* 1995. IOS Press, Amsterdam, Oxford, Washington DC, OHMSA, Tokyo, Osaka, Kyoto.

Jau, B. M. 1992. Man-equivalent telepresence through four fingered human-like hand system, *Proc. IEEE Int. Conf. Robotics and Automation,* 843–848, Nice, France.

Jau, B. M., Lewis, M. A., and Bejczy, A. K. 1994. Anthropomorphic telemanipulation system in terminus control mode, *Proc. ROMANSY '94,* Gdansk, Poland.

Kim, W. S. 1993. Graphical operator interface for space telerobotics, *Proc. IEEE Int. Conf. Robotics and Automation,* p. 95, Atlanta, GA.

Kim, W. S., 1994. Virtual reality calibration for telerobotic servicing, *Proc. IEEE Int. Conf. Robotics and Automation,* 2769–2775, San Diego, CA.

Kim, W. S. and Bejczy, A. K. 1991. Graphics displays for operator aid in telemanipulation, *Proc. IEEE Int. Conf. Systems, Man and Cybernetics,* 1059–1067, October, Charlottesville, VA.

Kim, W. S. and Bejczy, A. K. 1993. Demonstration of a high-fidelity predictive preview display technique for telerobotics servicing in space, *IEEE Trans. Robotics and Automation, Oct. 1993, Special Issue on Space Telerobotics,* 698–702; also Kim, W. S., Schenker, P. S., Bejczy, A. K., Leake, S., and Ollendorf, S. 1993. An advanced operator interface design with preview/predictive displays for ground-controlled space telerobotic servicing, *SPIE Conference No. 2057; Telemanipulator Technology and Space Telerobotics,* Boston, MA.

Lee, P., Hannaford, B., and Wood, L. 1990. Telerobotic configuration editor, *Proc. IEEE Int. Conf. Systems, Man and Cybernetics,* 121–126 Los Angeles, CA.

Proc. 1st Int. Symp. Medical Robotics and Computer Assisted Surgery, Sept. 22–24, 1994. Shadyside Hospital, Pittsburgh, PA.

Rovetta, A. and Sala, R. 1995. Execution of robot-assisted biopsies within the clinical context, *Proc. 9th World Congress on the Theory of Machines and Mechanisms,* August 29–September 2, Milan, Italy.

Sala, R. 1995. Construction of a new automatic telemeter for medical applications and robotic telesurgery, *Proc. 9th World Congress on the Theory of Machines and Mechanisms,* August 29–September 2, Milan, Italy.

Sheridan, T. B. 1992 *Telerobotics, Automation, and Human Supervisory Control,* The MIT Press, Cambridge, MA, London, England.

Further Information

Backes, P. G. 1992. Ground-remote control for space station telerobotics with time delay, *Proc. AAS Guidance and Control Conference,* AAS Paper No. 92–052, February 8–12, Keystone, Co.

Backes, P. G. 1992. Supervised autonomous control, shared control, and teleoperation for space servicing, *Proc. Space Operations, Applications, and Research Symposium,* August 4–6, Houston, TX.

Baron, S. and Kleinmann, D. C. 1969. The human as an optimal controller and information processor, *IEEE Trans. Man-Machine Systems,* MMS–10(1):9–17.

Bejczy, A. 1980. Kinesthetic and graphic feedback for integrated operator control, *Proc. 6th Annual Advanced Control Conf.*, 137–147, April 28–30, Purdue University, West Lafayette, IN.

Bejczy, A. 1980. Sensors, controls, and man-machine interface for advanced teleoperation, *Science,* 208:1327–1335.

Bejczy, A. K., Brooks, T. L., and Mathur, F. P. 1981. *Servomanipulator Man-Machine Interface Conceptual Design,* JPL Report No. 5030–507, August U.S. Dept. of Energy.

Brooks, T. 1979. Superman: *A System for Supervisory Manipulation and the Study of Human Computer Interactions,* Master's thesis, Man-Machine Systems Laboratory, MIT.

Ferrell, W. R. and Sheridan, T. B. 1967. Supervisory control of remote manipulation, *IEEE Spectrum,* October, 4:81–88.

Ferrell, W. R. 1973. Command language for supervisory control of remote manipulation, *Remotely Manned Systems,* E. Heer, ed., 369–373, California Institute of Technology, Pasadena, CA.

Funda, J., Lindsay, T. S., and Paul, R. P. 1992. Teleprogramming: toward delay-invariant remote manipulation, *Presence,* 1(1):29–44.

Groome, R. C. 1977. *Force Feedback Steering of Teleoperator System,* MS thesis, MIT.

Handlykken, M., and Turner, T. 1980. Control systems analysis and synthesis for a six-degree-of-freedom universal force-reflecting hand controller, *Proc. 19th IEEE Conf. Decision and Control,* 1197–1205, Albuquerque, NM. December 10–12.

Hayati, S., and Venkataraman, S. T. 1989. Design and implementation of a robot control system with traded and shared control capability, *Proc. IEEE Int. Conf. Robotics and Automation,* 1310–1315.

Hill, J. W., and Sword, A. 1979. Manipulators based on sensor directed control: an integrated end effector and touch sensing system, *Proc. 17th Annual Conf. Human Factors,* Washington, D.C. October.

Hill, J. W. 1979. Study of modeling and evaluation of remote manipulation tasks with force feedback, *Final Report For JPL, SRI Project 7696 JPL Contract 95–5170,* March.

Jagacinski, R. J. and Miller, R. A. 1978. Describing the human operator's internal model of a dynamic system, *Human Factors,* 20(4):425–433.

Jelatis, D. C. 1975. Characteristics and Evaluation of master-slave manipulators, *Performance Evaluation of Programmable Robots and Manipulators, NBS Special Publication 459,* 141–145, October.

Johnson, E. G. and Corliss, W. R. 1971. *Human Factors Applications in Teleoperator Design and Operation,* John Wley and Sons, New York, NY.

Kohler, G. W. 1981. *Manipulator Type Book,* Verlag Karl Thiemig, Munchen, FRG.

Nevins, J. L., Sheridan, T. B., Whitney, D. E., and Woodin, A. E. 1973. The multi-model remote manipulator system, in *Remotely Manned Systems,* E. Heer, ed., 173–187, California Institute of Technology, Pasadena, CA.

Setzer, W. and Vossius, G. 1981. On the stability problem of human arm and hand movements controlling external load systems, *Proc. 1st Annual European Conf. Manual Control,* 243–253, Delft University of Technology, Delft, The Netherlands, May 25–27.

Sheridan, T. B. 1983. Supervisory control of remote manipulators, vehicles and dynamic processes: experiments in command and display aiding, *M.I.T. Man-Machine Systems Laboratory Report,* March Cambridge MA.

Shultz, R. E., Tesar, D., and Doty, K. L. 1979. Computer augmented manual control of remote manipulator, *Proc.1978 IEEE Conf. Decision and Control,* San Diego, CA. January.

Spiger, R. J., Farrell, R. J., and Tonkin, M. H. 1982. Survey of Multi-Function Display and Control Technology, *NASA Report No. CR-167510,* Boeing Co.

Stark, L. and Ellis, S. S. 1981. Revisited: cognitive models in direct active looking, *Eye Movement, Cognition and Visual Perception,* Fisher, Monty, and Senders, eds., 193–226, Album Press, New Jersey.

Starr, G. P. 1981. Supervisory control of remote manipulation: a preliminary evaluation, *Proc. 17th Annual Conf. Manual Control,* 95–107, University of California, Los Angeles, CA., June 16–18.

Starr, G. P. 1979. A comparison of control modes for time-delayed remote manipulation, *IEEE Trans. Systems, Man, and Cybernetics,* SMC–9(4):241–246.

Stassen, H. G. 1976. Man as controller, *Introduction to Human Engineering,* Koln: Verlag TUV, Rheinland.

Vertut, J. 1975 Experience and remarks on manipulator evaluation, *Performance Evaluation of Programmable Robots and Manipulators,* NBS Special Publication 459, 97–112, October.

Vertut, J. and Coiffet, P. 1983. *Teleoperations,* Vol. 3 in Robot Technology Series, Prentice Hall, Englewood Cliffs, NJ.

Vykukal, H. C., King, R. F., and Vallotton, W. C. 1973. An anthropomorphic master-slave manipulator system, *Remotely Manned Systems,* E. Heer, ed.,199–205, California Institute of Technology, Pasadena, CA.

Wagner, E. and Hanett, A. 1978. MINIMAC—the remote-controlled manipulator with stereo TV viewing at the SIN accelerator facility, *Trans. Am. Nuclear Society,* 30:759–760.

White, T. N. 1981. Modeling the human operator's supervisory behavior, *Proc. 1st European Annual Conf. Human Decision Making and Manual Control,* 203–217, Delft University of Technology, Delft, The Netherlands, May 25–27.

Yoerger, D. R. 1982. Supervisory control of underwater telemanipulators: design and experiment, M.I.T., *Man-Machine Systems, Laboratory Report,* August, Cambridge, MA.

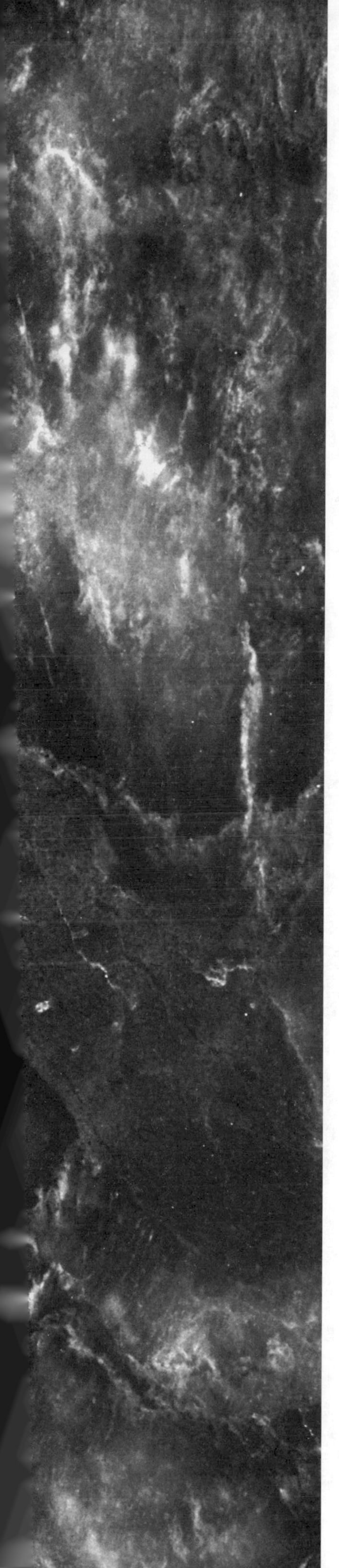

PART 2

INTELLIGENT ELECTRONICS and EMERGING TECHNOLOGIES

VII

Expert Systems and Neural Networks

Expert Systems

54 Current Applications of Expert Systems in Industrial Electronics *Mary Lou Padgett and Robert Shelton* 805
Emerging Trends for Expert Systems in Industrial Electronics • Defining Terms • Resources

55 Expert Systems Methodology *Gary Riley* 808
Capturing Human Expertise in a Program • Rule-Based Programming • Truth Table Simplification Program

56 Expert Systems and Their Use in Complex Engineering Systems *Robert E. Uhrig and Lefteri H. Tsoukalas* 824
Introduction • Definition of Expert Systems • Characteristics of Expert Systems • Components of an Expert System • Knowledge Representation and Inference • Uncertainty Management • State of the Art of Expert Systems • Use of Expert Systems • Potential Implementation Issues for Expert Systems • Legal Aspects of Expert Systems • Use of Expert Systems in Nuclear Power Plants

Neural Networks

57 Strategies and Tactics for the Application of Neural Networks to Industrial Electronics *Mary Lou Padgett, Paul J. Werbos, and Teuvo Kohonen* 835
Computational Intelligence Connections and Future • Engineering Intelligent Electronics Applications • Summary of Basic Modeling Concepts • Applications • Future • Defining Terms • Resources

58 The Basic Ideas in Neural Networks *David E. Rumelhart, Bernard Widrow, and Michael Lehr* 853
Introduction • Learning by Example • Generalization • Hints for Successful Applications

59 Neural Networks on a Chip *Clifford Lau* 858
Artificial Neural Network Technology Compared with Conventional • Examples of Chips • Comparisons of NN VLSI Microchips • Applications of Neural Network Technology • BMDO/IST Demonstration Project: 3-D ANN Silicon Neuron Seeker

60 Commercially Available Artificial Neural Network Chips *Seth Wolpert* 867
Introduction • Analog ANN Products • Digital ANN Products • Hybrid ANN Products • Discussion

61 Implementing Neural Networks in Silicon *Seth Wolpert and Evangelia Micheli-Tzanakou* . 874
Introduction • The Living Neuron • Neuromorphic Models • Neurological Process Modeling

62 An Avionics Application: MIMD Neural Network Processor *Richard Saeks* 885
NNP Architecture • Summary

63 Backpropagation to Neurocontrol *Paul J. Werbos* 888
Neurocontrol: Where It Is Going and Why It Is Crucial

64 CMAC Neural Networks and Color Correction *King-Lung Haung* 906
Introduction • High-Order CMAC Neural Networks for Color Correction • Experimental Result • Conclusion

65 Temporal Signal Processing *Simon Haykin* 910
Introduction • Temporal Neural Networks with Observable States • Temporal Neural Networks with Hidden States • Conclusions

66 **Feature Selection for Pattern Recognition Using Multilayer Perceptrons** *Dennis W. Ruck and Steven K. Rogers* 916
Introduction • Background • Methodology • Applications • Conclusions

67 **Wavelets for Pattern Recognition** *George W. Rogers, David J. Marchette, and Jeffrey L. Solka* 923
Wavelet Based Segmentation • Resistive Grid Local Averaging • Examples

68 **Fractals for Pattern Recognition** *George W. Rogers, Carey E. Priebe, and Jeffrey L. Solka* 933
A PDP Approach to Localized Fractal Dimension Computation with Segmentation Boundaries

69 **Multilayer Perceptrons with ALOPEX and Backpropagation** *Daniel A. Zahner and Evangelia Micheli-Tzanakou* 942
Introduction • The Backpropogation Algorithm • The ALOPEX Algorithm • Multilayer Perception Network • ALOPEX in VLSI • Discussion

70 **Supervised Neural Networks for Handwritten Digit Recognition in Industrial Processing** *WooGon Chung and Evangelia Micheli-Tzanakou* 951
Introduction • Preprocessing of Handwritten Digit Images • Zernike Moments (ZM) to Characterize Image Patterns • Dimensionality Reduction • Analysis of Prediction Error Rates from Bootstrapping Assessment • Summary

71 **Neocognitron** *Kunihiko Fukushima* 966
Neocognition • Selective Attention Model (SAM)

72 **Studies of Pattern Recognition with Self-Learning Layered Neural Networks** *Faiq A. Fazal and Evangelia Micheli-Tzanakou* 975
Abstract • Introduction • Neocognitron and Pattern Classification • Objectives • Methods • Study A • Study B • Summary and Discussion

73 **Analog 3-D Neuroprocessor for Fast Frame Focal Plane Image Processing** *Tuan A. Duong, Sabrina Kemeny, Taher Daud, Anil Thakoor, Chris Saunders, and John Carson* 990
Introduction • Neural Network Architecture • Neural Network Design and Operation • Experimental Results • Cascade-Backpropagation (CBP) • Six-Bit Parity Problem • Conclusions

74 **Simulated Annealing, Boltzmann Machine, and Hardware Annealing** *Tony H. Wu and Bing J. Sheu* 1003
Simulated Annealing • Botzmann Machine • Hardware Annealing on Hopfield Networks for Optimization • Hardware Annealing on Cellular Neural Networks

75 **Radial Basis Function (RBF) Neural Networks** *Thomas Lindblad, Clark S. Lindsey, and Åge Eide* 1014
Introduction • Topology • Operation • Training • Summary • Defining Terms

76 **Hardware Implemented Radial Basis Function (RBF): The IBM Zero Instruction Set Computer** *Thomas Lindblad, Clark S. Lindsey, and Åge Eide* 1019
Introduction • The ZISC036 VLSI Chip • Processing and Training • Implementing the Chip • Summary and Extrapolations

77 **The RCE Neural Network** *Douglas L. Reilly* 1025
Introduction • Training the RCE Network • RCE Network Responses • Practical Guides to RCE Network Training and Use • Applications of RCE to Pattern Recognition • RCE Network on a Commercially Available Neural Network Chip

78 **Probabilistic Neural Networks Model** *Donald F. Specht* 1038
Basic PNN • Adaptive PNN • High-Speed Classification • Other Considerations • Summary

79 **General Regression Neural Network Model** *Donald F. Specht* 1047
GRNN • Adaptive GRNN • Summary

80 **Classifiers: An Overview** *WooGon Chung and Evangelia Micheli-Tzanakou* 1055
Introduction • Criteria for Optimal Classifier Design • Categorizing the Classifiers • Classifiers • Neural Networks • Comparison of Experimental Results • System Performance Assessment • Analysis of Prediction Rates from Bootstrapping Assessment

54

Current Applications of Expert Systems in Industrial Electronics

Mary Lou Padgett
Auburn University

Robert Shelton
NASA/Johnson Space Center

54.1 Emerging Trends for Expert Systems in Industrial Electronics.. 805
54.2 Defining Terms 805
54.3 Resources 807

Capturing human expertise in a program is a goal of expert systems. Methodologies for using knowledge and heuristics to solve problems are described in Expert Systems Methodology by Riley (Chapter 55). This material covers expert system environments which make the use of expert systems in industrial electronics practical. Rule based programming and the rete algorithm are described and compared to traditional procedural approaches. In industrial electronics, monitoring a series of sensors for overheating is often necessary. Example programs are provided to show the simplicity of a CLIPS implementation compared to a C-language code segment. CLIPS is a forward chaining rule-based programming language developed at NASA/JSC. The CLIPS Object Oriented Language (COOL) is recommended for simulating circuit components for such tasks as a truth table simplification.

Continuing the process monitoring theme, the material on Expert Systems and Their Use in Complex Engineering Systems by Uhrig and Tsoukalas (Chapter 56) details methods designed to reduce the uncertainty inherent in operator decisions. Embedding expert advise in a rapidly accessible database promises improvements in safety and reliability and efficiency of operations in many arenas. Futures for such systems include the incorporation of computational intelligence modules (NN, FZ, EC) and virtual reality.

54.1 Emerging Trends for Expert Systems in Industrial Electronics

The use of expert systems in industrial electronics can enhance the reliability, safety and efficiency of operations. Since expert systems are functional only within very narrow limitations of performance, combining these systems with neural, fuzzy and/or evolutionary/genetic systems components enhances performance. Such combinations are covered in the chapter on computational intelligence and hybrid systems in industrial electronics and in the other chapters in the intelligent electronics section.

54.2 Defining Terms

The following terms are taken from the articles in this chapter, in some cases verbatim. They were also placed in the public domain prior to being submitted to the CI Standards News, an IEEE New Opportunities in Standards activity (Padgett, 1995).

Actions of a rule. Part of a rule to fire or take place when the rule applies. Synonyms: consequent, then portion or right-hand-side (RHS) of a rule.

Antecedent: Conditions of a rule.

Artificial intelligence: Branch of computer science that attempts to emulate certain mental processes of humans by using computer models.

Backward chaining (modus tollens): Inference moves from conclusions to facts that might support that conclusion.

Case-based reasoning: Seeks a solution by finding the problem in the case base most similar to the current problem. Useful when heuristics or exhaustive list of inputs is not available, but a large set of cases or examples exists. Contrast with rule induction.

Certainty factors: Measure of belief of the user that a piece of evidence is true (not probabilities).

Conditions of a rule: Portion of a rule tested to see if rule applies. Synonyms: antecedent, conditional part, pattern part, if portion or left-hand-side (LHS).

Conflict resolution strategy: Inference engine's process of selecting which rule to fire among many that apply.

Consequent: Actions and conclusions of a rule.

0-8493-8343-9/97/$0.00+$.50

Control structure: Collection of rules assigned to select appropriate rule clusters at a given time.

Cycle: Process of the inference engine selecting and executing a rule. Also called recognize-act cycle, select-execute cycle and situation-response cycle.

Data-driven: Changes in data determine which rules apply, as in forward-chaining rule-based programs.

Decision trees: Represent knowledge by encoding a problem solution as a series of questions that determine the path to be taken through a tree where the leaves of a tree represent solutions.

Demons: Set of rules which may be present to function outside the control structure for immediate action.

Domain: Scope of the knowledge contained within the data base.

Domain expert: Person familiar with the details specific to the application

Expert systems: Computer programs which exhibit proficiency in problem solving typically exhibited only by human experts. Apply substantial knowledge of specific areas of expertise to solve finite, well-defined problems. Functions include: reasoning using formal logic, seeking information from a variety of sources including data bases and the user, and interacting with conventional program to carry out tasks. Limitations include inability to solve problems outside their domain of expertise, and frequent inability to detect the limitations of their domain.

Expert systems environments: User-friendly suite of expert system tools and components for easy application construction.

Expert system "shell": A computer program used to develop an expert system.

Expert systems tools: Individual tools not encapsulated in a shell. More flexible than a shell, but not as easy to use. Provide paradigms such as rule-based programming, object-oriented programming and/or procedural programming; features such as hypothetical reasoning, temporal reasoning, truth maintenance and/or explanation/justification systems.

Explanation/justification systems: Features for explaining why decisions were reached.

Facts: See working memory.

Firing: Selection and execution of rules by the inference engine.

Forward chaining (modus ponens): Inference moves from facts to conclusions supported by these facts.

Heuristics: Human expertise obtained directly from humans, or indirectly from books, codes, standards, or data bases, as well as general and specialized knowledge that pertains to the specific situation.

Hypothetical reasoning: Features for "what-if" situations.

If portion: Conditions of a rule.

Interface: Tools or environment helpful in translating user input into computer language, allowing changes to the knowledge base, and presenting conclusions and explanations to the user in written or graphical form.

Inference: Process through which new facts are derived from existing facts and facts which are no longer true are removed.

Inference engine: Component of an expert system which controls the process of inference, e.g. through the selection and firing of rules.

Knowledge-based systems: See expert systems.

Knowledge base: The expertise embodied by an expert system, usually stored in the form of rules or frames.

Knowledge engineer. A programmer familiar with the techniques of eliciting and converting a domain expert's knowledge to a knowledge representation format for use in an expert system.

Left-hand-side (LHS): Conditions of a rule.

Pattern matching: Process where inference engine automatically matches data against patterns and determines which rules apply. Continues selecting, applying a rule and updating the data until no more rules apply.

Pattern part: Conditions of a rule.

Procedural programming: Traditional sequential execution of program statements.

Productions: See rules.

Production systems: See rule-based systems.

Rete Pattern Matching Algorithm: Rules are checked only when facts that affect that rule are added or removed. Speeds execution at the cost of storing data on rule states.

Right-hand-side (RHS): Actions and conclusions of a rule.

Rules: Representation of knowledge. Consist of a series of patterns which specify the conditions for applying the rule, and a set of actions to perform (or conclusions to reach) when the rule applies. Whenever-then statements rather than if-then statements.

Rule base: Database of rules for an expert system.

Rule-based programming: Rules executed through forward-chaining, backward-chaining or both.

Rule clusters: A group of rules in the rule base dedicated to a certain task or subtask.

Rule induction: Technique where rules are created from examples supplied by the domain expert. Especially good when all input parameters and output results can be enumerated in a table. Contrast with case based reasoning.

Rule interpreter: Mechanism through which rules are selected to be fired.

Rules of thumb: See heuristics.

Shells: Programs designed for specific types of applications and providing a framework for developing an expert system by just adding domain knowledge to the shell.

Subtask: Fraction of a task assigned to a group of rules.

Temporal reasoning: Features for time-based events.

Temporal redundancy: Property where most of the rules that apply at one cycle, still apply at the next cycle since the actions of a rule change very few facts for the next cycle.

Then portion: Actions and conclusions of a rule.

Truth maintenance: Features for maintaining the consistency of facts.

Uncertainty management: Representing or combining uncertain data and drawing reliable inferences from it.

Working memory: Global database holding input data, inferred hypotheses and internal information about the program.

54.3 Resources

Periodicals: Based on a recent article in *IEEE Expert* (Cheng, Holsapple, and Lee, 1994) the 10 leading periodicals for expert systems research are

IEEE Expert
Artificial Intelligence
AI Magazine
Expert Systems
International Journal of Man-Machine Studies
Machine Learning
Communications of the ACM
Knowledge Acquisition
AI Expert
Expert Systems and Applications

Internet: Information about access to CLIPS is found in the Appendix of the article by Riley in this chapter, but more detail is included in AI on the Internet (Hengl, 1995). The latter covers expert and knowledge-based systems:

World Wide Web Sites
FTP Sites
America OnLine Resources
CompuServe Resources
UseNet
Vendor Information.

Since internet information changes so rapidly, Hengl recommends strategies for finding the latest applicable sites.

References

Cheng, C. H., Holsapple, C. W., and Lee, A. 1994. "The Impact of Periodicals on Expert Systems Research," *IEEE Expert,* 9:(6), December.

Hengl, T. 1996. *PC AI Magazine Presents AI on the Internet,* Knowledge Technology, Inc.

Padgett, Mary L., ed. 1995. *Computational Intelligence Standards News,* an electronic newsletter approved by the IEEE New Opportunities in Standards Committee in 1995 (m.padgett@ieee.org).

55 Expert Systems Methodology

55.1 Capturing Human Expertise in a Program 808
55.2 Rule-Based Programming 809
The Rete Algorithm • Comparison of a Procedural Approach to a Rule-Based Approach
55.3 Truth Table Simplification Program 811
Representing the Circuit • Updating the Circuit Components • Finding the Source and LED Components • Iterating through the Source Component Value Combinations • Saving the Truth Table Results • Simplifying the Truth Table • Printing the Simplified Truth Table • The Completed Program • Additional Comments

Gary Riley
NASA/Johnson Space Center

55.1 Capturing Human Expertise in a Program

Expert systems, also referred to as knowledge-based systems, are computer programs which exhibit proficiency in problem solving typically exhibited only by human experts. Expert systems differ from more ubiquitous programs (such as accounting systems) in that they utilize knowledge and heuristics rather than procedures or algorithms to solve problems. Expert systems are typified by problems that require the processing of symbolic or conceptual data rather than numeric data. Expert systems have been developed which span a broad range of human expertise in fairly specific domain areas. Well known expert system applications include MYCIN (Davis et al., 1977), a program for diagnosing bacterial infections; CRYSALIS (Englemore and Terry, 1979), a program for determining protein structure; PROSPECTOR (Duda et al., 1979), a program for interpreting geologic data for minerals; and XCON (McDermott and Bachant, 1984), a program for configuring DEC computer systems.

The expertise embodied by an expert system is referred to as the knowledge base. Facts, also referred to as working memory, are the data on which an expert system operates. Facts represent the body of information that the expert system believes to be true. Inference is the process through which new facts are derived from existing facts and facts which are no longer true are removed. The inference engine is the component of an expert system which controls the process of inference.

Early developers of expert systems found themselves building the same types of software components to support the development of their knowledge-based systems. There is a saying that "if you only have a hammer, then all of your problems start to look like nails." Even though an expert system is classified by what it does rather than how it was built, expert system developers found that the use of procedural programming languages such as C made the development of an expert system much more difficult. Developers wanting to represent knowledge as heuristics and "rules of thumb" encountered difficulties in translating their conceptual model of solving a problem with expert system techniques to a procedural programming paradigm which required knowledge to be represented as procedures and step by step algorithms. Developers began to create software components to layer on top of the procedural languages which allowed them to model knowledge in a form similar to their conceptual model of a problem solution. As more and more expert systems were developed, it became apparent that special purpose programs specifically designed for the construction of expert systems could drastically reduce the amount of time and effort it took to create an expert.

Many of the early programs used to build expert systems were shells. These programs were designed for specific types of applications and provided a framework for developing an expert system by just adding domain knowledge to the shell. An early example of this type of program is the EMYCIN (empty MYCIN) shell which was developed by removing the medical knowledge contained in the MYCIN expert system. Shells lack the generic capabilities of a programming language, but require little programming knowledge to develop an expert system. This allows a domain expert to create an expert system without the need of a knowledge engineer, a programmer familiar with the techniques of eliciting and converting a domain expert's knowledge to a knowledge representation format suitable for use in an expert system.

Many shells represent knowledge as rules—what we call heuristics. A rule is generally composed of a set of conditions and a list of actions to perform if the conditions are true. Some shells simplify the creation of rules using rule induction. With this technique, rules are created from examples supplied by the domain expert. This technique is particularly useful when all of the input parameters and associated output results can be

0-8493-8343-9/97/$0.00+$.50

enumerated in a tabular format. Case-based reasoning is a particularly useful technique when it is difficult to acquire heuristics or specify all possible inputs, but a large body of examples or cases exist. Under these circumstances, a case based reasoning system would attempt to determine a solution by finding the most similar problem in the case base to the current problem. Cased-based reasoning is particularly well suited to help desk applications since problems and solutions are typically archived and it's virtually impossible to enumerate all potential problems. Decision trees represent knowledge by encoding a problem solution as a series of questions that determine the path to be taken through a tree where the leaves of the tree represent possible solutions. For example, the first question asked by a car diagnosis expert system might be, "Does the engine start?" If the answer is no, the next question might be, "Does the engine turn over?" If the answer is yes, the next question might be, "Does the engine make any noise?" The questions would continue until the problem was diagnosed.

Expert system tools are more difficult to use than shells, but provide more flexibility. Tools usually provide the same level of generic programming capabilities provided by other programming languages, although they are generally not designed for solving algorithmic or procedural problems. The most common programming paradigm provided by expert system tools is rule-based programming. Some tools provide additional programming paradigms such as object-oriented programming and procedural programming. Since many expert systems contain components best represented in different paradigms, the availability of multiple paradigms within a tool allows the developer to choice the most appropriate paradigm for each component. Tools can also provide a variety of additional features including hypothetical reasoning (for what-if situations), temporal reasoning (for time-based events), truth maintenance (for maintaining the consistency of facts), and explanation/justification systems (for explaining why decisions were reached).

Expert system environments extend the development capabilities of tools by providing suites of tools and components that make the creation of end-user applications much easier. Components commonly provided by environments include sophisticated debugging/browsing utilities, database access libraries, graphical user interface builders, and data acquisition routines.

55.2 Rule-Based Programming

In rule-based programming, rules are used to represent heuristics, or "rules of thumb." Rules are also referred to as productions and rule-based systems as production systems. A rule consists of a series of patterns which specify the conditions which must exist for the rule to be applicable and a set of actions to perform or conclusions reached when the rule is applicable. The conditions of a rule are often referred to as the antecedent, conditional part, pattern part, if portion, or left-hand-side (LHS) of the rule. The actions and conclusions of a rule are often referred to as the consequent, then portion, or right-hand-side (RHS) of a rule.

The inference engine controls the selection and execution (also referred as firing) of rules. Rule-based inference engines generally allow rules to be executed through a forward chaining process, backward chaining process, or a combination of the two. In forward chaining, inferencing occurs from facts to the conclusions supported by those facts. For example, if the oil light on your car dashboard is illuminated, you'd conclude that you're running low on oil. In backward chaining, inference occurs from conclusions to the facts that support those conclusions. For example, if your doctor thinks you have a bacterial infection, he might order blood tests to support that conclusion.

In forward chaining systems, the inference engine automatically matches data against patterns and determines which rules are applicable in a process called pattern matching. During execution, the inference engine repeatedly selects a rule from the list of applicable rules and performs its actions (which may affect the list of applicable rules by adding or removing data) until there are no applicable rules remaining. The process of selecting one rule among many to execute is handled by the inference engine's conflict resolution strategy. The process of the inference engine selecting and executing a rule is called a cycle and has been referred to by many names including the recognize-act cycle, select-execute cycle, and situation-response cycle. Figure 55.1 illustrates the process performed by the inference engine.

Rules are often described as being similar to the if then statements familiar to programmers who have used procedural programming languages such as C. This analogy is misleading, however, because a procedural if-then statement is considered only when the flow of execution reaches the statement. In contrast, a rule is always ready to be executed when its conditions are satisfied since pattern matching always occurs whenever changes are made to facts. Because of this characteristic, forward chaining rule-based programs are said to be data-driven. Thus, rules should be thought of as whenever-then statements rather than if-then statements. A rule acts in many ways like a software interrupt where the inference engine selects the interrupt with the highest priority to be executed.

The Rete Algorithm

Most forward chaining rule-based program exhibit a property called temporal redundancy. Typically, the actions of a rule

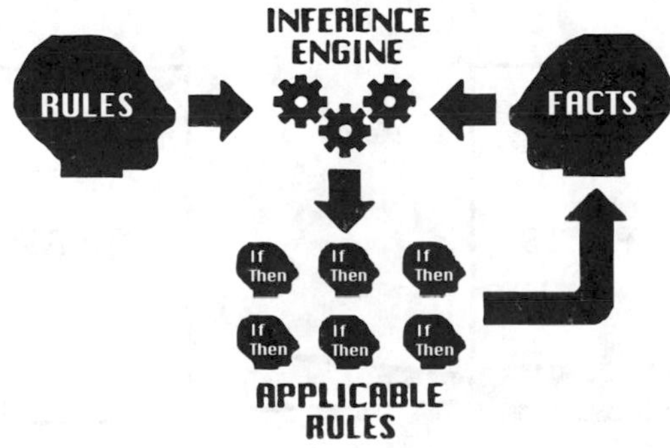

Figure 55.1 Execution of a rule-based program.

change very few facts from one cycle of execution to the next. Because of this, most of the facts present in the current cycle of execution will be present in the next cycle of execution. If rules search for the facts that satisfy their conditions after every cycle as illustrated by the Figure 55.2, then there are a large number of unnecessary computations performed if the shaded area represents the facts that have actually changed.

The Rete pattern matching algorithm (Forgy, 1982), utilized by many forward chaining rule-based languages, improves execution performance by remembering state information about which patterns have already been matched from one cycle to the next. The Rete algorithm increases program execution speed at the cost of the additional memory needed to store the state information of the patterns that have already been matched. Rules are checked to see if their conditions are satisfied only when facts that affect the rule are added or removed. Figure 55.3 illustrates the reduced number of computations that must be made from one cycle to the next when the facts update the rules that are searching for them.

Comparison of a Procedural Approach to a Rule-Based Approach

To illustrate the advantages of rule-based programming, consider the problem of monitoring a series of sensors. The following example program written in the C programming language illustrates how these sensors could be monitored using a procedural programming paradigm to determine if any two of the sensors have *bad* values (which a hypothetical expert indicates represents an overheated device).

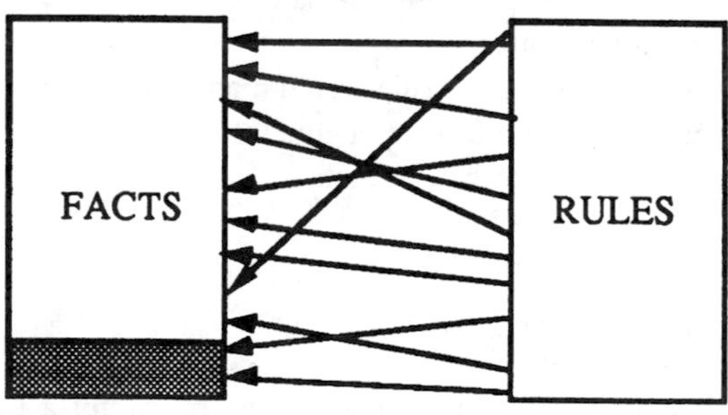

Figure 55.2 Unnecessary computations when rules search for facts.

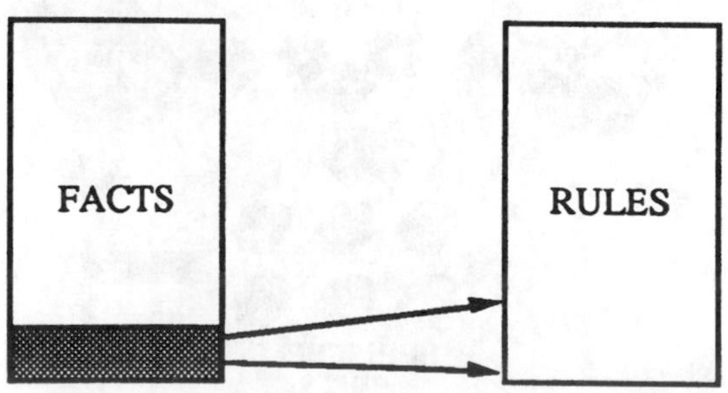

Figure 55.3 Facts searching for rules.

```
#define BAD 0
#define GOOD 1
#define DEVICE_OVERHEATED 0
#define DEVICE_NORMAL 1

int CheckSensors(sensorValues,numberOfSensors)
  int sensorValues[ ];
  int numberOfSensors;
  {
   int firstSensor, secondSensor;

   for (firstSensor = 1;
        firstSensor <= numberOfSensors;
        firstSensor++)
      {
       for (secondsensor = 1;
            secondSensor <= numberOfSensors;
            secondSensor++)
      {
       if ((firstSensor != secondSensor) &&
           (sensorValues[firstSensor] == BAD) &&
           (sensorValues[secondSensor] == BAD))
         { return(DEVICE_OVERHEATED); }
      }
   }
  return(DEVICE_NORMAL);
}
```

The *CheckSensors* function is implemented by storing the values of the sensors as integers in an array and then using two *for* loops to compare all combinations to determine if any two sensors have *bad* values. This function is relatively efficient if the sensors only need to be checked once. However, if this check is performed each time a sensor's value is changed, then all possible combinations are rechecked which is inefficient. In addition, the programmer has the responsibility for calling this function whenever an update is made to a sensor's value. An additional function could be written to check only one sensor against all other sensors, however, this increases the burden on the programmer. For contrast, the equivalent CLIPS code for a rule which performs the same task is shown following (CLIPS will be extensively discussed in the next section).

```
(defrule Two-Sensors-are-Bad
   (Sensor (ID-number ?id) (status Bad))
   (Sensor (ID-number ~?id) (status Bad))
   =>
   (assert (Device (status Overheated))))
```

The first line of the rule contains the keyword *defrule* which indicates that a rule is being defined. The symbol *Two-Sensors-are-Bad* is the name of the rule. The next two lines beginning with the symbol *Sensor* are the patterns that form the *if* portion of the rule. Essentially, the first pattern searches for any *Sensor* fact that contains a *status* value of *Bad* and the second pattern searches for another *Sensor* fact with a *status* value of *Bad* that does not have the same *ID-number* as the *Sensor* fact matching the first pattern. The => symbol serves to separate the *if* portion of the rule from the *then* portion of the rule. Finally, the assert command in the *then* portion of the rule creates a new fact which indicates that the device has overheated.

Because of the overhead associated with the inference engine and the generality provided through pattern matching, a rule-based program generally does not execute as quickly as a procedural program. However, significantly less code is required and the programmer does not have to explicitly check for applicable rules when sensor values are changed. Rules are always looking for new facts which satisfy their conditions. Indeed, careless implementation of pattern matching capabilities in a procedural language may result in a program which runs much less efficiently than its rule-based counterpart.

55.3 Truth Table Simplification Program

To demonstrate the techniques of rule-based programming, we'll develop a rule-based program which will simplify the truth table of an electronic circuit consisting of Boolean components. CLIPS (Giarranto and Riley, 1994; NASA, 1993), a forward chaining rule-based programming language developed at NASA's Johnson Space Center, will be used to build the program. Figure 55.4 illustrates the circuit that we'll use as a sample. It has two inputs, *Source #1* and *Source #2,* and two outputs, *LED #1* and *LED #2.* There are three other Boolean components to the circuit: *Not Gate #1, Or Gate #1,* and *And Gate #1.* The output of a *not gate* is the inverse of its single input. The output of an *or gate* is *True* if either of its inputs are *True,* otherwise its output is *False.* The output of an *and gate* is *True* if both of its inputs are *True,* otherwise its output is *False.*

Since the output of *Not Gate #1* is the inverse of *Source #1, Or Gate #1* will always have an output of *True* (or *on* or *1*) since it receives input directly from *Source #1* and also receives input from *Not Gate #1.* The output of *LED #1* is the input from *Or Gate #1.* Since the output of *Or Gate #1* is always *True,* the output of *LED #1* is also always *True.* Since the input to *And Gate #1* from *Or Gate #1* is always *True,* the output from *And Gate #1* will be *True* if the input from *Source #2* is *True,* otherwise the output from *And Gate #1* will be *False.* Table 55.2 shows the complete truth table for the circuit shown in Figure 55.4.

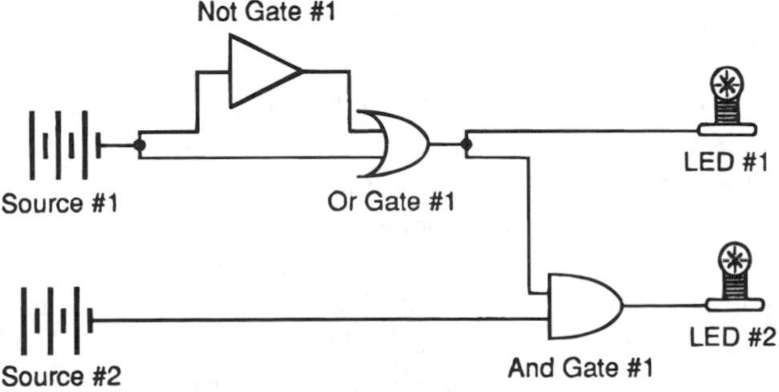

Figure 55.4 An electronic circuit.

Table 55.1 Truth Table for Circuit in Figure 55.4

Source #1	Source #2	LED #1	LED #2
False	False	True	False
False	True	True	True
True	False	True	False
True	True	True	True

Table 55.2 Simplified Truth Table for Circuit in Figure 55.4

Source #1	Source #2	LED #1	LED #2
*	False	True	False
*	True	True	True

An examination of Table 55.2 shows that the output values of the *LED* components are completely unrelated to the values for *Source #1.* Because of this the truth table from Table 55.1 can be simplified as shown in Table 55.2 (where the * indicates that it does not matter whether the value is *True* or *False*).

Representing the Circuit

The purpose of the program we will develop will be to derive the information shown in Table 55.1 from the Boolean circuit and then generate the information contained in Table 55.2. To start, we'll need data structures for representing the components of the circuit. CLIPS provides a programming construct called a *deftemplate* which is similar to a record in Pascal or a structure in C. The deftemplate for representing a circuit component is shown following.

```
(deftemplate component
   (slot name)
   (slot type)
   (slot value (default FALSE))
   (slot input-1 (default no-input))
   (slot input-2 (default no-input))
   (slot output-1 (default no-output))
   (slot output-2 (default no-output)))
```

The *deftemplate* keyword indicates to CLIPS the type of construct being defined. Constructs are the basic building blocks of a CLIPS program. Following the construct type is the name of the construct, in this case *component,* which can be used in the CLIPS interpreted environment to manipulate the construct. Slots in which values can be stored are specified by the *slot* keyword followed by the name of the slot. CLIPS is a weakly typed language, so it is not necessary to specify the type of value being stored in the slot (such as a float or integer type), although a slot can be strongly typed if desired. Several of the slots have been given specific default values using the *default* keyword. When *component* facts are created, these slots will be assigned the specified default value if a value is not supplied.

The *name* slot will be used to store a short-hand name for each component (e.g., *S#1* for *Source #1*). The *type* slot will indicate the type of component (e.g., *AND-GATE* for an *And Gate*). The *value* slot will store the component's Boolean output value (which is sent to the components specified in the *output* slots). By default, we'll store the value *FALSE* in the slot. Each component can have up to two input connections, *input-1* and *input-2*, and up to two output connections, *output-1* and *output-2*. Note that the *value* slot contains the component's Boolean value based on its inputs. The *output* slots are only used to propagate this value to other components.

With the *component* deftemplate defined, it's now possible to represent the circuit shown in Figure 55.4 as a collection of facts. The *deffacts* construct allows a collection of facts to be defined that will be automatically created whenever the **reset** initialization command is executed. The following deffacts represents the circuit shown in Figure 55.1.

```
(deffacts circuit
  (component (name S#1) (type SOURCE)
  (output-1 N#1) (output-2 O#1))
  (component (name S#2) (type SOURCE)
             (output-1 A#1))
  (component (name N#1) (type NOT-GATE)
             (input-1 S#1)
             (output-1 O#1))
  (component (name O#1) (type OR-GATE)
             (input-1 N#1) (input-2 S#1)
             (output-1 LED#1) (output-2 A#1))
  (component (name A#1) (type AND-GATE)
             (input-1 S#2) (input-2 O#1))
             (component (name LED#1) (type LED)
             (input-1 O#1))
  (component (name LED#2) (type LED)
             (input-1 A#1)))
```

Again, the *deffacts* keyword indicates to CLIPS the type of construct being defined. The name of the deffacts is *circuit*. Following the name is a series of 7 facts representing the components in the circuit. Notice that many slot values are left unspecified. For example, the fact representing *Source #1* has no values specified for its *input* slots since a *Source* component has no inputs.

CLIPS provides an interpretive environment where commands and constructs can be entered at a command prompt. The deftemplate and deffacts shown previously could be entered directly at the "CLIPS>" command prompt or saved to a file and loaded in CLIPS using the load command. The following sequence of commands and their output illustrate how interaction occurs in the CLIPS environment. User inputs are shown in bold.

```
CLIPS> (load circuit.clp)
Defining deftemplate: component
Defining deffacts: circuit
TRUE
CLIPS> (facts)
CLIPS> (reset)
CLIPS> (facts)
f-0     (initial-fact)
f-1     (component (name S#1) (type SOURCE)
                   (value FALSE)
                   (input-1 no-input)
                   (input-2 no-input)
                   (output-1 N#1)
                   (output-2 O#1))
f-2     (component (name S#2) (type SOURCE)
                   (value FALSE)
                   (input-1 no-input)
                   (input-2 no-input)
                   (output-1 A#1)
                   (output-2 no-output))
f-3     (component (name N#1) (type NOT-GATE)
                   (value FALSE)
                   (input-1 S#1)
                   (input-2 no-input)
                   (output-1 O#1)
                   (output-2 no-output))
f-4     (component (name O#1) (type OR-GATE)
                   (value FALSE)
                   (input-1 N#1)
                   (input-2 S#1)
                   (output-1 L#1)
                   (output-2 A#1))
f-5     (component (name A#1) (type AND-GATE)
                   (value FALSE)
                   (input-1 S#2)
                   (input-2 O#1)
                   (output-1 no-output)
                   (output-2 no-output))
f-6     (component (name L#1) (type LED)
                   (value FALSE)
                   (input-1 O#1)
                   (input-2 no-input)
                   (output-1 no-output)
                   (output-2 no-output))
f-7     (component (name L#2) (type LED)
                   (value FALSE)
                   (input-1 A#1)
                   (input-2 no-input)
                   (output-1 no-output)
                   (output-2 no-output))
For a total of 8 facts.
CLIPS>
```

The **load** command loads the file "circuit.clp" which contains the *component* deftemplate and *circuit* deffacts. As the file is loaded, informational messages are displayed indicating which constructs are being loaded. Some commands have return values. For the **load** command, the return value *TRUE* indicates that the file was successfully loaded. The **facts** command is used to display the list of facts in the CLIPS knowledge base. Loading the *circuit* deffacts does not automatically create the facts specified in the construct. Hence the first **facts** command indicates that there are currently no facts in the knowledge base. When the

reset initialization command is issued, the facts in the *circuit* deffacts are created. The subsequent **facts** command verifies this (note that the output from the **facts** command has been hand formatted with additional spacing and carriage returns to improve readability). The symbols f-1, f-2, etc. are a shorthand notation used to refer to the facts. In addition to the *component* facts defined by our *circuit* deffacts, an *initial-fact* fact has been added to the list of facts. This fact is automatically added by CLIPS when a **reset** command is issued. It is useful for starting program execution by creating rules which match this fact, although we won't use this technique for the program we're developing.

Updating the Circuit Components

Now that the knowledge base contains facts describing the circuit, these facts can be manipulated. The first change needed is to update a component's *value* slot based on its inputs. For example, Not Gate #1 (represented by fact f-3) receives an input value of *FALSE* from *Source #1,* fact f-1. This means that its *value* slot should be *TRUE,* however, its *value* slot contains FALSE. The *value* slot of a *Not Gate* can be updated by adding a rule which checks for an output value that is inconsistent with its input values.

```
(defrule not-gate
   ?c <- (component (type NOT-GATE)
                    (input-1 ?i1)
                    (value ?v))
   (component (name ?i1)
              (value ?v1))
   (test (neq ?v ?v1))
   =>
   (modify ?c (value (not ?v1))))
```

The *not-gate* rule has three patterns. The first pattern will only match *Not Gate* components. It binds the name of the component's input component to the variable ?i1 and the output value of the component to the variable ?v. The syntax "?c <- . . ." assigns the value of the fact matching the pattern to the variable ?c. This value can then be used later to specify a fact to be deleted or modified. The next pattern matches the component that provides the input value for the *Not Gate* component. It stores the output value of the input component in the variable ?v1. The third and final pattern is used to determine if the *Not Gate* component's output value is inconsistent with its inputs. The *test* keyword indicates that a function call is to be executed. The *test* pattern is satisfied if the return value of its function call is any value other than *FALSE.* The function neq is used to compare the values stored in the variables ?v and ?v1. The neq function returns *FALSE* if its arguments are the same value, otherwise it returns *TRUE.* Note that variable bindings made in one pattern restrict the value of the variable when subsequently used in the same or other patterns. For example, in order for both the first and second patterns to be successfully matched, the value of the variable ?i1 bound in the *input-1* slot of the first pattern must be the same as the value of the variable ?i1 bound in the *name* slot of the second pattern.

The action of the rule changes the output value of the *Not Gate* using the **modify** command to change the output value of the fact referred to by the variable ?c. The **not** function inverts the input value of the *Not Gate* component. The third pattern in the *not-gate* rule prevents the rule from being reconsidered once the output value has been changed to the correct value. If this pattern was not included, the rule would continually fire in an endless loop. (Even if the **modify** command changes a slot value to its current value, CLIPS treats the "non-modification" as if the fact has been changed.)

Once the *not-gate* rule has been added to the knowledge base (either by loading it from a file or entering it at the command prompt), it is activated by the existing circuit facts as shown in the following command sequence.

```
CLIPS> (agenda)
0       not-gate: f-3, f-1
For a total of 1 activation.
CLIPS> (watch rules)
CLIPS> (watch facts)
CLIPS> (run 1)
FIRE    1 not-gate: f-3, f-1
<== f-3      (component (name N#1) (type NOT-GATE)
                        (value FALSE)
                        (input-1 S#1)
                        (input-2 no-input)
                        (output-1 O#1)
                        (output-2 no-output))
==> f-8      (component (name N#1) (type NOT GATE)
                        (value TRUE)
                        (input-1 S#1)
                        (input-2 no-input)
                        (output 1 O#1)
                        (output-2 no-output))
CLIPS>
```

The **agenda** command displays the list of rules that are activated (have their patterns satisfied and are ready to fire). The output of the command shows the salience (to be discussed later) of each activation, the name of the rule associated with the activation, and the list of facts that matched the rule creating the activation. For the *not-gate* rule the fact f-3 matched its first pattern and the fact f-1 matched its second pattern. The third pattern of the rule just evaluates an expression and has no fact associated with it, so none is displayed in the activation output. The **watch** command is used to provide debugging information. Watching *rules* will display informational messages whenever a rule is fired (executed). Watching *facts* will display information messages whenever a fact is added, removed, or modified. The addition of facts is denoted by ==> and the removal is denoted by <==. The modification of a fact is implemented by removing the fact and then adding a new fact with the specified modifications. The **run** command begins execution of rules. It accepts an optional argument which indicates the number of rules to be fired before stopping. In the preceding command sequence, the *not-gate* rule is fired with the result that the *Not Gate #1* component fact has its value slot changed from *FALSE* to *TRUE*.

Now that *Not Gate #1* has an output value of *TRUE, Or Gate #1* needs to have its output value updated as well. This in turn will require *LED #1's value* slot to be updated as well. Similar to the *not-gate* rule, the following three rules will update the *value* slots of each of the component types.

```
(defrule and-gate
  ?c <- (component (type AND-GATE)
                   (input-1 ?i1) (input-2 ?i2)
                   (value ?v))
  (component (name ?i1)
             (value ?v1))
  (component (name ?i2)
             (value ?v2))
  (test (neq ?v (and ?v1 ?v2)))
  =>
  (modify ?c (value (and ?v1 ?v2))))
(defrule or-gate
  ?c <- (component (type OR-GATE)
                   (input-1 ?i1) (input-2 ?i2)
                   (value ?v))
  (component (name ?i1)
             (value ?v1))
  (component (name ?i2)
             (value ?v2))
  (test (neq ?v (or ?v1 ?v2)))
  =>
  (modify ?c (value (or ?v1 ?v2))))

(defrule LED
  ?c <- (component (type LED)
                   (input-1 ?i1)
                   (value ?v))
  (component (name ?i1)
             (value ?v1))
  (test (neq ?v ?v1))
  =>
  (modify ?c (value ?v1)))
```

Once the preceding three rules have been added to the knowledge base, the *or-gate* rule is activated. Once this rule has fired, the output value of *Or Gate #2* is changed which causes the activation of the *LED* rule for *LED #1*. The output of the following command sequence shows this. The *activations* argument passed to the **watch** command causes informational messages to be displayed whenever a rule is activated or deactivated.

```
CLIPS> (watch activations)
CLIPS> (agenda)
0        or-gate: f-4, f-8, f-1
For a total of 1 activation.
CLIPS> (run)
FIRE    1 or-gate: f-4, f-8, f-1
<== f-4      (component (name O#1) (type OR-GATE)
                        (value FALSE)
                        (input-1 N#1)
                        (input-2 S#1)
                        (output-1 L#1)
                        (output-2 A#1))
==> f-9      (component (name O#1) (type OR-GATE)
                        (value TRUE)
                        (input-1 N#1)
                        (input-2 S#1)
                        (output-1 L#1)
                        (output-2 A#1))
<== Activation 0      LED: f-6, f-9
FIRE 2 LED: f-6, f-9
==> f-6      (component (name L#1) (type LED)
                        (value FALSE)
                        (input-1 O#1)
                        (input-2 no-input)
                        (output-1 no-output)
                        (output-2 no-output))
<== f-10     (component (name L#1) (type LED)
                        (value TRUE)
                        (input-1 O#1)
                        (input-2 no-input)
                        (output-1 no-output)
                        (output-2 no-output))
CLIPS>
```

Finding the Source and LED Components

To generate the truth table for the circuit, we will need to iterate through all possible output value settings for the *Source* components. We'll also have to retrieve the values of each *LED* component for a given combination of *Source* component values. To facilitate these goals, the following constructs will create facts containing the list of *Source* components and *LED* components.

```
(deffacts list-info
   (LED-list)
   (source-list))

(defrule Find-LEDs
   ?f <- (LED-list $?list)
   (component (name ?n) (type LED))
   (test (not (member$ ?n ?list)))
   =>
   (retract ?f)
   (assert (LED-list ?n ?list)))

(defrule Find-Sources
   ?f -> (source-list $?list)
   (component (name ?n) (type SOURCE))
   (test (not (member$ ?n ?list)))
   =>
   (retract ?f)
   (assert (source-list ?n ?list)))
```

The *list-info* deffacts creates two facts containing the list of *LED* and *Source* components. Initially there are no items in either list. The *LED-list* and *source-list* facts appear different from the *component* facts defined using a deftemplate in that they have no slot names or corresponding *deftemplate* construct. In addition to deftemplate facts, CLIPS also allows ordered facts, such as the *LED-list* and *source-list* facts, which are just a sequence of values without any slot names.

The first pattern in the *Find-LEDs* rule matches against the *LED-list* fact which contains the current list of *LED* components. The multifield variable $?list is assigned the value of all the *LED* components in the *LED-list.* (A single field variable such as ?n must match a single value. A multifield variable can match any number of values including none at all so long as other constraints in the pattern are satisfied.) The second pattern matches any *LED* component. The third pattern is satisfied if the *LED* component matching pattern two is not contained in the *LED-list* that matched the first pattern. The **member$** function determines if a specified value is contained within a list of values. Note that the $ is not required before a multifield variable if the variable is passed to a function. It is only required when pattern matching to indicate that zero or more values may match the pattern. The actions of the *Find-LED* rule deletes the old *LED-list* fact using the **retract** command and creates a new *LED-list* fact containing an additional *LED* component using the **assert** command.

The *Find-Sources* rule is identical to the *Find-LEDs* rule except that it adds *Source* components to the *source-list* fact instead of *LED* components to the *LED-list* fact. The following command sequence shows the creation of the *LED-list* and *source-list* facts. The initially empty *LED-list* and *source-list* facts are asserted manually since the *list-info* deffacts was added after the initial **reset** command was executed. Executing a **reset** command now would remove all existing facts in addition to adding the facts contained in the deffacts constructs. This would cause the rules that had previously fired to be reactivated and we would have to go through the rule firing sequence again that would bring us to the current state.

```
CLIPS> (assert (LED-list))
==> f-11       (LED-list)
==> Activation 0     Find-LEDs: f-11, f7
==> Activation 0     Find-LEDs: f-11, f10
<Fact-11>
CLIPS> (assert (source-list))
==> f-12       (source-list)
==> Activation 0     Find-Sources: f-12, f-
1
==> Activation 0     Find-Sources: f-12, f-
2
<Fact-12>
CLIPS> (run)
FIRE     1 Find-Sources: f-12, f-2
<== f-12       (source-list)
<== Activation 0     Find-Sources: f-12, f-
1
==> f-13       (source-list S#2)
==> Activation 0     Find-Sources: f-13, f-
1
FIRE     2 Find-Sources: f-13, f-1
<== f-13       (source-list S#2)
==> f-14       (source-list S#1 S#2)
FIRE     3 Find-LEDs: f-11, f-10
<== f-11 (LED-list)
<== Activation 0     Find-LEDs: f-11, f-7
==> f-15       (LED-list L#1)
==> Activation 0     Find-LEDs: f-15, f-7
FIRE     4 Find-LEDs: f-15, f-7
<== f-15       (LED-list L#1)
==> f-16       (LED-list L#2 L#1)
CLIPS>
```

With the execution of the *Find-LEDs* and *Find-Sources* rules we now have complete lists of all the *LED* and *Source* components in the knowledge base.

```
f-14     (source-list S#1 S#2)
f-16     (LED-list L#2 L#1)
```

Iterating through the Source Component Value Combinations

The next step in generating a simplified truth table is to generate the unsimplified truth table. To do this, it's necessary to iterate through all of the possible value combinations for the *Source* components. For each value combination, the resultant value of the *LED* components needs to be determined. Because of the opportunistic way in which rules are activated, changes in the *Source* components are only propagated through the circuit for those components that need their values changed. For example, changing the *Source #2* component has no effect on the *Not Gate #1, Or Gate #1,* or *LED #1* components.

Table 55.3 shows one approach for iterating through all possible source value combinations using the standard binary representation for positive integers. Notice that in moving from iteration 2 to 3 that both the *Source #1* and *Source #2* components have their values changed. Indeed, as more *Source* components are added, the number of iterations in which more than one *Source* component has its value changed approaches half the number of total iterations (if a standard binary representation is used).

Table 55.3 Standard Binary Representation

Iteration	Source #1	Source #2	Binary Representation
1	FALSE	FALSE	00
2	FALSE	TRUE	01
3	TRUE	FALSE	10
4	TRUE	TRUE	11

To limit the number of *Source* component changes between iterations and thus reduce the amount of computation that is required to update the final *LED* component values, we'll use a Gray code representation (NASA, 1993) to iterate through the possible *Source* component values. The useful property of a Gray code representation, that is of interest for this problem, is that successive representations of integer values differ by only one binary digit. Table 55.4 shows the values for iterating through two *Source* components using a Gray code representation.

To iterate through all possible combinations, we will first need to determine the maximum number of iterations, set up a fact indicating which iteration we are processing, and create another fact which stores the current Gray code representation being processed. The following rule creates all of these facts.

```
(defrule max-iterations-and-gray-code
  (source-list $?list)
  (not (and (component (name ?n) (type SOURCE))
            (test (not (member$ ?n ?list)))))
  =>
  (bind ?max (integer (** 2 (length$ ?list))))
  (assert (max-iterations ?max))
  (assert (iteration 1))
  (bind ?code (create$))
  (loop-for-count (length$ ?list)
    (bind ?code (create$ FALSE ?code)))
  (assert (gray-code ?code)))
```

The first pattern of the *max-iterations-and-gray-code* rule gets the list of *Source* components generated by the *Find-Sources* rule. The next two patterns, the *component* and *test* patterns, are nested within *not* and *and* Boolean pattern operators. The *not* keyword surrounding the group of patterns indicates that the second pattern is satisfied if the patterns within are not satisfied. The *and* keyword immediately following, groups the two remaining patterns together (syntactically the *not* pattern can only be followed by a single pattern so the *and* pattern allows two or more patterns to be grouped within). These two patterns nested with the *not*/*and* pattern insure that the *source-list* fact matching the first pattern is indeed the entire list of *Source* components and not just a partial list containing only some of the *Source* components. The actions of the rule then compute all of the appropriate facts to begin iteration through the value combinations. A number of functions are called to help compute these facts. Briefly, the **bind** function is used to assign a variable a new value; the **length$** function determines the number of values in a multifield value (in our case the number of *Source* components); the ** function is for exponentation (the number of source value iterations is 2^N where N is the number of *Source* components); the **integer** function converts a numeric argument to an integer (since the ** function returns a floating point number); the **create$** function appends values together to create multifield values; and the **loop-for-count** function provides a procedural mechanism for iterating through a set of actions N times (where N is the first argument supplied to the function).

Table 55.4 Gray Code Binary Representation

Iteration	Source #1	Source #2	Binary Representation
1	FALSE	FALSE	00
2	FALSE	TRUE	01
3	TRUE	TRUE	11
4	TRUE	FALSE	10

Adding the *max-iterations-and-gray-code* rule to our knowledge base and then executing it produces the following output.

```
CLIPS> (agenda)
0      max-iterations-and-gray-code: f-14,
For a total of 1 activation.
CLIPS> (run 1)
FIRE    1 max-iterations-and-gray-code: f-14,
==> f-17      (max-iterations 4)
==> f-18      (iteration 1)
==> f-19      (gray-code FALSE FALSE)
CLIPS>
```

Since there are two *Source* components, it will require 4 iterations to go through all of the possible combinations. We're starting with iteration 1 and the first set of output values for each of the *Source* components is *FALSE* (which is consistent with their default values).

The next step in iterating through the Gray codes is to add a function which indicates which bit (*Source* component) to change when starting the next iteration. CLIPS provides several constructs for procedural programming. We'll use the following *deffunction* construct (which is similar to functions in languages such as C and Pascal) to determine which *Source* component needs to have its value toggled.

```
(deffunction change-which-bit (?it)
  (bind ?i 1)
  (while (and (evenp ?it) (<> ?it 0)) do
    (bind ?it (div ?it 2))
    (bind ?i (+ ?i 1)))
  ?i)
```

Again, the *deffunction* keyword indicates to CLIPS the type of construct being defined. The name of the deffunction is *change-which-bit.* Following the name is the parameter list for the function. This function has a single parameter indicated by the variable ?it. This variable should contain the current iteration. The value returned by the function is the variable ?i because it is the last expression evaluated by the deffunction. This value indicates the position of the bit in the Gray code that needs to be modified for the next iteration. We'll treat the operation of this function as a black box magically returning the position of the correct bit to change. For those interested in understanding the function's behavior: the **while** function evaluates the actions contained in its body so long as its first argument evaluates to any value other than *FALSE;* the **evenp** function returns a Boolean value (*TRUE* if the argument is an even integer, otherwise *FALSE*); the <> function tests for numeric inequality (returning *TRUE* if its arguments are not equal, otherwise *FALSE*); and the **div** function performs integer division.

Our next step is to define a *defrule* construct that uses the *change-which-bit* function to change the *iteration* and *gray-code* facts when an appropriate point has been reached. This rule is shown following.

```
(defrule next-gray-code
  (declare (salience -20))
  ?f1 <- (gray-code $?g)
  (max-iterations ?m)
  ?f2 <- (iteration ?i)
  (test (< ?i ?m))
  =>
  (retract ?f1 ?f2)
  (bind ?bit (change-which-bit ?i))
  (bind ?new-value (not (nth$ ?bit ?g)))
  (bind ?new-code (replace$ ?g ?bit ?bit ?new-value))
  (assert (gray-code ?new-code))
  (assert (iteration (+ ?i 1))))
```

The second line of the *next-gray code* rule, "(declare (salience −20))", is not a pattern, but a rule declaration. In this case, the salience value of the rule is declared to be −20. Salience allows priorities to be assigned to rules. If two or more rules are activated, the rules with higher salience values are executed before the rules with lower salience values. By default, rules have a salience of 0. Since all the previous rules we've defined have the default salience value of 0, the *next-gray-code* rule will only be allowed to execute after all activations of these rules have fired. Indeed, this is the exact behavior that we want—all of the changes to the *Source* component output values should be propagated through the circuit and the results saved before we proceed to the next iteration in deriving the truth table.

The first pattern in the *next-gray-code* rule retrieves the *Source* component settings for the current Gray code iteration. The second pattern retrieves the value of the maximum number of iterations to be performed. The third pattern retrieves the value of the current iteration. The fourth pattern checks that the current iteration is less than the maximum number of iterations to be performed. This prevents the rule from firing when all the possible combinations have already be processed.

The actions of the *next-gray-code* rule remove the previous iterations of the *iteration* and *gray-code* facts using the retract command. The position in the *gray-code* multifield value bound to the variable ?g to be changed is determined by a call to the *change-which-bit* deffunction (for convenience, the left-most component value in the *gray-code* fact will be treated as the first bit—it doesn't matter what order the bits are changed as long as we iterate through all possible values and only change one at a time). The **nth$** function is used to extract the current value of the *Source* component to be changed. The **not** function inverts this value which is then bound to the variable ?new-value. The **replace$** function is then used to replace the old-value of the *Source* component in the gray-code multifield value bound to the variable ?g. The resultant multifield value with the new Gray code *Source* component settings is then assigned to the variable ?new-code. Finally, new *gray-code* and *iteration* facts are asserted with the appropriate values for the next iteration.

Once the Gray code *Source* component values for the next iteration have been determined, the values must be copied to the *Source* components that are being changed. The following rule accomplishes this task.

```
(defrule change-source-from-gray-code
  (gray-code $?b1 ?v $?)
  (source-list $?b2 ?s $?)
  (test (= (length$ ?b1) (length$ ?b2)))
  ?f <- (component (name ?s) (type SOURCE)
    (value ~?v))
  =>
  (modify ?f (value ?v)))
```

The first pattern in the *change-source-from-gray-code* rule binds one of the values in the current *gray-code* fact to the variable ?v. The values preceding this value are bound to the multifield variable $?b1. Similarly, the second pattern in the rule binds one of the values in the *source-list* fact to the variable ?s and the values preceding this value are bound to the multifield variable $?b2. Note that there is a one-to-one positional correspondence between the Gray code values stored in the *gray-code* fact and the *Source* component names stored in the *source-list* fact. The third pattern in the rule enforces this correspondence by checking that the number of values stored in the variables ?b1 and ?b2 are the same. The fourth pattern limits changes to only those *Source* components with values that do not match the next iteration's Gray code values. There should only be one such component in each iteration (since only one binary digit should change in the Gray code representation). The fourth pattern also prevents the endless loop that would occur if the *Source* component fact were modified to contain the same value they had in the last iteration (e.g. changing the value from *FALSE* to *FALSE*). The ~?v contained in the *value* slot of the fourth pattern checks that the value stored in the *value* slot is not equal to the value stored in the variable ?v. The only action of the *change-source-from-gray-code* rule is to change the value of the *Source* component to its new setting.

With the *next-gray-code* and *change-source-from-gray-code* defrule constructs and the *change-which-bit* deffunction construct added, we can now step through all of the Gray code iterations.

```
CLIPS> (agenda)
-20 next-gray-code: f-19, f-17, f-18
For a total of 1 activation.
CLIPS> (run 1)
FIRE    1 next-gray-code: f-19, f-17, f-18
<== f-19     (gray-code FALSE FALSE)
<== f-18     (iteration 1)
==> f-20     (gray-code TRUE FALSE)
```

```
==> Activation 0 change-source-from-gray-code: f-20,
f-14, f-1
==> f-21     (iteration 2)
==> Activation -20      next-gray-code: f-20, f-17,
f-21
CLIPS>
```

The *next-gray-code* rule makes the appropriate changes to proceed from the first iteration to the second iteration. The *change-source-from-gray-code* rule is activated to process the change of the *gray-code* fact from (gray-code FALSE FALSE) to (gray-code TRUE FALSE), and the *next-gray-code* rule is activated (with a lower salience) so that when we're finished with processing the second iteration, we'll move on to the third iteration.

```
CLIPS> (run 1)
FIRE  1 change-source-from-gray-code: f-20, f-14, f-1
<== f-1    (component (name S#1) (type SOURCE)
                      (value FALSE)
                      (input-1 no-input)
                      (input-2 no-input)
                      (output-1 N#1)
                      (output-2 O#1))
==> f-22    (component (name S#1) (type SOURCE)
                       (value TRUE)
                       (input-1 no-input)
                       (input-2 no-input)
                       (output-1 N#1)
                       (output-2 O#1))
==> Activation 0    not-gate: f-8, f-22
CLIPS>
```

The *change-source-from-gray-code* rule changes the output value of the *Source #1* component from *FALSE* to *TRUE*. This affects the *Not Gate #1* component so the *not-gate* rule is activated to change the value of this component.

```
CLIPS (run 1)
FIRE 1 not-gate: f-8, f-22
<== f-8     (component (name N#1) (type NOT-GATE)
                       (value TRUE) (input-1 S#1)
                       (input-2 no-input)
                       (output-1 O#1)
                       (output-2 no-output))
==> f-23    (component (name N#1) (type NOT-GATE)
                       (value FALSE)
                       (input-1 S#1)
                       (input-2 no-input)
                       (output-1 O#1)
                       (output-2 no-output))
CLIPS> (agenda)
-20     next-gray-code: f-20, f-17, f-21
For a total of 1 activation.
CLIPS>
```

The *not-gate* rule is allowed to fire updating the value of the *Not Gate # 1* component. Since the change to this component's output value doesn't affect any other components, no new rules are activated and we are ready to fire the *next-gray-code* rule to change from the second iteration to the third iteration.

Having gone through one iteration watching facts and activations, we will process all remaining iterations by just watching the rules fire.

```
CLIPS> (unwatch facts)
CLIPS> (unwatch activations)
CLIPS> (run)
FIRE    1 next-gray-code: f-20, f-17, f-21
FIRE    2 change-source-from-gray-code: f-24, f-14,
f-2
FIRE    3 and-gate: f-5, f-26, f-9
FIRE    4 LED: f-7, f-27
FIRE    5 next-gray-code: f-24, f-17, f-25
FIRE    6 change-source-from-gray-code: f-29, f-14,
f-22
FIRE    7 not-gate: f-23, f-31
CLIPS>
```

The **unwatch** command is used to disable the printout of debugging information enabled with the **watch** command. Each new iteration begins with *next-gray-code* rule updating the Gray code values. The *change-source-from-gray-code* rule then transfers the new values to the *Source* components. Finally, the *not-gate, or-gate, and-gate,* and *LED* rules are allowed to fire when necessary to update the output values of the components affected by the *Source* component change.

Saving the Truth Table Results

We have iterated through all of the possible combinations for the *Source* components and determined the values for the *LED* components, however, this information for each of the iterations was lost as soon as the next iteration was begun. We'll need to save the information for each iteration before moving on to the next iteration. The following two rules will perform this task.

```
(defrule start-retrieve-result
  (declare (salience -10))
  (iteration ?)
  (source-list $?s)
  (LED-list $?1)
  ==>
  (assert (result ?s -> ?1)))

(defrule fill-result
  ?f <- (result $?b ?n $?e)
  (component (name ?n) (value ?v))
  ==>
  (retract ?f)
  (assert (result ?b (if ?v then T else F)
                  ?e)))
```

The *start-retrieve-result* rule initiates the task of saving the truth table information. It's given a salience of −10 so it will fire after the rules used to propagate changes through the circuit, but before we move on to the next iteration (by firing the *next-gray-code* rule which has a salience of −20). The first pattern in the *start-retrieve-result* rule causes the rule to be retriggered for each iteration by matching the *iteration* fact. The ? used in the pattern indicates that a value must exist (in this case the iteration number), but it does not matter what the value is. If this pattern wasn't present, the rule would only fire once for the first iteration since the *source-list* and *LED-list* facts aren't changed after they are initially determined. The action of the rule asserts a *result* fact which contains all of the *Source* component names, followed by a −> symbol, followed by all of the *LED* component names.

The *fill-result* rule replaces all of the component names in the *result* fact created by the *start-retrieve-result* rule with the actual values of the components for this iteration. The first pattern will match each of the fields in the *result* fact binding each field to the variable ?n and binding the variables $?b and $?e to the fields preceding and following the field bound to ?n. For our example circuit, since there are two *Source* components, two *LED* components, and one −> symbol used in each newly asserted *result* fact, the *fill-result* pattern can match each newly asserted *result* fact in five different ways. The second pattern matches the *component* fact corresponding to the component name bound to ?n in the first pattern. Since the −> symbol isn't a component name, the *fill-result* rule won't be activated for the case where ?n is bound to the −> symbol. The actions of the rule replace the component name in the *result* fact with the actual value of the component. To make the output of the simplified truth table more readable, we replace the values *TRUE* and *FALSE* with *T* and *F* using the (if ?v then T else F) in the **assert** command. Once the component names have been replaced with their values in the *result* fact, the *fill-result* rule won't be reactivated for these fields since there are no components named *T* or *F*.

Since we have already iterated through all of the gray codes we will need to restart the program so that the *start-retrieve-result* and *fill-result* rules get an opportunity to fire at the end of each iteration. We'll load the constructs from a file and stop at the point where the *start-retrieve-result* rule is activated.

```
CLIPS> (clear)
CLIPS> (unwatch compilations)
CLIPS> (load circuit.clp)
%$****$***!****
CLIPS> (watch rules)
CLIPS> (reset)
CLIPS> (run 8)
FIRE    1 Find-SOURCES: f-9, f-2
FIRE    2 Find-SOURCES: f-10, f-1
FIRE    3 max-iterations-and-gray-code: f-11,
FIRE    4 Find-LEDs: f-8, f-7
FIRE    5 Find-LEDs: f-15, f-6
FIRE    6 not-gate: f-3, f-1
FIRE    7 or-gate: f-4, f-17, f-1
FIRE    8 LED: f-6, f-18
CLIPS> (agenda)
-10   start-retrieve-result: f-13, f-11, f-16
-20   next-gray-code: f-14, f-12, f-13
For a total of 2 activations.
CLIPS>
```

The **clear** command deletes all existing constructs from the CLIPS environment. The **load** command loads the constructs contained in the file *circuit.clp*. The characters following the **load** command indicate the types and number of constructs loaded (% for deftemplates, $ for deffacts, * for defrules, and ! for deffunctions). The single characters are printed rather than the "Defining . . ." message because of the preceding **unwatch compilations** command which disables the lengthier compilation messages.

Allowing the *start-retrieve-result* rule to fire creates a *result* fact representing the current iteration.

```
CLIPS> (watch facts)
CLIPS> (watch activations)
CLIPS> (run 1)
FIRE    1 start-retrieve-result: f-13, f-11, f-16
==> f-20     (result S#1 S#2 -> L#1 L#2)
==> Activation 0      fill-result: f-20, f-7
==> Activation 0      fill-result: f-20, f-19
==> Activation 0      fill-result: f-20, f-2
==> Activation 0      fill-result: f-20, f-1
CLIPS>
```

The assertion of the *result* fact causes four activations of the *fill-result* rule—one for each component contained in the *result* fact. Allowing the first of the four activations to fire replaces the S#1 component name with its actual value.

```
CLIPS> (run 1)
FIRE    1 fill-result: f-20, f-1
<== f-20     (result S#1 S#2 -> L#1 L#2)
<== Activation 0      fill-result: f-20, f-2
<== Activation 0      fill-result: f-20, f-19
<== Activation 0      fill-result: f-20, f-7
==> f-21     (result F S#2 -> L#1 L#2)
==> Activation 0      fill-result: f-21, f-7
==> Activation 0      fill-result: f-21, f-19
==> Activation 0      fill-result: f-21, f-2
CLIPS>
```

Retracting the *result* fact removes the remaining activations of the *fill-result* rule. When the new *result* fact is asserted containing the value of the S#1 component rather than its name, three new activations of the *fill-result* rule are created—one for each of the remaining components in the result fact.

Allowing the *fill-result* rules to continue firing until no activations of the *fill-result* remain will replace all of the component names in the *result* fact with each component's value.

```
CLIPS> (unwatch activations)
CLIPS> (run 1)
FIRE    1 fill-result: f-21, f-2
<== f-21      (result F S#2 -> L#1 L#2)
==> f-22      (result F F -> L#1 L#2)
CLIPS> (run 1)
FIRE    1 fill-result: f-22, f-19
<== f-22      (result F F -> L#1 L#2)
==> f-23      (result F F -> T L#2)
CLIPS> (run 1)
FIRE    1 fill-result: f-23, f-7
<== f-23      (result F F -> T L#2)
==> f-24      (result F F -> T F)
CLIPS> (agenda)
-20     next-gray-code: f-14, f-12, f-13
For a total of 1 activation.
CLIPS>
```

Issuing a **run** command and allowing the existing rules to fire to completion will compute and store the results of each iteration. Shown following are the *result* facts present in the list of facts after the program has been allowed to run to completion. A list of all existing facts can be seen by using a **facts** command.

```
f-24    (result F F -> T F)
f-33    (result T F -> T F)
f-43    (result T T -> T T)
f-60    (result F T -> T T)
```

Simplifying the Truth Table

Notice that facts 24 and 33 and facts 43 and 60 can be merged to simplify the truth table. In both cases, the value of the first *Source* component has no effect on the final value of the *LED* components. Facts 24 and 33 can be merged to the following where the * indicates that the value does not matter.

```
(result * F -> T F)
```

Similarly, facts 43 and 60 can be merged to the following.

```
(result * T -> T T)
```

The following rule will be used to merge similar truth table values.

```
(defrule merge-responses
  ?f1 <- (result $?b ?x $?e -> $?response)
  ?f2 <- (result $?b ~?x $?e ->
  $?response)
  ==>
  (retract ?f1 ?f2)
  (assert (result $?b * $?e -> ?response)))
```

The first pattern matches a *result* fact and binds the variable ?x to one of the values of a *Source* component. The variable $?b and $?e are bound to the *Source* components preceding and following the *Source* component value bound to ?x. The variable $?response is bound to the values of the *LED* components. The second pattern will match a *result* fact that is identical to the fact matching the first pattern with the exception that the *Source* component corresponding to the variable ?x from the first pattern has the opposite value in the second pattern. The ~ contained in the expression ~?x is a negation operator and in this case means the value can not be the value stored in the variable ?x. If both patterns of the rule are satisfied, then the facts matching the patterns can be merged to simplify the truth table. The actions of the *merge-response* rule replaces the two *result* facts with a single *result* fact containing a * for the *Source* component to indicate that the value of the specified *Source* component does not affect the response of the *LED* components.

If the *merge-response* rule is added to the knowledge base, four activations of this rule are immediately placed on the agenda.

```
CLIPS> (agenda)
0       merge-responses: f-60, f-43
0       merge-responses: f-43, f-60
0       merge-responses: f-33, f-24
0       merge-responses: f-24, f-33
For a total of 4 activations.
CLIPS>
```

Notice that there are two activations for each of the fact pairs that we want merged: facts 24 and 33 and facts 43 and 60. This occurs since all four result facts match the first pattern of the *merge-responses* rule and the second pattern then matches the corresponding fact. The extra activation for each of the pairs will be removed when the pair of facts are merged by the other activation.

Issuing a **run** command will now allow the facts to be merged.

```
CLIPS> (watch facts)
CLIPS> (watch activations)
CLIPS> (watch rules)
CLIPS> (run)
FIRE    1 merge-responses: f-60, f-43
<== f-60      (result F T -> T T)
<== Activation 0      merge-responses: f-43, f-60
<== f-43      (result T T -> T T)
==> f-61      (result * T -> T T)
FIRE    2 merge-responses: f-33, f-24
<== f-33      (result T F -> T F)
<== Activation 0      merge-responses: f-24, f-33
<== f-24      (result F F -> T F)
<== f-62      (result * F -> T F)
CLIPS>
```

Once the *result* facts are merged, the remaining *result* facts represent the simplified truth table. The remaining *result* facts for our example are shown following.

```
f-61 (result * T -> T T)
f-62 (result * F -> T F)
```

Printing the Simplified Truth Table

The final step for our program is to print the simplified truth table. The following rule initiates the printing of the truth table.

```
(defrule print-header
  (declare (salience -30))
  (source-list $?s)
  (LED-list $?1
  =>
  (assert (print-results))
  (progn$ (?x ?s) (format t "%3s" ?x))
  (printout t " | ")
  (progn$ (?x ?1) (format t "%3s" ?x))
  (format t "%n")
  (progn$ (?x ?s) (printout t "-----"))
  (printout t "-+-")
  (progn$ (?x ?1) (printout t "-----"))
  (format t "%n"))
```

The *print-header* rule initiates printing by asserting the *print-results* fact and printing the header which lists the *Source* and *LED* components. Since we only want the truth table printed after all *result* facts have been merged, the rule is given a salience of −30 so that it fires only after all combinations of *Source* component values have been processed. The *print-header* rule matches against the *source-list* and *LED-list* facts, so it will be activated very early in program execution when the component lists are determined and remain active until it is allowed to fire after all other rules have executed. The assertion of the *print-results* fact will trigger the activation of the *print-result* rule to be discussed next. This rule will iterate through all of the *result* facts, printing them in the appropriate order beneath the table header. The **progn$** function is used to iterate over a multifield value performing an action for each field in the multifield. In the *print-header* rule, this function is used to iterate through each of the *Source* components captured in the variable $?s and the *LED* components captured in the variable $?1. The iteration through the components is used to print the component names and a line beneath all of the component names. The **printout** function is used for the simple printing of strings to the screen or a file. The **format** function provides formatted printing to the screen or a file and is similar to the **printf** function in C.

Adding this rule to the knowledge base and allowing it to fire produces the following output.

```
CLIPS> (unwatch all)
CLIPS> (agenda)
-30 print-header: f-11, f-16
For a total of 1 activation.
CLIPS> (run)
 S#1   S#2 |  L#1   L#2
-----------+-----------
CLIPS>
```

The final rule for the knowledge base prints the contents of the *result* facts beneath the truth table header.

```
(defrule print-result
  (print-results)
  ?f <- (result $?input -> $?response)
  (not (and (result $? input-2 -> $? response-2)
            (test (< (str-compare (implode$
            ?response-2)
                                        (implode$
                          ?response))
                          0))))
  =>
  (retract ?f)
  (progn$ (?i ?input) (printout t " " ?i " "))
  (printout t " | ")
  (progn$ (?i ?response) (printout t " " ?r " "))
  (printout t crlf))
```

The first pattern in the *print-result* rule matches against the *print-results* fact. Thus, this rule will not be activated until all processing is completed and the *print-header* rule has fired. The second pattern matches against any of the *result* facts generated by the program. The third pattern, preceded with the keyword *not* and containing another group of patterns, is rather complex and requires some explanation. Since rules act opportunistically, we can't rely on the *result* facts matching the second pattern to fire in a specific order. For example, depending upon the order in which the activations of the *merge-responses* rule fired, the final output of our program could be either

```
 S#1   S#2 |  L#1   L#2
-----------+-----------
  *     F  |   T     F
  *     T  |   T     T
```

or

```
 S#1   S#2 |  L#1   L#2
-----------+-----------
  *     T  |   T     T
  *     F  |   T     F
```

It's desirable to have the table printed in a specific order so that it's easy to look up entries. Specifically, if we view the *Source* component values as the binary representation of a positive integer, the entries should be printed in ascending order. The source combination "F F" represents 0 and should be printed first; the source combination "F T" represents 1 and should be printed second; and so forth. Since the * character can also be included as part of the source combination, it's inconvenient to convert the source combination to a number, but if the combination is converted to a string, a string comparison can be performed instead of a numeric comparison. For example, using ASCII values the following string relationships exist:

```
"**" < "*F" < "*T" < "F*" < "FF" < "FT" < "T*" < "TF" < "TT"
```

Thus, if the source combination values can be converted to strings, a comparison can be made to determine the order in which the *result* facts should be printed. This brings us back to the third pattern of the *print-result* rule. In plain English, the first two patterns of the rule say "If we want to print the results and there is a result to be printed, then . . ." The third pattern adds the condition "and there aren't any results that should be printed before the result we want to print." Referring back to the truth table results of our example circuit shown following, we don't want to print the results of fact f-61 until the results of fact f-62 have been printed since "*F" is less than "*T".

```
f-61    (result * T -> T T)
f-62    (result * F -> T F)
```

Recalling from the *max-iterations-and-gray-code* rule, the *not* keyword surrounding the group of patterns indicates that the third pattern is satisfied if the patterns within are not satisfied. The *and* keyword immediately following groups the two remaining patterns together. The *result* pattern binds the variables $?input-2 and $?response-2 to the *Source* and *LED* component values of a *result* fact (which could be the same fact matching the second pattern). The remaining *test* pattern, however, determines if the new matched *result* fact should be printed before the *result* fact matching the second pattern. The **implode$** function converts the multifield variable values representing the *Source* component values to strings. The **str-compare** function is used to compare the resulting two strings. A return value of less than zero indicates the first string passed to the function is less than the second string passed to the function. The < function is used to determine if the return value of the function was less than zero. The resulting effect of all the portions of the third pattern is to prevent a *result* fact from being printed while other *result* facts exist that should be printed before it.

The first action of the *print-result* rule is to retract the fact that it is printing. Removal of this fact allows the next fact to be printed to activate the *print-result* rule by removing the order restriction imposed by the third pattern. The **progn$** function is again used with the **printout** function to display the values of the *Source* and *LED* components.

Adding the *print-result* rule to the knowledge base and tracing its execution generates the following output.

```
CLIPS> (agenda)
0      print-result: f-63, f-62,
For a total of 1 activation.
CLIPS> (watch rules)
CLIPS> (watch activations)
CLIPS> (watch facts)
CLIPS> (run 1)
FIRE     1 print-result: f-63, f-62,
<== f-62          (result * F -> T F)
==> Activation 0       print-result: f-63, f-
61,
*   F   |   T   F
CLIPS> (agenda)
0      print-result: f-63, f-61,
For a total of 1 activation.
CLIPS> (run 1)
FIRE 1      print-result: f-63, f-61,
<== f-61          (result * T -> T T)
*   T   |   T   T
CLIPS> (agenda)
CLIPS>
```

The first execution of the *print-result* rule retracts fact f-34, causing the activation of the *print-result* rule for fact f-62, and then prints the *Source* and *LED* component values associated with fact f-34. The second execution of the rule similarly retracts fact f-62 and prints its associated component values. Since there are no other *result* facts, no new activations are generated.

The Completed Program

With all the rules completed, the entire program can now be loaded and executed. The complete execution is shown as follows.

```
CLIPS> (unwatch all)
CLIPS> (load circuit.clp)
%$****$***!*******
TRUE
CLIPS> (reset)
CLIPS> (run)
 S#1    S#2  |  L#1   L#2
 - - - - - + - - - - -
  *      F   |   T     F
  *      T   |   T     T
CLIPS>
```

Additional Comments

The original program from which this example was derived made use of the CLIPS Object Oriented Language (COOL) to represent and simulate the various circuit components. Because of the interconnections and similarities of the components, using object-oriented programming techniques to represent the circuit would have been highly desirable, however, describing both the object-oriented and rule-based features of CLIPS would have required considerably more background information, and explanation. In addition, the use of rules to propagate component values throughout the circuit is a good example for illustrating the data-driven nature of a rule-based system.

References

Davis, R., Buchanan, B. G., and Shortliffe, E. H. 1977. Production systems as a representation for a knowledge-based consultation program, *Artificial Intelligence,* 8(1):15–45.

Duda, R., Gaschnig, J. and Hart, P. 1979. Model design in the PROSPECTOR Consultant System for Mineral Exploration, *Expert Systems in the Micro-Electronic Age,* Michie, D., ed., 153–167, Edinburgh University Press, Edinburgh.

Englemore R. and Terry A. 1979. Structure and function of the CRYSALIS systems, *IJCAI,* 250–256.

Forgy, C. 1982. Rete: a fast algorithm for the many pattern/many object pattern match problem, *Artificial Intelligence,* 19:17–37.

Giarratano, J. and Riley, G. 1994. *Expert Systems: Principles and Programming,* 2d ed., PWS Publishing, Boston, MA.

Levy, L. S. 1980. *Discrete Structures of Computer Science,* pp. 9–12, John Wiley & Sons, New York, NY.

McDermott, J., and Bachant, J., 1984. R1 revisited: four years in the trenches, *AI Magazine,* (3):21–32, Fall.

NASA 1993. *CLIPS Reference Manual,* Version 6.0, NASA document JSC–25012, Houston, TX.

Further Information

The CLIPS 6.0 UNIX, PC, and Macintosh distribution packages are available by anonymous ftp from ftp.cs.cmu.edu and can be found in the user/ai/areas/expert/systems/clips directory. The distribution packages are also available by anonymous ftp from eecs.nwu.edu and can be found in the /pub/CLIPS directory.

An anonymous ftp site maintained by the CLIPS development team can be accessed at hubble.jsc.nasa.gov. Bug fixes and other CLIPS related information including a Frequently Asked Questions (FAQ) list are contained in the /pub/clips directory. Inquiries about CLIPS can be sent by electronic mail to stbprod@fdr.jsc.nasa.gov. A CLIPS World Wide Web page can be accessed using the URL http://www.jsc.nasa.gov/~clips/CLIPS.html.

56

Expert Systems and Their Use in Complex Engineering Systems

56.1 Introduction 824
56.2 Definition of Expert Systems 824
56.3 Characteristics of Expert Systems 825
56.4 Components of an Expert System 825
56.5 Knowledge Representation and Inference 826
56.6 Uncertainty Management 828
56.7 State of the Art of Expert Systems 830
56.8 Use of Expert Systems 830
56.9 Potential Implementation Issues for Expert Systems 831
General Implementation Issues • Specific Implementation Issues That Need to be Addressed
56.10 Legal Aspects of Expert Systems 832
56.11 Use of Expert Systems in Nuclear Power Plants 833

Robert E. Uhrig
University of Tennessee and Oak Ridge National Laboratory

Lefteri H. Tsoukalas
Purdue University

56.1 Introduction

In the operation of complex engineering systems, great quantities of numeric, symbolic, and quantitative information are handled by the system operators even during routine operation. The sheer magnitude of the number of process parameters and systems interactions poses difficulties for the operators, particularly during abnormal or emergency situations. Recovery from an upset situation depends upon the facility with which available raw data can be converted into, and assimilated as, meaningful knowledge. In operating a complex engineering system, people are sometimes affected by fatigue, stress, emotion, and environmental factors that may have varying degrees of influence on their performance. Expert systems provide a method of removing some of the uncertainty from operator decisions by providing expert advice and rapid access to a large information base.

Application of artificial intelligence (AI) technologies, particularly expert systems, to the control room activities in a complex engineering system has the potential to reduce operator error and improve plant safety and reliability. Furthermore, in a large number of nonoperating activities (e.g., testing, routine maintenance, outage planning, equipment diagnostics, fuel or feedstock management, etc.) expert systems may increase the efficiency and effectiveness of overall plant and corporate operations.

56.2 Definition of Expert Systems

Artificial intelligence is a branch of computer science that attempts to emulate certain mental processes of humans by using computer models. As the field developed, a number of specialized areas evolved including natural language processing, natural vision, image recognition, automatic learning, robotics, and expert systems. In expert systems, one of the primary objectives is to mimic human judgment using a computer program by applying substantial knowledge of specific areas of expertise to solve finite, well-defined problems. These computer programs contain human expertise (called heuristic knowledge) obtained either directly from human experts or indirectly from books, publications, codes, standards, or data bases, as well as general and specialized knowledge that pertains to the specific situation. Expert systems have the ability to reason using formal logic, to seek information from a variety of sources including data bases and the user, and to interact with conventional programs to carry out a variety of tasks including sophisticated computation.

0-8493-8343-9/97/$0.00+$.50

56.3 Characteristics of Expert Systems

A number of characteristics of expert systems are unique and generally advantageous (Feigenbaum, et al., 1988; Van Horn, 1986):

1. Experts need not be present for a consultation; expert systems may be delivered to remote locations where expertise may not be available.
2. Expert systems do not suffer from some of the shortcomings of human beings (e.g., they do not get tired or careless as the work load increases), but, when properly used, continue to provide dependable and consistent results.
3. The techniques inherent in the methodology of expert systems minimize the recollection of information by requesting only relevant data (i.e., data encountered in the reasoning path) from the user or appropriate data bases.
4. Expert knowledge is saved and readily available because the expert system can become a repository for undocumented knowledge that might otherwise be lost (e.g., through retirement).
5. The development of expert systems forces documentation of consistent decision-making policies. The clear definition of these policies makes the overall decision-making process transparent and the implementation of policy changes instant and simultaneous at all sites.

On the other hand, expert systems have disadvantages that affect their use:

1. They usually deal only with static situations.
2. They must be kept up to date as conditions change.
3. They often cannot be used in novel or unique situations.
4. Results are very dependent on the correctness of the knowledge incoporated into the expert system.
5. Perhaps most important, they do not benefit from experience except through updating of the knowledge base (based on human experience).
6. Expert systems are unable to solve problems outside their domain of expertise and in many cases are unable to detect the limitations of ther domain (Swartout and Smoliar, 1987; Ricker, 1986).

The domain of an expert system refers to the scope of the knowledge contained within the knowledge base. If the expert system operates outside its domain, it is possible that it may generate incorrect results by utilizing nonapplicable, irrelevant knowledge while searching for a solution. The inability of expert systems to recognize the limitations of their knowledge has been identified as a serious shortcoming.

Expert systems can, under certain circumstances, deal with imprecise or "fuzzy" information, missing information, and even a certain amount of conflicting information through the use of "certainty factors" and Bayesian probabilities (Kaplan, et al., 1987). Certainty factors represent a measure of belief of the user that a piece of evidence is true. They are not probabilities but rather simply a subjective judgment on the degree of truth or validity of an assertion. Some of the information used in development and application of an expert system may not be absolutely certain, and the use of certainty factors allows this subjective evaluation to be incorporated into the expert system. The final results in these cases may be the "most probable" solution or the "best" solution, but there is no absolute guarantee that the solution is the "correct" solution.[1]

A comparison of human and artificial expertise will help convey the strengths and weaknesses of expert systems (Van Horn, 1986). Human expertise is perishable and difficult to transfer, whereas artificial expertise is permanent and easy to transfer. Human expertise is not always consistent, whereas artificial expertise is consistent. (If you give an expert system the same problem on two occasions you will get the same answer unless stochastic processes are involved; this is not necessarily true of human expertise.) On the other hand, human expertise is creative and has a broad focus, whereas artificial expertise is uninspired and usually has a very narrow focus. Above all, human expertise is adaptive and demonstrates common sense, characteristics usually lacking in expert systems because the knowledge is entirely technical or objective in nature.

56.4 Components of an Expert System

The principal components of an expert system are the inference engine, the knowledge base, and the interface between the expert system and humans (users, knowledge engineers, and experts). The inference engine gathers the information needed from the knowledge base, from associated data bases, or from the user; guides the search process in accordance with a programmed strategy; uses rules of logic to draw inferences or conclusions for the processes involved; and presents the conclusions (where warranted) with explanations or bases for the conclusions. The knowledge base consists of information stored in retrievable form in the computer, usually in the form of rules or frames. The correctness and completeness of the information within the knowledge base is the key to obtaining correct results or solutions using expert systems. The interface between the human and the expert system must translate user input into the computer language, and it should present conclusions and explanations to the user in written or graphical form. It should also include an editor to assist in adding to or changing the knowledge base.

[1] Recent work incorporating "fuzzy logic" and "reasoning under uncertainty" into expert systems has significantly improved the performance of expert systems when dealing with complex systems. These topics are covered in a later chapter dealing with hybrid systems that combine two or more artificial intelligence methodologies (in this case, expert systems and fuzzy logic). (Tsoukalas and Uhrig, 1996)

One of the major breakthroughs in development of expert systems came in the mid 1970s with the expert system MYCIN, a diagnostic system for infectious diseases of the blood. The MYCIN architecture completely separated the knowledge base from the inference engine, which permitted modification of the knowledge base without any influence on the inference engine. Hence, it was possible to start with a simple expert system and incrementally add features and complexity as needed.

The knowledge base of an expert system contains the expertise (facts and heuristics) collected from experts, books, publications, and other sources and encoded into rules, frames, or other computer representations of knowledge. This information describes a methodology for solving the problem as a human expert would solve it. Collecting adequate knowledge from experts and translating it into computer code (a process called "knowledge acquisition") has proven to be a very difficult task. All too often, experts really do not understand the processes by which they reason or solve problems. In other cases, experts are reluctant to give up their expert knowledge because they perceive that the availability of an expert system with their expertise may lessen their value to their employer or clients. Because an expert system is only as good as its knowledge base, proper collection and representation of knowledge is critical for the successful implementation and operation of expert systems.

In addition to a knowledge base, an expert system includes a user interface to perform data collection, editing functions, and consultations. This interface almost always uses a written format to facilitate presentation of system knowledge, processor explanations, and results.

Some expert systems contain a degree of self-awareness or self-knowledge that allows them to reason about their own operation and to display inference chains and traces of the rationale behind their results (Waterman, 1986). These abilities (the explanation facilities) have been recognized as one of the most valuable features of expert systems. The user can take advantage of explanation facilities to request a complete trace for a consultation, request an explanation on how a particular goal or subgoal was inferred, or request an explanation of why a particular piece of information is needed. These facilities can be used to obtain information on the status of a system. Explanation-generating facilities are also of great use in debugging expert systems and may play a key role in verification and validation of expert systems.

The performance of mature expert systems has shown that the reliability of an expert system in a given subject area asymptotically approaches the reliability of the expert as the knowledge base approaches the expert's knowledge in that area. In some cases the reliability of an expert system exceeds the reliability of the expert, not because the expert system is "smarter" than the expert, but rather because the expert system does not forget anything contained in the knowledge base and is capable of rapidly carrying out analytic and mathematical operations.

An expert system "shell" is a computer program used to develop an expert system. Early shells were expert systems from which the domain-specific knowledge bases had been removed, and the mechanism for creating a new knowledge base of the user's choice had been made "user friendly." Often a shell also has provisions for changing the reasoning processes of the inference engine to adapt to the specific problem. The first shell was EMYCIN (Essential MYCIN) in which the knowledge base on infectious diseases of the blood was removed from MYCIN and knowledge bases on cancer treatment and pulmonary diseases were used to create new expert systems (ONCONIN and PUFF, respectively) to assist doctors in these fields. The pioneering efforts of Stanford University on EMYCIN paved the way for virtually all modern expert system shells. Indeed, only in the last few years have expert system shells begun to deviate significantly from the overall structure developed for MYCIN.

Expert system shells today differ significantly from each other and offer the user a wide variety of capabilities. Some have sacrificed size of knowledge base to improve ease of updating the knowledge base, and vice versa. Certain expert systems (e.g., 1ST CLASS, and VP EXPERT) have the ability to derive the knowledge base from a series of examples by induction. The ability to extract information from databases and experimental results is one of the strengths of artificial neural networks. Hence, the use of a hybrid consisting of an artificial neural network in the knowledge base of an expert system is feasible and often advantageous.[2] Recently, an expert system shell was introduced with HYPERTEXT as part of the knowledge base. Selection of an expert system to fit a specific need is almost a research project in itself and could be a topic for an expert system.

56.5 Knowledge Representation and Inference

There is a variety of approaches to encode human expertise in expert systems, the most common one being IF-THEN rules. Semantic networks, frames, and logical expressions are alternative paradigms of knowledge representation, yet, the majority of industrial expert systems uses the rule-based paradigm (for a good introduction to the subject see Gonzalez and Dankel, 1993).

The three basic constituents of a rule-based expert system are *rule base, working memory* and *rule interpreter.* As seen in Figure 56.1, the rule base is often partitioned into groups of rules, called *rule clusters.* Each rule cluster encodes the knowledge required to perform a certain *task* or a fraction of a task, usually referred to as a *subtask.* There may also be rules for internal control purposes, e.g., to signal which rule cluster to select as holding potentially relevant knowledge at a given time. Collectively, these rules are referred to as the *control structure* of an expert system. Another class of rules, called *demons,* may be present; they are designed to function outside the control structure of the program for the purpose of enhancing its ability to respond quickly to the occurrence of an event requiring some immediate action. Demons address inefficiency issues that may arise from excessive control over the rule base (Cooper and Wogrin, 1988).

[2] A hybrid system in which an artificial neural network is imbedded in the knowledge base of an expert system is discussed in a later chapter.

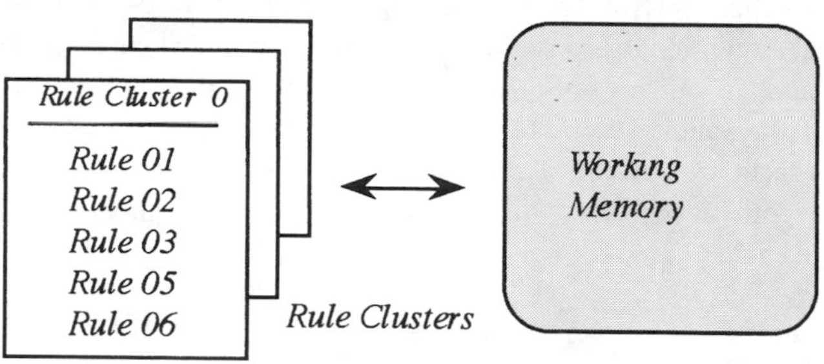

Figure 56.1 In a rule based expert system rules are often grouped into *Rule Clusters* operating on a global database called *Working Memory.*

Working memory is a database holding input data, inferred hypotheses and internal information about the program. In an on-line expert system with monitoring functions, for example, the state of working memory at any given time reflects changes occurring in the process being monitored as well as internal changes due to the reasoning process of the program itself.

The mechanism through which rules are selected to be fired is called the *rule interpreter.* It is based on a pattern matching algorithm whose main purpose is to associate at any given time the state of the system (input data, inferred hypotheses, etc.) with applicable rules from the rule base.

The inference engine of an expert system is in charge of manipulating the data presented to the system and arriving at a conclusion. In expert system technology, the two most widely used reasoning techniques are forward chaining (forward reasoning) and backward chaining (backward reasoning). In forward chaining the system reasons forward from a set of known facts and tries to infer the conclusions or goals. Design of a complex system is a forward chaining application where the expert system starts with the known requirements, investigates the very large array of possible arrangements, and makes a recommendation based on criteria specified by the user. In backward chaining the system works backward from the conclusions or goals and attempts to find supporting evidence to verify their correctness. Solving a crime is a backward chaining application where the expert system identifies the possible suspects, looks for evidence indicating the guilt and innocence of each suspect and makes a recommendation regarding which suspect is the most likely criminal. In many cases, backward-chaining systems are more efficient than true forward-chaining systems because they tend to reduce the search space and arrive at a conclusion more quickly. Many advanced expert systems use a combination of both forward and backward chaining. Different search strategies, such as "depth first" or "breadth first" may be incorporated into either backward or forward chaining.

Data enters an expert system either through a user interface or from other programs such as databases, data acquisition systems, simulation packages etc., and forms the initial facts (or assertions or evidence) available to the rules. From the input data conclusions are drawn in a process called inferencing. The two basic inferencing strategies, *forward chaining* and *backward chaining* are also referred to as *modus ponens* and *modus tollens*, respectively. In *modus ponens,* if we have the rule

IF *A is true*
THEN *B is true*

and we know that "*A is true,*" then we can infer that "*B is true.*" Most expert systems use this powerful inferencing strategy which due to familiarity looks very much commonplace if not trivial. In *modus tollens,* if we know that the rule is true and we also know that "*B is false,*" then we can infer that "*A is false.*" We often simply write *A* instead of "*A is true*" and *NOT A* instead of "*A is false.*" The requirement for an exact match between input data and what is stated in a rule is relaxed in fuzzy expert systems where fuzzified versions of the basic inferencing strategies have been developed.

To illustrate the nature of forward and backward chaining let us consider the simple rule base shown in Figure 56.2 where six rules are used to relate seven facts $A_1, A_2, \ldots, A_7$ to conclusion D_1. For example, *Rule 1* says

IF A_1 *AND* A_2 *AND* A_3
THEN B_1

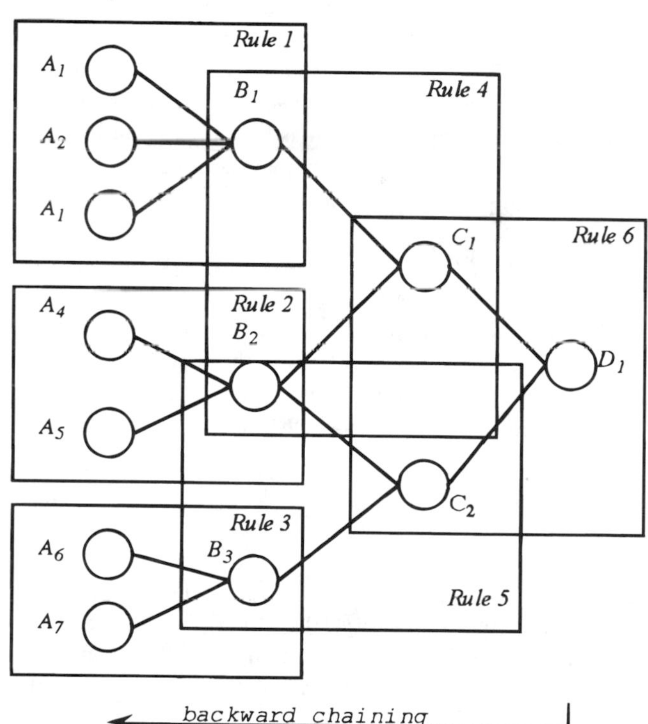

Figure 56.2 In this simple rule-base six rules are used to relate $A_1, A_2, \ldots, A_7$ to D_1.

When input data to the system matches all A_1,A_2,A_3 the given rule fires and hypothesis B_1 is added to working memory. Similarly, B_1 in conjunction with hypothesis B_2 produce C_1 through *Rule 4;* and hypothesis C_1 in conjunction with C_2 produces hypothesis D_1 through *Rule 6*. Hypothesis D_1 may be considered the *conclusion* drawn through this particular reasoning process. In *forward chaining,* the input data is matched against facts A_1, . . ., A_7 and, after intermediate hypotheses B_1,B_2,B_3,C_1,C_2 generated by rules *Rule* 1, . . ., *Rule 5* are drawn, the final hypothesis (conclusion) D_1 would be reached through *Rule 6*. In *backward chaining,* on the other hand, the opposite direction is taken as seen in figure 2. The inference process uses D_1 as the point of departure; intermediate hypotheses B_1,B_2,B_3,C_1,C_2 are sought to support D_1, and ultimately one or more of the facts A_1, . . ., A_7 will be reached, in order to explain D_1.

56.6 Uncertainty Management

An important issue in expert systems is uncertainty management. Representing or combining uncertain data and drawing reliable inferences from it has been extensively investigated over the past two decades and several theories of uncertainty have provided tools for solving uncertainty problems (see Cooper and Wogrin, 1988 for a comprehensive treatment of the subject). Probability theory and certainty factors which will be discussed in this section, have been frequently used; but, also possibility (fuzzy) theory (which will be discussed in another chapter), Dempster-Shafer belief measures, Cohen's theory of endorsements, and subjective Bayesian methods such as the one used in the expert system Prospector, are uncertainty management paradigms that have found important applications.

Bayesian Probabilities

Of the several approaches that have been taken to the problem, the oldest has been to ascribe probabilities to facts and rules and to use Bayes' rule. We recall that Bayes' rule says that the conditional probability $p(H|E)$ of a hypothesis H being valid when evidence E has been observed is given by

$$p(H|E) = \frac{p(E|H) \cdot P(H)}{p(E)} \tag{56.1}$$

where $p(H|E)$ is the probability that the evidence E was observed given the hypothesis H, and $p(H)$ and $p(E)$ are the probabilistic uncertainties associated with the strength of the hypothesis H and of the evidence E, respectively. Equation 56.1 is often written in the equivalent form

$$p(H|E) = \frac{p(E|H) \cdot p(H)}{p(E|H) \cdot p(H) + p(E|\neg H) \cdot p(\neg H)} \tag{56.2}$$

where $\neg H$ denotes the negation of hypothesis H. It can be observed by comparing Equations 56.1 and 56.2 that their denominators are equal, i.e., $p(E) = p(E|H) \cdot p(H) + p(E|\neg H) \cdot p(\neg H)$, indicating that the probability for evidence E to be observed equals the sum of the conditional probability that evidence E is observed when the hypothesis H is valid and the conditional probability that E is observed when H is not valid; in other words the probability that E occurs regardless of the validity of H. More generally, when m hypotheses and n facts are present, a Bayesian measure of the uncertainty of hypothesis H_i based on observing evidences E_1,E_2, . . ., E_n is given by

$$\begin{aligned} p(H_i|E_1 E_2 \cdots E_n) &= \frac{p(E_1 E_2 \cdots E_n|H_i) \cdot p(H_i)}{p(E_1 E_2 \cdots E_n)} \\ &= \frac{p(E_1|H_i) \cdot p(E_2|H_i) \cdot \cdots \cdot p(E_n|H_i) \cdot p(H_i)}{\sum_{k=1}^{m} p(E_1|H_k) \cdot p(E_2|H_k) \cdot \cdots \cdot p(E_n|H_k) \cdot p(H_k)} \end{aligned} \tag{56.3}$$

In Equation 56.3 Bayes' rule assumes that the pieces of evidence E_1,E_2, . . ., E_n are all conditionally independent given some hypothesis, an assumption which is not always valid. For example, in equipment diagnosis two symptoms E_1,E_2 might each independently indicate that some fault is likely with a certain probability. Yet, taken together, these symptoms may contradict or reinforce each other and the assumption of independence may not be valid.

A serious drawback of the probabilistic approach to uncertainty is not being able to distinguish between *absence of belief* and *doubt* or to represent *ignorance* related to the lack of knowledge. In addition, it requires a rather large amount of statistical data to construct the various probabilities in the knowledge base (Kruse et al., 1991).

EXAMPLE 56.1 Diagnosis Using Probabilistic Uncertainties

To illustrate how uncertainty is propagated using Bayes' rule, consider an expert system using *backward chaining* to decide whether an electronic product is manufactured according to technical specifications. The system receives electrical measurements from an automated test station and arrives at a diagnosis by combining probabilities relating evidence and hypotheses. Suppose that in its knowledge base there is a rule relating the quality of the product to the value of the *signal-to-noise-ratio (SNR)*, e.g.,

IF *quality is low*
THEN *SNR is low* (with probability $p = 0.8$)

We treat here the proposition "*SNR is low*" as evidence E and the *proposition "quality is low"* as a hypothesis H. Thus, the probability given for the rule is the conditional probability $p(E|H) = 0.8$. Since we are using backward chaining for diagnosis we want to derive the certainty with which we can believe that quality is low when the only thing known is that low *SNR* has been observed and the above rule is valid. In other words, we would like to compute $p(E|H)$. Equation 56.2 indicates that the probability "quality is low given that *SNR* is low" is the ratio of the probability that both quality and

SNR are low to the probability that *SNR* is low. The probability of *SNR* being low is the sum of the conditional probability that *SNR* is low when quality is low and the conditional probability that quality is low when *SNR* is not low, e.i., the probability that quality is low regardless of whether the *SNR* is low. Suppose that from past experience in this particular production line it is known that

$$p(H) = p(\textit{quality is low}) = 0.05, \text{ hence,}$$
$$p(\neg H) = p((\textit{quality is not low}) = 0.95$$
$$p(E|H) = p(\textit{SNR is observed low} \mid \textit{quality is low}) = 0.8$$
$$p(E|\neg H) = p(\textit{SNR is observed low} \mid \textit{quality is not low}) = 0.1$$

The probability of quality being low given that *SNR* is low is computed substituting values from above to Equation 56.2

$$p(H|E) = \frac{p(E|H)\cdot p(H)}{p(E|H)\cdot p(H) + p(E|\neg H)\cdot p(\neg H)}$$
$$= \frac{(0.8)\cdot(0.05)}{(0.8)\cdot(0.05) + (0.1)\cdot(0.95)} = 0.30$$

Thus, we may say that there is a 30% chance of having a defective product on the basis of only one piece of evidence, that is, measurement of low *SNR*. Additional rules may be called to contribute to a final estimate of the certainty with which the product may be found of low quality and a final decision can be made on the basis of a cutoff value. Typically, additional evidence is offered by inspecting the electronic product for physical defects through a vision system. Thus, two conditionally independent pieces of evidence are available: E_1, *SNR is low;* and, E_2 *physical defect is present.* In the rule base, three mutually exclusive hypotheses may then be considered: H_1, *quality is low;* H_2, *quality is average;* and H_3, *quality is good.* If the prior probabilities $p(H_i)$ $(i = 1,2,3)$ and the conditional probabilities $p(E_1|H_i)$ and $p(E_2|H_i)$ are known then Equation 56.3 may be used to determine the likely quality of the product.

Certainty Factors

To overcome the difficulties of Bayesian probabilities, e.g., requiring large volume of data or distinguishing between *absence of belief* and *doubt,* the developers of MYCIN came up with an alternative approach using certainty factors. In the certainty factor (*CF*) formalism, knowledge is expressed as a set of rules having the form

IF E
THEN H with *CF(H|E)*

where, *E* is the *evidence,* i.e., one or more facts known to support the *hypothesis H, and CF(H|E)* is the certainty factor for the rule, a measure of belief in *H* given that *E* has been observed. The value of *CF* ranges from -1 to $+1$. When $CF = -1$ the *hypothesis H* is totally denied, while *at* $CF = +1$, the *hypothesis H* is totally confirmed.

Certainty factors are derived from two further measures: *measures of belief, MB(H,E),* and *measures of disbelief MD(H,E),* both taking values between 0 and 1. A measure of belief *MB(H,E)* represents the degree to which the belief in hypothesis *H* is supported by observing evidence *E,* and is computed by

$$MB(H, E) = 1, \quad \text{if} \quad p(H) = 1, \text{ else}$$
$$= \frac{p(H|E) - p(H)}{1 - p(H)} \tag{56.4}$$

A measure of disbelief MD(H,E), on the other hand, represents the degree to which the disbelief in hypothesis *H* is supported by observing evidence *E.* It is computed by

$$MD(H, E) = 1, \quad \text{if} \quad p(H) = 1, \text{ else,}$$
$$= \frac{p(H) - p(H|E)}{p(H)} \tag{56.5}$$

The certainty factor *CF* is defined in terms of *MB(H,E)* and measure of disbelief *MD(H,E)*

$$CF \equiv \frac{MB(H, E) - MD(H, E)}{1 - \min[MB(H, E), MD(H, E)]} \tag{56.6}$$

During the execution of a knowledge base, multiple rules are typically capable of deriving the same hypothesis or conclusion, resulting in modification of the *CF*'s involved. Consider, for example, a case where two different evidences E_1 and E_2 lead to the same hypothesis *H.* In such cases, certainty factors of the same or opposite signs can be combined directly by the following formulas (Gonzalez and Dankel, 1993; Kruse, et al., 1991).

Case 1: When both $CF(H|E_1)$, $CF(H|E_2) > 0$:

$$CF(H|E_1, E_2) = CF(H|E_1) + CF(H|E_2) - CF(H|E_1)\cdot CF(H|E_2)$$

Case 2: When $-1 < CF(H|E_1) \cdot CF(H|E_2) \le 0$:

$$CF(H|E_1, E_2) = \frac{CF(H|E_1) + CF(H|E_2)}{1 - \min(|CF(H|E_1)|, |CF(H|E_2)|)}$$

Case 3: When $CF(H|E_1) \cdot CF(H|E_2) = -1$:

$$CF(H|E_1, E_2) = \text{undefined}$$

Case 4: When both $CF(H|E_1)$, $CF(H|E_2) < 0$:

$$CF(H|E_1, E_2) = CF(H|E_1) + CF(H|E_2) + CF(H|E_1)\cdot CF(H|E_2)$$

The assumption under which the above equations are based is that we have absolute confidence in the evidence of premises used to derive these values. If, however, we have the typical situation that exists in expert systems where a hypothesis from a rule is used as the evidence of another we may not actually have absolute confidence in the evidence, and the certainty factor approach does not materially contribute to the final results. Another drawback of certainty factors seems to be the complexity of maintaining them. When, for example, new knowledge is added or deleted from the knowledge base, the certainty factors of existing knowledge changes as well, making the maintenance of the system rather complicated. For these reasons, and others, use of fuzzy set theory in reasoning under uncertainty is more commonly encountered today.

56.7 State of the Art of Expert Systems

The impact of expert systems technology has been felt in many areas of science, education, and industry. In the past decade a great many applications have been initiated, and many are now operational or in the prototype stage. The extent of the potential application of this technology is not yet known, as expert systems in the future may be used in completely new settings to solve quite different problems. This is particularly true of expert systems that use fuzzy rules.

It is very difficult to gain a true picture of just how widespread the use of expert systems has become. In many cases, organizations are using expert systems internally. Even the fact that they are used, let alone the details of the expert systems, are treated as proprietary for the simple reason that the company or organization wants to gain competitive advantage. By one analyst's estimate, about half of the companies listed in the Fortune 500 are developing expert systems (Coates, 1988). General Motors is reportedly insisting that manufacturers supply diagnostic expert systems with the equipment they provide. The Boeing Aircraft Company has indicated that their future prosperity may be tied to innovative applications of artificial intelligence in their aerospace systems (Hertz, 1988). To prepare for this, they selected some 80 individuals from throughout the company and provided them with an intensive training program in the use of artificial intelligence with heavy emphasis on expert systems.

A very different and innovative program was initiated by the DuPont Company (Feignbaum, et al., 1988). After an extensive study, it concluded that expert systems could be very useful in individual situations but that no one could predict where those applications might occur. The company then provided all interested employees with low-cost expert system shells and short (2-day) training sessions on the use of these shells. The individual users then determined whether or not there were any potential applications of expert systems in their work. Experience with the first 200 expert systems at DuPont has been that an average of one person-month of effort is expended on each expert system and the payback has averaged about $100,000.

Expert systems may change the manner in which many organizations operate, and they could change the work place in general. In large organizations such as government, big corporations, and associations, one expert predicts that 60 to 90% of all jobs are candidates for augmentation, displacement, or replacement by expert systems (Coates, 1988). He further predicts that by about the turn of the century the capabilities of expert systems will have grown to such a degree that their impact will be felt throughout most occupations and workplaces.

56.8 Use of Expert Systems

Generally, but not always, problems that are amenable to a numerical solution should be solved using conventional computer programs. However, there are many situations in which expert systems offer unique advantages over conventional programs. Most applications of expert systems today can be classified into the following five categories:

1. Monitoring systems.
2. Control systems.
3. Configuring systems.
4. Planning and scheduling systems.
5. Diagnostic systems.

Monitoring systems. Monitoring systems are dedicated to data collection and analysis over a period of time. The collected values are compared against expected performance, and if discrepancies are identified the expert system generates recommendations and/or notifies the operator.

Control systems. Control systems are monitoring systems in which action (e.g., opening a valve, turning on a heater, etc.) is taken as a result of the discrepancy identified.

Configuring systems. Configuring systems address problems in which a finite set of components is to be arranged in one of many possible patterns. The classical example in this category is XCON, an expert system used by a large computer manufacturer to configure its equipment in accordance with user specifications.

Scheduling and planning systems. Scheduling and planning expert systems coordinate the capabilities or components within an organization to optimize production and/or increase efficiency. The difference between planning and scheduling systems is that the components for a task are not always known in planning systems.

Diagnostic systems. Diagnostic systems analyze and observe data and map the analysis results to a set of problems. Once the problems have been identified, the system recommends a solution based on facts in its knowledge base and on the other information it can acquire.

Expert systems have been used to solve many different problems in a variety of fields. Some of these areas are listed in Table 56.1, which is intended to give a brief overview of the breadth of applications that has developed. One area in which there has been extensive efforts to utilize expert systems in the nuclear power field, many of which could affect safety and safety-related systems. The scope of these applications has been well documented by Bernard and Washio (1989).

Table 56.1 Applications of Expert Systems

Field	Use
Design and engineering	Collecting and storing knowledge of best designers speeding the design process
Computer applications	Configuring experiment to user specifications
	Diagnosing problems with computer equipment
Manufacturing	Managing human and machine resources
	Facilitating factory automation
Finance	Decision support tools
	Providing tax and other business advice
	Processing loan and mortgage applications
	Analyzing financial risk
Science and medicine	Providing medical advice in hospitals
	Providing diagnostic assistance to medical personnel
	Patient monitoring
Geological applications	Advising regarding mineral deposit and oil locations
	Advising drillers regarding stuck bits
Training	Interface for computer-aided instruction
	Assisting in computer-based training

56.9 Potential Implementation Issues for Expert Systems[3]

Potential problems in implementing expert systems in complex engineering systems can be projected from past experience with the introduction of new and innovative systems.

General Implementation Issues

1. Most complex engineering systems, as presently built and operated, are considered by the operators to be safe enough. With the possible exception of the severe accidents (i.e., Three Mile Island, Bhopal, etc.), expert systems are not perceived to be needed to provide additional safety functions.
2. Introduction and use of an expert system must not introduce a new operational or safety problem. A thorough analysis of what could go wrong and what effect it could have on the plant and its safety system would be essential. The ultimate criterion in judging any new system is whether its failure can, in any way, lead to a challenge of existing safety systems.

Specific Implementation Issues That Need to be Addressed

A number of specific issues regarding the implementation of expert systems in complex engineering systems need to be addressed. These include, but are not limited to, the following.

1. Quantitative and Objective Performance Guidelines for Expert Systems. The introduction of expert systems into the operation of complex engineering systems has the promise of significant contributions to improved operation and safety. These applications may occur naturally with plant upgrades and perhaps with plant life extension. Alternately, the introduction may be driven by the productivity concerns of the industrial organizations.

   The primary concern about the introduction of any new system into a complex engineering system appear to be the impact it can have on the safety system when something goes wrong. The ultimate question in judging any new system must be "Can the failure of the expert system lead to a challenge of the existing safety systems?" Above all, replacement of an existing system with an expert system must not introduce new unresolved issues (i.e., new unreviewed safety hazards).

   Introduction of a new system must not lead to confusion of operators or other plant personnel. New tools may be needed to evaluate and measure the performance of expert systems and the impact of these systems on human performance. Objective criteria that are quantitative in nature are needed.
2. Validation and Verification (V&V). In conventional software programming, verification and validation have well-established meanings; verification is a determination that software has been developed in a formally correct manner in accordance with a specified software engineering methodology, and validation means demonstrating that the completed program performs the functions in the requirements specification and is usable for its intended purposes. However, expert systems go beyond the procedures of conventional software engineering, and a modularized, top-down, hierarchically decomposed design that makes conventional V&V possible may not be achievable. Expert systems, especially those operating under uncertainty or with incomplete data, may have so many states as to make exhaustive testing unfeasible. Hence, new approaches to V&V are needed for expert systems.

   A major issue in the use of expert systems undoubtedly will be the adequacy of the validation and verification provided. (Some industries seem to be waiting to see the regulatory requirements in this area before considering expert systems for use in safety-related systems.) The inference engine may be considered simply as another digital computer program, and its V&V can be dealt with in the same way as with other digital computer programs (e.g., IEEE-6.4.3.2). The real problem is the adequacy of the knowledge base—the qualifications of the expert whose expertise is incorporated into the knowledge base, the method used for acquisition of this expertise, and the method used to represent this expertise in the knowledge base. Except for relatively simple expert systems, exhaustive testing of the expert

[3] These perceptions regarding the use of expert systems are those of the authors.

system or its knowledge base to cover all likely situations may not be adequate or feasible.

Generally, as a matter of policy, V&V should always be carried out by a group completely independent of the group that developed the expert system. Because V&V in expert systems is so intimately related to the design, true independence may extremely difficult to achieve. To the extent possible, the independence of the group that does V&V should be ensured by quality assurance procedures and organization policy.

3. Human Factors. A primary human factors concern is that the expert system should present information to the user in a way that is comprehensible and understandable. Information must mesh well with the perspectives used by the human, and the way in which the information is displayed should correspond to the user's mental model of the plant. The user should be able to understand the expert system's behavior.

   Another concern is user reaction to the expert system. Will they like the system and accept it? Will they be comfortable with an expert system and use it when needed? Will they believe that the system will work and that it is useful? Above all, will they trust and have confidence in the information presented by the expert system? On the other hand, the user could become too dependent upon the guidance of an expert system and ignore other indications that might not agree with the conclusion of an expert system.

   The function allocation and division of responsibility between the expert system and the human is another important issue. Humans should be assigned those functions that they are most capable of performing and that utilize their abilities. Expert systems should relieve some of the physical and cognitive workload on users and not overload them. The system should make human jobs more efficient. The expert system should be integrated with the other hardware, software, and tools in the user's work environment. Clearly users should be involved in this analysis.

4. Accident Management. One of the major potential applications of expert systems is accident management, especially those extremely rare accidents that involve an unusual combination of events and have severe consequences. It is reasonable to expect plant operators to handle all sorts of upset conditions, but it may not be reasonable to expect them to handle all sorts of "beyond design basis" events that are beyond the scope of most operator training. Expert systems could be the preferred method of preparing for low-probability, high-damage events. They could provide the expertise of the world's experts on severe accidents to an ice-bound chemical plant in Wisconsin or to a power plant in the middle of a hurricane on a barrier island off the coast of Florida.

   An expert system could also be very helpful under severe accident conditions. For instance, an expert system might be used for containment assessment—there presently are a limited number of experts in the entire nation who are capable of assessing the status of a containment under accident conditions.

56.10 Legal Aspects of Expert Systems

Perhaps the major cloud that hangs over expert systems is the issue of product liability (Warner, 1988). What will be the reaction, for example, if a glitch in the software causes an expert system to specify the wrong action to a nuclear power plant operator? Lawyers warn that before expert systems become commonplace they will probably become ensnared in the widening web of product liability litigation. According to the Brookings Institute, the number of computer product liability lawsuits increased eightfold from 1974 to 1986. Should an expert system user suffer damage, there will be no shortage of parties to blame. Experts say possible lawsuit targets may extend from the user to the programmer, to the supplier, and even to the expert whose knowledge went into the program. In determining who is at fault, the lawyers indicate that a great deal will hinge on whether the system lets the user make the final decision. Another indicator of blame, lawyers say, is whether the software has been found to have a "bug" (programming error). Specialists in the field tend to believe that bugs, not an expert's error, pose the greatest potential for litigation against any type of software (Warner, 1988).

Apart from the liability issue, expert systems also raise the question of who owns the knowledge in them. A New Jersey specialist who provided the expertise for an expert system as part of his job (and who has since lost that job) wants to receive royalties for the system he created. He is claiming that, in the absence of any specific contract addressing the issue of ownership of knowledge and expert systems, the system is his intellectual property (Warner, 1988).

The U.S. government is already enacting regulations to control the use of expert systems. In some cases, government agencies have applied the same measures they use to regulate human experts. In 1986, for example, the Internal Revenue Service (IRS) began treating income tax advisory software the same way it deals with human tax consultants (i.e., if the program gives "substantive instructions" for completing a tax return and makes a mistake, it is liable). Another U.S. government agency, the Securities and Exchange Commission, refused to rule that a company's expert system would be exempt from registration as a financial advisor—at least until the program was completed. The result was that an expert system that could make specific financial recommendations was scrapped. The Food and Drug Administration (FDA) has so far been the most aggressive expert systems regulator because of its role in approving medical products. The FDA says the number of computerized medical products is increasing at an astronomical rate—so fast that the agency specified the kinds of software were not subject to review (e.g., spreadsheets used in medical offices). Software that makes treatment decisions, however, still must receive FDA approval. Ironically, the IRS, the Federal Bureau of Investigation, and the Environmental Protection Agency are all developing or applying expert systems for in-house use but cannot be sued for expert system errors (Warner, 1988).

In the case of expert systems as applied to nuclear power plants, the issue of liability is further complicated by the role played by Price-Anderson liability insurance. Who is protected by Price Anderson against liability claims in the case of an expert system that provides a wrong recommendation with serious consequences? The vendor that wrote the expert system shell? The expert who provided the expertise? The knowledge engineer who prepared the knowledge base? The architect-engineer who installed the expert system? The utility that used the expert system? The regulatory agency that approved installation of the expert system? Some undefined "third party"? The "public"? Litigants will probably attempt to sue all of these parties.

56.11 Use of Expert Systems in Nuclear Power Plants

As an example of the application of expert systems, let us consider their use to support control room activities in a nuclear power plant where they may reduce operator error and increase plant safety and reliability. Beyond the applications discussed here, there are a large number of nonoperating activities (testing, routine maintenance, outage planning, equipment diagnostics, and fuel management) in which expert systems can increase the efficiency and effectiveness of overall plant and corporate operations. Table 56.2 presents a number of potential applications of expert systems in the nuclear power field.

The Appendix presents a list of several expert systems typical of those now in operation or under development in the United States, Canada, France, Great Britain, Germany, Japan, and Sweden. All of these applications are advisory in nature and, except for the expert system presenting emergency operating procedures (in the Kuosheng Nuclear Plant in Taiwan), deal with nonsafety-related systems. U.S. utilities appear to be reluctant to introduce expert systems into safety-related systems of their nuclear plants until the various implementing issues are clarified.

Demands by the safety and environmental regulatory authorities for increased safety margins and lower environmental impacts and those by the economic regulatory authorities and the financial community for increased efficiency in operation (fewer trips, higher availability, plant investment protection) inevitably lead to more sophisticated plants with additional systems that must be controlled and/or automated. Hence, expert systems seem to be a natural addition to the control and instrumentation systems of the next generation of nuclear power plants. Indeed, integration of expert systems into the safety, control, and management systems of power plants is an integral part of the automation process that is evolving.

Table 56.2 Potential Utility Applications of Expert Systems in the Nuclear Power Field

Field	Function of Expert System
Diagnostics and monitoring	Predicting incipient failure of components plant behavior
	Monitoring for long-term gradual deterioration
	Diagnosing equipment malfunctions
Outage planning	Optimal sequencing of refueling activities
	Optimal fuel handling
	Minimizing radiation exposure
Compliance with specifications	Complying with technical specifications
	Complying with limiting conditions of operation
	Proper classification of emergencies
	Resolving ambiguous situations
Operational advisor	Analyzing plant "trips"
	Tracking emergency procedures
	Guiding and monitoring plant maneuvers
	Ensuring the ability to remove residual heat
	Monitoring "bypassed/inoperable" equipment
	Data logging and interpretation
	Emergency management
Nuclear personnel	Intelligent computer-aided instruction and training
Mitigation of accident consequences	Real-time management of evacuation
	Fast-time prediction of plume travel
	Minimization of radiation exposure
Reactor safety	Gives "big picture" to NRC safety team assessment system
	Monitors and projects core conditions, containment conditions, and fission product barriers
Reviewer aid	Provides a consistent framework and interactive process for reviewing licensing applications submitted by power plant licensees

Acknowledgment

The review of expert systems that resulted in this paper was supported, in part, by the U.S. Nuclear Regulatory Commission, Office of Nuclear Regulatory Research, under DOE Interagency Agency Agreement 1886–8085–2B, NRC FIN No. B0852. This support is gratefully acknowledged. The views and opinions expressed here are those of the authors and do not necessarily reflect the criteria, requirements, or guidelines of the NRC.

References

Bernard, J. and Washio, T. 1989. *Expert Systems in the Nuclear Power Reactors,* ANS Publishing, La Grange Park, II.

Coates, J. 1988. Artificial intelligence: observations on applications and control, *Computer Security Journal,* V(1).

Cooper, T. A., and Wogrin, N. 1988. *Rule-based Programming with OPS5,* Morgan Kaufmann, San Mateo, CA.

Expert-EASE Systems 1987. Seminar Notebook of Seminar. Expert Systems Applications in Power Plants, prepared by

Expert-EASE Systems, Inc., for the Electric Power Research Institute, Palo Alto, CA, May 27–29, Boston, Ma.

Feigenbaum, E., McCorduck, P., and Nii, H. P. 1988. *The Rise of the Expert Company,* Times Books, New York, NY.

Gonzalez, A. J., and D. D. Dankel, *The Engineering of Knowledge-Based Systems,* Prentice-Hall, Englewood Cliffs, NJ.

Hertz, D. B. 1988. Boeing has high hopes for AI, *AI Week,* July 1.

Kaplan, S., Frank, M. S., Bley, D. C., and Lindsay, D. G. 1987. Outline of COPILOT, expert system for reactor operational assistance using a Bayesian diagnostic module, *Proc. Int. Post SMiRT-9 Seminar on Accident Sequence Modeling: Human Actions, System Response Intelligent Decision Support,* Munich, August 24–25.

Kruse, R., Schwecke, E., and Heinsohn, J. 1991. *Uncertainty and Vagueness in Knowledge Based Systems,* Springer-Verlag, Berlin.

Ricker, M. 1986. An evaluation of expert system development tools, *Expert Systems,* 3(3).

Sackett, J. I. ed. 1987. *Proc. ANS Topical Meeting on Artificial Intelligence and Other Innovative Computer Applications in the Nuclear Industry,* August 31–September 2, Snowbird, UT.

Swartout, W., and Smoliar, S. 1987. On making expert systems more like experts, *Expert Systems,* 4(3).

Uhrig, R. E. 1987. Application of artificial intelligence in the U.S. nuclear industry, *Proc. ANS Topical Meeting on Artificial Intelligence and Other Innovative Computer Applications in the Nuclear Industry,* August 31–September 2, Snowbird, UT.

Uhrig, R. E. 1988. Applications of artificial intelligence in nuclear power plants, *POWER Magazine,* June.

Van Horn, M. 1986. *Understanding Expert Systems,* Bantam Books, New York, NY.

Warner, E. 1988. Expert systems and the law, *High Technology Business,* October.

Waterman, D. 1986. How do expert systems differ from conventional programs, *Expert Systems,* 3(1).

APPENDIX

LIST OF APPLICATION OF EXPERT SYSTEMS TO NUCLEAR POWER PLANTS

Current development and use of expert systems in activities associated with nuclear power are well documented (Bernard and Washio, 1989; Uhrig, 1988, 1987; Expert-EASE Systems, 1987; Sackett, 1987) and are listed in this Appendix. The applications presented here are typical of those in use or being developed in the nuclear power industry today.

1. Reactor Emergency Alarm Level Monitor
2. Computerized Tracking System for Emergency Operating Procedures
3. Clones of Experts at Fast Flux Test Facility
4. Trip Buffer Expert System
5. Technical Specification Monitor
6. On-Line Generator Diagnostic System
7. Intelligent Eddy Current Data Analyzer
8. Motor-Operated Valve Expert System
9. Expert Systems for Training
10. OECD Halden Reactor Project Expert Systems

    DISKETT—a rule-based diagnosis system to aid operators in analysis of plant disturbances.
    EARLY FAULT DETECTION (EFD)—a computer-based operator aid designed to assist operators in diagnosis of feedwater system faults.
    COPMA—a computer-based procedure system for use by plant operators.

11. Alarm Diagnosis and Filtering
12. Improving Nuclear Emergency Response with an Expert System
13. Reactor Safety Assessment System
14. DYSIS, a Real-Time Diagnostic/Control System
15. Residual Heat Removal Expert System
16. Accident Diagnosis and Prognosis Aide
17. Transient Analysis of Multiple Failure Simulations
18. COPILOT, an Expert System Advisor for Nuclear Power Plants
19. Handling Potentially Invalid Sensor Data
20. Spare Parts Inventory Control
21. Plant Status Monitor System
22. Search Procedure for Fuel Shuffler
23. BWR Fuel Channel Tracking System
24. ATHENA Code Input Model Preparation
25. Operational Control of PWR Cores
26. Diagnostics Using Model Base Reasoning
27. Nuclear Plant Technical Specification Tracking
28. Fault Tree Analysis in Expert Systems
29. Use of PRA in Expert Systems
30. Outage Planning
31. Heat Rate Improvement
32. Diagnostics for Instruments and Equipment
33. Welding Rod Selection Advisor
34. Generating Welder Procedures That Comply with NRC Codes
35. Signal Validation
36. Condensate Feedwater Monitor
37. Radwaste Processing System Advisor
38. Bypass-Inoperable Status Indicator System
39. Sequencing BWR Control Rods after Maneuvering
40. Pressure-Temperature Control During Startup
41. Water Chemistry Control
42. Real-Time Emergency Evacuation Planning
43. Real-Time Radiation Exposure Management

57

Strategies and Tactics for the Application of Neural Networks to Industrial Electronics[1]

Mary Lou Padgett
Auburn University

Paul J. Werbos
NSF

Teuvo Kohonen
Helsinki University of Technology

57.1 Computational Intelligence Connections and Future 835
The Problem • Neural Networks, Artificial Versus Biological
57.2 Engineering Intelligent Electronics Applications 836
Approaches • Methodology • Refinement • Applications • Implementations • Tools
57.3 Summary of Basic Modeling Concepts 846
57.4 Applications 846
57.5 Future 846
57.6 Defining Terms 847
Controls • Modeling • Single Neuron and Neural Networks
57.7 Resources 851

57.1 Computational Intelligence Connections and Future

The Problem

Ideally, intelligent electronics for industrial applications should incorporate "common sense" in autonomous operations. Instead of blindly following predetermined algorithmic procedures, electronics for use in industry should be responsive to changing system requirements, and should guard against taking actions with adverse consequences. Hardware and software should be able to optimize performance, over time, in an adaptive fashion. This kind of performance was difficult to achieve in the past, but recent advances in computer hardware and other technology have substantially expanded the possibilities. The material in the chapters on neural networks, fuzzy systems, evolutionary systems and computational intelligence (CI) in general will suggest approaches to the problem of incorporating "common sense" capabilities into industrial electronics. Examples of current and future applications will be discussed, and further resources will be identified. Careful incorporation of currently available computational intelligence techniques into existing systems can strengthen and stabilize performance. (See Figure 57.1.) Such successes then point the way to future improvements The material following is intended to assist a practicing engineer in expanding the horizons of intelligent industrial electronics.

Neural Networks, Artificial Versus Biological

A few years ago, the connection between the artificial (ANN) and biological neural network (BNN) was questionable at best because of the much greater complexity to be found in the BNN. The work on BNNs, on the one hand, and the development of software tools and hardware exploiting ANNs, on the other hand, were carried out by two different cultures with different goals. In recent years, some convergence has been developing between biological and artificial designs, particularly as more complex artificial control designs have appeared. Nevertheless this biologically motivated work is important mainly to those engineers developing more advanced designs through fundamental research. For engineers focused on today's control challenges, it is better to begin by mastering the engineering properties of the artificial designs already in existence. That by itself is no trivial accomplishment, given the variety of designs which have recently appeared (Figure 57.2).

With industrial applications we thus prefer talking of strategies and tactics based on complete algorithmic tools (that live their own life) rather than discussing the biological accuracy of related neuron models.

[1] This work was funded in part by AF Contract AF F 8530-94-1-0002 and AFOSR Contract SREP F49620-C-0063.

0-8493-8343-9/97/$0.00+$.50

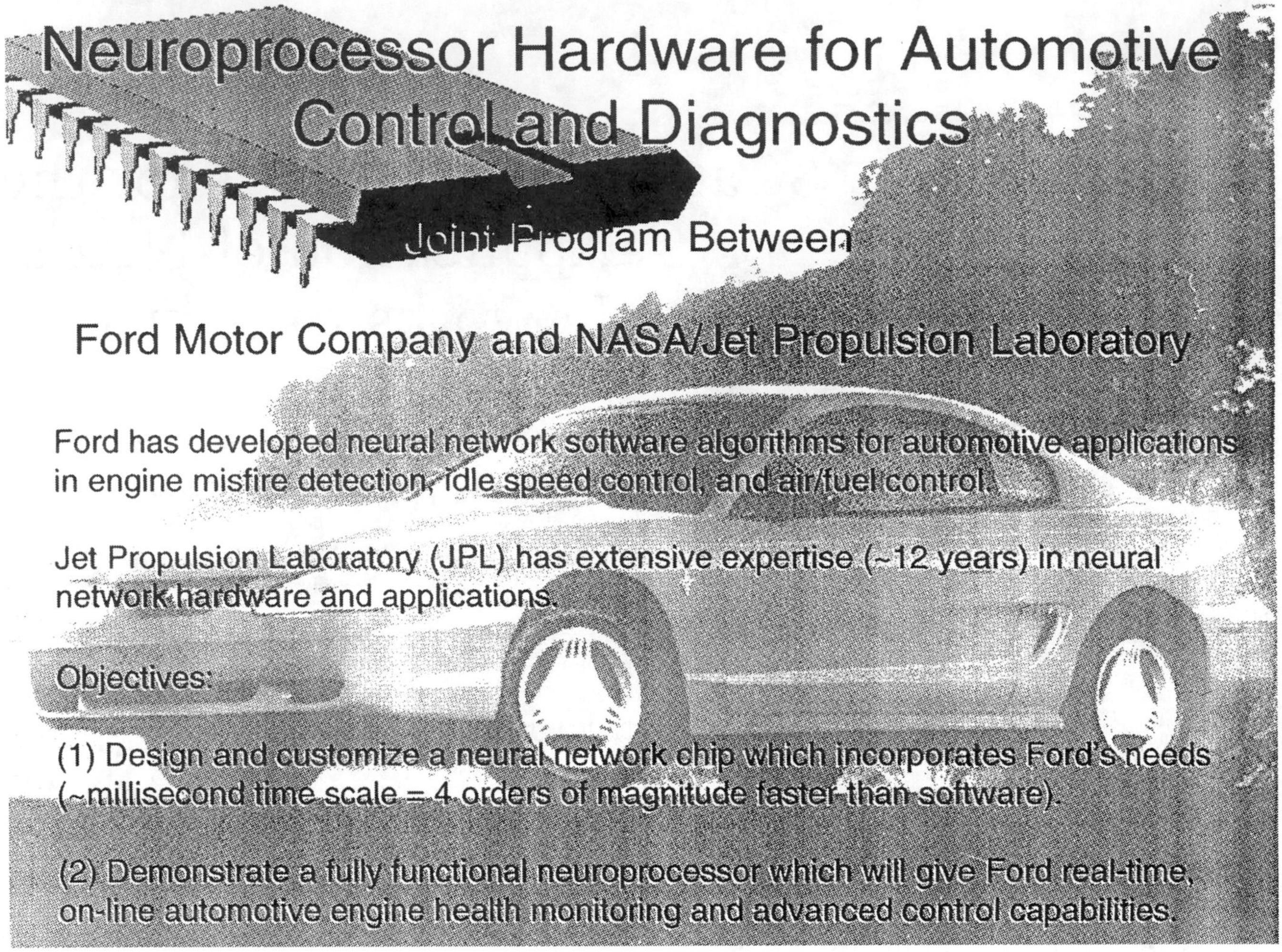

Figure 57.1 Neuroprocessor hardware for automotive control and diagnostics.

57.2 Engineering Intelligent Electronics Applications

Approaches

Strategies and tactics are two vital elements of application development. A good contract monitor will ask, "Did you answer the right question?" Engineering strategy must determine what the real problem is, and work toward solving it. Early efforts to apply artificial neural networks often suffered from a limited world view. If the only tool available is a hammer, the entire world begins to look like nails. It is important to consider new ways of specifying the tasks to be done, rather than falling back on old habits which force the mindless reuse of old tools.

A *metadesign* for the problem should be developed (Figure 57.2). This is a high-level design where the components are CI tools. The metadesign should have two separate levels. One is application into macrotasks which macrolevel CI designs (or other tools) can perform. The other is to fill in the components of these microdesigns with microlevel CI components, suitable for the specific application. The first may be thought of as strategy for solving the application-level problem. The second comprises the tactics for implementing the solution.

CI System Design Strategies and Tactics

Strategies: APPLICATION → MACRO/SYSTEM
Tactics: MACRO/SYSTEM → MICRO

In terms of traditional hardware-in-the-loop simulation, validation requires examining the real-world relevance of the simulation in question. In implementing artificial neural networks for industrial electronics applications, the same considerations and evaluation techniques apply. (See Tables 57.1 and 57.2.) Validation includes supplying an answer to the problem, and tailoring the detail to conserve resources while preserving accuracy. Early establishment of performance measures as validation benchmarks can help in the design of sensitivity analysis for elimination of unnecessary detail and complexity (Padgett and Padgett, 1995). Performance or utility measures actually constitute a mathematical definition of what the system is actually being designed to do; thus it is crucial to be careful in articulating such definitions, based on the real needs of the client, and to update these definitions on the basis of practical experience (Werbos, 1990).

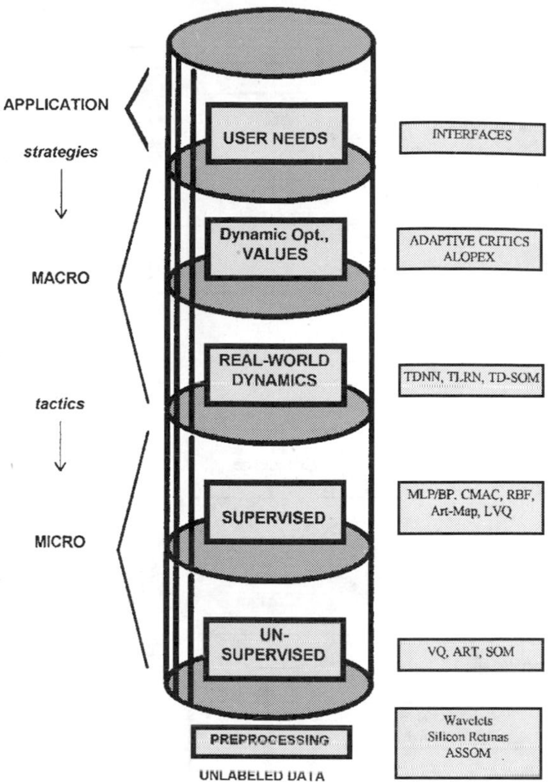

Figure 57.2 Levels of aggregation of artificial neural network applications. The sole purpose of each block is to serve the levels above it.

Planning ahead is essential, but foresight is always limited in a new application. A *flexible plan* should include a range of options to consider. First, consider where the current application lies in the research/feasibility/proof-of-concept/rapid-prototype/commercial production continuum. Then estimate where it may lead and examine the constraints and resources involved. Once a flexible global strategy has been outlined, tactics for implementation can be more closely examined. Here, another ladder of possibilities must be constructed. A particular component of a system can be implemented in many different ways. Selection of the easiest way that might work is usually a good start. It is important not to become discouraged when the first trial does not perform perfectly (or is very wrong). The flexible metaplan should consider this probable outcome and outline a ladder of options for improvement. Shifts in ladder level may go down to reduce complexity enough to produce feasibility. They may go up to increase the intelligence or other capabilities of the component. The important thing is to consider in advance that such shifts are nearly always a part of applications, and plan to make them easy. To achieve such flexibility, it is crucial to aim for simplicity and modularity of design, even if this results in a temporary increase in computational costs.

Modification and reuse of software and hardware is essential for most real-world applications. This of course means development of modular software and hardware, and keeping a bank of easy test problems for analysis of strange or problematic results. Easy test problems which work on a spreadsheet are very useful. Shrink-wrapped software packages can help build intuition and skill at using the older neural system types. These should be mastered before venturing into the more advanced types. Since the state of the art is advancing so rapidly, the commercial products lag behind available techniques quite a bit. For a price, in money and programmer skill, shareware and public domain software under development by the scientific community can increase the power and flexibility available. Dupont developed its own software system for applications of neural systems in the chemical industry. This was extremely successful. A similar complete package for industrial control applications has been developed at 3M (by Mr. Esa Vilkama). Ford has successfully implemented an on-vehicle control and diagnostics system designed to avoid the "*Fail-Dumb*" syndrome. If all is "OK", there is no decision. Otherwise, an operator decision is needed, and a red, yellow or green light may provide fuzzy information to the driver. (See Figure 57.1.) After off-line training, the neuron information is transferred to the Recurrent Network Chip designed for Ford by JPL (Wunsch and Prokhorov, 1995; Puskorius and Feldkamp, 1994; Feldkamp et al., 1995). The NEUROCLASS simulator of Thomson CSF in France is an example of European industrial neural software.

Although it is necessary to know the basic solutions contained in generally used software packages in detail, it is often profitable to tailor the software tools in-house, to fit them better to company traditions and policy.

Approaches to intelligent electronics use and/or development vary. Many successes in industry cited by Carver Mead follow the design procedures recommended by the 1990–1995 IEEE Standards study groups on computational intelligence. Researchers in the field, such as Carver Mead, Bernard Widrow, Laurence Fogel, Michio Sugeno and Lotfi Zadeh, recommend *clearly stating objectives and values for each possible outcome*, then amending them as knowledge of the application under development increases. Traditional industrial applications illustrating these concepts abound. (See the following chapters on intelligent electronics and the chapter on factory automation in this handbook.)

State Objectives. Successful industry applications of intelligent electronics are *tuned to their application-level objectives.* These projects may be aiming for micro-, systems- or applications-level solutions (see Figure 57.2 and the paper by Werbos in this chapter). Tiny (micro-) modules of intelligent electronics make useful tools when combined into task-oriented systems-level problem solutions. These task-oriented solutions in turn can be incorporated into multisystems or applications-level projects which are increasingly combining the strengths of neural (NN), fuzzy (FZ) and evolutionary systems (EC) to solve real-world problems facing industrial electronics engineers. Embedding elements of NN, FZ and EC into expert systems and into traditional systems is a design task which can be strongly aided by interactions with intelligent virtual reality systems (VR).

For example, robotics (and other applications) involve many systems and can be successfully approached using a *top-down/bottom-up design* incorporating neural, fuzzy and genetic systems. Work by Fukuda and Shimojima (1995) illustrates the success of this technique. Advances planned by this group will further

Table 57.1 Suggested Design Considerations for the MLP Approach

PARADIGM SPECIFICATION MODULES				
OBJECTIVES and PERFORMANCE MEASURES: Value of each Possible Outcome to the Application *(Ends)*				
Measure of Performance	Criteria: Accuracy, Efficiency, etc.	Optimality given Multiple Criteria	Criticality Degree	% of Time Important
MODELS and FUNCTIONS *(Strategies)*				
	(Biological)	Mathematical	Engineering	
TOOLS and RESOURCES Ready and Available *(Means)*				
Hardware	Software	Personnel	Experts	Experience
INPUT/OUTPUT *(Interfaces)*				
Data Sources	Characteristics	Interfaces	Dynamics	Validation
ARCHITECTURES and CONNECTIONS *(Structures)*				
Data Paths: Direction	Node Groups: Layers, Slabs	Node Number per Group	Cycles / Loops Feedforward	Data Paths: Timing
COMPONENT FUNCTIONS *(Functions)*				
Weight Precision	Activation Function	Local Memory (in Node)	Bias Node	Other
VARIABLE PARAMETERS: Adaptive Mechanism, Data Flow *(Functions cont.)*				
Bias Values	Initial Weights	Moment	Gain	Other
RECALL, LEARNING, UPDATE MECHANISMS *(Tactics)* *(Functions cont.)*				
	Batch	Sequential	Other	

expand use of these concepts. In this and other applications, global system objectives are met by proper selection and control of lower-level modules. It is necessary to be familiar with the capabilities of the lower-level modules before formalizing global system objectives. Planning to dynamically monitor and adapt these relationships should be part of specification of global objectives. Global performance measures should be developed to communicate the value of all possible system outcomes with respect to the global objectives (Fogel, 1995).

Data visualization helps suggest approaches and may clarify problems (Kohonen, 1996; Mead, 1994). Expanding the scope of interaction with the computer model to allow extensive real-time interaction between humans and computers can bring all of the human's senses to bear on a problem. Tracking an expert's reactions and responses can provide insight and numerical input to an intelligent electronics industry application. (See the papers on control by Werbos (cloning and shaping) and the paper by Sugeno et al. in the fuzzy systems chapter.) Often the best way to train an ANN to control a robot is by first asking a human being to control that same robot, in a virtual reality mode. If the human succeeds, the resulting data can be used for the initial training of an ANN controller; if not, the design of the robot itself may be reconsidered.

Visualizing machine states can be accomplished using a variety of techniques. For example, consider power transformer analysis using self organizing maps (SOM) (Kohonen et al., 1996). In general, the SOM and variations of it can convert on-line measurements into a simple and easily understandable display which *preserves the relationships of the system states despite the dimensionality reduction.* Advantages of this display include the following:

1. System operators can visually monitor the changing system states during development.
2. Estimation of future system states can be stimulated by better understanding of data.
3. Fault identification can be made possible by display of current or predicted faults.
4. System control can be based on state analysis.

Table 57.2 Suggested Design Considerations for the Competitive Learning Approach

PARADIGM SPECIFICATION MODULES				
OBJECTIVES and PERFORMANCE MEASURES				*(Ends)*
MODELS and FUNCTIONS				
	(Biological)	Mathematical	Engineering	
TOOLS and RESOURCES				*(Means)*
Hardware	Software	Personnel	Experts	Experience
INPUT/OUTPUT				*(Interfaces)*
Data Sources	Characteristics	Interfaces	Dynamics	Validation
ARCHITECTURES				*(Structures)*
Hierarchical / Nonhierarchical	Topology	Structures and Sizes of Codebooks	Buffers for Sequential Data	
COMPONENT FUNCTIONS				*(Functions)*
Distance Metric	Neighborhood Function	Winner Search	Weight Precision	Conscience Learning
VARIABLE PARAMETERS				*(Functions cont.)*
	Initial Weights	Neighborhood Function Parameters	Learning Rate Function	
LEARNING MECHANISMS				*(Functions cont.)*
	Unsupervised / Supervised	Teachers	Passing Results Between Algorithms	

Applications enhanced by this technique include: preprocessing and feature extraction; process and systems analysis (visualization of machine states and fault identification); statistical pattern recognition; telecommunications; measuring and evaluation techniques; design and testing methods; and robotics (parameterized SOM for control; Ritter, 1995; Kohonen et al. 1996). (See the Defining Terms section of this article.)

The careful engineering of a computationally intelligent (CI) system targets objectives and selects methodological variations to approach these objectives in a manner which can be validated and verified. A system with a meaningfully high machine IQ (MIQ) (Zadeh, 1995) has a system IQ (SIQ) (Padgett and Padgett, 1995) based on the "soft" application of computational intelligence: use it when needed, bypass it when appropriate. The capability of a system to self-diagnose and self-correct is economically valuable. Make the CI system meet the needs of the particular application and keep industry management satisfied.

CI System Objectives

- Know the target market (Mead, 1994).
- Design for solutions.
  - Develop performance measures.
  - Assign values to all potential outcomes (Fogel, L., 1995).
- Develop a flexible metaplan for strategy and tactics.
  - Application- to macro/system-level solutions *strategy* map.
  - Macro/system- to micro-level solutions *tactics* map.
- Clarify difference between mathematical and computer models (Mead, 1994)—and biological models if of interest at the research stage
- Realize and clearly state limitations due to platforms (hardware, software).
- State constraints due to resource availability.
- Consider skill level, man-hours, environmental restrictions.
- Achieve a high system IQ (Zadeh, 1995).

Methodology Refinement

Once the initial objectives have been formulated, methodologies can be selected, and adjusted to fit the particular application. The paper by Rumelhart, Widrow and Lehr in the neural networks chapter discusses issues frequently addressed in the design of NN applications. These considerations apply to all the CI modeling strategies. The list below mentions some of the considerations and options for methodology refinement. These are also depicted for types of neural systems in Tables 57.1 and 57.2.

CI System Design Considerations

- Appropriate application goals and performance measures—application → system/macro → micro.
- Preliminary mathematical model—learning strategies.
- Implementation constraints-affordability and availability (Fogel, L., 1995).
- Architectures—block diagrams, components and connections.
- Learning strategy details—parameters, functions, timing.
- Modeling data and links to system—scaling, sources.
- Validation/verification—generalization, performance.
- Future modifications—modularity, environments, user interactions, data hooks.
- Interactive visualization and intelligent VR.

Traditionally, artificial neural networks have been implemented in most industry applications with a multilayer perception structure and a backpropagation learning scheme. There are now many variations of neural networks which are very successful, and should be considered as potential solutions. Figure 57.2 illustrates a cylinder made up of the various levels of detail which need to be considered in ANN applications. The sole purpose of each block is to serve the levels above it. Flow of control begins at the top, application level of user needs. Strategy maps the application level to the macro-or system level of dynamics optimization, values and real-world dynamics. After system tasks have been determined, tactics for implementation map the macro- to the microlevel of supervised and unsupervised modules. Specification of a neural network involves setting the structure, or possible data paths and separately specifying the actions on the data. Structure and adaptive mechanisms are the essential elements. These tactical implementations are only effective if they meet the needs of the user. Block labels are: unsupervised, supervised, real-world dynamics, values or dynamic optimization, and user needs. These bridge the gap between top level interfaces with the larger system or human user, the multimodule tasks keyed to values, the real-world dynamics problems using forecasting and the supervised solutions which match the observed output to a target output. Unsupervised networks serve as preprocessors to the other, higher rungs on the ladder, operating on UNLABELED DATA. In practice, design upgrades may occur on all levels of the system, in parallel, based on experience at neighboring levels.

Applications

Types of Problems

Every application has a strategy and a tactical *level of design.* Some groups of engineers use existing systems (or refine them) to solve complex real-world problems. Others work on producing systems for use by various application domain specialists. Many efforts are directed toward producing generalizable microlevel modules which can be used in many systems. One of the first steps in an application is to decide which level is of current interest. Higher-level applications and systems need to consider use of existing tools, but also plan to modify them if necessary.

Engineering design methodologies are selected and refined to address certain types of problems. At the *microlevel.* CI paradigms exist which perform specific operations on input of a certain nature. Neural learning from data may be supervised, unsupervised, or based on immediate reinforcement feedback. (Reinforcement learning over time, however, requires macro-level designs.) Fuzzy modules, or even classical AI, can be used to translate rules provided by a human expert into working computational algorithms. However, unlike traditional AI, fuzzy logic performs in a more smooth and analog fashion permitting a more comfortable interface with other continuous variable techniques. Modules fuzzify, process, then defuzzify at an appropriate time (Jang and Sun, 1995). Genetic modules process binary strings, performing operations such as selection, crossover or mutation (Goldberg, 1994). Evolutionary modules operate top-down instead of bottom-up, evolving behavior traits of an individual, instead of genes along a chromosome. Evolutionary programming evolves behavior traits of a species instead of an individual. The process can be a continuous function optimization (Fogel, D., 1995).

Neural modeling of data, fuzzy modeling of human judgment and evolutionary/genetic modeling of exploratory change are powerful building blocks. Combinations and variations of these elements abound, and can combine the strengths of all the approaches. Neural learning can adjust fuzzy rules (see the article by Werbos in the fuzzy systems chapter) or enhance the selection procedures guiding evolution (Fogel, D., 1995). Fuzzifying a neural module can strengthen communication back to the expert or larger system. Fuzzy genetic algorithms may add logic to genetic or evolutionary strategies. Conversely, evolutionary/genetic explorations can help free a neural module from a local minimum, and genetic explorations can evolve fuzzy rules. Combining neural, fuzzy and genetic properties in a top-down and bottom-up method may use the advantages of each technique to overcome the disadvantages of the others (see the paper by Shibata, Fukuda, and Tanie in the computational intelligence chapter and Fukuda et al., 1995).

At the *system level,* these elementary modules can be combined to address tasks such as pattern recognition, controls, or decision support. Fault detection, identification and recovery (FDIR) and

system identification are frequent task-targets in CI design (Padgett and Padgett, 1995). Control systems have a hierarchy of possible tasks:

1. Cloning (of a human or automated expert).
2. Tracking (of a set-point or trajectory)
3. Optimizing performance measures or goal satisfaction over some planning horizon.

A complex control system application such as autonomous flight may combine a vision system, FDIR capabilities, system identification and a range of control tasks. Guidance may be initialized as a clone, tuned to track a trajectory, and modified to keep the system approaching acceptable performance levels such as fuel consumption or responsiveness to commanded maneuvers (see the articles by Werbos in the neural networks and the fuzzy systems chapters and Padgett, et al. in the emerging technologies chapter).

At the *application level,* general purpose or application-specific systems may combine in a hierarchy or be used concurrently to produce a product. Elastic neural networks, soft computing, computational intelligence and intelligent VR emphasize the strengths of NN, FZ, EC and VR to solve an application problem.

CI Problem Levels

- Micro—Modules.
- System/Macro—Combination of modules for tasks.
- Application—Combination of general-purpose and application-specific systems to produce a product.

Mixing NN, FZ, EC and VR. Recent successes mixing NN, FZ, EC, and VR with other expert and/or traditional systems are discussed and illustrated in the computational intelligence chapter. Intelligent VR is explored in the emerging technologies chapter.

Implementations

The design of complete signal processing and/or control systems can be based on many alternative categories or "brands" of algorithms. (See Figure 57.2.) Implementations evolve from unlabeled data to dollar-based applications meeting user needs. They should be driven by the *application strategies* using a detailed knowledge of the possible *tactics for implementation.*

Issues in implementing these and other applications using CI systems are addressed in this handbook for neural, fuzzy and evolutionary/genetic systems. Expert systems and hybrid systems are covered at the beginning of the chapter, and variations of neural systems follow this article. The discussion below therefore centers on description of common types of neural systems to introduce these concepts. The first step in many applications is preprocessing, often in the form of feature extraction. Next, unsupervised classification and clustering are often considered. Here they are described in terms of vector quantization (VQ) and self organizing maps (SOM). (See Table 57.2.) Next supervised learning is addressed (Table 57.1). The classical multilayer perceptron (MLP) trained by backpropagation is described and related to biological neurons. Finally, supervised classification by learning vector quantization (LVQ) is discussed. In summary, hybrid approaches combining neural, fuzzy and evolutionary systems are recommended as successful approaches for solving applications-level problems, such as those encountered in industrial electronics.

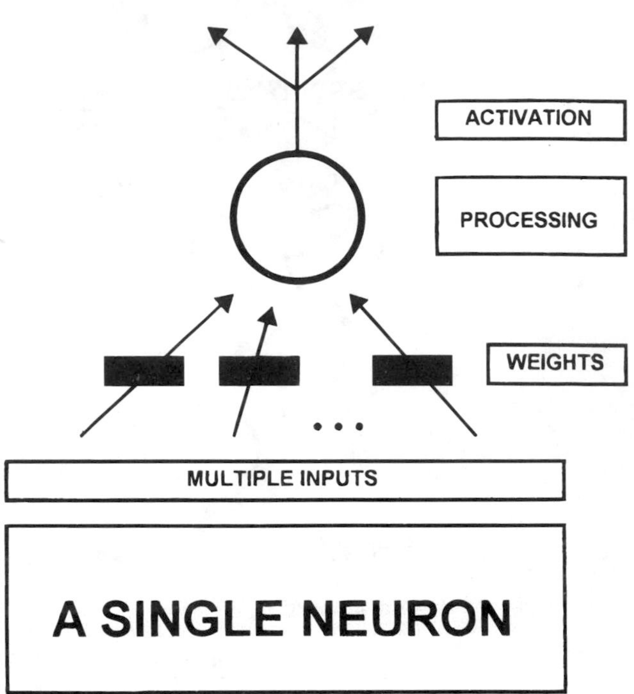

Figure 57.3 Single neuron of a multilayer perceptron.

Feature extraction

Industrial measurements can seldom be used directly as inputs to neural-network algorithms. At a minimum they should be normalized to common scales. In pattern recognition, especially, the natural variations in the data sets due to movements of the objects, varying illuminating conditions, etc. may be eliminated or compensated for. For instance, in image analysis, one should first extract a number of features from the primary variables that are as invariant with respect to these variations as possible. Classification is then performed on the basis of these new features. The two standard approaches to feature extraction are:

1. If it is possible to standardize the images by positioning, orientation, scaling and compensation for varying illumination, then various eigenfunctions (such as the principal components) of the input data vectors can be used as features.

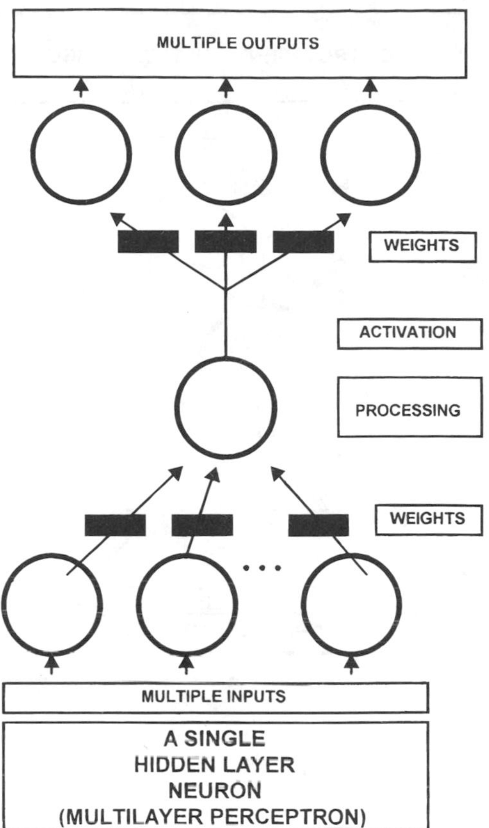

Figure 57.4 Single hidden layer neuron of a multilayer perceptron.

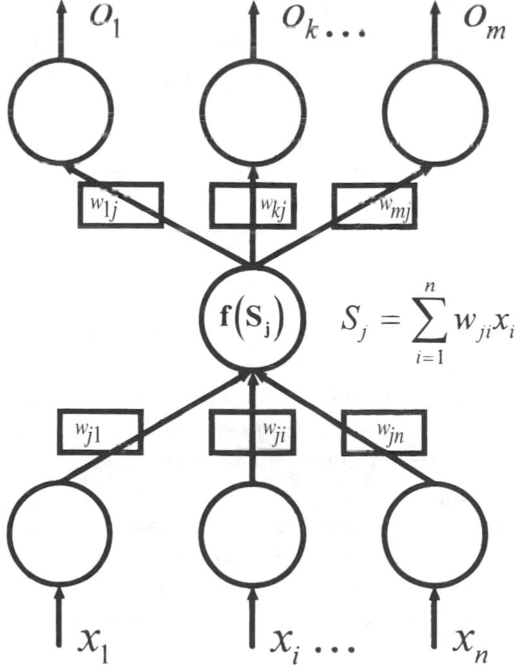

Figure 57.5 Typical multilayer perceptron (MLP), recall state (Zurada, 1994).

2. If the objects are moving, or cannot be tracked exactly, it is a more effective strategy to extract from the primary observations a set of local features that are as invariant as possible with respect to the natural transformation groups of patterns.

For instance, the wavelet and Gabor transforms are examples of elementary feature filters. An alternative approach to the fixed preprocessing of images is to use a "silicon retina" such as those developed by Carver Mead or by Stephen Grossberg.

In image analysis one has constructed such invariant-feature filters heuristically, by postulating the mathematical form of the filter functions and optimizing their parameters. It has recently transpired (Kohonen 1996, 1995) that such invariant feature filters can emerge in a new type of competitive, adaptive learning process called the adaptive-subspace SOM (ASSOM). It is characteristic of the ASSOM that it learns the transformations from sequences of almost general patterns such as noise patterns, photographic images, etc.

In general, ASSOM is one example of a growing tendency to use learning-based methods in place of fixed preprocessors, when preprocessing is necessary. Usually, such methods involve some form of unsupervised learning. The following section will describe VQ and SOM, which are among the most important such methods. Other such methods would include various varieties of encoder/decoder design, such as the deterministic designs of Hinton, Werbos and Cottrell, or the stochastic encoder/decoder/predictor (White and Sofje, 1992, Chapter 13), a variety of biologically inspired designs, and others (Kohnen, 1990; Kosko, 1988; Werbos, 1988; Grossberg, 1987). For more detail and additional references, see the articles by Werbos and by Carpenter and Grossberg in this handbook.

VQ and SOM for Unsupervised Classification

On the level of analysis that deals with the primary data, one usually starts with unsupervised classification (often the same as clustering), which means that the input data can still be unlabeled. Acquisition of unlabeled or unclassified data is usually much less expensive than preparation of the corresponding amount of carefully preclassified and labeled data. This phase is sometimes called exploratory data analysis or "data mining." One purpose of this phase is to map the raw data into categories or clusters that can be described with variables of low dimensionality.

The so-called competitive-learning algorithms (VQ, ART, SOM) are unsupervised classification algorithms that may accept very-high-dimensional input vectors (say, with hundreds of components). These vectors are usually regarded as elements of a real, metric vector space. This means that the input variables, say elementary measurements or signals that constitute the vector components, ought to represent the respective component variables in scales as similar as possible. As the signals or measurements may originally be defined in very different scales, it is a reasonable strategy to normalize the scales first, for instance by subtracting the mean from each variable and transforming the variables into new scales in which each variable has the same variance.

For instance, in the classical vector quantization, clustering of input data results if a number of codebook vectors is placed into the input signal space in such a way that the average expected quantization error over all input data is minimized. This condition may be expressed in terms of the error function E,

$$E = \int \|x - w_c\|^2 p(x)dx$$

where x is the input vector, w_c is one of the codebook vectors as specified below, $p(x)$ is the probability density function of x, and the integral is taken over the whole signal space. The vector w_c is the closest codebook vector to x in the signal space:

$$c = \arg\min_i \|x - w_i\|$$

In many practical cases, especially if the input vectors are high-dimensional, it thereby becomes possible to use the closest w_l instead of x to approximate it. Since the number of codebook vectors, w_i, is predetermined and their values are optimized by minimization of E, this usually results in big savings in representation, transmission, and classification of information.

The self-organizing map (SOM) is a vector quantization method, too. The codebook vectors in it are organized as a regular array (say, a two-dimensional, rectangular or hexagonal array) in such a way that the distance of two array elements in any direction roughly corresponds to the dissimilarity (distance, absolute value of difference) of the corresponding two codebook vectors in the original signal space. This correspondence is evidently best if the mapping of signal domains onto the array elements is ordered so that neighboring signal domains correspond to neighboring elements in the array. It is a characteristic property of the SOM algorithm that it creates this ordered map ping in an unsupervised way, on the basis of raw data (after normalization).

The images of input signals on the SOM are clustered. The basic difference with respect to the classical VQ is that the codebook vectors can also be represented by the coordinates of the SOM array, which is a very compressed encoding method. The benefit achievable by using the SOM instead of the traditional VQ methods is that since the mapping of input data into array coordinates is ordered, sensitivity of the mapping to errors is low. Noise in the signal space changes the array coordinates gradually.

Multilayer Perceptron (MLP) Trained by Backpropagation

Next described is a design to perform supervised learning. At present, the vast majority of ANN applications are based on supervised learning. Even in the future, it is expected that supervised learning systems will be crucial components of more complex systems to perform more sophisticated, more macro tasks.

Most supervised learning applications today are based on the multilayer perceptron (MLP).

Many design variations have followed the early work of Werbos (1988) and Widrow and Hoff (1960). "Global" network designs like the MLP and the simultaneous recurrent network (SRN) (White and Sofje, 1992, Chapter 3) typically offer good accuracy in approximating a function of many variables, and good generalization, but there are other designs—"local" designs—which typically offer faster learning, especially in real-time applications.

Local designs may also offer greater accuracy when the data tend to fall into clusters, or when there are relatively few input variables and a reasonably dense covering of the input space. In essence, local designs predict the desired output or classification based on a nearest-neighbor kind of principle. Among the most useful local designs are LVQ, radial basis function (RBF) and cerebellar model articulation controllers (CMAC) (Albus, 1975) and possibly—in the future—elastic fuzzy logic. In the future, there are good prospects for research to combine the advantages of both types of design, global and local (Werbos, 1993).

The structure of a neural network and the adaptation technique chosen are two important elements of the complete design or paradigm. The MLP architecture combined with adaptation by backpropagation (BP) form the paradigm commonly called basic backpropagation. Other combinations and modifications can be selected, depending on the focus of the project and its particular tasks. The focus may be microlevel, system-level or application-level. The particular tasks to be addressed should steer the decisions involving formulation and articulation of real-world applications of neural systems. Suppose a task at the system level is identified. It may be pattern classification, function approximation, dynamic system modeling, static or dynamic optimization, tracking, cloning, data compression, clustering or some other common application of neural systems. For a given task, there is a ladder of implementation possibilities ranging from quick but coarse or low-level solutions to complicated but highly intelligent, complex, perhaps slow solutions. Modular development of the actual implementation should foresee the progression from rapid-prototype feasibility studies to more elaborate proof-of-concept and possibly even commercial implementations. Planning ahead to ease the transition from stage to stage is vital. Software and hardware modularization and reusability are critical. Modeling error types and artifacts due to processing algorithms and platforms is essential. Maintaining a battery of simple examples in a spreadsheet is a very good way to examine a problem when things go wrong. There is a critical need to know what the bag of tools available does:

> What is the essence of the capabilities of each tool-type?
> Which are needed for this particular task?

Consider the structure and function of a typical neuron. First, a single neuron is illustrated in Figure 57.3. This neuron is shown in Figure 57.4 embedded in a multilayer perceptron as its single hidden layer neuron. Input layer and output layer neurons have connections outside the network, but a "hidden" neuron does not. Large numbers of such hidden layer neurons may be present. The number and their connections are design parameters.

The MLP adapted by backpropagation has two states: learning and recall. In use, the recall state operates to give a set of outputs for any selected input. The learning state operates on a comprehensive "training set" of input/output data pairs. As learning progresses, the error between observed and desired output is minimized, for all inputs in the training set. This is done by adjusting the weights to change those responsible for the most output error. An appropriately trained MLP will, in recall state, correctly estimate the desired output for inputs of a similar nature to those in the training set. If generalization has been achieved, recall results are accurate for a large number of inputs not identical to those in the training set. Many articles below discuss ways to improve the quality of generalization. Figure 57.5 shows a typical activation function for a multilayer perceptron's recall state.

The relationship of the artificial neuron to biology is worth noting. Engineering applications do not depend on knowledge or study of biology. The artificial neuron/biological neuron relationship is more meaningful to researchers trying to produce new neural networks based on ideas from biology. In Figure 57.3, it is shown that multiple inputs are accepted from the environment, from other neurons (or from earlier output of the neuron itself). The inputs are modified by weights representing the resistance encountered in a biological neuron at its synapses (connections between neurons). On arrival at the node (cell body), the weighted input signals are accumulated over time. Local information stored in the node may impact the reaction to these signals. Figure 57.5 illustrates the summation of the weighted input signals. When conditions in the node are right (perhaps based on a threshold), the neuron "fires" by outputting a signal which is a function, say f, of the sum of the weighted inputs, say S. This signal departs the "axon hillock" of the firing neuron. The output connection of the neuron represents an axon which branches, to meet input connections of other neurons (or the originating neuron). The input connections, representing dendrites, are branched. A potentially unique weight on each input connection represents the physical gap, or synapse, between axon fibers and dendrites. Signals traveling down axons are thought to move quickly and change little compared to the rapid dissipation of a signal over time and space as it travels down a dendrite toward the cell body. Learning is represented by the positive or negative values of the weights as they are adjusted to represent the excitation or inhibition of neural interconnections based on experience. A bias weight with a fixed input of 1 may be added to each neuron to represent its resting state, or to adjust the range of the activation function. Momentum terms, based on past history of activation may also be added to activation functions to help slow oscillations. Many such modifications of basic paradigms are in widespread use. Their effectiveness depends on the particular application and the skill of the system designers in adjusting parameter values. Techniques for intelligent manipulation of such internal workings of the neuron are illustrated in the chapters on fuzzy systems, evolutionary systems and computational intelligence.

As a number of neurons are assembled and massively interconnected, interesting properties are observed. The collective behavior of small, medium and large neural networks varies distinctly. (Here size is based on number of neurons and number of connections.) Nature provides biological models which are drastically different from actual physiology, but form the basis for ideas for artificial neural systems. These abstractions from biology give rise to mathematical models for artificial neural systems. Moving from a mathematical model to an implementation also changes many things. There are significant impacts attributable to the analog, digital or hybrid implementation choices. Careful examination of design objectives can lead to successful applications. Ignoring implementation constraints and assumptions leads to unpredictable results. Most successful applications are very focused, and build heavily on traditional methods of control, pattern recognition and decision aides. Material in Part 2 of this handbook is intended to guide the construction of neural systems applications in intelligent industrial electronics.

LVQ for Supervised Classification

The term *learning vector quantization* (LVQ) signifies a family of competitive-learning algorithms for supervised classification. The LVQ algorithm is closely related to VQ and SOM, and can therefore easily be combined with them; they can operate on common files. The complete implementation tactics of the MICRO levels can therefore also be based on these algorithms only.

The LVQ needs well-initialized values for its codebook vectors. It is therefore recommended to start with the unsupervised SOM algorithm, to define a set of clustered codebook vectors first. These SOM vectors are then assigned to different classes or categories of the input signals by calibrating the SOM units with labeled samples of input data. Each class of the input signals becomes thus represented by one or several codebook vectors, and the class borders are defined as pieces of midplanes between closest pairs of codebook vectors having different classification.

The characteristic property of the LVQ algorithms is that they fine-tune the codebook vectors optimally. The class borders between adjacent codebook vectors are moved into positions such that the average rate of misclassification errors is minimized. The LVQ thus defines near-Bayesian decision borders for supervised classification.

Hybrid Approaches

Many successful applications of pattern recognition and controls combine neural and fuzzy systems. These are discussed in the fuzzy systems and soft computing chapter. Fuzzy extensions of LVQ are covered by Karyiannis, elastic fuzzy logic and neurocontrol by Werbos, and fuzzy adaptive resonance theory (ART) by Carpenter and Grossberg. The newer concepts in evolutionary systems are added following the suggestions of (Fogel, D., 1995). These evolutionary computing and genetic algorithms approaches to CI are explained in the evolutionary systems and computational intelligence chapter. Top-down/bottom-up approaches to design are illustrated in the article by Fukuda et al. The strengths of each CI category (NN, FZ, EC) are utilized to construct working industrial applications.

In each CI category, task oriented issues are addressed by applications illustrating these microlevel configurations. Preprocessing by wavelets, comparisons to existing statistical pattern recognition techniques, and modifications to controls system approaches illustrate the problems and suggest some solutions. It is difficult to precisely categorize neural network systems, since so many variations exist. For example, if a net is adapted by derivative of error methods, feedback is a vector of derivatives. This is not exactly *supervised* learning, but it is a micro module (see Figure 57.2).

An overview of integrated applications and possible futures for the field is presented in the chapter on emerging technologies as a summary and a link with other emerging technologies. Intelligent virtual reality applications which integrate neural, fuzzy, evolutionary with VR and its precursors are covered by Padgett et al. The migration of aerospace simulators toward intelligent VR systems is an example of the combination of neural, fuzzy and evolutionary systems to solve applications problems beyond the scope of the usual classic techniques. Design considerations for implementing such systems are a key feature of this handbook [Part 2].

Tools

As diagrammed in Tables 57.1 and 57.2, identification of available tools and resources is an important step in applications development. Many articles in Part 2 of this handbook address these topics and issues in tool selection.

Hardware

For neural, fuzzy and fuzzy/neural applications, hardware implementations are discussed. Specialized neural electronics for neural networks vary from those designed to emulate biological systems to high-tech aerospace applications borrowing ideas from biology (see papers by Lau, Wolpert, Saeks and Daud et al. in the neural networks chapter). Wavelets aid image analysis (see Szu, 1995 and papers by Rogers et al., and Ruck and Rogers in the neural networks chapter). RBF hardware aids computer communications (see the papers by Lindblad, et al. in the neural networks chapter). Fuzzy hardware solutions range from very specialized to general purpose. Recent work cites successes in use of general purpose hardware for fuzzy solutions (see the paper by Padgett in the fuzzy systems chapter). A practical example of neural fuzzy hardware follows in the paper by Lindblad and Lindsey. Many competing companies are producing neural hardware which works. Details are proprietary, having a high economic value. Ford Motor is using a chip recently developed by JPL (Figure 57.1). There are many chips on the market, which are very different in quality. It is difficult to evaluate these products.

NSF is encouraging efforts which evaluate products by doing a check, performing benchmark tests. It is not obvious which products are best. For a particular application, there are not yet sufficient real-world comparisons of different boards on the same application. Thus, it is helpful to perform benchmark testing as described below:

BENCHMARK TESTING
- *Problem Selection:*
  - Premature, but
  - Control problems and NN designs that solve them exist
- *Platform:*
  - PC or MAC due to market
- *Criteria:*
  - Accuracy:
    - will it work on a PC or a MAC?
  - Speed:
    - how fast will it run?
  - Cost in Skill and Time to Implement:
    - how hard is it to program?
  - Scaling:
    - is it feasible to scale up programs to use more variables
      - (on PC or MAC)?
      - without trouble?
  - Flexibility:
    - is it modular
      - to support modifications, reuse, validation and verification?
  - Reliability:
    - is it bullet-proof?
    - is it hard to get to the system?
    - is the system development in a state of flux?
  - Embedding:
    - is it a rigid single-level model?
      - is this adequate for the application?
    - does is allow embedding boxes in boxes
      - for dynamic linking?
    - how heavy are the costs for dynamic linking?

Software

Software packages available are discussed in some of the applications articles, and internet sources for code and/or comments are described as further resources. Many very simple shrink-wrapped packages are available, and they meet the needs of some applications. It is important to start by understanding the engineering aspects of the popular, commercially available solutions. It is equally important to realize that these products are dated in capabilities compared to those still under development, or those tailored in-house to meet specific needs. For example, for a *cost,* Dupont has implemented thousands of real world applications. They built their own highly modular package which lots of chemical engineers use. Many groups are developing and maintaining their own inexpensive, shareware or freeware packages. Peterson at Clemson has such a generalized MLP package (Werbos, 1994). Extensive SOM and LVQ packages are available on the Internet at the address cochlea.hut.fi (Kohonen, 1995).

In summary, it is useful to obtain a deep understanding of simple neural networks. Some programs are free, some are more expensive, but more *powerful.*

Other tools

Environments for CI development abound. Obtaining good source code for rapid prototyping is a consideration for any engineering team. Statistical analysis tools and graphical interfaces are critically important, also. Articles describing methods, hardware, dynamic systems and pattern recognition applications provide reliable sources for the combination of micro- and systems-level CI tools useful packages for research and development of more applications in intelligent industrial electronics.

57.3 Summary of Basic Modeling Concepts

For the past five years, many of the authors of the section on intelligent electronics and the articles on virtual reality in the emerging technologies section of this handbook have contributed to a series of IEEE Standards study groups relating to computational intelligence (Padgett et al., 1994; Padgett and Karplus, 1993). The focus of the initial effort was to develop a consensus for a glossary of terms for artificial neural networks (ANNs). This quickly centered on developing a modular chart of components of an ANN system to be specified when defining a neural paradigm. (See Tables 57.1 and 57.2). The indicated material presented in this article was a public domain contribution to these pre-Standards efforts, but is not to be considered a draft or a standard. Other groups have also submitted contributions.

Due to the multidisciplinary nature of the study of ANNs, the theory has evolved from diverse sources, couched in the language familiar to its originators. Experts in electrical engineering, controls, pattern recognition, psychology, biology and microelectronics joined others making early outstanding contributions to the field. Their choice of terminology was designed to reach specific segments of the ANN community. Efforts to produce a dictionary have centered on providing links for people in one discipline to the work of experts in other areas, rather than on insistence on use of a particular set of terms and symbols.

In order to provide a meaningful organization to the task of specifying of an ANN paradigm, the components of a typical implementation were outlined. Enough detail was requested to allow the user to duplicate the work of others and potentially to make modifications and improvements. Evaluation criteria were also sought.

Consensus was quickly obtained on the importance of clearly stating objectives and assumptions. Successful applications of ANNs are application-specific and meet a need of the users. Carver Mead, in particular, supported the necessity for early marketing research (or just a series of conversations with potential users) in the design and development of a neural paradigm and its application. A very effective handwriting analysis tool will not sell if it is awkward to use, or there is simply no market for it. Because of the importance of the target use of industrial electronics applications, Mead's points are emphasized here.

As the design of an ANN paradigm for an application progresses, the project objectives and performance measures should be refined. New information obtained during design and testing must be incorporated, and the user should be kept informed of changing costs and options. Performance measures satisfactory to the potential customer and meaningful to the ANN designer are not easy to state. Iterative improvement of early attempts at performance measure is usually necessary.

Many of the surprising developments during the design and testing phase of ANN development stem from lack of precision in the statement of assumptions. There are major differences in the physiological facts of neural processing, biological models of these processes and mathematical models. The differences do not end here, but explode as the discrepancies between elegant closed form solutions and their digital or analog implementations mount. Sources of variation should be anticipated, and identified as needed. The literature of hardware-in-the-loop simulation abounds with very practical techniques for handling these situations. Careful use of the scientific method and traditional simulation validation and verification schemes is highly recommended. Again, finding these solutions and case studies means searching the literature in fields different from those of any one individual's experience.

Another realm of assumptions has to be documented and dealt with in an orderly manner: modeling decisions made by engineers and/or domain experts. Decision theory amplified into soft computing helps quantify subjective and qualitative aspects of the system model. Carefully used, these can help make the system more robust and sensible.

The diagram in Table 57.1 suggests modules of importance to the specification of an ANN paradigm using the multilayer perceptron approach, and Table 57.2 does the same for the competitive learning approach. These modules reflect some important points in stating the ANN specification clearly enough to facilitate its evaluation, use, re-use, and potential modification.

A chart similar to the ones illustrated, but tailored to the specific application can provide the basis for a practical check-list during the planning and development of a neural system paradigm. Use of concurrent engineering in computational intelligence implementations is promoted by early consultation by all parties of a cooperatively developed set of design charts and check-lists. (See the chapter on factory automation.)

57.4 Applications

Following this article, basic modeling concepts and hardware implementations are covered, then dynamic systems techniques and applications are addressed. Pattern recognition techniques and applications papers complete this chapter. The material is intended to be basic enough to be easily understood. Practical applications and examples of use of the techniques introduced are included to help in the implementation of similar projects. The contributions of Werbos give suggestions for future research into more brainlike control systems, and point out some prevalent errors in current literature and practices. These suggestions are continued in his contributions to the chapter on fuzzy systems and soft computing.

57.5 Future

The future of applications of neural networks in industrial electronics seems very positive. Articles in the following chapters will detail successes gained from evolution of traditional neural systems to neural/fuzzy and fuzzy/neural systems. Further expansion including evolutionary systems is explored, and combinations of all these techniques is amplified in the chapter on computational intelligence.

Particular note should be taken of the promise of neural/fuzzy applications to controls and pattern recognition listed below. The following chapter on fuzzy systems and soft computing contains articles addressing these concepts.

Controls: Neural/Fuzzy

- Reinforcement
- Adaptive Critic
- Sugeno Method

Pattern Recognition: Neural/Fuzzy

- Self-Organizing Map and LVQ Extensions
- Fuzzy ART

Details about these and other intelligent electronics topics mentioned in this paper can be found elsewhere in the handbook, under headings such as:

- INTELLIGENT ELECTRONICS TOPICS
  - EXPERT SYSTEMS
  - NEURAL SYSTEMS
    - Basic Modeling Concepts
    - Hardware Implementations
    - Dynamic Systems and Applications
    - Pattern Recognition Techniques and Applications
  - FUZZY AND SOFT SYSTEMS
    - Fuzzy Systems
      - Basic Modeling Concepts
      - Hardware Implementations
      - Techniques and Applications
    - Fuzzy Neural (FN)/Neural Fuzzy (NF) Systems
      - Basic Modeling Concepts
      - Hardware Implementations
      - Dynamic Systems and Applications
      - Pattern Recognition and Applications
- EVOLUTIONARY/GENETIC SYSTEMS
- COMPUTATIONAL INTELLIGENCE AND HYBRID SYSTEMS
- INTELLIGENT VIRTUAL REALITY

Computational intelligence in industrial electronics should continue to expand. Many forecasts imply that engineering development environments which are interactive, use and monitor all the human senses and in fact include virtual reality systems, will open the way to engineering practical intelligent electronics for industry applications.

57.6 Defining Terms

The following terms (except for the quoted material) were placed in the public domain prior to being submitted to the *CI Standards News*. In several cases, there are no commonly used terms for concepts which are crucial to neural network system development. The terms below are suggestions designed to stimulate comment. They are not intended to be a draft or a standard.

Controls

Closed loop: standard control theory term for any system responding to external feedback. This may be simple feedback, adaptive or learning control. See also: feedback, adaptive, learning control.

Feedback (ordinary or simple) control: a particular form of feedback control system involving an immediate response to an observed state variable, such as in PID control, e.g., response of a thermostat to current measurement of temperature. See also adaptive control, learning control.

Adaptive control: as in adaptive control, the technical use of the word adaptive is far less broad than the intuitive use many readers are familiar with. An adaptive controller is a controller which responds to feedback on the estimated values of some unknown parameters in the plant to be controlled. Typically these are slowly changing but familiar parameters such as mass or friction. Typically this kind of adaptation should occur on a time-scale longer than that of simple feedback, but shorter than that of true learning. Some conventional adaptive control systems use direct inverse designs which implicitly have similar properties See also feedback and learning.

Learning control: a form of feedback control in which the system builds up cumulative knowledge about the dynamics and the functional relationships of the plant to be controlled, information which should remain valid on a relatively long time scale. For example, it is possible to build controllers which learn off line to find the optimal gain to be used in an adaptive controller which will then be operated in real-time mode. While there is no formal mathematical distinction between an adaptive controller and a learning controller, there is great practical utility in maintaining a level of learning above the level of adaptation. See also feedback and adaptation.

Modeling

Note: All levels of analysis make sense for an application.

Neural network paradigm or design: a network has two components: functional form and learning rule. Put the two together to form a complete system. Paradigm or design = functional form + learning rule

Network synthesis: use first principles knowledge to come up with weights (sometimes useful).

Application level: level of design or analysis which employs multiple macrolevel ANN designs and other computational tools to meet the needs of a specific real-world application.

Microlevel: development and understanding of generic NN designs to perform relatively low-level generic tasks such as supervised learning, or other tasks which can contribute to supervised learning, such as feature extraction, clustering, etc. Even though microlevel components are at a lower level of the system in terms of scale, there is still a great variety of microlevel systems varying from simple pattern classification designs through to very complex hybrid brainlike approaches to performing similar tasks.

Macrolevel: development and understanding of generic NN learning designs which perform high-level generic tasks such as dynamic system identification or dynamic optimization. Such designs typically take the form of block diagrams or main programs which allow a choice of different NN paradigms in the microlevel components or blocks of the design.

Artificial neural network: a mathematical design (algorithm or architecture) designed to achieve some sort of generic information processing capability within the constraints of "sixth generation computer hardware." These constraints involve use of elementary processing elements in a massively parallel structure which should be able to scale efficiently to large numbers of inputs and outputs, and should include the human brain, in principle, as an example of neural like hardware. (Werbos, 1996.)

Biological neuron: nerve cell, serving as fundamental processing unit of the nervous system.

Bottom-up NN: models observations of structure and function of individual nerve cells.

Top-down NN: groups of neurons which model outward observations of capability and behavior, using simplified models of individual neurons.

Conflict: differing requirements for identification and control (exploration and exploitation).

Exploration: probe of environment to obtain information about how to change its behavior.

Exploitation: need to change behavior in order to improve performance based on current estimates.

Design levels for artificial neural networks: driven by user needs, each rung exists solely to meet the needs of the levels above it. The levels range from micro (unsupervised, supervised) to macro (real-world dynamics, values) to application (user needs).

Paradigms: various implementations of ANNs, specified by documenting the combination of network structure and adaptation technique selected.

Performance measure: a function defined over the set of possible external behaviors of the learning system, visualized as a surface. To improve performance, or to minimize error, move to high or low point of surface.

Subsystem functions: pattern recognition or neuroidentification, for sensor fusion, diagnostics, etc.

Cloning control: task or operation of building an ANN or other intelligent system which mimics the input output behavior of a human expert or other controller. See also tracking and optimization.

Tracking control: a form of control which attempts to make a plant or environment adhere to some predetermined setpoint (as in a thermostat) or to follow some predetermined trajectory (as in a robot arm).

Optimization control: in neurocontrol, a processes maximizing throughput, minimizing energy use, maximizing goal satisfaction or utility over *many* time periods into the future.

Single Neuron and Neural Networks

Activation function: operation on the neuron inputs giving neuron output. Models the impact of the neuron on the neural system, e.g., binary hard-limiter, linear-graded threshold, sigmoid.

Activity level: model of the state of polarization of the neuron. Implementation based on input from all input connections, with magnitude modified by magnitude of input signal and connection weight, and direction by the sign of the weight.

Anti-Hebbian learning: process where repeated stimulation of a synapse is inhibitory only.

Architecture: number and connections of neurons.

Associative memory: recall of contents of a storage location based on its associations with the contents of other storage locations. For example, recall of one pattern from a particular time (or location) triggers recall of another. Patterns from one time can also be linked with patterns from a different time.

ASSOM: adaptive-subspace self-organizing map meant for the detection of invariant features. The neural units decode signal subspaces rather than signal patterns.

Attractor (NN): one of a set of stable states of a feedback system. See also basin of attraction.

Basin of attraction (NN): set of all possible initial states from which a feedback system will converge to a particular attractor.

Batch processing: learning system where update of parameters takes place after processing of all training data. See also epoch learning.

Bias value: internal resting level of the neuron, modeled as a fixed weight to help bound the activations.

Binary hard-limiter: activation function where output is all or none, e.g., McCullouch and Pitts model.

Bootstrap learning (NN): given a finite population, training samples are selected at random (with replacement) to produce an arbitrarily large set of sequentially statistically independent samples from the population.

Classification: process of assigning a label to the input feature vector.

Cluster (NN): set of data points with some common attribute, such as a set of pattern or codebook vectors within a neighborhood with respect to the distance measure of the dataset.

Codebook vector: vector of parameters used as a reference vector in vector quantization (VQ, SOM and LVQ).

Competitive learning: supervised learning where output layer neurons compete for the right to fire by producing the output signal closest to the target. Sometimes reinforcement learning is added.

Computational intelligence (CI): neural, fuzzy, evolutionary, and/or virtual reality aspects of systems.

Computational intelligence (CI) systems: use of software and hardware tools based on neural, fuzzy, evolutionary, and/or virtual-reality systems to solve complex application problems without human intervention.

Correlation matrix memory (NN): Input vectors have a correlation matrix stored in a distributed content-addressable memory.

Cups (NN): connection updates (or weight value modifications) per second.

Delta rule: (Widrow-Hoff, Adaline, least mean square) iterative adjustments of weights to minimize delta (difference) between target and observed outputs. (See chapter by Rumelhart et al.).

Discriminating ability: number of patterns an ANN can distinguish.

Dynamic model: includes a self-relaxation term that insures that the cell output potential decays to 0 when all dendritic inputs are zero. May be implemented using a leaky integrator which allows for the duration or persistence of input signals to be controlled. (See chapters by Lau, Wolpert.)

Dynamically expanding context (DEC): input-output mappings have a variable context. Conflicts in input-output relations with short context impact length modifications.

Elemental features (biology/vision): lines, shapes, colors, relative locations—things recognized by different layers of visual cells.

Epoch learning: processing of a finite sequence of input patterns with parameter updates occurring as each pattern is presented. Contrast with batch learning.

Error function: expression of the expected mean of the error in some optimization task.

Error signal: in supervised learning, the difference between target and observed responses.

Excitatory: positive input signal, (depolarizing).

Feature extraction (data reduction): process designed to produce a feature vector, or to eliminate irrelevant information and generate a set of features that are invariant to unimportant factors but sensitive to those producing the class label of the class.

Feature saliency: attribute is based on sensitivity analysis of the outputs of the MLP to its inputs. Important features (salient) cause changes in the MLP outputs, e.g., noise feature often implemented as repetitions of MLP training to get sensitivity estimate accurately. (See paper by Ruck and Rogers.)

Frequency based neuronal models: designed to perform temporal and spatial summation of an arbitrary number of dendritic inputs and their own activity state. Their activation *functions* depend on type of cell and network. They cannot form a new action potential until the old one fires. (See chapters by Wolpert.)

Global: all weights impact the output of the neural network at all times.

Gradient descent rule: iterative adjustment of weights to minimize delta (difference) between target and observed outputs, where weight adjustment is governed by the rate of change of error with respect to the value of the weight. Intended to avoid local minima.

Gradient: an error vector (performance vs desired or target action).

Hard limiter: activation function which is linear, bounded and discontinuous (not simply differentiable).

Hebbian learning: repeated stimulation of a synapse increases its weight (or decreases it for a repeated negative stimulation).

Heuristic: intuitively based selection or design.

Hypermap: SOM or LVQ where a searching phase determines a winner by sequentially searching subsets of candidates determined by a previous phase.

Inhibitory: negative input signal, (hyperpolarizing).

Input activation: neuron inputs, often the sum of the values of all the input connections, where each is computed as the product of the input signal to the connection, and the connection weight, e.g., input activation S, for neuron i, with input connections x_j, and weights w_{ij} is,

$$S = \sum_{j=1}^{n} W_{ij} x_j$$

Interconnect: connection, often variable and adaptive, between neurons that perform the elementary computation in ANN.

IPS: interconnects per second, a measure of speed in nonlearning mode.

Iteration (NN): processing each element of a sequence. For example, in epoch training, processing and updating each vector in the training set completes one epoch, or iteration of the training procedure.

K-means clustering: vector quantization where K codebook vectors values are updated to become the mean of Voronoi sets based on the old codebook vectors (Kohonen, 1995)

Label: name or discrete symbol for a class or cluster.

Learning equations: equations for learning rules.

Learning rule: process for adjusting synaptic weights.

Learning subspace method: "supervised competitive-learning method in which each neuron is described by a set of basis vectors, defining a linear subspace of its input signals." (Kohonen, 1995, p. 266.)

Learning vector quantization (LVQ): "supervised-learning vector quantization method in which the decision surfaces, relating to those of the Bayesian classifier, are defined by nearest-neighbor classification with respect to sets of codebook vectors assigned to each class and describing it." (Kohonen, 1995, p. 266.)

Linear-graded threshold: neurons have variable output levels applied to distinctly designated excitatory and inhibitory inputs, e.g., Hopfield networks.

Linearly separable: ANN can separate input patterns into groups divided by a straight line.

Local: a given weight used only over a small part of the input space (not necessarily a localized region).

Magnification factor (SOM): percent of available area occupied by the representation of an input item.

McCulloch-Pitts model: feedforward stage of a single neuron in a typical MLP.

Momentum factor: scalar multiplier of the previous weight change. Designed to compute a momentum term to add to the current weight change to possibly damp oscillations and speed convergence.

Multilinearly separable: ANN which can separate input patterns into groups divided by a set of straight lines.

Neighborhood (NN): for a given neuron, n, and distance, d, the d-neighborhood of n is the set of all neurons in the network less than d from n.

Network capacity: number of pattern signals which can be recalled.

Neurocontrol: use of neural networks—artificial or natural—to directly control actions intended to produce a physical result in a world which changes over time.

Neurodynamics: time-varying properties of ANNs: learning, recall, association, comparative evaluation of new information, classification of new information, formation of new classes (Kartalopoulas, 1996).

Nonlinearly separable: ANN can separate input patterns into groups, but not with straight lines.

Nonlinear transfer function: modulation of input signal received by neuron.

Output: may be modeled as the value of the activation function, or as is current in form of $C\ dS/dt$, which features I_B, a fixed biasing current that represents a baseline level of activity, and S = aggregated sum of weighted inputs which must exceed a threshold in order for cell i to respond. (See output value.)

Output connections: the axon and its branches.

Output value: signal produced by the activation function or firing rate of the neuron. Models the value or frequency of the output of the axon hillock of the biological neuron.

Pattern recognition (PR): process of assigning a class label to a signal (information reduction). Steps for automating on computers are: segmentation, feature extraction (data reduction), and classification. Given a set of N features, select a subset of $p < N$ features for use so that these p give better recognition performance than the N (Kohonen, 1995).

PSOM: parametrized self-organizing map used mainly in robotics. The PSOM defines a continuous weight manifold, described by some smooth analytical parametrized function $w(s)$. The selected value of parameter s is determined by the value of $w(s)$ closest to input x.

Perceptron: neuron with a feedforward or recall stage and a learning stage, where weights are adjusted in proportion to feedback of error between observed and target output.

Ramp: linear activation function allowing adjustment of slope.

Random specialization: supervised competitive learning.

Reference vector: see codebook vector.

Reinforcement learning: refers to a class of learning tasks and algorithms in which the learning system learns an associative mapping A by maximizing scalar evaluations (reinforcement of its performance from the environment (user)).

Resistive grid local averaging: dendritic processes can be modeled by one-dimensional resistive networks. The 2–d resistive network or grid has been a central component in a silicon retina. These are low level neural networks that are central to much of neural computation.

Segmentation: process designed to find regions of interest (ROI) using a coarse filtering scheme.

Self-organization: learning designed to modify both structure and parameters of a network at the same time.

Self-organizing map (SOM): "result of a nonparametric regression process that is mainly used to represent high-dimensional, nonlinearly related data items in an illustrative, often two-dimensional display, and to perform unsupervised classification and clustering." (Kohonen, 1995, p. 275.)

Sigmoid: popular activation function which is nonlinear, bounded, continuous, monotonic and has a simple derivative, e.g., logistic function, f, for neuron i, with input activation S

$$f_i(S) = \frac{1}{1 + e^{(-\text{activation}(S)/T)}}$$

where, T is the temperature or slope of f and $T > 0$.
Note: For $T = 0$, f reduces to the step function, with value 1 for $S > 0$ and 0 otherwise.

Spatial summation: sum of input from all neurons at the target neuron.

Spatiotemporal summations: output of neuron based on sum over time of input signals from all input neurons.

Stability-plasticity dilemma: need to generalize without error. Adapt to new input without forgetting how to treat old input.

Supervised learning: a task in which systems are given a database of input vectors and target vectors, where a nonlinear mapping from inputs to outputs is a function of both the inputs and a set of weights. The mapping is "learned" by adapting the weights in either real-time or off line, in "batch" mode.

Supervisor: see teacher.

Supervised learning systems (SLS): any system which learns a nonlinear *function* or static mapping from a vector X to a vector Y.

Synapse: connection (contact or junction) of axon to dendrite (or to cell body or other axons). Composed of presynaptic terminal, cleft, and postsynaptic terminal. Output signal from axon is significantly altered as it crosses the synapse.

Synaptic connection: either excitatory or inhibitory as level of activity varies. Often modeled as the weight for a neural input connection.

Synaptic weights: adjusted to store knowledge about responses to inputs.

Target response: in supervised learning, the ideal or desired response for a particular input.

TD-SOM: time-delay self-organizing map in which the statistical dependence between successive input samples is taken into account, either by representing sets of related signal samples as sample vectors, or including low-pass filters at the inputs and/or output of each neuron.

Teacher: mechanism for evaluation of progress of learning.

Temporal summation: sum over time of an impulse train of input signals.

Threshold: value of activation which must be exceeded to allow firing.

Topology-preserving map: see SOM.

Unsupervised learning: learning without specific target for output. Classification of inputs into groups with similar features and creation of new groups when needed. Some validation criteria must be provided for guidance for feature selection.

Validation: process of determining whether the correct problem was solved based on the needs of the upper layers of the hierarchy, e.g., did you solve the user's problem? (See Figure 57.2.)

Vector quantization: distribution of a set vectors represented by a set of reference or codebook vectors which minimize the expected mean quantization error for the space.

Verification: process of determining whether the right answer was obtained based on the specification of the model, e.g., check for typographical errors, loose connections, logic errors, etc.

Voronoi tessellation: segmentation of a space of data vectors such that for each codebook vector and its nearest neighbor, the midplane between them is determined, and the resulting hyperplane segments partition the space.

Weight: synaptic strength associated with each input connection. Positive (excites) or negative (inhibits) flow of information.

Winner-takes-all learning algorithm: competitive, unsupervised learning.

57.7 Resources

The number of resources available for artificial neural networks software, hardware and engineering information is large and expanding rapidly. A few key sources of reliable information and software are listed below. These are not all-inclusive, but will give a working engineer support and direction. Sources listed under other chapters in this section often have neural networks material included, since the trend in applications is to combine these techniques. To participate in discussion about possible standardization of neural networks applications, contact the *CI Standards News* (m.padgett@ieee.org).

Internet:

Cascade-correlation: ftp://ftp.cs.cmu.edu//aft/cs.cmu.edu/project/connect/code/supported/cascor*

Learning vector quantization (LVQ): and self-organizing maps (SOM): Helsinki Univ. of Technology

Internet address: cochlea.hut.fi

ftp: cochlea.hut.fi subdirectory:/pub/Ivq pak or /pub/som pak

Multilayer perceptron: Aspirin/MIGRAINES Neural Network Simulation Environment: ftp://ftp.cognet.ucla.edu//pub/alexis/am6.*

Multilayer perceptron: NETS—contact COSMIC.

Newsgroups such as comp.ai.neural-nets also abound.

Publications:

AI/Expert, Miller-Freeman.

IEEE Transactions on Neural Networks, IEEE Press

Neural Networks, Pergamon Press.

Simulation, SCS Press, San Diego, CA.

Acknowledgment

The suggestions of Lotfi Zadeh and Walter Karplus played a key role in the selection and arrangement of the material in this chapter. Their comments and advice are appreciated.

References

Albus, J. 1975. A new approach to manipulator control: the cerebellar model articulation controller, *Trans. ASME J. Dyn. Syst. Meas. and Control*, 97:220–227.

Feldkamp, L. A., Marko, K. A., and Wunsch D. 1995. Personal communications regarding Ford Motor Co. and JPL Recurrent NN Chip.

Fogel, D. B. 1995. *Evolutionary Computation,* IEEE Press, New York, NY.

Fogel, L. 1995. The valuated state space approach and evolutionary computation for problem solving, *Sym. Computational Intelligence,* November, IEEE Press, NY, NY. Perth, WA.

Fukuda, T. and Shimojima, K. 1995. Intelligent control for robotics, *Sym. Computational Intelligence,* 202–219. Perth, WA., IEEE Press, NY, NY.

Goldberg, D. 1994. Genetic and evolutionary algorithms come of age, *Comm. ACM,* 37(3):113–119.

Grossberg, S. 1987. Competitive learning from interactive activation to adaptive resonance, *Cognitive Science,* 11:23–63.

Jang, I-S, and Sun, C-T. 1995. Neuro-fuzzy modeling and control, *Proc. IEEE,* 83(3).

Kartalopoulas, S. 1996. *Understanding Neural Networks and Fuzzy Logic,* IEEE Press, New York, NY.

Kohonen, T., Oja, E., Simula, O., Visa, A., and Kangas, J. 1996. Engineering applications of the self-organizing map, *Proc. IEEE,* (in press).

Kohonen, T. 1995. *Self-Organizing Maps.,* Springer-Verlag, New York, NY.

Kohonen, T. 1990. The self-organizing map, *Proc. IEEE.,* September.

Kosko, B. 1988. Bidirectional associative memories, *IEEE Trans.,* SMC-18:49–60.

Mead, C. 1994. (personal communications.)

Padgett, M. L. and Karplus, W. 1993. Neural network basics: applications, examples and standards, *IJCNN '93,* Tutorial Text, p. 385–413, Nagoya, Japan.

Padgett, M. L., Karplus, W. J., Deiss, S., and Shelton, R. 1994. Computational intelligence standards: motivation, current activities and progress, *Computer Standards and Interfaces,* 16(1994):185–203.

Padgett, M. L. and Padgett, W. D. 1995. Simulation and computational intelligence in real-world applications, *Simulation,* 65(1):5–9.

Puskorius, G. and Feldkamp, L. 1994. Neurocontrol of nonlinear dynamical systems with kalman filter trained recurrent networks, *IEEE Trans. Neural Networks,* 5(2):279–297.

Ritter, H. 1995. Parameterized self-organizing maps for vision learning tasks, *Proc. ICANN,* 94(II):803–810.

Szu, H. 1995. Wavelet dynamics, *The Handbook of Brain Theory and Neural Networks,* Arbib, M. A. ed. 1049–1053. MIT Press, Cambridge, MA.

Werbos, P. J. 1988. Backpropagation, past and future. *Proc. ICNN.*

Werbos, P. J. 1996. Neurocontrol and Elastic Fuzzy Logic: Capabilities, Concepts, and Applications, and Neurocontrol: Where It Is Going and Why It Is Crucial, *Intelligent Control Systems: Theory and Applications,* Gupta, M. M. and Sinha, N. K., eds., IEEE Press, New York, NY.

Werbos, Paul J. 1990. Rational approaches to identifying policy objectives, *Energy: The International Journal,* 15:(3/4).

Werbos, Paul J. 1993, Supervised learning: can it escape from its local minimum?, *WCNN Proceedings,* Erlbaum,

Werbos, P. J. 1994. Backpropagation through time: what it does and how to do it, Chapter 8, *The Roots of Backpropagation,* John Wiley & Sons, New York, NY.

White, D. A. and Sofje, D. A. 1992. *Handbook of Intelligent Control: Neural, Fuzzy and Adaptive Approaches,* Chapter 3, 10 and 13, Van Nostrand Reinhold, New York, NY.

Widrow, B. and M. Hoff 1960. Adaptive switching circuits, *IRE WESCON Convention Record,* 96–104.

Wunsch, D. and Prokhorov, D. 1995. Adaptive critic approaches, *Computational Intelligence,* 98–107, Perth, WA.

Zadeh, L. 1995. (Personal communications.)

Zurada, J. 1994. Learning algorithms in neural networks, *WCCI Tutorial Series,* Orlando, FL., IEEE Press, NY, NY.

58

The Basic Ideas in Neural Networks*

David E. Rumelhart
Stanford University

Bernard Widrow
Stanford University

Michael Lehr
Stanford University

58.1 Introduction 853
58.2 Learning By Example 855
58.3 Generalization 856
58.4 Hints for Successful Applications 857

58.1 Introduction

The last several years have witnessed a remarkable growth in interest in the study of neural networks. This effort has variously been characterized as the study of brain style computation, connectionist architectures, parallel distributed processing systems, neuromorphic computation, artificial neural systems and other names as well. The common theme to all of these efforts has been an interest in looking at the brain as a model of a parallel computational device very different from that of a traditional serial computer. The strategy has been to develop simplified mathematical models of brain-like systems, and then to study these models to understand how various computational problems can be solved by such devices. The work has attracted scientists from a number of disciplines: neuroscientists who are interested in making models of the neural circuitry found in specific areas of the brains of various animals; physicists who see analogies between the dynamical behavior of brain-like systems and the kinds of nonlinear dynamic systems familiar in physics; computer engineers who are interested in fabricating brain-like computers; workers in artificial intelligence who are interested in building machines with the intelligence of biological organisms; engineers interested in solving practical problem; psychologists who are interested in the mechanisms of human information processing; mathematicians who are interested in the mathematics of such neural network systems; philosophers who are interested in how such systems change our view of the nature of mind and its relationship to brain; and many others. The wealth of talent and the breadth of interest has made the area a magnet for bright young students.

Although proposals differ in the detail, the most common models take the neuron as the basic processing unit. Each such processing unit is characterized by an activity level (representing the state of polarization of a neuron), an output value (representing the firing rate of the neuron), a set of input connections, (representing synapses on the cell and its dendrite), a bias value (representing an internal resting level of the neuron) and a set of output connections (representing a neuron's axional projections). Each of these aspects of the unit are represented mathematically by real numbers. Thus, each connection has an associated weight (synaptic strength) which determines the effect of the incoming input on the activation level of the unit. The weights may be positive (excititory) or negative (inhibitory). Frequently, the input lines are assumed to sum linearly yielding an activation value for unit i at time t, given by

$$a_i(t) = \sum_j w_{ij} x_j(t) + \beta_i, \tag{58.1}$$

where w_{ij} is the strength of the connection from $unit_j$ to $unit_i$, β_i is the unit's bias value, and x_j is the output value of unit j.

Note the effect that the output of a particular unit has on the activity of another unit is jointly determined by its output level and the strength (and sign) of its connection to that unit. If the sign is negative, it lowers the activation, if the sign is positive it raises the activation. The magnitude of the output and the strength of the connection determines the amount of the effect. The output of such a unit is normally a nonlinear function of its activation value. A typical choice of such a function is the sigmoid. The logistic,

$$y_i(t) = \frac{1}{1 + e^{-a_i(t)/T}}, \tag{58.2}$$

* Reprinted from Rumelhart, D., Widrow, B., and Lehr, M. 1994. The basic ideas in neural networks, *Commun. ACM* 37(3):87–92, Mar. © 1994 ACM. With permission.

The Basic Idea

Inputs

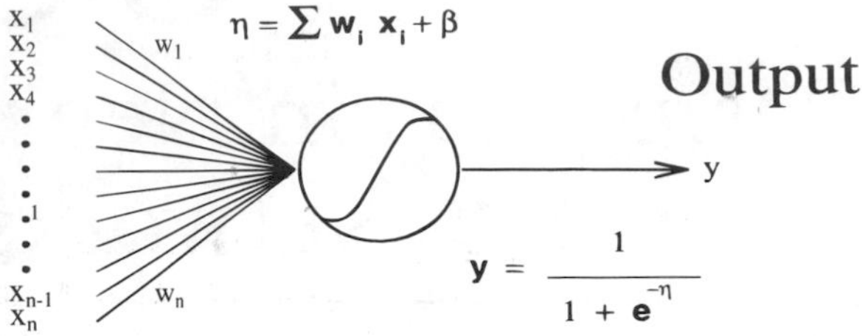

Figure 58.1 A simple sigmoidal output function.

illustrated in Figure 58.1, will be employed in the examples illustrated below. The parameter of the logistic T, yields functions of differing slopes. As T approaches zero the logistic becomes a simple logical threshold function which takes on the value of 1 if the activity level is positive and zero otherwise.

A brain style computational device consists of a large network of such units, richly connected to one another. In real brains there are tens of billions of such units and tens of trillions of such connections. Such a network is a general computing device. The function it computes is determined by the pattern of connections. Thus, the configuration of connections is the analogue of a program. The goal is to understand the kinds of algorithms that are naturally implemented by such networks.

Although there has been a good deal of activity recently, the study of brain style computation has its roots over 50 years ago in the work of McColluch and Pitts (1943) and slightly later in Hebb's famous *Organization of Behavior* (1949). The early work in artificial intelligence was torn between those who believed that intelligent systems could best be built on computers modeled after brains (cf. Selfridge, 1955; Rosenblatt, 1962, Widrow and Hoff, 1960), and those like Newell and Simon (1956), McCarthy (1959), and Minsky and Papert (1969) who believed that intelligence was fundamentally symbol processing of the kind readily modelled on the von Neumann computer. For a variety of reasons, the symbol processing approach became the dominant theme in artificial intelligence. The reasons for this were both positive and negative. On the one hand, the stored program digital computer became the standard of the computer industry. Such computers were easy to design and easy to program. The symbol processing/logic based approach to AI is well suited for such an architecture. On the other hand, the fundamentally parallel neural network systems, such as Rosenblatt's perceptron system, were not well suited to implementation on serial computers. Moreover, the perceptron turned out to be rather more limited than first expected (cf. Minsky and Papert, 1969) and this discouraged both scientists and funding agencies. Although work continued throughout the 1970s by number of workers including Amari, Anderson, Arbib, Fukishima, Grossberg, Kohonen, Widrow and others, and although a number of important results were obtained during this period, the work received relatively little attention.

The 1980s showed a rebirth in interest. There seem to be at least five reasons for this. Three of the reasons are essentially pragmatic and two theoretical. First, on the more pragmatic side:

1. Today's computers are much faster than those of the 1950s and 1960s. It is thus possible to use conventional computers to simulate and experiment with much larger and more interesting networks than ever before.
2. Everyone believes that the future for faster computers must be in parallel computation. Unfortunately, there is no generally accepted paradigm for parallel computation. It is generally easier to build parallel computers than to find algorithms that are efficient for them. There is a hope that algorithms which prove efficient and effective on brain style computers may prove a useful general paradigm for parallel computation.
3. The basic empirical tools of neuroscience are expanding and we are learning more and more about how the neuron functions and how neurons communicate with one another, but little is known about how to go from this information about specific neurons to a theoretical account of how large networks of such neurons might function. It is hoped that the theoretical tools developed in the study of neural network computational systems will allow for the modelling of real neural networks.

In addition to these three reasons, there have been two theoretical results which have now been developed and appreciated.

1. The first of these results is due to Hopfield (1982) and provides the mathematical foundation for understanding the dynamics of an important class of networks. In particular, Hopfield pointed out that recurrent networks with symmetric weights have a point-attractor dynamics which makes their behavior relatively simple to understand and analyze. This observation has been extended and applied by Hinton and Sejnowski (1986), Grossberg and Kohn (1984), Smolensky (1986) and a number of others to provide us with a useful mathematical understanding of how networks such as these might be configured to solve important optimization problems.
2. The second result is an extension of the work of Rosenblatt (1962) and Widrow and Hoff (1960), to deal with learning in complex, multilayer networks and thereby provide an answer to one of the most severe criticisms of the original perceptron work. In this case, it was observed that by selecting differentiable, nonlinear functions (such as the sigmoid described above) it was possible to use the gradient search methods of Widrow and Hoff (1960) for nonlinear multilayer networks. In this was a technique by which multilayer perceptron-like devices could be reliably trained. This procedure, known as the backpropagation learning algorithm has had a major impact on the field and is the primary method employed in most of the applications we will discuss (cf. Rumelhart, Hinton and Williams, 1986).

In this paper we focus on the learning results since they have had the greatest influence on applications.

58.2 Learning by Example

The problem of learning in neural networks is simply the problem of finding a set of connection strengths which allow the network to carry out the desired computation. In this section we focus on backpropagation, currently the most popular form of learning system and the one on which virtually all of the applications are based. The usual network architecture is illustrated in Figure 58.2.

There is a set of input units which are connected, through a set of so-called *hidden units,* to a set of output units. In the general case, there may be any number and configuration of hidden units and connections among the units. (For simplicity, we will restrict discussion here to the case of *feedforward* networks in which the activity of a given unit cannot influence, even indirectly, its own inputs.) The network is provided with a set of example input/output pairs (a training set) and is to modify its connections so as to approximate the function from which the input/output pairs have been drawn. The networks are then tested for ability to generalize. The error correction learning procedure is simple enough in conception. The procedure is as follows. During training an input is put into the network and it flows through the network generating a set of values on the output units. Then, the actual output is compared with the desired target and a match is computed. If the output and target match, no change is made to the net. However, if the output differs from the target a change must be made to some of the connections. The problem is to determine which connections in the entire network were at fault for the error—this is called the *credit assignment* (or perhaps better the blame assignment) problem. Although the solution to this problem for the case of networks without hidden layers has been known for some time, this is, in general, a difficult problem and the lack of a satisfactory solution was a major factor in the earlier loss of interest in neural networks systems. The 1980s has led to the development of a rather simple, yet powerful, solution to this problem. The basic idea is to define a measure of the overall performance of the system and then to find a way to optimize that performance. In this case, we can define the performance of the system as

$$E = \sum_{p,i} (t_{ip} - y_{ip})^2 \qquad (58.3)$$

where i indexes the output units, p indexes the I/O pairs to be learned, t_{ip} indicates the target for a particular output unit on a particular pattern, y_{ip} indicates the actual output for that unit on that pattern, and E is the total error of the system. The goal, then, is to minimize this function. It turns out, if the output functions are differentiable, then this problem has a simple solution—namely, we can assign a particular unit blame in proportion to the degree to which changes in that unit's activity leads to changes in the error. That is, we change the weights of the system in proportion to the derivative of the error with respect to the weights. This simple procedure works remarkably well on a wide variety of problems. The problem of learning is thus reduced to the problem of parameter estimation.

A key advantage of neural network systems is that these simple, yet powerful learning procedures can be defined which will allow the systems to adapt to their environments. It was work on the learning aspect of these neurally-inspired models which first led to an interest in them (c.f. Rosenblatt, 1962), and it was the conjecture that learning procedures for complex networks could never be developed that contributed to the loss of interest (c.f. Minsky and Papert, 1969). Although the *perceptron convergence procedure* and its variants had been around for some time, these learning procedures were limited to simple one-layer networks involving only *input* and *output* units. There were no *hidden units* in these cases and no *internal representation.* The coding provided by the external world had to suffice. Nevertheless, these networks have proved useful in a wide variety of applications. Perhaps the essential character of such networks is that they map similar input patterns to similar output patterns. This is what allows these networks to make reasonable generalizations and perform reasonably on patterns that have never before been presented. The similarity of patterns in a connectionist system is determined by their overlap. The overlap in such networks is determined outside the learning system itself *by whatever produces the patterns.*

The constraint that similar input patterns lead to similar outputs can lead to an inability of the system to learn certain mappings from input to output. Whenever the representation provided by the outside world is such that the similarity structure of the input and output patterns is very different, a network without internal representations (i.e., a network without hidden units) will be unable to perform the necessary mappings.

In a multilayer network, the information coming to the input units is *recoded* into an internal representation and the outputs are generated by the internal representation rather than by the original pattern. If we have enough connections from the input units to a large enough set of hidden units, we can always find a representation that will perform any mapping from input to output through these hidden units.

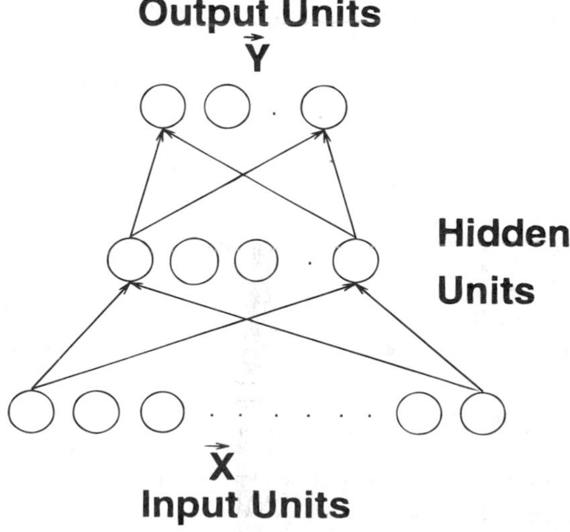

Figure 58.2 A simple multilayer network. The key to the effectiveness of the multilayer network is that the hidden units learn to rerepresent the input variables in a task-dependent way.

The existence of multilayer networks illustrates the potential power of hidden units and internal representations. The problem, as noted by Minsky and Papert (1969), is that whereas there is a very simple guaranteed learning rule for all problems that can be solved without hidden units, namely, the perceptron convergence procedure (or the variation due originally to Widrow and Hoff, 1960), there has been no equally powerful rule for learning in multi-layer networks. We are thus not assured of optimal solutions—local minima are always a possibility. Nevertheless, the backpropagation procedure is sufficiently robust that local minima rarely turn out to be serious limitations.

It is clear that if we hope to use these connectionist networks for general computational purposes, we must have a learning scheme capable of learning its own internal representations. This is just what the back propagation learning procedure is. It is a simple generalization of the perceptron learning procedure, and the Widrow-Hoff learning procedure which allows the system to learn to compute arbitrary functions. The constraints inherent in networks without self-modifying internal representations are no longer applicable. The basic weight update procedure is a two stage process. First, an input is applied to the network. After the system has responded to this input, certain of the units of the network are informed of the values they ought to have at this time. If they have attained the desired values, the weights are not changed. If they differ from the target values then the weights are changed according to the difference between the actual value the units have attained and the target for those units. This difference becomes an error signal. This error signal must then be sent back to those units which impinged on the output units. Each such unit receives an error measure which is equal to the error in all of the units to which it connects multiplied by the weight connecting it to the output unit. Then, based on this error, the weights into these "second layer" units are modified after which the error is passed back another layer. This process continues until the error signal reaches the input units or until it has been passed back for a fixed number of times. Then a new input pattern is presented and the process repeats. Although the procedure may sound difficult, it is actually quite simple and easy to implement within these nets. Moreover, it can be shown that this system will work for any network whatsoever.

Although the learning results do not *guarantee* that we can find a solution for all solvable problems, our analyses and simulation results have shown that as a practical matter, the backward error propagation scheme leads to solutions in virtually every case.

58.3 Generalization

The backpropagation learning procedure sketched above has become, perhaps, the single most popular method for training networks. The procedure has been used to train networks in problem domains including character recognition, speech recognition, sonar detection, mapping from spelling to sound, motor control, analysis of molecular structure, diagnosis of eye diseases, prediction of chaotic functions, playing backgammon, the parsing of simple sentences, and many many more areas of application. Perhaps the major point of these examples is the enormous range of problems to which the backpropagation learning procedure can usefully be applied. In spite of the rather impressive breadth of topics, and the success of some of these applications, there are a number of serious open problems. The theoretical issues of primary concern fall into four main areas:

1. The learning problem—can the network learn how to solve the problem at hand?
2. The architecture problem—are there useful architectures, beyond the standard three layer network employed in most of these areas, which are appropriate for certain areas of application?
3. The scaling problem—how can we cut down on the substantial training time that seems to be involved for the more difficult and interesting problem application areas?
4. The generalization problem—how can we be certain that the network trained on a subset of the example set will generalize correctly to the entire set of exemplars?

The original efforts were focused on the first of these problems. The primary applications of our learning algorithms were to see if a network could learn some complex non-linear function. Thus we focused on such problems as parity, exclusive-or and other similar analytically defined problems. We found that with a sufficiently large network we could learn essentially any function. The initial worries about the role of local minima and similar such problems turned out to be much less serious than we originally thought. However, we have come to understand that the "generalization" problem is much more serious than we might have thought. This, of course, is just the mirror image of the learning problem. The more general our learning procedure, the less constraints we have on the way the network actually solves the problem and therefore the less certain we can be about the network's ability to properly generalize to new cases. In the statistics literature this in known as the "overfitting" problem. Models of many parameters can fit essentially any function in many different ways. Our problem is to fit the function in such a way that it maximizes its ability to generalize to an as yet unseen collection of data. There have been essentially two strategies in the connectionist literature to deal with this problem.

The first strategy is a version of "Occam's Razor"—i.e., the notion that the simplest hypothesis consistent with the data is the one that should be chosen. In the world of connectionist networks this involves the view that the simplest network consistent with the data should be chosen. There are a number of measures of simplicity in a network. We, for example, have suggested that the following variables covary with simplicity: number of weights, number of units, number of symmetries among the weights, number of bits per weight, etc. It is possible to define cost functions which lead to minimal complexity networks as measured by any or all of these measurements. Generally we find that minimal networks offer better generalization performance than more complex networks.

The second basic scheme for training networks and, in fact, the most commonly used scheme is a version of cross-validation. In this scheme, the data is divided into three parts. One part is used for training, one part is used to evaluate the generalization performance and is set aside for a final test, and one part of the data is used for cross-validation. The procedure is as follows: following each training epoch the performance of the network is evaluated on the validation set. As long as the network continues to improve on the validation set training is continued. If over-fitting is occurring, the network will at some point begin to show poorer performance on the validation data. At that point we stop training and select the weights which give optimal performance on the validation set for testing against the "test-set" and the performance on this set is used as a measure of the quality of the generalization. This method is reasonably powerful and simple and often leads to good results. In particular, the results are nearly as good for this method as for the more complex method described above and the training time is generally much less.

58.4 Hints for Successful Applications

Although some authors have suggested that neural networks are simple black boxes that can be applied without much consideration of the details of the problem, most successful applications require great care in approaching the problem at hand. Below are a number of considerations that have proven useful in a number of areas of application.

1. Be certain to have enough data to constrain your model sufficiently for the problem at hand.
2. Carefully design appropriate input data. This will often require theory-based data reduction of the number of input variables.
3. Build known symmetries (often through weight linking) into your network wherever possible.
4. Build a probabilistic model of the task. Make use of "forward models" to map from a representation of the input that you want to discover to a target set that is easy to construct.
5. Use the network to solve problems that it is good at, but feel free to combine the network with other statistical methods. Making certain that you can offer a clear probablistic/Bayesian interpretation of the behavior of network will facilitate in interfacing the network with other statistical methods.

References

Grossberg, S. 1976. Adaptive pattern classification and universal recording: Part I Parallel development and coding of neural feature detectors, *Biological Cybernetics,* 23, 121–134.

Hebb, D. O. 1949. *The Organization of Behavior,* John Wiley & Sons, New York, NY.

Hinton, G. E. and Sejnowski, T. 1986. Learning and relearning in Boltzmann machines. In D. E. Rumelhart, J. L. McClelland and the PDP Research Group, 2 *Parallel Distributed Processing: Explorations in the Microstructure of Cognition. Vol. 1: Foundations,* MIT Press/Bradford Books, Cambridge, MA.

Hopfield, J. J. 1982. Neural networks and physical systems with emergent collective computational abilities, *Proceedings of the National Academy of Sciences, USA,* 79, 2554–2558.

Minsky, M. and Papert, S. 1969. *Perceptrons,* MIT Press, Cambridge, MA.

McCulloch, W. S. and Pitts, W. 1943. A logical calculus of the ideas immanent in nervous activity, *Bulletin of Mathematical Biophysics,* 5:115–133.

Rosenblatt, F. 1962. *Principles of Neurodynamics.* Spartan, New York, NY.

Rumelhart, D. E., Hinton, G. E., and Williams, R. J. 1986. Learning internal representations by error propagation. In D. E. Rumelhart and J. L. McClelland and the PDP Research Group, *2 Parallel Distributed Processing: Explorations in the Microstructure of Cognition. Vol. 1: Foundations,* MIT Press/Bradford Books, Cambridge, MA.

Rumelhart, D. E., McClelland, J. L. and the PDP Research Group 1986. *2 Parallel Distributed Processing: Explorations in the Microstructure of Cognition. Vol. 1: Foundations,* MIT Press/ Bradford Books, Cambridge, MA.

Selfridge, O. G. 1955. Pattern recognition in modern computers, *Proc. Western Joint Computer Conference.*

Smolensky, P. 1986. Information processing in dynamical systems: foundations of harmony theory. In D. E. Rumelhart, J. L. McClelland and the PDP Research Group, *Parallel Distributed Processing: Explorations in the Microstructure of Cognition. Vol. 1: Foundations,* MIT Press/Bradford Books, Cambridge, MA.

Widrow, B. and Hoff, M. E. 1960. Adaptive switching circuits, Institute of Radio Engineers, *Western Electronic Show and Convention, Convention Record, Part 4,* 96–104.

59

Neural Networks on a Chip

59.1 Artificial Neural Network Technology Compared with Conventional 858

59.2 Examples of Chips 858

Artificial Retina • Cellular Neural Network • CCD Neural Processor Chip • SIMD Numerical Array Processor (SNAP) Chip • Connectionist Neural Adaptive Processor (CNAP) Chip • NI1000 Recognition Accelerator Chip

59.3 Comparisons of NN VLSI Microchips 864

59.4 Applications of Neural Network Technology 864

59.5 BMDO/IST Demonstration Project: 3-D ANN Silicon Neuron Seeker 864

Clifford Lau

Office of Naval Research

59.1 Artificial Neural Network Technology Compared with Conventional

Neural networks are alternative ways of doing signal processing and pattern recognition, based on the idea of interconnecting large numbers of simple processors. Figure 59.1 compares conventional signal processing to the neural network approach.

59.2 Examples of Chips

Several different neural network chips will be discussed:

- Artificial retina (C. Mead, Caltech).
- Cellular neural network (L. Chua, UC Berkeley).
- CCD neural processor chip (A. Chiang, Lincoln Lab).
- SIMD Numerical Array Processor (SNAP) chip (R. Hecht-Nielson, HNC, Inc.).
- Connectionist Neural Adaptive Processor (CNAP) chip (D. Hammerstrom, Adaptive Solutions, Inc.
- Ni1000 Recognition Accelerator chip (M. Glier/M. Holler, Nestor/Intel).

A comparison of the number of processing elements, connectivity, precision, algorithms, commercial applications and defense applications of each of these chips will be made, then summarized in Table 59.4.

Artificial Retina

The artificial retina developed by Carver Mead of Caltech has a number of notable advantages:

- Parallel
- High speed
- Low power
- High dynamic range
- Analog, real-time
- Scaleable to larger size

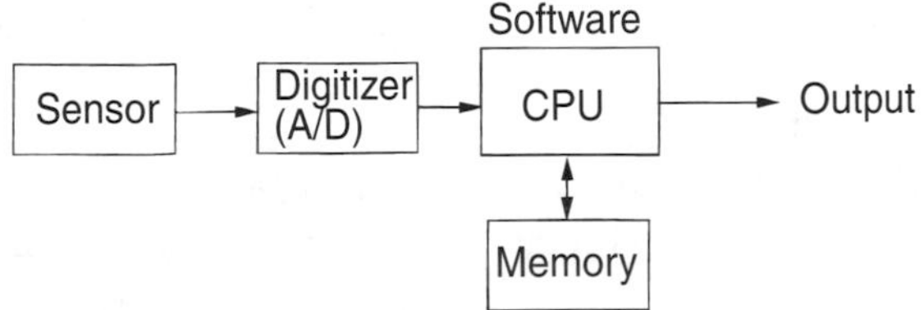

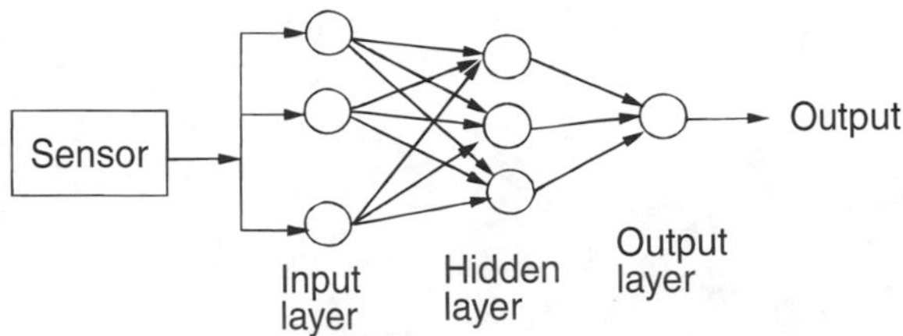

Figure 59.1 Conventional signal processing compared to neural network approach.

0-8493-8343-9/97/$0.00+$.50

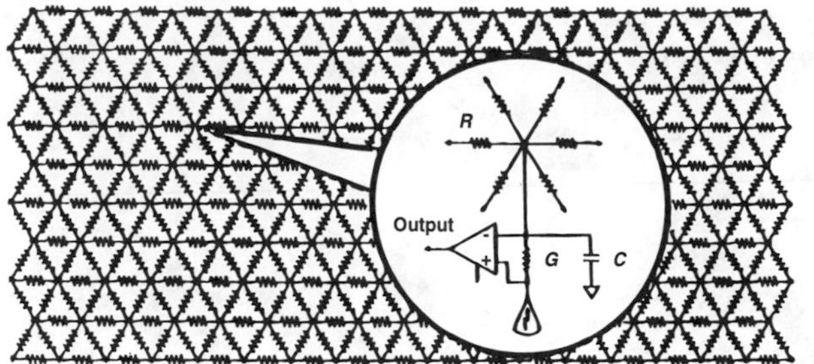

Figure 59.2 Electronic retina chip: schematic wiring diagram (Carver Mead, Caltech).

The schematic wiring diagram is illustrated in Figure 59.2. The chip's function uses a "Gaussian" convolution for edge enhancement. This chip has 50 × 60 (=3,000) pixels.

Assuming a 5 × 5 kernel for the current spread in the grid, the number of computations is 25 × 25 × 3000 = 1,875,000 in less than 1 μs. (A 25 MIPS machine would take 7.5 ms.)

Cellular Neural Network

A cellular neural network (CNN) is a locally connected uniform nonlinear analog processing array Its system equations are as follows:
State equation:

$$C\frac{dv_{xij}}{dt} = -\frac{1}{r_i}v_{xij}(t) + \sum_{C(k,l)\in N_r(i,j)} A(i,j;k,l)v_{jkl}(t) + \sum_{C(k,l)\in N_r(i,j)} B(i,j;k,l)v_{ukl} + I$$

Input equation:

$$v_{uij} = E_{ij}$$

Output equation:

$$v_{yij}(t) = \frac{1}{2}(|v_{xij}(t)+1| - |v_{xij}(t)-1$$

The connected component detection chip architecture is displayed in Figure 59.3. The CNN Chip applications include the following:

- US and Canadian currency recognition.
  - US: Simple convex object size recognition.
  - Canadian: Color decomposition and merging.
- Handwriting recognition.
- Medical image processing.
  - Ultrasound.
  - Computerized tomography.
  - Mammogram.
- Data compression and decompression.
- Color half-toning.
- Flexible image processing: scratch removal, noise removal

The analogic scratch removal algorithm is illustrated in Figure 59.4 and Figure 59.5.

CCD Neural Processor Chip

The CCD neural net classifier architecture is displayed in Figure 59.6. Its design characteristics are as follows:

- Fully connected two-layer net—6144 connections/chip.
- 5×10^9 connections/cm^2-s.
- Weighted sum computation.
- 50-dB dynamic range.

The status of the classifier is final design completed, prototype fabricated and tested, and functionality fully demonstrated.

SIMD Numerical Array Processor (SNAP) Chip

The SIMD numerical array processor (SNAP) is a floating-point parallel array processor based on an HNC designed ASIC VLSI gate array (1 micron CMOS manufactured by LSI Logic). By leveraging state-of-the-art gate array technology, the SNAP delivers an unparalleled price/performance ratio (−$40 per megaflop). This low price/performance ratio enables the development and delivery of many computationally intensive information processing solutions. The SNAP's SIMD architecture is well-suited to applications in signal processing, image processing, neural network processing and general matrix algebra.

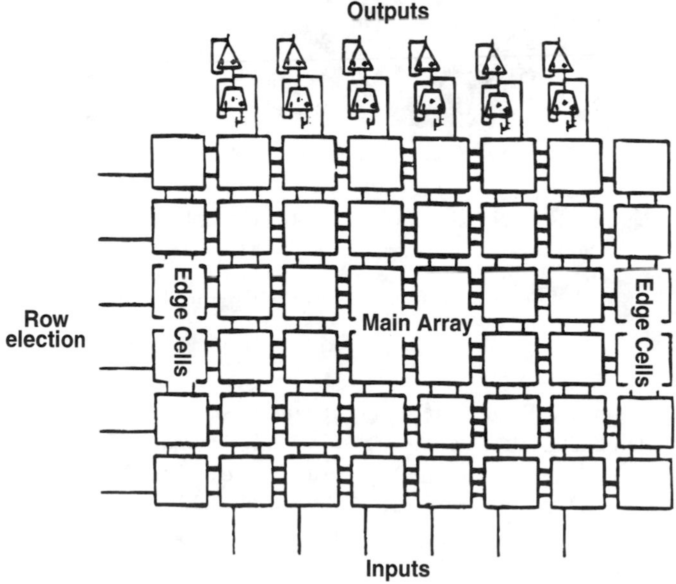

Figure 59.3 Connected component design (CNN) chip: architecture (Chua, UC Berkeley).

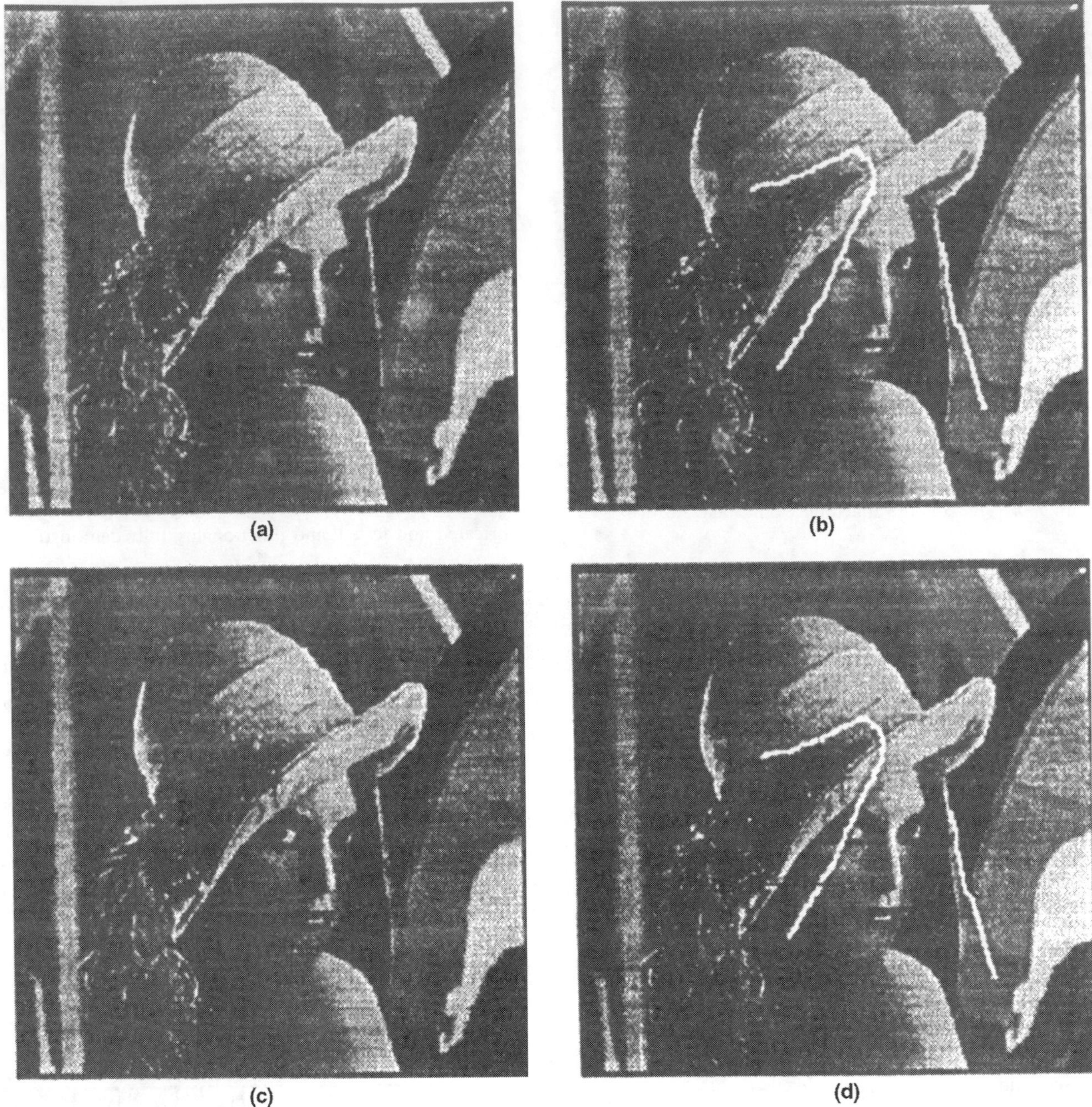

Figure 59.4 CNN chip analogic scratch removal algorithm. (a) The original image. (b) The scratched image. (c) The restored image. (d) The boxed area is expanded in Figure 59.5, which shows the intermediate steps in the process.

The SNAP hardware architecture and a single SNAP processing element block diagram are shown in Figures 59.7 and 59.8, respectively. This architecture performs the LINPACK benchmark very well compared with a Cray Y-MP8 (Table 59.1).

In 1993, SNAP was the IEEE Computer Society Gordon Bell Prize Winner for the best price/performance on a complete supercomputing application. It improved on the 1992 winner by almost 600%.

The SIMD numerical array processor (SNAP) is a single-board system with 640 Mflops, 25 watts, 6U VME single slot. The Software Development Toolkit for the Sun SPARCStation features a Neural Network Library with backpropagation, probabilistic neural network (PNN), radial basis function (RBF) and competitive learning (Kohonen) options. The Math Library allows FFT, FIR filter, SVD and many matrix and vector operations. The performance of SNAP-16 is compared with that of other processors in Table 59.2.

Applications include signal processing, image processing,

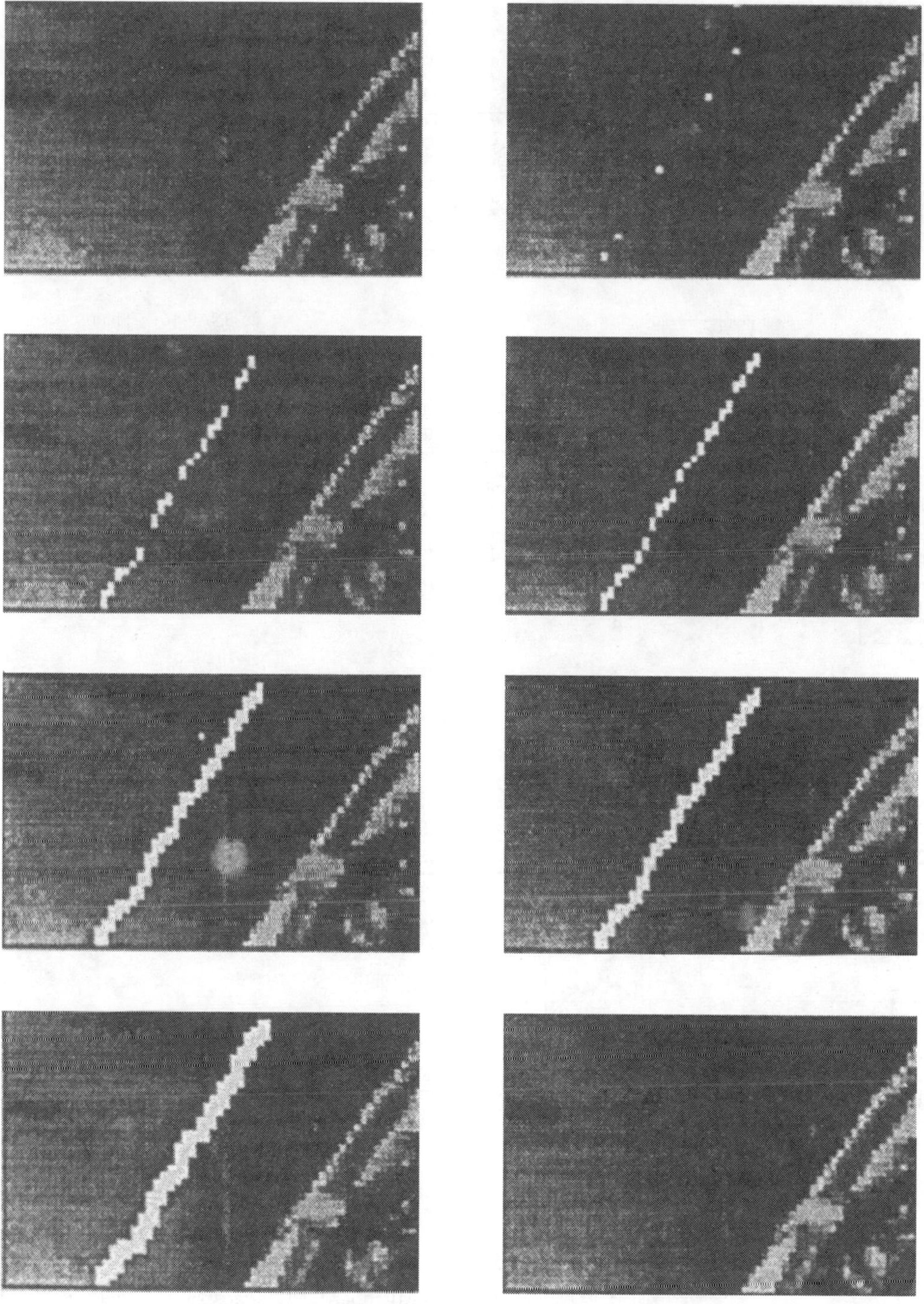

Figure 59.5 CNN chip analogic scratch removal algorithm. The frame on the top left shows the original image. The frame on the top right shows this image scratched. Following from left to right and top to bottom are successive steps in the removal process, finally resulting in the restored image on the bottom right.

mathematical modeling, linear algebra, many-body problems, sonar and radar, tomography and neural networks. As mentioned above, paradigms implemented include: multilayer perceptron with backpropagation training, radial basis function network with backpropagation or SVD training, probabilistic neural network, and Kohonen network.

Connectionist Neural Adaptive Processor (CNAP) Chip

The CNAPS Digital VLSI Neural Network Chip, made by Adaptive Solutions (Dan Hammerstrom), has a simple DSP-like configuration (SIMD). There are 64 PNs with 4K bytes local weight

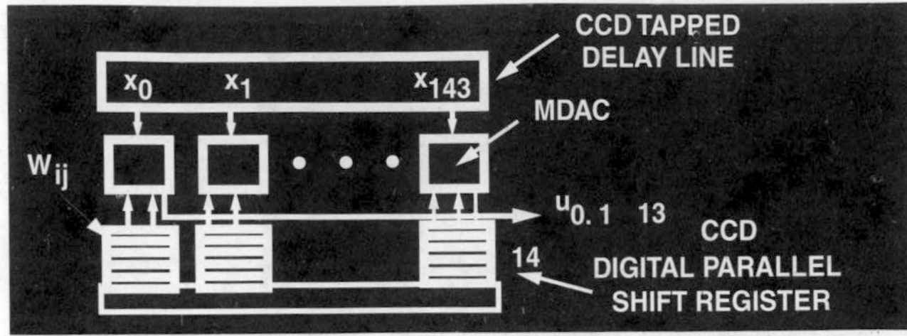

Figure 59.6 CCD neural net classifier (Chiang, Lincoln Lab).

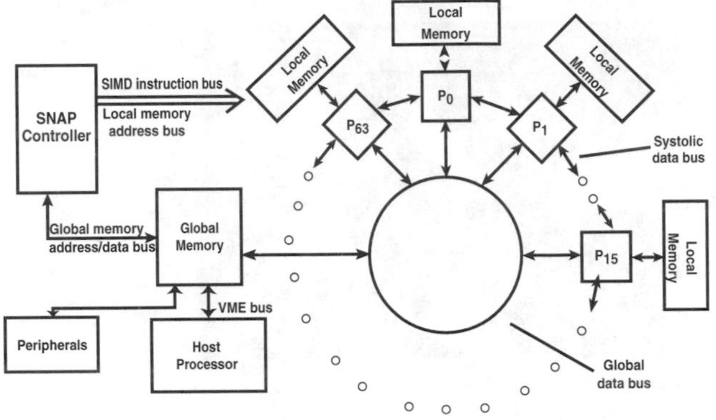

Figure 59.7 SNAP hardware architecture (HNC).

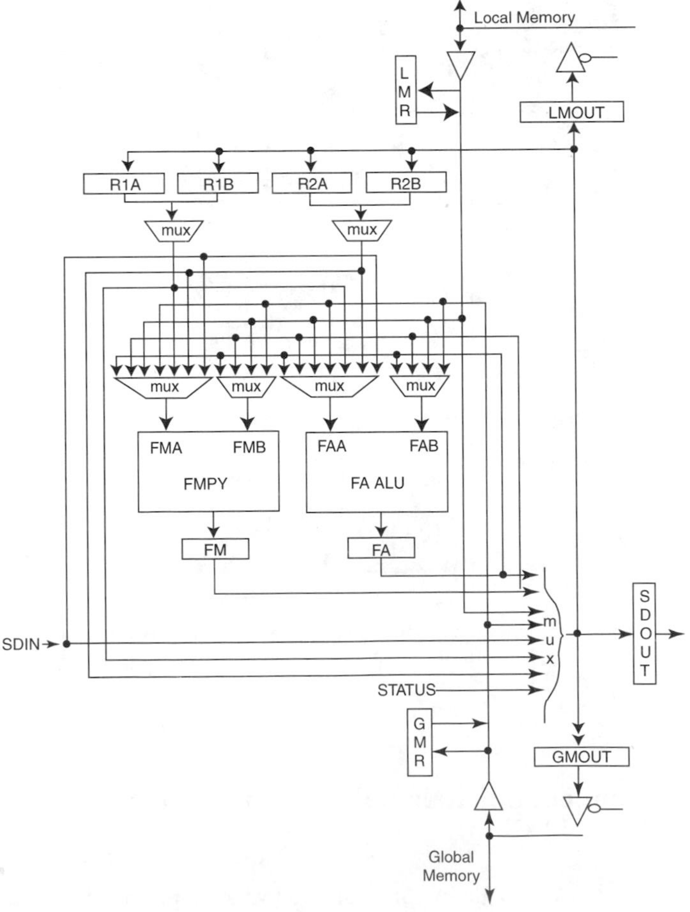

Figure 59.8 Single SNAP processing element block diagram (HNC).

Table 59.1 SNAP Price/Performance Benchmark

Price/Performance	SNAP-64	Cray Y-MP8
LINPACK R(max)	2.158 Gflops	2.144 Gflops
Price	$89,500	$15,000,000

Table 59.2 SNAP-16 Performance Comparison

	HNC SNAP-16	SKY Shamrock	CSPI SC-4XL	Mercury NC860VB
Number of processors	16	4 i860s	8 i860s	1 i860
Peak megaflops	640	320	640	80
16K element single precision vector ADD	155 μs	440 μs	1568 μs	1318 μs
16K element single precision dot product	104 μs	319 μs	1120 μs	864 μs
16K element single precision square root	412 μs	1266 μs	5120 μs	6240 μs
512 × 512 complex 2-D FFT	104 μs	127 μs	220 μs	510 μs

storage. Multiple virtual nodes can be mapped to Ps for large neural nets. Other properties include: broadcast instruction and input data bus, 128K (16 bit) to 2M (1 bit) connections/chip, and at 25 Mhz, 1.6 billion CPS for 8 × 8 or 8 × 16. Thirty different neural network architectures can be mapped onto the chip. Signal and image processing algorithms can also be performed:

- FFT, Wavelets, Gabor filters.
- Dynamic time warping.
- Image processing.
- Compression/decompression.
- OCR.

The CNAPS chip is available as 64 PN chip, 16 PN chip, VME board, PC board or workstation with extensive software development tools. A design study for CNAPS-2 is underway: 32 bit, FP? Bus? Applications?

The architecture, which allows chip boundaries to be ignored by the programmer, is shown in Figure 59.9. A single CNAPS processor node is shown in Figure 59.10, and its image processing performance is diagrammed in Figure 59.11.

For neural networks, the learning performance is shown in Table 59.3. Back-propagation learning speed, connections updated per second is recorded.

Successful commercial applications include image processing and commercial graphics. OCR for printed kanji is used by Mitshubishi. Details are listed below:

Image Processing and Commercial Graphics

- Image column stored each PN.
- 512 PN assumed for 512 × 512 image.
- 3 × 3 convolution = 1.9 ms, 7 × 7 convolution = 5.7 ms.
- 2D FFT = 240 ms.
- Capable real-time JPEG video.
- Adobe Photoshop NuBus board developed.
- −20X speed improvement over existing accelerators for same cost.
- LNK Inc. developed complete image processing library.
- Terrain classification: processes 6K × 8K.
- LandSat Image in 8 minutes vs. 8 hours on Sparc 1.
- Combines feature extraction (wavelets, Gabor filters) and neural net classification on same chip.

OCR, Printed Kanji (Customer: Mitsubishi)

- 4 CNAPS chips on VME board.
- 3500 characters in 15 fonts at 350 cps.
- greater than 99.5% accuracy (vs. 96% current).
- This product order of magnitude improvement in performance/price. 2X speed at 1/5 price!

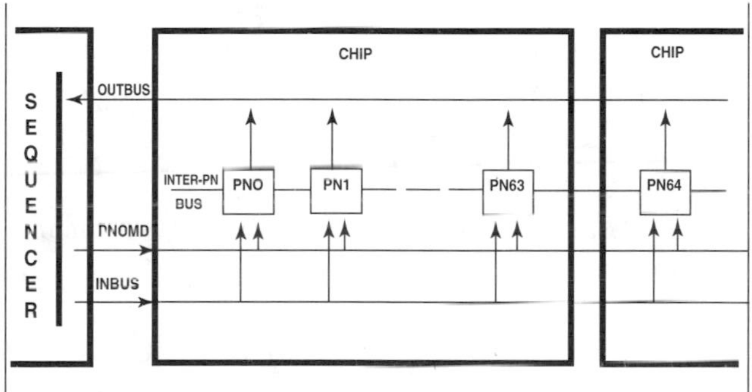

Figure 59.9 CNAPS array (Adaptive Solutions, Inc.).

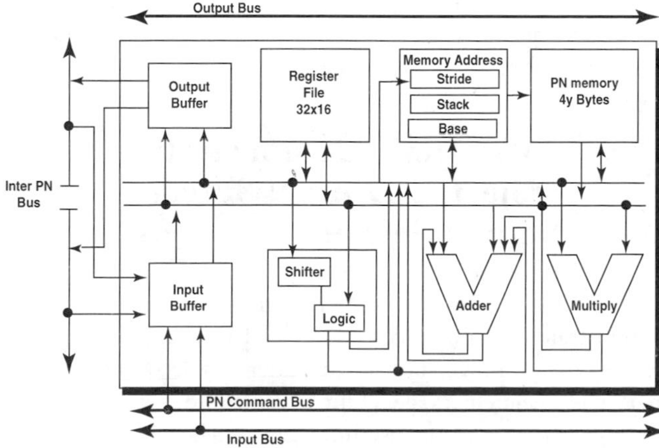

Figure 59.10 CNAPS processor node (Adaptive Solutions, Inc.).

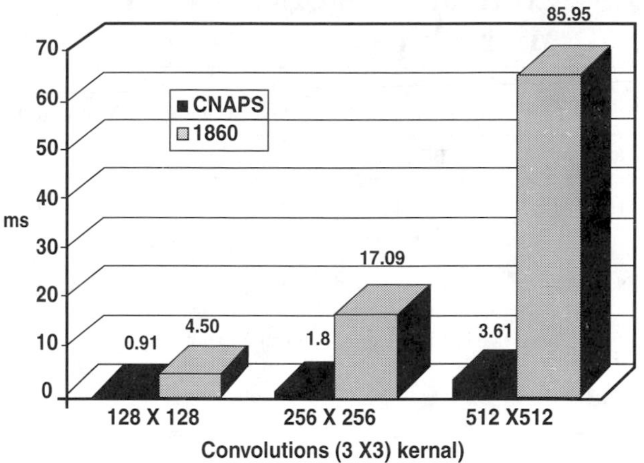

Figure 59.11 CNAPS image processing (Adaptive Solutions, Inc.).

Table 59.3 Neural Network Learning Performance

CNAPS (8 chips, 25 Mhz)	2380 M
Convex C1	0.06M
Cray 2	7M
Connection machine	40M
Warp	20M

NI1000 Recognition Accelerator Chip

The Intel/Nestor NI1000 Neural Network Chip is a digital neural network chip optimized for rapid computation of distance measures in Cooper's RCE Networks and related networks such as Radial Basis Function (RBF) and Probabilistic neural Networks (PNN). The architecture uses multiple pipelines. There are 14 processors per chip, scaleable for multi-chip boards. One version having 4 chips per board (PCI bus) was developed by Nestor for NestorReader handwriting recognition. Another having 9 chips per board (VME) was developed by Lockheed for implementation of PNN nets for sonar signal classification. The performance is 30 GOPS.

A block diagram of the architecture is found in Figure 59.12, and specifications are listed below:

Clock frequency	40 Mhz
DCU performance	20 GOPS
FPU performance	160 MFLOPS
16b Microcontroller	20MIPS
Array access bandwidth	10 Giga-word/s (5-bit words)
Prototype vector size	256 dimensions × 5 bits
Number of prototypes	1024
Array flash EEPROM	2.6 M trs.
SRAM	57k × 6 trs.
Total transistors	3.5M trs.
Die size	15.8 mm × 13.7 mm
Power dissipation	5W@5V, 0.5W@12V
Process	0.8μ CMOS, 2 metal, 2 poly, flash ET
Package	168-pin PGA

59.3 Comparisons of NN VLSI Microchips

Table 59.4 Comparisons of Neural Network VLSI Microchips

Chip	PE/Chip	Connectivity	Precision	Algorithms	Commercial	Defense
Silicon Retina	1250	Local	Analog	Image convolution	Check reader, face recognition	Oil debris monitor
CNN-Berkeley	500	Local	Analog	Flexible image processing	Medical imaging	Motion detection
CCD-LL	144 in 14 out	Sequentially fully connected	Hybrid	NN: vector matrix multiplier	Speech recognition	Sonar processing
SNAP-HNC	4	Systolic ring	32 bit FP	FFT, general NN	Tomography, image processing, modeling	Sonar 6.3, mechanical diagnostics 6.3
CNAPS-ASI	64	Broadcast fully Connected	16 bit Integer	FFT, general NN, graphics, JPEG video	Graphics, OCR, medical, image recognition	Sonor 6.3, missile seeker 6.3
Ni 1000-Intel	14	Multiple pipeline	5 bit (cascadable to 10 bits)	NN: RCE, radial basis, PNN	Handwriting recognition	Sonar target recognition

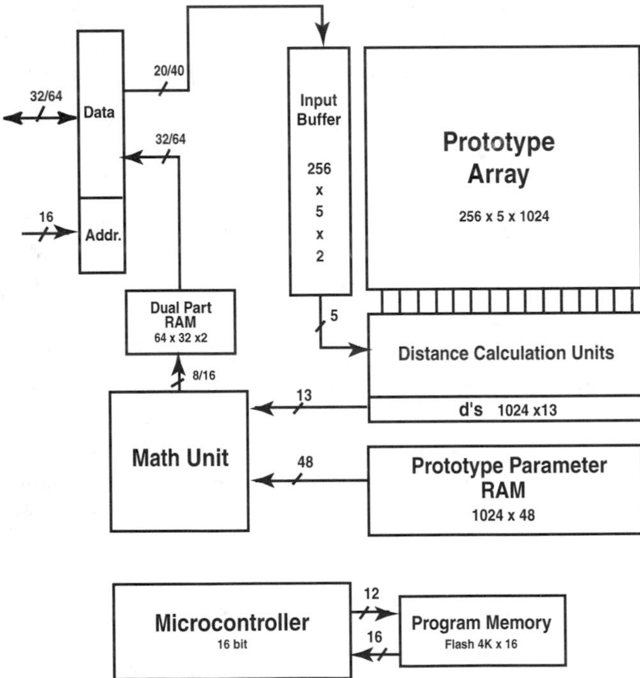

Figure 59.12 Ni1000 recognition accelerator chip block diagram (Nestor/Intel).

59.4 Applications of Neural Network Technology

Applications of neural network technology are listed below:

- Automatic target recognition.
- Detection of target in a clutter environment.
- Image processing, e.g., segmentation and feature extraction.
- Sonar signal detection and classification.
- Speech processing, voice recognition.
- Helicopter gear-box diagnosis.
- Integrated diagnostics for condition-based maintenance.
- Multitarget tracking and weapon-to-target assignment.
- Robotic and autonomous vehicle control.
- Ocean wakes detection in SAR images.
- Battle management, command, control, and communications.

Summary

Potential of artificial neural network technology can be realized only with special purpose VLSI hardware. Many neural network chips are already available on the market. Neural network chips can be used for real-time image processing and pattern recognition. Many new applications, such as multimedia and two-tone FAX, are waiting to be explored.

59.5 BMDO/IST Demonstration Project: 3-D ANN Silicon Neuron Seeker

The objective of the BMDO/IST Demonstration Project is to demonstrate detection, acquisition, and tracking of a multiple number of fast moving target in a clutter and decoy environment in a small satellite package. Using a frame rate greater than 1000 Hz, real-time acquisition discrimination and homing, and human level recognition, Irvine Sensors Corporation's 3DANN FPA implements many neural network architectures and algorithms real time. Figure 59.13 illustrates the 3DANN-1 FPA and the

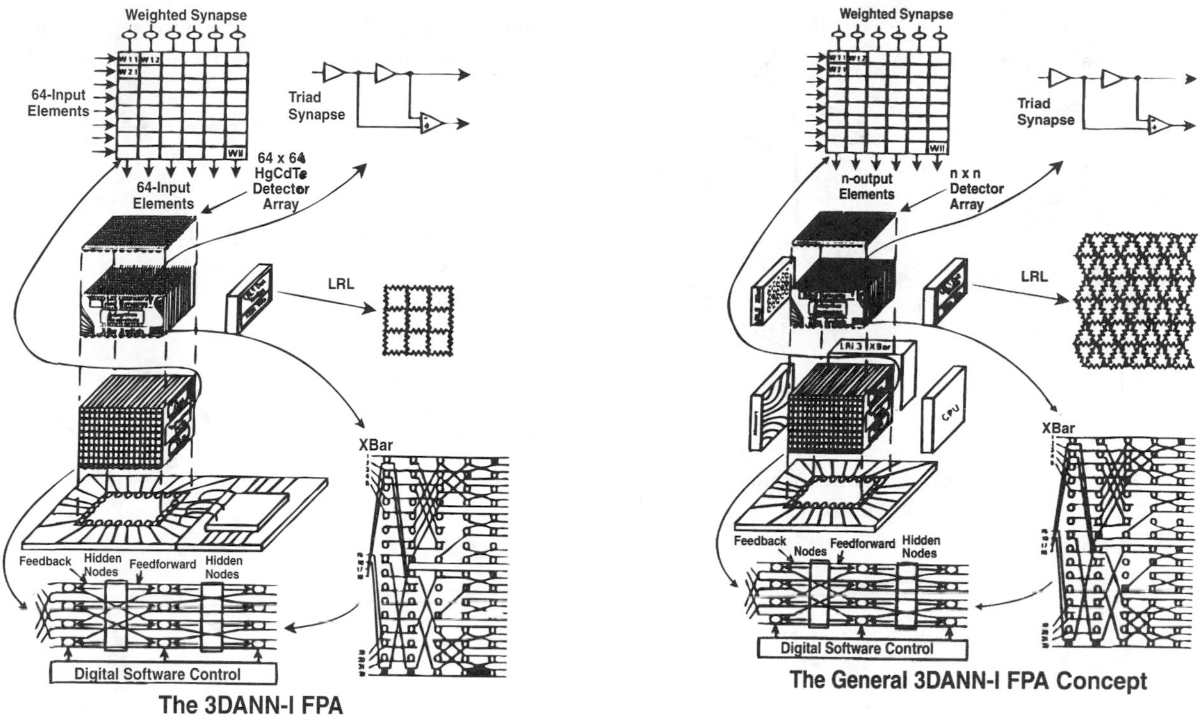

Figure 59.13 3-D ANN silicon neuron seeker: the 3DANN FPA implements many neural network architectures & algorithms real-time (Irvine Sensors Corp. BMDO/IST Demonstration Project).

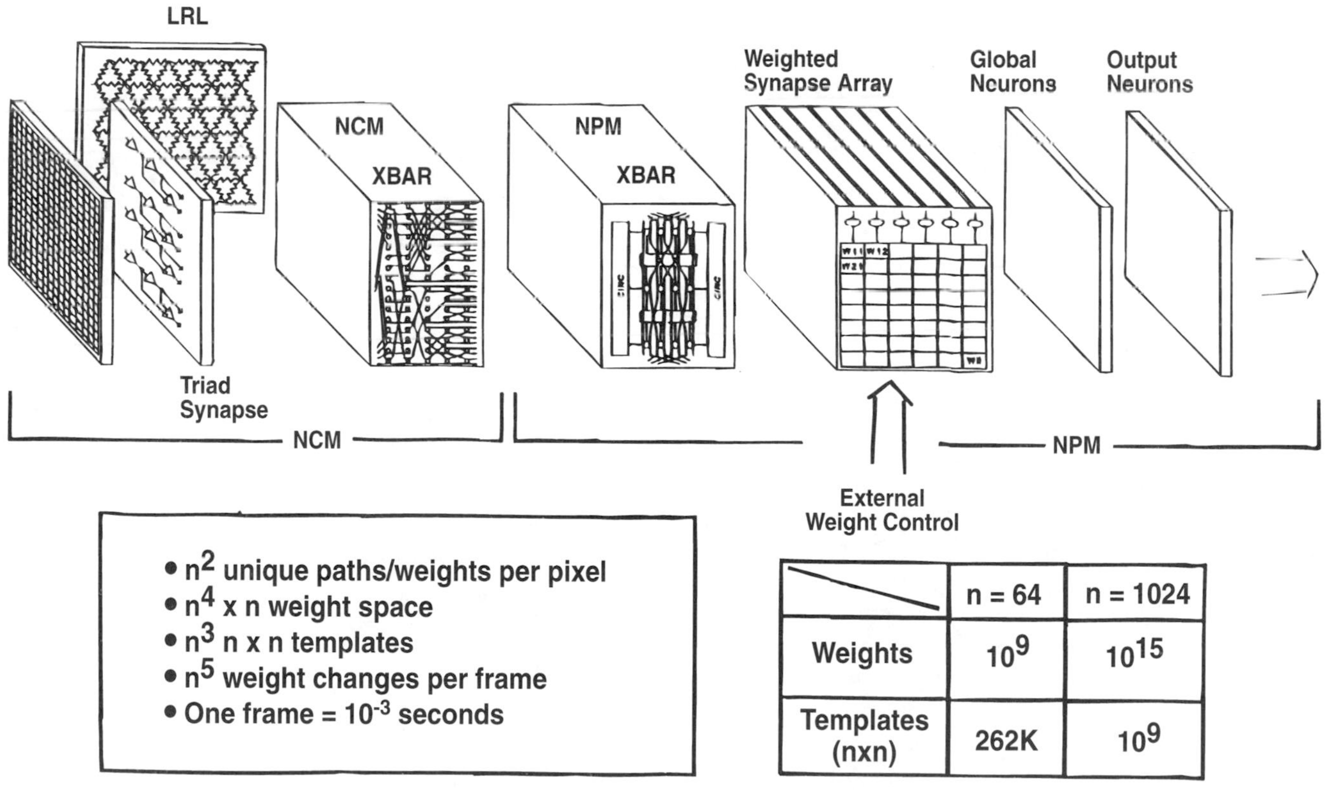

	n = 64	n = 1024
Weights	10^9	10^{15}
Templates (nxn)	262K	10^9

Figure 59.14 3-D ANN architecture (Irvine Sensors Corp. BMDO/IST Demonstration Project).

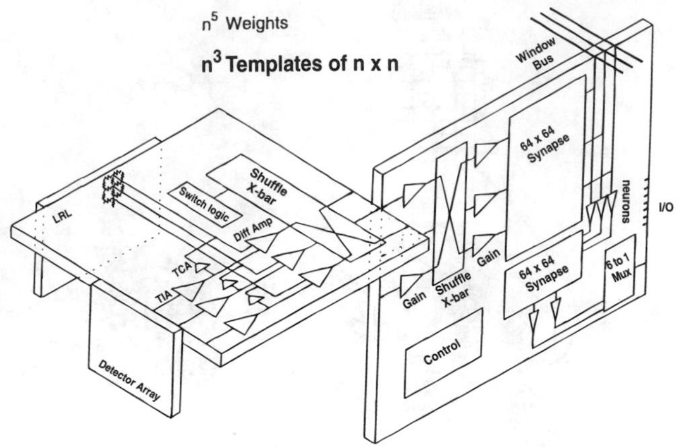

Figure 59.15 3-D ANN baseline perfect shuffle architecture (Irvine Sensors Corp. BMDO/IST Demonstration Project).

General 3DANN FPA concept. Figure 59.14 diagrams the 3DANN architecture, and Figure 59.15 shows the baseline perfect shuffle architecture.

Expected results are as follows: Hardware and software seeker in a can ready to put on a satellite

- >1000 frames/second
- 64 × 64 pixels per frame, expandable to 128 × 128
- 2″ cylinder by 5″ long (including optics)
- 1/4 cubic inches for NCM and NPM
- 1 watt power

HW/SW will demonstrate real-time multiple-target recognition and tracking, and weapon-to-target assignments.

60

Commercially Available Artificial Neural Network Chips

Seth Wolpert
Penn State University at Harrisburg

60.1 Introduction 867
60.2 Analog ANN Products 867
60.3 Digital ANN Products 869
60.4 Hybrid ANN Products 871
60.5 Discussion 872

60.1 Introduction

The widespread application of artificial neural networks (ANNs) to problems in pattern recognition, prediction, computation, and categorization is testimony to the great power, versatility, and fault-tolerance they offer in comparison to more conventional computational and statistical methods. These advantages have inspired numerous ANN applications to specific problems in a variety of disciplines within both academic and industrial research settings. Results from these applications are encouraging in both the short and long term. In addition, several commercial ventures have aspired to tap into the great wealth of more generalized problems to which ANNs may be applied.

The suitability of VLSI for ANN implementation was described early on by John Hopfield of Cal-Tech (1982), and has been pursued since that time by many researchers. One pioneer in the field is Robert Hecht-Nielsen (1990), who implemented a digital neurocomputer in 1982. It was designed using Zilog Z8000-based single-board computers, each with 32 KB of RAM and a 4 MHz clock speed. By 1984, this design had evolved into the Mark III by Todd Gutschow and Hecht-Nielsen at TRW, Inc., under DARPA sponsorship. The updated system had eight Motorola 68010 processors with 512 KB of RAM and a 12 MHz clock speed on a VME bus. This system could accommodate 8000 neuronal processing elements and 480,000 synaptic connections, and learned at a rate of 380,000 connections per second. Its successor, the Mark IV, employed similar processors, but a more sophisticated design and a novel pipelining scheme to accommodate 262,144 processing elements, 5.5 million connections, and a learning rate of 5 million connections per second. These techniques are employed as a matter of course in the digital ANN implementations on the market today.

Other pioneers in artificial neural networks with commercial application include the analog visual image processor described by Sivilatti, Mahowald, and Mead (1987) of Cal-Tech, the reduced coulomb energy network (Johnson and Brown, 1988), conceived by Leon Cooper, Nobel laureate from Brown University and founder and chairman of Nestor, Inc., one of the most prominent of today's commercial ANN producers. In conjunction with Charles Elbaum, also of Brown University, he developed a multilayered perceptron, which was patented in 1982, and has since been applied to a number of pattern recognition problems in research, government, and industry. Other significant efforts include the ANNA chip by Boser et al. (1992) and Sackinger et al. (1992) of AT&T Bell Laboratories, the CNAPS chip by Hamerstrom (1990) of Adaptive Solutions, the digital neural signal processor by Ramacher (1992) of Siemens, the BNU chip by Arima et al. (1992) of Mitsubishi, the Hitachi digital ANN by Watanabe et al. (1993), the Boltzmann/field learning chip by Alspector et al. (1989) of Bellcore, and the ETANN chips by Intel corporation. These products, some from these large corporations and others from a number of smaller fledgeling operations, offer a variety of ANN architectures, capacities, and functions. Some of these devices are implemented using analog circuitry, while others are realized totally in digital circuitry. A third category, the hybrid ANNs, makes appropriate use of both modes. A discussion of the comparative advantages of each circuit type along with descriptions of two commercially available examples of each type will follow.

60.2 Analog ANN Products

Many of the ANN products available now and in the past have made use of analog VLSI circuitry. Analog circuitry, although transistor-for-transistor, is more difficult to design, it can generally realize ANN circuitry in far fewer components than a digitally implemented counterpart. A lower component count affords

0-8493-8343-9/97/$0.00+$.50

higher circuit density, smaller chip area, and greater power efficiency. Such circuits use dedicated circuitry, with minimal serial multiplexing of data and control signals. This affords faster circuit operation, as well. In addition, such circuits offer operation that is more configurable and consistent with biologically based waveforms and algorithms. Finally, analog waveforms offer better comparative resolution than discretized signals, and small differences between vying signals are easier to distinguish by mutual inhibition or similar methods, especially since most digital systems are constrained by sheer circuit bulk to a limited number of bit widths for signals and synaptic weights.

There are also a number of disadvantages to analog implementation of ANNs. The principle one is information storage. ANNs require local memory for extensive and frequent storage of nodal values, synaptic weights, and other network management parameters. Stability of these stored values is of key importance to the training and function of the network. To date, several methods of analog voltage memory have been explored, including on and off-chip storage capacitors and floating gates, but these methods impose constraints on the stability, convenience, and precision afforded for user-friendly ANN application. Analog circuits, while structurally simpler than their digital counterparts, are far more difficult to simulate and synthesize. Other disadvantages of analog ANNs are thermal variability in on-chip resistances, MOSFET threshold and transconductance, junction leakage, and other circuit parameters. Thermal and induced noise also affect signals whose fine resolution is critical to ANN training and function.

ANNs are highly interconnected, but extensive interconnect on-chip introduces significant and nonuniform capacitive delay and crosstalk between adjacent signal routes. Analog CMOS processes rely upon distributions on a molecular scale, and are inherently inconsistent from one wafer to another. This makes resistor and transistor behavior difficult to precisely control *a priori*. In addition, the values of on-chip resistors and capacitors are limited in range and accuracy by the manufacturing process. Finally, the packages in which such ANNs are housed offer limited pin capacity, which constrains the number of on-chip signals that may be accessed from off-chip. This limits the size of the network on-chip, necessitating analog multiplexing or loss of access to internal nodes.

A wealth of circuitry is available for realizing analog ANNs in CMOS VLSI. Many of these circuits are detailed in the book *Analog VLSI and Neural Networks* by Carver Mead (1989). In it, a number of fundamental CMOS circuit implementations of a variety of mathematical processes are given. These processes include addition and subtraction, multiplication in two or four quadrants, integration, differentiation, nonlinear activation function transformations, logarithmic and exponential compression, temporal delay, voltage thresholding amplification, frequency filtering, absolute value, and mutual inhibition. From these processes, the majority of ANN behaviors may be realized in a minimal number of circuit components. From these circuits, a number of analog ANN implementations have been achieved. Two such implementations will be described here.

Intel Corporation has been developing both analog and digital ANN ICs for several years. Their devices are denoted as ETANNs (electrically trainable ANNs). One of their analog devices is the M64 ETANN (Maren et al., 1990). This IC contains an array of 64 rows by 128 columns of synaptic units, each of which contains a pair of floating gate storage elements, one for excitatory strength, and one for inhibitory strength. Complementary synapses were utilized to improve linearity and thermal stability in overall synaptic coupling, as were specialized biasing circuits to keep the multipliers linear.

The circuit accepts 64 parallel analog bimodal input signals, and applies them to the first 64 columns of the synaptic weight matrix. For each row, the 64 resulting products are summed. The 64 bipolar sums of products that result are then fed into the second 64 columns of the synaptic weight matrix. The 64 outputs from the second half are then threshold and 64 binary comparison results are outputted off-chip. This organization is depicted in Figure 60.1.

Synaptic weights are stored in capacitors as quantities of charge on-chip, and must naturally be periodically refreshed. Each synapse has a unique binary address. They are modified one at a time from off-chip, which can take up to several milliseconds, depending on the magnitude of the change in weight, i.e., amount of charge to be injected or removed. The multipliers are specified to be linear to within several percent, and the overall multiplication process is accurate to seven bits of resolution, although stored coefficients are only accurate to four bits of resolution, with a facility for minute changes to be effected, as well. Data transfer on or off-chip takes about 3 μS, which affords a a learning rate of 2.4 billion connections per second. This circuit formed the basis for Intel's ultimate analog ETANN IC, the 80170.

The Intel 80170 was introduced in 1991, intended for commercial application to real-time neurocomputation, including recognition of optical, electrical, tactile, and acoustic patterns, robotic and nonlinear process control, and signal processing. The 80170 is housed in a 208-pin package, and has a capacity of 64 analog inputs, 64 neurons and 10,240 synaptic weights. Each neuron computes a sum of products of signal inputs with 32 fixed bias weights and 128 variable weights. A block diagram of the 80170

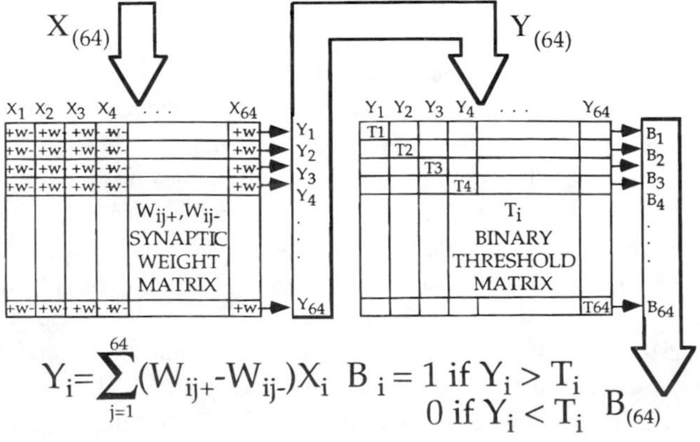

Figure 60.1 Internal organization of the Intel M64 ETANN.

is given in Figure 60.2. Like the M64 chip, it stores each weight as a complementary pair of excitatory and inhibitory synaptic coupling coefficients stored on floating gates MOSFETs, and multiplied by a voltage difference derived from the analog input voltages, converted to a current differential, summed, mapped back to a sigmoidal voltage distribution, and outputted. Floating gate voltages are modified by means of high-voltage pulses applied from off-chip.

The 80170 may be trained in two different ways. Weights may be derived in a virtual environment, totally within a host personal computer, using either Intel-supplied development system software, or user software. Training takes place according to user-specified algorithms, input-output spaces, and operating limits. Once training is completed, the weights are downloaded to the chip, which then evaluates input patterns in real-time. The other training option involves use of the chip to gauge response to a set of inputs and weights. Modifications to weights are formulated by the host computer and downloaded during the training process. Each synaptic weight is derived from a six-bit digitally converted value, and is stored in a floating gate EEPROM-like cell, which may be written and erased many times. In order to form larger or multilayered networks, 80170s may be networked either serially or in parallel to form arrays of up to 1024 neurons and 81,920 synaptic weights. Intel's approach to design of the 80170 provides maximum throughput once weights are computed, but cannot alter those weights without interruption to the system.

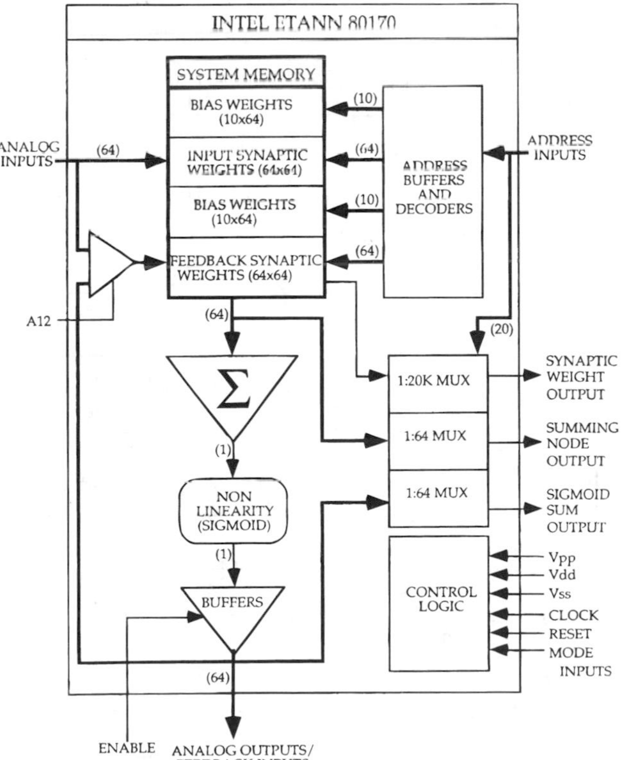

Figure 60.2 Internal organization of the Intel 80170 analog ETANN.

60.3 Digital ANN Products

As algorithms for structure and learning in ANNs are developed from mathematical models, digital implementations may be derived directly from the formulas that express their relationships. Digital circuitry offers a number of advantages over analog realizations of ANNs. Digital circuitry may be made as precise as is required by simply providing a sufficient number of bits for input and output data and synaptic weights. Analog circuits, on the other hand, are generally empirical approximations of mathematical relationships, and their precision is limited. Digital circuitry is inherently noise-immune, while analog circuitry requires compensation for thermal, parametric, and supply-voltage variation, and noise from a variety of sources. Digital circuits may be designed for fast operation and power efficiency, being implementable in a variety of MOS and gallium arsenide technologies, while analog circuitry has less flexibility in how it is implemented. Digital circuitry is far easier to interface to host processors, and resident signals are more easily multiplexed and demultiplexed than analog circuitry, economizing on interconnect. Digital circuits may be designed to be more user-programmable and configurable than analog circuits, offering greater flexibility and efficiency. The digital realm supports the use of design and synthesis software for fast custom VLSI design, while analog VLSI must still be done manually. Finally, accurate, long-term storage of digital data is easily managed through a diverse and well-developed memory technology, while accurate and convenient long-term storage of analog signals on-chip remains elusive. Disadvantages of digital ANN designs are that they require far more circuitry than analog designs, and they require A/D and D/A conversion in order to interface with analog input signals, synaptic weights, and output responses.

A wealth of circuitry for digital ANNs has already been developed for existing applications to microprocessors, memory, and logic. Combinatorial logic gates are widely used as components for larger scale digital modules. ANN and X-OR gates double as minimalistic 1-bit multipliers and adders respectively. Shift registers are used to rapidly double or halve a binary value. A host of contiguous adders and multipliers has been developed for basic mathematical operations. The ultimate in flexibility for non-linear scaling and conversion of signals and weights is afforded by look-up tables. RAM and ROM are available with a number of performance and efficiency levels. Counters monitor repetitive algorithms and processes. These circuits are already available in the form of cell libraries, offering automated design of dedicated ANN ICs tailored to a variety of applications and constraints.

One example of a digital ANN is the Intelligent Pattern Recognition Memory, IPRM, produced by Oxford Computer, Oxford, CT. Functionally, this circuit is used to recreate a network centered around a matrix-vector multiplier and a 2-dimensional convolver realized by repetitive shift, multiply, and add operations. Individual IPRMs are assembled into large storage arrays in a bit-slice approach, where each IC stores one bit of each element in a weight matrix of 1024 rows and 64 columns and several bits of each element of a 64-element input vector. Each IPRM stores its input vectors in a 512-by- 8-bit dual port static RAM and its weight matrices in a dual-port 64K-by-1- bit static RAM. Data from the weight matrix are downloaded, either in forward sequence, or through a moving 8-by-8-byte window by the shifter module to the processor. Data from the vector memory are also shifted and applied to the central processor. The processor, which is of a 64-bit slice architecture, then either searches for an exact match or computes a binary product, and outputs its result to a 64-bit adder. The sum of products is then outputed. Logical organization of an IPRM is given in Figure 60.3.

The IPRM is fabricated in a 1-micron CMOS technology, and is housed in a 132-pin PGA package. It operates at clock frequencies of up to 20 MHz, powered from a single-ended 5-volt supply, and is TTL-compatible. It is able to accommodate matrices of varying widths and lengths, and may be cascaded to provide as many bits of resolution as desired. Accessory ICs are available from Oxford to perform summation of partial products from IPRM ICs and apply nonlinear transformations to those sums. Each chip is able to compute a dot-product of two 64-element vectors in 50 ns. Oxford has begun planning on a next generation IC with a 256-bit architecture, and clock frequencies of up to 40 MHz. A multichip module (MCM) containing eight such ICs, plus summing and nonlinear transforming accessory ICs is expected to be able to perform 1.28 billion 8-bit-by-8-bit multiplications per second.

Another example of a digitally implemented ANN IC is the CNAPS (connected network of adaptive processors) 1064 Digital Parallel Processor, sold by Adaptive Solutions, Inc. of Beaverton, Oregon. The CNAPS 1064 is organized around a linear array of 64 synchronized processors, all of which are coordinated by a CNAPS sequencer chip (CSC), with which they share a command bus, an input data bus, and an output data bus, as shown in Figure 60.4. The CSC stores, fetches, and distributes input data, synaptic weights, and processor commands, and manages all input/output interactions by the system. Each of the 64 processors on a 1064 is organized in a 16-bit architecture, and contains an 8-bit-by-16-bit 2's complement multiplier, a 32-bit adder, a shifter and logic unit, a 32-word register file, a 12-bit weight address register, and 4 KB of storage for individual synaptic weights and signal strengths in either 16-bit or signed 8-bit mode. The internal organization of a processor is given in Figure 60.5. Instructions from the sequencer are issued to all processors, and executed simultaneously. Each processor can complete one multiply and add operation in each cycle of the clock.

The CNAPS 1064 consists of 14 million transistors fabricated in a 0.8-micron low-power CMOS technology. It is able to learn, with an on-board storage capacity of 2 million 1-bit synaptic weights, or 256K 8-bit weights. With a clock rate of 20 MHz, it is able to perform 960 million multiply and accumulate operations per second. For a system of eight CNAPS chips, this figure is 10 billion. The CSC requires separate static RAM for storage of commands and data, and interfaces to a host controller as a memory-mapped coprocessor via a digital control processor. This processor can be any moderate-high performance digital microprocessor, such as a Motorola 68030. The organization of such a system is shown in Figure 60.6. A CNAPS system is suitable

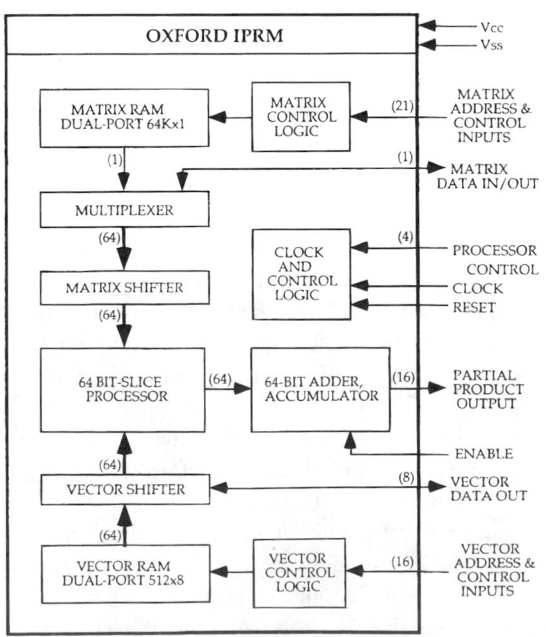

Figure 60.3 Organization of the Oxford IPRM ANN IC.

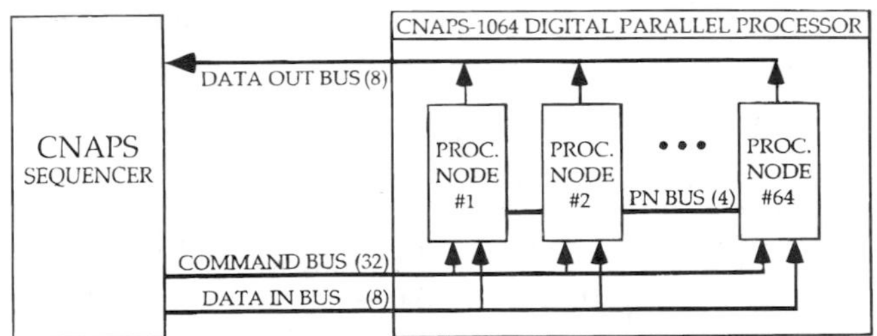

Figure 60.4 Organization of the CNAPS digital parallel processor system.

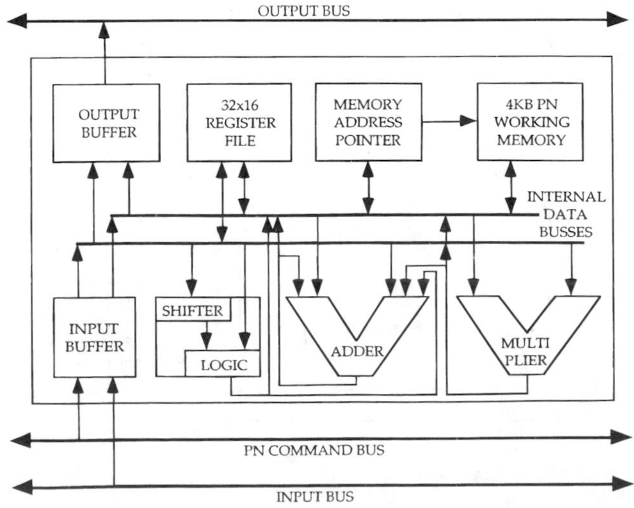

Figure 60.5 Organization of a processor node of the CNAPS 1064 ANN IC.

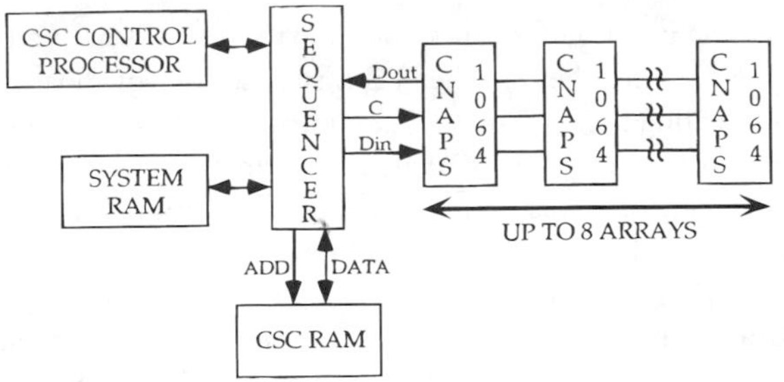

Figure 60.6 Interface of a CNAPS digital parallel processor to a host system.

for implementation of trainable ANNs, such as multilayered back-propagation networks, and a variety of real-time image processing tasks. Adaptive solutions also offers board-level processors for PC and VME bus architectures, as well as a parallel C compiler and foundation library for development of CNAPS algorithms, and the CNAPS Server II, a high-performance computer platform for developing parallel processing algorithms.

60.4 Hybrid ANN Products

As detailed earlier in this chapter, analog and digital circuit implementations each offer distinct advantages and disadvantages in the design, realization, and use of ANNs. Ideally, a circuit that made appropriate use of both circuit modes would take maximal advantage of both circuit types. Although hybrid circuits necessitate more complex mixed-mode design and fabrication technology, they offer the designers and users of ANN ICs the best of both worlds. Analog circuits would perform waveshaping and simple mathematical and logical processes with maximal speed, power efficiency, comparative resolution, and compactness, while digital circuits would provide a noise-immune environment for host interfacing and performing precise calculation and long-term storage of network signal strengths and synaptic weights. This has been the approach taken in some of the most successfully and comprehensively implemented ANN integrated circuits to date.

One of the original and most successfully applied hybrid commercial ANN ICs is the AT&T ANNA (analog neural network arithmetic-logic unit), developed by Boser and Sackinger of AT&T Bell Laboratories (Sackinger et al., 1992). It was designed to realize a multilayer perceptron model, and features a level of resolution that is configurable to the demand in various points in the ANN. The circuit realizes perceptron behavior by loading an input vector into a barrel shifter, and simultaneously evaluating the inner products of the shifted input vector with eight weight vectors. The eight scalar values are then transformed according to a sigmoidal nonlinearity and outputted off-chip. Weight vectors are stored in a 4096-location on-chip dynamic RAM, which must be externally refreshed. One complete cycle of an evaluation process takes four clock cycles, or 200 ns. The architecture of the chip is reconfigurable so that it can accommodate any combination of input and weight vectors 64, 128, or 256 bits wide, and neurons with 64, 128, or 256 synaptic inputs.

The barrel shifter, which preprocesses input vectors performs convolution of input data, as well as facilitating pipelining of input data for greater throughput and ease of multiplexing of input data. If all neurons on chip are configured to have 256 synapses, the ANNA chip is capable of evaluating 10 billion connections per second. This organization is illustrated in Figure 60.7.

The ANNA chip was implemented in 0.9 micron CMOS technology, and contains 180,000 transistors on a 4.5 by 7 mm die. On chip, there are 4096 analog synapses, each with a resolution of 6 bits for 64 possible levels. Neuronal input/output signals have resolutions of 3 bits each with eight possible analog levels, and each neuron has a 4-bit multiplicative scale factor to allow its dynamic range to cover a 16-fold output range. It is installed on a VME board in a host computer, which supples input data, downloads output data, configures the topology of the network, and refreshes analog weights, stored on capacitors on-chip. The ANNA has been successfully applied as an engine for high-speed optical recognition of alphanumeric characters by researchers at AT&T Bell Laboratories.

Another hybrid ANN IC is the Ni1000 Recognition Accelerator IC developed by Nestor, Inc., of Providence, R.I. The Ni1000 is sold as a stand-alone IC, or as the centerpiece of a PC-based development system, which includes a 16-bit ISA plug-in board, a microcontroller code assembler, Ni1000 hardware and assembler libraries in C, and programming environments for MS-DOS and Windows-based systems. Ni1000 ICs are available with clock speeds of 10 MHz without pipelined operation, or at 33 MHz with both pipelined and unpipelined operation. The Ni1000-33 with pipelined operation, is able to process 8.2 billion connections, and evaluate over 32,000 patterns per second, where each pattern contains 256 5-bit elements. Organized according to the internal 16-bit architecture shown in Figure 60.8, the Ni1000 consists of internal data, address, and control busses accessed by three modules: the microcontroller, the classifier, and the host interface.

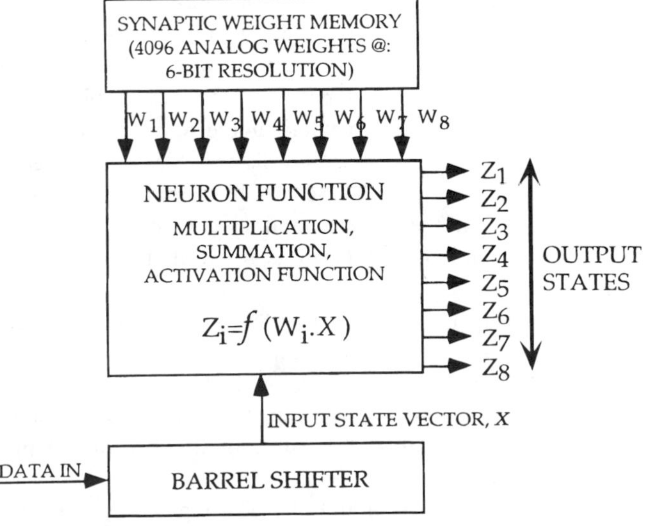

Figure 60.7 Organization of the AT&T Bell Laboratories ANNA IC.

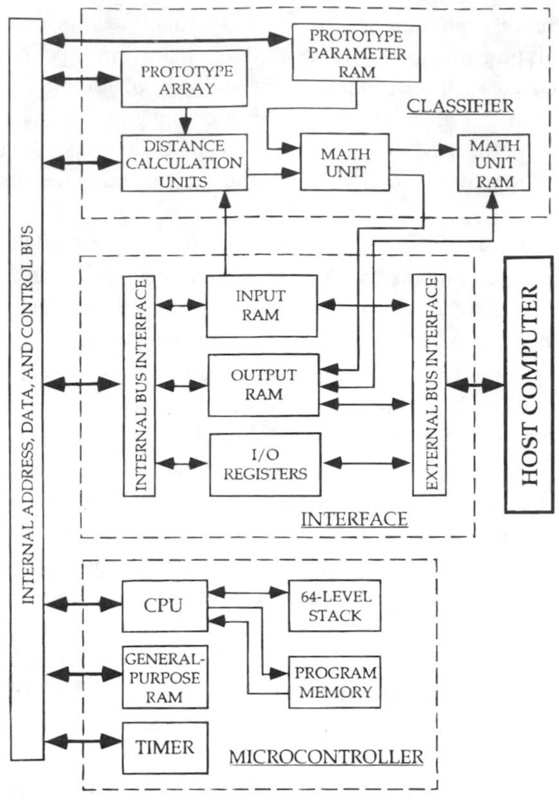

Figure 60.8 Organization and interfacing of the Nestor Ni1000 ANN IC.

The microcontroller is of a 16-bit wide Harvard-type that manages the learning process, communications with the host system, and a number of on-chip maintenance tasks. It includes a 4K-by-16-bit flash memory for microcontroller programs, a 64-level stack, 256 words of general-purpose RAM, and a 32-bit timer. Using a radial basis function paradigm, it includes code for learning according to restricted coulomb energy (RCE), Probabilistic RCE, and probabilistic neural network algorithms, and may be user-programmed for other algorithms.

The classifier module includes a digital prototype array, distance calculation unit, a math unit with RAM, and a prototype parameter RAM. The prototype array stores 256 5-bit features for each of 1024 patterns, reconfigurable to as many as 8000 patterns of 32 5-bit features. The distance classifier correlates input vectors with prototype vectors by computing city-block distances between their corresponding components. It has a capacity of up to 512 comparisons between the input vector and corresponding components of each pattern stored in the prototype array. The math unit interprets correlation results, calculating probability densities for each prototype. It employs six stages of pipelining and a sixteen bit floating point architecture for data. The prototype parameter RAM stores RBF radius, smoothing factor, count, and classification number for each prototype in the array.

The interface contains input buffer RAM, output buffer RAM, and 16 I/O control registers. It is compatible with host system data busses 32 or 64 bits wide, and can transfer data in burst mode. The input buffer can accommodate two vectors at a time. Output data may be in either IEEE standard 32-bit floating or 16-bit floating point formats. The control registers are accessible by the host, and by the Ni1000 system, and provide an avenue for exchange of status and command information between the host and the Ni1000. For applications where larger prototype vocabularies are required, the Ni1000 circuits may be cascaded in parallel, sharing access to the host system, as shown in Figure 60.9. The Ni1000 is fabricated in a 0.8 micron CHMOS technology and housed in a PGA package with 168 pins, of which over a third are redundant power and ground connections. The Ni1000, as well as the hardware, software, and technical support available from Nestor represents the culmination of fifteen years of development by a highly capable and motivated team of mathematicians, computer scientists, and engineers.

60.5 Discussion

In describing the history and variety of commercially available ANN implementations in VLSI, a number of factors are apparent. First and foremost is the astonishing computational power and efficiency they offer. This level of performance is a product of the high clock speeds afforded by state-of-the-art wafer fabrication technology and the parallel and simultaneous nature by which calculations are conducted. The majority of contemporary ANN realizations are virtual, taking place in computer-based simulations. The majority of computers used in these simulations utilize a Von Neumann, or inherently serial, architecture and cannot begin to approach the billions of connections per second featured in hardware-based implementations. One should note that these figures are quoted as absolute capacities, derived under conditions of ideal delivery and offloading of data. In cases where the application does not exactly fit the capacity of the ANN, or where the host processor cannot supply data at commensurate rates, this figure will be lower.

Mention of constraints on the type of problem for which VLSI ANNs may reach their full computational potential should not be misconstrued with those for which ANNs may successfully be applied. In fact, commercial ANN ICs have been used with success by the U.S. Postal Service, the U.S. Internal Revenue Service, and thousands of institutions with applications in finance, commerce, education, and the government.

The sheer computational power and versatility offered by the standalone hardware-based ANN, however, is offset by the up-front expertise and resources required to design, manufacture,

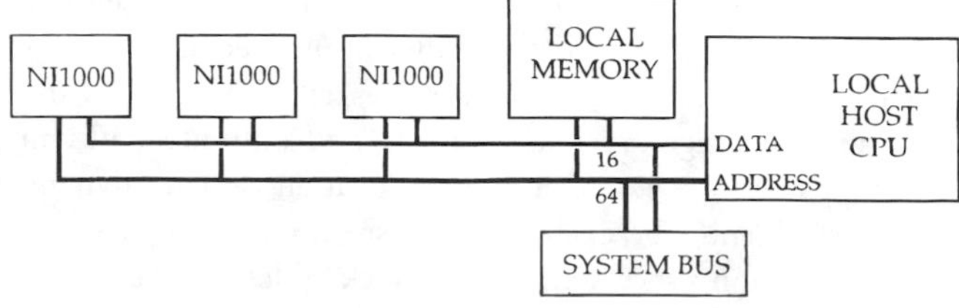

Figure 60.9 Organization of a Nestor Ni1000 recognition accelerator system.

and support products that are expected to compete in today's market. End-users getting started in the products described in this chapter should expect to invest thousands of dollars, and mobilize expertise in ANNs, computer hardware, and programming. Contemporary ANN ICs are fabricated exclusively in sophisticated sub-micron CMOS technologies, with many millions of gates per chip and complex designed-in circuit redundancy.

The products described in this chapter represent a mere handful of survivors of a promising, yet demanding market. A number of fledgeling enterprises, as well as large corporations have dropped out of this market. Synaptics, a company that was co-founded by Carver Mead once offered the OCR (optical character recognition) IC. It is no longer commercially available. Siemens, an international corporation with considerable human and financial resources for design and manufacturing published a digital neural signal processor IC in 1992 (Ramacher, 1992), but has not brought it to market. Similarly, Mitsubishi's BNU digital ANN IC was described in a 1992 paper (Arima, et al., 1992), and plans to make it available, but is not offering product information. Intel has since dropped its line of ETANNs, deferring this technology to Nestor. ANNs in VLSI is clearly a volatile and demanding market.

Finally, the potential of IC ANNs for image processing, pattern recognition, and computation becomes staggering. Able to process billions of interconnections per second for hundreds of neuronal units, each with hundreds of connections, these products offer a processing rate that dwarfs that of living neuronal systems, which operate in the millisecond range. The potential to dramatically impact such a broad range of technical pursuits can hardly be overstated.

References

Alspector, J., Gupta, B., and Allen, R. 1989. *Performance of a Stochastic Learning Microchip, Advances in Neural Information Processing Systems I,* Touretzky, D., ed, 748–760, Morgan Kaufmann, San Mateo, CA.

Arima, Y., Mashiko, K., Okada, K., Yamada, T., Maeda, A., Nontani, H., Kondoh, H., and Kayano, S. 1992. 336-neuron 28K synapse self-learning neural network chip with branch-neuron-unit architecture, *IEEE J. Solid-State Circuits,* 26:1637–1644.

Boser, B., Sackinger, E., Bromley J., LeCun, Y., and Jackel L. 1992. Hardware requirements for neural network pattern classifiers, *IEEE Micro,* 12:32–40.

Hamerstrom, D., 1990. A VLSI architecture for high-performance, low-cost, on-chip learning, *IEEE-INNS Int. Joint Conf. Neural Networks,* 2:537–544, San Diego, CA.

Hecht-Nielsen, R. 1990. *Neurocomputing,* Addison-Wesley, Reading, MA.

Hopfield, J. 1982. Neural networks and physical systems with emergent collective computational ability, *Proc. Nat. Academy of Sciences of the U.S.A.,* vol 79:2554.

Johnson, R., and Brown, C. 1988. *Cognizers: Neural Networks and Machines That Think,* John Wiley & Sons, New York, N.Y.

Maren, A., Harston, C., and Pap, R. 1990. *Handbook of Neural Computing Applications,* Academic Press, San Diego, CA.

Mead, C. 1989. *Analog VLSI and Neural Systems,* Addison-Wesley, Reading, MA.

Ramacher, U. 1992. SYNAPSE—A neurocomputer that synthesizes neural algorithms on a parallel systolic engine, *J. Parallel and Distributed Computing,* 14:306–318.

Sackinger, E., Boser, B., and Jackel, L. 1992. A neuocomputer board based on the ANNA neural network chip, *Advances in Neural Information Processing Systems 4,* Moody, J. E., Harrison, S. J., Hanson, S. J., and Lippmann, R. P., eds, 773–780, Morgan Kaufman, San Mateo, CA.

Sivilatti M., Mahowald, M., and Mead C. 1987. Real-time visual computations using analog CMOS processing arrays *Advanced Research in VLSI: Proceedings of the 1987 Stanford Conference,* 295–312, Losleben, ed., MIT Press, Cambridge, MA.

Watanabe, T., Kimura, K., Aoki, M., Sakata, T., and Ito, K. 1993. A single 1.5-V digital chip for a 10^6 synapse neural network, *IEEE Trans. Neural Networks,* 4:387–393.

Zurada, J. 1992. *Introduction to Artificial Neural Systems,* West, St. Paul, MN,

61

Implementing Neural Networks in Silicon

Seth Wolpert
Penn State University at Harrisburg

Evangelia Micheli-Tzanakou
Rutgers University

61.1 Introduction 874
61.2 The Living Neuron 874
61.3 Neuromorphic Models 875
61.4 Neurological Process Modeling 881

61.1 Introduction

In spite of dramatic increases in the capacity and throughput of automated systems, a number of descriptively simple, yet highly desirable, tasks remain elusive. These tasks are associated with the process known as pattern recognition. If machines were able to identify patterns in electrical, visual, mechanical, acoustic, or chemical signals as quickly and reliably as living systems, our world would be a very different place. A number of tedious operations could be performed tirelessly and accurately. We would no longer need locks on our automobiles and homes, or keyboards on our computers. For many years, engineers and mathematicians have worked to perform computer-based pattern recognition using geometric and statistical methods, but levels of accuracy commensurate with those of human operators have been difficult to obtain. To address the overwhelming utility to perform these tasks, engineers have begun to take cues from biological systems, the simplest of which are able to perform pattern recognition with relative ease and high reliability, as a matter of their very survival.

To deal with the sheer magnitude of living nervous systems, in-roads have been taken to understand their workings by "top-down" and "bottom-up" approaches. The top-down approach is based on outward observations of capability and behavior. This approach has given rise to the field of Artificial Neural Networks, or ANNs. ANNs are based upon simplified models of individual neurons, which are highly interconnected via an array of variably coupled transmission units known as synapses. Such systems are generally implemented as computer models, and have been most effective when configured and controlled when tailored to a specific pattern, method of assimilation and processing, and identification criteria. Generally implemented as a computer simulation, these networks acquire data, train themselves, and evaluate possible solutions serially, and therefore, require an inordinate amount of time and computational resources to function as well as traditional non-ANN pattern recognition methods. Clearly, these ANN methods have the potential to easily surpass conventional methods, but to do so, they must be transplanted from the virtual environment within a serial computer to a dedicated hardware platform, where they may be implemented in a parallel and simultaneous manner. This is consistent with theories of how living nerve circuits operate so quickly and reliably, and introduces the motivation for pursuing bottom-up approaches, based on observations of the structure and function of individual nerve cells.

61.2 The Living Neuron

Living nerve cells have always been studied and modeled to the limits of available electronic technology. As early as the 19th century, electrical models of processes observed in living nerve cells have undergone ardent development. The justification has been that since machines were unable to achieve the same tasks that living systems could do so easily, perhaps we could emulate living systems and put them to work in a number of endeavors. After all, the brain of an Einstein or a Shakespeare is not significantly different in structure or composition than the average human brain. The computational potential of the average human brain, then, must be remarkable, and an electronic model, which functions 100,000 times faster would hold great potential as a computer for a variety of applications. The problem with this objective, however, has been in the sheer magnitude of the machinery. Consisting of 100 billion nerve cells, many of which have many thousands of interconnections, the human brain is a machine far beyond the analysis, design and manufacturing capacities of any existing human technology. If such a "machine", or even a small part of it were to be replicated, however, such an effort must begin on the cellular level. To describe these efforts, a review of nerve cell structure and function is in order.

A typical living cell is depicted in Figure 61.1. Physically, it may be described as a tentacled elastic sac, whose interior and exterior are bathed in different conductive fluids separated from each other by the cell's outer material known as *cell membrane*. This membrane draws upon the cell's metabolic processes to

0-8493-8343-9/97/$0.00+$.50

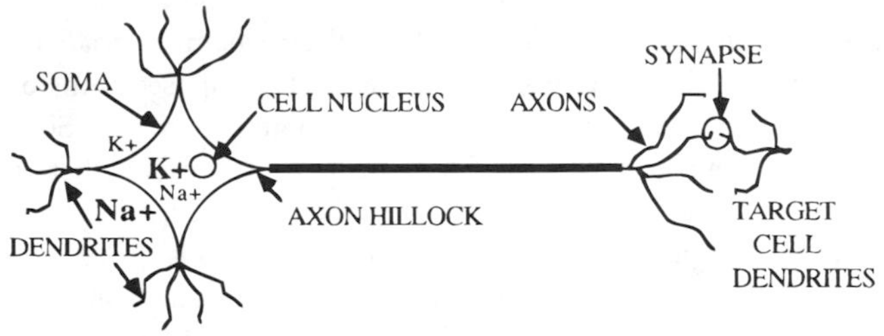

Figure 61.1 Depiction of a typical neuron showing physical structure and distributions of sodium and potassium ions inside and outside the cell membrane.

supply the energy it requires to accumulate specific ions against concentration gradients. Two key ions, potassium, which is accumulated inside the cell, and sodium, which is ejected to the exterior of the cell, have been identified as having the strongest role in the function of the nerve cell. The imbalance in distribution of these ions and several others forms the basis for an electrical potential within the cell relative to the conductive environment outside. This potential, known as *cell membrane potential,* rests nominally at 60–80 mV negative with respect to its external environment. Fluctuations in this cell membrane potential are the means by which neurons express their activity and communicate with sensory receptor cells, muscle cells, and other neurons.

The tentacles that branch off from the neuron's body, or *soma,* are known as *dendrites.* Dendrites, which may number from zero to well into the thousands, branch off to other cells and collect sensory input signals based on those *cells'* levels of activity. These signals appear as electrical transients in membrane potential, which are accumulated over time and space, with the resultant sum appearing in the cell soma. Emanating from the soma is another singular tentacle known as the *axon.* Typically larger, longer, and better insulated than the dendrites, the axon conveys the output of its cell over long or short distances to target nerve and muscle cells. The point at which the axon attaches to the cell body is known as the *axon hillock.* There, the accumulated cell membrane potential is compared against a cellular *threshold* potential. When that threshold potential is exceeded, a separate mechanism in the axonal membrane gives rise to a single impulse, typically 80–100 mV in amplitude, and 1 ms in duration. This impulse is then propagated down the axon to its remote terminus, where individual fibers branch off and adjoin target nerve or muscle cells. While such an impulse is being generated, the cell enters a temporary state of total inexcitability, where no amount of stimulation can cause a second impulse to be superimposed over the first. This state soon elapses, and the cell gradually returns to an excitable condition. This phenomenon is known as *refraction* and the interval of inexcitability is known as the *refractory period.*

Between the axonal terminus and the soma or dendrite of the target cell, a fluid gap forms a *synapse,* a physical discontinuity from one cell to the next. At the axonal terminal, a packet of chemicals is released into the synaptic gap, where it will migrate to the target cell, and induce transient impulses in that target *cell's* membrane potential. These chemicals are known as neurotransmitter, of which over twenty different types have been identified. Neurotransmitters that induce negative transients in target cell membrane potential are known as inhibitory, while those that induce positive transients in target cell membrane potential are known as excitatory. Inhibitory stimuli suppress activity in target cells, while excitatory stimuli facilitate activity in target cells. Cells that induce large transients are said to have high synaptic weights, while those inducing little or no transients in target cells are said to have low synaptic weights. The magnitude or the duration of that transient may be affected by the synaptic weight, and changes in synaptic form the basis for training of ANNs, as well as learning in living nervous systems. The synapse also prevents reflection of impulses back to source cells, which would cause unbridled chaos to engulf the entire nervous system in a very short time.

Orchestrating the modification of synaptic weight in a network of cells learning to perform a new task or recognize a sensory image is the basis for top-down neuronal study. For bottom-up study, two other aspects of the operation of living neurons are of particular interest to those modeling its function: formulating the threshold of a nerve cell in terms of the spatial and temporal distribution of stimuli directed toward it and the relationship between conductivity of the membrane to the ions giving rise to membrane potential, present membrane potential magnitude, and time. These two aspects have been the bases for modeling individual nerve cells from two schools of thought.

61.3 Neuromorphic Models

Since the era of the vacuum tube, a multitude of neuronal models composed of discrete components and off-the-shelf ICs have been published. Similar efforts in custom VLSI, however, are far fewer in number. A good introduction to a number of neuronal attributes, however, was presented by Linares-Barranco et al. of Texas A & M University (1990). CMOS compatible circuits for approximating a number of mathematical models of cell behavior are described. In its simplest form, this model represents the cell membrane potential in the axon hillock as nothing more than a linear combination of an arbitrary number, *n,* of dendritic inputs, *X,* each of which is weighted by a unique multiplier, *W,* summed, and processed by a non-linear range-limiting operator, *f.* The mathematical equation for this relationship is:

$$Y_k = f\left\{\sum_{i=1}^{n} W_i X_i\right\} = f\{S_k\} \tag{61.1}$$

and this relationship is realized in the circuit model shown in Figure 61.2a, and the CMOS circuit implementation in Figure 61.2b. This circuit is totally static, and makes no provision for time-courses of changes in input or output signals, or intracellular relationships. In the implementation of Figure 61.2b, the *operational transconductance amplifier,* OTA, as described in Wolpert and Micheli-Tzanakou (1993) and depicted in Figure 61.3, is used in lieu of operational amplifiers for this and most

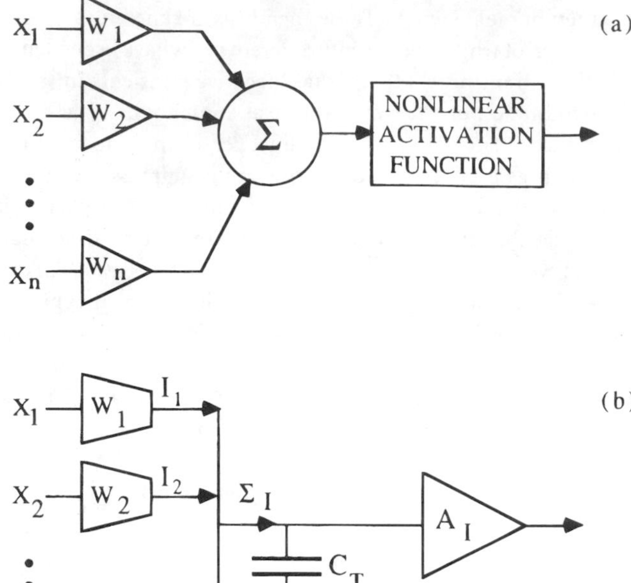

Figure 61.2 Circuit organization of a general purpose neuronal model (a), and a CMOS VLSI circuit implementation of such a model (b).

other VLSI neural network applications. Highly compatible with CMOS circuit technology, it is structurally simple and compact, realizable with only nine transistors, and provides reasonable performance. The only consideration it warrants is that its transfer function is a transconductance. As such, operations performed on its output signals must be oriented to its current, rather than its voltage. When driving high load impedances, as is usually the case with CMOS circuitry, this is only a minor inconvenience, necessitating buffering for lower load impedances. In fact, under some circumstances, such as when algebraic summation is performed, a current output actually may be an advantage, allowing output nodes to be simply tied together.

The nonlinear range-limiting operator, f, mentioned earlier is necessitated by the observation that, for a given biological neuron, there are limits on the strength of the electrochemical gradients that the *cell's* ionic pumps can generate. This imposes limits on how positive and negative cell membrane potential may go. Since a neuron may receive inputs from many other neurons, there is no such limit on the aggregate input voltage applied. As a result, an *activation function,* a nonlinearity of the relationship between aggregate input potential and output potential of a neuron, must be imposed. This is done typically in one of three different ways: the binary hard-limiter, which assumes one of only two possible states-active or inactive; the linear-graded threshold, which assumes a linear continuum of active states between its minimal and maximal values; the sigmoid, which assumes a sigmoidal distribution of values between its negative minimal and positive maximal output values. All three of these relationships are shown in the graphs of output potential versus input potential of Figures 61.4a, b, and c, respectively.

Which type of activation function is employed depends on the type of artificial neuron and network in which it is implemented. In networks where cell outputs are all-or-none, such as McCullouch and Pitts models (1943), the binary threshold model is used. In networks where neurons are theorized to have variable output levels applied to distinctly designated excitatory and inhibitory inputs, such as Hopfield networks, the linear threshold model is used. In networks where a synaptic connection must be both excitatory and inhibitory, depending on the level of activity, the sigmoid threshold is used. In either of the latter two activation functions, the slope of the overall characteristic can be varied to suit the sensitivity of the cell in question.

The basic neuron cell model shown in Figure 61.2a was designed for primitive neuronal models and learning algorithms. It performs linear summation of independently weighted synaptic inputs applied to a single node, and discriminates according to a binary threshold of zero, as shown in Figure 61.4a. Although the linear combination is an easy process to comprehend, its fidelity in the face of biological nerve behavior is restricted. To improve the applicability of such models, several improvements must be made to their mathematical descriptions. The first such improvement is the dynamic model. Like the model described by Equation 61.1, it includes linear combination, summation and a nonlinear operator—in this case, the sigmoidal activation function. Consistent with transconductance amplifiers, however, its output is now expressed as a current in the form of CdS/dt. It also features I_B, which represents a fixed biasing current that determines a baseline level of activity, or threshold. This activity

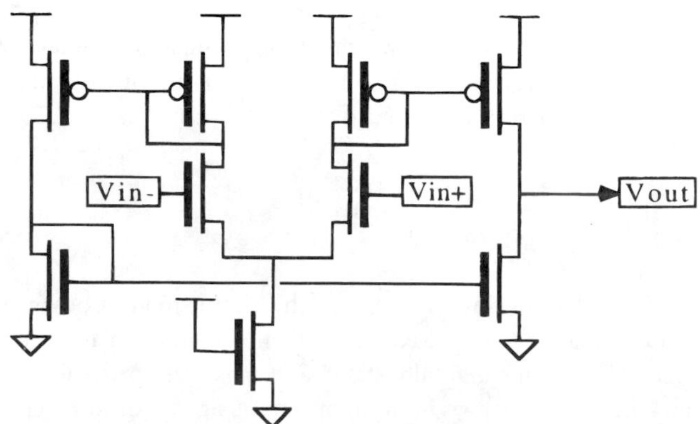

Figure 61.3 CMOS implementation of an operational transconductance amplifier, OTA, widely used for realization of neuronal models and networks in VLSI.

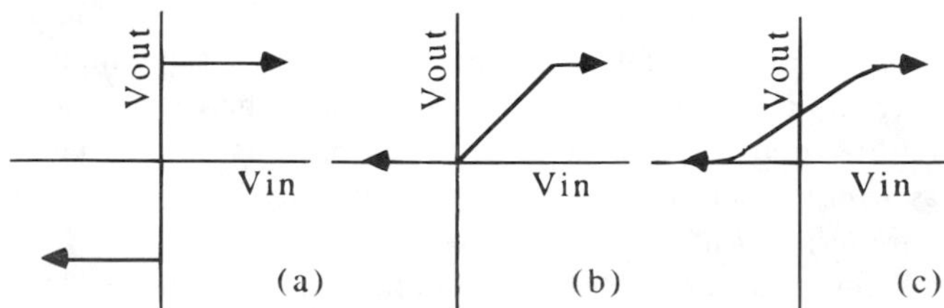

Figure 61.4 Nonlinear activation functions imposed on outputs of nerve cell models. Shown are the binary bipolar hard limiter (a), the linear graded potential (b), and the sigmoid potential (c).

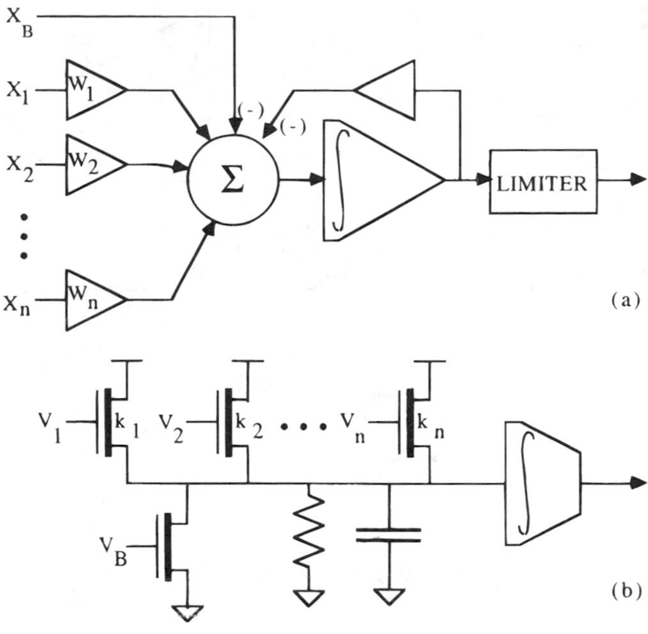

Figure 61.5 Circuit organization of a dynamic cell model (a) and a CMOS VLSI implementation of such a model (b).

level represents a threshold that must be surpassed by the aggregate sum of weighted inputs in order for cell k to respond. Different cells may be assigned different thresholds, so that their responsiveness may be tuned to the demands of the network in which it resides. Finally, the dynamic model includes R, a self-relaxation term that insures that S, the cell output potential will decay to zero when all dendritic inputs, X, are zero. The dynamic model is implemented using a "leaky integrator", which allows for the duration, or persistence of input signals to be controlled. The equation for this behavior is:

$$C\frac{dS_k}{dt} = I_{B\bar{k}}\frac{S_k}{R_k} + \sum_{i=1}^{n} W_{ik}f(S_i) \qquad (61.2)$$

A mathematical model of this equation is given in Figure 61.5a, and the CMOS implementation in Figure 61.5b.

More comprehensive features to facilitate functioning in a large population of nerve cells have been incorporated into the generalized model described by Gail Carpenter and Stephen Grossberg of Boston University (Grossberg, 1988). This model includes the features of the dynamic model, as well as a more comprehensive facility for temporal summation with the self-forgetting, or persistence term, A_k. The H and L coefficients allow for the fixed and output voltage-dependent levels of activity in the network to be controlled. This keeps the network's signals from saturating at too high or low an overall level of activity. The E coefficient represents a fixed applied bias signal analogous to the I_B term of Equation 61.2. The Z coefficient represents a synaptic coupling analogous to W of Equation 61.2. Mathematically, the equation for the generalized model is given as:

$$\frac{dS_k}{dt} = -A_kS_k + (H_k - L_kS_k)\left\{E_k + \sum_{i=1}^{n} Z_{ik}f(S_i)\right\} \qquad (61.3)$$

and the model, along with a CMOS implementation are given in Figures 61.6a and 61.6b, respectively.

This model is comprehensive enough to be appropriate for implementing artificial neural networks that realize adaptive resonance theory (ART), as well as Hopfield networks, and McCulloch and Pitts networks, but still lacks one vital attribute of the living neuron. All of the cell models presented so far portray neurons as simplified cells whose output is expressed as a DC level that reflects some nonlinear function of the aggregate sum of input signals. This forms the basis for most ANN implementations. There is also a class of cell models whose output is is a train of similar pulses whose frequency is varied, rather than a variable DC potential. For these frequency modulated models, there are also a series of circuit implementations.

Frequency-based neuronal models are similar to those already presented, in that they perform temporal and spatial summation of an arbitrary number of dendritic inputs, as well as their own current state of activity. They will also have activation functions assigned depending on the type of cell and network. Unlike the activation functions of voltage-based models, these are imposed in recognition of the fact that, for a given biological neuron, an action potential cannot be elicited during the formation of its predecessor. This manifests itself as a limit on how close together in time two action potentials may occur from a given cell, and therefore, a limit on the maximal frequency at which a neuron can generate pulses. A neuron may receive inputs from many other neurons. While each of those inputs has a similar upper limit on its frequency, there is no such limit on the number of inputs, and therefore no limit on the overall input frequency. As a result, a nonlinearity of the relationship between input frequency and output frequency of a neuron must be imposed. In most cases, this type of behavior may be brought about with the simple addition of a voltage-controlled oscillator, or VCO to the

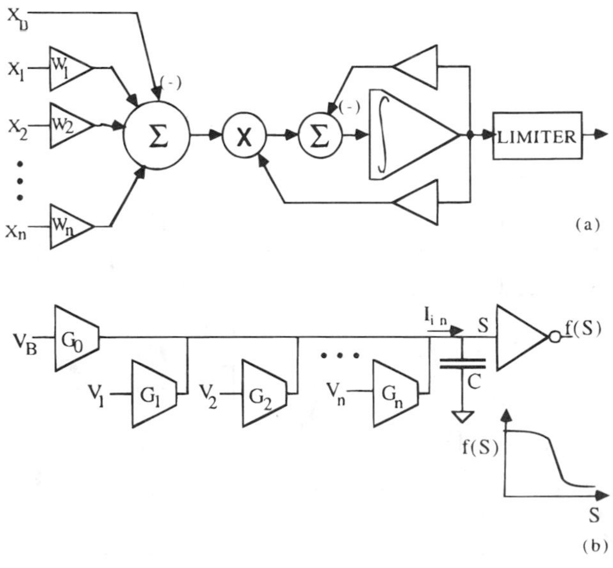

Figure 61.6 Circuit organization of a dynamic cell model providing unconstrained assimilation of excitatory and inhibitory inputs (a), and a CMOS implementation of that model (b).

output stage of one of the previously defined models with an activation function operator.

There are two CMOS VLSI implementations of oscillatory models of note, both of which are derived from the system of differential equations formulated by Hodgkin and Huxley (1952). In the course of producing an action potential, the neuronal cell membrane exhibits conductances to Sodium and Potassium ions that were found to be mathematical functions of time and of cell membrane potential. The Hodgkin-Huxley equations were derived to describe those time and voltage relationships. A popular circuit approach to realizing the oscillatory behavior required to synthesize a single pulse from one control input is to employ a hysteretic output stage. The organization of such a system is shown in Figure 61.7a, along with a CMOS circuit implementation in Figure 61.7b. It is apparent that this is a simple adaptation of the dynamic model shown in Figure 61.5. The other approach to recreating such a circuit instability is in a Hodgkin-Huxley derivative known as the Fitzhugh-Nagumo model (Fitzhugh, 1961). Based on a mutually antagonistic relationship between two cells and an I-V characteristic outwardly similar to that of a tunnel diode, the Fitzhugh-Nagumo model is somewhat more complex, but still realizable in conventional CMOS VLSI subcircuits. The model for this circuit is given in Figure 61.8, and it formed the basis for one of the more successful CMOS VLSI implementations of single-neuron models.

In 1991, Bernabe' Linares-Barranco et al. (1991) of Texas A & M University fabricated and characterized a circuit whose

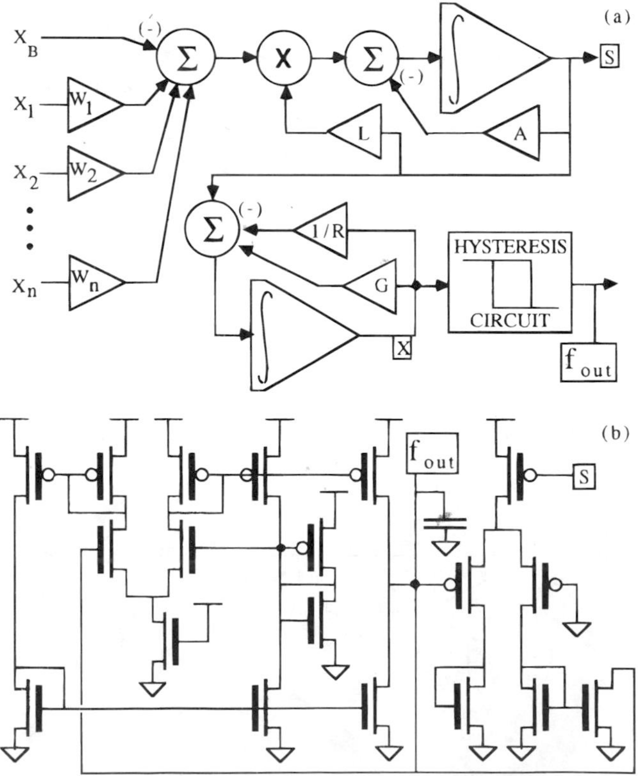

Figure 61.7 Circuit organization for a dynamic cell model producing variable-frequency output signals by the application of a hysteresis subcircuit (a), and a CMOS implementation of that model (b).

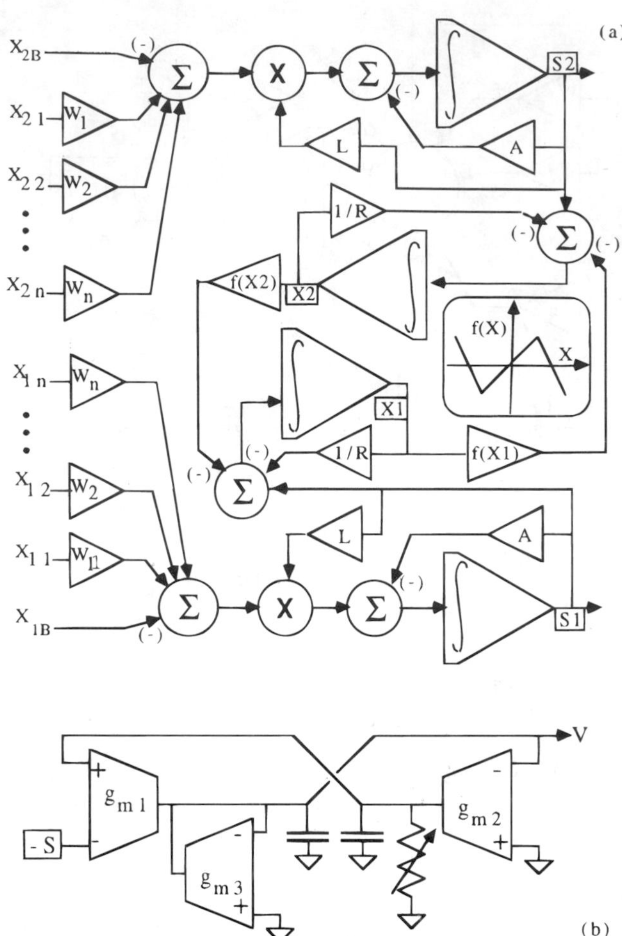

Figure 61.8 Circuit organization of a model of the Fitzhugh-Nagumo equation (a) and a CMOS implementation of the model (b).

behavior is based on the Fitzhugh and Nagumo equations. The variability membrane conductance characterized by the Hodgkin-Huxley equation was recreated as a piece-wise linear model, which was realized empirically using the circuit of Figure 61.8b. A series of OTAs whose transconductances correspond to the membrane ionic conductances over specified input voltage ranges were used to realize the transients in membrane conductance that give rise to the action potential. Fabricated prototypes were demonstrated to replicate several types of behavior commonly seen in living nerve cells, i.e., free-running sustained oscillation in a single cell, and on-and-off, or *bursting* oscillation, as seen in a pair of mutually antagonistic cells. For both circuit configurations, oscilloscope photographs appear similar, albeit less noisy than intracellular recordings from live nerve cells.

Along similar lines, another CMOS implementation was developed by Misha Mahowald of the Computation and Neural Systems Laboratory at Cal-Tech, and Rodney Douglas, of Oxford University in 1988 (Mahowald and Douglas, 1991). In this model, the time course of sodium and potassium currents are recreated empirically, by virtue of fundamental similarities between ionic conductivity in neural membrane and that of appropriately biased MOSFETs. Structurally simple, yet elegant circuits shown in Figure 61.9a, b, and c recreate the time and voltage courses of

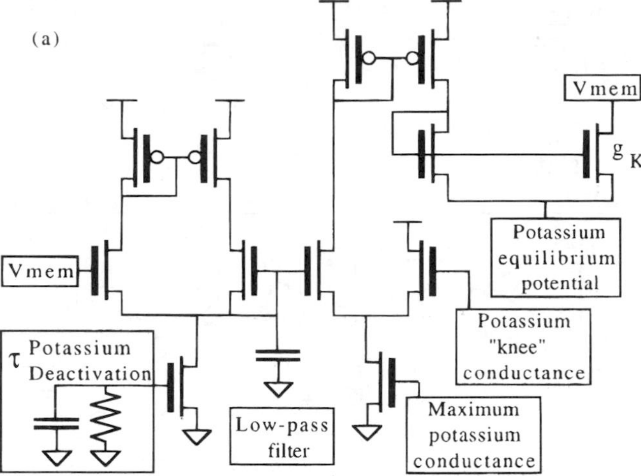

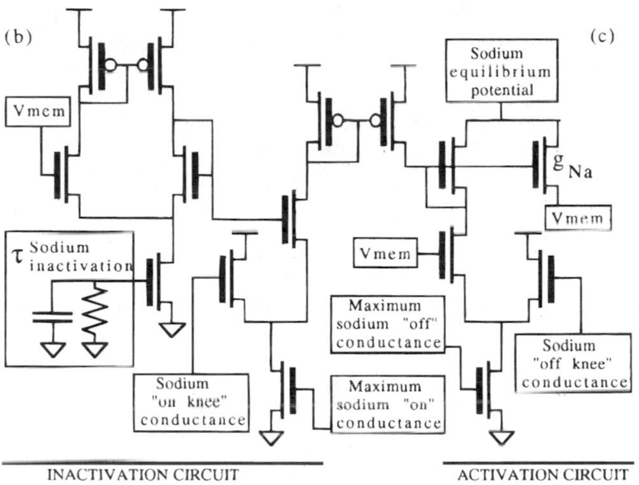

Figure 61.9 Circuits described by Mahowald and Douglas to realize sodium and potassium currents in active membrane. The potassium activation circuit is shown in part (a), the sodium activation is shown in part (b), and the sodium inactivation is shown in part (c).

potassium activation, sodium activation, and sodium inactivation respectively in neural membrane. Rectangular current pulses of various amplitudes applied to the circuit show a striking similarity to similar impulses applied to living nerve cells. The circuit is highly compatible with larger scale applications, requiring minimal off-chip support, occupying under 0.1 mm^2 of chip area, consuming under 60 μW of power, and able to operate a million times faster than their biological counterparts. With the incorporation of a dendritic array, networks of several hundred nerve cell analogs on a single chip have been envisioned.

Another well-executed implementation of VLSI-based nerve cells complements the Mahowald-Douglas model, concentrating less on overall nerve cell behavior, and more specifically on how inputs to a neuron combine over time and space to affect a target cell (Elias, 1993; Northmore and Elias). Temporal and spatial summation, and some topical applications have been modeled extensively in CMOS VLSI by John Elias and David Northmore of the University of Delaware. Recognizing that the strength, duration, and delay of a neuronal stimulus depend strongly on the physical location to which that stimulus is applied, Elias and Northmore recreate a linearly arrayed multicompartmental silicon dendrite in which each segment, or compartment, has a specific capacitance to the *cell's* exterior, *Cm,* impedance of the internal fluid, or, cytoplasm, *Ra,* and impedance of a leakage path to the *cell's* exterior, *Rm.* Implemented using on-chip switched-capacitor analog networks, the authors demonstrate impulses that can persist millions of times longer that the impulses from which they originated. They also showed a mechanism by which a target cell's sensitivity may be keyed to any of a wide range of impulse shapes, durations, latencies, directional velocities, and repetition frequencies, as applied at various locations along a dendritic tree (topographic connection), or across a dendritic tree (laminar connection). The design of distributed compartments and their incorporation into a dendritic tree is shown in Figure 61.10a.

The facility of such networks to recognize specific spatial and temporal frequencies in arbitrary images was then applied to a VLSI-based system for recognition of binarized two-dimensional visual images. Due to the large number of possible input sites to a dendritic tree contained in a 40-pin IC package, a multiplexed approach was taken to transmission of data on and off-chip. For the two-dimensional input images, one dimension is applied to topographic connections of the dendritic tree, and the other

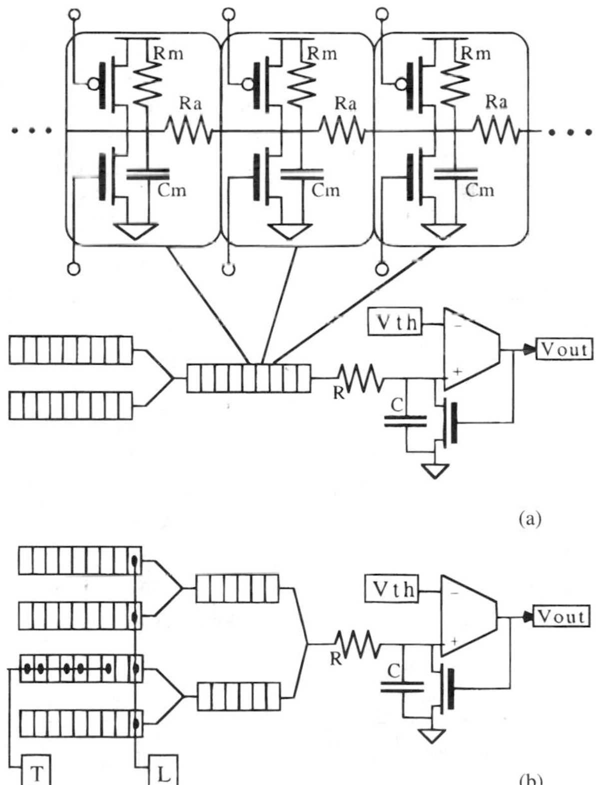

Figure 61.10 Circuits used to realize dendritic trees by Elias and Northmore. Dendritic compartment circuitry and the organization of compartments into a dendritic tree are shown in part (a), and the application of laminar and topographic summation in a dendritic tree are shown in part (b).

dimension is applied to laminar connections of the tree. As the image is scanned into the dendritic tree, spatial summation of the laminar inputs and temporal summation of the topological inputs results in a synchronized response unique to the pattern of the input image. Depictions of topographic and laminar connections to a dendritic tree are given in Figure 61.10b. The remainder of the circuitry in the implementation is associated with encoding and transferring data and synaptic coefficients. Dendritic trees of higher dimensions may be used to recognize images of higher dimension, and lateral inhibition and other real-time image processing operations are highly compatible with this method.

Another well-developed implementation of individual artificial nerve cells is the one by Wolpert and Tzanakou (1996, 1986) of Rutgers University. While most neuromorphic models are based on the Hodgkin-Huxley equations, this one uses a sequencer to synthesize the action potential in three distinct phases. It also employs a different formulation for cell membrane and threshold potentials known as an integrate-and-fire model, presented and implemented in discrete components by French and Stein (1988) in 1974. It uses the aforementioned leaky integrator, and provides off-chip control over the response and persistence of stimuli assimilated into membrane potential. The model affords similar controls over the resting level and time constant of the cell threshold potential, and allows for refraction to be recreated. This organization also affords control over the shape, resting level, and duration of the action potential, and produces a TTL-compatible pulse in parallel with the action potential. These controls, all of which are continuously and precisely adjustable, make this model ideal for replicating the behavior of a wide variety of individual nerve cells, and it has been successfully applied as such. The organization for the French and Stein model is shown in Figure 61.11, and the Wolpert and Tzanakou VLSI circuit was implemented as shown in Figure 61.12.

The Wolpert and Tzanakou model is organized around three critical nodes, the somatic potential, the axonal potential, and the threshold potential. Each of these nodes is biased off-chip with an R-C network so that its resting level and time constant are independently and continuously controllable. Stimuli to the cell are buffered and standardized by truncation into 10 μs impulses. Synaptic weight inputs on the excitatory and inhibitory pathways allow for this value to be increased or decreased from off-chip. The impulses are then applied to somatic potential by a push-pull MOSFET stage, and compared to threshold potential by an OTA acting as a conventional voltage comparator. When

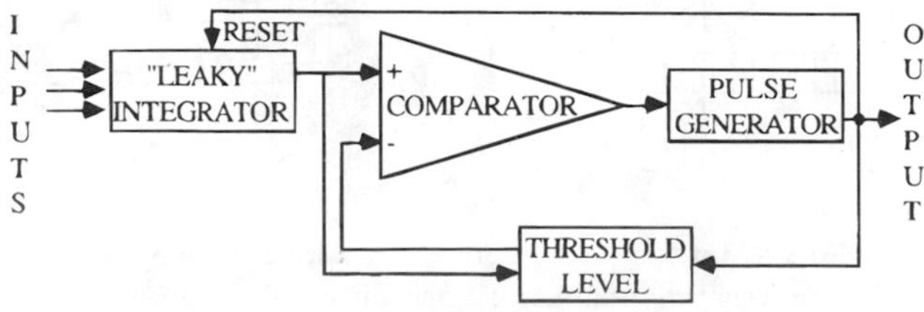

Figure 61.11 Organization of the "integrate and fire" model of neuronal behavior described by French and Stein.

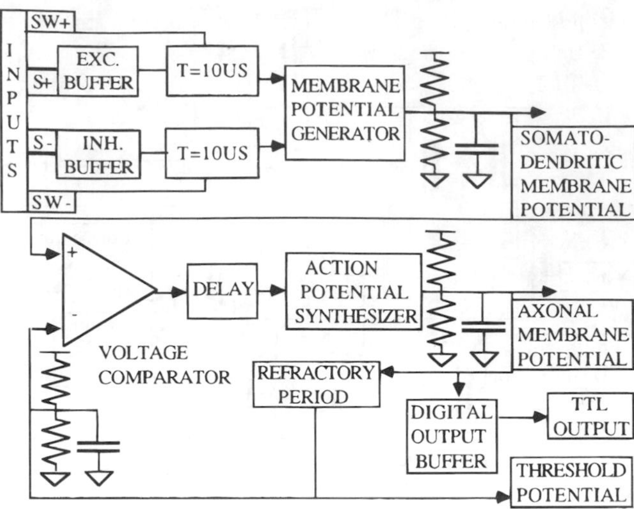

Figure 61.12 Implementation of the French and Stein model used for the VLSI prototype of the artificial nerve cell described by Wolpert and Tzanakou.

threshold is exceeded, an action potential is synthesized and outputted. This waveform is then binarized and buffered to form a binary-compatible output pulse. Also at the same time, threshold is elevated to form the refractory period. The circuit consists of approximately 130 transistors plus a few on chip and discrete resistors and capacitors, and was implemented in a conventional CMOS technology, requiring a single-ended DC supply of 4–10 volts DC, and occupying 0.6 mm^2 of chip area.

With its critical nodes bonded out off-chip, the Wolpert-Tzanakou neuromime's rate of operation may be accelerated from a biologically compatible time frame over several orders of magnitude. This model was first implemented in 1986, and is intended as a flexible and accurate aesthetic, rather than a mathematical model of cell behavior. Since then, it has been used successfully to recreate a number of networks from well-documented biological sources. Waveforms obtained in these recreations have shown a striking similarity to intracellular recordings taken from their biological counterparts. It has also been applied successfully to problems in robotics and rehabilitation.

Another well-conceived VLSI-based model of neuronal response is a hybrid neural processing element, PE described by DeYong, Findley, and Fields (1992) of New Mexico State University. Running at nominal CMOS VLSI speeds, and having no need for internal nodes representing membrane and threshold potentials, this implementation requires far fewer components and is therefore much more appropriate for large-scale implementations in VLSI. In this model, each of the synaptic types, excitatory, inhibitory, and shunting, is implementable using seven transistors or less, and variability in synaptic weight costs an additional five transistors per synapse. The accumulated somatic potential is then applied to an axon hillock circuit, which performs threshold discrimination, and generates an action potential pulse from under twenty transistors. This circuit has many of the features of the Wolpert-Tzanakou model, including an arbitrary number of excitatory, inhibitory, and shunting inputs, a tangible threshold potential node, and biologically aesthetic waveforms, even though their durations and amplitudes are oriented to conventional analog and digital circuitry. The circuit is used to realize a one-by-four-celled laterally inhibited winner-take-all network, which is of particular interest in pattern recognition operations, where the known pattern that is most similar to the unknown image is singled out over the remainder of less secure match candidates. Finally, models of neuronal function may be radically simplified to a voltage-controlled oscillator. This function may be realized in large quantity using a minimalist model known as the NTC, or neural-type cell.

Recognizing that a neuron may be described as a voltage-driven pulse generator, Moon and Zaghloul of George Washington University and Newcomb (1992) of the University of Maryland have been developing and applying NTCs to various problems in artificial neural networks. The description of a neuron as a VCO is one that can be implemented as a small circuit of three MOSFETs, three resistors, and a capacitor, as shown in Figure 61.13. Although the circuit does not oscillate over a wide range of frequencies, and its output frequency is not linearly related to its input level, its simplicity, small number of outward connections and compact size make the NTC appropriate for implementation in large quantities. With the replacement of R6 with a voltage-controlled variable resistor, this circuit is able to assimilate variable synaptic weight, as manifested by a variable duty cycle on its output waveform. This circuit may also be tuned to function over a wide range of operating frequencies, as controlled by R6 and C. The NTC and the other VLSI circuits presented so far have all been conceived with the intent or replicating one or more aspects of nerve cell behavior. There are also many efforts directed at modeling cell-to-cell interactions, as theorized and observed in living nervous systems.

61.4 Neurological Process Modeling

The modeling of interaction between nerve cells has been pursued most widely with respect to problems in image processing and computation. Image processing applications were pioneered mostly in VLSI form by Carver Mead of Cal-Tech. Dr. Mead is one of the world's leading educators and implementers of VLSI; the models of vision and audition he has developed focus on the simultaneous and immediate preprocessing of sensory images that is believed to take place before interpretation. The most common processing step is known as lateral, reciprocal, or mutual inhibition; modeling of sensory processes that use lateral inhibition has been foremost in neural process modeling. Computation, on the other hand, encompasses system control, pattern recognition, clustering, and prediction. The latter three of these topics will be discussed in more detail in a future chapter. The former has several VLSI applications, one of which is as a general-purpose servo element.

DeWeerth and Mead of Cal-Tech, and Nielsen and Astrom of the Lund Institute of Technology in Sweden, have implemented a simple servo controller in custom (DeWeerth et al., 1991). The authors recognize that human tissues possess friction, elasticity, and internal damping, yet are capable of precise positioning and movement due to the presence of copious feedback and redundancy. Such a precise control system surely can provide excellent positional and motional resolution to electromechanical systems, as well. The OTA of Figure 61.3 was modified with the addition of a second, parallel output, whose current is the complement of the primary output. In addition, the biasing transistor, whose gate was depicted as being tied to V_{DD} in Figure 61.3, now has its gate tied to a DC reference input voltage, V_b. This input serves as an overall gain control for each OTA.

To implement the servo system, a number of these OTAs are connected with their corresponding outputs in parallel, as shown in Figure 61.14. As such, they represent a number of independently weighted synaptic inputs, whose outputs saturate as they approach maximal and minimal levels, forming a sigmoidal activation function. The aggregate complementary output currents are then pulse width modulated. The complementary pair of variable duty cycle pulse trains that result are then buffered and applied directly to the terminals of a bidirectional DC motor. When system conditions demand motion in the positive direction, its synapses approach positive output currents, and the pulse train to the positive terminal of the motor approaches 100% while the duty cycle of the pulse train to the negative terminal approaches zero. This effects rotation in the positive

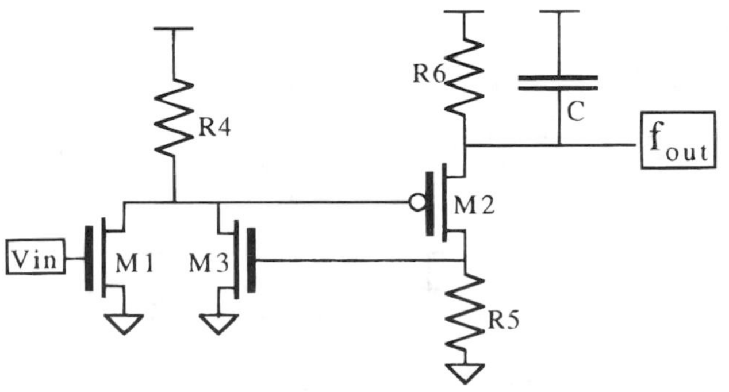

Figure 61.13 Schematic of the neural-type cell described by Moon, Zaghloul, and Newcomb.

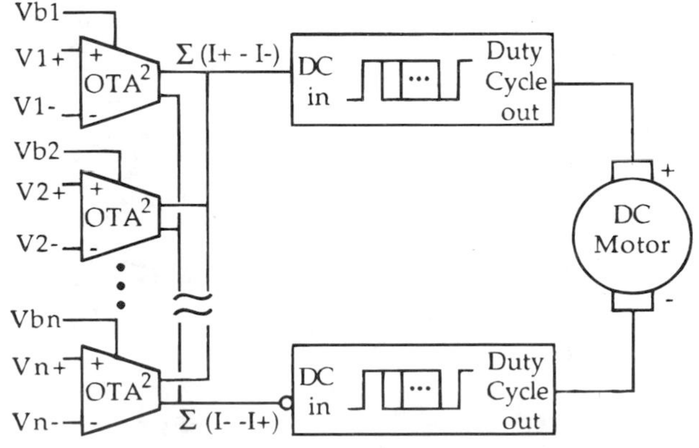

Figure 61.14 Organization of the VLSI neuron servo. System control directives are inputted to OTAs fitted with complementary outputs. Aggregate positive and negative motion directives are pulse width modulated, buffered, and fed to the input terminals of a reversible DC motor.

direction. When full positive and negative motion are invoked, the motor turns rapidly, yet when the positive and negative pulse trains are roughly equal, there is a very fine resolution of motor control. In the servo system, the complementary outputs assume an agonistic/antagonistic relationship, where one signal exists at the expense of the other, and both signals cannot coexist simultaneously. This mutually inhibitory relationship is a frequently recurring theme in a wide variety of living nervous systems in a variety of organisms, and has been modeled by a number of researchers.

Lateral inhibition is the process in which a cell containing some level of information encoded as its output level acts to inhibit, and is inhibited by a similar adjoining cell, as depicted in Figure 61.15. For many years, this process has been observed with striking regularity in both one- and two-dimensional arrays of sensory receptors in a variety of systems, in a variety of organisms. In numerous morphological, mathematical, and circuit studies, it has been identified as a key image preprocessing step which optimizes a sensory image in order to facilitate fast and accurate recognition in subsequent operations. Lateral inhibition accomplishes this by amplifying differences, enhancing image contrast, lending definition to its outward shape and isolating the image from its background. While a digital computer would accomplish this process one pixel at a time, biological systems manage it in a manner that is both immediate and simultaneous.

Laterally inhibited behavior has been observed in pairs of cells implemented in hardware and software models by many researchers, but in dedicated VLSI by only a few. Notable among them, Nabet of Drexel University, and Pinter and Darling of University of the Washington have studied extensively the stability and effectiveness of both pairs and linear strings of mutually inhibiting cells in CMOS VLSI and have obtained results well correlated with biological data (Nabet and Pinter, 1991). This line of work has been explored in two dimensions in another series of VLSI-based models by Wolpert and Micheli-Tzanakou (1993) of Rutgers University. Arrays of mutually inhibiting cells that inhibit via continuously active connections and cells that inhibit by dynamic, or strobed controls were both found to offer stable and variable control over the degree of inhibition. Arrays of hexagonally interconnected cells were found to be more stable than the square array, which tended to "checkerboard" when significant levels of inhibition were attempted. Feedback inhibition, where one array is used to store both the initial and inhibited images was found to be as effective, but less convenient to access than feed-forward inhibition, where separate input and inhibited images are maintained.

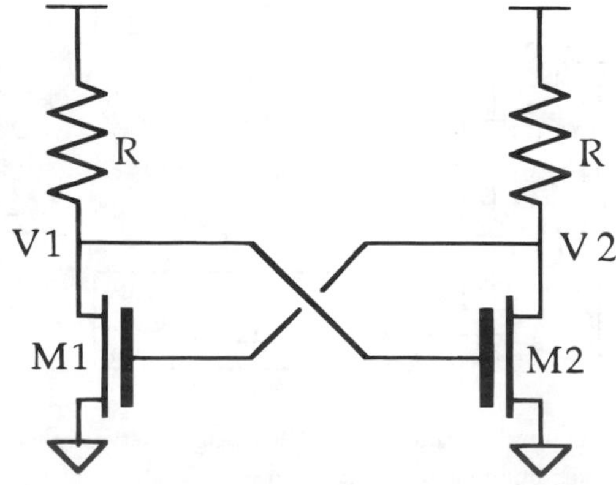

Figure 61.15 A pair of mutually inhibiting circuit nodes.

Characterization of lateral inhibition in the context of a more specific biological model has been pursued in another noteworthy effort by Andreou of Johns Hopkins University and Boahen of Cal-Tech. Multiple facets of cell-cell interactions, including both mutual inhibition and leakage of information between adjoining cells were implemented in VLSI, as a model of early visual processing in the mammalian retina (Andreou et al., 1991). There, adjacent cells on the photoreceptor layer intercommunicate through gap junctions, where their cell membrane potentials couple through a resistive path. Simultaneously, optical information from the photoreceptor cells are downloaded to corresponding cells of the horizontal layer, which have been shown to have mutually inhibitory connections. This interaction is illustrated in Figure 61.16. One-dimensional arrays, and subsequently, two-dimensional models of these relationships were implemented in analog VLSI and tested. Although little numerical data were published from these arrays, the two-dimensional array was demonstrated to produce a number of optical effects associated with the human visual system, including Mach bands, simultaneous contrast enhancement, and the Herman-Herring illusion, all of which are indicative of the real-time image processing known to occur in the mammalian retina.

Finally, the definitive VLSI implementation of a two-dimensional array is the well-known silicon retina devised by Carver Mead of Cal-Tech, and described in his text *Analog VLSI and Neural Systems* (Mead, 1991). In addition to presenting a comprehensive treasury of analog VLSI circuits for a variety of mathematical operations necessary to implement neural networks in VLSI, the book presents several applications of analog ANNs, culminating in an auditory model of the cochlea and a visual model of the retina.

The "silicon retina" is built around a 48-by-48-cell array of photosensors on a microchip. This array is then overlaid with a

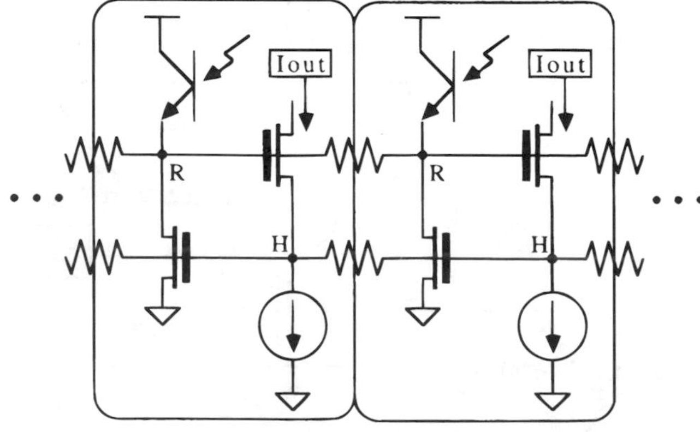

Figure 61.16 A mutually inhibitory pair, as implemented in analog VLSI by Andreou et al. (1991). This model represents the highly interconnected photoreceptor and horizontal cells of the retina, as indicated by the R and H nodes, respectively.

grid of resistors that replicates the gap junctions of the cells of the photoreceptor layer. Also incorporated into the array is a network of amplifiers, whose inputs are drawn from each adjoining node. The output of these amplifiers is an image that represents the Laplacian of the image, which replicates the mutual inhibition inherent in the horizontal cell layer. Because there are more pixels in the array than pins on the IC package that houses it, individual pixel data must be conveyed off-chip by an analog decoder/multiplexer. In tests, the circuit was shown to possess temporal and spatial response similar to those of living retinas, as evidenced by its recreation of a number of optical illusions associated with human vision. Since its initial description in 1988, many interesting modifications to the silicon retina have been implemented by Mead's students in the Computation and Neural Systems Laboratory at Cal-Tech.

An on-chip photoreceptor capable of transducing visual light over six orders of magnitude was implemented and published by Delbruck. They also developed a motion-sensitive silicon retina, which reacts to moving, rather than stationary objects. This phenomenon has been observed many times in living retinas in a variety of organisms. Directional sensitivity was then applied to this principle by Delbruck and Benson. Velocity-sensitivity was later implemented by Delbruck, as well as the facility to optimize the focus of an image onto the surface of a chip by means of a distributed system of differentiators, a maximizer, and a servo mechanism to control positioning of an optical lens over the chip. This, along with intrinsic electronic control over contrast and brightness constitutes a crucial first step in implementing a totally parallel visual system. This same objective has been brought to fruition in auditory system modeling, which has resulted in a number of commercial products now on the market.

A custom CMOS VLSI model of the human middle and inner ear has been implemented by Liu, Andreou, and Goldstein (1992) of Johns Hopkins University. The eardrum and bones of the middle ear are modeled as a fifth-order low-pass filter with a second order pole at 15 kHz, and a third order pole at 100 kHz. The cochlea is modeled as a bank of thirty second-order band-pass filters, whose Q and center frequency are tuned by on-chip resistors and DC bias voltages. The hair cells of the cochlea perform nonlinear transformations and dynamic range compression, and are modeled by a series of 128 analog switches, voltage dividers, and voltage comparators.

The voltage comparison in this model is accomplished with the use of the OTA shown in Figure 61.3. The active filters, on the other hand, required an amplifier with a higher output impedance. This is because the circuits were operating in their subthreshold region. Subthreshold operation is used when the Vgs of a MOSFET is varied between zero and its specified threshold voltage. Transistors biased in this region pass small, although coherently exponential drain-source currents. These currents are necessitated by the extremely long time constants of the human audio spectrum and the limited value of on-chip capacitors. A schematic for the modified subthreshold OTA with heightened output impedance is shown in Figure 61.17. Electrical signals in the VLSI implementation measured in response to sinusoidal tones correlate quite well to signals recorded under similar conditions in the auditory nerve of the cat. The overall organization of the system is shown in Figure 61.18. The technology for cochlear modeling has resulted in dramatic progress, not only in research efforts, but in commercial endeavors, as well.

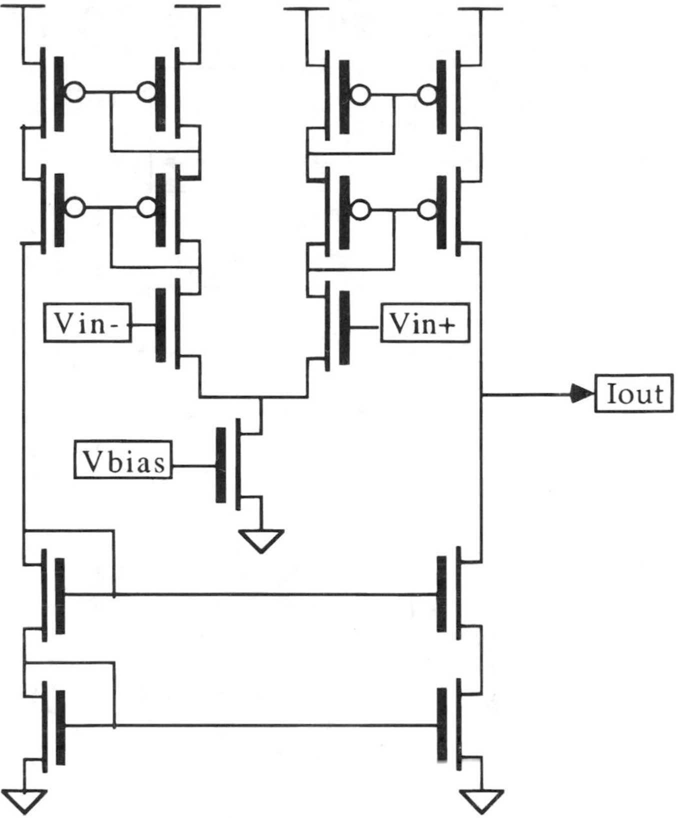

Figure 61.17 Operational transconductance amplifier adapted for use in subthreshold mode and elevated output impedance.

A good deal of commercial success has been made in the area of cochlear implants. Deafness in humans is the result of a number of possible pathologies. In cases where deafness occurs due to physical damage to the structures of the outer and middle ear, those structures may be augmented by a surgically implantable microchip that decomposes incoming sounds into the fundamental and harmonic frequencies of which they are composed. The output of the device is then a multipolar electrode that applies various frequency signals to various points on the basilar membrane of the intact cochlea. In devices such as the implant manufactured by Cochlear Corp., a microphone is worn on the outside of the ear in much the same way as a hearing aid. The microphone then decomposes the sound wave into frequency bands, and transmits them through the skin to the implanted device in the form of a modulated radio frequency signal. The implanted device then rectifies the information signal to derive the power required to run its internal circuitry and to provide electrical stimulation of the appropriate region of the cochlear membrane. This eliminates the need for internal batteries, which may pose a health hazard due to their chemical contents or the surgical procedures required to install and replace them.

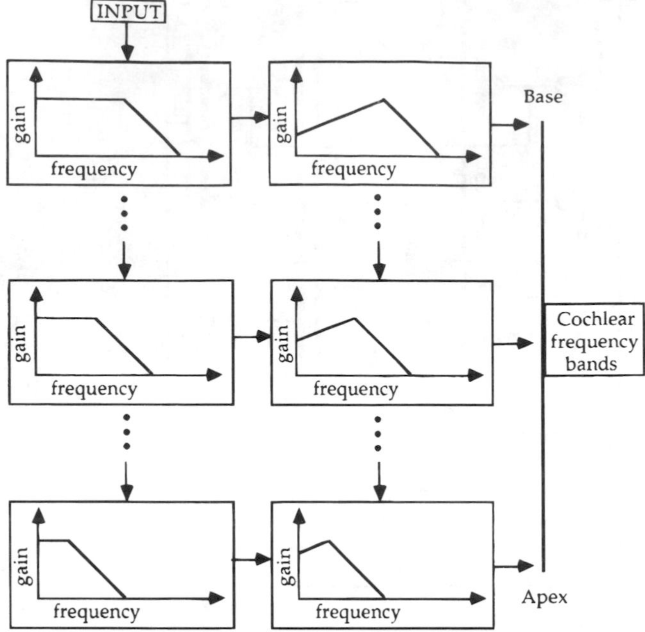

Figure 61.18 Organization of a silicon-based model of the auditory periphery, as described in Liu, et al., 1992. Input sound waves are low-pass filtered to simulate the eardrum and bones of the middle ear, and then bank-filtered to simulate the function of the cochlea. Finally, their dynamic range is compressed to simulate the hair cells.

While cochlear implants do not restore total hearing, they do impart the ability to receive sounds which aid in lipreading and overall awareness of the auditory environment. The cochlear implant also represents the leading edge in cybernetic implants-blazing the path for devices to augment hearing, vision, and sensation and movement, both visceral and somatic.

Clearly, the living nervous system has been the inspiration for a substantial amount of engineering development. Astounded and reassured by the utter throughput, reliability, and robustness of living sensory systems, engineers, mathematicians, neuroscientists, and computer scientists have doggedly endeavored to understand the workings of living nervous systems. Much progress has been made in understanding and recreating the structure and function of individual nerve cells. From morphological and electrophysiological study, progress has also been made in understanding the structure of small nerve circuits. Ahead lies the most profound frontier—that of the algorithms and control over activity in the brain.

References

Andreaou, A. G., Boahen, K. A., Pouliquen, P. O., Pavasovic, A., Jenkins, R. E., and Strohbehn, K. 1991. Current-mode Sb threshold MOS circuits for analog VLSI neural systems, *IEEE Trans. Neural Systems,* vol. 2, no. 2, March.

DeWeerth, S. F., Nielsen, L., Mead, C. A., and Astrom, K. J. 1991. A simple neuron servo, *IEEE Trans. Neural Networks,* vol. 2, no. 2, March.

DeYoung, M. R., Findley, R. L., and Fields, C. 1992. The design, fabrication, and test of a new VLSI hybrid analog-digital neural processing element, *IEEE Trans. Neural Networks,* vol. 3, no. 3, May.

Elias, J. G. 1993. Artificial dendritic trees, *Neural Computation,* 5:648–663.

Fitzhugh, R. 1961. Impulses and physiological states in theoretical models of nerve membrane, *Biophys. J.,* vol. 1.

Grossberg, S. 1988. Nonlinear neural networks: principles, mechanisms, and architectures, *Neural Networks,* 1:17–61.

Hodgkin, A. L. and Huxley, A. F. 1952. A qualitative description of membrane current and its application to conducting and excitation in nerves, *J. Phys.,* 177:500–544.

Linares-Barranco, B., Sanchez-Sinencio, E., and Rodriguez-Vazquez, A. 1990. CMOS circuit implementations for neuron models, *Proc. IEEE Int. Symp. Circuits and Systems,* 3:2421–2424, May, New Orleans, LA.

Linares-Barranco, B., Sanchez-Sinencio, E., Rodriguez-Vazquez, A., and Huertas, J. L. 1991. A CMOS implementation of Fitzhugh-Nagumo model, *IEEE J. Solid-State Circuits,* 26(7), July.

Liu, W., Andreaou, A. G., and Goldstein, M. H. 1992. Voiced-speech representation by an analog silicon model of the auditory periphery, *IEEE Trans. Neural Networks,* vol. 3, no. 3, May.

Mahowald, M. and Douglas, R. 1991. A silicon neuron, *Nature,* 354:515–518.

McCullouch, W. S. and Pitts, W. 1943. A logical calculus of the ideas imminent in nervous activity, *Bulletin of Mathematical Biophysics,* 5:115–133.

Mead, C. A. 1989. *Analog VLSI and Neural Systems,* Addison-Wesley, Reading, MA.

Moon, G., Zaghloul, M. E., and Newcomb, R. W. 1992. VLSI implementation of synaptic weighting and summing in pulse coded neural-type cells, *IEEE Trans. Neural Networks,* 3(3):394–403.

Nabet, B. and Pinter, R. B. 1991. *Sensory Neural Networks: Lateral Inhibition,* CRC Press, Boca Raton, FL.

Northmore, D. P. and Elias, J. G. Evolving synaptic connections for a silicon neuromorph, *Proc. IEEE World Congress on Computational Intelligence.*

Wolpert, S. and Micheli-Tzanakou, E. 1986. An integrated circuit realization of a neuronal model, *Proc. IEEE Northeast Bioeng. Conf.,* March 13–14, New Haven, CT.

Wolpert, S. and Micheli-Tzanakou, E. 1993. Silicon models of lateral inhibition, *IEEE Trans. Neural Networks.*

Wolpert, S. and Micheli-Tzanakou, E. A neuromime in VSLI, *Trans. Neural Networks,* submitted.

62

An Avionics Application: MIMD Neural Network Processor

Richard Saeks
Accurate Automation Corp.

62.1 NNP Architecture 885
62.2 Summary 887

The accurate automation neural network processor (NNP)[1] shown in Figure 62.1 is an MIMD parallel processor designed around a "linked list-like" parallel processing architecture which allows one to implement virtually any neural network paradigm or training methodology with high efficiency. A full MIMD parallel processing architecture is implemented with one processing element per NNP and up 8 NNPs per system. The neural network processor supports:

- 8K 16 bit neurons; 32K 16 bit connection weights and 4 user defined 14-bit-by-16-bit transfer functions per processor,
- Each processor runs at 35 MHz. performing 140,000,000 connections (8 bit multiply/accumulates) per processor per second for a total of more that a billion connection per second in an 8-processor system.

The NNP is designed to operate as a coprocessor with a standard microprocessor or DSP chip pre- and postprocessing the neural network data and serving as the master processor for the NNP.

The accurate automation neural network processor is available for PC/ISA and VME bus systems. The PC/ISA board is designed to use the PCs 80×86 processor as its master processor and includes one built-in NNP with sockets for 7 additional NNPs. The VME board includes built-in DSP support (two TI TMS320-C40s) and is designed to support up to 3 NNPs with either one of DSP processors or a VME host processor serving as master.

The NNP is supported by a full suite of software tools including an assembler which implements a special purpose neural network design language and a subroutine library (in C or ADA) to facilitate communications between the master processor and the NNP system. Additionally, a (UNIX or DOS) neural network toolbox is provided to facilitate the off-line design and training of neural networks.

62.1 NNP Architecture

A schematic diagram of the AAC neural network processor is shown in Figure 62.2 while the details of its interprocessor bus architecture are shown in Figure 62.3. The program and weight memories in Figure 62.2 combine to form a 32K-by-32-bit instruction store which defines the architecture of the neural network paradigm being implemented and the required connection weights. The computational heart of the processor is the 16-bit-by-16-bit MAC (multiplier/accumulator) while the 64K-by-16-bit-function memory holds four 16K by 16 bit lookup tables which can be programmed to define four neural network transfer functions. Indeed, by defining the neural network paradigm and its transfer functions in software the NNP can be readily programmed to implement any of the standard neural network paradigms and/or hybrids thereof.

The key to the NNPs parallel processing performance is the interprocessor bus and the 8K-by-16-bit neuron and buffer memories shown in Figure 62.3. Indeed, each processor has two data memories, a neuron memory which operates in a read-only mode and a buffer memory which operates in a write-only mode, both of which are replicated in every processor. On any given layer (or iteration) of a network the processors each read from their local copy of the neuron memory, compute the activations and neuron values for the next layer, and then simultaneously write the result to the buffer memories of all processors via the FIFO and the interprocessor bus.

Assuming that the neuron and buffer memories in the various processors are initialized identically, they remain identical throughout the process, since the neuron memory operates in a read-only mode while all writes to the buffer memory are done simultaneously via the interprocessor bus to the buffer memories of all processors. As such, perfect memory coherence is maintained. Once all computations on the given layer are

[1] The development of the Accurate Automation Corp. Neural Network Processor was supported by the U.S. Office of Naval Reserves under Phase II SBIR contract N00014–91–C–0268.

0-8493-8343-9/97/$0.00+$.50

Figure 62.1 Accurate automation PC/ISA neural network processor.

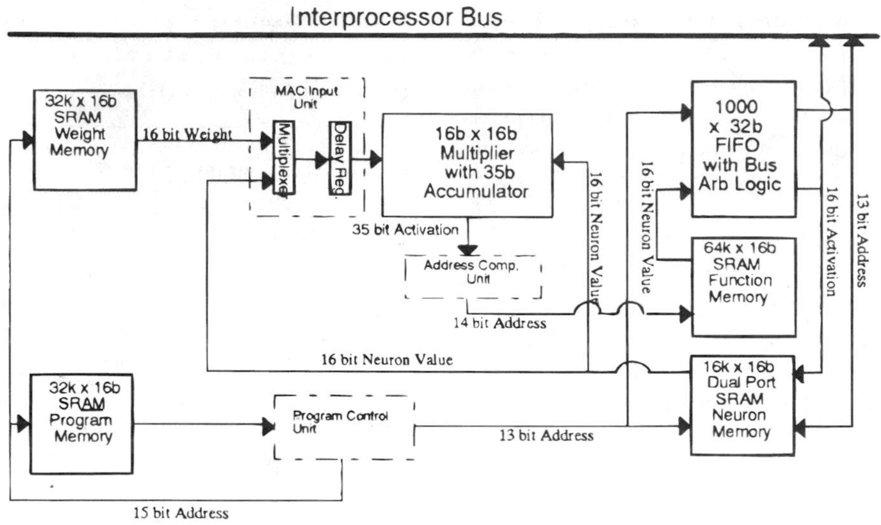

Figure 62.2 Schematic diagram of the AAC neural network processor.

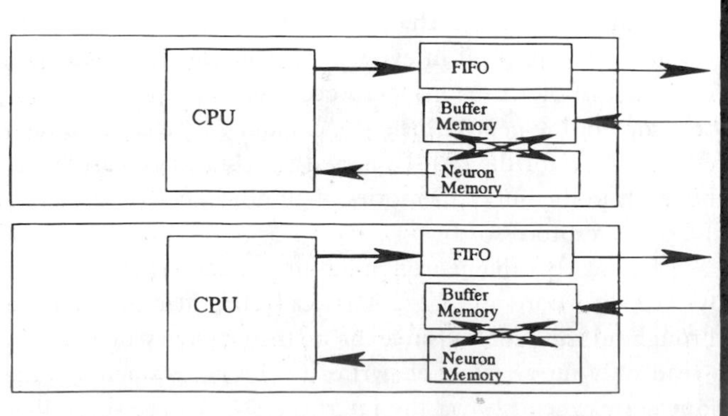

Figure 62.3 NNP interprocessor bus architecture.

completed the processor executes an instruction to interchange the neuron and buffer memories after which it pauses. When all processors have executed such an instruction, the memories in all processors are interchanged simultaneously and computation resumes using the new data now stored in the neuron memory, i.e., the data that was stored in the buffer memories before the interchange.

Although this bus architecture is seemingly expensive, requiring duplicate neuron and buffer memories in each processor, it guarantees perfect memory coherence and simultaneously minimizes bus contention. Indeed, since every processor has a local read-only copy of the neuron memory, no contention whatsoever is generated by the read operations which typically dominate the bus activity in a neural network processor. On average each processor only requests access to the bus once every $f + 1$ cycles (where f is the average fan-in of the neural network), whereas on a worst case basis it takes 4 clock cycles for a processor to gain control of the bus and write a neuron value to the buffer memories. As such, to a first order approximation, one can efficiently employ $p = (f + 1)/4$ processors without encountering bus contention as illustrated in Figure 62.4.

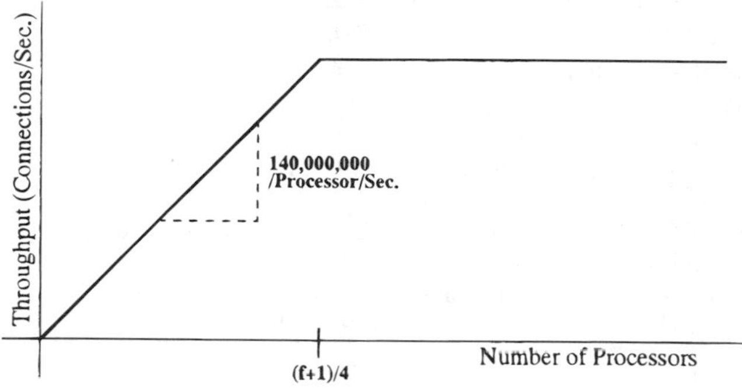

Figure 62.4 NNP throughput as a function of the number of processors.

62.2 Summary

The accurate automation neural network processor (NNP) may be compared with chips covered in the article by Lau. The AAC-NNP has the following attributes:

- PE/Chip: 1
- Connectivity: broadcast fully connected
- Precision: 16-bit integer
- Algorithms: neural networks, vector and sparse matrix
- Commercial applications: fault diagnostics and control
- Defense applications: radar signal processing

Wind tunnel tests recently validated the design of an 8-foot-long model of a neural-controlled hypersonic aircraft. The jet-powered, 24-foot-long full-sized aircraft is scheduled for testing in 1997. Ten of the extensible AAC-NNP chips are mounted on a VMEbus board capable of 1.4 billion cps to fly the hybersonic aircraft. The chip and supporting software are commercially available. Numerous applications of neural technology can be implemented with the AAC-NNP.

63

Backpropagation to Neurocontrol

Paul J. Werbos
*National Science Foundation**

63.1 Neurocontrol: Where It Is Going and Why It Is Crucial 888
Basic Definitions • Why Neurocontrol Is Crucial to Intelligence • A Basic Roadmap of Neurocontrol • What's New in Cloning (Supervised Control) • Tracking Methods: What's New and What's What • Backpropogating Utility • Adaptive Critics/Approximate Dynamic Programming • Practical Advantages of Brainlike Control • Recent Accomplishments in Brainlike Intelligent Control • Supervised Learning: What It Does, Applications, and How to Do It • Neuroidentification: What It Is, Applications, and How to Do It

Recent progress in neurocontrol will be surveyed in the first part of this article, *Neurocontrol: Where It Is Going and Why It Is Crucial.* More detail will be provided in *Supervised Learning: What It Does, Applications, and How to Do It.* Some of the latter material is written mainly for the beginner, but points will be clarified which have confused even the experts. Next, *Neuroidentification: What It Is, Applications and How to Do It* will be covered. All this material leads into the chapter *Neurocontrol and Elastic Fuzzy Logic: Capabilities, Concepts, and Applications* in the Fuzzy Systems and Soft Computing section of this Handbook.

63.1 Neurocontrol: Where It Is Going and Why It Is Crucial (Werbos, 1992)

In the past few years, enormous progress has been made by a relatively small group of researchers in developing and understanding new kinds of neural network designs which show real promise in explaining and replicating "intelligence" as we see it in biological organisms. These new designs come from the emerging field of neurocontrol. Already, there has been substantial real-world success in the control of robot arms (including the main arm of the space shuttle), in chemical process control, and in the continuous production of high-quality composite parts. New benchmark problems and early successes suggest that the neurocontrol may become crucial to the development of hypersonic flight, which in turn may be crucial to the cost-effective settlement of outer space and to the use of hydrogen instead of oil in aviation. (Hypersonic craft may be able to reach orbit as airplanes, at low cost). Environmental and automotive applications now appear very important. On the other hand, there have been many failures and many reinventions of the wheel due to inadequate appreciation of what has already been done and how it relates to control theory. This paper will discuss the goals of neurocontrol and then describe some applications in the context of a roadmap stretching from the past through to new opportunities to build truly intelligent systems in the future.

Basic Definitions

Neurocontrol is the use of neural networks—artificial or natural—to directly control actions intended to produce a physical result in a world which changes over time.

In many cases, neural networks are thought of entirely as artificial neural networks (ANNs) designed to perform a task called supervised learning. They are thought of as systems which are always given a database of input vectors, $\boldsymbol{X}(t)$, and target vectors, $\boldsymbol{Y}(t)$, for $t = 1$ to T. Neural networks are thought of as systems which implement a non-linear mapping, $\boldsymbol{Y}(t) = f(\boldsymbol{X}(t), W)$, which is "learned" by adapting the weights W either in real-time (t) or in batch mode, offline. "Basic research" is thought of as the development of new supervised learning designs or the analysis of such designs. Control is thought of as one of many applications areas. However, neurocontrol is an area of basic research in its own right. It calls for new types of ANNs, operating at a different level, performing different kinds of tasks.

Many control theorists have been impressed by the many theorems showing that neural networks used in supervised learning can approximate any well-behaved function f to an arbitrary degree of accuracy. They have been impressed by theorems saying that one can approximate even ill-behaved functions, like those required in some control applications, if they contain two hidden

* The views herein are those of the author, not the views of NSF. This article is a modification of Werbos (1992)—which was in the public domain—updated on the author's own personal time. Some parts of the article come from Werbos (1995a), and others come from Werbos (1991a)

0-8493-8343-9/97/$0.00+$.50

layers or simultaneous recurrence (which can emulate two hidden layers as a special case). (See Chapter 91.) But it is argued—with justice—that real-world control problems (like those addressed by biological organisms) involve a time dimension. Instead of seeking the optimal mapping from $\boldsymbol{X}(t)$ to $\boldsymbol{Y}(t)$, one needs to seek optimal maps of the form:

$$\hat{\boldsymbol{Y}}(t) = f_Y(\boldsymbol{X}(t), \boldsymbol{Y}(t-1), \boldsymbol{R}(t-1), W)$$
$$\boldsymbol{R}(t) = f_R(\boldsymbol{X}(t), \boldsymbol{Y}(t-1), \boldsymbol{R}(t-1), W) \qquad (63.1)$$

where f_Y and f_R represent two vector outputs of a single network f, and where $\boldsymbol{R}$ (for "recurrent" or "reality") is a kind of vector of memories. This is equivalent to the problem of system identification in control theory where we try to adapt a model network which predicts $\boldsymbol{X}(t+1)$:

$$\hat{\boldsymbol{X}}(t+1) = f_X(\boldsymbol{X}(t), \mathbf{u}(t), \boldsymbol{R}(t), W)$$
$$\boldsymbol{R}(t+1) = f_R(\boldsymbol{X}(t), \boldsymbol{u}(t), \boldsymbol{R}(t), W) \qquad (63.2)$$

where $\boldsymbol{X}(t)$ represents sensor data observed at time t and $\boldsymbol{u}(t)$ represents actions we take after observing $\boldsymbol{X}(t)$. The vector $\boldsymbol{R}$ essentially estimates the extra information we need (beyond $\boldsymbol{u}$ and $\boldsymbol{X}$) to specify the system state at time t.

Supervised learning is defined as the task of adapting neural networks to learn a static mapping from $\boldsymbol{X}(t)$ to $\boldsymbol{Y}(t)$. (Again, there may be dynamics or recurrence inside the network used to output a prediction of $\boldsymbol{Y}(t)$ as a function of $\boldsymbol{X}(t)$, but no dependence of $\boldsymbol{X}(t')$ on earlier times t'.) *Neuroidentification* may be defined as the effort to adapt neural networks of the form shown in Equation 63.1 or 63.2 (with the possibility of additional time lags and noise models). Neuroidentification is not a special case of supervised learning, for many reasons; for example, the vectors $\boldsymbol{R}(t)$ are not known in advance. Nevertheless, neuroidentification *is* a special case of system identification as defined in control theory. (Thus many of the concepts in White and Sofge (1992, Chapter 10) apply to both neural and nonneural systems.) *Neurocontrol* may be defined as the effort to build (or formulate) systems which include an adapted action network:

$$\boldsymbol{u}(t) = \boldsymbol{A}(\boldsymbol{X}(t), \boldsymbol{R}(t), W) \qquad (63.3)$$

Systems such as Equations 63.1 or 63.2 are a variety of "recurrent network," but they are very different from classical recurrent networks such as Cohen-Grossberg or Hopfield nets. Much of the conventional wisdom about recurrent networks is unreliable because it fails to distinguish between these very different kinds of recurrence and the many different ways of implementing each (White and Sofge, 1992, Chapter 3; Werbos and Pang, 1996).

Why Neurocontrol Is Crucial to Intelligence

The human brain, as a whole system, is clearly not a supervised learning system. It clearly is a "computer," an information processing system. The function of any computer, as a whole system, is to compute its outputs. The outputs of the brain as a whole system are actions. Therefore, the human brain as a whole system fits the definition of a neurocontroller, as given above. To understand the brain as a whole system, one must first understand neurocontrol; one must understand how it is possible to build (or to exist) a neurocontroller with the kinds of capabilities that the brain possesses. Clearly, these include very sophisticated capabilities, such as planning and problem-solving and foresight and the like. (The logic of this paragraph does not tell one what kind of neurocontroller the brain is; in fact, it tends to remind one that the field of neurocontrol still contains some unknown territory in need of greater research.)

Within the brain, one knows that there are subsystems and phenomena such as memory, pattern recognition, etc. But one cannot really hope to understand these subsystems until we know what their functions are. One cannot understand the functions of a subsystem until one knows how it fits in to the design of the whole system; therefore, once again, an understanding of neurocontrol is a prerequisite.

From the viewpoint of control, the brain is living proof that it is possible to build a generalized controller which takes full advantage of parallel distributed hardware, which can handle many thousands of actuators (muscle fibers) in parallel, which can handle noise and nonlinearity, and which can achieve goals or optimize over a long-range planning horizon. This proves that neural net designs of some kind (known or unknown) could achieve substantially better performance than classical controllers today.

In summary, to understand or replicate true brain-like intelligence, the primary challenge to our community is to climb the ladder of ever more sophisticated neurocontrol designs. It will be argued that one now can see the next steps of the ladder, far enough up to replicate all the capabilities mentioned in this section. (See White and Sofge, 1992; Werbos, 1994a, 1994b, 1996a; for more details of the argument.)

A Basic Roadmap of Neurocontrol

A paper this short cannot give all the equations of all the basic designs, let alone all the applications. Neurocontrol systems in the real world can be understood at three levels of analysis:

1. At the micro level, which discusses individual supervised learning modules (multilayer perceptions, radial basis functions, CMAC, etc.) or other low-level modules *within* a control architecture.
2. At the middle level, which describes how these modules are put together to build a general-purpose system or methodology.
3. The application level, which describes how general purpose systems are used in stages, and in combination with application-specific modules, to generate a product.

This may be compared to the three levels of building chips, putting chips together to make a computer, and figuring out how to use computers.

This paper will generally focus on the middle level. For a more complete discussion of the design (including equations, flow charts, and subroutine structures) options at all three levels see Werbos (1994a). For a discussion of how to use neurocontrol designs with fuzzy systems as modules, see Chapter 91. Many of these designs are subject to a patent pending in the author's name through Scientific Cybernetics, Inc. of Boca Raton, Florida. Several reviewers have reported that the introductory material in Werbos (1994b) is important to implementing the more complex designs in White and Sofge (1992).

At the middle level, ANN designs may be classified according to what kinds of generic tasks are performed. ANNs have performed four kinds of useful functions in control:

1. *Subsystem* functions such as pattern recognition or neuroidentification, for sensor fusion or diagnostics, etc.
2. *Cloning* functions, such as copying the behavior of a human being able to control the target plant.
3. Tracking functions, such as making a robot arm follow a desired trajectory or reference model, or making a chemical plant stay at a desired setpoint.
4. *Optimization* functions, such as maximizing throughput or minimizing energy use or maximizing goal satisfaction or utility over many time periods into the future.

Again, these are all functions which the neural network learns to perform.

The first of these functions can be extremely useful in practical applications (e.g., see some of the applications in Rauch, 1994); however, it does not meet the definition of neurocontrol given previously. (Also, the diversity of possibilities is too great to be reviewed here.) ANNs for the second function are called "supervised controllers". They have been reinvented many times, usually by people who use supervised learning and base their system on a database of correct actions (often without specifying how "correct actions" are determined). The third function—tracking—is performed by "direct inverse controllers" and by "neural adaptive controllers". Some authors seem to assume that following a trajectory is the only interesting problem in control; however, the human brain is not a simple trajectory follower, and real-world engineering faces many other tasks as well. The fourth group of designs is clearly the only working group with any chance of replicating brain-like capabilities. Within the fourth group itself, there are two useful subgroups—the "backpropagation of utility" (i.e., direct maximization of future utility) and the "adaptive critic family" (broadly defined); only the latter has a serious chance of someday replicating true brain-like capability (Werbos 1994a, 1994b). Within the adaptive critic family, one faces a similar ladder of designs, from simple methods which learn slowly except on small problems, through to moderate-scale methods, through to large-scale methods requiring a neuroidentification component, through to methods capable of true "planning" and "chunking" but requiring the use of simultaneous-recurrent modules (White and Sofge, 1992, Chapters 3 and 13; Werbos and Pang, 1996; Werbos 1996a).

In summary, there is a ladder here, starting from straightforward designs, easy to implement today, which can take one step by step to a true understanding of intelligence ... if only one has the will to climb higher.

What's New in Cloning (Supervised Control)

When it is decided to "simply use" ANNs in control, one often builds up a database of sensor inputs ($\boldsymbol{X}$(t)) and "correct actions" ($\boldsymbol{u}^*$(t)), and uses supervised learning to try to learn the mapping from $\boldsymbol{X}$ to $\boldsymbol{u}$. Widrow's pole balancer in the 1960s was based on this principle (Widrow and Smith, 1964). The intellectual challenge here is in building the database of "correct actions," which usually comes from a human being already able to solve the control problem.

Supervised control can be very useful when humans or computer programs are already able to compute an adequate control, but are too slow or too imperfect to meet the needs of the application. A neural net clone (especially with neural net hardware) can solve the problem. For example, the National Aerospace Plane (NASP) was designed to fly at a speed too fast for a human to stabilize in flight; under NSF support, Pap of Accurate Automation developed a supervised control system on Silicon Graphics which can replicate the reactions of humans in controlling a slowed-down version of a NASP simulator. In actual flight, the ANN could be run at electronic speed, or it could be used to provide the initial weights for a more sophisticated neurocontrol design (Pap, 1992). (Good initial weights can be very valuable when using complex optimization designs, because of stability issues and because of the local minimum problem.) Jorgensen and Schley controlled an F-15 simulator years ago in a similar way (Miller et al., 1990). Pap used a slight generalization of this approach, learning the map from $\boldsymbol{u}(t-1)$ and $\boldsymbol{X}(t)$ to $\boldsymbol{u}^*(t)$. The success of this preliminary work, and the good human-machine interface, led to a large follow-on project supported by the NASP program (Cox, Loflyte et al., 1993). AAC is now the prime contractor for Loflyte, the follow-on to NASP.

McAvoy has used a similar approach to try to "clone" good chemical plant operators. (The best operators are both rare and expensive.) McAvoy's Neural Network Club includes 25 paying corporate members, mainly large chemical process companies like Texaco. They have reported large savings from already-fielded applications of ANNs. Cloning the good operators is only one of many such applications. (See White and Sofge, 1992, Chapter 10 for a review of McAvoy's applications.) The differences between the best human operators and the worst may be worth thousands or millions of dollars because of their ability to maintain efficiency when the plant is taken through transitions. However, good operators—like good adaptive controllers—pay attention to past trends, not just to $\boldsymbol{X}(t)$; therefore, to capture their abilities, it is important to treat this as a problem in neuroidentification, not as a problem in supervised learning. One tries to predict the operators with a dynamic neural net model, not just a static map. Robust methods for neuroidentification are discussed in White and Sofge, 1992, Chapter 10. *Improved efficiency in chemical processing translates directly into reduced waste and large potential reductions in environmental pollution;* work on such applications could be enormously valuable to human society.

Supervised control is similar to expert systems in philosophy, but it copies what an expert *does,* not what an expert *says.* It has similarities to "pendant" systems for training robots, but pendants do not learn how humans respond to different input vectors $\boldsymbol{X}(t)$.

In many applications, in cloning an existing controller or plant model, it is possible—using backpropagation—to obtain selected *derivatives* of the controller or model; in that case, one can use gradient-augmented learning to adapt an ANN to match both the output and the derivatives. The details and the pseudocode are available through Scientific Cybernetics.

Tracking Methods: What's New and What's What

Classical adaptive control—championed by Narendra (Narendra and Annaswamy, 1989) and Astrom (Astrom and Wittenmark, 1989)—builds linear controllers whose parameters are adapted in real time to control linear plants whose parameters are unknown, so as to make the plants follow a reference model. Tracking a trajectory or staying at a setpoint are special cases of tracking a reference model. Even when a stable controller is adapted in this way, the interaction between the plant, the controller, and the adaptation process can cause instabilities and breakdown; the crowning achievement of Narendra and others here was the development of whole system stability proofs showing that this cannot happen for certain controllers (Narendra et al., 1980). (See Narendra and Annaswamy, 1989 for a survey of later stability proofs under various sets of assumptions, such as the special case of "total stability," which assumes state-dependent disturbances.) There are a few proofs for nonlinear systems as well, but they are difficult and limited in scope. Narendra has put major efforts into neural adaptive control, with NSF support, in order to achieve a more general nonlinear capability which, in simulations, breaks down far less often than linear control does on realistic problems (White and Sofge, 1992).

In neural tracking, as in classical adaptive control, there are two major design alternatives—the direct inverse approach and the indirect approach. Kawato has also developed a third approach (Miller et al., 1990), feedback error learning (FEL), which is essentially a hybrid neural/expert approach; it presupposes the existence of a stable classical feedback controller, which is then used in training the ANN.

Direct inverse control fits the biologists' notion of learning the mapping from spatial coordinates to motor coordinates—a subject often discussed by Grossberg, Eckmiller and many others (Bullock, 1994; Hakala et al., 1990; Pellionisz and Llinas, 1985). For example, given a two-degree-of-freedom robot arm, controlled by changing the joint angles θ_1 and θ_2, we may try to move the arm to the point with coordinates $x_1(t)$ and $x_2(t)$. If the mapping from θ_1 and θ_2 to x_1 and x_2 is one-to-one, then there will exist an inverse map from (x_1,x_2) to (θ_1,θ_2). Given a desired point $\boldsymbol{x}(t)$, we can use that inverse map to tell us the angles which send the arm to that point. We can learn the inverse map by first flailing the arm around at random, and building up a database of actual $\boldsymbol{\theta}(t)$ and $\mathbf{x}(t)$; we can then use supervised learning on this database to learn the map from $\boldsymbol{x}$ to $\boldsymbol{\theta}$.

Direct inverse control has many limitations. Neural applications to robotics typically have 3–4% error, far too much for practical use. J. Walter has done better (Walter et al., 1991), but only by using a highly accurate supervised learning method which limits the possibility of real-time readaptation. Miller (White and Sofge, 1992; Miller et al., 1990) has done well by modifying the approach. One still uses $\boldsymbol{u}(t)$ as the target output, but one uses $\mathbf{X}(t)$ *and* $\mathbf{X}(t-1)$ as the input, using a differentiable version of CMAC as the supervised learning method. This approach reduced error to a fraction of a percent in using a real Puma robot to push an unstable cart around a figure-eight track; even more impressively, it changed the weight on the cart, and the system readapted completely within three loops around the track. Nevertheless, it should be possible to readapt much faster than this to familiar disturbances, like a change in weights, if one uses time-lagged recurrent networks (as in Equations 63.1 and 63.2) instead of supervised learning here (Werbos, 1992b).

This example from robotics raises a very important issue—the difference between learning and adaptation, and the possibility of "learning off-line to be adaptive on-line." In many applications—like robotics—there is a great practical need to adapt to changes in parameters like weight or friction coefficients which tend to drift a lot in normal plant operation. In direct inverse control, people try to solve this adaptation problem by using learning—by continually relearning the dynamics of the plant, as if one were expecting a totally new plant. (There is an analogy here to the learning strategy used by primitive, submammalian species in response to "pattern reversals" (Bitterman, 1965).) A more sophisticated approach is to use time-lagged recurrent neurons, in effect, to detect changes in these kinds of parameters. This is equivalent to adapting the parameters of the adaptation rules themselves, so as to tune them to the needs of the specific application. For example, one can build up a database of plant operation, in which these parameters fluctuate as they normally do; one can then train a time-lagged recurrent network (TLRN) off-line to perform the desired task, taking care to insure robustness (White and Sofge, 1992, Chapter 10). The author proposed this approach in 1990 (Werbos, 1990a), and called it "learning off-line to be adaptive on-line." Lee Feldkamp of Ford (Feldkamp et al., 1994 and 1996) has achieved real-world success with an elaborated version of this approach, called "multi-streaming," applied to an optimization task, crucial to Ford's ability to meet new, strict air quality standards. Mammalian brains clearly combine the capabilities of real-time learning and time-lagged recurrence, in order to track their environment on multiple time-scales (Wilson, 1975; Bitterman, 1965).

In another recent application under NSF support, AAC used direct inverse control to replace the dynamic joint controllers *within* a more classical, hierarchical control design developed by Seraji (Adkins et al., 1992). In simulation, this led to the first controller fast enough and robust enough to control the main arm of the space shuttle, under gross supervision by humans.

(Millions of dollars have been spent, unsuccessfully, to use classical methods and AI in that application, as an alternative to the present slower, more manual approach.) After those simulations, the NASA project officer stated that this should increase productivity of these activities in space ten-fold. Since then, robustness was increased still further by putting a neural optimizer (a critic network developed in cooperation with NSF) on top of the structure, and success has been reported in controlling a physical arm, an early version of the main arm which was not used because it was harder to control than the real thing. The U.S. Navy is supporting a large follow-on in underwater robotics (Davis and Schaper, 1994). AAC has also made an arrangement with a major U.S. robot manufacturer to market this technology.

Nevertheless, direct inverse control is not powerful enough to explain human arm movement. Uno and Kawato et al. (Miller et al., 1990) have performed many experiments proving that human arms do include an optimization capability. This can only be explained, in the author's view (Werbos, 1994a, 1994b), by assuming that they are based on an indirect design.

In the indirect approach, one tries to minimize a utility function, U, defined as $(\boldsymbol{X}(t) - \boldsymbol{X}^*(t))^2$ plus terms for energy consumption, etc. The tracking application tells one something about the form of U but, beyond that, we simply move on to one of the optimization methods of the next two sections. In past applications by Jordan, Kawato (Miller et al., 1990), Narendra (White and Sofge, 1992), and others, the backpropagation of utility was used, but biological systems presumably use adaptive critics instead, for reasons discussed in the section on backpropagating utility. Narendra has shown that the indirect approach is more powerful than the direct approach, and has even proven a whole-system stability theorem for a simple version of it (White and Sofge, 1992). The theorem can probably be generalized substantially.

Farrell of MIT Draper Labs (under NSF and Air Force support) has used classical adaptive control together with a neural net parameter predictor to control a simulated F-15 from an AIAA control challenge; however, the control did poorly when noise was added (Farrell, 1992). This highlights the weakness of classical adaptive control with respect to noise—a problem solved by adaptive critic designs.

More recent work comparing neurocontrol to classical control theory has provided even stronger evidence that the stablest form of control—classical or neural—involves a proper use of optimization over time, similar in spirit to the work of Feldkamp et al. (Werbos 1996b).

Backpropagating Utility

After we have a deterministic model of the plant to be controlled, the sum of utility U over all future time can be expressed as a function of our actions $\boldsymbol{u}$ (past and future) or as a function of the weights in our action network. The task of maximizing future utility can then be treated as a straightforward problem in function maximization. Some people solve this problem by a purely random search or by Hopfield nets (as in earlier work by Kawato), but one can do much better by exploiting gradient information. To get this information, we can use the generalized form of backpropagation, the form which the author first applied in 1974 (Werbos, 1974, 1981), which works on any sparse nonlinear structure (not only on the so-called backpropagation networks, which are properly called multilayer perceptrons). To use the gradient well, we can use adaptive learning rates or sophisticated numerical methods, both of which are much faster than steepest descent. (See Chapter 91.)

Actually, there are three ways to calculate the gradient of utility:

1. Backpropagation through time (BTT), which is highly efficient even for large systems, but is basically an off-line or batch method.
2. The conventional or forward perturbation method (Werbos, 1981), which works in real time but grows in cost as N^2 where N measures the network size.
3. The truncation method, which simply ignores certain cross-time connections.

All three are implausible as models of biology (though truncation may exist in lower organisms), because the brain operates in real-time, at less than N^2 cost, and can account for cross-time connections. Fortunately, the adaptive critic designs do provide a biologically plausible alternative (Werbos, 1994a, 1994b).

BTT was first applied in 1974 (Werbos, 1974), and by 1988 there were four working examples in control: Jordan's robot controller (Jordan, 1989), Kawato's cascade robot controlled (Miller et al., 1990), Widrow's truck-backer-upper (Miller et al., 1990), and an official Department of Energy model of the natural gas industry used in their *1987 Annual Energy Outlook.* By now, there are many others, including McAvoy's (real-time) model predictive controller for chemical plants (White and Sofge, 1992, chapter 10), which has many imitators. Narendra's work uses forward perturbation. Many authors have reinvented truncation, which is useful only for the simplest tracking problems. Even though BTT is mainly an off-line method, the possibility of using time-lagged recurrent networks lets us train off-line a net which appears adaptive on-line, due to the recurrence, as discussed in the previous section. Among the more interesting recent successes of backpropagating utility are those reported by Hrycej of Daimler-Benz (Hrycej, 1992) and by Feldkamp et al. (1994) of Ford.

The failures of backpropagating utility have mainly been due to inadequate models of the plant to be controlled; such models are often based on random perturbations or on supervised learning. McAvoy and the author have used a more sophisticated neuroidentification method to solve this problem, and reduced prediction errors by orders of magnitude on real-world data from a refinery and a wastewater treatment plant (White and Sofge, 1992 Chapter 10) This is a crucial area for future research.

The use of forward perturbation in neuroidentification has given the false impression that time-lagged recurrent nets are expensive to use, at least if big. BTT is much cheaper, and new critic-based designs should permit low-cost real-time adaptation (White and Sofge, 1992, Chapter 13).

Adaptive Critics/Approximate Dynamic Programming

For a full explanation of the "ladder" of adaptive critic designs, see (White and Sofge, 1992 Chapters 3 and 13). Adaptive critics are often seen as a type of reinforcement learning system, but they are more powerful than conventional reinforcement learning can be. Critic networks may be defined as networks trained to approximate the evaluation function (J) of dynamic programming, or something very close to J. The simplest useful systems include a critic adapted by heuristic dynamic programming (HDP) and an action net adapted by Barto's Arp method. They have worked well in many applications, but grow very slow on medium-sized problems. Klopf and Baird have shown that drive-reinforcement theory modified to incorporate action-dependent HDP (ADHDP) explains a variety of animal behavior experiments which were intractable to all other attempted models (Baird and Klopf, 1993). (The reader should be warned, however, that there were many other old papers on "reinforcement learning" which did not consider dynamics, and are not relevant here.)

The next step up the ladder involves "advanced adaptive critics," which combine generalized backpropagation and adaptive critics in a unified way in a fully real-time system. In 1988, there was theory by the author (Miller et al., 1990), but no examples (and a few typographical errors). Now there are at least four working systems; both AAC and a group in Russia are also far along in developing them.

The most striking example comes from White and Sofge (1992), when they were at McDonnell-Douglas (McAir). McAir was a world leader in making high-quality composite materials, which are stronger and lighter than other structural materials (for which the U.S. market is circa $400 billion/year). Their market was limited because of high costs due to the lack of a continuous production process. For obvious reasons, McAir and others had spent millions of dollars on this problem, using the best classical and AI approaches, to no avail. After reading Miller et al., (1990), White tried neurocontrol. Direct inverse control did not work. Simple adaptive critics worked on a small test version, but learned too slowly for the real thing. Using an advanced adaptive critic (really just the second step on the ladder, using ADHDP and backpropagation), he and Sofge developed the first workable system, which has been used to make real parts in St. Louis.

Essentially the same design was used in reconfigurable control for the F-15—a controller which adapts to conditions like a wing being shot off, and reduces plane losses by a factor of two. In simulations (White and Sofge, 1992, Chapter 11) White found that a critic design using derivatives (backpropagation) to adapt an action network could adapt in real time in two seconds to a totally new aircraft configuration. Jim Urnes of McDonnell-Douglas reports that 10 seconds to a minute are required, when this design is applied to a physical test vehicle in a wind tunnel; however, this is still enough to cut plane losses in half. Flight tests on real F-15s have been scheduled in stages, from September 1994 through 1997 (Urnes and Jorgensen, 1994). Even better performance may be possible by blending adaptive critic approaches with time-lagged recurrence (Section 5) and insights derived from Rauch's approach to this application (Rauch, 1994).

White and Sofge—now at MIT and Neurodyne—claim that a prototype thermal controller for NASP, based on the same approach, looks very promising and may well be the only way to improve efficiency and reduce weight enough to allow NASP to reach orbit. A benchmark test problem representing NASP, developed in September 1992 by White, by NASA Ames and by McDonnel-Douglas is in White and Sofge (1992, Chapter 11). Neurodyne has submitted a Phase I final report on a Small Business Innovation Research (SBIR) contract from NSF, which indicates success in preliminary efforts to develop improved thermal control for NASP. NASA has indicated that this thermal control technology may have many other applications as well.

In 1994, Donald Wunsch of Texas Tech and his student, Danil Prokhorov, reported success (Prokhorov et al., 1995) in using an ADHDP design very similar to that used by White and Sofge, in finding a clean solution to the bioreactor benchmark test problem in Miller et al. (1990), a problem which is very difficult for many older adaptive critic and adaptive control designs. Shibata and Okabe (1994) obtained good results using a similar design on a simulated robot motion problem.

Also in 1994, John Jameson of Palo Alto, California tried both a second-level critic design (similar to the Neurodyne design) and a third-level design (using backpropagation through a neural net emulator of the plant—BAC (White and Sofge, 1992, Chapter 3))—to control a simple-looking single-link robot arm, in simulation. The robot arm model was formulated in a way which ended up being non-Markovian. After much effort, Jameson found that the third-level design could work on this simple but tricky problem, while the second-level design could not. Adding new inputs to the controller could have made the process Markhovian, and might have solved the problem here; however, the best way to do that is usually to generate new inputs by carrying out the neural-net version of Kalman filtering, which is based on building a neural net model of the plant (White and Sofge, 1992, Chapter 10). Thus, a neural-net model of the plant is needed in any case. The details are given in the final report from Jameson Robotics to ARPA.

Even more recently, Robert Santiago and Werbos reported a few brief simulations of Dual Heuristic Programming (DHP), which showed faster learning than the third-level system in simple simulations of an inverted pendulum, when parameters of the pendulum were changed in unexpected and massive ways. A very simple SRN-based critic performed best (Santiago and Werbos, 1994; Werbos and Santiago, 1993). A generalized implementation of this method was tested by Santiago at BehavHeuristics. A level 4 critic was more robust than a level 1 system, even in the simple pole-balancing problem used by BSA in their classical work. More recently, Santiago has developed generic, industrial-grade software to implement some of these designs (including those now used for revenue management at USAir and a more accurate variation thereof). He has also collaborated with Wunsch and Prokhorov (Prokhorov et al., 1995).

S. N. Balakrishnan of the University of Missouri-Rolla has also reported substantial success in using DHP in simulated target interception problems, based on standard models from the aerospace sector. Balakrishnan (who consults at times with McDonnell-Douglas) compared a level 4 critic against half a dozen classical methods normally used in the missile interception problem. (Part of his system used parametrized models instead of neural networks, but the adaptation methods given in White and Sofge, 1992, are generic.) His system showed a very substantial improvement in performance. This is quite interesting, insofar as this is a well-studied problem of rather extreme interest to the military. A version of this work may be forthcoming in *Neural Networks.*

The fourth level and fifth level use more powerful techniques to adapt the critic-dual heuristic programming (DHP) and globalized DHP (GDHP). These two techniques explicitly minimize the error in the derivatives. The first published successful implementation of a fifth-level system was presented by Prokhorov in WCNN95 (Prokhorov and Wunsch, forthcoming). More details about the upper levels and their practical advantages are discussed below in text taken from Werbos (1995a).

Practical Advantages of Brainlike Control

This section only discusses neurocontrol proper (where the control signals themselves come from a neural net). As mentioned above, every useful example of neurocontrol to date rests on a generic capability to perform one or more of three basic tasks:

1. Cloning of a human or other expert.
2. Tracking a set-point or desired reference trajectory.
3. Dynamic optimization, maximization of a performance measure over time, accounting for the impact of present actions on performance many periods into the future.

Cloning is still quite useful as a way to initialize neural nets. (It is very popular in adaptive fuzzy control, but losing popularity in neurocontrol.) In practical applications, tracking error or performance is the real objective; it is better to use cloning designs as a starting point, and then adapt them further to do better, using tracking or optimization approaches. Often there are better ways to do the initialization.

Tracking is now mainly done by using neural nets instead of matrices in classical model-based adaptive control designs. Narendra of Yale–who pioneered stability theorems for classical adaptive control–has proven similar theorems for the neural versions, and many others have followed him. (He has now begun–step by step, carefully–to move on to more brain-like approaches.) In essence, these designs use some form of backpropagation to train an action network to output those actions, $\boldsymbol{u}(t)$, which maximize a measure of tracking error at time $t+1$; however, some forms of optimal neurocontrol may be interpreted as constructive methods to find such an error measure, which is normally quite difficult (Werbos, 1995a).

A second problem is that generic real-time learning is a slow way to adapt to changes in familiar parameters like mass and friction; it results in unnecessarily long transient responses and unnecessarily weak performance during transients. A better approach is to "learn off-line to be adaptive on-line", as the author proposed in Maren (1990), to tune the adaptation parameters themselves, in effect to the specific parameters. This requires the use of optimization over time (which could be done in real-time as well) applied to a time-lagged recurrent network used as an action network, exactly as described in the plenary talk by Feldkamp of Ford at WCNN95. (See also Feldkamp 1996). As Feldkamp has stressed, it is critical to know how to calculate the required derivatives correctly here, and the literature is now pervaded by inaccurate shortcuts and unnecessarily expensive methods for doing this. (See Werbos, 1994b, for a tutorial and White and Sofge, 1992, Chapter 10 for more advanced methods for exact derivative calculation.)

Third, if we wish to directly optimize performance measures like fuel consumption, mass ratios and pollution over time in a highly dynamic system, we must move on to the designs for optimization over time. (Some people try to optimize performance by use of hand-tweaking here, or the equivalent, but this is not as effective as an automated, rigorous approach directly addressing the nonlinear dynamic optimization problem.) These kinds of performance metrics are absolutely critical in many applications, paticularly in the automotive, aerospace and chemical sectors.

A few researchers still perform model-free tracking based on "learning the mapping from spatial to motor coordinates." There have even been a couple of designs which achieved useful, practical levels of performance—the early work by Miller et al. (1990) and recent work by Gaudiano and Grossberg. But the direct approach has many limitations relative to the indirect approach, as discussed by many authors, including Narendra, Kawato, Jordan, and myself, (Werbos, 1995b, White and Sofge, 1992; Miller et al., 1990).

In optimization over time, there are two dominant practical approaches: (1) an explicit model-based approach (like model predictive control (MPC), using backpropagation through time (as defined in Werbos, 1994b) to calculate the derivatives; (2) an indirect approach, sometimes (loosely) called "reinforcement learning," "adaptive critics" or "approximate dynamic programming." The first approach—the basis of Widrow's famous truck-backer-upper (Miller et al., 1990)—was first spelled out in an example in the author's 1974 Ph.D. thesis (reprinted in Werbos, 1994b). The second approach was first implemented in neural networks by Widrow et al. (1973), who invented the term "*critic*."

Strictly speaking, these approaches are not mutually exclusive in engineering. For example, one could use MPC to look ahead 30 time steps, and use a critic network to initialize the backwards derivative calculations. (Werbos, 1995a, proposes this for the first time. . . .) In effect, the critic would try to approximate the derivatives which would have been calculated at time $t + 30$, if we could have afforded to compute all the way from t to $t + \infty$ in MPC. (The critic may actually be more accurate than an explicit calculation would have been, if uncertainty or noise tend to grow over long time intervals.) For example, in battery control, a critic might be trained to assess conditions which affect the future lifetime and performance of the battery (in effect), while

MPC could be used to optimize some combination of current performance and battery damage over the coming 10–60 seconds.

The MPC approach clearly is not plausible as part of any model of the brain, because of the structure of the derivative calculations, no matter how the derivatives are calculated. But in engineering, using fast chips, it does have some advantages—not least of them, exactness. Still, it cannot address noise or uncertainty in a numerically efficient manner, and the cost of the computations can become a problem, especially when millisecond sampling times are required.

The adaptive critic approach—broadly defined—is the only type of design which anyone has ever formulated, in engineering or biology or elsewhere, with any hope of explaining the generic kinds of capabilities we see in the brain. But the adaptive critic approach, like neurocontrol in general, is a complex field of study, with its own "ladder" of designs from the simplest and most limited all the way up to the brain itself. (Please bear in mind that the adaptive critics are not intended to be an alternative to backpropagation in simple pattern classification problems; they are systems for solving a different type of problem, an optimal control problem over time!)

Roughly speaking, level zero of this ladder (described above) is the original Widrow design (Widrow, et al., 1973), which no one uses any more. Level one is the 1983 Barto-Sutton-Anderson (BSA) design, described by Barto in White and Sofge (1992) and Miller et al., (1990), which uses a global reward system ("Arp") to train an Action network and "TD" methods to adapt the critic. It learns very slowly in medium-sized problems involving continuous variables, but it is very robust. It is still extremely popular among computer scientists, who often deal with a smaller number of action variables, all of them binary rather than continuous. "TD" is a special case of heuristic dynamic programming (HDP), a method which the author first published in 1977 (discussed in White and Sofge, 1992).

From 1990 to 1993, many people in the community climbed up one step in the ladder to level 2, which was once called "advanced adaptive critics." The idea was to use an action-dependent adaptive critic (ADAC), which I first defined in 1989 (Werbos, 1989a), and discussed in several other places, culminating in White and Sofge (1992). In ADAC, the Critic sends derivative signals back to the action network, so that backpropagation (but not supervised learning!) can be used to adapt the Action network. The rich feedback to the action network makes it possible to control more action variables, more effectively. ADAC was the basis for the numerous practical applications by White and Sofge, also discussed in Hrycej (1992), ranging from carbon-carbon composite parts fabrication, through to rapid stabilization of a simulated F-15 after battle damage, through to recent work in semiconductor manufacturing which has achieved great visibility in that industry. The basic equation for "J" given in Werbos (1989) is identical to that for "Q" in Watkins' 1989 Ph.D. thesis; however, Watkins' "Q learning" (level 2? level 1?) used an explicit enumeration and evaluation of alternative action vectors rather than an Action network adapted by backpropagation. In recent years, several people have reinvented ADAC as a "modified form of Q-learning," sometimes (e.g., in a recent workshop) replicating whole chunks of equations already published with White and Sofge (1992). Still, these designs are all a step up from the 1983 BSA design.

In criticizing this entire literature, Grossberg has explained again and again that an "expectations system" is essential to explaining the wide range of experiments in "classical conditioning." (Grossberg has formulated models of conditioning which meet this test, though without the link to engineering functionality; see Levine and Elsberry (1995) for important concepts in that literature.) Likewise, there are good engineering-based reasons to believe that an expectations subsystem is crucial to functionality, in coping with very complex control problems (Werbos, 1995b; Narendra, 1994; Werbos, 1994; White and Sofge, 1992). Unfortunately, some computer scientists seem to believe that it is "cheating" to use a model of the external environment or plant (even a neural network model!). Yet in many practical applications, industrial people would actually prefer to use their own model, with off-line adaptation, in developing a controller. (The "noise wrapper" techniques used by Feldkamp are an important part of making this work.) From a research point of view, there are many advantages to accepting this preference for the time being, in part of our work, in order to learn more about critic and action networks without the complications caused by concurrent model adaptation. Again, such approaches are more brainlike than the model-free approaches at lower levels.

Brainlike control represents level 3 and above on the ladder. Level 3, as mentioned above, is to use HDP to adapt a Critic, and backpropagate through a model to adapt the action network. (See Werbos 1994b and White and Sofge 1992, Chapter 10 for how to backpropagate through a nonneural model.) Levels 4 and 5 respectively use more powerful techniques to adapt the critic—dual heuristic programming (DHP) and globalized DHP (GDHP). These last two techniques explicitly minimize the error in the derivatives which would be passed back in the battery example which the author above. In 1981 and 1987 (Werbos, 1981), I proposed a 3-network system (critic, action, model) based on GDHP as a strawman model of the brain. From late 1993 to March 1995, several groups developed the capability to build such system—including Prokhorov's level 5 system mentioned above. (See Prokhorov and Wunsch, forthcoming.)

As a technical matter, note that Werbos (1994b) and White and Sofge (1992) are important prerequisites to success in this kind of work. In White and Sofge (1992), Chapter 13, equation 10 has a typo which some have found a problem: it uses λ-hat where it should be λ^* on the left of Equation 63.10. Likewise, in Equation 63.1 "s_x" should be "x_j". In Figure 63.6, the middle block should be labeled "Model." The pseudocode in Miller et al. (1990) has much more serious typos.

Recent Accomplishments in Brainlike Intelligent Control

At the 1994 NASA Ames Workshop on neural networks for flight control, organized by Jorgensen and Pellionisz, whose proceedings may be published by IEEE Press, Wunsch and Prokhorov reported on their efforts to use a well-tuned classical controller

(PID), a level 2 critic, and a level 3 critic on the bioreactor and autolander test problems in Miller et al. (1990), problems which are extremely difficult for less powerful methods. They solved both problems cleanly with a level 2 critic, and solved the autolander using PID, even in the "noisy" version of the problem. However, when they added still more noise, and drastically shortened the runaway, both the PID and the level 2 critic crashed the airplane 100% of the time. The level 3 critic was able to land the plane (more or less, using tolerant criteria) 80% of the time. With stringent criteria, it was only 40%. (See Prokhorov et al., 1995 for this and more recent extensions.) Their paper for WCNN95 shows still more accurate control as one climbs up to level 5 of the ladder. Naturally, there is a great deal of research still to be done in optimizing the use of such designs; by analogy, research in how to use backpropagation was not all complete after it was shown to work on the XOR problem.

Also at Ames, AAC reported work much closer to real-world application—use of a level 4 critic to control the first physical prototype of "LoFlite" (built by AAC and subcontractors), the first US prototype of a class of airplanes able to reach earth orbit as an airplane, at airplane-like costs. AAC reported in conversation that this level of capability was absolutely essential to solving this very difficult control problem, and that the results have been uniformly successful in actual wind-tunnel tests to date. AAC's neural chip is discussed in the article "An Avionics Example," by Saeks (1996) in this Handbook.

Issues like exploration, learning speed and "persistence of excitation" suggest that the next step up will do still better, if robust hybrid designs and robust neuroidentification are used (Miller et al., 1990). For true planning problems, like "Star Wars" or robot navigation through *novel* cluttered space, the fourth rung of the ladder, given in Prokhorov et al., may be necessary. Naturally, it is this highest rung which now seems to fit mammalian brains (Werbos, 1994b; Miller et al., 1990).

Supervised Learning: What It Does, Applications, and How To Do It

The following subsection is written mainly for the beginner, though it does define some important notation and terminology—terminology which is sometimes confused. The third subsection will mention a few points which have confused even the experts. The material for this section and the next is taken from Werbos (1991a).

What is Supervised Learning?

Several dozen designs now exist for ANNs which perform supervised learning. Figure 63.1 shows what the task of supervised learning is.

In supervised learning, one tries to adapt a neural network so that its actual outputs ($\hat{Y}$) come close to some target outputs (Y) for some training set which contains T patterns. The goal is to adapt the parameters of the network so that it performs well for patterns from outside of the training set. The parameters of the network are usually called "weights."

In actuality, there are two forms of the supervised learning task, called the real-time and non-real-time forms. (Unfortunately, the word "off-line" has been misused by many researchers, so it is avoided here.) Begin with an example of non-real-time supervised learning.

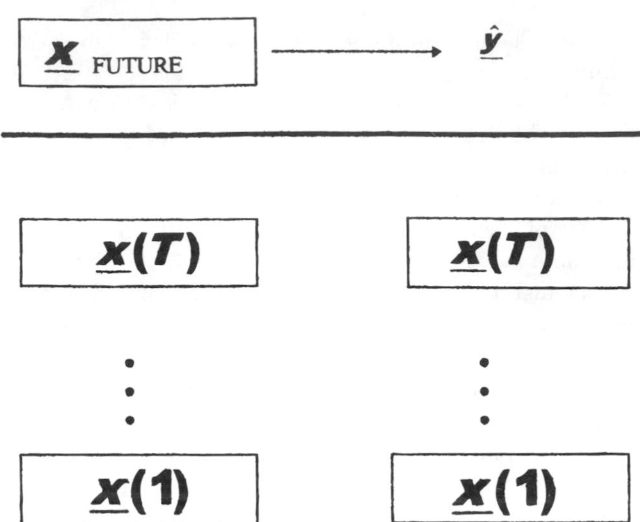

Figure 63.1 Illustration of the supervised learning task

The main use of supervised learning today lies in pattern recognition work. For example, suppose that one is trying to build an ANN which can learn to recognize handwritten ZIP codes. Assume that one already has a camera and preprocessor which can digitize the image, locate the five digits, and provide a 19 by 20 grid of ones and zeros representing the image of each digit. One wants the ANN to input the 19 by 20 image, and output a classification; for example, one might ask the network to output four binary digits which, taken together, identify which decimal digit is being observed. (In other applications, of course, there may be grey-scale patterns, in which the inputs are not all ones and zeros; this presents no problem for most ANN designs.)

In this example begin by building up a database of correct classifications. For example, build a database consisting of 20,000 examples of correct classifications. In that case, $T = 2000$. Give each example a label t between 1 and 2000. For each sample t, we have a record of the input pattern and the correct classification. Each input pattern consists of 380 numbers, which may be viewed as a vector with 380 components; call this the vector $\mathbf{X}(t)$. The desired classification consists of four numbers, which may be treated as a vector $\mathbf{Y}(t)$. The actual output of the network will be $\hat{\mathbf{Y}}(t)$, which may differ from the desired output $\mathbf{Y}(t)$, especially in the period *before* the network has been adapted.

Even when using a simple network design, the vectors $\mathbf{X}(t)$ and $\mathbf{Y}(t)$ need not be made up of ones and zeros. They can be made up of any values which the network is capable of inputting and outputting. Denote the components of $\mathbf{X}(t)$ as $X_1(t) \ldots X_m(t)$ so that there are m inputs to the network. Denote the components of $\mathbf{Y}(t)$ as $Y_1(t) \ldots Y_n(t)$ so that there are n outputs. Throughout this paper, the components of a vector will be represented by the same letter as the vector itself, in the same case; this convention turns out to be convenient because $\mathbf{x}(t)$ will represent a different vector, very closely related to $\mathbf{X}(t)$.

Figure 63.5 illustrates the supervised learning task in the general case. Given a history of $X(1) \ldots X(T)$ and $Y(1) \ldots Y(T)$, one wants to find a mapping or function from X to Y which will perform well when encountering new vectors X outside the training set. The index "t" may be interpreted either as a time index or as a pattern number index.

Once again, there are two forms of supervised learning. In non-real-time supervised learning, the order of the patterns is not assumed to be meaningful. One simply cycles through the training set as often as one likes, adjusting the weights through finer and finer tuning, until the estimates of the weights converge. In real-time learning, the data from times t' with $t' < t$ disappears forever after the data from time t becomes available; one can cycle through the data one observation at a time, and one has to take the data as they come, as one would in a true real-time control application, adapting the weights on a real machine in real operation.

Some researchers try to make it seem as if they are doing real-time learning when they are doing non-real-time learning. This has led to some serious confusion. There are actually two ways to solve a non-real-time learning problem. In one strategy—called batch learning—one begins each iteration with an estimate W of the weights. One goes through all of the patterns in the training set, calculating such things as the derivatives of error with respect to the weights. Then, after reexamining every pattern, one updates the weights and starts all over again. (See Figures 63.2 and 63.3.) In another strategy—called pattern learning—one begins with an estimate of the weights, and then one examines the first pattern. One adjusts the weights based on that individual pattern, and then one moves on to the next pattern, and so on. Most people simply cycle through the entire set of patterns in the training set, in the order given, but there are many variations on this.

Some authors call batch learning "off-line learning" and they call pattern learning "on-line learning," even though it is not true real-time learning. Still, there are similarities between pattern learning and real-time learning which make it interesting from a research point of view.

Applications of Supervised Learning

Approximately 70 to 80 percent of the real-world applications of ANNs involve supervised learning in some form.

$$E(t) = \frac{1}{2}\sum (Y_i(t) - \hat{Y}_i(t))^2$$

Minimize $E = \sum E(t)$

With Respect to $\{W_{ij}\}$

$$W_{ij} = W_{ij} - Learning\_rate \bullet \frac{\partial^+ E}{\partial W_{ij}}$$

Figure 63.2 Basic backpropagation (BATCH).

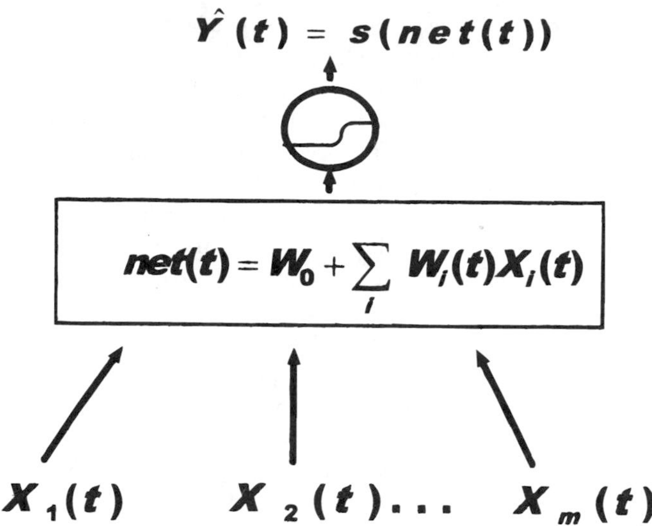

Figure 63.3 The McCullock-Pitts "neuron"—the building block of MLP systems.

There are two major applications of supervised learning systems: applications in areas like static pattern recognition, where supervised learning is the right way to formulate the problem; applications in areas like system identification, where supervised learning can provide a crude *initial* solution, where it can be useful as a component of a more complete solution, and where the lessons learned from building supervised learning systems can carry over to the more complex systems.

In the first category, ZIP code recognition has already been mentioned. Automated ZIP code recognition is a major concern of the U.S. Postal Service, where costs have grown over time and reduced costs could be worth billions of dollars. Because this is a major problem, the post office has been carrying out an ongoing competition to come up with better systems. Many private companies have claimed to lead this competition; however, former Post Office officials have told me that AT&T's system and one other system based on backpropagation have the best success on single digit recognition. (See Werbos and Pang, 1996, for "windowing"—a crucial extra trick they use.) Pattern recognition is also important in areas like target recognition and sonar recognition, areas where DARPA in the U.S. Department of Defense has concentrated much of its research effort; Barbara Yoon, the director of that effort, stated in Seattle at IJCNN91 that ANNs typically improve performance in such pattern recognition tasks by 50–100%. One hears many reports of important installed applications of ANNs in the U.S. Department of Defense, but the details are classified in some cases. Takehito Tanak of Fujitsu (in Kawasaki), with collaborators, has described a system to recognize hot spots in continuous casting in steel mills, which will soon be installed in all the steel mills of Nippon Steel Company and in other plants as well. Tom McAvoy, of the University of Maryland, has used ANNs to input spectral data on sample chemicals (which are available in real time in many chemical plants) and predict properties of the chemicals (which are known for sure only after lengthy assays, available with a time-delay of hours or days). Likewise, McDonnell-Douglas has used supervised learning to perform nondestructive evaluation of composite materials in real time, using real-time sensor data, in effect, to predict a later more precise assay; this real-time quality assessment plays a crucial role in their real-time control system (White and Sofge, 1992; Sofge and White, 1991).

The main advantage of ANNs over more traditional pattern recognition is the possibility of automated feature discovery. This, in turn, allows less human effort to be required. In many applications, the availability of high-performance special-purpose neural chips and boards is also a crucial factor, especially when real-time classifications are required.

In the second area, supervised learning systems are sometimes used to classify time-varying patterns, to diagnose time-varying systems for possible failures (Werbos, 1990b), to predict dynamic systems, and to serve as controllers. In principle, these problems all require more complex ANNs, as will be discussed later and in Chapter 91; however, a simple static mapping can be a useful first approximation, and it is often good enough for many applications. Siemens, for example, has developed an ANN which inputs the noise from a truck motor (recorded through acoustic sensors) and outputs an evaluation of the motor quality; this can be useful in quality control in motor production. Ford once worked with the Carnegie Group in Pittsburgh to develop a system which again can listen to the sound of an engine, and help diagnose what is wrong with it; a highly skilled human mechanic can do the same thing, but the supply of good mechanics is very scarce. Many speech recognition systems are being built by use of simple supervised learning systems, where the input to the network includes speech data across multiple time periods; Waibel of Carnegie-Mellon and NTT/ATR has used this kind of "time-delay neural network" quite often.

One should not underestimate the importance of having a good first approximation when developing a more complex system. The first approximation will be limited in its capabilities. However, it will give you experience in using ANN computer programs and/or hardware, which is important as a first step. It will give you values of weights which can be transferred over and used as initial weights in parts of a more complex design. Of course, the supervised learning system itself may receive its initial weights from another analysis, such as a fuzzy logic analysis this is discussed in Chapter 91.

Alternative Designs for Supervised Learning Systems (SLS)

Because there are dozens of designs for supervised learning, described at length in books like Maren (1990), this article cannot explain all the details of how to implement all of them. Instead, it will provide a condensed overview, with references to more detailed treatment. It will provide the basic equations for basic backpropagation, which accounts for perhaps two-thirds or more of the present-day applications. Hopefully this overview will be useful even to the expert. There are individual sentences in this material which can make the difference between success and failure in some applications.

Broadly speaking, there are two types of SLS which are useful in practice: SLSs based on minimizing error—minimizing some measure of $Y - \hat{Y}$ across the training set; SLSs based on forecasting by analogy—comparing a new input vector X against earlier inputs $X(t)$ and predicting that the new value of Y will equal an average of $Y(t)$ for the closest past analogues. Designs in class 2 are technically called "heteroassociative memories." In addition, there are many designs available for preprocessing the input vector, most of which involve: fixed preprocessing, such as Fourier transforms or breaking the data up into regions (as in CMAC, radial basis functions (RBF) and much of Kuperstein's work); unsupervised clustering or feature extraction methods, such as Grossberg's ART, Kohonen's topological maps, Cooper's mean-field projection pursuit system, etc. Many of these methods are described at length in the book by Maren (1990), in the chapters not dealing with neurocontrol. Systems based on CMAC (White and Sofge, 1992; Miller et al. 1990) and RBF display many of the same capabilities as heteroassociative memories, and are sometimes even discussed as if they were in the same group of methods.

Some authors advocate a third type of SLS: systems with an initial layer of decorrelation, followed by a classification layer based on simple correlations between the inputs and the targets. They like this approach because it fits the well-known theories of Hebb, which are considered to represent biological truth; however, the other approaches are also biologically plausible, and there are serious problems with the correlation approach (Werbos 1991a). The author is not aware of any realistic applications. Many of the associative memory designs—which are more useful—are also consistent with the theories of Hebb.

See Werbos and Pang (1996) for a fourth type of SLS, which is more difficult to use but more powerful.

The vast majority of SLSs used in engineering are based on the concept of minimizing error. Most often, they are based on the concept of minimizing square error, illustrated in Figure 63.2. However, in true classification applications, statistical theory says that we should minimize the well-known Bernoulli measure of error; if we do, then the output of our network can be interpreted as a probability of the classification. (In the Bernoulli measure, the error; $E_i(t) = -\log Y_i(t)$ in cases where $Y_i(t)$ equals one, and it equals $-\log(1 - Y_i(t))$ when $Y_i(t)$ equals zero.)

In real-time learning, it is now most common to use a complex preprocessor network, so that there is only one layer of artificial neurons which actually learn over time. This makes learning easier and faster; it allows a learning rate near 1.0 in practice; and it avoids the need to use the backpropagation algorithm to calculate the derivatives of error (E) with respect to the weights (W_{ij}). For example, the usual CMAC system (Miller et al., 1990) consists of a complex preprocessor, with a single layer of adapted neurons for the upper layer.

In non-real-time applications, practical people usually use a particularly type of ANN called a "multilayer perceptron" (MLP), which is illustrated in Figures 63.3 through 63.5. This allows an adaptation at the same time of the upper layer and of the "hidden layers"—the preprocessors, if you will. Because the entire structure is optimized, this approach generally gives more accurate results (i.e., better generalization to new examples), if it is done

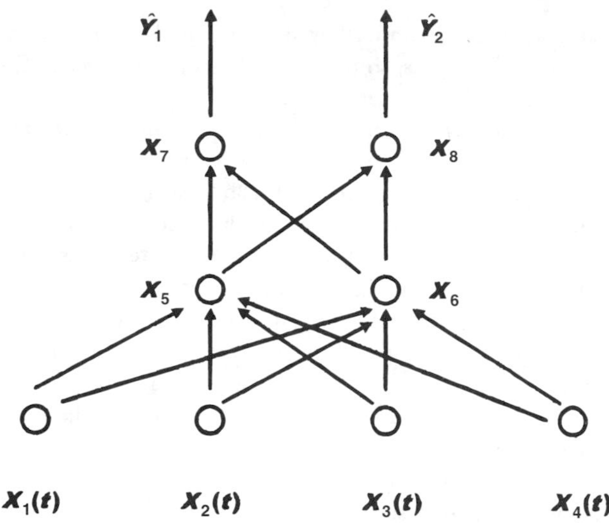

Figure 63.4 The most common MLP design.

for i = M+1 up to N+n,

$$\begin{cases} net_i(t) = \sum_{j<i} W_{ij} x_j(t) \\ x_i(t) = s(net_i(t)) \end{cases}$$

$$\hat{Y}_i(t) = x_{i+N}(t) \qquad (x_0 = 1)$$

Figure 63.5 Equations of a more general MLP design.

properly. The adaptation of the hidden or intermediate layers provides the "automatic feature extraction" capability. In non-real-time applications, the need for accuracy in generalization and improved feature extraction is usually the dominant issue.

In order to adapt a multilayer network, based on the derivatives of error (as in Figure 63.2), one needs an efficient derivative-calculation procedure. Backpropagation is essentially just that: an efficient, exact algorithm for calculating derivatives. Figure 63.7 shows how to do it, for other measures of error; it is not restricted to the case of MLPs, or even to the case of neural networks. Werbos (1994b; 1989b) describes how to apply back-propagation, in general, to feed-forward systems which are not necessarily made up of model neurons. Figures 63.8 and 63.9 go even further, by illustrating how to apply backpropagation to implicit systems or interative, relaxation systems; the concepts were described (with the details of a 1981 application in energy modeling) in Werbos (1989a), White and Sofge (1992) discuss the concepts more simply. (Figures 63.8 and 63.9 are taken from White and Sofge, 1992.) Good quality testing and debugging are crucial to making this kind of system work properly (White and Sofge, 1992; Werbos,1989a,b). It is also extremely important

$$F\_\hat{Y}_i = \hat{Y}_i - Y_i$$

For i=N+n down to m+1

$$\begin{cases} F\_x_i = F\_\hat{Y}_{i-N} + \sum_{j>i} W_{ji} F\_net_j \\ F\_net_i = s'(net_i(t)) F\_x_i(t) \end{cases}$$

$$F\_W_{ij} = F\_net_i * x_j$$

Figure 63.6 Equation of backpropagation for an MLP.

(White and Sofge, 1992, Chapter 10) to delete unnecessary connections from the network, to keep it from growing too big.

In actuality, there is no reason why a multilayer network cannot be adapted in real time. For example, one can still use a high learning-rate on the upper layer, and a slow learning rate (which is better than no learning at all) on the lower layers. To do this, it helps to have a formula which can be used, separately, on each layer of the network. This approach worked very well in speeding up the official U.S. Department of Energy model of the natural gas industry (Werbos 1989) in 1987, and it is described more concisely by Werbos in White and Sofge (1992). There is some additional value (White and Sofge, 1992) in scaling the current gradient vector, F_W′ to make sure that it is not very large compared to gradients in other, recent examples with similar overall error. In batch learning, F_W is the gradient of error from the previous iteration for weights in the current layer; in real-time learning, it would be a filtered average of that gradient in previous iterations. In batch learning examples, the author has chosen values of "a" on the order of 95 and values of "b" on the order of 0.1; for real-time learning, one would presumably want "a" closer to 1 and "b" closer to zero, to prevent too much volatility in the learning rate. Another simpler approach is to adapt a multilayer network off-line, then adapt only its top layer in real time (White and Sofge 1992). McDonnell-Douglas (Sofge and White, 1991) reported that the fastest learning it obtained, in a high-speed real-time control application, came with an upper network based on a differentiable CMAC system, and a lower layer based on an MLP, adapted by backpropagation in real time.

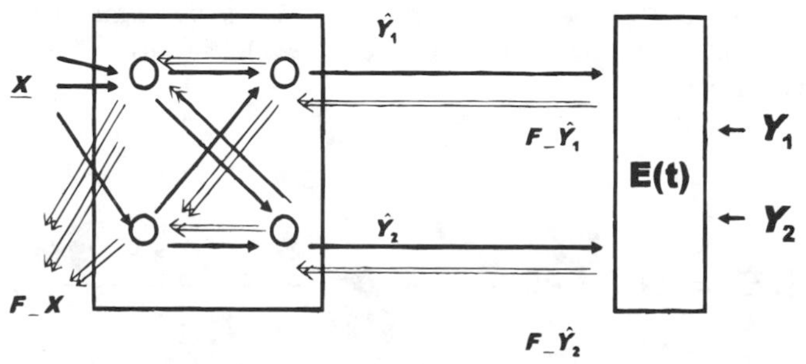

Figure 63.7 Flowchart for equation in Figure 63.6.

$F\_\hat{Y}$ inputs

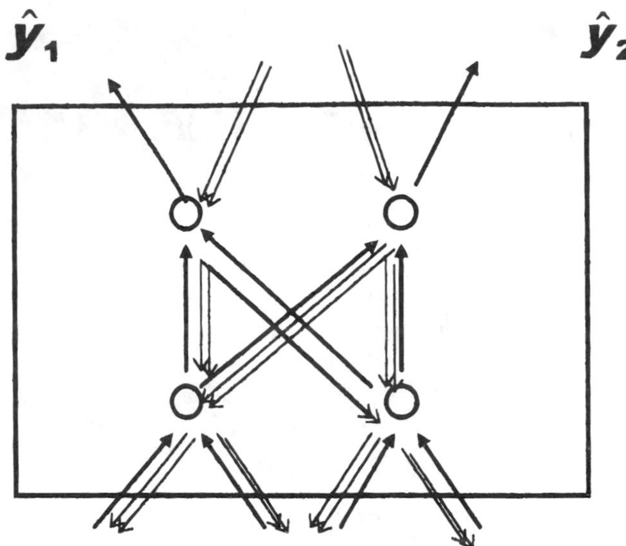

F_X outputs

Figure 63.8 Backpropagation through a network: An important concept. A subroutine was programmed for an extraction—part of Figures 63.5 and 63.6. From *Handbook of Intelligent Control Neural, Adaptive and Fuzzy Approaches*. 1992. White, D. and Sofge, D. (eds.). Van Nostrand. With permission.

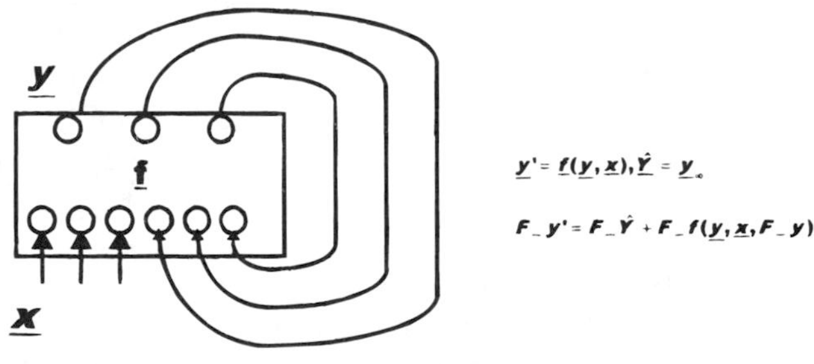

$$\underline{y}' = f(\underline{y}, \underline{x}), \hat{\underline{Y}} = \underline{y}_.$$

$$F\_y' = F\_\hat{Y} + F\_f(\underline{y}, \underline{x}, F\_y)$$

Figure 63.9 Backpropagation through a recurrent or relaxation network. From *Handbook of Intelligent Control: Neural, Adaptive and Fuzzy Approaches*. 1992. White, D. and Sofge, D. (eds.). Van Nostrand. With permission.

$$LR' = a * LR + \left\{ 1 + b * \frac{F\_W' \bullet F\_W}{|F\_W|^2} \right\}$$

Figure 63.10 Adaptive Learning Rate.

For fast adaptation with backpropagation, it also helps to scale the inputs and outputs, so that the weights in any layer tend to be of similar orders of magnitude.

Some researchers claim that pattern learning is faster than batch learning, even for non-real-time learning. However, it is the author's experience (Werbos, 1988b) that this is only true for large, highly redundant databases like the AT&T zip code database. For nonredundant databases, it is better to use batch learning together with highly efficient optimization methods instead of steepest descent (in Figure 63.2). This does not mean the use of Polak-Ribiere or Fletcher-Reeves conjugate gradient methods; it requires the use of more sophisticated methods (as in Shannon's chapter in Miller et al., 1990.) Someday, when more research has been done on pattern learning, it may become a more efficient competitor.

In the future, one suspects that a closer marriage between associative memory approaches and error minimization approaches will be needed before this becomes real. The existing forms of supervised learning are still good enough for a very wide range of applications.

Neuroidentification: What It Is, Applications, and How To Do It

What is Neuroidentification?

Broadly speaking, neuroidentification is the design and adaptation of neural networks which provide predictions of dynamic systems or act as models of dynamic systems. It is the ANN version of system identification (the engineers' term) or of the estimation of time-series models (the statisticians' term). It is also directly applicable to the classification and diagnosis of dynamic patterns over time. Unlike supervised learning, it is not static.

As with supervised learning, there is a batch version of the neuroidentification task, and a real-time version. One can mix and match, and use all the same learning tricks discussed above and in the references. Also, there is a conventional formulation of the neuroidentification problem, and a stochastic version as well.

In the conventional version, one is given a vector of inputs, $\boldsymbol{X}(t)$, at every time t. One is also given a vector of exogenous variables or control variables, $\boldsymbol{u}(t)$. A network may be built which inputs information about $\boldsymbol{X}$, $\boldsymbol{u}$ and its own internal state from times t' earlier than time t. Some authors like to describe such networks in differential equation form; however, this author finds that it works much better to use a difference equation formulation, which refers to data at times $t - 1$, $t - 2$, and so on. (For example, this formulation avoids confusing the idea of short-term memory with the idea of relaxation networks as described in the previous section.) Our job is to output $\hat{\boldsymbol{X}}(t + 1)$, a prediction of $\boldsymbol{X}(t + 1)$.

In the stochastic version—which was defined in Werbos (1977) but only recently translated into an explicit, detailed design with consistency results available (White and Sofge, 1992)—the job is to output a simulated version of $\boldsymbol{X}(t + 1)$, instead of a prediction. It is allowed to use random numbers in the network. The task is to simulate $X(t + 1)$ in such a way that the probability of outputting a particular value, X_0 is equal to the actual probability of observing X_0, at time $t + 1$, in the real system, conditional upon what has been observed at earlier times.

Applications

System identification is a necessary prerequisite to many control designs, including the most powerful neurocontrol designs discussed above. In control, the challenge is to use the best possible system identification, but then to put "outer loops" around the prediction system so as to make the controller as robust as possible with respect to errors in the predictor. The author's choice of notation above is based on the control application. In the very most powerful neurocontrol designs, it is important that the memory go back only one time interval; some authors think of this as a limitation, but it is easy to see that a lag-one system can easily and parsimoniously represent arbitrary time lags (e.g., via a bucket brigade of neurons or cascaded exponential decay neurons).

In 1991, the vast majority of failures in the field of neurocontrol could be traced back to inadequate system identification. The most general and most powerful forms of control require, in principle, a system for stochastic identification, in order to locate an optimal control strategy; however, this is probably not important to most engineering applications, where it is good enough to focus on the "best guess" or the "nominal" trajectory.

Neuroidentification is also useful in prediction problems, such as predicting the stock market. Actually, there are several firms which have made rather large profits, using very simple supervised learning techniques. The author has seen some of the details, but the best, large-scale examples are highly proprietary. (It should be obvious, however, that these firms have worked very hard to make nonneural methods work as well.) Presumably, however, more accurate forecasting methods would lead to still better results, especially over long time intervals. Stochastic neuroidentification would help the trader do a better job of hedging against the inevitable uncertainties. P. Treleaven, of the University of London, has developed a neural network club with a number of large clients in the financial sector. Nestor, Inc. of Rhode Island claims a number of financial clients as well.

Neuroidentification is also important in diagnostic applications (Werbos, 1990b). In diagnostic applications, there is often a huge database on normal (good) performance, but only a few observations of a few of the possible failure modes. In that situation, a dynamic model of the normal plant can be doubly useful: big deviations from the predictions of that model can be a good sign of abnormality, which can catch even novel modes of failure which were never seen in the database; the model can be used to predict ahead many steps, in case the normal operation is leading to an undesirable state. Of course, a complete diagnostic system would combine all these capabilities and a few others besides (Werbos, 1990b). Better automated diagnostics are crucial for improved safety and cost factors in nuclear plants and in aerospace vehicles (like the shuttle or the National Aerospace Plane) where conventional, all-human monitoring has led to astronomical costs and frequent anomalies. Of course, there are applications to quality control in other sectors as well. Good but preliminary work is going on in all these applications.

Neuroidentification can also be applied to pattern classification, with a minor change in where we obtain the database. The vector $X(t + 1)$ can be used to represent the desired classification and the observed sensor data a time $t + 1$. (In other words, the two sets of information can be combined). The vector $u(t)$ can be used to represent the sensor data at time t. The network can be forbidden from inputting the vector X itself. (In other words, we zero out all weights which input information from X, for simplicity.) If the network still contains a memory of its own past state, then this scheme can produce a very substantial improvement over what is possible with an SLS system—if it is used effectively. For example, in image processing, this kind of ANN can learn the optimal relaxation algorithms and exploit "optical flow"—which is extremely useful in applications like machine vision and robotics. In submarine detection, it can yield a rich network, with many useful hidden nodes, because it exploits the huge volume of data on what is heard from the ocean; in effect, it has the ability to subtract out the normal noise of the ocean, because it has a dynamic understanding of that noise. In speech recognition, the "memory" in such an ANN can develop what amounts to a model of speaker characteristics, which in turn permits true speaker-independent recognition systems. At present, however, the author is not aware of any projects which begin to exploit the true potential in these applications. The key problem here is the lack of intuitive understanding (White and Sofge, 1992; Miller, et al., 1990) by most researchers of how these capabilities work, or of how to implement them in an effective, flexible way.

Alternative Designs for Neuroidentification

In 1991, all of the published applications of neuroidentification were based on the conventional formulation, rather than the stochastic version. The publication of White and Sofge (1992) has since provided more complete descriptions of the stochastic version to encourage its wider use. The 1991 published applications were all based on two alternative approaches: the supervised learning approach, using "time-delay neural networks" (TDNN); the adaptation of more complex ANNs, called time-lagged recurrent networks (TLRN). (A few authors sometimes use the abbreviation TDNN to refer to both designs, but this is not done here.)

In the TDNN approach, one uses a simple supervised learning system, with a very big network. The vector $X(t + 1)$ is used as the target vector for the network (what was called $Y(t)$ above). The input vector is constructing by combining $X(t)$, $u(t)$, $X(t - 1)$, $u(t - 1)$ and so on, down through $X(t - k)$ and $X(t - k)$ and $u(t - k)$, for some k, into one very large vector. Because the input vector is very large, the network has to be very large. Everything else being equal, a larger network will lead to slower learning and poorer generalization; that is one serious disadvantage of this approach. (Large networks are not bad when they are very sparse, or when there is a large database on the target variables; however, to really use all these many inputs here, one needs a lot more weights in the network.) Furthermore, for a reasonable value of k, the network will display no ability to account for information before time $t - k$. The chief advantage of the TDNN approach is the ease of implementing it.

In the TLRN approach, one attempts to adapt an ANN like the example shown in Figure 63.11. (One can allow multiple time-lags very easily (Werbos, 1990c), but, for reasons described above, this is usually not a good thing to do.) Figure 63.11 was drawn for people interested in dynamic pattern recognition, but the key concept here applies to all forms of neuroidentification. The key concept is that we are asking the network to output a new vector, $\boldsymbol{R}$ which it then uses as an input in the next time-period. One may think of $\boldsymbol{R}$ as a kind of internal blackboard. One may think of $\boldsymbol{R}$ and $\boldsymbol{x}$ together as *R*epresentation or *R*econstruction of *R*eality, using a *R*ecurrent network. In control theory, $\boldsymbol{R}$ by itself is a vector of "observables," while $\boldsymbol{x}$ and $\boldsymbol{R}$ together provide a kind of estimated or filtered state vector. The vector $\boldsymbol{R}$ may also learn to represent slowly varying system parameters—something of great importance to adaptive control.

One way to implement the idea of Figure 63.11 is by using a multilayer perception, like that of the equations in Figure 63.12, but with memory of the past state of the network. One can replace the equation for the "net" in Figure 63.5 by the top equation of Figure 63.12. When this is done, the entire state of the network—the entire vector $\boldsymbol{x}$—serves in effect as the vector $\boldsymbol{R}$. One can still adapt this network using steepest descent (or other gradient methods) if it is known how to calculate the derivatives of error with this new kind of network. To calculate the derivatives, one can still use the same equations as before—shown in Figure 63.6—but must replace the equation for $\boldsymbol{F\_x}$ with a new version, shown in the bottom of Figure 63.12. The neuroidentification method shown in Figure 63.12 is discussed at greater length in (Werbos, 1990c). (To adapt this kind of network, it is important to use adaptive learning rates, initially set to small values, and to initialize the weights to values which came from adapting simpler versions of the network—e.g., versions which hold the W' to zero.) Notice that the cross-time weights W' in Figure 63.12 are just as rich as the same-time weights W.

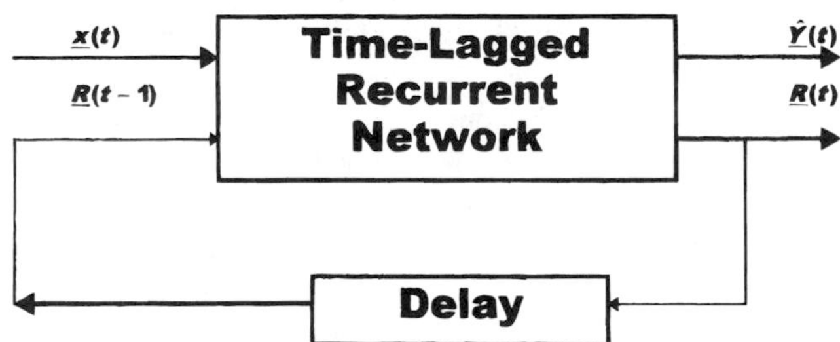

Figure 63.11 What a TLRN does (assuming lag 1).

$$net_i(t) = \sum_{i>j} W_{ij} x_j(t) + \sum W_{ij}' x_j(t-1)$$

$$F\_x_i(t) = F\_\hat{Y}_{i-N}(t) + \sum_{j>i} W_{ji} F\_net_j(t)$$

$$+ \sum_{all\ j} W_{ji}' F\_net_j(t+1)$$

Figure 63.12 The most common form of TLRN. From *Handbook of Intelligent Control: Neural Adaptive and Fuzzy Approaches.* 1992. White, D. and Sofge, D. (eds.). Van Nostrand. With permission.

The design in Figure 63.12 is probably the best state of the art, as used by all but a few researchers. Unfortunately, it is still not good enough for most control applications. To improve performance, there are four possible improvements, which are logically independent of each other:

1. Change the error measure to achieve more robustness.
2. Change the functional form to permit greater stability.
3. Use a real-time adaptation method.
4. Consider the stochastic case.

In principle, all four improvements can be combined.

The first of these improvements is the most important, by far, to 1991 applications. The conventional least-squares error measure, based on a prediction for time $t + 1$ which uses actual data from time t, tends to be very nonrobust, especially when you try to forecast multiple periods into the future. This in turn leads to poor multiperiod control. Every 1991 working application of neurocontrol that the author knows of, with a system identification component, uses one of the three following tricks:

1. Use an existing differentiable model of the process instead of an ANN (this works, because one can still backpropagate through any differentiable model (Werbos, 1989b)).
2. Use the method in Figure 63.12, but use $\boldsymbol{X}(t + 1) - \boldsymbol{X}(t)$ rather than $\boldsymbol{X}(t + 1)$ as the target vector.
3. Use the pure robust method, described in White and Sofge (1992) and Werbos (1990a).

The first two methods are purely *ad hoc;* therefore, it is hoped that more researchers will use the pure robust method and improved general-purpose methods of the same flavor. The pure robust method was used in Werbos (1994b) and Werbos (1977) in nonneural applications, and by Kawato (in his "first pass cascade method" in Miller, 1990) in a neural network robotics application. McAvoy, Werbos and Su (White and Sofge, 1992) used it to predict real-world data from several chemical plants, and found that it reduced forecast error by orders of magnitude compared with conventional methods of adapting ANNs. Unfortunately, the pure robust method is not really suitable for all applications; Werbos (1990a) and White and Sofge (1992) describe the ANN form of a more general "compromise" method, which has worked well in nonneural applications, but still leaves room for further improvement. Schmidhuber has also developed some ideas of relevance here.

A second direction for improvement is to consider alternative functional forms, to improve stability. For example, in Figure 63.12, one can have W' multiply net$(t - 1)$ instead of $x(t - 1)$;

when W' is close to the identity matrix, this yields a very stable underlying process (i.e., no decay of memory). Specifications of this sort are discussed in Werbos (1990a), along with the corresponding backpropagation equations (which are basically straightforward). This kind of "sticky neuron" is actually more plausible as a model of biology than is the usual McCulloch-Pitts model, as discussed in Werbos (1990a) and verified in recent conversations with Karl Pribram. Specifications of this sort are particularly important in adaptive control applications, where we want the hidden nodes to have the ability to represent slowly varying system parameters. On the other hand—like other TLRNs but more so—the adaptation of networks with sticky neurons requires multiple stages of learning, phasing in from simpler forms of the network; it helps to start out with small values of the sticky weights. It is possible that forcing the W'_{-ij} weights on the output layer to be closer to one will have advantages similar to those of the pure robust method; experiments on this are currently ongoing at NSF, to the extent that their schedules allow.

The third direction for improvement concerns real-time learning. Backpropagation through time is essentially a batch method. One can partially overcome this limitation by ad hoc tricks. For example, one can use backpropagation through time in an off-line mode, and then freeze the W′ weights; one can then adapt the upper part of the network in real-time using backpropagation. Alternatively, one can sample a "string" of data from time t to time $t + T$, adapt the network to that string of data, and then move on to $t + T$ through $t + 2T$. Clearly the human brain is not based on these kinds of tricks (This was used in Feldkamp, 1996.).

There are three approaches now used to achieve real-time learning with TLRNs. The crudest method is one denoted "truncation"; the network in Figure 63.12, for example, would be adapted as if $\boldsymbol{x}(t-1)$ were fixed, and did not depend on the weights. In other words, one simply uses supervised learning, and one treats $\boldsymbol{x}(t-1)$ as one more thing to be added to the input vector. This approach does not really minimize error, since it does not account for the effect of weights in changing what $\boldsymbol{x}(t-1)$ was; however, it can be useful if simple applications, like trajectory following, where memory across multiple time-periods is not really necessary. In applications like adaptive control—where one wants to represent slowly varying process parameters—it is likely to be very poor.

The second approach now used for real-time adaptation of TLRNs is to calculate derivatives in forward-time, using a method called the "conventional perturbation" method (Werbos, 1982). Williams and Zipser (see the Williams chapter in Miller, 1990) have applied this method to a particular class of TLRNs, in simulation studies, and call it the "Williams-Zipser method." Narendra and Parthasarathy (1990) have applied it to a more general class of TLRNs, in control applications, and have called it "dynamic backpropagation." This method is not a brainlike method, because the cost of calculating derivatives in forwards time is much greater than the cost of doing so in backward time; the cost is greater by a factor of N_W, the number of weights. For a very large system, this additional cost becomes prohibitive; however, for a small-scale applications, the method can be very useful.

The third approach, described briefly in Werbos (1988a) but spelled out at length in White and Sofge (1992), is to turn the neuroidentification problem into a control problem. This allows consistent real-time operation without paying a large additional cost. There is reason to believe that this is how the human brain adapts networks of this kind (Werbos 1991b). The control problem is then solved using the adaptive critic approach, described in Werbos (1996). The author calls this approach the "Error Critic" approach. Schmidhuber has done preliminary experiments on these lines, but a more powerful Critic design (as described in White and Sofge, (1992) will be needed to achieve good results. This is a complex approach needed only for very demanding applications—such as understanding or replicating the human brain.

Finally, the fourth possible improvement of Figure 63.12 comes from solving the stochastic version of the neuroidentification problem. Figure 63.13 shows a design for doing this, which is explained at greater length in White and Sofge (1992). Note that there are three component ANNs here—the encoder, the decoder and the predictor—as well as error calculations, a random number generator and the block where $\tilde{R}'$ is calculated. To adapt the three ANNs and the sigma's, one minimizes the error function:

$$E = \sum_i (X_i - \hat{X}_i(\tilde{R}'(x, \sigma^T, e^R))^2/\sigma_i^X)^2 + \sum_j (\hat{R}_j - \tilde{R}_j(x))^2 + \sum_i \log(\sigma_i^X)^2 + \sum_j ((\sigma_j^R)^2 - \log(\sigma_j^R)^2)$$

For all practical purposes, the predictor and decoder networks are simple TLRNs, adapted to solve the conventional neuroidentification problem; one can apply the usual methods to design them

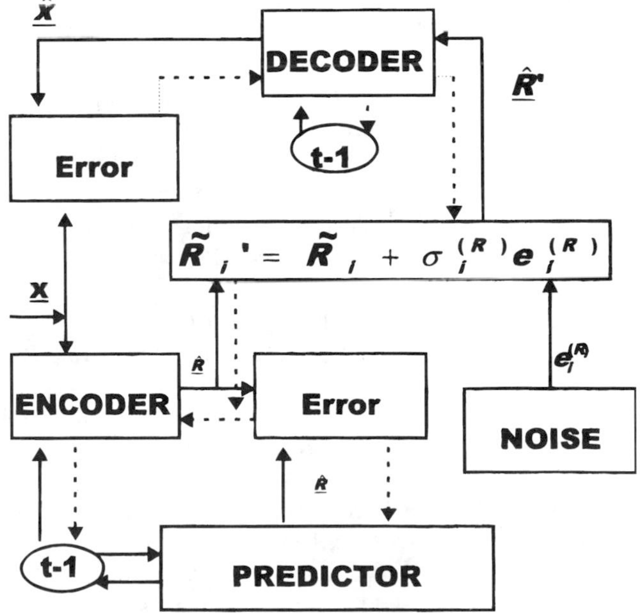

Figure 63.13 Stochastic encoder/decoder/predictor.

as subsystems of this larger scheme. For the encoder network, however, a direct use of backpropagation is essential. One may cite Werbos (1991b) as the source for this very limited description here; however, White and Sofge (1992) contains far more detail, including consistency results for the general multivariate linear case.

Many of the concepts introduced in this article will be extended in the fuzzy logic and soft computing chapter in this Handbook, where my article, Werbos (1996), discusses neurocontrol and elastic fuzzy logic.

References

Adkins, M., Cox, Pap, R., Thomas, and Saeks, R. 1992. Neural joint control for space shuttle remote manipulator system, *Proc. 1992 IEEE/RSJ Int. Conf. Intelligent Robots.*

Astrom, K. and Wittenmark, B. 1989. *Adaptive Control,* Prentice Hall, Englewood Cliffs, NJ.

Baird, L. and Klopf, E. H. 1993. Extensions of the associative control process (ACP): Heuristics and provable optimality, *Proc. 2nd Int. Conf. on Simulation of Adaptive Behavior, (Hawaii 1992),* MIT Press, Cambridge, MA.

Bitterman, M. E. 1975. Comparative analysis of learning, *Science,* 188:699–709. See also Bitterman, The evolution of intelligence, *Scientific American,* January 1965.

Bullock, D. 1994. Flexible motor control by forebrain, cerebellar and spinal circuits, *WCNN94 Proceedings.*

Cox, C., Pap, R., Saeks, R. and Mach, K., 1993. Neurocontrol of a hypersonic aircraft, AIAA-93-5155, *AIAA/DGLR Int. Aerospace Planes and Hypersonics Technologies Conf.,* Washington, D.C.

Davis, J. and Schaper, V., eds., 1994. *1994 Navy Workshop on Neural Networks (Washington, D.C.),* Office of Naval Research, Arlington, VA.

Farrell, 1992. The source for this statement was a technical report by Farrell submitted to Dr. E. H. Klopf of the Wright Research and Development Center at Wright-Patterson Air Force Base in Ohio, and used in a 1992 program review.

Feldkamp, L., Puskorius, G. V., Davis, L. I., and Yuan, F. 1994. Enabling concepts of neurocontrol, *Proc. of the 8th Yale Workshop on Adaptive and Learning Systems,* New Haven, CT. This material was also presented by Feldkamp orally at WCNN94.

Feldkamp, L., 1996, On-vehicle training for engine idle speed control, *Int'l Conf. Neural Networks* (ICNN96), IEEE Press, Piscataway, NJ, Plenary volume.

Hakala, J., Stein, R., and Eckmiller, R. 1990. Quasi-local solution for inverse kinematics of a redundant robot arm, *Proc. IJCNN90.*

Hrycej, T. 1992. Model-based training method for neural controllers, *Artificial Neural Networks II,* Aleksander, I. and Taylor, J., eds., North Holland.

Jordan, M. 1989. Generic constraints in underspecified target trajectories, *IJCNN89 Proceedings.*

Levine, D. and Elsberry, W., eds. 1995. *Optimality in Biological and Artificial Networks,* Lawrence Erlbaum Associates, Hillsdale, NJ.

Maren A., ed. 1990. *Handbook of Neural Computing Applications,* Academic Press, San Diego, CA.

Miller, W, Sutton R. and Werbos, P. 1990. *Neural Networks for Control,* MIT Press, Cambridge, MA.

Narendra, K. and Annaswamy, 1989. *Stable Adaptive Systems,* Prentice Hall, Englewood Cliffs, NJ.

Narendra, K. and Parthasarathy, 1990. Identification and control of dynamical systems using neural networks, *IEEE Trans. Neural Networks,* 1 March.

Narendra, K., ed. 1994. *Proc. 8th Yale Workshop on Adaptive and Learning Systems,* New Haven, CT.

Narendra, K., Lin, Y. H., and Valvani, L. 1980. Stable adaptive controller design—part II: proof of stability, *IEEE Trans. Automatic Control,* 25:440–448.

Pap, R. 1992. Design of neurocontroller to operate active flight surfaces, Technical Report to NSF Grant ECS-9147774, Accurate Automation Corporation, Chattanooga, TN.

Pellionisz, A. and Llinas, R. 1985. Tensor network theory of the metaorganization of functional geometries in the central nervous system, *Neuroscience,* 16:245–273.

Prokhorov, D. and Wunsch, D., Adaptive critic designs, *IEEE Trans. Neural Networks,* forthcoming.

Prokhorov, D., Santiago, R., and Wunsch, D. 1995. Adaptive critic designs: a case study for neurocontrol, *Neural Networks,* Vol. 8, No. 9.

Rauch, H. E. 1994. Adaptive control and fault analysis using neural networks, *Proc. of the 8th Yale Workshop on Adaptive and Learning System,* New Haven, CT.

Saeks, R. 1996. An avionics applications, Chap. 62, *Handbook on Industrial Electronics,* Irwin, J. D., ed., CRC Press, Boca Raton, FL.

Santiago, R. and Werbos, P. 1994. New Progress towards truly brain-like control, *WCNN94 Proceedings,* Erlbaum.

Shibata, K. and Okabe, Y. 1994. A robot that learns an evaluation function for acquiring of appropriate motions, *WCNN94 Proceedings,* Erlbaum.

Sofge, D. and White, D. 1991. Neural network based process optimization and control, *Proc. 29th IEEE Conf. Decision and Control,* IEEE Press, New York, NY.

Urnes, J. 1994. Personal communication, Jim Urnes (McDonnell-Douglas, St. Louis) and Charles Jorgensen (NASA Ames, Palo Alto), July 6, 1994.

Walter, J. A., Martinez, and Schulten, 1991. Industrial robot learns visuo-motor coordination by means of neural-gas network, *Artificial Neural Networks,* Kohonen, T., et al., eds. North Holland.

Werbos, P. 1974. *Beyond Regression: New Tools for Prediction and Analysis in the Behavioral Sciences,* Ph.D. thesis, Harvard University. Reprinted in *The Roots of Backpropagation: From Ordered Derivatives to Neural Networks and Political Forecasting,* John Wiley & Sons, New York, NY.

Werbos, P. 1977. Advanced forecasting methods for global crisis warning and models of intelligence, *General Systems Yearbook,* (an annual journal).

Werbos, P. 1981. Applications of advances in nonlinear sensitivity analysis, *System Modeling and Optimization (Proceedings of IFIP 1981),* Drenick and Kozin, eds., Springer-Verlag, New York, NY. Reprinted in *The Roots of Backpropagation: From Ordered Derivatives to Neural Networks and Political Forecasting,* John Wiley & Sons, New York, NY.

Werbos, P. 1982. Applications of advances in nonlinear sensitivity analysis, *Systems Modeling and Optimization: Proc. Int. Federation for Information Processing,* Drenick, R. and Kozin, eds., Springer-Verlag, New York, NY.

Werbos, P. 1988a. Generalization of backpropagation, *Neural Networks,* October. When simultaneous recurrence is not present, the calculations are much simpler, as in Jordan, M., Generic constraints on underspecified target trajectories in *IJCNN Proceedings,* June, 1989, New York, NY.

Werbos, P. 1988b. Backpropagation: past and future, *Proc. 2nd International Conf. Neural Networks,* New York. A transcript of the actual talk, with slides, is available from the author, and covers some additional topics.

Werbos, P. 1989a. Neural networks for control and system identification, *IEEE CDC,* IEEE Press, New York, NY.

Werbos, P. 1989b. Maximizing long-term gas industry profits in two minutes in Lotus using neural network methods, *IEEE SMC,* March/April.

Werbos, P. 1990a. Neurocontrol and related techniques, *Handbook of Neural Computing Applications,* Maren, A., ed, Academic Press, San Diego, CA.

Werbos, P. 1990b. Making diagnostic work in the real world: a few tricks, *Handbook of Neural Comp. Applications,* Maren, A., ed., p. 332, Academic Press, San Diego, CA.

Werbos, P. 1990c. Backpropatioan through time: what it does and how to do it, *Proc. IEEE,* October.

Werbos, P. 1991a. Neural network based control, *IECON '91 Tutorial Text,* IEEE Press, Piscataway, NJ.

Werbos, P. 1991b. The cytoskeleton: why it may be crucial to human learning and neurocontrol, *Nanobiology,* in press. See also Werbos, P. Neurocontrol, biology and the mind: new connections and developments, *IEEE SMC,* 1991.

Werbos, P. 1991c. Links between artificial neural networks (AN) and statistical pattern recognition, *Artificial Neural Networks and Statistical Pattern Recognition: Old and New Connections,* Sethi, I. and Jain, eds., Elsevier Science Publishing, New York, NY.

Werbos, P. 1992. Neurocontrol: Where It Is Going and Why It Is Crucial, *Artificial Neural Networks II,* Aleksander, I. and Taylor, J., eds, North Holland.

Werbos, P. 1994a. The brain as a neurocontroller: new hypotheses and experimental possibilities, *Origins: Brain and Self-Organization,* Pribram, K., ed., INNS Press, Lawrence Erlbaum Associates, Hillsdale, NJ.

Werbos, P. 1994b. *The Roots of Backpropagation: From Ordered Derivatives to Neural Networks and Political Forecasting,* John Wiley & Sons, New York, NY.

Werbos, P. 1995a. Optimal Neurocontrol: Practical Benefits, New Results and Biological Evidence, *WCNN95 Proceedings,* Erlbaum.

Werbos, P. 1995b. Why neural networks, Ch. A.2, and Control, Ch. F.1.10, *Handbook of Neural Computation,* Fiesler, E. and Beale, R., eds., Oxford University Press New York, 1995.

Werbos, P. 1996a. Learning in the Brain! an engineering interpretation. In N. Pribram, ed., *Learning as Self-Organization,* Erlbaum.

Werbos, P. 1996b. Necrocontrol, biological intelligence and engineering applications: evaluation and prognosis. In *Statusseminar de BMBF: Neuroihformatik und Kuenstliche Intelligenz* (Munich). Projehtträger Informations technik der BMBF, DLR, Berlin. (Marius.v.d.Meer@dlr.de).

Werbos, P. and Pang, X., 1996. Generalized maze navigation: SRN Critics learn what feed forward or Hessian nets cannot, *Proc. Int. Conf. SMC,* IEEE, Piscataway, NJ.

White D. and Sofge, D., eds. 1992. *Handbook of Intelligent Control: Neural, Adaptive and Fuzzy Approaches,* Van Nostrand Reinhold, New York, NY.

Werbos, P. and Santiago, R. 1993. Neurocontrol, *Above Threshold,* 2(2):INNS Press, Washington, D.C.

Widrow, B. and Smith, F. W. 1964. Pattern-recognizing control systems, *Computer and Information Sciences (COINS) Proceedings,* Washington, D.C.

Widrow, B., Gupta, N., and Maitra, S. 1973. Punish/reward: learning with a critic in adaptive threshold systems, *IEEE SMC 1973,* 5:455–465.

Wilson, E. O. 1975. *Sociobiology: The New Synthesis,* Ch. 7, Harvard University Press, Cambridge, MA.

64
CMAC Neural Networks and Color Correction

King-Lung Huang
OES/ITRI

64.1 Introduction 906
64.2 High-Order CMAC Neural Networks for Color Correction 907
64.3 Experimental Result 907
64.4 Conclusion 908

64.1 Introduction

The cerebellar model articulation controller (CMAC) was proposed by Albus to formulate the processing characteristics of the cerebellum. This model has the ability to learn arbitrary nonlinear relationships existing between input and output data. Unlike the backpropagation-styled neural networks, CMAC is characterized by the feature of local weight updating. Due to this feature, it has the advantage of fast learning and a high convergence rate in function approximation problems. Besides, in a hardware implementation, CMAC is easily implemented by the table look-up technique.

A CMAC neural network is a perceptron-like associative memory with overlapping receptive fields that is capable of multi-dimensional nonlinear functions. Representation of nonlinear functions,

$$y = f(x) \tag{64.1}$$

is accomplished by CMACs using two primary mappings,

$$S: X \Rightarrow A \tag{64.2}$$

and

$$P: A \Rightarrow Y \tag{64.3}$$

where X is a continuous s-dimensional input space, A is an N_A-dimensional association cell space, and Y is a one-dimensional output space. The schematic of the basic CMAC architecture is illustrated in Figure 64.1. The function S is usually fixed and maps each point in the input space, X, onto an association vector $\alpha \epsilon A$ that has N_L nonzero elements. These are the active association cells which will affect the value of output. $A^*(x)$ is defined as a set of those active association cells activated by x. In Figure 64.1, $N_L = 4$ and $A^*(x) = \{\alpha_3, \alpha_4, \alpha_{10}, \alpha N_A - 1\}$.

The $N_A \times 1$ association cell vector is defined as

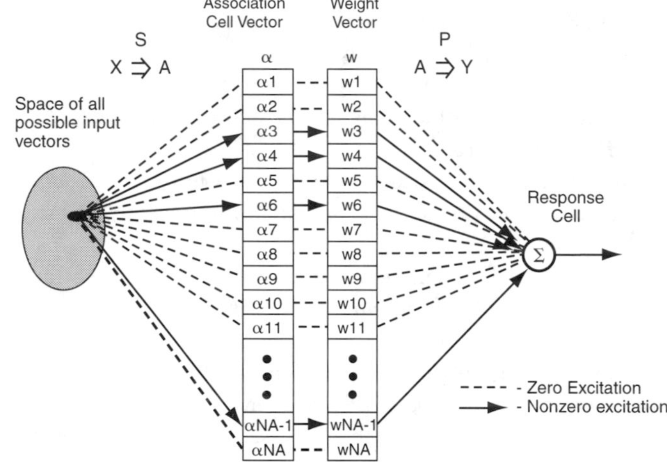

Figure 64.1 Schematic of the basic CMAC architecture.

$$\alpha = S(x) \tag{64.4}$$

The function $P(\alpha)$ computes a scalar output, y, by projecting the association vector determined by $S(x)$ on a $N_A \times 1$ vector, w, whose components are attached to their corresponding association cells, that is,

$$y = yP(\alpha) = \alpha^T w = \sum_{i=1}^{N_A} w_i S_i(x) \tag{64.5}$$

The mapping $S(x)$ of Equation 64.4 can be characterized by the following three submappings

$$R: X \Rightarrow M \tag{64.6}$$

$$Q: M \Rightarrow L \tag{64.7}$$

$$E: L, M \Rightarrow A \tag{64.8}$$

where R is a receptive field function, Q is a quantization function, and E is an embedding function. M is the matrix of receptive

0-8493-8343-9/97/$0.00+$.50

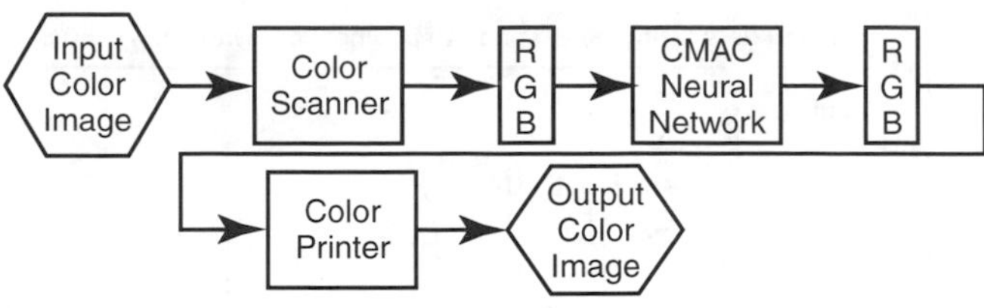

Figure 64.2 Schematic diagram of color image reproduction systems involving CMAC neural network for color correction.

field activation values, and L is an array containing column vectors for identifying the locations of maximally activated receptive fields along each input dimension.

For a standard CMAC neural network, the receptive filed activation value of each components of M is equal to either 0 or 1. Thus, R maps the input X to a binary state activation matrix, $m \in M$. The both quantization and embedding functions are used to assign m to its expected association cell vector. Generally, the receptive field functions are limited to be the rectangular functions in terms of high-order polynomials, i.e., Cubic B__Splines may be considered in the CMAC-based function approximation. The details including the learning algorithm were described in Rosenblatt (1962).

64.2 High-Order CMAC Neural Networks for Color Correction

We adapt a CMAC to learn and record the nonlinear relationships existing in the color feature values of input scanning devices and those of output printing devices.

The High-Order B__Spline CMAC architecture for color correction is shown in Figure 64.3. To obtain the generalized inverse for both scanner function $\varnothing_s$ and printer function $\varnothing_p$, the configuration of the learning system is arranged as shown in Figure 64.4. The system output d is used as neural network input. The system's input, X, is the desired response of the CMAC network. Thus, the error vector of network training is computed as X-0, where 0 is the output of neural network. The error to be minimized through learning is therefore $E = \|X - 0\|$. Once the network has been successfully trained to mimic the system inverse, the generalized inverse for both $\varnothing_s$ and $\varnothing_p$ are also obtained $f^{-1} = (\varnothing_s \circ \varnothing_p)^{-1} = \varnothing_p^{-} \circ \varnothing_p^{-1}$, $\circ$ denotes functional composition operator. The inverse model can be inserted between $\varnothing_s$ and $\varnothing_p$ so that the composed system results in an identity mapping between desired response and the controlled system output.

Due to its superiority in function approximation, CMAC is suitable for being embedded in a color image reproduction system as shown in Figure 64.2. In such a system, to make sure that the printed output images can faithfully reproduce the original input images, CMAC is used to overcome the non-linear mapping problem existing between image scanning device and image printing device.

64.3 Experimental Result

A laboratory setup has been used to test the CMAC neural network color correction system. The Microtek MSF-300Z color scanner and Canon color bubble jet printer BJC-820 serve as the image input and output device. Notice that the color correction system in Figure 64.2 is a transformation between additive RGB color space in scanner and subtractive CMY color space in printer. But, in the experiment, there is a problem that the CMY signals cannot drive Canon BJC-820 bubble jet printer directly. Therefore, the input drive signals of the Canon printer are expressed as in terms of RGB values. Thus the color correction mechanism is a transformation from RGB data to R′G′B′ data. Overall 512 training data points are generated from the sampling algorithm.

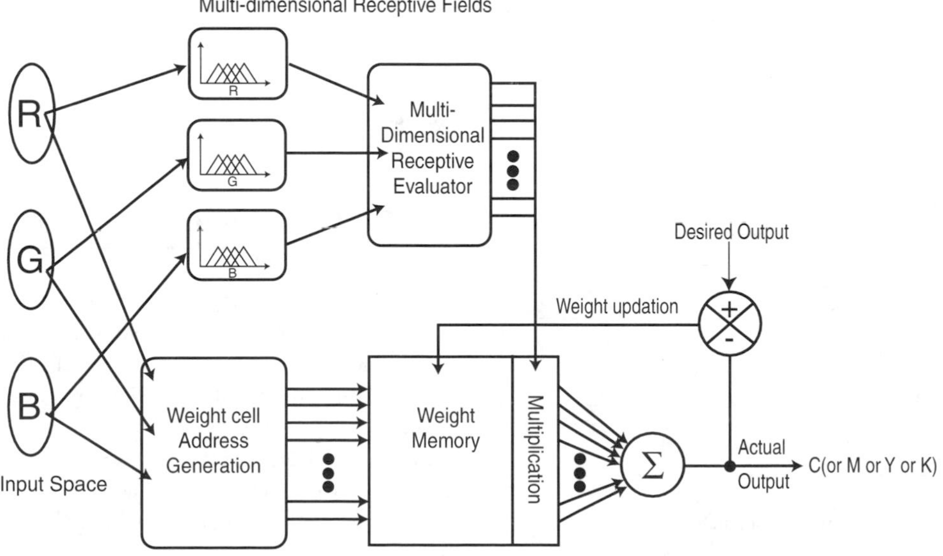

Figure 64.3 The functional schematic diagram of the developed CMAC neural network.

To perform the B__Spline CMAC learning algorithm, the related parameters are chosen as association cells number $|A^*| = N_L = 64$, G = 4, physical table size, $|A^P| = 64\ K$ for each variable, learning rate $\eta = 0.3$. The fully connected weights addressing method is used to deal with the addressing of association cells in the three-dimensional RGB space.

An average RGB color error is used to evaluate the performance of training the generalized inverse plant by multioutput CMAC and given by

$$\Delta E_{RGB} = \frac{1}{N}\sum_{k=1}^{N} \sqrt{(\Delta R)^2 + (\Delta G)^2 + (\Delta B)_k^2} \qquad (64.9)$$

where N is the number of training patches.

Figure 64.4. shows an example of the time evolution of the reduction of ΔE_{RGB} performed the multi-output CMAC. ΔE_{RGB} decreases rapidly and reaches a steady value at about the 40th iteration. Figure 64.5, Figure 64.6, and Figure 64.7 illustrate the tracking performance of the CMAC along R, G, and B components by inputting a staircase signal consisting of 512 training patches when the steady value of its associated $\Delta E_{RGB} = 0.5$, respectively.

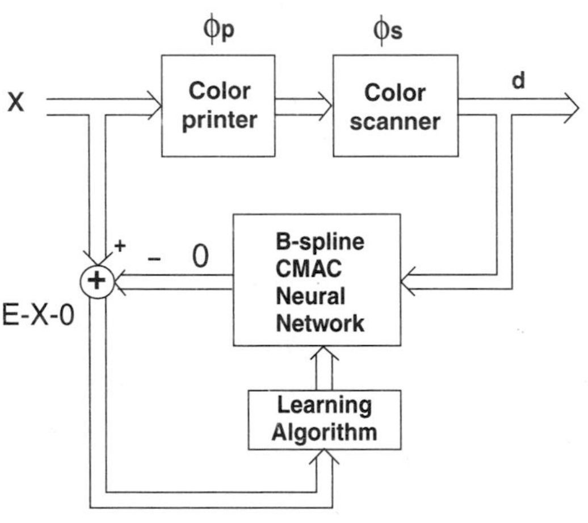

Figure 64.4 The time evolution of the reduction of ΔE_{RGB} in B-spline CMAC neural network.

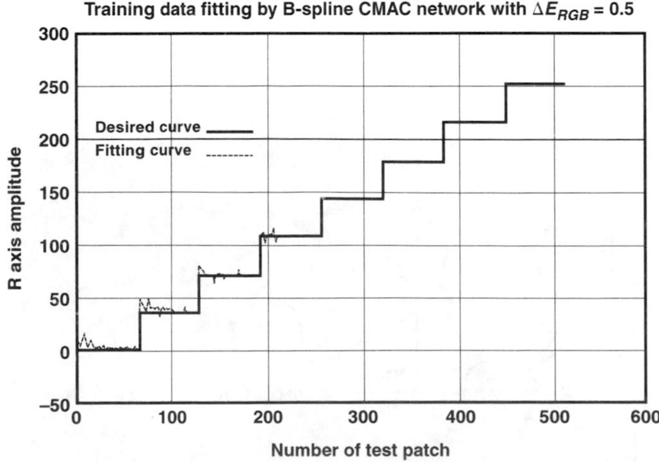

Figure 64.5 The R channel tracking performance of B-spline CMAC neural network.

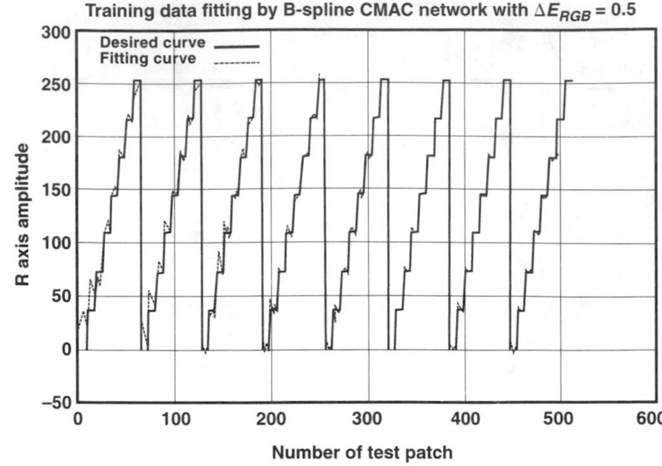

Figure 64.6 The G channel tracking performance of B-spline CMAC neural network.

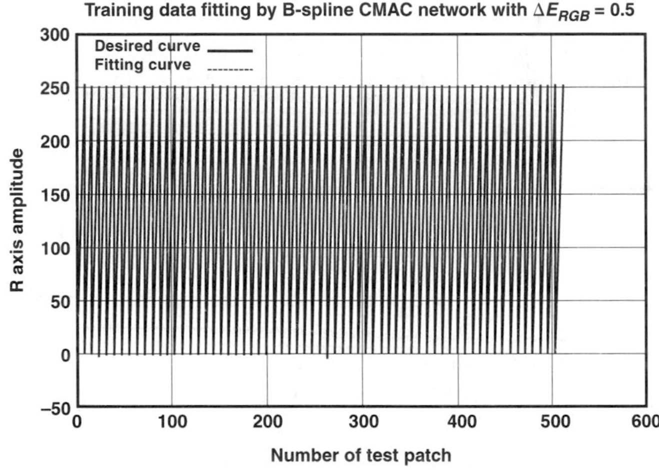

Figure 64.7 The B channel tracking performance of B-spline CMAC neural network.

64.4 Conclusion

The CMAC neural network color correction system is proposed. The CMAC-based neural network algorithm can eliminate the total color error due to the mismatch and the nonlinearity of both scanner and printer. The combination of both scanner and printer is recognized as a composite system plant. The strategy of the generalized inverse plant control is to add a plant inverse between two target plants. It acts as a preequalizer in front of the printer and an equalizer behind the scanner. Therefore, the resulting system becomes an identity plant. In other words the color errors have been reduced significantly. Experiment shows that the colors of several test samples are corrected effectively.

References

Albus, J. S. A New Approach to Manipulator Control: The Cerebellar Model Articulation Controller (CMAC), *Transactions on ASME, Jouranl of Dynamic Systems, Measurement and Control,* Vol. 97, pp. 220–227, Sep. 1975.

Albus, J. S. Data Storage in the Cerebellar Model Articulation Controller (CMAC), *Transaction on ASME, Journal of Dynamic Systems, Measurement and Control,* Vol. 97, pp. 228–233, Sep. 1975.

Lane S. H. et al., Theory and Development of High-Order CMAC Neural Networks, *IEEE Control Systems Magazine,* pp. 23–30, Apr. 1992.

Rosenblatt, F., Principles of Neurodynamics, *Spartan Books,* New York, 1962.

65
Temporal Signal Processing

65.1 Introduction 910
65.2 Temporal Neural Networks with Observable States 910
Narendra-Parthasarthy Models • Gamma Model
65.3 Temporal Neural Networks with Hidden States 912
FIR Multilayer Perceptrons • Time-Delay Neural Networks (TDNN) • Multilayer Perceptrons with Local Feedback • Recurrent Neural Networks
65.4 Conclusions 914

Simon Haykin
McMaster University

65.1 Introduction

Temporal signal processing problems of various kinds are commonly encountered in practice; they include the following:

- Prediction of a nonstationary signal.
- Modeling of an unknown dynamic system.
- Encoding of an information-bearing signal for the purpose of data compression.
- Equalization of an unknown communication channel.

And the list goes on. To use an artificial neural network (hereinafter referred to simply as a neural network) as a tool for solving temporal signal processing problems, the representation of *time* has to be built into the constitution of the network. Regardless how this objective is attained, the neural network must exhibit a *dynamic* behavior, which, in turn, means that the network must have *memory.* In other words, the terms *time, dynamic,* and *memory,* are all intimately related to each other.

In this article, we describe some important techniques for temporal signal processing, based on neural networks. These techniques may be grouped under two broadly defined classes (Giles, 1994):

1. Temporal neural networks with observable states.
2. Temporal neural networks with hidden states.

The term *state,* as used herein, refers to a vector of time-dependent variables that are involved in a partial or full description of the neural network's dynamic behavior.

Temporal neural networks with observable states include the *Narendra-Parthasarathy models* (Narendra and Parthasarathy, 1990), and their extensions using the *gamma memory* (deVries and Principe, 1992). In these models, a "static" multilayer perceptron is given memory and therefore made dynamic by the addition of a tapped-delay-line (or gamma memory) at the input end, output end, or both ends of the network.

Temporal neural networks with hidden states include finite-duration impulse-response (FIR) multilayer perceptron (Wan, 1990), time-delay neural network (Lang and Hinton, 1988), multilayer perceptrons with local feedback acting around each neuron of the network (Back and Tsoi, 1991), recurrent neural networks (Williams and Zipser, 1989), and pipelined recurrent neural networks (Haykin and Li, 1995). These models are all characterized by the presence of hidden neurons that exhibit dynamic behaviors of their own. Thus, although both classes of temporal neural networks do contain hidden neurons, the class of networks with hidden states differs from the class of networks with observable states in that in the former case the individual neurons are dynamic (i.e., they have memory), whereas in the latter case they are all static (i.e., memoryless).

In the next two sections we describe the aforementioned examples of these two classes of temporal neural networks in some detail; the discussion also includes their applications. The article concludes with some final remarks.

65.2 Temporal Neural Networks with Observable States

Narendra-Parthasarthy Models

Figure 65.1 shows four different single-input single-output (SISO) discrete-time models for solving nonlinear difference equations (Narendra and Parthasarathy, 1990). The nonlinear functions $g(\cdot)$ and $f(\cdot)$ are approximated by standard multilayer perceptrons under fairly weak conditions. The boxes labeled z^{-1} represent unit-delay elements. Among the four Narendra-Parthasarathy models described here, model IV is analytically the least tractable. Accordingly, for practical applications (e.g., system identification), models I to III are the preferred ones. In particular, model II is well suited for control applications.

To train the multilayer perceptrons [i.e., the approximators of $f(\cdot)$ and $g(\cdot)$] in these models, the most straightforward procedure is to use the standard back-propagation algorithm, which

0-8493-8343-9/97/$0.00+$.50

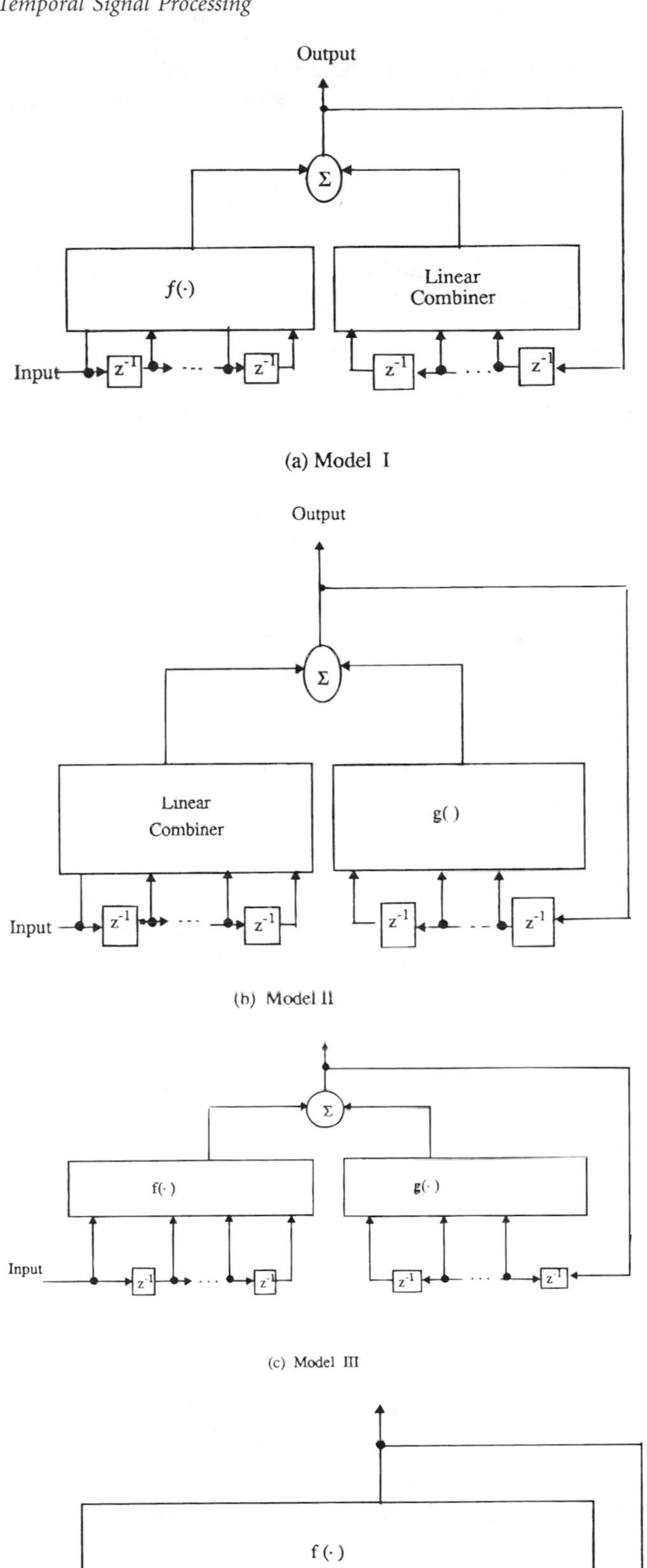

Figure 65.1 The Narendra-Parthasarathy models.

is based on the method of steepest descent. Such a procedure is said to be a first-order method, in that it involves instantaneous estimation of *gradient* of the error performance surface with respect to the free parameters of the multilayer perceptron. Feldkamp et al. (1994) have extended the dynamic gradient formalism of the Narendra-Parthasarathy models to include the training of recurrent neural controllers by second-order methods to include the training of recurrent neural controllers by second-order methods for real-world applications.

It is also of interest to note that model I includes the nonlinear one-step predictor as a special case. Figure 65.2 shows the block diagram of such a predictor. The structure of Figure 65.1a reduces to that of Figure 65.2 simply by omitting the feedback paths involving the output signal, and omitting the connecting link to the input.

Gamma Model

The memory structure built into the Narendra-Parthasarathy models of Figure 65.1 and the non-linear predictor of Figure 65.2 consists of a tapped-delay-line, that is, a cascade of ideal delay units denoted by the operator z^{-1}. A natural extension of the tapped-delay-line is obtained by substituting a leaky integrator for each delay unit, as depicted in Figure 65.3, where μ is an adjustable parameter. The resulting memory structure is called the *gamma model* (memory) (deVries and Principe, 1992). The primary function of the gamma memory is to provide a representative part of the signal's past behavior for use in current processing. The gamma memory derives its name from the fact that

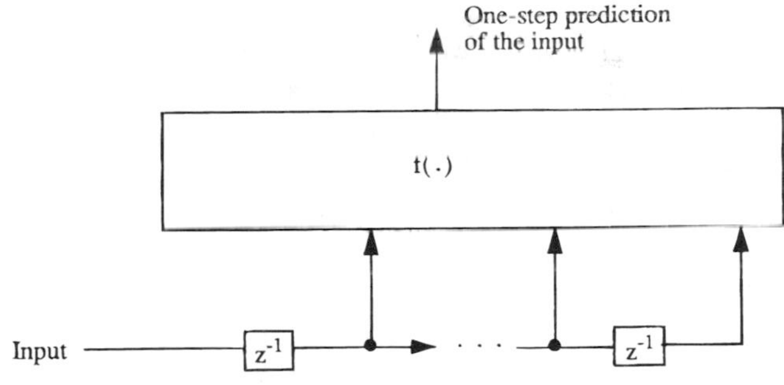

Figure 65.2 One-step predictor.

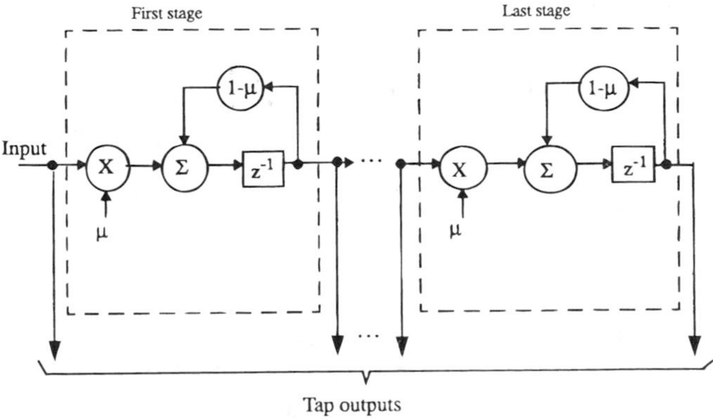

Figure 65.3 Gamma memory.

its impulse response, measured from the input up to some tap inside the structure, is the integrand of the gamma function (Principe et al., 1993).

In the context of a temporal neural network, the advantage of gamma memory is that the memory depth can be controlled by adapting the time constant of the leaky integrator, using gradient descent on the output mean-square error. So with a fixed number of taps, the memory depth can be adapted to find, in time, the best region of support to solve the processing task. The gamma model has been applied to time series prediction (Principe et al., 1992), and echo cancellation (Palkar and Principe, 1994). Recently, it has been also extended to two-dimensional signal processing, and applied to automatic target recognition.

65.3 Temporal Neural Networks with Hidden States

FIR Multilayer Perceptrons

In the FIR multilayer perceptron, each synapse of a neuron is represented by a discrete-time FIR filter as shown in Figure 65.4, where the coefficients $w_0, w_1, \ldots, w_p$ represent adjustable parameters. Thus, finite memory is built right into the synapse of an artificial neuron, much like the way it is with a biological synapse (Scott, 1977). In any event, with the synapse of each artificial neuron of a multilayer perceptron modeled as in Figure 65.4, the resulting neural network structure extends the capability of a standard multilayer perceptron by including memory that is distributed throughout the network. The result is a dynamic model with hidden states.

To train the FIR multilayer perceptron, we may use a supervised learning algorithm in which the actual response of each (visible) neuron in the output layer is compared with a target (desired) response at each time instant. The free parameters of the network are then adjusted using the standard back-propagation algorithm. To begin with, the network is unfolded in time by starting at the input (sensory) layer and then moving forward through the network, layer by layer. The result of this forward unfolding in time process (Haykin, 1994) is a "static" network, to which the standard back-propagation algorithm is applied in the usual way. However, this rather straightforward approach is handicapped by the following negative attributes (Wan, 1990):

- The unfolded structure is highly redundant.
- There is a loss of a sense of symmetry between the forward propagation of states and the backward propagation of terms needed to calculate instantaneous error gradients.
- There is no nice recursive formula for propagating the error terms.
- There is need for global bookkeeping, in that we have to keep track of which "static" parameters are actually the same in the unfolded-in-time network that is equivalent to the original FIR multilayer perceptron.

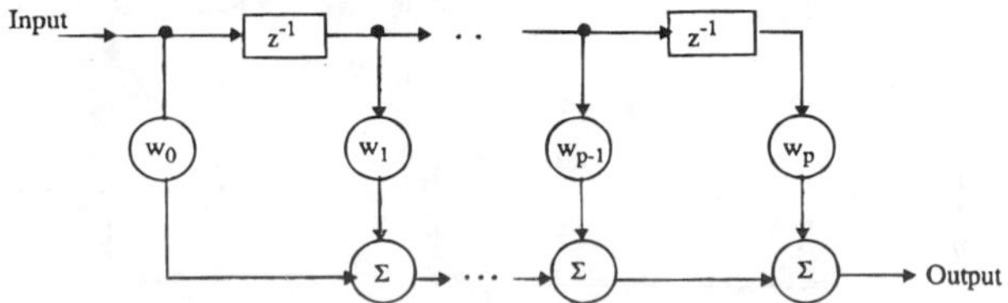

Figure 65.4 FIR model of a synapse.

To overcome these limitations, it is recommended that we use the temporal back-propagation algorithm devised by Wan (1990). For the derivation of this algorithm (Wan, 1990; Haykin, 1994) it is assumed that the free parameters of the FIR multilayer perceptron are fixed for all gradient calculations. Clearly, this is not a valid assumption during actual adaptation. Accordingly, discrepancies in performance will arise between the temporal back-propagation algorithm applied directly to the FIR multilayer perceptron and the standard back-propagation algorithm applied to the unfolded-in-time equivalent network. However, these discrepancies are usually of a minor nature (Wan, 1990). Indeed, for small values of the learning-rate parameter used in computing the adjustments to the free parameters, the differences in the learning characteristics of these two algorithms are negligible for all practical purposes.

Wan (1994) has used an FIR multilayer perceptron to perform nonlinear prediction on an actual time series representing the "chaotic" intensity pulsations of an NH_3 laser. The generalization performance of the network was tested using the method of recursive (iterated) prediction; this is a standard procedure in chaotic signal processing (Haykin and Li, 1995). Specifically, after the completion of network training, recursive prediction proceeds by initializing the network with enough samples of test data not seen before. Then, after each prediction is made, the predicted sample is fed back to the input layer of the network, thereby dispensing with an original data sample. This procedure is continued until the data vector applied to the input layer of the network consists entirely of predicted samples. Thereafter, the network operates in an autonomous fashion. Using such a procedure, Wan (1994) was able to demonstrate a superior performance of the FIR multilayer perceptron for modeling the chaotic time series produced by the intensity pulsations of an NH_3 laser, compared to a linear autoregressive (AR) model (of corresponding order) using a standard least-squares method for computing its coefficients.

Time-Delay Neural Network (TDNN)

The TDNN, first described in (Lang and Hinton, 1988), is a multilayer feedforward network whose hidden neurons and output neurons are repeated across time. It is indeed a simplified version of the FIR multilayer perceptron, as illustrated in Figure 65.5 for the example of a TDNN with the input applied to the synapse of a neuron is repeated several times. Simply stated, the adjustable coefficients of the discrete-time synaptic filter of each neuron in the TDNN are all assigned a common value.

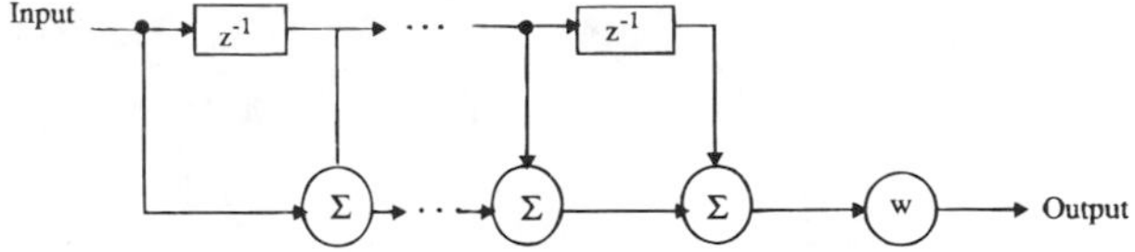

Figure 65.5 Model of a synapse in the TDNN network.

The TDNN has been applied to speech recognition (Waibel et al., 1989), and sonar signal processing (Smith et al., 1992).

Multilayer Perceptrons with Local Feedback

In this type of temporal neural network, the synaptic filters are modeled as infinite-duration impulse response (IIR) filters, which means that each synapse of the neuron has local feedback around it, as indicated in Figure 65.6. To be more specific, the neuron model may include local output feedback as in Figure 65.6a, or local activation feedback as in Figure 65.6b. For more details, see Back and Tsoi (1991), Frasconi et al. (1992), Leighton and Conrath (1991), and Podder and Unnikrishnan (1991).

Recurrent Neural Networks

The recurrent neural network devised by Williams and Zipser (1989) is a fully connected neural network with abundant feedback, as depicted in Figure 65.7 for the example of four processing units. The network has two important properties:

- It contains hidden neurons.
- It has arbitrary dynamics.

Of particular interest is the fact that the network has the ability to deal with a time-varying input or output through its own temporal operation.

The algorithm used to train such a recurrent neural network is called the real-time recurrent learning (RTRL) algorithm (Williams and Zipser, 1989); the term "real-time" is used to signify the fact that the learning process is performed on a continuous basis while the input signal is being processed. The algorithm is based on the method of steepest descent applied to the error surface.

Unfortunately, the computational complexity of the RTRL algorithm increases exponentially with the number of neurons in the network, which limits the scope of its practical applications. One way of overcoming this limitation is to use the following two structural properties (Haykin and Li, 1995).

- Modularity, whereby a number of modules are cascaded together, with each module consisting of a recurrent neural network.
- Weight sharing, whereby all the modules share the same synaptic weight matrix

The resulting structure is called a pipelined recurrent neural network (PRNN). An application of the PRNN to nonlinear adaptive differential pulse-code modulation (ADPCM) of speech signals is described in (Haykin and Li, 1993).

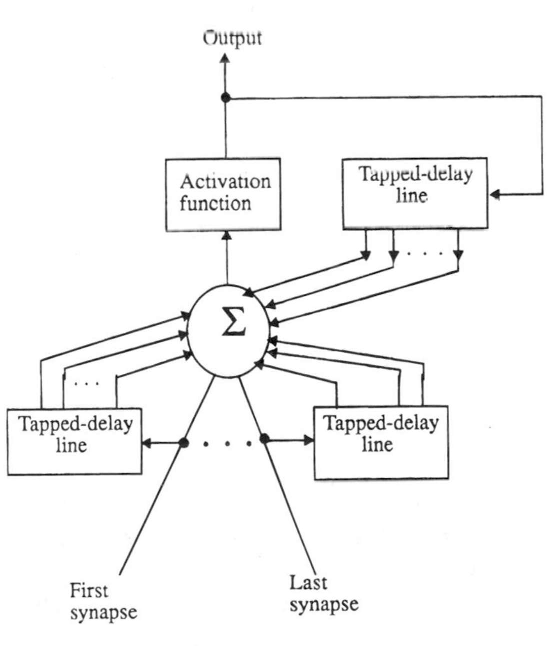

(a) Local output feedback

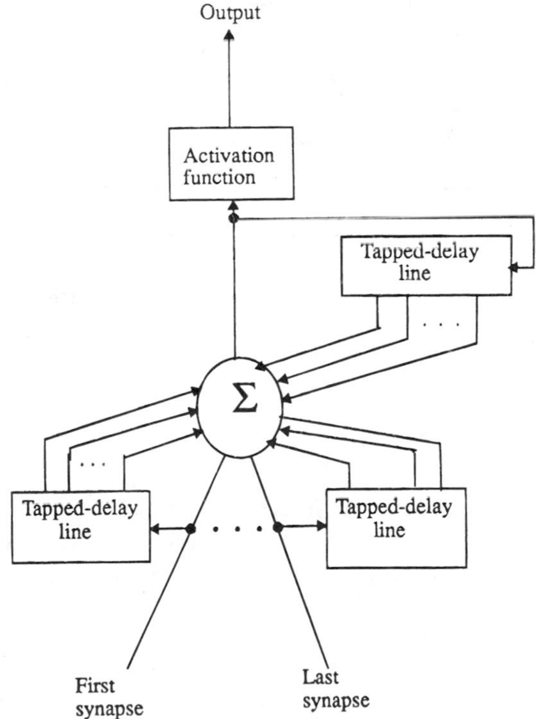

(b) Local activation feedback

Figure 65.6

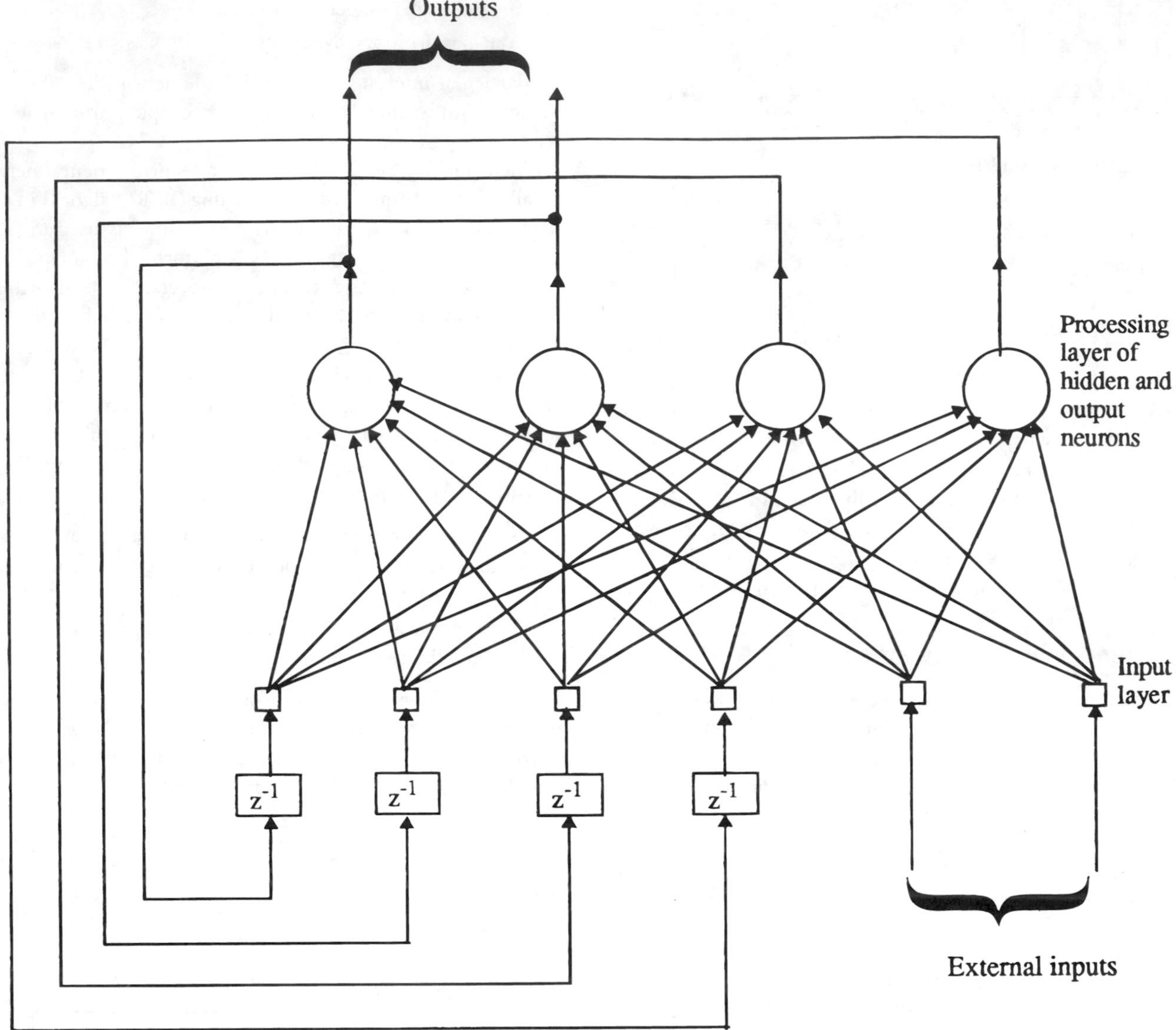

Figure 65.7 Recurrent network.

65.4 Conclusions

Temporal neural networks, incorporating time into their design, play a key role in signal processing applications where the source responsible for signal generation is governed by two properties:

- Nonlinearity.
- Nonstationarity.

These properties are indeed exhibited by many of the signals encountered in real-life applications.

The dynamic characteristics of the underlying physical mechanisms of signal generation can be learned in one of two ways:

- By having the network undergo a training session with input-output examples of the environment of interest, as in the Narendra-Parthsarathy models and FIR multilayer perceptrons.
- By performing continuous learning on the network, as in the standard recurrent neural network and its extension, namely, the PRNN.

Both of these approaches have their own specific areas of application.

For additional information on temporal signal processing, the interested reader is referred to Chapter 13 of the book by Haykin (1994), and the *Special Issue of IEEE Transactions on Neural Networks,* March 1994, devoted to dynamic recurrent neural networks under the editorship of C. L. Giles, G. M. Kuhn, and R. J. Williams.

References

Back, A. D., and Tsoi, A. C., 1991. FIR and IIR synapses, a new neural network architecture for time series modelling, *Neural Computation,* 3:337–350.

deVries, B., and Principe, J., 1992. The gamma model—a new neural network for temporal processing, *Neural Networks,* 5:565–576.

Feldkamp, L. A., Puskorius G. V., Davis, Jr., Davis, L. I., and Yuan, F., 1994. Enabling concepts for application of neurocontrol. *Proc. Eighth Yale Workshop on Adaptive and Learning Systems,* pp. 168–173, Yale University.

Frasconi, P., Gori, M., and Soda, G. 1992. Local feedback multilayered networks, *Neural Computation,* 4:120–130.

Giles, C. L. 1994. Dynamically-driven recurrent neural networks: models, training algorithms, and applications, Tutorial Notes, *WCNN-94,* San Diego, CA.

Haykin, S., and Li, L. 1995. Nonlinear adaptive prediction of nonstationary signals, *IEEE Trans. Signal Processing,* 43:526–525.

Haykin, S., and Li, X. B. 1995. Detection of signals in chaos, *Proc. IEEE,* 83:95–122.

Haykin, S. 1994. *Neural Networks: A Comprehensive Foundation,* Macmillan, New York, NY.

Haykin, S., and Li, L. 1993. 16 kb/s adaptive differential pulse-code modulation of speech, *Applications of Neural Networks to Telecommunications,* Alspector J, Goodman, R., and Brown, T. X., ed. 132–137, Lawrence Erlbaum, Hillsdale, NJ.

Lang, K. J., and Hinton, G. E. 1988. The development of the time-delay neural network architecture for speech recognition, Technical Report CMU-CS-88–152, Carnegie-Mellon University, Pittsburgh, PA.

Leighton, R. R., and Conrath, B. C. 1991. The autoregressive backpropagation algorithm, *Int. Joint Conf. on Neural Networks,* 2:369–377, Seattle, WA.

Narendra, K. S., and Parthasarathy, K. 1990. Identification and control of dynamical systems using neural networks, *IEEE Trans. Neural Networks,* 1:4–27.

Palkar, M., and Principle, J. 1994. Echo cancellation with the gamma filter, *Proc. ICASSP-94,* 3:369–372, Adelaide, Australia.

Podder, P., and Unnikrishnan, K. P. 1991. Nonlinear prediction of speech signals using memory neuron networks, *Neural Networks for Signal Processing,* in B. H. Juang, S. Y. Kung, and C. A. Camm, editors. *Proc. 1991 IEEE Workshop,* 395–404.

Principe, J. C., deVries, B., and de Oliveira, P. G., 1993. The gamma filter—a new class of adaptive IIR filters with restricted feedback, *IEEE Trans. Signal Processing,* 41:649–656.

Principle J., deVries, B., Kuo, J-M., and de Oliveira, P. 1992. Modeling applications with the focused gamma network, in *Advances of Neural Information Processing Systems* 4, pp. 143–150, Morgan Kaufmann.

Scott, A. C. 1977. *Neurophysics,* John Wiley, & Sons, New York, NY.

Smith, D. J., Bailey, T. C., and Munford, G. 1993. Robust classification of high-dimensional data using artificial neural networks, *Statistics and Computing,* 3:71–81.

Waibel, A., Hanazawa, T., Hinton, G., Shikamo, K., and Long, K. 1989. Phoneme recognition using time-delay neural networks, *IEEE Trans. Acoustics, Speech, and Signal Processing,* 37:328–339.

Wan, E. A. 1994. Time series prediction by using a connectionist network with internal delay lines, in *Time Series Prediction: Forecasting the Future and Understanding the Past,* A. S. Weigend and N. A. Gershenfeld, eds., 195–217, Addison-Wesley, Reading, MA.

Wan, E. A. 1990. Temporal backpropagation for FIR neural networks, *Int. Joint Conf. on Neural Networks,* 1:575–580, San Diego, CA.

Williams, R. J., and Zipser, D. 1989. A learning algorithm for continually running fully recurrent neural networks, *Neural Computation,* 1:270–280.

66

Feature Selection for Pattern Recognition Using Multilayer Perceptrons

Dennis W. Ruck
Air Force Institute of Technology

Steven K. Rogers
Air Force Institute of Technology

66.1 Introduction 916
Pattern Recognition Overview • The Feature Selection Problem • Feature Saliency
66.2 Background 918
MLP Architecture • MLP Training
66.3 Methodology 918
Feature Saliency Method • Automatic Selection of Features • Computation of Partial Derivatives
66.4 Applications 920
XOR Problem • Breast Cancer Detection
66.5 Conclusions 921

66.1 Introduction

Pattern Recognition Overview

Pattern recognition (PR) is the process of assigning a class label to a signal. It is an information reduction process. For example, you are performing PR as you read this article. Each letter on the page represents a signal to which you assign a class label, namely, the identification of the character.[1] The process of listening to a person speak and understanding the spoken word is another example of pattern recognition. When a radiologist views a breast mammogram looking for telltale signs of cancer, he is performing a high level of PR.

We attempt to automate the process of PR using computers. The basic process of performing PR using a computer is schematized in Figure 66.1. The signal to be recognized is first *segmented* from the input. The process of segmentation is necessary so that the entire input signal need not be processed. For example, in breast cancer detection the segmentation process will find regions of interest (ROI), perhaps masses or regions containing microcalcifications, using a coarse filtering scheme. The input image may be 1000 × 2000 pixels and the extracted ROIs may be 32 × 32 pixels. Even if 100 ROIs are extracted from the image, only 10 percent of the image pixels need to be processed further.

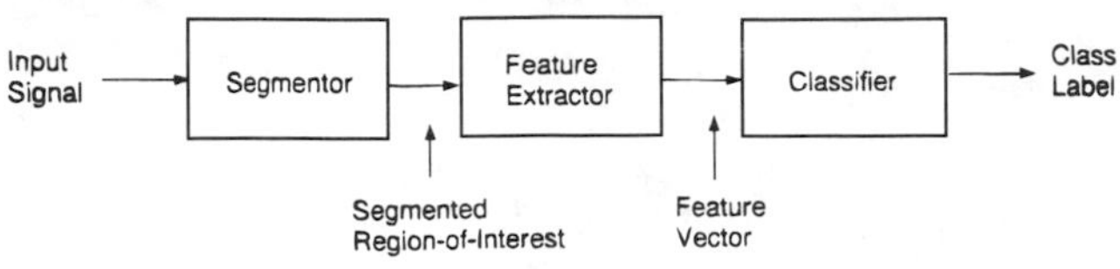

Figure 66.1 The pattern recognition process.

The next step in the PR process is *feature extraction.* The purpose of feature extraction is again data reduction. Continuing with the breast cancer detection example, each ROI contains 32^2 pixels, which is too many to use as input to a classifier. Additionally, the pixel data possesses information that is irrelevant to the determination of class. For example, the absolute brightness, orientation and the position of the mass within the ROI is unimportant. The feature extraction process is designed to generate a set of features that are invariant to these unimportant factors yet sensitive to those properties which help determine the class label of the mass. The output of the feature extractor is a *feature vector* which is the input to the final stage.

The final stage in the PR process is *classification.* The purpose of classification is to assign a label to the input feature vector. In the breast cancer example, the label is either benign or malignant, a two class problem. A wide variety of techniques are available for design of the classifier. Traditionally, statistical approaches have been used for classifier design (Duda and Hart, 1973). The classifier implemented in this article, the multilayer perceptron (MLP), is loosely patterned after the connectivity of neurons observed in the brain. Since the only existent proof that PR can be performed is the human brain, it makes sense to take clues from the architecture of the brain to design a PR system. More importantly, from a PR perspective it can be shown that

[1] Note: Some researchers have argued that reading involves *word* recognition and not *character* recognition (O'Hair, 1990).

0-8493-8343-9/97/$0.00+$.50

the MLP generates an approximation to the Bayes optimal discriminant function (Ruck et al., 1990).

The Feature Selection Problem

The classifier needs to be trained using a set of feature vectors representing the classes to be recognized. Developing such a training set involves collecting the input data, segmenting it, extracting features, and labeling each feature vector as to its class. This process of collecting labeled data is time consuming and costly; hence, training sets are normally limited in size.

No single feature will allow error-free labeling of input signals. As more features are added, we expect better discrimination between the classes. However, we cannot increase the number of feature vectors without limit due to the curse of dimensionality (Bellman, 1961) which says that as the number of features increases the amount of training data must increase exponentially for accurate classifier design. Other work shows that the number of feature vectors in the training set should grow proportionally to the number of features being used. Foley (1972) states that as a general rule of thumb at least $3N$ training vectors per class should be used for classifier design where N is the number of features. Foley derived this result for the special case using Gaussian data; hence, his result should be considered a lower bound on the number of training vectors required. Tou and Gonzalez (1974) suggest that the number of training vectors be greater than $10W$ where W is the number of free parameters in the classifier which is directly related to the number of features. A similar result for feedforward neural networks has been reported by Baum and Waussler (1989).

We have a conflict here between a desire to use a large number of features to achieve accurate classification and the need to limit the number of features because training data are limited and accurate classifier design requires a large number of training vectors relative to the number of features. Hence, we must find a way to limit the number of features.

Problem Statement

Given a set of N features for a PR problem, select a subset of $p < N$ features for use in the system design that yield increased recognition performance over the original feature set.

It is generally not practical to find the subset that provides the optimal classification performance since that entails examining every possible subset of size p. For a problem of any practical size, the number of subsets to be examined is prohibitive. For example, if $N = 100$ and $p = 10$, then the number of subsets to be examined is

$$C(N, p) = \frac{N!}{(N-p)!p!} = 10^{13}$$

There are several methods for selecting the subset of features. We will briefly review these methods. The probability of error method (Roggemann, 1989) computes the probability of error, (P_e), achieved by the classifier using only one feature at a time. The features are then ranked from lowest to highest resulting probabilities of error and the p features yielding the lowest individual error rates are chosen. This method completely ignores any interaction between features. It is entirely possible that the top two features may be nearly perfectly correlated, therefore the second feature provides no additional information.

Two related methods of feature selection that consider some feature interaction are the Add-on and Knock-out methods (Parsons, 1987). In the Add-on method (Goldstein, 1976), the classifier is initially designed using each feature individually and P_e is measured. The feature resulting in the lowest P_e is selected as the first feature. Next each of the $N - 1$ remaining features is combined with the first selected feature and a classifier is designed. The classifier resulting in the best P_e is used to select the second feature. Add-on builds the feature set in this way until p features have been selected. The Knock-out method (Sambur, 1975) is similar except that the feature set is sequentially pruned down to p features. Neither of these methods considers the interaction of all features simultaneously and, hence, are necessarily suboptimal. They may choose poor feature sets. The method described in this article considers the interaction of all features simultaneously.

Feature Saliency

The Feature Saliency method described in this article determines the feature subset directly from the weights in a trained multilayer perceptron. (See also Belue and Bauer, 1995; Priddy et al., 1993; Ruck, 1990; Ruck et al., 1990 for more information.) Because the weights in the trained MLP encode the processing of the features for classification of the input, it is logical to examine the weights for clues as to the relative importance of the features. Feature saliency is based on sensitivity analysis of the outputs of the MLP to its inputs. A feature is termed important or salient if the MLP outputs are sensitive to changes in the features values. If a feature can change drastically and cause few changes in the output, then that feature is unimportant and should be discarded. As an example, consider a feature which is totally independent of the label assigned to the feature vector (this defines a noisy input or noise feature). When the MLP is trained, this feature should have little if any effect on the outputs of the network since it conveys no information regarding the class label. That is, the MLP through observation of numerous training samples where the noise feature is changing without correlation to the class label will contain weights that give little emphasis to this feature. This property will be used later to determine salient and non-salient features. An advantage of feature saliency is that since it is based on the trained MLP it should take into account the interactions between features. Also, it does not require training of numerous classifiers; however, to obtain reliable feature saliency values we train the given classifier several times.

The next section provides some background information necessary to understand the methodology presented in the section on methodology. Several real-world example applications are discussed in the section on applications. Conclusions are presented in the final section.

66.2 Background

This section describes the multilayer perceptron architecture and the training methodology used in this article.

MLP Architecture

The multilayer perceptron is shown in Figure 66.2. The input to the MLP is the feature vector extracted from the signal. The first layer of nodes are simple fan-out units that distribute the inputs to all nodes in the first hidden layer. The *i*th component of the feature vector is designated x_i. The *j*th output of the MLP is denoted z_j. The MLP shown can be denoted in several different ways: a single hidden layer network, a two-layer network (referring to the number of layers of connecting weights, and a three-layer network (referring to the input layer, hidden layer, and output layer). In this article, we use the first method of describing the net in terms of the number of hidden layers.

Although only a one hidden layer network is depicted, the algorithm is valid for networks with an arbitrary number of hidden layers. We number the layers as follows: the input layer is the zeroth layer; the hidden layer is the first layer; the output layer is the second layer. The weights connecting the input layer to the hidden layer are the first layer weights, w^1_{ji}; the weights connecting the hidden layer to the output layer are the second layer weights, w^2_{kj}. The output of the *i*th node in the *l*th layer is designated x^l_i. The weight connecting this node to the *j*th node in the $l+$ 1st layer is designated w^{l+1}_{ji}. The outputs of the *l*th layer can be formed into a vector, $\mathrm{x}^l = (x^l_1, x^l_2, \ldots, x^l H^l)^T$ where H^l is the number of nodes in the *l*th layer and ^{T} is the transpose operator. The collection of weights from the $l-$ 1st layer to the next layer is a matrix, W^l, where *j*th row of the matrix contains the weights connecting nodes in the $l-$ 1st layer to the *j*th node in the *l*th layer. Thus, if the $l-$ 1st layer contains H^{l-1} nodes and the next layer contains H^l nodes, then x^l is an H^l-dimensional vector and W is $H^l \times H^{l-1}$.

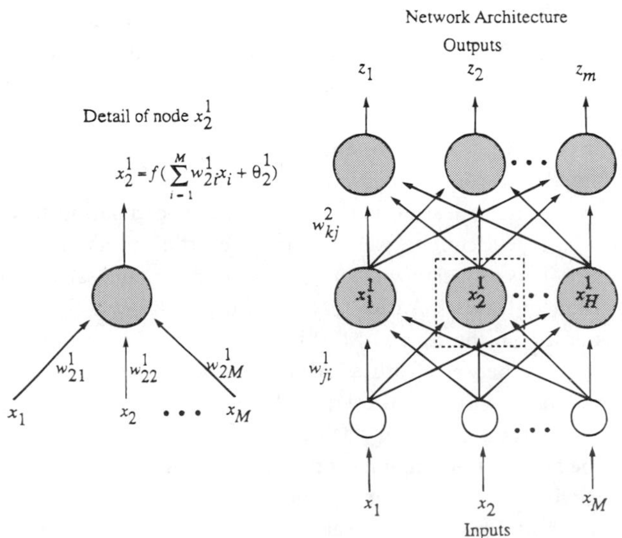

Figure 66.2 Multilayer perceptron.

The output of a node is computed from the weighted sum of the inputs to the node from the previous layer. The weighted sum of inputs is termed the activation and is denoted by *a*. Each node has an associated activation function, $f(a)$. The output of the node is the result of applying the activation to this function; hence, $x^l_i = f(a^l_i)$. The activation function is generally chosen from one of three possibilities:

$f(a) = a$, linear activation

$f(a) = \tanh(a)$, hyperbolic tangent or symmetric sigmoid activation

$f(a) = \dfrac{1}{1+e^{-a}}$, sigmoid activation

The derivative of the activation function with respect to its input is required in the feature saliency computation. Using the above activation functions, the derivative can be easily computed as follows:

$f'(a) = 1$, linear activation derivative

$f'(a) = 1 - f^2(a)$, hyperbolic tangent or symmetric sigmoid activation derivative

$f'(a) = f(a)[1 - f(a)]$, sigmoid activation derivative

Cybenko (1989) has shown that a single hidden layer network with linear output nodes and sigmoidal hidden layer nodes can approximate arbitrarily closely any continuous function on the unit hypercube. Hence, we use only single hidden layer networks in this article. It is possible in certain instances, however, for a two-hidden layer network to train more quickly using fewer nodes (Chester, 1990), so it is always wise to consider using more than one hidden layer.

MLP Training

The MLP is trained using the standard backpropagation training algorithm that was originally developed by Werbos (1974) and later popularized by Rumelhart et al. (1986). (See Lippmann, 1987, for a description of this algorithm.) As part of this algorithm, we need to initialize the weights in the network to random values. Typically, a uniform distribution on [−0.5, 0.5] is used. For more information on the use of multilayer perceptrons for pattern recognition, see Rogers and Kabrisky (1991).

66.3 Methodology

This section describes the method of computing feature saliency and gives the specific implementation equations. Feature saliency is based on the idea that once the MLP is trained, that the weights by encoding a classification rule for the input feature vector necessarily represent the relative importance of the input features. Since the network is trained using all the input features, we expect that feature saliency yields a ranking based on the interaction of the weights. A feature is termed salient if the outputs of the MLP are sensitive to changes in the feature. The following subsections describe the basic feature saliency method, the automatic selection of features using noise, and the computation of the partial derivatives.

Feature Saliency Method

Feature saliency is computed as follows:

1. Train an MLP using the given training data.
2. Compute the derivative of each output with respect to each input, $\frac{\delta z_j}{\delta x_i}$.
3. Compute the saliency of each feature

$$\Lambda_i = \sum_p \sum_j \left|\frac{\partial z_j}{\partial x_i}\right|$$

where j ranges over all outputs and p ranges over all feature vectors in the training set.

We will see in the section on the computation of the partial derivatives that $\frac{\delta z_j}{\delta x_i}$ is dependent on the current activations in the network; hence, the summation over all the vectors in the training set. Also, we are only interested in the overall sensitivity of the MLP to its inputs; hence, we sum over all outputs the absolute value of $\frac{\delta z_j}{\delta x_i}$.

Since the computation of feature saliency is dependent on the trained weights in the network and the trained weights are dependent on the randomly chosen initial weights, then the actual computed saliency for each feature will be a random variable. Therefore, we train our MLP a number of times where each trained MLP uses a different set of initial weights. In other words, we are performing a Monte Carlo analysis. Let Λ_i^n represent the saliency of feature x_i computed from the nth MLP. Then the overall feature saliency for x_i is the average saliency over all the networks trained, i.e.,

$$\Lambda_i = \frac{1}{T} \sum_{n=1}^{T} \Lambda_i^n$$

where T is the total number of MLPs trained (typically, 10 or more).

The features are now ranked from highest average feature saliency to lowest. If a set of p features is desired, then the p features with the highest average feature saliency are selected. In the next section, we describe a method for determining a threshold on average feature saliency to automatically determine which features to retain.

The feature saliency method has been shown to select features consistent with minimizing the Bayes risk (Priddy et al., 1993).

Automatic Selection of Features

Once the feature saliencies Λ_i are calculated, we need to determine which features to keep and which to discard. Through the introduction of noise as a feature, we can set a threshold for separating salient from unimportant features (Belue, 1992; Belue and Bauer, 1995).

In Belue's method (Belue and Bauer, 1995), the feature vector is augmented by a noise variable that is simply a random variable, x_{noise}, with a distribution that is independent of the label associated with the feature vector. The new feature vector is

$$\mathbf{x}_{\text{new}} = (\mathbf{x}_{\text{old}}^T, x_{\text{noise}})^T.$$

Typically, x_{noise} is a uniform or Gaussian distribution. The type selected depends on the normalization method used for the data. If the data are normalized such that each feature lies in the range [0,1], then a uniform distribution on that same range should be selected. If the data are normalized such that each feature has zero mean and unit variance, then a Gaussian distribution with those parameters should be selected. The network is now trained T times with the noise augmented feature set.

The feature saliency is computed as before for each feature including the noise feature. Additionally, the distribution of the noise feature's saliency is determined. A threshold can then be set using the pth percentile of the distribution. Any feature with an average saliency below this level is considered to be unimportant to the classification task.

Computation of Partial Derivatives

All that remains to complete the feature saliency method is to derive the partial derivatives, $\frac{\delta z_j}{\delta x_i}$, used in the computations. These partials are easily derived using the ordinary chain rule of differential calculus. The derivations will not be shown here only the resulting implementation equations. The implementation equations will be presented for an MLP with an arbitrary number of layers in both a scalar and vector form. Also, the specific equations for the most common MLP, the single hidden layer architecture, will be presented.

Scalar Form

The computation of $\frac{\delta z_j}{\delta x_i}$ proceeds as follows:

1. For each layer starting with the output layer, compute the following

$$\begin{aligned} \gamma_{jm}^l &= \delta_{jm} f'(a_m^l), && \text{output layer} \\ &= f'(a_m^l)\Sigma_k \gamma_{jk}^{l+1} w_{km}^{l+1}, && \text{hidden layers} \end{aligned}$$

where δ_{jm} is the Kronecker delta. The above computation is performed for each output, j, and each node, m, in the lth layer. The quantity γ_{jm}^l is the partial derivative of the jth output with respect to the mth node's activation α_m^l, in the lth layer.

2. Then compute for the input layer

$$\frac{\partial z_j}{\partial x_i} = \sum_m \gamma_{jm}^1 w_{mi}^1$$

where the sum is over all nodes in the first hidden layer. Note that the computation of $\delta z_j/\delta x_i$ depends not only on the weights in the network but also the current state of the network as indicated by the presence of the activation values in the formulas; hence, feature saliency, Λ_i, involves a summation over the training set data.

Vector Form

The above computations can be stated more cleanly in a vector-matrix notation for implementation in a high-level language such as MATLAB®. Consider implementation for an MLP with $L-1$ hidden layers

1. For the output layer, compute

$$\Gamma^L = \text{diag}[f'(\mathbf{a}^L)]$$

where $\mathbf{a}^L$ is the vector of activations for the output layer nodes, $f'(\mathbf{a}^L)$ is the vector of derivatives of the outputs for this layer, and the "diag" operator creates a diagonal matrix where the elements along the diagonal are the vector that the operator takes as input.

2. For each hidden layer, compute

$$\Gamma^{l-1} = \{\Gamma^l * \mathbf{W}^l\} .* \{\text{ones}(H^l, 1) * f'(\mathbf{a}^{l-1})^T\}$$

where H^l is the number of nodes in layer l, ones$(H^l, 1)$ is an H^l element vector of all ones, "*" denotes the usual matrix multiplication, ".*" denotes element by element multiplication of the two matrices, and ^{T} denotes the transpose operator.

3. For the input layer, compute

$$\Gamma^0 = \Gamma^1 * \mathbf{W}^1$$

Then

$$\Gamma^0_{ji} = \frac{\partial z_j}{\partial x_i}$$

Hence, the ith column of Γ^0 represents the derivatives of the outputs with respect to the ith feature.

Single Hidden Layer Implementation Equations

We can now present the implementation equations for the single hidden layer network. The derivative is calculated using (scalar form)

$$\frac{\partial z_j}{\partial x_i} = \sum_{k=1}^{H^1} f'(a_j^2) w_{jk}^2 f'(a_k^1) w_{ki}^1$$

For a typical single hidden layer network with linear output nodes and sigmoidal activation functions on the hidden layer nodes, the previous equation becomes

$$\frac{\partial z_j}{\partial x_i} = \sum_{k=1}^{H^1} w_{jk}^2 f(a_k^1)[1 - f(a_k^1)] w_{ki}^1$$

In vector form,

$$\Gamma^0 = [(\text{diag}(f'(\mathbf{a}^2)) * \mathbf{W}^2) * \{\text{ones}(J, 1) * f'(\mathbf{a}^1)^T\}] * W^1$$

where J is the number of output nodes. For the MLP with linear output nodes and sigmoidal hidden layer nodes, the above becomes

$$\Gamma^0 = [\mathbf{W}^2 .* \{\text{ones}(J, 1) * [f(\mathbf{a}^1) .* \{1 - f(\mathbf{a}^1)\}]^T\}] * \mathbf{W}^1$$

66.4 Applications

In this section, we examine the application of the feature saliency method for two different data sets. The first data set is the traditional XOR problem that has become a *de facto* test case for artificial neural networks. The second example is a breast cancer detection application.

XOR Problem

The XOR problem is a two class problem and a two-dimensional feature vector. Figure 66.3 shows a plot of a typical training set. The data are uniformly distributed over the square $(-1, 1) \times (-1, 1)$ in 2-space. For this problem four additional features $x_3, \ldots, x_6$ are added as useless features where each feature has a uniform distribution on (0, 1) and is independent of all other features and the class label. Belue and Bauer (1995) computed the feature saliency for this feature set using the method of introducing a noise feature to determine a threshold. The results

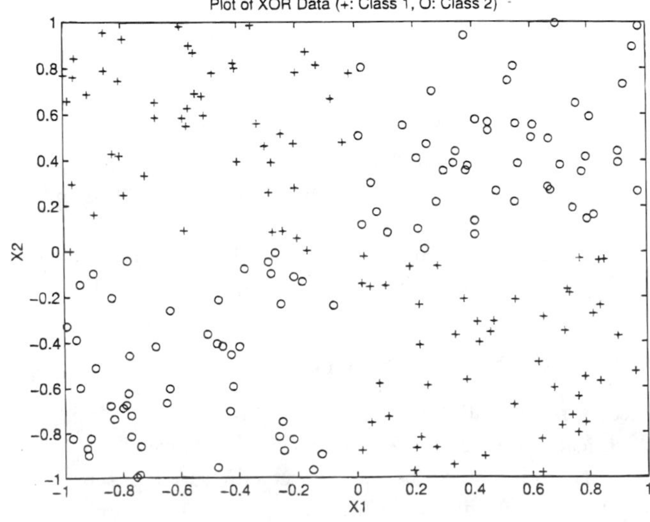

Figure 66.3 XOR data.

are shown in Table 66.1. It is immediately apparent from the feature saliencies that features x_3 through x_6 are useless and equivalent to noise for this classification problem. Assuming a Gaussian distribution for the noise saliency, the 95th percentile threshold for saliency is 0.4158. Clearly, all irrelevant features are eliminated by the threshold and only relevant features are retained. Figure 66.4 shows a plot of the assumed distribution for the noise saliency and the computed saliencies. Figure 66.5 shows the classification error rate of the MLP as a function of training using both the original feature set and the feature set selected using the saliency measure. We see that the network not only trains more quickly but also more accurately. This is not an unusual result.

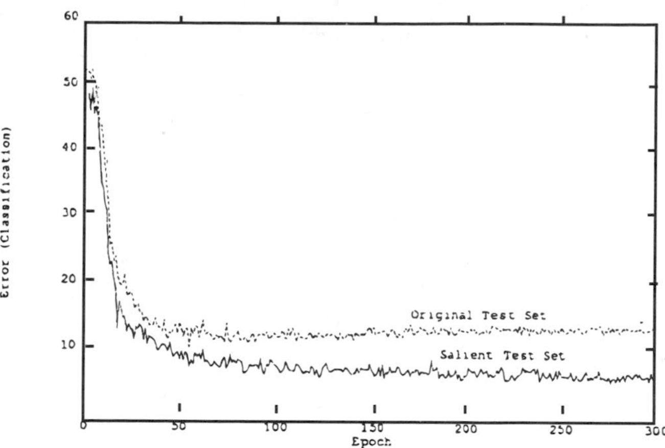

Figure 66.5 XOR training performance (from Belue and Bauer, 1995).

Breast Cancer Detection

The breast cancer database consists of 94 feature vectors: 31 malignant and 63 benign. There are 21 features that are derived from a biorthogonal wavelet decomposition applied to an ROI of size 32 × 32. A leave one out testing procedure (Fukunaga, 1990) was used for this data. For each feature vector in the database, an MLP was trained using all feature vectors except the one under consideration. Then the network is tested using the retained feature vector. This process is repeated until all feature vectors have been tested.

Table 66.1 XOR Problem—Feature Saliencies

	Λ_i	
Feature	Mean	Std Dev
x_1	2.2987	0.0927
x_2	2.2158	0.1110
x_3	0.2157	0.0954
x_4	0.3299	0.0825
x_5	0.2254	0.1062
x_6	0.3267	0.0961
x_7 (noise)	0.2934	0.0742

From Belue and Bauer, 1995.

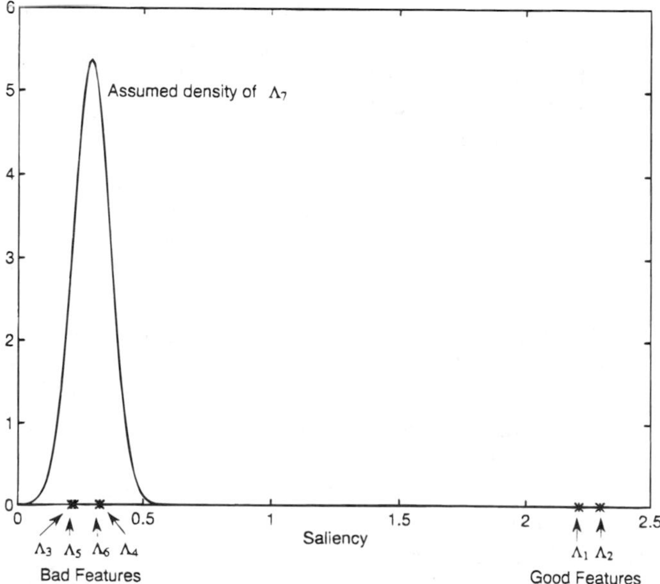

Figure 66.4 Plot of noise saliency density function and computed saliencies.

Table 66.2 Confusion Matrix on Breast Cancer Database Using All Features

Actual Class/ Assigned Class	Malignant	Benign	Accuracy Rate
Malignant	13	18	42%
Benign	7	56	89%
Overall	—	—	73%

Prior to feature saliency computation, the optimal performance on the complete feature set was found to be 73% classification accuracy using the leave out method. The confusion matrix is shown in Table 66.2.

The results of applying the feature saliency algorithm for this data are shown in Table 66.3. Using the 95th percentile on the noise saliency distribution yields a threshold of 9.85 that eliminates all but seven of the original 21 features. Training and testing an MLP as described above using the reduced feature set results in an increase in classification accuracy to 78%. The confusion matrix is shown in Table 66.4.

66.5 Conclusions

In this article, we have presented a method for selecting a subset p of features from an initial set of N features where $p < N$. This method, feature saliency, is based on a sensitivity analysis of the MLP outputs to the individual features. A method for calculating the derivatives of the outputs of the MLP with respect to the inputs was given for MLPs with an arbitrary number of layers. The equations were given in both matrix and scalar form. The specific equations for the single hidden layer network were also given.

The method was applied in two examples. The first example used the standard XOR problem with four added features as distractors. By adding an additional noise feature for calibration,

Table 66.3 Feature Saliency Results for Breast Cancer Data

	Λ_i	
	Feature	
	Mean	Std Dev
x_1	3.0003	3.3714
x_2	3.3215	3.9034
x_3	5.9368	7.2793
x_4	13.2007	17.9494
x_5	9.5691	15.9225
x_6	3.0207	3.6357
x_7	15.1136	19.7882
x_8	12.7782	17.1474
x_9	5.9238	8.2493
x_{10}	3.5143	4.3446
x_{11}	11.1934	15.0193
x_{12}	6.6860	8.5942
x_{13}	9.1896	13.6759
x_{14}	3.7308	4.3588
x_{15}	4.9961	5.6624
x_{16}	10.0983	13.1330
x_{17}	6.6545	9.0446
x_{18}	3.9698	5.1856
x_{19}	13.2059	14.9680
x_{20}	7.1637	12.3904
x_{21}	17.3424	23.0641
x_{noise}	3.6877	3.7352

Table 66.4 Confusion Matrix on Breast Cancer Database using Seven Features Selected Using Feature Saliency

Actual Class/Assigned Class	Malignant	Benign	Accuracy Rate
Malignant	15	16	48%
Benign	5	58	92%
Overall	—	—	78%

it was shown that the four useless features were identified as such. In the second example, the method was applied to a real world database for breast cancer detection. A set of wavelet features was examined using feature saliency. We found that the feature set could be reduced by two-thirds and at the same time increase classification accuracy.

Feature saliency is a useful tool for determining which features should be included in a pattern recognition system design.

Acknowledgments

The authors would like to thank William Polakowski for his help implementing feature saliency in MATLAB and applying it to the breast cancer data.

References

Baum, E. and Haussler, D. 1989. What size net gives valid generalization? *Neural Computation,* 1(1):151–160.

Bellman, R. E. 1961. *Adaptive Control Processes: A Guided Tour.* Princeton University Press, Princeton, NJ.

Belue, L. M. 1992. *Multilayer Perceptrons for Classification,* Master's thesis, Air Force Institute of Technology, Wright-Patterson, AFB, OH.

Belue, L. M. and Bauer, Jr. K. W. 1995. Determining input features for multilayer perceptrons. *Neurocomputing,* 7:111–121.

Chester, D. 1990. Why two layers are better than one, *Proc. Joint Int. Conf. Neural Networks,* 265–268, Hillsdale, NJ.

Cybenko, G. 1989. Approximations by superpositions of sigmoidal functions, *Mathematics of Control, Signals, and Systems,*

Duda, R. O. and Hart, P. E. 1973. *Pattern Classification and Scene Analysis.* John Wiley and Sons, New York, NY.

Foley, D. H. 1972. Considerations of sample and feature size, *IEEE Trans. Information Theory,* IT-18:618–626.

Fukunaga, K. 1990. *Introduction to Statistical Pattern Recognition,* 2d ed., Academic Press, San Diego, CA.

Goldstein, U. 1976. Speaker-identifying features based on formant tracks, *JASA,* 59(1):176–182.

Lippmann, R. P. 1987. An introduction to computing with neural nets, *IEEE ASSP Magazine,* vol. 4.

O'Hair, M. A. 1990. *A Whole Word and Number Reading Machine Based on Two Dimensional Low Frequency Fourier Transforms,* Ph.D. thesis, Air Force Institute of Technology, Wright-Patterson AFB, OH.

Parsons, T. W. 1987. *Voice and Speech Processing,* McGraw-Hill, New York, NY.

Priddy, K. L. Rogers, S. K., Ruck, D. W., Tarr, G. L., and Kabrisky, M. 1993. Bayesian selection of important features for feedforward neural networks, *Neurocomputing,* 5:91–103.

Rogers, S. K. and Kabrisky, M. 1991. *An Introduction to Biological and Artificial Neural Networks for Pattern Recognition,* SPIE Optical Engineering Press, Bellingham, WA.

Roggemann, M. C. 1989. *Multiple Sensor Fusion for Detecting Targets in FLIR and Range Images,* Ph.D. thesis, Air Force Institute of Technology, Wright-Patterson AFB, OH.

Ruck, D. W. 1990. *Characterization of Multilayer Perceptrons and their Application to Multisensor Target Detection,* Ph.D. thesis, Air Force Institute of Technology, Wright-Patterson AFB, Ohio.

Ruck, D. W. Rogers, S. K., and Kabrisky, M. 1990. Feature selection using a multilayer perceptron, *J. Neural Network Computing,* 2(2):40–48.

Ruck, D. W., Rogers, S. K., Kabrisky, M., Oxley, M. E., and Suter, B. W. 1990. The multilayer perceptron as an approximation to a Bayes optimal discriminant function, *IEEE Trans. Neural Networks,* 1(4):296–298.

Rumelhart, D. E. and McClelland, J. L. 1986. *Parallel Distributed Processing,* vol. 1: Foundations, MIT Press, Cambridge, MA.

Sambur, M. 1975. Selection of acoustic features for speaker identification, *IEEE Trans. Acoustics, Speech and Signal Proc.,* ASSP-23(2):176–182.

Tou, J. T., and Gonzalez, R. C. 1974. *Pattern Recognition Principles.* Addison-Wesley, Reading, MA.

Werbos, P. J. 1974. *Beyond Regression: New Tools for Prediction and Analysis in the Behavioral Sciences,* Ph.D. thesis, Harvard University, Cambridge, MA.

67

Wavelets for Pattern Recognition

George W. Rogers
Naval Surface Warfare Center

David J. Marchette
Naval Surface Warfare Center

Jeffrey L. Solka
Naval Surface Warfare Center

67.1 Wavelet-Based Segmentation 923
The 1-D Wavelet Transform • Wavelet Maxima and Contrast Boundaries • The 1-D Maxima Detection Wavelet • 2-D Wavelet Transform and Maxima

67.2 Resistive Grid Local Averaging 925

67.3 Examples 928
Local Features • Boundary Gating • Coefficient of Variation (CoV) • Simple Boundary Gating • Examples

67.1 Wavelet-Based Segmentation

The 1-D Wavelet Transform

We define the convolution of two functions written $f^*\, g(x)$ as

$$f^* g(x) = \int_{-\infty}^{\infty} f(u) g(x - u) du. \tag{67.1}$$

Let us denote the dilation of a function $\psi(x)$ by a factor s as $\psi_s(x)$ where $\psi_s(x)$ is given by

$$\psi_s(x) = \frac{1}{s}\psi\left(\frac{x}{s}\right). \tag{67.2}$$

We may then write the wavelet transform (WT) of a function at scale s and position x as the convolution product $W_s f(x) = f^*\, \psi_s(x)$. Although the parameter s can in practice vary continuously one is often interested in the case where s is sampled along the dyadic sequence $(2^j)_{j\in Z}$. The wavelet transform at scale 2^j is given by

$$W_{2^j} f(x) = f^* \psi_{2^j}(x). \tag{67.3}$$

We can assure that the whole frequency axis is covered by the wavelet by imposing the additional constraint that

$$\sum_{j=-\infty}^{\infty} |\hat{\psi}(2^j\omega)|^2 = 1. \tag{67.4}$$

Any wavelet that satisfies Equation 67.4 is known as a dyadic wavelet.

Let us next define a function $\phi(x)$ whose Fourier transform is given by:

$$|\hat{\phi}(\omega)|^2 = \sum_{j=1}^{\infty} |\hat{\psi}(2^j\omega)|^2. \tag{67.5}$$

It can be shown that the energy of $\hat{\phi}$ (ω) is concentrated in the low frequencies. Hence, we call $\phi(x)$ a smoothing function and we denote the smoothing operator S_{2^j} by:

$$S_{2^j} f(x) = f^* \phi_{2^j}(x), \tag{67.6}$$

where

$$\phi_{2^j} = \frac{1}{2^j}\phi\left(\frac{x}{2^j}\right).$$

Wavelet Maxima and Contrast Boundaries

Contrast boundaries which occur in images often separate regions of disparate textures. For this reason it is often important to be able to accurately detect these edges. Once detected these edges can be used either as a fundamental type representation of the image along the lines of Marr (1982) or as inputs to an averaging process that makes use of them.

There are many different schemes of edge detection. There is the Canny edge detector (Canny, 1986) that is essentially the first order derivative of a Gaussian. Alternatively one can use the Marr-Hildreth edge detector (Marr and Hildreth, 1980) that is the Laplacian of a Gaussian. The method that we use for the detection of contrast boundaries within an image is based on the wavelet transform.

Consider a wavelet that is the p-th-order derivative of a smoothing function $\phi(x)$. In the case where the wavelet is the first-order derivative of the scaling function, the local extrema of the wavelet transform obtained from such a wavelet represent sharp variations in the signal at multiple scales. Alternatively if the wavelet is the second-order derivative of the scaling function then the zero crossings of the wavelet transform convey this information. Through the use of some calculus these intuitive arguments can be made somewhat more rigorous.

Let $\psi_1(x)$ and $\psi_2(x)$ be the first and second derivative of $\phi(x)$, respectively. Hence we write

$$\psi_1(x) = \frac{d\phi(x)}{dx} \tag{67.7}$$

and

$$\psi_2(x) = \frac{d^2\phi(x)}{dx^2} \tag{67.8}$$

where $\phi(x)$ is the smoothing function mentioned previously. The dyadic WTs of the function $f(x)$ with these two wavelets is given by

$$W^1_{2^j}f(x) = f*\left[2^j \frac{d\phi_{2^j}}{dx}\right](x) \tag{67.9}$$

and

$$W^2_{2^j}f(x) = f*\left[2^{2j} \frac{d^2\phi_{2^j}}{dx^2}\right](x). \tag{67.10}$$

Using the fact that differentiation and convolution commute we may rewrite these two equations as

$$W^1_{2^j}f(x) = 2^j \frac{d}{dx}(f*\phi_{2^j})(x) = \left(2^j \frac{df}{dx}\right)*\phi_{2^j}(x) \tag{67.11}$$

and

$$W^2_{2^j}f(x) = 2^{2j} \frac{d^2}{dx^2}(f*\phi 2^{2j})(x) = \left(2^{2j} \frac{d^2 f}{dx^2}\right)*\phi_{2^j}(x). \tag{67.12}$$

It is important to note that Equations 67.11 and 67.12 can be cast as differentiation of the signal followed by smoothing at multiple scales. This is the design philosophy inherent in the Canny and Marr-Hildreth edge detector mentioned previously. As discussed in Mallat (1989) and Mallat and Zhong (1992) the local minima and zero crossings of the WTs given in Equations 67.11 and 67.12 do not correspond to sharp variation points in the signal but are instead inflection points of $f* \phi_2^{\,j}(x)$. The local maxima of $|W_{2^j}^1 f(x)|$ are, however, sharp variation points in the signal observed at multiple scales. Following Mallat and Zhong we denote the local maxima of $|W_{2^j}^1 f(x)|$ as modulus maxima and interpret them as multiscale edges.

It is helpful to construct a pedagogical example to explain these concepts. In Figure 67.1 we represent the action of hypothetical smoothing and wavelet functions on a piecewise linear signal which contains two jump discontinuities and an inflection point. We state the obvious when we note that x_0, x_1, and x_2 are all zero crossings of $|W_{2^j}^2 f(x)|$, but x_0 and x_2 correspond to sharp variation points in the signal while x_1 corresponds to an inflection point. We contrast this with the fact that the maxima of $|W_{2^j}^1 f(x)|$ clearly correspond to points of sharp variation in the signal.

In the case of the 1-D wavelet transform we denote the set of wavelet maxima as follows:

$$A_{2_j} = \{(x_i, W_{2^j}f(x_i)): \; |W_{2^j}f(x)| \text{ has local maxima at } x = x_i\}. \tag{67.13}$$

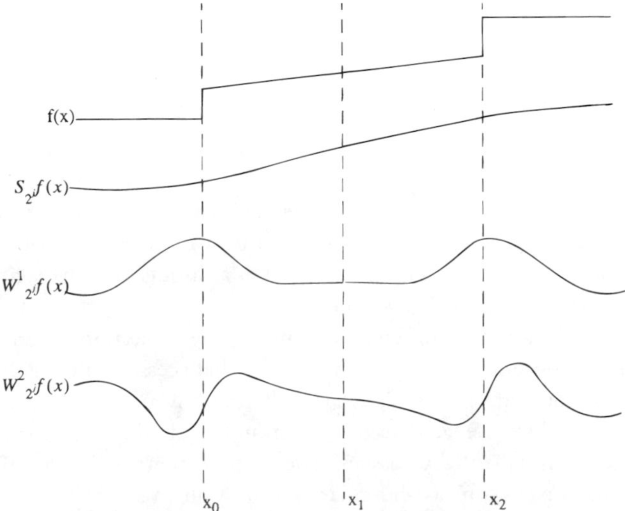

Figure 67.1 The local extrema of $|W^1{}_{2^j}f(x)|$ and zero-crossings of $|W^2{}_{2^j}f(x)|$. x_0 and x_2 are local maxima while x_1 is a local minima.

The 1-D Maxima Detection Wavelet

Previously Mallat (1989) and Mallat and Zhong (1992) have discussed a family of wavelets that have the desired edge detection properties discussed above. We choose to use wavelets that are similar in structure to theirs. In our case the father wavelet or scaling function ϕ takes the form of a spline approximation to a Gaussian type curve and the mother wavelet or ψ takes the form of a spline approximation to the derivative of this function. Representative father and mother wavelets are plotted in Figure 67.2.

2-D Wavelet Transform and Maxima

It is beyond the scope of our discussions to provide algorithmic details for the implementation of the 2-D wavelet transform. Let it suffice to say however that we employ two-dimensional wavelets that are separable into the tensor product of one-dimensional wavelets. In addition, we require that the two one-dimensional

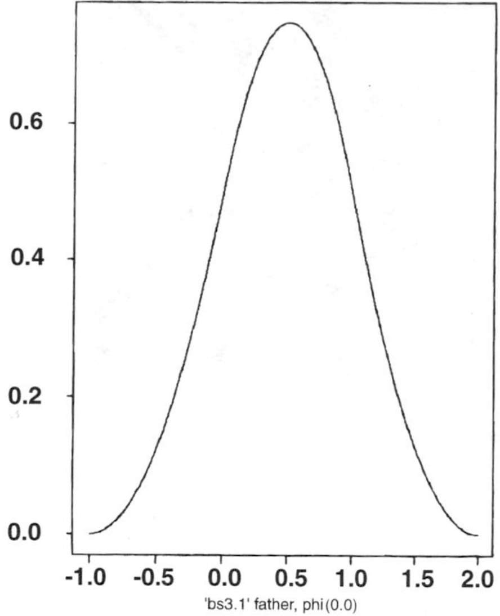

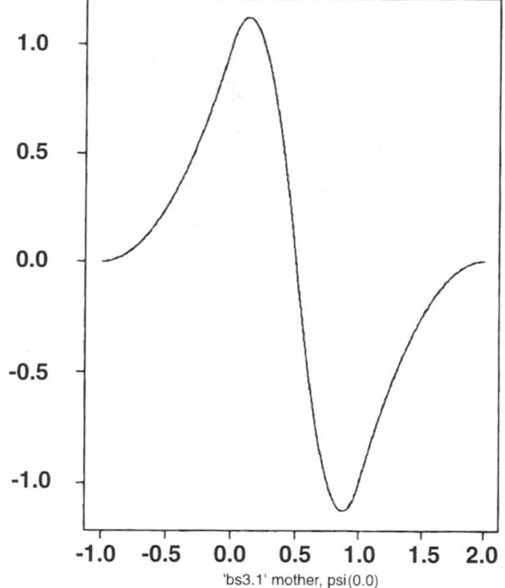

Figure 67.2 Spline scaling function and wavelet for edge detection in 1-D signals.

wavelets take the form of the partial derivatives of the 2-D scaling function. Specifically we write:

$$\psi^1(x, y) = \frac{\partial}{\partial x}\phi(x, y), \tag{67.14}$$

and

$$\psi^2(x, y) = \frac{\partial}{\partial y}\phi(x, y). \tag{67.15}$$

We may then cast the 2-D WT as the gradient of a function smoothed by the scaling function $\phi(x,y)$ at multiple scales. We may write this as the "multiscale gradient"

$$\nabla_{2^j} f(x, y) = \{W^1_{2^j} f(x, y), W^2_{2^j} f(x, y)\} = \frac{1}{2^{2j}}(\nabla f) * \phi(x, y). \tag{67.16}$$

We may characterize this gradient by its magnitude and phase. The magnitude and phase are given below:

$$\rho_{2^j} f(x, y) = \sqrt{(W^1_{2^j} f(x, y))^2 + (W^2_{2^j} f(x, y))^2} \tag{67.17}$$

and

$$\theta_{2^j} f(x, y) = \operatorname{atan}\left[\frac{W^2_{2^j} f(x, y)}{W^1_{2^j} f(x, y)}\right]. \tag{67.18}$$

The edges that we utilize in the procedures described later in this paper are extracted from the local maxima of Equation 67.17.

67.2 Resistive Grid Local Averaging

Resistive networks perform level normalization, or equivalently local averaging, in many neural systems. Dendritic processes can be modeled by one-dimensional resistive networks (Mead, 1989) while a two-dimensional resistive network or grid has been used as a central component in a silicon retina (Mead, 1989). Resistive networks can thus be considered as low level neural networks that are central to much of neural computation. In this section, we examine the modeling and use of a two-dimensional resistive grid for performing local spatial averaging of a signal. The grid is based on a regular square lattice suitable for use in image processing. The signal can be the image intensity at each pixel as in the silicon retina or a feature derived from the intensity at each pixel such as the coefficient of variation.

In this method of averaging, each pixel is modeled as a node connected to its four nearest neighbors by fixed conductances and to a voltage source through a resistance. Both the overall structure and the detailed circuitry associated with each node are shown in Figure 67.3.

Mead (1989) has demonstrated an analog VLSI implementation of a resistive grid. One of the artifacts of the analog VLSI implementation is a nonlinearity in the conductances connecting lattice sites. This nonlinearity can lead to an automatic segmentation effect. Rather than use this effect for segmentation, we are using linear elements with fixed conductance values and relying on segmentation from a separate source such as the wavelet-based method in the previous section. All conductances connecting to a node in the grid corresponding to a segmentation boundary are set to zero. Then with a continuous boundary, and all conductances that connect to the boundary set to zero, no current and hence no information can flow across the boundary. Since the resistive grid performs a local averaging

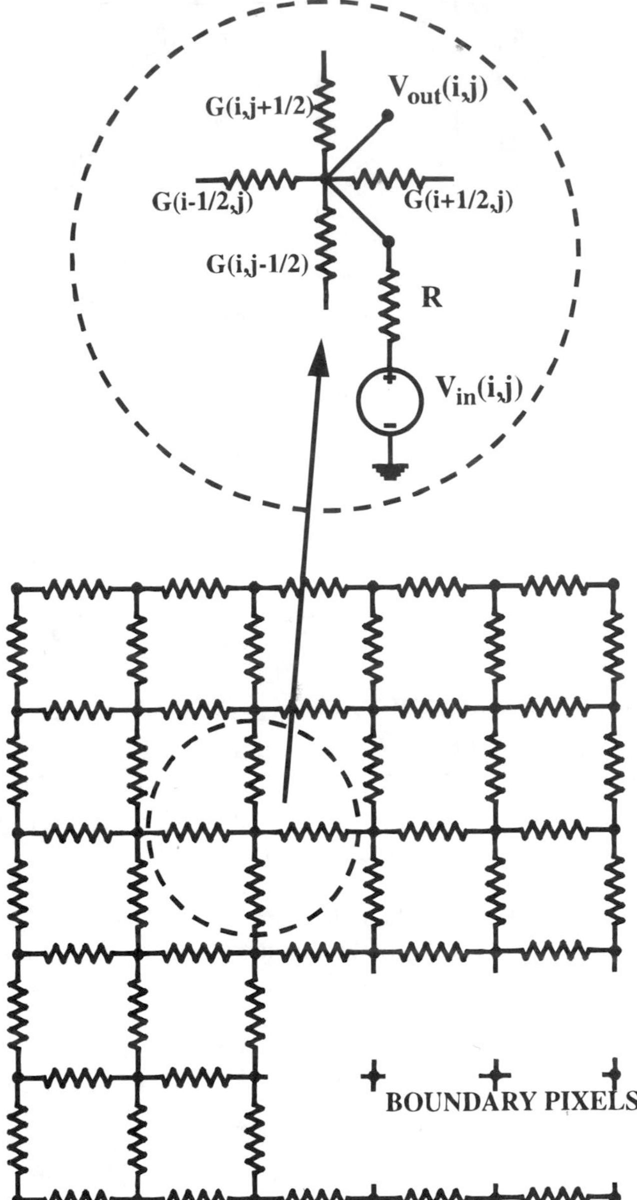

Figure 67.3 Detailed circuitry associated with each pixel in the image along with the overall resistive grid structure. Conductances connecting boundary pixels are set to zero.

computation, the effect of a boundary is to preclude any averaging across the boundary.

With fixed conductances/resistances, this circuit constitutes a linear system. Hence it makes sense to think in terms of the impulse response function at each node as an effective kernel that is used to convolve the input image or signal. In the absence of edge effects and boundaries, the effective kernel at each node will be identical. In a one-dimensional network this effective kernel is exponential. In our two-dimensional grid it falls off slightly faster than exponentially. However, in the presence of boundaries, the effective kernel at each node will in general be different as it locally adapts to the boundary configuration.

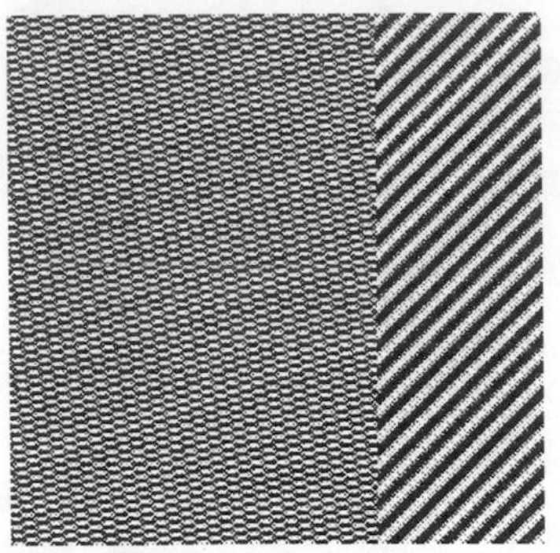

Figure 67.4 Example textures.

The circuit equations corresponding to the circuit depicted in Figure 67.3 can be easily written as

$$\begin{aligned} V_{\text{out}}(i, j) = \{ & V_{\text{in}}(i, j)/R + V_{\text{out}}(i-1, j)G(i-1/2, j) \\ & + V_{\text{out}}(i+1, j)G(i+1/2, j) \\ & + V_{\text{out}}(i, j-1)G(i, j-1/2) \\ & + V_{\text{out}}(i, j+1)G(i, j+1/2)\}/ \\ & \times \{(1/R) + G(i-1/2, j) + G(i+1/2, j) \\ & + G(i, j-1/2) + G(i, j+1/2)\}, \end{aligned} \quad (67.19)$$

where the indexing scheme is that of Figure 67.3 and V_{in} is the quantity to be locally averaged.

This presents us with a coupled set of difference equations that model a discrete anisotropic diffusion problem. Standard relaxation methods (Golub and Ortega, 1993) can be used to obtain a solution. Two updating methods are commonly used. The Jacobi method updates each value based on the values from the previous iteration while the Gauss-Seidel method sequentially updates each value based on the current mix of new/old values. While the Gauss-Seidel method is the method of choice for a uniprocessor computer, the Jacobi method corresponds to a massively parallel implementation where each value is to be updated synchronously. The successive overrelaxation (SOR) version of the Jacobi method is

$$\begin{aligned} V_{\text{out}}^{k+1}(i, j) = (1-\omega)V_{\text{out}}^{k}(i, j) + \omega\{ & V_{\text{in}}(i, j)/R \\ & + V_{\text{out}}^{k}(i-1, j)G(i-1/2, j) \\ & + V_{\text{out}}^{k}(i+1, j)G(i+1/2, j) \\ & + V_{\text{out}}^{k}(i, j-1)G(i, j-1/2) \\ & + V_{\text{out}}^{k}(i, j+1)G(i, j+1/2)\}/\{(1/R) \\ & + G(i-1/2, j) + G(i+1/2, j) \\ & + G(i, j-1/2) + G(i, j+1/2)\}, \end{aligned} \quad (67.20)$$

where the superscripted index corresponds to the iteration number and $0 < \omega \leq 1$ is the relaxation parameter. This method is less efficient than the Gauss-Seidel method but readily lends itself

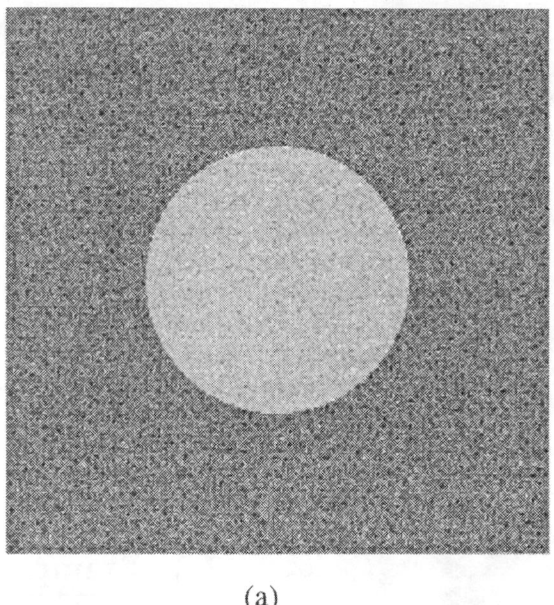

(a)

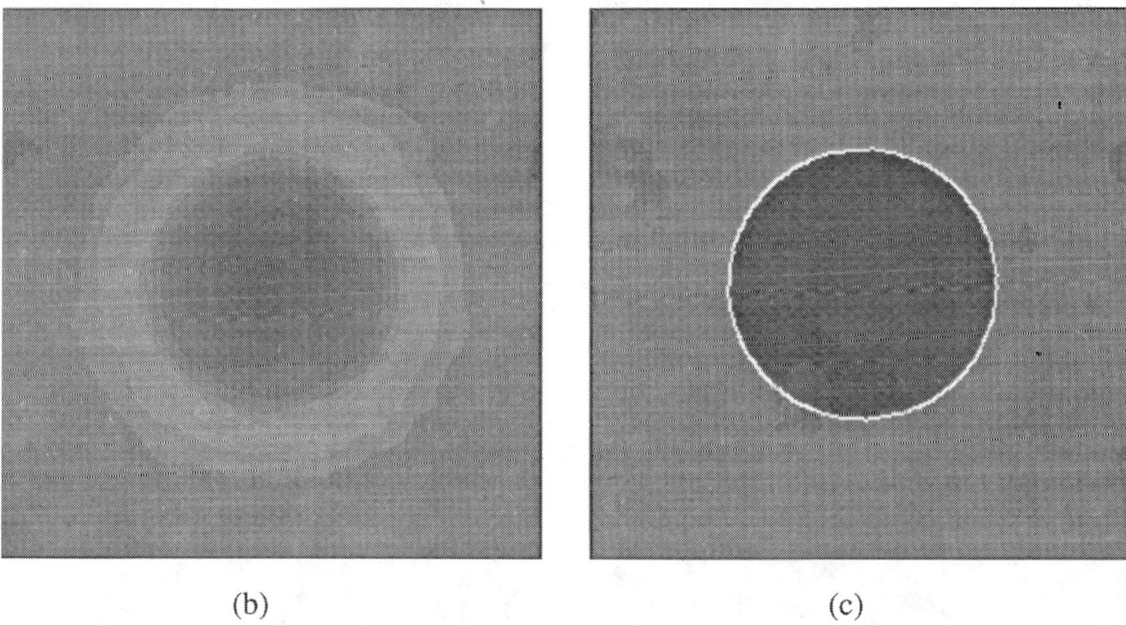

(b) (c)

Figure 67.5 (a) A simple synthetic image. (b) CoV of the image with no boundary gating and (c) with boundary gating.

to a massively parallel implementation. The SOR version of the Gauss-Seidel method is given by

$$\begin{aligned} V_{out}^{k+1}(i,j) = (1-\omega)V_{out}^{k}(i,j) &+ \omega\{V_{in}(i,j)/R \\ &+ V_{out}^{k+1}(i-1,j)G(i-1/2,j) \\ &+ V_{out}^{k}(i+1,j)G(i+1/2,j) \\ &+ V_{out}^{k+1}(i,j-1)G(i,j-1/2) \\ &+ V_{out}^{k}(i,j+1)G(i,j+1/2)\}/\{(1/R) \\ &+ G(i-1/2,j) + G(i+1/2,j) \\ &+ G(i,j-1/2) + G(i,j+1/2)\}. \end{aligned} \quad (67.21)$$

It differs from the Jacobi method in that current neighboring values are used in the updating so that the $(k+1)$ update for the node at (i,j) makes use of the $(k+1)$ value at node $(i-1,j)$, but uses the old (k) value at the $(i+1,j)$ node (which has not been updated yet). This is an inherently sequential algorithm well suited for uniprocessor machines. This method converges for $0 < \omega < 2$ and typically has a faster rate of convergence on a given problem.

An electronic implementation of the circuit has the advantage

Figure 67.6 Aerial image used in the examples.

of settling to the solution in just a few time constants of the circuit. This would allow for real time operation as, for example, in the silicon retina. For applications where the lack of precision attendant to analog computation is not an issue, the electronic implementation is the most efficient.

There are many instances where a locally weighted average is desired. In neural systems, vision provides a good example. It is normal for the illumination level in a visual field of view to vary from one region to another by several orders of magnitude. If a global average were used, both exceptionally bright and dark areas would lose all detail due to the limited dynamic range of the neurons involved. The solution is to use a locally weighted average as a local reference to the signal.

In statistical pattern recognition, features are often defined in terms of a local spatial average. The local mean has already been mentioned. The coefficient of variation is another example which is illustrated in the next section. In general, any feature that is computed using a window for averaging may be computed using the resistive grid averaging.

The advantages to this method of computing local averaging are the ability to locally adapt to any known boundaries produced by any desired segmentation algorithm and the massively parallel nature of the algorithm, including its suitability for analog VLSI implementation. The disadvantages are the fixed nature of the effective kernel (approximately exponential) and the inefficient nature of the algorithm on a uniprocessor digital computer.

67.3 Examples

Local Features

Most image processing tasks related to the detection and classification of regions in the image rely on local features. A local feature is one in which only pixels within a contiguous region are used in the calculation of the feature. Often one centers a rectangular window on a pixel and assigns to the pixel the feature calculated on the pixels within the window. A simple example of this is a median filter, where the feature returned is the median value of the pixels in the window.

When computing a local feature, one is making the assumption that the pixels within the region are homogeneous. For example, if we are trying to filter out noise, we assume that all the pixels in the region have been "corrupted" by the same noise process. This assumption may be valid if one is restoring an image which

(a)

(b)

Figure 67.7 (a) Sobel filter on aerial image. (b) Sobel filter thresholded to obtain a boundary map.

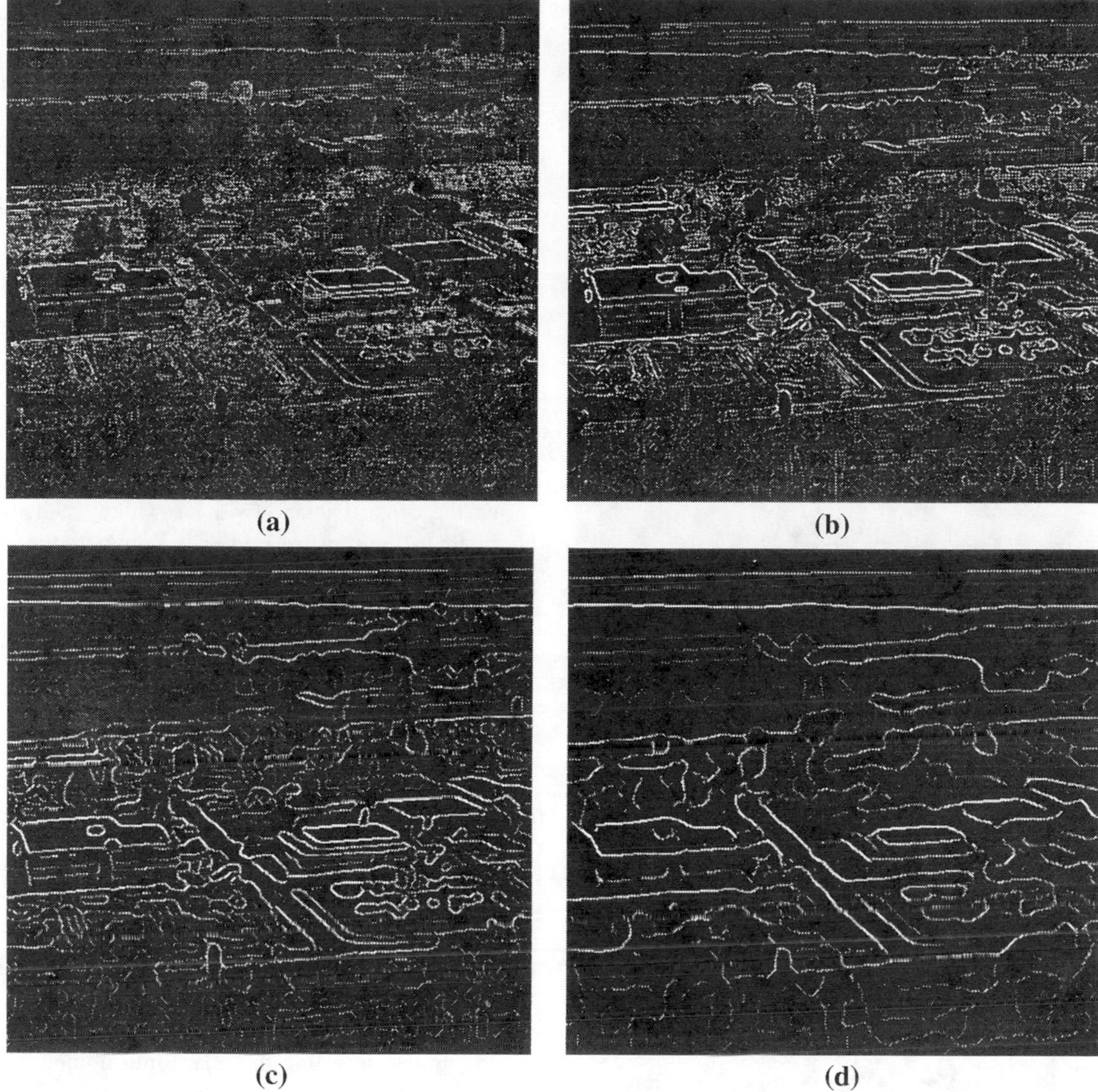

Figure 67.8 Wavelet maxima of the aerial image at scales (a) 1, (b) 2, (c) 3, and (d) 4.

has been corrupted uniformly (or in some known manner) across the image. However, in most situations the assumption is clearly not valid. Imagine trying to characterize the texture of an image. If the image consists of more than one texture, which is typical of real world images, the regions which contain more than one texture will be given a feature which does not well characterize the local texture. What is needed is a method for segmenting the image into homogeneous regions.

Boundary Gating

To calculate a local feature within a given window it is thus necessary to determine which pixels within that window should be used in the calculation. Figure 67.4 gives an example. This figure shows a window in which two distinct textures are evident. The center pixel clearly should not use pixels from the right texture in calculating its local feature since these are from a different texture. If we used a boundary map to separate the two regions we could determine which pixels are of the same texture type as the center pixel and extract a feature which better represents the texture surrounding the center pixel. This is the idea behind boundary gating.

If the boundaries are always closed it is a simple problem to determine which pixels to use, and there are quite efficient fill routines available for this purpose. However, there are few reliable methods of constructing closed boundaries which are computationally tractable, and flexible enough to use in a wide range of situations. We consider (below) a very simple version of boundary gating which will work reasonably well with any boundary map, whether it produces simple closed regions or not.

Note that in Figure 67.4 a boundary map based on intensity is unlikely to work well, particularly on the rightmost texture. Thus the boundary detection algorithm used must be related to the feature desired, in the sense that the boundaries should do

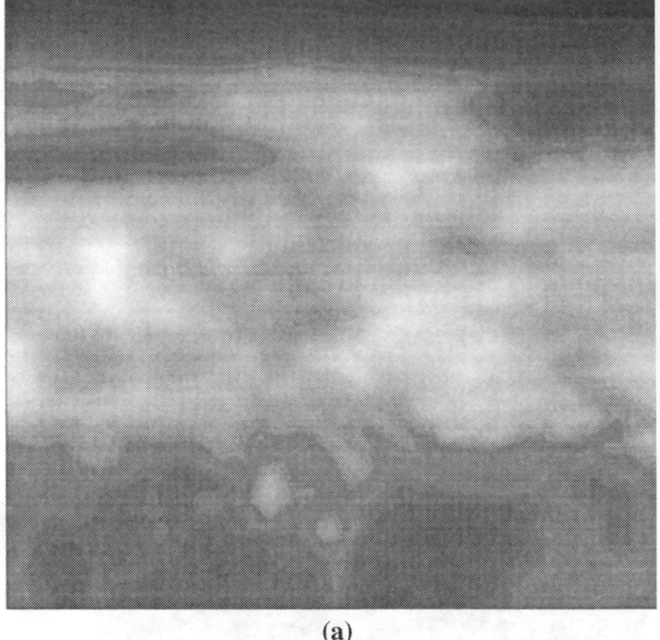

(a)

(b)

Figure 67.9 CoV for the aerial image (a) without boundary gating, and (b) with boundary gating.

a good job of delineating the regions which are homogeneous for the desired feature. It is not always easy to do this, and so one usually uses intensity based boundaries as they are the simplest to implement.

Coefficient of Variation (CoV)

The simplest and most familiar local features are the gray scale mean and variance of a window on the image. The problem with the mean is that it needs to be normalized across images because images that are darker overall will have lower mean values. Similarly, in general, small differences in intensity are significant to the variance if the overall intensity is low while these same differences tend not to be significant if the overall intensity is high. To alleviate these problems, one sometimes uses the coefficient of variation. The coefficient of variation is defined to be the standard deviation divided by the mean,

$$CoV = \frac{\sigma}{\mu}. \tag{67.22}$$

This feature tries to treat the variance in high- and low-intensity regions appropriately. We define the local coefficient of variation to be the *CoV* of the pixels within a given region computed using the resistive grid as follows. First, the mean μ is computed by using the original image (with or without a boundary map) as input to the resistive grid. The output is the resistive grid local mean

$$\mu = RG(\text{raw image}) \tag{67.23}$$

The resistive grid standard deviation is then computed via

$$\sigma = \sqrt{RG[((\text{raw image}) - \mu)^2]} \tag{67.24}$$

so that using the resistive grid to compute *CoV* requires using the resistive grid twice, first to find the local mean and then to find the local variance.

We will be focusing on the coefficient of variation throughout this paper for three reasons. First, it is easy to understand and very simple to compute. Second, it is a useful feature in many pattern recognition tasks. Finally, it has a strong relationship to intensity, and so intensity based boundaries will tend to do a fairly good job of segmenting the image into homogeneous regions.

Simple Boundary Gating

The simplest form of boundary gating uses what we call the light source algorithm. Imagine a light source at the center pixel and think of the boundaries as opaque walls. The pixels used in the calculation are those that are not shadowed by the boundaries. This is a very crude method which will omit pixels which are merely hidden from the center by a corner or singleton edge pixel. However, it will not select any pixels which require crossing an edge and it is extremely simple to implement. It also does not assume the boundaries are closed, and so can work with any edge detection algorithm.

Other algorithms are certainly possible. For example, if the edges are assumed closed, there are efficient fill routines to mark all the pixels in the region. However, the edges extracted using the wavelet transform are not guaranteed to produce closed regions, and so these algorithms cannot be used unless some edge closing algorithm is first applied.

Another more involved method of boundary gating would be to weight the pixels according to their distance from the center pixel. One way to do this is to weight each pixel p with the weight:

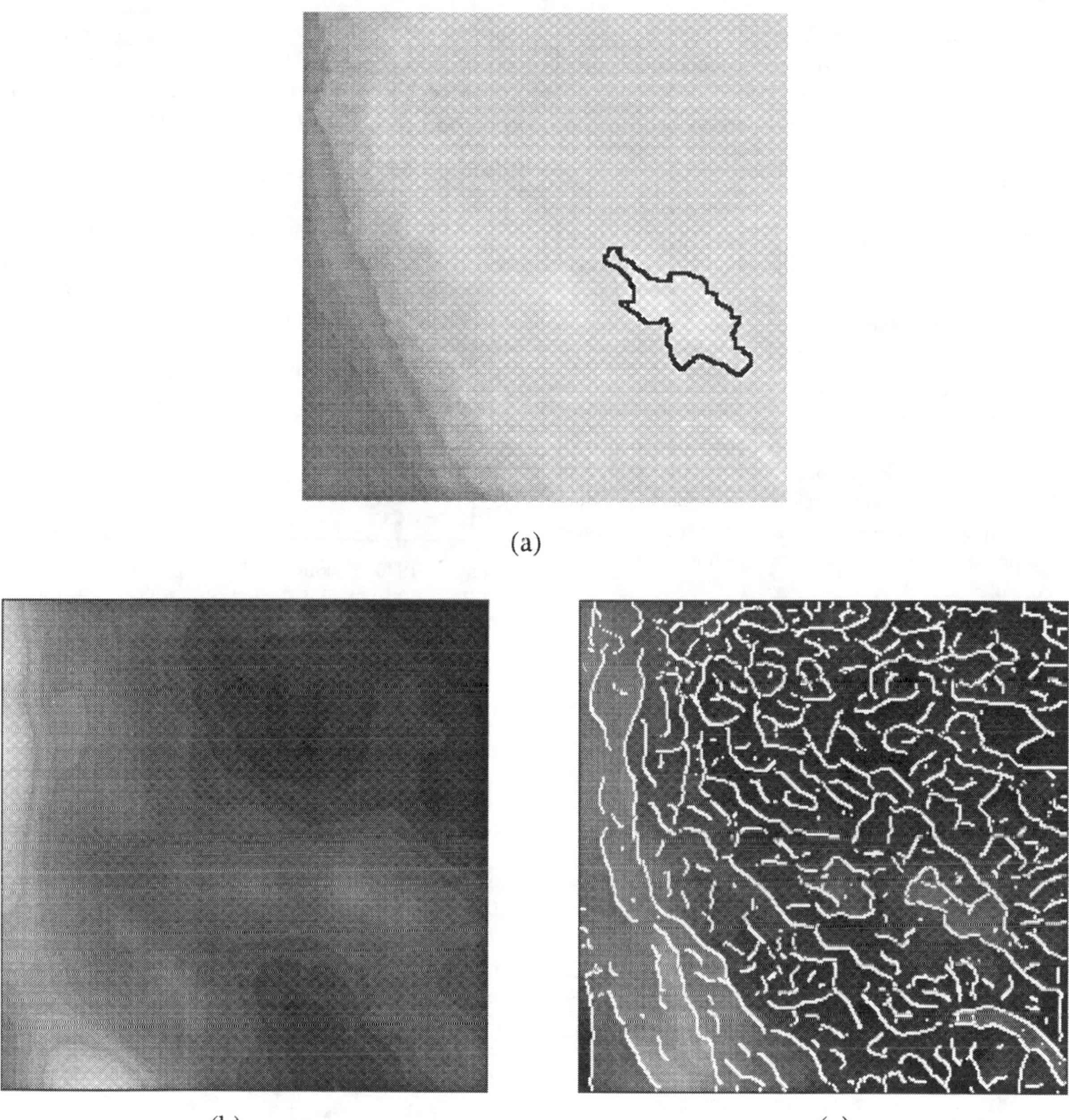

Figure 67.10 (a) Mammogram with tumor outlined in black. (b) CoV of mammogram with no boundary gating. (c) CoV of mammogram, with boundary gating.

$$w(p) = \frac{Edist(c, p)}{Bdist(c, p)} \tag{67.25}$$

where c is the center pixel. *Edist* is the usual Euclidean distance and *Bdist* is the length of the shortest path between p and c which does not cross a boundary, defined to be infinity if no such path exists.

The resistive grid method, where the presence of a boundary "breaks" the circuit at the boundary, is the method that will be used in the following examples. In all of the examples, a decay length of 16 pixels was used in the resistive grid computations.

Examples

The first example we consider is a simple generated image consisting of a background of Gaussian noise with mean 100 and variance 400, with a center circle whose mean has been increased to 175. This image is pictured in Figure 67.5a. We wish to compute the local coefficient of variation, which should be 0.2 for the background and $20/175 \approx 0.11$ inside the circle.

Figure 67.5b shows the *CoV* with no boundary gating. It is easy to see that the edge of the circle causes a region one window width wide of boundary effects. The values of the *CoV* are correct as long as one stays away from the boundary of the circle. If one is interested in pixels near the boundary of the circle, the boundary effect will clearly cause a difficulty in obtaining good values. Thus the need for boundary gating is clear.

Figure 67.5c shows the boundary gated version of the *CoV*. The boundary used is obtained using the wavelet transform as described above. The edge is indicated by the white circle in the image. This is a much cleaner image with the pixels near the boundary having very nearly the correct values. The only potential problem with this approach, given good boundaries as were obtained here, is that pixels near a boundary effectively use fewer points in computing the feature than pixels whose effective kernel/window does not intersect the boundary. We will not pursue this further in this paper.

The example above is a fairly trivial one, used merely to indicate the methodology. We now turn to real images. The first is an aerial image (Figure 67.6). Figure 67.8 shows 4 scales of wavelet maxima derived from this image. Scale 2 is thresholded, producing a boundary image which is then used in the boundary gated *CoV* calculation, and contrasted with the no-boundary calculation in Figure 67.9.

Figure 67.9b shows clearly the advantage of using the boundary gating. It also shows some of the effects of using unconnected boundaries. "Bleed through" is evident through holes in the edges, which can only be avoided by closing the edges. For some of these, such as along the tree line, an edge closing algorithm is appropriate and could be used. In other regions of the image edge closing algorithms are less applicable.

Most edge detectors essentially give edges at all scales at once. Figure 67.7 shows a sobel filter applied to an aerial image, along with a thresholded version. Note the extra noise due to the fine detail which persists through the thresholding operation. This results in a noisier boundary map than the wavelet approach, which gives one argument for considering the wavelet approach.

The final example is one from digital mammography. In the interest of space we show (Figure 67.10) only the original image, and the *CoV* before and after boundary gating with a wavelet, as was done above. The boundaries are evident as white curves in the boundary gated *CoV* image (Figure 67.10c). The image contains a benign calcification and a malignant tumor (biopsy proven). A radiologist drawn boundary around the tumor is shown on the original image, however this boundary is not used in the *CoV* calculations.

Once again we see that the boundary gated version is much smoother, and less sensitive to changes in intensity, which the ungated version picks up rather dramatically. The mammogram example is a good one from the perspective that the boundaries in it are far from obvious to the human eye, and in fact a Sobel filter does rather poorly on this image. However the wavelet boundaries do seem to be obtaining boundaries between regions of different *CoV* values, and so is doing a good job for the problem at hand.

References

Canny, J. 1986. A computational approach to edge detection, *IEEE Trans. Pattern Anal. and Machine Intell.*, PAMI-8:679–698.

Golub, G. and Ortega, J. M. 1993. *Scientific Computing An Introduction with Parallel Computing.* Academic Press, New York, NY.

Mallat, S. G. 1989. A complete signal representation from multiscale edges, Technical Report, No. 483, New York University, New York, NY.

Mallat, S. G. and Zhong, S. 1992. Characterization of signals from multiscale edges, Technical Report No. 592, New York University, New York, NY.

Marr, D. and Hildreth E. 1980. Theory of edge detection, *Proc. R. Soc. Lond.* B-207:187–217.

Marr, D. 1982. *Vision.* W. H. Freeman, San Francisco, CA.

Mead, C. 1989. *Analog VLSI and Neural Systems,* Addison-Wesley, New York, NY.

68
Fractals for Pattern Recognition

George W. Rogers
Naval Surface Warfare Center

Carey E. Priebe
The Johns Hopkins University

Jeffrey L. Solka
Naval Surface Warfare Center

68.1 A PDP Approach to Localized Fractal Dimension Computation with Segmentation Boundaries 933
Introduction • Formulation of the Covering Method with Boundary Incorporation • Adaptive Mixture Model Probability Density Estimation • Kullback-Leiber Information for Density Comparison • Results • Conclusions

68.1 A PDP Approach to Localized Fractal Dimension Computation with Segmentation Boundaries

A parallel distributed processing approach to the computation of localized fractal dimension values in imagery is presented. This approach is a further development of the covering method which requires only nearest neighbor communication. A major benefit of our approach is the ability to readily incorporate any boundary information that may be available. Many fractal textures or surfaces are fractal only in distribution. With this in mind, we show that comparison of the fractal dimension distributions via Kullback-Leibler can give an improved texture discrimination capability over comparison of computed fractal dimension. Results are presented for a set of textures.

Key Words: Parallel Distributed Processing, Fractal Dimension, Covering Method, Kullback-Leibler

Introduction

An automated texture recognition/classification capability is an important component of any artificial vision system. One approach to this segment of the vision problem which has shown promise is based on the computation of fractal dimension (fd) and/or related power law features (Stein, 1987; Peli, 1990; Solka et al., 1992; Peleg et al., 1993).

A two-dimensional grayscale image can be thought of as a manifold embedded in a three dimensional space. From this viewpoint, we can consider the image to have a fd that is somewhere between the images topological dimension of two and the dimension of the embedding space, resulting in a value between two and three. The fd is thus a characterization of roughness. For our purposes, the defining equation for the fd of an image is Richardson's law (Mandelbrot, 1977). This law describes the manner in which a measured property of a fractal varies as a function of the scale of the measuring device. It is given by

$$M(\epsilon) = K\epsilon^{(d-D)}, \qquad (68.1)$$

where $M(\epsilon)$ is the measured property of the fractal at scale ϵ. K is a constant of proportionality, d is the topological dimension, and D is the fractal dimension. The measured property varies as a power of the scale. Hence, this is a power law relationship and we term features derived from Equation 68.1 as "power law features."

If we take the logarithm of Equation 68.1 we obtain a linear relationship between the $\log(M(\epsilon))$ and D,

$$\log[M(\epsilon)] = (d - D)\log(\epsilon) + \log(K), \qquad (68.2)$$

which is the equation of a straight line with slope $(d - D)$ from which D can be recovered. While exact fractals will, in principle, conform to Richardson's law, texture analysis has as its idealization, the concept of "statistical fractals" (Hayes et al., 1993). For statistical fractals the straight line relationship given in Equation 68.2 holds only in distribution and a least squares regression can be used to find the best linear fit to a particular set of observations $\{\epsilon_i, M(\epsilon_i)\}$. The fd is estimated by the slope provided by this linear regression. In addition to the slope, regression also yields the y-intercept, $\log(K)$. Furthermore, for objects that do not strictly obey Equation 68.2, a measure of the goodness of fit based on an F-test provides a third useful feature. The fd, the y-intercept, and the F-test constitute three fractal dimension based power law features that are useful features for texture discrimination (Solka et al., 1992).

Methods of estimating fd from Richardson's law use the measured property as a function of scale to estimate D. Two statistical fractal objects may have the same fd but different statistical, distributions of the estimate of D. Similarly, two fractal objects may have the same fd but different values of K, or the same mean K values but different distributions. Finally, most objects are either non-fractal or fractal-like (obey Richardson's law) only over some limited range of scales, which makes the F-test value/distribution an important feature for texture discrimination, similar in concept to the so-called lacunarity (Wu et al., 1992). To

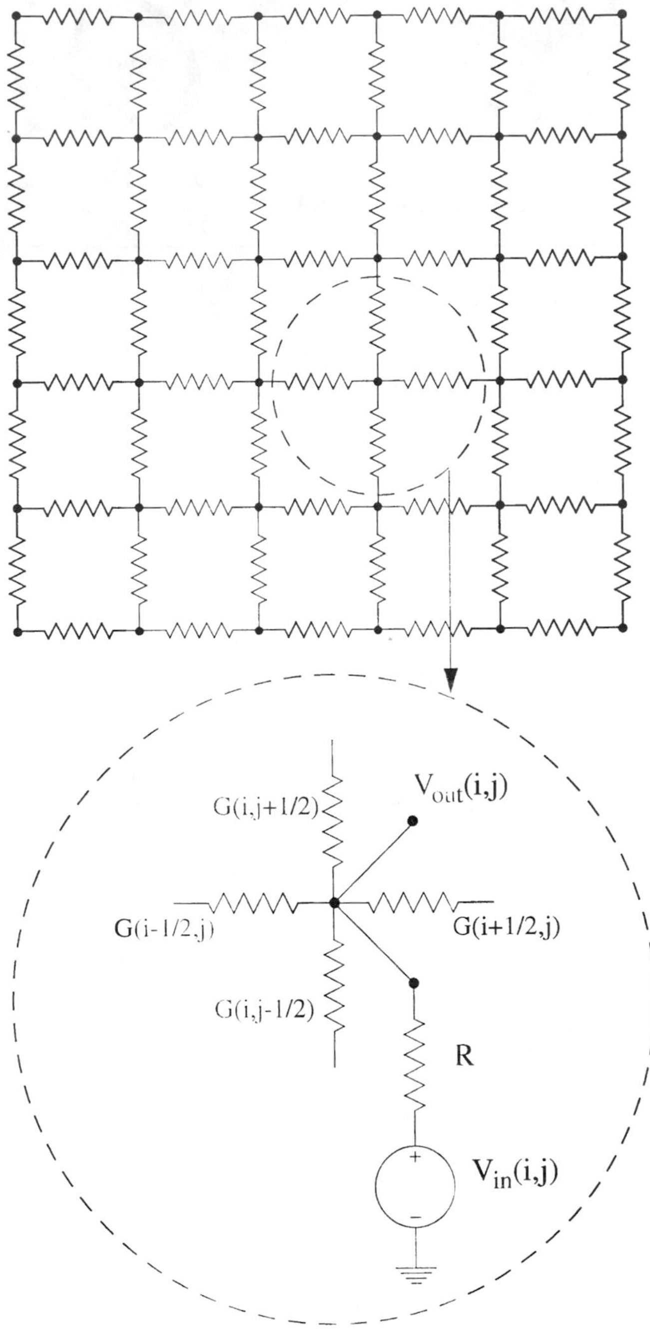

Figure 68.1 The resistive grid along with the detailed circuit model associated with a single pixel.

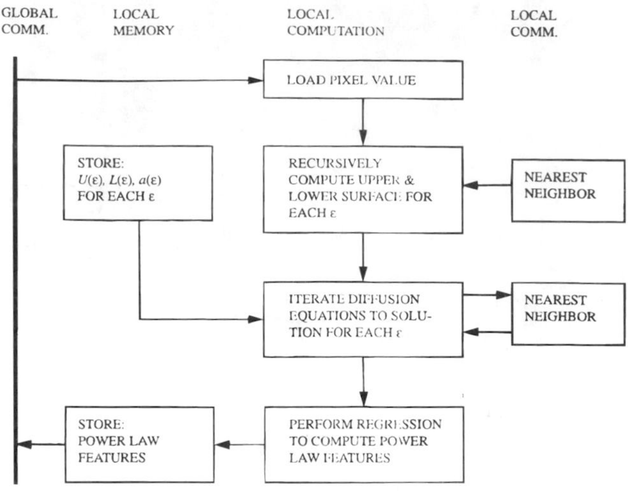

Figure 68.2 A block diagram of the power law feature computation at each pixel/processor.

most fully characterize textures in terms of Richardson's law, one should compare the computed distributions of the power law features instead of just the mean values. With this in mind, we present an alternate method of computing fd derived power law features that generalizes the covering method approach of Peli (1990) while requiring only nearest neighbor communication at all stages of the computation. We also present methods of estimating and comparing the resulting distributions.

We are motivated for this alternate approach by the need to perform texture computations in near real time as well as to be readily able to incorporate any *a priori* segmentation or boundary information that may be available. This has led us to develop a formulation suitable for a massively parallel implementation.

The bounding surfaces used in the covering method of computing fd are computed at each point (pixel) based on the characteristics of the image in some neighborhood of that point. The presence of a boundary between textures in this neighborhood can result in a greatly perturbed set of bounding surfaces, especially if there is a large difference in luminance between the two textures. This in turn can substantially perturb the computed fd. Segmentation schemes generally include the goal of detecting boundaries between texture types. This raises the question of how to incorporate any segmentation information that may be available into the texture feature computation so as to avoid basing features on both textures (as well as the luminance difference) near a boundary.

In this paper we present a formulation based on work by Rogers et al. (1993) that addresses both of these concerns simultaneously. We begin by presenting this formulation in the next section. We next describe the statistical methods we use to analyze the fractal based features extracted using this new method. This is followed by a section showing some results based on this approach, with the final section devoted to some concluding remarks.

Formulation of the Covering Method with Boundary Incorporation

In this section we develop our implementation of the covering method. By basing the implementation on local (nearest neighbor) computations at each step, we have a method that easily incorporates any *a priori* boundary information. The method results in a distribution of values for each of the power law features.

To use Richardson's law to estimate the fd of an image, we follow the covering method of Peli (1990) to estimate the surface area of the image in a window about a given pixel. This

Figure 68.3 The quilt of textures used in this paper.

method makes use of dilation and erosion operators which act recursively to bound the surface above and below at progressively larger scales. This results in a set of volume approximations at different scales which allow us to obtain estimates of the surface area as a function of scale. To take a potentially irregular segmentation boundary into account, we introduce the following modification. Assume that a segmentation map $M(i, j)$ is given where $M(i, j) = 0$ on a boundary and $M(i, j) = 1$ for all pixels not on a boundary. Let the dilation and erosion operators for the pixel (i, j) at scale ϵ be denoted by $U(i, j; \epsilon)$ and $L(i, j,; \epsilon)$, respectively. Then we introduce the new recursion relations

$$
\begin{aligned}
U(i, J; \epsilon + 1) = \max\{&U(i, j; \epsilon) + 1, [U(i + 1, j; \epsilon)M(i + 1, j) \\
&+ U(i, j; \epsilon)(1 - M(i + 1, j))], [U(i - 1, j; \epsilon)M(i - 1, j) \\
&+ U(i, j; \epsilon)(1 - M(i - 1, j))], [U(i, j + 1; \epsilon)M(i, j + 1) \\
&+ U(i, j; \epsilon)(1 - M(i, j + 1))], \\
&[U(i, j - 1; \epsilon)M(i, j - 1) \\
&+ U(i, j; \epsilon)(1 - M(i, j - 1))],\}
\end{aligned}
\tag{68.3}
$$

and

$$
\begin{aligned}
L(i, j; \epsilon + 1) = \min\{&L(i, j; \epsilon) - 1, [L(i + 1, j; \epsilon)M(i + 1, j) \\
&+ L(i, j; \epsilon)(1 - M(i + 1, j))], [L(i - 1, j; \epsilon)M(i - 1, j) \\
&+ L(i, j; \epsilon)(1 - M(i - 1, j))], [L(i, j + 1; \epsilon)M(i, j + 1) \\
&+ L(i, j; \epsilon)(1 - M(i, j + 1))], [L(i, j - 1; \epsilon)M(i, j - 1) \\
&+ L(i, j; \epsilon)(1 - M(i, j - 1))],\},
\end{aligned}
\tag{68.4}
$$

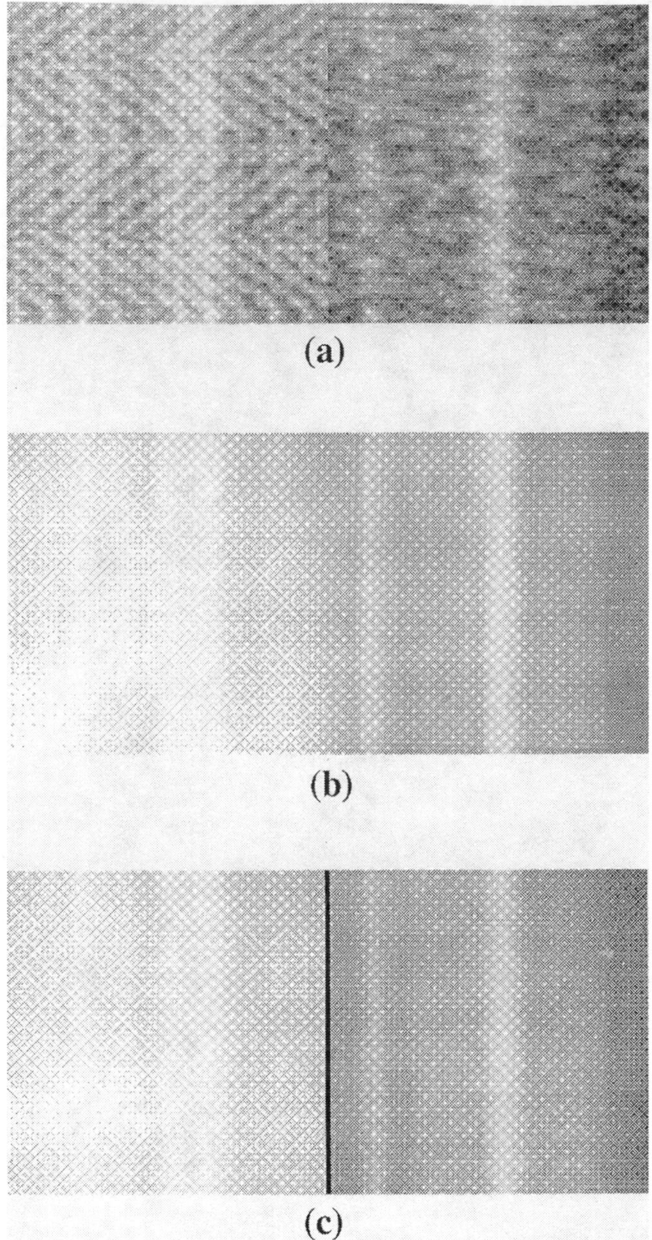

Figure 68.4 The two textures (#9 & #10) used in the first example are depicted in (a). The area map is shown with no boundary (b) and with boundary (c). In (c), the boundary is shown in black.

where the zero level recursions are defined as $U(i, j; 0) = L(i, j{:}\ 0) = G(i, j)$ for an original image grayscale value $G(i, j)$. In Equations 68.3 and 68.4, the segmentation map prevents values on the boundary from being used outside the boundary. For example, if the $(i + 1, j)$ pixel is on a boundary, while the (i,j) pixel is not, the $U(i, j; \epsilon)$ value is substituted for the $U(i + 1, j, \epsilon)$ value. Hence, the $U(i + 1, j, \epsilon)$ value will not affect the computation of $U(i, j, \epsilon + 1)$. Since the operators only involve nearest neighbors, we are guaranteed that no information will cross a continuous segmentation boundary at this stage of the computation. With $\mathrm{M}(i, j) = 1$ for all pixels, these expressions reduce to the previous versions (Solka et al., 1992; Peli, 1990).

Many segmentation approaches yield boundaries with an associated strength or degree of certainty. Instead of thresholding such a segmentation map to obtain a binary segmentation map we can simply use the continuous values if they are mapped to the unit interval. The continuous segmentation map can then be used for M in Equations 68.3 and 68.4 as well as in the steps to follow.

Once the upper and lower surfaces at a given scale have been computed, the bounding area is given by

$$a(i, j; \epsilon) = [U(i, j; \epsilon) - L(i, j; \epsilon)]/2\epsilon. \tag{68.5}$$

This method can only be expected to give the correct fd in distribution. Hence it has been customary to average over the entire patch of texture (Peleg et al., 1984) or over a window (Solka et al., 1992; Peli, 1990) as in

$$A(i, j; \epsilon) = \sum_{(l,m) \in W(i,j)} c\left[\frac{U(l, m; \epsilon) - L(l, m; \epsilon)}{2\epsilon}\right]. \tag{68.6}$$

where $W(i, j)$ is some window about (i, j) and c is a constant for all scales that can be used for normalization, if desired. The fd can then be estimated on a pixel-by-pixel basis as a regression of $\log[A(\epsilon)]$ verses $\log[\epsilon]$ as indicated by Richardson's law Equation 68.1.

The method of computing the areas embodied in Equation 68.6 using a fixed window does not readily accommodate the incorporation of segmentation boundaries. If used with boundaries, it will average across the boundary with attendant undesired effects. This problem has led us to introduce a method of averaging that is based on the physics of a resistive grid.

In this method, we model each pixel as though it is connected to its four nearest neighbors by fixed conductances and to a voltage source through a resistance. This is depicted in Figure 68.1. To account for a regular or irregular segmentation boundary we set all conductances to zero that connect to a pixel on the boundary. The circuit simulation then can be viewed as adapting the effective kernel associated with each pixel to account for the segmentation boundaries.

The circuit equations can be easily written as

$$\begin{aligned} V_{\text{out}}(i, j) = \{&V_{\text{in}}(i, j)/R + V_{\text{out}}(i-1, j)G(i-1/2, j) \\ &+ V_{\text{out}}(i+1, j)G(i+1/2, j) \\ &+ V_{\text{out}}(i, j-1)G(i, j-1/2) \\ &+ V_{\text{out}}(i, j+1)G(i, j+1/2)\}/\{(1/R) \\ &+ G(i-1/2, j) + G(i+1/2, j) \\ &+ G(i, j-1/2) + G(i, j+1/2)\}, \end{aligned} \tag{68.7}$$

where the indexing scheme is that of Figure 68.1 and V_{in} is given by

$$V_{\text{in}}(i, j) = a(i, j; \epsilon). \tag{68.8}$$

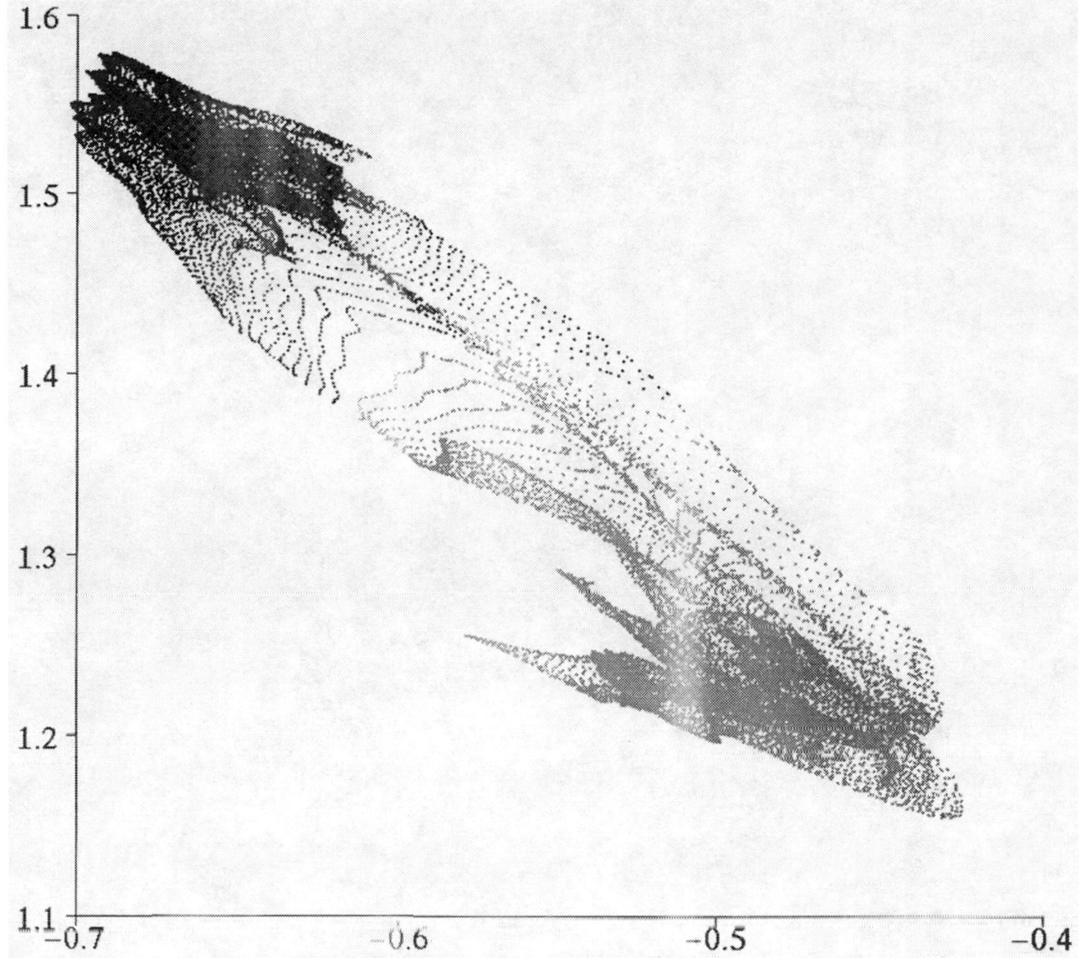

Figure 68.5 The y-intercept is plotted against the slope for each pixel of the two preceeding texture patches with no segmentation.

This presents us with a coupled set of difference equations that model a discrete anisotropic diffusion problem. Standard relaxation methods (Golub and Ortega, 1993) can be used to obtain a solution. Two updating methods are commonly used. The Jacobi method updates each value based on the values from the previous iteration, whereas the Gauss-Seidel method sequentially updates each value based on the current mix of new/old values. Although the Gauss-Seidel method is the method of choice for a uniprocessor computer, the Jacobi method corresponds to a massively parallel implementation where each value is to be updated synchronously. We chose to solve the equations using a successive overrelaxation (SOR) (Golub and Ortega, 1993) version of the Jacobi method. This method simulated a massively parallel implementation. For a relaxation parameter value of (1/2), convergence to several significant figures always occurred in less than 2000 steps for an effective decay length of $L=(GR)^{1/2} = 16$. For smaller decay lengths the convergence is faster. After convergence, we set

$$A(i, j; \epsilon) = V_{\text{out}}(i, j) \tag{67.9}$$

for scale ϵ. Thus, the computation defined by the set of Equations 68.7–9 replaces the computation embodied in Equation 68.6.

The final step in obtaining fd or the associated power law features is to perform a regression on $\log[A(i, j; \epsilon)]$ against $\log[\epsilon]$ to find the slope for each pixel. This is a local computation (no communication required). The quantity (2 − slope) gives the fd while the y-intercept and the F-test of the regression provide the two additional power law features (Solka et al., 1992).

The computation is depicted schematically in Figure 68.2 for a massively parallel implementation where one processor is dedicated to each pixel. Global communication is only required to distribute the image and the segmentation map, and to output the results. As shown in Figure 68.2, each pixel is acted on independently. During the course of the computation only nearest neighbor communication is required and local memory requirements are minimal. The same operation is performed in parallel on every element in the data set, making this algorithm appropriate for massively parallel architectures.

A massively parallel implementation (Hayes et al., 1993) of the power law feature computation (Solka et al., 1992) has been

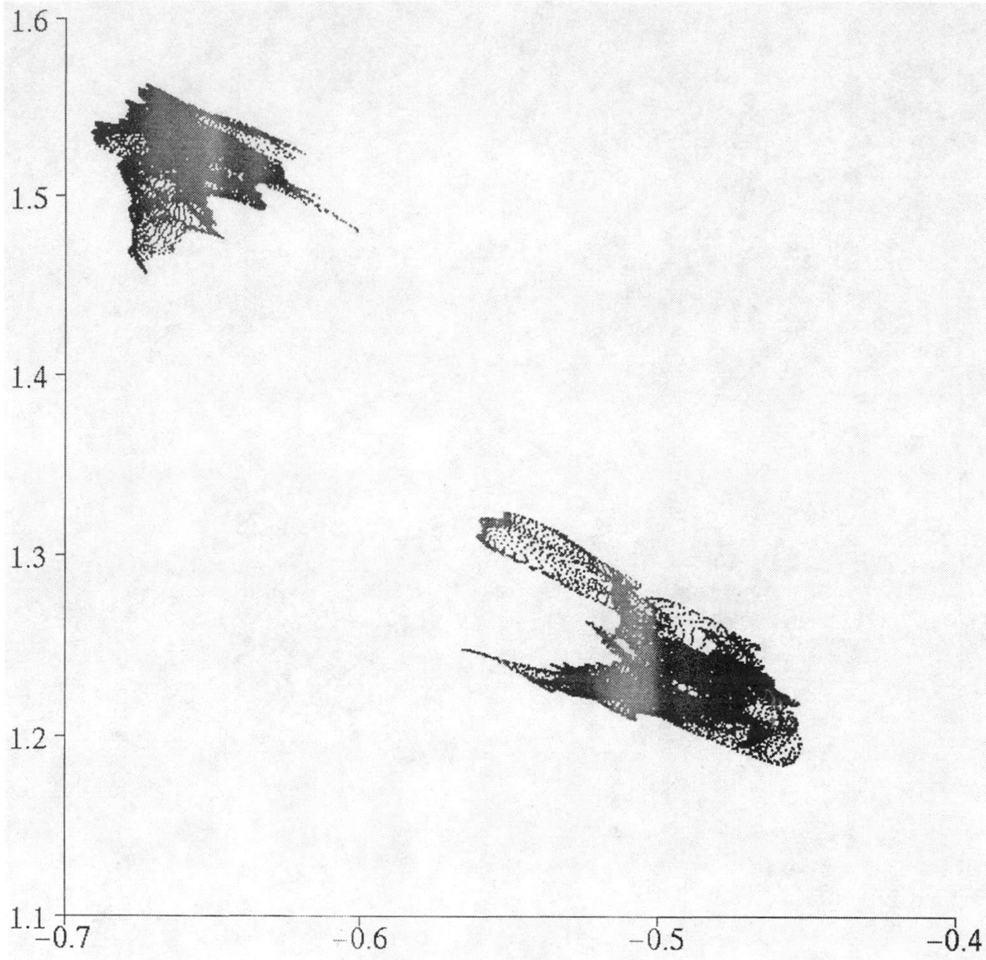

Figure 68.6 The y-intercept is plotted against the slope for each pixel of the two texture patches with segmentation.

conducted on a connection machine. This study yielded real time performance on that method. Based on this result, it is anticipated that the entire power law algorithm presented here could be processed in a similar parallel environment yielding results in a subsecond processing time for an image.

Adaptive Mixture Model Probability Density Estimation

Adaptive mixtures (Priebe, 1994) is a recursive nonparametric algorithm for the estimation of probability density functions (pdfs). It estimates the density of the data as a mixture of normals, and chooses the number of components in a data driven manner.

The adaptive mixtures estimator is related to both the kernel estimator and to finite mixture models. For the kernel estimator, a fixed kernel is placed at each point in the data, in effect convolving the data with the kernel. One of the problems with the kernel estimator is the computational and storage requirements, particularly for large data sets. Finite mixture models consist of a fixed number of (usually Gaussian) kernels where the parameters (means, variances, etc.) are updated via the EM (for Expectation/Maximization) algorithm which is based on maximum likelihood. The major problems associated with finite mixture models are the fixed, parametric nature and the susceptibility of the EM algorithm to get trapped in local maxima.

The adaptive mixtures estimator addresses these problems by adding terms within the mixture model framework in a data driven manner. The ability to add terms allows the escape from local likelihood maxima. The creation of new terms occurs at a much slower rate than with the kernel estimator. This results in a robust density estimator with nonparametric capability and only moderate complexity.

Adaptive mixtures process each observation sequentially and either updates the parameter values according to a recursive version of the EM algorithm, or adds a new term to the model. The decision to add a new component rather than to update the old one is based on the determination that the latest observation is not well explained by the current mixture. If a point has a low likelihood for each component, that is, is an outlier of each component, then the algorithm adds a component centered at the new data point. This approach results in a robust but generally overdetermined mixture model estimate of the density, giving it the flexibility of a nonparametric approach but with the complexity of a parametric one.

Kullback-Leibler Information for Density Comparison

There are many approaches to determining how different two densities are. The one used in this work, the Kullback-Leibler (KL) information (Kullback, 1959), is designed to give a measure of discriminatory power. The formula for the *KL* information of density *f* compared to density *g* is:

$$KL(f, g) = \int f(x)\log\left(\frac{f(x)}{g(x)}\right)dx \qquad (68.10)$$

The log of the likelihood ratio, $\log((f(x))/(g(x)))$, is the information at x for discrimination in favor of *f* against *g*. Thus, the *KL* information is the expected discriminatory information and hence a measure of the overall discriminatory power of *f* against *g*. This gives a single number to use in comparing the two densities and comparing different estimates of a single density.

Since the *KL* information is nonsymmetric (in general, $KL(f, g) = KL(g, f)$), we actually get two numbers from the *KL* processing. As described above, $KL(f, g)$ is the information in *f* for discriminating against *g*. Similarly, $KL(g, f)$ is the information in *g* for discriminating against *f*.

The *KL* information gives us a tool for comparing *fd* probability density estimates corresponding to different texture patches.

Results

In this section we examine the application of the diffusion based averaging feature extraction method to do texture analysis. Figure 68.3 shows a quilt of textures (Brodatz, 1966) used for this analysis. The numbering scheme for the patches used here is the column plus four times the row index. Thus, the texture numbers run from zero to fifteen.

We begin by analyzing the discrimination information in features extracted from two textures in the quilt, textures #9 and #10. These textures are replicated, with a linear boundary, in Figure 68.4a.

Figures 68.4b and 68.4c demonstrate the difference in the computed bounding area (at scale $\epsilon = 5$) with and without segmentation, respectively. The segmentation boundary, which was assigned *a priori*, is denoted by the black line in Figure 68.4c. Clearly, there is a significant blending of the computed area in the region about the texture interface in Figure 68.4b which is not experienced in Figure 68.4c.

In Figures 68.5 and 68.6 we present scatter plots of slope vs. y-intercept. Figure 68.5 shows results with no segmentation. The effect of including segmentation is readily apparent in the greater feature space separation of the textures in Figure 68.6. It is obvious that the information available for discrimination between these two textures (or the detection of a spatial change point) is much greater for the case presented in Figure 68.6 which has had the segmentation information incorporated into the feature extraction algorithm.

Figure 68.7 shows the results of the adaptive mixtures probability density estimate procedure for *fd* features extracted from all 16 textures from Figure 68.3. The pdfs are based on the *fd* estimates for those pixels at least one decay length (16 pixels) away from the boundaries of the texture so as to minimize edge effects. Exact segmentation boundaries between textures were used. The nonnormality of these pdfs is noteworthy. In particular, it should be noted that there are textures which might have a very similar mean *fd* (say, textures #3 and #4) but which have quite distinct pdfs and could, therefore, be discriminated if one used the densities rather than mean values.

To support the conjecture that the current approach provides results that are comparable with those previously reported in the literature (Table 3 of Sarkar and Chaudhuri, 1992) we also show in Figure 68.7 the calculated *fd* for those textures for which Sarkar and Chaudhuri report values using five different fractal calculation methods. These reported results lie, for the most part, in the support of our probability density estimates.

We now turn our attention to comparison of the fractal dimension pdfs. The overlap between the probability density estimates $\hat{f}_i$ and $\hat{f}_j$ for each pair of textures is measured via the Kullback-Leibler information. The smaller the overlap of the density estimates the greater the *KL* number. Figure 68.8 gives a pictorial representation of *KL* (texture #i, texture #j). The values are zero (light gray) on the diagonal, and the darker values indicate large *KL* numbers corresponding to more discriminable pairs of textures.

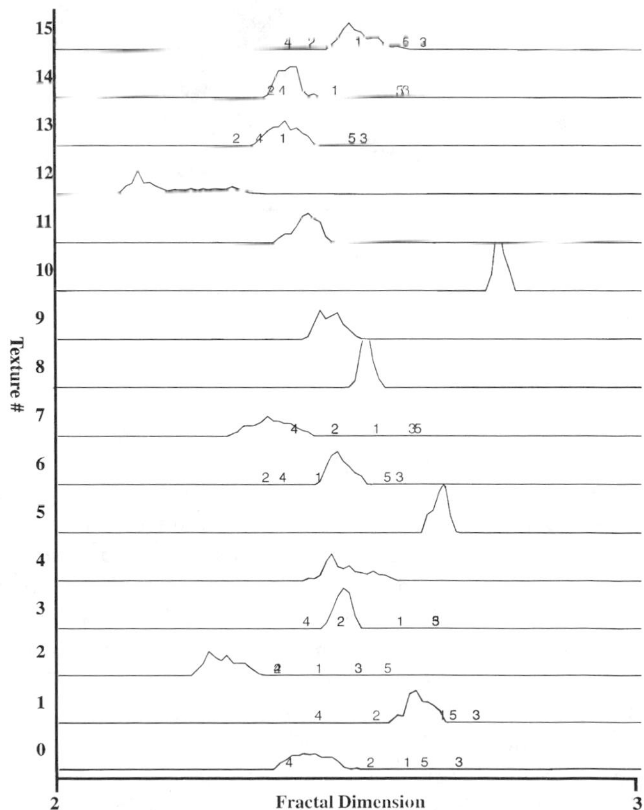

Figure 68.7 The probability density functions (pdfs) for the fractal dimension (fd) for all sixteen textures.

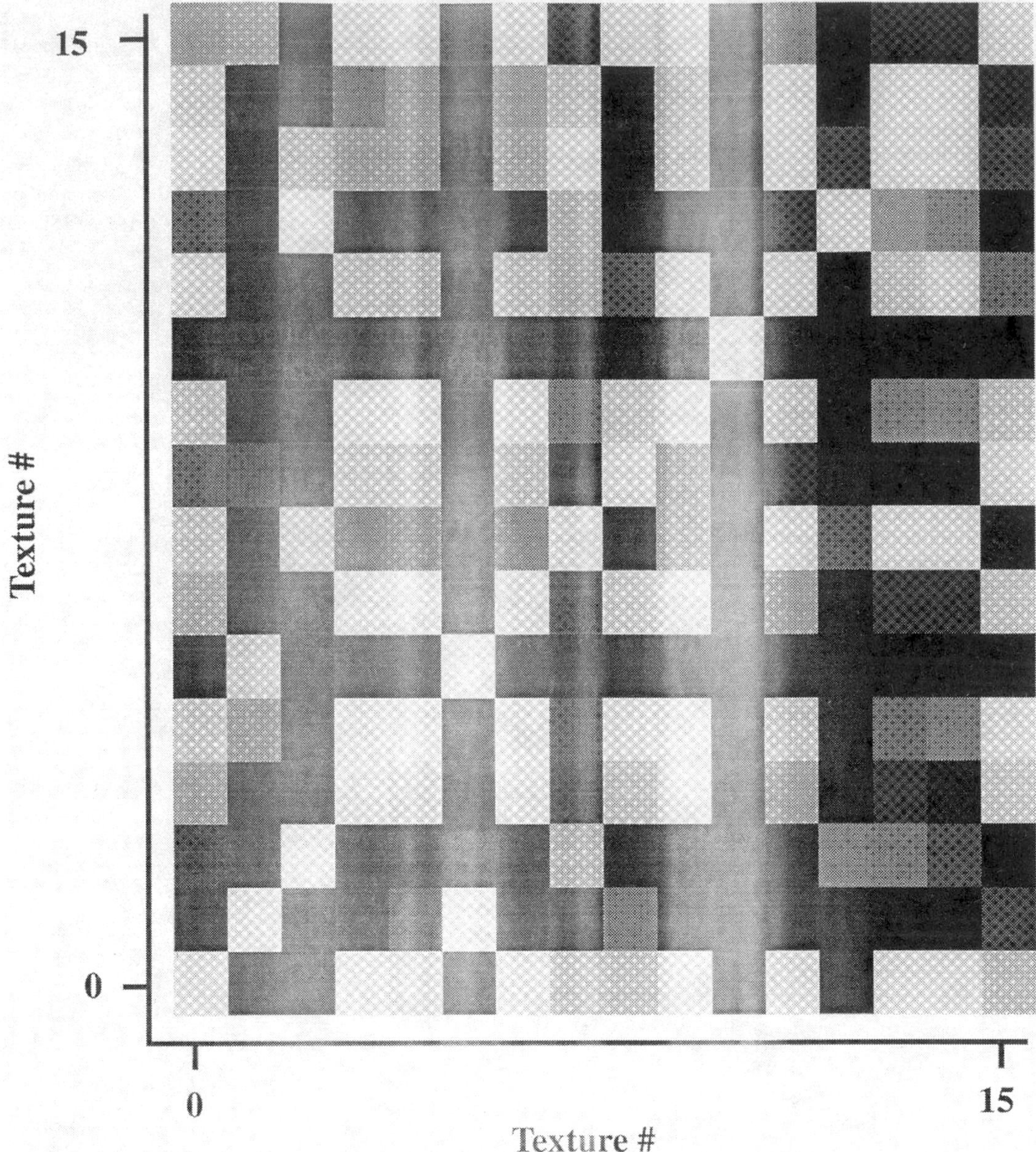

Figure 68.8 Kullback-Leibler number grayscale matrix plot that shows the relative Kullback-Leibler numbers comparing different textures.

Figure 68.9 considers a more detailed analysis of boundary effects on texture #10. This figure shows both the edge effects and the boundary effects for the texture with the largest pdf differences due to these effects. For each segmentation case, the "entire" pdf shows the density based on the computed fd from all pixels of the texture. The "interior" pdf shows the density based on the computed fd from only those pixels at least one decay length (sixteen pixels) from the texture boundaries. Figure 68.9a shows the pdfs for the full segmentation scheme. While there are differences between the "entire" and "interior" pdfs, they are reasonably small. For the partial (grayscale) segmentation shown in Figure 68.9b we see that the pdfs have been degraded significantly from the full segmentation version, with the pdf built from the entire patch preserving even less of the structure than that built on just the interior observations. The no segmentation results depicted in Figure 68.9c show even more degradation. The large tails for these latter two estimates, especially on the left, mean the discrimination capabilities between this texture and the others will be significantly reduced. Thus, the calculation of the texture features with boundaries incorporated can be a major advantage.

Conclusions

The diffusion equation based covering method described in this paper constitutes a boundary gated fd algorithm that produces estimates that are in general agreement with previously published values (Sarkar and Chaudhuri, 1992). The largest discrepancies occur for textures #0, #2, and #7 of the quilt. All three of these textures tend to vary significantly only at scales larger than the

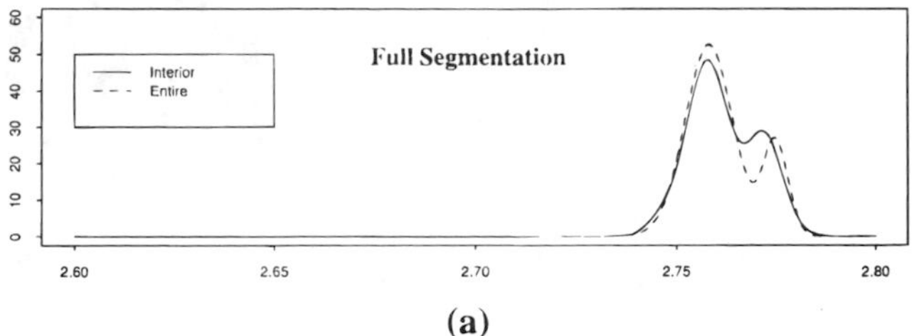

(a)

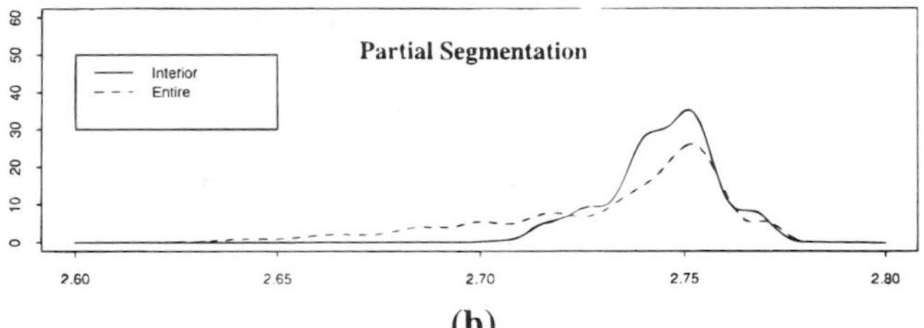

(b)

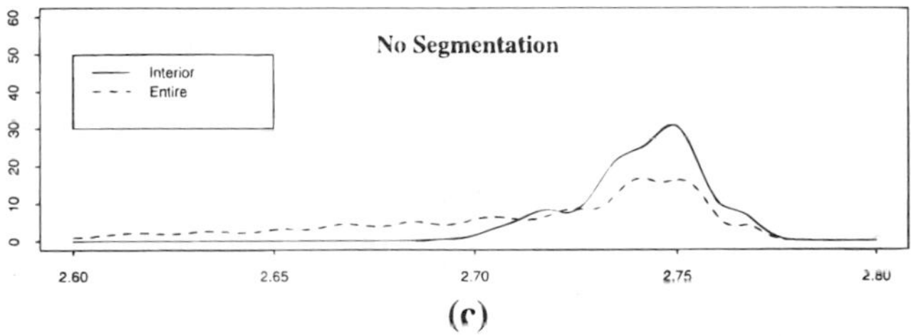

(c)

Figure 68.9 Edge and boundary effects for fractal dimension probability density functions (pdfs) for texture #10. The cases are (a) full segmentation, (b) partial segmentation and (c) no segmentation.

scales of 3, 4, and 5 pixels used in this work to estimate the *fd*. Thus, these three textures should all yield low estimates of the *fd* at these scales due to a relatively large scale of variation. In the other cases (and including texture #0) the mean of the fractal dimension pdf is within the range of *fd* values reported in earlier work.

If the averaging step is left out, the density functions are much more spread out and discrimination between them becomes much more difficult. The greater the degree of local averaging, the greater the compactness of the resulting density estimates until, in the limit of a global average, the estimate reduces to a single value. Thus, by varying the decay length and hence the degree of local averaging, we will affect the computed densities. This in turn gives us the freedom to choose a decay length that will enhance differences in the densities corresponding to different textures.

Kullback-Leibler information numbers, computed from the texture fractal dimension pdfs give a convenient means for quantifying the similarity or difference between pdfs. For non-normal distributions, pdf comparison is a more powerful tool for pattern recognition than simply comparing the distribution means.

The incorporation of known segmentation boundaries can lead to a dramatic reduction in the tail and variance of the *fd* distribution. This in turn can lead to dramatic improvements in classification accuracy whether in terms of mean *fd* or distribution via Kullback-Leibler numbers.

In the future we will report on ongoing work in adaptive kernel based averaging methods for the covering method as well as on applications to medical computer aided diagnosis.

Acknowledgments

This work was supported in part by the Office of Naval Research (R&T #4424314) and the NSWCDD Independent Research program. The authors gratefully acknowledge Halford Hayes for help with the initial programming and Richard Lorey for his careful reading of the paper and many helpful suggestions.

References

Brodatz, P. 1966. *Texture: A Photographic Album for Artists and Designers.* Dover, New York, NY.

Golub, G. and J. M. Ortega. 1993. *Scientific Computing An Introduction With Parallel Computing,* Academic Press, San Diego, CA.

Hayes, H. I., Solka, J. L., and Priebe, C. E. 1993. Parallel Computation of Fractal Dimension, *Adaptive and Learning Systems II,* Sadjadi, F. A., ed., Proc. SPIE 1962: 219–230.

Kullback, S. 1959. *Information Theory and Statistics.* John Wiley & Sons New York, NY.

Mandelbrot, B. 1977. *The Fractal Geometry of Nature.* W. H. Freeman and Company, New York, NY.

Peleg, S., Naor, J., Hartley, R., and Avnir, D. 1984. Multiple resolution texture analysis and classification. *IEEE Trans. Pattern Anal. Mach. Intell.,* PAMI-6: 518–523.

Peli, T. 1990. Multiscale fractal theory and object characterization, *J. of the Optical Soc. of America* A, 7(6):1101–1112.

Priebe, C. E. 1994. Adaptive mixtures, *J. Am. Statist. Assoc.,* 89:796–806.

Rogers, G. W., Priebe, C. E., Hayes, H., and Solka, J. L. 1993. A parallel distributed processing algorithm for power law features which requires only nearest neighbor communication, *Proc. Fifth Workshop on Neural Networks,* SPIE 2204:269–275.

Sarkar, N., and Chaudhuri, B. B. 1992. An efficient approach to estimate fractal dimension of textural images, *Pattern Recognition,* 25(9):1035–1041.

Solka, J. L., Priebe, C. E., and Rogers, G. W. 1992. An initial assessment of discriminant surface complexity for power law features, *Simulation,* 58(5):311–318.

Stein, M. C. 1987. Fractal image models and object detection, *Visual Comm. and Image Processing II,* SPIE-845:293–300.

Wu, C.-M., Chen, Y.-C., and Hsieh, K.-S. 1992. Texture features for classification of ultrasonic liver images, *IEEE Trans. Medical Imaging,* 11(2):141–152.

69

Multilayer Perceptrons with ALOPEX and Backpropagation

Daniel A. Zahner
Rutgers University

Evangelia Micheli-Tzanakou
Rutgers University

69.1 Introduction 942
69.2 The Backpropagation Algorithm 943
69.3 The ALOPEX Algorithm 944
69.4 Multilayer Perceptron Network 945
69.5 ALOPEX in VLSI 947
69.6 Discussion 949

69.1 Introduction

Common network architectures consist of multiple layers of neurons, where each neuron in layer 1 is connected to all neurons in layer 2. A three-layer feedforward architecture is shown in Figure 69.1. The input layer receives external stimuli while the output layer generates the output of the network. The hidden layer and all the interconnections, are responsible for the neurocomputation. The number of neurons in each layer and the number of layers necessary is problem dependent. Generally, as the number of nodes or neurons increases, problem complexity increases, as does the time to train the network. For linear activation functions, one layer is all that is necessary for linear separability. Additional layers are redundant (Minsky and Papert, 1969) when linear activation functions are employed. In the nonlinear cases, using two layers increases the nonlinear separability. Figure 69.1 shows a three-layer network, the simplest multilayer perceptron network.

The optimal number of hidden neurons needed to perform an arbitrary mapping is a subject of much debate. Methods used in practice are mainly intuitive determination or are found by trial and error. Mathematical derivation proves that a bound exists on the number of hidden nodes, m, needed to map a k element input set. The formulation is that $m = k - 1$, is an upper bound (Huang and Huang, 1991). These results are consistent with the optimal number of hidden neurons, determined empirically in Huang et al. (1988). Others believe the number of hidden nodes necessary, to be a function of the number of the separable regions needed as well as the dimension of the input vector (Mirchandani, 1989).

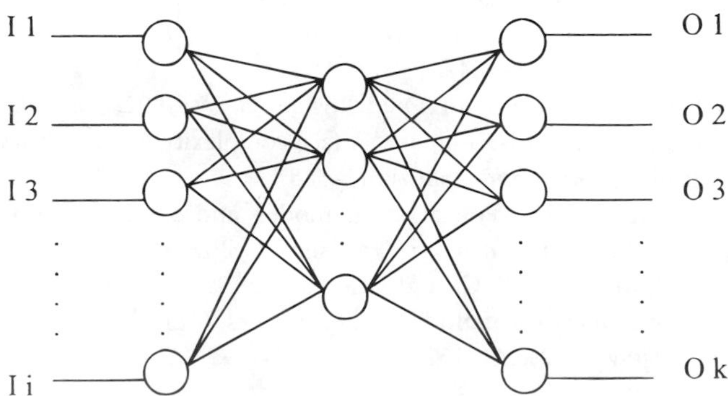

Figure 69.1 Multi-layer perceptron network. (*Source*: Zahner, D. and Micheli-Tzanakou, E. 1995.)

For most artificial neural networks there is an initial training phase where the interconnection strengths are adjusted until the network has a desired output. Only after training is the network capable of performing the task it was designed to do. The training phase can be either supervised or unsupervised. In supervised learning, there exists information about the correct or desired output for each input training pattern presented (Moore, 1992). In unsupervised learning no *a priori* information exists and training depends on the properties of the patterns. Unsupervised learning is highly dependent on the training data, and information about the proper classification is often lacking (Moore, 1992). For this reason, most neural network training is supervised.

The computational power of neuron-like networks was first demonstrated in 1943 by McCulloch and Pitts. After this landmark paper much effort was given to developing networks that could learn. In 1949, Donald Hebb proposed the strengthening of connections between pre-synaptic and post-synaptic units when both were active simultaneously (Hebb, 1949). The next major advancement in neural networks was by Frank Rosenblatt, who developed simple networks called perceptrons (Rosenblatt, 1958, 1959, 1962). In 1960, Widrow and Hoff proposed a model, called the adaptive linear element or ADALINE, which learns by modifying variable connection strengths, minimizing the square of the error in successive iterations (Widrow and Lehr, 1990). This

0-8493-8343-9/97/\$0.00+\$.50

error correction scheme is now known as the Least Mean Square (LMS) algorithm, and it has found widespread use in digital signal processing.

There was great interest in neural network computation until Minsky and Papert published a book in 1969, criticizing the perceptron. Minsky's book contained a mathematical analysis of perceptron-like networks, pointing out many of their limitations. They proved that the single-layer perceptron was incapable of performing the XOR mapping. The single-layer perceptron was severely limited in its capabilities. For linear activation functions, multilayer networks were no different from single layer models. Minsky and Papert pointed out that multilayer networks with nonlinear activation functions could perform complex mappings. However, the lack of any training algorithms for multilayer networks made their use impossible.

It was not until the discovery of multilayer learning algorithms that interest in neural networks resurfaced. The most widely used training algorithm, called backpropagation, was initially discovered by Werbos (1974), although it went virtually unnoticed until 1985 when Parker (1985) rediscovered it. In 1986, Rumelhart, Hinton, and Williams (1986) rediscovered the algorithm, again, and called it the delta rule. Their main contribution was not the discovery of the algorithm but their popularization of the algorithm, which has led to a renewed interest in neural networks. Another algorithm used for multilayer perceptron (MLP) training is the ALOPEX algorithm. ALOPEX was originally used for receptive field mapping by Tzanakou and Harth in 1973 (Harth and Tzanakou, 1973; Tzanakou and Harth, 1974; Tzanakou et al., 1979), and has since been applied to a wide variety of optimization problems. These two algorithms are explained in detail below.

69.2 The Backpropagation Algorithm

The backpropagation algorithm (Werbos, 1974; Parker, 1985; Rumelhart et al., 1986) is a learning scheme where the error is backpropagated layer by layer and used to update the weights. The algorithm is a gradient descent method that minimizes the error between the desired outputs and the actual outputs calculated by the multilayer perceptions (MLP). Let

$$= \frac{1}{2} \sum_{i=1}^{N} (T_i - Y_i)^2 Ep \tag{69.1}$$

be the error associated with template p. N is the number of output neurons in the MLP, T_i is the target or desired output for neuron i and Y_i is the output at neuron i calculated by the MLP. Let $E = \Sigma E_p$ be the total measure of error. The gradient descent method updates an arbitrary weight, w, in the network by the following rule:

$$w(n+1) = w(n) + \Delta w(n) \tag{69.2}$$

where

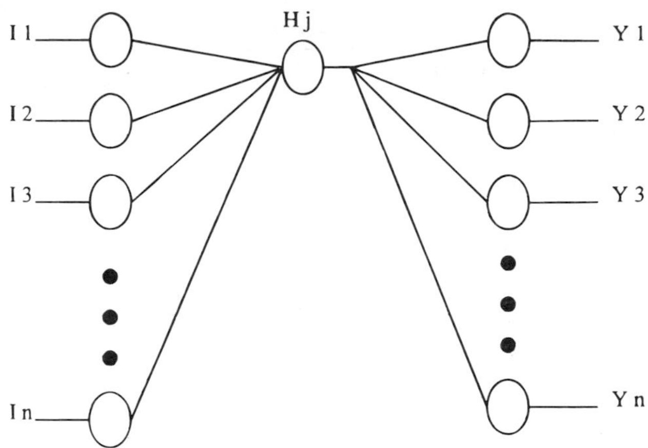

Figure 69.2 Hidden layer connections.

$$\Delta w(n) \propto -\eta \frac{\partial E}{\partial w(n)} \tag{69.3}$$

where n denotes the iteration number and η is a scaling constant. Thus, the gradient descent method requires the calculation of the derivatives $\partial E/\partial w(n)$ for each weight, w, in the network.

Figure 69.2 shows an arbitrary hidden layer neuron. Its output, H_j, is a nonlinear function f of the weighted sum of all its inputs (net_j).

$$H_j = f(\text{net}_j) \tag{69.4}$$

The function f is called the activation function. The most commonly used activation function is the sigmoid function given by

$$f(x) = \frac{1}{1 + e^{-x}} \tag{69.5}$$

The explanation of the calculation of the derivative $\partial E/\partial w_{ij}$ for any layer is now given. Using the chain rule, we can write

$$\frac{\partial E}{\partial w_{ij}} = \frac{\partial E}{\partial \text{net}_j} \cdot \frac{\partial \text{net}_j}{\partial w_{ij}} \tag{69.6}$$

and since

$$\text{net}_j = \sum_{j=1}^{n} w_{ij} I_i \tag{69.7}$$

we have

$$\frac{\partial \text{net}_j}{\partial w_{ij}} = I_i \tag{69.8}$$

and thus Equation (69.6) becomes

$$\frac{\partial E}{\partial w_{ij}} = \frac{\partial E}{\partial \text{net}_j} \cdot I_i \tag{69.9}$$

$$\frac{\partial E}{\partial \text{net}_j} = \sum_{k=1}^{m} \frac{\partial E}{\partial \text{net}_k} \cdot \frac{\partial \text{net}_k}{\partial H_j} \cdot \frac{\partial H_j}{\partial \text{net}_j} \tag{69.10}$$

recalling that

$$\text{net}_k = \sum_{j=1}^{n} w_{jk} H_j \tag{69.11}$$

it follows that

$$\frac{\partial \text{net}_k}{\partial H_j} = w_{jk} \tag{69.12}$$

also

$$\frac{\partial H_j}{\partial \text{net}_j} = f'(\text{net}_j) \tag{69.13}$$

Therefore,

$$\frac{\partial E}{\partial \text{net}_j} = f'(\text{net}_j) \cdot \sum_{k=1}^{n} \frac{\partial E}{\partial \text{net}_k} \cdot w_{jk} \tag{69.14}$$

Assuming f to be the sigmoid function of Equation 69.5, then

$$f'(\text{net}_j) = Y_i \cdot \mathcal{M} - Y_i \tag{69.15}$$

Equation 69.14 gives the unique relation that allows the backpropagation of the error to all hidden layers. For the output layer

$$\frac{\partial E}{\partial \text{net}_j} = \frac{\partial E}{\partial H_j} \cdot f'(\text{net}_j) \tag{69.16}$$

$$\frac{\partial E}{\partial H_j} = -(T_i - Y_i) \tag{69.17}$$

The backpropagation algorithm is now summarized. First, the output Y_i for all the neurons in the network is calculated. The error derivative needed for the gradient descent update rule of Equation 69.2 is calculated from

$$\frac{\partial E}{\partial w} = \frac{\partial E}{\partial \text{net}} \cdot \frac{\partial \text{net}}{\partial w} \tag{69.18}$$

If j is an output neuron then

$$\frac{\partial E}{\partial \text{net}_j} = -(T_i - Y_i) \cdot Y_i(1 - Y_i) \tag{69.19}$$

If j is a hidden neuron then the error derivative is backpropagated by using Equations 69.14 and 69.15. Substituting, we get

$$\frac{\partial E}{\partial \text{net}_j} = Y_i(1 - Y_i) \cdot \sum_{k=1}^{m} \frac{\partial E}{\partial \text{net}_k} \cdot w_{jk} \tag{69.20}$$

Finally the weights are updated as in Equation 69.2

There are many modifications to the basic algorithm that have been proposed to speed the convergence of the system. Convergence is defined as a reduction in the overall error below a minimum threshold. It is the point at which the network is said to be fully trained. One method (Rumelhart and McClelland, 1986) used for this thesis is the inclusion of a momentum term in the update equation such that

$$w(n+1) = w(n) - \eta \frac{\partial E}{\partial w(n)} + \alpha \Delta w(n) \tag{69.21}$$

η is the learning rate and is taken to be 0.25. α is a constant momentum term which determines the effect of past weight changes on the direction of current weight movements.

Another approach used to speed the convergence of backpropagation is the introduction of random noise (Holmstrom and Koistinen, 1992). It has been shown that while inaccuracies resulting from digital quantization are detrimental to the algorithm's convergence, analog perturbations actually help improve convergence time.

Backpropagation has achieved widespread use as a training algorithm for neural networks. Its ability to train multilayer networks has lead to a resurgence of interest in the field. Backpropagation has been successfully used in applications such as adaptive control of dynamical systems (Venugopal et al., 1993) and in many general neural network applications.

69.3 The ALOPEX Algorithm

The ALOPEX process is an optimization procedure which has been successfully demonstrated in a wide variety of applications. Originally developed for receptive field mapping in the visual pathway of frogs, ALOPEX's usefulness combined with its flexible form have increased the scope of its applications to a wide range of optimization problems. Since its development by Tzanakou and Harth (1973) ALOPEX has been applied to real time noise reduction (Ciaccio and Tzanakou, 1990), pattern recognition (Dasey and Micheli-Tzanakou, 1982), adaptive control systems (Venugopal et al., 1992), and multilayer neural network training to name a few.

Optimization procedures, in general, attempt to maximize or minimize a function $F()$. The function $F()$ is called the cost function and its value depends on many parameters or variables. When the number of parameters is large, finding the set ($x1$ $x2$ ... x_N) that corresponds to the optimal (maximal or minimal) solution is exceedingly difficult. If N were small an exhaustive search of the entire parameter space could be performed to find the "best" solution. As N increases intelligent algorithms are needed to quickly locate the solution. Only an exhaustive search can guarantee a global optimum is found, however, near optimal solutions are acceptable because of the tremendous speed improvement over exhaustive search methods.

As previously described, backpropagation, being a gradient descent method, often gets stuck in local extremes of the cost function. The local stopping points often represent unsatisfactory convergence points. Techniques have been developed to avoid the problem of local extrema, with simulated annealing (Kirkpatrick et al., 1983) being the most common. Simulated annealing incorporates random noise, which acts to dislodge the process from local extremes. Crucial to the convergence of the process is that the random noise be reduced as the system approaches the global optimum. If the noise is too large, the system will never converge and can be mistakenly dislodged from the global solution.

ALOPEX is another process which incorporates a stochastic element to avoid local extremes in search of the global optimum of the cost function. The cost function or response is problem dependent and is generally a function of numerous parameters. ALOPEX iteratively updates all parameters simultaneously based on the cross-correlation of local changes, ΔX_i, and the global response change ΔR, plus an additive noise. The cross-correlation term $\Delta X_i \Delta R$ helps the process move in a direction that improves the response. Table 69.1 shows how this can be used to find a global maximum of R.

All parameters X_i are changed simultaneously at each iteration according to:

$$X_i(n) = X_i(n-1) + \gamma \Delta X_i(n) \Delta R(n) + r_i(n) \quad (69.22)$$

The basic concept is that this cross-correlation provides a direction of movement for the next iteration. For example, take the case where $X_i\downarrow$ and $R\uparrow$. This means that the parameter X_i decreased in the previous iteration, and the response increased for that iteration. The product $\Delta X_i \Delta R$ is a negative number, and thus X_i would be decreased again in the next iteration. This makes perfect sense because a decrease in X_i produced a higher response. If you are looking for the global maximum, then X_i should be decreased again. Once X_i is decreased and R also decreases, then $\Delta X_i \Delta R$ is now positive and X_i increases.

These movements are only tendencies since the process includes a random component which will act to move the weights unpredictably, avoiding local extremes of the response. The stochastic element of the algorithm helps it to avoid local extreme at the expense of a slightly longer convergence or learning period.

The general ALOPEX updating equation (Equation 69.22) is explained as follows. $X_i(n)$ are the parameters to be updated, n is the iteration number, and $R()$ is the cost function, of which the "best" solution in terms of X_i is sought. γ is a scaling constant,

Table 69.1

	ΔX		ΔR	$\Delta X \Delta R$
$X\uparrow$	+	$R\uparrow$	+	+
$X\uparrow$	+	$R\downarrow$	−	−
$X\downarrow$	−	$R\uparrow$	+	−
$X\downarrow$	−	$R\downarrow$	−	+

$r_i(n)$ is a random number from a Gaussian distribution whose mean and standard deviation are varied, and $\Delta x_i(n)$ and $\Delta R(n)$ are found by:

$$\Delta X_i(n) = X_i(n-1) - X_i(n-2) \quad (69.23)$$

$$\Delta R(n) = R(n-1) - R(n-2) \quad (69.24)$$

The calculation of $R()$ is problem dependent and can be easily modified to fit many applications. This flexibility was demonstrated in the early studies of Harth and Tzanakou (1974). In mapping receptive fields, no *a priori* knowledge or assumptions were made about the calculation of the cost function, instead a "response" was measured. By using action potentials as a measure of the response (Harth and Tzanakou, 1974; Tzanakou et al., 1979; Micheli-Tzanakou, 1983,1984) receptive fields can be determined by using the ALOPEX process to iteratively modify the stimulus pattern until it produced the largest response.

It should be stated that due to its stochastic nature, efficient convergence depends on the proper control of both the additive noise and the gain factor γ. Initially, all parameters X_i are random. The additive noise is of Gaussian distribution with mean 0, and standard deviation, σ, is initially large. The standard deviation, σ, decreases as the process converges to ensure a stable stopping point. Conversely, γ increases with iterations. As the process converges, ΔR becomes smaller and smaller. An increase in γ is needed to compensate for this. Figures 69.3–69.5 show the response, gamma, and sigma with iterations for a typical ALOPEX run.

Additional constraints include a maximal change permitted for X_i, for one iteration. This bounded step size prevents the algorithm from drastic changes form one iteration to the next. These drastic changes often lead to long periods of oscillation, during which the algorithm fails to converge.

69.4 Multilayer Perceptron Network

A three-layer perceptron is trained for pattern recognition using both backpropagation and ALOPEX. The network was trained

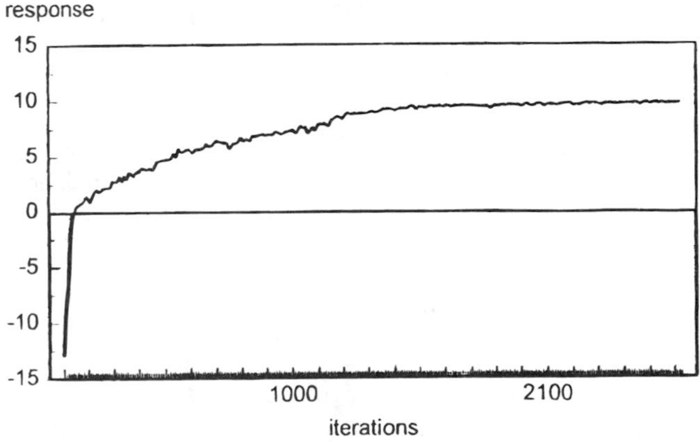

Figure 69.3 Response vs. iterations.

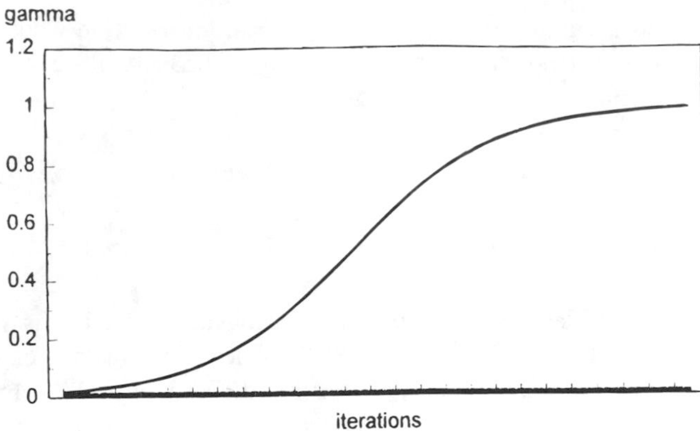

Figure 69.4 Gamma vs. iterations.

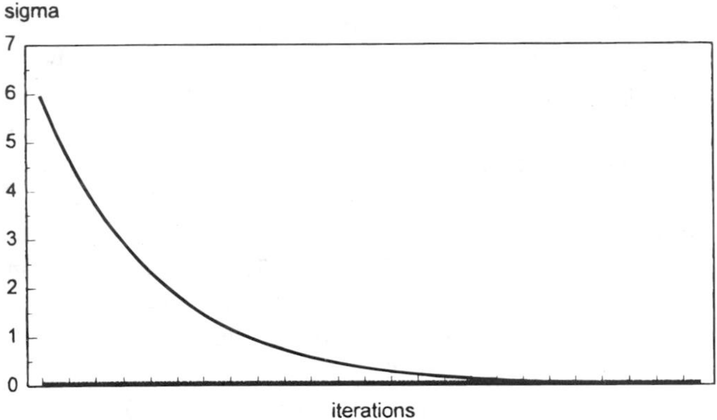

Figure 69.5 Sigma vs. iterations.

to recognize ten 5 × 7 templates corresponding to the ten digits 0–9. Backpropagation training occurs as previously described. How ALOPEX is implemented in this application is described below.

For each 5 × 7 input pattern there exists a desired output vector $O^{des}{}_k$. The observed output, $O^{obs}{}_k$, is found by a single feed forward pass through the fully interconnected layers of the network. Neurons or nodes in the hidden and output layers incorporate a nonlinear activation function, called a sigmoid.

A response is calculated for the jth input pattern based on the observed and desired output

$$R_j(n) = O^{des}{}_k - (O^{obs}{}_k(n) - O^{des}{}_k)^2 \tag{69.25}$$

where $O^{obs}{}_k$ and $O^{des}{}_k$ are vectors corresponding to O_k for all k. The total response for iteration n, is the sum of all the individual template responses, $R_j(n)$.

$$R(n) = \sum_{j=1}^{m} R_j(n) \tag{69.26}$$

In Equation 69.26, m is the number of templates used as inputs. ALOPEX iteratively updates the weights using both the global response information and local weight histories, according to the following:

$$W_{ij}(n) = r_i(n) + \gamma \Delta W_{ij}(n) \Delta R(n) + W_{ij}(n-1) \tag{69.27a}$$

$$W_{jk}(n) = r_i(n) + \gamma \Delta W_{jk}(n) \Delta R(n) + W_{ik}(n-1) \tag{69.27b}$$

where γ is an arbitrary scaling factor, $r_i(n)$ is an additive Gaussian noise, ΔW represents the local weight change and ΔR represents the global response information. These values are calculated by:

$$\Delta W_{ij}(n) = W_{ij}(n-1) - W_{ij}(n-2) \tag{69.28a}$$

$$\Delta W_{jk}(n) = W_{jk}(n-1) - W_{jk}(n-2) \tag{69.28b}$$

$$\Delta R(n) = R(n-1) - R(n-2) \tag{69.28c}$$

After training the network, it was tested for correct recognition using incomplete or noisy input patterns. The results show the robustness of the system to noise corrupted data. It should be noted, that regardless of which training procedure was used, backpropagation or ALOPEX, the recognition ability of the system was the same. The only difference being in how the response grew with iterations. Two response curves are shown in Figure 69.6.

It can be seen from Figure 69.6 that backpropagation converges faster than ALOPEX, particularly in the early periods of training. The networks were trained to 99% of maximal response; backpropagation converged in 1910 iterations, whereas ALOPEX took 2681 iterations to reach the same level.

The neural networks robustness is derived from its parallel architecture, and depends on the network topology not the learning scheme used to train. The network used, was a three-layer feedforward network with 35 input nodes, 20 hidden nodes, and 10 output nodes. The network's recognition ability was tested with noisy input patterns. Each 5 × 7 digit of the training set was subjected to noise of varying Gaussian distribution, and tested for correct recognition. The original training templates were binary (0 or 1) images. The results, demonstrating the networks robustness, are shown in Figure 69.7. Note that even when the standard deviation approaches one, the network correctly recognizes over 50% of the trained templates.

Artificial neural networks have shown a limited ability at solving problems that conventional computers are unable to resolve.

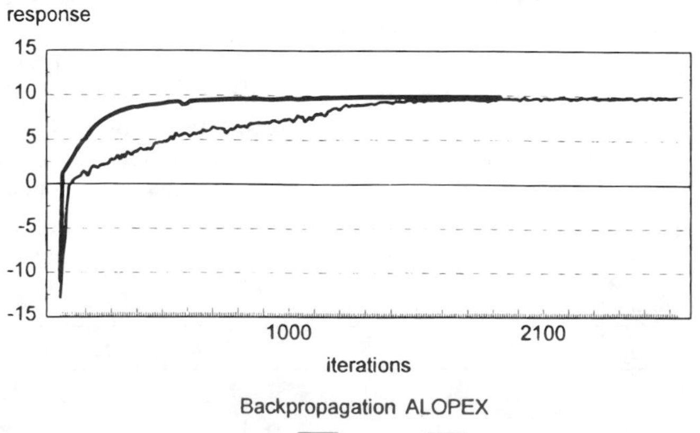

Figure 69.6 ALOPEX and backpropagation response curves.

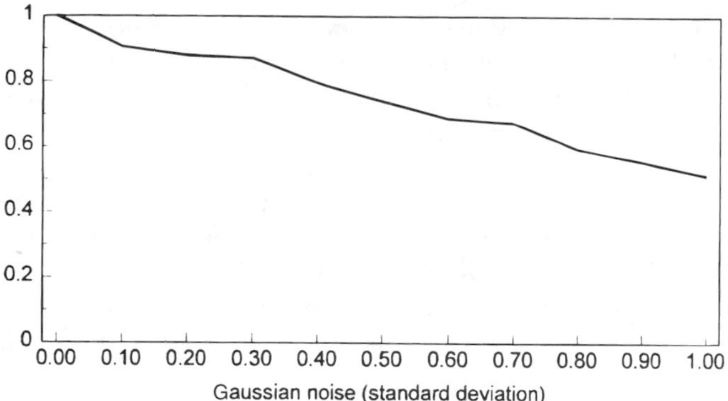

Figure 69.7 Recognition of noisy images by a trained MLP.

Image and speech recognition, motor control, and other such tasks which human brains perform well are stumbling blocks for the serial architecture. Artificial neural networks (ANN) were derived from a conscious effort to mimic brain functions and are models of their biological counterparts. Although ANNs are modeled after the human brain, they are far from repeating the brain's behavior. Severe limitations still exist, especially in terms of size and speed of the networks, and in the understanding of the biological system.

69.5 ALOPEX in VLSI

Artificial neural networks have existed for many years, yet because of recent advances in technology they are again receiving much attention. Major obstacles in ANN, such as a lack of effective learning algorithms, have been overcome in recent years. Training algorithms have advanced considerably, and now very large scale integration (VLSI) technology may provide the means for building superior networks. In hardware, the networks have a greater degree of speed, allowing for significantly larger architectures.

The tremendous advancement in technology during the past decades, particularly in VLSI technology, has renewed interest in ANN. Hardware implementation of neural networks are motivated by a dramatic increase in speed over software models. The emergence of VLSI technology has and will continue to lead neural network research in new directions. VLSI has advanced considerably over the last few years. Chips are now smaller, faster, contain larger memories, and are becoming cheaper and more reliable to fabricate.

Neural network architectures are varied, with over fifty different types being explored in research (Hect-Nielsen, 1988). Hardware implementations can be either electronic, optical, or electro-optical in design. A major problem in hardware realization, is often not due to the network architecture but to the physical realities of the hardware design. Optical computers, while they may eventually become commercially available, suffer far greater problems than do VLSI circuits. Thus, for the immediate and near future, neural network hardware designs will be dominated by VLSI.

Much debate exists as to whether digital or analog VLSI design is better suited for neural network applications. In general, digital designs are easier to implement and better understood methodologically. Also, in digital designs, computational accuracy is only limited by the chosen word length. While analog VLSI circuits are less accurate, they are smaller, faster, and consume less power than digital circuits (Mead and Ismail, 1989). For these reasons, applications that do not require great computational accuracy are dominated by analog designs.

Learning algorithms, especially backpropagation, requires high precision and accuracy in modifying the weights of the network. This has led some to believe that analog circuits are not well suited for implementing learning algorithms (Ramacher and Ruckert, 1991). Analog circuits can achieve high precision, at the cost of increasing the circuit size. Analog circuits with high precision (8 bits) tend to be equally large as their digital counterpart (Graf and Jackel, 1989). Thus, high precision analog circuits lose their size advantage over digital circuits. Analog circuits are of greater interest in applications requiring only moderate precision.

Early studies show that analog circuits can realize learning algorithms, provided that the algorithm is tolerant to hardware imperfections such as low precision and inherent noise. In a paper by Macq et al. (1993) a fully analog implementation of a Kohonen map, one type of neural network, with on-chip learning is presented (Macq et al., 1993). Because analog circuits have been proven capable of the computational accuracy necessary for weight modification, they should continue to be the choice of neural network research.

Size, speed, and power consumption are areas where analog circuits are far superior to digital circuits, and it is these areas that constrain most neural network applications. To achieve greater network performance, the size of the network must be increased. The ability to implement larger, faster networks is the major motivation for hardware implementation, and analog circuits superior in these areas. Power consumption is also of major concern as networks become larger (Andreou et al., 1991). As the number of transistors per chip increases, power consumption becomes a major limitation. Analog circuits dissipate less power than digital circuits, thus permitting larger implementations.

Besides its universality to a wide variety of optimization procedures, the nature of the ALOPEX algorithm makes it suitable for VLSI implementation. ALOPEX is a biologically influenced optimization procedure that uses a single value global response feedback to guide weight movements toward their optimum. This single value feedback, as opposed to the extensive error propagation schemes of other neural network training algorithms, makes ALOPEX suitable for fast VLSI implementation.

Recently, a digital VLSI approach to implementing the ALOPEX algorithm was undertaken by Pandya et al. (1990). Results of their study indicated that ALOPEX could be implemented using a single instruction multiple data (SIMD) architecture. A simulation of the design was conducted in software, and good convergence for a 4×4 processor array was demonstrated.

The importance of VLSI to neural networks has been demonstrated. For neural networks to achieve greater abilities, larger and faster networks must be built. In addition to size and speed advantages, other reasons including cost and reliability make VLSI implementations the current trend in neural network research. The design of a fast analog optimization algorithm, ALOPEX, is covered below.

ALOPEX is an optimization procedure where the "best" value of a cost function or response is sought. The process uses a stochastic element (added Gaussian white noise) to avoid local extremes of the response. In other words, the added noise helps the procedure to find the global maximum or minimum value of the response. ALOPEX is an iterative procedure where a large number of pixels are simultaneously changed by small amounts and then a new response is computed. The changes in the pixels are determined from the change in the response, the change in the pixel from the previous two iterations, plus the additive noise.

Let us assume that we have an array of 64 pixels which we call $I_i(n)$ where n represents the iteration. The additive Gaussian white noise is denoted by $r_i(n)$ and $R_j(n)$ is the response (or cost function) of the jth template at iteration n. The parameter $I_i(n)$ can then be found by the following equation:

$$I_i(n) = r_i(n) + \gamma \Delta I_i \Delta R + I_i(n-1) \tag{69.29}$$

where γ is an arbitrary scaling constant and ΔI_i and ΔR are found from the following:

$$\Delta I_i(n) = I_i(n-1) - I_i(n-2) \tag{69.30}$$

$$\Delta R(n) = R(n-1) - R(n-2) \tag{69.31}$$

$$i = 1, 2, 3, \ldots, 64$$

Let us assume that there are 16 templates to choose from, each with 64 pixels. The ALOPEX process is run on each of them with the objective being to recognize (converge to) an input pattern. Due to the iterative behavior, if allowed to run long enough ALOPEX will eventually converge to each of the templates. However, a "match" can be found by choosing that template which took the least amount of time to converge.

By convergence we mean finding either the global maximum or minimum of the response function. This response function can be calculated many different ways, dependent on the application. To allow this chip to be general enough to handle many applications, the response will be computed off the chip. A PROM will be used to compute the response based on the error between the input, $I_i(n,)$ and the template. The PROM will enable the response function to be changed to meet the needs of the application.

Although the chip design is limited to only 64 ALOPEX subunits, the parallel nature of ALOPEX will enable many chips to be wired together for larger applications. Parallel implementations are made easy since each subunit receives a single global response feedback, that governs its behavior. Backpropagation, on the other hand, requires dense interconnections and communication between each node. This flexibility is a tremendous advantage when it comes to hardwired implementations.

Originally the ALOPEX chip was designed using digital VLSI techniques. Digital circuitry was chosen over analog because it is easier to test and design. Floating point arithmetic was used to insure a high degree of accuracy. The digital design consisted of shift registers, floating point adders, and floating point multipliers. However, after having done much work toward the digital design, it was abandoned in favor of an analog design. The performance of the digital design was estimated and was found to be much slower than an analog design. The chip area of the digital design was much larger than an analog design would be. Also, the ALOPEX algorithm would be tolerant of analog imperfections due to its stochastic nature. For these reasons, it seemed clear that a larger, faster network could be designed with analog circuitry.

The analog design needed components similar to the digital design to implement the algorithm. Mainly, there needed to be an adder, multiplier, difference amplifier, a sample and hold mechanism, and a multiplexing scheme. These cells each perform a specific function and are wired together in a way that implements the ALOPEX process.

The chip is organized into 64 ALOPEX subunits, one for each pixel in the input image. They are stacked vertically, wiring by abutment. Each subunit is made from smaller components that are wired together horizontally and contains the following cells: a group selector, demultiplexor, follower aggregator, multiplier, transconductance amplifier, multiplexor, and another group selector.

The Gaussian white noise required for the ALOPEX process is added to the input before it reaches the chip. This will allow precise control of the noise, which is very important in controlling the stability of the algorithm. If there is too much noise the system will not converge. If there is too little noise, the system will get stuck in local minima of the cost function. By controlling the noise during execution, using a method similar to simulated annealing (Kirkpatrick et al., 1983) where the noise decays with time, it has been shown that the convergence time can be improved (Dasey and Micheli-Tzanakou, 1989). Also, by having direct control of the added noise the component and functional testing can be done with no noise added, greatly simplifying the testing.

The addition, multiplication, and subtraction required by the ALOPEX algorithm are performed by the follower aggregator, Gilbert multiplier (Mead, 1989), and the transconductance amplifier, respectively. To understand how these units implement the equations of the ALOPEX process, let us rewrite Equation 69.1 as follows:

$$I_i(n) = r_i(n) + \text{bias}(n) \tag{69.32}$$

where $r_i(n)$ is Gaussian white noise and bias(n) is defined as:

$$\text{bias}(n) = \gamma \Delta I \Delta R + \text{bias}(n-1) \tag{69.33}$$

The follower aggregation circuit computes the weighted average of its inputs. By weighing the inputs equally, the circuit computes the average of the two inputs. The average was chosen instead of the sum since the circuit is more robust, in that the output never has to exceed the supply voltage. A straight summer is more difficult to design because voltages greater than the supply voltage could be needed. The output of the follower aggregator, $I_i(n)$ is sent to the multiplier where a C-switch acts as a sample and hold, to store the value of the previous iteration, $I_i(n-1)$. The difference between these signals is ΔI and is one input to the multiplier. The previous two responses, calculated off chip, are the other two inputs, representing ΔR. The output of the multiplier is $\gamma\Delta I\Delta\ R$, where γ is the gain of the multiplier and is controlled by the control signal gamma.

The output of the multiplier is a current equal to $\gamma\Delta I\Delta R$. The current can be either positive or negative depending on the signs of ΔI and of ΔR. The output node acts as a capacitor that holds the bias voltage from Equation 69.33. This bias is then adjusted by an amount, $\gamma\Delta I\Delta R$, after each iteration. This bias is one of the inputs to the follower aggregator, the other being the input stimulus with added Gaussian white noise. The follower aggregator implements Equation 69.32, except that it computes the average or the sum divided by two.

The error signal is computed by the transconductance amplifier. The error is simply equal to the difference between $I_i(n)$ and the template to which you are matching.

However, since $I_i(n)$ is equal to the sum divided by two, the template values must be halved before being multiplexed onto the chip. The error signal is computed then multiplexed with $I_i(n)$ and sent off the chip. The error is used to compute the response $R(n)$. $I_i(n)$ is sent off the chip so that the operator can see the image as the algorithm converges, by sending the signal to some sort of display.

The power is supplied to the chip by four pins, two each for VDD and GND. The purpose of having two pins of the same signal is so that by placing them on opposite sides of the chip, and by proper wiring, the resistive drop can be reduced.

In designing the chip, much effort was made in making it controllable and testable, while making the chip general enough that it could be used in a wide variety of applications. This is why the Gaussian white noise is added off chip, and also why the error signal is taken off chip for the computation of the response. This not only allows the response function to be changed to meet the requirements of the specific application but it also provides the operator with accessible test points.

Despite the decrease in operating speed by a factor of four, due to time division multiplexing at both the input and outputs, the chip still operates at over 7,000,000 complete iterations per second. This speed may not even be attainable, given possible interfacing bottlenecks, and much slower support hardware that is necessary for operation. Support hardware necessary for chip operation include circuitry for the response calculation as well as memory to store template. Depending on the application A/D and D/A converters may be necessary. If this is the case then 7 Mhz operation speed is more than adequate.

While backpropagation is the most widely used software tool for training neural networks, it is less suitable for VLSI hardware implementation than ALOPEX for many reasons. While backpropagation converges quickly, due to its gradient descent method, it can often get stuck in local extrema. ALOPEX tends to avoid local extrema by incorporating a random noise component, at the expense of slightly longer convergence times.

The major differences arise when hardware implementation is discussed. Backpropagation is computationally taxing, due to the error computation needed for each node in the network. Each error is a function of many parameters (i.e., all the weights of the following layer). In hardware, very complex interconnections between all nodes are required to compute this error.

ALOPEX is ideal for VLSI implementation for a couple of reasons. First, the algorithm is tolerant to small amounts of noise, in fact, noise is incorporated to help convergence. Secondly, all parameters change based on their local history and a single value global response feedback. This single valued feedback is much simpler to implement than the error propagation used in backpropagation.

69.6 Discussion

We have presented a MLP architecture trained by ALOPEX and backpropagation for recogntion of digital numerals. The binary templates of these digits were presented as inputs to the MLP first as clean images and secondly as contaminated by noise and results were compared. For this particular application, BP was a little faster than ALOPEX. This fact as well as the ALOPEX characteristics *per se* made us think of putting together a VLSI design and implementation of ALOPEX. The basics behind this design are presented and the requirements for such a chip are discussed.

References

Andreou, A., et al. 1991. VLSI neural systems, *IEEE Trans. Neural Networks,* 2(2):205–213.

Ciaccio, E. and Tzanakou, E. 1990. The ALOPEX process: Application to real-time reduction of motion artifact, *Annual Int. Conf. of IEEE EMBS,* 12(3):1417–1418.

Dasey, T. J., and Micheli-Tzanakou, E. 1989. A pattern recognition application of the Alopex process with hexagonal arrays, *Int. Joint Conf. Neural Networks,* II:119–125.

Graf, H. P. and Jackel, L. D. 1989. Analog electronic neural network circuits, *IEEE Circuits and Devices Mag.,* 44–55, July.

Harth, E. and Tzanakou, E. 1974. Alopex: a stochastic method for determining visual receptive fields, *Vision Research,* 14:1475–1482.

Hebb, D. 1949. *The Organization of Behavior,* John Wiley & Sons, New York, NY.

Hect-Nielsen, R. 1988. Neurocomputing: picking the human brain, *IEEE Spectrum,* 36–41.

Holmstrom, L. and Koistinen, P. 1992. Using additive noise in backpropagation training, *IEEE Trans. Neural Networks,* 3(1):24–38.

Huang, S. and Huang, Y. 1991. Bounds on the number of hidden neurons in multilayer perceptrons, *IEEE Trans. Neural Networks,* 2(1):47–55, Jan. 1991.

Kirkpatrick, S., Gelatt, C. D., and Vecchi, M. P. 1983. Optimization by simulated annealing, *Science,* 220:671–679.

Kung, S. Y., Hwang, J., and Sun, S. 1988. Efficient modeling for multilayer feedforward neural nets, Proc. *IEEE Conf. Acoustics, Speech Signal Processing,* 2160–2163, New York, NY.

Macq, D., Verlcysen, M. Jespers, P. and Legat, J. 1993. Analog implementation of a Kohonen map with on-chip learning, *IEEE Trans. Neural Networks,* 4(3):456–461.

McCulloch, W. C. and Pitts, W. 1943. A logical calculus of the ideas immanent innervous activity, *Bulletin of Mathematical Biophys.,* 5:115–133.

Mead, C. 1989. *Analog VLSI and Neural Systems,* Addison-Wesley Publishing, New York, NY.

Mead, C. and Ismail, eds., 1989. *Analog VLSI Implementation of Neural Systems,* Kluwer Academic Publishers, Boston, MA.

Micheli-Tzanakou, E. 1983. Methods and designs: visual receptive fields and clustering., *Behav. Res. Methods and Instrum.,* 15(6):553–560.

Micheli-Tzanakou, E. 1984. Non-linear characteristics in the frog's visual system, *Biol. Cybern.,* 51:53–63.

Minsky, M. and Papert, S. 1969. *Perceptrons: An Introduction to Computational Geometry,* MIT Press, Cambridge, MA.

Mirchandani, G. 1989. On hidden nodes for neural nets, *IEEE Trans. Circuits and Sys.,* 36(5):661–664.

Moore, K. 1992. Artificial neural networks: weighing the different ways to systemize thinking, *IEEE Potentials,* 23–28.

Pandya, A. S., Shandar, R., and Freytag, L. 1990. An SIMD architecture for the Alopex neural network, *Parallel Architectures for Image Processing,* SPIE-1246:275–287.

Parker, D. B. 1985. *Learning Logic,* Technical Report, Center for Computational Research in Economics and Management Science, MIT Cambridge, MA.

Ramacher and Ruckert, eds. 1991. *VLSI Design of Neural Networks,* Kluwer Academic Publishers, Boston, MA.

Rosenblatt, F. 1959. *Principle of Neurodynamics,* Spartan Books, New York, NY.

Rosenblatt, F. 1958. The perceptron: a probabilistic model for information storage and organization in the brain, *Psychological Review,* 65:386–408.

Rosenblatt, F. 1962. *Principles of Perceptrons,* Spartan Press, Washington, DC.

Rumelhart, D. E. and McClelland, J. L., eds., 1986. *Parallel Distributed Processing,* MIT Press, Cambridge, MA.

Rumelhart, Hinton, and Williams. 1986. Learning Internal Representations by Error Propagation, *Parallel Distributed Processing, Vol. 1: Foundations,* Rumelhart and McClelland, eds., MIT Press, Cambridge, MA.

Tzanakou, E. and Harth, E. 1973. Determination of visual receptive fields by stochastic methods, *Biophys. Journal,* 15:(42a).

Tzanakou, E., Michalak, R., and Harth, E. 1979. The ALOPEX process: visual receptive fields by response feedback, *Biol. Cybern.,* 35:161V174.

Venugopal, K., Pandya, A. and Sudhakar, R. 1992. ALOPEX algorithm for adaptive control of dynamical systems, *Proc. IJCNN 1992,* II:875–880.

Venugopal, V., Sudhakar, R., and Pandya, A. 1993. An improved scheme for direct adaptive control of dynamical systems using backpropagation neural networks, *Circuits, Systems, and Signal Processing.*

Werbos, P. J. 1974. *Beyond Regression: New Tools for Prediction and Analysis in the Behavioral Sciences,* Ph.D. Thesis, Harvard University, Cambridge, MA.

Widrow, B. and Lehr, M. A. 1990. 30 Years of adaptive neural networks: perceptron, madaline, and backpropagation, *Proc. IEEE,* 78(9):1415–1442.

Zahner, D. A. and Micheli- Tzanakou, E. 1995. Artificial neural networks: definitions, methods, applications. In *The Biomedical Engineering Handbook,* J. D. Bronzino, ed. CRC Press, Boca Raton, FL.

70

Supervised Neural Networks for Handwritten Digit Recognition in Industrial Processing

WooGon Chung
Sung Kyun Kwan University

Evangelia Micheli-Tzanakou
Rutgers University

70.1 Introduction 951
70.2 Preprocessing of Handwritten Digit Images 951
70.3 Zernike Moments (ZM) to Characterize Image Patterns 955
Introduction • Reconstruction by Zernike Moments • Features from Zernike Moments
70.4 Dimensionality Reduction 960
Principal Component Analysis • Discriminant Analysis
70.5 Analysis of Prediction Error Rates from Bootstrapping Assessment 962
70.6 Summary 964

70.1 Introduction

Visual pattern recognition has long been an interesting problem both from the application and technical aspects. We have designed a system that understands characters, words, and even sentences. Handwritten digit recognition is one of the most challenging problems. Its applications are extensive: automatic document processing, banking systems, etc. Depending on the writer's environment, the writing style differs and this causes the difficulty in the system design even though the fundamental assumption in writing communications is that differences between characters are more significant than differences among the same character.

Handwritten digit recognition has a long history and many researchers have proposed different models (Tappert et al., 1990; Bitchell and Gillies, 1989; La Cun et al., 1989; Mantas, 1986; Suen et al., 1984). These are mostly *model-based.* The developed model is usually specific to the given data set and its applicability for a different data set is rather restricted. These methods find local properties, or primitives, e.g., arcs, lines, starting/end points, and the rules that combine the individual properties from the skeletonized images. Painstaking processes to fine tune the properties are some of the difficulties and variabilities of the resulting systems.

A simple and important image pattern (Arabic numerals, for example) analysis is carried out to demonstrate that a simple *model-free* strategy, via global moments with proper statistical analysis, renders a quite acceptable result. The moment calculation for features is model-free since no other information of the data set than the group label is required in order to design the pattern recognition system. All the groups of data are treated the same way to extract the global features, while the model-based methods are required to describe each different digit by a certain list of properties.

70.2 Preprocessing of Handwritten Digit Images

The images are passed through a sequence of preprocessing steps before the Zernike moments calculation, a global feature extraction method which will be described in the section on Zernike moments. A block diagram for the sequence of preprocessing procedures and the intermediate results of digit images are shown in Figure 70.1.

The objectives and the methods for each preprocessing are described in the following paragraphs. The major objective in the preprocessing stage of the pattern recognition system is in getting unique features from the same group of patterns.

Noise due to acquisition or transmission is reduced by a *smoothing* operation with neighboring pixel values which generally is low-pass filtering. Smoothing substitutes the value of the pixel in the center of a window with the average of pixels in the window. Such an operation has the effect of suppression of the distortions in the gray values caused by sensor noise or transmission errors. Edges in an object are typical changes in

0-8493-8343-9/97/$0.00+$.50

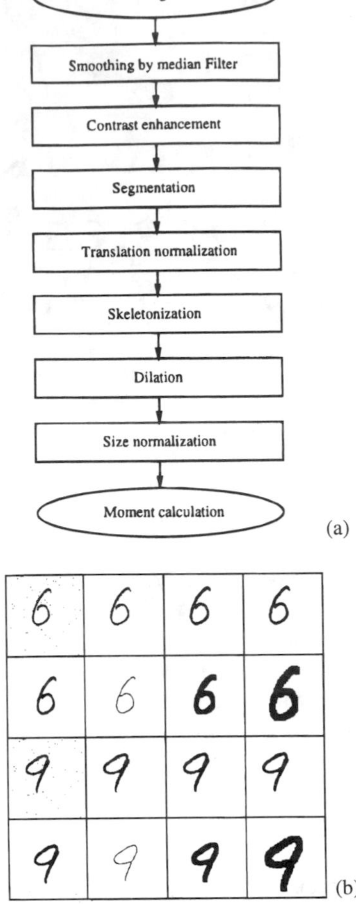

Figure 70.1 (a) Sequence of the preprocessing. (b) Two original images (9,6) and their preprocessed results. Starting with the original images, the results of "smoothing", "contrast enhancement", "thresholding", "centering", "skeletonization", "dialization", and "size normalization" are presented from left to right and top to bottom of the figure.

the gray levels. Thus *smoothing* and *edge detection* are contradictory. In image analysis, however, one likes to smooth without distorting the edges.

Median filtering, which is a nonlinear operation, is a well known method for noise removal while preserving the edges (Jähne, 1991) rendering a solution to this contradiction. Since the binary noise (i.e., shot noise) is the noise type to be removed, we apply *median* filtering to our data. The pixels in the window (usually a 3 × 3-matrix) are sorted and a robust median value is chosen to replace the pixel value. Since the binary noise, like the shot noise, completely changes the gray level value, it is very unlikely to be the median value in the window. Thus, the median of the pixel values in the window is used to estimate its gray level value.

Due to variations in the acquisition systems, e.g., cameras and scanners, reflection angle, etc., recorded pixel values are not exactly what objects really are. Thus the smoothed images are further processed for gray-scale modifications to enhance contrast.

The contrast of an image in a given gray level range can be increased by stretching the range of the gray levels in the image. The brightest and the darkest pixel values are found and they are assigned to white and black, i.e., 255 and 0 in an 8-bit representation. This is an affine transformation taking the acquisition value and changing it to the full gray levels. Some benefits from the contrast enhancement (usually known as histogram equalization) are:

- The elimination of the irregular acquisition effects.
- The enhancement of contrast.

The enhanced contrast not only helps in viewing, but also in building more confidence in finding the threshold in order to separate an object from the background. Segmentation of an image into parts is an important stage in image analysis. It uses a clustering of pixels by their values. An ideal clustering would result in homogeneity in the distribution of pixels in a cluster, segmenting the images into parts by their pixel values.

In digit recognition, we have only one object to be segmented from the background. For this purpose, simply taking the midpoint as the threshold of the gray level in the histogram will result in good binary images.

Another preprocessing step is done for the varied positions of the centroids of the digits as seen in Figure 70.2. This translational variance of the images is interpreted as the camera movement in a direction perpendicular to the optical axis. The centroid of an image $f(x,y)$ is given by

$$\bar{x} = M_{1,0}/M_{0,0}, \qquad \bar{y} = M_{0,1}/M_{0,0}$$

where

$$M_{p,q} = \sum \sum x^p y^q f(x, y)$$

is the $(p + q)$-th order moment. The image is translated to the center of the frame by moving the centroid to that point.

Depending upon the writing instruments and writer's habits, stroke widths are different as it can be seen in the sample digits of Figure 70.2. *Skeletonization*[1] is used to find an approximation to the medial axis of planar objects.

The basic requirements in the skeletonization algorithms are end point preservation and pixel connectivity (Zhang and Suen, 1984; Pavlidis, 1980). The algorithm used for our study is that of Zhang and Suen (1984). Eight neighbor pixel values, either 0 or 1, are usually compared and a decision is made as to whether to delete the center pixel or not. The eight neighbors are denoted as $(p_2, p_3, \ldots p_9)$ as shown in the Figure 70.3(a). Using the eight neighbor values we test for four conditions in order to decide for the removal of the center pixel, p_1.

The algorithm works in two directions. The conditions for the two directions are:

[1] Some other terms, like shrinking and thinning, appear in the literature and are used interchangeably (Zhang and Suen, 1984; Pavlidis, 1980).

Figure 70.2 Some digits from the training data. Five people are involved in writing digits on a grid and of one inch square. We assume that the digits are well separated, that is *interaction* and *occlusion* problems are already solved. Different sizes and widths of writing styles are notable.

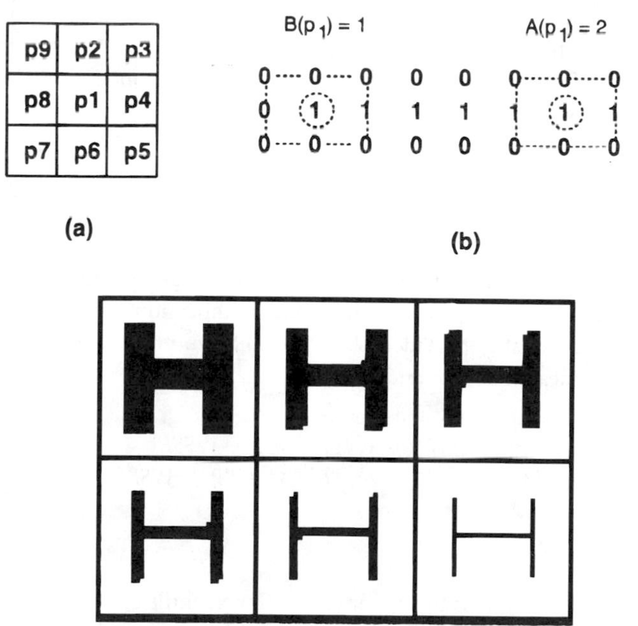

Figure 70.3 (a) Neighboring pixels and (b) preventing end points and middle points from deletion. A series of skeletonized patterns next to the original pattern. Starting from upper left, original pattern, 1st, 2nd 3rd, 4th, and 5th (the last one) are displayed. As the procedure goes, it peels off the boundary points and an opposite corner point, then it does the same from the opposite direction. In the first peeling-off all the N/W boundary points and a S/E corner point are deleted.

$2 < B(p_1) \leq 6$	$2 \leq B(p_1) < 6$	(70.1)
$A(p_1) = 1$	$A(p_1) = 1$	(70.2)
$p_2 * p_4 * p_6 = 0$	$p_2 * p_4 * p_8 = 0$	(70.3)
$p_4 * p_6 * p_8 = 0$	$p_2 * p_6 * p_8 = 0$	(70.4)

where the first two conditions of the second set are the same as the ones in the first set of conditions. $B(p_1)$ is the sum of all the eight neighboring pixels, that is, $B(p_1) = p_2 + p_3 + \ldots + p_9$, and $A(p_1)$ represents the number of the (0,1) patterns around the neighboring pixels (Figure 70.3).

The conditions of Equation 3 and Equation 4 in the first set above, are satisfied when $p_4 = 0$ or $p_6 = 0$ or ($p_2 = 0$ and $p_8 = 0$). So point p_1, which has been removed, might be an East/South boundary point or a North-West corner point. This set of conditions is valid for East/South boundary point or North-West corner point deletion. The conditions of Equation 1 and Equation 2, protect the end points from being deleted (Figure 70.3): the first loop at the left end point has $B(p_1) = 1$ which does *not* meet the condition of Equation 1, and the second loop shows that $A(p_1) = 2$, meaning that the middle point cannot be deleted. A set of skeletonized patterns and the original is also displayed in Figure 70.3. Note that the procedures take turns in both directions as the algorithm passes the two sub-iterations with the corresponding conditions.

After segmentation by thresholding, the binary images are skeletonized to obtain the invariance of the stroke width that resulted from different writing styles and writing instruments.

For global moment calculation a dilation process is desired. Pen path-width standardization by dilation is proven to be important for that purpose. (This will be indirectly seen later in Figure 70.6, where the reconstruction of patterns is progressively done for some font images. In the reconstruction, the narrow strokes are less prominent compared to the wider width parts of the fonts).

Another reason for the path-width standardization is that the moment values obtained from the skeletonized images (width of one pixel) are more vulnerable to perturbation by a little change in the location of the skeletonized pixels (Figure 70.4). Therefore a certain width in a given image size is desired in order:

- To stabilize the moment values against the variation of the skeletonized patterns.
- To build tighter clusters in the same group and larger separations between the clusters of the different classes.

Nonlinear morphological processing as opposed to the linear processing (e.g., convolution) achieves certain effects such as dilation, erosion, opening, closing, and boundary extraction (Haralilck and Shapiro, 1992; Jähne, 1991; Serra, 1982).

Let $\mathcal{F}$ be the set of all the pixels of the matrix which are not zero and $\mathcal{M}$ the set of the non-zero mask pixels. With $\mathcal{M}_p$ we denote the mask shifted or centered on this reference point to the pixel p.

The dilation is defined with a set operation as follows:

$$\mathcal{F} \oplus \mathcal{M} = \{p: \mathcal{M}_p \cap \mathcal{F} \neq \emptyset\}$$

that is, dilation operation produces the points on which the mask $\mathcal{M}$ and the image $\mathcal{F}$ have at least one non-zero pixel in common. Erosion is defined as

$$\mathcal{F} \ominus \mathcal{M} = \{p: \mathcal{M}_p \subseteq \mathcal{F}\}$$

that is, erosion produces the points for which the mask is a

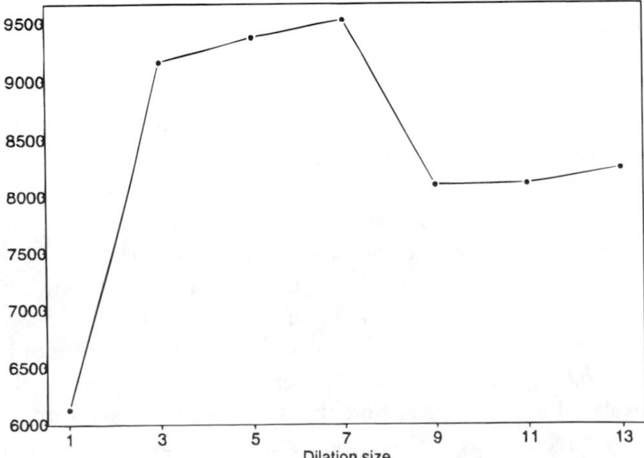

Figure 70.4 Bartlett statistic against dilation matrix size. Dilation increases the statistic as K increases and starts decreasing after size $2K + 1 = 7$ with the image frame of size 129×129.

subset of the original image. These are equivalent to the regular binary operations for dilation and erosion, respectively:

$$f_{xy} = \bigvee_{k=-K}^{K} \bigvee_{k=-K}^{K} (M_{k,l} \wedge f_{x-k,y-l}) \tag{70.5}$$

and

$$f'_{xy} = \overline{\vee_{k=-K}^{K} \vee_{k=-K}^{K} M_{k,l} \wedge \overline{f_{x-k,y-l}}} \tag{70.6}$$

where the $\vee$ and $\wedge$ denote the logical OR and AND operations, respectively. The binary image f is convolved with a symmetric $(2K + 1) \times (2K + 1)$ mask M. The erosion has to be done as shown in the Equation 70.6 since the all-zero mask $\mathcal{M}$ would have no meaning in a binary AND operation. In other words, the erosion operation is done by first dilating with the background and then inverting the result to get the erosion effect.

Optimal Size of the Mask for Dilation

The intuition for the dilation operation is justified via a simulation to find an optimal dilation matrix of size $2K + 1$. The strategy is, given a size of the image frame, to find the size of the dilation matrix of size $2K + 1$ which gives a larger separation between group means (or higher confidence in order to reject the null hypothesis of MANOVA model), in comparing J population mean vectors. The MANOVA model and the modified Wilks' statistic (or Bartlett statistic) (Johnson and Wichern, 1988) are used to measure the separation. Leaving the details to Chung (1994) we introduce its definition as well as results from a simulation study.

Bartlett Statistic is the modified Wilks' lambda statistic, given by

$$\Lambda^* = \frac{|WSSP|}{|BSSP + WSSP|} \tag{70.7}$$

where the $WSSP$ and $BSSP$ are the within and between sum of squares and cross-products. A simple modification results in the Bartlett statistic, provided that the null hypothesis (i.e., same group means) is true and $N = \Sigma_{j=1}^{J} n_j$ is large:

$$-\left(N - 1 - \frac{(p + J)}{2}\right) \ln\left(\frac{|WSSP|}{|BSSP + WSSP|}\right) > \chi^2_{p(J-1)}(\alpha) \tag{70.8}$$

where $\chi^2_{p(J-1)}(\alpha)$ is the upper (100α)th percentile of a chi-square distribution with $p(J - 1)$ degrees of freedom and J is the number of classes while p represents the dimensionality of the covariate.

The size of the digital images used in this study is about 128×128 because the moment approximation by digital calculation requires high resolutions. This fact is partly studied for lower moment invariants (Teh and Chin, 1986) requiring the image size to be larger than 60×60 pixels. For the higher order moments, a higher resolution may be required. With the image size fixed (129 × 129), Bartlett statistics (or modified Wilks' lambda Λ^*) are calculated for different dilation matrix size, $2K + 1$, as in Equation 70.5.

For the simulation study, an image pattern "A" is preprocessed in the same way except for the size of dilation. The skeletonized image is dilated with dilation matrix sizes $2K + 1 = 1,3,5,7,9,11,13,15$. A set of Zernike moments are obtained for different dilation sizes and the Bartlett statistics (Equation 70.8) are calculated and plotted against the size of the dilation matrix $2K + 1$ (Figure 70.4). The null hypothesis (that is, all the mean vectors are the same) test is obviously rejected in all K values at the significance level $\alpha = 0.01$.

From Figure 70.4, the statistic with $2K + 1 = 7$ is the highest. In fact results using size 7 look the best (Figure 70.1) for an image of size around 129 × 129, which is the size we have chosen.

It is worth noting the assumption made on the statistics. The statistics of Equation 70.8 assume that the error term follows the multinormal distribution $\epsilon_{ij} \sim N(0, \Sigma)$ in the one-way classification model

$$X_{ij} = \mu_X + \mu_j + \epsilon_{ij}$$

where $i = 1,2, \ldots, n_j$ and $j = 1,2, \ldots, J$. μ_X is an overall mean and μ_j represents the jth treatment effect (or jth group mean) with $\sum_{j=1}^{J} n_j\mu_j = 0$.

Furthermore, the statistic does not necessarily measure the separation between multi-group mean vectors where $J > 2$. For example, with a scalar statistic in a two-dimensional three-group setting, a large statistic may also result from the case that any two mean vectors are unacceptably close, but the other mean vector is far from the two. However, the more ideal separation among the groups is in the case when the three mean vectors are equilateral in distance. The Bartlett statistic in this sense gives little insight on how well the mean vectors are separated, however it still gives some feeling about the separation.

After the translation invariance has been obtained by the translation standardization stage in Figure 70.1, size standardization follows. The radius of an image function $f(x,y)$ can be defined as (Dudani et al., 1977)

$$r = (\mu_{20} + \mu_{02})^{1/2} \tag{70.9}$$

where μ_{20} and μ_{02} are the moments of order 2 after the centralization, and represent the variance in x- and y-directions of the ellipsoidal approximation of the image. In the stage of size standardization the desired radius r^s, after normalization, is fixed to be 60% of one half the smaller side of the image frame:

$$r^s = 0.6 * \min\{ncol/2, nrow/2\} \tag{70.10}$$

where $ncol \times nrow$ is the size of the image frame. All the object pixels are scaled in such a way that, the radius r^s of the scaled object becomes the prescribed value. The 60% restriction can be thought of as a control parameter that contains all the scaled objects inside the frame. This prevents the scaled objects from spilling outside the frame and it corresponds to the coordinate normalization in the Zernike moment calculation which will be treated in the following section. It should be that, the radius in Equation 70.9 is neither the principal axis length a nor the secondary principal axis length b of an ellipsoid approximation of the image function $f(x,y)$, but that it is directly related to a and b; the area of an ellipse of parameters (a,b) is equal to πab. Digits such as '1' have larger major principal axis but smaller secondary principal axis, whereas the digit '0' and '4' give relatively equal principal and secondary principal axes a and b. The effect of the size normalization with the control constant 0.6 in Equation 70.10 is shown in Figure 70.1.

70.3 Zernike Moments (ZM) to Characterize Image Patterns

Introduction

The complex Zernike moments of order n with repetition l are defined as

$$A_{n,l} = \frac{n+1}{\pi} \int_0^{2\pi} \int_0^{\infty} [V_{nl}(r, \theta)]^* f(r \cos\theta, r \sin\theta)\, r dr d\theta \tag{70.11}$$

where $n = 0,1,2, \ldots, \infty$ and l takes on positive and negative integer values such that

$$n - |l| = \text{even}, \qquad |l| \leq n. \tag{70.12}$$

The Zernike polynomials (Bhatia and Wolfe, 1954) given by

$$V_{nl}(r \cos\theta, r \sin\theta) = R_{nl}(r)\exp(il\theta) \tag{70.13}$$

are a complete set of complex-valued orthogonal functions on a unit disk $x^2 + y^2 \leq 1$:

$$\int_0^{2\pi} \int_0^1 [V_{nl}(r, \theta)]^* V_{mk}(r, \theta)\, r\, dr\, d\theta = \frac{\pi}{n+1} \delta_{nm}\delta_{kl} \tag{70.14}$$

In Figure 70.5 the luminance of gray images represents the real part of the polynomials which are in $[-1,1]$ and 256 gray levels are assigned to the discrete level of the polynomials. The periodicity in Equation 70.13 being equal to $2\pi/l$ relates the polynomial image to an l-fold symmetric image.

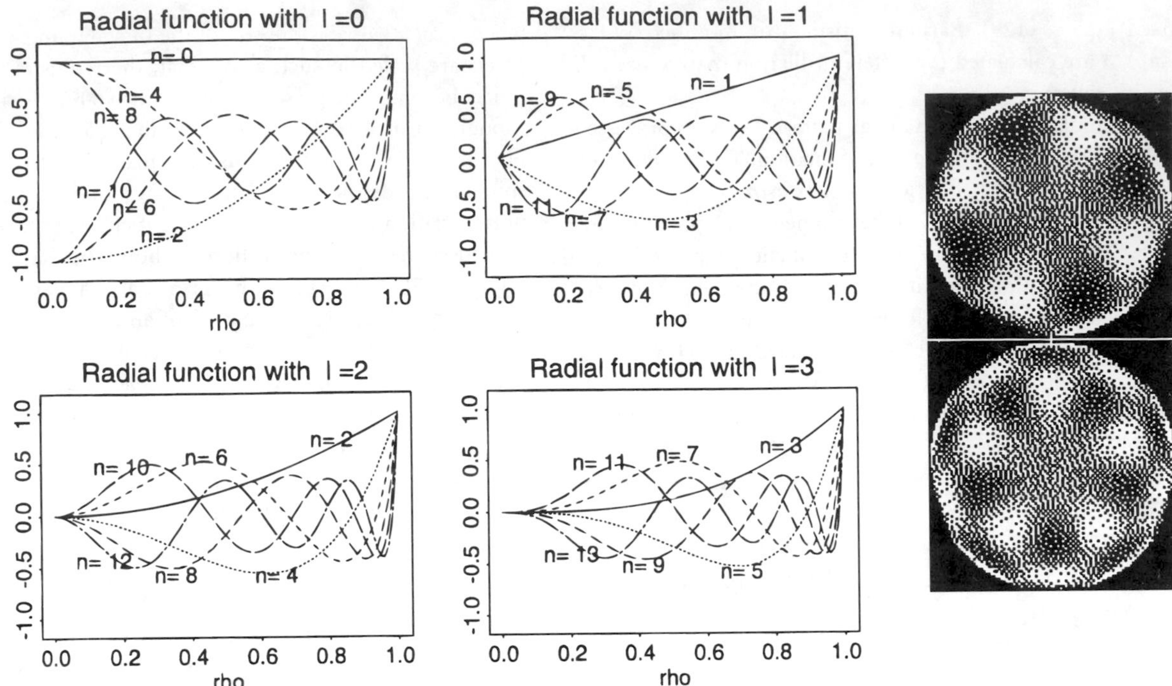

Figure 70.5 Radial and Zernike polynomials $R_{nl}(r)$ for different orders for a given azimuthal repetition l. Two real parts of Zernike polynomials with (n,l) = (6,4) and (n,l) = (9,5) are also shown.

The real-valued radial polynomial shown in Figure 70.5 and represented by Equation 70.13 satisfies the following condition:

$$\int_0^1 R_{nl}(r)R_{mk}(r)r\,dr = \frac{1}{2(n+1)}\delta_{mn} \tag{70.15}$$

and is defined as

$$R_{nl}(r) = \sum_{s=0}^{(n-|l|)/2} (-1)^s \frac{(n-s)!}{s!\left(\frac{n+|l|}{2}-s\right)!\left(\frac{n-|l|}{2}-s\right)!} r^{n-2s}$$

$$= \sum_{\substack{k=|l| \\ n-k=even}} B_{n|l|k} r^k \tag{70.16}$$

where the $B_{n|l|k}$ is the new expression (by changing the variable) for the coefficient part of the radial polynomial:

$$B_{n|l|k} = (-1)^{n-k/2} \frac{\left(n - \frac{n+k}{2}\right)!}{\left(\frac{n-k}{2}\right)!\left(\frac{k+l}{2}\right)!\left(\frac{k-l}{2}\right)!}$$

The orthogonality of the Zernike polynomials enables a given $f(x,y)$ to be expressed in terms of the polynomials

$$f(x, y) = \sum_{n=0}^{\infty} \sum_{\substack{|l|\le n \\ n-|l|=even}} A_{nl}V_{nl}(x, y) \tag{70.17}$$

where the Zernike moments A_{nl} are computed over the unit disk $x^2 + y^2 \le 1$:

$$A_{nl} = \frac{n+1}{\pi} \int\int_{x^2+y^2\le 1} [V_{nl}(r, \theta)]^* f(x, y)dxdy$$

$$= [A_{n,-l}]^*$$

This is obtained simply by the orthogonality property of the Zernike polynomials in Equation 70.13. The second equal sign holds because $f(x,y)$ is real and the radial polynomials satisfy $R_{n,l} = R_{n,-l}$. $A_{n,l}$ and it can be interpreted as the projection, correlation, or proximity of a given image onto each complex valued polynomial. Thus the set of Zernike moments are the collection of the projections of a given image onto the set of the Zernike polynomials of order n and azimuthal repetition l.

In practice, we cannot have an infinite limit in the summation of Equation 70.17. Instead the finite order of N is used:

$$\hat{f}(x, y) = \sum_{n=0}^{N} \sum_{\substack{|l|\le n \\ n-|l|=even}} A_{nl}V_{nl}(x, y). \tag{70.18}$$

This approximation with the finite order N is the optimal among all the other representations of $f(x,y)$ expressed by moments due to the orthogonality property.

The Zernike moments can be represented by the regular geometric moments (GM) by expressing the terms r^k in Equation 70.16 and $\exp(-il\theta)$ in Equation 70.13 in terms of x and y:

$$r^k = (x^2 + y^2)^{k/2}$$

$$\exp(-\theta) = (x^2 + y^2)^{-1/2}(x - iy)$$

$$\exp(-il\theta) = \exp(-i\theta)^l = (\cos\theta - i\sin\theta)^l \quad (70.19)$$

$$= (x^2 + y^2)^{-1/2}(x - iy)^l$$

$$= (x^2 + y^2)^{-1/2} \sum_{m=0}^{l} \binom{l}{m} (-i)^m x^{l-m} y^m$$

The resulting expression for the A_{nl} is

$$A_{nl} = \frac{n+1}{\pi} \iint R_{nl}(r)\exp(-il\theta) f(x, y)dxdy$$

$$= \frac{n+1}{\pi} \sum_{\substack{k=|l| \\ n-k=even}} \sum_{j=0}^{q} \sum_{m=0}^{|l|}$$

$$w^m \binom{q}{j} \binom{|l|}{m} B_{n|l|k} M_{k-(2j+m),2j+m} \quad (70.20)$$

where $w = -i, +i$ for $l > 0$, $l \leq 0$, respectively, and $q = 1/2\,(k - |l|)$.

Reconstruction by Zernike Moments

In designing a pattern recognition system one should be concerned with what constitutes the feature elements. What is the best set (if any at all) of the possible features for the classification purpose? How does one get it? A trade-off is to be made between representability and complexity of the system that resulted from the selected set of the global features.

The order of the ZM to be included can be found by the reconstruction process. Due to the orthogonality of the Zernike polynomials (Equation 70.14), we are able to reconstruct the image $\hat{f}(x,y)$ by its finite order representation (Equation 70.18) of the original image $f(x,y)$. In order to illustrate the reconstruction process and to find the optimal order to be used, we revisit Equation 70.18 and simplify it in terms of real-valued functions (Khotanzad and Hong, 1990)

$$\hat{f}(x, y) = \sum_{n=0}^{N} \sum_{l<0} A_{nl}V_{nl}(\rho, \theta) + \sum_{n=0}^{N} \sum_{l\geq 0} A_{nl}V_{nl}(\rho, \theta)$$

$$= \sum_{n=0}^{N} \sum_{l>0} A_{n,-l}V_{n,-l}(\rho, \theta) + \sum_{n=0} \sum_{l\geq 0} A_{nl}V_{nl}(\rho, \theta)$$

$$= \left[\sum_{n=0}^{N} \sum_{l>0} [A_{nl}^{*}V_{nl}^{*}(\rho, \theta) + A_{nl}(V_{nl}(\rho, \theta)] \right]$$

$$+ \sum_{n=0}^{N} A_{n0}V_{n0}(\rho, \theta)$$

$$= \sum_{n=0}^{N} \left[\sum_{l>0} (C_{nl}\cos l\theta + S_{nl}\sin l\theta)R_{nl}(\rho) + \frac{C_{n0}}{2} R_{n0}(\rho) \right]$$

with

$$C_{nl} = 2\mathrm{Re}(A_{nl}) = \frac{2(n+1)}{\pi} \iint_{x^2+y^2\leq 1} f(x, y)$$

$$R_{nl}(\rho)\cos l\theta\, dx\, dy$$

$$S_{nl} = 2\mathrm{Im}(A_{nl}) = \frac{-2(n+1)}{\pi} \iint_{x^2+y^2\leq 1} f(x, y)$$

$$R_{nl}(\rho)\sin l\theta\, dx\, dy$$

In the introduction to this section, the azimuthal index l is limited by the condition:

$$n - l = \text{even and } n \geq l \quad (70.21)$$

Two digits of Times-bold 14 font were reconstructed from the ZM. The reconstruction is done up to a certain high order, say 15; the order up to 15 renders a total of 72 moments:

$$72 = \sum_{n=0}^{15} \left\{ \left\lfloor \frac{n}{2} \right\rfloor + 1 \right\}$$

Figure 70.6 shows the original image and its reconstruction by ZM. It is evident that lower order ZMs capture gross shape information and that the more fine structures are filled in by higher order moments. Each digit consists of 16 small frames which are the original, top-left, and its reconstruction in the direction from left to right and top down for orders 1 to 15. Most of the digits are well reconstructed by an order around 11 ~ 15, except for digit '4'. We conjecture that the handwritten digits with various writing styles need orders up to 15 for the reconstruction to be close enough to the original images. The possible redundant variables included by higher moments will be removed via PCA (see the following section).

Order 15 was chosen to be the cut-off point for our handwritten digit data through visual inspection of Figure 70.6. In this way we have resolved the question of how large the feature set should be.

Features from Zernike Moments

The advantage of ZMs for pattern recognition has been reported in terms of noise immunity, discrimination power (Abu-Mostafa and Psaltis, 1984) and image representation ability, noise insensitivity and information relevance (Teh and Chen, 1988). These are considered a basic theoretical support for ZM. A simulation study that supports the theoretical work can be found in Chung and Micheli-Tzanakou (1994). The application of ZM for pattern recognition is also in favor of the ZMs compared to others (Belkasim et al., 1991).

Functions of the Zernike moments, called Zernike Moment Invariants (ZMIs), are introduced in order to get the rotational invariance from different orders m and azimuthal indices h for

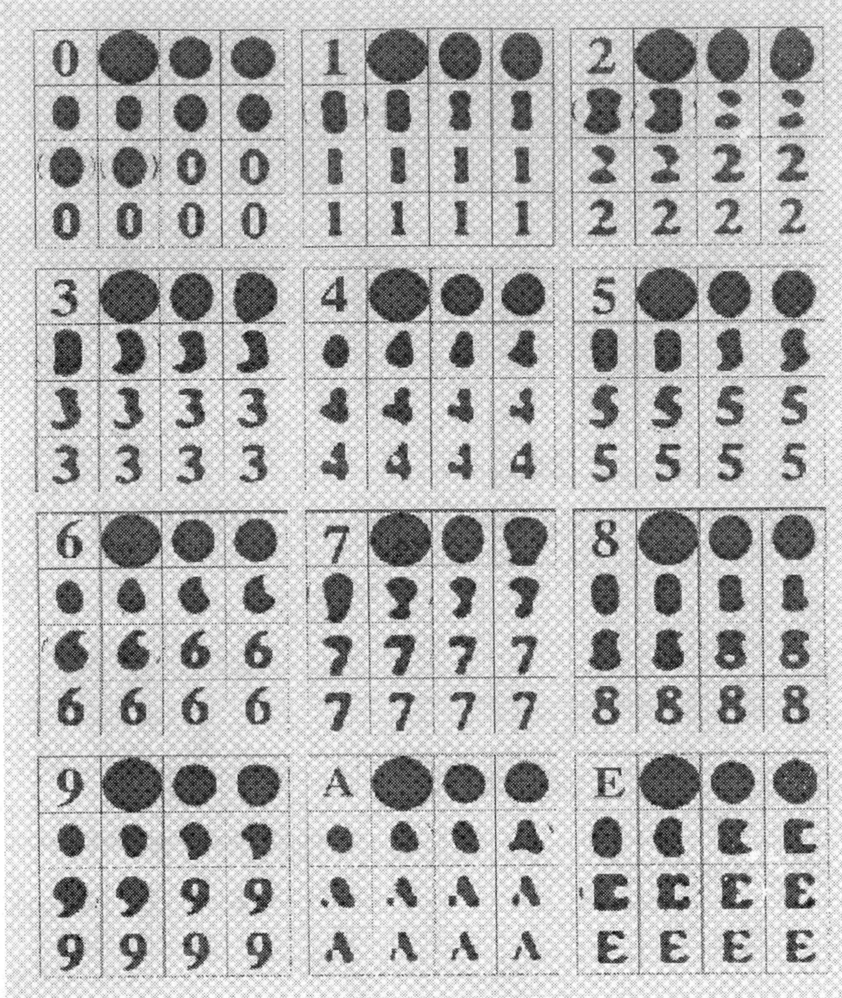

Figure 70.6 Reconstruction via ZM. The original image and the reconstruction by 1st to 15th order of moments shows the effects of the orders in the reconstruction.

a given order n and l. Teague (1980) introduced a form of rotational invariance

$$ZMI_{n0} = A_{n0}; \qquad ZMI_{nl} = |A_{nl}|^2 \tag{70.22}$$

$$ZMI_{nz} = [A^{\star}_{nl}(A_{mh})^p] \pm [A^{\star}_{nl}(A_{mh})^p]^{\star} \tag{70.23}$$

where the integers m, n, h, l and positive integer p are constraints such as

$$m = \text{any integer} \qquad p = \frac{l}{h} \text{ with } l \bmod h = 0$$

$$h \le l \qquad z = p + l + h \text{ for index}$$

The first two invariants in Equation 70.22 are called primary invariants and the third in Equation 70.23 secondary invariants. The number of the primary invariants for a given order n is $\lfloor n/2 \rfloor + 1$, due to the constraint $n - |l| =$ even in Equation 70.12 of the ZM definition. The secondary invariants are found by forcing the exponential term to be 1, thus to become independent of the angle θ,

$$\begin{aligned}
&[A^{\star}_{nl}(A_{mh})^p] + [A^{\star}_{nl}(A_{mh})^p]^{\star} \\
&= R_{nl}(r)\exp(-jl\theta)R^p_{mh}(r)\exp(jph\theta) \\
&\quad + R_{nl}(r)\exp(jl\theta)R^p_{mh}(r)\exp(-jph\theta) \\
&= R_{nl}(r)R^p_{mh}(r)[\exp\{j(ph - l)\theta\} + \exp\{-j(ph - l)\theta\}] \\
&= R_{nl}(r)R^p_{mh}(r)\cdot\cos(ph - l)\theta
\end{aligned} \tag{70.24}$$

with the constraint on p,h and l ensuring the (cos) term to be one, thus resulting in $R_{nl}(r)R^p_{mh}(r)$ being independent of the angle θ. Since there is no restriction on the order m of the secondary invariant we could have an infinite number of invariants by varying m while satisfying Equation 70.23. However, by the definition of the functional independence of the invariants only $n + 1$ number of invariants are functionally independent. The moment invariants are functionally independent if the invariants can be solved for the moments which form them (Belkasim et al., 1991). $n + 1$ is the number of the independent moments from the definition of ZM (Equation 70.11) and its constraints on the indices (Equation 70.12).

Another set of Zernike moment invariants has been introduced recently (Belkasim et al., 1991). The idea is the same as that of Teague's in Equation 70.22 and Equation 70.23, and is given by

$$ZMI'_{n0} = A_{n0}; \qquad ZMI'_{nl} = |A_{nl}| \tag{70.25}$$

$$ZMI'_{nz} = [A^{\star}_{mh}(A_{nl})^{p}] \pm [A^{\star}_{mh}(A_{nl})^{p}]^{\star} \tag{70.26}$$

where

$$m \leq n \qquad p = \frac{h}{l} \text{ with } 0 \leq p \leq 1$$

$$h \leq l \qquad z = \frac{l}{h} \text{ for index}$$

The difference of this formulation from the original ones (Equation 70.22 and Equation 70.23) is that the modulus values are taken instead of their squares and the constraints on the indices are rational power multiplications rather than integer power. The first constraint $m \leq n$ ensures that only combinations of moments of orders lower than n are used to form the secondary invariants. The factor p ranges between 0 and 1. This constraint tends to decrease the magnitudes of the secondary invariants since p decreases as l increases. This magnitude decreasing property of the new invariants ZMI' (Equation 70.26) is desirable, and was not present in the original ZMI of Equation 70.23.

The secondary parts of the ZMI and ZMI' (Equation 70.23 and Equation 70.26) are the additional ($\lfloor n/2 \rfloor$) rotational invariant values that are obtained from the power multiplication of the higher order moments or lower order moments, respectively.

As shown in the ZMI and ZMI' the rotational invariance is obtained in various ways by forcing the phase information of complex-valued ZMs to be one. Using only radial information means that all the points of a circle of radius r, in the complex domain, are the same. In addition, in digit recognition, 180-degree rotation conflict digits such as 9 and 6 are not taken care of.

Khotanzad and Hong (1990) used the modulus value of the complex-valued ZM, the primary invariant, to eliminate the rotational problem. Their argument is based on the fact that the ZM for a rotated image $f^r(x,y)$ due to the rotation by θ, results in a simple phase shift:

$$A^r_{nl} = \frac{n+1}{\pi}\int_{\phi=0}^{2\pi}\int_{r=0}^{1} f(r, \phi - \theta)R_{nl}(r)\exp(-il\phi)rdrd\phi$$

$$= \frac{n+1}{\pi}\int_{\phi^{\star}=0}^{2\pi}\int_{r=0}^{1} f(r\,\phi^{\star})R_{nl}(r)\exp(-il(\phi^{\star}+\theta))rdrd\phi^{\star}$$

$$= A_{nl}\exp(-il\theta) \tag{70.27}$$

The original function $f(x,y)$ and the rotated one $f^r(x,y)$ results in the same modulus value:

$$|A_{nl}| = |A^r_{nl}|$$

As a remedy to this problem we have included skewness information into the modulus value of all the variables used (up to order 15). The skewness of a 2-dimensional function $f(x,y)$ is obtained for each variable x and y. The skewness for an image function $f(x,y)$ is given by

$$S_x = \frac{\mu_{3,0}}{(\mu_{2,0})^{3/2}} \tag{70.28}$$

$$S_y = \frac{\mu_{0,3}}{(\mu_{0,2})^{3/2}} \tag{70.29}$$

Two more new variables for the skewness information are added to the modulus values of the ZM order from 2 to 15. The 0th and 1st order are deleted since the image has been preprocessed to be size standardized and to be centered by the centroid. The new moment moduli with the skewness values added are now not only rotation invariant, but also free of the 180-degree rotation conflict.

The section on the analysis of prediction error rates from Bootstrapping assessment includes the results from both the modulus values of ZM called 'V' and the modulus values of ZM with skewness information added, called 'V1'.

An argument is developed here to justify the use of only the real components of the ZM. The 180-degree rotation conflict problem is taken care of by the third order moments $\mu_{0,3}$ and $\mu_{3,0}$ of Equation 70.28 and Equation 70.29. This skewness information is contained in the real part of the phase components of the lower orders of ZM ($A_{3,l}$ and $A_{2,l}$). We call 'R' the real part of the ZM. The number of the real part of the ZM for a given order n is $\lfloor n/2 \rfloor + 1$ and is obtained with $m =$ even from Equation 70.20. That is, the real part of ZM is given by

$$C_{nl} = 2\text{Re}[A_{nl}] = 2\,\frac{n+1}{\pi}\iint_{x^2+y^2\leq 1} R_{nl}(r)\cos(l\theta)f(x,y)dx\,dy$$

$$= 2\,\frac{n+1}{\pi}\sum_{\substack{k=|l|\\ n-k=even}}\;\sum_{j=0}^{q}\;\sum_{\substack{m=0\\ m=even}}^{|l|} (-1)^{m}\binom{q}{j}\binom{|l|}{m}B_{n|l|k}M_{k-(2j+m),2j+m} \tag{70.30}$$

with $q = 1/2(k - |l|)$.

The rotational invariance by the modulus operation of ZM or moment invariants has been successful with the patterns that have no 180-degree rotation conflict, such as printed English alphabets, the aerial views of the four Great Lakes, aircraft recognition tasks, etc.

The circular symmetry property of the Zernike polynomials seems to handle the rotational variance of the patterns well. The Zernike polynomial $V_{nl}(r,\theta)$ is circularly symmetric in periods of $2\pi/l$ (Equation 70.13) and has a wedge shape implying the rotational variance of patterns.

If the patterns from a group vary within a certain orientation range (as is the case with handwritten digits) the modulus operation or the ZMI costs too much for the rotational invariance. The range of the modulus operation of the complex-valued ZM is only the distance, represented by a radius in a complex domain. The real part of ZM, however, has a range twice as large as that of the modulus value; it explains more than the radius does.

The modulus value (called 'V') or squared modulus of the complex-valued ZM is the primary part of the Zernike moment invariants [ZMI] (Equation 70.22 and Equation 70.25). The secondary part of the ZMI shown in Equation 70.23 and Equation 70.26 are not included in our features because the secondary invariants are simply the power multiplication that adds another ($\lfloor n/2 \rfloor$) number of the orientation independent values. Instead, we have followed the strategy of including the primary invariants of all the moments that have been included by the reconstruction process in finding the finite number of moments for the given patterns.

70.4 Dimensionality Reduction

The subject of dimensionality reduction in pattern recognition is concerned with mathematical tools for reducing the size of the features. The most revealing facts with dimensionality reduction are discussed in Kittler (1986) and summarized below:

- Reduction of the physical system complexity as required by feasibility limitations of either a technical or economical nature.
- It ensures the reliability of the decision-making procedure by removing the redundant and irrelevant information which has a derogatory effect on the classification process.
- More importantly, the dimensionality is strongly related to the size of the sample used for training: as the dimensionality increases the size of the training is required to grow exponentially. Neural networks, however, train well regardless of the dimensionality, except that the networks require more time to learn and result in poor convergence.

Two stages are employed for this purpose: Principal Component Analysis (PCA) followed by Discriminant Analysis (DA) which are eigen analyses on the covariance-type matrices. These eigen analyses can be interpreted as finding $p < q$ directions on which the projections of the data result in some interesting properties, such as large variance, or separation among the group means under a set of constraints.

Principal Component Analysis

Principal component analysis of a multivariate random sample can be viewed as finding an axis optimizing a criterion in a geometric sense. Illustration with projection of a simulated two-dimensional data points is shown in Figure 70.7. Pearson (1901) looked for a new axis on which the projection gives the *least sum of squares* of d_i. Hotelling (1933) was interested in finding a new axis on which the *maximum variance* of the projection values is obtained (see Granadesikan, 1977). Even though the approaches are different and opposite, the resulting axis from the two different approaches is the same. The optimal axis for minimal sum of the squares of d_i's is the same as the one with the axis in which maximum variance of z_i is obtained.

$$\max_{P_Y} \sum \frac{z_i^2}{N-1} \Leftrightarrow \min_{P_Y} \sum d_i^2 \qquad (70.31)$$

where P_Y is the projection operator defined by a projection axis.

The idea of the PCA is to find a rotational transformation (i.e., an orthogonal transformation) matrix $R_{q \times q}$ such that the sample variances of the new rotated variables are in decreasing order of magnitude (Granadesikan, 1977). Thus the first principal component is such that the projections of the given points onto it have maximum variance among all possible linear coordinates; the second principal component has maximum variance subject to being orthogonal to the first; and so on.

The PCA is done on the sample version of total covariance matrix $T_{q \times q}$ of the handwritten data matrix, $Y_{N \times q}$, where the dimensionality $q = 70$ and the size $N = \Sigma_{j=1}^{J} n_j = 1000$, where $J = 10$ is the number of classes.

The lower curve of the plot in Figure 70.8 is called *scree plot* and represents the variance information contained in the new derived variates. The upper curve represents the accumulated version of the lower scree curve; which is the total variance of the newly obtained variables from the first to the corresponding variable indices. The 95% and 99% of the accumulated variance are indicated by the two broken lines. The 99% explanation of the variance information is obtained by the first 35 newly obtained variables.

Since the dimensionality of the original data $q = 70$ was too large, we reduced the dimensionality using PCA of the total covariance. With the new data set which is supposed to be uncorrelated (or less correlated) we are ready to do more statistical treatment in order to find multidimensional outliers for robust analysis and reduce the *heteroscedacity* (and as a by-product enhance the multi-normality, if possible at all).

A strategy we follow for such large dimensionality, is a two step dimensionality reduction. First, principal component analysis on the total sample covariance matrix, T, is carried out. Then discriminant analysis follows, in order to reduce the dimensionality even further to $J - 1$.

Even though the PCA is well known to be sensitive to outliers (Ammann, 1993; Devlin et al., 1981) we argue that the whole data set is preserved, as much as we want, in a lower dimensional space, provided that the explanation of the variance information is over, say 99%. The whole data set as a single batch from the different clusters of different classes is decorrelated via principal component analysis. Now the lower $p = 35$ dimensional space is processed by discriminant analysis for further dimensionality reduction.

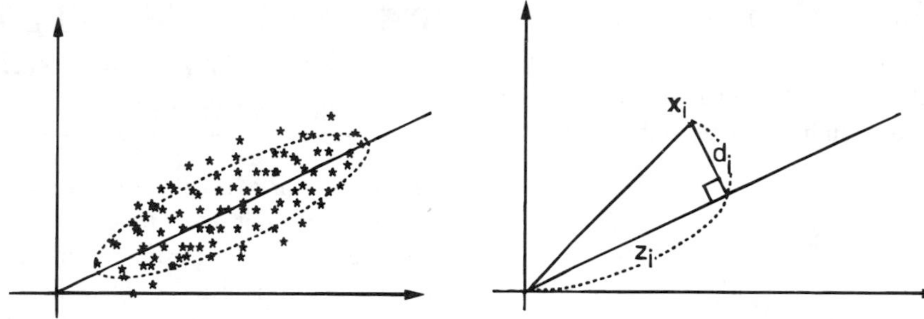

Figure 70.7 Illustration of projection of a vector point y_i onto the principal axis. z_i represents the projected value of x_i onto the axis and d_i the error component of the projection. $z_i^2 + d_i^2 = const$ confirms the equivalence of the two motivations for finding the optimal axis.

Discriminant Analysis

Suppose that we wish to find a linear transformation matrix F, which maximizes some distance criterion d defined over a sample of random vectors in a new transform space. Two interesting pairwise distance measures are the *intraset* and the *interset* distances (Kittler, 1975). The intraset distance, or averaged within-class distance, between the kth variable of all pattern vectors in one class, averaged over all classes:

$$d_W^{(k)} = \frac{1}{2}\sum_{i=1}^{J} P(w_i)\frac{1}{n_i^2}\sum_{j=1}^{n_i}\sum_{l=1}^{n_i} \mathbf{f}_k^t(\mathbf{y}_{ij} - \mathbf{y}_{il})(\mathbf{y}_{ij} - \mathbf{y}_{il})^t\mathbf{f}_k \qquad (70.32)$$

where n_i is the number of vectors $\mathbf{y} \in w_i$ and $\mathbf{f}_k$ is the kth column of the transformation matrix F.

The interset distance, or between-class distance, of the kth direction in the new transform space is defined as:

$$d_B^{(k)} = \sum_{i=2}^{J} P(w_i)\sum_{h=1}^{i-1} P(w_h)\frac{1}{n_i n_h}\sum_{j=1}^{n_i}\sum_{l=1}^{n_h} \qquad (70.33)$$

$$\mathbf{f}_k^t(\mathbf{y}_{ij} - \mathbf{y}_{hl})(\mathbf{y}_{ij} - \mathbf{y}_{hl})^t\mathbf{f}_k.$$

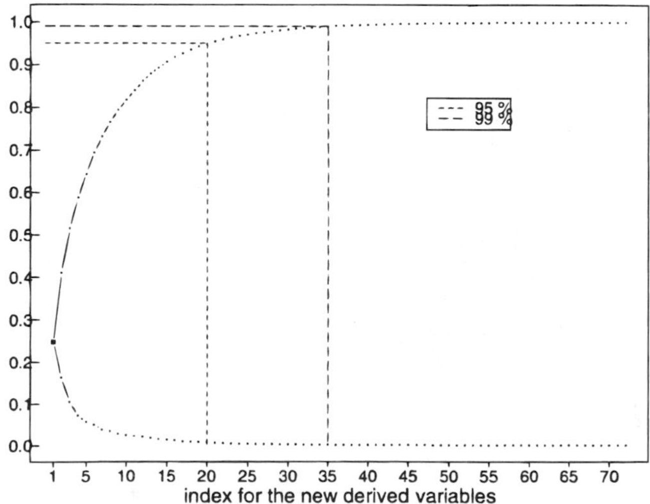

Figure 70.8 PCA on the sample total covariance matrix of the handwritten data set 'R'.

The first two summation indices hold for $N(N - 1)/2$ interpoint distances.

These averaged distance measures are expressed in terms of sample within-groups covariance matrix W and between-groups matrix B (Kittler and Young, 1973), defined as:

$$W = \frac{1}{N - J}\sum_{i=1}^{J}\sum_{j=1}^{n_i}(\mathbf{y}_{ij} - \bar{\mathbf{y}}_i)(\mathbf{y}_{ij} - \bar{\mathbf{y}}_i)^t$$

$$B = \frac{1}{J - 1}\sum_{i=1}^{J} n_i(\bar{\mathbf{y}}_i - \bar{\mathbf{y}})(\bar{\mathbf{y}}_i - \bar{\mathbf{y}})^t$$

respectively, where $N = \sum_{i=1}^{J} n_i$.

Using the definition of W and B, the distance measures $d_W^{(k)}$ (Equation 70.32) and $d_B^{(k)}$ (Equation 70.33) can be written, in terms of W and B as follows:

$$d_W^{(k)} = \mathbf{f}_k^t W \mathbf{f}_k$$

$$d_B^{(k)} = \mathbf{f}_k^t B \mathbf{f}_k$$

Now, we are interested in maximizing a distance measure $d_B^{(k)} = \mathbf{f}_k^t B\mathbf{f}_k$ with respect to the transformation vector $\mathbf{f}_k$ subject to a constraint, e.g., holding a distance measure, (i.e., $d_W^{(k)} = \mathbf{f}_k^t W\mathbf{f}$) constant. The constraints are usually chosen to be irrelevant for maximization of $d_B^{(k)}$ while guaranteeing a unique solution $\mathbf{f}_k$, i.e., $\mathbf{f}^t W\mathbf{f} = 1$. The solution for this kind of optimization problem can be obtained by the method of Lagrange multipliers. Maximization of $d_W^{(k)}$, subject to $d_W^{(k)}$ constant, has the form:

$$\max_{\mathbf{f}_k}\{J = d_B^{(k)} - \lambda(d_W^{(k)} - \text{const}) = \mathbf{f}_k^t B\mathbf{f}_k - \lambda(\mathbf{f}_k^t W\mathbf{f}_k - \text{const})\}$$

Setting the first derivative of J with respect to $\mathbf{f}_k$ equal to zero yields:

$$\frac{\partial J}{\partial \mathbf{f}_k} = B\mathbf{f}_k - \lambda W\mathbf{f}_k = 0$$

$$(B - \lambda W)\mathbf{f}_k = 0.$$

If we premultiply the above by W^{-1}, it results in an eigenvalue problem, i.e.:

$$(W^{-1}B - \lambda I)\mathbf{f}_k = 0 \qquad (70.34)$$

The traditional disCRIMinant COORDinate system (or CRIMCOORD) is interpreted as finding functions that maximize the quadratic forms:

$$d_B^{(k)} = \mathbf{f}_k^t B \mathbf{f}_k,$$

with respect to $\mathbf{f}_k$, subject to the constraint of

$$d_W^{(k)} = \mathbf{f}_k^t W \mathbf{f}_k = 1 \qquad (70.35)$$

resulting in the solution of Equation 70.34.

Two consecutive linear transformations by R (via PCA) followed by F (via DA) are represented by a linear transformation matrix $F\,R$ of dimension $J - 1 \times q$, for example, 9×70 for our data set. Figure 70.9 shows two-dimensional projections of 30 randomly selected patterns from each group on the first five discriminate variates (CRIMCOORD) with corresponding digit representation. Remarkably, some distinction of the digits is clear from the figures, implying that the discriminant variates discriminate among the different groups.

70.5 Analysis of Prediction Error Rates from Bootstrapping Assessment

Prediction error is usually a good measure of the performance of pattern recognition systems. In practice, a random sample, called training data set from an unknown population described by distribution F is given. Any statistic $\theta(F)$ requires a distribution F, but in practice, F is not known and is difficult to estimate. An empirical distribution $\hat{F}$ from the given sample from an unknown distribution F is defined in a bootstrap setting, by giving an equal probability mass $1/N$ on each of the values x_i. A bootstrap sample is a random sample from the empirical distribution:

$$X_1^*, X_2^*, \ldots, X_N^* \sim \hat{F}.$$

Each x_i^* is drawn independently *with replacement* and with equal probability from the sample, i.e., training data:

$$\mathcal{X} = \{x_i\}_1^N = \{(v_i, y_i)\}_1^N$$

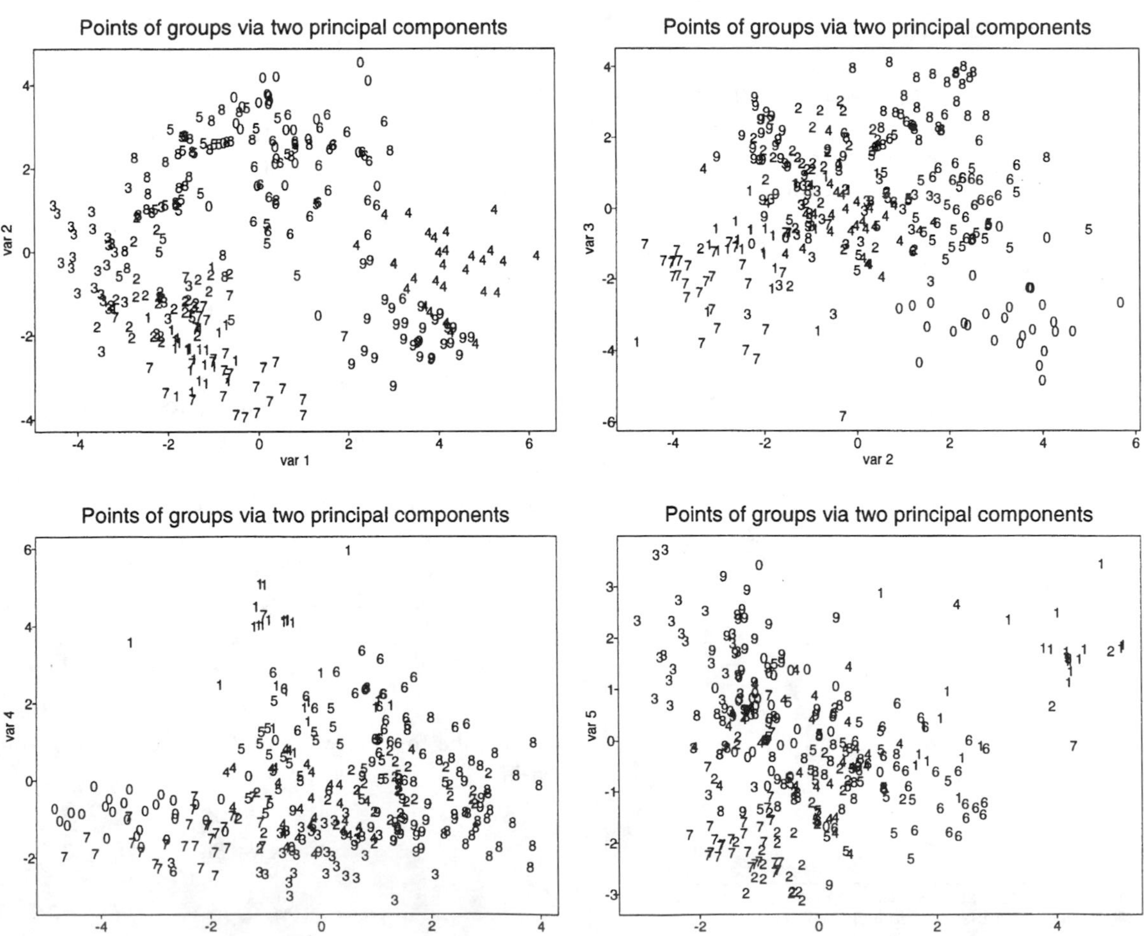

Figure 70.9 Two-dimensional projections of the handwritten data with the first five discriminant variates.

Standard error and bias estimation using Bootstrap resampling techniques can be found in Efron and Tibshirani (1993) and Efron and Gong (1983). Here we introduce the algorithms for estimation of the standard error and the bias for prediction error estimation, leaving the technical details in the references above.

The Monte Carlo bootstrapping algorithm proceeds in three steps.

1. Using a random number generator, independently draw a large number $50 \leq B \leq 200$ of bootstrap samples, $\{F^{*b}\}_{b=1}^{B}$,
2. For each bootstrap sample F^{*b}, evaluate the statistic of interest, $\hat{\phi}^*(b) = \hat{\theta}(F^{*b})$ for $b \in \{1,2,\ldots, B\}$ from the training data $\mathscr{X}$,
3. Calculate the sample standard deviation of $\hat{\theta}^*(b)$ values

$$\hat{\sigma}_B = \left(\frac{1}{B-1}\sum_{b=1}^{B}\{\hat{\theta}^*(b) - \hat{\theta}^*(\cdot)\}^2\right)^{1/2}, \quad (70.36)$$

$$\hat{\theta}^*(\cdot) = \frac{1}{B}\sum_{b=1}^{B}\hat{\theta}^*(b)$$

Standard errors are crude but useful measures of statistical accuracy (Efron and Tibshirani, 1986). An approximated confidence interval for an unknown parameter θ is given by

$$\theta \in \hat{\theta} \pm \hat{\sigma} z^{(\alpha)}, \quad (70.37)$$

where $z^{(\alpha)}$ is the $100 \cdot \alpha$ percentile point of a standard normal variate, e.g., $z^{(0.95)} = 1.64485$. The standard error approximation (Equation 70.37) for a confidence interval bears the assumption that:

$$\frac{\hat{\theta} - \theta}{\hat{\sigma}} \sim N(0, 1)$$

Bias about an estimator $\hat{\theta}$ is the next to consider. Bootstrap bias estimation is an estimation of the *optimistic bias op* resulting from using the same training data for prediction, e.g., via the resubstitution method. One way to estimate the system performance from the given sample is to correct the *apparent error rate* (or resubstitution error rate) by the estimation of the optimistic (or positive) bias. The optimistic bias is defined as

$$op(\mathscr{X}; F) = \theta - \theta_{app}$$

where θ is the true error rate for the unknown distribution F and θ_{app} for the apparent error rate. Since we do not know the bias $op(X, F)$ the bootstrap estimate of the bias, op_{boot} is found instead and the optimistic θ_{app} is corrected by adding the estimated bias:

$$\hat{\theta} = \theta_{app} + op_{boot} \quad (70.38)$$

Let $\eta(v,\mathscr{X})$ be a decision rule based on the training set $\mathscr{X}$ and let $Q[y_i, \eta(v_i,\mathscr{X})]$ be an indication of mis-classification of v_i by $\eta(\ )$:

$$Q[y_i, \eta(v_i, \mathscr{X})] = \begin{cases} 1 & \eta(v_i, \mathscr{X}) \neq y_i, \\ 0 & \text{otherwise.} \end{cases}$$

Thus, $Q[\cdot] = 1$ indicates the misclassification of a training observation from the system designed by the training data.

The bootstrap procedure for estimating the bias, op_{boot}, follows:

1. Select $50 \leq B \leq 200$ bootstrap samples from the empirical distribution $\hat{F}$.
2. From each bootstrap sample compute bias w_b

$$w_b = \sum_{i=1}^{N}\left(\frac{1}{N} - P_i^{*b}\right)Q[y_i, \eta(v_i, F^{*b})]$$

with P_i^{*b} indicating the proportion of the bootstrap sample on x_i, i.e.,

$$P_i^{*b} = \text{Cardinality of } \{j \mid x_j^{*b} = x_i\}/N$$

and $\eta(v_i, F^{*b})$ being the prediction of v_i from the system trained by F^{*b}.

3. Repeat step 2 to get $\{w_1, w_2, \ldots, w_B\}$.

Then the bootstrap bias op_{boot} is estimated by

$$op_{boot} = \frac{1}{B}\sum_{b=1}^{B} w_b$$

thus, the bootstrap error estimate $\hat{\theta}$ (Equation 70.38) is obtained.

The E0 prediction error estimation is equivalent to counting the number of patterns that are not included in the bootstrap samples and normalize the misclassification count of the samples (Efron, 1983) by the total number of the training patterns not selected in the bootstrap samples. Thus, E0 uses the testing set which is asymptotically 36.8% of the original training, according to the argument that follows: In a typical bootstrap sample, about 63% of the original observations are likely to be chosen. This is easily seen since the probability that an observation does not belong to a bootstrap sample is

$$(1 - 1/N)^N = 1/e.$$

Thus, an observation x_i will be in the bootstrap sample with about $1 - 1/e = 63.2\%$ chances.

Let $A_b = \{i \mid P_i^{*b} = 0\}$ denote the index set of training patterns which do not appear in the F^{*b}, then the prediction error θ_o estimated by the E0 estimator is defined by

$$\theta_0 = \frac{\sum_{b=1}^{B}\sum_{i \in A_b} Q[y_i, \eta(v_i, F^{*b})]}{\sum_{b=1}^{B} \text{Cardinality of } \{A_b\}}.$$

This E0 estimator is a form of cross-validation in that the testing data has not been used in training. The difference from the cross-validation is that the E0 separates the training and the testing data randomly while the cross-validation selects the testing pattern sequentially such that all the training patterns are used for testing.

The testing patterns used in the apparent error rate obtained by the resubstitution method, are too close or the distance is 'zero' from the training patterns while the test patterns for E0 estimator are 'too far' from the training set. From that the asymptotic probability argument that a pattern will not be included in a bootstrap sample is 0.368, the weighted average of θ_{app} and θ_0 involves patterns at the 'right' distance from the training set in estimating the error rate (Efron, 1983):

$$\theta_{632} = 0.368 * \theta_{app} + 0.632 * \theta_0 \qquad (70.39)$$

The E632 was shown to be optimal in terms of least variance and bias from comparison study for various estimators (Jain et al., 1987; Efron, 1983) among cross-validation, ordinary bootstrap bias correction (Equation 70.38) and E632 (Equation 70.39). We used the E632 prediction error as a standard performance measure. The bootstrap package[2]

```
bootstrap.funs
```

contains various resampling techniques and is available via *anonymous* ftp to

```
statlib@lib.stat.cmu.edu.
```

The boxplots in the figure (Figure 70.10) represent the E632 estimator superimposed to the distribution of the $B = 100$ bootstrap sample errors, $\hat{\theta}^*(b)$'s (Equation 70.36). The median value

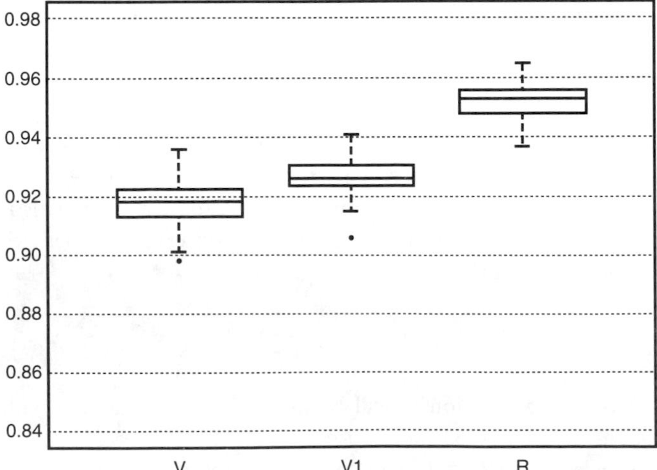

Figure 70.10 Boxplots from nnet with hidden layer size = 15 for data sets of V, V1, R. E632 prediction error rate and $B = 100$ bootstrap samples are used.

of the B error rates is replaced by the E632 estimate, thus the B bootstrap errors are shifted according to the E632 estimate. For ease of display and understanding the system performance, the recognition rates, $1 - \hat{\theta}'s$, are plotted.

The height of the box is the interquartile range which is the difference between the upper quartile and the lower quartile, and is considered to be a robust estimation of the scalar multiple of the dispersion. The median of the batch is represented by the line in the box. The bars, represented by the vertical dotted lines, are extended up to the points 1.5 times of the inter quartile range. Outliers are represented by the individual dots to signify their existence. The boxplots display the distribution very simply but well enough, especially when many different batches are to be compared.

The correct recognition rates from a three-layer feed-forward neural networks with the Broyden, Flecher, Goldfarb and Shanno (BFGS) algorithm (in Peressini et al., 1988) are displayed in the boxplots for each data set obtained from the different treatment in section 3.3. The classifiers[3] used in this study can be obtained via *anonymous* ftp to

```
statlib@lib.stat.cmu.edu.
```

Each boxplot shows the distribution of the recognition rate of 100 systems designed by 100 bootstrap samples.

70.6 Summary

A simple model-free feature extraction by the two-dimensional Zernike polynomials was shown to be a powerful pattern recognition system, (via the correct recognition rate $\gtrsim 95\%$ via E632 prediction error measure) for handwritten digits. The images are preprocessed before ZM calculations take place and the dimensionality of the feature vectors is reduced by PCA followed by DA.

For the 180-degree rotation conflict data, addition of the skewness variables improves the performance of the system. Simply by taking the real (or imaginary) part of the complex-valued Zernike moments, one obtains more information than losing by the rotational invariance operation for the rotational variance of the patterns, which is inherent in handwritten digit data. The rotational variance of the patterns seems to be observed by the *wedge* type Zernike polynomials.

The skewness information addition (V1) to the modulus value (V) of the complex-valued ZM generally improves the correct recognition rate by 2–3%, while the real part (R) generally yields 3–4% improvement over the modulus value (V). The wedge shape of the polynomial also possesses an important property that the variation, at around the outer region of the patterns, results in less variance than the one from the Cartesian coordinates, such as the regular moments and their invariants.

[2] The bootstrap was contributed by Efron and Tibshirani.

[3] The package nnet was contributed by Ripley.

Acknowledgments

The authors wish to thank Dr. R. Gnanadesikan for insightful discussions and Dr. G. Kontaxakis for help with the final version of the manuscript.

References

Abu-Mostafa, Y. S., and Psaltis, D. 1984. Recognitive aspects of moment invariants, *IEEE Trans. Pattern Analysis and Machine Intelligence*, 6(6):698–706.

Ammann, L. P. 1993. Robust singular value decompositions: a new approach to projection pursuit, *J. Am. Stat. Assoc.*, 88(422):505–514.

Belkasim, S. O., Shridhar, M., and Ahmadi, M. 1991. Pattern recognition with moment invarinats: a comparative study and new results, *Pattern Recognition*, 24(12):1117–1138.

Bhatia, A. B. and Wolf, E. 1954. On the circle polynomials of zernike and related polynomials orthogonal sets, *Proc. Camb. Phil. Soc.*, 50:40–48.

Bitchell, B. T. and Gillies, A. M. 1989. A model-based computer vision system for recognizing handwritten zip codes, *Machine Vision and Applications*, 2:231–243.

Chung, W. 1994. *A Strategy for Visual Pattern Recognition*, PhD thesis, Electrical and Computer Engineering, Rutgers University, The State University of New Jersey.

Chung, W. and Micheli-Tzanakou, E. 1994. A simulation study for different moment sets, in *Document Recognition, IS&E/SPIE 1994 Int. Symp. Electronic Imaging*, February.

Devlin, S., Gnanadesikan, R., and Kettenring, J. 1981. Robust estimation of dispersion matrices and principal components, *J. Am. Stat. Assoc.*, 76:354–362.

Dudani, S. A., Breeding, K. J., and McGhee, R. B. 1977. Aircraft identification by moment invariants, *IEEE Trans. Computers*, 26(1):39–45.

Efron, B. 1983. Estimating the error rate of a prediction rule: improvement on cross-validation. *J. Am. Stat. Assoc.*, 78(382):316–331.

Efron, B. and Gong, G. 1983. A leisurely look at the bootstrap, the jackknife, and cross-validation, *The American Statistician*, 37(1):36–48.

Efron, B. and Tibshirani, R. 1986. Bootstrap methods for standard errors, confidence intervals, and other measures of statistical accuracy, *Statistical Science*, 1(1):54–77.

Efron, B. and Tibshirani, R. J. 1993. *An Introduction to the Bootstrap.*, Chapman & Hall, New York, NY.

Gnanadesikan, R. 1977. *Methods for Statistical Data Analysis of Muitivariate Observations*, John Wiley & Sons, New York, NY.

Haralilck, R. M. and Shapiro, L. G. 1992. *Computer and Robot Vision, vol. 1*, Addison-Wesley, Reading, MA.

Jähne, B. 1991. *Digital Image Processing: Concepts, Algorithms and Scientific Applications.* Springer-Verlag, New York, NY.

Jain, A. K., Dubes, R. C., and Chen, C. C. 1987. Bootstrap techniques for error estimation, *IEEE Trans. Pattern Analysis and Machine Intelligence*, 9(5):628–633.

Johnson, R. A. and Wichern, D. W. 1988. *Applied Multivariate Statistical Analysis*, Prentice Hall, Englewood Cliffs, NJ.

Khotanzad, A. and Hong, Y. H. 1990. Invariant image recognition by zernike moments, *IEEE Trans. Pattern Analysis and Machine Intelligence*, 12(5):489–497.

Kittler, J. 1975. Mathematical methods of feature selection in pattern recognition, *Int. J. Man-Machine Studies*, 7:609–637.

Kittler, J. 1986. Feature selection and extraction, in *Handbook of Pattern Recognition and Image Processing*, ch. 3, pp. 59–83, Academic Press, San Diego, CA.

Kittler, J. and Young, P. C. 1973. A new approach to feature selection based on the Karhunen-Loeve expansion, *Pattern Recognition*, 5:335–352.

Le Cun, Y., Jacket, L. D., Boser, B., Denker, J. S., Graf, H. P., Guyon, I., Henderson, D., Howard, R. E., and Hubbard, W. 1989. Handwritten digit recognition: applications of neural network chips and automatic learning, *IEEE Communications Magazine*, pp. November, 41–46.

Mantas, J. 1986. An overview of character recognition, *Pattern Recognition*, 19(6):425–430.

Pavlidis, T. 1980. A thinning algorithm for discrete binary images, *Computer Graphics and Image Processing*, 13:142–157.

Peressini, A. L., Sullivan, F. E., and Uhl, Jr. J. J. 1988. *The Mathematics of Nonlinear Programming*. Springer-Verlag, New York, NY.

Serra, J. 1982. *Image Analysis and Mathematical Morphology*, Academic Press, San Diego, CA.

Suen, C. Y., Berthod, M., and Mori, S. 1980. Automatic recognition of handprinted characters—the state of the art, *Proc. IEEE*, 68(4):469–487.

Tappert, C. C., Suen, C. Y., and Wakahara, T. 1990. The state of the art in on-line handwriting recognition, *IEEE Trans. Pattern Analysis and Machine Intelligence*, 12(8):787–808.

Teague, M. R. 1980. Image analysis via the general theory of moments, *J. Opt. Soc. Am.*, 70(8):920–930.

Teh, C-H. and Chin, R. T. 1986. On digital approximation of moment invariants, *Computer Vision, Graphics, and Image Processing*, 33:318–326.

Teh, C-H. and Chin, R. T. 1988. On image analysis by the methods of moments, *IEEE Trans. Pattern Analysis and Machine Intelligence*, 10(4):496–513.

Wang, C. H. and Srihari, S. N. 1988. A framework for object recognition in a visually complex environment and its application to locating address blocks on mail pieces, *Int. J. Computer Vision*, 2:125–151.

Zhang, T. Y. and Suen, C. Y. 1984. A fast parallel algorithm for thinning digital patterns, *Communication of ACM*, 27(3):236–239.

71

Neocognitron

71.1 Neocognitron 966
Network Architectures • Deformation-Resistant Recognition • Unsupervised Learning • Handwritten Character Recognition

71.2 Selective Attention Model (SAM) 969
Network Architecture • Gate Signals • Segmentation • Threshold Control • Attention Focusing by Gain Control • Search Area • Attention Switching • Performance Test • Connected Character Recognition

Kunihiko Fukushima
Osaka University

71.1 Neocognitron

The "neocognitron", proposed by Fukushima (1988, 1980) is a neural network model for deformation-resistant visual pattern recognition. It is a hierarchical network consisting of many layers of neuron-like cells. There are forward connections between cells in adjoining layers. Some of these connections are variable and can be modified by learning.

The neocognitron can acquire the ability to recognize patterns by learning. Since it has a large power of generalization, presentation of only a few typical examples of deformed patterns (or features) is enough for the learning process to be successful. It is not necessary to present all the deformed versions of the patterns that might appear in the future. After learning, the neocognitron can recognize input patterns robustly, with little effect from deformation, changes in size, or shifts in position. It is even able to correctly recognize a pattern that has not been presented before, provided it resembles one of the training patterns.

The network architecture was suggested by physiological data on the visual systems of the brain. In the network, simple features are first extracted from a stimulus pattern, then integrated into more complicated ones. In this hierarchy, a cell in a higher stage generally has a larger receptive field, and is more insensitive to the position of the stimulus.

Network Architecture

The neocognitron has a multilayered architecture as shown in Figure 71.1, in which each rectangle represents a two-dimensional array of cells. Each cell receives its input connections from only a limited number of cells situated in a small area on the preceding layer. The density of cells in each layer is designed to decrease with the order of the stage.

The lowest stage of the hierarchical network is an input layer U_0, consisting of a two-dimensional array of receptor cells. Each succeeding stage has a layer U_S consisting of "S-cells" followed by another layer U_C consisting of "C-cells". Thus, in the whole network, layers of S-cells and C-cells are arranged alternately.

Each layer of S-cells or C-cells is divided into subgroups, called "cell-planes", according to the features to which they respond. The cells in each cell-plane are arranged in a two-dimensional array. Each rectangle drawn with heavy lines in Figure 71.1 represents a cell-plane. The connections converging to the cells in a cell-plane are homogeneous and topographically ordered. In other words, the connections have a translational symmetry, such that each of the cells of a cell-plane shares the same set of input connections. This condition of translational symmetry holds for both fixed and variable connections. The modification of variable connections is always done under this condition.

S-cells are feature-extracting cells. They resemble simple cells in the visual cortex in their response. Connections converging to these cells may be modified by learning. After learning, S-cells are able to extract features from input patterns. In other words, an S-cell is activated only when a particular feature is presented in its receptive field. The features extracted by the S-cells are determined during the learning process. Generally speaking, local features, such as lines in particular orientations, are extracted in the lower stages. More "global" features, such as parts of a training pattern, are extracted in higher stages.

C-cells, which resemble complex cells in the visual cortex, are inserted in the network to allow for positional errors in the features of the stimulus. The connections from S-cells to C-cells are fixed and invariable. Each C-cell receives signals from a group of S-cells that extract the same feature, but from slightly different positions (Figure 71.2). The C-cell is activated if at least one of these S-cells is active. Even if the stimulus feature is shifted in position and another S-cell is active instead of the first one, the same C-cell keeps responding. Therefore, the C-cell's response is less sensitive to shifts in the position of the input pattern.

The layer of C-cells at the highest stage is the recognition layer: the response of the cells in this layer is the final result of pattern recognition by the neocognitron.

0-8493-8343-9/97/$0.00+$.50

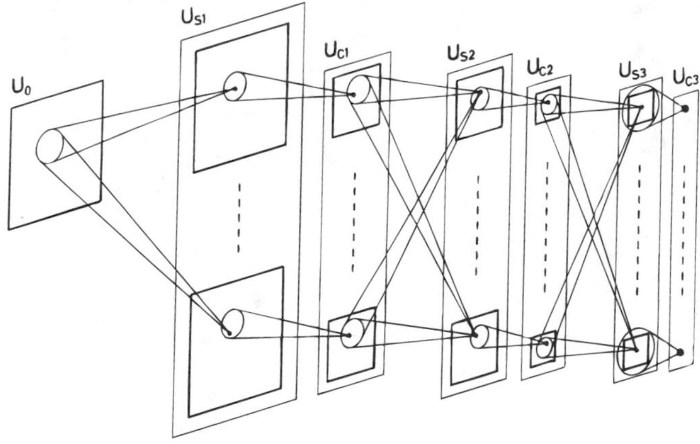

Figure 71.1 The network architecture of the neocognitron (modified from Fukushima, 1980). Each rectangle drawn with heavy lines represents a "cell-plane". The cells in each cell-plane are arranged in a two-dimensional array.

Deformation-Resistant Recognition

In the whole network, with its alternate layers of S-cells and C-cells, the process of feature-extraction by the S-cells and toleration of positional shift by the C-cells is repeated. During this process, local features extracted in lower stages are gradually integrated into more "global" features. Finally, each C-cell of the recognition layer at the highest stage integrates all the information of the input pattern, and responds only to one specific pattern. Figure 71.2 illustrates this situation schematically.

Tolerating positional error a little at a time at each stage, rather than all in one step, plays an important role in endowing the network with the ability to recognize even distorted patterns. Figure 71.3 illustrates this situation. Let an S-cell in an intermediate stage of the network have already been trained to extract a global feature consisting of three local features of a training pattern "A", as shown in Figure 71.3a. The cell tolerance a positional error of each local feature if the deviation falls within the dotted circle. Therefore, the S-cell responds to any of the deformed patterns shown in Figure 71.3b. The tolerance of positional errors should not be too large at this stage. If large errors are tolerated at one step, the network may come to respond erroneously, such as by recognizing a stimulus like Figure 71.3c as an 'A' pattern.

Since errors in the relative position of local features are thus tolerated in the process of extracting and integrating features, the same C-cell responds in the recognition layer at the highest stage, even if the input pattern is deformed, changed in size, or shifted in position.

Unsupervised Learning

The neocognitron can be trained to recognize patterns through either unsupervised or supervised learning. This section introduces the former process.

In the case of unsupervised learning, the self-organization of the network is performed using two principles. The first principle is a kind of "winner-take-all" rule: among the cells situated in a certain small area, only the one responding most strongly has its input connections reinforced. The amount of reinforcement of each input connection to this maximum-output cell is proportional to the intensity of the response of the cell from which the relevant connection leads.

Figure 71.4 illustrates this process of reinforcement, showing only the connections converging to an S-cell. The S-cell receives variable excitatory connections from a group of C-cells of the preceding stage. The cell also receives a variable inhibitory connection from an inhibitory cell, called a V-cell. The V-cell receives fixed excitatory connections from the same group of C-cells as does the S-cell, and always responds with the average intensity of the output of the C-cells.

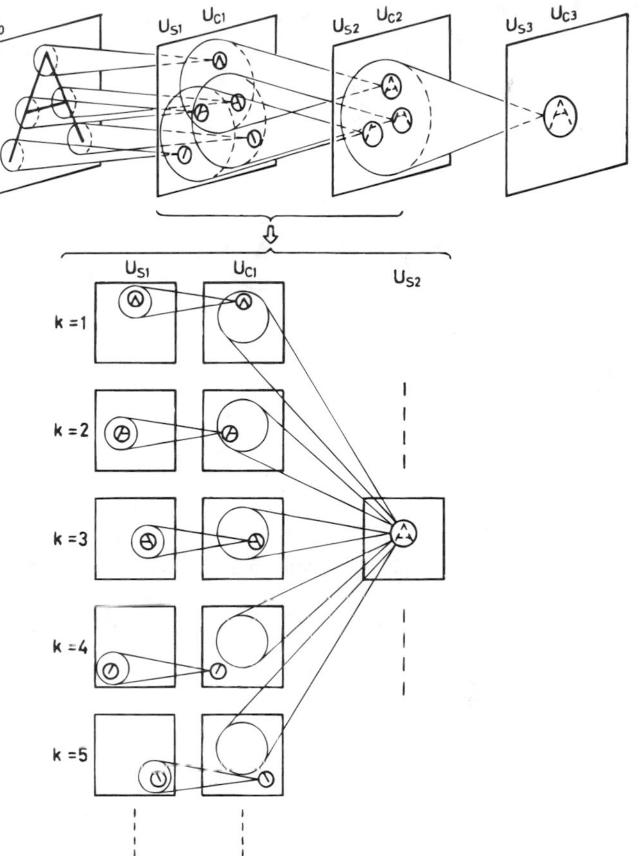

Figure 71.2 Illustration of the process of pattern recognition in the neocognitron (modified from Fukushima, 1980). As shown in the upper half of the figure, local features extracted in lower stages are gradually integrated into more "global" features. The lower half of the figure is an enlarged illustration of a part of the network. The cell-plane with $k = 1$ in layer U_{S1} consists of S-cells that extract ∧-shaped features. Since the stimulus pattern 'A' contains the ∧-shaped feature at the top, an S cell near the top of this cell-plane is active. A C-cell in the succeeding cell-plane ($k = 1$) in U_{C1} has excitatory input connections from S-cells situated in the circle, and is activated if one of these S-cells is active. Only one cell-plane is shown in U_{S2} in this enlarged illustration. Each S-cell in this cell-plane detects the existence of features $k = 1,2,3$ in U_{C1} and at the same time the absence of features $k = 4,5$ as well.

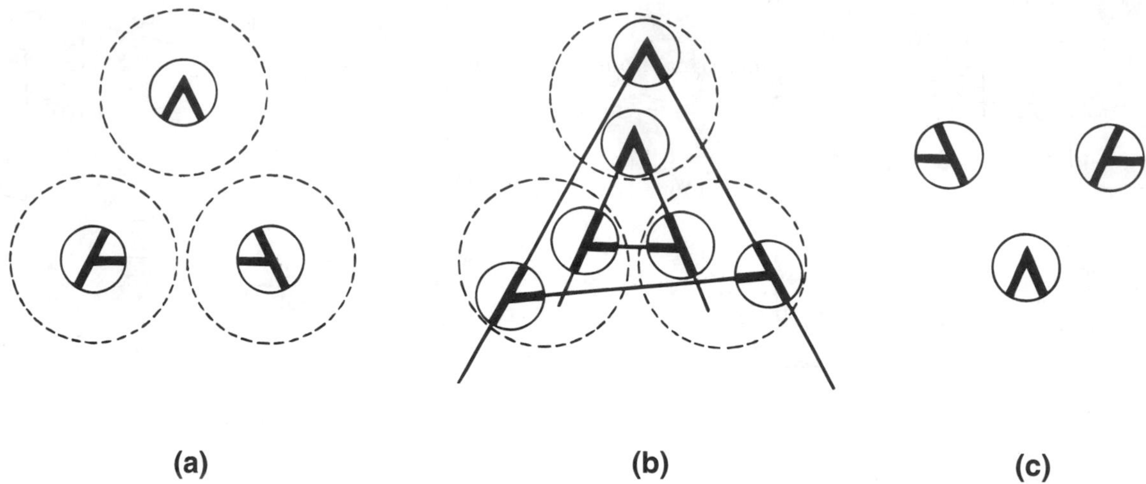

Figure 71.3 Illustration of the principle for recognizing deformed patterns (modified from Fukushima, 1988a). An S-cell, which has already been trained to extract a global feature consisting of three local features as shown in (a), tolerates a positional error of each local feature if the deviation falls within the dotted circle. Therefore, the S-cell responds to any of the deformed patterns shown in (b). The toleration of positional errors should not be too large at this stage. If large errors are tolerated at any one step, the network may come to respond erroneously, such as by recognizing a stimulus like (c) as an 'A' pattern.

The initial strength of the variable connections is very weak and nearly zero (Figure 71.4a). Suppose the S-cell responds most strongly of the S-cells in its vicinity when a training stimulus is presented (Figure 71.4b). According to the winner-take-all rule described above, variable connections leading from active C- and V-cells are reinforced, as shown in Figure 71.4c. The variable excitatory connections to the S-cell grow into a "template" that exactly matches the spatial distribution of the response of the cells in the preceding layer. The inhibitory variable connection from the V-cell is also reinforced at the same time, but not strongly, because the output of the V-cell is not as large.

After the learning, the S-cell acquires the ability to extract a feature of the stimulus presented during the learning period. Through the excitatory connections, the S-cell receives signals

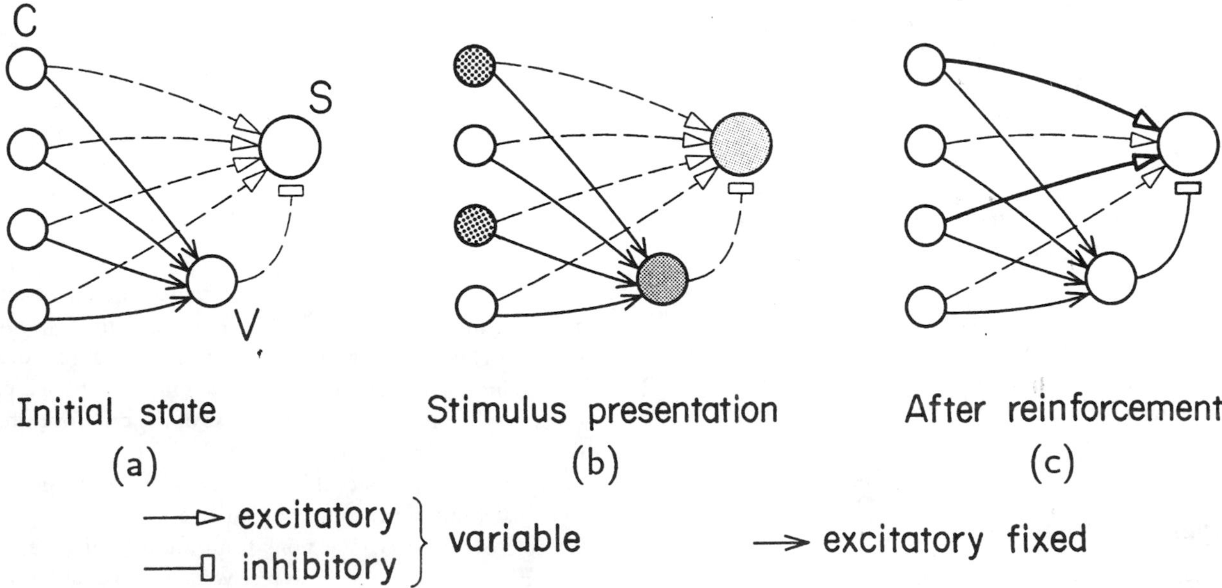

Figure 71.4 The process of reinforcement of the forward connections converging to a feature-extracting S-cell (modified from Fukushima, 1988a). The density of the shadow in the circle represents the intensity of the response of the cell. (a) Shows the initial state before training. (b) Shows stimulus presentation during the training. (c) Shows the connections after reinforcement.

indicating the existence of the relevant feature to be extracted. If an irrelevant feature is presented, the inhibitory signal from the V-cell becomes stronger than the direct excitatory signals from the C-cells, and the response of the S-cell is suppressed (Fukushima, 1989).

Once an S-cell is thus selected and reinforced to respond to a feature, the cell usually loses its responsiveness to other features. When a different feature is presented, a different cell usually yields the maximum output and has its input connections reinforced. Thus, a "division of labor" among the cells occurs automatically.

The second principle for the learning is introduced in order that the connections being reinforced always preserve translational symmetry. The maximum-output cell not only grows by itself, but also controls the growth of neighboring cells, working, so to speak, like a seed in crystal growth. To be more specific, all the other S-cells in the cell-plane, from which the "seed cell" is selected, follow the seed cell, and have their input connections reinforced by having the same spatial distribution as those of the seed cell.

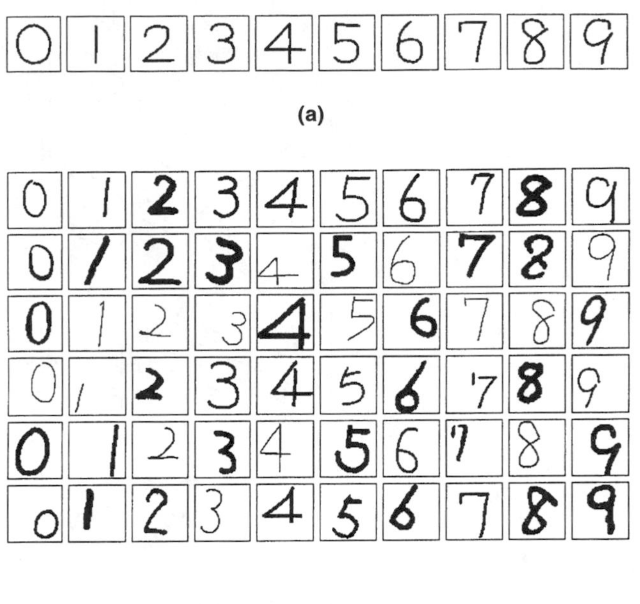

Figure 71.5 A neocognitron trained to recognize handwritten numeric characters (Fukushima and Wake, 1992). (a) Training pattern set used for unsupervised learning. (b) Some examples of deformed characters that the network recognized correctly.

Handwritten Character Recognition

The principle of the neocognitron can be used in various kinds of pattern recognition systems, such as systems recognizing handwritten characters (Fukushima, 1988; Fukushima and Wake, 1991).

Let us show an example applied to ten numeric character recognition (Fukushima and Wake, 1992). The network was trained by an unsupervised learning using the training pattern set shown in Figure 71.5a. An extremely short training time, compared to other learning algorithms such as backpropagation, is another advantage of the neocognitron. In this particular example, only three presentations of this training set consisting of one training pattern from each category was sufficient to train the network. No deformed version of these patterns has been presented to the network during the training phase.

Figure 71.5b shows some examples of deformed numeric characters that the network recognized correctly after finishing the learning. As can be seen from the figure, the network recognizes input patterns robustly, with little effect from deformation, changes in size, shifts in position, or changes in thickness of the lines.

71.2 Selective Attention Model (SAM)

Although the neocognitron has considerable ability to recognize deformed patterns, it does not always recognize patterns correctly when two or more patterns are presented simultaneously. The "selective attention model" (SAM) has been proposed to eliminate these defects (Fukushima, 1988a, 1987, 1986). In the SAM, backward (i.e., top-down) connections were added to the conventional neocognitron-type network, which had only forward (i.e., bottom-up) connections.

When a composite stimulus, consisting of two patterns or more is presented, the SAM focuses its attention selectively to one of the patterns, segments it from the rest, and recognizes it. After the identification of the first segment, the SAM switches its attention to recognize another pattern. The SAM also has the function of associative recall. Even if noise or defects affect the stimulus pattern, the SAM can recognize it and recall the complete pattern from which the noise has been eliminated and defects corrected.

These functions can be successfully performed even for deformed versions of training patterns, which have not been presented during learning—in other words, not only the recognition of the patterns, but also the filling-in process for defective parts of imperfect input patterns works on the deformed and shifted patterns themselves. The SAM can repair the deformed pattern without changing the basic shape and location of the deformed input pattern. The deformed patterns themselves can be repaired at their original locations, thus preserving their deformation.

Network Architecture

Figure 71.6 illustrates the network architecture of the SAM schematically. In this diagram, the layers in the forward paths are denoted by U, and those in the backward paths by W.

If we consider the forward paths only, the model has almost the same structure and function as the neocognitron. The signals through forward paths manage the function of pattern recognition. Layer U_{C0} at the lowest stage is the input layer, to which

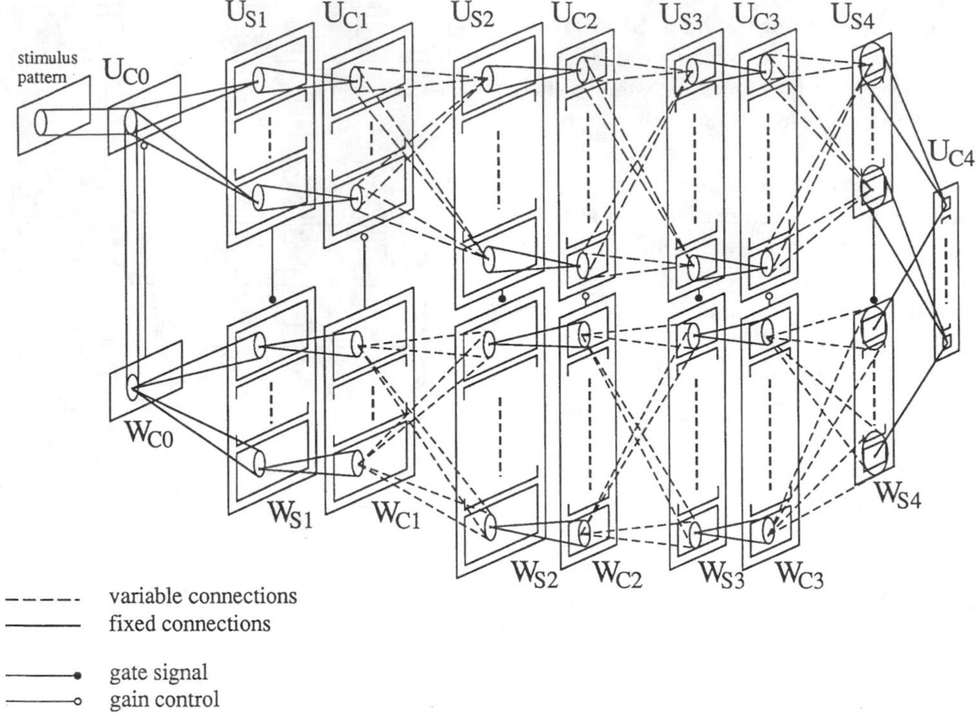

Figure 71.6 The network architecture of the SAM (modified from Fukushima and Imagawa, 1993). U_{C0} is the input layer, and U_{C4} is the recognition layer. Layer W_{C0} at the lowest stage of the backward paths is the recall layer, in which the result of associative recall or the result of segmentation appears.

stimulus patterns are presented. Layer U_{CL} at the highest stage ($L = 4$ in this diagram) is the recognition layer, where the result of the pattern recognition appears.

The cells in the backward paths are arranged in the network in a mirror image of the cells in the forward paths. The forward and the backward connections also make a mirror image to each other, but the directions of signal flow through the connections are opposite.

The signals through backward paths manage the function of selective attention, pattern-segmentation and associative recall. The output signal of the recognition layer U_{CL} is sent to lower stages through the backward paths, and reaches the recall layer W_{C0} at the lowest stage of the backward paths. In the recall layer, the result of associative recall appears. We can also interpret the output of the recall layer as the result of segmentation. The response of the recall layer is fed back positively to the input layer U_{C0}.

The forward and backward signals interact in the network. The forward signals gate backward signal flow, and, at the same time, the backward signals modulate forward signal flow. The process of these interactions is discussed in more detail below.

Gate Signals

The network is so designed that the backward signals, which are sent back from the recognition layer, flow retracing the same route as the forward signals. The route control of the backward signals is made by the gate signals from cells of the forward paths.

Since the backward connections have been reinforced to make a mirror image of the forward paths, the backward signals from an arbitrary backward S-cell will retrace the same route as the forward signals if they are simply transmitted through backward connections.

As for the backward signals from backward C-cells, however, the mirror-imaged network structure alone is not enough. Corresponding to the fixed forward connections that converge to a forward C-cell, many backward connections diverge from the corresponding backward C-cell towards many backward S-cells. However, we do not want all the backward S-cells receiving excitatory backward signals to be activated for the following reason: to activate a forward C-cell, the activation of at least one preceding S-cell is sufficient. Usually only a small number of preceding S-cells are actually active. To elicit a similar response from the backward S-cells, the network is synthesized in such a way that each backward S-cell receives not only excitatory backward signals from backward C-cells but also a gate signal from the corresponding forward S-cell. Guided by the gate signals from the forward paths, the backward signals retrace the same route as the forward signals.

Thus, the backward signals finally reach the recall layer W_{C0} at the lowest stage, and at exactly the same positions as the stimulus pattern presented to the input layer.

Segmentation

Now let's consider the case in which a stimulus consisting of two or more patterns is presented to the input layer. Sometimes, two or more cells may be active in the recognition layer. However, all of these cells but one, stop responding because of competition by lateral inhibition between cells in the forward paths, and also because of the process of focusing attention, mentioned later.

Since the backward signals are sent only from the active recognition cell, only the signal components corresponding to the recognized pattern reach the recall layer, W_{C0}. Even if the stimulus pattern that is now recognized is a deformed version of a training pattern, the deformed pattern is segmented and emerges with its deformed shape. Therefore, the output of the recall layer can be interpreted as the result of segmentation, where only components relevant to a single pattern are selected from the stimulus.

From the pattern emerged at the recall layer, noise and blemishes have been eliminated, because no backward signals are returned for components of noise or blemishes in the stimulus. Thus the segmentation of patterns can be successful, even if the input patterns are incomplete and contaminated with noise. Components of other patterns that are not recognized at this time are also treated as noise.

Threshold Control

Take, for example, a case in which the stimulus contains a number of incomplete patterns that are contaminated with noise and have several parts missing. Even when the pattern recognition in the forward path is successful, and only one cell is active in the recognition layer U_{CL}, it does not necessarily mean that the segmentation of the pattern is also completed in the recall-layer W_{C0}.

When some part of the input pattern is missing and the feature that is supposed to exist there fails to be extracted in the forward paths, the backward signal flow is interrupted at that point and cannot proceed any further because no gate signals are received from the forward paths.

There are monitoring cells in the network that always watch for failures of feature extraction. A monitoring cell responds when it detects a situation, in which a backward C-cell is active but forward S-cells around it are all silent. If this situation is detected, the monitoring cell sends threshold-control signals to the forward S-cells around that area, and decrease the threshold for feature-extraction. Thus, the forward S-cells are made to respond even to incomplete features, to which, in the normal state, no cell would respond. In other words, the SAM is forced to extract even vague traces of the undetected feature.

Once a feature is thus extracted in the forward paths, the backward signals can then be further transmitted to lower stages through the paths unlocked by the gate signals from the newly activated forward cells. Therefore, a complete pattern, in which defective parts are interpolated, emerges in the recall-layer W_{C0}.

If all the recognition cells are silent, the no-response detector in the network is activated and sends another threshold-control signal to the forward S-cells of all stages and decreases their threshold for feature extraction. The value of the threshold-control signal increases until at least one recognition cell becomes active.

Attention Focusing by Gain Control

Forward cells receive gain-control signals from corresponding backward cells. More specifically, the gain of each forward C-cell is increased by the signal from the corresponding backward C-cell. Therefore, forward signal flow is facilitated only in paths in which backward signals flow.

The gain-control signal plays the role of focusing attention. Let's consider the case in which a stimulus consisting of two or more patterns is presented to the input layer. Let one of the recognition cells be active, and one of the patterns of the stimulus be recognized. Only the forward signal flows relevant to this pattern, which is now recognized, are facilitated by the gain-control signals, because the backward signals flow from that recognition cell only. This means that attention is selectively focused on one of the patterns of the stimulus.

A forward C-cell is fatigued if it receives a strong gain control signal. It can maintain high gain only when it is receiving a large gain-control signal. Once the gain control signal disappears, the gain of the forward C-cell drops rather rapidly, and cannot recover for a long time. This fatigue is effectively used for switching attention to another pattern. It prevents the model from recognizing the same character twice.

Search Area

In an improved version of the SAM that is used for connected character recognition (Fukushima and Imagawa, 1993), a search controller is introduced in order to restrict the number of patterns to be processed simultaneously. The search controller moves a small "search area", and the SAM mainly processes the patterns contained in the area. The search controller sends gain-control signals of another type, and decreases the gains of the forward C-cells situated outside the search area.

The position of the search area is shifted to the place in which a larger number of cells extracting lower features (e.g, line extracting cells) are active. The search area has a size somewhat larger than the size of one character. The boundary of the search area is not sharply restricted: the gain of the forward C-cells is controlled to decrease gradually around the boundary.

It is not necessary to control the position and the size of the area accurately because the original SAM, which does not have a search controller, can segment and recognize patterns by itself, provided the number of patterns present is small. The only requirement is that the search area covers at least one pattern. It does not matter if it covers a couple of patterns simultaneously. Also, it does not matter if the center of the area happens to be placed between two characters, provided that at least one complete character is contained in the area.

Attention Switching

Once a character has been recognized and segmented, the attention is automatically switched to recognize another pattern. To be more exact, there is a detector in the network that determines the timing of attention switching (Fukushima and Imagawa, 1993). The detector monitors the following two conditions: whether the number of active recognition cells is only one, and whether the response of the network has reached a steady state. When both of these conditions are simultaneously satisfied, the detector sends a command to switch attention.

The fatigue of the cells is effectively used in the SAM for switching attention to another pattern. Once a command to switch attention is given to the network, the backward signal flow is cut off for a short period. Since the gain control signals from the backward cells disappear, the gains of the forward cells drop if they had been earlier kept high by strong gain-control signals. Therefore, signals corresponding to the previous pattern now have difficulty in flowing through the forward paths, and another pattern will be recognized.

The search controller again seeks a place in which a larger number of feature-extracting cells are responding, and shifts the search area to the new place. However, if all of the responses from the cells are small enough because of fatigue, the SAM stops working, assuming that all characters in the input string have already been processed, and that no more characters are left unrecognized. In order to prevent the model from recognizing the same character twice, the fatigue of the cells after attention switching is made to continue until all characters in a string have been processed.

Performance Test

We will first show some results of performance test for the original SAM simulated on a computer (Fukushima, 1988a, 1987).

The variable connections were reinforced by unsupervised learning. Figure 71.7a shows the five training patterns, which were repeatedly presented to the network during the learning period. The training patterns were presented in this shape only; deformed versions were not presented at all.

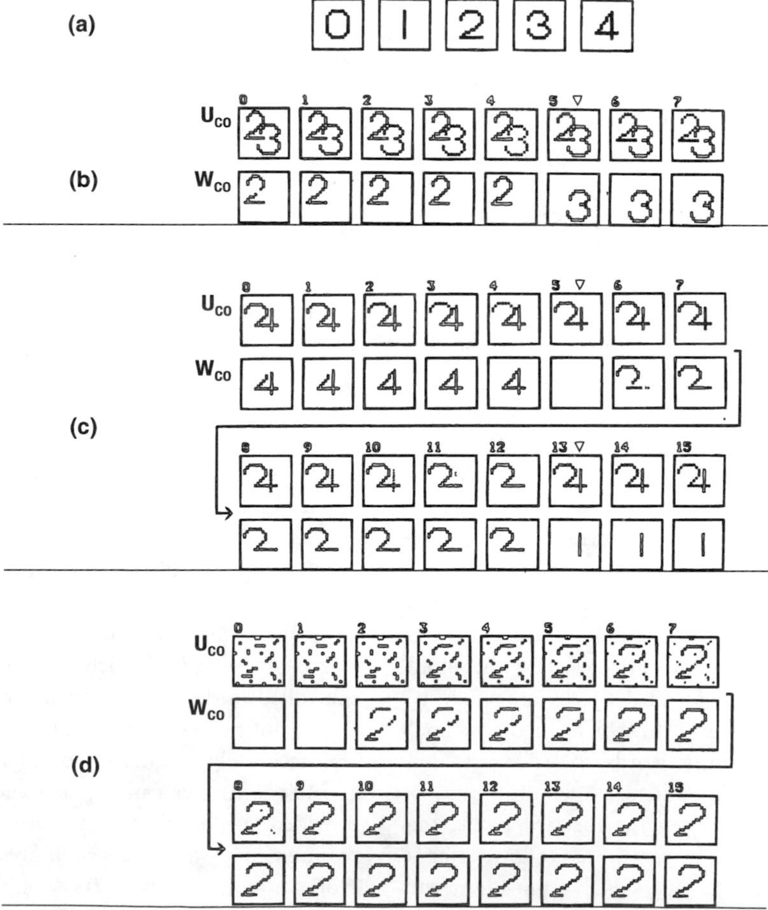

Figure 71.7 Some examples of the response of the SAM (Fukushima, 1988a, 1987). (a) Five training patterns used for learning. (b) An example of the response to juxtaposed patterns. (c) An example of the response to superimposed patterns. (d) An example of the response to an incomplete distorted pattern with noise.

Figure 71.8 Computer simulation of cursive word recognition (Shouno and Fukushima, 1994). (a) Training patterns. (b) Test patterns. Most of the characters have been recognized correctly, but a few of them are erroneously recognized or failed to be correctly segmented. The letter written below the image of an input pattern indicates how the corresponding character in the word was erroneously recognized. No such letters are written for characters recognized correctly.

Figures 71.7b–71.7d show the behavior of the SAM that has finished the learning process. In these figures, the responses of the cells in the input layer U_{C0} and the recall layer W_{C0} are shown in time sequence. The numeral to the upper left of each pattern represents time t after the start of stimulus presentation. The stimulus pattern presented to this network is identical to the response of the input layer at $t = 0$, shown in the upper left of each figure. (Note that the input pattern appears directly in layer U_{C0} at $t = 0$, because no response has been elicited from layer W_{C0} at $t < 0$).

Figure 71.7b shows the response to a stimulus consisting of two juxtaposed patterns, "2" and "3". In the recognition layer, not shown in this figure, the cell corresponding to pattern "2" happens to be active first. This signal is fed back to the recall layer through backward paths, but the middle part of the segmented pattern "2" is missing because of interference from the closely adjacent "3". However, the interference soon decreases and the missing part recovers, because the signals for pattern "3", which is not being attended to, are gradually attenuated without receiving facilitation by gain-control signals.

At $t = 5$, the backward signal-flow is interrupted for a moment to switch the attention. The mark ▼ denotes this operation. Since the gain-control signals from the backward cells stop, the forward paths for pattern "2", which have so far been facilitated, now lose their conductivity. The recognition cell for pattern "3" is now active. Since backward signals are fed back from this newly activated recognition cell, pattern "3" is segmented and emerges in the recall layer W_{C0}.

Recognition and segmentation of individual patterns can thus be successful even if the input patterns are deformed or shifted from the training patterns.

Figure 71.7c shows an example of the response to a stimulus consisting of superimposed patterns. The pattern "4" is isolated first, the pattern "2" next, and finally pattern "1" is extracted. The recalled pattern "4" initially has one part missing, but the missing part is soon restored to resemble the training pattern.

Figure 71.7d shows the response to a greatly deformed pattern with several parts missing and contaminated by noise. Because of the large difference between the stimulus and the training pattern, no response is elicited from the recognition layer (not shown in the figure) at first. Accordingly, no feedback signal appears at the recall layer W_{C0}. The no-response detector detects this situation, and a threshold-control signal is sent to all feature-extracting cells in the network, which makes them respond more easily even to incomplete features. Thus, at time $t = 2$, the recognition cell for "2" becomes active, and backward signals are fed back from it. Noise has been completely eliminated from the pattern now sent back to the recall-layer W_{C0}, and some missing parts have begun to be interpolated. This partly interpolated signal, namely the output of the recall layer W_{C0}, is again fed back positively to input layer U_{C0}. The interpolation continues gradually while the signal circulates through the feedback loop, and finally the missing part of the stimulus is completely filled in. The missing part is interpolated quite smoothly, despite a considerable difference in shape between the stimulus and the training pattern. In other words, the style of writing of the stimulus pattern is kept as faithful as possible, and only indispensable missing parts are restored.

Connected Character Recognition

The principles of the SAM can be extended to be used for several applications: for example, the recognition and segmentation of connected characters in cursive handwriting of English words (Shouno and Fukushima, 1994; Fukushima and Imagawa, 1993), the recognition of Chinese characters (Fukushima, et al., 1991), and the recognition of faces.

As an example of these applications, a result of computer simulation of a new system for cursive word recognition (Shouno and Fukushima, 1994) is presented below.

The system has been trained using ten alphabetical characters shown in Figure 71.8a. Although we used ten characters instead of twenty-six because of the limitation of the computer power, we chose characters whose shapes are similar to each other and difficult to be segmented when they are connected in handwriting. In other words, a character set difficult to discriminate have been chosen intentionally so that the performance can be tested with a small number of test patterns.

Figure 71.8b shows how the characters in cursive words have been recognized and segmented. Most of the characters have been recognized correctly, but few of them are erroneously recognized or failed to be correctly segmented. In Figure 71.8b, the letter written below the image of an input pattern indicates how the corresponding character in the word was erroneously recognized by the system. No such letters are written for characters recognized correctly. When one character in a word was recognized twice by mistake, the two results are indicated by letters enclosed in parentheses. A question mark shows the character could not be recognized.

As can be seen from Figure 71.8b, most of the characters were recognized and segmented correctly. Even in the words in which some characters were erroneously recognized, the rest of the characters were usually recognized correctly.

References

Fukushima, K. 1980. Neocognitron: a self-organizing neural network model for a mechanism of pattern recognition unaffected by shift in position, *Biological Cybernetics,* 36(4):193–202.

Fukushima, K. 1986. A neural network model for selective attention in visual pattern recognition, *Biological Cybernetics,* 55(1):5–15.

Fukushima, K. 1987. Neural network model for selective attention in visual pattern recognition and associative recall, *Applied Optics,* 26(23):4985–4992.

Fukushima, K. 1988. Neocognitron: a hierarchical neural network capable of visual pattern recognition, *Neural Networks,* 1(2):119–130.

Fukushima, K. 1988a. A neural network for visual pattern recognition, *IEEE Computer,* 21(3):65–75.

Fukushima, K. 1989. Analysis of the process of visual pattern recognition by the neocognitron, *Neural Networks,* 2(6):413–420.

Fukushima, K., Imagawa, T., and Ashida, E. 1991. Character recognition with selective attention, *Int. Joint Conf. Neural Networks,* July 8–12, 1991 Seattle, WA. I:593–598.

Fukushima, K., and Imagawa, T. 1993. Recognition and segmentation of connected characters with selective attention, *Neural Networks,* 6(1):33–41.

Fukushima, K. and Wake, N. 1991. Handwritten alphanumeric character recognition by the neocognitron, *IEEE Trans. Neural Networks,* 2(3):355–365.

Fukushima, K. and Wake, N. 1992. Improved neocognitron with bend-detecting cells, *Int. Joint Conf. Neural Networks,* June 7–11, 1992, Baltimore, MD. IV: 190–195.

Shouno, H. and Fukushima, K. 1994. Connected character recognition in cursive handwriting using selective attention model with bend processing, *Trans. IEICE D-II,* J77-D-II(5):940–950, in Japanese.

72

Studies of Pattern Recognition with Self-Learning Layered Neural Networks

Faiq A. Fazal
AT&T Network Systems

Evangelia Micheli-Tzanakou
Rutgers University

72.1 Abstract 975
72.2 Introduction 975
72.3 Neocognitron and Pattern Classification 976
72.4 Objectives 978
72.5 Methods 978
72.6 Study A 979
Network Description • Results from Study A
72.7 Study B 985
Results from Study B
72.8 Summary and Discussion 989

72.1 Abstract

Neurocomputing principles are being increasingly applied to the task of pattern recognition. This paper analyzes the mechanics of pattern recognition by a self-learning layered neural network in terms of the classical principles of pattern recognition. It also reports on the simulation-based study and analysis of the performance of a layered neural network when applied to the task of character recognition. Practical applications and future work are briefly discussed.

72.2 Introduction

The basic task of any pattern recognition system is to decide on the class membership of the current input pattern to the system. One approach is to make use of *decision functions*. If the input pattern has n items describing it, then each instance of a pattern can be viewed as a point (or vector) in an n-dimensional Euclidean space. Consider, for example, the two-dimensional cases depicted in Figure 72.1. We note that in Figure 72.1a the input patterns can be put into two classes, *C1* and *C2*, and that a linear decision function *D1* exists such that for any pattern, p, $D1(p) > 0$ if p belongs to C1 and $D1(p) < 0$ if p belongs to C2. Figure 72.1b shows a more complicated case of clustering which requires three decision functions to establish a pattern's membership. For more involved classification schemes one may have to turn to a nonlinear *decision surface*. For example, Figure 72.1c shows pattern classes separated by a circle. A detailed and mathematically rigorous discussion on this topic can be found in Tou and Gonzalez (1974). It is apparent that the success of this scheme depends on two factors: the form of the decision function and the ability to determine its coefficients.

Often, decision functions are not prewired into pattern classifiers, but heuristically develop as the classifier experiences input patterns during the training period. This is referred to as *clustering*. Several methods of clustering exist (Tou and Gonzalez, 1974) and have found a variety of applications (Micheli-Tzanakou, 1983; Chon and Micheli-Tzanakou 1990). For example, the first of the input patterns during the training period forms a class of its own and becomes the initial prototype for the class. If the second pattern is *similar* to the first pattern, it is put in the first class and the prototype for the class is adjusted so that the difference between it and the two patterns in the class is minimized. If the second pattern is not *similar* to the first pattern, then it forms a new class of its own, and becomes the initial prototype for that class. This process is repeated for each of the patterns, forming a new class only if the pattern does not *match* the prototypes of the existing classes. Several measures exist for *similarity*, and include finding the following two: the minimum distance between the prototype and the pattern, the dot product between the prototype and the pattern. A pattern belongs to a class represented by the prototype, if this measure is less than or exceeds a specified threshold, respectively.

Numerous questions can be asked regarding the quality of the clustering mechanism, such as:

0-8493-8343-9/97/$0.00+$.50

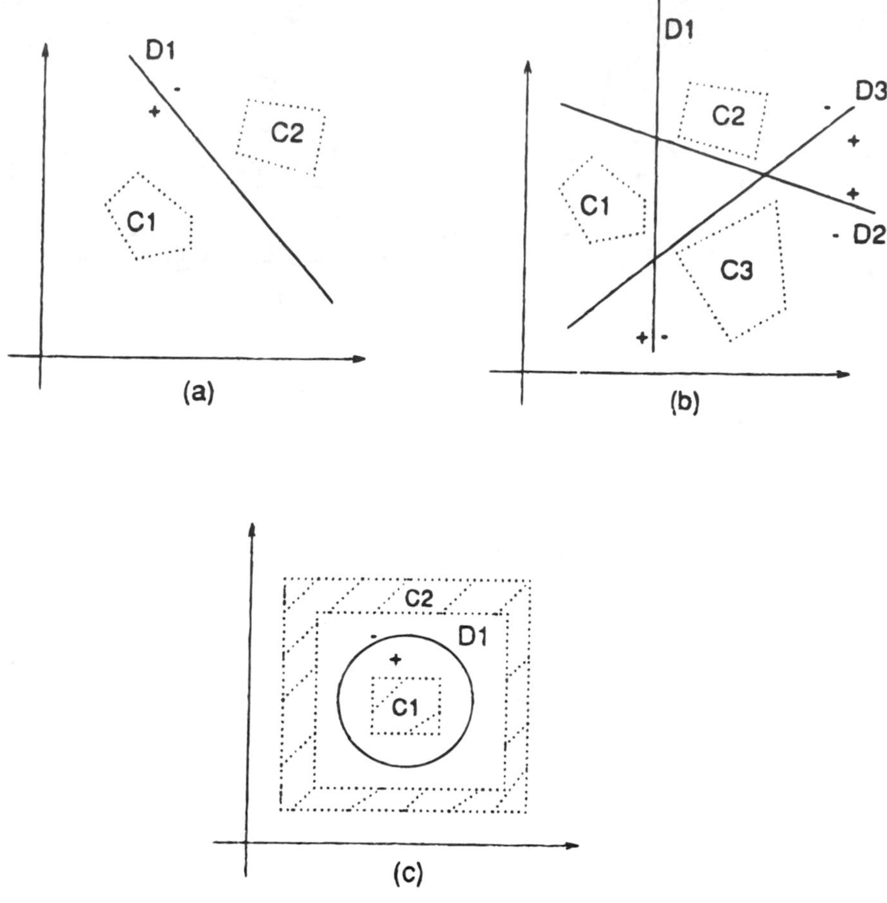

Figure 72.1 Decision functions and pattern classifications.

- How distinct or redundant are the prototypes for the various classes? For this estimation, the dot product between prototype pairs could be used.
- How much of an "overlap" exists between the classes?
- How correlated are the samples within each class? This can serve as a measure of the selectivity of the clustering procedure.
- Are there lots of prototypes with very few samples in them? This reflects on the sensitivity of the clustering mechanism to noise.
- Is the clustering mechanism dependent on the order in which the patterns are applied?
- Is the clustering mechanism dependent on the rate at which clusters are formed?

In this paper, we study the pattern clustering performance of a well known neural network (NN) model, namely the layered NN (Neocognitron) of Fukushima (1982). This model is a self-organizing (implying unsupervised learning) classifier of input patterns, which is capable of tolerating shifts in position and a certain degree of deformity of the input pattern. The following section reviews the Neocognitron model in terms of the classical pattern recognition techniques without getting into the details. This will help provide insight into the underlying mechanisms of the Neocognitron, suggest quantitative measures for its performance, and encourage experimentation with techniques not discussed by Fukushima (1982). A simplified version of the Neocognitron is also described in Deutsch and Micheli-Tzanakou (1987).

Note that the present study excludes aspects of the Neocognitron which deal with tolerance to deformity and shifts in position. This is because the underlying mechanisms for the functionality are not necessary to, or explicitly integrated into the more difficult task of unsupervised pattern classification. In addition, the functionality is hard-wired and does not involve learning.

72.3 Neocognitron and Pattern Classification

Figure 72.2 conceptualizes the pattern classification model embodied in the Neocognitron. The first thing we note is the *distribution* of the *decision functions* involved in pattern classification. Instead of having a set of decision functions which operate over the entire input field, the Neocognitron architecture distributes the decision mechanism over several levels. The decision functions at the first level work over very small portions of the input representation and, accordingly, decide over the existence

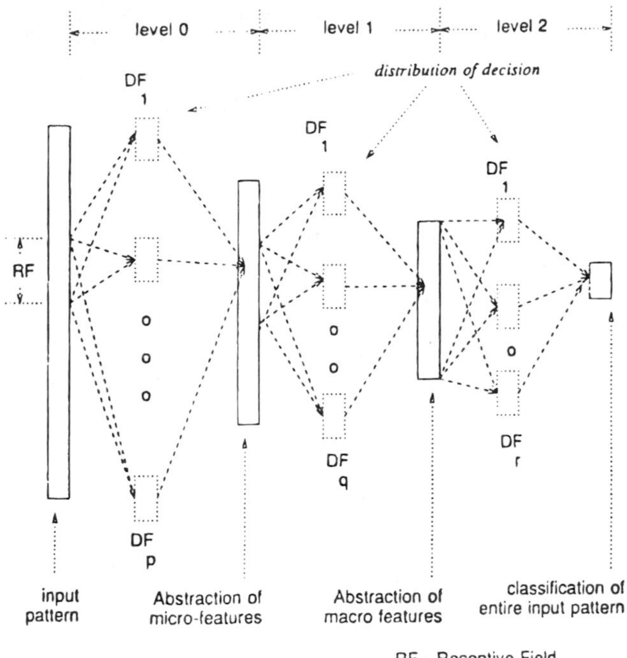

Figure 72.2 Pattern classification model embodied in the Neocognitron.

of low-level features in the various parts of the input field. Thus, given the pixel-input representation of Figure 72.3a, the first level decision function may collectively map it into a representation involving corners and line-segments as depicted in Figure 72.3b. The mapped representation, which must preserve the *spatial relationship* of the higher-level features, now serves as input to the next level of decision functions.

This level, in turn, produces a topographic map of the primary input in terms of more complex features. The process continues to the top-most level whose decision functions collectively decide on the correct classification of the entire input to the network. Thus, in Figure 72.3c the top-most level of a two-level network will put the topographic map of Figure 72.3b into a class which could be labeled A.

This approach to classification can be viewed as a *divide-and-conquer* technique to solving the problem. However, this distribution implies that at the lower level decisions are made on a local basis without taking the entire picture into account. In the presence of noise in the input pattern this scheme may not have performed as well as a single-level scheme. This is also true even when such distributed schemes take top-down expectations into account. After all, top-down expectation can only be initiated after receiving some initial evidence from bottom-up. Also, in image reconstruction through top-down excitation this scheme may cause convergence problems.

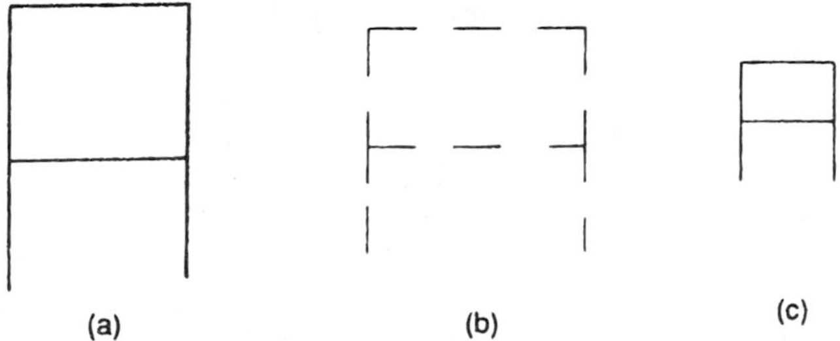

Figure 72.3 Example of distributed classification.

The basic decision mechanism is the same at all levels. Each level has a set of decision functions which work in *parallel* and in *competition* in order to decide on the features present in the different sections of the input-field. This process is conceptualized in Figure 72.4. All of the decision functions for level l work in parallel on their respective receptive field in order to decide which feature is present in it. The decision functions also compete with one another to decide on the *winning* feature for each section (also referred to as *competition area*). The decision for each competition-section collectively forms the input to the next level of decision functions. It should be noted that the output of each decision-function, which is shown as a single point in Figure 72.4, is typically represented by the states of a collection of units in the actual neural model. In the Neocognitron, the decision functions at each level are implemented by the **a** vector associated with each plane in the feature-detecting layer (known as the S-layer) of that level. The number of planes in each S-layer, thus places an upper limit on the number of classifications that can be made at the corresponding level. Each unit of a given S-plane attempts to decide if the prototype feature represented by the plane's **a** vector is presented in the unit's receptive field in the previous layer. Mathematically, this decision is specified by the following discriminant:

$$df = r\Phi\left[\frac{\sum_i a_i u_i \quad kb\sqrt{\sum_i c_i u_i^2}}{1 + kb\sqrt{\sum_i c_i u_i^2}}\right] \tag{72.1}$$

where

$$\Phi(x) = \begin{bmatrix} 0, & x \leq 0 \\ x, & x \geq 0 \end{bmatrix}$$

and

$$k = \frac{r}{r+1}; \qquad (r = \text{Inhibition Factor})$$

While the vector **a** is the same for each of the units in a plane, vector **u**, which represents the feature currently present in a unit's receptive field, may be different for each unit. Vector **c** and scalar b are used to compute the average excitation in a unit's receptive field. From Equation 72.1 it can be seen that df is a decision function of the quadratic form, based on the equations below:

$$\sum_i a_i u_i > kb\sqrt{\sum_i c_i u_i^2} \tag{72.2}$$

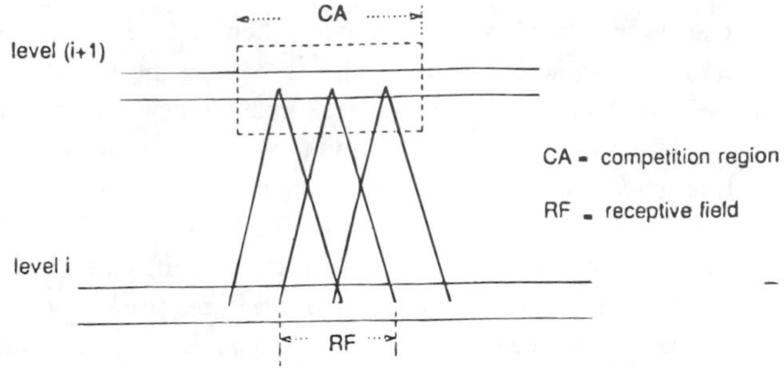

Figure 72.4 Parallel and competitive execution of decision functions.

$$\left(\sum_i a_i u_i\right)^2 - k^2 b^2 \left(\sum_i c_i u_i^2\right) > 0 \tag{72.3}$$

In the Neocognitron, the values of **a** and b are developed during the training period, while **c** is a constant vector associated with each plane. All planes at a given level have the same **c.** In the Neocognitron, **c** follows an exponentially decreasing function over the receptive field with the constraint that:

$$\sum_i c_i = 1 \tag{72.4}$$

The learning (or clustering) mechanism of the Neocognitron can be described by the procedure given below:

Training Algorithm

1. Apply the next training pattern.
2. For each level in the network perform the following, bottom-up for each section in the input to this level:

   a. determine the plane whose **a** vector has the closest match with the feature contained in this section,
   b. update **a** and b as follows:

$$\begin{bmatrix} \Delta a_i = q c_i u_i \\ \Delta b = q v \end{bmatrix}$$

   where q is the learning rate and v is the average inhibitory excitation computed using **u** and **c.**

72.4 Objectives

The objectives of the simulation experiments include an attempt to understand the issues about the clustering mechanism that were raised in the Introduction. Specifically, we are interested in the dependence of these issues on the form and the parameters of Equation 72.1 and Equation 72.4. The varied parameters are the following:

- Inhibition-factor, r, from Equation 72.1.
- Learning-rate, q from Equation 72.4.
- The form of vector **c,** i.e., exponentially decreasing vs. uniform.
- The initial values for vector **a,** i.e., random vs. primed.
- Thresholding the selection of winning units in the competition area with a *threshold factor,* θ. Experimentation has shown that such threshold can reduce the development of noisy or redundant features. Essentially, only those units whose activation exceeds θ**Average-activation* are selected for a weight update.

In relation to the distributed nature of clustering and the fact that the Neocognitron has a prefixed limit on the number of clusters that can be formed at each level, the following issues are also investigated:

- How should the inhibition-factor and learning rate vary from one level to another?
- How many applicants of the training patterns are necessary for learning to develop?
- Does it help to intermix the patterns from the different classes?

72.5 Methods

The applied input stimulus consisted of an array of pixels whose values are set to 0 or 1 in order to create different types of patterns. A facility provided by the simulation environment allowed creation of noisy patterns from the originals. The noise introduced by this facility is random. However, the user can control the Hamming-distance[1] of the noisy pattern from the original one. In the two studies reported in the next section, the mix of the original (i.e., non-noisy) pattern to noisy patterns was 2:1:1, where the three numbers refer to the proportion of original, 1-Hamming noisy and 2-Hamming-noisy patterns, respectively. In these simulations there exist several fixed and variable parameters as listed below:

The *fixed* parameters are the number of levels, layers, planes and units, and the size of the receptive field. The *variable* parameters are the learning rate, q, the inhibition factor, r, and the form of the vector **c.**

Two cases are considered for vector **c,** namely, a uniform distribution of connection weights, and an exponential distribution of connection weights. In addition, the initial value for vector **a** assigned to each of the learning planes is a variable parameter. For **a,** two cases of initial values are considered, namely, *random* assignment of the weights, and primed assignments, mixed with random assignments. *Primed* assignment of the initial value to **a** vector, gives it a slight bias to certain types of patterns in its receptive field. The pattern types included horizontal, vertical, and diagonal lines (Figure 72.5).

[1] If the pixels composing a pattern are viewed as elements of a vector then the Hamming distance between two patterns is equal to the number of pixels in which they differ.

Another variable parameter is θ, the threshold factor used in deciding the winning unit in each competition area.

The *Network Performance* is evaluated in terms of its capability to *learn* and *recall* after learning is over.

Learning is evaluated in terms of the number of planes used at each level, and how orthogonal the **a** vectors are for the planes that are used in learning. The *dot-product* between the **a** vectors is used for this purpose. After learning is completed, for each level the dot-product between the **a** vectors is computed, along with the minimum, average, and maximum values. In the results presented here (Figures 72.6–72.10 and 72.14–72.16), the learned **a** vector for each plane is shown in a two-dimensional grid format to make obvious the correspondence between the vector and the feature that it detects. Each number within the grid represents the relative sensitivity of the learned pattern to excitation occurring at the location of the number. For clarity, zero sensitivity is represented by a blank space. *Recall* is evaluated in terms of the activation states of the units at the various levels.

Two different studies, Study *A* and *B*, were performed. The results and their implications are presented next.

72.6 Study A

Study A was made with a 2-level network having an input level and a recognition level.

Network Description

The table below summarizes the description of the network used in this study. The entry *NA* stands for Not Applicable.

Table 72.1 Description of Network Used in Study A

	Level 0	Level 1	
	Layer INP	Layer Vc	Layer S
# of planes	1	1	10
Units per plane	5 × 5	1	1
Receptive field size	NA	5 × 5	5 × 5

The input patterns consists of a 5 × 5 array of pixels. Figure 72.5 shows the patterns that are used in Study A. If we consider patterns 5 and 6 as being similar to 4 and 1, respectively, then we expect the network to cluster them into four classes. This should result in four planes being used in level 1.

Results from Study A

Several observations are made, and are listed below:

A1 *The larger the inhibition factor, the more discriminatory is the clustering process, which in turn results in larger numbers of clusters.*

This can be seen in Figures 72.6a, 72.7a, and 72.8a for different inhibition factors. These figures show that there is an increase in the number of planes used up in the clustering process as the inhibition factor increases.

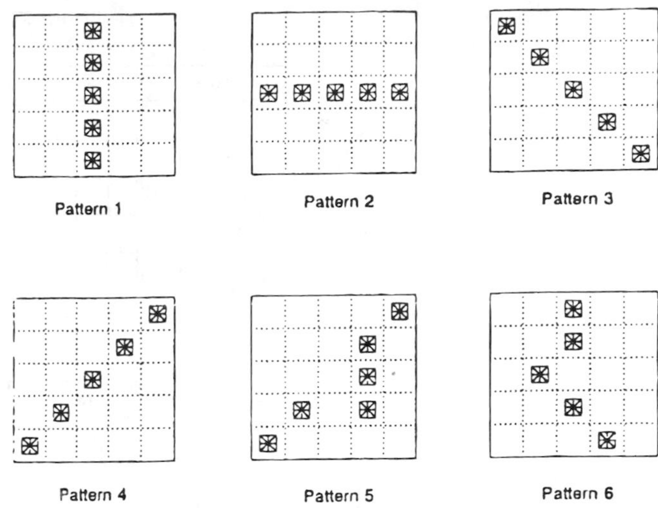

Figure 72.5 Input patterns for Study A.

A2 *The greater discrimination resulting from increased inhibition may cause the development of redundant planes.*

This can be seen by comparing the connections in Figures 72.6a, 72.7a, and 72.8a. For example, in Figure 72.8a, which shows the results for the highest inhibition, we see similar patterns (4 and 7, 2 and 6) represented by different planes. However, in Figure 72.6a, which corresponds to the least inhibition, only four planes develop. This is about the number that we would expect based on a human inspection of the stimulus shown in Figure 72.5. This redundancy is also reflected in the increase of the dot-product of the **a** vectors of the developed planes, as seen in Figures 72.6b, 72.7b, and 72.8b.

A3 *Decreasing the learning rate has little effect (actually, negative if any) on the results for the one level classification attempted in study A.*

This can be inferred by comparing the results in Figure 72.8 with those in Figure 72.9. In the latter, the learning rate was decreased to 0.75 and the number of vectors increased appropriately so that the same amount of learning occurred. As a result of a decrease in the learning rate, *q*, we note that the number of clusters increased by 1 and the dot-product also increased slightly.

A4 *Using the exponential form of* **c** *only seems to worsen the classification in the one-level case.*

This can be seen by comparing the results shown in Figure 72.6 with those shown in Figure 72.10. They differ in the form of vector **c.** The exponential form of **c** results in less clusters than are actually required. For the exponential case, we also note an increase in the value of the dot-products between clusters.

A5 *A high inhibition factor causes a sharp drop in the response of a feature detecting plane even with a single missing element in the feature. This could be a problem if the input consists of a macro-feature which contains many instances of this feature.*

a - CONNECTIONS for level 1 plane

PLANE 0					PLANE 1					PLANE 2					PLANE 3				
49				11						11		86							
	49											73	11						
		48		11						11	48	48							
		11	49									99							
				49								48	48						

PLANE 4					PLANE 5					PLANE 6					PLANE 7				
															11		11		98
																		98	
																	49	48	
															11	86	11	49	
															98	11			

PLANE 8					PLANE 9					PLANE 10					PLANE 11				
		11																	
48	36	49	49	49															
				11															

b - DOT_PRODUCT of CONNECTIONS after LEARNING

PLANE_NO	0	1	2	3	4	5	6	7	8	9
0	1.00		0.21					0.29	0.31	
1										
2			1.00					0.16	0.31	
3										
4										
5										
6										
7								1.00	0.23	
8									1.00	
9										

**minimum = 0.16 maximum = 0.31 average = 0.25

Figure 72.6 Results from Study A with INPUT__STIMULUS: stimulus 1 IF: 1.0 LR: 3.0 type of c: UNIFORM Initialization of a: RANDOM.

a - CONNECTIONS for level 1 plane

PLANE 0					PLANE 1					PLANE 2					PLANE 3				
56				13	13				13	13		56					41		
	56							13				41	13				41		
		56		13			13					56			13	56			
		13	56									56					56		
				56	13							56						56	

PLANE 4					PLANE 5					PLANE 6					PLANE 7				
																	13		98
																		98	
																	41	56	
															13	98	13	56	
															99	13			

PLANE 8					PLANE 9					PLANE 10					PLANE 11				
		13																	
56	41	56	56	56															
				13															

b - DOT_PRODUCT of CONNECTIONS after LEARNING

PLANE_NO	0	1	2	3	4	5	6	7	8	9
0	1.00	0.45	0.31	0.06				0.26	0.31	
1		1.00	0.32	0.01				0.70	0.21	
2			1.00	0.52				0.20	0.27	
3				1.00				0.06	0.27	
4										
5										
6										
7								1.00	0.22	
8									1.00	
9										

****minimum = 0.16 maximum = 0.31 average = 0.25**

Figure 72.7 Results from Study A with INPUT__STIMULUS: stimulus 1 IF: 2.0 LR: 3.0 type of c: UNIFORM Initialization of a: RANDOM.

a - CONNECTIONS for level 1 plane

PLANE 0					PLANE 1					PLANE 2					PLANE 3				
99				24			24		98			74					49		
	98							99				73	23				74		
		98		24				98				73				73			
		24	98			98	24	98				74					73		
				99	98	24						73						74	

PLANE 4					PLANE 5					PLANE 6					PLANE 7				
23				24						24		24							73
			24															73	
		24										23					74		
												24			24	73			
24												24			74				

PLANE 8					PLANE 9					PLANE 10					PLANE 11				
		24					24												
98	73	98	98	98	24	24													
							24												
				24				24											

b - DOT_PRODUCT of CONNECTIONS after LEARNING

PLANE_NO	0	1	2	3	4	5	6	7	8	9
0	1.00	0.24	0.25	0.06	0.45		0.44	0.25	0.31	0.06
1		1.00	0.15	0.08	0.54		0.10	0.71	0.20	0.10
2			1.00	0.56	0.27		0.79	0.26	0.26	0.40
3				1.00	0.01		0.36	0.00	0.30	0.78
4					1.00		0.41	0.79	0.21	0.01
5										
6							1.00	0.21	0.27	0.41
7								1.00	0.21	0.01
8									1.00	0.42
9										1.00

**** minimum = 0.00 maximum = 0.79 average = 0.30**

Figure 72.8 Results from Study A with INPUT__STIMULUS: stimulus 1 IF: 4.0 LR: 3.0 type of c: UNIFORM Initialization of a: RANDOM.

a - CONNECTIONS for level 1 plane

PLANE 0					PLANE 1					PLANE 2					PLANE 3				
86		5	5	5	5		11	5	92		5	92	5				86		11
	86		5					99			5	86	11	5			92	5	5
5		86		5	5			98	5			92				86	11		
		5	86		5	98	11	98			5	92	5				92	5	
				80	98	5	5					92		5		5		92	

PLANE 4					PLANE 5					PLANE 6					PLANE 7				
5				5						5		5				5		5	86
			5														11	86	
		5			11	11	5	11	11			5					86		5
									11			5			5	86			
5								5				5			86	5		5	

PLANE 8					PLANE 9					PLANE 10					PLANE 11				
5		5		5			5												
				5															
86	80	86	86	86	5	5													
5							5												
5				5				5											

b - DOT_PRODUCT of CONNECTIONS after LEARNING

PLANE_NO	0	1	2	3	4	5	6	7	8	9
0	1.00	0.23	0.28	0.08	0.45	0.14	0.44	0.24	0.27	0.07
1		1.00	0.11	0.10	0.57	0.22	0.12	0.72	0.24	0.10
2			1.00	0.62	0.26	0.11	0.81	0.27	0.22	0.42
3				1.00	0.10	0.31	0.45	0.11	0.23	0.80
4					1.00	0.13	0.42	0.81	0.27	0.05
5						1.00	0.14	0.13	0.85	0.50
6							1.00	0.23	0.26	0.44
7								1.00	0.25	0.05
8									1.00	0.42
9										1.00

**** minimum = 0.05 maximum = 0.85 average = 0.31**

Figure 72.9 Results from Study A with INPUT__STIMULUS: stimulus 1 IF: 4.0 LR: 0.75 type of c: UNIFORM Initialization of a: RANDOM.

a - CONNECTIONS for level 1 plane

PLANE 0					PLANE 1					PLANE 2					PLANE 3				
												12							
								5				29							
								16			14	37							
						5	1	5				33							
					1							5	2						

PLANE 4					PLANE 5					PLANE 6					PLANE 7				

PLANE 8					PLANE 9					PLANE 10					PLANE 11				
1				1															
	5	1	6																
5	13	99	14	6															
	5	1	5																
1																			

b - DOT_PRODUCT of CONNECTIONS after LEARNING

PLANE_NO	0	1	2	3	4	5	6	7	8	9
0										
1		1.00	0.08	0.55					0.30	
2			1.00	0.18					0.64	
3				1.00					0.23	
4										
5										
6										
7										
8									1.00	
9										

**** minimum = 0.08 maximum = 0.64 average = 0.31**

Figure 72.10 Results from Study A with INPUT__STIMULUS: stimulus 1 IF: 1.0 LR: 3.0 type of c: EXPONENTIAL Initialization of a: RANDOM.

Figure 72.11 illustrates this phenomenon for a plane detecting a diagonal line. With a high inhibition-factor (in this case equal to 4) the decrease in activation is significantly greater than with a low inhibition factor (in this case equal to 1).

Figure 72.12 illustrates the problem this could cause in the network's response to a macro feature containing contiguous instances of a micro-feature (feature 1). With a high inhibition factor, the one missing pixel may cause the activations of units at level $l + 1$ to decrease significantly as to prevent the feature's detection at that level.

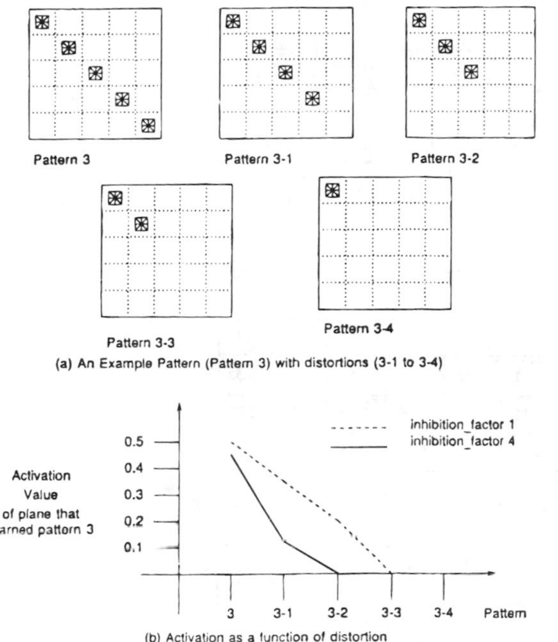

Figure 72.11 Neocognitron's sensitivity to distortion in learned features.

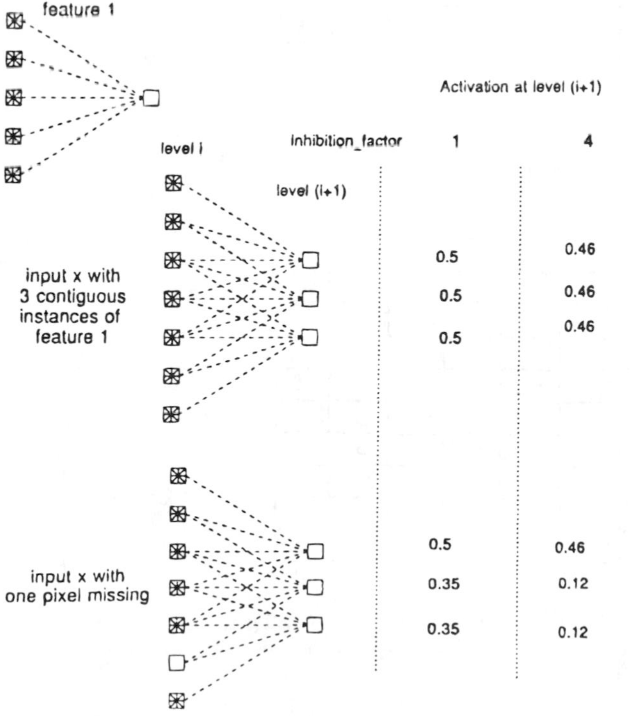

Figure 72.12 Adverse effect of high inhibition.

72.7 Study B

This study was made with a 3-level network: an input level, a micro-feature recognition level, and a level recognizing the total input pattern. The table summarizes the description of the network used in this study. The entry *NA* stands for not applicable. The input patterns consisted of a 9 × 9 array of pixels. The patterns are shown in Figure 72.13.

Table 72.2 Description of network used in Study B

	Level 0	Level 1		Level 2	
	Layer INP	Layer Vc	Layer S	Layer Vc	Layer S
# of planes	1	1	12	1	12
units per plane	9 × 9	7 × 7	7 × 7	1	1
receptive field size	NA	3 × 3	3 × 3	7 × 7	7 × 7
competition area	NA	NA	2 × 2	NA	3 × 3

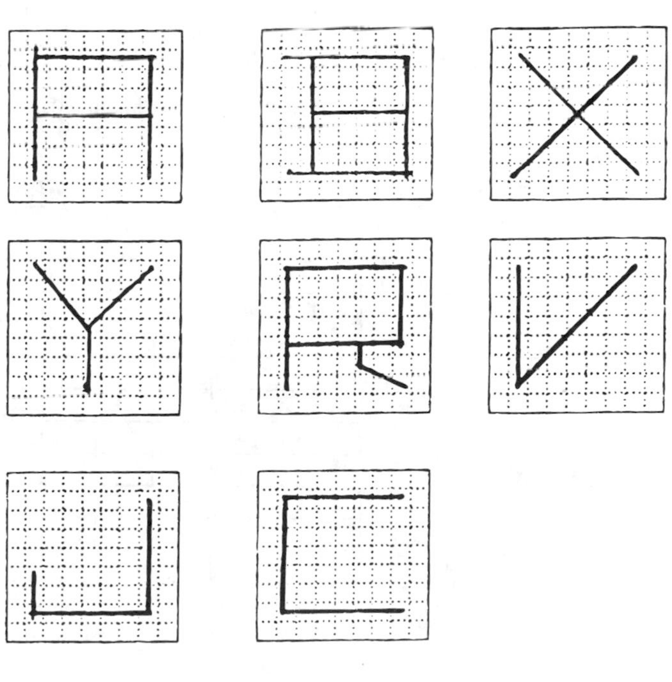

Figure 72.13 Input patterns for Study B.

a - CONNECTIONS for planes in Level 1 after Learning

PLANE 0			PLANE 1			PLANE 2			PLANE 3			PLANE 4			PLANE 5		
				99		3	1	4	49								48
1	1	1		99	67		4			49						48	
		1		99		4		4			49				48		

PLANE 6			PLANE 7			PLANE 8			PLANE 9			PLANE 10			PLANE 11		
32	32	32			65	71			48	48	49					60	
					63	71					48	77	77	77	60	60	
			65	65	65	71					49					29	

b - DOT_PRODUCT of CONNECTIONS for Level 1

PLANE_NO	0	1	2	3	4	5	6	7	8	9	10	11
0	1.00	0.45	0.53	0.56		0.40	0.11	0.52	0.29	0.52	0.85	0.55
1		1.00	0.34	0.31		0.31	0.31	0.40	0.00	0.40	0.52	0.75
2			1.00	0.75		0.78	0.60	0.61	0.49	0.66	0.26	0.35
3				1.00		0.33	0.33	0.26	0.33	0.52	0.33	0.32
4												
5						1.00	0.33	0.52	0.33	0.26	0.33	0.32
6							1.00	0.26	0.33	0.77	0.00	0.32
7								1.00	0.26	0.60	0.26	0.13
8									1.00	0.26	0.33	0.32
9										1.00	0.26	0.25
10											1.00	0.64
11												1.00

**** minimum = 0.00 maximum = 0.85 average = 0.40**

c - ACTIVATION of PLANES in Level 2

	PLANES										
PATTERNS	0	1	2	3	4	5	6	7	8	9	10
A				.31							
B		.27									
X						.18					
Y							.20				
R										.30	
V	.21										
J											.15
C					.34						

Figure 72.14 Results from Study B referred to in observation.

a - CONNECTIONS for planes in Level 1 after Learning

PLANE 0			PLANE 1			PLANE 2			PLANE 3			PLANE 4			PLANE 5		
29	98		50	38	50			58	27	15	27	42	42	42			8
	99							56		27				42	68	68	68
	98					58	58	58	13		27			42			

PLANE 6			PLANE 7			PLANE 8			PLANE 9			PLANE 10			PLANE 11		
70			15		15					42		58	58	56			37
70	70	70		13		42			42	27		58					24
42				15			42			42		58					37

b - DOT_PRODUCT of CONNECTIONS for Level 1

PLANE_NO	*0*	*1*	*2*	*3*	*4*	*5*	*6*	*7*	*8*	*9*	*10*	*11*
0	1.00	0.38	0.26	0.50	0.33	0.33	0.36	0.64	0.40	0.82	0.33	0.00
1		1.00	0.28	0.71	0.77	0.05	0.30	0.64	0.00	0.26	0.77	0.40
2			1.00	0.53	0.60	0.29	0.34	0.46	0.32	0.26	0.40	0.76
3				1.00	0.75	0.31	0.52	0.69	0.00	0.32	0.64	0.60
4					1.00	0.29	0.43	0.46	0.00	0.24	0.60	0.76
5						1.00	0.83	0.30	0.41	0.52	0.30	0.30
6							1.00	0.46	0.34	0.44	0.56	0.20
7								1.00	0.37	0.44	0.46	0.33
8									1.00	0.76	0.32	0.00
9										1.00	0.50	0.00
10											1.00	0.28
11												1.00

**** minimum = 0.00 maximum = 0.83 average = 0.41**

c - ACTIVATION of PLANES in Level 2

	PLANES										
PATTERNS	*0*	*1*	*2*	*3*	*4*	*5*	*6*	*7*	*8*	*9*	*10*
A					.38						
B									.41		
X								.19			
Y						.12					
R				.42							
V										.12	
J		.26									
C							.44				

Figure 72.15 Results from Study B referred to in observation.

a - CONNECTIONS for planes Level 1 after Learning

PLANE 0			PLANE 1			PLANE 2			PLANE 3			PLANE 4			PLANE 5		
28	98		43	43	43			56	1		1	41	41	41			
	99							56		1				41	41		
	98					56	56	56	1		1			41		41	

PLANE 6			PLANE 7			PLANE 8			PLANE 9			PLANE 10			PLANE 11		
69			15					41		26		57	57	56	41		
69	69	70	15		15		41		26	25		57				41	
41			15	15		41				26		57					41

b - DOT_PRODUCT of CONNECTIONS for Level 1

PLANE_NO	0	1	2	3	4	5	6	7	8	9	10	11
0	1.00	0.43	0.26	0.33	0.33	0.40	0.35	0.33	0.33	0.85	0.33	0.43
1		1.00	0.26	0.49	0.77	0.00	0.28	0.26	0.33	0.29	0.77	0.33
2			1.00	0.62	0.60	0.32	0.34	0.62	0.52	0.23	0.40	0.26
3				1.00	0.59	0.00	0.54	0.41	0.80	0.23	0.58	0.76
4					1.00	0.00	0.43	0.42	0.26	0.23	0.60	0.52
5						1.00	0.34	0.63	0.00	0.71	0.32	0.00
6							1.00	0.77	0.44	0.48	0.56	0.55
7								1.00	0.26	0.45	0.60	0.29
8									1.00	0.28	0.51	0.33
9										1.00	0.45	0.28
10											1.00	0.26
11												1.00

**** minimum = 0.00 maximum = 0.85 average = 0.41**

c - ACTIVATION of PLANES in Level 2

	PLANES										
PATTERNS	0	1	2	3	4	5	6	7	8	9	10
A					.45						
B									.45		
X			.36								
Y	.32										
R							.34				
V											.28
J								.17			
C				.34							

Figure 72.16 Results from Study B referred to in observation.

Results from Study B

Several observations are made, and are listed below:

B1 *With appropriate values for the inhibition factor and the learning rate, the Neocognitron seems to extract appropriate micro-features at the first level, which are then used in recognizing the different letters at the second level.*

This observation is substantiated by noticing that the **a** vectors developed for level 1 (Figure 72.14a) correspond to features that are apparent through visual inspection of the letters (Figure 72.13).

Also, after learning has occurred, each letter is associated with a response from only one plane in level 2 (Figure 72.14c). The responding plane is unique to that letter thus signifying the recognition of the input letters by the trained network.

For the experiment on which this observation is based, the variable parameters are set as follows:

a. For pass 1 in which the network is trained with 4 instances of each pattern:
—the inhibition factor is set at 5 for level 1, and 8 for level 2.
—The learning rate is set for 0.5 for level 1, and 1 for level 2.

b. For pass 2 in which the network is trained with 7 instances of each pattern:
—The inhibition factor is set to 5 for level 1 and 8 for level 2.
—The learning rate is set to 2.0 for level 1 and 9 for level 2.

B2 *The high inhibition factor required to distinguish between letter A and R resulted in the network being very sensitive to missing features in the input.*

B3 *Lowering the inhibition factor to reduce this sensitivity resulted in the failure to distinguish between A and R.*

B4 *Not thresholding the selection of winning features in each competition area resulted in the development of redundant micro features.*

This is evidenced by the development of connections for plane 8 in level 1 as shown in Figure 73.15a.

B5 *Primed, instead of totally random initialization of* ***a*** *resulted in a better clustering at level 1.*

This is evidenced by comparing the connection tables for level 1 in Figure 72.14a and Figure 72.16a (the unprimed case). We note that the unprimed case resulted in more clusters, with increased dot-product between them. However, this does not seem to affect the capability of the network to distinguish between the letters, as evidenced by the distinct activation of the planes in level 2 (Figure 72.16c).

72.8 Summary and Discussion

The Neocognitron is analyzed in terms of classical pattern recognition techniques. Its ability to recognize characters is demonstrated through simulations. Useful observations are made about the performance of this task. The most critical factors for the process appear to be the selection of the learning rate and the inhibition factor. The Neocognitron seems to provide a viable approach for optical character recognition. Several copies of the type of network used in study B could be used in parallel to recognize items like zip codes or social security numbers.

Future work will consider the hardware implementation of this type of Neocognitron. The local decision functions at each level could be implemented with simple processors with small amounts of local memory which could store the values of **a, c,** b and other parameters. Finally, it would be interesting to perform studies of the type reported in this paper on some of the other (Kohonen, 1984, Carpenter and Grossberg, 1987, Linsker, 1988, Widrow and Winter, 1988) pattern clustering approaches which are based on neuro-computing paradigms.

References

Carpenter, G. A. and Grossberg, S. 1987. A massively parallel architecture for a self-organizing neural pattern recognition machine, *Computer Vision, Graphics, and Image Processing,* 37:54–115.

Chon, T.-S. and Micheli-Tzanakou, E. 1990. Pattern and feature extraction, *Proc. IASTED—Int. Symp. Machine Learning and Neural Networks,* October, pp 14–17.

Deutsch, S. and Micheli-Tzanakou, E. 1987. *Neuroelectric Systems,* New York University Press, New York, NY.

Fukushima, 1982. Neocognitron: A new algorithm for pattern recognition tolerant of deformations and shifts in position, *Pattern Recognition,* 15(6):455–469.

Kohonen, T. 1984. *Self-Organization and Associative Memory,* Springer-Verlag, Berlin, Heidelberg, NY, Tokyo.

Linsker, R. 1988. Self-organization in a perceptual network, *IEEE Computer,* 21(3):105–117.

Micheli-Tzanakou, E. 1983. Visual receptive fields and clustering, *Behavior Research—Methods and Instrumentation,* 15(6): 553–560.

Tou, J. T. and Gonzalez, R. C. 1974. *Pattern Recognition Principles,* Addison-Wesley, Reading, MA.

Widrow, B. and Winter, R. 1988. Neural nets for adaptive filtering and adaptive pattern recognition, *IEEE Computer,* 21(3):25–39.

73

Analog 3-D Neuroprocessor for Fast Frame Focal Plane Image Processing

Tuan A. Duong
California Institute of Technology

Sabrina Kemeny
California Institute of Technology

Taher Daud
California Institute of Technology

Anil Thakoor
California Institute of Technology

Chris Saunders
Irvine Sensors Corporation

John Carson
Irvine Sensors Corporation

73.1 Introduction 990
73.2 Neural Network Architecture 991
73.3 Neural Network Design and Operation 991
Synapse Design • Neuron Design
73.4 Experimental Results 994
73.5 Cascade-Backpropagation (CBP) 995
Mathematical Model • Quantization of Weight Space • Procedure for Learning in Hardware • Weight Update Issues
73.6 Six-Bit Parity Problem 999
Cascade Backpropagation (CBP) Simulations
73.7 Conclusions 999

73.1 Introduction

Pattern recognition is computationally intensive and for many defense and commercial applications, real time response requires a hardware implementation with neural network's inherent parallelism. Neural networks, configured in software, have been reported for such applications but are slow (IEEE, 1994). VLSI-implemented neural network chips have been utilized to reduce processing time by orders of magnitude and are useful in a variety of applications (Eberhardt et al., 1991, 1992). However, the size of the VLSI networks is often limited by available silicon area (constrained by increasing cost and decreasing reliability as die size increases). Silicon area can be increased through the use of wafer-scale integration, multichip modules, or die stacking. Recent advances in die stacking are particularly attractive since they provide an extremely compact realization. A cube, constructed from many (e.g., 64) thinned die, would occupy approximately the same footprint as a single die. In addition to the tremendous processing power afforded by such a dense IC cube, mating of a 3-D IC stack to an image sensor array would enable spatially parallel signal processing to be performed on image data at extremely high speed. An architecture has been developed which combines the focal plane array with a spatially parallel 3-D neural processing cube, promising, for the first time, tremendous speed and problem size enhancements over conventional VLSI techniques (Duong et al., 1994).

A particularly challenging application that requires processing capability afforded by an integrated neural image processing cube is the missile seeker functionality which requires spatio-temporal recognition of both point and resolved targets at extremely high speeds (milliseconds). A reconfigurable neural network architecture, trained properly (loaded with appropriate weights), may discriminate targets from clutter or classify targets once resolved. By mating a 64×64 image sensor to a stack of 64 neural net ICs, each with different weights, a variety of image processing tasks could be performed in parallel at extremely high speeds and in an extremely small package (≈ 1 inch cube). The simultaneous requirements dictated by such an application on the neuroprocessing cube are:

1. Cold temperature operation, IC stack being mated to the infrared (IR) imager required to operate at $\sim$90°K.
2. Low power dissipation of 1.5 to 2.0 watts because of the need to maintain cold temperatures and overall power budget economy.
3. High speed operation approaching 1000 frames per second, which would translate into a 4 MHz pixel image processing rate, and hence a $<$250 nanoseconds signal processing speed.

These requirements have led to the design of low power analog circuits for implementation of the VLSI neural network ICs. Use

This article originally appeared in *Simulation* 65(1):11–25, July 1995. Reprinted by permission of Simulation Councils, Inc.

of analog circuitry (as opposed to digital) enables very compact, low power neural network realizations (Carson, 1991; Hopfield, 1990). In addition, for the 3-D stack coupled to an image array, the spatially parallel input to the neural networks is in analog form. Digital neural processing would require at least one (high speed) or up to 64 (moderate speed) analog to digital converters on each IC, impractical for the low power requirement of the proposed stack. This paper focuses on the analog neural network portion of the 3-D architecture. Test results of the neuron and synapse circuits are presented. In addition, since analog processing limits the synaptic resolution (5 to 10 bits) well below 32 or 64 bits available with software simulation, a new learning algorithm of cascade backpropagation (CBP) that is simpler to train and tolerant of limited synaptic resolution is described. Further, using this architecture, simulation results are given for solution of a nonlinear 6-bit parity problem as an illustrative example.

73.2 Neural Network Architecture

The stacked architecture promises a practical realization of three dimensional electronic circuitry, offering unprecedented computational power in such a compact package (Carson, 1991). Figure 73.1 illustrates an emulation of the silicon architecture. Sixty-four thinned VLSI chips are stacked to form a three-dimensional "sugarcube." Using the bump bonding technique, an imager is then mated to the IC stack so that each row in the imager array is directly attached to the inputs of one IC. Thus, an individual connection is made for each and every pixel. Communication among the stacked ICs is made possible by providing meta bus lines running across side planes.

Neural network input could be controlled by a sequencer circuit termed "window grabber," that controls signal flow along 64 common bus lines. The novel window grabber circuit, currently under development, is a switching matrix that would: select a desired window (e.g., 8 × 8) from the imager; convert the 2-dimensional window to a single-line vector; shuffle the individual vector elements to obtain them in the right sequence; and provide this vector as input to one or more ICs in the stack. In addition, the switching matrix could be designed to allow window rotation, selected adjacent-pixel averaging, etc. for signal preprocessing when required (Duong, T. A., Mapping Pixel Windows to Vectors for Parallel Processing, NASA Tech. Briefs, Vol. 20, No. 3, pp. 4a–6a, 1996).

Each IC in the stack would be identical, but could be programmed with different weights and/or different input windows. This architecture allows tremendous flexibility in the cube functionality. For instance, each of the 64 neural networks could look at the same window with differing weight templates, or at the other extreme, the entire image could be divided into 64 windows with different 8 × 8 windows input to each of the 64 neural networks. Any combination between these extremes could also be programmed. By processing the image data in parallel, enormous throughput can be achieved. For instance, the stack of 64 of the envisioned multilayer perceptrons operating at 4 MHz, (required for the 1000 frames per second data rate) potentially results in a connectivity of 10^{13} connections per second.

A reconfigurable feed-forward neural network, consisting of 7-bit programmable multiplying synapses, variable gain sigmoidal neurons, and control and addressing circuitry, enables a variety of neural architectures and algorithms such as multilayer perceptron, cascade backpropagation, and inner-product scheme with winner-take-all (WTA) to be realized. Shown schematically in Figure 73.2, the network can accommodate 64 parallel inputs plus a bias line, convenient for processing an 8 × 8 image window (kernel).

The input signals are broadcast as input to the 70 × 65 (a 64 × 65 and a 6 × 65) synaptic matrix at left (Figure 73.2) for processing (A) through the gang-switches (i) as a multilayer perceptron achitecture (switch positions as shown) to the hidden layer neurons, or (ii) as an inner-product scheme with WTA neurons for template matching (gang-switches in the lower position), or (B) first directly to the lower 6 × 65 synaptic matrix leading to the output neurons to the right and additionally through the upper 64 × 65 synaptic matrix on to the hidden neurons as required for such algorithms as cascade correlation [Fahlman and Lebiere, 1990)] when hidden units can be added one by one. It may be noted that the half-populated middle synaptic matrix allows signals from the previous hidden units to be fed to the newly added hidden units. For multilayer perceptron, up to 64 of the synapse rows can be connected to 64 hidden neurons. These hidden neurons would be connected to a 64 × 6 synapse array for 6 outputs. The synaptic matrix on the right is a 70 × 6 matrix out of which the top 64 × 6 matrix would take inputs from the 64 hidden units and provide them after proper weighting to the 6 output neurons. The lower 6 × 6 matrix of synapses is basically providing a linear connection to the 6 outputs with a weighting of unity. Thus, a direct input to output neuron connection through an input synapse is made for such constructive architectures as CBP algorithm (Duong et al., 1995).

73.3 Neural Network Design and Operation

The missile defense application imposes severe speed, computation, and mass requirements on any implementation. The 3-D architecture to be operated at ~90°K introduces an additional ultra low power requirement on the circuitry. To minimize heating, the power requirement is to be restricted to ≤2 watts for the 64 IC cube with the 64 × 64 imager operating at 1000 frames per second. To ensure that any one of the neural ICs can process the entire image at a 1 ms frame rate, a feed-forward pass through the network must be accomplished in 250 ns (4 MHz). To facilitate the multiplexing and processing of different blocks of data, it is assumed that only 16 of the 64 neural net ICs will be on at any given time, when operated at this high speed. Driven by the high-speed low-power requirements of the missile defense application implemented with a 3-D neural image processing cube, new synapse and neuron circuits have been designed.

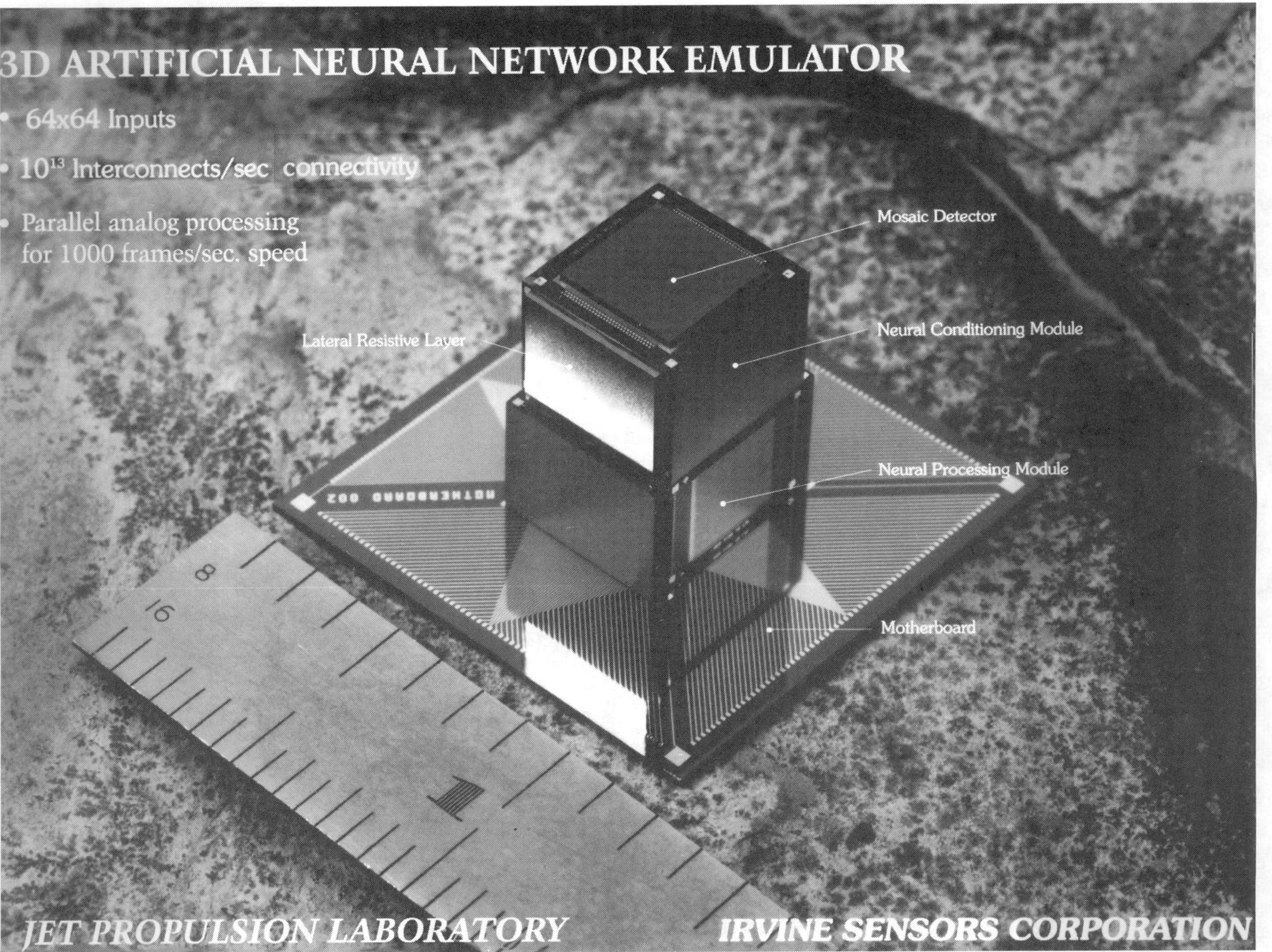

Figure 73.1 A 3-dimensional "sugarcube" architecture emulation in silicon that would combine an infrared focal plane array mated for full parallelism with a stacked multichip cube, each chip with preprocessing and neural network circuitry.

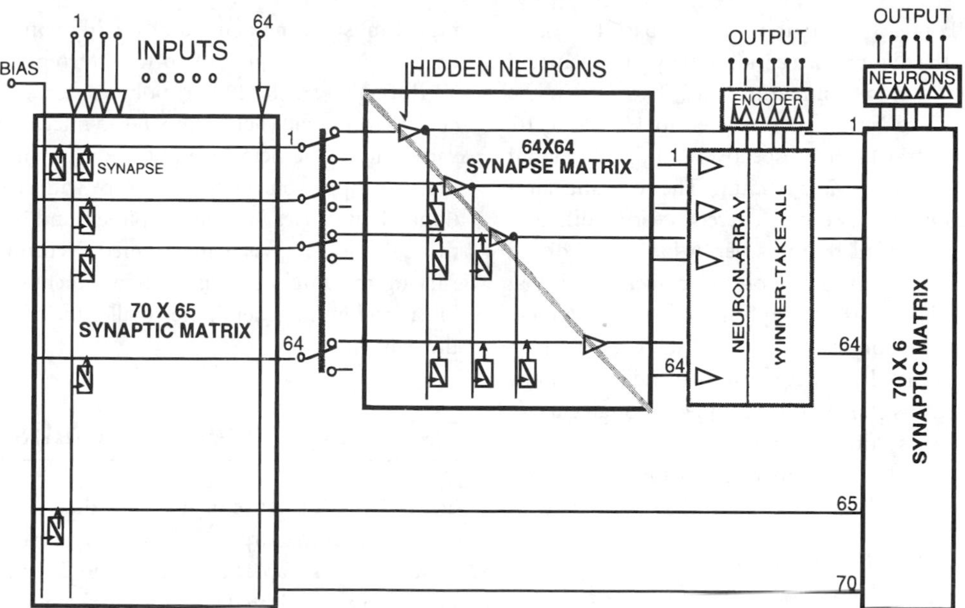

Figure 73.2 A schematic of a multi-circuit neural network architecture with 64 inputs and 6 outputs. This architecture can be configured either as a multilayer perceptron, a cascade backpropagation, or an inner-product based feed-forward circuit.

As the starting point for the new designs, our successfully demonstrated analog synapse and neuron designs, described in detail in Duong et al. (1995) were utilized. Modifications to these designs have increased speed by over an order of magnitude and potentially reduced power consumption by approximately an order of magnitude (Duong et al., 1994). Several global changes that affect both the neuron and synapse designs were implemented. The operating voltages were reduced from 8 to 5 volts, lowering power dissipation at the cost of reduced dynamic range. However, since the swing level is reduced, this decrease in voltage range has the advantage of higher speed operation. In addition to decreasing on-chip power consumption, the 5 volt power supply decreases system power consumption and complexity by simplifying and reducing the digital interface circuitry (e.g., no level shifters were required). Another reduction in power was realized by operating the network with smaller currents, nano-amperes to micro-amperes as opposed to up to tens of micro-amperes. Finally, a speed enhancement was obtained by fabricating the circuits in 1.2 μm design rules as opposed to 2 μm design rules.

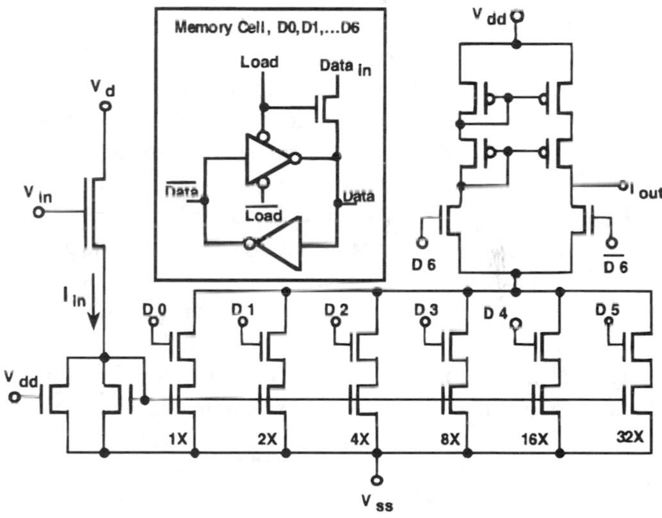

Figure 73.3 Synapse circuit containing a voltage-to-current input stage, a 7-bit multiplying digital to analog converter, and a 7-bit digital memory. (*Source*: *Fuzzy Logic and Neural Network Handbook*, Ch. 27. 1996. McGraw-Hill, NY. With permission.)

Synapse Design

The synapse circuit, shown in Figure 73.3, consists of a voltage-to-current converter at the input, a 7-bit multiplying digital to analog converter (MDAC), and a 7-bit digital memory. The synapse memory is randomly accessed through row and column decoders located adjacent to the array. This type of synapse, with its on-chip storage of digital weights, allows a very simple digital interface (as opposed to the need to refresh circuitry to update volatile analog storage of weights) and has been successfully incorporated into a number of implementations (Duong, et al., 1994, 1996; Eberhardt et al., 1992). The current realization utilizes single transistor current mirrors rather than the cascade current mirrors of previous designs. This difference results in higher speed and a more compact design at the cost of a possible decrease in circuit robustness.

Operation of a synapse cell is as follows: An input transistor, biased in the linear region ($Vdrain < Vgate - V_t$), converts an input voltage (V_{in}) applied to its gate into a drain current (I_{in}) which is almost linearly proportional to V_{in}. This input current is then multiplied by the stored digital word (weight) to produce the desired output current (I_{out}). Multiplication is accomplished by conditionally scaling the input current I_{in} by a series of current mirror transistors. For each current mirror, a pass transistor controlled by 1 bit of the digital word conditionally allows current to be placed on a common summation line. The bits in the digital word from LSB to MSB are connected to 1, 2, 4, 8, 16, and 32 current mirror transistors, respectively so that the input current is scaled by the appropriate amount. The resulting summation current is unipolar. However, a current steering differential transistor pair, controlled by the seventh bit of the digital word, determines the direction of the output current, such that two-quadrant multiplication is accomplished (−63 to +63 levels). The 7-bit digital memory consisting of 7 static latches provides programmable, nonvolatile weight storage and is randomly accessible. One input transistor circuit is coupled through current mirrors to all the synapses along one column in the input synapse matrix (or row for output synapse matrix) because the current I_{in} is required as input to all the synapses in that column (rows for the output).

Neuron Design

Neurons which produce a sigmoid activation function are usually based on transconductance amplifiers that have been modified to optimize performance (e.g., increase the input voltage range). The neuron circuit, shown in Figure 73.4, consists of a very simple variable gain transconductance operational amplifier with no feedback (compensation capacitor) connection. The stability conventionally achieved with a feedback capacitor is not needed in the present architecture which contains no feedback connections (Eberhardt et al., 1992).

The elimination of the capacitor greatly enhances speed and eliminates the need for a separate (power hungry) gain control circuit. In addition, the new simple design is operated with smaller currents and is much more compact than previous implementations (threefold reduction in area if utilizing the same design rules). For proper operation of the MDACs and the operational amplifier, the voltage on the row summation line (neuron input) must be maintained at a fairly constant voltage. To this end, the neuron input is tied to a reference voltage (Vref = 2.5 V) through a small (200 Ω) poly resistor. Since the synapse row currents are small (nano-amperes to micro-amperes), the voltage variations at the neuron input node are small (< millivolt), so that the input voltage range is kept within the required operational boundaries of the amplifier and uniform conduction through the synapses is maintained. Neuron gain is achieved by varying the amplifier bias current which alters the slope of the linear region. In general, the smaller the bias current, the higher the gain.

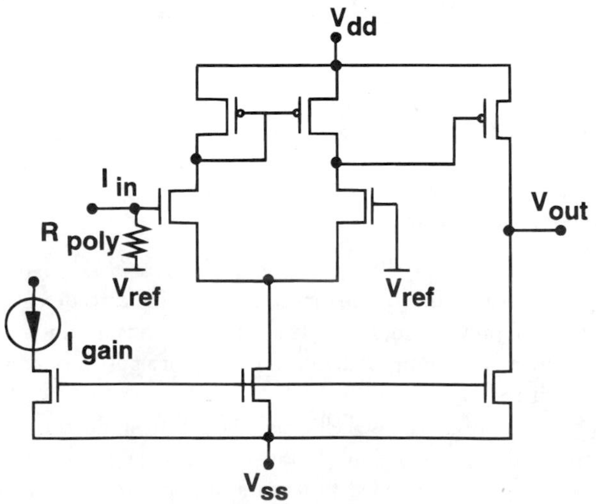

Figure 73.4 Circuit diagram of a wide range, variable gain sigmoidal neuron. (*Source: Fuzzy Logic and Neural Network Handbook*, Ch. 27. 1996. McGraw-Hill, NY. With permission.)

73.4 Experimental Results

To verify performance of the individual circuits prior to implementation of the large IC, a tiny chip (2.4 mm × 2.4 mm die) containing a synapse, a neuron, and a synapse connected to a neuron was fabricated in a 1.2 μm single poly n-well CMOS process. A full custom layout resulted in ultra compact cells (synapse occupies 119.4 μm × 100.8 μm and neuron occupies 75.6 μm × 61.2 μm) enabling the realization of the large network in a reasonable, albeit large chip size (< 10 mm^2). The outputs of the circuits were buffered through source followers that introduced about a 25 ns delay and a 0.2 signal attenuation. The tiny chip was mounted on a test board and interfaced to a laboratory PC for automated data acquisition. For cold temperature (77 °K) measurements, the chip was mounted on a separate fixture which was then immersed in liquid nitrogen. All circuits were found to be operational and exceeded the design speed specification of 4 MHz, i.e., 64 new inputs presented every 250 ns. The circuits were modeled with the PSPICE circuit simulation tool and experimental results correlate closely with simulation. Simulation results indicate an average power consumption of less than 30 mWatts/chip for operation of the network at 4 MHz.

The 7-bit MDAC synapse is nearly monotonic in its characteristics both at room and at cold temperatures. These characteristics are shown respectively in Figures 73.5(a) and (b). The neuron outputs the desired sigmoidal activation function and range of gain variations. Figure 73.6 shows the response of the neuron receiving input from a synapse as its digital weights are being ramped from −63 to +63 and its input is held constant at *Vin* = 5 volts. The three curves, illustrating gain variation, were obtained by adjusting the neuron bias current (Duong et al., 1994, 1995).

A synapse-neuron pair was utilized to measure the speed response of a single layer of the network. All bits of the digital synapse weight were set to one, and a 5 volt square wave was applied to the synapse input, while the neuron output voltage was monitored through a source follower. As shown in Figure 73.7, the rise and fall times respectively for the neuron output were 150 and 117 ns respectively at room temperatures. The delay figures at cold temperature were even better at 94 and 81 ns. The measurements were performed while driving the large 20 pF oscilloscope load. Thus, the 4 MHz design specification for operation of the neural network was satisfied.

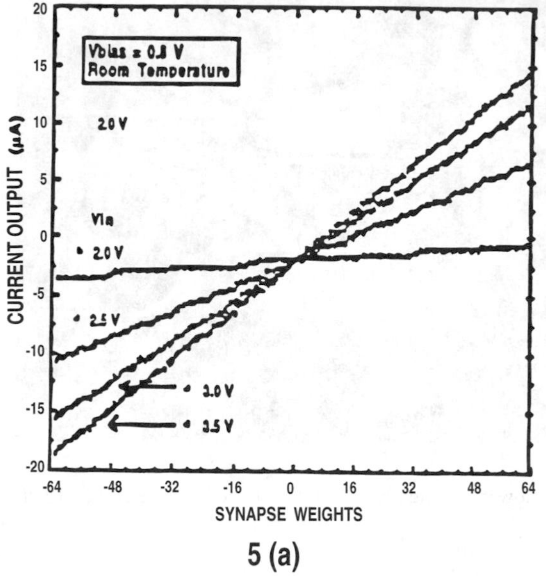

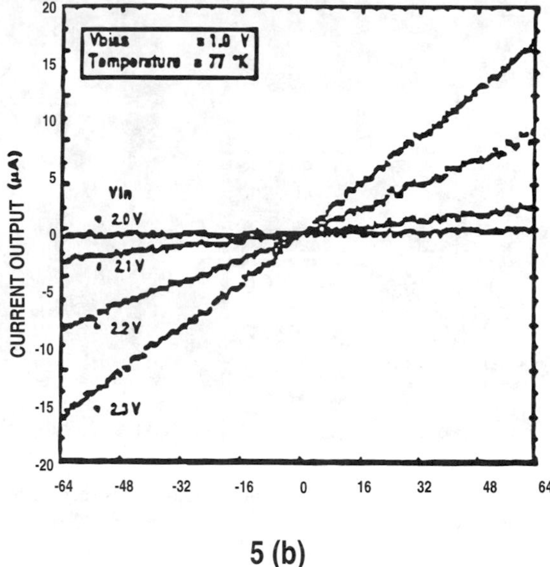

Figure 73.5 Synapse characteristics at (a) room, and (b) cold temperatures, with current output in micro-amperes as a function of digital weight variations from −63 to +63.

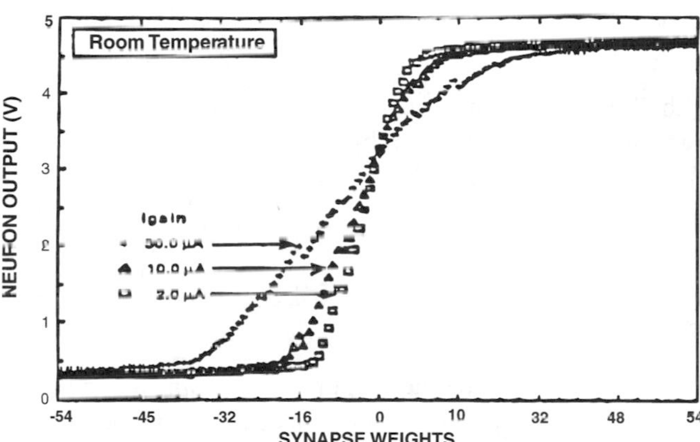

Figure 73.6 Synapse-neuron transfer characteristics at three gain settings. Neuron is receiving input from a synapse as its weights are being ramped from −63 to +63 and its input is held constant at 5 volts.

73.5 Cascade-Backpropagation (CBP)

In this section we develop a new self-evolving architecture that is highly efficient with respect to hardware implementations, and demonstrate its capability to learn with reduced synaptic weight dynamic-range. This new learning architecture of CBP is shown in Figure 73.8. In comparison with the error backpropagation (EBP) algorithm, CBP was designed with a clear motivation to avoid the arbitrary and *a priori* assignment of hidden units, and thus avoid identical subspaces in weight-space that may cause convergence problems (Chen and Hecht-Neilson, 1991). In addition, CBP's most important feature is its learning efficiency even with limited weight resolution, unlike EBP which is particularly costly to implement in hardware (Hollis et al., 1990). Further, the theory of self-evolving architecture shows that each added hidden unit potentially reduces the energy level and hence moves the network continuously toward minimum energy level (Duong, T., Cascade Error Projection: An Efficient Learning Algorithm, Ph.D., 1995, University of California, Irvine).

CBP uses the stochastic gradient-descent technique and the self-evolving architecture (Duong et al., 1995; Fahlman and Lebiere, 1990). The process of adding a new hidden unit is based on a number of fixed iterations. Learning is required for the synaptic weights that are related to the new hidden unit and the output bias weights only. However, in this study, we have not optimized the number of iterations that may be required to learn the input-output relationship for the particular problem.

Mathematical Model

We first define some variables as follows:

p is the variable for the number of training patterns, where $p = \{1, \ldots P\}$;
o is the variable for the output components with $o = \{1, \ldots O\}$;
x_0 is the bias input which is kept fixed at 1;
x_j is the input signal with $j = \{1, \ldots, Ni\}$; and
$x_h(l)$ is the output from hidden unit l with $l = \{1, \ldots, n\}$.

Here, Ni represents the input dimension, O the output dimension, and n is the number of added hidden units (or the expanded input space). The energy function can, therefore, be written as:

$$E = \sum_{p=1}^{P} E^p = \sum_{p=1}^{P} \sum_{o=1}^{O} (t_o^p - y_o^p)^2 \tag{73.1}$$

Let T be the target matrix, with a column for each input target pattern, given by:

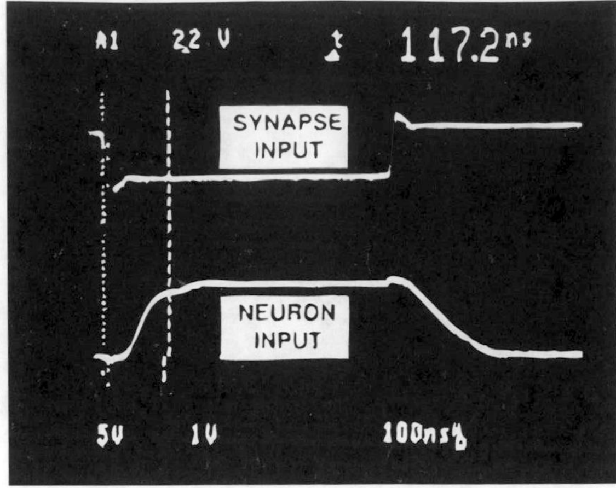

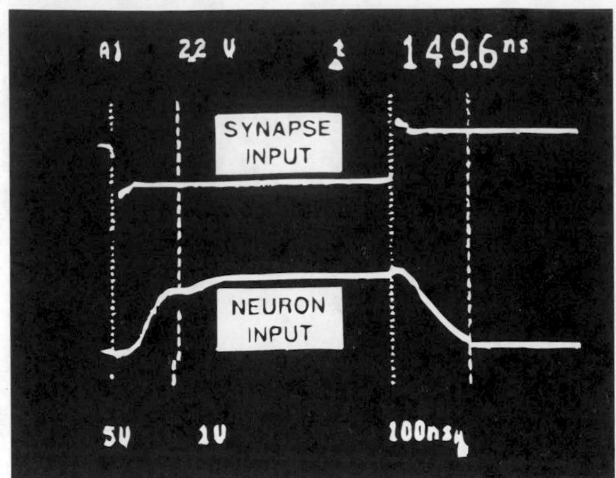

Figure 73.7 Delay through a synapse-neuron circuit as seen through a source follower driving a 20 pF oscilloscope load for rise and fall pulses of the synapse inputs.

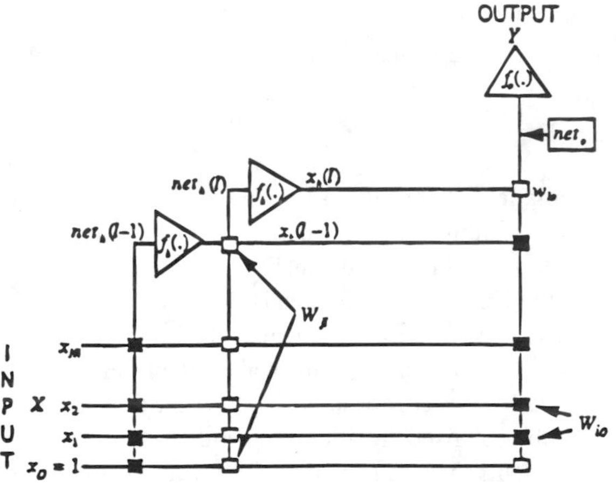

Figure 73.8 A schematic diagram of a cascade backpropagation (CBP) architecture showing added hidden units. Synaptic weights are shown as small rectangles where the filled rectangles signify that they are frozen after completion of training when the next hidden unit is added. (*Source: Fuzzy Logic and Neural Network Handbook*, Ch. 27. 1996. McGraw-Hill, NY. With permission.)

$$T = \begin{bmatrix} t_1^1 t_1^2 \cdots t_1^P \\ \vdots \\ t_o^1 t_o^2 \cdots y_o^P \end{bmatrix}$$

and, the corresponding actual output matrix is:

$$Y = \begin{bmatrix} y_1^1 y_1^2 \cdots y_1^P \\ \vdots \\ y_o^1 y_o^2 \cdots y_o^P \end{bmatrix}$$

Then, with no hidden units in the network, one can calculate the output as:

$$Y = F(WX) \tag{73.2}$$

where $W = W_{io}$ is the set of weights between input and output neurons. The best estimation weight-set of the given energy function (Equation 73.1) in affine space is calculated as:

$$W_{io} = F^{-1}(T)X^{+} \tag{73.3}$$

with X^{+} as the pseudo inverse of X[14], and F^{-1} as an inverse transformation matrix, given by:

$$F^{-1}(T) = \begin{bmatrix} f^{-1}(t_1^1) f^{-1}(t_1^2) \cdots f^{-1}(t_1^P)) \\ \vdots \\ f^{-1}(t_o^1) f^{-1}(t_o^2) \cdots f^{-1}(t_o^P)) \end{bmatrix}$$

The set of weights W_{io} is then kept frozen. Assume that n hidden units are added, and the output is calculated as follows:

$$y_o = f(net_o) \tag{73.4}$$

where

$$net_o = \sum_{l=1}^{n} x_h(l) w_{lo} + \sum_{j=0}^{Ni} x_j w_{jo};$$

and f is the transfer function of the output neuron (termed f_o). Further, for an arbitrary number of hidden units l added,

$$x_h(l) = f(net_h(l)) \tag{73.5}$$

where f is the transfer function of the current hidden neuron (termed f_h and is equal to f_o),

$$net_h(l) = \sum_{k=1}^{l-1} x_h(k) w_{kl} + \sum_{j=0}^{Ni} x_j w_{jl};$$

and

$$x_i = \begin{Bmatrix} 1 & \text{if } i = 0 \\ x_j & \text{if } i < Ni + 1 \\ x_h(l) & \text{if } i \geq Ni + 1 \end{Bmatrix}$$

Let us define:

$$f_h'(l) = \frac{df(net_h(l))}{dnet_h(l)};$$

and

$$f_o' = \frac{df(net_o)}{dnet_o}$$

With η as the learning rate, the stochastic gradient-descent gives the weight update as:

$$\Delta w_{ij} = -\eta \frac{\partial E^P}{\partial w_{ij}} \tag{73.6}$$

where, i and j denote the starting node i and the destination node j. Applying the chain rule to Equation 73.6 for the weights between the hidden and the output neurons, and the bias synapses connected to the output, we get:

$$\frac{\partial E^P}{\partial w_{ij}} = \frac{\partial E^P}{\partial y_o^P} \frac{\partial y_o^P}{\partial net_o^P} \frac{\partial net_o^P}{\partial w_{ij}}$$

which can be written as, (we are only interested in the newly added hidden unit),

$$\frac{\partial E^P}{\partial w_{ij}} = -2(t_o^P - y_o^P) f_o'^P x_h^P(n) \tag{73.7}$$

Using Equation 73.7, we can rewrite Equation 73.6 explicitly with a first order and a second order term (Parker, 1987) as:

$$\begin{aligned} \Delta w_{no}(k) &= \eta x_h^P(n)(t_o^P - o_o^P) f_o'^P \\ &\quad - \alpha \Delta w_{no}(k-1) \qquad \text{with } 0 < \alpha < 1 \end{aligned} \tag{73.8}$$

which gives the weight updates for the synaptic components between the currently added hidden unit n and the output o as shown in Figure 73.8. Similarly, the updates for the weights between the inputs (including expanded inputs and the bias weight at the currently added hidden unit) and the current hidden unit are given by:

$$\frac{\partial E^P}{\partial w_{ij}} = \frac{\partial net_h^P}{\partial w_{ij}} \frac{\partial f_h(net_h^P)}{\partial net_h^P} \sum_{o=0}^{0} \frac{\partial net_o^P}{\partial f_h(net_h^P)} \frac{\partial y_o^P}{\partial net_o^P} \frac{\partial E^P}{\partial y_o^P}$$

which then can be written as,

$$\frac{\partial E^P}{\partial w_{ij}} = -2x_i^P f_h'^P(n) \sum_{o=0}^{0} w_{no} f_o'^P(t_o^P - y_o^P)$$

This equation is similarly written in an explicit form with a first order and a second order term as:

$$\begin{aligned} \Delta w_{in}(k) &= \eta x_i^P f_h'^P(n) \sum_{o=0}^{O} w_{no}(t_o^P - y_o^P) f_o'^P \\ &\quad + \alpha \Delta w_{in}(k-1) \end{aligned} \tag{73.9}$$

with $0 < \alpha < 1$ for the weight components. The change in learning rate after addition of each new hidden unit is given by:

$$\eta_{new} = \eta_{old} - \frac{c}{\#\ of\ iterations} \tag{73.10}$$

with η_{new} as the current learning rate, η_{old} as the previous learning rate, and c as a constant. When α is zero, we obtain the first order gradient descent and if α is a non-zero constant, then the two terms in both the Equations 73.8 and 73.9 contribute to the weight updates, and the second order gradient-descent is obtained.

Quantization of Weight Space

Because of the limited quantization of weight space, the value ΔW_{ij} of the weight update will have to be modified to ΔW_{ij}^* to fit with the available quantization. The closeness between them will depend on the weight resolution available. Let *nbit* be the bit resolution of the weight space. Then the maximum level of weight space will be MAXLEVEL $= 2^{(nbit)} - 1$. We define *stepsize*(n) to be a step size for the weight space of a hidden unit n. The *stepsize*(n) can be generated from a constant *stepsize*(0) which is fixed before starting the learning process. The *stepsize*(n) is obtained as follows:

$$stepsize(n) = \beta E_{n-1} \tag{73.11}$$

where $\beta = stepsize(0)/E_0$ is a constant and E_0 is the energy of the network with no hidden units and bias inputs added (includes only the input-to-output weights calculated using pseudo-inverse technique), i.e.,

$$E_0 = \sum_{p=1}^{P} E^P(W_{io}; X^P, Y^P)$$

and E_{n-1} is the energy of the network with $n - 1$ hidden units added. There are two ways to obtain number of steps for ΔW_{ij}:

one is the round-off technique where the number of steps is calculated as follows:

$$\#stepi = \left\{ \begin{array}{l} (int)\left(\dfrac{\Delta W_{ij}}{stepsize(n)} + 0.5\right) \text{ if } \Delta W_{ij} \geq 0 \\ (int)\left(\dfrac{\Delta W_{ij}}{stepsize(n)} - 0.5\right) \text{ if } \Delta W_{ij} < 0 \end{array} \right\} \tag{73.12}$$

and the other is the truncation technique where number of steps is calculated as given below:

$$\#stepi = (int)\left(\frac{\Delta W_{ij}}{stepsize(n)}\right) \tag{73.13}$$

Before updating the candidate weight using ΔW^*_{ij}, one must ensure that the final quantized weight will not exceed the limit provided as MAXLEVEL. Therefore, first the previously stored weight is converted to an equivalent number of *steps,* which is given by:

$$\#stepa = (int)\left(\frac{W_{ij}}{stepsize(n)}\right) \tag{73.14}$$

Then,

$$\Delta W^*_{ij} =$$

$$\left\{ \begin{array}{ll} 0 & \text{if } |\#stepi + \#stepa| > \text{MAXLEVEL} \\ stepsize(n)(\#stepi) \text{otherwise} & \end{array} \right\} \tag{73.15}$$

Procedure for Learning in Hardware

A clear procedure for the learning algorithm, used later for solution of a 6-bit parity problem as an illustrative example, is now presented. Based on the mathematical analysis of the EBP learning algorithm (Parker, 1987), the weight update (consisting of the first and second order terms) can be performed by incorporating either the first order term only; or the summation of the two terms to obtain the second order effect as well. The idea of this development effort, of course, is to make the algorithm implementable in hardware given the limited synaptic weight resolution.

When considering the transfer characteristics of a neuron, either a mathematical equivalent of the sigmoid such as a logistic function is considered, or a look-up table is constructed. A look-up table requires step updates and hence a quantization of the values. It has been shown that such a neuron quantization is not as sensitive as a synaptic quantization for the convergence properties of the circuit (Hoehfeld and Fahlman, 1992; Hollis et al., 1990). In addition, the number of synapses on a chip is much higher than the number of neurons. Thus, it is important to keep the synapse quantization as high as possible, commensurate with proper learning. Therefore, in our study, the effect of neuron quantization has not been considered. On the other hand, synaptic weight quantization is known to affect the sensitivity of learning to a larger extent, and the synaptic weights in hardware may be limited in their resolution anywhere from 5 to 10 bits.

Weight Update Issues

The weight update Δw_{ij} is obtained as an analog number. However, the weight space is discrete in a hardware, based on hybrid digital-analog synapse designs as is the case with our MDAC approach described earlier. Therefore, to update the weight, the value Δw_{ij} must be converted into the number of steps by which the weight is to be updated. The conversion from an analog level to the respective limited discrete level would, in general, result in a partial loss of precision.

As noted earlier, in A/D conversion techniques, there are typically two conversion schemes. One is the "round-off" technique and the other is the "truncation." In our simulation, we have compared these two schemes for their effectiveness in learning of the selected 6-bit parity problem. We find that the two schemes provide different results as illustrated later in this paper. During the learning phase, the constraint of the maximum value of the discrete level limited by the available weight resolution must also be considered. For example, with an 8-bit synapse, the total number of discrete levels should not exceed 255.

Some of the salient features of our new learning algorithm are:

1. The step size is dynamically changed after addition of each hidden unit. The change is based on the level of energy left over with the previous hidden unit (as a ratio of the original level of energy). In general, with the addition of a new hidden unit and subsequent training of the respective weights, the energy of the network decreases, resulting in smaller step size for the next stage of the added hidden unit. Even though, in the present simulation for the 6-bit parity problem, the maximum number of hidden units added was limited to twenty irrespective of whether each additional hidden unit decreased the energy or not, or whether the network converged to the right solution, the algorithm is not limited in that respect. However, it may be noted that the more hidden units are added, the slower will be the signal propagation.
2. The input to a neuron can be adjusted using two variables beside the input to the synapse itself. One is the weight value which can be updated during training, and the other is the bias voltage applied to the synapse input transistor (V_d in Figure 73.3) It is this latter feature that allows for easy adjustment of the step size and, more importantly, promotes convergence with lower quantization of synaptic weights. Furthermore, this new design will provide independent, programmable, bias voltages to rows of synapses connected to each hidden unit.

73.6 Six-Bit Parity Problem

To assess the effectiveness of our methodology, and for easy comparison with other work reported in the literature, we selected for study the 6-bit parity problem using our new CBP learning algorithm. The 6-bit parity problem has 64 discrete patterns to be classified. The neural network architecture has 6 inputs and one bias line, directly connected to one output line through seven programmable weights. The procedure used for training is as follows:

1. Calculate the six weight values for the input-output connection weights using the pseudo-inverse relationship. In this particular case, the solution of the pseudo-inverse calculation is very close to zero. Therefore, we have arbitrarily set all the weights to 0.5. These weights are then kept frozen throughout.
2. Provide the input patterns (with bias weights not connected) and evaluate the respective output errors and calculate the energy E(0). If the errors are within a given tolerance, then the training is complete. If not, proceed further.
3. Set a learning rate, $\eta = 3.5$, and $\alpha = 0$ for first order effects and $\alpha = 0.9$ for second order effects, and a weight step size given by:

$$stepsize(0) = 0.015 * 2^{(8-nbit)};$$

where *nbit* = synaptic weight resolution in bits.

4. Add a new hidden unit along with randomly selected input and output weights, including the bias weights. These weights have to be converted to quantized levels of weights where each weight = *stepsize*(*n*)*(*#stepa*.) Further, *#stepa* should be an integer given by either the round-off or the truncation method.
5. Again provide the inputs, measure the outputs, and evaluate the new error values for all the input patterns to ascertain if training is complete. Otherwise, continue the training process.
6. $\eta = \eta - 3.5/10{,}000$, and *stepsize*(*n*) is given by Equation 73.11.
7. Apply a random input pattern to the network.
8. Calculate the change in weights ΔW_{ij} using Equations 73.8 and 73.9.
9. Calculate the number of steps required, *#stepi* using Equations 73.12 and 73.13.
10. The total number of steps, *#step*(*total*) = *#stepa* + *#stepi*. If the absolute value | *#step*(*total*) | > MAXLEVEL then set ΔW^{*}_{ij} (= *#stepi* * *stepsize*(*n*)) to 0. Otherwise, update W_{ij} and *#stepa*.

This procedure will update all the weights for the added hidden neurons and the output bias weights.

11. Go to 7, until the required application of number of iterations of the random patterns is completed. The number of iterations can be decided depending upon the requirement of the problem and the time available. In our case, we used 6000 iterations as an outer loop, and 64 iterations as an inner loop for each pattern.
12. Calculate the error for all the patterns and evaluate for completion of training. If complete, stop training, otherwise, calculate the energy E(n) and go to 4.
13. If the number of added hidden units is greater than 20, give up and quit.

Cascade Backpropagation (CBP) Simulations

Using the above procedure, simulations for hardware were performed and the mean error and the standard deviation of the error were obtained for the four cases, two with only the first order term, with both round-off and truncation methods of conversion, and similarly the other two with the second order term included. As expected, the simulation showed that including the second order term made the errors go down considerably compared to that with just the first order term. As a result, this led to an acceptable solution with reduced synaptic weight resolutions. The mean error and the standard deviation curves for these four cases as an average of 10 runs are shown in Figures 73.9 (a-d) and 73.10 (a-d), respectively. Overall, the method showed tremendous tolerance to reduced weight resolution and that with second order term included, the hardware with ≥7-bit resolution performed as well as that with full floating point accuracy with about 12 neurons added as hidden units. In addition, with the second order term included, the results with 6- and 7-bit resolution had close to 100% correctness, and even 5-bit resolution weights provided 80 to 90% correctness. Table 73.1 summarizes these results of weight quantization and the correctness of the solution in the four cases (out of 64 patterns) (Duong et al., 1995).

73.7 Conclusions

In conclusion, demonstration of high-speed low-power neuron and synapse circuits enables the realization of a 3-D neural network architecture for high speed pattern recognition. Measured

Table 73.1 Percent of correct CBP learning runs for the 6-bit parity problem with variation of synaptic weight resolution, using first order and second order terms in learning algorithm, with round-off (RO) and truncation (Tr) modes of weight value conversion

Weight Resolution	Percent correct, First order (RO)	Percent correct, First order (Tr)	Percent correct, Second o.(RO)	Percent correct, Second o.(Tr)
5-bit	40%	10%	90%	80%
6-bit	90%	80%	100%	90%
7-bit	100%	80%	90%	100%
8-bit	100%	100%	100%	100%
9-bit	100%	100%	100%	100%
Floating point	100%	100%	100%	100%

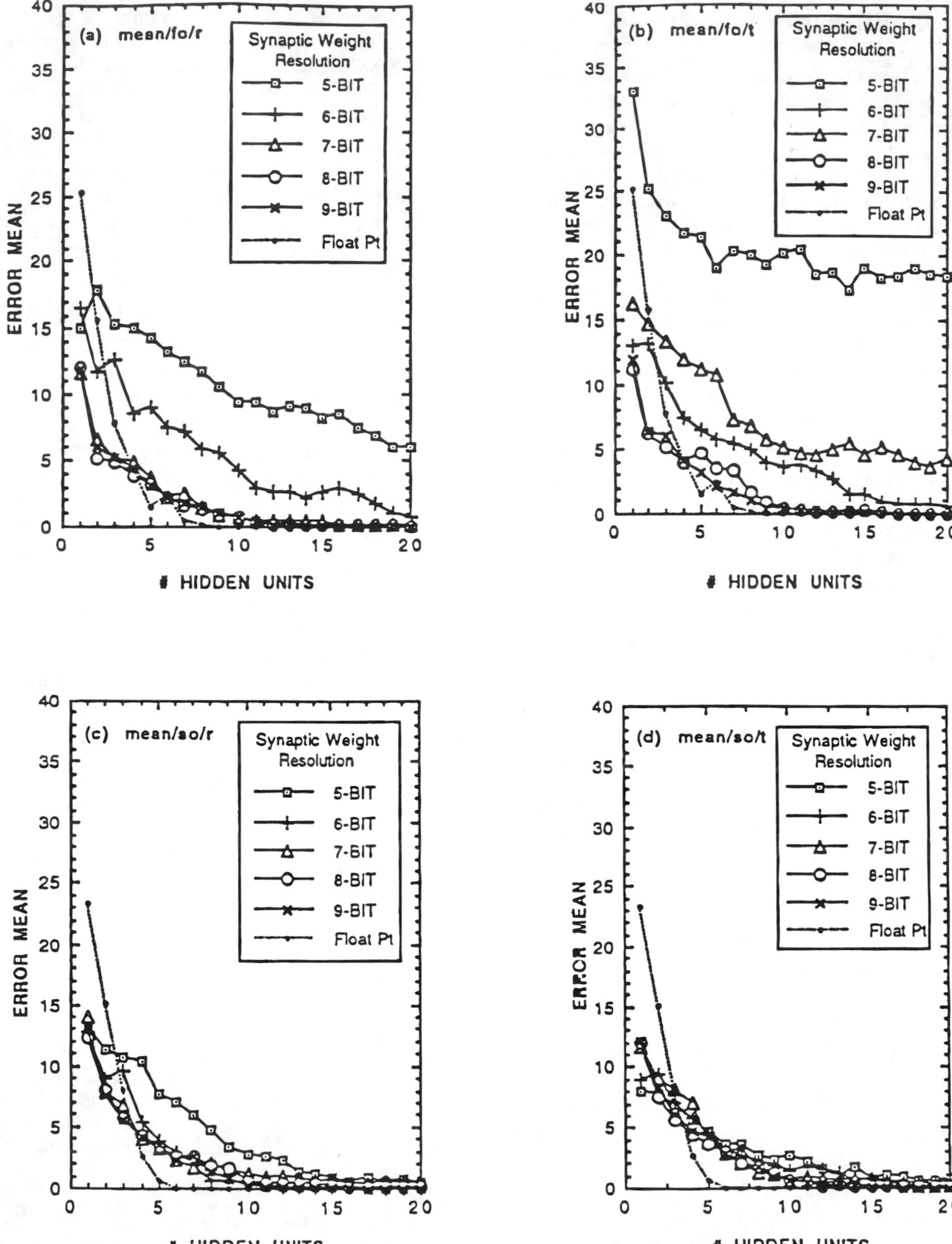

Figure 73.9 Mean error as a function of added hidden units with synaptic weight resolution as a parameter for the cascade backpropagation (CBP) learning simulation for hardware with (a) only the first order weight update term and round-off (*fo*/*r*) conversion; (b) first order term and truncation (*fo*/*t*) conversion; (c) including second order term and round-off (*so*/*r*) conversion; and (d) second order term and truncation (*so*/*t*) conversion methods. (*Source*: *Fuzzy Logic and Neural Network Handbook*, Ch. 27. 1996. McGraw-Hill, NY. With permission.)

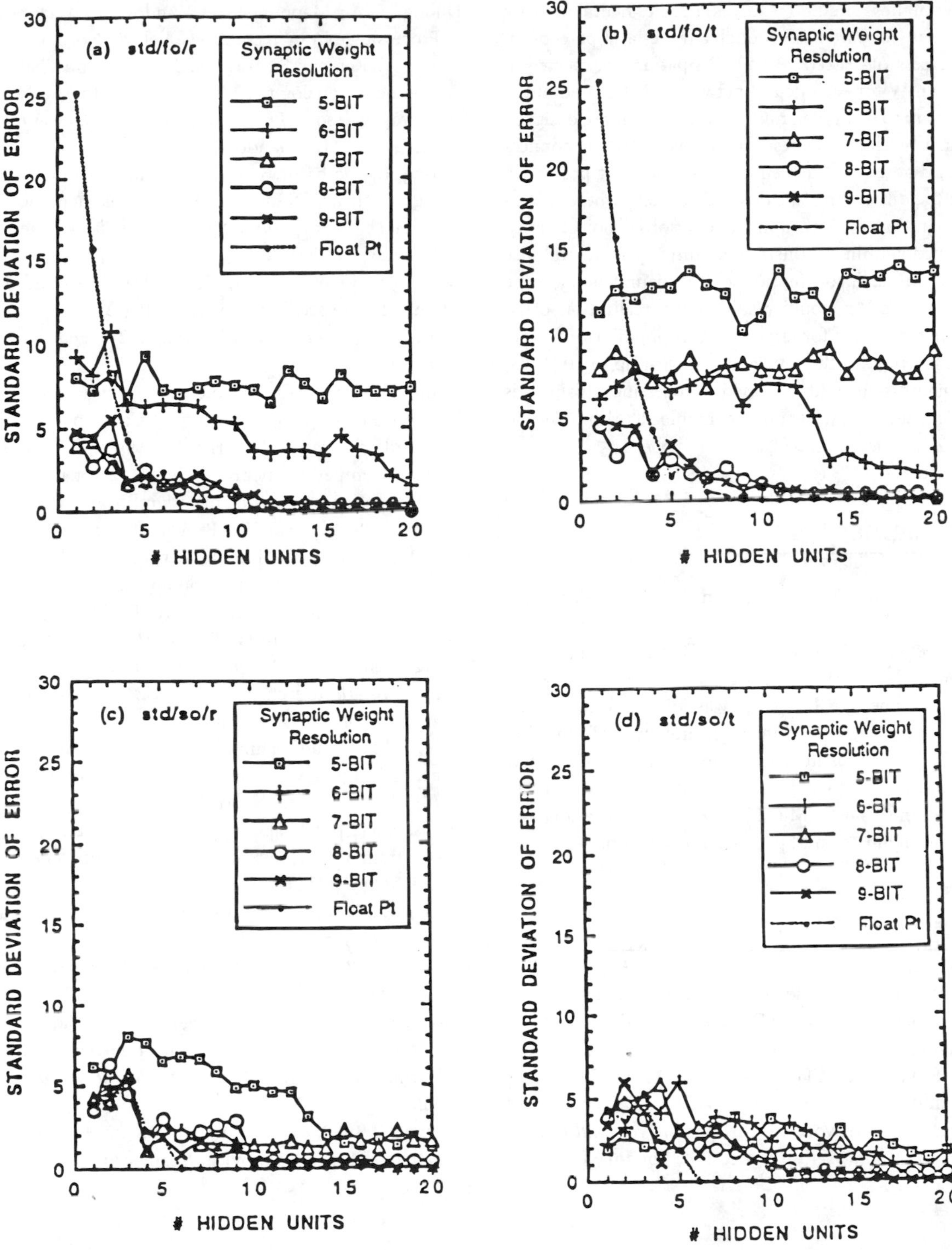

Figure 73.10 Standard deviation as a function of added hidden units with synaptic weight resolution as a parameter for the cascade backpropagation (CBP) learning simulation for hardware with (a) only the first order weight update term and round-off (*fo*/*r*) conversion; (b) first order term and truncation (*fo*/*t*) conversion; (c) including second order term and round-off (*so*/*r*) conversion; and (d) second order term and truncation (*so*/*t*) conversion methods. (*Source*: *Fuzzy Logic and Neural Network Handbook*, Ch. 27. 1996. McGraw-Hill, NY. With permission.)

response times of these circuits ($\leq$150 ns) implies neural network processing speeds (forward pass through a three-layer perceptron network) in excess of 4 MHz. Power dissipation is expected to be less than 30 mW for each neural network IC. The neural ICs with 64 inputs, up to 64 hidden units, and 6 outputs are designed for stacking in a 64-chip cube, resulting in over 1 trillion connections per second potential. Such enormous processing power can be applied to the difficult missile defense problem where spatio-temporal recognition of both point and resolved objects must be accomplished within a constrained time, power, and size budget. The eventual incorporation of the neural image cube into a fast frame seeker would result in the realization of an extremely compact system for target acquisition, discrimination, tracking, and homing. In the future, stacking of up to 1024 ICs, with a subsequent multi-fold increase in processing capability is anticipated. The new hardware implementable algorithm will aid in more efficient processing of data at high speed.

Acknowledgments

The authors wish to thank Dr. D. Duston, Mr. L. Lome, and Dr. C. Lau for encouragement and useful discussions, and gratefully acknowledge technical discussions with David Ludwig, Harry Langenbacher, and Tim Shaw during the course of this work. The research described in this paper was jointly performed by the Center for Space Microelectronics Technology, Jet Propulsion Laboratory, California Institute of Technology, and the Irvine Sensors Corporation, and was jointly sponsored by the Office of Naval Research, the Ballistic Missile Defense Organization, and the National Aeronautics and Space Administration.

References

Carson, J. 1991. On-focal plane array feature extraction using a 3-D artificial neural network (3DANN), *Proc. SPIE*, vol. 1541, *Infrared Sensors: Detectors, Electronics, and Signal Processing*, Jayadev, T. S. J., ed., Part I:141–144, Part II:227–231.

Chen, A. M. and Hecht-Neilson, R. 1991. On the geometry of feedforward neural network weight spaces, *Proc. 2nd IEE Int. Conf. Artificial Neural Networks*, pp. 1–4, London, UK.

Duong, T. A., Mapping Pixel Windows to Vectors for Parallel Processing, NASA Tech. Briefs, Vol. 20, No. 3, pp. 4a–6a, 1996.

Duong, T., Kemeny, S., Tran, M., Daud, T., and Thakoor, A. 1994. Low power analog neurosynapse chips for a 3-D "sugarcube" neuroprocessor, *Proc. IEEE Int. Conf. Neural Networks*, III:1907–1911, Orlando, FL.

Duong, T., Eberhardt, S., Daud, T., and Thakoor, A. 1996. Learning in Neural Networks: VLSI Implementation Strategies, *Fuzzy Logic and Neural Network Handbook*, Chen, C. H., ed., McGraw-Hill, NY.

Duong, T., Kemeny, S., Tran, M., Daud, T., and Thakoor, A. 1995. High speed low power analog ASICs for a 3-D neuroprocessor, *Proc. SPIE/IST*, vol. 2424, February 4–10, San Jose, CA.

Duong, T., Cascade Error Projection: An Efficient Learning Algorithm, Ph.D., 1995, University of California, Irvine.

Eberhardt, S., Daud, T., Kerns, D., Brown, T., and Thakoor, A. 1991. Competitive neural architecture for hardware solution to the assignment problem, *Neural Networks*, 4(4):431–442.

Eberhardt, S. P., Tawel, R., Brown, T., Daud, T., and Thakoor, A. 1992. Analog VLSI neural networks: implementation issues and examples in optimization and supervised learning, *IEEE Trans. Industrial Electronics*, 39(6):552–564.

Fahlman, S. and Lebiere, C. 1990. The Cascade Correlation Learning Architecture, *Advances in Neural Information Processing Systems—II*, Touretzky, D., ed., 524–532, Morgan Kaufman, San Mateo, CA.

Hoehfeld, M. and Fahlman, S. 1992. Learning with limited numerical precision using the cascade correlation algorithm, *IEEE Trans. Neural Networks*, 3:602–611.

Hollis, P., Harper, J., and Paulos, J. 1990. The effects of precision constraints in a backpropagation learning network, *Neural Computation*, 2:363–373.

Hopfield, J. 1990. The effectiveness of analogue "neural network" hardware, *Network*, 1(1):27–40.

IEEE 1994. *Proc. IEEE Int. Conf. Neural Networks*, Vols. I–VII, Orlando. FL.

Parker, D. B. 1987. Optimal algorithms for adaptive networks: second order backpropagation, and second order Hebbian learning, *Proc. IEEE 1st Int. Conf. Neural Networks*, II:593–600, San Diego, CA.

Strang, G. 1988. *Linear Algebra and its Applications*, 3rd ed., Harcourt Brace Jovanovich, San Diego, CA.

74

Simulated Annealing, Boltzmann Machine, and Hardware Annealing

Tony H. Wu
University of Southern California

Bing J. Sheu
University of Southern California

74.1 Simulated Annealing 1003

74.2 Boltzmann Machine 1004
Characteristics of Boltzmann Machine • Learning Algorithm

74.3 Hardware Annealing on Hopfield Networks for Optimization 1005
Starting Voltage Gain for the Cooling Schedule • Final Voltage Gain for the Cooling Schedule

74.4 Hardware Annealing on Cellular Neural Networks 1007
Application of Hardware Annealing • Simulation Results

Optimization is an important subject in solving many scientific and engineering problems. One conventional searching technique for finding the global minimum is to use gradient descent, which finds the direction for the next iteration from the gradient of the objective function. For complicated problems, the gradient descent technique often gets stuck at a local minimum where the objective function has surrounding barriers. In addition, the complexity of most combinatorial optimization problems increases dramatically with the problem size and makes it very difficult to obtain the globally optimal solution in a reasonable computational time. Complicated optimization problems could also be solved by using the branch-and-bound searching method, or by relaxing some constraints and solving the simplified problems. In this section, several attractive methods, such as simulated annealing, Boltzmann machine, and hardware annealing are described and they have been reported to help the solutions of the optimization problems to escape from local minima (Lee and Sheu, 1991; Aarts and Korst, 1989).

74.1 Simulated Annealing

Simulated annealing can be applied to the software computation of artificial neural networks with the following procedure (Lee and Sheu, 1991):

Step 1. Start with a high temperature and an initial state as the current state.

Step 2. Choose at random a new state from the neighborhood of the current state.

Step 3. Calculate the difference in energy, ΔE, between the new state and the current state.

Step 4. If ΔE is less or equal to zero, replace the current state by the new state. Otherwise, accept the new state only if $e^{-\Delta E/kT}$ is greater than a random number drawn from a uniform distribution on [0, 1].

Step 5. If equilibrium is not closely established at this temperature, go to Step 2.

Step 6. If the system is frozen, terminate the annealing process. Otherwise, decrease the temperature and return to Step 2.

Since it usually takes a lot of time at "Step 2" to compare all possible states, a specific perturbation rule is often used. The perturbation is usually called artificial noise in software computation.

Although simulated annealing is quite slow, it is easy and simple to apply to new problems. The simulated annealing technique can help recursive neural networks such as Hopfield nets to escape from local minima by replacing the transfer function of the neuron from a sigmoid function to the Boltzmann distribution function. In software simulation, the Hopfield network operation is described at two consecutive time steps during each iteration cycle. At the first time step, input signals to the neurons are summed up; while at the second step, the neuron outputs are updated. The update rule for the original Hopfield networks is

$$V_i = g(u_i), \tag{74.1}$$

and that for the Boltzmann machine is

$$V_i = \frac{1}{1 + e^{-u_i/T}} \tag{74.2}$$

0-8493-8343-9/97/$0.00+$.50

Here, $g(\cdot)$ is the input-output transfer function of the neuron, T is the effective temperature for Boltzmann distribution and includes the effect of the Boltzmann constant, as shown in Figure 74.1. At the steady state, the relative probability of state $S1$ to state $S2$ in Boltzmann distribution is determined by energy differences of the two states,

$$\frac{P_{S1}}{P_{S2}} = e^{-(E_{S1}-E_{S2})/T}. \tag{74.3}$$

Here, E_{S1} and E_{S2} are the energy levels for states $S1$ and $S2$, respectively. This update rule allows the network to escape from local minima in the energy well.

This algorithm was mainly developed for the case where the space D is countable. It had recently been extended to the case where D is the n-cube $(-1, 1)^n$ and this method is called the diffusion machine (Wong, 1989). The diffusion machine can be viewed as a continuous-state, continuous-time Hopfield network and noise is injected at each node of the network. Thus, the simulated annealing can be achieved by a suitable cooling schedule. An analog realization of the diffusion machine would be faster than a simulated version and would not involve a discrete-time approximation of a diffusion. There has been an attempt to realize the diffusion machine using MOSFET's operating in their subthreshold region.

Simulated annealing is one of the popular approaches which is widely applicable to the combinatorial optimization problems. Simulated annealing is a generalization of iterative improvement methods in that it accepts, with a non-zero but gradually decreasing probability, deterioration in the cost function of an optimization problem. A solution from simulated annealing is close to the global minimum within a polynomial upper bound for the computational time and is independent of the initial condition. Mean field annealing is a deterministic approximation to simulated annealing. It analytically approximates the relevant Boltzmann distribution and can execute at approximately 50 times faster than simulated annealing for some problems. The quality of the solutions produced by mean field annealing is highly dependent on the choice of the initial temperature, the cooling speed, and the final temperature. So far, simulated annealing has been successfully reported in layout generation of VLSI circuits, test pattern generation, integrated-circuit delay reduction, image segmentation, and noise filtering in image processing. Since the number of iterations at a given temperature and the cooling rate of the temperature should be compromised in order to speed up the convergence process, a very large computation time is usually required in software simulation.

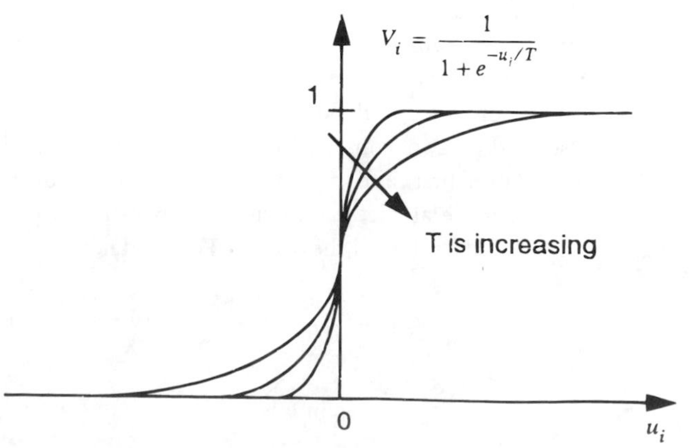

Figure 74.1 Analogy between the annealing temperature of a Boltzmann machine and the gain of an electronic neuron.

74.2 Boltzmann Machine

The Boltzmann machine is a parallel network which can solve the problem of constraint satisfaction tasks with a large number of weak constraints (Rumelhart et al., 1989). The machine updates the connection weight values so that it can produce the examples with the same probability distribution as the shown examples. It computes the best possible solution or the lowest cost function even though some constraints are violated. The quality of a solution is then determined by the total cost of all violated constraints. The distributed computation abilities of the Boltzmann machine make it suitable for realization by standard VLSI technologies and by future technologies such as molecular electronics and photonics.

Characteristics of Boltzmann Machine

The Boltzmann machine consists of the processing elements, *unit*, and the bidirectional *links* which connects two units. The unit takes one of the two states such as *on* or *off* and the weight of the link is a real value of either polarity. The weight value of the link is symmetric so that the weight value w_{ij} connecting from the unit j to the unit i is same as w_{ji} which connects from the unit i to j. Units are divided into two different kinds: visible units and hidden units. As shown in Figure 74.2, the visible units are used for interfacing the network operation and the

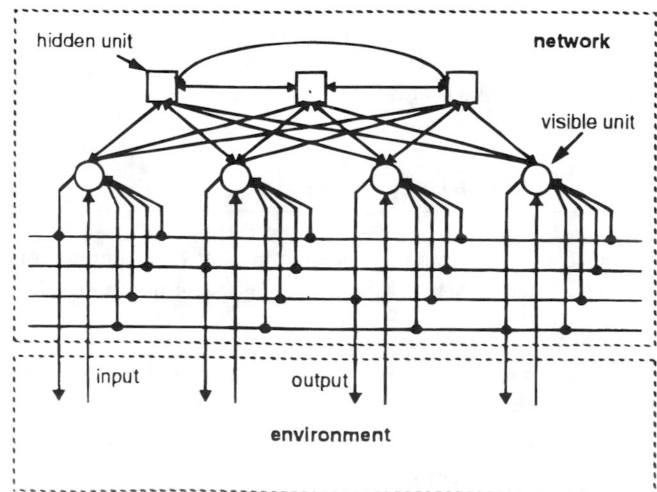

Figure 74.2 Conceptual block diagram of a Boltzmann machine.

environmental inputs and outputs. The hidden units are internally used to build the characteristics of the underlying constraints for the network. By using the hidden units, a higher-order constraints satisfying problems can be reduced to first- and second-order constraints throughout the whole set of units. The input neuron state is always known while the output neuron state is known only in the learning phase.

The energy of the global configuration is defined as,

$$E = -\frac{1}{2}\sum_{i=1}^{N}\sum_{j=1,j\neq i}^{N} w_{ij}s_i s_j + \sum_{i=1}^{N}\theta_i s_i, \tag{74.4}$$

where s_i and θ_i are the state and the threshold of the unit i, respectively. If the unit i is on, s_i is 1. Otherwise, s_i is 0. The unknown states are determined by searching for the global minimum of the energy function. Simulated annealing techniques can be applied because the problem is of combinatorial optimization nature. Since the weights are symmetric, the energy difference between the on-state and off-state of the unit k can be locally determined by,

$$\Delta E_k = \sum_i w_{ki}s_i - \theta_k. \tag{74.5}$$

Thus, the state should be the on-state if the summing input from other units is larger than the threshold value, in order to decrease the energy. The probability of unit k being on is

$$p_1 = \frac{e^{-E_1/T}}{(e^{-E_0/T} + e^{-E_1/T})} = \frac{1}{1 + e^{-\Delta E_k/T}} \tag{74.6}$$

while the probability of unit k being off is $p_0 = 1 - p_1$. In addition, the relative probability of two global states, α and β, has the form of the Boltzmann distribution such as,

$$\frac{P_\alpha}{P_\beta} = e^{-(E_\alpha - E_\beta)/T}, \tag{74.7}$$

where $P_{\alpha(\beta)}$ $E_{\alpha(\beta)}$ are the probability and the corresponding energy of the global state $\alpha(\beta)$, respectively. Notice that the equilibrium state is only determined by the energy difference regardless of the path followed in reaching the equilibrium.

Learning Algorithm

The learning algorithm for the Boltzmann machine can be performed with a cost function that is based on the information theoretic measure of disagreement between the internal model of the network and the environment (Rumelhart et al., 1989) as follows,

$$G = \sum_\alpha P^+(V_\alpha)\ln\left(\frac{P^+(V_\alpha)}{P^-(V_\alpha)}\right), \tag{74.8}$$

where $P^+(V_\alpha)$ is the probability of the state α of the input and output neurons when their states are clamped by the environment, and $P^-(V_\alpha)$ is the corresponding probability for the freely running network when the output neurons are not clamped. The cost function G is greater than 0 except when $P^+(V_\alpha)$ equals $P^-(V_\alpha)$. Through minimization of G by the gradient descent method, the weight update rule can be derived as follows. From (74.4) and (74.7), the partial derivative of G with respect to w_{ij} can be expressed as,

$$\frac{\partial G}{\partial w_{ij}} = -\frac{1}{T}(p_{ij}^+ - p_{ij}^-), \tag{74.9}$$

where p_i^+, and p_{ij}^- are the average probability, measured as the equilibrium, of both units i and j being on-state with and without environmental driving, respectively. The weight updating rule is the application of a Hebbian and an anti-Hebbian rule. Thus, the amount of weight update is,

$$\Delta w_{ij} = \eta(p_{ij}^+ - p_{ij}^-) + noise, \tag{74.10}$$

where η is the learning rate constant. The gradient descent technique does not guarantee the finding of the global minimum. The added noise will help to reach the global minimum solution. Instead of using the update amount proportional to the probability difference, Ackely et al. [1985], used a constant value which can be decremented or incremented according to the sign of difference of two probabilities.

74.3 Hardware Annealing on Hopfield Networks for Optimization

The parallel annealing technique for a Boltzmann machine is most suitable for VLSI neurocomputing. Changing the temperature of the probability function for a Boltzmann machine is equivalent to varying the voltage gain of the neurons. Thus, the cooling process in a Boltzmann machine is equivalent to the neuron gain increase process in an analog neural network. The neuron gain in electronic neural circuits can be updated continuously, while the annealing temperatures in software computation on digital computers are always updated in the discrete fashion. Hence, the final results of electronic neural circuits after hardware annealing can be guaranteed to be the optimal solutions, which are in sharp contrast to the results of simulated annealing with approximated convergence on digital computers.

Starting Voltage Gain for the Cooling Schedule

At a very high temperature, all metal atoms lose the solid phase so that they position themselves randomly according to statistical mechanics. An important quantity in metallurgic annealing is the lowest temperature that still could provide enough energy to completely randomize the metal atoms; equivalently, the highest neuron gain to make an electronic neural network escape from local minima. The Hopfield neural-based analog-to-digital (A/D) converter (Tank and Hopfield, 1986) was used for the illustration purpose because the optimal solutions are always known. Figure 74.3 shows the circuit schematic diagram of a Hopfield neural network for a 4-bit analog-to-digital conversion. Electronic neurons are made of operational amplifiers while synapse weights are implemented with equivalent resistances. If the neurons consist of high-gain amplifiers, their outputs will saturate at 0 V and -1 V due to the positive feedback nature of the network. However, given a very low neuron gain, the network could lose the well-defined state.

Assume that all neurons operate in the linear region. By Kirchoffs Current Law (KCL), the governing equation for the ith neuron in the Hopfield network is given as

$$C_i \frac{du_i(t)}{dt} + T_i u_i(t) = \sum_{j=1, j\neq i}^{N} T_{ij} v_j(t) + I_i(t), \quad (74.11)$$

where T_{ij} is the conductance between the ith and j^{th} neurons and T_i and C_i are the equivalent input conductance and input capacitance of the ith neuron. The input bias current $I_i(t)$ is provided by the inner products of the reference voltage V_R and the input voltage V_S with the corresponding weight conductances. Here, $u_i(t)$ is the input voltage to the ith neuron and $v_j(t)$ is the output voltage of the jth neuron.

By taking the Laplace transform, Equation 74.11 becomes

$$(sC_i + T_i)U_i(s) = \sum_{j=1, j\neq i}^{N} T_{ij} V_j(s) + I_i(s) + P_i, \quad (74.12)$$

where P_i is a constant and $U_i(s)$, $V_i(s)$, and $I_i(s)$ are transformed variables of $u_i(t)$, $v_i(t)$, and $I_i(t)$, respectively. If all neurons are assumed to operate in the nonsaturated region with the transfer function being $A(s)$ and to have the bandwidth much larger than T_i/C_1, then

$$V_i(s) = A_i(s) \cdot U_i(s). \quad (74.13)$$

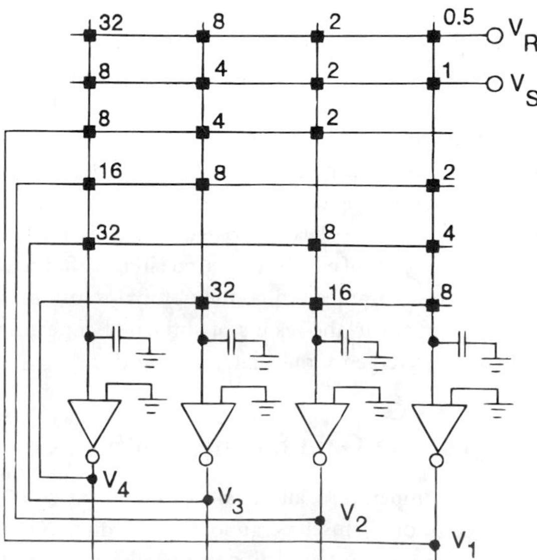

Figure 74.3 Circuit schematic diagram of a Hopfield neural network for a 4-bit analog-to-digital conversion.

The system equation can be expressed as $\mathbf{BV} = \mathbf{F}$ with the matrix $\mathbf{B}$ being an $N \times N$ matrix and $\mathbf{V}$ and $\mathbf{F}$ being $N \times 1$ vectors,

$$\mathbf{B} = \begin{pmatrix} -\frac{sC_1 + T_1}{A_1} & T_{12} & T_{13} & \cdots & T_{1N} \\ T_{21} & -\frac{sC_2 + T_2}{A_2} & T_{23} & \cdots & T_{2N} \\ \vdots & \vdots & \cdots & \cdots & \vdots \\ T_{N1} & T_{N2} & \cdots & \cdots & -\frac{sC_N + T_N}{A_N} \end{pmatrix}, \quad (74.14)$$

$$\mathbf{V} = [V_1, V_2, \ldots, V_N]^T, \quad (74.15)$$

and

$$\mathbf{F} = [-I_1 - P_1, \ldots, -I_N - P_N]^T. \quad (74.16)$$

The sum and product of eigenvalues of the system matrix $\mathbf{B}$ are

$$\sum_{i=1}^{N} \lambda_i = -\sum_{i=1}^{N} \frac{sC_i + T_i}{A_i} \quad (74.17)$$

and

$$\prod_{i=1}^{N} \lambda_i = \det(\mathbf{B}), \quad (74.18)$$

respectively, where det ($\mathbf{B}$) is the determinant of matrix $\mathbf{B}$. If the voltage gain of the neurons is sufficiently large, which is the same condition used in Hopfield's analysis, Equations 74.17 and 74.18 become

$$\sum_{i=1}^{N} \lambda_i \approx 0 \quad (74.19)$$

and

$$\prod_{i=1}^{N} \lambda_i = \det(\mathbf{B}) \neq 0. \quad (74.20)$$

With the constraint that $T_{ij} = T_{ji}$, all eigenvalues will lie on the real axis of the s-plane. Thus, at least one positive real eigenvalue exists. It makes the neuron outputs saturated at extreme values of 0 V or -1 V.

Figure 74.4 shows the radius of the eigenvalues determined from the Gerschgorin Theorem,

$$\left|\lambda + \frac{sC_i + T_i}{A_i}\right| = \sum_{j=1, j\neq i}^{N} |T_{ij}| \quad \text{for all } i. \tag{74.21}$$

Notice that A_i and T_i are always positive. To assure that the network contains positive feedback action, there should be at least one eigenvalue whose real part is positive. The lowest neuron gain (A_N) which satisfies the above condition can be determined from

$$A_N = \max\left[\frac{T_i}{\sum_{j=1, j\neq i}^{N} |T_{ij}|} \quad \text{for } 1 \leq i \leq N\right]. \tag{74.22}$$

The subscript N denotes the number of neurons operating in the nonsaturated region. The above derivation is based on the condition that the eigenvalue lies on a circle. Since validity of this condition is dependent upon the maximum real value of the eigenvalues of matrix **B** determined by resistive network $\{T_{ij}\}$, the above gain requirement is a sufficient condition that the neurons in the Hopfield network stay in the positive feedback action. With the amplifier gain less than A_N, all output states of the Hopfield network become legal for any input signal level. Thus, A_N is the maximum amplifier gain (equivalently lowest annealing temperature) that can randomize the neuron outputs.

Final Voltage Gain for the Cooling Schedule

Some neurons start to be biased in the saturation region for a given input voltage if the voltage gain of the neurons is increased above A_N. Let's assume that only the kth neuron output is saturated at a digital value V_k. The governing equations of the network are

$$C_i \frac{du_i(t)}{dt} + T_i u_i(t) = \sum_{j=1, j\neq i,k}^{N} T_{ij} v_j(t) + I_i(t) + T_{ik} V_k \quad \text{for any } i \neq k. \tag{74.23}$$

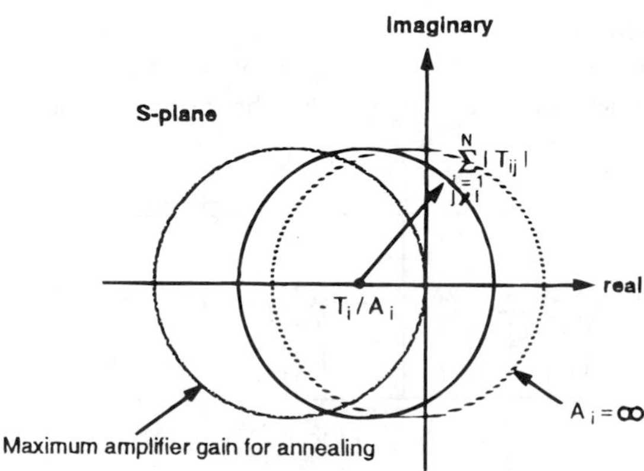

Figure 74.4 Radii of eigenvalues with different neuron gains.

The corresponding system matrix can be formed with the kth column and the kth row being deleted from Equation 74.14. Therefore, the lowest neuron gain which makes the network stay in the positive feedback action is determined by

$$A_{N-1} = \max\left[\frac{T_i}{\sum_{j=1, j\neq i,k}^{N} |T_{ij}|} \quad \text{for } 1 \leq i \leq N\right]. \tag{74.24}$$

The same T_i is used in Equations 74.22 and 74.24. The neuron gain for the positive-feedback action increases as the number of neurons which operate in the nonsaturated region decreases. A critical case is when only two neurons operate in the nonsaturated region. Let's assume that the outputs of the pth and q^{th} neurons are not saturated, the system matrix $\mathbf{B}_{pq}$ can be expressed as

$$\mathbf{B}_{pq} - \begin{pmatrix} -\dfrac{sC_p + T_p}{A_p} & T_{pq} \\ T_{qp} & -\dfrac{sC_q + T_q}{A_q} \end{pmatrix}, \tag{74.25}$$

Since the resistive network $\{T_{ij}\}$ is symmetrical, $T_{pq} \times T_{qp}$ is always positive. The neuron gain which makes two neurons operate in the nonsaturated region is

$$A_2 = \max\left[\sqrt{\frac{T_p T_q}{T_{pq} T_{qp}}} \quad \text{for every } p \text{ and } q \text{ with } p \neq q\right]. \tag{74.26}$$

When the voltage gain of the neurons is increased slightly above A_2, only one neuron will operate in the nonsaturated region. Even though the neuron output is an analog value, the digital bit can be easily decided using the middle value of the neuron output range as a reference. The corresponding logical state of the remaining neuron can then be determined. The neuron gain during the hardware annealing process should start from a value smaller than A_N and stop at a value larger than A_2.

74.4 Hardware Annealing on Cellular Neural Networks

A cellular neural network (CNN) is a continuous-time artificial neural network that features a multi-dimensional array of neuron cells and local interconnections among the cells. The basic CNN proposed by Chua and Yang (1988) is an n-by-m rectangular-grid array where n and m are the numbers of rows and columns, respectively. Each cell in a CNN corresponds to an element of the array.

The r-th neighborhood cells $N_r(i, j)$ of a cell $C(i, j)$, $1 \leq i \leq n$, $1 \leq j \leq m$, are defined as the cells $C(k, l)$, $1 \leq k \leq n$, $1 \leq l \leq m$, for which $|k - i| \leq r$ and $|l - j| \leq r$. The cell $C(i, j)$ has the direct interconnections with $N_r(i, j)$ through two kinds of weights, i.e., the feedback weights $A(k, l; i, j)$ and $A(i, j; k, l)$

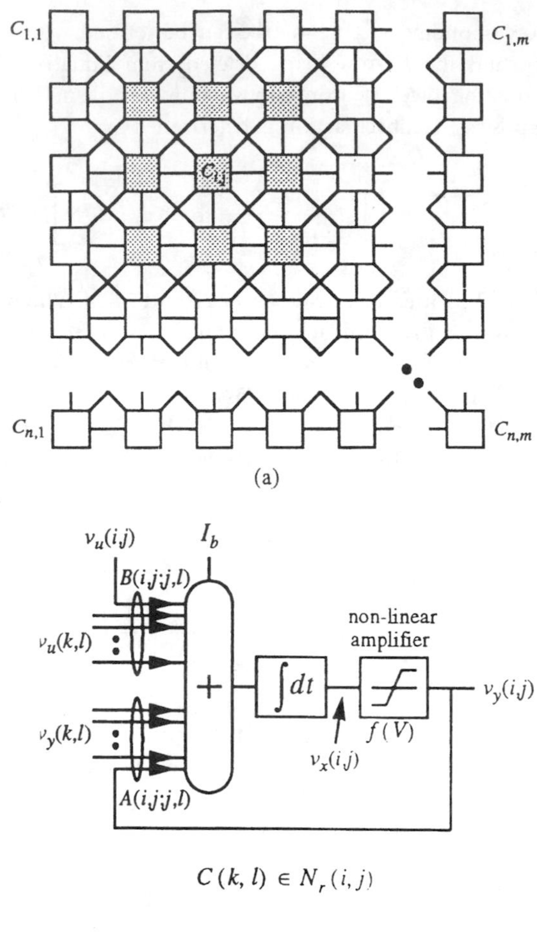

Figure 74.5 Cellular neural network (CNN). (a) An n-by-m cellular neural network on rectangular grid (shaded boxes are the neighborhood cells of C(*i,j*)). (b) Functional block diagram of neuron cell.

and feedforward weights $B(k, l; i, j)$ and $B(i, j; k, l)$, where the index pair $(k, l; i, j)$ represents the direction of signal from $C(i, j)$ to $C(k, l)$. The cell $C(i, j)$ communicates directly with its neighborhood cells $C(k, l) \in N_r(i, j)$. Since the cells $C(k, l)$ have their neighborhood cells, it also communicates with all other cells indirectly. Figure 74.5 (a) shows an n-by-m CNN with $r = 1$. The cells filled with dashed lines represent the neighborhood cells $N_1(i, j)$ of $C(i, j)$, including $C(i, j)$ itself.

The block diagram of a cell $C(i, j)$ is shown in Figure 74.5. The external input to the cell is denoted by $v_{uij}(t)$ and typically assumed to be constant $v_{uij}(t) = v_{uij}$ over an operation interval $0 \leq t < T$. The input is connected to $N_r(i, j)$ through the feedforward weights $B(i, j; k, l)$'s. The output of the cell, denoted by v_{yij}, is coupled to the neighborhood cells $C(k, l) \in N_r(i, j)$ through the feedback weights $A(i, j; k, l)$'s. Therefore, the input signals consist of the weighted sum of feedforward inputs and weighted sum of feedback inputs. In addition, a constant bias term is added to the cell. If the weights represent the transconductance values among the cells, the total input current i_{xij} to the cell is given by

$$i_{xij}(t) = \sum_{C(k,l) \in N_r(i,j)} A(i, j; k, l) v_{ykl}(t) + \sum_{C(k,l) \in N_r(i,j)} B(i, j; k, l) v_{ukl}(t) + I_b, \quad (74.27)$$

where I_b is the bias current. The equivalent circuit diagram of a cell is shown in Figure 74.6, where R_x and C are the equivalent resistance and capacitance of the cell, respectively. For the simplicity of illustration, I_b, R_x, and C_x are assumed to be the same for all cells throughout the network. All inputs are represented by dependent current sources and summed at the state node. Due to the capacitance C_x and resistance R_x, the state voltage v_{xij} is established at the summing node and satisfies a set of differential equations

$$\begin{aligned} C_x \frac{dv_{xij}(t)}{dt} &= -\frac{1}{R_x} v_{xij}(t) + i_{xij}(t) \\ &= -\frac{1}{R_x} v_{xij}(t) + \sum_{C(k,l) \in N_r(i,j)} A(i, j; k, l) v_{ykl}(t) \\ &\quad + \sum_{C(k,l) \in N_r(i,j)} B(i, j; k, l) v_{xkl}(t) + I_b; \\ &\qquad 1 \leq i \leq n,\ 1 \leq j \leq m, \end{aligned} \quad (74.28)$$

The cell contains a nonlinearity between the state node and the output and its input-output relationship is represented by $v_{yij}(t) = f(v_{xij}(t))$. The nonlinear function used in a CNN can be any differentiable, nondecreasing function $y = f(x)$, provided that $f(0) = 0$, $df(x)/dx \geq 0$, $f(+\infty) \to +1$ and $f(-\infty) \to -1$. Two widely used nonlinearities are the piecewise-linear and sigmoid functions as given by

$$y = f(x) = \begin{cases} \frac{1}{2}(|x + 1| - |x - 1|) & \text{piecewise-linear,} \\ \frac{1 - e^{-\lambda x}}{1 + e^{-\lambda z}} & \text{sigmoid.} \end{cases} \quad (74.29)$$

Here, the parameter λ is proportional to the gain of the sigmoid function. For a unity neuron gain at $x = 0$, $\lambda = 2$ may be used for the sigmoid function. The gain of neurons in a Hopfield neural network is very large so that the steady-state outputs are

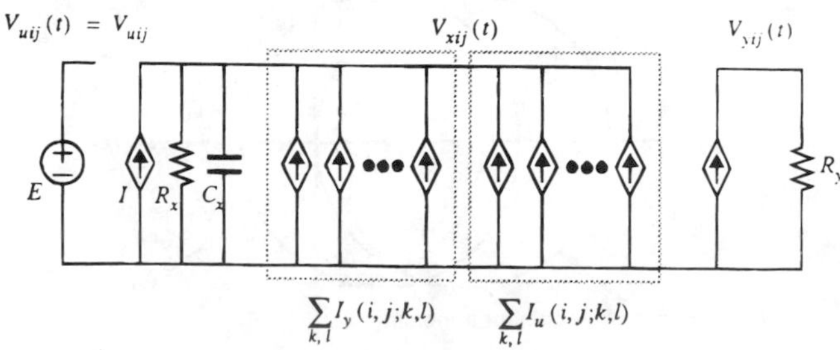

Figure 74.6 Equivalent circuit diagram of one cell.

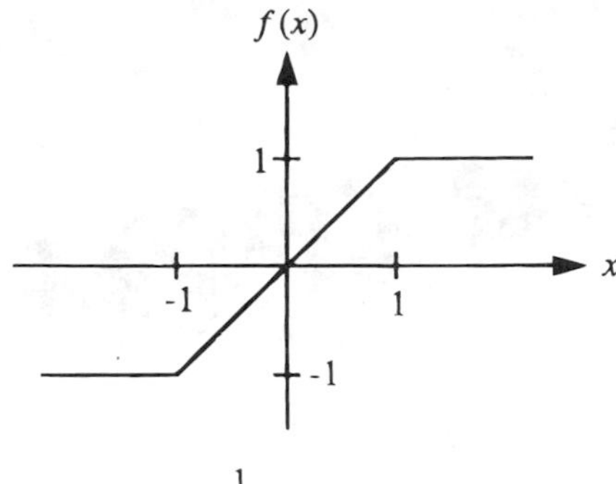

$$f(x) = \frac{1}{2}(|x+1| - |x-1|)$$

Figure 74.7 Piecewise-linear function.

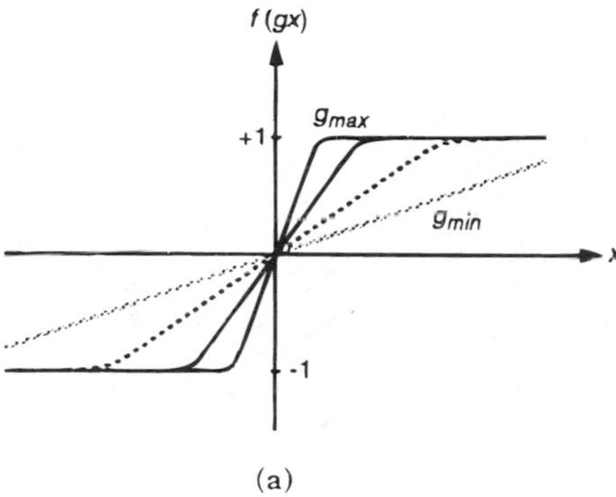

(a)

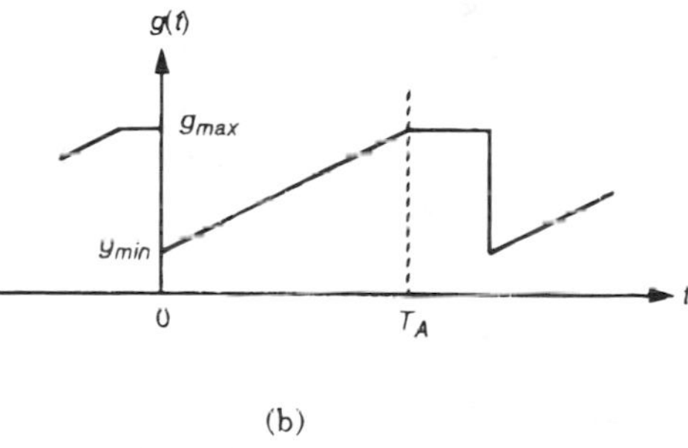

(b)

Figure 74.8 Modified neuron cell for hardware annealing in the CNN. (a) Transfer characteristics of nonlinearity for several gain control parameters. (b) Gain control function $g(t)$.

all binary-valued. However, if the positive feedback in the CNN cell is so strong that the feedback factor is greater than one, the gain of the cell needs not to be large for guaranteed binary output in the steady state. Typically, a unity gain $df(x)/dx|_{x=0} = 1$ is used in CNNs. The transfer characteristics of the piecewise-linear function are shown in Figure 74.7.

Application of Hardware Annealing

The hardware annealing is performed by the neuron-gain control $g(t)$, which is assumed to be the same for all neurons throughout the network for simplicity of analysis and illustration. The initial gain at time $t = 0$ can be set to an arbitrarily small, positive value such that $0 \leq g(0) << 1$, and after the annealing process for t_A seconds the final gain $g(t_A) = 1$ is maintained until the next operation. When the hardware annealing is applied to a cellular neural network (CNN) by increasing the neuron gain $g(t)$, the transfer function can be described by

$$v_y = f(gv_x) = \begin{cases} +1, & \text{if } v_x > +1/g \\ gV_x, & \text{if } -1/g \leq v_x \leq +1/g \\ -1, & \text{if } v_x < -1/g \end{cases}. \quad (74.30)$$

Figure 74.8a shows the transfer characteristics of the piecewise nonlinearity for several gain control parameters g.

Note that the saturation level is still $y = \pm 1$ and only the slope of $f(x)$ around $x = 0$ varies. In Fig.74.8b, the gain control function $g(t)$ with constant slope is plotted. In each annealed operation, $g(t)$ increases linearly from $g_{min} = g(0)$ to $g_{max} = 1$ for $0 \leq t \leq T_A$. Then, the maximum gain is maintained for $T_A < t \leq T$, during which the network is stabilized and the initialization operation $\mathbf{x} = \mathbf{x}(0)$ may take place.

Two gain values are very important in the annealing operation. First, the critical gain g_c represents the value at which the phase of search for an optimal solution is completed. In relation to this critical gain, two conditions must be satisfied:

- The initial gain g_0 is a small positive number less than g_c
- A sufficient amount of time is allowed for the network state to reach the basin of attraction to which the global minimum of E belongs, before the gain reaches g_c.

Having one or more eigenvalues with positive real parts is a sufficient condition for the system instability in the sense of bounded-input/bounded-output (BIBO). Therefore, for the condition $g > g_c$, at least one neuron output is saturated as time elapses. However, it does not guarantee binary-valued outputs for all neurons in the steady state. In accordance with the method to determine g_c, a saturation gain g_s can be defined as the annealing gain beyond which all saturated binary outputs are guaranteed.

In summary, the hardware annealing process may consist of the following operation phases:

1. Search Phase ($g_0 \leq g < g_c$): The annealing process forces the network state to move to the basin of attraction to which the global minimum of E belongs, i.e., $y(g_o) \in \mathbf{D}^N \rightarrow y(g_c) \in \Omega^N$ where $\mathbf{D}^N$ and Ω^N are N-dimensional hypercube and manifold, respectively.
2. Attraction Phase ($g_c \leq g < g_s$): Once $y(g_c) \in \Omega^N$, a further gain increase makes the output approach the global minimum rapidly. In addition, the energy barriers begin to stand out so that Ω^N becomes disjointed from others and accidental jumps to local minima can be prevented.
3. Completion Phase ($g_s \leq g \leq 1$): In this interval, the formation of the original energy landscape is complete and all saturated binary outputs are obtained.

Thus, the annealing schedule must be chosen such that $g_{min} = g_0 < g_c = g(t_c)$ and the time period t_c for search phase is long enough.

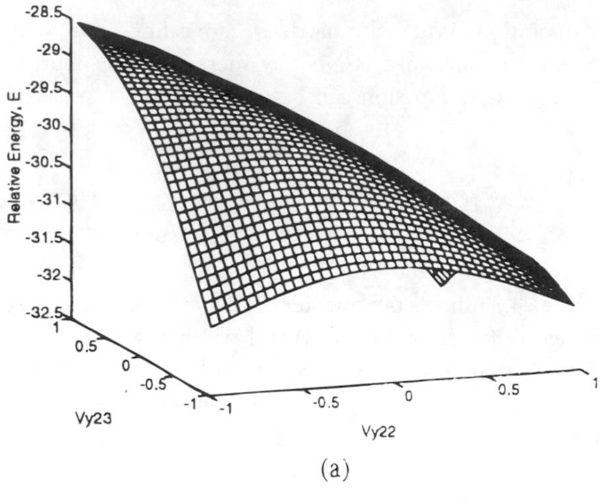

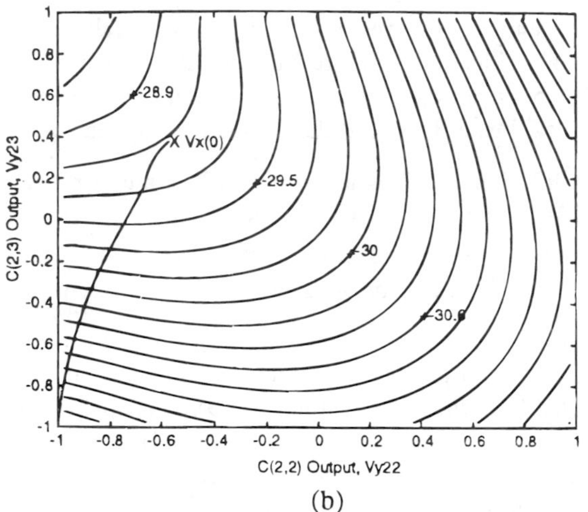

Figure 74.9 Two-dimensional energy surface of a 4 × 4 CNN. (a) Three minima over $[v_{y22}, v_{y23}]$ plane. (b) Trajectory of initial state $v_x(0)$ toward a local minimum.

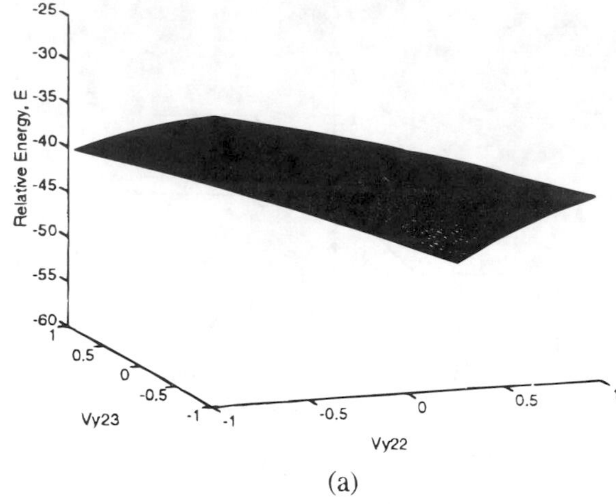

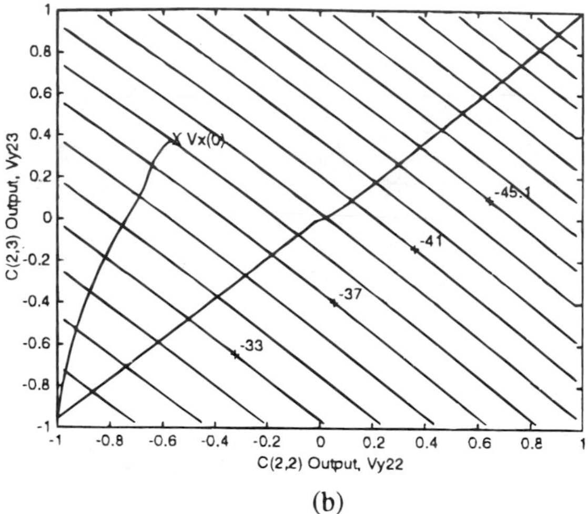

Figure 74.10 Two-dimensional energy surface of a 4 × 4 CNN with hardware annealing. (a) One global minimum over $[v_{y22}, v_{y23}]$ plane. (b) Trajectories of outputs from local minimum to global minimum during annealing process.

Simulation Results

Figure 74.9a shows the energy function of a 4 × 4 CNN in the steady state as the functions of v_{y22} and v_{y23}, where the CNN uses the bipolar sigmoid function as the neuron nonlinearity and has the parameters of $R_x = 10^3\Omega$ and $I_b = 0$; and the cloning templates of

$$\mathbf{T_A} = 10^{-3} \times \begin{pmatrix} 1 & 1 & 1 \\ 1 & 2 & 1 \\ 1 & 1 & 1 \end{pmatrix}, \qquad (74.31)$$

and

$$\mathbf{T_B} = 10^{-3} \times \begin{pmatrix} 0 & 0 & 0 \\ 0 & 1 & 0 \\ 0 & 0 & 0 \end{pmatrix}, \qquad (74.32)$$

For these parameters, the matrix **M** has six negative eigenvalues and each of three distinct eigenvalues has the multiplicity of 2. In the figure, all neurons other than $C(2, 2)$ and $C(2, 3)$ have fixed, steady-state output values determined by the network operation. Notice that, because $\mathbf{y}_0 \in \mathbf{D}^{16}$, there are three minima at the locations $[v_{y22}, v_{y23}] = [-1, -1],[+1, -1]$, and $[+1, +1]$.

Depending on the value of $\mathbf{v} = (0)$ and $\mathbf{v}_u$, the output $[v_{y22}, v_{y23}]$ will be attracted to one of vertices $[-1, -1]$, $[+1, -1]$, and $[+1, +1]$ in the steady state. The points $[v_{y22}, v_{y23}] = [-1, -1]$ and $[+1, -1]$ are the local minima and $[v_{y22}, v_{y23}] = [+1,+1]$ is the global minimum with the lowest energy value. Figure 74.9b shows the contours of the energy function and the trajectory of output values during the network operation. The figure shows that the initial output values $[v_{y22}(0), v_{y23}(0)]$ indicated by X is attracted to the point $[-1, -1]$, which is one of the local minima.

Figure 74.10 shows the results of hardware annealing for the

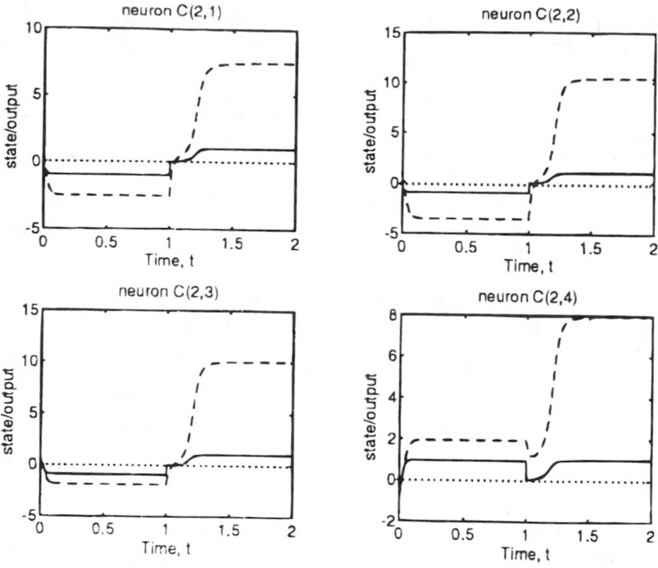

Figure 74.11 Plots of states and outputs of 4 neurons. All neuron outputs except $C(2.4)$ are toggled as the result of annealing.

same network parameters including the initial state as in Figure 74.9, and the annealing schedule used is $g(t) = \alpha t$, where α is a constant. In Figure 74.10a, the local minima do not exist and the output $[+1, +1]$ which was the global minimum before annealing, became the only minimum that can be reached after annealing. However, this does not mean that the local minima were removed by the hardware annealing. As a matter of fact, the local minima still exist at the locations they do all the time, and as a result of the annealing process the two-dimensional subspace $\mathbf{D}^2$ spanned by $[v_{y22}, v_{y23}]$ moved to a different location in $\mathbf{D}^{16}$ at which the global minimization of E can be achieved.

In Figure 74.11, the corresponding waveforms for the states and outputs of four cells C (2, 1) $-C(2, 4)$ are shown. The steady-state value υ_{y24} is not affected by the annealing but the polarities of the other outputs changed.

Figure 74.12 shows how often the local-minima problems are encountered, and how effectively the hardware annealing solves them in 50 independent experiments with the parameters given earlier. In Figure 74.12a, each row denotes an experiment and each neuron is numbered as a column. In each experiment, v_x (0) and v_u are chosen randomly in $\mathbf{D}^{16}$ so that the constraint

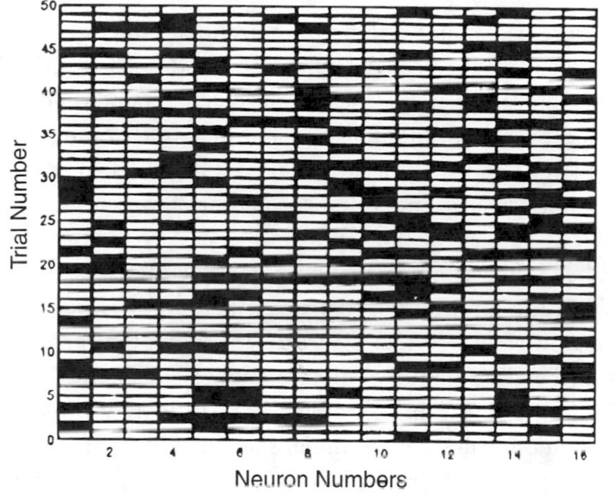

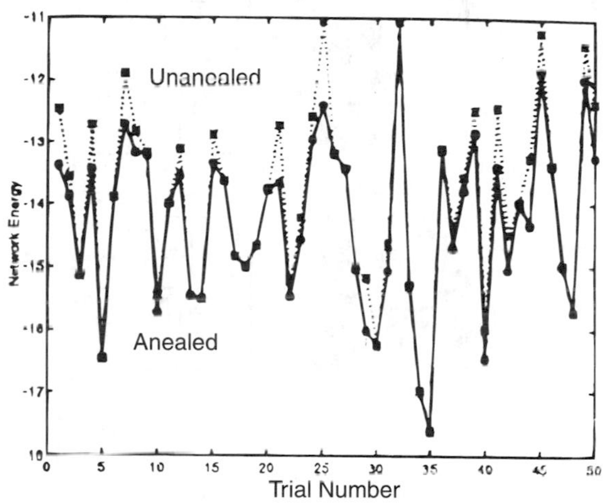

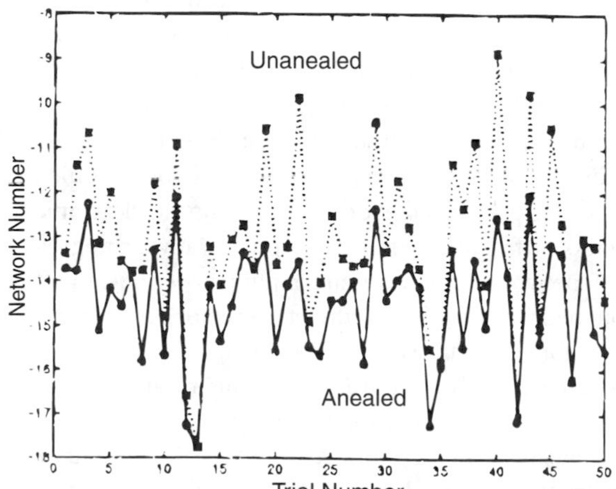

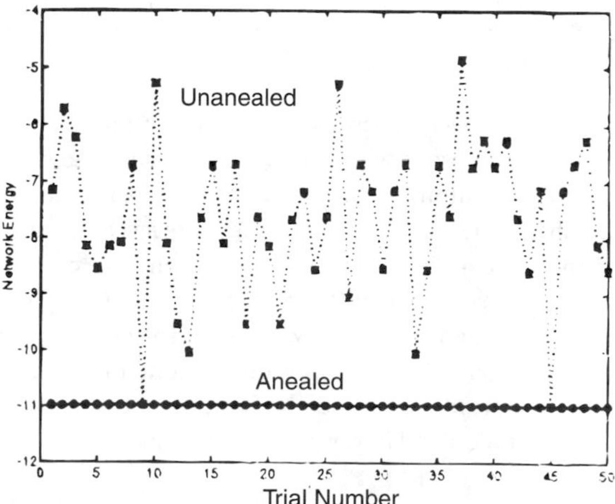

Figure 74.12 Effectiveness of hardware annealing in 50 independent experiments. (a) Changes of neuron outputs by annealing, showing the occurrence of optimal solutions. (b) Our proposed future compact vision with optical-input CNN wafers.

Table 74.1 Computational speed of hopfield-network 4-bit A/D conversion

Method / Item	Hardware computation using IC	Software computation			
		Adaptive time step	Fixed-time step		
			1×10^{-2}	1×10^{-3}	1×10^{-4}
Computational time (sec)	1.8×10^{-3}	314	182	1697	16147
CPU-time factor	1	1.7×10^5	1.0×10^5	9.4×10^5	9.0×10^6
Quality of solution	Good	Good	Bad	Poor	Good

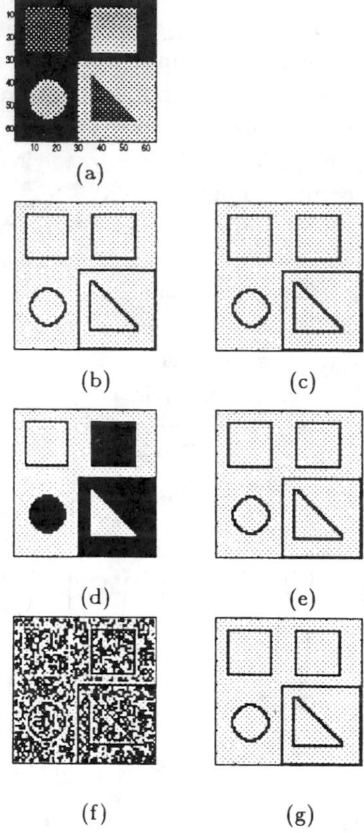

Figure 74.13 Edge detection results of the CNN. (a) A 64-by-64 original gray-scale image. (b) $\mathbf{v}_x = \mathbf{0}_1$ $\mathbf{v}_u$ = input image and without annealing. (c) $\mathbf{v}_x = \mathbf{0}$, $\mathbf{v}_u$ = input image and with annealing. (d) $\mathbf{v}_x = \mathbf{v}_u$ = input image and without annealing. (e) $\mathbf{v}_x = \mathbf{v}_u$ = input image and with annealing. (f) $|\mathbf{v}_{xij}(0)| > 1$, all i, j and without annealing. (g) $|\mathbf{v}_{xij}(0)| > 1$, all i, j and with annealing.

conditions and the independency of the experiments are satisfied. In the figure, neurons with the black color indicate the changes in steady-state outputs as the result of hardware annealing. Figure 74.12b shows the comparison of the energy levels for unannealed and annealed conditions. Similar experiments are conducted for two special cases when $\mathbf{v}_x(0) = \mathbf{0}$ and $\mathbf{v}_u = \mathbf{0}$. Figure 74.12c shows the energy levels when $\mathbf{v}_x(0) = \mathbf{0}$. In this case, the initial value of E is zero. This energy value is quite high compared to those in the steady state and the basin of attraction can be as large as a whole $\mathbf{D}^N$. However, it can be seen that

$$E_{max} = \frac{1}{2}\mathbf{b}^T\mathbf{M}^{-1}\mathbf{b} \geq E(0) = 0, \tag{74.33}$$

for positive definite M. Correspondingly, the network still can stay at the nearest local minimum in the steady state and the zero initial condition does not always provide the optimal solution. On the other hand, the annealing process not only increases the value of the energy function to near zero initially, but also forces it to reach $E_{\max}$ during the relaxation process as described in the previous section. In Figure 74.12d, the external forcing function $\mathbf{b}$ is set to zero in which case the energy is solely determined as $E = -(1/2)\mathbf{y}^T\mathbf{M}\mathbf{y}$. The figure shows that the hardware annealing provides the globally optimal solution in each experiment. By using the exact steady-state values $|y_k| = 1, \forall k$, the computer-generated solution for the global minimum of E is shown to be -10.8. Note that for the piecewise-linear function this value is -14.

Figure 74.13 shows the edge detection results of a 64-by-64 CNN for unannealed and annealed conditions. Figure 74.13a shows the original gray-scale image. The sigmoid function is used as the neuron nonlinearity and the same cloning templates as in (Chua and Yang, 1988) are used. Figures 74.13c and 74.13d show the CNN outputs due to the annealing effect. Here, the image is applied to the input only and $\mathbf{v}_x = \mathbf{0}$. Two networks resulted in the same, correct outputs. When the image is applied to both the input and initial state, the output is not correct as shown in Figure 74.13e. Figure 74.13f shows the correct output when the annealing operation is applied. In Figure 74.13g and 74.13h, the constraint condition $|v_{xtj}(0)| \leq 1, \forall i, j$ is removed and random values of the initial state between 1 and 5 are used. As a result, the output in Figure 74.13g contains many neurons that are not able to toggle the states. However, as shown in Figure 74.13h, the hardware annealing provides enough stimulation to those frozen neurons caused by such ill-conditioned initial states.

Since the number of iterations at a given temperature and the cooling rate of the temperature should be compromised in order to speed up the convergence process, a very large computation time is usually required in software implementations of simulated annealing. However, recent advances in microelectronic technologies make possible the design of compact electronic neural networks with hundreds of neurons. Many analog electronic neural network processors are equipped with gain-adjustable output neurons which allow the execution of hardware annealing. Applications of hardware annealing in back-propagation networks are explicitly supported by the neural network processor. Tremendous speed-up factor can be achieved with the hardware annealing. It is the key to achieve optimal solutions in a very short period of time because the annealing process is carried out in parallel and the voltage gain of the neuron is changed continuously. The speed comparison between the integrated circuit implementation and a software computation is listed in

Table 74.1 (Lee and Sheu, 1991). The neural-based A/D converter used in the previous subsection helps to demonstrate the superiority of hardware annealing.

Contribution by Bang W. Lee and Sa H. Bang (Sheu and Choi, 1995; Lee and Sheu, 1991) to the quick search of optimal solutions in recursive neural networks is highly appreciated.

References

Aarts, E. H. L. and Korst, J. 1989. *Simulated Annealing and Boltzmann Machines,* John Wiley & Sons, Chichester, Great Britain.

Ackley, D. H., Hinton, G. E., and Sejnowski, T. J. 1985. A learning algorithm for Boltzmann machines, *Cognitive Science,* 9:147–168, Ablex Publishing, Norwood, NJ.

Chua, L. O. and Yang, L. 1988. Cellular neural networks: theory, and applications, *IEEE Trans. Circuits and Systems,* 35:1257–1272; 1273–1290.

Lee, B. W. and Sheu, B. J. 1991. *Hardware Annealing in Analog VLSI Neurocomputing,* Kluwer Academic Publishers, Boston, MA.

Rumelhart, D. E., McClelland, J. L. and the PDP Research Group 1989. *Parallel Distributed Processing,* vol. 1, The MIT Press, Cambridge, MA.

Sheu, B. J. and Choi, J. 1995. *Neural Information Processing and VLSI,* Kluwer Academic Publishers, Boston, MA.

Tank, D. W. and Hopfield, J. J. 1986. Simple neural optimization networks: an A/D converter, signal decision circuit, and a linear programming circuit, *IEEE Trans. Circuits and Systems,* 33(5):533–541.

Wong, E. 1989. Stochastic neural networks, ERL Memo. UCB/ERL M89/9, University of California at Berkeley, Berkeley, CA.

75
Radial Basis Function (RBF) Neural Networks

Thomas Lindblad
Royal Institute of Technology

Clark S. Lindsey
Royal Institute of Technology

Åge Eide[1]
Royal Institute of Technology

75.1 Introduction 1014
75.2 Topology 1014
75.3 Operation 1015
75.4 Training 1015
75.5 Summary 1017
75.6 Defining Terms 1017

75.1 Introduction

Although radial basis function (RBF) architectures are generally included in the concept of neural networks, they are really very simple nets with no direct relation to biological structures. They are normally intended for classification problems. Indeed, they have been used for pattern classification without being referred to as neural networks (Batchelor) or RBF (Powell, 1986). When implemented in hardware, they provide a network that does fast forward processing and, perhaps most importantly, also trains quickly. Generally, the RBF is considered as a way of modeling an input-output mapping as a linear combination of radially symmetric functions. It is frequently found that as such a device, it has a poor ability to generalize. However, this may depend to some extent on the learning method, as will be demonstrated here. It is also well known that the RBF network can separate complicated patterns for which other types of feedforward/back-propagation do not perform well. The famous example here is the double-spiral.

75.2. Topology

Using the neural network concepts, one may refer to the RBF net as a very simple three-layer feedforward network. A typical architecture is shown in Figure 75.1. The architecture of the RBF net is of the multilayer (normal) feedforward type, i.e., each layer is only connected to adjacent layer(s). The input layer is made of source nodes and, in this case, the input vector has the dimension $N = n + 1$. The "hidden layer" may be referred to as the prototype layer. Although referred to as a hidden layer, it serves

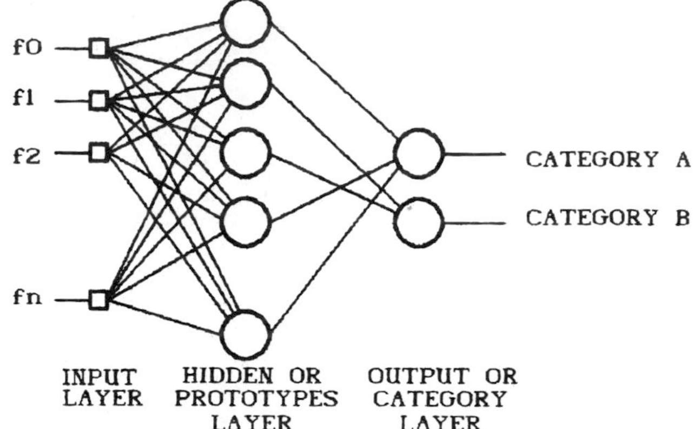

Figure 75.1 A typical radial basis function (RBF) neural network. In the present architecture each neuron in the hidden layer is connected to one and only one neuron in the output layer. Note the alternative names of these layers.

a different purpose than a perceptron multilayer network. Generally, there are many neurons in this layer. The output layer supplies the response of the network to the input vector. It has as many neurons as there are categories. In this architecture, all inputs are connected to all neurons in the "hidden" layer, each neuron in this layer is only connected to one neuron in the "output" layer. This is typical for, e.g., the IBM ZISC036 neural network chip (Le Bouqoin, 1994). Other chips, like N*i*1000 (Intel) may each have hidden neuron connected to all neurons in the output layer. In this case, a more "weighted" output is obtained. Note that in the case of an RBF net there can only be one hidden layer (contrary to the perceptron neural net). Another difference is that the hidden layer RBF network is nonlinear (and the output layer is linear). Also note that the hidden neurons have activation functions with arguments that calculate *distances* rather than *inner products* as in the case of perceptrons.

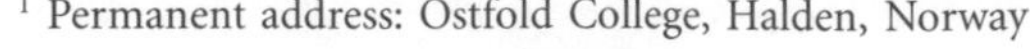

[1] Permanent address: Ostfold College, Halden, Norway

0-8493-8343-9/97/$0.00+$.50

75.3. Operation

In simplest terms, a RBF network determines *distances* (d) between stored vectors, i.e., the *prototypes,* and the input vector. The *activation* of each hidden (prototype) neuron is determined by the *type* of radial basis function $f(d)$, typically a Gaussian funcion, i.e., $\exp(-d^2/c^2)$, or a quadratic functions, e.g., $\sqrt{d^2 + c^2}$ or its inverse, etc. One can also use a simple step function centered on the prototype vector point in hyperspace. If the activation exceeds a threshold, then that prototype is "ON". The *category* assigned to it in turn is activated. As mentioned, the prototype neurons correspond to the "middle" or hidden layer and the category neurons correspond to the output layer. The category neurons can be simple threshold neurons. Figure 75.2 shows a two-dimensional mapping performed by, e.g., a network of the architecture outlined in Fig. 75.1. More elaborate types of RBF networks, such as the probablistic neural network (PNN), have more complicated output neurons whose output values are interpreted as *probabilities* for the categories. The RBF network is, thus, a simple classification device and its task consists of evaluating whether an N-dimension input vector lies within the influence field of any *prototype* stored in the network. Each prototype has an *area of influence* (AIF) value within which the neuron is ON, if it is greater than the distance, d between the prototype and the input vector. The influence field may be said to represent a part of the N-dimensional space around the prototype, where generalization occurs. A prototype is thus simply a *training vector* and the RBF "hidden" neuron stores this prototype and creates a connection to an output. Each prototype is thus associated with a category and an influence field. During operation the *distance* between the input vector and each stored prototype is calculated and compared to the thresholds of each prototype neuron. One easily separates three cases:

1. If an input vector does not lie within any influence field, it is not recognized.
2. If it lies within the influence fields of one or more prototypes of the same category, it will be classified as belonging to that category.
3. If the influence field is associated with two or more prototypes of different categories, it will be recognized but not identified.

There are several distance norms. The most common ones include Euclidian distance ($d = \| V - P \|$), where V is the input vector and P is the prototype vector, the *Manhatten block* distance ($d = \text{sum}|V_i - P_i|$) and maximum distance ($d = \max|V_i - P_i|$), where sum and max run from 1 to n for an n element vector. The latter two norms are typically used in hardware since no slow multiplication operations are involved.

One of the major problems with the RBF networks is that when the vector dimension grows, a significantly higher number of prototypes are required. This is referred to as the *curse of dimensionality.* Another drawback of common RBF nets is the poor capability to generalize. However, this depends strongly on the applied training paradigm.

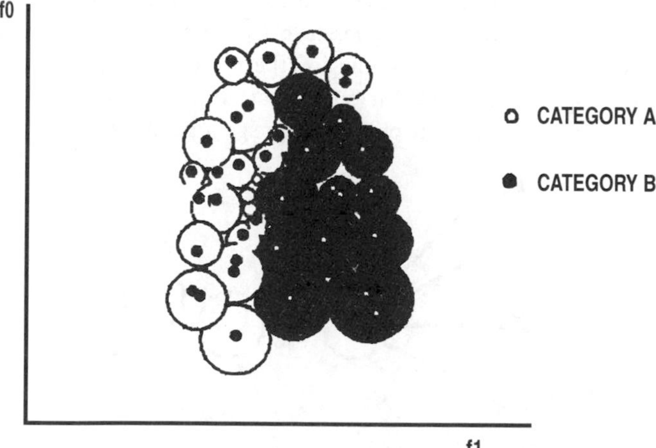

Figure 75.2 Typical performnce of an RBF-network with architecture as shown in Figure 75.1. The white and the shaded areas represents the two categories in the classification problem.

75.4 Training

The training of a RBF network is local, i.e., no information is stored in multiple neurons. The learning task consists of mapping the space by prototypes and adjusting the influence fields according to all neighbors. Presenting new prototypes results in either of three actions:

1. No change.
2. One or more influence fields are modified (reduced) or,
3. A new prototype neuron is created in the net.

There are several algorithms used in connection with the training of RBF nets. The Region of Interest or ROI (Batchalor) is frequently used (e.g., with the IBM ZISC036 on-chip learning implementation). The RCE method is very similar and is due to Reilly, Cooper, and Elbaum (1982). Here, as is the case in its probabilistic extension, P-RCE (Scofield and Reilly, 1991), one takes advantage of a growing structure in which hidden units are only introduced when necessary. The training using these methods reaches stability much faster than is the case for gradient-descent algorithms. The dynamic decay adjustment (DDA) paradigm (Berthold and Diamond, 1995) uses two threshold parameters and a Gaussian to overcome the shortcomings of P-RCE in connection with conflicting categories and low confidence levels (the standard RCE algorithm may not always commit new prototypes in the areas of conflict). Although fairly easy to implement, the DDA algorithm performs extremely well (Székely et al., 1995).

Like the P-RCE, the DDA uses Gaussian response functions. The key to the algorithm is the use of *two thresholds* denoted θ_{pos} and θ_{neg}, such that for every pattern x of class c, we have

$$\theta_{pos} \leq R_j^c(\mathbf{x}) \quad \text{and} \quad \theta_{neg} > R_j^k(\mathbf{x}), \qquad (75.1)$$

for all $k \neq c$ and $1 \leq j \leq m_e$, and where R is the response function and m_o is the prototype of class c. Only when the activations of proper classes are all below θ_{pos}, will a new prototype be added. The following pseudo code (Aizerman et al., 1964) shows what the training for one new pattern x of class c looks like:

1. find $i : 1 \leq i \leq m_c \wedge R_i^c(x) = \max\{R_j^c(x), 1 \leq j \leq m_c\}$
2. if $R_i^c(x) \geq \theta_{pos}$
3. [$A_i^c + = 1.0$
4. else
5. [add new prototype p^c with:
6. [$r^c = x$
7. [$\sigma^c = \max\{\sigma: k \neq c \wedge 1 \leq j \leq m_k \wedge R^c(r_j^k) < \theta_{nag}\}$
8. [$A^c = 1.0$
9. $\forall\ k \neq c, \leq j \leq m_k : \sigma_j^k = \max\{\sigma : R_j^k(x) < \theta_{neg}\}$

The choice of the two new parameters does not appear to be critical, but some default values can be used (they need to be different or the P-RCE like situation occurs).

Figure 75.3 shows the results obtained for a "noisy character set." In this case, the threshold parameters were set equal to 0.45 and 0.1, respectively. Included in this figure are also the results from a hardware (ZISC036) rún using the ROI algorithm. As the figure shows, we obtain better results with the software net trained by the DDA paradigm, than for the ZISC with on-chip ROI learning. The RBF/DDA network manages surprisingly well with noisy data (A noise level of "30" in our character independent definition here corresponds roughly to 50% noise, i.e., every second pixel changed). Indeed, the capability to generalize is not all that poor.

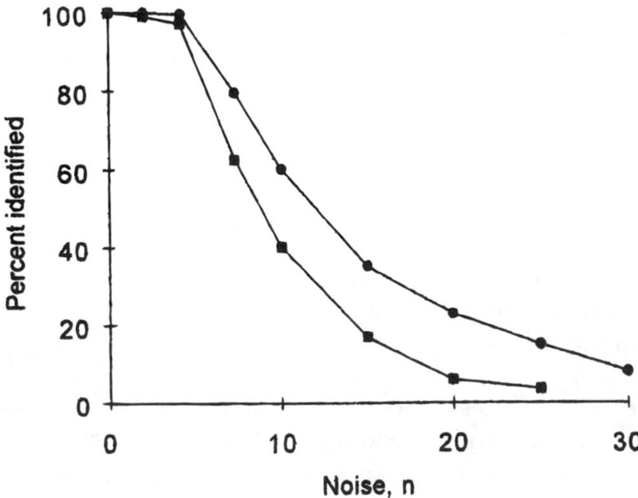

Figure 75.3 Results of "noisy character recognition" test. Recognition vs. *noise is plotted here. The upper curve represents the results using the DDA algorithm, while the lower one is obtained with the IBM ZISC036/ ROI hardware. In both cases, 52 prototypes were used. The noise is defined to be character independent and the value of 30 corresponds to 50% noise in average.*

Another approach to the present type of network is the O-algorithm (Eide and Lindblad, 1992). This is a classification method with limits obtained from chi-square cuts. When applied, the artificial neural network can be used as a measuring device. In such a topology, the node function (Figure 75.4) is assumed to be

$$f(\alpha) = 1/[1 + \exp(-\alpha)], \qquad (75.2)$$

where α is the input to the node. It is furthermore assumed that the set of points defining each class, m_j, are obtained as average values and has a standard deviation, σ_j. In other words, we choose α to be

$$\alpha = a + b \sum [(x_j - m_j)^2/\sigma_j^2] = a + b\chi^2, \qquad (75.3)$$

where x_j is the measured value for the point j. The value will be the input to node j and if there are N such points, then N input nodes are required. As pointed out (Berthold and Diamond, 1995), the constant $b < 0$ is chosen in such a way that the desired maximum value of χ^2 result in a value of $f(\alpha)$ greater than a chosen (desired) limit. The constant $a > 0$ controls the sensitivity of the system, i.e., if χ^2 is "too large" then $f(\alpha)$ will indicate a "no" response.

Although this network was originally designed to be used as a measuring device, it can also be used for ordinary pattern classification (Eide and Lindblad, 1992). Since the network provides a measure of how similar the input patterns are, a network using this "o-algorithm" may sometimes be preferred. In the present example, we have chosen to let the node input α be a linear function of χ^2. Depending on the actual situation and how close together the different output categories are located, one may choose a nonlinear dependence, e.g., a parabola. Finally, it

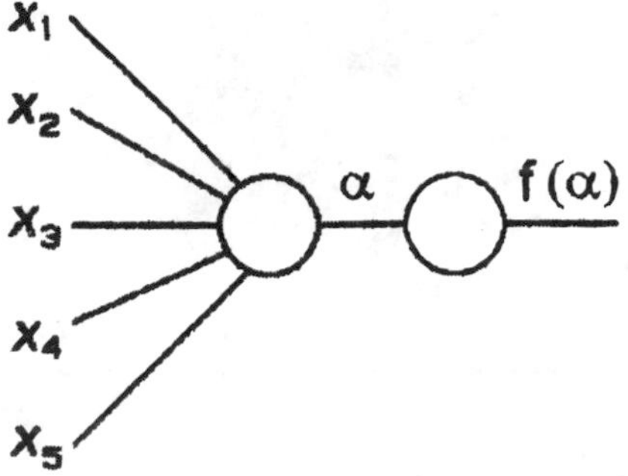

Figure 75.4 The neuron used in neural networks design as measuring devices. The operations performed are given by Equations 75.2 and 75.3.

should be stressed that this network measures how similar the input patterns are, which makes it directly applicable as a triggering device with a certain confidence level.

75.5 Summary

RBF are simple devices used mainly for classification and pattern recognition. According to Wasserman (Wasserman, 1993), early applications of RBF for classification using a trained output layer were published in 1964 to 1968 (Specht, 1968; Baskirov, 1964). Today, the performance of RBF (software) nets have been "boosted" using DDA algorithms (Berthold and Diamond, 1995), while at least two hardware devices (LeBouquin, 1994; Intel, 1994) are available today. The renewed interest in RBF is most likely due to the increased interest in pattern recognition and the need for massively parallel devices operating in real time applications. When the neural network is to be used as a measuring device, or the confidence level is needed, then the "o-algorithm" should be very useful.

75.6 Defining Terms

Here we describe some important definitions and concepts related to the radial basis functions, their architecture, neuron function, learning paradigms, etc. In addition to the references given above, we suggest Lindblad et al. (1995), Eide et al. (1994), Specht (1990), Broomhead and Lowe (1987), Moody and Darken (1986), Botros and Atkeson, Brown and Harris, Haykin (1994), and Lindsey et al. (1994) for additional reading.

Basis function: The activation of the neuron (processing element) is determined by its basis function. This function, $f(d)$, may be a simple step function, a Gaussian, etc.

Influence field: Each prototype (see below) has an area of influence (AIF) value within which the neuron is ON, if it is greater than the distance, d (see below) between the prototype and the input vector. The influence field may be said to represent a part of the N-dimensional space around the prototype, where generalization occurs.

Training and training algorithms: RCE, DDA, PRCE, PNN, ROI, O-algorithm: The training of a RBF network is local, i.e., no information is stored in multiple neurons. The learning task consists of mapping the space by prototypes and adjusting the influence fields according to all neighbors.

Presenting new prototypes results in either of three actions: (1) no change, (2) one or more influence fields are modified (reduced), or (3) a new prototype neuron is created in the net. There are several algorithms used in connection with RBF nets. The region of interest or ROI is frequently used (e.g., with the IBM ZISC036 on-chip learning implementation) although in the early book by Bruce Batchelor neither neural nets nor RBF is mentioned. The RCE method is very similar and is due to Reilly, Cooper and Elbaum (1982). Here, as is the case in its probabilistic extension, P-RCE, advantage can be taken of a growing structure in which hidden units are only introduced when necessary. The training using these methods reaches stability much faster than is the case for gradient-descent algorithms. The dynamic decay adjustment (DDA) paradigm uses two threshold parameters and a Gaussian to overcome the shortcomings of P-RCE in connection with conflicting categories and low confidence level. The choice of the two new parameters does not appear to be critical, but some default values can be used (they need to be different or the P-RCE like situation occurs). The O-Algorithm is a classification method with limits obtained from chi-square cuts. When applied, the artificial neural network can be used as a measuring devices.

Classification: This task consists of evaluating whether a N-dimension input vector lies within the influence field of any prototype stored in the network. To do so, the distance between the input vector and all stored prototypes are calculated and compared to the thresholds of each prototype neuron. Clearly, if an input vector does not lie within any influence fields, it is not recognized. If it lies within the influence fields of one or more prototypes of the same category, it will be classified as belonging to that one. However, if the influence field is associated with two or more prototypes of different categories, it will be recognized but not identified.

Prototype neuron: A prototype is a training vector and the RBF neuron simply stores this prototype and creates a connection to an output. Each prototype is associated with a category and an influence field.

Distance norm: There are several distance norms, the most common ones include Euclidian distance ($d = \|V - P\|$), where V is the input vector and P is the prototype vector, the Manhatten block distance (d = sum $|d_1 - P_1|$) and maximum distance (d = max$|V_1 - P_1|$), where sum and max run from 1 to n for an element vector. The latter two norms are typically used in hardware since no slow multiplication operations are involved.

Category neuron: The prototypes each belong to a given category defined by the classification problem. There are, of course, more prototypes than categories and generally many prototypes are associated with the same category.

Regularization theory: This is a method due to Tikhonov and is referred to as the regularization for solving ill-posed problems. The basic idea of regularization is to stabilize the solution by means of auxiliary non-negative functions. Regularization networks are very similar to RBF networks and may simply have a hidden layer of Green's functions.

Curse of dimensionality: Refers to the high number of prototypes required as the vector dimension of the RBF network grows.

Cover's theorem: This theorem refers to the separability of patterns. It states that complex pattern-classification problems cast in high-dimensional space are more likely to be linearly separable than in a low-dimensional space.

Separability patterns: Linear separability may be generalized to polynomial separability. Given a vector x, representing a set of patterns in an input space of (arbitrary) dimension N, we can generally find a nonlinear mapping $f(x)$ of high enough dimension M, such that we have linear separability in the f-space.

Acknowledgment

This work is supported by the Swedish Research Council for Engineering Sciences, which is gratefully acknowledged.

References

Aizerman, M. A., Braverman, E. M., and Rozonoer, L. I. 1964. Theoretical foundations of the potential function method in pattern recognition learning, *Automat. Remote Control,* 25:821–837.

Baskirov, O. A., Braverman, E. M., Muchnik, I. B. 1964. Potential function algorithms for pattern recognition learning machines, *Automat. Remote Control,* 25:629–631.

Batchelor, B. *Practical Approach to Pattern Classification,* Plenum Press, London, UK.

Berthold, M. R. and Diamond, J. 1995. Boosting the Performance of the RBF Networks with Dynamic Decay Adjustment, *Advances in Neural Information Processing Systems 7,* Tesauro, G., Touretzky, D. S., and Leen, T. K., eds., MIT Press, Cambridge, MA.

Botros, S. M. and Atkeson, C. G. Generalization Properties of Radial Basis Functions, *Advanced Neural Information Processing Systems,* Touretzky, D. S., ed., Kaufmann.

Broomhead, D. S. and Lowe, D. 1987. Multivariable functional interpolation and adaptive network, *Commun. Statist. Simulat.,* 16(1):263–297.

Brown, M. and Harris, C. *Neurofuzzy Adaptive Modeling and Control,* Prentice Hall, Englewood Cliffs, NJ.

Eide, Å, and Lindblad, T. 1992 Artificial neural networks as measuring devices, *Nucl. Inst. Meth.,* A317:607,–609.

Eide, Å, Lindblad, T., Lindsey, C. S., Minerskjöld, M., Sekhniaidze, G., and Sźkely, G. 1994. An implementation of the zero instruction set computers (ZISC036) on a PC/ISA bus card, *WNN/FNN WDC,* p. 30.

Haykin, S. 1994. *Neural Networks, A Comprehensive Foundation,* IEEE Press, Piscataway, NJ.

Intel Corporation, 1994. *Ni1000 Data Sheet,* Intel Corp., Santa Clara, CA (and Nestor Inc., Providence, RI); Chin, Park, Buckman, K., Diamond, J., Santoni, U., The, S.-C., Holler, M. Glier, M., Scofield, C. L., and Nunez, L., A radial basis function neural network with on-chip learning.

LeBouquin, J-P. 1994. IBM microelectronics ZISC, zero instruction set computer, preliminary information, *Poster Show WCNN,* San Diego, CA and addendum to conf. proc.; Lebouquin, J-P. and Grandguillot, M. IBM Microelectronics, Essones, G. Laurans and B. ZISCO36/PCMCIA card, GIAT Industries, Toulouse and G. Paillet, Neuroptics, Monpellier, private communication.

Lindblad, T., Lindsey, C. S., Minerskjöld, M., Sekhniaidze, G., Szëkely, G., and Eide, Å. 1995. The IBM ZISC036 zero instruction set computer, *(CERN) ONLINR,* 10:36–38; Implementing the new zero instruction set computer (ZISC036) from IBM for a Higgs search, *Nucl. Instr. Meth.,* A-357:192–194.

Lindsey, C. S., Lindblad, T., Minerskjöld, M., and Sekhniaidze, G. 1994. Experience with the IBM ZISC Neural Network Chip, *AI and Software Engineering in High Energy and Nuclear Physics.*

Moody, J. and Darken, C. 1986. Fast learning in networks of locally tuned processing units, *Neural Computation,* 1(2):181–200.

Powell, J. D. 1986. Radial basis functions for multivariable interpolation: a review, *Algorithms for Approximation,* Masona, J. C. and Cox, M. G., eds., Clarendon Press, Oxford, UK.

Reilly, D. L., Cooper, L. N., and Elbaum, C. 1982. A neural model for category learning, *Biol. Cybernet.,* 45:35–41.

Scofield, C. L. and Reilly, D. L. 1991. Into silicon: real time learning in a high density RBF network, *Proc. IEEE Conf. Neural Networks,* July, Seattle, WA.

Specht, D. F. 1968. *A Practical Technique for Estimating General Regression Surfaces,* LMSC–6–79–68–6, Lockheed Missiles and Space, Palo Alto, CA.

Specht, D. F. 1990. Probabilistic neural networks, *Neural Networks,* 3:18.

Székely, G., Lindblad, T., and Lindsey, D. 1995. *Evaluation of a RBF/DDA neural network, AIHEP,* Pisa, Italy.

Wasserman, P. D. 1993. *Advanced Methods in Neural Computing,* Van Nostrand Reinhold, New York, NY.

76

Hardware Implemented Radial Basis Function (RBF): The IBM Zero Instruction Set Computer

Thomas Lindblad
Royal Institute of Technology

Clark S. Lindsey
Royal Institute of Technology

Åge Eide[1]
Royal Institute of Technology

76.1 Introduction 1019
76.2 The ZISC036 VLSI Chip 1019
76.3 Processing and Training 1020
76.4 Implementing the Chip 1021
76.5 Summary and Extrapolations 1022

76.1 Introduction

There are several good reasons for implementing neural network techniques for software or hardware. Generally, neural networks are systems that learn how to process data, rather than being programmed, as is the case with von Neumann computers. This means that neural networks are of interest whenever a suitable algorithm is not easily available, when a highly nonlinear problem is at hand, when a certain slack in operation is desired, or when the capability to generalize is required.

Since neural nets are massively parallel in their intrinsic structure, they should also be fast. Of course, this is not the case if a neural network is executed on a serial computer. For the sake of gaining speed, this implies implementation in hardware.

The radial basis function (RBF) neural network is very simple in its topology, as well as in its operation. The reader is referred to Chapter 75 for an overview of the state-of-the-art networks. The IBM ZISC036 (Le Bouquin, 1994) is a *digital* implementation with 36 neurons and 2304 synapses in each chip. Operating at 20 MHz, it compares to biological systems as shown in Figure 76.1. A well-known analog implementation is the Intel ETANN 80170 having 64 neurons and 10,240 synapse weights (Intel Corp.). We have included this VLSI implementation in Figure 76.1, to show the state-of-the-art of commercial hardware in June 1995. Using today's technology, analog implementations are generally faster than digital ones.

Although only a few commercial implementations of neural network hardware are found today, the field is receiving considerable attention and a lot of prototype chips are available (Lindsey and Lindblad, 1995; Heemskerk). The aim of this paper is to present some experience (Lindsey et al., 1995; Lindblad, 1995; Lindblad et al., 1995, 1994; Eide et al., 1994) with the ZISC036 chip, compare it to other systems, and suggest future applications.

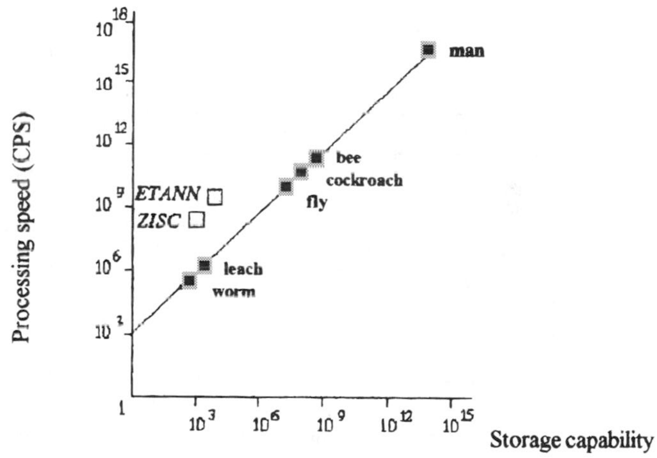

Figure 76.1 Although it is perhaps not fair to compare biological and silicon neural networks, we here plot speed vs. storage capability for a few biological systems and for the digital ZISC036 from IBM and the analog ETANN 80170 from Intel. The precision of the latter is about 1%, corresponding to 6 or 7 bits.

76.2 The ZISC036 VLSI Chip

This chip is the first one in a series of building blocks from IBM. It has been implemented in a fairly standard technology, 1 μ

[1] Permanent address: Ostfold College, Halden, Norway

0-8493-8343-9/97/$0.00+$.50

CMOS and is available in a 144 pin surface mount package. It is a highly cascadable chip, which means that systems with more than 36 neurons are easily obtained. Also, the software is independent of whether you are using one or several chips. The architecture of the ZISC036 is shown in Figure 76.2 and 76.3. Each neuron has a register file for prototype storage, as well as a unit for evaluation of distances. The upper limit of the number of *categories* is equal to 16K. It could be mentioned here that the Ni1000 RBF-chip (Intel, 1992) has 256 5-bit inputs, 1024 hidden and 64 output neurons. The 1K prototypes are stored in a 1.3 Mb flash EPROM.

In the case of the ZISC036, the hidden or prototype neurons use simple stepfunctions (Ni1000 uses Gaussians). When a new vector is presented to the network, each RBF neuron calculates the distance d between its prototype vector and the input vector. If the neuron fires, it will activate the output or category neuron to which it is connected. In the ZISC architecture, each hidden neuron is only connected to one output neuron (category). In other RBF topologies, all neurons in these layers may be interconnected.

The *distances* (d) can be calculated according to two norms, L_1 or L_{sup}

$$L_1 = \sum |V_i - P_i|, \qquad (76.1)$$

where V_i and P_i represents the input vector and the stored prototype elements, respectively. The summation runs from 1 to 64

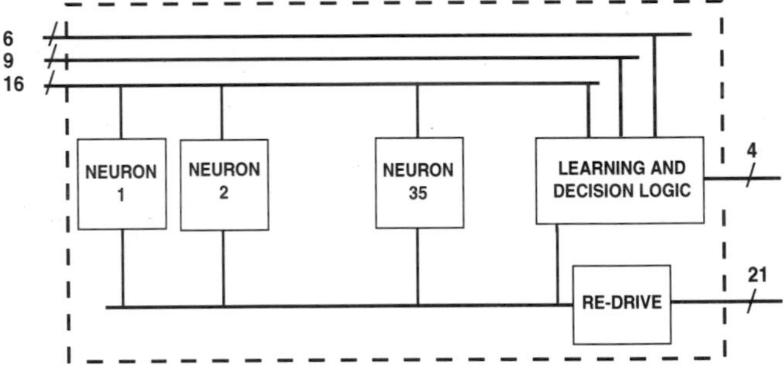

Figure 76.2 ZISC036 block diagram. The top part shows the address (6-bit), control (9-bit) and I/O data (16-bit) buses; to the right is shown the decision bus (4-bit) and the inter-ZISC communication bus (21-bit)

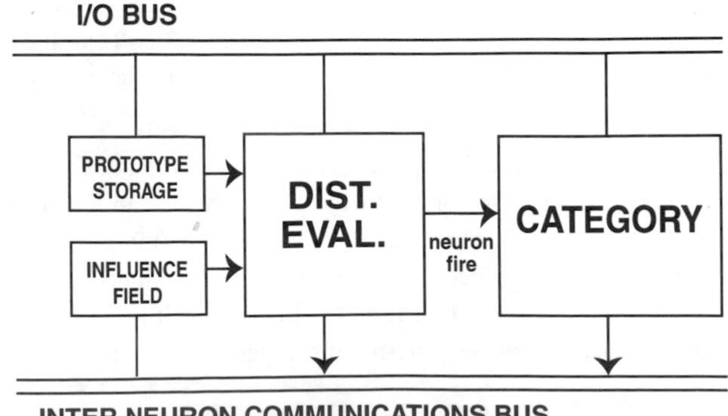

Figure 76.3 Block diagram of one of 36 neurons in the ZISC036 circuit. Connections are bidirectional except when shown using arrows.

and each component of the vector is coded as an 8-bit number. This is referred to as the Manhattan block distance and the distance calculations are carried out using a 14-bit accumulator. Alternatively, we can have

$$L_{sup} = \max |V_i - P_i|, \qquad (76.2)$$

with $i = 1, n$, for an n element vector. This is referred to as a "hypercube field." The components of the vectors are fed in sequence and processed in parallel by each neuron. This means that if the ZISC036 is operated at 20 MHz, 64 components can be fed and processed in 3.2 μs. The evaluation is obtained in 0.5 μs (or ten o'clock cycles) after the feeding of the last component, which corresponds to a quarter of a million evaluations per second on a 2000 MIPS von Neumann processor.

The ZISC036 has two registers holding the values of the minimum and the maximum *influence fields* associated with the *prototypes* (`MIF` and `MAF`, respectively). These values may both have values between 0 and 16,383, of course with `MIF` < `MAF`. The CLEAR signal will set the values to default values of 2 and 4096, respectively. As discussed in the processing and training section, if an input vector falls outside (inside) the influence field(s) of any prototype(s) associated with a particular *category*, it is not recognized (declared belonging to that category). However, if the input vector lies within the influence field of two or more prototypes associated with different categories, it is declared as recognized but *not formally identified.*

Other features of the neural network include a K-nearest neighbors mode (KNN), which returns the distances between a vector feature and stored prototypes in ascending order of distance and a "save/restore mode," which makes it possible to read out the current network or download a new one.

76.3. Processing and Training

In the forward or reference processing, the following steps occur when a vector is presented to the input:

1. Each existing prototype neuron calculates, in parallel, the distance of its prototype vector to the input vector according to either of the two distance norms mentioned above (Equations 76.1 or 76.2).
2. Each prototype uses a neuron area of influence (NAIF) value such that if the distance is smaller than this value the neuron is "ON."
3. If the *only* prototypes that fire are of the *same* category, then the input vector is declared to be *identified* (ID).
4. If prototypes of *different* categories fire, the input vector is declared to be *unclassified* (UNC)
5. If *no* prototype fires, then the input vector is declared to be *unidentified* (NID).

If the ID signal occurs, the category of the neuron can be read. The status signals ID, NID, and UNC are available both from a status register and on three pins. In the KNN mode, the procedure is similar, except the distances to all prototypes are made available in *ascending* order, along with the corresponding category of each prototype.

However, the ZISC employs a ROI-like method and to initiate learning, the 14-bit category value of the input vector is applied to the chip immediately following the reference processing. Now, either of three things may happen:

1. If *unidentified,* a new prototype neuron is created. The vector is equal to the input vector. The NAIF is set to MAIF or to the distance to the closest prototype, whichever is the smallest.
2. If *unclassified,* or identified with the category differing from the input category, the NAIF for all prototypes of the *wrong* category neurons are reduced to the distance to the input vector. If there are *no* correct prototypes firing, a *new* neuron will be created. This neuron will have NAIF equal to the distance to the closest neuron.
3. If the input is *identified* then no further processing occurs.

Using this method the training pattern must be presented several times before there is no further increase in the number of prototypes and changes to NAIF[2] values have ceased.

Neural networks with RBF architectures generally employ learning paradigms of the ROI (Batchelor) or RCE (Reilly et al., 1982) type, or possibly some probabalistic extension (Scofield and Reilly, 1991). These paradigms all have some shortcomings. To overcome these problems, a dynamic decay algorithm (DDA) has recently been introduced (Berthold and Diamond, 1995). It uses two different thresholds, one of which must be overtaken by an activation of a prototype of the same class so that the new prototype is added, and a second threshold which is the upper limit for the activation of conflicting areas. This results in better classification and confidence in areas where the training patterns did not result in new prototypes. The DDA method also seems to generate a network with the possibility to generalize much better than nets trained using the RCE paradigm (Székely et al., 1995).

76.4. Implementing the Chip

A first implementation of the ZISC036 was made on an ISA bus using two ZISC chips (Lindblad et al., 1995). Later implementations include 4 to 40 chips on a VME card (Lindblad et al.) and 16 chips on an ISA board (Mårtensson, 1995) and recently 3 chips on a PCMCIA card (Figure 76.4) (Aware).

As most neural networks are tested for that ability to recognize characters, the ISA implementation with two ZISC036 was used for this purpose (Lindblad et al.). In this case, we started with a network in which a simple picture can be drawn by using "." for a picture element that is off, and "#" for a picture element that is on. A set of 8 * 8-byte patterns of the capital and lower-case letters (26 + 26) was used, and presented to the chip together with their pertinent ASCII codes. The parameters used during learning were:

1. Number of components of input vectors: 64.
2. Number of categories: 52(or 36).
3. Norm used: L1.

Noise was added to the input picture where noise was defined as random pixels set fully on or off (Lindblad et al., 1995). As a reference it could be kept in mind that a noise value of 30 corresponds to, on the average, 50% noise or every second pixel changed.

As shown in Figure 76.5, the performance of the ZISC036 depends strongly on the value of the maximum influence field (MAF), while it is insensitive to changes in the corresponding minimum value. Very good results are obtained for MAF > 2000. However, for still larger values, no improvement is seen. The performance of the ZISC for this benchmark is indeed significantly better than corresponding results for popular software RBF-codes. However, still better results are obtained using the new DDA algorithm (Berthold and Diamond, 1995) as pointed out in Székley, et al. (1995). The present results may be suggestive for implementation of this algorithm in forthcoming versions of the ZISC.

In another test referred to as "a Higgs search" (Lindblad, 1995; Lindblad et al., 1995, 1994), simulated data Figure 76.6, was used to train and test an RBF network with 8 inputs (the total and the transversal momentum of the four leading particles in the reaction $H^0 \rightarrow Z^0Z^0 \rightarrow \mu^-\mu^+\mu^-\mu^+$ and background reactions) and two category outputs (i.e., only Higgs particle or background reaction). In such an application, the ZISC could be used both for identification of Higgs and rejection of the background. The ZISC performance as reported by Lindsey et al. (1995) is 73% and 79%, respectively, for the RBF mode. In the KNN mode, a slightly better value, or 85%, for the rejection rate is obtained. Several ways of preprocessing/normalizing the data did not improve these values significantly.

It is well known that RBF/ROI/RCE networks do not generalize as well as other types of neural networks for similar numbers of neurons. As mentioned in Lindsey et al., 1995, an 8–5–1 analog feedforward network trained in 100 k to 400 k iterations with *backpropagation* generalizes better than the ZISC by 10% to 15%. A similar test (for the Higgs search mentioned above) in software using the DDA training method, showed a constant value (with increasing number of training vectors) of identification just below 90% (Székely et al., 1995). The background rejection, however, improved with increasing number of vectors. This is reasonable, since the background events surely may look much more different than the Higgs signals. We may conclude that a ZISC with the DDA algorithm would probably do as well as a feedforward/backpropagation net, but train much faster.

[2] The NAIF has selectable minimum and maximum values between 0 and 16,383. The choice of the maximum influence field may have a strong influence on how the network generalizes.

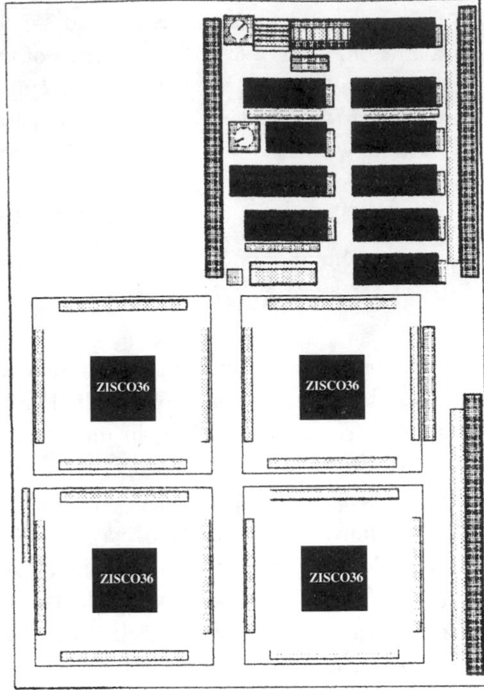

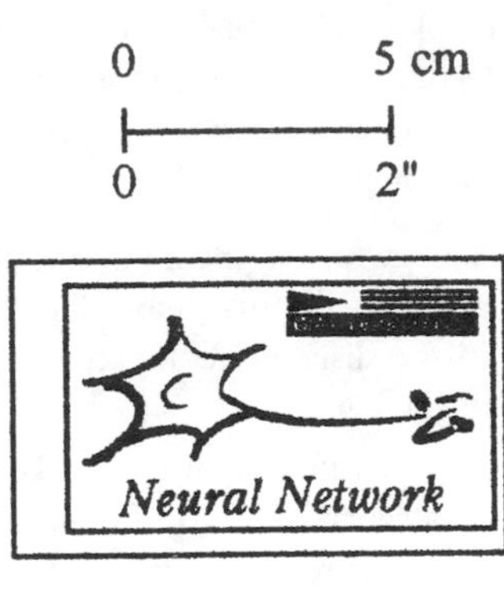

Figure 76.4 Schematic layout of the VME/ZISC036 board (left) (Lindblad et al., 1994). The lower part shows the piggy-back area that can hold 1–40 ZISC chips (by stacking), the upper right part holds the VMEBus interface and the upper left is user free. The VME card is of double Eurocard size (6U). The card for the PCMCIA bus is shown (right) for comparison. It holds three ZISC036 chips, a ZILOG bus interface chip with pertinent EEPROM and a 20 MHz clock. It is type I, i.e., 3.3 mm thick.

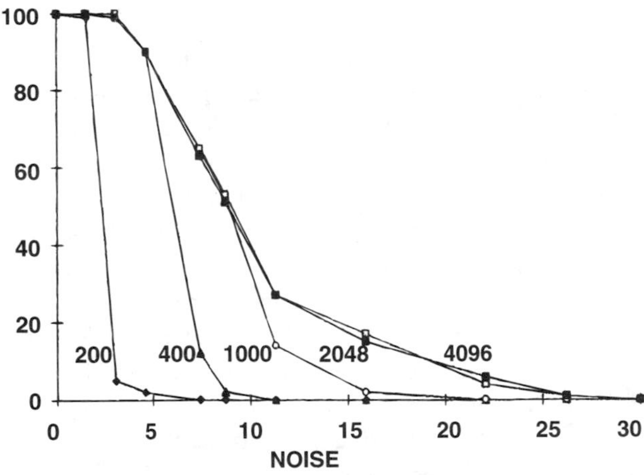

Figure 76.5 Recognized characters as a function of the noise. The parameter here is the maximum influence field (MAF). Values larger than MIF > 2048 all yield similar curves. The minimum influence field was kept constant in all cases and equal to the default value (2).

76.5 Summary and Extrapolations

The ZISC036 represents an implementation of a simple RBF neural network with on-chip learning. Although manufactured in standard IBM CMOS technology it is fairly fast at 20 MHz and has the feature of being easily cascadable. Since almost all pins of the chip are to be connected in such a case, it is tempting to build ZISC-towers, as mentioned in connection with the VME-implementation. One may also consider using the multi chip module (MCM) technology as described by Mårtensson (1995). Some of the benefits with this type of integration of several dice in the same package would be: reduced size, reduced number of solder points to the PCB, capability to handle high speed signals, and interconnection density. It is quite clear that for many experiments one needs a hundred or more neurons and thus several ZISC036 will be required. The architecture of the ZISC036 is such that the programming is independent of whether there are several chips or only one.

The fact that the ZISC036 is said to be the first in a series of neural network chips and the fact that there is a "0" before "36" may suggest that the next chip will carry more neurons (at least 100). Using the MCM technology on this chip would be very interesting. If using a more advanced technology than the present 1 μ CMOS, which yields a silicon area of about 1 sq. cm, MCMs with 4, 6, 8 or 9 ZISCs could probably be made.

The MCM technology is also very interesting when it comes to integrating two or more different dice, interconnected on the same substrate. Neural networks may frequently require preprocessing of the input data. This preprocessing may be for noise reduction or feature extraction or both. Figure 76.7 shows an example of a system with a wavelet transform chip (Aware) and a neural net (Lindblad, 1995). Wavelet analysis is fairly new to mainstream signal processing, although the mathematics have been known for many years. The technique is a projective one similar to the Fast Fourier Transform (FFT). However, rather

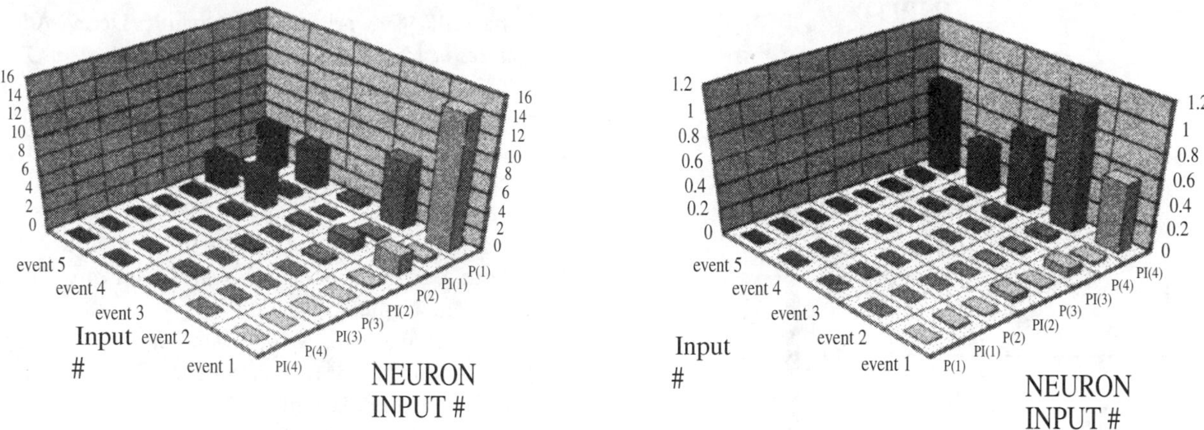

Figure 76.6 Ten examples of inputs to the neural network used for a "Higgs search". The left figure shows five Higgs and the right figure shows five background events. The eight inputs are p *and* p^T for the μ-particles 1 to 4.

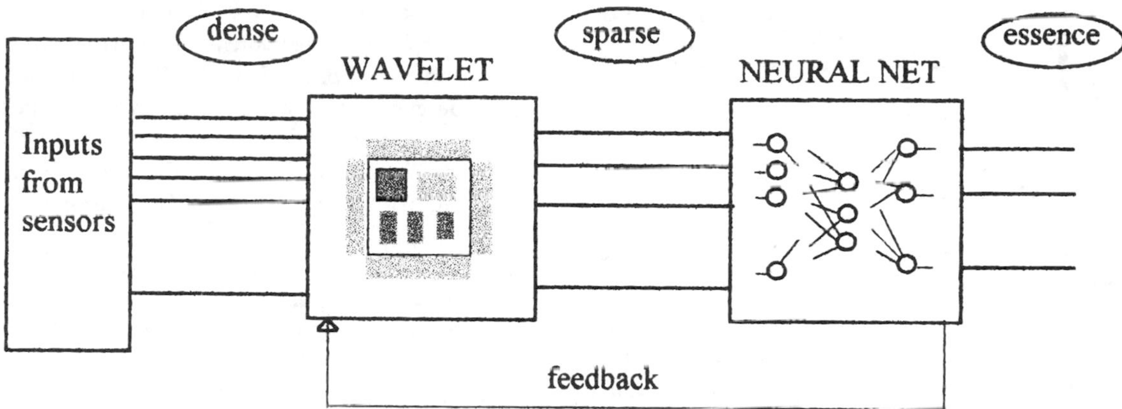

Figure 76.7 A combination of a wavelet preprocessor and a neural network may handle large volumes of primary data input. The data may be sparse after the feature extraction and the neural network will be trained to find the essential information.

than simply decomposing the signal into sinusoids of varying frequency, the data is represented as projections onto the affine group translations and dilations, of a fixed (basis) function called the *mother wavelet.* In a very general sense one may say that while the FFT is a technique used to locate regularities in a data set, wavelet analysis is useful for processing singularities and other nonstationary signals. Different wavelet transforms (Haar, Morlet, Meyer, etc.) may be used to extract different features of the original input signal and the MCM module may be tailored for different purposes. Inputs from pattern recognition systems have a tendency to be very dense. Hopefully, such a preprocessing system could yield a more sparse input to the ZISC in order to produce the essence in the final output.

Another interesting MCM module would be one that has one or more ZISCs combined with a von Neumann CPU (Figure 76.8). We feel that to make full use of the concept of neural networks in connection with intelligent front-end systems, one should perhaps learn from the interplay of the *hippocampus/ neocortex* in the human brain. The former is a kind of "quick and dirty" system used to "recall where I parked the car," while the latter is the system "used when learning HOW to drive a car." We think that consolidated neural nets will be very useful in the future in a wide area of research as well as in specific applications. If a military aircraft is hit, it may not be easy for the pilot to learn how to control the damaged plane. A neural network might learn faster than a human pilot can control a plane that suddenly behaves differently than the way it was designed. The damaged fighter, would instead have a guidance system with a consolidated neural network, that in many cases could restore its internal variables to proceed with the action of estimation as if nothing had happened.

The ZISC systems could be the *hippocampus* and the *neocortex* could be a feedforward/backpropagation network arranged as the interplay shown in Figure 76.8. The system is, of course, very useful when "unforseen things" happen. Hence, the coupling to front-end sensors. Supervising sensors may detect malfunctioning systems very early and complex physics experiments may not risk losing unforseen new data containing "novel physics".

Another way to reduce the number of interconnections when cascading ZISC would be to develop a bit-serial version. This technology seems to become more and more popular today.

The implementation on a card for the PCMCIA bus (Anderson) is also quite interesting. This bus is predicted, by several journals, including the *IEEE Spectrum*, to become very popular

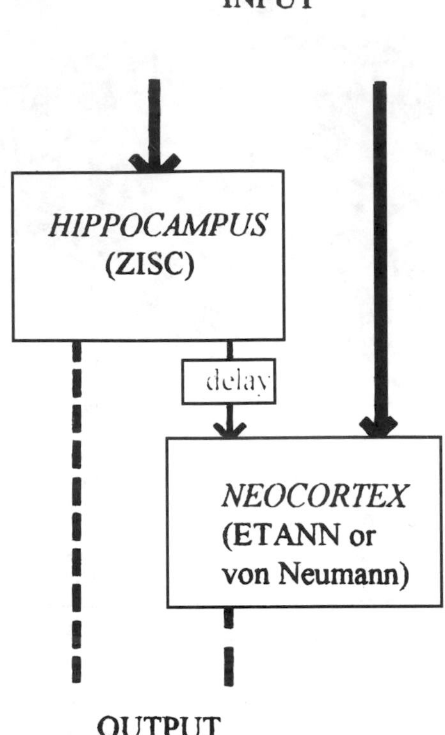

Figure 76.8 A consollidated neural network. Hippocampus and neocortex are replaced by "silicon."

in the future, and within a few years most desk-top computers will be equipped with this bus. This lightweight system will most likely also find applications in airborne systems, satellites, and space probes (Lindblad et al., 1995).

Acknowledgments

The present work was conducted under contract with the Swedish Research Council for Engineering Sciences (TFR), which is gratefully acknowledged. A special grant from the C. Trygger Foundation for studying front-end electronics is also very much appreciated. We would also like to acknowledge the support and comments of Drs. J-P LeBouquin and M. Grandguillot at IBM Microelectronics, Essonnes, G. Paillet of Neuroptics Consulting, Montpellier, and G. Laurens and B. Villiers of GIAT Industries, Toulouse.

References

Anderson, D. *PCMIA System Architecture,* Mind Share.

Aware. *Wavelet Transform Processor P/N 22500,* AD920505, Aware, Inc., Cambridge, MA.

Batchelor. *Practical Approach to Pattern Classification,* Plenum Press, London, UK.

Berthold, M. R. and Diamond, J. 1995. Boosting the Performance of the RBF Networks with Dynamic Decay Adjustment, *Advances in Neural Information Processing Systems 7,* Tesauro, G., Touretzky, D. S., and Leen, T. K., eds., MIT Press, Cambridge, MA.

Eide, Å., Lindblad, T., Lindsey, C. S., Minerskjöld, M., Sekhniaidze, G., and Székely, G. 1994. An implementation of the zero instruction set computers (ZISC036) on a PC/ISA bus card, *WNN/FNN 95 WDC,* p. 30; to be published by *SPIE.*

Heemskerk, J. N. H. *Overview of Neural Hardware,* University of Leiden, The Netherlands.

Intel Corporation. 1992. *Ni1000 Data Sheet,* Intel Corp., Santa Clara, CA (and Nestor Inc., Providence, RI); Chin, Park, Buckman, K., Diamond, J., Santoni, U., The, S.-C., Holler, M. Glier, M, Scofield, C. L., and Nunez, L., A radial basis function neural network with on-chip learning.

LeBouquin, J-P. 1994. IBM microelectronics ZISC, zero instruction set computer, preliminary information, *Poster Show WCNN,* San Diego, CA and addendum to conf. proc.; Lebouquin, J-P. and Grandguillot, M. IBM Microelectronics, Essones, G. Laurans and B. ZISC036/PCMCIA card, GIAT Industries, Toulouse and G. Paillet, Neuroptics, Monpellier, private communication.

Lindblad, T. 1995. Implementing the new zero instruction set computer from IBM, *The Hardware Designer,* 13:4–7, Älvsjö, Sweden.

Lindblad, T. Lindsey, C. S., Minerskjöld, M., Sekhniaidze, G., Székely, G., and Eide, Å. 1995. Implementing the new zero instruction set computer (ZISC036) from IBM for a Higgs search, *Nucl. Instr. Meth.,* A-357:192–194.

Lindblad, T., Lindsey, C. S., Minerskjöld, M., Eide, Å., Lindén, T., and Shelton, R. S. 1995. Attitude control systems for spacecrafts using neural networks and fuzzy logic, *AIHEP,* Pisa, Italy; Bockman, B., Eide, Å., Remme Johansen, S-E., Knutsen, M., Lindén, T., Lindblad, T., Lindsey, C. S., Minerskjöld, M., and Shelton, R. S. Star tracking using neural networks, submitted.

Lindblad, T., Lindsey, C. S., Minerskjöld, M., Sekhniaidze, G., Székely, G., and Eide, Å. 1994. The IBM ZISC036 zero instruction set computer, *(CERN) ONLINE,* 10:36–38.

Lindsey, C. S. and Lindblad, T. 1995. Survey of neural network hardware, *SPIE,* Orlando, FL.

Lindsey, C. S., Lindblad, T., Sekhniaidze, G., Székely, and Minerskjöld, M. 1994. Experience with the IBM ZISC036 Neural Network Chip, *AHIEP,* April, Pisa, Italy; to be published by *World Scientific.*

Mårtensson, E. 1995. Multi chip module technology, *The Hardware Designer,* 13:8–9.

Reilly, D. L., Cooper, L. N., and Elbaum, C. 1982. A neural model for category learning, *Biol. Cybernet.,* 45:35–41.

Scofield, C. L. and Reilly, D. L. 1991. Into silicon: real time learning in a high density RBF network, *Proc. IEEE Conf. Neural Networks,* July, Seattle, WA.

Székely, G., Lindblad, T., and Lindsey, D. 1995. Evaluation of a RBF/DDA neural network, *AIHEP,* Pisa, Italy.

77

The RCE Neural Network

77.1 Introduction 1025
Background • Network Description • Relation to Other Neural Networks • Advantages of RCE Network

77.2 Training the RCE Network 1027
Pattern Classification • Prototype Cell Responses During Training • The RCE Training Algorithm

77.3 RCE Network Responses 1032
Fast Response Mode • Output Probabilities Mode • RCE Network Responses on the Ni1000

77.4 Practical Guides to RCE Network Training and Use 1033
Statistically Reliable Training Set • Choice of Representation Features • Choice of Values for RCE Network Parameters

77.5 Applications of RCE to Pattern Recognition 1034
Character Recognition • Image Analysis Applications • Decision Making in Financial Services

77.6 RCE Network on a Commercially Available Neural Network Chip 1035

Douglas L. Reilly
Nestor, Inc.

77.1 Introduction

Background

The RCE neural network was designed as a general-purpose, adaptive pattern classification engine. Following a patent application submitted in 1980, a U.S. patent was granted for the RCE network in 1982 (Cooper et al., 1982). The first description of the network to appear in a technical journal was published in 1982, with a later elaboration appearing in 1987 (Scofield et al., 1988; Reilly et al., 1982).

As an adaptive pattern classification engine, the RCE network can solve pattern recognition problems in which data classes are represented by disjoint class distributions, linearly and nonlinearly separable class distributions, as well as nonseparable classes whose class distributions overlap. In this latter case, the RCE network outputs local probability density information that, along with known or assumed information on *a priori* class probabilities, can be used to compute an optimal pattern classification decision.

Network Description

The RCE network consist of three layers of "neuron cells", with a full set of connections (each represented by a connection weight) between the first and second layers, and a partial set of connections between the second and third layers (Figure 77.1). Each input layer cell represents a feature (a measurable characteristic) of an incoming pattern (an input signal) that the network assigns to some pattern class (category). The input signal is sometimes referred to as the pattern of activity (or activation pattern) of the input layer cells. The choice of input features is made based upon the nature and complexity of the pattern recognition problem.

The middle-layer cells are called prototype cells. Each prototype cell contains information about an example of a learned pattern class that occurred in the training data. The connections between a prototype cell and the input layer cells store the feature values of the class exemplar associated with the prototype.

Each cell on the output layer corresponds to a different pattern

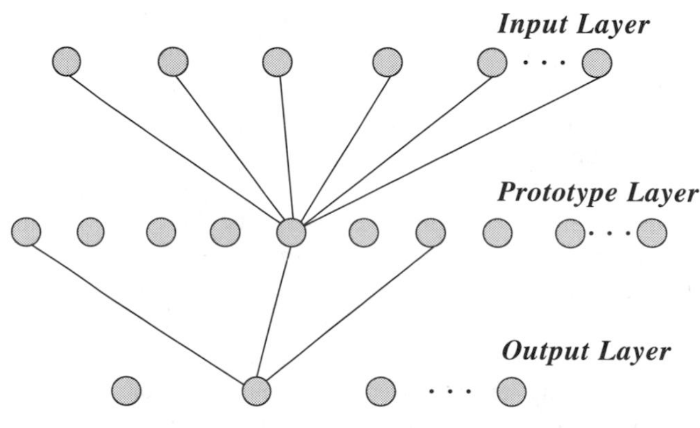

Figure 77.1 RCE network.

0-8493-8343-9/97/$0.00+$.50

class represented in the training data set. Prototype cells are class-specific. This class affiliation is represented structurally in the network by the single connection that a prototype cell makes to one and only one output cell (Figure 77.1). However, more than one prototype can be associated with the same pattern class. This means that an output cell can be connected to more than one prototype cell.

The knowledge about a class of patterns is stored in the network as a set of reference examples (prototypes) and the capability to generalize from these examples to new class instances. The RCE network applies a procedural, supervised training algorithm to grow the numbers of prototype and output cells, and to define values for their network connections, in order to perform pattern classification.

Relation to Other Neural Networks

RCE prototype cells use an exemplar-based function to compute their responses to a pattern of activity on the input layer. In most cases, this function computes either the Euclidean or city-block distance between the signal on the input layer and the vector of weights (the prototype vector) associated with the cell. Because of this, the RCE network is related to the class of radial basis function networks (RBF's) introduced in 1988 (Moody and Darken, 1989).

The most commonly used training algorithm for neural networks is currently the backpropagation of errors algorithm, first described in 1974 and, independently, in 1986 (Rumelhart et al., 1986; Werbos, 1974). The many variations of this algorithm all involve modifying network weights based upon a gradient descent approach to minimizing an error term. The error term is defined as a function of the difference between the desired and actual network responses to a pattern of activity on the network input layer. Whereas the backpropagation of errors technique has the advantage of being able to train neural networks with arbitrary numbers of cell layers, it has the disadvantage of training very slowly, requiring many passes (epochs) through the training set before the network weights converge on a final set of values. Further, the training algorithm can occasionally result in the network becoming "stuck" in a condition that prevents further changes to the weights, but without having arrived at an accurate solution to the pattern recognition problem.

By contrast, the RCE network uses a procedural training algorithm that avoids the long training times and problems of false convergence that can occur with backpropagation. Because it does not employ the gradient descent approach to minimizing an error function, RCE training offers guaranteed convergence, completing its training usually in 3–4 training passes through the data. Unlike backpropagation with its ability to train any neural network regardless of structure, the RCE training algorithm can only be applied to networks having three layers of cells. For pattern recognition applications, this is not a serious limitation since researchers have shown that any pattern classification problem can be solved by a neural network having at most, three layers (Irie and Miyake, 1988). Additionally, systems have been constructed with multiple component RCE neural network modules, each of which learns to solve portions of a pattern recognition task and which, together, cooperate to provide an integrated solution to the overall classification problem (Reilly and Cooper, 1990).

The RCE training algorithm grows the number of middle and output layer cells used by the network to solve the pattern recognition problem. RCE training differs from that employed in the related Probabilistic Neural Network (PNN), in that it allocates a new middle-layer cell only when the existing set of prototype cells is insufficient to correctly classify a pattern in the training set (Specht, 1988). PNN allocates a middle-layer cell for each exemplar in the training set.

Advantages of the RCE Network

The distinctive capability of the RCE network to automatically size itself during training solves a design issue for its users. By controlling the allocation of prototype and output layer cells, the RCE training algorithm eliminates the need to know in advance how many cells to specify in the middle layer of the network. This choice is a critical design parameter for users of networks trained with backpropagation. Choosing a number of middle layer cells that is either too small or too large can prevent such networks from training to a good solution for a pattern classification problem.

During RCE training, the middle-layer prototype cells develop expertise in classifying input signals that occur within their neighborhood of feature space. Information about a pattern class is represented among a subset of these prototype cells. Because of this, and because of the ability of the network to commit such cells dynamically, it is possible to incrementally train a previously trained network on new data examples without having to re-present the entire training set to the network.

During the course of training on new data, the RCE network will produce incorrect network responses that will guide the developer in deciding which kinds of previously trained data needs to be represented. As an example, a network trained to recognize handwritten numbers will confuse 2's with examples of Greek α's when examples of Greek handwritten letters are first presented. In this case, only examples of 2's need be represented to the network while it is being trained on the new class α. This dynamic category learning can be important for applications where in-field training is required and representation of an entire initial training set is not possible or practical.

The RCE network is a relatively simple network to understand, in terms of its training procedure and its mechanism for classification. The procedural aspect of the training function lends itself toward a straightforward description in terms of feature space diagrams. This, together with the relatively simple mathematics employed by the network, makes the RCE network intuitively easy to apply to a pattern recognition problem.

There are a number of variations of the RCE network that have been implemented for pattern classification problems. The following description characterizes the RCE network training algorithm and output response modes as they are executed in a commercially available chip (the Ni1000 Recognition Accelerator™) that implements RCE along with other radial basis function networks.

77.2 Training the RCE Network

Pattern Classification—Learning Territories in Feature Space

The clearest description of RCE network functions makes use of feature space diagrams. It is helpful to begin by introducing the term "feature space."

An input signal to the RCE network consists of a set of feature values, each value represented as the activation of a particular input cell. The set of features chosen to characterize an input signal for the network defines a feature space. The number of features in the set is referred to as the dimensionality (the number of axes) of the space.

The feature values describing a particular input signal locate the signal as a point in the feature space. The feature space itself is the set of all possible feature value combinations; i.e., it is the set of all possible points in the space. To measure the closeness or similarity between two input signals, a distance may be computed between their corresponding points in the feature space.

The correspondence between an input signal and a point in the feature space implies that a class of patterns is represented by a region or territory (i.e., a set of points) of the feature space. In general, the shape of the territory associated with a given class of patterns may be arbitrarily complex. A class of patterns may even consist of a collection of disconnected (disjoint) regions (Figure 77.2).

The solution to a pattern recognition problem requires an accurate description of the relevant class territories in feature space. With such a description, the class of an input signal can be identified by determining if the signal is contained within any of the feature space regions associated with that class.

The challenge in solving a pattern recognition problem is to accurately characterize the shapes of class territories that may be arbitrarily complex. It is useful to distinguish between two kinds of problems. In the case of simple (or separable) class regions, each point in the feature space belongs to one and only one category of patterns. This means that there is no overlap between the territories of any classes, although their shapes may be arbitrarily complex and disjoint class regions are allowed (Figure 77.3a).

Pattern classes whose regions overlap are said to have non-separable (or overlapping) class territories (Figure 77.3b). Any point in their shared feature space regions is associated with more than one class. In such cases, a probability of class membership must be estimated for a given point in the overlap regions.

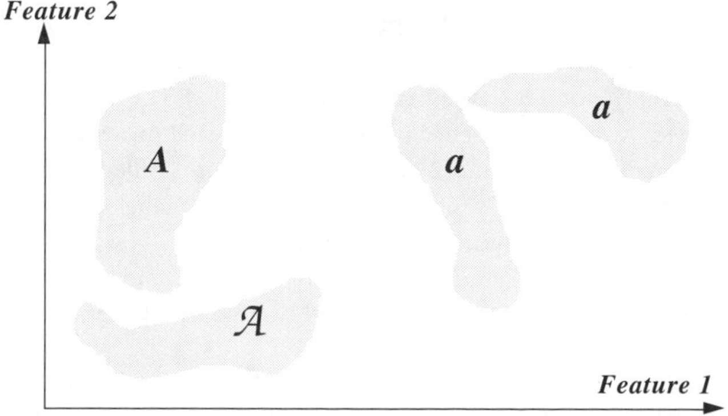

Figure 77.2 Disjoint regions for the class of patterns corresponding to the letter "A" in a hypothetical 2-D feature space.

Prototype Cell Responses During Training

When the RCE network trains on data, it learns the shapes of class territories in the feature space. These characterizations are developed by and stored in prototype cell parameters. An RCE prototype cell is characterized by five elements: its class, χ, its weight vector, ω, its cell threshold, λ, its pattern count, κ, and its smoothing factor, σ. During training, all but the smoothing factor play a role in prototype cell development.

The prototype cell weight vector, ω, represents the set of weighted connections between the prototype cell and each of the

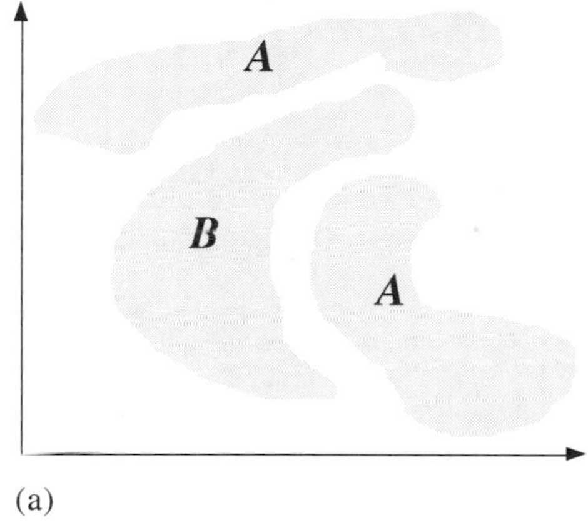

(a)

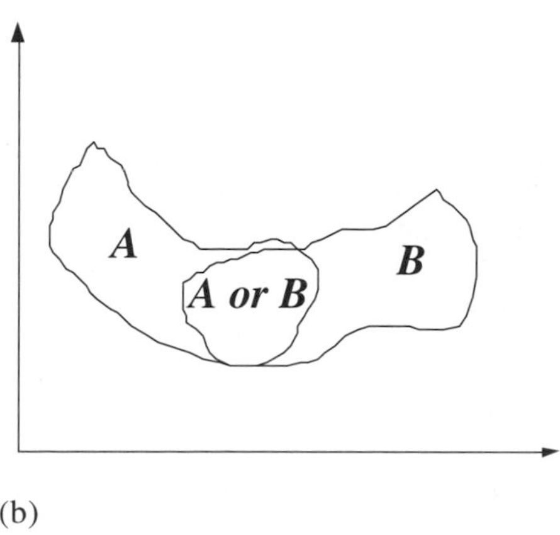

(b)

Figure 77.3 a. Pattern classes A and B are separable (A consists of two disjoint regions; b. Pattern classes. A and B are overlapping, sharing a region of points that could belong to either class A or B.

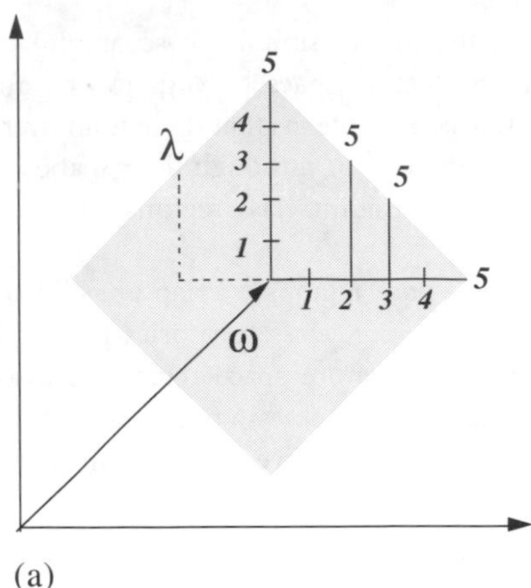

(a)

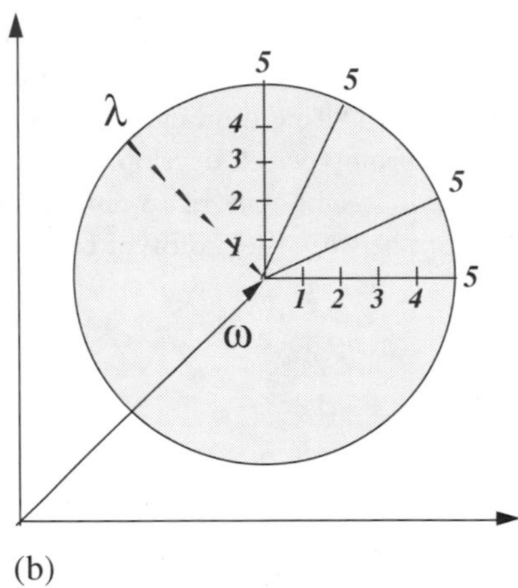

(b)

Figure 77.4 Examples of differently shaped prototype influence fields, each of a size $\lambda = 5$, in a hypothetical 2-D feature space. The ω marks the location of the weight vector. Use of the city-bloc distance for the prototype-to-pattern comparison yields the diamond-shaped influence field (a); use of the Euclidean distance function yields the circular influence field (b).

input layer cells. Because each prototype cell has one connection with each input cell, the prototype weight vector has the same dimensionality as the input signal. Just as the input signal defines a point in the feature space, so a prototype cell weight vector defines a point in the same feature space (Figure 77.4).

In response to a signal on the input layer, each prototype cell computes a distance between the input signal and the prototype vector stored in its weights. When the "city-block" function is used for this distance, it is computed by the ith prototype cell as

$$d_i = \sum_{j=1}^{N_D} |\omega_{ij} - x_j| \quad \text{(city block distance)} \qquad (77.1)$$

where ω_{ij} = the weight connecting the ith prototype cell to the jth input cell

x_j = the activity of the jth input cell (i.e., the jth feature value of the vector x)

and N_D = the number of input cells (i.e., the dimensionality of the feature space)

During training, a prototype cell will become active if the prototype-to-pattern distance, d, is less than the cell threshold, λ; if the distance d is greater than or equal to the cell threshold λ, then the prototype will not respond to the input signal. Referring to the output of the ith prototype cell as p_i,

$$p_i = 1 \quad \text{if} \quad d_i < \lambda_i \quad \text{(prototype fires)} \qquad (77.2a)$$

$$p_i = 0 \quad \text{if} \quad d_i \geq \lambda_i \quad \text{(prototype inactive)} \qquad (77.2b)$$

The cell threshold, together with the city-block distance function, describes a "region of influence" around the prototype cell in the feature space. During training, a prototype cell will fire for any input signal whose corresponding feature space location lies within the prototype's influence field. In the two-dimensional feature space illustrated in Figure 77.4a, the city-block distance function creates an influence field that looks like a diamond-shaped area centered on the point defined by the prototype weight vector.[1] As indicated earlier, it is also possible to choose a Euclidean distance function for prototypes.

$$d_i = \left[\sum_{j=1}^{N_D} (\omega_{ij} - x_j)^2\right]^{1/2} \quad \text{(Euclidean distance)} \qquad (77.3)$$

In this case, the influence field of a prototype in a two-dimensional feature space looks like a circular disk as shown in Figure 77.4b. Prototypes will be pictured with circular influence fields in the diagrams referred to in the following discussion.

The RCE Training Algorithm

Each prototype cell represents some local information (i.e., information in the small neighborhood of the feature space defined by its influence field) about the nature of the pattern class with which it is associated. During training, the RCE network will allocate prototype cells, positioning and sizing their corresponding influence fields so as to cover the feature space regions for each class of patterns present in the training data.

Before any training occurs, the RCE network can be pictured as consisting of a set of input cells and a set of unallocated prototype and output cells. By unallocated, we mean that they are simply not yet "wired into" the network.[2] The network is trained through a sequence of input signals, each presented with its correct classification. (A set of such patterns is called a labeled training set. A training algorithm that requires a labeled training set is called a supervised learning algorithm.)

The training procedure makes use of three mechanisms: prototype cell commitment, prototype threshold modification and

[1] In n dimensions, the influence field is an n-dimensional tetrahedron.

[2] In the commercial chip that implements the RCE, there are a total of 1000 unallocated prototypes and 64 unallocated output cells available for training purposes.

prototype pattern count modification. The process is illustrated for the pattern recognition problem shown in Figure 77.5, which portrays two nonlinearly separable pattern classes, C_1 and C_2, in a hypothetical two-dimensional feature space.

Prototype Cell Commitment

Let the first pattern presented to the network be an input signal, x_1, belonging to class C_1. Presentation of this pattern causes a new prototype cell to become committed (i.e., wired up) in the network. The influence field of this new prototype will be centered on the pattern x_1 (Figure 77.6). In the process of wiring up a prototype cell, several changes are made to the network.

First, the input signal is loaded into the prototype weight vector:

$$\omega_1 \leftarrow x_1 \tag{77.4}$$

This means that the influence field of the newly committed prototype will be centered on the pattern that caused the cell to be committed to the network. In effect, the prototype cell is "memorizing" a class exemplar from the training set.

Secondly, the prototype cell is assigned a cell threshold, λ. This assignment creates an influence field around the prototype.

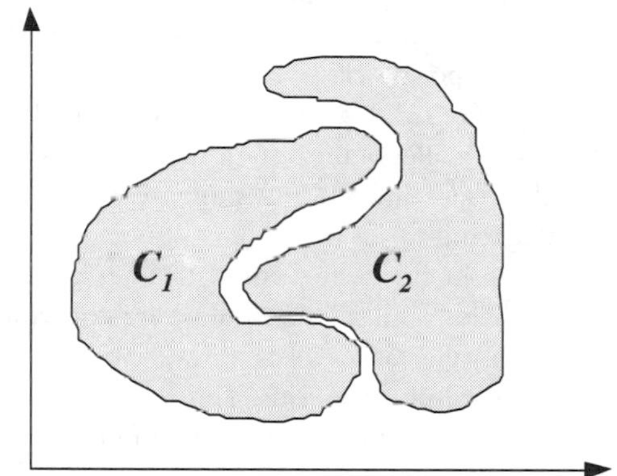

Figure 77.5 Hypothetical two-class recognition problem.

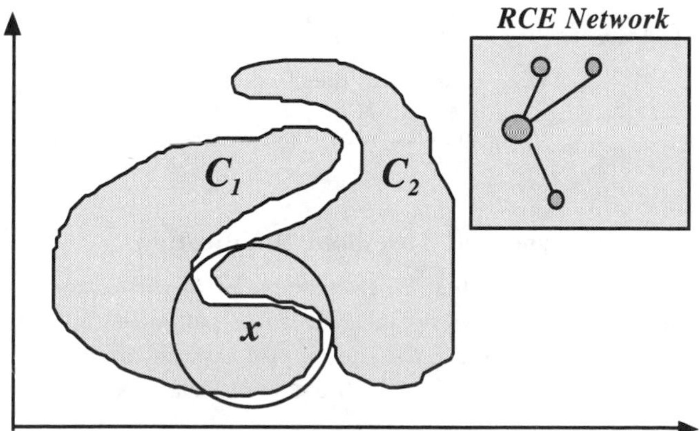

Figure 77.6 Prototype for C_1 class committed as a result of exemplar x. Pictured at upper right is diagram of RCE network, with newly committed prototype and output cell for C_1.

The prototype will use its influence field to determine how much it can generalize to respond to novel patterns that are similar to the memorized exemplar. In the case of the first prototype to be committed, the prototype is assigned the cell threshold λ_{max}, a user-specified parameter that defines the largest size that any prototype influence field can ever have:

$$\lambda_1 \leftarrow \lambda_{max} \tag{77.5}$$

Prototypes committed after this prototype will have their cell thresholds set either at λ_{max} or at some value less than λ_{max}, based upon their position with respect to other prototypes already present in the network.

Thirdly, a connection is made between the prototype cell and the output cell belonging to the class of the current input signal. This assigns a pattern class to the prototype:

$$\chi_1 \leftarrow C_1 \tag{77.6}$$

In this case, since no previous examples of this class (or any other) have been seen, a new output cell is committed to the network. Output cells are committed simply by establishing a connection to the newly committed prototype cell. The connection between the prototype cell and its associated output cell will carry a counter (the pattern counter, κ) that will store the number of times this prototype has correctly fired in response to a pattern belonging to its associated class. For a newly committed prototype, the pattern counter is set to one:

$$\kappa_1 = 1 \tag{77.7}$$

When the next input signal is presented to the network, the prototype activation is computed according to Equation 77.1 and 77.2. If the input falls within the prototype's influence field, the prototype cell will fire; this, in turn, triggers the corresponding output cell to fire. If the input signal is an example of class C_1, the network output will correctly classify the pattern. In effect, the network uses the prototype to generalize to recognize this new instance of the pattern class.

As long as subsequent input signals belonging to this class fall within the influence field of the prototype representing this class, no additional prototype cells are committed and no changes occur to the influence field of the prototype. However, each time an input falls within the prototype's influence field and matches the prototype's class, the prototype's pattern counter is incremented in order to keep a count of the number of "correct-class" patterns that have occurred within the prototype's influence field.

If

$$\text{Class}(x) = \text{Class}(\omega_i) \text{ AND } p_i = 1,$$

then

$$\kappa_i \leftarrow \kappa_i + 1 \tag{77.8}$$

The first occurrence of an input signal that belongs to this class but falls outside the influence field of the existing class prototype causes a second prototype to be committed for the class (Figure 77.7). The same commitment process occurs as described above: the input signal is loaded into the weight vector of the new prototype, the prototype cell threshold is set to λ_{max} and a connection is made between this new prototype and the output classification cell. The counter stored in this connection is initialized to 1.

As successive examples of this class are presented during training, each prototype cell determines its response by computing its distance to the input signal according to Equation 77.1 and comparing that distance to the prototype cell threshold stored with each prototype. A new prototype is committed in the RCE network only when an input signal does not fall within the influence field of any existing prototype belonging to the input signal's class.

Suppose an input signal belonging to a new pattern class, C_2, is presented to the network. Assume it falls outside the influence fields of any of the existing C_1 prototypes. This input will cause a new prototype cell to be committed; the input signal values will be loaded into the new prototype weight vector and the prototype will be assigned a cell threshold equal to λ_{max} (Figure 77.8). Because this is the first example of a new class of patterns, a new output cell is committed as well, representing the class C_2. The counter connection between the new prototype cell and the C_2 output cell is initialized to 1.

As illustrated in Figure 77.8, the influence field of a newly committed prototype may overlap the influence fields of existing prototypes belonging to different pattern classes. During training, influence fields are only tentative hypotheses about the class membership of the feature space points they contain. As discussed in the next section, future training examples may cause prototype influence fields to be reduced in size. If the class affiliation of influence fields is still uncertain during training, what is certain is that the central point of a prototype influence field (as defined by the weight vector, ω) must have a non-zero probability of belonging to the class of the prototype. (There is at least one

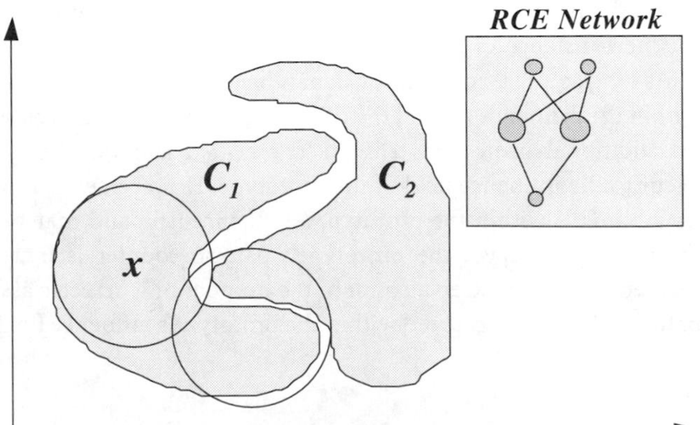

Figure 77.7 Second prototype for C_1 committed as a result of example pattern that is too dissimilar for initial prototype to classify. Upper right picture shows additional prototype cell being committed and connected to output cell for class C_1.

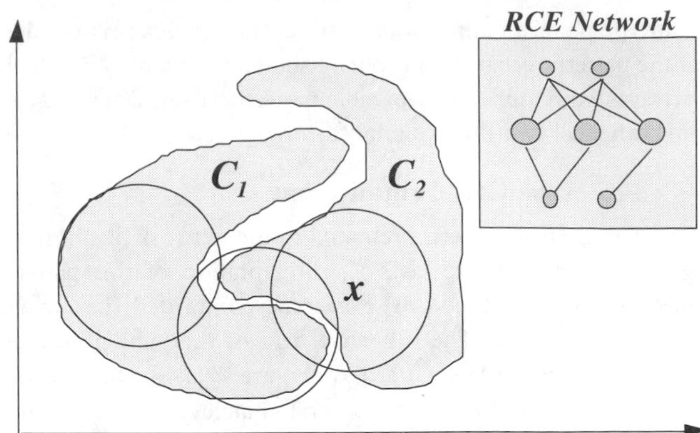

Figure 77.8 Example for new class C_2 causes new prototype cell (and output cell for C_2) to become committed in network. Note overlap between newly committed prototype and previous prototype for C_1.

input training pattern located at that point in the feature space that belonged to that class; this is the pattern that gave rise to the prototype.) Thus, the training algorithm allows a newly committed prototype to overlap the influence fields of other prototypes (in the chance that further training may yet revise their current thresholds to yield smaller influence field sizes), but it will not allow the influence field of a newly committed prototype to be so large as to contain the central point of a prototype for an opposing class.[3]

Thus, the influence field size of a newly committed prototype is chosen to be the smaller of the distance to the closest prototype of any other class (i.e., different from the class of the prototype to be committed) and λ_{max}.

As we shall see in the next section, there is a value below which influence fields (and cell thresholds) are not reduced; this value is λ_{min}. The value of λ_{min} sets a lower bound on the size of newly committed influence fields. Thus, the full specification for influence field determination for newly committed prototypes is the following:

$$\begin{aligned}&\text{Initial threshold of newly committed prototype}\\&\quad = \text{the smaller of } [\lambda_{max}, \text{ the larger of}\\&\qquad (\lambda_{min}, \text{ distance to the closest}\\&\qquad \text{opposing class prototype})] \qquad (77.9)\end{aligned}$$

Prototype Cell Threshold Modification

Now suppose that a new example of the first class C_1 is presented to the network, and that it falls within the influence field of a prototype for C_2. The C_2 prototype incorrectly fires, causing the output cell for C_2 to fire (Figure 77.9a). The network's

[3] A prototype that is to be committed with an influence field size of λ_{min} (defined as the smallest value an influence field can have) may contain the central point of prototypes that do not belong to its class.

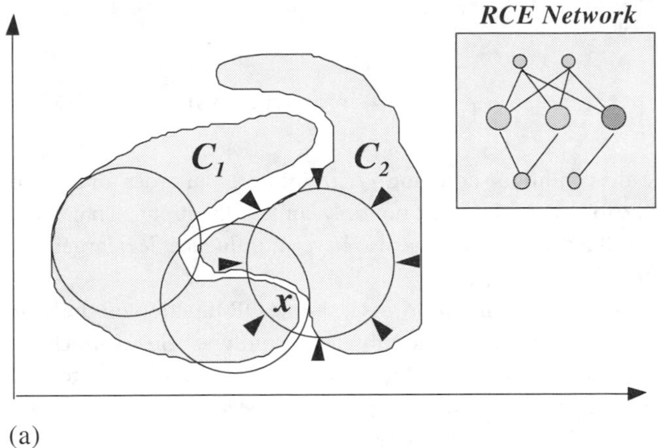

(a)

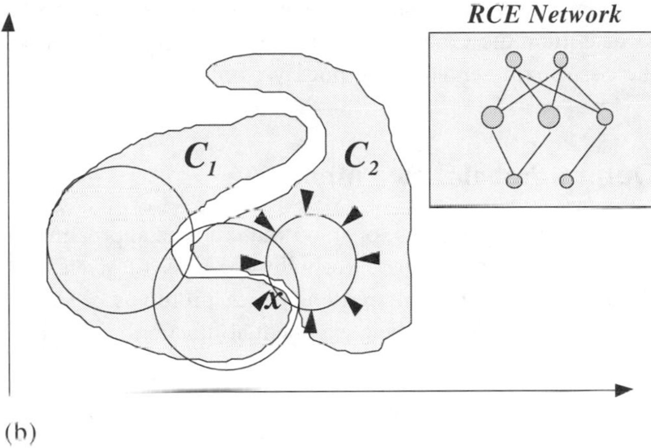

(b)

Figure 77.9 Example of class C_1 causes prototype for C_2 to fire. The training algorithm identifies the incorrectly active prototype (shown as the darkened cell in network diagram at upper right of (a) and reduces the cell threshold so that the influence field just excludes the input signal (b). Prototype cell with modified threshold is pictured with a smaller circle in the network prototype layer in the upper right of (b).

response is corrected during training by reducing the value of the threshold for the C_2 prototype to a point where the influence field of the C_2 prototype just excludes the input signal (Figure 77.9b). (Remember that the size of the prototype influence field is controlled by the value of the cell threshold.)

Just as there is a user-specified parameter which determines the maximum size of a prototype threshold (and correspondingly, the maximum influence field size), so there is a user-specified minimum threshold, λ_{min}, beyond which prototype thresholds are not reduced. As we shall see in the section on RCE Responses, prototypes that have reached this minimum size participate differently than prototypes that are above this threshold value in generating the network response to a pattern.

Since prototypes cannot be reduced below this value of λ_{min}, the value of λ_{min} sets a lower bound on the size of the influence field of a newly committed prototype. The new threshold for the C_2 prototype is

$$\lambda_{\text{modified}} = \text{larger of } [\lambda_{\text{min}}, \text{distance between input signal and prototpye for } C_2] \tag{77.10}$$

If none of the influence fields for C_1 prototypes contain the input signal, then a new prototype is committed for C_1 based on the input signal. Otherwise, the reduction in the incorrect prototype cell's threshold is sufficient to correct the response of the network and correctly classify the input.

Prototype Pattern Counts

In addition to prototype (and possibly output) cell commitment and prototype cell threshold modification, the only other mechanism involved in the RCE training procedure is the incrementing of prototype pattern counts. The pattern count of a prototype is incremented for every pattern that falls within the prototype influence field and that belongs to the same class as the prototype.

Prototype pattern counts are used by the network to approximate the local probability density values for a given class in a particular region of the feature space. This is important in those problems where the class territories are nonseparable.

In RCE training, the last training epoch is one in which no new prototypes are committed and no prototypes have their cell thresholds modified. (Further training with the given training data set would cause no changes to the network parameters.) At the beginning of every training epoch, all prototype pattern counts are initialized to zero. The pattern counts that develop during the last training epoch are those that are finally stored with the prototypes.

As noted before, the reduction in the influence field size of a prototype can alter the subset of correct class training patterns that lie within its new influence field size. Obviously, the introduction of a new prototype during the course of a training pass can also result in a smaller pattern count for this prototype than would occur if the prototype existed at the beginning of the training epoch. Thus, the pattern counts that develop during the last training pass are the most accurate estimators of probability density values because they develop for a set of prototypes whose number and influence field sizes have remained unchanged during the training pass.

Guaranteed Rapid Convergence of RCE Training

Figure 77.10 shows that, for separable pattern class problems, this simple RCE training procedure will result in coverings of the pattern class regions that correctly approximate their shape, regardless of the complexity of the shape and regardless of the number of disjoint territories that may comprise the definition of a pattern class.

The RCE network requires only a small number of presentations of the training set before it converges to a final solution. More than one training pass is required because a reduction in the size of a prototype's influence field during training may result in its failure to identify patterns which, when initially presented, fell within its formerly larger influence field. In such cases, these patterns may give rise to additional prototypes. Eventually, however, a training pass will occur in which no new prototypes are committed and no prototypes have their influence fields reduced. At this point, the RCE training has converged on a solution of the pattern recognition problem. This convergence is guaranteed to occur, and usually occurs in no more than 3–4 passes through the training set.[4]

77.3 RCE Network Responses

The RCE network can generate responses in either of two response modes. The first mode is geared toward providing a rapid identification of a pattern class that is separable from all other classes in the training set. However, if this mode does not provide a unique class identification, a second output mode can be invoked to provide an estimate of pattern class probabilities for the input signal.

Fast Response Mode

In this first mode of response, the network computes prototype cell activities for each prototype cell in the network by computing the pattern-to-prototype distances and comparing these with the threshold values stored with each prototype. In this mode of response, prototype cells use a modified version of the activation function used during network training.

If

$$(d_i < \lambda_i \text{ AND } \lambda_i > \lambda_{\min})$$

then

$$p_i = 1 \qquad \text{(prototype fires)} \qquad (77.11)$$

As the condition (Equation 77.10) indicates, in order for a prototype to be active, it must not only contain the input signal within its influence field, it must also have an influence field larger than the minimum size.

Each output unit performs a simple OR function on the input signals arriving from the subset of prototype cells to which it is connected. Thus, the output cell functions as a detector to indicate if any of its associated prototype cells that are above minimum-influence-field size are responding to the input signal.[5]

A single responding cell on the output layer of the network indicates an unambiguous identification of the input signal with that pattern category. If multiple output cells are active, or if none are active, then a second mode of response can be invoked to determine the probabilities that the input signal belongs to the classes represented by output layer cells.

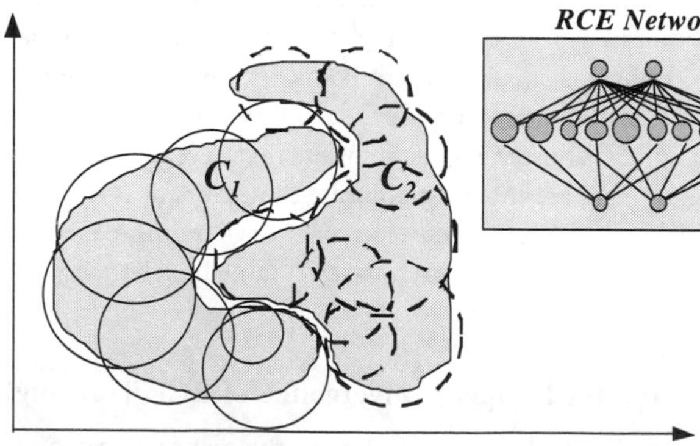

Figure 77.10 Fully trained network has committed sufficient prototype cells and modified their cell thresholds so that the prototypes for C_1 (shown as solid line circles) cover the class territory for C_1, while the prototypes for C_2 (shown as dashed circles) cover the class territory for C_2.

Output Probabilities Mode

In this mode, the response of an output cell is an approximation to $p(C|x)$, the conditional probability of class C, given input signal x. To compute this response, each prototype cell uses a radially symmetric, decaying exponential function of the form

$$p_i = e^{-\sigma_i d_i} \qquad (77.12)$$

where σ_i, the prototype smoothing factor, controls the rate at which the term decays as a function of d_i, given by Equation 77.1 as the distance between the input signal and the ith prototype. Each output cell then computes a weighted sum of the activations of the prototypes to which it is connected. In the case of the kth output cell, these are the prototypes associated with class C_k. In the activation sum, the activation of the ith prototype is weighted by the pattern count for that prototype, κ_i. Thus, the response of the output cell is given by

$$o_k = \sum_{P_i \in C_k} \kappa_i p_i \qquad (77.13)$$

The actual conditional probability $P(C_k|x)$ is computed by dividing o_k by N_f, a normalizing factor which is simply the sum of the activations of the output cells for all classes:

[4] The number of training epochs required is sensitive to the ordering of class examples in the training set. Faster convergence occurs for a randomly ordered training set as opposed to a set in which all examples of one class are presented, followed by all examples of the next, etc.

[5] When operating in this network response mode, the Ni1000 is designed to generate a list of classes represented among the "minimum-influence-field" prototypes that have been activated by the input signal. If no prototypes of any influence field size are active, this class list will be empty. In this case, host logic can produce a response of "Unidentified."

$$N_f = \sum_{k=1}^{N_C} o_k, \tag{77.14}$$

where N_c is the number of output cells.

$$P(C_k \mid x) = \frac{o_k}{N_f} \tag{77.15}$$

RCE Network Responses on the Ni1000

To achieve very high operating speed targets and to satisfy the objective of a scalable pattern recognition architecture, certain design modifications were made to the implementation of the RCE network on the Ni1000 Recognition Accelerator chip.

For scalability, the Ni1000 implementation of the RCE's probability response mode requires that the final normalization of output cell responses (i.e., the computation of Equations 77.14 and 77.15) be done off-chip, by the host processor. This enables a pattern recognition task to be distributed among a number of Ni1000 processors, working in parallel. In such an application, each chip computes its output cell terms, o_k. Host logic computes the N_f term, based upon the sum of all output cell activities for all chips. In the computation of the class specific output term, o_{class}, this logic may need to combine the output terms of different output cells for different chips. (A given class can be represented by the kth output cell on one chip and the mth output cell on another.)

To enhance operating speed, the Ni1000 implementation of the RCE network uses a particular form of a decaying exponential activation function for the prototype cell layer that is more naturally supported in silicon. By implementing an exponential decay function in base 2 as opposed to base e, the Ni1000 avoids unnecessary and time-consuming computational overhead. Specifically, on the Ni1000, the expression (Equation 77.12) for prototype activation is replaced by the following:

$$p_i = 2^{-\sigma_i d_i} \tag{77.16}$$

77.4 Practical Guides to RCE Network Training and Use

Like any other neural network or statistical learning algorithm, the performance of the RCE network is dependent upon the nature of the problem, the effectiveness of the input signal representations (i.e., the feature set) and the choices made for values of the network internal parameters that govern training and output response generation.

Statistically Reliable Training Set

For pattern recognition systems to develop a good solution to a recognition problem, the training set must be chosen in a way that represents the problem. Training sets that are composed from unrealistic or biased sampling will not have the same statistics as real world data. These statistics determine, after all, the location of pattern class territories, and, in the case of overlapping classes, the relative probabilities for different classes in such territories. To the extent that the statistics (i.e., class distributions in feature space) of the training set do not accurately reflect the statistics of real-world data, the network performance on the training set will not be predictive of its performance on "live" data. In such cases, what the network learns from the training set will not allow it to perform well in the real world.

Nonetheless, in some cases, it is possible to convert the output probabilities of a network trained on one sample of data to those that should be produced when the network is applied to a second data sample whose statistics are different from that of the training set. This need arises in those problems in which some classes have extremely low probabilities of occurrence. In composing a training set, the more likely occurring pattern classes are undersampled in order to avoid creating training sets that are excessively large.

Suppose the true *a priori* probabilities of a set of classes are represented by $P(C_1), \ldots, P(C_N)$, while their *a priori* probabilities in the training set are given by $P'(C_1), \ldots, P'(C_N)$. If, in response to an input signal x, the network produces a class probability $P(C_i|x)$ on the training set, the actual probability, in the context of the real world data, can be approximated by scaling $P(C_i|x)$ by the ratio $P(C_i)/P'(C_i)$. This is an approximation, and is useful only if the sampling of the training set has been random within each class of patterns.

Choice of Representation Features

The shape of the pattern class territory is very dependent on the selection of features chosen to characterize the input patterns. Omitting features from the representation that are critical in distinguishing one class of patterns from another will result in separable classes appearing as nonseparable, overlapping territories. At the same time, the inclusion of features that have no relevance to a pattern recognition problem can result in class territories occupying larger volumes in the pattern space than they would otherwise, resulting in RCE networks with large numbers of prototype cells.

Even relevant features, if too "low-level," will result in a need for large numbers of training patterns because the class territories that arise may consist of a large number of disjoint regions scattered throughout the pattern space. The RCE network is sensitive to the effects of too low-level a representation because its training does not generate new feature representations that can be used to re-engineer the feature space, rearranging pattern class territories. Within the feature space defined by the input signals, the RCE works to accomplish the best separation of input pattern classes and estimates of their class probabilities. A complex feature space class distribution that consists of numerous, unrelated, disjoint territories will result in a large number of prototype cells being committed by the RCE network.

In nearly all cases, an understanding of the problem domain creates the opportunity to engineer higher-level representations for presentation to the network. As an example, the best representations for a character recognition task are not pixel-based, but employ, instead, higher-level groupings of pixels that reflect some structures in the image (e.g., straight lines of different orientations, corners, intersections, etc.). Use of such higher-level representations can yield substantial benefits in terms of fewer prototypes committed, higher accuracies and better performance of the network in generalizing to correctly classify novel examples outside of the training set.

Choice of Values for RCE Network Parameters

The RCE parameters that most affect performance on a given problem are λ_{max}, λ_{min} and σ, the prototype smoothing factor.[6] In no case can choices be made for any of these values which would result in the RCE training failing to converge. The convergence guarantee is independent of the values chosen for λ_{max} and λ_{min}.[7] However, choices for these parameters can affect the numbers of prototypes committed during training and, to a lesser extent, recognition accuracy.

Smaller values for λ_{max} will result in the commitment of more prototypes, simply because the network will require more prototypes to cover pattern class territories. More specifically, if the size of λ_{max} is small relative to the average size of the class territories in the feature space, more prototypes will be required to solve the pattern recognition problem.

On the other hand, the effect of choosing larger values of λ_{max} will be a tendency by the system to generalize more aggressively when presented with novel input patterns. This effect will be most noticeable if the initial training set does not fully capture the statistics of the data that the network will process in the future. This effect will be particularly observable if the network is trained incrementally in the field through dynamic category addition, where, as often happens with in-field training, wholly new pattern classes are introduced at later times.

Similarly, choices for λ_{min} can have very noticeable effects on network performance. The larger the value for λ_{min}, relative to the average size of pattern class territories, the more likely that a separable pattern class problem will be treated as if it is non-separable. The smaller the value for λ_{min}, the more likely the network will commit a large number of prototypes for problems in which there are overlapping class territories that are large compared to the value chosen for λ_{min}. This is easy to see in the extreme case of choosing λ_{min} so small that only "point-value" influence fields are allowed. (The influence field is large enough to contain only the prototype weight vector.) In this case, training will cause a prototype to be committed for every distinct example of a pattern class contained in the overlap region.

Finally, although the value chosen for σ does not in any way affect the training of the RCE network (the training procedure is independent of σ), it can affect the probability responses generated by the network. As an example, choosing a value of $\sigma_i = 0$ for all prototypes (surely an extreme choice) would make the class probability estimate generated by the network independent of the actual value of the input signal. All prototypes would have activations of value 1, and the output cell responses (computed as normalized probabilities) would simply be equal to the *a priori* probabilities of each class, as computed from the training data set.

In general, the further the input signal is from the prototype, the less that prototype should contribute to the estimate of the probability of the input signal's belonging to the given prototype's pattern class. As an example, choosing $\sigma_i = 1/\lambda_i$ ensures that the contribution of the ith prototype to an output cell's (unnormalized) probability response falls to a value of $1/e$ of its contribution for input signals at the edge of its influence field as compared to those signals that fall at its field center.

[6] The value of σ, the smoothing factor, is not used during RCE training.

[7] The values for λ_{max} and λ_{min} must be properly chosen; i.e., $\lambda_{max} \geq \lambda_{min} > 0$.

77.5 Applications of RCE to Pattern Recognition

As in other neural network systems for pattern recognition, the mechanics of RCE network training and classification are completely independent of the meaning of the input signals presented to the network. This makes the network applicable to a broad range of pattern recognition tasks. The network has been applied to a wide variety of pattern recognition problems, including character recognition, image recognition, and a range of decision-making tasks in the area of financial services. The following discussion highlights the approach to feature generation and network training taken in several of these applications.

Character Recognition

The network has been applied to the problem of recognizing unconstrained handwritten characters, described either by image-based information, as might be available from scanning devices, or by stroke-based information, as might be available from devices that capture handwriting information online. Different features are defined for these two different contexts.

In the case of image based character recognition, one set of possible feature values corresponds to the registration of feature templates positioned at different locations over an image box containing the pattern to be recognized (Scofield et al., 1991). As an example, if the pattern is represented by a grid, 256 × 256 pixels in dimension, templates can be defined that are 16 × 16 pixels, for line segments oriented at 0°, 30°, 45°, 60° and 90°. Additional templates can be defined for different corner combinations and intersection styles (*T*'s, +'s, and *X*'s).

A given template is moved across the image to different sampled locations, and at each location, a function is computed which measures how well the template matches the pixel values in that area of the image. This produces a feature that measures the degree to which the given template is present at the particular image location. The set of all such feature values for each template at every sampled image location produces an input signal for the RCE network to use in classifying the image.

In the case of online character recognition, features can be defined that are based upon the sequence of points that occur as the pattern is being drawn. A stroke sequence of points can then be characterized by the magnitude of successive motions in the x and y directions, along with information on the rate of curvature change at various positions along the stroke. Such a representation will mean that very different feature input signals will be generated if a pattern is drawn using one sequence of strokes versus another. The RCE network will accommodate this by creating additional prototypes to learn these stroke variations.

Image Analysis Applications

One example of the application of the RCE network to an image analysis problem is vehicle detection on roadways (Bullock et al., 1993). Accurate, automatic vehicle detection systems can be used for a variety of traffic engineering applications, including queue length measurement and traffic disruption detection. A detection system must process a gray-scale image of a roadway view in order to determine the number of vehicles present in the scene.

The input features chosen for the problem convert the high-resolution gray-scale image provided by the video camera into a coarser representation for the neural network. An $a \times b$ pixel tile is defined whose value is computed from the average of the pixel gray-scale values it contains. Converting the image from an $m \times n$ array of gray-scale pixels to a $p \times q$ array of tiles reduces the dimensionality of the data. (Tiles do not overlap.) The coarser representation of the image is still a low-level representation; it makes use of no features corresponding to structural primitives. Although this representation does not carry enough information to enable the network to solve the problem of vehicle identification (e.g., deciding the particular make of automobile in the scene), it preserves enough information for the comparatively less demanding task of vehicle detection. This image analysis application illustrates the point that the complexity of the pattern recognition problem influences the level of complexity required in the input feature set.

In a similar spirit, the Ni1000 has been applied to the problem of classifying a fingerprint image in terms of component orientation maps (Shmurun et al., 1994). The maps, also known as ridge direction maps, are used in many fingerprint identification systems. The RCE network is trained to store in each of its prototype cells, a weight vector that corresponds to pregenerated templates for specific ridge directions. A fixed window is moved along the fingerprint image. To the center pixel of each window location, the network assigns the closest matching ridge classification. These local image classifications can then be provided as input to a second classification process to identify the fingerprint image.

Decision Making in Financial Services

Outside the realm of image data or general signal processing tasks, the RCE network has also been applied to pattern recognition problems that are at the heart of risk assessment problems in financial services. In particular, RCE networks have been trained to provide accept/decline decisions as made by underwriters on residential mortgage applications (Collins et al., 1993). In this case, the features used to create input signals for the network are computed from information available on the loan application. Such applications contain information on the borrower (e.g., length of employment, salary, etc.), the borrower's credit history (number of trade lines open, number of foreclosures, number of times 30 days late, 60 days late, etc.) and various ratios that underwriters consider in making their decision (loan to value ratio, etc.). By training on samples of accepted and declined mortgages, the network can learn to emulate the quality of decision-making capabilities of the mortgage underwriters as reflected in the training data set.

Another application in the financial services area involves the use of the network to detect fraudulent activity in credit card usage (Reilly, 1995; Ghosh and Reilly, 1994). Here the network is used to assign to each credit card transaction a score that reflects the likelihood of the transaction's being fraudulent. The input features that characterize a transaction are defined from characteristics of the transaction itself (amount of the transaction, location of the transaction, type of goods or service being purchased), the characteristics of recent purchase activity on the card (number of transactions made in the past several days, weeks and months, average dollar value of purchases in the past several days, weeks or months, etc.) as well as general information available about the cardholder's account (amount of available credit, how long the account has been open, elapsed time since a new credit card was mailed to the customer, etc.) All of these features, taken together, provide a picture of the current transaction in the context of the normal use of the card. By presenting the network with examples of both good and fraudulent transactions, each characterized by these features, the network is able to learn to identify a significant portion of fraudulent activity.

The above examples illustrate the broad applicability of the RCE network to pattern recognition problems. The principal requirements for applying the network to such problems are access to a reliably labeled training data set of sufficient examples and a means of characterizing data examples in terms of features that are relevant and appropriate for the problem domain.

77.6 RCE Network on a Commercially Available Neural Network Chip

As is the case with other neural networks, there is a high degree of parallelism in the computations that the RCE network performs for both training and pattern classification. Recently, a special purpose neural network chip, the Ni1000 Recognition Accelerator, has been designed and developed to implement in truly parallel fashion many of the operations performed by the RCE network and other networks of similar structure.[8] Significantly, the Ni1000 has been designed so as to perform not only

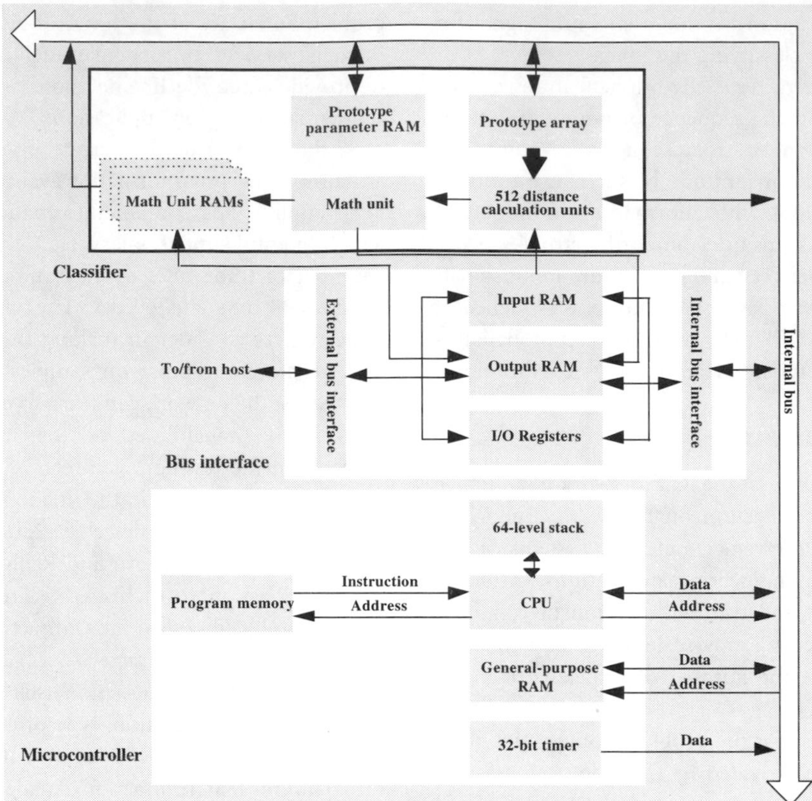

Figure 77.11 Block diagram of Ni1000 recognition accelerator chip.

RCE recognition, but also RCE training operations with on-chip logic (Holler et al., 1992). This makes it ideally suited for applications that require rapid, real-time, in-field trainability.

The Ni1000 Recognition Accelerator supports classification of over 32,000 patterns per second, with real-time adaptation. The chip is compatible with commonly used radial basis function paradigms, including RCE and PNN networks. The Ni1000 is designed to accept input vectors with a maximum of 256 features, each with 32 levels of resolution, and produces up to 64 classes and/or probabilities. High-speed parallel processing units compute the city-block distance between an input vector and up to 1000 stored prototypes. A block diagram of the Ni1000 Recognition Accelerator appears in Figure 77.11. The on-chip, custom, 16-bit microcontroller has separate program and data memories. The 4K × 16-bit nonvolatile FLASH EPROM memory can hold training algorithms, chip maintenance utilities and other software required by the application. A general purpose 256 × 16-bit RAM is also available to the microcontroller.

The microcontroller can enable an automatic classification mode in which a series of logic blocks, arranged as a pipeline, process data and output results to a host. The classification pipeline consists of input buffers, distance calculation units, a large FLASH prototype array that stores the results from the training process, a mathematical unit and its output memories, and an output buffer. At 33 MHz, the pipeline can classify over 32,000 input vectors per second, in which each input vector has up to 256 features with 5-bit resolution for each feature. The performance is made possible by the Ni1000 parallel architecture, which executes up to 16.5 billion operations per second. A typical Von Neumann machine would need to execute more than 65 billion instructions per second to approach the processing rate achieved by the Ni1000 Recognition Accelerator.

The Ni1000 makes it practical to apply the RCE network to numerous pattern classification tasks that have extremely high throughput requirements, or that require real-time or near real-time performance.

[8] The PNN network, as well as other radial basis function networks can also be implemented by the chip.

Acknowledgments

The author gratefully acknowledges the very careful editorial review and helpful suggestions made by Linda Mensinger Nunez, Michael Glier, Mark Laird and Christopher Bray in the preparation of this paper.

References

Bullock, D., Garrett, Jr., J., and Hendrickson, C. 1993. A neural network for image-based vehicle detection, *Transportation Research—C,* 1(3):235–247.

Collins, E., Ghosh, S., and Scofield, C. L. 1993. An Application of A Multiple Neural Network Learning System to Emulation of Mortgage Underwriting Judgments, *Neural Networks in Finance and Investing,* Trippi, R. T. and Turban, E., eds., 305–311, Probus Publishing, Chicago, IL.

Cooper, L. N., Elbaum, C., and Reilly, D. L. 1982. Self-organizing general pattern class separator and identifier, U.S. Patent No. 4,326,259, Apr. 1982.

Ghosh, S. and Reilly, D. L. 1994. Credit card fraud detection with a neural network, *Proc. Twenty-Seventh Hawaii Int. Conf. System Sciences,* IEEE Computer Society, 621–630, January, Wailea, HI.

Holler, M., Park, C., Diamond, J., Santoni, U., The, S. C., Glier, M., and Scofield, C. L. 1992. A high performance adaptive classifier using radial basis functions, *Proc. Gov. Microcircuit Applications Conf.,* 261–264.

Irie, B. and Miyake, S. 1988. Capabilities of three-layered perceptrons, *Proc. IEEE 2nd Int. Conf. Neural Networks,* I:641–648, July, San Diego, CA.

Moody J. and Darken, C. 1989. Learning with localized receptive fields, *Proc. 1988 Connectionist Models Summer School,* Touretzky, D. S., Hinton, G. E., and Sejnowski, T. J., eds., 133–143, Morgan Kaufman Publishers, San Mateo, CA.

Reilly, D. L. 1995. Neural Network Fraud Control in the Bank Card Industry, *Artificial Intelligence in the Capital Markets,* R. Freedman, R., Klein, R., and Lederman, J., eds., 437–468, Probus Publishing, Chicago, IL.

Reilly, D. L. and Cooper, L. N. 1990. An Overview Of Neural Networks: Early Models To Real World Systems, *An Introduction to Neural and Electronic Networks,* Zornetzer, S. F., Davis, J. L., and Lau, C., eds., 227–248, Academic Press, San Diego, CA.

Reilly, D. L., Cooper, L. N., and Elbaum, C. 1982. A neural model for category learning, *Biol. Cybern.,* 45:35–41.

Rumelhart, D. E., Hinton, G. E., and Williams, R. J. 1986. Learning representations by back-propagating errors," *Nature,* 332:533–536.

Scofield, C. L., Reilly, D. L. Elbaum, C., and Cooper, L. N. 1988. Pattern class degeneracy in an unrestricted storage density memory, *Neural Information Processing Systems,* Anderson, D. Z., ed., 674–682, American Institute of Physics, New York, NY.

Scofield, C. L., Kenton, L., and Chang, J.-C. 1991. Multiple neural net architectures for character recognition, *COMPCON Spring '91 Digest of Papers,* pg. 487–491.

Shmurun, A., Bjorn, V., Tam, S., and Holler, M. 1994. Extraction of fingerprint orientation maps using a radial basis function recognition accelerator, *Proc. WCCI,* 1186–1190, June, Orlando, FL.

Specht, D. F. 1988. Probabilistic neural networks for classification, mapping, or associative memory, *Proc. IEEE Second Int. Conf. Neural Networks,* I:525–532, July, San Diego, CA.

Werbos, P. J. 1974. *Beyond Regression: New Tools For Prediction And Analysis In The Behavioral Sciences,* Ph.D. dissertation, Harvard University.

78

Probabilistic Neural Networks Model

Donald F. Specht
Lockheed Martin Missiles & Space
Palo Alto, California

78.1 Basic PNN 1038
Limiting Conditions as $\sigma \to 0$ and $\sigma \to \infty$ • Estimating A Posteriori Probabilities • Probabilistic Neural Networks Using Alternative Estimators of $f(X)$ • Summary of Basic PNN

78.2 Adaptive PNN 1041
Comparisons with Backpropagation • Summary of Adaptive PNN

78.3 High-Speed Classification 1042
Hardware • Clustering • Maximum Likelihood Training

78.4 Other Considerations 1044
Detection—A Single Category Problem • Different Kernels for Different Categories • Provision for Unknown Categories

78.5 Summary 1046

78.1 Basic PNN

The Probabilistic Neural Network (PNN) is a network of parallel nodes (neurons) which can classify unknown patterns into categories based on their similarities to labeled training patterns, and which is trained by directly applying the principles of statistics. The "training" technique for the basic PNN is obtained by combining the Bayes strategy with a nonparametric estimator for probability density functions (Specht, 1988, 1990). Training of the basic PNN is very fast; advanced PNNs require iterative training but provide networks which are fast for evaluating new patterns and which can be used for automatic feature selection.

The Bayes strategy for pattern classification consists of making the classification decision which minimizes the expected risk. Risk, in turn, is the probability of making a wrong classification multiplied by the loss associated with making the wrong decision.

Consider the two-category situation in which the state of nature θ is known to be either θ_A or θ_B. If it is desired to decide whether $\theta = \theta_A$ or $\theta = \theta_B$ based on a set of measurements represented by p-dimensional vector $\boldsymbol{X}^t = [X_l \ldots X_j \ldots X_p]$, the Bayes decision rule becomes

$$d(\mathbf{X}) = \theta_A \quad \text{if} \quad h_A \ell_A f_A(\mathbf{X}) > h_B \ell_B f_B(\mathbf{X}) \qquad (78.1)$$
$$d(\mathbf{X}) = \theta_B \quad \text{if} \quad h_A \ell_A f_A(\mathbf{X}) < h_B \ell_B f_B(\mathbf{X})$$

where $f_A(\boldsymbol{X})$ and $f_B(\boldsymbol{X})$ are the probability density functions for categories A and B, respectively; ℓ_A is the loss associated with the decision $d(\boldsymbol{X}) = \theta_B$ when $\theta = \theta_A$; ℓ_B is the loss associated with the decision $d(\boldsymbol{X}) = \theta_A$ when $\theta = \theta_B$ (the losses associated with correct decisions are taken to equal zero); h_A is the *a priori* probability of occurrence of patterns from category A; and $h_B = 1 - h_A$ is the *a priori* probability that $\theta = \theta_B$.

A similar decision rule can be stated for the many-category problem:

$$d(\boldsymbol{X}) = \theta_k \quad \text{if} \quad h_k \ell_k f_k(\boldsymbol{X}) > h_q \ell_q f_q(\boldsymbol{X}) \quad \text{for all} \quad q \neq k \qquad (78.2)$$

where ℓ_k is the loss associated with the decision $d(\boldsymbol{X}) \neq \theta_k$ when $\theta = \theta_k$. For complete generality, ℓ should be defined as a matrix with different losses assigned for misclassification of a θ_k pattern to each of the incorrect categories. This is not usually necessary in practical problems, with one exception. If it is desired to classify uncertain patterns as "unknown," rather than risk the wrong classification, the loss associated with the unknown is less than that for a hard decision to classify $\boldsymbol{X}$ into the wrong category (Washburne et al., 1993). See subsection "Provision for Unknown Categories."

The key to using Equation 78.1 or Equation 78.2 is the ability to estimate PDFs based on training patterns. Often the *a priori* probabilities are known or can be estimated accurately, and the loss functions require subjective evaluation. However, if the probability densities of the patterns in the categories to be separated are unknown, and all that is given is a set of training patterns (training samples), then it is these samples that provide the only clue to the unknown underlying probability densities.

Parzen (1962) showed that a class of PDF estimators asymptotically approaches the underlying parent density, provided only that it is continuous. Cacoullos (1966) extended Parzen's results to cover the multivariate case. In the particular case of the Gaussian kernel, the multivariate estimates can be expressed as

0-8493-8343-9/97/$0.00+$.50

$$f_k(\mathbf{X}) = \frac{1}{(2\pi)^{p/2}\sigma^p}\,\frac{1}{m}\sum_{i=1}^{m}\exp\left[-\frac{(\mathbf{X}-\mathbf{X}_{ki})^t(\mathbf{X}-\mathbf{X}_{ki})}{2\sigma^2}\right] \qquad (78.3)$$

where
k = category
i = pattern number
m = total number of training patterns
X_{ki} = *ith* training from category k
σ = "smoothing parameter"
p = dimensionality of measurement space.

Note that $f_k(X)$ is simply the sum of small multivariate Gaussian distributions centered at each training sample. However, the sum is not limited to being Gaussian. It can, in fact, approximate any smooth density function.

Figure 78.1 shows a neural network organization for classification of input patterns X into two categories. In Figure 78.1, the input units are merely distribution units that supply the same input values to all of the pattern units. Each pattern unit measures a distance between the input pattern vector $\boldsymbol{X}$ and a stored vector $\boldsymbol{X}^i$. The *activation* of the pattern unit is a non-linear function of this distance. Two equivalent types of pattern units are shown in Figure 78.2 and Figure 78.3 The pattern unit of Figure 78.2

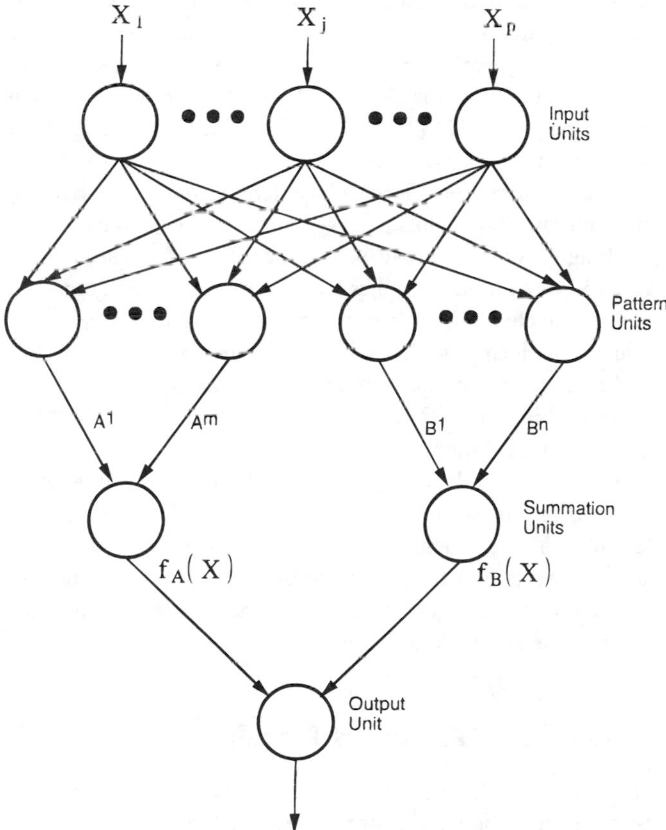

Figure 78.1 Organization for classification of patterns into categories.*

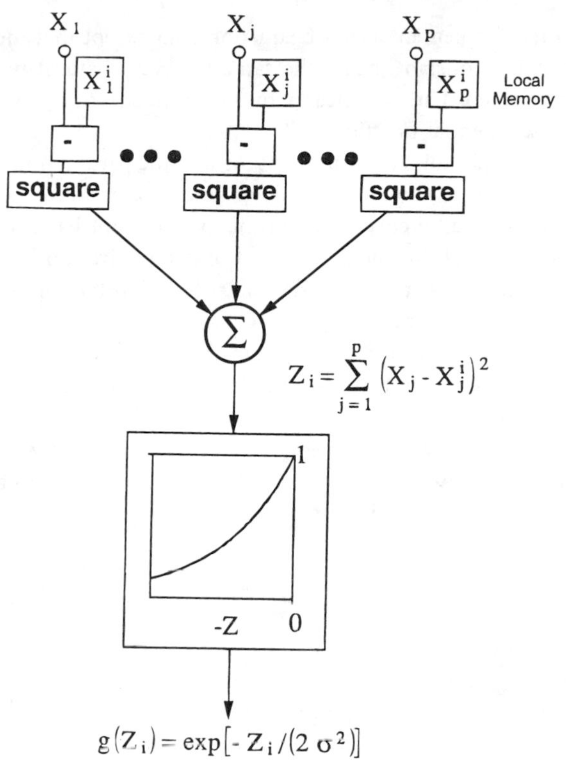

Figure 78.2 A pattern unit (Euclidian distance form).*

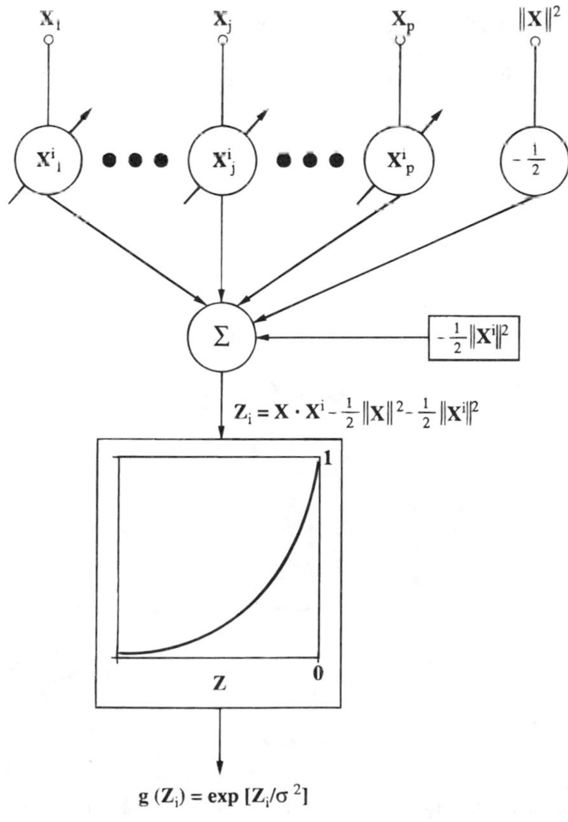

Figure 78.3 A pattern unit (dot product form).*

* (*Source*: Specht, D. F. 1992. Enhancements to probabilistic neural networks. In *Proc. IEEE Int. Joint Conf. Neural Networks*, Baltimore, MD. (c) 1992 IEEE. With permission.)

is a direct implementation of Equation 78.3 except that the pre multiplying constant, being the same for every pattern unit, is not computed. The nonlinear activation function of the neuron is the exponential function.

Figure 78.3 shows a dot product form of the pattern unit. This is very similar in form to the hidden units in a backpropagation network. The differences are that this pattern unit has two bias terms supplementing the dot product, and the activation function is exponential instead of the sigmoidal function typical of back propagation networks.

The summation units simply sum the inputs for the pattern units that correspond to the category from which the training pattern was selected.

For a two-category problem, the output units are two-input neurons, as shown in Figure 78.4. These units produce binary outputs. They have only a single variable weight,

$$C = -\frac{h_B \ell_B}{h_A \ell_A} \cdot \frac{n_A}{n_B} \tag{78.4}$$

where

n_A = number of training patterns from category A
n_B = number of training patterns from category B,

Note that C is the ratio of *a priori* probabilities divided by the ratio of samples and multiplied by the ratio of losses.

For a multiple category problem, the outputs of the summation units need to be multiplied by $h_k \ell_k / n_k$ and then the output unit becomes a maximum detector.

The network is trained by setting the $\boldsymbol{X}^i$ weight vector in one of the pattern units equal to each of the $\boldsymbol{X}$ patterns in the training set and then connecting the pattern unit's output to the appropriate summation unit. A separate neuron (pattern unit) is required for every training pattern (in basic PNN).

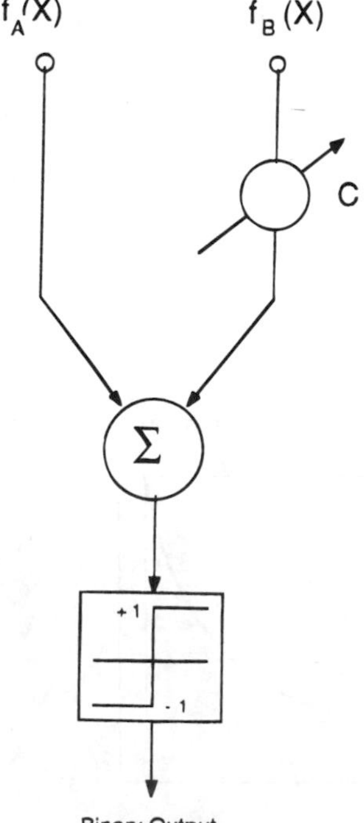

Figure 78.4 An output unit. (*Source*: Specht, D. F. 1990. Probabilistic neural networks. *Neural Networks* 3: 109–118. (c) 1990 Pergamon Press. With permission.)

For basic PNN, the weights A^j and B^j shown in Figure 78.1 all equal 1, and have no effect. It becomes necessary to have weights other than one when clustering is incorporated.

The best pattern unit to use for a particular application will depend on hardware availability. Implementation technologies that lend themselves to vector subtraction will be best used with the distance measuring form, and those that lend themselves to vector multiplication (such as those using digital signal processing (DSP) chips, and optical computers) will be best used with the dot-product form. Fixed-point computation is well suited for the distance measuring forms, whereas floating-point computation can be used equally well with either form.

Limiting Conditions as $\sigma \to 0$ and as $\sigma \to \infty$

It has been shown that the decision boundary defined by Equation 78.1 varies continuously from a hyperplane when $\sigma \to \infty$ to a very nonlinear boundary representing the nearest neighbor classifier when $\sigma \to 0$. In general, neither limiting case provides optimal separation of the two distributions. A degree of averaging of nearest neighbors, dictated by the density of training samples, provides better generalization than basing the decision on a single nearest neighbor. The PNN network is similar in effect to the K-nearest neighbor classifier.

In the typical classification problem involving overlapping distributions, if one plots a curve of accuracy vs. value of σ, the resulting curve has a broad peak with values of σ between 0.05 and 0.5 (when input variables have been normalized). Since the peak is usually broad, it is not difficult to find experimentally a value of σ which yields classification accuracies acceptably close to the peak. The only parameter to be adjusted in basic PNN is the smoothing parameter, σ, which is the same for every pattern unit and every dimension.

In order for each input variable to have equal influence on the decisions of the network, it is necessary to prescale the input variables to have roughly the same range or standard deviation. Standard deviations for each dimension should be computed by subtracting the category mean from each pattern vector, and pooling the data from all categories.

Estimating *A Posteriori* Probabilities

The outputs $f_k(\mathbf{X})$ of the summation units can also be used to estimate the a posteriori probabilities that $\boldsymbol{X}$ belongs to category $k, P[k|X]$. If pattern $\boldsymbol{X}$ belongs to one and only one of c categories θ_1 through θ_c, we have, from the Bayes theorem

$$P[k|\mathbf{X}] = \frac{h_k f_k(\mathbf{X})}{\sum_{j=1}^{c} h_j f_j(\mathbf{X})} \tag{78.5}$$

Probabilistic Neural Networks Using Alternative Estimators of $f(X)$

Many kernel functions besides Equation 78.3 satisfy Parzen's requirements to insure that the PDF estimators asymptotically approach the underlying parent density (Specht, 1990). Perhaps the most interesting of these is that of Equation 78.6, because it is easy to implement in parallel hardware or easy to compute in software simulation.

$$f_k(\mathbf{X}) = \frac{1}{n(2\lambda)^p} \sum_{i=1}^{n} \exp\left[-\frac{1}{\lambda} \sum_{j=1}^{p} |X_j - X_{kij}|\right] \quad (78.6)$$

Summary of Basic PNN

Operationally, the most important advantage of the PNN is that training is easy and instantaneous. PNN can be used in real time because, as soon as one pattern representing each category has been observed, the network can begin to generalize to new patterns. As additional patterns are observed and stored into the network, the generalization will improve and the decision boundary can become more complex.

Other advantages of PNN are:

a. The shape of the decision surfaces can be made as complex as necessary or as simple as desired, by choosing the appropriate value of the smoothing parameter σ.
b. The decision surfaces can approach the Bayes optimal.
c. Erroneous samples are tolerated.
d. Sparse samples are adequate for network performance.
e. σ can be made smaller as n gets larger, without retraining.
f. For time-varying statistics, old patterns can be overwritten with new patterns.

The major disadvantage of basic PNN stems from the fact that it requires one node or neuron for each training pattern. For large databases, this presents a computational and storage problem that can be overcome by using one of various types of clustering algorithms. All result in each PNN node representing a cluster center rather than an individual pattern. The use of clustering in conjunction with basic PNN is discussed later.

78.2 Adaptive PNN

An important improvement to PNN, called adaptive PNN, is obtained by adapting separate smoothing parameters for each measurement dimension. This often greatly improves the generalization accuracy. The dimensionality of the problem and the complexity of the network can usually be simultaneously reduced. Adaptive PNN can be used for automatic feature selection. The price paid for these improvements is increased training time.

Equation 78.3 is used to estimate a PDF as the sum of Gaussian kernels which all have the simple covariance matrix, $\sigma^2 I$, where I is the identity matrix. PDFs can also be estimated as the sums of Gaussians with a full covariance matrix. However, the complexity of a full covariance matrix is not justified for most problems. Also, the use of a full covariance matrix does not lend itself to automatic feature selection as does the method to be described.

We have found that the simpler technique of *adapting* separate σ's for each dimension greatly improves generalization accuracy. Adaptation is accomplished by perturbing each σ a small amount to find the derivative of the optimization criterion with respect to each sigma. Then conjugate gradient descent (Press et al., 1992) (ascent) is used to find iteratively the set of σ's that maximize the optimization criterion. Brent's method (Press et al., 1992, Chapter 10), used for finding a maximum along a gradient line, is modified to constrain the σ's to positive values (Specht, 1995).

The optimization criterion used in the examples presented in Table 78.1 emphasizes improvements in category separation only between categories where misclassifications occurred. When patterns from category k are misclassified as members of category q, the likelihood ratios, $LR = f_k(\boldsymbol{X})/f_q(\boldsymbol{X})$, are calculated for all category k patterns, using the hold-one-out validation method. The hold-one-out validation method consists of evaluating $f_k(\boldsymbol{X})$ using Equation 78.3 except that the training pattern $\boldsymbol{X} = \boldsymbol{X}_{ki}$ is not used in the summation. The mean log likelihood ratios of misclassified and correctly classified patterns are calculated separately, and their ratio is taken. This ratio is summed over all cross categories where misclassifications have occurred. The following criterion was then maximized:

$$\sum_k \sum_{q \neq k} \frac{\text{Mean log LR for misclassified patterns } (\text{CAT}_k/\text{CAT}_q)}{\text{Mean log LR for correctly classified patterns } (\text{CAT}_k/\text{CAT}_q)}$$

This adaptation, with a criterion of separating classes rather than simply estimating PDFs, not only finds a separate smoothing parameter, σ, for each variable, but also discovers variables that are poorly correlated with the desired output. It will be noted that variables with a large σ have a relatively small effect on the estimation of PDFs. After adaptation, as described above, has progressed for several passes, resulting in some variables being almost irrelevant to the classification decisions, these variables are removed one at a time and are left out if the resulting classification accuracy is improved or the same. Thus, adaptation of σ's can also be used for feature selection and dimensionality reduction.

As examples of the improvements which can be obtained, the comparative results of basic PNN and adaptive PNN are shown for fifteen databases in Table 78.1. All of the databases represent real problems with overlapping distributions and measurements contaminated with noise. Databases A through E and J through O came from sensor measurements with naturally occurring noises; simulated noise was used for F through I.

Adapting just the p smoothing parameters (one for each measurement) does not increase the complexity of the trained network, because the usual preprocessing needed for PNN requires division of each variable by its standard deviation or range to ensure that the numerical ranges of all input variables are comparable. Since

Table 78.1 Comparative Accuracy: Adapted Smoothing Parameters per Feature vs Standard PNN Accuracy

Database	Number of Patterns	Original Number of Features	Basic PNN Accuracy (%)	Adapted Number of Features	Accuracy with Adapted σ's (%)
A Drawings of 3-D parts	73	21	74	5	93
B Aircraft health monitoring	90	16	78	6	95
C Automatic targeting	270	9	93	4	97
D Automatic targeting	792	9	94	5	98
E 97 categories	4187	3	72	3	74
F 17 categories active sonar	1530	9	95	7	95
G Missile track discrimination	498	10	92	3	95
H Missile track discrimination	1684	10	97	2	100
I Missile track discrimination	3570	10	97	4	99.4
J Multispectral imagery	756	12	98.68	6	100
K Voice grade	3037	10	97.05	8	99.44
L Thematic mapper sensor	628	12	92.25	6	96.26
M Aviris hyperspectral imagery	648	209	94.33	122	100
N Engine misfire detection	2520	4	77.54	3	87.40
O Propellant pressure	68	51	97.3	21	100

(*Source*: Specht, D. F. and Romsdahl, H. 1994. Experience with adaptive probabilistic neural networks and adaptive general regression neural networks. In *Proc. IEEE Int. Conf. Neural Networks*, Orlando, FL. (c) 1994 IEEE. With permisison.)

the smoothing parameters are used subsequently but in the same way, they can be combined with the preprocessing divisors.

The adaptive PNN described here, while often finding a reduced feature set and greatly increased accuracy, is iterative and trains much more slowly than basic PNN. The advantages of adaptive PNN depend on the underlying distributions in the database. For Database E, for example, no dimensionality reduction was possible and the adapted σ's were almost equal; in this one case, the advantage of adaptive PNN was insignificant.

Comparisons with Backpropagation

It would be useful to know that PNN always produces networks that generalize better to new patterns than networks trained using backpropagation (BP), or vice versa. Unfortunately, depending on the underlying statistics of the database, one or the other may prove to produce the better results. However, the National Institute of Standards and Technology performed studies with basic PNN on two important databases, hand-printed character recognition (Grother and Candela, 1993) and fingerprint classification (Candela and Chellappa, 1993). Both of these reports, and the published journal article (Blue et al., 1994), conclude that PNN had the highest generalization accuracy of any techniques they tried. Its error rate was about half that of BP for hand-printed character recognition, but was only marginally better for the fingerprints. Given the demonstrated improvements of adaptive PNN over basic PNN, we infer that adaptive PNN would, in turn, be superior to BP.

Summary of Adaptive PNN

Adaptive PNN usually greatly outperforms basic PNN in terms of generalization accuracy. Adaptive PNN differs from basic PNN in that a separate smoothing parameter is adapted for each input feature. PNN is usually orders of magnitude faster in training than BP. Since adaptive PNN incorporates the incremental adaptation characteristic of BP, the learning speed advantage has been sacrificed in favor of an accuracy advantage. The user thus has a choice of fast learning with basic PNN or superior accuracy with adaptive PNN. However, once adaptive PNN has been used to select features and set σ's, the resulting network has the real-time learning ability of basic PNN.

Adaptive PNN also provides for automatic feature selection. Because large values of the smoothing parameter imply that the corresponding input feature has little influence on the classification, the algorithm tests for deletion of features. The reduction of the dimensionality of feature space actually leads to increased generalization accuracy with finite training sets.

Gradient descent is not the only technique which could be used for discovering separate smoothing parameters for each dimension. Genetic algorithms have been used very effectively for this same purpose. When using genetic algorithms, there is no need to take derivatives of the optimization criterion. Therefore, the criterion to be minimized could be simply $\sum_k$ (number of category k patterns misclassified) $h_k l_k / n_k$.

78.3 High-Speed Classification

The major disadvantage of PNN stems from the fact that it requires one node or neuron for each training pattern. Although training is extremely fast, classification of large numbers of new patterns can be slow because the amount of computation required to classify a new pattern is proportional to the number of neurons in the network.

Special-purpose parallel hardware has been developed to speed up classification. One example is the DARPA/Nestor/Intel Ni1000 chip, which has 512 parallel processors that perform kernel computations common to PNN, GRNN, RCE, P-RCE, and RBF paradigms. Another is the Adaptive Solutions, Inc., CNAPS chip, which is a more general, single-instruction multiple-data architecture.

An approach to speeding up classification in a dedicated application is to simplify the network. Several researchers have suggested various types of clustering techniques to overcome the limitation. These techniques yield a smaller number of cluster centers, so that each node represents a group of training patterns.

Any standard clustering technique, such as K-means clustering (Tou and Gonzales, 1974) or ISODATA, can be used for this purpose. Burrascano (1991) has advocated using Kohonen's learning vector quantization (LVQ) technique to find representative exemplars to be used for PNN.

A soft clustering technique in which PDFs are estimated as mixtures of Gaussians using a maximum likelihood criterion yields a PNN network which usually has far fewer Gaussian nodes than training patterns.

The following description is adapted from Streit and Luginbuhl (1994):

Let p denote the dimension of the input vector X, and let M denote the number of different class labels in the training set τ of size T. For $j = 1, \ldots, M$, let $G_j \geq 1$ denote the total number of different components in the j-th class mixture PDF. Let $p_{ij}(X)$ denote the multivariate PDF of the i-th component in the mixture for class j, and let π_{ij} denote the proportion of the component i in class j. The "within-class" mixing proportions π_{ij} are nonnegative and satisfy the equations

$$\sum_{i=1}^{G_j} \pi_{ij} = 1, j = 1, \ldots, M. \tag{78.7}$$

The PDF of class j, denoted by $f_j(X)$, is approximated by a general mixture PDF, denoted by $g_j(X)$, that is,

$$f_j(X) \approx g_j(X) = \sum_{i=1}^{G_j} \pi_{ij} p_{ij}(X), j = 1, \ldots, M \tag{78.8}$$

In Streit and Luginbuhl (1994), only multivariate homoscedastic Gaussian mixtures are considered, hence $p_{ij}(X)$ has the form

$$p_{ij}(X) = (2\pi)^{-p/2} |\Sigma|^{-1/2} \exp \left\{ -\frac{1}{2}(X - \mu_{ij})^t \Sigma^{-1} (X - \mu_{ij}) \right\} \tag{78.9}$$

where μ_{ij} is the mean vector and Σ is the positive definite covariance matrix of $p_{ij}(X)$, and where superscript t denotes transpose. The covariance matrix Σ is chosen independent of the class index j and the component index i. $|\Sigma|$ denotes the determinant of matrix Σ.

Let h_l denote the *a priori* probability of class *1*. Let ℓ_{jl} denote the loss associated with classifying an input vector X into class j when the correct decision should have been class *1*. The risk $\rho_j(X)$ of classifying the input X into class j is the expected loss, so that

$$\rho_j(X) \approx \sum_{l=1}^{M} \ell_{jl} h_l f_l(X) \tag{78.10}$$

The decision risk $\rho_j(X)$ is thus approximated by a mixture of Gaussian PDFs, as is seen by substituting Equation 78.8 into Equation 78.10. The minimum risk decision rule is to classify X into that class j having the minimum risk, that is, $j = \arg\min \{\rho_j(X)\}$. The decision j is the optimum Bayesian classification decision provided the approximation in Equation 78.8 is an equality.

The PDFs are estimated by a maximum likelihood method (rather than maximizing classification accuracy directly as in adaptive PNN). A brief description of the training algorithm for finding the mixtures of Gaussians to be implemented in the nodes of PNN is given here; the mathematical justification is given in Streit and Luginbuhl (1994).

The first step of the estimation process is somewhat arbitrary. It is necessary to specify in advance how many Gaussian nodes will be assigned to each category, and to give them starting centers and a common starting covariance matrix. The PDF of each category is estimated as the weighted sum of each of the Gaussian densities (with the restriction that the sum of the weights must equal 1). Once the conditional PDFs are estimated, classification proceeds as in basic PNN.

Unlike hard clustering, each training sample for category j is considered to belong partially to every cluster node which comprises the estimate of the PDF for category j. Assignment of a sample X to each cluster i of category j is made in proportion to the likelihood $p_{ij}(X)$, and is designated $\omega_{ij}(X)$.

The sum of the proportions of sample X assigned to each component in its class must equal unity so that each training sample is assigned 100% to its category, although less than (or at most equal to) 100% to each component.

Once this is done, the component means must be recomputed. The component mean μ_{ij} is the weighted average of all of the training vectors in category j (weighted by the proportion of each sample in that component). The weight of the component in the estimation of the conditional PDF for the category α_{ij} is the sum of the proportions for each of the training samples, divided by the total number of training samples in category j.

Next, the covariance is recomputed using all of the training samples from all of the categories, but the μ_{ij} of the appropriate category j and component i is subtracted from each training vector before being used in the computation:

$$\Sigma^{(n+1)} = T^{-1} \sum_{j=1}^{M} \sum_{i=1}^{G_j} \sum_{k=1}^{T_j} \omega_{ij}^{(n)}(X_{kj})(X_{kj} - \mu_{ij}^{(n+1)})\,(X_{kj} - \mu_{ij}^{(n+1)})^t \tag{78.11}$$

The summation is over all training vectors X_{kj}, each multiplied by the computed proportions to be assigned to each category j and component i. T is the total number of training samples for all categories, and T_j is the number of samples with class label j.

The computations indicated in the last two paragraphs are repeated for n iterations until a stopping criterion is satisfied.

A typical stopping criterion is to stop when the likelihood function, $\Im$, as a function of iteration number stops increasing at a sufficient rate. The likelihood function is the sum of the log likelihoods for all patterns in the training sets,

$$\Im(\tau) = \sum_k \log \sum_{l=1}^{m} h_l g_l(X_k) \qquad (78.12)$$

Since Σ is positive definite, matrix L^{-1} can be chosen such that $\Sigma^{-1} = (L^{-1})^t L^{-1}$. If L^{-1} is chosen to be the Cholesky factor of Σ^{-1}, then L^{-1} is lower triangular. The Cholesky factor (sometimes referred to as the "square root" of the matrix) can be computed easily using the algorithm in Press et al (1992, Section 2.9). Substituting into Equation 78.9 yields

$$\begin{aligned} p_{ij}(X) &= (2\pi)^{-p/2} |\Sigma|^{-1/2} \\ &\quad \times \exp\left\{-\frac{1}{2}(X - \mu_{ij})^t (L^{-1})^t L^{-1}(X - \mu_{ij})\right\} \\ &= (2\pi)^{-p/2} |\Sigma|^{-1/2} \\ &\quad \times \exp\left\{-\frac{1}{2}\|L^{-1}X - L^{-1}\mu_{ij}\|^2\right\} \end{aligned} \qquad (78.13)$$

where $\|\cdot\|$ is the usual Euclidean norm on R^N.

Alternatively, the matrix L can be chosen so that it characterizes the discrete Karhunen-Loeve transformation corresponding to Σ, that is $L^{-1} = \Lambda^{-1/2} U^t$, where $\Sigma = U \Lambda U^t$. However, the Cholesky decomposition requires less computation to determine L^{-1}, and less computation in the evaluation of Equation 78.10, since L^{-1} is then lower triangular.

A neural network topology for implementation of classification using the mixture of Gaussians technique is shown in Figure 78.5. It is similar to, but more general than, the PNN topology of Figure 78.1. Between the input units and the pattern units are now placed p L^{-1} transform units, which perform the function of rotating the measurement space to a new set of axes. The Gaussian mixture components are identical to the pattern units, except for the method used for training and the π coefficients, which give them weight in proportion to the number of training samples represented.

The summation units are unchanged. The Risk units implement the minimum risk strategy for decision making for multiple categories, and are equally appropriate for basic PNN with multiple categories.

Estimation of PDFs by a mixture of Gaussians can proceed iteratively in the same way with or without constraints on covariance matrices. If the covariance matrices for all nodes are constrained to be the same and diagonal, the procedure will simultaneously find a set of prototype vectors and the set of variances. When this is done, the L^{-1} transform units of Figure 78.5 can be eliminated if the input units perform a simple division of the raw input by the square root of the corresponding variance term.

From this diagram, it is clear that the benefit of restricting mixtures to homoscedastic kernels is that only one L^{-1} transform has to be performed on each input (pattern) vector. After that, the pattern units, of which there may be large numbers, can be as simple as those described in Figure 78.2 or 78.3.

Note that the adaptive PNN can be used very effectively to reduce the dimensionality p of the patterns before any of the clustering techniques are applied.

78.4 Other Considerations

Detection—A Single Category Problem

When it is necessary to classify a pattern as the category of interest versus everything else, it may be impractical to get a sufficient number of training samples for "everything else." In this case, it is important to train a PDF on just one category, and then establish a threshold on that PDF.

The PDFs for one category by itself can be estimated by using maximum likelihood to establish the values of the smoothing parameters to be used. Referring back to Equation 78.3, which is the likelihood of pattern X belonging to category k, the likelihood (LH) that all patterns X_{ki} belong to category k is the sum of the logs of $f_k(X)$ evaluated at each pattern (using the holdout method to avoid an artificial maximum at $\sigma = 0$).

$$\begin{aligned} \text{Log } LH = \sum_{i=1}^{m} \log \sum_{\substack{j=1 \\ j \neq i}}^{m} \frac{1}{(2\pi)^{p/2}\sigma^p} \\ \times \exp\left(\frac{-\|X_{ki} - X_{kj}\|^2}{2\sigma^2}\right) \end{aligned} \qquad (78.14)$$

The best σ for the one-category case can be found as the value that maximizes the log likelihood.

The log likelihood can also be maximized with more than one free parameter, such as a separate smoothing parameter for each dimension, or the mixture of Gaussians procedure of the previous section.

In any of these cases, the threshold on the estimated PDF will have to be determined on the basis of the number of false positive detections or misses that can be tolerated.

Different Kernels for Different Categories

In many problems, it is clear that the underlying probability distributions are quite different for different categories. Again, the maximum likelihood technique can be used separately for each category (with or without the mixture of Gaussians technique). It is also possible to optimize classification accuracy by adapting separate σ's for each category.

The choice between selection of σ based on classification accuracy or selection based on maximum likelihood depends on the problem to be solved. Selection based on classification accuracy optimizes the value of σ at the decision boundary, with little concern for estimating the shape of the pdf in other regions. On the other hand, estimation based on maximum likelihood is better when categories have widely differing variances, and therefore one size estimating kernel is not appropriate for all categories,

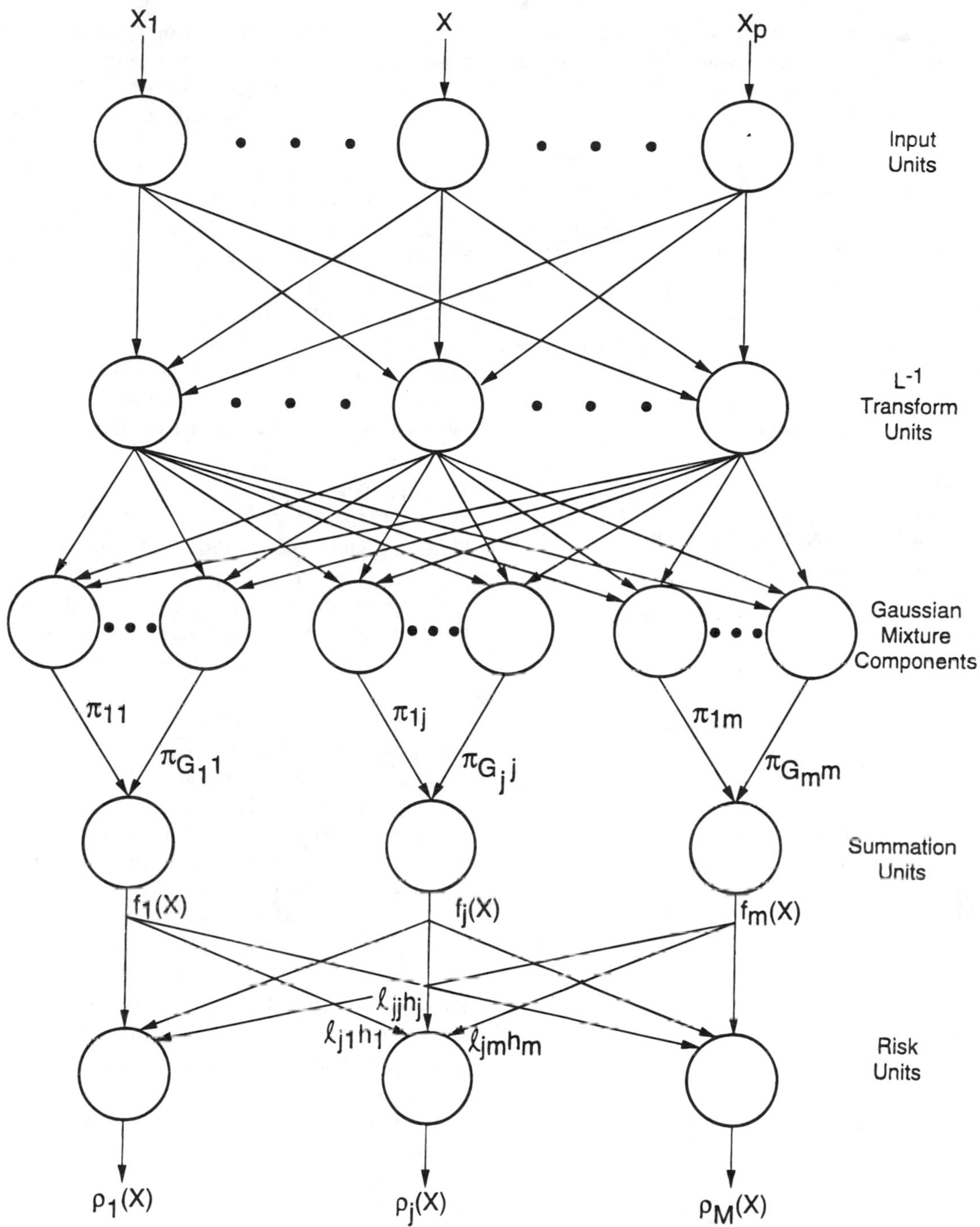

Figure 78.5 Probabilistic neural network using mixture of Gaussians estimation of PDF (after Streit and Luginbuhl, 1994). (*Source*: Specht, D. F. 1996. In *Fuzzy Logic and Neural Network Handbook*, Ch. 3. C. H. Chen, ed. (c) 1996 McGraw-Hill, NY. With permission.)

or when the PDF estimates are to be used for probability estimation using Equation 78.9 in addition to or instead of finding the decision boundary.

Provision for Unknown Categories

Most neural network classifiers are based on the assumption that all patterns belong to one of a fixed set of possible categories. When examples of a new category, for which there have been no training examples, first appear, it is important not to classify them into one of the known categories. Instead, the classifier should recognize novelty in the new patterns and establish a new category (Washburne et al., 1993).

This can be accomplished within the framework of PNN by simply postulating an unknown category, θ_u, and then assigning lower values of loss to misclassification of a pattern from a known category as unknown than the loss associated with misclassifying the pattern into the wrong known category. To simplify the following analysis, ℓ_i is defined as the loss associated with misclassification of a category *i* pattern as belonging to any other known category. In addition to supplying values for the ℓ_i's, the user must supply rough approximations for the following values:

- ℓ_{uk}, the loss associated with misclassification of a pattern from an unknown category into a known category.
- ℓ_{ku}, the loss associated with misclassification of a pattern from any of the *N* known categories as an unknown.
- h_i, the *a priori* probability of observing a pattern from category *i*.
- h_u, the *a priori* probability of observing a pattern from a new category (unknown).
- $f_u(X)$, the pdf for unknowns—assumed uniform over the range of measurement variables.

The loss values ℓ_i are always larger than $\ell_{ku} \cdot \ell_{ku}$ may or may not be equal to ℓ_{uk}.

The risk associated with classifying a pattern *X* into one of the *N* known categories (θ_j) is then:

$$\text{Risk}(\theta = \theta_j) = \ell_{uk} h_u f_u(X) + \sum_{\substack{i=1 \\ i \neq j}}^{N} \ell_i h_i f_i(X) \qquad (78.15)$$

The risk associated with classifying a pattern *X* into the unknown category is:

$$\text{Risk}(\theta = \theta_u) = \ell_{ku} \sum_{i=1}^{N} h_i f_i(X) \qquad (78.16)$$

The decision to classify the pattern *X* into one of the *N* known categories is:

$$d(\theta = \theta_j) \text{ if } \ell_j h_j f_j(X) > \ell_i h_i f_i(X) \text{ for all } i \text{ not equal to } j$$

and

$$\ell_{uk} h_u f_u(X) + \sum_{\substack{i=1 \\ i \neq j}}^{N} \ell_i h_i f_i(X) < \ell_{ku} \sum_{i=1}^{N} h_i f_i(X) \qquad (78.17)$$

The decision to classify the pattern *X* as an unknown is made if the second condition is not true.

These decision rules replace the original PNN decision rule of Equation 78.2.

78.5 Summary

This chapter has described probabilistic neural networks (PNNs), which have broad applicability to classification problems based on learning from labeled samples.

Basic PNN is based on Parzen window estimation of probability density functions and the Bayes strategy for decision making. Further developments which are also described are adaptive PNN, clustering to reduce the number of nodes, mixture of Gaussians clustering, discovery of unknown categories using PNN, and detection of a single category. Adaptive PNN is a technique which can be used to simultaneously reduce dimensionality and improve generalization accuracy.

In all, many different variations of PNN have been discussed. The basic form has the important advantage of instant learning. It is ideal for exploring new databases and preprocessing algorithms because frequent retraining is often required in these circumstances. The Adaptive version, being iterative, requires long training times but offers (usually) much improved generalization accuracy relative to the basic form and also relative to backpropagation networks. The clustering versions, particularly the mixture of Gaussians form, provide compact networks that are the fastest for evaluating new patterns after training is complete.

References

Blue, J. L., Candela, G. T., Grother, P. J., Chellappa, R., and Wilson, C. L. 1994. Evaluation of pattern classifiers for fingerprint and OCR applications, *Pattern Recognition,* 27(4):485–501.

Burrascano, P. 1991. Learning vector quantization for the probabilistic neural network, *IEEE Trans. Neural Networks,* 2:458–461.

Candela, G. T. and Chellappa, R. Comparative performance of classification methods for fingerprints, National Institute of Standards and Technology Report NISTIR 5163, April.

Cacoullos, T. 1966. Estimation of a multivariate density, *Ann. Inst. Statist. Math. (Tokyo),* 18(2):179–189.

Grother, P. J. and Candela, G. T. 1993. Comparison of handprinted digit classifiers, National Institute of Standards and Technology Report NISTIR 5209, June.

Parzen, E. 1962. On Estimation of a probability density function and mode, *Ann. Math. Statist.,* 33:1065–1076.

Press, W. H., Teukolsky, S. A., Vetterling, W. T., and Flannery, B. P. 1992. *Numerical Recipes in C: The Art of Scientific Computing, Second Edition,* Cambridge University Press, Cambridge, MA.

Specht, D. F. 1988. Probabilistic neural networks for classification, mapping or associative memory, *Proc. IEEE Int. Conf. Neural Networks,* 1:525–532.

Specht, D. F. 1990. Probabilistic neural networks, *Neural Networks,* 3:109–118.

Specht, D. F. 1995. PNN: From fast training to fast running, *Computational Intelligence, a dynamic system perspective,* 246–260, M. Palaniswami et al., eds., IEEE Press, New York.

Streit R. L. and Luginbuhl, T. E. 1994. Maximum likelihood training of probabilistic neural networks, *IEEE Trans. Neural Networks,* 5:764–783.

Tou, J. T. and Gonzales, R. C. 1974. *Pattern Recognition Principles.* Addison Wesley, Reading, MA.

Washburne, T. P., Specht, D. F., and Drake, R. M. 1993. Identification of unknown categories with probabilistic neural networks, *Proc. IEEE Int. Conf. Neural Networks,* 1:434–427, March 28–April 1, San Francisco, CA.

79

General Regression Neural Network Model

Donald F. Specht
Lockheed Martin Missiles & Space
Palo Alto, California

79.1 GRNN 1047
General Regression • Normalization of Input and Selection of the Value of the Smoothing Parameter • Clustering and Adaptation to Nonstationary Statistics • Comparison with Other Techniques • Neural Network Implementation of GRNN • A One-Dimensional Example • Adaptive Control Systems • Summary of Basic GRNN
79.2 Adaptive GRNN 1052
Adaptation of Kernel Shapes • Results Using Adaptive GRNN
79.3 Summary 1053

79.1 GRNN

This chapter describes a function approximation neural network called the General Regression Neural Network (GRNN). The GRNN provides estimates of continuous variables and converges smoothly to the underlying (linear or nonlinear) regression surface (Specht, 1991). Like PNN, the GRNN features instant learning and a highly parallel structure.

Even with sparse data in a multidimensional measurement space, the GRNN provides smooth transitions from one observed value to another. The mathematical form can be used for any regression problem in which an assumption of linearity is not justified. The parallel network form can be used with parallel processors in high-speed applications such as learning the dynamics of a system (modeling) for prediction or control.

Regression is the least mean squares estimation of the value of a variable based on examples. The term "general regression" implies that the regression surface is not restricted to being linear. If the variables to be estimated are future values, the network is a predictor. If they are dependent variables related to input variables in a process, plant, or system, the network can be used to model the process, plant, or system. Once the system is modeled, a control surface can be defined in terms of samples of control variables that, given a state vector of the system, improve the output of the system. If a GRNN is taught these samples, it can estimate the entire control surface, and it becomes a controller. A GRNN can be used to map from one set of sample points to another. If the target space is the same dimension as the input space, and if the mapping is one to one, an inverse mapping can easily be formed using the same examples. GRNN can also be used as an interpolator in multidimensional space, with no requirement that the sample data be regularly spaced.

In all cases, basic GRNN instantly adapts to new data points. An adaptive version described in section 79.2 provides for automatic feature selection and improved generalization accuracy.

General Regression

Assume that $f(\mathbf{x},y)$ represents the known joint continuous probability density function of a vector random variable, $\mathbf{x}$, and a scalar random variable, y. Let $\mathbf{X}$ be a particular measured value of the random variable $\mathbf{x}$. The conditional mean of y given $\mathbf{X}$ (also called the regression of y on $\mathbf{X}$) is given by

$$E[y|\mathbf{X}] = \frac{\int_{-\infty}^{\infty} y f(\mathbf{X}, y)\,dy}{\int_{-\infty}^{\infty} f(\mathbf{X}, y)\,dy} \tag{79.1}$$

When the density $f(\mathbf{x},y)$ is not known, it must usually be estimated from a sample of observations of $\mathbf{x}$ and y. For a nonparametric estimate of $f(\mathbf{x},y)$, we will again use the class of consistent estimators proposed by Parzen. These estimators are a good choice for estimating the probability density function, f, if it can be assumed that the underlying density is continuous and that the first partial derivatives of the function evaluated at any $\mathbf{x}$ are small. The probability estimator of Equation 78.3, Chapter 78 is expanded to estimate the joint density $\hat{f}(\mathbf{X},Y)$ based on sample values $\mathbf{X}^i$ and Y^i of the random values $\mathbf{x}$ and y:

$$\hat{f}(\mathbf{X}, Y) = \frac{1}{(2\pi)^{(p+1)/2}\sigma^{(p+1)}} \cdot \frac{1}{n}\sum_{i=1}^{n} \exp\left[-\frac{(\mathbf{X}-\mathbf{X}^i)^t(\mathbf{X}-\mathbf{X}^i)}{2\sigma^2}\right] \cdot \exp\left[-\frac{(Y-Y^i)^2}{2\sigma^2}\right] \tag{79.2}$$

where

0-8493-8343-9/97/$0.00+$.50

$n =$ is the number of sample observations and
$p =$ is the dimension of the vector variable **x**.

A physical interpretation of the probability estimate $\hat{f}(\mathbf{X},Y)$ is that it assigns sample probability of width σ for each sample of $\mathbf{X}^i$ and Y^i, and the probability estimate is the sum of those sample probabilities. Substituting the joint probability estimate $\hat{f}$ in Eq. 79.2 into the conditional mean, Equation 79.1, gives the desired conditional mean of y given **X**. In particular, combining Equations 79.1 and 79.2 and interchanging the order of integration and summation yields the desired conditional mean:

$$\hat{Y}(\mathbf{X}) = \frac{\sum_{i=1}^{n} \exp\left[-\frac{(\mathbf{X}-\mathbf{X}^i)^t(\mathbf{X}-\mathbf{X}^i)}{2\sigma^2}\right] \int_{-\infty}^{\infty} y \exp\left[-\frac{(y-Y^i)^2}{2\sigma^2}\right] dy}{\sum_{i=1}^{n} \exp\left[-\frac{(\mathbf{X}-\mathbf{X}^i)^t(\mathbf{X}-\mathbf{X}^i)}{2\sigma^2}\right] \int_{-\infty}^{\infty} \exp\left[-\frac{(y-Y^i)^2}{2\sigma^2}\right] dy} \quad (79.3)$$

Defining the scalar function D_i^2,

$$D_i^2 = (\mathbf{X}-\mathbf{X}^i)^t(\mathbf{X}-\mathbf{X}^i) \quad (79.4)$$

and performing the indicated integrations yields:

$$\hat{Y}(\mathbf{X}) = \frac{\sum_{i=1}^{n} Y^i \exp\left(-\frac{D_i^2}{2\sigma^2}\right)}{\sum_{i=1}^{n} \exp\left(-\frac{D_i^2}{2\sigma^2}\right)} \quad (79.5)$$

Because the particular estimator Equation 79.3, is readily decomposed into x and y factors, the integrations were accomplished analytically. The resulting regression, Equation 79.5, which involves summations over the observations, is directly applicable to problems involving numerical data.

The estimate $\hat{Y}(\mathbf{X})$ can be visualized as a weighted average of all of the observed values, Y_i, where each observed value is weighted according to its Euclidean distance from **X**. When the smoothing parameter σ is made large, the estimated density is forced to be smooth, and in the limit becomes a multivariate Gaussian with covariance $\sigma^2 I$. On the other hand, a smaller value of σ allows the estimated density to assume non-Gaussian shapes, but with the hazard that wild points may have too large an effect on the estimate. As σ becomes very large, $\hat{Y}(\mathbf{X})$ assumes the value of the sample mean of the observed Y^i, and σ goes to 0, $\hat{Y}(\mathbf{X})$ assumes the value of the Y^i associated with the observation closest to **X**. For intermediate values of σ, all values of Y^i are taken into account, but those corresponding to points closer to **X** are given heavier weight.

When the underlying parent distribution is not known, it is not possible to compute an optimum σ for a given number of observations, n. It is therefore necessary to find σ on an empirical basis. This can be done easily when the density estimate is being used in a regression equation, because there is a natural criterion that can be used for evaluating each value of σ, namely, the mean squared error between Y^j and the estimate $\hat{Y}(\mathbf{X}^j)$. For this purpose, the estimate in Equation 79.5 must be modified so that the jth element in the summation is eliminated. Thus each $\hat{Y}(\mathbf{X}^j)$ is based on inference from all the observations except the actual observed value at $\mathbf{X}^j$. This procedure is used to avoid an artificial minimum error as $\sigma \to 0$ that results when the estimated density is allowed to fit the observed data points. Overfitting of the data is present in the least-squares estimation of linear regression surfaces, but it is not as severe there because the linear regression equation has only $p + 1$ degrees of freedom. If $n >> p$, the phenomenon of overfitting is commonly ignored.

Y and $\hat{Y}$ can be vector variables instead of scalars. In this case, each component of the vector **Y** would be estimated in the same way and from the same observations (**X**,**Y**), except that Y is now augmented by observations of each component. Note, from Equation 79.5, that the denominator of the estimator and all of the exponential terms remain unchanged for vector estimation.

Normalization of Input and Selection of the Value of the Smoothing Parameter

As a preprocessing step, it is usually necessary to scale all input variables such that they have approximately the same ranges or variances. The need for this stems from the fact that the underlying probability density function is to be estimated with a kernel that has the same width in each dimension. This step is not necessary in the limit as $n \to \infty$ and $\sigma \to 0$, but it is very helpful for finite data sets. Exact scaling is not necessary, so the scaling variables need not be changed every time new data are added to the data set.

After rough scaling, the width of the estimating kernel, σ, must be selected. A useful method for selecting σ is the holdout method. For a particular value of σ, this method consists of removing one sample at a time and constructing a network based on all of the other samples. The network is then used to estimate Y for the removed sample. By repeating this process for each sample and storing each estimate, the mean-squared error can be measured between the actual sample values Y^i and the estimates. The value of σ giving the smallest error should be used in the final network. Typically, the curve of mean-squared error versus σ exhibits a wide range of values near the minimum, so it is not difficult to pick a good value for σ without a large number of trials.

Finally, the Gaussian kernel used in Equation 79.2 could be replaced by any of the Parzen windows. Again, the kernel of Equation 78.6, Chapter 78 is attractive from the point of view of computational simplicity. Using this kernel results in the estimator

$$\hat{Y}(X) = \frac{\sum_{i=1}^{n} Y^i \exp\left(-\frac{C_i}{\sigma}\right)}{\sum_{i=1}^{n} \exp\left(-\frac{C_i}{\sigma}\right)} \quad (79.6)$$

where

$$C_i = \sum_{j=1}^{p} |X_j - X_j^i| \quad (79.7)$$

Clustering and Adaptation to Nonstationary Statistics

For some problems, the number of observations (**X**,Y) may be small enough so all the data obtainable can be used directly in the estimator of Equation 79.5 and Equation 79.6. In other problems, the number of observations obtained may be large enough so that it is no longer practical to assign a separate node (or neuron) to each sample. Various clustering techniques can be used to group samples so that the group can be represented by only one node, which measures distance of input vectors from the cluster center. Burrascano (1991) has suggested using learning vector quantization to find representative samples to use for PNN to reduce the size of the training set. This same technique also can be used for the current procedure. K-means averaging (Tou and Gonzales, 1974), adaptive K-means (Moody and Darken, 1989), one-pass K-means clustering (Tseng, 1991) or the clustering technique used by Reilly et al. (1982) for the restricted Coulomb energy (RCE) network could be also used. However the cluster centers are determined, let us assign new variables, B^i and A^i, to indicate respectively the number of samples and the sum of the observed values that are represented by the ith cluster center. Equation 79.5 can then be rewritten as

$$\hat{Y}(X) = \frac{\sum_{i=1}^{m} A^i \exp\left(-\frac{D_i^2}{2\sigma^2}\right)}{\sum_{i=1}^{m} B^i \exp\left(-\frac{D_i^2}{2\sigma^2}\right)} \quad (79.8)$$

where

$$\begin{Bmatrix} A^i(k) = A^i(k-1) + Y^j \\ B^i(k) = B^i(k-1) + 1 \end{Bmatrix} \quad (79.9)$$

are incremented each time a training observation Y^j for cluster i is encountered. $m < n$ is the number of clusters.

The method of clustering can be as simple as establishing a single radius of influence, r. Starting with the first sample point (**X**,Y), establish a cluster center, $\mathbf{X}^i$, at **X**. All future samples for which the distance $|\mathbf{X} - \mathbf{X}^i|$ is less than the distance to any other cluster center and is also $\leq r$ would update Equations 79.9 for this cluster. A sample for which the distance to the nearest cluster $>r$ would become the center for a new cluster. The numerator and denominator coefficients are completely determined in one pass through the data; no iteration is required to improve the coefficients.

Since the A and B coefficients can be determined using recursion equations, it is easy to add a forgetting function. This is desirable if the network is being used to model a system with changing characteristics. If Equations 79.9 are written in the form

$$\begin{Bmatrix} A^i(k) = \frac{\tau - 1}{\tau} A^i(k-1) + \frac{1}{\tau} Y^j \\ B^i(k) = \frac{\tau - 1}{\tau} B^i(k-1) + \frac{1}{\tau} \end{Bmatrix}$$

new simple assigned to cluster i

$$\begin{Bmatrix} A^i(k) = \frac{\tau - 1}{\tau} A^i(k-1) \\ B^i(k) = \frac{\tau - 1}{\tau} B^i(k-1) \end{Bmatrix}$$

new sample assigned to a cluster $\neq i$

(79.10)

then τ can be considered the time constant of an exponential decay function (where τ is measured in update samples rather than in units of time). If all of the coefficients were attenuated by the factor $(\tau - 1)/\tau$, the regression Equation 79.8 would be unchanged; however, the new sample information will have an influence in the local area around its assigned cluster center.

For practical considerations, there should be a lower threshold established for B^i, so that when sufficient time has elapsed without update for a particular cluster, that cluster (and its associated A^i and B^i-coefficients) would be eliminated. In the case of dedicated neural network hardware, these elements could be reassigned to a new cluster.

When the regression function of Equation 79.8 is used to represent a system that has many modes of operation, it is undesirable to forget data associated with modes other than the current one. To be selective about forgetting, one might assign a second radius, $\rho >> r$. In this case, Equations 79.10 would be applied only to cluster centers within a distance ρ of the new training sample.

Higher moments can also be estimated with y^q substituted for y in Equation 79.1. Therefore variance of the estimate and standard deviation can also be estimated directly from the training examples.

Comparison With Other Techniques

Conventional nonlinear regression techniques involve either a priori specification of the form of the regression equation with subsequent statistical determination of some undetermined constants, or statistical determination of the constants in a general regression equation, usually of polynomial form. The first technique requires that the form of the regression equation be known a priori or guessed. The advantages of that approach are that it usually reduces the problem to estimation of a small number of undetermined constants and that the values of these constants when found may provide some insight to the investigator. The disadvantage is that the regression is constrained to yield a "best fit" for the specified form of equation. If the specified form is a poor guess and not appropriate to the database to which it is applied, this constraint can be serious. Classical polynomial regression is usually limited to polynomials in one independent variable or low order, because high-order polynomials involving multiple variates often have too many free constants to be determined using a fixed number, n, of observations ($\mathbf{X}^i,\mathbf{Y}^i$). A classical polynomial regression surface may fit the n observed points very closely, but unless n is much larger than the number of coefficients in the polynomial, there is no assurance that the error for a new point taken randomly from the distribution $f(x,y)$ will be small.

With the regression defined by Equations 79.5 or 79.6, however, it is possible to let σ be small, which allows high-order curves if they are necessary to fit the data. Even in the limit as σ approaches 0, Equation 79.5 is well behaved. It estimates $\hat{Y}(\mathbf{X})$ as being the same as the Y^i associated with the $\mathbf{X}^i$ that is closest in Euclidean distance to $\mathbf{X}$ (nearest neighbor estimator). For any $\sigma > 0$, there is a smooth interpolation between the observed points (as distinct from the discontinuous change of Y from one value to another at points equidistant from the observed points when $\sigma = 0$). Other methods used for estimating general regression surfaces include the back propagation of errors neural network (BP), radial basis functions (RBFs) (Broomhead and Lowe, 1988), the method of Moody and Darken (1989), CMAC (Miller, et al., 1990), and the polynomial ratio approximation to Equation 79.5 (Specht, 1968).

The principal advantages of GRNN are fast learning and convergence to the optimal regression surface as the number of samples becomes very large. GRNN is particularly advantageous with sparse data in a real-time environment because the regression surface is instantly defined everywhere, even with just one sample. The one-sample estimate is that $\hat{Y}$ will be the same as the one observed value regardless of the input vector $\mathbf{X}$. A second sample will divide hyperspace into high and low halves, with a smooth transition between them. The surface becomes gradually more complex with the addition of each new sample point.

The principal disadvantage of the technique of Equation 79.5 is the amount of computation required of the trained system to estimate a new output vector. The version of Equation 79.8–79.10 using clustering overcomes this problem to a large degree.

Finally, GRNN can be combined with linear techniques. When linear regression explains most of the data in a database, a linear equation can be used for first-order estimation, leaving GRNN to model only the deviations from linear. This concept applies also to extended Kalman filters which are typically used for nonlinear control problems. Fisher and Rauch (1994) have demonstrated in one class of applications that the use of GRNN in combination with an extended Kalman filter greatly outperforms the EKF alone.

Neural Network Implementation of GRNN

Figure 79.1 is the overall block diagram of neural network topology implementing GRNN in its adaptive form, represented by Equation 79.8. The input units are merely distribution units, which provide all of the (scaled) measurement variables X to all of the neurons on the second layer, the pattern units. It turns out that the first two layers, the input and pattern units, are

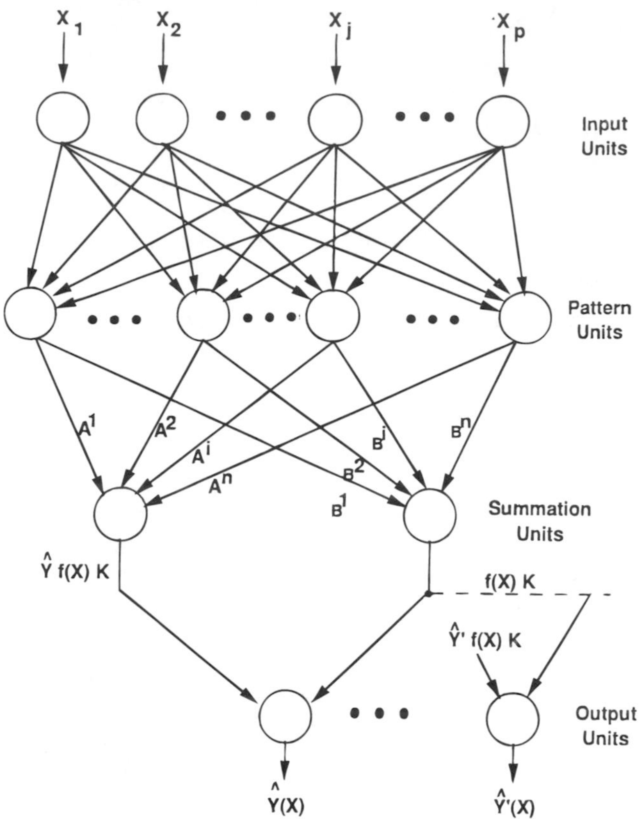

Figure 79.1 GRNN block diagram. (*Source*: Specht, D. F. 1991. A general regression neural network. *IEEE Trans. Neural Networks* 2(6), Nov. (c) 1991 IEEE. With permission.)

identical to those for PNN. The pattern unit outputs are passed on to the summation units.

The summation units perform a dot product between a weight vector and a vector composed of the activations from the pattern units. The summation unit that generates an estimate of $f(\mathbf{X})K$ sums the outputs of the pattern units weighted by the number of observations each cluster center represents. When using Equation 79.10, this number is also weighted by the age of the observations. K is a constant determined by the Parzen window used, but is not data-dependent and does not need to be computed. The summation unit that estimates $\hat{Y}f(\mathbf{X})K$ multiplies each value from a pattern unit by the sum of the samples Y^j associated with cluster center $\mathbf{X}^i$. The output unit merely divides $\hat{Y}f(\mathbf{X})K$ by $f(\mathbf{X})K$ to yield the desired estimate of Y.

To estimate a vector, $\mathbf{Y}$, each component is estimated using one extra summation unit, which uses as its multipliers sums of samples of that component of the vector $\mathbf{Y}$ associated with each cluster center $\mathbf{X}^i$. There may be many pattern units (one for each exemplar or cluster center); however, the addition of one element in the output vector requires only one summation neuron and one output neuron.

What is shown in Figure 79.1 is a feedforward network that can be used to estimate a vector $\mathbf{Y}$ from a measurement vector $\mathbf{X}$. Because they are not interactive, all of the neurons can operate in parallel. Not shown in Figure 79.1 is a microprocessor that assigns training patterns to cluster centers and updates the coefficients A^i and B^i.

A One-Dimensional Example

A simple problem with one independent variable will serve to illustrate some of the differences between the techniques that have been discussed. Suppose that a regression technique is needed to model a "plant" which happens to be an amplifier that saturates in both polarities and has an unknown offset. Its input/output (I/O) characteristic is shown in Figure 79.2. With enough sample points, many techniques would model the plant well. However, in a large measurement space, any practical data set appears to be sparse. The following illustration shows how the methods work on this example with sparse data, namely, five samples at $X = -2, -1, 0, 1,$ and 2. When polynomial regression using polynomials of first, second, and fourth order was tried, the results were predictable; the polynomial curves are poor approximations to the "plant" except at the sample points. In contrast, Figure 79.3 shows the input/output characteristic of this same plant as estimated by GRNN. Since GRNN always estimates using a (nonlinearly) weighted average of the given samples, the estimate is always within the observed range of the dependent variable. In the range from $x = -4$ to $x = 4$, the estimator takes on a family of curves depending on σ. Any curve in the family is a reasonable approximation to the plant of Figure 79.2. The curve corresponding to $\sigma = 0.5$ is the best approximation. Larger values of σ provide more smoothing, and lower values provide a close approximation to the sample values plus a "dwell" region at each sample point. When the holdout method was used to select σ, $\sigma = 0.3$ was selected (based on only four sample points at a time).

In this case, GRNN was used to interpolate among only 5 samples which were noise free. GRNN can also interpolate using noisy samples, but then the data set must be larger and σ must be large enough so that several samples are averaged for each estimate. The same method for finding σ applies with either noise-free or noisy data; namely, find a value which minimizes mean squared error.

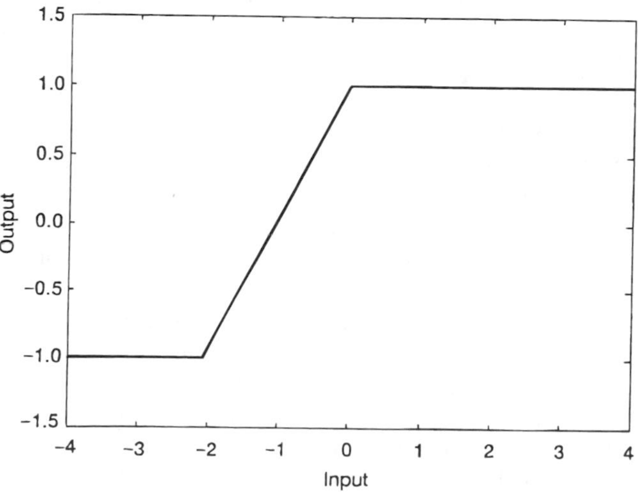

Figure 79.2 Input/output characteristics of simple "plant". (*Source*: Specht, D. F. 1991. A general regression neural network. *IEEE Trans. Neural Networks* 2(6), Nov. (c) 1991 IEEE. With permission.)

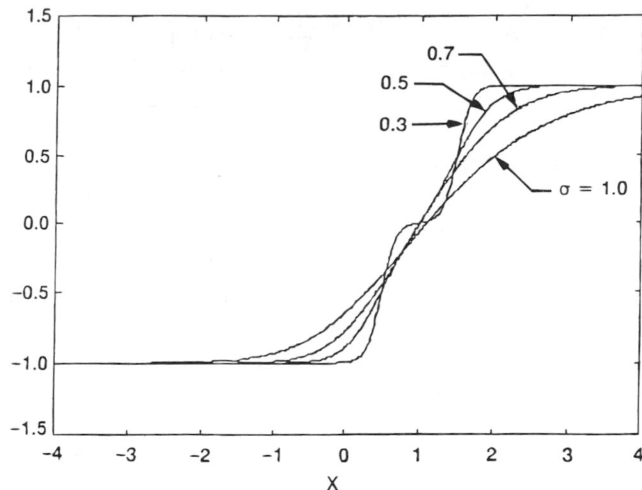

Figure 79.3 Input/output characteristics of Figure 79.2 as estimated by GRNN based on sample points at $X = -2, -1, 0, 1,$ and 2. (*Source*: Specht, D. F. 1991. A general regression neural network. *IEEE Trans. Neural Networks* 2(6), Nov. (c) 1991 IEEE. With permission.)

Adaptive Control Systems

The fields of nonlinear control systems and robotics are particularly good application areas that can use the potential speed of neural networks implemented in parallel hardware, the adaptability of instant learning, and the flexibility of a completely nonlinear formulation. A straightforward technique can be used. First, model the plant as in Figure 79.4. The GRNN learns the relationships between the input vector (the input state of the system and the control variables) and the simulated or actual output of the system. Control inputs can be supplied by a nominal controller (with random variations added to explore inputs not allowed by the nominal controller) or by a human operator. After the model is trained, it can be used to determine control inputs by an automated "what if" strategy or by finding an inverse model. Modeling involves discovering the association between inputs and outputs, so an inverse model can be determined from the same database as the forward model by assigning the input variable(s) to the function of the desired output, **Y**, in Figure 79.1, and the state

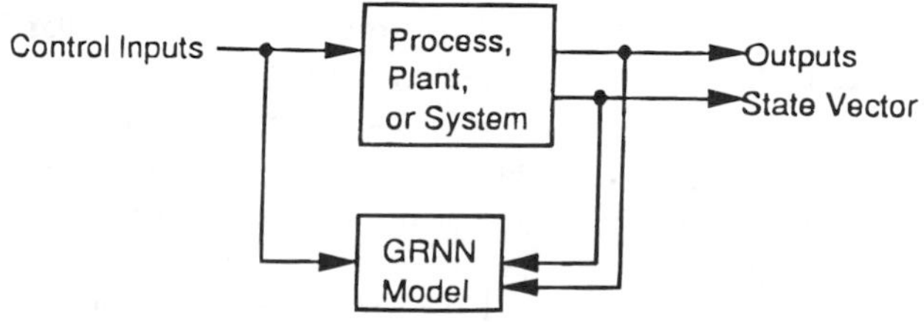

Figure 79.4 Modeling the system. (*Source*: Specht, D. F. 1991. A general regression neural network. *IEEE Trans. Neural Networks* 2(6), Nov. (c) 1991 IEEE. With permission.)

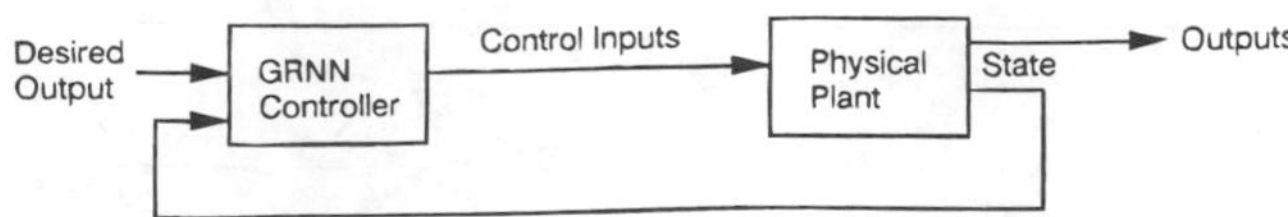

Figure 79.5 A GRNN controller. (*Source*: Specht, D. F. 1991. A general regression neural network. *IEEE Trans. Neural Networks* 2(6), Nov. (c) 1991 IEEE. With permission.)

vector and other measurements are considered components of the **X** vector in Figure 79.1. One way the neural network could be used to control a plant is illustrated in Figure 79.5.

Adaptive inverse neural networks can be used for control purposes in either the feedforward path or the feedback path. Atkeson et al. (1990) used an adaptive inverse in the feedforward path with positional and velocity feedback to correct for residual error in the model. They noted that the feedback had less effect as the inverse model improved from experience. They also used a content-addressable memory as the inverse model and reported good results. Interestingly, the success reported was based on using only single nearest neighbors as estimators. Their paper mentions the possibility of extending the work to local averaging.

Farrell et al. (1990) used both sigmoidal processing units and Gaussian processing units as neural network controllers in a model reference adaptive control system. They note that the Gaussian processing units have an advantage in control systems, because the localized influence of each Gaussian node allows the learning system to refine its control function in one region of measurement space without degrading its approximation in distant regions. The same advantage would hold true when using Equation 79.8 as an adaptive inverse.

Narendra and Parthasarathy (1990) separate the problem of control of nonlinear dynamical systems into an identification of system (modeling) section and a model reference adaptive control (MRAC) section. In Specht (1991) it is shown that GRNN can be used effectively for the identification section. Whereas Narendra and Parthasarathy (1990) used back propagation for both sections, in Specht (1991) it was shown, for one of the more difficult examples, that GRNN achieved approximately the same error between the model and the plant as did back propagation; however, the GRNN model required only 1000 time steps to achieve this degree of accuracy, compared with 100,000 time steps required for the back-propagation model.

Summary of Basic GRNN

The general regression neural network (GRNN) is similar in form to the probabilistic neural network (PNN). Whereas PNN finds decision boundaries between categories of patterns, GRNN estimates values for continuous dependent variables. Both do so through the use of nonparametric estimators of probability density functions.

The advantages of GRNN relative to other nonlinear regression techniques are as follows:

1. The network "learns" in one pass through the data and can generalize from examples as soon as they are stored.
2. The estimate converges to the conditional mean regression surfaces as more and more examples are observed; yet, as indicated in the examples, it forms very reasonable regression surfaces based on only a few samples.
3. The estimate is bounded by the minimum and maximum of the observations.
4. The estimate cannot converge to poor solutions corresponding to local minima of the error criterion (as sometimes happens with iterative techniques).
5. A software simulation is easy to write and use.
6. The network can provide a mapping from one set of sample points to another. If the mapping is one to one, an inverse mapping can easily be generated from the same sample points.
7. The clustering version of GRNN, Equation 79.8, limits the numbers of nodes and (optionally) provides a mechanism for forgetting old data.

The main disadvantage of GRNN (without clustering) relative to other techniques is that it requires substantial computation to evaluate new points. There are several ways to overcome this disadvantage. One is to use the clustering versions of GRNN. Another is to take advantage of the inherent parallel structure of this network and design semiconductor chips to do the computation. The two in combination provide high throughput and rapid adaptation.

79.2 Adaptive GRNN

Just as adapting a separate smoothing parameter for each measurement dimension leads to greatly improved generalization accuracy for PNN, the same technique can be applied to the PDF estimation kernel for GRNN to greatly improve its accuracy. This change results in Adaptive GRNN. Like Adaptive PNN, Adaptive GRNN can be used for automatic feature selection. Again, the price paid for these benefits is increased training time.

Adaptation of Kernel Shapes

Adapting separate σ's for separate dimensions is a bit simpler for Adaptive GRNN than for Adaptive PNN (Chapter 78) because the primary criterion to be minimized is inherently continuous. This criterion is the mean squared error between the GRNN estimate and the desired response measured by the holdout method.

Adaptation is accomplished by perturbing each σ a small amount to find the derivative of the optimization criterion. Then conjugate gradient descent (Press, et al., 1992) is used to find iteratively the set of σ's that minimize the criterion. Brent's method (Press, et al., 1992), modified to constrain the σ's to positive values, is used to find the minimum along each gradient line.

After adaptation has progressed for several passes, some σ's will usually become so large that their corresponding inputs are almost irrelevant to the estimation of the dependent variables. These inputs are tentatively removed one at a time. If the resulting

Table 79.1 Comparative Accuracy; Mean Squared Error Rate for Basic GRNN and Adaptive GRNN

Database	Original Number of Patterns	Original Number of Features	Basic GRNN Error Rate (MSE × 10,000)	Adapted Number of Patterns	Adapted Number of Features	Adaptive GRNN Error Rate (MSE × 10,000)	Improvement Ratio
Pressure predictor	17450	17	3	17450	8	2	1.5
Stock forecast	372	9	9334	98	6	6936	1.4
Sales forecast	64	17	4186	37	8	1117	3.8
Sales forecast	416	9	5381	236	9	1410	3.8
Phase diversity 1	543	245	1498	265	39	349	4.3
Phase diversity 2	543	245	1118	293	43	225	5.0
Phase diversity 3	543	245	147	280	50	56	2.6
Sim. Active Sonar 1	910	10	79	418	8	14	5.6
Sim. Active Sonar 2	910	10	1047	344	8	277	3.8
Sim. Active Sonar 3	910	10	467	401	9	119	3.9
Sim. Active Sonar 4	910	10	274	370	8	72	3.8
Sim. Active Sonar 5	910	10	126	343	5	12	10.5
Sim. Active Sonar 6	910	10	531	369	10	64	8.3

(*Source:* Specht, D. F. and Romsdahl, H. 1994. Experience with adaptive probabilistic neural networks and adaptive general regression neural networks. In *Proc. IEEE Int. Conf. Neural Networks*, Orlando, FL. (c) 1994 IEEE. With permission.)

regression accuracy is improved or left the same, the input is left out.

Results Using Adaptive GRNN

Although basic GRNN has been found to be very valuable for interpolation and extrapolation of multivalued functions, the accuracy obtained with Adaptive GRNN is usually better and often greatly improved. Table 79.1 shows comparative results for 13 databases of 5 distinct types (Specht and Romsdahl, 1994). "Pressure predictor" is prediction of pressure profiles in a rocket motor. "Phase diversity" refers to an optical wavefront sensor based on image data at two focal planes. The estimated wavefront can be used to correct for optical aberrations by controlling a deformable mirror. GRNN is used to estimate the piston positions needed to bring the object into focus (Kendrick, et al. 1994). For the active sonar databases, GRNN was used to infer aspect angles of six different bodies.

The accuracy criterion in Table 79.1 is the mean squared error normalized by the variance of the predicted variable. Adaptive GRNN achieved significant reduction in the error rate in all cases. In addition, the numbers of features and of prototypes required were almost always reduced. Clustering, which was not used here, could further reduce the number of prototypes. Prototype pruning was not attempted on the pressure predictor database. An equivalent criterion, the multiple coefficient of determination (R Squared) can be obtained by dividing the MSE shown by 10,000 and subtracting the result from 1.0. The improvement ratio, which is the ratio of the error rate for GRNN to that of Adaptive GRNN, varies from a minimum of 1.4:1 to better than 10:1 for these databases.

In the experimental work reported here, adaptation of the σ vector was accomplished using conjugate gradient descent. Other techniques for discovering the best combination of σ's are possible. Ward Systems Group, Frederick, Maryland, uses genetic algorithms for this purpose.

79.3 Summary

This chapter has described General Regression Neural Networks (GRNN), which estimate values of continuous variables such as future position, future values, or multivariate interpolation. There is no linearity constraint on the relationship between inputs and outputs.

The principal advantage of basic GRNN is that the network "learns" in one pass through the data and can generalize from examples as soon as they are stored. This characteristic makes it ideal for exploring new databases and preprocessing algorithms because frequent retraining is often required in these circumstances.

Besides basic GRNN, there are versions that use clustering to reduce the complexity of the networks, and versions that can follow nonstationary statistics.

Adaptive GRNN is a technique that can be used to simultaneously reduce dimensionality and improve generalization accuracy. Adaptive GRNN usually greatly outperforms basic GRNN in terms of estimation accuracy, but at the cost of increased training time.

Adaptive GRNN provides for automatic feature selection by adapting a separate smoothing parameter for each input feature. Because large values of the smoothing parameter imply that the corresponding input feature has little influence on the output estimates, the algorithm tests for deletion of features. The reduction of the dimensionality of feature space leads to increased generalization accuracy with finite training sets.

References

Atkeson, C. G. and Reinkensmeyer, D. J. 1990. *Neural Networks for Control,* Ch. 11, Miller, T. et al., eds., MIT Press, Cambridge, MA.

Broomhead, D. S. and Lowe, D. 1988. Multivariable functional interpolation and adaptive networks, *Complex Systems,* 2:321–355.

Burrascano, P. 1991. Learning vector quantization for the probabilistic neural network, *IEEE Trans. on Neural Networks,* 2:458–461.

Farrell, J. et al. 1990. Connectionist learning control systems: submarine depth control, *Proc. 29th IEEE Conf. Decision and Control,* December.

Fisher, W. A. and Rauch, H. E. 1994. Augmentation of an extended Kalman filter with a neural network, *Proc. IEEE Int. Conf. Neural Networks,* II:1191–1196, June, Orlando, FL.

Kendrick, R. L., Acton, D. S., and Duncan A. L. 1994. Phase-diversity wave-front sensor for imaging systems, *Applied Optics,* Vol. 33:6533–6546.

Miller III, W. T. Glanz, F. H., and Kraft III, L. G. 1990. CMAC: an associative neural network alternative to backpropagation, *Proc. IEEE,* Oct. 78:1561–1567.

Moody, J. and Darken, C. 1989. Fast learning in networks of locally tuned processing units, *Neural Computation,* 1:281–294.

Narendra, K. S. and Parthasarathy, K. 1990. Identification and control of dynamical systems using neural networks, *IEEE Trans. Neural Networks,* Mar 1:4–27.

Press, W. H., Teukolsky, S. A., Vetterling, W. T., and Flannery, B. P. 1992. *Numerical Recipes in C: The Art of Scientific Computing,* 2nd ed. Cambridge University Press, Cambridge, MA.

Reilly, D. L. et al. 1982. "A Neural Model for Category Learning," *Biol. Cybern.,* 45:35–41.

Specht, D. F. 1968. A Practical Technique for Estimating General Regression Surfaces, LMSC-6-79-68-6, June, Lockheed Missiles & Space Company, Inc., Palo Alto, CA; also available as Defense Technical Information Center AD-672505 or NASA N68-29513.

Specht, D. F. 1991. A general regression neural network, *IEEE Trans. Neural Networks,* 2:568–576.

Specht, D. F. and Romsdahl, H. 1994. Experience with adaptive PNN and adaptive GRNN, *Proc. IEEE Int. Conf. Neural Networks,* II:1203–1208, June 28–July 2, Orlando, FL.

Tou, J. T. and Gonzales, R. C. 1974. *Pattern Recognition Principles,* Addison-Wesley, Reading MA.

Tseng, Ming-Lei 1991. Integrating neural networks with influence diagrams for multiple sensor diagnostic systems, Ph.d. Dissertation, University of California at Berkeley, August.

80
Classifiers: An Overview

80.1 Introduction 1055
80.2 Criteria for Optimal Classifier Design 1055
80.3 Categorizing the Classifiers 1056
Bayesian Optimal Classifiers • Exemplar Classifiers • Space Partition Methods • Neural Networks
80.4 Classifiers 1057
Bayesian Classifiers • Bayesian Classifiers with Multivariate Normal Populations • Learning Vector Quantizer • Nearest Neighbor Rule
80.5 Neural Networks 1062
Introduction • Feed-Forward Networks • Error Backpropagation • Issues in Neural Networks • Enhancing Convergence Rate and Generalization of an Optimization Machine • Two-group Regression and Linear Discriminant Function • Multi-response Regression and Flexible Discriminant Analysis • Optimal Scoring • Canonical Correlation Analysis • Linear Discriminant Analysis • Translation of Optimal Scoring Dimensions into Discriminant Coordinates • Linear Discriminant Analysis via Optimal Scoring • Flexible Discriminant Analysis by Optimal Scoring
80.6 Comparison of Experimental Results 1075
80.7 System Performance Assessment 1076
Classifier Evaluation • Analysis of Prediction Error Rates from Bootstrapping Assessment
80.8 Analysis of Prediction Rates from Bootstrapping Assessment 1080

WooGon Chung
Sung Kyun Kwan University

Evangelia Micheli-Tzanakou
Rutgers University

80.1 Introduction

One way to better understand a subject is to classify or categorize it among related subjects. Many classifiers result from different approaches to classification problems. The purpose of this article is to categorize the well-known classifiers in the literature according to those approaches.

Lippmann's (1989) tutorial paper described various classifiers as well as neural networks in detail after his first discussion (Lippmann, 1987) on the general application of neural networks. Another general overview on this subject is found in a recent paper by Hush and Horne (1993) in which neural networks are reviewed in the broad dichotomy of stationary versus dynamic networks. Weiss and Kulikowski's book (1997, Ch. 6) generally touches the classification and prediction methods from the point of view of statistics, neural networks, machine learning, and expert systems.

The purpose of this article is not to give a tutorial on the well-developed networks and other classifiers, but to introduce another branch in the growing classifier tree, that of nonparametric regression approaches to classification problems. Recently, Hastie, Tibshirani, and Buja (1993) introduced the Flexible Discriminant Analysis (FDA) in the applied statistics literature, after the unpublished work by Breiman and Ihaka (1984).

Canonical Correlation Analysis (CCA) for two sets of variables is known to be a scalar multiple equal to the Linear Discriminant Analysis (LDA). Optimal Scaling (OS) is an alternative to the CCA, where the classical Singular Value Decomposition (SVD) is used to find the solutions. OS brings the *flexibility* obtained via nonparametric regression and introduces this flexibility to discriminant analysis, hence the name *Flexible Discriminant Analysis.*

A number of recently developed multivariate regressions are used for classification in addition to other groups of classifiers for a data set obtained from handwritten digit images. The software is contributed mainly from the authors or active researchers in this area. The sources are described in later sections after the description of each classifier.

80.2 Criteria for Optimal Classifier Design

We start with a general description of the classification problem and then proceed to a discussion of simpler cases in which assumptions are made. Which criterion should be used is application specific. Expected Cost for Misclassification (*ECM*) is applied to problems in which the cost of misclassification differs among the cases. For example, one may expect to assign a higher cost

0-8493-8343-9/97/$0.00+$.50

for misdiagnosing a patient with a serious disease as healthy than for misdiagnosing a healthy person as unhealthy. If a meteorologist forecasts fine weather for the weekend but a heavy storm strikes the town, the cost of the misclassification will be much more than if the opposite situation occurs.

Sometimes we do not care about the resulting cost of misclassification. The cost for misclassification for a pattern recognition system to misclassify pattern "A" as pattern "B" may be considered the same as the cost for misclassifying pattern "B" as pattern "A." In this situation, we can disregard the cost information, or assign the same cost to all cases. An optimal classification procedure might also consider only the probability of misclassification (from conditional distributions) and its likelihood to happen among different classes (from the *a priori* probabilities). Such an optimal classification procedure is referred to as the Total Probability of Misclassification (*TPM*). The *ECM*, however, requires three kinds of information, that is, the conditional distribution, the *a priori* probabilities and the cost for misclassification.

In the simplest case, we also ignore the *a priori* probabilites or assume that they are all equal. In this case we only wish to reduce misclassification for all the classes without considering the class proportion of the given data. It should be noted, however, that it is relatively simple to estimate the *a priori* probabilities from the sample at hand by the frequency approximation. Thus the *TPM* is often the choice as a criterion in which the class conditional distribution and *a priori* probabilities are considered.

80.3 Categorizing the Classifiers

Bayesian Optimal Classifiers

Bayesian classifiers are based on probabilistic information on the populations from which a sample of training data is to be drawn randomly. Randomness in sampling is assumed and it is necessary for a better representation of the sample of the underlying population probability function. An optimal classifier would be one that minimizes the criterion, *ECM*, which consists of three probabilistic types of information. Those are, the class conditional probabilities $p_i(\mathrm{x})$, *a priori* probabilities P_i, and cost for misclassification $C(i|j)$, $i \neq j$ for $i \in \mathcal{G}$. Another criterion for an optimal Bayesian classifier is ignoring the cost for different misclassifications or using the same cost for all the different misclassifications. Then the probabilistics information used is $p_i(\mathrm{x})$ and P_i for $i \in \mathcal{G}$. This minimum *TPM* classifier is the *Maximum A Posterior* classifier which may be familiar. This will be shown in section on Bayesian Classifiers. For the minimum *ECM* and *TPM* optimal classifiers, we need to estimate the class conditional densities for different classes which is usually difficult for $q \gtrsim 2$. This difficulty in density estimation is related to the *curse of dimensionality* caused by the fact that a high-dimensional space is mostly empty.

A simplified Bayesian classifier can be obtained by assuming a normal distribution for the class conditional density functions. With the normal distribution assumption, the conditional density functions are parameterized by the mean vector μ_i and the covariance matrices Σ_i for $i \in \mathcal{G}$ where $\mathcal{G}$ is the set of class labels. Depending on the assumption of the covariance matrices we have a *quadratic discriminant classifier* or a *linear discriminant classifier.*

Exemplar Classifiers

The most simple-minded non-parametric classifier is to use the label information of the training data to allocate the unknown input **x**. The idea is to find the distribution of the labels in a neighborhood of a new observation **x** in the training sample and pick the label whose occurence is maximum. The well known classifier in this group is the *K*-nearest neighbors (KNN) classifier. This classifier is justified either via nearest-neighbor density estimation, or using the nearest-neighbor nonparametric regression (Hardle, 1991).

Practical issues in the KNN include the choice of a metric to measure the distance between the *K* nearest points and the unknown pattern point, and fast searches for neighbors. Advanced data structures such as K-D trees (Omohundro, 1987) are suggested for faster searches at the expense of complications in training and adaptation.

Other examples are the feature-map classifier (Huang and Lippmann, 1986), Learning Vector Quantization (LVQ) (Kohonen, 1990), Adaptive Resonance Theory (ART) classifier (Carpenter and Grossberg, 1987) and others which are found in the survey paper by Lippmann (1989).

Vector Quantization (VQ) (Gray, 1984; Linde et al., 1980) is another classical representative exemplar finding algorithm that has been used in communications engineering for the purpose of data reduction for storage and transmission. The exemplar classifiers (except for the KNN classifier) cluster the training patterns via unsupervised learning followed by supervised learning or label assignment. A Radial Basis Function (RBF) network (Powell, 1984) is also a combination of unsupervised and supervised learning. The basis function is radial and symmetric around the mean vector which is the centroid of the clusters formed in the unsupervised learning stage, hence the name radial basis function. The RBF networks are two-layer networks in which the first layer nodes represent radial functions (usually Gaussian). The second layer weights are used to linearly combine the individual radial functions, and the weights are adapted via a linear least squares algorithm during the training by supervised learning. Figure 80.1 depicts the structure of the RBF networks.

The LMS algorithm (Widrow and Hoff, 1960), a simple modification for the linear least squares, is usually used during training for the output layer weights. Any unsupervised clustering algorithm such as the K-means algorithm (i.e., LBG algorithm, Linde et al., 1980) or Self-Organizing Map (Kohonen, 1990) may be used in the first clustering stage.

The most common basis is a Gaussian kernel function of the form:

$$\theta_i = \exp\left[-\frac{(\mathbf{x}-\mathbf{m}_j)^t(\mathbf{x}-\mathbf{m}_j)}{2\sigma_j^2}\right] \qquad j = 1, 2, \ldots, n \tag{80.1}$$

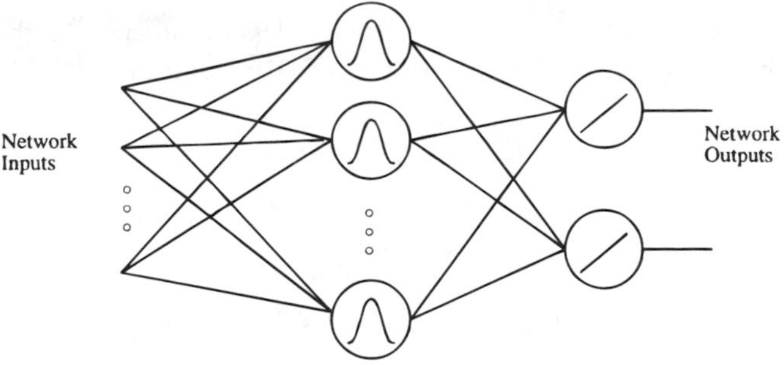

Figure 80.1 RBF network. Two layer network with first layer node being any radial functions imposed on different locations and second layer node being linear.

where $\mathbf{m}_j$ is the mean vector of the jth cluster found from a clustering algorithm, and $\mathbf{x}$ is the input pattern vector. The σ_j^2 is the normalization factor which is a spread measure of the points in a cluster. The average squared distance of the points from the centroid is the common choice for the normalization factor:

$$\sigma_j^2 = \frac{1}{M_j} \sum_{\mathbf{x} \in w_j} (\mathbf{x} - \mathbf{m}_j)^t (\mathbf{x} - \mathbf{m}_j) \tag{80.2}$$

where w_j is the set of the points in the jth cluster and M_j is the number of the points in the jth cluster.

A generalization of the radial function utilizes the variance of an individual variable and covariance among the variables in the training sample. The *Mahalanobis distance* in the Gaussian kernel has the form:

$$\theta_j = \exp[-(\mathbf{x} - \mathbf{m}_j)^t \Sigma_j^{-1} (\mathbf{x} - \mathbf{m}_j)] \qquad j = 1, 2, \ldots, n \tag{80.3}$$

where Σ_j is the covariance matrix in the jth cluster. The localized distribution function is now ellipsoidal rather than a radial function. A more extensive study on the RBF networks can be found in (Hush and Horne, 1993).

Space Partition Methods

The input space $\mathscr{X}$ is recursively partitioned into children subspaces such that the class distributions of the subspaces become as *impure* as possible: impurity of class distribution in a subspace measures the partitioning of the input space by classes.

There are a number of different schemes for estimating trees. Quinlan's *ID*3 (1986) is well known in the machine learning literature. The citations for some of its variants can be found in a review paper by Ripley (1994). The most well known partitioning method is the Classification and Regression Tree (CART) (Breiman, et al., 1984) which is used to build a binary tree partitioning the input space. At each split of the subspace each variable is considered with a separating value, and the separating variable with the best separating value is chosen to split the subspace into two children subspaces.

The main issue in this CART algorithm is how to 'grow' to fit the given training data well and 'prune' it to avoid over-fitting, i.e., to improve the regularization.

Neural Networks

Neural networks are popular and there are numerous text books and journals devoted to the topic. Lippmann (1987) is recommended for a general overview of neural networks for classification and (auto) associative memory applications. A statistician's view on using neural networks for multivariate regression and classification purposes is found in extensive review papers by Ripley (1993, 1994). Different learning algorithms with historical aspects in learning can be obtained from a reference by Hinton (1989).

In this paper we are mainly interested in multivariate regression and classification properties of neural networks, usually in the form of feed-forward multilayer perceptrons.

80.4 Classifiers

Bayesian Classifiers

For simplicity, we would like to start with a two-class classification problem and develop it for multi-class cases in a straight-forward way. Three kinds of information for an optimal classification design procedure in Bayesian sense are denoted as

$C(2\|1)$, $C(1\|2)$	cost of misclassification
P_1, P_2	*a priori* probabilities
$p_1(\mathbf{x})$, $p_2(\mathbf{x})$	class conditional probability density functions

where $C(i|j)$ is the cost for misclassification of j as i. With the notations introduced, the probability that an observation is misclassified as w_2 is represented by the product of the probability that an observation comes from w_1 but falls in w_2 and the probability that the observation comes from w_1:

$$P(\text{misclassified as } w_2) = P(\mathbf{X} \in R_2)P(w_1) = P(2|1)P_1 \tag{80.4}$$

where the regions R_2 and $P(2|1)$ (i.e., the integration of p_1 (x) in the region R_2) are depicted in Figure 80.2.

R_i, $i \in \{1,2\}$ is an optimum decision region in the input space such that minimum error results are obtained. $P(i|j)$, $i \neq j \in \{1,2\}$ is the integration of the conditional probability function in the region of the other class, thus measuring the possibility of error due to the regions and the conditional probability functions.

Minimum *ECM* Classifier

When the criterion is to minimize the *ECM* (Expected Cost for Misclassification), the optimal resulting classifier is called

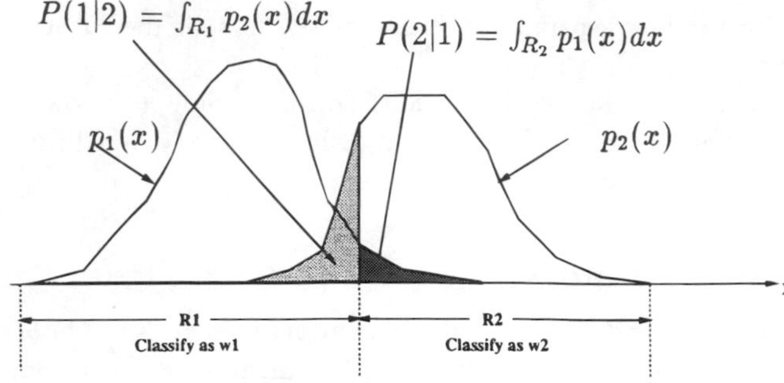

Figure 80.2 Misclassification probabilities and decision regions R_1 and R_2.

a *minimum ECM classifier.* The cost for correct classification is usually set to zero and positive numbers are used for misclassification costs. The whole supporting region is the input space X and is divided into 2 exclusive and exhaustive subregions: $\mathscr{X} = R_1 \cup R_2$.

By the definition of the minimum *ECM* classifier for class 1 the following is formed:

$$ECM = C(2|1)P(2|1)P_1 + C(1|2)P(1|2)P_2 \tag{80.5}$$

$$= C(2|1)P_1 \int_{R_2} p_1(\mathbf{x})d\mathbf{x} + C(1|2)P_2 \int_{R_1} p_2(\mathbf{x})d\mathbf{x}$$

$$= C(2|1)P_1\left(1 - \int_{R_1} p_1(\mathbf{x})d\mathbf{x}\right) + C(1|2)P_2 \int_{R_1} p_2(\mathbf{x})d\mathbf{x}$$

$$= \int_{R_1} (C(1|2)P_2 p_2(\mathbf{x}) - C(2|1)P_1 p_1(\mathbf{x}))d\mathbf{x} + C(2|1)P_1 \tag{80.6}$$

with all the individual quantities being positive. The minimization is achieved as close to zero as possible by having the integration in Equation 80.6 to be equal to a negative quantity. Thus the *ECM* is minimized if the region R_1 includes those values $\mathbf{x}$ for which the integrand becomes as negative as possible with which the absolute value is equal to the last quantity $C(2|1)P_1$:

$$\{C(1|2)P_2 p_2(\mathbf{x}) - C(2|1)P_1 p_1(\mathbf{x})\} \le 0 \tag{80.7}$$

and excludes those $\mathbf{x}$ for which this quantity is positive. That is, R_1, the decision region for class 1, must be the set of points $\mathbf{x}$ such that

$$C(1|2)P_2 p_2(\mathbf{x}) \le C(2|1)P_1 p_1(\mathbf{x}) \quad \text{or} \tag{80.8}$$

$$\frac{p_1(\mathbf{x})}{p_2(\mathbf{x})} \ge \frac{C(1|2)}{C(1|2)} \frac{P_2}{P_1} \tag{80.9}$$

Here we have chosen to express the region as the set of solution $\mathbf{x}$ of the inequality. The fractional form of Equation 80.9 for the region R_1 is the preferred format, since it reduces to a simple form (which will be shown) when the conditional distribution function $p_i(\mathbf{x})$, $i = 1,2$ is assumed to be normal (and thus assuming the same covariance matrix for the two conditional distributions) for simple Bayesian classifiers.

Assuming the same cost for each misclassification reduces the criterion *ECM* to Total Probability of Misclassification (*TPM*):

$$P_2 p_2(\mathbf{x}) \le P_1 p_1(\mathbf{x}) \quad \text{or} \tag{80.10}$$

$$\frac{p_1(\mathbf{x})}{p_2(\mathbf{x})} \ge \frac{P_2}{P_1} \tag{80.11}$$

from Equation 80.9. Due to the Bayes theorem:

$$P(w_k|\mathbf{x}) = \frac{P_k p_k(\mathbf{x})}{\sum_{i=1}^{2} P_i p_i(\mathbf{x})} \quad \text{for all } k \in \{1, 2\} \tag{80.12}$$

the corresponding decision rule (Eq. 10) becomes the Maximum A Posteriori (MAP) criterion, that is to allocate $\mathbf{x}$ into w_1 if

$$P(w_2|\mathbf{x}) \le P(w_1|\mathbf{x}). \tag{80.13}$$

Multi-class Optimal Classifiers

The boundary regions of the minimum *ECM* optimal classifier for a multi-class classifier are obtained in a straight-forward manner from Equation 80.6 by minimizing

$$ECM = \sum_{i=1}^{J} P_i\left[\sum_{\substack{k=1\\k\ne i}}^{J} P(k|i)C(k|i)\right] \tag{80.14}$$

The probability of misclassification of $\mathbf{x} \in w_i$ into w_k is represented as

$$P(k|i) = \int_{R_k} p_i(\mathbf{x})d\mathbf{x}. \tag{80.15}$$

The optimal regions $\{R_i\}$ that minimize the *ECM* are the set of the points $\mathbf{x}$ for which the allocation of $\mathbf{x}$ to a group w_k, $k = 1, 2, \ldots, J$ results in the least cost. It can be shown that an equivalent form of Equation 80.14 can be represented without the integral term $P(k|i)$. The equivalent minimizing ECM' is interpreted intuitively[1] as:

> "The minimizing *ECM* is equivalent to minimizing the *a posteriori* probabilities for the wrong classes with the corresponding costs."

That is, the equivalent ECM' has the form

$$ECM' = \sum_{\substack{k=1\\k\ne i}}^{J} P(k|\mathbf{x})C(i|k)$$

$$= \sum_{\substack{k=1\\k\ne i}}^{J} \frac{P_k p_k(\mathbf{x})}{\sum_{j=1}^{J} P_j p_j(\mathbf{x})} C(i|k) \tag{80.16}$$

[1] The fact that *ECM* and *ECM'* are equivalent is shown analytically in the text (Hinton, 1989).

and since the denominator is a constant independent of the indices j, this can be further simplified as

$$ECM' = \sum_{\substack{k=1 \\ k \neq j}}^{J} P_k p_k(\mathbf{x}) C(i|k) \qquad (80.17)$$

In other words, the optimal minimum *ECM* classifier assigns **x** to w_k such that Equation 80.17 is minimized. The minimum *ECM* (*ECM'*) classifier rule determines mutually exclusive and exhaustive classification regions R_1, R_2, R_J such that Equation 80.14 (Equation 80.17) is a minimum.

If the cost is not important (or the same for all misclassifications) the minimum *ECM* rule becomes minimum *TPM*. The resulting classifier is, again as in the two-class case, a MAP classifier.

Assign unknown **x** to w_k:

$$\mathbf{x} \in w_k = \arg\min_{i \in \mathcal{G}} \sum_{\substack{i=1 \\ i \neq k}}^{J} P_i p_i(\mathbf{x}) \qquad (80.18)$$

$$= \arg\max_{i \in \mathcal{G}} P_i p_k(\mathbf{x}) \qquad (80.19)$$

$$= \arg\max_{i \in \mathcal{G}} P(w_k|\mathbf{x}) \qquad (80.20)$$

The Bayesian classification rule, which is based on the conditional probability density functions for each class, $p_i(\mathbf{x})$, is the optimal classifier in the sense that it minimizes the cost of the probability of error (Fukunaga, 1990). However, the class conditional probability density function $p_i(\mathbf{x})$ needs to be estimated. The density estimation is realizable and efficient if the dimensionality is low, such as $1 \sim 2$ or 3, at most. The parametric Bayesian classification, even if it renders the optimal result in the sense that probability of error is minimized, is difficult to realize in practice. Alternatively, we look for other simple approximations using a normality assumption on the class conditional distributions.

Bayesian Classifiers with Multivariate Normal Populations

If the conditional distribution of a given class is assumed to be p-dimensional multivariate normal:

$$p_i(\mathbf{x}) = \frac{1}{(2\pi)^{p/2}|\Sigma_i|^{1/2}} \exp\left(-\frac{1}{2}(\mathbf{x} - \mu_i)^t \Sigma_i^{-1} (\mathbf{x} - \mu_i)\right), \quad i = 1, 2, \ldots, J \qquad (80.21)$$

with mean vectors μ_i and covariance matrices Σ_i, then, the resulting Bayesian classifiers are easily realized.

Quadratic Discriminant Score

With the assumption of having the same cost for all misclassifications added to the multivariate normality, we get a simple classification rule directly from Equation 80.19. Then the minimum *TPM* decision rule can be expressed as follows:

Allocate **x** to the class w_k:

$$\mathbf{x} \in w_k = \arg\max_{i \in \mathcal{G}} \{\ln P_i p_i(\mathbf{x})\}$$

$$= \arg\max_{i \in \mathcal{G}} \{d_i^q(\mathbf{x})\} \qquad (80.22)$$

where the quadratic discriminant score is defined as

$$d_i^q(\mathbf{x}) = -\frac{1}{2}\ln|\Sigma_i| - \frac{1}{2}(\mathbf{x} - \mu_i)^t \Sigma_i^{-1}(\mathbf{x} - \mu_i) + \ln P_i \qquad (80.23)$$

and consists of contributions from the generalized variance $|\Sigma_i|$, the *a priori* probability P_i, and the squared distance from **x** to the population class mean μ_i. Note that $d_i^q(\mathbf{x})$ is the quadratic form of the unknown **x**.

Linear Discriminant Score

If we further assume that the population covariance matrices Σ_i are all the same, we can simplify the quadratic discriminant score Equation 80.23 into the linear discriminant score:

$$d_i(\mathbf{x}) = \mu_i^t \Sigma_i \mathbf{x} - \frac{1}{2}\mu_i^t \Sigma^{-1} \mu_i + \ln P_i \qquad (80.24)$$

Then the optimal minimum *ECM* classifier with the assumptions that

1. The multivariate normal distribution in the class conditional density function is $p_i(\mathbf{x})$.
2. We have equal misclassification cost (thus a minimum *TPM* classifier).
3. And that we have equal covariance matrices Σ_i for all classes, reduces to the simplest form with a linear discriminant score as follows:

$$\mathbf{x} \in w_i = \arg\max_{i \in \mathcal{G}} \left\{ d_i(\mathbf{x}) = \mu_i^t \Sigma_i \mathbf{x} - \frac{1}{2}\mu_i^t \Sigma^{-1} \mu_i + \ln P_i \right\} \qquad (80.25)$$

where **x** was assigned to class w_k.

As the name indicates, the linear discriminant score $d_i(\mathbf{x})$ for a class i used in the special case of the minimum *TPM* classifier Equation 80.25 is a *linear* functional of the input **x**. The boundary regions $R_1, R_2, \ldots, R_J$ are hyper-linear, e.g., lines in 2-dimensional, planes in 3-dimensional input space, etc. However, the minimum *TPM* classifier with different covariances for the classes, is given by the *quadratic* form of x as in Equation 80.23.

Linear Discriminant Analysis and Classification

The Fisher's discriminant function is basically for description purposes. With new lower dimensional discriminant variables, multidimensional data may be visualized to find some interesting structures; hence, the linear discriminant analysis is exploratory. The objective of this section is to relate the linear discriminant analysis to Bayesian optimal classifiers based on *normal theory.*

The linear transform by which the discriminant variates are obtained is defined by the $q \times q$ matrix F in the transform:

$$\mathbf{x} = F\mathbf{y} \tag{80.26}$$

where q is the dimensionality of vector x and the matrix F consists of $s = \min\{q, J - 1\}$ eigenvectors of $W^{-1}B$ whose corresponding eigenvalues are nonzero. This result is obtained by maximizing the quadratic form of matrix B with respect to the constraint in the form of the quadratic expression of matrix W. W and B are the sample versions of pooled within and between covariance matrices, respectively defined as:

$$W = \frac{1}{N - J} \sum_{i=1}^{J} \sum_{j=1}^{n_i} (\mathbf{y}_{ij} - \bar{\mathbf{y}}_i)(\mathbf{y}_{ij} - \bar{\mathbf{y}}_i)^t$$

$$B = \frac{1}{J - 1} \sum_{i=1}^{J} n_i(\bar{\mathbf{y}}_i - \bar{\mathbf{y}})(\bar{\mathbf{y}}_i - \bar{\mathbf{y}})^t$$

where $N = \Sigma_{i=1}^{J}$ is the size of the sample and J is the number of classes.

In the transformed domain, or in the discriminant coordinate space (CRIMCOORD) the class mean vectors are given by:

$$\mu_{i,x} = [\mu_{i,x_1}, \mu_{i,x_2}, \ldots, \mu_{i,x_s}]^t = F\mu_{i,y}$$

for $x \in w_i$, and by the definition of the LDA $\text{cov}(X) = I$. Thus it is appropriate to consider a Euclidean distance in order to measure the separation of the discriminant variates. The classification rule from the discriminants is now to allocate x into class w_k:

$$\mathbf{x} \in w_k = \arg\min_{i \in \mathcal{G}}\{\|\mathbf{x} - \mu_{i,x}\|^2\} \tag{80.27}$$

Here the dimensionality of **x** is $s \leq \min\{q, J - 1\}$. The dimensionality of the transformed variables, i.e., the discriminant variates, becomes s and the classification rule needs only s variables in the linear discriminant classification rule (Equation 80.27).

The reason for only s variables needed for this classification purpose follows. The sample pooled *within* covariance matrix W and the *between* covariance matrix B have full ranks, hence the $W^{-1}B$, $(q \times q)$-matrix, has full rank. The number of nonzero eigenvalues should not be greater than the full rank:

$$s \leq q. \tag{80.28}$$

And the class mean vectors span a multi-dimensional space with dimensionality:

$$p \leq J - 1 \tag{80.29}$$

which is obvious since by definition $\Sigma_{i=1}^{J}(\mu_i - \bar{\mu}) = 0$. From Equation 80.28 and Equation 80.29 we can conclude that $s = \min\{q, J - 1\}$. The remaining $(q - s)$-dimensional subspace is called the *null* space of the linear transformation represented by the matrix F and consists of all the vectors y that are mapped into 0 by the linear transformation of Equation 80.26.

Equivalence of LDF to Minimum *TPM* Classifier

It is interesting to observe the equivalence of the linear discriminant classification rule Equation 80.27 with that of the minimum *TPM* classification rule, with the assumption that all covariances $\Sigma_i = \Sigma$ are the same for all classes $i \in \mathcal{G}$.

The argument of the minimization quantity of Equation 80.27 becomes

$$\begin{aligned} \|F(\mathbf{y} - \mu_{i,y})\|^2 &= \|\mathbf{x} - \mu_{i,x}\|^2 \\ &= (\mathbf{y} - \mu_{i,y})^t\Sigma^{-1}(\mathbf{y} - \mu_{i,y}) \\ &= -2d_i(\mathbf{y}) + \mathbf{y}^t\Sigma^{-1}\mathbf{y} + 2\ln P_i \end{aligned} \tag{80.30}$$

where the last equation is due to:

$$d_i(\mathbf{y}) - \frac{1}{2}\mathbf{y}^t\Sigma^{-1}\mathbf{y} = -\frac{1}{2}(\mu_{i,y}^t\Sigma^{-1}\mu_{i,y} - 2\mu_{i,y}^t\Sigma^{-1}\mathbf{y} + \mathbf{y}^t\Sigma^{-1}\mathbf{y}) + \ln P_i \tag{80.31}$$

The minimization of the squared distance in the Fisher's discriminant variate domain is equivalent to the maximization of the linear discriminant score $d_i(\mathbf{y})$, which results in the equivalence of the 'linear discriminant classification rule' to the 'minimum *TPM* optimal classifier.' (Johnson and Wichern, 1988).

This is an interesting observation or justification of Fisher's LDF. Even though the derivation of the Fisher's discriminant functions do not require the 'multivariate normality' assumption, the same classification rule is obtained from the minimum *TPM* criterion Bayesian classification rule in which normality is assumed.

Learning Vector Quantizer

Learning Vector Quantization (LVQ) is a combination of the self-organizing map and of supervised learning (Kohonen, 1990). The self-organizing map is a typical competitive learning method and results in a number of new vectors, called *codebook vectors,* $\mathbf{m}_i$, $i = 1,2, \ldots, L$. The codebook vectors represent an input vector space with a small number of representative vectors (codebook $\mathcal{M}$). It is a quantization of the given data set $\{x_i, g_i\}_1^N$ to get a quantized codebook $\{\mathbf{m}_i, g\}_1^L$.

Competitive Learning

Given a training vector $\{x_i, g_i\}_1^N$ and a size L of a randomly chosen codebook $\{\mathbf{m}_i\}_1^L$, an input of time instance k, $\mathbf{x}^{(k)}$, is compared to all the code vectors, $\mathbf{m}_i$, in order to find the closest one, $\mathbf{m}_c$, by a distance measure such that:

$$d(\mathbf{x}^{(k)}, \mathbf{m}_c) = \min_l\{d(\mathbf{m}_l, \mathbf{x}^{(k)})\} \tag{80.32}$$

L_2-norm is a common choice and the competitive learning with this measure utilizes the *steepest descent gradient* step optimization (Kohonen, 1990). Once the closest code vector $\mathbf{m}_c$ is found, the competitive learning (or, the steepest descent gradient optimization) updates the closest code vector, m_c but it does not change the other code vectors, $\mathbf{m}_l$ $l \neq c$.

$$\mathbf{m}_c^{(k+1)} = \mathbf{m}_c^{(k)} + \alpha(k)(\mathbf{x}^{(k)} - \mathbf{m}_c^{(k)}) \tag{80.33}$$

$$\mathbf{m}_l^{(k+1)} = \mathbf{m}_l^{(k)} \quad \text{for} \quad l \neq c \tag{80.34}$$

with $\alpha(k)$ being a suitable constant $0 < \alpha < 1$, or monotonically decreasing sequence, $0 < \alpha(k) < 1$. for which the optimization LVQ (or OLVQ that will be discussed later) is concerned with.

Self-organizing map is an algorithm for finding a codebook $\mathcal{M}$ (or a set of feature-sensitive detectors) in the input space χ. It is known that the internal representations of information in the brain are generally spatially organized, and the self-organizing map mimics the spatial organization of the cells (Kohonen, 1990) in its structure. A self-organizing map enforces the logically inspired network connections, with "*lateral inhibition*" in a general way by defining a neighborhood set N_c; a time-varying monotonically decreasing set of code vectors:

$$N_c^{(k)} = \{\mathbf{m}_l^{(k)} \mid d(\mathbf{m}_l^{(k)}, \mathbf{m}_c^{(k)}) \leq r(\mathrm{k})\} \tag{80.35}$$

where $\gamma(k)$ represents the radius of the $N_c^{(k)}$. Once the winning code vector (or cell) is found from Equation 80.32 all the code vectors in the neighborhood N_c which is centered on the winning code vector $\mathbf{m}_c$ are updated and the others remain untouched. It has been suggested (Kohonen, 1990) that the $N_c^{(k)}$ be very wide in the beginning and shrink monotonically with time as $\gamma(k)$ is a function of time, k.

Thus the updating has a similar form to simple competitive learning as in Equation 80.33,

$$\mathbf{m}_l^{(k+1)} = \begin{cases} \mathbf{m}_l^{(k)} + \alpha(k)(\mathbf{x}^{(k)} - \mathbf{m}_l^{(k)}) & \text{if } \mathbf{m}_l \in N_c^{(k)} \\ \mathbf{m}_l^{(k)} & \text{if } \mathbf{m}_l \notin N_c^{(k)} \end{cases} \tag{80.36}$$

where $\alpha(k)$ is a scalar-value "adaptation gain" $0 \leq \alpha(k) \leq 1$.

Learning Vector Quantization

If we now have a codebook that represents the input vector space χ by a set of quantized vectors, i.e., a codebook $\mathcal{M}$, then the *nearest neighbor* rule can be used for classification problems, provided that the codebook vectors m_l have their labels in the space to which each codebook vector belongs. The labeling process is similar to the K-nearest neighbor rule in which (a part of) the training data are used to find the majority labels among the K closest patterns to a codebook vector m_l. Thus the LVQ, a form of supervised learning, follows the unsupervised learning, self-organizing map, as shown in Figure 80.3.

The last two stages in the figure are called LVQ and researchers (Kohonen, et al., 1992; Kohonen, 1990) have come up with different updating algorithms (LVQ1, LVQ2, LVQ3, OLVQ1) from different methods of updating the codebook vectors. The LVQ1 and its optimization version OLVQ1 are considered in the next sections.

LVQ1 is similar to the simple competitive learning (Equation 80.33), except that it includes the pushing of any wrong closest codebook vector in addition to the pulling operations (Equation 80.33 and Equation 80.36).

Let $\mathcal{L}(\mathrm{x}^{(k)})$ be an operation to get the label information, then the codebook updating rule LVQ1 has the form (Figure 80.4)

$$\mathbf{m}_c^{(k+1)} = \mathbf{m}_c^{(k)} + \alpha(k)(\mathbf{x}^{(k)} - \mathbf{m}_c^{(k)}) \text{ for } \mathcal{L}(\mathbf{x}^{(k)}) = \mathcal{L}(\mathbf{m}_c) \tag{80.37}$$

$$\mathbf{m}_c^{(k+1)} = \mathbf{m}_c^{(k)} - \alpha(k)(\mathbf{x}^{(k)} - \mathbf{m}_c^{(k)}) \text{ for } \mathcal{L}(\mathbf{x}^{(k)}) \neq \mathcal{L}(\mathbf{m}_c)$$

$$\mathbf{m}_i^{(k+1)} = \mathbf{m}_i^{(k)} \text{ for } i \neq c \tag{80.38}$$

Here, $0 < \alpha(k) < 1$ is a gain, which is decreasing monotonically with time, as in the competitive learning, (Equation 80.33). The authors suggest a small starting value, i.e. $\alpha(0) = 0.01$ or 0.02.

Optimized LVQ1(OLVQ1)

For fast convergence of the LVQ1 algorithm in Equation 80.37 and Equation 80.38, an optimized learning rate for the

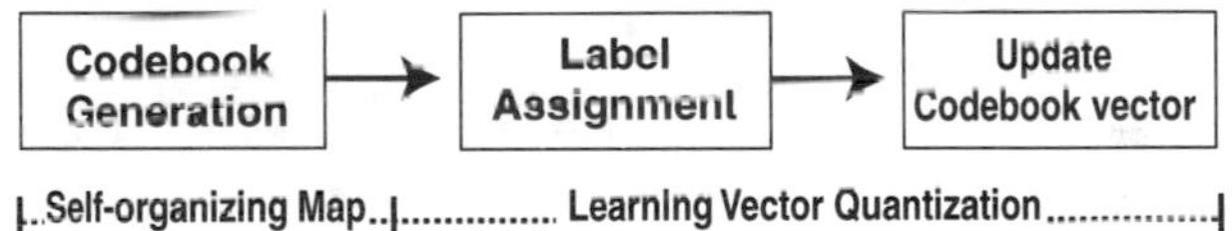

Figure 80.3 Block diagram for a system of self-organizing map and learning vector quantization.

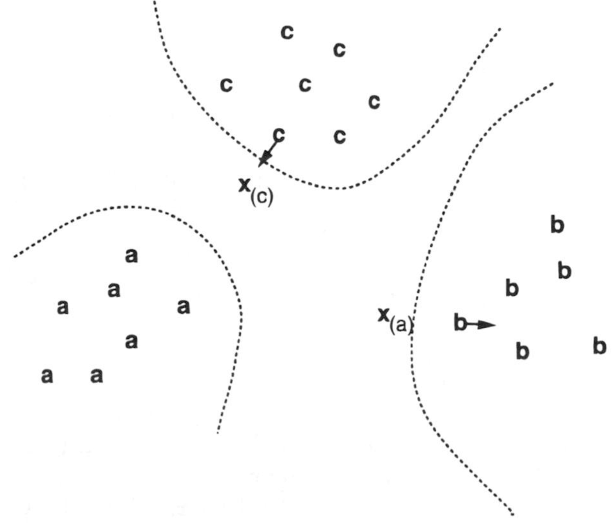

Figure 80.4 LVQ1 learning, or updating the initial codebook vectors **a, b, c.**

LVQ1 is suggested (Kohonen, et al., 1992). The objective is to find an optimal learning rate $\alpha_l(k)$ for each codebook vector $\mathbf{m}_l$, so that we have individually optimized learning rates:

$$\mathbf{m}_c^{(k+1)} = \mathbf{m}_c^{(k)} + \alpha_c(k)(\mathbf{x}^{(k)} - \mathbf{m}_c^{(k)}) \text{ for } \mathcal{L}(\mathbf{x}^{(k)}) = \mathcal{L}(\mathbf{m}_c) \quad (80.39)$$

$$\mathbf{m}_c^{(k+1)} = \mathbf{m}_c^{(k)} - \alpha_c(k)(\mathbf{x}^{(k)} - \mathbf{m}_c^{(k)}) \text{ for } \mathcal{L}(\mathbf{x}^{(k)}) \neq \mathcal{L}(\mathbf{m}_c)$$

$$\mathbf{m}_l^{(k+1)} = \mathbf{m}_l^{(k)} \text{ for } l \neq x \quad (80.40)$$

Equation 80.39 and Equation 80.40 can be stated with a new sign term $s(k) = 1$ or -1 for the right class and the wrong class, respectively, as follows:

$$\mathbf{m}_c^{(k+1)} = [(1 - s(k)\alpha_c(k))]\mathbf{m}_c^{(k)} + s(k)\alpha_c(k)\mathbf{x}^{(k)} \quad (80.41)$$

It can be seen that m_c is directly independent but is recursively dependent on the input vector x from Equation 80.41.

The argument on the learning rate (Kohonen, 1990) is that:

> "Statistical accuracy of the learned codebook vectors $\mathbf{m}_c^{(*)}$ is optimal if the effects of the corrections made at different times are of equal weight."

The learning rate due to the current input $\mathbf{x}^{(k)}$ is a $\alpha_c(k))$ from Equation 80.41 and due to the previous input $\mathbf{x}^{(k-1)}$ the current learning rate is $(1 - s(k)\alpha_c(k)) \cdot \alpha_c(k-1)$. According to the argument, the effects to the learning rates are to be the same for two consecutive inputs $\mathbf{x}^{(\kappa)}$ and $\mathbf{x}^{(\kappa-1)}$:

$$\alpha_c(k) = [1 - s(k)\alpha_c(k)]\alpha_c(k-1). \quad (80.42)$$

If this condition is to hold for all k, by induction, the learning rates from all the earlier $\mathbf{x}^{(\kappa)}$, for $k = 0, 1, \ldots, k$ should be the same. Therefore, due to the argument, the optimal values of learning rate $\alpha_c(k)$ are determined by the recursion from Equation 80.42 for the specific code vector $\mathbf{m}_c$ as:

$$\alpha_c(k) = \frac{\alpha_c(k-1)}{1 + s(k)\alpha_c(k-1)} \quad (80.43)$$

with which the OLVQ1 is defined as in Equation 80.39 and Equation 80.40.

Nearest Neighbor Rule

The nearest neighbor classifier, a nonparametric exemplar method, is the natural classification method one can first think of. Using the label information of the training sample, an unknown observation $\mathbf{x}$ is compared with all the cases in the training sample. N distances between a pattern vector $\mathbf{x}$ and all the training patterns are calculated and the label information, with which the minimum distance results, is assigned to the incoming pattern $\mathbf{x}$. That is, the NN rule allocates the $\mathbf{x}$ to w_κ if the closest exemplar $\mathbf{x}_c$ is with the label $k = \mathcal{L}(\mathbf{k}_c)$:

$$\mathbf{x}_c = \arg\min_i\{d(\mathbf{x}_0, \mathbf{x}_i)\}, \; i = 1, 2, \ldots, N$$

$$\mathbf{x}_0 \in w_k = \mathcal{L}(\mathbf{x}_k) \quad (80.44)$$

The distance measure between the unknown and the training sample has a general quadratic form:

$$d(\mathbf{x}, \mathbf{x}_k) = (\mathbf{x}_0 - \mathbf{x}_k)^t M(\mathbf{x}_0 - \mathbf{x}_k) \quad (80.45)$$

With $M = \Sigma^{-1}$, the inverse of the covariance matrix in the sample, the result is the *Mahanalobis* distance. Euclidean distance is obtained when $M = I$, i.e., the identity matrix. Another choice may be the measure considering only the variance for which $M = \Lambda$, where Λ is a diagonal matrix with its elements (λ_i) $1/2 = \text{var}(x_i)$ and $\mathbf{x} = (x_1, x_2, \ldots, x_p)^t$.

The K-Nearest Neighbor (KNN) rule is the same as the NN rule except that the algorithm finds K nearest points within the points in the training set, from the unknown observation χ and assigns the class of the unknown observation to the majority class in the K points.

Recent VLSI technology advances have made memory cheaper than ever, thus the KNN rule is becoming feasible. Some modified versions of the original KNN rules are reported in what follows. These approaches interpolate between outputs of nearest neighbors stored during training to form complex nonlinear mapping functions (Wolpert, 1988; Farmer and Sidorowich, 1988). Much of the work with the modified KNN rules is in designing effective distance metrics (Lippmann, 1989). Some modified KNN are developed for parallel machine implementation, called the connectionist machine (Stanfill and Waltz, 1986), as well as for serial computing (Farmer and Sidorowich, 1988).

80.5 Neural Networks

Introduction

Neural networks have been a much-publicized topic of research in recent years and are now beginning to be used in a wide range of subject areas. One of the strands of interest in neural networks is to explore possible models of biological computation. Human brains contain about 1.5×10^{15} neurons of various types, with each receiving signals through 10 to 10^4 synapses. The response of a neuron is known to be happening in about $1 \sim 10$ milliseconds (Ripley, 1994). Yet we can recognize an old friend's face and call him in about 0.1 seconds. This is a complex pattern recognition task which must be performed in a highly parallel way, since the recognition is done in about $100 \sim 1000$ steps. This suggests that highly parallel systems can perform pattern recognition tasks more rapidly than current conventional sequential computers. As yet, our VLSI technology, which is essential planar implementation with at most 2 or 3 layer cross connections, is far from achieving these parallel connections that require 3-dimensional interconnections.

Artificial Neural Networks

Even though originally the neural networks were intended to mimic a task specific subsystem of a mammalian or human brain, recent research has been mostly concentrated on the *artificial neural networks* which are only vaguely related to the biological system. Neural networks are specified by the net topology, node characteristics, and training or learning rules.

Topological consideration of the artificial neural networks for different purposes can be found in review papers (Hush and Horne, 1993; Lippmann, 1987). Since our interests in the neural networks are in classification, only the feed-forward multilayer perceptron topology is considered, leaving the feedback connections to the references.

The topology describes the connection with the number of the layers and the units in each layer for feed-forward networks. Node functions are usually nonlinear in the middle layers but can be linear or nonlinear for output layer nodes. However, all of the units in the input layer are linear and have fan-out connections from the input to the next layer.

Each output y_j, is weighted by w_{ij} and summed at the linear combiner represented by a small circle in Figure 80.5. The linear combiner thresholds its inputs before it sends them to the node function ϕ_j. The unit functions are (nonlinear, monotonically increasing and bounded functions as shown on the right of Figure 80.5.

Usage of Neural Networks

One use of a neural network is *classification*. For this purpose each input pattern is forced, adaptively, to output the pattern indicators which are part of the training data; the training set consists of the input covariate **x** and the corresponding class labels. *Feed forward* networks, sometimes called multilayer perceptrons (MLP), are trained adaptively to transform a set of input signals, $\mathcal{X}$, into a set of output signals, $\mathcal{G}$. Feedback networks start with an initial activity state of a feedback system, and after state transitions have taken place the asymptotic final state is identified as the outcome of the computation. One use of the feedback networks is the case of *associative memories:* on being presented with a pattern near a prototype X it should output pattern X', and as *autoassociative memory* or *contents-addressable memory* by which the desired output is completed to become X.

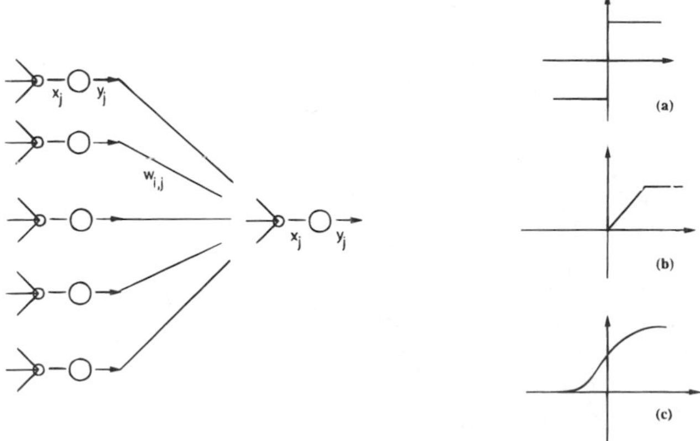

Figure 80.5 (I) The linear combiner output $x_j = \sum_{i=1}^{n} y_i w_{ik}$ is input to the node function ϕ_j to give the output y_j. (II) Possible node functions. Hard limiter (a), threshold (b), and sigmoid (c) nonlinear functions.

In all cases the network *learns* or *is trained by* the repeated presentation of patterns with known required outputs (or pattern indicators). Supervised neural networks find a mapping $f: \mathcal{X} \rightarrow \mathcal{G}$ for a given set of input and output pairs.

Other Neural Networks

The other dichotomy of the neural networks family is *unsupervised learning*, that is, clustering. The class information is not known or it is irrelevant; the networks find the groups of the similar input patterns. The neighboring code vectors in a neural network compete in their activities by means of mutual lateral interactions, and develop adaptively into specific detectors of different signal patterns. Examples are the self-organizing map (Kohonan, 1990), and the Adaptive Resonance Theory (ART) (Carpenter and Grossberg, 1987) networks. ART is different from other unsupervised learning networks in that it develops new clusters by itself; the network develops a new code vector if there exist sufficiently different patterns. Thus the ART is truly adaptive, whereas others require the number of clusters to be specified in advance.

Feed-Forward Networks

In feed-forward networks the signal flows only in the forward direction; no feed-back exists for any node. This is perhaps best seen graphically in Figure 80.6. This is the simplest topology and has been shown to be good enough for most practical classification problems (Ripley, 1993).

The general definition allows more than one hidden layer, and also allows 'skip-layer' connections from input to output. With this skip-layer, one can write a general expression for a network output y_k with one hidden layer,

$$y_k = \phi_k\left(b_k + \sum_{i \rightarrow k} w_{ik} x_i + \sum_{j \rightarrow k} w_{jk} \phi_j\left(b_j + \sum_{i \rightarrow j} w_{ij} x_i\right)\right) \qquad (80.46)$$

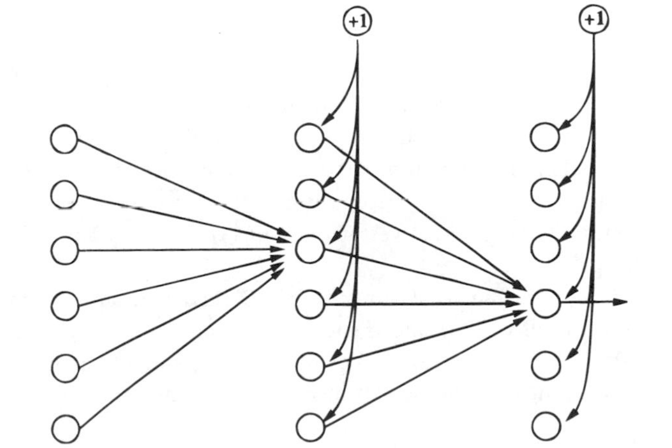

Figure 80.6 A generic feed-forward network with a single hidden layer. For bias terms the constant input with 1 are shown and the weights of the constant inputs are the bias values which will be learned as training proceeds.

where the b_j and b_k represent the thresholds for each unit in the jth hidden layer and the output layer, which is the kth layer. Since the threshold values b_j, b_k are to be adaptive, it is useful to have a threshold for the weights for constant input value of 1 as in Figure 80.6. The function $\phi()$ is almost inevitably taken to be a linear, sigmoidal ($\phi(x) = e^x/(1+e^x)$) or threshold function ($\phi(x) = \ (x > 0)$).

Rumelhart, Hinton and Williams (1986) showed that the feed-forward multilayer perceptron networks can learn using gradient values obtained by an algorithm, called *error backpropagation.*[2] This contribution is a remarkable advance since 1969, when Minsky and Papert (1969) claimed that the nonlinear boundary, required for the XOR problem, can be obtained by a multilayer perceptron. The learning method was unknown at the time.

Since Rosenblatt (1959) introduced the one layer, single perceptron learning method, called the *perceptron convergence* procedure, the research on the single perceptron had been widely active until the counter example of the XOR problem was introduced which the single perceptron could not solve.

In multilayer network learning, the usual objective or error function to be minimized has the form of a squared error:

$$E(\mathbf{w}) = \sum_{p=1}^{P} \|\mathbf{t}^p - \mathbf{f}(\mathbf{x}^p; \mathbf{w})\|^2 \tag{80.47}$$

that is to be minimized with respect to **w**, the weights in the network. Here p represents the pattern index, $p = 1, 2, \ldots, P$, and $\mathbf{t}^p$ is the target (or desired) value when $\mathbf{x}^p$ is the input to the network. Clearly this minimization can be obtained by any number of unconstrained optimization algorithms: gradient methods or stochastic optimization are possible candidates.

The updating of weights has a form of the steepest descent method:

$$w_{ij} \leftarrow w_{ij} - \eta \frac{\partial E}{\partial w_{ij}}, \tag{80.48}$$

where the gradient value $\partial E/\partial w_{ij}$ is calculated for each pattern being present; the error term $E(\mathbf{w})$ in the on-line learning is not the summation of the squared error for all the P patterns.

Note that the gradient points are in the direction of maximum increasing error. In order to minimize the error, it is necessary to multiply the gradient vector by minus one and by a learning rate η.

The updating method (Equation 80.48) has a constant learning rate η for all weights and is independent of time. The original Method of Steepest Descent has the time-dependent parameter, η_k, hence η_k needs to be calculated as iterations progress.

[2] A comment on the terminology 'backpropagation' is given in **error backpropagation.** There, the backpropagation is interpreted as a method to find the gradient values of a feedforward multilayer perceptron network, not as a learning method. A pseudo-steepest descent method is the learning mechanism used in the network.

Error Backpropagation

The backpropagation was first discussed by Bryson and Ho (1960), later by Werbos (1974), and Parker (1985) but was rediscovered and popularized later by Rumelhart, Hinton and Williams (1986). Each pattern is presented to the network, and the input x_j and output y_j is calculated as in Figure 80.7. The partial derivative of the error function with respect to weights is

$$\nabla E(t) = \left[\frac{\partial E(t)}{\partial w_1(t)}, \ldots, \frac{\partial E(t)}{\partial w_n(t)}\right]^T \tag{80.49}$$

where n is the number of weights, and t is the time index representing the instance of the input pattern presented to the network.

The former indexing is for the 'on-line' learning in which the gradient term of each weight does not accumulate. This is the simplified version of the gradient method that makes use of the gradient information of all training data. In other words, there are two ways to update the weights by Equation 80.49:

$$w_{ij}^{(p)} \leftarrow w_{ij}^{(p)} - \eta \left(\frac{\partial E}{\partial w_{ij}}\right)^{(p)} \quad \text{temporal learning} \tag{80.50}$$

$$w_{ij} \leftarrow w_{ij} - \eta \sum_{p} \left(\frac{\partial E}{\partial w_{ij}}\right)^{(p)} \quad \text{epoch learning} \tag{80.51}$$

One way is to sum all the P patterns to get the sum of the derivatives in Equation 80.51, and the other way (Equation 80.50) is to update the weights for each input and output pair temporally without summation of the derivatives. The temporal learning, also called on-line learning (Equation 80.50), is simple to implement in a VLSI chip because it does not require the summation logic and storing each weight, while the epoch learning in Equation 80.51 does require one to do so. However, the temporal learning is an asymptotic approximation version of the epoch learning which is based on minimizing objective function (Equation 80.47).

With the help of Figure 80.7 the first derivatives of E with respect to a specific weight w_{jk} can be expanded by the chain rule:

$$\frac{\partial E}{\partial w_{jk}} = \frac{\partial E}{\partial x_k}\frac{\partial x_k}{\partial w_{jk}} = \frac{\partial E}{\partial x_k} y_j = \phi_k'(x_k)\frac{\partial E}{\partial y_k} y_j \tag{80.52}$$

$$= \frac{\partial \phi_k(x_k)}{\partial x_k}\frac{\partial E}{\partial y_k} y_j = \delta_k y_j \tag{80.53}$$

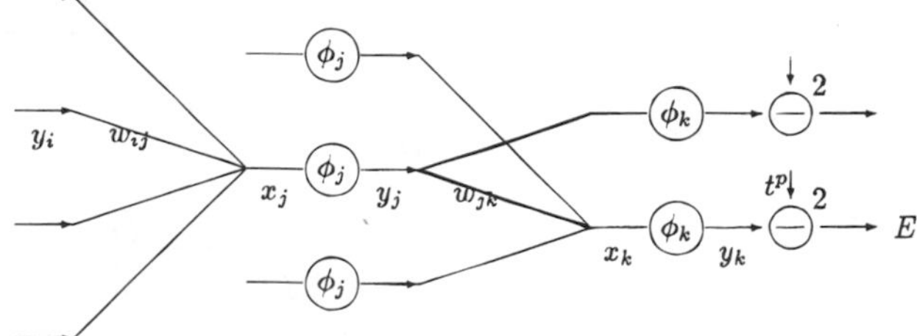

Figure 80.7 Error backpropagation. The δ_j for weight w_{ij} is obtained, δ_k's are then backward propagated via thicker weight lines w_{jk}'s.

For output units, $\partial E/\partial_{yk}$ is readily available, i.e., $2(y_k - t^p)$, where y_k and t^p are the network output and the desired target value for input pattern $\mathbf{x}^p$. The $\phi_k' x_k$ is straight forward for the linear and logistic nonlinear node functions; the hard limiter on the other hand is not differentiable.

For the linear node function:

$$\phi'(x) = 1 \quad \text{with} \quad y = \phi_x = x$$

and for the logistic unit the first order derivative becomes

$$\phi'(x) = \frac{e^x(1 + e^x) - (e^x)^2}{(1 + e^x)^2} \tag{80.54}$$

$$= y(1 - y) \quad \text{when} \quad \phi(x) = \frac{e^x}{1 + e^x} \tag{80.55}$$

The derivative can be written in the form

$$\frac{\partial E}{\partial w_{ij}} = \sum_p y_i^p \delta_j^p \tag{80.56}$$

which has become known as the generalized delta rule.

The δs in the generalized delta rule, Equation 80.56, for output nodes, therefore becomes

$$\delta_k = 2y_k(1 - y_k)(y_k - t^p) \quad \text{for a logistic output unit}$$

$$\delta_p = 2(y_p - t^p) \qquad \text{for a linear output unit} \tag{80.57}$$

The interesting point in the backpropagation algorithm is that the δs can be computed from output to input through hidden layers across the network. δs for the units in earlier layers can be obtained by summing the δs in the higher layers. As shown in Figure 80.7, the δ_j are obtained as

$$\delta_j = \phi_j'(x_j)\frac{\partial E}{\partial y_j}$$

$$= \phi_j'(x_j) \sum_{j \to k} w_{jk} \frac{\partial E}{\partial x_k}$$

$$= \phi_j'(x_j) \sum_{j \to k} w_{jk}\delta_k \tag{80.58}$$

The δ_ks are available from the output nodes. As the updating (or learning) progresses backwards, the previous (or higher) δ_k are weighted by the weights w_{jk}s and summed to give the δ_js. Since Equation 80.58 for δ_j only contains terms at higher layer units, it is clear that it can be calculated backwards from the output to the input of the network; hence the name backpropagation.

Madaline Rule III for Multilayer Network with Sigmoid Function

Widrow took an independent path in learning as early as in the 1960's (Widrow and Lehr, 1990, Widrow, 1962). After some 20 years of research in adaptive filtering, Widrow and colleagues returned to the neural network research (Widrow and Lehr, 1990), and extended the Madaline I with the goal of developing a new technique that could adapt multiple layers of adaptive elements, using the simpler hard-limiting quantizer. The result was Madaline Rule II (or simply MRII), a multilayer linear combiner with a hard-limiting quantizer.

Andes (1988, unpublished) modified the MRII by replacing the hard-limiting quantizer resulting in MRIII by a sigmoid function in the Adaline, i.e., a single layer linear combiner with a hard-limiting quantizer. It was proven later that MRIII is in essence equivalent to backpropagation. The important difference from the gradient based backpropagation method is that the derivative of the sigmoid function is not required in this realization; thus the analog implementation becomes feasible with this MRIII multilayer learning rule.

A Comment on the Terminology Backpropagation. The terminology 'backpropagation' has been used differently from what it should mean. To get the partial derivatives of the error function (at the system output node) with respect to the weights of the units in lower than the output unit, the δ terms in the output unit are propagated backward as in Equation 80.58. However, the network (actually the weights) learns (or weights are updated) using the pseudo steepest descent method, (Equation 80.48); it is *pseudo* because a constant term is used whereas the steepest descent method requires an optimal learning rate for each weight and time instance, i.e., $\eta_{ij}(k)$. The error back-propagation is indeed to find the necessary gradient values in the updating rule. Thus it is not a good idea to call the back propagation a learning method; the learning method is a simple version of the steepest descent method, which is one of the classical minimizer finding algorithms. Backpropagation is an algorithm to find the gradient ∇E in a feed-forward multilayer perceptron network.

Optimization Machines with Feed-forward Multilayer Perceptrons. Optimization in multilayer perceptron structures can be easily realized by gradient-based optimization methods with the help of back propagation. In the multilayer perceptron structure the functions can be minimized/maximized via any gradient-based unconstrained optimization algorithm, such as Newton's method or steepest descent method.

The description of the optimization machine has the functional description depicted in Figure 80.8 and consists of two parts: gradient calculation and weight (or parameter) up-dating.

The gradient ∇E of the multilayer perceptron network is

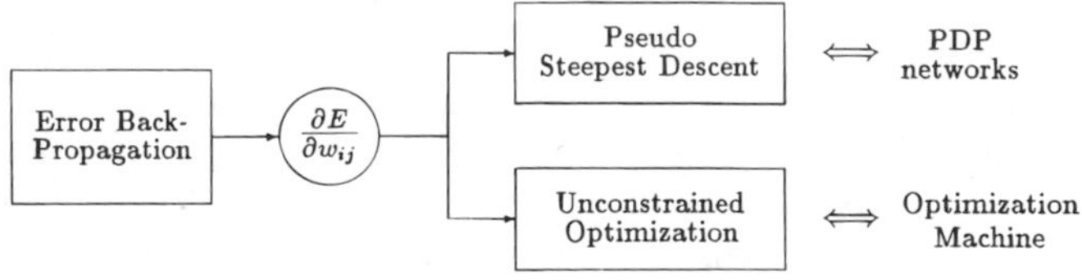

Figure 80.8 Functional diagram for an optimization machine.

obtained by error back propagation. If this gradient is used in an on-line fashion with the constant learning rate η as in Equation 80.48, then this structure is the neural network used earlier (Rumelhart et al., 1986). This on-line learning structure possesses a desirable feature in VLSI implementation of the algorithm since it is temporal: no summation over all the patterns is required but the weights are updated as the individual pattern is presented to the network. It requires little memory but sometimes the convergence is too slow.

The other branch in Figure 80.8 shows unconstrained optimization of the nonlinear function. The optimization machine gets the gradient information as before, but various and well developed unconstrained optimizations can be used for finding the optimizer. The unconstrained nonlinear minimization is divided basically into two categories; gradient methods and stochastic optimization. The gradient methods are deterministic and use the gradient information to find the direction for the minimizer. Stochastic optimization methods such as ALOPEX are discussed in Chapter 69 of this handbook as well as in Micheli-Tzanakou, 1995 and Zahner and Micheli-Tzanakou, 1995. Comparisons of ALOPEX with backpropagation are shown in Chapter 69 and Micheli-Tzanakou et al. (1995).

Justification for Gradient Methods for Non-linear Function Approximation. Getting stuck in local minimizers is a well-known problem for gradient methods. However, the size of the weights (or the dimensionality of the weight space in the neural networks) is usually much larger than the dimensionality of the input space: $X \subset R^p$ that we like to search for optimization. The employed redundant degrees of freedom in the ways to find the better minimizer is a good reason or the justification for the gradient methods used in neural networks.

Another justification for the gradient method in optimization may be due to the approximation by the Taylor expansion of highly nonlinear functions (Ripley, 1994) where the first and second order approximation, i.e., a quadratic approximation to the nonlinear function, is used. The quadratic function in a covariate $\mathbf{x}$ has a unique minimum or maximum.

Training Methods for Feedforward Networks

There exist two basic ways to train the feedforward networks. They are gradient based learning and stochastic learning. Training or learning is essentially an unconstrained optimization problem. Abundant algorithms in optimization can be applied to the function approximated by the network in a structured way defined by the network topology.

In the gradient based methods, the most popular learning is the steepest descent/ascent method with the error backpropagation algorithm to get the required gradient of the minimizing/maximizing error function with respect to the weights in the network (Rumelhart et al., 1986a, 1986b). Another method using the gradient information is Newton's Method, which is basically used for zero finding of a nonlinear function. The function optimization problem is the same as the zero finding of the first derivative of the function, hence the Newton's method is valid.

All the deterministic (as opposed to stochastic) minimization techniques are based on either or both the steepest descent and Newton's method. The objective function to be optimized is usually limited to a certain class in the network optimization. The square of the error $\|t - \hat{y}\|^2$ and the information theoretic measure, the Kullback-Leibler distance, are objective functions used in the feedforward networks. This is due to the limitation in calculating the gradient values of the network utilized by the error backpropagation algorithm.

The recommended 'method of optimization' due to Broyden, Fletcher, Goldfarb, and Shanno (BFGS) is the well known Hessian matrix update in the Newton's method of unconstrained optimization (Peressini et al., 1988). It requires gradient values. For the optimization machine of Figure 80.8 the feedforward network with backpropagation provides the gradients and the Hessian approximation is obtained by the BFGS method.

The other dichotomy of the minimization of an unconstrained nonlinear multivariate function is grouped into, the so-called 'stochastic optimization.' The representative algorithms are simulated annealing (Kirkpatrick, 1983), Boltzman machine learning (Hinto and Sejnowski, 1986) and ALgorithm Of Pattern EXtraction (ALOPEX) (Unnikrishnan and Venugopal, 1994; Harth and Tzanakou, 1974). Simulated annealing (Kirkpatrick, 1983) has been successfully used in combinatoric optimization problems, such as the traveling salesman problem, VLSI wiring and VLSI placement problems. An application of feedforward network learning has been reported (Engel, 1988) with the weights being constrained to be integers or discrete values rather than continuum of the weight space.

Boltzman machine learning by Hinton and Sejnowski (1986) is similar to simulated annealing except that the acceptance of randomly chosen weights is possible even when the energy state has decreased. In simulated annealing, the weights yielding the decreased energy state are always accepted; but in the Boltzman machine, probability is used in accepting the increased energy states.

The simulated annealing and the Boltzman machine learning (a general form of Hopfield Network (Hopfield and Tank, 1985) for the associative memory application) are mainly for combinatoric optimization problems with binary states of the units and the weights. Extension from binary to M-ary in the states of the weights has been reported for classification problems (Engel, 1988) in simulated annealing training of the feedforward perceptrons.

ALOPEX was originally used for construction of the visual receptive field but with some modifications was later applied to the learning of any type of network, not restricted to multilayer perceptrons. It is a random walk process in each parameter in which the direction of the constant jump is decided by the correlation between the weight changes and the energy changes (Unnikrishnan and Venugopal, 1994). Since the stream of this chapter consists of the gradient-based optimization methods and the scope of the stochastic optimization is examined in Chapter 69 of this handbook, we do not include the other important optimization stream of stochastic methods in this chapter.

Issues in Neural Networks

Universal Approximation

In the introduction section of the article by Hornik, Stinchcombe and White (1989) previous work about the approximation capability of multilayer perceptrons is summarized and is referenced here. More than 20 years ago, Minsky and Papert (1969) showed that simple two layer (no hidden layers) networks cannot approximate the nonlinearly separating functions (e.g., XOR problems), but a multilayer neural network could do the job. Many results on the capability of the multilayer perceptron have been reported. Some theoretical analysis for the network capability of the universal approximator are listed below and are extensively discussed by Stinchcombe and White, 1989.

Kolmogorov (1957) tried to answer the question of Hilbert's 13th problem, i.e., the multivariate function approximation by a superposition of the functions of one variable. The superposition theory sets the upper limit of the number of hidden units to $2n + 1$ units, where n is the dimensionality of the multivariate function to be approximated. However, the functional units in the network are different for the different functions to be approximated, while one would like to find an adaptive method to approximate the function from the given training data at hand. Thus Kolmogorov's superposition theory says nothing about the capability of a multilayer network nor which method is to be used.

More general views were reported. Le Cun (1987) and Lapedes and Farber (1988) showed that monotone squashing functions can be used in the two hidden layers to approximate the functions. Fourier series expansion of a function is realized by a single layer network by Gallant and White (1988) with cosine functions in the units. Further related results using the sigmoidal (or logistic) units are shown by Hecht-Nielsen (1989). Hornik, Stinchcombe and White (1989) presented a general approximation theory of one hidden layer network using arbitrary squashing functions such as cosine, logistic, hyperbolic tangent, and etc., provided that sufficiently many hidden units are available. However the number of hidden units are not considered to attain any given degree of approximation in Hornik et al. (1989).

The number of hidden units obviously depends on the characteristics of the training data set, i.e., the underlying function to be estimated. It is intuitive to say that the more complicated functions to be trained, the more hidden units are required.

For the number of the hidden units, Baum and Haussler, 1989, limit the size of general networks (not necessarily the feedforward multilayer perceptrons) by relating it to the size of the training sample. The authors analytically showed that if the size of the sample is N, and we want to correctly classify future observations with at least a fraction $1 - \epsilon/2$ correctly, then the size of the sample has a lower bound given by

$$N \geq O\left(\frac{W}{\epsilon}\log\frac{N}{\epsilon}\right)$$

where W is the number of the weights and N the number of the nodes in a network. This, however, does not apply to the interesting feedforward neural networks and the given bound is not useful for most applications.

There seems to be no rule of thumb for the number of hidden units (Ripley, 1993). Finding the size of the hidden units can usually be done usually by cross-validation or any other resampling methods. Usual starting value for the size is suggested to be about the average of the number of the input and output nodes (Ripley, 1993). Failure in learning, can be attributed (Hornik, Stichcombe, and White, 1989) to three main reasons:

- Inadequate learning.
- Inadequate number of hidden units.
- Or, presence of a stochastic rather than a deterministic relation between input and target in the training data, i.e., noisy training data.

Enhancing Convergence Rate and Generalization of an Optimization Machine

While the steepest descent method used originally with the backpropagation algorithm, (Equation 80.48), can be an efficient method for obtaining the weights that minimize an error measure, error surfaces frequently possess properties that make this procedure slow convergence. There are at least two reasons (correlated in a sense as will be seen below) for this slow rate of convergence (Jacobs, 1988).

1. The magnitude of a gradient may be such that modifying a weight by a constant proportion, η as in Equation 80.48, of that gradient will yield too little reduction in the error measure. There are two cases for this situation. When the error surface is fairly smooth (or nearly flat), the gradient magnitude is small, and consequently the convergence is too slow. The other situation involves the case where the error curve is too wiggly. Even a small change in the weight space may result in 'overshooting' which may produce a small reduction of the error measure. Oscillating over a local minimum can happen with this error function.
2. The second reason for the slow convergence is that the negative gradient may not point to the actual minima, as is usually the case. Figure 80.9 shows an example of an error function of the two parameters with the elliptic curves representing the contour of the error function. With the given weight point $w(t)$ at time t, the negative gradient does not point to the real minima which is represented by a bullet in the center of the inner contour. Given the negative gradient the magnitude in the direction of the major axis x_1 is too small, whereas the component in the minor direction x_2 is too large.

Suggestions for Improving the Convergence

Jacobs(1988) summarized four heuristics proposed in the literature for increasing the rate of convergence:

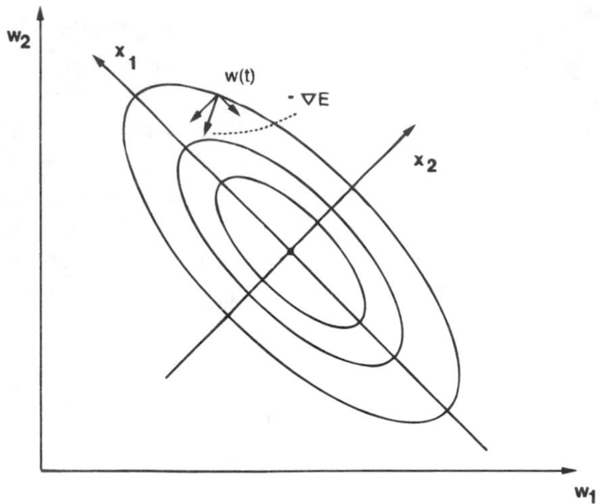

Figure 80.9 Error surface with contours over a two-dimensional weight space.

1. Every parameter of the performance measure to be minimized should have its own individual learning rate, η_{ij}.
2. Every learning rate should be allowed to vary over time, $\eta_{ij}(k)$.
3. When the derivative of a parameter possesses the same sign for several consecutive time steps, the learning rate for that parameter should be increased.
4. When the sign of the derivative of a parameter alternates for several consecutive time steps, the learning rate for that parameter should be decreased.

Note that from Figure 80.9 by providing different learning rates for each parameter dimension, the current point in the weight space is not modified in the direction of the negative gradient, but toward the real minima.

Another cause for the slow convergence comes from the sigmoidal units $\phi()$'s that are used to impose the network with nonlinearity. The derivative of the nonlinear unit function has been shown to be in the form of Equation 80.55. The logistic units may become 'stuck' at a round value, either 0 or 1, since $\phi'(x) = y(1 - y)$ (Equation 80.55) gives a very small value for an output $\simeq$ 0 or 1:

$$\phi'(x) = y(1 - y) \simeq 0 \quad \text{for } y \simeq 0 \text{ or } 1 \tag{80.59}$$

Unfortunately, any saturating unit function is bounded, resulting in the property: near the saturation points the derivative vanishes. With nonlinear units with the backpropagation learning and the general objective function $E = \|t - y\|^2$ giving the $\partial E/\partial w = y(1 - y)$, the convergence of a network is known to be slow as discussed earlier.

In the original work of Rumelhart, Hinton and Williams (1986a) a 'momentum' term was added, that is an exponential smoothing was applied to the correction term, so that

$$w_{ij} \leftarrow w_{ij} - \eta\left[(1 - \alpha)\frac{\partial E}{\partial w_{ij}} + \alpha(\Delta w_{ij})\right] \tag{80.60}$$

They also considered the 'on-line' version of Equation 80.60, that is

$$w_{ij} \leftarrow w_{ij} - \eta' y_i^p \delta_j^p + \alpha'(\Delta w_{ij}) \tag{80.61}$$

and updated the weights as each pattern was presented to the network.

Quick Prop

Some other interesting ideas to speed up the convergence have been introduced. *Quickprop* (Fahlmann, 1989) used a second-order method, based loosely on Newton's method. Quickprop is based on two risky assumptions: that the error vs. weight graph for each weight can be approximated by a parabola with one minimum value; that the change in the slope of the error curve, as seen by each weight, is not affected by all the other weights that are changing at the same time.

Everything else proceeds as in standard back-propagation, but for each weight w_{ij} a set of information for the previous time update is retained to get a second order approximation. The steps to follow are: (1) find the error derivative $S_{ij}(t - 1) = \partial E(t - 1)/\partial w_{ij}(t - 1)$, and (2) update $\Delta w_{ij}(t - 1) = w_{ij}(t) - w_{ij}(t - 1)$. The computation for the next step size of a found direction according to the heuristics above is then given by:

$$\Delta w(t) = \frac{S(t)}{S(t - 1) - S(t)} \Delta w(t - 1) \tag{80.62}$$

where $S(t)$ and $S(t - 1)$ are the current and previous values of $\partial E/\partial w$. This is a crude approximation to the optimal minima. The fraction portion η in each parameter w_{ij} is adaptively adjusted using the Equation 80.62.

To get around this pitfall, Fahlmann (1989) suggested also using an offset in order for the *delta* (as in Equation 80.57) to be at least 0.1, i.e., $\phi'(x) = 0.1 + y(1 - y)$.

Kullback-Leibler Distance

A more interesting treatment for the problem with the classical gradient descent method has been shown in the literature (van Ouyan and Niehhuis, 1992; Golden, 1988; Solla et al., 1988). A relative (or cross) entropy of target t with respect to output y is defined and interpreted as Maximum A Posteriori (MAP) estimation for the optimal minima of the weight space,

$$E = \sum_p \sum_k \left[t_k^p \log \frac{t_k^p}{y_k^p} + (1 - t_k^p) \log \frac{1 - t_k^p}{1 - y_k^p} \right] \tag{80.63}$$

This entropy measure becomes the measure of 'maximum likelihood' if the targets t_k are $t_k \in \{0,1\}$ [20], and may be called the 'Kullback-Leibler' distance, one of the probabilistic distances.

The interpretation of the output vectors with this distance measure is that the output vector represents the conditional probability of target t, given the input pattern $\mathbf{x}$. A binary random variable B_k associated with the kth output unit describes the presence($B_k = 1$) or absence($B_k = 0$) of the kth output attribute.

For a given input pattern $\mathbf{x}^p$, the activity y^p reflects the conditional probabilities

$$P\{B_k = 1|\mathbf{x}^p\} = y^p, \text{ and} \tag{80.64}$$

$$P\{B_k = 0|\mathbf{x}^p\} = 1 - y^p. \tag{80.65}$$

With this distance measure the δ value in the generalized delta rule, (Equation 80.56), becomes simpler and linear with the error $(t^p - y_k)$:

$$\begin{aligned}\delta_k &= \phi_k'(x_k)\frac{\partial E}{\partial y_k}\\ &= y_{jk}(1-y_k)\left(\frac{-t^p}{y_k}+\frac{1-t^p}{1-y_k}\right)\\ &= y_k(1-y_k)\frac{y_k - t^p}{y_k(1-y_k)}\\ &= y_k - t^p\end{aligned} \tag{80.66}$$

Thus the error signal propagates towards the inner layers backwards and the pitfall problem (Equation 80.59) no longer exists for this distance measure.

Weight Decay. Another way to avoid saturation is to discourage large weights and hence large inputs (Hinton, 1989); ones with large deviations from the data set are used for training. One can modify the error function to obtain the regularization effects by adding an extra term which penalizes the over-fitting. Also, the discouragement of unusual inputs (e.g., outlier patterns) works as *robust* learning. This generalization in learning is related to the *bias-variance* trade-off in the scatter plot smoothing.

A new error to be minimized is the sum of the squared error:

$$E' = E + \lambda \sum_{ij} w_{ij}^2 \tag{80.67}$$

where the λ is the *weight decay* parameter. The weight update rule, (Equation 80.48), turns out to be (with the penalty term)

$$w_{ij} \leftarrow w_{ij} - \eta \sum_p y_i^p \delta_j^p - 2\eta\lambda w_{ij} \tag{80.68}$$

This is the gradient (or steepest) descent learning method with a new error term.

Two effects from the weight decay can be realized. One is the *generalization* obtained by the shrinkage effect of the weight decay. This shrinkage method is the same idea as ridge regression in statistics, which may be written in a modified linear regression form as:

$$(X^tX + \Lambda)\hat{\beta} = X^tY \tag{80.69}$$

where Λ is a non-negative diagonal matrix. This is motivated by a prior on β or as a penalty term or a device to avoid large parameter values in nearly collinear problems (Ripley, 1993). It is also known that weight decay helps the numerical stability of optimization algorithms, especially in avoiding almost flat regions in iterative methods, such as in Equation 80.48.

The extra penalty term in Equation 80.68, weight growing, is equally discouraged; there is no discrimination of the weights by their hierarchical position in a multilayer network. With the help of Figure 80.7, the weights $\{w_{ij}\}$ relate the system inputs $y_i = x_i$ and x_j, the input to the next layer units, but the weights $\{w_{jk}\}$ are in between y_j and x_k. To give the same penalty for all the weights evenly, (Equation 80.68), the input vector x to the system should have the same range as the y_js. Thus it is more sensible that the system inputs have the same range as the intermediate values y_js, by scaling so that the input $\{x^p\}$ will be in [0,1], approximately.

For the decay parameter λ, Ripley (1994) suggested $\lambda \simeq 10^{-4} \sim 10^{-2}$ for the sum of the squares criterion (Equation 80.68), and $0.01 \sim 0.1$ for the entropy measure criterion, (Equation 80.63).

If regression and classification are to be considered in a unified frame, the distinguishing characteristic is in the interpretation and use of the response variable. Regression is a method of model fitting for the given data point pairs. Regression has the continuous response variable, representing outputs of the estimating function $\hat{f}(\cdot)$, and usually continuous in the region of the function $f(\cdot)$. One likes to find or estimate the underlying function that relates input and output pairs, $\{(x_i, y_i)\}_1^N$, for many reasons. *Prediction* for future observations x_0, *inference* on the estimated function f, and *interpretation* of the function of covariate x_i are the principal objectives. Neural networks are a new surge in this regression paradigm although research for regression purposes is not as active as is for classification problems. Classification is meant to analyze different group data and to represent the group data well so that future observations could be classified as correctly as possible. The response variable can be considered as a categorical variable taking the value from a finite set of class labels.

The difference between regression and classification is whether the response variable is the continuous region of the function or the categorical variable respectively.

Regression Methods for Classification Purposes. The recent success and popularity of neural networks motivated some applied statisticians to look for similar methodologies in the statistical literature and to develop methods to use the existing nonparametric regression techniques (Hastie et al., 1993) for classification. The classification problem is recasted in the form of a regression problem. To establish relationship between regression and classification, the two-class linear discriminant function can be shown to be the scalar (not a constant) multiple of the least square regression function in the next section.

Generalization for multiple group settings is given in the section "Multi-response Regression and Flexible Discriminant Analysis." A number of recently developed adaptive regression methods are studied. Those are Classification And Regression Tree (CART) (Breiman et al., 1984), BRUTO (Hastie, 1989), and Multivariate Adaptive Regression Splines (MARS) (Friedman, 1991) and incorporated with a bridging tool FDA (Flexible Discriminant Analysis) (Hastie et al., 1993) for classification purposes.

Two-Group Regression and Linear Discriminant Function

The linear discriminant function for two-group classification has been viewed by Fisher (1936) alternatively in a regression context. (See pp. 212–213 of Anderson (1984)). The linear projector $W^{-1t}(\bar{x}^{(2)} - \bar{x}^{(1)})$ in the linear discriminant function $(\bar{x}^{(2)} - \bar{x}^{(1)})^t W^{-1}\mathbf{x}$ is actually a scalar multiple of the linear regression function.

A dummy variate is introduced for two class response values. Let the two variables be

$$y_i^{(1)} = \frac{n_2}{n_1 + n_2}, \qquad i = 1, 2, \ldots, n_1, \tag{80.70}$$

$$y_i^{(2)} = \frac{-n_1}{n_1 + n_2}, \qquad i = 1, 2, \ldots, n_2. \tag{80.71}$$

The regression function $\mathbf{b}^t\mathbf{x}$ is obtained by minimizing the sum of squared residual (SSR)

$$\sum_{j=1}^{2}\sum_{i=1}^{n_i} [y_i^{(j)} - \mathbf{b}^t(\mathbf{x}_i^{(j)} - \bar{\mathbf{x}})]^2.$$

where $\mathbf{x}^{(j)}$ is the ith observation from group j, $j = 1{,}2$ and $\bar{x}$ is the overall mean of the training data.

The normal equations are obtained by taking the derivative of the SSR with respect to b, the newly defined unknown coefficients of the two-group regression, and set it equal to zero:

$$\begin{aligned}\sum_{j=1}^{2}\sum_{i=1}^{n_i} (\mathbf{x}_i^{(j)} - \bar{\mathbf{x}})(\mathbf{x}_i^{(j)} - \bar{\mathbf{x}})^t\mathbf{b} &= \sum_{j=1}^{2}\sum_{i=1}^{n_i} y_i^{(j)}(\mathbf{x}_i^{(j)} - \bar{\mathbf{x}}) \qquad (80.72)\\ &= \frac{n_1 n_2}{n_1 + n_2}[(\bar{\mathbf{x}}^{(1)} - \bar{\mathbf{x}}) - (\bar{\mathbf{x}}^{(2)} - \bar{\mathbf{x}})]\\ &= \frac{n_1 n_2}{n_1 + n_2}(\bar{\mathbf{x}}^{(1)} - \bar{\mathbf{x}}^{(2)}) \qquad (80.73)\end{aligned}$$

The outer product in the LHS of Equation 80.72 is the total covariance of the predictor variables and can be decomposed in the form of within-covariance and between-covariance matrix combination as

$$\sum_{j=1}^{2}\sum_{i=1}^{n_i} (\mathbf{x}_i^{(j)} - \bar{\mathbf{x}})(\mathbf{x}_i^{(j)} - \bar{\mathbf{x}})^t \tag{80.74}$$

$$\begin{aligned}&= \sum_{j=1}^{2}\sum_{i=1}^{n_i} (\mathbf{x}_i^{(j)} - \bar{\mathbf{x}}^{(j)})(\mathbf{x}_i^{(j)} - \bar{\mathbf{x}}^{(j)})^t\\ &\quad + n_1(\bar{\mathbf{x}}^{(1)} - \bar{\mathbf{x}})(\bar{\mathbf{x}}^{(1)} - \bar{\mathbf{x}})^t + n_2(\bar{\mathbf{x}}^{(2)} - \bar{\mathbf{x}})(\bar{\mathbf{x}}^{(2)} - \bar{\mathbf{x}})^t\\ &= \sum_{j=1}^{2}\sum_{i=1}^{n_i} (\mathbf{x}_i^{(j)} - \bar{\mathbf{x}}^{(j)})(\mathbf{x}_i^{(j)} - \bar{\mathbf{x}}^{(j)})^t\\ &\quad + \frac{n_1 n_2}{n_1 + n_2}(\bar{\mathbf{x}}^{(1)} - \bar{\mathbf{x}}^{(2)})(\bar{\mathbf{x}}^{(1)} - \bar{\mathbf{x}}^{(2)})^t \qquad (80.75)\end{aligned}$$

Thus Equation 80.72 is rewritten as

$$\begin{aligned}&\frac{n_1 n_2}{n_1 + n_2}(\bar{\mathbf{x}}^{(1)} - \bar{\mathbf{x}}^{(2)})\\ &= \left[\sum_{j=1}^{2}\sum_{i=1}^{n_i} (\mathbf{x}_i^{(j)} - \bar{\mathbf{x}}^{(j)})(\mathbf{x}_i^{(j)} - \bar{\mathbf{x}}^{(j)})^t \right.\\ &\quad \left. + \frac{n_1 n_2}{n_1 + n_2}(\bar{\mathbf{x}}^{(1)} - \bar{\mathbf{x}}^{(2)})(\bar{\mathbf{x}}^{(1)} - \bar{\mathbf{x}}^{(2)})^t\right]\mathbf{b} \qquad (80.76)\end{aligned}$$

If we define the within-group SSP(sum of squares and products) as W:

$$W = \sum_{j=1}^{2}\sum_{i=1}^{n_i} (\mathbf{x}_{i=1}^{(j)} - \bar{\mathbf{x}}^{(j)})(\mathbf{x}_{i=1}^{(j)} - \bar{\mathbf{x}}^{(j)})^t,$$

the normal equation Equation 80.76 has the form

$$W\mathbf{b} = \frac{n_1 n_2}{n_1 + n_2}(\bar{\mathbf{x}}^{(1)} - \bar{\mathbf{x}}^{(2)}) - \frac{n_1 n_2}{n_1 + n_2}(\bar{\mathbf{x}}^{(1)} - \bar{\mathbf{x}}^{(2)})(\bar{\mathbf{x}}^{(1)} - \bar{\mathbf{x}}^{(2)})^t\mathbf{b} \tag{80.77}$$

$$= (\bar{\mathbf{x}}^{(1)} - \bar{\mathbf{x}}^{(2)})\left[\frac{n_1 n_2}{n_1 + n_2} - \frac{n_1 n_2}{n_1 + n_2}(\bar{\mathbf{x}}^{(1)} - \bar{\mathbf{x}}^{(2)})^t\mathbf{b}\right]. \tag{80.78}$$

Since the whole bracket is a scalar, the solution b of Equation 80.78 is proportional to the projection vector $W^{-1}(\bar{\mathbf{x}}^{(1)} - \bar{\mathbf{x}}^{(2)})$ of the linear discriminant function.

Multi-Response Regression and Flexible Discriminant Analysis

Multi-response linear/nonlinear regression can also be used for classification. The most simple and common way is to transform the categorical variable $j \in \{1,2, \ldots, J\}$ in the form of $(N \times J)$ matrix $Y_{N\times J}$ such that an element y_{ij} has a value 1 in the jth column if the observation is in class j. The multi-purpose multivariate regression is carried onto the predictors $\mathbf{x}$. A new observation $\mathbf{x}_0$ is fitted with the J fits and is classified by the class having the largest fitted value, i.e., $\hat{Y}_j$.

Since we cannot expect the regression fit $\hat{y}_k = f_k(\mathbf{x}^o)$, the kth regression fit, to be in the region [0,1], the indicator matrix Y whose elements are either 0 or 1 is not a good way of introducing dummy response variables. Optimal scoring, which will be studied in the next section, transforms a categorical variable to real line R such that, linear regression of the transform is best regressed on the predictor variables $\mathbf{x}$.

Powerful Nonparametic Regression Methods for Classification Problems

Recently, Hastie, Buja, and Tibshirani (1993) introduced a new treatment of regression methods to be used for classification problems. They showed that the discriminant analysis could be tackled via optimal canonical correlation analysis (CCA), especially its asymmetric version, optimal scoring (OS). The idea is based on the fact that CCA is equivalent to linear discriminant analysis (LDA) and that the OS results to CCA, via various nonparametric regression methods.

Linear discriminant analysis in Bayesian classifiers with multivariate normal populations of multi-group has been the traditional choice in classification and discriminant analysis. The robustness and the simplicity of LDA (Gnanadesikan and Kettenring, 1989) in implementation and interpretation are responsible for its popularity. Recently a group of applied statisticians found and developed ways of using regression techniques for classification applications. Breiman and Ihaka (1984) noticed that the regression approach to the classification problem can be extended from the two-group to a multi-group setting via *scaling* and ACE. This idea has been adapted by Hastie, Tibshirani, and Buja and developed to render the Flexible Discriminant Analysis (FDA). (Hastie et al., 1993).

The basic concept is that the LDA, CCA, and OS are equivalent. One can find the discriminant variates via either CCA or OS. Since this equivalence is so critical some space is devoted here to the understanding of this property. The generalization of the LDA to nonlinear flexible discriminant analysis is due to the fact that an OS solution can be obtained by any linear/nonlinear regression method. This has the important consequence that we can simply use the tools for nonparametric regression to perform nonparametric discriminant analysis, which the authors termed as Flexible Discriminant Analysis (FDA). This section is a somewhat concise version of section 3 of Hastie, Buja, and Tibshirani's unpublished paper (Hastie et al., 1993).

It is known that discriminant variates are the same as the so-called 'canonical variates' which result from an associated canonical correlation analysis (CCA), and often the latter term is used interchangeably with discriminant variates. Somewhat less known is that an asymmetric version of canonical correlation analysis, here called optimal scoring (OS), well known in correspondence analysis, can also yield a set of dimensions which coincide with those of LDA and CCA. Each of the three techniques (OS, CCA, LDA) to be explained has an associated criterion and constraints under which the criterion is to be optimized. The equivalence of LDA, CCA, and OS follows as each of them are briefly described.

Optimal Scoring

Optimal scoring is used to turn categorical variables into quantitative ones by assigning scores to classes (groups, categories). Suppose $\theta:\mathcal{T}\mapsto\mathcal{R}$ is a function that assigns scores to the classes, such that the transformed class labels are optimally predicted by linear regression on X. This produces a one-dimensional separation between the classes. More generally, we can find K sets of independent scorings for the class labels, $\{\theta_1\theta_2, \ldots, \theta_K\}$, and K corresponding linear maps $\eta_k(X) = X^t\beta_k$, $k = 1,2 \ldots K$, chosen to be optimal for multiple regression in R^K. Thus the OS problem is to find the two sets of unknown functions to minimize a certain criterion.

Let $(\mathbf{x}_i, g_i)$ $i = 1,2 \ldots, N$, be the training sample, then the scores $\{\theta_k(g)\}_1^K$ and the maps $\{\beta_k\};_1^K$ are chosen to minimize the average squared residual (ASR):

$$ASR = \frac{1}{N}\sum_{k=1}^{K}\sum_{i=1}^{N}(\theta_k(g_i) - \mathbf{x}_i^t\beta_k)^2$$

In the criterion ASR above, $\theta(g)$ assigns a real number, θ_j, to the jth label of g, the categorical response variable. With the matrix notation, given a J-vector of such scores θ_k, a N-vector $Y\theta$ is a vector of scored training data which one may try to regress onto the predictor matrix H, the $N \times p$-matrix.

For simple notational purposes we proceed with a single solution only. The multi-response multivariate regression can be thought of as simply the K duplicates for the single response multivariate regression. Thus a single solution pair (θ, β) is used in the following instead of the series of solution (θ_k, β_k), $k = 1, 2 \ldots, K$ to simplify the notation.

Definition: The optimal scoring problem is defined by the criterion

$$ASR(\theta, \beta_{OS}) = \min_{\beta}\left\{\frac{1}{N}\left(\sum_{i=1}^{N}[\theta(g_i) - h(\mathbf{x}_i)^t\beta]^2\right)\right\} \quad (80.79)$$

$$= \min_{\beta}\frac{1}{N}\|Y\theta - H\beta\|^2 \quad (80.80)$$

which is to be minimized (or made stationary) under the constraint $N^{-1}\|Y\theta\|^2 = 1$ which is a unique solution for θ.

A unified view for the three similar but equivalent techniques (OS, CCA, and LDA) can be conveniently achieved by rewriting the ASR in Equation 80.80 in a quadratic form:

$$ASR(\theta, \beta) = \theta^t\Sigma_{11}\theta - 2\theta^t\Sigma_{12}\beta + \beta^t\Sigma_{22}\beta \quad (80.81)$$

where the matrices Σ are defined as:

- $\Sigma_{11} = 1/N\ Y^tY$, a diagonal matrix with the class proportions $p_j = n_j/N$ in the diagonal,
- $\Sigma_{22} = 1/N\ (H^tH)$, the total covariance matrix of the predicator variables,
- $\Sigma_{22} = 1/N\ (Y^tH)$, $\Sigma_{21} = \Sigma_{12}^t$.

If all considered classes are in the sample, i.e. $n_j > 0$, Σ_{11} is invertible.

Partially Minimized ASR

If we assume that the score vector θ is fixed, the minimizing β for the OS problem is obtained by the least squares estimate of β:

$$\beta_{OS} = (H^tH)^{-1}H^tY\theta = \Sigma_{22}^{-1}\Sigma_{21}\theta \tag{80.82}$$

The linear regression of $Y\theta$ onto the design matrix H with the least square criterion gives the following results. From Equation 80.80 and Equation 80.82:

$$\min_{\beta} ASR(\theta, \beta)$$

$$= \frac{1}{N}\|Y\theta\|^2 - \frac{1}{N}((Y\theta)^tS^tSY\theta) = 1 - \frac{1}{N}(Y\theta)^tS(Y\theta) \tag{80.83}$$

$$= 1 - \frac{1}{N}\theta^tY^tSY\theta = 1 - \theta^t\Sigma_{12}\Sigma_{22}^{-1}\Sigma_{21}\theta \tag{80.84}$$

where $S = H(H^tH)^{-1} H^t$ denotes that 'hat' or 'smoother' matrix of predictor matrix H, which is the result of the least square linear regression.

The same equation on the $ASR(\theta\beta)$ has a matrix form as

$$\begin{aligned}\min_{\beta} ASR(\theta, \beta) &= \frac{1}{N}\|Y\theta\|^2 - \frac{1}{N}((Y\theta)^tS^tSY\theta)\\ &= \frac{1}{N}\|Y\theta\|^2 - \frac{1}{N}(SY\theta)^t(SY\theta)\\ &= \frac{1}{N}(Y\theta - SY\theta)^t(Y\theta - SY\theta)\\ &= \frac{1}{N}\{(Y\theta)^t(I - S)(I - S)Y\theta\}\\ &= \frac{1}{N}\{\theta^tY^t(I - P_H)Y\theta\}\end{aligned} \tag{80.85}$$

with a new notation for the projection matrix, P_H, based on the predictor design matrix H for the least square linear regression

$$P_H = S = H(H^tH)^{-1}H^t$$

With the assumption of fixed θ we have reached the partially minimized ASR where the minimizing β was obtained via the least square linear regression. Now, we need to find the θ that transforms the indicator matrix to yield the scalings $Y\theta$ such that the linear regression yields the best fit to the new scalings. The question then is given in Equation 80.85, what θ gives the least possible *ASR?*

It is the quadratic form of the symmetric matrix $Y^t(I - P_H)Y$ that we like to look for the vector θ, that results in the minimum quadratic value. Minimizing θ for the whole matrix $Y^t(I - P_H)Y$ is the same as maximizing θ for the matrix $Y\hat{Y} = YP_HY$ provided that the regression fit $\hat{Y} = P_HY$ is *shrunk,* which is a property of linear smoothers (Buja et al., 1989). The projection operation P_H is a linear smoother. Therefore, the minimizing θ in Equation 80.85 is the eigenvector corresponding to the largest eigenvalue of $Y\hat{Y} = YP_HY$.

This is the point at which nonlinear nonparametric regressions come into play for classification application of regression. Direct calculation of the projector matrix P_H of the expanded predictor space $h(\mathbf{x})$, or spanned by the columns of the matrix $\hat{H}$ is possible, but the fact that any regression can calculate the fitted value $\hat{Y}$ allows various linear/nonlinear regressions to be used.

Canonical Correlation Analysis

Canonical Correlation Analysis (CCA) seeks to identify and quantify the associations between two sets of variables. The correlation of two linear combinations of the two sets of variables is to be maximized.

Definition: The canonical correlation problem is defined by the criterion

$$COR(\theta_{CCA}, \beta_{CCA}) = \max_{\theta,\beta}\{\theta^t\Sigma_{12}\beta\} \tag{80.86}$$

which is to be maximized under the constraints

$$\theta^t\Sigma_{11}\beta = 1, \qquad \text{and} \quad \beta^t\Sigma_{22}\beta = 1. \tag{80.87}$$

The Σ's are the same as in the previous section for optimal scoring. The criteria of the optimal scoring $ASR(\theta, \beta)$ and canonical correlation analysis $COR(\theta, \beta)$ are related to each other by Equation 80.81 and the two *CCA* constraints:

$$ASR = 2 - 2\ COR$$

which means that the OS and the CCA differ only in the additional constraint on β through Equation 80.87.

The partially maximizing β_{CCA} with the θ for both the OS and the CCA is obtained by minimizing β_{OS} with the constraint of the β_{CCA} in Equation 80.87:

$$\beta_{CCA} = \beta_{OS}/\sqrt{\beta_{OS}^t\Sigma_{22}\beta_{OS}}. \tag{80.88}$$

The maximizer β_{CCA} representation in terms of the minimizer β_{OS} in the above equation (Equation 80.88) and the definition of the CCA (Equation 80.86) entails the identity in the fixed linear coefficients θ in the *OS* and *CCA:*

$$\begin{aligned}\max_{\substack{\theta^t\Sigma_{11}\theta=1\\ \beta^t\Sigma_{22}\beta=1}} COR(\theta, \beta) = \theta^t\Sigma_{12}\beta &= \frac{\theta^t\Sigma_{12}\beta_{OS}}{\sqrt{\beta_{OS}^t\Sigma_{22}\beta_{OS}}}\\ &= \left(\frac{\theta^t\Sigma_{12}\beta_{OS}\beta_{OS}^t\Sigma_{21}\theta}{\beta_{OS}^t\Sigma_{22}\beta_{OS}}\right)^{1/2}\\ &= (\theta^t\Sigma_{12}\beta_{OS}\beta_{OS}^{-1}\Sigma_{22}^{-1}\beta_{OS}^{t-1}\beta^t\Sigma_{21}\theta)^{1/2}\\ &= (\theta^t\Sigma_{12}\Sigma_{22}^{-1}\Sigma_{21}\theta)^{1/2}\end{aligned} \tag{80.89}$$

which verifies the identity of θ in that the minimizer in Equation 80.84 is the same as the one in the maximizer in Equation 80.86. With the identity of the θ for both the OS and the CCA as just shown and the relationship between the βs (Equation 80.88) verifies that the OS is essentially the same as the CCA with the constraint on the β_{CCA}.

Linear Discriminant Analysis

Linear Discriminant Analysis (LDA) is a standard tool for classification and dimension reduction purposes. The LDA is a special case of the Bayesian Classifier (see the section on this topic) where the group conditional distributions are assumed to be multivariate normal, have a common covariance matrix, and have different mean vectors for the different classes.

LDA Revisited

The optimizing problem of the multi-class data is to find the $K \leq J - 1$ linear combinations which separate the class means $\mathbf{m}_j$ as much as possible in the K dimensional subspace satisfying the constraint that the linear combinations are to be spherical, i.e., uncorrelated and with unit variance, with respect to Σ_W, the within-class covariance. The columns of the matrix U of LDA vectors $\mathbf{u}_k$ are the eigenvectors corresponding to the K largest eigenvalues of the matrix of $\Sigma_B^{-1}\Sigma_W$. The procedure of the LDA is first to sphere $\mathbf{x}$ with respect to the common within-groups covariance matrix, project these data onto the $J - 1$ dimensional subspace spanned by the J group mean vectors $\mathbf{m}_j$'s, and then classify the new discriminant covariate, $U\mathbf{x}_0$, vector to the class corresponding to the closest centroid.

Following the notations of the two sets of variables as in optimal scoring, the matrix M of mean vectors, Σ_B, and Σ_W have the following simple form with $P_Y = Y(Y^tY)^{-1}Y^t$ the projector onto Y-column space:

- $M = \Sigma_{11}^{-1}\Sigma_{12}$, a $J \times p$-matrix whose rows are the class means $\mathbf{m}_j = \text{avg}\{\mathbf{h}_i;\ i \in \text{Class } j\}$: $M = (\mathbf{m}_1, \mathbf{m}_2, \ldots, \mathbf{m}_J)^t$.
- $\Sigma_B = \dfrac{1}{N}(P_YH)^t(P_YH) = \Sigma_{21}\Sigma_{11}^{-1}\Sigma_{12} = M^t\Sigma_{11}M.$
- $\Sigma_W = \dfrac{1}{N}[((I - P_Y)H)^t(I - P_Y)H] = \Sigma_{22} - \Sigma_B.$

The matrix M consists of rows of class mean vectors $\mathbf{m}_j$. The between-class covariance Σ_B is the covariance of H regressed onto Y, or, equivalently, the class-weighted covariance of the class means. The within-class covariance is the left of the subtraction of the Σ_B from the total covariance Σ_{22}.

The criterion of the linear discriminant problem is the maximization problem of the between-class variance under a constraint on the within-class variance.

Definition: The criterion of the linear discriminant problem to be maximized is the between-class variance:

$$BVAR(\beta_{LDA}) = \max_{\beta}\{\beta^t\Sigma_B\beta\} \tag{80.90}$$

with the constraint:

$$WVAR(\beta_{LDA}) = \beta_{LDA}^t\Sigma_W\beta_{LDA} = 1 \tag{80.91}$$

Translation of Optimal Scoring Dimensions into Discriminant Coordinates

It is convenient to use CCA as a link between OS and LDA. CCA is a generalized singular value problem for Σ_{12} with regard to the metrics given by Σ_{11} and Σ_{22}. Remember that it is the maximizing problem, Equation 80.86 in which the generalized quadratic form is used, hence it is called the generalized singular value problem.

The associated singular value decomposition (SVD), essentially a collection of stationary solutions of the CCA problem, takes on the form:

$$\Sigma_{11}^{-1}\Sigma_{12}\Sigma_{22}^{-1} = \Theta D_\alpha B^t \tag{80.92}$$

$$\Theta^t\Sigma_{11}\Theta = I_L \tag{80.93}$$

$$B^t\Sigma_{22}B = I_L \tag{80.94}$$

where $L = \min(J,p)$, Θ is a $J \times L$ matrix whose columns θ_k are left-stationary vectors. B is a $p \times L$ matrix whose columns β_k are right-stationary vectors, and D_α is a diagonal matrix of size $L \times L$ with non-negative diagonal elements α_k sorted in descending order.

A simple(non-generalized) SVD of the form $A = UDV^t$ entails the trivial consequences:

$$A = UDV^t$$

$$AV = UD$$

$$A^tU = VD$$

$$U^tAV = D$$

$$V^tA^tAV = D^2$$

$$U^tAA^tU = D^2.$$

These are translated to the generalized SVD as follows. The left column is for the regular SVD and right column for the generalized SVD.

$A = UDV^t$	$\Sigma_{11}^{-1}\Sigma_{12}\Sigma_{22}^{-1} = \Theta D_\alpha B^t$	(80.95)
$AV = UD$	$\Sigma_{11}^{-1}\Sigma_{12} = \Theta D_\alpha B^t\Sigma_{22}$	
	$\Sigma_{11}^{-1}\Sigma_{12}B = \Theta D_\alpha$	(80.96)
$A^tU = VD$	$\Sigma_{22}^{-1}\Sigma_{21}\Sigma_{11}^{-1} = BD_\alpha\Theta^t$	
	$\Sigma_{22}^{-1}\Sigma_{21}\Theta = BD_\alpha\Theta_t\Sigma_{11}\Theta = BD_\alpha$	(80.97)
$U^tAV = D$	$\Theta^t\Sigma^{-1}11\Sigma_{12}\Sigma_{22}^{-1}B = D_\alpha$	
	$\Theta^t\Sigma_{12}B = D_\alpha$	(80.98)
$V^tA^tAV = D^2$	$\Theta^t\Sigma_{12}\Sigma_{22}^{-1}\Sigma_{21}\Theta = D_\alpha{}^2$	(80.99)
$U^tAA^tU = D^2$	$B^t\Sigma_{21}\Sigma_{11}^{-1}\Sigma_{12}B = D_\alpha{}^2$	(80.100)

In particular, Equation 80.98 implies $COR(\theta_k, \beta_k) = \alpha_k$.

As noted before from Equation 80.84 and Equation 80.89 the

stationary θ vectors of OS and CCA are the same, while the B vectors of OS and CCA are related according to Equation 80.82 and Equation 80.97 by

$$B_{OS} = BD_{\alpha}^{t}, \quad (80.101)$$

B_{OS} being a matrix of OS-stationary column vectors $\beta_{OS,k}$. From Equation 80.84 and Equation 80.89 it follows that $ASR(\theta_k, \beta_k) = 1 - \alpha_k^2$.

To link CCA and LDA, we rewrite Equation 80.100 using the expression of the $\Sigma_b = \Sigma_{21}\Sigma_{11}^{-1}\Sigma_{12}$ as:

$$B^t\Sigma_B B = D_{\alpha^2} \quad (80.102)$$

and

$$B^t\Sigma_W B = B^t(\Sigma_{22} - \Sigma_B)B \quad (80.103)$$

$$= I_L - D_{\alpha^2} = D_{1-\alpha^2} \quad (80.104)$$

These two equations, (Equation 80.102 and Equation 80.104), show that B diagonalize both Σ_B and Σ_W. If we define,

$$B_{LDA} = BD_{(1-\alpha^2)^{1/2}},$$

we get a matrix whose columns $\beta_{LDA,k}$ are stationary solutions of the LDA problem:

$$B_{LDA}^t\Sigma_W\Sigma_{LDA} = I_L, \quad (80.105)$$

$$B_{LDA}^t\Sigma_B B_{LDA} = D_{\alpha^2(1-\alpha^2)}. \quad (80.106)$$

Finally, the relation between the LDA and the OS solutions is given by

$$B_{LDA} = B_{OS}D_{[\alpha^2(1-\alpha^2)]^{-1/2}}. \quad (80.107)$$

Linear Discriminant Analysis via Optimal Scoring

The minimization criterion, average squared residual(ASR), for a multi-response optimal scoring has the form

$$ASR = \frac{1}{N}\sum_{k=1}^{K}\sum_{i=1}^{N}(\theta_k(g_i) - \mathbf{x}_i^t\beta_k)^2$$

$$= \frac{1}{NH}\|Y\Theta - XB\|^2 \quad (80.108)$$

with a constraint $N^{-1}\|Y\Theta\|^2 = 1$ for a unique solution Θ.

If Θ is fixed, we get the transformed value $\Theta_{N\times}^{*}K = Y_{N\times J}\Theta_{J\times K}$.

With a new notation for the projection matrix, P_H, and smoothing operation S, based on the predictor design matrix H for the least square linear regression $P_H = S = H(H^tH)^{-1}H^t$, the partially minimizing ASR with the Θ^* fixed becomes

$$ASR(\Theta, B) = \frac{1}{N}\|Y\Theta\|^2 - \frac{1}{N}((Y\Theta)^tS^tSY\Theta)$$

$$= \frac{1}{N}\|Y\Theta\|^2 - \frac{1}{N}(SY\Theta)^t(SY\Theta)$$

$$= \frac{1}{N}(Y\Theta - SY\Theta)^t(Y\Theta - SY\Theta)$$

$$= \frac{1}{N}\{(Y\Theta)^t(I - S)(I - S)Y\Theta\}$$

$$= \frac{1}{N}\{\Theta^tY^t(I - P_H)Y\Theta\} \quad (80.109)$$

If we set the constraints on the Θ^* of zero mean and being unit variance and uncorrelated:

$$\frac{1}{N}\sum_{i=1}^{N}\Theta_i^* = 0 \qquad \frac{1}{N}\Theta^{*t}\Theta^* = I_K$$

the minimizing Θ is obtained from Equation 80.109 by the K largest eigenvectors Θ of Y^tP_HY with the constraint $\Theta^tD_p\Theta = K_K$ and with $D_p = Y^tY/N$.

A direct approach for such optimal score Θ would be by explicitly building the project (or hat) matrix P_X and doing eigen analysis via singular value decomposition,

$$P_X = X(X^tX)^{-1}X^t$$

$$Y^tP_XY = \Theta\Lambda\Theta^t.$$

A more convenient approach avoids the explicit calculation P_X and takes advantage of the fact that P_X computes linear regression: $\hat{Y} = P_XY$.

An algorithmic approach to compute the usual canonical variates by OS provides an equivalent procedure to get the LDA by OS.

LDA via OS

As the equivalence of OS and LDA from Equation 80.107, the algorithm for LDA via OS is:

1. Initialize: form $Y_{N\times J}$, the indicator matrix, whose index y_{ij} is 1 if the ith observation belongs to the jth group, otherwise is 0.
2. Linear multivariate regression: find the linear regression

$$\hat{Y} = P_XY = SY$$

and by the linear least squares, set B such that

$$\hat{Y} = XB.$$

3. Optimal scores: find the eigenvector matrix Θ of rank $K \le J$ matrix $Y^t\hat{Y}$ via SVD

$$Y^t\hat{Y} = \Theta\Lambda\Theta^t \text{ with } \Theta^tD_p\Theta = I_J.$$

4. Update the coefficient matrix of the linear combination matrix B obtained in step 2.

$$B \leftarrow B\Theta$$

The final coefficient matrix B_{OS} is, up to a diagonal scale matrix, the same as the LDA coefficient matrix B_{LDA} obtained from Equation 80.107.

$$B_{LDA} = B_{OS}D$$

where the diagonal matrix D has the elements

$$d_{kk} = [\alpha_k^2(1 - \alpha_k^2)]^{-1/2}$$

and α_k is the kth element of the diagonal matrix Λ, in the spectral decomposition of the rank $K \leq J$ matrix $Y^t\hat{Y}$ via SVD:

$$Y^t\hat{Y} = \Theta^t\Lambda\Theta.$$

Flexible Discriminant Analysis by Optimal Scoring

If we apply nonparametric regression $\hat{Y} = S(\hat{\lambda})Y$, in step 2 above, we can reduce the *flexibility* of the nonparametric regression into a classification problem. Here the smoothing parameter $\hat{\lambda}$ controls the fitness of the regression $\hat{Y}$ to Y, and is thus the control parameter.

The nonparametric multivariate regression in $\hat{Y} = S(\hat{\lambda})Y$ comes into play in two ways: (Hastie et al., 1993)

- The regularization property by bias-variance control is obtained.
- And, a model selection (i.e., variable selection) and inter action between variables may be exploited in the multivariate regression.

There exist many powerful nonparametric multivariate regression methods and more are expected to be developed. The most recently developed are

1. Projection Pursuit Regression(PPR) (Friedman and Stuetzle, 1981).
2. Alternate Conditional Expectation(ACE) (Breiman and Friedman, 1985).
3. Additivity and Variance Stabilization(AVAS) (Tibshirani, 1988).
4. Additive Model(AM)(Freidman, 1991).
5. Multivariate Adaptive Regression Splines(MARS) (Breiman, 1991a).
6. π-method (Wahba, 1990).
7. Interaction spline method (Breiman, 1991b).
8. Hinging-hyperplanes.
9. Neural networks.

The FDA by OS method is similar to the algorithmic LDA by OS of the previous section. The steps to follow are:

1. Initialize: Choose an initial score matrix Θ_0 satisfying the constraints

$$\Theta^t_{K\times J}D_p\Theta_{J\times K} = I_K$$

and get the scoring matrix $\Theta_0^m = Y\Theta_0$. The Θ_0 may be obtained by a contrast matrix.[3]

2. Multivariate nonparametric regression: Fit a multi-response, adaptive nonparametric regression of Θ_0^m on X by one of the nonparametric regressions listed above.

$$\hat{\Theta}_0^* = S(\hat{\lambda})\Theta_0^* = \eta(\mathbf{x})$$

where $\eta(\mathbf{x})$ is the vector of fitted regression functions.

3. Optimal Scores: Obtain the eigenvector matrix Φ of $\Theta_0^{*t}\hat{\Theta}_0^*$ and hence the optimal scores $\Theta_{J\times K} = \Theta_0\Phi$.
4. Update the final model from step 2 using the optimal scores:

$$\eta(\mathbf{x}) \leftarrow \Theta^t\eta(\mathbf{x}).$$

It is worth noting step 3 in both procedures in order to distinguish the way of obtaining the optimal scores. For the first procedure for the LDA via OS, the indicator matrix Y is regressed onto X. But in the second procedure for FDA via OS, the transformed score data, $\Theta_0^* = Y\Theta_0$ are regressed onto X by any of the various nonparametric regression methods. The optimal score Θ is thus updated as $\Theta = \Theta_0\Phi$.

For a J class problem, it is known from the discriminant analysis that the vector of canonical variates or functions $\eta(\mathbf{x})$ has at most $K = J - 1$ components. If $\bar{\eta}^j = \Sigma_{gi=j}\eta(\mathbf{x}_i)/n_j$ denotes the fitted centroid of the jth class in this space of canonical variates, the discrimination rule has the form of a (weighted) nearest centroid rule:

$$\mathbf{x} \in j = \arg\min_k\{\|D(\eta(\mathbf{x}) - \bar{\eta}^k)\|^2\} \qquad (80.110)$$

D is the diagonal matrix of scale factors that convert optimally scaled fits to discriminant analysis variables.

80.6 Comparison of Experimental Results

In general, any pattern recognition system consists of two basic subsystems: Feature extraction and classifier design. In this study,

[3] The contrast matrix is the $K - 1$ linear combinations of a factor variable with K levels. It is an encoding method of the factor variable such that the linear combination of the levels becomes linearly independent. There exist the Helmert, polynomial contrasts and others (Chambers and Hastie, 1991, ch. 2).

however, we are mainly interested in classifiers. There are many different classifiers from the simple and powerful nonparametric KNN rule to the recently popularized neural networks, as well as the newly developed multivariate regression methods. Eleven classifiers which are all explained in the introduction are experimented with the same data set obtained by Zernike moments, a global feature extraction method (Chung, 1994; Khotanzad and Hong, 1990; Teague, 1980)

A new branch in the growing tree of the classifiers has been developed in applied statistics (Hastie et al., 1993; Breiman and Ihaka, 1984) and is by now popularized. It is based on the fact that optimal scoring (OS) is equivalent to the linear discriminant analysis (LDA) (Equation 80.107) and the OS can be obtained by various regression techniques which are well researched in statistics (see the section on the translation of optimal scoring). The multivariate regression methods were used for classification and the results were proven to be competitive to the classical statistical methods.

Table 80.1 describes the classifiers in a simple format with control parameters, learning and operation processes. Details on the classifiers are given in the Introduction.

The core part of the software for the classifiers used in the study has been obtained from contributed software. They are written mostly by originators or some active researchers in the area. The archive `package classif` is a collection contributed by B. Ripley and is maintained in the `statlib@lib.stat.cmu.edu` which is accessible by *anonymous ftp*. It can be found under the S directory of the maintainer. This `classif` library also contains LDA, OLVQ1, KNN and others that we did not experiment with.

Hastie and Tibshirani contributed the programs that were recently developed by themselves and A. Buja. The package fda contains the Flexible Discriminant Analysis (FDA) which is a way of using the Optimal Scoring by nonparametric regression for classification problems. The library `fda` comes with POLYREG, BRUTO, MARS and BRUTO which are the recently developed multivariate regression methods. MARS can also be obtained from the directory general in the same maintainer, `statlib@lib.stat.cmu.edu.` The CART and PPREG can also be found from the S directory of the same maintainer. These are also available in function `type tree()` and `ppreg()` from the commercial package Splus.[4]

Table 80.1 List of the Classifiers Used

Classifiers	Control parameters	Learning	Operation
LDA		F, μ_i	$\arg\min_{i\in\mathcal{G}} \|F(x-\mu_i)\|^2$
OLVQ1	Codebook	find$\{\mathbf{m}_i\}_1^L$	find $d_i(\mathbf{x},\mathbf{m}_i)$
KNN	$k = 1,3,5$		$\arg\min_i\{d_i(\mathbf{x},\mathbf{x}_i)\}$
NNET	$h = 15$ $\lambda = 0.005$	Minimize $\Sigma(\hat{y}_i - t_i)^2 + \lambda\Sigma W^2$	$\arg\max_j\{P(j\|\mathbf{x})\}$
CART		find $B_m(x) =$ $\Pi_{k=1}^{Km} H[S_{km}(x_{v(k\cdot m)} - t_{km})]$	$[\arg\max_m\{B_m\|x\}_1^M]$
LREG	deg = 1	$P_x = X(X'X)^{-1}X'$	$\hat{y} = P_X\mathbf{y}$
POLY	deg = 2	$P_H = H(H'H)^{-1}H'$	$\hat{y} = P_H\mathbf{y}$
PPREG	min = 9 max = 15	Minimize $\Sigma(y_i - \Sigma(\beta_m\psi_m(\alpha_m^t x_i)))^2$	$\hat{y} = \Sigma\beta_m\phi_m(\alpha^t\mathbf{x})$
BRUTO	cost = 2.5	Backfitting	$\hat{y} = \Sigma f_i(x_i)$
MARS	cost = 2 deg = 1	TURBO	$\hat{y} = \Sigma f_i(x_i)$
Nnet	h =15	Minimize $\Sigma(\hat{\theta}(j) - \theta(j))^2 + \lambda\,\Sigma\,W^2$	$\arg\min_j\{(\theta(j) - \hat{\theta}(j))^2\}$

The NNET neural networks written by Ripley are different from the original ones (Rumelhart et al., 1986) in that he uses the modified Newton's optimization algorithm with BFGS algorithm (the most popular Hessian matrix update algorithm). The description is depicted in Figure 80.8. The NNET has been very reliable in experiments and yields a better convergence to better minima than any other software that has been tested for the feedforward multilayer neural network study with backpropagation.

80.7 System Performance Assessment

In practice we are given a data set and required to design a system for a certain objective. The system is a realization of the function of an unknown input space D. If we know all the necessary characteristics of the input space D, it is fairly easy to design an optimal system for the objective, such as the Bayesian classification rule with class conditional distributions and *a priori* probabilities for the classes. We, however, usually do not know the underlying generating function that generates the sample we

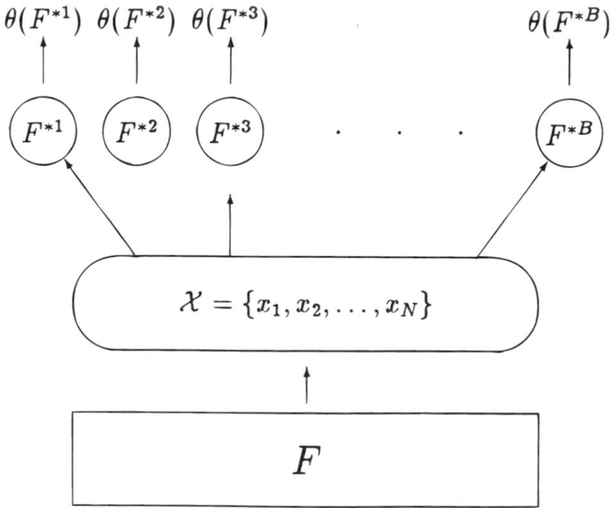

Figure 80.10 Illustration of the Bootstrap sampling.

[4] The commercial version of S (Beeker, Chambers and Wilks, 1988) which is developed in AT&T Bell Lab. Splus is an extended version of S from Statistical Sciences, Inc. Seattle, WA., USA.

have at hand. Instead, from the sample we like to find the underlying generating function, i.e., the population distribution. This is the *inference* problem.

Let us say that the input space is fully described by a certain distribution function $F(\cdot)$. The system we are interested in can be represented as a functional θ that takes the population distribution F: $\theta(D) = \theta(F)$. The functional θ is known, but the distribution F is not. θ could be any statistic or a complicated error rate in a classification problem.

The distribution is usually estimated parametrically or nonparametrically, thus providing the input argument to the system functional $\theta()$ in order to estimate the system's functional of the real population distribution F. Thus we have an estimation for $\theta(F)$:

$$\theta(F) \simeq \theta(\hat{F}).$$

With this estimation strategy the next question is how accurate $\hat{\theta}$ as the estimator of θ is.

Classifier Evaluation

Once we have designed a classifier, we like to know how accurately the system can do the job or to quantify the quality of the system performance. Prediction error is the criterion that we like to employ to see how good the designed system is. For both regression and classification system design, the usual system performance measure is its prediction error. In the context of regression, prediction error refers to the expected squared difference between the response value and its prediction from the model:

$$PE = E(y - \hat{y})^2. \tag{80.111}$$

The expectation operation refers to the repeated sampling from the true underlying population distribution.

Prediction error also arises in classification problems, where the response falls into one of J not ordered classes. The prediction error is commonly defined as the probability of an incorrect classification

$$PE = \text{Prob}(\hat{y} \neq y) \tag{80.112}$$

which is called mis-classification rate.

How to assess the system performance is an important issue in order to better quantify the designed system in terms of a criterion, e.g., error rate.

Hold-out Method

If the data set at hand is large we may divide it in two parts; use one for training and hold-out the other for testing, hence the name *hold-out* method. This is a popular method to assess the system's performance. In most cases the data is limited in size, thus a hold-out method is ad-hoc in the sense of which subset is held-out for testing. The performance evaluation via this method depends on how the data is separated.

***K*-fold Cross Validation.** A natural compromise to the hold-out above is the so-called K-fold cross-validation method. The given data is divided evenly into K parts. One or more of the K parts is used to test the designed system by the remaining parts of the data. An average among the results is called the K-fold cross-validation estimate of the true error rate. An extreme case results to the *leave-one-out* method, in which one observation, (y_k, X_k), is left out and the rest $N - 1$ cases, $\{(y_i, \mathbf{x}_i\}_{i \neq k}\}$ are used for training. The prediction error, PE (Equation 80.111) from the leave-one-out method is the average of the N errors

$$PE_{cv} = \frac{1}{N}\sum_{i=1}^{N}(y_i - \hat{f}^{-i}(\mathbf{x}_i))^2 \tag{80.113}$$

where $\hat{f}^{-i}(\mathbf{x}_i)$ is the estimation of the response of $f(\mathbf{x}_i)$ based on the system trained with the data in which the $\mathbf{x}_i$ is missing. In general, with a notation w_i being the index group in which the index i falls, the cross-validation has a form of prediction error in regression:

$$PE_{cv} = \frac{1}{N}\sum_{i=1}^{N}(y_i - \hat{y}^{-(w_i)}(\mathbf{x}_i))^2$$

and in classification setting:

$$PE_{cv} = \frac{1}{N}\sum_{i=1}^{N}[y_i \neq \hat{y}^{-(w_i)}(\mathbf{x}_i)]. \tag{80.114}$$

Other than cross validation for estimation, some modification of the apparent error, the sum of squared residuals (SSR)

$$\frac{1}{N}\sum_{i=1}^{N}(y_i - \hat{f}(\mathbf{x}_i))^2$$

has also been used (Efron and Tibshirani, 1993); such as $SSR/(N - p)$, $SSR/(N - 2p)$, and $C_p = SSR/N + 2p\hat{\sigma}^2/N$. Leaving these modifications of SSR aside (since they are beyond the scope of our interest) we like to use the *Bootstrap* estimate of prediction error, which is also used for the performance analysis of our classification system.

Bootstrapping Method for Estimation

Bootstrapping is a method of nonparametric estimation of statistical errors, which are the bias and the standard error of an estimator. The nonparametric techniques known to date are the Bootstrap, the Jackknife, and the cross-validation. Nonparametric methods for testing the accuracy of an estimator all have some common desirable features: they require very little in the way of modeling, assumptions, or analysis, and can be applied in an automatic way to any statistics, no matter how complicated these are (Efron and Gong, 1983).

To see what they are, a simple statistic, the sample mean $X_{\bar{}}$, is employed to assess the accuracy of the estimation for the true mean μ. We consider the available data set as a random sample of size N from an unknown distribution F in the sense that it represents the population F relatively well. As shown in Figure 80.10, a random sample is drawn from an unknown probability distribution F,

$$X_1, X_2, \ldots, X_n \sim F. \tag{80.115}$$

With a sample from F, we compute the sample average $\bar{x} = \Sigma_1^N x_i/N$ as an estimate of the expectation of F, $E_F(X)$. For this special statistic (sample average), we can get more information about the estimator $\bar{x}$. The accuracy of the estimator is represented by the standard deviation of $\bar{x}$:

$$\begin{aligned}\hat{\sigma}(F; N, \bar{x}) &= \{\mathrm{var}(\overline{X})\}^{1/2} \\ &= \{\mathrm{var}(X)/N\}^{1/2} \\ &= \left\{\frac{1}{N(N-1)}\sum_{i=1}^{N}(x_i-\bar{x})^2\right\}^{1/2} \qquad (80.116)\\ &\simeq \left(\frac{\mu_2(F)}{N}\right)^{1/2} \qquad (80.117)\end{aligned}$$

where $\mu_2(F)$ is the central moment of the population with distribution F. This standard error formula with the raw sample realization of Equation 80.115, does not extend to the other statistics, such as median, correlation, or prediction error. This is the point where computer methods, such as the resampling techniques for accuracy estimation, come into play.

Jackknife estimation. Let $\bar{x}_{(i)}$[5] be defined as:

$$\bar{x}_{(i)} = \frac{1}{N-1}\sum_{j\neq i} x_j = \frac{N\bar{x}-x_i}{N-1} \tag{80.118}$$

with $N-1$ points, be the sample average of $N-1$ points for all $i = 1, 2, \ldots, N$. Then the jackknife estimate of standard error is represented by

$$\hat{\sigma}_J(F; N, \bar{x}) = \left[\frac{N-1}{N}\sum_{i=1}^{N}(\bar{x}_{(i)}-\bar{x}_{(\cdot)})^2\right]^{1/2}. \tag{80.119}$$

The $x_{(\cdot)} = \Sigma_i^N \bar{x}_i/N$ is the average among the $N\bar{x}_{(i)}$s. This can be proved to be equal to the standard error for sample average of Equation 80.117 by substituting Equation 80.118 onto the Equation 80.119.

The jackknife standard error estimation of any statistic θ may have the form of Equation 80.119 to get the accuracy information of the estimator, $\hat{\theta}_J$. The advantage with the estimate of standard error for a statistic is to use Equation 80.119 where any statistic $\hat{\theta}_{(i)} = \hat{\theta}(X_1, \ldots, X_{i-1}, X_{i+1}, \ldots, X_N)$ is replaced by $\bar{x}_{(i)}$ and $\hat{\theta}_{(i)} = \frac{1}{N}\Sigma_{i=1}^N \hat{\theta}_{(i)}$ for $\bar{x}_{(i)}$.

[5] Note the change of the notation in the deletion statistic from the usual superscript with negative sign, e.g., $\hat{f}^{-i}(x_i)$ in Equation 80.113.

Bootstrap method. Bootstrap generalizes Equation 80.117 in an apparently different way. Any statistic $\theta(F)$, which is a functional, requires the distribution F. But in practice F is not known and is difficult to estimate. An empirical distribution $\hat{F}$ from the given sample from an unknown distribution F is defined in a bootstrap setting by giving an equal probability mass $1/N$ to each of the values x_i, and draw a sample from the *empirical distribution* $\hat{F}$:

$$X_1^*, X_2^*, \ldots, X_N^* \sim \hat{F}.$$

Each x_i^* is drawn independently *with replacement* and with equal probability from the set $\{x_1, x_2, \ldots, x_N\}$. Then the standard error of sample mean $\bar{X}^* = \Sigma_{i=1}^N X_i^*/N$ is given as

$$\begin{aligned}\sigma(\hat{F}; N, \bar{x}^*) &= \left(\frac{1}{N}\mu_2(\hat{F})\right)^{1/2} = \left(\frac{1}{N}\sum_{i=1}^{N}\frac{1}{N}(x_i-\bar{x})^2\right)^{1/2} \\ &= \left(\frac{1}{N^2}\sum_{i=1}^{N}(x_i-\bar{x})^2\right)^{1/2}. \qquad (80.120)\end{aligned}$$

where $\mu_2(\cdot)$ is the second order central moment of a given distribution. Comparing this standard error for bootstrap sample average with Equation 80.117 we note that they are almost the same. Thus the jackknife (Equation 80.119) and the bootstrap (Equation 80.120) standard error for sample average (a simple statistic as an example) are shown to be nearly equal to Equation 80.117; a special statistic that is the sample average as an estimate for mean has an explicit form. Formulas like Equation 80.117 do not exist for most statistics.

This is where the computing intensive jackknife and bootstrap estimations are used. It turns out[6] that we can always numerically evaluate the bootstrap estimate for standard error $\hat{\sigma} = \sigma(\hat{F})$, without a simple expression like Equation 80.117.

Analysis of Prediction Error Rates from Bootstrapping Assessment

Prediction error is an often used measure of performance for pattern recognition systems. In practice, a random sample, called training data, from an unknown population described by distribution F is given. Any statistic $\theta(F)$ requires a distribution F, but in practice, F is not known and difficult to estimate. An empirical distribution $\hat{F}$ from the given sample from an unknown distribution F is defined in a bootstrap setting by giving an equal probability mass $1/N$ to each of the values x_i. Let the distribution $\hat{F}$ be:

$$X_1^*, X_2^*, \ldots, X_N^* \sim \hat{F}.$$

Each x_i^* is drawn independently *with replacement* and with equal probability from the sample, i.e., training data:

[6] The proof can be found in reference (Efron and Tibshirani, 1993.)

$$\mathcal{X} = \{x_i\}_1^N = \{(v_i, y_i)\}_1^N$$

Standard error and bias estimation using the Bootstrap resampling technique can be found from references (Efron and Gong, 1993; Efron and Tibshirani, 1993). Here we introduce the algorithms for estimation of the standard error and the bias for prediction error estimation, leaving the technical details in the references above.

The Monte Carlo bootstrapping algorithm proceeds in three steps.

1. Using a random number generator, independently draw a large number $50 \leq B \leq 200$ of bootstrap samples, $\{F^{*b}\}_{b=1}^{B}$,
2. For each bootstrap sample F^{*b}, evaluate the statistic of interest, $\hat{\theta}(b = \hat{\theta}(F^{*b})$ for $b \in \{1, 2, \ldots, B\}$ from the training data χ,
3. Calculate the sample standard deviation of $\hat{\theta}^*(b)$ values

$$\hat{\sigma}_B = \left(\frac{1}{B-1}\sum_{b=1}^{B}\{\hat{\theta}^*(b) - \hat{\theta}^*(\cdot)\}^2\right)^{1/2}, \qquad (80.121)$$

$$\hat{\theta}^*(\cdot) = \frac{1}{B}\sum_{b-1}^{B}\hat{\theta}^*(b)$$

Standard errors are crude but useful measures of statistical accuracy (Efron and Tibshirani, 1986). An approximated confidence interval for an unknown parameter θ is given by

$$\theta \in \hat{\theta} \pm \hat{\sigma} z^{(\alpha)}, \qquad (80.122)$$

where $z^{(\alpha)}$ is the $100 \cdot \alpha$ percentile point of a standard normal variate, e.g., $z^{(0.95)} = 1.64485$. The standard error approximation Equation 80.122 for confidence interval bears an assumption:

$$\frac{\hat{\theta} - \theta}{\hat{\sigma}} \sim N(0, 1)$$

Bias about an estimator $\hat{\theta}$ is the next to consider. Bootstrap bias estimation is an estimation of the *optimistic bias op* resulting from using the same training data for prediction, e.g., via the resubstitution method. One way to estimate the system performance from the given sample is to correct the *apparent error rate* (or resubstitution error rate) by the estimation of the optimistic (or positive) bias. The optimistic bias is defined as

$$op(\mathcal{X}; F) = \theta - \theta_{app}$$

where θ is the true error rate for the unknown distribution F and θ_{app} for the apparent error rate. Since we do not know the bias $op(\chi, F)$ the bootstrap estimate of the bias, op_{boot} is found instead and the optimistic θ_{app} is corrected by adding the estimated bias:

$$\hat{\theta} = \theta_{app} + op_{boot} \qquad (80.123)$$

Let $\eta(\upsilon, \mathcal{X})$ be a decision rule based on the training set $\mathcal{X}$ and let $Q[y_i, \eta(\upsilon_i,\mathcal{X})]$ be indication of misclassification of υ_i by $\eta()$:

$$Q[y_i, \eta(v_i, \mathcal{X})] = \begin{cases} 1 & \eta(v_i, \mathcal{X}) \neq y_i, \\ 0 & \text{otherwise.} \end{cases}$$

Thus $Q[\cdot] = 1$ indicates the misclassification of a training observation from the system designed by the training data.

The bootstrap procedure for estimating the bias, op_{boot}, follows:

1. Select $50 \leq B \leq 200$ bootstrap samples from the empirical distribution $\hat{F}$.
2. From each bootstrap sample compute bias w_b

$$w_b = \sum_{i=1}^{N}\left(\frac{1}{N} - p_i^{*b}\right)Q[y_i, \eta(v_i, F^{*b})]$$

with P_i^{*b} indicating the proportion of the bootstrap sample on x_i, i.e.,

$$P_i^{*b} = \text{Cardinality of } \{j \mid x_j^{*b} = x_i\}/N$$

and $\eta(\upsilon_i, F^{*b})$ being the prediction of υ_i from the system trained by F^{*b}.

3. Repeat step 2 to get $\{w_1, w_2, \ldots, w_B\}$.

Then the bootstrap bias op_{boot} is estimated by

$$op_{boot} = \frac{1}{B}\sum_{b=1}^{B} w_b$$

thus, the bootstrap error estimate $\hat{\theta}$ (Equation 80.123) is obtained.

E0 prediction error estimation is to count the number of patterns that are not included in the bootstrap samples and normalize the misclassification count of the samples (Efron, 1983) by the total number of the training patterns not selected in the bootstrap samples. Thus, E0 uses the testing set which is asymptotically 36.8% of the original training, according to the argument that follows. In a typical bootstrap sample, about 63% of the original observations are likely to be chosen. This is easily seen since the probability that an observation does not belong to a bootstrap sample is

$$(1 - 1/N)^N = 1/e.$$

Thus an observation x_i will be in the bootstrap sample with about $1 - 1/e = 63.2\%$ chances.

Let $A_b = \{i \mid P_i^{*b} = 0\}$ denote the index set of training patterns which do not appear in the F^{*b}, then the prediction error θ_0 estimated by the E0 estimator is defined by

$$\theta_0 = \frac{\Sigma_{b=1}^{B} \Sigma_{i \in A_b} Q[y_i, \eta(v_i, F^{*b})]}{\Sigma_{b=1}^{B} \text{Cardinality of } \{A_b\}}.$$

This E0 estimator is a form of crossvalidation in that the testing data has not been used in training. The difference from the crossvalidation is that the E0 separates the training and the testing data randomly, while the crossvalidation selects the testing pattern sequentially such that all the training patterns are used for testing.

The testing patterns used in apparent error rate obtained by the resubstitution method are too close or the distance is 'zero' from the training patterns while the test patterns for E0 estimator are 'too far' from the training set. From that, the asymptotic probability argument that a pattern will not be included in a bootstrap sample is 0.368, the weighted average of θ_{app} and θ_0 involves patterns at the 'right' distance from the training set in estimating the error rate (Efron, 1983):

$$\theta_{632} = 0.368 * \theta_{app} + 0.632 * \theta_0 \qquad (80.124)$$

The E632 was shown to be optimal in terms of least variance and bias from a comparison study for various estimators (Jain et al., 1987; Efron, 1983) among crossvalidation, ordinary bootstrap bias correction (Equation 80.123) and E632 (Equation 80.124). We used the E632 prediction error as a standard performance measure. The bootstrap package `bootsrap.funs`[7] contains various resampling techniques and is available via *anonymous* ftp to `statlib@lib.stat.cum.edu.`

80.8 Analysis of Prediction Rates from Bootstrapping Assessment

The boxplots in Figure 80.11 represent the E632 estimator superimposed by the distribution of the $B = 100$ bootstrap sample

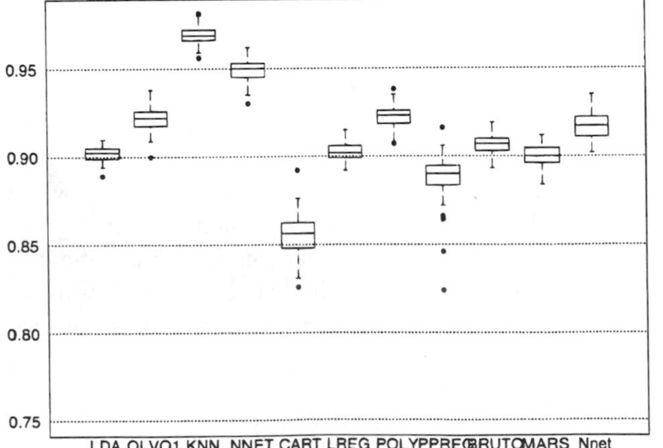

Figure 80.11 Boxplots for different classifiers for data set R. 100 bootstrap samples are used to assess each classifier.

[7] The bootstrap is contributed by Efron and Tibshirani.

errors, $\hat{\theta}^*(b)$'s in Equation 80.121. The median value of the B error rates are replaced by the E632 estimate, thus the B bootstrap errors are shifted according to the E632 estimate. For ease of display and understanding the system performance the recognition rates, i.e., $1 - \hat{\theta}'$, are plotted.

The *generality* issue of the designed system is related to its reliability in terms of standard error of the estimator for the prediction error. To make the analysis simpler, we assume the symmetry of the system performance of the classifiers in the boxplot figures. Then the standard error of the prediction rule by the mean (or median simply from the boxplots) is relatively approximated by the inter quartile range of the boxplots.

The mean value is the bootstrap sample estimate $\hat{\theta}$ of the true statistic $\theta = 1 - PE.$. The standard deviation implicitly represents the reliability of the estimate, i.e., standard error of the estimate. From the result of the classifiers considered in this study, (Figure 80.11), the 95% confidence interval of the estimate $\hat{\theta} = 0.955$ is given by Equation 80.122:

$$\hat{\theta} - \hat{\sigma} \times 1.645 \leq \theta \leq \hat{\theta} + \hat{\sigma} \times 1.645$$

$$0.941 \leq \theta \leq 0.96$$

where the multiple factor 1.645 is the 95% percentile point of the standard normal variate, $N(0,1)$.

The graphical display seems to reveal more for the comparison study of the classifiers and different treatments of the data. The Boxplot display of a batch is a very simple and useful way to show the distribution of the sample. The Inter Quartile Range (IQR) which is the difference between the upper quartile and the lower quartile is considered to be the robust estimation of the scalar multiple of the dispersion. The height of the box is the IQR. The median of the batch is represented by the line in the box. The whiskers represented by the dotted lines are extended up to the points in which the 1.5 times of the IQR contains. Outliers are represented by the individual dots to signify their existence. The boxplot, thus, displays the distribution very simply but well enough, especially when many different batches are to be compared.

The correct recognition rates from 11 classifiers are displayed with the boxplots for each data set obtained from the different treatments. Each boxplot shows the distribution of the recognition rate of the 100 systems designed by $B = 100$ bootstrap samples. The corresponding figures for the data are in Figure 80.11.

The results from LDA and LREG (via linear regression) would have been the same due to the equivalence of the LDA and OS (Equation 80.107 and Equation 80.110) if the same bootstrap sample were used for both classifiers; the bootstrap samples used to train the classifiers are different for no reason.[8]

The best performance of the optimization machine with the feedforward neural network structures can be observed (Figure

[8] If the different classifiers were trained with the same B bootstrap samples, then the classification by the linear regression method and the LDA would have been the same.

80.11). This is seen with the mean values for the estimation of the correct recognition error. Note that we do not consider the KNN classifier as a learning mechanism so it is not of concern. It does not learn but performs by the exemplars, i.e., the computation in the operation phase is the largest, which is inappropriate in real-time processing applications.

Acknowledgments

The authors wish to thank Dr. G. Kontaxakis for help with the final version of the manuscript.

References

Anderson, T. W. 1984. *An Introduction to Multivariate Statistic Analysis,* John Wiley & Sons, New York, NY.

Baum, E. and Haussler, D. 1989. What size net gives valid generalization?, *Neural Computation,* 1:151–160.

Becker, R. A., Chambers, J. M., and Wilks, A. R. 1988. *The New S Language,* Wadsworth, Pacific Grove, CA.

Breiman, L. 1991a. The π-method for estimation multivariate functions from noisy data, *Technometrics,* 33(2):125–160.

Breiman, L. 1991b. Hinging Hyperplanes for Regression, Classification and Function Approximation, Technical Report 324, University of California, Berkeley.

Breiman, L. and Friedman, J. H. 1985. Estimating optimal transformations for multiple regression and correlation, *J. Am. Statistical Association,* 80(391):580–619.

Breiman, L. and Ihaka, R. 1984. Non-linear Discriminant Analysis via Scaling and Ace, Technical report, Tech. Report, Univ. of California, Berkeley.

Breiman, L., Friedman, J. H., Olshen, R. A., and Stone, C. J. 1984. *Classification and Regression Trees.* Wadsworth and Brooks/Cole, Belmont, CA.

Bryson, A. E. and Ho, Y. C. 1969. *Applied Optimal Control,* Bleisdell.

Buja, A., Hastie, T., and Tibshirani, R. 1989. Linear smoothers and additive models, *The Annals of Statistics,* 17(2):453–555.

Carpenter, G. A. and Grossberg, S. 1987. Art2: self-organization of stable category recognition codes for analog input patterns, *Applied Optics,* 26:4919–4930.

Chambers, J. M. and Hastie, T. J. 1991. Statistical Models. *Statistical Models in S,* Chambers, J. M. and Hastie, T. eds., Wadsworth & Brooks, New York.

Chung, W. 1994. *A Strategy for Visual Pattern Recognition,* PhD thesis, Electrical and Computer Engineering, Rutgers University.

Engel, J. 1988. Teaching feed-forward neural networks by simulated annealing, *Complex Systems,* 2:641–648.

Efron, B. 1983. Estimating the error rate of a prediction rule: improvement on cross-validation, *J. Am. Statistical Association,* 78(882):316–331.

Efron, B. and Gong, G. 1983. A leisurely look at the bootstrap, the jackknife, and cross-validation, *The American Statistician,* 37(1):36–48.

Efron, B. and Tibshirani, R. 1986. Bootstrap methods for standard errors, confidence intervals, and other measures of statistical accuracy, *Statistical Science,* 1(1):54–77.

Efron, B. and Tibshirani, R. J. 1993. *An Introduction to the Bootstrap,* Chapman & Hall, New York, NY.

Farmer, J. D. and Sidorowich, J. J. 1988. Exploiting Chaos to Predict the Future and Reduce Noise, Technical report, Los Alamos National Laboratory, Los Alamos, NM.

Friedman, J. H. 1991. Multivariate adaptive regression splines, *The Annals of Statistics,* 19(1):1–141.

Fukunaga, K. 1990. *Introduction to Statistical Pattern Recognition,* 2d ed., Academic Press, Inc., San Diego, CA.

Friedman, J. H. and Stuetzle, W. 1981. Projection pursuit regression. *J. Am. Statistical Association,* 76(376):817–823.

Gallant, A. R. and White, J. 1988. There exists a neural network that does not make avoidable mistables, *IEEE 2nd Int. Conf. Neural Networks,* I:657–664, San Diego, CA.

Golden, R. M. 1988. A unified framework for connectionist systems, *Biological Cybernetics,* 59:109–120.

Gnanadesikan, R. and Kettenring, J. 1989. Discriminant analysis and clustering, *Statistical Science,* 4(1):34–69.

Gray, R. M. 1984. Vector quantization, *IEEE ASSP Magazine,* 1:4–29.

Härdle, W. 1991. *Smoothing Techniques With Implementation in S.* Springer-Verlag, New York.

Harth, E. and Tzanakou, E. 1974. Alopex: a stochastic method for determining visual receptive fields, *Vision Res.,* 14:1475–1482.

Hastie, T. 1989. Discussion in flexible parsimonious smoothing and additive modeling, *Technometrics,* 31(1):23–29.

Hastie, T., Buja, A., and Tibshirani, R. 1993. Penalized discriminant analysis; can be obtained from \netlib\ stat via ftp to netlib.att.com., July.

Hastie, T., Tibshirani, R. and Buja, A. 1993. Flexible discriminant analysis by optimal scoring, can be obtained from \netlib\ stat via ftp to netlib.att.com., February 1993.

Hecht-Nielsen, R. Theory of the back propagation neural network, *Proc. Int. Joint Conf. Neural Networks,* I:593–608, San Diego, CA.

Hinton, G. E. 1989. Connectionist learning procedures. *Artificial Intelligence,* 185–234.

Hinton, G. E. and Sejnowski, T. J. 1986. Learning and Relearning in Boltzmann Mahcines, *Parallel Distributed Processing: Explorations in the Microstructure of Cognition, I: Foundations,* chapter 7, Rumelhart, D.E. McClelland, J.L, and the PDP Research Group, eds., MIT Press, Cambridge, MA.

Hopfield, J. J. and Tank, D. W. 1985. Neural computation of decisions in optimization problems, *Biological Cybernetics,* 52:141–152.

Hornik, K., Stichcombe, M., and White, H. 1989. Multilayer feedforward networks are universal approximators, *Neural Networks,* 2:359–366.

Huang, W. and R. Lippmann, R. 1986. Neural net and traditional classifiers. In D. Anderson, editor, *Neural Info. Processing Syst.,* pages 387–396. NY: American Institute of Physics, 1986.

Hush, D. and Horne, B. 1993. Progress in supervised neural networks, *IEEE Signal Processing Magazine,* January, 8–39.

Jacobs, R. A. 1988. Increased rates of convergence through learning rate adaptation, *Neural Networks,* 1:295–307.

Jain, A. K., Dubes, R. C., and Chen, C. C. Bootstrap techniques for error estimation, *IEEE Trans. Pattern Analysis and Machine Intelligence,* 9(5):628–633.

Johnson, R. A. and Wichern, D. W. 1988. *Applied Multivariate Statistical Analysis,* Prentice Hall, Englewood Cliffs, NJ.

Khotanzad, A. and Hong,Y. H. 1990. Invariant image recognition by zernike moments, *IEEE Trans. Pattern Analysis and Machine Intelligence,* 12(5):489–497.

Kirkpatrick, S., Gelatt Jr., C. D., and Vecchi, M. P. 1983. Optimization by simulated annealing. *Science,* 220(4598): 671–680.

Kohonen, T. 1990. The self-organizing map, *Proc. IEEE,* 78(9): 1464–1480.

Kohonen, T., Kangas, J., Laaksonen, and Torkkola, K. 1992. Lvq-pak: The Learning Vector Quantization Program Package, Technical report, Helsinki University of Technology, Laboratory of Computer and Information Science; Ivq-pak is available for anonymous ftp user at the Internet site cochlea.hut.fi(130.233.168.48).

Kolmogorov, A. N. 1957. On the representation of continuous functions of many variables by superposition of continuous functions of one variable and addition, *Doklady Akademii Nauk SSR,* 114:953–956.

Lapedes, A. and Farber, R. 1988. How Neural Networks Work, Technical Report, Los Alamos National Laboratory, Los Alamos, NM.

le Cun, Y. 1987. *Medeles connexionistes de l'apprentissage,* PhD thesis, Universite Pierre et Marie Curie.

Linde, Y., Buzo, A., and Gray, R. M. 1980. An algorithm for vector quantization, *IEEE Trans. Communications,* COM-8:84–95.

Lippmann, R. P. 1987. An introduction to computing with neural nets, *IEEE ASSP Magazine,* April, 4–22.

Lippmann, R. P. 1989. Pattern classification using neural networks, *IEEE Communications Magazine,* November, 47–64.

Micheli-Tzanakou, E., Uyeda, E., Sharma, A., Ramanujan, K. S., and Dong, J. 1995. Face recognition: comparison of neural networks algorithms, *Simulation,* 64(1):37–51.

Micheli-Tzanakou, E. 1995. Neural Networks in Biomedical Signal Processing, *The Biomedical Eng. Handbook,* ch. 60, 917–931, Bronzino J., ed., CRC Press, Boca Raton, FL.

Minsky, M. and Papert. 1969. *Perceptrons: An Introduction to Computational Geometry,* MIT Press, Cambridge, MA.

Omohundro, S. M. 1987. Efficient algorithms with neural network behavior, *Complex Systems,* 1:273–347.

Parker, D. B. 1988. Learning-logic, Technical Report TR-47, Center for Comp. Res. in Econ. and Man., MIT, Cambridge, MA, April 1985.

Peressini, A. L., Sullivan, F. E., and Uhl Jr, J. J. 1988. *The Mathematics of Nonlinear Programming,* Springer-Verlag, New York, NY.

Powell, M. J. D. 1985. Radial Basis Functions for Multivariate Interpolations, Technical Report DAMPT 1985/NA12, Dept. of Appl Math. and Theor. Physics, Cambridge University., Cambridge, England.

Quinlan, J. R. 1986. Induction of decision tree, *Machine Learning,* 1:81–106.

Ripley, B. D. 1994. Neural networks and related methods for classification, PS file is available by anonymous ftp from markov.stats. ox.ac.uk (192.76.20.1) in directory pub/ neural/ papers.

Ripley, B. D. 1994. Neural networks and flexible regression and discrimination; PS file is available by anonymous ftp from markov.stats.ox.ac.uk (192.76.20.1) in directory pub/neural/ papers.

Ripley, B. D. 1993. Statistical aspects of neural networks, *Chaos and Networks: Statistical and Probabilistic Aspects,* Barndorff-Nielsen, O. E., Cox, D. R., Jensen, J. L., and Kendall, S. S., eds., Chapman & Hall, London, UK.

Rosenblatt, R. 1959. *Principles of Neurodynamics,* Spartan Books, New York, NY.

Rumelhart, D. E., Hinton, G. E., and Williams, R. J. 1986a. Learning internal representations by error backpropagation, *Parallel Distributed Processing: Explorations in the Microstructure of Cognition, I: Foundations,* ch. 8, Rumelhart, D. E., McCleland, J. L, and the PDP Research Group eds., MIT Press, Cambridge, MA.

Rumelhart, D. E., Hinton, G. E. and Williams, R. J. 1986b. Learning representation by back-propagating errors, *Nature,* 323:533–536.

Solla, S. A., Levin, E., and Fleisher, M. 1988. Accelerated learning in layered neural networks, *Complex Systems,* 2:625–640.

Stanfill, C. and Waltz, D. 1986. Toward memory-based reasoning., *Commun. of the ACM,* 29(12):I:213–228.

Teague, M. R. 1980. Image analysis via the general theory of moments, *J. Opt. Soc. Am.,* 70(8):920–930.

Tibshirani, R. 1988. Estimation optimal transformations for regression via addivity and variance stabilization, *J. Am. Statistical Association,* 83:394–405.

Touretzky, D., Hinton, D., and Sejnowski, T., eds. 1989. *Faster-learning Variations on Back-propagation: An Empirical Study,* Morgan Kaufmann, San Mateo, CA.

Unnikrishnan, K. P. and Venugopal, K. P. 1994. Alopex: A correlation-based learning algorithm for feed-forward and recurrent neural networks, *Neural Computation,* June.

van Ooyen, A. and Niehhuis, B. Improving the convergence of the back-propagation algorithm, *Neural Networks,* 5:465–471.

Wahba, G. 1990. *Spline Models for Observational Data,* SIAM, Philadelphia, PA.

Weiss, S. M. and Kulikowski, C. A. 1991. *Computer Systems that Learn: Classification and Prediction Methods From Statistics; Neural Nets, Machine Learning, and Expert Systems,* Kaufmann Publishers,

Werbos, P. J. 1974. *Beyond Regression: New Tools for Prediction and Analysis in the Behavioral Sciences,* PhD Thesis, Harvard University, Cambridge, MA.

Widrow, B. 1962. Generalization and information storage in networks of adaline 'neurons', *Self-Organizing Systems.* Yovitz, M., Jacobi, G., and Goldstein, G., eds., 435–461. Spartan Books, Washington, DC.

Widrow, B. and Hoff, M. 1960. Adaptive switching circuits, *1960 IRE WESCON Convention Record,* New York, NY, 96–104.

Widrow, B. and Lehr, M. 1990. 30 years of adaptive neural networks: perceptron, madaline, and backpropagation, *Proc. IEEE,* 78(9):1415–1442.

Wolpert, D. 1988. Alternative generalizers to neural nets, *Neural Networks,* 1, 1988. Abstracts of 1st Annual INNS Meeting, Boston, MA.

Zahner, D. and Micheli-Tzanakou, E. 1995. Artificial Neural Networks: Definitions, Methods and Applications, *The Biomedical Eng. Handbook,* ch. 184, 2689–2705, J. Bronzino, ed., CRC Press, Boca Raton, FL.

VIII

Fuzzy Systems and Soft Computing

81 **Applications of Fuzzy Systems and Soft Computing in Industrial Electronics**
Mary Lou Padgett 1087
Introduction • From Basic Implementations to New Research

82 **Fuzzy Numbers: The Application of Fuzzy Algebra to Safety and Risk Analysis**
J. Arlin Cooper 1091
Background • Analytical Processing of Input Data • Fuzzy-Algebra Background • Fuzzy-Algebra Depiction of Uncertainty • Example Applications

83 **Fuzzy Systems** *Mo-yuen Chow* 1096
Brief Description of Fuzzy Logic • Qualitative (Linguistic) to Quantatitive Description • Fuzzy Operations • Fuzzy Rules, Inference • Fuzzy Control

84 **Fuzzy Hardware** *Mary Lou Padgett* 1103
Introduction • Challenges and Rewards • Approaches • Futures • Defining Terms

85 **Fuzzy Modeling and Applications: Controls, Visions, Decisions** *Mary Lou Padgett* 1112
Introduction • Engineering Approaches • Futures

86 **Fuzzy Logic Control: Basics and Applications** *Robert N. Lea, Yashvant Jani, and Joseph A. Mica* 1116
Introduction • A Simple Example of Fuzzy Logic Control • The Example of the Inverted Pendulum • Remote Manipulator System • Collision Avoidance • Summary

87 **Development of an Intelligent Unmanned Helicopter Based on Fuzzy Systems**
Michio Sugeno, Howard A. Winston, Isao Hirano, and Satoru Kotsu 1127
Introduction • Helicopter Hardware System • Software System for Helicopter Control • Results • Conclusions

88 **Fuzzy and Neural Modeling** *Mary Lou Padgett* 1139
Introduction • Engineering Approaches and Applications • Futures

89 **NeuFuz: A Combined Neural Net/Fuzzy Logic Tool** *Thomas Lindblad and Clark S. Lindsey* 1143
Introduction • Working with the Neural Network of NeuFuz4 • Working with the Fuzzy Logic Part of NeuFuz4 • Working with the Code Generator Part of NeuFuz4 • Summary

90 **Neural Network Learning in Fuzzy Systems** *Yashvant Jani and Robert N. Lea* 1147
Introduction • Reinforcement Learning • Architecture of ARIC • ARIC and 6 DOF Space Operations • GARIC and Attitude Control • Six Degree-of-Freedom Proximity Operations Trajectory Controller

91 **Neurocontrol and Elastic Fuzzy Logic: Capabilities, Concepts, and Applications**
Paul J. Werbos 1157
Introduction • Neurocontrol in General • Basic Principles of Design • Supervised Learning for Neurocontrol • Elastic Fuzzy Logic: Principle and Subroutines • Current Designs in Neurocontrol: A Roadmap • Appendix

92 **Integrated Health Monitoring and Control in Rotocraft Machines** *Gary G. Yen* 1182
Introduction • Artificial Neural Networks • Fuzzy-Based Feedforward Neural Networks • FDIA Architecture • Simulation Study • Conclusions

93 Autonomous Neural Control in Flexible Space Structures *Gary G. Yen* 1192
Learning Control System • Adaptive Time-Delay Radial Basis Function Network • Eigenstructure Bidirectional Associative Memory • Fault Detection and Identification • Reconfigurable Control • Simulation Studies • Conclusion

94 Fuzzy Pattern Recognition *Witold Pedrycz* 1207
Introductory Remarks—Pattern Recognition in the Framework of Fuzzy Sets • The General Methodological Structure of Fuzzy Modeling • Formation of the Feature Space • Implicit and Explicit Knowledge Representation in Pattern Recognition • From Supervised to Unsupervised Pattern Recognition—A Continuum of Classification Models • Fuzzy Neural Structures • Supervised Learning • Implicitly Supervised Pattern Recognition • Unsupervised Learning

95 Neural Fuzzy Systems in Handwritten Digit Recognition *Timothy J. Dasey and Evangelia Micheli-Tzanakou* 1231
Introduction • System Design • Application to Handwritten Digits • Discussion • Summary

96 Fuzzy Algorithms for Learning Vector Quantization *Nicolaos B. Karayiannis* 1264
Introduction • Learning Vector Quantization • Generalized Learning Vector Quantization • Fuzzy Learning Vector Quantization Algorithms • GLVQ-F and FLVQ Algorithms • Fuzzy Algorithms for Learning Vector Quantization • The FALVQ 1 Family of Algorithms • The FALVQ 2 Family of Algorithms • The FALVQ 3 Family of Algorithms • Competition Measures • Alternative FALVQ Algorithms • Experimental Results • Discussion and Concluding Remarks

97 Adaptive Resonance Theory *Gail A. Carpenter and Stephen Grossberg* 1286
Match-Based Learning and Error-Based Learning • ART and Fuzzy Logic • ART Dynamics • Fuzzy ART • Fuzzy ARTMAP • Fuzzy ART Algorithm • Fuzzy ARTMAP Algorithm • ART Applications

98 Future Directions for Fuzzy Systems and Soft Computing in Industrial Electronics *Mary Lou Padgett and Lotfi A. Zadeh* 1299

81

Applications of Fuzzy Systems and Soft Computing in Industrial Electronics

81.1 Introduction 1087

81.2 From Basic Implementations to New Research 1087

Basic Fuzzy Modeling Concepts • Fuzzy Hardware Implementations • Fuzzy Techniques and Applications • Fuzzy Neural Modeling • Fuzzy Neural Hardware • Fuzzy Neural/Neural Fuzzy Dynamic Systems and Applications • Fuzzy Neural/Neural Fuzzy Pattern Recognition and Applications

Mary Lou Padgett

Auburn University

81.1 Introduction

Use of fuzzy systems and soft computing for industrial electronics is an economically viable concept which can add to the robustness of an existing, traditional system. There are many aspects of fuzzy modeling to be addressed, especially when considering expansion of the traditional concepts to include soft computing. For example, merging neural network systems with fuzzy systems can add a degree of adaptability to the fuzzy system. Fuzzifying neural networks can contribute sensible, generalizable interpretations of neural output, or repeatable algorithms for training neural systems. There seems to be no end to the variations and combinations possible, and improvements continue to surface.

The material included in this chapter is intended to guide an engineer interested in building applications of fuzzy systems for industrial electronics applications. The organization of the topics and selection of authors was guided in large part by comments from Walter Karplus, Paul Werbos, Harold Szu, Lotfi Zadeh, Gail Carpenter and Thomas Lindblad. Witold Pedrycz and Nicolas Karayiannis contributed significant segments of very pertinent material. The support of Michio Sugeno and his students was extremely helpful, and provided a large boost to the CI Standards News (contact: m.padgett@ieee.org). The news will feature debates about fuzzy systems and soft computing, along with contributions by individuals suggesting definitions, design procedures, lessons learned and documentation tips suitable for future standardization.

Topics in this chapter are arranged to first feature electronic implementations suitable for commercial use, then move into areas of promising research development.

81.2 From Basic Implementations to New Research

Basic Fuzzy Modeling Concepts

Applications of fuzzy systems range from risk analysis; fault detection, identification and recovery through controls and all types of pattern recognition. A straight-forward application of fuzzy numbers is risk analysis, in use by Cooper of Sandia National Labs. This very practical technique is useful as a decision aide in the design of computational intelligence systems. Padgett and Padgett (1995) suggests subtracting a fuzzy number representing *cost* of adding an intelligent component to a system from a fuzzy number representing the *benefit* of the component. Taking J. Cooper's concept further, use of fuzzy number subtraction for trend analysis at the top management level of an aerospace manufacturer provides the basis for Padgett (1995). The latter article discusses simulation experiments with the use of this risk analysis method as part of the real-time analysis of a missile control system. An ideal or historical window of observations gives the base-line fuzzy number, and the current window of observations gives the current fuzzy number. A large difference provides early warning of actuator failure. A window of observations can be used as input to a neural network to produce a fuzzy number. There are a number of ways that very simple neural and fuzzy modules can be combined to solve practical problems.

Fuzzy Hardware Implementations

Several recent IEEE special issues cover fuzzy hardware excellently. They contain a number of design and specification suggestions. They list some of the factors to be considered and

0-8493-8343-9/97/$0.00+$.50

documented during the iterative process of system design and validation. Use of these factors as part of a concurrent engineering process tends to reduce the number of iterations needed (see the chapter on factory automation for more details). There are many tradeoffs between digital systems which heavily support design modifications and interfaces with nonfuzzy algorithms and analog systems which can be tuned to specific applications for top speed and efficiency. There are a variety of implementations ranging between these options. One of the most promising included in these special issues is Pedrycz's presentation which features fuzzy and neural capabilities. *Flexibility* and *speed* are prime goals of the fuzzy hardware implementations discussed. Defining terms and further resources are included.

Fuzzy Techniques and Applications

A fuzzy system specification moves beyond statement of objectives and models, through selection of implementation tools and resources. Decisions about complexity and variability to be allowed are vital. System components and variables may need to be adaptable, or a fixed application specific design may be appropriate.

Examining the nature of the modules for *fuzzification, fuzzy inference* and *defuzzification* modules is important. Prior to selecting methods for implementing these three tasks, the characteristics of the system to be modeled should be assessed. For example, in fuzzification, the range and dynamics of variables must be determined, and the nature of various membership functions explored (Kartalopolous, 1996). For fuzzy inference, the TSK or Sugeno method for determining fuzzy system structure has proven very effective for recently reported applications. It simplifies the defuzzification task, but caution must be used with regard to scaling.

System identification in a fuzzy system can help plan ahead and guide the selection of modeling techniques, then help fine-tune the emerging system. Two elements of system identification commonly considered are *structure identification,* and *parameter identification.* The first is amenable to the Sugeno method, and the second is receptive to neural systems techniques. Paul Werbos suggests some novel approaches to consider, and many of the other articles in this chapter offer detailed suggestions for approaching these problems with particular applications in mind.

The recent work by R. Lea, J. Mica, and Y. Jani involves experiences with real-world applications and the stress of real-time *development* of a control system which potentially has to operate without access to domain experts dangling in space around the gigantic robot arm frantically tuning PID controllers. NASA working engineers found a solution using fuzzy logic which can be tweaked into place in a practical manner. In early 1995, a videotape of the working (ground-based) arm was aired. This application of fuzzy controls to full-sized remodeled construction equipment is of interest to practical engineers. There may be other, more theoretically esoteric ways to implement this system, but given the time and resource constraints for development and the target operating environment, this RMS arm presents a practical way to solve an engineering problem.

One of the most respected fuzzy applications world-wide is the autonomous helicopter developed by M. Sugeno's laboratory. His helicopter can, for example, be used to dust crops, freeing humans from a hazardous task. His team uses a hierarchical fuzzy control approach so successfully implemented.

Fuzzy Neural Modeling

Moving from fuzzy systems to soft computing applications combining fuzzy and neural approaches, some of the strategies in common use for structure determination and parameter identification are documented in the work of Jang and Sun (1995). Their neuro-fuzzy modeling techniques have been very successful. Specification of fuzzification, fuzzy inference, and defuzzification modules can be done in such a manner that integration with neural systems techniques is encouraged and carefully documented. Combining fuzzy and neural approaches in commercial and research-level projects is highly productive.

Fuzzy Neural Hardware

One commercial solution to the combination of neural and fuzzy systems is *NeuFuz,* a combined neural networks/fuzzy logic tool. Other solutions are also being developed. As these mature, the use of fuzzy algorithms in combination with nonfuzzy techniques should increase.

Fuzzy Neural/Neural Fuzzy Dynamic Systems and Applications

Moving into the area of interesting applications, Neural Network Learning in Fuzzy Systems is one of the most promising approaches for learning. Reinforcement learning in particular can enhance existing systems and expand their capabilities.

The advantages and disadvantages of reinforcement learning and the combination of temporal elements with traditional and nontraditional models are discussed in depth in *Elastic Fuzzy Logic (ELF) Neural Networks in Fuzzy Systems for Controls* by Paul J. Werbos. In a tutorial-level appendix, many basic concepts are described and diagrammed. Particular attention should be paid to the recommended methods for calculating derivatives. Suggestions for research proposals to NSF are also included. Werbos' suggestions for Elastic Fuzzy Logic and combinations with neural systems can be compared to the techniques of Yager and Filev (1995). These procedures build on the material covered by Jani and Lea in this chapter.

Werbos presents a roadmap of basic designs and concepts, building from basic to future research suggestions. The articles by Werbos in the neural networks and the fuzzy systems chapters also discuss these concepts, and mention some of the most advanced work currently in progress. Werbos explains cloning, tracking and dynamic optimization and includes basic principles of design. Supervised learning for neurocontrol is described and illustrated. Then, elastic fuzzy logic is covered. Approaches to adapting fuzzy systems are covered, including adaptation of the membership function (which can cause problems) or use of a multilayer perceptron as a membership function. Many suggestions and variations are presented. Some of them are outlined below.

Problems with Linguistics

Varying meaning of words
Need freedom to adapt the rules, especially when same input word appears in many rules

Suggestions:

Werbos: *Use MLP as a membership function, train with many examples of words. Each expert can estimate degree of membership of the word. Use supervised learning to estimate the expert's true meaning of word.*
Yager: *Use MLPs in final action stage as part of a more flexible approach, but this limits degree of communication back to expert.*
Mendel and Wang: *Use another similar approach.*

Levels of Analysis

Micro level

Individual supervised learning modules:
MLP, RBF, CMAC, etc. within a control architecture

Middle level

How these modules are put together to build a general purpose system or methodology
What kinds of generic tasks they perform, e.g., in control

1. Pattern recognition or neuroidentification for sensor fusion or diagnosis (not control)
2. Cloning function, such as copying the behavior of a human being able to control the target plant supervised learning: should tell how to know correct actions
3. Tracking, e.g., robot arm trajectory or chemical plant setpoint direct inverse controllers or neural adaptive controllers
4. Optimization functions maximize throughput minimize energy use, or maximize goal satisfaction
Possible Brain-Like Capability

Application level

Describe how general purpose systems are used in stages, and in combination with application-specific modules, to generate a product.
These three levels are likened to building chips, putting chips together to build a computer, vs. figuring out how to use a computer.
The tutorial appendix describes neural networks and fuzzy systems in supervised learning. This is followed by discussions and comparisons of supervised control, direct inverse control, neural adaptive control, backpropagating utility, adaptive critics (2-net), BAC + DHP and other variations. Applications and examples are provided. Six tools Werbos would like to see developed are listed:

1. Simple sensitivity analysis tool.
   Input: utility function or target function.
   Output: the 10 or 100 most important inputs based on user selection of largest derivatives of utility with respect to all the inputs largest elasticities, or the largest weighted derivatives.
2. Flowchart of important connections (generated as intermediate information by backpropagation).
3. Extended version of tools 1 and 2 above, using low-cost second derivatives.
4. Backpropagating utility—a full-fledged version.
5. Model calibration tool based on backpropagation of error and robust estimation.
6. Integrated nonlinear version of existing linear diagnostic tools for identifying how output parameter estimates and rules were influenced by different cases in the input dataset.

In summary, creating tools aiming for effective two-way man-machine communication is a primary goal. Tools such as those for sale by Jang and under development by others are addressing these goals.

Applications of combined neural and fuzzy systems include Integrated Health Monitoring and Control in Rotorcraft Machines and Autonomous Neural Control in Flexible Space Structures. Study of concrete examples such as these should help the working engineer apply the concepts covered to other applications.

Fuzzy Neural/Neural Fuzzy Pattern Recognition and Applications

Recent conferences and technical publications have featured promising new ideas and results based on fuzzy pattern recognition. Combining fuzzy and neural pattern recognition techniques seems particularly fruitful. As with all practical applications of modeling, it is vital to keep the validation and verification of the model an intrinsic part of the developmental process so that the resulting system will have the appropriate amount of detail: enough to satisfy the system goals without wasting resources.

Fuzzy pattern recognition and its recent developments are making excellent contributions to the field. The rapid development of this technology is precipitating breakthroughs in applications. Facets of this topic include:

- Pattern recognition in the framework of fuzzy sets.
- Methods for fuzzy modeling in pattern recognition.
- Formation of the feature space.
- Knowledge representation.
- A continuum of classification models.
- Fuzzy neural structures.

- Supervised learning.
- Implicitly supervised pattern recognition.
- Unsupervised learning.

Neural fuzzy systems are often tested on handwritten digit recognition problems. Applications of digit recognition in industrial processing include processing sales orders, reading income tax forms and addresses on envelopes.

The learning vector quantization (LVQ) approach originated by Kohonen has been in use for years, and can be very effective if its limitations are realized and measures taken to compensate for them. Recent developments in LVQ applications have thrived and expanded as the theory advances on a daily basis. Fuzzy algorithms for LVQ and all their variations are worth monitoring and studying for potential industrial electronics applications.

The well-known work of Carpenter and Grossberg has evolved. Combination of the innovative Adaptive Reasonance Theory (ART) techniques with fuzzy systems provides the basis applying some of the pioneering work accomplished by this team.

Future directions for fuzzy systems and soft computing in Industrial Electronics should include attempts to formalize the properties of intelligent systems which are most valuable to users in specific situation. As suggested by Lotfi Zadeh, performance measures related to machine intelligence (MIQ) could help direct the design process, and suggest directions for further research.

Acknowledgments

Discussions with Lotfi Zadeh, Paul Werbos, James Bezdek, Ronald Yager, and Walter Karplus guided the selection of material for this chapter.

Fuzzy Systems and Soft Computing Chapter, Intelligent Electronics Section, *CRC Handbook on Industrial Electronics,* J. D. Irwin, ed. in chief, CRC Press and IEEE Press, 1996.

References

Bezdek, J. 1994. An Introduction to Fuzzy Logic, *WCCI Tutorial #7,* IEEE Press, Piscataway, NJ.

Padgett, M. L. 1995. A practical filter for conflicting or subjective data, *ANZIIS 95,* Perth, Australia.

Padgett, M. L. and Padgett, W. D. 1995. Simulation and computational intelligence in real-world applications, *Simulation,* July, pp. 1–7.

Kartalopoulos, S. V. 1996. Time-Dependent Fuzzy Logic, *Understanding Neural Networks and Fuzzy Logic: Basic Concepts and Applications,* 130–152, IEEE Press, Piscataway, NJ.

Jang, J.-S. R. and Sun, C.-T. 1995. Neuro-fuzzy modeling and control, *Proc. IEEE, Special Issue on Engineering Applications of Fuzzy Logic,* 83(3):378–349.

Yager, R. R. and Filev, D. P. 1994. *Essentials of Fuzzy Modeling and Control,* John Wiley & Sons, New York, NY.

82

Fuzzy Numbers: The Application of Fuzzy Algebra to Safety and Risk Analysis

J. Arlin Cooper
Sandia National Laboratories

82.1 Background 1091
82.2 Analytical Processing of Input Data 1091
82.3 Fuzzy-Algebra Background 1091
82.4 Fuzzy-Algebra Depiction of Uncertainty 1092
82.5 Example Applications 1093

82.1 Background

Safety (risk) analyses describe the potential for system safety failure or weigh decision cost/benefit tradeoffs. Many of these analyses depend on uncertain inputs and on mathematical models chosen from various alternatives, making the meaningful portrayal of output uncertainty imperative. Since analysis results can be a major contributor to safety-measure or other decision processes, risk management depends on relating uncertainty to the information available (Klir, 1993; Morgan and Henrion, 1990).

Safety analyses are frequently based on probabilities (e.g., probabilistic risk assessments). This approach almost always depends on models using logic structures (e.g., fault trees, event trees). The input uncertainty may be due to variability of potential input values, interpolation or extrapolation, human or measurement error, disagreements in interpretation, problem specification language vagueness or ambiguity, assumptions, simplifications or approximations, instrumentation resolution limits, sampling variability, etc. These uncertainties are basically either stochastic (probabilistic) or subjective in nature.

The stochastic approach to describe input uncertainty is to use probability density functions. These are most appropriate when the data uncertainties are well defined. When data uncertainty is less well defined such as in abnormal (outside the normal operating specifications) environments, more subjective approaches based on expert opinion are appropriate. One of the most useful of these is fuzzy-(approximate numbers) algebra analysis.

82.2 Analytical Processing of Input Data

Analytical processing of input data requires a mathematical model (Figure 82.1) and a technique for incorporating uncertainty. Uncertain output data can be presented based on what is known about input data uncertainty. Standard probabilistic processing utilizes probabilistic calculus, or Monte-Carlo/Latin Hypercube sampling techniques. Subjective data can be processed using fuzzy algebra.

The description of logical and algebraic operations on stochastic variables using integral calculus[1] (Sveshnikov, 1978) is a probabilistic tool. For example, the sum of the density functions for two independent random variables is:

$$f(y) = \int_{-\infty}^{\infty} f_1(x) f_2(y - x)\, dx. \qquad (82.1)$$

The density function of the product of two random non-negative (e.g., probabilities) variables is:

$$f(y) = \int_{0}^{\infty} \frac{f_1(x) f_2(y/x)\, dx}{x}. \qquad (82.2)$$

Monte Carlo and Latin hypercube sampling techniques simulate these mathematical processes. Numerous other mathematical operations are possible using this approach. In the following sections, fuzzy-algebra processing will be described, accounting for subjective uncertainty.

82.3 Fuzzy-Algebra Background

Fuzzy logic first emphasized set membership (Zadeh, 1965). "Crisp" (conventional) set membership is fixed (an element is either a member of a set or it isn't). However, an entity can have some of the characteristics of more than one set description (e.g.,

[1] Transform techniques can be similarly used.

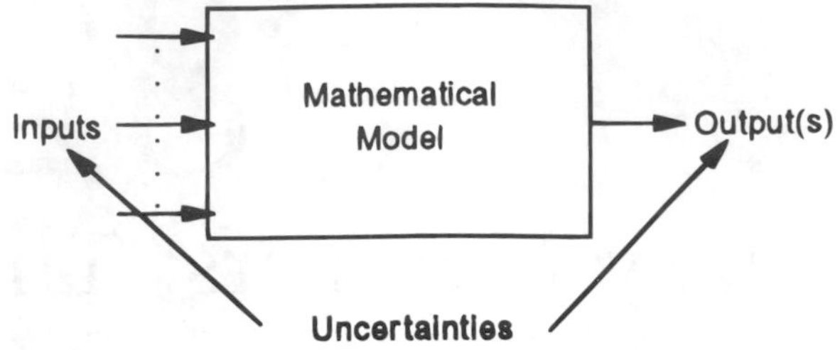

Figure 82.1 Analytical modeling with data uncertainty.

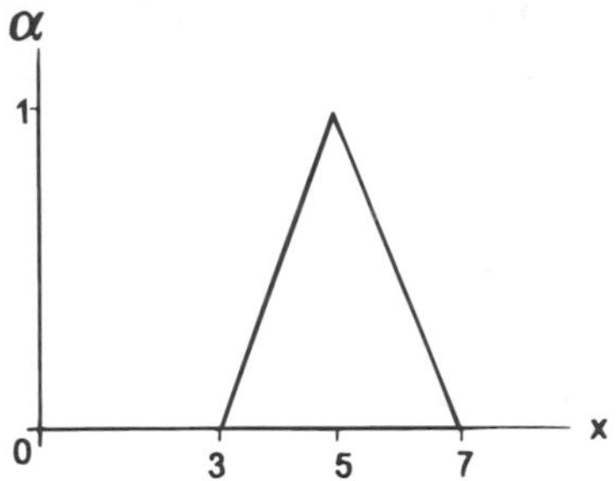

Figure 82.2 An example of a "fuzzy" number (e.g., "approximately" five).

a person's hair may be somewhat black and somewhat gray). Like probabilistic calculus, fuzzy algebra also can be applied to introduce variability to fixed parameters. For example, an uncertain parameter can have some of the characteristics of more than one number (e.g., "approximately" five may indicate a range of real numbers including, but not limited to, five). Fuzzy models can therefore be applied to represent uncertainty of parameters in probability analysis (Tanaka et al., 1983), and this has some similarity to strictly probabilistic descriptions. However, fuzzy algebra differs from probabilistic calculus both mathematically and in concept. Some mathematical background is helpful.

A fuzzy number (formally a convex and normal fuzzy set) (Figure 82.2) can be represented mathematically (Kaufmann and Gupta 1991) as:

$$A^{\alpha}(x) = A^{\alpha} = [a_1^{\alpha}, a_2^{\alpha}], \tag{82.3}$$

where the a_1 and a_2 values on x represent the lower and upper limits, respectively, of the variation possible for the parameter as a function of x, and α is a "level of presumption." The level of presumption represents a collection of subjective judgments[2] about the range specified. One must be more presumptuous in order to specify a narrower variable range (maximum level of presumption is presumption of minimum uncertainty). The "normal" restriction fixes the maximum level of presumption at 1 and the minimum to 0. An example of fuzzy number (over the real numbers) is shown in Figure 82.2.

[2] Preferably from "experts," preferably based on data (even if limited), and possibly weighted according to expertise.

Fuzzy addition is specified[3] as:

$$A^{\alpha} + B^{\alpha} = [a_1^{\alpha} + b_1^{\alpha}, a_2^{\alpha} + b_2^{\alpha}]. \tag{82.4}$$

Fuzzy subtraction is:

$$A^{\alpha} - B^{\alpha} = [a_1^{\alpha} - b_2^{\alpha}, a_2^{\alpha} - b_1^{\alpha}]. \tag{82.5}$$

Fuzzy multiplication, for nonnegative numbers, is:

$$A^{\alpha} \times B^{\alpha} = [a_1^{\alpha} b_1^{\alpha}, a_2^{\alpha} b_2^{\alpha}]. \tag{82.6}$$

It should be noted that there are definitions of multiplication that are slightly more complex, and which allow negative portions.

Multiplication is a nonlinear operation. Note that the above fuzzy algebra operations only utilize ranges of values, and make no use of or assumptions about relationships between parameters, or of independence between parameters.[4] The applicability of the operations shown is useful for parameters for which relative probabilities and independence are not well known (a common situation). On the other hand, probabilistic operations are limited to parameters for which these characteristics *are* well known (a less common situation).

In situations such as abnormal environment safety assessment, the inputs are often not well known, and the relations between possible values are not well known. The independence or dependence between inputs may also not be well known. However, we usually have access to expert judgment, along with limited data, which can be applied (to the appropriate extent) using fuzzy algebra (Cooper, 1985).

82.4 Fuzzy-Algebra Depiction of Uncertainty

Figure 82.3 depicts two basic types of uncertainty. On the left is a continuous representation of stochastic uncertainty (which

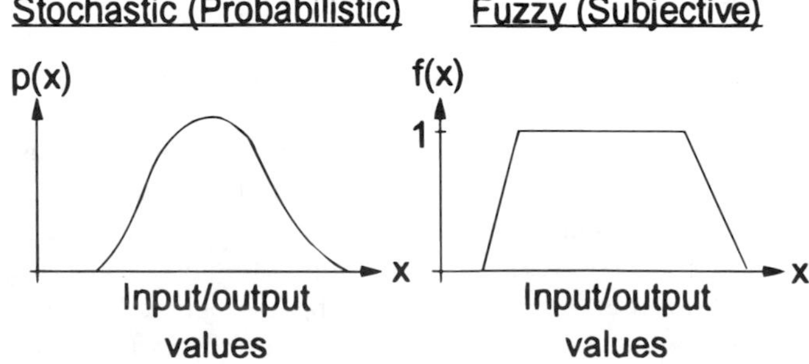

Figure 82.3 Two basic different models of uncertainty.

[3] Fuzzy arithmetic can be derived using the "extension principle" (Tanaka et al., 1983; Kaufmann and Gupta, 1991).

[4] However, treatment of independence/dependence properties is not precluded.

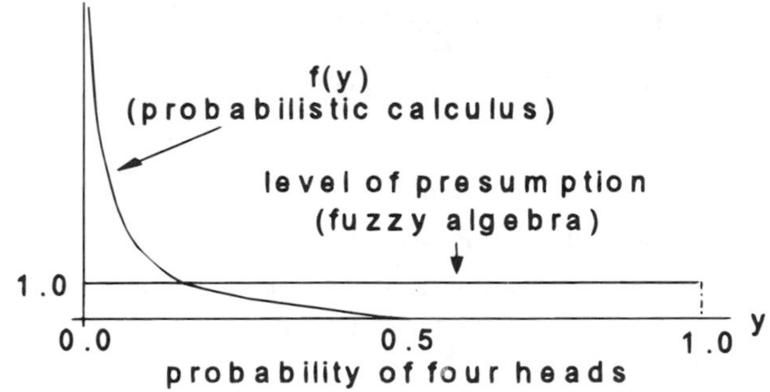

Figure 82.4 Probabilistic assessment example (four heads; deformed coins).

also could reflect discrete values, e.g., for tossing dice). Here the abscissa represents variability in a parameter, x; the ordinate represents relative probability of each x value. There is an overall requirement that the integral of the variability function be one (convention for probability). On the right is a subjective fuzzy representation of x variability. The ordinate represents a level of presumption, or strength of opinion (variable from zero to one) about the horizontal spread. The mathematical requirement for fuzzy numbers is that the maximum level of presumption be one. The two functions appear similar, but the meaning of the two is only weakly related.

Situations that require aspects of both types of uncertainty can be handled through hybrid number analysis (Kaufmann and Gupta, 1991).

82.5 Example Applications

EXAMPLE 82.1: Four Coin Tosses

A straightforward example game uses familiar concepts. The probability of throwing four heads in four coin tosses is an undesirable outcome (you would immediately lose the game). Traditional analysis says that the probability is 1/16, and this might be viewed as an acceptably small risk.

Suppose the coins have been subjected to an abnormal environment and may well be significantly warped or deformed. The best guess requires that we account for this new information. One (futile) approach that attempts to expand on the discrete probability of 1/2 is to assume that each coin has a "uniform distribution" of probabilities of heads ranging between zero and one. Using a calculus solution for the probability of getting four heads based on this information, a composite distribution can be derived which shows a continuum of possible values for the probability of four heads, but which has a concentration around very small (e.g., 1/16) values, because the function calculated is $-(\ln y)^3/6$ (derived from Equation 82.2, and illustrated in Figure 82.4). In other words, this approach gives results qualitatively similar to those obtained for "fair" coins, although we removed the "fair" condition. The possibility that all four coins could be deformed so that heads was likely for each is not well recognized, even though such extremes might be likely, and even though they are of prime concern. The concentration in probabilistic assessment arises from the assumptions of equal likelihood for probabilities and independence of coin tossing. These would be impossible to assure for general abnormal environment cases. The implication for safety is that assuming more than is known can lead to unwarranted expectations.

Qualitative information can be gained by looking at the coins. The "point estimate" of the probability of heads for the four coins might shift from 1/2, 1/2, 1/2, 1/2 to something else (e.g., 3/4, 1/2, 1/2, 3/4) based on the observation. If this is done, the new estimate for the probability of four heads would be 9/64. This approach has the advantage of taking the observation into account, but suffers the disadvantage of not indicating the inherent uncertainty in the estimate.

"Expert" estimates combined with fuzzy algebra is an appropriate approach. This is intended to provide an effective tool for approaching poorly defined problems (e.g., abnormal-environment problems) such that the potential for the occurrence of extreme values is not overlooked.

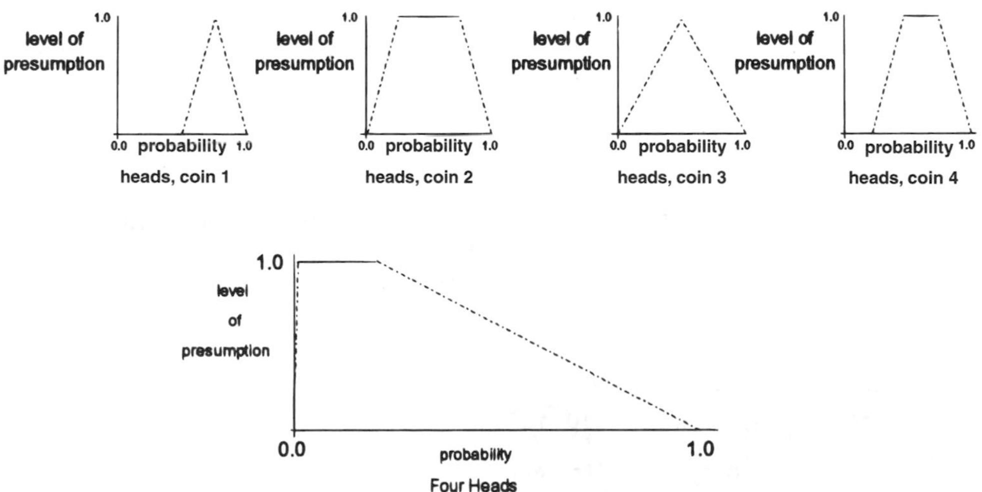

Figure 82.5 "Expert" inputs and fuzzy algebra output (example).

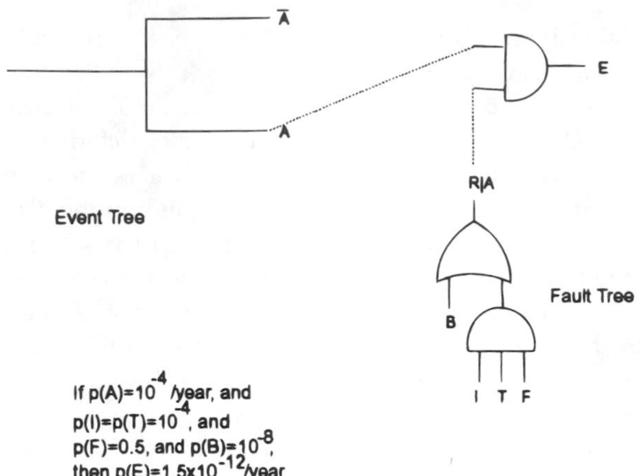

Figure 82.6 Example of described with an event tree/fault tree.

Fuzzy algebra is not limited to trivial bounds. Suppose that "experts" have a brief opportunity to view the deformed coins. Their experience may allow them to make qualitative judgments about the possibility of heads for each coin. Furthermore, each expert may have varying levels of presumption in his or her own judgment, and the level of expertise may vary from "expert" to "expert." The guidance provided for generating such descriptions is that the maximum level of presumption (one) corresponds to the smallest range of values the expert would judge plausible, and the lowest level of presumption (zero) corresponds to the largest range of values the expert would judge plausible. This allows for a characterization of the fuzzy input parameters by "level of presumption." Assume that the expert judgments have been consolidated (e.g., by a weighted combination) into the four graphs shown for the four coins in Figure 82.5. Multiplication of the fuzzy input variables (Equation 82.6, depicted linearized for simplicity) is appropriate only if the coins are independently tossed. This leads to the output shown in Figure 82.5.

But what if a detailed study of the coins (analytical or experimental) were possible? The fuzzy algebra approach can transition toward the probabilistic approach as the amount of knowledge increases. It is also possible to combine probabilistic variables and fuzzy variables, as well as to combine probabilistic and fuzzy characteristics in the same variable.

EXAMPLE 82.2: A System Safety Scenerio

A more practical problem involves logic for an undesired outcome.

- Call an unwanted event E.
- An incident, A, causes the undesired environment.
- An unwanted response, R, can be caused by failure of subsystem, I, subsystem, T, and subsystem F; or by bypass, B.

This word description results in the following logic and math description, where juxtaposition represents the "and" function and + represents the "or" function.:

$$E = A \text{ and } R(\text{given } A) \qquad E = A(R|A)$$
$$R(\text{given } A) = (I \text{ and } T \text{ and } F) \text{ or } B \qquad (R|A) = ITF + B$$
$$E = A(ITF + B)$$

The event tree and fault tree (Figure 82.6) for this problem are trivial, but help illustrate how an event tree leading to an incident and fault tree for the response to that incident can be coupled to describe an undesired incident. For illustration, some point estimates for the inputs are shown along with the trees, and the final result is computed. Because of input inaccuracy, the result is also inaccurate, so the uncertainty question needs to be addressed.

Assume input variability as shown in Figure 82.7. Fuzzy algebra is used to obtain the result shown in Figure 82.8.

There are many other example applications, notably those arising in risk analysis. Risk analysis typically involves monetary values, which may have substantial uncertainty. Even more subjective values arise when factors such as the values of human life, morale, litigation potential, etc., must be weighed (Cooper, 1989).

Defining Terms

Event tree: A logical procedure to inductively follow a progression of occurrences from an initiating event to a (usually undesired) consequence of that event.

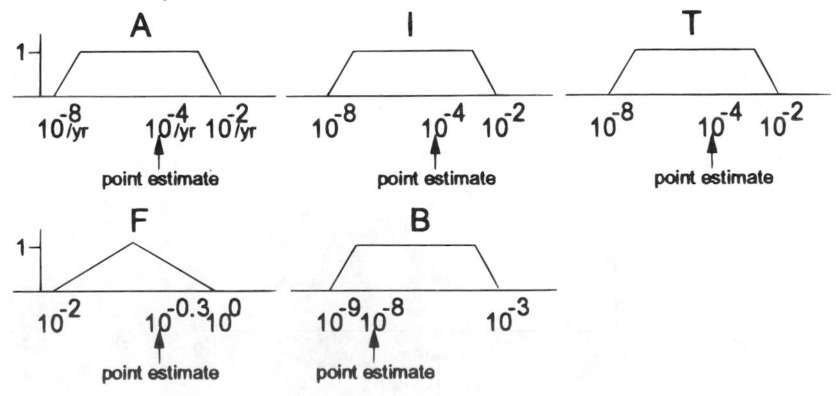

Figure 82.7 Fuzzy input variability.

Fault tree: A logical procedure for determining deductively how some unwanted event might occur.

Fuzzy algebra: A mathematical system based on variables whose membership includes a degree of belonging to some number or numbers.

Fuzzy logic: A logic of variables having set membership (possession of some characteristic) in a gradated rather than crisp manner.

Risk analysis: A formal technique for comparing the "cost" of some decision with the "benefit," to determine if the decision "pays off."

Stochastic variability: Variation due to statistically describable events, such as throwing dice or flipping coins.

Acknowledgments

This work was supported by the United States Department of Energy under Contract DE-AC04-94AL85000.

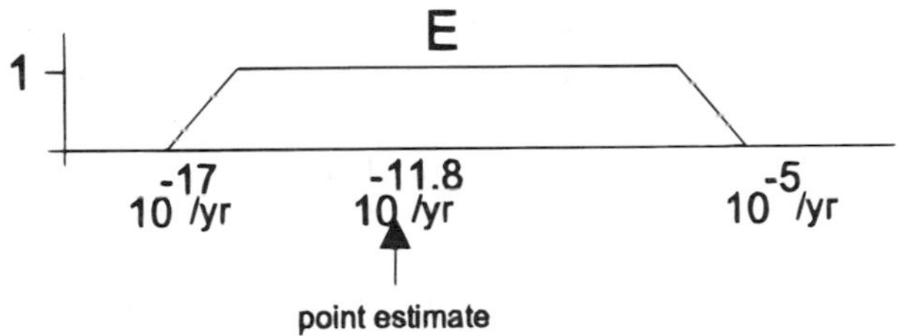

Figure 82.8 Fuzzy output variability.

References

Cooper, J. A. 1995. Fuzzy-algebra uncertainty analysis for abnormal-environment safety assessment, *J. Intelligent and Fuzzy Systems,* January. Vol. 2, Issue 4, 337–345.

Cooper, J. A. *Computer and Communications Security* McGraw-Hill, New York, NY, 1989.

Kaufmann, A. and Gupta, M. M. 1991. *Introduction to Fuzzy Arithmetic,* pp. 9, 82–91, Van Nostrand Reinhold, New York, NY.

Klir, G. 1993. Developments in Uncertainty-Based Information, *Advances in Computers,* Vol. 36, Academic Press, New York, NY.

Morgan, M. and Henrion, M. 1990. *Uncertainty,* Cambridge University Press, New York, NY.

Sveshnikov, A. A. 1978. *Problems in Probability Theory, Mathematical Statistics, and Theory of Random Functions,* pp. 121, 129, 409, Dover Publications, New York, NY.

Tanaka, H., Fan, L. T., Lai, F. S., and Toguchi, K. 1983. Fault tree analysis by fuzzy probability, *IEEE Trans. Reliability,* 453–457.

Zadeh, L. A. 1965. Fuzzy sets, *Information Control,* 8:338–353.

Further Information

Handbook of Intelligent Control—Neural, Fuzzy, and Adaptive Approaches, Edited by David A. White and Donald A. Sofge, Van Nostrand Reinhold, New York, NY.

Klir, G. J., Yuan, B. *Fuzzy Sets and Fuzzy Logic: Theory and Applications,* Prentice-Hall, Englewood Cliffs, NJ.

Ross, T., 1995. *Fuzzy Logic with Engineering Applications,* Timothy J. Ross, McGraw-Hill, New York, NY.

83
Fuzzy Systems

Mo-yuen Chow
North Carolina State University

83.1 Brief Description of Fuzzy Logic ... 1096
Crisp Set • Fuzzy Set
83.2 Qualitative (Linguistic) to Quantitative Description ... 1097
83.3 Fuzzy Operations ... 1098
Union • Intersection • Complement
83.4 Fuzzy Rules, Inference ... 1100
Fuzzy Relation/Composition/Conditional Statement • Compositional Rule of Inference • Defuzzification
83.5 Fuzzy Control ... 1101

83.1 Brief Description of Fuzzy Logic

Fuzzy logic can easily implement human experiences and preferences via *membership functions* and *fuzzy rules,* from a qualitative description to a quantitative description that is suitable for microprocessor implementation of the automation process. Fuzzy membership functions can have different shapes depending on the designer's preference and/or experience. The fuzzy rules, which describe the control strategy in a human-like fashion, are written as antecedent-consequent pairs of IF-THEN statements and stored in a table. Basically, there are four modes of derivation of fuzzy control rules (Chow and Menozzi, 1993).

1. Expert experience and control engineering knowledge.
2. Behavior of human operators.
3. Derivation based on the fuzzy model of a process.
4. Derivation based on learning.

These do not have to be mutually exclusive. In later sections, we will discuss more about *membership functions* and *fuzzy rules.*

Due to the use of *linguistic variables* and *fuzzy rules,* the fuzzy controller can be made understandable to a nonexpert operator. Moreover, the description of the control strategy could be derived by examining the behavior of a conventional controller. The fuzzy characteristics make it particularly attractive for control applications because only a linguistic description of the appropriate control strategy is needed in order to obtain the actual numerical control values. Thus, fuzzy logic can be used as a general methodology to incorporate knowledge, heuristics or theory into a controller.

In addition, fuzzy logic has the freedom to completely define the control surface without the use of complex mathematical analysis, as discussed in later sections. On the other hand, the amount of effort involved in producing an acceptable rule base and in fine-tuning the fuzzy controller is directly proportional to the number of quantization levels used, and the designer is left to choose the best tradeoff between being able to create a large number of features on the control surface and not having to spend much time in the fine tuning process. The general shape of these features depends on the heuristic rule base and the configuration of the membership functions (Zimmermann, 1991; Sugenon, 1985; Zadah, 1965). Being able to quantize the domain of the control surface using linguistic variables allows the designer to depart from the mathematical constraints (e.g., hyperplane constraints in PI control) and achieve a control surface which has more features and contours.

In 1965, L. A. Zadeh (1965) laid the foundations of *fuzzy set theory,* which is a generalization of conventional set theory, as a method of dealing with the imprecision of the real physical world. Bellman and Zadeh write: "Much of the decision-making in the real world takes place in an environment in which the goals, the constraints and the consequences of possible actions are not known precisely" (Bellman and Zadeh, 1970). This "imprecision" or fuzziness is the core of fuzzy logic. Fuzzy control is the technology that applies fuzzy logic to solve control problems.

This section is written to provide readers with an introduction of the use of fuzzy logic to solve control problems; it also intends to provide information for further exploration on related topics. This section, provides an overview of some fundamental fuzzy logic concepts and operations. The advantages of using fuzzy control is more substantial when applied to nonlinear and ill-defined systems. If the readers are interested in more details about fuzzy logic and fuzzy control, please refer to Chapter 39 and to the references at the end of this chapter.

Crisp Set

The basic principle of conventional set theory is that an element either is or is not a member of a set. A set that is defined in this way is called a *crisp* set, since its boundary is well defined. Consider the set, *W,* of motor speed operating range between 0

0-8493-8343-9/97/$0.00+$.50

rad/sec and 175 rad/sec. The proper motor speed operating range would be written as:

$$W = \{w \in W | 0 \text{ rad/sec} \leq w \leq 175 \text{ rad/sec}\}. \quad (83.1)$$

The set W could be expressed by its membership function $W(w)$, which indicates whether the motor is within its operating range:

$$W(w) = \begin{cases} 1; & 0 \text{ rad/sec} \leq w \leq 150 \text{ rad/sec} \\ 0; & \text{otherwise} \end{cases}. \quad (83.2)$$

Fuzzy set and it's membership function are often used interchangebly. A graphical representation of $W(w)$ is shown in Figure 83.1.

Fuzzy Set

However, in the real world sets are not always so crisp. Human beings routinely use concepts that are approximate and imprecise to describe most problems and events. The human natural language largely reflects these approximations. For example, the meaning of the word "fast" to describe the motor speed depends on the person who uses it and the context in which he/she uses it. Hence, various speeds may belong to the set "fast" to various degrees, and the boundary of the set operating range is not very precise. For example, if we wish to consider 150 rad/sec as fast, then 160 rad/sec is certainty fast. However, do we still say 147 rad/sec is fast, too? How about 145 rad/sec? Conventional set theory techniques have difficulties in dealing with this type of linguistic or qualitative problem.

Fuzzy sets are very useful in representing *linguistic variables*, which are quantities described by natural or artificial language (Zadeh, 1975) and whose values are linguistic (qualitative) and not numeric (quantitative). A linguistic variable can be considered either as a variable whose value is a fuzzy number (after *fuzzification*), or as a variable whose values are defined in linguistic terms. Examples of linguistic variables are fast, slow, tall, short, young, old, very tall, very short, etc. More specifically, the basic idea underlying the fuzzy set theory is that an element is a member of a set to a certain degree, which is called the *membership grade* (*value*) of the element in the set. Let U be a collection of elements denoted by $\{u\}$ which could be discrete or continuous. U is called the *universe of discourse* and u represents the generic element of U. A *fuzzy set A* in a universe of discourse U is then characterized by a *membership function* $A(.)$ that maps U onto a real number in the interval $[A_{min}, A_{max}]$. If $A_{min} = 0$ and $A_{max} = 1$, the membership function is called a *normalized* membership function, between 0 and 1, and A: $U \rightarrow [0,1]$.

For example, a membership value $A(u) = 0.8$ suggests that u is a member of A to a degree of 0.8, on a scale where zero is no membership and one is complete membership. One can then see that crisp set theory is just a special case of fuzzy set theory. A fuzzy set A in U can be represented as a set of ordered pairs of an element u and its membership value in A: $A = \{ (u, A(u)) | u \in U\}$. The element u is sometimes called the *support,* while $A(u)$ is the corresponding membership function of u of the fuzzy set A. When U is continuous, the common notation used to represent the fuzzy set A is

$$A = \int_U A(u)/u, \quad (83.3)$$

and when U is discrete, the fuzzy set A is normally represented as

$$A = \sum_{i=1}^{n} A(u_i)/u_i = A(u_1)/u_1 + A(u_2)/u_2 + \cdots + A(u_n)/u_n, \quad (83.4)$$

where n is the number of supports in the set. It is to be noted that, in fuzzy logic notation, the summation represents union, not addition. Also, the membership values are 'tied' to their actual values by the divide sign, but they are not actually being divided.

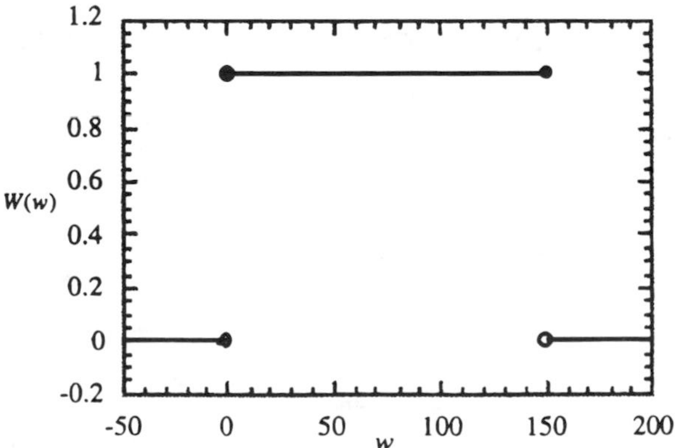

Figure 83.1 Crisp membership function for proper speed operating condition defined in Equation 83.1.

83.2 Qualitative (Linguistic) to Quantitative Description

In this section, we use *height* as an example to explain the concept of fuzzy logic. The word "tall" may refer to different heights depending on the person who uses it and the context in which he/she uses it. Hence, various heights may belong to the set "tall" to various degrees, and the boundary of the set tall is not very precise. Let's consider the linguistic variable "tall" and assign values in the set zero to one. A person seven feet tall may be considered "tall" with a value of one. Certainly, anyone over seven feet tall would also be considered tall with a membership of 1. A person six feet tall may be considered "tall" with a value of 0.5, while a person five feet tall may be considered tall with a membership of zero. As an example, the membership function *TALL*(*height*), which maps the values between four to eight feet into the fuzzy set "tall", is described continuously by Equation 83.5 and shown in Figure 83.2.

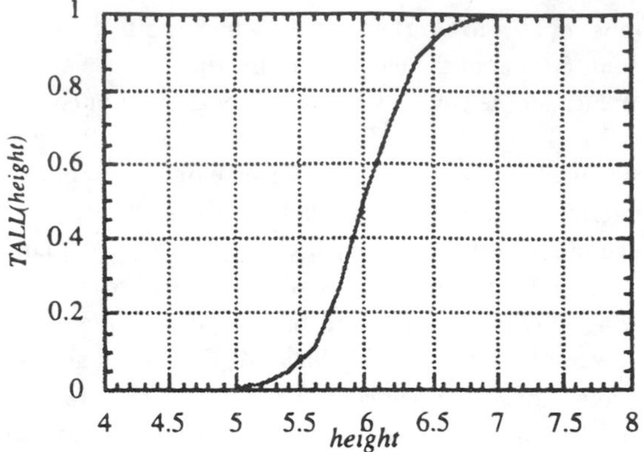

Figure 83.2 Continuous membership function plot for linguistic variable "tall" defined in Equation 83.5.

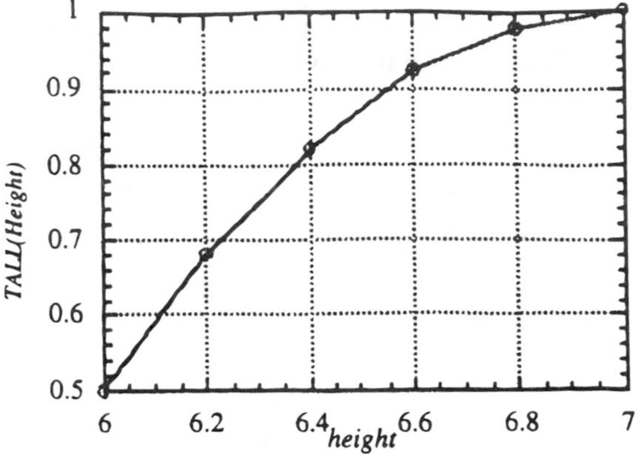

Figure 83.3 Discrete membership function plot for linguistic variable "tall" defined in Equation 83.6.

$$\text{TALL} = \int_{[4,8]} \frac{1}{1 + e^{-5(\text{height}-6)}} \Big/ \text{height} \qquad (83.5)$$

Membership functions are very often represented by the discrete fuzzy membership notation when membership values can be obtained for different supports based on collected data. For example, a five-player basketball team may have a fuzzy set "TALL" defined by:

$$\text{TALL} = 0.5/6 + 0.68/6.2 + 0.82/6.4 + 0.92/6.6 + 0.98/6.8 + 1.0/7, \qquad (83.6)$$

where the heights are listed in feet. This membership function is shown in Figure 83.3.

In the discrete fuzzy set notation, linear interpolation between two closest known supports and corresponding membership values is often used to compute the membership value for the u that is not listed. For example, if a new player joins the team and his height is 6.1 feet, then his *TALL* membership value based on the membership function defined in (6) is;

$$\text{TALL}(6.1) = 0.5 + \frac{6.1 - 6(0.68 - 0.5)}{6.2 - 6} = 0.59. \quad (83.7)$$

Note that the membership function defined in Equation 83.5 is normalized between 0 and 1 while the one defined in Equation 83.6 is not, because its minimum membership value is 0.5 rather than 0.

It should be noted that the choice of a membership function relies on the actual situation and is very much based on heuristic and educated judgment. In addition, the imprecision of fuzzy set theory is different from the imprecision dealt with by probability theory. The fundamental difference is that probability theory deals with randomness of future events due to the possibility that a particular event may or may not occur, whereas fuzzy set theory deals with imprecision in current or past events due to the vagueness of a concept—the membership or non-membership of an object in a set with imprecise boundaries (Zadeh, 1978).

83.3 Fuzzy Operations

The combination of membership functions requires some form of set operations. Three basic operations of conventional set theory are intersection, union, and complement (Sugenon, 1985; Lee, 1990; Kosko, 1992; d. Glas, 1982). The fuzzy logic counterparts to these operations are similar to those of conventional set theory. Fuzzy set operations such as union, intersection and complement are defined in terms of the membership functions. Let A and B be two fuzzy sets with membership functions A and B, respectively, defined for all $u \in U$. The *TALL* membership function has been defined in Equation 83.6, and a *FAST* membership function is defined in Figure 83.4:

$$\text{FAST} = 0.5/6 + 0.8/6.2 + 0.9/6.4 + 1/6/6 + 0.75/6.8 + 0.6/7. \qquad (83.8)$$

These two membership functions will be used in the next sections as examples for fuzzy operation discussions.

Union

The membership function of the union $A \cup B$ is pointwise defined for all $u \in U$ by

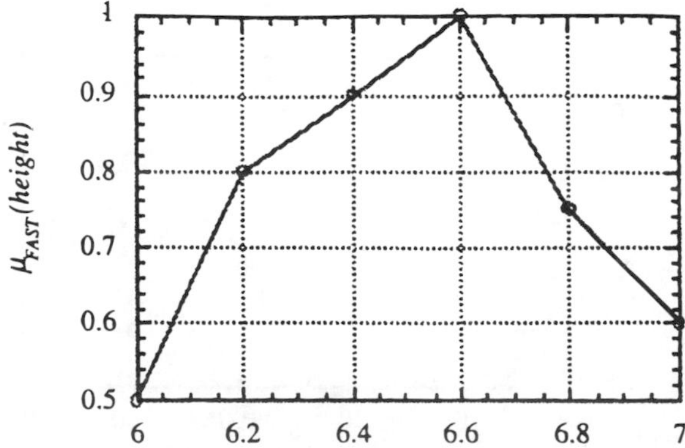

Figure 83.4 Discrete membership function plot for linguistic variable "fast" defined in Equation 83.8.

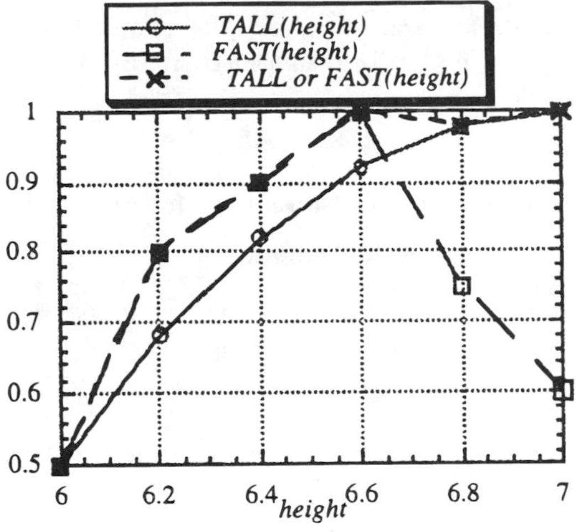

Figure 83.5 Membership function plot of *TALL*∪*FAST*(*height*) defined in Equation 83.10.

$$A \cup B(u) = \max\{A(u), B(u)\} = \vee_u \{A(u), B(u)\}, \quad (83.9)$$

where ∨ symbolizes the max operator. As an example, a fuzzy set could be composed of basketball players who are "tall *or* fast." The membership function for this fuzzy set could be expressed in the union notation as:

$$\begin{aligned} \text{TALL} \cup \text{FAST(height)} &= \max\{\text{TALL(height)}, \text{FAST(height)}\} \\ &= (0.5 \vee 0.5)/6 + (0.68 \vee 0.8)/6.2 \\ &\quad + (0.82 \vee 0.9)/6.4 + (0.92 \vee 1)/6.6 \\ &\quad + (0.98 \vee 0.98)/6.8 + (0.6 \vee 1)/7, \\ &= 0.5/6 + 0.8/6.2 + 0.9/6.4 + 1/6.6 \\ &\quad + 0.98/6.8 + 1/7. \end{aligned} \quad (83.10)$$

The corresponding membership function is shown in Figure 83.5

Intersection

The membership function $A \cap B$ is pointwise defined for all $u \in U$ by

$$A \cap B(u) = \min\{A(u), B(u)\} = \wedge_u \{A(u), B(u)\}, \quad (83.11)$$

where ∧ symbolizes the min operator. As an example, a fuzzy set could be composed of basketball players who are "tall *and* fast." The membership function for this fuzzy set could be expressed in the intersection notation as:

$$\begin{aligned} TALL \cap FAST(height) &= \min\{\text{TALL}(height), \text{FAST}(height)\}, \\ &= (0.5 \wedge 0.5)/6 + (0.68 \wedge 0.8)/6.2 \\ &\quad + (0.82 \wedge 0.9)/6.4 + (0.92 \wedge 1)/6.6 \\ &\quad + (0.98 \wedge 0.98)/6.8 + (0.6 \wedge 1)/7, \\ &= 0.5/6 + 0.68/6.2 + 0.80/6.4 \\ &\quad + 0.92/6.6 + 0.75/6.8 + 0.6/7. \end{aligned} \quad (83.12)$$

The corresponding membership function is shown in Figure 83.6

Complement

Also, the membership function $\overline{A}$ of complement of the fuzzy set A is pointwise defined for all $u \in U$ by

$$\overline{A}(u) = 1 - A(u). \quad (83.13)$$

The fuzzy set "short" could be considered to be the complement of the fuzzy set *TALL* expressed as:

$$SHORT(u) = \overline{Tall}(u) = 1 - TALL(u). \quad (83.14)$$

The *SHORT* fuzzy set correspondent to the *TALL* fuzzy set defined in Equation 83.5 is shown in Figure 83.7

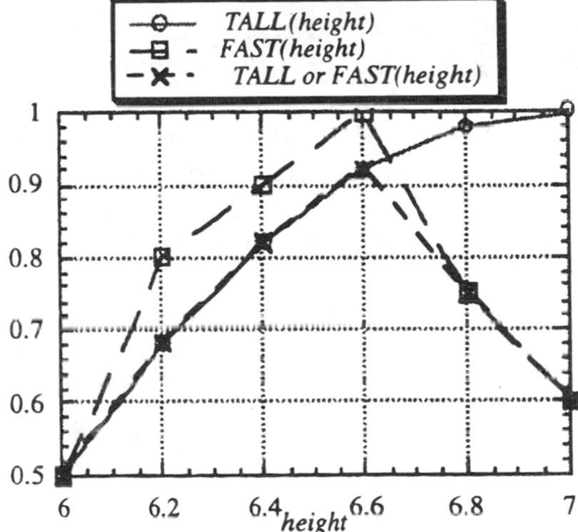

Figure 83.6 Membership function plot of *TALL*∩*FAST*(*height*) defined in Equation 83.12.

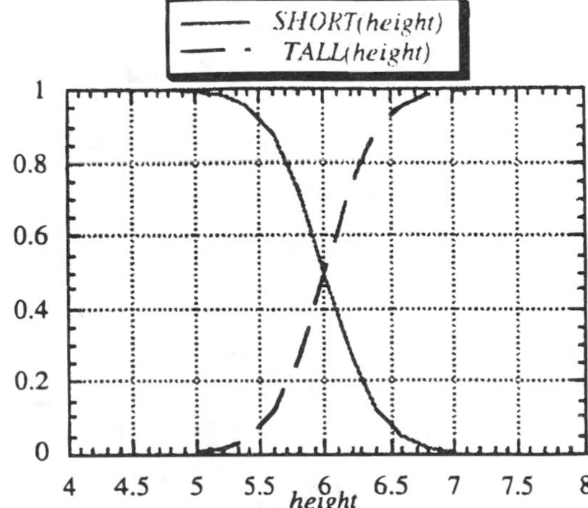

Figure 83.7 Membership functions *SHORT*(height) and *TALL*(height) defined in Equations 83.5 and 83.14.

The *SHORT* fuzzy set correspondent to the *TALL* fuzzy set defined in Equation 83.6 is shown in Figure 83.8.

Other operations exist for fuzzy sets. Different modifications of the union, intersection, and complement have also been proposed and used. For example, unions and intersections of various strengths can be achieved by using the Yager class (among others) to perform the fuzzy union and intersection operations (Klir and Folger, 1988). But the three operations described above represent the most popular ones.

83.4 Fuzzy Rules, Inference

Fuzzy Relation/Composition/Conditional Statement

Another concept of fuzzy logic is the fuzzy relation. Fuzzy logic techniques can be used to translate natural language into heuristic responses and also to combine fuzzy membership functions to formulate fuzzy rules. A *fuzzy relation R* from a set X to a set Y is defined as a fuzzy subset of the Cartesian product $X \times Y$, which is the collection of ordered pairs (x, y), where $x \in X$, and $y \in Y$ (Zadeh, 1973). In the same way that a membership function defines an element's membership in a given set, the fuzzy relation is a fuzzy set composed of all combinations of participating sets which determines the degree of association between two or more elements of distinct sets (Klir and Folger, 1988). It is characterized by a bivariate membership function $R(x,y)$, written as:

$$R_Y(x) = \int_{X \times Y} R(x, y)/(x, y) \tag{83.15}$$

in continuous membership function notation, and

$$R_Y(x) = \sum_{i=1}^{n} R(x_i, y_i)/(x_i, y_i) \tag{83.16}$$

in discrete membership function notation.

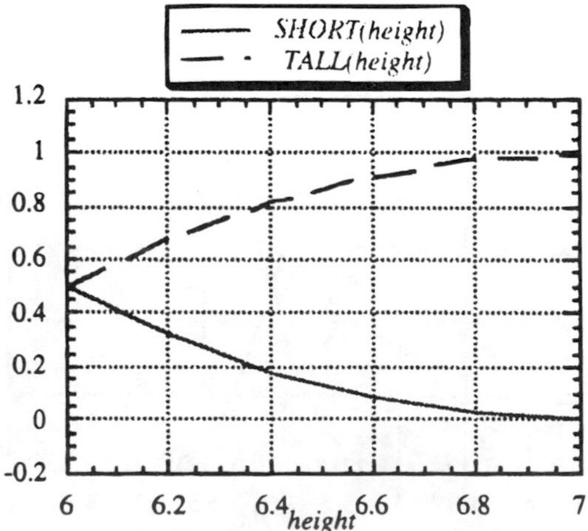

Figure 83.8 Membership functions *SHORT*(height) and *TALL*(height) defined in Equations 83.6 and 83.14.

Let's consider a fuzzy rule:

If a player is much taller than 6.4 feet, his scoring-average per game should be high.

Let the fuzzy sets:

$$\begin{aligned} X &= \text{much taller than 6.4 feet (with [6 feet, 7 feet]} \\ &\qquad \text{as the universe of discourse)} \\ &= 0/6 + 0/6.2 + 0.2/6.4 + 0.5/6.6 + 0.8/6.8 + 1/7, \end{aligned} \tag{83.18}$$

and

$$\begin{aligned} Y &= \text{scoring-average per game (with [0, 20]} \\ &\qquad \text{as the universe of discourse)} \\ &= 0/0 + 0.3/5 + 0.8/10 + 0.9/15 + 1/20, \end{aligned} \tag{83.19}$$

If we use the min operator to form the R (other operators such as "product" is another popular choice (Kosko, 1992; Sugenon, 1985), then the consequent relational matrix between X and Y:

$R_Y(x) =$

X\Y	0	5	10	15	20
6	$0 \wedge 0$	$0 \wedge 0.3$	$0 \wedge 0.8$	$0 \wedge 0.9$	$0 \wedge 1$
6.2	$0 \wedge 0$	$0 \wedge 0.3$	$0 \wedge 0.8$	$0 \wedge 0.9$	$0 \wedge 1$
6.4	$0.2 \wedge 0$	$0.2 \wedge 0.3$	$0.2 \wedge 0.8$	$0.2 \wedge 0.9$	$0.2 \wedge 1$
6.6	$0.5 \wedge 0$	$0.5 \wedge 0.3$	$0.5 \wedge 0.8$	$0.5 \wedge 0.9$	$0.5 \wedge 1$
6.8	$0.8 \wedge 0$	$0.8 \wedge 0.3$	$0.8 \wedge 0.8$	$0.8 \wedge 0.9$	$0.8 \wedge 1$
7	$1 \wedge 0$	$1 \wedge 0.3$	$1 \wedge 0.8$	$1 \wedge 0.9$	$1 \wedge 1$

$=$ (83.20)

X\Y	0	5	10	15	20
6	0	0	0	0	0
6.2	0	0	0	0	0
6.4	0	0.2	0.2	0.2	0.2
6.6	0	0.3	0.5	0.5	0.5
6.8	0	0.3	0.8	0.8	0.8
7	0	0.3	0.8	0.9	1

Note that in (20), min $R_y(x) = 0$, and max $R_y(x) = 1$. Therefore a player with height 6.8 feet and 10 scoring-average per game should have a fuzzy relation 0.8 in the scale between 0 to 1.

Compositional Rule of Inference

Compositional rule of inference, which may be regarded as an extension of the familiar rule of modus ponens in classical propositional logic, is another important concept. Specifically, if R is a fuzzy relation from X to Y, and x is a fuzzy subset of X, then the fuzzy subset y of Y that is induced by x is given by the composition of R and x, in the sense of Equation 83.12. Also, if R is a relation from X to Y and S is a relation from Y to Z, then the *composition* of R and S is a fuzzy relation denoted by $R \circ S$, defined by:

$$R \circ S \equiv \int_{X \times Z} \max_y(\min(R(x, y), S(y, z))) \Big/ (x, z). \tag{83.21}$$

Equation 83.21 is called the *max-min* composition of R and S.

For example, if a player is about 6.8 feet with membership function:

$$x = 0/6 + 0/6.2 + 0/6.4 + 0.2/6.6 + 1/6.8 + 0.2/7, \qquad (83.22)$$

and we are interested in his corresponding membership function of scoring-average per game, then

$$\mu_Y = \max_{y_i}\{\min(X \circ R_Y(x))\}$$

$$= \max_{y_i}\left\{ [0 \quad 0 \quad 0 \quad 0.2 \quad 1 \quad 0.2] \wedge \begin{bmatrix} 0 & 0 & 0 & 0 & 0 \\ 0 & 0 & 0 & 0 & 0 \\ 0 & 0.2 & 0.2 & 0.2 & 0.2 \\ 0 & 0.3 & 0.5 & 0.5 & 0.5 \\ 0 & 0.3 & 0.8 & 0.8 & 0.8 \\ 0 & 0.3 & 0.8 & 0.9 & 1 \end{bmatrix} \right\}$$

$$= \max_{y_i}\left\{ \begin{bmatrix} 0\wedge 0 & 0\wedge 0 & 0\wedge 0 & 0\wedge 0 & 0\wedge 0 \\ 0\wedge 0 & 0\wedge 0 & 0\wedge 0 & 0\wedge 0 & 0\wedge 0 \\ 0\wedge 0 & 0\wedge 0.2 & 0\wedge 0.2 & 0\wedge 0.2 & 0\wedge 0.2 \\ 0.2\wedge 0 & 0.2\wedge 0.3 & 0.2\wedge 0.5 & 0.2\wedge 0.5 & 0.2\wedge 0.5 \\ 1\wedge 0 & 1\wedge 0.3 & 1\wedge 0.8 & 1\wedge 0.8 & 1\wedge 0.8 \\ 0.2\wedge 0 & 0.2\wedge 0.3 & 0.2\wedge 0.8 & 0.2\wedge 0.9 & 0.2\wedge 1 \end{bmatrix} \right\},$$

$$= \max_{y_i}\left\{ \begin{bmatrix} 0 & 0 & 0 & 0 & 0 \\ 0 & 0 & 0 & 0 & 0 \\ 0 & 0 & 0 & 0 & 0 \\ 0 & 0.2 & 0.2 & 0.2 & 0.2 \\ 0 & 0.3 & 0.8 & 0.8 & 0.8 \\ 0 & 0.2 & 0.2 & 0.2 & 0.2 \end{bmatrix} \right\}$$

$$= \begin{bmatrix} 0\vee 0\vee 0\vee 0\vee 0 \\ 0\vee 0\vee 0\vee 0\vee 0 \\ 0\vee 0\vee 0\vee 0\vee 0 \\ 0\vee 0.2\vee 0.2\vee 0.2\vee 0.2 \\ 0\vee 0.3\vee 0.8\vee 0.8\vee 0.8 \\ 0\vee 0.2\vee 0.2\vee 0.2\vee 0.2 \end{bmatrix} = \begin{bmatrix} 0 \\ 0 \\ 0 \\ 0.2 \\ 0.8 \\ 0.2 \end{bmatrix}. \qquad (83.23)$$

Defuzzification

After a fuzzy rule base has been formed (such as the one shown in Equation 83.17 along with the participating membership functions (such as the ones shown in Equation 83.22 and 83.23), a defuzzification strategy needs to be implemented. The purpose of defuzzification is to convert the results of the rules and membership functions into usable values, whether it be a specific result or a control input. There are a few techniques for the defuzzification process (Kosko, 1992). One popular technique is the "center-of-gravity" method which is capable of considering the influences of many different effects simultaneously. The general form of the center of gravity method is:

$$f_j = \frac{\sum_{i=1}^{N} A_i \times c_i}{\sum_{i=1}^{N} A_i}, \qquad (83.24)$$

where N is the number of rules under consideration. μ_i and c_i are the membership and the control action associated with rule i, respectively, and f_j represents the j-th control output. Consider the rule base:

1.	if	X = tall	and Y = fast	then $C = 1$
2.	if	X = short	and Y = slow	then $C = 0$.

The "center-of-gravity" method applied to this rule base with $N = 2$ results in:

$$f_i = \frac{\min(TALL(X), FAST(Y)) \times 1 + \min(SHORT(X), SLOW(Y)) \times 0}{\min(TALL(X), FAST(Y)) + \min(SHORT(X), SLOW(Y))}, \qquad (83.25)$$

where the "min" operator has been used in association with the "and" operation.

83.5 Fuzzy Control

The concepts outlined above represent the basic foundation upon which fuzzy control is built. In fact, the potential of fuzzy logic in control systems was shown very early by Mamdani and his colleagues (Pappis and Mamdani, 1977). Since then, fuzzy logic has been used successfully in a variety of control applications. Since the heuristic knowledge about how to control a given plant is often in the form of linguistic rules provided by a human expert or operator, fuzzy logic provides an effective means of translating that knowledge into an actual control signal. These rules are usually of the form:

IF (a set of conditions is satisfied)

THEN (the adequate control action is taken),

where the conditions to be satisfied are the *antecedents* and the control actions are the *consequent* of the fuzzy control rules, both of which are all associated with fuzzy concepts (linguistic variables). Several linguistic variables may be involved in the antecedents or the consequents of a fuzzy rule, depending on how many variables are involved in the control problem. For example, let x and y represent two important state variables of a process, and let w and z be the two control variables for the process. In this case, fuzzy control rules have the form:

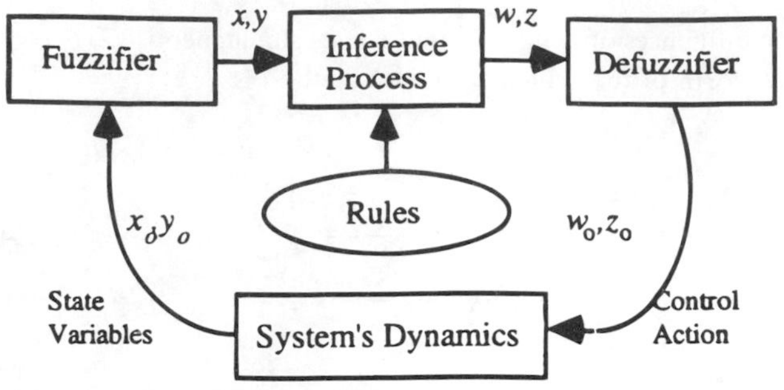

Figure 83.9 Block diagram representation of the concept of Fuzzy Logic control.

$$\begin{aligned} &Rule_1\text{: if } x \text{ is } A_1 \text{ and } y \text{ is } B_1 \text{ then } w \text{ is } C_1 \text{ and } z \text{ is } D_1,\\ &Rule_2\text{: if } x \text{ is } A_2 \text{ and } y \text{ is } B_2 \text{ then } w \text{ is } C_2 \text{ and } z \text{ is } D_2,\\ &\qquad \cdots\ \cdots\ \cdots\\ &\qquad \cdots\ \cdots\ \cdots\\ &Rule_n\text{: if } x \text{ is } A_n \text{ and } y \text{ is } B_n \text{ then } w \text{ is } C_n \text{ and } z \text{ is } D_n, \end{aligned}$$

where A_i, B_i, C_i and D_i are the linguistic values (fuzzy sets) of x, y, w, and z, in the universes of discourse X, Y, W, and Z, respectively, with $i = 1,2, \ldots, n$. Using the concepts of fuzzy conditional statements and the compositional rule of inference, w and z (fuzzy subsets of W and Z, respectively) can be inferred from each fuzzy control rule.

Typically, though, in control problems, the values of the state variables and of the control signals are represented by real numbers, not fuzzy sets. Therefore, to convert real information into fuzzy sets, and vice versa, it is necessary to convert fuzzy sets into real numbers. These two conversion processes are generally called *fuzzification* and *defuzzification*, respectively. Specifically, a fuzzification operator has the effect of transforming crisp data into fuzzy sets. Symbolically,

$$x = fuzzifier(x_0), \tag{83.26}$$

where x_0 is a crisp value of a state variable and x is a fuzzy set. Alternatively, a defuzzification operator transforms the outputs of the inference process (fuzzy sets) into a crisp value for the control action. That is,

$$z_0 = \text{defuzzifier}(z), \tag{83.27}$$

where z_0 is a crisp value and z is a fuzzy membership value. Referring back to the height example used previously, if a basketball player is about 6.8 feet, then after fuzzification, his height membership value is shown in Equation 83.22. Various fuzzification and defuzzification techniques are described in the references (Kosko, 1992; Zimmermann, 1991; Sugenon,1985). Figure 83.9 is a block diagram representation of the mechanism of the fuzzy control described above.

In Chapter 39, we described a motor fuzzy control design process to illustrate the application of fuzzy control to industrial electronics areas.

Acknowledgment

The author of this paper would like to thank Mr. Alberico Menozzi and Mr. Jason Teeter for their contributions to this article.

References

Bellman, R. E. and Zadeh, L. A. 1970. Decision-making in a fuzzy environment, *Management Science,* 17:41–164.

Chow, M.-y. and A. Menozzi, 1993. Design methodology of an intelligent controller using artificial neural networks, *IECON'93,* Maui, HA.

d. Glas, M. 1982. A Mathematical Theory of Fuzzy Systems, *Fuzzy Information and Decision Processes,* Gupta, M. M. and Sanchez, E., eds., 401–410, North-Holland Publishing New York, NY.

Klir, G. J. and Folger, T. A. 1988. *Fuzzy Sets, Uncertainty, and Information,* Prentice-Hall, Englewood.

Kosko, B. 1992. *Neural Networks and Fuzzy Systems: A Dynamical Systems Approach to Machine Intelligence,* Prentice Hall, Englewood Cliffs, NJ:

Lee, C. C. 1990. Fuzzy logic in control systems: fuzzy logic controller, *IEEE Trans. Systems, Man, and Cybernetics,* 20:404–435.

Pappis, C. P. and Mamdani, E. H. 1977. A fuzzy logic controller for a traffic junction. *IEEE Trans. Systems, Man, and Cybernetics,* 7:707–717.

M. Sugenon, 1985. An introductory survey of fuzzy control, *Information Sciences,* 36:59–79.

Zadeh, L. A. 1965. Fuzzy sets, *Information and Control,* 8:338–353.

Zadeh, L. A. 1973. Outline of a new approach to the analysis of complex systems and decision processes, *IEEE Trans. Systems, Man, and Cybernetics,* vol. 3.

Zadeh, L. A. 1975. The concept of a linguistic variable and its application to approximate reasoning, parts 1 and 2, *Information Sciences,* 8:199–249, 301–357.

Zadeh, L. A. 1978. Fuzzy sets as a basis for a theory of possibility, *Fuzzy Sets and Systems,* 1:3–28.

Zimmermann. H.-J. 1991. *Fuzzy Set Theory—and Its Applications,* Kluwer Academic Publishers, Norwell, MA.

84

Fuzzy Hardware

Mary Lou Padgett
Auburn University

84.1 Introduction 1103
84.2 Challenges and Rewards 1103
84.3 Approaches 1103
Objectives and Performance Measures • Models and Functions • Implementations: Tools and Resources • Input/Output • Architectures and Connections • Component Functions • Variable Parameters • Recall, Learning, Dynamics, Update Mechanisms
84.4 Futures 1110
84.5 Defining Terms 1110

84.1 Introduction

Soft computing solutions can be carefully designed to extend the capabilities of traditional computing, but are not intended to replace traditional methods completely. Some design issues are conventionally difficult to resolve, but make good targets for soft computing applications. Fuzzy systems and soft computing are particularly useful when the mathematical model is unknown or not easily implemented and when human intervention or experience can make solutions more reasonable or more rapid. These situations contrast with those in which the major objective is computational precision. A wide range of applications have facets where fuzzy solutions can improve the performance of the system. (See the paper by Pedrycz in this chapter, and Costa et al., 1995; Eichfeld et al., 1995; Kandel, 1995; Watanabe, 1993.)

84.2 Challenges and Rewards

Recent trends in fuzzy system hardware offer a wide range of choices between software for conventional processors, the inclusion of fuzzy logic function in the instruction set, fuzzy coprocessors, modular fuzzy chips (analog and digital) and application specific integrated circuits (ASICs).

Developers are challenged to produce flexible systems with fast through-put from crisp input to crisp output. The potential rewards are enormous. As the *March 1995 Special Issue of the IEEE Proceedings on Fuzzy Logic with Engineering Applications* indicates, fuzzy systems and soft computing offer practical (commercial) solutions to task-oriented control functions. Reliability and fault tolerance are also rising in commercial applications (Chand and Chiu 1995). Fuzzy pattern recognition is a rapidly-moving technology, discussed in the following articles of this chapter by Pedrycz and by Karayiannis. Fuzzy neural applications are very successful, and promise extended flexibility and commercial value (Zadeh, 1995). (See the articles in the Neural Networks and Fuzzy Systems chapters by Lau, Lindblad, Werbos, and Wolpert.)

84.3 Approaches

Approaches to the development of analog and digital fuzzy hardware frequently focus on fuzzy controllers. Many have a dedicated architecture aimed at high processing speed and efficient silicon usage (Kandel, 1995). Other approaches such as that of Pedrycz focus on flexibility. Deciding how to implement a fuzzy system implies making informed choices, which should be carefully documented for later evaluation.

Although every designer has favorite procedures for successful implementations, the following paragraphs describe some strategies suggested by various experts. Perhaps these charts and lists can help the reader tailor a set of check-lists to guide applications development. As an idea moves from concept to commercial product, the design team iterates through the specification of modules such as those in Table 84.1. As knowledge is gained, objectives and realizations are modified. Documentation of such transitions provides a solid basis of lessons learned, which may be helpful in the future. The literature is full of success stories, but paper and time constraints, as well as protective caution, often suppress the publication of, say, "the 137 ways that did not work before the following method was decided upon." A project team and its management can benefit from examining such lessons-learned for interesting phenomena and/or for future avoidance.

Suggestions for modification and improvement to the following charts may be sent to the CI Standards News (edited by the author) for public circulation and discussion via the Internet to m.padgett@ieee.org.

Table 84.1 outlines some of the paradigm specification modules suggested as design aides for fuzzy system implementation. These modules are:

0-8493-8343-9/97/$0.00+$.50

Table 84.1 Suggested Design Considerations for Fuzzy System Paradigms

PARADIGM SPECIFICATION MODULES				
OBJECTIVES and PERFORMANCE MEASURES				
MODELS and FUNCTIONS				
	Biological	Mathematical	Engineering	
TOOLS and RESOURCES for IMPLEMENTATION				
Hardware	Software	Personnel	Experts	Experience
INPUT/OUTPUT				
Data Sources	Characteristics	Interfaces	Dynamics	Validation
ARCHITECTURES and CONNECTIONS				
Number of Crisp and fuzzy inputs and outputs	Number and shape of membership functions	Number of Rules	Number of Antecedents / Consequents per rule	Precision (number of bits for discretizing
COMPONENT FUNCTIONS				
	Fuzzification: Computation- or Memory-oriented	**Fuzzy Inference Max-Min, max-dot, etc.**	**Defuzzification center of gravity mean of maxima centroid, etc.**	
VARIABLE PARAMETERS				
Inputs and Outputs: Number of Crisp and Fuzzy	Membership Functions: Number & Shape per I/O	Rules: Number	Antecedent / Consequent Number per rule	Precision e.g. bits for LOOKUP TABLE
RECALL, LEARNING, DYNAMICS, UPDATE MECHANISMS				
Interaction with Non-Fuzzy Algorithms	Policies	Teachers	Value Functions	Error Functions

1. Objectives and Performance Measures
2. Models and Functions
3. Tools and Resources for Implementations
4. Input/Output
5. Architectures and Connections
6. Component Functions
7. Variable Parameters
8. Recall, Learning, Dynamics and Update Mechanisms

Objectives and Performance Measures

The project objective should be clearly stated to the satisfaction of the customer and the engineers. This is a nontrivial task requiring skilled communication and documentation of ideas, intents, and desires. As a design progresses from conception (R&D research ideas) to proof of concept (demonstrations) to product (implementation), more knowledge is gained and this should be incorporated into the written objectives. Some of the best solutions occur when unexpected or "interesting" phenomena are encountered during product development. It is helpful to keep iterating though the assumptions and specifications, updating and clarifying them as needed. Many working engineers and project managers keep check-lists of potential problem areas, items needing to be documented or explained and notes on lessons learned. The following paragraphs and charts are intended to suggest starting places for a practical (iterative) design specification and documentation structure. Real utility of the suggestions will come as the specification modules and lists are tailored by the user to the application at hand.

Typical Objectives

Level. Objectives vary according to the focus of the project (see the papers by Werbos in the Neural Networks and in the Fuzzy Systems chapters). A micro-level project might aim to construct a chip to be used in a larger system or class of systems. Combining chips into boards and assembling them is a middle-level project. Learning to use the assembled equipment for a particular application is the higher level project type.

Micro
Middle (System/Task)
Application

Design Phase. The design phase also impacts the project objectives. It may be adequate to put together a rapid-prototype digital simulation of a fuzzy system to explore its capabilities before investing in more complicated or time-consuming phases of production. Design tools to help this type of endeavor are becoming more prevalent.

Research and Development
Proof of Concept
Commercial Product

Later articles in this chapter by Padgett will discuss design trade-offs for fuzzy modeling and for fuzzy-neural modeling.

Speed Required

Measurement of speed required is defined by [Costa et al. 1995] as follows:

> Speed parameters: measure performance in terms of system response time (time crisp out—time crisp in).

This contrasts with the commonly used but ambiguous term, FLIPS:

FLIPS: fuzzy logic inferences per second

Caution: This term can be confusing or misleading because:

- The fuzzy inference meaning depends on assumptions regarding the numbers and types of inputs/outputs, precision, rule format, etc.
- The outputs per second; processor throughput time, may be misleading, e.g., feedback loops where the output depends on future inputs so have to prime the pump before get meaningful output—pipelining may be processing inputs before necessary outputs are generated to be fed-back.

Real-time Constraints

Application-specific constraints place demands on the system dynamics. Some rough generalizations are given in Table 84.2, Fuzzy System Implementations.

Complexity Supported

Flexibility regarding complexity supported may be critical. On the other hand, a fixed application may be optimized for speed by carefully tuning a particular selection of capabilities to that application.

Inputs: number of crisp and fuzzy.
Outputs: number of crisp and fuzzy.
Membership functions: number and shape (per input/output).
Rules: number.
Antecedent/Consequence: number per rule.
Precision: number of bits for discretizing the span and the alpha values.
Algorithm variants: membership computation method, rule inferencing method, defuzzification method, etc.

Interaction with Nonfuzzy Algorithms

Interfaces may be external, impacting the timing, availability and characteristics of input and output to the fuzzy system. Internal interfaces may be used during fuzzification, fuzzy inference, or defuzzification. Sometimes fuzzy processing alternates with other algorithms, so switching mechanisms are needed. Planning ahead for potential interactions is desirable, but the actual impact of the algorithmic hybridization usually will stimulate modifications and tuning during the design cycle. Later in this chapter, Lindblad, Werbos, and Pedrycz offer interesting insights into the combination of neural and fuzzy systems for controls, pattern recognition and other applications.

Table 84.2 Fuzzy System Implementations (Miki and Yamakawa 1995; Costa et al., 1995)

CATEGORY	General Purpose (Regular Digital Microprocessor or DSP)	Fuzzy Logic Operations in Microcode	CPU Supported by Fuzzy Coprocessor	CPU with Fuzzy Operation at Hardware Level (Fuzzy CPU,Core)	Fuzzy Chip (Exclusive Fuzzy Inference Engine) Digital or Analog	Fuzzy ASIC's capable of standalone operations
FACTOR						
Processing Speed	SLOW (but FASTER with higher precision) LOOKUP TABLE fast but high storage, low complexity	MAY be faster than Regular Micro-processors	FAST Fuzzy Operations, SLOW Data Handling	FASTER Fuzzy Operation	FASTEST Fuzzy Operation	FASTEST, for a fixed application Inference Speed. 100 kHz-1MHz
Complexity Supported	LOW-8 b to HIGH- 32 b (complexity slows performance) LOOKUP TABLE 2-3 in, 1-2 out, low res. (4-8b)	HIGH (but slows performance)	MODERATE Usually- Triangular M, Flexible Rule, Max-Min Infr, 8-b precision, centroid defuz.	HIGH	DIGITAL-limited by storage space for LOOKUP TABLE ANALOG- less accuracy	Medium (15-20 gates)
R-T Constraints (Dynamics)	Relatively Slow (0 1-1 kHz)	1-10 kHz (CISC) 1-50 kHz (RISC)	10-100 kHz	FAST	FAST	
Interaction with Non-Fuzzy Algorithms	HIGH	HIGH	LOW also need Reg. Microprocessor	HIGH	VARIES	LOW also need Reg. Microprocessor
Flexibility	HIGH	HIGH	SOME	MODERATE	SOME	NONE
Cost of Equipment	LOW	LOW	HIGH	MODERATE	HIGH	LOW
Silicon Area					COMPACT	LOW COST (5-10K gates)
Modularity			MODULAR		MODULAR	Block in a Single-chip control system
Manpower/ Expertise Required	EASY	EASY			HIGH COST	EASY
Design Phase: R&D	EASY	EASY			HARD	
Prototyping	EASY	EASY			HARD	FAST (Field Programmable Gate Array (FPGA)
Low Volume Production						
High Volume Production						COST EFFECTIVE
Maintenance, System Support, Development Systems	HIGH	HIGH	SOME	HIGH		HIGH
Level	Micro to Full Application	Micro to Full Application	Middle - System / Task	Micro to Full Application	Micro-Modules	
EXAMPLES						
	FIDE (Aptronix)	Fuzzy-166 Firmware (Inform-CISC)	F^{2}RU-8 (Fujitsu)	Reconfig. Fz. Proc RFP Pedrycz et al	Miki & Yamakawa	
	fuzzyTECH (Inform)	R&D (RISC)	81C99 (Siemens)		Others	
	TIL Shell (Togai InfraLogic)		Omron FP series			
			ST WARP SGS-THOMSON(ST)			
			Togai Infra Logic FCA			
			Toshiba T/FC150			
		FLORA (DIBE& ST-RISC)	WARP (ST)			VHDL (DIBE)

Flexibility Needed—Which Are Variable, Which Are Fixed

Flexibility needed varies with design stage (concept, prototype or production), as well as, with the intended use for the product. Some considerations are listed below:

- Programmability
- Algorithm variants
- Precision, etc.

Comparison of general purpose versus specialized systems strengths should provide significant direction to the project:

Dedicated hardware plus: structure and relative simplicity of fuzzy processing algorithms.

General purpose hardware plus: flexibility and developmental support.

Models and Functions

Having determined (for the moment) the project objectives and performance measures, appropriate models and functions can be addressed. In the article on neural network design by Padgett, the developer was cautioned to distinguish between biological, mathematical and engineering models (Mead, 1994). For fuzzy system design, the biological model shifts from the inner workings of neural tissue to conceptual decisions, higher level brain activity.

Integration of human experience and decision-making patterns into a computerized system is a worthwhile challenge. Tight documentation of assumptions, transformations and data collection mechanisms is essential to the validation of these techniques, and to the acceptance of the resulting product. Fuzzy hardware design should ensure that these items are properly recorded, and that decisions to implement the model in one manner or another reflect the nature of the data and the designer's intent.

Methods such as the Analytical Hierarchy approach of Saaty (1980) have been in use for years. The author has worked since 1982 with data collection and computer implementation of decision aides, extrapolating this approach for government laboratories. Computation of eigenvalues based on unpredictable and voluminous human input is undesirable. Obtaining accurate input for a large number of paired comparisons is difficult, especially when the input is needed from busy executives and experts. Currently, the evaluative processes are being recorded by management as fuzzy numbers: best guess, low range and high range for current performance vs. best guess, low range and high range for subjectively gathered historical benchmark data. These human inputs are recorded as linguistic variables ranging from very low to very high. Five to nine subdivisions are recommended, based on extensive experimentation with human capabilities for distinguishing categories. The fuzzification process transforms these linguistics into numbers suitable for computer processing. The origin and accuracy of these numbers should be a recorded part of the modeling process, so that resources are not wasted. Savings in computational resources and speed can be realized by reducing the precision of calculations based on rough estimates and hunches. These resources can then be made available elsewhere in the model. Defuzzification reverses the transformation from computerized data to linguistics which can be interpreted by human experts, or by other facets of the application external to the fuzzy system. Most top management decision-makers want output that can be scanned in 30 seconds to tell them where to focus attention next. This summary/focusing preference for output applies to engineering applications as well. The defuzzification system selected should give a concise summary of conditions, flagging areas needing attention. A key to success is that after the need for attention is *detected*, enough tracking mechanisms should exist so that the area needing more attention can be *identified*, and then thoroughly *analyzed* so that *recovery* or the next appropriate action can be instigated. The existence of data records and mechanisms for this type of summary/focusing action should be an integral part of the design process.

Models should record the higher level, decision-type goals, the mathematical processes desired, and the difference between an abstraction such as a closed form solution to a system requiring integration or summation to infinity, etc., and the reality of an implementation in analog or digital hardware. Remembering that developing a fuzzy system is just enhancing a traditional engineering design should guide the precise development and documentation of the models. Table 84.1 suggests some of the design considerations which may help in the specification of a fuzzy system which can be validated and which will meet the user's needs. Tools and resources, input/output, architectures and connections, component functions, variable parameters, and algorithmic mechanisms can be chosen to implement the models based on the objectives and performance measures. The examples given below help illustrate these concepts.

Implementations: Tools and Resources

Given objectives, performance measures, models and desired functions, there are a number of possible implementations to consider. Suggested classes are listed below.

Classes (Costa et al., 1995; Watanabe, 1993)

General purpose components.

Regular Microprocessor
DSP

General purpose processors with specialized instructions dedicated fuzzy coprocessors.

Off-the-shelf fuzzy logic chips.
Special purpose microprocessor.
Parameterized module generator: semi-custom layout, full-custom layout, integrated CAD system.

Stand-alone fuzzy application-specific ICs (ASICs).

Semi-custom.
Full-custom.

Issues: General Purpose vs. Dedicated

As discussed above, the need for flexibility should be balanced against the need for speed, structure and relative simplicity:

Flexibility needed: programmability, membership functions, inferencing method, defuzzification method, rule format, precision, etc. to be supported in a dedicated fuzzy device

Dedicated hardware plus: structure and relative simplicity of fuzzy processing algorithms

As more detail about the needs of the project surfaces, decisions can be made about analog versus digital or hybrid implementations. Table 84.2 suggests some factors to investigate and consider carefully. It is a table of categories of implementation vs. performance measures and objectives. As with any set of generalizations, the groupings suggested in Table 84.2 are somewhat arbitrary, and exceptions exist. As this rapidly moving technology improves, new solutions to the problems shown in the table are found. The examples listed are detailed in the 1995 special issues of *IEEE Proceedings* and *IEEE Micro.* The interested reader can find detailed schematic diagrams, complexity supported and other information in these issues.

Input/Output

For the purposes of this fuzzy hardware article, the input/output stage of specification will be summarized by the list below:

- Data Sources.
- Characteristics.
- Interfaces.
- Dynamics.
- Validation.

Further suggestions regarding modeling of fuzzy and neural fuzzy systems are found in other articles in this chapter.

Architectures and Connections

Fuzzy architectures and connections include:

- Number of crisp and fuzzy inputs.
- Number of crisp and fuzzy outputs.
- Number and shape of membership functions per input/output.
- Number of rules.
- Number of antecedents/consequents per rule.
- Number of bits for discretizing, number of bits in digital chip, or precision of analog implementation.

These elements may be fixed or flexible, but they bind the project to a subset of possible implementations. High fan-in and fan-out can be traded for other considerations in various implementations. A considerable amount of detail about the architectures of various examples is provided in the special issues referenced above. In particular, Costa, (1995) and the article by Pedrycz in this chapter are helpful.

Component Functions

The traditional components of fuzzy systems are those dedicated to fuzzification, fuzzy inference, and defuzzification. Variations of fuzzy systems incorporating neural systems and other non-fuzzy algorithms abound. These will be addressed by other articles in this chapter. Examples of fuzzification, inference and defuzzification follow. These examples should not be taken to be an exhaustive list of possibilities, but just examples and suggestions.

Fuzzification

Fuzzification is the process of converting the crisp data value of a linguistic variable to a membership degree, or alpha value. The membership degree is related to the shape of the membership function. Figure 84.1 illustrates a triangular membership function implemented in a digital (discrete) manner and in an analog (continuous) manner (Kandel, 1995). Tradeoffs for digital/analog/hybrid implementations are discussed by Wolpert with regard to neural chips. With fuzzy systems, input values are inherently imprecise, so the lower precision and higher speed of analog implementations is more attractive. On the other hand, rapid-prototyping flexibility in digital systems and choice of complexity may be more important. In the early stages of a project, digital implementations may be more practical. Hardware-in-the-loop simulations are a traditional way of moving from digital explorations to very specialized analog implementations.

Fuzzy Inference

The fuzzy inference module of the fuzzy system inputs the alpha values computed or recalled by the fuzzification step. The rule base is used to deduce the fuzzy output. Two common methods are the max-min method and the max-dot method of fuzzy inference (Miki and Yamakawa, 1995).

Fuzzy Logic Functions.

Digital: Commonly used Min and Max operations may be implemented with a comparator and a multiplexer (Miki and Yamakawa, 1995), but are limited to two inputs. Increasing the number of inputs could be accomplished with a slower serial-mode circuit or a large-area tree-structure circuit.

Analog: Implementing an analog Max or Min circuit can be done easily, using the physical characteristics of transistors. Large fan-out and multiple inputs are easily accommodated (Miki and Yamakawa, 1995).

Defuzzification

The fuzzy inference engine produces a fuzzy value which must be interpreted. Some common defuzzification methods include (Miki and Yamakawa, 1995).

- Center of Gravity (COG).
- Centroid.
- Mean of Maxima.

Membership Function Example: Memory vs. Computation (Miki and Yamakawa 1995)

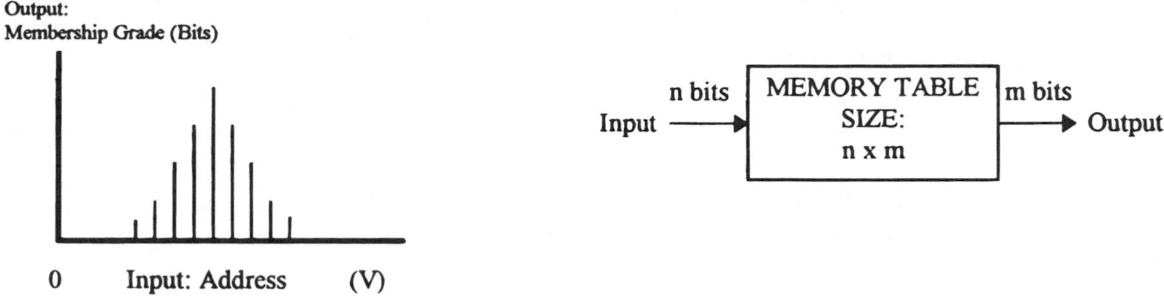

DIGITAL MEMBERSHIP FUNCTION: discrete function, table lookup, memory-oriented approach

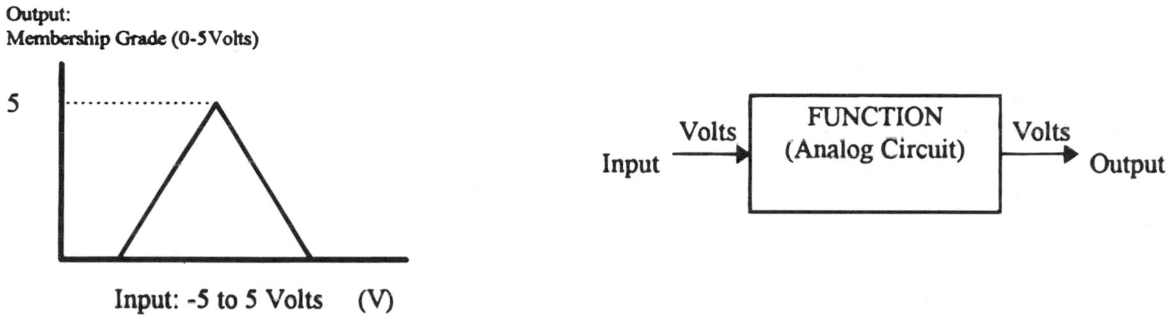

ANALOG MEMBERSHIP FUNCTION: continuous function, computational-oriented approach

Figure 84.1 Design trade-offs: digital vs. analog fuzzification example. Design Tradeoffs: Digital offers high resolution, complex function shapes and fast recall, but may use large amounts of memory. Analog offers a compact circuit, fast processing, and stores only function parameters, but choice of function shape is limited (Miki and Yamakawa, 1995). Please note that digital implementations may also use the computation-oriented approach.

For the COG method, input is the weight and membership grade of each element. The sum over all elements of the product of the element's weight in the universe of discourse and its grade of membership is divided by the sum over all elements of the membership grades This calculation slows as resolution (bit count) increases, and needs a larger circuit.

Substituting integrals for summations give the continuous version of this operation. The analog implementation is small and fast. It lacks the precision often desirable in crisp values, but special measures can be taken to increase this precision if justifiable.

Variable Parameters

The most obviously variable parameters are those impacting the complexity of the implementation, listed above. Other variable parameters internal to the component functions of fuzzification, fuzzy inference and defuzzification will depend on the particular choice of function. Sometimes these variable parameters are set in a rather arbitrary manner. Usually it is helpful in producing a generalizable design to annotate these choices (even if arbitrary). It is even better to design some type of experiment to test the sensitivity of the system to variations of these parameters.

In neural networks, training can be enhanced greatly by algorithmic adjustment of parameters such as gain and step size during training. Fuzzy systems can be used to implement the rules of thumb a designer frequently uses. Such an implementation produces a documentable and reproducible method of "tweaking" system parameters. The prospect of adding to the body of knowledge about the process in also enhanced. Likewise, neural learning of the reactions of a fuzzy system to its use and interactions with the environment can help "tune" the fuzzy rules, and suggest other systematic and sensible variations of system parameters.

Current research into "Soft Learning Vector Quantization" (Karayiannis, 1996) is an example of the rapid developments in this field.

Recall, Learning, Dynamics, Update Mechanisms

As often noted by Paul Werbos, proper calculation of derivatives and keeping track of time steps carefully is essential for successful soft computing. In particular, when fuzzy systems interface with other systems, the dynamics must be considered carefully.

Combining reinforcement learning with fuzzy systems is an exciting horizon in the development of commercial applications.

A few possibilities for such interactions are described in this chapter by Werbos. Pattern recognition futures are covered by Pedrycz. Further extensions of fuzzy hardware and interfaces with non-fuzzy algorithms are summarized in articles by Padgett.

84.4 Futures

Flexibility and speed from crisp input to crisp output are the prime criteria mentioned by most fuzzy hardware experts and users. Hardware fuzzy systems of the future may offer a set of rapid-prototype design tools to help developers move from digital simulations to hardware-in-the-loop simulations to working prototypes. Commercial applications will vary from flexible designs for systems needing to respond to changing conditions and subject to upgrade, to fixed applications for fast, specialized fuzzy processing. The combination of neural learning and fuzzy systems is very promising. Pedrycz's article in this chapter provides more detail about his fuzzy neural hardware suggestions.

84.5 Defining Terms

Application-specific project: Combining micro and system/task level elements to solve a particular problem or class of problems.

Antecedent: First part of a fuzzy rule, concerning the inputs or premises (IF part).

Consequent: Second part of a fuzzy rule, concerning outputs or conclusions (THEN part).

Crisp set: Membership is binary. Elements have zero membership or total membership.

Fuzzification: A process that inputs the crisp data value of a linguistic variable and converts it to a membership degree (alpha value).

Fuzzy rules: Statements having an if/then structure and using linguistic values for input and output variables. See also, antecedent, consequent.

Fuzzy set: Membership is continuous. Elements may have membership ranging from zero to total.

Fuzzy system: Process unusually having three steps: fuzzification, fuzzy inference, defuzzification

Fuzzy inference: Process that deduces a fuzzy output value using the rule base and alpha values. May evaluate the entire rule base, with regard to the contribution of the single rule. See also fuzzy system.

Defuzzification: Process that converts fuzzy output to crisp output. In controls, this is usually the value of the control action.

Membership: Degree to which an element belongs to a set. A crisp set has binary membership. A fuzzy set allows partial membership, ranging from none to total.

Micro-level project: Having a generic purpose, say to create a building block with a simplistic function.

Rule base: The whole set of rules involved in a fuzzy application.

System/task-level project: Having a supervisory, decision-making role (most commercial controls applications currently fall in this category.)

Further Information

Many examples of fuzzy systems implemented in hardware can be found described in detail in Watanabe (1993), Costa (1995), and Kandel (1995). Some of these are listed and briefly described below:

Examples of Fuzzy Chips in Watanabe (1993)

Products

- Togai InfraLogic.
- Omron Corp.
- Oki Electric.
- Fujitsu Device.
- Olympus Camera.
- NeuraLogic.

R&D (Technical Papers Published)

- TI.
- Seimens.
- Mitsubishi Electric.
- Thomson (in Italy).
- Toshiba.
- Nissan Motors.
- Tokyo Gas (analog circuits).

Examples in the Micro Special Issue Kandel (1995)

- Analog chip.
- A processor.
- Digital systems.
- Rule-based systems for use on a general-purpose processor.
- A controller.

Specific Chips Discussed

Yamakawa.

- 1980 bipolar fabrication, discrete components.
- 1984–85 Yamakaw and Miki first fuzzy logic chip using PMOS and CMOS.

Togain and Watanabe 1984–85.

- First fuzzy inference engine on a VLSI chip.
- 250,000 rules per second.

Pedrycz et al.

- Reconfigurable fuzzy processor (RFP) for aggregative (Or, And) and referential (Matching, Difference, Dominance and Inclusion) operations.
- Flexible structure and parametric.
- RFP: 3 main units.

1 The RFP.
2 A learning unit with fuzzy back-propagation learning.
3 A local memory unit.

- Fuzzy Neural Network as a bidirectionally linked series of shared buses gives a modular and scaleable design, easy to interface.

Benefits: simple, versatile new high-density, user-programmable logic devices, powerful CAD tools, gives high performance fuzzy systems.

Hung. Dedicated digital hardware gives improved performance over systems based on general-purpose computing machines (Hung).

Surmann and Ungering.

- General-purpose processors: concepts for fast processing on general-purpose processors.
- Speed is very important for interfacing with neural networks and genetic algorithms.
- Analyze bottleneck in terms of I/O.

Vidal-Verdu and A. Rodruguesz Vazquez. Parallel architecture for fuzzy controllers and a methodology for their realization in the form of analog CMOS chips:

1 Learn with adaptation of electrically controllable parameters.
2 Use a dedicated hardware-compatible learning algorithm.
3 Simplify at the circuit level for increasing process complexity and operating speed.
4 Use a 3 input four rule controller chip in 1.5 micron CMOS, single-poly double-metal technology.

IEEE Proceedings Special Issue Costa et al. (1995)

Topics Discussed

- Targeting of different approaches to different application domains or market areas.
- Design trade-offs.
- Examples of dedicated fuzzy coprocessors, RISC processors with specialized fuzzy support and application specific fuzzy ASICS.
- Example applications: control, decision making tasks based on medical diagnosis, business forecasting, image processing and computer vision, more.

Companies Referenced in Costa et al. (1995)

Aptronix Inc., FIDE Application Notes, 1992, San Jose, CA, USA.
Inform Software Corp., Evanston, IL 60201, USA.
INFORM Inc., Fuzzy-166 Processor—The Chip for Flexible High-Performance Fuzzy Solutions. INFORM GmbH, Aachen, Germany.
Motorola Inc., Motorola Semiconductor Technical data, 1992.
Neuralogic Inc., Neuralogic Data Sheets, 1992.
OMRON Corp., Tokyo 140, Japan.
SGS-THOMSON Inc., WARP Users Manual, Mar. 1994.
Siemens GmbH., Siemens Data Sheets, May 1993.
Togai InfraLogic Inc., Irvine, CA 92718.
Toshiba Corp.

References

Chand, S. and Chiu, S. L. 1995. Scanning the issue: Special issue on fuzzy logic with engineering applications, *Proc. IEEE,* 83(3):343–344.

Chiaberge, M. L. and Reyneri, M. 1995. Cintia: a neuro-fuzzy real-time controller for low-power embedded systems, *IEEE Micro,* 15(3):40–47.

Costa, A., De Gloria, A., Faraboschi, P., Pagni, A., and Rizzotto, G. 1995. Hardware solutions for fuzzy control, *Proc. IEEE,* 83(3):422–434.

Eichfeld, H., Klimke, M., Menke, M., Nolles, J., and Kunemund, T. 1995. A general-purpose fuzzy inference processor, *IEEE Micro,* 15(3):12–17.

Kandel, A. 1995. The fuzzy boom, *IEEE Micro,* Special Issue, August, pp. 6–7.

Mead, C. 1994. Personal communication.

Miki, T. and Yamakawa, T. 1995. Fuzzy inference on an analog fuzzy chip, *IEEE Micro,* Special Issue, August, pp. 7–18.

Saatay, T. L. 1980. *The Analytic Hierarchy Process.* McGraw-Hill, New York, NY.

Watanabe, H. 1993. *Hardware Systems for Fuzzy Inference,* Second IEEE FUZZ Tutorial #7, IEEE Press, Piscataway, NJ.

Zadeh, L. 1995. Personal communication.

85

Fuzzy Modeling and Applications: Controls, Visions, Decisions

Mary Lou Padgett
Auburn University

85.1 Introduction 1112
85.2 Engineering Approaches 1112
System Identification in a Fuzzy System • Dynamics
85.3 Futures 1115

85.1 Introduction

Fuzzy modeling and applications are a natural combination. In 1965, when the author was a student of Mathematical Topology, Lotfi Zadeh's non-boolean logic premises formed an intriguing part of the theorem-proving course. His concepts were considered a natural extension of the set theory being studied, but no practical use was extrapolated. Years later, as a working engineer studying the simulation of industrial processes, the extremely practical nature of fuzzy systems and soft computing became apparent to the author. Attempting to model a real-world forest harvesting process based on data, observations and conversations with operators of feller bunchers and skidders illuminated the fallacies of many traditional models based on assumptions of Gaussian or other distributions. As the work progressed, many heated discussions took place among the researchers and the domain experts regarding the validity of standard assumptions and modeling techniques. The resulting compromise took the form of a study to determine the level of detail needed in different modules of the simulation. The most perplexing part of the data analysis involved the seemingly erratic time-to-shear a tree. No standard distribution fit the empirically derived histogram for this data, and the sensitivity analysis indicated that maximum detail was essential for this part of the model. Monte Carlo simulation of the empirical histogram was therefore employed, even though its inclusion was expensive in terms of system resources and time. Lack of generalizability was a concern. The solution to this strange histogram's origin was found on a field trip and interviews with operators of feller-bunchers. If an attempt to shear was successful on the first try, the time-to-shear was short (usually the case). Longer time-to-shear was associated with second attempts by the feller buncher operator to grab a tree with its collecting arms, and pinch it off at the base. Sometimes a third attempt was necessary. Factors like swampiness of the ground causing very thick tree bases, and percentage of hardwood vs. pine in the collecting arms already (blocking the operator's vision and resisting the addition of a new tree) increased the frequency of second and third attempts to shear. Once these rules-of-thumb, known all the time to the feller buncher operators, became part of the recorded data history, the patterns in the empirical histogram became very explainable. If the Monte Carlo simulation of the empirical histogram had been replaced by a fuzzy system based on operator knowledge, a much faster and more generalizable simulation would have been produced. Instead of a set of heuristics embedded in an enormous simulation code, a rule base would have existed. This rule base and its application to the system could then have been used to validate the model and provide information for potentially improving the harvesting process, or at least giving better estimates of time to harvest for the timber manager (Webster et al., 1984, 1983; Padgett et al., 1988; Hines et al., 1982).

Many industrial models can be improved by restating heuristics embedded in the code as fuzzy systems. Careful design and analysis of such systems is a practical solution to addressing problems of high complexity where human knowledge should be incorporated into the electronics.

85.2 Engineering Approaches

Engineering approaches to fuzzy modeling benefit from the application of traditional scientific methods and design strategies developed for software engineering and for design of complicated (layered, hierarchical) circuits. The differences in approach result from the nature of the problems being addressed. Complexity and human decisions are not as easy to fold into a computerized system as a closed form solution to a computational problem might be. Instead of focusing simply on the computational aspects of a problem, the designer of a fuzzy system should carefully investigate the nature of data and assumptions.

Building on the material presented in the earlier article on Fuzzy Hardware by Padgett (in this chapter) this article will focus on other aspects of specification. (See Table 85.1)

0-8493-8343-9/97/$0.00+$.50

Table 85.1 Suggested Design Considerations for Fuzzy System Paradigms

PARADIGM SPECIFICATION MODULES				
OBJECTIVES and PERFORMANCE MEASURES				
MODELS and FUNCTIONS				
TOOLS and RESOURCES for IMPLEMENTATION				
INPUT/OUTPUT				
Data Sources	Characteristics	Interfaces	Dynamics	Validation
ARCHITECTURES and CONNECTIONS				
Number of Crisp and fuzzy inputs and outputs	Number and shape of membership functions	Number of Rules	Num.of Antecedents / Consequents per rule	Precision (number of bits for discretizing
COMPONENT FUNCTIONS				
	Fuzzification: Computation- or Memory-oriented	**Fuzzy Inference Max-Min, max-dot, etc.**	**Defuzzification COG, Mean of Max., Centroid, etc.**	
VARIABLE PARAMETERS				
Inputs and Outputs: Number of Crisp and Fuzzy	Membership Functions: Number & Shape per I/O	Rules: Number	Antecedent / Consequent Number per rule	Precision e.g. bits for LOOKUP TABLE
	RANGE of each variable	**Membership Profile for each Variable Range**	**Rules and Actions Needed**	**[Kartalopolous 96]**
RECALL, LEARNING, DYNAMICS, UPDATE MECHANISMS				
Interaction with Non-Fuzzy Algorithms	Policies	Teachers	Value Functions	Error Functions
DYNAMICS of Membership Grade	**DYNAMICS of Membership Function**	**TIME INTERVAL Definition**	**TEMPORAL Fuzzy Operator Specification**	**[Kartalopolous 96]**

Suppose the design has progressed to examination of the Input/Output, Architectures and Connections and Component Functions.

As detailed in Yaver and Filev, (1995), expert opinions can be gathered and stated as rules. Instead of relying only on these rules of thumb, experimental data can be used to validate and tune these rules. Use of input-output data can be regarded as a form of system identification.

System Identification in a Fuzzy System

1 Structure identification.

- I/O values.
- Structure of the rules.
- Number of rules in the rule base.
- Partitioning of the input and output variables into fuzzy sets (Sugeno and Yasukawa, 1993).

2 Parameter identification.

- Estimation of the membership functions of the fuzzy sets, or
- Estimation of the fuzzy relation associated with the fuzzy model.
- Learning fuzzy model parameters with neural networks techniques.
- **Very successful** technique: Takagi-Sugeno-Kang (TSK) method (Takagi and Sugeno, 1985).

Structure Identification

The structure identification part of system identification may be based on template linguistic values, or by clustering the input-output space. In the first method, expert knowledge and data are combined. The expert provides linguistic values used to postulate potential rules for the system. The weights or credibilities of the rules are obtained from the data. Tong and Kosko originated techniques like this.

When expert templates are not available, clustering the input-output space based on available data is in order. Here, clustering and especially Yaver's mountain clustering technique may aid in estimation of the relationship between variables, rough estimates of the membership functions and number of rules and their importance.

Parameter Identification

Recent presentations by Bezdek and others reiterate the success found when using the T-S-K method of parameter identification. A good tutorial presentation of its use is found in Yaver and Filev (1995).

Using neural network systems to aid in parameter identification is also a promising technique. It is worth noting that even those combining neural and fuzzy approaches applaud the success of the very simple T-S-K technique.

Dynamics

The proper identification of fuzzy system dynamics is of critical importance (Werbos, 1996). Errors in derivative calculation have plagued developmental efforts for many. The works of Werbos, White and Sofje (1992) and Gupta and Sinha (1996) offer suggestions and explicit diagrams. A new text by Kartopoulas (1996) provides a concise summary of temporal fuzzy logic and suggestions for its implementation.

85.3 Futures

Many current applications summaries report success with the T-S-K method compared to some others.

Future developments in fuzzy modeling may emphasize the integration of fuzzy systems with neural systems and evolutionary systems. Soft computing holds the promise of letting the applications engineer combine the strengths of many techniques.

Standardization of design procedures, or at least development of check-lists for documentation of such procedures is being discussed vigorously in the CI Standards News. To contribute, contact m.padgett@ieee.org.

References

Bezdek, J. 1994. An Introduction to Fuzzy Logic, *WCCI Tutorial #7,* IEEE Press, Piscataway, NJ.

Galicher, S. and Roulloy, L. 1995. Fuzzy controllers: synthesis and equivalences, *IEEE Trans. Fuzzy Systems,* 3(2):140–148.

Gupta, M. M. and Sinha, N. K. 1996. *Intelligent Control Systems: Theory and Applications,* IEEE Press, Piscataway, NJ.

Hines, G. S., Padgett, M. L., Webster, D. B., and Sirois, D. L. 1982. Statistical methodology for forest harvesting model development, *Wood Science,* 14(4):178–187.

Jani, Y. 1995. Fuzzy logic in control, *SPIE's AeroSense '95,* SC53, Orlando, FL.

Kartalopoulos, S. V. 1996. Time-Dependent Fuzzy Logic, *Understanding Neural Networks and Fuzzy Logic: Basic Concepts and Applications,* 130–152, IEEE Press, Piscataway, NJ.

Kaufmann, A. and Gupta, M. M. 1991. *Introduction to Fuzzy Arithmetic,* Van Nostrand Reinhold, New York, NY.

Padgett, M. L., Webster, D. B., Hines, G. H., and Sirois, D. L. 1983. Southern forest timber harvesting computer simulation model: feller-buncher module, *Simulation,* 40(1):39.

Padgett, M. L. ed., 1993. *Simulation: Special Issue on Neural Networks—Model Development for Applications,* SCS Press. San Diego, CA.

Padgett, M. L. 1993. *Simulation: Special Issue on Computational Intelligence in Simulation Application,* SCS Press, San Diego, CA.

Padgett, M. L. 1993. Neural Network Basics: Applications, Examples and Standards, *IJCNN Nagoya Tutorial Book,* 383–412, IEEE Press, Piscataway, NJ.

Padgett, M. L., Karplus, W. J., Deiss, S., and Shelton, R. 1994. Computational intelligence standards motivation, current activities and standards, *Computer Standards and Interfaces,* 16:185–203.

Padgett, M. L. and Padgett, W. D. 1995. Simulation and Computational Intelligence in Real-World Applications, *Simulation,* July, 1995.

Sandri, S., Dubois, A. D., and Kalfsbeek, H. W. 1995 Elicitation, assessment, and pooling of expert judgments using possibility theory, *IEEE Trans. Fuzzy Systems,* 3(3):313–335.

Sugeno, M. and Yasukawa, T. 1993 A fuzzy-logic-based approach to qualitative modeling, *IEEE Trans. Fuzzy Systems,* 17–31.

Takagi, T. and Sugeno, M. 1985. Fuzzy identification of systems and its application to modeling and control, *IEEE Trans.,* SMC (15):116–132.

Terano, T., Asai, K., and Sugeno, M. 1989. *Applied Fuzzy Systems,* Academic Press, San Diego, CA.

Terano, T., Asai, K., and Sugeno, M. 1989. *Fuzzy Systems Theory and Its Applications,* Academic Press, San Diego, CA.

Webster, D. B., Padgett, M. L., and Sirois, D. L. 1983. Features of a felling module using a feller buncher in a timber harvesting computer simulation model, *Forest Products Journal,* 33(6):11–16.

Webster, D. B., Padgett, M. L., Hines, G. S., and Sirois, D. L. 1984. Determining the level of detail in a simulation model—a case study, *Computers in Industrial Engineering,* 8(3/4):215–255.

White, D. A. and Sofge, D. A. 1992. *Handbook of Intelligent Control: Neural, Fuzzy, and Adaptive Approaches,* Van Nostrand Reinhold, New York, NY.

Yaver, R. R. and Filev, D. P. 1994. *Essentials of Fuzzy Modeling and Control,* John Wiley & Sons, New York, NY.

Yen, J., Langari, R., and Zadeh, L. A. 1995. *Industrial Applications of Fuzzy Logic and Intelligent Systems,* IEEE Press, Piscataway, NJ.

Zadeh, L. A. 1995. Fuzzy logic and calculi of fuzzy rules, fuzzy graphs and fuzzy probabilities, *WCNN95 Tutorials,* INNS.

Zadeh, L. A. 1965. *Fuzzy sets, Inform, and Control,* 8:338–353.

86

Fuzzy Logic Control: Basics and Applications

Robert N. Lea
Ortech Engineering, Inc.

Yashvant Jani
Hitachi America Ltd.

Joseph A. Mica
NASA/Goddard Space Flight Center

86.1 Introduction 1116
86.2 A Simple Example of Fuzzy Logic Control 1117
86.3 The Example of the Inverted Pendulum 1118
System Components and Control Objectives • Controller Definition- • Conventional Approach • Fuzzy Logic Approach
86.4 Remote Manipulator System 1122
86.5 Collision Avoidance 1123
86.6 Summary 1124

86.1 Introduction

Fuzzy logic deals with the imprecision of natural world data which cannot be described by precise equations and traditional logic. It is constructed around the idea of a set in which the membership is a matter of degree rather than yes or no. Though known for more than 30 years, the value of fuzzy logic has been appreciated only recently through applications mainly in consumer appliances in Japan. Fuzzy logic principles have been investigated at Johnson Space Center in the late 80s, for several space applications such as translational and rotational control of spacecraft (Lea and Jani, 1992b, Lea et al. 1991a, 1991b), camera tracking systems (Lea et al., 1992c; Lea and Jani, 1992a, 1995a), auto focusing of microscopes (Tren and Weiss, 1991), temperature control in a payload module (Lemback, 1992), tether length control (Lea, et al., 1992d; Stefano, 1991; Copeland et al., 1991), and robotic manipulator control (Lea et al., 1995b; Lea et al., 1993). Since then fuzzy logic has also been explored at several other NASA centers, including the Goddard Space Flight Center, Ames Research Center, and the Jet Propulsion Laboratory. Currently many United States industries are considering the application of fuzzy logic in the development of consumer goods. Some of these include Westinghouse, General Electric, General Motors, and Ford Motor Company, as well as Hitachi America and Ortech Engineering which was formerly the Technology Systems Division of Togai Infralogic. Other government agencies, such as the Environmental Protection Agency, the Department of Defense, and the Department of Commerce, are also funding, or have funded, fuzzy logic research.

Neural Networks and fuzzy logic present the next step in computerizing the human thought processes. Rule based expert systems allow one to develop computer programs in a manner that relate rules to numbers. Fuzzy logic takes the next step by relating rules to fuzzy sets. This fuzzy logic attribute allows the capture of the human thought processes in an optimal manner for automation. For example, if a car is going too fast and the driver finds it necessary to slow down, braking control consists of fuzzy sets defined over a graded range of decelerating braking speeds. Fuzzy logic provides the ability to handle control problems where there is uncertainty due to complex dynamics of an environment. Fuzzy logic can be used to intuitively control the system as long as insight about the system behavior exists in the operators mind. Neural networks and fuzzy logic jointly provide the ability to address the mechanization in computer software and hardware of very difficult control and pattern matching problems in a more natural and optimal manner.

Lotfi Zadeh first developed the concepts of fuzzy sets (Zadeh, 1965) in 1965 and established these concepts firmly into fuzzy logic during the 70s with other pioneers (Kosko, 1991; Klir and Folger, 1988; Zimmermann, 1991; Dubois and Prade, 1980; Zadeh, 1973, 1968). Contrary to its name, fuzzy logic is a precise subdiscipline in mathematics that enables mathematicians and engineers to utilize human like thinking in decision making processes. Handling imprecise information is easier in the fuzzy logic architecture than it is in conventional logic; however, it was not accepted in the U.S. for a long time simply because the word 'fuzzy' implied unclear. It was the impression that this logic explains decision making processes by using magic rather than sound mathematical principles. Control engineers argued that the desired control could be achieved using existing control theory principles until the triple inverted pendulum was balanced using fuzzy logic principles. There is not an adequate treatment of this problem using conventional logic as yet. The real utility of this logic was shown by Japanese engineers (Shingu and Nishimori, 1989; Tobi et al., 1989; Yasunobu and Miyamoto, 1985) when they applied these principles to subway control, automatic transmission control, camera focusing, and many more applications. These applications showed the most interesting aspect to U.S.

business people—using this logic saved developmental time and costs. It is only recently that U.S. business has taken any interest in this field:

- First Industrial Conference on Fuzzy Logic Systems sponsored by MCC, the consortium of corporations, June, 1990, Austin, TX.
- First and Second International Workshop on Industrial Applications of Fuzzy Control and Intelligent systems, sponsored by Texas A&M University, College Station, TX, November 1991 and December 1992.
- IEEE International Conference on Fuzzy Systems, *FUZZ-IEEE '92* and *'93*, sponsored by the IEEE Neural Network Council, San Diego and San Francisco, CA, March 1992 and 1993.

Investigations in the areas of fuzzy logic and neural networks have been underway since 1984–1985, as reported at Tech2000 conference (Lea and Jani, 1990b), at the Johnson Space Center (JSC). The utility of this logic in several space applications, especially in autonomous operations, has been shown utilizing high fidelity simulations (Lea and Jani, 1992b). The JSC objectives were to investigate new technologies for control and decision making processes, particularly, the feasibility of applying these technologies to space operations to achieve desired operational efficiency and reduce overall life cycle costs.

In the next section the basic philosophy of fuzzy logic control is discussed and in subsequent sections, example applications of fuzzy logic are described.

86.2 A Simple Example of Fuzzy Logic Control

Fuzzy logic is a rule-based approach to control which is particularly suited to complex systems where good mathematical models cannot be developed or require to much time to develop. The reason fuzzy logic based control works well, where classical rule based expert systems tend to fail, is due to the nature of fuzzy logic inferencing which computes degrees of existence, or criticality, of a problem, or the degree of necessity for a control action. It is a natural approach in the design of control systems to automate functions that historically have been performed by human beings based on their evaluations of information from sensors and information from other humans. People placed in this type of control situation develop rules that they follow, but these rules are usually of a fuzzy nature such as: if the temperature is *very high* and the pressure is *high* then open the release valve *significantly.*

The italicized words have fuzzy interpretations, i.e., *very high* typically does not mean precisely greater than or equal to a predetermined temperature, but more likely it implies a predetermined temperature about which the operator's concern shifts from slight to extreme that the temperature is *very high.* For example, consider the graph in Figure 86.1.

At the current indicated value of *T,* the temperature would be interpreted to be *high* to degree approximately 0.6, whereas it would be *very high* to degree 0.3. Consequently the rule would

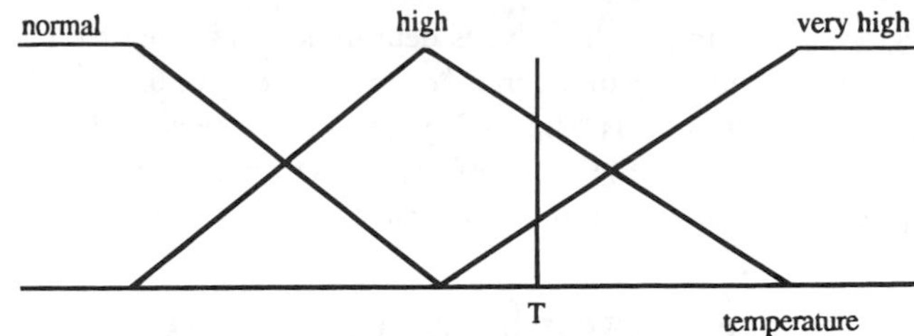

Figure 86.1

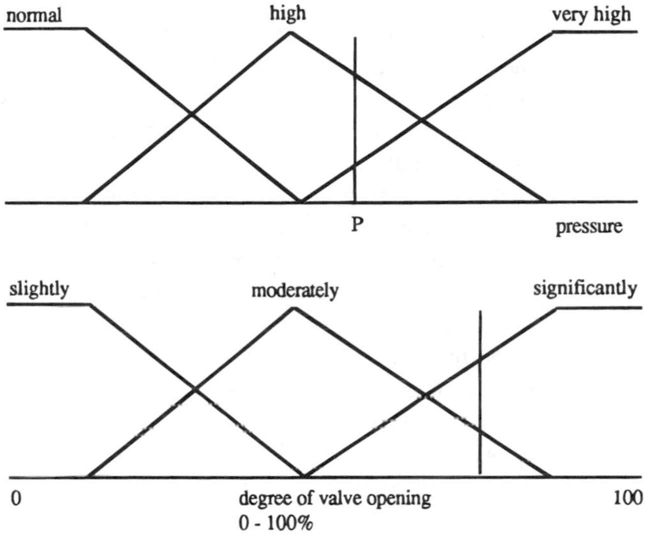

normal
high
very high
P
pressure

slightly
moderately
significantly
0
degree of valve opening
0 - 100%
100

Figure 86.2

fire at less strength than a rule, for example, that states, "if temperature is *high* and pressure is *high* then open the release valve *moderately*" where fuzzy sets representing pressure and percent of valve opening are given in Figure 86.2. Note here that the first rule and the second rule both have pressure, *P, high* to degree approximately 0.7, estimating from the graph, whereas the first rule has temperature, *T, very high* to degree approximately 0.3 and the second rule has temperature, *T, high* to degree 0.7. However, even though one rule has a weaker strength than the other, they both fire and contribute to the strength of the control action.

To illustrate how these rules are combined, we must have an intersection operator, *and,* and a union operator, *or.* We will take these to be the operators originally suggested by Zadeh in his seminal paper, although many others have been proposed and studied for different applications (Negoita, 1985). We choose Zadeh's operators since they are particularly simple to apply, and have been used successfully in many applications. In this definition the intersection of two fuzzy sets is taken to be the minimum of the degrees to which each of the fuzzy set conditions are satisfied, while the union is taken to be the maximum of the degrees. Therefore, the degree to which temperature $= T$ and pressure $= P$ satisfies temperature is *very high* and pressure is *high* is the minimum of $\{0.3, 0.7\}$ which is 0.3. On the other hand, the degree to which temperature is *high* and pressure is *high* is the minimum of the set $\{0.7, 0.7\}$ which is 0.7. Consequently the degree of opening of the valve is determined as in Figure 86.3, where the *moderately* function is "clipped" at degree 0.7 and the fuzzy set *significantly* is "clipped" at 0.3. Taking the *or* of the two functions and using the center of gravity defuzzification method we get a degree of opening as indicated at about 60%.

86.3 The Example of the Inverted Pendulum

In this section we discuss the familiar inverted pendulum mounted on a cart that moves in such a way as to balance a pole that is attached with a single in plane gimbal device. We will discuss the basic concept, system components, objectives of the control, parameters measured and used as input for the controller, and controller output.

The inverted pendulum is a classical control problem in the sense that it requires continuous control for all states except for the perfect one which is hard to achieve. The state which does not require control is not maintainable under normal operating conditions, and the slightest perturbation force to the state causes the pendulum to migrate into an unstable region. Our purpose in discussing this example is many-fold. First of all, it is well known, and many readers have had experience with this problem. Second, it is easy to understand the intuitive control process and generate a fuzzy rule-base using common sense. However, classical control techniques require a mathematical background that many users may not have or be inclined to deal with. Third, we want to show that the process of generating this rulebase can be based on experience with no detailed mathematical analysis necessary. Humans in their childhood can learn how to balance a stick in their palm without knowing state space equations. Only a few basic intuitive rules are used and tuned to perfect the balancing act.

System Components and Control Objectives

The inverted pendulum system (Figure 86.4) consists of a cart that can be pushed in two directions, and a pole that is situated in the center of the cart with a mass at the top. The objective is to balance the mass at the top of the pole by moving the cart from side to side. If the mass starts falling to the right side then push the cart to the right side to stop or reverse the falling rate. Typically, the cart is pushed so that the falling rate is reversed and an angle from the vertical which is close to zero is achieved. In terms of parametric values, the desired angle and the desired rate are zero. If the initial angle and initial rate, both, are zero then the balanced position of the pole is maintained in the absence of any other perturbation. However, if there is a slight perturbation, then, the force of gravity increases the rate and forces the angle to one side or the other of the vertical. If no control action is taken, the pole falls down, and the objective is not achieved. If the starting position is not zero angle and zero rate, then, control is required immediately.

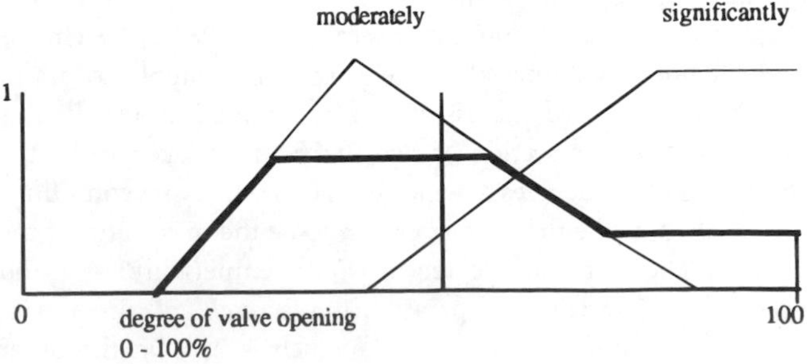

Figure 86.3

Once the pole is off-balance and control action starts, the system dynamics are such that continuous control actions are required. As we will see later the desired state of zero angle and zero rate can be achieved by adjusting the control actions continuously, however, the perturbations typically prohibit this.

Controller Definition

When a human balances the pole, he observes the state of falling and the direction in which the pole is falling. When any automatic system is controlling the pole balancing, we need to provide it with some intelligent inputs so that it can make decisions. Since the desired angle and rate both are zero, the state is defined in terms of angle and angle rate. Thus, the input to the controller are two measurements: angle and angular rate. Within the system, we can measure rate, and then, integrate it to obtain the angle. Thus, we can have a rate gyro as a sensor. If we have an angle measurement, then, we can differentiate the angles measured at two consecutive times and derive the rate measurement. Or, we may decide to use two different sensors and get angle as well as rate as independent information. In any case, the controller inputs are angle and rate. The controller output is the force given to the cart. This force results in the acceleration of the cart, which is coupled with the angular acceleration and thus correct force will provide desired angular acceleration to reduce the rate and angle. Thus, the controller definition is very simple, with angle and rate measurements as inputs and force as the output, regardless of the type of controller, as indicated by the following diagram (Figure 86.5).

Conventional Approach

The conventional approach based on state space theory is to write the equations of motion where the angle, angular rate, displacement x and velocity x-dot are coupled through gravitational force. The set of equations should also include modeling of the friction between the surface and the wheels. The mass of the cart and the mass at the top of the pole should be considered, however, the pole mass is usually assumed to be zero, or is assumed to be included in the mass at the top. Furthermore, the mass distribution is not considered, and point mass design is accepted for analysis. Otherwise the equations become even more complex. A set of nonlinear equations that represent the inverted pendulum model is

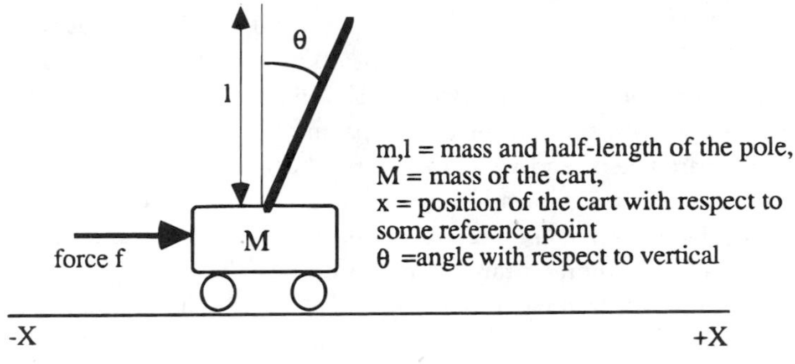

Figure 86.4 Inverted pendulum configuration.

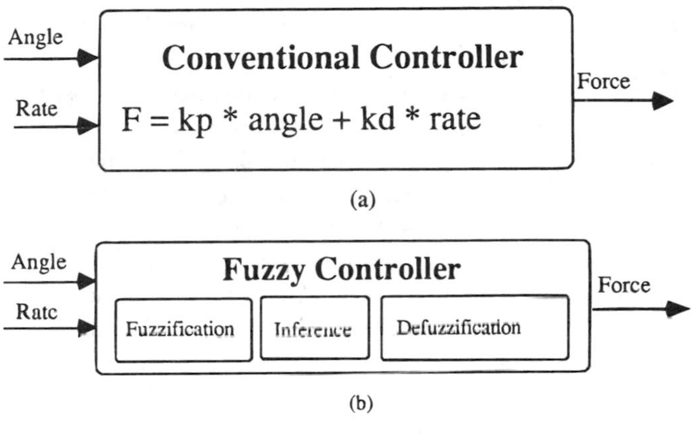

Figure 86.5

$$\ddot{\theta} = \frac{g \sin\theta + \cos\theta\left[\dfrac{-f - ml\dot{\theta}^2 + \mu_c \operatorname{sgn}(\dot{x})}{(M+m)}\right] - \dfrac{\mu_p\theta}{ml}}{l\left[\dfrac{4}{3} - \dfrac{m\cos\theta^2}{(M+m)}\right]}$$

$$\ddot{X} = \frac{f + ml[\dot{\theta}^2 \sin\theta - \ddot{\theta}\cos\theta] - \mu_c \operatorname{sgn}(\dot{x})}{(M+m)}$$

where μ_c and μ_p are coefficients of friction for the cart and pole, respectively. The other parameters are defined in the inverted pendulum diagram (Figure 86.4)

The next step is usually to linearize these equations by neglecting terms which are less significant, possibly neglecting the friction terms. Once we linearize the equations, we can write control equations and determine the gains. Gains are the constant values in the controller which are used to convert errors into control actions. For the inverted pendulum, the desired state is zero angle and zero rate. Therefore, any non-zero angle measured by a sensor automatically becomes the error in the angle. Similarly, the rate measured by a sensor is rate error. The following two sets of equations are linearized, the first set, including friction, is

$$\ddot{\theta} = \frac{g\theta + \left[\dfrac{-f + \mu_c \operatorname{sgn}(\dot{x})}{(M+m)}\right] - \dfrac{\mu_p\theta}{ml}}{l\left[\dfrac{4}{3} - \dfrac{m}{(M+m)}\right]}$$

$$\ddot{X} = \frac{f + ml[-\ddot{\theta}] - \mu_c \operatorname{sgn}(\dot{x})}{(M+m)}$$

and the second set ignoring friction is given as follows.

$$\ddot{\theta} = \frac{g\theta + \left[\dfrac{-f}{(M+m)}\right]}{l\left[\dfrac{4}{3} - \dfrac{m}{(M+m)}\right]} = \frac{3(M+m)g\,\theta - 3f}{4Ml + ml}$$

$$\ddot{X} = \frac{f + ml[-\ddot{\theta}]}{(M+m)}$$

At this point, we make a decision about what kind of controller we want. If we want a proportional-derivative-integral (PDI or PID) control, then we must determine three gains. The proportional gain modifies the angle, the derivative gain operates on the rate measurement, and the integral portion scales the integral of the angle. If we decided to build a PD controller, then, we must determine only two gains. The PID equation is given as

$$f = K_p * \theta + K_d * \dot{\theta} + K_i * \int \theta \, dt$$

Once we determine these gains, the controller design is complete. Therefore, we implement the controller with these gains and test it. Please note that the gains are functions of mass of the cart, mass at the top of the pole, and gravitational constant g. For the case with significant surface friction, the gains are also a function of the friction coefficient.

As the mass of the cart changes, or the mass at the top of the pole is changed, or the friction characteristics change, these gains will need modifications or recomputing. When we change these gains to match the system performance we usually call it gain tuning. Once the gains are tuned, the controller will perform good as long as the set of parameters do not change. The controller will work for small perturbations also, and will handle the state changes appropriately.

Fuzzy Logic Approach

The fuzzy logic approach to building a controller for the inverted pendulum is simple. As is the usual case, we must define membership functions for input and output parameters, and we must create a rulebase. Then, we must select a defuzzification scheme, and we are done. This seems easy, and in fact it is once we understand the process. It is not trivial and requires a through understanding and design of the logic of the controlling methodology. We will now examine the process in detail.

The two input parameters for our controller are angle and angle rate. This is very similar to the conventional controller. The maximum value of the angle is 90 degrees at which the pole lies flat on the right hand side. The minimum value of the angle is −90 degrees when the pole lies flat on the left hand side, thus, the universe of discourse for the angle parameter is from −90 to +90 degrees. If the angle is exactly 0.0, we will say that it belongs to the 'ZERO' membership function (or 'ZERO' set) with a belief value of 1.0, and as the angle increases, our belief that it belongs to 'ZERO' set decreases. (Membership functions are shown in the figure 86.6 below.) When the angle is 5.0 deg then our belief value will go to 0.0 and we will quit including it in this 'ZERO' set. So we have assigned a 0.0 belief value at that point. Similarly, when the angle starts to decrease toward a −5.0 deg value, our belief that it belongs to the 'ZERO' set will decrease. If the angle is 5.0 deg, we will say that it's a small positive angle (call it a 'PS' set for Positive Small) with a belief value of 1.0, and as the angle decreases toward 0.0, our belief value that it belongs to 'PS' will decrease to 0.0. If the angle increases toward 12.0 deg, then again our belief value that it belongs to the 'PS' set will decrease. After 12 deg, we will say that it belongs to a large positive set abbreviated as 'PL'. As the angle goes toward 90 deg, our belief value remains at 1.0, and does not decrease at all. However, as the angle decreases toward 5.0 deg, our belief value that it belongs to 'PL' set decreases. Similarly, we can define two more sets or membership functions 'NS' and 'NL' for the negative part of the universe of discourse. A total of five membership functions are defined for the angle parameter, which is our first input. It should be noted that we simply have assigned our belief values. These belief values can change from designer to designer, user to user, and person to person. There is no rule that this is the definition and nobody can change it. In fact, the definition provided here can be changed as we learn more about the inverted pendulum from experience.

The angle rate is estimated from the time it takes this pole to fall. Let us say that the time is about five seconds, thus the maximum rate is 90 divided by 5, or 18 deg/sec. We also know that the speed of the pole slowly increases as it falls down, and its speed begins with zero. Therefore, the maximum rate is somewhere between 0 and 18 deg/sec. Let us make a choice of 10 deg/sec, just for convenience. Note that we will be able to change this as we learn more about the behavior of the pole during its fall. Specifically, we will learn more about the speed and its behavior. Thus, the universe of discourse for rate is −10 to 10 deg/sec. When the rate is 0.0, we will say that it belongs to set 'ZERO' with a belief value of 1.0, and as the rate increases toward 1.0, our belief value will decrease to 0.0. Our belief value will similarly decrease to 0.0 as the rate goes to −1.0 deg/sec. Thus we have defined our 'ZERO' membership function for rate parameter. We follow similar arguments to define four more membership functions for rate parameters.

Now, we come to the "push" values (push is equivalent to force in figure 86.6). If a human is balancing the pole, we have to estimate how hard that person can push the pole from side to side. If a child is holding a relatively heavy pole, then, we know that the range of that "push" will be small. If a strong person is holding that pole then the range of "push" is large. Similarly, when the pole is placed on a cart that has a small power motor, it will move this cart with a small value of "push", or its universe of discourse will be small. If the motor that pushes this cart has large power, then, the range of "push" will be large. Thus, the universe of discourse is derived from the power of the motor or its range. We can say that we used hardware specifications to derive the range of universe of discourse. Let us say that a motor of 3/4 hp (roughly 1/2 kw power) is used for our cart-pole inverted pendulum. This will translate into some acceleration of the system. Actually, the acceleration of the system is force divided by the total mass of the system, and thus we only need to estimate the mass of the system. Let us say that this translates into maximum of 10 cm/sec of acceleration in the positive direction. It will also be the same in negative direction, because we can change the direction of this force. Thus, the universe of discourse for our output parameter "push" is the interval from −10 to 10 cm/sec. We go through a similar process as described in earlier paragraphs to define five membership functions for this "push" parameter.

Now, let us derive the rules for our fuzzy controller. It is very simple to derive the first rule. If the angle is 0.0, and rate is 0.0 then we do not want to push at all. Similarly, if the angle is 'ZERO' and rate is 'ZERO' then we do not want to have any push or we can say that "push" is 'ZERO'. Our first rule is:

"If angle is ZERO and rate is ZERO then push is ZERO"

where we have used three membership functions designated as 'ZERO' for each parameter.

Next, if the angle is 'ZERO' but it is falling toward the right side, then, we want to push the cart toward right side. If it is falling slowly, then we want to push slowly. Falling toward the right side slowly means the rate is 'PS', and we want to push slowly means 'push' is 'PS'. Thus, we have our second rule as:

"If angle is ZERO and rate is PS then push is PS"

If the rate was large, then we want to push the cart with a large force. Thus our third rule is:

"If angle is ZERO and rate is PL then push is PL"

Similarly, is the pole is falling on the left side, then we will have negative rate. If the rate is negative slow, then, we will push toward left slow. If the rate is negative large, then, we will push left with a large force. Thus, we have two more rules as follows:

"If angle is ZERO and rate is NS then push is NS"

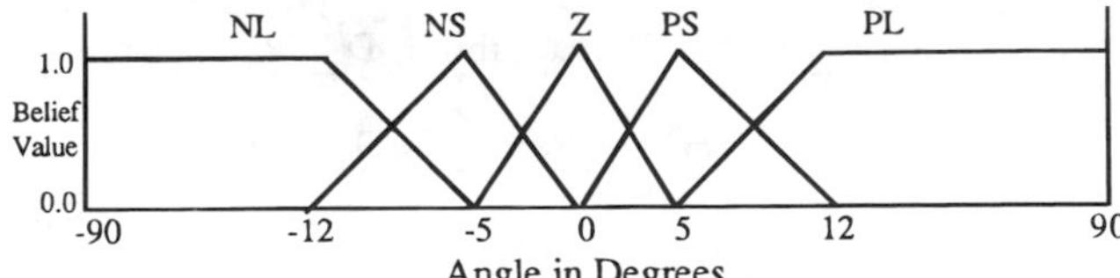

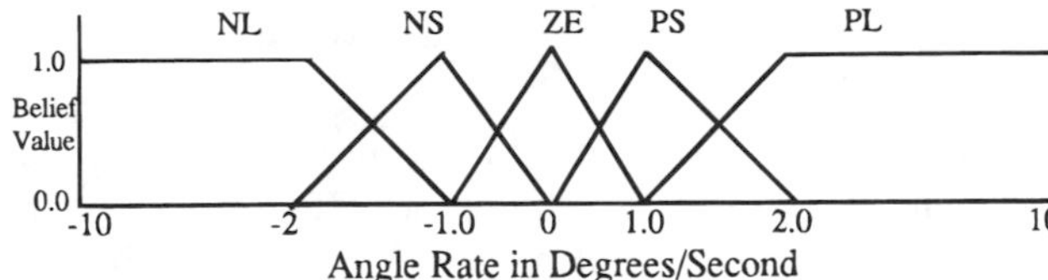

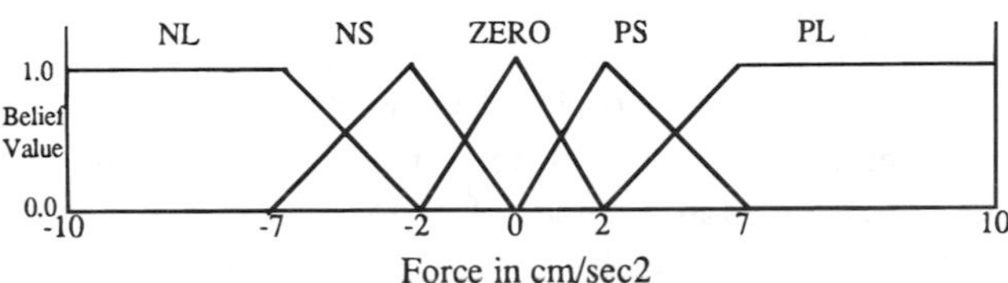

Figure 86.6 Membership functions for angle, angle rate, and force.

"If angle is ZERO and rate is NS then push is NL"

Now, let us take a situation where the angle is small positive, but it is decreasing toward zero. That means the rate is negative small. What we realize is that the angle is not zero but is coming toward zero, and therefore we do not want to push or do anything. Thus, the angle is PS but rate is NS then do not apply push. We have one more rule as:

"If angle is PS and rate is NS then push is ZERO"

However, if the rate at this point is zero then we want to push slowly. That means we can add one more rule as:

"If angle is PS and rate is ZERO then push is PS"

If the angle and rate both are positive slow then we realize that the angle is not what we want and furthermore it is increasing in the wrong direction. In this case, we definitely want to push, and we may want to push with a large force. This thinking provides a rule as:

"If angle is PS and rate is PS then push is PL"

If the angle is negative small but is coming toward zero, then again we do not want to push. This is similar to the situation where the angle was positive but the rate was negative. Thus we can add a rule as:

"If angle is NS and rate is PS then push is ZERO"

If the rate at that point (meaning angle is NS) is zero then we want to push slowly toward left (meaning toward zero angle or negative push). This translates into a rule as:

"If angle is NS and rate is ZERO then push is NS"

And, if the rate is increasing such that the pole is falling further left, then, we definitely want to push left. This translates into a rule as:

"If angle is NS and rate is NS then push is NL"

Similarly, we derive all other rules from our thinking relative to what action we will take to balance the pole. These rules will make sense intuitively and will also conform with the control actions and mathematics behind it. The rulebase so derived is given below.

	Angle				
Rate	NL	NS	Z	PS	PL
PL	ZERO	PS	PL	PL	PL
PS	NS	ZERO	PS	PL	PL
ZE	NL	NS	ZERO	PS	PL
NS	NL	NL	NS	ZERO	PS
NL	NL	NL	NL	NS	ZERO

Rulebase for the Inverted Pendulum

Figure 86.7 Rulebase for the inverted pendulum.

Since we have defined membership functions for input/output parameters and have derived all necessary rules, we can now select a fuzzy inference scheme and a defuzzification method. There are two inference schemes that are used more often than any others, the Max-Min and Max-Dot methods. Since the Max-Dot inference method provides a speed advantage, we will select that scheme. For the defuzzification method we will select the Centroid method. There are several methods the user can select depending on his/her choice.

Once we make these choices, we are now ready to implement our fuzzy controller using the TILShell (Togai Infralogic, 1993a). It should be noted that we have not used any equations in deriving rules or membership functions. The reason is that we did not need them. However, if a user feels comfortable using equations there is no restriction in using them. If the user understands the problem using mathematical equations, then, that insight into perfecting this controller can be very well utilized. If a user has other insight from other experience, that can be used also to enhance this controller. Also note that the rules are general and the user/designer has a choice of writing whatever rules he/she feels comfortable with. If the rulebase is not correct, then control of the inverted pendulum will not be achieved. Then, the user will have to change the rulebase. Similarly, the definition of membership functions can be changed to get the desired performance from this controller.

Such controllers have been built by a number of people, particularly Yamakawa in the late 1980s and Hamid Berenji in the early 1990s. These controllers have performed with remarkable success, being able to withstand perturbations to the system and recover. For example, Yamakawa built a pendulum with a platform at the top where he allowed a mouse to roam around, or where he placed a wine glass to which he added wine during the control process, by pouring from the bottle into the glass while it was being balanced on top of the pendulum. Berenji's system added rules to maintain a position on the track from which his system could always recover from perturbations.

86.4 Remote Manipulator System

In this section the system under development at the Goddard Space Flight Center for the control of a remote manipulator system is described. This system generates the joint rate commands required to move the point of resolution of a robotic arm based on commands generated by remote manual operation of translational and rotational hand controllers. A second function of the system is to control the joint rates at the required levels based on feedback from the joint angle encoders. This section concentrates on the closed loop control function.

The Goddard Space Flight Center (GSFC) Remote Manipulator System (RMS) consists of a six degree of freedom arm, a set of software modeled on the Shuttle RMS Flight Software Systems Requirements (FSSR) (Mica and Lea 1994), and a fuzzy logic controller for controlling the six joint rates to the required levels as generated by the FSSR software.

The controller developed is designed to control the rate of each of six joints to match the commanded rate as computed by the FSSR. This software accepts translational and rotational rates of the point of resolution (POR) of the arm (for this discussion we assume the POR is the end of the wrist roll joint) generated by a remote manipulator system operator who utilizes translational and rotational hand controllers. Related work on the concept of fuzzy logic for control of the point of resolution of robotic arms was developed and reported in Lea et al. (1993). However, that work concentrated on the problem of specifying an end position and orientation of the point of resolution and the fuzzy logic control of the transition. In this work the goal is to take commands from an operator that specifies a linear motion of the point of resolution, or rotations to achieve a new orientation of the end effector, convert them into joint rates required to achieve these end effector rates through the FSSR software, and to achieve and maintain the commanded rates through fuzzy logic control based on feedback from angle position encoders. The angle position encoder measurements and the time tag associated with the measurements are the only feedback to the control system.

This method controls the joint rates based on the feedback of the joint angle measurements. From these measurements estimated joint rates are computed. From the joint rates and the Rate__Cmd that are generated by the FSSR, joint rate errors, E, and changes in joint rate errors, DE, are computed. Next E and DE are treated as fuzzy input variables to a fuzzy rulebase that outputs a fuzzy variable D__Rate__Cmd which is defuzzified and added, initially to the Rate__Cmd, and then to the previous Rate command, to create a Rate__Cmd__Out to be sent to the joint. The equations are,

$$\begin{aligned}\text{Rate\_\_Cmd\_\_Out} &= \text{Rate\_\_Cmd} + \text{D\_\_ Rate\_\_Cmd}\\ &\quad + \text{D\_\_Rate\_\_CMD\_\_Prev}\end{aligned}$$

where

$$\begin{aligned}\text{D\_\_Rate\_\_Cmd\_\_Prev} &= \text{Rate\_\_Cmd\_\_Out}\\ &\quad - \text{Rate\_\_Cmd}.\end{aligned}$$

The rulebase (Table 86.1) and the fuzzy membership functions were built using the Togai Infralogic TIL__SHELL (Togai Infralogic 1993a). Seven levels of membership functions, NL, NM, NS, Z, PS, PM, and PL, were used for E, DE = D__Rate__Error, and D__Rate__Cmd.

An additional computation of a Delta, based on an integral like computation, was introduced to smooth the commanded rate. Experiments were used with varying lengths of history of actual rate errors and methods of performing the integration like operations. It was decided that actual integral operators, such as the trapezoidal or Simpson's rule, place too much emphasis on past history. This system is constantly changing and past information, although quite useful for trending, will cause too much of a delay in taking required action. It was decided that the best results were obtained by simply adding rate errors over the previous *k* cycles to determine if the actual rate is consistently high or low and then adding a delta to the rate command to compensate for the bias. It was also necessary to filter out abnormally large rate errors that occur when a new rate command has been issued by the FSSR. The parameter which was computed as input to the integral rule base was,

$$\text{Rate\_\_Error\_\_Integral} = \sum_{l=0}^{k} \text{Rate\_\_Error}_{n-i}$$

For the tuned Goddard RMS system *k* equal to six was chosen. Output from the system is a second delta to the Rate__Cmd, Delta, that is generated by a second fuzzy rulebase and fuzzy membership functions for the variables Rate__Error__Integral and Delta.

Table 86.1 Rulebase for Delta__rate__cmd

D_ Rate Cmd		E						
		NL	NM	NS	Z	PS	PM	PL
D Rate Error	NM	PL	PL	PM	PM	PS	Z	NL
	N	PL	PM	PS	PS	NS	NM	NL
	Z	PL	PM	PS	Z	NS	NM	NL
	P	PM	PS	PS	NS	NM	NL	NL
	PM	PS	Z	Z	NM	NL	NL	NL

The primary objective of the system is to control the motion of the point of resolution of the arm consistent with the operator's hand controller commands. For example, if he commands a translation in the x direction, the POR should move essentially along a linear path in the x direction. This will only occur if the commanded rates are achieved over approximately the same time span and with rate build-up in each joint proportional to joint rate change required for the joint. If some joints have slower response to rate commands, then the motion will not be in the desired direction and will require additional commands to achieve the desired arm position. Therefore, it is desirable to increase the acceleration of the lagging joints in order that they reach the commanded level at the same time as the other joints.

Since a scalar multiple of translational velocity is equivalent to the same scalar multiple of joint rates, if joint rates are increased or decreased proportionally to the joint rate errors, we will achieve motion in the correct direction although possible slower or faster than the actual commanded rates. To add this feature, inputs of all six angle positions are used to compute estimates of joint rates and then joint rates are used to compute rate errors. Next the maximum joint rate error, K, is computed, and all six ratios $K_i = E_i/K$ are computed. These K_i are then input to another rulebase that outputs a gain, G, that weights the Rate_ Cmd_Out proportional to the size of the errors in the six joint rates. Thus, those joints that are slow to respond are given priority. The final equation for computation of Rate_Cmd_Out is given by

$$\begin{aligned}\text{Rate\_Cmd\_Out} &= G^*(\text{Rate\_Cmd} \\ &+ \text{D\_Rate\_Cmd\_Prev} \\ &+ \text{D\_Rate\_Cmd} + \text{Delta})\end{aligned}$$

This software has been integrated into the hardware/software system at the Goddard Space Flight Center. This system has been tuned and demonstrated at the WNN/FNN '94 conference held in Washington, D.C. in December 1994 and reported by Lea et al. (1995b). It has been demonstrated to perform as good or better than a PID controller. Initial tests have indicated the need for other sensors, such as strain gauges or accelerometers, to indicate oscillations of the flexible links caused by accelerations and braking. Future work will investigate this problem in an effort to design a near optimal system that will significantly reduce the ringing effects of the arm during operations.

86.5 Collision Avoidance

Future unmanned missions to Mars (Kahl and Bailey, 1989) will investigate the terrain and collect soil samples in advance of manned missions. Path planning is a crucial element in the activities to be undertaken by an autonomous rover. Obstacles such as boulders or troughs may block the shortest path from the current position to the target position for the next sample acquisition. Collision avoidance algorithms have been developed that take fuzzified sensor data and generate short-term path decisions. These algorithms utilize range to obstacles, current velocity and orientation to generate steering and velocity commands as shown in Figure 86.8. For the rover trajectory control, the range parameter was limited to 20 meters, and the speed was limited to 5 meters per second. The azimuth was completely covered by taking -200 to 200 degree universe of discourse. These membership functions are shown in Figure 86.9. The steering angle range for the rover is limited to 30 degrees, and the increment in the delta velocity was limited to the same range as the velocity. These membership functions are shown in Figure 86.10. The rules for the collision avoidance are shown in Table 86.2, which fall into two groups because the range has two membership functions. The first group of rules is for the *Critical* membership function, while the second group is for the *Proximity* membership function. There are 20 rules in each group and initially all steering angle membership functions were used. However, as we performed the tests, we tuned the rulebase as well as

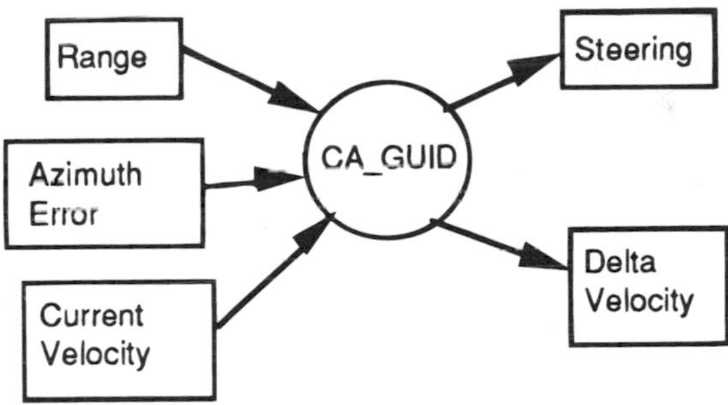

Figure 86.8 Collision avoidance guidance for Mars Rover.

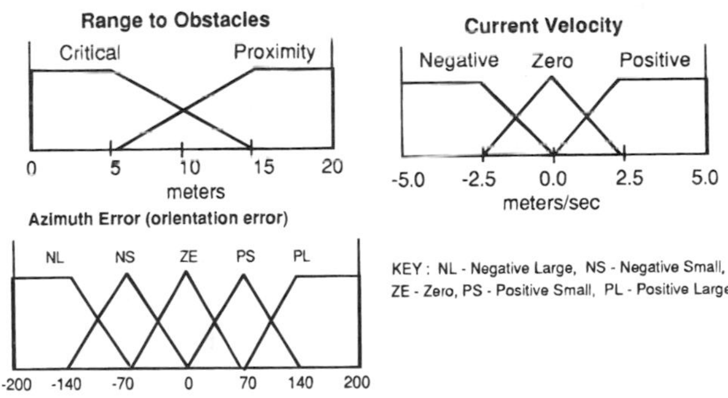

Figure 86.9 Membership functions for range to obstacles, current velocity and azimuth (orientation) error.

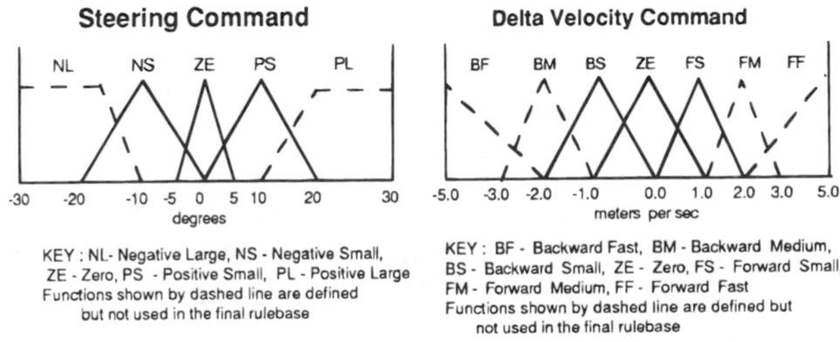

Figure 86.10 Membership functions for steering and delta velocity commands for Mars Rover.

Table 86.2 Rulebase for Collision Avoidance Guidance

When RANGE equals CRITICAL							When RANGE equals PROXIMITY						
	Azimuth Error (orientation error)							Azimuth Error (orientation error)					
	NL	NS	ZE	PS	PL			NL	NS	ZE	PS	PL	
Current Velocity							Current Velocity						
POS	NS	ZE	ZE	ZE	PS	STEERING	POS	NS	PS	ZE	NS	PS	STEERING
NEG	PS	ZE	ZE	ZE	NS		NEG	PS	NS	ZE	PS	NS	
POS	FS	ZE	ZE	ZE	FS	DELTA	POS	FS	FM	ZE	FM	FS	DELTA
NEG	BS	ZE	ZE	ZE	BS	VELOCITY	NEG	BS	BM	ZE	BM	BS	VELOCITY

KEY: NL—Negative Large, NS—Negative Small, ZE—Zero, PS—Positive Small, PL—Positive Large, POS—Positive, NEG—Negative, FS—Forward Small, BS—Backward Small, FM—Forward Medium, BM—Backward Medium

membership functions. As a result, we have only three membership functions for both, steering and delta velocity command. The membership functions that are eliminated in the tuning process are shown with a dash line.

Simulation testing is performed for a set of representative test cases, and performance of the guidance algorithms is evaluated in a variety of obstacle scenarios. We have designed five test cases to test the capabilities of the collision avoidance system. The first test case shows that an obstacle directly in the center can be avoided very easily by basic steering commands. The second test case was designed to test a large central obstacle, which was successfully avoided. The third and fourth test cases show that the obstacles can be successfully avoided one by one and the rover still reaches the desired destination. The fifth test case shows that the rover can avoid many obstacles even if they are lined up, leaving only a small opening.

Some important points of our results are as follows:

- Our trajectory guidance (Lea, 1990a; Lea et al., 1990c) includes not only position control but also orientation control. When the collision avoidance algorithms (Lea, et al., 1991c) are integrated with the trajectory control, the orientation control could not be managed properly without major modification to the system. Proper orientation can be achieved only if there is sufficient distance between the last obstacle avoided and the target point so that the rover can turn itself. Since this distance can not be guaranteed, the desired orientation of the rover at arrival was not addressed at this time.
- The back-off situation requires some knowledge about the obstacles just avoided, or some information about the obstacles in the path. Since our cameras were looking forward, we postponed the 'back-off' situation study.
- When the rover is going forward, and it encounters an obstacle that can be avoided only by going back, it must remember that the distance it must go back is a factor. Otherwise, it can get into a situation where it continues to go back and forth when the critical distance for collision avoidance is not altered. Thus, the transition from forward to backward requires special care.

It was found that a higher-level path planner is needed when the vehicle is caught in a back-off setting, that is, when it is not possible for the vehicle to pursue a "forward" path. It is significant to note that the method employed does not depend on object identification, but rather, detection of the degree to which an object (where present) or (more generally) an angular sector represents an obstacle. This is a significant relaxation over most collision avoidance schemes. Our simulation results and planned enhancements for the future point towards refinements in the algorithm, the possibility of adaptive tuning of the system and, as expected, the need for a higher-level path planner to handle cases that involve backoff, sensor fusion, positioning of the vehicle at the destination, moving obstacles, and other situations that involve radically changing environments of operation.

Fuzzy logic control together with straightforward algorithms yield an effective system for autonomous collision avoidance in an environment of uncertain information. The technique is robust and avoids complexity in initial stages where simple obstacle avoidance is a key element rather than involved object identification or mapping of a complete world model. This technique could be integrated with fast and sophisticated object identification algorithms if desired.

86.6 Summary

Fuzzy logic is simple, easy to understand and reflects human type thinking. Its architecture is very well suited for implementing heuristic knowledge or the knowledge gained through experience. For example, control of a processing plant typically performed by human operator can be easily automated using this framework in software. Several applications have shown that fuzzy logic based control is usually robust, nonlinear and comparatively stable. It provides an ability to combine seemingly unrelated parameters for higher order reasoning. Control of systems that are nonlinear and difficult to model is easily achieved using fuzzy logic principles. In our experience, fuzzy controllers can be easily designed using heuristics and experiential knowledge. Tuning of membership functions and modifications to the rule base is also simplified at a point where rapid implementation and testing is possible. Maintenance of the algorithms is minimal. Since the algorithms are in a graphical form, the knowledge transfer from one generation to another generation is very easy.

In the United States, the word "fuzzy" has a bad connotation, and therefore, industry is afraid that if their appliances are based on fuzzy logic, nobody will use them. The market share will be lost resulting in less profit. Even though it is a precise mathematical formulation, confusion is always there because "fuzzy" is used to imply imprecise in our daily life. Since it allows human like thinking, control of processes looks very easy, and thus engineers sometimes feel that since complexity is lost, somehow the importance of their work is reduced. They are also concerned about the question of stability. Due to the inherent design of fuzzy control systems based on human actions, and the lack of mathematical models of the process, stability analyzes of the control system can not be carried out in the traditional way. However, this is an important question and research work is continuing to develop stability criteria for fuzzy control.

From a logic and reasoning point of view, fuzzy logic provides symbolic as well as numerical processing ability. Thus, the algorithms are mixed. Rules are written using the logic form but are processed using precise numerical computations. This characteristic of fuzzy logic has provided a path for merging fuzzy with the CLIPS environment (C-Language Integrated Production System, developed at NASA Johnson Space Center in the 1980s) so developers can build hybrid expert systems (Togai Infralogic, 1993b). Fuzzy logic provides a framework to perform numerical as well as symbolic reasoning. As a result, it fills the gap between the expert systems and neural network processing. Fuzzy logic cannot map complex relationships unless the developer already has knowledge to create the rules. Fuzzy systems typically do not learn on the fly. When combined with neural systems, a fuzzy-neuro system can handle the control and decision making aspect as well as learn the changing environment, user requirements, or new demands for performance. Thus, fuzzy-neuro systems can adapt to changes and can provide the flexibility required in many applications. As we deal with more and more complex systems in the future, our experience indicates that the fuzzy-neuro based expert systems will be suitable to perform in varied environments and will meet the demands placed on these complex systems.

In our opinion, it is time to exploit these fields for decision making and expert system applications and enjoy the advantages offered by fuzzy logic, neural networks, and combined architecture for efficiency, and cost savings. In space operations, autonomy at a higher level can be easily achieved resulting in operational efficiency required for cost-effectiveness.

References

Coledan, S. 1991. Tethered satellite advances, *Space News,* 2(15):8.

Copeland, C., Lea, R., Jani, Y., and Villarreal, J. 1991. Fuzzy logic based tether control, *NAFIPS '91,* Columbia, MO.

Dubois, D. and Prade, H. 1980. *Fuzzy Sets and Systems—Theory and Applications,* Academic Press, New York, NY.

Kahl, R. and Bailey, S. 1989. Mars Rover sample return: project description and mission operations review, presented at MRSR Phase A Midterm Review, November, NASA JSC (New Initiatives Office), Houston TX.

Klir, G. J. and Folger, T. A. 1988. *Fuzzy Sets, Uncertainty, and Information,* Prentice Hall, Englewood Cliffs, NJ.

Kosko, B. 1991. *Neural Networks and Fuzzy Systems,* Prentice Hall, Englewood Cliffs, NJ.

Lea, R. N. 1990a. Fuzzy logic approach to Mars Rover guidance, presented at *Int. Conf. Fuzzy Logic and Neural Networks, IIZUKA '90,* July, Iizuka, Japan.

Lea, R. N. and Jani, Y. K. 1990b. Applications of fuzzy logic to control and decision making, *Proc. Tech. 2000 Conf.,* NASA Conference Publication 3109, vol. 2.

Lea, R. N., Walters, L., and Jani, Y. K. 1990c. A fuzzy logic approach to Mars Rover trajectory planning and control, *Proc. 1st ISMCR,* Session D.3, pp. D.3.1.1, June.

Lea, R. N., Hoblit, J., and Jani, Y. K. 1991a. A fuzzy logic based spacecraft controller for six degree of freedom control and performance results, *Proc. AIAA Guidance, Navigation and Control Conf.,* 3:1680.

Lea, R. N., Hoblit, J., and Jani, Y. K. 1991b. Performance comparison of a fuzzy logic based attitude controller with the shuttle on-orbit digital auto pilot, *NAFIPS '91 Workshop Proc.,* 291–295.

Lea, R. N., Murphy, M., and Walters, L. 1991c. Fuzzy logic control for autonomous collision avoidance, *NAFIPS '91 Workshop Proc.,* May, University of Missouri-Columbia, MO.

Lea, R. N. and Jani, Y. K. 1992a. Design and performance comparison of fuzzy logic based tracking controllers, presented at *AIAA Space Program and Technologies Conf.,* March, Huntsville, AL.

Lea, R. N. and Jani, Y. K. 1992b. Fuzzy logic in autonomous orbital operations, *Int. J. Approximate Reasoning,* 6(2):151–184.

Lea, R. N., Chowdhary, I., Jani, Y. K., and Shehadeh, H. 1992c. Design and performance of the fuzzy tracking controller in software simulation, *Proc. FUZZ-IEEE '92,* March 8–12, San Diego, CA.

Lea, R. N., Copeland, C., Jani, Y., and Villarreal, J. 1992d. Tether operations using fuzzy logic based length control, *Proc. FUZZ-IEEE '92,* p. 1335, San Diego, CA.

Lea, R. N., Jani, Y., and Hoblit, J. 1993. Fuzzy logic based robotic arm control, *Proc. 2nd Int. Conf. Fuzzy Systems,* 1:128–133, March 28–April 1, San Francisco, CA.

Lea, R. N. and Jani, Y. K. 1995a. Intelligent Sensor System for Space Operations, *Industrial Applications of Fuzzy Logic and Intelligent Systems,* chap. 11, Yen, J., Langari, R., and Zadeh, L., eds., IEEE Press, Piscataway, NJ.

Lea, R. N., Mica, J. A., Jani, Y., and Dohmann, E. L., 1995b. Fuzzy logic approach to enhance the performance of remote manipulator systems, *SPIE Applications of Fuzzy Logic Technology II,* April, pp. 84–93.

Lembeck, M. F. 1992. TES/CRIM fuzzy logic thermal control system, presented to Software Technology Branch/JSC, March.

Mica, J. A. and Lea, R. N., 1994. *Space Shuttle Orbiter Remote Manipulator System Simulator (RMMS) Software Program Description Document (PDD),* Goddard RMSS, NASA GSFC document, May 29.

Negoita, C. V. 1985. *Expert Systems and Fuzzy Systems,* 19??, Menlo Park, Reading, London, Amsterdam.

Shingu, T. and Nishimori, E. 1989. Fuzzy based automatic focusing system for compact camera, *Proc. IFSA '89,* 436–439.

Tobi, T., Hanafusa, T., Itoh, S., and Kashiwagi, N. 1989. Application of fuzzy control system to coke oven gas cooling plant, *Proc. IFSA '89,* 16–22.

Togai Infralogic, 1993a. *TilShell User Manual, V 3.0.0,* Togai Infralogic, Irvine, CA, USA.

Togai Infralogic, 1993b. *Fuzzy CLIPS User's Guide,* Togai Infralogic, Irvine, CA.

Tran, L. P. and Weiss, J. 1991. Fuzzy control system for a remote focusing microscope, *Proc. 5th Ann. Workshop on Space Operations, Automation and Robotics,* July 9–11, Johnson Space Center, TX.

Yasunobu, S. and Miyamoto, S. 1985. Automatic train operation system by predictive control, *Industrial Applications of Fuzzy Control,* 1–18, North Holland, Amsterdam.

Zadeh, L. 1965. Fuzzy sets, *Information and Control,* 8:338–353.

Zadeh, L. 1968. Fuzzy algorithms, *Information and Control,* 12:94–104.

Zadeh, L. 1973. Outline of a new approach to the analysis of complex systems and decision processes, *IEEE Trans. Syst., Man and Cyberns.,* SMC-3:28–44.

Zimmerman, H. J. 1991. *Fuzzy Set Theory and Its Applications,* Kluwer-Nijhoff Publishing, Boston/Dordrecht/London.

87

Development of an Intelligent Unmanned Helicopter Based on Fuzzy Systems

Michio Sugeno
Tokyo Institute of Technology

Howard A. Winston
United Technologies Research Center

Isao Hirano
Hitachi Ltd.

Satoru Kotsu
Tokyo Institute of Technology

87.1 Introduction 1127
Subject • Background • Contribution • Other Work
87.2 Helicopter Hardware System 1129
Body • Sensors • Controller • Command Transmitter • Image Processing System
87.3 Software System for Helicopter Control 1131
Lower Layer: Fuzzy Control Modules • Upper Layer
87.4 Results 1135
Simulations • Experiments
87.5 Conclusions 1136

Abstract

This paper describes recent results from the Sugeno Laboratory project on autonomous control of an unmanned helicopter. A fuzzy-logic based control system has been developed that enables single inputs (e.g., voice commands) to replace the aircraft's normal set of control inputs. As a result, a novice can use this system to control an unmanned helicopter without prior knowledge of the vehicle's flight dynamics. The fuzzy-controlled helicopter described in this paper can execute basic flight modes, such as forward flight, and can also blend them to execute more complex maneuvers, such as climbing turns. To accomplish this, the fuzzy controller is organized hierarchically with modules for primitive control inputs (e.g., rudder control) in a lower layer that can be activated by basic flight mode modules (e.g., forward flight) in an upper layer. Flight mode modules can, in turn, be used by higher level tasks such as *(1)* command sequence programs, *(2)* GPS-guided, and *(3)* image-guided control and navigation. This organization not only makes it easier to implement flight mode mixing, such as climbing turns, but is also ideally suited for active cross-coupling compensation and switching between different flight modes.

Reprinted with permission from "Intelligent Control of an Unmanned Helicopter Based on Fuzzy Logic" presented at the American Helicopter Society 51st Annual Forum, Fort Worth, TX, May 9–11, 1995.

87.1 Introduction

Subject

Stengel (1993) describes the benefits of designing control systems that emulate, in part, the functions of natural intelligence. One of the objectives mentioned for such systems is the enhancement of aircraft mission capabilities. In particular, Stengel states that

> "In the future, teleoperated or autonomous systems could find increasing use for missions that expose human pilots to danger."

The subject of this paper is the design of an intelligent control system for such an autonomous system in the form of an unmanned helicopter. To achieve this goal, fuzzy logic has been employed to represent the knowledge associated with and the execution of the reasoning processes of natural intelligence.

Background

The Sugeno Laboratory at the Tokyo Institute of Technology is developing an intelligent unmanned helicopter as a testbed for experimentation with new fuzzy logic-based technologies. The project focuses on and integrates three major areas of fuzzy logic research—control, natural language understanding, and computer vision.

In general, the unmanned helicopter is an example of an intelligent autonomous agent. This paper supports the belief that fuzzy logic is an important enabling technology for building these types of systems. The use of an aerial robot testbed is motivated by the challenges and opportunities associated with the autonomous control of an unmanned aircraft. Helicopters, in particular, are uniquely challenging because they are:

1. Nonlinear.
2. Cross-coupled.
3. Unstable.
4. Multivariate (i.e., there are many input-output and state variables.)
5. Sensitive to external disturbances and environmental conditions.
6. Used in many different flight modes (e.g., hover or forward flight), each of which requires different control laws.
7. Often used in dangerous environments (e.g., at low altitudes near obstacles).

Nevertheless, characteristics that make it difficult to automate the operation of a helicopter with conventional control present fewer challenges to the design of fuzzy control systems. For example, although helicopters are nonlinear plants, fuzzy controllers are capable of controlling them because they are also inherently nonlinear. The instabilities that result from time delays between plant input and output changes can be addressed with fuzzy control rules that capture the feedforward knowledge used by pilots to stabilize aircraft. Cross-couplings between control inputs can also be compensated for by implementing predictive feedforward fuzzy control rules. Finally, the environmental sensitivity of helicopters can be ameliorated in a fuzzy control system by incorporating control rules that describe the actions pilots take to adapt to changing external conditions.

Vision and natural language are important sensor and communication modalities, respectively, for intelligent agents. A goal of our project is to incorporate and integrate natural language and image processing with intelligent control. Important problems in each of these areas are being addressed by fuzzy logic methodologies, and, as a result, it should be possible to design a consistent architecture for autonomous rotorcraft based on fuzzy logic technologies.

Contribution

This paper contributes novel methodologies for the design of unmanned aerial vehicles (UAVs) in three areas—control, natural language command, and image-guided navigation (including the use of GPS information).

Control

The difference between fuzzy control systems[1] and conventional control systems stems from the effort to build systems that use representations of control knowledge similar to those employed by skilled humans (e.g., pilots) in the fuzzy logic case, and the effort to build systems based on a deeper analytical understanding of plant and control system physics in the conventional design case. The ability of humans to pilot manned aircraft, with only qualitative knowledge, is taken to be an existence proof that fuzzy logic-based controllers with similar capabilities can also be developed.

The fuzzy logic-based helicopter control system described here is organized hierarchically into two major divisions—a lower level subsystem containing modules for controlling basic flight modes such as hover, climb, forward, and turning flight and a higher level subsystem containing modules for blending basic flight modes (e.g., climbing turns), transitioning between basic flight modes (e.g., forward flight to hover), executing complex maneuvers (e.g., programmed figure eights and rectangular flight trajectories), and navigation (e.g., image- and GPS-guided flight).

In general, fuzzy logic controllers work by executing control rules represented in the form of fuzzy implication relations such as "If X is A and Y is B Then Z is C", where X and Y are error inputs and Z is a control output. In this example, A, B, and C are imprecise, but robust, linguistic predicates such as *small* or *large*. In this way, control rules can represent the linguistic knowledge used by pilots to control manned aircraft and found as verbal descriptions in pilot operating manuals, etc. As a result, they can endow a control system with the flexibility required to operate in complex, ambiguous, and unpredictable environments as opposed to the use of conventional crisp (i.e., nonfuzzy) control systems that, for example, have been used to optimize more narrowly defined objective functions. This motivation is a special case of the more general principle of incompatibility first expounded by Zadeh (1965) that expresses an inverse relationship between the precision and relevance of statements about increasingly complex systems.

Natural Language

As mentioned above, a fuzzy logic controller expresses control rules linguistically. As a result, it can handle verbal and numerical control inputs. A goal of the unmanned helicopter project is the demonstration that a consistent fuzzy logic-based flight control system architecture can be developed. The ability to interpret imprecise natural language commands, such as "Fly a little slower," is an important part of this demonstration.

Vision

As Keller (1993) points out, fuzzy logic can directly realize two of David Marr's principles for the design of vision algorithms (Marr, 1982), and this motivates considering the use of fuzzy logic to address problems in computer vision.

Least Commitment. Fuzzy logic supports the principle of least commitment in computer vision (i.e., information should be preserved as long as it might still be useful for future computations) because with it the results of computations can be represented in the form of membership functions that preserve

[1] Neural network-based controllers (Walker and Mo, 1994) can indirectly model human cognitive performance by emulating the biological processes underlying human skill acquisition. Fuzzy logic-based controllers directly model human cognitive performance by emulating the more transparent linguistic processes supported by this neural substrate.

more information about the outcomes of calculations than crisp numerical values.[2]

Graceful Degradation. Fuzzy logic supports the principle of graceful degradation in computer vision (i.e., degrading data will not prevent the delivery of at least some of the answer), because its interpolative nature facilitates the implementation of continuous processes.

Nishimori et. al. (1994) used visual information to control an autonomous automobile. The machine vision and vehicle control systems were based on fuzzy logic, and the authors concluded that simple fuzzy image processing techniques were sufficient to guide the fuzzy controller as opposed to the requirement for significantly more image processing power needed to guide classical automobile navigation and control systems.

Although we have not yet used fuzzy logic for image processing itself, a goal of the helicopter project is to develop machine vision design methods that will prove to be useful in other aerial or terrestrial unmanned vehicles.

Other Work

See Fagg et al., (1993) for an example of an alternative approach to real-time control of an autonomous flying vehicle based on a behavioral, or reactive, approach to control system design.

Other investigators have developed fuzzy logic flight controls. For example, in Phillips et al. (1994), Wade et al. (1994), Wade and Walker (1994) describe systems that include mechanisms for discovering and tuning fuzzy rules in adaptive controllers. Larkin (1984) describes a model of an autopilot controller based on fuzzy algorithms.

The research and development reported in this paper draws on several earlier investigations carried out in the Sugeno Laboratory. Issues related to fuzzy reasoning for control and fuzzy controller design can be found in Sugeno (1985). Prior work on verbal command-based control is described in Sugeno, et al (1989) in which spoken commands were used to control a model car. In Sugeno and Park (1993a, 1993b), an architecture was reported that supported learning control of a helicopter.

Figure 87.1 Yamaha R50.

A general description of the Sugeno Laboratory unmanned helicopter project can be found in technical reports (Sugeno, 1993; Sugeno et al., 1993a). In Sugeno et al., (1993b) and Sugeno et al., (1993c), the use of hierarchically structured control systems to facilitate the implementation and coordination of multiple flight modes was reported. Finally, prior work in the area of image feedback control was reported in Ozawa, 1994.

87.2 Helicopter Hardware System

Body

This research was conducted with an R50 unmanned helicopter manufactured by Yamaha Motor Corporation, shown in Figure 87.1. The R50 is widely used for agricultural spraying over rice paddy fields in Japan, and approximately 600 airframes have been sold to date.

Figure 87.2 shows the specifications of the R50 helicopter. It has four control inputs:

1. Longitudinal cyclic.
2. Lateral cyclic.
3. Collective pitch.
4. Rudder.

In the R50 helicopter, throttle and collective are mixed into a single collective pitch control.

Sensors

For telemetered control of the helicopter, we use the following set of sensors to measure 15 state variables:

- Laser Height Meter for altitude (Z).

YAMAHA R-50

Body :	
Overall Body Length	3.57 m
Main Rotor	3.07 m
Tail Rotor	0.52 m
Body Length	2.66 m
Width	0.7 m
Height	1.08 m
Empty Weight	44 kg
Performance :	
Payload	20 kg
Hovering Ceiling	100 m
Flying Hours	30 min
Engine (water-cooled 2 cycle) :	
Displacement	98 cc
Power Output	12 hp

Figure 87.2 Dimensions of Yamaha R50.

[2] This can also be accomplished with the use of distributions in statistical analysis.

- Radio Wave Speed Meters for velocities ($\dot{X}$, $\dot{Y}$, $\dot{Z}$).
- Magnetic Azimuth Sensor for heading direction (ψ).
- Integrated Gyro Sensor for rectilinear accelerations ($\ddot{X}$, $\ddot{Y}$, $\ddot{Z}$); roll, pitch, and yaw (ϕ, θ, ψ); and roll, pitch, and yaw velocities ($\dot{\phi}$, $\dot{\theta}$, $\dot{\psi}$).
- Differential Global Positioning System (DGPS) for horizontal position (X, Y).

A conventional differential GPS (DGPS) was modified to measure the horizontal position of the helicopter. The DGPS measures positions once every second. However, the sampling period of our control system is 100 milliseconds, so more frequently sampled position data is needed. As a result, the DGPS output data was interpolated by integrating sensor velocity data (i.e., $\dot{X}$ and $\dot{Y}$) to obtain position data every 50 milliseconds.

Controller

The Sugeno Laboratory has designed an on-board fuzzy flight controller, as shown in Figure 87.3. Figure 87.4 shows that it is equipped with four fuzzy inference chips that can store 1500 fuzzy control rules. It can receive radio transmissions encoding 8 bit input signals from a ground station. By using such signals, we can transmit 256 different commands to control the helicopter.

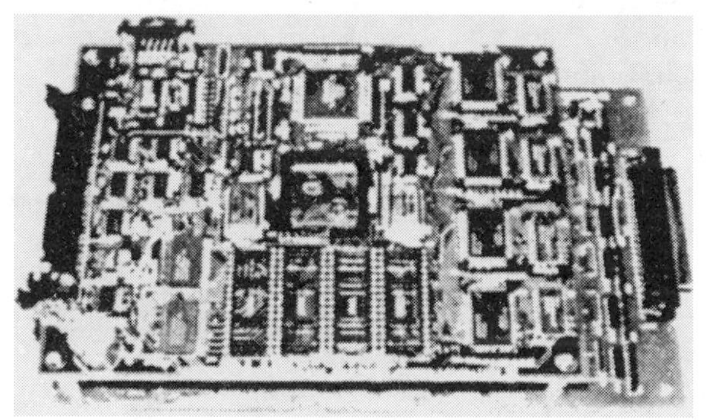

Figure 87.3 On-board fuzzy flight controller.

On-Board Fuzzy Flight Controller

CPU : MPU TMP68301
ROM : 2M Byte
RAM : 1M Byte
A/D : 12 Bit , 24 Channels
D/A : 8 Bit , 8 Channels
PWM Inputs : 8 Channels
Fuzzy Inference Module
Chip : FP3000 x 4 (1.5K rules)
SRAM : 32K Byte x 4
Sampling Period : 100 msec

FP3000 (OMRON)	
rule format	: 8 inputs - 4 outputs
number of rules	: 128 rules x 3
inference speed	: 1 msec / 60 rules

Figure 87.4 Specifications of on-board fuzzy flight controller.

Command Transmitter

The command transmitter is composed of a hand-held radio controller, a personal computer that executes a planning program, and a voice recognition device. It is shown in Figure 87.5. Operators can transmit their commands either through a computer keyboard or through a microphone.

Image Processing System

A color charge-coupled device (CCD) camera is installed in the helicopter, as shown in Figure 87.6. The camera can be panned left or right and tilted up or down with servo-motors by remote signals from the ground station. Figure 87.7 shows the image processing system used for image-guided flights. VTR signals from the CCD camera are sent to a ground-based image processor through a TV channel. A planning program receives telemetered height and attitude (i.e., roll, pitch, and yaw) information from the helicopter. From processed images and telemetered information, the planner gives guidance (by commands through the command transmitter) to the helicopter.

Figure 87.5 Command transmitter.

Figure 87.6 CCD camera.

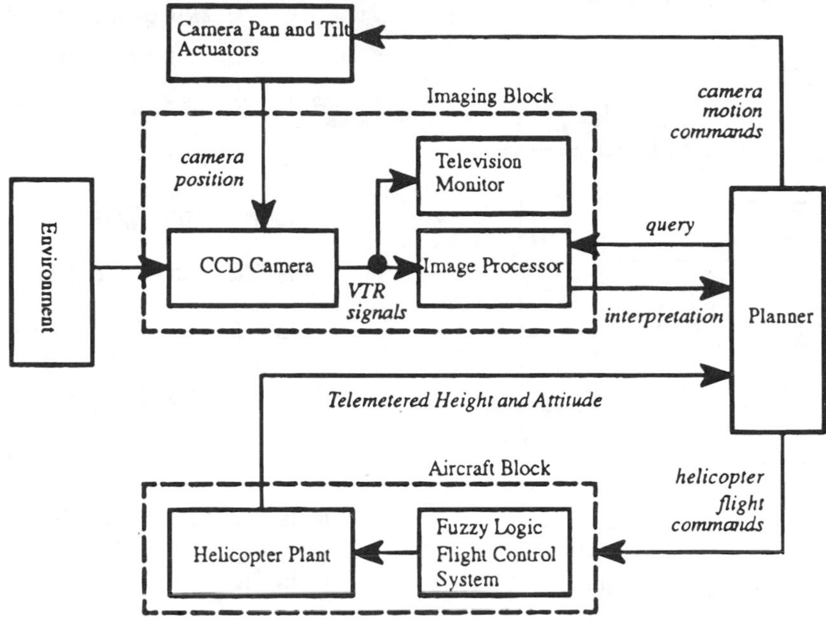

Figure 87.7 Image processing system.

87.3 Software System for Helicopter Control

Four years ago, the Sugeno Laboratory designed a two-layered hierarchical control system that has been and is still being improved every year. This year, its structure has almost been completed. Its total configuration is shown in Figure 87.8. The lower layer contains a number of fuzzy control modules corresponding to the four control inputs (i.e., longitudinal and lateral cyclic, collective pitch, and rudder pedals). The upper layer consists of a number of modules for command interpretation, navigation, flight mode management, and basic flight.

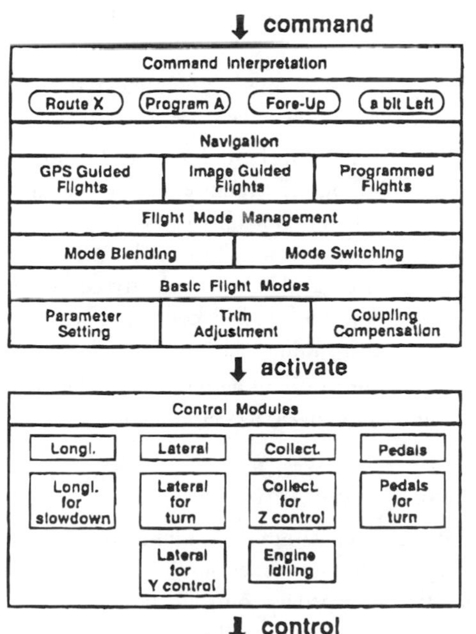

Figure 87.8 Hierachical control system.

Flight commands are first input into the upper layer. The upper layer then activates the lower layer according to a given command, and the lower layer gives actual control inputs to the helicopter.

Lower Layer: Fuzzy Control Modules

The lower layer contains fuzzy control modules corresponding to longitudinal, lateral, collective, and rudder controls. Each module consists of if-then fuzzy control rules.

Longitudinal Module Structure

The Longitudinal module has the control structure shown in Figure 87.9. Given a reference velocity $R_{\dot{x}}$, this module controls the forward velocity $\dot{X}$. From the reference $R_{\dot{x}}$ signal, the first controller (i.e., the controller labeled X-Velocity), computes a reference pitch angle R_θ signal for the second controller (i.e., the controller labeled Pitch). By referring to R_θ, the second controller provides the longitudinal control input. The parameters GDX, GDDX, GPIT, and GDPIT are the input gains of the controllers. GR_θ and GLON are the output gains. T_{pit} is an attitude trim, and T_{lon} is a control trim.

- The X-Velocity controller matrix represents the fuzzy control rule table shown in Figure 87.10, where PO, ZE, and NE are labels for the fuzzy sets *positive, zero,* and *negative,* respectively. Moreover, NB, NM, PM, and PB are labels for the fuzzy sets *negative big, negative medium, positive medium,* and *positive big,* respectively. For instance, the first rule says 'If the X acceleration ($\ddot{X}$) is *positive* and the

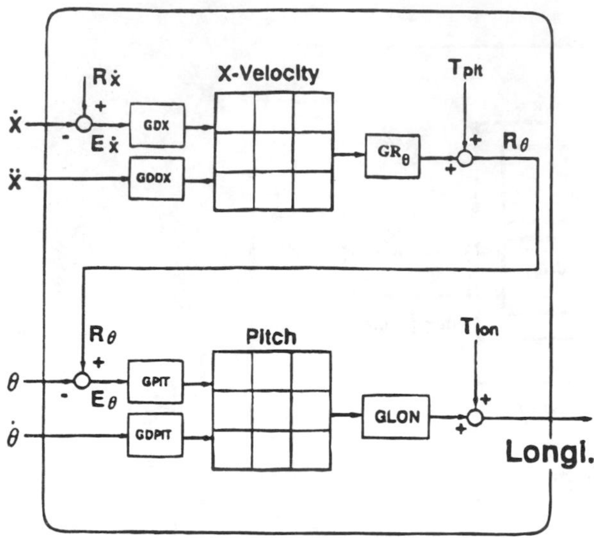

Figure 87.9 Structure of longitudinal control.

Longitudinal Module

(velocity : 9 rules)

$\ddot{X}$ is PO and $E_{\dot{X}}$ is PO → R_θ is NB

$\ddot{X}$ is PO and $E_{\dot{X}}$ is ZE → R_θ is NM

$\ddot{X}$ is PO and $E_{\dot{X}}$ is NE → R_θ is ZE

$\ddot{X}$ is ZE and $E_{\dot{X}}$ is PO → R_θ is NM

$\ddot{X}$ is ZE and $E_{\dot{X}}$ is ZE → R_θ is ZE

$\ddot{X}$ is ZE and $E_{\dot{X}}$ is NE → R_θ is PM

$\ddot{X}$ is NE and $E_{\dot{X}}$ is PO → R_θ is ZE

$\ddot{X}$ is NE and $E_{\dot{X}}$ is ZE → R_θ is PM

$\ddot{X}$ is NE and $E_{\dot{X}}$ is NE → R_θ is PB

Figure 87.10 Longitudinal module: velocity.

X velocity error ($E_{\dot{X}}$) is *positive,* then the desired pitch angle (R_θ) is *negative big.* Given actual $\dot{X}$ and $\ddot{X}$ flight data, the fuzzy controller infers R_θ based on fuzzy logic using these nine rules. We omit the details of fuzzy reasoning since it is widely known. These fuzzy reasoning calculations are performed by fuzzy inference engines (i.e., fuzzy chips installed in the fuzzy flight controller).

- The Pitch controller matrix in Figure 87.9 also consists of nine fuzzy control rules, as shown in Figure 87.11.

Lateral, Collective, and Pedals Module Structure

In addition to the Longitudinal control module, the lower layer contains Lateral, Collective, and Pedals control modules.

Longitudinal Module

(pitch : nine rules)

$\dot{\theta}$ is PO and E_θ is PO → Lon. is NB

$\dot{\theta}$ is PO and E_θ is ZE → Lon. is NM

$\dot{\theta}$ is PO and E_θ is NE → Lon. is ZE

$\dot{\theta}$ is ZE and E_θ is PO → Lon. is NM

$\dot{\theta}$ is ZE and E_θ is ZE → Lon. is ZE

$\dot{\theta}$ is ZE and E_θ is NE → Lon. is PB

$\dot{\theta}$ is NE and E_θ is PO → Lon. is ZE

$\dot{\theta}$ is NE and E_θ is ZE → Lon. is PM

$\dot{\theta}$ is NE and E_θ is NE → Lon. is PB

Figure 87.11 Longitudinal module: pitch.

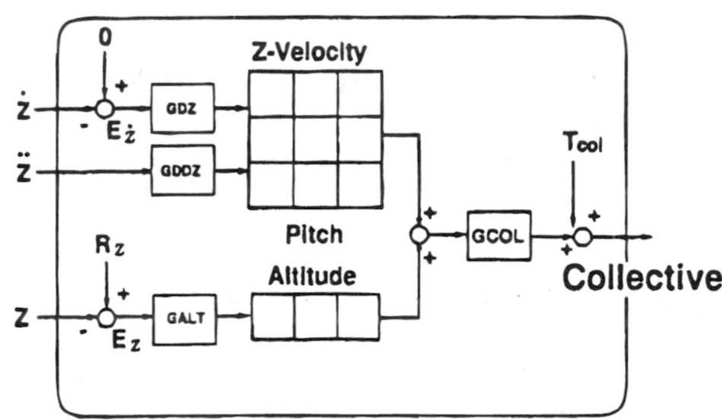

Figure 87.12 Structure of collective module for hovering.

- The Lateral control module is almost the same as the Longitudinal module.
- The Collective control module for the hovering flight mode has the control structure shown in Figure 87.12. In this case, the fuzzy controller consists of a (nine rule) Z-Velocity controller and a (three rule) Altitude controller connected in parallel. The Z-Velocity controller keeps $\dot{Z}$ and $\ddot{Z}$ zero, and the Altitude controller keeps the altitude at a reference R_z value.
- The Pedals control module sets the rudder control input with nine rules, given a reference yaw angle R_θ.

Control Module Activation

In order to achieve hovering flight, we set R_X and R_Y to zero, R_Z and R_ψ to certain values, and activate the longitudinal, lateral, collective, and pedals control modules. In this case, we use 57 fuzzy control rules altogether.

As is easily understood, the longitudinal module can also be used for forward or rearward flight, the lateral module can also be used for rightward or leftward flight, and the pedals module can also be used to make either hovering turns or rudder turns.

Fuzzy Control Design

It is widely believed that, in the area of fuzzy control, a fuzzy controller is designed using only an experienced operator's knowledge. However, we actually use all available knowledge sources. Generally speaking, there are two important sources of information that are needed to realize fuzzy control: knowledge of an objective system, and sensors to utilize appropriate information from the system.

In the helicopter control described in this paper, 5 different sensors are used to measure 15 state variables. For example, by taking the lateral-for-turn and the pedals-for-turn modules to achieve a coordinated turn, we briefly explain how to design a fuzzy controller. It is of crucial importance to find an appropriate control structure. (The structure's internal control parameters can subsequently be easily tuned through experiments and/or simulations.) Taking account of dynamic balance in a coordinated turn, we obtain the following relation:

$$\dot{\psi} = (g \tan \phi)/\dot{X} \tag{87.1}$$

This equation states that, during a coordinated turn, the yaw rate $\dot{\psi}$ is determined by a function of roll angle ϕ and forward velocity $\dot{X}$. This represents kinetic knowledge of the objective system. According to pilots and helicopter flight control manuals, we also know the following:

1. Tilt the lateral stick to keep the bank angle constant,
2. Do not use pedals for turning, and
3. Use pedals to control yaw rate and to get rid of any side-slip.

From these knowledge sources, we can find a control structure for making coordinated turns as shown in Figure 87.13, where R_ϕ is a given constant, $R_{\dot{Y}}$ is set to zero, and $R_{\dot{\psi}}$ is determined by Equation 87.1 according to ϕ and $\dot{X}$ during a turn.

The longitudinal-for-slowdown module is used for GPS-guided navigation. In this module, another fuzzy controller is added to the Longitudinal module in Figure 87.9 to infer a desired velocity $R_{\dot{X}}$ for slowdown based on a reference X position R_X. The lateral-for-Y control module is also designed for navigation in which the Y position is kept at a desired value. Hence, in this case, a Y-Position control is substituted for a Y-Velocity control in the lateral module. The collective-for-$\dot{Z}$ control module is constructed by removing the altitude block from the Collective module shown in Figure 81.12 (i.e., it consists of only the Z-Velocity block).

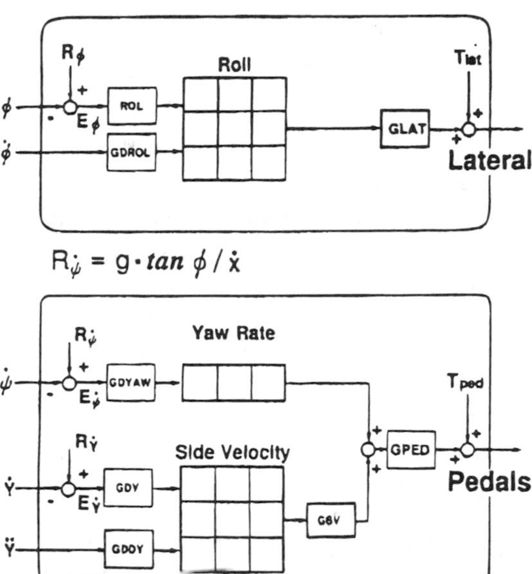

Figure 87.13 Structure of a coordinated turn.

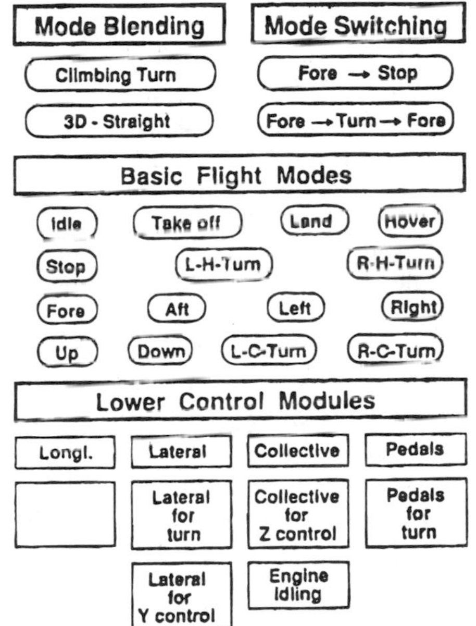

Figure 87.14 Flight mode management.

Upper Layer

Flight Mode Management Module

As seen in Figure 87.8, the flight mode management module consists of three submodules for basic flight modes, mode blending, and mode switching. The main task of this module is to select and appropriately activate four control modules in the lower layer according to a given command. Figure 87.14 shows the fine structure of the flight management module, including the lower control modules. We can manage 14 basic flight modes and an Engine-Idling mode. (Here, for example, L-H-Turn means "left hovering turn" and L-C-Turn means "left circling turn").

For example, in the case of the Hover basic flight mode, the flight management module assigns Hover mode to the Longitudinal (Longi.), Lateral, Collective and Pedals lower control modules with appropriate reference inputs and other necessary parameters.

Hover Mode

Longitudinal: = Hover
Lateral: = Hover
Collective: = Hover
Pedals: = Hover

In the case of the Forward (Fore) basic flight mode, it assigns Forward mode to the Longitudinal lower control module, and the other lower control modules are kept at Hover mode.

Forward Mode

Longitudinal: = Forward
Lateral: = Hover
Collective: = Hover
Pedals: = Hover

Moreover, in the case of the Left-Circling-Turn mode, it assigns Forward mode to the Longitudinal control module, the Left-Circling-Turn mode to the Lateral-for-Turn and Pedals-for-Turn control modules, while the Collective control module is maintained at Hover mode.

Left Circling Turn Mode

Longitudinal: = Forward
Lateral for Turn: = Left Circling Turn
Collective: = Hover
Pedals for Turn: = Left Circling Turn

In this way, many flight modes can be realized by combining a rather small number of lower-level control modules.

For mode blending, the flight mode manager can blend a few basic flight modes to achieve complex flight. For example, 3D-Straight mode blends together the Forward, Up, and Right basic flight modes. In this case, the Longitudinal control module is assigned Forward mode, the Lateral control module is assigned Right mode, the Collective-for-Z control module is assigned Up mode, and the Pedals control module is assigned Hover model.

3D Straight Mode

Longitudinal: = Forward
Lateral: = Right
Collective for Z: = Up
Pedals: = Hover

(We can blend physically compatible basic flight modes (e.g., Forward and Left, Up and Left-Circling-Turn, etc.). However, Right and Left, for example, cannot be blended.)

Mode switching refers to smoothly connecting two flight modes (not necessarily different) by inserting another flight mode between them. For example, in the case of Forward-Stop, we insert a Slowdown mode (by activating the Longitudinal-for-Slowdown lower control module) between the Forward and Stop basic flight modes as follows:

$$\text{Forward} \rightarrow \text{Slowdown} \rightarrow \text{Stop}$$

Moreover, in order to change flight direction, we execute mode switching of the form:

$$\text{Forward} \rightarrow \text{Turn} \rightarrow \text{Forward}$$

Parameter Setting Module

This module manages all parameters necessary to drive the fuzzy control modules in the flight mode manager. For example, when a mode shift is made from Left-Circling-Turn to Forward, the parameter setting module automatically transforms the final heading direction ψ of Left-Circling-Turn into a reference R_ψ for the Pedals control module under the Forward flight mode.

Trim Adjustment Module

This module adjusts attitude trims and control trims. For example, in Figure 87.9, T_{pit} is an attitude trim and T_{lon} is a control trim. Both of these are usually held constant. However, in general, when flight conditions change, trims need to be adjusted. (Detailed descriptions have not been included in this paper.)

Coupling Compensation Module

In the fuzzy control system described here, four inputs are controlled in a distributed manner. As a result, we have to compensate for some control couplings when the helicopter makes a significant change in its flight. As of the time of this writing, only the following couplings are compensated:

- Longitudinal to Lateral and Longitudinal to Pedal Couplings
- Lateral to Longitudinal and Lateral to Pedal Couplings
- Collective to Pedal Couplings

Conventional logic is used for this compensation according to a linear equation. It will be extended to rule-based logic in the future.

Navigation Module

The navigation module consists of three submodules for GPS-guided flights, image-guided flights, and programmed flights. In GPS-guided flights, the helicopter flies according to GPS signals from a point *A* to a point *B*, as indicated by a relative direction and distance. To achieve this, two new control modules were constructed, namely Longitudinal-for-Slowdown (to stop at a desired point) and Lateral-for-*Y*-Control (to maintain a desired route). In addition, the mode switching system is used to change flight direction at a waypoint.

In image-guided flights, the helicopter automatically searches for, flies to a landing strip, and then lands on a landmark with the aid of images from a CCD camera. Other visual guidance maneuvers, such as tracking and collision avoidance, will be studied in the near future.

In programmed flights, the helicopter can achieve a flight trajectory according to a prescribed plan. For instance, it can execute the sequence of events:

Takeoff → Up 10m → Forward 20m →
→ 45° Left Turn → Forward 50m → Land

Any such command sequence can be expressed as a flight program.

Command Interpretation Module

As explained in the previous hardware description, 256 commands can be transmitted from the ground station, where the helicopter's CCD camera signal is observed through a video monitor. Commands are organized into seven levels, as shown by the following examples.

- Level 7 commands (e.g., route X for navigation) are concerned with GPS-guided flight.
- Level 6 commands (e.g., automatic landing) are concerned with image-guided flight.
- Level 5 commands are concerned with programmed flights.
- Level 4 commands (e.g., "move a little to the left") are concerned with fuzzy commands.
- Level 3 commands are concerned with blended flight modes.
- Level 2 commands are concerned with mode switching.
- Level 1 commands are concerned with basic flight modes.

As for fuzzy commands, one can use fuzzily modified linguistic commands with respect to positions and velocities (e.g., "a bit forward", "a little faster", etc.).

87.4 Results

Simulations

The performance of the fuzzy controller was tested by computer simulations. A helicopter simulation program is installed in an IRIS Crimson graphic workstation. It models the BK117 Kawasaki Heavy Industries helicopter.

Figure 87.15 shows the altitude changes of the helicopter under a 5m/sec right crosswind. The solid line shows the changes with fuzzy control, while the dotted line shows the changes without fuzzy control. Figure 87.16 shows the effect of coupling compensation. The forward velocity $\dot{X}$ changes from 0 to 10m/sec. The solid line in Figure 87.16(a) shows the change in Y and that in Figure 87.16(b) shows the change in yaw angle when coupling is compensated. The dotted lines show the corresponding results without compensation. It can be seen that longitudinal to lateral/pedals coupling compensation works well.

Figure 87.17 shows the results of making a coordinated left turn. The helicopter flies with a speed of 5m/sec without a sideslip. As is seen in Figure 87.17(b), (c), and (d), it turns very smoothly.

Figure 87.18 shows the results of 3D-Straight flight. Figure 87.18(a) shows a desired velocity in 3D space. From this, we obtain the reference velocities $R_{\dot{x}}$, $R_{\dot{y}}$, and $R_{\dot{z}}$ that are assigned to the Longitudinal, Lateral, and Collective control modules, while $R_{\dot{\psi}}$ is set equal to zero. Figure 87.18(b) shows the trajectory in the X-Y plane, and Figure 87.18(c) shows the change of altitude with time. From these figures, it is apparent that the control system enables the helicopter to precisely realize a complex flight.

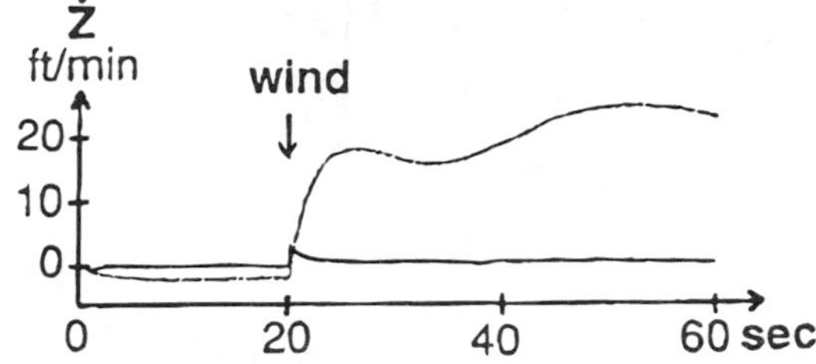

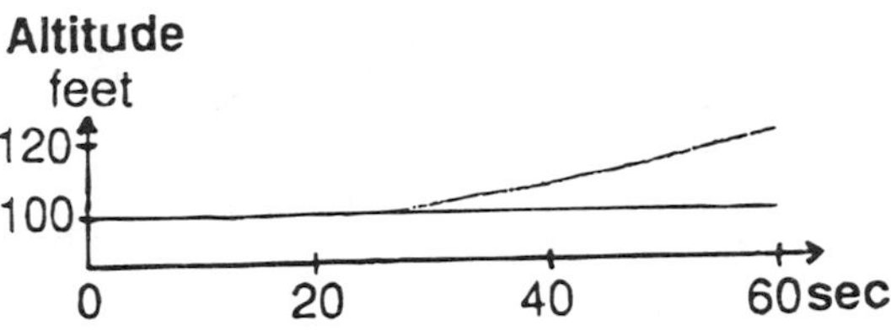

Figure 87.15 Hovering with right cross-wind of 5 meters per second.

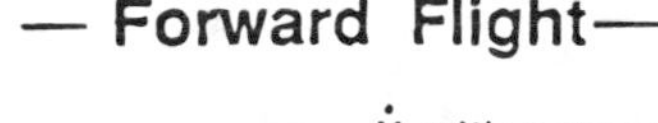

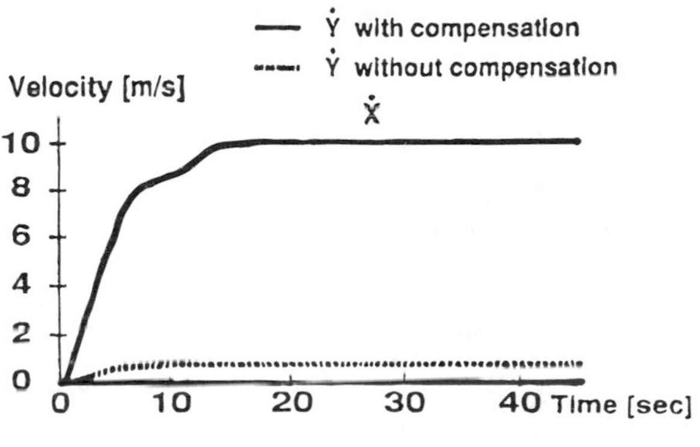

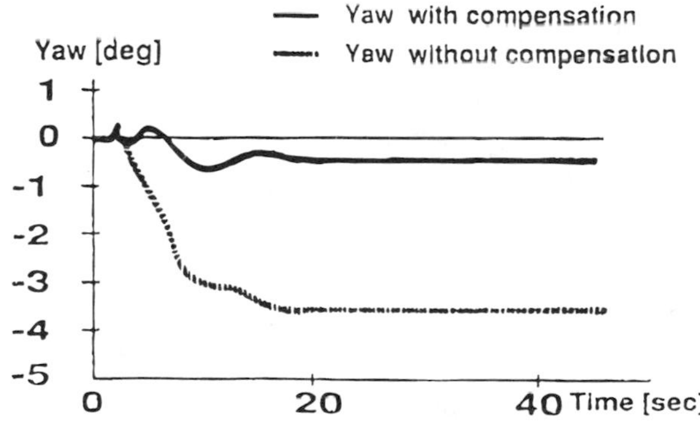

Figure 87.16 Effect of cross-coupling compensation.

Experiments

Experiments are performed in a 100m by 100m field. Figure 87.19 shows the velocity change that occurs as a result of switching

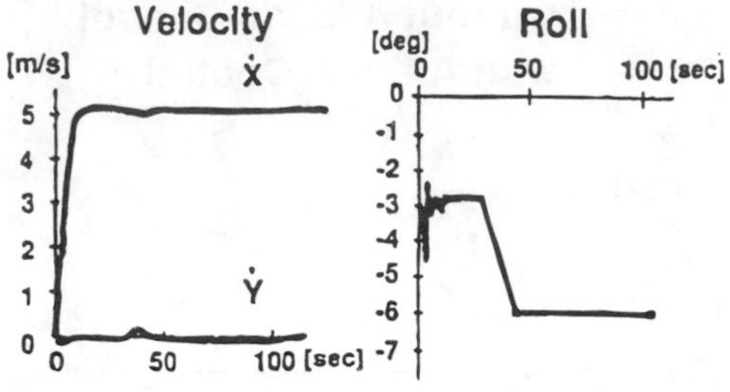

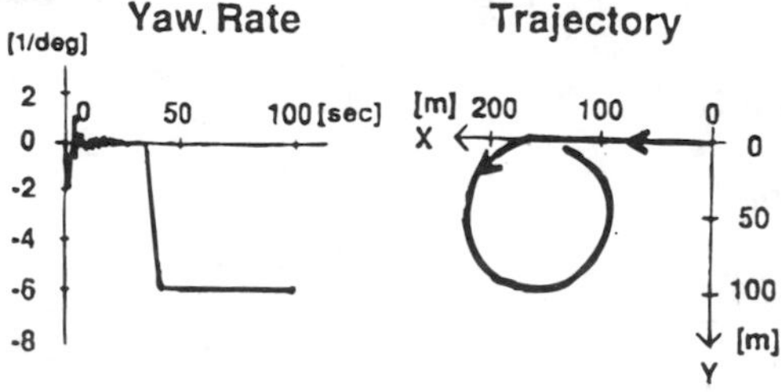

Figure 87.17 Coordinated left turn.

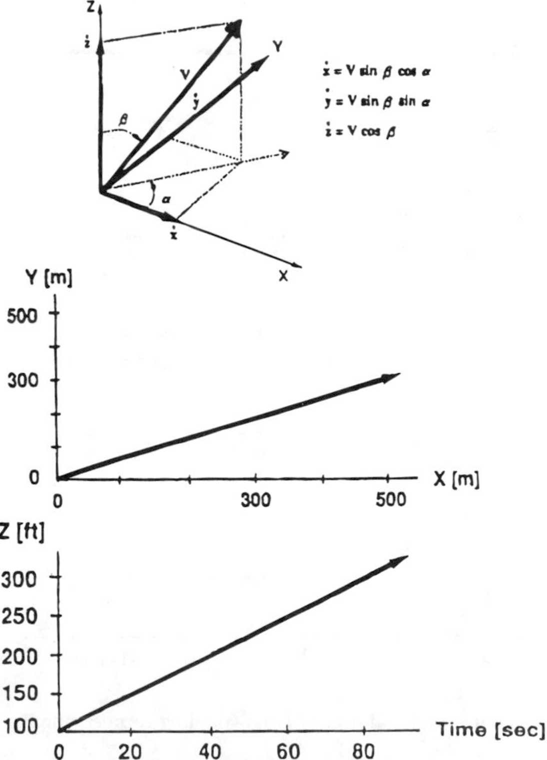

Figure 87.18 3D straight flight.

modes from Forward to Stop with slowdown control. The helicopter starts the Forward mode at 157 seconds and flies forward until 175 seconds. Then the Longitudinal-for-Slowdown control module is activated as a result of GPS position information. By referring to the GPS position signal, the helicopter slows down and stops at 183 seconds.

Figure 87.20 shows the result of GPS-guided flight. The vertical axis shows North-South position, and the horizontal axis shows East-West position. Starting from point *A*, the helicopter flies to point *D* via two waypoints *B* and *C*. As shown in Figure 87.20, the helicopter undergoes the following sequence of flight modes:

1. Hover at starting point *A*.
2. Forward to waypoint *B*.
3. 90° Left-Circling-Turn.
4. Forward to waypoint *C*.
5. 90° Right-Circling-Turn.
6. Forward to target point *D*.
7. Slowdown and Stop.
8. Hover at target point *D*.

For image-guided flight, one can test automatic landing based on image information derived from a landmark. A 5m by 10m white rectangular landing strip was established that contains a 1.8m diameter red circle. Using the on-board CCD camera, the helicopter successfully lands with commands produced by the planner shown in Figure 87.7. However, it was found to be very difficult to land the helicopter on the small red circular landmark.

87.5 Conclusions

Free flights of an unmanned helicopter in 3D space were successfully performed. The helicopter can fly with linguistic commands: either verbally expressed through a microphone or symbolically expressed through a computer keyboard. It can also fly guided by a GPS or by images. To achieve this goal, we have designed a sophisticated hierarchical control system with fuzzy logic. It has been found that fuzzy control techniques work sufficiently well to control a highly nonlinear and unstable system.

The most important results from this investigation are:

1. Multiple Flight Modes: Hierarchically structured control systems can facilitate the implementation of multiple flight modes because each high-level flight mode controller can take advantage of a common set of lower-level modules that control basic cyclic, collective, and rudder actuator inputs. Moreover, fuzzy control meta-rules can be used to implement flight mode switching.
2. Transient State Control: Fuzzy control systems can be used to implement transient state control (e.g., take-offs and landings).
3. Free 3D Flight: Fuzzy control systems can be used to implement free flight in three dimensional space. Specific control objectives (e.g., precise trajectories) do not have to be specified in the design of a fuzzy control system. Rather, a controller can be designed with knowledge about control strategy that only implicitly includes a control objective (e.g., the execution of circular flight without specifying a definite turning radius.).

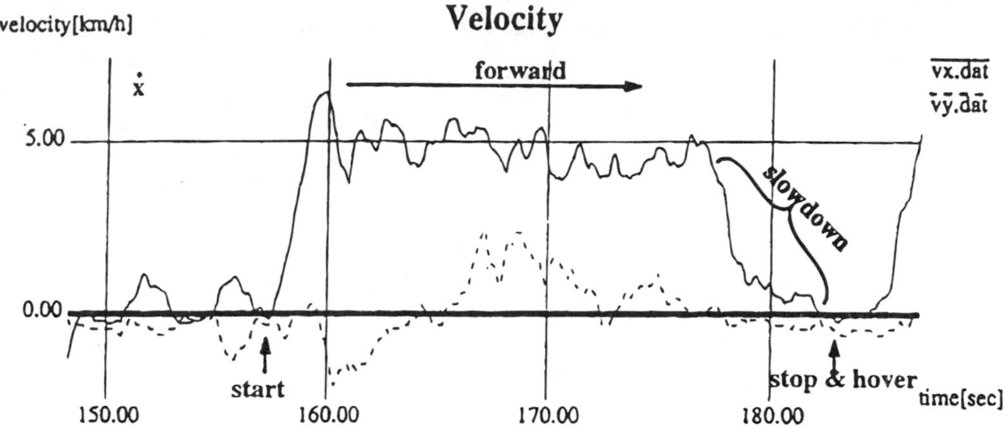

Figure 87.19 Forward to Stop Mode Switching with GPS.

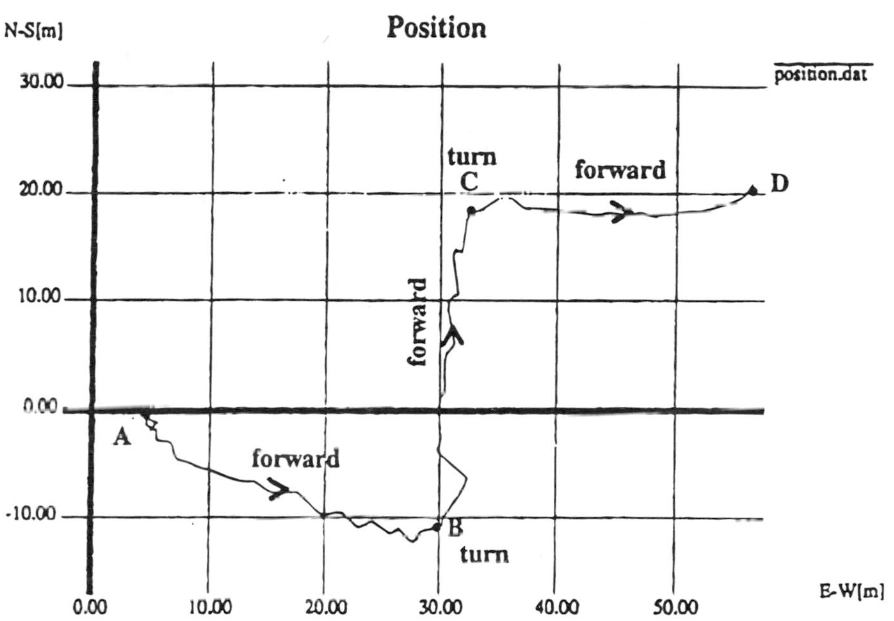

Figure 87.20 GPS-guided flight.

4. Fuzzy Commands: Fuzzy control systems can be designed to execute fuzzy instructions. Because the overall control system is written in terms of linguistic rules and has a mechanism to understand the meaning of fuzzy concepts, it is able to accept and interpret imprecise commands (e.g., "fly a little slower" or "fly to the right").
5. Image Feedback Control: Fuzzy control systems can be used to implement image feedback control. Because fuzzy control rules are compatible with both quantitative and qualitative information, visually-derived qualitative information can be used as input to fuzzy control systems.
6. Expert Control: Expert control has been demonstrated by developing voice- and keyboard-based linguistic controllers that enable novices to control the aircraft without any prior knowledge of helicopter dynamics.

For future study, we will continue to improve GPS-based navigation with visual collision avoidance and tracking systems. A fail-safe system and a monitoring system are particularly indispensable for an unmanned helicopter.

Acknowledgments

This research is supported by the Ministry of International Trade and Industry, Japan in cooperation with United Technologies Research Center and Yamaha Motor Corporation. The authors would like to thank S. Nakamura (Tokyo Institute of Technology) for his assistance in developing the helicopter's image-guided flight capabilities.

References

Fagg, A., Lewis, M., Montgomery, J., and Bekey, G. 1993. The USC autonomous flying vehicle: An experiment in real-time behavior-based control, *Proc. 1993 IEEE/RSJ Int. Conf. on Intelligent Robots and Systems,* July, Yokohama, Japan.

Keller, J. 1993. The impact of fuzzy set theory in computer vision, *Proc. 12th Annl. Mtg. of the North American Fuzzy Information Processing Society,* August, Allentown, PA.

Larkin, L. 1984. A fuzzy logic controller for aircraft flight control, *Proc. 23rd Conf. on Decision and Control,* December, Las Vegas, NV.

Marr, D. 1982. *Vision,* W. H. Freeman, San Francisco, CA.

Nishimori, K., Mataharu, O., Naganori, I., and Tokutaka, H. 1994. Vision system for fuzzy driving control of an automatic vehicle, *Proc. 3rd Int. Conf. Fuzzy Logic, Neural Nets and Soft Computing,* August, Iizuka, Japan.

Ozawa, Y. 1994. *Image—Guided Control of an Unmanned Helicopter,* Master's Thesis, Tokyo Institute of Technology, Tokyo, Japan.

Phillips, C., Karr, C., and Walker, G. 1994. A genetic algorithm for discovering fuzzy logic rules, *Proc. Int. Fuzzy Systems and Intelligent Controls Conf.,* March.

Stengel, R. 1993. Toward intelligent flight control, *IEEE Trans. Syst., Man, and Cyber.,* 23(6), November/December.

Sugeno, M. 1985. An introductory survey on fuzzy control, *Information Sciences,* vol. 36.

Sugeno, M. 1993. *Development of an Intelligent Unmanned Helicopter 1991–1993,* Technical Report, Tokyo Institute of Technology, Tokyo, Japan.

Sugeno, M. and Park, G. 1993a. An approach to linguistic instruction based learning and its application to helicopter flight control, *5th Int. Fuzzy Systems Assoc. Congress,* July, Seoul, Korea.

Sugeno, M. and Park, G. 1993b. An approach to linguistic instruction based learning, *Int. J. Uncertainty, Fuzziness and Knowledge-Based Systems,* vol. 1(1).

Sugeno, M., Murofushi, T., Mori, T., Tatematsu, T., and Tanaka, J. 1989. Fuzzy algorithmic control of a model car by oral instructions, *Fuzzy Sets and Systems,* 32:207–219.

Sugeno, M., Griffin, M., and Akagi, H. 1993a. *Fuzzy Logic Flight Control of an Unmanned Helicopter,* Technical Report, Tokyo Institute of Technology, Tokyo, Japan.

Sugeno, M., Griffin, M., and Bastian, A. 1993b. Fuzzy hierarchical control of an unmanned helicopter, *IFSA Congress,* Korea.

Sugeno, M., Griffin, M., Walker, G., and Bastian, A. 1993c. Issues for blending fuzzy controllers in hierarchical systems, *1st Asian Fuzzy Systems Symp.,* November, Singapore.

Wade, R. and Walker, G. 1994. Fuzzy logic adaptive controller—helicopter (FLAC-H): a multi-platform, rotary-winged aerial robotic control system, *19th Army Science Conf.,* June, Orlando, FL.

Walker, G. and Mo, S. 1994. Forward modeling of helicopter flight dynamics using recurrent neural networks, *19th Army Science Conf.,* June, Orlando, FL.

Wade, R., Walker, G., and Phillips, C. 1994. Combining genetic algorithms and aircraft simulations to tune fuzzy rules in a helicopter control system, *Advances in Modeling and Simulation Conf.,* April, Huntsville, AL.

Zadeh, L. 1965. Fuzzy sets, *Information and Control,* vol. 8.

88

Fuzzy and Neural Modeling

Mary Lou Padgett
Auburn University

88.1 Introduction 1139
88.2 Engineering Approaches and Applications 1139
Structure Determination • Parameter Identification • Dynamics - • Hardware Improvements for Pattern Recognition
88.3 Futures 1141

88.1 Introduction

The integration of fuzzy systems and neural networks systems is a natural way to capture the benefits of both techniques.

Fuzzy systems excel in the processing of nonstatistical, poorly defined information where some rule-of-thumb can be obtained from experts or deduced from data clustering or other methods. The flexibility of resulting systems is a major issue in their successful application. The addition of neural networks for adaptation of fuzzy systems parameters solves this problem, at the expense of designing a hybrid system.

Neural systems process data and give function approximations, classifications, and other useful generalizations about the data, even when no system model is known. The black box quality of the resulting neural system can be hard to interpret. Fuzzy systems can be used to help interpret these results to the supervisory system. Fuzzy decision-making procedures can also help train or adapt a neural system in an orderly manner. Heuristics often employed by designers can be captured in a meaningful and generalizable manner using fuzzy systems.

A neural fuzzy or fuzzy neural system can be much more robust than a traditional system or a simple neural or fuzzy system. Weaknesses of any one approach can be compensated by hybridization. Of course, care must be taken to avoid excessive and unnecessary complication of a system. A truly high System IQ is produced by a machine or system with the intelligence to cut to the core of the matter at hand and use the simplest solution appropriate (Padgett and Padgett, 1995; Zadeh, 1995).

88.2 Engineering Approaches and Applications

Many of the articles in this chapter deal with engineering approaches to fuzzy/neural modeling. Some of the most promising include soft LVQ (by Karayiannis) for pattern recognition and elastic fuzzy logic for controls (by Werbos). Pedrycz and Lindblad and Lindsay offer hardware solutions.

Current problems and possible solutions are concisely summarized by Jang and Sun (1995). Beginning with the modeling strategies discussed in the Padgett's previous article on Fuzzy Modeling, Jang considers structure determination and parameter identification.

Structure Determination

Design considerations for the determination of structure for a neuro-fuzzy system include the following (Jang and Sun, 1995).

- Manner of partitioning.
- Number of membership functions (MF's) for each input.
- Number of fuzzy if-then rules.
- And others.

Approaches to solving this problem include:

- Fuzzy CART—Jang.
- Reinforcement learning—Lin.
- Fuzzy k-d trees—Sun.
- Iterative method—Sugeno.
- Clustering algorithms—Chiu Khedkar and Wang.
- RECENT advances on constructive and destructive learning of neural networks.

Parameter Identification

Parameter identification by back-propagation gradient descent and the least-squares methods is addressed by (Jang and Sun 1995).

Design Considerations:

FAST: least-squares estimates for parameter identification.
SLOW: gradient descent.
NEW: new learning strategies are being sought, and reinforcement learning is very promising. (See papers by Jani et al and by Werbos in this chapter.)

0-8493-8343-9/97/$0.00+$.50

Dynamics

An example application of neuro-fuzzy modeling and control is given by Jang and Sun (1995), in their description of their Adaptive-Network-Based Fuzzy Inference System (ANFIS).

The technique employed follows a "minimum disturbance principle" rather than a sigmoidal approach. The addition of learning capability to a fuzzy inference system allows the design of a fuzzy controller using of all the design methodologies usually employed for neural networks controllers.

First, a fuzzy system is described. Next adaptive networks are described, giving back-propagation learning and radial basis function networks as examples. Then, the ANFIS is compared to backpropagation learning and design techniques for fuzzy neural controllers are recommended.

Fuzzy System

A fuzzy system operates on a fuzzy set. Such a set is distinguished from "crisp" sets by allowing a continuous range of membership, from zero to total. Such membership ranges may be discretized. Fuzzy membership is subjectively determined, having nothing to do with randomness.

Fuzzy sets and probability differ in useful ways. In probability, the area of the universe is considered to be 1, and subsets are crisp. An element is either totally a member of a set, or totally not in the set. The probability of a randomly selected element being in a particular subset of the universe depends on the area of the subset (also may be discretized). This offers an objective treatment of randomness contrasting with the subjective nonrandomness of fuzzy sets.

Fuzzification

Specifying a fuzzification system requires:

- Identifying a suitable universe (range of the variable, and its dynamics).
- Specification of an appropriate membership function (shape, temporal characteristics).

Fuzzy set operations are defined in many places and will not be detailed here. It is critical for system specification to state these plainly, however, because there are many subtle variations of operations in use. Some examples include:

- Containment (subset).
- Union.
- Intersection.
- Complement.
- Others: T-norm, T-conorm, min, max, . . .

Membership functions (MFs) (also frequently encountered and defined):

- Triangular.
- Trapezoid.
- Gaussian.
- Generalized Bell.
- Sigmoidal.
- Any type of continuous probability function.
- Other.

Fuzzy Inference

The fuzzification module inputs crisp values from the range of the universe and processes them using the fuzzy membership function for that variable to produce a fuzzy "alpha" value. The Fuzzy Inference module inputs these alpha values and processes them using Fuzzy If-Then rules to produce a fuzzy output for Defuzzification.

Fuzzy Rules (Rule Base): A fuzzy relation maps any alpha value input to the IF statement to a set of possible conclusions, or THENs.

Database or Dictionary: Membership functions used in the fuzzy rules.

Fuzzy Reasoning Mechanism: Given a set of fuzzy rules and their alpha inputs, a fuzzy reasoning process combines all these to produce a fuzzy output. Computation of this output becomes more complex when multiple antecedents (IF's) and/or multiple rules are present.

Examples of Fuzzy Inference Systems

Mandani Fuzzy Model: Operators may be the original max-min, a popular variation called product-max, or other variations such as T-norm or T-conorm. Variations are continually tried for improving performance for particular applications.

Sugeno Fuzzy Model (TSK fuzzy model) (Takagi, Sugeno, 1985, Sugeno, Kang, 1988): The TSK approach attempts to generate fuzzy rules from a representative set of input-output data (Jang and Sun, 1995). For example, a fuzzy rule might be

$$\text{if } x \text{ is } A \text{ and } y \text{ is } B \text{ then } z = f(x, y)$$

where

x and y = alpha values of two separate variables
A and B = fuzzy sets in the *IF* (antecedent) part of the rule (here the rule has two separate antecedents to be combined)
and, z = a crisp function in the *THEN* (consequent) part of the rule.

Bezdek reports significant success in using polynomial forms of z (personal communication). Any appropriate function may be used, but polynomials are favored.

In the TSK model, the order of the polynomial is important.

Zero order polynomial (f is a constant): Zero-order Sugeno fuzzy model

- Special case of Mamdani fuzzy inference system.
- Special case of the Tsukamoto fuzzy model.
- RADIAL BASIS FUNCTION (RBF)—functionally equivalent with minor constraints.

Output is a smooth function ONLY if the membership functions in the antecedent overlap enough

First-order Sugeno fuzzy model: First-order polynomial.

Output is crisp, so no defuzzification, just combination

Note: combination may be weighted average or weighted sum (if sum is nearly 1.).

Tsukamoto Fuzzy Model: The THEN part (antecedent) of each rule is a monotonical membership function which gives a crisp output. The total output is thus just the weighted average, avoiding headdefuzzification.

Defuzzification

Defuzzification takes a fuzzy inference output and converts it to a crisp value useful to the external environment. Techniques include:

- Centroid of area.
- Bisector of area.
- Mean of maximum.
- Largest of maximum, smallest of maximum.
- Other.

Partitioning Styles for Fuzzy Models. The role of the THEN or antecedent part of the rule is to partition the input space so that the localized regions of the input space have similar characteristics with respect to the particular antecedent. This is known as determining the preimage of a subset of the range of a relation. Three approaches are common: grid, tree and scatter. Grid works well for a small number of inputs, but not for a large number. With a tree partition, fewer rules are used, but more membership functions per input are needed, so clarity of linguistic meaning may be lost.

Neuro Fuzzy Modeling. Common practice is to use fuzzy modeling (and available data) to determine model structure, then move to neural techniques for parameter identification. Two techniques frequently used for this are the multilayer perceptron trained using backpropagation and the radial basis function network. Many of the following articles give specific examples of neuro-fuzzy modeling commercial applications (Lindblad and Lindsey) and research suggestions. The paper by Werbos gives ideas for approaches which are beyond current applications, but appear to offer significant (long-term) commercial advantage. The work of Yager and Filav (1994) is also worthy of note (Zadeh, 1995).

Hardware Improvements for Pattern Recognition

On-chip learning capability should expand the range of fuzzy neural applications in the areas of adaptive signal processing and controls. See papers by Pedrycz and Lindblad and Lindsay for a description of some of these hardware improvements underway.

88.3 Futures

Soft LVQ and modifications discussed by Pedrycz and Karayiannis in this chapter are one of the most promising avenues of research for pattern recognition. The paper on fuzzy ART by Carpenter and Grossberg reflects the similar trends, and describes promising extensions to their adaptive resonance theory.

For controls applications, reinforcement learning covered by Jani, et al. and extended by Werbos seems extremely promising.

More extensive incorporation of evolutionary programming techniques, expert systems, and virtual reality into the design and use of soft computing systems is indicated by trends exemplified in the articles in the Computational Intelligence chapter of this book.

References

Jang, J.-S. R, and Sun, C.-T. 1995. Neuro-fuzzy modeling and control, *Proc. IEEE, Special Issue on Engineering Applications of Fuzzy Logic,* 82(3):378–406.

Padgett, M. L. and Padgett, W. D. 1995. Simulation and computational intelligence in real-world applications, *Simulation,* July.

Sugeno, M. and Kang, G. T. 1988. Structure identification of fuzzy model, *Fuzzy Sets and Systems,* 28:15–33.

Takagi, T. and Sugeno, M. 1985. Fuzzy identification of systems and its application to modeling and control, *IEEE Trans.,* SMC-15:116–132.

Yager, R. R. and Filev, D. P. 1994. Essentials of Fuzzy Modeling and Control, John Wiley & Sons, New York, NY.

Zadeh, L. A. 1995. Personal Communications.

Further Resources

See the other articles by Padgett in this section for definitions, journals, and internet sources in addition to the material listed below.

Tutorials and Basics

Bezdek, J. 1994. An Introduction to Fuzzy Logic, *WCCI Tutorial #7,* IEEE Press, Piscataway, NJ.

Galicher, S. and Roulloy, L. 1995. Fuzzy controllers: synthesis and equivalences, *IEEE Trans. Fuzzy Systems,* 3(2):140–148.

Gupta, M. M. and Sinha, N. K. 1996. *Intelligent Control Systems: Theory and Applications,* IEEE Press, Piscataway, NJ.

Jang, J.-S. R. and Sun, C.-T. 1995. Neuro-fuzzy modeling and control, *Proc. IEEE. Special Issue on Engineering Applications of Fuzzy Logic,* 83(3):378–406.

Jani, Y. 1995. Fuzzy logic in control, *SPIE's AeroSense '95,* SC53, Orlando, FL.

Kartalopoulos. S. V. 1996. Time-Dependent Fuzzy Logic, *Understanding Neural Networks and Fuzzy Logic: Basic Concepts and Applications,* 130–152, IEEE Press, Piscataway, NJ.

Kaufmann, A. and Gupta, M. M. 1991. *Introduction to Fuzzy Arithmetic,* Van Nostrand Reinhold, New York, NY.

Padgett, M. L., ed. 1993. *Simulation: Special Issue on Neural Networks—Model Development for Applications,* May, *SCS Press,* San Diego, CA.

Padgett, M. L. ed. 1995. *Simulation: Special Issue on Computational Intelligence in Simulation Applications,* July, SCS Press, San Diego, CA.

Padgett, M. L. 1993. Neural Network Basics: Applications, Examples and Standards, *IJCNN Nagoya Tutorial Book,* 383–412, IEEE Press, Piscataway, NJ.

Padgett, M. L., Karplus, W. J., Deiss, S., and Shelton, R. 1994. Computational intelligence standards: motivation current activities and standards, *Computer Standards and Interfaces,* 16:188–203.

Padgett, M. L. and Padgett, W. D. 1995. Simulation and computational intelligence in real-world applications, *Simulation,* July.

Sugeno, M. and Kang, G. T. 1988. Structure identification of fuzzy model, *Fuzzy Sets and Systems,* 28:5–33.

Sugeno, M. and Yasukawa, T. 1993. A fuzzy-logic-based approach to qualitative modeling, *IEEE Trans. Fuzzy Systems,* 1:7–31.

Terano, T., Asai, K. and Sugeno, M. 1989. *Applied Fuzzy Systems,* Academic Press, San Diego, CA.

Terano, T., Asai, K. and Sugeno, M. 1989. *Fuzzy Systems Theory and Its Applications,* Academic Press, San Diego, CA.

White, D. A. and Sofge, D. A. 1992. *Handbook of Intelligent Control: Neural, Fuzzy, and Adaptive Approaches,* Van Nostrand Reinhold, New York, NY.

Yager, R. R. and Filev, D. P. 1994. *Essentials of Fuzzy Modeling and Control,* John Wiley & Sons, New York, NY.

Yen, J., Langari, R. and Zadeh, I. A. 1995. *Industrial Applications of Fuzzy Logic and Intelligent Systems,* IEEE Press, Piscataway, NJ.

Zadeh, L. A. 1995. Fuzzy Logic and Calculi of Fuzzy Rules, Fuzzy Graphs and Fuzzy Probabilities, *WCNN95 Tutorials,* INNS.

89

NeuFuz: A Combined Neural Net/Fuzzy Logic Tool

Thomas Lindblad
Royal Institute of Technology, Stockholm

Clark S. Lindsey
Royal Institute of Technology, Stockholm

89.1 Introduction 1143
89.2 Working with the Neural Network of NeuFuz4 1143
89.3 Working with the Fuzzy Logic Part of NeuFuz4 1145
89.4 Working with the Code Generator Part of NeuFuz4 1145
89.5 Summary 1146

89.1 Introduction

The NeuFuz4 from National Semiconductors (1993) is a software development tool for designing fuzzy logic engines for processors (mainly from National). The software tool uses a neural network to determine the fuzzy rules. This means that you first train a software neural network by presenting inputs and desired outputs and then a fuzzy logic module is produced. The "heart" of the software is the conversion of the neural net into an equivalent (at least functionally) fuzzy system. The fuzzy logic module is then coded into an embedded processor. This generally requires some approximation techniques, but even if the end result is an approximation, it is good enough for the application. The total procedure then looks like

Neural Network → Fuzzy Logic → Assembler (or C-) code.

It thus is a two step technique and the whole procedure must be repeated if continuous learning of the processor application is required. While the second step in the above procedure is fairly simple, the first one is more delicate and tricky. While a neural network is trained rather than programmed to solve a problem, fuzzy logic can be said to solve problems through a succession of increasingly accurate steps. This is accomplished by replacing the traditionally crisp sets by fuzzy sets.

Other methods of implementing solutions from neural network training may involve Boolean neural nets. One example is the hardware synthesis presented by Beagles (1992) in which the results from NASA "NETS" neural networks code is translated into the OCT hardware description language (BDS) using the "OCT" VLSI CAD tools from U.C. Berkely, and a dedicated chip can be produced. This method is quite interesting for cases where you do not expect any changes to the system and the number of embedded systems is large. For other situations, the advantages of the more flexible NeuFuz are rather clear.

A compromise between the two methods mentioned above could be the use of Field Programmable Gate Arrays, FPGA. Fast Boolean neural networks can be implemented using standard hardware from, e.g., Altera (Betin and Bouchard, 1993). Recently, Lundheim et al. (1995) showed a system developed by DEC, which could be used to implement a programmable active memory (PAM) of a neural network for a triggering circuit in connection with the Large Hadron Collider (LHC) at CERN laboratory in Geneva. Still the implementation of the algorithm is said to take one or two weeks. While this may be quite all right with a large longterm project like LHC/CERN, there are other applications where a fast implementation is required and where thus NeuFuz should be quite useful.

The NeuFuz approach may be quite "logical" since neural networks are not only redundant but also inherently exhibit a certain "slack" in operation, which may fit well into the approximation approach of fuzzy logic. However, since the implementation is on a (single) von Neumann processor, one will not benefit from the massively parallel structure of neural nets.

89.2 Working with the Neural Network of NeuFuz4

NeuFuz4 is a Windows program and is quite easy to use. A flow chart is shown in Figure 89.2 and for comparison we have included the "direct" method of Beagles (1992) in Figure 89.1.

There are a few limitations to the NeuFuz4. The number of inputs is limited to 4 and the number of outputs is a single output neuron. The network appears to be of a layered feedforward type with several hidden layers. It also appears to be trained using a backpropagation learning paradigm. The input file is a standard ascii-file with the input and the desired output on the same line (i.e., max 5 numbers). All numbers may be positive or negative integers or floating-point numbers and the maximum number of lines or patterns is 1200. An example

0-8493-8343-9/97/$0.00+$.50

from a high energy physics (HEP) experiment (Lindblad et al., 1996) is shown below.

```
14.47480 0.39768 2.18173 0.26454 1
 0.57865 0.02176 0.07198 0.01586 0
 7.36548 0.25612 1.39249 0.35353 1
 1.04654 0.04892 0.01344 0.00357 0
 1.84157 0.02293 0.00086 0.00026 0
 4.49228 0.46454 4.02087 0.42985 1
 1.03644 0.10334 0.01663 0.00743 1
 . . . .
```

Here the first four values of each line correspond to the total and the transversal linear momentum of two emerging particles, i.e., p^1, p_T^1, p^2, and p_T^2. The fifth number indicates if the decaying particle was a Higgs particle (1) or just a background event (2). The first values are in the range 0 to 17, while the other three input values all are in the range 0 to 0.7. Generally, Higgs events are recognized here by the first value being quite large. However, this is not always the case as can be seen from the last vector in the table above. In order to get good results, several thousand training vectors are required. However, for a crude separation of the two kinds of events, 600 of each type of training vector may be sufficient. Before learning is commenced, the user is requested to give the desired error (e), the learning rate and learning factor. Default values are 0.01, 0.1, and 0.1.

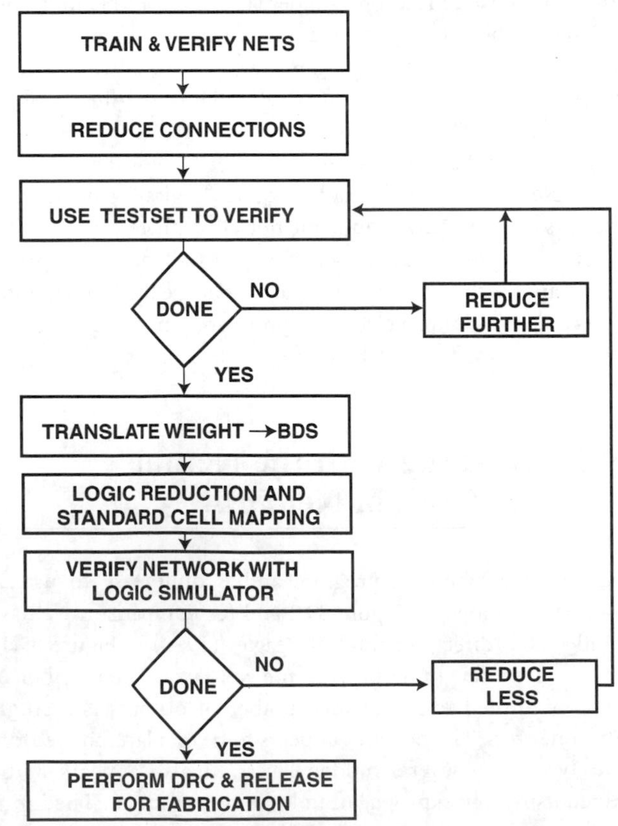

Figure 89.1 Application specific neural network synthesis process (Lundheim et al., 1995). The software neural network is directly translated to hardware language for fabrication (cf. Figure 89.2.)

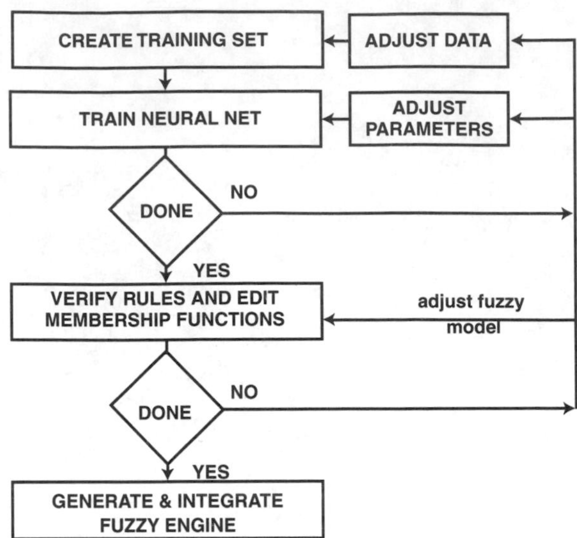

Figure 89.2 The *NeuFuz4* flowchart. The software neural network is translated into a fuzzy logic system, which in turn is converted to assembler or C-language for implementation on von Neumann processors.

The user is also requested to set the number of membership functions. Membership functions are inherent to fuzzy logic and will be used to represent each input variable. Valid values in *NeuFuz4* are 2 to 7 (or 0 if that input is not used). In order to get maximum dispersion in the above case, in which the momentum values can vary from zero to 17, we may choose 7 for all four inputs. One may try different values of the number of membership functions, but a change will reset the model and the net will have to relearn with the new values.

At this stage it may also be rewarding to consider the range of I/O values and the scaling. The input values must be scaled before being passed to the FUZZ routine (see below) of the fuzzy code. This scaling is performed using

$$I_{\text{scaled}} = \{(I_{\text{input}} - I_{\text{min}})256/(I_{\text{max}} - I_{\text{min}})\} - 1. \quad (89.1)$$

whereas the value from the sensor, I_{min} is the smallest input value and I_{max} is the largest input value. Note that this equation may be different for each input. In our Higgs example, input number 1 certainly spans over a wider region than the other inputs. A similar formula is used for the output value, or

$$\text{Out} = \text{Out}_{\text{scaled}}/128^{*}\text{weight} \quad (89.2)$$

i.e., you should apply this formula to remove the scale factor from the output value of the fuzzy code.

To check how the learning proceeds, you can look periodically at the maximum error/ϵ. As usual, this value increases, decreases or oscillates depending on the actual learning. Changing the learning rate is done manually in very much the same way as in other neural network software.

A test file (or data recall file) with other input data than those used above may be employed to test the trained network. The file looks similar to the file above but does not include the desired value. Hence testing is preferentially performed using one file with "Higgs" and one file with "background."

89.3 Working with the Fuzzy Logic Part of NeuFuz4

In this second step NeuFuz4 generates membership functions (overlapping bellshaped curves with a range of input values on the x-axis and the degree of membership on the y-axis as shown in Figure 89.3) and fuzzy rules for your application. Contrary to many other fuzzy logic codes, the membership functions here are non-linear curves rather than simple triangles. However, the fuzzy logic engine code uses a linear approximation and there is a possibility of editing the membership functions (cf. Figure 89.3). This is a simple operation using a graphical display window and the mouse.

The fuzzy rules (which describe a system's behavior) are used to calculate the output value. It is necessary to verify that these rules are working to an acceptable degree of accuracy. Marginally useful rules may also be eliminated in this validation and editing procedure. The procedure is referred to as "adjusting the deletion factor." The latter is a number between 0 and 1 that determines the number of rules that NeuFuz will extract from the neural net. The default value is 0.0 which means that no fuzzy rules are deleted. So why do you want to delete any fuzzy rules. The reason is obvious: fewer rules means smaller code, which in turn yields a faster execution of the code.

Generally, one would like to reduce the amount of memory required and *NeuFuz4* provides two methods. The first one is to increase the aformentioned deletion factor. It will reduce the number of fuzzy rules, which in turn will reduce the ROM memory needed. You can also reduce the number of membership functions (maximum 7), which reduces both the RAM and ROM needed by your application.

89.4 Working with the Code Generator Part of NeuFuz4

The system provides a code generator utility to translate the fuzzy rules and the membership functions of your application into assembly or C code. The required information is provided in a file called `model.neu`, which includes information on the fuzzy system based on the preceding neutral net training and information extraction. The code generator will also require a set of generic assembly language code modules. These are found in a binary file called `n4.lib`. When generating the code you may also specify the ROM and RAM memory sizes you plan to use for your application. You may also skip this part by specifying "No memory check." After clicking "Generate Code" you get a window with a message something like:

```
2401 rules, 4 input variables
12139 bytes of ROM
66 bytes RAM (inc 6 fast RAM, 3 register
RAM)
Rules "higgs.neu"
Log "higgs.log"
Code "higgs.asm"
```

This is what we get in the case of the Higgs search implementation mentioned before. Clearly, this code is not going to execute very fast on a von Neumann processor. Generally speaking most codes will take many milliseconds and generally one seldom finds applications which execute in less than one millisecond.

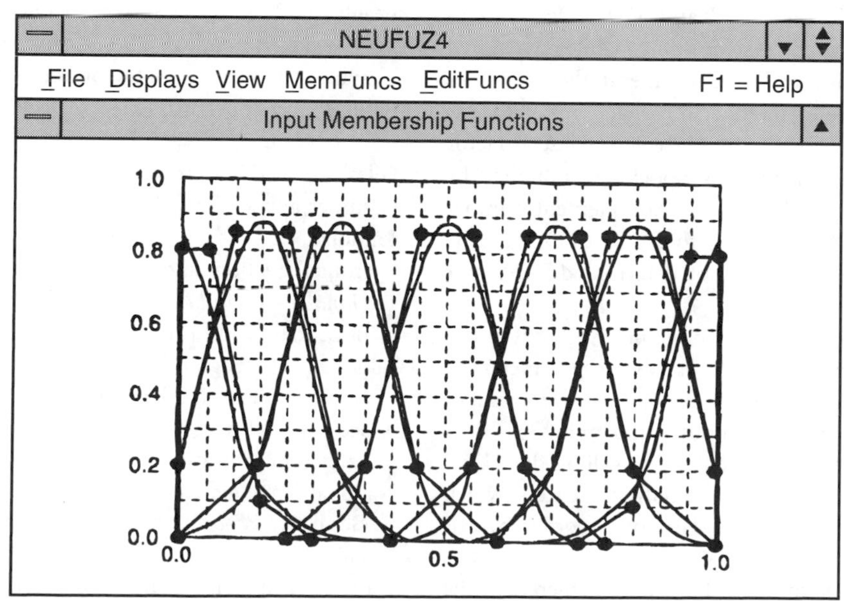

Figure 89.3 The bell shaped membership functions and the editing of the trapezoids associated with these is carried out stepwise one at a time using the mouse. In the case of the Higgs search, the x-axis associated with input 1 (2–4) would have a range from 0 to 17 (approx 0.7).

For comparison, we can mention that the standard case of an xor-gate with two inputs, results in the following "size" of the code:

```
49 rules, 2 input variables
844 bytes of ROM
50 bytes RAM (inc 6 fast RAM, 3 register
RAM)
Rules "exor.neu"
Log "exor.log"
Code "exor.asm"
```

The assembly code file contains a number of relocatable modules. Each module contains routines for one or two specific functions used in the fuzzification, rule evaluation and defuzzification processes. A list of code modules and their basic functions are given below:

Module Name	Routine	Function
FUZDAT	None	Variable storage
FAST	None	Variable storage
TOP	None	Variable storage
FUZZ	FUZZ	Mainloop
	GOFUZ	Mainloop; fuzzification
	MFDCALC	Calculate degree-of-membership for an input
MFCODE	MFPLUP	Table look-up of membership function data'
RULECODE	RULEVAL	Mainloop: rule evaluation
	COUT	Calculate output from each rule and sum results
	RDOM	Look-up DoM for each rule atencedent
RULE	RULEUP	Calculate location and rule storage
RULEn	LUPRSn	Table look-up of rules and rule storage
MATH	DIV	16-bit by 8-bit division
	MUL08	8-bit by 8-bit multiply
	MUL16	16-bit by 8-bit multiply
	MUL24	24-bit by 8-bit multiply

The routines contained in the FUZZ and MFCODE modules perform the fuzzification of the system inputs, while data stored in the MFTABLE module is used by the routines in these modules for calculating the degree-of-membership. Each fuzzy rule is processed individually. The final output is scaled (the scale factor is equal to the value given in the file model.log multiplied by 128, cf. above) and stored as 5-byte twos complement number in the data memory location assigned to OUT1 to OUT5. The absolute location of all variables is not determined until your assembled modules are linked.

The interfacing between the application and the "fuzzy code" is simple. A set of registers is defined for passing parameters to and from the fuzzy code. The number of input registers will, of course, be equal to the number of inputs (maximum 4). Obviously these registers should be loaded before calling the FUZZ routine in the fuzzy code. There are five output registers, which return a single 5-byte output result. The lowest order byte is stored in the lowest register (OUT1).

NeuFuz4 is originally designed to be used together with the COP8 family of processors from National Semiconductors. However, in the case of the C-code, there is little problem implementing the system on several other processors. Probably the most suitable processor is the COP88GW with hardware multiply and divide functions.

89.5 Summary

The *NeuFuz4* concept involves neural network training to generate a fuzzy system to be implemented on a standard von Neumann processor. It can take a maximum of four inputs and has a single output, but has several hidden layers. It appears as a rather standard feed forward network trained using backpropagation, and a fuzzy system is created. The latter is then automatically converted to assembler or C-code to be implemented on microprocessors. Typical applications with 2–4 inputs will execute in the ms to several hundred ms range.

It should be remembered that NeuFuz4 is a development tool mainly designed to support applications performing control and regulation tasks (Giles, 1992; National Semiconductor). This can be done in open or closed loop systems using micro- controllers/ processors. It is probably not very well suited for detailed pattern recognition, but rather in complex control situations, where it should be very useful and time saving for application engineers.

Acknowledgments

The present work is carried out in part under contract with the Swedish Research Council for Engineering Sciences, which is gratefully acknowledged. Support from the Carl Trygger foundation is also very much appreciated. Thanks are due to Mr. Bengt Andersson of National Semiconductors, Stockholm.

References

Beagles, G. P. 1992. *VLSI Synthesis of Digital Application Specific Neural Networks,* MS Thesis, Montana State University, Bozeman, MA, April; private communication.

Betin, P. and Bouchard, P. 1993. *DecPeRLe-1 Hardware Programmers Manual,* DEC-PRL, Paris, France.

Giles, M. 1992. Microcontroller based fuzzy solutions for consumer applications, *Northcon* '92, Santa Clara, CA.

Lindblad, T., Lindsey, C. S., and Eide, Å. 1996. Intelligent Electronics Section and Emerging Technologies Section, *CRC Handbook on Industrial Electronics,* Irwin, J. D., ed., CRC Press, Boca Raton, FL, IEEE Press, Piscataway, NJ.

Lundheim, L., Legrand, I., and Moll, L. 1995. A programmable active memory implementation of a neural network for the second level triggering in ATLAS, *4th AIHEP,* Pisa, Italy; private communication.

National Semiconductor, National Semiconductor-Inform GmBH, Germany: *PC-based Recycling Glass Classifier by Using the NeuFuz4 Neuro-Fuzzy Technology.*

National Semiconductor, *NeuFuz4 in a "Real World" System,* 0802/6602(H)08141/103389(W).

National Semiconductor, 1993. *NeuFuz4 User's Guide,* NSC Pub. No. 424421645–001PB, January.

90
Neural Network Learning in Fuzzy Systems

90.1 Introduction 1147
90.2 Reinforcement Learning 1147
90.3 Architecture of ARIC 1147
90.4 ARIC and 6 DOF Space Operations 1149
90.5 GARIC and Attitude Control 1150
90.6 Six Degree-of-Freedom Proximity Operations Trajectory Controller 1154

Yashvant Jani
Hitachi America Ltd.

Robert N. Lea
Ortech Engineering Inc.

90.1 Introduction

Fuzzy systems have provided a successful framework for the implementation of knowledge provided by skilled operators in the form of linguistic rules. The typical requirement of having an analytical model is thus replaced by emulating the performance of an expert operator. However, the process of tuning the fuzzy rules and/or membership functions remains a difficult task. As a first step, it is easier to tune the control rules to achieve the desired performance. Once the rules are more or less decided, the performance of a fuzzy system can be further improved by tuning the membership functions.

Neural networks play an important role in this tuning process and allows the learning of rules as well as membership functions. Typically, weights are attached to the rules, and the weights are modified during the learning process. If the value drops to zero, then, obviously the rule is not required. If the value remains very high, then, its importance is definitely high and the rule remains in the rulebase. Membership function tuning is slightly different. Because membership functions can be described by so many parameters, it is difficult to develop a generalized method for membership function update. Earlier approaches have been to use linear membership functions, such as Tsukomoto's one sided triangles, and attach a weight to the slope of the line. When the weight changes, the slope changes and thus the importance of the membership functions increase or decrease. More recently, an approach has been developed by which linear membership functions such as triangles or trapezoids can be tuned to redefine their points rather than their slopes.

Connectionist approaches distinguish three classes of learning: supervised learning, reinforcement learning and unsupervised learning. In supervised learning, a supervisor or teacher provides a measure of error at every time step by giving the expected value at that time. Thus, there is a necessity for an input-output set in supervised learning. In reinforcement learning, the supervisor's response is not always available. Further, it is not as direct or informative, but rather it serves as a critic to evaluate the state of the system. In unsupervised learning, no information about the correct output or an example of desired output is available.

90.2 Reinforcement Learning

Reinforcement learning techniques based on fuzzy control and multilayer neural networks have been successfully demonstrated by Berenji et al. (1991); Lee et al., (1989) at Ames Research Center (ARC) using the inverted pendulum as an example. As a joint project between the NASA Johnson Space Center (JSC) and ARC, a concept was developed for applying this technique to spacecraft docking operations (Berenji et al., 1990). Two architectures, approximate reasoning-based intelligent control (ARIC) and generalized ARIC (GARIC), have been developed and applied to space operations. In the following sections, a general description of these architectures are provided along with discussions of space applications and results.

90.3 Architecture of ARIC

Typical back propagation neural networks or space time neural networks undergo training processes during which a supervisor computes the error (or deviation) in the expected output and updates the weights. Desired outputs are assumed available from the supervisor in the form of supervised learning. In reinforcement learning, there is no supervisor to critically judge the control action chosen as the output by the neural network. The learning system is told indirectly about the effects of its chosen control action. First, the effect of the control action is observed and the amount of reinforcement is evaluated by applying a temporal

0-8493-8343-9/97/$0.00+$.50

difference method. Then, a reward or blame is distributed to the individual elements contributing to that performance. For control problems, such as cart-pole balancing, the state space is partitioned into nonoverlapping smaller regions and then the credit assignment is performed on a local basis. This is most appropriate when a state space based control is used. For a rule-based control, partitions should correspond to the regions pointed to by rules, and individual rules engaged in solving the problem must be assigned proper credit. In fuzzy reinforcement learning, these partitions can overlap leading to the use of fuzzy partitions in the antecedents and consequents of fuzzy rules. The reinforcements coming from the environment are then used to refine the definition of the fuzzy labels in the rules.

The ARIC architecture as shown in Figure 90.1 extends Anderson's method (Anderson, 1986) by including the prior control knowledge of expert operators in terms of fuzzy control rules. The two main elements in this hybrid architecture are: the action selection network (ASN) which includes a fuzzy controller based on the operators knowledge, and the action-state evaluation network (AEN), which acts as a supervisor (or a critic) and provides advice to the main controller. Both, the ASN and AEN, are multilayer neural networks with the output layer employing reinforcement learning and the hidden layer using a modified error backpropagation scheme.

The fuzzy controller implemented in the ASN consists of a fuzzifier, a rulebase and decision making logic, and a defuzzifier. The design of the rulebase is based on a hierarchical process and considers the interaction of multiple goals. The multilayer neural network consists of an input layer, hidden layer, and an output layer. The input layer has a fuzzifier whose task is to match the values of the input variables against the labels used in the preconditions of fuzzy control rules. The hidden layer in the ASN corresponds to the rules used in the controller and includes the decision making logic. Knowledge from an operator or an expert is implemented in this hidden layer and typically connects the input to output via label references. The output layer includes the decoding or defuzzification process. It combines the conclusion of the individual rules by using the center of area method.

Let $w(i)$ represent the degree that rule 'i' is satisfied by the input state variables in X,

$$w(i) = \text{Min}\{d_{i1}\mu_{i1}(x_1), \ldots, d_{in}\mu_{in}(x_n)\} \tag{90.1}$$

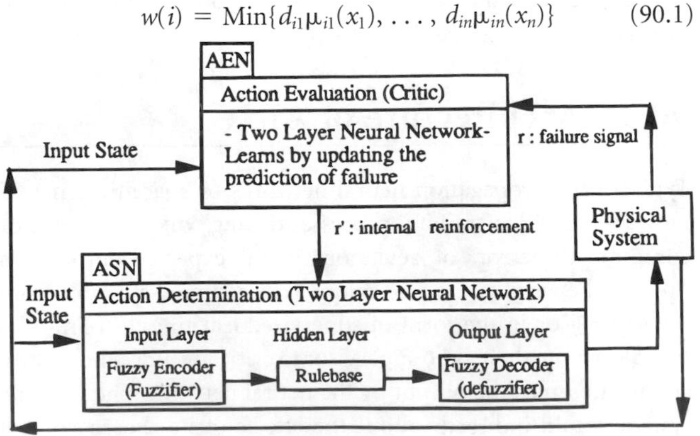

Figure 90.1 The ARIC architecture with fuzzy controller.

where $\mu_{ij}(x_j)$ represents the degree of membership of the input x_j in a fuzzy set which represents the label used in the first precondition of the rule i, d_{ij} represents the connection weights on the input from node j to a hidden layer node i, and n is the number of inputs. Please refer to Figure 90.2 to observe the relationship between the input layer and the hidden layer via connection weights d_{ij} presented as matrix D. Then, $m(i)$, which represents the results of applying $w(i)$ on the conclusion of rule i, is calculated implicitly from

$$w(i) = \mu C_i(m(i)) \tag{90.2}$$

where μC_i represents the monotonic membership function of the label used in the conclusion of rule i. The amount of the control action, $u(t)$, for the combined set of control rules is then calculated from

$$u(t) = \sum_{i=1}^{k} f_i * m(i) * w(i) \Big/ \sum_{i=1}^{k} w(i) * f_i \tag{90.3}$$

where, k is the number of nodes in the hidden layer which is equivalent to the number of rules used in the model. The variable f_i represents the connection weight on the link from node i of the hidden layer to the output node. Again, please refer to Figure 90.2 to see the relationship between the output layer and the hidden layer via connection weights f_i presented as matrix F. The inputs and outputs are also connected by a standard feedforward neural network with connection weights e_i presented as matrix E in Figure 90.2.

In the backpropagation scheme, weights for each link are updated using the error which is the difference between the actual output and the expected output. Again, it should be noted that the expected output is provided in the so-called "training" data. Since there is no supervisor (or critic) to provide a measure of error or deviation, the internal reinforcement r' (see Equation 90.7 in the following discussion) predicted by the AEN is used as a measure of error to update the weights of the links. Since another contributing factor to this error is the difference between the selected control action and its expected value given the current weights and the current state, the product of internal reinforcement and this difference is used as an error in backpropagating and updating the weights in the network. The learning algorithm in ARIC architecture is described in detail in (Berenji, 1992).

The action-state evaluation network (AEN) in ARIC basically apportions the blame for the failure among states and actions in the sequence leading to a failure. The only information received by this network is the state of the physical system (as shown in Figures 90.1 and 90.2) and a signal, r, indicating whether or not a failure has occurred. The state of the physical system is provided in terms of state variables ($x_1, x_2, \ldots\ldots x_n$). The AEN structure consists of m_h hidden units and n inputs from the environment. Each hidden unit receives $n + 1$ inputs and has $n + 1$ weights, while each output unit receives $n + 1 + m_h$ inputs and has $n + 1 + m_h$ weights. Here, the AEN plays the role of an adaptive critic element and constantly tries to predict reinforcements associated with different input states. The equations for calculating the output as well as updating the weights are as follows: if A, B, and C are the matrices of connection weights as shown in Figure 90.2, then,

$$y_i[t_1, t_2] = g\left(\sum_{j=1}^{n} a_{ij}[t_1]x_j[t_2]\right) \tag{90.4}$$

where, the y_i's are the output of the nodes in the hidden layer and

$$g(s) = 1/(1 + \exp(-s)) \tag{90.5}$$

The output of the evaluation network is

$$v[t_1, t_2] = \sum_{i=1}^{n} b_i[t_1]x_i[t_2] + \sum_{i=1}^{m_h} c_i[t_1]y_i[t_1, t_2] \tag{90.6}$$

where t_1 and t_2 are time points and double time dependencies are used to avoid instabilities in the updating of weights. Note that the weights in the above equation at time t_1 are multiplied by the x_i's at time t_2. If the same time index is used, then one cannot detect whether the change in v was caused by the change in the weights (i.e., b_i and c_i) or if it was caused by the change in the state of the system (i.e., x_i). Writing the equation in the manner above with different time steps allows comparison of different v's over time and to notice whether the system has moved to a more desirable state (i.e., higher reinforcement) or to a worse one (i.e., lower reinforcement).

The AEN evaluates the action recommended by the action selection network as a function of the failure signal and the change in state evaluation based on the state of the system at time $t + 1$. Let

$$r' = \begin{cases} 0 & \text{start;} \\ r[t+1] - v[t, t] & \text{failure;} \\ r[t+1] + \gamma v[t, t+1] - v[t, t] & \text{otherwise} \end{cases} \tag{90.7}$$

where $r'[t + 1]$ is called the heuristic reinforcement and plays the role of an error in the AEN's weight modifications. The discount rate γ is between 0 and 1, and v is the state evaluation in terms of predicted reinforcement.

90.4 ARIC and 6 DOF Space Operations

Translational and rotational fuzzy controllers developed at Johnson Space Center (JSC) have been implemented in the ARIC architecture (Figures 90.1 and 90.2) and several test cases have been performed using a high fidelity simulation of the shuttle (Jani, 1992a,b). The rotational or attitude controller developed for the shuttle has been implemented in the ARIC architecture (Figure 90.2) and test cases for attitude hold have been performed.

The space shuttle attitude controller is expected to perform four basic operations.

1. Attitude hold or maintaining the desired attitude within a small region of the desired value, typically known as a deadband.
2. Attitude maneuver or going from one attitude to another attitude.
3. Rate hold or maintaining a desired rate on a given axis.
4. Rate maneuver or going from one rate value to another rate value for a given axis.

Typical controllers based on the phase plane concept have angle errors and rate errors as input values. The controller output is a command for generating a correcting torque. For the space shuttle, the rotational corrective torques are generated by thrusters. The fuzzy logic attitude control rulebase described in the section on the six degree-of-freedom proximity operations trajectory controller with seven fuzzy labels (Figure 90.7) has been defined and implemented in the ARIC architecture. The rulebase has 31 rules (Table 90.1a) based on the attitude error and rate error. Figure 90.2 illustrates an ARIC model for the attitude controller of the space shuttle. The input layer in each network includes nodes which represent the angle error, angle rate error, and a bias node. Although ARIC does not require that the AEN and ASN have an equal number of hidden layer nodes, both have been modeled with 31 nodes in their hidden layers. Also, both networks have a single node at their output layer. It should be noted that the AEN requires a single node at the output layer

Table 90.1a Fuzzy Rulebase for Attitude Control

		Angle Error						
	NB	NM	NS	ZO	PS	PM	PB	
	NB	PM	PM	PS	Pm			
	NM	PM	PM	PS	PM			
Rate	NS	PS	PS	PS	PS			
Error	ZO	PS	PS	PS	ZO	NS	NS	NS
	PS				NS	NS	NS	NS
	PM				NS	NS	NS	NS
	PB				NM	NE	NM	NM

Table 90.1b Rulebase for Elevation Control

		Elevation Rate Error				
		NM	NS	ZO	PS	PM
	PM			NM	NM	NM
Elev.	PS		ZO	NS	NS	NM
Angle	ZO	PM	PS	ZO	NS	NS
Error	NS	PM	PS	PS	ZO	
	NM	PM	PM	PM		

Table 90.1c Rulebase for Azimuth Control

		Azimuth Rate Error				
		NM	NS	ZO	PS	PM
	PM			NM	NM	NM
Azimuth	PS			NS	NS	NM
Angle	ZO	PM	PS	ZO	NS	NM
Error	NS	PM	PS	PS		
	NM	PM	PM	PM		

Table 90.1d Rulebase for Range Control

		Range Rate Error						
		NB	NM	NS	ZO	PS	PM	PB
	PB					NB	NB	NB
	PM					NM	NB	NB
	PS					NS	NM	NB
Range	ZO	PM	PS	PS	ZO	NS	NS	NM
Error	NS	PB	PM	PS				
	NM	PB	PB	PM				
	NB	PB	PB	PB				

KEY: NB—Negative Big, NM—Negative Medium, NS—Negative Small, ZO—Zero, PS—Positive Small, PM—Positive Medium, PB—Positive Big.

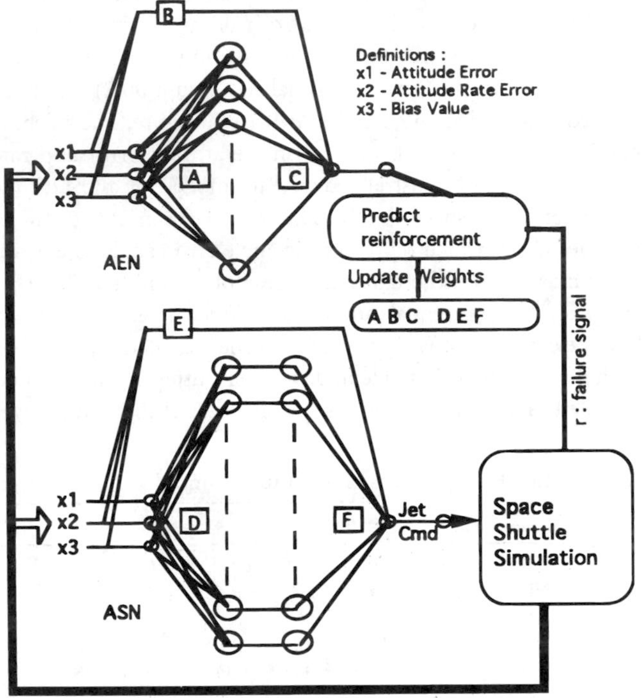

Figure 90.2 Space shuttle attitude control in ARIC.

since a single value for internal reinforcement is calculated there. However, the ASN places no restriction on the number of nodes to be placed at the output layer.

Several experiments, testing ARIC's performance using a high fidelity simulator called the orbital operations simulator (OOS) at JSC were run. For the AEN, the only reinforcement comes in the form of failure signal from the physical system. Therefore, a failure criteria was developed which uses the angular excursion as a parameter and is consistent with the control design goals. Results show that the fuzzy reinforcement learning technique has no problem in controlling the angles or angular rates (Berenji et al., 1993, 1991). It performs the attitude control properly meeting the control design goals and provides adequate learning ability during these operations.

Since the impact of fuel usage was not considered in the failure criteria, the fuel usage did not compare well with the nonlearning fuzzy control system which is designed to achieve efficient fuel usage. Our detailed analysis of the defuzzification method and the definitions of PS and NS membership functions shows that a small hysteresis exists during rate reversal. A small overlap of 0.01 degrees between these two membership functions removes the hysteresis and provides nearly the same fuel usage. Another way to improve this situation is to include the impact of fuel usage in the failure criteria to significantly improve the learning in the AEN and ARIC's performance in general. ARIC, which was being applied to the shuttle control problem at this time, could only alter slopes of membership functions. It had no capability of adjusting the end points of the membership functions. This capability is provided in the new GARIC architecture. Furthermore, triangular membership functions are required in ARIC if the output parameters are to add inertia for the no action case. As currently implemented, the ZO (zero) output membership function does not provide any control action to cause a slowing of the changes in the series of actions.

Translational control for parameters such as range and elevation angle have also been investigated. The rules given in Tables 90.1b–d were implemented with the fuzzy labels described in Figures 90.7b, 90.7c, and 90.7d. Several test cases were performed to evaluate the learning process and neural network performance. Failure criteria have been developed for range, elevation and azimuth control. Analysis of our results indicate that the failure criteria needs careful evaluation. Learning rates are slow in ARIC for very stable operations. If there are no failures in the system's performance, the learning ability of ARIC degrades significantly and it can not improve on definitions of fuzzy labels. Only monotonic membership functions are used in ARIC, and the fuzzy labels are adjusted locally within each rule. Two improvements needed in ARIC are the ability to use any shape membership functions and the capability of tuning the rules globally rather than locally. Such capabilities are provided in GARIC.

90.5 GARIC and Attitude Control

Many shortcomings of ARIC are taken care of in the new GARIC architecture and several enhancements have been developed and implemented. It is basically an extension of ARIC, but has new components and new methods in implementing the fuzzy controller to achieve global learning. The architecture of GARIC is schematically shown in Figure 90.3. It has three main components:

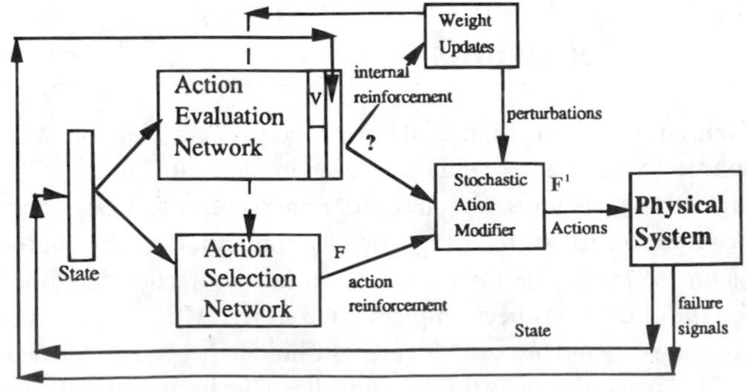

Figure 90.3 The architecture of GARIC.

1. The action selection network (ASN) maps a state vector into a recommended action F using fuzzy inference.
2. The action evaluation network (AEN) maps a state vector and a failure signal into a scalar score which indicates the state of goodness, and generates an internal reinforcement r'.
3. The stochastic action modifier using both F and r' produces an action F which is applied to the plant.

The ensuing state is fed back into the controller, along with a Boolean failure signal. Learning occurs by fine tuning the free parameters in the two networks: in the AEN, the weights are adjusted; in the ASN, the parameters describing the fuzzy membership functions change.

The AEN in the GARIC architecture is similar to that used in ARIC and no major modifications or enhancement have been implemented. It produces a prediction of future reinforcement for a given state. This prediction is used to update the weights and modify the fuzzy labels in the ASN. Changes in this prediction are provided to the stochastic action modifier to guide the selection of actions.

The ASN implements the fuzzy control rules in the form of five layer nodes, each layer performing one stage of the fuzzy inference process (Figure 90.4). The connections are fed forward with each node performing a local computation. However, this computation may be different from the conventional weighted sum of inputs and sigmoid function evaluations. Layer 1 consists of real-valued input variables and does not perform any computations. These variables are linguistic variables of interest whose values are defined in terms of labels. Layer 2 consists of nodes which correspond to one possible value of the linguistic variables in layer 1. For example, if HIGH is one of the values that an input x can take, a node computing $\mu_{\mathrm{HIGH}}(x)$ belongs to layer 2. This node will have exactly one input, and its output $\mu_{\mathrm{HIGH}}(x)$ is connected to all nodes in layer 3 where the clause 'if x is HIGH' is used in the "if" part of the rule.

The computational function is given by

$$m = f(x, c_V, s_{VL}, s_{VR}) \tag{90.8}$$

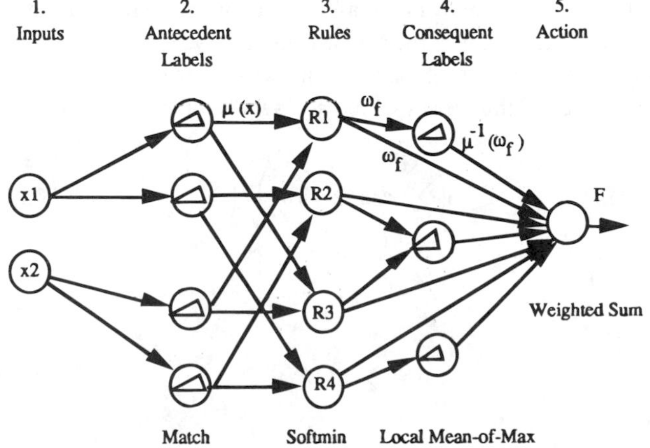

Figure 90.4 The action selection network with its five layers.

where, V indicates a linguistic value (e.g., HIGH), and c_V, s_{VL}, and s_{VR} correspond to the center, left and right spread of the fuzzy membership function of label V as shown in Figure 90.5. The value c_V serves as a reference point for the fuzzy label V and the spread on both sides permit asymmetry from this center point. More parameters may be included if desired, and their use could be defined in the function in Equation 90.8.

An instance of a smooth membership function is

$$\mu(x) = \frac{1}{1 + \left|\frac{x - c}{s}\right|^b} \tag{90.9}$$

where $s = s_{VL}$ or s_{VR} accordingly as $x < c$ or $x \geq c$ and b controls the curvature. For triangular shapes, this function is given by

$$\mu_{c,s_L,s_R}(x) = \begin{cases} 1 - |x - c|/s_R, & c \leq x \leq c + s_R \\ 1 - |x - c|/s_L, & c - s_L \leq x \leq c \\ 0 & \text{otherwise} \end{cases} \tag{90.10}$$

Triangular membership functions are easy to use, simple to define, and have been proven as sufficient in many applications. The center and spreads may be considered as weights on the input links, analogous to the approach taken in neural networks with the radial-basis-function units.

Layer 3 implements the conjunction of all the antecedent conditions in a rule. Each node in layer 3 corresponds to a rule in the rule base. Layer 2 nodes which participate in the "if" part of the rule supplies the inputs to the node in layer 3. The node itself in layer 3 performs the *softmin* operation which is a continuous and differentiable operation as follows.

$$O_{R3} = \omega_r = \left(\sum_i \mu_i \exp(-k\mu_i)\right) \Big/ \left(\sum_i \exp(-k\mu_i)\right) \tag{90.11}$$

Here, μ_i is the degree of match between a fuzzy label occurring as one of the antecedents of rule r and the corresponding input variable. The *softmin* operation computes the degree of applicability, ω_r, for a given rule r, and the parameter k controls the hardness of the *softmin* operation. As k tends to infinity, the usual minimum operator of fuzzy logic is recovered. However, for k finite, the *softmin* operation results in a differentiable function of the inputs making it convenient for calculating gradients

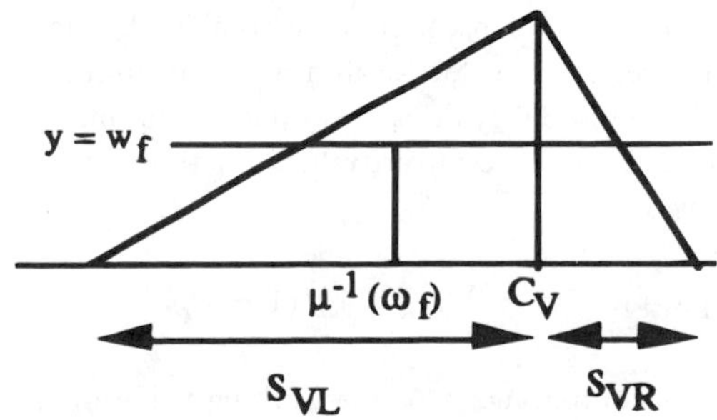

Figure 90.5 Triangular membership function.

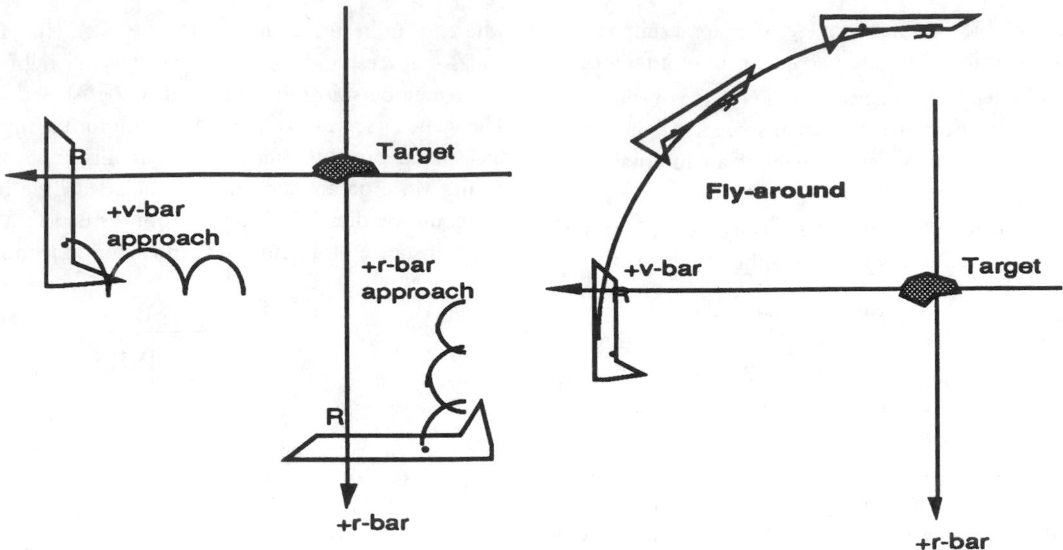

Figure 90.6 Proximity operations: v-bar, r-bar approaches and fly-around segment.

during the learning process. Experiments have shown that the choice of k is not critical.

Layer 4 nodes correspond to consequent labels. Each node receives its inputs from all rules which use this particular consequent label in the "then" part of the rule. Based on the input value ω_r, this node computes the corresponding output action as suggested by rule r using the inverse mapping written as

$$\mu^{-1} c_V, s_{VL}, s_{VR}(\mu_r), \tag{90.12}$$

where V is the specific fuzzy consequent label, and c, s_L and s_R are membership parameters as defined earlier.

Inverse mapping implies a suitable defuzzification procedure applicable to an individual rule. In general, the mathematical inverse of μ may not exist if the function is not strictly monotonic. A simple procedure very similar to the mean-of-maximum method is proposed to determine this inverse.

If w_f is the degree to which rule i is satisfied, then, $\mu_{\bar{V}}^{1}$ (wf), is the X-coordinate of the centroid of the set $\{x{:}\mu_V(x) \geq w_f\}$.

The difference between the proposed method and the mean-of-maximum method is when it is applied. The mean-of-maximum method is applied after all rule consequents have been combined, whereas the proposed method is applied locally to each rule before the consequents are combined. Therefore, the method is referred to as the local mean-of-maximum (LMOM) method. In Figure 90.5, the set, limited to the line $y = w_f$, is used to determine the centroid value. For triangular functions, LMOM gives,

$$\mu^{-1} c_V, s_{VL}, s_{VR}(w_r) = c_V + 0.5^*(s_{VR} - s_{VL})^*(1 - w_r) \tag{90.13}$$

If the rule output $w_r = 0$, the limiting value of the inverse $\mu^{-1}(w_r \rightarrow 0^+)$ is used, which is $c_V + 0.5^*(s_{VR} - s_{VL})$ for triangular functions. If the membership function is monotonic, then, $\mu^{-1}(w_r)$ is just the standard mathematical inverse with proper limiting values. Since sharing of consequent labels is allowed, a node in this layer may have multiple outputs carrying different values. For each rule, feeding it a non-zero degree, the node should produce a corresponding output action for the next layer. This unusual feature, even though it is useful, has not, as yet, been tried and experimented with in applications. In fact, for many classes of membership function, it can be eliminated. In particular, it can be eliminated for triangular shaped functions by generating the output as

$$O_{V4} = (c_V + 0.5^*(s_{VR} - s_{VL}))^*\left(\sum_r w_r\right) - (0.5^*(s_{VR} - s_{VL}))^*\left(\sum_r w_r^2\right) \tag{90.14}$$

Whenever $\mu^{-1}(x)$ is a polynomial, then one output is sufficient regardless of the number of inputs.

Layer 5 has as many nodes as there are output action variables. Each output node uses the rule strength as a weight and combines the recommended (defuzzified) values from each rule in the rule base. The weighted sum formula is very simple.

$$F = \left(\sum_r w_r\mu^{-1}(w_r))/(\right) \Big/ \left(\sum_r w_r\right) \tag{90.15}$$

or

$$F = \sum_V O_{V4} \Big/ \sum_R O_{R3} \tag{90.16}$$

where the inputs come from layer 4 and layer 3 and we have used formulas (Equation 90.14) and (Equation 90.11) in reducing

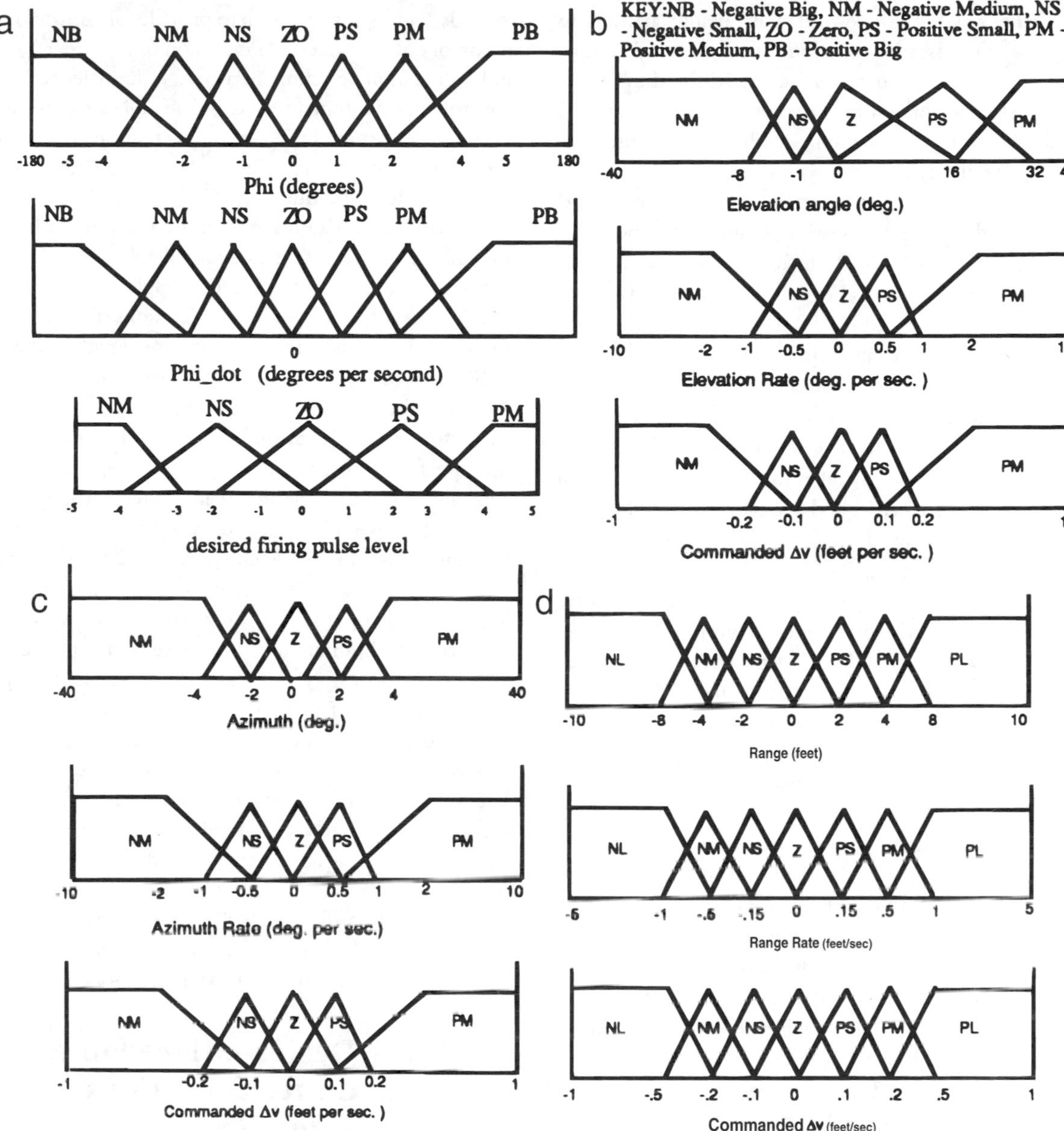

Figure 90.7 (a) Membership functions for Phi, Phidot and desired pulse level. (b) Membership functions for elevation angle control. (c) Membership functions for azimuth angle control. (d) Membership functions for range control.

the equation. The node simply sums up each set of inputs and forms a quotient. In this way, a continuous value of the output variable is delivered as an action selected by the ASN. When the input space is completely covered by the antecedent label functions, the output F is always defined. It should be noted that the weights on input links in layer 2 and 4 are modifiable and all others are fixed at unity. Therefore, the gradient descent procedure will work effectively on only two layers.

The stochastic action modifier generates an action F' based on the internal reinforcement r' from the previous time step and an action F recommended by the ASN. The action F', actually applied to the plant, is a stochastically generated value with mean F and standard deviation $\sigma(r'(t-1))$. This function $\sigma(\ )$ is a non-negative and monotonically decreasing, e.g., $\exp(-r')$. This stochastic perturbation in the action leads to a better exploration of the state space and better generalization ability. The difference between F and F' is large when the internal reinforcement r' is low, and small when it is high. A large random step away from the ASN value may occur when the last action performed is bad, but the controller will remain consistent with the fuzzy rules when the previous action selected is a good one. The actual form of $\sigma(\ )$, especially its scale and rate of decrease, should take the units and range of variation of the output into account. The perturbation at each time step is simply the normalized deviation from the ASN-recommended action and contributes as a learning factor in the ASN.

Weight updating in the AEN is based on the reward/punishment scheme. If positive internal reinforcements are received, the values of the weights are rewarded by being changed in the direction that increases its contribution to the total sum. If reinforcements are negative, then weights are adjusted to reduce its contribution to the total sum. Learning in GARIC and its application to the cart-pole balancing problem is discussed in detail in Berenji, et al. (1992).

The space shuttle attitude control membership functions and rules described in the following section, Figure 90.7 and Table 90.6, were implemented in GARIC format within the OOS at Ames Research Center and several experiments were performed to understand the learning in GARIC. For the ASN, there are two inputs, attitude error and rate error, each with 7 fuzzy labels (NB, NM, NS, ZE, PS, PM, PB). The single output of the ASN, jet firing commands, has five fuzzy labels (NM, NS, ZE, PS, PM). A total of 31 rules control the state space. Each rule has two antecedents, the first uses a fuzzy label for error and the second uses a fuzzy label for rate error, and one consequent that uses a fuzzy label for the jet firing command. Hence, the network has 2, 14, 31, 5, and 1 neurons in its five layers. For the AEN, a biased unit was also included as an input. Thus, there are three inputs, error, rate error and a bias, and one output. Since there were 31 rules in the rulebase, the hidden layer has 31 nodes. Hence the network has 3, 31, and 1 neurons in its three layers.

The goal for the attitude control is to keep the angle and rate deviations within a certain deadband. In a learning experiment, a failure occurs when the value of this deviation exceeds the allowed deadband. Every time a failure occurred in a GARIC execution, the control was shifted to a supervisory control routine to bring the state of the system back to within the deadband. Indeed, a requirement of these architectures, is that a restart of the learning process occurs after each failure is detected. The original fuzzy controller with a small modification in the NS membership function was implemented as a supervisory controller. The center c_V of the NS membership function was shifted to -3 degrees from its original value of -2 degrees. This modification was sufficient to bring back the error to within its deadband of 0.4 degrees after a failure, but not sufficient to hold the error there. The original fuzzy controller, developed at JSC in the late 1980's and discussed in the following section of this chapter, holds the errors within 0.5 degrees but not within 0.4 degrees. After a small number of trials (less than 10) GARIC learns to do this task by revising the membership functions. A similar experiment was performed to train the new controller to hold the errors within 0.3 degrees, and GARIC learned this new task within 5 trials. The fuel consumption for each test case was about 222 lb which is in the same range as the non-adaptive fuzzy controller and less than the conventional controller in OOS. Since the conventional controller has more constraints on its design (such as hardware concerns about the jet life which is improved by using longer, but as few, firings as possible), this fuel comparison was done just to show that the GARIC performance is within an acceptable range of the conventional controller.

Although GARIC is very similar in structure to that proposed by Anderson (1986), the action selection network is a synthesis of fuzzy logic and neural networks. The fuzzy controller in GARIC extends Anderson's approach to provide for continuous representation of output values and inclusion of the expert's control rules into the action selection network. A knowledge base of fuzzy control rules is implemented in the ASN as the starting point. Tuning via learning in a continuous manner is then achieved in the neural networks in GARIC. Thus, the goal is not restricted to strategy learning only.

The stochastic action modifier, even though it has similarities to Gullapalli's method, uses a completely different approach for defining internal reinforcement. Single layer neural networks require identification of trace functions for keeping track of the visited states and their evaluations. Further, the single-layer networks typically consider only the generation of output values, and leave the preconditioning of fuzzy rules as is. The approach suggested in GARIC is different in both respect. First, it does not use the trace functions. The new representation of fuzzy rules allows faster development and faster learning. Second, based on the reinforcement received from the environment, both the preconditions and the conclusions of the rules can be modified.

In summary, a new method of designing and tuning fuzzy logic controllers has been proposed in the ARIC and GARIC architectures. Both architectures have been fully developed with applications to cart pole balancing as well as the space shuttle attitude control. The knowledge acquired through experience and used by an operator in controlling a process can now be modeled and implemented using linguistic terms. The rules can later be refined through the process of learning from experience. Neural networks are configured in a particular fashion so that the reinforcement signals can be used to update the weights for improvement in the fuzzy controller's performance. The architecture of GARIC provides a well-balanced method for combining the experiential knowledge in the form of fuzzy rules and the learning power of artificial neural networks.

90.6 Six Degree-of-Freedom Proximity Operations Trajectory Controller

A six degree of freedom (6 DOF) controller for a spacecraft has been designed and tested (Lea et al., 1991a,b) in a shuttle simulation for proximity operations (Figure 90.6). When the shuttle approaches the satellite in an orbit, a local vertical local horizontal frame (Lea and Jani, 1992) is used as a reference frame. The x-axis of this frame generally along the velocity vector of the satellite is called v-bar, and the z-axis pointing towards the center of the orbit is called r-bar. When the shuttle approaches along the x-axis, the approach is known as a v-bar approach. This approach being in the same orbit saves fuel and provides operational flexibility. An approach along the z-axis is known as an r-bar approach as shown in Figure 90.6. During the station-keeping phase, a constant distance is maintained from the shuttle to the satellite to correct for any relative position and to establish the ground communication for information update. The preferred axis for station-keeping is v-bar due to fuel savings. A fly-around phase consists of the shuttle moving from the v-bar to negative r-bar to properly inspect the satellite and find out the position of the grapple mechanism. These operations performed manually by a crew member requires extensive training in the real-time mission simulators to achieve proficiency.

The 6 DOF controller uses sensor measurements such as range, elevation and azimuth angles directly as input and generates the commands for the jet select logic to null out the errors. For each degree of freedom, there is a rulebase that requires the parameter error and its rate as input. For example, the pitch attitude is controlled by a rulebase of 31 rules given in Table 90.1a. This rulebase requires pitch error and error rate whose membership functions are given in Figure 7a. The same rulebase and membership functions are used for the two other attitude (roll and yaw) control. The elevation, azimuth and range control are performed by the rules given in Table 90.1b, 90.1c, and 90.1d. Each parameter has different rules tuned to achieve desired performance. Corresponding membership functions are given in Figures 90.7a–d. It should be noted that these membership functions are tuned for the shuttle operations specifically. Therefore, application of this software to other spacecraft operations requires minor modifications.

For a given mission profile, the controller maintains a proper range and range rate. The elevation and azimuth angle measurements are used in conjunction with the angular rates to follow a desired trajectory. For example, during the v-bar approach, the controller maintains zero elevation and azimuth angles, desired range and range rate. The attitude is maintained by the rotational part of the controller. If the range is smaller than the desired range, the controller will slow down accordingly. If the range rate is slower than the desired rate then the controller will increase the speed. In keeping the elevation and azimuth angles close to zero, the controller adjusts its actions based on the angle errors, rate errors as well as current pitch rate and roll rate.

The performance of the 6 DOF controller in terms of simultaneous relative trajectory and attitude control is very good and robust. The controller is also very responsive and maintains flight profiles within the expected and desired envelope. The controller holds the proper elevation and azimuth angles during all proximity operations test cases, and performs proper range and range rate control. It transitions from v-bar and r-bar approaches to station keeping operation in a very natural manner without having any discontinuity in its performance. It performs fly-around operations very well and continuously maintains proper range deadbands for the expected trajectory. Again, it transitions from the fly-around to station-keeping operation without any loss of control. Since the membership functions and rule base are shuttle specific, the controller primarily uses single jet firings for translational corrections.

Table 90.2a Proximity Operations Test Cases Flown Manually by Experienced Pilots

RUN #	Manuever	Distance	Rates	Arrival time	Fuel used
RUN 1	*V* bar approach	400 ft–50 ft	zero rates	1410 sec	31 lbs
RUN 2	*R* bar approach	400 ft–50 ft	zero rates	1560 sec	65 lbs
RUN 3	1/4 Fly around	@ 200 ft,	0.2 rate	1380 sec	50 lbs
RUN 4	Station keep	@200 ft	zero rates	1800 sec	36 lbs

Table 90.2b Test Results with 6 DOF Fuzzy Controller for the Same Test Cases

RUN #	Manuever	Distance	Rates	Arrival time	Fuel used
RUN 1	*V* bar approach	400 ft–50 ft	zero rates	1436 sec	33 lbs
RUN 2	*R* bar approach	400 ft–50 ft	zero rates	1686 sec	58 lbs
RUN 3	1/4 Fly around	@200 ft	0.2 rate	1800 sec	47 lbs
RUN 4	Station keep	@200 ft	zero rates	1800 sec	12 lbs

Proximity operations flight profiles and fuel usage during these operations have been excellent. The comparison of the fuzzy logic controller with mission planning data for fuel usage is shown in Table 90.2. For mission planning, experienced pilots fly the mission segments in a simulator. Table 90.2a shows fuel usage for four proximity operations segments flown by these pilots. Table 90.2b shows the corresponding fuel usage by the 6 DOF fuzzy controller. Fuel usage for v-bar, r-bar and fly around test cases is comparable, but the fuzzy controller uses significantly less fuel for the station keeping task. Further tests of this 6 DOF controller should compare its performance with other manual crew procedure test cases flown in the shuttle mission simulator, and possibly flight data.

Acknowledgments

The authors very much appreciate the encouragement provided by Chris Culbert, Bob Savely, and Robert Shelton of Software Technology Branch at the Johnson Space Center and Ken Kristie and Steve Imsen of Hitachi America Limited. The assistance provided by Hamid Berenji and Anil Malkani at Ames Research Center, Jeff Hoblit at LinCom Corporation, Luc P. Tran and Jonathan Weiss at McDonnell Douglas Space Systems Company, Jack Aldridge at Advantex Inc., and Edgar Dohmann at Ortech Engineering Inc. is also very much appreciated.

References

Anderson, C. W. 1986. *Learning and Problem Solving with Multilayer Connectionist Systems,* Ph.D. Thesis, University of Massachusetts.

Berenji, H. R., Lea, R. N., and Jani, Y. 1990. Fuzzy logic controller with reinforcement learning for proximity operations and docking, *Presentation at the 5th IEEE Int. Symp. Intelligent Control,* 2:903.

Berenji, H. R. 1991. Strategy learning in fuzzy logic control, *Proc. North American Fuzzy Information Processing Society Workshop,* May 14–17, Columbia, MO.

Berenji, H. R. 1992. An architecture for designing fuzzy controllers using neural networks, *Int. J. Approximate Reasoning,* 6(2):267–292.

Berenji, H. R. and Khedkar, P. 1992. Learning and tuning fuzzy logic controllers through reinforcements, *IEEE Trans. Neural Networks,* 3(5).

Berenji, H. R., Lea, R. N., and Jani, Y. 1991. Approximate reasoning-based learning and control for proximity operations and docking in space, *Proc. AIAA Guidance, Navigation and Control Conf.* 3:1707.

Berenji, H. R., Lea, R. N., Jani, Y., Malkani, A., Khedkar, P., and Hoblit, J. 1993. Space shuttle attitude control by reinforcement learning and fuzzy logic, *Proc. FUZZ-IEEE '93 Conf.* sponsored by IEEE Neural Network Council, San Francisco, CA. 2:1396

Lea, R. N. and Jani, Y. 1992. Fuzzy logic in autonomous orbital operations, *Int. J. Approximate Reasoning,* 6(2):151–184, 1992.

Lea, R. N., Hoblit, J., and Jani, Y. 1991a. Performance comparison of a fuzzy logic based attitude controller with the shuttle on-orbit digital auto pilot, *NAFIPS '91 Workshop Proc.,* 291–295.

Lea, R. N., Hoblit, J., and Jani, Y. 1991b. A fuzzy logic-based spacecraft controller for Six Degree of Freedom control and performance results, *Proc. AIAA Guidance, Navigation and Control Conf.,* 3:1680.

Lee, C. C. and Berenji, H. R. 1989. An intelligent controller based on approximate reasoning and reinforcement learning, *Proc. IEEE Int. Symp. Intelligent Control,* Albany, NY.

Jani, Y. 1992a. *Application of Fuzzy Logic—Neural Network Based Reinforcement Learning to Proximity and Docking Operations: Attitude Control Results,* Report No. 2 for the Research Activity No. AR.06 under Cooperative Agreement NCC 9-16, The research Institute for Computing and Information Systems (RICIS), University of Houston at Clear Lake, TX.

Jani, Y. 1992b. *Application of Fuzzy Logic—Neural Network Based Reinforcement Learning to Proximity and Docking Operations: Translational Controller Results,* Report No. 3 for the Research Activity No. AR.06 under Cooperative Agreement NCC 9-16, The Research Institute for Computing and Information Systems (RICIS), University of Houston at Clear Lake, TX.

91

Neurocontrol and Elastic Fuzzy Logic: Capabilities, Concepts, and Applications

Paul J. Werbos
National Science Foundation

91.1 Introduction 1157
91.2 Neurocontrol in General 1158
The Availability of High-Throughput Chips and Boards • Universal Approximation Theorems • Ease of Use and Teaching • Maintaining the Link to the Brain
91.3 Basic Principles of Design 1159
91.4 Supervised Learning for Neurocontrol 1160
Basics of Supervised Learning • Supervised Learning: Recent Results • A Simple Example of the Implementation of Supervised Learning • Basic Subroutines for the 3-Layer MLP
91.5 Elastic Fuzzy Logic: Principle and Subroutines 1162
Background and General Concepts • Approaches to Adapting Fuzzy Systems • Basic Subroutines for the ELF Net
91.6 Current Designs in Neurocontrol: A Roadmap 1165
91.7 Appendix (Tutorial Level Background Information): Neurocontrol and Fuzzy Logic 1166

Abstract

In recent years, enormous progress has been made in *neurocontrol*—the use of neural nets as controllers. Designs which originated in neurocontrol can also be used with a wide variety of *nonneural* systems. This article will try to facilitate *both* types of applications. For example, the article will show how elastic fuzzy logic (ELF) nets make it possible to combine the capabilities of expert systems with the *learning* capabilities of neural nets at a high level. Still, artificial neural network (ANN) implementations have advantages in terms of hardware implementation, ease of use, generality, and links to the brain, which is still the only true intelligent controller available to us.

Neurocontrol is useful in cloning experts, in tracking trajectories or setpoints, and in optimization (e.g., approximate dynamic programming). There has been substantial success in controlling robot arms (including an arm built for the space shuttle), in chemical process control, in continuous production of high-quality parts, and other aerospace applications. This article will provide a tutorial or roadmap of the basic designs and concepts, with reference both to applications and future research opportunities.

91.1 Introduction

This article will provide an overview of the new "toolbox" of control designs which has been developed in the last two or three years in the field of neurocontrol. Applications of these tools have generally involved artificial neural networks (ANNs), but the underlying mathematics is quite general. The same tools could be applied to more conventional controllers or models (or to fuzzy logic systems or PDE codes) as well. This article will describe how, very briefly.

The article will begin by describing the goals and advantages of neurocontrol in broad terms. Then, in Section 91.3, it will describe basic concepts which are used in building neurocontrol systems. Section 91.4 will describe the current status of *supervised learning*, a fundamental area which neurocontrol draws on very heavily. It will also provide examples of *modular design* which is crucial to the engineering implementation of neurocontrol designs. Section 91.5 will build on Section 91.4, by describing

* The views and ideas expressed in this article are those of the author, not those of NSF. This article is an updated version, written on personal time, of Werbos (1993a). Since Werbos (1993a) was written, many of the more advanced techniques described here—including ELF—have been included in a patent pending through Scientific Cybernetics, Inc., of Coral Springs, Florida.

0-8493-8343-9/97/$0.00+$.50

elastic fuzzy logic, a new approach to intelligent control and to the synthesis of fuzzy logic and neural nets. Finally, Section 91.6 will describe the three basic tasks which neurocontrollers can perform, which have proven useful in many applications in the real world (described in more detail in Chapter 63):

1. Cloning—transferring the expertise of a human expert or of a complex automatic controller into an ANN. Unlike expert systems, the ANN copies what a person *does,* not what he *says.* The ANN is usually faster and cheaper than what it copies—which is crucial in many applications.
2. Tracking—making a system or plant follow a desired trajectory, adhere to a desired setpoint or follow a reference model. This is essentially just a nonlinear extension of conventional adaptive control, but with additional capabilities involving learning and efficient hardware implementation.
3. Dynamic optimization—systems designed to maximize utility or performance or profits, etc., *over time,* based on learning. Even when solving a simple tracking problem, one can use optimization methods to minimize a *combination* of tracking error plus energy use or the like.

Within neurocontrol we have a ladder of designs, going up from the simplest cloning designs to the most complex optimization designs. New researchers are urged to begin with the simplest designs, to build up expertise, credibility, and software; however, they should be warned that many applications will be resistant to the simpler designs. Advanced groups, like those seeking funding from my program at the National Science Foundation (NSF), need to push up the ladder as high as possible, in order to build up to true general-purpose intelligent systems. Even within the optimization area, there are ladders on top of ladders of ever more sophisticated designs.

91.2 Neurocontrol in General

Neurocontrol is defined as the use or study of well-specified neural networks as *controllers*—as networks which output a vector of control signals, $u(t)$, as a function of time t.

Hundreds of papers have been published so far on neurocontrol. Because the literature is so complex, many people find it difficult to develop a "map" of what has been done. As a result, many researchers have reinvented the same basic designs dozens of times, leaving the most interesting designs and powerful applications to a smaller group of researchers. This article will try to help alleviate this problem by providing a fairly complete roadmap of the field.

Back in October 1988, NSF sponsored a small workshop in New Hampshire to try to review the existing work in neurocontrol (Miller et al., 1990). At that time, all the existing useful work was based on *five* underlying designs. *None* of the five involved any kind of black magic; all five could be understood completely within the framework of classical control theory. They were not an alternative to control theory, but a subset of control theory. All five could be applied to all kinds of large, sparse nonlinear controllers or models, although most of the designs require that the controller or model be differentiable. The classification of designs given here is somewhat broader and easier to understand than the classification used in Miller et al. (1990), but these basic points remain valid.

When they are reformulated as general purpose methods, the designs now used in neurocontrol offer an important advantage over more conventional methods: a generalized capability for learning. By using learning instead of tweaking models by hand, one can save an enormous amount of effort in system development. Both in aviation and in robotics applications, industrial engineers have told me that 80% of the project development costs tend to be in the tweaking which occurs *after* the basic design has been "finalized." Cutting those costs can be crucial, especially in applications where cost overruns have a good chance of getting a product line canceled altogether. Learning in real time has many other benefits as well, in enabling a smoother, faster adaptation to new tasks, new products, etc. (In the automotive industry, the time required to bring out a new product can be crucial to corporate competitiveness.) All of these benefits can be had, in principle, by applying the methods of neurocontrol to classical models and controllers, or to fuzzy logic systems (as described below).

For pure tracking problems, where there is no need to minimize energy use or the like, classical *adaptive* control provides capabilities very similar to those of certain designs used in neurocontrol. However, Chapter 63 describes how learning techniques, used at a global level, can be used to adjust the adaptation rate parameters used in local adaptation circuits, so as to improve transient response, even in pure tracking applications. The distinction between learning and adaptation (discussed in other articles of this book) is difficult to define in a precise, mathematical way, but it has substantial practical implications.

In practice, almost all of the applications of the methods which originated in neurocontrol have used ANNs as models or controllers. So far as I know, my own early work using backpropagation (e.g., Werbos, 1989, 1988, 1974) has been the main exception. Why have people chosen to limit themselves to the case of ANNs as such? There are four major reasons, which vary greatly in importance from application to application.

The Availability of High-Throughput Chips and Boards

Neuroengineering has sometimes been defined as the effort to develop general-purpose nonlinear algorithms which are transparently suitable for implementation on massively parallel, analog, fixed-instruction computing systems. That kind of implementation permits orders of magnitude greater throughput, in principle, than the best digital parallel computers, which typically require an entire chip to hold a processor; a neural chip can hold thousands or millions of effective processors *on a chip.* (Of course, applications-specific chips—ASICs—suitable for implementing ANNs, offer similar capabilities when they are used in this way.)

Neural chips are available today from a variety of vendors. David Andes of the U.S. Navy China Lake has estimated that one handful of the Intel neural chips has more computational power, for what it does, than all of the Crays and Amdahls in the world put together. Adaptive solutions have used advanced neural chips to build a turnkey workstation which substantially outperforms conventional supercomputers on a number of benchmark problems. They have also sold less expensive add-on boards offering similar capabilities in PCs and Macs. Motorola has provided software simulators to assist developers in using its new chip, which is expected to become available at truly mass-market quantities and prices. And there are many other important products in process.

The implications of all this in real-world control are enormous. There are many algorithms which perform very well in machine vision or control when they are simulated at slow speed on a Cray. But they are often not used, because it is not practical to carry two Crays around on an aircraft, or to install a Cray on every workstation in a factory. If a neural net *clone* can be developed, these controllers can suddenly become useful in the real world, for a much larger market.

Universal Approximation Theorems

Just like Taylor series, ANNs can approximate any well-behaved function to any desired degree of accuracy. This makes ANNs useful as a way of working with nonlinear functions, just as Taylor series are useful.

From an engineering point of view, this means that a neural net chip or board can be "rewired" by electrical signals to represent any nonlinear function. Other function approximators may be just as useful, at times, on conventional computers or conventional massively-parallel supercomputers, however, the ability to "rewire" a chip or a board is critical to many applications.

Numerous theorems have been proven showing that commonly used ANNs can represent any well-defined function, both in its value and in its derivatives up to any finite order. There have even been proofs that they can represent the *ill-behaved* functions often encountered in control theory, in tracking applications (Sontag, 1990). Barron (1993) has proven that the most popular class of ANN (the MLP) can approximate smooth functions with less growth in complexity, as the number of inputs grows, than with the linear basis-function methods—like Taylor series—which have been used in virtually all practical nonneural practical applications. In very complex applications, such as optimal planning over time, one sometimes encounters functions which are hard to approximate by *any* conventional means, including the most popular ANNs. However, there is a class of neural networks—the simultaneous recurrent network (SRN)—which can cope with such problems (Werbos and Pang, 1996).

Ease of Use and Teaching

Realistically, this is the major reason why neurocontrol designs have mainly been used with ANNs. Many papers, going back many years, explain in great detail how to use neurocontrol methods with *arbitrary* nonlinear models (White and Sofge, 1992; Werbos, 1989, 1974); however, these techniques require that the user *apply new concepts in calculus* to the equations of the model, and that special-purpose computer subroutines specific to that model be written. (Actually, Griewank of Argonne National Laboratories and M. Iri in Japan have begun to develop computer programs which could automate this process; this could eventually lead to some very useful software packages.) For now, it is much easier to copy existing equations directly for specific forms of ANNs, or to borrow existing programs. After all, why bother to write your own programs when the standard ANNs can approximate any function anyway?

This same advantage carries over to teaching (and inspiring) new students in the field. It explains the great success of books like Rumelhart and McClelland (1986), which simplified and popularized neural net methods, and communicated some of the basic ideas to a nontechnical audience.

Maintaining the Link to the Brain

The effort to understand the human brain is still a vital source of inspiration in developing new neurocontrol designs. The human brain is living proof that it *is* possible to build modular control systems which take full advantage of massively parallel analog circuitry, and which learn in real time to accomplish difficult tasks over time in an unknown noisy, nonlinear, and complex environment, requiring *millions* of actuators. Until we have built artificial control systems which combine *all* these capabilities, we will have good reason to respect, emulate, and reverse-engineer the brain.

This is a two-sided coin. Because the brain *as a whole system* is a neurocontroller, we need to understand the mathematics of neurocontrol before we can hope to understand it. So long as neurophysiologists limit themselves to species of mathematics which are unable to replicate these higher-order control capabilities, they will not be able to understand the basic essentials of human intelligence. Thus, the effort to develop working neurocontrol designs is crucial not only for engineering but for biology and psychology as well. The goal is to develop *general* mathematics which remain valid, *regardless* of the material (or immaterial) substratum used to implement the phenomenon of intelligence.

In neurocontrol, we have already developed the basic concepts which—in principle—should permit us to achieve true brain-like intelligence as discussed above (Werbos, 1994a, 1994b, 1974). Tremendous progress has been made in moving *some* of these concepts into real-world industrial applications; however, there is still a great deal to be done.

91.3 Basic Principles of Design

Years ago, some neural net researchers felt that they could summarize everything useful about their work, including all of the equations necessary to implement their designs, in a handful of pages. Modern neurocontrol—like the brain itself—is more complex.

Modern neurocontrol may be compared with modern computing. In computing, there are at least three important levels of research and analysis. At the lowest level, people build chips. At a middle level, people combine chips to make computers. At the highest level, people study how to use computers to solve practical problems.

There are three main levels of analysis in neurocontrol as well. At the lowest level, people try to build supervised learning systems (SLS). An SLS is any system which *learns* a nonlinear function or static mapping from a vector X to a vector Y. At the middle level, in neurocontrol proper, we build complex systems made up of SLS components, and other similar components; we try to develop general-purpose designs to perform the tasks of cloning, tracking, or dynamic optimization mentioned above. Finally, in applications research, we *use* the neurocontrol systems in combination with other systems to build complex systems for specific applications. (For example, we may use fuzzy logic to initialize a network trained to be a clone, which can then be used to initialize an optimizer.) All three levels of research need to respect each other and work with each other.

Section 91.4 of this article describes the current state of the *lowest level* of research—the effort to build supervised learning systems. My Chapter 63, in the section on neural networks, outlines the designs which exist at the *middle level,* and discusses some applications. Section 91.4 provides the equations for a couple of very basic, popular designs and explains the concept of *modular design* in more detail. Using modular design, it is actually very easy to program the complex designs mentioned in the later part of this book as surprisingly short programs, *calling on subroutines* like those discussed in Section 91.4; however, for reasons of space, these programs will not be displayed here. (See White and Sofga, 1992, and the introductory background material in Werbos, 1974.)

Figure 91.1 gives a more complete picture of the research fields important to neurocontrol.

At a lower level, we need to build both supervised learning systems *and* "backpropagation learning" systems. (In the backpropagation learning task, as defined in Figure 91.1, we are *not given* the values of the target variables $Y_i(t)$ at time t; instead, we are given only the error derivatives, $\partial E/\partial Y_i(t)$.) At an intermediate level, many neurocontrol designs require *neuroidentification* components; in other words, they need components which are capable of predicting or modeling dynamic systems. There are other aspects of the neural network field—including methods for clustering, feature extraction, combinatorial optimization, and associative memory—which could be useful as subsystems or sources of inspiration for the key tasks indicated in Figure 91.1.

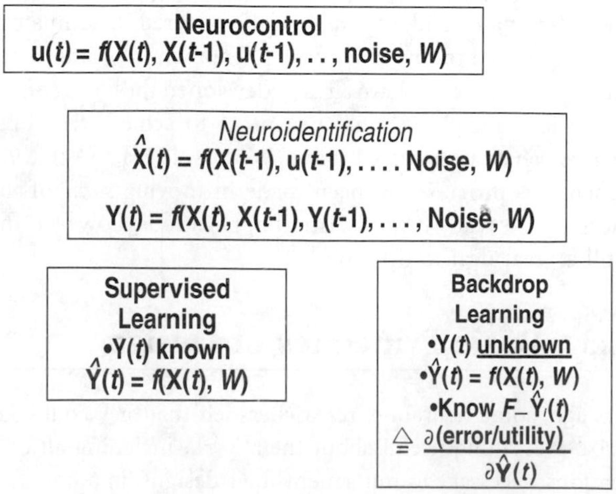

Neurocontrol
$u(t) = f(X(t), X(t\text{-}1), u(t\text{-}1), \ldots, \text{noise}, W)$

Neuroidentification
$\hat{X}(t) = f(X(t\text{-}1), u(t\text{-}1), \ldots \text{Noise}, W)$
$Y(t) = f(X(t), X(t\text{-}1), Y(t\text{-}1), \ldots, \text{Noise}, W)$

Supervised Learning
•$Y(t)$ known
$\hat{Y}(t) = f(X(t), W)$

Backdrop Learning
•$Y(t)$ unknown
•$\hat{Y}(t) = f(X(t), W)$
•Know $F\_\hat{Y}_i(t) \triangleq \frac{\partial(\text{error/utility})}{\partial \hat{Y}(t)}$

Figure 91.1 Research areas critical to neurocontrol.

91.4. Supervised Learning for Neurocontrol

Basics of Supervised Learning

In mathematical terms, supervised learning is the task of learning a *mapping* from a vector $X(t)$ to a vector $Y(t)$, based on a database of training examples, where each example is labelled by some value of t.

For example, suppose that we want to train a neural net to learn to recognize handwritten numbers. We can create a database of 2000 handwritten numbers. In other words, we can build a training set which goes from $t = 1$ to $t = 2000$. For each example, t, we need to obtain an image, $X(t)$, and we need to record the correct classification, $Y(t)$, of that image. Suppose that the image consists of a 19-by-20 array of pixels; then the vector $X(t)$ will have 380 components, $X_1(t)$ through $X_{380}(t)$. Suppose that the correct classification consists of a 1 ("yes") or 0 ("no") for each of 10 possible digits, 0...9; then the vector $Y(t)$ will have 10 components, $Y_1(t)$ through $Y_{10}(t)$.

To solve this problem, our first job is to choose a *functional form,* a *particular* network design or topology. For any type of neural network, we can always write the output of the network as:

$$\underline{\hat{Y}}(t) = \underline{f}(\underline{X}(t), W), \tag{91.1}$$

where $X(t)$ is the input to the network, where the function f represents what the network does, and W is a set of adjustable weights. (For example, an elastic fuzzy network meets this definition.) There are many forms of neural net design which are popular in control. Perhaps the most popular are the multilayer perceptron (MLP), the CMAC, and the radial basis function (RBF) networks. The MLPs are sometimes called "backpropagation networks," but this is not accepted terminology because it is very misleading.

After we have defined a network, we need to *adapt* the network. In other words, we need to find a *learning procedure,* a procedure for how to *change the weights* W so that Equation 91.1 will do a good job of approximating $Y(t)$ over the training set. The most common procedure (with MLPs *and* CMAC *and* RBF) is to use steepest descent:

$$\text{new } W_{ij} = \text{old } W_{ij} - LR^* \frac{\partial}{\partial W_{ij}}((\underline{Y}(t) - \underline{\hat{Y}}(t))^2), \tag{91.2}$$

where LR is a constant called the "learning rate." *Backpropagation* is simply an efficient technique for calculating all of the derivatives required in Equation 91.2 in one sweep through the network. (This refers to the more general form of backpropagation which was proven valid in my Ph.D. thesis in 1974, and discussed in a neural net context in the 1981 IFIP Proceedings.) Backpropagation can be applied to *any* differentiable distributed system, *not just to MLPs.* (See White and Sofge, 1992, 1989, 1974.)

Simple supervised learning already has many applications in the real world. For example, the U.S. Post Office has a large research program to try to develop automated digit recognition systems, because this could save them billions of dollars in sorting mail. Already, Post Office officials have told me directly that all the very best recognition systems use supervised learning as *part* of their system. One of the very best systems is the AT&T system, developed by Guyon et al. (1993) based completely on a sophisticated use of backpropagation and MLPs. (Many other individuals who have never worked for the Post Office have made strong claims about the performance of their systems—neural or non-neural—which need to be considered with caution.) It is remarkable that such a new technology should already be proven superior in a complex recognition problem, where the existing methods had a long history of extremely capable and insightful research.

Supervised Learning: Recent Results

Most people now distinguish two kinds of neural net designs for supervised learning—global designs like MLPs, and *local* designs like CMAC and RBF. With global designs, all of the weights affect the output of the network at all times. With local designs, a given weight is used only over a small part of the input space (which may or may not be a localized region) (Werbos, 1993c). These three designs—MLP, CMAC, and RBF—are the dominant designs in use today in neurocontrol; thus, most of the articles of White and Sofge (1992) contain some examples of their use. All three are *feedforward* designs: they are easy to calculate, from inputs to intermediate or hidden variables, to outputs, without any need to solve nonlinear equations or (equivalently) to iterate through a nonlinear relaxation process.

Most people say that *global* designs lead to slower learning, but a better ability to predict new points very different from the training set. Thus, global designs are good to use when you have a fixed database, and you can afford to go over your training examples again and again. Local designs are used more often in real-time control. In theory (Werbos, 1993c), there is an approach called *syncretism* which could give us the benefits of *both* approaches; however, research is still necessary to implement this approach. A similar approach, called "continuous learning," has been used by Rosalind Picard of M.I.T. in machine learning.

Actually, there are limits to the capabilities of all feedforward networks. These are related to some of the early works of Minsky, where he showed that feedforward MLPs cannot represent concepts like a "connected path" very well. For maximum performance, in control, we need to use *recurrent* networks, networks which allow a neuron to input its own output, etc. There are two types of recurrent network important here: time-lagged recurrent networks (TLRNs) and simultaneous-recurrent networks (SRN). It is possible to combine *both* kinds of recurrence in a single network. For complex problems—like helping a robot find a path through a novel cluttered workspace, or like strategic defense—it may be necessary to use SRNs instead of feedforward nets, even if one is simply trying to "clone" a dynamic programming solution. Backpropagation can be used efficiently with *both* kinds of recurrence, and even with networks which combine both types of recurrence together. (See articles 3 and 10 of White and Sofge, 1992, for straightforward ways to combine both kinds of recurrence; see Werbos, 1993c, and the paper by Gupta also in *WCNN93* for additional ideas on improving the convergence and stability of SRNs.) Simultaneous-equation models in econometrics can be treated as *nonneural* SRNs; in fact, the very first application of simultaneous backpropagation, in 1981, was to the sensitivity analysis of an econometric model used by the Department of Energy for natural gas policy analysis (Werbos, 1988). For a recent neural application, see Werbos and Pang (1996).

There have been rumors that recurrent nets are very expensive to adapt as networks grow larger; however, those rumors are based on certain particular designs which scale as N^2. See White and Sofge (1992) and Werbos (1974) for designs which scale as N, just like backpropagation for the simple 3-layer net. Likewise, issues of stability depend critically on how the derivatives calculated by backpropagation are actually *used* within a larger system; the stability properties vary greatly from design to design, and are beyond the scope of this article.

Simple steepest descent (Equation 91.2) can be a slow way to adapt these networks. For realtime learning, one can do much better by using an adapted learning rate (Werbos, 1993e, Chapter 3). For offline learning, there are techniques in numerical analysis—like the methods of Nocedal and perhaps Karmarkar—which can speed things up much faster. For true classification problems (where $Y_i(t)$ is always 1 or 0), it is usually better to minimize the Bernoulli measure of error *instead* of square error:

$$L(\underline{Y}, \underline{\hat{Y}}) = \sum_i Y_i(t) \log \hat{Y}_i(t) + (1 - \hat{Y}_i(t)) \log (1 - \hat{Y}_i(t)) \quad (91.3)$$

Again, there is no problem in using backpropagation with such a modified error function. Modified error functions have also been used to help in "growing" or "pruning" connections in a neural net design; in other words, the design itself can be adapted over time.

A Simple Example of the Implementation of Supervised Learning

In neurocontrol, it is very important to build modular program libraries. It is important to build main programs, which call on subroutines to do a lot of the work. It is important to have many, alternative forms of these subroutines available. (In fact, we need subroutines which call subroutines which call subroutines, etc.)

This section will give an example of a main program, which calls on two subroutines, to perform supervised learning. This will be a very simple example, which ignores many of the advanced points in the previous section.

Suppose that we have a neural network design which implements some function f as in Equation 91.1. Before we can run the main program, we must first code up two subroutines for that network:

1. A *forwards* subroutine, f, which inputs X and W, and outputs Y.
2. A *backwards* or *dual* subroutine, F_f, which inputs X and W *and* the derivatives of error with respect to the variables Y_i, *and outputs* the derivatives we need to adapt the weights W.

The next subsection will show how to code up these subroutines, for one example of neural-net designs—the classic three-layer MLP. Section 91.5 will show how to code up the same two subroutines for the elastic fuzzy logic (ELF) network I have recently proposed (Werbos, 1993b). First, however, a simple main program, which can be used for *both* of these examples and for a wide variety of other designs too, will be described.

The main program will perform supervised learning in real time. In other words, it will start with a given set of weights W; it will then input a new input vector X and a new target vector Y, and then modify its weights to adapt to this new observation.

The "program" consists of the following five steps, to be performed in order:

1. Input X and Y.
2. Call on the forwards subroutine to calculate:

$$\underline{\hat{Y}} = f(\underline{X}, W) \tag{91.4}$$

3. Calculate the derivatives of error with respect to Y:

$$F\_\hat{Y}_i = \frac{\partial}{\partial \hat{Y}_i}((\underline{\hat{Y}} - \underline{Y})^2) = 2(\hat{Y}_i - Y_i) \tag{91.5}$$

4. Call on the dual subroutine to calculate the set of derivatives we need:

$$F\_W = F\_f_w(\underline{X}, W, F\_\hat{Y}) \tag{91.6}$$

5. Adapt the weights by steepest descent:

$$\text{new } W_{ij} = \text{old } W_{ij} - LR * F\_W_{ij} \tag{91.7}$$

Step 4 is the backpropagation step. Calling on the dual subroutine is the same as "backpropagating through" the network.

The *Handbook of Intelligent Control* (White and Sofge, 1992, Chapters 3, 10, 13) contains "main programs" of this sort for all the many designs used in neurocontrol. It also provides general-purpose instructions for how to program dual subroutines for all kinds of networks or models or programs, neural or nonneural. Unlike some of the material in Miller et al. (1990), this pseudocode is more complete, based on simulations, applications, and analyses of convergence. Correct pseudocode for GDHP, which may be even more powerful, is currently available only from BehavHeuristics, which has a patent pending on several of these designs. (The reader should be warned, however, that Chapter 3 of White and Sofge (1992) does contain a flowchart where the middle box is labeled "Critic" but should be labeled "Model." Chapter 10 has "s_x" instead of "x_i" in the equation explaining dual subroutines, and a missing term at the end of one equation which is part of a hypothetical simple example of how to generate a dual subroutine.)

Basic Subroutines for the 3-Layer MLP

The classic fully-connected 3-layer MLP, popularized by Rumelhart et al. (1986), is defined by the following equations, which can be coded up directly into a forwards subroutine:

$$v_i^- = \sum_{j=0}^{m} W_{ij}^- X_j \qquad i = 1 \cdots h \tag{91.8}$$

$$x_i = s(v_i^-) = 1/(1 + e^{-v_i^-}) \qquad i = 1 \cdots h \tag{91.9}$$

$$v_i^+ = \sum_{j=0}^{h} W_{ij}^+ x_j \qquad i = 1 \cdots n \tag{91.10}$$

$$\hat{Y}_i = \sum_{j=0}^{h} W_{ij}^+ x_j \qquad i = 1 \cdots n \tag{91.11}$$

where we adopt the convention that $X_0 = x_0 = 1$, where n and m are the number of components of the vectors Y and X, and where h is the number of "hidden units." (Any value of h is allowed, but Werbos (1993c) describes a few of the methods now available which try to pick the best value of h.) The letter "v" refers to the level of "voltage" exciting the membrane of a neuron. The set of weights, W, is made up of the two subsets, W^+ and W^-.

The dual subroutine of this network is defined by the following equations:

$$F\_v_i^+ = F\_\hat{Y}_i * s'(v_i^+) = F\_\hat{Y}_i * (\hat{Y}_i)(1 - \hat{Y}_i) \qquad i = 1 \cdots n \tag{91.12}$$

$$F\_x_j = \sum_{i=1}^{n} F\_v_i^+ * W_{ij}^+ \qquad j = 1 \cdots h \tag{91.13}$$

$$F\_W_{ij}^+ = F\_v_i^+ * x_j \tag{91.14}$$

$$F\_v_j^- = F\_x_j * s'(v_j^-) = F\_x_j * (x_j)(1 - x_j) \qquad j = 1 \cdots h \tag{91.15}$$

$$F\_W_{ij}^- = F\_v_i^- * X_j \tag{91.16}$$

Equations 91.14 and 91.16 define the set of values $F\_\_W$ used in our main program above.

In coding up the 3-layer MLP for a *general* range of applications, it can be important to add an extra equation, to output $F\_\_X$ as well as $F\_\_W$.

91.5 Elastic Fuzzy Logic: Principles and Subroutines

Background and General Concepts

The elastic fuzzy logic (ELF) network is a new design, described in Werbos (1993b), with a patent pending through Scientific Cybernetics, Inc. The goal is to permit a complete synthesis of the best capabilities of fuzzy logic and of neural networks, including neurocontrol.

The idea of ELF grew out of a view of intelligent control shown in Figure 91.2. That view, in turn, emerged from numerous discussions about the idea of intelligent control, which has been the subject of NSF workshops, IEEE conferences, and so on. In early years, the goal of intelligent control was to merge the capabilities of artificial intelligence (AI) with the capabilities of control theory. However, as shown in Figure 91.2, such efforts were limited in their success because of the intrinsic incompatibility of *linear* control designs and *forms* of AI based on 1/0 Boolean variables. In recent years, however, there has been tremendous growth in two *subsets* of AI and control theory—fuzzy logic and neurocontrol, respectively—which overcome this compatibility problem. These two subsets both focus on variables which vary *continuously,* in nonlinear systems, between *limits* such as 0 and 1 or −1 and 1.

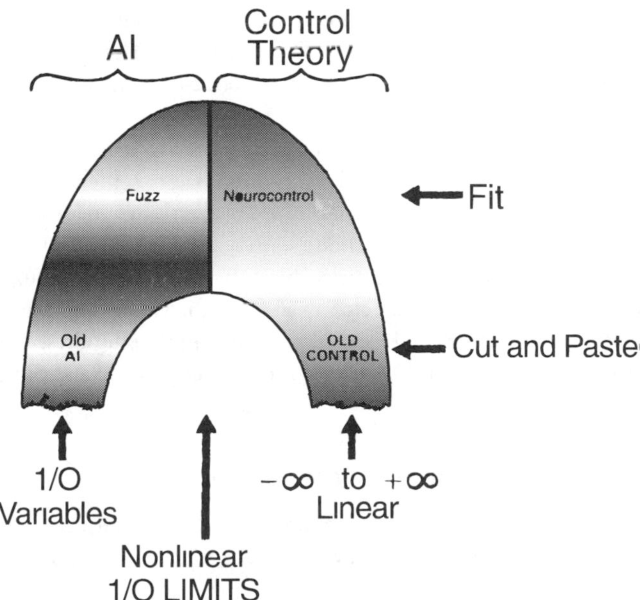

Figure 91.2 A view of intelligent control. *Source:* White D. and Sofge, D, eds., 1992. *Handbook of Intelligent Control,* Van Nostrand, New York, NY.

Based on this view, I proposed in 1990 (Werbos, 1992) that we could try to build a *single* nonlinear controller, using *both* classes of tools—fuzzy tools and neural tools—to complete the design. More precisely, I proposed that we could use fuzzy logic to *initialize* a controller—based on the words of a human expert—and then use neurocontrol methods to adapt the controller to improve its performance in response to real-time data. After adaptation, the methods of fuzzy logic could be used to explain to the human expert what the controller was doing. The obvious way to do this was to use a "fuzzy logic" controller—making the human interface straightforward—and using the methods of neurocontrol (which are mathematically quite general) to adapt the controller. Many varieties of this were discussed.

Many people have followed up on this approach, but it has turned out to have a number of difficulties. ELF is intended to overcome these difficulties.

To begin with, the classical forms of fuzzy logic led to controllers which were not differentiable (Kosko, 1992). The most powerful neurocontrol methods require differentiability. However, the overwhelming bulk of practical applications of fuzzy logic have used a very simple form of fuzzy logic, applied to control problems. Japanese researchers (e.g., as reported at the Iizuka conference organized by Yamakawa of Kyushu University) have done a number of comparative studies showing that a simple, *differentiable* version works best in practice. (This is not yet ELF, of course, but a step in that direction.)

In this version of fuzzy logic, an expert gives us a set of n_R "rules" in a simple if-then form. Each rule takes the form, "if A_1 and A_2 . . . then do B." The expert must also supply membership functions $\mu_A(X)$ to indicate the *degree* of applicability of word A to any vector of observables X. In theory, the expert also provides a membership function $\mu_B(\boldsymbol{u})$ to describe how well a given vector of actions $\boldsymbol{u}$ fits the action-word B; however, in practice, the expert can simply tell us $\boldsymbol{u}_r$ the vector of actions which *best fits* the action word B in rule number r.

The fuzzy system designer receives these rules and membership functions from the expert, and then implements the following equations in the actual controller:

$$x_i = \mu_{A_i}(\underline{X}) \qquad i = 1 \cdots n_A \tag{91.17}$$

$$R_r = x_{r,1} * x_{r,2} * \cdots * x_{r,n_r} \qquad r = 1 \cdots n_R \tag{91.18}$$

$$R_\Sigma = 1/\sum_{r=1}^{n_R} R_r \tag{91.19}$$

$$\underline{u} = R_\Sigma * \sum_{r=1}^{n_R} R_r \underline{u}_r \qquad (91.20)$$

where n_A is the number of input words, where n_r is the number of input words in rule number r, and where $x_{r,j}$ is an abbreviation for that value of x_i which corresponds to the jth input word appearing in rule number r. Intuitively, R_r represents the *degree* to which rule number r applies in the current situation X.

Approaches to Adapting Fuzzy Systems

Following Werbos (1992), we might try to adapt Equations 91.17 through 91.20 *directly,* using neural network methods. The big problem is this: Equations 91.17 through 91.20 do not contain *weights* or *parameters* to be adapted. A number of authors have encountered this problem and taken a variety of approaches to solve it. All of these approaches work, but they do not live up to our original goals.

The most common approach is to adapt the membership functions, μ_A. For example, the membership functions might be set up as simple neural nets like MLPs. Or the value of μ_A in selected areas of space might be treated as a parameter to be adapted. In either case, the membership function is adapted so as to improve the overall performance of the controller.

This approach has two problems: communication back to the human expert is impaired when the *definitions* of words (A) are *changed* so that they mean something else from what the expert intended: there is not freedom here to adapt the rules themselves—especially when the same input word appears in many rules. Intuitively, a better control strategy would often require better rules.

Actually, there is another way to use simple MLPs as membership functions, which does not have those problems. If a word A is so complex that an expert finds it hard to define the word, you can present many examples of possible situations X. You can ask the expert to estimate the degree to which the word A applies in each case. You can then use supervised learning to estimate what the expert really means by the word A. This approach is good for understanding very tricky words, but it does not help in making a controller work better.

Yager has developed an alternative approach which is far more flexible, but includes the use of MLPs in the final action stage (Equation 91.20). This, in turn, may limit the degree of communication back to the human expert. More recently, Yager has helped provide additional interpretation to ELF, and has developed some theoretical ideas relevant to understanding the significance of the approach (Yager, 1993).

Mendel and Wang (1992) are said to have developed another approach, which is essentially a subset of ELF. In ELF, I replace Equation 91.18 by:

$$R_r = \gamma_{r,0} * x_{r,1}^{\gamma_{r,1}} * \cdots * x_{r,n_r}^{\gamma_{r,n_r}} \qquad r = 1 \cdots n_R \qquad (91.21)$$

The gamma parameters are called "elasticities" (whence the term "elastic" fuzzy logic). Note that *all* the methods discussed in this section could be described as "adaptable fuzzy logic," but that I reserve the term *elastic* fuzzy logic for systems using Equation 91.21 or the equivalent.

In using ELF, the words of the human expert are translated into Equations 91.17 through 91.20, exactly as before: this is equivalent to setting all the elasticities to one in Equation 91.21. Then, in adaptation, the elasticities *and* the $\mathbf{u}_r$ vectors can be adapted using neurocontrol methods. After adaptation, the expert can be shown the new $\mathbf{u}_r$ vectors (which the expert should understand) *and* the elasticities. The γ_{r0} elasticities represent the degree of strength or validity of the rule. The $\gamma_{r,0}$ elasticities represent the degree of importance of each input condition. (For example, Yager has pointed out to me that a $\gamma_{r,i}$ of 2 would be equivalent to having the word A_i appear *twice* in the rule. Large elasticities allow a rule to approach the classic "min" rules of classic fuzzy logic, *if* this improves real-time control.) These elasticities are *exactly* equivalent to "elasticities" as used in economics; if economists can understand elasticities (after a brief introduction), then expert engineers should be able to as well.

ELF permits us to use all the techniques from neurocontrol (White and Sofge, 1992) to add and delete weights and units (rules). Thus, if an elasticity goes to zero in adaptation, that weight or rule could be deleted. Also, the adaptive routine might experiment with new words j in rule r (or even new rules), and *initialize* the relevant elasticity to zero; this would bring in the new word or rule significantly only if adaptation really called for a nonzero value. Thus, neurocontrol techniques could be used to *change* the rules, as well as adapt them.

ELF gives one the full flexibility of a (local learning) neural system, limited only by the initial vocabulary (words A_i) provided by the expert. To test whether that limitation is a problem (whether the expert needs to increase his or her vocabulary), you could compare the resulting performance against classical neural nets, against Yager's design, or against an extended form of ELF which also allows the computer to add new words and adapt their definitions (membership functions). The latter kind of test might give hints to the expert about the kinds of words which are needed.

Once we know how to program an ELF net *as a neural network* and how to program its dual subroutine, we can then use the techniques in White and Sofge (1992) *directly* to adapt fuzzy controllers, fuzzy predictive models, fuzzy performance measures or critics, fuzzy value networks, fuzzy planning systems, fuzzy model-based controllers, etc. Even though White and Sofge (1992) do not discuss ELF itself, one can simply plug these two critical subroutines into the "programs" which White and Sofge (1992) do provide.

Basic Subroutines for the ELF Net

The ELF network can be programmed as a neural network with two hidden layers (defined by Equations 91.17 and 91.21) and an output layer defined by Equation 91.19 and:

$$\underline{\hat{Y}} = R_\Sigma * \sum_{r=1}^{n_R} R_r \underline{Y}_r \qquad (91.22)$$

The equations for the dual subroutine come from differentiating this system. They are:

$$F\_\underline{Y}_r = F\_\underline{\hat{Y}} * R_r * R_\Sigma \qquad r = 1 \cdots n_R \tag{91.23}$$

$$F\_R_r = R_\Sigma * ((F\_\underline{\hat{Y}} \cdot \underline{Y}_r) - (F\_\underline{\hat{Y}} \cdot \underline{\hat{Y}})) \qquad r = 1 \cdots n_R \tag{91.24}$$

$$F\_\gamma_{r,j} = F\_R_r * \log x_{r,j} * R_r \tag{91.25}$$

$$F\_\gamma_{r,0} = F\_R_r * R_r/\gamma_{r,0} \tag{91.26}$$

where these equations clearly can be calculated more efficiently by calculating $F\_\_R_i{}^*R_i$, only once for each i, and by calculating $\log x_i$ for each input word i only once. These equations, together, yield the set of values $F\_\_W$ used in our main program. For a complete dual subroutine (as required in some neurocontrol designs), one would also need to include the equations:

$$F\_x_{r,j} = F\_R_r * \gamma_{r,j}/x_{r,j} \tag{91.27}$$

$$F\_x_i = \sum_{(r,j)=i} F\_x_{r,j} \tag{91.28}$$

$$F\_X_i = \sum_j \frac{\partial \mu_j}{\partial X_i} F\_x_j \tag{91.29}$$

where Equation 91.27 is bypassed by a lengthier version when x_{rj} is very close to zero, and Equation 91.28 is summed over combinations r,j such that x_{rj} is an abbreviation for x_i. (More precisely, combinations such that input word number i is the jth input word to appear in the list of input rules in rule number r.) Usually Equation 91.29 will be very parsimonious and simple; however, if μ_j is itself a neural network, it requires us to plug $F\_\_x_j$ into the dual subroutine for that network as part of minimizing the cost of invoking this equation.

For more complex fuzzy inference structure which vary on a case-by-case basis and include recurrences, one can still use the techniques in White and Sofge (1992) to develop the required dual subroutines (e.g., see Werbos, 1993 and 1992 and the section on simultaneous-recurrent nets in chapter 3 of White and Sofge, 1992), but the details are complicated.

91.6 Current Designs in Neurocontrol: A Roadmap

This section will provide a very brief roadmap of the capabilities available today in neurocontrol. For more details, and examples of important applications, see Chapter 63.

At the middle level of analysis, as discussed in Section 91.3. ANN designs may be classified according to *what kinds of generic tasks* they perform. ANNs have performed four kinds of useful functions in control:

1. Subsystem functions such as pattern recognition or neuroidentification, for sensor fusion or diagnostics, etc.
2. Cloning functions, such as *copying* the behavior of a human being able to control the target plant.
3. Tracking functions such as making a robot arm follow a desired trajectory or reference model, or making a chemical plant stay at a desired *setpoint.*
4. Optimization functions, such as maximizing throughput or minimizing energy use or maximizing goal satisfaction over the entire future.

The first of these functions does not qualify as neurocontrol. ANNs for the second function are called "supervised controllers." They have been reinvented many times, usually by people who use supervised learning and base their system on a database of "correct actions" (often without telling us how they know what the "correct actions" are). The third function—tracking—is performed by "direct inverse controllers" and by "neural adaptive controllers." Some authors seem to assume that following a trajectory is the *only* interesting problem in control; however, the human brain is *not* a simple trajectory follower, and real-world engineering faces many other tasks as well. The fourth group of designs is clearly the only working group with any chance of replicating brain-like capabilities (White and Sofge, 1992; Werbos, 1974). Within the fourth group itself there are two useful subgroups—the "backpropagation of utility" (i.e., *direct* maximization of future utility) and the "adaptive critic family" (broadly defined); only the latter has a serious chance of someday replicating true brain-like capability (Miller et al., 1990; Werbos, 1974). Within the adaptive critic family, we face a similar ladder of designs, from simple methods which learn slowly except on small problems, through to moderate-scale methods, through to large-scale methods requiring a neuroidentification component, through to methods capable of true "planning" and "chunking" but requiring the use of simultaneous-recurrent modules (Barron, 1993; Werbos, 1993c).

In summary, we have a ladder here, starting from straightforward designs, easy to implement today, which can take us up step by step to a true understanding of intelligence . . . if only we have the will to climb higher.

References

Barron, A. R. 1993. Universal approximation bounds for superpositions of a sigmoidal function, *IEEE Trans. Info. Theory,* 39(3):930–945.

Guyon, et al., 1989. *IJCNN Proc.,* June, Washington, DC.

Kosko, B. 1992. *Neural Networks and Fuzzy Systems,* Prentice-Hall, Englewood Cliffs, NJ.

Mendel, J. M. and Wang, L. X. 1992. Back-propagation fuzzy

system as nonlinear dynamic system identifier, *Proc. of Int'l Conf. Fuzzy Systems (FUZZ-IEEE).*

Miller, W. T., Sutton, R., and Werbos, P. eds., 1990. *Neural Networks for Control,* MIT Press, Cambridge, MA.

Rumelhart, D. and McClelland, 1986. *Parallel Distributed Processing,* MIT Press, Cambridge, MA.

Sontag, E. 1990. *Feedback Stabilization Using Two-Hidden-Layer Nets,* SYSCON-90-11, Rutgers University Center for Systems and Control, New Brunswick, NJ.

Werbos, P. 1974. *Beyond Regression,* Harvard U. Ph.D. thesis; reprinted in Werbos, P. 1994. *The Roots of Backpropagation: From Ordered Derivatives to Neural Networks and Political Forecasting,* John Wiley & Sons, New York, NY.

Werbos, P. 1988. Generalization of backpropagation with application to a recurrent gas market model, *Neural Networks,* October.

Werbos, P. 1989. Maximizing long-term gas industry profits in two minutes in Lotus using neural network methods, *IEEE Trans. SMC,* March.

Werbos, P. 1992. Neurocontrol and fuzzy logic: connections and designs, *Int. J. Aprox. Reasoning,* 6(2); this was an upgraded version of the paper in Lea R. and Villareal, J., *Proc. 2nd Joint Technology Workshop on Neural Networks and Fuzzy Logic,* (Houston, April 1990), NASA Conference Pub. 10061.

Werbos, P. 1993a. Neurocontrol and elastic fuzzy logic: capabilities, concepts and applications. *IEEE Trans. Indus. Electronics,* April; appeared in Gupta. M. M. and Sinha, N. K. 1995.

Werbos, P. 1993b. Elastic fuzzy logic: a better way to combine neural and fuzzy capabilities, *WCNN93 Proc.,* Erlbaum.

Werbos, P. 1993a. Supervised learning: can it escape from its local minimum, *WCNN93 Proc.* Erlbaum.

Werbos, P. 1994a. Control circuits in the brain: Basic principles, and critical tasks requiring engineers, *Proc. of Eighth Yale Workshop on Adaptive and Learning Systems,* Yale University.

Werbos, P. 1994b. The Brain as a Neurocontroller: New Hypotheses and Experimental Possibilities, Pribram, K., ed., *Origins: Brain and Self-Organization,* INNS Press, Erlbaum.

Werbos, P. and Pang, X. Z. 1996. Generalized maze navigation: SRN Critics solve what feedforward or Hebbian nets cannot. *Proc. Int'l. Conf. Systems, Man and Cybernetics,* IEEE.

White, D. and Sofge, D. eds., 1992. *Handbook of Intelligent Control,* Van Nostrand, New York, NY.

Yager, R. 1993. Toward a unified approach to aggregation in fuzzy and neural systems, *WCNN93 Proc.,* Erlbaum.

91.7 Appendix (Tutorial Level Background Information): Neurocontrol and Fuzzy Logic

The material here is extracted, with little change, from my paper, "Neurocontrol and Fuzzy Logic: Connections and Designs," originally scheduled for the May 1991 issue of the *International Journal for Approximate Reasoning.* Starting in Section A.4, it provides an overview of neurocontrol which does not require prior knowledge of fuzzy logic. It will also mention why neurocontrol is vital to understanding the brain, although for detailed information on that topic see Werbos (1991).

A.1 Abstract

Artificial neural networks (ANNs) and fuzzy logic are complementary technologies. ANNs extract information from systems to be learned or controlled, while fuzzy techniques most often use verbal information from experts. Ideally, both sources of information should be combined. For example, one can learn rules in a hybrid fashion, and then calibrate them for better whole-system performance. ANNs offer universal approximation theorems, pedagogical advantages, *very* high-throughput hardware, and links to neurophysiology. Neurocontrol—the use of ANNs to directly control motors or actuators, etc.—uses five *generalized* designs, related to control theory, which can work on fuzzy logic systems as well as ANNs. These designs can: copy what experts *do* instead of what they *say;* learn to track trajectories; generalize adaptive control; maximize performance or minimize cost over time, even in noisy environments. Design tradeoffs and future directions are discussed throughout. The final section mentions a few new ideas regarding reasoning, planning and chunking, with biological parallels.

A.2 Introduction

This paper will mainly discuss neurocontrol—the use of neural networks (artificial or neural) to *directly* control motors, actuators, muscles, or other kinds of overt physical action. It will also discuss the relation *between* artificial neural networks (ANNs) and fuzzy logic, and how best to combine them. It will begin by discussing the most basic and most popular application of ANNs—to learning a mapping from a vector X to a vector Y. Then it will discuss neurocontrol, and the central importance of neurocontrol to understanding intelligence. The final section will include some thoughts about reasoning and planning, directed more towards future research.

This paper will take the position that fuzzy logic and neurocontrol are complementary technologies. In many applications, the best approach is to use the two together, rather than decide which technology is "best." This complementarity is based in part on their common emphasis on using continuous variables, which also allows a high degree of complementarity with nonlinear control theory and a new generation of analog computer hardware.

Precisely *because* they are complementary technologies, there are certain semantic problems which arise in *defining* which technology is which (i.e., in defining the *boundaries* between neural nets and fuzzy logic). There are cases where neural networkers and fuzzy logicians would use the exact same mathematics to solve a specific problem, but would give the mathematics different names, and would refer back to different sources. In cases like this, it is particularly absurd to try to decide which technology is "better," even for a specific problem; it is more realistic to lay out a diverse inventory of techniques in concrete terms, while trying to exploit *both* traditions. This paper will survey the techniques which have been developed in neurocontrol, in the hope that this will be useful to both communities.

Section A.4 will make a crucial point which underlines the relevance of Sections A.6 through A.9 to fuzzy control: that the learning methods described in these sections can all be applied to fuzzy inference structures (or directly to fuzzy rules, in some cases); even though those later sections *talk* about block diagrams filled in with neural networks, one can use the same block diagrams by plugging in fuzzy structures instead, using the various options discussed in Section A.2. Some readers might prefer to have this substitution worked out explicitly, across the entire range of options, and labeled explicitly as "fuzzy learning control designs;" however, by describing neurocontrol on its own terms, the hope is to make this paper accessible to a wider audience, starting with Section A.4.

This paper will cite a few examples and surveys of learning methods developed *within* the fuzzy logic community; however, there is no claim that these surveys are complete, and no effort is made to provide a complete crosswalk between those learning methods and neurocontrol. Such a crosswalk would be useful, but it could probably be done better by someone more familiar with all the many strands of thought within fuzzy control, in addition to the material described here.

The paper *will* try to be relatively comprehensive in laying out the inventory of designs used in the neurocontrol field. It will not describe any *one* application in extensive detail; however, the references point to papers which do this, for a wide variety of designs and applications. Section IX will briefly discuss one example of a hybrid fuzzy/neural system now being worked on by a group in Washington, DC.

A.3 ANNs and Fuzzy Logic in Supervised Learning

Neurocontrol is still a small part of the greater neural network community. Most people use ANNs for applications like pattern recognition, diagnostics, risk analysis, and so on. They mostly use ANNs to learn static mappings from an "input vector," *X* to a "target vector," *Y*. For example, *X* might represent the pixels which make up an image, while *Y* might represent a classification of that vector. Given a training set made up of pairs of *X* and *Y*, the network can "learn" the mapping by adjusting its weights so as to perform well on the training set. In the example, it would learn to input the image and output the classification.

This kind of learning is called "supervised learning." There are many forms of supervised learning used by different researchers, but the most popular is basic backpropagation (Werbos, 1990a). Basic backpropagation is simply a unique implementation of least squares estimation. In basic backpropagation, one uses a special efficient technique to calculate the derivatives of square error with respect to all the weights or parameters in an ANN; then, one adjusts the weights in proportion to these derivatives, iteratively, until the derivatives go to zero. The components of *X* and *Y* may be 1's and 0's, or they may be continuous variables in some finite range.

Fuzzy logic is also used, at times, to infer well-defined mappings. For example, if *X* is a set of data characterizing the state of a factory, and *Y* represents the presence or absence of various breakdowns in the factor, then fuzzy rules and fuzzy inference may be used to decide on the likelihood that one of the breakdowns maybe present, as a function of *X*.

Which method is better to use, and when?

The simplest answer to this question is as follows: since ANNs extract knowledge from empirical databases used as "training sets", and fuzzy logic usually extracts rules from human experts, we should simply decide which source of knowledge we trust more, in the particular application. (When in doubt, we can try both and try for an evaluation after the fact.) In principle, empirical data represents the real bottom line while expert judgment is only a secondary source; however, when the empirical data is too limited to allow us to learn complex relations, expert judgment may be all we have.

In many applications, there are some parts of the problem for which we have adequate data, and others for which we do not. In that case, the practical approach is to divide the problem up, and use ANNs for part and fuzzy logic for another part. For example, there may be an intermediate proposition R which has an important influence on *Y*; we may build a neural net to map from *X* to R, and a fuzzy logic system to map *X* and R into *Y*, or vice-versa. Amano et al. (1989), for example, have built a speech recognition system in which ANNs detect the features, and a fuzzy logic system goes on to perform the classification. Many people building diagnostic systems have taken similar approaches (Schreinemakers and Touretzky, 1990).

In the current literature, many people are using fuzzy logic as a kind of organizing framework, to help them subdivide a mapping from *X* to *Y* into simpler partial mappings. Each one of the simple mappings is associated with a fuzzy "rule" or "membership function." ANNs or neural network learning rules are used to actually learn all of these mappings. There are a large number of papers on this approach, reviewed in Takagi (1990). Kosko's work in this area is particularly famous. Because these are typically very simple mappings—with only one or two layers of neurons—we can choose from a wide variety of neural network methods to learn the mappings; however, since the ANNs only minimize error in learning the individual rules, there is no guarantee that they will minimize error in making the overall inference from *X* to *Y*. This approach also requires the availability of data in the training set for all of the intermediate variables (little *R*) used in the partial mappings. Strictly speaking, this approach is a special case of the previous paragraph, in the general case, some rules can be learned while others come from experts.

Many people in fuzzy logic might say that fuzzy logic is more than just rules and inference. There is also such a thing as fuzzy learning. In fact, much of the neural network literature on learning (like backpropagation [Werbos 1990a]) applies directly to any well-behaved nonlinear network. It can be applied directly to the inference structures used in fuzzy logic. We could easily get into a situation where fuzzy logic people and neural network people use the exact same mathematical recipe for how to adapt a particular network, and use different names for the same thing.

Here we focus on the generalized mathematical learning rules, so that we can speak a more universal language, and avoid distinctions without a difference.

There are some problems which can not be easily subdivided into expert-based parts and learning-based parts. For example, there are theories of international conflict which involve a rich structure, continuing a large number of parameters known with varying degrees of confidence; it is important to expose the entire structure to the discipline of historical testing ("backcasting" and "calibration"). In situations like that, the best procedure is to combine fuzzy logic and learning. (In Bayesian terms, one would regard this as a convolution of prior and posterior knowledge, to determine the correct conditional probabilities, conditional upon available information). For example, we can use fuzzy logic and interviews with experts to derive an initial structure and estimates of uncertainty. Then, one can use generalized backpropagation directly to adjust the weights (or uncertainty levels or other parameters) in that network. We can even use backpropagation to minimize an error measure like:

$$E = \sum_i (Y_i - \hat{Y}_i)^2 + \sum_j C_j(W_j - W_j^{(0)})^2 \qquad (91.22)$$

where C_j is the prior degree of certainty about parameter W_j and $W_j^{(0)}$ is the prior estimate of the parameter. This kind of convolution approach could also be applied, of course, to the learning of independent rules or membership functions, as described in Takagi (1990). In a recent meeting to discuss long-term strategic planning issues, I suggested a two-stage approach: build up an initial inference system or model using conventional techniques, which adapt individual rules or equations; then—after assessing degrees of certainty—adjust all of the weights in a "calibration" phase, using backpropagation to make sure that the overall structure adequately fits the overall structure in historical data.

To the best of my knowledge, the idea of applying backpropagation to a fuzzy logic network was first published in 1988 (Werbos, 1988a). Matsuba of Hitachi, in unpublished work, first proposed the use of Equation 91.22. Backpropagation is important in this application, because it can adapt multilayer structures.

Backpropagation cannot be used to adapt the weights in a more conventional, Boolean-logic network. However, since fuzzy logic rules are differentiable, fuzzy logic and backpropagation are more compatible. Strictly speaking, it is not necessary that a function be everywhere differentiable to use backpropagation; it is enough that it be continuous and be differentiable almost everywhere. Still, one might expect better results from using backpropagation with modified fuzzy logics, which avoid rigid sharp corners like those of the minimization operator.

One reason for liking fuzzy logic, after all, is that it can do a better job than Boolean logic in representing what actually exists in the mind of a human expert. This being so, modified fuzzy logics—which are even smoother—may be even better. Fu (1990) has gotten good results applying backpropagation to simple fuzzy logic structures (using special rules to handle the corner points), while Hsu et al., (1990) have proposed a modified logic. Presumably the fuzzy logic literature itself includes many examples of smooth, modified fuzzy logics. Among the obvious possibilities are to use simple ANNs themselves in knowledge representation; to use functional forms similar to those used by economists, in production functions and cost functions, with parameters to reflect the importance, the complementarity, and the substitutability of different inputs.

Fuzzy logic has the advantage that it can be applied in a flexible way, using a different inference structure for each case in the training set. This inference structure may contain logic loops, which go beyond the capability of what ANN people call "feedforward" networks. The inference structure maybe a "simultaneously recurrent" network. Nevertheless, backpropagation can be used on such inference structures, using the memory-saving methods in Werbos (1988b) to calculate the derivatives of error with respect to every parameter, at a cost less than the cost of invoking the inference structure a single time. Thus, one can use backpropagation here as well. Hybrid systems like this may be too expensive to justify for unique applications, but they make considerable sense in generalized software system.

When complex inference is required, in fuzzy logic as in conventional logic, the design of an inference engine can be very tricky. Neurocontrol systems may be used, in essence, as inference engines. In fact, it can be argued that this is precisely how the human brain does inference—that the true "deep structure" of language is a collection of neural nets which learn, through experience, how to perform more and more effective inference (in a non-Boolean environment). Inference may be more difficult the other forms of control problem; however, there are parallels between neurocontrol systems and existing inference engines which suggest some real possibilities here.

Stinchcombe and White have proven (IJCNN, 1989) that conventional ANNs can represent essentially any well-behaved nonlinear mapping. Sontag (1990) has extended this result to some of the ill-behaved mappings one sometimes encounters in control designs. Nevertheless, in applications of ANNs, many researchers have begun to encounter the limitations of *any* static mapping. In recognizing dynamic patterns (Werbos, 1990a), like speech or moving targets, or in real-world diagnostics (Werbos, 1990c), it is often necessary to add memory of the past. As one adds such memory, it becomes more and more important to build up robust *dynamic models* of the system to be analyzed or controlled. Neural networks can do this (Werbos, 1990d), in part by adapting intermediate features and developing representations which an expert might not have thought of.

A.4 Neurocontrol in General

In 1988, neurocontrol was just beginning a major period of growth. At that time, NSF sponsored a workshop on neurocontrol at the University of New Hampshire, chaired by W. Thomas Miller (Miller et al., 1990), who brought together a small, mixed group of neural network people, control theorists and experts in substantive applications areas. In the very early part of that workshop, a few people echoed the old arguments about who is better—control theorists or neural networkers. Within a very short time, however, it became apparent that this issue was utterly meaningless. It was meaningless because it revolved about a distinction without a difference. The reason for this is illustrated in Figure 91.3.

Figure 91.3 is a Venn diagram, telling us that neurocontrol is a subset *both* of neural network research and of control theory. In the course of the workshop, it became apparent that the existing work in neurocontrol could be reduced to five fundamental design strategies, each of which occurred over and over again, with variations, in numerous papers. (Individual papers tend to highlight their unique aspects, of course.) *All five* turned out to be *generic* approaches which could be applied to *any* large, sparse network of differentiable functions or to an even larger class of networks. One may call these "functional networks," as opposed to neural networks. All five methods could be fully understood as generic methods *within* control theory. By remembering that neurocontrol is a subset of *both* disciplines, we are in a position to draw upon both disciplines in developing more advanced designs and applications.

This situation is particularly important to fuzzy logicians because the inference structures of fuzzy logic are themselves functional networks. In this paper, I will present numerous boxes labeled as "neural networks," but every such box could just as easily be filled in with a fuzzy inference structure varying over time. In other words, every one of the five "neurocontrol" methods can also be applied directly to fuzzy learning as well. In practice, one would often want to fill in different boxes with different things—perhaps an ANN for one box, a hybrid neural/fuzzy map (as described in the previous section) for another, and a conventional fixed algorithm for a third. This kind of mixing and matching is quite straightforward, once the basic principles are understood.

Why should we be interested at all in the special case where the functional network is built up from the traditional kinds of artificial neurons? Why should we be interested in functional forms close to the conventional form used in ANNs (Werbos, 1990a):

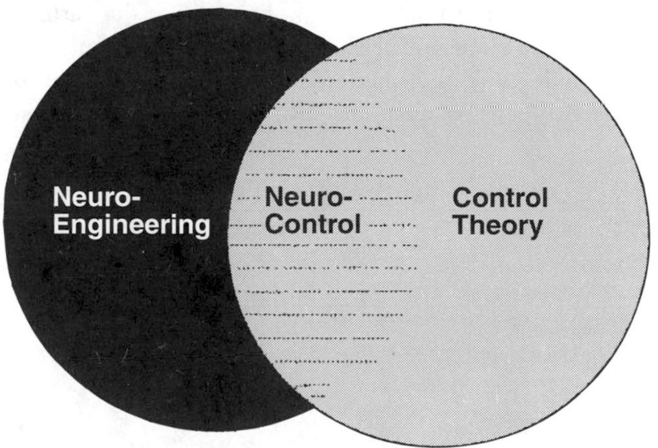

Figure 91.3 A VENN diagram.

$$x_i = s(\sum W_{ij}x_j) \tag{91.23}$$

where:

$$s(z) = \frac{1}{1 + e^x} \tag{91.24}$$

(Here, x_j represents the "output" or "activation" of a model neuron, while W_{ij} represents a "weight" or "parameter" or "connection strength" or "synapse strength.")

There are at least four reasons for paying attention to the special case represented by neural networks:

1. The universal mapping theorems of Sontag (1990) and White and others.
2. The availability of special purpose computer hardware.
3. The pedagogical value of the special case.
4. The link to the brain.

The theorems of White and others have excited great interest in the control community because they show that conventional ANNs do something very similar to what Taylor series do—provide a basis for approximating an arbitrary nonlinear function. As with Taylor series, the nonlinearity is very simple, offering a hope of workable practical tools.

The availability of special purpose computer hardware is a decisive factor in favor of ANNs. There are many cases where a task can be done equally well using conventional sequential methods or neural nets, and where both approaches involve a similar degree of computational complexity. (For example, there are cases where an ANN can simply be trained to mimic the input-output behavior of an existing algorithm.) In such cases, ANNs may have a decisive advantage in real-world implementation because of the hardware.

Fuzzy logic chips have also been developed. However, because of the complexity of fuzzy logic, as normally practiced, these chips cannot take advantage of parallel distributed architecture as much as neural chips do. At the May 1990 conference in Houston on neural nets and fuzzy logic, the Japanese developer of one of the leading fuzzy chips stated unequivocally that one could expect far more computational throughput from a neural chip than from a fuzzy chip.

Harold Szu of the Naval Surface Warfare Center has often argued that digital parallel computers constitute the real "fifth generation" of computers, as far beyond current PCs as the PCs are beyond the old LSI mainframes. In a similar vein, he argues that fixed-function, analog distributed hardware—either VLSI or optical—represents a sixth generation. The NSF program in neuroengineering got its start when people like Carver Mead (1989)—often viewed as the father of all VLSI—and people like Psaltis and Farhat and Caulfield (famous in optical computing) argued that this sixth generation could achieve a thousand-fold or million-fold improvement in throughput over even the fifth generation. The challenge was to find a way to use this hardware in a truly general-purpose way. That is the goal which led to the neuroengineering program at NSF. Similar considerations have been crucial to the neural networks program at the Air Force Office of Scientific Research, which has also begun to stress neurocontrol. Some engineers would simply define an ANN as a general-purpose system capable (in principle) of efficient implementation in such hardware.

A third reason for being interested in neural networks as such is their pedagogical value. The importance of this should not be underestimated. For example, when I first published backpropagation as a generalized method for use with any functional network, it received relatively little attention, in part because the mathematics were unfamiliar and difficult. Later, when several authors (including myself) presented it as a method for use with simplified ANNs—with interesting interpretations, with nice flow charts using circles and lines, and with easy-to-use software packages (exploiting the simplicity which comes from giving the user no choice of functional form)—the method became much better known (Werbos, 1982). Even now, for many people, it is easier to learn how to use a new design in the ANN special case, and then generalize the knowledge, than it is to start with the purest, most general mathematics. The explosion of interest in neural networks has also been very useful in motivating a new generation of graduate students, with diverse backgrounds, to learn the relevant mathematics.

A fourth reason for being interested in neurocontrol is the desire to be explicit about the link to the human brain. This link can be useful in both directions—from engineering to biology, and from biology to engineering.

The output of the human brain as a whole system is the control over muscles (and other actuators), as illustrated in Figure 91.4. Therefore the function of the brain as a whole system is control over time, so as to influence the physical environment in a desired direction. Control is not part of what goes on in the brain, it is the function of the whole system. Even though lots of pattern recognition and reasoning and so on occur within the brain, they are best understood as subsystems or phenomena within a neurocontroller. To understand the subsystems and phenomena, it is most important to understand their function

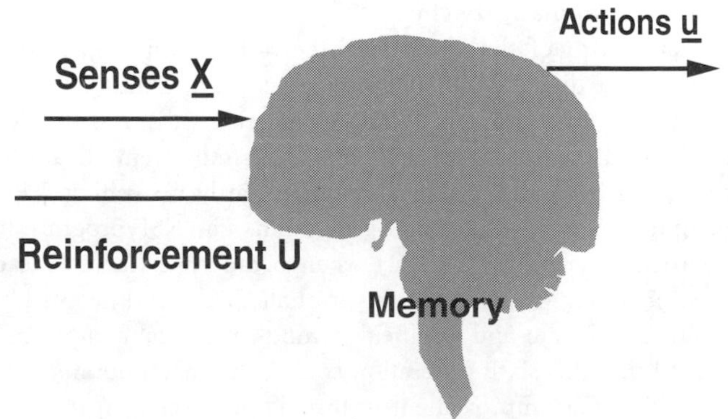

Figure 91.4 The function of the brain as a whole system is control over time to influence the physical environment in a desired direction.

within the larger system. In short, a better understanding of neurocontrol will be crucial, in the long-term, to a real understanding of what happens in the brain. (For a more concrete discussion of this, see Werbos, 1990d). Because the mathematics involved are general mathematics, they should be applicable to chips, to neurons, and to any other substrate we are capable of imagining to sustain intelligence.

The brain is living proof that it is possible to build an analog, distributed controller which is capable of effective planning (long-term optimization) under conditions of noise, qualitative uncertainty, nonlinearity, and millions of variables to be controlled at once, all with a very low incidence of falling down or instability. Control at such a high level necessarily includes pattern recognition and systems identification as subsystem. Table 91.1 compares the five major design strategies now used in neurocontrol against the four most challenging capabilities of the brain of engineering importance.

Table 91.1 was developed in 1988 (Miller et al., 1990), but it still applies to all the recent research that I am aware of (except that a few clever researchers like Narendra have developed interesting ways to combine some of these approaches). Supervised control is the strategy of building a neural network which imitates a pre-existing control system; this is like expert systems, except that we copy what a person says instead of what he does, and can operate at higher speed. Direct inverse control builds neural nets which can follow a trajectory specified by a user or a higher level system. Neural adaptive control does what conventional adaptive control does, but it uses neural networks for the sake of nonlinearity and robustness; for example, an ANN may learn how to track an external reference model (as in conventional MRAC design). Backpropagating utility and adaptive critics are two techniques for optimal control over time—to maximize utility or performance, or to minimize cost, over time. All five will be discussed in more detail in later sections. (A similar taxonomy has been published by Sugeno (1985) for fuzzy control approaches.)

Table 91.1 does suggest that we are now on a well-defined path to duplicating the most important capabilities of the human brain. However, the human brain is more than just a set of cells and learning rules. For the next few years, it may be better to think of ANNs as artificial mice (at best) rather than artificial humans. Mice are magnificent at some very difficult control and even planning tasks, but they are not very good at calculus (or is it that they don't pay attention?). Artificial humans are certainly

Table 91.1 Neurocontrol Versus Brain Capabilities

	Many motors	Noise	Long-term optimization planning	Real-Time learning
Supervised Control	X	X		X
Direct Inverse Control	(X)	X		X
Neural Adaptive Control	X	?		?
Backpropagating Utility	X		X	
Adaptive Critics 2-Net		X	X	X
BAC + DHP, etc.	X	X	X	X

possible, in my view, but there are many reasons to move ahead one step at a time. Personally, I find myself most interested in the last group of methods, because of its importance to understanding true intelligence, however, there are many engineering applications where it pays to use a simpler approach and the brain itself may be a hybrid of many approaches.

A.5 Areas of Application

Four major areas have been discussed at length (Werbos 1990d; Miller et al., 1990) for possible applications of neurocontrol:

- Vehicles and structures.
- Robots and manufacturing (especially of chemicals).
- Teleoperation and aid to the disabled.
- Communications, computation and general-purpose modeling (e.g., economics).

This paper cannot describe all these areas in-depth, but a few words may be in order.

In vehicles and structures, the aerospace industry has been a leader in applying these concepts. Unfortunately, the most exciting applications remain proprietary. NSF has been mainly interested in sponsoring high-risk applications which in turn serve as risk-reducers in high-risk projects of economic importance. Risk reduction comes from an alternative, back-up approach to solving very difficult problems which conventional techniques may or may not be adequate to solve. The National Aerospace Plane is a prime example. The goal is not to replace humans in space, but to improve the economics required to make the human settlement of space a realistic possibility.

As this paper goes to press, McDonnell-Douglas (Sofge and White, 1991) has revealed some of the details of one important application of neurocontrol, involving the manufacture of thermoplastic composite materials. Composite materials are both lighter and stronger than metals, and would have enormous benefits throughout the economy if only they could be manufactured more cheaply. McDonnell-Douglas has used neurocontrol to solve problems in the continuous, lower-cost manufacture of these materials, problems which had previously proven resistant to conventional methods (including expert systems). These problems involving the integration of propulsion, steering, and thermal control will be published in the proceedings of the October 1990 workshop (White and Sofge, 1992). It is hoped that successful solutions of those problems, incorporating noise and uncertainty, will produce greater confidence that a vehicle like NASP is actually feasible, even without allowing for future progression areas like propulsion technology, etc.

By the time of this tutorial, there is evidence that a classical thermal control system for the NASP would weigh so much that the vehicle could not possibly reach escape velocity; however, a neurocontrol system based on an adaptive critic seems to be adequate to do the job (White and Sofge, 1992).

The chemical industry has also been quite active in neurocontrol. Major sessions have been held at the American Control Conference and at the annual meetings of the chemical societies on this topic. The Chemical Reaction Processes program at NSF held a workshop in January 1991, focusing on neurocontrol, and laying the groundwork for expanded activity. The Bioengineering and Aid to the Disabled program has recently held a broad workshop to prepare for its approved initiative in the general area. McAvoy at Maryland (White and Sofge, 1992) has developed a neural network club, with membership from at least 25 Fortune 500 firms, which have developed numerous applications to improve efficiency and so on at chemical plants.

All of these new activities were motivated by interest expressed in the engineering community itself. There are many cases where industry or industry-oriented researchers are coping with fundamental issues which mainstream academics are barely beginning to address.

A.6 Supervised Control and Conventional Fuzzy Control

In the usual expert systems approach, a control strategy is developed by asking a human expert how to control something. Supervised control is essentially the ANN equivalent of that approach.

In supervised control, the first task is to build up a training set—a database—which consists of sensor inputs (X) and desired actions (u). Once this training set is available, there are many neural network designs and learning rules (like basic backpropagation) which can learn the mapping for X to u. Once the training set has been set up, the rest of this method is extremely simple.

Usually, the training set is built up by asking a human expert to perform the desired task, and recording what the human sees (X) and what the human does (u). There are many variations of this, of course, depending on the task to be performed. (Sometimes the input to the human, X, comes from electronics sensors, which are easily monitored; at other times, it may be necessary to develop an instrumented version of the task, using teleoperation technology, as a prelude to building the database.) The goal is essentially to "clone" a human expert.

Supervised control has two other applications besides cloning a human expert. First, it can generate a controller which is faster than the expert. For example, a human might be asked to fly a slowed down simulated version of a new aircraft. The ANN could then be implemented on a neural net chip, which allows it to operate a at a higher speed—higher than what a human could keep up with. Second, it can be used to create a compact, fast version of an existing automated controller, developed from expert system or control theory, which was too expensive or too slow to use in real-time, on-board applications. Supervised control is similar, in a way, to the old "pendant" system used to train robots; however, unlike the pendant system, it learns how to respond to different situations, based on different sensor input.

When should we use supervised control with ANNs (or other networks), and when should we use fuzzy knowledge-based control?

Knowledge-based control is like following what a person says, while supervised control is like copying what the person does. Parents of small children may remember the famous plea: "Do what I say, not what I do." Knowledge-based systems obey this injunction. Supervised controllers do not.

There are many tasks where it is not good enough to ask people what they do, and follow those rules. For example, if someone asked you how to ride a bicycle, and coded those rules up into a fuzzy controller, the controller would probably fall down a lot. Your system would be like a child, who just started riding a bicycle, based on rules he learned from his mother. The problem is that your knowledge of how to ride a bicycle is stored "in your wrists," in your cerebellum, and in other parts of your brain which you can't download directly into words. A supervised controller can imitate what you do and thereby achieve a more mature, complete, and stable level of performance. (This may be one reason why children have evolved to be so imitative, whether their parents like it or not.) Other forms of ANN control can go further and learn to do better than the human expert; however, it may be best to initialize them by copying the human expert, as a starting point, in applications where one can afford to do so.

The example here does not tell us that neurocontrol should be preferred over fuzzy logic in all cases. As with the problem of learning a mapping, discussed above, the theoretical optimum is to combine knowledge-based approaches and ANN approaches. As a practical matter, the theoretical optimum is often unnecessary and too expensive to implement. However, there are tasks which are too difficult to do in any other way.

As an example, consider the problem of learning how to do touch typing. Even a human being cannot learn to do touch-typing simply by hunting and pecking, and gradually increasing speed. In a technical sense, we would say that the problem of touch typing is fraught with "local minima," such that even the very best neural network—the human brain—can get stuck in a suboptimal pattern of behavior. To learn touch-typing, one begins with a teacher, who explicitly conveys rules using words. Then one fine-tunes the behavior, using neural learning. Then one learns additional rules. Only after one has initialized the system properly—by learning all the rules—can one rely solely on practice to improve the skill. Morita et al. (1990) have shown how a two-stage approach—knowledge-based control followed by backpropagation-based learning—can improve performance, in certain supervised control problems. There are other ways to deal with local minima, but they complement the use of symbolic reasoning, rather than compete with it.

In actuality, practical users of fuzzy control often tweak their rules and assumptions to get good control after the fact (e.g., see Morita et al., 1990). Morita's approach may be seen as a way of replacing that tweaking stage with something more objective, more automatic, and more suitable for larger and more confusing problems. At the October 1990 workshop mentioned above, Robicon Systems of Princeton, New Jersey reported successful results using a four-stage-strategy, which carries Morita's approach still further, consistent with ideas in Werbos (1990d).

Advanced practitioners of supervised control no longer think of supervised control as a simple matter of mapping $X(t)$, at time t, onto $u(t)$. Instead, they use past information as well to predict $u(t)$. They think of supervised control as an exercise in "modeling the human operator." The best way to do this is by using neural nets designed for robust modeling or "system identification," over time. There is a hierarchy of such ANN designs, the most robust of which has yet to be applied to supervised control (Werbos, 1990d).

Supervised control with an ANN was first performed by Widrow (1963). Kawato, in conversation, has stated that Fuji has widely demonstrated working robots based on supervised control. Many other applications have been published.

A.7 Direct Inverse Control

Direct inverse control is a highly specialized method used to make a plant (like a robot arm) follow a desired trajectory, a trajectory specified by a human being or by a higher-order planning system. The underlying idea is illustrated in Figure 91.5.

Let us suppose, for example, that we had a simple robot arm, controlled by two joints. One joint controls the angle θ_1, and the other determines θ_2. Our goal is to move the robot hand to a point in two-dimensional space, with coordinates X_1 and X_2. We know that X_1 and X_2 are functions of θ_1 and θ_2. Our job, here, is to go *backwards*—for *given* (*desired*) X_1 and X_2, we want to calculate the θ_1 and θ_2 which move the hand to that point. If the original mapping from X_1 to X_2 is invertible (i.e., if a unique solution always exists for θ_1 and θ_2), then we can try to learn this inverse mapping directly.

To do this, we simply wiggle the robot arm about for a while, to get examples of θ_1, θ_2, and the resulting X_1 and X_2. Then we adapt a neural network to input X_1 and X_2 and output θ_1 and θ_2. To *use* the system, we plug in the *desired* X_1 and X_2 as input.

Miller et al. (1990) has used direct inverse control to achieve great accuracy (error less than 0.1%) in controlling an actual, physical Puma robot. Morita (1990) has used direct inverse control with a fuzzy network, but with an ANN learning rule, and claims that this is better than supervised control for the same problem. Grossberg and Bullock and Grossberg and Kuperstein

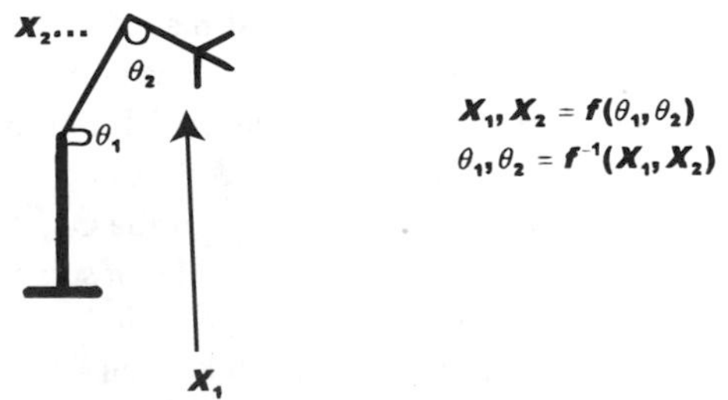

Figure 91.5 Direct inverse control.

have given many talks arguing that neural networks which implement direct inverse control are a good model for biological phenomena like hand-arm coordination and visual tracking.

In direct inverse control, as in supervised control, it works better to think of the mapping problem in a dynamic context (Werbos, 1990d), to get better results. This may explain why Miller has gotten better accuracy than many other researchers using this method. (For example, some authors report positioning errors of 4% of the work space. Miller's method may be like getting 4% error in reducing the remaining gap between the desired position and the actual position. As that gap is reduced from one time step to the next, it should go to zero quite rapidly.) Because he uses a highly appropriate supervised learning rule (Narendra, 1990), Miller reports that he can get a robot to adapt in real time to changing parameters. For example, in pushing an unstable cart around a figure-8 track, his robot arm demonstrates highly accurate tracking after three loops around the track, after the weight of the cart is changed. If Miller went further, by using a full-fledged system identification network with memory (Werbos, 1990d), using fast learning in the upper layer of the network, I would expect that his robot could adapt to changing weights even more rapidly than it now does.

Direct inverse control does not work when the original map from θ to X is not invertible. For example, if the degrees of freedom of the control variables (like T) are more or less than the degrees of freedom of the observable (like X), there is a problem. Eckmiller (1989) has found a way to break the tie, in cases where there are excess control variables, however, methods of this sort do not fully exploit the value of additional motors in achieving other desirable goals such as smooth motion and low energy consumption.

Kawato's "cascade method" (Werbos, 1990d; Jordan, 1989) describe more general ways of following trajectories, which do achieve these other goals, by rephrasing the problem as one of optimal control. They define a cost function as the error in trajectory following, plus a term for jerkiness or torque change. Then they adapt a neural network to minimize this cost function. To do this, they use the backpropagation of utility—a different ANN design, to be discussed later on. Kawato also argues that optimizing networks of this kind fit more recent experiments better than direct inverse control can.

Earlier, Kawato developed a special-purpose inverse control design called "feedback error learning" (also in Werbos, 1990d), which requires starting off from a known feedback controller of adequate quality. SAIC (San Diego, California) has widely distributed a videotape demonstration of a working vibration suppressor (applied to glasses on a table), which may be seen as a special case of feedback error learning when the known feedback controller happens to be an identity map.

A.8 Neural Adaptive Control

Neural adaptive control tries to do what conventional adaptive control does, using ANNs instead of the usual linear mappings. Because there are many tools used in conventional adaptive control, this is a complex subject (Werbos, 1990d; Narendra, 1990; Narendra and Parthasarathy, 1990; Narendra and Annaswamy, 1989).

One common tool in adaptive control is model reference adaptive control, where a controller tries to make a system follow specifications laid down in a reference model. In the conference on neural networks and fuzzy logic in Houston in 1991, Narendra described a straightforward way to do this with ANNs. One can simply define a cost function to equal the *gap* between the output of the reference model and the actual trajectory, and then minimize this cost function exactly as Jordan and Kawato did—by backpropagation utility. In actuality, one does not *have* to use the backpropagation of utility to minimize this cost function; one could also use adaptive critic methods here (Werbos, 1990d).

In adaptive control, the goal is often to cope with slowly varying hidden parameters. There are two different ways of doing this with ANNs, which are complementary. One is by *real-time* learning—where an ANN, like a biological neural network, adapts its weights in real time in response to experience. Another is by adapting *memory* units which are capable of estimating the hidden parameters. Even without real-time learning, it is possible to train an ANN *off-line* so that is will be adaptive *in real-time,* because of this memory (Werbos, 1990d). Ideally, one would want to combine both kinds of adaptation, but there is a price to be paid in so doing. The main price is that backpropagation through time must be replaced by adaptive critics (Werbos, 1990d) both in control and in system identification. The tradeoffs involved will be discussed in the next section.

In conventional, linear adaptive control it is often possible to prove stability algebraically in advance by specifying a Liapunov function (Narendra and Annaswamy, 1989). In nonlinear adaptive control, it is far more difficult (Narendra, 1990). In actuality, however, the "Critic" networks to be discussed below function very much like Liapunov functions (especially in the BAC design). For many complex, nonlinear problems, it may be necessary to *adapt* a Liapunov function after the fact, and verify its properties after the fact, rather than specify it in advance.

A.9 Backpropagation of Utility and Adaptive Critics

General Concepts

Backpropagation of utility and adaptive critics are two general-purpose designs for *optimal* control, using neural networks. In both cases, the user specifies a utility function or performance index to be maximized, or a cost function to be minimized. In both cases, these designs will always have *more than one* ANN component. Different components are adapted by different learning rules, aimed at minimizing or maximizing different things.

There will always be an Action network, which inputs current state information (and perhaps other information), and outputs the actual vector of control, $u(t)$. The utility function itself can also be thought of as a network (the Utility network), even though it is not adapted. (Some earlier papers talked about "reinforcement learning," which is logically a special case of utility maximization—Werbos, 1990d; Miller et al., 1990). In most cases, there will also be a Model network, which inputs a current description of reality, $R(t)$, and the action vector $u(t)$; which outputs a forecast of $R(t + 1)$ and of $X(t + 1)$, the vector of sensor inputs at time $t + 1$. (In some cases, the Model network can be a stochastic network, which outputs simulated values rather than forecasts.) Finally, in the case of critic designs, there will be a Critic network, which inputs $R(t)$ and possibly $u(t)$, and outputs something like an estimate of the sum of future utility across all future times.

The real challenge in maximizing utility over time lies in the problem of linking present action to future payoffs, across all future time periods. There are really only two ways to address this problem, in the general case. One is to take a proposed Action network, and explicitly work out its future consequences, for every future time period. This is exactly what the calculus of variations does, in conventional control theory, and it is also what the backpropagation of utility does. The backpropagation of utility is equivalent to the calculus of variations, but—because derivatives are calculated efficiently through large sparse nonlinear structures—one may hope for less expensive implementation. A second approach is to adapt a network which predicts the optimal future payoff (over all future times) starting from a given value for $R(t + 1)$, and to use that network as the basis for choosing $u(t)$. This requires that we approximate the payoff function, J^* of dynamic programming. This is the adaptive critic approach.

Backpropagating Utility

The backpropagation of utility through time is illustrated in Figure 91.6.

In the backpropagation of utility, we must *start* with a model network which has *already* been adapted, and a utility network which has already been specified. Our goal is to *adapt* the weights in the action network. (In practice, of course, we can adapt both the action net and the model net concurrently; however, when we adapt the action net, we treat the model net *as if* it were fixed.) To do this, we start from the initial conditions, $X(0)$ and use the *initial* weights in the action network to predict $X(t)$ at all future times t. Then we use generalized backpropagation to calculate the derivatives of *total* utility, across all future time,

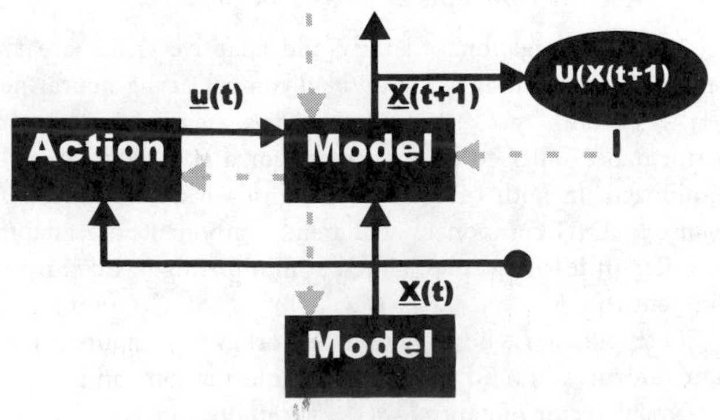

Figure 91.6 Backpropagating utility through time.

with respect to all of the weights in the action network. This involves backwards calculations, following the dashed lines in Figure 91.6. Then we adjust the weights in the action network in response to these derivatives, and start (Werbos, 1990a) all over again. We iterate until we are satisfied. The mechanics are described in more detail in 91.6, but Figure 91.6 really tells the whole story. One can implement Figure 91.6 simply by a few vector additions and subroutine calls, if one codes up a subroutine to backpropagate through the model network. (See Section 3.)

The backpropagation of utility was first proposed in 1974 (Werbos, 1974). By 1988, there were four working examples. There was the truck-backer-upper of Nguyen and Widrow, and the "cascade" robot arm controller of Kawato, both published in Werbos, 1990d. There was Jordan's robot arm controller (Jordan, 1989), and my own official DOE model of the natural gas industry (Werbos, 1989a). Recently, Narendra and Hwang have reported success with this method. Widrow has recently shown videotapes of trucks with double trailers doing complex loops to avoid obstacles while backing up, all based on the same methodology. McAvoy has developed variations which may also be seen as generalizations of control methods which have been widely used in the chemical industry (Donat et al., 1990).

The backpropagation of utility is a very straightforward and exact method. Unfortunately, there have been few reported successes this past year. This may be due in part to a lack of straightforward tutorials (though Werbos, 1989a and Werbos, 1990a should help). The biggest problem in practical applications may be the difficulty of adapting a good model network. In some applications, it may be good enough to build a model network which inputs $X(t)$ and $u(t)$, which uses $X(t + 1)$ as its target, and contains time-lagged memory units, as described in Werbos (1990a), to complete the state vector description; however, in some applications, it is crucial to go beyond this, and insert special "sticky neurons"—designed to represent slowly-varying hidden parameters—and elements of robust estimation (Werbos, 1990d).

Kosko (1990), in comparing a fuzzy truck-backer-upper against a truck-backer-upper based on backpropagation utility, has reported results with the latter quite inferior to what Widrow claimed. Clearly, there must be some difference in the two implementations here, which puts into doubt any conclusions about fuzzy logic versus neurocontrol and the like. The most obvious explanation would be differences in how the model network is adapted—something quite crucial to success with this method—but there are other possibilities as well, which need to be investigated.

The biggest limitation of backpropagating utility is the need for a forecasting model, which cannot be a true stochastic model. In fuzzy logic, this is not so bad, because the variable being forecasted may itself be a measure of likelihood or probability. In some applications, however, like stock market portfolio optimization, more explicit treatment of probabilities and scenarios may be important. There are tricks which can be used to represent noise, even when backpropagating utility, but they are somewhat ad hoc and inefficient (Werbos, 1990d).

Another problem in backpropagation utility is the need to

learn in an off-line mode. The calculations backwards through time require this. Various authors have devised ways to do backpropagation through time in a time-forwards direction (e.g., Werbos, 1982), but those techniques are either very approximate or do not scale well with large problems or both. In any case, Narendra and Parthasarathy (1990) has questioned the stability of such methods. Nevertheless, even if we backpropagate utility in an off-line mode, we can still develop a network which adapts in real-time to changes in slowly varying parameters; we can "learn off-line to be adapted on-line" (Werbos, 1990d). This should be very attractive in many applications, because true real-time learning is more difficult.

Adaptive Critics

Adaptive critic methods, by contrast, do permit true real-time learning and stochastic models, but only at a price: they lack the exactness and simplicity of backpropagating utility. One reason for their lack of simplicity is the wide variety of designs available—from simple 2-Net structures, which work well on small problems, to complex hybrids, which hopefully encompass what goes on in the human brain (Werbos, 1990d; Miller et al., 1990).

Adaptive critic methods may be defined, in broad terms, as methods which attempt to approximate dynamic programming as first described in Werbos (1977). Dynamic programming is the *only* exact and efficient method available to control actions or movements over time, so as to maximize a utility function in a noisy, nonlinear environment, without making highly specialized assumptions about the nature of that environment. Figure 91.7 illustrates the trick used by dynamic programming to solve this very difficult problem.

Dynamic programming requires as its input a utility function U and a model of the external environment, F. Dynamic programming *produces*, as its major output, *another* function, J, which I like to call a secondary or *strategic* utility function. The key insight in dynamic programming is that you can maximize the function U, in the *long-term, over time,* simply by maximizing this function J in the immediate future. After you know the function J and the model F, it is then a simple problem in function maximization to pick the actions which maximize J. The notation here is taken from Raiffa (1968), whose books on decision analysis may be viewed as a highly practical and intuitive introduction to the ideas underlying dynamic programming.

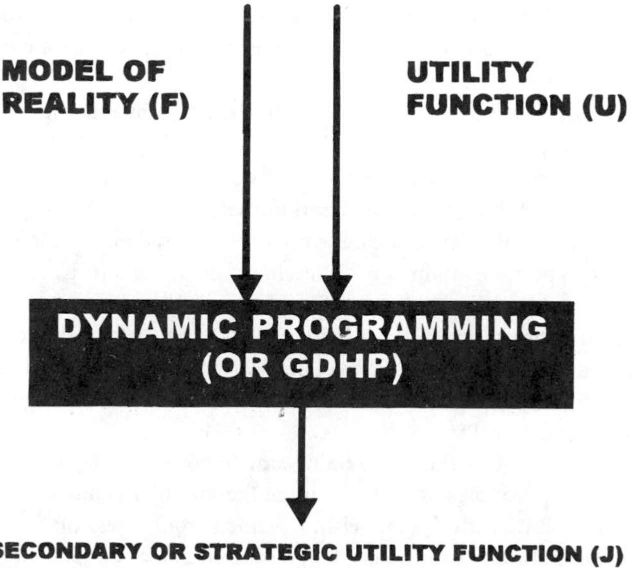

Figure 91.7 Inputs and outputs of dynamic programming.

Unfortunately, we cannot use dynamic programming *exactly* on complicated problems, because the calculations become hopelessly complex. (Bayesian inference sometimes entails similar complexities). However, it *is* possible to *approximate* these calculations by using a model or *network* to estimate either the J function or the J' function of Lukes et al. (1990) and Werbos (1989c). Adaptive critic methods may be defined more precisely as methods take this approach.

If this kind of design were truly fundamental to human intelligence, as I would claim, one might expect to find it reflected in a wide variety of fields. In fact, notions like U and J do reappear in a wide variety of fields, as illustrated in Table 91.2 (taken from Werbos, 1991). Please note that the last entry in Table 91.2, the entry for Lagrange multipliers, corresponds to the *derivative* of J, rather than the value of J itself. In economic theory, the prices of goods are supposed to reflect the *change* in overall utility which would result from *changing* your level of consumption of a particular good. Likewise, in Freudian psychology, the notion of emotional charge associated with a *particular object* corresponds more to the *derivatives* of J; in fact, the original inspiration for backpropagation (Werbos, 1968) came from Freud's theory that emotional charge is passed *backwards* from object to object, with a strength proportionate to the usual *forwards* association between the two objects (Yankelovitch and Barrett, 1971). The backpropagated adaptive critic (BAC) design reflects that theory

Table 91.2 Examples of J and U

Domain	Basic Utility (U)	Strategic Utility (J)
Chess	Win/Lose	Queen = 9 points, etc.
Business theory	Current profit cash flow	Present value of all assets (performance measures)
Human thought	Pleasure/pain hunger	Hope/fear reaction to job loss
Behavioral psychology	Primary reinforcement	Secondary reinforcement
Artificial intelligence	Utility function	Static position evaluator (Simon evaluation function (Hayes-Roth)
Government finance	National values, long-term goals	Cost/benefit measures
Physics	Lagrangian	Action function
Economics	Current value of product to you	Market price or shadow price ("Lagrange multipliers")

very closely. The word "pleasure" in Table 91.2 should not be interpreted in a narrow way; for example, it could include such things as parental pleasure in experiencing happy children.

To build an adaptive critic controller, we need to specify two things:

1. How to adapt the action network in response to the critic.
2. How to adapt the critic network.

The most popular adaptive critic design by far is the 2-network arrangement of Barto, Sutton, and Anderson (1983), illustrated in Figure 91.8. In this design, there is no need for a model of the process to be controlled. The estimate of J is treated as a gross reward or punishment signal.

This design has worked well on a wide variety of real-world problems, including robotics (Franklin, 1988), autonomous vehicles, and fuzzy logic system. Williams (Narendra and Annaswamy, 1989) has reported some interesting new results on convergence. Unfortunately, this approach becomes very slow as the number of control variables or state variables grows to 10 or 100. The reason for this is very straightforward: knowing J is not enough to tell us which actions were responsible for success or failure, and it does not tell us whether we need more or less of any component of the action vector. This design is like telling a student that he or she did "well" or "poorly" on an exam without pinpointing which answers were right or wrong; it is a lot harder for a student to improve performance when he or she has no specific idea of what to work on.

Fortunately, there are alternative designs which can overcome this problem. Note that it is critical to modify *both* the Action network *and* the Critic network, to permit learning at an acceptable speed when the number of variables is large (as in the human brain). There are also some other tricks which can help, discussed by myself, by Barto, and by Sutton (Werbos, 1990d; Miller et al., 1990; Narendra and Parthasarathy, 1990).

To speed up learning in the Action network, for *large* problems, there are now two major alternatives:

1. The backpropagated adaptive critic (BAC), shown in Figure 91.9.
2. The action-dependent adaptive critic (ADAC), shown in Figure 91.10.

The BAC design is closer to dynamic programming than is the 2-net design, because there is a more explicit attempt to pick $u(t)$ so as to maximize $J(t+1)$, based on the use of generalized

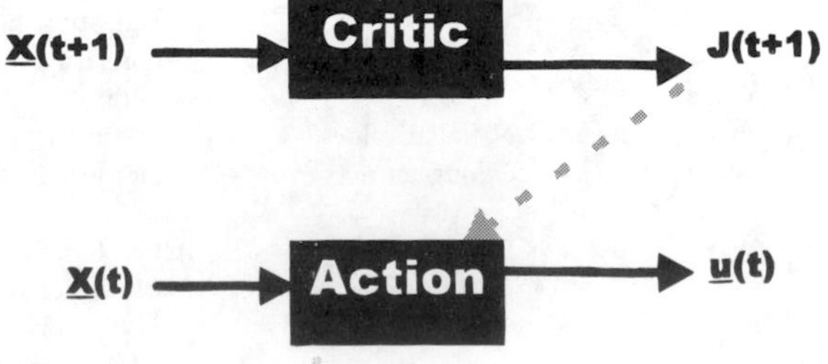

Figure 91.8 The two-network design of Barto, Sutton, and Anderson.

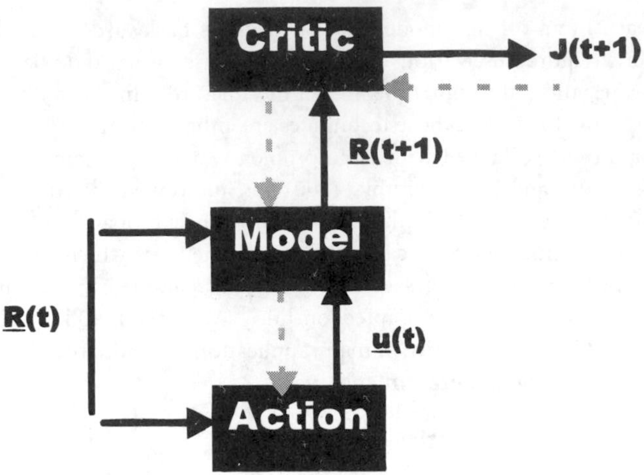

Figure 91.9 Backpropagated adaptive critic (BAC).

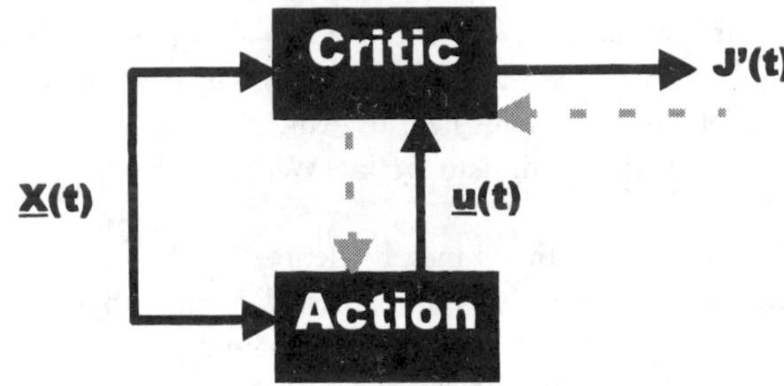

Figure 91.10 Action-dependent adaptive critic (ADAC). *Source:* Fu, L. 1990. Backpropagation in neural networks with fuzzy conjunction units, *Proc. IJCNN,* January, Erlbaum, Hillsdale, NJ; Amano et al. 1990. *Proc. IJCNN,* June, New York, NY.

backpropagation to calculate the derivatives of $J(t+1)$ with respect to the components of $u(t)$. The dashed lines in Figure 91.9 represent the calculation of derivatives. (Usually we adapt the *weights* in the action network in proportion to these derivatives, rather than adapting u(t) itself.) The cost of BAC is that we need to develop a model network, as we do when backpropagating utility. The adaptation of a good dynamic model can be a challenging task at times (Werbos, 1990d).

ADAC (Werbos, 1989b; Lukes et al., 1989) avoids the need for an explicit model, but the Critic network in Figure 91.8 would have to represent the *combination* of the critic and model in Figure 91.9. Jordan, in conversation, has stated that he adapted an action-dependent critic network in 1989, based on an independent paper by Watkins on "Q learning" (discussed in Narendra and Parthasarathy, 1990), but found the resulting Critic network to be rather complex. A variety of ADAC, with a few additional features proposed in (Miller et al., 1990), was the basis for the McDonnell-Douglas success with composite materials (Sugeno, 1985) discussed in Section A.4.

In an ideal world, one would want to combine the BAC and ADAC approaches, so as to combine the modularity and cleanliness of BAC with the model-independent robustness of ADAC; however, BAC may be good enough by itself in many applications. Jameson has reported some preliminary results with BAC (Jameson, 1990), and other aerospace-oriented researchers may have dealt with larger applications; however, more work is needed. Whatever the details, the adaptation of the action network in large-scale problems is clearly central to the future of this discipline and of our ability to understand organic intelligence.

In adapting the critic networks, few people have gone beyond simple, scalar methods which are more or less equivalent (Werbos, 1990a) and which have severe scaling problems. There are two alternatives which should scale much better:

1. Dual heuristic programming (DHP), which outputs estimates of the *derivatives* of *J*.
2. Globalized DHP (GDHP), which outputs an estimate of *J* (or its components), but which adapts the critic by minimizing error in the implied derivatives as well as the estimate of *J*.

These methods were first proposed in the 1970s (Werbos, 1982, 1977), but are described in more modern language in Werbos, 1990d and Miller et al., 1990. Both methods *require* the existence of a model network. Hutchinson of BehavHeuristics has claimed real-world commercial success in applying such methods, but many of the details are proprietary. See White and Sofge (1992) for *critical* design features and new consistency results.

Most neural networks researchers have adapted Model networks to *predict* rather than simulate the plant to be controlled. One can build up a stochastic simulation model from a prediction mode, simply by measuring the errors in prediction, and generating random numbers to simulate errors of the measured magnitude; however, this assumes that these errors at the point of prediction are uncorrelated with each other. It now seems possible to develop neural networks capable of simulating an unknown plant, in a way which fully accounts for correlations between errors across time and space (Ch. 13, White and Sofge, 1992). To prove that this can work is an area for future research. It is not obvious, *a priori*, that human brains have this kind of capability at the neuronal level; in other words, it is conceivable that human brains use fuzziness rather than true probabilities to handle uncertainty.

A.10 Example of a Hybrid System

In 1988, a friend of mine asked how to use these methods to assist in some very complex social decision problems, well beyond the scope of this paper. Given the nature of this application, a very conservative approach is recommended for the time being. As a first stage, obtain a conventional sort of modeling system, capable of storing and analyzing time-series data, and capable of manipulating forecasting models built up from any of three methodologies: econometric-style equations, fuzzy logic, and ANNs. Look for a *linkage* capability so that models of specific sectors (built up from different methodologies and often revised) could be combined together to yield composite streams of forecasts. Then build a general purpose "dual compiler." The dual compiler would input a sectoral model (in text form or parsed into a tree), and output a "dual subroutine" (like those in White and Sofge, 1992), so as to facilitate the use of generalized backpropagation. Then implement a whole set of tools using backpropagation.

Tool number one would be a simple sensitivity analysis tool. The user would type in a utility function or target function. The tool would then calculate the derivatives of utility with respect to all of the inputs—initial values, policy variables, and parameters—which affected the original forecast, in one quick sweep through the process. It would report back the ten or the hundred most important inputs. (There is a scaling problem here in deciding which input is most important; the user could be given a choice, for example, between looking for the biggest derivatives, the biggest elasticities, or the biggest derivatives weighted by some other variables.) The user could go on to make plans to *change* these inputs, so as to increase utility, *or* he could first evaluate in detail whether he believes that the inputs are really important. (Tests of this sort can in fact be very useful in pinpointing *weaknesses* of an integrated modeling system (Werbos, 1988b), or real-world uncertainties which require more analysis). The *cost* of a comprehensive sensitivity analysis is the key issue here; using more conventional tools one must often wait a long time and spend a lot of money to get even a partial sensitivity analysis, and the results are usually out of date.

Tool number two would help in reassessing the importance of the key inputs. For any given input, it would use the *intermediate* information generated by backpropagation (as in Werbos, 1988b) to identify the path of connections which really made that input important. It could even display this information as a kind of tree or flow chart. This would be similar in purpose to the inference sequences printed out as "explanations" by many expert systems.

Tool number three would be an extended version of tools one or two. Instead of first derivatives, it would provide information based on low-cost second derivatives (as described in Werbos, 1982, based on calculations like those in Werbos, 1988; Miller et al., 1990). For example, the sensitivity of utility to dollars spent in 1992 may be a key measure of policy effectiveness; it may be useful to see how that measure, in turn, would be changed by other factors (such as diminishing returns or complementary variables). At the optimum, the first derivative of utility with respect to any policy variable will be zero; the derivatives of *that* derivative give information about why the policy variable should be set at a particular level.

Tool number four would be a full-fledged version of backpropagating utility. The user could flag certain variables or parameters as policy variables, and the computer would be asked to suggest an optimal *improvement* upon current plans, so as to maximize utility. The resulting suggestion may be a local minimum, but it should at least be better than the starting plans.

Tool number five would be a model calibration tool, based on the backpropagation of error, and robust estimation concepts like those of Werbos (1990d). At a minimum, this would be a relatively quick and objective way to calibrate a model as a whole system to fit the past; it could replace the rather elaborate and ad hoc "tweaking" which usually goes into the most complex models in the real world for calibration purposes.

Tool number six would go back and identify how the resulting parameter estimates or rules were influenced by different cases in the input dataset; this would provide an integrated, nonlinear version of the highly respected linear diagnostic tools developed by Belsley, Kuh, and Welsh (1980).

These six tools are the most obviously needed tools, exploiting backpropagation, but a host of other tools are possible involving estimation diagnostics, decision diagnostics, and convergence tools. Also, there is no need to develop the six tools in the order of my discussion.

In principle, one can even build a strategic assessment or stochastic planning tool, based on adaptive critic methods but permitting user-specified assessment models, as described in Werbos (1991a).

To bring all these tools together in a general-purpose modeling package capable of running on desktop workstations would not be a trivial task. However, these are important applications, and some work has begun in this direction. All of these tools aim at effective *two-way* man-machine communication, so as to exploit the capabilities of both forms of intelligence.

A.11 Reasoning, Planning, and Chunking: Thought for Future Research

Experts in traditional artificial intelligence (AI) often ask two questions about neurocontrol or about neural networks in general:

1. How could such low-level architectures be extended to large-scale planning systems, capable of a long-term planning horizon and capable of structuring very complex decision processes?
2. Where does symbolic reasoning fit into this picture?

This section will mainly focus on the first of these questions. There are other mammals besides humans that are clearly capable of complex forms of problem solving, without using symbolic reasoning as such. At the neural level, it seems very clear that the brains of other mammals are really quite similar to those of humans. Early speculations (and hopes) that language is based on fundamentally unique kinds of neural structures have not held up well in recent research (Segalowitz, 1983). Formal symbolic reasoning as we practice it today is relatively recent, even within the history of the human species, and—at its best—it uses the entire structure of strategic planning of the brain to learn complex rules for manipulating those actions we call speech or writing. One should not expect to reduce the actual content of these rules and strategies to a simple, modular structure transmittable by the genes. In other words, the rules of symbolic reasoning as such may not be hard-wired in the brain, but learned. To learn such complex rules, however, the brain must have a great inborn capacity for what is called "planning" in AI.

The problem of hierarchical or multilayer planning occurs not only in neural networks, but in AI and control theory as well. For example, automatic control for the main arm of the space shuttle presents a severe challenge to classical control theory. Seraji developed a hierarchical control scheme which worked on a Puma robot arm to some degree, but was computationally intensive and never deployed. This past year, the joint controllers used by Seraji were replaced by neurocontrollers (using direct inverse control and computationally affordable), with a substantial improvement in performance, at least in simulations (Parten et al., 1990). Tests on the real arm on the ground have just started up.

At any one level of abstraction or aggregation, the higher order adaptive critic architectures described in Section 8 are very similar in spirit to the planners used in AI, as Table 91.2 would suggest, except for their learning ability. The emphasis is on refining the evaluation function, rather than performing better tree searches; however, studies of human abilities in games like chess and Go suggest that a high-quality evaluation function is the real root of those abilities. One would therefore expect neurocontrol designs to be reasonable for planning problems at this level.

The greater difficulty lies in how to handle *multiple* levels of abstraction or aggregation—a problem which is often called "chunking" in the literature of AI. As in AI, some neural network researchers have developed elaborate hierarchies to break down complex decision problems, both to allow chunking and to prevent any one neurocontroller in the system from being overloaded with too many inputs. With adaptive critic controllers, for example, one can use $J(t+1) - J(t) - U(t)$ as calculated by a higher-level controller as the intrinsic utility function $U(t)$ to be input to a subordinate controller (Werbos, 1990d).

Unfortunately, these kinds of designs tend to require prior knowledge about how to structure the hierarchy. True hierarchies are not consistent with the more modular or "heterarchical" structure of the brain, and they limit the possibility of rich feedback between layers. Fortunately, a hierarchical array of neural networks performing similar functions can usually be represented as a *single* neural network with restrictions on which neuron inputs from which. Thus, one can build a *single* network which initially reflect one's prior knowledge (if that prior knowledge points towards a hierarchy), and then allow the network to make and break new connections in all directions, based on adaptive rules for making and breaking connections. This allows us to recover modularity and flexibility, without losing the crucial benefits of a sparsely connected network.

This sparsity approach can work very well if the various networks in the original hierarchy all work on a common cycle time. For example, in the human limbic system, which appears to function like a critic network (Werbos, 1987), there is a standard cycle time of about a quarter of a second, which corresponds to the classical theta rhythm of brainwaves. This example leads to a fundamental question which has yet to be answered: can an adaptive critic system, based on a uniform cycle time shorter than a second, *effectively* plan over very long intervals of time? If we adapt such networks by use of new learning rules, based on the best that numerical analysis has to offer (as proposed in Jameson, 1990), will the approximation to dynamic programming be fast enough to compete with brute force methods like AI planning? The brute force approach tends to fit poorly with the neural network approach, because it does not have a natural way of accommodating complex, noisy, nonlinear environments in the general case. However, the use of adaptive critics is still relatively new, and the possibilities for faster convergence on larger problems with sophisticated learning rules have hardly been tested at all.

One way to extend the foresight capability of adaptive critics—and thus help explain the foresight of biological brains—would be to use complex types of networks which somehow lend themselves better to foresight. However, simple feedforward networks can learn to represent almost any function (Sontag, 1990). *All* of the known methods for adapting critic networks are based on supervised learning (Werbos, 1990d; Miller et al., 1990), with properly calculated inputs and targets. How could it help to use a more complex network design when solving a supervised learning problem, if simple feedforward networks are certain to be good enough anyway? There are many ways that this can happen, especially if the structure of a complex network fits what we know about the problem to be solved; even though a feedforward network can do the job, eventually, it may be possible to *learn* the job much faster with a more appropriate network design, a design which requires fewer weights or parameters to fit the problem at hand.

In the special case of planning problems, there is an interesting argument to suggest that *recurrent* networks might in fact work better (Werbos, 1987 and Werbos and Pang, 1996).

In conventional planning, one tries to build networks or models which represent well-defined *tasks.* The descriptions of these tasks are like Model networks which input a description of the task initiation state A (at time t), and immediately output the state B which would be achieved (at time $t + T$). In a stochastic world, this approach becomes increasingly difficult as the complexity and the noise increase; it becomes ever more difficult to predict the final state. In fact, the whole notion of *deterministic* planning becomes less and less valid as the time horizon grows. One way to deal with this problem is to rely on *evaluation* networks rather than models to jump over multiple time intervals. Since evaluation functions (j) are based on dynamic programming, the paradigm remains valid even at very high noise levels. The challenge is to build an evaluation or critic network which can translate a *value* or *goal* B for time $t + T$ into a value or goal A for time t, in a single step of calculation. The obvious way to do this is to build a Critic network in which the value weight on variable B is represented by a neuron which is then used as input to the neuron which evaluates A. Once this connection is learned, then *changes* in the value placed on B (which are not really time-indexed) should lead *immediately* into changes in the value placed on A. This kind of approach would only work with the more sophisticated types of Critic architecture, and it is critical to *avoid* treating the recurrent links as if they were memories of the external environment at earlier times.

An interesting aspect of this approach is that the Critic network could look more like the traditional Hopfield network which is known to be highly effective in solving combinatorial optimization problems. Designs of this sort would probably be ideal in applications like SDI, where dynamic control is required but the critic must somehow accomplish calculations which are qualitatively similar to combinatorial optimization. Recurrent networks can take time to settle down to an optimum or equilibrium output; this, in turn, is consistent with our subjective understanding of how humans sometimes take time to perform evaluations in games like Go or Tetris. Feedforward nets can learn to *emulate* recurrent nets, but only with greater complexity, which requires greater learning time; this is consistent with the way in which humans can learn to do well in games like Tetris at slow speed, and then only slowly learn to perform the evaluations in a faster, more reflex mode.

In summary, the best solution to the long-term planning problem, in a noisy environment, may well be an adaptive critic structure with a recurrent Critic network. The best short-term motor control may use an adaptive critic design with a feedforward critic. Nevertheless, one may still argue that the ultimate learning control system would still take advantage of more deterministic task-oriented planning in the mid-term—at time intervals short enough that defined tasks have a high probability of being accomplished. One can imagine a three-level control system, with a long-term controller directing a mid-term controller, and both of those two controlling a short-term optimization system. Each of the three would be as complex as the BAC system described in Section 5.

As this tutorial goes to press, evidence has begun to accumulate suggesting that the human brain does indeed use a Critic network and a Model network based on a recurrent relaxation network as described in Section 3 of Werbos (1991). One can plug in such networks in place of MLPs in all the designs of this section, without changing anything, *so long as* one can program a subroutine which backpropagates through a recurrent network. Section 3 did describe how to program such a subroutine, using a figure taken from White and Sofge (1992).

From an engineering point of view, we are a long way from needing that degree of complexity, and we may never really need it. However, these notions do have intriguing biological parallels. There does exist a kind of midterm control structure in the brain (a part of the basal ganglia (Brooks, 1986; Marsden, 1982)) which does seem to generate a kind of task representation. The basal ganglia are clearly subordinate to the longer-term planning system, made up of the limbic system and the cerebral cortex proper (neocortex), and superior to the short-term motor control system, made up of the brain stem and cerebellum. As evolution has progressed, the basal ganglia appear to have grown less important; however, the cerebral cortex itself has important links, both in anatomy and in evolution, with the basal ganglia.

How could one build such a midterm task-oriented planning system out of analog distributed hardware-like neurons? The biological literature may offer some valuable clues, or we may need to work with the biologist to help design new experiments which proved the kind of clues we need for our purposes. One practical possibility among many would be to design an adaptive critic system, in which the output of a midlevel critic network would consist of two vectors—R' (to be interpreted as a short-term goal state) and w- with the estimate of J estimated implicitly as:

$$J(\underline{R}) = \sum_j w_i(R_i^* - R_i)^2 \qquad (91.25)$$

Despite its unique structure, the Critic network could still be adapted by HDP or DGHP, or perhaps a learning rule could be found which works better for this specific case. If goal states tend to be attained, under the proper conditions, then the part of the critic network which generates R' as a function of R could then be used as a kind of higher-level model network, jumping from one time to a much later time. This provides for only one level of temporal chunking, but the biological literature (Brooks, 1986; Marsden, 1992) suggests that is all we need. The success of such a scheme would depend critically on the ability of the system to learn new vector components R_t, which tend to represent the degree of progress towards accomplishing particular tasks. In this connection, it is interesting that the basal ganglia appear to have less of a distributed, holographic architecture than many other parts of the brain; a single cell may well play a crucial role in triggering an entire task. For more recent ideas, see Werbos (1996).

Considerable research will be needed to fully exploit and understand all the higher order options described in this section. One may hope that neural networks which can learn complex planning tasks will also be able to learn symbolic reasoning, without any hard-wiring of the basic concepts, just as humans appear to do. Research on those lines is promising (Shastri, 1990), but has only just begun.

A.12 Conclusions

Neurocontrol and fuzzy logic are complementary, rather than competitive, technologies. There are numerous ways of combining the two technologies. Which combination is best depends very heavily on the particular application; there is always a tradeoff between "general syntheses"—which combine everything but require the expense of *implementing* everything—and direct, simple designs tuned to particular concrete problems. Given the natural human tendency towards inertia, it is critical to be aware of a wide variety of options, and to as "Why not?" when considering new approaches. Even within neurocontrol, there is a wide variety of designs available, ranging form simple off-the-shelf technologies (easily applied to fuzzy logic networks) through to areas where fundamental research is still needed and vital to our understanding of real intelligence.

Note: The author's article, Chapter 63 of this Handbook, provides some updates on recent research.

References

Amano, A. et al., 1989. On the use of neural networks and fuzzy logic in speech recognition, *Proc. Int. Joint Conf. Neural Networks (IJCNN)*, June, New York, NY.

Barto, A., Sutton, and Anderson. Neuron-like adaptive elements that can solve difficult learning control problems, *IEEE Trans. SMC*, SMC-13:834–846.

Besley, D., Kuh, and Welsch. 1980. *Regression Diagnostics Identifying Influential Data and Sources of Collinearity*, John Wiley & Sons, New York, NY.

Brooks, V. B. 1986. *The Neural Basis of Motor Control*, Oxford University Press, New York, NY.

Donat, J., Bhat, and McAvoy, T. 1990. Optimizing neural net based predictive control, *American Control Conf*, IEEE, New York, NY; White, D. and Sofge, D. 1992. *Handbook on Intelligent Control*, Van Nostrand, New York.

Eckmiller, R. et al. 1989. Neural kinematics net for a redundant robot arm, *Proc. IJCNN*, June, IEEE, New York, NY.

Franklin, J. 1988. Reinforcement of robot motor skills through reinforcement learning, *IEEE/CDC Proc.*, IEEE, New York, NY.

Fu, L. Backpropogation in neural networks with fuzzy conjunction units, *Proc. IJCNN*, January, Erlbaum, Hillsdale, NJ.

Hsu, L. et al. 1990. Fuzzy logic in connectionist expert systems, *Proc. IJCNN*, January, Erlbaum, Hillsdale, NJ.

Jameson, J. 1990. A neurocontroller based on model feedback and the Adaptive Heuristic Critic, *Proc. IJCNN*, June, San Diego, CA.

Jordan, M. 1989. Generic constraints on underspecified target trajectories, *Proc. IJCNN*, June, IEEE, New York, NY.

Kosko, B. 1990. Comparison of fuzzy and neural truck backer-upper control systems, *Proc. IJCNN*, New York, NY.

Lokendra, S., ed., 1990. *Connectionism Meets AI: Workshop Proc.*, University of Pennsylvania, Philadelphia, PA.

Lukes, G., Thompson, B., and Werbos, P. 1990. Expectations driven learning with an associative memory, *Proc. IJCNN*, Washington, DC, Erlbaum, Hillsdale, NJ.

Marsden, C. D. 1982. The mysterious motor function of the basal ganglia, *Neurology (NY)*, 32:514–539.

Mead, C. 1989. *Analog VLSI and Neural Systems*, Addison-Wesley, Reading, MA.

Miller, W. T. 1990. *Neural Networks for Control*, Sutton and Werbos, eds., MIT Press, Cambridge, MA.

Morita et al. 1990. Fuzzy knowledge model of neural network type, *Proc. IJCNN*, Erlbaum, Hillsdale, NJ.

Narendra, K. and Annaswamy. 1989. *Stable Adaptive Systems*, Prentice Hall, Englewood Cliffs, NJ.

Narendra, K. and Parthasarathy. 1990. Identification and control of dynamical systems using neural networks, *IEEE Trans. Neural Networks*, 1(1).

Narendra, K., ed. 1990. *Proc. 6th Yale Workshop on Adaptive Learning Control*, Yale, New Haven, CN.

Parten, C., Pap, R., and Thomas, C. 1990. Neurocontrol applied to telerobotics for the space shuttle, *Proc. Int. Neural Network Conf.*, Paris.

Raiffa, H. 1968. *Decision Analysis: Introductory Lectures on Making Choices Under Uncertainty*, Addison-Wesley, Reading, MA.

Schreinemakers, J. and Touretzky, D. 1990. Interfacing a neural network with a rule-based reasoner for diagnosing mastis, *Proc. IJCNN*, January, Erlbaum, Hillsdale, NJ.

Segalowits, S. J., ed., 1983. *Language Functions and Brain Organization*, Academic Press, New York, NY.

Sofge, D. and White, D. 1991. Neural network based process

optimization and control, *Proc. 29th IEEE Conf. Decision and Control,* January, New York, NY.

Sontag, E. 1990. *Feedback Stabilization Using Two-Hidden-Layer Nets,* SYCON-90-11, October, Rutgers University Center for Systems and Control, New Brunswick, NJ.

Sugeno. 1985. *Industrial Applications of Fuzzy Control,* North-Holland, Amsterdam.

Takagi, H. 1990. Fusion technology of fuzzy theory and neural networks, *Proc. Fuzzy Logic and Neural Networks,* Izzuka, Japan.

Werbos, P. Links Between Artificial Neural Networks (ANN) and Statistical Pattern Recognition, *Artificial Neural Networks and Statistical Pattern Recognition: Old and New Connections,* Sethi, I. and Jain, eds., Elsevier, New York, NY.

Werbos, P. Neural networks for prediction and system identification, *IEEE Expert.*

Werbos, P. 1968. Elements of intelligence, *Cybernteica* (Namur), No. 3.

Werbos, P. 1974. *Beyond Regression: New Tools for Prediction and Analysis in the Behavioral Sciences,* Ph.D. thesis to Harvard University Committee on Applied Mathematics.

Werbos, P. 1977. Advanced forecasting methods for global crisis warning and models of intelligence, *General Systems Yearbook.*

Werbos, P. 1982. Applications of Advances in Non-Linear Sensitivity Analysis, *Systems Modeling and Optimization: Proc. Int. Federation for Information Processing,* Drench, R. and Cozen, eds., Springer-Verlag, New York, NY.

Werbos, P. 1987. Building and understanding adaptive systems: a statistical/numerical approach to factory automation and brain research, *IEEE Trans. SMC,* January/February.

Werbos, P. 1988. Backpropagation: past and future, *Proc. 2nd Int. Conf. Neural Networks,* New York, NY.

Werbos, P. 1988. Generalization of backpropagation, *Neural Networks,* October.

Werbos, P. 1989a. Maximizing long-term gas industry profits in two minutes in Lotus using neural network methods, *IEEE Trans. SMC,* March/April.

Werbos, P. 1989b. Neural Networks for control and system identification, *IEEE CDC Proc.,* IEEE, New York, NY.

Werbos, P. 1990a. Backpropagation through time: what it does and how to do it, *Proc. IEEE,* October.

Werbos, P. 1990b. Consistency of HDP applied to a simple reinforcement learning problem, *Neural Networks,* 3:179–189.

Werbos, P. 1990c. Making Diagnostics Work in the Real World: A Few Tricks, *Handbook of Neural Computer Applications,* Maren, A., ed., p. 337, Academic Press, New York, NY.

Werbos, P. 1990d. Neurocontrol and Related Techniques, *Handbook of Neural Computer Applications,* Maren, A., ed., p. 337, Academic Press, New York, NY.

Werbos, P. 1991a. The cytoskeleton: why it may be crucial to human learning and neurocontrol, *Nanobiology;* see also Werbos, P. 1991b., Neurocontrol, biology and the mind: new connections and development, *IEEE Proc. SMC.*

Werbos, P. 1996, Learning in the brain: an engineering interpretation. In K. Pribram, ed., *Learning As Self-Organization,* Erlbaum, Hillsdale, NJ.

White, D. and Sofge, D. 1992. *Handbook on Intelligent Control,* Van Nostrand, New York.

Widrow, B. and Smith 1963. Pattern-recognizing control systems, *1963 Computer Information Science (COINS) Symp. Proc.,* Spartan, Washington, DC.

Yankelovitch, D. and Barrett. 1971. *Ego and Instinct The Psychoanalytic View of Human Nature—Revised,* Vintage, New York, NY. This is an unusually clear presentation, though my own work would suggest it is too pessimistic in its conclusions.

92

Integrated Health Monitoring and Control In Rotorcraft Machines

Gary G. Yen
Oklahoma State University

92.1 Introduction ... 1182
92.2 Artificial Neural Networks ... 1184
92.3 Fuzzy-Based Feedforward Neural Network ... 1185
92.4 FDIA Architecture ... 1187
Detection • Feature Extraction • Fault Isolation • Fault Severity Estimation • Hierarchical Structure
92.5 Simulation Study ... 1189
92.6 Conclusions ... 1190

Abstract

In this paper we propose to design and evaluate an on-board intelligent health assessment tool for rotorcraft machines, which is capable of detecting, identifying, and accommodating various failure modes in rotorcraft machines under an adverse operating environment. A fuzzy-based neural network paradigm with an on-line real-time learning algorithm is developed to perform expert advising to the ground maintenance crew. A hierarchical fault diagnosis architecture is advocated to fulfill the time-critical and on-board needs in different levels of structural integrity over a global operating envelope. The research objective is to demonstrate the feasibility and flexibility of the proposed health monitoring procedure through numerical simulations of bearing faults in helicopter transmissions. The proposed fault detection, identification, and accommodation architecture is applicable to various generic rotorcraft machines. In a similar spirit, the proposed technology can be easily transferred to damage and identification of any general purpose dynamic systems, e.g., aeropropulsion engines, underwater vehicles, chemical processes, nuclear power plants, and manufacturing production equipments. The proposed system will greatly reduce the operational and developmental costs and serve as an essential component in an autonomous control system.

92.1 Introduction

Modern engineering technology is leading to increasingly complex rotorcraft machines with ever more demanding performance criteria. However, currently used transmission diagnostic systems which primarily rely on magnetic chip detectors, torque meters, oil pressure sensors, and temperature indicators to provide warning, caution, and maintenance information to the pilot and ground crew is capable of identifying only a few failure modes. For example, the debris rate monitoring sensor is developed to detect the debris generating failures. But fatigue cracks in gears and shafts are far more difficult to identify. Experience indicates that these fatigue cracks may propagate slowly or even drastically. To increase the ability to detect, predict, and diagnose problems in the complex interrelated network of rotating components that makes up the rotorcraft, vibration sensors were added to the list. Figure 91.1 schematically outlines the main sensor configuration in the tail rotor drive shaft and bearing of the MH-53J PAVE LOW helicopter.

Initially, vibration measurements were used to indicate the dynamic imbalances of the main and tail rotors of rotorcrafts. Later Fast Fourier Transform were included to measure vibration spectrums of various rotating components. This capability was made realistic by hand-carrying the recording equipment on the aircraft. Based on this need, the on-board vibration monitoring System (VMS) was developed and utilized on various commercial and military helicopters. The present VMS system typically gathers 60 bands of vibration data every 2 minutes which after 10 hours of flight time results in 18,000 bands of data to analyze. After landing, the data is then transferred to a ground based PC for analysis on VibraLog, which is a trend monitoring software being used worldwide. A typical reading is given in Figure 92.2 for reference. This screening process calls upon a human expert to visually analyze the frequency bands, and to heuristically reason the causes. Obviously, this open-loop iterative process of health assessment is unable to deliver a time-critical decision and is often subject to unpredicted human factors.

0-8493-8343-9/97/$0.00+$.50

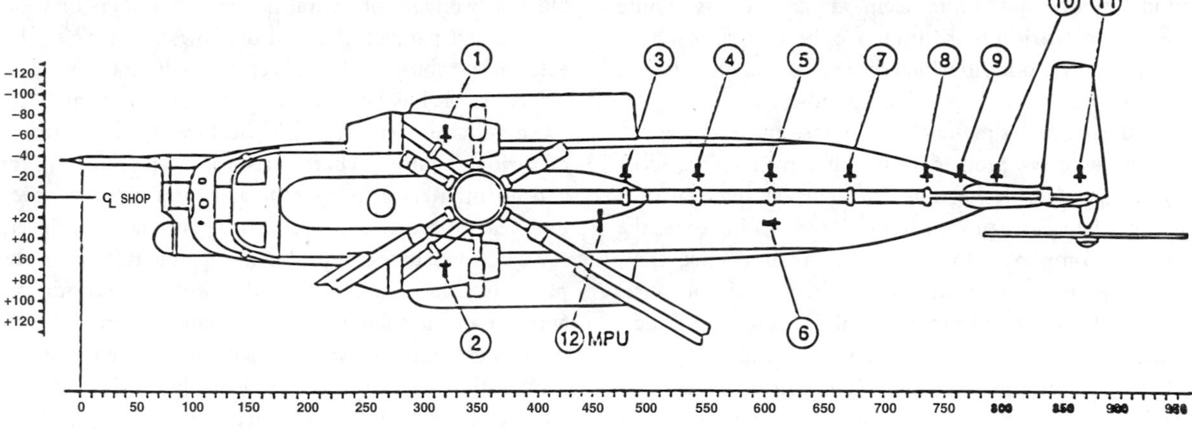

Figure 92.1 Sensor configuration in the tail rotor drive shaft and bearing of MH-53J helicopter.

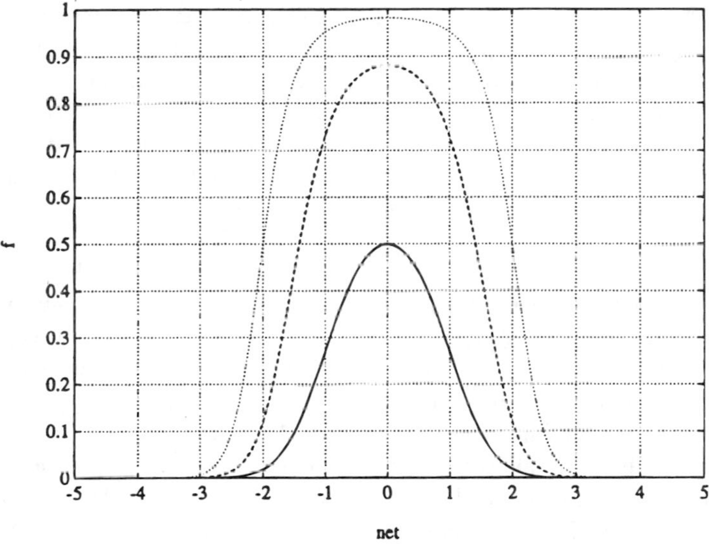

Figure 92.2 A typical reading of VibraLog screen.

The ultimate pursuit of a higher degree of autonomous behavior that provides constant health monitoring and fault tolerance for rotorcrafts with minimum human intervention has high priority in order to achieve a successful flight mission (Yen, 1994a). In addition, the next generation rotorcraft machines would inevitably necessitate an intelligent fault detection, identification, and accommodation (FDIA) system as well as autonomous controller reconfigurable mechanism (Yen, 1994e) which can respond in real-time under adverse thermal and aerodynamic conditions. As a result, the U.S. Army is currently conducting a seeded fault testing on gearboxes, and the Office of Naval Research has initiated a health-monitoring program (i.e., the condition-based maintenance program) (Rock et al., 1993). The NASA-Marshall Space Flight Center is also engaged in a program on autonomous fault detection of the space shuttle's main engine (Cikanek, 1985). In addition, the U.K. and Norway will impose a mandatory requirement of all rotorcrafts flying over the North Sea to have health and usage monitoring (HUM) technology on-board by 1995 (Rock et al., 1993).

Fault assessment has long been identified as a key technology within an intelligent control system framework (Rauch, 1994). Advanced rotorcrafts which call for a highly sophisticated controller to ensure that demanding performance be met can be extremely difficult to design due to factors such as high dimensionality, multiple inputs and outputs, operational constraints, as well as the unexpected failures of different sensors, actuators or other components. (Randell, 1982) This further complicates the synthesis of a given health monitoring procedure (Yen and Kwak, 1993) The conventional damage detection methods often depend upon the off-line observation and destructive tests or computationally intensive finite element simulations. Quite often, these *ad hoc* algorithms are limited to the analysis of a fixed structural concept or model, in which the loadings, materials, and design constraints need to be specified in advance.

Because of the need for time-critical response and autonomous operations, available symptom data is either misinterpreted or unused by the pilot, often leading to incorrect removal of a system's components. This process has unavoidably increased the cost, time, and complexity to the ground maintenance crew. *Hardware redundancy* is often utilized to achieve fault tolerance. This approach is assumed to be impractical, because time-critical response and additional cost, volume, weight and complexity of redundant hardware are needed. One example of hardware redundancy is multiple sensors measuring the same quantity. The best estimate can be obtained by a majority vote. Most of the current research is based on the concept of *analytical redundancy.* Multiple sensor measurements are processed analytically, which relate mathematical estimations of observed parameters. The appeal of the analytical redundancy lies in the fact that information processing techniques using powerful computing tool can be employed to create the necessary redundancy. An effective way to obtain statistical estimates is to operate on the measurements directly rather than processing through a Kalman filter bank (Chin et al., 1993). The neural network approach pursued in this study is designed to provide the required analytical redundancy with respect to the approaches reported in literature (e.g., detection filter, innovation test (Mehra and Pershon, 1971)), parity space method (Deckert et al., 1971) and the parameter estimation technique (Kitamura, 1980) which assume a high fidelity mathematical model of the rotorcraft dynamics.

The rationale of utilizing artificial neural networks in the fault detection, identification, and accommodation (FDIA) framework will first be validated in Section 92.2. In Section 92.3, an innovative fuzzy based feedforward neural network will be proposed, providing a justification to achieve real-time performance. A generic fault detection, identification and accommodation architecture for condition based machinery maintenance will be developed in Section 92.4 to analyze vibration signature in MH-53J helicopter tail rotor derive shaft bearings. The proposed FDIA advisor will provide a trouble shooting tree with confidence probability rating to greatly facilitate ground crew to perform a time-critical maintenance. A reduced scope of simulation study will then be presented in Section 92.5 while some pertinent comments will be drawn in Section 92.6.

92.2 Artificial Neural Networks

A connectionist system consists of a set of interconnected processing elements and is capable of improving its performance based on past experimental information. An artificial neural network (herein referred to as simply "neural network") is a connectionist system which was originally proposed as a simplified model of the biological nervous system. Neural networks have been shown to provide an efficient means of learning concepts from the past experience, abstracting features from uncorrelated data, and generalizing solutions from unforeseen inputs. Other advantages of neural networks are their distributed data storage and parallel information flow which cause them to be extremely robust with respect to malfunctions of individual devices as well as being computationally efficient.

There have been many architectures (i.e., schema consisting of various neuronic characteristics, interconnecting topologies, and learning rules) proposed for neural networks over the last five years (last count over 200). Simulation experience has revealed that success is problem-dependent. Some networks are more suitable for pattern classification whereas others are more appropriate in adaptive control, signal filtering, or associative searching. Neural networks which employ the well-known back-propagation learning rule are capable of classifying any non-convex regions with an arbitrary degree of accuracy (Hornik et al., 1989). Although the back-propagation algorithm proves its effectiveness in many instances, it is generally known that it takes considerable time to train the neural network and the network may get trapped into local minima. Radial basis function network (Moody and Darken,1989) which employ locally distributed function (e.g., Gaussian function or *erf* function) and hybrid learning rule, on the other hand, is promised to serve as an efficient universal approximator (Hartman, et al., 1990). These model-free neural network paradigms are more effective in solving pattern classification problems than conventional statistical approaches which require vast memory usage and intensive computation. Neural network, on the other hand, overcome common memory and intensive computation. Neural network, on the other hand, overcome common memory intensive problems and yet provide a sufficiently generalized solution space.

Neural network-based approaches are attractive in FDIA realization for several arguments, including their potential to reduce complexity, to increase speed, and to minimize cost. Neural networks reduce complexity by providing generic and reusable software and hardware modules applicable in a wide variety of modeling and decision-making applications. Neural networks are not dependent on *a priori* analytic models or statistics, and can significantly simplify the process by which models are synthesized. Neural network can readily address modeling problems that are analytically difficult and for which conventional approaches are not practical, including complex physical processes having nonlinear, high-order and time-varying dynamics, and those for which analytic models do not yet exist (Yen, 1994d). They can increase fault tolerance through adaptation and can be self-modifying over life-cycle maturation to compensate. Reduced complexity is critical because experience shows that complex conventional computer systems fail 100 times more often than the main helicopter gearboxes that they are intended to diagnose.

A second reason for using neural network approach is speed. Appropriately synthesized neural networks can be used to implement accurate and fast on-line pattern recognition systems, due to their ability to fuse numeric information from multiple and possibly disparate channels nearly instantaneously. Thus, data from the powertrain, drivetrain, structure, rotor, and oil systems may be brought jointly to bear for diagnostic and prognostic purposes. This fusion of data may also take into full consideration both the linear and nonlinear interactions among the channels, given the ability of neural networks to handle nonlinearity as readily as linearity (Parker, et al., 1993).

For rotorcraft condition-based machinery maintenance applications, the hardware upon which neural network paradigms execute will largely be determined by program and system requirements. Neural network algorithms can, of course, be implemented on many types of hardware, including microprocessors, digital signal processors (e.g., TMS320C31), and neural processors (e.g., Ni1000). In general, special purpose hardware is only necessary when high throughput is required. The trade-off between the cost and performance is essential. Too often, neural processors are used to provide an alternative for overcoming the well-known performance deficiencies of backpropagation and related learning algorithms when fitting training data. With properly tuned neural network, such problems can often be avoided, and relatively low-cost systems developed which offer compromised accuracy and speed, can deliver a satisfactory solution with less computational efforts. Taken to the extreme, the computational unit in a helicopter transmission FDIA system may itself be collocated with the vibration sensor and allowed to communicate with neighboring sensors, forming an integrated diagnostic/prognostic system with distributed components and dynamic architecture to address sensor reliability problems. Either way, hardware implementations of the proposed on-board FDIA system are feasible and can even be built into the on-board VMS system.

The third reason for using neural network implementations is reducing cost, resulting both from reduced manpower requirements necessary to design functional diagnostic/prognostic systems and from the opportunity to implement systems that can learn and adapt in the field, rather than requiring analyst development and programming. The domain expert of reading the vibration signatures can be kept off the maintenance loop. Unnecessary removal of healthy components can also be prevented. The developed FDIA system can be easily transferred to the applications of various rotorcraft machines, including military and civilian usage. Neural networks offer the opportunity to replace explicit parametric models with empirically derived implicit models, thereby greater automaticity in the system design process and reducing engineering development time, and of course, the cost.

92.3 Fuzzy-Based Feedforward Neural Network

Among all the failure modes that appear in transmission, an experienced expert can usually tell that an erratic frequency response of the bearing sensor may indicate the degradation of the transmission gearbox. Often, the well trained technician can identify and locate the delamination of composite material from the frequency response readings of the collocated sensor. The capability of incorporating this huge amount of expert knowledge into the artificial neural network becomes essential in solving the FDIA problem. The existing feedforward neural network paradigms, which learn and generalize all nonlinear mappings from scratch, didn't provide such a mechanism.

A fuzzy-based feedforward neural network is developed in this paper to incorporate the existing expert knowledge and heuristic experience on the structural domain accumulated over years into the design constraints. The proposed neural network can set aside part of its structure to incorporate the existing expert knowledge for a particular problem domain by assigning the connecting weights involving neurons in different layers. After the network is well trained, the possibility of interpreting the underlying meanings of decisions made by the fuzzy neural network will provide a clear judgment for the ground-base crew to perform routine maintenance.

The proposed fuzzy based feedforward neural network utilizes *guadratic sigmoid function* as activation function (see Figure 92.3). The quadratic sigmoid function which exhibits the second order characteristics as Equation 92.1, is a localized basis function and possesses all attributes radial basis function has. The locality of representation and linearity of the fuzzy learning rule have made the training extremely efficient to curve fit the high-dimensional training data. The *incremental learning* is also made possible by including new fuzzy neurons in different layers of network to incorporate the novel rule. The incremental learning is defined as the ability of including additional rules into the database without the necessity of re-configuring the existing neural network architecture. The fuzzy back-propagaton network (see Figure 92.4), equipped with four layers of neurons, is contructed according to the architecture of fuzzy expert systems as described in the following.

$$f(net, \theta) = \frac{1}{1 + \exp(net^2 - \theta)} \tag{92.1}$$

Input Neurons

Each input neuron represents an input fuzzy variable, and is used to distribute the input to its membership-function neurons.

Membership-Function Nuerons

Each membership-function neuron represents one membership function asociated with a particular fuzzy variable. A membership-function neuron has a unique input connection from an input neuron, z_h, and its output links may be connected to several fuzzy-AND neurons. The output of the *i*th membership-function neuron, z_i is described by

$$z_i = \frac{1 + \exp^{-s_i w_i}}{1 + \exp^{s_i[(z_h - c_l)^2 - w_i]}}, \tag{92.2}$$

where c_l is the centroid, w_i controls the width, and s_i affects the slope of the normalized trapezoidal membership function.

Fuzzy-AND Neurons

Each fuzzy-AND neuron represents the IF-part of some fuzzy rules. A fuzzy-AND neuron may have several input connections from membership-function neurons, and its output may

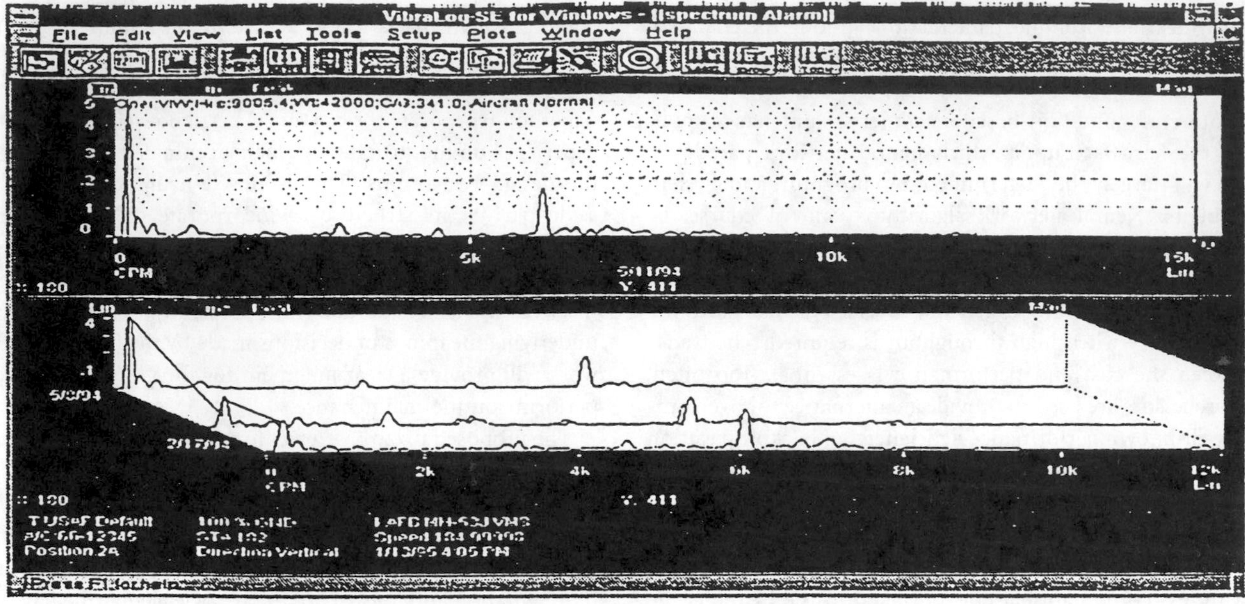

Figure 92.3 Quadratic sigmoid function with different θ.

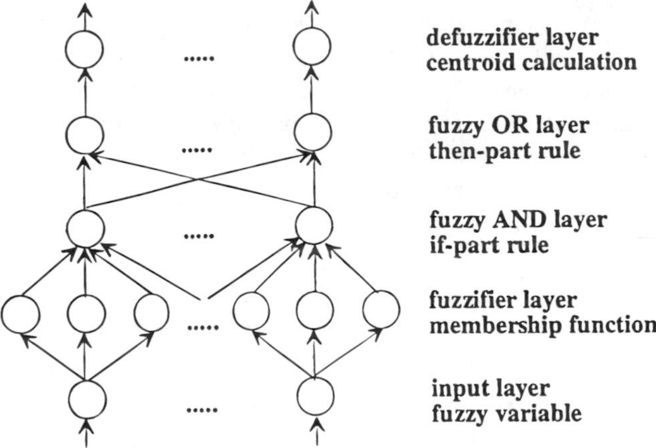

Figure 92.4 Fuzzy based feedforward neural network.

be linked to several fuzzy-OR neurons. The output of the *j*th fuzzy-AND neuron, z_j is described by

$$z_j = \gamma_j \min_{i \in P_j} (z_i) + (1 - \gamma_j) \frac{\sum_{i \in P_j} (z_i)}{|P_j|}, \tag{92.3}$$

where $\gamma_j \in [0,1]$, and P_j denotes the set of membership-function neurons which has its output connected to the *j*th neuron in the fuzzy-AND layer. $|P_j|$ is defined as the cardinal number of a finite set P_j. The weights of the input connections of the fuzzy-AND neuron are chosen to be unity.

Fuzzy-OR Neurons

Each fuzzy-OR neuron represents the THEN-part of some fuzzy rules. A fuzzy-OR neuron may have several input connections from fuzzy-AND neurons, and its output is linked to exactly one defuzzification neuron. The output of the *k*th fuzzy-OR neuron, z_k is described by

$$z_k = \gamma_k \max_{j \in P_k} (z_j w_{kj}) + (1 - \gamma_k) \frac{\sum_{j \in P_k} (z_j w_{kj})}{|P_k|}, \tag{92.4}$$

where $\gamma_k \in [0,1]$, and P_k denotes the set of fuzzy-AND neurons which has its output connected to the *k*th neuron in the fuzzy-OR layer. $|P_k|$ is defined as the cardinal number of a finite set P_k. The weight of the input connections of the *k*th fuzzy-OR neuron from the *j*th fuzzy-AND neuron is w_{kj}.

Defuzzification Neurons

Each defuzzification neuron represents an output variable, and performs the defuzzification of all the related membership functions of the variable. A defuzzification neuron may have several connections from fuzzy-OR neurons, and its output represents an output variable. The output of the defuzzification neuron, z_t is described by

$$z_l = \frac{\sum_{k \in P_l} (z_k a_{lk} c_{lk})}{\sum_{k \in P_l} (z_k a_{lk})}, \tag{92.5}$$

where a_{lk} and c_{lk} are the area and centroid of the membership function related to the *k*th fuzzy-OR neuron. The weights of the input connections of the defuzzification neuron are chosen to be unity.

The gradient descent learning rule can be used to minimize the mean square error,

$$E = \frac{1}{2} \sum_{k=1}^{M} (T_l - z_l)^2, \tag{92.6}$$

where M is the number of output fuzzy variables, and T_l and z_l are the desired and the actual output values. Let N_i be the set of indices of neurons which has an input link from neuron i. The mathematical derivation of the delta values for each type of neuron are as listed below.

The deal value of a defuzzication neuron:

$$\delta_l = T_l - z_l. \tag{92.7}$$

The delta value of a fuzzy-OR neuron:

$$\delta_k = \delta_l \frac{a_{lk}\left[c_{lk} \sum_{k' \in P_l} (z_{k'} a_{lk'}) - \sum_{k' \in P_l} (z_{k'} a_{lk'} c_{lk'}) \right]}{\left[\sum_{k' \in P_l} (z_{k'} a_{lk'}) \right]^2}. \tag{92.8}$$

The delta value of a fuzzy-AND neuron:

$$\delta_j = \sum_{k \in N_j} \delta_k \cdot \begin{cases} \gamma_k w_{kj} + (1 - \gamma_k) \dfrac{w_{kj}}{|P_k|}, & \text{if } z_j w_{kj} = \max\limits_{j' \in P_k} (z_{j'} w_{kj'}) \\ (1 - \gamma_k) \dfrac{w_{kj}}{|P_k|}, & \text{otherwise} \end{cases} \tag{92.9}$$

The delta value of a membership-function neuron:

$$\delta_l = \sum_{j \in N_l} \delta_j \cdot \begin{cases} \gamma_j + \dfrac{1 - \gamma_j}{|P_j|}, & \text{if } z_l = \min\limits_{l' \in P_j} (z_{l'}) \\ \dfrac{1 - \gamma_j}{|P_j|}, & \text{otherwise} \end{cases} \tag{92.10}$$

92.4 FDIA Architecture

The rotorcraft transmission health assessment is often approached from a pattern classification perspective. Neural networks can be designed to provide a failure index which is dependent upon the structural response to given payloads, so that perturbations in structural geometry and material properties can be identified by the outputs of the neural network. The output of the neural network indicates the damage index of prespecified transmission components or a complex failure scenario.

The generic fault detection, identification, and accommodation (FDIA) system is schematically given in Figure 92.5. The estimated model is a constantly updated mathematical/analytical representation of the transmission system. To detect any drastic changes in the system dynamics, the estimated model is compared to a nominal system model. A residual generator provides a measure of the deviation between the estimated and the nominal model. Based on the residual vectors, a decision is made as to whether a failure has occurred. If a failure is detected, a parallel decision logic mechanism compares the characteristics of the failure, as given by the residual vector, with the signature of any known failure modes. Signature of known failures, which we refer to as anticipated failures, are stored in a post-failure model bank that can be implemented via associated memory (Yen, 1994; Yen and Michel, 1992). The information can then be fed to the control system to invoke an effective controller to achieve autonomous control reconfiguration as given in Figure 92.6. This information could also generate corrective maintenance actions or guidance. The controller solution space is stored within an associative memory as opposed to a look-up table and therefore offers the capabilities of real-time reconfiguration and generalization. A look-up table approach would only provide *discrete* controller solutions in a lengthy and sequential search.

Health monitoring may be decomposed into five general tasks: detection, feature extraction, fault isolation, and fault severity estimation. Post-processing operations, such as multiple-look strategies, expert systems, data fusion, etc. may also be included as a fifth task. To achieve a completely robust diagnostic/prognostic system, all of the above aspects will, in general, require implementation.

To realize optimal and reliable fault detectors, isolators, and severity estimators, it is essential to explicitly recognize the direct dependence of fault detection, isolation, and accommodation signal processing on an adequate understanding of the basic characteristics of the data. This can be achieved by characterizing the statistical properties of the vibration signatures that most readily distinguish them from background vibration and noise. This can be accomplished experimentally by analyzing the second- and higher-order spectral characteristics of the data, both those containing fault signals and those that are fault-free. These characteristics should be examined in both the time and frequency domains, through the use of quadratic and high-order time-frequency representations. Because the helicopter transmission environment is reverberant, additional techniques, such as nonlinear homomorphic deconvolution, may also provide useful information concerning vibration signals. Here, reverberation represents a distortion of the basic signal whose recovery is desired. Where applicable, the added dimension of space should also be examined to discern whether greater use of spatial characteristics may be possible.

To achieve spatial selectivity, it is necessary to receive and process multivariate signals arising from an array comprised of two or more independent and spatially separated sensors. Whether this is done implicitly, by having the classifier perform the data fusion or explicitly, through analytic beamforming, the addition of a spatial dimension into the signal processing introduces new capabilities. For example, if a signal being monitored is corrupted by interfering signals that occupy the same temporal frequency band, spatial separation can sometimes be exploited; signal-to-noise ratios may be improved through the reduction or elimination of directional interference by adaptive canceling or nulling; adaptive arrays can be steered to isolate signals without

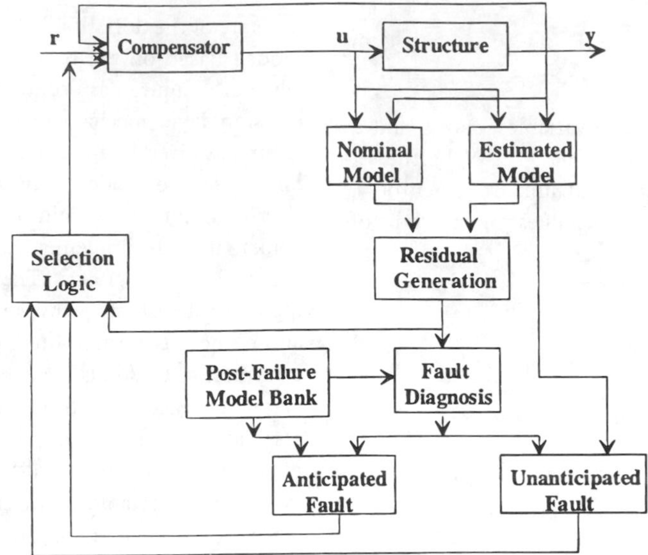

Figure 92.5 FDIA system architecture.

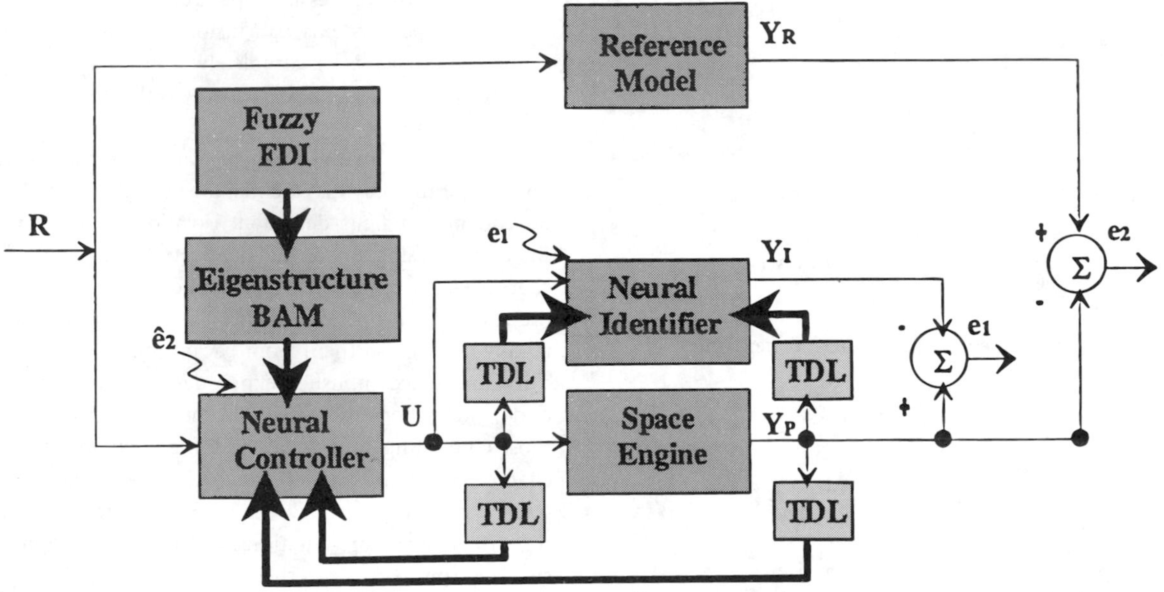

Figure 92.6 Reconfigurable control system architecture.

a priori knowledge of their direction of arrival; adaptive receiving arrays can be made sensitive to signals originating nearby and insensitive to signals originating at a distance, and vice versa. They can also be tuned to infrequent transient signals at the expense of frequent stationary signals and vice versa. Explicit beamforming provides the advantage that the three-dimensional space inside a gearbox can be scanned selectively in range, azimuth, elevation angle, and temporal frequency to detect and isolate fault. The following subsections discuss the subtasks involved to facilitate the process of FDIA system.

Detection

Detection is often the first step to separating potential fault characteristic signals from background vibration and noise. This step may also include pre-processing operations, such as use of a whitening filter to decorrelate the signals of interest from background noise, interference, and sensor characteristics. Effective performance of the detection task can greatly enhance subsequent health diagnosis/prognosis tasks by freeing these algorithms from spurious detection and by eliminating low-amplitude vibration signatures from further consideration. A certain level of vibration occurs even in normal transmissions due to manufacturing tolerances and to the effects of maneuvering and atmospheric turbulence on the drivetrain. Vibration in any component is radiated to all other components.

The fault signals of interest in transmission health diagnosis/prognosis are generated by complex, possibly nonlinear, vibrations. Such signals typically exhibit non-Gaussian statistical behavior (Hite, 1993). For vibration signatures of a purely random nature, deterministic signal analyses based on matched-filtering (for the coherent case) or banks of matched filters (for the incoherent case) generally do not work. For many applications, such signals can be detected adequately using conventional radiometric techniques, which do not exploit the distribution characteristics of the signals. Radiometry assumes that there is no structure to either the vibration signal or the noise and that both the signal and the noise have Gaussian amplitude distributions. It involves comparing the received normalized energy to a threshold, which may be set and adapted as a function of flight conditions. The nominal threshold can be learned online, for example, by continually reducing it to a level where many false alarms are observed. Then the actual threshold used is set to some function of the recurring false-alarm threshold. Radiometry, however, may be suboptimal for this application; it may be possible to improve significantly upon it by exploiting the more detailed statistical behavior of the signals of interest. For example, the use of higher-order spectral information can potentially result in significant improvements in the early detection capabilities of subtle fault signatures. More sophisticated detectors and Gabor and wavelet representations may also be useful in detecting transient helicopter transmission fault signals when the signal is non-white. Such improvements may be particularly crucial to the early detection of subtle signals, such as those generated by hard faults.

Feature Extraction

Feature extraction involves preliminary processing of sensor measurements to obtain suitable parameters that, in linear and/or nonlinear combination, reveal whether or not a bearing fault is developing. This process also serves to compress and to filter the huge amount of vibration signals generated constantly. All potential algorithms will, in general, require windowing of the time-series vibration data to form data segments on which linear, bilinear, or nonlinear transformations are applied. Some relevant feature extraction techniques are given in Table 92.1

All of these algorithms have potential merits and deficiencies for vibration signal feature extraction. Parameters extracted from these representations can be used as input vectors to detection and classification neural networks. Spectral and recursive techniques make the assumption that the signal is not white, but that the noise is. Nonlinear and higher-order techniques (e.g., cumulants, polyspectra) are especially valuable when the vibration signatures and/or noise are non-Gaussian. In such cases, techniques based on second-order statistics (e.g., coherence, spectral analysis), and linear (e.g., wavelet) and bilinear (e.g., Wigner) transformations are suboptimal. All of the above approaches may be considered in a three-dimensional context (e.g., amplitude vs. frequency vs. time), and where applicable combined with spatial selectivity derived from implicit or explicit multi-channel sensor beamforming, to provide distinct and representative spatiotemporal patterns.

Table 92.1 High-Order Statistical Spectral Algorithms

• spectral analysis,	• Prony's method,
• moment analysis,	• higher moments (e.g., kurtosis),
• generalized hypercoherence,	• phase-domain averaging,
• homomorphic filtering,	• cumulants,
• time-frequency analysis,	• time-scale analysis,
• recursive estimation method,	• bispectrum, trispectrum,
• hypercoherence filtering,	• nonstationary analysis
• heuristic features.	

Fault Isolation

Isolation is a diagnostic classification process, whose goal is to reveal the type of fault condition. Multiclass neural network classifiers discussed above (i.e., probabilistic neural network (Specht, 1990), adaptive time-delay radial basis function network (Yen, 1994c) and fuzzy-based feedforward neural network) will yield statistically optimal performance on classification task. The output of these networks are estimates of the probabilities of class memberships.

Fault Severity Estimation

Fault severity estimation is a prognostication task, designed to yield information regarding the severity of faults and to predict the time to failure of the machine, thereby providing additional opportunity for the fault condition to reflect whether or not emergency procedures are required. Estimation of neural networks can be either static or dynamic (i.e., recurrent). Generally, a constrained, square-error fitting criterion is appropriate for synthesizing estimators.

Hierarchical Structure

The Rotorcraft transmission FDIA system may be implemented in a hierarchical and distributed manner. At the bottom level, damage detection may be designed for parts, such as bearings, shafts, cables, sensors, or actuators. Once the appropriate failure scenarios are available, a high-level decision maker is employed to perform a proper control action.

92.5 Simulation Study

In this proof-of-concept study, pre-processing tasks will be kept to a minimum. For example, Figure 92.7 illustrates the signal processing incorporated within the neural network architecture.

The time series vibration data is recorded from twelve accelerometers, mounted radically at the points indicated in Figure 92.1. A 256-point short-term Fourier transform (STFT) is used to generate the spectral components from twelve time domain channels. After obtaining the spectra, principal component analysis (PCA), which is also known as the Karhunen-Loéve transform, will be applied to generate a set of features that accounted for most of the variation in the data. PCA reduces the dimensionality

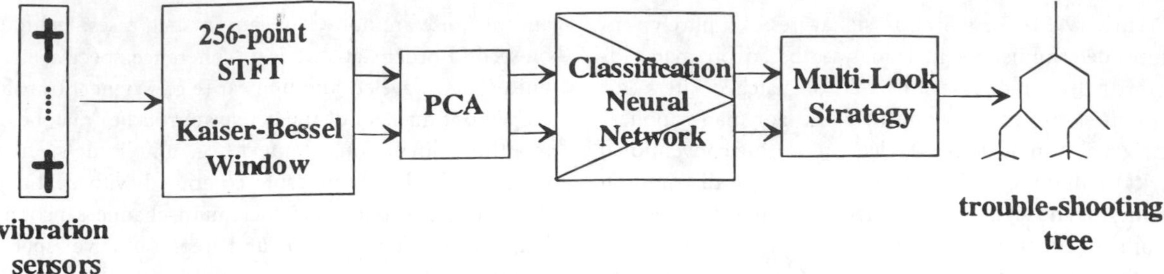

Figure 92.7 Rotorcraft transmission bearing fault classification signal processing diagram.

of the feature space through the use of linear combinations of the input features (power spectra) to account for an analyst-selected percentage of the cumulative variance in the data. The procedure involves generating a reduced-order transformation matrix (whose rows are the most significant eigenvectors of the data covariance matrix) that is used to reduce the dimension of the original training and evaluation feature sets, which are then used by the classification networks.

The fuzzy based feedforward network is simulated based on fifteen IF-THEN rules, a reduced scope of the entire scenario. The fifteen failure modes are quantified as: no fault, twelve bearing faults and two combinations of multiple faults. Twelve linguistic variables and twelve defuzzifier neurons are used in this setting. In generating the vibration signals, faults were introduced under controlled conditions, to emulate the characteristics of actual in-flight faults. Approximately ten minutes of sample data were provided for each fault type. Based on expert's experience, only a subset of the available frequency spectrum (i.e., 4k–11k Hz) is used. Frequencies below 4 kHz are dominated by high energy noise, while frequencies above 11 kHz are negligible, since the response of both accelerometers drops off at approximately 10 kHz. A multi-look post-processing strategy can be used to eliminate false alarms while still allowing the models to have high sensitivity. By tracking the network outputs over time and majority vote policy, intermittents are averaged out of the output of the overall system. The effect is that the overall system reports *trends* in network outputs, removing intermittent misclassifications. If the neural network consistently reports a bearing fault, for instance, then the pilot may be alerted. Because proprietary information is involved, we limit the discussion up to this scale.

Using the voting scheme described above, the system is able to achieve 99.4% fault classification with a zero false alarm rate. The system had difficulty discriminating between multiple bearing faults. This result, partly due to highly similar vibration spectra, can be improved by increasing the training sets, or by expanding the frequency domain of interest. The near future study will focus on resolving this issue.

92.6 Conclusions

Rotorcraft safety, survivability, and mission effectiveness highly depend on the structural integrity of dynamic components. The need to develop an on-board, constant vibration diagnostic system to detect and prognostic faults in these components (e.g., bearings, gears, and shafts) prior to failures is essential. This paper addresses a generic fault detection, identification and accommodation (FDIA) architecture for condition based machinery maintenance applications. The innovative fuzzy based neural network used for fault pattern classification is developed to analyze normal and defect vibration signatures in helicopter transmissions. The proposed FDIA advisor provides a trouble shooting tree with confidence probability rating to greatly facilitate ground crew to perform a time-critical maintenance.

References

Beard, R. V. 1971. Failure accommodation in linear systems through self-organization, *Technical Report MVT-71-1,* February Man Vehicle Lab, MIT, Cambridge, MA.

Chin, H., Danai, K., and Lewicki, D. G. 1993. Fault detection of helicopter gearboxes using the multi-valued influence matrix method, *NASA Technical Memorandum 106100,* March.

Cikanek, H. A. 1985. SSME failure detection, *Proc. American Control Conf.,* June, 282–286.

Deckert, J. C., Desai, M. N., Deyst, J. H., and Willsky, A. S. 1977. Dfbw sensor failure identification using analytic redundancy, *IEEE Trans. Automatic Control,* 20:795–809.

Hartman, E. H., Keeler, J. D., and Kowalski, J. M. 1990. Layered neural networks with Gaussian hidden units as universal approximations, *Neural Computation,* Summer, 2:210–215.

Hite, S. W. 1993. An algorithm for determination of bearing health through automated vibration monitoring, *Technical Report AEDC-TR-93-19,* December USAF Arnold Engineering Development Center.

Hornik K., Stinkchcombe, M., and White, H. 1989. Multilayer feedforward networks are universal approximators, *Neural Networks,* 2:359–366.

Kitamura, M. 1980. Detection of sensor failures in nuclear plant using analytic redundancy, *Trans. American Nuclear Society,* 34:581–583.

Mehara, R. K. and Peshon, I. 1971. An innovations approach to fault detection and diagnosis in dynamic systems, *Automatica,* 7:637–640.

Moody, J. and Darken, C. J. 1989. Fast learning in networks of locally-tuned processing units, *Neural Computation,* Summer, 1:281–294.

Parker, B. E., Nigro, T. M., Carley, M. P., Barron, R. L., Ward, D. G., Poor, H. V., Rock, D, and DuBois, T. A. 1993. Helicopter gearbox diagnostics and prognostics using vibration signature analysis, *Proc. SPIE Conf. Applications of Artificial Neural Networks IV,* April, 531–542.

Randall, R. B. 1982. A new method of modeling gear faults, *ASME J. Mech. Design,* 14:259–267.

Rauch, H. E. 1994. Intelligent fault diagnosis and control reconfiguration, *IEEE Control Systems Magazine,* 14:6–12.

Rock, D., Malkoff, D., and Stewart, R., 1993. AI and aircraft health monitoring, *AI Expert,* 8:28–35.

Specht, D. F. 1990. Probabilistic neural networks, *Neural Networks,* pp. 3:109–118.

Yen, G. G. 1994a. Autonomous neural/fuzzy control in precision space structures, *IFAC Control engineering Practice,* April, 3:471–483.

Yen, G. G., 1994b. Eigenstructure bidirectional associative memories: a computationally efficient synthesis procedure, *IEEE Trans. Neural Networks.* June 1994, 1038–1043; to appear; also in *Pro. IEEE Conf. Neural Networks.*

Yen G. G. 1994c. Adaptive time-delay neural control in space structural platforms, *Proc. IEEE Conf. Neural Networks,* June, 2622–2624.

Yen, G. G. 1994d. Identification and control of large structures using neural networks, *Computers and Structures,* 52:859–870.

Yen, G. G. 1994e. Reconfigurable learning control in large space structures, *IEEE Trans. Control Systems Technology,* December, 362–370.

Yen, G. G. and Michel, A. N. 1992. A learning and forgetting algorithm in associative memories: the eigenstructure method, *IEEE Trans. Circuits and Systems, Part II: Analog and Digital Signal Processing,* 39:212–225.

Yen, G. G. and Kwak, M. K. 1993. Neural network approach for the damage detection of structures, *Proc. AIAA/ASME/ASCE/AHS/ASC Structures, Structural Dynamics, and Material Conf.* April, 1549–1555.

93

Autonomous Neural Control in Flexible Space Structures

Gary G. Yen
Oklahoma State University

93.1 Learning Control System 1192
93.2 Adaptive Time-Delay Radial Basis Function Network 1194
93.3 Eigenstructure Bidirectional Associative Memory 1195
93.4 Fault Detection and Identification 1198
93.5 Reconfigurable Control 1199
93.6 Simulation Studies 1202
93.7 Conclusion 1205

93.1 Learning Control System

Contemporary control design methodologies (e.g., robust, adaptive, and optimal controls) face limitations for some of the more challenging realistic systems. In particular, flexible space structural platforms, which may be highly nonlinear, time-varying, and poorly modeled, pose serious difficulties for all currently advocated methods as summarized in White and Sofge (1992). These control system design difficulties arise in a broad spectrum of aerospace applications; e.g., surveillance satellites, military robots or space vehicles. The ultimate autonomous control, intended to maintain above acceptable performance over an extended operating range, can be especially difficult to achieve, due to factors such as high dimensionality, multiple inputs, and outputs, complex performance criteria, operational constraints, imperfect measurements, as well as the unavoidable failures of various actuators, sensors, or other components. Indeed, an iterative and time-consuming process is required to derive a high fidelity model to capture all of the spatiotemporal interactions among the structural members effectively. Therefore, the controller needs either to be exceptionally robust or adaptable after deployment. Also, catastrophic changes to the structural parameters, due to component failures, unpredictable uncertainties, and environmental threats, require that the controller be reconfigurable.

In the present paper, we investigate a hybrid connectionist system as a means of realizing a learning controller (Fu, 1970) with reconfiguration capability. The proposed control system integrates adaptive time-delay radial basis function (ATRBF) networks, an eigenstructure bidirectional associative memory (EBAM), and a fuzzy based back-propagation network (FBPN). A connectionist system consists of a set of interconnected processing elements and is capable of improving its performance based on past experimental information. An artificial neural network (herein referred to as simply "neural network") is a connectionist system which was originally proposed as a simplified model of the biological nervous system. Neural networks have been shown to provide an efficient means of learning concepts from past experience, abstracting features from uncorrelated data, and generalizing solutions from unforeseen inputs. Other advantages of neural networks are their distributed data storage and parallel information flow which cause them to be extremely robust with respect to malfunctions of individual devices as well as being computationally efficient. Neural networks have been successfully applied to the control of many dynamical systems, including aerospace and underwater vehicles (Troudet et al., 1991), (Venugopal et al., 1992), nuclear power plants (Guo and Uhrig, 1992), chemical process facilities (Watanabe et al., 1989), and manufacturing production lines (Leem and Dreyfus, 1992).

There have been many architectures (i.e., schema consisting of various neuronic characteristics, interconnecting topologies, and learning rules) proposed for neural networks over the last five years (least count over 200). Simulation experience has revealed that success is problem-dependent. Some networks are more suitable for adaptive control, whereas others are more appropriate for pattern recognition, signal filtering, or associative searching. Neural networks which employ the well known back-propagation learning algorithm are capable of approximating any continuous functions (e.g., nonlinear plant dynamics and complex control laws) with an arbitrary degree of accuracy (Hornik et al., 1989). Similarly, radial basis function networks (Moody and Darken, 1989) are also shown to be universal approximators (Hartman et al., 1990). These model-free neural network paradigms are more effective at memory usage in solving control problems than conventional learning control approaches. An example is the BOXES algorithm, a memory intensive solution, which partitions the control surface in the form of a look-up table (Michie and Chambers, 1968).

0-8493-8343-9/97/$0.00+$.50

Our goal is to approach structural autonomy by extending the control system's operating envelope, which has traditionally required vast memory usage. Connectionist systems, on the other hand, deliver less memory intensive solutions to control problems and yet provide a sufficiently generalized solution space. In vibration suppression problems, we utilize the adaptive time-delay radial basis function network as a building block to allow the connectionist system to function as an indirect closed-loop controller. Prior to training the compensator, a neural identifier based on an ARMA model is utilized to identify the open-loop system. The horizon-of-one predictive controller then regulates the dynamics of the nonlinear plant to follow a prespecified reference system asymptotically as depicted in Figure 93.1 (i.e., the model reference adaptive control architecture). The reference model, which is specified by an input-output relationship (R, Y_R), describes all desired features associated with a specific control task, e.g., a linearly and highly damped system to suppress the structural vibration. As far as trajectory slewing problems are concerned, the generalized learning controller synthesized by the adaptive time-delay radial basis function network compensates the non-linear flexible structure in a closed-loop fashion in order to follow the motion specified by the command outputs as given in Figure 93.2. Tapped delay lines (TDL) are incorporated to process the time-varying structural parameters as suggested in (Narendra and Parthasarthy, 1990).

The function of the neural controller is to map the states of the system into corresponding control actions in order to force the plant dynamics (Y_P) to match a certain output behavior which is specified either by the reference model (Y_R) or command output (Y_D). However, we cannot apply the optimization procedure (e.g., gradient descent, conjugate gradient, or Newton-Raphson method) to adjust the weights of the neural controller because the desired outputs for the neural controller are not available. In Psaltis et al., (1988), a specialized learning algorithm which treats the plant as an additional unmodifiable layer of network is proposed. The output error, e_2, is back-propagated through the plant to derive the neural controller output error $\hat{e}_2$. However, the authors fail to suggest a reliable way to compute $\hat{e}_2$. In Elsey (1988), the inverse Jacobian of the plant is used to estimate $\hat{e}_2$ at each weight update, which results in a complicated and computational expensive learning procedure. Moreover, since the plant is often not well-modeled because of modeling uncertainties, the exact partial derivatives cannot be determined. In Saerens and Soquet (1989), a 'dynamic sign approximation' is used to determine the direction of the error surface, assuming the qualitative knowledge of the plant. This is not necessarily

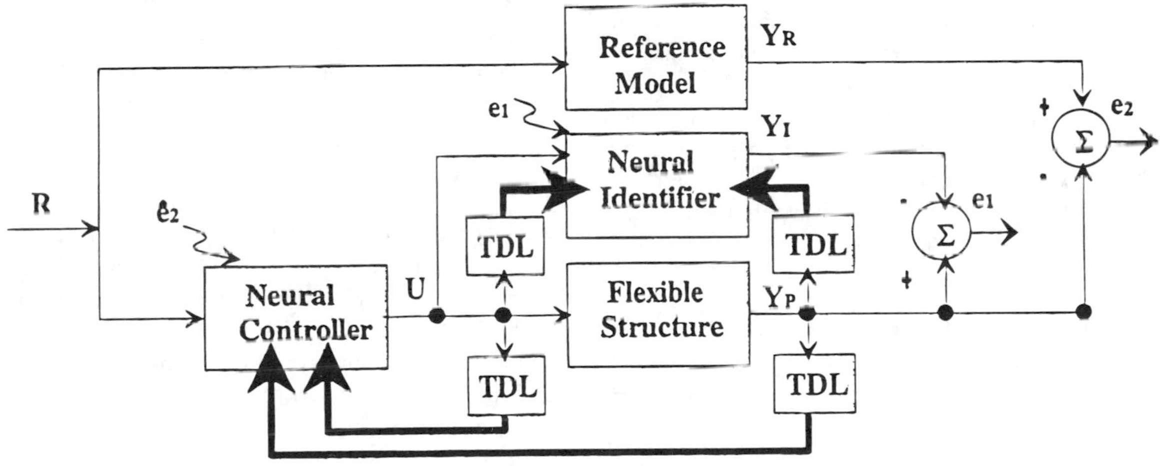

Figure 93.1 Vibration suppression learning control architecture.

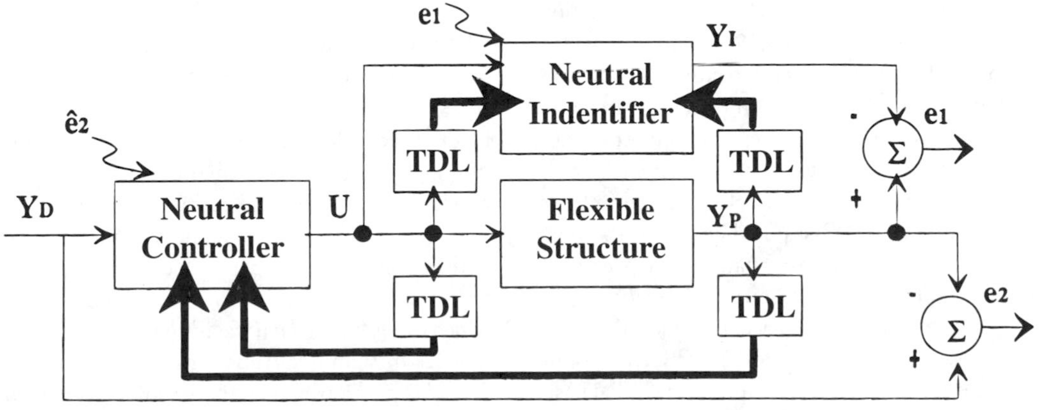

Figure 93.2 Trajectory slewing learning control architecture.

the case in space structure applications which are often equipped with highly correlated parameters. To achieve the true gradient descent of the square of the error, we use 'dynamic back propagation' to accurately approximate the required partial derivatives. A single-layer ATRBF network is first trained to identify the open-loop system. The resulting neural identifier then serves as extended unmodifiable layers to train the compensator (i.e., another single-layer ATRBF network). If the structural dynamics are to change as a function of time, the neural identifier would require the learning algorithm to periodically update the network parameters accordingly.

The proposed architecture for autonomous neural control includes identification-dedicated and control-dedicated neural networks, structural health component assessment, and controller association retrieval and interpolation. For the purpose of system identification and dynamic control of flexible space structures, an adaptive time-delay radial basis function network which serves as a building block is discussed in Section 93.2, providing a justification to achieve real-time performance. A novel class of bidirectional associative memories synthesized by the eigenstructure decomposition algorithm is covered in Section 93.3 to fulfill the critical needs in real-time controller retrieval. This is followed by introducing a class of back-propagation networks incorporated with fuzzy linguistics in Section 93.4 for fault detection and identification of structural failures. The integration of various components in an intelligent manner is then presented to achieve the structural reconfigurable learning control in Section 93.5. Specific examples in space structural testbeds are used in Section 93.6 to demonstrate the effectiveness of the proposed neural control architecture. The paper is concluded with a few pertinent observations regarding potential commercial applications in Section 93.7.

93.2 Adaptive Time-Delay Radial Basis Function Network

Biological studies have shown that variable time-delays do occur along axons due to different conduction time and different lengths of axonal fibers. In addition, temporal properties such as temporal decays and integration frequently occur at synapses. Inspired by this observation, the time-delay back-propagation network was proposed by Waibel et al. (1988) for solving the phoneme recognition problem. In their architecture, each neuron takes into account not only the current information from all neurons of the previous layer, but also a certain amount of past information from those neurons due to delay on the interconnections. However, a fixed amount of time-delay throughout the training process has limited the usage possibly due to the mismatch of the temporal location in the input patterns. To overcome this limitation, Lin et. al. (1992) has developed an adaptive time-delay back-propagation network to better accommodate the varying temporal sequences, and to provide more flexibility for optimization tasks. In a similar spirit, adaptive time-delay radial basis function network is proposed to take full advantages of temporal pattern matching and learning/recalling speed.

A given adaptive time-delay radial basis function network can be completely described by its interconnecting topology, neuronic characteristics, temporal delays, and learning rule. The individual processing unit performs its computations based only on local information. A generic radial basis function network is a two layer neural network whose outputs form a linear combination of the basis functions derived from the hidden neurons. The basis function produces a localized response to input stimulus as do locally tuned receptive fields in our nervous systems. The Gaussian function network, a realization of an RBF network using Gaussian kernels, is widely used in pattern classification and function approximation. The output of a Gaussian neuron in the hidden layer is defined by

$$u_j^1 = \exp\left(-\frac{\|x - w_j^1\|^2}{2\sigma_j^2}\right), \qquad j = 1, \ldots, N_1 \tag{93.1}$$

where u_j^1 is the output of the jth neuron in the hidden layer (denoted by the superscript 1), x is the input vector, w_j^1 denotes the weighting vector for the jth neuron in the hidden layer (i.e., the center of the jth Gaussian kernel), σ_j^2 is the normalization parameter of the jth neuron (i.e., the width of the jth Gaussian kernel), and N_1 is the number of neurons in the hidden layer. Equation 93.1 produces a radially symmetric output with a unique maximum at the center dropping off rapidly to zero for large radii. The output layer equations are described by

$$y_j = \sum_{i=1}^{N_1} w_{ji}^2 u_i^1, \qquad j, = 1, \ldots, N_2 \tag{93.2}$$

where y_j is the output of the jth neuron in the output layer, w_{jI}^2 denotes the weight from the ith neuron in the hidden layer to the jth neuron in the output layer, u_i^1 is the output from the ith neuron in the hidden layer, and N_2 is the number of linear neurons in the output layer. Inspired by the adaptive time-delay back-propagation network, the output equation of ATRBF networks is described by

$$y_j(t_n) = \sum_{i=1}^{N_1} \sum_{i=1}^{L^{ji}} w_{ji,1}^2 u_i^1(t_n - \tau_{ji,1}^2), \qquad j = 1, \ldots, N_2 \tag{93.3}$$

where $w_{ji,l}^2$ denotes the weight from the ith neuron in the hidden layer to the jth neuron in the output layer with the independent time-delay τ_{jil}^2, u_i^1 $(t_n - \tau_{jil}^2)$ is the output from the ith neuron in the hidden layer at time $t_n - \tau_{jil}^2$, L_{ji} denotes the number of delay connections between the ith neuron in the hidden layer and the jth neuron in the output layer. Shared with generic radial basis function networks, adaptive time-delay Gaussian function networks have the property of undergoing *local* changes during training, unlike adaptive time-delay back-propagation networks which experience *global* weighting adjustments due to the characteristics of sigmoidal functions. The localized influence of each Gaussian neuron allows the learning system to refine its functional approximation in a successive and efficient manner. The hybrid learning algorithm (Moody and Darken, 1989) which employs the K-means clustering for the hidden layer and the least mean square (LMS) algorithm for the output layer further ensures a faster convergence and often leads to better performance and generalization. The combination of locality of representation and linearity of learning offers tremendous computational efficiency to achieve real-time adaptive control compared to the back-propagation network which usually takes considerable time to converge. K-means algorithm is perhaps the most widely known clustering algorithm because of its simplicity and its ability to produce good results. The normalization parameters, σ_j^2, are obtained once the clustering algorithm is complete. They represent a measure of the spread of the data associated with each cluster. The cluster widths are then determined by the average distance between the cluster centers and the training samples,

$$\sigma_j^2 = \frac{1}{M_j} \sum_{x \in \Theta_j} \|x - w_j^1\|^2, \quad (93.4)$$

where Θ_j is the set of training patterns belonging to jth cluster and M_j is the number of samples in Θ_j. This is followed by applying a LMS algorithm to adapt the time-delays and interconnecting weights in output layer. The training set consists of input/output pairs, but now the input patterns are pre-processed by the hidden layer before being presented to the output layer. The adaptation of the output weights and time delays are derived based on error back-propagation to minimize the cost function,

$$E(t_n) = \frac{1}{2} \sum_{j=1}^{N_2} (d_j(t_n) - y_j(t_n))^2, \quad (93.5)$$

where $d_j(t_n)$ indicates the desired value of the jth output neuron at time t_n. The weights and time-delays are updated step by step proportional to the opposite direction of the error gradient respectively,

$$\Delta w_{ji,1}^2 \equiv -\eta_1 \frac{\partial E(t_n)}{\partial w_{ji,1}^2}, \qquad \Delta \tau_{ji,1}^2 \equiv -\eta_2 \frac{\partial E(t_n)}{\partial \tau_{ji,1}^2}, \quad (93.6)$$

where η_1 and η_2 are the learning rates. The mathematical derivation of this learning algorithm is straightforward. We summarize the learning rule given as follows.

$$\Delta w_{ji,1}^2 = \eta_1(d_j(t_n) - y_j(t_n)) \sum_{l=1}^{L^{ji}} u_i^1(t_n - \tau_{ji,1}^2), \quad (93.7a)$$

$$\Delta \tau_{ji,1}^2 = -\eta_2(d_j(t_n) - y_j(t_n)) \sum_{l=1}^{L^{ji}} w_{ji,1}^2 u_i^{1'}(t_n - \tau_{ji,1}^2). \quad (93.7b)$$

93.3 Eigenstructure Bidirectional Associative Memory

Based on the failure scenario determined by a fault diagnosis network (to be covered in Section 93.5), an eigenstructure bidirectional associative memory will promptly retrieve a corresponding controller configuration from a continuous solution space. This controller configuration in the form of weighting parameters will then be loaded into the neural controller block to achieve controller reconfiguration.

Bidirectional associative memory (BAM) (Kosko, 1987) is a two-layer nonlinear feedback neural network. Unlike the Hopfield network, bidirectional associative memory is a hetero associative memory that provides a flexible nonlinear mapping from input data to output data. However, bidirectional associative memory does not guarantee that a network will necessarily store the desired vectors as equilibrium points. Furthermore, experience has shown that BAM networks synthesized by 'correlation encoding' (Kosko, 1987) can effectively store only up to $p < \min(m,n)$ arbitrary vectors as equilibrium points, where m and n denote the number of neurons in each of the two layers. In Yen (1994), we have shown that the BAM network can be treated as a variation of a Hopfield network. Under appropriate assumptions, we have demonstrated that the present class of continuous BAM is a gradient system with the properties of *global stability* (i.e., for any initial condition, the trajectories of solution will tend to some equilibrium.) and *structural stability* (i.e., stability persists under small weight perturbations).

The qualitative and quantitative results (i.e, equilibrium condition, asymptotical stability criteria, and the estimation of trajectory bounds) which we have developed for Hopfield-type networks (Michel et al., 1991; Yen and Michel, 1991; Yen and Michel, 1992) can then be extended to the BAM networks through a special arrangement of interconnection weights. Based on these results, we investigate a class of discrete-time BAM networks defined on a closed hypercube of the state space. For the present model, we establish *stability analysis* which enables us to generalize the solutions of discrete-time systems, and to characterize the set of system equilibria. In addition, we develop an efficient *synthesis procedure* utilizing the eigenstructure decomposition method for the present class of neural networks. The synthesized networks are capable of *learning* new vectors as well as *forgetting* learned vectors without the necessity of recomputing all interconnection weights and external inputs. The resulting network can easily be implemented in digital hardware. Furthermore, when simulated by a serial processor, the present system offers extremely efficient means of simulating discrete-time BAM (modeled by a system of difference equations) compared to the computational complexity involved to approximate the dynamic behavior of the continuous system.

We now consider a class of neural networks described by a pair of difference equations DN_i which are defined on a closed hypercube of the state space, given by

$$x_i(k+1) = sat\left(\sum_{j=1}^{n} W_{ij} y_j(k) + I_i\right),$$

$$i = 1, \ldots, m, \; k = 0, 1, 2, \ldots \quad (DN_i a)$$

$$y_j(k+1) = sat\left(\sum_{i=1}^{m} V_{ji} x_i(k) + J_j\right),$$

$$j = 1, \ldots, n, \; k = 0, 1, 2, \ldots \quad (DN_i b)$$

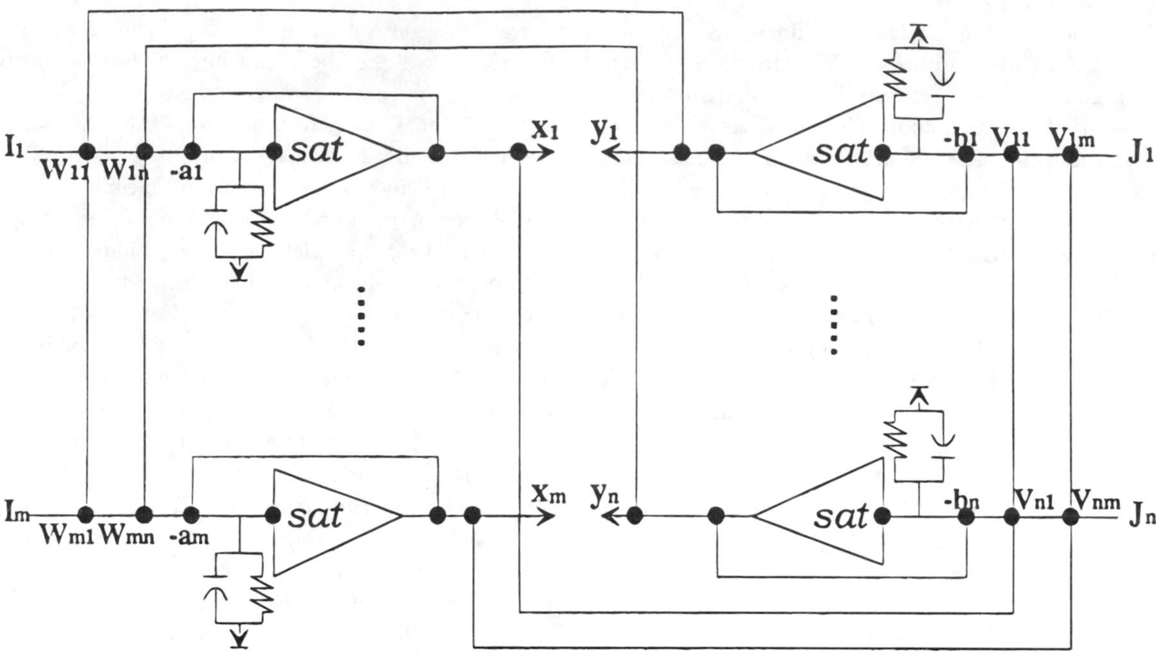

Figure 93.3 An implementation of eigenstructure bidirectional associative memory.

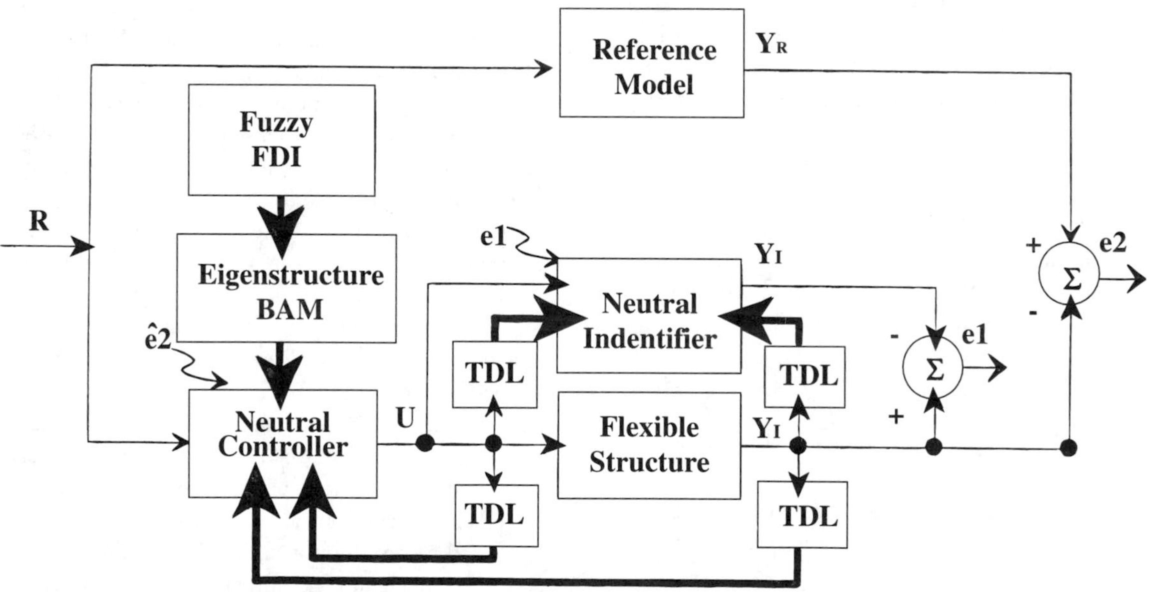

Figure 93.4 Autonomous reconfigurable learning control system architecture.

where the saturation function $sat(\theta_i) = 1$, if $\theta_i \geq 1$, $sat(\theta_i) = \theta_i$, if $-1 < \theta_i < 1$, and $sat(\theta_i) = -1$, if $\theta_i \leq -1$. The *sat* function serves as the model for all neurons. Figure 93.3 depicts an analog implementation of the eigenstructure BAM network. In our application, vector x refers to the failure index while vector y points to the weighting parameters for the retrieving controller configuration. Equation DN_i can be put into a compact form DN,

$$x(k+1) = sat(Wy(k) + I), \qquad k = 0, 1, 2, \ldots \text{ (DNa)}$$

$$y(k+1) = sat(Vx(k) + J), \qquad k = 0, 1, 2, \ldots \text{ (DNb)}$$

where $sat(\theta)$ is defined componentwisely, W and V are matrices denoting the interconnection weights, and I and J are vectors representing the external inputs. In contrast to the usual system defined on open subsets of $\Re^{m+n}$, system DN is described on a closed hypercube,

$$D = \{x \in \Re^m, y \in \Re^n: -1 \leq x_i \leq 1, -1 \leq y_i \leq 1, i = 1, \ldots, m, j = 1, \ldots, n\}. \qquad (93.8)$$

The results which we establish for system DN fall into one of two categories. One type of result addresses the *stability analysis* of system DN while the other type pertains to *synthesis procedure* for system DN. In Yen (1994), we conducted a thorough and complete qualitative analysis of system DN. Among other aspects, this analysis discusses the distribution of equilibrium points in the state space, the qualitative properties of the equilibrium points, global stability and structural stability properties of system DN, and the like. For the completeness of this paper, we briefly summarize the synthesis procedure given below (Yen, 1994).

Synthesis Problem

Given p pairs of vectors in B^{m+n}, say $(x^1, y^1), \ldots, (x^p, y^p)$, we wish to design a system DN such that

1. $(x^1, y^1), \ldots, (x^p, y^p)$ are asymptotically stable equilibrium points of system DN.
2. The system has no periodic solutions.
3. The total number of asymptotically stable equilibrium points of DN in the set B^{m+n} is as small as possible.
4. The domain of attraction of each (x^i, y^i), $i = 1, \ldots, p$ is as large as possible.

We summarize the synthesis procedure given below (called the *eigenstructure decomposition method*).

Synthesis Procedure

Suppose we are given p pairs of vectors, $\Lambda = \{(x^1, y^1), \ldots, (x^p, y^p)\}$ as desired library vectors to be stored as asymptotically stable equilibrium points for system DN. We proceed as follows.

1. Form the vectors

$$\mu^i = \begin{bmatrix} x^i \\ y^i \end{bmatrix}, \qquad i = 1, \ldots, p.$$

2. Compute the matrices

$$S^p = [\mu^1 - \mu^p, \ldots, \mu^{p-1} - \mu^p] = [s^1, \ldots, s^{p-1}],$$

where $s^i = \mu^i - \mu^p$, $i = 1, \ldots, p - 1$, and the superscript p for matrix S^p denotes the number of vectors to be stored in the BAM network.

3. Perform a singular value decomposition on matrix S^p

$$S^p = U\Sigma V^T,$$

where U and V are left and right singular matrices, respectively, and where Σ is a diagonal matrix with the singular values of S^p on its diagonals. (i.e., This can be accomplished by standard computer routines (e.g., *LSVRR* in IMSL, Singular Values in Mathematica, and *svd* in MATLAB or Matrix$_x$)). Let

$$L = \text{Span}(s^1, \ldots, s^{p-1}), \qquad L^a = \text{Aspan}(\mu^1, \ldots, \mu^p).$$

Then L is the linear subspace spanned by the vectors $\{s^1, \ldots, s^{p-1}\}$ and $L^a = L + \mu^p$ denotes the affine subspace generated by the vectors $\{\mu^1, \ldots, \mu^p\}$.

4. Decompose the matrix

$$U = [U^+ \quad U^-],$$

where $U^+ = \{u_1, \ldots, u_k\}$, $U^- = \{u_{k+1}, \ldots, u_{m+n}\}$, and $k = \text{rank}(\Sigma) = \dim(L)$. From the properties of singular value decomposition, we know that $\{u_1, \ldots, u_k\}$ is an orthonormal of L and $\{u_{k+1}, \ldots, u_{m+n}\}$ is an orthonormal basis of $L^\perp$ (i.e., $L^\perp$ denotes the orthogonal complement of space L).

5. Compute the matrices

$$T^+ = \sum_{i=1}^{k} u_i u_i^T = U^+ U^{+T}, \quad T^- = \sum_{i=k+1}^{m+n} u_i u_i^T = U^- U^{-T}.$$

6. Choose parameters $\tau_1 > 1$ and $-1 \leq \tau_2 < 1$, and compute

$$T_\tau = \tau_1 T^+ - \tau_2 T^-, \qquad K_\tau = \tau_1 \mu^p - T_\tau \mu^p.$$

7. Decompose matrix T_τ and vector K_τ by

$$T_\tau = \begin{bmatrix} A_1 & W_\tau \\ V_\tau & A_2 \end{bmatrix} \begin{matrix} \}m \\ \}n \end{matrix}, \qquad K_\tau = \begin{bmatrix} I_\tau \\ J_\tau \end{bmatrix} \begin{matrix} \}m \\ \}n \end{matrix}.$$

Then all vectors in Λ will be stored as asymptotically stable equilibria of the synthesized system DN_τ,

$$x(k+1) = sat(W_\tau y(k) + I_\tau), \quad k = 0, 1, 2, \ldots \quad (DN_\tau a)$$

$$y(k+1) = sat(V_\tau x(k) + J_\tau), \quad k = 0, 1, 2, \ldots \quad (DN_\tau b)$$

8. With $\tau_1 > 1$ fixed, choose parameter τ_2 as large as possible. The set of asymptotically stable equilibrium points will be approximately equal to $L^a \cap B^{m+n}$.

The eigenstructure decomposition method developed above possesses several advantages since it is possible by this method to exert control over the number of spurious states, since it is possible to estimate the extent of the basin of attraction of the stable memories, and since it is possible, under certain circumstances, to store by this method a number of desired stable vectors which by far exceeds the order of the network.

In synthesizing bidirectional associative memory, we usually assume that all desired vectors (i.e., fault scenarios) to be stored are known *a priori*. However, in the large space structure applications, this is usually not the case. Sometimes, we are also required to update the stored vectors (i.e., controller configurations) dynamically to accommodate new scenarios (e.g., when a novel fault condition is identified.). In a similar spirit of development as Yen and Michel, (1992), we have successfully incorporated the *learning* and *forgetting* capabilities into the present synthesis algorithm, where learning refers to the ability of adding vectors to be stored as asymptotically stable equilibria to an existing set of stored vectors in a given network, and where forgetting refers to the ability of deleting specified vectors from a given set of stored equilibria in a given network. The synthesis procedure is capable of adding an additional pattern as well as deleting an existing pattern without the necessity of *recomputing* the entire interconnection weights, i.e., W and V, and external inputs, i.e., I and J.

Making use of the updating algorithm for singular value decomposition (Bunch and Nielsen, 1978), we can construct the required orthonormal basis set i.e., $\{u^1, \ldots, u_{m+n}\}$, for space L where $L = Span(s^1, \ldots, s^{p-1})$ in accordance with the new configuration. The detailed development of the learning and forgetting algorithms can be found in Yen (1994). Furthermore, the *incremental learning and forgetting* algorithm is proposed to improve the computational efficiency of the eigenstructure decomposition method by taking advantage of recursive evaluation (Yen, 1994).

93.4 Fault Detection and Identification

Detection of structural failures in large scale systems has been an interesting subject for many decades. The existing damage detection methods highly depend on off-line destructive tests or computationally intensive finite element simulations. Quite often, these heuristic algorithms are limited to the analysis and design of a fixed structural concept or model, where the loadings, materials, and design constraints need to be specified in advance. Because of the need for time-critical response in many situations, available symptom data is either misinterpreted or unused, often leading to the incorrect removal of a system's components. Fault tolerance issues have usually been ignored or have been assumed to be handled by a simple strategy such as triple modular redundancy.

To date, relatively little *systematic* work has been pursued in connection with damage detection, isolation, and identification. Literature surveys have shown a promising potential in the application of artificial neural networks to quantify structural failures (Chin *et al.*, 1993; Chou and Kuo, 1993). It has become evident that neural networks can also be trained to provide failure information based on the structural response to given payloads, so that perturbations in structural geometry and material properties can be identified by the outputs of the neural network. This information can then be fed back to the bidirectional associative memory to invoke an effective neural controller before the structure breaks down. In addition, the neural-network based fault diagnosis system developed for a certain structural component can also be used in a hierarchical manner where the same structural component is used in several places on large space structures.

We approach the damage detection of flexible structures from a pattern classification perspective. In doing so, we classify the loading to structures and the output response to such a loading as an input pattern to the neural network. The output of the neural network indicates the damage index of structural members or scenarios. Neural networks trained with a back-propagation learning rule have been used for various problems (Napolitano et al., 1993). Simulation results show that the neural network is capable of performing fault detection and identification. Although the back-propagation algorithm proves its effectiveness in this case, it is generally known that it takes considerable time to train the neural network and the network may get trapped into local minima. In addition, existing expert knowledge and heuristic experience on specific problem domain need to be built into the design constraints. Conventional back-propagation network won't be sufficient to handle this requirement. The proposed fuzzy back-propagation network is dedicated to satisfy this need (Kosko, 1992; Carli, 1994).

The fuzzy back-propagation network, equipped with four layers of neurons, is constructed according to the architecture of fuzzy expert systems as described in the following.

Input Neurons. Each input neuron represents an input fuzzy variable, and is used to distribute the input to its membership-function neurons.

Membership-function Neurons. Each membership-function neuron represents one membership function associated with a particular fuzzy variable. A membership-function neuron has an unique input connection from an input neuron, z^h, and its output links may be connected to several fuzzy-AND neurons. The output of the ith membership-function neuron, z_l is described by

$$z_i = \frac{1 + \exp^{-s_i w_t}}{1 + \exp^{s_i[(z_h - c_i)^2 - w_i]}}, \tag{93.9}$$

where c_i is the centroid, w_l controls the width, and s_l affects the slope of the normalized trapezoidal membership function, respectively.

Fuzzy-AND Neurons. Each fuzzy-AND neuron represents the IF-part of some fuzzy rules. A fuzzy-AND neuron may have several input connections from membership-function neurons, and its output may be linked to several fuzzy-OR neurons. The output of the jth fuzzy-AND neuron, z_j is described by

$$z_j = \gamma_j \min_{i \in P_j} (z_i) + (1 - \gamma_j) \frac{\sum_{i \in P_j} (z_i)}{|P_j|}, \tag{93.10}$$

where $\gamma_j \in [0,1]$, and P_j denotes the set of membership-function neurons which has its output connected to the jth neuron in the fuzzy-AND layer. $|P_j|$ is defined as the cardinal number of a finite set P_j. The weights of the input connections of the fuzzy-AND neuron are chosen to be unity.

Fuzzy-OR neurons. Each fuzzy-OR neuron represents the THEN-part of some fuzzy rules. A fuzzy-OR neuron may have several input connections from fuzzy-AND neurons, and its output is linked to exactly one defuzzification neuron. The output of the kth fuzzy-OR neuron, z_k is described by

$$z_k = \gamma_k \max_{j \in P_k} (z_j w_{kj}) + (1 - \gamma_k) \frac{\sum_{j \in P_k} (z_j w_{kj})}{|P_k|}, \qquad (93.11)$$

where $\gamma_k \in [0,1]$, and P_k denotes the set of fuzzy-AND neurons which has its output connected to the kth neuron in the fuzzy-OR layer. $|P_k|$ is defined as the cardinal number of a finite set P_k. The weight of the input connections of the kth fuzzy-OR neuron from the jth fuzzy-AND neuron is w_{kj}.

Defuzzification neurons. Each defuzzification neuron represents an output variable, and performs the defuzzification of all the related membership functions of the variable. A defuzzification neuron may have several connections from fuzzy-OR neurons, and its output represents an output variable. The output of the defuzzification neuron, z_l is described by

$$z_l = \frac{\sum_{k \in P_l} (z_k a_{lk} c_{lk})}{\sum_{k \in P_l} (z_k a_{lk})}, \qquad (93.12)$$

where P_l denotes the set of fuzzy-OR neurons which has its output connected to the lth neuron in the defuzzification layer. a_{lk} and c_{lk} are the area and centroid of the membership function related to the kth fuzzy-OR neuron. The weights of the input connections of the defuzzification neuron are chosen to be unity.

The back-propagation learning rule is used to minimize the error,

$$E = \frac{1}{2} \sum_{jl=1}^{M} (T_l - z_l)^2, \qquad (93.13)$$

where M is the number of the output fuzzy variables, and T_l and z_t are the desired and the actual output values. Let N_i be the set of the indices of neurons which has an input link from neuron i. The mathematical derivation of the delta values for each type of the neurons are listed as below.

The Delta Value of a Defuzzification Neuron.

$$\delta_l = T_l - z_l; \qquad (93.14)$$

The Delta Value of a Fuzzy-OR Neuron.

$$\delta_k = \delta_l \frac{a_{lk}\left[c_{lk} \sum_{k' \in P_l} (z_{k'} a_{lk'}) - \sum_{k' \in P_l} (z_{k'} a_{lk'} c_{lk'})\right]}{\left[\sum_{k' \in P_l} (z_{k'} a_{lk'})\right]^2}; \qquad (93.15)$$

The Delta Value of a Fuzzy-AND Neuron.

$$\delta_j = \sum_{k \in N_j} \delta_k \cdot \begin{cases} \gamma_k w_{kj} + (1 - \gamma_k) \dfrac{w_{kj}}{|P_k|}, & \text{if } z_j w_{kj} = \max_{j' \in P_k} (z_{j'} w_{kj'}) \\ (1 - \gamma_k) \dfrac{w_{kj}}{|P_k|}, & \text{otherwise} \end{cases}; \qquad (93.16)$$

The Delta Value of a Membership-Function Neuron.

$$\delta_l - \sum_{j \in N_l} \delta_j \cdot \begin{cases} \gamma_j + \dfrac{1 - \gamma_j}{|P_j|}, & \text{if } z_l = \min_{l' \in P_j} (z_{l'}) \\ \dfrac{1 - \gamma_j}{|P_j|}, & \text{otherwise} \end{cases} \qquad (93.17)$$

The proposed fuzzy based back-propagation network can set aside part of structure to incorporate the existing knowledge base for a particular problem domain by assigning the parameters of involving neurons in different layers. After the network is well trained, the possibility of interpreting the underling meanings of decision made by the fuzzy neural network is currently under investigation. Fault detection may be implemented in a hierarchical and distributed manner. At the bottom level, damage detection may be designed for parts, such as bearings, shafts, cables, sensors, or actuators. Once the appropriate failure scenarios are available, a high level decision maker can be employed to perform a proper control action. Incorporated with learning and forgetting capabilities of associative memory, a robust FDI system can be designed to detect, isolate, and identify evolutionary variations as well as catastrophic change of large structures on a real-time basis.

93.5 Reconfigurable Control

Critical to autonomous system design is the development of a control scheme with globally adaptive and reconfigurable capabilities. *Reconfiguration* refers to the ability of retrieving a workable controller from the solution space (created prior to the failure). The motivation is to strive for a high degree of structural autonomy in flexible space platforms, thereby severing the dependence of the dynamical system on *a priori*

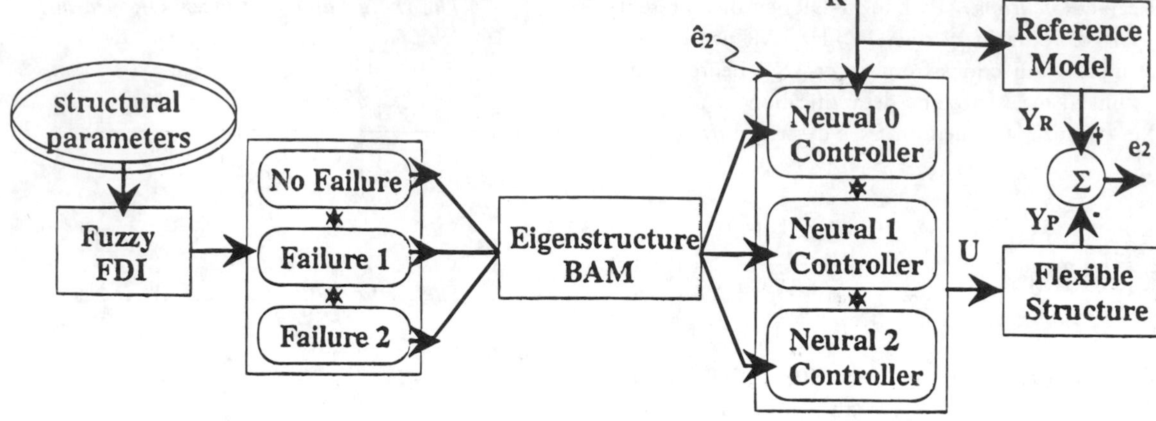

Figure 93.5 Functional diagram for controller association retrieval performed by EBAM.

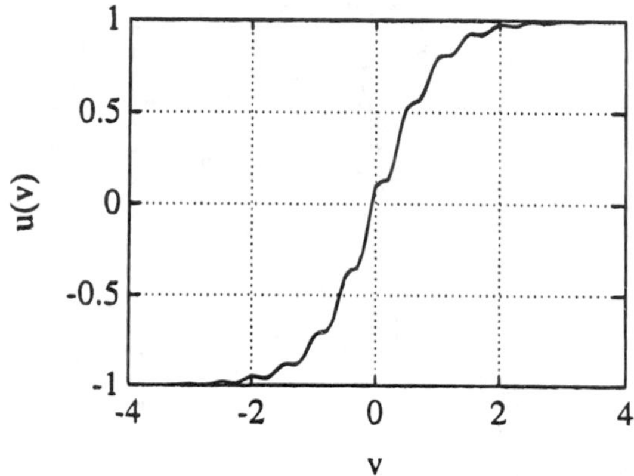

Figure 93.6 Simulated actuator input/output relationship.

programming, perfect communications, as well as the flawless operation of the system components, while maintaining a precision pointing capability.

Existing reconfigurable control techniques often rely on computationally intensive simulations (e.g., finite element analysis) or simple strategies such as gain scheduling (Åström and Wittenmark, 1989) and triple modular redundancy (Barron et al., 1990). In the present paper, we achieve controller reconfiguration capability by integrating an eigenstructure bidirectional associative memory into a model reference adaptive control framework. In a similar spirit, eigenstructure bidirectional associative memory can be applied to the control of slewing flexible multibody. The proposed architecture is expected to maintain stability for extended periods of time without external intervention, while possibly suffering from unforeseeable perturbations. The architecture of a reconfigurable control system is given in Figure 93.4.

The adaptive control framework handles slowly varying system parameters, which commonly occur on structures exposed to the adverse environment (e.g., increased thermal and aerodynamic load). Subsequently, as experience with the actual plant is accumulated, the learning system would be used to anticipate the appropriate control or model parameters as a function of the current plant operating condition. Catastrophic changes to the system dynamics are compensated for by retrieving an acceptable controller from a *continuous* solution space, which is created beforehand and reflects a host of various system configurations. The solution space is stored within a EBAM network as opposed to a look-up table and therefore offers the capabilities of *real-time reconfiguration* and *generalization* (see Figure 93.5). A look-up table approach would only provide

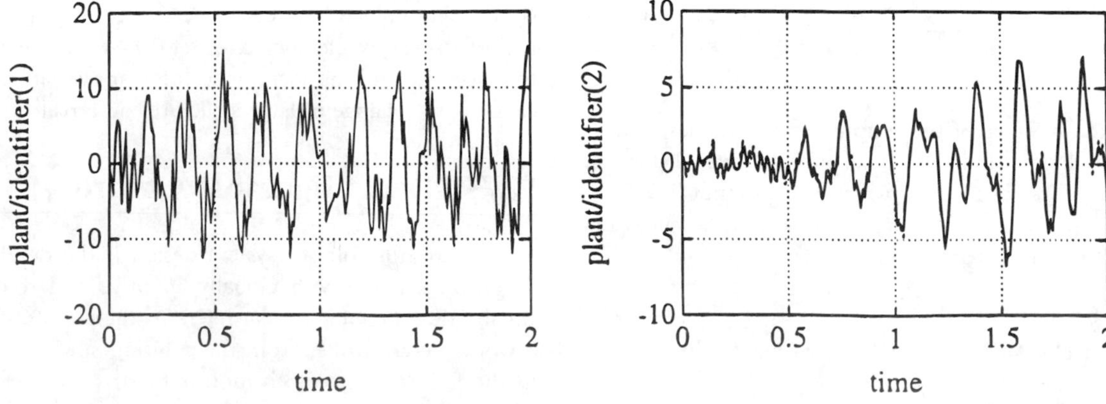

Figure 93.7 Open-loop responses (neural identifier vs. nonlinear plant).

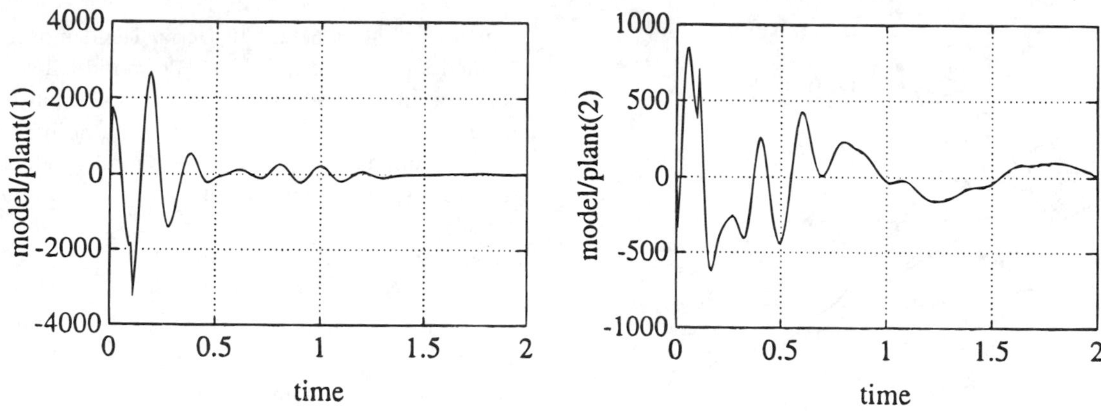

Figure 93.8 Closed-loop responses (reference model vs. nonlinear plant).

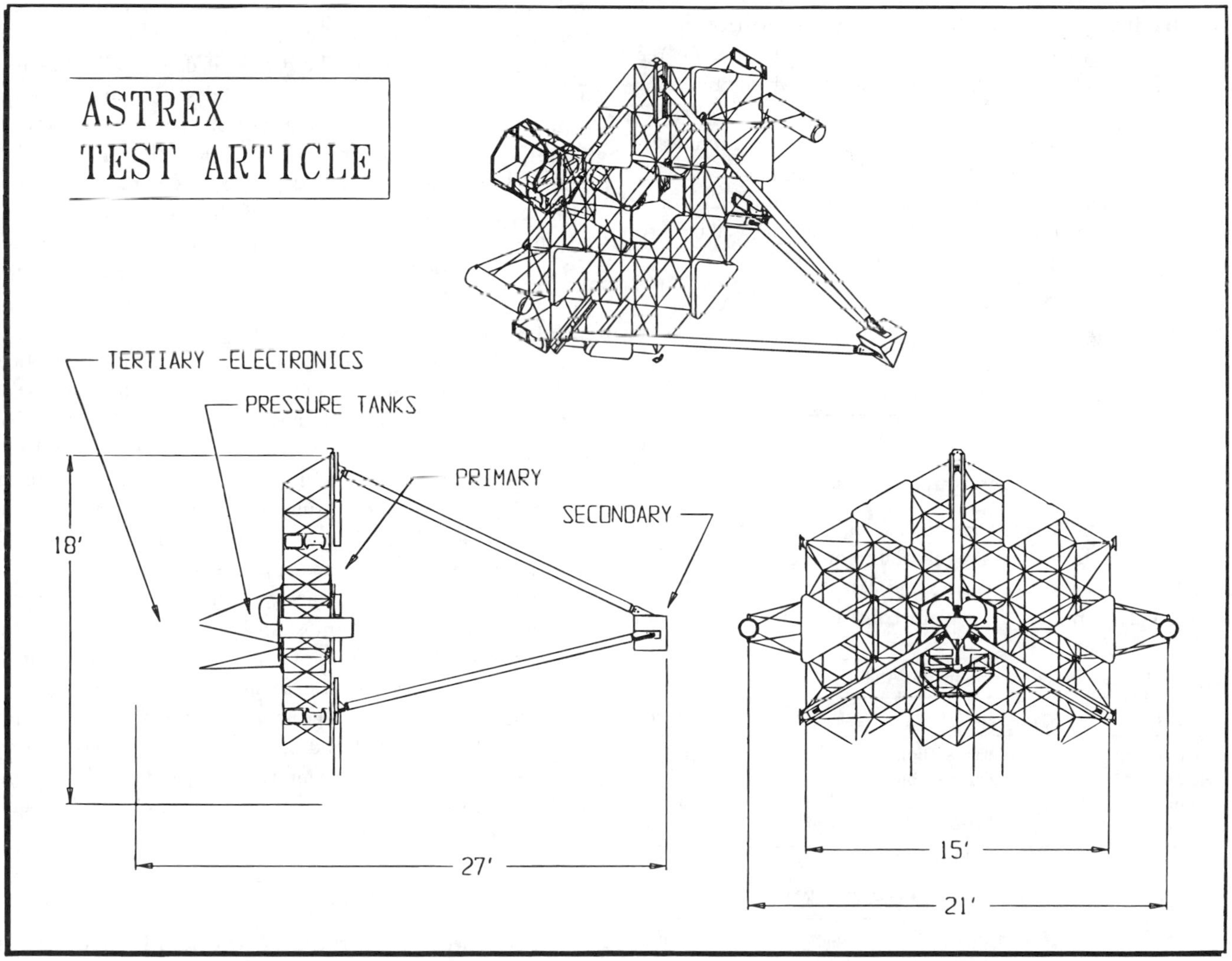

Figure 93.9 The Advanced Space structures Technology Research EXperiments (ASTREX) test article.

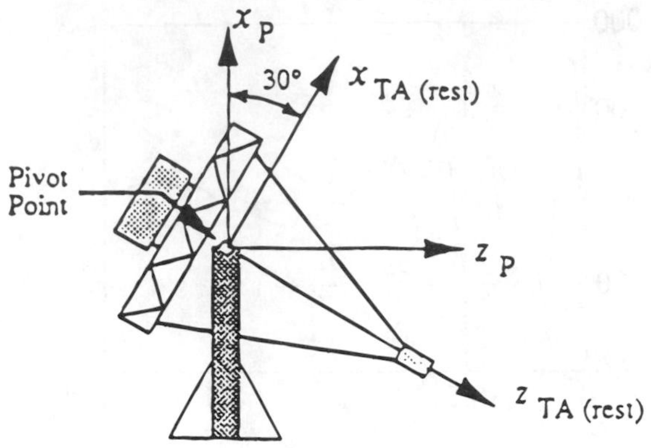

Figure 93.10 Reference frame for rigid-body motion model.

discrete controller solutions in a lengthy and sequential search. Reconfiguration capability proposed entails the design of a health monitoring procedure in detecting, isolating and identifying adverse conditions.

To achieve successful reconfiguration capabilities, we devise a reliable bidirectional associative memory synthesized by the eigenstructure decomposition method. As pointed out in, Yen and Michel (1992), the eigenstructure method, which utilizes the energy function approach, guarantees the storage of a given set of desired fault scenarios/weight configurations as asymptotically stable equilibria in the state space. The assumption is made that an acceptable fault detection and identification algorithm synthesized by the fuzzy back-propagation network will be used for health monitoring to provide the required information to the eigenstructure bidirectional associative memory (Yen and Kwak, 1993).

93.6 Simulation Studies

EXAMPLE 93.1 Generic large space structure

To simulate the characteristics of a large space structure, the plant is chosen to process low natural frequencies and damping as well as high modal density, and the actuators are chosen to be highly nonlinear. The plant consists of five modes with frequencies: 1, 4, 5, 6, and 10 Hertz. The damping ratio for all five modes is selected to be 0.15% of critical. Two sensors, two actuators, and ten states are used in this multi-input multi-output system. The eigenvectors are arbitrarily selected under the condition that they remain linearly independent. The actuators are chosen to exhibit a combination of saturation and exponentially decaying ripple. The input/output relationship is shown in Figure 93.6 and is given below

$$u(\nu) = \tan h(2\nu) + 0.1 \times e^{-|\nu|} \times \cos(4\pi\nu). \quad (93.18)$$

A compensator is trained so that the closed-loop system containing the nonlinear actuators and lightly damped plant emulates the linear, highly damped reference model. The five natural frequencies of the reference model were set equal to those of the plant. This is realistic in a practical sense because in many cases natural frequencies of spacecraft structures can be identified with reasonable accuracy by modal testing. However, it is much more difficult to accurately identify the eigenvectors (corresponding to the mode shapes). Therefore, the eigenvectors of the reference model were chosen arbitrarily and they were different from the eigenvectors of the plant. The degree of damping is chosen to be 10% of critical for each of the five modes. Prior to training the compensator, an adaptive time-delay Gaussian function network consisting of 40 hidden neurons with learning rates equal to 0.001 is trained to identify the open-loop system. The resulting neural identifier assists the training of the compensator (another adaptive time-delay Gaussian function network with 40 hidden neurons) by translating the plant output error to compensator output error. There are chosen to possess 4 time delays from each hidden neuron to each output neuron.

Figure 93.7 presents the performance of the neural identifier with respect to sensors 1 and 2, respectively, in response to random inputs for 2 seconds after training for 100 trials. Mean square error converged to 0.01. Within the scale of vertical axis, the plant output and the neural identifier output are indistinguishable. The simulation results show that the neural identifier has successfully emulated the structural dynamics of this generic space structure. Although the neural identifier learned to match the open-loop system very quickly, the neural compensator with learning rate 0.001 took almost an hours to converge to mean square error 0.01. The choice of a smaller learning rate ensures a monotonically decreasing mean square error in the LMS training. Figure 93.8 displays the closed-loop performance for 2 seconds with respect to sensors 1 and 2, respectively, in response to an impulse. Again, the reference model output and the plant output are indistinguishable. The neural controller has learned to damp out the vibration.

Table 93.1 Sensor Locations in the ASTREX Testbed

Type	Location	Node	Direction
accelerometer 1	secondary section	1	(1,0,0)
accelerometer 2	secondary section	1	(0,1,0)
accelerometer 3	tripod	1525	(1,0,0)
accelerometer 4	tripod	3525	(0,1,0)

Table 93.2 Actuator Locations in the ASTREX Testbed

Type	Location	Node	Direction
shaker	primary truss	62	(0.5,0,0.86)
proof mass 1	secondary section	462	(0.86,0.5,0)
proof mass 2	secondary section	461	(−0.86,−0.5,0)
proof mass 3	secondary section	459	(0,1,0)

EXAMPLE 93.2 ASTREX plant

The Advanced Space structures Technology Research EXperiments (ASTREX), currently located at the Phillips Laboratory, Edwards Air Force Base, is a testbed equipped with 3-mirror Space-based laser beam expander to develop, test and validate control strategies for large space structures (Abhyankar et al.,

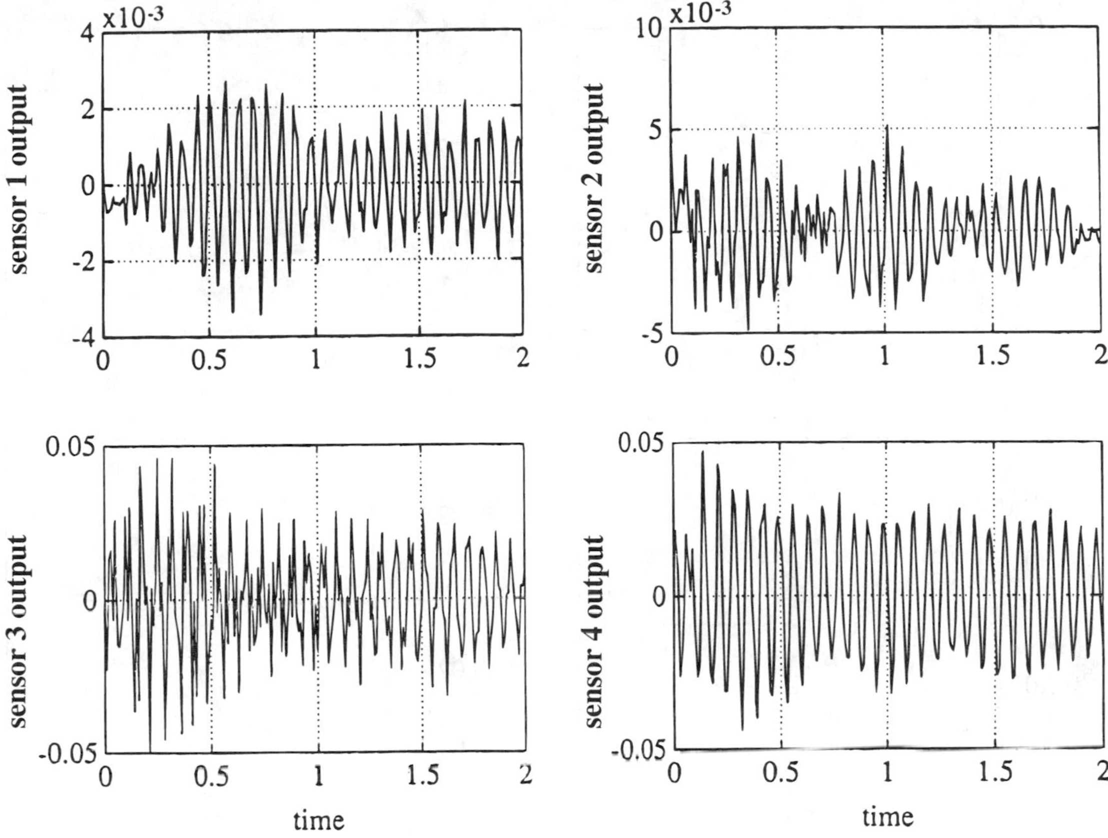

Figure 93.11 Open-loop responses of sensors 1, 2, 3, and 4 (neural identifier vs. nonlinear plant).

1992). The unique features of the experimental facility include a three-axis large angle slewing maneuver capability and active tripod members with embedded piezoelectric sensors and actuators. The slewing and vibration control can be achieved with a set of reaction control thrusters, a reaction wheel, active members, control moment gyros, and linear precision actuators. The test article allows three degrees of rigid body freedom, $\pm 20^\circ$ in pitch and roll and $\pm 180^\circ$ in yaw. A dedicated control and data acquisition computer is used to command and control the operations. This test article has provided a great challenge for researchers from academia and industry to implement the control strategies to maneuver and to achieve retargeting or vibration suppression. The test structure is shown in Figure 93.9.

The test article itself consists of three major sections:

1. *The Primary Structure* is a 5.5-meter diameter truss constructed of over 100-cm diameter graphite epoxy tubes with aluminum end fitting that are attached to star node connections. The primary structure includes six sets of steel plates mounted on its surface to simulate the primary mirror and two cylindrical masses mounted on its sides to simulate tracker telescopes. A pair of 30 gallon air tanks are attached inside the hub directly above the air-bearing system.
2. *The Secondary Structure* is a triangular structure which houses the reaction wheel actuators and the mass designed to simulate the secondary mirror. It is connected to the primary truss by a tripod arrangement of three 5.1 meter graphite epoxy tubes manufactured with embedded sensors and actuators.
3. *The Tertiary Structure* is a structure designed to hold the electronics and power supply for the data acquisition and control system and other masses to balance the secondary mirror.

The finite element model (FEM) of the entire testbed consists of approximately 615 nodes and over 1000 elements. Even though the FEM has been constantly modified based on the detailed modal survey, it is not considered as an accurate dynamic model. The complicated factors in this control design problem are lack of an accurate dynamic model, nonlinear thruster characteristics, and nonlinear aerodynamic effects. In the rigid-body motion model, two reference frames are employed. The base pedestal axis is an inertially fixed reference frame which points in the true vertical and true horizon plane. The ASTREX rest position is pitch down 30° in this coordinate system. The test article axis is the body-fixed reference frame. As shown in Figure 93.10, the origin for both systems is the pivot point, the location where the test article is attached to the base pedestal at the air bearing. Modeling of the physical structure is implemented by a FEM formatted as a NASTRAN data deck. The dynamical modal equation is given by

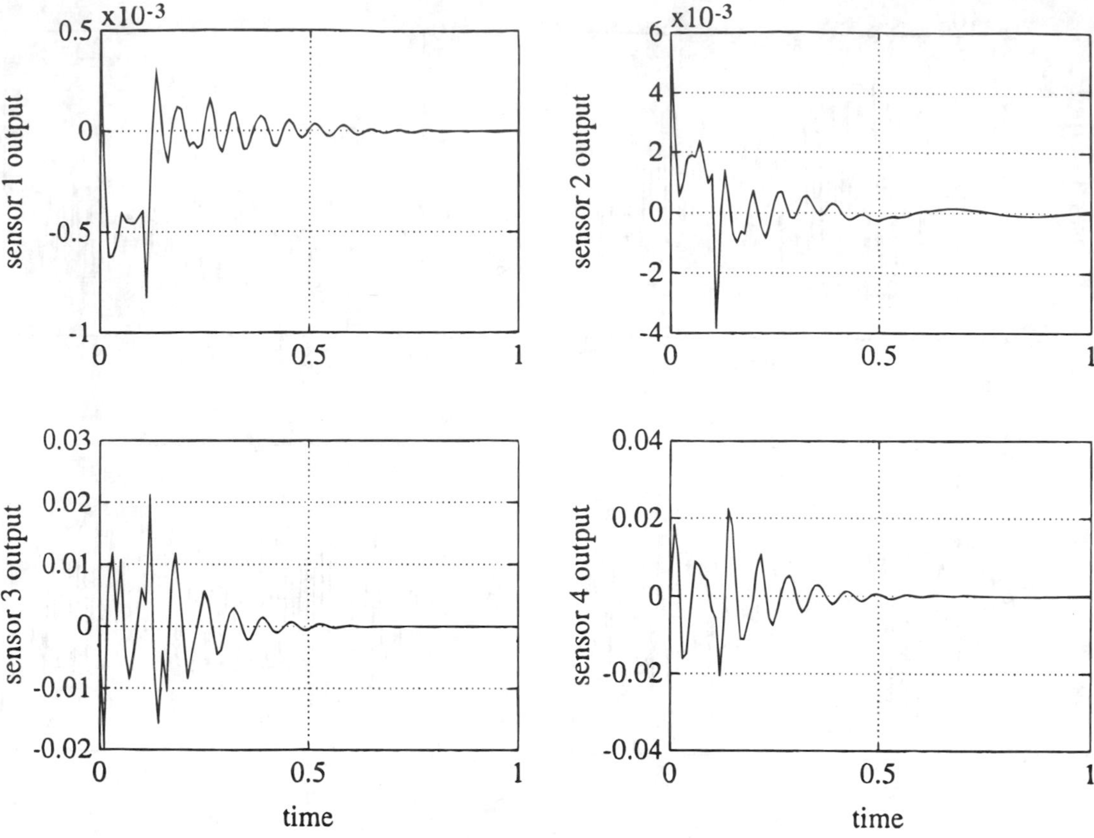

Figure 93.12 Closed-loop responses of sensors 1, 2, 3, and 4 (reference model vs. nonlinear plant).

$$M\ddot{x} + E\dot{x} + Kx = f, \tag{93.19}$$

where M is the mass matrix, E denotes the viscous damping matrix, K is the stiffness matrix, x is a vector representing the physical degrees of freedom, and f is the force vector applied to structure. Through a mass normalization procedure on the modal matrix, the state space model of ASTREX can be obtained

$$\dot{x} = Ax + Bu + Dw, \tag{93.20a}$$

$$y = Cx, \tag{93.20b}$$

$$z = Mx + Hu, \tag{93.20c}$$

where A, B, C, D, M, and H are constant matrices, and x, u, w, y and z denote state, input, noise, output, and measurement vectors, respectively. The data required for the system identification is obtained from accelerometers and thrusters through finite element analysis simulations. The locations for accelerometers are carefully selected based on expectations of capturing all the relevant structural modes. For simplicity, only four accelerometers and four actuators as described in Tables 93.1 and 93.2 are used for this preliminary study. Three linear precision proof mass actuators are situated on the front of the secondary structure, in the direction parallel to the edges of test article axis.

System identification is simulated by an adaptive time-delay Gaussian function network with 100 hidden neurons, while vibration suppression is performed by another adaptive time-delay Gaussian function network with 100 hidden neurons. The closed-loop controller regulates the dynamics of the ASTREX structure to follow a linearly and highly damped reference model in which the degree of damping is chosen to be 10% of critical for all modes. The five natural frequencies of the reference model were determined based upon modal test results. The eigenvectors of the reference model were arbitrarily selected under the condition that they remain linearly independent. Both the neural identifier and the neural controller with learning rate 0.01 took roughly five hours to converge to mean square error 0.01. Six time delays are used in each pair of neurons from the hidden layer to the output layer. Open-loop responses of sensors 1, 2, 3, and 4 for random inputs are given in Figure 93.11, while the closed-loop performance of sensors 1, 2, 3, and 4 are displayed in Figure 93.12 in response to an impulse.

Three possible configurations are simulated based on different fault scenarios (i.e., no fault, fault condition 1, and fault condition 2). A fault diagnosis system synthesized by a fuzzy back-propagation network is performed by mapping patterns of input sensors to damage indices of line-of-sight errors that represent fault conditions. Angular rate sensors are used at different locations for line-of-sight error measurements where failure scenarios may be evolutionary varying or catastrophically changing. Figure 93.13 shows that for each fault condition, the outputs exhibit distinct thresholds crossing from the no fault region to fault

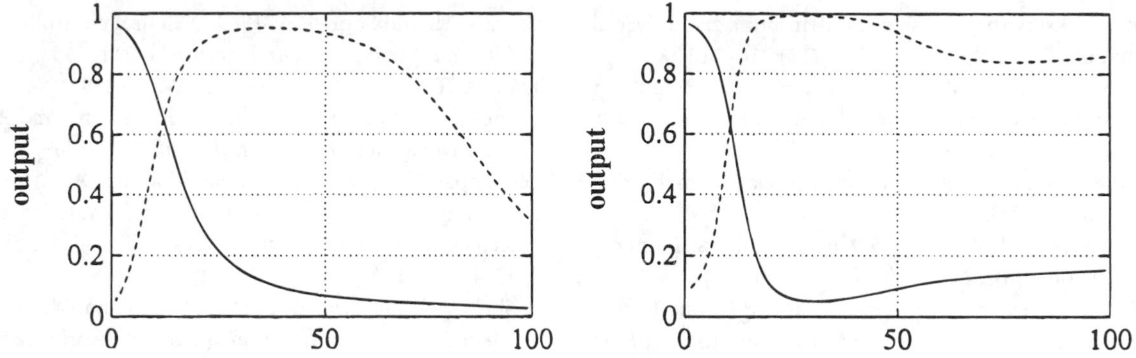

Figure 93.13a Fault condition 1 test (—: no fault; ...: fault 1). **Figure 93.13b** Fault condition 2 test (—: no fault; ...: fault 2).

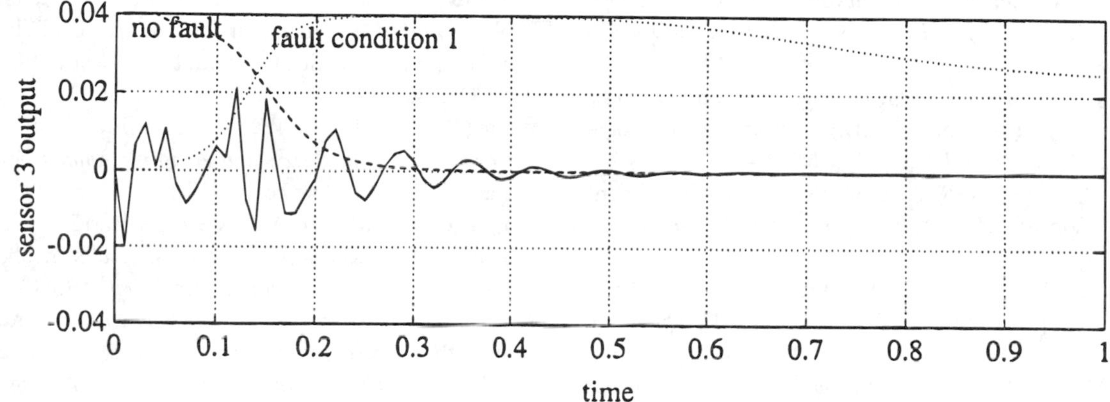

Figure 93.14 Neural reconfigurable control with fault detection and identification.

regions. The eigenstructure bidirectional associative memory, which is created prior to dynamic simulation, provides a probability for decision making based on the information derived from the fuzzy FDI network. Figure 93.14 displays, the closed-loop reconfiguration performance of sensor 3 when the neural controller switches from the no fault region to fault condition 1.

93.7 Conclusion

The architecture proposed for autonomous neural control successfully demonstrates the feasibility and flexibility of connectionist learning systems for flexible space structures. The salient features associated with the proposed reconfigurable learning controls are discussed. In addition, a real-time autonomous control system is made possible to accommodate uncertainty through on-line interaction with the nonlinear structures. In a similar spirit, the proposed architecture can be extended to the dynamic control of aeropropulsion engines, underwater vehicles, chemical processes, power plants, and manufacturing scheduling.

References

Abhyankar, N. S., Ramakrishnan, J., Byun, K. W., Das, A., Cossey, F., and Berg, J. L. 1992. Modeling, system identification and control of ASTREX, *Proc. NASA/DoD Control Structures Interaction Technology Conference,* 727–750.

Åström, K. J. and Wittenmark, B. 1989. *Adaptive Control,* Addison-Wesley, Reading, PA.

Barron, R. L., Cellucci, R. L., Jordan, P. R., Beam, N. E., Hess, P., and Barron, A. R. 1990. Applications of polynomial neural networks to FDIE and reconfigurable flight control, *Proc. IEEE Nat. Aerospace and Electronics Conf.,* 507–519.

Bunch, J. R. and Nielsen, C. P. 1978. Rank-one modification of the symmetric eigenproblem, *Numerische Mathematik,* 31:31–40.

Chin, H., Danai, K., and Lewicki, D. G. 1993. Fault Detection of Helicopter Gearboxes using the Multi-valued Influence Matrix Method, *NASA Technical Memo 106100.*

Chou, C. L. and Kuo, R. J. 1993. On-line diagnosis for turbine blade faults using neural network, *Proc. INNS World Congress on Neural Network,* IV–55–58.

De Carli, A. 1994. NeurO-fuzzy control, *Control Engineering Practice,* 2(1):1–153.

Elsey, R. 1988. A learning architecture for control based on back-propagation neural network, *Proc. IEEE Int. Conf. Neural Networks,* 587–594.

Fu, K. S. 1970. Learning control systems—review and outlook, *IEEE Trans. Auto. Control,* 15(2):210–221.

Guo, Z. C. and Uhrig, R. E. 1992. Using Modular Neural Networks to Monitor Accident Conditions in Nuclear Power Plants, *Proc. SPIE Conf. Applications of Artificial Neural Networks III,* 505–516.

Hartman, E. J., Keeler, J. D., and Kowalski, J. M. 1990. Layered neural networks with Gaussian hidden units as universal approximations, *Neural Computation,* 2(2):210–215.

Hornik, K., Stinchcombe, M., and White, H. 1989. Multilayer feedforward networks are universal approximators, *Neural Networks,* 2(5):359–366.

Kosko, B. 1987. Adaptive bidirectional associative memories, *Applied Optics,* 26(23):4947–4960.

Kosko, B. 1992. *Neural Networks and Fuzzy Systems,* Prentice-Hall, Englewood Cliffs, NJ.

Leem, C. S. and Dreyfus, S. E. 1992. Learning input feature selection for sensor fusion in tool wear monitoring, *Proc. Artificial Neural Networks in Engineering Conf.,* 815–820.

Lin, D. T., Daynoff, J. E., and Ligomenides, P. A. 1992. Adaptive time-delay neural network for temporal correlation and prediction, *Proc. SPIE Conf. Biological, Neural Net, and 3-D Methods,* 170–181.

Michel, A. N., Si, J., and Yen, G. G. 1991. Analysis and synthesis of a class of discrete-time neural networks described on hypercubes, *IEEE Trans. Neural Networks,* 2(1):32–46.

Michie, D. and Chambers, R. A. 1968. BOXES: an Experiment in Adaptive Control, *Machine Intelligence,* Dale, E. and Michie, D. eds. 137–152.

Moody, J. and Darken, C. J. 1989. Fast learning in networks of locally-tuned processing units, *Neural Computation,* 1(2):81–294.

Napolitano, M. R., Chen, C. I., and Naylor, S. 1993. Aircraft failure detection and identification using neural networks, *AIAA J. Guidance, Control, and Dynamics,* 16(6):999–1009.

Narendra, K. S. and Parthasarthy, K. 1990. Identification and control of dynamical systems using neural network, *IEEE Trans. Neural Networks,* 1(1):4–27.

Psaltis, D., Sideris, A., and Yamamura, A. A. 1988. A multilayered neural network controller, *IEEE Control Systems Magazine,* 8(3):17–21.

Saerens, M. and Soquet, A. 1989. A neural controller, *Proc. IEEE Int. Conf. Artificial Neural Networks,* 211–215.

Troudet, T., Garg, S., and Merrill, W. C. 1991. Neural network application to aircraft control system design, *Proc. AIAA Guidance, Navigation, and Control Conf.,* 993–1009.

Venugopal, K. P., Sudhakar, R., and Pandya, A. S. 1992. On-line learning control of autonomous underwater vehicles using feedforward neural networks, *IEEE J. Oceanic Eng.,* 17(4):308–319.

Waibel, A., Hanazawa, T., Hinton, G., Shikano, K., and Lang, K. 1988. Phoneme recognition: neural networks versus hidden markov models, *Proc. IEEE Int. Conf. Acoustics, Speech and Signal Processing,* 107–110.

Watanabe, K., Matsuura, I., Abe, M., Kubota, M., and Himmelblau, D. M. 1989. Incipient fault diagnosis of chemical processes via artificial neural networks, *AIChE J.,* 35(11):1803–1812.

White, D. A. and Sofge, D. A. 1992. *Handbook of Intelligent Control—Neural, Fuzzy, and Adaptive Approaches,* Van Nostrand Reinhold, New York, NY.

Yen, G. G. and Michel, A. N. 1991. A learning and forgetting algorithm in associative memories: results involving pseudo inverses, *IEEE Trans. Circuits and Systems,* 38(10):1193–1205.

Yen, G. G. and Michel, A. N. 1992. A learning and forgetting algorithm in associative memories: the eigenstructure method, *IEEE Trans. Circuits and Systems, Part II: Analog and Digital Signal Processing,* 39(4):212–225.

Yen, G. G. and Kwak, M. K. 1993. Neural network approach for the damage detection of structures, *Proc. AIAA/ASME/ASCE/AHS/ASC Structures, Structural Dynamics, and Material Conf.,* 1549–1555.

Yen, G. G. 1994. Stability analysis and synthesis algorithm of discrete-time bidirectional associative memories, *Proc. IEEE Conf. Neural Networks,* 1038–1043.

94

Fuzzy Pattern Recognition

94.1 Introductory Remarks—Pattern Recognition in the Framework of Fuzzy Sets 1207
94.2 The General Methodological Structure of Fuzzy Modeling 1208
94.3 Formation of the Feature Space 1209
Main Properties of the Input Interface • Representing and Processing Uncertainty in Patterns • Optimization of the Linguistic Feature Space
94.4 Implicit and Explicit Knowledge Representation in Pattern Recognition 1212
94.5 From Supervised to Unsupervised Pattern Recognition—A Continuum of Classification Models 1213
94.6 Fuzzy Neural Structures 1213
Aggregative and Referential Functions of the Fuzzy Neurons • Fuzzy Neurons with Feedback • Learning in the Fuzzy Neural Networks
94.7 Supervised Learning 1218
Logic Processors as a Generic Classification Architecture • Pseudomedian Filtering • Modeling Spatial Relationships in High-Level Computer Vision • Fuzzy Perceptron • Nearest—Neighbor Classification Algorithm Realized as a Fuzzy Neural Network • Hierarchical Pattern Classifiers
94.8 Implicitly Supervised Pattern Recognition 1223
Problem Statement • The General Architecture • The Design of the Classifier
94.9 Unsupervised Learning 1225
FUZZY ISODATA Clustering Algorithm • Directional Clustering and Its Role in System Identification

Witold Pedrycz
University of Manitoba, Winnipeg

94.1 Introductory Remarks—Pattern Recognition in the Framework of Fuzzy Sets

It is perceivable that fuzzy sets have put pattern recognition into a new and a broader perspective by developing an innovative methodological and algorithmic framework to cope with complex and ill-defined systems. As lucidly pointed out by L. A. Zadeh in one of his early papers on pattern recognition (Zadeh, 1977), the fundamental role of fuzzy sets is to make the *opaque* classification schemes, as usually being used by a human, *transparent* by developing a formal, computer—implementable framework. In this context, it is indispensable to identify a general role which fuzzy sets play in the design of intelligent systems. In general, the primary role of fuzzy sets is to help deploy a qualitative domain knowledge about a classification task onto the relevant algorithmic structure (Bellman et al., 1966). For instance, a vast number of successful applications of fuzzy controllers (Pedrycz, 1992, 1995) hinge on the premise that a substantial amount of experiential knowledge could be efficiently downloaded onto the rule-based architecture of the controller. In pattern recognition, as many of its classification algorithms are highly interactive, the role of the domain knowledge and its successful utilization becomes vital to the success of the classifier. The qualitative domain knowledge, if properly represented, constitutes a fundamental ingredient of any advanced and successful topology of the classifier. This does not imply that numerical data (patterns) are not essential. To the contrary; the main point is that all pieces of classification knowledge should be utilized as efficiently as possible by putting their processing and utilizing in the most appropriate information processing framework.

What is the most visible and profound role of fuzzy sets in pattern recognition? The pertinent general features can be enumerated accordingly:

- Fuzzy sets offer an important possibility to develop *explicit* rather than *implicit* classification schemes; quite often the classifiers designed with the aim of fuzzy set technology are logic-oriented, naturally reflecting the conceptual layout of the classification problem. Interestingly enough, owing to the logic-based format of these classifiers, fuzzy system modelling and dynamic pattern recognition become uniform to a high extent.

0-8493-8343-9/97/$0.00+$.50

- The degrees of class membership are no longer binary values (inducing a straightforward yes-no quantification) and can be treated as membership degrees distributed within the unit interval. This conceptual extension definitely expands the range of the classification problems that naturally fall under this conceptual umbrella. Intermediate membership degrees help identify the most "unclear" (doubtful) patterns; this aspect becomes of a primordial importance in unsupervised learning. In fact, many methods of fuzzy clustering vigorously exploit this feature.
- As a consequence of the use of continuous grades of membership, a traditional dichotomy of supervised versus unsupervised classification is substantially enriched. This rises by admitting a series of schemes employing *implicit* rather than *explicit* pattern labeling or allowing for a portion of the patterns to be labeled. This crucial observation about a continuous transition between supervised and unsupervised pattern recognition will be reflected in the presentation of the overall material of the paper.

From a methodological standpoint, an exposition of the material is top-down, meaning that we will proceed first with the general concepts afterwards converting them into more operational classification models. The exposition of the material concentrates primarily on the applied and algorithmic aspects of fuzzy pattern recognition; the more theoretical material can be readily accessed through a series of easily available references (Bezdek and Pal, 1992; Pedrycz, 1990; Pal and Dutta Majumder, 1986; Kandel, 1982). We start off by placing fuzzy classification in a general frame of fuzzy modelling which allows us to emphasize the key aspects of fuzzy information processing. A notion of a linguistic feature space operating at the level of logical variables is discussed first. The discussion is augmented by some design methods. By contrasting between mechanisms of implicit and explicit knowledge representation, we are able to point out a main role played by the fuzzy set technology in pattern recognition. Fuzzy neural networks regarded as generic computational structures will be then used in various classification schemes including pseudomedian filtering and nearest neighbor classifier. The mechanisms of unsupervised learning are concentrated on FUZZY ISODATA and its enhancements such as clustering under partial supervision. The algorithm of implicitly supervised pattern recognition comes as a representative example of all classification models spread between the models coping with situations of completely supervised and completely unsupervised learning. The organization of the paper reflects the way in which fuzzy sets have been utilized in the realm of pattern recognition. Some of the ideas that are well-known in the literature, are exposed here in a new unified framework. Some other concepts including those of implicit learning are novel. The paper is self-contained and does not assume any in-depth fuzzy sets prerequisites. Some necessary mathematical background dealing primarily with the calculus of triangular norm is summarized in Appendix A.

94.2 The General Methodological Structure of Fuzzy Modeling

Before getting into the concepts of pattern recognition, it is worth reviewing the general methodology of modeling with fuzzy sets as this applies equally well to this area. The key observation is that the resulting algorithms (such, as for instance, fuzzy classifiers or fuzzy controllers) have a lot in common as all of them operate at the abstract set-theoretic level rather than that exhibiting a purely numerical ground of precisely quantified variables. Bearing in mind that quite often the modeling environment generates an abundance of numerical data and requires numerical actions to affect it, we will concentrate our discussion on a general framework of fuzzy modeling equipped with the suitable interfaces. This general three-phase scheme supporting the functions of knowledge representation and knowledge processing is schematically shown in Figure 94.1, cf also Pedrycz (1995).

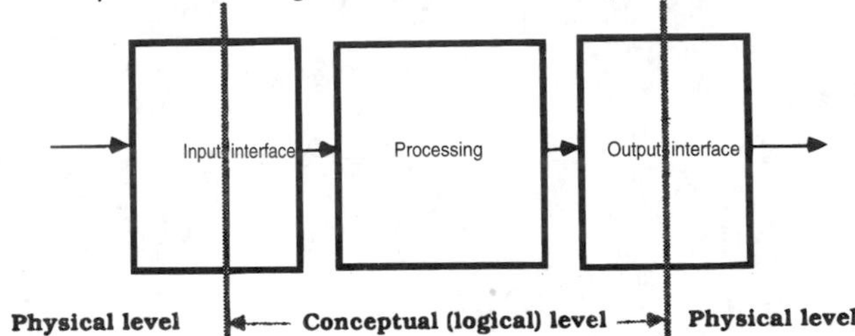

Figure 94.1 A general scheme of fuzzy modeling.

The overall scheme works accordingly. The collected data and facts about a certain application (environment) are transformed (elevated) from the usual numerical level into a conceptual (logical) level formed by fuzzy sets. This phase is completed with the aid of the input interface. Fuzzy sets (linguistic labels) included here, play a crucial role in placing all the available pieces of information in a certain suitable cognitive perspective (processing context). The algorithmic part of the scheme (being the second module of the architecture) is responsible for all necessary processing faculties. It should be noted that this processing is carried out at the level of fuzzy sets. Finally, the results produced by this functional block are converted by the output interface.

Let us briefly summarize the roles and functions of the main components of this scheme in more detail. As said, the aim of the input interface is to convert the input data into a suitable format required for the internal model. More specialized tasks to be realized there pertain to:

- Specification of model requirements aimed at achieving a rational generality-relevance trade-off within the model that complies with the general guidelines of the principle of incompatibility.
- Development of a relevant frame of cognition (fuzzy partition) with a special emphasis focused on a proper assignment of information granularity.
- Conversion of heterogenous input data (including numerical quantities, intervals etc.) into a coherent internal format.

- Expressing data incompleteness.
- Performing spatial and temporal fuzzy filtering.

The role of the processing block is in the revealing of the relationships between the linguistic entities of the system's variables. The tasks to be solved there include:

- Definition of a structure of the model.
- Parameter estimation of the structure of the model.
- Model verification.

Finally, the role of the output interface is to transform the results of modeling being available in their internal format characteristic for the processing phase, to the format that is acceptable at the environment level; quite often the results are transformed down to plain numerical quantities.

The underlying tasks include:

- Linguistic-numerical processing.
- Linguistic approximation.

In pattern recognition, the same general architecture can be employed (see Figure 94.2.) The processing module becomes a classifier itself operating at the logical level. The input interface is linked with the phase of feature formation. The output interface is mainly reduced to the methods of class membership interpretation.

94.3 Formation of the Feature Space

Main Properties of the Input Interface

The knowledge about the patterns as well as the perspective from which one is interested to take a look at them, is articulated with the aid of linguistic labels. The linguistic labels constitute generic pieces of knowledge which are conceived by the user as essential in describing a correspondence between the patterns and classes. The information granularity derived in this fashion implies that the use of more detailed pieces of information becomes superfluous in representing knowledge about the classification problem.

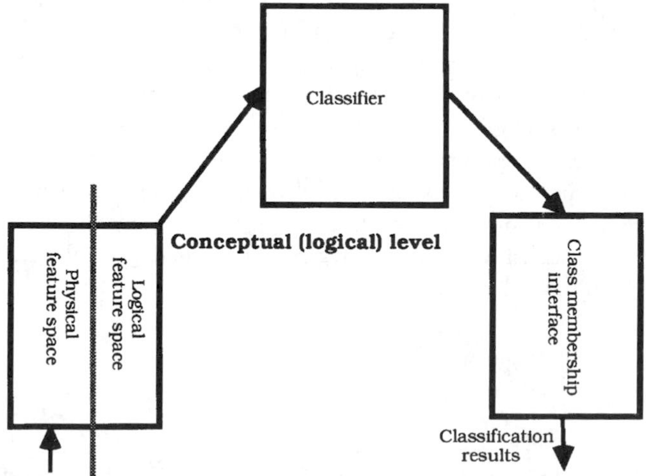

Figure 94.2 General topology of classification with fuzzy sets.

The linguistic labels are treated as fuzzy sets. As demonstrated in Zadeh (1978, 1983), they can be also viewed as elastic constraints defined over the feature space. Their main function is to highlight the elements with the highest degrees of compatibility with the specified linguistic term. Quite often the linguistic labels are also referred to as information granules (Pedrycz, 1992a). All the information granules defined in the same feature space and analyzed together constitute a frame of cognition of this variable (Pedrycz, 1992b). More formally, the family of fuzzy sets

$$A = \{A_1, A_2, \ldots, A_c\}$$

(where A_i: $X \to [0, 1]$) constitutes a frame of cognition A if the following properties of semantic integrity are fulfilled:

- A "covers" universe X, namely each element of the universe is assigned to at least one granule with a nonzero degree of membership.

  $$\forall x \in X \, \exists i \quad A_i(x) > 0$$

  As it will become obvious in the course of the discussion, the coverage property assures that any piece of information (new feature value, no matter whether numerical or nonnumerical) defined in X can be sufficiently represented in terms of A_is.
- The elements of A are unimodal fuzzy sets. By stating that we identify several regions of $\mathbf{X}$ (one for each A_i) that are highly compatible with the labels (i.e., with significantly high grades of membership in A_i). The regions defined in this way are characterized by a well-defined semantics.

Overall, we can contemplate the frame of cognition as a certain systematic way of looking at the classification problem from the most suitable perspective, leading in this way towards customizing the fuzzy classifier by the individual user to fully accommodate her/his modelling requirements.

From a pragmatic point of view, we will distinguish two distinct approaches commonly found while determining the components of the frame of cognition.

1. The linguistic labels can be specified by studying the problem and recognizing relevant information granules being necessary in its description. In this way the subjective evaluation of the membership functions completed by the user of the model becomes essential to the problem of knowledge elicitation. It is the user who provides relevant membership functions for the variables of the system reflecting his individual cognitive perspective. In this regard some standard methods of membership function estimation are fully applicable (Pedrycz, 1992).
2. The second approach, which could be helpful in most cases when some records of numerical data are available, takes advantage of fuzzy clustering techniques and FUZZY ISODATA. We will study these ideas in Section 94.9.

Quite often, the frame of cognition A can be also referred to as a fuzzy partition of X[1].

The requirements pertaining to the frame of cognition as listed above give rise to the following characterization:

- Specificity of the frame of cognition. We say that the frame of cognition A' is more specific than A if all the elements of A' are more specific than the elements of A. The measure of specificity of fuzzy sets can be introduced in several ways; refer e.g., to Dubois and Prade (1988). Higher specificity of A' entails that the number of elements of A' is greater than the number of the labels in A.

For instance, the frame

$$A = \{\text{negative, zero, positive}\}$$

distinguishing between three linguistic landmarks of the feature is less specific than the frame

$$A' = \{\text{Negative Large, Negative Medium, Negative Small, Zero, Positive Small, Positive Medium, Positive Large}\}$$

in which the same feature takes on more levels of the linguistic quantification. The partition A' is less general than the previous one. Or put this equivalently, the information granularity of A' is finer than that of A.

- Information hiding of the frame of cognition refers to each element of A. The essence of this property is such that some elements of X are made nondistinguishable (equivalent) as characterized by the same level of membership in the partition.

By defining the membership function and its specific alpha-cuts, we selectively hide the information about the elements situated within the corresponding interval, see Figure 94.3. In other words, in the context formed by this granule A, there is no further distinction between elements a_1 and a_2, $a_1, a_2 \in X$, as far as both of them are contained in the same alpha-cut.

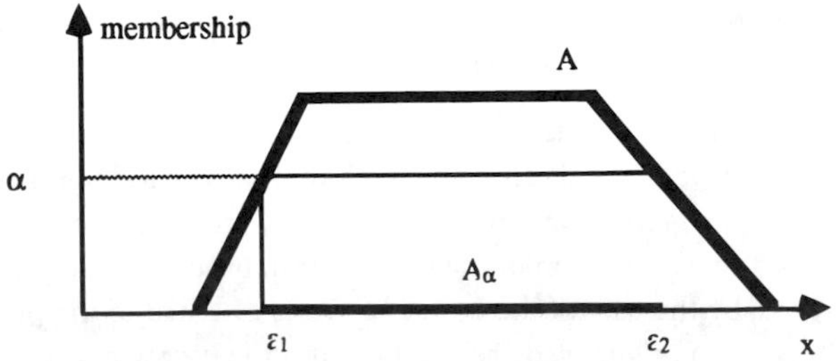

Figure 94.3 Fuzzy set A, its alpha-cuts and their role in information hiding.

Information hiding is completed on purpose so that all the computations carried out by the classifier do not involve details emerging below the conceptual level defined by the threshold α. Obviously, this significantly reduces a potential computational burden caused by handling too many details at the level of precise numerical processing. Fuzzy sets parameterize the concept of information hiding by admitting various grades of membership in quantification of this process. In other words, a certain α-cut completes information hiding at this particular level.

Representing and Processing Uncertainty in Patterns

Situations could emerge in which some of the patterns might not be fully specified or could be described with a certain precision. To express this fact, the input interface should be designed in a way it is capable of quantifying the factor uncertainty. The use of possibility and necessity measures is one of the alternatives worth studying.

Let A denote a fuzzy set viewed as a reference. Any input datum X (despite its character) is then "translated" into an internal logical format by computing the possibility and necessity measures computed with respect to A (Dubois and Prade, 1988; Zadeh, 1978).

$$\text{Poss}(X|A) = \sup_{x \in X} [\min(X(x), A(x))]$$

$$\text{Nec}(X|A) = \inf_{x \in X} [\max(1 - X(x)), A(x))]$$

The basic properties of these measures have been thoroughly studied in the literature; for more details the reader can refer to Dubois and Prade (1988). Note that the possibility measure evaluates a degree of overlap of X and A while the necessity measure is involved in expressing a degree of inclusion of X in A, see Figure 94.4.

In particular, the fundamental inequality states that Poss $(X/A) \geq \text{Nec}(X/A)$. These two generic definitions can be immediately generalized by replacing the lattice (max and min) operators used above by triangular norms. This approach is useful in capturing a global aspect of evaluation of the above matching properties. Observe that due to the sup and inf operations, the previous expressions are noininteractive and the final numerical results produced these depend solely upon a single element of the universe of discourse—thus, the aggregation operations are rather insensitive with this regard. The sound generalization would be of the type (Pedrycz, 1993)

$$\text{Poss}(X|A) = \underset{x \in X}{S} (X(x)\ tA(x)$$

$$\text{Nec}(X|A) = \underset{x \in X}{T} ((1 - X(x))\ sA(x))$$

that involves the s-t and t-s composition, respectively. One can also study the sup-t or inf-s aggregation which implies interactivity at a "local" level of the individual elements of the universe of discourse. This gives rise to the expressions

[1] Notice that the fuzzy partition satisfies an additional property stating that $A_1(x) + A_2(x) + \cdots + A_c(x) = 1$ which holds for any x; this constraint is automatically preserved by the FUZZY ISODATA, however the first method used in the formation of A may violate this requirement.

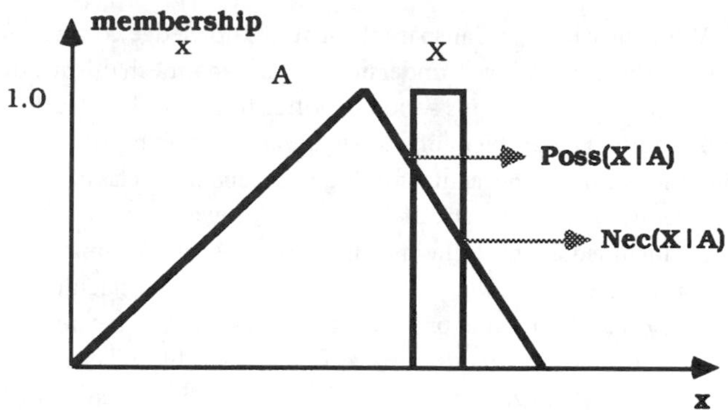

Figure 94.4 Fuzzy sets A and X and their possibility and necessity measures.

$$\text{Poss}(X|A) = \sup_{x \in X} (X(x)\ tA(x))$$

$$\text{Nec}(X|A) = \inf_{x \in X} ((1 - X(x))\ sA(x))$$

The possibility and necessity measures processed together can be useful in handling uncertainty, in particular the aspects of ignorance and conflict manifested in the available value X of the feature. Again, these two notions are context-dependent and as such should be analyzed with respect to the given fuzzy set A. Furthermore, the aspect of context-dependency implies that the numerical qualifications of these phenomena depend upon the environment (the frame of cognition) within which they are embedded. Let us define two indices

$$\lambda = \text{Poss}(X|A)$$

$$\xi = 1 - \text{Nec}(X|A)$$

used to express the relationships between X and A. For a pointwise (numerical) character of X, $X = \{x_0\}$, the quantities λ and ξ are linked together via a straightforward relationship

$$\lambda + \xi = 1$$

In general, when the feature is of a nonpointwise character, then we end up getting one of these inequalities

$$\lambda + \xi < 1, \qquad \lambda + \xi > 1$$

These cases are worth studying since they tackle the situations including information *ignorance* and *conflict:*

- Let $\lambda + \xi > 1$ that can be expressed as $\lambda + \xi = 1 + \gamma$ where $\gamma \in [0,1]$. The higher the value of γ, the higher the level of conflict emerging out of X evaluated in the context of A; γ denotes the level of conflict.
- The case in which $\lambda + \xi < 1$, with $\lambda + \xi = 1 - \gamma$, $\gamma \in [0, 1]$, articulates a situation of ignorance arising from expressing X via A. More precisely, γ is utilized to express this level of ignorance.

Optimization of the Linguistic Feature Space

In addition to the previous information-theoretic aspects of the linguistic labels discussed in the first part of Section 94.3, an intriguing question arises concerning the use of the labels in the optimization of the feature space. Additionally, the optimization should be general enough to be valid regardless of the specific functional form of the classifier being used in the classification problem. The issue raised now is different from a well known standard problem of feature selection as encountered in the existing literature (Fukunaga, 1991; Devijer and Kitler, 1982; Tou and Gonzalez, 1974; Duda and Hart, 1973). While the problem there deals with feature selection or aggregation, we are focused on the specification of the linguistic labels to be defined in the given (physical) feature space with a clear intent of supporting the design of the classifier itself. The potential is there: as already highlighted very clearly, fuzzy sets could be looked at as a vehicle furnishing with a nonlinear feature normalization (realizing a mapping from reals to the unit interval). The optimization problem leading to the formation of the feature space is guided by structure-free criterion (namely, the criterion should not be confined to any particular form of the classifier). Consider a collection of N labeled patterns distributed, in R^n; the pattern involve "c" classes. The collection of the linguistic labels is viewed as fuzzy relations defined again in R^n, A_i, $i = 1, 2, \ldots, p$. The underlying optimization criterion is motivated by the following interpretation. Let be two patterns, x and x', along with their class assignment ω and ω', respectively. Roughly speaking, when considering the pair of these patterns with conjunction of their class membership, four qualitatively distinct situations could occur as summarized below,

Class Membership

Features of Patterns	Similar	Different
Similar	+	?
Different	+	+

All the entries of the table but one (marked by ?) do not require any modification of the feature space as the classifier could easily cope with these cases. For this specific entry, two pattern with (almost) the same values of the features are very distinct classwise. The role of the linguistic labels is to make the logical images of the physical features (via their transformation using the linguistic labels) as distinct as possible. In other words, the intent of the optimization is to make x and x' disjoint as much as possible. To make this requirement formal and manageable from a numerical perspective, let us introduce the following index,

$$V = \sum_{k,l=1}^{N} \left\{ \frac{1}{p} \sum_{i=1}^{p} [A_i(x_k) \equiv A_i(x_l)] \rightarrow \frac{1}{c} \sum_{j=1}^{c} [\omega_{kj} \equiv \omega_{lj}] \right\}$$

where the symbol $\equiv$ stands for the degree of equality (matching) between two objects situated either in the logical feature space or class space, namely

$$a \equiv b = 0.5[(a\varphi b) \wedge (b\varphi a) + (\bar{a}\varphi\bar{b}) \wedge (\bar{b}\varphi\bar{a})]$$

where φ is the pseudocomplement (inclusion) induced by some t- norm.

Analogously, the implication operator is induced by one of the triangular norms, see Appendix A for more details. Owing to the properties of the implication operator, the above table is well captured as the entry with the question mark implies low values of V. Obviously, our goal is to maximize V. This is done parametrically by adjusting the values of the membership values of A_is. Note that in this optimization, one has to assure that the fuzzy sets represent ("cover") the entire feature space thus maintaining the completeness of the developed classifier. An additional observation resulting directly from the format of the maximization of V is that by augmenting the feature space, the value of V can be easily increased.

94.4 Implicit and Explicit Knowledge Representation in Pattern Recognition

One among many taxonomies we can establish in the genuine abundance of diverse schemes of pattern classification can be settled by studying how a given classification scheme represents knowledge about the patterns. Most of the existing algorithms, primordially in the realm of statistical pattern recognition (Fukunaga, 1991), imply an implicit character of knowledge that resides within linear or nonlinear discriminant functions, Bayesian decision regions, potential functions, or connections of neural networks. The knowledge accumulated in this manner cannot be directly translated (converted) into a format easily comprehended by the user (such as e.g., rules, frames, two-valued or multivalued predicates, etc). In all the above situations, the classification paradigm can be elicited in the following obvious static relationship,

$$F(\text{pattern}) \rightarrow \text{class membership}$$

where "F" is used to denote a general assignment rule (classifier) associating features with classes. In this respect the exclusive processing domain of the classification knowledge is purely numerical. The learning capabilities of this number-based approach are usually significant and can be additionally enhanced by expanding the classes of the discriminant functions taken under consideration (e.g., by moving from linear to quadratic classifiers), expanding architectures of the neural networks, etc.

On the other hand, the knowledge given in an explicit manner can be utilized in designing symbol-oriented classifiers. In these classifiers, the knowledge is conveyed in terms of symbolic entities being structured into logical expressions and plug in as a part of the classification schemes. One can refer e.g., to knowledge-based classification systems as one among available possibilities. Quite commonly these systems encapsulate knowledge about relationships between the features of the patterns and the classification outcomes as a series of "if-then" statements.

While providing a transparent form of knowledge coding and intuitively clear and well-understood reasoning mechanisms, the knowledge-based systems seriously suffer from a lack of learning capabilities. The problem of knowledge acquisition becomes visible and tends to be acute for high-dimensional classification problems. Due to the lack of numerical knowledge that has not been included in the scheme, the processes of learning could be prolonged and quite inefficient. Since both the faculties of numerical and symbolic processing are present in pattern recognition, it is advantageous to reveal their possible relationships and try to hybridize them into a form of a single classification structure. On second thoughts, the direct transformation between objects (patterns) and classification results, as outlined above, splits now into the two conceptual stages that involve:

- Constructing the logical features. The choice of the linguistic entities should be legitimate in the context of the classification domain. They should be material pieces of knowledge that are actively engaged in the classification activity. This, in fact, pertains to the input interface.
- Expressing logical relationships between the symbolic entities and the results of classification.

The more or less visible conceptual regularities arising in most of the classification problems are enhanced by the knowledge-oriented schemes by their direct expression in terms of logical (or geometrical relationships) occurring between the entities and classes. One can refer e.g., to Minsky and Pappert (1988) where many well-known classification tasks have been articulated in the language of some basic geometric constructs.

Conceptually, the three-phase classification scheme resulting within this environment can be envisioned accordingly.

$$\text{patterns} \rightarrow \text{descriptors} \rightarrow \text{symbolic classification}$$

The descriptors are viewed as symbols describing several geometric regions distinguished in the problem. The classification rules are given in the form,

pattern x belongs to class ω if

$$\psi_1(\phi_1, \phi_2, \ldots, \phi_r) \text{ is true}$$

or

$$\psi_2(\phi_1, \phi_2, \ldots, \phi_r) \text{ is true}$$

or

$$\vdots$$

$$\psi_p(\phi_1, \phi_2, \ldots, \phi_r) \text{ is true} \tag{94.1}$$

where $\psi_1, \psi_2, \ldots, \psi_p$ are two-valued predicates defined over logical variables ϕ_j's while (94.1) ϕ_i denotes truth value of satisfaction of the given descriptor of the pattern, say

$$\phi_i = \begin{Bmatrix} 1, \text{ if pattern } x \text{ satisfies the } i\text{-th descriptor} \\ 0, \text{ otherwise} \end{Bmatrix}$$

An example classification rule would read as pattern x in class ω if

$$\psi_1(\phi_1, \phi_2)$$

or

$$\psi_2(\phi_3, \phi_4)$$

where $\psi_1(\phi_1, \phi_2) = \phi_1$ and ϕ_2 and $\psi_2(\phi_3, \phi_4) = \phi_3$ or ϕ_4. The terms descriptor and predicate are usually used interchangeably.

When getting to fuzzy sets, the classification rules (1) exhibit a series of extensions:

- The logical terms $\phi_1, \phi_2, \ldots, \phi_r$ are fuzzy variables (Zadeh, 1978). Any pattern to be classified matches them to a certain degree. The degrees become aggregated in a logical way with the aid of the predicate ψ_i.
- The truth value of the complete classification statement (rule) is derived based on the truth values of the individual predicates $(\psi_1, \psi_2, \ldots, \psi_p)$.

Bearing these generalizations in mind, the straightforward extension of (1) reads as

truth_value (pattern x belongs to class ω) =

$$\text{truth\_value}(\psi_1(\phi_1, \phi_2, \ldots, \phi_r))$$
$$\Box$$
$$\text{truth\_value}(\psi_2(\phi_1, \phi_2, \ldots, \phi_r))$$
$$\Box$$
$$\vdots$$
$$\text{truth\_value}(\psi_p(\phi_1, \phi_2, \ldots, \phi_r))$$

where $\Box$ is used to denote aggregation of the truth values of the predicates included in the rule. (the detailed meaning of this aggregation will be clarified in further sections).

94.5 From Supervised to Unsupervised Pattern Recognition—A Continuum of Classification Models

The classic taxonomy of supervised and supervised learning commonly encountered in pattern recognition (Duda and Hart, 1973; Tou and Gonzalez, 1974) is a useful yet somewhat simplified categorization of many real-world classification problems. In the setting of fuzzy sets, the issue of a fundamental dichotomy of supervised versus unsupervised learning need to be carefully addressed and substantially revised. Let us recall that in the existing literature the methods of unsupervised learning viewed as two-valued or fuzzy clustering (Bezdek and Pal, 1992; Bezdek, 1981) are usually contrasted with the algorithms of supervised learning (Boolean or fuzzy classifiers). The supervised learning is based on a training set with each pattern being labelled. For the unsupervised learning, this class assignment is not available. These two are very distinct modes of learning. More realistically, there could also be many intermediate scenarios in which the training set is composed of a relatively small population of the labeled patterns and a vast majority of the unlabeled objects. This mixture of the patterns calls for the algorithms of the fuzzy clustering carried under partial supervision, the idea introduced in (Pedrycz, 1985). Depending upon the sizes of the respective populations of the labeled and unlabeled patterns the effect of partial supervision can be diminished or strengthened. Another aspect of partially supervised learning is the one in which the information about classes is given in an implicit, rather than explicit, format. A convincing example arises in the realm of referential classification where the training set includes the patterns arranged in pairs along with their similarity levels (e.g., patterns x and y are λ—*similar* with $\lambda \in [0, 1]$ standing for this degree of similarity). This classification outcome is definitely less detailed (more generalized or synthesized) than that achieved when dealing with complete vectors of membership values. The problem exhibiting this format of information about the classes will be referred to as an implicit (or implicitly-supervised) classification. The panoply of the possible learning situations is schematically visualized in Figure 94.5.

The taxonomy is based upon two fundamental criteria: a percentage of the labeled patterns and a level of generality of the classification information. The first criterion is quite easy to quantify: the level of partial supervision is arranged on the basis of a percentage of the labelled patterns (f%). The second criterion is more difficult to illustrate graphically. The attempt documented in Figure 94.5 (vertical axis) constitutes a somewhat simplified view. In general, we cannot impose a linear order; in fact, any linear ordering should be viewed as a useful and concise, yet obviously simplified illustration.

94.6 Fuzzy Neural Structures

Fuzzy neural structures constitute an interesting computational structure aimed at the efficient numerical processing of fuzzy sets in a certain logical setting (Pedrycz, 1995; Pedrycz and Rocha, 1993; Hirota and Pedrycz, 1991). By this class of networks we

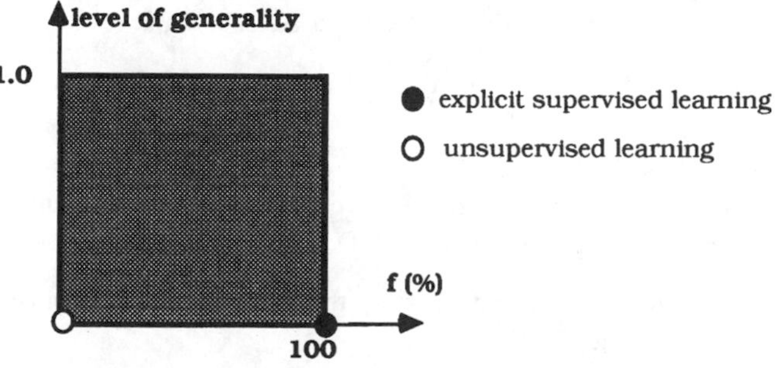

Figure 94.5 A continuum of learning schemes in pattern recognition.

mean distributed and parallel computing structures heavily employing logic operations existing in the theory of fuzzy sets. As opposed to standard neural networks, the networks emerging within this framework are usually heterogeneous, i.e., they consist of neurons of a different conceptual and numerical nature. When put together the neurons exhibit diverse functional characteristics and play quite distinct roles in the network. Firstly, we introduce some basic models of the neurons (aggregative and reference ones) and afterwards concentrate on their main properties that will be of interest when building neural networks. Owing to the fuzzy set operations being used in the defining the neurons, they will be also referred to as fuzzy neurons. Subsequently, the detailed learning algorithms will be dealt with.

Aggregative and Referential Functions of the Fuzzy Neurons

The logic-based neurons aggregate input signals $x_1, x_2, \ldots, x_n \in [0, 1]$ using fuzzy set operators. The two basic logical connectives AND and OR give rise to so-called AND and OR neurons. The AND neuron combines the input signals $x_1, x_2, \ldots, x_n$ ANDwise

$$y = x_1 \text{ AND } x_2 \text{ AND} \cdots \text{AND } x_n \tag{94.2}$$

The OR neuron is described as follows,

$$y = x_1 \text{ OR } x_2 \text{ OR} \cdots \text{OR } x_n \tag{94.3}$$

Both the AND and OR logical connectives are represented as triangular norms (t- and s-norms); refer to Appendix A. While the above expressions are very basic and do not allow for any flexibility of the neuron, a natural extension to the above formulas could include weights (connections) associated with the inputs. Following that, Equations 94.2 and 94.3 are translated into the following formulas:

AND neuron

$$y = (x_1 \text{ OR } w_1) \text{ AND } (x_2 \text{ OR } w_2) \text{ AND} \cdots \text{AND } (x_n \text{ OR } w_n) \tag{94.4}$$

OR neuron

$$y = (x_1 \text{ AND } w_1) \text{ OR } (x_2 \text{ AND } w_2) \text{ OR} \cdots \text{OR } (x_n \text{ AND } w_n) \tag{94.5}$$

$w_i \in [0, 1], i = 1, 2, \ldots, n$

The weights are used to enhance or eliminate an influence of x_i's on the output y. For the AND neuron we observe:

- The lower the value of w_i, the more evident influence of x_i on y.
- Higher values of w_i's enhance the importance of the associated input (x_i).

In limit cases ($w_i = 0$ for the AND neuron and $w_i = 1$ for the second one taken for all the input variables) formulas (94.4)–(94.5) reduce to (94.2) and (94.3), respectively.

The fuzzy neurons can be functionally enhanced in two different ways, that is by including the complements of the input signals and equipping them with an additional nonlinear transformation:

- The complemented input signals, $\bar{x}_i = 1 - x_i$, allow to realize an inhibitory behavior of the neuron while still preserving the unit interval as a relevant range of coding. By choosing appropriate values of the connections the neuron can easily exhibit inhibitory and excitatory characteristics.
- Despite a well-defined semantics of the neurons stemming directly from the logical operations applied therein, the main concern one may raise about these constructs occurs at the numerical side. Once the connections (weights) are set (that happens after learning), each neuron realizes an "in" mapping between the inputs and the output. From a formal point of view, this means that the values of z for all possible inputs cover a subset of the unit interval. More specifically, for the OR neuron this yields the containment $y \in [0, S_{i=1}^{n} w_i]$. Similarly, the achievable range of the values the dual neuron equals $[T_{i=1}^{n} w_i, 1]$. In both the situations, the dynamics of the neurons have been reduced. The improvement may arise from augmenting the neuron by a nonlinear element placed in serial with the previous logical component. Thus the neurons formalized as follows

$$y = \psi\left(\underset{i=1}{\overset{n}{S}} (x_i t w_i)\right)$$

$$y = \psi\left(\underset{i=1}{\overset{n}{T}} (x_i s w_i)\right)$$

where ψ: $[0, 1] \to [0, 1]$ forms a nonlinear monotonic mapping. In contrast to the standard nonlinearity commonly encountered in neurocomputations, we admit both monotonically increasing as well as decreasing functions. A useful sigmoidal nonlinearity takes on the form

$$z = \psi(u) = \frac{1}{1 + \exp\{-(u - m)\sigma\}}$$

$m \in [0, 1]$ $\sigma \in R$

The neurons described so far are of an aggregative character. The second class of the fuzzy neurons realizes referential computations. Given is a fuzzy reference point, say r. Now the neuron is not driven directly by the input signals x_i but by a collection of degrees of matching achieved for the individual coordinates of x and r. These degrees are afterwards processed by OR-ing them (including eventually their weighting by connections w_i). The formalism of this neuron is expressed by the formula:

$$y = \overset{n}{\underset{i=1}{S}} [(x_i \equiv r_i) t w_i] \tag{94.6}$$

where $x_i \equiv r_i$ is the degree of matching achieved between x_i and r_i. The matching operator provides an input signal to the OR neuron. We can write it down symbolically as y = MATCH (x, r, w). Bearing this in mind one can express the neuron described by Equation 94.6 as a serial structure of the reference block

$$f_i = x_i \equiv w_i$$

followed by the OR neuron

$$y = (f_1 \text{ AND } w_1) \text{ OR } (f_2 \text{ AND } w_2) \text{ OR } \cdots \text{ OR } (f_n \text{ AND } w_n)$$

While Equation 94.6 is referred to as ORwise aggregation of the matching results, the conjunctive form of the neuron reads as

$$y = \overset{n}{\underset{i=1}{T}} [(x_i \equiv r_i) s w_i]$$

The neural network can also be formulated in such a way that some other logic-oriented relationships between x and r are captured. With this regard, we can enumerate two interesting dependencies:

- Constraint inclusion. The neuron Equation 94.6 is modified by combining x and r with the use of the relationship OR__INCLUDED:

$$y = \text{OR\_INCLUDED}(x, r, w) \tag{94.7}$$

namely,

$$y = \overset{n}{\underset{i=1}{S}} [(x_i \to r_i) t w_i]$$

The dual form of the neuron reads as

$$y = \overset{n}{\underset{i=1}{T}} [(x_i \to r_i) s w_i]$$

In contrast to the previous equality relationship the level of membership of the output y becomes elevated once x is strongly included in the relevant constraints characterized by r. Similarly as before, Equation 94.7 is converted into a coordinatewise notation and reads as

$$y = [(x_1 \text{ INCLUDED\_IN } r_1) \text{ AND } w_1] \text{ OR}$$
$$[(x_2 \text{ INCLUDED\_IN } r_2) \text{ AND } w_2] \text{ OR}$$
$$\cdots \text{ OR}[(x_n \text{ INCLUDED\_IN } r_n) \text{ AND } w_n]$$

- The opposite type of dependency induces constraint covering. The grade of membership of resulting y increases as x "covers" r (in other words r is included in x). This fact is expressed as

$$y = \text{OR\_COVER}(x, r, w)$$

where the relationship "COVER" is a dual to the predicate INCLUDED IN, namely

$$y = [(r_1 \text{ INCLUDED\_IN } x_1) \text{ AND } w_1] \text{ OR}$$
$$[(r_2 \text{ INCLUDED\_IN } x_2) \text{ AND } w_2] \text{ OR} \cdots$$
$$\text{OR } [(r_n \text{ INCLUDED\_IN } x_n) \text{ AND } w_n]$$

Briefly,

$$y = \overset{n}{\underset{i=1}{S}} [(r_i \to x_i)\ t\ w_j]$$

Similarly,

$$y = \overset{n}{\underset{i=1}{T}} [(r_j \to x_i)\ t\ w_i]$$

It is worth noting that the inclusion and coverage properties when summarized as in Figure 94.6, give rise to the matching neuron.

Fuzzy Neurons with Feedback

The neurons studied so far are essentially not capable of expressing dynamical (time-implied) relationships existing in the classification task. This time-dependency aspect could be, however, essential in a proper description of the classifier. Consider, for instance, a classification of faults based upon a collection of symptoms. More specifically, one of the system's sensors provides information about abnormal (elevated) temperature. Observe

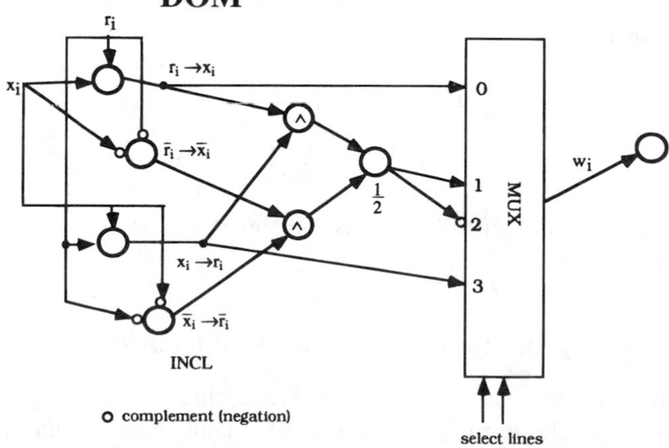

Figure 94.6 Fuzzy matching neuron as an aggregation of inclusion and dominance (coverage) neurons.

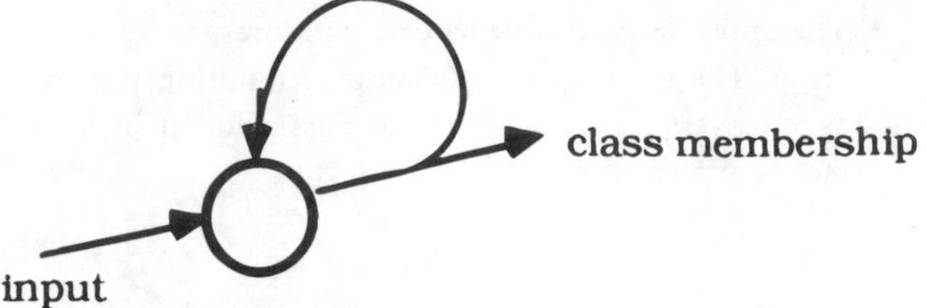

Figure 94.7 Fuzzy neuron with a feedback loop.

that the duration of this phenomenon (raise of temperature) has a primordial impact on expressing the confidence about the failure (class). If the elevation of the temperature prolongs, our confidence about the failure rises up. It might well be that some short temporary temperature elevations (spikes) reported by the sensor could be almost ignored. To properly capture this effect in the classifier, one has to equip the basic logic neuron with some feedback link. A straightforward extension of this nature is schematically illustrated in Figure 94.7.

The neuron is described accordingly

$$\omega(k+1) = [b \text{ OR } x(k)] \text{ AND } [a \text{ OR } \omega(k)] \qquad (94.9)$$

$a, b \in [0, 1]$.

Its dynamics is uniquely defined by the feedback connection. The initial condition of class membership, $\omega(0)$, expresses a priori confidence associated with the failure (class) ω. The level of accumulation of evidence as well as a speed at which this accumulation takes place is defined by the value of the feedback connection (a). For sufficiently long period of time, $\omega(k+1)$ could take on higher values in comparison to the level of the original evidence being available at the input. A proper selection of the connections of the neuron in the feedback loop leads to a very high flexibility of its characteristics, as clearly visible in Figure 94.8. In these situations, the neuron modelling a first-order dynamics in class membership, is realized using the product and probabilistic sum.

Any higher-order dynamical dependencies to be accommodated by the network, if necessary, have to be taken care of via a feedback loop consolidating several pieces of temporal information e.g.,

$$\omega(k+2) = [b \text{ OR } x(k)] \text{ AND } [a_1 \text{ OR } \omega(k)] \text{ AND } [a_2 \text{ OR } \omega(k+1)] \qquad (94.10)$$

One can also refer to Equations 94.9 and, 94.10 as fuzzy difference equations.

Learning in the Fuzzy Neural Networks

The discussed neural networks or even a single neuron require learning. The learning procedures are mainly of a parametric nature and deal with a series of suitable adjustments of the weights (connections) of the networks. The learning is carried out on the basis of a learning set of input-ouput patterns (x_k, y_k), $k = 1, 2, \ldots, N$, and is driven by a specified performance index Q. Usually Q is given as a sum of squared errors (MSE criterion) measuring distances between y_k's and the output of the network being driven by x_k, say $N(x_k)$, where $N(\cdot)$ stands for a general notation of the output of the neural network

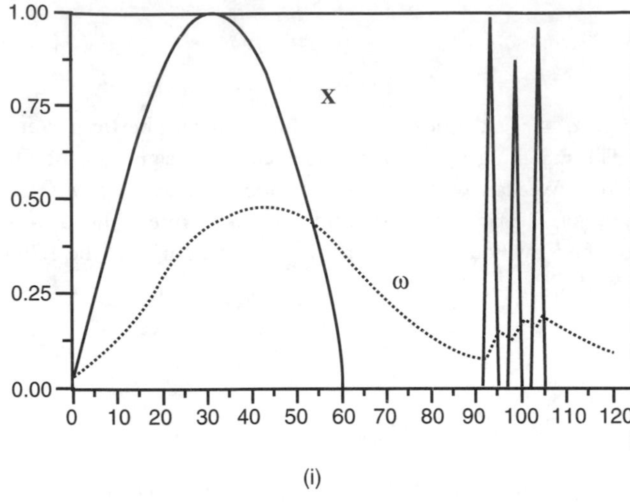

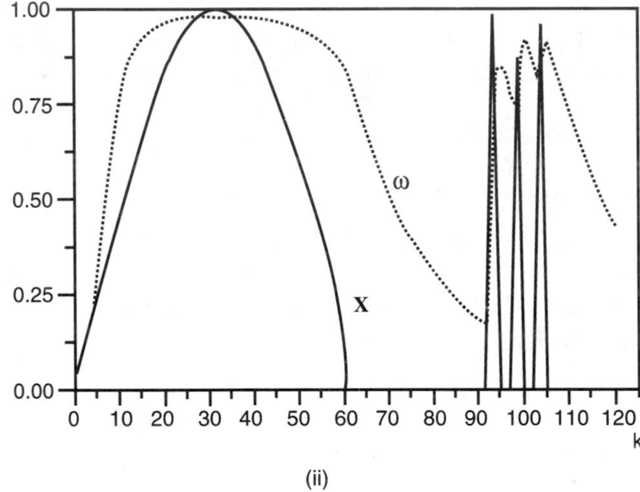

Figure 94.8 Class memberships $\omega(k)$ in successive time moments for a given input $x(k)$ (i) $a = 0.05$, $b = 0.95$ (ii) $a = 0.70$, $b = 0.95$.

$$Q = \sum_{k=1}^{N} (y_k - N(x_k))^2$$

The adjustments of the connections are worked out by a standard Newton-like method. Generally speaking, the abbreviated form of the update scheme looks as follows

$$(\text{connections})_{\text{new}} = (\text{connections}) - \alpha \frac{\partial Q}{\partial(\text{connections})} \qquad (94.11)$$

while α specifies a learning rate situated in the unit interval. The choice of the learning rate implies a particular pace of learning. In general, too high values of α could result in oscillations of the performance index, while too small rates could slow down the learning.

The general learning formula can be applied to different networks upon specification of all the details of the structure (including its topology, types of the neurons, initial values of the connections, triangular norms, etc.). Below we summarize the detailed learning scheme for the basic architecture with a single hidden layer. We assume that the dimension of the hidden layer is fixed (we will comment later on the selection of this parameter). The triangular norms are also specified in advance, namely we treat the *t*-norm as the product operation while the *s*-norm will be given as the probabilistic sum. The connections to be modified are those between the input and the hidden layer (w_{ij}) and between the hidden layer and the output node (v_j). The general formula (Equation 94.11) is translated into the following expressions;

$$w_{ij} = w_{ij} - \alpha \frac{\partial Q}{\partial w_{ij}}$$

$$v_j = v_j - \alpha \frac{\partial Q}{\partial v_j}$$

$j = 1, 2, \ldots, p$, $i = 1, 2, \ldots, n$.

The computations of the above derivatives are straightforward. Below we will discuss an on-line type of the learning scheme, in which each pair of the training set successively modifies the connections

$$\frac{\partial Q}{\partial v_j} = \frac{\partial}{\partial v_j}(y_k - N(x_k))^2 = -2(y_k - N(x_k))\frac{\partial N(x_k)}{\partial v_j}$$

The inner derivative $\partial N(x_k)/\partial v_j$ is equal to

$$\frac{\partial}{\partial v_j}\left(\underset{l=1}{\overset{p}{S}} v_l t z_l\right) = \frac{\partial}{\partial v_j}[A + v_j z_j - A v_j z_j] = z_j - A z_j = z_j(1 - A)$$

where

$$A = \underset{l \neq j}{S}\, v_l z_l$$

To obtain $\partial Q/\partial w_{ij}$ we proceed accordingly,

$$\frac{\partial}{\partial w_{ij}}(y_k - N(x_k))^2 = -2(y_k - N(x_k))\frac{\partial N(x_k)}{\partial w_{ij}}$$

$$\frac{\partial N}{\partial w_{ij}} = \sum_{l=1}^{p} \frac{\partial y}{\partial z_l}\frac{\partial z_l}{\partial w_{ij}} = \frac{\partial y}{\partial z_j}\frac{\partial z_j}{\partial w_{ij}}$$

Finally

$$\frac{\partial y'}{\partial z_j} = \frac{\partial}{\partial z_j}\left(\underset{l=1}{\overset{p}{S}} v_l t z_l\right) = \frac{\partial}{\partial z_j}[A + v_j z_j - A v_j z_j] = v_j(1 - A)$$

$$\frac{\partial z_j}{\partial w_{ij}} = \frac{\partial}{\partial w_{ij}}\left[\underset{l=1}{\overset{2n}{T}} (w_{lj} s x_i)\right]$$

$$= \frac{\partial}{\partial w_{ij}}[B(w_{ij} + x_i - w_{ij}x_i)]\; B(1 - x_i),$$

where

$$B = \prod_{l \neq i} (w_{lj} s x_l')$$

In this topology, the size of the hidden layer "p" determines its representation capabilities, i.e., uniquely specifies a number of minterms (maxterms) of the logic function the network is capable of handling (approximating).

The learning capabilities of the fuzzy neural network used as a classifier are illustrated with the aid of a two-class binary

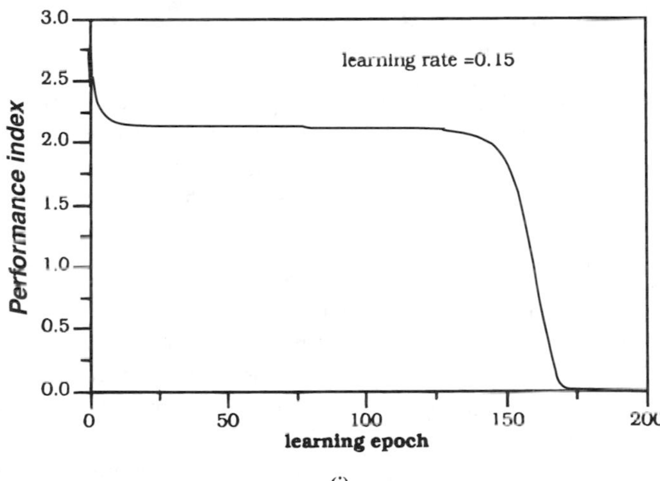

(i)

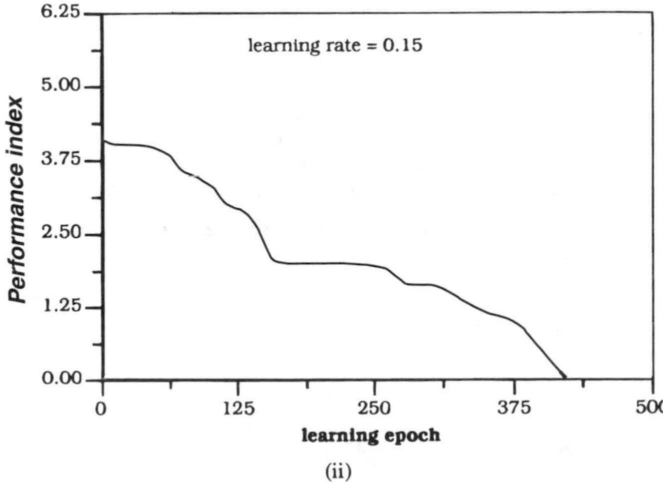

(ii)

Figure 94.9 Performance index in successive learning epochs for two (i) and three (ii) dimensional patterns.

problem. The first class includes all pattern that satisfy an Exclusive-Or (XOR) predicate. The second one consists of the binary patterns satisfying the predicate of equivalence. The problem is solved for two- and three dimensional patterns; the learning is summarized in terms of the performance index monitored through successive steps of learning, Fig. 94.9.

The determination of the size of the hidden layer (p) is out of the stream of the above parametric learning. Its choice should be directed by the values of Q. Usually by increasing "p," the corresponding values of Q get lower.

For the reference neuron and the logic processor driven by the reference preprocessor, the reference points can be either provided in advance or may be learned as well as a part of the overall parametric learning.

94.7 Supervised Learning

In this section we review a variety of fuzzy classifiers whose training is based on a series of labelled patterns.

Logic Processors as a Generic Classification Architecture

The AND and OR neurons can be put together in a form of a so-called logic processors (LP). The role of the logic processors, which are heterogeneous neural networks, is to realize (approximate) any fuzzy mapping from the logical feature space to the space of class membership. Essentially, there are two basic structures of the processor:

- The first one consisting of the three layers comes as a sum of products (SOM). The input layer consists of "$2n$" nodes and includes both x_i's as well as their complements ($\bar{x}_i$). The hidden layer includes "p" AND nodes. The output layer has a single OR node.

  The formal notation for this architecture looks as follows:
- The hidden layer forms "p" minterms z_j

$$z_j = \overset{2n}{\underset{i=1}{T}} (w_{ij} s x_i') \tag{94.12}$$

  $j = 1, 2, \ldots,$ p, where x' is an extended vector of "$2n$" inputs including direct and complemented values of all x_i's.
- Output layer. The minterms are combined by taking the OR operation on z_j's:

$$y = \overset{p}{\underset{j=1}{S}} (v_j t z_j) \tag{94.13}$$

- The dual structure of the logical processor computes y by considering a product of minterms and combining the results produced by the hidden layer in the AND form. We will be referring to this structure as a product of maxterms (POM). Its formal model is described accordingly,
- Hidden layer

$$z_j = \overset{2n}{\underset{i=1}{S}} (w_{ij} t x_i') \tag{94.14}$$

  $j = 1, 2, \ldots, p$
- Output layer

$$y = \overset{p}{\underset{j=1}{T}} (v_j s z_j) \tag{94.15}$$

The architecture of the network mimics the general topology of the classification problem. The development of the classifier involves three phases:

1. The phase of feature matching returns truth values of the fuzzy variables that is truth values $\phi_1, \phi_2 \ldots, \phi_m$.
2. The geometric regions are built by forming intersections of the fuzzy variables. Within this step one generates a conjunctive from of the truth values $\phi_1, \phi_2, \ldots, \phi_m$ as well as their complements $\bar{\phi}_1, \bar{\phi}_2, \ldots, \bar{\phi}_m$. This step is attained by applying a series of AND neurons, see Figure 94.10. The role of the complements of these truth values is to facilitate a generation of the most condensed logical characteristics of the classifier within its learning.
3. The final class description is formed by OR-ing the outputs of the neurons situated at the level of the classification regions. This phase corresponds to the merging the geometrical regions. The use of the OR operation is legitimized by the fact that several disjoint regions can contribute to the formation of the same class.

The detailed formulas of the network follow the topology of the network:

- The hidden layer pertaining to "p" classification regions and emitting activation signals z is described as

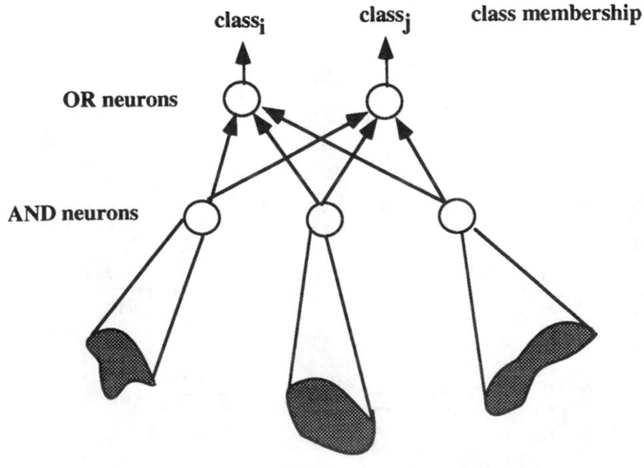

Figure 94.10 Fuzzy classifier implemented as a Logic Processor.

$$z = \text{AND}(\phi, w)$$

with

$$z = [z_1 \; z_2 \cdots z_p],$$
$$w = [w_{ji}],$$
$$j = 1, 2, \ldots, p,$$
$$i = 1, 2, \ldots, m$$

- The output layer aggregating z_j's is given by

$$y = \text{OR}(z, v)$$

with the connections

$$v = [v_1 v_2 \cdots v_p] \in [0, 1]^p$$

Pseudomedian Filtering

Quite often in filtering procedures one is interested in determining median as a suitable representative of a certain collection (window) of data. We will concentrate on a sequence of numerical entries $\mathcal{X} = \{x_i, x_j, \ldots, x_n\} \in [0, 1]^n$ forming an n-point window. Overall, median filtering (Pratt, 1991; Gonzalez and Woods, 1992) is computationally greedy as it requires generation and processing a significant number of strings of the data set in order to determine this value. For instance, a five-element data set ($n = 5$) calls for the max-min operations of the form,

$$\begin{aligned}\text{Median}(\mathcal{X}) = \text{Max}[&\text{Min}(x_1, x_2, x_3), \text{Min}(x_1, x_2, x_4),\\ &\text{Min}(x_1, x_2, x_5), \text{Min}(x_1, x_3, x_4),\\ &\text{Min}(x_1, x_3, x_5), \text{Min}(x_1, x_4, x_5),\\ &\text{Min}(x_2, x_3, x_4), \text{Min}(x_2, x_3, x_5),\\ &\text{Min}(x_2, x_4, x_5), \text{Min}(x_3, x_4, x_5)]\end{aligned}$$

or, alternatively, when utilizing the dual min-max composition one gets,

$$\begin{aligned}\text{Median}(\mathcal{X}) = \text{Min}[&\text{Max}(x_1, x_2, x_3), \text{Max}(x_1, x_2, x_4),\\ &\text{Max}(x_1, x_2, x_5), \text{Max}(x_1, x_3, x_4),\\ &\text{Max}(x_1, x_3, x_5), \text{Max}(x_1, x_4, x_5),\\ &\text{Max}(x_2, x_3, x_4), \text{Max}(x_2, x_3, x_5),\\ &\text{Max}(x_2, x_4, x_5), \text{Max}(x_3, x_4, x_5)]\end{aligned}$$

In total, for any n-element data set we require to process $n!/m!\,(n - m)!$ sequences where $m = n + 1/2$.

Noticing this evident drawback, the method proposed in Pratt (1991), Section 10.3.2, realizes an operation of a so-called pseudomedian filtering where accuracy is traded for a higher computational efficiency. That is to say that the number of the sequences becomes limited; the proposed formulas uses an average of the max-min and min-max compositions of some "sliding" sequences of the elements. Again for $n = 5$ we obtain,

$$\begin{aligned}&\text{Pseudomedian}(\mathcal{X})\\ &= \frac{1}{2}\text{Max}[\text{Min}(x_1, x_2, x_3), \text{Min}(x_2, x_3, x_4), \text{Min}(x_3, x_4, x_5)]\\ &+ \frac{1}{2}\text{Min}[\text{Max}(x_1, x_2, x_3), \text{Max}(x_2, x_3, x_4), \text{Max}(x_3, x_4, x_5)]\end{aligned}$$

The max-min and min-max compositions produce upper and lower bound of the median, so the average of the two of them can cancel or reduce the biases produced by these extremities. Interestingly enough, one can also look at the pseudomedian filter as a network of weightless (Boolean) OR and AND neurons with the connections set either to 0 or 1, Figure 94.11.

It is worth noting that only the specific subsequences of the elements aggregated by the min and max operations are included in (or excluded from) the successively applied operations.

The resulting architecture is not equipped with any modifiable elements that could eventually improve the estimates of the median by gathering more specific and data-dependent information that could be eventually accommodated in the network (Hirota and Pedrycz, 1994). One among possible extensions would be then to treat the corresponding subsequences more individually by considering additional weights added to the resulting network. In sequel, the proposed architecture is illustrated in Figure 94.12.

Modeling Spatial Relationships in High-Level Computer Vision

An interesting utilization of the fuzzy neural networks as logic-oriented classification algorithms can be encountered in the description of spatial relationships occurring at high level tasks of computer vision (Krishnapuram, et al., 1993; Huntsberger, et al., 1986). Any scene analysis and understanding requires analysis of mutual distribution of the objects that form this particular scene. Each image (object) constitutes a certain region in the

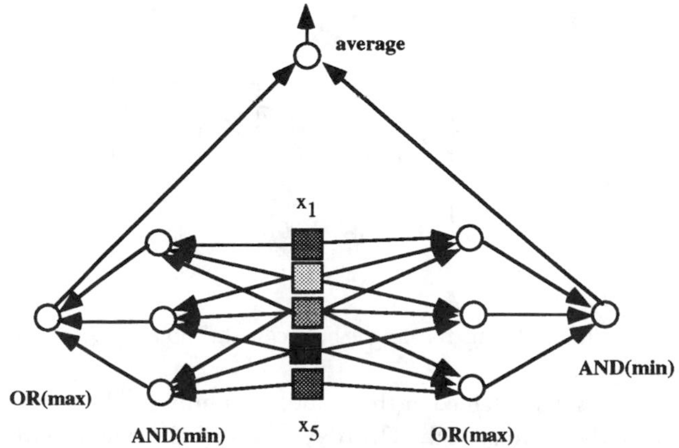

Figure 94.11 Pseudomedian filter as a weightless fuzzy neural network (only significant connections indicated).

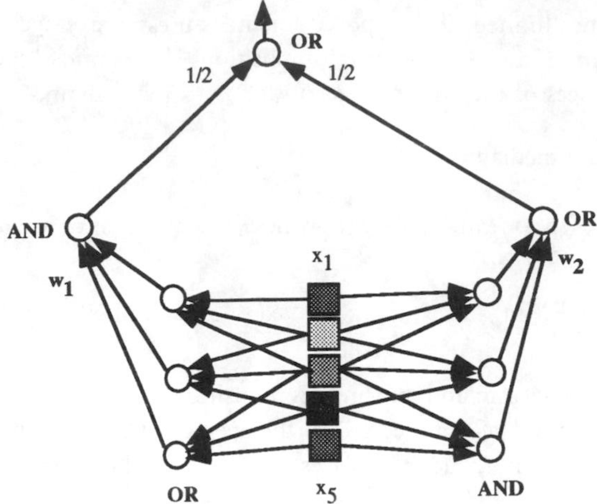

Figure 94.12 Fuzzy neural network realization of a pseudomedian filter.

x-y plane. The relationships between the objects could be conveniently treated as fuzzy relations (like *above, below, inside,* etc.), see Krishnapuram et al., (1993). The architecture of the network, as illustrated in Figure 94.13, exhibits several layers:

- The first layer consists of the processing units that are associated with each pixel and generate a degree of satisfaction of the given spatial relationship by the considered regions *A* and *B*. Essentially, each unit acts rather as an AND node in the AND graph with $r_L(x_j, x_j)$ standing for the satisfaction of the spatial relationship tied with the body of the unit rather than its individual inputs as found in the neurons. This yields

$$x_{ij} = \min[A(z_i), B(z_j), r_L(\psi(z_i, z_j))]$$

for each pair of pixels z_j and z_j. The function ψ depends on the spacial relationship to be described between the objects. For example, the relationship *left of*, Figure 94.14, is defined as,

$$\text{left}(\vartheta) = \left\{ \begin{array}{c} 1, \text{ if } |\vartheta| < \dfrac{a\pi}{2} \\ \dfrac{\dfrac{\pi}{2} - |\vartheta|}{\dfrac{\pi}{2}(1-a)}, \text{ if } \dfrac{a\pi}{2} \le |\vartheta| \le \dfrac{\pi}{2} \\ 0, \text{ if } |\vartheta| > \dfrac{\pi}{2} \end{array} \right\}$$

with ϑ being the angle between the objects (A and B), and "a" in [0,1].

The results obtained at this node are summarized in the OR and AND neurons of the OR/AND neuron. The output $\mathcal{D}_L(A, B)$ describes a degree of satisfaction of the spatial relationship holding for *A* and *B*. The respective connections (v_1 and v_2) are modified accordingly in order to assure that the network follows

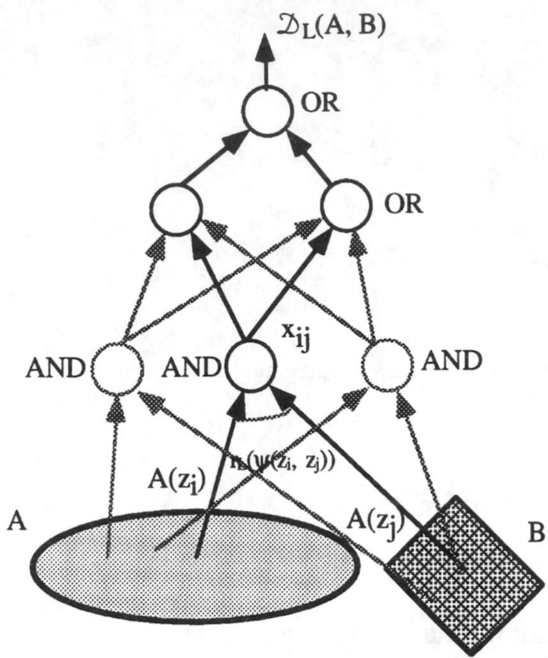

Figure 94.13 General network architecture for modelling spatial relationships.

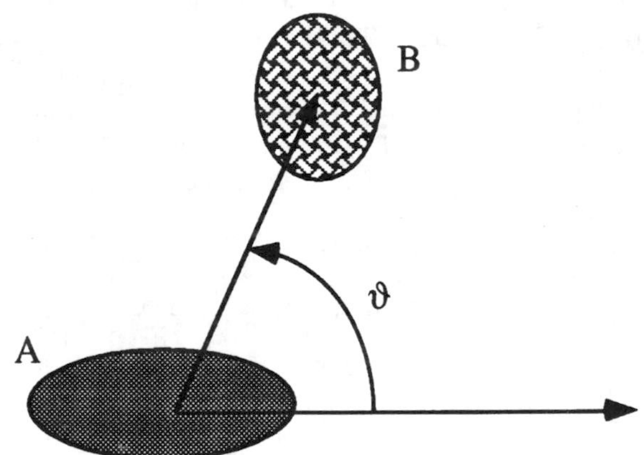

Figure 94.14 An example of a spatial relationship left of.

the elements of the training set. In contrast, the scheme discussed in Krishnapuram, et al. (1993) is less selective as it uses α-cuts of *A* and *B*, accumulates the results and finally produces their average.

Fuzzy Perceptron

The perceptron as originated by Rosenblatt (see also Minsky and Pappert, 1988) is one of the earliest gradient-descent models of supervised pattern recognition. In its simplest version, the algorithm deals with two classes of patterns and produces a hyperplane in $\mathbf{R}^n$ such that all patterns belonging to class ω_1 lie on one side of the hyperplane while the others (class ω_2) are situated on the opposite side of the hyperplane. More formally, the perceptron $\mathcal{P}$ treated as a mapping from $\mathbf{R}^{n+1}$ to $\mathbf{R}$ is a linear mapping such that

$$\mathcal{P}(\mathbf{x}) = \mathbf{w}^T\mathbf{x}$$

such that

$$\mathcal{P}(\mathbf{x}) < 0 \text{ for all } x \in \omega_1 \text{ and } \mathcal{P}(\mathbf{x}) > 0 \text{ for } x \in \omega_2.$$

Where $\mathbf{x} = [1\ x_1\ x_2 \ldots x_n]$ and $\mathbf{w} = [w_0\ w_1\ w_2 \ldots w_n]$. The above relationship defines a hyperplane called also a linear discriminant function.

The key theoretical finding states that if two classes of patterns are linearly separable (namely, there exists a hyperlane separating the patterns), then the perceptron $\mathcal{P}$ is guaranteed to find it in a finite number of steps. The algorithm leading to the construction of $\mathcal{P}$ is surprisingly compact:

- Let $x_1, x_2, \ldots, x_N$ bethe given labelled patterns in a two-class classification problem.
- Pick up an arbitrary weight vector $\mathbf{w}$ in $\mathbf{R}^{n+1}$ (initialization).
- Repeat

  cyclically present the patterns $\mathbf{x}_k$, $k = 1, 2, \ldots, N$, to the classifier; if $\mathbf{x}_k$ has been misclassified by the perceptron, modify its weights according to the formula,

$$\mathbf{w}(\text{new}) = \mathbf{w} + a\mathbf{x}_k,$$

  $a > 0$.
- Until $\mathbf{w}$ remains unchanged.

While the perceptron guarantees a nonzero classification error in the linearly separable case, it states nothing about the performance of the algorithm itself—a finite number of iteration could indeed be a very large number. The method proposed by Keller and Hunt (1985) attempts to improve the performance of the perceptron by making a distinction between the patterns which should have a primordial impact on the updates of the weight vector and those whose impact on the modifications of **w**'s should be very much reduced. The underlying idea is to associate membership values to each of the patterns, say u_{k1} and u_{k2}, such that $u_{k1} + u_{k2} = 1$, $k = 1, 2, \ldots, N$. More specifically, as the patterns have been already labelled, this assignment indicates a degree to which one can regard the pattern as being prototypical to the given class. Then the original update scheme of the perceptron becomes generalized by taking the values of u_{k1} and u_{k2} as a part of the update formula.

$$\mathbf{w}(\text{new}) = \mathbf{w} + a|u_{k1} - u_{k2}|^2\mathbf{x}_k,$$

where $a > 0$, $p > 1$. The above algorithm is also referred to as fuzzy perceptron.

Note that:

1. The influence of uncertain patterns (i.e., those for which $u_{k1} \approx u_{k2}$) in the update scheme is very much reduced. In particular, for the pattern x_k with $u_{k1} = u_{k2} = 1/2$, the correction effect is totally ignored.
2. If $u_{k1}, u_{k2} \in \{0,1\}$, then the fuzzy perceptron reduces to the previous two-valued version.

The selection of the membership values can be done in many different ways. As we are, in fact, concerned with the labeled patterns, these membership values should be assigned based upon the level of *prototypicality* of the patterns. One can study a similarity measure between the patterns and the means of the classes. The membership class assignment proposed in Keller and Hunt (1985) reflects this observation as they are defined accordingly,

$$u_{ik} = 0.5 + \frac{e^{-f(d_1 - d_2)/d} - e^{-f}}{e^f - e^{-f}}$$

with f standing for the scaling factor being used to control the rate at which the membership level decreases towards 0.5. Similarly, d_1 and d_2 denote the distance function between x and the means (averages) of the classes m_1 and m_2, that is $d_1 = \| x - m_1 \|$ and $d_2 = \| x - m_2 \|$. It should be stressed that the two fuzzy sets u_1 and u_2 capture an additional domain knowledge so that the perceptron learning can benefit from this preprocessing effort.

Nearest—Neighbor Classification Algorithm Realized as a Fuzzy Neural Network

The nearest neighbor (NN) classification model is one among the most popular classification schemes (Cover and Hart, 1967). The underlying idea is to classify the pattern based on a distance between it and prototypes of the classes. The class assignment is done based upon the best matching of the pattern and the prototype. In comparison to the NN schemes operating in the physical feature space, here we are concerned with the matching carried out in the logical feature space. Let $p_1, p_2, \ldots, p_c$ be the prototypes of the classes while x describes a pattern to be classified. The result of matching is obtained using the AND type referential neuron in the architecture shown in Figure 94.15.

In fact, Figure 94.15 shows only the portion of the classifier carrying out the matching for the i-th prototype; the complete system calls for "c" subsystems as portrayed in Figure 94.16. It is also remarkable that the different matching neurons exhibit their own set of connections. This implies a heterogeneous feature space with several local similarity measures associated with the individual prototypes.

The output y_i describes a level of matching occurring between x and p_i. The final classification is driven by the maximum of the level of matching (winner takes all strategy), namely

$$\text{assign } \mathbf{x} \text{ to class } \omega_{i_0} \text{ where } y_{i_0} = \max_{i=1,2,\ldots,c} y_i$$

It is worth underlining that the connections of the AND neuron can be easily adjusted to improve the performance of the classifier. The modifications of the connections are guided by the classification error. Let i_0 denotes the winner takes all class identified by the classifier, while j_0 is the index of the class coming from the training set. If these two indices are different, the

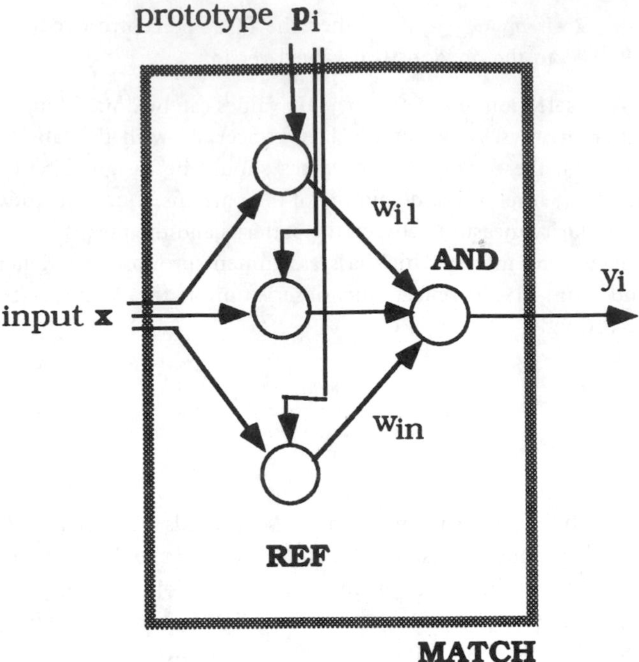

Figure 94.15 NN classifier for i-th class.

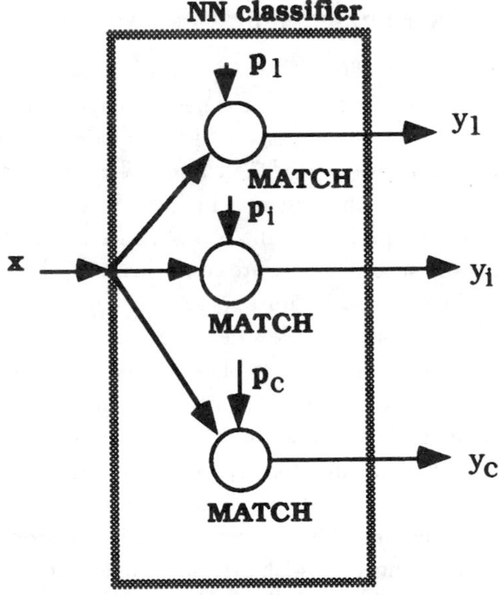

Figure 94.16 A complete NN classifier for "c" classes.

classifier needs to be improved by changing its connections so that the correct class (j_0) becomes identified. Let $I = \{i_1, i_2, \ldots, i_r\}$ be a set of indices for which the corresponding levels of matching assume values higher than that for y_{i0}. These class assignments should be lowered, while the one for y_{i0} requires elevation. More specifically, the changes in the values of y_is allowing y_{i0} to succeed in the competition is to select an increment Δ such that

$$\Delta = \max_{i \in I}(y_i - y_{i_0})$$

and modify the connections so that all the MATCH neurons with the outputs in I are set up accordingly,

$$y_i(\text{new}) = y_i - \Delta, \qquad i \in I$$

whereas the output of the i_0-th to be increased,

$$y_{i_0}(\text{new}) = y_{i_0} + \Delta$$

E.g., consider $c = 5$ and assume that $i_0 = 2$ while the outputs of the neurons are equal to [0.1 0.3 0.5 0.4]; definitely the classifier identifies the third class as the winner in this competition. In our case $I = \{3, 4\}$ and $\Delta = \max(0.5-0.3, 0.4-0.3) = 0.2$. The target values are set accordingly,

$\text{target}_2 = 0.3 + \Delta = 0.5$
$\text{target}_3 = 0.5 - \Delta = 0.3$
$\text{target}_4 = 0.3 - \Delta = 0.2$

(more precisely, the target values should always be maintained in the unit interval).

Hierarchical Pattern Classifiers

As fuzzy neural networks are definitely domain-oriented models of classification schemes, their efficiency becomes most profound in all these situations when the qualitative knowledge about the classification problem is made accessible. Very often, the tasks of pattern recognition lend themselves to the situations where we are confronted with a mixture of qualitative knowledge and extensive logs of numerical patterns. Consider, for instance, computer vision. One can easily distinguish between low-level recognition tasks that are primarily identified and handled at the level of thousands of individual pixels and the ones taking place at the level of some conceptual primitives of an image (such as regions, lines, etc.). At the pixel level, when the processing is numerically intensive and the domain knowledge is practically nonexistent, one can apply standard statistically-based methods. The domain knowledge becomes crucial at the higher level of information processing. Thus, the low level computer vision (image processing) constitutes a realm of numerical processing while the high end tasks call for fuzzy set technology. In this sense we can talk about a symbiosis between numerical neural networks and fuzzy neural networks realized in multiple neural network architecture, cf. Figure 94.17.

From the conceptual point of view, we can consider classification situation in which instead of a single global classifier, the problem becomes solved through a series of local, eventually linear, classification algorithms. By doing that, we can make the local classifiers more flexible and capable of coping with some local peculiarities of the classification data. The idea of such hierarchical classification could be realized in many different ways; below we elaborate on two of them.

The classification rules relate specific regions in the feature space with the corresponding local classifiers,

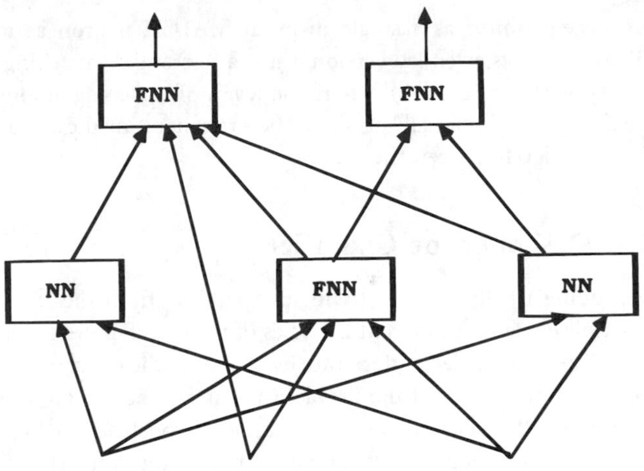

Figure 94.17 Cooperation between fuzzy neural networks (FNN) and neural networks (NN) in a multiple network topology.

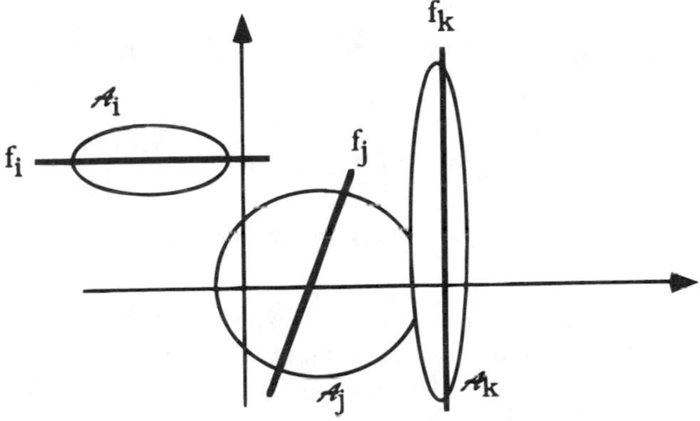

Figure 94.18 Classification through a series of local classifiers.

$$\text{if features } \mathbf{x} \text{ are in } \mathcal{A}_1 \text{ then the } \omega = \mathbf{f}_1(\mathbf{x}, \mathbf{a}_1)$$
$$\text{if features } \mathbf{x} \text{ are in } \mathcal{A}_2 \text{ then the } \omega = \mathbf{f}_2(\mathbf{x}, \mathbf{a}_2)$$
$$\dots$$
$$\text{if features } \mathbf{x} \text{ are in } \mathcal{A}_q \text{ then the } \omega = \mathbf{f}_q(\mathbf{x}, \mathbf{a}_q)$$

where $\mathbf{f}_i$s given as the mappings from R^n to $[0, 1]^c$ describe the classifiers associated with the predefined regions (relationships) $\mathcal{A}_j$ in the feature space and $\omega = [\omega_1 \ \omega_2 \ \dots \ \omega_c]$ stands for the vector of membership values in the corresponding classes, refer also to Figure 94.18.

The classifier f_i itself could be either linear or non-linear. It is important to note that f_i acts only as a local classifier so that it represents a classification formula whose validity is restricted only to $\mathcal{A}_i$. The aggregation of the classification outcomes produced by the specific local classifiers and completed as

$$\omega_{\text{aggr}} = \frac{\sum_{i=1}^{q} \mathbf{f}_i(\mathbf{x}, \mathbf{a}_i)\mathcal{A}_i(\mathbf{x})}{\sum_{i=1}^{q} \mathcal{A}_i(\mathbf{x})}$$

gives rise to a smooth switching between the local classifiers.

It is well known that the *NN* classification rules become the most effective in all these regions of the feature space in which the classes overlap, see Figure 94.19. It is therefore highly justifiable to aggregate two types of distinct classifiers:

- A linear (or non-linear) classifier is used for classifying those patterns x that are far from the regions of class overlap, and
- A NN classifier operates in the regions of the feature space exhibiting a visible overlap between the classes.

This combination means that we are confined to the two types of the classification rules,

$$\text{if } \mathbf{x} \text{ is in far from the region of class overlap then } \omega = \mathbf{f}(\mathbf{x}, \mathbf{a})$$
$$\text{if } \mathbf{x} \text{ is in the region of high class overlap then } \omega = NN(\mathbf{x})$$

In particular, the classification rules pertinent to the situation visualized in the same figure read as

$$\text{if } \mathbf{x} \text{ is in } \mathcal{A}_2 \text{ or } \mathcal{A}_3 \text{ then } \omega = \mathbf{f}(\mathbf{x}, \mathbf{a})$$
$$\text{if } \mathbf{x} \text{ is in } \mathcal{A}_1 \text{ then } \omega = NN(\mathbf{x})$$

94.8 Implicitly Supervised Pattern Recognition

Problem Statement

We are now concerned with the logical feature space formed by n dimensional vectors of the unit hypercube, that is $x \in [0, 1]^n$. Let us recall that in explicit learning, the training set of patterns consists of the classification results that are provided as the membership vectors of an c-dimensional classification space; each coordinate of ω is interpreted as a degree of membership in the corresponding class, $\omega_i \in [0, 1]$. The learning set, called $\mathfrak{D}$—training set, comes as a collection of the feature—class assignment pairs,

$$(\mathbf{x}_1, \omega_1)(\mathbf{x}_2, \omega_2), \dots, (\mathbf{x}_M, \omega_M)$$

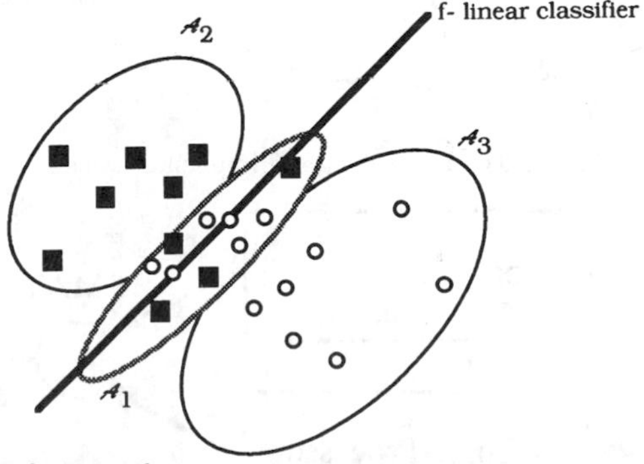

Figure 94.19 Classification for two overlapping and three regions distinguished in the feature space.

On the other hand, the implicit format of the available classification results implies that the m-dimensional classification space is usually reduced to a unit interval the values of whose represent the referential characteristics of the patterns. For the matching (similarity) property we are concerned with the pairs of the patterns $\mathbf{x}_k$, $\mathbf{x}'_k$ and their associated similarity levels; altogether these constitute a so-called $\mathfrak{R}$—training set,

$$((\mathbf{x}_1, \mathbf{x}'_1), \text{sim}_1)((\mathbf{x}_2, \mathbf{x}'_2), \text{sim}_2) \cdots ((\mathbf{x}_N, \mathbf{x}'_N), \text{sim}_N)$$

where $\text{sim}_i \in [0, 1]$ describes a similarity between the corresponding pairs of the patterns. Essentially, each sim_i could be regarded as an aggregation of the class membership ω_i and $\boldsymbol{\psi}'_i$, namely

$$\psi : [0, 1]^c \rightarrow [0, 1]$$

being treated as the referential transformation. Graphically, one can portray the results of the explicit and implicit classification as shown in Figure 94.20.

One can also assume (that is intuitively appealing) that $M >> N$ as the assignment of the detailed class membership values is far more demanding than the scalar appointment done at the higher level of some general properties of the classes (such as e.g., similarity, difference, etc.). The same family of the training examples may also include instances coming from the $\mathfrak{D}$—as well as $\mathfrak{R}$—training set.

The General Architecture

The overall architecture of the classifier, Figure 94.21, consists of the two main functional blocks. The first one is responsible for the direct classification ($\mathfrak{D}$—training set). It is followed by the module of the referential classification that is developed using the $\mathfrak{R}$—training set.

The direct classification is realized through a logic processor (LP). The referential part of the classifier treats the vectors of class membership as its inputs and returns the values of their referential computations. The simplest version of this module can be envisioned as a single m-input MATCH neuron as in Figure 94.21. Its extended version is just a referential (matching) processor (Pedrycz, 1995), where the levels of matching among the classes are additionally processed before generating the overall similarity level between the classes.

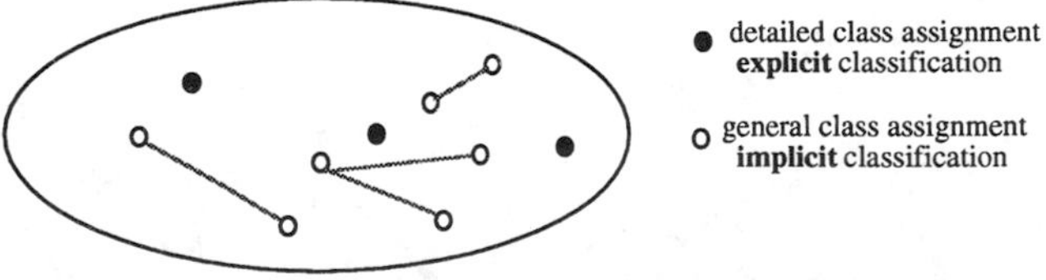

Figure 94.20 Explicit and implicit class assignment.

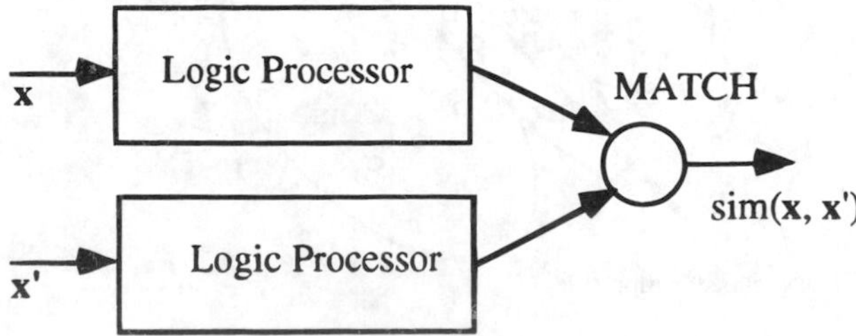

Figure 94.21 General architecture of a classifier.

The Design of the Classifier

Depending whether we deal with the patterns from the $\mathfrak{D}$ or $\mathfrak{R}$—training family, different modules of the classifier are developed. The fully labeled patterns are used to train the logic processor. The formulation of the problem is fairly standard: given is a learning set $(x_1, \omega_1), (x_2, \omega_2), \ldots, (x_M, \omega_M)$, construct the logic processor (including both its architecture and the connections) minimizing the given performance index Q,

$$\min_{LP} Q$$

where

$$Q = \sum_{k=1}^{M} (\omega_k - LP(\mathbf{x}_k, \text{connections}))^T \times (\omega_k - LP(\mathbf{x}_k, \text{connections}))$$

The structural learning usually involves additions and/or removals of the individual processing units (units). The parametric learning is driven by the gradient of the above performance index, $-\partial Q/\partial\text{connections}$. It is important to note here that the cardinality of the labeled patterns ($\mathfrak{D}$—training set) is usually very low in comparison to the overall training set. Therefore, one should not be concerned too much about a plain memorization carried out by the processor; let us stress that the primary objective is to minimize Q even at an expense of a significant architectural expansion of the processor. While at this phase the vast portion of the learning activities take place, some fine tuning can be done later in conjunction with the learning the referential part of the classifier. For the referential phase of learning (*viz* the learning exploiting the elements in the $\mathfrak{R}$ training set), the two identical copies of the previously constructed logic processor are used in the configuration shown in Figure 94.21. The training set used there comprises of the triples $((x_k, x_k), \text{sim}_k)$, $k = 1, 2, \ldots N$, constituting the dominant part of the entire training set, $M << N$, while sim_k denotes a degree of similarity reported for x_k and x_k (generally speaking, the discussed learning scheme works well for any other referential operation). The patterns x_k as well as x'_k are propagated through the logic processors and produce two vectors of class assignment, say ω_k and ω'_k. These are presented at the inputs of the matching neuron. The new training set comes in the form of the ordered triples $(\omega_k, \omega'_k, \text{sim}_k)$, $k = 1, 2, \ldots, N$. The parameters of the MATCH neuron are updated based on a squared error between sim_k and the output of the matching neuron MATCH (ω_k, ω'_k, w). The performance index Q is defined as the sum taken over the $\mathfrak{R}$-training set,

$$Q = \sum_{k=1}^{N} (\text{sim}_k - \text{MATCH}(\omega_k; \omega'_k, \mathbf{w}))^2$$

The gradient-driven scheme of updating is obvious,

$$\mathbf{w}(\text{new}) = \mathbf{w} - \xi \frac{\partial Q}{\partial \mathbf{w}}$$

$\xi \in [0, 1]$. In an on-line learning mode the modifications of w occur after presenting an individual input-output triple of data, $(\omega, \omega', \text{sim})$ (the subscript "k," as irrelevant in this scheme, has been left out)

$$\frac{\partial Q}{\partial \mathbf{w}} = -2(\text{sim} - \text{MATCH}(\omega, \omega', \mathbf{w})) \frac{\partial \text{MATCH}(\omega, \omega', \mathbf{w})}{\partial \mathbf{w}}$$

Its scalar notation leads to the expression,

$$\frac{\partial \text{MATCH}(\omega, \omega', \mathbf{w})}{\partial w_i} = \frac{\partial Q}{\partial w_i} \left\{ \mathop{T}_{j=1}^{m} (\omega_j \equiv \omega'_j) s w_j \right\}$$

$$= \frac{\partial Q}{\partial w_i} \{At[(\omega_i \equiv \omega'_i) s w_i]\}$$

and

$$A = \mathop{T}_{j \neq 1}^{m} [(\omega_j \equiv \omega'_j) s w_j].$$

Once the learning of the referential part is over but the value of the performance index is still not acceptable, two conceptually distinct remedial steps could be sought:

- An expansion of the referential module of the classifier by replacing the single MATCH neuron by the referential processor.
- An incremental retraining the logic processor already constructed with the aid of the $\mathfrak{B}$—training set. A special caution should be exercised, though, as this part of the classifier has been already trained and too radical modifications of its connections done at this stage could easily wash away the previous structure (due to the relatively small size of the $\mathfrak{B}$—training set). Hence, the learning should be vigilantly monitored. The corresponding learning schemes are shown in Figure 94.22. One can eventually consider two learning rates: the higher one is applied to the matching neuron while the lower rate guides the updates of the logic processors.

The connections of these two logic processors are updated simultaneously so that they are always maintained equal. Let x, y, and sim be provided. In view of this learning policy, the derivative of Q reads as

$$\frac{\partial Q}{\partial \text{ connection}} = \frac{\partial Q}{\partial \omega_x} \frac{\partial \omega_x}{\partial \text{ connection}} + \frac{\partial Q}{\partial \omega_y} \frac{\partial \omega_y}{\partial \text{ connection}}$$

(note that we have started with the two identical copies of the logic processor).

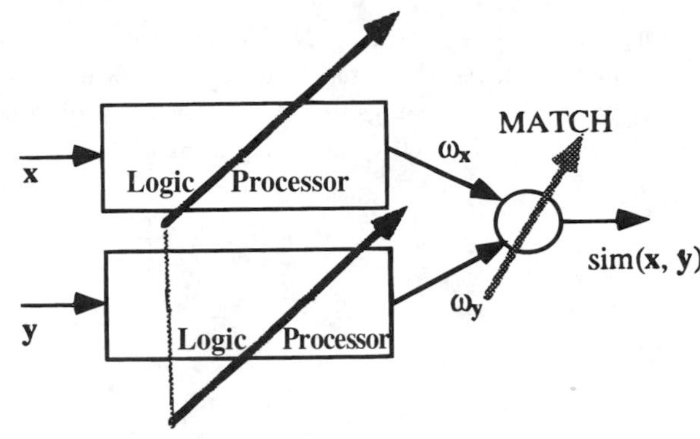

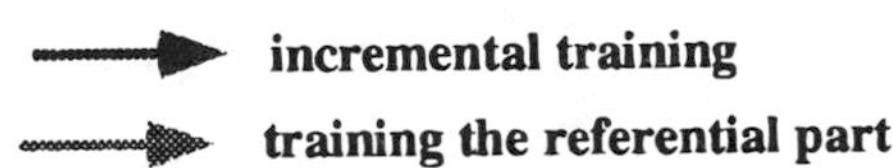

Figure 94.22 Learning modes in the classification scheme.

94.9 Unsupervised Learning

The mode of unsupervised learning in pattern recognition, known as clustering, is essentially carried out without teacher, namely the training set has no labeling associated with the patterns. Thus the main objective is to determine class membership of the patterns through revealing relationships (similarities) between them. As these dependencies could, in general, provide an approximate partition of the family of the patterns, it is highly desirable to equip the algorithm with some self-flagging mechanisms the aim of whose is to evaluate the quality of the clusters (Bezdek and Hathaway, 1992; Roubens, 1982; Backer, 1978; Ruspini, 1970).

FUZZY ISODATA Clustering Algorithm

Among several categories of fuzzy clustering (*viz.* the algorithms employing the notion of fuzzy sets in their development) those methods based on minimization of a certain objective (performance) function are evidently dominant (Bezdek, 1981). Their widespread utilization is due to the following reasons:

- Firstly, the results of clustering are usually appealing and very easy to interpret. Furthermore, the existing extensions and generalizations of the basic method, allow to handle a broad variety of problems and topologies conveyed by the data sets,
- Secondly, the underlying numerical scheme of clustering is well established as one has at his disposal quite efficient iterative schemes of optimization along with their sound theoretical properties.

Being more specific, let us consider a finite set of patterns viewed as elements in an n-dimensional feature space of reals R^n, $X = \{x_1, x_2, \ldots x_n\}$. Introduce also a so-called partition matrix describing in an unique manner a split of these patterns into c-classes,

$$U = [u_{ik}]$$

$i = 1, 2 \ldots c$, $k = 1, 2, \ldots, N$. The (i, k)-th element of the partition matrix, u_{ik}, characterizes a degree to which the k-element (pattern) of X belongs to the i-th cluster. Additionally, we impose the following self-evident requirements

$$\text{(i)} \sum_{i=1}^{c} u_{ik} = 1 \qquad k = 1, 2 \cdots N$$

and

$$\text{(ii)}\ 0 < \sum_{k=1}^{N} u_{ik} < N, \qquad \forall\ i = 1, 2 \cdots c$$

The objective function to be optimized within the clustering process is given in the form

$$Q = \sum_{i=1}^{c} \sum_{k=1}^{N} u_{ik}^{p} \|\mathbf{x}_n - \mathbf{v}_i\|$$

where $\| \,.\, \|$ stands for a distance function. The role of parameter p, $p > 0$, is to control a "fuzziness" of the clusters. The minimization of Q proceeds with respect to the partition matrix and the centroids of the clusters, v_1. $v_2, \ldots, V_c$. This leads to the formal statement of the form,

$$\min_{\mathbf{v}_1, \mathbf{v}_2, \ldots, \mathbf{v}_c} Q$$

subject to

$$U \in \mathfrak{U}$$

where $\mathfrak{U}$ denotes a family of the partition matrices satisfying (i) and (ii).

The well-known optimization scheme as described in Bezdek (1981) constitutes a sequence of iterations that leads to a local minimum of Q. The detailed algorithm is given below.

FUZZY ISODATA—Generic Version

1. Fix the number of clusters (c), select the distance function $\|.\|$, and initialize partition matrix U
2. Calculate centers (prototypes) of the clusters

$$\mathbf{v}_i = \frac{\sum_{k=1}^{N} u_{ik}^{p} \mathbf{x}_k}{\sum_{k=1}^{N} u_{ik}^{p}}$$

$i = 1, 2, \ldots, c$

3. Update partition matrix

$$u'_{ij} = \frac{1}{\sum_{l=1}^{c} \left[\frac{d_{ij}}{d_{lj}}\right]^{1/1-p}}$$

where

$$d_{ik} = (\mathbf{x}_k - \mathbf{v}_i)^T (\mathbf{x}_k - \mathbf{v}_i)$$

4. Compare U' to U, if $\|U - U'\| < \partial$ (with ∂ being a tolerance limit) then stop, else go to (2) with $U = U'$

The results of the algorithm is self-explanatory as the entire information about clusters are conveyed by the computed partition matrix. The higher the membership value u_{ik}, the stronger cohesion of the k-th pattern in the i-th cluster. One can also envision u_{ik} as a discrete membership function of this cluster defined over X.

As an illustrative example, let us discuss a two-dimensional data set shown in Figure 94.23—note that two clusters are visible with a single pattern whose membership to any of these clusters could raise some hesitation. In fact, the FUZZY ISODATA was able to reflect this phenomenon very well through the entries of the partition matrix—the assignment of this pattern is not that decisive in comparison to the rest of the patterns:

class	membership	
1	0.0123	0.9877
2	0.0246	0.9754
3	0.0197	0.9803
4	0.0256	0.9744
5	0.4205	0.5795
6	0.9981	0.0019
7	0.9982	0.0018
8	0.9858	0.0142
9	0.9691	0.0309
10	0.9745	0.0255

Figure 94.24 summarizes the values of the minimized objective function—in fact the minimization did not take more than a few iterations.

Even though the clustering algorithm essentially forms clusters without any supervision, a certain amount of a priori information about the structure one is looking for, becomes imperative and, in fact, is provided in advance. This aspect of unsupervised learning is not substantially emphasized, however a decision made with this regard has far reaching consequences concerning the clusters being obtained. The two crucial structural parameters are the form of the objective function and the number of clusters-both of them are very much domain (pattern) dependent. The

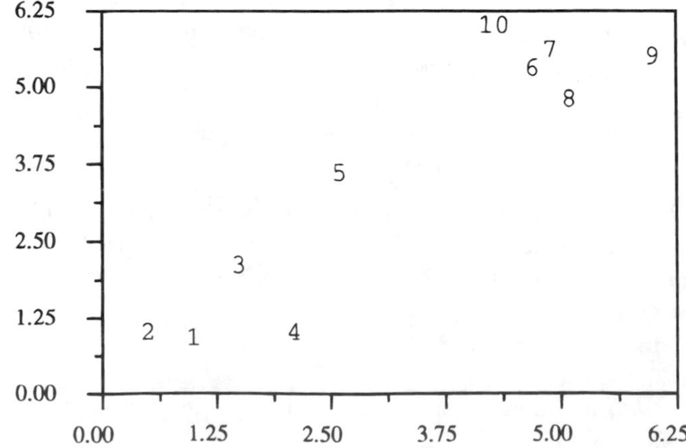

Figure 94.23 Two-dimensional synthetic data set.

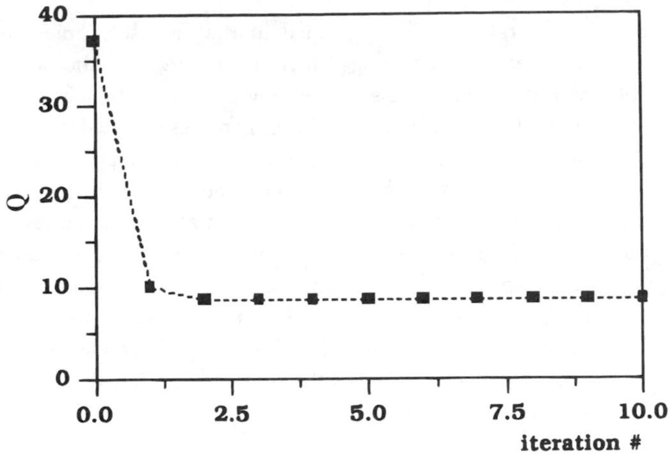

Figure 94.24 The values of Q in successive learning epochs.

number of clusters should be estimated prior to running the optimization procedure; usually one specifies their lower and upper bound, say c_{min}, c_{max}, respectively and repeats the clustering procedure for the number of clusters situated within these bounds. Some auxiliary cluster validity indices might be helpful in this situation (see e.g., Xie and Beni, 1991 and Windham, 1982). Unfortunately, one should be aware that the universality of these indices could be questioned in many instances.

The form of the distance function (metrics) $\|.\|$ becomes vital to the geometrical form of the clusters being formed. In essence, the type of the distance function predetermines the shape of the clusters. By admitting specific distance, the method is well equipped to "detect" and capture a certain quite limited category of these geometrical constructs. To a certain extent, one can claim that in this sense, the clustering method becomes "sensitized" to these generic geometric objects and looks at the patterns from this somewhat limited, if not biased, processing perspective. The objects favored by the different distance functions could be various constructs like spheres, hyperellipsoides, linear varieties, etc. The generic FUZZY ISODATA relies on the Euclidean distance defined as

$$\|\mathbf{x} - \mathbf{v}_i\| = (\mathbf{x} - \mathbf{v}_i)^T(\mathbf{x} - \mathbf{v}_i)$$

which naturally favors spherical groups of the patterns. The obvious extension would be the one accepting the Mahalanobis distance,

$$\|\mathbf{x} - \mathbf{v}_i\| = (\mathbf{x} - \mathbf{v}_i)^T\mathbf{A}_i^{-1}(\mathbf{x} - \mathbf{v}_i)$$

where $\mathbf{A}_i$ is a positive definite square matrix. Some essential generalizations include those of linear varieties and *c*-shells (Bezdek and Hathaway, 1992; Dave, 1992; Dave and Bhaswan, 1992).

Referring to the first property of the family of the partition matrices, one can note it has a certain probabilistic flavor by requiring that the sum of the membership values equals one. In fact the most "unclear" (indeterminate) clustering situation happens for the *k*-th pattern when all the grades of membership are equal, say $u_{1k} = u_{2k} = \ldots = u_{ck} = 1/c$. The membership value is dependent upon the number of clusters. To alleviate this restriction an extension proposed in (Krishnapuram and Keller, 1993) is concentrated as a so-called possibilistic clustering in which the probabilistic requirement (*i*) is replaced by the "coverage" condition expressed as

$$\max_{i=1,2,..c} u_{ik} > 0$$

Directional Clustering and Its Role in System Identification

The clustering techniques have been intensively utilized in construction of fuzzy models. The existing approaches have been concentrated either on the development of linguistic labels (fuzzy sets) of the system's variables or formation of clusters within the available numerical data.

More formally, we will be interested in clustering the patterns (data) (x,y) regarded as discrete fuzzy sets distributed within the corresponding unit hypercubes. By concatenating the elements of $[0, 1]^n$ and $[0, 1]^m$, that is x and y, say,

$$\mathbf{z} = [\mathbf{x}|\mathbf{y}]$$

one can look at the clustering method as being applied to these new objects distributed now within the $[0, 1]^{n+m}$ hypercube. By not introducing any *a priori* knowledge about the functional dependencies between the variables, the existing clustering methods are essentially aimed at discovering *relational* rather than *functional* dependencies between the variables. N.B. one can make the distinction between the character of these dependencies even more transparent by contrasting between the use of functions in procedural languages and predicates (relations) in declarative languages such as e.g., PROLOG. While this clustering might constitute an interesting task per se, this formulation does not reflect well the very nature of the function as a *directional* construct holding between the variables. In contrast, the enhanced clustering method taking into account the direction of these dependencies (viewed essentially as a mapping from $[0, 1]^n$ to $[0, 1]^m$) applying this to data analysis will be referred to as a *directional* clustering. The required behavior of the clustering mechanism will be accomplished by defining a suitable direction-sensitive objective function guiding the formation of the clusters. Principally, the property of directionality requires that the clustering criterion Q is asymmetrical with its arguments, namely,

$$Q(\mathbf{x}, \mathbf{y}) \neq Q(\mathbf{y}, \mathbf{x})$$

Any method of directional clustering should assuredly be dwelled upon this observation.

The objective function (clustering criterion) used to describe an extent to which two pairs of patterns (x_k, y_k) and (x_l, y_l) could be regarded as the elements belonging to the same cluster should comply with the following qualitative observations:

(i) (x_k, y_k) and (x_l, y_l) treated as two candidates to be included in the same cluster should be similar coordinatewise, namely the corresponding coordinates of x_k and x_l as well as y_k and y_l should be *similar*.

(ii) The "directionality" component of the performance index should contemplate the functional direction to be discovered within the data (recall that the mapping $y = f(x)$ denotes that "x implies y" but not other way around) and this fact should be reflected by the character of the elements assigned to the cluster. In such a sense, with x_k almost equal to x_l but quite different y_k and y_l, these two patterns should fall into the same cluster. In the opposite case when x_k and x_l differ significantly while simultaneously having similar y_k and y_l, these patterns should be definitely allocated to the two distinct clusters.

The directionality aspect to be included in the objective function is incorporated through the relationship,

$$(\mathbf{y}_k \equiv \mathbf{y}_l) \rightarrow (\mathbf{x}_k \equiv \mathbf{x}_l)$$

where

$$\mathbf{x}_k \equiv \mathbf{x}_l = \frac{1}{n}\sum_{i=1}^{n} (x_{ki} \equiv x_{li})$$

and

$$\mathbf{y}_k \equiv \mathbf{y}_l = \frac{1}{m}\sum_{j=1}^{m} (y_{kj} \equiv y_{lj})$$

As the clusters have to be formed on a basis of (i) and (ii), the multiplicative aggregate of these two expressions gives rise to the formula.

$$Q = \frac{1}{2}\,[(\mathbf{x}_k \equiv \mathbf{x}_l) + (\mathbf{y}_k \equiv \mathbf{y}_l)][(\mathbf{y}_k \equiv \mathbf{y}_l) \rightarrow (\mathbf{x}_k \equiv \mathbf{x}_l)]$$

that from now on will be used as the clustering objective function.

The clustering procedure is applied successively to the individual elements of the data set. The process is carried out bottom-up in an aglomerative manner: we proceed with N clusters each consisting of a single pair of the input-output patterns and merge them successively based on the values of the objective function produced via this combination. Starting from "N" single-element clusters $\{(x_1, y_1)\}\ \{(x_2, y_2)\}, \ldots, \{(x_N, y_N)\}$, the new two-element cluster $\{(x_{i0}, y_{i0}), (x_{j0}, y_{j0})\}$ is formed in such a way that this cluster leads to a maximum of Q determined over all possible mergings of the available data points. Subsequently, the clusters to be expanded (merged) at the successive stages are guided by the maximal value of the performance index Q averaged over the corresponding cluster. This leads to the general merging rule:

Merge clusters $\mathcal{X}$ and $\mathcal{X}'$ for which the sum

$$\frac{1}{\mathrm{card}(\mathcal{X})\,\mathrm{card}(\mathcal{X}')} \sum_{(x_k,y_k)\in\mathcal{X}} \sum_{(x_l,y_l)\in\mathcal{X}'} Q(k, l)$$

attains a maximal value among all possible mergings.

The fundamental question that usually emerges when using any clustering technique is the one about a "plausible" number of the clusters to be distinguished in the data set. Since the proposed method is of an agglomerative nature, one should be able to control the process of merging by terminating it when the produced clusters cannot sufficiently represent the data. The deficient representation phenomenon occurs due to an excessive variety of the objects placed within the same cluster. This in turn calls for a formal definition of the representation capabilities of the clusters. More precisely, we will be interested in expressing how well the prototypes represent the elements of the generated clusters. Let us introduce the following notion. A prototype $p = (p_x, p_y) \in [0, 1]^n \times [0, 1]^m$ of cluster $\mathcal{X}$ is an element of $\mathcal{X}$, say $(\mathbf{x}_{i_0}, \mathbf{y}_{i_0})$ that maximizes the discussed objective function computed with regard to this cluster.

$$\mathbf{p}_x = \mathbf{x}_{i_0} \text{ and } \mathbf{p}_y = \mathbf{y}_{i_0} \text{ if } \max_{(\mathbf{x}_l,\mathbf{y}_l)} \sum_{k=1}^{\mathrm{card}(\mathcal{X})} Q(k, l) = \sum_{k=1}^{\mathrm{card}(\mathcal{X})} Q(k, i_0)$$

(Note that the indices in Q are used to emphasize the patterns being discussed). The resulting value of Q, say $Q(\mathcal{X})$, is used as a measure of the representation capabilities of the prototype taken with respect to $\mathcal{X}$. For the single-element clusters, $\mathrm{card}(\mathcal{X}) = 1$, the elements of the clusters are obviously ideal prototypes and the above expression always equals 1. In sequel, the global sum of Q taken over all the clusters "c",

$$V(c) = \sum_{\text{all clusters}} Q(\mathcal{X})$$

could be admitted as an indicator of representativeness of the data conveyed by their clusters (more precisely, their prototypes). For $c = N$ one has $V(c) = N$. Generally speaking, $V(c)$ is a non-decreasing function of the number of clusters, namely $V(c_1) \leq V(c_2)$ for $c_2 > c_1$. The analysis of the behavior of $V(c)$ being plotted versus c could be used to detect the most "plausible" number of clusters: the minimal value of c, say, c^*, that does not lead to a substantial and abrupt decrease in V can be accepted as a viable candidate for the structure in this set of patterns.

Acknowledgment

Support from the Natural Sciences and Engineering Research Council of Canada, MICRONET and University Research Grants Program (URGP) of the University of Manitoba is gratefully acknowledged.

References

Backer, E. 1978. *Cluster Analysis by Optimal Decomposition of Induced Fuzzy Sets,* Delft University Press, Delft, the Netherlands.

Bellman, R. E., Kalaba, R., and Zadeh, L. A. 1966. Abstraction and pattern classification, *J. Math. Anal. Appl.,* 13:1–7.

Bezdek, J. C. 1981. *Pattern Recognition with Fuzzy Objective Function Algorithms,* Plenum Press, New York, NY.

Bezdek, J. C. and Pal, S. K., 1992. *Fuzzy Models for Pattern Recognition,* IEEE Press, New York, NY.

Bezdek, J. C. and Hathaway, R. J. 1992. Numerical convergence and interpretation of the fuzzy c-shells clustering algorithm, *IEEE Trans. Neural Networks,* 3:787–793.

Cover, T. and Hart, P. 1967. Nearest neighbor pattern classification, *IEEE Trans. Information Theory,* 13:21–27.

Dave, R. N. 1992. Boundary detection through fuzzy clustering, *IEEE Int. Conf. Fuzzy Systems,* 127–134, March 89–12. Sand Diego, CA.

Dave, R. and Bhaswan, K. 1992. Adaptive fuzzy c-shells clustering and detection of ellipses, *IEEE Trans. Neural Networks,* 3:643–662.

Devijver, P. and Kittler, J. 1982. *Pattern Recognition: A Statistical Approach,* Prentice-Hall, Englewood Cliffs, NJ.

Dubois, D. and Prade, H. 1988. *Possibility Theory—An Approach to Computerized Processing of Uncertainty,* Plenum Press, New York, NY.

Duda, R. and Hart, P. 1973. *Pattern Classification and Scene Analysis,* John Wiley & Sons. New York, NY.

Fukunaga, K. 1991. *Statistical Pattern Recognition,* Academic Press, San Diego, CA.

Gonzales, R. C. and Woods, R. E. 1992. *Digital Image Processing,* Addison-Wesley, Readings, MA.

Hirota, K., and Pedrycz, W. 1991. Fuzzy logic neural networks: design and computations, *Int. Joint Conf. Neural Networks,* 18–21 Nov. 1991, 152–157 Singapore.

Hirota, K. and Pedrycz, W. 1994. OR/AND neuron in modeling fuzzy set connectives, *IEEE Trans. Fuzzy Systems,* 2:151–161.

Huntsberger, T., Rangarajan, C., and Jayaramamurthy, S. N. 1986. Representation of uncertainty in computer vision using fuzzy sets, *IEEE Trans. Computers,* C-35:145–156.

Kandel, A. 1982. *Fuzzy Techniques in Pattern Recognition,* John Wiley, & Sons New York, NY.

Keller, J. M. and Hunt, D. J. 1985. Incorporating fuzzy membership functions into the perceptron algorithm, *IEEE Trans. Pattern Analysis and Machine Intelligence,* PAMI-7:693–699.

Krishnapuram, R. and Keller, J. M. 1993. A possibilistic approach to clustering, *IEEE Trans. Fuzzy Systems,* 2:98–110.

Krishnapuram, R., Keller, J. M., and Ma, Y. 1993. Quantitative analysis of properties and spatial relations of fuzzy image regions, *IEEE Trans. Fuzzy Systems,* 1:222–233.

Menger, K. 1942. Statistical metric spaces, *Proc. Nat. Acad. Sci.,* 28:535–537, USA.

Minksy, M. L. and Papert, S. A. 1988. *Perceptrons (Expanded edition),* MIT Press, Cambridge, MA.

Pal, S. K. and Dutta-Majumder, D. K. 1986. *Fuzzy Mathematical Approach to Pattern Recognition,* John Wiley & Sons, New York, NY.

Pedrycz, W. 1985. Algorithms of fuzzy clustering with partial supervision, *Pattern Recognition Letters,* 3:13–20.

Pedrycz, W. 1990. Direct and inverse problem in comparison of fuzzy data, *Fuzzy Sets and Systems,* 34:223–236.

Pedrycz, W. 1990. Fuzzy sets in pattern recognition, *Pattern Recognition,* 2/3:121–146.

Pedrycz, W. 1991. Neurocomputations in relational systems, *IEEE Trans. Pattern Analysis and Machine Intelligence,* 13:289–296.

Pedrycz, W. 1991. Fuzzy logic in development of fundamentals of pattern recognition, *Int. J. Approximate Reasoning,* 5(3):251–264.

Pedrycz, W. 1992. Selected issues of frame of knowledge representation realized by means of linguistic labels, *Int. J. Intelligent Systems,* 7:155–170.

Pedrycz, W. 1992. Fuzzy neural networks with reference neurons as pattern classifiers, *IEEE Trans. Neural Networks,* 3:770–775.

Pedrycz, W. 1992. *Fuzzy Control and Fuzzy Systems,* 2d extended ed., Research Studies Press/J. Wiley, Taunton/New York.

Pedrycz, W. 1993. Fuzzy neural networks and neurocomputations, *Fuzzy Sets and Systems,* 56:1–28.

Pedrycz, W. 1995. *Fuzzy Sets Engineering,* CRC Press, Boca Raton, FL.

Pedrycz, W. and Rocha, A. F. 1993. Fuzzy-set based models of neurons and knowledge-based networks, *IEEE Trans. Fuzzy Systems,* 1:254–266.

Pratt, W. K. 1991. *Digital Image Processing,* 2d ed. John L Wiley & Sons, New York, NY.

Roubens, M. 1982. Fuzzy clustering algorithms and their cluster validity, *European J. Operations Research,* 10:294–301.

Ruspini, E. 1970. Numerical methods for fuzzy clustering, *Information Sciences,* 2:319–350.

Tou, J. T. and Gonzales, R. C. 1974. *Pattern Recognition Principles,* Addison-Wesley, Reading, MA.

Xie, X. L. and Beni, G. A. 1991. Validity measure for fuzzy clustering, *IEEE Trans. Pattern Analysis and Machine Intelligence,* 3:841–846.

Windham, M. P. 1982. Cluster validity for the fuzzy c-means clustering algorithm, *IEEE Trans. Pattern Analysis and Machine Intelligence,* 4:357–362.

Zadeh, L. A. 1977. Fuzzy Sets and Their Applications to Classification and Clustering: *Classification and Clustering,* 251–299. van Ryzin, J., ed., *Academic Press,* New York, NY.

Zadeh, L. A. 1978. Fuzzy sets as a basis for a theory of possibility, *Fuzzy Sets and Systems,* 1:3–28.

Zadeh, L. A. 1979. Fuzzy Sets and Information Granularity, *Advances in Fuzzy Set Theory and Applications,* 3–18, M. M., Gupta, R. K., Ragade, and R. R., Yager, eds., North Holland, Amsterdam.

Zadeh, L. A. 1983. The role of fuzzy logic in the management of uncertainty in expert systems, *Fuzzy Sets and Systems,* 11:199–227.

Appendix A

The appendix briefly summarizes the main ideas of triangular norms (Menger, 1942) as being used as a vehicle for representing logic operations (logic connectives) on fuzzy sets.

Definition 94.1: A *t*-norm is a function

$$t\colon [0, 1] \times [0, 1] \rightarrow [0, 1]$$

Satisfying the following conditions:

1. monotonicity

$$x < x' \text{ and } y < y' \text{ implies that } xty \leq x'ty'$$

2. commutativity

$$xty = ytx$$

3. associativity

$$(xty)tz = xt(ytz)$$

4. boundary conditions

$$xt0 = 0 \qquad xt1 = x$$

Definition 94.2: An *s*-norm (called also *t*-conorm) is a function

$$s: [0, 1] \times [0, 1] \rightarrow [0, 1]$$

fulfilling the following properties:

1. monotonicity

$$x < x' \text{ and } y < y' \text{ implies that } xsy \leq x'sy'$$

2. commutativity

$$xsy = ysx$$

3. associativity

$$(xsy)sz = xs(ysz)$$

4. boundary conditions

$$xs0 = x, \; xs1 = 1$$

It is also worth noting that with any s-norm one can associate a *t*-norm satisfying the relationship,

$$asb = 1 - (1 - a)t(1 - b)$$

or equivalently,

$$1 - asb = (1 - a)t(1 - b)$$

$a, b \in [0, 1]$.

Considering that the complement of A, Ā, is defined in a usual way,

$$\overline{A}(x) = 1 - A(x)$$

the above dependency is in fact nothing but De Morgan identity stating that

$$\overline{A \cup B} = \overline{A} \cap \overline{B}$$

Equivalently we may write,

$$\overline{A \cap B} = \overline{A} \cup \overline{B}$$

The *t*- and *s*-norm satisfying this condition are called dual. An important operation of implication associated with any continuous t-norm (more precisely, we may relax this requirement by admitting left continuity of *t*-norms) is defined as follows,

$$a\varphi b = a \rightarrow b = \sup\{c \in [0, 1] | atc \leq b\}$$

A list of commonly used triangular norms is given below:

t-norm	*s*-norm
$\min(x, y) = x \wedge y$	$\max(x, y) = x \vee y$
xy	$x + y - xy$
$1 - \min\left[1, [(1-x)^p + (1-y)^p]^{\frac{1}{p}}\right]$	$\min\left(1, (x^p + y^p)^{\frac{1}{p}}\right), p \geq 1$
$\dfrac{xy}{\gamma + (1-\gamma)(x + y - xy)}$	$\dfrac{xy(\gamma - 2) + x + y}{xy(\gamma - 1) + 1}, \gamma \geq -1$
$\max[0, (\lambda + 1)(x + y - 1) - \lambda xy]$	$\min[1, x + y + \lambda xy], \lambda \geq -1$
$\log_p [1 + \dfrac{(p^x - 1)(p^y - 1)}{p - 1}$	$1 - \log_p [1 + \dfrac{(p^{1-x} - 1)(p^{1-y} - 1)}{p - 1}], p>0, p \neq 1$
x, if $y = 1$	x, if $y = 0$
y, if $x = 1$	y, if $x = 0$
0, otherwise	1, otherwise

95

Neural Fuzzy Systems in Handwritten Digit Recognition

95.1 Introduction 1231
Pattern Recognition • Optimization • Artificial Neural Networks
95.2 System Design 1240
Feature Extraction • Clustering
95.3 Application to Handwritten Digits 1248
Introduction to Character Recognition • Data Collection • Preprocessing • Center of Mass Adjustment • Results
95.4 Discussion 1256
95.5 Summary 1258

Timothy J. Dasey
MIT Lincoln Labs

Evangelia Micheli-Tzanakou
Rutgers University

95.1 Introduction

This chapter is divided into four components. In this section, the concepts and background relevant to pattern recognition, some typical optimization techniques, including ALOPEX, and a tutorial on the ideas and early works in neural networks are discussed. The danger in this presentation is that these fields might be construed as disjoint problems. The truth is that a large amount of overlap exists between these conceptual divisions. Pattern recognition has benefited from the application of neural networks and optimization. Neural networks commonly use optimization routines to guide their training, and have achieved many of their greatest successes in pattern recognition applications. These relationships should be kept in mind during the reading. The last section of this introduction includes a philosophical discussion explaining the rationale for this work.

Pattern Recognition

Theory and Applications

To most individuals, a pattern recognition task involves an ability of the brain to assign labels to objects, sounds, feelings or ideas and discriminate one from another. Most of us are extremely adept at this processing task, while being unaware of the precise mechanism that provides us with this power. In fact, it is through the scientific field of pattern recognition, which relegates this task to machines, that the methods of our brain may be fully realized. Yet, there has not been a machine ever designed which has our capability to be a general-purpose pattern recognition machine.

Regardless of the limitations, machines perform quite well in the grouping and labeling of patterns from certain problem sets. Machines excel when the recognition task is confined to a specific application. An extensive body of literature describes the recent attempts at relegating many pattern recognition tasks to machines as the explosive growth of information overworks the human classifiers (Oja, 1983; Fu, 1982; Young, 1974). The use of automated pattern recognition machines has now touched nearly every field in an enormous variety of working places.

The special nature of each pattern recognition task requires selection of the best approach (Highleyman, 1962). Heuristic approaches, which rely on the designer's intuition and familiarity with the problem, are often sufficient to provide excellent solutions to many problems. Linguistic (syntactic) approaches are often useful when numerical measurements are not sufficient to describe the problem. Many pattern recognition problems can be solved through several mathematically substantiated techniques, using statistical variability between patterns or certain pattern similarity measures (Devijver and Kittler, 1982).

When confronting a typical pattern recognition task, three particular problems must be addressed by the designer. The first is the representation of the input data which the system will use in its classifications. When determined, these comprise the pattern vector x as

$$x = (x_1, x_2, x_3, \ldots, x_n) \tag{95.1}$$

where n is the total number of parameters needed for analysis. In many mathematical pattern recognition problems, it is often convenient to envision each parameter x_i as describing an axis

0-8493-8343-9/97/$0.00+$.50

in n-dimensional space (n-space), where each pattern then comprises a point in that space, as in the two parameter space depicted in Figure 95.1.

The second problem concerns the extraction of certain characteristic attributes from the pattern vectors and a reduction in the dimensionality (from n to m) of those vectors. This is usually termed the preprocessing and feature extraction problem. The attributes of features to be selected vary with the application, but involve the selection of pattern attributes which can best be used for the discrimination among the patterns. The feature extraction process can be thought of as an intermediate formulation of the more prominent goal of pattern recognition: the compression of large numbers of attributes to a small number of class determinants (Watanabe, 1972).

The third problem involves the determination of optimum decision procedures, which are used for the identification and classification process. Many such procedures involve the separation of n-space (or m-space) into clusters, much as Figure 95.1 includes pattern points which are grouped into three similar categories.

Although mathematical formulations of pattern recognition methods have been available for several decades, many prominent problems still must be solved before great theoretical improvements can be made (Kohonen, 1988b, Lerner, 1972). One issue, that of properly estimating the classification performance of a machine, has been largely agreed upon (Toussaint, 1974). It is generally thought that two pattern populations are needed to ensure that a machine pattern recognizer can generalize. One group is reserved for the training of the machine (determination of the decisions involved in making a classification), and the other is used for post-training testing. This helps to ensure that the decisions formulated for the training group also apply to the similar but distinct non-training group.

Feature Extraction

A feature of a given parameter set refers to an attribute described by one or more elements of the original pattern vector.

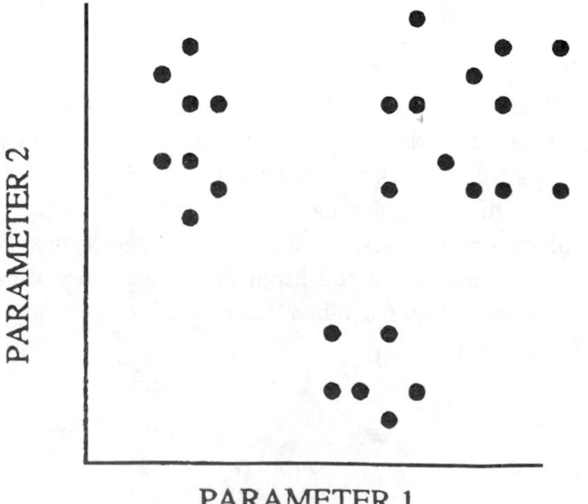

Figure 95.1 A two parameter space. Each point in the space corresponds to aninputr pattern vector.

In an application to imaging the elements are each pixel, and a feature may be selected as a subset of the pixel intensity values. More commonly, a feature describes some combinations of the original pixels, as in a Fourier expansion or a spatial filtering operation. The precise meanings of a preprocessing operation and a feature extraction process overlap, but in general a feature extraction operation involves the reduction in dimensionality of the pattern vector. The primary reason for such a transformation is to provide a set of measurements with more discriminatory information, and less redundancy, to the classifier.

The precise choice of features is perhaps the most difficult task in pattern processing. In order to know the most successful set of features for a particular problem, the accuracy of the classifier must be known. Yet the classifier depends on the information from a feature extraction device, and thus cannot normally provide that information without a completely designed feature extractor. This enigma remains the primary reason for the difficulty in evaluating competing feature extraction methods. It has prompted many researchers to subjectively select features from an educated guess of what will be most important to the classifier. These techniques can be effective, but are increasingly more difficult to ascertain as the complexity of the patterns increases, and are always subjected to personal bias.

Many common mathematical parameters are used as feature measurements (Levine, 1969). A set of n-space Euclidean distance measurements is a very common example. In other situations where the identity of the patterns are known, transformation matrices to minimize intraset pattern entropy (Watanabe, 1988) or intraset pattern dispersion, and functional approximation methods are commonly used. Several orthogonal expansions are also used, including the Fourier expansion and the Karhunen-Loève (K-L) expansion (Fukunaga, 1970). The K-L expansion offers certain optimal properties, and will be reviewed in more detail later in this chapter. Other common measurements, such as moment invariants (Teh, 1986), are used because of their constancy under many common pattern transformations. The number of different schemes for feature extraction, even in similar applications (such as the processing of handwritten characters), is typically enormous.

Clustering

A clustering operation may connote different meanings to different people, even in the pattern recognition community. This article will use a terminology commonly accepted by many scientists. A clustering operation involves the grouping of like patterns with one another without any knowledge of pattern identity beforehand (an unsupervised classification operation). Classification problems generally have this information. In the clustering problem, patterns must be separated solely on their specific attributes, whereas the classification problems have access to error signals which can be generated to guide the decision making of the machine.

Continuing with the geometrical analogy outlined in the theory and applications section, let us envision each pattern as a point in n-space, much as we see each star as a point in the sky. If we are asked to group the stars in the sky, what measurement do we use for this determination? The exact formulation of this answer often depends on the application (Romesburg, 1984, Scoltock, 1982). In some instances, the distances between patterns can be used to separate patterns. In other cases, pattern density in regions of space are used to indicate locations where patterns likely are drawn from the same class, the techniques known as histogram approaches (Leboucher and Lowitz, 1978).

It should be clear already that no clustering operation can ever be guaranteed to operate without error. The successful operation of the clustering method relies on the separability of the data from the attributes used as the pattern vectors. If two or more classes overlap in n-space, they will never be perfectly separated. All clustering problems rely on the fidelity of this input data and generally are based on the separation of highly dense regions of patterns from one another. Each of these "modes" of the distribution is assigned a particular class label at a later time.

A large class of problems rely on a hierarchical grouping of pattern data (Dante and Sharma, 1985). The procedures used usually have the disadvantage of a phenomenon called "chaining", where small errors in grouping at the extremes of the tree accentuate at later levels. Patterns in this scheme can be arbitrarily given a classification by choosing to "cut" at a particular level of the tree, but recent thinking is that there is significant information in leaving the class identity of patterns "fuzzy". Fuzzy clustering refers to assigning grades of membership to patterns, and is currently a widely touted method (Gath and Geva, 1989; Davis and Economou, 1984).

Optimization

Theory and Objectives

In many situations it is desired to find the values of a set of parameters that best define the solution to a particular problem. As a rule, it is always possible to perform an exhaustive search over the entire parameter space, choosing the parameter values that are closest to the desired operation of the system. In most cases, a measure can be formalized to assess the degree of fit of the proposed solutions to the ideal. This measure is usually termed a cost, energy, Hamiltonian, or objective function. Although an exhaustive search through all allowable combinations of system parameters is always theoretically possible, it is generally not feasible for even a moderately high number of system parameters. In fact, the number of possible choices (N) explodes exponentially with the number of parameters (q) involved in the space as

$$N = n_1 n_2 n_3 \cdots n_q \tag{95.2}$$

where n_j is the number of samples of parameter j, or as

$$N = n^q \tag{95.3}$$

when $n_1 = n_2 = \ldots n_q$. As an example, if the system is composed of 3 parameters and the search is conducted by sampling the parameters every 0.1 in the interval [0,1], then there are 11^3 or 1,331 search items. With the same sampling and 10 parameters, the search list includes 11^{10} or 2.6×10^{10} items.

It is obvious that this scheme is untenable for all but simple problems, and is certainly impossible for the implementation of a dynamic system. It is for this reason that optimization procedures have received attention for a long period. The goal is to find the optimal (or at least close to optimal) solution with a shorter search time than the exhaustive search method. One of the conceptual means of achieving this uses a hyperdimensional geometrical visualization of the cost function as it varies with each of the (presumably uncoupled)[1] system parameters. This parameter space is usually widely variant, and the search over that space involves the extraction of a global minimum (or maximum, depending on the cost function used) from among all of the local minimum. In truth the global extremum is rarely consistently attainable for realistic situations in finite time, but usually a very close approximation is both achievable and sufficient.

Two means for adjusting the exhaustive search technique readily come to mind. The first involves sampling the entire parameter space at a low resolution and finding the lowest (in the case of a minimization) region. Then that subregion can be sampled at a higher resolution ad infinitum. This procedure has occasionally been adapted, but makes the major assumption that the global extremum is contained in a larger depression about it (and that the boundary of the global minimum is at least approximately funnelling into that extremum). This is a gross oversimplification, and application of these methods can result in a solution far from the best choice. In many other schemes, referred to as gradient descent techniques, the effect of a parameter change on the cost function is calculated, and the parameter is adjusted so that it is moving downhill toward a better solution. This results in a rapidly converging iterative procedure, but the technique is fortuitous if the solution arrived at is a global, not a local, minimum. All good optimization routines work on the concept that a short-term deleterious move, moving uphill as well as downhill, is necessary to ensure the possible escape from local minima and arrive near, or most preferably at, the global extremum.

Background

Much of the literature on optimization deals with the analogy between optimization and statistical mechanics. Perhaps the first to draw this comparison was the technique which has become known as the Metropolis algorithm (Metropolis, et. al., 1953). The system was originally written as a means for investigating such macromolecular properties as the states of substances at the level of a set of N individual molecules. At any particular time, the potential energy of the system can be found as

$$E = \frac{1}{2} \sum_{i=1}^{N} \sum_{j=1}^{N} V(d_{ij}) \forall i \neq j \tag{95.4}$$

[1] Two parameters are considered coupled if the location of the minimum in one variable is affected strongly by the value of the other.

where V is the potential between molecules and d_{ij} the minimum distance between molecules i and j. The problem consists of optimizing the positions of the particles in space (in this case 2-D space) to arrive at the lowest potential energy of the system. Starting with random positions of the particles, each particle is moved a random amount in a random direction. The new energy of the configuration is checked. If the energy (E) is decreased, the move is allowed. However, if the move results in a higher potential energy, the move is allowed with a probability $P(\Delta E)$

$$P(\Delta E) = e^{-\Delta} E/kT, \tag{95.5}$$

which is the Boltzmann distribution. Notice that this is no longer a gradient descent technique, but rather there is always a finite probability that the system can move uphill, out of a local minimum. In this way, the equilibrium states of the set of molecules could be analyzed, and it was seen that the system settled in configurations which also conformed to a Boltzmann distribution. It turns out that this method is a simple modification of a Monte Carlo scheme, where instead of choosing configurations randomly and weighting those configurations with a Boltzmann factor, the configurations are chosen with a Boltzmann distribution (evident through the simulations) and weighted evenly.

The analogy between this statistical mechanics problem and optimization was explored even further with the introduction of the "*simulated annealing*" procedure (Kirkpatrik et al., 1983). If we examine the Boltzmann update from the Metropolis algorithm, it is clear that the higher the temperature, the more likely that an uphill move will be accepted. Conversely, at zero temperature all uphill moves will be denied and the system will fall to an energy minimum. To ensure a ground state configuration (without crystal imperfections) in a material, the system must be carefully annealed, a process where the substance is first melted and then slowly cooled, with extra time spent near the vicinity of the phase transition. With the analogy to the optimization problem, a "ground state" (global minimum) of the system may be found by starting off at a high value of temperature. This corresponds to melting the system so that uphill moves are nearly equiprobable to downhill moves, and the system randomly wanders in parameter space. By slowly lowering the temperature (and thus reducing the probability of uphill moves), the system can slowly settle in to a minimum. It has been shown that the simulated annealing procedure can find the global minimum under certain conditions with probability 1.0, but that finding may take an inordinate amount of time. An analogous calculation to the specific heat of the system can be used to signal phase transitions in the optimization. The simulated annealing procedure has been applied to a wide variety of pattern recognition tasks (Xu, 1988). The primary emphasis of experimentation with the algorithm has been the adjustment of the cooling schedule, the process of lowering the temperature (Rutenbar, 1989). The simulated annealing procedure has received great attention over the last decade, but is burdened by the application dependant optimum cooling method and the necessity of a large number of iterations for convergence.

Another has been dubbed Mean Field Annealing (Snyder, et. al., 1989, Peterson and Hartman, 1989). In this scheme, the cost function $H(x)$, which may have many local minima and in other ways be "ugly", is replaced by another function $H(x,m)$ which "resembles" $H(x)$ but has components which are much easier to minimize (they could be convex functions with only one minimum). To make the two functions resemble each other, the set of parameters m_l must be estimated. To perform this, another technique is borrowed from statistical mechanics, the mean field approximation. The details are too intensive to consider in this synopsis, but by estimating each of the parameters m_i, the problem is reduced to a series of gradient descents at each value of temperature (note that this theory also utilizes a Boltzmann probability distribution).

Several researchers have noted that each of the above methods involves a local search about the current point in parameter space. That is, even with the capability to move uphill, all operations are still local. The odds of crossing a wide gap to a region of "better" minima is low, and thus more global search methods have been proposed. One commonly used technique is to run multiple trials on a given data set, saving the best result. Given a high number of random starts, the hope is that the global optimum will be among the optima identified. Galar (Galar, 1989) proposed a similar optimization routine to that of Eigen's theory of macromolecular evolution (Eigen, 1971) which was more capable of crossing wide gaps. This method has many striking similarities, at least in concept if not in the method of application, to the ALOPEX process discussed in detail in the next section. Galar describes a two term parameter update, one of which is a modified Markov chain and the other a random walk component. He claims that the resulting "biased random walk" is more capable of crossing wide gaps between local extrema than procedures like simulated annealing.

Another recent approach has been deemed the dynamic tunneling algorithm (Levy and Montalvo, 1985). This routine uses gradient descent to go to a local minimum, at which time the system "tunnels" through the surrounding hill (using an appropriately defined tunneling function) for the purpose of finding a point, other than the last minimum, which when gradient descent is continued will arrive at a point lower than the last minimum. The calculations are quite intensive, but the algorithm converges relatively often to the global minimum, and may be more effective for problems with high density of local minima.

ALOPEX

The optimization routine ALOPEX (*AL*gorithms *O*f *P*attern *EX*traction) presents an alternative to the previously reviewed algorithms. It was originally applied to the measurement of the visual receptive fields of cells in the adult frog tectum (Tzanakou et al., 1979; Harth and Tzanakou, 1974). In the original application of the method, the cost function was referred to as the response function R.

The method normally updates the model parameters (the pixel intensity values in the original application) as

$$P_i(n+1) = P_i(n) + B_i(n) + r_i(n+1), \tag{95.6}$$

where $B_i(n)$ represents the influence of a term due to historical bias, and $r_i(n)$is a random noise component. The bias term is calculated as

$$B_i(n) = B_i(n-1) + \gamma \Delta P_i(n) \Delta R(n) \tag{95.7}$$

where $\Delta P_i(n)$ represents the previous change in the ith parameter value $P_i(n)$ as

$$\Delta P_i(n) = P_i(n) - P_i(n-1), \tag{95.8}$$

and $\Delta R(n)$ indicates the similar change in the response function as

$$\Delta R(n) = R(n) - R(n-1). \tag{95.9}$$

The two terms in the modification of Equation 95.6 provide different influences on the optimization. The first term is a bias term which tends to move the parameter in the direction which has been successful in the past. It is actually an aggregation of the biases to that point in the simulation, where the direction of the latest addition to the bias is determined by the change in the response function due to the last move. The second term is a random number, generated for each parameter at each iteration, which provides the opportunity for the parameter to move against the direction of recent success. As mentioned earlier, this capability to move "uphill" is what provides a good optimizer with the ability to escape local extrema[2]. The term $r_{ij}(n)$ in the ALOPEX update equation is typically implemented as a gaussian random number with zero mean and standard deviation s.

The accumulation of the biased terms in Equation 95.7 must be controlled in order to prove helpful. Without this regulation, the magnitude of B_i due to past iterations may overpower the relatively smaller change from the current iteration. In this scenario, the system has effectively gained "mass", so that the "momentum" of the movement in one direction will not allow the system to stop quickly enough at the sites of the extrema. In all simulations in this work, the magnitude of B_i is constrained to the limits $[-a,a]$. The first two iterations of the simulation supply random numbers for the forthcoming update statements. The responses are found for each, and the update Equations 95.6–95.9 are applied to all of the parameters. This process repeats itself until the simulation is finished. Note that the ALOPEX process provides for the simultaneous update of all parameters at once, which the simulated annealing algorithm does not. This generally makes the ALOPEX process more time conservative. In addition, the magnitude of the random component does not depend on the amount by which that component raises or lowers the response (there is a dependence via the Boltzmann distribution is simulated annealing). This makes it easier for parameters to traverse wide gaps between the extrema.

Even with the differences previously indicated, there are some analogies between the parameters of simulated annealing and those of ALOPEX. If the magnitude of the random component is much higher than that of the biasing component, then the parameters will be overwhelmingly driven by randomness, a situation analogous to the "melting" process in simulated annealing. Conversely, with no noise, the ALOPEX process simplifies to a gradient descent[3]. This indicates that the choices of γ and s are critical for controlling the speed and accuracy of the convergence.

The suspicion that the ALOPEX process could be run under similar conditions of "annealing" was confirmed by earlier (Tzanakou, 1978) and later work (Dasey and Micheli-Tzanakou, 1989a) which showed that slowly shrinking the magnitudes of both the noise and bias components in the update of the parameters could result in a great improvement in both the speed and accuracy of the optimization. In this work, the values of g and s were initially high and were lowered during the course of the simulation by the schedule

$$\gamma(n) = (\gamma_0 - \gamma_{infinity})e^{-n/t} + \gamma_{infinity}, \tag{95.10}$$

and

$$S(n) = (\sigma_0 - \sigma_{infinity})e^{-n/t} + \sigma_{infinity}, \tag{95.11}$$

where t was used to control the rate of the "cooling" and the initial and final parameter values are user entered.

Many other improvements have also been suggested, including a parallel implementation of the algorithm (Mellissaratos and Micheli-Tzanakou, 1989), averaging between multiple ALOPEX processes (Tzanakou, 1977) and an interleaved formulation of the algorithm to work on multiple response functions (Chon and Micheli-Tzanakou, 1989). Recent work has used distributed ALOPEX processes working on overlapping "fields" of an image to enhance convergence speed (Marsic and Micheli-Tzanakou, 1990). In a situation where the algorithm is used for noise removal or the correction of pattern imperfections, and there exist a set of templates to guide the optimization, it has been shown (Mellissaratos and Micheli-Tzanakou, 1989) that multiple response functions from each of the m templates $R_j(n)$ can be used to get a single response function as

$$R' = \sum_{j=1}^{m} \left(\frac{R_j^2(n)}{\sum_{k=1}^{m} R_k(n)} \right) \tag{95.12}$$

A similar function, with the inversion of both the numerator and denominator of Equation 95.12, was used for minimizing a particular response function (Dasey and Micheli-Tzanakou, 1989b).

---

[2] The form of Equation 95.6 is correct for the maximization of the response function R. For minimization, the sign of the bias term should be changed.

[3] Note that there is not complete freedom of movement of the parameters with no noise. This is due to the fact that only one response function is used to update a large set of parameters.

The ALOPEX process has been successfully applied to many application areas since its introduction, in large part because of its general and flexible form. The ALOPEX process is interesting in that the pattern recognizer can be converted to a pattern extractor (Deutsch and Micheli-Tzanakou, 1987; Micheli-Tzanakou, 1984). Other applications include curve fitting to waveforms such as Visual Evoked Potentials (Wang and Micheli-Tzanakou, 1990; Micheli-Tzanakou and O'Malley, 1985), crystal growth (Harth et al., 1988), the traveling salesman problem (Harth and Pandya, 1986), and pattern recognition applications (Dasey and Micheli-Tzanakou, 1989b). Using ALOPEX in perceptual tasks has also been addressed (Harth et al., 1986). Recent applications include the use of ALOPEX in reconstructing compressed images (Micheli-Tzanakou and Dasey, 1990), reducing motion artifacts (Ciaccio and Micheli-Tzanakou, 1990), and use of the VEP as a generator through ALOPEX of patterns for stimulation (Tezzi et al., 1990).

Artificial Neural Networks

Foundations of Neural Network Research

In the pioneering years of neurocomputing, the field was dominated by physiologists and cognitive psychologists eager to place the accumulated neurological and psychophysical data into a mathematical framework. The 1943 paper of McCullogh and Pitts was perhaps the most instrumental work of the early years (McCullogh and Pitts, 1943), the influence of which is still seen in present day publications. In their minds the most important attribute of neuron behavior was what was seen as the "all-or-none" response of the cell to its inputs. With this motivation, model neurons were attributed a binary character, where during each time quantum, neurons responded to the accumulated activity of its synapses with an active response (firing level 1), or an inactive behavior (firing level 0). Each of the synaptic inputs to a cell were added and compared to a threshold. If the sum exceeded the threshold, the cell was made active. Otherwise, it was kept inactive. Additionally, the model allowed the existence of absolute inhibitory synapses, which if active at any time would ensure the neuron was inactive regardless of the other inputs. In this way, each cell was performing a simple threshold gated logic operation, and it was illustrated that any finite logical expression could be realized by a set of McCullogh and Pitts neurons. This was perhaps the first true connectionist model, in which simple computing elements, working partly in parallel, could perform powerful computations with appropriately constructed connection strengths. Their idea was that the neuron was simple, deriving its power by embedding them in a nervous "system". Additionally, they speculated about the possibility of using continuous changes in threshold to enact learning and adaptation. A good introduction to McCullogh-Pitts neurons in the context of later work is given in Minsky's 1967 book (Minsky, 1967). Ironically, there are indications that Von Neumann's use of gated logic in digital computing was at least partly influenced by McCullogh and Pitts (Von Neumann, 1982, 1945), while McCullogh and Pitts themselves used a spatially distributed *analog* computation in their later model of the superior colliculus (McCullogh and Pitts, 1947).

Although this type of gated model can perform many interesting computations, many of them, such as the one proposed by McCullogh and Pitts, fell under harsh criticism. Particularly tough on such systems was the developer of the perceptron (Rosenblatt, 1958), a model which dominated neural network thinking for a decade. A number of variations of Rosenblatt's perceptron were proposed in the original paper, but the primary focus was the organization of a "layered" system, in which one layer of model neurons projects to another layer of such cells by way of parallel bundles. The first layer, receiving information from the environment, was called the retinal, or *S*-layer. It connected in a random but localized way to an association, or *A*-layer. In many later applications of the perceptron, the *A*-layer was omitted. Then the *A*-layer connected reciprocally to a final response, or *R*-layer. The feedback connections from the *R*-layer to the *A*-layer were organized so that each *R* unit inhibits each *A* unit not connected to it. In this way, activation of a particular *R* unit tends to decrease the activation of all other *R* units. This "winner-take-all" philosophy would be reincarnated many times in the subsequent years (Feldman and Ballard, 1982; Kohonen, 1982). The model neurons were linear thresholded devices, where values above threshold varied linearly with the sum of inputs. The goal of such a system was the development of a "learning associator", which could make classification responses to stimuli. Initially, training in the perceptron used simple reinforcement. If an *A*-unit was active, its activity was increased further, so that the next stimulation of that pattern would produce an even greater response. The perceptron was poor at discriminating between arbitrary random patterns, but good at separating items in a "differentiated environment" where "each response is associated to a distinct class of mutually correlated ... stimuli".

A more explicit training procedure was later proposed for a two-layer system (without the A-layer, or hidden layer), in which the desired response supplied by an external "teacher" was available to generate an error for which to direct the training of the weights (Rosenblatt, 1959, 1962). Such a system could be proven to converge (the perceptron convergence theorem) for any set of linearly separable inputs (Block, 1962), meaning the connection weights could learn to classify correctly, if classification was possible. The problem was that the convergence may take a very long time, and that oscillations in the decision line could occur when linear separation was not absolute. Work by Widrow and Hoff (Widrow, 1962; Widrow and Hoff, 1960) was related to perceptrons, although the model neurons were threshold logic units with a binary result. A teacher provided an error between the network and the desired solution. It was shown that the mean square error (MSE) is a simple quadratic energy surface with only one, global, minimum. The gradient descent method they proposed provided a very fast solution. The Widrow-Hoff system is usually termed ADALINE (ADaptive Linear NEuron) in the neural network field and the algorithm of MSE reduction is called the LMS algorithm in signal processing. Many variations and applications of the LMS procedure have been enacted since this time (Widrow and Stearns, 1985).

The perceptron dominated the succeeding era of neurocomputing research, largely because of the proof of the perceptron convergence theorem. Although the limitations of the perceptrons and other analogous models were already becoming apparent, Minsky and Papert's book on the subject (Minsky and Papert, 1969) is considered to be largely responsible, along with the emergence of the field of Artificial Intelligence, for the stagnation in research in neurocomputing for several years thereafter. The book was a discussion of the limits of the perceptron structures. As mentioned, some of these were already known: the restriction to linear separability and an inability of the networks to acquire extensive generalizing capabilities. It was also apparent that small systems were much more effective than larger systems. Minsky and Papert based their attacks on the proofs that preceptrons could not compute two of what they considered to be fundamental operations: connectedness (patterns which can be drawn without lifting a pen from paper) and parity (counting the number of points). The mathematical analysis was brilliant, but the sound attacks on single-layer (only one set of connections) perceptrons were tainted with the subsequent conjectures on future extensions of such systems. As they put it

> "... we consider it to be an important research problem to elucidate ... our intuitive judgement that the extension [to multilayer systems] is sterile"(p.232).

That is, why bother with more complex systems, since they are bound to have the same problems?

The back-propagation algorithm, reemerging in 1986, uses a multi-layered perceptron with a somewhat different training routine. Although a similar algorithm was described in a Ph.D. thesis by Werbos (1974), it apparently went unnoticed until the nearly simultaneous independent solutions proposed by Parker (1985), LeCun (1986) and Rumelhart et al. (1986). Minsky and Papert recognized that multilayered perceptrons (with at least one layer of association, or hidden, units) were capable of much more complicated mappings from input to output. With such systems, decision surfaces of arbitrary complexity could be generated (Lippmann, 1987). The problem as they saw it, was the difficulty in training those hidden layers, when the source of error was only known at the output units.

The back-propagation learning algorithm addressed this doubt. The architecture for the network utilizes a multi-layered perceptron with non-localized connections, as shown in Figure 95.2. With the knowledge of the desired outputs of the network,

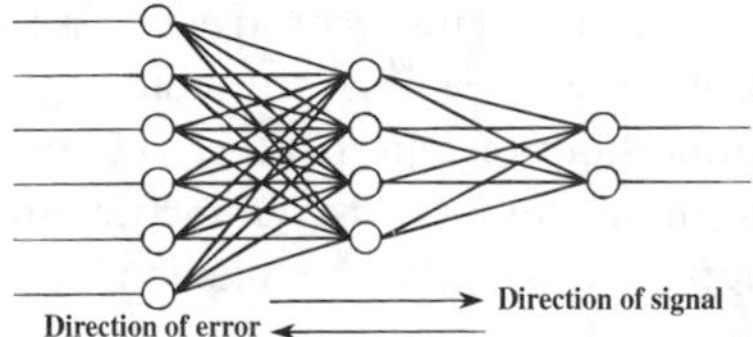

Figure 95.2 The conventional architecture for the multi-layered perceptron used with the backpropagation training method.

a simple extension to the Widrow-Hoff LMS rule could be used to train the hidden units. In this enactment, the weight from i to j at iteration t $(C_{ij})(t))$ could be adjusted by

$$C_{ij}(t+1) = C_{ij}(t) + \eta\delta_j O_i \tag{95.13}$$

where δ_j is an error term for unit j, η is a constant, and O_i is either the output of the ith hidden node (if the jth cell is an output unit), or an input (if the jth cell is a hidden node). If node j is an output unit, then

$$\delta_j = O_j(1 - O_j)(d_j - O_j) \tag{95.14}$$

where O_i is the actual output of node j and δ_j is the desired output of that node. If however the node j is a hidden node, then the difference from actual to desired activity is replaced by the weighted errors from the output layer as

$$\delta_j = O_j(1 - O_j) \sum_k \delta_h C_{jk} \tag{95.15}$$

In other words, the hidden nodes change their connection vectors based on the contributions they make toward the output unit errors. The process consists of feeding the inputs forward in the net to determine the output error, then propagating this error back to earlier layers of the net. Generally, the convergence can be made faster by adding a momentum term to the update rule as

$$C_{ij}(t+1) = C_{ij}(t) + \eta\delta_j O_i + \alpha(C_{ij}(t) - C_{ij}(t-1)) \tag{95.16}$$

where $0 < \alpha < 1$. The additional non-linearity introduced by the use of a sigmoidal firing of the hidden units to an integrated input x, given by

$$f(x) = \frac{1}{1 - e^{-r}} \tag{95.17}$$

provided the system with an extra degree of flexibility in the formulation of the interclass boundaries.

The algorithm has by far been the most extensively used neural network method since its introduction, and a good introduction to the theory of back-propagation has recently been written (Hecht-Nielson, 1989). Many attempts have been made to speed up the convergence (Stometta and Huberman, 1987; Cater, 1987), understand the information handling of the hidden layers (Mirchandani, 1989), and apply the network to a wide variety of classification tasks, including several medical applications (Marconi et al., 1989). Although there still exists no proof of convergence, as existed for the single-layer perceptron, it cannot be disputed that the back-propagation algorithm is a powerful classification technique for supervised pattern recognition problems.

It is clear that there exists a need for good supervised pattern recognition tasks, such as back-propagation. However, this type of problem is not the sort which originally attracted neurophysiologists to the field, nor is it the sort which can readily be applied to many real world applications, where a priori knowledge about the classes of patterns is unavailable, or difficult to obtain. There exists a separate class of neural network research devoted to the training of networks based only on information in the input patterns, often referred to as self-organizing neural networks. The vast majority of such systems use extensions on the concept proposed by D. O. Hebb in 1949 (Hebb, 1949).

> "When an axon of cell *A* is near enough to excite a cell *B* and repeatedly or persistently takes part in firing it, some growth process or metabolic change takes place in one or both cells such that *A*'s efficacy, as one of the cells firing *B*, is increased." (p. 50).

Note that this is not a precise mathematical formulation, and although many researchers enjoy arguing about how it should be expressed mathematically, a number of quite different training rules can reasonably be called Hebbian. In addition to this construct, he proposed the idea that there could exist temporarily stable activity patterns that have important functions in mental activity. This idea has often reappeared in the various neural network "attractor" models (Hopfield, 1982 and Grossberg, 1976). From the earliest realizations of the Hebbian rule, it was apparent that some additional constraints were necessary for numerical simulation, including some form of normalization of connection values to control unlimited growth, and a means for the reduction of connection values as well as the growth (Rochester et al. 1956). Several methods have been proposed to realize this idea (Bienenstock, 1982; Stent, 1973). Stent was one of the first to introduce spatial competition between the synapses of a cell. An increase in a synaptic efficacy according to the Hebb hypothesis was balanced by a compensatory decrease in other synapses to the same cell. In Bienenstock's experiments, whether the synaptic strength increases or decreases depends on the cell's output magnitude as compared to a threshold. They described the used of this learning threshold along with a nonlinear modification equation as providing a means of temporal competition between patterns for synapse strengths, as opposed to the aforementioned spatial competition schemes.

The type of problem generally attacked by the Hebbian systems are usually the construction of systems for the storage of associations or consistent transformations. The key difference between these systems and the forms used by the aforementioned schemes is in the representation of the output data. Those systems were designed to have a "grandmother" cell which was, ideally, the only cell firing for each pattern or class of patterns. In the associative models, the output representation of an input pattern is assumed to be *distributed*, where a set of cell activities describe the result. An example is a corrupted image input to a system which outputs a noiseless version of that image. In that case, the association is between the noisy input and the "clean" output pattern.

The strong interest in this form of storage was encouraged by the concurrent work by Kohonen and Anderson in 1972 (Kohonen, 1972; Anderson, 1972). In Kohonen's model, the model neurons were "linear associators", with no thresholding or gating applied to the weighted sum of input connections. Removing this restriction allows for the use of linear algebra on the dynamics of the system. Kohonen's generalization of the Hebb synapse allowed for change in connections as

$$\Delta C_{ij} = \alpha O_i O_j, \tag{95.18}$$

where α is a growth constant. In Anderson's model, the change was proportional to the correlation between input and output. As it turns out, these implementations are different means to the same result. With either form of training, the connection matrix becomes the outer product between the input and output vectors. With such a system, the only perfect associations can be formed when the input patterns are orthogonal. If that is not the case, the outputs are somewhat corrupted by the "crosstalk" between the stored transformations. Kohonen described two variations on this architecture, the heteroassociative system, where the inputs and outputs are distinct, and the autoassociative system, where the outputs are fed back into the model as inputs. Moreover, it was possible to use rapid gradient descent techniques such as the Widrow-Hoff LMS rule in training. Although the crosstalk was generally described as a limitation of the model, Cooper suggested that there may be ways to use crosstalk to generate "weaker" associations in addition to the main associations (Cooper 1973). The most complete analysis of this class of models is found in Kohonen's books (Kohonen, 1977 and Kohonen, 1984).

Many variations have been constructed to the associative models. Aman used a recurrent net (outputs feedback to the inputs) which received input from a teacher to organize his "concept forming nets" (Amari, 1977). Anderson et al. used a recurrent autoassociative network with positive feedback, using limits on the firing rates of the individual neurons to ensure stability of the system (Anderson et al., 1977). After several time periods, the system settled to a stable state (all output firings remain constant), corresponding to a solution at the corner of a hypercube. Thus the system is often referred to as the Brain-State-in-a-Box (BSB) model. Other variations have been used to illustrate the organization of receptive field characteristic reminiscent of cells in the mammalian visual system (Von der Marlsburg, 1973; Nass and Cooper, 1975; Cooper et al., 1979, Linsker, 1986; Ruff et al., 1987; Kammen, 1988).

Even with all of these interesting presentations, it was left to Hopfield in 1982 to give mass credibility to the neurocomputing field (Hopfield, 1982). This paper is usually credited with the modern resurgence in neural networks. He implemented the stable state, or attractor, concept with a network which could perform error correction and reconstruct missing information. His model neurons were simple threshold logic units and he used global connections (except that each cell was not connected to itself). Using the assumption of a symmetric connection matrix and the development of an energy function

$$E = \frac{1}{2} \sum_{i \neq j} \Sigma C_{ij} O_i O_j \tag{95.19}$$

which was minimized during the evolution of the network, he was able to illustrate the connection with Ising spin models in physics, thus encouraging the entry of many physical scientists into the field. The similarity with Equation 95.4, describing the energy function used by Metropolis (Metropolis et. al., 1953), is readily visible. It is no wonder that simulated annealing has later appeared in the training of Hopfield nets (Foo and Takefuji, 1988; Levy and Adams, 1987; Lee and Sheu, 1989). The connections in the Hopfield net were hard wired (by computation beforehand), and the system evolved through a transition in output state space. The number of memories which could be stored averages near 15% of the dimensionality (the number of cells) of the system. This result was later extended for linear cells with sigmoid nonlinearities (Hopfield, 1984). Many complex computational problems were later solved using a heuristic choice of connections, which often could be inferred from the structure of the problem (Hopfield and Tank, 1986, 1985; Tank and Hopfield, 1986). That this could be done was proposed many years earlier (Marr and Poggio, 1976). Perhaps more importantly, practical implementations were highlighted, and VLSI and optical implementations were soon to follow (Graf et al., 1988, Farhat et al., 1985). The laboratory of Carver Mead has also produced some interesting implementations of neural networks in VLSI circuitry (Sivilotti et al., 1987, Mead, 1989).

The use of optimization methods in the training of neural networks was also used with the introduction of Boltzmann machines (Ackley et al., 1985; Hinton and Sejnowski, 1987), an extension of the simulated annealing procedure to neural networks. In this case, cells would fire probabilistically according to the Boltzmann distribution. The reasoning for the use of optimization routines in neural networks is the same as the logic for their use in other applications: the avoidance of local extrema. Many training algorithms are manipulated so that only one extrema exists (such as the Widrow-Hoff LMS training), but other applications, especially those which cannot use an external "teacher", are destined to have complex energy surfaces with many extrema.

Recent trends in neural network research have attempted to consolidate rule-based expert systems with neurocomputing systems (Fu, 1989), or to include emerging pattern recognition methods, such as the concept of fuzzy clustering (Amano et al., 1989), into the networks. Other future directions involve the construction of hybrid machines, capable of both supervised and unsupervised learning under various circumstances (Holdaway, 1989). The use of optical hardware, which can expand the capabilities of network connections to holographic storage, appears promising in the future (Psaltis et al., 1990).

In pattern classification tasks, particularly when the training is unsupervised, there is a high likelihood for multiple extrema in the cost function. We have used the ALOPEX optimization algorithm to alter the connection strengths of a neural network. In addition to the avoidance of local extrema, in certain cases the use of such an algorithm will reduce the time required for a designer to create a useful network. All that is required from the designer is the selection of an appropriate energy function to perform the global operations of a network, without worrying about the long-term global effects of local Hebbian operations. This procedure is applied to both the feature extraction and classification portions of a pattern recognition system, further illustrating the diversity of the method. The following section will give a more in depth rationale for such a system.

Computer bound pattern recognition schemes have been often utilized in medical settings over the past several years. The artificial intelligence community developed one of the first expert systems to act as a tool in medical diagnoses for physicians (Sandell and Bourne, 1985). In the past couple of years, the interest in artificial neural networks has resulted in several applications of neural network classifiers for medical diagnoses (Marconi et al., 1989). These systems have relied most heavily on the backpropagation algorithm for the training of the networks.

In both expert system and neural network applications, there is a great reliance on either heuristic information provided by the specialist experts or accurate diagnoses of several example cases. Consequently, when the systems have completed development, replication of that information is the highest attainable performance. In neither case are the computerized systems likely to provide any additional information to the physician than what was already available to them.

An unsupervised pattern recognition system may supersede this limitation, in that the decisions that it makes are not based on maximizing agreement with clinical diagnoses. In this respect a properly designed unsupervised pattern recognition scheme is capable of a more valuable decision, because the decision itself is formed from an entirely separate set of assumptions. It may also be capable of stronger generalizations to non-training data than supervised schemes. The backpropagation algorithm, as an example, requires a large number of example patient data for accurate generalization. This is rarely available to the average developer of medical systems. The result is an algorithm which can demonstrate "near perfect" decisions for the 10–50 patients it was trained for, but is nearly worthless for reliable decisions on other patients. The cause of this is the dependance of the algorithm on an agreement with the clinician. As long as there is this agreement, the network can make the decision under any rule set it chooses.

This problem is pictorially demonstrated in Figure 95.3. The points in the figure are representative of the data collected from individual patients, and their distances from each other denote to some degree the similarity of the data. The x's are diseased patient data (for purposes of illustration), and the o's are control subjects. A supervised scheme (such as backpropagation) may make the decision boundaries as shown in Figure 95.3(a), while the unsupervised method could choose that of Figure 95.3(b). Clearly the supervised scheme is more accurate in its groupings when compared to the actual diagnoses, but its groupings destroy the similarity geometry presented by the data. Assuming that the data is pertinent and is entirely sufficient for classification (an assumption which can never be asserted with complete confidence), it is less likely that this supervised scheme can generalize to additional patient information. Accurate generalization is the single most crucial factor in applying any trained pattern recognition system to its expected surroundings.

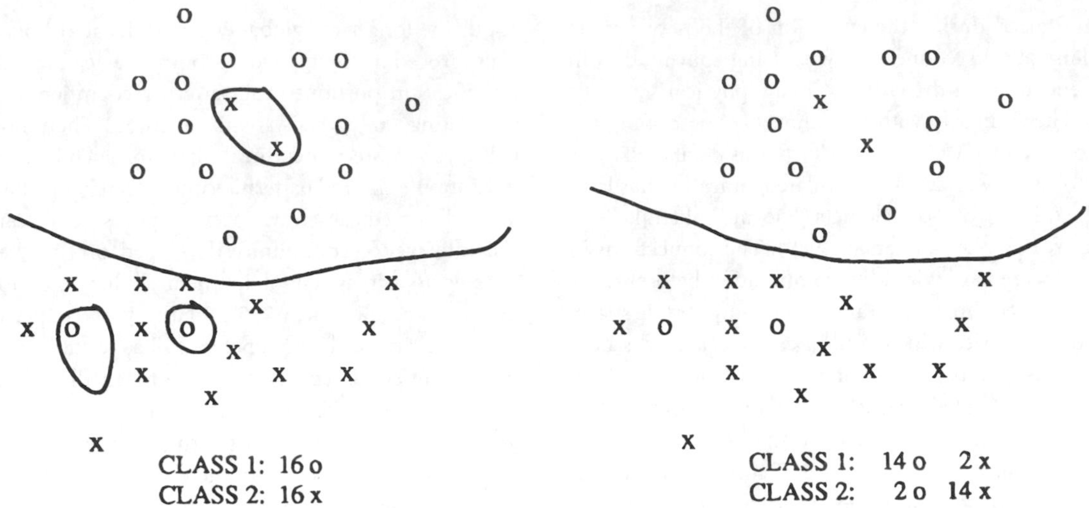

Figure 95.3 Hypothetical decision surfaces (solid lines) generated by (a) a supervised and (b) an unsupervised classifier. The x's represent disease populations and the o's normal subjects.

Based on these speculations, attention should be paid to the formulation of unsupervised pattern recognition schemes for medical applications. Little attention seems to have been given to this subject area. The first reason is a matter of interpretation and marketing. Clearly the evaluation of the success of an unsupervised scheme in performing a classification is much more difficult, and probably less quantitative. This is because any quantitative assessment must utilize the known pattern identities, and that contests the motives for using an unsupervised method. Without information about the success of the algorithm, the developer has little opportunity to improve and fine tune the method.

What we describe next is an attempt to develop an unsupervised pattern recognition system that can be executed on a parallel architecture. The primary application of the system is to the analysis of the Visual Evoked Potential (VEP) signals of control and Multiple Sclerosis patients. In this enactment, both the signal features and the classification decisions are based solely on the statistics of the data set. This chapter discusses the development and fine-tuning of the system by application to the classification of unconstrained handwritten digits. This is a common task of pattern recognition methods, allowing the development of benchmarks (even though supervised schemes have a natural competitive advantage) against other proven methods. The classes of the digits are known, allowing a reasonable assessment of the performance of the algorithm, but are not utilized in the training of the system.

There are other features of this pattern recognition system that deserve to be distinguished. Through the study of the artificial neural network field, limitations imposed on such systems (usually due to an implied adherence to the biological systems and/or a mathematical or computational convenience) have excluded viable options in the training and calculation of the network units. The training of unsupervised systems has centered on the Hebbian rule, in which changes in weightings is performed through a *local* analysis of input-output correlations. It seems more reasonable that the entire pattern of input weightings to each unit be changed through mutual understanding of the influences of the other weights in the pattern as well as the local input and output to each individual connections. If the pattern of weightings is considered as a whole, the informational retention of the units can be altered to suit the needs of the application. This is possible only with an algorithm which can alter many parameters (the weights) with one feedback element (the output), a situation ideally suited to optimization routines.

The optimization routine ALOPEX is used to dynamically alter the weights during the training period. The cost function used to alter the weights can be extracted through a statistical evaluation of the outputs of the unit to the training patterns (as is used in the feature extraction phase of the system), or it can be supplied as an additional input from an external evaluator (as in the clustering portion of the network, which must analyze the performance of a group of cells). This use of ALOPEX in the training, along with some unconventional computing and storage facilities attributed to the cells, may cause some purists to discount this system as a "true" neural network. That is perhaps not so important, since it is certain that the system can at least retain the advantages of the broader category of parallel distributed computing systems.

95.2 System design

A pattern recognition system is comprised of four essential components, as labeled in Figure 95.4. The preprocessing module is an application dependant stage. The feature extracting and a clustering modules are the trainable commodities in this scheme and comprise the bulk of the discussion of this chapter. As indicated by Figure 95.4, these modules are under the training control of the ALOPEX process, although the depiction of ALOPEX as a control external to the individual module is merely used as a convenience. A more accurate depiction would place

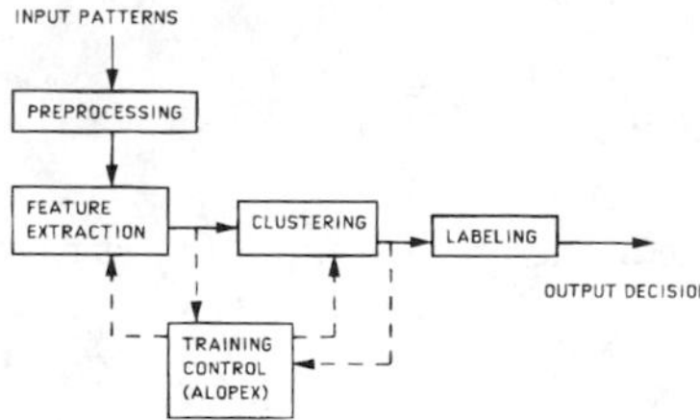

Figure 95.4 A component diagram of the pattern recognition system used in this research. The dotted lines indicate control signal input to the modules, whereas the solid lines denote transfer of data.

an independent ALOPEX process within each stage. The final module is a labeling stage, in which the clusters formed in the previous stage are assigned an identity.

Feature Extraction

A feature of an input pattern refers to any measurement from a set of pattern measurements which characterizes some attribute of that pattern. It was previously mentioned that a good feature extraction routine will compress the input space to a lower dimensionality while still maintaining a large portion of the information contained in the original pattern space. Although this is the most often cited advantage of feature extraction, it is also true that an appropriate choice of features can help eliminate redundant and irrelevant information from the data set, thereby reducing the overhead for the classifier (Tou and Gonzalez, 1974). Most feature extraction routines can significantly aid in the performance of a classifier.

In unsupervised situations, the only information available to the feature extraction module is the statistical distribution of the patterns. In such a scenario, it is impossible to quantitatively analyze the effectiveness of a feature extraction routine in improving pattern classification. However, there are operators designed to maintain high information content in the features (as compared to the original measurement pattern space) with a minimal number of dimensions.

The Karhunen Loève Expansion

Perhaps the most widely used feature extraction routine with some of these information conserving properties is the Karhunen-Loève (K-L) expansion (Karhunen, 1947; Watanabe, 1965; Chien and Fu, 1967). If it is desired to represent the kth N-dimensional input pattern x with an M-dimensional feature pattern y, then an MxN matrix Φ can be chosen so that

$$\overline{y}^k = \Phi \overline{x}^k \qquad (95.20)$$

The matrix Φ is actually a set of M orthonormal vectors (meaning they lie perpendicular to one another in N-dimensional space and have unit magnitude) constructed by taking the M eigenvectors corresponding to the M largest eigenvalues of the covariance matrix C constructed by the input patterns. This matrix C is formed by

$$C = \sum_{k=1}^{P} \sum_{j=1}^{N} (\overline{x}_j^k - \overline{\mu})^T (\overline{x}_j^k - \overline{\mu}) \qquad (95.21)$$

where Φ is the number of patterns in the set and the vector μ is the mean vector of the pattern set as

$$\overline{\mu} = E(\overline{x}) \qquad (95.22)$$

It has been shown that the eigenvectors of this covariance matrix exhibit certain optimal properties as a feature extractor when they are ordered with their correspondence to the eigenvalues of the matrix from highest to lowest (Chien and Fu, 1967). One of these properties is that the mean square representation error is the minimum for any choice of M orthogonal vectors, meaning that the approximation error (e) of reconstructing the original pattern space with only M features

$$\epsilon = \sum_{k=1}^{P} \sum_{i=1}^{N} \left(x_i^k - \sum_{j=1}^{M} \Phi_{ij} y_j^k \right) \qquad (95.23)$$

is the smallest for any choice of M vectors in Φ. This means the expansion answers a key requirement in the information compression problem of feature extraction. It's other optimum property is that this choice of vectors associates with the coefficients of the expansion a minimum measure of entropy or dispersion. The borrowed concept of entropy is often used in the pattern recognition field as a clustering measure (Tou and Heydorn, 1967, Watanabe and Kaminuma, 1988), and so this minimum entropy property characterizes the K-L expansion as likely containing clustering transformational properties. The crux of the theory is that the features contain the most information (without knowing the pattern identities) for the price and probably retain the existing pattern groups in the population.

An implementation of the K-L expansion as a feature extractor (Fukunaga and Koontz, 1970, Kittler and Young, 1973, Kirby and Sirovich, 1990) generally proceeds in the opposite direction to the above analysis as the following steps:

1. Calculate the covariance matrix in Equation 95.21 using all available patterns and the mean vector from Equation 95.22.
2. Find the eigenvalues and eigenvectors of that covariance matrix.
3. Select the eigenvectors corresponding to the M largest eigenvalues and store them in the matrix Φ.
4. Find the feature values for each of the patterns via Equation 95.20.

The primary advantage in using the K-L expansion for selection of features is that it requires no previous knowledge of pattern labels and thus is perfectly suited to unsupervised tasks. Many people confuse the aforementioned optimum properties of the expansion with an assumption that the features generated for the expansion provide optimum performance of the classifier. This is certainly not the case. A feature extractor can never provide optimum classifier information without information from the classifier about its historical performance. Nevertheless, the K-L expansion is useful in situations where that information is not reliable or available. The K-L expansion is also a linear operation, and considerable evidence suggests that other nonlinear features can often provide more useful information to the classifier (Cover, 1974).

There is an additional point which should be mentioned here. It turns out that the primary eigenvector points in such a direction so that the variance of the patterns in that feature space is maximum for all vectors. Subsequent eigenvectors find other locally maximum variance features in orthogonal directions. Furthermore, the eigenvalue representing each eigenvector is exactly equivalent to the variance of projected patterns onto the corresponding eigenvector (Devijver and Kittler, 1982). It is this realization which provides the impetus for the enactment of the K-L expansion onto a neural network, as described in the next section.

Application by a Neural Network

The linear projection of a pattern vector onto one of the K-L expansion vectors is a simple inner product operation, as denoted by Equation 95.20. Conveniently, this is the same operation which is commonly given to units of an artificial neural system, (although many artifical neurons use a nonlinear transformation subsequent to this inner product). In concept then, an artificial neuron can use its connection weightings to act as one of the eigenvectors contained in the matrix Φ in the last section. It only remains to consider the method of training a cell to retain that specific pattern of connections. After training, the output of the neuron is a real number corresponding to the feature value from the input pattern.

A hint has already been given about the means for training a cell to retain the "maximum" eigenvector. It was mentioned that the primary K-L expansion vector had the property that the output features generated by it contained the maximum variance of any features generated by any other choice of vectors. In direct relation to the neural network, if the variance of the output of the cell for all training patterns is maximized during training, the connection vector retained after training corresponds to the primary K-L vector. The scheme for training any one cell follows the steps outlined below:

1. Set the connections to the cell to random values.
2. Calculate the output value of the cell for every training pattern.
3. Find the variance of the output over all patterns used in step 2.
4. Update the connection weights to the cell via an ALOPEX update equation.
5. Normalize the connection weights to the cell to a vector magnitude of 1.0.
6. Go to step 2 until a convergence criterion has been met.

In particular, the output y of the jth cell from the input of the kth N-dimensional pattern x is found as

$$O_j^k = \sum_{j=1}^{N} C_{ij} x_i^k \tag{95.24}$$

and the variance of the jth cell output (V_j) is calculated over all P patterns by

$$V_j = \frac{\sum_{k=1}^{P} (O_j^k - \mu_j)^2}{P}, \qquad \mu_j = E[O_j]. \tag{95.25}$$

The ALOPEX update equation uses the changes in the variance (ΔV_{ij}) and connections (ΔC_{ij}) from the current iteration (n) to the previous iteration ($n - 1$) to change the connections by

$$C_{ij}(n+1) = C_{ij}(n) + \Delta B_{ij}(n) + r_{ij}(n) \tag{95.26}$$

where

$$\Delta B_{ij}(n) = \gamma \Delta V_j(n) \Delta C_{ij}(n) \tag{95.27}$$

(see Equations 95.6–95.9). The term $r_{ij}(n)$ is a gaussian random number with zero mean and standard deviation s. The factors γ and s are adjusted as in Equations 95.10–95.11.

To illustrate the performance of the algorithm, a simple 2-dimensional pattern space was constructed and one neuron was trained on 60 patterns in this space using ALOPEX output variance maximization. The performance of the algorithm can be easily tested, since the "ideal" maximum is known (it can be found by the analysis of the K-L expansion section). The resultant vectors of the ALOPEX method and the K-L expansion are shown in Table 95.1. Also included in that table are two other commonly used methods for unsupervised neural network training. The first is a simple normalized Hebbian scheme, where the connections are changed during each iteration by

$$\Delta C_{ij} = \eta O_j x_i \tag{95.28}$$

where h is a gain factor. The second is a method employed by Oja (Oja, 1982), which is a variant of the Hebbian proposal and includes output feedback as

$$\Delta C_{ij} = \eta O_j (x_i - O_j C_{ij}) \tag{95.29}$$

Table 95.1 shows some clear results, the first of which is that the variance maximization scheme with ALOPEX was able to very accurately mimic the optimal K-L vector. Secondly, it clearly points out a weakness in the Hebbian proposal. As further experiments will show, the Hebbian training (and the Oja training, which is nearly identical in most real world situations) cannot optimize to the "best" vector because the input patterns are not zero-mean centered (they do not share a center of mass at the origin of the coordinate system). Thus essentially the Hebbian mechanism is unable to compensate for DC offsets, a situation which must be tolerated in nearly all real world applications. Obviously this weakness can be compensated for by centering

Table 95.1 Converged Connection Vectors for the Two-Dimensional Sample Data Space Shown in Figure 95.5. The First Row in the Table Refers to the First Connection Value in the Neuron, While the Second Row Denotes the Second Connection Value.

Principal K-L vector	ALOPEX variance maximization	Normalized Hebbian	Oja scheme
0.2011	0.1920	0.6305	0.6299
0.9796	0.9810	0.7760	0.7770

the data before it enters a Hebbian training module, but this requires additional post-training computation and is impractical in a highly connected system. Oja claims that this scheme will result in the cell retaining the principal component of the input pattern distribution [4].

Figure 95.5 shows the pattern space which was used for the training results in Table 95.1. Clearly the space is composed of three quite distinct clusters. However, not all choices of vectors allow all of those clusters to be clearly evident at the output of the feature cell. The illustrations of Figure 95.6 makes it obvious why the K-L and ALOPEX trained connections are superior choices to the Hebbian scheme for this pattern set. The diagrams in Figure 95.6 are histograms of the output values of the cell for each of the different vectors in Table 95.1. It is easy to see why the K-L expansion and ALOPEX trained vectors are preferable to the Hebbian and Oja schemes: the increased range of the neuron output levels increases the information content of the output by allowing all three clusters to be evident on the output line. So the concept of variance maximization indirectly promotes the retention of cluster forming information. Note that this is not a proof of superiority of the methods, but is a valid observation with an example training set.

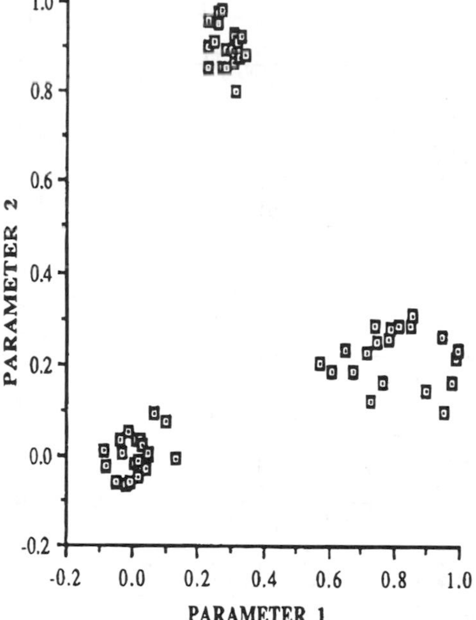

Figure 95.5 The input pattern space used for the training of the vectors in Table 95.1. There are 60 patterns in the space, with 20 grouped in each cluster.

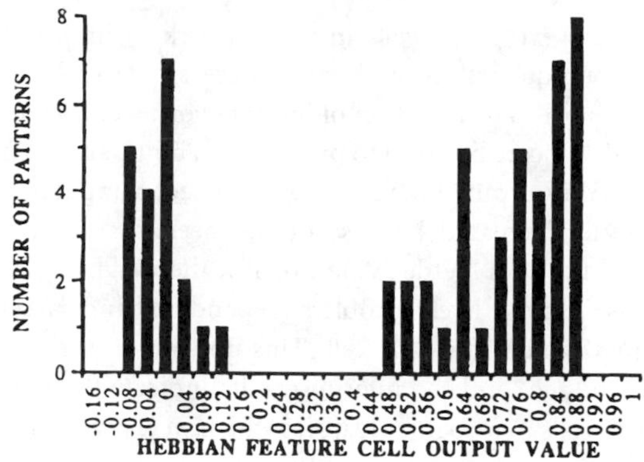

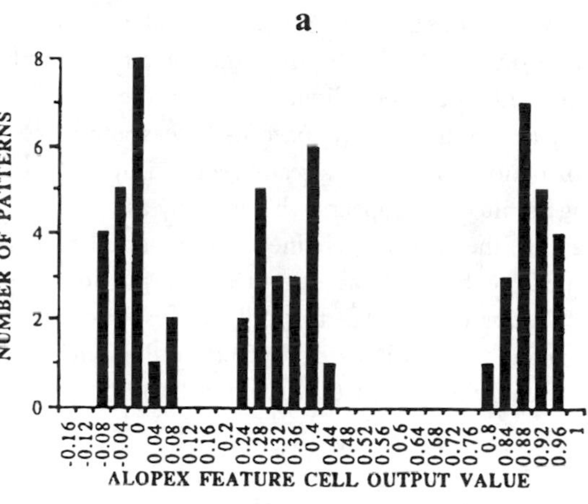

Figure 95.6 The feature cell output value histograms for the space of Figure 95.5 for (a) the normalized Hebbian training and (b) the ALOPEX variance maximization scheme.

With the one cell implementation described thus far, there is still the question of how a network of cells is to optimize to other vectors in the K-L expansion. Since each cell is searching for the "optimal" vector, all cells in a network will arrive at or near the same vector when using the same input pattern set. There are several possible means for forcing other neurons to optimize to other K-L vectors when more than one feature output is necessary. The training routine can be altered so that the optimization is not a global search. This will force each cell to arrive at a local optimum that differs depending on the initial conditions, but this does not guarantee that redundant cells will not be formed, and imposes a strong likelihood that some cells will arrive at local maxima which are not K-L expansion vectors. A second possibility is for an additional term to be added to the ALOPEX cost function which uses feedback connections between

[4] The principal component is identical to the first K-L expansion vector when the center of mass of the patterns in data space is at the origin.

other feature extracting cells in the network to impose a constraint on uniqueness of each cell. There are several problems with this choice. The insertion of internetwork feedback requires that additional care be taken to prevent instability, and even when stability of the output of the cells is guaranteed, computation time is drastically increased because of the need to wait until the outputs of the cells "settle". Additionally, the weighting between the terms is most likely problem dependant, and is certainly optimally different for each cell. This possibility was examined in detail though, and the interim results indicate the possibility of training cells to recognize features which were highly nonlinear and yet often retained some of the optimal aspects required by the problem. The work on this possibility was suspended when the training times became exorbitant. Instead, a more viable solution was obtained. The actual method used imposes the constraints on the ALOPEX optimization in a more implicit way. Instead of adding extra qualifying terms to the ALOPEX update equation, the input pattern space itself was altered to exclude the information which was already retained by other cells. The resulting architecture appears schematically as in Figure 95.7. In this method, the cells are "chained" to one another in series by a weighted feedback of the output of the previous cell to the input of the current cell. The first cell receives the original pattern space unaltered, and will perform exactly as the single cell simulations shown previously. Subsequent cells receive the original pattern space (x) minus the component of that space extracted by the previous cells as

$$\hat{x}_i = x_i - C_{i,j-1}O_{j-1} \tag{95.30}$$

for the ith input to the jth cell in the chain. The output of the jth cell is then found in the same way as before, only now using the modified input space as

$$O_j = \sum_{i=1}^{N} C_{ij}\hat{x}_i \tag{95.31}$$

Using this method, only the first cell, which receives no interference from other cells in the network, can optimize to the principal vector, while all others are relegated to finding secondary, yet potentially important, vectors. When a network such as this is trained, it can be seen that the first cell optimizes to the first Karhunen-Loève eigenvector, and the other cells locate in order the remaining eigenvectors. Sanger published an architecture similar to this scheme (Sanger, 1989), but his training relies on the gradient descent method of Oja (Oja, 1982).

To illustrate the performance of this architecture, let us use the pattern space employed in the one neuron simulation in Figure 95.5. Since the first cell in the chain architecture above does not receive any influence from other cells, it will optimize in the same

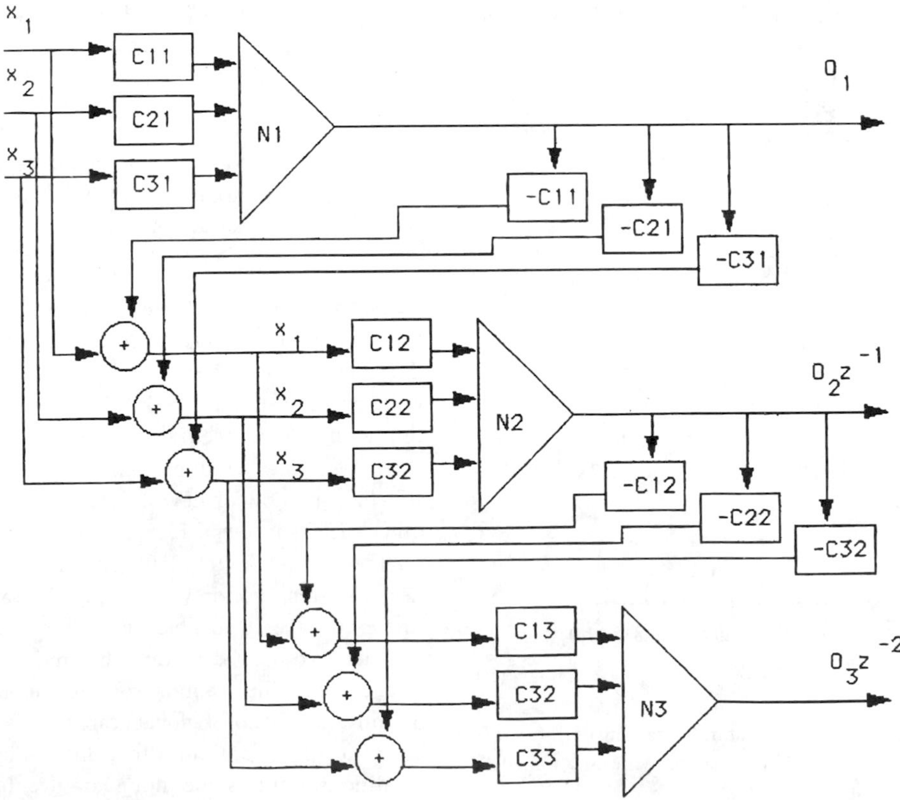

Figure 95.7 The feature extraction architecture for 3 feature cells. After the first cell, all inputs are modified by a subtracted weighted input from the outputs of the previous cells as in Equation 95.30. The triangle symbol represents the neuron integrator, while the symbol C12 indicates the connection weight from the first input to the second cell.

way as in Table 95.1. If we now find the new pattern space to provide to the second cell according to Equation 95.31, we get the pattern space shown in Figure 95.8. It can be clearly seen that all information in the direction of the first cell vector of Table 95.1 has been removed from the pattern space given to the second cell. It is then a trivial matter for the second cell to optimize on this pattern space. Note that we were, in this example, extracting two features from what was originally a two dimensional pattern space. This is not a compression scenario. In a more general case where the number of features is less than the number of pattern dimensions, the input space would never compress to a line as in Figure 95.8. Rather, with each cell in the chain, there is essentially a virtual removal of a dimension from the pattern space in the direction of the previously optimized vectors (not a physical removal, since the feature extraction is still performed over the same number of dimensions). The application of this feature extraction method to a more complicated pattern space is shown in subsequent chapters, where it can be clearly seen that this architecture and training allows for the network cells to converge on the K-L expansion vectors in order.

Clustering

The Fuzzy c-Means (FCM) Clustering Algorithm

The concept of fuzzy logic was first introduced by Zadeh, whose classic paper has become the philosophical bible in the field (Zadeh, 1965). The concept is simple: set membership, and indeed reasoning of any sort, carries more information when there are a continuum of grades of membership. The reasoning is based on Zadeh's Principle of Incompatibility, which maintains that high precision is incompatible with high complexity. The suggestion is that the complexity of a system and the precision with which it can be analyzed bear a roughly inverse relation to one another. He asserted that since real world ideas appear to be fuzzy in nature, there is reasonable cause for adapting this approach to machines. Since that time, the number of applications to decision making, and pattern clustering in particular, have been numerous (Davis and Economou, 1984).

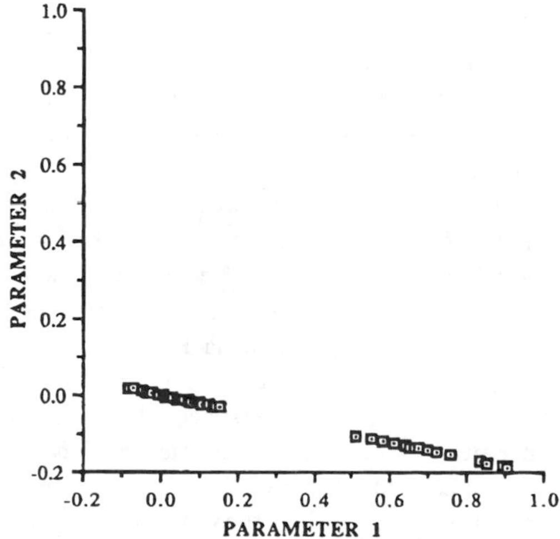

Figure 95.8 The revised pattern space of Figure 95.5 as seen by the second cell in a feature extraction network chain.

One of the first to apply fuzzy reasoning to pattern recognition was James Bezdek (Bezdek, 1973). The method which he and his colleagues have introduced, the Fuzzy c-Means clustering (FCM) algorithm (Bezdek and Dunn, 1975; Bezdek, 1981; Bezdek et al., 1984), has seen great popularity as a flexible and easily implemented method. The method itself is actually a spinoff of the venerable ISODATA algorithm (Ball and Hall, 1965). The ISODATA clustering method is one of a set of techniques which assumes that the optimal cluster partitioning is described as the minimum (or maximum) of an objective function. For the ISODATA algorithm and others like it, which use a set of c prototype "centers" of clusters around which patterns are grouped by their resemblance to these centers, the most common choice of objective function J is of the form

$$J = \sum_{k=1}^{P} \sum_{i=1}^{c} u_{ik} d_{ik} \tag{95.32}$$

where u_{ijk} is the membership strength of pattern k in cluster i and d_{ijk} is the squared distance from pattern k to cluster center i in m-dimensional feature space. The ISODATA algorithm is normally used to generate *hard* partitions of the data. A hard partition is one in which each pattern is allocated entirely to one cluster or another, so that the membership strengths take on the values of zero or one.

In most scenarios, the assignment of patterns to any one cluster prototype in exclusion of all others is a gross simplification of the complexity of the pattern space. Bezdek used the concept of fuzzy logic, where decisions are made through analog weightings, and applied it to this objective function J. In doing so, J was defined as

$$J = \sum_{k=1}^{P} \sum_{i=1}^{c} (u_{ik})^q d_{ik} \tag{95.33}$$

It is easiest to think of this objective function as representing the sum of the errors (the distances) in representing the patterns by a set of c cluster centers, weighted by the membership of the patterns to those clusters. The exponent q controls the sharpness of the decision boundaries, so that when $q = 1$, hard clusters are constructed, and when $q = \infty$ all patterns share the same membership to each cluster.

Most importantly for the mathematical analysis of this function, the use of continuous memberships means that the decision space is now continuous for all $q > 1$. It now becomes possible to examine the conditions for minimization of this function. It was demonstrated that J could be locally optimal for any one q only if

$$\overline{v}_i = \frac{\sum_{k=1}^{P} (u_{ik})^q \overline{x}_k}{\sum_{k=1}^{P} (u_{ik})^q}; \qquad 1 \le i \le c, \tag{95.34}$$

and

$$u_{ik} = \sum_{j=1}^{c} \left(\frac{d_{ik}}{d_{jk}}\right)^{-2/q-1}; \qquad 1 \le k \le P, \qquad 1 \le i \le c \tag{95.35}$$

where v_i is the ith cluster center and x_k the kth pattern. By iterating through these conditions, Bezdek (Bezdek, 1981) claimed that a local minimum of the function J would be achieved. It was later seen that this iteration could only guarantee stationary points and not necessarily local minima (Tucker, 1987); nevertheless the FCM method was found widely useful in practically achieving rapid (usually < 25 iterations) and "good" clusterings of data from many application areas (Bezdek et al., 1985; Bezdek and Fordon, 1984; Granath, 1982; Cannon et al., 1978).

There are two factors in the use of the FCM procedure which still require discussion. One of these points is the optimal selection of the parameter q. To this point, there is no automated way of selecting the best value of q for any one pattern set, but most applications seem to find reasonable values as lying somewhere between 1.2 and 4.0 (Bezdek, 1981).

The second factor is the way in which the distance d_{ijk} is calculated. In general, the distance d can be calculated through the quadratic form

$$d_{ik} = \|\overline{x}_k - \overline{v}_i\|_A^2 = (\overline{x}_k - \overline{v}_i)^T A(\overline{x}_k - \overline{v}_i) \qquad (95.36)$$

which is termed the A-norm distance. If the matrix A is chosen to be the identity matrix, then the distance is the squared Euclidean distance from pattern x_k to center v_i. This causes the FCM algorithm to form roughly hyperspherical clusters. This makes convergence simpler (since each cluster shape is identical), but is not optimal for most data sets with clusters of unequal shape. If the matrix A is chosen as C^{-1}, where C is found as the fuzzy covariance matrix

$$\overline{C}_i = \frac{\sum_{k=1}^{P} (u_{ik})^q (\overline{x}_k - \overline{v}_i)(\overline{x}_k - \overline{v}_i)^T}{\sum_{k=1}^{n} (u_{ik})^q} \qquad (95.37)$$

then the axes of the cluster are effectively scaled according to the distribution of the data points within those clusters. This is a modification of the Mahalanobis distance measure for "hard" data sets (Tou and Gonzalez, 1974). Other forms of the matrix A are also popular, including a diagonal matrix of the eigenvalues of the matrix C which Equation 95.37 calculates (Bezdek et al., 1984).

Using $A = C^{-1}$, and with the same calculation of membership strengths as in Equation 95.35, clusters of essentially hyperellipsoidal shape can now be found. Furthermore, the cluster shapes can be variant from one cluster to another, since each cluster has its own covariance matrix C. The incidence of local optima in the use of a variant A matrix such as this has been shown to rise drastically, affecting almost every problem, even with small data sets (Bezdek, 1981). To compensate for this, elaborate means of choosing initial conditions have been used, with unproven ability to guarantee global success (Gath and Geva, 1989).

Finally, every clustering algorithm must develop a proven means for determining the optimal number of clusters in the data set, and whether a converged set of clusters is a "good" clustering. This is usually termed the *cluster validity* problem, and there are at least as many opinions as to what are the best set of parameters to provide this information as there are clustering routines.

The originators of the FCM routine usually use an entropy measure to characterize the effectiveness of the clustering operations which is given by

$$H_c = -\left(\frac{1}{P}\right) \sum_{i=1}^{c} \sum_{k=1}^{P} u_{ik} \log_a(u_{ik}), \quad a \in (1, \mathit{INFINITY}), \qquad (95.38)$$

where $H_c = 0$ for hard partitions and $H_c = \log_a(c)$ for an entirely "blurred" (or indecisive) clustering. Another related parameter, termed the partition coefficient (F), is found by

$$F = \left(\frac{1}{P}\right) \sum_{j=1}^{c} \sum_{k=1}^{P} (u_{ik})^2 \qquad (95.39)$$

Both of these parameters rely on one of the major paradoxes of fuzzy logic. That is, although the pretext of fuzzy clustering is to incorporate more information via using analog decision criteria, heuristically the "best" clustering is one in which the resultant clusters are hard (have binary membership strengths). This idea is the basis behind the use of the entropy (H) and partition coefficient (F) measures, and assumes that the optimal number of clusters is the choice for which H is minimized and F is maximized.

The FCM algorithms already fit many of the requirements for ALOPEX training. There is an explicit cost function which determines the "optimal" choice of clustering, a requirement for an optimization routine such as ALOPEX. Further, there is really only one set of independent parameters which must be varied in order to minimize the cost function of Equation 95.33: the set of cluster centers (c times m parameters in total). All other information for the determination of membership strengths results from the specification of cluster centers. The distances (Equation 95.36) are found in reference to the cluster centers and the membership strengths are based entirely on distance information (Equation 95.35).

It remains only to justify the use of ALOPEX to this application. ALOPEX will reduce the likelihood of arriving at a locally optimum solution at the price of increased computation time. However, most situations demand accuracy, even when having to sacrifice increased computation time. This is especially true when you consider that the decision of these clustering partitions are usually needed only once, after which those partitions are used to make rapid decisions about new data.

The danger of locally optimal solutions becomes especially apparent when clusters of nonhyperspherical shape are assumed. The distance measurements often become very local in certain directions from the cluster center. The result is a much more localized cost function, which is therefore much more volatile. The resulting distances generated can often exceed the real number range of most software languages, and special care must be taken to ensure the stability of the algorithm as it iterates.

As mentioned before, all of the FCM family of algorithms share the danger of locally optimal solutions. Even with a Euclidean distance measurement, it is easily apparent that multiple runs of the FCM algorithm arrive at different solution points for pattern spaces of reasonable complexity. A more complete example of this behavior is shown in the context of the classification of handwritten characters.

To incorporate a global optimizer into the fuzzy *c*-means family, ALOPEX is used to adjust the cluster centers iteratively in the steps outlined below.

1. Randomly choose initial cluster centers.
2. Find squared distances between patterns and centers.
3. Calculate membership strengths via Equation 95.35.
4. Find the current cost function J from Equation 95.33.
5. Use ALOPEX to update the centers based on recent change in the cost function and the centers.
6. Go to step 2 until a convergence criterion is met.

The performance of this routine, and that of the feature extractor, is illustrated in the context of two application domains. These are described in detail in the next section.

Most pattern recognition schemes need to consider the assignment of labels, or pattern identities, to decision codes generated by the pattern recognition system. Most commonly, this consideration is important for supervised schemes, in which the pattern identities are known without ambiguity. In unsupervised methods, the notion of pattern labeling is somewhat self-defeating. That is, if the identities of the patterns used in the training were known before training, then a supervised method would have been more productive. If however the pattern labels are suppositions or decisions with an amount of uncertainty, it would be more useful to assign labeling based only on cluster membership strengths.

In the two application areas analyzed by this method, there are two distinct goals. The application to the clustering of Visual Evoked Potential (VEP) signals is interested in understanding how well the pattern distribution allows clustering which is in agreement with the physician's diagnosis. To this extent, it is not important whether the pattern labels are in agreement with the clinician's conclusions.

The labeling of this application area is performed by the clustering module and is comprised of the membership strengths of the patterns in each of the formed clusters. If a non-fuzzy labeling is desired, the labeling can be performed by assigning the pattern to the cluster with which it shares the maximum membership strength.

The classification of handwritten digits by this unsupervised system is an unusual task. As mentioned before, it is motivated by a desire to improve the algorithm without biasing the answer toward concurrence with the medical diagnoses. Paradoxically, the labels of each of the characters used in the study (which were never subjected to a segmentation process) are known unambiguously before training, but the knowledge is only used in the assignment of pattern labels after training is completed. This allows for a quantitative description of the performance of the algorithm.

The labeling of such a scenario is performed in this research by analyzing the constituent

$$\Omega_{ij} = \frac{\sum_{k=1}^{P} \{u_{ik}: \forall k \in PATTERN\ TYPEj\}}{\sum_{i=1}^{c}\sum_{k=1}^{P} u_{ik}}, \quad 1 \leq i \leq c, 1 \leq j \leq R \tag{95.40}$$

memberships of each clusters into an array W where W_{ij} is the percentage of cluster i membership from pattern type j (for R pattern types in the simulation, i.e., 10 digits in the character recognition problem), and there were c clusters formed from the P training patterns. Then the degree (y_{kj}) to which pattern k belongs to pattern type j is calculated as

$$\psi_{kj} = \frac{\sum_{i=1}^{c} u_{ik}\Omega_{ij}}{\sum_{l=1}^{R}\sum_{i=1}^{c} u_{ik}\Omega_{ij}}, \quad 1 \leq k \leq P, \quad 1 \leq j \leq R \tag{95.41}$$

The label of pattern k (L_k) is then the maximum of the degrees of memberships to the pattern types as

$$L_k = \max\{\psi_{kj}\} \tag{95.42}$$

In this way the labeling is performed not only on the membership strengths of the patterns given by the clustering module but also on the specificity to a single pattern type demonstrated by the clusters. That is, the labeling of a pattern with a high membership strength to a cluster with a high population of more than one pattern type will downplay that cluster membership strength in favor of other clusters with more "pure" pattern types.

Most neural network decisions formulate their decisions in a highly intertwined and complicated way. Even if the network is purely feedforward (as is the multilayer perceptron used in the backpropagation algorithm), there is usually only a limited idea of the criteria that the network used in making its decision.

The neural network just presented is an intriguing exception to this category of systems. The primary finding of the clustering module is the set of "centers" around which the cluster boundaries are formed. Since the coordinates of this center reside in the same space as the feature vectors of the input patterns, the cluster center coordinates can be thought of as the feature values which would have been extracted if there were a corresponding input pattern.

The primary question is, knowing these feature values, can we find out what the input pattern would have looked like? The answer is a resounding yes! The feature extracting neural network implements the K-L expansion, as we have already mentioned. Since the K-L expansion is really a linear expansion of the input pattern, and since the K-L expansion is used both as a feature extractor and as a data compression method, the feature vector can be reconstructed to find the input pattern with the knowledge of the K-L vectors that derived the features. This is essentially the same concept which was used to *remove* information from subsequent cells in the feature extraction network applied to a different task.

Given an m-dimensional feature vector y and a desired representation of the input pattern x, we can reconstruct an approximation to the input pattern (x) as

$$\hat{x}_i = \sum_{j=1}^{m} C_{ij} y_j,$$

where x_t is the ith input of the vector x and C_{ij} is the network connection strength from input i to feature extracting cell j. The vector x is an approximation to m terms of the K-L expansion the original input pattern x.

The realization that the cluster solutions (the centers) can be reconstructed into the corresponding input pattern (with hopefully a small error) allows the system to be used in an entirely new light. Not only can the system provide unsupervised classifications of a set of patterns, but through the reconstruction of the input pattern, a glimpse of the reasoning of the decisions can be made. This is made more apparent when the reconstructions of specific applications are displayed, as is done in the next section with Figure 95.16.

95.3 Application to Handwritten Digits

Introduction to Character Recognition

Among the widely varied applications of pattern recognition techniques, perhaps none has been more intensively studied than the machine recognition of character data (Ullmann, 1982). The number of potentially profitable uses for such systems are nearly limitless, since so much of the information resident in today's industrial society is textual. This is one reason that an application of the methods developed in this thesis is devoted to the character recognition arena. Also, the nature of the task permits the experimenter a concrete success formulation since the correct classes can be determined unambiguously. This is a much more desirable environment for the development of a new method than the less clearly formulated class memberships of medical data.

The industrial application of character recognition (CR) systems fall into several broad categories. One area is certainly that of data entry of handwritten information into conventional computer systems. Such arenas are typically constrained to data sets with limited character sets and constrained paper format (i.e., banking). This overlaps with the text entry area, which is more concerned with the input of typewritten characters into a word processing or publishing environment. These systems can only recognize characters of certain fonts, but with very high success (>99.9%). Other character recognition systems use the deciphered information to control a process, as would happen in a post office branch with a CR system which sorts mail. A final application area deals with providing an interface with the visually impaired, which often involves both a recognition procedure and a translator into speech.

Any comparison of character recognition tasks must be cautiously approached, since the difficulty of the task is largely determined by the constraints imposed on the data and the information available to the machine. It is certainly much easier for a machine to recognize typewritten characters than handwritten characters, since the typewritten characters would usually follow more standard guidelines and be less variable. Similarly, a signature verification system would likely be more successful if the machine had access to the pen pressure, velocity, and acceleration information at the time of the writing as well as the shape characteristics of the signature.

The recognition of handwritten characters is a subset of the much more extensive optical character recognition (OCR) problem, as shown in Figure 95.9. In short, it deals with the recognition of single hand-drawn characters of an alphabet which is unconnected. It must be differentiated from script recognition, which is concerned with the recognition of handwritten characters which may be connected and cursive. In this sense the developers of handwritten character recognition schemes do not need to concern themselves with the extremely challenging task of segmenting the characters (Taxt et al., 1989). Still, handwritten character recognition is not as simple a task as it may appear, since some claim (Suen et al., 1977) that even human beings can make up to 4% of mistakes when reading certain characters in the absence of context. Errors in reading handprinted characters, in addition to deriving from the algorithm and scanning methods, can also arise because of variations in shape due to the habits, style, mood, health, and other conditions of the writer (Suen, 1973).

The recognition of handwritten characters must consider at least two problems: the means of scanning the image, and the method for its recognition. The choice of a scanning device is not considered in this discussion, but the methodology of the recognition has been categorized as (Mantas, 1986, Gaillat and Berthod, 1979):

1. Point by point global comparison with stored images (Golshan and Hsu, 1970).
2. Global Transformations such as Karhunen-Loeve (Gudeson, 1976; Krause et al., 1974), Fourier (Persoon and Fu, 1977; Granlund, 1972), Walsh (Andrews, 1971), Moments of inertia (Tucker and Evans, 1974) and others (Niemann, 1976; Ott, 1974).
3. Extraction of the local properties such as endpoints, line crossings, and angles (Kwon and Lai, 1976; Spanjersberg, 1974; Beun, 1973).
4. Use of curvature and stroke information for analysis (Iwata et al., 1978; Hosking, 1972; Toussaint and Donaldson, 1970).

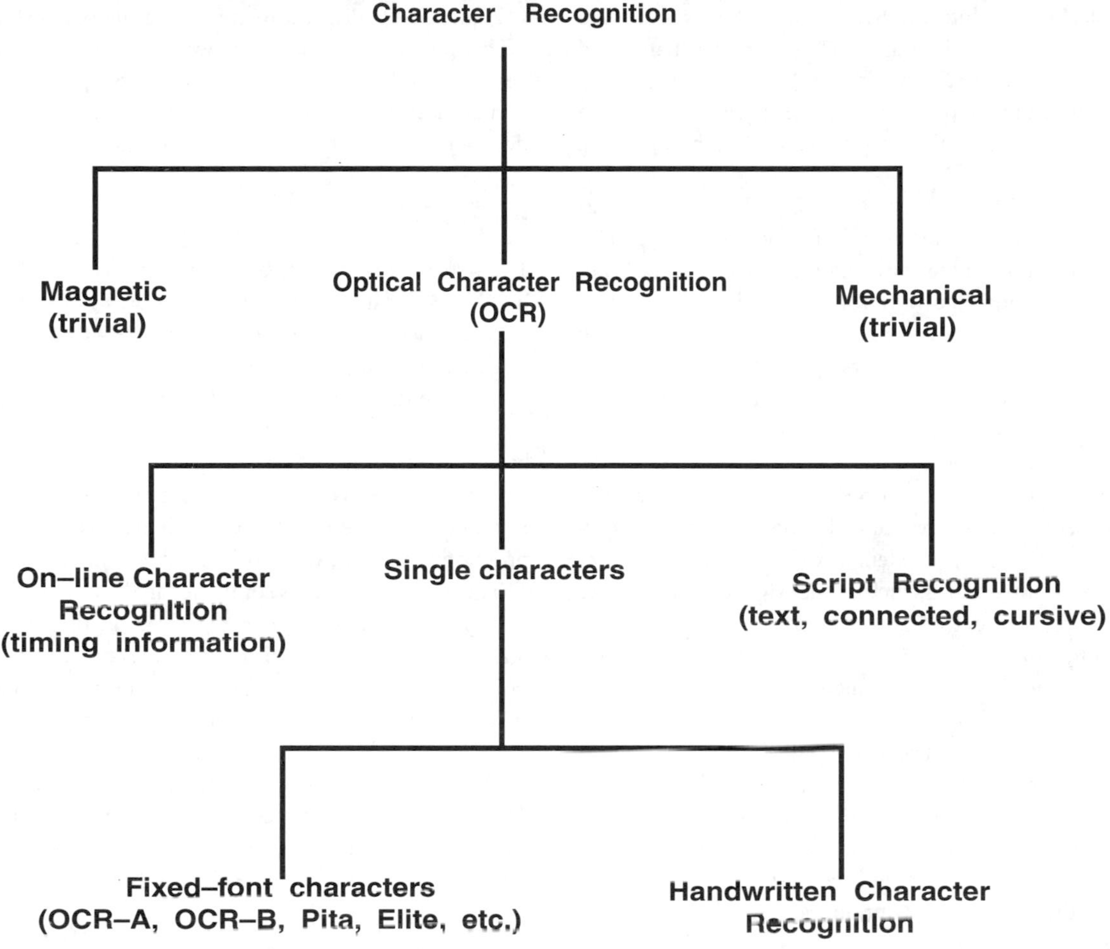

Figure 95.9 The different areas covered under the more general term "character recognition" (Mantas, 1986).

5. Structural methods, including decomposition of the character into graphs or other constituent elements (Sue and Chen, 1976; Watt and Beurle, 1971).

Many techniques contain portions which overlap between these categories. Each technique must be assessed by their ability to "ignore" deformation of the image caused by noise, translation, rotations, style variations and other distortions as well as practical considerations of the implementation such as speed and complexity.

The work of Grimsdale et al. represents one of the earliest attempts at character recognition (Grimsdale et al., 1958). In this scheme, each digitized pattern is analyzed for shape by a computer which extracts heuristic features and compares them to feature values stored on the computer. A few years later the notion of "analysis-by-synthesis" was presented by Murray Eden (Eden, 1961, 1968). He initially proposed that all Latin characters could be formulated by only 18 strokes, which in turn could be generated by a subset of four strokes, called segments. More generally, the concept was that handwritten characters are formed by a small, finite number of schematic features, which, when known, can be used for recognition of character data.

Perhaps more than Latin character recognition, the study of the Chinese alphabet is a stringent test for any algorithm (Mori and Masuda, 1980). One of the first attempts at this problem was made by Casey and Nagy at IBM (Casey and Nagy, 1966). A step-by-step approach was used for this large character set, in which the first stage grouped similar characters, then "group masks" and "individual masks" were employed to further specify the character. This type of method, in which a hierarchical decision process is employed, is characteristic of several OCR schemes (Parks, 1974).

Other researchers implemented a more mathematically formulated process for their systems. Tou and Gonzalez used a two stage system, the first stage performing a series of measurements for subgroup separation and the second extracting a set of specialized features (Tou and Gonzalez, 1972). Pavlidis and Ali used a "split-and-merge" algorithm to produce polygonal approximations of the characters which could provide enough information for decision making (Pavlidis and Ali, 1975), while others used clustering procedures on the task (Suen, 1982).

The review paper of Suen et al. (1978) discusses the efforts on the recognition of handprinted numerals. The best classifications of over 30 studies ranged from 85% to 99.79%, but direct comparison of methods is rarely feasible. This is due to the large discrepancy in the experimental setups. Not surprisingly, the 85% success rate used a realistic data set collected from the U.S. Postal

Service (Neill, 1969), while the 99.79% accuracy was derived after training writers to write numerals in specified shapes and sizes (Masterson et al., 1962). In addition to this widely varying data quality, the number of training patterns was different in each case, and some studies never reserved any patterns for testing of the system after training.

The arrival of neural network concepts into the pattern recognition field has spurned some wonderful successes, and great disappointments, in the character recognition field. Most studies use the backpropagation algorithm for the training and network architecture (Rajavelu et al., 1989). The study at AT&T Bell Laboratories is one of the more notable projects, in which postal zips code numerals were trained with a modified backpropagation algorithm (LeCun et al., 1989). Using this challenging data set of 7291 training patterns, they were able to show 0.14% error on the training set and 5% on the 2007 pattern test set. Fukushima's Neocognitron network also has demonstrated an invariance to recognizing characters of different rotations, sizes and translations. Other works, such as the ART topologies, have very limited success in detecting *any* discrepancies among patterns (Carpenter, 1989).

The testing of newly established algorithms often relies on a realistic data base for the development of the method. Several popular data sets are widely available for this purpose, the most popular of which are those created by Highleyman (Highleyman, 1961), Munson (Munson, 1968), and Suen (Suen et al., 1978). The Munson data set seems to be the most popular, chiefly because of its difficulty. A more recent large data set was created in which the optimal writing style for recognition was also examined (Shinghal and Suen, 1982). Still, many researchers choose to construct their own data sets, and this is the strategy used in this work.

Data Collection

Digits were collected from 13 subjects who were instructed to write several of each of the digits on a clean sheet of paper. No limitations were imposed as to the style, size, clarity, thickness, or slant of the digits. Each of the digits was segmented by hand, scanned at 300 DPI with 16 gray levels, and saved in separate files. A total of 1500 digits were collected in this manner, approximately 150 of each of the 10 digits types.

There are only a few assumptions which were imposed on the data. Among these are:

1. The digits were to be clearly segmentable from one another. That is, a rectangular box could be drawn around each digit so that the entire content of one digit resided inside that box, and no portion of any other digits were contained in that region.
2. The background was relatively noiseless so that there exists a clear threshold between background and digit intensities.
3. No character was rotated more than 45° from what is normally considered its upright position (the character was not upside down).

Figure 95.10 displays many of the digits collected with this process. It is clear that they were written without regard to neatness, and in fact some of the digits appear ambiguous to human classifiers. This variety was encouraged to provide a realistic environment for the training process.

Preprocessing

The networks used for feature extraction and classification are highly dependant on spatial overlap of digits of the same class for their success. The original digits were written without regard to this constraint, and so it was necessary to process the digits to alleviate differences in size, thickness, rotation, location, and intensity. This was not expected to destroy the recognition capabilities of either humans or machine, since the information content of the digits is largely contained in the form of the digits. In addition, the resolution of the digits was reduced to prevent prohibitive training time for subsequent modules.

The preprocessing was conducted in the following steps: intensity thresholding to remove noise, center of mass adjustment, line thinning, simultaneous rotation to standard axis and translation to standard center of mass, size determination and fixation, reduction in resolution, and smoothing of digits as a form of anti-aliasing. This sequence is depicted in Figure 95.11 and the methods for these steps are described in the following paragraphs. The inputs to the preprocessing stages come from the digitized characters from the digital scanner, while the outputs of the preprocessing feed into the inputs of the feature extraction network.

Noise Thresholding

A threshold was applied to each pixel of the original digits, creating a binary image for further processing. The threshold served a dual purpose. It eliminated weak and extraneous information from the digit, thereby aiding a separation from the background. Secondly, it eliminated intensity variability from within the contour of the digit. Each pixel of the digit was checked against the threshold value. If it was lower, it was set to zero. If the pixel value was equal to or higher than the threshold, it was set to a maximal value (assigned to be 2). An effective threshold was found to be at a gray level of 4, and this value was used for the processing of all digits. A more flexible approach would have been to use an adaptive threshold, whereby the deciding value is based on the content of each digit by analyzing an intensity histogram. In part because of the controlled lighting conditions of the digital scanner, and also due to the assumption of a clear separation of the digit from its background, it was felt that this additional computation was not necessary. This analysis was confirmed by the high quality of the digits after thresholding was applied.

Center of Mass Adjustment

Particular problems were encountered when some digit types (nines, eights, sixes) had small loops or regions of high density

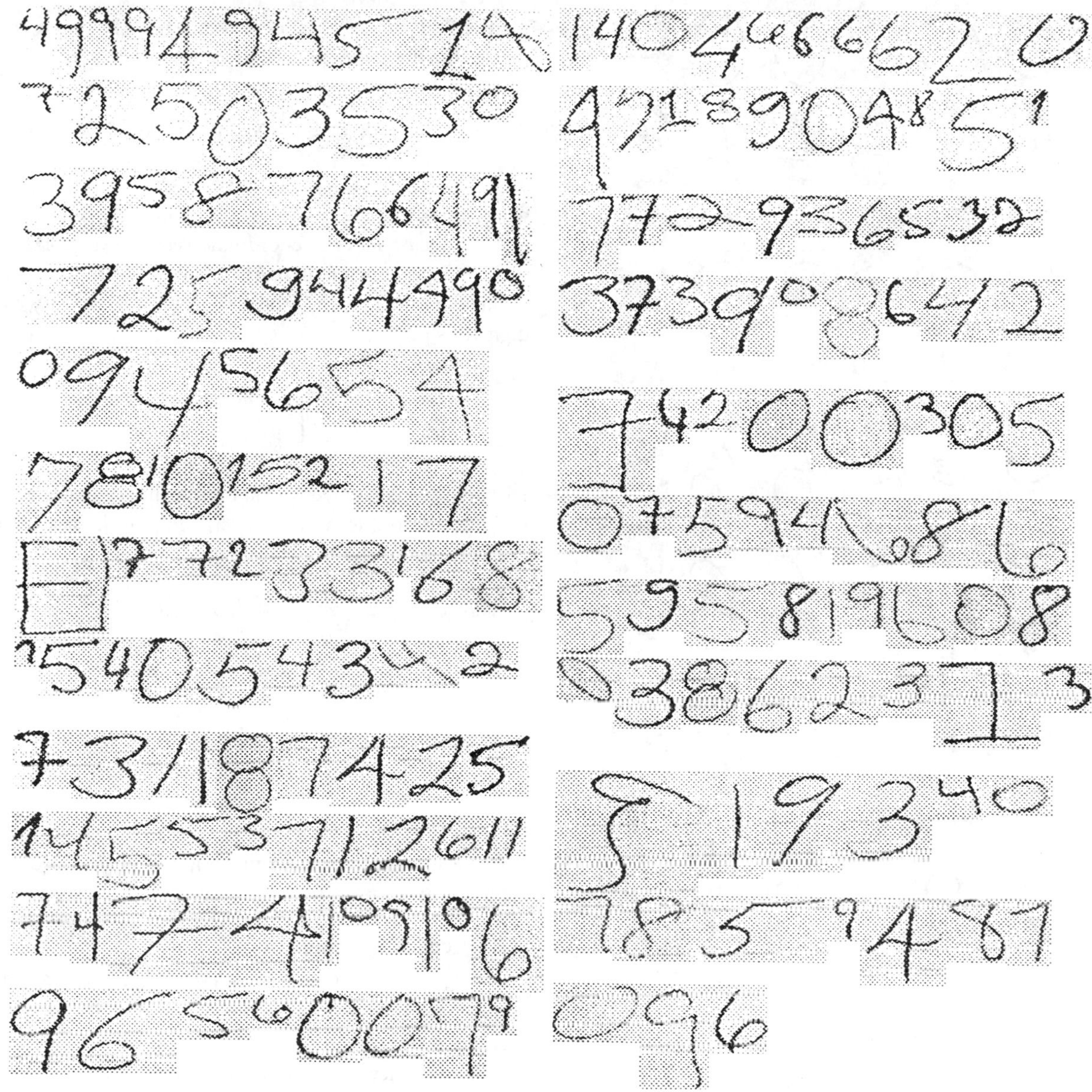

Figure 95.10 Random samples of the original unprocessed characters used in this study.

of pen marks. In such instances, the center of mass of the digit was highly skewed toward that region, and overlap with similar digits were often small. This also often resulted in an abnormal rotation when that routine was applied. An adjustment was applied to each digit to expand small regions as this and so move the center of mass toward the absolute center of the digit. In this method, the center of mass ($CM = [x_c\ y_c]^T$) was located and the digit split into quadrants about this point. Each quadrant was then mapped into its corresponding quadrant in absolute space (using the absolute center $AC = [x_a\ y_a]^T$) by scaling each of the regions as

$$x' = \frac{(x - x_c)(x_{\max} - x_c)}{(x_{\max} - x_c)} \tag{95.43}$$

where the old x coordinate is mapped to the new location x'. The y coordinate is changed in the same way.

Line Thinning

A thinning routine was used on the binary level digits to reduce the effects of line thickness. The method used is familiar in the literature (Pavlidis, 1980; Beun, 1973). Basically, the algorithm pares away all boundary points in the digit until it is left with only skeletal pixels which must be kept in order to preserve the integrity of the digit contour. A pixel is considered to be a skeletal pixel if it is part of the digit (has a non-zero value), one of its four neighbors is zero valued, and passes either of two conditions as is described thoroughly in a previous article (Beun, 1973). Several passes of this procedure are necessary to reduce the image to one consisting of only skeletal pixels, since the above criteria

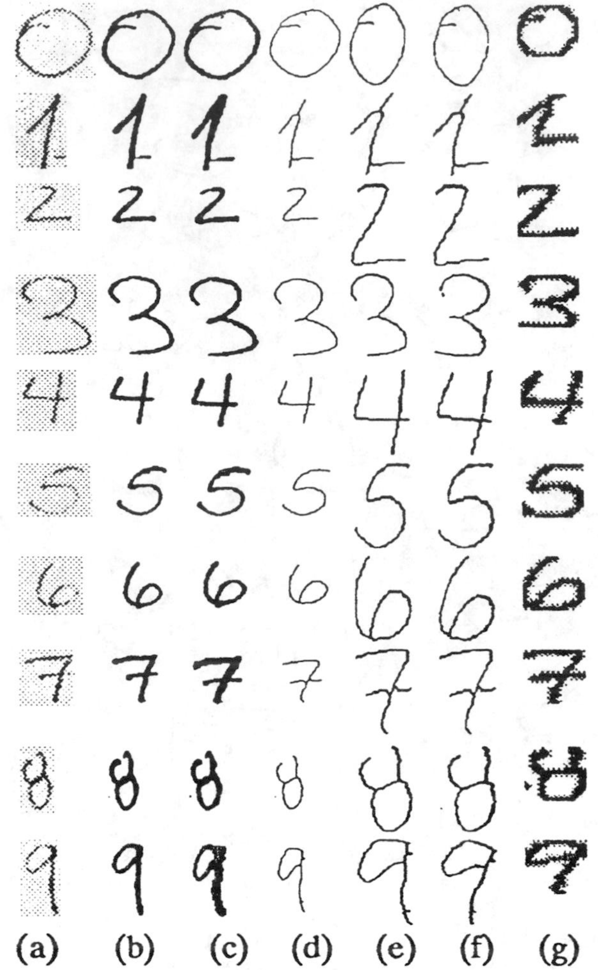

Figure 95.11 The sequence of steps in the preprocessing of the handwritten digits.

will remove only boundary pixels with each pass through the digit.

Fixing to Size

Prior to the use of the rotation routine, the image is fixed to a standard size (60 by 100 pixels). This is necessary to avoid errors in the calculation of the digit principle axes caused by distortions in portions of the digit. To perform this task, the corners of the digit are located and scaled to the new size. Pixels are mapped into the nearest pixel after the scaling factor has been applied. The operation is performed in the same way as Equation 95.48a,b, where the x coordinate magnifier is 60.0 and the y magnifier is 100.0.

Rotation

The rotation algorithm uses the coordinates of each of the non-zero valued image pixels to find a principal vector of the image. This vector specifies the angle of the principal axis of the digit in 2-D space, which can then be manipulated to create a transformation matrix which will rotate the image.

Each of the digits is rotated and translated in this space to a standard location and primary axis. The center of mass of the digit ($M = [m_x\ m_y]^T$) is located and used to find a correlation matrix for the digit, calculated as

$$C = \left(\frac{1}{r}\sum_{i=1}^{r} \overline{PP^T}\right) - \overline{MM^T} \tag{95.44}$$

where the summation is over all r nonzero pixels in the image and the vector P is the coordinate vector of the pixel ($P = [p_x\ p]^T$). An eigenvector matrix E is calculated from the 2×2 matrix c. encoding the angle (f) through which the primary axis of the digit runs as

$$E = \begin{bmatrix} \cos(\phi) & -\sin(\phi) \\ \sin(\phi) & \cos(\phi) \end{bmatrix} \tag{95.45}$$

Each digit was rotated so that the primary axis lies vertically (90°). Each pixel location $\mathbf{P}' = [p_x\ p_y]^T$ of the rotated image is calculated from the original image as

$$\overline{P}' = E'(\overline{P} - \overline{M})^T \tag{95.46}$$

where

$$E' = \begin{bmatrix} \cos(90° - \phi) & -\sin(90° - \phi) \\ \sin(90° - \phi) & \cos(90° - \phi) \end{bmatrix} \tag{95.47}$$

and the vector $\mathbf{P}$ contains the original pixel coordinates. The elements of $\mathbf{P}'$ are rounded to the nearest integer locations.

Reducing Resolution

The corners of the rotated image (smallest rectangle which completely encloses the digit) are found and used to scale the digit to a new resolution of 16×16. This is performed by calculating a new pixel coordinate $[x'\ y']$ by

$$x' = nint\left(\frac{x^*16}{x_{\max} - x_{\min}}\right) + 1 \tag{95.48a}$$

and

$$y' = nint\left(\frac{x^*16}{y_{\max} - y_{\min}}\right) + 1 \tag{95.48b}$$

where the *n*int( ) operation nearest integer takes the nearest integer of the resultant division. A pixel in the new 16×16 digit is assigned a value of 2 if *any* of the positive valued pixels of the higher resolution are mapped into that location. An alternative is to assign an additional threshold to turn on a pixel in the lower resolution image if the number of original pixels mapping into that location exceeds the threshold. The resultant 16×16 images were considered to be generally of good enough character (by subjective analysis) to avoid this additional complication.

Blurring

As was mentioned previously, one of the assumptions fundamental to the success of the subsequent neural network processors is that of a high degree of spatial overlap of similar digits. That is, because of the hard wiring of neural inputs to image locations, the neural networks are not position invariant. The aforementioned preprocessing steps can aid in creating an invariance, but is by no means invincible in this task. To assist in the overlap of the digit contours of similar digits, a simplified smoothing operation was applied to the 16×16 images. This operation can also be thought of as an anti-aliasing operation. Basically, if a zero valued pixel has one or more of its four primary neighbors with a nonzero value, that pixel is turned on with a value of 1 (it should be remembered that the pixels on the contour were given values of 2).

Results

The feature extraction routine (ALOPEX variance maximization of network node outputs using the architecture of Figure 95.7) was applied to each of the digits in the data set. A random 1000 digits were selected for the training of this module, and 32 features were extracted from each 256 dimensional input image[5]. Since the feature extraction module has an architecture reminiscent of a pipeline, it was more efficient computationally to allow each neuron in the module to complete training before any subsequent nodes were altered. Training appeared most efficient with ALOPEX parameters of $\gamma_0 = \alpha_0 = 5.0 \times 10^{-3}$, $\sigma_0 = 7.5 \times 10^{-3}$, $\gamma_\infty = \alpha_\infty = 5.0 \times 10^{-5}$, $\sigma_\infty = 7.5 \times 10^{-5}$, and $\tau = 1000$, and typically required between 8,000 and 12,000 iterations per node for a good convergence, as seen by the response curve of Figure 95.12. The vast bulk of the processing time is due to the large number of patterns (1000) used in the training, since each pattern must be presented to the neuron during each iteration.

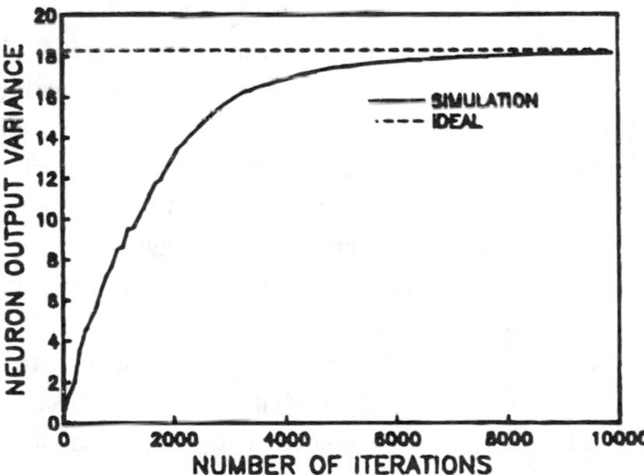

Figure 95.12 A representative sample of the effects of each of the preprocessing stages on the digit types, where (a) is the original image, (b) thresholded, (c) center of mass adjusted, (d) thinned (e) fixed for size 100×60 pixels, (e) rotated to standard axis, and (g) resolution reduced and low pass filtered.

The number of features to retain was calculated by plotting the eigenvalues in descending order as generated from the conventional K-L expansion, as in Figure 95.13. The magnitude of the eigenvalues is identical to the optimum variances of the cell outputs in the FE network, and it is convenient to relate the magnitude of the eigenvalue with the amount of information the corresponding K-L vector carries. The number of features to extract (32) was subjectively obtained from Figure 95.13 as the point in which the information given by an additional eigenvector reduces to near zero. Another way of finding the optimum number of neurons in the feature extracting network is to set a threshold. If a neuron optimizes to an output variance below this threshold, the node is not logically added to the network and the training simulation is stopped. In this way both the extent of the network and its connectivities can be adaptable in the training.

Figure 95.14 depicts the feature cell vectors as they would appear in image form. Each connectivity strength (C_{ij} from Equation 95.13) is given a corresponding intensity (relative to the strength of the connectivity) in the spatial position where the input to the connection arose. Very high intensities (white) indicate large positive connections, and large negative connections are shown as a low intensity value (black). Some of the feature "filters" have regions of high contrast which remind us of features in the character data set. It is clear that the last few feature images are quite "noisy", and this is consistent with their low information content. Moreover, it is very obvious that these "optimal" vectors would be very difficult to specify heuristically.

The clustering operation needed to incorporate some understanding of the number of clusters necessary to accurately describe the data. Since the ALOPEX optimization for the clustering operation was quite time consuming, the standard FCM algorithm with Euclidean distance measurements was used to find the cluster validity measures described in section 2.3.2 for as few as 2 and as many as 40 clusters. These simulations typically required no more than 30 iterations for convergence.

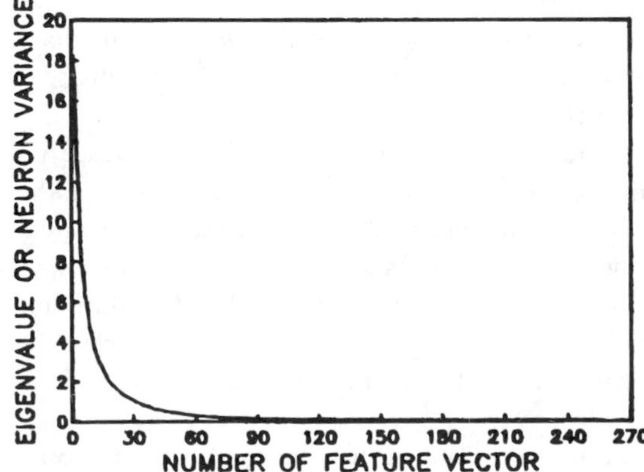

Figure 95.13 The eigenvalue as a function of the number of the K-L expansion vector. The eigenvalue can also be thought of as the optimum nodal output variance for the ALOPEX trained feature cell.

[5] Note that the neural networks in this study have global inputs and are not spatially interdependent. This means that the preprocessed 16 by 16 digits are viewed as a 256 dimensional vector by the networks.

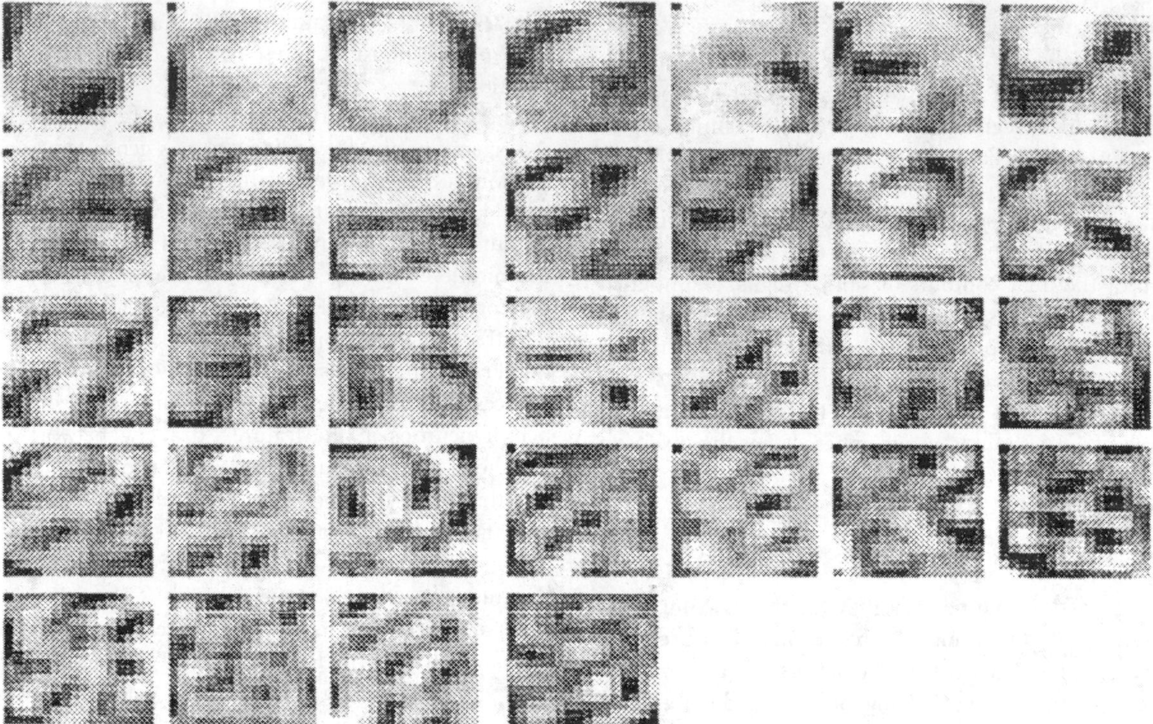

Figure 95.14 The 32 feature cell connection vectors displayed in image form. The vectors are shown in descending order of their output cell variance as you view from left to right and top to bottom.

In the section on FCM clustering algorithm, the determination of the number of clusters necessary for any given data space was discussed (the cluster validity problem), and the cluster validity measures *F* and *H* were introduced. Figures 95.15(a)–(b) illustrate the change in the validity measures *F* and *H* for the converged clusterings of the FCM algorithm from 2 to 40 clusters ($q = 1.2$). It is hoped that a distinct minimum in the entropy (H) measure, and a distinct maximum in the partition coefficient (F) measure will present themselves definitively around a certain value of *c*. This is not the case in any of the plots of Figures 95.15. In fact, the data space created by these digits appears rather homogeneous in nature, with few well separated regions for simple cluster identification. This may be an artifact of the high number of patterns used in the cluster formation, which may "fill in" many of the less dense regions of feature space used for simple cluster identification.

The credibility to the notion that the data space is highly uniform is enhanced by the extremely low value of *q* which was necessary to form clusters. At $q = 1.2$, the clusters are formed with very sharp decision boundaries. When a more commonly used value ($q = 2$) was used, the FCM algorithm converged every cluster center to the same point, so that the class memberships were entirely fuzzy and no distinguishing information was provided.

There are two regions in the curves of Figure 95.15 in which it is reasonably safe to assume that there are relatively "better" clusterings than for other *c* values. The first is for the value of $c = 2$, which the curves of Figure 95.15(a) and 95.15(b) show as locally optimal. For the purposes of this study, the value of $c = 2$ had to be rejected simply because of the understanding that there are at least 10 clusters desired. This is because of the 10 digits types (zero through nine) used in the data set.

The second region occurs for values of $c > 30$. In this region of the curves of Figure 95.15, there begins a "plateau" region, beyond which a mental extrapolation of the curves would anticipate little improvement for a much higher number of clusters[6] The region from $c = 30$ to $c = 40$ is heuristically an acceptable region, and still maintains an adequate number of average samples (25–33) per cluster. The heurisitic basis for the credibility of this range of *c* values resides in the belief that each of the digit types can be written in, on average, 3 to 4 different styles. For example, a one can be written as a single vertical line, or additions of an upper diagonal line alone or with an accompanying lower horizontal line.

It is interesting to note that there is indication that the FCM algorithm was not finding the globally optimal solution to the clustering problems it was presented with. One evidence of this was that the cost function value of the converged solution was often higher than one of the intermediate solutions through which the simulation has passed. But by far the simplest determination of locally optimal solutions is to run the program several times with the same parameter set and the same patterns. When this was performed at $c = 30$, the FCM algorithm obtained different solutions each time, as evidenced by discrepancies in

[6] This was partially confirmed with a simulation performed at $c = 50$, which indicated a continuation of this trend.

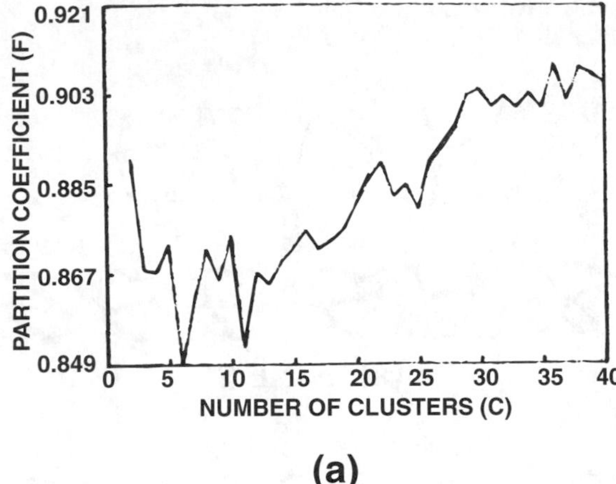

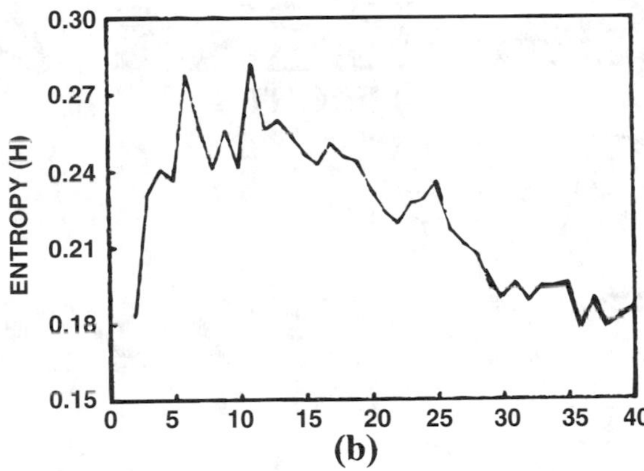

Figure 95.15 The variation in the (a) partition coefficient (F) and (b) the entropy (H) for choices for the number of clusters (c) from 2 to 40.

the cost function value[7] and cluster membership distributions (the array Ω in equation 95.40). This lends further credence to the use of an optimizer in the FCM routines.

The ALOPEX trained FCM algorithm was trained on 30 clusters for the same 1000 training patterns. In order to reduce the computational overhead, the simulation was started by using the center coordinates converged upon by the standard FCM algorithm. The simulation typically required between 1,000 and 2,000 iterations for a "good" convergence, and seemed to perform best with ALOPEX parameters of $\gamma_o = a_0 = 0.2$, $\sigma_0 = 0.3$, $\gamma = a_\infty = 2.0 \times 10^{-3}$, $\sigma_\infty = 3.0 \times 10^{-3}$, and $\tau = 2500$. Primarily for computational reasons, the Euclidean distance measure was used in the ALOPEX trained FCM algorithm. The use of a non-Euclidean measure would have, for this application, resulted in exorbitant execution times, since the calculation of the covariance matrices results in substantial computational overhead. Another reason for not using the fuzzy-covariance matrices in the formulation of a non-Euclidean distance metric was that later simulations showed that such a selection to result in an unusually hard membership assignment, which may be disadvantageous for medical applications in particular.

Table 95.2 shows the classification results for the ALOPEX modified FCM scheme with the labeling method described in section 2.4. The total classification accuracy is 86.3% for the 1000 training digits, and 86.0% for the 500 post-training digits, as indicated in Table 95.3.

Figure 95.16 depicts the cluster centers as images (using the method of section 2.4), to give us a flavor for the aspects of the characters which each cluster emphasizes. As Figure 95.16 shows, most clusters have fields which are strongly reminiscent of one of the digit types, but there are a few clusters which are blends of portions of several types.

Figure 95.17a,b shows the misclassified digits as they appeared in their original unprocessed form, grouped by the digit type they were incorrectly identified with. Some of the misclassifications can be directly connected with preprocessing problems (i.e. improper rotations, noise in the image retained), while other are probably due to the strong overlap between certain characteristics of the digit types.

The data set was also tested with the backpropagation neural network training algorithm. This technique was described thoroughly in section 95.1 (see especially figure 95.2 and Equations

Table 95.2 A Comparison of the Classification Results of the ALOPEX Trained Network with the Actual Pattern Identities of the Digits Used in the Training. Of the 1000 of These Training Digits, 863 were Correctly Classified (86.3%)

	ASSIGNED DIGIT CLASS									
	0	1	2	3	4	5	6	7	8	9
0	96	0	0	0	0	0	3	0	1	0
1	0	85	3	0	6	1	0	2	3	0
2	0	2	88	0	1	0	0	6	2	1
3	0	0	1	86	0	2	0	3	7	1
4	0	0	0	0	84	6	1	0	0	9
5	0	0	0	0	0	96	2	0	2	0
6	1	1	0	0	2	14	81	0	1	0
7	0	0	0	0	11	0	0	89	0	0
8	0	1	0	5	1	1	1	0	90	1
9	2	0	0	5	12	3	0	5	5	68

Table 95.3 A Comparison of the Classification Results of the ALOPEX Trained Network with the Actual Pattern Identities of the Digits Not Used in the Training. Of the 500 of these Digits, 430 were Correctly Classified (86.0%)

	ASSIGNED DIGIT CLASS									
	0	1	2	3	4	5	6	7	8	9
0	46	0	0	0	1	0	1	0	0	0
1	0	41	3	0	3	0	1	1	1	0
2	0	6	44	0	0	0	0	2	0	0
3	0	0	0	41	0	2	0	2	4	1
4	0	0	0	0	41	2	1	0	0	4
5	0	0	0	0	0	49	1	0	0	0
6	0	0	0	0	0	5	47	0	0	0
7	0	0	1	0	2	0	0	46	0	1
8	0	0	0	2	1	1	0	0	48	0
9	2	0	0	3	12	1	0	1	2	27

[7] When the FCM algorithm was run twice at $c = 30$, final cost function values of $J = 52422.17$ and $J = 56321.15$ were obtained.

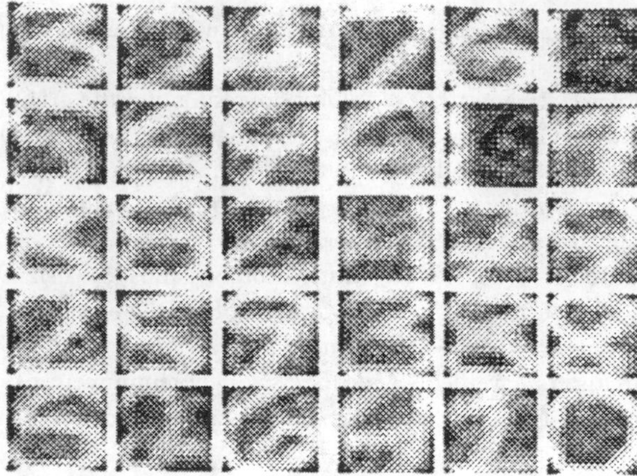

Figure 95.16 The 30 cluster centers displayed in image form.

95.13–95.17. A direct comparison of the backpropagation results with the ALOPEX trained network developed in this study is not equitable, since the backpropagation algorithm is a supervised technique. However, since backpropagation is so widely used, and since it has been used in the specific application of character recognition, the results which it provides can give a calibration of the difficulty of the data set. These results can also demonstrate the degree of additional accuracy which can be extracted by knowing the pattern identities a priori.

The training was conducted with a network comprised of 256 input nodes, 100 hidden nodes, and 10 output nodes on 1000 input patterns (consisting of the same preprocessed character training set as was used in the ALOPEX trained system). The desired low value of the output lines was set at 0.1, and the desired high value at 0.9. The network was trained for 300 epochs with values of $\eta = 0.1$ and $\alpha = 0.75$. Upon the completion of training, the training pattern classification error was determined by assigning it the class identity of the output node with the highest activity. For the 1000 training patterns, the backpropagation network correctly classified all but two of them, for an accuracy of 99.8%. The 500 patterns not used in the training were classified with 93% accuracy, as shown in Table 95.4 below.

95.4 Discussion

Our primary interest in this application is to be able to fine tune the training algorithm so that it is of maximum efficiency and accuracy for subsequent medical applications. In this regard, the classification of handwritten digits tests the limits of the applicability of the method. This is because the large number of clusters, features, and patterns stress the algorithm to its maximum load. The computing times for all phases of the ALOPEX trained algorithm were significant, but the accuracy of the optimization was nearly ideal in the feature extraction training. For the clustering module, the ALOPEX simulation for $c = 30$ provided a moderate improvement over the standard FCM algorithm. Clearly a much more substantial computational demand is caused by the use of a non-Euclidean distance metric, particularly when the calculation of a fuzzy covariance matrix is required. For a

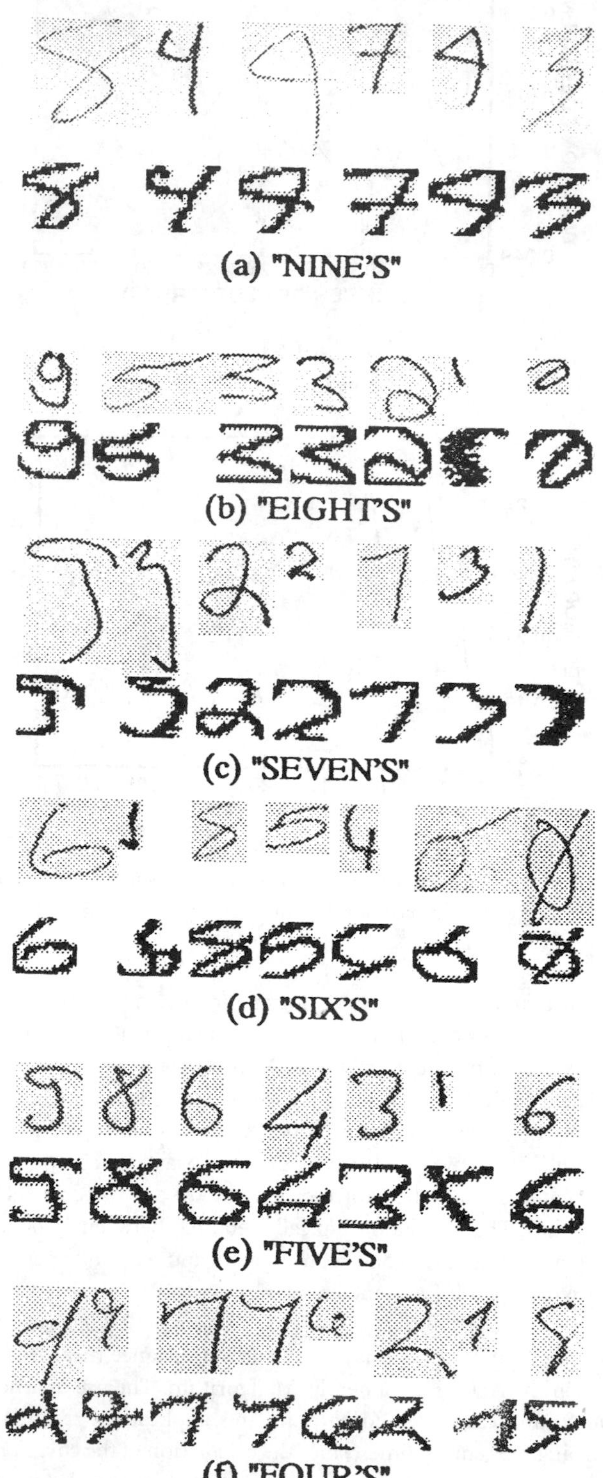

Figure 95.17a A sample of some of the characters misclassified as (a) nine's, (b) eight's, (c) seven's, (d) six's, (e) five's, (f) four's, (g) three's, (h) two's, (i) one's, and (j) zero's. Each character is shown in its unprocessed form, with the preprocessed character shown directly beneath it.

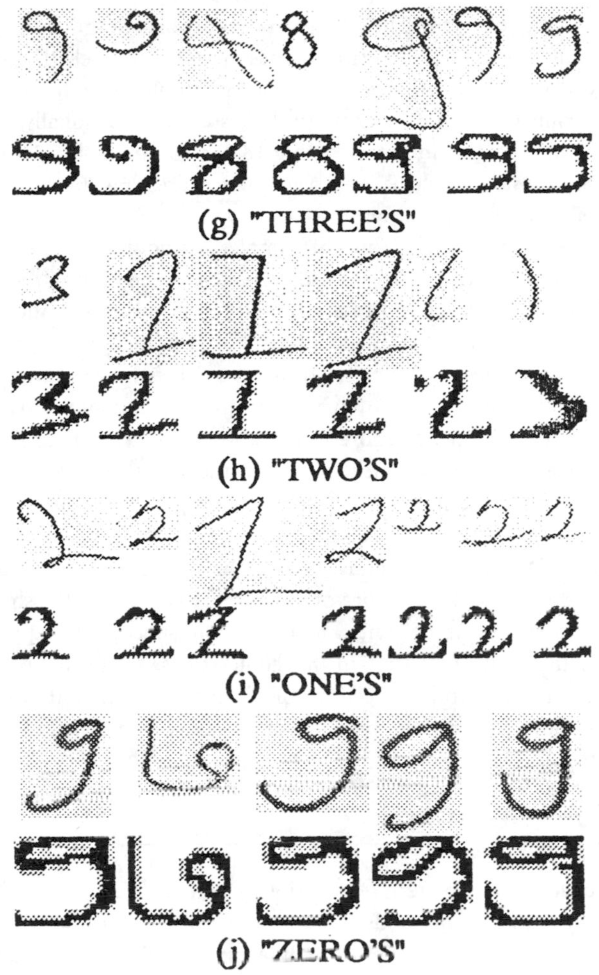

Figure 95.17b See Figure 95.17a caption.

Table 95.4 A Comparison of the Classification Results of the Backpropagation Trained Network with the Actual Pattern Identities of 500 Digits Not Used in the Training. Of the 500 of these Digits, 465 were Correctly Classified (93%).

	0	1	2	3	4	5	6	7	8	9
0	46	0	0	0	1	0	1	0	0	0
1	0	42	2	1	0	0	1	3	1	0
2	0	5	46	1	0	0	0	0	0	0
3	1	1	0	44	0	0	0	1	0	3
4	0	0	0	0	47	0	1	0	0	0
5	0	0	0	0	0	47	1	0	1	1
6	0	0	0	0	0	0	51	0	1	0
7	0	1	1	0	0	0	0	47	0	1
8	0	0	0	1	0	1	0	0	49	1
9	0	0	0	0	1	0	0	0	1	48

large cluster, large pattern set application such as this, the Euclidean metric becomes one of the only feasible possibilities.

One of the largest problems appears to be the determination of the number of clusters necessary for an accurate depiction of the data space. Both of the cluster validity measures we used, along with about a half dozen others not included in this document, were not able to give us a definitive idea of the proper number of clusters. There is a natural tendency for all of the measures to drift toward their ideal values as the number of clusters increases, since when the number of clusters equals the number of patterns, we have a trivial but perfect set of clusters. Whether the plateau region of the curves of Figure 95.15 is an artifact from this tendency is unknown, but since the number of clusters was still substantially lower than the number of patterns ($30 << 1000$), the assumption is that the saturation of the validity measures after $c = 30$ contains real information.

As was mentioned in the results section, the low value of q, and the lack of a distinct local extremum in the cluster validity measures is indicative of a data space with few, if any, well separated and compact clusters. The data space appears to be naturally "fuzzy". It is curious that the FCM algorithm requires a rather hard decision to partition this fuzzy data space, while the theoretical reasoning for fuzzy logic presumes a fuzzy algorithm as ideal for this type of decision making. In any event, it is unclear whether the ALOPEX trained FCM algorithm really arrives at a good clustering of the data space, or whether there is just a large enough number of cluster centers to "fill in" the data space.

This low value of q may account for many of the classification errors presented in Figure 95.17a,b. Many of the misclassified digits are visually clearly a member of another class. The sharp decision boundaries created by the low value of q can push the cluster membership strengths toward their limits (zero and one). Thus even though a character lies near the boundaries between clusters, it is given a strong membership to a cluster. This pushes a marginal pattern (i.e., a pattern with 50% similarity to each of two clusters) to become decisively incorrect! With this information misgiven in the clustering module, no labeling scheme can reclaim a correct classification.

The errors in the classification of the handwritten digits appears to arise from multiple sources. A few of the characters are of uncertain identity to myself and several others who have viewed the data, and so it is unreasonable to expect that the computer algorithm should perform any better. Much of the erroneous classifications can be traced to the preprocessing of the characters. This is perhaps the most crucial and controllable portion of the system, and yet very difficult to design and improve. Alterations to any of the algorithms seemed to improve the operations on some characters to the detriment of several others. There are some adjustments which can and should be made to any implementable system, including the addition of thresholds to the resolution reduction, and a contour tracing program to alleviate noisy elements which cannot be removed by threshold analysis and yet contribute to improper rotations.

Even with these contributions, the most significant source of error is from the decisions made by the classification system itself. The question becomes: how can these erroneous decisions be reduced? The answer seems to lie in the selection of the training set and other parameters of the simulation, including the selection of the resolution of the digit representation. Subjectively, using an 8 by 8 image would probably result in even larger classification errors, as there is barely enough resolution to unambiguously represent most of the digits with a 16 by 16 image. A slightly higher resolution would probably provide some relief, but was not feasible computationally with the speed and memory of the machines available.

Most importantly, the composition of the data set seems to be in question. Most other studies with digit recognition have used a much larger training set, up to 10 times the size. This larger training set helps to generalize the networks trained with supervised algorithms. The data set used in this study may have more than a "normal" share of unusual and exceptional characters, as several of the providers of the data (notably the authors) intentionally wrote the characters using various writing styles, slants, and sizes.

The use of the backpropagation algorithm allows a calibration of this method with a more widely used strategy. It appears that this data set is of comparable difficulty to the postal zip code data used by LeCun, et al. (1989), since both data sets were tested with the backpropagation technique and performed similarly. Both data sets performed nearly perfectly in classifying the training set, and the untrained set classified 93% accurately in this study, as opposed to 95% accuracy for that AT&T Bell Labs group[8].

Given that the unsupervised system developed in this study was not privy to the class identities during training, the 86.3% accuracy of the method is, in my opinion, outstanding. Even more striking is the ability of the unsupervised system to generalize its decision capabilities more easily than the supervised system, as indicated by the 86.0% error in classifying the untrained data. The inherent ability of unsupervised decision making to generalize more readily, and with fewer training patterns, was one of the motivations for the construction of the system. It is clear, at least for this application, that the decisions made by the unsupervised classifier are more general than that of backpropagation, in that there is little loss of information after training. It is also clear that the inherent accuracy, whether with regard to the training set or test set, is greater for the backpropagation system.

This understanding gives credence to a recent trend to incorporate unsupervised and supervised decision making in the same system. Each scenario is beneficial at different times. It may be particularly useful to use an unsupervised decision until the time that there are enough training patterns to construct supervised decisions. I would expect that if the training set for this study was reduced (say from 1000 patterns to 500 patterns) that the unsupervised method would have classification accuracy on the untrained data closer to the supervised decision accuracy on the same data set.

There is one other important aspect which should be considered in the comparison of these two techniques: the amount of hardware resources necessary to implement these methods in a true parallel form. For the backpropagation system, the decision were made with 110 computing nodes and 26,600 connections, versus 93 nodes and 17,178 connections for the ALOPEX trained unsupervised system. The savings in connections for the unsupervised method is the most critical, since this is the most challenging aspect of implementing these systems in parallel (the hardware interconnects require a great deal of space). Additionally, the unsupervised system requires only 18,015 additions and 17,388 multiplications, while backpropagation requires 26,490 additions and 26,600 multiplications.

This discussion would not be complete without some speculation about the strategy used for the character recognition process. It is my belief that the primary limitation of the classification is the assumption that the spatial form of the digit is the most important aspect of character recognition. It is more likely that an analysis of the stroke curvatures and other contour based principals will provide more information to the classifier. As an example, notice how much confusion there was (Figure 95.17) in classifying the number two from the way many people write the number two. As difficult as this was for the software, it is remarkably easy for humans, even though the only real distinction is the smooth curvature of the number two versus the sharp vertices of the number one. The most promising systems appear to be those which can retain this contour based information of curvature and stroke direction, or if shape information is still important, preserve the pixel neighborhoods in the analysis.

95.5 Summary

An unsupervised pattern recognition system has been introduced, largely based on methods originating elsewhere, but bound by their application to a parallel architecture and their ability to be trained by a single optimization algorithm. The method has been tested in two widely varying application domains; in the classification of handwritten digits and the diagnosis of Visual Evoked Potential signals of normal and abnormal subjects. The scope of the second application is beyond the interest of this handbook.

The ALOPEX trained system has proven itself capable of strong generalizations, as evidenced by the application to handwritten digits, and is able to extract a significant amount of information without the advice of an omnipotent instructor. When properly tuned, the clustering module can make decisions of an analog nature, so that an understanding of the certainty of its decision can be analyzed.

To the artificial neural network purists, this architecture can only loosely be referred to as a neural network. In the sense that the system is trained by example, highly parallel, and comprised of highly interconnected elements performing simple computations, the neural network label is quite fitting. If however, the label also connotes a system trained through local information sharing, and nodal units based only on inner product variations[9], then a more general label should be applied. In the truest sense, the backpropagation algorithm is not a "local" training regime,

[8] The slightly higher accuracy of their study is probably attributable to the slightly variant neural architecture they used to accentuate certain inherent aspects of the digits.

[9] The clustering module differs from this in that some of the units perform squaring and difference operations.

since the errors propagate through the layers imaginatively, and do not actually reside on any signal lines.

The reason for mentioning this small labeling problem addresses the direction of the neural network community as a whole. Only recently has the field begun to merge with other more well established disciplines, partly because of limited utility alone and partly because of a reluctance to share in the spotlight. In the context of parallel processing systems in general, the neural network extension has a good deal to gain by attributing more computing power to the processing elements. From a hardware perspective alone, the high connectivity of "pure" neural network systems has been a technological stumbling block. Providing more power to the "neuron" means releasing the tight relationship to neurobiology which many researchers rely upon.

One of the fields which has long been tightly interwoven with the neural network field is that of combinational optimization. The usual hope is that the optimization schemes, such as the ALOPEX technique used in this chapter will provide a higher probability of reaching a globally optimal solution. That ALOPEX, in this network construction, can provide this is certain. Whether it is always computationally necessary is less certain[10]. More overlooked, but of primary importance, is the utility of a proven optimization scheme as a flexible and reliable design tool. Regardless of the architecture, if the information which the user desires to retain in the network after training is expressible through the minimization or maximization of a function of the network, then ALOPEX can be used to find that information. In our laboratory, ALOPEX has been used as an alternative to the backpropagation training on the multilayer perceptron architecture, as well as in the training of Hopfield nets. It seems reasonable that specialized problems can be solved with this general tool. In this sense ALOPEX can perform, as it has in this study, as a conversion of the desired information onto a parallel architecture.

As hoped, the conclusion of this study has produced more questions and promising directions than answers. The character recognition arena was an interesting demonstration, but it is unlikely that this scheme can ever compete equally with the supervised methods. Still, the blending of unsupervised and supervised training methods at varying times in the learning process is intriguing, and probably beneficial if a suitable application is available.

References

Ackley, D. H., Hinton, G. E., and Sejnowski, T. J. 1985. A learning algorithm for boltzmann machines, *Cognitive Sci.,* 9:147–169.

Amano, A., Aritsuka, T., Hataoka, N., and Ichikaua, A, 1989. On the use of neural networks and fuzzy logic, *Proc. J. Intl. Conf. on Neural Networks,* 1:301–306 Washington, DC.

[10] The ALOPEX simulation results for the clustering module in the application to the VEPs performed no better than the standard FCM simulation.

Amari, S.-I. 1977. Neural theory of association and concept formation, *Biological Cybern.,* 26:175–185.

Anderson, J. A. and Rosenfeld, E. 1988. *Neurocomputing: Foundations of Research,* MIT Press: Cambridge, MA.

Anderson, J. A. 1972. A simple neural network generating an interactive memory, *Mathematical Biosciences,* 14:197–220.

Anderson, J. A., Silverstein, J. W., Ritz, S. A., and Jones, R. S. 1977. Distinctive features, categorical perception, and probability learning: some application of a neural model, *Psychol. Review,* 84:413–451.

Andrews, H. C. 1977. Multi-dimensional rotation in feature selection, *IEEE Trans.,* SMC-7:537–541.

Ball, G. H. and D. J. Hall, 1967, A Clustering Technique for Summarizing Multivariate Data, *Behavior Science,* 12:153–155.

Beun, M. 1977. A flexible method for automatic reading of handwritten numerals, *Phillips Tech. Rev.,* 33:89–101, 130–137.

Bezdek, J. C. 1973. *Fuzzy Mathematics in Pattern Classification,* PhD Dissertation, Cornell University, Ithaca, N.Y.

Bezdek, J. C. 1988. Pattern Recognition with Fuzzy Objective Function Algorithms, 65–85, Plenum Press, New York, NY.

Bezdek, J. C. and Dunn, J. C. 1975. Optimal fuzzy partitions: a heuristic for estimating the parameters in a mixture of normal distributions, *IEEE Trans. Computers,* August.

Bezdek, J. C., Ehrlich, R., and Full, W. 1984. FCM: The fuzzy c-means clustering algorithm, *Computers and Geosciences,* 10(2):191–203.

Bezdek, J. C. and Fordon, W. A. 1978. Analysis of hypertensive patients by the use of the fuzzy ISODATA algorithm, *Proc. JACC,* 3:349–256.

Bezdek, J. C., Trevedi, M., Ehrlich, R., and Full, W. 1982. Fuzzy clustering: a new approach for geostatistical analysis., *Int. J. Sys., Meas., and Decisions.*

Bienenstock, E. L., Cooper, L. N., and Munro, P. W. 1982. Theory for the development of neuron selectivity: orientation specificity and binocular interaction in visual cortex, *J. Neurosci.,* 2(1):32–48.

Block, H. D. 1962. The perceptron: a model for brain functioning I, *Reviews of Modern Physics,* 34:123–135.

Cannon, R. L., Dave, J. V., Bezdek, J. C., and Trivedi, M. M. 1985. Segmentation of thematic mapper image data using fuzzy c-means clustering, *Proc. 1985 IEEE Workshop on Languages for Automation,* IEEE Computer Society:93–97.

Carpenter, G. A. 1989. Neural network models for pattern recognition and associative memory, *Neural Networks,* 2:243–257.

Casey, R. and Nagy, G. 1966. Recognition of printed chinese characters, *IEEE Trans. Elec. Comput,* 15:91–101.

Chien, Y. T. and Fu, K. S. 1967. On the generalized Karhunen-Loève expansion, *IEEE Trans. Inform. Theory,* 15:518–520.

Chon, T. and Micheli-Tzanakou, E. 1989. A probabilistic approach to the ALOPEX process using moment invariants of images, *Proc. Int. Joint Conf. Neural Networks,* II: 611.

Ciaccio, E. J. and Micheli-Tzanakou, E. 1990. The ALOPEX process: application to real-time reduction of motion artifact, *Proc. 12th Ann. Int. Conf. IEEE/EMBS,* 12:1417–1418.

Cooper, L. N. 1973. A possible organization of animal memory and learning, *Proc. Nobel Symp. Collective Properties of Physical Systems,* 252–264.

Cover, T. M. 1974. The best two independent measures are not the two best, *IEEE Trans.,* SSC-6: 33.

Dante, H. M. and Sharma, V. V. S. 1985. Optimum decision tree classifiers for classification in large populations, *Proc. IEEE Int. Conf. Cybernetics and Society,* 559–563.

Dasey, T. and Micheli-Tzanakou, E. 1989a. A pattern recognition application of the ALOPEX process on hexagonal images, *Proc. Int. Joint Conf. Neural Networks,* 2:119–125.

Dasey, T. and Micheli-Tzanakou, E. 1989b. Efficiency exploration of ALOPEX based recognition of hexagonalized images, *Proc. 15th Ann. Northeast Biomedical Engineering Conf.,* 177–178.

Davis, J. C. and Economou, C. E. 1984. A review of fuzzy clustering methods, *Adv. Eng. Software,* 6(4).

Deutsch, S. and Micheli-Tzanakou, E. 1987. *Neuroelectric Systems,* NYU Press, New York, NY.

Devijver, P. A. and Kittler, J. 1982. *Pattern Recognition: A Statistical Approach,* Prentice-Hall, Englewood Cliffs, NJ.

Eden, M. 1961. On the Formalization of Handwriting, *Structure of Language and its Mathematical Aspect,* 83–88, American Mathematical Society, Providence, RI.

Eden, M. 1968. Handwriting Generation and Recognition, *Recognizing Patterns,* 38–154, Kolers and Eden, M., eds., MIT Press, Cambridge, MA.

Eigen, M. 1971. Self-organization of matter and the evolution of biological macromolecules, *Naturwissenschaften,* 58:465–523.

Farhat, F. A., Psaltis, D., Prata, A., and Paek, E. 1985. Optical implementation of the Hopfield model, *Applied Optics,* 24:1469–1475.

Feldman, J. A. and Ballard, D. H. 1982. Connectionist models and their properties, *Cognitive Science,* 6:205–254.

Foo, Y.-P. and Takefuji, Y. 1988. Stochastic neural networks for solving job-shop scheduling (series), *Proc. IEEE Int. Conf. Neural Networks,* 1:275–282, 283–290, San Diego, CA.

Fu, K. S. 1982. *Application of Pattern Recognition,* CRC Press. Boca Raton, FL.

Fu, L.-M. 1989. Building expert systems on neural architecture, *Proc. IEEE 1st Int. Conf. Neural Networks,* 221–225, London.

Fukunaga, K. and Koontz, W. L. G. 1970. Application of the Karhunen-Loeve expansion to feature selection and ordering, *IEEE Trans. Computers,* C-19(4):331–318.

Gaillat, G. and Berthod, M. 1979. Panorama des Techniques d'extraction de Traits Caracteristiques en Lecture Optique des Caracteres, *Revue Tech,* Thomson—CSF 11:943–959.

Galar, R. 1989. Evolutionary search with soft selection, *Biol. Cybem,* 60:357–364.

Gath, I. and Geva, A. B. 1989. Unsupervised optimal fuzzy clustering, *IEEE Trans. PAMI,* 11(7):773–781.

Golshan, N. and Hsu, C. C. 1970. A recognition algorithm for handprinted arabic numerals, *IEEE Trans. Syst. Sci. Cybem.,* 6:246–250.

Graf, H. P., Jackel, L. D., and Hubbard, W. E. 1988. VLSI implementation of a neural network model, *IEEE Computer,* 21(3):41–49, March.

Granath, G. 1984. Application of fuzzy clustering and fuzzy classification to evaluate provenance of glacial till, *Math. Geol.,* 16:283–301.

Granlund, G. H. 1972. Fourier processing for hand print character recognition, *IEEE Trans. Comput.,* 21:195–201.

Grimsdale, R. L., Sumner, F. H., Tunis, C. J. and Kilbum, T. 1958. A system for the automatic recognition of patterns, *Proc. IEEE,* 106B:210–251.

Grossberg, S. 1976. Adaptive pattern classification and universal recoding: I. Parallel development and coding of neural feature detectors, *Biological Cybemetics,* 23:121–134.

Gudeson, A. 1976. Quantitative analysis of preprocessing techniques for the recognition of handprinted characters, *Pattern Recognition,* 8:219–227.

Harth, E. and Tzanakou, E. 1974. A stochastic method for determining visual receptive fields, *Vision Res.,* 12:1475–1482.

Harth, E., Kalogeropoulos, T., and Pandya, A. S. 1988. ALOPEX: a universal optimization network, *Proc. Special Symp. Maturing Technology and Emerging Horizons in Biomed. Eng.,* 97–107.

Harth, E. and Pandya, A. S. 1988. Dynamics of the ALOPEX Process: Applications to Optimization Problems, *Biomathematics and Related Computational Problems,* Ricciardi, L., ed., 459–471, Kluwe Academy, Norwell, MA.

Harth, E., Pandya, A. S., and Unnikrishnan, K. P. 1986. Perception as an optimization process, *Proc. IEEE Computer Soc. Conf. on Computer Visual and Patt. Recog.,* 662–665, Washington, DC.

Hebb, D. 1949. *The Organization of Behavior. A Neurophysiological Theory.* John Wiley & Sons, New York, NY.

Hecht-Nielson, R. 1988. Neurocomputing: picking the human brain, *IEEE Spectrum,* March, 36–41.

Hecht-Nielson, R. 1989. Theory of the back-propagation neural network, *Proc. IEEE Joint Int. Conf. Neural Networks,* 1:593–606, Wash. DC.

Highleyman, W. H. 1961. An analog method for character recognition, *IRE Trans. Elec. Comput.,* 502–512.

Highleyman, W. H. 1962. The design and analysis of pattern recognition experiments, *Bell System Tech. J.,* 41:723–744.

Hinton, G. E. and Seinowski, T. J. 1986. Learning and Relearning in Boltzmann Machines, *Parallel Distributed Processing, Vol. 1,* Rumelhart D. E., and McClelland, J. L., eds., 194–281, MIT Press. Cambridge, MA.

Holdaway, R. M. 1989. Enhancing supervised learning algorithms via self-organization, *Proc. IEEE Joint Int. Conf. Neural Networks,* II:523–530, Washington, DC.

Hopfield, J. J. 1982. Neural networks and physical systems with emergent collective computational abilities, *Proc. Nat. Acad. Sci.,* 79:2554–2558.

Hopfield, J. J. and Tank, D. W. 1988. "Neural" computation of decisions of optimization problems, *Biol. Cybern.,* 52:141–152.

Hopfield, J. J. and Tank, D. W. 1986. Computing with neural circuits: a model, *Science,* 233:625.

Hosking, K. H. 1972. A Contour Method for the Recognition of Handprinted Characters, *Machine Perception of Patterns and Pictures,* 19–27, The Institute of Physics, London, England.

Iezzi, R., Jr., Micheli-Tzanakou, E., and Cottaris., N. 1990. Effects of pattern convergence and orthogonality on visual evoked potentials, *Proc. 12th Ann. Int. Conf. IEEE/EMBS,* 12:897–898.

Iwata, K., Yoshida, M., and Tokunaga, Y. 1978. High speed OCR for handprinted characters, *Proc. 4th Int. Joint Conf. Pattern Recognition,* 826–828.

Kammen, D. M. and Yuille, A. L. 1988. Spontaneous symmetry-breaking energy functions and the emergence of orientation selective cortical cells, *Biol. Cybern.,* 59:23–31.

Karhunen, K. 1947. Uber lineare Methoden in der Wahrscheinlichkeitsrechnung, *Ann. Acad. Sci. Fennicae,* Ser. A137 (translated by I. Selin in "On Linear Methods in Probability Theory", T-131, The RAND Corp., Santa Monica, Ca., 1960).

Kirby, M. and Sirovich, L. 1990. Application of the Karhunen-Loève procedure for the characterization of human faces, *IEEE Trans.,* PAMI-12(1):103–108.

Kirkpatrick, S., Gelatt, C. D., and Vecchi, M. P. 1983. Optimization by simulated annealing, Science, 220:671–679.

Kittler, J. and Young, P. C. 1973. A new approach to feature selection based on the Karhunen-Loève expansion, *Pattern Recognition,* 5:335–352.

Kohonen, T. 1972. Correlation matrix memories, *IEEE Trans. Computers,* C-21:353–359.

Kohonen, T. 1977. *Associative Memory—A System Theoretic Approach,* Springer-Verlag, Berlin.

Kohonen, T. 1982. Self-organized formation of topologically correct feature maps, *Biol. Cybern.,* 43:59–69.

Kohonen, T. 1984. *Self-organization and Associative Memory,* Springer-Verlag, Berlin.

Kohonen, T. 1988a. An introduction to neural computing, *Neural Networks,* 1:3–16.

Kohonen, T. 1988b. Problems in practical pattern recognition, *Neural Networks,* 1(suppl.):29.

Krause, P., Schwerdtman, W., and Paul, D. 1974. Two modifications of a recognition system with pattern series expansion and bayes classifier, *Proc. 2nd. Int. Joint Conf. Pattern Recognition,* 215–219.

Kwon, S. K. and Lai, D. O. 1976. Recognition experiments with handprinted numerals, *Proc. Joint Workshop on Pattern Recognition and Artificial Intelligence,* 74–83.

Leboucher, G. and Lowitz, G. E. 1978. What a histogram can really tell the classifier, *Pattern Recognition,* 10:351–357.

LeCun, Y. 1986. Learning Processes in an Asymmetric Threshold Network, *Disordered Systems and Biological Organization,* Bienenstock, E., Fogelman Souli, F., and Weisbuch, G. eds., Springer-Verlag. Berlin.

LeCun, Y., Boser, B., Denker, J. S., Henderson, D., Howard, R. E., Hubbard, W., and Jackel, L. D. 1989. Backpropagation applied to handwritten zip code recognition, *Neural Computation,* 1:541–551.

Lerner, A. 1972. Crisis in the Theory of Pattern Recognition, *Frontiers of Pattern Recognition,* Wanatabe, S., ed., 367–372, Academic Press, New York, NY.

Levine, M. D. 1969. Feature extraction: a survey, *Proc. IEEE,* 57(8):1391–1407.

Levy, A. C. and Montalvo, A. 1985. The tunneling algorithm for the global minimization of functions, *SIAM J. Sci. Stat. Comput.,* 6(1):15–29.

Linsker, R. 1986. From Basic Network Principles to Neural Architecture (series), *Proc. Na. Acad. Sci.* 83:7508–7512, 8390–8394, 8779–8783, USA.

Lippman, R. 1987. An introduction to computing with neural nets., *IEEE ASSP Magazine,* April, 4–22.

Mantas, J. 1986. An overview of character recognition methodologies, *Pattern Recognition,* 19(6):425–430.

Marconi, L., Scalia, F., Ridella, S., Arrigo, P., Mansi, C., and Mela, G. S. 1989. Application of back propagation to medical diagnosis, *Proc. Int. Joint Conf. on Neural Networks,* II: 577.

Marr, D. and Poggio, T. 1976. Cooperative computation of stereo dispanty, *Science,* 194:283–287.

Marsic, I. and Micheli-Tzanakou, E. 1990. Distributed optimization with the ALOPEX algorithms, *Proc. 12th. Int. Conf. IEEE/EMBS,* 12:1415–1416.

Masterson, J. L. and Hirsch, R. S. 1962. Machine recognition of constrained handwritten arabic numerals, *IRE Trans. Human Factors Electron.,* 3:62–65.

McCullogh, W. S. and Pitts, W. 1943. A logical calculus of ideas immanent in nervous activity, *Bull Math Biophys.,* 5:115–133.

Mead, C. A. 1989. *Analog VLSI and Neural Systems,* Addison-Wesley, Readiag, MA.

Mellissaratos, L. and Micheli-Tzanakou, E. 1989. The parallel character of the Alopex process, Proc. 15th Ann. *Northeast Biomedical Engineering Conf.,* 179–180.

Metropolis, N., Rosenbluth, A. W., Rosenbluth, M. N., and Teller, A. H. 1953. Equation of state calculations by fast computing machines., *J. Chemical Physics,* 21(6):1087–1092.

Micheli-Tzanakou, E. 1984. Non-linear characteristics in the prog's visual system, *Biol. Cybern.,* 51:53–63.

Micheli-Tzanakou, E. and Dasey, T. J. 1990. Pattern recognition with neural networks on compressed images, *6th IASTED Int. Conf. Expert Systems and Neural Networks,* 9–11.

Minsky, M. 1967. *Computation: Finite and Infinite Machines,* Prentice-Hall, Englewood Cliffs, NJ.

Minsky, M. and Papert, S. *Perceptrons: An Introduction to Computational Geometry,* MIT Press, Cambridge, MA.

Mirchandani, G. 1989. On hidden nodes for neural nets., *IEEE Trans. Circuits and Systems,* 36(5):661–664.

Mori, K. and Masuda, J. 1980. Advances in recognition of chinese characters, *Proc. 5th Int. J. Conf. Pattern Recognition,* 692–702, Miami, FL.

Munson, J. H. 1968. Experiments in the recognition of handprinted text: part I—character recognition, *Proc. AFIPS,* 33:1125–1138.

Nass, M. N. and Cooper, L. N. 1975. A theory for the development of feature detecting cells in the visual cortex, *Biol. Cybern.,* 19:1–8.

Neill, J. 1969. Numeric script mail sorter, *Proc. Automat. Pattern Recognition,* 49–65.

Niemann, H. 1976. A comparison of classification results in character recognition by man and by machine, *Proc. 3rd. Int. Joint Conf. Pattern Recognition,* 144–147.

Oja, E. 1982. A simplified neuron model as a principal component analyzer, *J. Math. Biol.*

Oja, E. 1983. *Subspace Methods of Pattern Recognition,* Research Studies Press, Letchworth.

Parker, D. 1985. *Learning Logic,* Technical Report TR-87, Center for Computational Research in Economics and Management Science, MIT, Cambridge, MA.

Parks, J. R. 1974. An articulate recognition procedure applied to handprinted numerals, *Proc. 2nd Int. J. Conf. Pattern Recognition,* 416–420. Copenhagen.

Pavlidis, T. 1980. A thinning algorithm for discrete binary images, *Computer Graphics and Image Processing,* 13:142–157.

Pavlidis, T. and Ali, F. 1975. Computer recognition of handwritten numerals by polygonal approximation, *IEEE Trans.,* SMC-5:610–614.

Persoon, E. and Fu, F. S. 1977. Shape discrimination using fourier descriptors, *IEEE Trans.,* SMC-7:170–179.

Peterson, C. and Hartman, E. 1989. Explorations of the mean field theory learning algorithm, *Neural Networks,* 2:475–494.

Psaltis, D., Brady, D., Gu, X.-G., and Lin, S. 1990. Holography in artificial neural networks, *Nature,* 343:325–330.

Rajavelu, A., Musavi, M. T., and Shirvaikar, M. V. 1989. A neural network approach to character recognition., *Neural Networks,* 2:387–393.

Romesburg, H. C. 1984. *Cluster Analysis for Researchers,* Lifetime Learning Publications, London.

Rosenblatt, F. 1958. The perceptron: a probabilistic model for information storage and organization in the brain, *Psychological Review,* 65:386–408.

Rosenblatt, F. 1959. *Principles of Neurodynamics,* Spartan Books. New York, NY.

Rosenblatt, F. 1962. *Principles of Perceptrons,* Spartan Press, Washington, DC.

Ruff, P. I., Rauschecker, J. P., and Palm, G. 1987. A model of direction-selective "simple" cells in the visual cortex based on inhibition asymmetry, *Biol. Cybern.,* 57:147–157.

Rumelhart, D. E., Hinton, G. E., and Williams, R. J. 1986. Learning Internal Representations by Error Propagation, *Parallel Distributed Processing: Explorations in the Microstructure of Cognition,* Vol 1. Foundations, Rumelhart, D. E. and McClelland, J. L., eds., Cambridge, MIT Press, 318–362.

Rutenbar, R. A. 1989. Simulated annealing algorithm: an overview, *IEEE Circuits and Devices Magazine,* 5(1):19–26.

Sandell, H. S. H. and Bourne, J. R. 1985. Expert systems in medicine: a biomedical engineering perspective, *CRC Critical Reviews in Biomedical Engineering,* 12(2):95–129.

Sanger, T. D. 1989. Optimal unsupervised learning in a single-layer linear feedforward neural network, *Neural Networks,* 2:459–473.

Scoltock, J. 1982. A survey of the literature of cluster analysis, *The Computer Journal,* 25(1):130–134.

Shingal, R. and Suen, C. Y. 1982. A method for selecting constrained hand-printed character shapes for machine recognition, *IEEE Trans.,* PAMI-4(1):74–78.

Sivilotti, M. A., Emerling, M. R., and Mead, C. A. 1986. VLSI architectures for implementation of neural networks, *Proc. Conf. Neural Networks for Computing,* 151:408–413.

Sivilotti, M. A., Mahowald, M. A., and Mead, C. A. 1987. Real-time visual computations using analog CMOS processing arrays, *Advanced Research in VLSI: Proc. of 1987 Stanford Conf.,* 295–312.

Snyder, W., Bilbro, G., and Van den Bout, D. 1989. *New Techniques in Optimization: A Tutorial,* Technical Report NETR-89-12, Center for Communication and Signal Processing. North Carolina State University, Raleigh, NC.

Spanjersberg, A. A. 1974. Combinations of different systems for the recognition of handwritten digits, *Proc. 2nd Int. Joint Conf. Pattern Recognition,* 208–209.

Stent, G. S. 1973. A physiological mechanism for Hebb's postulate of learning, *Proc. Natl. Acad. Sci. USA,* 70:997–1001.

Stornetta, W. and Huberman, S. 1987. Improved three-layer back propagation algorithm, *Proc. IEEE 1st Conf. Neural Networks,* II:637–643.

Sue, T.-J. and Chen, Z. 1976. Skeleton chain code approach to recognition of handwritten numerals, *Proc. Nat. Computer Symp.,* 15–4.26.

Suen, C. Y. 1973. Factors affecting the recognition of handprinted characters, *Proc. Int. Conf. Cybernetics and Society,* 174–175.

Suen, C. Y. 1982. The role of multi-directional loci and clustering in reliable recognition of characters, *Proc. 6th Int. Joint Conf. Pattern Recognition,* 1023–1026. Munich.

Suen, C. Y., Berthod, M., and Mori, S. 1978. Advances in recognition of handprinted characters, *Proc. 4th Int. Conf. Pattern Recognition,* 30–44, Kyoto.

Suen, C. Y., Shinghal, R., and Kwan, C. C. 1977. Dispersion factor: a quantitative measurement of the quality of handprinted characters, *Proc. Int. Conf. Cybernetics and Society,* 681–685.

Tank, D. W. and Hopfield, J. J. 1986. Simple "neural" optimization networks: an A/D converter, signal decision circuit, and a linear programming circuit, *Trans. IEEE Circuits Syst.,* CAS-33:533.

Taxt, T., Flynn, P. J., and Jain, A. K. 1989. Segmentation of document images, *IEEE Trans.,* PAMI-11(12):1322–1329.

Teh, C.-H. and Chin, R. T. 1986. On digital approximation of moment invariants, *Computer Vision, Graphics, and Image Processing,* 33:318–326.

Tou, J. T. and Gonzalez, R. C. 1972. Automatic recognition of handwritten characters via feature extraction and multilevel decision, *Int. J. Comp. Inf. Sci.,* 1:43–65.

Tou, J. T. and Gonzalez, R. C. 1974. Pattern Recognition Principles, 243–246, Addison-Wesley, Reading, MA.

Tou, J. T. and Heydom, R. P. 1967. Some Approaches to Optimum Feature Extraction, *Computer and Information Sciences—II,* Tou, J. T., ed., Academic Press, New York, NY.

Toussaint, G. T. 1974. Bibliography on estimation of misclassification, *IEEE Trans. Information Theory,* IT-20(4):472–479.

Toussaint, G. T. and Donaldson, R. W. 1970. Algorithms for recognizing contour-traced handprinted characters, *IEEE Trans. Comput.,* 19:541–546.

Tucker, N. D. and Evans, F. C. 1974. A two-step strategy for character recognition using geometrical moments, *Proc. 2nd Int. Joint Conf. Pattern Recognition,* 223–225.

Tucker, W. T. 1987. Counterexamples to the Convergence Theorem for the Fuzzy C-Means Clustering Algorithms, *Analysis of Fuzzy Information,* Vol. III—*Applications in Engineering and Science,* 109–121, CRC Press: Boca Raton, FL.

Tzanakou, E. 1977. *Principles and Design of the ALOPEX Device: A Novel Method od Mapping Visual Receptive Fields,* Doctoral Dissertation, 1977. International Publication No. 77–30, 771.38/8, 1978.

Tzanakou, E, Michalak, R., and Harth, E. 1979. The ALOPEX process: visual receptive fields by response feedback, *Biol. Cybern.* 35:161–174.

Von der Malsburg, C. 1973. Self-organization of orientation sensitive cells in the striate cortex, *Kybemetik,* 14:85–100.

Von Neumann, J. 1945/1982. First Draft of a Report on the EDVAC, *The Origins of Digital Computers: Selected Papers,* 3rd ed., Randall, B. ed., Springer Verlag, Berlin.

Wang, J.-Z. and Micheli-Tzanakou, E. 1990. The use of the ALOPEX process in extracting normal and abnormal visual evoked potentials, *IEEE-EMBS Magazine Special Issue on DSP,* 9(1):44–46.

Watanabe, S. 1965. Karhunen-Loève expansion and factor analysis—theoretical remarks and applications, *Proc. 4th Conf. Information Theory,* Prague.

Watanabe, S. 1972. Pattern Recognition as Information Compression, *Frontiers of Pattern Recognition,* Watanabe, S., ed., 561–567, Academic Press, New York, NY.

Watanabe, S. and Kaminuma, T. 1988. Recent developments of the minimum entropy algorithm, *9th Int. Conf. Pattern Recognition,* 536–540.

Watt, A. H. and Beurle, R. L. 1971. Recognition of handprinted numerals reduced to graph-representable form, *Proc. 2nd Int. Joint Conf. Artificial Intelligence,* 322–332.

Werbos, P. J. 1974. *Beyond Regression: New Tools for Prediction and Analysis in the Behavioral Sciences,* Ph.D. thesis, Harvard University, Cambridge, MA.

Widrow, B. and Hoff, M. E. 1960. Adaptive Switching Circuits. 1960 IRE WESCON Convention Record, IRE, 96–104, New York, NY.

Widrow, B. 1962. Generalization and Information Storage in Networks of Adaline "Neurons", *Self-Organizing Systems 1962,* Yovitz, M. C., Jacobi, G. T., and Goldstein, G., eds., Spartan Books, 435–461, Washington, DC.

Widrow, B. and Stearns, S. D. 1985. Adaptive Signal Processing, Prentice-Hall. Englewood Cliffs, NJ.

Xu, L. 1988. Some application of simulated annealing to pattern recognition, *Proc. In. Conf. Pattern* Recognition, 1040–1042, Rome.

Young, T. T. and Calvert, T. W. 1974. *Classification, Estimation, and Pattern Recognition,* American Elsevier, New York, NY.

Zadeh, L. 1965. Fuzzy sets, *Information and Control,* 8:338–353.

96

Fuzzy Algorithms for Learning Vector Quantization

Nicolaos B. Karayiannis
University of Houston

96.1 Introduction 1264
96.2 Learning Vector Quantization 1265
96.3 Generalized Learning Vector Quantization 1266
The GLVQ Algorithm • GLVQ-F: Improved GLVQ Algorithms
96.4 Fuzzy Learning Vector Quantization Algorithms 1268
96.5 GLVQ-F and FLVQ Algorithms 1269
96.6 Fuzzy Algorithms for Learning Vector Quantization 1270
96.7 The FALVQ 1 Family of Algorithms 1272
96.8 The FALVQ 2 Family of Algorithms 1274
96.9 The FALVQ 3 Family of Algorithms 1275
96.10 Competition Measures 1277
96.11 Alternative FALVQ Algorithms 1280
Harmonic FALVQ 1 • Geometric FALVQ 1 • Arithmetic FALVQ 1
96.12 Experimental Results 1282
96.13 Discussion and Concluding Remarks 1284

96.1 Introduction

Vector quantization can be seen as a mapping from an n-dimensional Euclidean space into a finite set of prototypes (Gray, 1984). Vector quantization has strongly influenced the structure and training of a special class of artificial neural networks, known as self-organizing feature maps. The self-organizing feature map (SOFM) is a sheet-like neural network, the cells of which become specifically tuned to various input signal patterns or class of patterns through an unsupervised learning process (Kohonen, 1988, 1989, 1990a–c Karayiannis and Venetsanopoulos, 1993; Hertz et al., 1992). During the training, or ordering, of a feature map, its cells are not updated independently but as topologically related subsets. The selection of the subset of cells to be updated at each learning step requires the definition of a center cell and also a topological neighborhood around the center cell, which must be very wide when the ordering process begins and shrink monotonically with time (Kohonen, 1989, 1990a, Karayiannis and Venetsanopoulos, 1993). A wide neighborhood in the initial steps guarantees a rough global ordering of the weight vectors assigned to the cells of the map, while the shrinking of the neighborhood during the ordering improves the spatial resolution of the map.

Despite the common belief, the self-organizing feature map was not intended to perform pattern classification tasks. Rather, the self-organizing feature map attempts to find topological structure hidden in the input data and display it in one or two dimensions (Pal et al., 1993). If the feature map is used as classifier, the cells must be grouped into subsets which correspond to discrete classes. In this case, the unsupervised ordering of the map alone is not enough. After its original unsupervised ordering, the feature map must be trained to function as a classifier using a training set of feature vectors whose classification is already known. Kohonen proposed supervised learning algorithms for the fine tuning of the map, which follows its original ordering, known as the LVQ1, LVQ2, and LVQ3 algorithms (Kohonen, 1990a–c). If a feature map functions as a classifier, several weight vectors may represent the same class. Since the identity of each weight vector within a certain class is not particularly important, the main objective of these algorithms is to determine near-optimal boundaries between the classes.

The use of the name learning vector quantization (LVQ) for the supervised algorithms proposed by Kohonen for the fine tuning of the map was a source of confusion among researchers. By definition, vector quantization is an unsupervised process which maps a feature vector into a certain vector, called the codeword or prototype. In contrast, Kohonen's LVQ1, LVQ2, and LVQ3 algorithms are attempting to create clusters of weight vectors which represent certain classes of input data that have been labeled by an external teacher. The weight vectors resulting from such a process can hardly be considered as prototypes, given that more than one weight vectors may represent the same class.

0-8493-8343-9/97/$0.00+$.50

In addition to the conceptual problems of Kohonen's LVQ1, LVQ2, LVQ3 algorithms, there are also practical problems associated with their application. Pal et al. identified various problems associated with the unsupervised learning rules for feature maps, which include their dependence on strategies for gradually reducing the learning rate and shrinking the topological neighborhood, the lack of optimality criteria behind these learning rules, and the reliability of the criteria for the termination of the learning process (Pal et al., 1993). The unsupervised ordering process proposed by Kohonen requires a map of a reasonable size for its implementation, because of the shrinking topological neighborhood associated with the learning process. Since Kohonen's algorithms are applied on an ordered feature map, they involve the adaptation of many redundant weight vectors. Such a strategy is computationally expensive, given the large number of iterations typically required for the unsupervised ordering of feature maps.

Bezdek et al. suggested that LVQ can be implemented through an unsupervised and competitive learning process (Bezdek, 1992; Pal et al., 1992, 1993). According to their formulation, LVQ can be achieved by minimizing a loss function which measures the locally weighted error of the input vector with respect to the winning the prototype, that is, the prototype that is closest to the input vector in the Euclidean distance sense. This formulation resulted in the generalized learning vector quantization (GLVQ) algorithm (Bezdek, 1992; Pal et al., 1992, 1993). The GLVQ algorithm was designed to realize competitive learning, since all the prototypes compete to match every input vector. However, the weights assigned to the non-winning prototypes depend directly on the distance between the corresponding input vector and the prototypes. Thus, the performance of the GLVQ algorithm can be affected by the magnitude of the input vectors and the number of prototypes. The GLVQ-F algorithms were developed recently in an attempt to overcome the scaling problems associated with the original GLVQ algorithm (Bezdek et al., 1995; Karayiannis et al., 1996).

Huntsberger and Ajjimarangsee attempted to establish a connection between feature maps and fuzzy clustering by modifying the learning rule proposed by Kohonen for the SOFM (Huntsberger and Ajjimarangsee, 1989). Their learning rule was obtained by replacing the learning rate involved in Kohonen's learning rule by the membership function associated with fuzzy *c*-means algorithms (Dunn, 1973; Bezdek, 1981; Karayiannis, 1994, 1995). Nevertheless, the input vectors are presented to the map sequentially while the prototypes that are attracted by each input belong to a topological neighborhood centered at the winning prototype. As in Kohonen's SOFM, the topological neighborhood shrinks with time and provides the basis for the termination of the learning process. According to this scheme, the adaptation of the prototypes within a particular topological neighborhood of the map is not uniform. Rather, the adaptation of each prototype depends on its distance from the input vector. This hybrid learning scheme can only be seen as a first attempt to merge fuzzy clustering and feature maps because of the lack of theoretical foundation, formal derivation and clear objectives. Bezdek et al. argued that the sequential presentation of the input vectors to the map is one of the main disadvantages of the learning rule proposed by Huntsberger and Ajjimarangsee (Bezdek, 1992; Pal et al., 1992; Tsao et al., 1994). They also proposed batch learning schemes that update all the prototypes with respect to all input vectors. These learning schemes will be referred to throughout this section as the FLVQ algorithms.

Karayiannis and Pai proposed a framework for the development of fuzzy algorithms for learning vector quantization (FALVQ). (Karayiannis and Pai, 1994a–d, 1995a). The development of FALVQ algorithms was based on the minimization of a loss function formed as the weighted sum of the squared Euclidean distances between an input vector, which represents a feature vector, and the weight vectors, which represent the prototypes. The distances between each input vector and the prototypes are weighted by a set of membership functions, which regulate the competition between various prototypes for each input and, thus, determine the strength of attraction between each input and the prototypes during the learning process. According to this formulation, the design of specific FALVQ algorithms reduces to the selection of membership functions that satisfy certain properties. This section reviews the formulation that led to the development of the FALVQ 1, FALVQ 2, and FALVQ 3 families of algorithms (Karayiannis and Pai, 1994c–d). In addition, this section introduces two competition measures that can be used by the user to control the competition between the winning and non-winning prototypes during the learning process. Finally, this section presents the development of alternative FALVQ algorithms, which allow the nonwinning prototypes to be more competitive during the learning process (Karayiannis and Pai, 1994a and c).

96.2 Learning Vector Quantization

The objective of LVQ algorithms is the representation of a set of vectors $\mathbf{x} \in \mathcal{X} \subset \mathfrak{R}^n$ by a set of prototypes $\mathcal{V} = \{\mathbf{v}_1, \mathbf{v}_2, \ldots, \mathbf{v}_c\} \subset \mathfrak{R}^n$. LVQ is associated with a neural architecture whose structure is roughly illustrated in Figure 96.1. The LVQ network consists of an input layer and an output layer. Each node in the input layer is connected directly to the cells, or units, in the output layer. A weight vector, also referred to as the prototype, is assigned to each cell in the output layer.

Consider the finite set $\mathcal{X}$ formed by M feature vectors from an n-dimensional Euclidean space, that is, $\mathcal{X} = \{\mathbf{x}_1, \mathbf{x}_2, \ldots, \mathbf{x}_M\}$, $\mathbf{x}_i \in \mathfrak{R}^n \; \forall i = 1, 2, \ldots, M$. Learning vector quantization is frequently based on the minimization of the following functional (Bezdek, 1992; Pal et al., 1992, 1993)

$$L = \frac{1}{M} \sum_{i=1}^{M} \sum_{j=1}^{c} u_{ij} \, \|\mathbf{x}_i - \mathbf{v}_j\|^2, \tag{96.1}$$

where $u_{ij} = u_j(\mathbf{x}_i)$, $j = 1, 2, \ldots, c$ is a set of weights assigned to the prototypes $\mathbf{v}_j$, $j = 1, 2, \ldots, c$. The functional in Eq. (96.1) can be interpreted as the mean of

$$L_{\mathbf{x}} = \sum_{j=1}^{c} u_j(x) \, \|\mathbf{x} - \mathbf{v}_j\|^2 \tag{96.2}$$

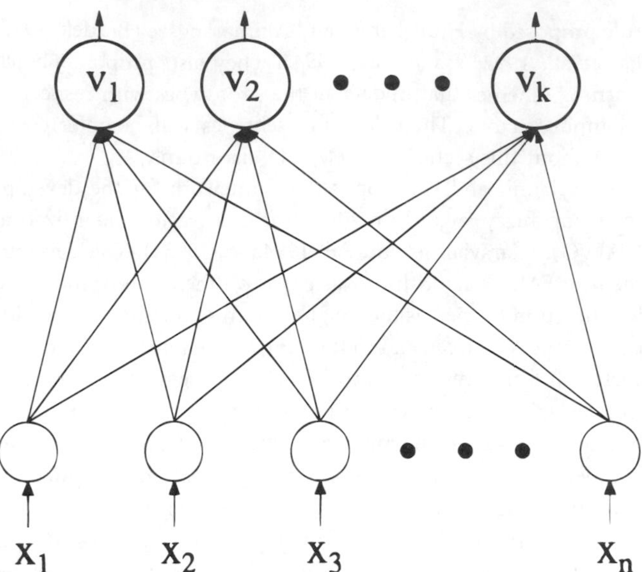

Figure 96.1 The LVQ network.

evaluated over the finite set $\mathcal{X}$. The weights $u_j(\mathbf{x})$, $j = 1, 2, \ldots, c$ regulate the competition between the prototypes $\mathbf{v}_j$, $j = 1, 2, \ldots, c$ for the input vector $\mathbf{x}$. The loss function in Eq. (96.2) is often defined with respect to the winning prototype. Assuming that $\mathbf{v}_i$ is the winning prototype corresponding to the input vector $\mathbf{x}$, that is, the closest prototype to $\mathbf{x}$ in the Euclidean distance sense, the weights $u_j(\mathbf{x})$, $j = 1, 2, \ldots, c$ can be of the form

$$u_j(\mathbf{x}) = \begin{cases} 1 & \text{if } \|\mathbf{x} - \mathbf{v}_j\|^2 = \|\mathbf{x} - \mathbf{v}_i\|^2 \\ u_j(\mathbf{x}, \mathbf{v}_l \in \mathcal{V}) < 1 & \text{otherwise.} \end{cases} \tag{96.3}$$

A rather trivial case of the above formulation is the minimization of the functional in Eq. (96.1) under the assumption that the weights $u_j(\mathbf{x})$, $j = 1, 2, \ldots, c$ are obtained according to the nearest neighbor condition as

$$u_j(\mathbf{x}) = \begin{cases} 1 & \text{if } \|\mathbf{x} - \mathbf{v}_j\|^2 = \|\mathbf{x} - \mathbf{v}_i\|^2 \\ 0 & \text{otherwise.} \end{cases} \tag{96.4}$$

Under this assumption, the functional in Eq. (96.1) becomes

$$L = \frac{1}{M} \sum_{i=1}^{M} \min_{1 \le j \le c} \{\|\mathbf{x}_i - \mathbf{v}_j\|^2\} \tag{96.5}$$

The functional in Eq. (96.5) represents the average of the squared Euclidean distances between the inputs $\mathbf{x}_i$ and their closest neighbors among the prototypes $\mathbf{v}_\ell$, $\ell = 1, 2, \ldots, c$. The minimization of the functional in Eq. (96.5) implies that each input attracts only the winning prototype, while the non-winning prototypes are not allowed to compete with the winning prototype for it.

The minimization of the functional in Eq. (96.1) can produce batch LVQ algorithms. However, the minimization of Eq. (96.1)using the gradient descent method is a difficult task if the loss function L_x is defined with respect to the winning prototype (Karayiannis and Pai, 1994a, 1995a). The obvious reason is that the winning prototype must be determined with respect to each input vector $\mathbf{x}_i \in \mathcal{X}$. The gradient of L can be approximated by the gradient of the loss function L_x (Pal et al., 1993). In fact, this approach implies the sequential adaptation of the prototypes with respect to the input vectors $\mathbf{x}_i \in \mathcal{X}$ and is frequently used in the development of learning algorithms (Karayiannis and Venetsanopoulos, 1993; Tsypkin, 1973). Under certain conditions regarding the learning process, the adaptation of the prototypes in an iterative fashion along the direction of the gradient of L_x is expected to result in a set of prototypes $\mathbf{v}_\ell$, $\ell = 1, 2, \ldots, c$ which approximate the prototypes that minimize L.

96.3 Generalized Learning Vector Quantization

The formulation of the LVQ problem presented previously provided the basis for the development of a broad variety of LVQ algorithms, including the original GLVQ algorithm (Bezdek, 1992; Pal et al., 1992, 1993) and the GLVQ-F algorithms (Karayiannis et al., 1996; Bezdek et al., 1995). The GLVQ and GLVQ-F algorithms are briefly reviewed here.

The GLVQ Algorithm

Pal et al. developed the GLVQ algorithm by minimizing a loss function which measures a locally weighted error of the input with respect to the winning prototype (Bezdek, 1992; Pal et al., 1992, 1993). Assuming that $\mathbf{x}$ is the input vector and $\mathbf{v}_i$ is the winning prototype, i.e., the closest prototype to $\mathbf{x}$ in the Euclidean distance sense, the loss function minimized in this approach can be written as (Pal et al., 1993)

$$L_\mathbf{x}(\mathbf{v}_r, r = 1, 2, \ldots, c) = \sum_{r=1}^{c} u_{ir} \|\mathbf{x} - \mathbf{v}_r\|^2, \tag{96.6}$$

where the weights $u_{ir} = u_{ir}(\mathbf{x})$ are defined in terms of $S = S(\mathbf{x}) = \Sigma_{l=1}^{c} \|\mathbf{x} - \mathbf{v}_\ell\|^2$ as

$$u_{ir} = \begin{cases} 1 & \text{if } r = i \\ \dfrac{1}{S} & \text{if } r \neq i. \end{cases} \tag{96.7}$$

The GLVQ algorithm was derived by minimizing the loss function defined by Eqs. (96.60) and (96.7) using the gradient descent method. If $\mathbf{x}$ is the input vector, the winning prototype $\mathbf{v}_i$ can be updated by (Pal et al., 1993)

$$\Delta \mathbf{v}_i = \alpha\, (\mathbf{x} - \mathbf{v}_i) \frac{S^2 - S + \|\mathbf{x} - \mathbf{v}_i\|^2}{S^2}, \tag{96.8}$$

while the non-winning prototypes $\mathbf{v}_j \neq \mathbf{v}_i$ can be updated by (Pal et al., 1993)

$$\Delta \mathbf{v}_j = \alpha\,(\mathbf{x} - \mathbf{v}_j)\,\frac{\|\mathbf{x} - \mathbf{v}_i\|^2}{S^2}, \tag{96.9}$$

where $\alpha \in [0, 1]$ is the learning rate.

The update of the non-winning prototypes can be studied by observing that

$$\frac{S^2}{\|\mathbf{x} - \mathbf{v}_i\|^2} = S\,\frac{\sum_{l=1}^{c} \|\mathbf{x} - \mathbf{v}_l\|^2}{\|\mathbf{x} - \mathbf{v}_i\|^2} = S\left(1 + \sum_{l \neq i}^{c} \frac{\|\mathbf{x} - \mathbf{v}_l\|^2}{\|\mathbf{x} - \mathbf{v}_i\|^2}\right). \tag{96.10}$$

Since $\|\mathbf{x} - \mathbf{v}_\ell\|^2/\|\mathbf{x} - \mathbf{v}_i\|^2 > 1\ \forall \ell \neq i$,

$$\sum_{l \neq i}^{c} \frac{\|\mathbf{x} - \mathbf{v}_l\|^2}{\|\mathbf{x} - \mathbf{v}_i\|^2} > c - 1. \tag{96.11}$$

The combination of Eqs. (96.10) and (96.11) results in

$$\frac{\|\mathbf{x} - \mathbf{v}_i\|^2}{S^2} < \frac{1}{cS}. \tag{96.12}$$

According to Eq. (96.12), the adaptation of the non-winning prototypes depends on the value of the sum S relative to 1 and also on the number c of the prototypes. As the value of kS increases, the adaptation of the non-winning prototypes becomes negligible. Given that the value of c is fixed for a given application, it could be argued that the value of S could be modified by normalizing the input data. However, in such a case, the performance of the algorithm would depend rather strongly on the normalization scheme chosen.

The update of the winning prototype depends on the term

$$\frac{S^2 - S + \|\mathbf{x} - \mathbf{v}_i\|^2}{S^2} = 1 - \frac{1}{S} + \frac{\|\mathbf{x} - \mathbf{v}_i\|^2}{S^2}. \tag{96.13}$$

According to the previous analysis,

$$\frac{S^2 - S + \|\mathbf{x} - \mathbf{v}_i\|^2}{S^2} < 1 - \frac{1}{S} + \frac{1}{cS} = 1 - \frac{1}{S}\left(1 - \frac{1}{c}\right). \tag{96.14}$$

Since $c > 1$, then $1 - 1/c < 1$. In particular, $1 - 1/c$ approaches 1 for sufficiently large of c. In any case, the adaptation of the winning prototype is strongly affected by the size of the sum S relative to unity. As S increases, $1/S$ becomes negligible and $(S^2 - S + \|\mathbf{x} - \mathbf{v}_i\|^2)/S^2$ approaches 1. Given that in this case the term $\|\mathbf{x} - \mathbf{v}_i\|^2/S^2$ is close to zero, the behavior of the GLVQ algorithm is closer to that of a crisp LVQ algorithm.

GLVQ-F: Improved GLVQ Algorithms

The strengths and weaknesses of the GLVQ algorithm motivated the search for alternative objective functions that can provide the basis for learning vector quantization. This search resulted in the GLVQ-F algorithms, an improved version of the GLVQ algorithm that were developed by considering an alternative formulation of the learning vector quantization problem, (Karayiannis et al., 1996; Bezdek et al., 1995). According to this formulation, learning vector quantization is based on the minimization of the loss function

$$L_{\mathbf{x}}(\mathbf{v}_r, r = 1, 2, \ldots, c) = \sum_{r=1}^{c} u_r \|\mathbf{x} - \mathbf{v}_r\|^2, \tag{96.15}$$

where $u_r = u_r(\mathbf{x})$ are the membership values obtained from the fuzzy c-means algorithm as (Bezdek, 1981, Bezdek et al., 1984, Karayiannis, 1995, 1994)

$$u_r = u_r(\mathbf{x}) = \left(\sum_{j=1}^{c}\left(\frac{\|\mathbf{x} - \mathbf{v}_r\|^2}{\|\mathbf{x} - \mathbf{v}_j\|^2}\right)^{1/(m-1)}\right)^{-1}. \tag{96.16}$$

This loss function is not defined with respect to the winning prototype, since the weights u_r, $r = 1, 2, \ldots, c$ are all defined by Eq. (96.16). If $\mathbf{v}_i$ is the winning prototype, then $\|\mathbf{x} - \mathbf{v}_i\|^2 < \|\mathbf{x} - \mathbf{v}_r\|^2\ \forall\ v \neq i$ and also $u_r(\mathbf{x}) < u_i(\mathbf{x}) < 1\ \forall\ r \neq i$. Thus, this approach favors the winning prototype. Nevertheless, the bias toward the winning prototype is affected by the distances between the non-winning prototypes and the input $\mathbf{x}$.

The gradient of $L_{\mathbf{x}} = L_{\mathbf{x}}(\mathbf{v}_r, r = 1, 2, \ldots, c)$ with respect to any prototype $\mathbf{v}_j$ is given by (Karayiannis et al., 1996; Bezdek, 1995)

$$\frac{\partial L_{\mathbf{x}}}{\partial \mathbf{v}_j} = -f_j(m)(\mathbf{x} - \mathbf{v}_j)\ \forall\ j = 1, 2, \ldots, c, \tag{96.17}$$

where

$$f_j(m) = \frac{2}{m-1}\,u_j\left[(m-2) + u_j\left(\sum_{r=1}^{c}\left(\frac{\|\mathbf{x} - \mathbf{v}_j\|^2}{\|\mathbf{x} - \mathbf{v}_r\|^2}\right)^{(1/m-1)}\right)^{2-m}\right]. \tag{96.18}$$

Thus, the prototypes can be updated according to the gradient descent method by

$$\Delta \mathbf{v}_j = -\eta\,\frac{\partial L_{\mathbf{x}}}{\partial \mathbf{v}_j} = \eta\, f_j(m)(\mathbf{x} - \mathbf{v}_j)\ \forall\ j = 1, 2, \ldots, c, \tag{96.19}$$

where η is the learning rate, $f_j(m)$ is given in Eq. (96.18) and $m \in (1,00)$. It was shown that the GLVQ-F algorithms are not subject to the scaling problems associated with the GLVQ algorithm (Karayiannis et al., 1996; Bezdek et al., 1995).

Eq. (96.19) describes a family of infinitely many competitive LVQ algorithms, while the parameter m can be selected to regulate the competition between the prototypes during the learning process. The behavior of Eq. (96.19) was studied in the case where $m = 2$ and in the limiting cases where m approaches 1 and ∞.

For $m = 2$, $f_j(2) = 2\,c\,u_j^2$ and Eq. (96.19) reduces to

$$\Delta \mathbf{v}_j = \eta(2\,c\,u_j^2)(\mathbf{x} - \mathbf{v}_j)\ \forall\ j = 1, 2, \ldots, c. \quad (96.20)$$

In this case, u_j is inversely proportional to $\|\mathbf{x} - \mathbf{v}_j\|^2$. Thus, Eq. (96.20) guarantees that the attraction between the input vector and each prototype becomes stronger as their distance decreases.

As $m \to \infty$, all prototypes are updated by an equal amount according to (Karayiannis et al., 1996; Bezdek et al., 1995)

$$\Delta \mathbf{v}_j = \frac{2\eta}{c}(\mathbf{x} - \mathbf{v}_j)\ \forall\ j = 1, 2, \ldots, c. \quad (96.21)$$

The behavior of the GLVQ-F algorithm in this case is in full agreement with the limiting behavior of the membership values Eq. (96.16), which all approach $1/c$ as m approaches infinity (Bezdek, 1981, Bezdek et al., 1984). As m approaches infinity, Eq. (96.19) results in the least selective or maximally fuzzy adaptation of the prototypes, since Eq. (96.21) treats all prototypes equally regardless of their distance from the input vector **x**.

As m approaches 1 from the right, the GLVQ-F algorithm approaches a crisp LVQ algorithm. More specifically, in the limit where $m \to 1_+$ the winning prototype $\mathbf{v}_i$ is updated according to (Karayiannis et al., 1996; Bezdek et al., 1995)

$$\Delta \mathbf{v}_i = (2\eta)(\mathbf{x} - \mathbf{v}_i), \quad (96.22)$$

while the non-winning prototypes are not attracted by the input vector **x**. This is also consistent with the behavior of the membership function in Eq. (96.16). As m approaches 1 from the right, the membership in Eq. (96.16) becomes an indicator function that identifies the nearest neighbor to the input **x** among the prototypes (Bezdek, 1981, Bezdek et al., 1984)

96.4 Fuzzy Learning Vector Quantization Algorithms

Fuzzy learning vector quantization (FLVQ) algorithms were proposed by Bezdek et al. (Bezdek, 1992; Tsao et al., 1994) as an alternative to Huntsberger and Ajjimarangsee' attempt to establish a connection between the feature maps and fuzzy clustering (Huntsberger and Ajjimarangsea, 1989). The update equations associated with the FLVQ algorithms also involve the membership functions associated with fuzzy *c*-means algorithms, which are used to determine the strength of attraction between each prototype and the input vectors. This learning scheme was not the result of a formal derivation. Nevertheless, Bezdek et al. validated their update equation by pointing out its close relationship with fuzzy *c*-means and crisp *c*-means algorithms (Bezdek, 1992; Tsao et al., 1994).

Given a finite set of feature vectors $\mathbf{x}_1, \mathbf{x}_2, \ldots, \mathbf{x}_M$, the prototypes of the map can be updated by (Pal et al., 1992; Bezdek, 1992, Tsao et al., 1994)

$$\Delta \mathbf{v}_j = \eta_j \sum_{i=1}^{M} (u_{ij})^m (\mathbf{x}_i - \mathbf{v}_j)\ \forall\ j = 1, 2, \ldots, c \quad (96.23)$$

where $u_{ij} = u_j(\mathbf{x}_i)$ are the membership values obtained from fuzzy *c*-means algorithms as (Bezdek 1981, Bezdek, et al., 1984; Bezdek)

$$u_{ij} = \left(\sum_{\ell=1}^{c} \left(\frac{\|\mathbf{x}_i - \mathbf{v}_j\|^2}{\|\mathbf{x}_i - \mathbf{v}_\ell\|^2} \right)^{(1/m-1)} \right)^{-1} \quad (96.24)$$

Despite some superficial similarities between the update equation proposed by Bezdek et al. and Kohonen's learning rule for self-organizing feature maps, their structures, properties and objectives are different. According to the update equation proposed by Bezdek et al., each prototype is updated with respect to all inputs (batch learning). According to Kohonen's learning rule, the prototypes are updated with respect to each input vector while the input vectors are presented to the map in a sequential fashion (sequential learning).

The close relationship between fuzzy *c*-means and FLVQ algorithms can be established by considering the explicit formula that relates the prototypes in two successive iterations of FLVQ algorithms. Let $\mathbf{v}_{j,\nu-1}$, $j = 1, 2, \ldots, c$ be the set of prototypes obtained after the $(\nu - 1)$th iteration. According to Eq. (96.23), a new set of prototypes $\mathbf{v}_{j,\nu}$, $j = 1, 2, \ldots, c$ can be obtained according to the formula

$$\mathbf{v}_{j,\nu} = \mathbf{v}_{j,\nu-1} + \eta_{j,\nu} \sum_{i=1}^{M} \alpha_{ij,\nu} (\mathbf{x}_i - \mathbf{v}_{j,\nu-1})\ \forall\ j = 1, 2, \ldots, c, \quad (96.25)$$

where

$$\alpha_{ij,\nu} = \left(\sum_{\ell=1}^{c} \left(\frac{\|\mathbf{x}_i - \mathbf{v}_{j,\nu-1}\|^2}{\|\mathbf{x}_i - \mathbf{v}_{\ell,\nu-1}\|^2} \right)^{(1/m-1)} \right)^{-m}. \quad (96.26)$$

Clearly, $\alpha_{ij,\nu}$ are evaluated in terms of the membership values obtained from fuzzy *c*-means after the $(\nu - 1)$ iteration as $\alpha_{ij,\nu} = (u_{ij,\nu})^m$. The close relationship between Eq. (96.25) and the 'centroid' formula for the prototypes associated with fuzzy *c*-means can be established by rewriting Eq. (96.25) as

$$\mathbf{v}_{j,\nu} = \left(1 - \eta_{j,\nu} \sum_{i=1}^{M} \alpha_{ij,\nu} \right) \mathbf{v}_{j,\nu-1} + \eta_{j,\nu} \sum_{i=1}^{M} \alpha_{ij,\nu}\, \mathbf{x}_i \quad (96.27)$$

$$\forall\ j = 1, 2, \ldots, c.$$

The term $(1 - \eta_{j,\nu} \sum_{i=1}^{M} \alpha_{ij,\nu})\mathbf{v}_{j,\nu}-1$ represents the direct effect of the prototype $\mathbf{v}_j = \mathbf{v}_{j,\nu-1}$ available after the $(\nu - 1)$th iteration on the evaluation of its updated version $\mathbf{v}_j = \mathbf{v}_{j,\nu}$. The term $\eta_{j,\nu} \sum_{i=1}^{M} \alpha_{ij,\nu}\, \mathbf{x}_i$ represents the cumulative effect of the input vectors $\mathbf{x}_i \in \mathcal{X}$ on the adaptation of the prototype $\mathbf{v}_j$ during the νth iteration and depends indirectly on the prototypes $\mathbf{v}_j$, $j = 1, 2, \ldots, c$ as indicated by the definition of $\alpha_{ij,\nu}$ in Eq. (96.26). According to Eq. (96.27), $\mathbf{v}_{j,\nu}$ can be evaluated only in terms of the input vectors $\mathbf{x}_i \in \mathcal{X}$ if the direct effect of $\mathbf{v}_{j,\nu-1}$ diminishes. This can be accomplished if $1 - \eta_{j,\nu} \sum_{i=1}^{M} \alpha_{ij,\nu} = 0$ or, equivalently,

$$\eta_{j,\nu} = \frac{1}{\sum_{i=1}^{M} \alpha_{ij,\nu}}. \tag{96.28}$$

The formula provided by fuzzy c-means algorithms for the evaluation of the prototypes at each iteration can easily be obtained by substituting the learning rates defined in Eq. (96.28) in Eq. (96.27) as

$$\mathbf{v}_{j,\nu} = \frac{\sum_{i=1}^{M} \alpha_{ij,\nu}\, \mathbf{x}_i}{\sum_{i=1}^{M} \alpha_{ij,\nu}} = \frac{\sum_{i=1}^{M} (u_{ij,\nu})^m\, \mathbf{x}_i}{\sum_{i=1}^{M} (u_{ij,\nu})^m}. \tag{96.29}$$

The algorithms described by Eq. (96.25) are identical with fuzzy c-means algorithms if m is fixed during the learning process and the learning rates are evaluated at each iteration according to Eq. (96.28). In this case, the algorithm described by Eq. (96.25) approaches asymptotically the crisp c-means algorithm as m approaches asymptotically unity from the right. Nevertheless, m is not constant during the learning process. Bezdek et al. suggested that m can be evaluated during the learning process as a linear function of the iteration number ν as

$$m = m(\nu) = m_i + \nu \frac{m_f - m_i}{N}, \tag{96.30}$$

where m_i and m_f are the initial and final values of m, respectively, and N is the total number of iterations (Bezdek, 1992). A variable m can affect the competition between the prototypes for each input of the feature map during the learning process. If m is evaluated according to Eq. (96.30) and $m_f > m_i$, then the algorithm guarantees the gradual transition from a maximum uncertainty or minimum selectivity phase, where all prototypes compete for all inputs, to a minimum uncertainty or maximum selectivity phase, where each prototype is attracted by a particular set of inputs which are clustered together.

The FLVQ algorithms can be summarized as follows:

1. Select c, m_i, m_f, ϵ; fix N; set $\nu = 0$; generate an initial set of prototypes $\mathbf{V}_0 = \{\mathbf{v}_{1,0}, \mathbf{v}_{2,0}, \ldots, \mathbf{v}_{c,0}\}$.
2. Set $\nu = \nu + 1$.
3. Calculate $m = m_i + \nu[(m_f - m_i)/N]$.
   - $\alpha_{ij,\nu} = (\sum_{l=1}^{c} (\|\mathbf{x}_i - \mathbf{v}_{j,\nu-1}\|^2 / \|\mathbf{x}_i - \mathbf{v}_{l,\nu-1}\|^2)^{1/(m-1)})^{-1}$ $\forall i,j$.
   - $\eta_{j,\nu} = (\sum_{i=1}^{M} \alpha_{ij,\nu})^{-1}$ $\forall j$.
   - $\mathbf{v}_{j,\nu} = \mathbf{v}_{j,\nu-1} + \eta_{j,\nu} \sum_{i=1}^{M} \alpha_{ij,\nu} (\mathbf{x}_i - \mathbf{v}_{j,\nu-1})$ $\forall j$.
   - $E_\nu = \sum_{j=1}^{c} \|\mathbf{v}_{j,\nu} - \mathbf{v}_{j,\nu-1}\|^2$.
4. If $E_\nu > \epsilon$ and $\nu < N$, then go to step 3.
5. Calculate $E_\nu = \sum_{j=1}^{c} \|\mathbf{v}_{j,\nu} - \mathbf{v}_{j,\nu-1}\|^2$.
6. If $E_\nu > \epsilon$ and $\nu < N$, then go to step 2.

96.5 GLVQ-F and FLVQ Algorithms

Both GLVQ and GLVQ-F algorithms are sequential, since the prototypes are sequentially updated with respect to the input vectors which are presented to the LVQ network one by one. Since the loss function in Eq. (96.15) that resulted in the GLVQ-F algorithms is not defined with respect to the winning prototype, a variety of batch LVQ algorithms can be obtained by slightly modifying the formulation that led to GLVQ-F algorithms. This modified formulation is studied here in the case where $m = 2$, in an attempt to establish a relationship between GLVQ-F and FLVQ algorithms.

Given a finite set of M feature vectors $\mathbf{x}_1, \mathbf{x}_2, \ldots, \mathbf{x}_M$, consider the minimization of the functional

$$L(\mathbf{v}_r, r = 1, 2, \ldots, c) = \frac{1}{M} \sum_{i=1}^{M} \sum_{r=1}^{c} u_{ir} \|\mathbf{x}_i - \mathbf{v}_r\|^2, \tag{96.31}$$

where $u_{ir} = u_r(\mathbf{x}_i)$ are the membership values obtained from the fuzzy c-means algorithms with $m = 2$, that is,

$$u_{ir} = u_r(\mathbf{x}_i) = \left(\sum_{j=1}^{c} \frac{\|\mathbf{x}_i - \mathbf{v}_r\|^2}{\|\mathbf{x}_i - \mathbf{v}_j\|^2} \right)^{-1} = \frac{1}{\|\mathbf{x}_i - \mathbf{v}_r\|^2} \left(\sum_{j=1}^{c} \frac{1}{\|\mathbf{x}_i - \mathbf{v}_j\|^2} \right)^{-1}. \tag{96.32}$$

Combining Eqs. (96.31) and (96.32) gives

$$\begin{aligned} L(\mathbf{v}_r, r = 1, 2, \ldots, c) &= \frac{c}{M} \sum_{i=1}^{M} \left(\sum_{j=1}^{c} \frac{1}{\|\mathbf{x}_i - \mathbf{v}_j\|^2} \right)^{-1} \\ &= \frac{1}{M} \sum_{i=1}^{M} D_H(\mathbf{x}_i), \end{aligned} \tag{96.33}$$

where $D_H(\mathbf{x}_i)$ is the harmonic mean of the distances between $\mathbf{x}_i$ and the prototypes $\mathbf{v}_j$, $j = 1, 2, \ldots, c$, defined as

$$\frac{1}{D_H(\mathbf{x}_i)} = \frac{1}{c} \sum_{j=1}^{c} \frac{1}{\|\mathbf{x}_i - \mathbf{v}_j\|^2}. \tag{96.34}$$

The functional in Eq. (96.33) is the average harmonic mean of the distances between the feature vectors and the prototypes. It is shown here that the minimization of the functional in Eq. (96.33) provides the basis for a formal derivation of the FLVQ algorithm that corresponds to $m = 2$.

The gradient of L with respect to each prototype $\mathbf{v}_j$ can be evaluated as

$$\frac{\partial L}{\partial \mathbf{v}_j} = \frac{c}{M}\sum_{i=1}^{M}\frac{\partial}{\partial \mathbf{v}_j}\left(\sum_{r=1}^{c}\frac{1}{\|\mathbf{x}_i - \mathbf{v}_r\|^2}\right)^{-1}$$

$$= -\frac{2c}{M}\sum_{i=1}^{M}\left(\sum_{r=1}^{c}\frac{\|\mathbf{x}_i - \mathbf{v}_j\|^2}{\|\mathbf{x}_i - \mathbf{v}_r\|^2}\right)^{-2}(\mathbf{x}_i - \mathbf{v}_j)$$

$$= -\frac{2c}{M}\sum_{i=1}^{M}(u_{ij})^2(\mathbf{x}_i - \mathbf{v}_j) \quad (96.35)$$

where u_{ij} are the membership values defined in Eq. 96.32.

The prototypes can be updated according to the gradient descent method as

$$\Delta\mathbf{v}_j = -\eta_j'\frac{\partial L}{\partial \mathbf{v}_j} = \eta_j\sum_{i=1}^{M}(u_{ij})^2(\mathbf{x}_i - \mathbf{v}_j) \quad (96.36)$$

where $\eta_j = 2c/M\eta_j'$. Thus, the minimization of the functional L defined in Eq. 96.33 results in Eq. 96.36 which is associated with FLVQ algorithms in the case where $m = 2$. Since the implementation of FLVQ algorithms assumes that m is not constant during the learning process, the derivation of the update equation for $m \neq 2$ is a necessary step toward the formal development of FLVQ algorithms. This problem is currently under investigation.

96.6 Fuzzy Algorithms for Learning Vector Quantization

Fuzzy algorithms for learning vector quantization were developed by minimizing a loss function formed as the weighted sum of the squared Euclidean distances between an input vector, which represents a feature vector, and the weight vectors of the LVQ network, which represent the prototypes (Karayiannis and Pai, 1994a–d, 1995a). Assuming that $\mathbf{x}$ is the input to the LVQ network and $\mathbf{v}_i$ is the winning prototype, the loss function can be formed as

$$J_{\mathbf{x}} = J_{\mathbf{x}}(\mathbf{v}_j, j = 1, 2, \ldots, c) = \sum_{r=1}^{c} u_{ir}\|\mathbf{x} - \mathbf{v}_r\|^2, \quad (96.37)$$

where $u_{ir} = u_{ir}(\mathbf{x})$, $r = 1, 2, \ldots, c$ is a set of membership functions, which regulate the competition between the prototypes $\mathbf{v}_r$, $r = 1, 2, \ldots, c$ for the input $\mathbf{x}$. In fact, the specific form of the membership functions determines the strength of attraction between each input and the prototypes during the learning process (Karayiannis and Pai, 1994a–d, 1995a).

The development of competitive learning vector quantization algorithms requires the selection of the membership functions assigned to the prototypes (Karayiannis and Pai, 1994a–d, 1995a). A fair competition among the prototypes is guaranteed if the membership function assigned to each prototype: (i) is invariant under uniform scaling of the input vectors, (ii) is equal to unity if the prototype is the winner, (iii) takes values between 1 and 0 if the prototype is not a winner, (iv) approaches zero if the prototype is not a winner and its distance from the input vector approaches infinity.

The development of fuzzy learning vector quantization algorithms can be achieved by selecting membership functions of the form

$$u_{ir} = \begin{cases} 1 & \text{if } r = i \\ u\left(\dfrac{\|\mathbf{x} - \mathbf{v}_i\|^2}{\|\mathbf{x} - \mathbf{v}_r\|^2}\right) & \text{if } r \neq i. \end{cases} \quad (96.38)$$

Under this assumption, the loss function in Eq. 96.37 can also be written as

$$J_{\mathbf{x}} = \sum_{r=1}^{c} u_{ir}\|\mathbf{x} - \mathbf{v}_r\|^2 = \|\mathbf{x} - \mathbf{v}_i\|^2 + \sum_{r\neq i}^{c} u_{ir}\|\mathbf{x} - \mathbf{v}_r\|^2. \quad (96.39)$$

Assuming that $\mathbf{v}_i$ is the winning prototype, each non-winning prototype $\mathbf{v}_r \neq \mathbf{v}_i$ contributes to the loss function $J_{\mathbf{x}}$ through the term $u_{ir}\|\mathbf{x} - \mathbf{v}_r\|^2$. According to the properties of u_{ir}, the contribution of $\|\mathbf{x} - \mathbf{v}_r\|^2$ to the loss function $J_{\mathbf{x}}$ decreases as the distance between the non-winning prototype $\mathbf{v}_r$ and the corresponding input $\mathbf{x}$ increases. The contribution of the winning prototype $\mathbf{v}_i$ is represented by the term $\|\mathbf{x} - \mathbf{v}_i\|^2$. The search for admissible membership functions can be facilitated by requiring that the loss function Eq. 96.39 is a weighted version of $\|\mathbf{x} - \mathbf{v}_i\|^2$. According to Eq. 96.39, this requirement is satisfied by membership functions of the form (Karayiannis and Pai, 1994c–d)

$$u_{ir} = u\left(\frac{\|\mathbf{x} - \mathbf{v}_i\|^2}{\|\mathbf{x} - \mathbf{v}_r\|^2}\right) = \frac{\|\mathbf{x} - \mathbf{v}_i\|^2}{\|\mathbf{x} - \mathbf{v}_r\|^2}\,p\left(\frac{\|\mathbf{x} - \mathbf{v}_i\|^2}{\|\mathbf{x} - \mathbf{v}_r\|^2}\right) \quad \forall\, r \neq i. \quad (96.40)$$

Under this assumption, Eq. (96.39) becomes

$$J_{\mathbf{x}} = \|\mathbf{x} - \mathbf{v}_i\|^2\left(1 + \sum_{r\neq i}^{c} p\left(\frac{\|\mathbf{x} - \mathbf{v}_i\|^2}{\|\mathbf{x} - \mathbf{v}_r\|^2}\right)\right). \quad (96.41)$$

The selection of generalized membership functions of the form described by Eq. 96.40 implies that the winning prototype $\mathbf{v}_i$ is updated with respect to the input $\mathbf{x}$ by minimizing a weighted version of the squared Euclidean distance $\|\mathbf{x} - \mathbf{v}_i\|^2$. More specifically, the term $\Sigma_{r\neq i}^{c} p(\|\mathbf{x} - \mathbf{v}_i\|^2/\|\mathbf{x} - \mathbf{v}_r\|^2)$ represents the cumulative effect of the non-winning prototypes $\mathbf{v}_r \neq \mathbf{v}_i$ on the attraction of the winning prototypes $\mathbf{v}_i$ by the input $\mathbf{x}$. In the trivial case where $p(x) = 0\ \forall\ x \in (0, 1)$ the membership function in Eq. (96.38) corresponds to the nearest prototype condition, which was the basis for the development of crisp LVQ algorithms. According to Eq. (96.41), if $p(x) = 0\ \forall\ x \in (0, 1)$ the non-winning prototypes have no effect on the attraction of the winning prototype by the input $\mathbf{x}$. In this case, LVQ is based on the minimization of the squared Euclidean distance $\|\mathbf{x} - \mathbf{v}_i\|^2$ between each input vector $\mathbf{x}$ and the winning prototype.

The form of the loss function Eq. (96.41) can be the basis for selecting admissible functions $p(\cdot)$. The development of competitive algorithms requires that the non-winning prototypes which are sufficiently close to $\mathbf{v}_i$ have a minimal positive effect, no effect, or negative effect on the attraction of the winning prototype $\mathbf{v}_i$ by the input $\mathbf{x}$. As a result, $p = p(\|\mathbf{x} - \mathbf{v}_i\|^2/\|\mathbf{x} - \mathbf{v}_r\|^2)$ must attain its minimum value when $\|\mathbf{x} - \mathbf{v}_r\|^2 = \|\mathbf{x} - \mathbf{v}_i\|^2$ or, equivalently, $\|\mathbf{x} - \mathbf{v}_i\|^2/\|\mathbf{x} - \mathbf{v}_r\|^2 = 1$. The competition between $\mathbf{v}_i$ and the non-winning prototype $\mathbf{v}_r$ for the input vector $\mathbf{x}$ is required to decrease as $\|\mathbf{x} - \mathbf{v}_r\|^2$ increases from a value slightly higher than $\|\mathbf{x} - \mathbf{v}_i\|^2$ to infinity. Thus, the value of $p = p(\|\mathbf{x} - \mathbf{v}_i\|^2/\|\mathbf{x} - \mathbf{v}_r\|^2)$ must increase and approach 1, its maximum value, as $\|\mathbf{x} - \mathbf{v}_i\|^2/\|\mathbf{x} - \mathbf{v}_r\|^2$ decreases from 1 to 0. The previous discussion indicates that any function defined in Eq. (96.40) is an admissible generalized membership function provided that $p(\cdot)$ is a differentiable function which satisfies the following conditions: (i) $0 < p(\mathbf{x}) < 1$ $\forall\ x \in (0, 1)$, (ii) $p(x)$ approaches 1 as x approaches 0, (iii) $p(x)$ is a monotonically decreasing function in the interval (0, 1), and (iv) $p(x)$ attains its minimum value at $x = 1$.

The investigation of the loss function in Eq. (96.41) indicated that the term $1 + \Sigma_{r\neq i}^{c}\, p(\|\mathbf{x} - \mathbf{v}_i\|^2/\|\mathbf{x} - \mathbf{v}_r\|^2$ that multiplies the squared Euclidean distance $\|\mathbf{x} - \mathbf{v}_i\|^2$ between the input $\mathbf{x}$ and the winning prototype $\mathbf{v}_i$ increases during the learning process from $1 + (c - 1)\, p(1)$ to c (Karayiannis and Pai, 1994d). A similar investigation indicated that the weights u_{ir} that multiply the squared Euclidean distances between the input vector $\mathbf{x}$ and the non-winning prototypes $\mathbf{v}_r \neq \mathbf{v}_i$ decrease from $p(1)$ to 0 during the learning process. As the learning process progresses the winning prototype is increasingly attracted by the input $\mathbf{x}$, which has a gradually decreasing effect on the non-winning prototypes. In fact, the non-winning prototypes are attracted to another set of input vectors. In conclusion, the minimization of $J_{\mathbf{x}}$ implies that all prototypes compete to be attracted by the input $\mathbf{x}$, although the form of the proposed loss function favors the winning prototype $\mathbf{v}_i$. Moreover, the loss function $J_{\mathbf{x}}$ guarantees the gradual transition from a maximum uncertainty or minimum selectivity phase, where all prototypes compete for all inputs, to a minimum uncertainty or maximum selectivity phase, where each prototype is attracted by a particular set of inputs which are clustered together.

The derivation of fuzzy algorithms for learning vector quantization is based on the minimization of the loss function in Eq. (96.41) using the gradient descent method. The gradient of $J_{\mathbf{x}}$ with respect to the winning prototype $\mathbf{v}_i$ is

$$\begin{aligned}\frac{\partial J_{\mathbf{x}}}{\partial \mathbf{v}_i} &= \frac{\partial}{\partial \mathbf{v}_i}\left(\|\mathbf{x} - \mathbf{v}_i\|^2 + \sum_{r\neq i}^{c} \|\mathbf{x} - \mathbf{v}_i\|^2\, p\left(\frac{\|\mathbf{x} - \mathbf{v}_i\|^2}{\|\mathbf{x} - \mathbf{v}_r\|^2}\right)\right) \\ &= -2(\mathbf{x} - \mathbf{v}_i) - 2(\mathbf{x} - \mathbf{v}_i)\sum_{r\neq i}^{c} p\left(\frac{\|\mathbf{x} - \mathbf{v}_i\|^2}{\|\mathbf{x} - \mathbf{v}_r\|^2}\right) \\ &\quad + \sum_{r\neq i}^{c} \|\mathbf{x} - \mathbf{v}_i\|^2 \frac{\partial}{\partial \mathbf{v}_i} p\left(\frac{\|\mathbf{x} - \mathbf{v}_i\|^2}{\|\mathbf{x} - \mathbf{v}_r\|^2}\right).\end{aligned} \tag{96.42}$$

Using the chain rule,

$$\begin{aligned}\frac{\partial}{\partial \mathbf{v}_i} p\left(\frac{\|\mathbf{x} - \mathbf{v}_i\|^2}{\|\mathbf{x} - \mathbf{v}_r\|^2}\right) &= p'\left(\frac{\|\mathbf{x} - \mathbf{v}_i\|^2}{\|\mathbf{x} - \mathbf{v}_r\|^2}\right)\frac{\partial}{\partial \mathbf{v}_i}\left(\frac{\|\mathbf{x} - \mathbf{v}_i\|^2}{\|\mathbf{x} - \mathbf{v}_r\|^2}\right) \\ &= -2(\mathbf{x} - \mathbf{v}_i)\frac{1}{\|\mathbf{x} - \mathbf{v}_r\|^2}\, p'\left(\frac{\|\mathbf{x} - \mathbf{v}_i\|^2}{\|\mathbf{x} - \mathbf{v}_r\|^2}\right).\end{aligned} \tag{96.43}$$

Substituting Eq. (96.43) in Eq. (96.42) gives

$$\begin{aligned}\frac{\partial J_{\mathbf{x}}}{\partial \mathbf{v}_i} &= -2(\mathbf{x} - \mathbf{v}_i)\left(1 + \sum_{r\neq i}^{c}\left[p\left(\frac{\|\mathbf{x} - \mathbf{v}_i\|^2}{\|\mathbf{x} - \mathbf{v}_r\|^2}\right)\right.\right. \\ &\quad \left.\left. + \frac{\|\mathbf{x} - \mathbf{v}_i\|^2}{\|\mathbf{x} - \mathbf{v}_r\|^2}\, p'\left(\frac{\|\mathbf{x} - \mathbf{v}_i\|^2}{\|\mathbf{x} - \mathbf{v}_r\|^2}\right)\right]\right).\end{aligned} \tag{96.44}$$

Since $u(x) = xp(x)$, $u'(x) = p(x) + xp'(x)$. Thus, the update equation for the winning prototype $\mathbf{v}_i$ can be obtained from Eq. (96.44) as

$$\Delta\mathbf{v}_i = -\eta\frac{\partial J_{\mathbf{x}}}{\partial \mathbf{v}_i} = \eta\,(\mathbf{x} - \mathbf{v}_i)\left(1 + \sum_{r\neq i}^{k} w_{ir}\right), \tag{96.45}$$

where

$$\begin{aligned}w_{ir} &= p\left(\frac{\|\mathbf{x} - \mathbf{v}_i\|^2}{\|\mathbf{x} - \mathbf{v}_r\|^2}\right) + \frac{\|\mathbf{x} - \mathbf{v}_i\|^2}{\|\mathbf{x} - \mathbf{v}_r\|^2}\, p'\left(\frac{\|\mathbf{x} - \mathbf{v}_i\|^2}{\|\mathbf{x} - \mathbf{v}_r\|^2}\right) \\ &= u'\left(\frac{\|\mathbf{x} - \mathbf{v}_i\|^2}{\|\mathbf{x} - \mathbf{v}_r\|^2}\right).\end{aligned} \tag{96.46}$$

Similarly, the gradient of $J_{\mathbf{x}}$ with respect to the non-winning prototypes $\mathbf{v}_j \neq \mathbf{v}_i$ is

$$\begin{aligned}\frac{\partial J_{\mathbf{x}}}{\partial \mathbf{v}_j} &= \frac{\partial}{\partial \mathbf{v}_j}\|\mathbf{x} - \mathbf{v}_i\|^2\left(1 + \sum_{r\neq i}^{c} p\left(\frac{\|\mathbf{x} - \mathbf{v}_i\|^2}{\|\mathbf{x} - \mathbf{v}_r\|^2}\right)\right) \\ &= \|\mathbf{x} - \mathbf{v}_i\|^2\frac{\partial}{\partial \mathbf{v}_j} p\left(\frac{\|\mathbf{x} - \mathbf{v}_i\|^2}{\|\mathbf{x} - \mathbf{v}_j\|^2}\right).\end{aligned} \tag{96.47}$$

Using the chain rule,

$$\begin{aligned}\frac{\partial}{\partial \mathbf{v}_j} p\left(\frac{\|\mathbf{x} - \mathbf{v}_i\|^2}{\|\mathbf{x} - \mathbf{v}_j\|^2}\right) &= p'\left(\frac{\|\mathbf{x} - \mathbf{v}_i\|^2}{\|\mathbf{x} - \mathbf{v}_j\|^2}\right)\frac{\partial}{\partial \mathbf{v}_j}\left(\frac{\|\mathbf{x} - \mathbf{v}_i\|^2}{\|\mathbf{x} - \mathbf{v}_j\|^2}\right) \\ &= 2(\mathbf{x} - \mathbf{v}_j)\frac{\|\mathbf{x} - \mathbf{v}_i\|^2}{(\|\mathbf{x} - \mathbf{v}_j\|^2)^2}\, p'\left(\frac{\|\mathbf{x} - \mathbf{v}_i\|^2}{\|\mathbf{x} - \mathbf{v}_j\|^2}\right).\end{aligned} \tag{96.48}$$

Substituting Eq. (96.48) in (96.47) gives

$$\frac{\partial J_{\mathbf{x}}}{\partial \mathbf{v}_j} = 2(\mathbf{x} - \mathbf{v}_j)\left(\frac{\|\mathbf{x} - \mathbf{v}_i\|^2}{\|\mathbf{x} - \mathbf{v}_j\|^2}\right)^2 p'\left(\frac{\|\mathbf{x} - \mathbf{v}_i\|^2}{\|\mathbf{x} - \mathbf{v}_j\|^2}\right). \tag{96.49}$$

Since $u'(x) = p(x) + xp'(x)$, $x^2p'(x) = xu'(x) - xp(x) = x\,u'(x) - u(x)$. Thus, the update equation for the non-winning prototypes $\mathbf{v}_j \neq \mathbf{v}_i$ can be obtained from Eq. (96.49)

$$\Delta \mathbf{v}_j = -\eta \frac{\partial J_{\mathbf{x}}}{\partial \mathbf{v}_j} = \eta\,(\mathbf{x} - \mathbf{v}_j)\,n_{ij}, \tag{96.50}$$

where

$$\begin{aligned} n_{ij} &= -\left(\frac{\|\mathbf{x} - \mathbf{v}_i\|^2}{\|\mathbf{x} - \mathbf{v}_j\|^2}\right)^2 p'\left(\frac{\|\mathbf{x} - \mathbf{v}_i\|^2}{\|\mathbf{x} - \mathbf{v}_j\|^2}\right) \\ &= u\left(\frac{\|\mathbf{x} - \mathbf{v}_i\|^2}{\|\mathbf{x} - \mathbf{v}_j\|^2}\right) - \frac{\|\mathbf{x} - \mathbf{v}_i\|^2}{\|\mathbf{x} - \mathbf{v}_j\|^2}\,u'\left(\frac{\|\mathbf{x} - \mathbf{v}_i\|^2}{\|\mathbf{x} - \mathbf{v}_j\|^2}\right) \\ &= u_{ij} - \frac{\|\mathbf{x} - \mathbf{v}_i\|^2}{\|\mathbf{x} - \mathbf{v}_j\|^2}\,w_{ij}. \end{aligned} \tag{96.51}$$

The adaptation of the prototypes during the learning process depends on the learning rate $\eta \in [0, 1]$, which is a monotonically decreasing function of the number of iterations ν. The learning rate can be a linear function of ν defined as $\eta = \eta(\nu) = \eta_0(1 - \nu/N)$, where η_0 is the initial value of the learning rate and N is the total number of iterations predetermined for the learning process.

According to Eq. (96.45), the adaptation of the winning prototype $\mathbf{v}_i$ is affected by all the non-winning prototypes $\mathbf{v}_r \neq \mathbf{v}_i$, while w_{ir} represents the interference from the non-winning prototype $\mathbf{v}_r$ to the winning prototype $\mathbf{v}_i$. In fact, the term $\Sigma_{r\neq i}^{c} w_{ir}$ represents the cumulative effect of the non-winning prototypes on the attraction of the winning prototype by the input vector $\mathbf{x}$. In contrast, Eq. (96.50) indicates that the adaptation of each non-winning prototype $\mathbf{v}_j \neq \mathbf{v}_i$ is affected only by the winning prototype $\mathbf{v}_i$. In this case, n_{ij} represents the interference from the winning prototype $\mathbf{v}_i$ to the adaptation of the non-winning prototype $\mathbf{v}_j$.

The effect of the non-winning prototypes on the adaptation of the winning prototype can be investigated by studying the properties of the function $u_{ir} = u(\|\mathbf{x} - \mathbf{v}_i\|^2/\|\mathbf{x} - \mathbf{v}_r\|^2)$ for $r \neq i$. If $u(x)$ is a monotonically increasing function over the interval (0, 1), then $u'(x) > 0\ \forall\ x \in (0, 1)$. Since in this case $w_{ir} > 0$, each non-winning prototype $\mathbf{v}_r$ has an excitatory effect on the attraction of the winning prototype $\mathbf{v}_i$ by the input $\mathbf{x}$, which depends on the ratio $\|\mathbf{x} - \mathbf{v}_i\|^2/\|\mathbf{x} - \mathbf{v}_r\|^2$. If there exists an $x \in (0, 1)$ such that $u'(x) = 0$, then $u(x)$ possesses a maximum between 0 and 1. In this case, $u'(x)$ can take nonnegative as well as negative values. Depending on the value of the ratio $\|\mathbf{x} - \mathbf{v}_i\|^2/\|\mathbf{x} - \mathbf{v}_r\|^2$, each non-winning prototype $\mathbf{v}_r$ can have either an inhibitory or excitatory effect on the adaptation of the winning prototype $\mathbf{v}_i$.

The properties of $p(\cdot)$ can be the basis for evaluating the effect of the interference functions n_{ij} on the adaptation of the non-winning prototypes. Since $p(x)$ is required to be a decreasing function over the interval (0,1), $p'(x) < 0$, and thus, $-x^2 \cdot p'(x) > 0$. According to Eq. (96.51), $n_{ij} > 0\ \forall\ j \neq i$. Therefore, the winning prototype $\mathbf{v}_i$ has an excitatory effect on the adaptation of each non-winning prototype $\mathbf{v}_j$, which depends on the value of the ratio $\|\mathbf{x} - \mathbf{v}_i\|^2/\|\mathbf{x} - \mathbf{v}_j\|^2$.

Finally, the selection of admissible membership functions can be facilitated by examining the relationship between the form of $p(\cdot)$ and the competition among the prototypes in two extreme situations. If $p(x) = 0\ \forall\ x \in (0, 1)$, then $w_{ir} = 0\ \forall\ r \neq i$ and $n_{ij} = 0\ \forall\ j \neq i$. In this case, the winning prototype $\mathbf{v}_i$ is updated according to

$$\Delta \mathbf{v}_i = \eta\,(\mathbf{x} - \mathbf{v}_i), \tag{96.52}$$

while the non-winning prototypes $\mathbf{v}_j \neq \mathbf{v}_i$ remain unchanged. If $p(x) = 1\ \forall\ x \in (0, 1)$, then $u(x) = x\ \forall\ x \in (0, 1)$. Since $p'(x) = 0\ \forall\ x \in (0, 1)$, Eq. (96.46) indicates that $w_{ir} = 1\ \forall\ r \neq i$ and $\Sigma_{r\neq i}^{c}\, w_{ir} = c - 1$. According to Eq. (96.45), the winning prototype $\mathbf{v}_i$ is updated by

$$\Delta \mathbf{v}_i = \eta\, c(\mathbf{x} - \mathbf{v}_i). \tag{96.53}$$

Under the same assumption, $n_{ij} = 0\ \forall\ j \neq i$. Thus, the non-winning prototypes $\mathbf{v}_j \neq \mathbf{v}_i$ are not updated with respect to the input $\mathbf{x}$ regardless of their distance from the winning prototype $\mathbf{v}_i$.

The resulting FALVQ algorithms can be summarized as follows:

1. Select c; fix η_0 N; set $\nu = 0$; randomly generate initial codebook $\mathcal{V}_0 = \{\mathbf{v}_{1,0}, \mathbf{v}_{2,0} \ldots, \mathbf{v}_{k,0}\}$
2. Calculate $\eta = \eta_0\,(1 - \nu/N)$.
3. Set $\nu = \nu + 1$.
4. For each input vector $\mathbf{x}$:
   - find i such that $\|\mathbf{x} - \mathbf{v}_{i,\nu-1}\|^2 < \|\mathbf{x} - \mathbf{v}_{j,\nu-1}\|^2\ \forall\ j \neq i$
   - calculate $u_{ir,\nu} = u(\|\mathbf{x} - \mathbf{v}_{i,\nu-1}\|^2/\|\mathbf{x} - \mathbf{v}_{r,\nu-1}\|^2) \forall\ r \neq i$.
   - calculate $w_{ir,\nu} = u'(\|\mathbf{x} - \mathbf{v}_{i,\nu-1}\|^2/\|\mathbf{x} - \mathbf{v}_{r,\nu-1}\|^2) \forall\ r \neq i$.
   - calculate $n_{ir,\nu} = u_{ir,\nu} - (\|\mathbf{x} - \mathbf{v}_{i,\nu-1}\|^2/\|\mathbf{x} - \mathbf{v}_{r,\nu-1}\|^2) w_{ir,\nu}$ $\forall\ r \neq i$.
   - update $\mathbf{v}_i$ by $\mathbf{v}_{i,\nu} = \mathbf{v}_{i,\nu-1,1} + \eta(\mathbf{x} - \mathbf{v}_{i,\nu-1})(1 + \Sigma_{r\neq i}^{c}\, w_{ir,\nu})$
   - update $\mathbf{v}_j \neq \mathbf{v}_i$ by $\mathbf{v}_{j,\nu} = \mathbf{v}_{j,\nu-1} + \eta(\mathbf{x} - \mathbf{v}_{j,\nu-1}) n_{ij,\nu}$
5. If $\nu < N$, then go to step 2.

96.7 The FALVQ 1 Family of Algorithms

The development of the FALVQ 1 algorithm was based on the minimization of the loss function in Eq. (96.37) with (Karayiannis and Pai, 1994a–d)

$$u_{ir} = \begin{cases} 1 & \text{if } r = i \\ \left(1 + \dfrac{\|\mathbf{x} - \mathbf{v}_r\|^2}{\|\mathbf{x} - \mathbf{v}_i\|^2}\right)^{-1} & \text{if } r \neq i. \end{cases} \tag{96.54}$$

For $r \neq i$, u_{ir} can be written as

$$u_{ir} = \frac{\|\mathbf{x} - \mathbf{v}_i\|^2}{\|\mathbf{x} - \mathbf{v}_r\|^2}\left(1 + \frac{\|\mathbf{x} - \mathbf{v}_i\|^2}{\|\mathbf{x} - \mathbf{v}_r\|^2}\right)^{-1}. \qquad (96.55)$$

Thus, the membership function in Eq. (96.54) is of the form described by Eq. (96.40) with $p(x) = (1 + x)^{-1}$. Figure 96.2a plots $u_{ir} = u(\|\mathbf{x} - \mathbf{v}_i\|^2/\|\mathbf{x} - \mathbf{v}_r\|^2)$ as a function of $\|\mathbf{x} - \mathbf{v}_i\|^2/\|\mathbf{x} - \mathbf{v}_r\|^2$. Clearly, u_{ir} decreases from a value close to 1/2 to 0 as $\|\mathbf{x} - \mathbf{v}_r\|^2$ increases from a value slightly higher than $\|\mathbf{x} - \mathbf{v}_i\|^2$ to infinity. Therefore, the effect of the non-winning prototypes on the loss function in Eq. (96.39) diminishes as their distance from the input vector $\mathbf{x}$ increases.

The FALVQ 1 algorithm can be derived by minimizing the loss function defined by Eq. (96.41) with $p(x) = (1 + x)^{-1}$. If $\mathbf{x}$ is the input vector, the winning prototype $\mathbf{v}_i$ can be updated by Eq. (96.42) with

$$w_{ir} = (1 - u_{ir})^2 = \left(1 + \frac{\|\mathbf{x} - \mathbf{v}_i\|^2}{\|\mathbf{x} - \mathbf{v}_r\|^2}\right)^{-2}, \qquad (96.56)$$

while the non-winning prototypes $\mathbf{v}_j \neq \mathbf{v}_i$ can be updated by Eq. (96.44) with

$$n_{ij} = u_{ij}^2 = \left(\frac{\|\mathbf{x} - \mathbf{v}_i\|^2}{\|\mathbf{x} - \mathbf{v}_j\|^2}\right)^2\left(1 + \frac{\|\mathbf{x} - \mathbf{v}_i\|^2}{\|\mathbf{x} - \mathbf{v}_j\|^2}\right)^{-2}. \qquad (96.57)$$

The effect of each non-winning prototype $\mathbf{v}_r$ on the attraction of the winning prototype $\mathbf{v}_i$ by the input vector $\mathbf{x}$ is illustrated in Figure 96.2b, which plots w_{ir} as a function of $\|\mathbf{x} - \mathbf{v}_i\|^2/\|\mathbf{x} - \mathbf{v}_r\|^2$. Consider a non-winning prototype $\mathbf{v}_r$ for which $\|\mathbf{x} - \mathbf{v}_r\|^2 \gg \|\mathbf{x} - \mathbf{v}_i\|^2$. Under this assumption, $u_{ir} \approx 0$ and, therefore, $w_{ir} = (1 - u_{ir})^2 \approx 1$. Since the maximum value of w_{ir} is 1, the existence of a non-winning prototype $\mathbf{v}_r$ for which $\|\mathbf{x} - \mathbf{v}_r\|^2 \gg \|\mathbf{x} - \mathbf{v}_i\|^2$ results in the maximum possible reinforcement of the adaptation of the winning prototype $\mathbf{v}_i$. If a non-winning prototype $\mathbf{v}_r$ is very close to the winning prototype $\mathbf{v}_i$, that is, if $\|\mathbf{x} - \mathbf{v}_r\|^2 \approx \|\mathbf{x} - \mathbf{v}_i\|^2$, then $u_{ir} \approx 1/2$. In this case, $w_{ir} = (1 - u_{ir})^2$ approaches 1/4, which is its minimum value. Clearly, the adaptation of the winning prototype $\mathbf{v}_i$ is reinforced to a lesser degree by the non-winning prototypes sufficiently close to it. In this sense, this set of non-winning prototypes compete with $\mathbf{v}_i$ to match the input $\mathbf{x}$. The attraction of each non-winning prototype $\mathbf{v}_j$ by the input $\mathbf{x}$ is determined by the corresponding interference function n_{ij} defined in Eq. (96.57), which is plotted in Figure 96.2c as a function of $\|\mathbf{x} - \mathbf{v}_i\|^2/\|\mathbf{x} - \mathbf{v}_j\|^2$. Clearly, n_{ij} increases from 0 to 1/4 as $\|\mathbf{x} - \mathbf{v}_i\|^2/\|\mathbf{x} - \mathbf{v}_j\|^2$ increases from 0 to 1. Thus, Eq. (96.44) reinforces the attraction between each input vector $\mathbf{x}$ and the non-winning prototypes which are close to the winning prototype $\mathbf{v}_i$, while the adaptation of the non-winning prototypes $\mathbf{v}_j$ for which $\|\mathbf{x} - \mathbf{v}_j\|^2 \gg \|\mathbf{x} - \mathbf{v}_i\|^2$ is negligible.

The FALVQ 1 family of algorithms can be extended by introducing the membership function

$$u_{ir} = \begin{cases} 1 & \text{if } r = i \\ \dfrac{\|\mathbf{x} - \mathbf{v}_i\|^2}{\|\mathbf{x} - \mathbf{v}_r\|^2}\left(1 + \alpha\dfrac{\|\mathbf{x} - \mathbf{v}_i\|^2}{\|\mathbf{x} - \mathbf{v}_r\|^2}\right)^{-1} & \text{if } r \neq i, \end{cases} \qquad (96.58)$$

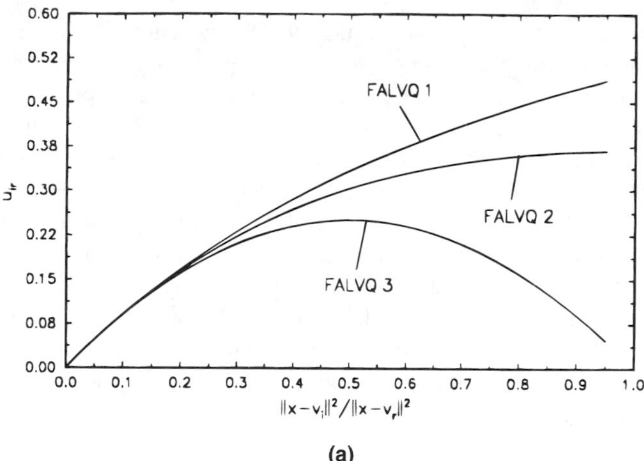

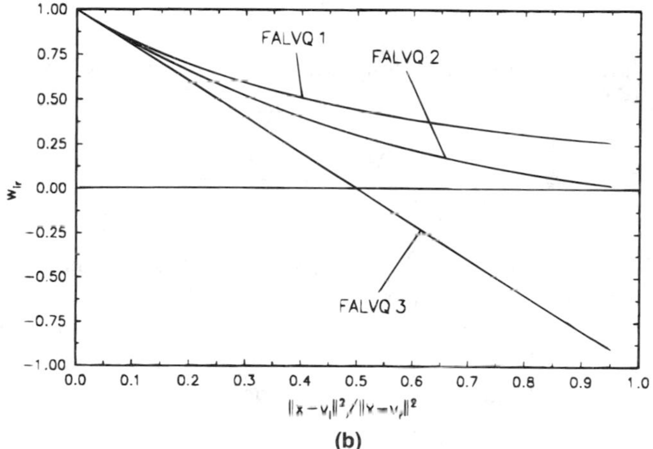

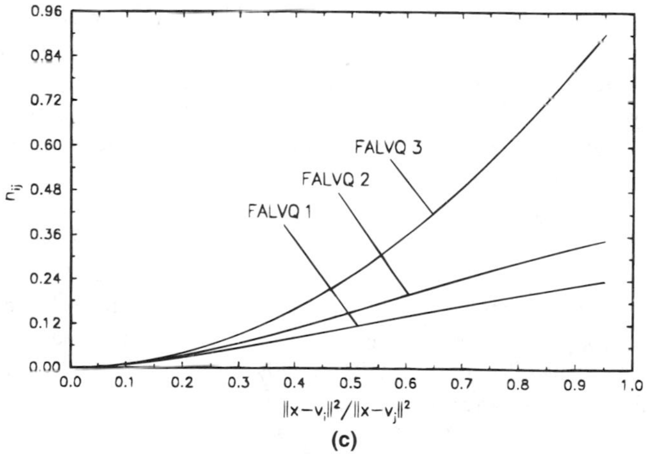

Figure 96.2 Membership function u_{ir} and interference functions w_{ir}, n_{ij} corresponding to FALVQ 1, FALVQ 2, and FALVQ 3 algorithms: (a) u_{ir} as a function of $\|\mathbf{x} - \mathbf{v}_i\|^2/\|\mathbf{x} - \mathbf{v}_r\|^2$, (b) w_{ir} as a function of $\|\mathbf{x} - \mathbf{v}_i\|^2/\|\mathbf{x} - \mathbf{v}_r\|^2$, and (c) n_{ij} as a function of $\|\mathbf{x} - \mathbf{v}_i\|^2/\|\mathbf{x} - \mathbf{v}_j\|^2$. (*Source: IEEE Trans. Neural Networks* 7(5): 1197–1211. © IEEE. Used with permission.)

where $\alpha > 0$. For $r \neq i$, Eq. (96.58) indicates that u_{ir} is a monotonically increasing function as $\|\mathbf{x} - \mathbf{v}_i\|^2/\|\mathbf{x} - \mathbf{v}_r\|^2$ increases from 0 to 1, regardless of the value of α. For values of α close to 0, u_{ir} increases almost linearly from 0 to 1 as $\|\mathbf{x} - \mathbf{v}_i\|^2/\|\mathbf{x} - \mathbf{v}_r\|^2$ spans the interval (0,1). For sufficiently large values of α, u_{ir} is almost constant and approaches $1/\alpha$ as $\|\mathbf{x} - \mathbf{v}_i\|^2/\|\mathbf{x} - \mathbf{v}_r\|^2$ increases from 0 to 1. The role of the parameter α is also shown in Figure 96.3a, which plots $u_{ir} = u_{ir}(\alpha)$ as a function of $\|\mathbf{x} - \mathbf{v}_i\|^2/\|\mathbf{x} - \mathbf{v}_r\|^2$ for different values of α.

The FALVQ 1 family of algorithms can be derived by minimizing the loss function defined by Eq. (96.41) with $p(x) = (1 + \alpha x)^{-1}$. If $\mathbf{x}$ is the input vector, the winning prototype $\mathbf{v}_i$ can be updated by Eq. (96.42) with

$$w_{ir} = (1 - \alpha u_{ir})^2 = \left(1 + \alpha \frac{\|\mathbf{x} - \mathbf{v}_i\|^2}{\|\mathbf{x} - \mathbf{v}_r\|^2}\right)^{-2}, \qquad (96.59)$$

while the non-winning prototypes $\mathbf{v}_j \neq \mathbf{v}_i$ can be updated by Eq. (96.44) with

$$n_{ij} = \alpha\, u_{ij}^2 = \alpha \left(\frac{\|\mathbf{x} - \mathbf{v}_i\|^2}{\|\mathbf{x} - \mathbf{v}_j\|^2}\right)^2 \left(1 + \alpha \frac{\|\mathbf{x} - \mathbf{v}_i\|^2}{\|\mathbf{x} - \mathbf{v}_j\|^2}\right)^{-2}. \qquad (96.60)$$

For sufficiently large values of α, w_{ir} approaches 0. In this case, the non-winning prototypes have almost no effect on the adaptation of the winning prototype. This is clearly shown in Figure 96.3b, which plots $w_{ir} = w_{ir}(\alpha)$ as a function of $\|\mathbf{x} - \mathbf{v}_i\|^2/\|\mathbf{x} - \mathbf{v}_r\|^2$ for different values of α. For sufficiently large values of α, n_{ij} behaves like $1/\alpha$. Thus, the adaptation of the non-winning prototypes is negligible regardless of their distance from the winning prototype. This is shown in Figure 96.3c, which plots $n_{ij} = n_{ij}(\alpha)$ as a function of $\|\mathbf{x} - \mathbf{v}_i\|^2/\|\mathbf{x} - \mathbf{v}_j\|^2$ for different values of α. In summary, for large values of α the non-winning prototypes are not allowed to compete with the winning prototype to match the corresponding input. As the value of α decreases, the non-winning prototypes become more competitive. As α approaches 0, w_{ir} approaches 1 regardless of the value of the ratio $\|\mathbf{x} - \mathbf{v}_i\|^2/\|\mathbf{x} - \mathbf{v}_r\|^2$. Thus, all non-winning prototypes contribute to the attraction between the input and the winning prototype by the same amount and $1 + \sum_{r \neq i}^{c} w_{ir}$ approaches c. As α approaches 0, the non-winnings prototypes are not substantially updated because n_{ij} approaches 0. In conclusion, the algorithms included in the FALVQ 1 family are genuinely competitive for a certain range of values of α.

96.8 The FALVQ 2 Family of Algorithms

The membership function in Eq. (96.54) which resulted in the FALVQ 1 algorithm corresponds to $p(x) = 1 + x)^{-1}$. An alternative admissible function $p(\cdot)$ can be obtained by observing that $(1 + x)^{-1}$ and $\exp(-x)$ are very similar if $|x| << 1$. In fact, $p(x) = \exp(-x)$ is an interesting candidate since $\exp(-x) < (1 + x)^{-1}$ as x approaches 1. For this particular choice, u_{ir} is of the form

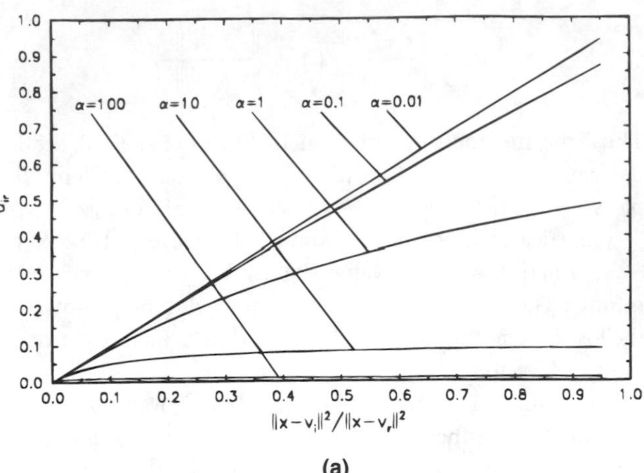

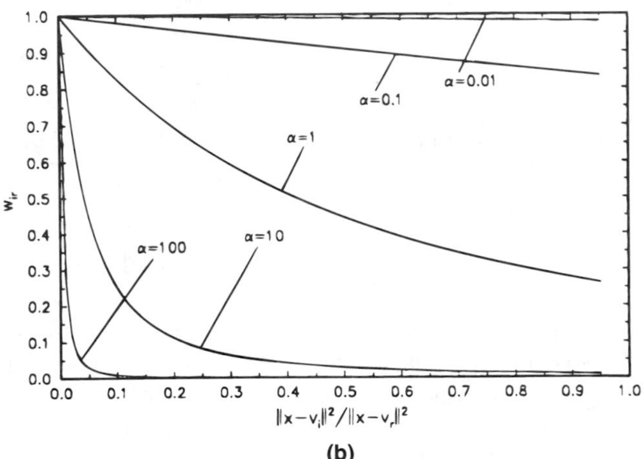

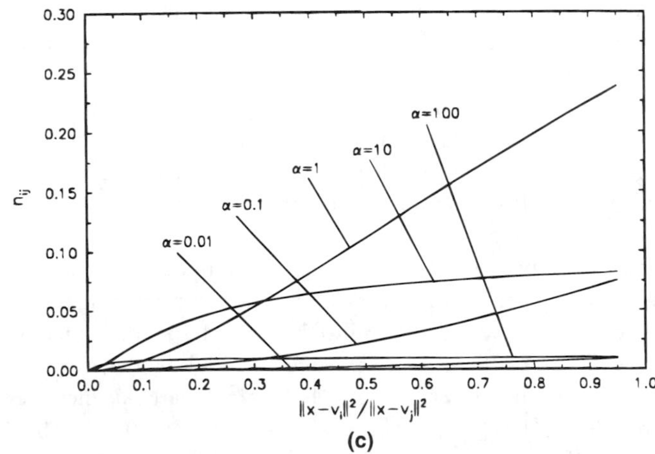

Figure 96.3 Membership function u_{ir} and interference functions w_{ir}, n_{ij} corresponding to the FALVQ 1 family of algorithms: (a) $u_{ir} = u_{ir}(\alpha)$ as a function of $\|\mathbf{x} - \mathbf{v}_i\|^2/\|\mathbf{x} - \mathbf{v}_r\|^2$ for different values of α, (b) $w_{ir} = w_{ir}(\alpha)$ as a function of $\|\mathbf{x} - \mathbf{v}_i\|^2/\|\mathbf{x} - \mathbf{v}_r\|^2$ for different values of α, and (c) $n_{ij} = n_{ij}(\alpha)$ as a function of $\|\mathbf{x} - \mathbf{v}_i\|^2/\|\mathbf{x} - \mathbf{v}_j\|^2$ for different values of α. (*Source: IEEE Trans. Neural Networks* 7(5): 1197–1211. © IEEE. Used with permission.)

$$u_{ir} = \begin{cases} 1 & \text{if } r = i \\ \dfrac{\|\mathbf{x} - \mathbf{v}_i\|^2}{\|\mathbf{x} - \mathbf{v}_r\|^2} \exp\left(-\dfrac{\|\mathbf{x} - \mathbf{v}_i\|^2}{\|\mathbf{x} - \mathbf{v}_r\|^2}\right) & \text{if } r \neq i. \end{cases} \quad (96.61)$$

The function u_{ir} defined in Eq. (96.61) decreases from a value slightly lower than e^{-1} to 0 as $\|\mathbf{x} - \mathbf{v}_r\|^2$ increases from a value slightly higher than $\|\mathbf{x} - \mathbf{v}_i\|^2$ to infinity. The difference between the functions defined in Eqs. (96.54) and (96.61) is clearly exhibited in Figure 96.2a, which plots both as functions of $\|\mathbf{x} - \mathbf{v}_i\|^2/\|\mathbf{x} - \mathbf{v}_r\|^2$.

The FALVQ 2 algorithm can be derived by minimizing the loss function in Eq. (96.41) with $p(x) = \exp(-x)$. If $\mathbf{x}$ is the input vector, the winning prototype $\mathbf{v}_i$ can be updated by Eq. (96.42) with

$$w_{ir} = \left(1 - \frac{\|\mathbf{x} - \mathbf{v}_i\|^2}{\|\mathbf{x} - \mathbf{v}_r\|^2}\right)\exp\left(-\frac{\|\mathbf{x} - \mathbf{v}_i\|^2}{\|\mathbf{x} - \mathbf{v}_r\|^2}\right), \quad (96.62)$$

while the non-winning prototypes $\mathbf{v}_j \neq \mathbf{v}_i$ can be updated by Eq. (96.44) with

$$n_{ij} = \left(\frac{\|\mathbf{x} - \mathbf{v}_i\|^2}{\|\mathbf{x} - \mathbf{v}_j\|^2}\right)^2 \exp\left(-\frac{\|\mathbf{x} - \mathbf{v}_i\|^2}{\|\mathbf{x} - \mathbf{v}_j\|^2}\right). \quad (96.63)$$

Figure 96.2b plots the interference function w_{ir} defined in Eq. (96.62) as a function of $\|\mathbf{x} - \mathbf{v}_i\|^2/\|\mathbf{x} - \mathbf{v}_r\|^2$. Although the interference function Eq. (96.62) is similar with that corresponding to the FALVQ 1 algorithm, it approaches 0 as $\|\mathbf{x} - \mathbf{v}_i\|^2/\|\mathbf{x} - \mathbf{v}_r\|^2$ approaches 1. In this case, the non-winning prototypes which are sufficiently close to the winning prototype do not affect its adaptation. Compared with the FALVQ 1 algorithm, the FALVQ 2 allows this set of non-winning prototypes to compete stronger with the winning prototype to match the input vector $\mathbf{x}$. This can also be verified by comparing the function n_{ij} corresponding to the FALVQ 1 and FALVQ 2 algorithms, shown in Figure 96.2c. Clearly, as $\|\mathbf{x} - \mathbf{v}_i\|^2/\|\mathbf{x} - \mathbf{v}_j\|^2$ approaches 1 the function n_{ij} corresponding to the FALVQ 2 algorithm approaches e^{-1}, which is higher than the value 1/4 attained by n_{ij} in the same case when the FALVQ 1 algorithm is used.

A broad variety of fuzzy learning vector quantization algorithms can be developed by modifying the membership function in Eq. (96.61) as

$$u_{ir} = \begin{cases} 1 & \text{if } r = i \\ \dfrac{\|\mathbf{x} - \mathbf{v}_i\|^2}{\|\mathbf{x} - \mathbf{v}_r\|^2} \exp\left(-\beta \dfrac{\|\mathbf{x} - \mathbf{v}_i\|^2}{\|\mathbf{x} - \mathbf{v}_r\|^2}\right) & \text{if } r \neq i, \end{cases} \quad (96.64)$$

where $\beta > 0$. The parameter β has a rather significant effect on the competition between winning and non-winning prototypes. For $r \neq i$, u_{ir} reaches its maximum value $(\beta e)^{-1}$ at $\|\mathbf{x} - \mathbf{v}_i\|^2/\|\mathbf{x} - \mathbf{v}_r\|^2 = 1/\beta$. If $\beta \leq 1$, then $1/\beta \geq 1$ and u_{ir} is a monotonically increasing function in the interval (0,1). If $\beta > 1$, then u_{ir} increases as $\|\mathbf{x} - \mathbf{v}_i\|^2/\|\mathbf{x} - \mathbf{v}_r\|^2$ increases from 0 to $1/\beta < 1$ and decreases as $\|\mathbf{x} - \mathbf{v}_i\|^2/\|\mathbf{x} - \mathbf{v}_r\|^2$ increases from $1/\beta$ to 1. The effect of the parameter β is also shown in Figure 96.4a, which plots $u_{ir} = u_{ir}(\beta)$ as a function of $\|\mathbf{x} - \mathbf{v}_i\|^2/\|\mathbf{x} - \mathbf{v}_r\|^2$ for different values of β.

The FALVQ 2 family of algorithms can be derived by minimizing the loss function in Eq. (96.41) with $p(x) = \exp(-\beta x)$. The winning prototype $\mathbf{v}_i$ can be updated by Eq. (96.42) with

$$w_{ir} = \left(1 - \beta \frac{\|\mathbf{x} - \mathbf{v}_i\|^2}{\|\mathbf{x} - \mathbf{v}_r\|^2}\right)\exp\left(-\beta \frac{\|\mathbf{x} - \mathbf{v}_i\|^2}{\|\mathbf{x} - \mathbf{v}_r\|^2}\right), \quad (96.65)$$

while the non-winning prototypes $\mathbf{v}_j \neq \mathbf{v}_i$ can be updated by Eq. (96.44) with

$$n_{ij} = \beta \left(\frac{\|\mathbf{x} - \mathbf{v}_i\|^2}{\|\mathbf{x} - \mathbf{v}_j\|^2}\right)^2 \exp\left(-\beta \frac{\|\mathbf{x} - \mathbf{v}_i\|^2}{\|\mathbf{x} - \mathbf{v}_j\|^2}\right). \quad (96.66)$$

Figure 96.4b plots the interference function $w_{ir} = w_{ir}(\beta)$ as a function of $\|\mathbf{x} - \mathbf{v}_i\|^2/\|\mathbf{x} - \mathbf{v}_r\|^2$ for various values of β. As $\|\mathbf{x} - \mathbf{v}_i\|^2/\|\mathbf{x} - \mathbf{v}_r\|^2$ increases from 0 to 1, w_{ir} decreases from 1 to $(1 - \beta)e^{-\beta}$. If $b \leq 1$, then $(1 - \beta)e^{-\beta} \geq 0$. If $\beta > 1$, then the value of w_{ir} at $\|\mathbf{x} - \mathbf{v}_i\|^2/\|\mathbf{x} - \mathbf{v}_r\|^2 = 1$ is $(1 - \beta)e^{-\beta} < 0$. In this case, there is a zero-crossing for w_{ir} at $\|\mathbf{x} - \mathbf{v}_i\|^2/\|\mathbf{x} - \mathbf{v}_r\|^2 = 1/\beta < 1$. This implies that the non-winning prototypes $\mathbf{v}_r$ for which $\|\mathbf{x} - \mathbf{v}_i\|^2/\|\mathbf{x} - \mathbf{v}_r\|^2 > 1/\beta$ inhibit the attraction of the winning prototype $\mathbf{v}_i$ by the input vector $\mathbf{x}$. Thus, the winning prototypes that are close to $\mathbf{v}_i$ become more competitive for values of β above 1, while the algorithm becomes increasingly biased toward the wining prototype as the value of β decreases below 1. For $\beta > 1$, w_{ir} is not necessarily a monotonically decreasing function of $\|\mathbf{x} - \mathbf{v}_i\|^2/\|\mathbf{x} - \mathbf{v}_r\|^2$. It can easily be verified that w_{ir} has a minimum at $\|\mathbf{x} - \mathbf{v}_i\|^2/\|\mathbf{x} - \mathbf{v}_r\|^2 = 2/\beta$. Thus, w_{ir} is a monotonically decreasing function over the interval (0,1) if $2/\beta \geq 1$ or, equivalently, $\beta \leq 2$. The effect of β on the adaptation of the non-winning prototypes is illustrated in Figure 96.4c, which plots $n_{ij} = n_{ij}(\beta)$ as a function of $\|\mathbf{x} - \mathbf{v}_i\|^2/\|\mathbf{x} - \mathbf{v}_j\|^2$ for different values of β. As β increases from 0 to 2, n_{ij} has an increasingly excitatory effect on the adaptation of the non-winning prototypes, especially those which are sufficiently close to the winning prototype. The attraction between the input vector and the non-winning prototypes begins decreasing as the value of β increases above 2.

96.9 The FALVQ 3 Family of Algorithms

The FALVQ 2 algorithm was derived by selecting a function $p(x)$ whose minimum value at $x = 1$ is lower than that of the function $p(x) = (1 + x)^{-1}$, which resulted in the FALVQ 1 algorithm. The approximation $(1 + x)^{-1} \approx 1 - x$, which is valid for $|x| << 1$, leads to another admissible function $p(x) = 1 - x$ which attains the value of 0, its minimum value, at $x = 1$. The function $p(x) = 1 - x$ also relates to the membership functions that resulted in the FALVQ 2 algorithm, since $\exp(-x \approx 1 - x$ if $|x| << 1$. Since $p(x) = 1 - x$ satisfies the conditions presented above, a new membership function can be obtained as

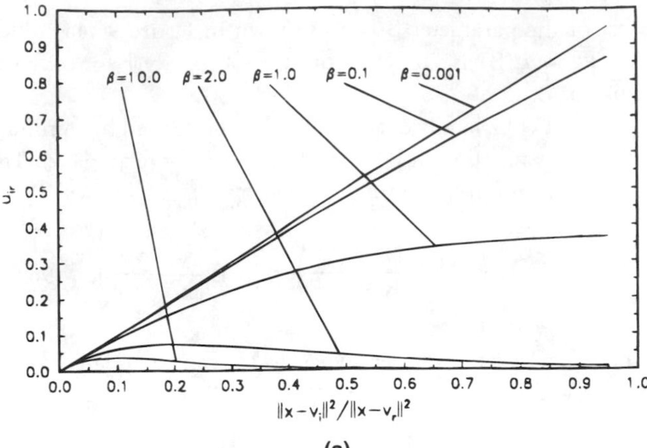

(a)

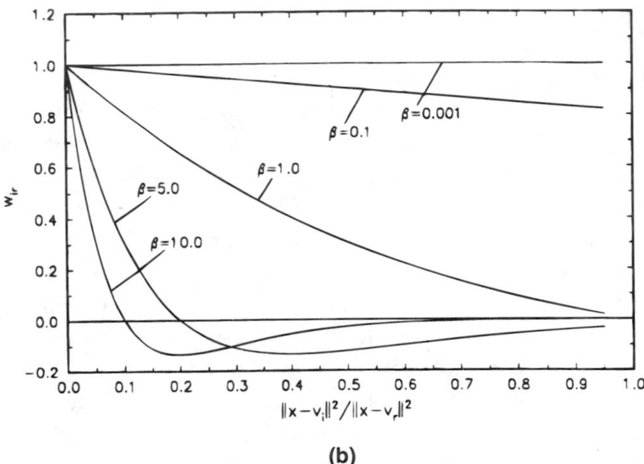

(b)

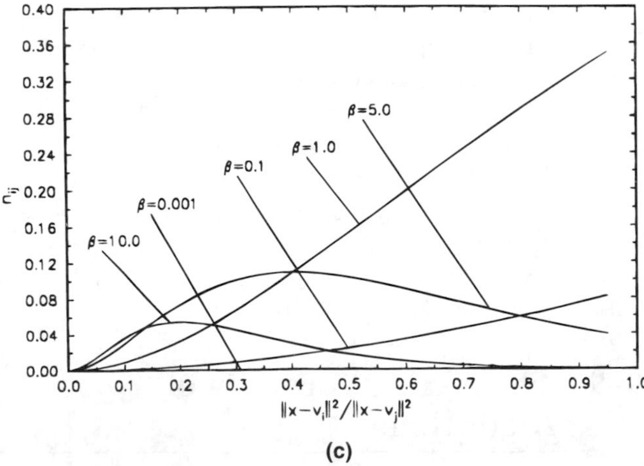

(c)

Figure 96.4 Membership function u_{ir} and interference functions w_{ir}, n_{ij} corresponding to the FALVQ 2 family of algorithms: (a) $u_{ir} = u_{ir}(\beta)$ as a function of $\|\mathbf{x} - \mathbf{v}_i\|^2/\|\mathbf{x} - \mathbf{v}_r\|^2$ for different values of β, and (c) $n_{ij} = n_{ij}(\beta)$ as a function of $\|\mathbf{x} - \mathbf{v}_i\|^2/\|\mathbf{x} - \mathbf{v}_j\|^2$ for different values of β. (*Source*: *IEEE Trans. Neural Networks* 7(5): 1197–1211. © IEEE. Used with permission.)

$$u_{ir} = \begin{cases} 1 & \text{if } r = i \\ \dfrac{\|\mathbf{x} - \mathbf{v}_i\|^2}{\|\mathbf{x} - \mathbf{v}_r\|^2}\left(1 - \dfrac{\|\mathbf{x} - \mathbf{v}_i\|^2}{\|\mathbf{x} - \mathbf{v}_r\|^2}\right) & \text{if } r \neq i. \end{cases} \tag{96.67}$$

Figure 96.2a plots u_{ir} as a function of the ratio $\|\mathbf{x} - \mathbf{v}_i\|^2/\|\mathbf{x} - \mathbf{v}_r\|^2$. For $r \neq i$, u_{ir} approaches 0 as $\|\mathbf{x} - \mathbf{v}_i\|^2/\|\mathbf{x} - \mathbf{v}_r\|^2$ approaches 0 and is an increasing function of $\|\mathbf{x} - \mathbf{v}_i\|^2/\|\mathbf{x} - \mathbf{v}_r\|^2$ in the internal (0, 1/2). When $\|\mathbf{x} - \mathbf{v}_i\|^2/\|\mathbf{x} - \mathbf{v}_r\|^2 = 1/2$, u_{ir} reaches the value of 1/4, its maximum value. Finally, u_{ir} is a decreasing function of $\|\mathbf{x} - \mathbf{v}_i\|^2/\|\mathbf{x} - \mathbf{v}_r\|^2$ in the interval (1/2, 1) and approaches 0 as $\|\mathbf{x} - \mathbf{v}_i\|^2/\|\mathbf{x} - \mathbf{v}_r\|^2$ approaches 1. It must be emphasized that in this case u_{ir} is not a monotone function of the ratio $\|\mathbf{x} - \mathbf{v}_i\|^2/\|\mathbf{x} - \mathbf{v}_r\|^2$ in the interval (0, 1). This is a major difference between the functions u_{ir} used in this formulation and those that led to the FALVQ 1 and FALVQ 2 algorithms. The direct implication of this formulation is that the contribution of the non-winning prototype $\mathbf{v}_r$ to the loss function in Eq. (96.39) is almost 0 when $\|\mathbf{x} - \mathbf{v}_r\|^2$ is sufficiently close to $\|\mathbf{x} - \mathbf{v}_i\|^2$. Thus, the loss function obtained by selecting a membership function of the form described by Eq. (96.67) favors strongly the winning prototype in situations where there is almost a tie.

The FALVQ 3 algorithm can be derived by minimizing the loss function in Equation (96.41) with $p(x) = 1 - x$. If $\mathbf{x}$ is the input vector, the winning prototype $\mathbf{v}_i$ can be updated by Eq. (96.42) with

$$w_{ir} = 1 - 2\frac{\|\mathbf{x} - \mathbf{v}_i\|^2}{\|\mathbf{x} - \mathbf{v}_r\|^2}, \tag{96.68}$$

while the non-winning prototypes $\mathbf{v}_j \neq \mathbf{v}_i$ can be updated by Eq. (96.44) with

$$n_{ij} = \left(\frac{\|\mathbf{x} - \mathbf{v}_i\|^2}{\|\mathbf{x} - \mathbf{v}_j\|^2}\right)^2. \tag{96.69}$$

According to Figure 96.2b, w_{ir} decreases linearly from 1 to 0 as $\|\mathbf{x} - \mathbf{v}_i\|^2/\|\mathbf{x} - \mathbf{v}_r\|^2$ increases from 0 to 1/2. For $\|\mathbf{x} - \mathbf{v}_i\|^2/\|\mathbf{x} - \mathbf{v}_r\|^2$ between 1/2 and 1, w_{ir} takes negative values. Clearly, the non-winning prototype $\mathbf{v}_r$ has an excitatory effect on the adaptation of the winning prototype $\mathbf{v}_i$ if $\|\mathbf{x} - \mathbf{v}_r\|^2 > 2\|\mathbf{x} - \mathbf{v}_i\|^2$. If $\|\mathbf{x} - \mathbf{v}_r\|^2 = 2\|\mathbf{x} - \mathbf{v}_i\|^2$, then $\mathbf{v}_r$ does not affect the adaptation of $\mathbf{v}_i$. Finally, the adaptation of the winning prototype $\mathbf{v}_i$ is inhibited by the existence of a non-winning prototype $\mathbf{v}_r$ for which $\|\mathbf{x} - \mathbf{v}_i\| \leq \|\mathbf{x} - \mathbf{v}_r\|^2 < 2\|\mathbf{x} - \mathbf{v}_i\|^2$. Since the interference coefficients w_{ir} can take negative values, the FALVQ 3 algorithm allows the non-winning prototypes to be more competitive. This is also illustrated in Figure 96.2c, which plots n_{ij} as a function of $\|\mathbf{x} - \mathbf{v}_i\|^2/\|\mathbf{x} - \mathbf{v}_j\|^2$. According to Figure 96.2c, n_{ij} increases from 0 to 1 as $\|\mathbf{x} - \mathbf{v}_i\|^2/\|\mathbf{x} - \mathbf{v}_j\|^2$ increases from 0 1. Compared with the FALVQ 1 and FALVQ 2 algorithms, the winning prototype $\mathbf{v}_i$ has an increasingly excitatory effect on the adaptation of $\mathbf{v}_j$, especially for values of $\|\mathbf{x} - \mathbf{v}_j\|^2$ sufficiently close to $\|\mathbf{x} - \mathbf{v}_i\|^2$.

The competition between the winning and the non-winning

prototypes can be regulated by slightly modifying the membership function (Equation 96.67) as

$$u_{ir} = \begin{cases} 1 & \text{if } r = i \\ \dfrac{\|\mathbf{x} - \mathbf{v}_i\|^2}{\|\mathbf{x} - \mathbf{v}_r\|^2}\left(1 - \gamma \dfrac{\|\mathbf{x} - \mathbf{v}_i\|^2}{\|\mathbf{x} - \mathbf{v}_r\|^2}\right) & \text{if } r \neq i, \end{cases} \qquad (96.70)$$

where $0 < \gamma < 1$. For $i \neq r$, u_{ir} attains its maximum value $1/(4\gamma)$ for $\|\mathbf{x} - \mathbf{v}_i\|^2/\|\mathbf{x} - \mathbf{v}_r\|^2 = 1/(2\gamma) \leq 1$ or, equivalently, for $\gamma \geq 1/2$. As the value of γ increases from 1/2 to 1, the value of $\|\mathbf{x} - \mathbf{v}_i\|^2/\|\mathbf{x} - \mathbf{v}_r\|^2$ for which u_{ir} attains its maximum value decreases. If $\gamma < 1/2$, u_{ir} is a monotonically increasing function of $\|\mathbf{x} - \mathbf{v}_i\|^2/\|\mathbf{x} - \mathbf{v}_r\|^2$ in the interval (0, 1). As γ approaches 0, u_{ir} approaches a linear function. The effect of γ on the shape of u_{ir} is illustrated in Figure 96.5a, which plots $u_{ir} = u_{ir}(\gamma)$ as a function of $\|\mathbf{x} - \mathbf{v}_i\|^2/\|\mathbf{x} - \mathbf{v}_r\|^2$ for different values of γ.

The FALVQ 3 algorithms corresponding to the membership functions in Eq. (96.70) can be derived by minimizing the loss function in Eq. (96.41) with $p(x) = 1 - \gamma x$. Since $p'(x) = -\gamma$, the winning prototype $\mathbf{v}_i$ can be updated by Eq. (96.42) with

$$w_{ii} = 1 - 2\gamma \frac{\|\mathbf{x} - \mathbf{v}_i\|^2}{\|\mathbf{x} - \mathbf{v}_r\|^2}, \qquad (96.71)$$

while the non-winning prototypes $\mathbf{v}_j \neq \mathbf{v}_i$ can be updated by Eq. (96.44) with

$$n_{ij} = \gamma \left(\frac{\|\mathbf{x} - \mathbf{v}_i\|^2}{\|\mathbf{x} - \mathbf{v}_j\|^2}\right)^2. \qquad (96.72)$$

Figure 96.5b plots $w_{ir} = w_{ir}(\gamma)$ as a function of $\|\mathbf{x} - \mathbf{v}_i\|^2/\|\mathbf{x} - \mathbf{v}_r\|^2$ for different values of γ. The interference function w_{ir} can take negative values if there is a zero-crossing between 0 and 1. Since $w_{ir} = 0$ for $\|\mathbf{x} - \mathbf{v}_i\|^2/\|\mathbf{x} - \mathbf{v}_r\|^2 = 1/(2\gamma)$, w_{ir} takes nonnegative values if $1/(2\gamma), \geq$ or $\gamma \leq 1/2$. The values of w_{ir} are negative if $1/(2\gamma), < 1$ or $\gamma > 1/2$. Since the non-winning prototypes can inhibit the attraction of the winning prototype $\mathbf{v}_i$ by the input $\mathbf{x}$ if $\gamma > 1/2$, increasing the value of γ above 1/2 intensifies the competition between the winning and non-winning prototypes. This argument is also supported by Figure 96.5c, which plots $n_{ij} = n_{ij}(\gamma)$ as a function of $\|\mathbf{x} - \mathbf{v}_i\|^2/\|\mathbf{x} - \mathbf{v}_j\|^2$ for different values of γ. Clearly, the maximum value of n_{ij} is γ and is reached when $\|\mathbf{x} - \mathbf{v}_j\|^2 = \|\mathbf{x} - \mathbf{v}_i\|^2$. Thus, increasing the value of γ results in a stronger attraction between the input vector $\mathbf{x}$ and the non-winning prototypes, especially those which are close to the winning prototype.

96.10 Competition Measures

The performance of FALVQ algorithms mainly depends on the properties of the corresponding membership functions. This section establishes a direct relationship between the properties of the membership functions used and the performance of the

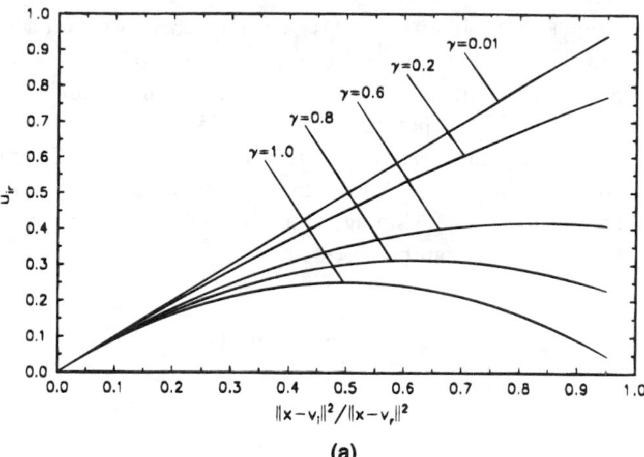

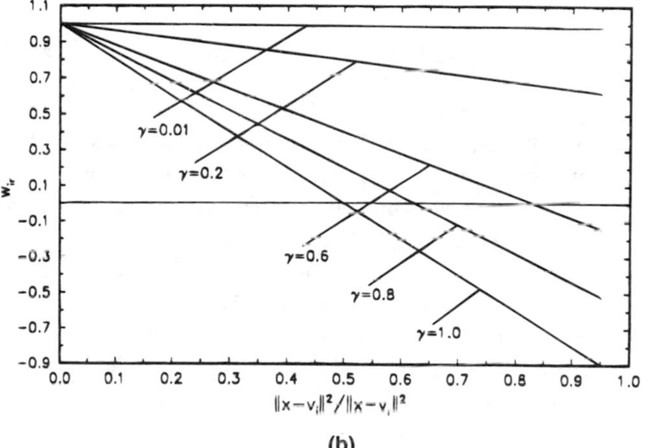

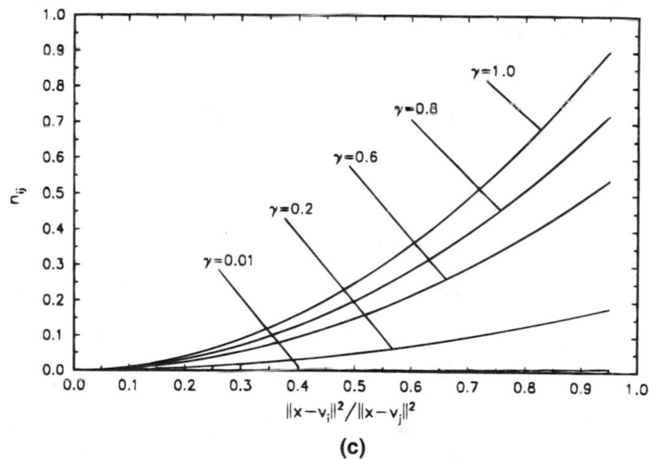

Figure 96.5 Membership function u_{ir} and interference functions w_{ir}, n_{ij} corresponding to the FALVQ 3 family of algorithms: (a) $u_{ir} = u_{ir}(\gamma)$ as a function of $\|\mathbf{x} - \mathbf{v}_i\|^2/\|\mathbf{x} - \mathbf{v}_r\|^2$ for different values of γ, (b) $w_{ir} = w_{ir}(\gamma)$ as a function of $\|\mathbf{x} - \mathbf{v}_i\|^2/\|\mathbf{x} - \mathbf{v}_r\|^2$ for different values of γ, and (c) $n_{ij} = n_{ij}(\gamma)$ as a function of $\|\mathbf{x} - \mathbf{v}_i\|^2/\|\mathbf{x} - \mathbf{v}_j\|^2$ for different values of γ. (*Source: IEEE Trans. Neural Networks* 7(5): 1197–1211. © IEEE. Used with permission.)

resulting FALVQ algorithms. This is accomplished by introducing two competition measures, which relate the form of the membership functions with the competition between the winning and non-winning prototypes during the learning process.

According to the formulation that resulted in the FALVQ families of algorithms, the non-winning prototypes are not allowed to compete with the winning prototype if $u(x) = x$ or $u(x) = 0 \ \forall \ x \in (0, 1)$. It can be observed that

$$\int_0^1 u(x)\, dx = \begin{cases} \frac{1}{2} & \text{if } u(x) = x \\ 0 & \text{if } u(x) = 0. \end{cases} \tag{96.73}$$

For any other membership function selected according to the proposed admissibility conditions,

$$0 < \int_0^1 u(x)\, dx < \frac{1}{2}. \tag{96.74}$$

Thus, the area $A_u = \int_0^1 u(x)\, dx$ can be used as a measure of the competition between the winning and non-winning prototypes. The development of genuinely competitive FALVQ algorithms requires that $A_u \in (0, 1/2)$. Moreover, the non-winning prototypes become increasingly competitive as A_u moves from 0 or 1/2 to the center of the interval (0, 1/2). This measure can be used to evaluate the membership functions that resulted in the FALVQ 1, FALVQ 2 and FALVQ 3 families of algorithms by investigating the effect of the parameters involved in their definition on the competition between the winning and non-winning prototypes during the learning process.

The FALVQ 1 family of algorithms is generated by membership functions of the form $u(x) = x(1 + \alpha x)^{-1}$. Thus,

$$A_u(\alpha) = \int_0^1 \frac{x\, dx}{1 + \alpha x} = \frac{1}{\alpha^2}(\alpha - \ln(1 + \alpha)). \tag{96.75}$$

If α approaches zero,

$$\lim_{\alpha \to 0} A_u(\alpha) = \lim_{\alpha \to 0} \frac{1}{2(1 + \alpha)} = \frac{1}{2}. \tag{96.76}$$

As α approaches infinity,

$$\lim_{\alpha \to \infty} A_u(\alpha) = 0. \tag{96.77}$$

This is a clear indication that the competition between the winning and non-winning prototypes during the learning process diminishes as α approaches 0 or infinity. Figure 96.6a plots the measure $A_u = A_u(\alpha)$ as a function of α. According to Figure 96.6, $A_u(\alpha)$ attains values very close to 1/2 for small values of α. In this case, the non-winning prototypes are not allowed to compete with the winning prototype to match the input vector. As α increases, the value of $A_u(\alpha)$ decreases very slowly to 0, the other extreme value of this competition measure which indicates

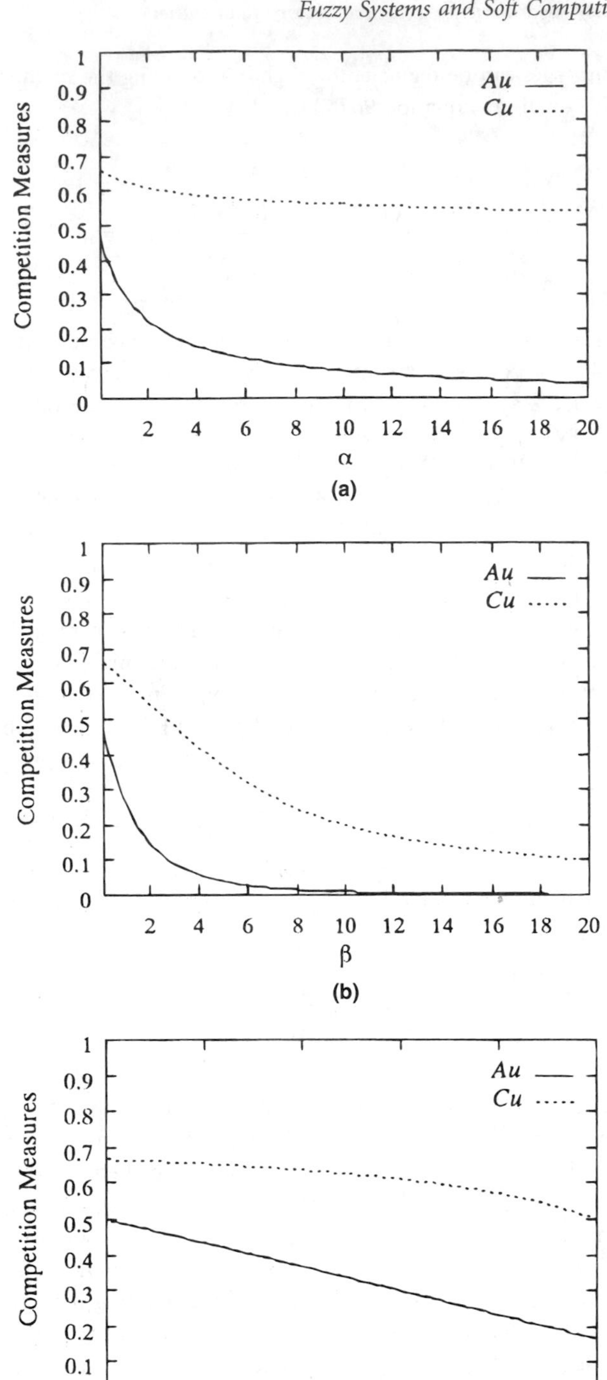

Figure 96.6 The competition measures (a) $A_u(\alpha)$ and $C_u(\alpha)$ corresponding to the FALVQ 1 family of algorithms as a function of α, (b) $A_u(\beta)$ and $C_u(\beta)$ corresponding to the FALVQ 2 family of algorithms as a function of β, and (c) $A_u(\gamma)$ and $C_u(\gamma)$ corresponding to the FALVQ 3 family of algorithms as a function of γ.

that there is no competition between the prototypes during the learning process.

The FALVQ 2 family of algorithms is generated by membership functions of the form $u(x) = x \exp(-\beta x)$. In this case,

$$A_u(\beta) = \int_0^1 x \exp(-\beta x)\, dx = \frac{1}{\beta^2}(1 - (1 + \beta)e^{-\beta}). \tag{96.78}$$

As β approaches zero,

$$\lim_{\beta \to 0} A_u(\beta) = \lim_{\beta \to 0} \frac{1}{2} e^{-\beta} = \frac{1}{2}. \tag{96.79}$$

As β approaches infinity,

$$\lim_{\beta \to \infty} A_u(\beta) = 0. \tag{96.80}$$

The FALVQ 2 algorithms become increasingly competitive as β moves away from the extremes 0 and infinity. Figure 96.6b plots $A_u = A_u(\beta)$ as a function of β. According to Figure 96.6b, $A_u(\beta)$ decreases quickly to values close to 0 as the value of β increases. Thus, the competition between the winning and non-winning prototypes diminishes quickly as the values of β exceeds a certain threshold.

The FALVQ 3 family of algorithms is generated by membership functions of the form $u(x) = x(1 - \gamma x)$. In this case,

$$A_u(\gamma) = \int_0^1 x(1 - \gamma x)\, dx = \frac{1}{6}(3 - 2\gamma). \tag{96.81}$$

Figure 96.6c plots $A_u = A_u(\gamma)$ as a function of $\gamma \in (0, 1]$. Clearly, $A_u(\gamma)$ attains its maximum value 1/2 for $\gamma = 0$, which corresponds to no competition, and decreases linearly from 1/2 to 1/6 as γ spans the interval (0, 1).

The area A_u alone is not sufficient to establish a relationship between the form of the membership function and the competition between the winning and non-winning prototypes during the learning process. This can be accomplished by considering the area A_u in conjunction with the 'centroid' or 'center of gravity' of the membership function $u(\cdot)$. Assuming that $A_u = \int_0^1 u(x)\, dx \neq 0$, the centroid of $u(x)$ over the interval $x \in (0, 1)$ is defined as

$$C_u = \frac{\int_0^1 xu(x)\, dx}{\int_0^1 u(x)\, dx}. \tag{96.82}$$

The centroid defined in Eq. (96.82) is a useful source of information regarding the shape of $u(\cdot)$ and, thus, the bias of the resulting FALVQ algorithm toward the winning prototype. In the extreme case where $u(x) = x$, $C_u = 2/3$. If $u(\cdot)$ is an admissible membership function, then $C_u < 2/3$. Since the selection of $u(x) = x$ implies that there is no competition between the winning and non-winning prototypes, the development of competitive FALVQ algorithms requires a membership function that corresponds to a centroid value lower than 2/3. Nevertheless, the non-winning prototypes become increasingly competitive as the centroid C_u decreases below 1/2. If the value of C_u is sufficiently close to 0, there is no competition between the winning and non-winning prototypes.

The centroid of the membership function that resulted in the FALVQ 1 family of algorithms can be obtained from Eq. (96.82) with $u(x) = x\,(1 + \alpha x)^{-1}$ as

$$C_u(\alpha) = \frac{1}{2}\left(\frac{\alpha}{\alpha - \ln(1 + \alpha)} - \frac{2}{\alpha}\right). \tag{96.83}$$

If α approaches zero, then $C_u(\alpha)$ approaches its maximum value, i.e.,

$$\lim_{\alpha \to 0} C_u(\alpha) = \frac{2}{3}. \tag{96.84}$$

This is consistent with the fact that if α approaches zero, then $u(x) = x(1 + \alpha x)^{-1}$ approaches x. It can also be verified that

$$\lim_{\alpha \to \infty} C_u(\alpha) = \frac{1}{2}. \tag{96.85}$$

Figure 96.6a plots $C_u = C_u(\alpha)$ as a function of α. As the value of α increases from zero to infinity, $C_u(\alpha)$ decreases asymptotically from its maximum value of 2/3 to 1/2, its lower bound. Thus, with the exemption of values of α sufficiently close to 0, the area $A_u = A_u(\alpha)$ is a more reliable competition measure the FALVQ 1 family of algorithms.

The centroid of the membership function that resulted in the FALVQ 2 family of algorithms can be obtained from Eq. (96.82) with $u(x) = x \exp(-\beta x)$ as

$$C_u(\beta) = \frac{1}{\beta}\frac{2 - (\beta^2 + 2\beta + 2)e^{-\beta}}{1 - (\beta + 1)e^{-\beta}}. \tag{96.86}$$

If β approaches zero, then $C_u(\beta)$ approaches its maximum value, i.e.,

$$\lim_{\beta \to 0} C_u(\beta) = \frac{2}{3}. \tag{96.87}$$

It can also be verified that

$$\lim_{\beta \to \infty} C_u(\beta) = 0. \tag{96.88}$$

Figure 96.6b plots $C_u = C_u(\beta)$ as a function of β. Clearly, the centroid $C_u(\beta)$ can take positive values significantly lower than 1/2 for large values of β. Such values of $C_u(\beta)$ indicate that there is practically no competition between the prototypes during the learning process. In conjunction with the area $A_u = A_u(\beta)$, $C_u(\beta)$ can be used to select the range of values of β that guarantee the competition between the winning and non-winning prototypes.

The centroid of the membership function that resulted in the FALVQ 3 family of algorithms can be obtained from Eq. (96.82) with $u(x) = x(1 - \gamma x)$ as

$$C_u(\gamma) = \frac{1}{2}\frac{4 - 3\gamma}{3 - 2\gamma}. \tag{96.89}$$

Figure 96.6c plots $C_u = C_u(\gamma)$ as a function of $\gamma \in (0, 1)$. Clearly, $C_u(\gamma)$ decreases from 2/3 to 1/2 as the value of γ increases from 0 to 1. Since $C_u(n, \gamma)$ takes values higher than 1/2 as γ spans the interval (0, 1), $C_u(\gamma)$ is no particularly informative competition measure in this case. Thus, the area $A_u = A_u(\gamma)$ can be used for selecting the values of γ that result in genuinely competitive FALVQ 3 algorithms.

96.11 Alternative FALVQ Algorithms

According to the formulation that led to the development of FALVQ algorithms, $u_{ir} = 1$ if $\mathbf{v}_r = \mathbf{v}_i$, where $\mathbf{v}_i$ is the winning prototype. Since $u_{ir} < 1\ \forall\ \mathbf{v}_r \neq \mathbf{v}_i$, the membership function u_{ir} favors the winning prototype. Although all prototypes are allowed to compete for each input vector $\mathbf{x}$, the bias inherent in the definition of u_{ir} guarantees that each input $\mathbf{x}$ has a more significant effect on the winning prototype.

The non-winning prototypes can be made more competitive by introducing a new set of membership functions u_{ir}, defined as (Karayiannis and Pai, 1994a,c)

$$u_{ir} = \begin{cases} 1 & \text{if } r = i \\ \left(1 + \dfrac{\|\mathbf{x} - \mathbf{v}_r\|^2}{D(\mathbf{x})}\right)^{-1} & \text{if } r \neq i, \end{cases} \tag{96.90}$$

where $D(\mathrm{x}) = D(\|\mathbf{x} - \mathbf{v}_j\|^2, \mathbf{v}_j \in \mathcal{V})$ is a differentiable function of $\|\mathbf{x} - \mathbf{v}_j\|^2$, $\mathbf{v}_j \in \mathcal{V}$ such that

$$D(\mathbf{x}) \geq D_{\min}(\mathbf{x}) = \min_{\mathbf{v}_j \in \mathcal{V}}\{\|\mathbf{x} - \mathbf{v}_j\|^2\} \tag{96.91}$$

Clearly, the function u_{ir} that resulted in the FALVQ 1 algorithm is a special case of Eq. (96.90) which corresponds to $D(\mathbf{x}) = D_{\min}(\mathbf{x})$. If $D(\mathbf{x}) > D_{\min}(\mathbf{x})$, then the membership functions defined in Eq. (96.90) are less biased toward the winning prototype. This can easily be verified by observing that, for $r \neq i$, Eq. (96.90) can also be written as

$$u_{ir} = \frac{1}{1 + \dfrac{\|\mathbf{x} - \mathbf{v}_r\|^2}{D_{\min}(\mathbf{x})}\dfrac{D_{\min}(\mathbf{x})}{D(\mathbf{x})}} = \frac{1}{1 + \dfrac{\|\mathbf{x} - \mathbf{v}_r\|^2}{\|\mathbf{x} - \mathbf{v}_i\|^2}\dfrac{D_{\min}(\mathbf{x})}{D(\mathbf{x})}}. \tag{96.92}$$

For $D_{\min}(\mathbf{x})/D(\mathbf{x}) = 1$, Eq. (96.92) gives the membership function in Eq. (96.38), which resulted in the FALVQ 1 algorithm. For a given ratio $\|\mathbf{x} - \mathbf{v}_r\|^2/\|\mathbf{x} - \mathbf{v}_i\|^2$, u_{ir} assigns a higher weight to the non-winning prototype $\mathbf{v}_r$ as the ratio $D_{\min}(\mathbf{x})/D(\mathbf{x})$ decreases. If $\|\mathbf{x} - \mathbf{v}_r\|^2/\|\mathbf{x} - \mathbf{v}_i\|^2$ is slightly higher than 1, u_{ir} approaches unity as the ratio $D_{\min}(\mathbf{x})/D(\mathbf{x})$ decreases. If $D_{\min}(\mathbf{x})/D(\mathbf{x})$ is sufficiently smaller than 1, each input affects almost equally the winning prototype and the non-winning prototypes which are sufficiently close to it. The increased level of competition between the winning and some non-winning prototypes might affect the selectivity of the learning vector quantization process, thus resulting in a highly uncertain partition.

The previous discussion indicates that the search for alternative fuzzy learning vector quantization algorithms involves the search for differentiable functions $D(\mathbf{x}) = D(\|\mathbf{x} - \mathbf{v}_j\|^2, \mathbf{v}_j \in \mathcal{V})$ such that $D(\mathbf{x}) > D_{\min}(\mathbf{x})$. Among the differentiable functions worth investigating are the *harmonic mean* $D_H(\mathbf{x}) = D_H(\|\mathbf{x} - \mathbf{v}_j\|^2, \mathbf{v}_j \in \mathcal{V})$, defined as

$$\frac{1}{D_H(\mathbf{x})} = \frac{1}{c}\sum_{j=1}^{c}\frac{1}{\|\mathbf{x} - \mathbf{v}_j\|^2}, \tag{96.93}$$

the *geometric mean* $D_G(\mathbf{x}) = D_G(\|\mathbf{x} - \mathbf{v}_j\|^2, \mathbf{v}_j \in \mathcal{V})$, defined as

$$D_G(\mathbf{x}) = \left(\prod_{j=1}^{c}\|\mathbf{x} - \mathbf{v}_j\|^2\right)^{1/c}, \tag{96.94}$$

and the *arithmetic mean* $D_A(\mathbf{x}) = D_A(\|\mathbf{x} - \mathbf{v}_j\|^2, \mathbf{v}_j \in \mathcal{V})$, defined as

$$D_A(\mathbf{x}) = \frac{1}{c}\sum_{j=1}^{c}\|\mathbf{x} - \mathbf{v}_j\|^2. \tag{96.95}$$

The investigation of the three averages defined above allows for a rough a priori prediction of some qualitative properties of the resulting learning vector quantization algorithms. In particular, the competition between winning and non-winning prototypes can qualitatively be predicted by combining the results of the above analysis with the well-known inequality

$$D_{\min}(\mathbf{x}) \leq D_H(\mathbf{x}) \leq D_G(\mathbf{x}) \leq D_A(\mathbf{x}). \tag{96.96}$$

A variety of alternative FALVQ 1 algorithms can be derived by minimizing the loss function $J_\mathbf{x}$ defined in Eq. (96.41), where u_{tr} is given in Eq. (96.90) and $D = D(\|\mathbf{x} - \mathbf{v}_j\|^2, \mathbf{v}_j \in \mathcal{V})$ is a differentiable function of $\|\mathbf{x} - \mathbf{v}_j\|^2$, $\mathbf{v}_j \in \mathcal{V}$. The gradient of $J_\mathbf{x}$ with respect to the winning prototype v_i is

$$\begin{aligned}\frac{\partial J_\mathbf{x}}{\partial \mathbf{v}_i} &= \frac{\partial}{\partial \mathbf{v}_i}\left(\|\mathbf{x} - \mathbf{v}_i\|^2 + \sum_{r \neq i}^{c} u_{ir}\|\mathbf{x} - \mathbf{v}_r\|^2\right) \\ &= -2(\mathbf{x} - \mathbf{v}_i) + \sum_{r \neq i}^{c}\|\mathbf{x} - \mathbf{v}_r\|^2\frac{\partial u_{ir}}{\partial \mathbf{v}_i}. \end{aligned} \tag{96.97}$$

From the definition of u_{ir} in Eq. (96.90), for $r \neq i$,

$$\frac{\partial u_{ir}}{\partial \mathbf{v}_i} = -u_{ir}^2 \|\mathbf{x} - \mathbf{v}_r\|^2 \frac{\partial}{\partial \mathbf{v}_i}\left(\frac{1}{D}\right) = u_{ir}^2 \frac{\|\mathbf{x} - \mathbf{v}_r\|^2}{D^2} \frac{\partial D}{\partial \mathbf{v}_i}. \tag{96.98}$$

Substituting Eq. (96.98) in (96.97) gives

$$\frac{\partial J_{\mathbf{x}}}{\partial \mathbf{v}_i} = -2(\mathbf{x} - \mathbf{v}_i) + \sum_{r \neq i}^{c} u_{ir}^2 \left(\frac{\|\mathbf{x} - \mathbf{v}_r\|^2}{D}\right)^2 \frac{\partial D}{\partial \mathbf{v}_i}. \tag{96.99}$$

Since $1 - u_{ir} = u_{ir}\,(\|\mathbf{x} - \mathbf{v}_r\|^2/D)$, Eq. (96.99) can also be written as

$$\frac{\partial J_{\mathbf{x}}}{\partial \mathbf{v}_i} = -2(\mathbf{x} - \mathbf{v}_i) + \sum_{r \neq i}^{c} (1 - u_{ir})^2 \frac{\partial D(\mathbf{x})}{\partial \mathbf{v}_i}. \tag{96.100}$$

The gradient of $J_{\mathbf{x}}$ with respect to the non-winning prototypes $\mathbf{v}_j$, $j \neq i$ is

$$\begin{aligned}\frac{\partial J_{\mathbf{x}}}{\partial \mathbf{v}_j} &= \frac{\partial}{\partial \mathbf{v}_j}\left(u_{ij}\,\|\mathbf{x} - \mathbf{v}_j\|^2 + \sum_{r \neq i,j}^{c} u_{ir}\,\|\mathbf{x} - \mathbf{v}_r\|^2\right)\\ &= -2(\mathbf{x} - \mathbf{v}_j)\,u_{ij} + \|\mathbf{x} - \mathbf{v}_j\|^2 \frac{\partial u_{ij}}{\partial \mathbf{v}_j}\\ &\quad + \sum_{r \neq i,j}^{c} \|\mathbf{x} - \mathbf{v}_r\|^2 \frac{\partial u_{ir}}{\partial \mathbf{v}_j}.\end{aligned} \tag{96.101}$$

According to the definition of u_{ij} in Eq. (96.90),

$$\begin{aligned}\frac{\partial u_{ij}}{\partial \mathbf{v}_j} &= u_{ir}^2 \frac{\partial}{\partial \mathbf{v}_j}\left(\frac{\|\mathbf{x} - \mathbf{v}_j\|^2}{D}\right)\\ &= 2(\mathbf{x} - \mathbf{v}_j)\frac{u_{ij}^2}{D} + u_{ij}^2 \frac{\|\mathbf{x} - \mathbf{v}_j\|^2}{D^2} \frac{\partial D}{\partial \mathbf{v}_j}.\end{aligned} \tag{96.102}$$

Similarly, for $r \neq j$,

$$\frac{\partial u_{ir}}{\partial \mathbf{v}_j} = -u_{ir}^2 \frac{\partial}{\partial \mathbf{v}_j}\left(\frac{\|\mathbf{x} - \mathbf{v}_r\|^2}{D}\right) = u_{ir}^2 \frac{\|\mathbf{x} - \mathbf{v}_r\|^2}{D^2} \frac{\partial D}{\partial \mathbf{v}_j}. \tag{96.103}$$

Substituting Eqs. (96.102) and (96.103) in Eq. (96.101) gives

$$\begin{aligned}\frac{\partial J_{\mathbf{x}}}{\partial \mathbf{v}_j} &= -2(\mathbf{x} - \mathbf{v}_j)\,u_{ij} + 2(\mathbf{x} - \mathbf{v}_j)\,u_{ij}^2 \frac{\|\mathbf{x} - \mathbf{v}_j\|^2}{D}\\ &\quad + \sum_{r \neq i}^{c} u_{ir}^2 \left(\frac{\|\mathbf{x} - \mathbf{v}_r\|^2}{D}\right)^2 \frac{\partial D}{\partial \mathbf{v}_j}.\end{aligned} \tag{96.104}$$

Since $1 - u_{ij}\,(\|\mathbf{x} - \mathbf{v}_j\|^2/D) = u_{ij}$ and $1 - u_{ir} = u_{ir}\,(\|\mathbf{x} - \mathbf{v}_r\|^2/D)$, Eq. (96.104) can also be written as

$$\frac{\partial J_{\mathbf{x}}}{\partial \mathbf{v}_j} = -2(\mathbf{x} - \mathbf{v}_j)\,u_{ij}^2 + \sum_{r \neq i}^{c} (1 - u_{ir})^2 \frac{\partial D(\mathbf{x})}{\partial \mathbf{v}_j}. \tag{96.105}$$

The FALVQ 1 algorithm can be obtained as the special case of the above formulation which corresponds to

$$D(\mathbf{x}) = D_{\min}(\mathbf{x}) = \|\mathbf{x} - \mathbf{v}_i\|^2 = \min_{\mathbf{v}_j \in \mathcal{V}}\{\|\mathbf{x} - \mathbf{v}_j\|^2\} \tag{96.106}$$

The gradient of $D_{\min}(\mathbf{x})$ with respect to the winning prototype $\mathbf{v}_i$ is

$$\frac{\partial D_{\min}(\mathbf{x})}{\partial \mathbf{v}_i} = \frac{\partial}{\partial \mathbf{v}_i}(\|\mathbf{x} - \mathbf{v}_i\|^2) = -2(\mathbf{x} - \mathbf{v}_i). \tag{96.107}$$

The gradient of $D_{\min}(\mathbf{x})$ with respect to each non-winning prototype $\mathbf{v}_j \neq \mathbf{v}_i$ is

$$\frac{\partial D_{\min}(\mathbf{x})}{\partial \mathbf{v}_j} = \frac{\partial}{\partial \mathbf{v}_j}(\|\mathbf{x} - \mathbf{v}_i\|^2) = 0. \tag{96.108}$$

The update equations for the winning and non-winning prototypes associated with the FALVQ 1 algorithm can easily be obtained by substituting Eqs. (96.107) and (96.108) in Eq. (96.100) and Eq. (96.105), respectively.

Harmonic FALVQ 1

Consider that learning vector quantization is based on the minimization of the loss function in Eq. (96.41), where u_{ir} is given by Eq. (96.90) with $D(\mathbf{x}) = D_H(\mathbf{x})$, that is,

$$u_{ir} = \begin{cases} 1 & \text{if } r = i \\ \left(1 + \dfrac{1}{c}\displaystyle\sum_{j=1}^{c} \dfrac{\|\mathbf{x} - \mathbf{v}_r\|^2}{\|\mathbf{x} - \mathbf{v}_j\|^2}\right)^{-1} & \text{if } r \neq i. \end{cases} \tag{96.109}$$

The Harmonic FALVQ 1 algorithm can be obtained under the assumption that $D = D(\mathbf{x})$ is the harmonic mean, defined in Eq. (96.93). The gradient of $D_H(\mathbf{x})$ with respect to any prototype $\mathbf{v}_i$ is

$$\begin{aligned}\frac{\partial D_H(\mathbf{x})}{\partial \mathbf{v}_i} &= \frac{\partial}{\partial \mathbf{v}_i}\left(\frac{c}{\displaystyle\sum_{l=1}^{c} \frac{1}{\|\mathbf{x} - \mathbf{v}_l\|^2}}\right)\\ &= -\frac{2}{c}(\mathbf{x} - \mathbf{v}_i)\left(\frac{D_H(\mathbf{x})}{\|\mathbf{x} - \mathbf{v}_i\|^2}\right)^2\end{aligned} \tag{96.110}$$

The update equation for the winning prototype $\mathbf{v}_i$ can be obtained by substituting Eq. (96.110) in Eq. (96.100) as

$$\Delta \mathbf{v}_i = \eta\,(\mathbf{x} - \mathbf{v}_i)\left(1 + \frac{1}{c}\sum_{r \neq i}^{c} u_{ir}^2 \left(\frac{\|\mathbf{x} - \mathbf{v}_r\|^2}{\|\mathbf{x} - \mathbf{v}_i\|^2}\right)^2\right) \tag{96.111}$$

The update equation for the non-winning prototypes $\mathbf{v}_j \neq \mathbf{v}_i$ can be obtained by substituting Eq. (96.110) in Eq. (96.105) as

$$\Delta \mathbf{v}_j = \eta\,(\mathbf{x} - \mathbf{v}_j)\left(u_{ij}^2 + \frac{1}{c}\sum_{r \neq i}^{c} u_{ir}^2 \left(\frac{\|\mathbf{x} - \mathbf{v}_r\|^2}{\|\mathbf{x} - \mathbf{v}_j\|^2}\right)^2\right) \tag{96.112}$$

The adaptation of the winning prototype $\mathbf{v}_i$ can be investigated by studying the term $1/c\Sigma_{r\neq i}^{c}\, u_{ir}^2(\|\mathbf{x} - \mathbf{v}_r\|^2/\|\mathbf{x} - \mathbf{v}_i\|^2)^2$, which represents the effect of the non-winning prototypes. Assume that $\mathbf{v}_r$ is a non-winning prototype such that $\|\mathbf{x} - \mathbf{v}_r\|^2 \gg \|\mathbf{x} - \mathbf{v}_i\|^2$. According to the definition of u_{ir} in the case where $r \neq i$,

$$u_{ir}^2\left(\frac{\|\mathbf{x} - \mathbf{v}_r\|^2}{\|\mathbf{x} - \mathbf{v}_i\|^2}\right)^2 = \left(\frac{\|\mathbf{x} - \mathbf{v}_r\|^2}{D_H(\mathbf{x}) + \|\mathbf{x} - \mathbf{v}_r\|^2}\,\frac{D_H(\mathbf{x})}{\|\mathbf{x} - \mathbf{v}_i\|^2}\right)^2. \tag{96.113}$$

Since it can reasonably be assumed that $\|\mathbf{x} - \mathbf{v}_r\|^2 \gg D_H(\mathbf{x})$, then for $r \neq i$,

$$u_{ir}^2\left(\frac{\|\mathbf{x} - \mathbf{v}_r\|^2}{\|\mathbf{x} - \mathbf{v}_i\|^2}\right)^2 \approx \left(\frac{D_H(\mathbf{x})}{\|\mathbf{x} - \mathbf{v}_i\|^2}\right)^2. \tag{96.114}$$

Since $D_H(\mathbf{x}) > \|\mathbf{x} - \mathbf{v}_i\|^2$, then $u_{ir}^2\,(\|\mathbf{x} - \mathbf{v}_r\|^2/\|\mathbf{x} - \mathbf{v}_i\|^2)^2 > 1$. If the non-winning prototype $\mathbf{v}_r$ is close to the winning prototype $\mathbf{v}_i$, i.e., $\|\mathbf{x} - \mathbf{v}_r\|^2/\|\mathbf{x} - \mathbf{v}_i\|^2 \approx 1$, then for $r \neq i$,

$$\begin{aligned} u_{ir}^2\left(\frac{\|\mathbf{x} - \mathbf{v}_r\|^2}{\|\mathbf{x} - \mathbf{v}_i\|^2}\right)^2 &\approx \left(\frac{1}{1 + \dfrac{\|\mathbf{x} - \mathbf{v}_r\|^2}{\|\mathbf{x} - \mathbf{v}_i\|^2}\,\dfrac{\|\mathbf{x} - \mathbf{v}_i\|^2}{D_H(\mathbf{x})}}\right)^2 \\ &\approx \left(\frac{1}{1 + \dfrac{\|\mathbf{x} - \mathbf{v}_i\|^2}{D_H(\mathbf{x})}}\right)^2. \end{aligned} \tag{96.115}$$

Since $0 < \|\mathbf{x} - \mathbf{v}_i\|^2/D_H(\mathbf{x}) < 1$, then $u_{ir}^2\,(\|\mathbf{x} - \mathbf{v}_r\|^2/\|\mathbf{x} - \mathbf{v}_i\|^2)^2 < 1$. In summary, if $\|\mathbf{x} - \mathbf{v}_r\|^2 >> \|\mathbf{x} - \mathbf{v}_i\|^2$ the term $u_{ir}^2\,(\|\mathbf{x} - \mathbf{v}_r\|^2/\|\mathbf{x} - \mathbf{v}_i\|^2)^2$ tends to increase the attraction of the winning prototype by the input $\mathbf{x}$. Conversely, the effect of the input $\mathbf{x}$ on the winning prototype $\mathbf{v}_i$ is inhibited by the term $u_{ir}^2\,(\|\mathbf{x} - \mathbf{v}_r\|^2/\|\mathbf{x} - \mathbf{v}_i\|^2)^2$ if $\|\mathbf{x} - \mathbf{v}_r\|^2 \approx \|\mathbf{x} - \mathbf{v}_i\|^2$.

The comparison between the update equations for the winning and non-winning prototypes is based on the observation that since $\|\mathbf{x} - \mathbf{v}_j\|^2 > \|\mathbf{x} - \mathbf{v}_i\|^2$ and $u_{ij}^2 < 1\ \forall_j \neq i$, then

$$\begin{aligned} u_{ij}^2 + \frac{1}{c}\sum_{r\neq i}^{c} u_{ir}^2\left(\frac{\|\mathbf{x} - \mathbf{v}_r\|^2}{\|\mathbf{x} - \mathbf{v}_j\|^2}\right)^2 &< 1 \\ &+ \frac{1}{c}\sum_{r\neq i}^{c} u_{ir}^2\left(\frac{\|\mathbf{x} - \mathbf{v}_r\|^2}{\|\mathbf{x} - \mathbf{v}_i\|^2}\right)^2. \end{aligned} \tag{96.116}$$

Clearly, the input vector $\mathbf{x}$ has a more significant effect on the winning prototype $\mathbf{v}_i$. Also, Eq. (96.116) indicates that the attraction of each non-winning prototype by $\mathbf{x}$ depends on the value of the ratio $\|\mathbf{x} - \mathbf{v}_j\|^2/\|\mathbf{x} - \mathbf{v}_i\|^2$ relative to 1.

Geometric FALVQ 1

Consider that learning vector quantization is based on the minimization of the loss function in Eq. (96.41), where u_{ir} is given by Eq. (96.90) with $D(\mathbf{x}) = D_G(\mathbf{x})$, that is,

$$u_{ir} = \begin{cases} 1 & \text{if } r = i \\ \left(1 + \left(\prod_{j=1}^{c}\dfrac{\|\mathbf{x} - \mathbf{v}_r\|^2}{\|\mathbf{x} - \mathbf{v}_j\|^2}\right)^{1/c}\right)^{-1} & \text{if } r \neq i \end{cases} \tag{96.117}$$

The Geometric FALVQ 1 algorithm can be obtained under the assumption that $D = D(\mathbf{x})$ is the geometric mean, defined in Eq. (96.94). The gradient of $D_G(\mathbf{x})$ with respect to any prototype $\mathbf{v}_i$ is

$$\frac{\partial D_G(\mathbf{x})}{\partial \mathbf{v}_i} = \frac{\partial}{\partial \mathbf{v}_i}\left(\prod_{l=1}^{c}\|\mathbf{x} - \mathbf{v}_l\|^2\right)^{1/c} = -\frac{2}{c}\,(\mathbf{x} - \mathbf{v}_i)\,\frac{D_G(\mathbf{x})}{\|\mathbf{x} - \mathbf{v}_i\|^2}. \tag{96.118}$$

The update equation for the winning prototype $\mathbf{v}_i$ can be obtained by substituting Eq. (96.118) in Eq. (96.100) as

$$\Delta \mathbf{v}_i = \eta\,(\mathbf{x} - \mathbf{v}_i)\left(1 + \frac{1}{c}\sum_{r\neq i}^{c} u_{ir}(1 - u_{ir})\,\frac{\|\mathbf{x} - \mathbf{v}_r\|^2}{\|\mathbf{x} - \mathbf{v}_i\|^2}\right). \tag{96.119}$$

The update equation for the non-winning prototypes $\mathbf{v}_j \neq \mathbf{v}_i$ can be obtained by substituting Eq. (96.118) in Eq. (96.105) as

$$\Delta \mathbf{v}_j = \eta\,(\mathbf{x} - \mathbf{v}_j)\left(u_{ij}^2 + \frac{1}{c}\sum_{r\neq i}^{c} u_{ir}(1 - u_{ir})\,\frac{\|\mathbf{x} - \mathbf{v}_r\|^2}{\|\mathbf{x} - \mathbf{v}_j\|^2}\right). \tag{96.120}$$

The update equations corresponding to the Harmonic FALVQ 1 and Geometric FALVQ 1 algorithms are very similar. The only significant difference is that the term $u_{ir}^2(\|\mathbf{x} - \mathbf{v}_r\|^2/\|\mathbf{x} - \mathbf{v}_i\|^2)^2$ appearing in Eq. (96.111) and (96.112) is replaced in Eqs. (96.119) and (96.120) by $u_{ir}(1 - u_{ir})(\|\mathbf{x} - \mathbf{v}_r\|^2/\|\mathbf{x} - \mathbf{v}_i\|^2)$. Nevertheless, the comparison between the adaptation of the winning and non-winning prototypes presented above for the Harmonic FALVQ 1 algorithm is also valid when $D(\mathbf{x}) = D_G(\mathbf{x})$.

Consider here the update equations of the Harmonic FALVQ 1 and Geometric FALVQ 1 algorithms in the case where $\|\mathbf{x} - \mathbf{v}_r\|^2 \approx \|\mathbf{x} - \mathbf{v}_i\|^2$. Under this assumption

$$\left(\frac{\|\mathbf{x} - \mathbf{v}_r\|^2}{\|\mathbf{x} - \mathbf{v}_i\|^2}\right)^2 \approx \frac{\|\mathbf{x} - \mathbf{v}_r\|^2}{\|\mathbf{x} - \mathbf{v}_i\|^2}. \tag{96.121}$$

Since $D_G(\mathbf{x}) > D_H(\mathbf{x})$, the weight u_{ir} which corresponds to $D(\mathbf{x}) = D_G(\mathbf{x})$ is higher than that corresponding to $D(\mathbf{x}) = D_H(\mathbf{x})$. In addition, $u_{ir}(1 - u_{ir})$ is an increasing function of u_{ir} if $0 < u_{ir} < 1/2$, attains its maximum value at $u_{ir} = 1/2$, and is a decreasing function of u_{ir} if $u_{ir} > 1/2$. Since $u_{ir} > 1/2$ when $\|\mathbf{x} - \mathbf{v}_r\|^2$ is sufficiently close to $\|\mathbf{x} - \mathbf{v}_i\|^2$, $u_{ir}(1 - u_{ir})$ decreases as the non-winning prototype $\mathbf{v}_r$ approaches the winning prototype $\mathbf{v}_i$. In contrast, u_{ir}^2 is a monotonically increasing function function of u_{ir}. Thus, the non-winning prototypes close to the winning prototype $\mathbf{v}_i$ result in a stronger inhibition of its adaptation when the Geometric FALVQ 1 algorithm is used.

Compared with the Harmonic FALVQ 1 algorithm, the Geometric FALVQ1 algorithm results in a stronger competition between the winning prototype and the non-winning prototypes which are sufficiently close to it.

Arithmetic FALVQ 1

Consider that learning vector quantization is based on the minimization of the loss function in Eq. (96.41), where u_{ir} is given by Eq. (96.90), with $D(\mathbf{x}) = D_A(\mathbf{x})$, that is,

$$u_{ir} = \begin{cases} 1 & \text{if } r = i \\ \left(1 + \left(\frac{1}{c}\sum_{j=1}^{c} \frac{\|\mathbf{x} - \mathbf{v}_j\|^2}{\|\mathbf{x} - \mathbf{v}_r\|^2}\right)^{-1}\right)^{-1} & \text{if } r \neq i. \end{cases} \tag{96.122}$$

The Arithmetic FALVQ 1 algorithm can be obtained under the assumption that $D = D(\mathbf{x})$ is the arithmetic mean, defined in (Eq. (96.95). The gradient $D_A(\mathbf{x})$ with respect to any prototype $\mathbf{v}_i$ is

$$\frac{\partial D_A(\mathbf{x})}{\partial \mathbf{v}_i} = \frac{\partial}{\partial \mathbf{v}_i}\left(\frac{1}{c}\sum_{l=1}^{c} \|\mathbf{x} - \mathbf{v}_l\|^2\right) = -\frac{2}{c}(\mathbf{x} - \mathbf{v}_i). \tag{96.123}$$

The update equation for the winning prototype $\mathbf{v}_i$ can be obtained by substituting Eq. (96.123) in Eq. (96.100) as

$$\Delta\mathbf{v}_i = \eta\,(\mathbf{x} - \mathbf{v}_i)\left(1 + \frac{1}{c}\sum_{r \neq i}^{c}(1 - u_{ir})^2\right). \tag{96.124}$$

The update equation for the non-winning prototypes $\mathbf{v}_j \neq \mathbf{v}_i$ can be obtained by substituting Eq. (96.123) in Eq. (96.105) as

$$\Delta\mathbf{v}_j = \eta\,(\mathbf{x} - \mathbf{v}_j)\left(u_{ij}^2 + \frac{1}{c}\sum_{r \neq i}^{c}(1 - u_{ir})^2\right). \tag{96.125}$$

The difference between the adaptation of the winning and non-winning prototypes depends on the relative size of the terms $1 + (1/c)\,\Sigma_{r\neq i}^{c}\,(1 - u_{ir})^2$ and $u_{ij}^2 + (1/c)\,\Sigma_{r\neq i}^{c}\,(1 - u_{ir})^2$, which appear in Eq. (96.124) and (96.125), respectively. Since $u_{ij} < 1$ $\forall\, j \neq i$,

$$u_{ij}^2 + \frac{1}{c}\sum_{r \neq i}^{c}(1 - u_{ir})^2 < 1 + \frac{1}{c}\sum_{r \neq i}^{c}(1 - u_{ir})^2. \tag{96.126}$$

This latter inequality indicates that each input $\mathbf{x}$ has a stronger effect on the winning prototype. However, the difference between u_{ij}^2 and 1 is not significant, especially when the non-winning prototype $\mathbf{v}_j$ is close to the winning prototype $\mathbf{v}_i$. As a result, the Arithmetic FALVQ 1 algorithm is not capable of discriminating between prototypes which are similar. This disadvantage of the algorithm is a consequence of the fact that the outliers in the set $\|\mathbf{x} - \mathbf{v}_r\|^2$, $\mathbf{v}_r \in \mathcal{V}$ have a significant effect on the arithmetic mean $D_A(\mathbf{x})$. In contrast, the harmonic mean and, to a lesser degree, the geometric mean are not significantly affected by the outliers in the set $\|\mathbf{x} - \mathbf{v}_r\|^2$, $v_r \in \mathcal{V}$.

96.12 Experimental Results

The FALVQ algorithms presented in this section were tested using Anderson's IRIS data set, which has extensively been used for evaluating the performance of pattern clustering algorithms (Anderson, 1939). This data set contains 150 feature vectors of dimension 4, which belong to 3 classes representing different IRIS subspecies. Each class contains 50 feature vectors. One of the three classes is well separated from the other two, which are not easily separable due to the existence of similar vectors. The performance of the algorithms tested on this data set is usually evaluated by counting the number of clustering errors, i.e., the number of feature vectors that are assigned to a wrong cluster by the algorithms. Unsupervised clustering of the IRIS data typically results in 12–17 clustering errors (Pal et al.,1993).

Table 96.1 shows the number of clustering errors recorded when the IRIS data were clustered by the FALVQ 1 algorithms with different values of α. The total number of iterations was $N = 100$ and the initial value of the learning rate η_o varied from 0.1 to 0.9. Table 96.1 also shows the values of the competition measures $A_u(\alpha)$ and $C_u(\alpha)$ that correspond to the FALVQ 1 algorithms tested was poor for very small values of α and improved as the value of α increased above 0.1. In fact, the minimum number of clustering errors was achieved by the algorithms when the value of α was between 0.1 and 10.0. This experimental outcome is consistent with the analysis presented in this section, which indicated that the non-winning prototypes become more competitive as the value of α increases above zero. According to Table 96.1, the performance of the algorithms degraded for values of α higher than 10.0. This is an experimental verification that the non-winning prototypes are not allowed to compete with the winning prototype as the value of α increases above a certain threshold.

Table 96.2 shows the number of clustering errors recorded when the IRIS data were clustered by the generalized FALVQ 2 algorithms with different values of the parameter β. The initial

Table 96.1 Number of Clustering Errors Recorded When the IRIS Data were Clustered by the FALVQ 1 Family of Algorithms with Different values of α.

α	$A_u(\alpha)$	$C_u(\alpha)$	$\eta_0 = 0.1$	$\eta_0 = 0.3$	$\eta_0 = 0.5$	$\eta_0 = 0.7$	$\eta_0 = 0.9$
0.001	0.499	0.666	62	63	16	100	100
0.01	0.497	0.666	62	63	16	100	100
0.1	0.469	0.661	63	16	16	16	16
0.5	0.378	0.644	64	16	16	16	16
1.0	0.306	0.629	64	16	16	16	16
2.0	0.225	0.609	65	16	16	16	16
5.0	0.128	0.597	65	16	16	16	16
10.0	0.076	0.558	65	16	16	16	16
100.0	0.009	0.514	61	62	17	17	17

Table 96.2 Number of Clustering Errors Recorded When the IRIS Data were Clustered by the FALVQ2 Family of Algorithms with Different Values of β.

β	$A_u(\beta)$	$C_u(\beta)$	$\eta_0 = 0.1$	$\eta_0 = 0.3$	$\eta_0 = 0.5$	$\eta_0 = 0.7$	$\eta_0 = 0.9$
0.001	0.499	0.666	62	63	16	100	100
0.01	0.496	0.666	62	63	16	100	100
0.1	0.468	0.661	63	64	16	16	100
0.5	0.361	0.638	65	16	16	16	16
1.0	0.264	0.608	16	16	16	16	16
2.0	0.148	0.544	17	16	16	16	16
5.0	0.038	0.365	51	43	26	26	53
10.0	0.009	0.199	53	45	61	65	50
100.0	0.000	0.020	75	72	76	75	71

Table 96.3 Number of Clustering Errors Recorded When the IRIS Data were Clustered by the FALVQ2 Family of Algorithms with Different Values of γ.

γ	$A_u(\gamma)$	$C_u(\gamma)$	$\eta_0 = 0.1$	$\eta_0 = 0.3$	$\eta_0 = 0.5$	$\eta_0 = 0.7$	$\eta_0 = 0.9$
0.001	0.499	0.666	62	63	16	100	100
0.01	0.497	0.666	62	63	16	100	100
0.1	0.467	0.661	63	16	16	16	100
0.2	0.433	0.654	63	16	16	16	16
0.4	0.367	0.636	65	16	16	16	16
0.6	0.300	0.611	16	16	16	16	16
0.8	0.233	0.571	16	16	16	16	16
1.0	0.167	0.500	72	16	16	16	16

value of the learning rate η_0 used in these experiments varied from 0.1 to 0.9. The total number of iterations was $N = 100$. Table 96.2 also shows the values of the competition measures $A_u(\beta)$ and $C_u(\beta)$ that correspond to the FALVQ 2 algorithms tested in these experiments. According to Table 96.2, the number of clustering errors decreased as the value of β increased from 0.001 to 0.1. In fact, the algorithms tested in this experiment achieved satisfactory performance for values of β in the interval [0.5, 2.0]. According to the analysis presented in this section, the non-winning prototypes compete stronger with the winning prototype to match each input vector as the value of β increases above 1. Thus, the algorithms corresponding to such values of β are expected to be more successful in the clustering of the IRIS data set, which consists of two clusters that are not well-separated. Nevertheless, the performance of the algorithms degraded for values of β higher than 5.0 due to the lack of competition between the winning and non-winning prototypes.

Table 96.3 shows the number of clustering errors recorded when the IRIS data were clustered by the generalized FALVQ 3 algorithms with values of γ between 0 and 1. The total number of iterations was $N = 100$ while the initial value of the learning rate η_0 varied from 0.1 to 0.9. Table 96.3 also shows the values of the competition measures $A_u(\gamma)$ and $C_u(\gamma)$ that correspond to the FALVQ 3 algorithms tested in these experiments. According to Table 96.3, the number of clustering errors decreased consistently as the value of γ increased above 0.1. The performance of the algorithms tested was not affected by the initial value of the learning rate for values of γ greater than 1/2. The performance of the algorithms tested is consistent with the analysis presented in this section, which indicated that the use of values of γ from the interval [1/2, 1] intensifies the competition between the winning and non-winning prototypes.

96.13 Discussion and Concluding Remarks

This section presented a review of some competitive LVQ schemes, which included the GLVQ and GLVQ-F algorithms. In addition to these sequential LVQ algorithms, this section presented a family of batch FLVQ algorithms and established their close relationship with fuzzy *c*-means clustering algorithms. The formulation that resulted in the sequential GLVQ-F algorithms was modified to produce a batch LVQ algorithm in the special case where $m = 2$. Since this formulation resulted in the corresponding FLVQ algorithm, it can be seen as the first step toward a formal derivation of the family of FLVQ algorithms that were proposed on the basis of intuitive arguments. This section also presented a new formulation of learning vector quantization that provided the basis for the development of a broad variety of FALVQ algorithms. The FALVQ algorithms presented in this section allow the non-winning prototypes to compete with the winning prototype for each input during the learning process. This competition can be controlled by the user by simply modifying a single parameter. This section introduced two quantitative measures that establish a relationship between the formulation that led to the FALVQ algorithms and the competition between the prototypes during the learning process. These competition measures can be used for selecting the parameters that lead to genuinely competitive FALVQ algorithms. Various algorithms from the FALVQ 1, FALVQ 2 and FALVQ 3 families were experimentally tested on the IRIS data set. The application of the algorithms on the IRIS data set evaluated the effect of there parameters on the performance of the algorithms. The validity of the proposed competition measures was tested in the limit where the FALVQ algorithms allow only the winning prototype to the updated in order to match the input vector. The low computational requirements of the FALVQ algorithms make them effective tools in applications involving large number of feature vectors of high dimensionality and LVQ of large size, such as image compression based on vector quantization (Karayiannis and Pai, 1995b; Gersho and Gray, 1992; Nasrabadi and King, 1988; Gray, 1989). The application of the proposed algorithms in codebook design for image compression exhibited their ability to design high quality vector quantizers for non-trivial tasks (Karayiannis and Pai, 1994a–d).

References

Anderson, E. 1939. The IRISes of the Gaspe Peninsula, *Bulletin of the American IRIS Society,* 59:2–5.

Bezdek, J. C. 1981. *Pattern Recognition with Fuzzy Objective Function Algorithms,* Plenum, New York, NY.

Bezdek, J. C. 1992. Integration and generalization of LVQ and c-means clustering, *Proc. SPIE, Intelligent Robots and Computer*

Vision XI: Biological, Neural Net, and 3-D Methods, 1826:280–299.

Bezdek, J. C., Ehrlich, R., and Full, W. 1984. FCM: The fuzzy c-means clustering algorithm, *Computers and Geosciences,* 10:191–203.

Bezdek, J. C., Pal, N. R., Hathaway, R. J., and Karayiannis, N. B. 1995. Some new competitive learning schemes, *SPIE Proc. vol. 2492: Applications and Science of Artificial Neural Networks,* 2487:538–549, April 17–21, Orlando, FL.

Bezdek, J. C., and Pal, S. K. eds., 1992. *Fuzzy Models for Pattern Recognition: Models That Search for Structures in Data,* IEEE Press, New York, NY.

Duda, R. O., and Hart, P. E. 1973. *Pattern Classification and Scene Analysis,* John Wiley & Sons, New York, NY.

Dunn, J. C. 1973. A fuzzy relative of the ISODATA process and its use in detecting compact well-separated clusters, *J. Cybernetics,* 3:32–57.

Gersho, A. and Gray, R. M. 1992. *Vector Quantization and Signal Compression,* Kluwer Academic Publishers, Boston, MA.

Gray, R. M. 1984. Vector quantization, *IEEE ASSP Magazine,* 1:4–29.

Hertz, J., Krogh, A., and Palmer, R. G. 1990 *Introduction to the Theory of Neural Computation,* Addison-Wesley, Redwood City, CA.

Huntsberger, T., and Ajjimarangsee, P. 1989. Parallel self-organizing feature maps for unsupervised pattern recognition, *Int. J. General Systems,* 16:357–372.

Karayiannis, N. B. 1994. Generalized Fuzzy *k*-means Algorithms, *Systems, Neural Nets, and Computing Technical Report* No. 94–14, University of Houston, December 1994.

Karayiannis, N. B. 1995. Generalized fuzzy *k*-means algorithms and their application in image compression, *SPIE Proceedings vol. 2493: Applications of Fuzzy Logic Technology II,* 2493:206–217 April 17–21, Orlando, FL.

Karayiannis, N. B., Bezdek, J. C., Pal, N. R., Hathaway, R. J., and Pai, P.-I 1996. A new family of competitive learning schemes, submitted to *IEEE Trans. Neural Networks,* in press.

Karayiannis, N. B., and Pai, P.-I 1994a. A Family of Fuzzy Algorithms for Learning Vector Quantization, *Systems, Neural Nets, and Computing Technical Report No. 94–07,* University of Houston, July 1994.

Karayiannis, N. B., and Pai, P.-I 1994b. A Fuzzy Algorithm for Learning Vector Quantization, *Proc. IEEE Int. Conf. Systems, Man and Cybernetics,* October, 126–131, San Antonio, TX.

Karayiannis, N. B., and Pai, P.-I 1994c. A Family of Fuzzy Algorithms for Learning Vector Quantization, *Intelligent Engineering Systems Through Artificial Neural Networks,* vol. 4, Dagli, C. H. et al., eds. 219–224 ASME Press, New York, NY.

Karayiannis, N. B., and Pai, P.-I 1994d. Fuzzy Algorithms for Learning Vector Quantization: Generalizations and Extensions, *Systems, Neural Nets, and Computing Technical Report No. 94–13,* University of Houston, December 1994.

Karayiannis, N. B., and Pai, P.-I 1995a. Fuzzy algorithms for learning vector quantization: generalizations and extensions, *SPIE Proc. Applications and Science of Artificial Neural Networks,* 2492:264–274. April 17–21, Orlando, FL.

Karayiannis, N. B. and Pai, P.-I. 1995b. Fuzzy vector quantization algorithms and their application in image compression, *IEEE Trans. Image Processing,* 4(9):1193–1201.

Karayiannis, N. B. and Venetsanopoulos, A. N. 1993. *Artificial Neural Networks: Learning Algorithms, Performance Evaluation, and Applications,* Kluwer Academic Publishers, Boston, MA.

Kohonen, T. 1988. An Introduction to Neural Computing, *Neural Networks,* vol. 1:3–16.

Kohonen, T. 1989. *Self-Organization and Associative Memory,* 3rd ed., Springer-Verlag, Berlin.

Kohonen, T. 1990a. The Self-Organizing Map, *Proc. IEEE,* 78:1464–1480.

Kohonen, T. 1990b. Improved versions of learning vector quantization, *Proc. Int. Joint Con. Neural Networks,* 1:545–550, June, San Diego, CA.

Kohonen, T. 1990c. Statistical pattern recognition revisited, *Advanced Neural Computers,* 137–144.

Linde, Y., Buzo, A., and Gray, R. M. 1980. An algorithm for vector quantizer design, *IEEE Trans. Communications,* 28:84–95.

Nasrabadi, N. M., and King, R. A. 1988. Image coding using vector quantization: A review, *IEEE Trans. Communications,* 36:957–971.

Pal, N. R., Bezdek, J. C., and Tsao, E. C.-K. 1992. Improving convergence and performance of Kohonen's self-organizing scheme, *Proc. SPIE, Science of Artificial Neural Networks,* 1710:500–509, April 21–24, Orlando, FL.

Pal, N. R., Bezdek, J. C., and Tsao, E. C.-K. 1993. Generalized clustering networks and Kohonen's self-organizing scheme. *IEEE Trans. Neural Networks,* 4:549–557.

Tsao, E. C.-K., Bezdek, J. C., and Pal, N. R. 1994. Fuzzy Kohonen clustering networks, *Pattern Recognition,* 27(5):757–764.

Tsypkin, Y. Z. 1973. *Foundations of the Theory of Learning,* Academic Press, New York, NY.

97

Adaptive Resonance Theory

97.1 Match-Based Learning and Error-Based Learning 1287
Match-Based Learning and Stable Coding • Boeing Neural Information Retrieval System • Error-Based Learning

97.2 ART and Fuzzy Logic 1288

97.3 ART Dynamics 1288

97.4 Fuzzy ART 1290
Fast-Learn Slow-Recode and Complement Coding Options

97.5 Fuzzy ARTMAP 1290

97.6 Fuzzy ART Algorithm 1292
ART Field Activity Vectors • Normalization by Complement Coding • Fuzzy ART Stable Category Learning

97.7 Fuzzy ARTMAP Algorithm 1294
ART_a and ART_b

97.8 ART Applications 1296

Gail A. Carpenter
Boston University

Stephen Grossberg
Boston University

Abstract

Adaptive Resonance Theory (ART) models are real-time self-organizing neural networks for category learning, hypothesis testing, pattern recognition, and nonstationary prediction. ART networks combine properties of production systems, neural networks, and fuzzy logic into a unified computational framework. The unique computational properties of these systems have been found useful in many types of industrial applications.

Unsupervised fuzzy ART and supervised fuzzy ARTMAP synthesize fuzzy logic and ART networks by exploiting the formal similarity between the computations of fuzzy subsethood and the dynamics of ART category choice, search, and learning. Fuzzy ART self-organizes stable recognition categories in response to arbitrary sequences of analog or binary input patterns. It generalizes the binary ART 1 model, replacing the set-theoretic intersection ($\cap$) with the fuzzy intersection ($\wedge$), or component-wise minimum. A normalization procedure called complement coding leads to a symmetric theory in which the fuzzy intersection and the fuzzy union ($\vee$), or component-wise maximum, play complementary roles. Complement coding preserves individual feature amplitudes while normalizing the input vector, and prevents a potential category proliferation problem. Adaptive weights start equal to one and can only decrease in time. A geometric interpretation of fuzzy ART represents each category as a box that increases in size as weights decrease. A matching criterion controls search, determining how close an input and a learned representation must be for a category to accept the input as a new exemplar. A vigilance parameter (ρ) sets the matching criterion and determines how finely or coarsely an ART system will partition inputs. High vigilance creates fine categories, represented by small boxes. Learning stops when boxes cover the input space. With fast learning, fixed vigilance, and an arbitrary input set, learning stabilizes after just one presentation of each input. A fast-commit slow-recode option allows rapid learning of rare events yet buffers memories against recoding by noisy inputs.

Fuzzy ARTMAP unites two fuzzy ART networks to solve supervised learning and prediction problems. Because it is a self-organizing architecture, fuzzy ARTMAP can operate in either a supervised or unsupervised mode. A Minimax Learning Rule controls ARTMAP category structure, conjointly minimizing predictive error and maximizing code compression. Low vigilance maximizes compression but may therefore cause very different inputs to make the same prediction. When this coarse grouping strategy causes a predictive error, an internal match tracking control process increases vigilance just enough to correct the error by triggering a search, or bout of hypothesis testing, for a better or new category. ARTMAP automatically constructs a minimal number of recognition categories, or "hidden units," to meet accuracy criteria. An ARTMAP voting strategy improves prediction by training the system several times using different orderings of the input set. Voting assigns confidence estimates to competing predictions given small, noisy, or incomplete training sets.

0-8493-8343-9/97/$0.00+$.50

97.1 Match-Based Learning and Error-Based Learning

Match-Based Learning and Stable Coding

A stable learning system needs to incorporate crucial new data into an existing memory system without destroying old memories. We effortlessly remember that a dog is still a dog, even as we learn that this particular dog is a Dalmatian named Spot. In a complex world, new information often complements the old, but both are important and correct.

An ART (Adaptive Resonance Theory) network constructs new memories based on the success or failure of old memories, as they guide the system in the world. As the network encounters examples, some categories become coarse (dog) (Figure 97.1a) or fine (Dalmatian), as needed. When we expect to hear "dog" but the answer is "Dalmatian," we are surprised into paying attention to features that had previously been ignored (Figure 91.1b). When we learn to recognize a Dalmatian as a breed, we do not forget that it is also a dog. Similarly, ART memories encode attended features, rather than the entire set of features that happen to be present at the moment. This is the basis for the stability of ART learning.

ART memories are stable in a complex world because the learning process is *match-based.* Memories are refined when attended portions of the external world provide a good enough match with our internal expectations. When more novel events occur that fail to match an ART network's expectations or predictions, a search process activates a new category. The new category represents a new hypothesis about what is important in the present environment. Match-based learning is a key characteristic of ART networks.

Boeing Neural Information Retrieval System

Learned code stability in response to fast incremental learning of a nonstationary environment is one of the main reasons that ART networks are selected for applications. One example of such a technology transfer is the Boeing Neural Information Retrieval System (NIRS) (Caudell et al., 1994; Caudell, 1993; Smith et al., 1993) in which ART networks are the critical system components. NIRS encodes an inventory of airplane parts in the form of 2-D and 3-D drawings. The system creates a compressed but stable memory structure for later retrieval by design engineers. The resulting neural database reduces inventory size by a factor of nine, thus alleviating a severe memory proliferation problem and permitting efficient reuse of stored designs. NIRS has moved from beta testing to implementation in CAD systems for design of the Boeing 777, and for manufacturing of the Boeing 747 and 767 planes. Other industrial applications are summarized in Section 97.8.

Error-Based Learning

Match-based learning generates a stable recognition code in a large, complex, evolving environment. A match-based learning system is thus well suited to problems such as the Boeing CAD neural database, which creates its own expert system as a function of experience. However, qualitatively different types of learning problems also exist. For example, as we grow, our eyes and limbs need to learn, or adapt, to their own internal changes so that we can pick up a pencil as an adult as well as we could at age

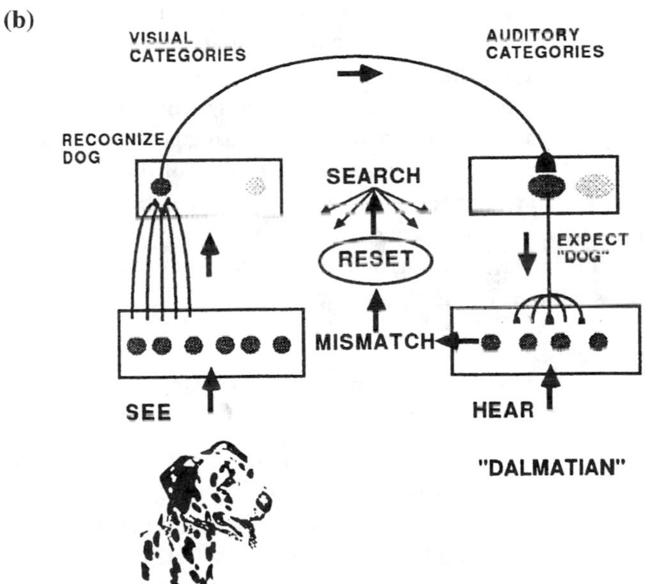

Figure 97.1 (a) An ART network creates a coarse category ("dog") by a two-step code compression process. First, dogs are grouped into visual recognition categories on the basis of their shared features. Then, these dissimilar categories learn the shared prediction "dog." (b) An ART network that makes an incorrect prediction learns a new specific category identification ("Dalmatian") but simultaneously preserves the coarse category representation ("dog").

two. As adults we have no need for the sensory-motor maps that we learned as babies. These codes need not, therefore, be stable in the sense that large knowledge systems—such as visual recognition, language, and database retrieval systems—need to be stable. Layers of old motor maps would most likely be a great nuisance.

Neural networks that employ *error-based* learning are well suited to adaptive sensory-motor control problems. Error-based learning systems include the perceptron (Rosenblatt, 1962, 1958), multilayer perceptrons such as back propagation (Rumelhart, Hinton, and Williams, 1986; Parker, 1982, Werbos, 1974), and vector associative map, or VAM systems (Bullock, Grossberg, and Guenther, 1993; Grossberg, Guenther, Bullock, and Greve, 1993; Gaudiano and Grossberg, 1991). In these systems, an error causes memories to change so that the same input, seen again, would give an answer that was closer to the "correct" one. If we see a dog and know it is a dog, but are then told that it is a Dalmatian, an error-based network would shift its learned weights in such a way that the next response would be toward Dalmatian, away from dog. If this happened several times in a row, the system would learn to respond "Dalmatian," but would completely forget that a dog is still a dog. Error-based learning is, hereby, subject to "catastrophic forgetting." This kind of forgetting is desirable, however, if the error signal registers that we have reached too far to touch a pencil and thereby recalibrates the sensory-motor transformations that enable us to reach correctly.

Catastrophic forgetting in error-based systems is typically controlled in applications to recognition, language, and database retrieval problems by running the system off-line in a slow-learning mode, restricting the size of the database, and using approximately stationary data. ART systems are designed to permit on-line fast learning of arbitrarily large nonstationary databases without enduring catastrophic forgetting.

97.2 ART and Fuzzy Logic

Stephen Grossberg (1976) introduced adaptive resonance as a theory of human cognitive information processing. The theory has led to an evolving series of real-time neural network models for unsupervised and supervised category learning and pattern recognition. These models form stable recognition categories in response to arbitrary input sequences with either fast or slow learning. Unsupervised ART networks include ART 1 (Carpenter and Grossberg, 1987a), which stably learns to categorize binary input patterns presented in an arbitrary order; ART 2 (Carpenter and Grossberg, 1987b) and fuzzy ART (Carpenter, Grossberg, and Rosen, 1991a), which stably learn to categorize either analog or binary input patterns presented in an arbitrary order; and ART 3 (Carpenter and Grossberg, 1990), which carries out parallel search, or hypothesis testing, of distributed recognition codes in a multi-level network hierarchy. Many of the ART papers are collected in the anthology *Pattern Recognition by Self-Organizing Neural Networks* (Carpenter and Grossberg, 1991).

A supervised network architecture, called ARTMAP, self-organizes arbitrary categorical mappings between m-dimensional input vectors and n-dimensional output vectors using a pair of ART networks joined together by an associative mapping network and an internal controller. ARTMAP's internal control mechanisms create stable recognition categories of optimal size by maximizing code compression while minimizing predictive error in an on-line setting. A pair of binary ART 1 models is used in the first ARTMAP network (Carpenter, Grossberg, and Reynolds, 1991), which therefore learns binary maps. Fuzzy ART directly generalizes ART 1 to learn stable recognition categories in response to analog and binary input patterns. When fuzzy ART replaces ART 1 in an ARTMAP system, the resulting fuzzy ARTMAP architecture (Carpenter, Grossberg, Markuzon, Reynolds, and Rosen, 1992) rapidly learns stable categorical mappings between analog or binary input and output vectors. Fuzzy ARTMAP learns to classify inputs by a fuzzy set of features, or a pattern of fuzzy membership values between 0 and 1, that indicate the extent to which each feature is present. Whereas set-theoretic operations may be used to describe ART 1 dynamics, fuzzy set-theoretic operations (Kosko, 1986; Zadeh, 1965) describe fuzzy ART dynamics.

97.3 ART Dynamics

Fuzzy ART incorporates the basic features of all ART systems, notably pattern matching between bottom-up input and top-down learned prototype vectors. This matching process leads either to a resonant state that focuses attention and triggers stable prototype learning or to a self-regulating parallel memory search. If the search ends with the selection of an established category, then the category's prototype may be refined to incorporate new information in the input pattern. If the search ends by selecting a previously untrained node, then the ART network establishes a new category.

Figure 97.2 illustrates the main components of an ART 1 network and Figure 97.3 illustrates an ART search cycle. During ART search, an input vector **I** registers itself as a pattern **X** of activity across level F_1 (Figure 97.3a). Multiple converging and diverging $F_1 \rightarrow F_2$ adaptive filter pathways multiply the vector **S** by a matrix of adaptive weights, or long term memory (LTM) traces, to generate a net input vector **T** to level F_2. The internal competitive dynamics of F_2 contrast-enhance vector **T**, generating a compressed activity vector **Y** across F_2. In ART 1, strong competition selects the F_2 node that receives the maximal $F_1 \rightarrow F_2$ input. Only one component of **Y** is nonzero after this choice takes place. Activation of such a *winner-take-all* node defines the category, or symbol, of the input pattern **I**. Such a category represents all the inputs **I** that maximally activate the corresponding node.

Activation of an F_2 node may be interpreted as "making a hypothesis" about an input **I**. An F_2 vector generates a signal vector **U** sent top-down through the $F_2 \rightarrow F_1$ adaptive filter. After multiplication by the adaptive weight matrix of the top-down filter, a vector **V** becomes the $F_2 \rightarrow F_1$ input (Figure 97.3b). Vector **V** plays the role of a learned top-down expectation. Activation of **V** by **Y** may be interpreted as "testing the hypothesis" **Y**, or "reading out the category prototype" **V**. The ART

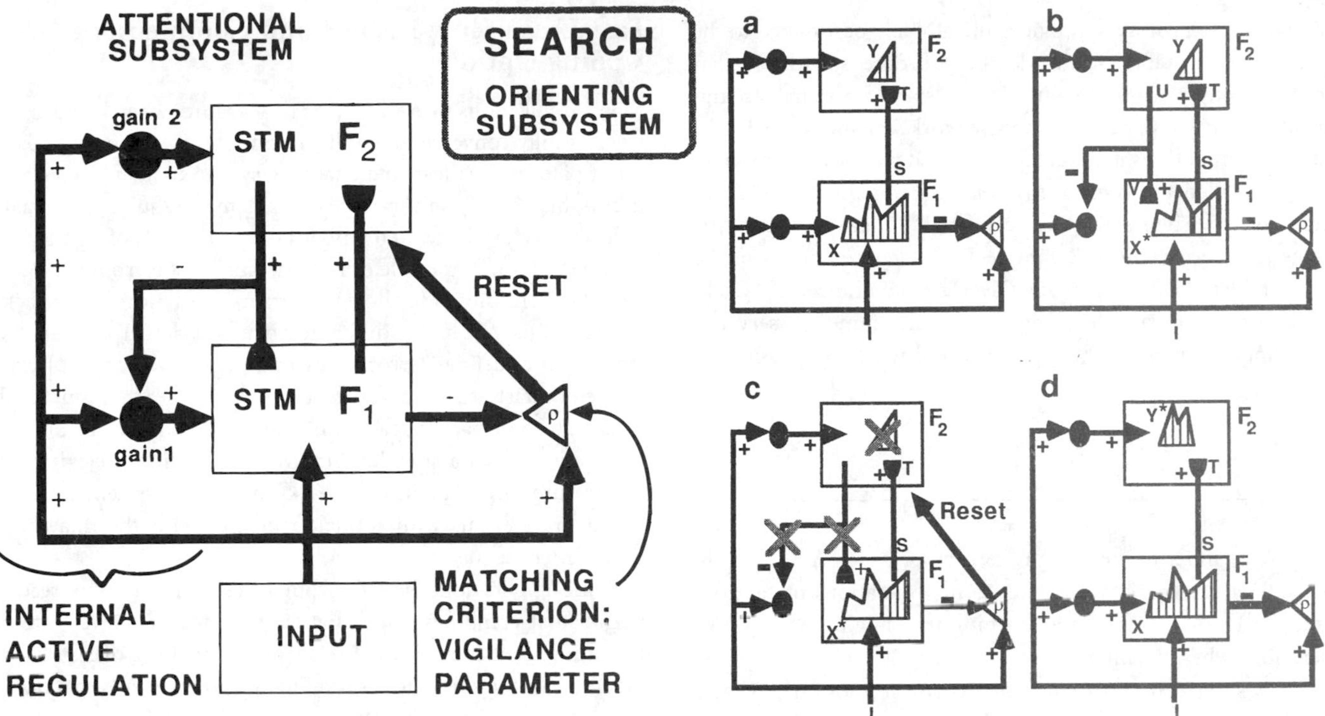

Figure 97.2 Typical ART 1 neural network (Carpenter and Grossberg, 1987a).

Figure 97.3 ART search for an F_2 code: (a) The input pattern **I** generates the specific STM activity pattern **X** at F_1 as it nonspecifically activates the orienting subsystem *A*. Pattern **X** both inhibits *A* and generates the output signal pattern **S**. An adaptive filter transforms the signal pattern **S** into the pattern **T**, which activates the STM pattern **Y** across F_2. (b) Pattern **Y** generates the signal pattern **U**, and a top-down adaptive filter transforms **U** into the prototype pattern **V**. If **V** mismatches **I**, then F_1 registers a new STM activity pattern **X***. The resulting reduction of total STM reduces the total inhibition from F_1 to *A*. (c) If the ART matching criterion fails, *A* releases a nonspecific signal that resets the STM pattern **Y** at F_2. (d) Since reset inhibits **Y**, it also eliminates the top-down prototype signal **V**, so **X** can be reinstated at F_1. Enduring traces of the prior reset allow **X** to activate a different STM pattern **Y*** at F_2. If the top-down prototype due to **Y*** also mismatches **I** at F_1, then the search for an F_2 code that satisfies the matching criterion continues.

network matches the "expected prototype" **V** of the category against the active input pattern, or exemplar, **I.**

This matching process may change the F_1 activity pattern **X** by suppressing activation of all features in **I** that are not confirmed by **V.** The resultant pattern **X*** encodes the pattern of features to which the network "pays attention". If the expectation **V** is close enough to the input **I,** then a state of *resonance* occurs, with the matched pattern **X** defining an attentional focus. The resonant state persists long enough for learning to occur; hence the term *adaptive resonance* theory. ART learns prototypes rather than exemplars because weights encode the attended feature vector **X,*** rather than the input **I** itself.

A dimensionless parameter called *vigilance* defines the criterion of an acceptable match. Vigilance weighs how close the input exemplar **I** must be to the top-down prototype **V** in order for resonance to occur. In ARTMAP, vigilance becomes an internally controlled variable, rather than a fixed parameter. Because vigilance can vary across learning trials, a single ART system can encode widely differing degrees of generalization, or morphological variability. Low vigilance leads to broad generalization, coarse categories, and abstract prototypes. High vigilance leads to narrow generalization, fine categories, and specific prototypes. In the limit of very high vigilance, prototype learning reduces to exemplar learning. Varying vigilance levels allow a single ART system to recognize both abstract categories of faces and dogs and individual faces and dogs.

ART memory search, or hypothesis testing, begins when the top-down expectation **V** determines that the bottom-up input **I** is too novel, or unexpected, to satisfy the vigilance criterion. Search leads to selection of a better recognition code, symbol, category, or hypothesis to represent input **I** at level F_2. An *orienting subsystem A* controls the search process. The orienting subsystem interacts with the attentional subsystem, as in Figures 97.2c and 97.2d, to enable the attentional subsystem to learn about novel inputs without risking unselective forgetting of its previous knowledge.

ART search prevents associations from forming between **Y** and **X*** if **X*** is too different from **I** to satisfy the vigilance criterion. The search process resets **Y** before such an association can form. If the search ends upon a familiar category, then that category's prototype may be refined in light of new information carried by **I.** If **I** is too different from any of the previously learned prototypes, then the search ends upon an uncommitted F_2 node, which begins a new category.

An ART *choice parameter* controls how deeply the search proceeds before selecting an uncommitted node. As learning self-stabilizes, all inputs coded by a category access it directly and search is automatically disengaged. The category selected is, then, the one whose prototype provides the globally best match to the input pattern. Stable on-line learning proceeds with familiar inputs directly activating their categories and novel inputs triggering adaptive searches, until the network's memory reaches its capacity. Simulations illustrate fuzzy ART dynamics in a parameter range called the *conservative limit.* In this limit, the choice parameter α (Figure 97.4) is very small. Then an input first selects a category whose weight vector is a fuzzy subset of the input, if such a category exists. Given such a choice, no weight change occurs during learning; hence the name conservative limit, since learned weights are conserved wherever possible.

97.4 Fuzzy ART

Fuzzy ART inherits the design features of other ART models. Figure 97.4 summarizes how the ART 1 operations of category choice, matching, search, and learning translate into fuzzy ART operations when the intersection operator (∩) of ART 1 replaces the fuzzy intersection, or component-wise minimum, operator (∧). Despite this close formal homology, this chapter summarizes fuzzy ART as an algorithm, rather than as a locally defined neural model. Carpenter, Grossberg, and Rosen (1991b) describe a neural network realization of fuzzy ART. For the special case of binary inputs and fast learning, the computations of fuzzy ART are identical to those of the ART 1 neural network.

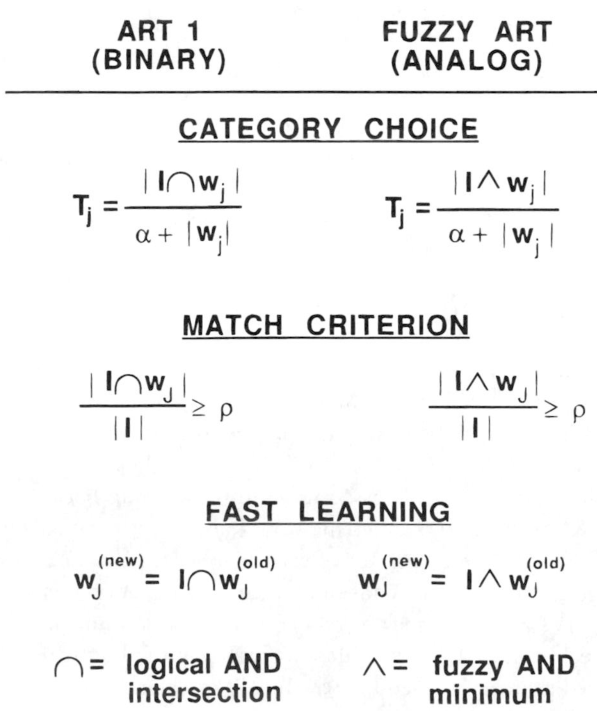

Figure 97.4 Analogy between ART 1 and fuzzy ART. In ART 1 $\mathbf{w}_j$ denotes the index set of top-down LTM traces that exceed a prescribed positive threshold value.

Fast-Learn Slow-Recode and Complement Coding Options

Many applications of ART 1 use fast learning, whereby adaptive weights fully converge to equilibrium values in response to each input pattern. Fast learning enables a system to adapt quickly to inputs that occur only rarely but that may require immediate accurate performance. Remembering many details of an exciting movie is a typical example of fast learning. Fast learning destabilizes the memories of feedforward, error-based models like backpropagation. When the difference between actual output and target output defines "error", present inputs drive out past learning, since fast learning zeroes the error on each input trial. This feature of backpropagation restricts its domain to off-line applications with a slow learning rate. In addition, lacking the key feature of competition, a backpropagation system tends to average rare events with similar frequent events that may have different consequences.

Some applications benefit from a fast-commit slow-recode option that combines fast initial learning with a slower rate of forgetting. Fast commitment retains the advantage of fast learning, namely, the ability to respond to important inputs that occur only rarely. Slow recoding then prevents features in a category's prototype from being erroneously deleted in response to noisy or partial inputs. Only a statistically persistent change in a feature's relevance to an established category can delete it from the prototype of the category.

Complement coding is a preprocessing step that normalizes input patterns. Complement coding solves a potential fuzzy ART category proliferation problem (Carpenter, Grossberg, and Rosen, 1991a; Moore, 1989). In neurobiological terms, complement coding uses both on-cells and off-cells to represent an input pattern, preserving individual feature amplitudes while normalizing the total on-cell/off-cell activity. Functionally, the on-cell portion of a prototype encodes features that are critically present in category exemplars, while the off-cell portion encodes features that are critically absent. Small weights in both on-cell and off-cell portions of a prototype encode as "uninformative" those features that are sometimes present and sometimes absent. In set theoretic terms, complement coding leads to a symmetric ART theory in which the fuzzy intersection (∧) and the fuzzy union (∨) play complementary roles. Complement coding allows a geometric interpretation of fuzzy ART recognition categories as box-shaped regions of input space. Fuzzy intersections and unions iteratively define the corners of each box. Simulations in this section illustrate fuzzy ART geometry for an example where inputs are two-dimensional, so boxes are rectangles.

97.5 Fuzzy ARTMAP

Each ARTMAP system includes a pair of Adaptive Resonance Theory modules (ART_a and ART_b) that create stable recognition categories in response to arbitrary sequences of input patterns (Figure 97.5). During supervised learning, ART_a receives a stream $\{\mathbf{a}^{(p)}\}$ of input patterns and ART_b receives a stream $\{\mathbf{b}^{(p)}\}$ of

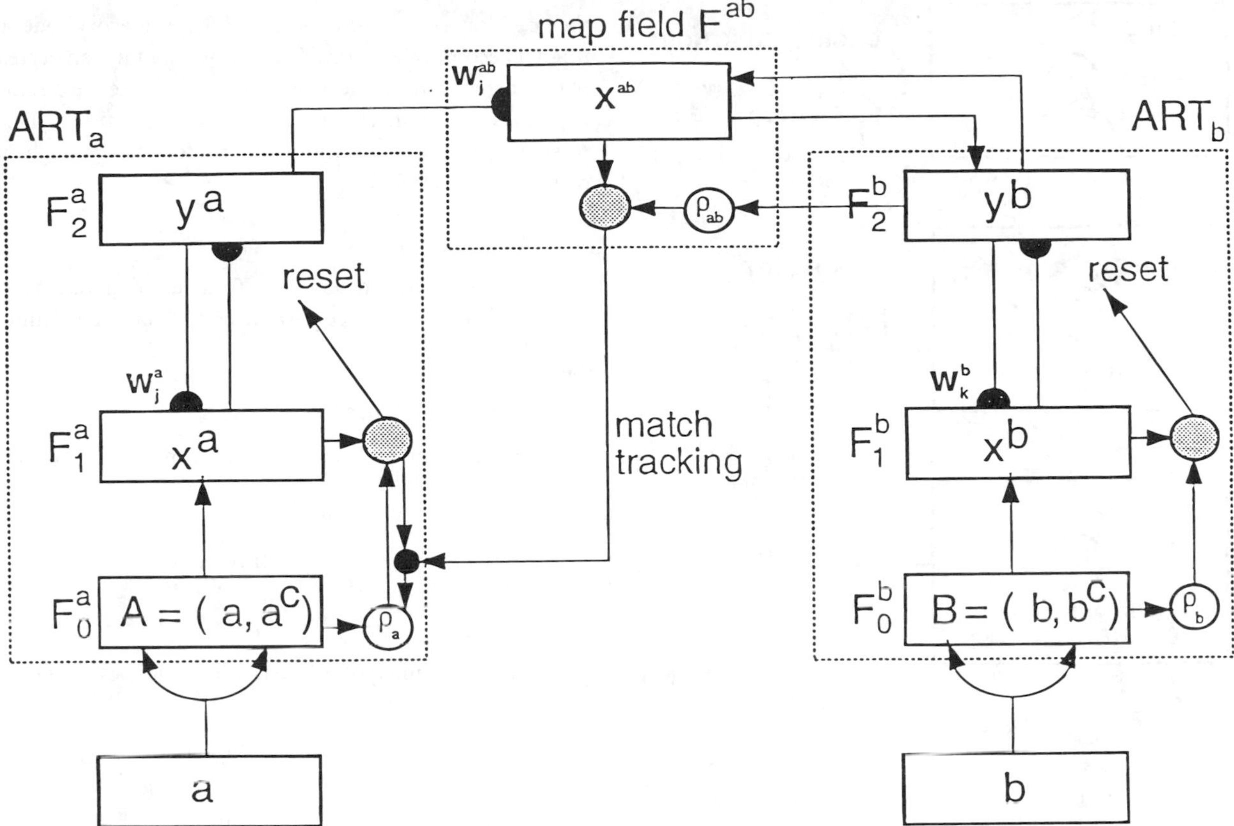

Figure 97.5 Fuzzy ARTMAP architecture. The ART_a complement coding preprocessor transforms the M_a-vector **a** into the $2M_a$-vector $\mathbf{A} = (\mathbf{a}^c)$ at the ART_a field F_0^a. **A** is the input vector to the ART_a field F_1^a. Similarly, the input to F_1^b is the $2M_b$-vector $(\mathbf{b},\mathbf{b}^c)$. When ART_b disconfirms a prediction of ART_a, map field inhibition induces the match tracking process. Match tracking raises the ART_a vigilance (ρ_a) to just above the F_1^a-to-F_0^a match ratio $|\mathbf{x}^a|/|\mathbf{A}|$. This triggers an ART_a search which leads to activation of either an ART_a category that correctly predicts **b** or to a previously uncommitted ART_a category node.

input patterns, where $\mathbf{b}^{(p)}$ is the correct prediction given $\mathbf{a}^{(p)}$. An associative learning network and an internal controller link these modules to make the ARTMAP system operate in real time. The controller creates the minimal number of ART_a recognition categories, or "hidden units," needed to meet accuracy criteria. A Minimax Learning Rule enables ARTMAP to learn quickly, efficiently, and accurately as it conjointly minimizes predictive error and maximizes code compression. This scheme automatically links predictive success in ART_b to category size in ART_a on a trial-by-trial basis using only local operations. It works by increasing the ART_a vigilance parameter (ρ_a) by the minimal amount needed to correct a predictive error at ART_b.

An ART_a *baseline vigilance* parameter $\overline{\rho_a}$ calibrates the minimum confidence needed for ART_a to accept a chosen category, rather than search for a better one through automatically controlled search. Lower values of $\overline{\rho_a}$ enable larger categories to form, maximizing code compression. Initially, $\rho_a = \overline{\rho_a}$. During training, a predictive failure at ART_b increases ρ_a by the minimum amount needed to trigger ART_a search, through a feedback control mechanism called *match tracking* (Carpenter, Grossberg, and Reynolds, 1991). Match tracking sacrifices the minimum amount of compression necessary to correct the predictive error. Due to match tracking, the vigilance parameter increases until it just exceeds the measure $|\mathbf{A} \wedge \mathbf{w}_J|/|\mathbf{A}|$ of how well the input vector **A** matches the weight vector $\mathbf{w}_J$ of the chosen category J; see Figure 97.4. Once vigilance exceeds the match value, hypothesis testing is triggered and leads to selection of a new ART_a category, whose prototype focuses attention on a new cluster of $\mathbf{a}^{(p)}$ input features that is better able to predict $\mathbf{b}^{(p)}$. With fast learning, match tracking allows a single ARTMAP system to learn a different prediction for a rare event than for a cloud of similar frequent events in which it is embedded.

A DARPA benchmark simulation circle-in-the-square (Wilensky, 1990) illustrates fuzzy ARTMAP dynamics. The simulation task is learning to identify which points lie inside and which lie outside a circle. During training, components of the ART_a input **a** are the x- and y-coordinates of a point in the unit square; and ART_b input equals 0 or 1, identifying **a** as inside or outside the circle. As fuzzy ARTMAP learns on-line, or incrementally, test set accuracy increases from 88.6% to 98.0% as the training set increases in size from 100 to 100,000 randomly chosen points. With off-line learning, the system needs from 2 to 13 epochs to learn all training set exemplars to 100% accuracy, where an epoch is one cycle of training on an entire set of input exemplars. Test

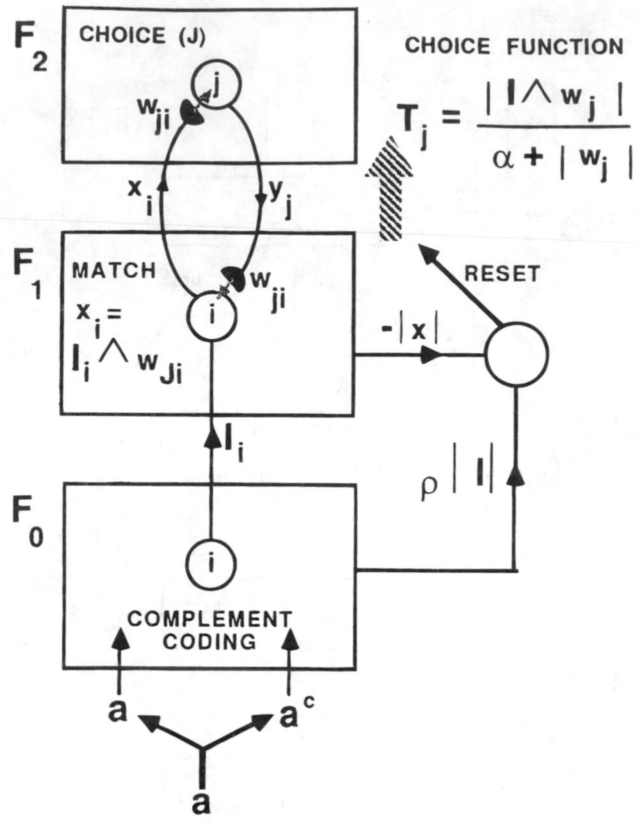

Figure 97.6 Fuzzy ART notation. In the fuzzy ART algorithm, $\mathbf{w}_j$, equals both the bottom-up weight vector and the top-down weight vector (Figure 97.2).

set accuracy then increases from 89.0% to 99.5% as the training set size increases from 100 to 100,000. Application of a voting strategy improves an average single-run accuracy of 90.5% on five runs to a voting accuracy of 93.9%, with each run trained on a fixed 1,000-item set for one epoch.

97.6 Fuzzy ART Algorithm

ART Field Activity Vectors

Each ART system includes a field F_0 of nodes that represent a current input vector and a field F_1 that receives both bottom-up input from F_0 and top-down input from a field F_2 that represents the active code, or category (Figure 97.6). Vector $\mathbf{I} = (I_1, \ldots, I_M)$ denotes F_0 activity, with each component I_i in the interval [0, 1], for $i = 1, \ldots, M$. Vector $\mathbf{x} = (x_1, \ldots, x_M)$ denotes F_1 activity and $\mathbf{y} = (y_1, \ldots, y_N)$ denotes F_2 activity. The number of nodes in each field is arbitrary.

Weight Vector

Associated with each F_2 category node $j (j = 1, \ldots, N)$ is a vector $\mathbf{w}_j \equiv (w_{j1}, \ldots, w_{jM})$ of adaptive weights, or long-term memory (LTM) traces. Initially

$$w_{j1}(0) = \cdots = w_{jM}(0) = 1; \tag{97.1}$$

then each category is *uncommitted.* After a category codes its first input it becomes *committed.* Each component w_{ji} can decrease but never increase during learning. Thus each weight vector $\mathbf{w}_j(t)$ converges to a limit. The fuzzy ART weight, or prototype, vector $\mathbf{w}_j$ subsumes both the bottom-up and top-down weight vectors of ART 1 (Figure 97.2).

Parameters

A choice parameter $\alpha > 0$, a learning rate parameter $\beta \in [0, 1]$, and a vigilance parameter $\rho \in [0, 1]$ determine fuzzy ART dynamics.

Category Choice

For each input $\mathbf{I}$ and F_2 node j, the *choice function* T_j is defined by

$$T_j(\mathbf{I}) = \frac{|\mathbf{I} \wedge \mathbf{w}_j|}{\alpha + |\mathbf{w}_j|}, \tag{97.2}$$

where the fuzzy intersection $\wedge$ (Zadeh, 1965) is defined by

$$(\mathbf{p} \wedge \mathbf{q})_i \equiv \min(p_i, q_i) \tag{97.3}$$

and where the norm $|\cdot|$ is defined by

$$|\mathbf{p}| \equiv \sum_{i=1}^{M} |p_i|. \tag{97.4}$$

The system makes a *category choice* when at most one F_2 node can become active at a given time. The index J denotes the chosen category, where

$$T_J = \max\{T_j : j = 1 \cdots N\}. \tag{97.5}$$

If more than one T_j is maximal, the category with the smallest j index is chosen. In particular, nodes become committed in order $j = 1, 2, 3, \ldots$. When the Jth category is chose, $y_J = 1$; and $y_j = 0$ for $j \neq J$. In a choice system, the F_1 activity vector $\mathbf{x}$ obeys the equation

$$\mathbf{x} = \begin{cases} \mathbf{I} & \text{if } F_2 \text{ is inactive} \\ \mathbf{I} \wedge \mathbf{w}_J & \text{if the } J^{th}\ F_2 \text{ node is chosen.} \end{cases} \tag{97.6}$$

Resonance or Reset

Resonance occurs if the *match function* $|\mathbf{I} \wedge \mathbf{w}_J| / |\mathbf{I}|$ of the chosen category meets the vigilance criterion:

$$\frac{|\mathbf{I} \wedge \mathbf{w}_J|}{|\mathbf{I}|} \geq \rho; \tag{97.7}$$

that is, by Equation 97.6, when the Jth category becomes active, resonance occurs if

$$|\mathbf{x}| = |\mathbf{I} \wedge \mathbf{w}_J| \geq \rho|\mathbf{I}|. \tag{97.8}$$

Learning then ensues, as defined below. *Mismatch reset* occurs if

$$\frac{|\mathbf{I} \wedge \mathbf{w}_J|}{|\mathbf{I}|} < \rho; \tag{97.9}$$

that is, if

$$|\mathbf{x}| = |\mathbf{I} \wedge \mathbf{w}_J| < \rho|\mathbf{I}|. \tag{97.10}$$

Then the value of the choice function T_J is set to 0 for the duration of the input presentation to prevent the persistent selection of the same category during search. A new index J represents the active category, selected by 97.5. The search process continues until the chosen J satisfies the matching criterion 97.7.

Learning

Once search ends, the weight vector $\mathbf{w}_J$ learns according to the equation

$$\mathbf{w}_J^{(\text{new})} = \beta(\mathbf{I} \wedge \mathbf{w}_J^{(\text{old})}) + (1 - \beta)\mathbf{w}_J^{(\text{old})}. \tag{97.11}$$

Fast learning corresponds to setting $\beta = 1$. The learning law of the NGE system (Salzberg, 1990) is equivalent to Equation 97.11 in the fast-learn limit with complement coding.

Fast-Commit, Slow-Recode

For efficient coding of noisy input sets, it is useful to set $\beta = 1$ when J is an uncommitted node, and then to take $\beta < 1$ for slower adaptation after the category is already committed. The fast-commit, slow-recode option makes $\mathbf{w}_J^{(\text{new})} = \mathbf{I}$ the first time category J becomes active. Moore (1989) introduced the learning law Equation 97.11, with fast commitment and slow recoding, to investigate a variety of generalized ART 1 models. Some of these models are similar to fuzzy ART, but none uses complement coding. Moore describes a category proliferation problem that can occur in some analog ART systems when many random inputs erode the norm of weight vectors. Complement coding solves this problem, as follows.

Normalization by Complement Coding

Normalization of fuzzy ART inputs prevents category proliferation. The $F_0 \rightarrow F_1$ inputs are normalized if, for some $\gamma > 0$,

$$\sum_i I_i = |\mathbf{I}| \equiv \gamma \tag{97.12}$$

for all inputs **I**. One way to normalize each vector **a** is:

$$\mathbf{I} = \frac{\mathbf{a}}{|\mathbf{a}|}, \tag{97.13}$$

which requires the nonlinear operation of division and loses amplitude information. Complement coding represents both the on-response and the off-response to an input vector **a** (Figure 97.6). In its simplest form, **a** represents the on-response and $\mathbf{a}^c$, the complement of **a**, represents the off-response, where

$$a_i^c \equiv 1 - a_i. \tag{97.14}$$

The complement coded $F_0 - F_1$ input **I** is the 2M-dimensional vector

$$\mathbf{I} = (\mathbf{a}, \mathbf{a}^c) \equiv (a_1, \ldots, a_M, a_1^c, \ldots. a_M^c). \tag{97.15}$$

A complement coded input is automatically normalized, because

$$\begin{aligned} |\mathbf{I}| &= |(\mathbf{a}, \mathbf{a}^c)| \\ &= \sum_{i=1}^{M} a_i + (M - \sum_{i=1}^{M} a_i) \\ &= M. \end{aligned} \tag{97.16}$$

With complement coding, the initial condition

$$w_{j1}(0) = \cdots = w_{j,2M}(0) = 1. \tag{97.17}$$

replaces the fuzzy ART initial condition (Equation 97.1).

The close linkage between fuzzy subsethood and ART choice/search/learning forms the foundation of the computational properties of fuzzy ART. In the conservative limit, where the choice parameter $\alpha = 0^+$, the choice function T_j measures the degree to which $\mathbf{w}_j$ is a fuzzy subset of **I** (Kosko, 1986). A category J for which $\mathbf{w}_J$ is a fuzzy subset of **I** will then be selected first, if such a category exists. Resonance depends on the degree to which **I** is a fuzzy subset of $\mathbf{w}_J$, by Equations 97.7 and 97.9. When J is such a fuzzy subset choice, then the match function value is:

$$\frac{|\mathbf{I} \wedge \mathbf{w}_J|}{|\mathbf{I}|} = \frac{|\mathbf{w}_J|}{|\mathbf{I}|}. \tag{97.18}$$

Choosing J to maximize $|\mathbf{w}_j|$ among fuzzy subset choices, by Equation 97.2, thus maximizes the opportunity for resonance in Equation 97.7. If reset occurs for the node that maximizes $|\mathbf{w}_J|$, then reset will also occur for all other subset choices.

A geometric interpretation of fuzzy ART represents each category as a box in M-dimensional space, where M is the number of components of input **a**. Consider an input set that consists of 2-dimensional vectors **a**. With complement coding,

$$\mathbf{I} = (\mathbf{a}, \mathbf{a}^c) = (a_1, a_2, 1 - a_1, 1 - a_2). \tag{97.19}$$

Each category j then has a geometric representation as a rectangle R_j. Following Equation 97.19, a complement-coded weight vector $\mathbf{w}_j$ takes the form:

$$\mathbf{w}_j = (\mathbf{u}_j, \mathbf{v}_j^c), \tag{97.20}$$

where $\mathbf{u}_j$ and $\mathbf{v}_j$ are 2-dimensional vectors. Vector $\mathbf{u}_j$ defines one corner of a rectangle R_j and $\mathbf{v}_j$ defines the opposite corner (Figure 97.7a). The size of R_j is:

$$|R_j| \equiv |\mathbf{v}_j - \mathbf{u}_j|, \tag{97.21}$$

which is equal to the height plus the width of R_j.

In a fast-learn fuzzy ART system, with $\beta = 1$ in Equation 97.11, $\mathbf{w}_J^{(\text{new})} = \mathbf{I} = (\mathbf{a}, \mathbf{a}^c)$ when J is an uncommitted node. The corners of $R_J^{(\text{new})}$ are then $\mathbf{a}$ and $(\mathbf{a}^c)^c = \mathbf{a}$. Hence $R_J^{(\text{new})}$ is just the point $\mathbf{a}$. Learning increases the size of R_j, which grows as the size of $\mathbf{w}_J$ shrinks during learning. Vigilance ρ determines the maximum size of R_j, with $|R_j| \le 2(1 - \rho)$, as shown below. During each fast-learning trial, R_J expands to $R_J \oplus \mathbf{a}$, the minimum rectangle containing R_J and $\mathbf{a}$ (Figure 97.7b). The corners of $R_J \oplus \mathbf{a}$, are $\mathbf{a} \wedge \mathbf{u}_J$ and $\mathbf{a} \vee \mathbf{v}_J$, where the fuzzy intersection $\wedge$ is defined by Equation 97.3, and the fuzzy union $\vee$ is defined by:

$$(\mathbf{p} \vee \mathbf{q})_i \equiv \max(p_i, q_i) \tag{97.22}$$

(Zadeh, 1965). Hence, by Equation 97.21, the size of $R_J \oplus \mathbf{a}$ is:

$$|R_J \oplus \mathbf{a}| = |(\mathbf{a} \vee \mathbf{v}_J) - (\mathbf{a} \wedge \mathbf{u}_J)|. \tag{97.23}$$

However, before R_J can expand to include $\mathbf{a}$, reset and search chooses another category if $|R_J \oplus \mathbf{a}|$ is too large. With fast learning, R_j is the smallest rectangle that encloses all vectors $\mathbf{a}$ that have chosen category j without reset.

If $\mathbf{a}$ has dimension M, the box R_j includes the two opposing vertices $\wedge_j\mathbf{a}$ and $\vee_j\mathbf{a}$, where the ith component of each vector is:

$$(\wedge_j\mathbf{a})_i = \min\{a_i\text{:}\mathbf{a} \text{ has been coded by category } j\} \tag{97.24}$$

and

$$(\vee_j\mathbf{a})_i = \max\{a_i\text{:}\mathbf{a} \text{ has been coded by category } j\} \tag{97.25}$$

(Figure 97.8). The size of R_j is

$$|R_j| = |\vee_j\mathbf{a} - \wedge_j\mathbf{a}| \tag{97.26}$$

and the weight vector $\mathbf{w}_j$ is

$$\mathbf{w}_j = (\wedge_j\mathbf{a}, (\vee_j\mathbf{a})^c), \tag{97.27}$$

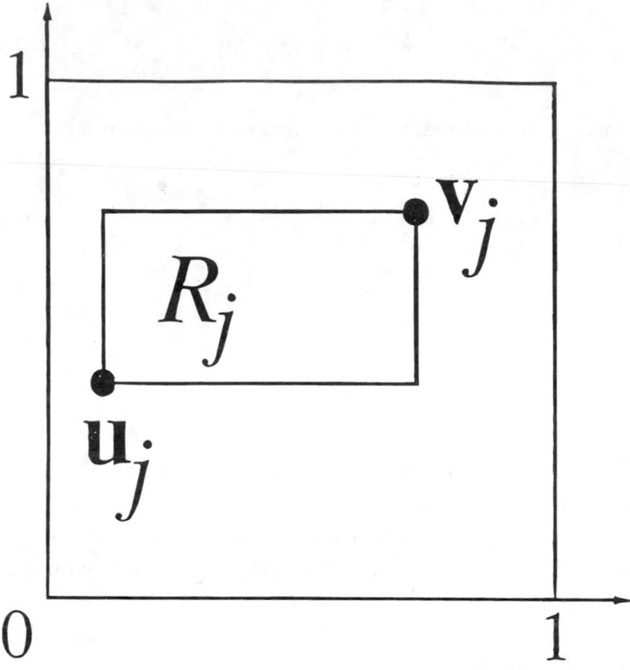

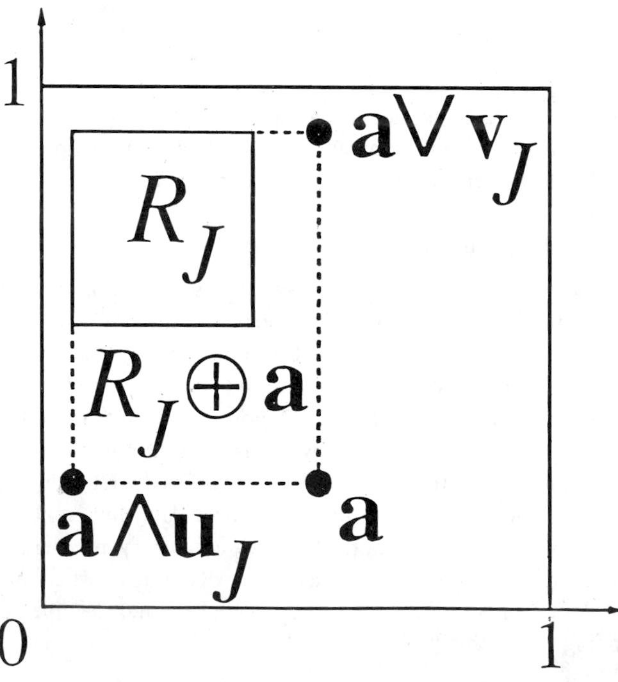

Figure 97.7 Fuzzy ART category boxes. (a) In complement coding form with $M = 2$, each weight vector $\mathbf{w}_j$ has a geometric interpretation as a rectangle R_j with corners $(\mathbf{u}_j, \mathbf{v}_j)$. (b) During fast learning, R_J expands to $R_J \oplus \mathbf{a}$, the smallest rectangle that includes R_J and $\mathbf{a}$, provided that $|R_J \oplus \mathbf{a}| \le 2(1 - \rho)$.

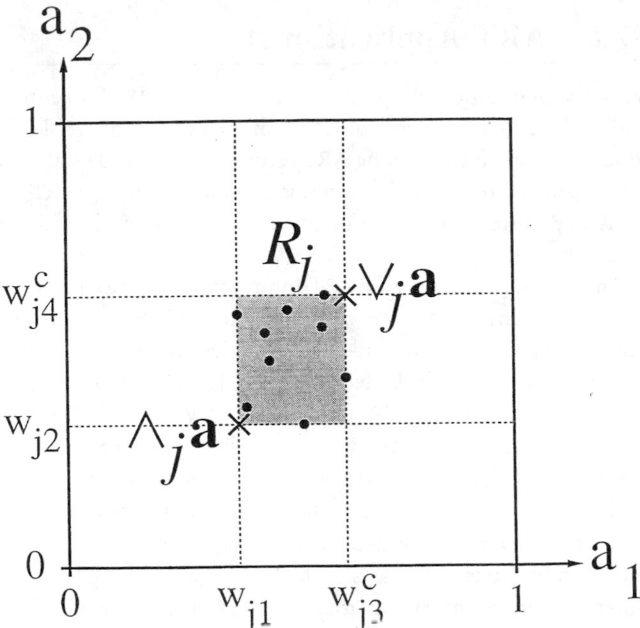

Figure 97.8 With fuzzy ART fast learning and complement coding, the jth category rectangle R_j includes all those vectors **a** in the unit square that have activated category j without reset. The weight vector $\mathbf{w}_j$ equals $(\wedge_j\mathbf{a}, (\vee_j\mathbf{a})^c)$.

as in Equations 97.20 and 97.21. Thus

$$|\mathbf{w}_j| = \sum_i (\wedge_j\mathbf{a})_i + \sum_i [1 - (\vee_j\mathbf{a})_i] = M - |\vee_j\mathbf{a} - \wedge_j\mathbf{a}|, \tag{97.28}$$

so the size of the box R_j is

$$|R_j| = M - |\mathbf{w}_j|. \tag{97.29}$$

How large can a box R_j grow due to fast learning? By Equations 97.8, 97.11, and 97.16,

$$|\mathbf{w}_j| \geq \rho M. \tag{97.30}$$

By Equations 97.29 and 97.30,

$$|R_j| \leq (1 - \rho)M. \tag{97.31}$$

Inequality (Equation 97.31) shows that high vigilance ($\rho \cong 1$) leads to small R_j while low vigilance ($\rho \cong 0$) permits large R_j. If j is an uncommitted node, $|\mathbf{w}_j| = 2M$, by Equation 97.17, so formally, $|R_j| \equiv -M$, by Equation 97.29. These observations are combined into the following summary of fuzzy ART dynamics.

Fuzzy ART Stable Category Learning

A fuzzy ART system with complement coding, fast learning, and constant vigilance forms categories that converge to limits in response to an arbitrary sequence of analog or binary input vectors. Category boxes can grow in each dimension, but never shrink. The size of a box R_j equals $M - |\mathbf{w}_j|$, where $\mathbf{w}_j$ is the corresponding weight vector. The size $|R_j|$ is bounded above by $M(1 - \rho)$. In the conservative limit, one-pass learning obtains such that no reset or additional learning occurs on subsequent presentations of any input. Moreover, if $0 \leq \rho < 1$, the number of categories is bounded, even if the number of exemplars in the training set is unbounded. Similar properties hold for the fast-learn slow-recode case, except that repeated presentations of each input may be needed before stabilization occurs, even in the conservative limit. The supervised learning of fuzzy ARTMAP can actively force the learning of new categories by changing the size of ρ from trial to trial.

97.7 Fuzzy ARTMAP Algorithm

Fuzzy ARTMAP incorporates two fuzzy ART modules ART_a and ART_b that are linked together via an inter-ART associative learning module F^{ab} called a *map field.* The map field forms predictive associations between categories and realizes the ARTMAP *match tracking rule.* Match tracking increases the ART_a vigilance parameter ρ_a in response to a predictive error, or mismatch, at ART_b. Match tracking reorganizes category structure so that subsequent presentations of the input do not repeat the error. An outline of the ARTMAP algorithm follows.

ART_a and ART_b

Inputs to ART_a and ART_b are complement coded. For ART_a, $\mathbf{I} = \mathbf{A} - (\mathbf{a}, \mathbf{a}^c)$; and for ART_b, $\mathbf{I} = \mathbf{B} = (\mathbf{b}, \mathbf{b}^c)$ (Figure 97.5). Variables in ART_a or ART_b are designated by subscripts or superscripts "a" or "b". For ART_a, $\mathbf{x}^a \equiv x^a_1 \ldots x^a_{2Ma}$ denotes the F^a_1 output vector; $\mathbf{y}^a \equiv y^a_1 \ldots y^a_{Na}$ denotes the F^a_2 output vector; and $\mathbf{w}^a{}_j \equiv w^a_{j1}, w^a_{j2}, \ldots w^a_{j,2M_a}$ denotes the jth ART_a weight vector. For ART_b, $\mathbf{x}^b \equiv x^b_1 \ldots x^b_{2M_b}$ denotes the F^b_1 output vector; $\mathbf{y}^b \equiv y^b_1 \ldots y^b_{N_b}$ denotes the F^b_2 output vector; and $\mathbf{w}^b_k \equiv (w^b_{k1}, w^b_{k2}, \ldots, w^b_{k,2M_b})$ denotes the kth ART_b weight vector. For the map field, $\mathbf{x}^{ab} \equiv x^{ab}_1, \ldots, x^{ab}_{N_b}$ denotes the F^{ab} output vector, and $\mathbf{w}^{ab}_j \equiv (w^{ab}{}_{j1}, \ldots, w^{ab}_{jN_b})$ denotes the weight vector from the jth $F^a{}_2$ node to F^{ab}. Vectors $\mathbf{x}^a$, $\mathbf{y}^a$, $\mathbf{x}^b$, $\mathbf{y}^b$, and $\mathbf{x}^{ab}$ are reset to **0** between input presentations.

Map Field Activation

The map field F^{ab} receives input from either or both of the ART_a or ART_b category fields. A chosen F^a_2 node J sends input to the map field F^{ab} via the weights $\mathbf{w}^{ab}_J$. An active F^b_2 node K sends input to F^{ab} via one-to-one pathways between F^b_2 and F^{ab}. If both ART_a and ART_b are active, then F^{ab} remains active only if ART_a predicts the same category as ART_b. The F^{ab} output vector $\mathbf{x}^{ab}$ obeys:

$$\mathbf{x}^{ab} = \begin{cases} \mathbf{y}^b \wedge \mathbf{w}_J^{ab} & \text{if the } J^{th}\ F_2^a \text{ node is active} \\ & \quad \text{and } F_2^b \text{ is active} \\ \mathbf{w}_J^{ab} & \text{if the } J^{th}\ F_2^a \text{ node is active} \\ & \quad \text{and } F_2^b \text{ is inactive} \\ \mathbf{y}^b & \text{if } F_2^a \text{ is inactive and } F_2^b \text{ is active} \\ \mathbf{0} & \text{if } F_2^a \text{ is inactive and } F_2^b \text{ is inactive.} \end{cases} \quad (97.32)$$

By (32), $\mathbf{x}^{ab} = \mathbf{0}$ if $\mathbf{y}^b$ fails to confirm the map field prediction made by $\mathbf{w}_J^{ab}$. Such a mismatch event triggers an ART_a search for a better category, as follows.

Match Tracking

At the start of each input presentation ART_a vigilance ρ_a equals a baseline vigilance parameter $\overline{\rho_a}$. When a predictive error occurs, match tracking raises ART_a vigilance just enough to trigger a search for a new F_2^a coding node. ARTMAP detects a predictive error when

$$|\mathbf{x}^{ab}| < \rho_{ab}|\mathbf{y}^b|, \quad (97.33)$$

where ρ_{ab} is the map field vigilance parameter. A signal from the map field to the ART_a orienting subsystem causes ρ_a to "track the F_1^a match." That is, ρ_a increase until it is slightly higher than the F_1^a match value $|\mathbf{A} \wedge \mathbf{w}_J^a||\mathbf{A}|^{-1}$. Then, since

$$|\mathbf{x}^a| = |\mathbf{A} \wedge \mathbf{w}_J^a| < \rho_a|\mathbf{A}|, \quad (97.34)$$

ART_a fails to meet the matching criterion, as in Equation 97.10, and the search for another F_2^a node begins. The search leads to an F_2^a node J with

$$|\mathbf{x}^a| = |\mathbf{A} \wedge \mathbf{w}_J^a| \geq \rho_a|\mathbf{A}| \quad (97.35)$$

and

$$|\mathbf{x}^{ab}| = |\mathbf{y}^b \wedge \mathbf{w}_J^{ab}| \geq \rho_{ab}|\mathbf{y}^b|. \quad (97.36)$$

If no such node exists and if all F_2^a nodes are already committed, F_2^a automatically shuts down for the remainder of the input presentation.

Map Field Learning

Weights w_{jk}^{ab} in $F_2^a \rightarrow F^{ab}$ paths initially satisfy

$$w_{jk}^{ab}(0) = 1. \quad (97.37)$$

During resonance with the ART_a category J active, $\mathbf{w}_J^{ab}$ approaches the map field vector $\mathbf{x}^{ab}$ as in Equation 97.11. With fast learning, once J learns to predict the ART_b category K, that association is permanent; i.e., $w_{JK}^{ab} = 1$ for all time.

97.8 ART Applications

Since the publication of the first ART network in 1987, scientists and engineers have applied these systems to a variety of problems. Researchers often cite unique ART features such as code stability, speed, and incremental learning as reasons for using ART or ARTMAP instead of an error-based neural network such as backpropagation.

The Boeing Company Neural Information Retrieval System (NIRS) has advanced from prototype to implementation in a state-of-the-art computer-aided airplane design system (Caudell, Smith, Escobedo, and Anderson, 1994; Caudell, 1993; Smith, Escobedo, and Caudell, 1993). Engineers now use NIRS for production of the Boeing 747 and 767 airplanes and for design of the Boeing 777. The Neural Information Retrieval System is a hierarchy of ART networks that form compressed content-addressable memories of 2-D and 3-D parts designs. The NIRS shows an engineer who has sketched a part on the CAD system other parts in inventory that may be similar. Inventory proliferation and design time are both saved. Working CAD systems that include the NIRS have already reduced parts inventories by a factor of nine, and Boeing estimates that this technology will save the company up to $80 million per year.

A trained ARTMAP system translates into a set of if-then rules at any stage of learning. This feature has made the network particularly useful in the analysis of large medical databases (Carpenter and Tan, 1993; Ham and Han, 1993; Harvey, 1993; Goodman, et al., 1992). Other ART medical applications include electrocardiogram wave recognition (Suzuki, Abe, and Ono, 1993). ARTMAP test set performance has proved superior to that of other neural networks in application domains such as diagnostic monitoring of nuclear plants (Keyvan, Durg, and Rabelo, 1993), land cover classification from remotely sensed data (Gopal, Sklarew, and Lambin, 1993), and the prediction of protein secondary structure (Mehta, Vij, and Rabelo, 1993). The ART-EMAP network adds to fuzzy ARTMAP spatial and temporal evidence accumulation capabilities (Carpenter and Ross, 1993). These new functions improve performance on both noisy and noise-free test sets, and expand the range of ARTMAP applications to spatio-temporal recognition problems such as 3-D object recognition and scene analysis. VIEWNET is an image processing architecture for invariant 3-D object recognition from sequences of 2-D ARTMAP view categories (Bradski and Grossberg, 1994a, 1994b). ART networks partition perceptual space to enable robot navigation in an unknown, cluttered environment (Dubrawski and Crowley, 1994). Researchers at MIT Lincoln Laboratory use ART systems for both spatial navigation (Bachelder, Waxman, and Seibert, 1993; Baloch and Waxman, 1991) and 3-D object recognition (Seibert and Waxman, 1992, 1991) by mobile robots, as well as for face recognition (Seibert and Waxman, 1993). The Macintosh commercial software Open Sesame! uses an unsupervised ART network to adapt the operating system to a user's work habits (Johnson, 1993). Other applications range from analyses of musical structure (Gjerdingen, 1990) and identification of airborne particles in scanning electron microscopy images for air quality monitoring (Wienke,

Xie, and Hopke, 1994) to military target recognition (Moya, Koch, and Hostetler, 1993) and multivariable optimization of high performance concrete mixes (Kasperkiewicz, Racz, and Dubrawski, 1994). Finally, applications of ART networks continue to include those of the original adaptive resonance theory: to organize, clarify, and predict neural and psychological data concerning learning, memory, recognition, and attention (Carpenter and Grossberg, 1993, 1991; Desimone, 1992; Gochin, 1990; Grossberg, 1988, 1987).

Acknowledgments

Supported in part by the Advanced Research Projects Agency (ONR N00014-92-J-4015), the National Science Foundation (NSF IRI 94-01659), and the Office of Naval Research (ONR N00014-91-J-4100) (G.A.C.).

Supported in part by the Advanced Research Projects Agency (ONR N00014-92-J-4015), the Air Force Office of Scientific Research (AFOSR F49620-92-J-0499), and the Office of Naval Research (ONR N00014-91-J-4100) (S.G.).

The authors wish to thank Cynthia E. Bradford and Diana J. Meyers for their valuable assistance in the preparation of the manuscript.

References

Bachelder, I. A., Waxman, A. M., and Seibert, M. 1993. A neural system for mobile robot visual place learning and recognition, *Proc. World Congress on Neural Networks (WCNN-93)*, I:512–517, Erlbaum, Hillsdale, NJ.

Baloch, A. A. and Waxman, A. M. 1991. Visual learning, adaptive expectations, and behavioral conditioning of the mobile robot MAVIN, *Neural Networks*, 4:271–302.

Bradski, G. and Grossberg, S. 1994a. A neural architecture for 3-D object recognition from multiple 2-D views, *Proc. World Congress on Neural Networks*, IV:211–219, San Diego, CA.

Bradski, G. and Grossberg, S. 1994b. Recognition of 3-D objects from multiple 2-D views by a self-organizing neural architecture, Cherkassky, V., Friedman, J. H., and Wechsler, H., eds., Springer-Verlag, New York, NY.

Bullock, D., Grossberg, S., and Guenther, F. H. 1993. A self-organizing neural model of motor equivalent reaching and tool use by a multijoint arm, *J. Cognitive Neuroscience*, 5:408–435.

Carpenter, G. A. and Grossberg, S. 1987a. A massively parallel architecture for a self-organizing neural pattern recognition machine, *Computer Vision, Graphics, and Image Processing*, 37:54–115.

Carpenter, G. A. and Grossberg, S. 1987b. ART 2: Stable self-organization of pattern recognition codes for analog input patterns, *Applied Optics*, 26:4919–4930.

Carpenter, G. A. and Grossberg, S. 1990. ART 3: Hierarchical search using chemical transmitters in self-organizing pattern recognition architectures, *Neural Networks*, 3:129–152.

Carpenter, G. A. and Grossberg, S., eds. 1991. *Pattern Recognition by Self-Organizing Neural Networks.* MIT Press, Cambridge, MA.

Carpenter, G. A. and Grossberg, S. 1993. Normal and amnesic learning, recognition, and memory by a neural model of cortio-hippocampal interactions, *Trends in Neurosciences*, 16:131–137.

Carpenter, G. A., Grossberg, S., Markuzon, N., Reynolds, J. H., and Rosen, D. B. 1992. Fuzzy ARTMAP: A neural network architecture for incremental supervised learning of analog multidimensional maps, *IEEE Trans. Neural Networks*, 3:698–713.

Carpenter, G. A., Grossberg, S. and Reynolds, J. H. 1991. ARTMAP: Supervised real-time learning and classification of nonstationary data by a self-organizing neural network, *Neural Networks*, 4:565–588.

Carpenter, G. A., Grossberg, S., and Rosen, D. B. 1991a. Fuzzy ART: Fast stable learning and categorization of analog patterns by an adaptive resonance system, *Neural Networks*, 4:759–771.

Carpenter, G. A., Grossberg, S. and Rosen, D. B. 1991b. *A Neural Network Realization of Fuzzy ART,* Technical Report CAS/CNS-TR-91-021, Boston University, Boston, MA.

Carpenter, G. A. and Ross, W. D. 1993. ART-EMAP: A neural network architecture for learning and prediction by evidence accumulation, *Proc. World Congress Neural Networks (WCNN-93)*, III:649–656, Erlbaum, Hillsdale, NJ.

Carpenter, G. A. and Tan, A.-H. 1993. Rule extraction, fuzzy ARTMAP, and medical databases, *Proc. World Congress on Neural Networks (WCNN-93)*, I:501–506. Erlbaum, Hillsdale, NJ.

Caudell, T., ed. 1993. *Adaptive Neural Systems*, 1992 IR&D Technical Report BCS-CS-ACS-93-008, The Boeing Company, Seattle, WA.

Caudell, T. P., Smith, S. D. G., Escobedo, R., and Anderson, M. 1994. NIRS: Large-scale ART 1 neural architectures for engineering design retrieval, *Neural Networks*, 7:1339–1350.

Desimone, R. 1992. Neural circuits for visual attention in the primate brain, *Neural Networks for Vision and Image Processing*, Carpenter, G. A., and Grossberg, S., eds., MIT Press, 343–364, Cambridge, MA.

Dubrawski, A. and Crowley, J. L. 1994. Learning locomotion reflexes: a self-supervised neural system for a mobile robot, *Robotics and Autonomous Systems*, 12:133–142.

Gaudiano, P. and Grossberg, S. 1991. Vector associative maps: unsupervised real-time error-based learning and control of movement trajectories, *Neural Networks*, 4:147–183.

Gjerdingen, R. O. 1990. Categorization of musical patterns by self-organizing neuronlike networks, *Music Perception*, 7:339–370.

Gochin, P. 1990. Pattern recognition in primate temporal cortex: but is it ART?, *Proc. Int. Joint Conf. Neural Networks (IJCNN-90)*, I:77–80, Erlbaum, Hillsdale, NJ.

Goodman, P. H., Kaburlasos, V. G., Egbert, D. D., Carpenter, G. A., Grossberg, S., Reynolds, J. H., Rosen, D. B., and Hartz, A. J. 1992. Fuzzy ARTMAP neural network compared to linear discriminant analysis prediction of the length of hospital stay in patients with pneumonia, *Proc. IEEE Int. Conf. Systems, Man, and Cybernetics,* I:748–753, Chicago, IL.

Gopal, S., Sklarew, D., and Lambin, E. 1993. Fuzzy-neural network classification of land cover change in the Sahel, *Proc. DOSES/EUROSAT Workshop on New Tools for Spatial Analysis,* Lisbon, Portugal, ECSC-EC-EAEC, 55–68, Brussels.

Grossberg, S. 1976. Adaptive pattern classification and universal recoding, II: feedback, expectation, olfaction, and illusions, *Biological Cybernetics,* 23:187–202.

Grossberg, S., ed. 1987. *The Adaptive Brain: Volumes I and II,* Elsevier, Amsterdam.

Grossberg, S., ed. 1988. *Neural Networks and Natural Intelligence,* MIT Press, Cambridge, MA.

Grossberg, S., Guenther, F. H., Bullock, D., and Greve, D. 1993. Neural representations for sensory-motor control, II: learning a head-centered visuomotor representation of 3-D target position, *Neural Networks,* 6:43–67.

Ham, F. M. and Han, S. W. 1993. Quantitative study of the QRS complex using fuzzy ARTMAP and the MIT/BIH arrhythmia database, *Proc. World Congress on Neural Networks (WCNN-93),* I:207–211, Erlbaum, Hillsdale, NJ.

Harvey, R. M. 1993. Nursing diagnosis by computers: an application of neural networks, *Nursing Diagnosis,* 4:26–34.

Johnson, C. 1993. Agent learns user's behavior, *Electrical Engineering Times,* June 28, pp. 43, 46.

Kasperkiewicz, J., Racz, J., and Dubrawski, A. 1994. HPC strength prediction using artificial neural networks, *ASCE Journal of Computing in Civil Engineering.*

Keyvan, S., Durg, A., and Rabelo, L. C. 1993. Application of artificial neural networks for development of diagnostic monitoring system in nuclear plants, *American Nuclear Society Conference Proceedings,* April 18–21.

Kosko, B. 1986. Fuzzy entropy and conditioning, *Information Sciences,* 40:165–174.

Mehta, B. V., Vij, L., and Rabelo, L. C. 1993. Prediction of secondary structures of proteins using fuzzy ARTMAP, *Proc. World Congress on Neural Networks (WCNN-93),* I:228–232, Erlbaum Hillsdale, NJ.

Moore, B. 1989. ART 1 and pattern clustering, *Proc. 1988 Connectionist Models Summer School,* 174–185, San Mateo, CA.

Moya, M. M., Koch, M. W., and Hostetler, L. D. 1993. One-class classifier networks for target recognition applications, *Proc. World Congress on Neural Networks (WCNN-93),* III:797–801, Erlbaum, Hillsdale, N.I.

Parker, D. B. 1982. *Learning-logic,* Invention Report 581–64, File 1, October, Office of Technology Licensing, Stanford University.

Rosenblatt, F. 1958. The perceptron: A probablistic model for information storage and organization in the brain, *Psychological Review,* 65:386–408; reprinted in Anderson, J. A. and Rosenfeld, E., eds., 1988. *Neurocomputing: Foundations of Research,* 18–27, MIT Press, Cambridge, M.A.

Rosenblatt, F. 1962. *Principles of Neurodynamics,* Spartan Books, Washington, DC.

Rumelhart, D. E., Hinton, G. and Williams, R. 1986. Learning internal representations by error propagation, *Parallel Distributed Processing,* McClelland, J. L., Rumelhart, D. E. eds., 318–362, MIT Press, Cambridge, M.A.

Salzberg, S. L. 1990. *Learning with Nested Generalized Exemplars,* Kluwer Academic, Hingham, MA.

Seibert, M. and Waxman, A. M. 1991. Learning and recognizing 3D objects from multiple views in a neural system. *Neural Networks for Perception,* vol. 1, Wechsler, H., ed., Academic Press, New York; NY.

Seibert, M. and Waxman, A. M. 1992. Adaptive 3D object recognition from multiple views, *IEEE Trans. Pattern Analysis and Machine Intelligence,* 14:107–124.

Seibert, M. and Waxman, A. M. 1993. An approach to face recognition using saliency maps and caricatures, *Proc. World Congress on Neural Networks (WCNN-93),* III:661–664, Erlbaum, Hillsdale, NJ.

Smith, S. D. G., Escobedo, R., and Caudell, T. P. 1993. An industrial strength neural network application, *Proc. World Congress on Neural Networks (WCNN-93),* I:490–494, Erlbaum, Hillsdale, NJ.

Suzuki, Y., Abe, Y., and Ono, K. 1993. Self-organizing QRS wave recognition system in ECG using ART 2, *Proc. World Congress on Neural Networks (WCNN'93),* IV:39–42, Erlbaum, Hillsdale, NJ.

Werbos, P. 1974. *Beyond Regression: New Tools for Prediction and Analysis in the Behavioral Sciences,* PhD Thesis, Harvard University, Cambridge, MA.

Wienke, D., Xie, Y., and Hopke, P. K. 1994. An Adaptive Resonance Theory based artificial neural network (ART 2-A) for rapid identification of airborne particle shpes from their scanning electron microscopy images, *Chemometrics and Intelligent Laboratory Systems.*

Wilensky, G. 1990. Analysis of neural network issues: Scaling, enhanced nodal processing, comparison with standard classification, *DARPA Neural Network Program Review,* October 29–30.

Zadeh, L. 1965. Fuzzy sets, *Information and Control,* 8:338–353.

98

Future Directions for Fuzzy Systems and Soft Computing in Industrial Electronics

Mary Lou Padgett
Auburn University

Lotfi A. Zadeh
University of California, Berkeley

Intelligent systems in the realm of industrial electronics are rapidly growing in number, visibility and importance. Much of the growth has taken place during the past few years, reflecting the advances in sensor technology; the availability of embedded systems which can process large volumes of data at high speed, high reliability and low cost; and the employment of newly developed soft computing methodologies for automated reasoning, learning and adaptation.

Basically, soft computing is a consortium or a partnership of methodologies which provide effective tools for the conception, design, and deployment of intelligent systems. The principal constituents of soft computing and fuzzy logic, neurocomputing and probabilistic reasoning, with the latter subsuming genetic algorithms, belief networks and chaotic systems. Within soft computing, the principal contribution of fuzzy logic is a methodology for dealing with imprecision, approximate reasoning, and computing with words; that of neurocomputing is a methodology for learning, adaptation and system identification; and that of probabilistic reasoning are methodologies for evidential reasoning and systematized random search. In the latter, the methodology of genetic algorithms plays a pivotal role.

Although there are some overlaps between fuzzy logic, neurocomputing and genetic algorithms, the underlying methodologies are, in the main, complementary rather than competitive. What this means is that in many cases better performance can be realized by employing a combination of fuzzy logic, neurocomputing and genetic algorithms than by the use of these methodologies in isolation. Such so-called hybrid systems are likely to grow in importance in coming years. A class of hybrid systems which are already in wide use is that of neuro-fuzzy systems. Currently, most neuro-fuzzy systems are basically fuzzy rule-based systems in which gradient programming techniques are employed for tuning and calibration. However, a trend which is becoming increasingly visible involves the use of fuzzy rule-based techniques for the purpose of tuning and optimization of basic algorithms in neurocomputing—especially the back-propagation algorithm. We are also beginning to see the use of neuro-fuzzy-genetic systems in industrial electronics, quality control and pattern matching. In this context, an important dimension of intelligence is a capability for self diagnosis and self-repair.

The concept of Machine Intelligence Quotient (MIQ) plays a central role in the assessment of performance of intelligent systems. So far, no attempts have been made to develop ways of measuring MIQ for particular products, e.g., intelligent battery chargers, washing machines, sorting machines, microwave ovens, etc. However, the time is approaching when the availability of industry-wide standards of MIQ assessment becomes a necessity. In anticipation of this need, you are invited to offer suggestions on (a) how MIQ could be measured for a specific class of systems or products; and (b) employed for purposes of guiding the design process in ways that would reflect users needs and preferences for systems which have a high MIQ.

To contribute a suggestion to the electronic CI Standards News, send e-mail to Mary Lou Padgett at m.padgett@ieee.org. To stay up to date on developments in fuzzy logic and soft computing, contact Professor L. A. Zadeh at zadeh@cs.berkeley.edu.

Future application of fuzzy logic and soft computing to industrial electronics will of necessity involve a wide range of methodologies and techniques. Some of these—and especially the methodologies relating to evolutionary computing and hybrid systems—are discussed in greater detail in the following chapters.

0-8493-8343-9/97/$0.00+$.50

Evolutionary Systems, Computational Intelligence, and Hybrid Systems Applications

Evolutionary Systems

99 **Applications of Evolutionary Systems in Industrial Electronics** *Mary Lou Padgett and V. Rao Vemuri* 1303
Introduction • From Basic Implementations to New Research • Defining Terms

100 **Evolutionary Computation** *Mary Lou Padgett* 1307
Introduction • Design of Evolutionary Systems • Applications • Summary

101 **Genetic Algorithms** *Mark G. Cooper and V. Rao Vemuri* 1316
Introduction • The Basic Genetic Algorithm • String Encoding • Evaluation • Test Fitness Functions • Premature Convergence • Selection • Replacement • Genetic Parameters

102 **Fuzzy Evolutionary and GA Systems** *Mary Lou Padgett* 1321
Introduction • Combining Evolutionary Systems and Fuzzy Systems • Summary

103 **Information Fusion by Fuzzy Set Operations and Genetic Algorithms** *Anna L. Buczak and Robert E. Uhrig* 1325
Information Fusion • Fuzzy Aggregation Connectives • Genetic Algorithms • Two Fuzzy-Genetic Fusion Techniques • Information Fusion for Object Classification • Vibration Monitoring • Results • Conclusions

104 **Neural Evolutionary and GA Systems and Applications** *Mary Lou Padgett* 1338
Introduction • Combining Evolutionary Systems and Neural Systems • Summary

Computational Intelligence and Hybrid Systems Applications

105 **Computational Intelligence Applications in Industrial Electronics** *Mary Lou Padgett and Robert Shelton* 1343
Introduction • Aerospace Applications of Computational Intelligence • From Basic Implementations to New Research

106 **Hybrid Artificial Intelligence Systems** *Lefteri H. Tsoukalas and Robert E. Uhrig* 1346
Introduction • Expert Systems and Fuzzy Logic Systems • Neural Networks and Expert Systems • Neural Networks and Fuzzy Logic Systems • Genetic Algorithms and Neural Networks • Genetic Algorithms and Fuzzy Systems • Discussion and Conclusions

107 **Application Techniques: Combining Fuzzy Logic, Artificial Neural Networks, and Probabilistic Reasoning—Soft Computing** *Okyay Kaynak* 1360
Combining Soft Computing Methodologies • Neurofuzzy Control • The Use of NNs in Consumer Products • The Fusion of GA and FS

108 **Synthesis of Fuzzy, Artificial Intelligence, Neural Networks, and Genetic Algorithm for Hierarchical Intelligent Control** *Takanori Shibata, Toshio Fukuda, and Kazuo Tanie* 1364
Introduction • Artificial Intelligence, Fuzzy, Neural Network, and Genetic Algorithm • Hierarchical Intelligent Control of Robotic Motion • Conclusions

109 Advanced Tools for Adaptive Nonlinear Modeling and Control of Power in Large Systems
Harold H. Szu and Brian A. Telfer 1369
Introduction • Modeling, Control, and Neural Networks • Wavelet and Adaptive Space-Frequency Techniques for Modeling and Control • Summary and Conclusions

110 Application of Model Reference Adaptive Control and Adaptive Time-Delay RBF Networks
Gary G. Yen 1372
Introduction • Dynamic Modeling of Flexible Multibody • Adaptive Time-Delay Radial Basis Function Network • Pace Simulation Study • Conclusions

99

Applications of Evolutionary Systems in Industrial Electronics

Mary Lou Padgett
Auburn University

V. Rao Vemuri
University of California at Davis

99.1 Introduction 1303
99.2 From Basic Implementations to New Research 1303
Basic Evolutionary System Modeling Concepts • Fuzzy/Evolutionary System Implementations • Neural/Evolutionary System Implementations • Future Trends
99.3 Defining Terms 1304

99.1 Introduction

Evolutionary systems are emerging as valuable assets in industrial electronics applications whenever exhaustive search is impractical. Success stories reported in this context include the extension of path-finding algorithms to enhance corner-turning capabilities (Fogel, 1995). Central to these evolutionary systems is the element of randomness in the search strategy. The generation of possibilities combined with planned reinforcement strategies can guide global searches through reduced dimensional spaces and offer novel solutions to problems. Such procedures do not always produce optimal results, but searches can be continued until satisfactory solutions are discovered, or a restart is indicated.

Engineering problems which may be approached profitably by evolutionary methods and genetic algorithms are typically optimization problems characterized by noncomplex, multimodal and/or noisy search spaces.

Some strategies start with the gene/chromosome level and look at the impact of mimicking such biological phenomena as linking, mutations and crossovers. A simulated new generation is rated and desirable attributes somehow reinforced. Design of efficient and appropriate mechanisms for rating and reinforcing is critical to the system's performance over time. Adjusting parameters to tune this performance is also important.

Other strategies take a population view and follow the statistical trends of the larger group. Both bottom-up and top-down manipulation are successfully employed in many applications. Many of the reported applications creatively combine neural systems with evolutionary systems or fuzzy systems with evolutionary systems. Some, of course, combine all three.

The articles in this chapter will describe the basic techniques for design and application of evolutionary systems which pertain to industrial electronics. Combination of evolutionary systems with neural systems or with fuzzy systems is also covered. The following chapter on computational intelligence and hybrid systems applications includes the combination of multiple techniques. Updates and discussions will be featured in the CI Standards News (contact: m.padgett@ieee.org). The news will feature debates about evolutionary systems with contributions by individuals suggesting definitions, design procedures, lessons learned and documentation tips suitable for future standardization.

As in the preceding chapters in this section, topics are arranged to first feature implementations suitable for commercial use, then move into areas of research development. The topics, articles, and authors are listed and discussed below.

99.2 From Basic Implementations to New Research

Basic Evolutionary System Modeling Concepts

Basic evolutionary system modeling concepts are an intrinsic part of evolutionary computation. The perspectives of [Fogel, 1995] on the effectiveness of evolutionary algorithms vs. genetic algorithms are presented, and their relationship is illustrated by biological examples. Some design considerations are given, to help industrial electronics engineers translate the evolutionary principles into working applications. Top-down, bottom-up design strategies combining elements of both evolutionary and genetic algorithms strategies are probably indicated for the construction of complex, applications level systems.

The introduction to evolutionary concepts is followed by a clear, tutorial description of bottom-up strategies commonly employed in engineering applications of genetic algorithms.

Fuzzy/Evolutionary System Implementations

Evolutionary systems can augment fuzzy systems by adding the ability to learn from data and optimize. EC techniques can be

0-8493-8343-9/97/$0.00+$.50

used to auto-tune structures and parameters in fuzzy systems, using historical data and adding an element of randomness for exploration. Conversely, fuzzy algorithms can provide real-time "soft" mathematical modeling based on human knowledge representations. Both EC and fuzzy systems are effective in the presence of nonlinearity, and the combination of their strengths provides a strong engineering basis for stable, flexible applications.

Fuzzy system properties present an effective methodology for optimizing decision fusion processes. When dealing with multiple sensors, fusion of their output is critically important in many industrial electronics applications. Information fusion by fuzzy set operations and genetic algorithms is a source of generalizable strategies for deciding which fuzzy aggregation methods to use in a particular application. This addresses a problem encountered by everyone constructing a fuzzy application, and offers examples of practical use.

Neural/Evolutionary System Implementations

Neural evolutionary/GA systems and their interrelationships spur many applications. Evolutionary systems add optimization capabilities to neural networks. These optimizations can be configured to give a sensitivity analysis on potential inputs and outputs to a NN, to auto-tune neural structures or parameters, to escape from local minima and, in general, to use historical data interspersed with randomness for exploration. Conversely, neural systems can serve as evaluation functions within an evolutionary system. Neural (and/or fuzzy) systems can also be an integral part of a global system validation process and the operator interface. Both EC and NN excel at learning from data and nonlinearity. Neither use a mathematical model, operator knowledge, nor knowledge representation. They are thus useful in automating data analysis without human involvement, and can be combined with fuzzy systems when such elements are needed.

Future Trends

Evolutionary systems, composed of evolutionary algorithms or genetic algorithms, provide valuable optimization tools for the industrial electronics engineer. Drawing ideas from biology, the articles in this chapter illustrate the types of engineering applications that are in current use, and suggest avenues for further development.

Effective strategies for increasing the flexibility and long-term utility of an application with an evolutionary computation (EC) component are presented. Key among these concepts are designing for parallelization, establishing effective step-wise validation procedures, and blending EC systems with neural and/or fuzzy systems.

Progress in this rapidly moving technology may be monitored by interacting with the IEEE Standards projects in the area. Contributions to the CI Standards News may be directed to m.padgett@ieee.org, or to r.vemuri@ieee.org.

99.3 Defining Terms

Some defining terms are listed below. Many of the definitions below have been taken from articles in this chapter, in some cases, verbatim. These definitions were placed in the public domain prior to being contributed to the CI Standards News (m.padgett@ieee.org). Updates can be found at website: http:\\www.mindspring.com\~pci-inc

alleles: (GA) possible values for a gene. May be finite or continuous, but frequently modeled as finite. (EA) abstracted behavioral or genetic traits of individuals (ES) or of an entire species (EP).

chromosome: (GA) abstraction of the biological grouping of genes into a connected string. Each individual has a collection of such strings with properties characteristic to the species. Frequently modeled as a finite, ordered string of p symbols from a fixed collection of n possible selections. (ES) components of the strings are frequently continuous in value, with Gaussian distributions, and may represent behavioral traits of an individual or species, as opposed to the expression of an attribute based on the chemical composition of a particular genetic site.

continuous function optimization: (EA) values assigned to genes may be selected from a continuous range.

crossover point: (GA) location along a string or chromosome where the physical characteristics of the adjacent genes allow twisting of adjacent chromosomes. When these twisted chromosomes separate during reproduction, the front of one may be attached to the former tail of the other (and vice versa). In biology, this allows genes from the mother, say M1 to M5, to combine with genes from the father, say P1 to P5, to produce a new chromosome: M1, M2, M3, P4, P5, and its partner, P1, P2, P3, M4, M5.

crossover: (GA) twisting and recombination of genetic material from adjacent chromosomes to form new chromosomes with genes from two different sources when the adjacent chromosomes separate during reproduction. See also: crossover point.

crowding selection: (GA) reproduction mates individuals in a randomly selected subset of the environment which are the most similar in phenotype or performance. This encourages mating of individuals from the same niche, and implements a multi-modal search.

dynamic GA's: variation of the parameters such as crossover or mutation probability occurs over time.

dynamic parameter encoding (DPE): parameter ranges are varied over time. When convergence appears near, the parameter window size may be reduced to allow focusing and greater precision when close to the desired optimal solution.

elitist strategy: (GA) reproduction is allowed only for the individuals with the highest fitness scores, assuming rating criteria are fixed over time. May prematurely restrict the homogeneity of the population, but good for quick unimodal search.

evaluation step: (GA) algorithm to rate the performance or acceptability of each string in the current population with respect to a desired target for such performance. (EC) algorithm to evaluate the behavior of each string with respect to the rest of the population, so that relatively poor solutions may be culled by blocking reproduction.

evolution: (GA) a reductionist or bottom-up process operating on the strings or chromosomes of an individual. (EA) a process operating on a population which impacts its heterogeneity (or diversity), and thus its ability to adapt.

evolutionary algorithms (EA): top-down optimization strategies based on population genetics focusing on individuals or species (evolutionary strategies (ES) and evolutionary programming (EP), respectively.)

evolutionary computation (EC): primarily algorithms for optimization, with mathematical models abstracted from biological evolution. Paradigms include: top-down evolution of populations (EA) focusing on the species level (EP) or the individual level (ES); and bottom-up evolution focusing on manipulation of individual genes (GA). See also: genetic algorithms (GA), evolutionary algorithms (EA), evolutionary strategies (ES) and evolutionary programming (EP).

evolutionary programming (EP): Simulated evolution of competing algorithms designed to develop artificial intelligence. Generation of a simulated population is followed by three steps: evaluation, selection and reproduction. Changes emphasize mutations such that the behavior of an entire species varies from generation to generation according to some distribution, such as Gaussian with zero mean and small variance. Contrast to EA's emphasis on individuals in the population, and GA's where population attributes are not considered.

evolutionary strategies (ES): Generation of a simulated population is followed by three steps: evaluation, selection, and reproduction. Evolution strategies use mutations such that the range of behavior from each individual parent to its offspring follows a selected distribution, such as Gaussian with zero mean and small variance. See also EC.

exploration: (GA) algorithmic pressure to search the entire space, avoiding premature convergence without unnecessarily jumping around the space. See also: exploitation.

exploitation: (GA) algorithmic pressure to reward the most beneficial traits without dangerously reducing population variance to premature homogeneity. A homogeneous population cannot respond to changing environmental pressures. See also: exploration.

fitness proportionate reproduction (FPR): (GA) probability of reproduction for a particular string is directly proportional to its non-zero fitness score.

fitness function: (GA) algorithm rating the performance or acceptability of a particular string. The function should converge to a target value as string performance improves.

fitness score: (GA) value of the rating of a particular string by the fitness function. (EC) rating of the behavior of a particular string or species with respect to the entire population.

gene: (GA) site on a chromosome corresponding to a particular attribute.

generation: (GA) group of chromosomes (representing individuals) subjected to a cycle of evaluation, selection and reproduction. (EC) population with behavioral traits usually varying from the parent population according to a Gaussian distribution.

genetic algorithm (GA): a stochastic search technique based on abstractions of the biological behavior of genes and chromosomes under pressure from the environment. Generation of a simulated population is followed by three steps: evaluation, selection and reproduction. Contrast with EC and EP, where population dynamics are of primary concern.

genotype: (GA) symbol(s) selected and placed at the site of a gene for a particular implementation.

homogeneous: (GA) population with small amount of genetic variability.

heterogeneous: (GA) population with a large amount of genetic variability.

hybrid GA's: individuals are gradually clustered into species with similar characteristics where reproduction is allowed only within a species.

intelligent behavior: (EP) the ability to 1) predict environmental states and 2) respond to the predictions based on a goal.

linking: (GA) genes with adjacent or very close sites on a chromosome may be linked due to physical proximity and chemical composition.

mutation: (GA) alteration of the material at a particular genetic site for an individual. The biological chemical changes to a gene are frequently modeled as bit inversion, addition or deletion. Some genes or expressions of a gene are highly prone to mutations (such as albinism). Others are not. The mutation algorithm usually includes a probability of mutation for each site. (EA) frequently modeled as samples from a normal distribution used to perturb vectors, producing a new generation with strong links to the parent generation.

niche: (GA) a subdomain of the search space.

phenotype: (GA) physical expression of the attribute governed by a gene, due to the selection of symbols for a particular implementation.

population: (GA) a simulated group of chromosomes representing potential individuals or trial solutions to the problem of survival of the fittest under environmental pressure, where the pressure is modeled as rating and reinforcement algorithms. Randomized information exchange offers hope for improvement over the best of the older generation.

ranking selection: (GA) FPR variant where reproduction for a particular string is computed based on its fitness rank compared to the rest of the population.

replacement: (GA) part of the reproduction algorithm governing the possible termination of parent chromosomes to make room for the next generation.

reproduction step: (GA) algorithm to produce offspring to form the next generation. Biological abstractions of genetic linking, mutations and crossovers are often included. (EC) algorithm to produce offspring for the next generation with variation due to mutations usually having Gaussian distribution with zero mean and small variance.

selection step: (GA) reinforcement of desirable attributes by designating a probability of reproduction to each individual (chromosome or string). A probability of zero would eliminate reproduction by that individual. (EA) punishment of undesirable population *behaviors* by culling relatively poor strings, usually limited so that the variation of the population behavior from generation to generation follows a Gaussian distribution with zero mean and low variance. In contrast to GA strategies which usually increase reproduction by favored individuals.

species: (GA) portion of the population having similar chromosomal characteristics. In biology, a group of individuals capable of producing fertile offspring.

tournament selection: (GA) FPR variant where groups of individuals are randomly selected then ranked according to fitness. The group winner reproduces, so that the most successful individuals in a population may reproduce many times. This may prematurely restrict the population, but is good for a short search for one peak.

References

Fogel, David B. 1995. *Evolutionary Computation,* IEEE Press, Piscataway, NJ.

100
Evolutionary Computation

100.1 Introduction 1307
100.2 Design of Evolutionary Systems 1307
Objectives and Performance Measures • Models and Functions • Tools and Resources for Implementations • Input/Output • Architectures and Connections • Component Functions • Variable Parameters • Recall, Learning, Dynamics and Update Mechanisms
100.3 Applications 1313
100.4 Summary 1315

Mary Lou Padgett
Auburn University

100.1 Introduction

Evolutionary systems are becoming increasingly practical as hardware for parallel and distributed computation becomes more sophisticated and economical. Current approaches for modeling biological evolution to produce computational intelligence vary. Evolutionary computation (EC) can be top-down or bottom-up. *Evolutionary algorithms* use top-down approaches based on dynamical behavior of populations of individuals or of species (evolutionary strategies (ES) or evolutionary programming (EP), respectively. *Genetic algorithm* approaches are bottom-up, modeling the phenotypic expression of attributes based on individual gene sites along an artificial chromosome (GA's) (Fogel, 1995). Each of these strategies has its strengths and appropriate applications. These are discussed in more detail in the following paragraphs.

100.2 Design of Evolutionary Systems

In each of the approaches to modeling evolutionary systems, generation of a simulated population is followed by three steps: evaluation, selection, and reproduction. See the paper by Cooper and Vemuri. Some design options and suggestions are discussed below. Table 1 outlines some of the paradigm specification modules suggested as design aides for evolutionary system implementation. These modules are:

1. Objectives and Performance Measures.
2. Models and Functions.
3. Tools and Resources for Implementations.
4. Input/Output.
5. Architectures and Connections.
6. Component Functions.
7. Variable Parameters.
8. Recall, Learning, Dynamics and Update Mechanisms.

Objectives and Performance Measures

Evolutionary systems are primarily implemented as optimization algorithms with convergence properties of interest. Applications of these systems vary. The selection of a top-down or bottom-up approach, or a combination of these will depend on the goals of the particular application. This paper, focuses on micro-level implementations of basic Evolutionary Computation paradigms, comparing alternatives. Later articles in this chapter will give more details about applications and combination of evolutionary systems with fuzzy or neural components.

The text below gives examples drawn from biology which illustrate the problems encountered in configuring an engineering solution to a problem with an EC component. Deciding whether to select an evolutionary algorithm or a genetic algorithm involves considerations such as those mentioned in the animal breeding and dermal ridge analysis examples below. The animal breeding methodologies (GA-like) are of concern to engineers studying the environmental impact of their policies and products. Some of the pro-offered solutions have ramifications that are self-defeating. These long-term effects can be simulated using EC technology. The dermal ridge analysis example provides details about the inheritance of dermal ridge transverseness. The FBI and other groups are currently studying fingerprints, trying to find ways of classifying and characterizing them. Wavelet analysis is being employed to try to determine transverseness, but it is hard to find the center of the fingerprint. Some of the approaches that would first occur to analysts in this area are inappropriate because of the properties of ridge transverseness inheritance. Study of population dynamics with regard to fingerprints is productive, but an individual inheritance strategy focusing on parent-child genotypes will have limited success. The properties of the dermal ridge/embryology intelligence differential relationships outlined may suggest design considerations necessary for successful implementation of related industrial electronics pattern recognition applications.

0-8493-8343-9/97/$0.00+$.50

In nature, evolution is considered to be a response to environmental pressures on the behavior patterns of species (top-down). For millennia, man has intervened in the breeding habits of animals and in plant reproduction (bottom-up). An individual shepherd may wish to produce sheep with solid white wool, and mate selected animals accordingly. If generation after generation of white sheep are interbred within a small population, and any black wool, striped or spotted sheep are culled from the breeding population, white sheep should be produced. Because albinism (absence of pigment) is frequently linked with other possibly undesirable traits, lethal or damaging genes may be preserved along with the desired "white wool" genes. Highly inbred dogs and cats frequently have this problem, also. If inbreeding has reduced the variability of the population in its attempt to produce homogeneity of the "white wool" gene, survival of the species is threatened. Modern day environmentalists are fighting to preserve viable gene pools for the animals of Africa, rare fish species threatened by dams, the rain forests of South America, and many other living things. These environmentalists look at population genetics and survival of the species, and have a more "top-down" approach than the ranchers raising sheep (bottom-up) near the habitat of endangered wolves. In order to practically implement the species survival plans, geneticists use bottom-up and top-down strategies to try to control breeding on a temporary, crisis-management basis. The population genetics, animal breeding and plant propagation practices of working geneticists are a matter of record, and can provide valuable insight in setting goals and performance measures for computer-based modeling of evolutionary systems. Only in the latter half of this century have these practices emerged from the secrecy of skilled breeders, to be placed in print. The advent of computers had aided record-keeping, and allowed the application of computationally intensive population genetics techniques to the toolbox of geneticists.

When the project on the population genetics of dermal ridges commenced, in 1970, population statistics were computed on Monroe calculators by teams of researchers, paired up for verification of tedious data entry. Discovery of an additive human gene governing fingerprint ridge alignment followed quickly after the advent of statistical analysis packages such as SAS and computers which worked with cards. Holes were punched to record data, and the same input dataset could be processed repeatedly, using different analytic techniques. A population study on the fingerprints of twins and their families showed trends indicative of additive inheritance of ridge variation from vertical alignment (Padgett, 1973). Racial and cultural patterns can be observed in this data, as niche-breeding apparently produced clusters of patterns. Trends within the entire population were easy to identify using regression analysis, but their high variability made analysis of individual families difficult. After some experience, the researchers learned to predict factors such as race, and to distinguish identical twins from fraternal twins when given a family unit or small selection of prints. This experience-based analysis was very difficult to characterize with regard to a few firm rules for making decisions. In a case like this, trying to selectively breed for slanted ridge alignment (thought to be linked to species intelligence, but not to individual intelligence) would be very difficult. On the other hand, keeping track of the population variability and comparing the statistical distributions of population clusters offers a very good way of tracking migrations, interbreeding and evolution of the species. (Many studies of monkeys, chimpanzees and other animals exist for comparative analysis. Their dermal ridges typically are highly whorled and the centers are vertically aligned)

Using this example may assist the working engineer in the planning of an evolutionary system model. According to popular reviews of evolutionary computation, some knowledge of the characteristics of the universe to be explored is very helpful. For example, some techniques can be successfully applied to explore a population having a smooth, unimodal distribution. Other strategies work best for multi-modal situations. Noise and non-convexity of the space to be explored also impact design strategies. The sooner in the design phase that such properties are recognized, the easier it is to adjust data analysis policies to cope with them.

To continue with the analogy above, an evolutionary computing strategy might input a population of ridge slopes, cull the nearly vertical slopes, and reproduce the rest. A number of pleasant and unpleasant outcomes could result. These will be discussed below. An engineer working in industrial electronics might be interested in finding a key feature associated with a desired performance or pattern. Physically, dermal ridge alignment depends on the swelling of the volar pad of the embryonic finger during a critical time period. Well-developed pads found on species and races with high intelligence are sloped. Lower animals and some human races have a prevalence of very symmetric, conical volar pads. Unfortunately, embyros with chromosomal aberrations (such as the extra chomosome found in Down's Syndrome) and some embryos with incipient heart deformities exhibit acute swelling of the volar pads which is very non-conical, and leads to extreme variation of the slopes from vertical.

In chemical processes in industry, similar conditions may be found. Suppose that a chemical process produces a product where interior bubbling is measurable and has an impact on final performance on stress tests. The "bubbling" measure may not have a clearly obvious correlation with all the test results. A desirable objective might be to find the best distribution of "bubbling" in a production lot with respect to performance on stress tests. How homogeneous does the population of "bubbles" need to be? What is an optimal "bubble" measure? How can a process be monitored over time so that bad production lots can be rejected or recycled early in the process? How can aberrations mimicking good results be avoided? What system parameters (such as temperature) can be allowed to vary over time and from vat to vat to encourage the exploratory production of slightly different "bubble" measures? These questions could be addressed by setting up parallel processes where temperature has some impact on "bubbling." Suppose at the end of the process, each vat is tested, and the contents rejected or accepted. Temperature variation and "bubble measure" are recorded as inputs and outputs of the process in addition to stress test results. There are many variations in evolutionary strategies which might be employed to investigate such a scenario.

As typical in computer programs or simulations, the automated evolutionary system does exactly as it is instructed, not necessarily what the analyst desired. Caution should be employed in any mathematical analysis of an EC application. Commonly used simplifying assumptions can drastically alter the convergence properties from those needed. Many mathematical properties of EC's can be illustrated by Markov Chains (Fogel, 1995), but active research programs are underway to extend the theoretical basis for EC.

For system evaluation, it is desirable to set a single performance criterion, or to express desired performance as a weighted combination of measures, and test the results for practicality (See Chapter 101 and Padgett, 1982.) All simulation designs need to be globally planned, then refined as the design iterates through stages of development and testing. This is particularly true of evolutionary systems.

Models and Functions

For the purpose of constructing a simple illustration, let us configure the dermal ridge alignment situation as a problem in evolutionary computation which is similar to situations encountered in industrial processes such as chip manufacture. The processes discussed below are modified from EP processes in (Fogel, 1995, p. 136).

1. Start with an initial population of, say n, randomly selected individuals each having 10 digits. The embryonic volar pad swelling of each individual serves as a base value for future finger ridge transverseness, or alignment of the central ridges of the print with the vertical axis of the finger. Arrangement of the ridges around the swollen pad is based on the shape of the pad, and an element of chance enters in. A conical, symmetric pad tends to produce concentric circles, with a vertical ridge in the center. Sloped pads tend to produce loops with non-vertical centers. Grossly swollen pads produce flattened arches with horizontal ridges.

   The volar pad shape for individual i can be considered to be a vector T_i. The influence of this vector is then perturbed by Uniformly distributed variations, (random ridge growth around the contours of the pad), giving the vector of outcomes, t_i. These outcomes give a continuous, quantitative measure of central ridge alignment with the vertical axis of the finger. Finger Transverseness (FT) is determined for each finger., and the average (AFT) is computed for each individual.
2. Evaluate each individual, say by performing an IQ test. The IQ test result may correspond to the fitness score, $G(F(t_i), v_i\}$ where F is the true fitness, and v represents an element of randomness or some other relation with t_i, and G is the score assigned by the algorithm. Thus, $G(C_i)$ is the *measured* IQ of child I. If there is any merit to the observation that vertically aligned, highly whorled prints are characteristic of lower primates and the more cultured races of humans have predominately slanted central dermal ridges (looped fingerprints), then this measure should produce strong pressure for selection of individuals with non-vertical central dermal ridges. Because abnormalities such as Down's Syndrome flatten fingerprints (and reduce variability in ridge alignment among an individual's digits), uniformly horizontally aligned ridges should be strongly selected against.
3. Begin a cycle of processing a generation by *reproduction*. Evolutionary systems vary at this step. (Fogel 1995, p. 136) recommends generating one new individual, $i + n$ from each parent, i, by adding Gaussian noise with zero mean and small, non-zero variance to each element of the parent vector. Care must be taken to sample from independent random number streams. Scaling is also a concern.

With respect to dermal ridge inheritance, this procedure requires strong assumptions about the similarity of parents of the resulting child, $i + n$. An alternate approach to generation of a child is for pairs of parents to generate children. This complication requires a procedure for selection of mates and number of children to generate. It focuses on reproduction at the individual level more than at the species level.

If individual-level, two-parent reproduction is chosen, there is a biological basis for generation of a new individual. As discussed above, an individual has a base-line volar pad swelling with variation between digits determined by perturbing the baseline measure for 9 digits with samples from a uniform distribution. This genetically determined baseline swelling cannot be directly measured, but can be estimated from the average dermal ridge angle of the mother (Mbar) and that of the father (Fbar) and the maximum and minimum values for the population, (Pmax and Pmin). The normalized baseline measure for an individual, C, is computed as

$$\text{Cbar} = 0.6\,[\text{Mbar}] + 0.4\,[\text{Fbar}],$$

$$\text{C} = [\text{Cbar} - \text{Pmin}] \,/\, [\text{Pmax} - \text{Pmin}].$$

to account for inheritance governed by an additive gene having some maternal effect due to prenatal environment (Padgett, 1973). Normalization gives the value, C, in terms of percent of maximum for the population. This normalized value is similar to the normalized value for the volar pad swelling. Dermal measures for the child are simulated by perturbing C with samples from nine Uniform distributions. (The selection of Uniform distribution versus Gaussian distribution here is somewhat arbitrary.)

If the above biological reproduction scheme is selected, there are no obvious restrictions on the variation of the population of children from the population of parents. This is an important difference between evolutionary systems. Convergence to local or global maxima is impacted by reproductive restrictions. Careful mathematical analysis of the particular problem being considered is essential for success!

Many evolutionary systems add variation to the reproduction step by introducing mutations and other simulated exchanges and modifications to genetic material or to entire chromosomes. Species-level systems focus on modeling mutations of individual genes, and tend to leave modification or exchange of large groups of genes (pieces of chromosomes) to the individual-level evolutionary systems.

4. *Evaluation* of the fitness of the new individuals is the next step. Suppose each new child is given an IQ test as its fitness score. This assumes that the test given is such that the score values move toward a target or ideal value (here a maximum) as the performance on a task improves (See Cooper and Vemuri's paper in this chapter.)

   Here, it also assumes that AFT is related to intelligence, so that movement of the population Mean AFT toward a peak value will be observed as intelligence scores increase. Mating individual parents with high IQ scores does not produce high IQ children, as a general rule. On the other hand, species and racial IQ averages are closely related to their mean AFT. Species-level algorithms thus seem to be appropriate mechanisms for exploring the relationship between mean AFT and IQ. It might be possible to estimate the IQ score of a child from the distribution and average of its finger transverseness in order to simulate the evolution of lower primate AFT to human AFT, but that would require a lot of assumptions, and is mentioned only as an interesting illustration of the types of problems encountered in applying evolutionary systems.

5. *Competition* is a valuable consideration in *selection* of individuals for reproduction. To rank the fitness of all the individuals in the population, a comparative value can be assigned. Here again, algorithms for doing this are varied. An exhaustive paired comparison can be reduced by uniformly sampling a subset of the possible pairs and ranking the individual compared to this sampled selection. Chapter 101 gives a clear and practical discussion of selection strategies. Fogel's algorithm, cited above, employs the most highly recommended of these to rank individuals with respect to the rest of the population, or with respect to a uniform sample from the rest of the population. Raw scores are used as tie-breakers for ranking, when needed.

6. *Selection* of some individuals to reproduce forces some type of elimination, to maintain a desirable population size. Selection strategies vary according to the needs of the particular application. Some are best for unimodal searches, others for multi-modal searches. See Chapter 101 for suggestions. In Fogal's algorithm, the n worst ranking individuals are culled from the population, and the remaining n form the next generation.

   Some variants to this procedure focus on elitism, making multiple copies of the best-performing solutions. In genetic algorithms, this strategy is often employed. As with human intervention in breeding schemes, inbreeding and premature homogeneity may become problems. The same solutions used by animal breeders are applied by evolutionary systems engineers to counter these problems.

7. The process repeats steps 3–6 until stopping criteria are met. A population mean close enough to the desired value may be achieved, or a time limit may stop the process. Recent developments in evolutionary systems feature dynamic modification of algorithmic parameters and explore the suitability of various implementation schemes to classes of applications.

Tools and Resources for Implementations

Evolutionary systems are strong candidates for parallel and distributed processing (Fogel, 1995). Serial simulation of evolutionary systems may be used, but thought should be given to migration to a parallel processing environment.

Resources needed include personnel with a practical knowledge of simulation of random processes. The need to check assumptions and assure independence of random number streams is acute. The type of models used in evolutionary systems lend themselves to tempting simplifications which then modify properties involving convergence.

Software tools to analyze data real-time and visually display the results are very helpful in evolutionary and other simulation models. The simulation literature and banks of software tools offer engineering-oriented "canned" tools which can greatly enhance the efficiency of development of an evolutionary system, and enhance its validity.

Continual validation of the appropriateness of evolutionary progress from generation to generation is vital. Resources should be allocated to make this reassurance easy and effective. The final utility of the system depends heavily on the faith that the users and project managers have in the results. Plan to reinforce their confidence in the correctness and appropriateness of the system.

Input/Output

Input and output data have characteristics governed by the application being developed. It is important to know something about the nature of the universe being explored. Input data should be representative of the system being characterized, or the results will not apply to the problems intended to be addressed.

Data representation may be real-valued or binary. For digital computer implementation, even the real-valued data must be discretized, but the precision may vary. Frequently, Evolution Strategies (ES) and Evolutionary Programming (EP) use real-valued representations. Genetic Algorithms (GA) use binary strings, but sometimes move to real, continuous valued implementations.

Factors such as modality, noise level, and character, convexity or concavity will impact the selection of implementation strategies at each step. As soon in the iterative design process as these factors become known (or change), this knowledge should be used to tune the algorithm. As with neural network systems, obtaining and analyzing the input and output to the system accounts for a large amount of the effort expended in developing the model.

Considerations include:

- Granularity.
- Precision.
- Range.
- Distribution.
- Test set with same properties as Universe of interest.
- Modality of Universe.
- Modality of current neighborhood.
- Noise level and character.
- Convexity or concavity of surface to be explored.

Architectures and Connections

In evolutionary systems, architectures and connections implement the flow and storage of data. These systems may be depicted as top-down or bottom-up. They may focus at the species or at the individual level. Mechanisms for evaluation, selection, and reproduction vary, as do capabilities for mutation and recombination. Data structures for these algorithms are not peculiar to evolutionary systems, but their combination is characteristic.

Population number and the structure of individuals are design decisions. Length, precision, and adaptive capabilities are involved. Structures may be adapted during evolution. The space of operation for the individual should be carefully defined and fully explored and tested at extrema. Strategy parameters for an individual (if any) are part of the structure. For example, the individual variables (say transverseness of each finger), the standard deviations or variances and the rotation angles or covariances must be accounted for in the individual's structure.

Component Functions

The main steps in evolutionary systems are considered to be:

- Evaluation
- Selection, and
- Reproduction with

mutation, and sometimes
recombination.

Evaluation

Evaluation involves assigning a fitness score which will improve as performance on a fixed task improves.

Test fitness functions may supplement the global performance measure specified above. The five functions below are typical test patterns used to check an algorithm's fit to its problem domain (See Chapter 101):

1. Function of a sine wave.
2. Product of an exponential function and the function above.
3. Shekel's Foxholes (25 optima).
4. Two far-distant global optima.
5. Five randomly and dynamically characterized optima.

Choose functions for testing to see whether the implementation will converge to a single local optimum, the global optimum, or find the k-best peaks, according to the system objectives discussed earlier. It may be that such objectives need to change over time from coarse estimates designed to quickly approach the neighborhood of the desired solutions, to finer-tuned explorations which assume that the correct solution is already nearby. In neural system training, parameters may be adapted over time to adjust step size and so forth. Similar techniques can be applied to tuning the performance of evolutionary systems over time.

Constraints on optimization problems may involve order, selection without replacement, and other restrictions. In evolutionary systems, constraints may be implemented in various ways, such as:

1. Penalty terms in the fitness function being optimized, good for a few, narrowly defined constraints; avoid production and testing of numerous illegal individuals.
2. Decoders which avoid building illegal individuals (Michalewicz, 1992).
3. Chromosome repair.

Comparative Fitness Functions

ES

- Top-down.
- *Individual*-level behavior (phenotype).
- Multiple effects for each gene (pleiotropy), multiple genes for each effect allowed (polygeny).
- Population fitness in an adaptive landscape.

EP

- Top-down.
- *Species*-level behavior (phenotype).
- Multiple effects for each gene (pleiotropy), multiple genes for each effect allowed (polygeny).
- Population fitness in an adaptive landscape.

GA (Fogel, 1996)

- Bottom-up building blocks of genes.
- *Gene* level.
- Fitness assigned to individual components based on average performance.
- Interactions hard to model.

Selection

Selection techniques are varied. (See Chapter 101; Baeck, 1994; Fogel, 1994.) They are intended to maintain or increase the fitness of a population. A "fit" population anticipates its environment, and moderates its behavior accordingly, successfully producing a proportionally large number of fertile offspring.

Comparative Selection Procedures

ES

- Deterministic, extinctive selection.
- Rank-based selection.
- No scaling required.
- High selective pressure.

EP

- Probabilistic, extinctive selection.
- Low selective pressure.

GA

- Probabilistic, preservation selection.
- Absolute fitness based selection.
- Scaling mechanisms needed.

Common Selection Techniques Include

1. Fitness proportionate reproduction (FPR). Need to scale fitness scores as homogeneity increases quickly locates a nearby peak and converges to it (careful with multimodal situations). May have premature convergence.
2. Ranking selection.

   Variant of FPR.
   Relative performance scales fitness scores.

3. Tournament selection.

   Variant of FPR.

4. Crowding selection.

   Multi-modal searching (niche mating).

5. Elitist strategy selection.

Exploration of the solution space to avoid premature convergence and homogeneity should be balanced with the *exploitation* of good attributes. Replacement eliminates some member of the population to make room for their offspring. This may focus on preserving desirable qualities (GA), or it may emphasize culling a percentage of undesirables (ES/EP). Adjustment of genetic parameters can also help balance the search.

Reproduction

Reproduction may be species-level or individual-level. Asexual or sexual reproduction may be simulated. Mating patterns and number of offspring are design parameters. Evolutionary strategies often maintain a linkage between generations such that the deviations from parent to child have a Gaussian distribution with 0 mean and small non-zero variance.

Variability in a new generation is often enhanced (on a trial basis) by simulated mutation and/or recombination of genetic material. Mutation includes the localized modification, addition or deletion of genetic material. Crossover or recombination is a relocation or swapping of a long string of adjacent genes. The top half of a mother's chromosome may wind up connected to the bottom half of a father's chromosome, and vice versa. Other sources of variation include errors in replication of the chromosome and environmental constraints (Fogel, 1994).

Mutation and crossover for GAs explained in the following article by Cooper and Vemuri. Some evolutionary algorithm conventions (Baeck, 1994) are listed below:

Mutations

Simple standard mutations:

Normal distribution.
Expected value zero.
Standard deviation must be adapted.

Alternate standard mutations:

Mutational stepsize control.
Momentum adaptation for nonzero expected values for gaussian random number.

Correlated mutations.

Recombination

Discrete, one source chosen.
Intermediate, combination of sources chosen.
Dual: 2 randomly chosen parents create one child.
Global: 1 parent randomly chosen for each component of the child
Recombination of the phenotype vector, the variance vector and the correlation vector are separate processes, and are usually different

Comparative Reproduction Procedures

ES

- Mutation is main operator.
- Normally distributed mutations:

  Distribution of new behaviors nearly continuous.
  Behavioral link from parent to offspring very strong.

- Recombination: discrete, intermediate.

EP

- Variation added as self-adapting variances.

- Normally distributed mutations:

  Distribution of new behaviors nearly continuous.
  Behavioral link between generations very strong.

- No recombination

GA

- Recombination as main operator.
- Crossover variations: uniform, k-point (and others).
- Mutations: bit reversal, addition, deletion (and others).

The lines between these evolutionary paradigms become indistinct as variations abound. One primary caution related to freely combining elements from different paradigms is to take care about the impact on convergence and on the relationship to the universe planned to be modeled.

Variable Parameters

Some typically found evolutionary parameters include:

Population size.
Crossover rates.
Mutation rates.

These are usually application-specific. Trends observed by Cooper and Vemuri include the following:

- Success improves with population size (assuming sufficient heterogeneity).
- For small populations, success improves with a high mutation rate and elitist selection.
- For premature convergence and homogeneity, success may improve with increased mutation rate.
- For exploring less successful areas of the string, success may improve with nonuniform mutation: increased mutation rates in the section to be explored, and lowered mutation rates in the successful area. (See Chapter 104.)
- Success may increase with adaptation of individual parameters coded as genes in the string.

Uniform, Gaussian, and exponential distributions abound in evolutionary systems. Random number streams and their independence are a major concern. The dynamic self-adaptation of standard deviations and variances occurs in (ES) and in (EP), but not in (GAs) (Baeck, 1994). Adaptation of many aspects of an evolutionary system is possible when breaks with the traditional models discussed here introduce fuzzy system and neural system interactions. These topics are covered in more detail in Buczak and Uhrig and in articles by Padgett in this chapter.

Recall, Learning, Dynamics and Update Mechanisms

Adaptation of structure and of parameters is possible. Granularity and string length may be permitted to evolve. Range and distribution may likewise adapt.

In Genetic Algorithms with dynamic parameter encoding (DPE), parameter ranges are varied over time. When convergence appears near, the parameter window size may be reduced to allow focusing and greater precision when close to the desired optimal solution.

In Evolutionary Algorithms, self-adaptation of strategy parameters recombines and/or mutates the strategy parameters without exogenous control, and exploits the link between fitness and useful internal model (Baeck, 1994). Conditions necessary for success include:

1. Generating more new individuals than needed to maintain population size, u.
2. Survival of the u best children.
3. Moderate to low selection pressure (elimination of about 6 of every 7 children).
4. Recombination of strategy parameters.
5. Intermediate recombination based on the average of two parents for each child is most recommended.

Active research and some of the most worthwhile new applications in evolutionary systems center on increasing the dynamic aspects of systems-level implementations and on integrating evolutionary systems with fuzzy and/or neural systems for practical applications. See papers by Padgett in this section, and Buczak and Uhrig in this chapter.

100.3 Applications

Evolutionary programming applications have recently been applied to a wide range of combinatorial optimization problems. Fogel (1995) cites examples of its use in path planning, training and design of neural networks, automatic control, gaming and general function optimization.

Evolution strategies have been explored as aides in the design of neural structures and in system identification. Parameter optimization (Baeck, 1995) applications include minimization of material losses, processing times, weight, deviation from a goal state and finding a feasible solution. The mathematical foundations of all these ideas are being investigated. Computational complexity and parallelism are issues for implementation.

Software tools for helping with applications are becoming more abundant (Lane, 1995; Goldberg, 1994). Parallel processing plays a strong role in applications such as design of laminated composite structures (Punch et al., 1995). An associative architecture for GA's is suggested by (Twardowski, 1994), and explained in detail. Punch and Fogel both emphasize the importance of parallelism and high-speed computing to the current success and future improvement of their applications.

Genetic algorithms applications have been used to solve optimization problems. They have also been used in rule-based classifier systems, and as a basis for artificial life simulations. Their popularity in applications is high because of the following (Goldberg, 1994):

Table 100.1 Suggested Design Considerations for Evolutionary System Paradigms

PARADIGM SPECIFICATION MODULES				
OBJECTIVES and PERFORMANCE MEASURES ***(Problem Definition and Values of Outcomes)***				
MODELS and FUNCTIONS				
Biological: Bottom-Up (GA) or Top-Down -- Species (EP) or Individual (ES)	GA: Rules and Functions that Work in Nature; Morphological Exhibitions	Mathematical Representation, Convergence	Engineering: Modality, Convexity or Concavity, Noise Precision, Real or Discrete Valued	Response Surface Jaggedness and Noisiness
TOOLS and RESOURCES for IMPLEMENTATION				
Hardware	Software	Personnel: Experts, Others	Experience	Readiness and Affordability
INPUT/OUTPUT				
Data Sources: Sample Size, Clean Data, Noise, Purpose, Scoring Time	Characteristics: Span the Space; Solution Set Rank Ordered;	Interfaces Database	Dynamics: Real World Representation	Validation: Solutions Match Objectives of ***Correct Problem*** and ***In Time***
ARCHITECTURES and CONNECTIONS				
Individual Structure Length	Individual Structure Precision	Ind. Str. Adaptive Capabilities	Individual Objects	Individual Strategy Parameters
COMPONENT FUNCTIONS ***(Structure)***				
Evaluation	**Selection**	**Reproduction**	**Exploratory Mech. (GA:Mutation)**	**GA:Recombination EC:Phenotype**
VARIABLE PARAMETERS ***(Adaptive)***				
Noise Distribution across Modes of Mutation	Population Size; Memory Length: Dynamically Evolve	Modes GA:Crossover Rates Mutation Rates	Objective Function (State Dependent - Not Static NN)	Objective Function Parameters
RECALL, LEARNING, DYNAMICS, UPDATE MECHANISMS ***(Adaptive)***				
Interaction with Neural and/or Fuzzy Algorithms	Strategies; Modes & Means for Adaptation	Structure: granularity, length, range, distribution memory,	DPE GA: Zoom Cost Granularity Level	Self-Adaptation of Strategy Parameters

1. GAs address hard problems.
2. Solutions are rapid and reliable (especially with parallel processing).
3. Interfaces with existing simulations and models are flexible and simple.
4. Extensions and modifications are abundant.
5. GAs are easy to combine with expert systems, neural systems and/or fuzzy systems.

In all of these evolutionary paradigms, mathematical research into their foundations and appropriate implementations is actively being pursued.

100.4 Summary

A detailed discussion of evolutionary computation and comparative benefits of evolutionary programming, evolution strategies and genetic algorithms can be found in Fogel (1995). A very practical tutorial design of genetic algorithms is provided in Cooper and Vemuri. Applications are illustrated in other articles in this chapter and the chapter on Computational Intelligence and Hybrid Systems Applications. Defining terms can be found in the introduction to this chapter.

The future of evolutionary system applications appears to lie in increased adaptation of structure and parameters, perhaps by using fuzzy and/or neural systems techniques.

Acknowledgments

The compilation of this material owes much to the work of Rao Vemuri and Robert Uhrig, and to Walter Karplus' advice recommending David Fogel's material. L. Fogel made many valuable suggestions for Table 100.1.

References

Baeck, T. 1994. Evolution Strategies: A Thorough Introduction, Tutorial #1, *IEEE Conf. EC, WCCI,* Orlando, FL.

Fogel, David B. 1994. An Introduction to Evolutionary Computation, Tutorial #3, *IEEE Conf. EC, WCCI,* Orlando, FL.

Fogel, David B. 1995. *Evolutionary Computation,* IEEE Press, Piscataway, NJ.

Goldberg, David E. 1994. Genetic and evolutionary algorithms come of age, *Comm. ACM,* 37(3), March.

Lane, A. 1995. The GA edge in analyzing data, *AI Expert,* June, 11–13.

Michalewicz, Z. 1992. *Genetic Algorithms + Data Structures = Evolution Programs,* Springer-Verlag, New York, NY.

Punch, W. F., Averill, R. C., Goodman, E. D., Lin, S.-Ch., and Ding, Y. 1995. Using genetic algorithms to design laminated composite structures, *IEEE Expert,* February 42–49.

Padgett, M. L. 1973. *Inheritance of Dermal Ridges in Humans,* M.S. Thesis, Auburn University, Alabama.

Padgett, M. L. 1982. *A Statistical Methodology for the Development of Computer Simulation Models and Forest Harvesting Applications,* M. S. Thesis, Auburn University, Alabama.

Twardowski, Kirk, 1994. An associative architecture for genetic algorithm-based machine learning, *IEEE Computer,* November 27–38.

101
Genetic Algorithms

Mark G. Cooper
University of California, Los Angeles

V. Rao Vemuri
University of California at Davis

101.1 Introduction 1316
101.2 The Basic Genetic Algorithm 1316
101.3 String Encoding 1317
101.4 Evaluation 1317
101.5 Test Fitness Functions 1317
101.6 Premature Convergence 1318
101.7 Selection 1318
101.8 Replacement 1320
101.9 Genetic Parameters 1320

101.1 Introduction

A *genetic algorithm* (GA) is a stochastic search technique based on the principles of biological evolution, natural selection, and genetic recombination, simulating "survival of the fittest" in a *population* of potential solutions or *individuals.* GAs are capable of globally exploring a solution space, pursuing potentially fruitful paths while also examining random points to reduce the likelihood of settling for a local optimum (Goldberg, 1989). They are conceptually simple yet computationally powerful, making them attractive for use in complex domains, and have been demonstrated on a wide variety of problems. We begin by describing the basic genetic algorithm framework and introducing the vocabulary of the field. In subsequent sections we examine in more detail the issues which govern genetic algorithm design decisions and the trade-offs which have given rise to variations on the basic algorithm.

101.2 The Basic Genetic Algorithm

All genetic algorithms begin with the premise that all of the information required to specify the evolved system can be encoded using a string of characters from a fixed-length alphabet. This string is referred to as a *chromosome.* Sites on the chromosome corresponding to specific characteristics of the encoded system are called *genes,* and may assume a set of possible values, or *alleles.* The instantiation of a string is its *genotype,* and the behavior pattern exhibited or *expressed* by the genotype is its *phenotype.* By analogy, each genotype represents a particular point in the solution space of the function being optimized. *Niches* are subdomains of the search space, and *species* are individuals with a common characteristic or set of characteristics.

The genetic algorithm begins with a population of strings generated either randomly or from some set of known specimens, and cycles through three steps—evaluation, selection, and reproduction (Figure 101.1). Each string is evaluated according to a given performance criterion or *fitness function,* and assigned a *fitness score.* The fitness function must be designed such that the fitness score of an individual or group of individuals moves toward an extremum as its performance on a given task improves. Once all of the individuals have been assigned a fitness score, a decision must be made as to which individuals will be permitted to produce offspring and with what probability—the *selection* step. One selection strategy, for instance, is for all individuals to reproduce with probability proportionate to their fitness score, and mating pairs would be selected randomly on this basis. In the third step, the *reproduction* of a pair of strings proceeds by copying bits from one string until a randomly triggered *crossover* point, after which bits are copied from the other string. As each bit is copied, there is also the probability that a *mutation* will occur. The most common mutation is the inversion of a bit, however, any function, including adding or deleting bits, can be labeled a mutation (Figure 101.2). An implicit component of reproduction is *replacement*—the determination of which members of the current population are forced to perish in order to make room for their offspring.

The cycle of evaluation, selection, and reproduction continues for a predetermined number of *generations,* or until an acceptable performance level is achieved. In addition, individuals in a population will become increasingly similar with each generation, converging upon a specific portion of the solution space due to the effects of repeated reproduction over a diminishing region. As the population becomes more homogeneous, the genetic operators become less effective. Therefore, a termination condition based upon the homogeneity of the population may also be specified.

Variations on the basic genetic algorithm differ in the details of how each of these steps are implemented. The specifics of a particular genetic algorithm implementation should be dictated

0-8493-8343-9/97/$0.00+$.50

by the nature of the given problem—a technique that works well in one situation may not be as appropriate in another. Since the components are essentially modular, each can be selected individually and assembled into the overall algorithm. In addition, hybrid methodologies (e.g. evolving a population broadly into a number of species, then further evolving each species more specifically separately) or dynamic algorithms (e.g. gradually changing the crossover and mutation probabilities) may be employed as required to solve the problem.

101.3 String Encoding

The two most critical factors in selecting a string encoding scheme are the range and granularity of each system component or parameter. Encoding a parameter in eight bits means that it can only assume 256 different values, regardless of the range. The system designer must assess the sensitivity of each parameter, and allocate enough bits to adequately cover the range to the desired sensitivity. Although lengthening the representation of a parameter increases the dimensionality of the search space and may make acceptable solutions statistically harder to find, we have found that this is often not the case, and that a solution may actually be converged upon more quickly. Of course, longer strings are also more computationally expensive to evolve, and the time required to complete the evolution should also be considered when designing the representation. In certain applications the length of the strings may be permitted to change during evolution, and all of the strings need not be the same length.

Typically, the system parameters are encoded into a string as the concatenation of their binary representations. Alternative representations include Gray codes which guarantee that similar strings represent similar systems, or codes representing specific actions (e.g., the bit strings 00, 01, 10, and 11 might represent "stop," "turn right," "turn left," and "go straight," respectively). Floating-point numbers or integers may be used as parameters instead of bit strings when a high degree of accuracy is required. Trees and tables are often selected to solve combinatorial optimization problems, data structures encoding conditional (IF-THEN-ELSE) information to facilitate expert system evolution, and tree structures to represent computer programs (genetic programming). Many of these techniques require specialized crossover and mutation operators, and are particularly useful when applied to specific problem domains.

101.4 Evaluation

Systems produced using genetic algorithms are sensitive to the selection pressures imposed by the system designer. The range and distribution of the test cases must be representative of those encountered by the evolved system. If the test cases are unevenly distributed, then the evolved systems will fail to learn the outliers correctly. It is generally preferable to maintain a consistent set of test cases throughout evolution so that their optimal distribution can be preserved.

The performance criterion selected is probably the most important factor in successful evolution. It is usually best to measure the performance of an individual with respect to a single measure. If performance is predicated on more than one measure, their relative weights must be carefully balanced. Individuals that perform well on one criterion, to the exclusion of all others, may be favored if their performance exceeds that of the more "well rounded" individuals. Eventually, sensitivity to all but that single criterion may be selected out of the population, and many times, the result is unintended. All other things being equal, genetic algorithms will tend to take the "path of least resistance," evolving the most superficial traits first. For example, in one of the authors' early experiments, individuals were awarded a bonus that was inversely proportional to the length of the string to encourage compact representations. Within two generations, this bonus superseded the performance of the individuals on the actual problem and the result was a population of single-bit individuals, none of whom could solve the problem. Genetic algorithms tend to implement systems *exactly* as mandated by the evaluation function-even when that is not necessarily what is expected.

Finally, fitness value scaling must be considered. For instance, with a range of payoff values [0, 100], the population could very quickly converge to a point where most individuals score in the range [99, 100]. At this point, the selection differentials will not be sufficient to select between individuals. Therefore, a payoff incentive must be provided to exploit such differentials. One way to accomplish this is to select individuals based on relative fitness within the population rather than the absolute raw fitness scores.

101.5 Test Fitness Functions

While the true test of a genetic algorithm is its performance with respect to a given problem domain, it is possible to evaluate a methodology using a set of test fitness functions. While there is no "standard" set of functions, the following five functions are commonly used to characterize the behavior of a genetic algorithm, to highlight its computational strengths and weaknesses, and to attempt to determine its appropriateness to a given problem domain. Functions $F_1(x)$ and $F_2(x)$ are shown in Figure 101.3 and are defined by:

$$F_1(x) = \sin^6(5.1\pi x + 0.5)$$

$$F_2(x) = \exp^{-4(\ln 2)(x-0.0667)^2/0.64} \sin^6(5.1\pi x + 0.5)$$

Function $F_3(x)$, called "Shekel's Foxholes" (Figure 101.4) has twenty-five optima. The peaks are located on the grid intersections formed by the values $x = [32, 16, 0, -16, -32]$ and $y = [32, 16, 0, -16, -32]$. All 25 peaks have the same base width. The height of peak i is given by the expression $0.002 + 1/i$, where the peaks are located at $(32, 32), (16, 32), \ldots, (-32, -32)$.

Function $F_4(x,y)$, on the left side of Figure 101.5, contains two global optima with the same height and width, but located far apart. Function $F_5(x,y)$, the sample on the right, contains five

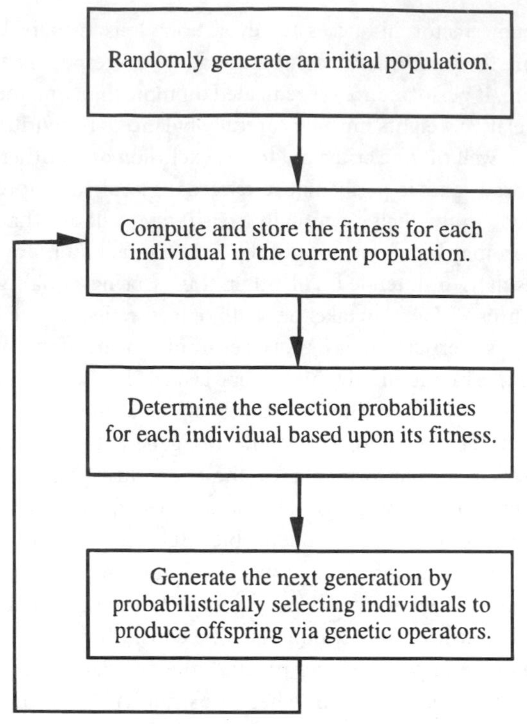

Figure 101.1 The Basic Genetic Algorithm.

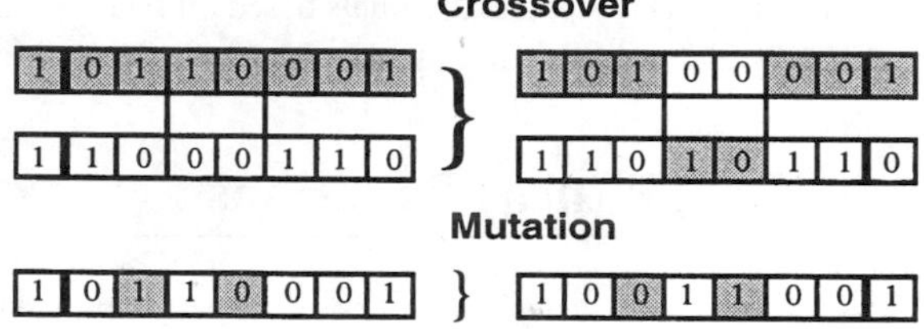

Figure 101.2 Genetic Operators.

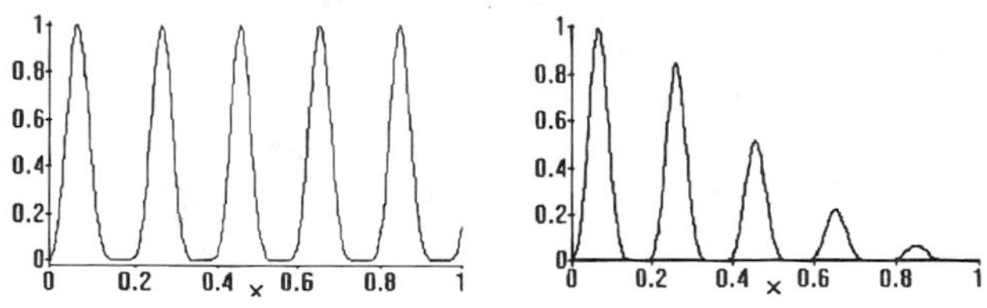

Figure 101.3 Five-Optima Sine Functions.

optima with height, width, and location chosen at random in every run. Both of these functions are defined by:

$$\sum_{i=1}^{p} \frac{A_i}{1 + W_i((x - X_i)^2 + (y - Y_i)^2)}$$

where p indicates the number of peaks in the function, (X_i, Y_i) the coordinates of peak i, A_i the height of peak i, and W_i the width of the base of peak i. Table 101.1 summarizes the parameters for functions F_4 and F_5 shown in Figure 101.5.

Of course, we cannot, in general, assume or conclude that a genetic algorithm will perform well by using the results of a different problem and it is usually best to select a test function that is related to the problem domain and that tests the genetic algorithm in extreme cases. The central idea behind the use of these functions is to determine whether a genetic algorithm method will tend to converge to a single local optimum, to the global optimum, or to several global optima simultaneously. If the objective is to design a genetic algorithm to find *all* of the peaks (or the k-best peaks), then it is best to select a test function having many peaks, each having a different height and each located at a random location. Conversely, if the global maximum is desired, the test function should be designed to have a single large peak surrounded by several "teaser" peaks that would attract local convergence.

101.6 Premature Convergence

As a stochastic search method, successful genetic algorithms must maintain a balance between the *exploration* of a wide area of the solution space so as to avoid premature convergence, and the *exploitation* of beneficial aspects of existing solutions in order to improve them. If weighted too far in favor of exploration, the algorithm essentially jumps randomly around the solution space. If weighted too far in favor of exploitation, the population becomes homogeneous and the genetic search degenerates to hill-climbing. Convergence and homogeneity are not typically problems in simulations that last only a few generations (up to 50 or 100), however in simulations which last hundreds or thousands of generations these issues become critical to the success of the experiment and must be considered in the design of the genetic algorithm.

One way to overcome this problem is to enforce constraints on reproduction which promote diversity. For instance, a string might be added to the population only if it differs from all other strings by some minimum distance metric, or individuals that are very similar may be prohibited from mating. The primary drawback of these approaches is that they significantly increase the time required to produce each generation. One encouraging approach is the use of crowding selection to promote niching and speciation (Cedeño and Vemuri, 1995). In this way, GAs can facilitate convergence to more than one extremum in a multimodal search space.

101.7 Selection

Once the population has been evaluated, individuals must be selected to produce the subsequent generation. The most common techniques are variations of *fitness proportionate reproduction* (*FPR*) or roulette wheel selection, where all of the individuals in the population are eligible for reproduction, and the probability with which a string is selected for reproduction is proportional to its fitness score. The most notable drawback of FPR is that as the population converges, fitness score differentials may not be sufficient to exert the necessary selection pressures to continue improvement. Nevertheless, FPR is the basis for the selection process in most GA applications.

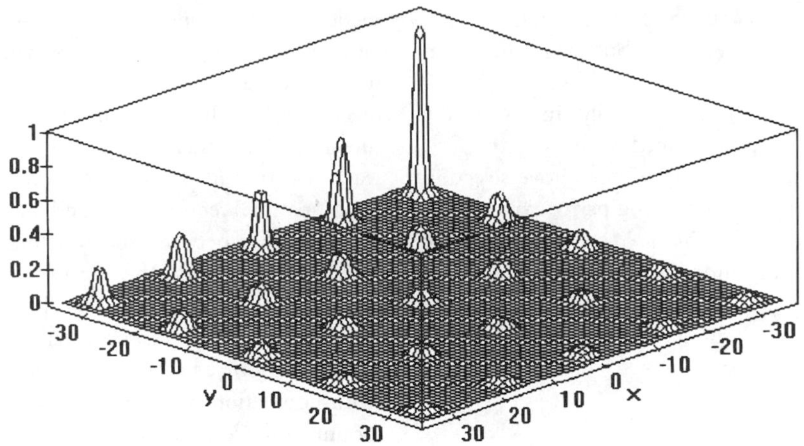

Figure 101.4 Shekel's Foxholes (Inverted).

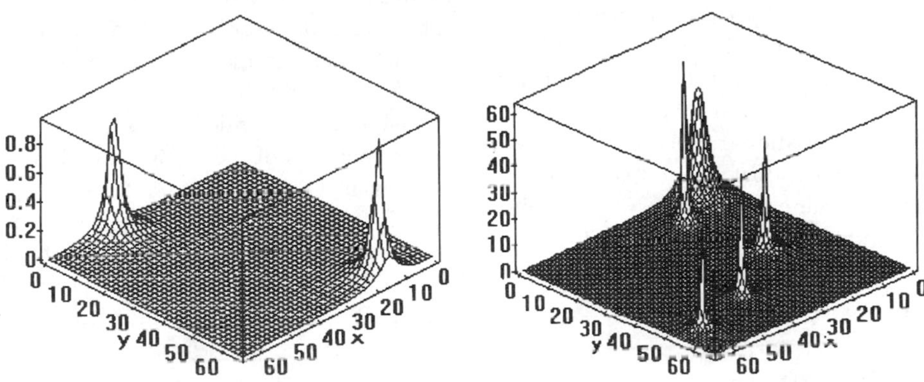

Figure 101.5 Functions $F_4(x,y)$ and $F_5(x,y)$.

Table 101.1 Parameters for Functions F_4 and F_5

Function	Peak Location	Width	Height
F_4	(45k, 2k)	0.0004	100
	(15k, 62k)	0.0004	100
F_5	(17.1, 34.4)	1.9	50.5
	(8.7, 4.1)	0.17	44.8
	(20.3, 11.0)	3.6	87.1
	(38.3, 47.9)	3.1	51.5
	(57.9, 54.1)	3.7	55.5

In *ranking selection,* the individuals in the population are sorted from best to worst according to their fitness values. Then, a number generated by a decreasing function is assigned to each sequentially. Finally, FPR is performed according to that assigned number. Ranking selection overcomes the fitness score differential problem by calculating probabilities based upon the *relative* performance of each individual within the population.

Another variation of FPR is *tournament selection.* Groups of individuals are chosen randomly from the population. The best-scoring individual from each group is selected to reproduce with the "winner" from another randomly selected group. This process is repeated until an entire new generation is produced. This method is closely related to ranking selection in that the number of times that an individual is selected for reproduction (i.e. wins a tournament) is roughly proportional to its rank in the population.

The most significant drawback of selection techniques based on FPR is the possibility of premature convergence. Once a candidate with an above average fitness score is identified, the FPR rule begins to favor that candidate from generation to generation until a better candidate emerges. This means that the FPR simply assigns an exponential number of mating chances to those members of the population that exhibit above average survival rates. This is analogous to the process of convergence to a local minimum from an initial guess in the neighborhood of that minimum. These techniques are therefore best used for quickly locating one peak and converging toward it; however, they will often fail to thoroughly explore a complex search space with multiple peaks.

One selection technique which does work well for multi-modal search is *crowding selection.* In this method, every individual in a population has the same probability of reproducing, but its mate is selected as the individual to which it is most similar (as measured by phenotypic distance or performance, rather than genotypic distance or bit-wise similarity) in a randomly selected subset of the population. Therefore, individuals from the same niche are most likely to mate. Here, the fitness function is not used to determine whether or how often, but rather to whom an individual mates.

Finally, an *elitist strategy* which automatically singles out the highest scoring strings and copies them directly into the next generation may be combined with any of the above selection techniques to achieve strictly non-decreasing performance maxima throughout the genetic search. This also guarantees that the highest scoring instance in the most recent generation is the best solution found by the genetic search (assuming that the performance criteria do not change from generation to generation).

101.8 Replacement

In implementations where individuals can only reproduce once, the pair of generated offspring usually replaces its parents. However, when an individual can participate in the generation of several offspring, it is most convenient to produce the entire next generation in a temporary buffer and then replace the current generation all at once. Other variations replace individuals with the lowest scores or the individuals that are most similar to each other. In some GA implementations where fitness is determined by the number of similar individuals rather than the fitness score of any given individual, a mating pair may be replaced by a large number of offspring.

101.9 Genetic Parameters

Besides the selection strategy, an acceptable balance between exploration and exploitation can also be accomplished through appropriate genetic parameter settings (population size, crossover, and mutation rates). Unfortunately, there is no way to definitively specify genetic parameters. Rather, they are usually set to "traditionally accepted values" or adjusted by trial and error based upon the experience of the system designer.

The first systematic study of genetic parameters was performed by DeJong (1975). In his classic experiment, he applied genetic algorithms using various parameter settings to several tasks and determined an "acceptable" set of parameters which is still used by many researchers today: crossover probability 60%, mutation probability 0.1%, and population size of 50 to 100. While these values are typically taken as the starting point for genetic experiments, they are often tuned by trial and error to improve performance, and some studies have found them to be inadequate. This should not be surprising since each parameter is responsible for controlling some aspect of the genetic search and the relative degrees of exploration and exploitation required may differ depending the topology of the solution space. Several other attempts have been made to formalize the relationship between population size, crossover rate, and mutation rate, but none have been consistently successful in practice. Ultimately, it is left to the individual system designer to set the genetic parameters since each selection involves tradeoffs between competing (and often incompatible) requirements and that success depends upon striking a balance between these factors. In fact, simulations which succeed quickly with the appropriate genetic parameter settings often fail completely when specified even slightly differently.

Some general trends, however, have been identified and a few generalizations can be made. Genetic search tends to improve with population size as long as diversity is maintained. In small populations, combining a high mutation rate with an elitist strategy tends to produce wide exploration while still "remembering" acceptable solutions. In most cases, the probability of mutation remains constant across all bits and over all generations. However, increasing the probability of mutation in later generations can sometimes overcome the effects of premature convergence, and varying the probability within each string may permit selected exploration of specific parts of the genome while leaving intact other portions of the string, possibly identified as being more successful. Overall, crossover plays an important role when construction and survival are required for good performance—when the population is diverse near the beginning of the simulation. Mutation, then, becomes more important later in the evolutionary process when a fairly homogeneous population seeks to make incremental improvements.

In our experiments (Cooper, 1994) we have found success in varying both the crossover and mutation rates within a predefined range. Another approach that has received recent attention is the evolution of the genetic parameters along with each individual in the population. In other words, each string is prefaced with bits that encode each of the genetic parameters for that individual.

References

Cedeño, W. and Vemuri, V. 1994. Multi-niche crowding in genetic algorithms and its application to the assembly of DNA restriction fragments, *Evolutionary Computation,* 2(4):321–345.

Cooper, M. G. 1994. *Genetic Design of Rule-Based Fuzzy Controllers,* PhD dissertation, Department of Computer Science, University of California, Los Angeles.

DeJong, K. A. 1975. *Analysis of the Behavior of a Class of Genetic Adaptive Systems,* PhD dissertation, Department of Computer and Communication Sciences, University of Michigan.

Goldberg, D. E. 1989. *Genetic Algorithms in Search, Optimization, and Machine Learning,* Addison-Wesley, Reading, MA.

102
Fuzzy Evolutionary and GA Systems

Mary Lou Padgett
Auburn University

102.1 Introduction 1321
102.2 Combining Evolutionary Systems and Fuzzy Systems 1321
Relationships • Fuzzy Evolutionary Systems • Evolutionary Fuzzy Systems • Application Examples
102.3 Summary 1323

102.1 Introduction

Fuzzy systems combined with evolutionary systems are proving practical and of high interest in engineering applications such as those useful in industrial electronics. Evolutionary systems modeling can be approached from a top-down perspective (evolutionary algorithms (EA)) or from a bottom-up perspective (genetic algorithms (GA)). Both evolutionary and genetic algorithms are covered by the evolutionary computation (EC) literature and conferences. These approaches will be compared in this paper, and suggestions made for incorporating fuzzy systems within them, and for inserting them into fuzzy systems.

102.2 Combining Evolutionary Systems and Fuzzy Systems

Relationships

Evolutionary systems and fuzzy systems can have a symbiotic relationship. They have compensating strengths which are ideal for practical applications combining the techniques. Nonlinear optimization problems without crisp mathematical models make good targets for this type of approach. Evolutionary computing offers *auto-tuning* of parameters, and can both *learn* from historical data and *explore* new solutions by adding randomness. Fuzzy systems offer operator *knowledge* and real-time implementation of soft mathematical models based on some type of knowledge representation. See Table 102.1 and Kaynak's paper in the computational intelligence chapter.

Fuzzy Evolutionary Systems

A fuzzy evolutionary system could help a user understand the progress of the algorithm, evaluating its approach toward broad system goals over time (See papers by Shibata and Fukuda et al and by Kaynak in the chapter on computational intelligence.) At the micro-level, evaluation of individuals and the effectiveness of the fitness function could be enhanced by fuzzy systems to replace the tedious, heuristic parameter adjustments a user is tempted to make. Fuzzy systems for enabling dynamic parameter adaptation could adjust system strategy parameters such as the mean and standard deviation of Gaussian mutation controllers (Baeck, 1994). Actual changes in type of implementation of the evaluation, selection and reproduction steps could be governed by a fuzzy controller monitoring the homogeneity of the population and convergence of the process (See the paper by Werbos in the fuzzy systems chapter.).

Some suggestions for monitoring the state of the evolutionary system and making adjustments are made in the paper by Cooper and Vemuri and by others. These could be molded into a set of fuzzy rules for GA model development. The following text summarizes some of the observations that could form the basis for a set of fuzzy rules:

- Success improves with population size (assuming sufficient heterogeneity).
- For small populations, success improves with a high mutation rate and elitist selection.
- For premature convergence and homogeneity, success may improve with increased mutation rate.
- For exploring less successful areas of the string, success may improve with non-uniform mutation: increased mutation rates in the section to be explored, and lowered mutation rates in the successful area. (See Chapter 103.)
- Success may increase with adaptation of individual parameters coded as genes in the string.

Adaptation of structure and of parameters is possible. Granularity and string length may be permitted to evolve. Range and distribution may likewise adapt.

In Genetic Algorithms with dynamic parameter encoding (DPE), parameter ranges are varied over time. When convergence appears near, the parameter window size may be reduced to

0-8493-8343-9/97/$0.00+$.50

Table 102.1 Relationships Between Evolutionary Systems and Fuzzy Systems

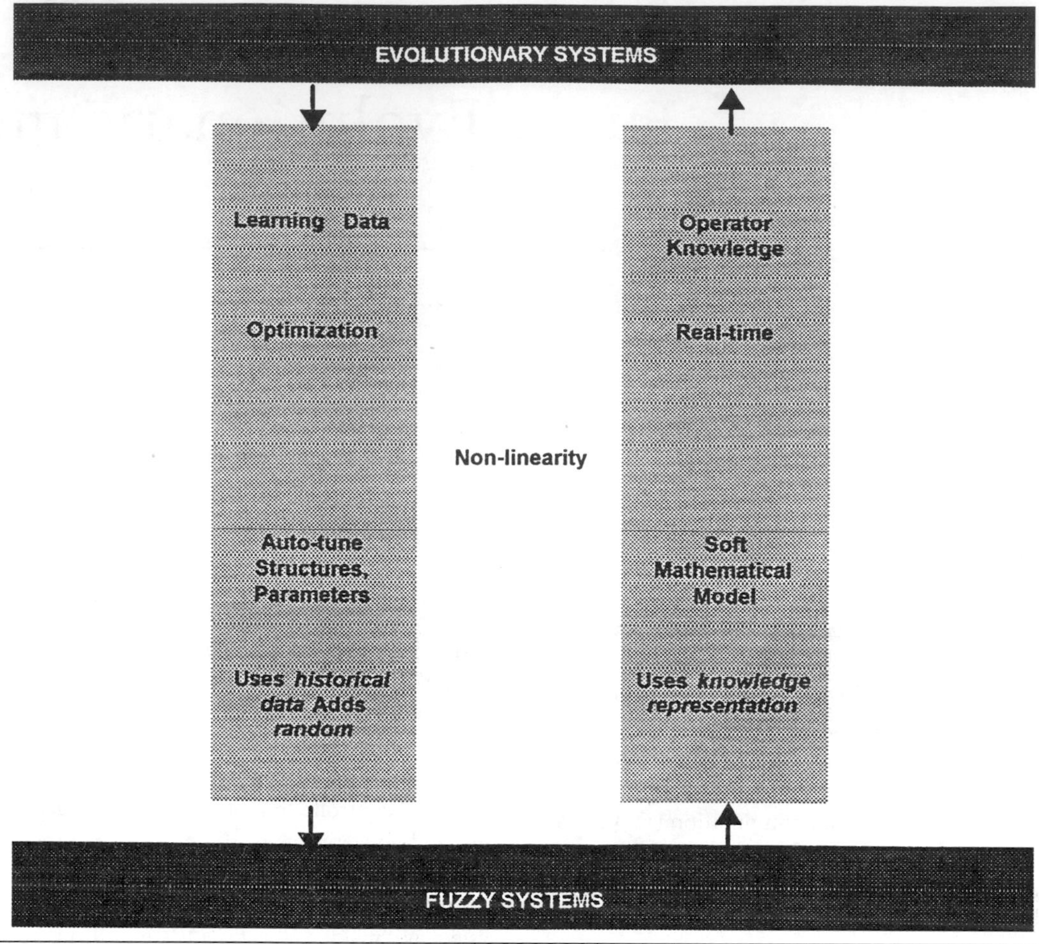

allow focusing and greater precision when close to the desired optimal solution.

In Evolutionary Algorithms, self-adaptation of strategy parameters recombines and/or mutates the strategy parameters without exogenous control, and exploits the link between fitness and useful internal model (Baeck, 1994). Conditions necessary for success include:

1. Generating more new individuals than needed to maintain population size, **u**
2. Survival of the **u** best children,
3. Moderate to low selection pressure (elimination of about 6 of every 7 children).
4. Recombination of strategy parameters.

Intermediate recombination based on the average of two parents for each child is most recommended.

Active research and some of the most exciting new applications in evolutionary systems centers on increasing the dynamic aspects of systems-level implementations and on integrating evolutionary systems with fuzzy and/or neural systems for practical applications. See papers by Padgett in this section and by Buczak and Uhrig in this chapter.

Evolutionary Fuzzy Systems

Starting with a collection of rules of thumb provided by domain experts, how can one configure a fuzzy system? Off-line, the design of the system can be roughed in structurally, then fine-tuned (usually off-line) with evolutionary system optimization of parameters. The rule-base matrix can be treated as a string and a population of solutions evolved. Undesirable or insignificant rules might be eliminated. The shape of fuzzy membership functions can be evolved by treating the parameters of the geometric shapes or distributions as strings.

Attention can be focused on solving the problem, rather than on whether the particular representation is adequate. Poor solutions, needing to be rethought, should quickly pop out as failures. Use of data visualization, even to the point of manipulating a virtual world, is helpful in this type of endeavor (Wiggins, 1992).

Application Examples

Fuzzy Evolutionary Systems

- FS as an *evaluation tool* for GA (See Chapter 108).

  Evaluation tools are hard to define—FS help.
  Human operator needs to understand—FS help.

- FS for *dynamic parametric* GA control (Lee and Takagi, 1993)

Evolutionary Fuzzy Systems

Examples of evolutionary fuzzy systems applications are listed below, with sources as indicated.

- EA for *shape and location of fuzzy membership functions* (Baeck, 1994).

  Vectors: Triangle(center, min x-intercept, max x-intercept); or, Gaussian(mean, variance)

- EC for *simultaneous design of fuzzy systems parameters* (Chapter 107 and Takagi, 1993).

  Shape of membership function.
  Number of rules.
  Consequent parameters.

- GA for *controls system parameters tuning.*

  Cart pole balancing system controller (Karr, 1991 in Baeck, 1994).
  On-line control of pH system in a laboratory (Gentry 1993; Karr, 1991). Chapter 107.
  Tuning parameters of a PID controller (Wang and Kwok 92). Chapter 107.
  System identification (Chapter 107).
  Dynamic control of robot arms (Chapter 107).
  Motion systems for multiple robots (Chapter 108).

- ES optimization for *time series prediction.*

  RBF networks with Gaussian membership functions (Wienholt, 1993 in Baeck, 1994).

- EC for *evolving rules for fuzzy control* of a cart (Thrift 1991; ICGA91, in Fogel, 1994).

  Vector: String representing a 5 × 5 matrix where,

  Columns: position x = {NM, NS, ZE, PS, PM or {__}

  Rows: velocity of x, v = {NM, NS, ZE, PS, PM or {__}

  N is negative, P positive, ZE zero, S small, M medium, and __ is no entry.

  If *x* is *NM* and *v* is *NM,* then the entry in the first row and column of the string, say *NM,* is a "gene expression" to be evolved into an optimal value.
  Defuzzification: weighted centroid.
  Mutation: up or down by one.
  Crossover: two-point.

In this example, the rule matrix was evolved to produce a sensible solution, enhancing the system.

- EP for *design of fuzzy* HVAC controller (Fogel, 1994).
- GA for *interpreting sensor data.*

  Controls problems: *docking a truck* (Wiggins, 1992).
  Vector: backing parameters.
  Graphics: visual aides and plotter essential.

  Fault Detection: *selecting the order of importance* of 9 sensor fault detectors (See the paper by Uhrig and Tsoukalas).
  Fault detectors often varied in diagnosis.
  Especially good for multiple faults.

- GA for *searching for consistent patterns* in financial databases (Wiggins, 1992).
- GA for *fuzzy relational descriptors.*
- *GA and gradient-based techniques* combined for learning purposes (Pedrycz, 1993, 1994).

  Initialize gradient based learning schemes.
  Partition universe into feasibility regions for further learning.
  GA initialization guides gradient-based learning by constructing a guarding zone, Ω.
  For the performance index or objective function, *Q,*

$$\text{connections (new)} = \left(\text{connections} - \alpha\frac{\partial Q}{\partial \text{connections}}\right)\Bigg|_{\Omega}$$

- GA for minimizing error instead of gradient methods or steepest descent (Chapter 107).

Evolutionary algorithms excel at optimizing strings of numbers. Only the imagination and experience of the designer limits the multitude of ways these techniques can be used in the practical, working environment of an industrial electronics engineer.

102.3 Summary

The relationships between fuzzy systems and evolutionary systems are typically similar to those in Figure 102.1 (Takagi, 1993). Many variations of this scenario exist, and more appear as better developmental tools and faster computers emerge. Some helpful references in addition to those mentioned above include Fogel, 1995, 1994; Fukuda et al., 1994, 1993; Kwok, 1993; Kristinsson and Dumon, 1992; Michalewicz, 1992; Davis, 1991; Karr, 1991a,b; Goldberg, 1989.

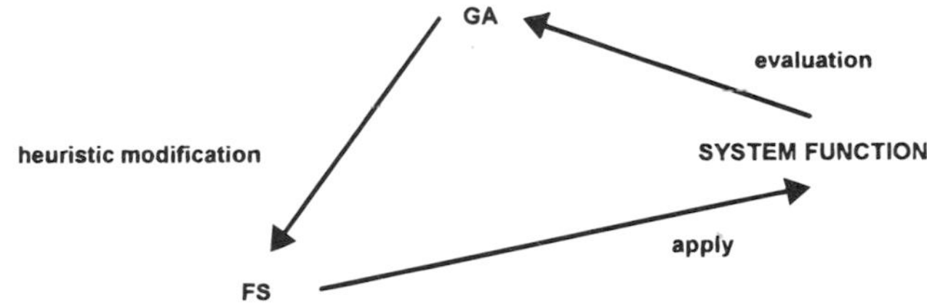

Figure 102.1 GA's tune FS parameters, FS aid evaluation of GA performance.

Acknowledgments

The advice of Paul Werbos and the slides provided by H. Takagi contributed significantly to this article.

References

Baeck, T. 1994. Evolution Strategies: A Thorough Introduction, Tutorial #1, *IEEE Conf. EC, WCCI,* Orlando, FL.

Davis, L., ed., 1991. *Handbook of Genetic Algorithms,* Van Nostrand Reinhold, NY.

Fogel, D. B. 1994. An Introduction to Evolutionary Computation, Tutorial #3, *IEEE Conf. EC, WCCI,* Orlando, FL.

Fogel, D. B. 1995. *Evolutionary Computation,* IEEE Press, Piscataway, NJ.

Fukuda, T., Shimojima, K., and Shibata, T. 1994. Fuzzy, neural network and genetic algorithm based control system, *Proc. of IECON'94. Int. Conf. on Ind. Electronics, Control and Instrumentation,* 2:1220–1225.

Fukuda, T, Ishigami, H., et al., 1993. Structure optimization of fuzzy neural networks using genetic algorithm; *Proc of IFSA.*

Goldberg, D. E. 1989. *Genetic Algorithms in Search, Optimization, and Machine Learning,* Addison-Wesley, Menlo Park, CA.

Karr, C. 1991. Genetic algorithms for fuzzy controllers, *AI Expert,* February, 26–33.

Karr, C. 1991. Applying genetics to fuzzy logic, *AI Expert,* March, 38–43.

Karr, C. and Gentry, E. J. 1993. Fuzzy control of pH using genetic algorithms, *IEEE Trans. FS,* 1:46–53.

Kristinsson, K. and Dumon, G. A. 1992. System identification and control using genetic algorithms, *IEEE Trans. SMC* 22:1033–1045.

Kwok, D. P., Leung, T. P. and Shang, F. 1993. Genetic algorithms for optima dynamic control of robot arms, *Proc. of IECON'93, Int. Conf. on Industrial Electronics, Control, and Instrumentation,* 1:380–385.

Lee, M. A. and Takagi, H. 1993. Integrating design stages of fuzzy systems using genetic algorithms, *FUZZ-IEEE'93,* 1:612–617.

Pedrycz, W. 1995. Genetic algorithms for learning in fuzzy relational structures, *Fuzzy Sets and Systems,* 69:37–52.

Michalewicz, Z. 1992. *Genetic Algorithms + Data Structures = Evolution Programs,* Springer-Verlag, New York, NY.

Pedrycz, W. 1994. Fuzzy Logic in Neurocomputations, Tutorial #1, *IEEE ICFS, WCCI,* Orlando, FL.

Takagi, H. 1993. Fusion techniques of fuzzy systems and neural networks, and fuzzy systems and genetic algorithms, *SPIE Proc. Tech. Conf. on Applications of Fuzzy Logic Technology,* v. 2061, Boston, MA.

Wang, P. and Kwok, D. P. 1992. Optimal design of PID process controllers based on genetic algorithms, *Proc. IFAC 11th World Congress,* 5:261–265.

Wiggins, R. 1992. Docking a truck: a genetic fuzzy approach, *AI Expert,* May, 26–35.

103

Information Fusion by Fuzzy Set Operations and Genetic Algorithms

Anna L. Buczak
Allied Signal

Robert E. Uhrig
University of Tennessee

103.1 Information Fusion 1325
103.2 Fuzzy Aggregation Connectives 1326
103.3 Genetic Algorithms 1328
Overview • Optimization Problems • Genetic Algorithms for Finding the Best Aggregation Parameters
103.4 Two Fuzzy-Genetic Fusion Techniques 1329
Fusion from Two Sensors in One Step • Fusion from All Sensors in One Step
103.5 Information Fusion for Object Classification 1331
103.6 Vibration Monitoring 1332
Background • Vibration Signatures
103.7 Results 1332
Data Flow • Fusion from Two Sensors in One Step • Fusion from All Sensors in One Step
103.8 Conclusions 1335

Abstract

This paper* describes novel multisensor information fusion methods based on fuzzy logic and genetic algorithms. Unlike most fuzzy logic-based systems that perform reasoning by fuzzy *IF-THEN* rules, the reasoning in this work takes place by means of fuzzy aggregation connectives. These connectives are capable of combining information not only by union and intersection used in traditional set theories but also by compensatory connectives that better mimic the human reasoning process. The particular connective used in this work for the purpose of data fusion is the generalized mean aggregation connective. The distinctive feature of this information fusion method is that the optimal parameters of the aggregation connective are automatically found by a genetic algorithm. Both elitist and nonelitist strategies for genetic algorithms are investigated.

Two different methods are developed. The first technique performs aggregation of evidence from two sensors in one step; if there are more sensors, information from the next sensor is fused with the data already aggregated. The second technique developed performs one step fusion from all the sensors available. The techniques devised are tested on a vibration monitoring problem and the results are described.

103.1 Information Fusion

Aggregation of evidence (fusion of information) from multiple sources is an increasingly important area of research and application. Multicriteria aggregation consists of a combination of different sources of information into one representational format (Luo and Kay, 1992). It is used whenever several sources of information exist in a system, in order to reduce the uncertainty and resolve the ambiguity present in the information from a single source. The problem of combining multiple sources of information is encountered in numerous applications, such as *decision making,* when opinions from individual decision makers are combined, and *multisensor fusion,* when data from different sensors are aggregated. Decision making is an area of operations research, while multisensor fusion is employed extensively in engineering disciplines.

Multisensor fusion is a process in which information from different sensors, representing the same modality or different modalities, is combined to derive meaningful information not available from any individual sensor. The range of sensor fusion applications includes robotics, automated manufacturing, autonomous navigation, target detection and computer vision. Multisensor fusion can take place at different levels, and depending on the level it is called signal fusion, feature fusion, or decision fusion. Fusion at all the levels is seldom possible in one system. For example, for fusion at the signal level there is a need for synchronized compatible sensors. Decision fusion is considered more robust than fusion at lower levels, because failure of one of the sensors in multisensor system does not signify total catastrophic failure of the entire system (Dasarathy, 1991).

The information to be fused can be *redundant* or *complementary.* When each sensor is perceiving the same features of the environment we deal with *redundant* information. Its fusion can reduce the overall degree of uncertainty. *Complementary* information allows us to perceive features that are impossible to perceive using just the information from each individual sensor operating separately.

We are mainly interested in the fusion of information from multiple sources for the purpose of pattern classification. The data fusion algorithms for classification can be divided into three major categories (Waltz and Linus, 1990):

- Physical models.
- Parametric classification techniques.
- Cognitive-based models.

Physical models strive to accurately model the data and estimate the identity by matching the observations predicted by the model with actual data. This category includes such techniques as simulation, Kalman filtering, and least-squares method.

Parametric classification techniques do not use physical models. Instead a direct mapping is made between the object features and the declaration of identity. These methods include statistical techniques, such as classical inference, Bayesian inference, Dempster-Shafer evidential reasoning, and information theoretic techniques such as clustering algorithms, artificial neural networks, voting methods, and entropy methods. Statistical techniques utilize *a priori* knowledge about the observation process to make inferences about object identity. Classical inference techniques describe the probability of observed data, given a hypothesis on the existence of the object. Bayesian inference updates the likelihood of a hypothesis, given a previous likelihood estimate and additional observations using the famous Bayes formula. Dempster-Shafer evidential reasoning allows each sensor to contribute information at its own level of detail. Clustering algorithms are used to group patterns into clusters, based on internal similarities among patterns. Neural networks are used for sensor fusion because of their robustness and noise tolerance, modeling ability, as well as possibility of fast parallel implementation. Voting schemes are based on the majority of individual sensor votes. The entropy methods use the concept of information entropy that assigns less significance to frequently appearing information, and more significance to infrequent ("surprising") information. All the information theoretic methods attempt to induce natural groupings in the data, associated later with classes of objects.

The third category of fusion algorithms for classification relies on cognitive-based models. These methods seek to mimic the human cognitive process in identifying entities from multiple sensor data. Techniques in this category encompass logical templates, expert systems and fuzzy logic methods. Templating is the process of matching a predetermined pattern against observed data to determine whether conditions have been satisfied, thereby allowing a logical inference to be made. Expert systems use production rules to symbolically represent the relation between sensory information and an attribute that can be inferred from the information. Fuzzy set theory is very advantageous for information fusion due to the fact that there are numerous ways of combining fuzzy sets in addition to the union and intersection used in traditional set theories. Since there is no unique extension of crisp theory to fuzzy set theory, there is a large amount of flexibility in the design of a fuzzy system.

103.2 Fuzzy Aggregation Connectives

Application of fuzzy set theory for the fusion process is advantageous, because the information that the fusion systems deal with is fuzzy rather than precise in nature, and fuzzy set theory allows us to model this imprecision appropriately and later permits us to reason in imprecise terms. In the fuzzy set theory several connectives (Krishnapuran and Lee, 1992; Dubois and Prade, 1992, 1980; Zimmermann, 1987; Kaeprzyk, 1983); can be used for the purpose of aggregation in addition to union and intersection, which are the only connectives allowed in the traditional set theory. The particular connective one chooses depends on the nature and relative importance of criteria, as well as the requirements imposed by the decision-making process. For instance, the requirement may be that all the criteria be satisfied, or that any one of the criteria be satisfied. In the first case an intersection connective should be used, and in the second case an union connective.

Aggregation operators, used to combine two or more fuzzy sets to produce a single set, are represented by a function f defined by

$$f.[0, 1]^n \rightarrow [0, 1], \qquad n \geq 2 \tag{103.1}$$

where n is the number of aggregation sources. The function f must satisfy the boundary conditions $f(0, 0 \ldots 0) = 0$ and $f(1, 1 \ldots 1) = 1$ and be monotonic and non-decreasing (Klir and Folger, 1988). The basic fuzzy set aggregation operators are *intersection* and *union.* In fuzzy logic, unlike in Boolean logic, several types of intersection and union operators exist, not just one type. The particular type one chooses depends on the properties that one would like the operator to satisfy. The intersection connective belongs to the general class of *T-norms* (triangular norms) and the union connective belongs to the class of *T-conorms. T-norms* are used when one requires that the aggregated value be high only when all the inputs are high. *T-conorms* are used when the aggregated value is required to be high whenever any one of the input values is high.

The *intersection* of n fuzzy sets $A_1, A_2 \ldots A_n$ with respective membership functions $\mu_{A1}(x), \mu_{A2}(x) \ldots \mu_{An}(x)$ is a fuzzy set C, written as $C = A_1 \cap A_2 \cap \ldots \cap A_n$, whose membership function is given by

$$\mu_c(x) = \min[\mu_{A_1}(x), \mu_{A_2}(x) \cdots \mu_{A_n}(x)], \quad x \in X \tag{103.2}$$

The min operator (Equation 103.2) proposed by Zadeh is the most popular intersection operator. The min operator is the most optimistic of all T-norms, because the aggregated confidence can never be lower than the lowest value among the inputs. The Yager intersection operator (Klir and Folger, 1988) allows different degrees of pessimism. It is defined as

$$\mu_c(x) = 1 - \min[1, ((1 - \mu_{A_1}(x))^{-p} + \cdots + (1 - \mu_{A_n}(x))^{-p})^{-1/p}], \quad x \in X \tag{103.3}$$

The Yager intersection operator allows us to obtain all the values between 0 and min by varying p from 0 to $-\infty$.

The *union* of n fuzzy sets $A_1, A_2 \ldots A_n$ with respective membership functions $\mu_{A1}(x), \mu_{A2}(x) \ldots \mu_{An}(x)$ is a fuzzy set C, written as $C = A_1 \cup A_2 \cup \ldots \cup A_n$, whose membership function is given by

$$\mu_c(x) = \max[\mu_{A_1}(x), \mu_{A_2}(x) \cdots \mu_{A_n}(x)], \quad x \in X \tag{103.4}$$

The max operator (Equation 103.4) proposed by Zadeh is the most popular union operator. The max operator is the most pessimistic of all T-conorms, because the aggregated output can never exceed the highest input. A generalization of the max operator is a Yager union introduced by Yager (Klir and Folger, 1988) which allows different degrees of optimism. Yager Union operator is defined as

$$\mu_c(x) = \min[1, (\mu_{A_1}^p(x) + \mu_{A_2}^p(x) + \cdots + \mu_{A_n}^p(x))^{1/p}], \quad x \in X \tag{103.5}$$

This operator allows obtaining all the values between max and 1 by varying p from ∞ to 0.

In many decision-making situations one wants to take a position between the two extremes of no compensation which is characterized by the intersection operator, and of full compensation which is characterized by the union operator. No compensation means that the information is complementary, and full compensation means that the information is redundant. When no compensation among different information sources exists, different features of the environment are perceived from each source. Usually in decision-making based on several criteria, a certain amount of compensation is desirable and therefore compensatory connectives will best describe the fusion process.

One of compensative operators proposed in the literature (Krishnapuran and Lee, 1992; Zimmermann, 1987; Kacprzyk, 1988; Dobois and Prade, 1980) is the mean operator. There are diverse kinds of mean operators:

- The weighted arithmetic mean.
- The geometric mean.
- The harmonic mean.
- The generalized mean.

The generalized mean is defined by:

$$g(x_1, x_2, \ldots, x_n; p, w_1, w_2, \ldots, w_n) = \left(\sum_{i=1}^{n} w_i x_i^p\right)^{1/p} \tag{103.6}$$

the w_i's can be thought as the relative importance factors for the different criteria, where

$$\sum_{i=1}^{n} w_i = 1 \tag{103.7}$$

The behavior of the generalized mean is presented in Figure 103.1. The attractive properties of the generalized mean are:

- min $(a, b) \leq$ mean $(a, b) \leq$ max (a, b).
- Mean increases with an increase in p; by varying the value of p between $-\infty$ and $+\infty$, one can obtain all values between min and max.

Therefore, in the extreme cases the generalized mean operator can be used as intersection or union. Furthermore, it can be shown that $p = -1$ gives the harmonic mean, $p = 0$ gives the geometric mean, and $p = 1$ gives the arithmetic mean. The rate of compensation for the generalized mean can be controlled by changing p; when using larger p, the partition becomes more fuzzy.

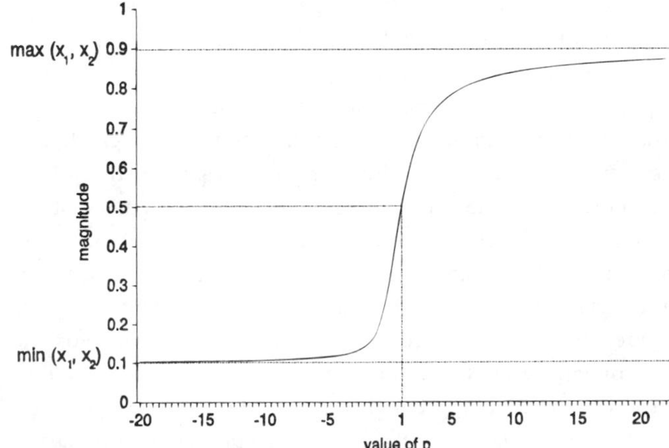

Figure 103.1 Behavior of generalized mean connective for $x_1 = 0.1$, $x_2 = 0.9$.

103.3 Genetic Algorithms

Overview

Genetic algorithms (Holland, 1992; Davis, 1991; Goldberg, 1989) are search algorithms based on the mechanics of natural selection and natural genetics. They efficiently utilize historical information to obtain new search points with expected enhanced performance. In every generation, a new set of artificial individuals is created, using the information from the best of the old generation. Genetic algorithms combine the survival of the fittest from the old population with a randomized information exchange which helps to form new individuals with higher fitness.

There are three basic genetic algorithm operators: *reproduction*, *crossover*, and *mutation*. Those operators, combined with the proper fitness function definition, constitute the main body of genetic algorithms. *Reproduction* is a process in which the mating pool for the next generation is chosen. Individual strings are copied into the mating pool according to their fitness function values. *Crossover* usually proceeds in two steps. First, members from the mating pool are mated at random. Second, each pair of strings undergoes crossover as follows: a position k along the string is selected uniformly at random from the interval $[1, l - 1]$, where l is the length of the string. Two new strings are created by swapping all characters between the positions k and l. *Mutation* is a random alteration of the value of a string position. In a binary coding, mutation means changing a zero to a one or vice versa. Mutation occurs with small probability and plays a secondary role in the operation of genetic algorithms. Mutation's primary role is not one of generating new structures—this role is filled by crossover. However, mutation is needed because occasionally during reproduction and crossover, some useful genetic material may be lost (like a one at a particular position). The mutation operator protects against such an irrecoverable loss.

Holland (1992) suggested that genetic algorithms should be used for identifying the high performance regions of the parameter space, and later the search should be performed by a local search routine to optimize the members of the final population. Fine local tuning capabilities of genetic algorithms can also be improved by *non-uniform* mutation (Michalewicz, 1992). The probability of a traditional mutation is the same regardless of the fact if the bit is located at the left or right portion of the chromosome. If the bit is located in the left fragment of the chromosome, its change results in a very significant change of the value of the variable. On the other hand, if the bit is located on the right hand side of the chromosome, its change results in a much smaller difference in the coded variable. Therefore, as the population ages, bits located further to the right of each chromosome should get higher probability of being mutated, while those on the left should have such a probability decreasing. *Non-uniform* mutation causes global search of the parameter space at the beginning of the iterative process, and an increasingly local search in later generations (Michalewicz, 1992).

In the reproduction strategy proposed so far, the best member of population may fail to produce offspring in the next generation. The *elitist* strategy solves this problem by copying the best individual from each generation into the next generation. The elitist strategy often appears to improve genetic algorithm performance but this strategy may increase the speed of domination of a population by a super individual, especially in small population genetic algorithms. De Jong (1975) found that on unimodal surfaces the elitist strategy improves the genetic algorithm's performance; however for multimodal surfaces, the elitist strategy degrades the performance. De Jong suggests that elitism improves local search at the expense of global perspective. Therefore caution should be used when applying elitist strategy.

Optimization Problems

The need to solve optimization problems arises in many fields and is especially dominating in the engineering environment. There are several analytic and numerical optimization techniques, but there are still large classes of functions which are not solvable by these techniques. Among functions which have difficulties being solved by classical methods are non-convex, multi-modal, and noisy functions (De Jong, 1992). As potential function optimizers, genetic algorithms have received a great deal of attention. Genetic algorithms proved to be very useful for this process because of their ability to efficiently use historical information to obtain new solutions with enhanced performance. Genetic algorithms are also theoretically and empirically proven to provide robust search in complex search spaces and they do not get trapped in local minima as opposed to gradient descent techniques.

Optimization problems often have constraints which make them more difficult to solve than problems without constraints. For example, in a traveling salesman problem the constraint usually used is that each of the cities should be visited once and only once. For a scheduling problem the constraints are usually that some of the processes have to be done on particular machines, and some processes cannot be performed before others. When optimizing an unconstrained function, any combination of bits in the chromosomes is legal. This means that the mutation and crossover operations will always produce a legal offspring. When the problem has some constraints, the crossover and mutation operations on legal chromosomes, may produce illegal offsprings.

One of the solutions for solving the constraint problem is to use penalty functions as an adjustment to the fitness function being optimized (Michalewicz, 1992). In this case a penalty term is introduced in the fitness function for each constraint violation. In this way the original constrained problem is transformed into an unconstrained one. If for the solved problem it is likely that an individual violating the constraint will be produced, it is also possible that the genetic algorithm will spend most of its time evaluating illegal individuals (Davis, 1991). This technique seems to work best for narrow classes of problems and for few constraints (Michalewicz, 1992). Another solution for resolving the constraint problem is to use decoders, which avoid building an illegal individual. The third approach consists of chromosome repair, which repairs the chromosome in such a way that it fulfills all the constraints. Both of those approaches are highly problem specific, but they seem to give good results.

Genetic Algorithms for Finding the Best Aggregation Parameters

The schemes that we are proposing for finding the best aggregation parameters are based on genetic algorithms and allow fusion of decisions from two or more sensors at a time. The aggregation function used is the generalized mean Equation 103.6. For n sensors, $n - 1$ weights (w_i) and one exponent (p) should be determined. The nth weight is determined by Equation 103.7. Our method can be described as follows:

- Code the exponent and the weights as bit strings.
- Generate a population with different exponents and weights.
- Perform fusion with the given parameters and calculate the fitness function.

The fitness function is the function which will be maximized. We define the fitness function as

$$\text{fitness} = 1 - \text{err1} \tag{103.8}$$

where err1, the error function, is given by

$$\text{err1} = \sqrt{\frac{\sum_{j=1}^{l}\sum_{k=1}^{m}(\text{desired}_{jk} - \text{actual}_{jk})^2}{l \cdot m}} \tag{103.9}$$

and

$$\text{actual}_j = \left(\sum_{l=1}^{n} w_i x_i^p\right)^{1/p} \tag{103.10}$$

The expression in Equation 103.9 under the square root is the sum of squared errors divided by the number of patterns l and the number of outputs m. The number of outputs (in case of a classification problem) is the number of classes. Equations 103.8 and 103.9 ensure that the fitness function will lie in the interval [0, 1]. The sum of the squared errors is computed for l patterns. The role of those patterns is the same as the role of the training patterns in a neural network. For those patterns we have the desired values. The goal is that the genetic algorithm adjusts the exponent and weight in such a fashion that the error for those training patterns is minimized in the least squares sense.

103.4 Two Fuzzy-Genetic Fusion Techniques

Two information fusion techniques based on fuzzy aggregation connectives and genetic algorithms were developed in this paper. The first method performs aggregation of evidence from two sensors at a time; if there are more sensors, information from the next sensor is fused with the data already aggregated. The second technique developed performs one-step fusion from all the sensors available.

Fusion from Two Sensors in One Step

When a large number of sensors exists, fusing information from all the sensors is perhaps not necessary. If the weight for one of the sensors is very small there is no real need to fuse the information from this sensor.

The proposed fusion scheme is comprised of several steps:

1. Choose two sensors for the fusion process.
2. Use the genetic algorithm to find the best fusion parameters for those sensors.
3. If one of the weights is very small, do not fuse the data from that sensor; otherwise perform fusion.
4. If data from more sensors are available, choose the next sensor for fusion with the previous fusion result, and repeat steps 2–4.
5. If data from all the sensors has been used this gives the aggregated decision.

The final decision has to be obtained from the aggregated decision by some defuzzification method. The method chosen was α-cuts. The *α-cut* of a fuzzy set A (denoted by A_α) is the crisp set of elements that belong to the fuzzy set A at least to the degree α

$$A_\alpha = \{x \in X, \mu_A(x) \geq \alpha\} \tag{103.11}$$

The membership function of a fuzzy set A can be expressed in terms of the characteristic functions of its α-cuts according to the formula.

$$\begin{aligned} \mu_A(x) &= 1 \quad \text{iff } x \in A_\alpha \\ \mu_A(x) &= 0 \quad \text{iff } x \notin A_\alpha \end{aligned} \tag{103.12}$$

Fusion from All Sensors in One Step

The method proposed in the previous section performs fusion of decisions from two sensors in each fusion step (multi-step method). The method described in this section allows fusion from any number of sensors in one fusion step (single-step method). Fusion is not carried out for the sensors for which the weight found by the genetic algorithm is zero.

When fusing information from n sources in one step, the genetic algorithm has $n + 1$ chromosomes. The first chromosome is used for coding the exponent p, the other n chromosomes are used for coding the weights. Each of the weights can have any value in the range [0, 1], but there is a constraint that the weights sum up to one (see Equation 103.7). Each crossover and mutation operation that cause the offspring's weights to not sum to one causes the offspring to be illegal. We use the chromosome repair in order to overcome this problem. After crossover and mutation take place, we use one of the two chromosome repair methods developed by us.

The first chromosome repair method adds all the weights, and if their sum exceeds one, a randomly chosen weight is set to zero. The process is repeated as long as the sum of weights is greater than one. This repair method leads to a relatively large number of weights set to zero. The second method after adding the weights, divides each of the weights by the sum of all weights, therefore normalizing the sum to one. This method involves chromosome recoding from the floating point value to the binary representation and therefore is more time consuming. The advantage of the second method is that it does not easily set the weights to zero as the first chromosome repair technique.

The convergence of the constrained genetic algorithm (for the fusion from all the sensors) was much less satisfactory than the convergence of the unconstrained algorithm (for the fusion from two sensors). To remedy the situation we introduced non-uniform mutation. Non-uniform mutation allows the mutation probability for the bits located on the left side of the chromosome to decrease as the population ages, and the mutation probability for the bits at the right side of the chromosome to increase. The function we developed to control the mutation rate is defined by

$$\text{mutation rate} = a \cdot (1 - b \cdot c)$$

where $a = e^{-\text{gen}/4 \cdot maxgen}$

$$b = \frac{1}{2} - \frac{\text{gen}}{\text{maxgen}}$$

$$c = \left(1 - \frac{\text{curchrom} - 1}{\text{lchrom}}\right)^3 \qquad (103.13)$$

where *gen* is the current generation number, *maxgen* is the maximum generation number, *curchrom* is the current chromosome number and *lchrom* is the chromosome length. To obtain the current mutation probability, the value of the function from Equation 103.13 should also be multiplied by the mutation probability, which we chose to be 0.0333 in all the experiments. The behavior of this function is presented by the three-dimensional plot in Figure 103.2.

To better describe the behavior of the mutation rate, let us consider Figures 103.3 and 103.4. Figure 103.3 depicts the behavior of the mutation rate as we change the current generation from 1 to the maximum generation. The behavior of the function for the rightmost chromosome is described by the solid line, and the behavior of the function for the leftmost is described by the dashed line. For the rightmost chromosome the mutation rate increases from 0.5 to 1.17 as the current generation increases from 1 to the maximum generation. For the leftmost chromosome the mutation rate decreases from 1.0 to 0.78 as the current generation increases from 1 to maximum generation. These values ensure that when the population is young, the rate of mutation is larger

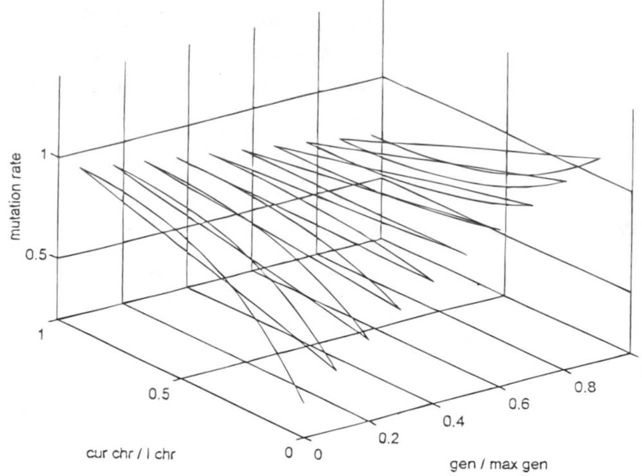

Figure 103.2 Three-dimensional plot of the mutation rate defined by Equation 103.13.

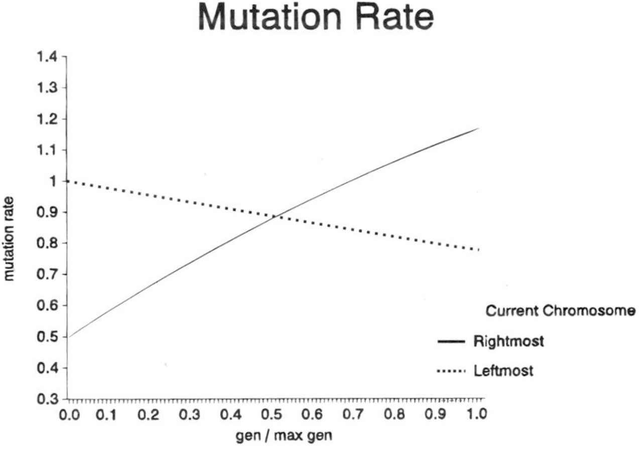

Figure 103.3 Mutation rate as the generation number increases.

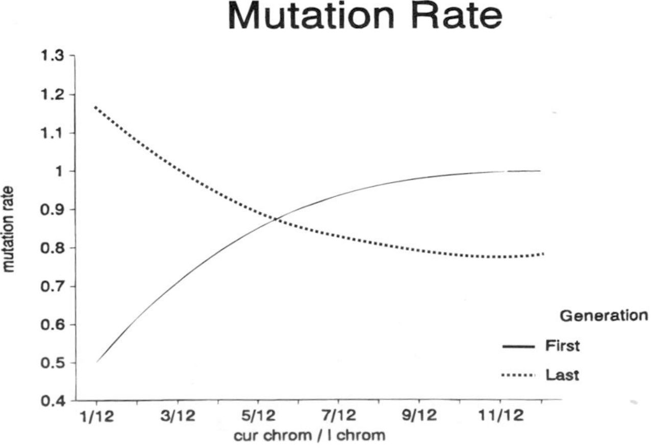

Figure 103.4 Mutation rate as the chromosome number increases.

for the bits located at the left side of the chromosome, and as the population ages the probability of mutation is larger for the bits at the right side of the chromosome. The functions for the leftmost and the rightmost chromosome are not symmetric. The reason behind this asymmetry is that when the population is young, much of the space is searched by crossover. As population ages, and individuals resemble each other more and more, only local tuning is performed and it is performed more efficiently by mutation than by crossover. Therefore, the mutation rate should be larger at the end of the run, than at the beginning of the run. Figure 103.4 describes the behavior of the mutation rate as we change the current chromosome from 0 to the last chromosome. The solid line describes the behavior of the function for the first generation, and it is a monotonically increasing function starting at 0.5 and ending at 1.0. The behavior for the last generation is described by the dashed line and it is a monotonically decreasing function starting at 1.17 and ending at 0.78. Non-uniform mutation allowed the genetic algorithm to perform more efficient local tuning, and in consequence it considerably improved the results.

103.5 Information Fusion for Object Classification

In the case of aggregation of information from multiple sensors for object classification, classifier modules can be structured and combined to form a complete system (Heading and Bedworth, 1991). Each of the sensors can attempt a classification, and then a fusion center makes an overall classification on the basis of information supplied to it. Each sensor classification can be obtained for example by a neural network classifier, and the decision fusion can be performed by a fuzzy logic module. When the classification problem involves multiple sensor readings, it is very convenient to break the problem into separate parts. This idea is illustrated on Figure 103.5 for two inputs. Each of the sensor processors S_i can produce a decision y_i based on the data from sensor i. The "fusion center" F, makes an overall classification based on the information supplied to it as proposed by Heading and Bedworth (1991).

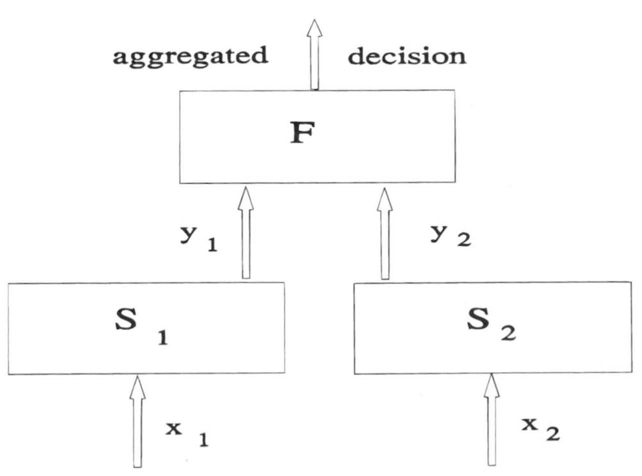

Figure 103.5 Model for data fusion.

The classification problem is addressed in three phases. Phase I includes the extraction of relevant features from the signatures for each sensor separately and the compression of the spectra using recirculation neural networks (RNN). Phase II includes the classification of compressed signatures using one backpropagation network (BPN) per sensor. Phase III, the fusion of decisions from individual classifiers, is performed by fuzzy fusion modules (FFM). The parameters of the fuzzy aggregation functions are determined by the genetic algorithms (GA). The schematic of the multi-step fusion method for object classification is depicted in Figure 103.6, and the diagram of a one-step method for object classification is depicted in Figure 103.7.

For each sensor i there is one recirculation neural network, which takes as input the signature registered by sensor i and compresses it. The compressed signature constitutes the input to BPN network. There is one backpropagation network per sensor. The backpropagation networks produce a decision as to

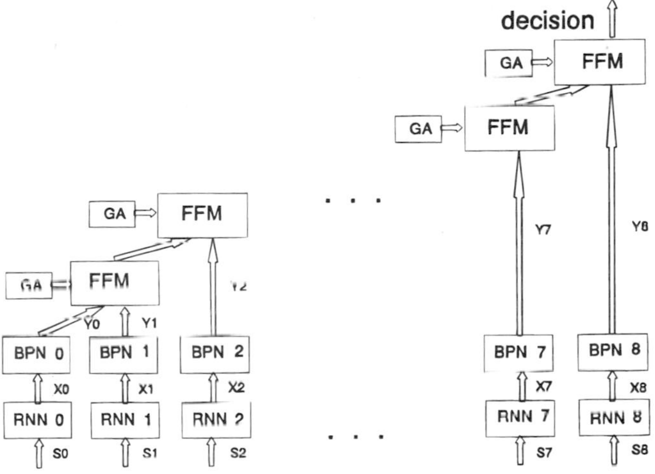

Figure 103.6 Architecture of the fuzzy-genetic aggregation system performing fusion step by step.

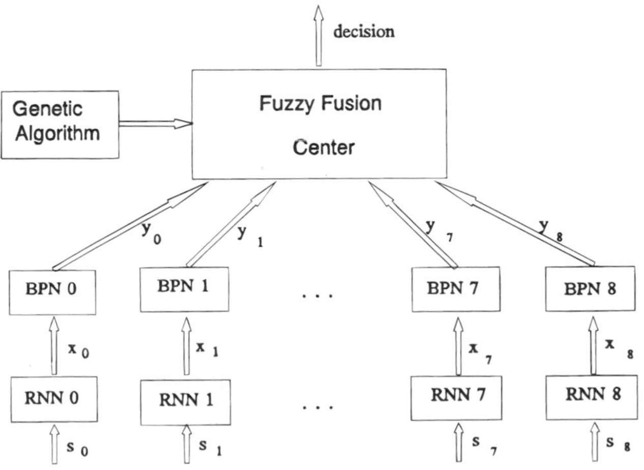

Figure 103.7 Architecture of the fuzzy-genetic aggregation system performing fusion from all the sensors at once.

what degree a given signature belongs to each of the classes. The outputs of BPN networks are the inputs to the fuzzy fusion module; its output is the fused decision. Recirculation and back-propagation neural networks are described in detail in references Hinton and McClelland (1988) and Rumelhart et al. (1986) respectively.

103.6 Vibration Monitoring

Background

The information fusion schemes developed are applied next to the problem of vibration monitoring. Vibration monitoring of components in manufacturing plants involves the analysis of vibration data obtained from vital components of the plant. The analysis leads to the identification of component failures and their causes, and makes it possible to perform efficient preventive maintenance. Each vibration frequency spectrum has a characteristic shape when the machine is operating properly, and it has a different shape when faults appear or some dynamic processes in the machine change. Recognition of faults can be accomplished in many cases by detecting features in the frequency spectrum which are known to be related to particular faults. The peaks in the frequency spectrum are related to a specific fault and its severity, and all vibration monitoring techniques are based fundamentally on the recording and quantification of these vibration impulses produced by the fault (Broch, 1984; Jackson, 1979). Generally, spectral features associated with specific defects are generated at frequencies that can be calculated from formulae; however, the task of recognizing a fault is complicated by a series of factors which affects the shape of the spectrum and which makes the problem specific to the component where failure occurs and to operating conditions. Among these factors are noise, presence of multiple faults, severity of the fault, machine geometry, and speed changes (Alguindigue et al., 1993).

For vibration monitoring, multiple sensors which record the vibration signals are located throughout the machines. Based on these sensor outputs, analysis techniques can be designed for early detection of faults. Fusion of information from multiple sensors is required in order to yield satisfactory performance levels. The goal is to design a diagnostic system, which will perform the fusion of decisions from multiple sensors in order to automate the interpretation of vibration spectra.

Vibration Signatures

Data used for this project consist of vibration signatures from "laminar flow" table rolls in a steel sheet manufacturing mill. Data are collected with sensors attached to the plant machinery at nine locations on each machine. Signals acquired from the nine sensors are correlated but not identical (see Figure 103.8). This is due to different vibration levels throughout the machine and to the fact that the faults which are particular to a bearing located near one sensor are not necessarily recorded by other sensors. The spectrum of each sensor output is generated using FFT (Fast Fourier Transform) techniques and the coefficients are stored in a database. Each spectrum contains 150 points. The data set contains signatures from 49 machines for which the type of fault has been identified. For some machines, one to three sensor readings are missing.

The data set reflects only faulty operating conditions such as misalignment (M), looseness (L), wear (W), outboard bearing damage (O), lubrication (C), and some double and triple faults. Double faults consist of misalignment and looseness (M&L), misalignment and wear (M&W), misalignment and outboard bearing damage (M&O), misalignment and lubrication (M&C), and looseness and outboard bearing damage (L&O). The triple fault encountered is the combination of misalignment, looseness, and outboard bearing damage (M&L&O).

We worked earlier on the fusion of decisions (Loskiewicz-Buczak and Uhrig, 1994) from this data set, but the parameters of the generalized mean were not obtained by genetic algorithms. Instead the weights for each sensor were set equal and some assumptions about the degree of redundancy/complementarity of data from the sensors allowed us to set the parameter p for each fusion step. The results were very satisfying. However the methods based on genetic algorithms are more general, since they do not need any assumptions about the data from the sensors.

103.7 Results

Data Flow

The range of the spectral data was [0.0, 1.0]. The 150-point input vectors to the recirculation networks were not further normalized by any method. The recirculation networks perform compression 3:1, and 50-point signatures constitute their output. The signatures compressed by RNNs form the input to each of the back-propagation networks. The collective output of BPN represents the degree to which each of the faults is present. In case of fusion from two sensors in one step, the output of two BPN networks is the input to the fuzzy fusion module. In case of fusion from all the sensors in one step, the output of all BPN networks is the input to the fuzzy fusion module. The output of the fuzzy fusion module is the aggregated decision. Fusion of decisions is performed for all the sensors for which the weight selected by the genetic algorithm is greater than 0.0.

The final decision is obtained from the aggregated decision by means of α-cuts. After fixing the value of α, an α-cut is performed on the aggregated decision. For each of the five faults (M, L, W, O, C) there is a corresponding α-cut (M_α, L_α, W_α, O_α C_α). Each of these sets includes all the patterns that are manifesting a given fault. If a pattern belongs only to one α-cut, it means that the final decision is that it is exhibiting only this fault (single fault pattern). If a given pattern belongs to more than one α-cut, it means that the final decision is that it is a multiple fault pattern, manifesting the faults, to which α-cuts the pattern belongs.

The value of α is determined as follows. A value of 0.5 is usually used as α. The reason is that the range of outputs is [0, 1] and therefore reserving half of the range to describe the presence of a given fault and the other half to describe the

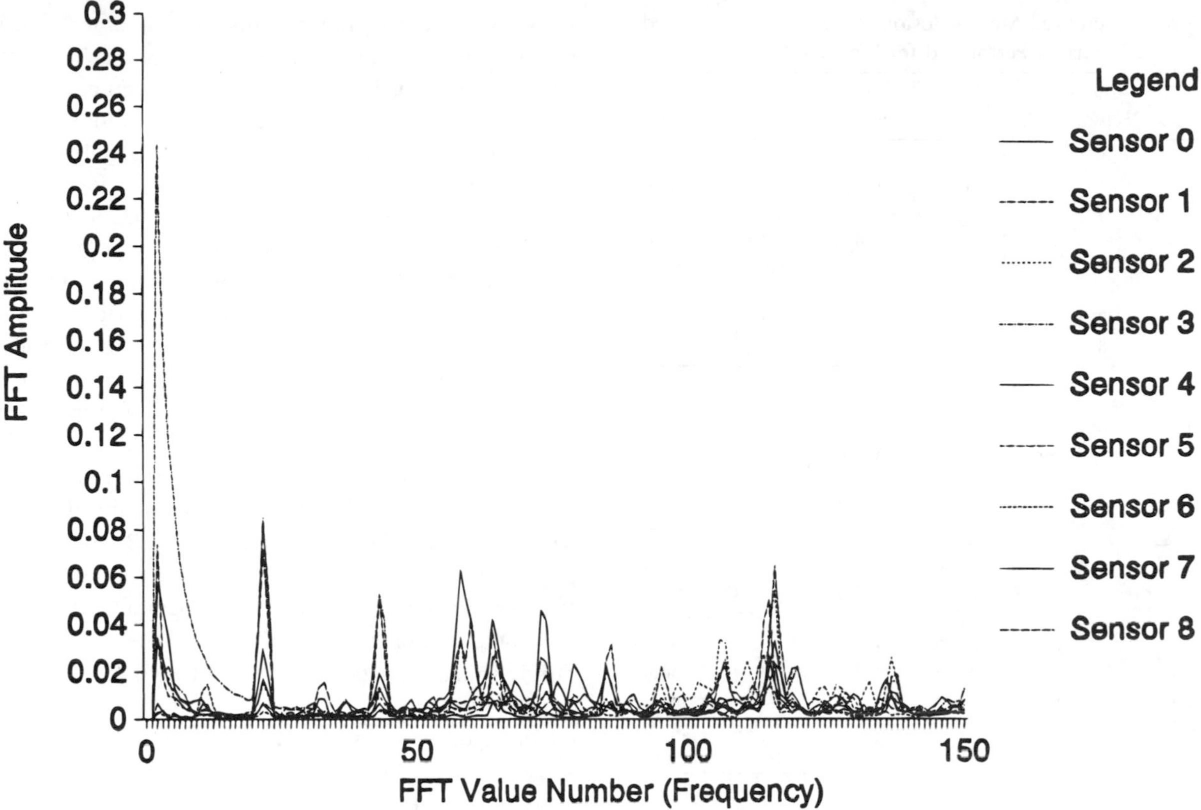

Figure 103.8 Vibration spectra from nine sensors on a machine with fault M.

absence of the fault, seems to be an unbiased decision. Sometimes, however, for one of the patterns the largest of the five outputs is smaller than 0.5, which would mean that the pattern is not classified in any category. To remedy this situation the value of the largest of the five outputs truncated to the first decimal place (for example 0.44784 is truncated to 0.4) is chosen as α for all the set. The method described is based on approximate reasoning for finding the parameters of the aggregation process. There is no reason to use a crisp number precise to the 5th decimal place (as in case of 0.44784) for the value of α; this number has to be truncated.

Fusion from Two Sensors in One Step

Results of compression by recirculation networks and classification based on one sensor information by backpropagation networks are described in detail in our earlier work in (Loskiewicz-Buczak and Uhrig, 1992, 1993). Here we will concentrate on decision fusion results.

For the fusion process the order of sensors chosen is that of consecutive numbers (0, 1, 2 . . . 8). Therefore, the first GA finds p and w for the fusion of decisions from sensors 0 and 1. The results for non-elitist strategy are presented in Table 103.1. The fusion is performed in all the cases, except when the weight for one of the sensors was 0.0. After 8 runs of the genetic algorithm, the parameters of all the fusion steps were determined. It is not necessary to perform the fusion of decisions from sensors 5, 6

Table 103.1 Generalized Mean's Fusion Parameters Determined by Nonelitist GA; Fusion Performed for All Weights

1st Sensor	2nd Sensor	p	w_1	w_2	Action
0	1	0.89	0.907	0.093	fused
0&1	2	1.998	0.441	0.559	fused
0&1&2	3	1.996	0.726	0.274	fused
0&1&2&3	4	0.02	0.982	0.018	fused
0&1&2&3&4	5	0.163	1.0	0.0	none
0&1&2&3&4	6	1.432	1.0	0.0	none
0&1&2&3&4	7	0.777	0.809	0.191	fused
0&1&2&3&4&7	8	0.91	1.0	0.0	none

and 8 for which the weights were 0.0. The α-cut is performed at 0.5. The fused decision is much more reliable than the decisions based on individual sensor information: 2% erroneous decisions compared with an average of 16% of such decisions for individual sensors. Also, 79.6% correct decisions were obtained while only 63.7% of such cases on the average were recognized when dealing with individual sensor classifiers.

The results for nonelitist strategy, when omitting the data from sensors for which the weight was less than 0, are shown in Table 103.2. In this case only decisions from four sensors (0, 2, 3, and 7) were fused. The results are not as good as in the previous case, giving 75.5% correct classifications, instead of 79.6% and 2% of misclassifications as previously. However, only four sensors are needed in this case, instead of six.

The results for the elitist strategy are presented in Table 103.3.

Table 103.2 Generalized Mean's Fusion Parameters Determined by Nonelitist GA; Fusion Performed for Weights ≥ 0.1

1st Sensor	2nd Sensor	p	w_1	w_2	Action
0	1	0.89	0.907	0.093	none
0	2	1.995	0.41	0.59	fused
0&2	3	1.994	0.738	0.262	fused
0&2&3	4	0.004	0.996	0.004	none
0&2&3	5	0.368	1.0	0.0	none
0&2&3	6	0.178	0.996	0.004	none
0&2&3	7	0.712	0.792	0.208	fused
0&2&3&7	8	0.338	0.993	0.07	none

Table 103.3 Generalized Mean's Fusion Parameters Determined by Elitist GA; Fusion Performed for All Weights

1st Sensor	2nd Sensor	p	w_1	w_2	Action
0	1	0.884	0.903	0.097	fused
0&1	2	2.0	0.452	0.548	fused
0&1&2	3	2.0	0.723	0.277	fused
0&1&2&3	4	0.003	0.993	0.007	fused
0&1&2&3&4	5	0.766	1.0	0.0	none
0&1&2&3&4	6	0.938	1.0	0.0	none
0&1&2&3&4	7	0.747	0.774	0.226	fused
0&1&2&3&4&7	8	0.174	1.0	0.0	none

The parameters chosen by GA are slightly different than in the case of non-elitist strategy, but the data from the same sensors is fused: 0, 1, 2, 3, 4 and 7. The results, however, are better: 81.6% correct classifications, 16.4% in which the most prominent fault was correctly recognized, and 2% misclassifications. When omitting in the fusion process the sensors for which the weight was less than 0 (Table 103.4), the result is 75.5% correct classifications, 22.5% most prominent faults correctly identified, and 2% misclassifications. In this case, only data from four sensors (0, 2, 3 and 7) were aggregated.

An important characteristic of the proposed system is that it works well even when some sensor readings are missing. In case of the elitist genetic algorithm performing fusion for all weights with the exception of zero weights, from the 8 cases in which only the most prominent fault was correctly identified 5 had some sensor readings missing. In such a situation, the output nodes of BPN classifiers are assumed to have activation of 0.5 each. These values are used as the contribution of the missing sensors to the aggregation function. The accuracy of our system on the patterns with no sensor readings missing is excellent: 89.7% correct classifications, 7.7% most prominent faults correctly identified, and 2.6% incorrect classifications.

Table 103.4 Generalized Mean's Fusion Parameters Determined by Elitist GA; Fusion Performed for Weights ≥ 0.1.

1st Sensor	2nd Sensor	p	w_1	w_2	Action
0	1	0.884	0.903	0.097	none
0	2	2.0	0.422	0.578	fused
0&2	3	1.999	0.745	0.255	fused
0&2&3	4	0.0	0.995	0.005	none
0&2&3	5	0.055	1.0	0.0	none
0&2&3	6	0.12	0.998	0.002	none
0&2&3	7	0.707	0.789	0.211	fused
0&2&3&7	8	0.331	0.996	0.004	none

The nonelitist genetic algorithm had a population size of 30 and was run for 70 generations, giving 2100 individuals which were tested. The elitist genetic algorithm had a population size of 41 and was run for 50 generations, giving 2050 individuals which were tested. The reason to use a larger population size in case of the elitist algorithm is that this algorithm may increase the speed of domination of the population by a super individual. When using elitist strategy, the best individual will produce at least two offsprings in almost each generation. Larger population size ensures that there will be more diversity in the population (more individuals will have the ability to reproduce). In both elitist and non-elitist algorithms, the probability of crossover was 0.6 and the probability of mutation 0.0333. We developed the genetic algorithm as a C program for executing on a 60 MHz Pentium computer. Each run of the program took approximately 20 seconds. Five of the patterns were used already in backpropagation learning; the other five were not used in the earlier learning process.

Our algorithm uses two chromosomes, not just one chromosome as most genetic algorithms do. Therefore, the question arises if crossover should be performed on all the chromosomes at once, or only on one randomly chosen chromosome. We tried both approaches and found out that when using multiple chromosome crossover, the fitness function grows much more rapidly than when using single chromosome crossover. Therefore, less generations were needed to achieve the same accuracy in case of multiple chromosome crossover than in the case of single chromosome crossover. Hence, we used multiple chromosome crossover exclusively.

The tremendous improvement of fused decision over the average of individual sensor decisions shows how well the generalized mean is suited for decision fusion and how well the parameters of the fusion process are determined by the elitist genetic algorithm. The method developed has the advantage that once the sensors from which decisions will be fused are determined, there is no need to process the data from the other sensors.

Fusion from All Sensors in One Step

Genetic algorithm with ten chromosomes (nine for weights and one for the exponent p) with a population size 101 was run for 500 generations. The first chromosome repair method, as described in the vibration signatures section is used. Using the second chromosome repair method, the weights are slowly moving toward zero and the method needs a significantly larger number of generations (about 5000) in order to reach the same value of the fitness function as the first chromosome repair technique. Both methods converged only when non-uniform mutation was introduced (as described in the vibration signatures section).

Table 103.5 Generalized Mean Fusion Parameters for One Fusion Step Determined by Elitist GA

p	w_0	w_1	w_2	w_3	w_4	w_5	w_6	w_7	w_8
1.397	0.266	0.0	0.313	0.151	0.0	0.001	0.0	0.269	0.0

The parameters found by the algorithm are presented in Table 103.5. Single step fusion performed with these parameters gave 79.6% correct classifications, 18.4% cases when the most prominent fault was correctly recognized and 2% incorrect classifications when five out of nine sensors were chosen for the fusion process: 0, 2, 3, 5 and 7. The accuracy of one step fusion on the patterns with no sensor readings missing is very good: 87.2% correct classifications, 10.2% most prominent faults correctly identified, and 2.6% incorrect classifications. The one step fusion method developed allows only one type of aggregation, as only one parameter p is determined. Therefore, unless the data from all sensors show the same degree of redundancy/complementarity, the results obtained may not be optimal. But this method finds the sensors that are essential for the final classification, and sensors which are not so important.

A significant fact was noticed about the 2% misclassifications obtained by each of the methods described: signature 12 was always wrongly classified. This fact made us investigate this problem in detail. Signature 12 was compared with the second signature in the set exhibiting the same fault (improper lubrication); these signatures were found to be completely different. For signature 12, spectra contain a lot of noise (see Figure 103.9): the peaks from different sensors are at very different frequencies and this implies a very severe problem during data acquisition. We believe signature 12 was misclassified initially. Either correcting this error or eliminating this signature would have given 0% inaccurate classifications for most cases.

103.8 Conclusions

In this paper, the problem of decision fusion from multiple sensors is solved by means of fuzzy aggregation connectives. The use of fuzzy aggregation connectives for the fusion process allows taking advantage of soft boundaries in fuzzy logic environments. In fuzzy logic, there is a possibility of using not only the union and intersection connectives for the purpose of aggregation, but also connectives with compensatory behavior which closely match the human decision-making process and therefore result in better classification than with classical aggregation connectives.

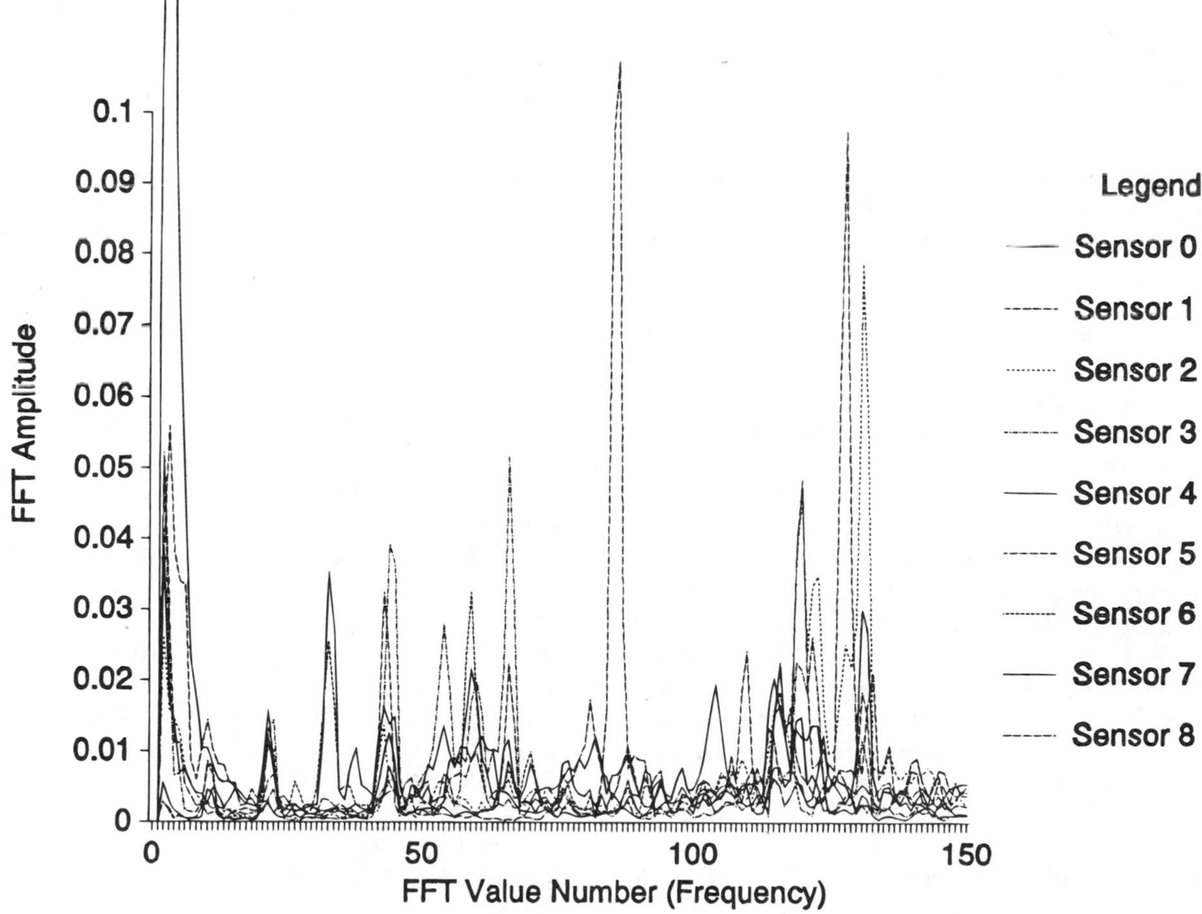

Figure 103.9 Vibration spectra from nine sensors on machine 12 with fault C.

The novel approach consists of finding the optimal parameters of the aggregation connectives by genetic algorithms. Genetic algorithms prove to be very useful for this process because of their ability to productively use historical information to obtain new individuals with enhanced performance. Genetic algorithms are also theoretically and empirically proven to provide robust search in complex search spaces, and they do not get trapped in local minima as opposed to gradient descent techniques. Genetic algorithms with elitist strategy proved to be more efficient than its nonelitist counterparts.

Two evidence aggregation techniques are developed in this paper. The first method allows fusion of evidence from two sensors at each step; if there are more sensors, information from the next sensor is fused with the data already aggregated. The second technique performs one step fusion from all the sensors available. When fusing information from more than two sensors in a single step, the problem of chromosome repair arises, because the weights, coded in the chromosomes, should sum up to one. Two chromosome repair techniques were developed, and the advantages and disadvantages of each of the techniques are described. Since multiple chromosome genetic algorithms have problems with convergence, non-uniform mutation is used. Non-uniform mutation allows the mutation probability for the bits located on the left side of the chromosome to decrease as the population ages, and the mutation probability for the bits at the right side of the chromosome to increase.

The use of genetic algorithm allows reduction of the number of sensors needed for the fusion process. This is an important issue, especially when the number of sensors employed is large. The algorithm is very useful in determining the sensors from which the data is unreliable; not choosing these data for the fusion process improves the classification accuracy. The performance of the system is very good when all the sensors are operating properly and is still good when some of the sensor readings are missing. This last property is especially useful for a real-time application, where we can expect some of the sensors being out of order some of the time.

For the vibration monitoring problem, the best results are obtained by elitist genetic algorithm performing consecutive fusion steps, from two sensors in each step for all the weights larger than 0.0. The results are very good given the fact that the multiple fault identification problem is a particularly difficult one. Experts often disagree when classifying multiple fault signatures. Sometimes one fault is very well pronounced, while the other is in very early stages of development and is almost not noticeable. In such cases experts tend to be inconsistent and classify one signature as, for example, misalignment and looseness, and another, very similar signature as misalignment only. This is why in a large percentage of cases the most prominent fault was correctly identified, while the second one was not (or vice versa).

The distinctive feature of the methods developed is that they can be applied to any problem involving fusion of decisions. The methods devised are very general and can be used not only for vibration analysis but also to combine evidence from arbitrary knowledge modules for arbitrary tasks. The task can be vibration monitoring, recognition of objects in a computer vision system, area surveillance systems for military purposes, or managerial decision making systems.

Acknowledgments

The authors wish to acknowledge Electricite de France (Paris, France) for partially supporting this work and Technology for Energy Corporation (Knoxville, TN) for providing the data.

References

Abidi, M. A. and Gonzalez, R. C. eds., 1992. *Data Fusion in Robotics and Machine Intelligence,* Academic Press, New York, NY.

Alguindigue, I. E., Loskiewicz-Buczak, A., and Uhrig, R. E. 1993. Monitoring and diagnosis of rolling element bearings using artificial neural networks, *IEEE Trans. Industrial Electronics: Special Issue on Applications of Intelligent Systems to Industrial Electronics,* 40(2):209–217.

Broch, J. T. 1984. *Mechanical Vibrations and Shock Measurements,* Bruel & Kjaer,

Dasarathy, B. V. 1991. Decision strategies in multisensor environments, *IEEE Trans. on Systems, Man and Cybernetics,* 21(5):1140–1154.

Davis, L. 1991. *Handbook of Genetic Algorithms,* Van Nostrand Reinhold, New York, NY.

De Jong, K. A. 1975. An Analysis of the behavior of a Class of Genetic Adaptive Systems, (Doctoral Dissertation, University of Michigan), *Dissertation Abstracts International,* 36(10):514B.

De Jong, K. A. 1992. Are Genetic Algorithms Function Optimizers?, *Parallel Problem Solving from Nature 2,* Manner, R. and Manderick, B. eds., North-Holland, Amsterdam.

Dubois, D. and Prade, H. 1980. *Fuzzy Sets and Systems: Theory and Applications,* Academic Press, New York, NY.

Dubois, D. and Prade, H. 1992. Combination of Fuzzy Information in the Framework of Possibility Theory, *Data Fusion in Robotics and Machine Intelligence,* Abidi, M. A. Gonzalez, R. C., eds., 481–505, Academic Pres New York, NY.

Goldberg, D. 1989. *Genetic Algorithms in Search, Optimization and Machine Learning,* Addison-Wesley, Reading, MA.

Heading, A. J. R. and Bedworth, M. D. 1991. Data fusion for object classification, *Proc. Int. Conf. Systems, Man and Cybernetics,* 837–840.

Hinton, G. E. and McClellend, J. L. 1988. Learning Representations by Recirculation, *Proc. IEEE Conf. Neural Information Processing Systems,* November.

Holland, J. H. 1992. *Adaptation in Natural and Artificial Systems—An Introductory Analysis with Applications to Biology, Control, and Artificial Intelligence,* The MIT Press, Cambridge, MA.

Jackson, C. 1979. *The Practical Vibration Primer,* Gulf Publishing, Houston, TX.

Kacprzyk, J. 1983. *Multistage Decision-Making under Fuzziness: Theory and Applications,* Verlang TUV Rheinland.

Klir, G. J. and Folger, T. A. 1988. *Fuzzy Sets, Uncertainty, and Information,* Prentice-Hall, Englewood Cliffs, NJ.

Krishnapuram, R. and Lee, J. 1992. Fuzzy-connective-based hierarchical aggregation networks for decision making; *Fuzzy Sets and Systems,* 46:11–27.

Loskiewicz-Buczak, A. and Uhrig, R. E. 1992. Probabilistic Neural Network for Vibration Data Analysis, *Intelligent Engineering Systems Through Artificial Neural Networks,* vol. 2, Dagli, C. H., Burke, L. I., and Shin, Y. C., eds., 713–718, ASME Press, New York, NY.

Loskiewicz-Buczak, A. and Uhrig, R. E. 1993. Aggregation of evidence by fuzzy set operations for vibration monitoring, *Proc. 3rd Int. Conf. Industrial Fuzzy Control and Intelligent Systems,* 204–209, Houston, TX.

Loskiewicz-Buczak, A. and Uhrig, R. E. 1994. Decision fusion by fuzzy set operation, *Proc. 3rd IEEE Int. Conf. Fuzzy Systems,* Vol. 2, June 26–29 Orlando, FL.

Luo, R. C. and Kay, M. G. 1992. Data Fusion and Sensor Integration: State-of-the-Art 1990s [*Data Cusion in Robotics and Machine Intelligence,* Abidi, M. A. and Gonzalez, R. C., eds., pp. 7–135, Academic Press, New York, NY. ABI-92], pp. 7–135.

Michalewicz, Z. 1992. *Genetic Algorithms + Data Structures = Evolution Programs,* Springer-Verlag, New York, NY.

Rumelhart, D. E., Hinton, G. E., and Williams, R. J. 1986. Learning Internal Representations by Error Propagation, *Parallel Distributed Processing,* Rumelhart, D. E. and McClelland, J. L., eds., MIT Press, Cambridge, MA.

Waltz, E. and Linas, J. 1990. *Multisensor Data Fusion,* Artech House, Norwood, MA.

Zimmermann, H. J. and Zysno, P. Latent connectives in human decision making, *Fuzzy Sets and Systems,* 4:37–51.

Zimmermann, H. J. *Fuzzy Sets, Decision Making, and Expert Systems,* Kluwer Academic Publishers, 1987.

104

Neural Evolutionary and GA Systems and Applications

Mary Lou Padgett
Auburn University

104.1 Introduction 1338
104.2 Combining Evolutionary Systems and Neural Systems 1338
Application Examples
104.3 Summary 1341

104.1 Introduction

Applications combining neural systems with evolutionary systems are emerging as successful, practical engineering solutions, and indicators of future developmental trends. As high-speed parallel processing becomes more accessible to engineers working in industrial electronics, and user-friendly software development tools evolve, systems level applications of evolutionary neural systems should flourish.

The material in the following paragraphs is intended to suggest strategies for applications in industrial electronics. Core ideas are outlined, and further references suggested as sources of details about applications using modifications of the basic ideas.

104.2 Combining Evolutionary Systems and Neural Systems

Both evolutionary systems and neural systems are effective at learning from data and modeling nonlinearity. They work when there is no mathematical model available. Operator knowledge and knowledge representation are not part of these systems, but can be incorporated using fuzzy algorithms or expert systems.

The symbiosis between evolutionary and neural systems comes from the EC strength in optimization and the NN capabilities for real-time evaluation. Figure 104.1 adapted from the paper by Shibata et al. and Table 104.1 based on the paper by Kaynak illustrate some of these relationships.

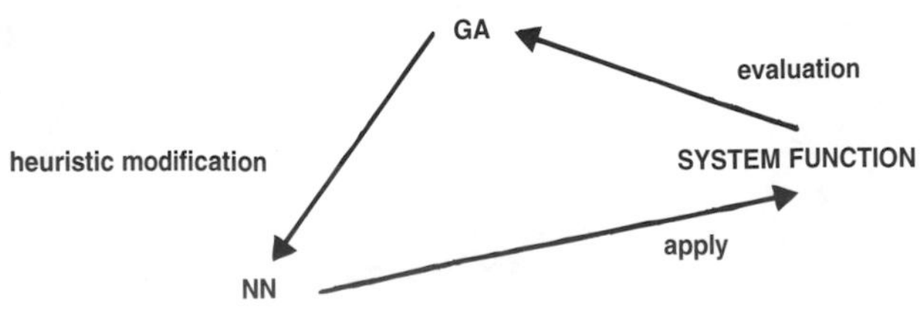

Figure 104.1 GAs tune NN structure, I/O, other parameters; NN aid evaluation of GA performance as user-tools or as evaluation functions.

As with fuzzy EC systems, incorporation of EC can augment neural systems by *auto-tuning* parameters. EC can both *learn* from historical data and *explore* new solutions by adding randomness. Optimization can serve as a sensitivity analysis, determining the most important of potential NN input and output factors. Several schema exist for auto-tuning neural structures. The paper by Shibata et al. mentions use of EC to optimize connection patterns in Multilayer Perceptrons (MLPs) and in Radial Basis Functions (RBFs). Success with RBFs is attributed to their more structured architecture. Takagi (1993, 1995) points out that GA determination of MLP connectivity is difficult to scale up. A grammar encoding method may provide an indirect approach to the problem. The ingenuity of the applications engineer is key to the successful configuration of an EC problem as a population of strings made up of parameters to optimize. The choice of coding for parameters and interpretation of the results is a worthwhile challenge in creativity and scientific method.

In two areas, the interpretation of EC progress can be aided by NN techniques. In the validation and verification of an EC system with respect to its overall goal, neural networks can help in approximation of a performance evaluation function that is complex and hard to specify. Fuzzy systems can also help—human operator interface is usually involved, and system results should be interpreted in terms appropriate for human communication.

The EC concepts discussed in earlier papers by Padgett in this chapter explore evolution of processes difficult to evaluate in terms of a simple fitness function. There are cases when a NN function approximation or classification scheme can ably assist in this endeavor. Again, fuzzy algorithms may help incorporate into the automated system any rules of thumb that are available for guidance. The blending of NN and EC techniques is less difficult than the addition of FS technology to either NN or EC. There are already many similarities between NN and EC systems. Both require intensive data processing resources. If a system already has the hardware and software for NN or EC, it is easy

0-8493-8343-9/97/$0.00+$.50

Table 104.1 Relationships Between Evolutionary Systems and Neural Systems

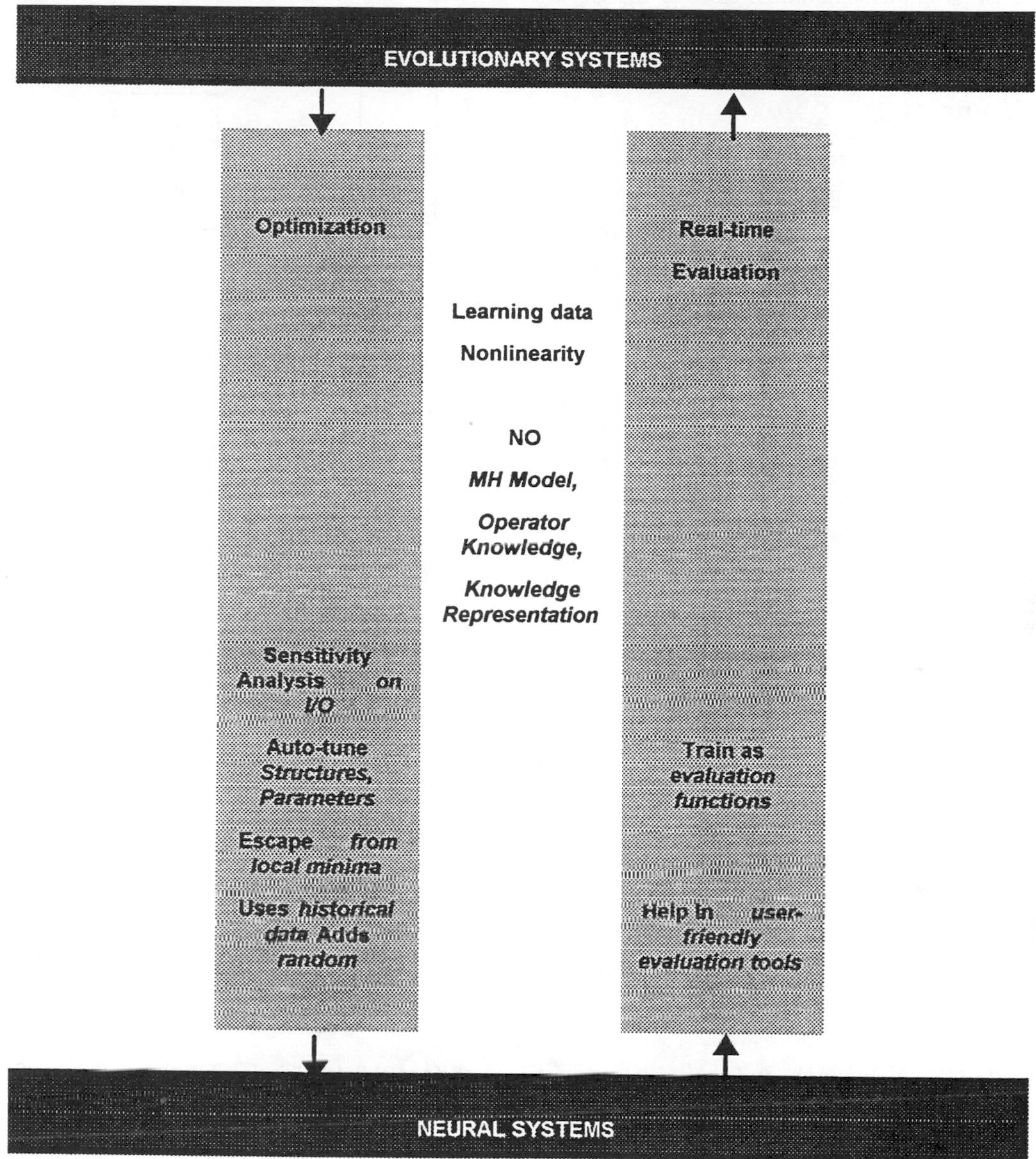

to expand to implement the other. FS, on the other hand, usually involves linguistics, knowledge bases, and other techniques different from pure number processing without knowledge representation.

The cost/benefit ratio for adding a new capability to an evolutionary system should be considered carefully before proceeding. Papers by Cooper and Vemuri (Chapter 101) and Padgett and Padgett (1995) discuss a very simple fuzzy number subtraction cost analysis procedure. Potential *benefit* from expanded capabilities is represented as one fuzzy number, and potential *cost* as another. Fuzzy subtraction gives a visual picture of the *risk* involved.

Based on current technology, the rate and direction of advances, the flexibility and more rigorous scientific modeling methodology inherent in multiple domain problem solutions, combining neural and evolutionary systems appears to be a cost effective way to proceed. The addition of fuzzy systems is supported by the same logic. Of course, the intelligence employed in any implementation will have an enormous impact on its practicality and longevity. A complex intelligent system should be flexible enough to bypass time and resource demanding elements when their contribution is unnecessary.

A system with a high machine IQ will appropriately allocate some tasks to off-line model development, and engineer other tasks to encourage evolution toward high-speed hardware solutions. An automatic intelligent focusing system, for example, should enable manual over-rides when more control is needed. On the other hand, if automatic intelligent processing consumes scarce resources, it should turn itself off or down when system states are steady or other indications show that the expensive option is not needed at the moment. These concepts are not unique to intelligent systems development. They are mentioned to remind the working engineer that "intelligence" in a system is only as effective as its implementation is close to the global needs of its users, and of the managers paying for its development.

Application Examples

Evolutionary Neural Systems

Application examples assembled by Takagi (1993, 1995) focus on work in progress at Berkeley, where soft computing has emerged as a technology which is rapidly producing more and more successful engineering results.

Neural Learning Using GAs. Comparisons of GA and BP optimization of synaptic weights in Takagi (1993, 1995) indicate that GAs may rapidly approach a near-optimal solution (low error) without reaching it in a feasible length of time. BP may stay at a high error rate for a long time, but reach the desired low error state faster than a GA. This suggests the use of GAs to initialize a training procedure to be finished by another method. Experience also indicates that adjustment of window size and algorithm parameters may be needed when approach to a solution plateaus, or appears trapped in a local minimum. It is possible to include window size and other parameters which respond to dynamic adaptation in a GA optimization plan. This type of system adjustment or auto-tuning offers automation and logic to replace "random", undocumented and unrepeatable perturbation of the system by the developer. The randomness in a GA is systematic as opposed to intuitive. Several neural strategies incorporate an element of randomness. GAs can lend this asset to other neural paradigms.

EXAMPLE 104.1: Climate Control (Takagi 1993, 1995)

Input:

Sensor inputs to NN (RCE model) and GA
(Room, Outside, Reference Temperature, Time)
Controller inputs to GA:
(Warm/Cool)

Interaction between NN and GA:

NN transmits current weights to GA
GA transmits new weights to NN

Output from NN:

Control Ref. Temp

As observed by Whitley (1993, 1994) use of GAs to train neural networks has similarities to standard backpropagation, but is slower than cascade correlation. In an adaptive critic NN, such as one used to control a cart-pole system, the evaluation net may be replaced by a GA and a NN builder algorithm.

EXAMPLE 104.2: Fully Recurrent Nets For Neurocontrol (Weiland, 1990 in Whitley, 1994)

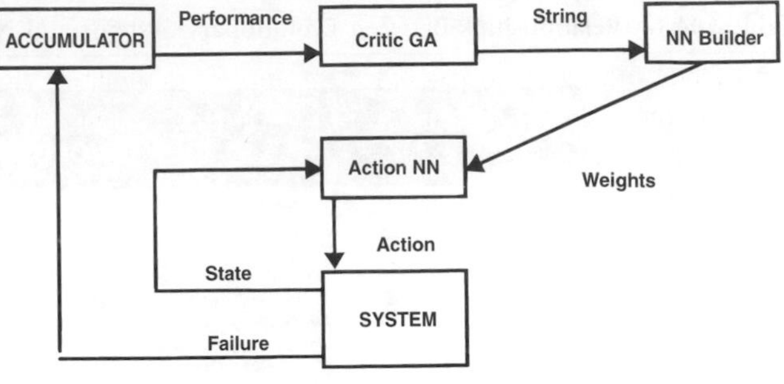

Figure 104.2 Evolutionary Critic (Whitley 1994).

Pole Balancing:

1. Fully recurrent NN.
2. Input: Pole and cart position only.
3. Output: Velocity.

EXAMPLE 104.3: Water Control for a Hydroponics System (Takagi, 1993, 1995)

Input to NN:

Water drainage and supply to plant
Carbon dioxide produced by plant

Output from NN:

Photosynthesis rate = Fitness Value

GA then tunes the water drainage and supply

EXAMPLE 104.4: Alcohol Fermentation Control (Takagi, 1993, 1995)

EXAMPLE 104.5: Job-Shop-Scheduling (Takagi 1993, 1995)

NN Pattern Recognition Using GAs

EXAMPLE 104.6: Genetic Sparse Distributed Memory (Das and Whitley, 1992 in Whitley, 1994)

NN: Counters are trained using perceptron-like learning
GA:

Individuals are strings of location addresses
Features are local minima (multiple)

NN Configuration by GAs. A direct approach to NN configuration by GAs is possible, but using GAs to encode a grammar appears to be a more flexible approach and faster. A neural network with highly structural properties may be easier to optimize with GAs than a fully connected, large multi-layer perceptron. (See Takagi, 1993, 1995; Weiland, 1990; in Whitley, 1994; and Chapter 108.)

Sensitivity Analysis

Determining the order of importance of variables in the potential input set of a NN can guide the reduction of complexity of the system. As implemented by Guo, and described by Uhrig and Tsoukalas, an input string consists of a set of possible input variables. Each variable has a value of 0 or 1 depending on whether it is selected as input. After training, the NN is evaluated by the GA fitness function, and a new generation simulated. Both reduction in number of inputs and decrease in training error produce higher values of the GA fitness function.

The resulting evolutionary neural system selects short strings which reduce training set error. This coding scheme eliminates variables which have a low impact on error reduction. Cooper and Vemuri point out the ease with which a GA can be encoded to produce a multitude of single variable strings, each of which is incapable of solving the problem. One way to monitor the validity of the implementation with respect to system goals is to place a test set in the loop. As Wiggins (1992) observes, visualization and operator interaction with the GA system during development allow rapid identification of problem configurations and exploration of new ideas.

Neural Evolutionary Systems

NN as a Fitness Function of a GA

EXAMPLE 104.7 Color Recipe Prediction (Takagi 1993, 1995)

Takagi's implementation of a neural fitness function combines the input of three subfunctions:

Input to all Subfunctions:
- GA population of genes for concentration of color pigment of different spectra.
- A string is a sequence of pigment concentrations for mixing a color. (This is the same as evolving a string of weights for an MLP.)

Function 1:
- Additional input:
  - Spectra of target color → NN → target pigment concentrations.
- Computation of output:
  - Distance of GA pigment string from NN pigment string.

Function 2:
- Additional input:
  - Rule-base for same color and complementary color.
- Computation of output:
  - Penalty calculation for mistakes.

Function 3:
- Additional input:
  - Characteristics of target color (numerical description).
- Computation:
  - NN estimate of numerical color characteristics based on GA string.
  - Distance of target characteristics from NN characteristics.

Three subfunctions are combined into the final fitness value, completing the Evaluation Step for the GA.

Challenges to researchers and engineers seeking to apply and improve NN/GA technology are given by Whitley (1993) as follows:

GA search slow—need parallelism.
Many EC varieties to choose from

- Need comparative performance on harder problems.
- Larger and more difficult test problems.
- More shared results.

104.3 Summary

The relationships between neural systems and evolutionary systems vary, but most can be categorized by the diagram in Figure 104.1 (Takagi, 1993, 1995). On a micro level, GAs can help in the design of NN structure and can assist in optimization during training. For more involved neural systems, such as those useful in control systems described by Werbos, GAs can serve as a component of a complex structure. A GA-based sensitivity analysis of input and/or output of a neural system can guide the elimination of unnecessary complexity and allow implementation of smaller, modular systems, which can be trained, validated, and verified more easily (Uhrig and Tsoukalas, 1996). Of course, as Fogel (1995) points out, an excessive amount of modularization can create problems with properly modeling interactions. The evolutionary (ES and EP) methods he advocates offer global solutions, less likely than neural systems, or even GAs, to be trapped by local minima, or to converge to an inappropriate solution. Neural evolutionary systems require value judgments by human operators developing these algorithms, and their internal evaluation functions can be difficult to configure. Using neural algorithms at the top level for validation, verification, and guidance during the cyclic evaluation of an EC process can be very helpful. Likewise, an internal neural network can serve to approximate a complex and hard-to-define EC evaluation function.

All of the above-mentioned suggestions for engineering more effective applications will be more successful if good visualization and user-interface tools are available (Wiggins, 1992). Emphasis on parallelization and techniques with reduced computational complexity should prove beneficial on a long-term basis (Fogel, 1995). The application of neural and evolutionary systems is limited only by the creativity and experience base of the engineer tuning these basic suggestions to fit specific industrial electronics applications. Useful resource for applications include the following: Padgett and Padgett, 1995; Baeck, 1994; Fogel, 1994; Fukuda et al., 1993, 1994; Pedrycz, 1994; Michalewicz, 1992; Wang and Kwok, 1992; Davis, 1991; Goldberg, 1989.

Acknowledgments

The work and advice of Toshio Fukuda guided the construction of this article. H. Takagi provided many of the application illustrations, and Paul Werbos directed attention to tight incorporation of neural and fuzzy algorithms within the most advanced intelligent control systems.

References

Baeck, T. 1994. Evolution strategies: A Thorough Introduction, Tutorial #1, *IEEE Conf. EC, WCCI,* Orlando, FL.

Davis, L., ed., 1991. *Handbook of Genetic Algorithms,* Van Nostrand Reinhold, New York, NY.

Fogel, D. B. 1994. An Introduction to Evolutionary Computation, Tutorial #3, *IEEE Conf. EC, WCCI,* Orlando, FL.

Fogel, David B. 1995. *Evolutionary Computation.* IEEE Press, Piscataway, NJ.

Fukuda, T., Shimojima, K., and Shibata, T. 1994. Fuzzy, neural network and genetic algorithm based control system, *Proc. IECON'94, Int. Conf. on Ind. Electronics, Control and Instrumentation,* 2:1220–1225.

Fukuda, T., Ishigami, H. et al. 1993. Structure Optimization of Fuzzy Neural Networks using Genetic Algorithm, *Proc. of IFSA.*

Goldberg, D. E. 1989. *Genetic Algorithms in Search, Optimization, and Machine Learning,* Addison-Wesley, Menlo Park, CA.

Michalewicz, Z. 1992. *Genetic Algorithms + Data Structures = Evolution Programs,* Springer-Verlag, New York, NY.

Padgett, M. L. and Padgett, W. D. 1995. Simulation and computational intelligence in real-world applications, *Simulation,* 65(1):5–9.

Pedrycz, W. 1994. *Fuzzy logic in neurocomputations,* Tutorial #1, *IEEE ICFS, WCCI,* Orlando, FL.

Takagi, H. 1993. Fusion techniques of fuzzy systems and neural networks, and fuzzy systems and genetic algorithms, *SPIE Proc. Tech. Conf. Applications of Fuzzy Logic Technology,* vol. 2061, Boston, MA.

Takagi, H. 1995. What Have We Learned from Experiences of Real World Applications in NN/FS/GA?, Tutorial Course at *WCNN'95.*

Wang, P. and Kwok, D. P. 1992. Optimal design of PID process controllers based on genetic algorithms, *Proc. IFAC 11th World Congress,* 5:261–265.

Whitley, D. 1993. A Genetic Algorithm Tutorial, Tutorial #10, *1993 IEEE-ICNN,* San Francisco, CA.

Whitley, D. 1994. Genetic Algorithms: Theoretical Foundations and Experimental Evaluation, Tutorial #6, *IEEE-ICEC, WCCI,* Orlando, FL.

Wiggins, R. 1992. Docking a truck: a Genetic fuzzy approach, *AI Expert,* May, pp. 26–35.

105

Computational Intelligence Applications in Industrial Electronics

Mary Lou Padgett
Auburn University

Robert Shelton
NASA/JSC

105.1 Introduction 1343
105.2 Aerospace Applications of Computational Intelligence 1343
105.3 From Basic Implementations to New Research 1344
Basic CI System Modeling Concepts • Computational Intelligence System Implementations • Future Trends of Computational Intelligence Applications in Industrial Electronics

105.1 Introduction

A form of machine intelligence that promises to play an increasing role in areas ranging from "smart weapons" to consumer electronics consists of a wide range of techniques all having the common feature that they appear to "learn" the desired action rather than requiring explicit programming. These methods are sometimes collectively termed "computational intelligence" (CI).

Motivation for the contents of this chapter was provided by applications examples such as the aerospace experiences discussed below.

105.2 Aerospace Applications of Computational Intelligence

As NASA is required to perform its mission with continually shrinking resources, the solution is seen to require massive automation of routine tasks, especially those connected with flight control. To this end, there has been significant progress in the application of rule-based systems for monitoring and control of systems amenable to symbolic reasoning. Beginning in early 1993, a new effort was undertaken to identify and exploit opportunities for computational intelligence in NASA's Mission Control Center at JSC. Several areas were identified and prototype applications were constructed to assess the effectiveness of various CI methods for the target applications. One early lesson learned from the initial prototypes was to build a reusable, extensible infrastructure for application development. This infrastructure—the Pattern Interpretation and Recognition Application Toolkit and Environment (PIRATE) was fabricated, and there are now three PIRATE applications running in the Control Center.

The most mature of these is now certified for official use by flight controllers monitoring the Electrical Generation and Integrated Loading (EGIL) system on the Space Shuttle orbiter. This application identifies start-up transients of electrical devices on the Space Shuttle AC electrical power buses. Formerly, EGIL flight controllers monitored this system by visual inspection of paper stripchart recorders. The AC signature recognition application uses a Bayesian statistical classifier developed at the Ames Research Center to identify features of the demand curves and typically achieves accuracy in the 99% range. Further, due to the Bayesian technique, a "confidence factor" is displayed so that the flight controllers can readily identify novel situations in which system performance is at variance with the nominal state characterized by the training examples.

The two more recent applications monitor accelerometers in the Shuttle Inertial Measuring Units (IMUs) and the combustion chamber pressure in the Auxiliary Power units (APUs). The IMU fault detection application uses wavelet transforms to magnify signals characterizing a certain kind of accelerometer failure. The APU application extracts features from the rapidly varying chamber pressure signal in order to facilitate analysis by flight controllers.

These and other computational intelligence applications are being developed at NASA/JSC and appropriate software is made available to the public through COSMIC.

The articles in this chapter will describe the basic techniques for design and application of CI systems which pertain to industrial electronics. Combination of multiple techniques such as neural, fuzzy and evolutionary systems is covered. Updates and discussions will be featured in the CI Standards News (contact: m.padgett@ieee.org). The news will feature debates about CI systems with contributions by individuals suggesting definitions, design procedures, lessons learned, and documentation tips suitable for future standardization.

As in the preceding chapters in this section, topics are arranged

0-8493-8343-9/97/$0.00+$.50

to first feature implementations suitable for commercial use, then move into areas of research development.

105.3 From Basic Implementations to New Research

Basic CI System Modeling Concepts

Hybrid systems composed of combinations of expert systems, neural networks, fuzzy logic systems, and/or genetic algorithms are discussed and illustrated with examples in Hybrid Artificial Intelligence Systems (Chapter 106). Design principles and lessons learned are covered clearly.

Fuzzy expert systems can solve problems with outstanding performance, and a large reduction in the number of rules for a standard expert system. Expert neural network systems can function more effectively with regard to interfaces to systems users and to quantified knowledge about the system than can neural networks working in isolation. Preprocessing and postprocessing of data for neural systems is critically important to their valid performance. This is a natural setting for embedding a neural system in an effective environment.

Neural networks (NN) and fuzzy logic (FL) both perform "fuzzy" mappings of nonlinear systems, but have useful differences. Neural systems can be embedded into fuzzy systems for inferencing, or fuzzy systems can help preprocess and postprocess NN data. A fuzzy logic controller example illustrates this. Fuzzy logic controllers tuned by neural networks are widely reported in the literature. In industrial applications, use of fuzzy and neural systems in sequence or in parallel is quite effective. Fuzzy control of parameter adjustments in backpropagation learning is discussed, and neural mapping of input variables to fuzzy set membership functions is explained. Neural fuzzy approaches to anticipatory control and intelligent combinations of fuzzy logic with neural networks are also cited.

Genetic algorithms and neural network combinations are illustrated with examples offering practical solutions to reducing the complexity of a neural network structure by genetically optimizing the number of inputs and the error level simultaneously. An example is given for detection of multiple faults using spectral analysis of nuclear power plant transients. Neural networks can conversely be used as GA fitness functions, when those functions are complex and hard to characterize.

A vital question for application of fuzzy systems is how to choose the best fuzzy aggregation method. A sensor fusion example is cited (see also Chapter 103.) The combination of GAs with gradient based learning and their use to partition problem space into feasibility regions is also mentioned. Hybrid system solutions are suggested as examples for providing ideas to help engineers solve "the problem at hand in a timely, reliable and cost-effective manner."

Chapter 107, Application Techniques: Combining Fuzzy Logic, Artificial Neural Networks, and Probabilistic Reasoning—Soft Computing, by O. Kaynak, discusses hard computing and the assets of soft computing. Various combinations of neural and fuzzy systems are discussed, and the relationships among neural networks, fuzzy logic, artificial intelligence, and control systems are charted with regard to mathematical model, learning data, operator knowledge, real time, knowledge representation, nonlinearity, and optimization strengths and weaknesses.

Computational Intelligence System Implementations

Combinations of CI tools in a complex, hierarchical system are illustrated in T. Shibata, T. Fukuda and K. Tanie in Synthesis of Fuzzy, Artificial Intelligence, Neural Networks and Genetic Algorithms for Hierarchical Control: Top-Down and Bottom-Up Hybrid Method (Chapter 108). Application strategies which have proven effective in many arenas are diagrammed, and a three layered scheme for autonomous robot control is discussed. The adaptation, skill and learning levels are implemented and linked using combinations of neural, fuzzy, GA and expert systems as needed.

Building blocks potentially useful for systematic compilation of effective control systems are being developed in a massive undertaking. H. Szu and B. Telfer discuss the concepts involved in Advanced Tools for Adaptive Nonlinear Modeling and Control of Power in Large Systems (Chapter 109). Using modeling strategies with some similarity, G. Yen presents a controls application of RBF networks, Application of Model Reference Adaptive Control and Adaptive Time-Delay RBF Networks (Chapter 110).

Future Trends of Computational Intelligence Applications in Industrial Electronics

Many areas of industrial electronics are heavily impacted by the addition of computational intelligence to the set of possible tools for design, development, and implementation of practical applications. Topics such as integrated design, machine vision, computer vision, education and training, machine control, piezomechanics, auronautical mechatronics, hardware/software codesign, design in textile engineering, balanced automation systems, control and algorithms, control of electrical drives, microelectromechanical systems, mobile robotics, sensors and measurement technology, nonlinear control, production automation, robotics and motion control, sensors and actuators in mechatronics, and vibration control are feasible settings for use of modern computational intelligence strategies. Isermann (1995) illustrates simultaneous engineering of design solutions for mechatronic systems. These techniques are discussed in terms of incorporation of neural networks, fuzzy systems and soft computing, evolutionary systems and/or expert systems in the intelligent electronics section of this Handbook. Mathematical models and their biological basis are introduced, and design strategies are outlined. The check-lists and modules cited in the design and modeling papers in this section are intended to be reviewed in the project planning stage and incorporated into validation strategies within each step of the developmental process. This look-ahead policy emphasizing planning, communications, checking of assumptions, and careful documentation is discussed in the section on factory automation in this Handbook and in Padgett (1982). The design modules prototyped at the beginning of the project are refined as the development of a practical product iterates through cycles of exploration and fine-tuning. The capabilities of neural systems for real-time learning of nonlinearities and complex functions from data without a mathematical model or operator knowledge can be applied to all of the above-mentioned areas. Pattern recognition, controls, fault detection, identification, and recovery can all be performed or enhanced by neural systems working tightly with conventional systems and with other CI augmented systems. Fuzzy systems provide real-time solutions to nonlinear problems and can incorporate operator knowledge and a database for knowledge representation. Control systems can profit from fuzzy supervisors, and fuzzy pattern recognition can also be very effective. Evolutionary systems learn from historical data and add an element of randomization. They are effective with nonlinear systems and optimization. Sensitivity analyses, fine-tuning of system parameters, escape from local minima, and the ability to turn corners distinguish these systems in the practical arena. Expert systems provide shells for interfacing these tools with the rest of the system.

Future directions for the incorporation of these CI elements into industrial electronics feature the integration of neural, fuzzy and evolutionary systems into top-down and bottom-up structural development of problem solutions. Increased speed for processing and reduction in computational complexity will help the merger of these techniques. Concurrent engineering of applications using computational intelligence can provide the basis for extensive use of these technologies. Carefully structured and recorded nontrivial experiments with designs and performance evaluation strategies can provide detailed guidance for implementation. As the lessons learned, the examples, and the successes are shared in detail, the field will grow and flourish. Development of standards for terminology and documentation can help encourage the recording of information for public use. The discussions and international forums accompanying this process may be even more beneficial than the final product. To participate in such discussions, contact m.padgett@ieee.org and inquire about the CI Standards News.

References

Isermann, R. 1995. Information Processing for Mechatronic Systems, *Int. Conf. Recent Advances in Mechatronics,* pp. 3–17, August 14–16, Istanbul, Turkey.

Padgett, M. L. 1982. *A Statistical Methodology for the Development of Computer Simulation Models and Forest Harvesting Applications,* M. S. Thesis, Auburn University.

Padgett, M. L., ed., 1995. *Simulation: Special Issue on Computational Intelligence,* SCS Press, San Diego.

106

Hybrid Artificial Intelligence Systems

Lefteri H. Tsoukalas
Purdue University

Robert E. Uhrig
University of Tennessee and Oak Ridge National Laboratory

106.1 Introduction 1346
106.2 Expert Systems and Fuzzy Logic Systems 1347
Fuzzy Rules in the Knowledge Base
106.3 Neural Networks and Expert Systems 1347
Neural Networks in the Knowledge Base of an Expert System
106.4 Neural Networks and Fuzzy Logic Systems 1347
Neural Network Mapping • Fuzzy Logic Mapping • Fuzzy Inputs to a Neural Network • Fuzzy Logic Controllers • Using Fuzzy Controllers with Neural Networks • Fuzzy Data Processing in Neural Networks • Fuzzy Representations of Variables that are Inputs and Outputs of Neural Networks • Fuzzy "One-of-n" Coding of Neural Network Inputs • Fuzzy Postprocessing of Neural Network Outputs • Fuzzy Control of Backpropagation Learning • Neural Network Based Fuzzy Logic Design System • Neural Fuzzy Approaches to Anticipatory Control • Intelligent Combinations of Fuzzy logic with Neural Networks
106.5 Genetic Algorithms and Neural Networks 1356
Selecting Most Important Inputs to a Neural Network
106.6 Genetic Algorithms and Fuzzy Systems 1357
Fuzzy Decisions of Sensor Fusion Parameters using Genetic Algorithms
106.7 Discussion and Conclusions 1357

106.1 Introduction

The term "artificial intelligence" (AI), in its broadest sense, encompasses a number of technologies that includes, but is not limited to, expert systems, neural networks, genetic algorithms, fuzzy logic systems, cellular automata, chaotic systems, anticipatory systems, and hypermedia databases. Interestingly, most of these technologies have their origins in biological or behavioral phenomena related to humans or animals, and crude analogues of these technologies exist in many human and animal systems. Hybrid intelligent systems generally involve two, three, or more of these individual AI technologies used either in series or integrated in a way to produce advantageous results through synergistic interactions. The combinations and permutations of these eight technologies, coupled with an equal or larger number of advanced conventional technologies (e.g., Dempster-Shafer theory, autoregression-moving average methods, etc.) indicate that the number of potential hybrid AI systems is the order of 10^{12}. Obviously, the most we can hope to explore in this chapter is a few of the more common hybrid AI systems in use today and to demonstrate the advantages of the hybrid combination.

In data and/or information processing, the objective is generally to gain an understanding of the phenomena involved and to evaluate relevant parameters quantitatively. This is usually accomplished through "modeling" of the systems, experimentally and/or analytically (using mathematics and physical principles). Most hybrid systems are used to relate experimental data to system models. Once we have a model of a system, we can carry out various procedures (e.g., sensitivity analysis, statistical regression, etc.) to gain a better understanding of the system. Such experimentally derived models give insight into the nature of the system behavior that can be used to enhance mathematical and physical models.

In this chapter, we shall confine our hybrid systems to a few examples of combinations of expert systems, neural networks, fuzzy logic systems, and genetic algorithms. Each of these technologies has been discussed elsewhere in this handbook, and a basic knowledge of these technologies by the readers is presumed. Therefore, we shall confine our discussions to the hybrid aspects of combining these technologies. However, when certain characteristics are important in examining hybrid system they will be introduced and discussed briefly.

0-8493-8343-9/97/$0.00+$.50

106.2 Expert Systems and Fuzzy Logic Systems

Fuzzy Rules in the Knowledge Base

Expert systems usually consist of computer systems that attempt to embed human and other expertise in a form that can be readily utilized by a non-expert. Most expert systems consist of two principal components plus the appropriate interfaces with humans and external databases: a "knowledge base" for storing information needed for the expertise and an "inference engine" that processes the information and infers conclusions based on reasoning and logic operations. One of the most popular methods of storing information in the knowledge base is through the use of *if/then* rules. Both the antecedent (the "*if*" part) and the conclusion or action (the "*then*" part) of the rules may have multiple statements connected by conjunctions and disjunctions such as "and" and "or."

For simple systems, the rules can be relatively simple and straight forward. If the individual components of a system are independent and follow a "logic tree" structure, the rules proceed in a monotonic manner, i.e., the inferring process always proceeds forward. However, if the components are interconnected, the logic trees interact with the result that the rules become longer (more qualifying conditions connected by conjunctions) and more complex. It becomes harder and harder to prevent rules from conflicting with each other. Indeed, it has been the experience of many investigators that when the number of complex rules gets beyond about 200, it is virtually impossible to write a meaningful rule that does not conflict with previously written rules. This paralysis of the knowledge base for complex systems caused interest in expert systems to decline in the middle to late 1980s. With the advent of fuzzy rules, based on fuzzy set technology, expert systems are again being introduced in high technology systems. For instance, an autonomous navigation system using sensor signals to navigate between moving objects was almost abandoned when 450 rules did not provide a satisfactory system. However, the replacement of the navigation system's 450 rules with 15 fuzzy rules provided a system with outstanding performance. Comparable results in the reduction in size of expert system knowledge bases by the introduction of fuzzy rules have been reported by many investigators (see Terano et al., 1994).

106.3 Neural Networks and Expert Systems

Neural Network in the Knowledge Base of an Expert System

Neural networks, in spite of their extraordinary capabilities, have relatively limited applicability. They are trained using available data, tested, and put into use. All they can do is recall an output when presented with an input consistent with the training data. They cannot reason, seek data from available databases to assist their operation, or provide an explanation for their outputs. They need a structured environment in which to operate, which can be provided by conventional software programming. However, recent experience indicates that the usefulness of a neural network can be enhanced significantly if an expert system is used to provide this operating environment. Indeed, an expert system can retrain a neural network to adapt this hybrid system to new situations, or it can intermittently update the training of the neural network to adapt to changing situations. Some recent work indicates that expert systems can be used to provide explanations for why a neural network gives the output it does.

Perhaps the most direct combination of these two AI technologies is the use of a neural network in the knowledge base of an expert system. This gives the expert system the ability to learn from data presented to it. The training may be on-line or performed during an initialization period. Multiple and/or modular neural networks may be incorporated into the knowledge base, and neural network outputs may be combined within the knowledge base. Control of the neural network is carried out by the inference engine in the same way that it seeks additional information from a database or initiates a logic reasoning step.

106.4 Neural Networks and Fuzzy Logic Systems

Neural Network Mapping

Both neural networks and fuzzy logic systems can be viewed as "mapping" systems. Neural networks typically have a multi-component vector as an input and a different multi-component vector (that usually has a different number of components) as an output. The mapping is carried out by adjusting the weights in the neural network until the error between the desired mapping and actual mapping is minimized in a least squares sense or is reduced below some specified value (Werbos, 1994a). The relationship between the input and output is usually expressed in terms of a weight matrix for linear mapping and a functional relationship for non-linear mapping. The use of nonlinear activation functions (typically sigmoidal functions) allows non-linear mapping to take place. However, since the mapping is not perfect (i.e., the error between desired and actual outputs is not zero), there is a degree of "fuzziness" in the matrix or operator relating the input and output (Kosko, 1993).

Fuzzy Logic Mapping

On the other hand, fuzzy logic mapping can be viewed as using a reversal of the order in neural network mapping. The components of the input vector are "fuzzified," then these fuzzy variables are operated on by "logic" operators, and the results are "defuzzified" to give a precise output. The fuzzification and defuzzification processes make it possible to handle nonlinear systems in a straight-forward (but complex) manner. Fuzzy logic attempts to model the imprecise modes of reasoning that are an essential part of the human ability to make rational decisions in an environment of uncertainty and imprecision.

Although neural networks and fuzzy logic both perform mappings, they are fundamentally different. Neural networks map a precise known input into a precise known output, typically with a minimization of least square error. The fact that this error is not zero introduces some fuzziness, the degree of which is dependent on the choice and quality of the data used in training. On the other hand, fuzzy logic systems map a known input into a possibly unknown output using a known, but fuzzy, logic processing procedure. Both the input and the output values are usually precise. (However, the input may not necessarily be a good representation of the phenomenon being studied.) A major difference between the two technologies is that the imprecision of the data is recognized and dealt with in the fuzzy system, whereas the illusion of precision in the relationship of the input and output data is retained in neural networks. Perhaps the greatest attribute of fuzzy logic systems is the ability to describe in everyday language how to carry out decisions and control actions without having to specify the process behavior in complex detail using mathematical equations. Furthermore, the fuzzy relationships (usually fuzzy rules) are readily available for examination and can be used to produce explanations about the decisions or actions that are taken.

The neural network and fuzzy logic technologies are very different, and each has unique capabilities that are useful in information processing. Yet they often can be used to accomplish the same results in different ways. For instance, they can speed the unraveling and specifying the exact mathematical relationships among the numerous variables in a complex dynamic process. Both can be used to control non-linear systems to a degree not possible with conventional linear control systems. They perform mappings with some degree of imprecision. However, their unique capabilities can also be combined in a synergistic way. It is this combination of two technologies (as well as combinations of other AI technologies) with the goal of gaining the advantages of both that is the focus of this chapter.

Neural networks and fuzzy logic systems can be integrated in a variety of ways. Certain neural network configurations can be directly embedded into systems for fuzzy inferrencing. The outputs of some neural networks represent the extent of activation for particular categories of process, i,e., a radial-basis-function network outputs classification scores in which each output value (0 for no activation to 1.0 for complete activation) represents the extent of activation for a particular category.

EXAMPLE 106.1: Fuzzy and Neural Mappings.

To illustrate what is meant by fuzzy and neural "mappings," consider the function $y = f(x)$ shown in Figure 106.1. The function is a particular kind of relation known as a *many-to-one* mapping where many x's can map to the same y but not vice versa (for example a_2, a_2', a_2'' and a_n all mapping to b_2, but b_2 cannot be mapped to a unique value of x). The same relation between the x's and the y's may be described in several alternative ways. For example, listing a sufficiently large number of (x,y) pairs or *points*, i.e.,

$$\begin{array}{c} (a_1, b_1) \\ (a_2, b_2) \\ \cdots\cdots \\ (a_i, b_i) \\ \cdots\cdots \\ (a_n, b_n) \end{array} \tag{106.1}$$

A point (a_i, b_i) can also be thought of as a crisp *if/then* rule, "*if x is a_i then y is b_i.*"

Hence an alternative way would be to express the relation as

$$\begin{array}{llll} \textit{if} & x \textit{ is } a_1 & \textit{then} & y \textit{ is } b_1 \\ \textit{if} & x \textit{ is } a_2 & \textit{then} & y \textit{ is } b_2 \\ & \cdots\cdots & & \\ \textit{if} & x \textit{ is } a_i & \textit{then} & y \textit{ is } b_i \\ & \cdots\cdots & & \\ \textit{if} & x \textit{ is } a_n & \textit{then} & y \textit{ is } b_n \end{array} \tag{106.2}$$

Intuitively, we expect this crisp linguistic rendition of $y = f(x)$ to become more accurate with increasing number of rules. Having 1000 crisp rules for $f(x)$ is preferable to let's say 10 rules. However, the number of crisp *if/then* rules needed to describe a relation actually depends on the specific nature of the relation as well as our tolerance for approximation error. Take for instance a linear function, a straight line going through the origin. In this case, one crisp *if/then* rule may suffice since one additional point on the *x-y* plane outside the origin uniquely identifies this straight line. On the other hand, a very "noisy" function with many "spikes" and slope-changes (e.g., a vibration spectrum with many peaks) will require considerably more rules. In practical terms, however, an approximate description of $y = f(x)$ may be acceptable, sometimes even preferable.

We are often interested in approximate associations of x's and y's such as

$$\textit{if } x \textit{ is 'about } a_i\textit{' then } y \textit{ is 'about } b_i\textit{'} \tag{106.3}$$

that is to say, we are interested not in a crisp point of $f(x)$ but a neighborhood around a point. This is illustrated in Figure 106.2 where instead of crisp point (a_i, b_i) we consider the neighborhood around (a_i, b_i) (circled area) which may be thought of as an *area-cum-point*, i.e., an area obtained from a point. Such an *area-cum-point* can be described by a fuzzy *if/then* rule. Let '*about a_i*'

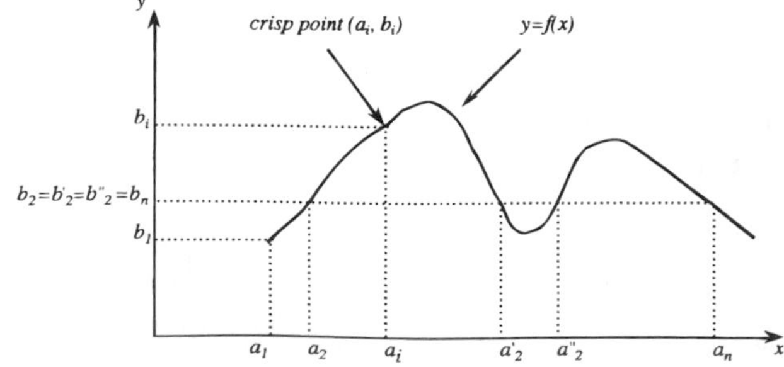

Figure 106.1 A function is a *many-to-one* mapping.

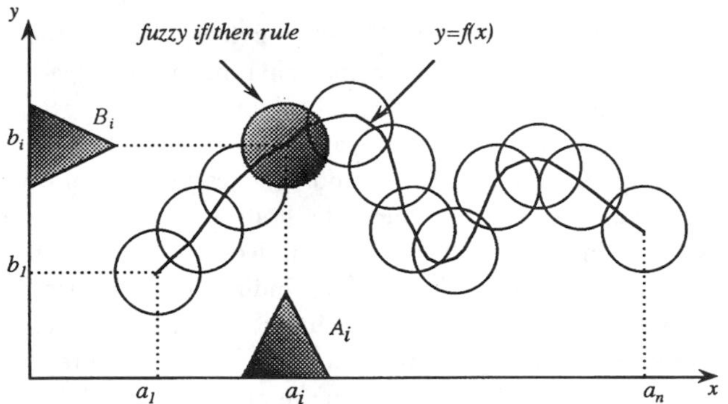

Figure 106.2 A more general relation called *many-to-many* mapping approximates a function and provides a basis for fuzzy and neural mappings.

be a fuzzy number A_i on the universe of discourse of the x's and *'about b_i'* a fuzzy number B_i on the universe of discourse of the y's. We define a fuzzy variable **X** as a variable whose arguments are fuzzy numbers on the x-axis, such as A_i, and a fuzzy variable **Y** taking as values fuzzy numbers on the y-axis, such as B_i. Hence the *area-cum-point* 'about (a_i, b_i)' (rule [Equation 106.3]) can be described by a fuzzy *if/then* rule

$$\textit{if}\quad \mathbf{X}\ \textit{is}\ A_i \quad \textit{then}\quad \mathbf{Y}\ \textit{is}\ B_i \tag{106.4}$$

Rule (Equation 106.4), like all fuzzy rules, has associated with it a fuzzy relation $R_i(x, y)$ called the *implication relation* of the rule. Given membership functions $\mu_{Ai}(x)$ and $\mu_{Bi}(y)$ for the fuzzy values in the left and right hand sides of the rule, the membership function for the fuzzy relation can be obtained as the product of the two $\mu(x, y) = \mu_{Ai}(y) \wedge \mu_{Bi}(y.)$ How this implication relation is obtained in general is a rather complicated issue which we will not examine here (see Lee, 1990, Mizumoto, 1988).

Thus, the relation originally described through a function $y = f(x)$ may be approximated as a collection of several fuzzy *if/then* rules, for example

$$\begin{array}{lllll} \textit{if} & \mathbf{X}\ \textit{is}\ A_1 & \textit{then} & \mathbf{Y}\ \textit{is}\ B_1 & \textit{ELSE} \\ \textit{if} & \mathbf{X}\ \textit{is}\ A_2 & \textit{then} & \mathbf{Y}\ \textit{is}\ B_2 & \textit{ELSE} \\ & \cdots\cdots & & & \\ \textit{if} & \mathbf{X}\ \textit{is}\ A_i & \textit{then} & \mathbf{Y}\ \textit{is}\ B_i & \textit{ELSE} \\ & \cdots\cdots & & & \\ \textit{if} & \mathbf{X}\ \textit{is}\ A_n & \textit{then} & \mathbf{Y}\ \textit{is}\ B_n & \end{array} \tag{106.5}$$

where $A_1, A_2, \ldots, A_i, \ldots, A_n$ are fuzzy numbers on the x-axis and, $B_1, B_2, \ldots, B_i, \ldots, B_n$ are fuzzy numbers on the y-axis. It is worth noting that the rules in Equation 106.4 are combined by the connective *ELSE*, which could be analytically modeled either as *union* or *intersection* depending on the *implication relation* of the individual rules. The collection of *if/then* rules in Equation 106.4 is called a *fuzzy algorithm,* and its analytical form is a relation $R_a(x,y)$ between the x's and the y's, called the *algorithmic relation.* As may be expected, the *algorithmic relation* depends on the *implication relation* of constituent rules.

Alternatively, the relation in Figure 106.1 can be encoded via a neural network as shown in Keller and Tahani (1992). A neural network is composed of fundamental processing elements called neurons such as the one shown in Figure 106.3. Each neuron performs a local computation that involves the incoming signals x_j (coming from the previous neuron) with their weighted sum minus a threshold $[net_j = \sum_j^n w_{ij}\, x_j - \vartheta_j\,]$ undergoing a nonlinear transformation $f(net_j)$ before producing output y_i propagated through its outgoing connections to other neurons in the network. During training a number of input-output pairs, called *examples,* are presented to a network of such neurons and the connection weights are adjusted until the network has "learned" the underlying relationship that the examples represent. This is called *supervised learning* and the process of weight adjustment is referred to as *training.* The most widely used algorithm for this kind of training is *backpropagation.* Figure 106.4 shows a typical neural network composed of three layers of neurons: *input, hidden (i.e., middle),* and *output.*

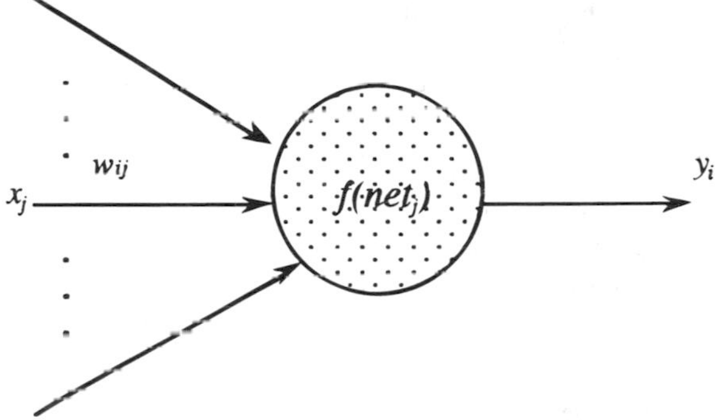

Figure 106.3 A typical processing unit of a neural network.

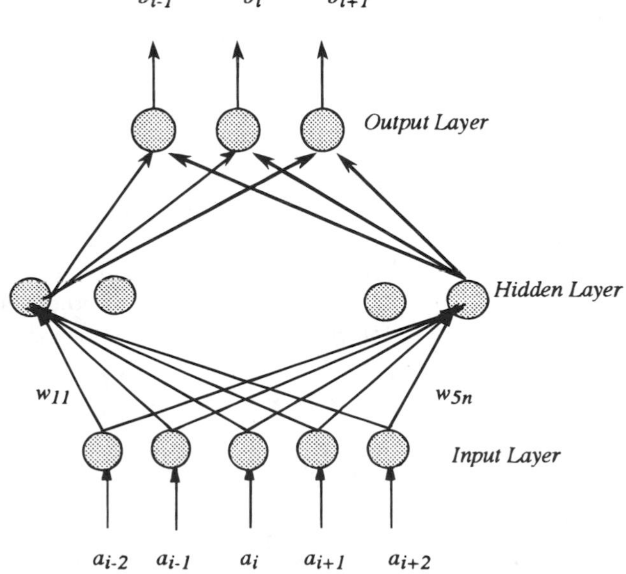

Figure 106.4 A neural network maps a number of inputs from the neighborhood of a_i onto a number of outputs in the neighborhood of b_i having learned the underlying relation.

The transition from conventional descriptions, such as $y = f(x)$, to fuzzy descriptions addresses the fact that functions are often overly restrictive mathematical idealizations. In most real-world problems, we do not have a $y = f(x)$ relation such as the one shown in Figure 106.1 but rather many-to-many mappings. For example, suppose that $y = f(x)$ is viewed as a control policy, i.e., a prescription recommending a control action y, for each state x. In many applications, the control system changes with time and conditions and it may manifest nonlinear and complex behaviors. Hence, the control policy may actually be a more many-to-many type mapping. Sometimes conventional descriptions, being overly idealized models of complex systems, may suffer from lack of robustness and exhibit undesirable side-effects.

Let us look again in Figure 106.2. We note that the transition from *points* to *area-cum-points* reduces the number of *if/then* rules needed to describe the relation. For example, we could approximate $f(x)$ with only 12 fuzzy *if/then* rules (circled areas) as shown in Figure 106.2. The rules are overlapping as are the various fuzzy numbers on the x and y-axes. Yet, we no longer have a function (a *many-to-one mapping*) but a more general relation $R_a(x,y)$ (a *many-to-many mapping*), and the obvious question is: how do we use such a relation? In conventional descriptions we evaluate functions by inputting a crisp value of x to $f(x)$ and obtain a unique crisp value of y as output. Something similar can be done with fuzzy systems as well. The process of evaluating a fuzzy linguistic description is called *fuzzy inference.* There are two important problems in fuzzy inference. First, given a fuzzy number A' as input to a fuzzy system we want to obtain a fuzzy number B' as its output, and second, given B' we want to obtain A' (the inverse problem). The first problem is addressed with an inferencing procedure called *generalized modus ponens (GMP)* and the second with another inferencing procedure called *generalized modus tollens (GMT).* Both *GMP* and *GMT* have their origin in the field of logic and approximate reasoning. Analytically, they involve the composition of fuzzy relations.

Hybrid systems in general and fuzzy in particular offer convenient tools for controlling the *granularity* of a description[1] in the sense that they facilitate the choice of appropriate precision levels, that is, levels that application-specific considerations call for. In terms of our example, when we use fuzzy numbers and fuzzy *if/then* rules to describe $y = f(x)$, we have at our disposal a mechanism for reducing the number of rules needed and, hence, control the *granularity* of this particular description and the overall cost of computation.

Fuzzy Inputs to a Neural Network

In another combination of these two technologies, the inputs to the neural networks are fuzzified. Travis and Tsoukalas (1992) have explored the concept of applying fuzzy logic to neural network inputs by utilizing membership functions to replace the input data to a neural network. The objective was to use operating data as the input to a backpropagation neural network to predict operating conditions. The methodology uses an artificial neural network architecture that couples both unsupervised and supervised learning neural networks in a unified structure. The major issue that emerged in this work was finding an appropriate way of representing the data from the actual system. The use of a fuzzy number representation scheme was found to be appropriate. After clustering the initial data using an unsupervised (Kohonen) neural network, the clustered data were transformed back into a set of fuzzy numbers (through the use of the centroid of the cluster) that are subsequently used as inputs to the backpropagation network.

Two different types of fuzzy variables were used in this project. The first consisted of seven fuzzy values evenly distributed over the universe of discourse from 0 to 1 that were applied to variables that were more or less uniformly distributed over the range 0 to 1. The second set of fuzzy variables consisted of five values evenly distributed over the universe of discourse that were applied to variables that tended to cluster around a few specific values. The centroids of these clusters were then used to train a backpropagation network to give a binary code representing the test conditions as outputs. Once trained, the input patterns used in clustering are then used to test the backpropagation neural network trained on the cluster centroids. This methodology showed improved results over other self-organizing neural networks investigated (i.e., competition, self-organizing map, and probabilistic neural networks).

Fuzzy Logic Controllers

A fuzzy logic controller consists of three parts: a fuzzifier, a rule base, and a defuzzifier. Each fuzzifier and defuzzifier contains a group of fuzzy sets that describe the universe of discourse, the fuzzy input and the fuzzy logic controller output. The fuzzifier takes a crisp input value and determines its degree of membership in a fuzzy set. We think of fuzzification as a mapping from the observed input to the fuzzy values involved in the *LHS* (left hand side) of the controller's rules. The rule base relates input sets to output sets and defines the characteristics of the fuzzy logic controller. It uses the degrees of membership from the fuzzifier in its calculations and produces an inferred value which is applied to one or more output sets leading to a resultant set out of which the defuzzifier produces a crisp output value (see Tsoukalas and Uhrig, 1997).

EXAMPLE 106.2: Fuzzy Logic Controller.

It is easy to see what is involved in fuzzy control by looking at Figure 106.5, where three rules are triggered at some time $t = t_k$ by two crisp inputs, *error* and *change-in-error.* The following three rules are the part of a fuzzy rule base for a first order system (used for controlling the level of a liquid in a tank) that

[1] By *granularity* we generally mean the coarseness of a description, the level of precision necessary to effectively represent a given system.

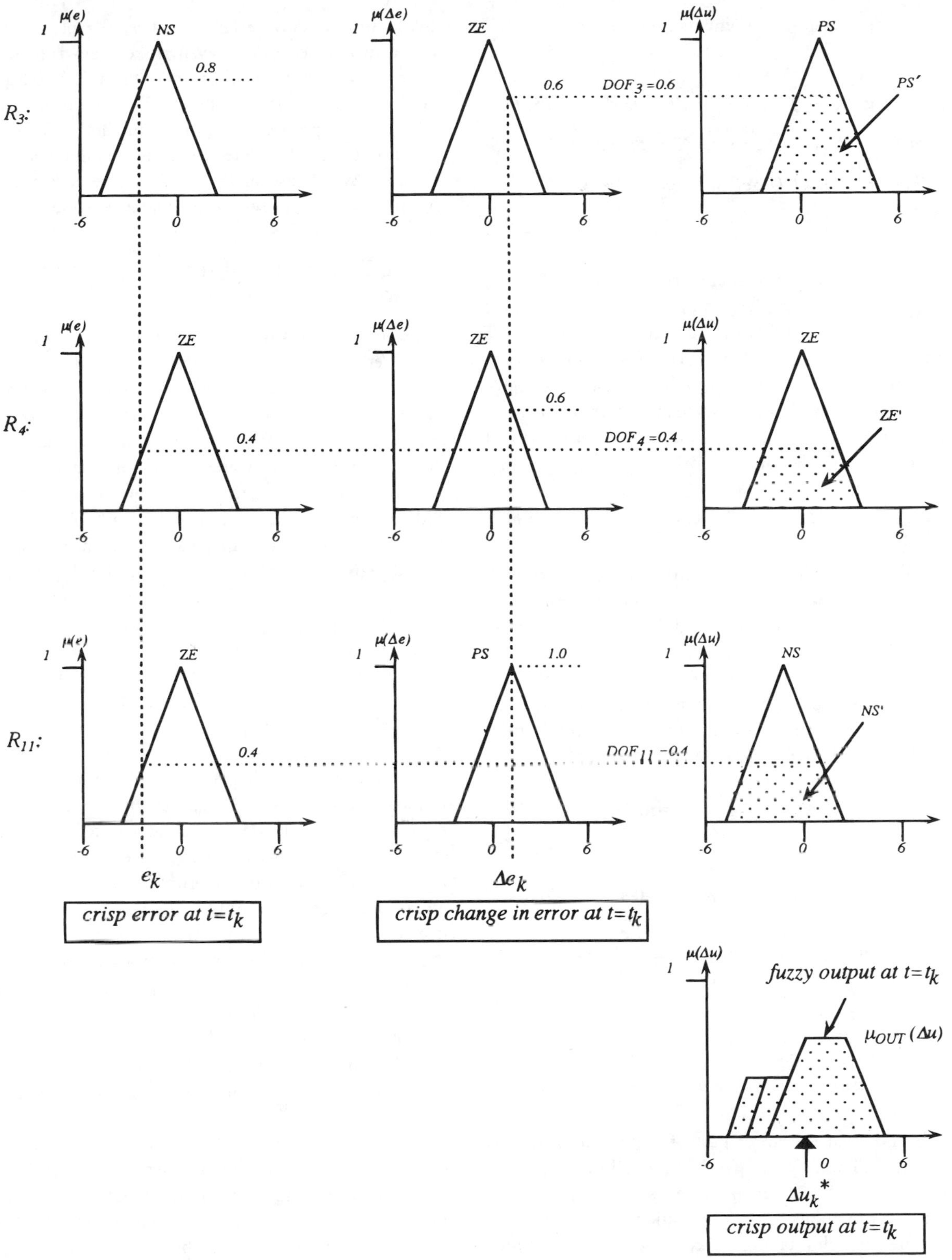

Figure 106.5 Rules from the rule base of a fuzzy control system controlling the level of a liquid in a tank, triggered at a certain time.

are triggered at some specific point in time (out of a total of 13 rules):

R_3: (w_3) *if* error is *NS AND* Δerror is *ZE* *then* Δu is *PS, ELSE*

R_4: (w_4) *if* error is *ZE AND* Δerror is *ZE* *then* Δu is *ZE, ELSE*

R_{11}: (w_{11}) *if* error is *ZE AND* Δerror is *PS* *then* Δu is *NS*

(106.6)

When at time $t = t_k$ crisp error e_k and crisp change in error Δe_k as shown in Figure 106.5 are given to these rules we say that the rules have "fired," provided that their degree of fulfillment *DOF* (the extent to which inputs match the LHS of a rule) is not zero. For example, in rule R_3 the crisp error e_k shown has a *0.8* degree of membership to *NS* while the crisp change in error Δe_k has a *0.6* degree of membership to *ZE*. Thus the degree of fulfillment of rule R_3 at this particular time is

$$DOF_3 = \mu_{NS}(e_k) \wedge \mu_{ZE}(\Delta e_k) = 0.8 \wedge 0.6 = 0.6 \qquad (106.7)$$

The *RHS* value *PS* will be transformed in accordance with DOF_3 in Equation 106.7. The nature of the transformation depends on the implication relation used. When the implication mentioned in Example 106.1 is used (called *Mamdani min*) the transformation amounts to clipping *PS* at the height of DOF_3 as shown in Figure 106.5. Thus R_3 contributes $\mu_{PS'}$ (Δu), the shaded part of the *RHS* value, towards a resultant fuzzy output. Similarly rules R_4 and R_{11} have degrees of fulfillment

$$DOF_4 = \mu_{ZE}(e_k) \wedge \mu_{ZE}(\Delta e_k) = 0.4 \wedge 0.6 = 0.4 \qquad (106.8)$$

$$DOF_{11} = \mu_{ZE}(e_k) \wedge \mu_{PS}(\Delta e_k) = 0.4 \wedge 1.0 = 0.4 \qquad (106.9)$$

and they contribute $\mu_{ZE'}$ (Δu) and $\mu_{NS'}$ (Δu) shown as shaded parts of the *RHS* values in Figure 106.5.

The total fuzzy output at $t = t_k$ is the *union* of the three outputs, i.e.,

$$\mu_{OUT}(\Delta u) = \mu_{PS'}(\Delta u) \vee \mu_{ZE'}(\Delta u) \vee \mu_{NS'}(\Delta u) \qquad (106.10)$$

$\mu_{OUT}(\Delta u)$ is shown at the lower part of Figure 106.5. At this point we need to defuzzify $\mu_{OUT}(\Delta u)$ and obtain a crisp value Δu_k^*, representative of $\mu_{OUT}(\Delta u)$, to be used as input to the process. Different methods for defuzzification may be used, the most common one being taking the centroid or center of area (COA) of the resultant fuzzy output. This is the value used to drive control actuators.

With each rule in a fuzzy rule base we can associate a weight as shown in rules (Equation 106.6). The weight is a value between 0 and 1 and indicates the strength or relevance of a given rule to a given situation. A neural network can be trained to adjust these weights for the purpose of tuning the rule base to any significant changes in the system under control (Blanco et al., 1995). An alternative arrangement is to replace the rule base with a neural network or even make them run in parallel in order to give a fuzzy logic controller the ability to learn through the learning capabilities of the neural network. The major problem with this approach is creating a method by which the target output values can be related to the desired inference values. This is solved by the use of the product inference method which multiplies an output set by the inference value.

Using Fuzzy Controllers with Neural Networks

Neurofuzzy systems are increasingly utilized in control technologies (Werbos, 1992). The hardest part in designing an ordinary fuzzy controller is selecting which fuzzy sets are best representing the controlled and controlling variables. Most fuzzy controllers are sensitive to the shapes of the membership functions, and as the number of rules increases, the use of "trial and error" tuning procedures become less and less feasible. A report in IEEE SPECTRUM magazine (Schwartz and Klir, 1992) describes work at Matsushita and Hitachi in Japan in which a backpropagation neural network learns the needed membership functions from a set of training examples (Hayashi et al., 1992). It is claimed that a tuning task that had previously taken six months was accomplished in one month.

There is a great number of reports in the literature where various types of fuzzy controllers are tuned and optimized via neural networks. A very promising hybrid system, however, from the viewpoint of industrial applications is when the two are used in series as shown in Figure 106.6. Here a neural network is trained to receive three measurements as inputs, for example, electrical and visual data from an automated test station testing electronic components for physical defects (O'INCA, 1994). This input is mapped to two numerical values that serve as input to a fuzzy rule base. The output of the neural module indicates the degree of the component's physical damage (through a number between 0 and 1) and the signal-to-noise ratio (through a number between 0 and 30). These two outputs are subsequently fed as inputs into a fuzzy rule base where they are involved in fuzzy variables in the left-hand-side of rules that map signal-to-noise ratio and physical damage information to the quality of the component. A decision can subsequently be made to accept the component or reject it.

There is a lot of potential, practical benefit to the use of hybrid combinations of neural and fuzzy systems such as the one shown in Figure 106.6. Numerical measurements may actually provide too much detail to be effectively used in an on-line manner (in addition to noise and other problems). Neural filtering, smoothing, and mapping of numerical measurements to a feature space (e.g., physical damage and signal-to-noise ratio) may facilitate quick action by a fuzzy controller.

Fuzzy Data Processing in Neural Networks

Inputs and outputs of neural networks are not always hard physical measurements; sometimes inputs are subjective responses or

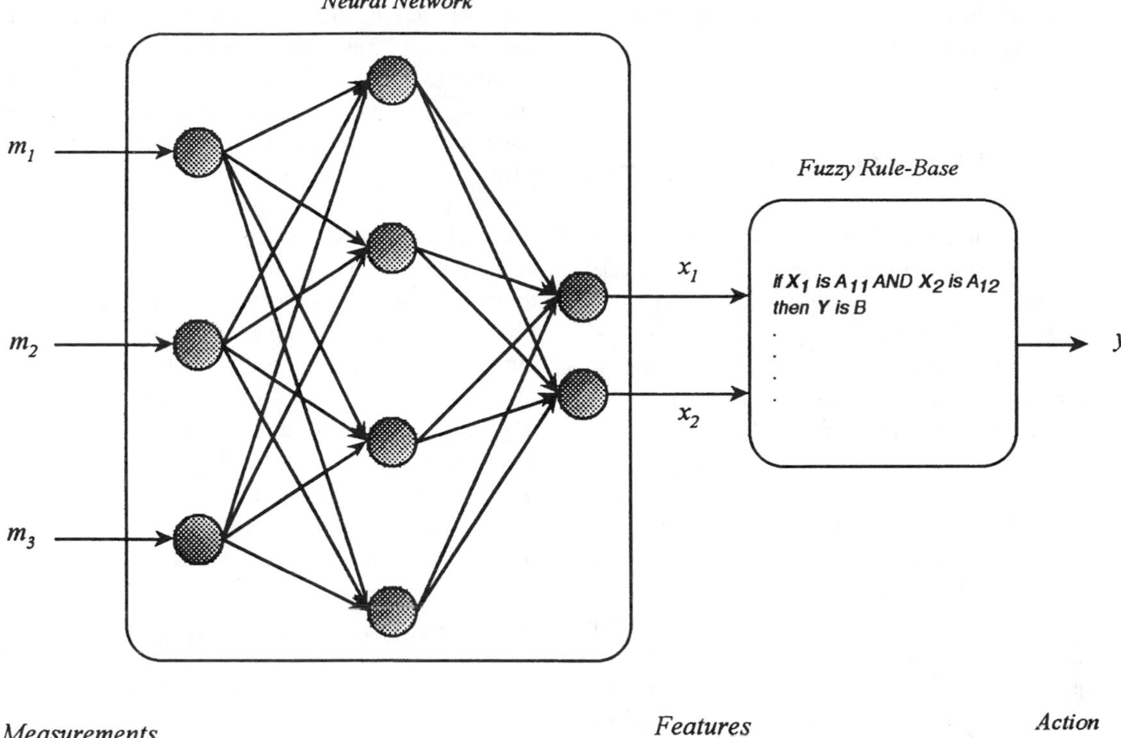

Figure 106.6 A hybrid system involving a neural network *in series* with a fuzzy rule base where measurements get mapped to features serving as inputs to the fuzzy system.

groupings into ill-defined categories, and outputs may not be a crisp number. If human judgments are involved, quantities are often described in linguistic terms, requiring special procedures to present such data to neural networks. The advantages of using fuzzy or linguistic notation is that fuzzy set operations are based on solid theoretical foundations, and although fuzzy sets deal with imprecise information, they do so in a precise and well defined way. This eliminates the need for user defined membership functions that are often arbitrary and can lead to erroneous results

Another advantage of fuzzy notation is that some contradictions in the data need not cause serious problems. Crisp systems have to deal with erratic data in a logical and precise way. Whereas rule-based systems that deal with probabilities must always sum to unity, fuzzy sets have membership functions that need not sum to unity. Hence, fuzzy systems can deal with a few outliers or some erratic data without significantly influencing the final result.

Fuzzy Representations of Variables that are Inputs and Outputs of Neural Networks

Sometimes dealing with all possible outputs of a neural network requires a large number of neurons, thereby increasing the complexity and training time. For instance, if we consider the temperatures between freezing and boiling of water, even on the Centigrade scale, there would be 100 integral values. The number can be reduced by grouping these 100 values into groups of 10 successive values and representing each group of 10 values with a single value (e.g., the 10 values in the range 21° and 30° degrees could be represented by 25°). Hence, the scale would become 5°, 15°, 25°, 35°, . . . 95°. Such groupings leads us to considering fuzzy or linguistic representation of the variable, where 0° to 10° might be "extremely cold," 10° to 20° might be "very cold," 20° to 30° might be "cold," 30° to 40° might be "slightly cold," etc. If one views these temperatures from the standpoint of human comfort, as opposed to the distance along a scale between the freezing and boiling points of water, a non uniform distribution with fewer values might be more appropriate, i.e., 0° to 15°, 16° to 20°, 21° to 23°, 24° to 30°, and 31° to 100°. In linguistic terms, these ranges might be designated too cold, cold, comfortable, hot, and too hot.

The sequence of events that are involved in utilizing fuzzy data in neural networks is as follows:

1. Crisp (or fuzzy) data are converted into membership functions or sets.
2. These memberships or sets are then subject to fuzzy logic operations.
3. The resultant sets are then defuzzified into crisp data that are presented to the neural network.
4. The network may also have other direct inputs that are crisp and do not need the fuzzy processing.
5. The output of the neural network is a crisp set that utilizes a membership function to convert it into a fuzzy variable.

6. This fuzzy output is then operated on by fuzzy logic.
7. The fuzzy logic output is then defuzzified to produce a crisp output.

Fuzzy "One-of-*n*" Coding of Neural Network Inputs

Typically, an input variable is represented by a single input node in the neural network. When an input variable has a special relationship with other variables over only a small portion of its range, the training process of the neural network is made especially difficult. Sometimes a nonlinear transformation is used to emphasize the particular region, but this is usually not a satisfactory process. This difficulty can be overcome by providing the neural network with neurons that focus on one region of the variables domain. The domain is divided into n regions (where n is typically 3, 5 or 7), and each is assigned a fuzzy set having a triangular membership function. (Of course, the lowest and highest sets have horizontal extensions starting at the minimum and maximum expected values respectively.) The membership value in each fuzzy set determines the activation level of its associated input neuron. This "one-of-*n*" coding expands the range of the variable into n network inputs, each covering a fraction of the domain. While the resulting specialization often facilitates learning, the increase in the number of neurons tends to slow down learning. This technique is advantageous only when the importance of the variable changes significantly across its domain.

There is a tendency to want more measurements of imprecise (or linguistic) data to compensate for lack of precision. Let us consider the case of two time signals that are to be sampled, digitized and fast-Fourier transformed so that one FFT is the input to a neural network and the other one is the desired output. If we have 100,000 simultaneously sampled data points for each variable and are dealing with spectra that have 128 points each (and another 128 points in the negative frequency range), dividing 100,000 points by 256 points per spectrum gives 390 complete spectra for each evaluation. The traditional approach with such FFTs is to average the 390 spectra to obtain an average spectrum with a high degree of confidence for each variable and to apply these two spectra to the neural network for training. A much better alternative would be to train the neural network using each of the 390 individual spectra, even though each of them is much less precise and would be considered "noisy" or perhaps "fuzzy." Subjecting the 128 components of the input and desired output vectors to "one-of-*n*" coding in the manner described above is another alternative that should be considered. Alternatively, grouping the 390 individual spectra into 39 spectra by averaging each successive group of 10 spectra would reduce the amount of computation involved.

Fuzzy Postprocessing of Neural Network Outputs

A neural network can be trained to produce the desired final product, but there are often advantages to training the network to present intermediate values with postprocessing to obtain the desired results. The advantages are that the neural network may be easier to train and the necessity for retraining if other outputs are desired can be avoided. An example of such a postprocessor might be control of the electrical output of a gas-fired plant when there is competition for the gas with residential users and industrial users, both of which have a higher priority for the gas. A neural network with such inputs as air temperature now and at several earlier times and at several locations, overall demand for industrial products now and several earlier times, the competitiveness of the products, plant efficiencies as a function of power output, etc. could be trained to predict the available gas. However, intermediate values such as future temperatures at several locations and future industrial output may be more appropriate since they can reasonably be obtained using an ordinary neural network. However, the relationship between the availability of gas and the intermediate network outputs are fuzzy and must be treated as such.

Fuzzy Control of Backpropagation Learning

Numerous methods of speeding up the learning in backpropagation neural networks have been attempted with varying degrees of success. One of the most common methods has been to adjust the learning rate during the training using an adaptive method that satisfies some index of performance. (The "delta-bar-delta" training procedure is such a method.) Wang and Mendel (1994) have shown that fuzzy systems may be viewed as a layered feedforward network and have developed a backpropagation algorithm for training this form of fuzzy system to match the input and desired output pairs of patterns or variables. Haykin has described a method in which the on-line fuzzy logic controller is used to adapt the learning parameters of a multilayer perceptron with backpropagation learning (Haykin, 1994). The system uses the classical four step fuzzy control process of (1) scaling and fuzzification of the crisp input, (2) development of a fuzzy rule base, (3) fuzzy inference using the fuzzy rule base, and (4) rescaling and defuzzification to give a crisp result or recommended action. The idea is to implement heuristics in the form of fuzzy *if/then* rules that are used for the purpose of achieving a faster rate of convergence. The heuristics (as is the case of almost all supervised training) are based on the behavior of the instantaneous sum of squared errors.

Neural Network Based Fuzzy Logic Decision System

Several investigators have proposed neural network based fuzzy systems (Kulkarni et al., 1994; Keller and Tahani, 1992). Kulkarni et al. have proposed a neural network model for fuzzy logic decisions that consists of six layers; the first three layers map the input variables to fuzzy set membership functions, and the last three layers implement the decision rules. Triangular shaped membership functions with five fuzzy values are used, and the model learns the decision rules using a supervised gradient descent procedure. The connection strengths between the last three layers encode the decision rules used in decision making. Layer 1 is the input layer that receives the input features. Layer 2 represents the linguistic term variables with five term variables (very low, low, medium, high, very high) for each input feature; hence, it has five times as many nodes as layer 1. Each node of layer 2 is connected with weights of plus and minus 1 to two nodes in layer 3 where the two nodes represent the left and right sides of the triangular membership functions. Each node in layer 4 combines the outputs of the corresponding two nodes in layer 3 so that it now represents the membership values, which is presented to layer 5. Layers 5 and 6 implement the inference engine process. Layers 4, 5, and 6 represent a simple three-layer feed-forward network with backpropagation learning. The number of nodes in the output layer is equal to the number of output decisions. During training, only the weights between layers 4, 5, and 6 are adjusted.

Kulkarni et al. used this system to recognize objects in multi-spectral satellite images, using data obtained from α thematic mapper sensor (a multispectral scanner that captures data in seven spectral bands). Five inputs to layer 1 were used, and layers 2, 3, and 4 contained 25, 50, and 25 nodes, respectively, since five linguistic term values were used in the universe of discourse in fuzzification. Layers 5 contained 35 nodes, and layer 6 contained 5 nodes representing output categories. Results obtained were virtually identical with results from a three layer conventional neural network classifier and a conventional maximum likelihood classifier. However, the conventional neural network took over 24 hours to train vs. about 25 minutes for the fuzzy neural system. Both the conventional and fuzzy neural network systems gave results very rapidly after training. In contrast, the conventional maximum likelihood classifier had to handle each pixel individually and sequentially; as a result conventional classifier took excessively long times for classification.

Neural Fuzzy Approaches to Anticipatory Control

In anticipatory control, a system uses predictions about the future to regulate its behavior at the present. Based on the assumption that future information is fuzzy in nature, that is, predicted values are imbued with fuzzy (not stochastic) uncertainty and that they can be modeled as fuzzy numbers, an approach has been suggested called *virtual measurement*, for measuring and predicting such values (Ikonomopoulos et al., 1993; Tsoukalas, et al., 1992). *Virtual measurement* offers considerable promise for the timely and reliable estimation of system parameters with functional or operational significance such as *performance, reliability,* or *availability.* It utilizes groups of artificially trained neural networks as shown in Figure 106.7. The networks $N_1, N_2, \ldots, N_n$ comprise the core of a program called *virtual instrument* being trained (in a process analogous to "calibration") with time series input vectors, and output vectors $\{o_1, o_2, o_3, o_4\}$ representing fuzzy numbers. Each network learns to map a constellation of input patterns to a particular fuzzy number. After the training of networks $N_1, N_2, \ldots, N_n$ is completed, they receive on-line

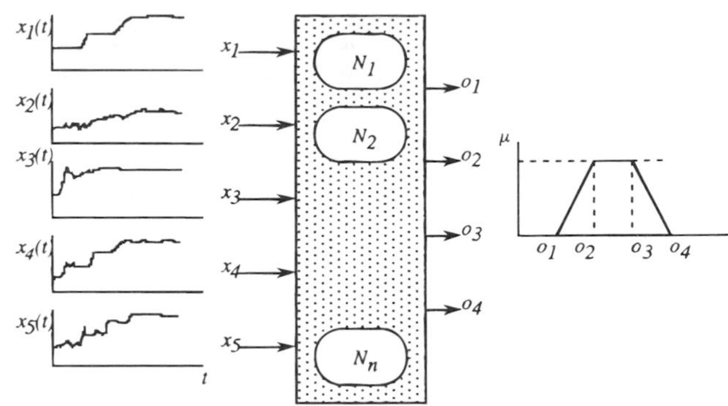

Figure 106.7 Each neural network in a virtual instrument maps a time-series input vector onto a vector $\{o_1, o_2, o_3, o_4\}$ representing a trapezoidal fuzzy number.

time signals as inputs and produce a set of membership functions as outputs.

A system that makes control decisions in the present on the basis of what may be happening in the future is considered different from conventional systems in three important respects: in the *language* used to formulate models of its behavior, in the manner used to *observe* its behavior, and in possessing a *built-in predictive capability* used to access future states. Research in these three areas has shown that hybrid systems are a promising approach for addressing the complexity of the problem.

Typically, the temporal behavior of a system is described by a set of difference (differential) equations of the form

$$x(t+1) = Ax(t) + Bu(t) + w(t); \qquad x(t_0) = x_0 \tag{106.11}$$

$$y(t) = Cx(t) + v(t)$$

where, $\{u(t)\}$ is an $r \times 1$ input sequence, $\{y(t)\}$ is an $m \times 1$ output sequence, $\{x(t)\}$ is an $n \times 1$ state sequence, *A, B,* and *C* are appropriate transition matrices, x_0 some initial state, and $w(t)$ and $v(t)$ are noise terms. It should be noted that it is rather difficult to include future information in the formalism of Equations 106.11 except by containing it within the noise terms as in the case of non-deterministic systems. Anticipatory control algorithms on the other hand are made through collections of fuzzy *if/then* rules pertaining to present and anticipated states. Thus, there are rules triggered by events at the present $(t - i)$ having the (simplified) form

$$\textit{if } \mathbf{X} \textit{ is } A \quad \textit{then} \quad \mathbf{Y} \textit{ is } B \tag{106.12}$$

as well as rules triggered by what is anticipated, that is predictions pertaining to $t = i + \Delta t$, and having the form

$$\textit{if } \mathbf{X} \textit{ will be } A \quad \textit{then} \quad \mathbf{Y} \textit{ is } B \tag{106.13}$$

Each rule is a *state/action* pair $s \rightarrow a$ where both *present* and *anticipated states are* considered in the *LHS,* while only *current action* is being prescribed through the *RHS.* The rules are clustered and structured to reflect temporal structures, that is, there are rules related to the state of the system at the present $t = i$, having the form $s(t = i) \rightarrow a(t = i)$, as well as rules that describe the possible state of the system sometime later, i.e., $s(t = i + \Delta t) \rightarrow a(t = i)$. Thus an anticipatory fuzzy algorithm can infer the current action $a(t = i)$ on the basis of the present state $s(t = i)$ as well as anticipated ones $s(t = i + \Delta t)$. Rules (Equations 106.12 and 106.13) describe relations of a more general type than that of functions, that is, *many-to-many* mappings, and can be designed on the basis of observing a system via neural network based virtual instruments.

Fuzzy *if/then* rules have been used in connection with a form of anticipatory control (rather similar to predictive control) where a predictive routine is used to anticipate the effect of the proposed decision on the system output (Yasunobu and Miyamoto, 1985). Additional rules may be called if the current decision will result in system behavior that is unacceptable; for example, the rule "*If the current decision* (u_c) *will cause the difference between the current and anticipated states to be big, then* $u = u_c\ (1 - \beta \cdot$ *bigt*)" where, β is a user-chosen parameter between 0 and 1 and *bigt* is the fulfillment function for the anticipated states. The parameter β may also be chosen by employing a predictive neural network (McCullough, 1993).

Intelligent Combinations of Fuzzy Logic with Neural Networks

National Semiconductors recently introduced a commercial product called NeuFuz that combines neural networks and fuzzy logic. It uses neural networks to learn system behavior from the sample system input/output data and automatically generates fuzzy logic rules and memberships. The rules and membership functions which are processed using fuzzy logic algorithms for defuzzification, rule evaluation, and antecedent processing are used for the implementation of the fuzzy system. These fuzzy logic algorithms replace conventional heuristic fuzzy logic algorithms and enable full mapping of neural networks to fuzzy logic systems. Full mapping provides an important key feature of generating fuzzy rules and membership functions to meet a prespecified accuracy level. It also claims significantly improved performance, improved reliability, reduced design time, and minimum system cost by optimizing the number of rules and membership functions. The system has many of the features of the Kulkarni et al. approach discussed above. However, *NeuFuz* does not use defuzzification since neural defuzzification is actually rule evaluations and the output of rules are nonfuzzy numbers. One disadvantage of this system is that it uses multiplication instead of the "minimum" operation in conventional fuzzy logic. Multiplication takes extra cycles and may be a problem unless dedicated hardware is used. However, software implementation of multiplication is probably acceptable in low cost systems, and the extra cost of dedicated hardware for multiplication is probably justified in more expensive systems or where speed is essential.

After learning is complete, *NeuFuz* actually provides two solutions for a problem: a neural network solution and a fuzzy logic solution. Hardware implementation of neural networks is usually expensive and software implementation may be too slow for many applications (e.g., control of rapidly changing systems). Since it is easier to implement fuzzy logic rules and membership functions in hardware at a lower cost while achieving a more rapid response, the fuzzy logic system (which is actually a hybrid system) is the solution of choice. Furthermore, a rule verifier and optimizer as well as the ability to generate assembly language code is important for practical use.

Other programming shells for developing hybrid neural-fuzzy systems are becoming increasingly popular in commercial applications, for example *O'INCA* by Intelligent Machines, *CubiCalc* by HyperLogic, and *TILShell* by Togai Infralogic. Most of them provide graphical user interfaces and a variety of testing, debugging, and simulation tools. An important issue in hybrid system development is assessing the reliability and completeness of a system before commercial operation. For this purpose some systems (for example O'INCA; see O'INCA, 1994) provide extensive simulation and validation utilities. Most systems, like O'INCA, also provide C code generation capabilities to facilitate the migration of a neural fuzzy system across different platforms.

106.5 Genetic Algorithms and Neural Networks

Selecting Most Important Inputs to a Neural Network

The most common use of genetic algorithms is to optimize a process or system. In recent work reported by Guo and Uhrig (1992a, b), genetic algorithms were used to select the most significant inputs to a neural network used to identify transients in nuclear power plants. Time records of some 25 variables from a high fidelity, full scope training simulator were sampled during a number of transients (e.g., loss of power, loss of coolant, offsite power failure, etc.) and presented to a recurrent type neural network. The network was then trained to identify the specific transient. The system was highly successful in that all the transients could be identified before the plant automatically shut down so that mitigating actions could be taken when appropriate. However, the recurrent neural network with 25 inputs was complex and hard to train. To overcome this problem, it was decided to use many "modular" neural networks (one for identifying each transient) in which only those inputs that were important to that transient were used as inputs to the modular network. Initially, sensitivity analysis was used on the recurrent neural network representing the nuclear plant dynamics in which the partial derivative of each output with respect to each input was evaluated, and only those inputs that were important to a particular transient (typically 4 to 6 inputs) were used. The result was a series of small, easy to train neural networks that typically had 5, 7, and 1 nodes in the input, hidden, and output layers, respectively. Training of the modular networks was very rapid, and the use of a series of small modular networks gave results that were equally as good as those obtained from the complex recurrent neural network. The problem was that the recurrent neural network had to be developed and trained in order to identify the important variables for each transient.

To overcome this problem, Guo (Guo and Uhrig, 1992b) used a genetic algorithm optimizing the type of procedure to identify the most important variables for each transient. The procedure involved the use of traditional binary bit strings as individuals to form a population for genetic algorithm search. Each position in a bit string is associated with an input variable, which may or may not be selected as an input for the neural network, depending upon the value of 1 or 0 in that position. Therefore, the number of bits in a string equals the number of variables that are candidate inputs. The *selection operator* used in this study was *stochastic remainder selection without replacement*, which was investigated by Goldberg (1989) who explained this process as

> *"Expected individual count values are calculated as the ratio of individual fitness and population's average fitness and integer parts are assigned. The fractional parts of the expected number values are treated as probabilities. One by one, weighted coin tosses (Bernoulli trials) are performed using the fractional parts as success probabilities. For example, a string with an expected number of copies equal to 1.5 would receive a single copy surely and another with a probability of 0.5. This process continues until the population is full."*

The fitness function used in this study contained three independent variables: the number of inputs selected, the network training error, and the generation indices. The fitness function value was designed to guide the search for fewer inputs, faster training, and more accurate recall by the neural network. Both fewer inputs and smaller training errors produce higher fitness function values.

When the results of the genetic algorithm optimization were compared with the results of the sensitivity analyses, the most important variable was the same in all cases, and the second most important variable was the same most of the time. Beyond this point, the comparison gave agreement only about half the time for the third, fourth, and fifth most important variables. However, when the effectiveness of determining the transient with modular networks with inputs selected by these two methods was compared, the results were virtually identical. Hence, the use of genetic algorithms made it possible to select the most important variables without going through the cumbersome process of training the recurrent neural network.

106.6 Genetic Algorithms and Fuzzy Systems

Fuzzy Decisions of Sensor Fusion Parameters using Genetic Algorithms.

Work reported by Loskiewicz and Uhrig (1994) involves the use of genetic algorithms to guide the fuzzy processing of sensor data from vibration measurements. Power spectral density measurements of accelerometer outputs at nine different locations on over 2000 identical rolling machines in a steel mill were processed using neural networks to identify the fault (or faults). Of the machines tested, 68 were identified as having one or more faults. However, the results using sensors at the 9 different locations on a machine did not always give the same fault diagnosis, and the less dominant of multiple faults were difficult to identify. A fuzzy sensor fusion methodology that involved starting with spectra from two sensors and adding spectra from other sensors, one at a time, subject to the limitation that the quality of the diagnosis improved with each addition, was developed. The procedure worked well as long as the process started using the most important of the 9 spectra and adding spectra in order of decreasing importance. A genetic algorithm was then used to select the order of importance of the spectra. This procedure was successful in improving the overall diagnosis of faults, especially when multiple faults were present.

An approach of integrated genetic learning employed in the construction of fuzzy relational descriptions has been reported by Pedrycz, where fundamental concepts of genetic algorithms are combined with gradient-based techniques for learning purposes (Pedrycz, 1995). It appears that genetic algorithms can be of benefit in initializing gradient based learning schemes and help with partitioning a problem's space into feasibility regions for the purpose of facilitating the supervision of further learning.

106.7 Discussion and Conclusions

An important question arising in connection with hybrid systems is "which one's are we to use in a given application?" Unfortunately, there is no straightforward answer. In a sense, all the various constituents and combinations of hybrid systems are tools whose merit is ultimately determined by their usefulness in solving particular problems. We can, however, make certain general statements.

All systems can be viewed in the manner shown in Figure 106.8a, that is, as relations between inputs and outputs (not necessarily functions but more general relations such as *many-to-many* mappings). In a highly idealized situation we can have one of two extremes: either we know exactly how the system should be working but we have no example of its input/output behavior (Figure 106.8b), or we know its input/output behavior but we know nothing of its internals (i.e., we have a black box) (Figure 106.8c). In the first case, it is best to write fuzzy *if/then* rules to prescribe the systems behavior at the appropriate level of precision. In the second case, it is convenient to use the available input/output data to train one or more artificial neural networks to model the internals of the system.

Of course, in real-world systems the crisp, highly idealized alternatives presented above are more fuzzy. We may have examples of a system's input/output behavior as well knowledge of what is inside the black (or better "gray") box. Hence we may use various hybrids of neural, fuzzy, and other tools to bear on successfully modeling the system. In final analysis, however, our choice of which ones to use will be guided not by a particular commitment to any given tool, but a commitment to solving the problem at hand in a timely, reliable, and cost-effective manner.

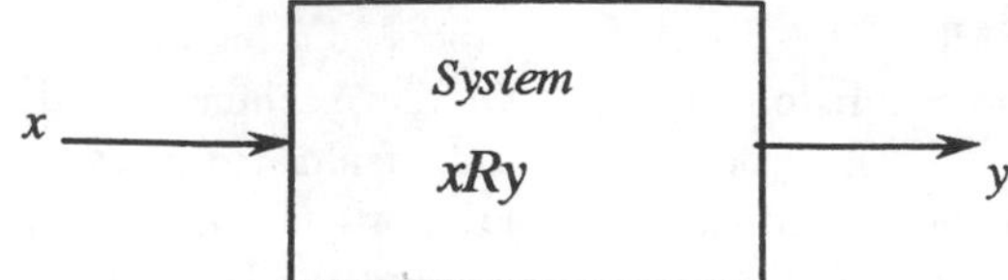

(*a*) *A System is a Relation Between Inputs and Outputs*

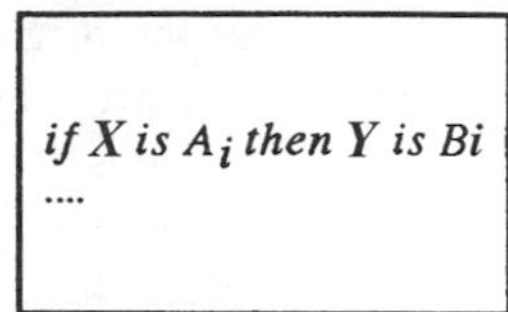

(*b*) *System Logic is Known: Use Fuzzy Rules*

(*c*) *System Input/Output is Known: Use Neural Network Model*

Figure 106.8 Depending on whether the internal relation or the input/output behavior of a system is known, fuzzy, neural, or hybrid tools may be chosen.

References

Blanco, A., Delgado, M., and Requena, I. 1995. A learning procedure to identify weighted rules by neural networks, *Fuzzy Sets and Systems,* 69:29–36.

Challoo R., Clark, D. A., Omar, S. I., and McLauchlin, R. 1993. *A Fuzzy Neural Hybrid System,* Internal report, Intelligent Controls Systems Laboratory, Texas A&M University, Kingsville, TX.

Fraleigh, S. 1994. Fuzzy logic and neural networks,, *PC-AI Magazine,* May/June.

Goldberg, D. E. 1989. *Genetic Algorithms in Search, Optimization, and Machine Learning,* Addison Wesley, New York, NY.

Guo, Z. and Uhrig, R. E. 1992a. Use of genetic algorithms to select inputs for neural networks, *Proc. Special Workshop on COGANN (Combination of Genetic Algorithms and Neural Networks), Int. Joint IEEE-INNS Conference on Neural Networks (IJCNN),* June 6, Baltimore, MD.

Guo, Z. and Uhrig, R. E. 1992b. Using modular neural networks to monitor accident conditions in nuclear power plants, *Proc. SPIE Technical Symposium on "Intelligent Information Systems,"* Application of Artificial Neural Networks III, April 20–24, Orlando, FL.

Hayashi, I., Nomura, H., Yamasaki, H., and Wakami, N. 1992. Construction of fuzzy inference rules by NDF and NDFL, *Int. J. Approximate Reasoning,* 6:241–266.

Haykin, S., 1994. *Neural Networks: A Comprehensive Foundation,* (IEEE Computer Society Press), Macmillan College Publishing Company, New York, NY.

Ikonomopoulos, A., Tsoukalas, L. H., and Uhrig, R. E. 1993. Integration of neural networks with fuzzy reasoning for measuring operational parameters in a nuclear reactor, *Nuclear Technology,* 104:1–12.

Ishibuchi, H., Fujioka, R., and Tanaka, H. 1993. Neural networks that learn from fuzzy if-then rules, *IEEE Trans. Fuzzy Systems,* 1(2):85–97.

Keller, J. M. and Tahani, H. 1992. Implementation of conjunctive and disjunctive fuzzy logic rules with neural networks, *Int. J. Approximate Reasoning,* 6:221–240.

Kosko, B. 1993. *Neural Networks and Fuzzy Systems,* Prentice Hall, Englewood Cliffs, NJ.

Kulkarni, A. D., Coca, P., Giridhar, G. B., and Bhatikar, Y. 1994 Neural network based fuzzy logic decision system, *Proc. World Congress on Neural Networks,* 1994 INNS Annual Meeting, June 5–9, San Diego, CA.

Lee, C. C. 1990. Fuzzy logic in control systems: fuzzy logic controller—Part-I, *IEEE Trans. Systems, Man and Cybernetics,* 20(2):404–418.

Loskiewicz-Buczak, A. and Uhrig, R. E. 1994. Determination of fuzzy decision fusion system parameters by genetic algorithms, *Proc. SPIE Conf. Applications of Neural Networks V,* Vol. 2243, Rogers, S. K. and Ruck, D. W. eds., April 5–8, Orlando, FL.

McCullough, C. L. 1993. Anticipatory neuro-fuzzy control: a powerful new method for real world control, *Proc. IEEE Int. Workshop on Neuro Fuzzy Control,* pp 267–272, March 22–23, Muroran, Japan.

Mizumoto, M. 1988. Fuzzy controls under various reasoning methods, *Information Sciences,* 45:129–141.

O'INCA Design Framework, 1994. *User's Manual,* Intelligent Machines, Sunnyvale, CA.

Pedrycz, W. 1995. Genetic algorithms for learning in fuzzy relational structures, *Fuzzy Sets and Systems,* 69:37–52.

Schwartz, D. G. and Klir, G. J. 1992. Fuzzy logic flowers in japan, *IEEE SPECTRUM,* July.

Terano, T., Asai, K., and Sugeno, M. 1994. *Applied Fuzzy Systems,* Academic Press, Boston, MA.

Travis, M. and Tsoukalas, L. H. 1992. *Application of Fuzzy Logic Membership Functions to Neural Network Data Representation,* Internal Report, University of Tennessee, Knoxville, TN, 1992; included as Appendix A in Uhrig, et al., *Application of Neural Networks,* EPRI Report TR-103443–P1–2, January 1994.

Tsoukalas, L. H., Ikonomopoulos, A., and Uhrig, R. E. 1992. Virtual measurements using neural networks and fuzzy logic, *Proc. American Power Conf.,* 54-II:1437–1442 April 13–15, 54th Annual Meeting, Chicago, IL.

Tsoukalas, L. H., Uhrig, R. E. 1997. *Fuzzy and Neural Approaches in Engineering,* John Wiley & Sons, New York (in press).

Wang, L. X. and Mendel, J. M. 1992. Backpropagation fuzzy systems as non-linear dynamic system identifiers, *IEEE Int. Conf. Fuzzy Systems,* 1409–1418, San Diego, CA.

Welstead, S. 1994a. *Neural Network and Fuzzy Logic Applications in C++,* John Wiley & Sons, New York, NY.

Werbos, P. J. 1992. "Neurocontrol and Fuzzy Logic: Connections and Designs," *International Journal of Approximate Reasoning,* 6:185–219.

Werbos, J. P. 1994b. *The Roots of Backpropagation,* John Wiley & Sons, New York, NY.

Yasunobu, S. and Miyamoto, S. 1985. Automatic Train Operation by Predictive Fuzzy Control, *Industrial Applications of Fuzzy Control,* Sugeno, M., ed., 1–18, North-Holland, Amsterdam.

107

Application Techniques: Combining Fuzzy Logic, Artificial Neural Networks, and Probabilistic Reasoning—Soft Computing

Okyay Kaynak
Bogazici University

107.1 Combining Soft Computing Methodologies 1361
107.2 Neurofuzzy Control 1361
107.3 The Use of NNs in Consumer Products........................... 1361
107.4 The Fusion of GA and FS 1362

In industrial applications, control engineers often have to deal with complex systems having multiple variable and multiple parameter models with perhaps nonlinear coupling. The conventional approaches for understanding and predicting the behavior of such systems based on analytical techniques can prove to be very difficult, even at the initial stages of establishing an appropriate mathematical model. The computational environment used in such an analytical approach is perhaps too categorical and inflexible in order to cope with the intricacy and the complexity of the real world physical systems. It turns out that in dealing with such systems, one has to face a high degree of uncertainty and tolerate imprecision. Trying to increase precision can be very costly.

In the face of difficulties summarized above, it may be more appropriate and advantageous to use a different approach to computation. Prof. Lotfi A. Zadeh has been stoutly promoting this notion through the center that he established and directs in the University of California, Berkeley; Berkeley Initiative in Soft Computing (BISC) (Kaynak, 1995; Zadeh, 1994). He separates hard computing based on binary logic, crisp systems, numerical analysis, and crisp software from soft computing based on fuzzy logic, neural nets, probabilistic reasoning, and genetic algorithms. The former has the attributes of precision and categoricity and the latter approximation and dispositionality as shown in Figure 107.1. Although in hard computing, imprecision and uncertainty are undesirable properties, in soft computing the tolerance for imprecision and uncertainty is exploited to achieve tractability, lower cost, high Machine Intelligence Quotient (MIQ), and economy of communication. Prof. Zadeh argues that soft computing, rather than hard computing, should perhaps be viewed as the foundation of artificial intelligence.

The principal constituents of soft computing are

- Fuzzy logic (FL).
- Artificial neural networks (ANN).
- Probabilistic reasoning (PR), including genetic algorithms (GA), chaos theory, and parts of learning theory.

Fuzzy logic is mainly concerned with imprecision and approximate reasoning, neural networks mainly with learning and curve fitting, and probabilistic reasoning mainly with uncertainty and propagation of belief. Table 107.1, constructed by Fukuda and Shibata (1993, 1994), gives a comparison of their capabilities in different application areas, together with those of control theory and artificial intelligence. It is seen that the approaches are complementary rather than competitive and there can be much to be gained in using them in a combined manner, rather than exclusively. For example, an integration of fuzzy logic and neurocomputing has already become quite popular (neurofuzzy control) with many diverse applications, ranging from chemical process control to consumer goods. Below, the fusion techniques of soft computing methodologies are discussed with particular

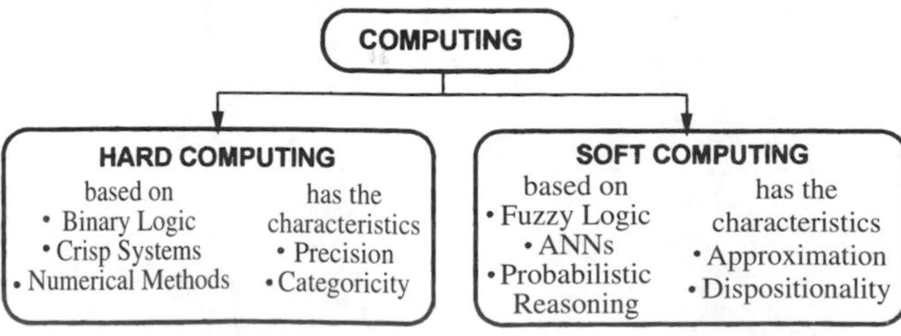

Figure 107.1 The characteristics of hard and soft computing.

0-8493-8343-9/97/$0.00+$.50

Table 107.1 Comparison of Fuzzy Logic, Neural Networks, Artificial Intelligence, and Genetic Algorithms

	Mathematical Model	Learning Data	Operator Knowledge	Real Time	Knowledge Representation	Non-Linearity	Optimization
Control Theory	●	×	▲	●	×	×	×
Neural Network	×	●	×	●	×	●	◖
Fuzzy Logic	◖	×	●	●	▲	●	×
Artificial Intelligence	▲	×	●	×	●	▲	×
Genetic Algorithms	×	●	×	▲	×	●	●

Explanation of Symbols: ● Good or suitable; ◖ Fair; ▲ Needs some other knowledge or techniques; × Unsuitable or does not require

emphasis on neurofuzzy control. The synergy of neural network and fuzzy reasoning follows naturally. They are the best couple to mimic the structure and the reasoning of human brain. Neural network accomplishes what a person does with data and fuzzy logic realizes what a person does with language. The resulting controller is a nonlinear one, suitable to overcome the difficulties involved in using linear controllers for (naturally) nonlinear systems.

107.1 Combining Soft Computing Methodologies

Fuzzy systems can very effectively handle explicit knowledge but have no capability for learning or adapting themselves. Neural networks, on the other hand, have the ability to learn so that the two techniques complement each other. Use of NNs for auto design of fuzzy controllers follows naturally. Figure 107.2 shows (Werbos et al., 1992) in a schematic form how neural networks

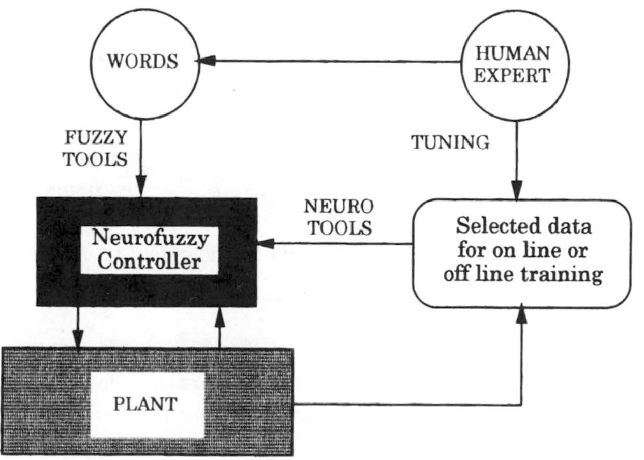

Figure 107.2 Combining neuro and fuzzy tools together with human expert knowledge in a neurofuzzy controller.

are combined with fuzzy logic expert knowledge and I/O data to achieve a desired performance.

The use of neurofuzzy approach in industrial systems has become very popular in recent years, especially in consumer goods. Japan is leading the technology in this respect; numerous products are available on the Japanese market, which are proudly promoted as incorporating a neurofuzzy controller.

The fusion of GA and FL, on the other hand, has attracted the attention of researchers more recently. Nevertheless, some encouraging practical applications are reported in the literature.

107.2 Neurofuzzy Control

In a neurofuzzy controller, the design of the fuzzy system is carried out automatically. Generally, the main purpose of the NN is the generation of membership functions (Kosko, 1992). There are two different approaches in this respect. In one, a neural network is used to represent the fuzzy systems, the weights being the parameters of the membership functions and the nodes the rules. Center position and width are commonly used shape parameters. The function of the NN is to modify these parameters so that the error between the output of the neurofuzzy controller and the supervised data is minimized. Gradient methods (commonly steepest descent) or GAs are employed in minimization.

In the second approach, NN is used to generate the multidimensional nonlinear membership functions. The approach is named NN-Driven Fuzzy Reasoning (Takagi and Hayashi, 1991, 1988) and is especially advantageous when a fuzzy system has inputs that are related. Consider, for example, humidity and temperature in an air conditioner. In a conventional fuzzy system (FS), independent membership functions are generated although the variables are dependent. In NN-driven fuzzy reasoning a nonlinear, multidimensional membership function is generated directly, the design steps being (1) clustering the given training data, (2) fuzzy partitioning the input space by NNs, and (3) designing the consequent part of each partitioned space.

Neural networks have also been used as inference engines and defuzzifiers. Kaynak et al. (1993) describe such an application in which the performance of a conventional fuzzy controller for a servo system is compared to that of a neurofuzzy controller.

107.3 The Use of NNs in Consumer Products

As is stated above, in recent years (since about 1990) we have seen a growing use of neurofuzzy controllers in consumer products. In general, NNs in such applications are not user trainable and their function is one of the four depicted in Figure 107.3 (Takagi, 1993). In Fig. 107.3.a, the use of NN is shown to be as a development tool. Designing and fine tuning of the membership functions are carried out by the NN, as is discussed above.

In some consumer products, NN is used completely independently (Figure 107.3b), mainly for nonlinear function interpolation. For example, in a Matsushita air conditioner, a NN is used,

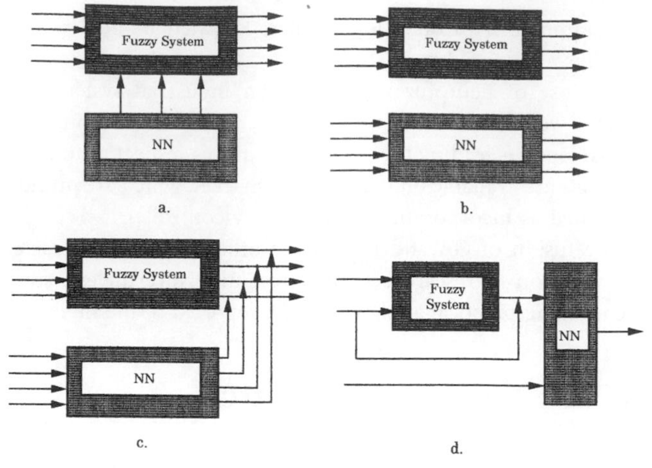

Figure 107.3 Different approaches to the use of NNs in combination with a FS.

completely independent of the fuzzy system that controls the heat pumping, to derive the Predictive Mean Vote as defined ISQ 773, as a measure of the comfort level. This requires a mapping from 6-D space to 6-D space, which is done by the NN.

In some other applications, the role of NN is corrective as shown in Fig. 107.3c. This is the approach used to improve the performance of an already marketed product. For example, the later models of a washing machine can incorporate a NN to handle some extra inputs that were not considered in the original fuzzy controller.

In cascaded systems (Figure 107.3d), FS accomplishes a part of the task and then passes it to a NN. For example in a Sanyo neurofuzzy fan, the fan is supposed to rotate towards the user, which requires the determination of the direction of the remote controller.

In some consumer products that have been marketed recently, the standard system modifies itself automatically, according to the usage style of the owner and/or the owner can modify the basic operational parameters of a system according to his/her personal preferences. The determination of the preheating time in a kerosene fan heater and the cooling level in an air conditioner can be cited as examples.

107.4 The Fusion of GA and FS

In recent years, we have seen an increasing number of propositions for the use of GAs in industrial systems. The application areas range from the relatively simple one of tuning the parameters of a PID controller (Wang and Kwok, 1992) to system identification (Kristinsson and Dumon, 1992) and dynamic control of robot arms (Kwok et al., 1993). The use of GAs in combination with the other components of soft computing is, however, relatively new. For example, in Karr and Gentry (1993), a fusion of GA and FS is proposed for the control of pH. In such applications, evolutionary computation is utilized for the auto-design of a FS (Fukuda et al., 1994; Leu and Takagi, 1993). The shape of the membership function, the number of rules, and the consequent parameters are determined by a GA. A detailed description of the issues involved can be found in Takagi (1993b).

In fusion of soft computing methodologies, the purpose of integration is generally the betterment of the performance or the easing of the design of a FS by the use of NNs or GAs, i.e. NN into FS or GA into FS. A fusion of all three constituents is not very common. It should be pointed out here that in any application, integration should not be understood just as an amalgamation but a synergetic combination should be sought.

In the above, only a limited number of fusion techniques has been discussed. Many other approaches have been proposed in the literature to benefit from the complementary capabilities of FL, NNs, and GAs. The fusion of FL into NNs or GAs (FL into NN or FL into GA) is not considered at all, although such approaches can yield equally propitious solutions.

Acknowledgment

The author acknowledges the grant provided by TUBITAK under the framework of DOPROG program and the grant of Bogazici University, Research Fund, Project No: 94A0235.

References

Fukuda, T. and Shibata, T. 1993. Hierarchical control system in intelligent robotics and mechatronics, *Proc. IECON'93, Int. Conf. Ind. Elec., Control, and Instrumentation,* 1:33–38.

Fukuda, T. and Shibata, T. 1994. Private communications.

Fukuda, T., Shimojima, K., and Shibata, T. 1994. Fuzzy, neural network and genetic algorithm based control system, *Proc. IECON'94, Int. Conf. Ind. Elec., Control, and Instrumentation,* 2:1220–1225.

Karr, C. and Gentry, E. J. 1993. Fuzzy control of pH using genetic algorithms, *IEEE Trans. Fuzzy Systems,* 1:46–53.

Kaynak, O. 1995. Viewpoint: mechatronics becoming state of art, *IEEE Spectrum,* 32(1):86.

Kaynak, O., et al. 1993. Motion control using a neuro-fuzzy approach, *Proc. IEEE Workshop on Neuro-Fuzzy Control,* 285–292, Mororan, Japan.

Kristinsson, K. and Dumon, G. A. 1992. System identification and control using genetic algorithms, *IEEE Trans. System, Man and Cybernetics,* 22:1033–1045.

Kusko, B. 1992. *Neural Networks and Fuzzy Systems: A Dynamical Systems Approach to Machine Intelligence,* Prentice Hall, Englewood Cliffs, NJ.

Kwok, D. P., Leung, T. P., and Shang, F. 1993. Genetic algorithms for optimal dynamic control of robot arms, *Proc IECON'93, Int. Conf. Ind. Elec., Control, and Instrumentation,* 1:380–385.

Lee, M. A. and Takagi, H. 1993. Integrating design stages of fuzzy systems using genetic algorithms, *Int. Conf. Fuzzy Systems (FUZZ-IEEE'93),* 1:612–617.

Takagi, H. 1993a. Cooperative System of Neural Networks and Fuzzy Logic and its Application to Consumer Products, *Industrial Applications of Fuzzy Control and Intelligent Systems,* Yen, J. and Langari, R., eds., Van Nostrand Reinhold, New York, NY.

Takagi, H. 1993b. Fusion techniques of fuzzy systems and neural networks, and fuzzy systems and genetic algorithms, *SPIE Proc. Technical Conf. Applications of Fuzzy Logic Technology,* v. 2061, Boston, MA.

Takagi, H. and Hayashi, I. 1991. NN-driven fuzzy reasoning, *Int. J. Approximate Reasoning (Special Issue of IIZUKA'88),* 5:191–212.

Wang, P. and Kwok, D. P. 1992. Optimal design of PID process controllers based on genetic algorithms, *Proc. IFAC 11th World Congress,* 5:261–265.

Werbos, P. J. et al. 1992. Foreword of *Handbook of Intelligent Control,* White, D. A. and Sofge, D. A., eds., Van Nostrand Reinhold, New York, NY.

Zadeh, I. 1994. Berkeley initiative on soft computing—BISC, *IEEE Ind. Elec. Soc. Newsletter,* 41(3):8–10.

108

Synthesis of Fuzzy, Artificial Intelligence, Neural Networks, and Genetic Algorithm for Hierarchical Intelligent Control

Takanori Shibata
Ministry of International Trade and Industry, and Massachusetts Institute of Technology

Toshio Fukuda
Nagoya University

Kazuo Tanie
Ministry of International Trade and Industry

108.1 Introduction 1364
108.2 Artificial Intelligence, Fuzzy, Neural Network, and Genetic Algorithm 1364
108.3 Hierarchical Intelligent Control of Robotic Motion 1366
108.4 Conclusions 1367

108.1 Introduction

Autonomous robots, which perform tasks without human operators, are required in many fields. The autonomous robots have to carry out tasks in various environments by themselves like human beings. They have to be intelligent to determine their own actions based on sensory information. In advance, human operators can give the robots their knowledge and skill to some extent in a top-down manner. However, when the robots perform tasks in an unknown environment, the knowledge may not be useful. In this case, the robots have to adapt to their environments and acquire new knowledge by themselves through learning. This process proceeds in bottom-up manner.

This paper introduces a control scheme for autonomous robots, which this paper refers to as *hierarchical intelligent control* scheme (Figure 108.1). The hierarchical intelligent control consists of three levels: adaptation level, skill level, and learning level. This scheme has two characteristics with respect to learning process: top-down approach and bottom-up approach. To link three levels and have such characteristics, the scheme uses artificial intelligence (AI), fuzzy logic, neural networks (NN), and genetic algorithm (GA) (Fukuda and Shibata, 1992; Goldberg, 1989; Zadeh, 1965). Each technique has advantages and disadvantages. To overcome the disadvantages, this paper introduces synthesis techniques of them. Those are key techniques for intelligent control of robots.

This paper describes advantages and disadvantages of each technique in Section 108.2. Section 108.3 explains how to construct the hierarchical intelligent control while using the advantages of the techniques and synthesis techniques. Section 108.4 concludes this paper.

108.2 Artificial Intelligence, Fuzzy, Neural Networks, and Genetic Algorithm

AI techniques have been used to synthesize knowledge-based systems as expert systems. For intelligent control, there were some examples of *symbolic control* which uses symbolic reasoning mechanisms for higher-level control. However, it is difficult to classify sensed data to map numerical data set into a symbolic data set for understanding process state. The signals are classified by using *if-then* rules as shown in Figure 108.2. On the other hand, the fuzzy logic is characterized as an extension of binary crisp logic. The fuzzy set is a class in which transition from membership to non-membership is gradual rather than abrupt as shown in Figure 108.3. Crisp sets allow only full membership or no membership at all, whereas fuzzy sets allow partial membership. Since the fuzzy set does not have learning capability, it is difficult for a human operator to tune the rules from the data set. The NN has capabilities of nonlinear mapping, parallel processing, and learning. The NN produces mapping rules from empirical training sets through learning, but the mapping rules

0-8493-8343-9/97/$0.00+$.50

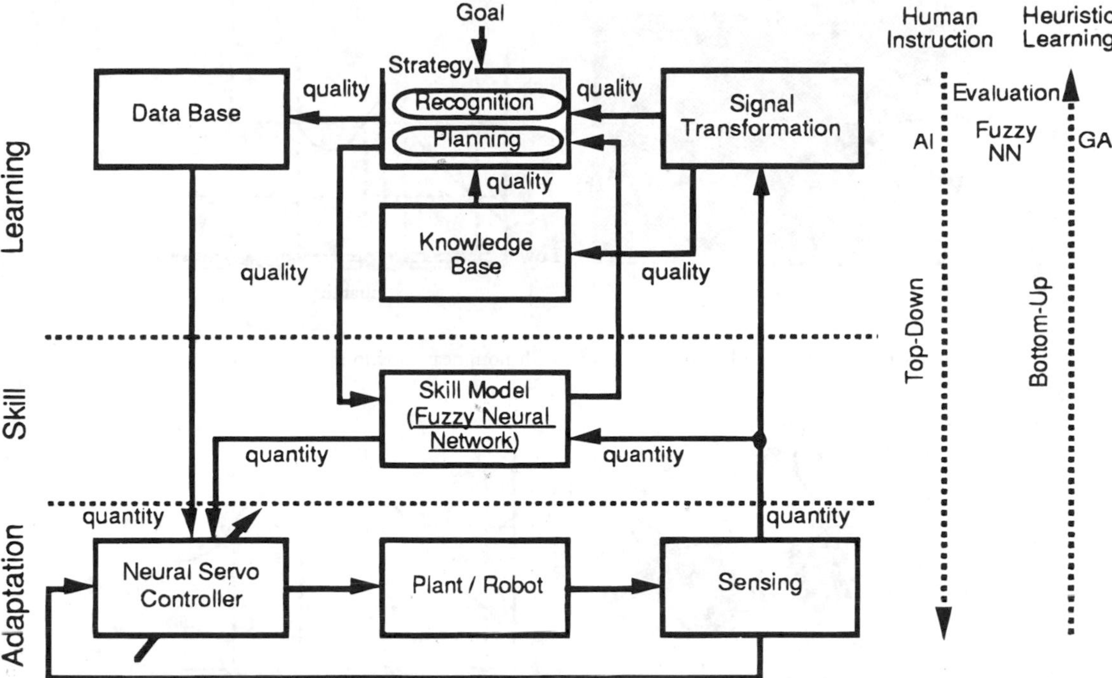

Figure 108.1 Hierarchical intelligent control system.

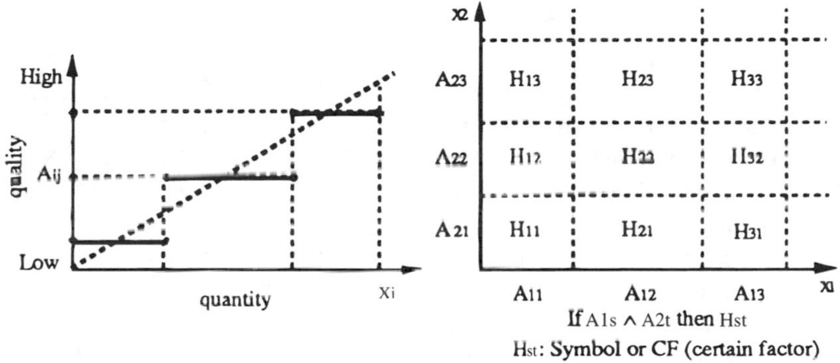

Figure 108.2 Classification by *if-then* rules.

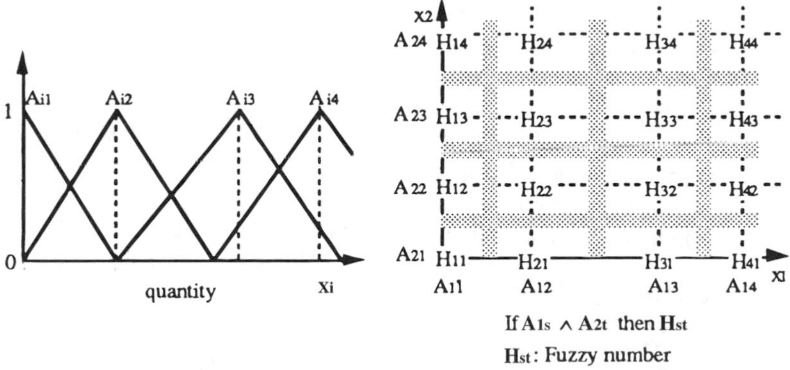

Figure 108.3 Classification by fuzzy logic with membership functions.

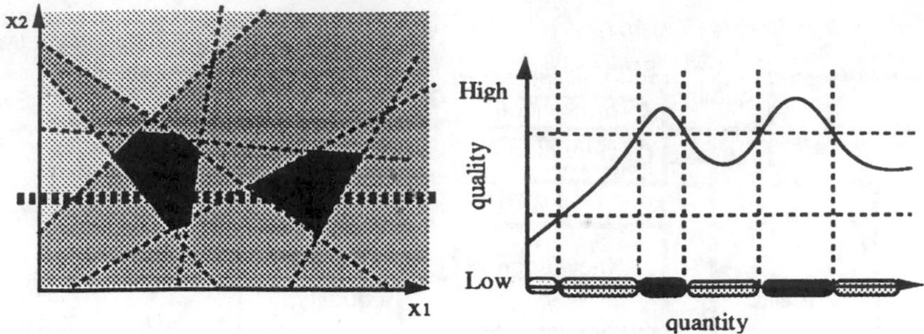

Figure 108.4 Classification by neural network with nonlinear functions, i.e., sigmoid function.

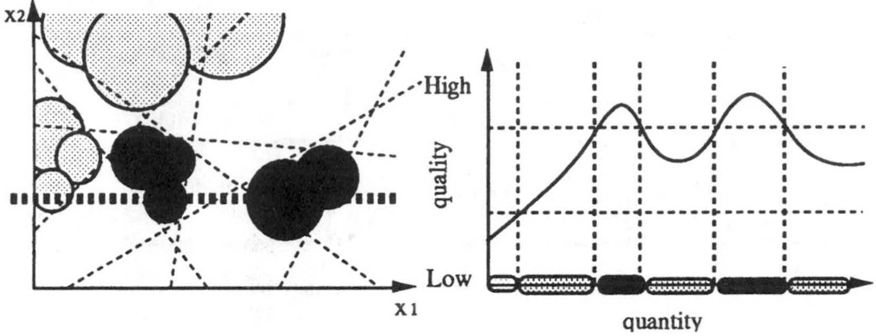

Figure 108.5 Classification by neural network with radial basis functions.

in the network are not visible and are difficult to understand as shown in Figure 108.4. Moreover, the structured neural network, i.e., neural network with radial basis function, has potential to learn more quickly and easily than the neural network with the sigmoid functions (Figure 108.5) (Fukuda, Shiotani et al., 1992). Therefore, human operators should give the neural network efficient structure if they have experiences. Or else, a heuristic approach for structure optimization is necessary.

GAs are search algorithms based on the mechanics of natural selection and natural genetics. They combine survival of the fittest among string structures with a structured yet randomized information exchange to form a search algorithm with some of the innovative flair of human problem solving. An occasional new part is tried for good measure. While randomized, GAs are no simple random walk. They efficiently exploit historical information to speculate on new search points with expected improved performance. The GA is a powerful tool for structure optimization of the fuzzy logic and the neural networks (Figures 108.6–108.8) (Fukuda et al., 1992). On the other hand, the fuzzy logic and the neural network can be an evaluation function for the GA (Shibata and Fukuda, 1993). It is difficult to define evaluation functions for complex optimization problems. However, while using the fuzzy logic and the neural network, human operators can transfer their criterion.

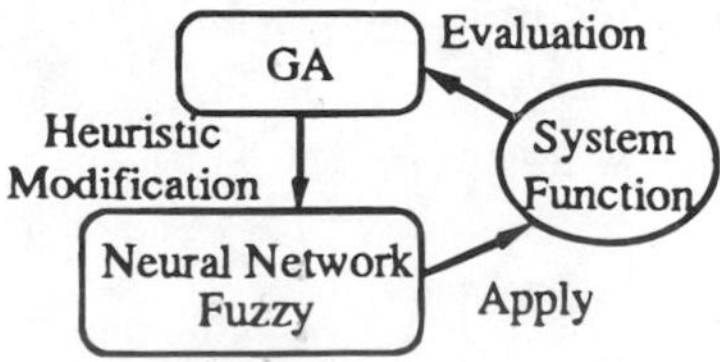

Figure 108.6 Structure optimization of fuzzy or neural network by genetic algorithm.

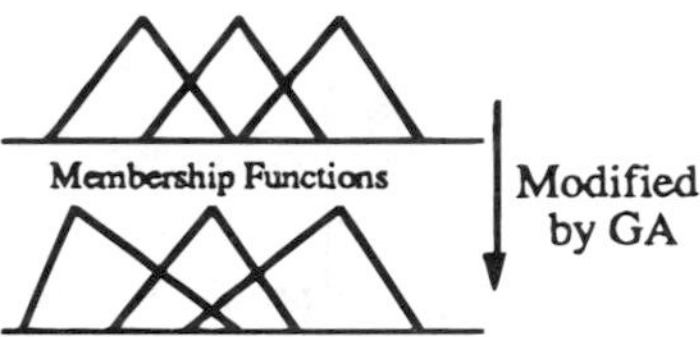

Figure 108.7 Structure optimization and learning of fuzzy logic by genetic algorithm.

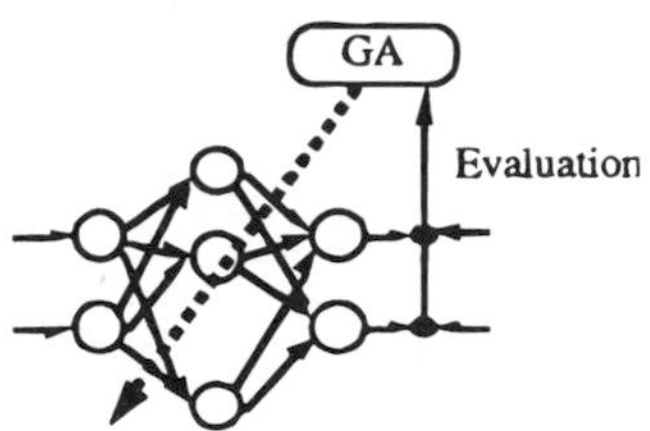

Figure 108.8 Structure optimization and learning of neural network by genetic algorithm.

108.3 Hierarchical Intelligent Control of Robotic Motion

The hierarchical intelligent control scheme comprises three levels: a *learning* level, a *skill* level, and an *adaptation* level as shown in Figure 108.1 (Shibata and Fukuda, 1992a). Therefore, there are three feed-back loops. The learning level is based on the expert system for a reasoning mechanism and has a hierarchical structure: recognition and planning to develop a control strategy. The recognition level uses neural networks and fuzzy logic combined with the neural network as nodes of a decision tree. In the case of the neural network, inputs are a numeric quantity sensed by some sensors, while outputs are a symbolic quality which indicates process states. A structured neural network for additional learning is effective to memorize new patterns (Fukuda, Shiotani et al., 1992). In the case of the fuzzy neural network, inputs and outputs are numeric quantities and the fuzzy neural network clusters input signals by using membership functions. That is, the fuzzy neural network transforms numerical quantity into symbolic quality by using membership functions. Both the neural network and the fuzzy neural network are trained with the training data of *a priori* knowledge obtained from human experts. As a result, the neural network and the fuzzy neural network can transform various sensed data from numerical quantities to symbolic qualities, and perform *sensor fusion* and production of *meta knowledge* at the learning level. The important information is sensed actively on using the knowledge base. The sensors of vision, weight, force, touch, acoustic, and others can be used as nodes of a decision tree for recognition of the environment.

Then, the planning level reasons symbolically for strategic plans or schedules of robotic motion, such as task, path, trajectory, force, and other planning in conjunction with the knowledge base. The system can include another *common sense* for robotic motion. The GA optimizes the planning of the motion heuristically (Shibata et al., 1992). The GA also optimizes structures of neural network and fuzzy logic. Thus, the learning level reasons an unknown fact from *a priori* knowledge and sensory information. Then, the learning level produces control strategies for skill level and adaptation level. Following the control strategy, the learning level selects an initial data set for a servo controller at the adaptation level from a database which maintains some gains and initial values of interconnection weights of the neural network in the servo controller. Moreover, the recent sensed information from the skill level and the adaptation level updates the learning level through long-term learning process with human instruction. Therefore, knowledge at the learning level is given by the human operator in top-down manner and by heuristics of the skill level and the adaptation level in bottom-up manner.

In the same task and different environments, it is necessary to change control references depending on the environment for the servo controller at the adaptation level. At the skill level, the fuzzy neural network is used for specific tasks following the control strategy produced at the learning level to generate appropriate control references. Input signals into the fuzzy neural network are numerical values sensed by some specific sensors and some symbols which indicate the control strategy produced at the learning level. Output of the fuzzy neural network is the control reference for the servo controller at the adaptation level. This output is based on the skill extracted from human experts through learning training sets obtained from them. At the same moment, the fuzzy neural network clusters the input signals in the shape of membership functions. These membership functions are used as the symbolic information for the learning level.

In the adaptation level, a neural network in the servo controller adjusts control law to current status of dynamic process (Fukuda et al., 1992). Particularly, compensation for nonlinearity of the system and uncertainties included in the environment must be dealt with by the neural network. Thus, the neural network in the adaptation process works more rapidly than that in the learning process. It is shown that the neural network-based controller, the Neural Servo Controller, is effective to the nonlinear dynamic control with uncertainties such as force control of a robotic manipulator. Eventually, the neural networks and the fuzzy neural networks connect neuromorphic control with symbolic control for hierarchical intelligent control while combining human skills.

The hierarchical intelligent control is applied not only to a single robot, but also to a multi-agent robot system. If there is no interaction between robots, each robot has to work optimally for its purpose so that the total task should be achieved optimally; that is, each robot should work selfishly, or else a conflict among the robots might occur when using a public resource. The competition may cause collisions and deadlock states among the robots in a local area. To avoid competition, it is necessary for the robots to communicate and to coordinate among themselves. The coordination among the robots is as important as selfishness. The GAs are applied hierarchically to balance selfishness with coordination for efficient motion planning (Shibata et al., 1992).

As results, synthesis of AI, fuzzy logic, NN, and GA is important for an intelligent system, depending on their characteristics. Hierarchical intelligent control using these techniques is effective for intelligent robot systems.

108.4 Conclusions

This paper introduced synthesis techniques of AI, fuzzy, neural network, and GA for intelligent systems. The approach is applied to a hierarchical intelligent control system of robots. The system is a hybrid system of top-down and bottom-up learning while synthesizing those techniques.

References

Fukuda, T. and Shibata, T. 1992. Theory and applications for neural networks for industrial control systems, *IEEE Trans. Ind. Elec.,* 39(6):472–489.

Fukuda, T., Ishigami, H., et al. 1993. Structure optimization of fuzzy neural network using genetic algorithm, *Proc. IFSA;* to appear.

Fukuda, T., Shibata, T., Tokita, M., and Mitsouka, T. 1992. Neuromorphic control—adaptation and learning, *Proc. Trans. Ind. Elec.,* 39(6):497–503.

Fukuda, T., Shiotani, S., et al. 1992. A new neuron model for additional learning, *Proc. IJCNN'92,* 1:938–943, Baltimore, MD.

Goldberg, D. E. 1989. *Genetic Algorithms in Search, Optimization, and Machine Learning,* Addison-Welsey, Reading, MA.

Shibata, T. and Fukuda, T. 1992a. Skill based control by using fuzzy neural network for hierarchical intelligent control, *Proc. IJCNN'92,* 2:81–86, Baltimore, MD.

Shibata, T. and Fukuda, T. 1992b. Hierarchical intelligent control of robotic motion, *Trans. Neural Networks,* to appear.

Shibata, T. and Fukuda, T. 1993. Fuzzy critic for robotic motion planning by genetic algorithm in hierarchical intelligent control, *Proc. IJCNN'93,* submitted.

Shibata, T., Fukuda, T., et al. 1992. Selfish and coordinative planning for multiple mobile robots by genetic algorithm, *Proc. 31st Conf. Decision and Control,* 3:2686–2691, Tucson, AZ.

Zadeh, L. A. 1965. Fuzzy sets, *Information and Control,* 8:28.

109

Advanced Tools for Adaptive Nonlinear Modeling and Control of Power in Large Systems

Harold H. Szu
Naval Surface Warfare Center, Dahlgren Division

Brian A. Telfer
Naval Surface Warfare Center, Dahlgren Division

109.1 Introduction 1369
109.2 Modeling, Control, and Neural Networks 1369
109.3 Wavelet and Adaptive Space-Frequency Techniques for Modeling and Control 1370
109.4 Summary and Conclusions 1371

109.1 Introduction

The Power Electronic Building Block (PEBB) will significantly alter power system generation, delivery, control, and uses. It creates the potential for a more electric ship (or plane, tank, or car). Aboard ship, it will reduce cost, number of people, and improve performance. Electrical distribution and control systems will replace or reduce the need for actuators controlled by compressed air, hydraulics, and mechanics. Energy distribution systems will consist of high-speed alternators, rectifiers, nonlinear regulators, solid-state circuit breakers, solid-state load transfer devices, adjustable speed motor drivers, pulse width modulated inverters and converters, and switch mode power supplies. Consequently, the distribution system will have many nonlinear loads, loads with $-180°$ phase shifts, and loads continually changing in response to process control requirements (NSWCCD, 1995). To address these nonlinear dynamic systems, mathematical techniques such as wavelets, adaptive time-frequency methods, artificial neural networks, and possibly chaos analysis tools will be needed for intelligent modeling, control, and diagnostics. Fuzzy logic also falls within the scope of intelligent modeling and control although we do not adress it here. Fuzzy logic techniques are typically integrated with neural networks to allow them to adapt to changing conditions.

These power system issues fit into a larger context of related needs of the Navy and the broader technological community, e.g., conditional maintenance, object recognition, etc. Like the new power systems, these needs also require automated techniques that can adapt to changing nonlinear conditions. All of these automation efforts provide leverage toward the automated techniques that are needed for new power systems.

109.2 Modeling, Control, and Neural Networks

For this and other complex applications, nonlinear models and controllers are needed that adapt to changing conditions. Artificial neural networks are ideally suited and have been demonstrated in a number of complex applications, e.g., chemical processes (McAvoy, 1992), the National Aerospace Plane (NASP) (Sofge and White, 1992), and in a number of power system applications (Sobajic, 1992). For example, in the NASP (with a goal of two-hour flight times between Washington, DC and Tokyo), neural networks are being studied for controlling non-uniform temperatures on the engine walls. This is a very complex problem that requires numerous sensors and controllers (a smart skin), and neural networks are felt to be the only technique that can solve it.

For more than two decades, distinct direct and indirect strategies have been employed for adaptive control. In direct control, the controller parameters are directly adjusted to reduce a norm of the output error. In indirect control, the current state of the plant is modeled by a set of parameters and the controller parameters are set according to the model parameters. The indirect approach points to the close relationship of modeling (or system identification) and control. These paradigms are depicted in Figure 109.1. If only modeling is needed and not control, then a controller need not be included in the system. The same type of nonlinear neural network architecture can be used for both system identification and control.

Neural network models are attractive for systems that are highly nonlinear and for which physical models are not known or too complex or changing, requiring adaptivity. Both of these are the case for the power systems we are considering. For example, switched-mode power supplies present a negative impedance,

0-8493-8343-9/97/$0.00+$.50

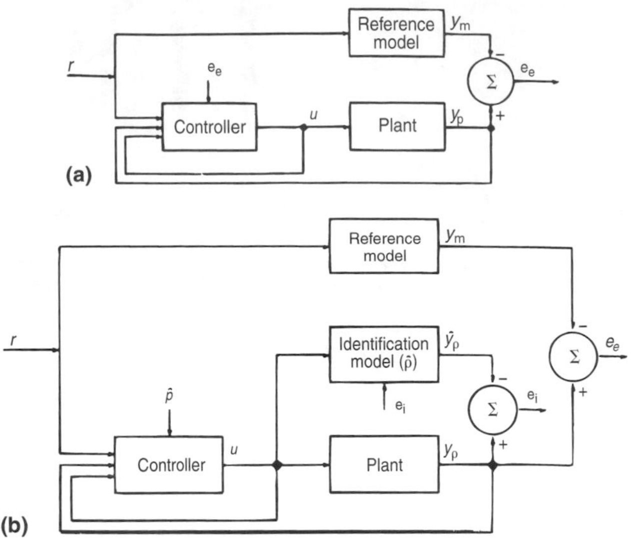

Figure 109.1 (a) Direct and (b) indirect control (from Narendra and Parthasarathy, 1990).

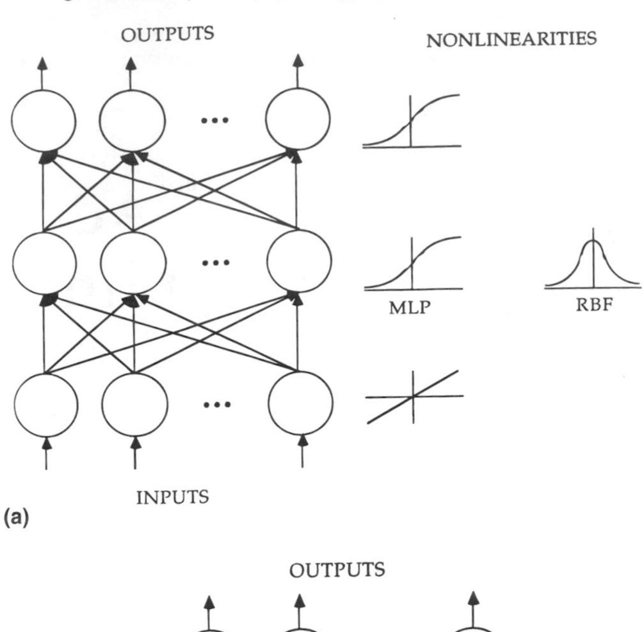

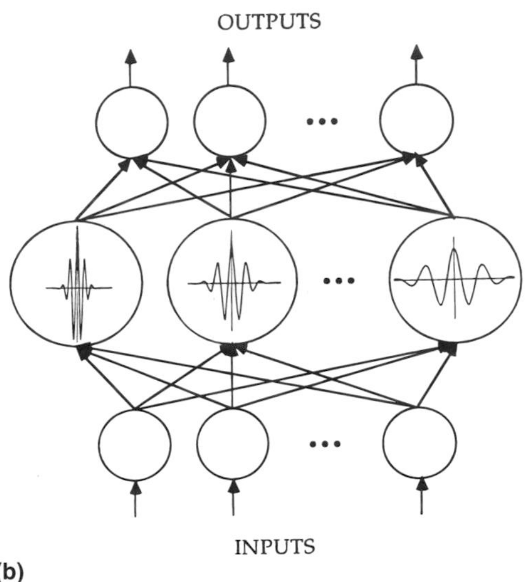

Figure 109.2 (a) Architecture of multilayer perceptrons (MLP) and radial basis function (RBF) networks: connections denote weights, nodes denote summation followed by nonlinearity. Non-linearities are given on right for MLP (global sigmoidal) and RBF (local Gaussian); (b) architecture of adaptive wavelet network.

for which PID controllers lack guaranteed stability (Doerry, 1995). Neural networks can also function well in conjunction with limited knowledge about the plant. For example, part of the operating region may be well characterized by a physical model, but the other regions may require a network.

Neural networks have even been demonstrated to model chaotic systems, in the case of the Mackey-Glass equations (Narendra, 1992). The Ott-Grebogi-York (OGY) algorithm has also been demonstrated to control chaotic signals, for the cases of controlling a chaotic magnetic ribbon (Ditto et al., 1990), a rabbit heart, etc. These methods are of interest because it is thought that nonlinear dynamics and chaos may play an important role in future power systems.

Two primary network architectures have been used for these applications: multilayer perceptrons (MLPs) (Narendra and Parthasarathy, 1990), and radial basis function (RBF) networks (Sanner and Slotine, 1992), shown in Figure 109.2a. Backpropagation through time has become an important algorithm for training networks to provide a response over time (Werbos, 1990). The nonlinearities in MLPs have global receptive fields, while the RBF networks have local receptive fields. Although many of the first neural network modeling and control demonstrations used MLPs, RBF networks have been used with increasing frequency because they can be simply synthesized constructively, i.e., by adding new basis functions to the network one at a time to minimize approximation error in local regions. This advantage is demonstrated by Sanner and Slotine (1992) using a network constructed from Gaussian RBFs.

109.3 Wavelet and Adaptive Space-Frequency Techniques for Modeling and Control

As pointed out in Sanner and Slotine (1992), wavelets appear more preferable as basis functions than Gaussian RBFs. This is because wavelets provide space-frequency information (they are bandpass filters while Gaussians are lowpass filters) at multiple scales or resolutions. Wavelets are being applied in numerous domains involving data approximation, compression, processing, noise reduction, and recognition (Szu et al., 1994, 1995). Their power arises from their multiresolution and local representation, and the flexibility to choose a wide variety of wavelet basis functions that meet the completeness conditions. Wavelets have been chosen to be orthogonal/biorthogonal, leading to fast O(N) wavelet transforms (Daubechies, 1992; Mallat, 1989).

Another attractive method for selecting wavelet functions is to optimize waveforms and weighting coefficients to best fit a particular application. This has been done by integrating wavelet functions with a neural network to form an adaptive wavelet network (Telfer et al., 1994; Szu et al., 1992), depicted in Figure 109.2b. The network is related to a RBF network, but adds the additional advantages of wavelets. The resulting wavelets are nonorthogonal, but can generally represent a function with fewer wavelets than required by orthogonal wavelets. This is important in practice to avoid overfitting data sets and to avoid poor generalization to unseen data. Mallat's matching pursuit has the same goal, and provides a fast O(NlogN) transform (Mallet and Zhang, 1993). Others have also demonstrated the advantages of wavelet network approaches (Zhang and Benveniste, 1992). We have applied adaptive wavelet networks to several signal processing domains (Telfer et al., 1994), and they have significant potential for power systems modeling and control.

As an example of wavelets already being applied to power systems, Robertson and Mayer have used wavelets as a feature space for classifying power transients (Robertson et al., 1994). Wavelets are ideally suited for characterizing transients because of their multiple-scale representation, as opposed to windowed Fourier transform techniques, which require a fixed window size. Features were extracted from wavelet representations of several types of simulated transients. The transient types were sufficiently well separated in feature space so that only a very simple classifier was needed to perfectly separate the classes. With more transient types and real transients, a more sophisticated neural network classifier would no doubt be required. An adaptive wavelet network would be particularly well suited to this application.

109.4 Summary and Conclusions

New PEBB-based power systems will require adaptive nonlinear techniques for modeling and control. Neural networks, wavelets, and integrated adaptive wavelet networks are ideal candidates. In addition to their desirable computational aspects, they fit well the need to have decentralized local control, since wavelet transforms and neural networks have already been implemented in hardware for real-time applications. They are also readily realized in software on the microprocessors and DSP chips to be incorporated in the PEBB.

Acknowledgment

Our understanding of the above issues has emerged from participating in a focus group on Advanced Network Theory, organized by John Joynes at NSWCCD-Annapolis. Some of the introductory statements are paraphrased from the group's Vision Statement. The group has had participants from NSWCCD, NSWCDD, Navy funding agencies, and several universities, companies, and labs.

References

Daubechies, I. 1992. *Ten Lectures on Wavelets,* SIAM, Philadelphia, PA.

Ditto, W., Rauseo, S., and Spano, M. 1990. *Physics Review Letters,* 65:3211.

Doerry, N., LCDR. Briefing to Advanced Network Theory Group, NSWCCD, February, unpublished.

Mallat, S. 1989. A theory for multiresolution signal decomposition: the wavelet representation, *IEEE Trans. PAMI,* 11:674–693.

Mallat, S. and Zhang, Z. 1993. Matching pursuit with time-frequency dictionaries, *IEEE Trans. Signal Processing,* December.

McAvoy, T., ed. 1992. *Special Session on Application of Neural Networks in the Chemical Process Industries, in Proc. World Congress Neural Networks,* Van Nostrand Reinhold, New York, NY.

Narendra, K. 1992. Adaptive Control of Dynamical Systems Using Neural Networks, *Handbook of Intelligent Control,* Sofge, D. and White, D. eds., p. 158, Van Nostrand Reinhold, New York, NY.

Narendra, K. and Parthasarathy, K. 1990. Identification and control of dynamical systems using neural networks, *IEEE Trans. Neural Networks,* 1:4–27.

NSWCCD. 1995. *Vision Statement of Advanced Network Theory Group,* NSWCCD, Joynes, J., organizer, March, unpublished.

Pati, Y. and Krishnaprasad, P. 1993. Analysis and synthesis of feedforward neural networks using discrete affine wavelet transforms, *IEEE Trans. Neural Networks,* 4:73–85.

Robertson, D., Camps, O., and Mayer, J. 1994. Wavelets and power system transients: feature detection and classification, *Proc. SPIE,* vol. 2242, April.

Sanner, R. and Slotine, J. 1992. Gaussian networks for direct adaptive control, *IEEE Trans. Neural Networks,* 3:837–863.

Sobajic, D., ed. 1992. *Proc. INNS/EPRI Workshop on Neural Network Computing for the Electric Power Industry,* Lawrence Erlbaum INNS Press, Hillsdale, NJ.

Sofge, D. and White, D., eds. 1992. *Handbook of Intelligent Control,* Van Nostrand Reinhold, New York, NY.

Szu, H., Telfer, B., and Kadambe, S. 1992. Adaptive wavelets for signal representation and classification, *Optical Engineering,* 31:1907–1916.

Szu, H., Crowley, J., Hewer, G., and Miceli, W., eds. 1994. *Proc. SPIE Wavelet Applications Conf.,* vol. 2242, April.

Szu, H., Crowley, J., Hewer, G., Meleis, H., and Miceli, W., eds. 1995. *Proc. SPIE Wavelet Applications Conf.,* vol. 2491, April.

Telfer, B., Szu, H., Dobeck, G., Garcia, J., Ko, H., Dubey, A., and Witherspoon, N. 1994. Adaptive wavelet classification of acoustic backscatter and imagery, *Optical Engineering,* vol. 33:2192–2203.

Werbos, P. 1990. Backpropagation through time: what it does and how to do it, *Proc. IEEE,* 78:1550–1560.

Zhang, Q. and Benveniste, A. 1992. Wavelet networks, *IEEE Trans. Neural Networks,* 3:889–898.

110

Application of Model Reference Adaptive Control and Adaptive Time-Delay RBF Networks

110.1 Introduction 1372
110.2 Dynamic Modeling of Flexible Multibody 1374
110.3 Adaptive Time-Delay Radial Basis Function Network 1376
110.4 Pace Simulation Study 1377
110.5 Conclusions 1379

Gary G. Yen
Oklahoma State University

Abstract

A distributive neural control system is advocated for flexible multibody structures. The proposed neural controller is designed to achieve trajectory slewing of structural member as well as vibration suppression for precision pointing capability. The motivation to support such an innovation is to pursue a *real-time* implementation of robust and fault tolerant structural controller. The proposed control architecture which takes advantage of the geometric distribution of piezoceramic sensors and actuators has provided a tremendous freedom from computational complexity. In the spirit of model reference adaptive control, we utilize adaptive time-delay radial basis function networks as a building block to controllers cooperatively regulates the dynamics of the nonlinear structure to follow the prespecified reference models asymptotically. The proposed control strategy is validated in the experimental facility, called the Planar Articulating Controls Experiment which consists of a two-link flexible planar structure constrained to move over a granite table. This paper addresses the theoretical foundation of the architecture and demonstrates its applicability via a realistic structural test bed.

110.1 Introduction

Modern engineering technology is leading to increasingly complex space structures with ever more demanding performance criteria. Specifically, precision pointing devices (e.g., robotic manipulators and surveillance satellites) are often made of lightweight composites and equipped with piezoelectric and/or piezoceramic sensors and actuators. These flexible multibody structures, which are likely to be highly non-linear with time-varying structural parameters and poorly modeled dynamics have posed serious difficulties for all currently advocated control methodologies (e.g., robust, adaptive, and optimal controls) (Antsaklis and Passino, 1992; White and Sofge, 1992). Furthermore, the ultimate pursuit of a higher degree of autonomous behavior, which calls for a highly sophisticated controller to ensure that demanding performance to be met, can be extremely difficult due to factors such as high dimensionality, multiple inputs and outputs, operational constraints, as well as the unexpected failures of sensors and actuators. Conventional control design approaches often depend upon the assumption of a high fidelity dynamic model containing identified system parameters. This hypothesis which inevitably necessitates an iterative process of finite element analysis and system identification is computationally expensive to validate. Consequently, design procedures to achieve the desired stability, robustness, and dynamic response for precision space structures with unknown parameters are incomplete.

Dynamic modeling, system identification, and vibration controls of a flexible multibody have been an active mission in the structures and controls division of the USAF Phillips Laboratory for years. At present, a critical need exists for the verification and comparison of various modeling and control theories based on an actual hardware experiment. To meet this need, Phillips Laboratory has constructed a flexible multibody structure which consists of two flexible beams connected in series with motors at both the hub and the elbow joint. Figure 110.1 shows the flexible multibody structure named the Planar Articulating Controls Experiment (PACE) (Denoyer and Kwak, 1993; Kwak et al., 1992). DC motors are mainly used to drive the PACE arms through the specified trajectory. Piezoceramic actuators and sensors have been chosen for vibration suppression on PACE. Because of their stiffness, good linearity, relative temperature

0-8493-8343-9/97/$0.00+$.50

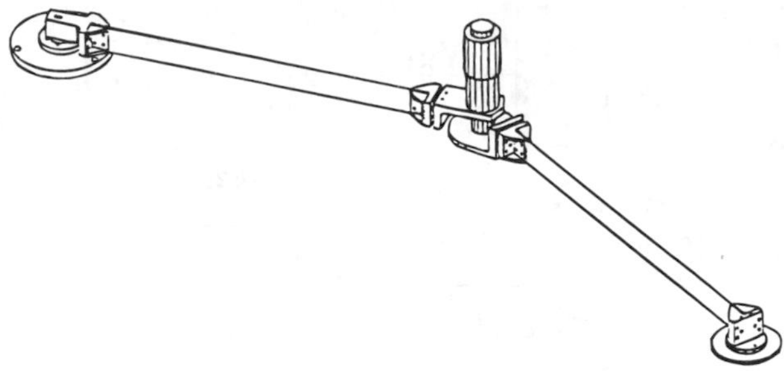

Figure 110.1 PACE test article.

insensitivity, and ease of implementation, piezoceramics such as lead zirconate titanate (PZT) have been determined to be a good candidate as actuators for many structural control applications (Garcia and Inman, 1991).

The mathematical formulation of a class of flexible multibodies based on equations of motion in terms of quasi-coordinates is derived for each substructure independently. The individual substructure is made to act as a single structure by means of a consistent kinematics synthesis. The resulting differential equations are nonlinear and hybrid, where the term 'hybrid' implies that the equations for the rigid-body translations and rotations are ordinary differential equations and those for the elastic motions are partial differential equations (Meirovitch, 1991). The advantage of this approach is that it yields equations of motion in terms of body axes, which are the same axes used for the control forces and torques.

In addition to modeling efforts, we propose to design and to validate a distributive neural control system which is capable of withstanding structural failures, component deviation, and unpredictable perturbations. Neural networks which employ the well known back-propagation learning algorithm are capable of approximating any continuous functions (e.g., non-linear plant dynamics and complex control laws) with an arbitrary degree of accuracy (Hornik et al., 1989). Similarly, radial basis function networks (Moody and Darken, 1989) are also shown to be universal approximators (Hartman et al., 1990). These model-free neural network paradigms are more effective at memory usage in solving control problems than conventional learning control approaches. A typical example is the BOXES algorithm, a memory intensive approach, which partitions the control law in the form of a look-up table (Michie and Chambers, 1968). Adaptive neural control system offers the capability of real-time adaptation and generalization while a look-up table approach would only provide discrete controller solutions in a lengthy and sequential search.

Our goal is to approach structural autonomy by extending the control system's operating envelope, which has traditionally required vast memory usage. Connectionist systems, on the other hand, deliver less memory intensive solutions to control problems and yet provide a sufficiently generalized solution space (Yen, 1994a). In vibration suppression/trajectory following problems, we utilize the adaptive time-delay radial basis function network as a building block to allow the connectionist system to function as an indirect closed-loop controller. Decentralized nature of control system provides a tremendous computation power to suppress the vibration modes that can be identified by the experimental modal testing or to follow a prespecified trajectory. Prior to training the compensator, a neural identifier based on an ARMA model is utilized to identify the open-loop system (see Figure 110.2). The m horizon-of-one predictive controllers then cooperatively regulate the dynamics of the nonlinear structure to follow a prespecified reference system (in terms of m linearized systems for each mode interested) asymptotically as depicted in Figure 110.3 (i.e., the model reference adaptive control architecture) (Yen, 1994b). m-1 backup copies of ATDRBF neural identifier are created to facilitate the training of ATDRBF neural controller. The reference models, which can be easily specified through an input-output relationship, described all desired features associated with the control task, e.g., a linear and highly damped model to suppress the vibration or a designate route to specify the desired trajectory.

Each control subsystem, which were designed dedicatedly for one set of PZT actuator and sensor, is utilized to suppress a specified vibration mode or to follow a designate trajectory, so that the computational load can be evenly distributed and executed on a real-time basis. The function of each ATDRBF neural controller is to map the system states into corresponding control actions in order to force the decomposed plant dynamics to match an output behavior which is specified by a linearized reference model. However, we cannot apply the energy minimization procedure (e.g., gradient descent, conjugate gradient or Newton-Raphson method) to adjust the interconnection weights of the neural controllers because the desired outputs of the neural controllers are not available. To achieve the true gradient descent of the square of the error, we use *dynamic back propagation* (Narendra and Parthasarthy, 1990) to accurately approximate the required partial derivatives. An adaptive time-delay radial basis function network is first trained to identify the open-loop system. The resulting neural identifier then serves as extended unmodifiable layers to train a set of neural controllers. If the structural dynamics are to change as a function of time, the neural identifier would require the learning algorithm to periodically update the network parameters accordingly (as well as the m-1 backup copies).

The proposed efforts address several issues to achieve a distributive fault tolerant control system in flexible multibody structures. In Section 110.2, the mathematical formulation and dynamic modeling of a class of flexible multibody are established.

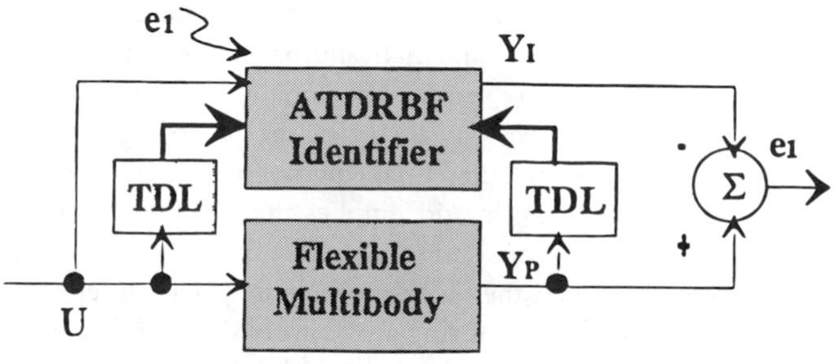

Figure 110.2 System identification of flexible multibody.

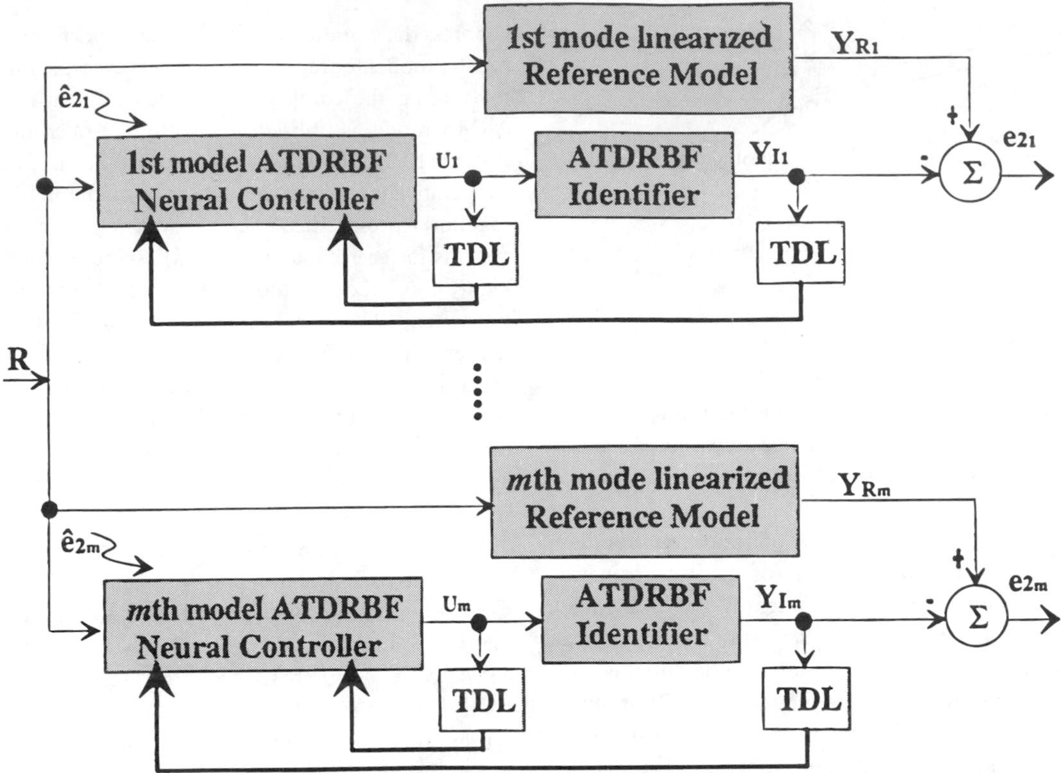

Figure 110.3 Decentralized model reference adaptive control.

In Section 110.3, adaptive time-delay radial basis function network is covered, providing an underlying issue pertaining to the learning algorithm. The proposed control strategy was validated in the PACE test article which consists of a two-link flexible planar structure constrained to move over a granite table in Section 110.4. The paper is concluded with a few pertinent observations in Section 110.5.

110.2 Dynamic Modeling of Flexible Multibody

A set of equations of motion suitable for the control task can be formulated by means of Lagrange's equations for flexible bodies in terms of quasi-coordinates. The advantage of this approach is that it yields equations in terms of body axes, which are the same axes as those used to express control forces and torques. In using the approach of Mierovitch (1991) to derive equations of motion for a chain of flexible multibody systems, it is convenient to adopt a kinematical procedure permitting the expression of the velocity vector of a nominal point in a typical body in terms of the velocity vector of the preceding body in the chain. The resulting differential equations are nonlinear and hybrid, where the term 'hybrid' implies that the equations for the rigid-body translations and rotations are ordinary differential equations and those for the elastic motions are partial differential equations. Because maneuvering and control design in terms of hybrid equations is not feasible, the partial differential equations must be transformed into sets of ordinary differential equations by means of a discretization-in-space procedure, such as the finite element method (Mierovitch, 1980) or a Rayleigh-Ritz based substructure synthesis (Mierovitch and Kwak, 1991). The resulting formulation consists of a high-order set of non-linear ordinary differential equations. A common approach to control design requires the solution of a two-point boundary value problem, which is not feasible for high-order systems, so that a different approach is advisable.

The non-linearity enters into the differential equations through the rigid-body motions. Indeed, the elastic motions tend to be small. In view of this, it appears natural to conceive of a *perturbation approach* whereby the rigid-body motions can be regarded as being of zero-order in magnitude and the elastic motions as being of first-order in magnitude. This approach permits dividing the problem into a low-dimensional set of non-linear zero-order equations for the rigid-body motions and a high-dimensional set of linear first-order equations for the elastic motions and the perturbations in the rigid-body motion, where the order is to be taken in a perturbation sense. Note that, because the zero-order solution enters into the first-order equations as a known function of time, the first-order equations represent a time-varying system. Moreover, the system is subjected to persistent disturbances. The perturbation approach just described was proposed in Mierovitch and Quinn (1987a) to maneuver and control flexible spacecraft.

The kinematical synthesis (Mierovitch and Kwak, 1990; Mierovitch and Quinn, 1987b) works quite well in the case in which the number of bodies in the chain is relatively small. When the number of bodies is larger than three, difficulties can be expected, so that a different approach is taken. In this paper, we consider a procedure whereby the equations of motion are derived first for each individual flexible body. Then, the sets of equations for the individual bodies are assembled into a global set by invoking the kinematical relations. In the process, the redundant coordinates and velocities resulting from considering the individual bodies separately are eliminated. It is convenient to carry out the kinematical synthesis on the zero-order problem and first-order problem separately. Implementation of the kinematical synthesis is based on recursive relations that lend themselves to ready computer coding. The resulting zero- and first-order global sets of equations are particularly suited for maneuvering and control design, respectively. The zero-order nonlinear equations govern the maneuver as if the system consisted of articulated rigid bodies where the maneuver amounts to driving the system from an initial state to a final state. The simplest approach is to carry out the maneuver by means of actuators that impart predetermined motions to the substructures relative to one another. The first-order equations govern the elastic vibrations and the perturbations in the rigid-body motions. They contain the zero-order solution as a known function of time. As a result, the system is time-varying. Moreover, it is subjected to persistent disturbances caused by the maneuver. The process can be likened to that in which the system must follow a reference state. In this case, the reference state is defined by the rigid-body maneuvering, which is characterized by zero elastic states. Then, the first-order equations are simply the equations in terms of the difference between the actual states and the reference states, where this difference can be identified as perturbations in the state variables.

The approach used in this paper is to derive equations of motion for the individual substructures separately and then impose kinematical relations of the type described earlier to obtain system's equations of motion. Although the approach is used for the case of a two-link flexible body system, the approach can be extend to the case of arbitrary N flexible multibody systems. Let us consider a typical flexible substructure moving on a horizontal surface (see Figure 110.4) and introduce the inertial axes XY with the origin at O and a set of body axes $x_s y_s$ with the origin at S and embedded in the undeformed substructure. Then, we can write the position vector of a typical point in the substructure with the spatial coordinates given symbolically as follows:

$$\overline{W}_s = \hat{n}^T(R_s + C_s^T u_s), \tag{110.1}$$

where $\hat{n} = [\bar{I}\ \bar{J}]^{\mathrm{T}}$ represents the column matrix consisting of the unit vectors in X- and Y-directions corresponding to inertial coordinates, $C_s = C(\theta_s)$ is the matrix of the direction cosines which is given by

$$C_s = \begin{bmatrix} \cos\theta_s & \sin\theta_s \\ -\sin\theta_s & \cos\theta_s \end{bmatrix}. \tag{110.2}$$

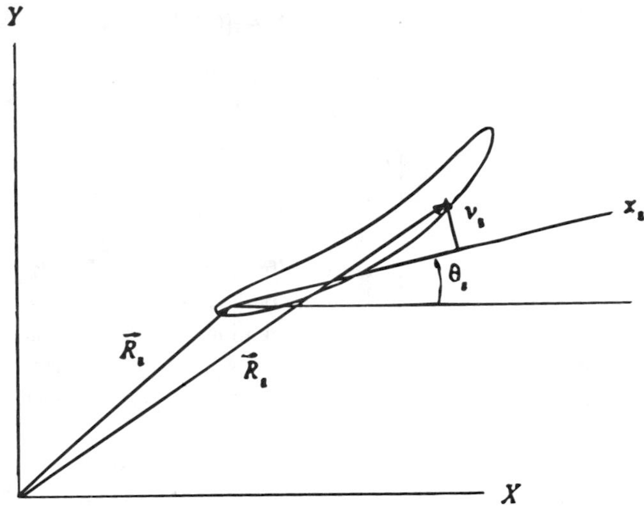

Figure 110.4 Flexible body.

In addition, R_s is the radius vector from I to S, and u_s includes the radius vector from S to a typical point in s and the elastic displacement vector of the same point relative to the body axes x_s y_s, respectively. Due to space limitation, the detailed mathematical derivation can be found in Yen and Kwak. The following equations represent the nonlinear ordinary differential equations for the single substructure;

$$m_s\ddot{R}_s - \ddot{\theta}_s D_s^T(LS_s + N\overline{\Phi}_s q_s) + C_s^T N\overline{\Phi}_s\ddot{q}_s - \dot{\theta}_s^2 C_s^T(LS_s + N\overline{\Phi}_s q_s) - 2\dot{\theta}_s D_s^T N\overline{\Phi}_s\dot{q}_s = 0, \tag{110.3a}$$

$$\ddot{R}_s^T D_s^T(LS_s + N\overline{\Phi}_s q_s) + I_s\ddot{\theta}_s + \tilde{\Phi}_s\ddot{q}_s = T_s, \tag{110.3b}$$

$$\overline{\Phi}_s^T I_s^T C_s\ddot{R}_s + \tilde{\Phi}_s^T\ddot{\theta}_s + M_s\ddot{q}_s + K_s q_s = 0, \tag{110.3c}$$

where

$$m_s = \int \overline{m}_s\, dx_s, \qquad S_s = \int \overline{m}_s x_s\, dx_s, \qquad I_s = \int \overline{m}_s x_s^2\, dx_s$$

in which $\overline{m}_s$ is the mass density per unit length. In addition,

$$\overline{\Phi}_s = \int \overline{m}_s\Phi_s\, dx_s,$$

$$\tilde{\Phi}_s = \int \overline{m}_s x_s\Phi_s\, dx_s$$

$$M_s = \int \overline{m}_s\Phi_s^T\Phi_s\, dx_s,$$

in which $\Phi_s(x_s)$ is the vector of admissible functions and q_s is the vector of generalized coordinates. Also

$$L = [1\ 0]^T, \qquad N = [0\ 1]^T,$$

$$D_s = \begin{bmatrix} \sin\theta_s & -\cos\theta_s \\ \cos\theta_s & \sin\theta_s \end{bmatrix},$$

$$\tilde{I} = \begin{bmatrix} 0 & -1 \\ 1 & 0 \end{bmatrix},$$

K_s represents the substructure stiffness matrix, and T_s is the torque applied to the origin of the body axis $x_s y_s$. Imposing the kinematical relations to the above equations to link two adjacent substructures is not an easy task. Thus, we propose the perturbation method to ease the kinematical synthesis and numerical calculations.

Designing the slewing and control for articulated systems of substructures is very difficult, especially if the design is to be optimal in some fashion. The difficulty can be traced to the fact that the system is nonlinear and of high order. The nonlinearity can be attributed to the rigid-body motions and the high order to the elastic motions. The perturbation approach is based on the simple observation that rigid-body motions tend to be large compared to the elastic motions. Consistent with this, let us *assume* that the translations and rotation of the body can be divided into zero-order terms and first-order terms in magnitude. Elastic displacements are assumed to be small so that the generalized displacements associated with elastic motion can be regarded as a first-order term. Thus, we may write

$$R_s = R_{s0} + R_{s1}, \qquad \theta_s = \theta_{s0} + \theta_{s1}, \qquad T_s = T_{s0} + T_{s1}. \tag{110.4}$$

Inserting Equation (4) into trigonometric functions yields the following relations,

$$\cos\theta_s = \cos\theta_{s0} - \sin\theta_{s0}\theta_{s1}, \qquad \sin\theta_s = \sin\theta_{s0} + \cos\theta_{s0}\theta_{s1}, \tag{110.5}$$

which lead to $C_s = C_{s0} - D_{s0}\theta_{s1}$, $D_s = D_{s0} + C_{s0}\theta_{s1}$. After mathematical manipulation, we finally obtain the first-order equations of motion,

$$m_s\ddot{R}_{s0} - D_{s0}^T L S_s \ddot{\theta}_{s0} - \dot{\theta}_{s0}^2 C_{s0}^T L S_s = 0, \tag{110.6a}$$

$$-S_s L^T D_{s0}\ddot{R}_{s0} + I_s\ddot{\theta}_{s0} = T_{s0}, \tag{110.6b}$$

as well as the first-order equations of motion,

$$\begin{aligned} &m_s\ddot{R}_{s1} - S_s D_{s0}^T L\ddot{\theta}_{s1} + C_{s0}^T N\overline{\Phi}_s q_s - 2\dot{\theta}_{s0} C_{s0}^T L S_s \dot{\theta}_{s1} \\ &\quad - 2\dot{\theta}_{s0} D_{s0}^T N\overline{\Phi}_s \dot{q}_s - (\ddot{\theta}_{s0} C_{s0}^T - \dot{\theta}_{s0}^2 D_{s0}) L S_s \theta_{s1} \\ &\quad - (\ddot{\theta}_{s0} D_{s0}^T + \dot{\theta}_{s0}^2 C_{s0}) N\overline{\Phi}_s q_s = 0 \end{aligned} \tag{110.7a}$$

$$\begin{aligned} &-S_s L^T D_{s0}\ddot{R}_{s1} + I_s\ddot{\theta}_{s1} + \tilde{\Phi}_s\ddot{q}_s - S_s L^T C_{s0}\ddot{R}_{s0}\theta_{s1} \\ &\quad - \ddot{R}_{s0}^T D_{s0}^T N\overline{\Phi}_s q_s = T_{s1}, \end{aligned} \tag{110.7b}$$

$$\begin{aligned} &\overline{\Phi}_s^T N^T C_{s0}\ddot{R}_{s1} + \tilde{\Phi}_s\ddot{\theta}_{s1} + M_s\ddot{q}_s - \overline{\Phi}_s^T L_s^T D_{s0}\ddot{R}_{s0}\theta_{s1} \\ &\quad + K_s q_s = -\ddot{\theta}_{s0}\tilde{\Phi}_s^T - \overline{\Phi}_s^T N^T C_{s0}\ddot{R}_{s0}. \end{aligned} \tag{110.7c}$$

The main advantage of using the perturbation method is that the zero-order equations can be solved independently of the first-order equations. The zero-order equations are nonlinear but in low order. Once the zero-order equations are solved, the solution of the zero-order equations enters into the first-order equations and complete the numerical calculations. Due to space limitation, we are concerned with the single substructure. Please refer to (Yen and Kwak) to see how to assemble individual equations into a global equation of motion by means of the kinematical synthesis.

110.3 Adaptive Time-Delay Radial Basis Function Network

Biological studies have shown that variable time-delays do occur along axons due to different conduction time and different lengths of axonal fibers. In addition, temporal properties such as temporal decays and integration occur frequently at synapses. Inspired by this observation, the time-delay back-propagation network was proposed by Waibel, et al. (1988) for solving the phoneme recognition problem. In this architecture, each neuron takes into account not only the current information from all the neurons of the previous layer, but also a certain amount of past information from those neurons due to delay on the interconnections. However, a fixed amount of time-delay throughout the training process has limited the usage mainly due to the mismatch of the temporal location in the input patterns. To overcome this limitation, Lin, et al. (1992) has developed an *adaptive* time-delay back-propagation network to better accommodate the varying temporal sequences, and to provide more flexibility for optimization tasks.

A given adaptive time-delay radial basis function network can be completely described by its interconnecting topology, neuronic characteristics, temporal delays, and learning rule. The individual processing unit performs its computations based only on local information. The basis function in the hidden layer produces a localized response to input stimulus as do locally-tuned receptive fields in our nervous systems. The Gaussian function network, a realization of the RBF network using Gaussian kernels, is widely used in pattern classification and function approximation. The output of a Gaussian neuron in the hidden layer is defined by

$$u_j^1 = \exp\left(-\frac{\|x - w_{1j}^0\|^2}{2\sigma_j^2}\right), \qquad i = 1, \ldots, N^1, \tag{110.8}$$

where u_j^1 is the output of the jth neuron in the hidden layer, x is the input vector, w_{1j}^0 denotes the weighting vector for the jth neuron in the hidden layer (i.e., the center of the jth Gaussian kernel), σ_j^2 is the normalization parameter of the jth neuron (i.e., the width of the jth Gaussian kernel), and N^1 is the number of neurons in the hidden layer. Equation 110.4 produces a radially symmetric output with a unique maximum at the center dropping off rapidly to zero at large radii. That is, it produces a significant nonzero response only when the input falls within a small localized region of the input space. Inspired by the adaptive time-delay back-propagation network, the output equation of ATDRBF networks is described by

$$y_j^2(t_n) = \sum_{i=1}^{N^1} \sum_{L=1}^{L^{ij}} w_{ij,l}^1 u_i^1(t_n - \tau_{ij,l}^1), \qquad i = 1, \ldots, N^2, \tag{110.9}$$

where $w_{ij,l}^l$ denotes the connection between the output of the ith neuron of the hidden layer and the input of the jth neuron of the output layer with an independent time-delay $\tau_{ij,l}^1$, $u_i^1(t_n - \tau_{ij,l}^1)$ is the output vector from the hidden layer at time $t_n - \tau_{ij,l}^1$, L_{ij}^1 denotes the number of delay connections from the ith neuron of the hidden layer to the jth neuron of the output neuron.

Shared with generic radial basis function networks, adaptive time-delay Gaussian function networks have the property of undergoing *local* changes during training, unlike adaptive time-delay back-propagation networks which experience *global* weighting adjustments due to the characteristics of sigmoidal functions. The localized influence of each Gaussian neuron allows the learning system to refine its functional approximation in a successive and efficient manner. The hybrid learning algorithm (Moody and Darken, 1989) which employs the K-means clustering for the hidden layer and the least mean square (LMS) algorithm for the output layer further ensures a faster convergence and often leads to better performance and generalization. The combination of locality of representation and linearity of learning offers tremendous computational efficiency in real-time adaptive control. K-means algorithm is perhaps the most widely known clustering algorithm because of its simplicity and its ability to produce good results. The normalization parameters, σ_j^2, are obtained once the clustering algorithm is complete. They represent a measure of the spread of the data associated with each cluster. The cluster widths are then determined by the average distance between the cluster centers and the training samples,

$$\sigma_j^2 = \frac{1}{M_j} \sum_{x \in \Theta_j} \|x - w_j^1\|^2, \tag{110.10}$$

where Θ_j is the set of training patterns belonging to jth cluster and M_j is the number of samples in Θ_j. This is followed by applying a LMS algorithm to adapt the time-delays and interconnecting weights in output layer. The training set consists of input/output pairs, but now the input patterns are preprocessed by the hidden layer before being presented to the output layer. The adaptation of the output weights and time delays are derived based on error back-propagation to minimize the cost function,

$$E(t_n) = \frac{1}{2} \sum_{j=1}^{N^2} (d_j(t_n) - y_j(t_n))^2, \tag{110.11}$$

where $d_j(t_n)$ indicates the desired value of the jth output neuron at time t_n. The weights and time-delays are updated step by step proportional to the opposite direction of the error gradient respectively,

$$\Delta w_{i,l}^2 \equiv -\eta_1 \frac{\partial E(t_n)}{\partial w_{ij,l}^2} = \eta_1 (d_j(t_n) - y_j(t_n)) u_i^1(t_n - \tau_{ij,l}^2), \tag{110.12a}$$

$$\Delta \tau_{ij,l}^2 \equiv -\eta_2 \frac{\partial E(t_n)}{\partial \tau_{ij,l}^2} = -\eta_2 (d_j(t_n) - y_j(t_n)) w_{ij,l}^2 u_i^{1'}(t_n - \tau_{ij,l}^2), \tag{110.12b}$$

where η_1 and η_2 are the learning rates.

110.4 Pace Simulation Study

The autonomous control of precision space structures requires a distributed computational architecture that provides the ability to perform on-line system identification and dynamic control after orbital deployment. A neural network based decentralized control system proposed in this paper provides an alternative way to reduce the need for *a priori* knowledge of structural qualitative behavior, although a minimum knowledge of modeling is assumed. The dynamics of the plant are assumed unknown. System identification is simulated by a single-layer ATDRBF network with 100 neurons to ensure the flexibility to approximate arbitrary non-convex regions. The control objective is to achieve trajectory slewing (i.e., θ_1: absolute angular displacement of the upper arm; $\dot{\theta}_1$: absolute angular velocity of the upper arm; β_2: relative angular displacement of the forearm; and $\dot{\beta}_2$: relative angular velocity of the forearm) as well as vibration suppression along the motion (i.e., t_1: tip displacement of the upper arm; $\dot{t}_2$: tip velocity of the upper arm; t_2: tip displacement of the forearm; and $\dot{t}_2$: tip velocity of the forearm) by applying the control forces (i.e., u_1: torque applied to the shoulder and u_2: torque applied to the elbow).

Our strategy is to relax the vibration suppression task for the first few seconds. The RMS errors are accumulated only based on the trajectory following states (i.e., θ_1, $\dot{\theta}_1$, β_2, and $\dot{\beta}_2$). Around 1.5 second, we begin to suppress the structural vibration states (i.e., t_1, $\dot{t}_1$, t_2, and $\dot{t}_2$) while maintaining the progress of trajectory following. However, by confining the desired responses to strictly zeros will deteriorate the network performance, even affecting the first four components. The way we proposed in this study is to setup an implicit exponential delaying envelop, so the network will smoothly catch up the requirements. Whenever the output falls within this exponential envelop, it indicates a zero error, meaning no weight adjustments will be taken. This process

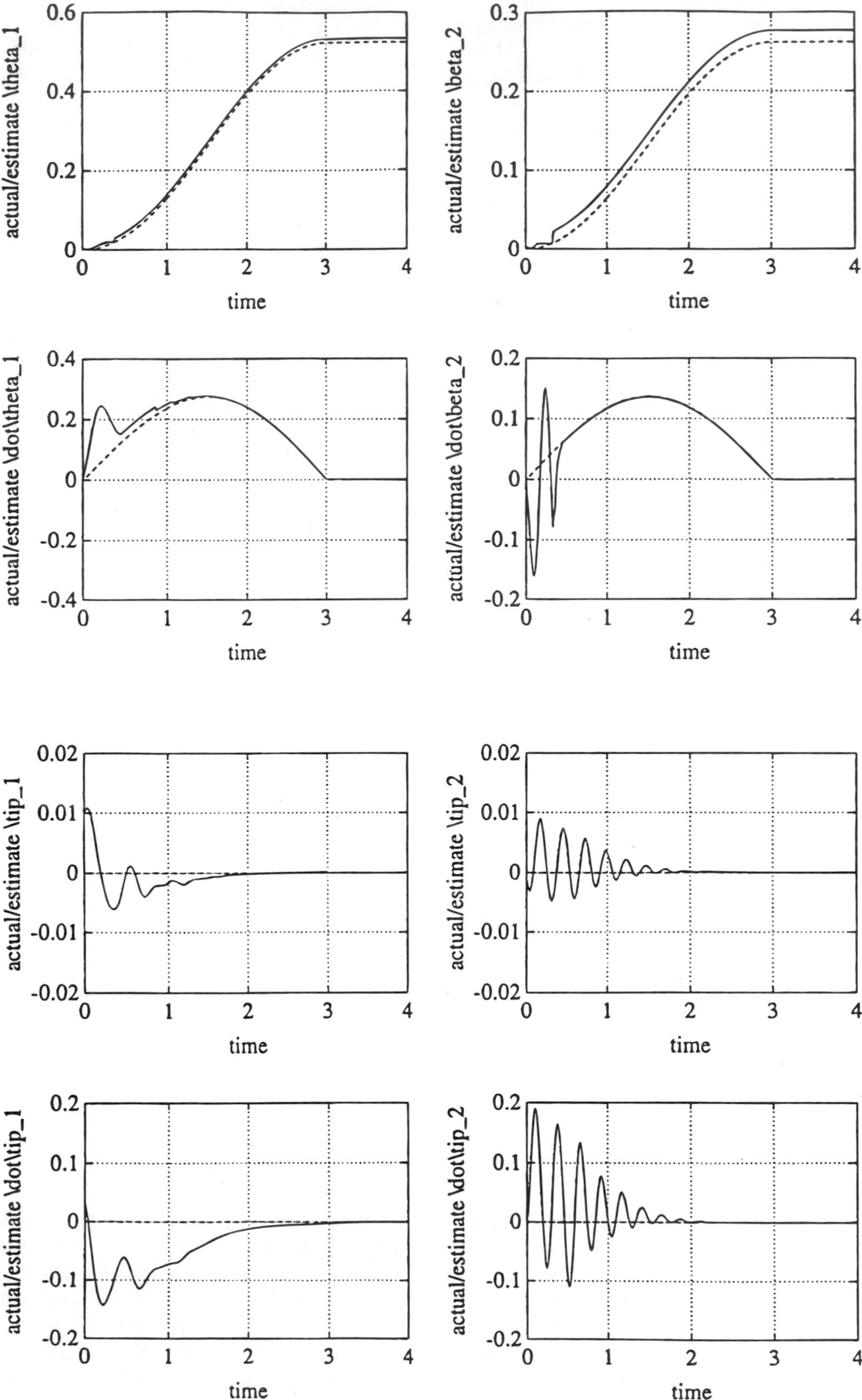

Figure 110.5 Closed response of distributive neural control of PACE test article.

indeed speeds up the training procedure significantly. Trajectory slewing/vibration suppression is performed by eight ATDRBF networks with 25 neurons each. The closed-loop controller regulates the dynamics of the PACE structure to follow the desired outputs as given below:

$$\theta_1(t) = \frac{\pi}{12}\left(1 - \cos\frac{\pi t}{3}\right), \qquad (110.13a)$$

$$\dot{\theta}_1(t) - \frac{\pi^2}{36}\sin\frac{\pi t}{3}, \qquad (110.13b)$$

$$\beta_2(t) = \frac{\pi}{24}\left(1 - \cos\frac{\pi t}{3}\right), \qquad (110.13c)$$

$$\dot{\beta}_2(t) = \frac{\pi^2}{72}\sin\frac{\pi t}{3}, \qquad (110.13d)$$

and tip displacements and their velocity are zeros, of course. Although the neural identifier learned to match the open-loop system in a reasonable time frame, the compensator took more than four days to converge to a reasonable accuracy, mean square error 0.000562. Figure 110.5 displays the closed-loop performance with respect to all output variables, respectively, in response to an impulse. The neural regulator learned to follow the specified trajectory and then damp out the structural vibration effectively.

110.5 Conclusions

The architecture proposed for distributed neural control system successfully demonstrates the feasibility and flexibility of our proposed solution for precision flexible multibodies. The salient features associated with the proposed control architecture are discussed. In a similar spirit, the proposed control structure can be extended to the dynamic control of aeropropulsion engines, underwater vehicles, chemical processes, power plants, and manufacturing scheduling. The applicability of the present methodology to various realistic CSI structural test beds will be pursued in our future research.

References

Antsaklis, P. J. and Passino, K. M. 1992. *An Introduction to Intelligent and Autonomous Control,* Kluwer Academic, Norwell, MA.

Denoyer, K. K. and Kwak, M. K. 1993. Dynamic modeling and vibration suppression of a slewing active structure utilizing piezoelectric sensors and actuators, *Proc. SPIE Conf. Smart Structures and Intelligent Systems,* February, 882–894.

Garcia, E. and Inman, D. J. 1991. Modeling of the slewing control of a flexible structure, *AIAA J. Guidance, Control and Dynamics,* 14(4):736–742.

Hartman, E. J., Keeler, J. D., and Kowalski, J. M. 1990. Layered neural networks with Gaussian hidden units as universal approximations, *Neural Computation,* 2(2):210–215.

Hornik, K., Stinchcombe, M., and White, H. 1989. Multilayer feedforward networks are universal approximators, *Neural Networks,* 2(5):359–366.

Kwak, M. K., Smith, M. J., and Das, A. 1992. PACE: a test bed for the dynamics and control of flexible multibody systems, *Proc. NASA/NSF/DoD Workshop on Aerospace Computational Control,* August, 100–105.

Lin, D. T., Dayhoff, J. E., and Ligomenides, P. A. 1992. Adaptive time-delay neural network for temporal correlation and prediction, *Proc. SPIE Conf. Biological, Neural Net, and 3-D Methods,* November, 170–181.

Meirovitch, L. 1980. *Computational Methods in Structural Dynamics,* Sijhoff & Noordhoff, Netherlands.

Meirovitch, L. 1991. Hybrid state equations for flexible bodies in terms of quasi-coordinates, *AIAA J. Guidance, Control and Dynamics,* 14(5):1008–1–13.

Meirovitch, L. and Kwak, M. K. 1990. Dynamics and control of a spacecraft with retargeting flexible antennas, *AIAA J. Guidance, Control, and Dynamics,* 13(2):241–248.

Meirovitch, L. and Kwak, M. K. 1991. A Rayleigh-Ritz based substructure synthesis flexible multi-body systems, *AIAA Journal,* 29(10):1709–1719.

Meirovitch, L. and Quinn, R. D. 1987a. Equations of motion for maneuvering flexible spacecraft, *AIAA J. Guidance, Control and Dynamics,* 10(5):453–465.

Meirovitch, L. and Quinn, R. D. 1987b. Maneuvering and vibration control of flexible spacecraft, *J. Astronautical Science,* 35(3):301–328.

Michie, D. and Chambers, R. A. 1968. BOXES: An Experiment in Adaptive Control, *Machine Intelligence,* Dale E. and Michie, D. eds., 137–152.

Moody, J. and Darken, C. J. 1989. Fast learning in networks of locally-tuned processing units, *Neural Computation,* 1(2):281–294.

Narendra, K. S. and Parthasarthy, K. 1990. Identification and control of dynamical systems using neural network, *IEEE Trans. Neural Networks,* 1(1):4–27.

Waibel, A., Hanazawa, T., Hinton, G., Shikano, K., and Lang, K. 1988. Phoneme recognition: neural networks versus hidden Markov models, *Proc. IEEE Conf. Acoustics, Speech and Signal Processing,* April, 107–110.

White, D. A. and Sofge, D. A. 1992. *Handbook of Intelligent Control—Neural, Fuzzy, and Adaptive Approaches,* Van Nostrand Reinhold, New York, NY.

Yen, G. G. 1994a. Identification and control of large structures using neural networks, *Computers and Structures,* 52(5): 859–870.

Yen, G. G. 1994b. Reconfigurable learning control in large space structures, *IEEE Trans. Control Systems Technology,* 1(4): 362–370.

Yen, G. G. and Kwak, M. K. The design of neural controller for flexible multibody systems, *AIAA J. Guidance, Control and Dynamics,* submitted.

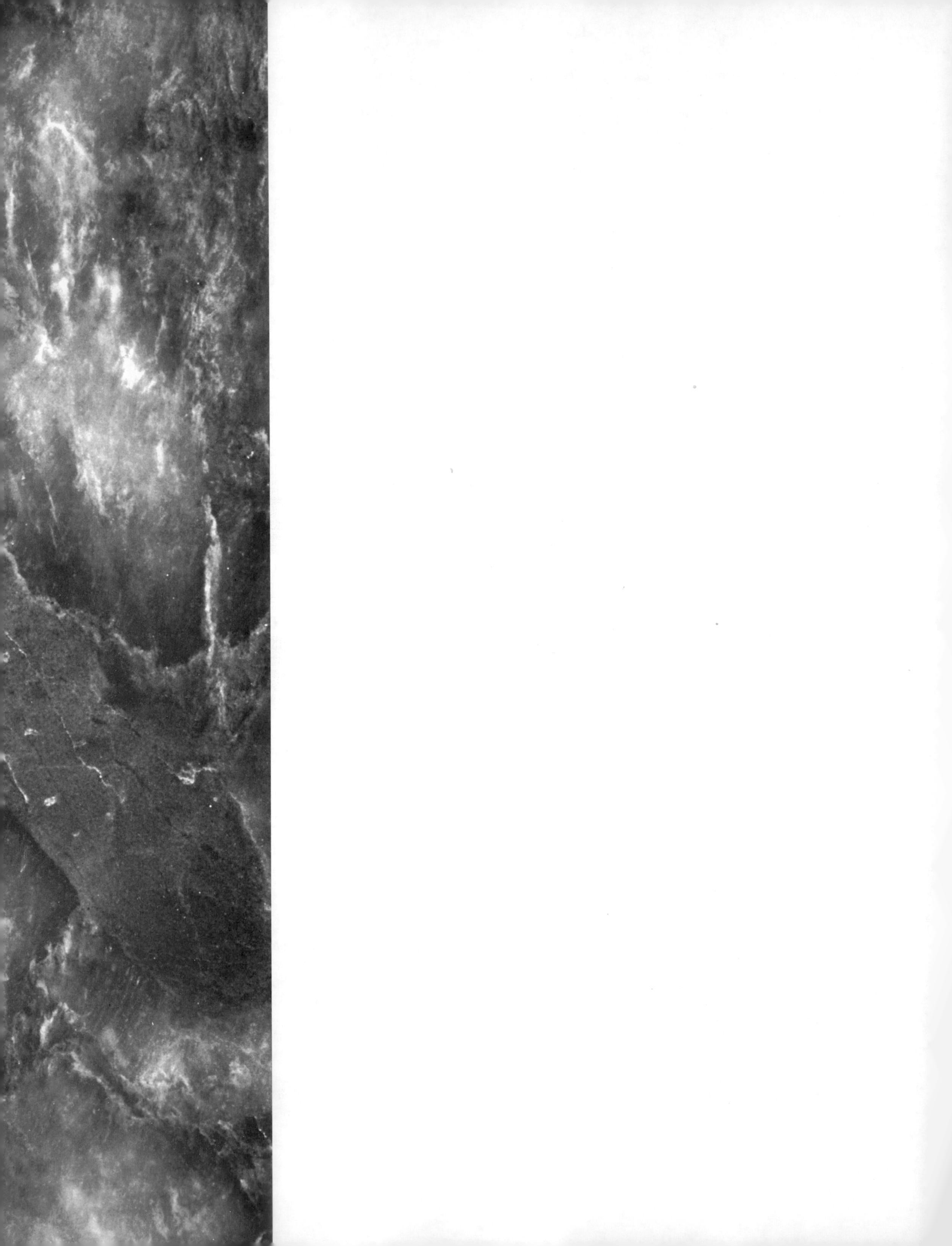

Emerging Technologies

Virtual Reality

111 **Virtual Reality** *Richard A. Blade, Mary Lou Padgett, Timothy Poston, Herschell Murry, Nadine Miner, Thomas Caudell, Johnny Evers, Charles R. White, and Hideyuki Takagi* 1383
Current Applications in Virtual Reality • The Virtual Workbench—A Path to Use for VR • Motion Tracking for Virtual Reality • Virtual Sound • Virtual Reality Systems • Fuzzy Logic Applications in Image Processing Equipment: Intelligent VR Futures

Asynchronous Transfer Mode for High-Speed Communication

112 **Asynchronous Transfer Mode Technology** *Thomas Lindblad* 1438
What is ATM Offering? • Why ATM? • What is ATM? • ATM Applications • The NEBULAS Project • Summary

113 **NEBULAS: High Performance Data-Driven Event Building Architectures Based on Asychronous Self-Routing Packet-Switching Networks** *Michele Costa, Jean-Pierre Dufey, Mike Letheren, Atsushi Manabe, Alessandro Marchioro, Christian Paillard, Denis Calvet, Kamel Djidi, P. Le Dû, I. Mandjavidze, P. Sphicas, Konstanty Sumorok, S. Tether, Leif Gustafsson, Klaudiusz Kobylecki, Kenneth Agehed, Solve Hultberg, Tawfik Lazrak, Thomas Lindblad, Clark S. Lindsey, Hannu Tenhunen, Martin DePrycker, B. Pauwels, Guido Petit, H. Verhille, and Michael Renard* 1444
Introduction • Technical Background • Computer Modeling • Event Building Protocols and Related Software Development • Hardware Development • Integration of Event Builder Demonstrators • Plan of Work

Micro Systems Technology

114 **Microelectromechanical Systems (MEMS)** *Yu-Chong Tai and Chang-Jin Kim* 1468
Introduction • Bulk Micromachining • Surface Micromachining • First Applications

115 **Micromachines** *Hiroyuki Fujita* 1472
Micromachines and the Scaling Effect • Difficulties in Miniaturization and Proposed Solutions • Microactuators • Architecture for MEMS: Autonomous Distributed Micromachines • Applications • Conclusion

116 **Selected Micromachining Fabrication Technologies** *A. Bruno Frazier, James Jara-Almonte, Noel C. MacDonald, M. T. A. Saif, S. A. Miller, Craig R. Friedrich, and Michael J. Vasile* ... 1489
Precision Metallic Micro Structures and Micro Molding Technologies • Nanotechnology • Precision Micromachining Technologies

117 **Microsensors** *Keith O. Warren and Antonio J. Ricco* 1515
Pressure Sensors and Accelerometers • Acoustic Wave-Based Chemical Sensors

118 **Micro Actuators and Energy Supply** *Toshi Fukuda and Fumihito Arai* 1526
Micro Actuators • Energy Supply Methods and Non-Contact Manipulation

119 **On-Board Power Supply and Remote Driving Mechanisms for Microelectromechanical Systems** *Jeong B. Lee* 1538
Power Requirements of Microelectromechanical Systems • On-Board Power Supply: Solar Cell Array • On-Board Power Supply: Microbattery • Remote Driving Mechanisms • Conclusions

120 **Si Micromachining in High-Frequency Applications** *Linda P. B. Katehi, Gabriel M. Rebeiz, Tom M. Weller, Rhonda F. Drayton, Stephen V. Robertson, and Chen-Yu Chi* 1547
Introduction • Applications • Fabrication Methodology • Membrane-Supported Distributed Circuits • Conformal Micromachined Packaging • Micromachined Lumped Elements • Conclusions

121 MEMS Integration—Technical and Economic Considerations *Janusz Bryzek* 1576
Introduction • Why MEMS Focus on Silicon • Market Growth Analogy: Transistors, Integrated Circuits and MEMS • Integrated MEMS Market Overview • To Integrate Or Not To Integrate • Mechanical On-Sensor-Chip Integration • Monolithic or Hybrid • Case Study: Lucas NovaSensor • Conclusions

Multisensor Fusion and Integration for Intelligent Systems

122 Multisensor Fusion and Integration for Intelligent Systems *Ren C. Luo, Michael G. Kay, Kota Takahashi, Hiro Yamasaki, Kazunori Umeda, Tamio Arai, Mark E. Kotanchek, James P. Helferty, W. Bosseau Murray, Charles Palmer, Zbigniew Korona, Mieczyslaw M. Kokar, Ryosuke Masuda, Michio Sasaki, and Karl Kluge* 1592
Introduction • Issues and Approaches of Multisensor Fusion and Integration • Audio-Visual Sensor Fusion System for Intelligent Sound Sensing • Industrial Vision System by Fusing Range Image and Intensity Image • Application of Data Fusion to Neonate Oxygenation Control • Multiresolution Multisensor Target Identification • Shaping Control of Plastic Object by Robot Hand with Sensor Fusion Processing • Multisensor System Integration for Autonomous Navigation Tasks • Future Trends for the Further Development in Multisensor Fusion and Integration

111
Virtual Reality

Richard A. Blade
University of Colorado, Colorado Springs

Mary Lou Padgett
Auburn University

Timothy Poston
Institute of Systems Science, National University of Singapore

Herschell Murry
Polhemus Inc.

Nadine Miner
Sandia National Laboratories

Thomas Caudell
University of New Mexico

Johnny Evers
USAF Armament Directorate

Charles R. White
Auburn University

Hideyuki Takagi
Kyushyu Institute of Design

111.1 Current Applications in Virtual Reality 1383
Elements of Virtual Reality (VR) • Enabling Technologies for VR • Current VR Applications • Future Directions

111.2 The Virtual Workbench—A Path to Use for VR 1390

111.3 Motion Tracking for Virtual Reality 1393
Overview • Magnetic Tracking Devices • Acoustic Tracking • Inertial Tracking Devices • Mechanical Tracking Devices • Optical Tracking Devices • Compass-Tilt Tracking

111.4 Virtual Sound 1397
Introduction • The Real-World Sound Process • Virtual Sound • Discussion and Conclusions

111.5 Virtual Reality Systems 1404
Introduction • What is Virtual Reality for Industrial Electronics? • Rendering Hardware • Software for Immersive Virtual Reality • Augmented Reality • Industrial Reality • Computational Intelligence (CI) Connections and Futures

111.6 Fuzzy Logic Applications in Image Processing Equipment: Intelligent VR Futures 1426
Introduction • Cameras • Camcorders • Photocopying Machines • Television Equipment • Codecs • Stepper Alignment in Semiconductor Manufacturing • Conclusion

111.1 Current Applications in Virtual Reality

Richard A. Blade and Mary Lou Padgett

Elements of Virtual Reality (VR)

The VR Workbench

Virtual reality near term can be realized by careful design and allocation of resources. Immersive interactive systems can be implemented in a variety of ways (L. and E. von Schweber, 1995). Styles of interactivity vary from fly-through VR to fully interactive VR. The former type of system allows six degree of freedom as the user "moves" through a world turning and looking in any direction. An example might be a walk-through of a CAD model for a planned product, say checking the views from the driver's and passenger's seats of a prototype car. An interactive VR system might allow the driver to adjust the rear-view mirror and side-view mirrors, adjust the bucket seat, and change the color of the upholstery. Levels of immersion also vary. The user can view the world through a window, be in the room, or be completely immersed in the world, from head to toe. A flatscreen monitor can supply a window, such as those used in popular flight simulator games. If glasses provide stereoscopic enhancements to give a feeling of depth, the VR style is called "in the room." A totally immersive VR system, may use some type of headmounted display (HMD) or a cave-like projection onto walls and ceiling to put the user visually into the world. A head-tracker communicates to the computer the head position. A display may change accordingly to whether floor, ceiling, or other surface should be in view.

A VR Workbench is the epitome of a practical implementation of very useful virtual objects. Reduction of detail, where it is unnecessary, and addition of useful capabilities is key to construction of VR systems which make optimal use of available resources. For example, a virtual tool needs a handle, not a graphical human hand holding it. The resistance needed for simulated tactile ability can be implemented. Haptic properties of VR systems add to realism and practicality as do odor and the more commonly used virtual sound and graphics.

Motion Tracking

Motion tracking is essential for interactive, immersive VR. The body position and orientation is needed for communication

0-8493-8343-9/97/$0.00+$.50

among the elements of the system. Graphical displays and 3-D sound projections can be altered to reflect the current, past, and predicted future states of the user with respect to other system elements.

Virtual Sound

Sound can greatly enhance the realism of a VR system, and extend the capabilities of the user. Complex information can be transmitted very quickly by triggering memories based on sound. Very good 3D simulations of sound can be achieved by filtering signals to adjust for differences between distance to the right or left ear, occlusions due to the head and upper body position, and even for ear shape. The shape and size of an individual's ears helps change the sound so that signals coming from the front are directed into the ear canal, but signals coming from the rear are muffled by the outside ear. Fine-tuning to an individual's ear size, shape, and hearing acuity can be enhanced by the addition of intelligent capabilities such as neural network algorithms.

Virtual Reality Systems: From Training Simulator to Intelligent VR

VR systems vary from enhanced training simulators to intelligent VR systems. The direction of emerging technologies for hybrid systems indicates that combining multiple algorithms and adding intelligence can help overcome current limitations. Time-critical computing and "graceful negotiated degradation" are key drivers (van Dam, 1995). Design-stage optimization (Sowrizal, 1995) by evolutionary systems techniques can improve speed and performance. Negotiated degradation is a strength of fuzzy neural control. Reinforcement learning often combines all the aspects of computational intelligence. Computer graphics, visual displays, body sensing, calibration and registration, non-visual sensory feedback display, and software tools are all important elements of VR (Caudell, 1994) which can be combined and enhanced in many ways. Scientific data visualization can be simple or elaborate. The principles of signal processing and analysis cannot be ignored in the process of evolving a virtual reality system from classical training simulator type applications. Moorhead and Zhu (1994) present many of the important considerations regarding sampling rate, reconstruction systems, and the human visual system. These need to be factored into selection of rendering hardware and VR software. Augmented reality, an interactive blend of graphical simulation and the reality, is very applicable to industry. Even simple, inexpensive software and hardware for PCs can be put together with little programming to provide a working engineer with good tools. These tools can be used in the design process or to develop a prototype for demonstration. Any engineer interested in learning about virtual reality can start with these tools and learn how to increase the interactive and immersive capabilities of computer applications today.

Enabling Technologies for VR

Telecommunications: ATM and NEBULAS at CERN

Developing high-speed telecommunications for virtual reality will help reduce the bottlenecks blocking ultra-realistic implementations of immersive virtual reality. The automotive industry, for example, would like to use virtual reality in the visualization of possible new car designs. The light reflecting off the car fender or its matte-finish dashboard should be accurate and believable (Adam, 1993). Medical applications such as virtual surgery require extreme accuracy. Other applications using communications from a technology center to remote locations or mobile teams also place heavy demands on telecommunications capabilities. The advances in progress at CERN are worth monitoring.

Fuzzy Logic Applications in Industry: Autofocus for Camcorders

The hybrid system, carefully engineered for flexibility, accuracy, speed, and economy, is the system of the future. Substantial improvements in performance with respect to these criteria can be achieved by using concurrent engineering design principles and incorporating fuzzy systems and soft computing. Many of the key design considerations in conservation of computer resources are determining the key points for focus and image registration. Commercially available camcorders use these techniques in their fuzzy autofocus systems. Integration of these techniques into intelligent virtual reality systems is critical to speedy enhancement of VR applications.

Current VR Applications

Factory Automation

Virtual reality (VR) in factory automation is expected to have a significant impact on all aspects of the discipline. Concurrent engineering is greatly assisted by the addition of interactive, immersive capabilities. In addition to the imagination of the visual implications of ideas, and to the verbal communication normally taking place in a design or manufacturing environment, VR allows manipulation of the environment. Design and data visualization, rapid prototyping, and validation and verification are enhanced by VR, which employs many senses. Possibilities range from use of a three-dimensional (3D) "joy-stick" or mouse to steer movement in a virtual environment to the addition of elaborate tracking capabilities. The "ding" and "bang" sounds of the common PC software are familiar, and provide instant alerts, warnings, and awards. This can be enhanced by the addition of realistic, even 3D, sound. Robotics offers an arena where practical addition of simulated human senses to an automated system can pay dividends. Robot vision, tactile sensing, and sense of smell are commercially useful attributes worth the investment of time and assets. Robot vision can be used for direction finding (controls) and pattern recognition (industrial inspection) in almost any application. Tactile sensing can help a robot (or an artificial hand) grip a can, lift a circuit board, or shake hands without crushing bones or dropping the object as readily. Texture and slippage detection are important assets for many applications. A sense of smell can detect obnoxious odors such as a selenium rectifier failure generates, or distinguish particles of pollution in the air too delicate for human detection. In the transportation industry drugs and explosives can be automatically detected in hard-to-access places without endangering human or dog lives. There are many instances where teleoperators perform tasks an unassisted human or hand-manipulated mechanism cannot do. The chapter on factory automation discusses many examples of the use of robots with simulated senses.

An example experienced by attendees of the 1994 Workshop on Neural Networks, Fuzzy Systems, Evolutionary Systems and Virtual Reality illustrates the emergence of intelligent VR as a practical tool. Using a large piece of equipment like the one illustrated in Figure 111.1 at Goddard Space Flight Center, a fuzzy joint controller lifted a dummy astronaut into position for simulated repairs to orbiting equipment. The scale of the virtual environment was full-size. Mechanical construction equipment was transformed into a fuzzy logic controlled device for positioning an astronaut strapped to the end of the jointed arm. The simulator was experienced by a visiting expert, Marilyn Panayi, who specializes in VR and training for the disabled. Panayi rated the experience as excellent. A description of the engineering design for the fuzzy controller can be found in the article by Lea, Jani, and Mica in the fuzzy logic and soft computing chapter.

Although the combination of computational intelligence with virtual reality is in its infancy, trends indicate that such integration of emerging technologies is the hallmark of the future.

Industrial Manufacturing

In Japan, where houses are "ordered" from a factory, VR has been used for some time to allow a walk-through of a potential house design (Kahaner, 1994; Kellar, 1993; et al., 1994). Other Japanese VR products include an exercise bicycle that moves through a virtual landscape, and a massage chair that provides relaxing images (Kahaner, 1994). A VR package in the US, called *Vegas*, also allows VR walk-throughs of building designs (Heichler, 1994), and a VR package called *Superscape* in the UK displays a model of a new housing development. Also in the UK, since April 13, 1993, the BBC Television News has been produced with a virtual studio in which there are no camera operators present (Hollingum, 1993).

VR has begun to find its way into advertising. *Virtual Voyage*, a 2.5 minute VR game promoting Cutty Sark scotch, was introduced May 14, 1994, at the National Restaurant Association show (Aho et al., 1994; Teinowitz et al., 1994). Nabisco Foods Group launched a Bubble Yum VR game in January, 1993 (Aho, 1994). A 30 second TV spot for Jolly Rancher goes after kids by simulating a VR experience (Fitzgerald, 1993). The *Design News* exhibit at the National Design Engineering trade show included a VR demonstration of an ornithopter and a lunar rover (Anonymous 5, 1993).

Scientific Visualization

A form of VR is called *scientific visualization.* Chemists are using VR to study molecular structures by generating stereographic images of molecules (Illman, 1994; Hann, 1993; Louchet, 1993). It is also used to visualize semiconductor lattice (Peterson, 1993), microcircuits, and solve various kinds of physics problems (Goldman, 1994; Hanson, 1992). At the University of North Carolina the "nanomanipulator" allows the observer to see him/herself floating over a surface as seen by a scanning electron microscope and even pick up atoms from the surface (Taylor, 1993).

Carver Mead's *Computation and Neural Systems Program* at CIT has worked extensively with scanners for visualizing activity of analog VLSI circuitry. For a tutorial description of how such scanners work and practical details of design and performance, see Mead and Delbruck (1991). Other work by this group is worth following. In discussions regarding directions for standards (Mead, 1994) the practical experience and sound design policies of Mead became obvious. He is a strong advocate of concurrent engineering and early marketing surveys to keep the needs of consumers in the forefront of product design.

Industry Applications

Aside from the airplane manufacturers mentioned before, there are other industries that are using VR in their design process. John Deer (Taubes, 1994), Ford Motor Co. (Keebler, 1993), Renault (Anonymous 6, 1993) are using VR in various aspects of their automobiles design. Also in the miscellaneous category, an advertising agency, Chiat/Day Inc., has a virtual office for its clients (Anonymous 7, 1994), the reactions of people to emergencies are being studied by fire researchers in the UK using VR (Geake, 1992), and fire fighters are being trained using virtual reality (Egsegian et al., 1993).

Power Industry

Detection of high-impedance faults on electrical distribution systems has been difficult to automate, but critical for control of operations. For example, when a "live wire" or energized conductor is ripped from its overhead position and dangles into the street or a yard, a severe safety hazard exists. The electric current in such instances is below threshold for normal detection and protection devices.

Research at Texas A&M University led to the recommendation for development of an integrated, multi-algorithm approach. Along with load current biased electromechanical relay, 3rd harmonic detection, mechanical catcher, and loss of voltage signaling approaches, neural networks are used to detect high-impedance faults. The most highly prized tool, discussed in Patterson (1995), is a digital model power system test facility which reproduces power system scenarios up to 30 minutes in length. Stored on optical disks, these scenarios enable the design optimization and evaluation of enhancements. Virtual reality makes testing intelligent fault detection, identification, and recovery systems feasible.

NASA

Probably the most ambitious projects in virtual reality exist in the general area of space exploration. VR has been successfully

Figure 111.1 Virtually walking in space to learn how to repair the Hubble Space Telescope.

applied to the training of astronauts at Goddard Spaceflight Center to the repair of the Hubble telescope. Development is moving ahead at Johnson Manned Spacecraft Center to replace the underwater simulation of zero gravity by virtual reality. Currently underwater divers provide the appropriate "pushes and pulls" on apparatus to provide the astronaut trainees with an approximation to the inertial properties of objects in outer space. Before the breakup of the Soviet Union there were plans for the late 1990's to launch a manned mission to orbit Mars for three years and explore the planet's surface via robots with a VR link to the orbiting spacecraft. The time delays for radio signals back to Earth make such a link with Earthbound explorers unfeasible. NASA Ames *Virtual Planetary Exploration* (VPE) testbed is developing VR simulators for Mars explorations (Hitchner, McGreevy, 1993; Hitchner, 1992), and there are discussions concerning the development of a *Telepresence Controlled, Remotely Operated Vehicle* (TROV) to roam the Martian surface with a VR link back to the base camp (Carlson, 1993; Pine, 1993). The use of VR is also being pursued by the European Space Agency (Bagiana and Mills, 1993), and the United Kingdom is exploring astronaut training using VR (Geake, 1992).

Aerospace Industry

The military and airlines industries continue to develop and use VR in flight simulators (Vince, 1993), and aircraft manufacturers are currently using VR to ensure maintenance access in the design of new aircraft (Esposito, 1992). In addition, Boeing is using enhanced reality (HMDs with see-through displays) to guide factory workers in laying out the wires in the enormous wiring harnesses used in aircraft. The HMD superimposes the correct path of each wire on the real wiring harness (Sims, 1994).

Military

Meanwhile, the military is increasing its use of VR in training. The army is using VR tank simulators that incorporate personal computer technology (Halle and Mariani, 1994), VR battlefield simulations are performed (Smith, 1994; Burdick, 1993), and future communications technology permits joint exercises worldwide in VR (Dennehy et al., 1994; Anonymous, 1994). Finally, hazardous operations training, such disassembling nuclear weapons, is done with VR (Kiernan, 1994).

Medical Instrumentation

The field of medicine continues to invite VR innovations. Surgical training, particularly that involving videoscopic surgery, increasingly uses simulators (Machlis, 1994; Merril et al., 1994; McGovern, 1994; Carroll, 1994; Dunkley, 1994; Anonymous 2, 1994; Sinclair, and Peifer, 1994). Probably the most exciting innovations are related to telemedicine. The Army is seriously studying remote battlefield monitoring of vital signs and telesurgery (Satava, 1993). Researchers at Ipswitch Hospital in England have developed and are currently testing a telesurgery system (Field, 1993). Therapeutic applications of VR include interactive talking cartoon characters on closed circuit TV for sick children (Mestel, 1993) and treatment of phobias using VR images for desensitization (Goddard, 1994; Williford et al., 1993).

Human Interfaces and Training

Training astronauts to repair the Hubble Space Telescope is the purpose of the NASA Johnson Space Center device pictured in Figure 111.1. Homan and Bell of the Automation and Robotics Division of JSC provided input by the *VR Special Report* article by Testa, "Virtually Walking in Space" (Testa, 1994). There are many challenges in simulation and training which have been met for years with hardware-in-the-loop simulators and a great deal of art and luck. High caliber VR simulators with proper speed and focusing can greatly enhance the comfort of the student and the probable survival of a pilot-in-training. At a simulation conference tour of a domed flight simulator facility (NAVY), test runs down a canyon were made by volunteer attendees. Each "flew" a strafing run down a canyon, trying to eliminate bridges. When the last volunteer used careful observations of the others' errors to fly straight down the gorge and HIT a bridge, the resulting explosion of bright light and sound was so startling that the "pilot" jumped and veered to the left, straight into the wall of the canyon. Success and then death! There is no way to imagine the flaring impact of a direct hit and explosion. This example of the effectiveness of virtual environments for training started the search for ways to make such reality attainable by the practicing engineer, learning on inexpensive home equipment.

There are many tools, gadgets, and inexpensive software and hardware pieces available to a working engineer interested in learning more about virtual reality without making a $50,000 investment. Suggestions for learning how to engineer a software-hardware system with some rudimentary virtual reality capabilities are given by Padgett et al., in this chapter.

Handicapped individuals are in the forefront of VR applications. VR wheelchair simulators are used to train patients (Anonymous 3, 1994; Anonymous 4, 1994; Machlis, 1992). A novel application of augmented reality puts a virtual image of regular spots on the floor of a Parkinson's disease victim. By focusing on those spots the victim is able to avoid the walking problems associated with the disorder (Dutton, 1994). A VR simulator is being used to perform virtual wheelchair tours of buildings in the design stage (Krumenaker, 1994), and studies and proposals abound to use VR for the disabled (Murphy, 1992).

Although industrial and biomedical virtual reality specialists prefer practical applications devoid of "hype", the existence of the entertainment industry and its explosive expansion cannot be ignored. Much of the arcade game type "VR" experiences are poor in quality. Nevertheless, the tremendous market for this type of toy is a stimulus to the economy. Manufacture of dual-use equipment or software can increase the market for a product so that the quick and dirty implementation focused on entertainment can finance speculative investments and careful research into use of virtual reality in industrial electronics.

Educational uses of VR outside the entertainment industry have been primarily focused on the handicapped as well (Sklaroff, 1994; McLellan, 1993; Holdsworth, 1993; McKeown, 1992). A particularly interesting innovation is a device being developed by Biocontrol of Palo Alto, California, that translates electrical signals from the head and face (EMG and EEG) to control a screen cursor or make musical sounds (Lusted et al., 1993). This provides severely impaired persons a control over their environment unequaled by other means. Connected to the arms and other parts of the body, it provides a person the ability to create music by body motions. There are also attempts to bring VR into the classroom for nonhandicapped students. An example is a VR system to do physics experiments that would be impossible in the real world (Yam, 1993). Additional applications being developed are a virtual university, a virtual library (Kniffel, 1993; Kurzweil, 1993; Oppenheim, 1993), and a networked multicultural virtual art museum (Loeffler, 1993).

There is great activity related to VR in the arcades and theme parks. Theme parks are discovering that VR rides are much less expensive than conventional rides and much safer and less expensive to build. Universal Studios has a *Back to the Future* flying automobile simulator; Disney World has changeable rides in its Pleasure Island (Asch, 1994); Luxor Hotel in Las Vegas has a futuristic aircraft ride (Patton et al., 1994), and companies like Sega are involved in developing a new generation of VR rides. Most of the rides involve large scale screens with images projected on them, in contrast to the head-mounted displays used in many of the arcade games.

Local shopping malls are importing VR technology into their arcades (Schuytema, 1993; McGrath, 1993; Corliss, 1993). Besides flight simulators and battle games, there are sports simulators like skiing (Lerman, 1993), baseball (Kinnaman, 1992), and golf (Akins, 1994; Puttre, 1993). Additional VR entertainment includes a VR simulation of Shakespeare's Globe Theater (Coughlan, 1994; MacRae, 1994), a VR museum (Teixeira, 1994), a simulated concert hall (Soviero, 1994), and architectural tours in VR (Johnston, 1994). A VR version of *Legend Quest* exists at a role-playing center in Nottingham, England (Whittington, 1992).

VR is moving into home entertainment with such things as Apple Computer's *Quicktime VR* (Lewis, 1994; Carlton, 1994; Lewis, 1994) and various platforms by Sega, Atari, 3DO, and others (Willcox, 1993). A new stereoscopic TV that requires no glasses will undoubtedly play a role in future home VR, as will plans to distribute VR simulations over telephone lines.

Future Directions

As this is being written, the number of commercial applications of VR becoming available is enormous mostly because of the increasing computational power available to the general public. However, many are being developed by small, undercapitalized companies which may not survive long, and the lack of standards in the VR industry continues to deter the development of a robust market. As noted in (Adam, 1993) an automotive industry-based report notes "there is a pressing need from potential industrial users for standardized measures of performance that would help select expensive equipment for a specific application." To contribute to discussions about appropriate standards, contact the *CI Standards News* (m.padgett@ieee.org or r.blade @ieee.org).

References

Adams, J. A. 1993. Virtual reality is for real, *IEEE Spectrum*, October.

Aho, D. 1994. What's hot in interactive marketing?, *Advertising Age*, 65(5):17.

Aho, D. and Teinowitz, I. 1994. Cutty Sark sets a virtual sail, *Advertising Age*, 65(20):16.

Akins A. S. 1994. Golfers tee off into the future, *Futurist*, 28(2):39–42.

Anonymous 5 1993. Magazine's booth exhibits engineering breakthroughs, *Design News*, 49(9):96–98.

Anonymous 6 1993. New-age transport: trains, planes and automobiles, Economist, 329(7843):96–98.

Anonymous 1 1994. DIS mixes real, virtual, *Aviation Week & Space Technology*, 140(19):73.

Anonymous 2 1994. Simulator for eye surgery, *R&D*, 36(2):146.

Anonymous 3 1994. Disabled people use virtual reality, *News for You*, 42(23):3.

Anonymous 4 1994. In virtual reality, tools for the disabled, *The New York Times*, April 13, Sec. C, 1, col. 1.

Anonymous 7 1994. Virtual office runs on telecommunications, *Managing Office Technology*, 39(6):57–58.

Asch, T. 1994. CyberTron: first permanent immersive VR systems installed at disney world, *Virtual Reality World*, 2(3):18–20.

Bagiana, F. and Mills, S. 1993. Virtual Reality for European Space Programmes, *Proc. 3rd Ann. Conf. Virtual Reality*, 138–43, Meckler, London, UK.

Burdick, C. D. 1993. Seamless simulation mixing live and virtual simulations, *1993 Winter Simulation Conf. Proc.*, 996–1002, IEEE, New York, NY.

Carlson, S. 1993. Virtual Mars? *Ad Astra (GADS)*, 5(1):59.

Carlton, 1994. Apple Unveils Technology that lets Users take 'Tours' of Places, Buildings, *The Wall Street Journal*, June 8, Section B, p. 6.

Caroll, L. 1994. Virtual Reality Shapes Surgeons' Skills, Medical World News, 35(2):26–27.

Caudell, T. P. 1994. The Application of Neural Networks to Virtual Reality, ICNN Tutorial Number 13, WCCI, Orlando, FL 1994.

Corliss, R. 1993. Virtual, Man, Time, 142(18):80–83.

Coughlan, S. 1994. Wherefore Art Thou, Super Romeo?, Times Educational Supplement, Issue 4052, February 25, p. SS17A.

Dennehy, M. T., Nesbitt, D. W., and Sumey, R. A. 1994. Real-Time Three-Dimensional Graphics Display for Antiair Warfare Command and Control, Johns Hopkins APL Technical Digest, 15(2):110–119.

Dunkley, P. 1994. Virtual Reality in Medical Training, Lancet, 343(8907):1218.

Dutton, G. 1994. Perpetual Pathway, Popular Science, 245(1):34.

Dysart, J. 1994. Wall Street Meets VR: Animated Investment Tracking, Virtual Reality World, 2(5):22–25.

Egsegian, R., Pittman, K., Farmer, K., and Zobel, R. 1993. Practical applications of virtual reality to firefighter training, *Proc. 1993 Simulationa Multiconf. Int. Emergency Management and Engineering Conf.* 155–60.

Enomoto, N., Nagamachi, M., Nomura, J., and Sawada, K. 1994. Virtual Kitchen System using Kansei Engineering, Human-Computer Interaction. Proceedings of the Fifth International Conference on Human-Computer Interaction, 2:657–562.

Esposito, C. 1992. Virtual Reality Research at Boeing, WESCON/92 Conference Record, 397–398.

Field, R. 1993. Surgeons Perform from a Remote Location, Medical World News, 34(2):35.

Fitzgerald, K. 1993. Jolly Rancher Takes Taste Message to New Level, Advertising Age, 64(11):44.

Flanagan, William G., and Contavespi, V. 1992. Cyberspace Meets Wall Street Forbes, 149(13):164–168.

Focardi, S. 1993. Virtually Reality in Scientific Computing Centers, Pixel, 14(5):18–21, (Italian).

Geake, E. 1992. Virtual Emergency Gives Clue to People's Behavior, *New Scientist (GNSC),* 136(1847):21.

Geake, E. 1992. Britain Urged to Coordinate Reality Research, New Scientist, 136(1851):18.

Goddard, A. 1994. Virtual Therapy Reaches New Heights, New Scientist, 142(1929):6.

Goldman, J. and Roy, T. M. 1994. The Cosmic Worm, IEEE Computer Graphics and Applications, 14(4):12–14.

Halle, R. and Mariani, D., 1994. Crewman's Associate Advanced Technology Demonstration, Proceeding of the SPIE, 2219:34–41.

Hann, M. and Hubbard, R. 1993. Molecular Visualization in Pharmaceutical Research, *Proc. 3rd Ann. Conf. Virtual Reality,* 122–125, Meckler, London, UK.

Hanson, A. J. 1992. Seeing the Right Picture: Graphics and Visualization for High Energy Physics, Proceedings of the International Conference on Computing in High Energy Physics '92, CERN, 90–95.

Heichler, E. 1994. Virtual Crowd's Add Dimension to Emergency Simulations, Computerworld, 28(10):82.

Hitchner, L. E. 1992. The NASA Ames Virtual Planetary Exploration Testbed, WESCON/92 Conference Record, Electron. Conventions Management, Ventura, CA, 376–81.

Hitchner, L. E. and McGreevy, M. W. 1993. Methods for User-Based Reduction of Model Complexity for Virtual Planetary Exploration, Proceedings of the SPIE, 1913:622–36.

Holdsworth, N. 1993. Driving to a New World, Times Educational Supplement, Issue 4014, June 4, p. SS17.

Hollingum, J. 1993. BBC Sets Daleks to Work, Industrial Robot, 20(5):26–28.

Illman, D. L. 1994. Researches make Progress in Applying Virtual Reality to Chemistry, Chemical & Engineering News, 72(12):22–25.

Johnston, S. J. 1994. Virtually Reality Takes Architectural Leap, Computerworld, 28(25):72.

Kahaner, D. 1994. Japanese Activities in Virtual Reality, IEEE Computer Graphics and Applications, 14(1):75–78.

Karlsson, K. 1991. Turbo-ISVAS: an Interactive Visualization System for 3D Finite Element Data, *Advances in Scientific Visualization,* 68–75, Springer-Verlag, New York, NY.

Keebler, J. 1993. Cyberspaced Out, Automotive News, 67(5523):8i.

Kellar, D. 1993. Virtual Reality, Real Money, Computerworld, 27(46):70.

Kiernan, V. 1994. Bomb-breakers Play Safe in a Fantasy World, New Scientist, 141(1916):9.

Kinnaman, D. E. 1992. Batter Up, Technology & Learning, 12(7):78.

Kniffel, L. 1993. Cal State U. Freezes Construction to Ponder Virtual Library, American Libraries, 24(8):692–694.

Krumenaker, L. 1994. Roll-Through Blueprints, Popular Science, 244(5):48.

Kurzweil, R. 1993. The Virtual Library, Library Journal, 118(5):54–55.

Lewis, P. H. 1994. Science Times: Virtual Reality Plans to Grow More Real, The New York Times, June 14, Section C 13.

Lerman, J. 1993. Virtue Not to Ski?, Skiing, 45(6):20.

Loeffler, C. 1993. The Networked Visual Art Museum, Bulletin of the American Society for Information Science, 19(1):13–14.

Louchet, J. 1993. Molecule Synthesis and Animation, *Informatique '93 (2nd Int. Conf. Interface to Real and Virtual Worlds.),* 425–432, in French.

Lusted, H. S., Knapp, R. B., and Lloyd, A. M. 1993. Applications for Biosignal Processing in Virtually Reality, Proceedings of the Third Annual Conference on Virtual Reality, Meckler, London, UK, 134–137.

Machlis, S. 1992. Computers Create a New Reality, Design News, 48(20):60–70.

Machlis, S. 1994. Virtual surgery: Computers Promise Better Training, Techniques, Design News, 49(11):44.

MacRae, A. C. 1994. The Virtual Globe Theatre, Virtual Reality World, 2(5):40–43.

Maples, C. and Peterson, C. 1995. MUSE (Multidimensional, User-oriented Synthetic Environment): A Functionality-Based, Human-Computer Interface, The International Journal of Virtual Reality, 1(1):2–9.

McGovern, K. T. 1994. The Virtual Clinic, a Virtual Reality Surgical Simulator, Virtual Reality World, 2(2):43–44. Also McGovern, Kevin T, 1994, Applications of Virtual Reality to Surgery, British Medical Journal, 308(6936):1054–1055.

McGrath, R. 1993. Virtuality Puts Retailer on New Plane, Advertising Age, 64(8):25.

McKeown, S. 1992. Learning Through a Looking-Glass Universe, Times Educational Supplement, Issue 3984, November 6, p. SS13.

McLellan, H. 1993. Virtual Reality Goes to School, Computers in the Schools, 9(4):5–12.

Mead, C. A. 1994., Personal communications.

Mead, C. A. and Delbruck, T. 1991. Scanners for Visualizing Activity of Analog VLSI Circuitry, Analog Integrated Circuits and Signal Processing 1, 93–106, 1991, Kluwer Academic Publishers, Boston, MA.

Merril, J., Allman, S., Merril, G. and Roy, R. 1994. Virtual Heart Surgery: Trade Show and Medical Education, Virtual Reality World, 2(4):55–57.

Mestel, R. 1993. Virtual Actors Help the Medicine Go Down, New Scientist, 139(1889):9.

Murphy, H. J. 1992. *Conference Proceedings: Technology and Persons with Disabilities,* California State Univ., Northridge, CA.

Oppenheim, C. 1993. Virtual Reality and the Virtual Library, Information Services & Use, 13(3):215–227.

Patterson, R. 1994. Signatures and Software Find High-Impedence Faults, IEEE Computer Applications in Power, V. 8, N. 3, July 1995, 12–15.

Patton, P., Britton, P. and Hutsko, J. 1994. Now Playing in the Virtual World, Popular Science, 244(4):80–85.

Peterson, I. 1993. Wandering into Virtual Physics, Science News, 143(14):220.

Pine, D. 1993. The Next Best Thing, Popular Science, 242(1):23.

Puttre, M. 1993. Teeing Off Indoors: Virtual Golf, Mechanical Engineering, 115(81):56–57.

Ribarsky, W., Bolter, J., Op den Bosch, A., van Teylingen, R. 1994. Visualization and Analysis Using Virtual Reality, IEEE Computer Graphics and Applications, 14(1):10–12.

Sadowsky, J. and Massof, R. W. 1994. Sensory Engineering: the Science of Synthetic Environments, Johns Hopkins APL Technical Digest, 15(2):99–109.

Satava, R. M. 1993. Surgery 2001: a Technologic Framework for the Future, *Proc. 3rd Ann. Conf. Virtual Reality,* 101–105, Meckler, London, UK.

Schuytema, P. 1993. Inside a Virtual Robot, Omni, 15(11):27.

Sims, D. 1994. New Realities in Aircraft Design and Manufacture, IEEE Computer Graphics and Applications, 14(2):91.

Sinclair, M. and Peifer, J. 1994. Socially Correct Virtual Reality: Surgical Simulation, Virtual Reality World, 2(4):64–66.

Sklaroff, S. 1994. Virtual Reality Puts Disabled Students in Touch, Education Week, 13(36):8.

Smith, R. D. 1994. Current Military Simulations and the Integration of Virtual Reality Technologies, Virtual Reality World, 2(2):45–50.

Soviero, M. M. 1994. Walk-through Mozart, Popular Science, 244(2):35.

Sowizral, H. A. 1995. Using a Rendering Pipeline Efficiently, ACM SIG-Graph Tutorial, 1995.

Taubes, G. 1994. Virtual Jack, Discover, 15(6):66–74.

Taylor, R. M. II, Robinett, W., Chi, V. L., Brooks, F. P., Jr., Wright, W. V., Williams, R. S., Snyder, E. J. 1993. The Nanomanipulator: a Virtual-Reality Interface for a Scanning Tunneling Microscope, Computer Graphics Proceedings, ACM, 127–134.

Teinowitz, I. and Aho, D. 1994. Cutty Sark Sails into High-Tech, Advertising Age, 65(10):1, 49.

Teixeira, K. 1994. Behind the Scenes at the Guggenheim, Virtual Reality World, 2(3):66–70.

Testa, B. 1994. Virtually Walking in Space, Virtual Reality Special Report, December 1994, 69.

van Dam, A. 1995. VR s a Forcing Function: Software Implications of a New Paradigm, SIG-Graph Tutorial Proceedings, 1995.

Vince, J. 1993. VR Impacts in Flight Simulation, Proceedings of the Third Annual Conference on Virtual Reality, Meckler, London, UK, 106–110.

von Schweber, L. and von Schweber, E. 1995. Virtual Reality: Virtually Here, PC Magazine. 168–183.

Whittington, Amanda, 1992. Fun with Eric the Spider, New Statesman & Society, 5(204):33.

Wilcox, J. K. 1993. Future Games, Popular Science, 170(12):108–109.

Williford, J. S., Hodges, L. F., North, M. M., and North, S. M. 1993. Relative Effectiveness of Virtual Environment Desensitization and Imaginal Desensitization in the Treatment of Acrophobia, Proceedings Graphics Interface '93, 162.

Yam, P. 1993. Surreal Science, Scientific American, 268(2):103–104.

111.2 The Virtual Workbench—A Path to Use for VR

Timothy Poston

The "standard model" of VR is immersive, with goggles and head motion detection. With present technology this suffers from the seriously low resolution of the goggles, their narrow field of view, and the heavy computational costs of continuously responding to the change of viewpoint. (For realistic "virtual cathedral tours," for instance, fine detail in distant carvings—necessary when they are approached—is wasted. Draw all those polygons anyway? Create elaborate "simplifier" routines? All alternatives are costly in CPU and/or programming time.) These costs have diverted computing resources from the features most needed in making VR a convenient tool. (After my own first experience of VR goggles, I was happy to remove them after a few minutes, despite being excited by the *potential* they represented. Routinely working in that visual environment is discomfort even to imagine.)

For generalized Dungeons and Dragons the sense of maneuverable personal presence is vital—chase, run, hide, look behind you . . . The field has been to some extent driven by the game players, as 1960s space efforts were by Space Race players. The moon voyages did not generate payoffs that kept them going; virtual games have a consumer side, but are they enough to pay for VR?

The less dramatic space economy has grown steadily. I wish to suggest a VR approach more analogous to communication satellites than to manned landings; more immediately practical with current technology, and much closer to monetary payoff, but still establishing a strong beach-head in (virtual) space, that can ripen to a comfortable full presence.

In many applications, a continuously mobile viewpoint is not necessary. The biologist working on a single cell, using a binocular microscope and micromanipulators, is effectively projected into the physical reality of interest, despite having a fixed viewpoint through the lenses. Watch a jeweller or sealmaker at work: how often does the head move? Similarly, one can look into a hands-on virtual space by an arrangement such as the Reality Box we have constructed at ISS, as shown schematically in Figure 111.2. High resolution and a wide field of view are easily achieved at a cost much less than for still-inadequate current goggle technology. Stereo is achieved by time-splitting the display with Crystal

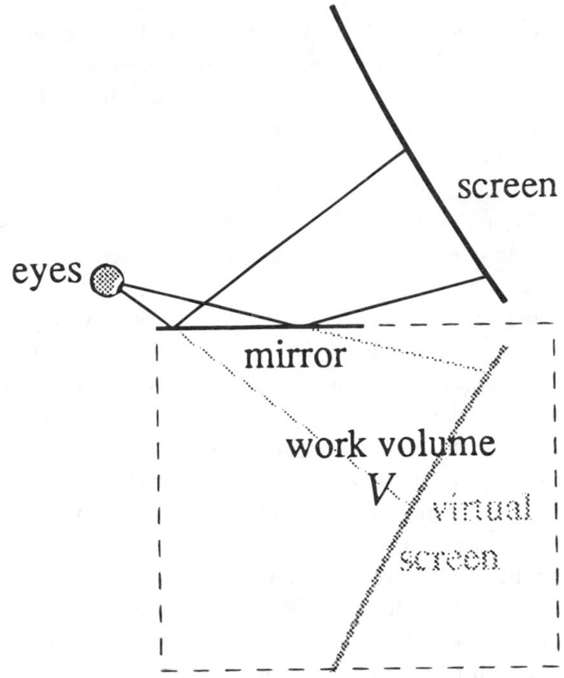

Figure 111.2 Virtual Reality Box.

Eyes™ glasses. This display configuration allows the real hands to move in the positions at which the eyes seem to look. Mammals have multiple 3D systems—visual, neuromuscular, aural—and human dexterity is based on unusually strong integration of the first two. Fine control depends on exploiting this.

The emphasis must be not merely on placing objects, but on transforming them (which is where the computing resources freed by fixing the viewpoint would now be dedicated). Virtual table-tennis etc., need only objects pre-defined by the game rules. It is productive crafting that will provide the true economic engine for VR. Virtual crafting—in CAD/CAM, plastic surgery, cloud painting and so on—will need far greater subtlety in the contents of the virtual space. This subtlety should be developing now, with feedback from now-possible hand-eye coordinated Virtual Tools; the combination of these tools with the Reality Box is the Virtual Workbench. When and if the immersive environments become livable for hours on end, there will then be more productive things to put in them than dragons and office furniture.

It is necessary to reach into this work volume and work, but it is *not* necessary to place a virtual hand there. From the craft viewpoint, a visible hand is a nuisance, masking the workpiece: it is vital to see where the tool is, and feel where the tool is, but why draw a picture of the hand that holds it? To a serious craft worker the tool is already an extension of the hand and the body, and need not be obscured. It is not merely a waste of ingenuity and CPU time to display the hand, it is a positive obstruction of effective work. Our approach is to use a universal "tool-handle" controlled by the hand, which in the Reality Box acquires different business ends; a blade, a brush, a tensor bender, and so on.

We model the handle, not the hand.

The next step, to force feedback, is far more practical with this approach. Pressure pads in a data glove cannot prevent the hand from moving through a virtual solid wall. Positional measurements cannot determine how hard the hand is pressing the wall, so what should the pressure be anyway? Only wild elaboration of muscle sensors along the arm could begin to estimate that. Even then the hand cannot be blocked; resistance is felt as much in the pushing muscles of the arm as it is in the pressure sensors of the fingers, and can only be made to feel right by truly preventing the physical hand from moving further. A wall must be represented by true resistive forces, mechanically delivered. A surgeon cutting flesh needs to know that the scalpel has reached bone by the fact that it cannot be forced further. To deliver resistance to every glove-fingertip is computationally hard, and intricate as micro-engineering to the point where the glove becomes very expensive, and very easily damaged. (Imagine a glove that can deliver the feeling of touching silk, then press your finger and thumb together, hard. Hear the crunch?)

To deliver computed forces to a six-degree-of-freedom handle is a far more reasonable task, computionally involving mainly a model of the workpiece and its elastic, etc. properties—where effort should anyway be concentrated—and engineering on a far less micro-fragile scale. The experienced tool user already feels as though there are nerve endings in the tooltip; you can sense textures through your fingernails, which are dead material. All the forces on individual fingers, the resistance experienced by arm muscles pushing the hand, and so on—the whole experience of "something solid there"—are provided gratis by Newtonian biomechanics, once the correct forces are delivered to the rigid toolhandle: six-degrees-of-freedom force feedback, not the innumerable degrees of freedom of skin, are enough. Simulated touch would be nice in virtual clay modelling and massage, but for most economically important uses of VR it is a hard-to-attain luxury, whose difficulty should harm VR as little as the problems of a Mars mission harm satellite TV.

Model the handle, not the hand.

Most small workbench tools can be described as 'handle plus business end.' Here, the business end would be purely virtual, chosen from a menu: scalpel, spatula, rasp, pressor, gripper, . . . or meta-tools, to be described below. A real handle is often a simple rod, sometimes shaped to the hand in ways convenient for transmission of the necessary forces. For delicacy of control, either a pen-style grip or a palm-fitting handle like a jeweller's graver, with a business end guided by thumb and forefinger, would be most convenient here. Even for tools without force feedback, the sense of fingertip control is much stronger through a rigid tool than through any glove too coarse to transmit texture. The hands can function at a natural position for benchwork, not uncomfortably raised as for a typical screen light-pen.

The possible uses for a Virtual Tool that can cut, pare down, press into shape, pierce, polish, *etc.* are obvious, since these are capabilities that humanity has had in its toolkit since the first experiments with flint and bone. A cave-dweller could quickly learn to use the virtual version, and to enjoy features such as *undo* to retract errors, forces experienced in logarithm to combine subtlety and strength, and *enlarge the workpiece by a factor of 10* to work in detail on a part that requires it. More subtle are the *tensor bender* tool options that change material properties *in situ* (and for which force feedback is not important, though the hand's sense of their location is). For example, a loop of stiff wire will naturally form a circle; apply a *softening* tool to reduce stiffness locally, and it relaxes to a new shape with a rounded corner, as in (Figure 111.3). Held close to the workpiece the tool would have a more local effect, producing a sharper corner. A jeweller does something similar with heat (which likewise can be adjusted to a narrow flame or to area heating, and which likewise involves no force feedback from the workpiece), and with certain special alloys can do the converse (form a shape by hand—gently, to avoid generating heat—and then fix it by applying a flame) but only with particular materials, and only isotropically. A Virtual Tool could adjust the properties of a virtual form that would later be physically manufactured in any material, once the shape was fixed; it could cause an area to flatten out by raising the virtual bending energy; it could "fibre" the surface, making it hard to bend along the stiff fibre direction, easy to bend transverse to that. These elasticity properties would be fully virtual, for shape control; the system should also be able to switch to answering the question, "If this object is physically made, by injection moulding/carbon fibre spinning/milling/ . . ., how will it elastically 'feel'?"

Note that the implied model of an object, in the above paragraph, is a variationally responsive one, which will adapt its shape *most* where modified, but *somewhat* everywhere unless specific constraints have been applied. This is computationally more expensive than, say, a piecewise spline surface that the user adjusts region by region, leaving untouched regions fixed. The paradigmatic CAD/CAM problem is to attach a side branching tube to a main one, smoothly, and much ingenuity has been spent on making an evenly rounded join; but the result is invariably ugly, because even if many derivatives match and the eye cannot see the exact curve where patches join, the two cylindrical tubes lack the harmony of a tree trunk and a branch that grows from it, each modifying the other in a way that fades with distance but does not wholly vanish. (Mathematically, the solution of such variational problems tends to produce an *analytic* form, for which any small part implies the whole, and the whole thus has a coherence and integrity rare in "piece-wise" assembled objects.) As this illustrates, the Virtual Workbench should have heavy, parallelized computing power available to devote to the properties of objects. To spend almost all resources on realistic display of simple, rigid objects is to let the tail of the reality wag the dog.

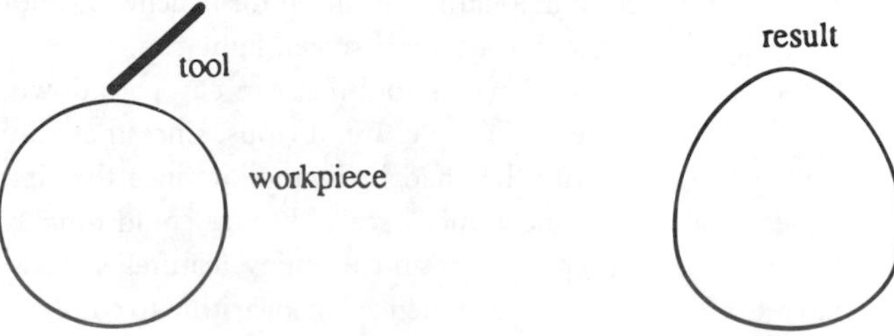

Figure 111.3 A 'springy' virtual loop reacts to local softening.

For each model of shape growth and deformation—triangulated surface with elastic properties, particle system, L-grammar, or whatever—the Virtual Work-bench would have matching Virtual Tools, capable of locally modifying the model parameters.

Let us consider some possible applications of the Virtual Workbench, illustrating the power of integrating VR with a serious model of an object that is to be made or modified.

Virtual surgery: Both the simulation of open surgery and the computer-aided design of prostheses such as knee and hip replacements would gain more from developments in the manipulability of the virtual objects than from superior maneuverability of the viewer. Hip bone prostheses, for example, are natural not quartic-patch forms; designing them needs good Virtual Tools operating on responsive shape models. Similarly, a 3D virtual shattered bone could be rebuilt using a transparent unbroken one as a template for the jigsaw work involved; but because no standard bone would be a perfect match, it should be adjustable to fit the evidence—as it grows—from the reassembled fragments. Pieces carried away by the accident or gunshot could be detected ahead of an operation, allowing planning for bone graft or prosthesis as necessary, and helping in their design.

It is precisely to achieve such fine-manipulative control in a medical context, over a practical timescale and at a practical cost, that we are doing Virtual Workbench development work at the Institute of Systems Science, as a part of our work on information-enhanced 3D medicine for use in the hospitals with whom we collaborate.

Minimally invasive surgery: This surgery is tele-immersive, but the motions of the camera in current and near term technology (move the guide tube forward and back, twist it) correspond poorly to head motions and cannot usefully be slaved to them. When a good viewpoint is found, the surgeon fixes it. The fixed viewpoint of the Reality Box is thus an appropriate match for both simulation and enhanced-view telepresence (for instance, seeing computer-added ribs "through" flesh makes navigation easier) in such fine-scale surgery.

Vehicle design: A car sells, above all, visually; the designer should be able to handle it in virtual form, like a sculptor. The variational approach described above could make it easier to produce convincing, coherent, elegant forms, and for this specialist and high investment purpose the modelling should include the air around it, which affect speed, fuel efficiency, stability in wind gusts, and so on. (The same applies, *a fortiori,* to other vehicles such as boats.) Virtual colouring tools could allow continuous variation, constrained to be realizable by paint-blend control in shopfloor painting robots, opening a whole new dimension for appearance and customization.

Plastic surgery: This should not work simply with a face surface; it should use a realistic model of the face, including skull, cartilage and muscle, and numerically predict the effects of growth—childhood defect removal often gives results that fall apart after a few years. To this one would add response to the 'pseudo-forces' exerted by Virtual Tools, changing the size and shape of these components, judging the effect, and carrying it into the future. This will allow much better long-term planning, effectively code possible faces directly by the surgical changes that would produce them, rather than leave an inverse problem of deducing the surgery from the required change, allow running "smile," 'frown' etc. routines through the model, so that the face could be considered in motion—the normal condition of a waking human face. While different humans would put different expressions into the same face, this would at least indicate the visual repertoire available to the modified person.

A similar, simpler system could be used in planning dental prostheses, and allow such dynamic features as bite tests.

Customized Clothes Design: Scan an individual's body shape, and input a "generalized" pattern, specifying (modifiable) shapes to be cut and sewn from flat cloth. The system would model the effects of body and gravity on the garment, while specialized Virtual Tools modify seams, add or remove darts and tucks, adjust cloth type, and so on. As in the case of plastic surgery, using tools that work through the right parameters, one produces full specifications for producing a desired effect in the non-virtual world in the very act of finding and choosing that effect in virtuality.

This is clearly a tiny subset of the range of commercially useful applications of the Virtual Workbench. Some of them would be enhanced by moving from the Reality Box to mobile-head immersive technology, once available at adequate price and resolution, but in none is immersion crucial. For some the current costs of a system would outweigh the benefits (could a dentist afford a Reality Engine and a massively parallel object modeller to design prostheses for individuals?), but for industrial mass-production applications such as car and crockery design the costs of a pioneer system would be easily amortized. With this solid initial market the Virtual Workbench would be set free to ride the price curve that has put into home computing the power that was once available only to superpower weapons establishments, and the mass market will open up access to tailors, dentists, plastic artists, and the creators of new arts such as 3D painting in volume-rendered shapes of pure light.

This is a clearer economic path to wide commercial availability of VR than the headset-based one, which will for some time yet be acceptable to enthusiasts and game-players, but frustrating and uncomfortable to the user who needs to spend multiple hours per day in the effort to create marketable products. In the Institute of Systems Science it is opening paths of collaboration both within the institution (for instance, between the Medical Imaging group, the Virtual Reality group, and the Accelerated Computing group, whose skills are required in numerically modelling organs and other objects), with the healthcare industry, and with other potential users of the Virtual Workbench for production.

References

Editorial Note: The papers below demonstrate the applicability and successful realization of the early ideas expressed in this paper. The design philosophy recommended received excellent reviews at the VRAIS 93 IEEE VR Standards Committee meeting, and deserves study by those interested in industrial electronics applications of virtual reality.

Lawton, W., Poston, T., and Serra, L. 1995. Time-lag reduction in a medical virtual workbench, *Virtual Reality and Its Applications,* Earnshaw, R., Jones, H., Vince, J., eds., BCS *Conf., Leeds '94,* Academic Press, San Diego, CA, 123–148.

Poston T. and Serra, L. 1994. The virtual workbench: dextrous VR, *Proc. ACM VRST'94,* 111–122.

Poston, T. and Serra, L. 1996. Dextrous virtual work, *Communications of the ACM,* 29, 5:37–45.

Poston, T., Serra, L., Lawton, W., and Chua, B. C. 1995. Interactive tube finding on a virtual workbench, *Medical Robotics and Computer Assisted Surgery, MRCAS 95,* Wiley-Liss, 119–123.

Serra, L., Poston, T., Ng, H., Heng, P. A., and Chua, B. C. 1995. Virtual space editing of tagged MRI heart data, *Proc., 1st Int. Conf. Computer Vision, Virtual Reality and Robotics in Medicine,* 70–76, April 3–5, Nice, France.

Serra, L. Poston, T., Ng, H., and Chua, B. C. 1995. Interaction techniques for a virtual workspace, *ICAT/VRST'95,* 221–230.

http://ciemed.iss.nus.sg/research/3dinterfaces/3dinterfaces.html.

111.3 Motion Tracking for Virtual Reality

Herschell Murry

Overview

The roots of motion tracking for the cyber world can be traced back to the early 70s and the Department of Defense. Primarily that use was in flight simulators, what we might call the original virtual reality game. This is still an important application where targeting and other symbology are moved with the pilot's head and in some instances the projector scenes are panned under head movement.

In the real world we have a scenario all about us that we view at will. But in a virtual environment, fabricating a spherical view in all directions is a daunting and expensive task. An alternative is to present a view directly in front of the eyes and then realistically scan that view in synchronism with a person's head. Immersed in this type of environment, one can achieve the sensation of an alternate world.

A way to track head motion is needed in order to tell the assisting machinery how to accomplish the movements needed in the computer scenario world. That is, how to navigate in cyberspace. And in order to totally define an object in space we must remember that both position and orientation are needed. Hence, the *xyz* position and the orientation angles of azimuth, elevation and roll are six parameters generally needed from a motion tracker, although some VR systems are designed to function with fewer parameters while the system scenario fleshes out the remainder of the action.

The earliest motion tracker was a mechanical apparatus rigged to perform head tracking. While mechanical approaches still have some important applications, such devices on the head were rather encumbering to the wearer. Inventors subsequently have devised magnetic, acoustic, inertial, mechanical, optical and perhaps other technologies for capturing motion and reporting it in electronic form to the controlling computer.

Of course the region over which head motion tracking is needed for a simulator is very limited, and such is also the case for many other applications such as current VR systems. So, primarily we will be concentrating our discussion here on motion tracking over a volume of a few cubic feet or in the range of a cubic meter, although extensions in some devices have been/are made well past these boundaries to a volumetric range of several square meters.

At the time of this writing the dominant motion tracking technology for Virtual Reality applications is magnetic. The primary reason for this is that magnetic fields penetrate most materials in our environment, and therefore the source of magnetic fields and the sensor using them to track do not need to be in line of sight with each other. This allows complete freedom of movement. Nevertheless, other technologies also can perform VR motion tracking, and these will be discussed along with magnetics.

Magnetic Tracking Devices

There are two basic types of magnetic tracking: pulsed DC fields (often called DC) and AC fields. Actually, a third form of tracking also has existed for a long time: use of the Earth's magnetic field. While this passive technique has been good for tracking, or navigating, over the surface of the globe, it is not very relevant to precise six parameter tracking over a small space about which we are concerned here (Note, however, the compass-tilt discussion in the last paragraph of the motion tracking section.).

Magnetic tracking is an active technology where three orthogonal fields are generated via three orthogonal coils and are sensed remotely by another set of three orthogonal coils. See Figure 111.4. Circuitry drives the magnetic field source and at the same time amplifies and processes the received signals to achieve position and orientation of the sensor relative to the coordinate system of the source coils.

A little more discussion about the terminology of active and passive trackers is in order here. An active tracker provides its own space reference and operating field (magnetic, light, sound)

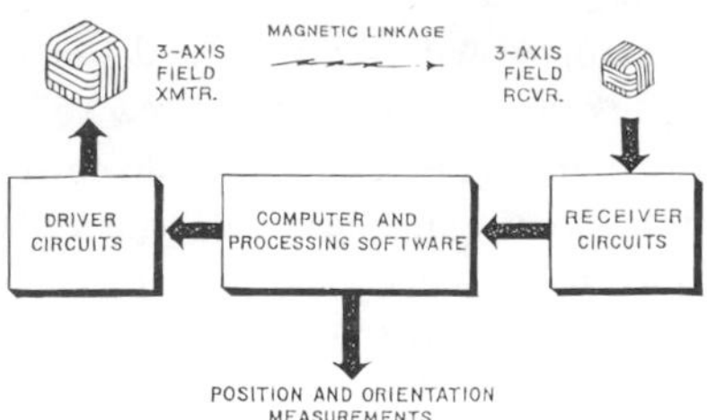

Figure 111.4 Magnetic Motion Tracker System (Courtesy of Polhemus Inc.).

so that the position of its sensors within that field can be determined at any time, including at start-up. Passive trackers using an ever-present physical phenomenon (inertia, earth's field) must be initialized to the space frame but afterward they track continuously. Each technique has its advantages and disadvantages. An inertial tracker, for instance, raises no concerns about signal-to-noise ratio from a signal source such as does magnetic nor about reflections, diffractions or distortions such as optical and acoustic techniques. On the other hand, besides needing initialization the inertial tracker must be serviced continuously to integrate for the position of the object it is tracking and for correcting for drift that inevitably occurs.

Pulsed DC Magnetic Tracking

Both the pulsed DC and AC techniques are active methods where they generate the magnetic fields in which they navigate. The pulsed DC and AC refers to the type of field environment created by their magnetic field source element. In the case of pulsed DC, during each cycle the DC magnetic field is brought up in each of the three cartesian axes to take readings at the sensor coils. Of course, the fields are generated by bringing up drive current on each of the X, Y and Z axes.

We know from EM theory that such a step in current/magnetic field will couple a transient not only into the sensor coils but also into any other conductors that may be in the vicinity. The transient in the sensor circuitry must be allowed to dissipate, which can be controlled in the design. Any current induced in metals in the locale, however, are not under control and must be allowed to die out because these currents will alter/distort the magnetic fields and therefore the tracking measurement. While it is true that a DC tracker may need slowing when encountering conducting metals (these products typically provide for such control via the host computer), in badly distorted environments this may be the only way to achieve magnetic tracking. In such cases, slowness may be preferable to no tracking at all.

For a different reason of biasing data, ferromagnetic materials which sustain permanent magnetic fields also must be avoided in the local environment of a pulsed DC tracker. Of course, there is one permanent magnet that cannot be avoided, the earth's magnetic field. The pulsed DC tracker must determine the local earth field bias and subtract that from the sensor readings. These sensor readings are a matrix of nine measurements where energizing each sensor axis XYZ yields three xyz sensor outputs The dipole fields produced by the source are known, so xyz measurements can be used to determine where on that field the sensor must be at each tracking point.

AC Magnetic Tracking

The AC magnetic tracker is somewhat like a transformer where the primary is the source and the secondary winding is the sensor. Since we know how to describe from basic EM theory the dipole fields generated and can measure the sensor outputs, we can compute the "structure" of this "air core transformer," which actually yields the tracking parameters.

Of course an AC magnetic field can induce currents in nearby conducting metals as well as the sensor coils. These eddy currents create their own fields so that the net field is distorted, causing tracking errors. Although there are techniques to combat such distortion, they can be involved and time consuming so that constructing components of a system from non-conducting (or poor conductors such as cast iron or some stainless steel types) to avoid distortion is the most prudent approach.

Using AC techniques accrue certain benefits that help offset the tendency to distortion. For one, very rapid operation is possible because there is no need to wait for pulse transients to die out nor to subtract out the earth's field. For another, the source can be much smaller because it is not competing with the Earth's field to create the same signal-to-noise ratio (SNR), which is important for some applications such as in aircraft cockpits, to create the same SNR, although one certainly can build large AC sources.

The AC magnetic tracker also can be made to operate faster because of the immediate induction of signals into the sensor and no need to wait for stable dc levels. Still another possibility not available to DC techniques is to drive each axis with a different frequency so that data on all three sensor axes can be collected simultaneously.

Another benefit of AC magnetic trackers for some applications is the ability to operate multiple AC trackers independently in the same environment by running them at different carrier frequencies (frequencies designed for optimal "frequency orthogonality" and usually synchronized to further guarantee this property). For instance, an aircraft with fore and aft cockpits cannot be provided helmet tracking from a single magnetic source to the accuracies and low noise levels required, but separate tracker systems operating at different frequencies can easily solve the problem.

Acoustic Tracking

Acoustic systems utilize an array, usually of three, sensors intercepting sound emitters, usually piezoelectric transducers, to determine direction and range. In other words, this also is an active motion tracking technology. Trigonometric relations then perform triangulation to determine position and orientation. In order to accomplish this the system either assumes a one atmosphere velocity of the sound waves (about 347 meters per second) or provides a means to more accurately determine this velocity. Pulses of acoustic energy can then be measured to basically determine angles in two orthogonal planes such that the intersection provides a loci of locations. Multiple sound sources can provide more solutions, and a fixed and known relationship between sounds allow computing position and orientation. Discounting the fact that the sensor array can be rather bulky, an acoustic tracker generally enjoys a cost advantage over more involved magnetic, optical or inertial trackers.

Unfortunately, many things in our environment, such as walls for reflections, can cause distortion or wavefront blurring as well as there often being many potential noise sources to contaminate the signals. Further, the signals can be attenuated by such things as clothing or largely blocked by solid objects in the environment. Consequently, acoustic trackers are considered to be LOS devices. While it is true that the sound waves may bend and reflect past some of these objects, the wave front then becomes more confused and very unlikely to yield accurate answers. All or these effects could be considered as noise in the environment such that acoustic trackers tend to make considerable demands on good SNR design.

Inertial Tracking Devices

Accelerometers and gyroscopes can be configured for tracking the navigation of objects in cyberspace for VR applications just as they have been used in large scale navigation in the real world. Because of the inertia of a mass or the rotation of a wheel, outputs can be obtained and integrated over time to report motion coordinates when we move a mass. In other words, we take advantage of Newton's laws of motion about the reluctance of masses to change their state of resisting movement or moving in a new direction or plane. Because these are basic properties of mass, no illuminating signal is required so that inertial techniques are categorized as being passive.

From one point of view, inertial techniques are a truer form of motion tracking than the other technologies discussed here. The logic of this statement comes from the fact that the device only outputs when there is movement. Optical and magnetic trackers, for instance, actually provide a series of high speed "snapshots" of where an object is located and then connecting the dots, so to speak, tracks the motion. But since the video for which the tracking is used also operates similarly by sequences of frames of pictures there is no incompatibility between these trackers and the application.

The early 90s have seen improvement in inertial devices for application to VR systems. Cost, complexity and size have been limiting factors for such applications. Now small rotating gyros, tuning fork accelerometers and other novel devices (see Figure 111.5) have become available where acceptable accuracy has been maintained for many applications as their size was reduced. Certainly compactness is a real issue in many VR applications where we want orientations and where an object is, up and down, as well as where it is in the two dimensions of classical navigation, creating the need for even more parts to obtain these degrees of

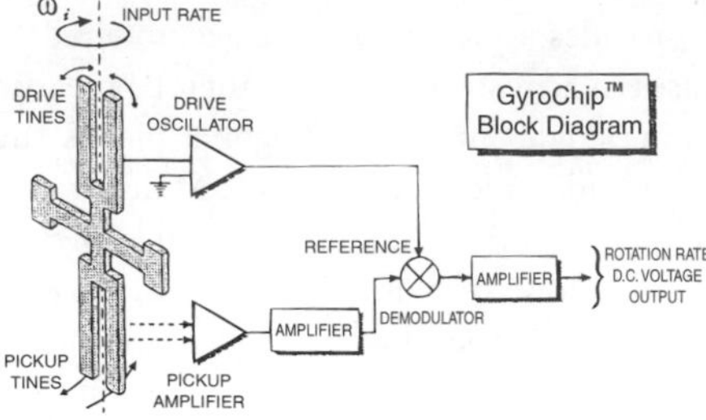

Figure 111.5 The GyroChip™ by BEI Systron Donner Inertial Div. Courtesy of BEI Systron Donner.

freedom in a small package. And achieving the goal of compactness flies in the face of the inertial technology. A gyro, for instance, becomes less massive as its size is reduced such that accuracy suffers, and friction/drag effects are more telling. The manufacture of very small moving elements has only recently become possible and still maintain acceptable tolerances, sometimes really primarily vibrating rather than rotating. Ever more capable microcircuits also are becoming available and cost effective so that the inertial tracker offers the potential of having practical solutions to most of its problems in the near future. Whether the cost performance gains will show adequate advantages compared to advances at the same time in other technologies remains for the VR system design engineer to evaluate.

Mechanical Tracking Devices

Various techniques are used to obtain linkage joint readings from which these angles and the known length of mechanical linkage elements can be used to track an object. As with any technology, mechanical techniques have strengths and weaknesses. A strength is that mechanical system accuracies vary little with range, unlike an electronic method where signals spread over distance and therefore degrade in signal-to-noise ratio. On the other hand, the mechanical structure must start from a very sturdy base which must guard against sag as the device reaches outward in range, and must have precise mechanical structures which tend to become costly with size although smaller ones can be quite economical. With a small reach they can possibly be the simplest and cheapest as long as all requirements of the application can be met. And they can operate without concern for the type of material they are near. Whether bright or dark, warm or cold, ferromagnetic or not, highly or less conductive or in an interfering signal field. Only solids/LOS, resolution and sometimes sag are an issue with a mechanical tracker.

Optical Tracking Devices

Tracking by light is another active technology that is accomplished in a number of ways. In each instance optical systems tend to be top performers but also are probably the most expensive. The extremely short wavelength of light of course offers a great deal to work with for resolution. One approach is to track spot reflectors of certain wavelengths of light while being observed by several video cameras. Processing from the known geometry of camera placement then tracks the spots and therefore the object of interest. Because the objects they track must always be in plain view (i.e., LOS limitation) there are times when the spots are obscured and upon reappearance are very difficult to sort as to which spot reappeared where, so that tracking may break down.

Another optical approach is to use an array of LEDs (light-emitting diodes) and focusing an image of this array on a camera where the direction to each element in the array can be computed and tracking accomplished. Sorting out the diode images can be aided by either energizing them in a known sequential pattern or pulsing/frequency multiplexing the various diodes. And, like many trackers, the tracking process can be accomplished symmetrically, That is, the LEDs can be tracked by a static camera (or cameras) or the camera can be mobile and the LEDs remain fixed.

As far as drawbacks besides LOS and high cost for optical tracking, there would seem to be little opportunity for degradation in a VR environment due to dust, smoke or vapor although these factors should be kept in mind. Also, interference by other light sources in any given installation is another factor for consideration.

There is another optical type of technology that should be mentioned here because it has been used for tracking at least in two dimensions and is generally cheaper but less accurate. This is using arrays of IR sensing diodes. An IR beam can be projected onto lenses, or portals, over an intersection of several sensor diodes such that the amount of incident infrared energy causes current output in proportion to the amount of sensor being illuminated. Through ratios of these outputs and simple trigonometry, the angle to the IR source can be computed. Several such sensors arrayed in the proper geometry can then be used to obtain tracking. An alternative to this type of tracking is to flood the area with IR light and track the shadows.

Compass-Tilt Tracking

Probably the cheapest way to perform tracking, even if only in three dimensions, is to marry a compass with a tilt sensing device. A compass provides orientation in a horizontal plane, and being able to sense inclination in the vertical plane can provide orientation in that plane. Coupling these two types of sensors in this orthogonal fashion can track in 3D orientation angles useful for some VR applications. Poor accuracy and resolution tend to offset the virtues of low cost and simplicity, especially when one considers that tilt sensing devices also are sensitive to movement accelerations just as they are to the acceleration due to gravity. In a dynamic environment this can be unacceptable, but such an approach has been employed in VR systems. Compass-tilt trackers are technically unsophisticated and have results not always very pleasing, but no discussion of motion trackers would be complete without their inclusion.

111.4 Virtual Sound

Nadine Miner and Thomas Caudell

Introduction

Sound is an integral part of the human perceptual process and is valuable for processing and understanding information. One of the goals of Virtual Reality (VR) is to intuitively convey information to a human participant. To this end, virtual sound may serve as an important sensory input tool. The term *virtual sound* refers to the processes of creating and displaying sounds in a virtual environment. This includes localization of the sound in the three-dimensional (3D) environment, environmental distortion modeling and sound source simulation. There are many difficulties involved in simulating the complex real-world sound process. Over the past decade, researchers have made significant progress towards achieving realistic sound simulations. However, techniques are still expensive in terms of equipment and time, forcing system developers to make difficult trade-off decisions.

Many VR researchers believe virtual sound may create a more compelling "virtual experience" by adding to the participant's sense of presence in the virtual world. Sound provides the ability to convey complex information to a participant. Sound is particularly useful when visual display of the information is ineffective. For example, the click of a robot gripper on a scalpel tells a remotely located surgeon that a grasp action has succeeded. Sounds, such as a fire alarm ringing or radiation sensors buzzing, inform a participant of an impending action or danger. In the absence of other sensory information, such as force or tactical feedback, sounds can indicate contact with a remote or virtual object. This may be particularly important for VR based training applications.

Sound in the form of voice feedback can greatly enhance the usability of a VR system. Voice feedback can provide real-time guidance as the user interacts with the virtual world through instructions and help utilities, confirmation of actions, and status information.

Traditionally, computer based information processing systems have emphasized the visual presentation of information and have tended to ignore the auditory dimension. The addition of sound to complex, multi-dimensional data sets may enhance the human's understanding of the data. Sound can be especially useful in applications that overload the human visual system. Techniques for mapping data to sound, otherwise known as data sonification, add an extra dimension to the data analysis process above a purely visual presentation.

Kramer summarizes the benefits of using virtual sound displays as: freeing hands and eyes, alerting (e.g. alarm sounds), orienting (telling the eyes where to look), back grounding (monitoring low priority sounds for significant changes), parallel listening (tracking several sounds simultaneously), monitoring broad dynamic range of data (range of few milliseconds to several thousand milliseconds), and discerning data relationships or trends (known as auditory gestalt) (Kramer, 1994).

To understand virtual sound, we will begin by discussing the real-world sound process including a description of the human auditory system and sound perception. Next, we will describe the components of a virtual sound system and present techniques for creating sounds and localizing them in 3D space. Finally, we will discuss the many open research questions and provide a list of resources and references for further reading.

The Real-World Sound Process

The goal of virtual sound is to convey perceptually meaningful information to a VR participant. Humans intuitively understand many real-world sounds. Thus, to better comprehend virtual sound, we will begin by describing the real-world sound process that we wish to emulate. The real-world sound process consists of several stages as summarized in Figure 111.6. A sound wave, created by a physical interaction, emanates through the environment to a human listener. The next sections describe each component of this process in more detail.

Origin of Sound

The first component of the real-world sound process is the sound source which describes the sound wave creation. A physical interaction causing a displacement of molecules results in the creation of the sound wave. The interaction can be any variety of things: the plucking of a string, two objects hitting each other, one object scraping against another, a solid transforming into a gas such as in an explosion, etc. In 1822, Jean Fourier showed that all waves are made up of the combination of many sinusoid signals of varying frequency and amplitude (Resnick, 1960). Multiple sound waves additively combine according to the superposition principle. The frequency of the wave is measured in cycles per second or hertz (Hz). Human audible sound waves range in frequency from about 20 Hz to 20,000 Hz.

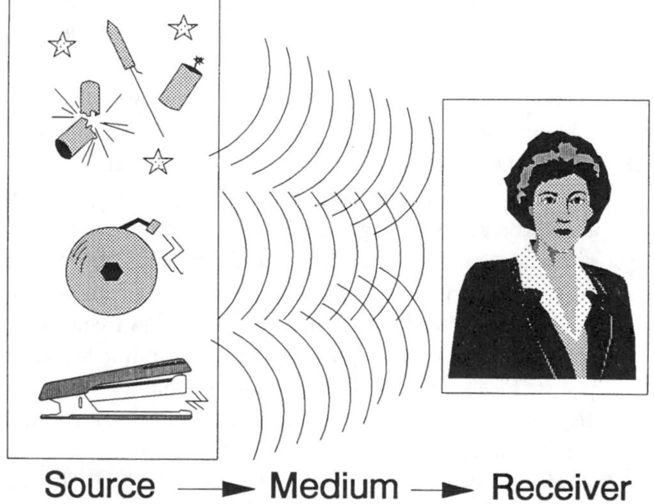

Figure 111.6 Illustration of the sound process. Physical world sound source creates waves that propagate through the environment (medium) to the human ear (receiver).

Sound Transmission and Environmental Effects

The next component of the real-world sound process is the transmission of the sound from the source through the medium or environment. Sound waves propagate outward from the source through a medium which can be solid, liquid or gas. If unimpeded, the sound wave spreads out spherically in all directions. The medium itself, and anything within the environment, can distort the sound wave due to effects of reflection, refraction, diffusion, or diffraction. Reverberation (multiple reflections of a sound wave within an enclosure) is also important for human sound perception. Reflection and refraction effects are frequency dependent and vary depending on the sound source location. Sound waves interacting with the listener's torso, head, and outer ear (pinnae) will also distort the sound. Thus, the environment significantly modifies the sound wave by the time it reaches the human listener. Furthermore, the sound is "customized" for a listener due to the individualized human distortion effects. This effect can be quite significant as we shall see in the next section.

Primary Localization Cues: the "Duplex Theory"

closer ear

time

ITD

Interaural Time Differences (ITDs): sources off to one side arrive sooner at the closer ear

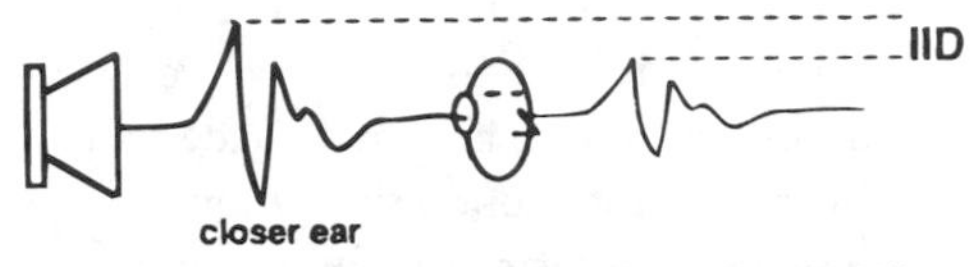

Interaural Intensity Differences (IIDs): sources off to one side are louder at the closer ear due to head-shadowing

Figure 111.7 Illustration of two primary auditory cues: Interaural Time Difference and Intensity Difference (Courtesy MIT Press. From Wenzel, E. 1992. Localization in virtual acoustic displays. *Presence* 1(1):80–103.)

The Human Auditory System as Receiver

The final component of the real-world sound process is the sound receiver. The human listener provides a very effective and efficient sound receiver. Sound source location, size, shape, density and velocity information can be perceived through the human's auditory processing system. In this system, processing of multiple, parallel sound sources occur with very little conscious effort. According to Georg Ohm, the human auditory system breaks the sound signal down into its basic frequency components of different phase and amplitude and performs a Fourier analysis on the sound to allow interpretation by the brain [Kandel 85]. The system is complex because of this combination of sophisticated physical structures and advanced auditory cognitive processing.

An important feature of the human auditory system is its ability to perceive sounds in three dimensions. Simply put, humans hear sounds spatially because we posses two ears located on either side of our head. Lord Rayleigh's "duplex theory" describes the role of two primary cues relating to sound localization: interaural time differences (ITD) and interaural intensity differences (IID) as depicted in Figure 111.7 (Rayleigh, 1945). The interaural time difference is the time difference between when a sound reaches the right ear of a listener and the left ear. For a stationary sound source, the sound appears to be located to the right of the listener if the sound reaches the listener's right ear first. The largest effect of time differences is at low frequencies, below 1500 Hz. At high frequencies, the time difference does not appear to be a significant factor due to the shadowing caused by the listener's head. The intensity difference of the same sound at each ear results in the IID. Intensity differences seem to be more of a factor at higher frequencies (above 5000 Hz). The use of only time and intensity difference measurements for a fixed listener and sound source may result in an increase of ambiguities in the sound location. The ambiguities are conical in nature and are thus often referred to as the *cone of confusion* in the literature (Wenzel, 1992; Mills, 1972). These ambiguities include front-back reversals and up-down, or elevation reversals. It is normal to have some ambiguities in sound location in our everyday world. To resolve the ambiguities, human listeners move their heads and use additional frequency and spectral cues.

Neither time differences nor intensity differences account for the vertical localization cues which humans receive. Today researchers believe direction cues (horizontal and vertical) are largely due to frequency and direction-dependent distortions of the sound wave as it intersects with the listener's head, torso, and pinnae (Wenzel, 1992; Gradecki, 1992). Humans are not particularly good at determining distances of sound sources because sound intensity dominates strongly over other factors. Thus, a sound that has very high intensity will be perceived as being closer to a listener than the same sound with lower intensity. The Doppler effect aids human listeners in judging the direction and velocity of moving sound sources. As predicted by J. Doppler (1803–1853), a sound moving towards a listener will result in a frequency increase, and a sound moving away from a listener will have a perceived frequency decrease (Resnick and Halliday, 1960).

It is difficult to quantify the effectiveness of the human auditory system because performance is individualized and depends on the combination of many variables, including: ambient noise in environment, quality of the sound, familiarity of the sound, stability of the sound source, participant's hearing sensitivity at particular frequencies, fatigue of the participant, and test conditions. One study performed by Wightman and Kistler played a static sound source over loudspeakers in an anechoic chamber to eight observers (Wightman and Kistler, 1989). Elevation judgments ranged in accuracy from 0.68 to 0.96 (median of 0.93), while azimuth judgments ranged from 0.96 to 0.99 (median of 0.98). Wightman and Kistler observed front-back reversals 3% to 12% (median of 5%) of the time. This data indicates humans are better at judging azimuth than elevation and front-back reversals normally occur in the real-world.

The physical components of the auditory system consist of the peripheral auditory system and the central auditory system. The ear and primary auditory neurons, known as the cochlear nerve, make up the peripheral auditory system. This system conditions the signal and acts as a spectral analyzer for the inner ear. The central auditory system is a parallel processing system made up of the nerve pathways and nuclei from the cochlear nerve inwards. This system processes the sound according to frequency range and sound function. For example, Barlow suggests that the processing of alarm sounds occur in different parts of the brain from speech sounds (Barlow and Mollon, 1982).

The human ear consists of three parts: the outer ear, middle ear, and the inner ear. Figure 111.8 shows a high-level anatomy diagram of the ear. Begault provides a brief description of the human auditory system (Begault, 1994). Sound is transformed by the pinnae (visible portion of the outer ear) and proximate parts of the body such as the shoulder and head. Following this are the effects of the meatus (or ear canal) that leads to the middle ear. The middle ear consists of the typmanic membrane (or ear drum), and the ossicles made up of the malleus, incus and stapes (or hammer-anvil-stirrup). The acoustical energy of the sound wave is converted to mechanical energy in the form of fluid pressure variations at the ossicles via motion at the oval window. The pressure variations result in frequency-dependent vibration patterns which bend the auditory "hair cells" in the inner ear, or cochlea. These in turn activate electrical action potentials within the neurons of the auditory system, which are combined at higher levels with information from the opposite ear. These neurological processes are eventually transformed into aural perception and cognition.

Human Sound Perception

Today we lack a complete understanding of the sophisticated perceptual and cognitive processing capabilities of the human auditory system. Individual differences and multiple stages of neurological processing make the area of psycho-acoustic research a challenging one. Ballas (1993), Gaver (1993), and Warren and Verbrugge (1984), among others, are working to understand the human auditory perceptual process.

Barlow outlines four basic perceptual parameters of sound: threshold, loudness, pitch, and timbre [Barlow and Mollon, 1982]. We present a brief description of each parameter here:

Threshold—A level of intensity above which a particular sound is distinguished from noise. Threshold varies between individuals and species (e.g., animals have a much different sound threshold than humans). Threshold levels vary across frequency and vary with age for an individual listener. Typically, the smaller the dimensions of the ear, the higher the threshold.

Loudness—The main factor is the level of the sound above threshold, although several other factors contribute to the quality of loudness. The magnitude of an acoustic signal is commonly expressed on a logarithmic power scale in decibels (dB). To judge sound loudness, humans interpret both the frequency and Sound Pressure Level (SPL in dB). For example, under certain test conditions, a 60 Hz tone at 60 dB SPL can seem to have the same loudness as a 1000 Hz tone at 40 dB SPL. The discomfort threshold or Loudness Discomfort Level (LDL) is at ~100 dB SPL. The threshold for pain is 130–140 dB SPL.

Pitch—The sound frequency determines the quality of pitch to a human listener. The human auditory system can categorize sound signals on a monotonic scale according to frequency.

Timbre—This perceptual quality of sound is difficult to define. Timbre refers to the spectral complexity of a signal.

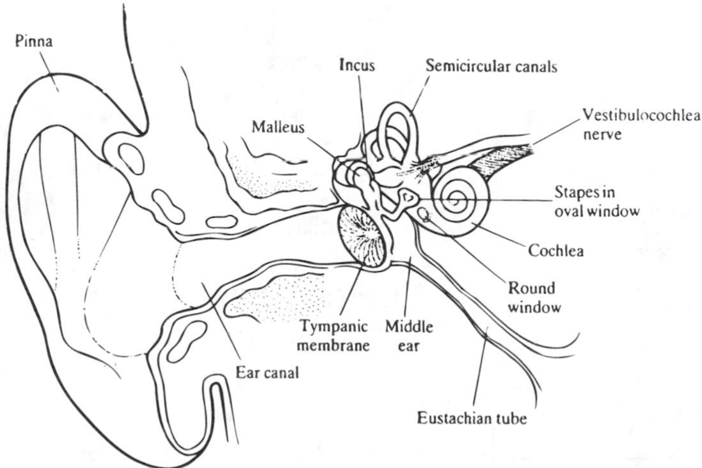

Figure 111.8 Anatomy of the ear. (*Source:* Kalawsky, 1993. *The Science of Virtual Reality and* Virtual *Environments,* p. 69, Addison Wesley, Wokingham, England. With permission.)

It creates the perceived qualities of richness, mellowness, or brightness. Timbre allows a human listener to distinguish between two signals with the same pitch and loudness. Along with attack (onset transients of the sound), timbre allows differentiation of two different instruments playing the same note.

Most researchers agree that interpretation of sound by the human's perceptual system includes processing of these auditory parameters; however, many details of this processing are still unknown. Kandel and Schwartz (1985) and, Barlow and Mollon (1982) provide a more in-depth treatment of the physiology of the human auditory system.

Virtual Sound

The goal of the virtual sound system is to emulate the real-world sound process as closely as possible. This requires creating sounds to convey the desired perceptual messages, locating the sounds in 3D space, distorting the sounds appropriately for the modeled virtual environment, and playing the sounds for the human listener. Figure 111.9 shows the parallel between the virtual sound system components and the real-world sound process. In an ideal system, all virtual sound processing would be accomplished with maximum fidelity, minimum cost and in "real-time." Real-time for sound processing can be defined as: within one frame of video, or less than one-thirtieth of a second. Achieving the ideal system is not possible today due to the many unresolved research questions which will be discussed in the last section.

Over the past ten years, researchers have made significant progress in the areas of sound creation and 3D sound localization. However, detailed physical models of object interactions and virtual environments require long processing times and prohibitively high system costs. Thus, system developers must trade-off sound fidelity, performance rate and system cost.

As in the real-world sound process, the first component of the virtual sound system is the sound source. Several components create the virtual sound source: host processor system, signal generator, and signal processor. The target application will largely determine the complexity and functionality of the sound source components. The host processing system controls the VR simulation, the sound signal equipment, and may perform some signal processing functions. The signal generator can include a synthesizer, a sound sampler, a tone generator or simply a large bank of memory for storage of predigitized sound samples. Situations where the sounds are not known *a priori* require dynamic sound generation. This is an active area of research. Today, obtaining even rudimentary dynamic sounds requires extensive equipment and software. Applications with a fixed set of known sounds typically use pre-digitized sounds and thus require minimal signal processing but vast amounts memory. A more detailed discussion of sound content creation follows in the next section. The signal processing element refers to the hardware and software required for 3D sound localization, sound mixing and sound source transformations. The 3D sound system individually localizes each sound source according to the user's current head position. Head position information is obtained via a position tracker placed on the user's head. The amount of signal processing equipment required increases as the number of participants and sound sources increase. A discussion of the techniques for creating 3D sound follows later in this section.

The environmental processing component refers to both the modeling of the virtual world and overcoming distortion effects

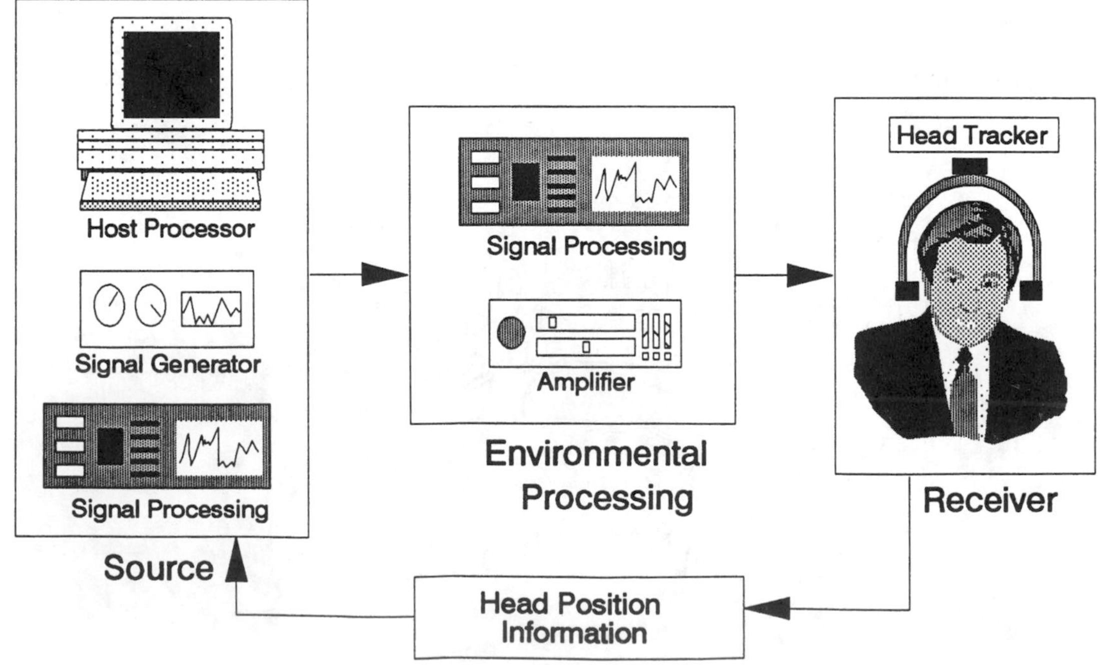

Figure 111.9 Configuration of a virtual sound system.

of the participant's physical environment. Accurate modeling of a virtual environment requires significant signal processing equipment. As explained earlier, environmental modeling is important due to the vast amounts of information humans obtain from environmental distortions. Development of techniques for simplifying environmental distortion modeling are on-going. It is possible, but difficult, to overcome sound wave distortions created by the participant's physical environment. Headphones are an easy way to minimize the distance between the source and receiver and thus minimize the environmental distortion. If the sound is displayed using speakers, real-world environmental distortions (which typically do not correlate with the virtual environment) will modify the sound. The types of headphones that fit inside the ear are often preferred for 3D sound systems because they minimize environmental distortions, have lower resonance than larger headphones that enclose the pinnae, and provide relatively good attenuation of external sounds (Durlach and Mavor, 1995). Active noise cancellation equipment can serve to further control the sound environment.

The last link in the virtual sound process is the human listener. Luckily, the human auditory system need not be simulated or modeled. However, determining the effectiveness of the simulated sounds requires at least a rudimentary understanding of human sound perception. This is being accomplished through extensive psycho-acoustic experiments. Significant on-going research continues in this area as the perceived importance of the auditory sensory system increases.

The cost of a virtual sound system relates directly to the system performance requirements. If the application requires high speed, and high quality sound rendering and localization, the equipment can be very expensive. For example, today sound mixing boards range from $100 to more than $10,000 for professional sound mixers. As of this writing, sound localization systems range from $2,000 to $10,000 per sound channel depending on localization quality. With current techniques, sound sources and individual reflections require separate sound channels; thus, the localization system cost increases directly with each modeled source or reflection.

Researchers in the field of virtual sound generation, 3D sound, auditory displays, and psycho-acoustics often require more extensive equipment. Additional equipment might include sound chambers to control the environmental effects. For example, soundproof rooms eliminate external noise from the environment and anechoic chambers reduce or eliminate sound wave reflections but are generally not sound proof. Depending on the quality and size of the environment, the price will vary from $10,000 to $200,000. Sound researchers may require a variety of auxiliary equipment: sound level meter, oscilloscope, audio filters, noise generators, equipment for testing the quality of human hearing, microphones, and recording devices. The reference section contains several sources which provide listings of state-of-the-art virtual sound hardware and software.

Creating Sound Content

The first issue in adding sound to a virtual environment is the creation of the virtual sound, or determining the sound content. We define sound content as the sampled sound signal which contains user interpretable sound information. This includes both verbal and non-verbal, or abstract, sounds.

For virtual environments, there are two basic sound content creation methods: using and manipulating pre-digitized sound sequences, or synthesizing the sound using some type of physical or spectral model. The sounds can be realistic sounds related to actions within the virtual environment, abstract sounds for information encoding, or verbal cues.

There are inherent advantages and disadvantages to using either pre-digitized or synthesized sound. The pre-digitized sound approach provides realistic, application specific sound feedback to the user. However, the sound sequences are static and thus cannot respond dynamically to users as they interact with the environment. Storage of sufficient pre-digitized sound sequences for a complex VR system requires significant amounts of memory. Alternatively, sound synthesis allows dynamic creation of sound sequences, providing that there is enough sophistication in the application software. Voice synthesis systems available today can translate text strings into voice sequences, however, the resulting voices tend to be unnatural. Currently, only limited hardware and software are available for synthesizing non-verbal sounds in real-time. Systems are available which provide tools for manipulating and editing sound parameters, but the end user is typically left to create their own application specific sounds.

The lack of hardware and software for creating dynamic, behavior oriented computer generated sound is a well-known problem in the VR community (Durlach and Mavor, 1995). Research is on-going to address this problem. Takala and Hahn developed an approach to sound rendering for producing synchronized sound tracks for animations (Takala and Hahn, 1992). Takala's techniques result in realistic real-world sounds, but are not rendered in real-time. Gaver experimented with different methods for creating impact and scraping sounds in real-time (Gaver, 1994). Gaver's method includes the ability to modify sound object parameters to increase the sound production variety; however, the sounds may not be realistic enough for some applications.

Today, the typical approach for adding sound to a VR or multi-media environment is to use pre-digitized sounds. Quality pre-digitized sounds can be very time consuming to create and often require sophisticated recording equipment and advanced sound manipulation tools. Thankfully, sound effect libraries used extensively by video, film and music professionals, are becoming more widely available to the general public.

Long sequences of sounds required for background sound effects present a different type of problem. High cost usually prohibits storage of lengthy sequences, especially considering that VR experiences can typically last many minutes. Thus, shorter background sound sequences are often replayed for the duration of a VR session. Background sequences with dramatic amplitude variations or distinct noises will be noticeably repetitive and seem artificial. Thus, short, non-repetitive sound sequences with distinctive sounds mixed in at random intervals are best for background sound. For example, to create the sound of a busy street, one could loop over a steady hum of traffic sound with random mixing of horns blowing, sirens blaring, or bells ringing. If the horn sound was embedded in the sequence with the traffic hum, only a very long sound sample would prevent human detection of repetition. The inclusion of sound in a VR experience can add a tremendous amount of realism if it's done properly. But, typically, initial system design does not include the sound elements. This serves to complicate the already expensive and time consuming process of sound creation and often leads to a less auditorially compelling experience.

Three-Dimensional Sound

Several different terms are synonymous with 3D sound: sound localization, spatialized sound, binaural audio or virtual acoustics. There are several compelling reasons for using 3D sound. As discussed earlier, the use of 3D sound is useful for discriminating between sounds and in directing the human's attention towards an urgent sound. Humans hear sounds spatially in the real world, thus in creating realistic simulations, sounds should be heard in the same way. This is especially important when simulating high-stress scenarios because often times the high stress is due to the intense, encompassing sound environment.

Over the past several years, techniques have been developed which allow sound placement in a 3-D environment around a human listener. Some sound localization systems on the market today use only time difference and intensity differences to locate a sound within a virtual environment without taking into consideration the distortions created due to the head, torso, and pinnae of the listener. As a result, these systems are lacking in horizontal direction accuracy, accurate vertical location ability and the externalization (out-of-the-head) sensation. Systems that additionally include digital filters to model the head, torso, and pinnae distortions are much more effective in achieving 3D sound simulation. These filters are often referred to as head-related transfer functions (HRTFs). By filtering a digitized sound source with the appropriate HRTF filter for the user's current head position (obtained from head position/tracker sensors), one can potentially place sounds anywhere in the virtual space about a listener.

HRTFs are created by inserting probe microphones at the opening of each ear canal and playing sounds from many source positions to the centrally located listener. The microphone allows measurement of the acoustical transfer functions in the form of finite impulse response (FIR) filters creating the HRTFs. These HRTFs are used as the basis of the digital filters for processing the synthesized sound. Figure 111.10 shows an illustration of the technique developed by Wenzel et. al (1992) to measure HRTFs and use HRTFs to localize virtual sound sources. For a more detailed description of 3D sound and the HRTF approach see, Wenzel (1992) and Begault (1994). HRTFs vary somewhat between listeners, because the shape and size of each person's head, torso and pinnae are unique. If a generic or standard HRTF is used, the variation is often times sufficient to create significant errors in perceived sound localization. A simplified method for measuring HRTFs recently developed by Crystal River Engineering reduces the measurement time to approximately 20 minutes.

To place a sound, the digital filter corresponding to the desired target location is convolved with the sound signal to be localized and current head position information. The convolution process requires multiplying the discrete fourier transform (DFT) of the original signal with the DFT of the digital filter. This process places a sound signal in the perceptual 3D space of the listener. This technique is usually used with headphones, rather than speakers, in order to minimize the environmental effects. The system processing capabilities and the high per channel cost typically limits the number of simulation channels in a system.

Although researchers have been developing and experimenting with the HRTF technique for some time, there is relatively little experimental data available to quantify the effectiveness of the approach. In the study mentioned earlier, Wightman and Kistler used the HRTF technique to localize static sound sources for eight listeners wearing headphones (Wightman and Kistler, 1989). Elevation judgments using the HRTF technique ranged in accuracy from 0.43 to 0.94 (median of 0.86) which compares favorably to the same participant's real-world elevation judgment scores (0.93 median). The azimuth judgments using the HRTF technique ranged from 0.95 to 0.99 (0.98 median). This also compares favorably to the participant's real-world azimuth judgment scores (0.98 median). Occurrences of front-back reversals using the HRTF technique increased with scores of 6% to 20% (10% median) as compared to real-world performance scores of 3% to 12% (5% median). Localization performance using the HRTF technique will vary depending on many variables, including: individual subject hearing abilities, the goodness of the HRTF measurements, and the environment and test conditions. The Wightman and Kistler study provide data to suggest 3-D sound localization using HRTFs is a valid and promising approach as long as increased front-back reversals can be tolerated or minimized.

Discussion and Conclusions

Virtual sound research is gaining in interest as an important component of VR and multi-media experiences. As a result, research in the area of virtual sound is increasing. In general, there are three ways in which virtual sound is expected to enhance a virtual environment:

1. Provide realistic sounds as events occur (sound source simulation and localization).
2. Provide voice output for user guidance and feedback.
3. Provide an additional dimension for representing information (data sonification).

Each of these areas contains many unresolved research questions.

In simulating sounds as events occur, how *realistic* do the sounds have to be to convey the required information? The answer will obviously vary depending on the application. The required realism of the virtual sound modeling, transmission

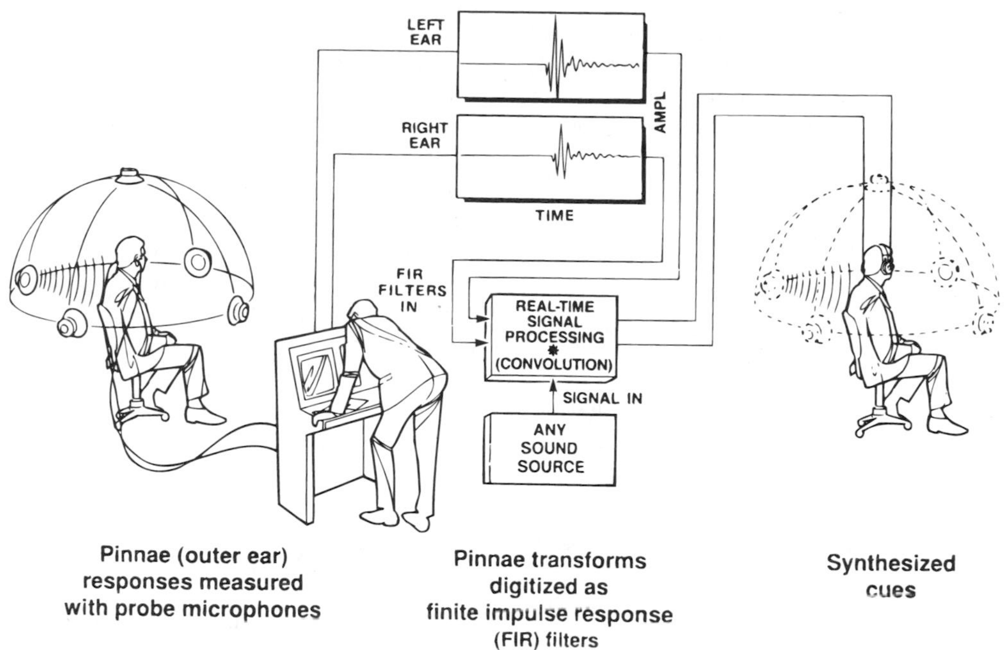

Figure 111.10 Synthesizing Virtual Acoustic Sources using HRTFs (Courtesy MIT Press. From Wenzel, E. 1992. Localization in virtual acoustic displays. *Presence* 1(1):80–103.)

and localization must be considered. For sound source modeling, is it necessary to create detailed physical models of the event in order to convince the participant of the perceptual intent of the sound? What model complexity can be used while still maintaining real-time performance? Will there be a *Foley Effect*, where the simulated sounds need to be "bigger than life," for events that are not directly viewed (e.g., a door closing behind the participant)? For sound localization, making a sound seemingly emit from a particular 3D location is a good start, but real-world sounds are strongly influenced by the environmental distortions (reflections, diffraction, diffusion, refraction) which occur during transmission. How many of these distortions should be modeled in order to create "realistic enough" sounds? Are there algorithmic techniques that can simulate these distortions without requiring the addition of expensive sound channels? Localizing 3D sounds for an individual listener is difficult without a cumbersome calibration procedure to account for individual differences in head, torso and pinnae size and shape. How can this process be generalized or simplified?

Several additional factors will influence sound simulation research. The synchronization of sounds with the visual event is an important consideration and will constrain the available processing time. The audio simulation researcher will need to determine the minimal information content and accuracy required to create high-fidelity, visually synchronized, real-time sounds. The complex and largely unexplored human auditory perceptual system further complicates this task.

Voice synthesis systems available today provide less than realistic human voice feedback. Researchers should consider the best way to present voice feedback in combination with other visual feedback. Often times, long lists of verbal commands are tedious and difficult to remember; however, text information is also not very effective in virtual environments. How can voice feedback systems most effectively be used?

In data visualization, determining the proper mapping between sound and data is of central importance. In general, mappings can be either analogic or symbolic. Analogic refers to an intrinsic, one-to-one correspondence between the data being represented and the sound (e.g., in a Geiger counter, the speed of clicks has a direct relation to the amount of radiation). Kramer (1994), Smith (1994), and Scaletti (1994), among others, have investigated analogic mapping approaches. Symbolic refers to an abstract mapping between sounds and the information the sound conveys (e.g., an alarm signal, speech). Cohen (1994) and Gaver (1989), among others, have investigated symbolic mappings. Details of the attempted approaches are beyond the scope of this section.

Several additional references will enable the curious reader to gain both a greater depth and breadth of information in this area. The National Research Council recently released a volume on Virtual Reality containing an in-depth review of the "Auditory Channel," including information on auditory scene analysis, human sound perception, sound localization, room-acoustics modeling and a summary of research needs. (Durlach, 1995). The proceedings of the first International Conference on Auditory Display (ICAD '92), edited by G. Kramer, provides an excellent synopsis of the state of the field at that time, and invaluable references to additional books, journals, and papers including an extensive annotated bibliography (Kramer, 1994). Begault provides an in-depth review of 3-D sound, and a resource chapter that presents a summary of manufacturers of 3-D audio systems, headphones, sound synthesis and analysis hardware and software (Begault, 1994). Additional sources for the latest in virtual sound technology can be found in the Computer Music Journal, Electronic Musician, Journal of the Acoustical Society of America, and Journal of the Audio Engineering Society (see reference section). Additional publications which often contain information on sound technology and recent research advances in virtual sound include: Presence, VR Special Report, VR News, and CyberEdge.

References

Ballas, J. A. 1993. Common factors in the identification of an assortment of brief everyday sounds, *J. Experimental Psychology: Human Perception and Performance,* 19(2):250–267.

Barlow, H. B. and Mollon, J. D., eds., 1982. *The Senses,* 239–332, Cambridge University Press.

Begault, D. R. 1994. *3-D Sound for Virtual Reality and Multimedia,* Academic Press, San Diego, CA.

Cohen, J. 1994. Monitoring Background Activities, *Auditory Display: Sonification, Audification, and Auditory Interfaces,* G. Kramer, ed., Santa Fe Institute Studies in the Sciences of Complexity, Proc. Vol. XVIII, 499–531. Addison-Wesley, Reading, MA.

Durlach, N. I, and Mavor, A. S., eds., 1995. *Virtual Reality: Scientific and Technological Challenges,* 134–160, National Academy Press, Washington, DC.

Durlach, N. I., Shinn-Cunningham, B. G., and Held, R. M. 1993. Supernormal auditory localization, I. general background, *Presence,* 2(2):89–103.

Gaver, W. 1989. The sonic finder: an interface that uses auditory icons, *Human-Computer Interaction,* 4(1):67–94.

Gaver, W. 1993. How do we hear in the world?: explorations in ecological acoustics, *Ecological Psychology,* 1993, 5(4):285–313.

Gaver, W. 1994. Using and Creating Auditory Icons, *Auditory Display: Sonification, Audification, and Auditory Interfaces, G. Kramer,* ed. Santa Fe Institute Studies in the Sciences of Complexity, Proc. Vol. XVIII, 417–446, Addison-Wesley, Reading, MA.

Gradecki, J., 1992. 3-D sound theory, *PCVR Journal,* 1(6):6–12.

Kalawsky, R. S. 1993. *The Science of Virtual Reality and Virtual Environments,* p. 69, Addison Wesley Publishers Limited, Wokingham, England.

Kandel, E. R. and Schwartz, J. H. 1985. *Principles of Neural Science, 2nd Edition,* 396–408, Elsevier Science, New York, NY.

Kramer, G., ed. 1994. *Auditory display: sonification, audification, and auditory interfaces,* Proc. *1st Int. Conf. Auditory Display (ICAD) '92,* Santa Fe Institute Studies in the Sciences of Complexity, Santa Fe, NM.

Mills, A. W. 1972. Auditory Localization. *Foundations of Modern Auditory Theory, Vol. II,* J. V. Tobias, ed., 301–345, Academic Press, New York, NY.

Rayleigh, J. W. S. 1945. *The Theory of Sound,* Volumes I and II, 433–478, Dover Publications, New York, NY.

Resnick, R. and Halliday, D. 1960. *Physics,* Part I, 3rd ed. 433–456, John Wiley & Sons, New York, NY.

Scaletti, C. 1994. Sound Synthesis Algorithms for Auditory Data Representations, *Auditory Display: Sonification, Audification, and Auditory Interfaces,* G. Kramer, ed. Santa Fe Institute Studies in the Sciences of Complexity, Proc. Vol. XVIII, 223–251, Addison-Wesley, Reading, MA.

Takala, T. and Hahn, J. 1992. Sound rendering, *Computer Graphics,* Vol. 26(2):211–220.

Warren, W. H. and Verbrugge, R. R. 1984. Auditory perception of breaking and bounding events: a case study in ecological acoustics, *J. Experimental Psychology: Human Perception and Performance,* 10(5):704–712.

Wenzel, E. M. 1992. Localization in Virtual Acoustic Displays. *Presence,* 1(1):80–107.

Wightman, F. L. and Kistler, D. J. 1989. Headphone simulation of free-field listening II, psychophysical validation. *J. Acoustical* Soc. *America,* 85:868–787.

Further Reading

Computer Music Journal, MIT Press Journals 55 Hayward Street, Cambridge, MA 02142-9902, USA.

CyberEdge Journal's Virtual Reality Products, #1 Gate 6 Road, Suite G, Sausalito, CA 94965, USA.

Electronic Musician, Mix Magazine, PO Box 41525, Nashville, TN 37204, USA.

Journal of the Acoustical Society of America, 500 Sunnyside Blvd., Woodbury, NY 11797, USA.

Journal of the Audio Engineering Society, 60 East 42nd Street, New York, NY, 10165-2520, USA.

International Journal of Virtual Reality (IJVR), IPI Press, 2608 N. Cascade Ave., Colorado Springs, CO 80907, USA.

Presence: Teleoperators and Virtual Environments, MIT Press Journals, 55 Hayward Street, Cambridge, MA 02142-9902, USA.

Virtual Reality Special Report, AI Expert, Miller Freeman Inc., 600 Harrison St., San Francisco, CA 94107, USA.

VR News, Cydata Limited, P.O. Box 2515, London N4 4JW, UK.

111.5 Virtual Reality Systems

Mary Lou Padgett,
Richard A. Blade, Johnny Evers, and
Charles R. White

Introduction

Virtual reality (VR) systems can be enhanced by integrating intelligent electronics components to add "common sense" to the VR system. Conversely, intelligent electronics applications and theory can be augmented by adding the interactive and immersive capabilities of VR systems to the development and final product platforms. Concurrent engineering looks forward to anticipate the needs of the system, its designers, and users from concept to production, sales, maintenance, and modifications. Coupling VR, neural networks, fuzzy systems, and evolutionary systems produces a mature, practical environment for research, development, production, and consummation in factory automation and other industrial electronics applications. As noted by Caudell (1994), VR is useful in government applications involving interactions, distribution of information, and crisis management. The *National Information Infrastructure* (NII) is a promising development. Applications in industry range from mathematics, engineering design, digital preassembly, interference testing, manufacturing, prototyping, maintainability, training, and telepresence to applications of telerobotics. The latter include space exploration, toxic waste cleanup, undersea exploration, and microscale exploration. The advent of fast, powerful computer systems and software tools to control them has opened the door to potential advances in industrial electronics applications in all areas. Aerospace training simulator technology provides a firm basis for extension into intelligent virtual reality systems. These range from simple simulations which perform well on portable PC's to elaborate networks of realistic scenarios. The other articles in this chapter describe many virtual reality applications.

Applications: Aerospace to Factory Automation

Examples described in the factory automation chapter cover machine vision and robots with rudimentary capabilities for vision, smell, touch, and hearing. Telepresence is illustrated. There are a multitude of scenarios where the human cannot directly manipulate the environment. Hostile conditions such as radiation, temperature, or weapon fire may exist. Physical limitations such as size may prevent a human from journeying down a VLSI chip. Vision limited to certain wavelengths may prevent inspection of IR images, or visualization of wind tunnel flow patterns.

Many of these applications intensively consume telecommunications resources. VR and multimedia applications which are networked benefit from *Asynchronous Transfer Mode* (ATM) capabilities being developed at CERN. Low cost chip sets are enabling advances in applications which require huge amounts of bandwidth. See sections by Lindblad and by Christiansen et al., in this chapter. Networking and distributed interactive simulation (DIS) also play key rolls in implementation of virtual reality. The *Institute for Simulation and Training* (IST) in Orlando, Florida, is working on standardization of DIS for military applications (Hofer and Loper, 1995). Virtual wargames played by people on different types of computer platforms are plagued with problems such as nonuniform level of detail. If a combatant fires at an opponent, appears to hit him, but he lives anyway, it can be very confusing. One explanation is that the "invincible" opponent hid behind a tree not displayed on the screen of the one who fired. In one sense, VR can be considered to be "modern" simulation: extremely interactive and immersive hardware-in-the-loop simulation with "modern" capabilities.

The very conservative pulp and paper industry labels its "intelligent" simulation presentations with the "modern" tag for marketing to its executives (Padgett and Karplus, 1993 TAPPI). Modern applications are mandatory. Intelligent applications are often plagued with "hype" and inflated promises. We shall attempt to suggest some practical approaches to modernizing simulation in a robust manner and to validating the work to the satisfaction of necessarily skeptical management.

Modern Simulation to Immersive, Interactive Systems

A taxonomy of techniques may be helpful in interpreting the literature and selecting appropriate strategies. Scientific data visualization is vitally important in understanding today's complicated systems and mathematics. This technology can be made interactive and immersive to expand the graphical displays into a virtual reality system. A VR system creates an artificial environment encouraging the user to feel a part of the "world" created, and to interact with it. An augmented reality (AR) system combines the real world with artificially generated objects which can be manipulated. Telepresence is immersive interactive communication with a completely real environment. Industry applications of augmented reality appear to be practical and popular.

Virtual reality and its variations are outlined in Caudell (1994). Immersive VR involves human perception. The human stereo vision system, with its illusions, motion detection, and foviation properties presents a challenge. See Kartalopoulas (1996) for neurobiology and Idesawa (1993) for 3D optical illusion in binocular viewing. Various SPIE proceedings and tutorial books deal with these issues. Other senses to consider include: 3D hearing, balance, smell, and touch. Human factors such as stress, ergonomics, task performance, simulator sickness, disbelief, communication, and teaming cannot be ignored. Industrial applications frequently select VR-related implementations such as see-though optical head-mounted displays for augmented reality because of safety and economy. Research on human factors is needed, but careful selection from existing (economical) techniques can avoid problems while more research is being done on elaborate extensions of VR.

VR technology touches on many fields of engineering (Burdea and Coiffet, 1994; Larijani, 1994; Gradecki, 1994; Loeffler and Anderson 1994; Jacobson, 1994, and Stampe et al., 1994). Caudell (1994) categorizes these as listed below:

- *Computer graphics*: Coordinate systems, polygons, models, transforms, surface representations, 3D rendering, color and texture maps, luminance, graphics hardware.
- *Visual displays*: Image vs non-image generation systems, CRT cathode ray tubes, flat panel displays: (liquid crystal, plasma discharge, polymers); head-mounts, color technology, optical systems for image relay or creation, variable acuity systems, foviation tracking, and projection displays.
- *Body sensing*: 6DOF position, orientation; local vs extended; theory and error analysis; tracking devices: (magnet, acoustic, beacons, imagine, fiber optics, stress gauges, boom encoders); other sensing.
- *Calibration and registration*: Visual registration with real world objects, virtual screen measurement; human eye location; registration fiducials; optimal methods for parameter selection.

- *Non-visual sensory feedback display*: Stereo visual vs 3D sound, touch, force reflection, olfaction, veracity vs computation issues.
- *Software tools*: Locomotion, navigation, transportation, orientation, collision detection, animations, visualization of information.

Extensive details on each of these areas can be found in the VR journals, *Presence* and the *International Journal of Virtual Reality*. The proceedings of the IEEE VRAIS conferences and ACM SIG-Graph conferences are also excellent sources for continual updates. These conferences feature technical papers and tutorial proceedings. To meet with VR vendors, Meckler conferences are recommended (Thompson, 1994).

Many VR and VR-related systems rely heavily on graphics and elaborate goggles or helmets and display screens to stimulate immersive experiences. Other senses can be brought into play to enhance these systems, or to compensate for the lack of expensive, cumbersome equipment. Blind computer users have an acute need for this type of VR system. Heavy emphasis on graphics to the exclusion of ASCII formatted text and other standardized methods of communication has cost leaders in all fields their jobs. Many people lose their visual acuity with age, but need to interact with their environment and with other people. Engineering expertise possessed by older, experienced practitioners and managers can be captured for reuse by enabling these people to continue accessing their computers. Other dedicated, intelligent computer users have limited control of their hands and other limbs. Accurate but forgiving input devices can be constructed with today's technology to allow victims of Parkinson's, cerebral palsy, or spinal injuries to control their computers. Rather than focus a specialized marketing effort at the "handicapped", who are often automatically assumed to be mentally retarded, it is more helpful to concentrate on developing tools useful to a busy executive or technician who needs to use eyes in one place, ears for another task, and hands for yet another. Aiming at a high volume market encourages the development of products which will be promptly updated, maintained, and supported by free technical help. Many specialized products are extremely expensive and rapidly outdated. Using the extensive entertainment market to finance the development of these systems is one way to raise money for research and development, but a sounder foundation may be attained by addressing the mass of people using PCs for business, word processing, financial, and database applications. At a recent IEEE VR Standards brainstorming session (MecklerMedia, 1994), exhibitors of VR products heavily endorsed addressing this market. Increasing the volume of sales stands to spur development tremendously. Spinoffs will also benefit customers, such as the automobile manufacturing industry, which has firmly expressed a need for elaborate, photorealistic VR systems and *Standardization* (Adams, 1993).

Performance measures are needed for evaluating products before making expensive purchases. It is always difficult to derive such performance measures. Given a test data set, or benchmarks, ingenious manufacturers can design to meet set criteria in an economical manner (for them). Drawing on the experience of the power industry, and standards developed for design and analysis of its complex systems, standards projects have been initiated for VR system design and analysis. These will initially consist of guidelines containing check-lists of considerations and issues based on lessons learned. Software engineering standards stress the same type of approach. In this multi-discipline area, it is helpful to have a list of important design criteria from each of the disciplines. Stating these lists in basic terms readily understood by nearly any working engineer can help with the concurrent engineering design process. Evaluation of potential purchases can be helped when the engineer reads a check-list of things that need to be documented about a system, and things that might go wrong. Asking a vendor to explain in writing how the product deals with certain design issues and how flexible it is can be very helpful. This motivates many of the discussions about system modules and design issues and trade-offs contained in this article and in others in this handbook.

Motion tracking and updating at speeds compatible with human capabilities without nausea is vital to VR. Flight simulation facilities at a military installation in Orlando once ran a study on equipment failures. Findings revealed that pilots subjected to an IR beam to the eye became ill, bored, or otherwise uncomfortable and reacted by disabling the equipment. This finding was used to focus the next set of design advances on relieving some of these problems. The section by Murry discusses implementation alternatives and design considerations for motion tracking in VR systems. Many of these technologies are rooted in military simulation technology and its training simulator capabilities. Compromises between factors, such as cost, resolution and robustness, to occlusion of line of sight and corruption by noise are best made by employing concurrent engineering strategies to try to anticipate needs of the system at all stages. Rapid prototyping and developmental stages may need one realization of motion tracking, whereas the final product may perform best with some other type.

Virtual sound simulates sound as it would be perceived by someone immersed in an environment. Enhancing the intuitive communication of information to a participant is a primary goal. Adding sound to an existing virtual environment is rarely as successful as preplanned integration. Two sources for sound are pre-recorded segments and simulated, or synthesized, sound. The prerecorded segments are fixed and limited in selection due to storage space. Synthesized sound lacks realism. For example, consider a synthesized concert versus a performance by a master. Something is missing. On the other hand, synthesized sound offers more variety in interactions with the user. See the section by Miner and Caudell.

Speed and conservation of resources are critically important for successful implementations. Poston's section discusses design choices and transformations. A virtual workbench contains tools. It models the tool handle, but not the hand. Why model the hand?

Many of the design decisions and interactive decision involve focusing on objects or areas of the space of possibilities. Choice of focal areas can be driven by fuzzy systems. For example, consider the auto-focus capabilities of camcorders explained by Takagi's section in this chapter. Selection of what is important in a scene identifies key areas. Establishment of these key areas in an image is an important research topic in virtual reality. Commercial camcorders have already solved part of this problem. Use of this existing technology can help enhance VR applications, and on VR applications can be "wired" to provide data for determining membership functions, training neural networks and so forth.

The material below will provide a brief summary of virtual reality as applied to industrial electronics, then draw from material in the chapters on neural, fuzzy, and evolutionary systems and on computational intelligence to suggest ways to integrate these technologies.

What is Virtual Reality for Industrial Electronics?

Virtual reality for industrial electronics has many components, and many variations which are practical, but less than completely immersive and interactive. Topics to be covered below are as follows:

- Scientific Data Visualization.
- Rendering Hardware.
- VR Software; interaction models, object oriented systems, application development frameworks.
- Augmented Reality.
- Industrial Reality.

Scientific Visualization

Scientific visualization for controls and for signal processing is becoming increasingly essential as the complexity of systems studied increases. It is becoming more achievable as the power and speed of computer hardware and software tools increases. Intelligent use of these new and needed capabilities can be enhanced by the incorporation of computational intelligence. Cautions and design considerations are provided by Moorhead and Zhu (1995) as a good introduction to the interaction of graphical considerations with signal processing technology. Emphasis on scientific visualization of data is focused on increasing the information content of the images presented to the viewer. Gaining understanding or insight a prime goal. Issues familiar to the signal processing community, but not often presented to the graphics community, have a substantial effect on data analysis and presentation. These include sampling rate, reconstruction filters, the human visual system. Techniques for aliasing and filtering, color theory and models, automatic feature extraction, data compression, and multiresolution visualization are extremely important.

The Visualization Process. An image may be considered as a mapping from discrete points, or a set of geometrical objects, each of which is a mapping from discrete points or smaller geometrical objects This construction from geometrical objects is reminiscent of the hierarchical analysis of control systems presented by Shibata et al. in the computational intelligence chapter and of the processing of images by the human eye (Kartalopoulas, 1996). Shape and color mappings are vital. Functional values such as temperature, pressure, humidity, salinity, density, velocity, or stress can be mapped to pixel intensity. The values can be scaled, thresholded, and smoothed in various ways to produce a shape which can be segmented or otherwise manipulated. Dilation and erosion can be employed; edge detection and other image processing and analysis techniques can also be employed. Simulated objects can be projected onto a background, and noise such as jitter can be added to turn a single static image into a series of images for analysis of the dynamics of the projected objects and their separability from the background (Padgett et al., 1985). Libraries of software subroutines for image and pixel manipulation can turn graphics equipment, a video camera, and a Puma robot into a rapid-prototyping lab for industrial applications. Moorhead and Padgett worked in such a laboratory at NCSU, where shape and color mappings were standard investigatory and rapid-prototyping tools. An industrial inspection for possibly defective parts or boards with cold solder joints was studied. Functional values were mapped into colors chosen to intuitively imply their importance or uniqueness in the image. For example, mapping the pressure readings of a plane fuselage described in Moorhead and Zhu (1995) shows a green stripe of lower pressure delimiting the high pressure area on the nose cone. These functional mappings are prime candidates for analysis using computational intelligence techniques. Graphical representation of shape with highlights and reflections based on ambient light and material qualities is a complementary technique, but usually described in a completely different set of literature. Generation of graphical images has been one field of study, whereas image processing and analysis has been another. Both sets of skills are needed in virtual reality. Graphical realism of a single frame or of a flic of images is one thing. Sensible construction and manipulation of synthetic and realistic images is another. It is becoming ever more critical to build a closely coordinated team of workers capable of addressing both issues.

Opacity mappings for two-dimensional functional values can be constructed so that the value at a coordinate on a plane is mapped to the intensity of the pixel at that location. A simulated x-ray effect can be achieved by mapping high functional values to low intensities, as though the high values belonged to opaque elements. Another useful representation of two dimensional functional values is to map the value of a set of coordinates in the plane to the height of a 3D histogram bar at that point. The resultant object can be rotated and viewed from various angles, revealing holes and rough places on a non-convex surface. One such rapid-prototyping experiment (Snyder, Padgett, Moorhead and others) involved automating assembly-line inspection of hexagonal nuts for rough surfaces and anomalies. Processing of images changing over time involves massive use of computer memory and data compression and restoration techniques. Processing can be passive, (just viewing a preprogrammed set of images) or interactive, where the viewer can explore the scenario and select actions real-time. Systems with such capabilities approach virtual reality environments.

Technical Issues with Signal Processing in Visualization. Aliasing and resultant artifacts are often *not clearly described in signal processing toolkits.* Standardization of VR documentation procedures could clarify this. In scientific visualization the presentation medium is usually a CRT, which is a sampled surface. Undersampling and the use of common filters such as triangular filters for exploratory visualization can result in artifacts being mistaken for objects. Moorhead illustrates with some sinusoidal functions which are depicted as having sharp peaks when not properly processed. Standards for virtual reality construction should pull together common knowledge from the many disciplines interacting to produce VR so that humans trained in one particular discipline can be cautioned to consider things like the Nyquist rate and optimal filter selection, which may be second nature to people trained in another facet of VR. Concurrent engineering is critically important here.

Two signal processing issues—Sampling and color space selection: Color space selection varies (Moorhead and Zhu, 1995).

- *RGB color model*: input and output based techniques, e.g., histogram equalization, image compositing good for sensors and displays based on Red/Green/Blue filters or phosphors
- *HSV color space*:
  - hue/saturation/value(brightness)
  - human visual perception model
  - direct volume visualization
- *Others*: luminance plus two chominance difference signals; TV standards and image compression

Survey of Visualization Techniques: Time-Invariant vs Time-Variant. Visualization can be real-time and exploratory, placing heavy demands on computer resources. The speed and flexibility constraints make visualization different from routine image processing. The need to conserve resources leads to processing only a sample of the data, so signal processing issues emerge: sampling, interpolation, reconstruction.

Some valuable techniques can be explained in terms of 2D, extendible to 3D. First consider time-invariant 2D data. Scalar visualization may use contours and/or shading techniques. Isocontours are a set of curves in a plane with constant functional values. Interpolation schemes vary. Techniques for shading within a region, or *rendering*, vary according to the need for quality versus speed. For example, flat shading can be done quickly, while Gourand shading (a form of bilinear interpolation) is more time-consuming. The computer graphics literature provides many details and suggestions for rendering. When a multivariate vector is being processed, arrows may be used to represent the magnitude and direction of the vector field. This approach often produces a cluttered look. The colorwheel mapping technique is a good alternative. Using a color-wheel with hue, saturation and value (HSV), it is effective to map vector direction to hue, and magnitude to both saturation and value. (S and V are hard for humans to distinguish.) Another approach is use of streamlines, which are lines everywhere tangential to the velocity field. Streamlining uses curve-fitting, starting anywhere. It must consider Nyquist sampling rate and interpolation theory: it needs two points on the curve in every cell. Speed and accuracy trade-offs exist. Structured grids may be mapped with bilinear interpolation, whereas irregular grids need scattered data interpolation, which is slower, but may be more accurate. In line interval convolution (LIC) the "output image is a smeared version of the input texture map" (Moorhead and Zhu, 1995). The smear direction may be determined by the vector field. Design success depends on selection of the best filter or convolution kernel. Evaluation requires viewing this technique in animation.

One of the most effective techniques is critical point analysis. At a critical point, the vector magnitude is zero. Types of critical points (based on sign of real and imaginary parts of the eigenvalues of the Jacobean) are given in Table 111.1.

Visualization of Time-Invariant 3-D Data. Projection of 3D data onto a 2D surface is ambiguous. Problems similar to this arise in simulating 3D sound. Some of the solutions are similar. Addition of motion perturbs the input and provides a way to learn more about the object (or sound) of interest. Visual cues to perspective include lines converging to a point in the far distance, shadows, highlights, size comparisons, and other things. Artificial animation of the data can also be revealing. Techniques suggested include varying a threshold, changing the color map, or moving the data slice shown up and down the image.

Visualization of Static Images

Scalar Data. Scalar Data (volume visualization or volume rendering) techniques include those listed below:

- Surface Fitting.
- Direct Volume Rendering (DVR).
- Frequency Domain Volume Rendering (Fourier volume rendering).
- Data Sampled on an Irregular Grid.

Surface fitting uses techniques such as the *Marching Cubes* (MC) *method.* Ambiguities which might create holes are smoothed, based on adjacent data. An ambiguous cube is treated as being undersampled.

In Direct Volume Rendering (DVR), mapping is directly from data to image without intermittent geometric objects. There are two different directions possible:

Map from image into data (feed-backwards).
Map from data into image plane (feed-forwards).

The process (Westwood, 1991 in Moorhead and Zhu 1995),

Table 111.1 Critical Points Useful in Data Visualization.

	Real			
	Negative	Zero	Positive	Mixed
IMAGINARY				
Non-zero	Attracting Focus	Center	Repelling Focus	
Zero	Attracting Node		Repelling Node	Saddle Point

begins with 1) discrete input samples, which are mapped to 2) a continuous volume function. Next, 3) the resulting image space, is shaded. This shading usually spreads the spectrum of the signal. Thus, 4) a low-pass filter, is used to "lower the signal's maximum frequency so that it is below one half the image resampling rate." Step 5) is to resample for image resolution, then 6) calculate the visibility function, and 7) generate the image. The usual procedure is to perform a quick and dirty operation for data orientation, then follow with non-interactive enhancements for physical and mathematically accurate image production.

Frequency Domain Volume Rendering (Fourier volume rendering), may make use of the Fourier Projection-Slice Theorem. This gives faster volume data images than spacial domain techniques. Complexity is $O(n^2 \log n)$ time for volume of size n^3. Like x-rays, these images lack depth information. This technique requires interpolation and reconstruction.

Data sampled on an irregular grid is a technique which is usually slower but more accurate. It eliminates Moiré patterns. Motion may be simulated by projections from different directions to reveal information about data's structure. Such data may be effectively visualized as 3D arrows or 3D streamlines.

Vector Data. Vector data presents a different set of problems: global vs local areas and projecting 3-D vectors onto a plane without clutter. Common techniques for global data representation involve virtual smoke, clouds, or flow volumes. For local data representation particle-based methods are used. The color sphere method is effective. It emphasizes and classifies critical points.

Visualization of Time-Varying Data

Visualization of time-varying data is a problem of another magnitude. One can use a sequence of 3D images, but temporal aliasing raises a problem. Using a constant sampling rate on time-varying data results in undersampling unless something is altered to compensate. For rotation, each rotation is sampled twice. Half a rotation per frame can be shown without aliasing. With translation, blurring, aliasing, and ringing must potentially be dealt with. Perception of continuous motion may be broken if movement magnitude is more than a few degrees in the eye's field of view (FOV) for a few milliseconds or more.

Applications include feature detection, tracking, and animation. Feature detection occurs within one time step, tracking connects the same features over time, and description involves abstracting in a meaningful, storable manner. Real-time techniques for doing this are described in the chapters on neural networks and on fuzzy systems.

Scalar data may involve feature definition using a specific data range or gradient, or using boundaries and edge detection. Features merge, evolve, and split over time. Some sort of dynamic pattern recognition technique is needed. System identification and reinforcement learning techniques, described by Werbos in this handbook, can help model and track this type of occurrence. Simulation of such changes might be implemented using genetic algorithms or evolutionary systems procedures to find optimal ways to generate or identify such processes.

Flow data can be pictured well by using critical points and regions around them. Tracking and animation monitor properties that vary continuously over time, but slowly between samples.

Multiresolution visualization and analysis offers a rapid economical estimate, with more resolution generated where needed. Wavelets offer this property. Work by Szu and Telfer (1995) and other sections in the neural networks chapter describe wavelet analysis and its successes. Szu chairs a wavelet application conference at SPIE in Orlando each year. The technical proceedings and the tutorial volume are of great value. There are many applications which need a fast progressive algorithm. Wavelet variations differ widely, so they need to be carefully tuned for use in different applications. The *Biorthogonal Wavelet Transform* gives fast, lossless coding, with helpful symmetries allowing good reconstruction. Adaptive wavelets also show great promise.

Rendering Hardware (Sowizral, 1995)

Extracting maximal performance from current generation rendering hardware will allow practical applications of virtual systems, but improvements are always needed in speed and capacity of hardware. Components of VR rendering hardware include the CPU, polygon processors (vertex processors or geometry engines), pixel processors (raster engines), and the display generator. It is important to optimize individual components and balance the flow of information to minimize bottlenecks. This is similar to optimal factory operations, where parallel assembly lines and sequential processes and supply of parts and failure rate all have to be planned and modeled. Factory automation techniques apply very well to high-speed parallel and distributed processing design. Study of available tools (software and hardware) is necessary to apply these simulation design techniques to VR-capable hardware. Design requirements for immersive VR surpass those required for simpler interactive graphics systems, and are thus worth discussion. Demands of speed and massive amounts of memory precipitate trimming designs in ways that have multiple side-effects. Careful choice of shortcuts can bring economy without disaster.

The CPU handles graphic commands, routing data down a rendering pipe(s). The command parser breaks data streams into sensible tasks for the geometry engines.
Polygon processing by geometry engines takes one of two forms:

SIMD (Single Instruction Multiple Data) Geometry Engines.
MIMD (Multiple Instruction Multiple Data) Geometry Engines.

Operations performed include:

Transform vertices and normals into world coordinates.
Scissor transformed triangles.
Light the triangles.
Apply the viewing transform.
Clip to the viewport.
Obtain perspective corrected vertices.
Generate triangle fragments with enough information for the rasterizer.

Pixel processors (raster engines) input triangle fragments and compute the location, color, and depth for each pixel.

The display generator is optimized according to mode. The display-list mode generators predefine their graphic objects, match data to underlying hardware, and use large amounts of memory. The immediate mode gives maximum flexibility.

Object simplification is of prime importance. It can be achieved by using simpler objects, fewer triangles, or by removal of texture. Preconditioning object information can also be extremely helpful. It is important to check determinants of transform matrices (+1) and to check the length of normals. In structuring objects, the designer should place transforms down at the leaves if possible, hoping to avoid unnecessary execution of these processes. It may also help to preprocess polygons or to decompose polygons into meshes.

Feeding the rendering engine is of prime importance. Consideration should be given to aspects such as culling, managing modes, ability to draw quickly, best level of detail, and level of detail substitution.

Some design considerations are managed by the VR application designer, others by the VR hardware designer. The principles of concurrent engineering apply. The most unique and innovative hardware product will not meet the needs of its users without an immense amount of preplanning and careful design.

Software for Immersive Virtual Reality (van Dam, 1995)

Immersive VR software issues differ from those of traditional graphics software. VR is designed to give the effect of interactive immersion in a computer-generated environment. It frequently uses displays with wide field of view (in head-mounted or head-coupled displays), stereoscopic images (LCD-shuttered glasses or twin displays), one or more 6DOF tracking devices for head and hand tracking, and a glove input device. Structure and object interpretation involve motion parallax from head tracking and binocular parallax of stereo. Conventional computer graphics techniques (object motion, perspective foreshortening and depth cueing) are supplemented by appeals to other senses using voice recognition, sound synthesis, and/IR haptic feedback devices. In the taxonomy of VR-related systems, true VR rates are nearly (1,1,1) when measured by (autonomy, interaction, presence). VR systems require a unique combination of qualities found in other systems. Coexistence of these in a single application force development of unique systems. There is a need for rapid update rates and minimal lag. Multiple input devices and human participants in parallel must be accommodated. Virtual environments may have many independent but interacting objects. This places a heavy demand on computer resources, so many hardware and software improvements are needed.

Interaction Models

Modes vary. In one mode a single user is in charge in passive environment (walkthroughs, CAD construction). Here extensions of conventional systems are acceptable. On the other hand, the system may be multimodal, parallel, with a high bandwidth input. For example, when a participant is moving, fixing his gaze, gesturing, and speaking a command such as "delete", the object to be deleted is selected based on a complicated function. This is termed by von Dam to be "probabilistic, context-dependent and hierarchical".

COMPUTATIONAL INTELLIGENCE/
REAL-TIME VERSION IS NEEDED

Techniques of merit in this situation may include reinforcement learning and controls, pattern recognition with wavelets, fuzzy LVQ in the future, and right now: fuzzy focus on commercial camcorders. For training simulators, or design-tool simulators, one can use fuzzy numbers to graphically display trends and risk analysis estimates.

Needs include a programming style where threads are cheap and numerous (thousands). Distributed process management software such as that by Ernest (portable C++ library from Ithaca Software) is recommended by von Dam. There is a need for unified interactors and application objects (e.g., for 10,000 simultaneous participants).

Object Oriented Systems

For object oriented systems, van Dam recommends a delegation-based graphics system: Examples mentioned are *Garnet* for 2D user interface construction (Myers et al., 1990); and *UGA* for 3D interactive illustrations and 3D widgets (Zeleznick et al., 1991). These promise the flexibility needed for highly interactive VR. Also needed are object systems that provide the ability to add either new objects or new gestures without modifying code.

Interactive Application Development Frameworks

Primitives such as graphics need to be augmented. Time-varying properties must be implemented to allow them to execute Newtonian physics or systems of rules. Faster than expert systems, fuzzy rules and fuzzy neural controllers offer such capabilities. These may be optimized off-line by graphic adapters. Neural networks can be trained with data, then used to fine-tune fuzzy systems. Estimates of time available must be given to objects so they can choose methods of rendering, simulated behavior, etc. that are feasible, and track and warn of consequences such as aliasing, etc. There is a need to communicate to other objects.

Measures to take for control include those used in a fast-moving factor automation problem.

CONCURRENT ENGINEERING IS NEEDED
WITH CI AND VR

Augmented Reality offers a useful and more achievable solution. In AR, time constraints include synchronization of graphics (and audio) with sampled data. The strategy moves from the graphics plan of just processing frames as fast as possible, to one similar to a queuing simulation, where tasks are scheduled and assigned a time. This can be handled as an extension of simulation queuing scheduling and the technology developed for integration of missile seeker images with guidance and control strategies. The most elaborate and accurate techniques in existence do no good if it takes so long to interpret the image that the target is long gone.

Reducing resolution or accuracy in favor of speed will have an application-specific cost. If a defending missile misses an incoming missile, people die. If a moving object in a walkthrough simulation is blurred a bit, the human perception of the event may be relatively unchanged. Acceptable miss distance, signal to noise ratio, probability of hit, etc., are design considerations which may be appropriately expressed as fuzzy numbers. In the words of an industrial statistician, errors can "fry the innocent" or "free the guilty". The possibilities and penalties associated with these conditions need to be dynamically assessed, acted upon, and recorded for future reference.

> *Time-critical computing and graceful negotiated degradation are ideal candidates for merger with neural, fuzzy and evolutionary systems using the principles of concurrent engineering design.*

An application framework cannot be monolithic. Instead, van Dam recommends a distributed object system of the type contemplated by the Object Management Group (1991).

As mentioned above, it is crucial to be aware of the division of responsibility between designers of application and designers of the application framework, and to use concurrent engineering design principles.

Augmented Reality (Azuma, 1995)

Augmented reality (AR) can use state of the art technology and is promising for industry applications. Augmented Reality can be thought of as a reality-in-the-loop interactive in a real time, immersive simulation which is registered in 3D. Azuma (1995) defines AR in terms of desired characteristics—not by specific technologies used to attain these (e.g., head-mounted displays are not required). This allows monitor-based interfaces, monocular systems, see-through HMDs, and combinations of these. The purpose is to enhance the user's perception and interaction with reality. With Augmented Reality, superman's x-ray vision becomes a reality.

Aerospace simulators deal with this type of sensor fusion in tracking, say in IR/MM wavelength target images. These are hard to register, but when fused, each sensor source provides information not in other one. Of course this also introduces two kinds of noise. The signal to noise ration (SNR) is an important consideration. One needs to be sure the process is increasing the INFORMATION content of the fused image. Here in AR, the purpose is to help a user perform a real-world task. AR can be used to map color-enhanced edges around an object to aid in identification, manipulation, or evaluation. Using another example from the NCSU laboratory, consider the GE-funded bottle-recognition project. The bottle was to be inspected for regularity and upright position on the assembly line. An AR approach might be to:

1. *Do the image processing*: Preprocess and segment the image, smooth the segments, use an edge detector, and then produce an image of brightly colored edges, outlining the base cylinder and cone of the bottle, and their intersection (or outlining an ellipse from the base of a fallen bottle).
2. *Project the edges onto the original bottle's edges.* Consider the common technique of superimposing the edge outline on the original photographic image of the bottle. Extend this process into AV by projecting the edges onto the original bottle.

Suppose this bottle-shaped piece is part of a complex machine. Misalignment can be quickly detected and adjusted. Grasping the cylindrical bottom of the bottle becomes easier for a robot. The grip needs to be well below the bottom of the cone. Insertion of the cone-shaped top of the object into its proper place, or polishing it, then replacing it, is easier, because the robot is not coping with fallen or inverted objects. Shining a light onto the object lets multiple observers see the same signal and work together.

A variation of this technique is teleoperation of a robot. Suppose there is a communication delay or some other access problem with directly manipulating a robot. A mock-up robot can be manipulated on-sight, with results mapped onto the real world. Once a strategy has been determined, the planned maneuvers can take place. Such planning and previewing capabilities remind one of the training of a truck-backer-upper neural network and the resulting extrapolation of a fuzzy ruleset. This type of model-based training might allow customized neural fuzzy systems to be tailored for awkward-to-manipulate or unusual physical systems in remote locations.

Assembly, maintenance, and repair of complex machinery can be done by people with less experience, or remotely directed, if 3D drawings are superimposed on the actual equipment. Step-by-step instructions can be provided. Some examples include a laser printer repair application by Steve Feiner's group from Columbia (Feiner, 1993). Another example is Boeing's wiring harness guide for technicians (Caudell, 1994). Promising for automobile engines (Adam, 1993). Another application is lighting the edges of objects in the shuttle bay, as seen in orbit. Still another is labeling. The lighted edges on the bottle might have segment labels and evaluations about equipment state displayed on them. A tracking device might be put on the bottle, so the computer knows where it is, and the label, BOTTLE, can appear on it continuously.

Assembly-line restructuring could be pre-visualized, with images of equipment to be relocated being displayed on the factory floor. Simulated machined parts could be displayed moving down an empty line at different rates, and bottle-necks could thus be observed in a realistic manner. A boat ride was planned in the construction of the lake in the international part of Epcot Center. There are four boat docks which continually move people back and forth. Simulation studies showed that three docks would work better than two or four. This was counter-intuitive, so a visual, real-time simulation was developed for the management presentation (and later presented orally at a simulation conference). An augmented reality version of this project might have used the architectural mockup of the lake with docks and displayed virtual boats moving across the lagoon in various patterns, including the optimal one. This would be much easier than building model boats with variable speeds and paths of motion. Even a simple red dot of light moving across the surface of the water would give the viewer a good idea of the speeds and patterns involved. As anyone who has seen the laser show on that lagoon knows, controlling the motion of a variety of lights skipping across the actual lake is a practical, spectacular capability available today.

Another application for consideration is stepper alignment in semiconductor manufacturing. See the article by Takagi in this chapter. Canon uses fuzzy logic to interpret the alignment error of chips on a wafer based on reflected laser light. An extension of this technique might display a red light on a chip determined to be too defective to correct, or too much out of line for automatic adjustment. A technician could then intercede.

The supply of such easily achieved applications is endless.

Characteristics and Issues of AV Systems.

Tradeoffs exist between optical and video systems. For industrial applications, safety and economy cause optical see-through systems to be favored. These do not blind the user during power failure and require less expensive camera equipment. There is a challenge beyond the registration of images: focus and contrast for human perception. Augmentation can add or remove objects from environment. It can clear out existing equipment to visualize an empty room, and it is effective even without detailed photorealism. Other senses can help. 3D sound with earphones, blended with environmental sound from mikes, could cancel some unwanted signals. Tactile feedback can be used in AR. Overlay the background feel of a hard desktop with the force associated with a virtual object "located" on the desktop.

Some design considerations for AR are given in detail in Azuma (1995). In particular, optical versus video solutions are discussed. Table 111.2 below summarizes the points raised.

Elements of an AR system to be specified include the scene generator, display device, and the tracking and sensing components. See Table 111.3.

Sources of registration error are both static and dynamic. Consider optical distortion. Regular distortion can be computed and compensated for using traditional techniques. Irregular distortion can be corrected by having a neural net learn the compensating function. Other errors include errors in the tracking system and mechanical misalignments. Errors in the output reported from the tracking system are difficult to characterize, but techniques exist for doing this. Perturbing the system and testing response against known elements is troublesome, but helpful. Mechanical misalignments are model vs reality deltas. Some can be calibrated and corrected. Others cause system rejection or re-engineering. Incorrect viewing parameters (FOV, tracker-to-eye position and orientation, interpupillary distance) are severe problems. For the HMD center of projection and viewport dimensions may be offset, with differences in both translation and orientation between the location of the head tracker and the user's eyes. This can cause systematic static errors which can be reduced by manual adjustments, direct measurement, or other methods. The recommended solution is to construct a set of view-based tasks that require positioning head and eyes with respect to objects in the environment, etc. Multiple measurements and use of optimizers can help to find the best-fit solution, or at least one that is good enough. For video systems and camera calibration, intelligent optimization schemes are commercially available.

Dynamic Errors

Dynamic errors in augmented reality systems are due to system delays or lags. The definition of end-to-end system delay given by Azuma is the time difference between the moment that the tracking system measures the position and orientation of the viewpoint, and the moment when the generated images corresponding to that position and orientation appear in the displays.

The difference in time from observation to display of calculation may be typically 100ms, but can be up to 250ms. Approaches are to reduce system lag, apparent lag, match temporal streams (video), or to predict one or more of the following:

- *System lag.* This sacrifices throughput for minimal latency, and is not practical to reduce enough using today's technology.
- *Apparent lag for systems using head orientation only image deflection.* This generates a large scene. Then at the last instant select a window near the current orientation. It won't work on translation, but one might try image morphing for small updates.
- *Temporal streams.* Dynamically adjust the video of real and graphical images to synchronize. This, however, means both real and graphical are delayed, which is a real problem for telepresence. It is NOT a problem for optical see-through HMDs.
- Inertial sensors and predictors work for short system delays, render scene with predicted locations

Vision based techniques close the loop. They detect features in the environment, and use these to force registration. Detection and matching takes place in real time and is robust. Special hardware and sensors may be used. For example, the placement of fiducials—LEDs or special markers in the environment—is very effective. Other template matching techniques for registration can be used, e.g., radial basis functions. Laser range finder and multiple sensors are another solution. The easiest, most practical recourse is to make the user do some things when automation breaks down, e.g., when a view is obscured. For sensing and tracking AR systems need a lot of input variety and bandwidth, accuracy and range.

Futures in AR

Trends in augmented reality favor the use of hybrid, multi-algorithmic approaches with fall-back strategies. Fuzzy decision strategies are in this category. The need for real-time systems requires planning ahead. Accurate time stamps are needed, since one cannot have arbitrary swap-out of processes. The technology is flight simulator-like. Multimodal enhancements, especially 3D

Table 111.2 Comparison of Optical and Video Displays for Augmented Reality (Azuma, 1995)

	Optical	**Video**
	HMD see-through version: reflect and transmit eg Head-Up Displays in military aircraft. percentage of light to allow from real world, and blending is a design problem	Monitor-based (optional stereo glasses), or HMD see-through version. closed-view HMD plus 1 or 2 head-mounted video cameras. Power dependent. graphic objects have strange colored background, which is detected and replaced by real world. Nicer to have depth perception as a threshold, real cover virtual and vice versa
SIMPLICITY	SIMPLE. ONE GRAPHIC STREAM. REAL WORLD VIEW NEARLY UNDISTORTED	SEPARATE VIDEO STREAMS FOR REAL AND VIRTUAL IMAGES. HARD TO SYNCHRONIZE. VIDEO DISTORTION: OPTICAL DISTORTION FROM FRONTS OF DISPLAY DEVICES. EXPENSIVE AND COMPLICATED CAMERAS AND COMBINERS
RESOLUTION	REAL WORLD RESOLUTION UNREDUCED	RESOLUTION OF DISPLAY DEVICES
SAFETY	NOT BLINDED IN POWER FAILURE	DANGEROUS IN POWER FAILURE
EYE OFFSET	NO EYE OFFSET	EYE OFFSET BY HEIGHT AND DISTANCE BETWEEN CAMERAS. MIRRORS CAN FIX
FLEXIBILITY	POOR OCCLUSION-GHOST-LIKE IMAGES	FLEXIBLE MERGER OF REAL AND VIRTUAL
WIDE FOV	NARROW FOV FOR GRAPHICS, WIDE FOR REALITY, WORKS WELL	CAN REMOVE DISTORTIONS WITH FILTERS (EG NN*** FILTERS)
TEMPORAL MISMATCHES	INSTANT REALITY, DELAYED GRAPHICS	CAN DELAY CAMERA IMAGE TO MATCH GRAPHICS, BUT THEN EVERYTHING LAGS AND PREDICTING FUTURE REALITY IS HARD *** FUZZY, FN CONTROLLERS
REGISTRATION STRATEGIES	HEAD TRACKER DEPENDENT	HAVE DIGITIZED IMAGE OF REALITY
MECHANICAL ASSEMBLY, REPAIR	CHEAP AND SAFE	MORE COMPLICATED
FOCUS		BOTH PROJECTED AT SAME DISTANCE, BUT VIDEO CAMERAS DEPTH OF FIELD AND FOCUS SETTINGS MAY MEAN PART OF REALITY NOT IN FOCUS ***FUZZY FOCUS. MATCH GRAPHICS RENDERING
CONTRAST	LARGE DYNAMIC RANGE IN REALITY. LESS IN GRAPHICS, EYE ADAPTATION CANNOT COPE WITH FULL RANGE CHANGE. SMALL FOV, REALITY IN PERIPHERY TOLERATE LOW RESOLUTION GRAPHICS BECAUSE HAVE HIGH RESOLUTION REALITY	BOTH HAVE LIMITED DYNAMIC RANGE
MEDICAL APPS		FLEXIBLE BLEND OF REAL AND VIRTUAL, NICER REGISTRATION
PORTABILITY	EASIER	

Table 111.3 Comparison of Augmented Reality and Virtual Environment Implementations

	AR	VE
SUBSYSTEM		
SCENE GENERATOR	supplementary few objects realism nice but not always needed	realistic replacement for reality
DISPLAY DEVICE	augmenting, not replacing, monochrome may be OK resolution of reality softens need for resolution in graphics, low graphics FOV OK due to peripheral reality	full color high resolution desirable
TRACKING AND SENSING	registration VIP	not as hard
	especially in medical apps	harder to detect can cause motion sickness, but less noticeable visually
	angular accuracy: can see differences of one minute of arc but HMD trackers and displays are not that accurate	visual capture vision dominates other senses of location can get used to systematic errors

sounds, are important. In many instances the human factors considerations in AR are easier to control than those in totally immersive VR, but still need careful consideration. Industry applications of AR should increase with time.

Industrial Reality

Real-world applications of computational intelligence can enhance the fault detection and identification capabilities of a missile guidance and control system. A simulation of a bank-to-turn missile demonstrates that actuator failure may cause the missile to roll and miss the target. Failure of one fin actuator can be detected using a filter and depicting the filter output as fuzzy numbers. The properties and limitations of artificial neural networks fed by these fuzzy numbers are explored. A suite of networks is constructed to detect a fault and determine which fin (if any) failed. Both the zero order moment term and the fin rate term show changes during actuator failure. Simulations address the following questions:

1. How bad does the actuator failure have to be for detection to occur.
2. How bad does the actuator failure have to be for fault detection AND isolation to occur.
3. Are both zero order moment and fin rate terms needed.

A suite of target trajectories are simulated, and properties and limitations of the approach reported. In some cases, detection and isolation of the failed actuator occurs within 0.1 sec. Suggestions for further research are offered.

What is the difference between the above-referenced missile simulation and *data visualization, interactive simulation, interactive immersive simulation?*

The missile simulation, (Wallis and Feeley, 1989), was intended to be a tool for design and analysis of missile guidance and control systems. It executes, showing the body and fins of the missile and the trajectories of both missile and target. Its speed is adequate for observation of fin dynamics and the effectiveness of the selected test algorithms. After a scenario is run, statistics saved can be graphed. Selected statistics are displayed during the run, but extensive data is available for post-run analysis. Graphical routines show two-dimensional plots, which can be printed using screen-capture.

Diagnostics such as the ones described above are needed to test the robustness of various control strategies in the face of actuator failure. Performing a set of simulator runs with statistical validity is a time-consuming and confusing job. The tedium involved promotes clerical errors due to lapses of attention and interruptions. Automation of the runs and reporting the results is critical to obtaining reliable results. Classical techniques for doing this involve pre-planning the set of tests and results and analytic procedures. Runs can be automated using DOS batch files and a series of data files to physically change the code for each run. Simple C programs can be used to make these procedures a little more elegant and flexible.

Improvements to this process can be made using programs such as AutoRun™ (Adagio, 1994) to time and execute a set of programs. AutoRun™ can be set to branch to different actions based on program output. AutoRun™ can also be set to allow user intervention. It can be suspended for unplanned intervention and alteration of program input or nearly any other exploratory action. This capability makes it easy for a domain expert in missile guidance and control to stop the runs or alter perturbation parameters to follow up on trends and questions that occur. Watching the results of a simulation run fly across the screen builds intuition and offers the chance for capturing the exploratory strategies of experts.

Watching and capturing these exploratory strategies provides potential input to computational intelligence strategies. Some of the analytic procedures considered easy to interpret for trend and risk analysis are based on fuzzy numbers, and can be displayed as part of a Lotus™ presentation, flashing onto the screen between simulation runs. This type of procedure can be effective when a simulation program of interest is available only in executable form or the source is proprietary for some reason. It requires almost no programming. Interactive simulation is within the grasp of any engineer with an interest and good background in statistics.

What does it take for a system to be immersive? Watching a train of residuals run across a screen for twenty minutes or so can be hypnotic. If the results are important, it can be engrossing to sit and "tweak" parameters to try and find patterns. Disregarding the realistic graphics problem for the moment, consider alternative enhancements. Building on experience with domed flight simulators, flashes of light and short bursts of sound can be added to the DOS batch programs to be triggered by simulation results. Prerecorded sounds can be selected from thousands of free or cheap sound effects, or specially recorded. Inexpensive sound accessories can be attached to the parallel port of a portable computer and used to record engine noise, people's voices, and so forth right on site. 16 bit stereo sound equipment and Koss speakers can produce pleasing sounds. Putting these into the simulation to trigger reflexes and instinctive feelings at appropriate times in the scenarios is a potent tool within the grasp of an unskilled programmer with a limited budget. Engineers with more hardware experience can try attaching a Mattel Power Glove™ to the portable computer for a 3D experience with unusual computer control. Books such as *Garage Virtual Reality* (Jacobson, 1994) suggest hardware and software solutions that may at first seem like toys, but Garage VR is published by SAMS and lives up to that group's reputation. *Rend 386* (Stampe, 1994) is a readily available software program that can put a good programmer into world-building in a hurry. Options for 3D with fresnel lens viewers open up real immersive VR opportunities on a "garage" level budget.

Are these items "serious" enough to rate time and attention? Think about the industry atmosphere and conservative management's attitudes toward emerging technologies. Learning to do some simple, straight-forward things with affordable tools can help an ambitious engineer work up a proof of concept demonstration for a practical application. Industrial applications must usually be straightforward and understandable and verifiable before they are considered reliable enough to implement. Existing technology can fit this mold and introduce neural, fuzzy and evolutionary algorithms, integrated with virtual systems into the industrial scene. Data visualization, interactive immersive simulation, and augmented reality are within grasp. Telepresence and elaborate, photorealistic virtual reality are worth reaching for in industrial electronics at many levels (Adam, 1993).

Construction of rules and membership functions for fuzzy systems is an art. Modeling and simulation have long needed "common sense" methods for preserving, imitating, and extending the abilities of domain experts and simulation experts in the realm of establishing parameters for their models. Thinking of rules and membership functions in a fuzzy system as modifiable elements of the system design, the need for systematic methods for formulating these elements becomes clear. How do intelligent system tools enter the picture? Articles in the chapters on neural, fuzzy, and evolutionary systems, and on computational intelligence are full of suggestions. Individualized data can be used to tune the shape of membership functions. Genetic algorithms can be used (off-line) to optimize parameter sets and to drop insignificant rules. These are numeric approaches to fuzzy system tuning which can be fed and tested by inserting "hooks" into virtual reality systems to monitor the techniques, successes and failures of expert and trainee users. A recent paper by Hackwood (1995) extends the scope of sources for rules beyond numerical or verbal descriptors. Illustrating the concepts with examples from law enforcement, the section provides ideas which are relevant for industrial security and trouble-shooting applications. Fuzzy controllers are extended to have multimedia input and output. An empirical but unbiased approach is taken to measuring distances in feature space based on human subjects. A distance metric is used as the membership function, where distance is basically the difference between the "concept," or centroid, and the human perception of the data sample location compared to the central concept. Quantification of the human perception is aided by interaction with the computer display of possible choices and interpretation of responses.

In the control of automated and robotic vehicles, verbal terms such as the following need to be translated into variables such as position and orientation coordinates for incorporation into applications such as training simulators:

Slightly behind.
Dangerous position.
Good view.
High-risk scenario.
Safely halt.

How does one direct the trainee or the situation simulator to move into such a position, or shift in response to such an evaluation? The chapter by Murry describes many motion tracker possibilities, which are configured to compare crisp position, rate and acceleration values. These "crisp" estimates have meaning relative to the global scenario and to the immediate localized objectives. Using techniques developed for missile system tracking systems, and guidance and control comparative analysis (Padgett, 1988), the immediate reaction is to "soften" these crisp values. A location may be interpreted as "close enough" to the target track, or as a "flag-raising" situation. Developing rules which govern concepts such as "close enough" or "flag-raising" involves study of expert evaluations. Interviewing the experts is not completely accurate because very few of them know the precise basis for their evaluations and reactions, beyond having an intuitive feel for what works. Scene-of-the-Crime Reconstruction described by Hackwood parallels the reconstruction of a mysterious event, such as a miss in a missile simulation, a high production line part rejection rate, or a power system failure (Patterson, 1995). Data input to such a problem solving scenario may be a mix of input from observers and measuring devices. Much is imprecise and highly subjective. Perception and recollection vary. Event description may be difficult to verbalize, subject to bias, or hampered by the observer's limited experience base or language skills.

The methodology proposed by Hackwood involved determining fuzzy rules from human input. The premise is that by presenting multimedia choices to humans, a continuous range of values can be selected from. The example discussed uses the verbal categorization of red as Light Pink, Pink, Coral, Peach or Red. It suggests the possibility of improvement in selection by a non-colorblind subject shown a graphical display on the computer screen. The subject can select from 25 equally spaced values of saturation of the red hue at a fixed brightness. Color concepts are extremely individualistic, but the graphical interface tool compensates for that, and was tested on 100 subjects. Extending this to virtual reality would possibly allow the user to turn color control knobs to tune the screen to match a memory or a preference. Interactions could be tested as well, by allowing the subject to further tune the brightness and add texture. Verbal descriptions of texture are nearly impossible to quantify. Similarly, sound could be tuned to match a memory or preference. Caution is advised by Hackwood with regard to drawing conclusions from user interaction with a complex environment. Factors such as sex, race, or age of a human in the scene may introduce bias. This can be detected by stimulating the scenario generator to vary factors such as sex, race, or age to look for correlations in reactions. Experiments of this nature are often done in traditional statistical analysis, and these experimental design techniques apply to virtual reality training and testing simulations as well.

Classical psychology indicates that humans can verbally distinguish about five to nine categories, say VeryVeryBad, VeryBad, Bad, Slightly Bad, Neutral, Slightly Good, Good, Very-Good, and VeryVeryGood (Saaty, 1980). Asking for a performance evaluation figure between 0 and 100% may produce answers such as a 57.39% rating for employee effectiveness. Answering this type of question with such precision is relatively meaningless, and has a negative impact on the perception of validity by the employee being rated, and by the aerospace industry managers which recently encountered this example. Backing up to the analytical hierarchy-type rating selected from nine categories makes more sense. Still, specifying one of the five to nine categories gives a crisp answer to a situation which may have a lot of variation. In recent work, (Padgett and Evers, 1996) propose a rating of a central value, a low range and a high range. These values can be used to graph a triangular fuzzy number. To track a trend or make an analysis, a fuzzy number representing the observed state can be subtracted from the fuzzy number representing the desired state. The resulting fuzzy number graphically portrays the relationship between expected and observed values. This visual display provides useful information about system dynamics in situations where assuming a particular probability distribution is unreasonable. Practical application of this technique has been made in aerospace industry management, and is being explored for use in missile system simulations. Actuator failures can be detected fairly rapidly using two sets of fuzzy numbers to describe variations between windows of measurements. The following example charts are reprinted from Padgett's ANZIIS presentation (1995).

For detecting an anomaly within 10 msec, missile fin positions and rates are monitored. Using data windows of 10 measurements in 10 msec, comparisons of the "centroids" and ranges of these windows gives an idea of the similarity of these two "clusters" of data. Similar windows are ignored, but differences in windows indicate a problem. If an actuator becomes sticky, response to acceleration commands slows. This immediately sets the zero-moment term of the control algorithm off balance See Figure 111.11. Fin rates are impacted by smaller-than-expected accelerations. See Figure 111.12. Other fins compensate by making wide swings and oscillating. Sometimes this action is enough to allow success of the mission, but often not. Since maneuvers also cause swings in the zero moment term and variations in fin rate among fins, care must be taken in assuming that cluster differences indicate failures. In this particular example, however, a missile can roll to move a questionable fin away from the sensitive rudder-control role, and continue to operate. Thus a fuzzy number measure of cluster difference seems appropriate as a decision aide to the automatic control system.

Fuzzy risk analysis compares two fuzzy numbers (fuzzy clusters) by subtracting their central points to obtain the risk central point, and taking the worst case and best case differences as risk end points. See Kaufmann and Gupta (1994) and the section by Cooper in the chapter on fuzzy systems.

Examples of these calculations and triangular fuzzy number diagrams illustrate the possibilities and cautions. See Figures 111.13–111.14. Numerous traditional methods for estimating a center and spread for a set of numbers exist. In the missile example presented, the data is fairly well-behaved.

A simple but effective use of fuzzy numbers is illustrated below. Actual figures have been modified to preserve the privacy of the source. Risk analysis is performed by subtracting the fuzzy number for observed expenses from the fuzzy benchmark for expenses. The fuzzy benchmark is taken from an industry source, with no calculation formula provided. The lower and upper ranges of the benchmark are the smallest and the largest yearly measurements provided by participating industries. This number has a dubious origin, but satisfies the industrial participants. No probabilities or distribution estimations can be extrapolated. The fuzzy expense observed by one industry in one month is expressed as this month's expense estimate as a central number. The lowest monthly expense of the year is the smallest point; the highest

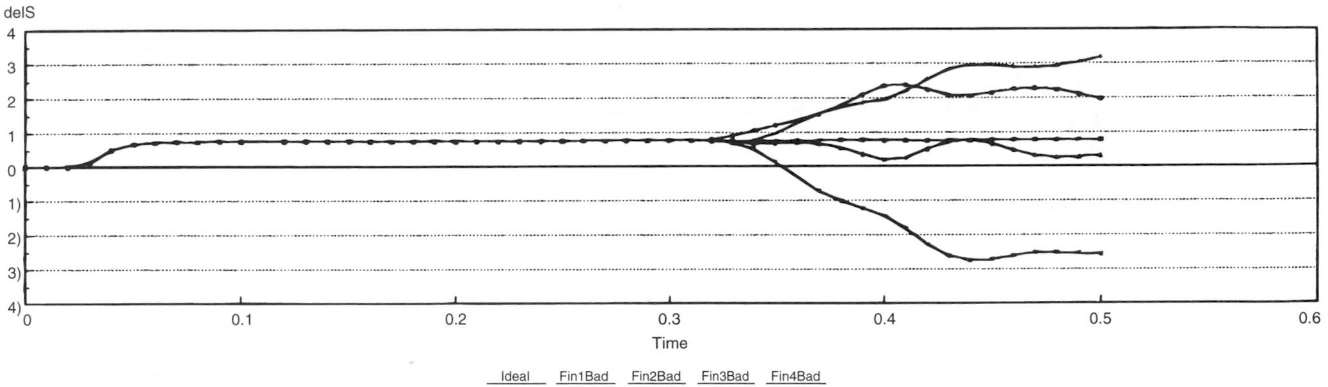

Figure 111.11 Zero moment term for fins.

monthly expense of the year is the largest point. These numbers are expressed as triangular fuzzy numbers where the central number is believed with a strength of 1, and the lower and upper bounds have a belief of 0. Values in between have a belief value based on the straight lines illustrated in Figure 111.15. The fuzzy risk, or fuzzy estimate of difference between observed and benchmark fuzzy numbers is calculated by fuzzy subtraction. (See Chapter 82). The central points are subtracted directly. The lower bound is the worst case, and the upper is the best case scenario. To compute the observation, A, minus the benchmark, B, consider the bounds of A to be (a1, a2) and the bounds of B to be (b1, b2). A–B has bounds (a1 b2, a2–b1) (Padgett and Padgett, 1995). Glancing at Figure 111.16 shows that the figure for this month is favorable compared to the benchmark. Range information displayed is also reassuring.

Another area of keen interest in fuzzy systems is the problem of how to aggregate fuzzy membership functions. Genetic algorithms may be used to test many possibilities and drop the least effective ones. In virtual reality scenarios, the time consumed by most genetic algorithm implementations precludes on line use. However, their incorporation into the development of training simulator test scenarios is quite practical. Buczak and Uhrig discuss this in the chapter on fuzzy systems.

A virtual environment for detection of high-impedance faults on distribution systems is describe in Patterson (1995). This virtual reality "digital model power system test facility is effectively a high-fidelity 70 KVA CD player with special control software." Detection of below-threshold high-impedance faults has been difficult to impossible, but is addressed quite effectively in this facility. A multi-algorithm approach couched in a virtual environment has proven effective. Neural networks algorithms are among the analytic approaches employed.

The relationship between neural networks and virtual reality systems is outlined by Caudell (1994). Capabilities of neural network systems which can be useful if integrated into a virtual reality system include signal processing, complex functional mappings, modeling and animation, pattern recognition, and speech processing. Control, autonomous agents, real-time knowledge acquisition, clustering, time series prediction, and vision are other roles for neural networks in VR systems. Specific applications include face animation, hand gesture recognition, head track prediction, laser fiber coding, force feedback models, adaptive intelligent agents, and design retrieval in Virtual Distributed Industry. Conversely, as discussed in other chapters, VR can be of great help in neural networks applications by allowing immersive interaction with neuroanatomy structure, network engineering, and state and dynamics visualization.

Computational Intelligence (CI) Connections and Futures

The Problem

Ideally, intelligent electronics for industry applications incorporate "common sense" in autonomous operations. Instead of blindly following predetermined algorithmic procedures, electronics for use in industry should be responsive to changing needs of the systems, and should guard against taking actions with adverse consequences. User interaction and immersive capabilities allow manipulation and exploration of scenarios, extending the realm of concurrent engineering, and enhancing many practical applications. Actual implementation of such sensible actions and capabilities is extremely difficult, but recent advances in computer hardware and other technology have brought some such implementations into the realm of the possible. The material following will suggest approaches to the problem of incorporating "common sense" capabilities and into industrial electronics. Examples of current and future applications will be discussed, and further resources will be identified. Careful incorporation of currently available computational intelligence techniques into existing systems can strengthen and stabilize performance. Such successes then point the way to future improvements. The material following is intended to assist a practicing engineer in expanding the horizons of intelligent industrial electronics into virtual reality.

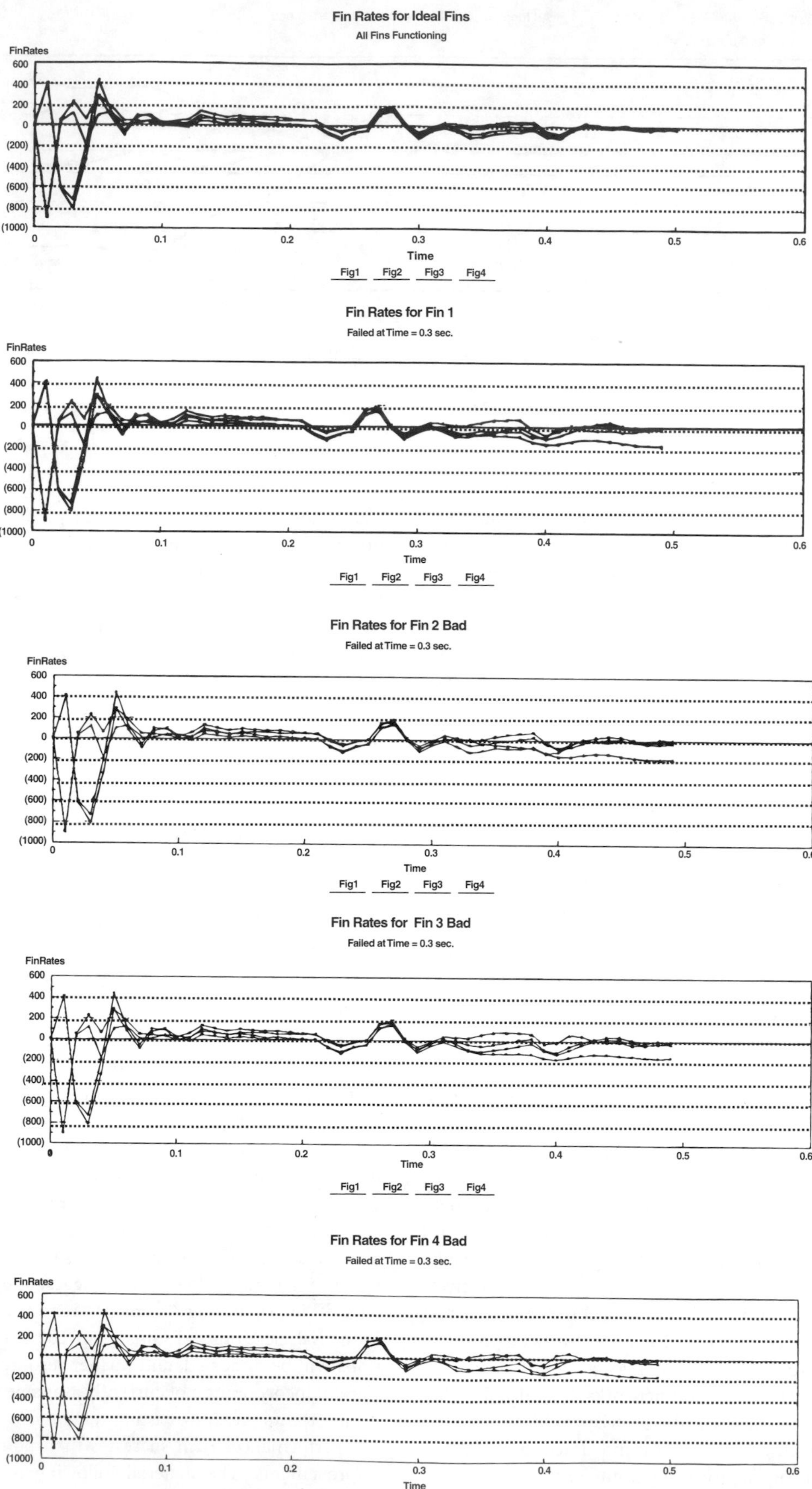

Figure 111.12 Fin rate for ideal fins and fin 1, 2, 3 or 4 bad.

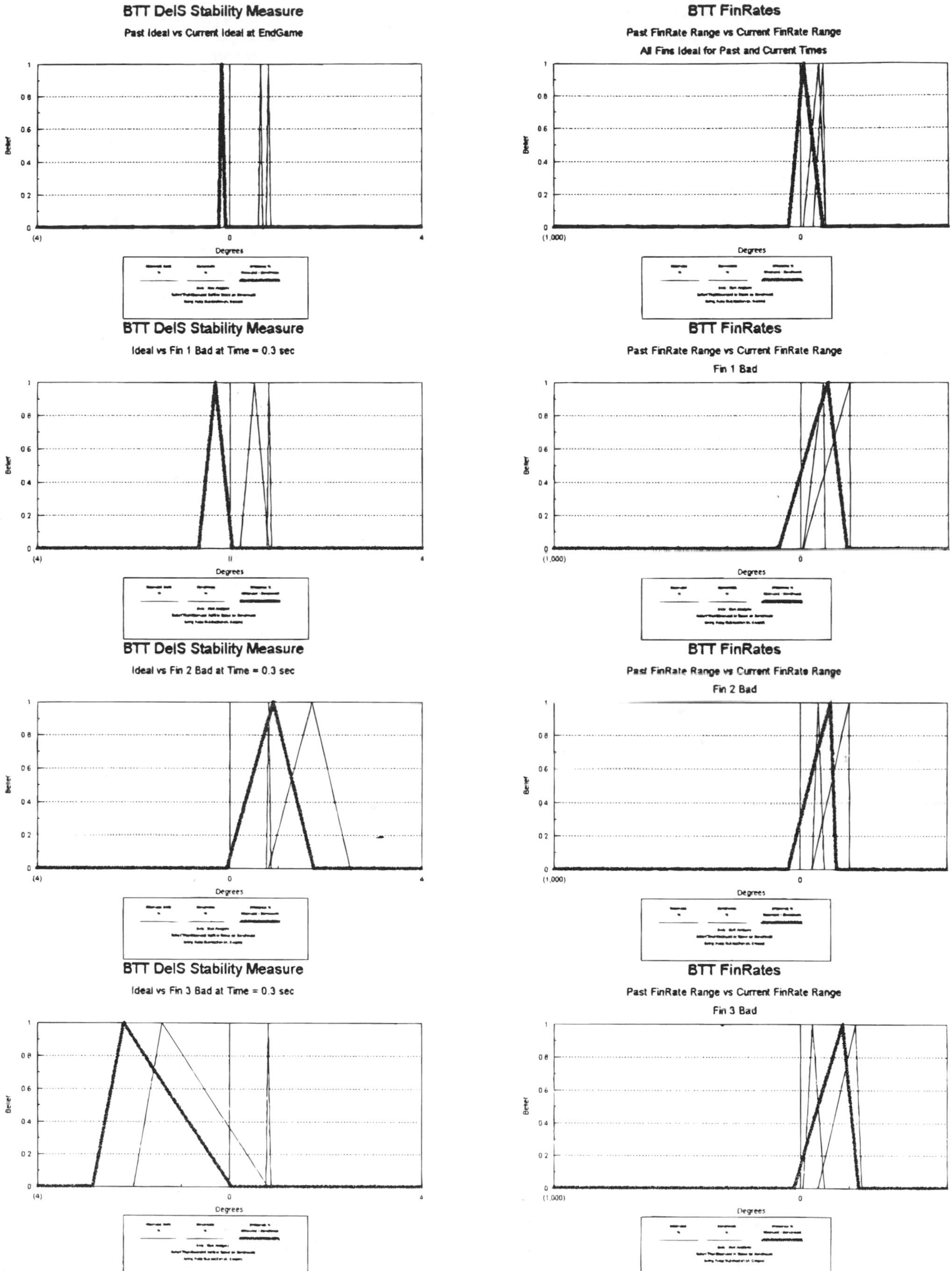

Figure 111.13 Fuzzy stability measure for zero moment term. DelS and for fin rates.

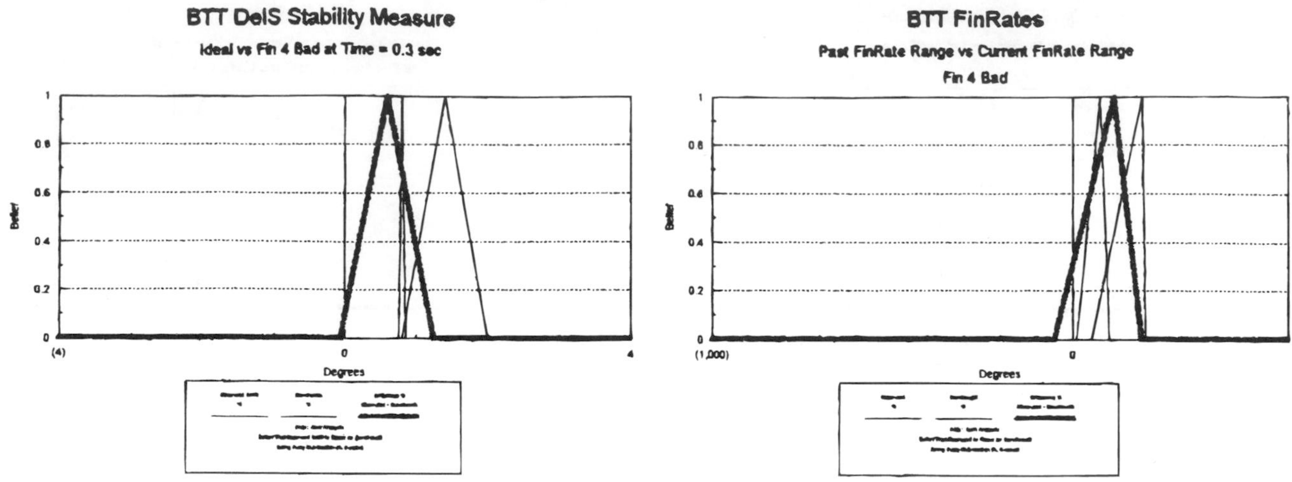

Figure 111.14 a. Fuzzy stability measure; b. Fuzzy fin rates.

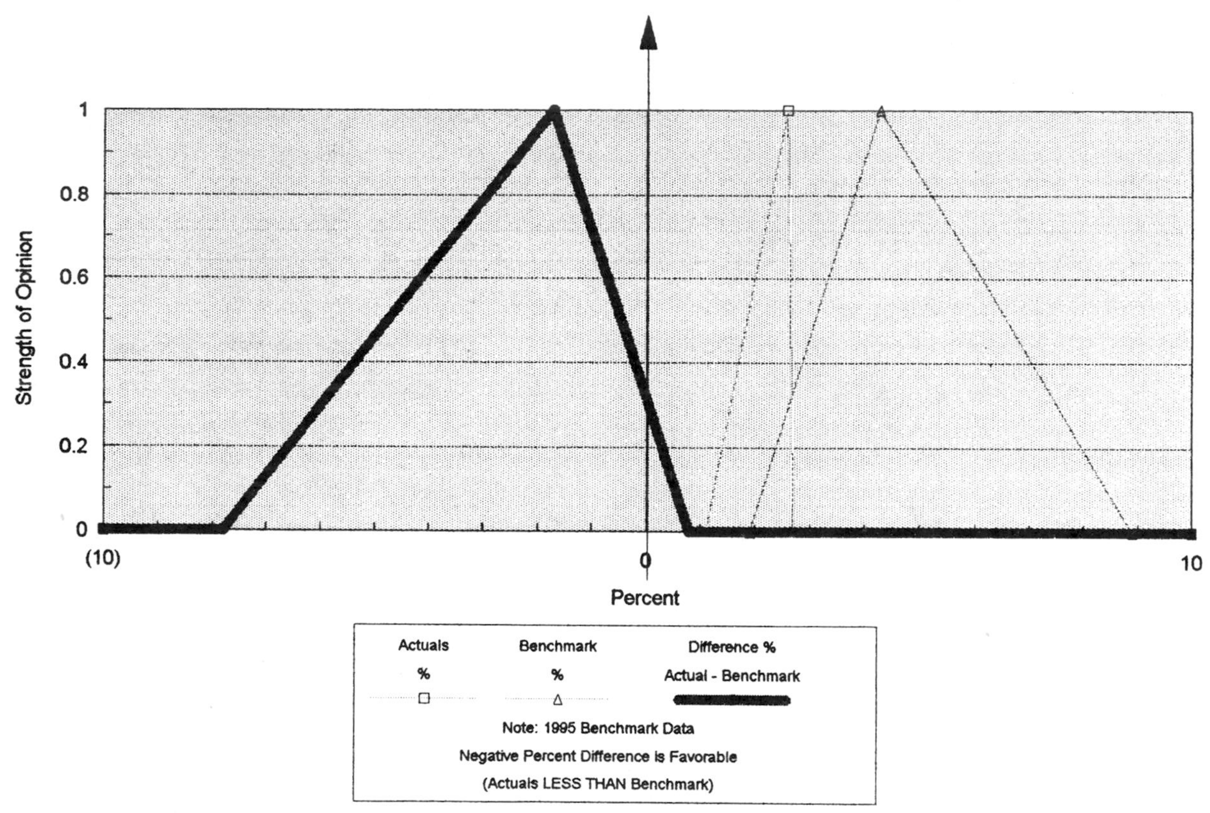

Figure 111.15 Fuzzy number calculation of operating expense.

Engineering Intelligent Electronics Applications

Approaches. Approaches to intelligent electronics use and/or development vary. Successes in industry cited by Carver Mead follow the design procedures recommended by the 1990–1995 IEEE Standards study groups on Computational Intelligence. Based on comments and suggestions from pioneers in the field, such as Carver Mead, Bernard Widrow, Paul Werbos, Teuvo Kohonen, Michio Sugeno, and Lotfi Zadeh, these suggestions recommend clearly stating objectives, and amending them as knowledge of the application under development increases. Traditional industrial applications illustrating these concepts abound. See the chapters on intelligent electronics include expert systems, neural networks, fuzzy systems, evolutionary systems and computational intelligence. The fusion of these technologies can be augmented dramatically with the support of virtual reality techniques fostering interaction and immersiveness. Conversely,

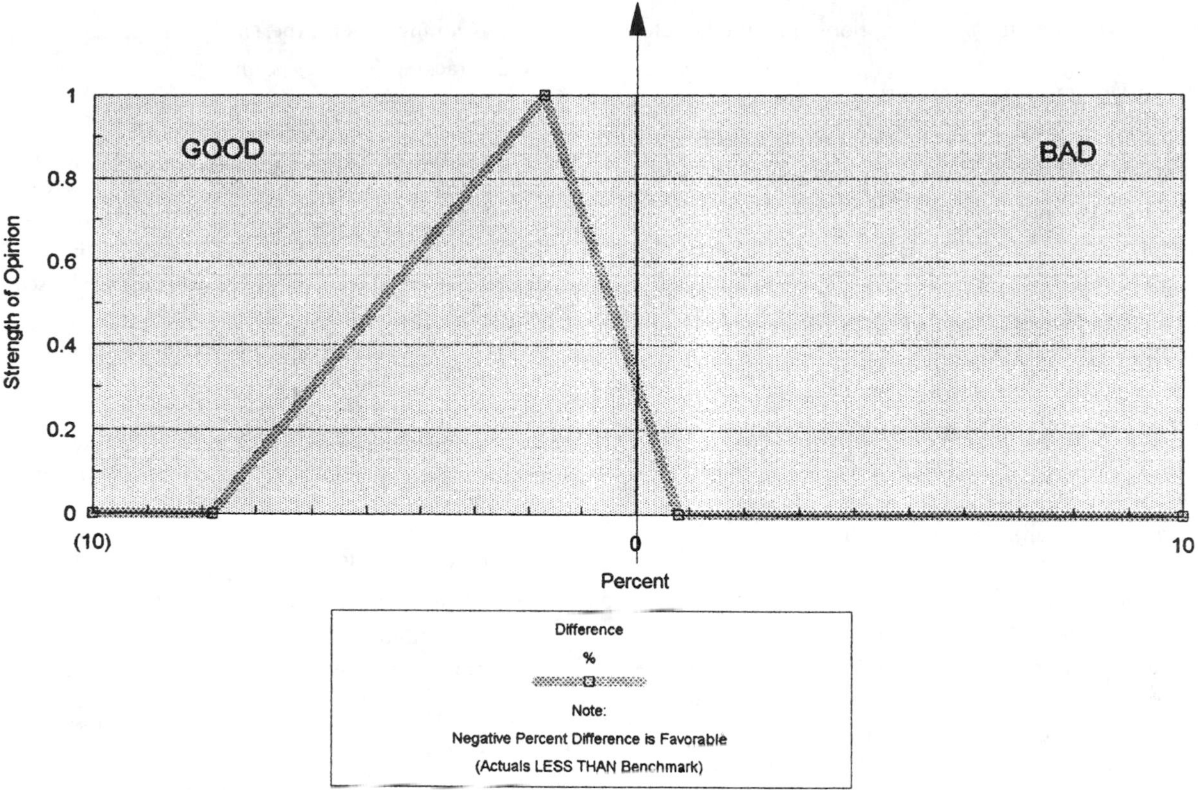

Figure 111.16 Operating expense performance expressed as a fuzzy number.

virtual reality systems can be made more automatic and more realistic with the incorporation of computational intelligence

State objectives. Successful industry applications of intelligent electronics are tuned to their objectives. These projects may be aiming for micro-, systems- or applications-level solutions (see the paper by Werbos in the neural networks chapter). Tiny (micro-) modules of intelligent electronics make useful tools when combined into task-oriented systems-level problem solutions. These task-oriented solutions in turn can be incorporated into multi-systems or applications-level projects which are increasingly combining the strengths of neural (NN), fuzzy (FZ) and evolutionary systems (ES) to solve real-world problems facing industrial electronics engineers. Embedding elements of NN, FZ and ES into expert systems and into traditional systems is a design task which can be strongly aided by interactions with intelligent virtual reality systems. Data visualization helps suggest approaches and may clarify problems (Mead, 94). Expanding the scope of interaction with the computer model to allow extensive real-time interaction between humans and computers can bring ALL of the human's senses to bear on a problem. Tracking an expert's reactions and responses can provide insight and numerical input to an intelligent electronics industry application. (See the chapter by Sugeno et al. in the fuzzy systems section). The careful engineering of a computationally intelligent (CI) system targets objectives and selects methodological variations to approach these objectives in a manner which can be validated and verified. A system with a meaningfully high Machine IQ (MIQ) (Zadeh, 1995) has a System IQ (SIQ) (Padgett and Padgett, 1995) based on the "soft" application of computational intelligence: use it when needed, bypass it when appropriate. Make the CI system meet the needs of the particular application and keep industry management satisfied. See the article by Padgett and Zadeh in the chapter on Fuzzy Systems and Soft Computing.

CI System Objectives:

- Aim for micro, systems or applications-level solutions.
- Know the target market (Mead, 1994).
- Clarify difference between biological, mathematical, and computer models (Mead, 1994).
- Realize and clearly state limitations due to platforms (hardware, software).
- State constraints due to resource availability.
- Consider skill level, man-hours, and environmental restrictions.
- Design for solutions.
- Achieve a high System IQ (Zadeh 1995).

Methodologies. Once the initial objectives have been formulated, methodologies can be selected and adjusted to fit the particular application. The paper by Rumelhart, Widrow, and Lehr in the neural networks chapter discusses issues frequently addressed in the design of NN applications. These considerations apply to all the CI modeling strategies. The list below mentions some of the considerations and options for methodology refinement.

CI System Design Considerations:

- Appropriate application goals and performance measures.
- Preliminary mathematical model: learning strategies.
- Implementation constraints.
- Architectures: block diagrams, components, and connections.
- Learning strategy details: parameters, functions, and timing.
- Modeling data and links to system: scaling, and sources.
- Validation/verification: generalization, and performance.
- Future modifications: modularity, environments, user interactions, and data hooks.
- Interactive visualizations and intelligent VR.

Applications.

Types of problems. Methodologies are selected and refined to address certain types of problems. At the *micro-level,* CI paradigms exist which perform specific operations on input of a certain nature. Neural learning from data may be supervised by reinforcement or unsupervised. Signal transfer, state transfer, and/or competitive learning may occur (Kohonen, 1995). Fuzzy modules translate linguistics or other imprecisely defined sets or classes into numerical quantities which can be manipulated by a computer. Modules fuzzify, process, then defuzzify at an appropriate time (Jang, 1995). Genetic modules process binary strings, performing operations such as selection, crossover, or mutation (Goldberg 1994). Evolutionary modules operate top-down instead of bottom-up, evolving behavior traits of an individual, instead of genes along a chromosome. Evolutionary programming evolves behavior traits of a species instead of an individual. The process can be a continuous function optimization (Fogel, 1995). Neural modeling of data, fuzzy modeling of human judgment, and evolutionary/genetic modeling of exploratory change are powerful building blocks. Combinations and variations of these elements abound, and can combine the strengths of all the approaches. Neural learning can adjust fuzzy rules (see Chapter 91) or enhance the selection procedures guiding evolution (Fogel 1995). Fuzzifying a neural module can strengthen communication back to the expert or larger system. Fuzzy genetic algorithms may add logic to genetic or evolutionary strategies. Conversely, evolutionary/genetic explorations can help free a neural module from a local minimum, and genetic explorations can evolve fuzzy rules. Combining neural, fuzzy, and genetic properties in a top-down and bottom-up method may use the advantages of each technique to overcome the disadvantages of the others (see Chapter 108).

At the *system-level,* these elementary modules can be combined to address tasks such as pattern recognition, controls, or decision support. Fault detection, identification and recovery (FDIR), and system identification are frequent task-targets in CI design (Padgett and Padgett, 1995). Control systems have a hierarchy of possible tasks:

1. Cloning (of an expert).
2. Tracking (of a set-point).
3. Approaching performance measures.

A complex control system application such as autonomous flight may combine a vision system, FDIR capabilities, system identification, and a range of control tasks. Guidance may be initialized as a clone, tuned to track a set-point, and modified to keep the system approaching acceptable performance levels (see Chapters 63 and 91).

At the *application-level,* general purpose or application-specific systems may combine in a hierarchy or be used concurrently to produce a product. Elastic neural networks, soft computing, computational intelligence, and intelligent VR emphasize the strengths of NN, FZ, ES, and VR to solve an application problem.

CI Problem Levels

- Micro—Modules.
- System—Combination of modules for tasks.
- Application—Combination of general-purpose and application-specific systems to produce a product.

Mixing NN, FZ, ES, and VR. Recent successes mixing NN, FZ, ES, and VR with other expert and/or traditional systems are discussed and illustrated in the computational intelligence chapter. Intelligent VR is explored in the emerging technologies chapter.

Implementations. Issues in implementing CI systems are addressed for each of NN, FZ, and ES. First, expert systems and potential combinations with NN, FZ, and/or ES are explored. In the next chapter, neural networks are covered. Micro-configurations of neural networks are discussed ranging from those with supervised learning schemes (multi-layer perceptrons and variations of backpropagation) to dynamic systems using reinforcement learning (time-delay neural networks (TDNN), time-lagged recurrent networks (TLRN), Adaptive Critics, Cerebellar Model Articulation Controllers (CMAC)), and modifications (see the neural networks chapter articles by Werbos and by Haykin); and on to self-organizing learning systems such as self-organizing maps (SOM) and ART. As pioneers such as Zadeh, Werbos, Kohonen, Grossberg and Carpenter have moved into the soft computing arena, variations of their work are presented next in the fuzzy systems and soft computing chapter. The newer approaches to evolutionary systems are added following the suggestions of Fogel (1995) to explain evolutionary computing and genetic algorithms approaches to CI in the evolutionary systems chapter.

In each category task oriented issues are addressed by applications illustrating these micro-level configurations. Preprocessing by wavelets, comparisons to existing statistical pattern recognition techniques, and modifications to controls system approaches illustrate the problems and suggest some solutions.

An overview of integrated applications and possible futures for the field is presented as a summary and a link with other emerging technologies.

Hardware. For neural, fuzzy, and fuzzy/neural applications, hardware implementations are discussed. Specialized neural electronics for neural networks vary from those designed to emulate biological systems to high-tech aerospace applications borrowing ideas from biology (see chapters by Lau, Wolpert, Saeks and Daud et al. in the neural networks section). Wavelets aid image analysis (see (Szu and Telfer, 1995) and chapters by Rogers et al. and Ruck and Rogers in the neural networks section). RBF hardware aids computer communications. (See the chapters by Lindblad et al. in the neural networks chapter). Fuzzy hardware solutions range from very specialized to general purpose. Recent work cites successes in use of general purpose hardware for fuzzy solutions (see the chapter by Padgett in the fuzzy systems chapter). A practical example neural fuzzy hardware follows in the chapter by Lindblad and Lindsey.

Software. Software packages available are discussed in some of the applications articles, and internet sources for code and/or comments are described as further resources.

Other tools. Environments for CI development abound. Obtaining good source code for rapid prototyping is a consideration for any engineering team. Statistical analysis tools and graphical interfaces are also critically important. Articles describing methods, hardware, dynamic systems, and pattern recognition applications provide reliable sources for the combination of micro- and systems-level CI tools into useful packages for research and development of more applications in intelligent industrial electronics.

Intelligent Electronics Topics Relevant to VR

Expert Systems
Neural Networks

Basic Modeling Concepts
Hardware Implementations
Dynamic Systems and Applications
Pattern Recognition Techniques and Applications

Fuzzy and Soft Systems

Fuzzy Systems: Basic Modeling Concepts, Hardware Implementations, Techniques, and Applications.
Fuzzy Neural (FN)/Neural Fuzzy (NF) Systems: Basic-Modeling Concepts, Hardware Implementations, Dynamic Systems and Applications, Pattern Recognition, and Applications.

Evolutionary/Genetic Systems
Computational Intelligence and Hybrid Systems

Future Directions

Many areas of industrial electronics are heavily impacted by the addition of computational intelligence to the set of possible tools for design, development, and implementation of practical applications. Topics such as integrated design, machine vision, computer vision, education and training, machine control, piezomechanics, aeronautical mechatronics, hardware/software codesign, design in textile engineering, balanced automation systems, control and algorithms, control of electrical drives, microelectro-mechanical systems, mobile robotics, sensors and measurement technology, nonlinear control, production automation, robotics and motion control, sensors and actuators in mechatronics, and vibration control are feasible settings for use of modern computational intelligence strategies. Isermann (1995) illustrates simultaneous engineering of design solutions for mechatronic systems. These techniques are discussed in terms of incorporation of neural networks, fuzzy systems and soft computing, evolutionary systems, and/or expert systems in the intelligent electronics section of this handbook. Mathematical models and their biological basis are introduced, and design strategies are outlined. The check-lists and modules cited in the design and modeling papers in this section are intended to be reviewed in the project planning stage, and incorporated into validation strategies within each step of the developmental process. This look-ahead policy emphasizing planning, communications, checking of assumptions, and careful documentation is discussed in the section on factory automation, and in Padgett (1982). The design modules prototyped at the beginning of the project are refined as the development of a practical product iterates through cycles of exploration and fine-tuning. The capabilities of neural systems for real-time learning of nonlinearities, and complex functions from data without a mathematical model or operator knowledge, can be applied to all of the above-mentioned areas. Pattern recognition, controls, fault detection, identification, and recovery can all be performed or enhanced by neural systems working tightly with conventional systems, and with other CI augmented systems. Fuzzy systems provide real-time solutions to non-linear problems and can incorporate operator knowledge plus a database for knowledge representation. Control systems can profit from fuzzy supervisors, and fuzzy pattern recognition can also be very effective. Evolutionary systems learn from historical data and add an element of randomization. They are effective with nonlinear systems and optimization. Sensitivity analyses, fine-tuning of system parameters, escape from local minima, and the ability to turn corners distinguish these systems in the practical arena. Expert systems provide shells for interfacing these tools with the rest of the system.

Future directions for the incorporation of these CI elements into industrial electronics feature the integration of neural, fuzzy, and evolutionary systems into top-down and bottom-up structural development of problem solutions. Increased speed for processing and reduction in computational complexity will help the merger of these techniques. Concurrent engineering of applications using computational intelligence can provide the basis for extensive use of these technologies. Carefully structured and recorded nontrivial experiments with designs and performance evaluation strategies can provide detailed guidance for implementation. As the lessons learned, the examples, and the successes are shared in detail, the field will grow and flourish. Development of standards for terminology and documentation can help encourage the recording of information for public use. The discussions and international forums accompanying this process may be even more beneficial than the final product. To participate in such discussions, contact m.padgett@ieee.org and inquire about the *CI Standards News.*

Comments on future directions for fuzzy systems and soft computing can be found in the chapter by Padgett and Zadeh in the fuzzy systems section. Combination of neural networks and virtual reality is encouraged by Caudell (1994), which lists research challenges for VR:

- Hardware and software tool development.
- Realism in virtual teleconferences.
- Human 3-D perception and reasoning.
- Human task performance measures and qualitative satisfaction measures; design considerations; impact on performance, types of tasks helped or hindered; system characteristics such as system time delays, positional errors, effective measures of performance, how to optimize them.
- See-through augmented reality (AR): how to get good visual fusion so virtual graphic seems part of the scene.
- Scientific visualization.

Computational intelligence in industrial electronics should continue to expand (Thompson, 1994). Many forecasts imply that engineering development environments which are interactive, use and monitor all the human senses, and, in fact, include virtual reality systems will open the way to engineering practical intelligent electronics for industry applications. http:\\www.mindspring.com\~pci-inc

Glossary

The following terms were placed in the public domain prior to being submitted to the *CI Standards News* (mpadgett@ieee.org or r.blade@ieee.org.). Additional VR glossary terms are found in the June, 1996 issue.

Acoustic tracker: device to receive sound and transform to position and orientation with respect to comparative coordinate systems. Bulky but relatively inexpensive. Subject to attenuation and distortions. SNR high unless LOS.

Acoustic tracking: active use of sound sensors, commonly piezoelectric transducers, to determining direction and range and transform to position and orientation.

Active tracker: generates its own operating field (magnetic, light, sound . . .) and coordinate system. Has a signal-to-noise ratio and various corruptions.

Binaural audio: 3D sound.

Compass-tilt tracking: computation of position and orientation based on integration of a compass and a tilt sensing device. Cheap but inaccurate.

Cone of confusion: symmetries based on duplicate explanations for zero ITD and IID cause ambiguity in location of stationary sound directly in back or front of an unmoving listener. Resolved by motion and additional sampling and by other sources of information.

Data sonification: relationship mapping data to sounds in order to convey information.

Doppler effect: motion of sound toward or away from listener will change the perceived frequency, raising it on approach, lowering it on retreat.

Head-related transfer functions (HRTF): digital filters designed to resolve ambiguities caused by the cone of confusion by modeling the individual's head, torso and outer ear.

Human audible sound range: about 20 Hz to 20k Hz (cycles per second).

Human sound perception: Measured by threshold, loudness, pitch and timbre, with interpretation impacted by variables such as environment, quality, familiarity, stability, frequency sensitivity, fatigue and test conditions.

Inertial tracker: device to compute motion based on properties of mass and inertia. Uses small gyros and accelerometers and microelectronic versions of these sensors.

Inertial tracking: passive use of the properties of mass and inertia to report motion.

Interaural intensity differences (IID): difference between right ear and left ear perception of intensity. Sensitive to high frequencies (above 5k Hz).

Interaural time differences (ITD): difference between right ear time of arrival and left ear time of arrival of a particular sound. Sensitive to low frequencies, below 1.5k Hz.

IR optical tracking: active computation of position and orientation based on current induced by illumination of sensor by infrared (IR) energy, or based on tracking shadows produced by IR light. Less expensive but less accurate than other optical tracking.

Line-of-sight (LOS). line from source to target without occurring objects.

Loudness: perception of the magnitude of the signal based on the level of the sound above threshold. Also impacted by frequency and Sound Pressure Level (SPL). Discomfort and pain thresholds exist.

Magnetic motion tracker: device to establish comparative orientation of source and target using magnetic fields, generated by one set of three orthogonal coils and detected by another such set.

Magnetic motion tracking: active use of magnetic fields and sensors to track the motion of the VR user without need for being in the line-of-sight (LOS). Use pulsed DC or AC fields.

Mechanical tracker: device to compute position and orientation from linkage joint readings and lengths. Subject to sag, resolution error, LOS.

Motion tracker: device for determining position and orientation coordinates as needed, transforming and communicating them to the controlling computer. May use mechanical, head mounted, magnetic, acoustic, inertial, mechanical, optical or other approaches.

Optical tracker: device to compute position and orientation based on location of spots of reflected light or based on an image of an array of LEDs. High resolution. Subject to LOS and interference. Expensive.

Optical tracking: active computation of position and orientation based on spot reflectors of specific wavelengths or based on an array of LEDs (light-emitting diodes), orientation coordinates: commonly azimuth, elevation and roll angles.

Passive tracker: uses some physical phenomenon (e.g., gravity, inertia) as an operating field and a coordinate system relative to that of the physical phenomena. Must be periodically update. Subject to drift.

Pitch: perception of the frequency of the signal.

Position coordinates: commonly in three-dimensional space (x,y,z).

Pre-digitized sound: manipulation of pre-recorded samples. Limited variety.

Propagation: transmission of sound waves through solid, liquid or gas, subject to distortion from reflection, refraction, diffusion or diffraction. Listener's upper body, head and outer ear distort sound.

Real-time sound: processing within one video frame, say 30 frames per second.

Reverberation: sound in an enclosure reflected multiple times.

Signal-to-noise ratio (SNR): measure of accuracy of transmission of information based on percentage of accurate information (signal) to corrupting factors (noise).

Sound localization: 3D sound.

Sound receiver: device for detecting sound source location, size, shape, density and velocity. Postulated by Ohm as a Fourier analysis by the human brain.

Sound source: device or physical reaction initiating a sound wave composed of a set of sinusoidal signals of varying frequency and amplitude.

Spatialized sound: 3D sound.

Synthesized sound: dynamic creation of artificial sounds. Often lacks realism.

Threshold (sound): threshold of distinction between noise and sound based on intensity. Varies with species, size of ear, individual, age and sound.

Timbre: perception of quality of sound based on spectral complexity of the signal. Synthesizers have great difficulty in simulating the timbre of great singers or the richness of sound produced by great musicians.

Virtual acoustics: 3D sound.

Virtual sound: creation and display of sound in a virtual environment which may be 3D. Conveys complex information to a participant in an intuitive manner.

Journals and Proceedings *International Journal of Virtual Reality,* Contact: r.blade@ieee.org.

References

Adagio Software, Inc. 1994. *AutoRun™ User's Guide,* Adagio Software, Inc. 2375 Troicana Ave. Ste. 322, Las Vegas, NV.

Adams, J. A. 1993. Virtual reality is for real, *IEEE Spectrum,* October.

Azuma, R. T. 1995. A Survey of Augmented Reality, *ACM 1995 SIG-Graph Tutorial Proceedings.*

Burdea, G. and P. Coiffet. 1994. *Virtual Reality Technology,* John Wiley & Sons, New York, NY.

Caudell, T. P. 1994. The Application of Neural Networks to Virtual Reality, ICNN Tutorial Number 13, WCCI, Orlando, FL 1994.

Feiner, S., MacIntyre, B., and Seligmann, D. 1993. Knowledge-based augmented reality, *Comm. ACM,* 36(7):53–62.

Fogel, D. B. 1995. *Evolutionary Computation,* IEEE Press, Piscataway, NJ.

Goldberg, D. E. 1994. Genetic and evolutionary algorithms come of age, *Comm. ACM,* March, 37(3):113–119.

Gradecki, J. 1994. *The Virtual Reality Construction Kit,* John Wiley & Sons, New York, NY.

Hackwood, S. 1995. Fuzzy Control and Multimedia with examples from Law Enforcement, SPIE Int. Symp. on Aerospace/ Defense Sensing and Control and Dual Use Photonics, April, Orlando, FL.

Hofer, R. C. and Loper, M. L. 1995. DIS today, *Proc. IEEE,* Special Issue on Distributed Interactive Simulation (DIS), August, Vol. 83, No. 8, pp. 1124–1138.

Idesawa, M. 1993. New types of 3-D optical illusion in binocular viewing, RIKEN Review No. 1 (April, 1993): Focused on Light Science and Technology.

Isermann, R. 1995. Information Processing for Mechatronic Systems, Int. Conf. Recent Advances in Mechatronics, 3–17, August 14–16, Istanbul, Turkey.

Jacobson, L. 1994. *Garage Virtual Reality: The Affordable Way to Express Virtual Worlds,* Prentice Hall Computer Publishers (SAMS),

Jang, I-S, and Sun, C-T. 1995. Neuro-Fuzzy Modeling and Control, *Proc. IEEE,* Vol. 83, No. 3, March, 1995.

Kartalopoulos, S. V. 1996. Time-Dependent Fuzzy Logic, Understanding Neural Networks and Fuzzy Logic: Basic Concepts and Applications. 130–152, IEEE Press, Piscataway, NJ.

Kaufmann, A. and Gupta, M. 1994. *Introduction to Fuzzy Arithmetic: Theory and Applications,* Van Nostrand Reinhold, New York, NY.

Kohonen, T. 1995. personal communications.

Krueger, M. W. 1991. *Artificial Reality II,* Addison-Wesley, Reading, MA.

Larijani, L. Casey. 1994. *The Virtual Reality Primer,* McGraw-Hill, New York, NY.

Loeffler, C. E. and Anderson, T. 1994. *The Virtual Reality Casebook,* VNR Computer Library,

Moorhead, R. J. II and Zhu, Z. 1995. Signal Processing Aspects of Scientific Visualization, *IEEE Signal Processing Magazine*, 12(5):20–41.

Myers, B. A. et al. 1990. Comprehensive support for graphical, highly-interactive user interfaces. *IEEE Computer*, 23(11): 71–85.

OMG 1991. *The Common Object Request Broker: Architecture and Specification*, OMG Document #91.12.1, Rev. 1.1, draft, 10 November 1991.

Padgett, M. L. 1982. A Statistical Methodology for the Development of Computer Simulation Models and Forest Harvesting Applications, M. S. Thesis: 1982, Auburn University

Padgett, M. L. 1988. Statistical Modeling and Analysis, Models for Midcourse and Terminal Guidance Simulations for the Guidance and Control for Strategic Defense Initiatives Project, prepared by the SPI for Eglin AFB., November 30, 59–92.

Padgett, M. L. 1993. Modern Simulation and Standards: Neural Networks, Fuzzy Systems, Evolutionary Systems and Virtual Reality, presentation to *TAPPI*, Atlanta, GA 1993; later published by request in *WNN* proceedings for 1995.

Padgett, M. L. 1995. A Practical Filter for Conflicting or Subjective Data, IEEE-ANZIIS 95, Perth, Australia.

Padgett, M. L. and Evers, J. 1996. Fault Detection and Identification in Missile System Guidance and Control: A Filtering Approach, SPIE's Int. Symp. Aerospace/Defense Sensing and Controls: Applications and Science of Artificial Neural Networks Conference, in press.

Padgett, M. L. and Karplus, W. 1993. Neural Network Basics: Applications, Examples and Standards, Tutorial Text, IJCNN '93, 385–413, Nagoya, Japan.

Padgett, M. L. and Padgett, W. D. 1995. Simulation and Computational Intelligence in Real-World Applications, *Simulation*, 65(1):5–9.

Padgett, M. L., Rajala, S. A., Snyder W. E. and Ruedger, W. H. 1985. Detection of Maneuvering Target Tracks, Proc. SPIE 29th Ann. Int. Tech. Symp. Optical and Electro-optical Engineering, August 18–23, San Diego, CA.

Patterson, R. 1994. Signatures and Software Find High-Impedence Faults, *IEEE Computer Applications in Power* 8(3)12–15, July.

Pimentel, K. and Teixeiera, K. 1993. Virtual Reality—Through the New Looking Glass, Intel, Windcrest, McGraw-Hill, New York, NY.

Rheingold, H. 1991. *Virtual Reality*, Simon and Schuster, New York, NY.

Saaty, T. L. 1980. *The Analytic Hierarchy Process.* McGraw-Hill, New York, NY.

Sowizral, H. A. 1995. Using a Rendering Pipeline Efficiently, ACM SIG-Graph Tutorial, 1995.

Stampe, D., B. Roehl and J. Eagan. 1993. Virtual Reality Creations: Explore, Manipulate, and Create Virtual Worlds on Your PC. The Waite Group.

Szu, H. and B. Telfer. 1995, Wavelet Dynamics, *The Handbook of Brain Theory and Neural Networks*, Arbib, M. A. ed. 1049–1053, MIT Press Cambridge, MA.

Thompson, J. 1993. *Virtual Reality: An International Directory of Research Projects*, Meckler, Westport, CT.

van Dam, A. 1995. VR s a Forcing Function: Software Implications of a New Paradigm, SIG-Graph Tutorial Proceedings, 1995.

Wallis, M. E. and J. J. Feeley. 1989. Bank-to-turn missile/target simulation on a desk-top computer, *Modeling and Simulation on Microcomputers*, 1989, SCS Press.

Zadeh, L. 1995. personal communications.

Zeleznick, R. C. et al. 1991. An object-oriented framework for the integration of interactive animation techniques, *Computer Graphics*, 25(4):15–112.

111.6 Fuzzy Logic Applications in Image Processing Equipment: Intelligent VR Futures

Hideyuki Takagi

Introduction

Since the inception of fuzzy logic three decades ago, fuzzy theory has been realized as a practical technology, and has appeared in widespread industrial and commercial applications. Japan has been the center of such applications, where since the 1980s fuzzy logic control has been applied in industry. One famous example is the Sendai subway system, which uses fuzzy logic to efficiently control its daily operations.

Other applications were presented at the Second IFSA Congress in Tokyo in 1987 and sparked further interest in the field. Since then electronics companies have incorporated fuzzy logic into consumer products such as showers (1988), air conditioner (1989), and camcorders (1989).

In 1990 fuzzy logic burst onto the consumer scene when the first washing machine explicitly advertising the use of *fuzzy* technology entered the market. Japanese consumers readily accepted the word *fuzzy* as a new AI technology since it previously did not exist in the Japanese language. This first washing machine used fuzzy logic in conjunction with sensors to measure dirt quantity and dirt composition for the purpose of determining washing time. These operational differences conveyed the impression of intelligence to the consumer. The combination of a new term and operational differences led to increased sales. The extremely high sales record for this machine prompted electronics companies to use and advertise *fuzzy* technology in other consumer products.

This survey was conducted while the author was a Visiting Industrial Fellow at the Computer Science Division, University of California at Berkeley.

This paper has been used in Hideyuki Takagi, *Survey of Fuzzy Logic Applications in Image Processing Equipment*, ed. by J. Yen, R. Langari, and L. Zadeh, in Industrial Applications of Fuzzy Control and Intelligent Systems, Ch. 4, pp. 69–92, IEEE Press, Piscataway, NJ, USA (1995).

Because of the increasing demand for fuzzy consumer products, many researchers and engineers entered the fuzzy technology field and, as a result both technical and application development grew. On the technical side, automatic fuzzy system design techniques developed rapidly. Many techniques proposing neural networks, gradient methods, genetic algorithms, and meta-rules appeared, and some have already become important tools in practical product development. New research efforts are moving toward the development of learning equipment capable of adapting to a particular user.

Applications that use fuzzy logic have expanded from control to signal processing, image processing, and non-engineering fields. This chapter will concentrate on surveying fuzzy applications in image processing equipment such as cameras, copy machines, and TVs. Like many other *fuzzy* consumer applications, fuzzy logic itself does not perform the low level control, but sets parameters for a conventional controller. The significant point is that fuzzy logic can model and emulate human decision making, which is outside the purview of conventional control theory. The image processing equipment detailed in the following sections provide examples of successful combinations of fuzzy inference and conventional controllers.

Cameras

Canon: Auto-Focus

Canon Inc. has applied fuzzy logic to determine the object of focus evaluate the field of view, and control the auto-focus mechanism to focus on that object (Shingu and Nishimori, 1989). Earlier cameras used the object centered in the field of view as the desired focus. This sometimes led to error, as in the case of two objects presented off-center. The problem had to be solved by the photographer, who focused on one object, locked the auto-focus, then reoriented the camera to get the desired shot.

To circumvent this awkward process, fuzzy reasoning has been introduced to guide the auto-focus mechanism. First, distances to three points in the field of view (shown in Figure 111.17) are measured. Using these locations and the relationships between them, fuzzy logic decides where the desired focus lies, and then focuses on that point. Fuzzy rules used to do this were obtained by an analysis of about 300 pictures taken by 8 people. Table 111.4 and Figure 111.18 show the rules used in the experimental simulations during development of the product. The final product may have more rules (Asami, 1989).

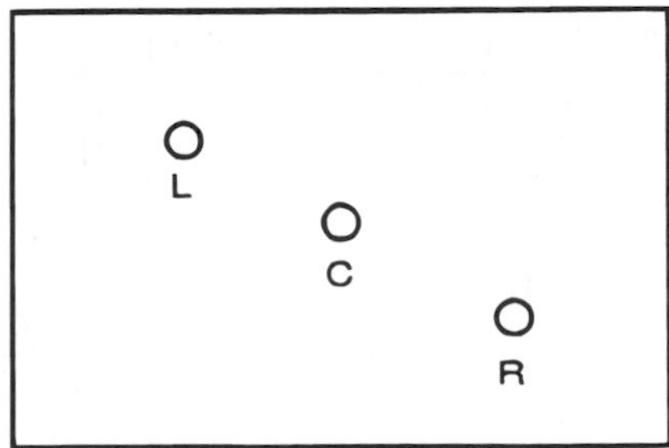

Figure 111.17 Three points to which distance is measured.

Table 111.4 Fuzzy Rules for the Canon Autofocus System: L, C, and R Denote Measures to the Three Points Shown in Figure 17. P_l, P_c, and P_r Denote Plausibility of Finding the Object of Focus There

Rule
#1: IF C is *near*, THEN P_c is *high*
#2: IF L is *near*, THEN P_l is *high*
#3: IF R is *near*, THEN P_r is *high*
#4: IF L is *far* and C is *medium* and R is *near*, THEN P_c is *very high*
#5: IF R is *far* and C is *medium* and L is *near*, THEN P_c is *very high*

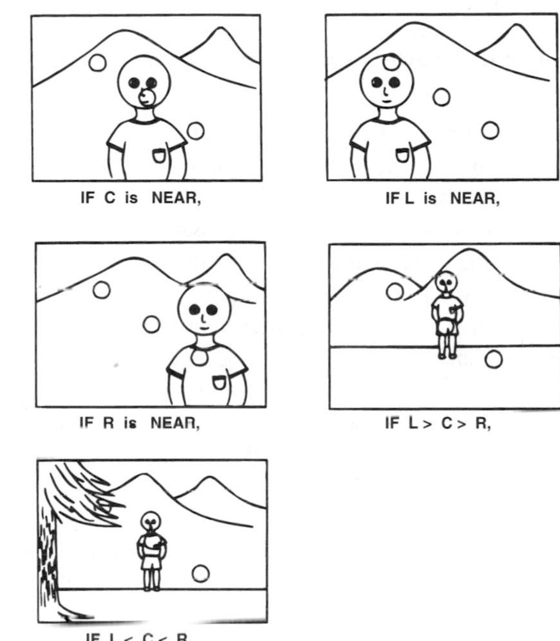

Figure 111.18 Five rules and positions of their main subject.

Comparing Figures 111.19a and 111.19b, we see that Figure 111.19a has the main subject on the left, whereas Figure 111.19b has the main subject at the center. However, the relationship is L < C < R in both cases, and both satisfy Rules #2 and #5. In the case the decision depends on the values of L, C, and R, and the comparison is done with the help of membership functions. It is very hard for binary logic rules to model this situation, and

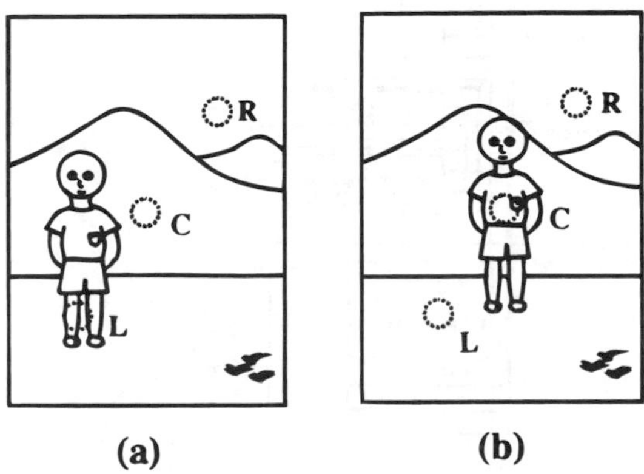

Figure 111.19 Example of scenes.

a large number of rules would be required. A few fuzzy rules, however, can easily deal with the problem.

The performance of this method has been evaluated by using 288 pictures taken by 8 people. The percentage of correctly focused pictures, with and without the fuzzy rules, is shown in Table 111.5. There is an increase of 22.9% in the focusing rate using the fuzzy rules.

The camera incorporating this technology was put on the market in 1989. The inference is realized on a 4-bit microcontroller with a 500-byte memory.

Minolta: Auto-Focus, Auto-Exposure, and Auto-Zoom

Minolta Camera Co., Ltd. has used fuzzy logic to combine the three mechanisms of focusing, zooming, and deciding exposure automatically. Figure 111.20 shows the whole system.

To implement auto-focusing it is necessary to locate the main subject. The fuzzy reasoning system for doing this uses six distance distributions, which are obtained by preprocessing the outputs of four auto-focus sensors, lens information, and one sensor which detects camera position. Seven fuzzy rules, obtained from the analysis of approximately 1000 pictures, determines the location of the main subject to focus on. These sensors are the same as those found in earlier models. Adding the fuzzy logic for decision-making leads to an improvement of 15% in the focus hit rate.

To implement auto-exposure fuzzy reasoning is used to determine exposure value and the best combination of shutter speed and aperture, depending on the type of scene being photographed. The exposure value is determined by three fuzzy inference modules, using brightness values obtained from 14 zones in the field of view and the position of the main subject (determined by the autofocus mechanism described above). The first fuzzy system uses the difference in brightness between the main subject and the background to output a measure of the amount of backlighting present. The second fuzzy system decides whether the exposure is to be focused only on the main subject or on the entire scene. The third system uses the outputs of these two fuzzy modules, weights three measures of exposures (an average, at the center, and at the main subject), then outputs the final exposure value.

Table 111.5 Performance of Auto-focus method

method	focusing rate
three measured distances + fussy inference	96.5%
distance to the center	73.6%

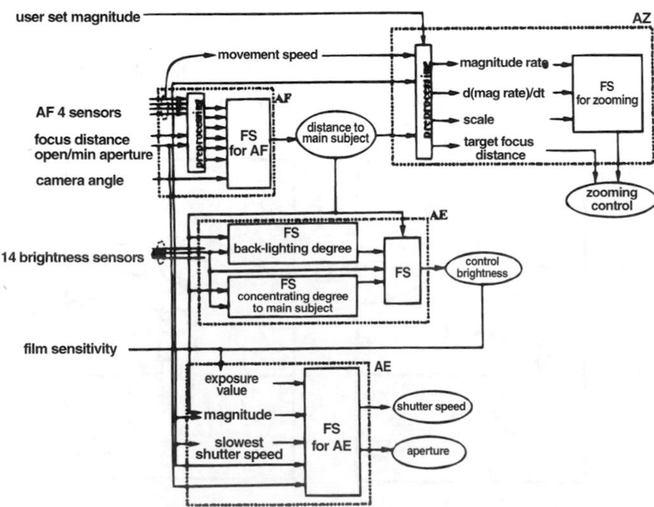

Figure 111.20 Auto-focus, auto-exposure, and auto-zooming of Minolta camera.

The optimal combination of shutter speed and aperture is determined by fuzzy inference using the type of scene and the lens being used. Figure 111.21 shows basic idea of inference rules. The type of scene (for example, snap, portrait, close-up or natural scenery) is determined by the distance to the main subject, depth of field, lens magnification, lowest allowable shutter speed, exposure value, etc. In a portrait photo the focus is solely on the subject. In a scenery shot the depth of field increases. All such detail and fine control can be automated using fuzzy logic.

To implement auto-zooming fuzzy reasoning needs to decide the speed to zoom the lens. When the main subject moves, the size of its image is held at a constant value by zooming appropriately to compensate for the movement. If the zooming is controlled by using the distance to the subject, the error in this value leads to hatching at the correct distance. When the distance information is smoothed, hatching can be prevented, but quick zooming cannot be realized. Fuzzy reasoning chooses the zooming speed by looking at the ratio of current lens magnification to that one unit time ago, as well as its rate of change. The rules change the speed of the lens depending on how the object moves.

Camcorders

Sanyo: Autofocus, Auto-Exposure, and Auto-White-Balancing

Sanyo Electric Co. Ltd. introduced fuzzy logic into the components of the camcorder (Kikuchi et al., 1991): auto-exposure (1989), autofocus (1990), and auto-white-balancing (1991). The brightness signal and color-difference signal obtained from the video signal (via a CCD) are filtered and integrated to give

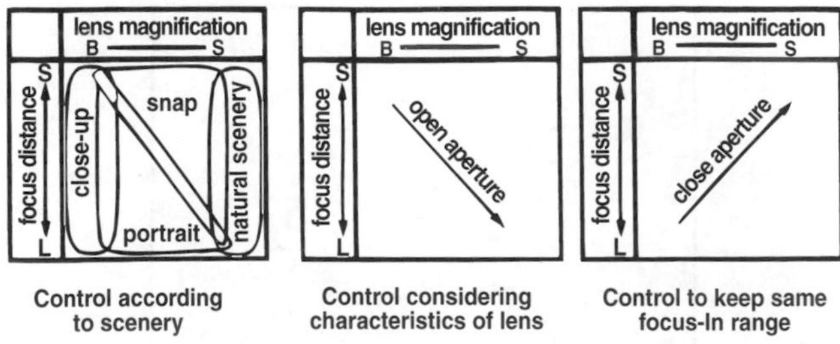

Figure 111.21 Basic concepts of inference rules for auto-exposure of Minolta Camera.

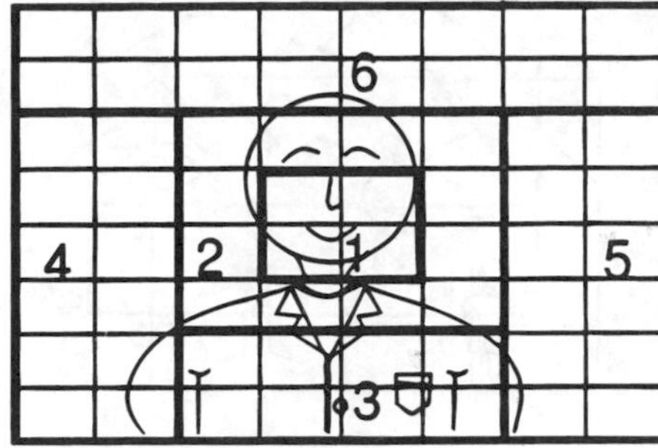

Figure 111.22 The six partitions of the image plane.

a set of 8 × 8 values as shown in Figure 111.26. These inputs are used to realize the automation of the components.

Autofocusing is based on the observation that the high frequency components of the signal increase as the focus comes close to its optimum value. This is easy to do if the area from which the high-frequency component is obtained is small. If the scene consists of moving objects, then focusing based on this idea becomes unstable. A stable focus, on the other hand, will be suboptimal if there is motion in the scene. To solve this problem a fuzzy system using eight signals is used: the high-frequency components are passed through two different bandpass filters, the brightness signal at the center of Figure 111.22 (zones 1 + 2), zooming information, aperture information etc. The system has 21 fuzzy rules and decides the focus area and the focus direction when the object is not in focus. An example of such a rule is:

RULE 111.1: IF brightness difference in center area is BIG and zooming is CLOSE to telephoto THEN $y = 0.8$,

where y is the degree to which the central area contributes in determining the focus.

The autoexposure system decides on the exposure value by evaluating brightness signals from six zones in the field of view (denoted by v_i, $1 \leq i \leq 6$) and assigning them different weights. This handles the problem of excessive lighting from the front or the back. Since the video camera deals with motion, this exposure value must be continuously changed, making the situation significantly different from a still camera. As the scene or the camera keeps moving, fuzzy rules keep matching the situation to different degrees. Since this degree is a real number, the changes in output are smooth responses. In contrast, crisp rules would match completely or not at all, and only one rule would be applicable at one time. This involves discontinuous shifts in the exposure value. There are eleven fuzzy rules, and the j-th rule produces a linear combination of the brightness values denoted by

$$E_j = w_{j1}v_1 + w_{j2}v_2 + \cdots + w_{j6}v_6$$

where w_{ji} is the weight given to the brightness value in area i. Each rule has a different set of weights. For instance, in cases of strong lighting, the rule may be:

RULE 111.2: IF maximum value of v_i is BIG, and average brightness is LOW, THEN those v_i which are smaller than the average are given BIG weights.

This rule emphasizes the importance of darker areas. The final value of the fuzzy system is a weighted combination of the v_i.

White balancing establishes a color reference by demonstrating the color white. This ensures that the recorded colors correspond to human visual senses. Usually this is done manually by showing the camera reference white patches. Auto-white-balancing has to arrive at the reference automatically. One way to do this is to assume that the image contains many colors and averages them to arrive at what the white color should look like. This causes an error because of large patches of specific colors. To solve this problem, the image is divided into 8 × 8 parts, and the color difference signal (R-Y, B-Y) is averaged after weighting each of these 64 areas differently. Large areas of one color or strong brightness are detected and given less weight. Then nine fuzzy rules use eight inputs (such as color phase information, high frequency component from the autofocus system, position of the main subject, zooming information, etc.) to output the reliability of the average color temperature computed in the previous step. For example, a rule might be:

RULE 111.3: IF the high-frequency component is STRONG and the lens is wide-angle and the main subject is at a LARGE distance, THEN reliability is HIGH.

The specification of the three fuzzy systems is shown in the Table 111.6. The three fuzzy systems are not implemented as table lookup, but inference is done in software running on an 8-bit microcontroller. The program size is about 1 KB. All three fuzzy systems use the max-min method for inference. The autofocus and auto-white-balancing systems use constants for rule consequents, whereas the auto-exposure system uses linear functions of its inputs in the consequent part (in the sense of Takagi-Sugeno-Kang model (Takagi and Sugeno, 1985)).

Canon: Autofocus System for Camcorders

Canon Inc. applied fuzzy logic for controlling the speed of the autofocus motor (Kaneda et al., 1990). Their focusing principle is the same as that mentioned in the previous section, namely, that high frequency components are strongest when the image is in sharp focus. To find the correct position quickly, the motor must work at a high speed when the high frequency components are weak (picture is blurred), but must slow down as the correct position is approached.

Table 111.6 Specification of fuzzy systems of Sanyo camcorder

	AF	AE	AWB
# of rules	21	11	9
input parameters	8	6	8
consequent	constant	linear equation	constant
reasoning by		software on 8 bit μ controller	
ROM		total about 1KB	
inference speed		several ms	

The problem of controlling this motor is that the control depends on both the main subject and the type of scene. For example, when the main subject is low contrast, the motor overshoots, whereas excessive brightness causes the motor to slow down prohibitively. Conventional autofocus motors are controlled by using absolute frequency thresholds, as shown in Figure 111.23.

To solve this problem Canon used the frequency values and their rate of change as inputs to a 10-rule fuzzy system to regulate the motor speed. An example of such a rule is:

RULE 111.4: IF the high-frequency component is WEAK, and the differential rate is SMALL, THEN the motor speed is HIGH.

The fuzzy inference used the max-min-gravity method with a simplified approximation for defuzzification. One cycle of inference takes at most 5 ms, and the program size is approximately 1.7 KB. The focusing time was reduced by 20% using this fuzzy technique.

Matsushita Electric: Image Stabilization for Camcorders

As consumer camcorders has become lighter and more portable, the problem of handshaking has become more acute. Shooting a movie while walking or riding in a vehicle exacerbates the problem. Matshushita Electric Industrial Co., Ltd., has applied fuzzy logic to determine whether movement of the image is due to shaking of the hand or of the object being photographed. A camera using this system was introduced in 1990. A customized LSI chip detects the motion vector, and a fuzzy system decides if the motion is due to hand-shaking, then compensates with digital signal processing.

There are two ways to detect jitter due to hand-shaking. One uses vibration sensors; the other uses pure signal processing. The Matsushita camcorder uses the latter method. The inputs to the fuzzy system are four motion vectors, each coming from one of the four regions into which the image has been divided (see Figure 111.24), plus their rates of change. Each of these four regions is further divided into 30 smaller areas. Two successive frames are compared to compute the spatial difference values for each area. These differences are summed over the 30 areas in a region to produce a net difference R_i for region i. For each region, one shift vector results in the smallest value of R_i. This minimizing shift is the motion vector v_i for that region.

Fuzzy reasoning uses the values v_i as inputs to detect shaking of the hand. When this is the case, and there is no moving object,

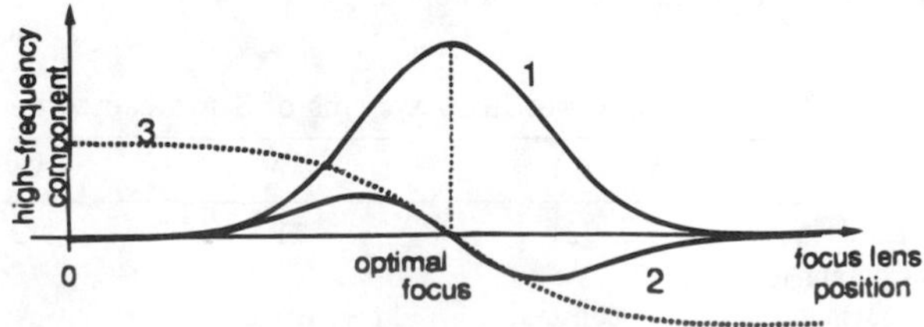

Figure 111.23 Characteristics of focus lens position: (1) high frequency component, (2) derivative of (1), (3) optimal speed of auto-focus motor.

Figure 111.24 Example of motion vectors.

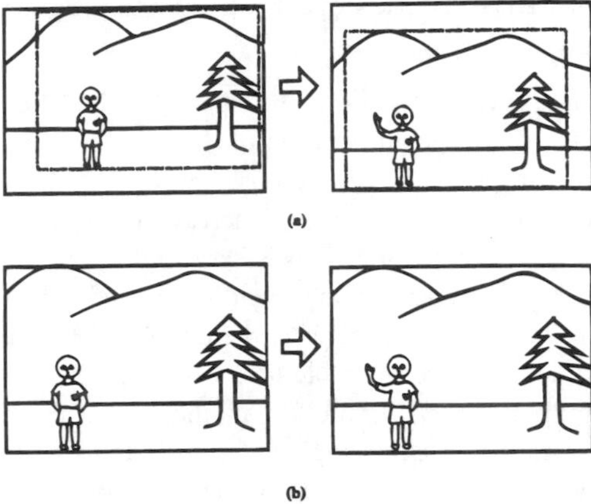

Figure 111.25 Image stabilizing by digital signal processing: (a) frames in memory at the time of t and $t + 1$, (b) recorded image at the time of t and $t + 1$ (Uomori et al., 1990)

then the minimum R_i is almost zero. If there is a small moving object in the image and the hand is steady, then the area corresponding to the moving object has a spatial difference value that is different from the surrounding areas. However, the net R_i values are small. When there is hand motion as well as object motion, then the minimum R_i is bigger than in the no-moving-object case.

The time derivatives of the R_i values are also used by the fuzzy system. This incorporates information about the size and motion of the moving object. A rule using these may be:

RULE 111.5: IF the four motion vectors are almost parallel, and their time differential is SMALL, THEN shaking of the hand is occurring, and the direction of the shaking is the direction of the moving vectors.

Once the direction of shaking is known, the frame in the buffer is shifted in the opposite direction to this motion, achieving stabilization. Figure 111.25 shows two images at the time of t and $t + 1$, depicting how digital signal processing realizes image stabilizing. This image processing is possible only because all signals used are digitized.

Photocopying Machines

Canon: Electrophotography Process

Canon Inc. used fuzzy technology in the electrophotography process of a copy machine put on the market in October 1990.

This process is a delicate one and is influenced by temperature, humidity, toner condition, the ratio of black to white on the original page, etc. In conventional machines, these variables are manually managed according to the environment of the machine. A new product by Canon automated this by using a temperature sensor and a potentiometer which picks up image density (the black/white ratio) from the charge distribution on the drum. The fuzzy system uses these two inputs and controls the charger so that the drum is given just the right amount of charge (Souma and Suda, 1991). Each of these inputs can attain the values high, medium, and low.

The fundamental process of a copy machine controls the charge imparted to the drum in several steps (shown in Figure 111.26):

1. Charge the photoconductor drum uniformly.
2. Erase the charge corresponding to the white areas by scanning.
3. Attach toner to the charged areas of the drum which correspond to the black portions.
4. Decrease the potential of the surface.
5. Transfer the toner to the copy paper by static electricity.
6. Remove the copy paper.
7. Fix the toner to the paper thermally.

Step 4, 5, and 6 are controlled by fuzzy logic.

For step 5, nine rules are used to produce the control value of the charge using three labels each for the two inputs, and three labels in the consequents. They are shown in Table 111.7. One such rule is:

RULE 111.6: IF temperature is HIGH and Image Density is LOW, THEN charge value is HIGH.

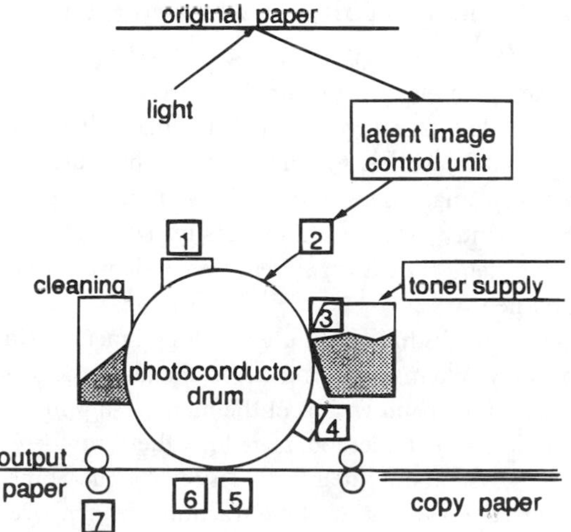

Figure 111.26 Diagram of electrophotography process: (1) charging the drum uniformly, (2) erasing the charge corresponding to the white areas, (3) attaching toner, (4) decreasing the potential of the drum, (5) transferring toner to the copy paper, (6) removing the copy, (7) fixing the toner to the paper.

Table 111.7 Fuzzy inference rules that decide charge value in the process of transferring toner to copy paper

	Image Density		
temperature	low	mid	high
low	low	low	mid
mid	mid	mid	high
high	high	high	high

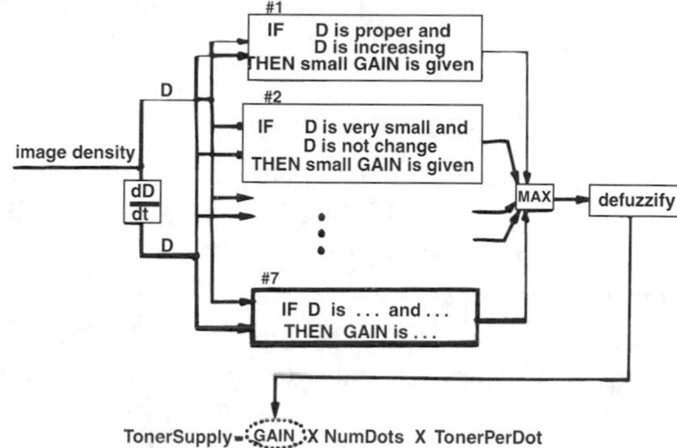

Figure 111.27 Toner supply fuzzy rules of Ricoh.

The inference method is max-min-gravity and uses a Mamdani-type fuzzy controller. The resulting system can adjust to a changing environment, eliminating the need for manual adjustment by maintenance personnel, reducing the number of paper jams, and maintaining a stable quality in copies.

Ricoh: Electrophotography Process

Ricoh Co., Ltd. applies neural networks and fuzzy logic to the electrophotography process to hold the quality of the image constant, even if the environment changes during copying (Morita et al., 1992). Fuzzy logic controls the toner supply (position 3 in Figure 111.26), whereas the neural networks provide correction in the latent image control unit (position 2 in Figure 111.27) when the environment fluctuates.

In order for the neural net to do its job, it must be trained in different environments. For this three different temperatures and three values of humidity were used, and one of five charging grid voltages and exposure voltage were supplied as corresponding output. The trained net accepts environmental conditions, fatigue parameters, and the uncorrected voltages, and outputs the corrected voltage values to be fed to the drum. Conventional control resulted in jumps in these voltages by as much as 50–60 volts. With the neural net, this has been reduced to 10 volts.

Fuzzy control realizes stable image quality by controlling toner supply (Figure 111.27). The supply keeps changing, and this may cause uneven contrast over the copy. The process has a time delay since a buffer is used for the toner delivery, and the response to any change in the supply is delayed because of the buffered quantity. Furthermore, the toner is sensitive to temperature and humidity. The fuzzy reasoning system developed to solve this problem uses the image density and its time differential as its inputs to regulate the toner supply. An example of the fuzzy rule here is:

RULE 111.7: IF contrast is OK and it is increasing, THEN stop toner supply quickly. IF contrast is light and it is steady, THEN supply more toner.

The reasoning method used is max-min-gravity. Figure 111.28 shows the Image Density when 500 copies are taken with 5 gray levels. The fuzzy system can reduce the fluctuation of the image density less than the same process that uses binary logic. For both analog and digital copying methods, this fuzzy system reduced the variation of image contrast to between 11% and 90% of that of conventional methods.

Sanyo: Toner Supply Control

Sanyo Electric Co., Ltd. has applied fuzzy logic to the toner supply controller (position 3 in Figure 111.26) to preserve high image quality (Nomura et al., 1991).

A magnetic sensor measures the ratio of toner to carrier (iron filings, etc.) at the position marked 3 in Figure 111.26, and more toner is supplied when the toner density becomes less than a certain level. Conventional control supplies a constant amount of toner which does not depend on the white/black ratio in the original image. This leads to unstable copy density which changes over time.

Sanyo has introduced a fuzzy controller which monitors not only toner density but also its time derivative, and changes the amount of toner supply accordingly. The main ideas used in the rules are:

RULE 111.8: (1) IF toner density is LOW (HIGH), THEN supply greater (lesser) amount of toner. (2) IF toner density DECREASES, THEN supply greater amount of toner. (3) IF toner density INCREASES, THEN supply greater amount of toner.

Toner density increases just after the toner is supplied. However, if a large part of the original image is black, toner density begins to decrease soon. Rule 111.8 attempts to guard against this situation. The actual rules are shown in Table 111.8.

Several output membership functions are used for the target toner density to realize precise control at this position. The inference method is max-min-gravity and uses a Mamdani-type fuzzy controller. Use of the fuzzy system reduced the variation of toner density to 0.94% from 1.53% for conventional methods. It also reduced the standard deviation in the image density of a half-tone copy from 0.0617 for conventional control to 0.0432, corresponding to a 30% improvement in the stability of the image quality.

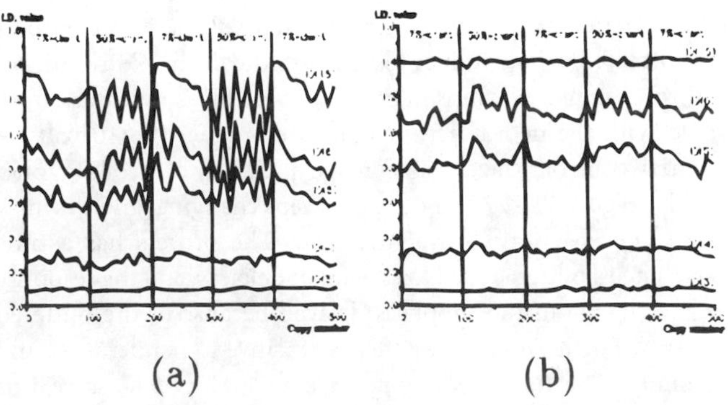

Figure 111.28 Image Density of 500 copies (5 gray levels) [9]: (a) binary logic, (b) fuzzy logic.

Table 111.8 Fuzzy inference rules for image quality control of toner supply. N: negative, P: positive, S: small, M: medium, L: large, and ZR: zero

		time derivative of toner density						
		NL	NM	NS	ZR	PS	PM	PL
toner density	NL	PL		PL	PM		PM	
	NS		PL		PS	ZR		PL
	ZR	PL		ZR	NS		NM	
	PS	PM		NM		NL		PS
	PL				NL			

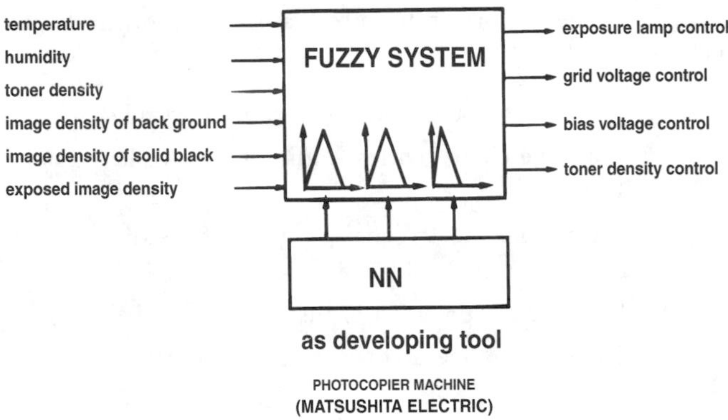

Figure 111.29 A neural net determines membership functions.

Matsushita Electric: Auto-Exposure and Toner Control

Matsushita Electric Industrial applied fuzzy system to control exposure lamp, grid voltage on drum, bias voltage, and toner density from the six sensor inputs of temperature, humidity, toner density, image density of back ground, image density of solid black, and exposure image density. Figure 111.29 shows the fuzzy system. This fuzzy system was designed automatically by neural networks.

Neural networks have been used to design membership functions of fuzzy systems used as decision making systems for control. The positions and widths of the membership functions are tuned by a gradient method to reduce the error between the actual fuzzy system output and the desired output. Some of such systems are the washing machine, vacuum cleaner, rice cooker, photocopying machine, etc. of Matsushita Electric Group (Takagi, 1994).

Sanyo: Color Copying

Sanyo Electric Co., Ltd. are considering using fuzzy sets to define certain natural colors correctly so that copy machine colors can be compared to this reference and the color temperature can be adjusted in case of deviation (Genno et al., 1990). Human subjective tests are used to set up the reference colors.

Television Equipment

Sanyo: Television Sets

Sanyo Electric Co., Ltd. used fuzzy inference for controlling the image quality of a TV reception set which appeared on the market in autumn 1990. The fuzzy system controlled contrast, brightness, velocity modulation, and sharpness. Velocity modulation is a technology which emphasizes sharpness by changing scanning velocity. The input parameters are the ambient brightness in the room and the distance of the viewer from the set. The rule is:

RULE 111.9: WHEN the room is brightly lit, and the viewer is far away, THEN the region boundaries on the picture should be sharper and clearer, whereas IF the viewer is close and in a darkened room, THEN the sharpness should be less as the high-frequency components are accompanied by noise.

See Table 111.9 for some of the other principles used.

The actual rules in the system are more detailed for finer control. There are four membership functions for brightness, three for distance, and seven for the output variables. The system implies that different sets of rules apply to ordinary TV programs and to movies. The inference is by the max-min-gravity method and is implemented by a look-up table in the final product. The table has 8 × 3 × 2 cells. From the two inputs and the two modes each cell containing a value for each of the four outputs.

Room brightness is computed from a light sensor to give one of eight values. Viewer distance is computed by locating the remote control, and has three possible values. The microcontroller uses these to consult the look-up table, and four output values are received. This process is repeated every 50 msec.

Others

Mitsubishi Electric Corp. has put similar television products on the market. Contrast and sharpness are controlled using brightness and viewer distance (Mitsubishi Electric Corp. 19??). The implementation of the fuzzy system is by table look-up. For satellite broadcasts, frequency characteristics for noise reduction and sharpness are controlled by fuzzy logic according to incoming signal power. For example, rain may reduce the signal-to-noise ratio, but noise can be reduced by suppressing high frequency components.

Table 111.9 Fuzzy Inference Rules for Image Quality Control of TV set

	room		from TV to viewer		
	bright	dark	far	mid	near
contrast	big	small	big	mid	small
brightness	big	small		same	
velocity modulation	big	small	big	mid	small
sharpasses	big	small		same	

Sony Corporation marketed a TV set in autumn 1989, in which image quality was continuously adjusted. There is also a product called AI-Television which uses fuzzy logic. Although the details of the fuzzy reasoning are not available, the inputs are brightness signal, color signal, beam current, and noise level. The system controls contrast, brightness, color, sharpness, noise reduction, etc.

Codecs

In April 1991 Mitsubishi Electric Corp. introduced fuzzy inference into the video codecs used by their video-conferencing system. Video-conferencing equipment enables various people to communicate via TV sets and cameras using high-speed bidirectional phone lines.

The coding essentially uses various data compression techniques depending on the extent of change in the picture. For instance, rapid change in the picture leads to finer quantization, and vice versa. According to their news release (Mitsubishi Electric Corp., 1991), the system compares successive frames to compute the changes. The fuzzy controller's inputs are the amounts of such changes now (x_0), in the recent past (x_1), in the distant past (x_3), and at some point in between (x_2) (Figure 111.30). The output controls the quantization process.

An example of a rule is:

RULE 111.10: IF x_3 is almost the same as, but x_0, x_1, and x_2 are below a threshold determined by the current quantization rate, THEN reduce the quantization rate.

The motion-tracking ability of the system increased by 30–50% as a result of using the fuzzy system.

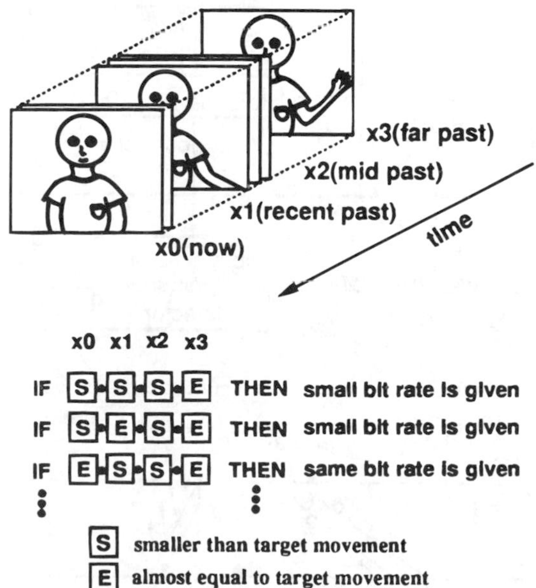

Figure 111.30 Rule examples of the codec for a video conference system.

Stepper Alignment in Semiconductor Manufacturing

Canon Inc. has used fuzzy logic to align semiconductor wafers in the process of producing chips (Imaizumi et al., 1990). In semiconductor device manufacturing the wafer must be located and aligned on the assembly-line very precisely. This has become increasingly important as chip complexity and density increases. Each wafer is equipped with an alignment mark, which is read automatically as a reference. However, since many different companies manufacture wafers, the structure and position of the alignment mark varies. The alignment process must know how to correct for this variation. Canon developed a system that corrects alignment error of a conventional alignment position detector, shown in Figure 111.31. Fuzzy logic is used for encoding and using the knowledge in the system.

The alignment mark is a sharp depression in the wafer surface (Figure 111.32). The aligning equipment uses the diffraction or reflection of a laser beam from the two edges of the mark to locate it. The peak location of this synthesized returning laser signal is fed into one fuzzy system, which outputs the correction to be applied to locate the mark. The alignment error between the position of the peak and the real position is shown in Figure 111.33.

Two rulebases are used to deal with the problem (Figure 111.34). The first rulebase relates the position of the peak (see Figure 111.33) to the actual center of the mark. Another rulebase stores the relationship between the alignment shot position on the wafer and the position of the mark. These two rulebases together determine the precise location of the mark. These relationships have been observed empirically, and the rules are constructed from data that has been experimentally obtained, analyzing 84 cases. The ensuing performance had an average error of zero, and the standard deviation was reduced by half.

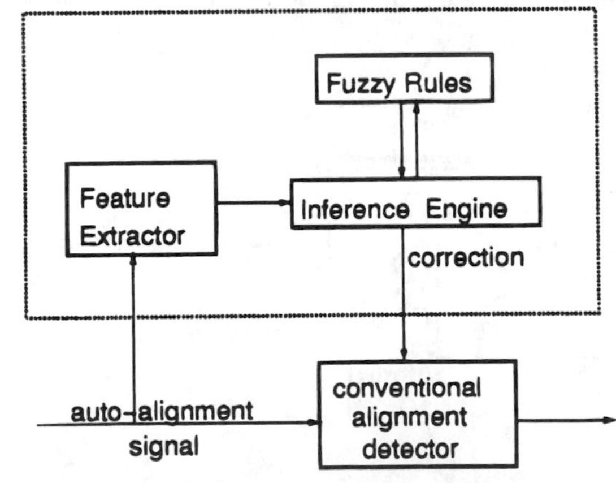

Figure 111.31 Fuzzy alignment correction system.

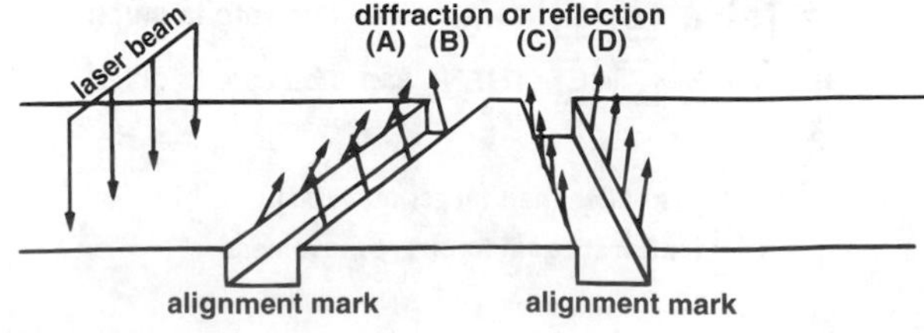

Figure 111.32 How to detect the position of the LSI chip on a wafer.

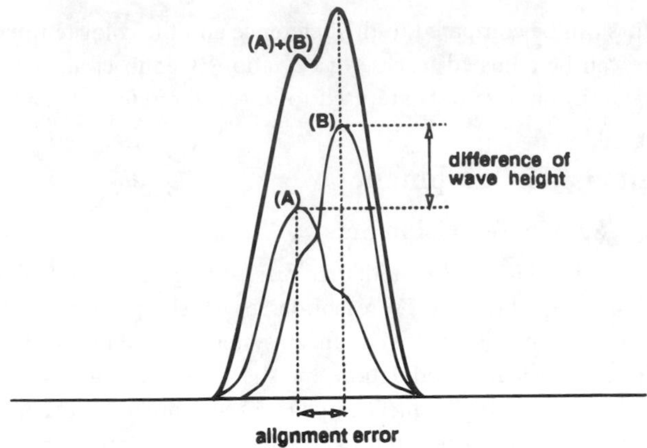

Figure 111.33 Correcting the alignment error.

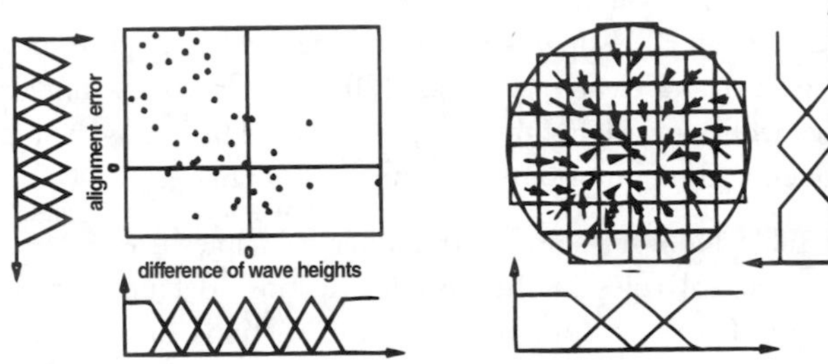

Figure 111.34 Alignment error on LSI wafer and fuzzy partitioning for rules.

Conclusion

Fuzzy logic has been widely applied to consumer home appliances following two major trends. One trend is to have learning ability in the equipment so that it can adapt itself to the user's environment and preferences (Takagi, 1994). Initial approaches did this by fusing neural networks and fuzzy logic. Current approaches add an emphasis on consumer safety in operating the product. The other trend focuses on applications to signal processing. Here, logic is used for decision-making and for controlling parameters and quality. The fuzzy logic applications to image processing equipment described here comprise the initial stage of research in this direction.

Acknowledgments

We would like to express thanks to various people for supplying the information used in this survey. Among these are Mr. T. Tachibana and Mr. K. Akahoshi of Minolta; Mr. T. Haruki and T. Nagata of Sanyo; Mr. M. Imaizumi of Canon; Ms. S. Kubo of

Matsushita Electric; Mr. T. Morita of Ricoh; Mr. Mitsuhashi, Ms. Kawase, and Mr. A. Murakami of Mitsubishi; and others.

References

Adagio Software, Inc. 1994. *AutoRun™ User's Guide,* Adagio Software, Inc. 2375 Troicana Ave. Ste. 322, Las Vegas, NV.

Asami, N. 1989. *Fuzzy Logic Comes To Home Appliances,* Nikkei Electronics 1989, 10.30, 485:167–171, in Japanese.

Azuma, R. T. 1995. A Survey of Augmented Reality, *ACM 1995 SIG-Graph Tutorial Proceedings.*

Ballas, J. A. 1993. Common factors in the identification of an assortment of brief everyday sounds, *J. Experimental Psychology: Human Perception and Performance,* 19(2):250–267.

Barlow, H. B. and Mollon, J. D., eds., 1982. *The Senses,* 239–332, Cambridge University Press.

Begault, D. R. 1954 *3-D Sound for Virtual Reality and Multimedia,* Academic Press, San Diego, CA.

Burdea, G. and P. Coiffet. 1994. *Virtual Reality Technology,* John Wiley & Sons, New York, NY.

Cohen, J. 1994. Monitoring Background Activities, *Auditory Display: Sonification, Audification, and Auditory Interfaces,* G. Kramer, ed., Santa Fe Institute Studies in the Sciences of Complexity, Proc. Vol. XVIII, 499–531. Addison-Wesley, Reading, MA.

Durlach, N. I, and Mavor, A. S., eds., 1995. *Virtual Reality: Scientific and Technological Challenges,* 134–160, National Academy Press, Washington, DC.

Durlach, N. I., Shinn-Cunningham, B. G., and Held, R. M. 1993. Supernormal auditory localization, I general background, *Presence,* 2(2):89–103.

Egusa, Y., Akahori, H., Morimura, A., and Wakami, N. 1992. An electronic video camera image stabilizer operated on fuzzy theory, *IEEE Int. Conf. Fuzzy Systems FUZZ-IEEE '92),* 851–858.

Feiner, S., MacIntyre, B., and Seligmann, D. 1993. Knowledge-based augmented reality, *Comm. ACM,* 36(7):53–62.

Fogel, D. B. 1995. *Evolutionary Computation,* IEEE Press, Piscataway, NJ.

Fujii, H., Ueda, H., Hayashi, K., and Akahoshi, K. 1992. Fuzzy Logic Application in α-7xi, *Minolta Techno Report, No. 9,* pp. 7–16, in Japanese.

Gaver, W. 1989. The sonic finder: an interface that uses auditory icons, *Human-Computer Interaction,* 4(1):67–94.

Gaver, W. 1993. How do we hear in the world?: explorations in ecological acoustics, *Ecological Psychology,* 1993, 5(4):285–313.

Gaver, W. 1994. Using and Creating Auditory Icons, *Auditory Display: Sonification, Audification, and Auditory Interfaces, G. Kramer,* ed. Santa Fe Institute Studies in the Sciences of Complexity, Proc. Vol. XVIII, 417–446, Addison-Wesley, Reading, MA.

Genno, H., Fujiwara, Y., Kano, H., and Fukushima, K. 1990. Human sensory perception oriented image processing, *1st Int. Conf. Fuzzy Logic and Neural Networks IIZUKA '90).*

Goldberg, D. E. 1994. Genetic and evolutionary algorithms come of age, *Comm. ACM,* March, 37(3):113–119.

Gradecki, J., 1992. 3-D sound theory, *PCVR Journal,* 1(6):6–12.

Gradecki, J. 1994. *The Virtual Reality Construction Kit,* John Wiley & Sons, New York, NY.

Hackwood, S. 1995. Fuzzy Control and Multimedia with examples from Law Enforcement, *SPIE Int. Symp. on Aerospace/Defense Sensing and Control and Dual Use Photonics,* April, Orlando, FL.

Hayashi, H. and Yoshida, K. 1991. Sanyo fuzzy-AI vision C-20ZS101, *Television Technics & Electronics,* 9:91–96, in Japanese.

Hofer, R. C. and Loper, M. L. 1995. DIS today, *Proc. IEEE,* Special Issue on Distributed Interactive Simulation (DIS), August, Vol. 83, No. 8, pp. 1124–1138.

Hongu, M, Amano, T., and Oda, O. 1991. Development of intelligent AV systems, *Inst. Television Engineers Annual Convention (ITEC '91),* s2–5:547–550, In Japanese.

Idesawa, M. 1993. New types of 3-D optical illusion in binocular viewing, RIKEN Review No. 1 (April, 1993): Focused on Light Science and Technology.

Imaizumi, M., Takakura, N., Tanaka, H., Nakai, A., and Uzawa, S. 1990. Advanced auto alignment system using approximate reasoning, *1st Int. Conf. Fuzzy Logic and Neural Networks (IIZUKA '90),* 119–122.

Isermann, R. 1995. Information Processing for Mechatronic Systems, Int. Conf. Recent Advances in Mechatronics, 3–17, August 14–16, Istanbul, Turkey.

Jacobson, L. 1994. *Garage Virtual Reality: The Affordable Way to Express Virtual Worlds,* Prentice Hall Computer Publishers (SAMS),

Jang, I-S, and Sun, C-T. 1995. Neuro-Fuzzy Modeling and Control, *Proc. IEEE,* Vol. 83, No. 3, March, 1995.

Kalawsky, R. S. 1993. *The Science of Virtual Reality and Virtual Environments,* p. 69, Addison Wesley Publishers Limited, Wokingham, England.

Kandel, E. R. and Schwartz, J. H. 1985. *Principles of Neural Science, 2nd Edition,* 396–408, Elsevier Science, New York, NY.

Kaneda, K., Homna, H., and Togai, M. 1990. A fuzzy auto-focus system for a portable video camera, *NAFIPS '90,* June.

Kartalopoulos, S. V. 1996. Time-Dependent Fuzzy Logic, Understanding Neural Networks and Fuzzy Logic: Basic Concepts and Applications. 130–152, IEEE Press, Piscataway, NJ.

Kaufmann, A. and Gupta, M. 1994. Introduction to Fuzzy Arithmetic: Theory and Applications, Van Nostrand Reinhold, New York, NY.

Kendall, G. and Rodgers C. 1981. The simulation of three-dimensional localization cues for head-phone listening, *Proc. 1981 Int. Computer Music Conf.,* 225–243.

Kikuchi, K., Haruki, T., Tsujino, K., Kitano, T., and Fujita, K. 1991. Video camera system using fuzzy logic, *Sanyo Technical Review,* 23(2):8–20, in Japanese.

Kohonen, T. 1995. personal communications.

Kramer, G., ed. 1994. *Auditory display: sonification, audification, and auditory interfaces,* Proc. *1st Int. Conf. Auditory Display (ICAD) '92,* Santa Fe Institute Studies in the Sciences of Complexity, Santa Fe, NM.

Krueger, M. W. 1991. *Artificial Reality II,* Addison-Wesley, Reading, MA.

Larijani, L. Casey. 1994. *The Virtual Reality Primer,* McGraw-Hill, New York, NY.

Loeffler, C. E. and Anderson, T. 1994. *The Virtual Reality Casebook,* VNR Computer Library,

Mills, A. W. 1972. Auditory Localization. *Foundations of Modern Auditory Theory, Vol. II,* J. V. Tobias, ed., 301–345, Academic Press, New York, NY.

Mitsubishi Electronics Corporation, 1992. Kyoto Works, personal communication.

Mitsubishi Electronics Corporation, 1991. Mitsubishi TV Meeting System: MELFACE 810, 850 Series Putting on the Market, news release, March 28.

Moorhead, R. J. II and Zhu, Z. 1995. Signal Processing Aspects of Scientific Visualization, IEEE Signal Processing Magazine, 12(5):20–41.

Morita, T., Kanaya, M., Inagaki, T., Murayama, H., and Kato, S. 1992. Electrophotography process control method based on neural network and fuzzy theory, *2nd Int. Conf. Fuzzy Logic and Neural Networks (IIZUKA '92),* 2:885–888.

Myers, B. A. et al. 1990. Comprehensive support for graphical, highly-interactive user interfaces. *IEEE computer,* 23(11):71–85.

Nomura, N., Asada, M., Kikuchi, N., Handa, Y., and Miyamoto, T. 1991. Toner Control of Plain Paper Copier Using Fuzzy Logic, *Sanyo Technical Review,* 23(2):82–88, in Japanese.

OMG 1991. *The Common Object Request Broker: Architecture and Specification,* OMG Document #91.12.1, Rev. 1.1, draft, 10 November 1991.

Padgett, M. L. 1982. *A Statistical Methodology for the Development of Computer Simulation Models and Forest Harvesting Applications,* M. S. Thesis: 1982, Auburn University

Padgett, M. L. 1988. Statistical Modeling and Analysis, Models for Midcourse and Terminal Guidance Simulations for the Guidance and Control for Strategic Defense Initiatives Project, prepared by the SPI for Eglin AFB., November 30, 59–92.

Padgett, M. L. 1993. Modern Simulation and Standards: Neural Networks, Fuzzy Systems, Evolutionary Systems and Virtual Reality, presentation to *TAPPI,* Atlanta, GA 1993; later published by request in *WNN* proceedings for 1995.

Padgett, M. L. 1995. A Practical Filter for Conflicting or Subjective Data, IEEE-ANZIIS 95, Perth, Australia.

Padgett, M. L. and Evers, J. 1996. Fault Detection and Identification in Missile System Guidance and Control: A Filtering Approach SPIE's Int. Symp. Aerospace/Defense Sensing and Controls: Applications and Science of Artificial Neural Networks Conference, in press.

Padgett, M. L. and Karplus, W. 1993. Neural Network Basics: Applications, Examples and Standards, Tutorial Text, IJCNN '93, 385–413, Nagoya, Japan.

Padgett, M. L. and Padgett, W. D. 1995. Simulation and Computational Intelligence in Real-World Applications, *Simulation,* 65(1):5–9.

Padgett, M. L., Rajala, S. A., Snyder W. E. and Ruedger, W. H. 1985. Detection of Maneuvering Target Tracks, Proc. SPIE 29th Ann. Int. Tech. Symp. Optical and Electro-optical Engineering, August 18–23, San Diego, CA.

Pimentel, K. and Teixeiera, K. 1993. Virtual Reality—Through the New Looking Glass, Intel, Windcrest, McGraw-Hill, New York, NY.

Rayleigh, J. W. S. 1945. *The Theory of Sound,* Volumes I and II, 433–478, Dover Publications, New York, NY.

Resnick, R. and Halliday, D. 1960. *Physics,* Part I, 3rd ed. 433–456, John Wiley & Sons, New York, NY.

Rheingold, H. 1991. Virtual Reality, Simon and Schuster, New York, NY.

Saaty, T. L. 1980. The Analytic Hierarchy Process. McGraw-Hill, New York, NY.

Scaletti, C. 1994. Sound Synthesis Algorithms for Auditory Data Representations, *Auditory Display: Sonification, Audification, and Auditory Interfaces,* G. Kramer, ed. Santa Fe Institute Studies in the Sciences of Complexity, Proc. Vol. XVIII, 223–251, Addison-Wesley, Reading, MA.

Shingu, T. and Nishimori, E. 1989. Fuzzy-based automatic focusing system for compact camera, *3rd IFSA Congress,* 436–439.

Smith, S., Pickett, R., and Williams, M. 1994. Environments for exploring auditory representations of multidimensional data. *Auditory Display: Sonification, Audification, Auditory Interfaces,* G. Kramer, ed. Santa fe Institute, Proc. Vol. XVIII, 167–183, Addison-Wesley, Reading, MA.

Souma, I. and Suda, T. 1991. Electrophotography process control for copy machine, *Elec. Eng.,* 33(1):48–51, in Japanese.

Stampe, D., B. Roehl and J. Eagan. 1993. Virtual Reality Creations: Explore, Manipulate, and Create Virtual Worlds on Your PC. The Waite Group.

Szu, H. and B. Telfer. 1995, Wavelet Dynamics, The Handbook of Brain Theory and Neural Networks, Arbib, M. A. ed. 1049–1053 MIT Press Cambridge, MA.

Takagi, H. 1994. Cooperative System of Neural Networks and Fuzzy Logic and Its Application to Consumer Products, *Industrial Applications of Fuzzy Control and Intelligent Systems,* IEEE Press, Piscataway, NY.

Takagi, T. and Sugeno, M. 1985. Fuzzy identification of systems and its applications to modeling and control, *IEEE Trans.,* SMC-15-1:116–132.

Takala, T. and Hahn, J. 1992. Sound rendering, *Computer Graphics,* Vol. 26(2):211–220.

Thompson, J. 1993. Virtual Reality: An International Directory of Research Projects, Meckler, Westport, CT.

Uomori, K., Morimura, A., Ishii, H., Sakaguchi, T., and Kitamura, Y. 1990. Automatic image stabilizing system by full-digital signal processing, *IEEE Trans. Consumer Electronics,* 36(3):510–519.

Wallis, M. E. and J. J. Feeley. 1989. Bank-to-turn missile/target simulation on a desk-top computer, Modeling and Simulation on Microcomputers, 1989, SCS Press.

Warren, W. H. and Verbrugge, R. R. 1984. Auditory perception of breaking and bounding events: a case study in ecological acoustics, *J. Experimental Psychology: Human Perception and Performance,* 10(5):704–712.

Wenzel, E. M. 1992. Localization in Virtual Acoustic Displays. *Presence,* 1(1):80–107.

Wightman, F. L. and Kistler, D. J. 1989. Headphone simulation of free-field listening II, psychophysical validation. *J. Acoustical* Soc. *America,* 85:868–787.

Zadeh, L. 1995. personal communications.

Zeleznick, R. C. et al. 1991. An object-oriented framework for the integration of interactive animation techniques, *Computer Graphics,* 25(4):15–112.

112

Asynchronous Transfer Mode Technology

112.1 What is ATM Offering? 1438
112.2 Why ATM? 1438
112.3 What is ATM? 1439
112.4 ATM Applications 1439
112.5 The NEBULAS Project 1440
An Example: • The Calorimeter
112.6 Summary 1442

Thomas Lindblad
Royal Institute of Technology, Stockholm

Several years ago the telecommunications industry started developing asynchronous transfer mode (ATM) techniques to be used as the roadbeds for "data superhighways." Today multimedia is receiving a great deal of attention and applications with appetites for consuming huge amounts of bandwith are already here. World Wide Web (WWW) browsers like Mosaic and Netscape are only two examples. Other bandwidth consumers are related to advanced physics training (APT) and experiments requiring high bandwidths by themselves. One example of such experiments is the next generation of high energy physics (HEP) experiments at the Large Hadron Collider at CERN in Geneva. Other examples are given by IBM in Figure 112.1. These demands could result in ATM products (chip sets) being available both at low cost and earlier than anticipated. The fact that the standard emmanates from the telecommunications industry should make the user confident that the components will be around for a long time.

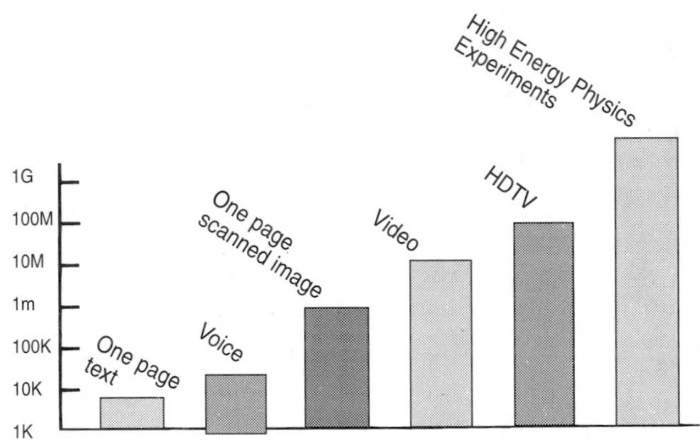

Figure 112.1 Bandwidth requirements.

112.1 What is ATM Offering?

The fixed size cell structure of the ATM, combined with a unique identifier, allows high-speed hardware techniques for switching. ATM works at speeds of 155 Mb/s and greater, thus offering advantages over most other networking technologies. However, other important features in network solutions are to obtain a scalable system (generally based on some quality of services, QOS) and a significantly small network delay to make the system work, e.g., video traffic. With new applications such as desktop video-conferencing and multi-media kiosks, advanced and updated distance learning will be possible with your new high-speed, point-to-point network.

Do you really need 155 Mb/s or 662 Mb/s? Well, just recall how the WWW started. Conceived as a unique way to give particle physicists easy access to their data wherever they worked, the WWW has grown into something much bigger. Today, everybody runs Mosaic or Netscape browsers. The WWW traffic on the NSF backbone has increased from 100,000 characters/s in January 1994 to 1,400,000 characters/s in January 1995. Born at CERN in 1989, WWW is now considered beyond the laboratory's mission, and the European Commission has asked the French National Institute for Research in Computer Sciences and Control (INDRA) to take up the "WebCore."

If you are familiar with WWW, then you know the importance of point-to-point connection, scalability, and bandwidth. Once you are in a large server system at MIT, clicking on a hypertext word may bring you to Ostfold University College in Halden. ATM is there to offer the "information superhighways."

112.2 Why ATM?

Local Area Networks (LAN) today generally rely on packet-based bridges and/or routers. The former are cheap and easy to implement, the latter offer a higher degree of LAN segmentation. So why ATM? Three reasons are already hinted above:

0-8493-8343-9/97/$0.00+$.50

1. Direct interconnection of 100 Mbps (or more) between workstations is required.
2. Scalability (power to increase bandwith without changing the information structure) is a key word and ATM cells can be carried on a number of transports.
3. Both bridges and routers introduce a variance in signals that the user sees as distortions in the picture or in the spoken word.

ATM has several advantages over FDDI. The most important one is perhaps that you can use the same type of network for both local and wide area networks. Campus networks are generally highly heterogeneous and essentially open environments and most computers run TCP/IP. At the University of Geneva, one is currently running a TCP/IP link over ATM and measures a performance of up to 80 Mb/s. Personal systems with IBM Turbo Ways adapters connects to concentrators at 25 Mb/s. This allows for application visualization of dentistry applications, where treatments are filmed using (volunteer) patients. Although ATM is not used to its maximum, it is here and can be used today.

112.3 What is ATM?

ATM is a standard developed for telecommunications by the ITU (formerly known as the CCITT) to form the basis for the future B-ISDN. The ATM technology has been designed to support a massive, low latency, non-blocking switching capacity that can be used for a wide range of traffic. ATM uses small, *fixed*-size (byte payload) packets referred to as cells. The size of the cell is pretty small: there is a 5-byte header and 48 bytes of information. The ATM standard specifies the connection between an end-station and the network (UNI) and between the sub-networks (NNI). The standard may be viewed as a three-layer implementation.

The ATM layer.
The ATM Adaptation Layer (AALn).
The physical layer.

The *ATM layer* involves the aforementioned small cells with payload and header. The latter includes 3 bytes that carry a label identifying a connection between the source and the destination. This label is used by the switching fabric hardware to route a cell to its destination (self-routing). The connection itself may be either set up permanently at hardware initialization or established/broken dynamically by a signalling protocol. Connections are not necessarily assigned a constant bandwith, but rather a variable one on demand. This is done based on information on peak and average characteristics. It should be stressed that ATM is a stochastically multiplexed technique, and several connections can be mixed on the same physical path, provided their aggregate characteristics do not exceed the physical ones. The connections are said to be virtual because the cell multiplexing and the switching technique obviates the need to reserve dedicated hardware paths.

The *adaptation layer* (AAL1 to AAL5) defines how to adapt the ATM layer requirements (voice, video, data). The AAL5 protocol specifies that data can be transferred at variable block length <64 kBytes. There is no header but rather a trailer that may include padding bytes to make sure the length of the package is a multiple of the ATM cell 48 bytes.

Finally, the *physical layer* specifies the transmission over a link. There are two sublayers:

The transmission convergence sublayer.
The physical medium sublayer.

The former defines the bit rates and the framing patterns, while the latter defines the physical support and the timings. There are several standards for different applications (SDH, SONET). SONET defines transmission rates which are multiples of 51.84 Mb/s. The SDH base bit rate of 155.52 Mb/s may be multiplied by factors of four to yield 622.08 Mb/s, 2.488 Gb/s, etc. Clearly, the overhead and control of the physical layer will lower the effective bit rate, but only by a few percent.

Finally, it should be stated that there is no "*guaranteed delivery*" in the ATM standard. This has been sacrificed by the requirements to have low-latency and non-blocking switches. If needed cell loss recovery must be implemented in higher level protocols.

112.4 ATM Applications

As mentioned above, the ATM concept was developed by the telecommunications industry, and it was quickly adapted by the computer manufacturers. However, this has reflected upon the types of ATM switches available today. Generally speaking, there are three classes:

Telecommunications industry
Switches with implemented flow control
Switches based on a shared medium

Cell loss occurs in a switching fabric when internal buffers overflow. Of course, the traffic may not always be random. As soon as there is a correlation between different virtual connections, the risk of congestion and overflow will increase. This may be overcome by including internal flow control, although the ATM standard does not include any such implementations for the lower levels. The shared medium switch is mostly used in LAN applications and is not expandable.

The three types of switches mentioned above hint at three different fields of application. However, ATM will probably be subject to further developments and changes to be the common technology for all types of communications. Movies distributed to homes on demand and interactive television programs are often envisaged as the features of the "information superhighways." Surely, the "entertainment" sector is important, but still the need for bandwidth is found in almost all areas. Hopefully the ATM will be implemented directly as a point-to-point network. Unfortunately, many vendors of data networks are trying to sell ATM only as a network interconnecting LANs. For anyone who

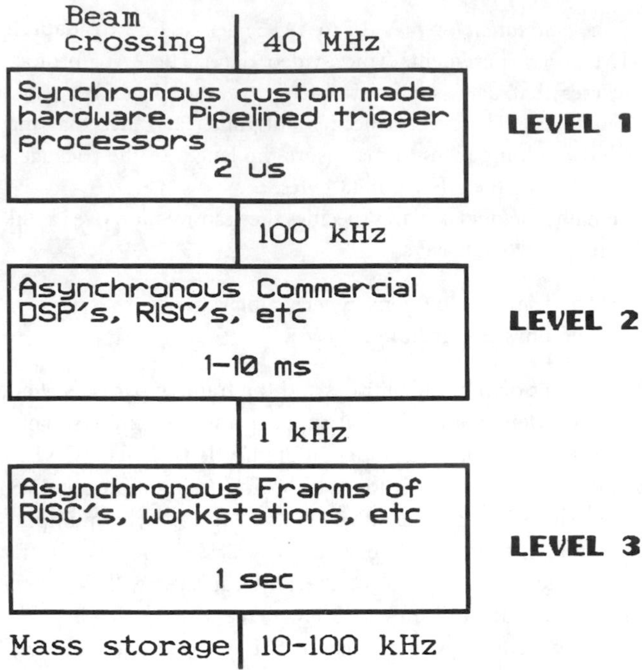

Figure 112.2 Schematic diagram of the trigger. The trigger will filter out "nonsense" data and select regions of interest in order to reduce the amount of data being stored on mass storage devices.

has used WWW, remote image processing, etc., it must be obvious that this is not true.

Indeed, ATM adapters are becoming readily available. There are adapters from IBM, SUN, etc. Olicom A/S in Denmark is presenting EISA-bus and PCI-bus adapters at full (155 Mb/s) speed. The design of the former is very similar to the VME-adapter developed for high-energy physics experiments at CERN (Figure 112.2). Both adapters use the Fujitsu chip set. The Olicom adapter has been tested with a software package and together these products should find a use not only as servers in qualified EISA PC, but also in a variety of client server applications.

112.5 The NEBULAS Project

Experiments at the Large Hadron Collider (LHC) at CERN have just been approved. In the present (LEP) circular tunnel, a second accelerator will be built to study collisions between particles (protons and anti-protons) at very high energies to search for new and more fundamental particles and to gain general insight in microcosmos. There are basically three different experiments conditionally approved and each experiment involves very large and modular detectors of various types (muon detectors, calorimeters, etc.). These different detector systems are all highly granular, which results in a vast number of detector outputs. These signals need to be amplified, filtered, digitized, filtered again, etc. before they can be "summed" together to form an "event", i.e., something describing what happened at that very collision. Since the signal-to-noise ratio in the most exciting experiments is very low or 1/10,000,000, one has to build special triggering systems. The reason is, of course, to provide a signal that enables interesting data to be sent to mass storage units, while rejecting "junk" data. A two- or three-level system is generally considered. In the case of the ATLAS, a three-level system is considered. Schematically, such a system looks like that shown in Figure 112.2.

The detector is large and modular, and consists of several subsystems. These subsystems will contribute to the final event in various degrees. A summary of the data volumes and their flow (bandwidths) at difference levels is given in Figure 112.1. Note that the total bandwidth at level 3 is more than 8 Gb/s.

The Trigger

The ATLAs trigger consists of three logical levels shown schematically in Figure 112.2. Beam crossing interactions occur at a rate of 40 MHz.

The scope of the level 1 trigger is to identify particles, high transversal momentum electromagnetic clusters, jets and muon tracks. Fixed algorithms will be used on raw data for pattern recognition and energy evaluation. The system will use hardwired pipelined processors and point-to-point links.

Flexible and programmable algorithms will be used as the second level for particle identification (electrons, photons, muons, etc.). A rejection factor of about 100 or greater is expected at level 2.

The scope of the third level is to investigate the physics process signature. This involves the "event builder", and the NEBULAS project concerns this part of the data acquisition system. It is realized as a software filter, using a farm of workstations fed by switching fabrics. A further event rejection of 10–20 times is expected. Events, accepted by the third level selection, will be stored on mass storage devices for subsequent off-line analysis.

Signals and Bandwidths

Each bunch crossing the signals from all subdetectors is locally stored in pipelined memories during the level 1 process. The accepted events are transferred via optical links to about 2000 read-out cards. The level 3 selection algorithm is expected to take 0.1–1 second, and hence, a farm of processors is needed in order to meet the 1 kHz level 2 rate. Only one processor is allocated per event and we have linear latency.

There are mainly three groups of detectors, and the estimated bandwidths for these three channels are given in the following table.

Subsystem	Event size kilobyte	Level 2 Gb/s	Level 3 Gb/s
Tracker	770	10	6
Calorimeter	400	6	3
Moon chamber	200	3	2

Thus, the aggregate bandwidth required for the triggering system is of the order of several tens of Gb/s. This data flow cannot be handled by conventional bus-based data systems. This fact, together with the requirement for compliances with adopted industrial standards, high MTBF, and low costs, suggests that the commercial ATM systems are used. ATM is the main candidate technology that is currently evaluated by the NEBULAS project.

Event Building and ATM

A "generic" event builder is shown in Figure 112.3. Here, the sources provide the event fragments. They are packed in ATM AAL5 packets. At this destination, the event fragments are received cell-by-cell and reassembled. The event generator decides when a new event has to be created. In software or demonstrators, the event rate is an important parameter; in an experiment it is determined by the collision frequency and the probability for the reactions. The destination assignment logic assigns the events to the destinations following various schemes. Finally, the switching network is built with a regular interconnection topology and can be either Banyan or Omega. Contention resolution in the switching elements can be selected from one of the following methods:

- Shared media switching with no link-level flow control (e.g., Fore Systems).
- Shared memory with no link-level flow-control (Alcatel/HSS).
- Output queueing with link-level flow-control (AT&T/Phoenix).
- Shared memory with link-level flow-control (IBM/Prizma).

The generic event builder described here can be used to simulate various sub-detector systems of the total event builder. Figure 112.3 shows a model which has been used to simulate the behaviour of the level 2 and level 3 selection systems of the total trigger. The data originates from the calorimeter. The global processors request the sources to send level 3 data (16 kByte per source) at an average rate of 1 kHz. The level 3 traffic adds some 20% load on the source-destination data path. If no precautions are taken, the level 3 traffic creates congestion in the switching fabric. This will delay not only the level 2 traffic but also the protocol traffic. At this moment traffic shaping and flow control is studied.

One of the tasks given to the experimentalists was to present demonstrators of event builders. One of these demonstrators is a prototype 8 × 8 with multipath, self-routing architectures from Alcatel Bell Telephone. A SUN Sparc station communicates with the embedded software via an Ethernet to transputer-link bridge. The switch supports the 155 Mb/s SONET UNI standard and the system will be used to test data acquisitions protocols and traffic shaping techniques. Commercial workstations supporting SONET can be incorporated.

A different demonstrator, based on the AT&T Phoenix ATM switch and using the internal flow control strategy, is planned to be set-up at Saclay this year. It will allow the verification of some of the behaviour patterns observed in the modeling of this type of switch.

Note that the architecture chosen results in a system which is scalable and has a linear latency as shown in Fig. 112.4. If the number of processors in the computer farm is not enough, one can simply add more workstations. Note that each processor is working with data fragments from one and only one beam crossing.

EXAMPLE: The calorimeter

The calorimeter is one of the major detector systems in the ATLAS detector. It will measure the deposited energy of several particles and to do so it is divided into several segments (each being highly granular). For each event accepted by the level 1 trigger, data are transmitted to read-out cards. Several of these cards form a "crate" and in the model example studied each crate has its own link to the level 2 and level 3 triggering systems.

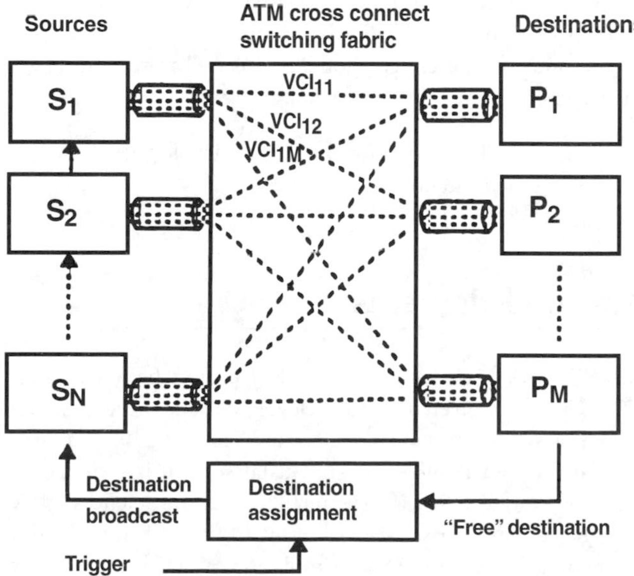

Figure 112.3 A generic event builder.

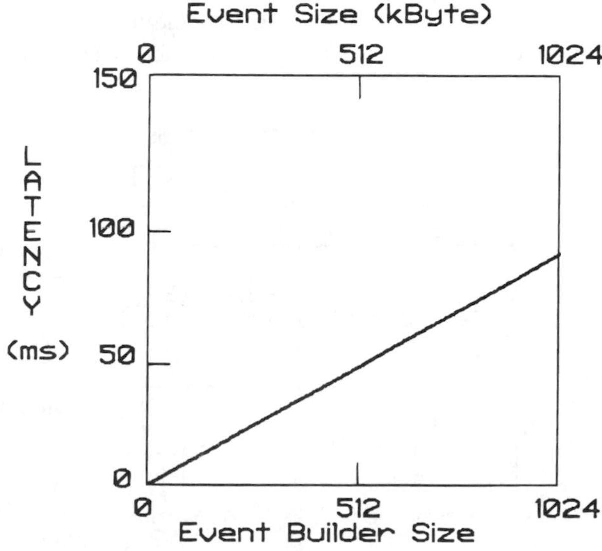

Figure 112.4 The latency of the suggested system shows a linear dependence.

A subsystem would then have 26 source modules, 16 farms of local processors, 14 farms of level 2 global and level 3 processors, and an ATM switching fabric with 64 bidirectional ports. A schematic drawing of the system is shown in Fig. 112.5.

An evaluation of the bandwidth requirements shows that for some sources we would need 622 Mbit/sec links, while other sources could do with 155 Mb/s links. High speed is required for the electromagnetic and hadronic calorimeter sections CEM and HAC in Fig. 112.5. However, at this point we assume all links to be 622 Mb/s.

In the case of the level 3 full event building, data fragments are roughly an order of magnitude larger than for the preceding triggering level. The funnelling of large data packets towards a destination processor seem to induce severe contention in the switching network. This can, of course, be avoided or reduced by an appropriated bandwidth allocation scheme. The scheme includes semi-permanent virtual connections (VC) associated with a high priority logical queue serviced in FIFO order at full link bandwidth and lower priority queues serviced when the other ones are empty. Rate control is used to limit the traffic on the VC so that the maximum bandwidth of all traffic does not exceed the available bandwidth at the output port. The peak level 3 bandwidth per source will be inversely proportional to a programmable parameter N (bigger or equal to the number of low priority logical queues; N times 0.68 μs will also be the period for servicing the logical queues).

During simulation each link is monitored with respect to bandwidth utilization. Depending on the subsystem, the electromagnetic (EM) sources are the "worst" ones showing an output link load of 60–80% at level 3 and 40% at level 2. The hadronic sources (HAC) require approximately 30% at both levels. However, the average input link load is fairly constant (and low) for all detectors. In fact, the HAC shows the highest load (5–6%), while all other subsystems including the EM sources are below 3%. On the *average 25% utilization* of the available switching fabric aggregate bandwidth is observed in these simulations.

To avoid blocking of less frequent data and high latencies at level 2, concurrent segmentation of packets belonging to different VCs are introduced. Referring to Figure 112.1, an equal bandwidth allocation will grant 24 Mb/s for all VCs. The EM barrel sources will top the occupancy by 75% load. However, one can try to distribute the available output link bandwidth among all sources proportionally to their contribution. Applying the rate division technique, a significantly lower internal buffer occupancy and contention in the fabric can be obtained. This is important since in the case of internal buffer overflow, cell loss will occur if no hardware link level flow-control system is present. At this stage the cell loss predicted by simulations is 10^{-8}.

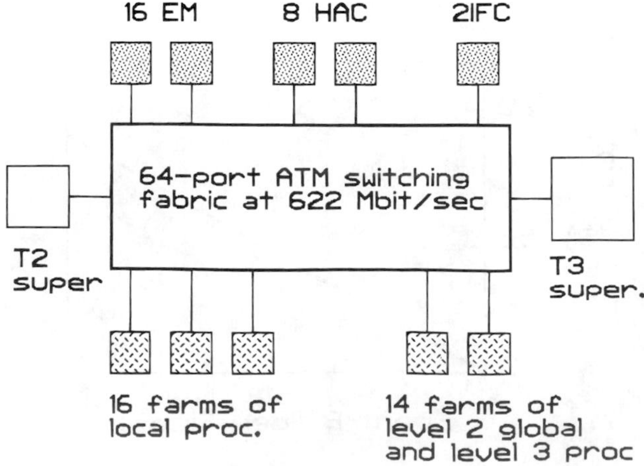

Figure 112.5 A subsystem for the electromagnetic sub-detectors.

112.6 Summary

The ATM packet switching network technology has been proposed as the interconnect for building high-performance data acquisition systems for physics experiments. For the ATLAS detector at LHC/CERN the network links several thousand front-end memories with pertinent processing elements and the data flow has been estimated as close to 50 GBits/sec. To build such a system, in particular if mainly standard components are to be used, is not a trivial task. It requires a good understanding of the industrial components still under development, modeling capability of SAR, flow control, traffic shaping, etc.

A detailed study suggests that including the bandwidth allocation technique (provided by ATM technology) makes an ATM network adequate to handle the ATLAS detector. Other detectors may have more sophisticated and tougher triggering systems pushing beyond the present ATM technology. The NEBULAS project is a part of a large European project (LHC/CERN) attempting to find out what micro cosmos looks like.

Maybe what is supposed to be the everyday technology for telecommunication will also be the tool to find those new particles and show us how they interact. In spite of the small cell size, it seems to have been accepted by the computer industry. Since last winter we have two Olicom ATM-EISA cards (in 386 PCs) running 155 Mb/s SONET/SDH. Within a few years ATM will be in everyone's computer, telephone, etc., as well as in sophisticated science experiments. Or as "Ny Teknik" expressed it:

> "The search for the Higgs particle goes through the switchboard."

Acknowledgments

The NEBULAS project is a joint R&D project between CERN, KTH, Uppsala University, and Saclay, Paris. The project includes the Physics Department and Electrum (Prof. H. Tenhunen) at KTH and Uppsala University (L. Gustafsson). It is supported in part by the Swedish Research Council for Natural Sciences (NFR). KTH has obtained a substantial grant from the Knut and Alice Wallenberg foundation for the simulations and the demonstrator (switch) parts of this project. Special thanks are due to Alcatel Bell, HP, Fujitsu, Olicom, and Novel.

Further Information

ATLAS, Letter of Intent (available from the Atlas secretary, CH-1211 CERN, Geneva).

Calvert, D., Djidi, K., Le Du, P., Mandjavidze, I., Costa, M., Dufey, J.-P., Letheren, M., Piallard, and Manabe, A. 1995. A study of performance issues of the ATLAS event selection system based on ATM switching network, *9th Conf. Rt Applications in Nuclear, Particle and Plasma Physics,* May, Michigan State University.

Christiansen, J., Dufey, J.-P., Letheren, M., Mandjavidze, I., Marchioro, A., Paillard, C., Agehed, K., Eide, Å., Hultberg, S., Lazrak, T., Lindblad, T., Lindsey, C. S., Minerskjöld, M., Tenhunen, H., Gustafsson, L. R., Pauwels, B., Petit, G., de Prycker, M., and Bernard, M. 1994. The NEBULAS Project: a study of ATM-based event building for future high-rate experiments, *RDT '94 (ESONE),* invited talk, July, Dubna, Moscow.

Costa, M., Dufey, J.-P., Letheren, M., Piallard, C., Calvet, D., Djidi, K., Ledu, P., Mandjavidze, I., Gustafsson, L., Lazrak, T., Lindblad, T., Tenhunen, H., Manabe, A., and Nomachi, M. 1994. *ATM-based Event Building,* ATLAS Internal Note DAQ-NO-024, 1 December, CERN, Geneva.

113

NEBULAS: High Performance Data-Driven Event Building Architectures Based on Asynchronous Self-Routing Packet-Switching Networks

Michele Costa, Jean-Pierre Dufey
CERN, Geneva

Mike Letheren, Atsushi Manabe
CERN, Geneva

Alessandro Marchioro, Christian Paillard
CERN, Geneva

Denis Calvet, Kamel Djidi
CEA DSM/DAPNIA, Saclay

P. Le Dû, I. Mandjavidze
CEA DSM/DAPNIA, Saclay

P. Sphicas, Konstanty Sumorok
MIT

S. Tether
MIT

Leif Gustafsson, Klaudiusz Kobylecki
University of Uppsala

Kenneth Agehed, Solve Hultberg
Royal Institute of Technology (KTH), Stockholm

Tawfik Lazrak, Thomas Lindblad
Royal Institute of Technology (KTH), Stockholm

Clark S. Lindsey, Hannu Tenhunen
Royal Institute of Technology (KTH), Stockholm

Martin DePrycker, B. Pauwels
Alcatel Bell Telephone, Antwerp

Guido Petit, H. Verhille
Alcatel Bell Telephone, Antwerp

Michael Benard
Hewlett Packard, Geneva

113.1 Introduction 1445
Executive Summary

113.2 Technical Background 1446
The Main Classes of ATM Switches

113.3 Computer Modeling 1447
A Generic Event Builder Model • Comparative Performance Evaluation of Traffic Shaping vs. Flow-Control Techniques • Parallel Simulation for Large Switches • A Custom-Designed Conical Switching Fabric • Modeling of the ATLAS Architecture • Modeling of the CMS Architecture

113.4 Event Building Protocols and Related Software Development 1454
The Layered Structure of the Event Builder Architecture Based on an ATM Switching Network • Protocol Traffic Transport via the Switch

113.5 Hardware Development 1460
ATM SONET Physical Layer Board • VME-ATM Adapter • ATM Data Generator

113.6 Integration of Event Builder Demonstrators 1464
The ALCATEL ATM-Based Event Builder Demonstrator • ATLAS AT&T ATM-Based Real-Time Demonstrator

113.7 Plan of Work 1466

0-8493-8343-9/97/$0.00+$.50

113.1 Introduction

The RD-31 proposal (Christiansen et al., 1992) was originally approved on 26 November 1992, and the first status report to the DRDC was presented in January 1994 (Christiansen et al., 1993). This document summarizes the work carried out by the collaboration since the previous status report. We recall here the milestones set by the DRDC for the second year of the project:

- Design and simulate full data acquisition protocol for the ATM-based event building, with "traffic shaping" and "internal flow-control" options.
- Demonstrate event building from VME microprocessor sources with ATM switch.
- In addition, it was stated that "the project might benefit from increased contacts with the LHC collaborations."

Two new groups have joined the collaboration: Saclay in the framework of ATLAS and MIT in the framework of CMS. Some further changes in individual collaborators are reflected by the updated list of signatures on the cover page.

In the course of the year it has been recommended that RD-31 should also study event building using other switching technologies, in particular Fibre Channel (Fibre Channel Assoc., 1995; ANSI X3T9.3 Committee), by applying a method of investigation similar to the one adopted for ATM-based event building. It has not been possible to carry out any significant work in the domain of Fiber Channel, partly because it is difficult to obtain detailed information about Fibre Channel switches from industry and partly due to a shortage of available manpower with the requisite skills. We have nevertheless prepared specifications, on request from a manufacturer, for simulation work that they proposed to carry out themselves (Dufey, 1995).

We do not give an overview of ATM technology in this report. A summary of those aspects that are relevant to the event building problem can be found in the previous status report (Christiansen et al., 1993). An ATM tutorial (LeBoudec, 1992) and the BJSDN standards (ITU) can be consulted by the interested reader.

Executive Summary

The goal of the RD-31 project is to demonstrate high-performance, parallel event building architectures that can satisfy the requirements for the level-2 and level-3 trigger systems of the LHC experiments. These architectures can be constructed around commercial or custom-designed parallel, multi-way switching fabrics. Many industrial switching fabrics are now available for switching traffic in broadband telecommunications networks or local area networks based on the asynchronous transfer mode (ATM) standard. High-speed switches for the interconnection of computers and peripherals, based on the Fibre Channel standard, are also becoming available. Within RD-31, event building architectures based on ATM switches have been studied extensively. Alternative architectures using custom-designed switching fabrics have also been explored. Investigations on the use of Fibre Channel switches are planned.

RD-31 was approved in November 1992 and the last status report was presented in January 1994. The DRDC assigned as milestones the tasks: "Design and simulate full data acquisition protocol for the ATM-based event building, with traffic shaping and internal flow-control options" and "Demonstrate event building from VME microprocessor sources with ATM switch." It was also recommended to increase contacts with the LHC experiments.

Simulation: The ongoing work on specific models of commercial switches has been complemented by a "generic" model. It includes many options of switching network components and event builder traffic control techniques which allows quick prototyping of models based on most of the technologies currently available or in preparation, and implementation of a large variety of architectures. The flexibility of this tool allows quick evaluation of new ideas and new products. In addition to the ATM switching technology, a custom-designed switching fabric, optimized for data acquisition, has been proposed and evaluated.

An important issue in using switching fabrics for event building is how to control the traffic patterns to avoid internal congestion (depending on the switch architecture, congestion may result in lost data, poor throughput, and scaling characteristics). An in-depth investigation of congestion control by the so-called "traffic shaping" and "internal link-level flow-control" techniques has been conducted. Interesting results about the switches with internal flow-control have been found and will need confirmation from a demonstrator test bench. The combination of both techniques has also been evaluated. Configurations with large switches (up to 1024 × 1024) have been studied, whenever possible. The implementation of a model on a parallel machine is underway and should permit study of switches as large as to 2048 × 2048 ports and to simulate longer real time sequences. A new traffic shaping scheme has been proposed and complements the three that had been proposed and studied earlier.

Most of the data acquisition protocols simulated belong to the class of "push" architecture where the sources, receiving the identifier of a destination push their data through the switch. An investigation of a "pull" architecture, where the destinations play an active role and collect the data selectively, has been initiated with an application to the RoI concept of ATLAS in view. The results are encouraging.

Collaboration with the LHC experiments: This has led to detailed investigations of their respective event building architectures. This work is continued by groups that are members of the experiments and at the same time members (or collaborators) of RD-31. For CMS, the "full read-out" and the "virtual level 2" architectures have been simulated. For ATLAS, a detailed study of the data flows based on physics simulations and detector read-out scenarios has lead to a proposal for the level 2 and level 3 event building for the calorimeter. An alternative approach, using a "pull" strategy, has been shown to be promising.

Event builder demonstrator hardware and software developments: A VME-ATM interface has been developed based on commercially available chip sets which implement the ATM protocols. A prototype has been successfully tested and operates correctly with standard equipment (an ATM switch from Alcatel, and a

SONET/ATM tester from HP). It has not been possible so far to reach the full performance expected, but we can still implement an event builder based on this interface (a small series is presently being manufactured). Simple data generators have been developed as a cheaper alternative and they can deliver data at full bandwidth through the switch. The software protocol layers and management functions, required for event building, have been developed and tested on the prototype. Some preliminary measurements have been made.

We expect the assembly of the demonstrator to be completed in the coming months, after which it will be possible to complete the implementation and measurement of various event building protocols and traffic shaping techniques. Another demonstrator, based on an internal flow-controlled switch (from AT&T), and using commercial interfaces is planned. Interfacing with the "intelligent" source memories has to be investigated. Studies of event builder management and control using standard ATM signaling protocols should be carried out in collaboration with the experiments in order to provide a user-friendly, self-regulating system. Simulation work should continue and new and more realistic traffic patterns should be studied using data from physics simulations and more detailed information about the detector read-out organization.

113.2 Technical Background

The principle of the parallel event builder architectures we are studying is the use of a switching fabric to interconnect the many front-end physics sources to the multiple "destinations" in which events are built for processing by the level 2 (L2) or level-3 (L3) trigger processor farms. Two standard, commercially available switching technologies seem promising candidates; these are ATM (ITU) and Fibre Channel (ANSI). Most of our effort has concentrated on commercially available technologies, and in particular on ATM-based solutions. Nevertheless, we have also investigated a custom-designed conical switching fabric architecture which has been optimized for overall DAQ system simplicity and cost.

If one excludes the switches based on shared media (busses), because they do not offer interesting scaling characteristics, the ATM switching fabrics are built of a number of elementary switching nodes interconnected in a web topology. The ideal switch would be an $N \times N$ cross-bar, allowing N independent paths to be established in parallel between the N inputs and the N outputs. The complexity of a cross-bar increases like N^2. The large ATM switching fabrics compromise by employing a network of elementary switching nodes in which the traffic of the N independent source-to-destination paths is packetized in ATM cells and asynchronously multiplexed over shared internal links. Contention for the internal links is resolved by introducing cell buffering in each elementary switching node. The complexity of these networks only grows as $N \cdot \log N$.

Internal links are not reserved for specific source-to-destination connections, but cells carry a label that allows them to be routed in hops between the switching nodes. The establishment of routing tables in the internal switching nodes allows the routing of cells according to their labels. Connections set up in this way are said to be *virtual connections,* and the label is called a *virtual connection identifier* (VCI). In principle, the number of virtual connections is only limited by the size of the VCI tables in the switching nodes. In summary, the traffic flowing on the virtual connections is statistically multiplexed onto the physical resources of the switch (bandwidth on demand), which makes for efficient use of the hardware when traffic on the individual virtual connections is fluctuating widely.

The Main Classes of ATM Switches

From our point of view there are two main classes of ATM switches which can be differentiated by the strategy they adopt when an internal buffer becomes full:

- Those designed for the telecommunications industry, where expandability to large dimensions, low-latency, and non-blocking characteristics are important. This class of switch simply drops incoming ATM cells whenever an internal buffer is full; therefore, delivery of data is not guaranteed. However, under the "random" traffic pattern resulting from the aggregation of the traffic of a large number of independent subscribers, the probability of data loss is acceptably small (of the same order as the loss probability in a long distance link). For random traffic, the switch's internal buffers are dimensioned to give a very low probability of loss, typically of the order of 10^{-10} or lower at 80% load on the switch (see for example Mandjavidze, 1993).
- Switches which implement a flow-control protocol on the internal links to guarantee lossless data transfer under all conditions. These are more likely to be used for LAN applications.

In switches of the first type, even if the network admission control system ensures that the connection characteristics do not exceed, on average, the resources of the switch, traffic burstiness and particular, traffic patterns (e.g., concentration of traffic) can still produce overflow in some of the internal buffers. Usually the telecom switches implement some mechanism to indicate internal congestion to the subscribers, who can then use a higher level signalling protocol to slow down traffic at the input. However, the reaction time of these higher level flow-control protocols is slow.

Whenever there is any correlation between the traffic flowing on different virtual connections, the probability of internal buffer overflow increases. In this case it is the task of the user-network interface (UNI) to regulate the traffic in such a way as to avoid congestion. This technique is called *traffic shaping* and can be used for event building over a telecom switch.

In switches of the second type, an *internal flow-control* protocol is used to prevent buffer overflows in the switch by holding up the traffic flowing towards a nearly full buffer until sufficient buffer space becomes available. In this way, no cells are lost in the switch. However, one must consider the case where the buffers of the fabric's first stage of switching elements overflow and the case where the destination user buffers overflow. The ATM standard does not specify an action in those cases, except for a higher level flow-control protocol which, as we mentioned above, might not react fast enough to prevent loss of data.

In all cases, there are techniques to limit the data losses to acceptable values. A careful evaluation of the switching network by means of simulation is necessary to properly dimension the network and the interface buffers.

113.3 Computer Modeling

Computer modeling is an indispensable method to investigate event builder architectures based on large switching networks. The size of the system, the variety of the technological solutions (available presently or within the time scale of the LHC projects), and the abundance of architectural options exclude full scale prototypes. However, practical experimentation on small scale prototypes is a necessary complement to computer modeling. It allows confirmation or correction to the understanding of the technology and serves to reveal some limitations that are not emphasized in the usual textbooks or even in the detailed documentation (if it exists at all!). The small prototypes can be used to verify the overall correctness of the simulation for that scale. But, we still depend on the correctness of the model when we use it to extrapolate to the performance of the full scale system. As a consequence, it is extremely important to develop high quality, accurate, and reliable models, and to cross-check results with independent models.

This section presents the progress accomplished in ATM modeling since the last report. At that time, the development of a detailed model of the Alcatel switch had permitted investigation of data losses and traffic shaping techniques. A first model of an event builder based on the flow-controlled AT&T Phoenix switch had also been developed.

Since then, we have recognized the need for a flexible modeling tool that would allow us to apply to the architectural research the variety of technologies that will be available in the time scale of the LHC projects, as well as the numerous methods of traffic control that can be envisaged. This has led to the development of a "generic model" described in the next section. This tool has been used to investigate large event builders with a variety of link speeds and switching element technologies, operated with various traffic shaping methods or with link-level flow-control. This model has revealed some interesting and unexpected behavior of the flow-controlled switches and we expect to be able to observe these effects on a real system soon.

The development of a switch model using the technique of parallel simulation is now underway. It promises to allow us to simulate very large switches (2048 × 2048) and to run much longer real-time sequences than can be achieved on sequential machines.

A detailed investigation of a custom switching fabric optimized for the event building task has been undertaken and has delivered very valuable results

The modeling of the proposed architectures for ATLAS and CMS has occupied a large fraction of our efforts, and has led to contributions to the technical proposals and to the publication of back-up reports. Several models of event builders realized in MODSIM have been our contribution to the global model of the ATLAS data acquisition system. In addition, a detailed investigation of the probable data flow scenarios for the ATLAS L2 trigger has led to the proposal of original architectural concepts. For CMS, the "virtual level 2" architecture has been studied and encouraging results have been obtained.

The use of standard simulation tools has been extended. Apart from μC++, models are available in C++ and MODSIM. As previously mentioned, cross-checking of simulation results obtained with independent models, possibly in different languages, has proved to be necessary to remove bugs and also to show that unexpected behavior was not due to modeling approximations adopted in a specific model.

A Generic Event Builder Model

The modeling activity aims at evaluating and optimizing the performance of a particular event building architecture. Poor performance is due to bottlenecks and, to obtain the desired performance, many parameters of the system can be adjusted. Each object constituting an event building system has a complex behavior which depends not only on its own architecture, but also on the behavior of the other objects. Examples of such subsystems are: the distribution of event data fragments among the sources, their size distribution, the discipline which sources follow while segmenting and sending the event fragments to destinations, the architecture of the switching network, the protocol which allows a destination to determine when the event building process for an event is finished, strategies for assigning events to destinations, etc. We need to understand how those parameters (and many others) influence the performance of the overall system.

Consequently, it is desirable to have flexible tools within the model which allow easy modification of the architecture and investigation of ways to optimize its performance. To this end, a generic event builder model, shown in Figure 113.1, has been developed. It is flexible enough and allows easy change of various system parameters via screen menu or from a parameter file. To facilitate the debugging process, it can update the statistics online, on the user screen, as the simulation task is executing. Below we give a description of each module with the options offered (*in italic*) by the generic model.

The *Event Generator* decides when a new event has to be created. The inter-trigger delay can be *constant* or follow a *negative exponential* or *geometric distribution*. The value of the mean inter-trigger delay parameter determines the average trigger rate. Also, the minimum time between two events can be specified. For each event, the event generator creates event data fragments and distributes them among the sources. The size of event data fragments can be *constant* or follow various distributions (*flat, exponential, erlang, normal*). The mean, min., max. and variance determine the shape of the distributions. When using one of the

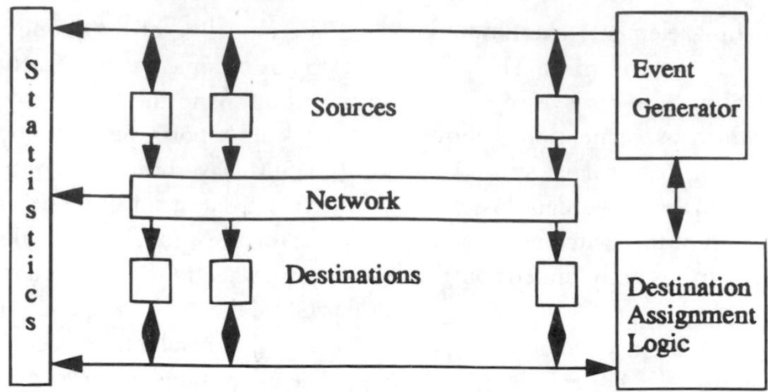

Figure 113.1 Block diagram of the generic event builder model.

above distributions, the sizes of the event data fragments in different sources are assumed to be uncorrelated. To study source correlation effects, the event generator can read event data *from a file,* which can contain, for example, events generated by Monte Carlo. From the destination assignment logic the event generator obtains a destination's identifier to which the event has to be sent and it passes this information to the sources.

The *Destination Assignment Logic* assigns events to destinations following one of various possible strategies: *simple periodic (sequential or butterfly), random* or a more complex assignment which *takes into account the status of the destinations* (current occupancy, number of events being scheduled to the destination, etc.). A suitable destination assignment strategy in conjunction with some traffic shaping scheme can significantly reduce the congestion probability in the switching networks (Mandjavidze, 1993).

In the *Source Modules* event, fragments are associated with the virtual connection to the assigned destination. If necessary, the sources provide the necessary buffering and queuing of the event fragments. The event fragments are packed in the ATM AAL5 packet format and then segmented into cells, which are injected in the switching fabric. In this way, the corresponding ATM and AAL5 overheads are modeled. The segmentation and cell injection strategies can be selected from *a simple FIFO* scheduling (no traffic shaping), or a *traffic shaping scheme* (*Cell Based Barrel Shifter* (Christiansen et al., 1993), *True Barrel Shifter* (Mandjavidze, 1993a) *or Randomizer* (Christiansen et al., 1993). Recently, a couple of other flow-control methods have been added to the source module behavior, such as *Static Rate Control* and *Dynamic Rate Control* techniques (their effect on the event builder performance is a subject for future studies). Of course, the number of source modules in the event building system is variable.

The *Destination Modules* receive, cell-by-cell, the event fragments sent by the sources, and they reassemble them. The event fragments are associated with the event structure to which they belong. Several events can be built simultaneously in a destination (especially, when source traffic shaping schemes are used to reduce contention in the fabric). One of the following protocols, which allow a destination to determine when the event building process is finished, can be chosen: *Known Sources, Empty Records* or *Time-Out* (Mandjavidze, 1994b). The number of destination modules in the event building system is variable.

The *Switching Network* is built with a regular interconnection topology of the switching elements that can be either *Banyan or Omega.* Switching elements can be variable size *(2 × 2, 4 × 4, 2 × 4, 8 × 4, etc).* Contention resolution in the switching elements can be selected from one of the following methods:

- Shared media switching element with no link-level flow-control (Fore Systems type, Fore Systems Inc.).
- Shared memory with no link-level flow-control (Alcatel/HSS type, Alcatel Data Systems).
- Output queueing with link-level flow-control (AT&T/Phoenix type, AT&T).
- Shared memory with link-level flow-control (IBM/Prizma type, IBM).

The switching element queue and buffer sizes are variable and can even be infinite. The switching element link rates are also variable and can be chosen to be either *160Mbit/s, 320Mbit/s, 640Mbit/s, 1.28Gbit/s or 2.56Gbit/s.* In the generic model, the switching elements operate on a cell, or transmission unit, of length 64 bytes, which carries 56 byte user payload. When studying ATM switching fabric implementations by particular vendors one finds, in most cases, that the internal switching elements operate on cells with a proprietary format (e.g., 55 byte cells for AT&T Phoenix). The size of the event builder switching fabric is variable.

During simulation runs various interesting statistics are gathered and stored in the form of distributions, tail distributions, or tables. The statistics can be visualized on-line or stored in ASCII files for off-line analysis. The source and destination buffer occupancies, the event building latency, the network load, and the switching elements' buffer occupancy are a few examples from the list of all the variables whose behavior is monitored and analyzed.

It is worth mentioning that the generic event building model is restricted to the data flow aspects and does not model possible control traffic due to the event building protocols. Presently, the generic model simulates one unique switching fabric that interconnects the source and destination modules. Recently, provisions have been made for allowing the event building switching network to be formed by several interconnected fabrics. The interoperability and performance of cascaded fabrics will be a subject for future study.

The generic model has been developed in μC++ (Buhr et al., 1992), an extension of the object oriented C++ language towards concurrency and simulation. When possible, the results derived from the simulations have been compared and successfully cross-checked with the results of other modeling activities which use C++ (Marchioro and Mandjavidze, 1994a) and MODSIM (Calvet, 1994) as the simulation environment. A C++ model is under development (Tether, 1995) and will provide easy portability, in addition to several other advantages.

Comparative Performance Evaluation of Traffic Shaping vs. Flow-Control Techniques

Traffic Shaping: The True Barrel Shifter

As mentioned elsewhere (Christiansen et al., 1993), the continuous strong concentration of data streams, typical of the event building traffic, creates congestion in switching fabrics. In the networks, which do not exploit link-level flow-control (WAN switches), unacceptably high cell losses occur due to buffers overflowing in the switching elements. Thus, the traffic originating from the sources has to be shaped before being injected into an event building switch. The effect of a well-chosen traffic shaping method is to reduce data loss probabilities in WAN-type networks to acceptably low values, to avoid unpredictably long event building latencies and to limit the required memory in the front-end source modules.

Traffic shaping has been described extensively (Christiansen et al., 1994; Letheren, 1994). To summarize, it consists of:

- Allocating an average bandwidth to all virtual connections between sources and destinations in such a way that the aggregate average bandwidth seen at each destination does not exceed the available bandwidth at the output port.
- Breaking the instantaneous time correlation between cells emitted from all the sources towards the same destination, as a result of the trigger.

The need for hardware traffic shaping, may preclude the use of commercial ATM interfaces. Thus, it is very important to find a traffic flow-control scheme which presents the following characteristics:

- Guarantees acceptable data loss probabilities due to residual congestion in the WAN ATM fabric.
- Results in low event building latencies and source/destination buffer occupancies.
- Needs a minimum of (or preferably no) specialized hardware to be added to a detector front-end source ATM interface.
- Requires minimum (or preferably no) centralized control of detector front-end sources.

Among the traffic shaping schemes that have been studied, one suffers from low bandwidth utilization (the event-based barrel shifter), another requires a strict synchronization of sources (the cell-based barrel shifter), while still another needs dedicated hardware (the Randomizer) thus excluding the use of commercial ATM interfaces in the sources.

A new, conceptually simple, traffic flow-control scheme, referred to as "True Barrel Shifter," has been proposed (Mandjavidze, 1994a). It presents almost all the desirable characteristics mentioned above. The scheme is very suitable for the level 3 event building processes. To summarize, the event building system operates as a barrel shifter which changes its states after a time period approximately equal to the emission time of an average event fragment size. A strict synchronization of the sources is not necessary. At the transition, and during a short period of time corresponding to the inaccuracy of the synchronization, it can happen that some destinations receive data from 2 sources. But this small bursty traffic can easily be smoothed out by an event building ATM switching fabric. Commercially available ATM interface chips offer all the necessary features to configure the source modules with the true barrel shifter requirements (hardware-maintained linked lists of virtual connections, activation-deactivation of virtual connections, segmentation processes, switching from one virtual connection to another with no overhead). It has been shown (Mandjavidze, 1994a) that event builder systems which use the "true barrel shifter" traffic shaping scheme, scale linearly to large dimensions (Figure 113.2a), while maintaining low cell loss probabilities (Figure 113.2b) shows that the probability of overflow the 2 kByte buffer of a switching element is less than $\sim 10^{-12}$).

It has also been shown that the performance of an unbalanced system (a system with sources generating event fragments with different average sizes) does not depend too much on the barrel shifting time period. As the event building latency remains the same whether the time period is equal to the smallest, the average or the biggest event fragment transition time (Mandjavidze, 1994a; Nomachi, 1993).

Use of Traffic Shaping in a Flow-Controlled Switch (AT&T Phoenix)

The combination of link-level flow-control and the traffic shaping technique has been studied for an event builder based on the AT&T Phoenix switch (Kumar et al., 1993). The switching element operates at 400 Mbit/s link speed and transports 55 Bytes long cells. Network adapters, based on the ALI chip (Oechslin et al., 1992), are used as inlets and outlets in order to match the external 155 Mbit/s link speed with the internal 400 Mbit/s rate, and to perform the conversion between the external ATM cell format and the internal proprietary cell format. For reasons of simplicity, the network adapters have been modeled as FIFO queues of cells. Backpressure, originating from the output network adapters or from the switching elements can propagate as far back as the input network adapters (as shown in Figure 113.3), but there is no hardware signal to transmit the backpressure to the source. If the backpressure persists for a long time, while cells are still delivered to the congested inlet, the buffer in the network adapter can overflow and the cells can be lost. Therefore, although the cell loss probability is zero in the fabric, cell losses can still occur in the network adapters. One could try to estimate,

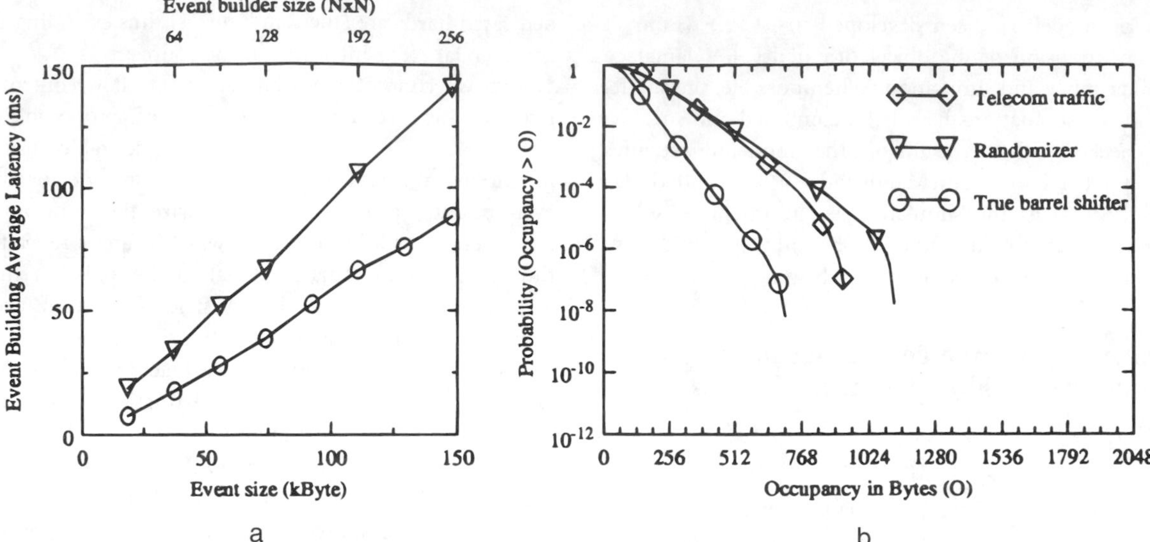

Figure 113.2 Simulation results for an event builder with true barrel shifter traffic flow-control scheme. Event builder based on the Alcatel 256 × 256 @ 155Mbit/s switch, 25 kHz trigger rate, 80% load. a. Scaling of events builder. b. Cell loss probability for the 256 × 256 event builder.

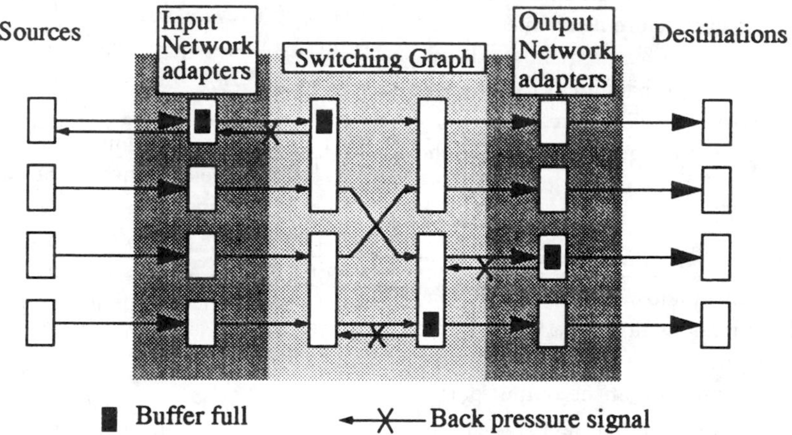

Figure 113.3 Buffer overflow and backpressure signals.

by simulation, the required buffer size in the network adapters to achieve low cell loss probabilities.

Simulations have been conducted for an event builder which operate at a 10 kHz trigger rate. Sources generate event data fragments with an average size of 1000 Bytes and a maximum size of 5000 Bytes. The resulting bandwidth utilization is 60% of the 155 Mbit/s links, but due to bandwidth expansion inside of the fabric, the core of the switching fabric operates at approximately 25% load. For the 1024 × 1024 event builder, the total event size is equal to 1024*1000 = 1 Mbyte.

Initially, no source traffic shaping has been applied. The source buffers were simple FIFO queues of event fragments injecting cells at full 155 Mbit/s speed. Event builder systems up to 256 × 256 have been modeled. Larger systems could not be simulated because of the required memory space and long execution times. The event building latency and the buffer space required in the input network adapter scales linearly with the size of the event builder and, for the 256 × 256 event builder, each network adapter had to buffer 5000 Cells (270 kBytes) in order to guarantee a cell loss probability lower than 10^{-10}. From the simulation results one could expect that a network adapter buffer size of 1 Mbyte would be necessary for the 1024 × 1024 event builder system. The effect of the rate control technique has not been studied and this is a subject for future investigation. We expect that rate control of sources will decrease the level of congestion in the fabric and, as a consequence, the network adapter occupancy.

When traffic shaping (randomizer) was applied at the sources, it became possible to simulate a 1024 × 1024 event builder. Figure 113.4 presents the simulation results. On average, 355 ms were necessary to build events of 1 Mbyte in a destination. This time also includes the queuing time of the event fragments in the source modules. The source modules have to buffer on average, 700 event fragments. A buffer space of 1 Mbyte in the sources (easy to provide) would guarantee a very low probability of buffer overflow. On the other hand, the network adapter occupancy charts (along with some other statistics) indicate that contention

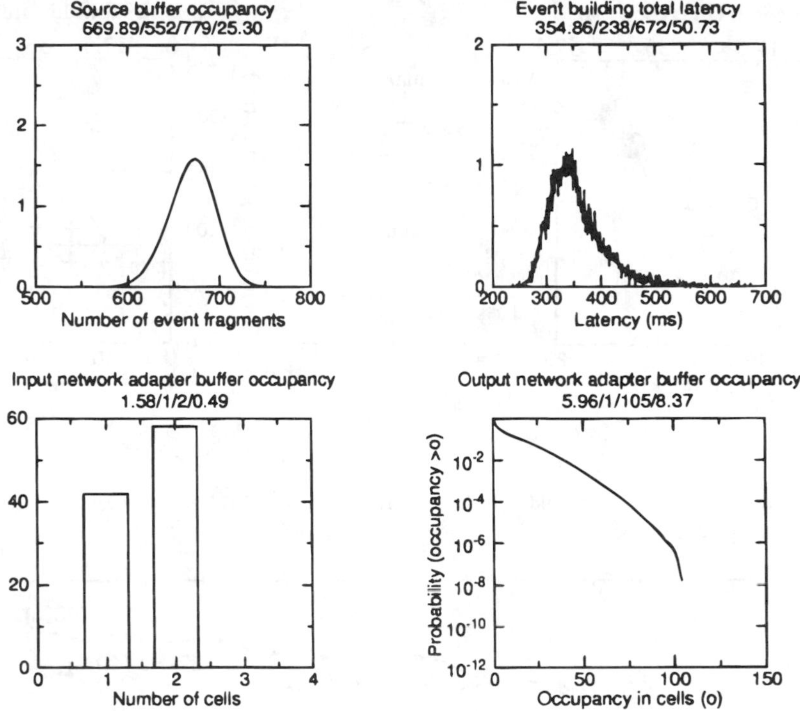

Figure 113.4 Simulation results for event builder which combines flow-control and traffic shaping techniques. 10 Khz trigger rate, 1 Mbyte event; 1024 × 1024 event builder, 155 Mbit/s link rate.

in the fabric never propagates back to the input network adapters (the input network adapter occupancy never exceeded two cells for approximately 350 × 10^6 cells injected in the fabric). From the tail distribution of the output network adapter buffer occupancy (see Figure 113.4) one can estimate that a buffer of 150 cells in this adapter will guarantee a cell loss probability much lower than 10^{-6} (today, network adapters can buffer up to 256 cells and more).

Study of the Flow-Control and Traffic Shaping Techniques Based on the Generic Event Builder Model

We have seen, in the previous section, that the simulation of large switches with pure flow-control (without traffic shaping) was not possible with the exact model of the Phoenix switch. Hence, the generic event builder model described in the section on a generic event build model has been used for this study. Sources generate on average 1 kByte of event fragment data. A 1024 × 1024 network was interconnecting source and destination modules and was constructed from 4 × 4 switching elements with 16 kByte of memory each. The fabric operates with 640 Mbit/s links, therefore, the aggregate bandwidth equals 640 Gbit/s. Figure 113.5 shows the event building latency versus trigger rate.

In one case, Figure 113.5a, no traffic shaping has been applied to the source modules, but backpressure, if necessary, could propagate as far back as the sources and prevent them from transmitting data. For trigger rates below 10 kHz, there is very weak dependency of the event building latency on the trigger rate. In the destination modules only one event is built at a time. For trigger rates in the range from 10 to 15 kHz there are always two events being built concurrently in a destination and event building latency stabilizes around 110 msec. The third state (if one can describe the system behavior in terms of states) is characterized by three events being built concurrently in a destination and is observed in the 17 to 20 kHz range. Event building latency continuously increased and system steady state could not be reached for the trigger rates above 22 kHz, even though the load on the switching fabric was not too high (around 40% of the available aggregate bandwidth). The same behavior has been observed for the event builders which operate at 2.56 Gbit/s link rates, the maximum reachable trigger rate for a stable system being 80 kHz. Those results were qualitatively confirmed by means of two other simulation programs, one in C++ (for conic event builders) and one in MODSIM (for the event builders based on the AT&T switching fabric).

Figure 113.6 shows how the event building latency changes with time, starting with an empty system. For low loads, one can see that the latency grows up to a maximum value and then drops to a lower value where it remains stable. This stable value grows with the load applied to the switch, but it grows by steps, and not in a continuous way. Beyond some value of load, the latency keeps growing with time: the system cannot stabilize and the event builder is not usable under this condition.

To better understand the observed behavior of flow-controlled systems, a more detailed study of a small (16 × 16) event builder has been carried out. It has shown that, due to backpressure, some kind of self-organization in the system leads to minimizing contention in the switching fabric. For example, by analyzing

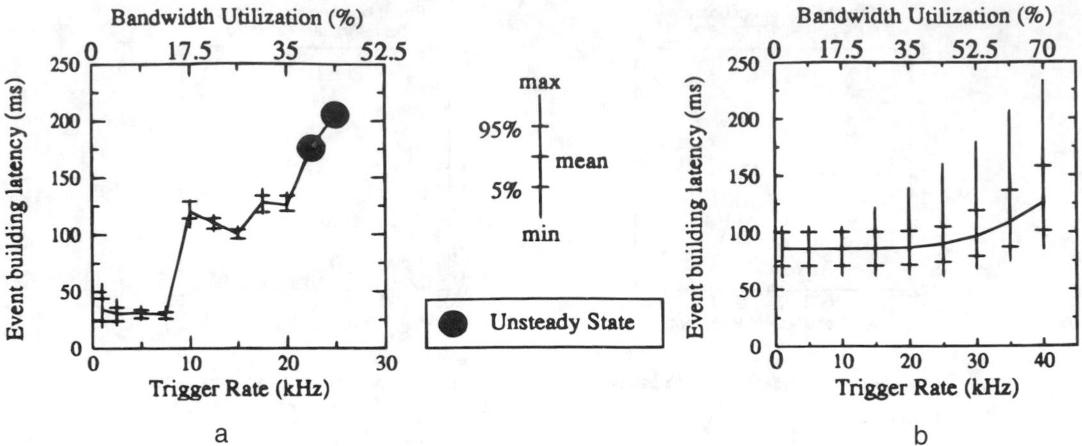

Figure 113.5 Simulation results for the generic event builder. 1 Mbyte average event size, 1024 × 1024 event builder, 640 Mbit/s link rate. a. Flow-control extended up to sources. b. Source traffic shaping.

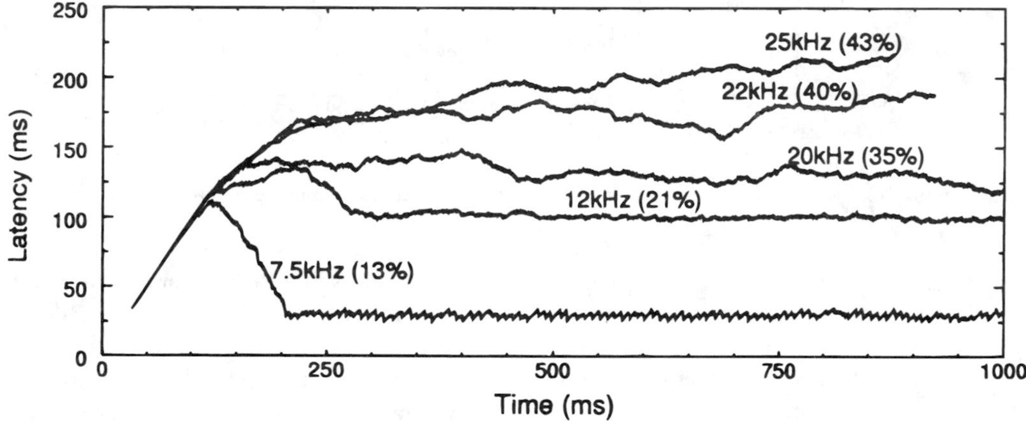

Figure 113.6 Latency profile in a fabric with link level flow-control. Event builder: 1024 × 1024 @ 640 Mbit/s. Switching element: 4 × 4 with 16 kByte memory. Back pressure: up to sources event size: 1 Mbyte.

the occupancy of sources it has been found that they are divided into groups which never send data to the same destination at a given time. A more detailed study of the observed phenomena could be a subject of future work.

Scaling characteristics of the event builders with flow-control, depend on the switching network architecture and event building traffic characteristics (trigger rate, event size, network load) (Mandjavidze, 1994c).

It should be mentioned that the effect of using the rate control technique on event builders with flow-control has not been studied and remains a subject for future investigation. One can expect, that rate control of sources will decrease the level of congestion in the fabric and, as a result, will lead to higher bandwidth utilization.

In the second case (Figure 113.5b) the randomizer traffic flow-control scheme was used to prevent congestion in the switching fabric. Available system bandwidth utilization up to 70% has been observed without significant performance degradation. These simulation studies confirm results obtained previously, which have been successfully cross-checked against queuing theory (Nomachi, 1993). It has also been shown that event builders which exploit traffic shaping techniques can be characterized by good scalability. Event building latency depends linearly on the system size (Mandjavidze, 1994c). Moreover, assuming that the traffic shaping guarantees the same level of cell loss probability for two different types of network architecture, the event building latency, the source/destination buffer occupancies do not depend on the particular network architecture, but are determined by the traffic shaping technique.

Parallel Simulation for Large Switches

As already mentioned, the simulation of event builders based on switches with internal flow-control requires a very large memory space and long execution times. Thus, our simulations of event builders based on the AT&T Phoenix switching element (when we do not apply traffic shaping) have been limited to fabrics not exceeding the 256 × 256 size.

We intend to use a parallel simulation environment, SIMA (Rajaei, 1992) to simulate large ATM switches on parallel computers. We expect to be able to simulate very large switches (up to 2048 × 2048) on a Convex multiprocessor system. The parallel approach will also allow runs simulating longer real time spans than those of the order of 1 second, which we typically achieve after many hours of computing on a high-performance uniprocessor workstation.

Custom-Designed Conical Switching Fabric

A custom-designed conical switching fabric, employing internal link-level hardware flow-control, has been proposed as an alternative architecture to those based on square commercial switching fabrics. The conical fabric has M inputs and N outputs, where $M > N$, and it has been proposed specifically as a simple, optimized solution for event building (Mandjavidze, 1993b; Marchioro and Mandjavidze, 1994a, 1994b; Christiansen, 1992). The conical fabric connects directly to the front-end modules via a large number of low speed input ports. It simultaneously performs the cell switching function and a data multiplexing function. It has the advantage of providing a homogeneous data acquisition system architecture, whereas in the case of the square commercial switches, the multiplexing function must be provided by specific dedicated hardware upstream of every input port.

A model of a proposed conic event builder (Mandjavidze, 1993b) for the Euroball experiment (Gerl and Lieder, 1993) has been developed. It was based on an optimized, partially interconnected, 6-stage Banyan network with 1536 inputs (connected directly to front-end modules) and 64 outputs. The proposed fabric could be constructed from very simple custom switching ASICs, each with 4 inputs and 2 outputs.

Modeling of the ATLAS Architecture

MODSIM Model of the Switches

A model of an ATM switch, based on the PHOENIX AT&T switching element (Kumar et al., 1991) has been developed (Calvet, 1994), using the MODSIM language (CACI Products Company, 1993). It can be plugged into the ATLAS DAQ model (Boggerts, A. et al. 1996). This work also had the goal of comparing the performances of switches with and without the use of internal flow-control. It has confirmed that, due to the bandwidth expansion inside of the switch, rather high loads (75%) can be applied to the Phoenix based switching fabric. Event building latencies are significantly lower than in the case of a switch operated with traffic shaping.

The MODSIM simulation allowed us to model switches up to size 128×128 and the μC++ program has confirmed the favorable scaling of latency up to a size of 256×256, but it has shown that the buffering space required in the network interfaces is probably too high.

Study of the ATLAS Calorimeter System

The Saclay group, within the RD-31 collaboration, has initiated a detailed study of the Atlas level-2 and level-3 triggering systems. Oriented towards the calorimeter subdetector part of the Atlas DAQ, it is nevertheless general enough to cover other subdetectors, as well.

According to the read-out scheme proposed in Costa et al. (1994), for each event accepted by level 1, data are transmitted from the calorimeter (PS, EM, HAC) front-end boards to the Intelligent Read-out Memories. The read-out of the calorimeter is organized in towers of 0.1×0.1 in the η, ϕ space. The detector consists of 64 ϕ by 60 η towers. In our model, we assume that 512 Intelligent Read-out Memories will be used to store the data during level 2 and level 3 decision latencies. The memories will be housed in 32 crates, 8 crates, for example, mapping the barrel part of the EM calorimeter, each one covering 1.4×1.6 in the η, ϕ space. One link per crate can be used to transmit the data from the Regions of Interest (RoI), required for the level 2 decision, into the feature extraction (local) processors via a switching network.

We started with studies of the data volumes and with an evaluation of the bandwidth required by the system. Starting from physics simulation data and, with simple numerical evaluation, we found that, for a trigger rate of 100 kHz from level 1, an aggregate bandwidth of 5 Gbit/s is required for collecting the calorimeter data for level 2 processing. Assuming that the event building traffic for the next selection level will use the same links, a total aggregate data bandwidth of approximately 7 Gbit/s is required. This results in a data throughput of 225 Mbit/s per crate. Assuming that standard ATM links at 622 Mbit/s are used to interconnect the read-out crates with the switching network, the links will be utilized at ~40% (including the overhead due to the ATM protocols), which is a reasonable value.

The simulated architecture is shown in Figure 113.7a. The local processors are grouped in farms. At the output of the switch, the level 2 data are delivered to each farm through a single link. Thirty-two processing farms have been used in our model. In this first modeling study, we are only interested in the latency due to the network (RoI collection latency). Therefore, the execution time of the Feature extraction (FEX) algorithm, though used in the model, has been chosen to be constant. We plan to introduce more realistic FEX algorithm time distributions at a later stage.

In our model, a generic ATM Multistage Interconnection Network (MIN), based on switching elements with a size that can be varied, provides a data path between the read-out crates and the farms of local processors (see the generic event builder model for more details about switching network).

Two completely independent simulation programs have been developed in concurrent object oriented languages: MODSIM (CACI Products, 1993) and μC++ (Buhr et al., 1992). The results derived from both programs have been compared in the same conditions and have shown to be in good agreement with each other.

From our initial simulation studies, we have found that the average crate occupancy (the probability that, for a given event, a part of at least one RoI will fall in a crate) amounts to 27%. For 100 kHz level 1 trigger rate this means that each crate must be able to collect, format and send level 2 data at a rate of 27 kHz. Up to four sources may contain data for a given RoI. Therefore one, two or four packets have to be collected in the

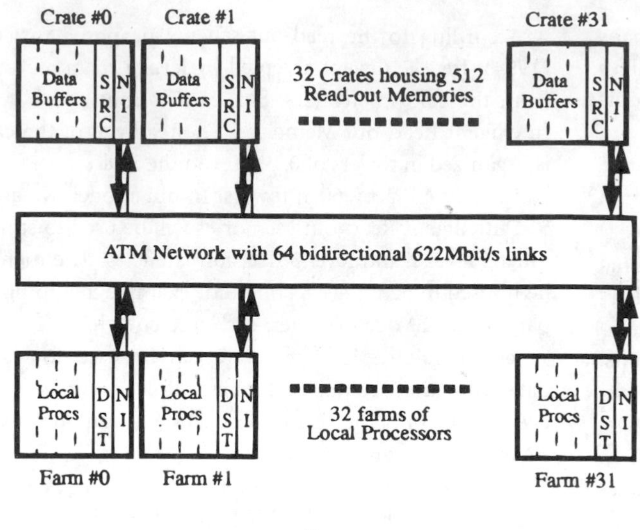

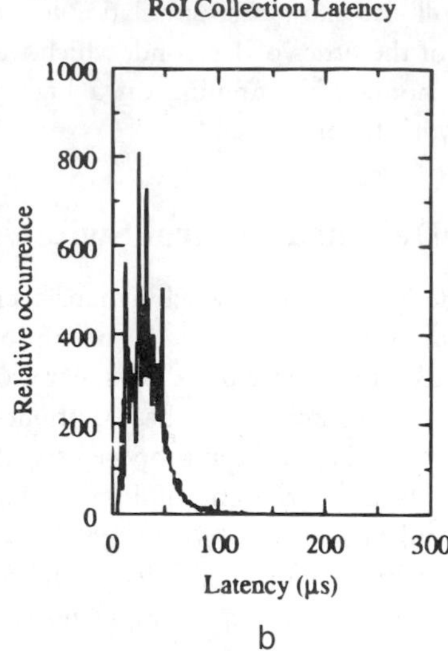

Figure 113.7 Initial simulation studies for the Atlas calorimeter L2 subsystem. a. Simulated architecture. b. RoI collection time distribution.

destination to reconstruct a RoI from the calorimeter. The distribution of the time necessary for this operation is shown on the "RoI Collection Latency" histogram in Figure 113.7b). The average RoI collection latency is ~50 μs. The different peaks observed on the time distribution correspond to the different types of RoI's and their distribution among the crates.

The simulation studies described here constitute a first approach and the status of our research is evolving rapidly. In the section protocol traffic transport via the switch, we introduce a possible scheme for the level 2 and level 3 event building of ATLAS, which uses the same switching network for data and control flows, and we present our latest simulation results.

Modeling of the CMS Architecture

The feasibility of using packet switching networks for the DAQ architecture of CMS has been studied. For events accepted by level 1, the CMS collaboration currently considers two types of event read-out schemes: full and partial (CERN, 1994b). The full read-out scheme corresponds to sending the full event from the front-end dual port memories to a destination (processing farm) via the switching network. The simulation results of the generic event builders, presented in the section on the study of the flow cost and traffic shaping techniques based on the generic event builder, are applicable to this scheme and will not be repeated here. This section will focus only on the partial read-out simulation architecture. A detailed description of the modeling for both architectures can be found in (Mandjavidze, 1995).

Partial Read-Out Architecture ("Virtual Level 2")

Partial read-out corresponds to transmitting only the information needed by the level-2 trigger for every event. In case of acceptance, the rest of the event data is sent to the destination processor farm. The model of the partial read-out scheme is shown in Figure 113.8.

The 1024 sources are interconnected with the 1024 destination farms via a switching network. Only 256 source modules (level 2 sources) participate in the level 2 decision, sending on average 400 bytes of data to a destination processing farm for every event accepted at level 1. The rest of the event data, which does not participate in the level 2 decision, are stored in the L2 buffers where they wait for completion of level 2 event building and decision. In the case of a negative decision, the event data fragments are flushed from the L2 buffers. In the case of a positive decision, the event data fragments are transferred to the level 3 sources, which then send them to the same destination farm that made the level 2 decision. The amount of data sent by each level 3 source is 1 kByte on average. Both level 2 and level 3 data fragment sizes follow normal distributions.

The same 1024 × 1024 network model, which was used for generic event builder studies, was used for the CMS partial read-out simulation. The network was constructed from 4 × 4 switching elements with 16 kByte of memory in each switching node. The fabric operates with 640 Mbit/s links, giving an available aggregate bandwidth of 640 Gbit/s. The source traffic shaping technique was applied to minimize congestion in the fabric. Figure 113.9 shows the event building latencies for level 2 and level 3 as a function of the trigger rate (Figure 113.9a) and as a function of the variance of the event fragment size distribution (Figure 113.9b). In the simulations, a level-2 rejection factor of 10 was assumed.

Compared to the full read-out scheme, the partial read-out approach substantially reduces (by a factor of 9) the data throughput. At a trigger rate of 100 kHz, only 30% of the available aggregate bandwidth is utilized. Both level 2 and level 3 latencies have approximately a flat distribution as a function of the trigger rate. Therefore, the safety factor of the system is high.

The event data generator, used in the simulations, is very simple and assumes uncorrelated front-end sources which generate normally distributed event fragments. As shown in Figure 113.9b, level 2 and level 3 latencies depend not only on the

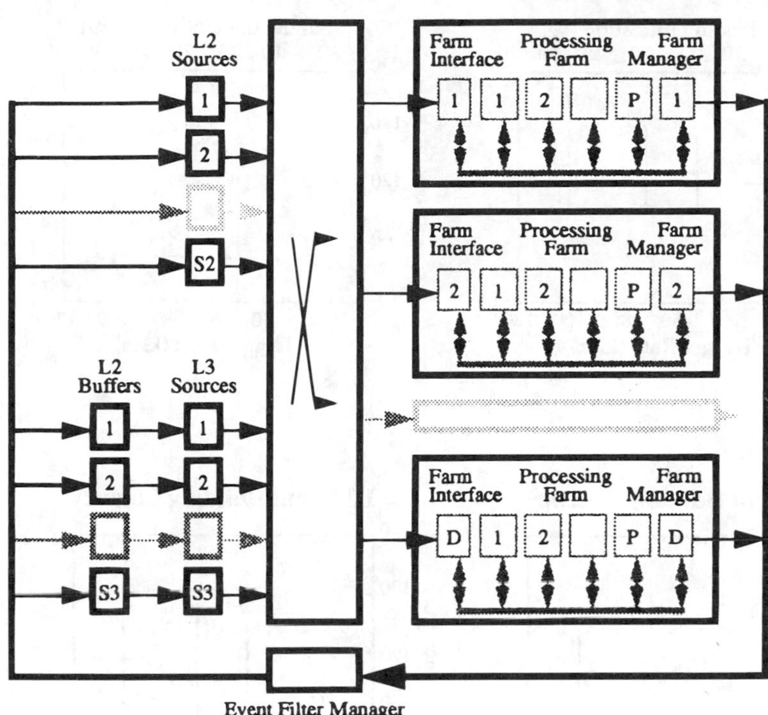

Figure 113.8 Simulation model of the CMS partial read-out architecture.

average size of the event data fragments, but also on the statistical distribution of the size: the larger the variation of the event fragment size, the longer it takes to build an event. What is even more important is that the tail distribution of the latencies also becomes longer. In the future, more realistic simulation studies should be based on the input data derived from physics simulations and from the detailed read-out architecture of the detectors. This will allow us to study the effects of the correlations between the sources.

113.4 Event Building Protocols and Related Software Development

The ATM-based event building process has been defined as a layered structure of protocols. These layers include the standard ATM layers and are complemented by higher-level layers that implement the event building functions. Most of the basic software required to run the full event building process has now been written. It is complemented by various programs which provide monitoring, event building control and data generation. This software has already been used extensively during the development of the VME-ATM board, the tests of interoperability with the switch and, currently, the development of the event builder demonstrator. A detailed presentation of the concepts and the software is given in Costa (1995).

We have defined additional layers, on top of the ATM protocol layers, to implement a full function event builder. See Costa (1995) for a detailed review. Most of the work has been done in the field of a "push" architecture, whereas some new work is starting now to investigate the possibilities of "pull" architectures, mainly for application to the ATLAS level-2 trigger (see Protocol Traffic Transport via the Switch).

The Layered Structure of the Event Builder Architecture Based on an ATM Switching Network

The Architecture of Protocol Layers

Figure 113.10 shows the layers of protocols as they are proposed and have been implemented. Layers 1 to 3 are the standard ATM layers (ITU). The top "event building" layer is sub-divided into 2 sublayers:

- The *Event fragment sublayer* ensures the independence of the layer(s) above it from the network-specific layers (1 to 3). It allows the handling of event-fragments which are longer than the maximum packet length defined by the underlying technology (in the AAL5/ATM case the maximum packet length is 64 kBytes).
- The *Event sublayer* has the task of linking together the received event-fragments to form an event (because they traverse different paths through the fabric, the event fragments reach the destination out of order). It must also recognize when an event is completely assembled. Several methods have been proposed for this latter task (Mandjavidze, 1994b) and can be selected according to the particular conditions of operation of an event builder. The event sublayer is only required in the destinations, where it must be able to build several events concurrently.

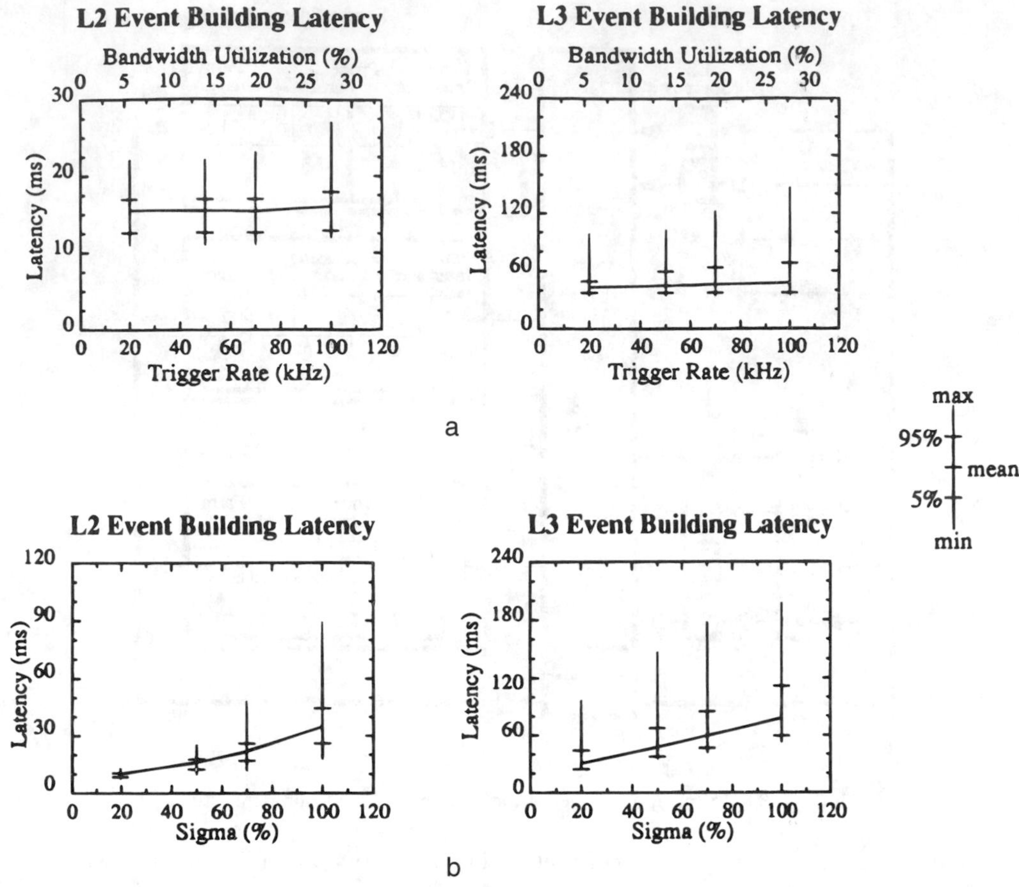

Figure 113.9 Simulation results of the CMS partial read-out architecture. a. Latency as a function of trigger rate (50% variation of event fragment size distribution) b. Latency as a function of event fragment size distribution (100 kHz trigger rate)

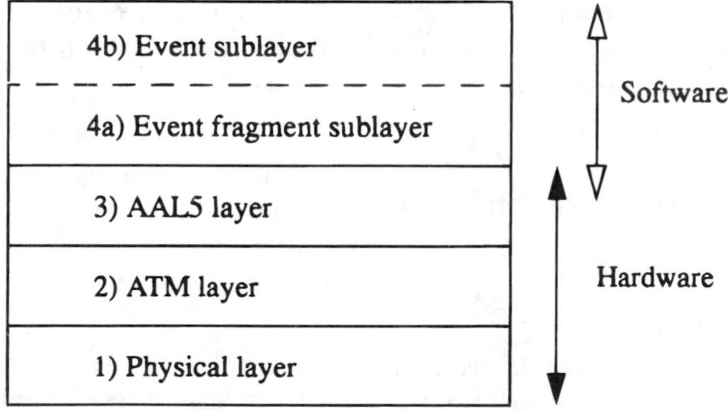

Figure 113.10 Protocol layer structure of the event builder architecture.

Of course, the event fragment sublayer introduces some overhead, but it is necessary in order to allow event fragments to be larger than the maximum 64 kByte AAL5/ATM packet. It is expected that the ALICE experiment will have event fragments much longer than 64 kByte. It will perhaps also be needed on other experiments when collecting calibration data. It can also optimize memory by allowing us to define the buffers in the memory interface to be smaller than the largest possible event fragment.

In the protocol stack described, there is no OSI transport layer. The CRC within an AAL5 packet provides error detection, but there is no mechanism to retransmit an errored packet. Actually, most transport protocols suppose that sources and destinations are both able to send and to receive. In the current implementation of the event builder demonstrator system (see The ALCATEL ATM-Based Event Builder Demonstrator), sources are emulated using traffic generators which can neither receive nor process data. This implies that no transport layer can be used. However, the use of a transport layer has to be evaluated, considering the trade-off between the overhead in the network interfaces and the increase of the traffic caused by retransmitted packets and the consequences on the reliability of the event builder. For limiting the protocol overhead the transport functionality could be included in the event fragment sublayer. The algorithms for avoiding further congestion due to the transport protocol have still to be investigated taking in account the recommendations of the ATM Forum.

The architecture proposed so far is a "push" architecture where the destinations have no means to direct the collection of data. It should be preferred in all cases where it is adequate because it is certainly simpler to implement than a "pull" architecture, in which the destination requests the data from the sources. However, we have just started investigating the "pull" architecture for applications which have a sparse distribution of data in sources (e.g., for systems collecting data from "Regions Of Interest", as discussed in ATLAS AT&T ATM-Based Real-Time Demonstrator). This might be a subject for further research.

Data Format, Data Structures

Figure 113.11 shows the data format for the event fragment sublayer. At the sending side, an event fragment PDU (Protocol Data Unit) is formed from a payload and a PCU (Protocol Control Unit) with information about event number, destination, and optional event building control information (e.g., the event sequence number which may be required for the event building completeness algorithm). It is segmented in AAL5 packets, each one being in turn complemented with a PCU containing the source number, a fragment sequence number and a segment type (continuation of message or last packet indication (LPI)). The PCUs are used in the destination to check for the completeness of every event fragment (Event Fragment SAR and AAL5 PCUs) and for the completeness of the event building (event fragment PCU).

The event building algorithms are implemented with linked lists of descriptors to keep track of the various segments received and possibly belonging to different events. The event building is performed keeping track of fragments using pointers and, as far as there are enough free buffers in the network interface memory, without moving data into the processor main memory.

Software Structure

The software has been designed taking into consideration its portability and reusability. These issues are particularly important because some aspects of the ATM standard as well as many other aspects of the event builder are not completely defined.

The structure of the software can be divided into three layers and two planes, as shown in Figure 113.12. *The Event Building Layer* implements the event building layer protocol and controls and manages the system in order to satisfy the requests from sources or destinations. The *Network Interface Layer* implements the communication protocol, manages the network interface resources and monitors the network performance. The *I/O Specific Library* provides a set of functions to access the interface.

The *Data Protocol* plane implements the event building protocol and the communication protocol; the *Management* plane manages the system resources, handles errors, and monitors the system performance. More details on the functions shown in Figure 113.12 can be found in Costa (1994).

In the current implementation, the software runs in stand-alone mode on a CES RIO module (Creative Electronics Systems, 1993). It works as a server executing tasks requested from a client controller program running on a CES RAID. The two programs communicate using a message passing mechanism via their FIFOs. Due to limitations in VME data transfer characteristics, no event building data is actually transferred between RIO and RAID.

Protocol Traffic Transport via the Switch

Until now, most of the performance studies of event builders have been focused on the data flow aspects. Recently, the collaboration has started to investigate various scenarios of control flow and data flow; in particular, we are evaluating the merits of

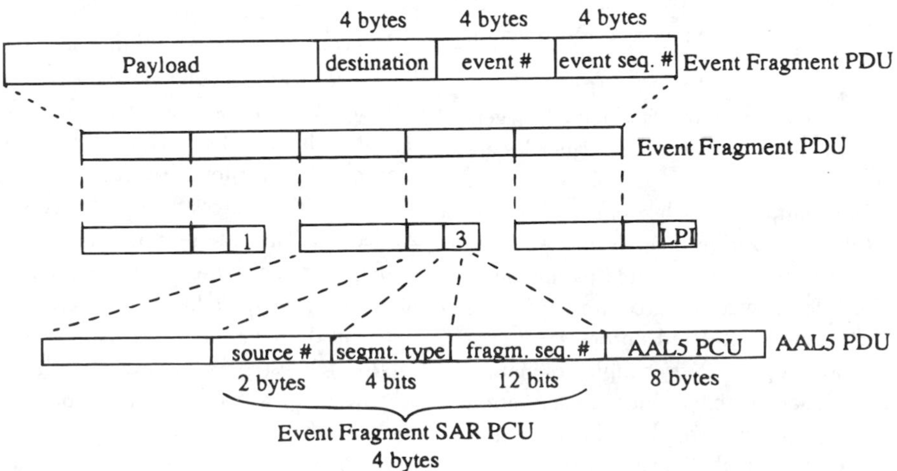

Figure 113.11 Event fragment protocol data unit (PDU) and its segmentation into AAL5 packets.

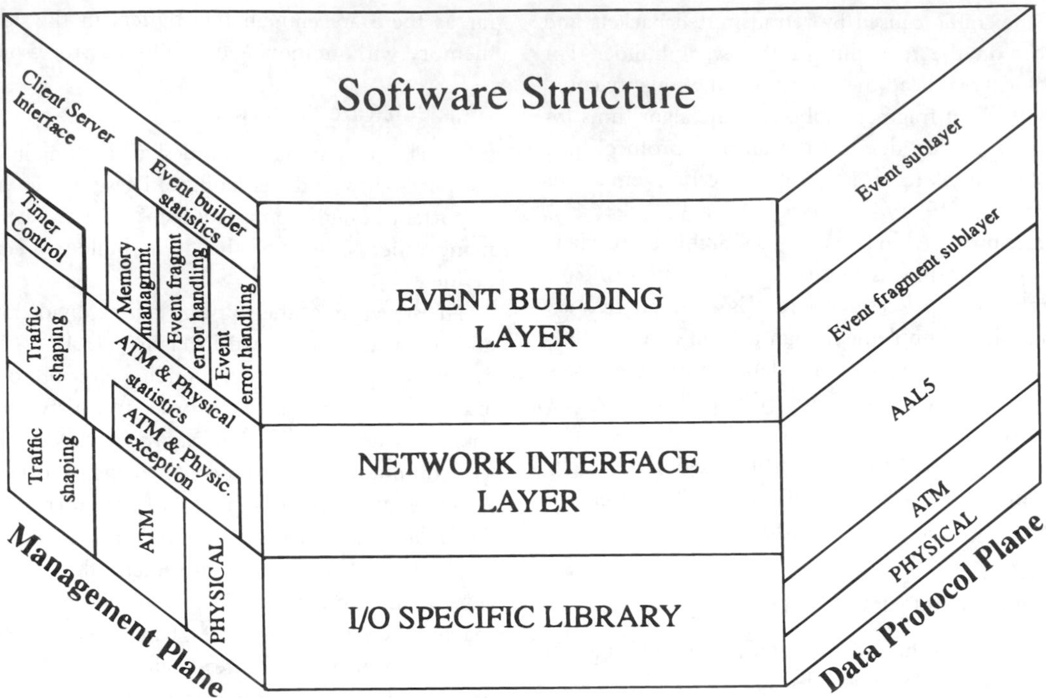

Figure 113.12 Software structure.

"Push" and "Pull" architectures. In any DAQ architecture control information should be exchanged between the various parts of the system, such as the data sources, the destination processors, etc. The control flow can use the same medium (network) as the one used for data transmission. The main advantages of this approach are:

- A unique switching network for all types of traffic (data and control).
- A single network adapter per node.
- Standard network protocols, available from industry. Therefore, assuming bidirectional control flows, and using bidirectional links one can take advantage of industry developments, thus greatly simplifying the error detection and recovery issues.

We propose a control scheme for the level 2 and level 3 triggers of ATLAS based on the considerations above. The principles are shown in Figure 113.13.

We assume that the information on each event accepted by the level 1 (number of RoIs and position in η, ϕ space) will be delivered to the trigger supervisor via a dedicated path. One of the tasks of the supervisor is to allocate resources for processing this event, e.g., assign a processor per RoI and a processor for the global decision. Currently, we propose a very simple destination processor assignment scheme, namely that a sequential allocation should be adequate. More sophisticated algorithms are not excluded. Each allocated (local or global) processor receives a notification message (one cell) from the supervisor (message flow (1) in Figure 113.13a). The message contains the Event ID, the RoI ID, the Global Processor ID, etc. We also envisage the possibility to replace flow 1b by flow 1c.

With this information, the global processor knows from which local processor it has to expect features data, and a local processor knows which sources contain data for a particular RoI. The local processor will then send a request message (one cell) to each source concerned (flow 2 in Figure 113.13b). In response, the sources send the requested data (flow 3). When all data for a given RoI have been delivered to the local processor, it executes the feature extraction algorithm. The result is then sent (flow 4) to the global processor. When all features of the event have been collected the global processor executes a global algorithm and generates the level 2 decision.

We consider two possibilities for the continuation after the level 2 decision. In one case, the source modules are notified only if the event has been accepted. The level 2 decision "Yes" is sent to the level 3 supervisor (flow 5 in Figure 113.13c), which then multicasts it to all sources (flow 6). No immediate action is taken in the sources for the events which didn't pass the level 2 selection. The oldest event is simply overwritten in the source buffer when a new event is being read from the front-end modules. This scheme is based on the consideration that the event buffer in the source modules has to be designed to be sufficient for the longest possible level 2 decision latency anyway. This scheme is attractive because it does not generate unnecessary traffic in the network (99% of the level 2 decisions are expected to be "No"), it simplifies the control logic of the data sources and it requires less actions in the system per event.

The other solution consists of sending either of the level 2

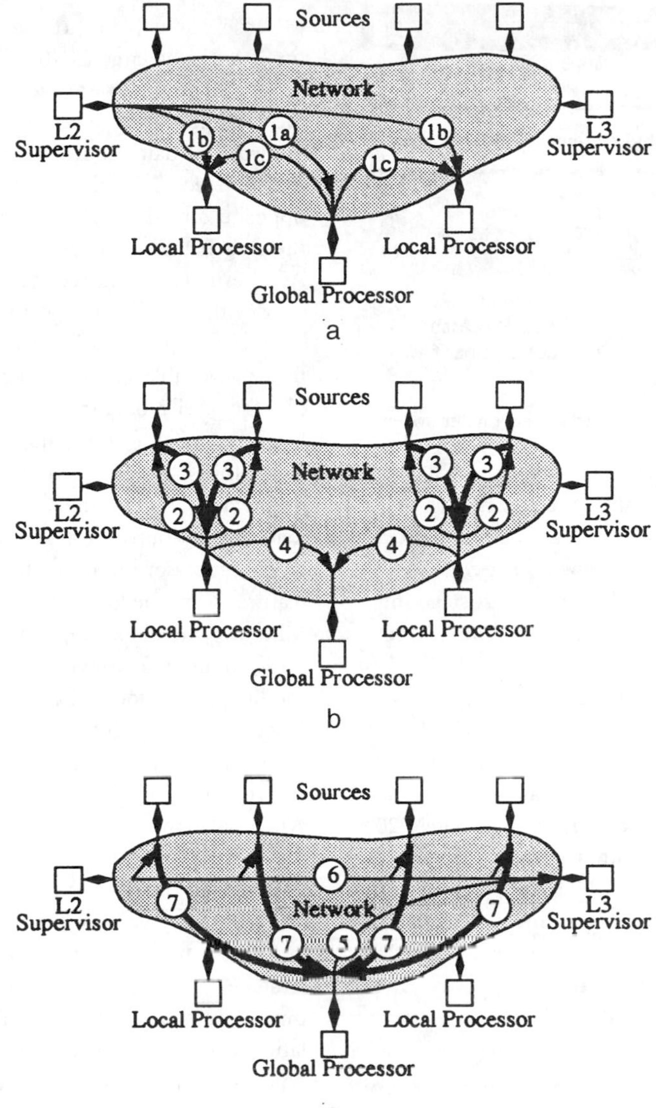

Figure 113.13 Possible scheme for the L2 and L3 levels of the ATLAS trigger system.

decisions, "Yes" or "No," to the level 3 supervisor (flow 5). To minimize the decision broadcast traffic, several (around 10) decisions are packed in one ATM cell which is then multicast to the sources (flow 6). If necessary, information about the accepted events is sent from the level 3 supervisor immediately, and only "No" decisions for consecutive events are packed together.

For the continuation of the work on the level 2 selection, we are studying two possibilities: parallel and sequential. The first one requires all RoI data to be sent to the local processors and examined in parallel for all subdetectors and the results to be combined in the global processor. In the second case, the RoI data for the particular subdetector are sent to the local processors only if they are required by the subsequent steps of level 2 selection algorithm (e.g., TRT data for a RoI is requested only if the decision based on the calorimeter data of the RoI was positive). The sequential flow of the level 2 selection can significantly reduce the aggregate bandwidth requirements for the switching network.

For level 3, it is not decided yet, whether the full event data is needed for the selection algorithm or whether partial event data will be sufficient. In the latter case the level 3 selection step will be followed by the event building. It is also possible that the same processor, which performs the level 2 global decision will continue to work on the level 3 selection for the same event (as it already possesses a substantial amount of information about this event). The Figure 113.13c represents a simplified scenario of the level 3 data flow, assuming that full event reconstruction happens in the same global processor. In principle, if necessary, the same steps, which have been performed for the level 2 decision, can be followed for the level 3 selection, namely a dedicated processor can be assigned for the event, which will request the necessary data from the sources, will perform either partial or full reconstruction of the event, will execute the level 3 algorithm, etc.

Figure 113.14 represents our model which allows us to simulate the behavior of the level 2 and level 3 selection systems described above. The studies have been performed for the Atlas calorimeter

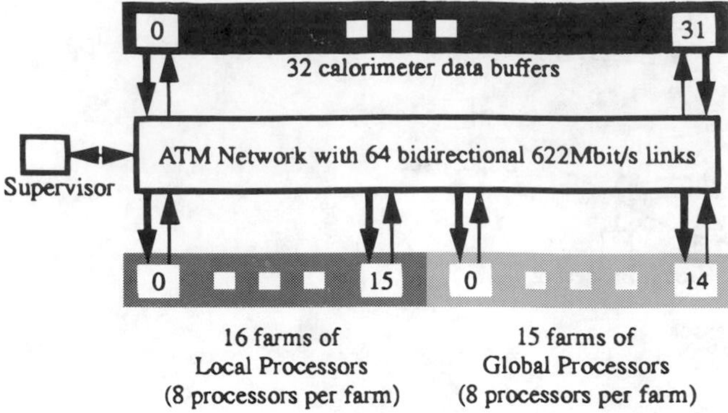

Figure 113.14 Simulation model for the Atlas calorimeter system.

subdetector. The 64 × 64 switching fabric interconnects a supervisor module, 32 front-end data sources, 16 farms of local processors and 15 farms of global processors. Each farm consists of 8 processing elements. The switching fabric is a multistage interconnection network of AT&T Phoenix-like switching elements and operates at 622 Mbit/s.

In our simulation studies, we used the same level 2 input data as described in Study of the Atlas Calorimeter System. In addition, the global processors requested the sources to send level 3 data (16 kByte per source, fixed) with an average rate of 1 kHz (level 2 acceptance rate). The level 3 traffic adds approximately 20% load on the source-destination data path. Compared to the average level 2 message size of the order of 500 bytes per source, the level 3 messages are long and can significantly delay level 2 packets in the sources, if level 2 and level 3 data are serviced in the same FIFO queue of the sources. Apart from that, if no precautions are taken, the level 3 traffic creates congestion in the switching fabric and therefore, delays not only the level 2 traffic, but also the protocol traffic. There are several methods which permit reduction of the congestion in the switching fabric and minimize level 2 and protocol traffic latencies. First, in the source modules each type of traffic can be serviced with different priorities. Protocol data will be sent prior to level 2 and level 3 data, while the level 2 packets will overtake level 3 packets. Another possibility is to apply the rate division technique to the level 3 traffic. Most existing SAR chips implement both the prioritized servicing and the rate division techniques. Some ATM switching fabrics support routing priorities. Also, different service classes can be used for protocol (CBR) traffic and data (VBR). All these techniques and their combinations are currently under investigation.

The simulation results obtained from the MODSIM and μC++ models have been compared and cross-checked against each other. The good agreement of the results is shown in Figure 113.15, which represents the protocol traffic latency tail distributions. On average, 12 μsec are necessary for the level 2 supervisor to provide the RoI information to the local processor (the "local processor notification latency" graph). The protocol traffic from the supervisor towards the local processors is disturbed by the level 2 data traffic from the sources toward the local processors. From the "local processor notification latency" tail distribution we observe that, due to the contention inside the fabric, this time can be as large as 40 μsec with a probability of the order of 10^{-3}. The time necessary for the RoI data request cell from the local processor to reach the source module is shown on the "source RoI data request latency" graph. In average it amounts to 7.5 μsec. Even though the data request stream travels in opposite direction to the data stream (Figure 113.13b), inside the switch they share internal links, which can be overloaded. As a result, the data request latency time can be as long as 30 μsec with a probability of the order of 10^{-3}.

During the simulation runs, the utilization of the switching fabric's available bandwidth has been monitored. Thus, for example, the level 2 data traffic (RoI) from the 32 sources to the 16 local processor farms require 25% of the source output links (622 Mbit/s) and 50% of the destination input links, respectively. As previously mentioned, level 3 data traffic adds another 20% load on the source output links and requires about 45% of the global processor input link bandwidth. The traffic which delivers features from the local processor to the global processor uses 19 Mbit/s of the global input links and increases their load up to 50%. To distribute the notification messages from the supervisor module to the local and global processor farms (40% of 622 Mbit/s rate) 250 Mbit/s are necessary. The RoI data request traffic requires 13 Mbit/s bandwidth on the input links of the sources (2% of the 622 Mbit/s rate). On average, 25% of the available switching fabric aggregate bandwidth utilization has been observed.

Based on the developed models and their future versions we are going to perform extensive studies of the relative merits and disadvantages of the "Push" and "Pull" data flow-control strategies. We plan to feed our simulations with more realistic input parameters derived from the physics simulations. The simulation code can be used for similar studies for other detector types, like the Muon detector, the TRT, etc.

113.5 Hardware Development

To set up an event builder demonstrator, we are developing a VME-ATM interface, traffic shaping hardware and an ATM data generator to produce flexible traffic patterns.

ATM SONET Physical Layer Board

An implementation of the ATM physical layer has been realized in the form of a piggy-back daughter card (Paillard, 1995) that plugs on top of the ATM SAR part of the ATM-VME adaptor card described below. This interface can be configured to comply either with STS-OC3 SONET (ANSI, 1991) or with STM-1 SDH (ITU, 1990), and it transmits and receives over serial optical-fibre at bit-rates of 155 Mbit/s.

Physical layer interface can be used to form the basis of an ATM data generator that will be integrated in the event builder demonstrator to emulate data sources.

Figure 113.16 shows the layout of the physical interface. It is

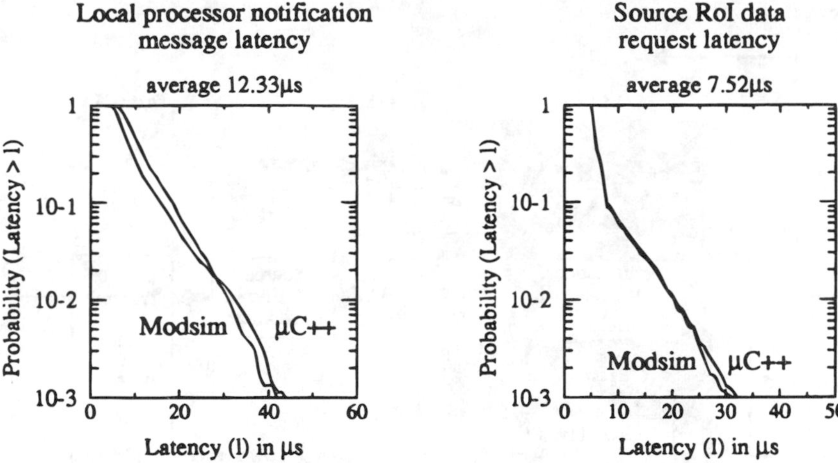

Figure 113.15 Tail distributions of the protocol traffic latency.

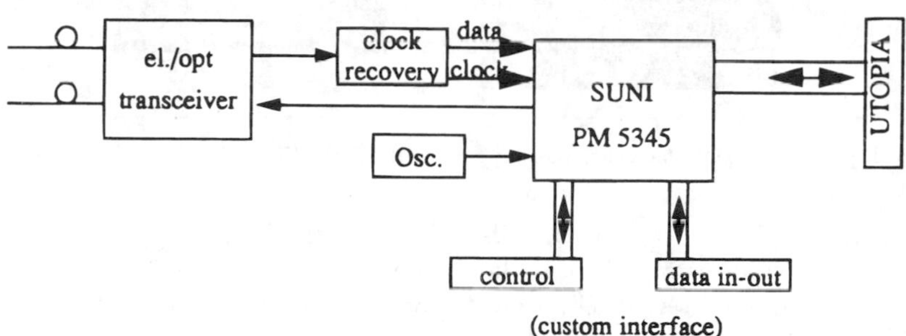

Figure 113.16 Layout of the physical interface.

built around the SUNI chip from PMC-Sierra (PMC-Sierra, Inc., 1993) which implements the CCITT standard 1.432 specification. The interface with the ATM layer part must include a bi-directional data path and a control path for SUNI set-up and for synchronization signals. The board implements 2 different interfaces to the ATM layer. One complies with the UTOPIA standard (UTOPIA specification,1993) and the other is custom defined and was adopted to simplify the design of our ATM board.

VME-ATM Adapter

The development of a VME-ATM interface has been undertaken to provide the source and destination modules for the event builder demonstator. This activity helped us to gain experience with ATM technology, and also to check if and how the functionality and performance needed for event building could be implemented using commercially available chip sets designed for building ATM host-interfaces. A custom development was necessary to integrate the Randomizer traffic shaping hardware (Lazraq et al., 1994) (a function specific to event building).

A prototype has been realized (Gustafsson et al., 1994) as a daughter board which plugs into a CES RIO module (Creative Electronics Systems, 1993). It has been tested successfully with regard to its functionality and interoperability with the Alcatel ATM switch (Henrion et al., 1992) and the HP broadband tester (Hewlett Packard, 1994). However, the theoretically achievable data transfer rates have not yet been reached, neither in the ATM interface itself, nor in the data transfers via the VME bus. Nevertheless, the performances obtained are adequate to proceed with the implementation of the event builder and we have launched the production of a small series of a printed circuit board version of the interface for this purpose. We have gained through this development effort, a deep knowledge of the technology and a very good understanding of the critical issues in ATM interface design.

Implementation of the VME-ATM Adaptor

The CES RIO module is used for the VME interface; it includes a 25 MHz RISC processor which will run the software implementing the higher layers of the protocol stack. The lower layers are implemented in hardware on a daughter board that communicates with the processor via the system bus. The architecture of the ATM adapter hardware is shown in Figure 113.17. During the prototyping phase we actually have three separate hardware plug-in modules. One implements the B-ISDN AAL5 and ATM protocol layers; the second implements the SONET physical layer (and was described in ATM SONET Physical Layer Board), and the third is an optional randomizer module that includes special hardware (see Lazraq et al., 1994 and Traffic Randomizing Hardware) to perform the traffic shaping required for event building over telecommunication switches.

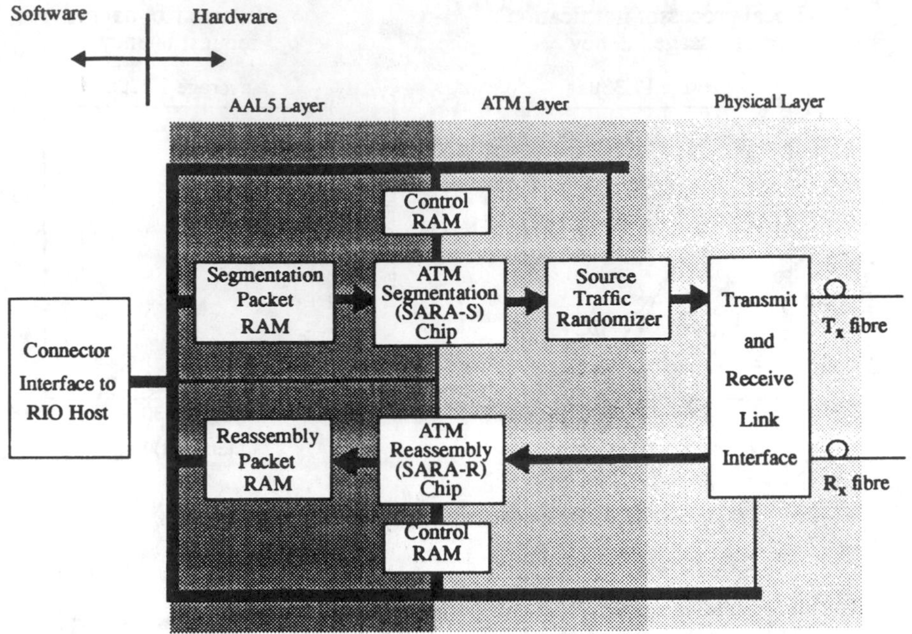

Figure 113.17 Block diagram of the interface hardware supporting the AAL, ATM and physical layers of the B-ISDN protocol.

A commercial chip set (Transwitch, 1992) performs in hardware the segmentation and reassembly of data packets, in the AAL5 format (up to 64 kByte long), into/from ATM cells. These segmentation and reassembly (SARA) chips require two dual-ported memories each. The first one, the packet memory, stores the actual data packet to be transmitted (or that has been received and reassembled). In order to sustain the full 155 Mbit/s rate, this memory is accessed by the SARA via a 32-bit port, and 12 memory accesses are required per ATM cell. The second port is also 32-bit wide and connects to the host's system bus. The port arbitration logic assigns equal priority to both ports. Currently we transfer data between VME bus and the packet memory using programmed I/O. Some improvements to the current design are required in order to be able to support block transfer mode between VME address space and the packet memory.

The second type of memory contains packet descriptors that point to the location of AAL5 packets in the packet memory and specify their length, the virtual connection index (VCI), and its associated traffic metering parameters. The segmentation chip implements sophisticated procedures to segment the packet when multiple VCIs are concurrently active. We measured that, for every ATM cell generated and passed to the physical layer, not less than 23 control memory accesses are required for this management. Each SARA chip can support up to 64k different VCIs, and can simultaneously segment/reassemble 8k packets, which is sufficient to construct very large event builders. The current design uses 512 kByte packet memories and 256 kByte control memories.

The physical layer hardware is included in Figure 113.17, and has already been described in the section on ATM Sonet Physical Layer Board. The interface between the physical layer board and the board with the AAL layer is a custom protocol rather than the standard UTOPIA (which would have been more difficult to implement). FIFOs of 4 cells on both sender and receiver paths are provided by the SUNI chip. Their role is to make the transition between the asynchronous ATM and the synchronous physical layers. The receive FIFO can also smooth out some burstiness in the cell rate, but the current size of 4 cells is not sufficient when the effective rate on the AAL layer is as low as it is in our case (see section on **Test and Performance Measurements** for more details).

Traffic Randomizing Hardware

Source traffic shaping can be used to control congestion within the switching fabric by regulating the bandwidth assignment to virtual connections, and by modulating the time at which cells are injected into the switch. Figure 113.18 shows the principle of the randomizer traffic shaping hardware developed for event building applications. Each source module in the event builder must maintain one logical FIFO queue of event data for each destination (in Figure 113.18 these logical FIFO queues are labeled with unique virtual connection identifiers associating them with a specific destination). The SARA segmentation chip services the packet queues in round robin, picking one cell from the head of each packet queue in each round robin cycle. Rate metering is effectively imposed by SARA applying a programmable delay between each service cycle.

The randomization of a cell injection time, which breaks the correlation between traffic from different sources and therefore minimizes congestion inside the fabric, is performed by the randomizer module (Lazraq et al., 1994). The randomizer contains two cell buffer memories (a "write" buffer and a "read" buffer). It operates by writing the ATM cells sent out by SARA during a segmentation cycle into pseudo-random locations in the write buffer. During the next segmentation cycle the write and read buffers are switched. The cells from the read buffer are always

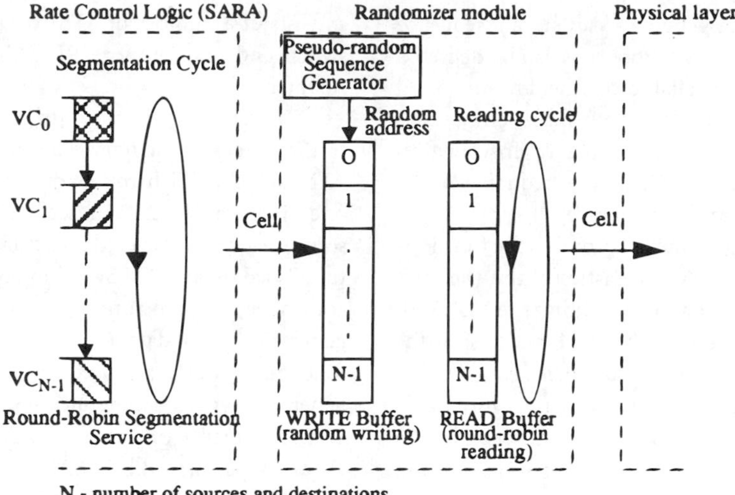

Figure 113.18 The principle of operation of the randomizer traffic shaping hardware.

read out by scanning the memory sequentially, thus effectively adding a random delay to the injection time of cells on a given VC. The algorithm guarantees that cell sequencing within each VC is preserved.

A modification to the original design of the randomizer had to be performed, due to the lower than expected bandwidth achieved by the interface. The new functional protocol between the AAL layer and the randomizer is described in Costa et al. (1995).

Tests and Performance Measurements

The VME-ATM prototype has been extensively tested to ensure that it worked properly with other ATM standard equipment, namely the Alcatel switch and the HP broadband tester. In addition, the various layers (ATM, AAL and physical) have been tested individually in loop-back mode. The randomizer part is still under development and has not yet been integrated in the global tests.

Currently we achieve transfer rates of 50 Mbits/s between VME bus and the packet memories using programmed I/O. The bit-rate on the optical fibre is 155 Mbit/s, but after subtracting SONET framing protocol overheads the theoretically available bandwidth is 149.7 Mbit/s. In loop-back mode, when packet data are transferred between packet memories (but not to the VME bus), we achieve a sustained effective data transfer rate of 95 Mbit/s. However, when the board sends data through the switch, we have to reduce the rate to 70 Mbit/s on the sender line because the switch output ports deliver the cells in bursts that cause the receive FIFOs in the physical layer card to overflow (i.e, when we send data at 95 Mbit/s bursts of up to 8 consecutive cells are delivered, which of course overflow the 4-cell deep FIFOs of the SUNI). Further optimization of the design is required in order to sustain the full bandwidth offered by the 155 Mbit/s bit-rate of the fibre optic transmission standard.

What We Have Learned

1. Link with VME: real DMA block transfer must be provided by the mother card. Future versions of RIO, implementing PCI, should solve this problem.
2. The AAL, ATM, and physical layer chip sets must be chosen in order to have the best match of their individual characteristics. Our choice was not optimized, mainly due to the limited availability of components when we started.
3. An ATM interface must be able, on the receiver side, to accept bursty traffic at full 155 Mbit/s bandwidth, even if its maximum average bandwidth performance is lower.

ATM Data Generator

A simple ATM data generator has been developed with the aim of providing a low cost source module for event builder demonstrators (Paillard, 1995). It is based on the ATM physical interface card, described in **ATM SONET Physical Layer Board,** to which a memory is attached. It is controlled through a connection to a PC (Figure 113.19).

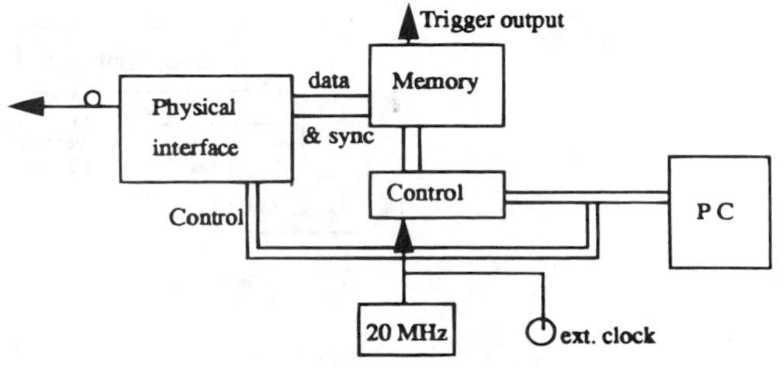

Figure 113.19 Data generator layout.

In its present version the data generator can store a sequence of 1230 cells in its 64 kByte memory. Those cells can be delivered continuously at the maximum available bandwidth supported by the 155 Mbit/s SONET standard. One can insert empty cells to reduce the ATM rate (e.g., to 77.5 Mbit/s by inserting an empty cell after every ATM cell. By varying the frequency of the external clock, any frequency can be achieved.

The control program running on the PC provides the functionalities to define the cells and the characteristics of the traffic. It includes an ATM cell editor which allows defining the VCI, VPI, PT, and CLP fields. It is used to control the SUNI and display the error messages. In addition, a general purpose program (running under UNIX) can generate ATM cell sequences according to global data and traffic characteristics and frame them within an AAL5 structure. A file is used as an intermediate storage medium (Costa, 1995).

Several of data generator modules as described above have been produced. It is intended to implement several enhancements in future versions in order to facilitate the use of the data generator in the event builder demonstrator. The new features will include an implementation in VME format, the possibility to use an external trigger to launch the emission of the next available AAL5 packet, and increased size of the memory to store longer data sequences.

113.6 Integration of Event Builder Demonstrators

The ALCATEL ATM-Based Event Builder Demonstrator

To test the largest configuration at minimum cost, we are planning to use the traffic generator (see ATM Data Generator) in the event builder demonstrator. The system (Figure 113.20, left) consists of a number of ATM traffic generator sources, an ATM switch and one or more RIO/ATM interfaces (see VME-ATM Adapter).

ATM cells containing event fragment data are stored in the traffic generator memory. The traffic generators send cells to an ATM switch, which routes data to dummy destinations and to one or more RIO/ATM interface which run the event building software (see The Layered Structure of the Event Builder Architecture Based on an ATM Switching Network). The traffic generator controller program running on a PC loads the memories of the generators with data read from a file. This file is generated following a two-step procedure (Figure 113.20, right): a fragment generator program generates AAL5 packets correspondent to a certain sequence of events; an AAL5 to ATM program simulates the work performed by a SAR chip and optionally can stimulate the true barrel shifter or randomizer traffic shaping mechanisms.

Performance Measurements

At the moment only one ATM interface is available so that hardware and software performance tests have been carried out with an optical loop-back and using the interface both as a source and as a destination. Figure 113.21a shows the software and hardware overheads for sending an Event Fragment PDU (protocol data unit). Most of the software overhead is due to the SAR chip control data structure initialization and network error checking. As the processor of the interface and the SAR chip work in parallel, a new packet can be initialized and submitted while the former packet is being transmitted.

Using the HP Broadband Test System, it has been possible to perform some measurements of the Alcatel switch. In Figure 113.21b the cell delay introduced by the Alcatel swicth is shown. When the speed gets close to the maximum ATM speed (149.7Mbit/s) the switch starts losing cells and the delay increases.

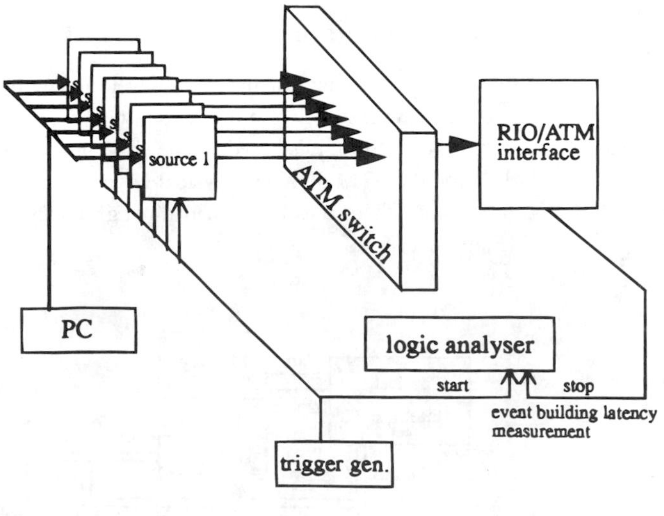

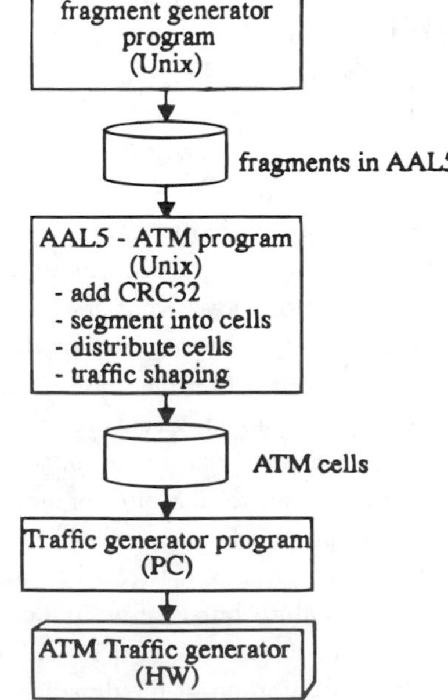

Figure 113.20 ATM event builder using a traffic generator.

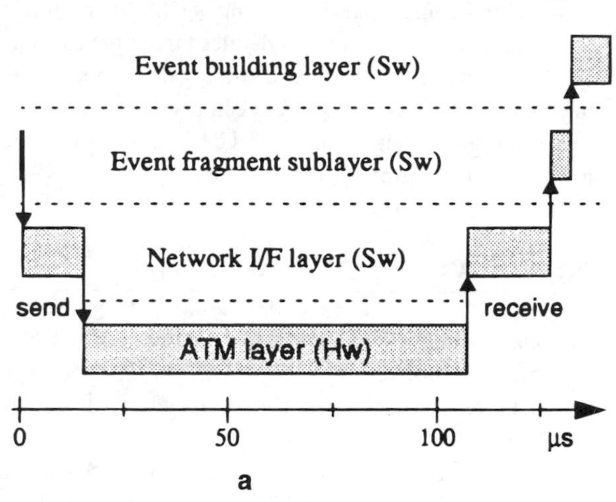

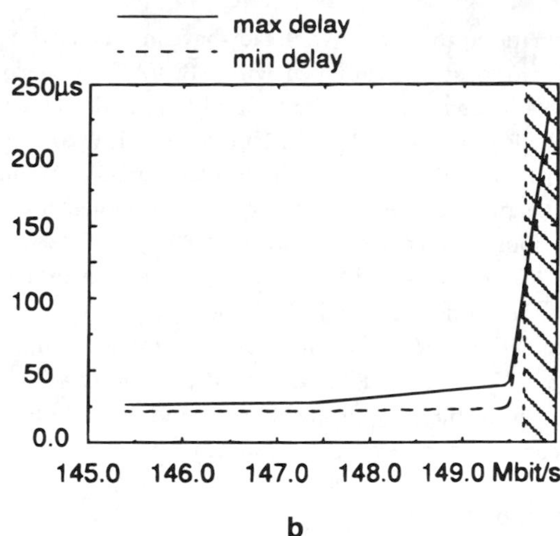

Figure 113.21 a. Sending and receiving an event fragment PDU of 1 AAL5 packet. Speed = 90 Mbit/s; hardware delay = 11 µs; packet size = 800 bytes. b. Cell delay measured between two Alcatel switch links.

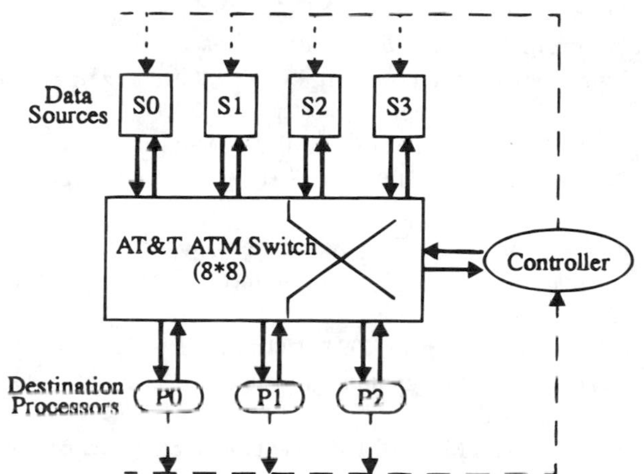

Figure 113.22 Schematic view of the Atlas ATM demonstrator based on the AT&T switch.

ATLAS AT&T ATM-Based Real-Time Demonstrator

A general purpose demonstrator based on the Phoenix AT&T switch (AT&T) is foreseen at Saclay. The aim of this demonstrator is to validate an ATM-based architecture as a possible level 2 solution for the ATLAS experiments. The foreseen test bench, shown in Figure 113.22, includes:

- Source data generators (developed by RD-31 and described in ATM Data Generator).
- An 8-port Phoenix-based ATM switching fabric (purchased from AT&T).
- Destination processors (workstation and/or VME processor board).
- Protocol software, e.g., data flow-control and error recovery mechanisms (currently under development within the RD-31 collaboration and described in Layered Structure of the Event Builder Architecture Base on an ATM Switching Network).
- Level 2 algorithms software.

This demonstrator will allow us to compare the measurements performed on real hardware against the results predicted by simulation and to refine the models. The test bench can be used to study the architecture of any level 2 subdetector. However, because Saclay is strongly involved in the Atlas calorimetry, in a first step we will consider the electromagnetic and hadronic calorimeter subsystems.

Components of the Demonstrator

Source data generator: At present, physics events have been generated by simulation using the ATRECON code (ATLAS Softman Group, 1994). Samples of events accepted by the level 1 selection algorithm are available. Those events will be used to produce traffic patterns for data generators. To achieve high sustained event data transmission rates, the event data fragments should be stored in the source modules. Assuming a few kByte event data fragments, the required source buffer size will amount to a few MBytes. The data generator is a VME format board developed within the RD-31 collaboration (Paillard, 1995 and ATM Data Generator). It can sustain the full 155Mbit/s rate.

Switching fabric: We expect that a 256 port switch running at 155 Mbit/s (or 64 ports at 622 Mbit/s) will be adequate for the Atlas level 2 and level 3 calorimeter subsystem (see Study of the ATLAS Calorimeter System and Protocol Traffic Transport via the Switch). At present, our system will use an 8 × 8 AT&T switching fabric with ports running at 155Mbit/s. We can implement a demonstrator which includes four data sources, three destination processors, and a controller (Figure 113.22).

Destination processors: We already have installed two Sparc-20 workstations equipped with SBus/ATM (Interphase, 1994a) interfaces and SunOs and Solaris device drivers. We also have a VME/ATM (Interphase, 1994b) interface card and CES RAID running under LynxOs. We plan to evaluate level 2 local and global selection algorithms on various platforms (e.g., PowerPc, C80). The TI C80 (TMS, 1994) multimedia video processor is currently under evaluation with TI emulation package.

Protocol software: The protocol software will implement both "Push" and "Pull" data flow-control. It will include the event building protocol layers and will address error detection and recovery mechanisms.

Current Status

We are evaluating commercial SBus/ATM Adapters in a LAN emulation mode. Two Sparc-20 stations running under SunOs are connected back-to-back via ATM. We measured data transmission performance using different classes of LAN protocols (UDP and TCP based UNIX sockets).

We are currently modifying the device driver software in order to work directly at the level of AAL5 protocol and to achieve higher transmission rates. We evaluate VME/ATM interfaces in real-time environment. The VME/ATM device driver, provided by the manufacturer, has to be ported to LynxOs. We expect the AT&T switching fabric to be delivered within a few months.

113.7 Plan of Work

The integration of the event builder demonstrators based on the Alcatel switch (at CERN) and the AT&T switch (at Saclay) is planned to continue (as was described in the previous section). These two demonstrators are complementary in that they investigate fundamentally different ATM switch architectures and congestion control methods. The Alcatel demonstrator is of a generic nature, whereas the AT&T-based demonstrator will be targeted more to the requirements of the Atlas level-2 trigger.

The generic event builder demonstrator at CERN will be used to study performance issues and evaluate various event building protocols and traffic shaping schemes. An important goal is to investigate methods of management and control of the event builder, perhaps based on high-level congestion control techniques developed by the ATM Forum (Jain, 1995), so as to make it a user-friendly system to operate and integrate into the overall DAQ.

The RD-31 "Atlas team" will continue the architecture design and simulation studies adapted to the Atlas level-2 trigger system, using the calorimeter sub-system as a realistic model. They will develop their demonstrator in conjunction with the development of an "intelligent", flexible dual-port source memory that can support the required data flows.

The CMS experiment will be actively investigating various switch-based event building schemes in the near future. Given the importance of this item to the CMS DAQ system, a CMS group is participating in RD-31 with the goals of developing architectural concepts and modeling of CMS event builders using more realistic input data flows deduced from physics simulations and from expected detector read-out organization, and investigating Dual-Ported Memory architectures as inputs to ATM-like switches (define, help simulate and design, and finally participate in the testing of prototype modules).

References

Alcatel. *Alcatel Data Networks,* Alcatel 1100 HSS, private communication.

ANSI. Fibre Channel Draft Proposed Standard, rev. 4.2, ANSI X3T9.3 Committee.

ANSI 1991. Digital Hierarchy—Optical Interface Rates and Formats Specifications (SONET), ANSI T1.105–1991.

AT&T. *Globe View-2000 Broadband System—System Description,* AT&T Network Systems, Red Bank, NJ.

ATLAS 1994. *ARTECOM Manual,* ATLAS Software Group, ATLAS Internal Not SOFT-NO-15.

ATM Forum 1993. *UTOPIA Specification: An ATM PHY Data Path Interface,* draft version 1.06 (October 1993), working paper of the ATM Forum, Foster City, CA.

Bogaerts, A. et al. 1994. ATLAS SIMDAQ, Modeling of the ATLAS Data Acquisition and Trigger System, ATLAS Internal Note, DAQ-NO-18.

Buhr, P. A. et al. 1992. μC++: concurrency in the object-oriented Language C++, *Software—Practice and Experience,* 22(2):137–172.

CACI Products Company 1993. *MODSIM II—The Language for Object-Oriented Programming,* January, CACI Products Company, La Jolla, CA.

Calvet, D. 1994. A MODSIM Model of the AT&T Phoenix Switching Fabric, RD-31 Internal Note 94-07, August.

CERN 1994a. ATLAS Technical Proposal, CERN/LHCC 94-43, LHCC/P2, 15 December.

CERN 1994b. The Compact Muon Solenoid (CMS), Technical Proposal, CERN/LHCC 94-38, LHCC/P1, December 15.

Christiansen et al., 1992. NEBULAS—a high performance data-driven building architecture based on an asynchronous self-routing packet-switching network, CERN/DRDC 92-14 and CERN/DRDC 92-47.

Christiansen et al., 1993. NEBULAS—a high performance data-driven building architecture based on an asynchronous self-routing packet-switching network, CERN/DRDC 92–55.

Costa, M. 1994. ATM Event Building Software, RD-31 Note 94-08, December 1994, revised February 1995.

Costa, M. 1995. An ATM Based Event Building Test System Using ATM Traffic Generators, RD-31 Note 95-05.

Costa, M. et al. 1994. ATM-Based Event Building, ATLAS Internal Note, DAQ-NO-024, December.

Costa, M. et al. 1995. Randomizer Protocol, RD-31 Note 95-01, February.

Creative Electronics Systems 1993. *RIO 8260 and MIO 8261 RISC I/O Processors—User's Manual,* ver. 1.1, March, Creative Electronics Systems, SA Geneva, Switzerland.

Dufey, J.-P. 1995. Problem Statement for Fibre Channel Event Builder Modeling, RD-31 Note 95-02, January.

Fibre Channel Association, 1994. Fibre Channel—Connection to the Future, ISBN 1-878707-19-1, The Fibre Channel Association, Austin TX.

Fore Systems. *ASX Family of ATM Switches,* Fore Systems, Inc., Pittsburgh, PA.

Gerl, J. and Lieder, R. M. 1993. Euroball III, European Gamma-Ray Facility, GSI Darmstadt.

Gustafsson et al. 1994. A 155 Mbit/s VME to ATM interface with special features for event building applications based on ATM switching fabrics, *Proc. Int. Data Acquisition Conf.,* October, Fermilab, to be published; also available as RD-31 NOTE 94-11.

Henrion, M. et al. 1992. Technology, distributed control and performance of a multipath self-routing switch, *Proc. 14th Int. Switching Symp.,* 2:2–6, October, Yokohama, Japan.

Hewlett Packard Corporation 1994. *Broadband Series Test System.*

IBM Corporation (IBM). Nways ATM Products, IBM Corporation

International Telecommunications Union (ITU). Recommendations G. 707, G. 708, G. 709, International Telecommunications Union, Geneva, Switzerland.

International Telecommunications Union (ITU). International Telegraph and Telephone Consultative Committee recommendations I. 150, I.211, I.311, I.321, I.327, I.361, I.362, I.363, I.413, I.432, I.610, ITU, Geneva, Switzerland.

Interphase Corporation 1994a. *S/ATM 4615 Adapter, User's Guide,* Interphase Corporation, June.

Interphase Corporation 1994b. *V/ATM 5215 Adapter, User's Guide,* Interphase Corporation, September.

Jain, R. 1995. Congestion control and traffic management in ATM networks: recent advances and a survey, draft version January 26, 1995, *Computer Networks and ISDN Systems,* submitted.

Kumar, V. P. et al. 1991. Phoenix: a building block for fault tolerant broadband packet switches, *Proc. IEEE Global Telecom. Conf.,* December, Phoenix, AZ.

Lazraq et al., 1994. ATM Traffic Shaping in Event Building Applications, RD-31 Note 94-09.

Le Boudec, J.-Y. 1992. The asynchronous transfer mode: a tutorial, *Computer Networks and ISDN Systems,* 24:279–309.

Letheren, M. et al. 1993. An asynchronous data-driven event building scheme based on ATM switching fabrics, *IEEE Trans. Nuclear Science,* 41(1), February; also available as CERN/ECP 93-14.

Mandjavidze, I. 1993a. Modeling and performance evaluation for event builders based on ATM switches, RD-31 Internal Note 93-06, December.

Mandjavidze, I. 1993b. A Data-Driven Event Building Scheme Based on Self-Routing Packet-Switching Banyan Network, RD-31 Note 93-07.

Mandjavidze, I. 1994a. A new traffic shaping scheme: the true barrel shifter, RD-31 Internal Note 93-03, February.

Mandjavidze, I. 1994b. Software protocols for event building switching networks, *Int. Data Acquisition Conf.,* October, Fermilab.

Mandjavidze, I. 1994c. Review of ATM, Fibre Channel and conical network simulations, *Int. Data Acquisition Conf.,* October, Fermilab.

Mandjavidze, I. 1995. Modeling of an ATM Implementation of the CMS Virtual Level 2 Architecture, RD-31 Note 95-3, February.

Marchioro, A. and Mandjavidze, I. 1994a. Pros and cons of commercial and non-commercial switching networks, *Proc. Int. Data Acquisition Conf.,* October, 1994, to be published; also available as RD-31 Note 94-12.

Marchioro, A. and Mandjavidze, I. 1994b. A Data-Driven Event Building Scheme Based on a Conic Self-Routing Packet-Switching Banyan Network, RD-31 Note 94-06.

Nomachi, M. 1993. *Event Builder Queue Occupancy,* SDC-93-566, August.

Oechslin, P. et al., 1992. ALI: a versatile interface chip for ATM systems, *Proc. IEEE Global Telecom. Conf.* '92, 1282–1287, December 6–9, Orlando, FL.

Paillard, C. 1995. An STS-OC3 SONET/STM-1 SDH ATM Physical Layer Implementation and Application to an ATM Data Generator, RD-31 Note 95-04, February.

PMC-Sierra Incorporated 1993. *The PMC5345 Saturn User Network Interface Manual,* PMC-Sierra, Incorporated.

Rajaei, H. 1992. SIMA, an environment for parallel discrete event simulation, *Proc. 25th Ann. Simulation Symp.,* April, Florida.

Tether, S. 1944. SchedSim: A Tool Kit for Building Scheduled-Event Simulations, private communication.

TMS320C80 Multimedia Video Processor, technical brief, TI, 1994.

Transwitch Corporation 1992. *SARA Chip Set, Technical Manual,* ver. 2.0, October, Transwitch Corporation, Shelton, CT.

114

Microelectromechanical Systems (MEMS)

Yu-Chong Tai
California Institute of Technology, Pasadena

Chang-Jin Kim
University of California, Los Angeles

114.1 Introduction 1468
114.2 Bulk Micromachining 1468
114.3 Surface Micromachining 1469
114.4 First Applications 1469

114.1 Introduction

The acronym, MEMS, was first used in the late 80s, but even now a consensus has not been reached on the definition of MEMS. In fact, people use different names, such as micromachines and microsystems, to describe this field. Nevertheless, a MEMS is generally recognized as a complete unit that contains both electrical and mechanical components (microstructures) with characteristic sizes ranging from nanometers to millimeters. Some people actually limit the overall size of a MEMS device to less than 1 cm^3. Even so, this definition is still vague and not complete. What is unique about typical MEMS devices is that when compared to normal machines, they have one or more of the following distinct features of miniaturization: component multiplicity, functional complexity, system integration and the ability to be mass produced. Currently, the worldwide market of MEMS is estimated to be about $2 billion (System Planning Corp., 1994). It is predicted that by the year 2,000, the market will reach nearly $14 billion and will have $100 billion worth of influence on other markets. As a result, MEMS has been recognized as one of the most promising fields of the future. The surge of MEMS is fueled by the development of silicon micromachining technology. Silicon micromachining technology has evolved mainly from silicon microfabrication technology, or more specifically, integrated circuit (IC) technologies, but it has also been enriched by other technologies such as LIGA and micro EDM. There are two important features of silicon micromachining which were inherited from IC technology; namely, small size and large quantity. These two traits make it impossible for conventional machining to compete with silicon micromachining. Finally, MEMS can be regarded as being truly interdisciplinary. Its broad applications have great potential for making significant contributions to other disciplines such as physics, chemistry, engineering, biology, and medicine. In the following sections, the origin and history of MEMS will be discussed by describing two main branches of silicon micromachining-bulk and surface micromachining, followed by examples of first commercial applications.

114.2 Bulk Micromachining

Bulk micromachining is a process for making microstructures from a starting substrate by selectively removing unwanted portions of the substrate. Theoretically, there are many choices for the starting substrate material and for the means of material removal. In the field of MEMS, however, bulk micromachining usually means that the substrate is single-crystalline silicon and the means of removing silicon is by isotropic or anisotropic chemical etching. To machine a specific geometry of a silicon microstructure, the selective etching of silicon is done through the use of lithographic masking in conjunction with techniques such as boron and electrochemical etch stops (Raley et al., 1984; Jackson et al., 1981; Bohg, 1971). As a result, many silicon microstructures, such as cantilevers, beams, diaphragms, channels, and nozzles can be successfully fabricated. These microstructures form the main elements of many MEMS devices.

The development of silicon chemical etchants has been documented since the 1950s, when researchers tried to find precise ways of etching silicon (Robbins and Schwartz, 1959). The aqueous silicon etchants that were developed were mainly isotropic and based on a mixture of hydrofluoric acid, nitric acid, and acetic acid (HNA). Anisotropic etchants for silicon were developed in the 1960s. The most studied etchants were potassium hydroxide (KOH) (Nature-Times, 1967), ethylene-diamine-pyrocatechol with water (EDP or EPW) (Finne and Klein, 1967), and hydrazine solutions (Lee, 1969; Pacific Semiconductors, 1962). The anisotropic feature of these etchants is that the etch rates in the <100> and <110> silicon crystallographic directions are much higher than in the <111> direction, which allows for the design of microstructures bound by {111} crystalline planes. This feature was broadly used, and these anisotropic etchants were further refined in the 1970s and 1980s (Kendall and De Guel, 1985;

0-8493-8343-9/97/$0.00+$.50

Reisman et al., 1979; Bassous, 1978; Bean, 1978). Recently, metal-exclusive anisotropic etchants of ammonium hydroxide (Schnakenberg et al., 1991a) and tetramethyl ammonium hydroxide (TMAH) (Schnakenberg et al., 1991b; Tabata et. al., 1991a) have been developed. As more and more MEMS devices are integrated with electronics, these etchants have become more important. Today, bulk micromachining using wet chemical etching can be considered a mature technology and many MEMS devices have been fabricated using this method. Interested readers should refer to some of the comprehensive review papers such as (Seidel, 1987; Kendall and De Guel, 1985; Petersen, 1982).

In addition to wet chemical etching, there has been research focusing on other etching processes including dry etching that can replace or even outperform wet ones. For example, laser etching, plasma etching, reactive-ion etching (RIE), ion beam etching and even micro electro-discharge-machining (EDM) have all been demonstrated (Bloomstein and Ehrlich, 1994, Miu et al., 1993; Linder et al., 1991). Bulk micromachining using dry etching is not limited to specific crystalline orientations, and there are no surface-tension-induced problems.

114.3 Surface Micromachining

As the name suggests, surface micromachining is performed on the surface of a substrate, which can be silicon, glass, alumina, or metal. The only function of the substrate is as mechanical support. The micromachining then involves combinations of thin-film deposition and patterning. At the end, selective etching is used to remove certain (sacrificial) layers and leave others (structural layers) free-standing.

According to the literature, the concept of surface micromachining was demonstrated in the 1950s (U.S. Patent, 1956). However, the actual use of this technology to make a complete MEMS device—the resonant-gate transistor—was realized much later, in 1967, by Nathanson et al. (1967). After Nathanson's paper, a series of surface-micromachined devices were documented including early work on a digital mirror display (Preston, 1972). However, it was not until 1982 that Howe and Muller (1982) used LPCVD polycrystalline silicon to make micro cantilevers and bridges. The significance of this polysilicon surface micromachining process is that it is compatible with IC technology. The work at Berkeley in the late 1980s (Kim et al., 1992; Tang et al., 1989; Tai and Muller, 1989; Fan et al., 1988a, b) then demonstrated the broad use of polysilicon micromachining for various devices such as micro pin-joints, sliders, micromotors, resonators, and tweezers. Today, polysilicon micromachining has established itself as one of the most important branches of surface micromachining. In addition to polysilicon surface micromachining, researchers have also explored many other materials such as aluminum (Storment et al., 1994; Sampsell, 1993) and silicon nitride (Tabata et al., 1991b). Surface micromachining continues to flourish as MEMS devices become smaller, lighter, faster and cheaper, and are integrated with ICs.

Lastly, there is the LIGA technology (a German acronym for Lithographie Galvanoformung Abformung) (Ehrfeld et al., 1988; Becker et al., 1986). The LIGA process uses X-ray lithography to generate a deep resist pattern on a substrate. The empty space in the X-ray resist is then electroplated from the bottom of the substrate to the top of the resist. This generates a negative replica of the resist mold. The use of X-ray lithography allows for the fabrication of structures with submicron resolution. These microstructures can be as thick as one millimeter and can still be built with excellent precision. There is no other micromachining technique which can match the aspect ratio (i.e., height to width) the LIGA can provide. However, the drawback is the cost for the high-energy synchrotron used as the X-ray source. Because of this drawback, LIGA may not be a popular technology, but it is useful for some special applications which require a fine spatial resolution and a high aspect ratio. The rarity of the light source and the cost of this technique has encouraged others to develop similar techniques using ultraviolet (UV) light sources (Frazier et al., 1992) which are available in any IC manufacturing laboratory. Light-heartedly termed as "poor-man's LIGA" or simply "cheap LIGA", this UV-based micro-electroplating has gained popularity in many areas such as motors, flaps, and channel fabrication (Liu et al., 1995; Joo et al., 1995; Hirano et al., 1993; Frazier et al., 1991.)

114.4 First Applications

Micro-pressure sensors currently represent the most successful application of bulk-micromachined devices. They have been used in a wide range of applications ranging from automobiles to biomedical instruments. Over the course, many different types of pressure sensors have evolved. Micromachined piezoresistive silicon-based pressure sensors were first introduced in 1958 by Kulite, Honeywell, and MicroSystems (Brysek et al., 1990). At that time, the devices were made with silicon piezoresistors glued to metal diaphragms. Today, the most widely used low-cost pressure sensors are made by anisotropic etching of silicon and require little hand assembly. Two examples of such devices are the fully-integrated Motorola pressure sensor (Fraden, 1993) and the silicon-fusion bonded millimeter-size pressure sensors by Lucas NovaSensor (Bryzek et al., 1990). Very recently, surface-micromachined pressure sensors have also become commercially available by SSI Technologies for automobile applications. These sensors are even smaller and contain integrated compensation circuits. It has been estimated that the total global pressure sensor market is currently about $0.5 billion (1995) but likely to increase to about $3.4 billion by the year 2000 (System Planning Corp., 1994).

The microaccelerometer is another prominent micromachined device. Acceleration is measured by piezoresistively reading the strain in the beam or by capacitively reading the displacement of the beam induced by the inertial force of the proof-mass. Following bulk-micromachined accelerometers (e.g., products from Lucas NovaSensor and IC Sensors) in the 1980s, surface-micromachined accelerometers have recently appeared on the market for automobile airbag deployment (e.g., ADXL50 of Analog Devices, MMAS40G by Motorola). These new generations of accelerometers incorporate complex electronic circuits for high functionality. Together with the navigation gyro, the inertial sensors market is estimated to be worth over $2.7 billion by the year 2,000 (System Planning Corp., 1994).

Another example of marketed micromachined devices include the printing mechanism in inkjet printers. Current inkjet printers use the pressure of vapor generated in microchambers to transfer ink onto paper. These microchambers are made either by electroforming (ThinkJet by Hewlett-Packard) or by glass bonding on photo-patterned resin structures (BubbleJet by Canon). Passive devices made of bulk-micromachined structures include scanning probe microscope (SPM) tips by Park Scientific and Nanoprobe. They utilize the directional etching of silicon wafers in anisotropic etchants.

As for microactuators, the most notable commercial product is the thermopneumatic valve, currently marketed by the start-up company Redwood Microsystems (Zdeblick et al., 1994). In this device, a diaphragm can close or open a nozzle depending on whether a micro-cavity underneath the diaphragm is pressurized or depressurized. A large pressure increase can be achieved by thermally generating a vapor bubble inside a cavity filled with liquid.

One last microactuator example is the Digital Mirror Display (DMD) devices developed by Texas Instruments. A 768 × 576 pixel display has been demonstrated using standard semiconductor processing on 6″ wafers (Sampsell, 1993). In the past, similar but less-dense devices have been used for military applications, but a commercial product is now being introduced. The applications of the DMD may include projection TVs and holographic displays.

References

Bassous, E. 1978. Fabrication of Novel three-dimensional microstructures by the anisotropic etching of (100) and (110) silicon, *IEEE Trans. Electron. Devices,* ED-25(10):1178–1185.

Bean, K. 1978. Anisotropic etching of silicon, *IEEE Trans. Electron. Devices,* ED-25(10):1185–1193.

Becker, E. W., Ehrfeld, W., Hagmann, P., Maner, A., and Munchmeyer, D. 1986. Fabrication of microstructures with high-aspect ratios and great structural heights by synchrotron radiation lithography, galvanoformung, and plastic moulding (LIGA Process), *Microelectronic Engineering,* 4:35–36.

Bloomstein, T. M. and Ehrlich, D. J. 1994. Laser stereo micromachining at one-half million cubic micrometers per second, Technical Digest, *Solid-State Sensors and Actuators Workshop,* 142–144, June, Hilton Head, SC.

Bohg, A. 1971. Ethylene diamine pyrocatechol water mixture shows etching anomaly in boron doped silicon, *J. Electrochem. Soc.,* 118:401–402.

Bryzek, J., Petersen, K., Mallon, J., Christel, L., and Pourahmadi, F. 1990. *Silicon Sensors and Microstructures,* NovaSensor.

Ehrfeld, W., Gotz, F., Munchmeyer, D., Schelb, W., and Schmidt, D. 1988. Process: sensor construction techniques via x-ray lithography, Technical Digest, *Solid-State Sensor and Actuator Workshop,* 1–4, June, Hilton Head, SC.

Fan, L. S., Tai, Y. C., and Muller, R. S. 1988a. Integrated movable micromechanical structures, *IEEE Trans. Electron. Devices,* 35:724–730.

Fan, L. S., Tai, Y. C., and Muller, R. S. 1988b. IC-processed electrostatic micromotors, Technical Digest, *IEEE IEDM Meeting,* 666–669, December, San Francisco, CA.

Finne, R. M. and Klein, D. L. 1967. A water-amine-completing agent system for etching silicon, *J. Electrochem. Soc.,* 114(9):965–970.

Fraden, J. 1993. *AIP Handbook of Modern Sensors,* American Institute of Physics, New York, NY.

Frazier, A. B. and Allen, M. G. 1992. High aspect ration electroplated microstructures using a photosensitive polyimide process, *Proc. Micro Electro Mechanical Systems Workshop,* 87–92, February, Travemünde, Germany.

Frazier, A. B., Babb, J. W., Allen, M. G., and Taylor, D. G. 1991. Design and fabrication of electroplated micromotor structures, ASME Winter Annual Meeting, *Micromechanical Sensors, Actuators, and Systems,* DSC-32:135–146, December, Atlanta, GA.

Hirano, T., Furuhata, T., and Fujita, H. 1993. Dry releasing of electroplated rotational and overhanging structures, *Proc. IEEE Micro Electro Mechanical Systems Workshop,* 278–283, February, Fort Lauderdale, FL.

Howe, R. T. and Muller, R. S. 1982. Polycrystalline silicon micromechanical beams, *1982 Spring Meeting of the Electromechanical Society,* May 9–14, Montreal, Canada.

Jackson, T. N., Tischler, M. A., and Wise, K. D. 1981. An Electrochemical P-N junction etch-stop for the formation of silicon microstructures, *IEEE Electron Device Letters,* EDL-2(2):44–45.

Joo, Y., Dieu, K., and Kim, C.-J. 1995. Fabrication of monolithic microchannels for IC chip cooling, *Proc. IEEE Micro Electro Mechanical Systems Workshop,* 362–367, January–February, Amsterdam, Netherlands.

Kendall, D. L. and De Guel, G. R. 1985. Orientations of The Third Kind: The Coming of The Age of (110) Silicon, *Micromachining and Micropackaging of Transducers,* Fung, C. D., Cheung, P. W., Ko, W. H., and Fleming, D. G., eds., Elsvier Science Publishers, Amsterdam, Netherlands.

Kim, C.-J., Pisano, A. P., and Muller, R. S. 1992. Silicon-processed overhanging micorgripper, *J. Microelectromechanical Systems,* 1(1):31–36.

Lee, D. B. 1969. Anisotropic etching of silicon, *J. Appl. Phys.,* 40(11):4569–4574.

Linder, C., Tschan, T., and Rooij, N. F. 1991. Deep dry etching techniques as a new IC compatible tools for silicon micromachining," Technical Digest, *Transducers '91,* 54–57, June, San Francisco, CA.

Liu, C., Tsao, T., Tai, Y.-C., Leu, T.-S., Ho, C.-H., Tang, W.-L., and Miu, D. 1995. Out-of-plane permalloy magnetic actuators for delta-wing control, *Proc. IEEE Micro Electro Mechanical Systems Workshop,* 7–12, January-February, Amsterdam, Netherlands.

Miu, D., Wu, S., Tatic, S., and Tai, Y.-C. 1993. Silicon micromachined microstructures for supercompact magnetic recording rigid disk drives, Technical Digest, *Transducers '93,* 771–773, June, Yokohama, Japan.

Nathanson, H. C., Newell, W. E., Wickstrom, R. A., and Davis, J. R. 1967. Resonant gate transistors, *IEEE Trans. Electron. Devices,* 14:117.

Nature-Times News Service, 1967. *The London Times,* 23, November, 1967.

Pacific Semiconductors, 1962. British Patent 896,669, Pacific Semiconductors Inc., May.

Petersen, K. E. 1982. Silicon as a mechanical material, *Proc. IEEE,* 70(5):420–456.

Preston, K. 1972. *Coherent Optical Computers,* McGraw-Hill, New York, NY.

Raley, N. F., Sugiyama, Y., and Van Duzer, T. 1984. (100) Silicon etch-rate dependence on boron concentration in ethylenediamine pyrocatechol water solutions, *J. Electrochem. Soc.,* 131(1):161–171.

Reisman, A., Berkenblit, M., Chan, S. A., Kaufman, F. B., and Green, D. C. 1979. the controlled etching of silicon in catalyzed ethylenediamine-pyrocatechol-water solutions, *J. Electrochem. Soc.,* 126(8):1406–1415.

Robbins, H. and Schwartz, B. 1959. Chemical etching of silicon, *J. Electrochem. Soc.,* 106(6):505–508.

Sampsell, J. B. 1993. The digital micromirror device and its application to projection display, Technical Digest, *Transducers '93,* 24–27, June, Yokohama, Japan.

Schnakenberg, U., Benecke, W., and Lange, P. 1991b. TMAHW etchants for silicon micromachining, Technical Digest, *Transducers '91,* 815–818, June, San Francisco, CA.

Schnakenberg, U., Benecke, W., Lochel, B., Ullerich, S., and Lange, P. 1991a. NH_4OH-based etchants for silicon micromachining: influence of additives and stability of passivation layers, *Sensors and Actuators,* A25–27:1–7.

Seidel, H. 1987. The mechanism of anisotropic silicon etching and its relevance for micromachining, Technical Digest, *Transducers '87,* 120–125, June, Tokyo, Japan.

Storment, C. W., Borkholder, D. A., Westerlind, V. A., Suh, J. W., Maluf, N. I., and Kovacs, G. T. A. 1994. Dry-released process for aluminum electrostatic actuators, Technical Digest, *Solid-State Sensor and Actuators Workshop,* 95–98, June, Hilton Head, SC.

System Planning Corporation, 1994. *MEMS Market Study,* System Planning Corporation (SPC), July, Arlington, VA.

Tabata, O., Asahi, R., Funabashi, H., and Sugiyama, S. 1991a. Anisotropic etching of silicon in $(CH_3)_4NOH$ solutions, Technical Digest, *Transducers '91,* 811–814, June, San Francisco, CA.

Tabata, O., Funabashi, H., Shimaoka, K., Asahi, R., and Sugiyama, S. 1991b. Surface micromachining using polysilicon sacrificial layer, Technical Digest, *2nd Int. Symp. Micromachine and Human Science,* October, Nagoya, Japan.

Tai, Y. C. and Muller, R. S. 1989. IC-processed electrostatic synchronous micromotors, *Sensors and Actuators,* 20:49–55.

Tang, W. C., Nguyen, T. H., and Howe, R. T. 1989. Laterally driven polysilicon resonant microstructures, *Sensors and Actuators,* 20:25–32.

U.S. Patent 2,749,598, Method of preparing electrostatic shutter mosaics, June. (1956).

Zdeblick, M., Anderson, R., Jankowski, J., Klineschoder, B., Christel, L., Miles, R., and Weber, W. 1994. Thermopneumatically actuated microvalves and integrated electro-fluid circuits, Technical Digest, *Solid-State Sensor and Actuator Workshop,* 251–255, June, Hilton Head, SC.

115
Micromachines

115.1 Micromachines and the Scaling Effect 1472
115.2 Difficulties in Miniaturization and Proposed Solutions 1473
Materials in Micromachining • Three-Dimensional Micromachining
115.3 Microactuators 1474
Electrostatic Micromotors • Utilization of Rolling Motion • Elastically Supported Actuators • Other Driving Principles
115.4 Architecture for MEMS: Autonomous Distributed Micromachines 1479
System with Micro Smart Modules • Two-Dimensional Conveyor • Arrayed Actuators
115.5 Applications 1483
Optics • Fluidics • Micro Magnetic Head • Electrostatic Handling of Biological Objects • Tunneling Current Detection and Control
115.6 Conclusion 1486

Hiroyuki Fujita
The University of Tokyo

Abstract

This paper provides a brief overview of micromachines and the fabrication technology involved. Its focus is on the IC-based micromachining.

115.1 Micromachines and the Scaling Effect

Making small machines such as artificial ants has been a human dream for a long time. Some of the possible applications of these machines are in medicine—micromachines cleaning a patient's veins; science—micromachines manipulating molecules or atoms; and in the environment—micromachines used to monitor or to clean a polluted area. It is important to understand the difference between the microscopic and macroscopic worlds to produce better designs resulting in more successful applications.

Engineering experience and common sense are essential for choosing the best alternatives. However, this may not necessarily be applicable to micromachine development. The reason is that rules describing the macroscopic and microscopic world may be quite different; the motion equation, for instance. Terms negligible for macromachines may become dominant for micromachines. The electrostatic force is a typical example.

The issue of micromachine dimension has to be considered first. This should be based on the requirements specification for a given application. The machine for handling bits in computers (a mass memory storage, for instance) can be made as small as the manufacturing technology allows. However, if a machine is to manipulate a physical object within a narrow space, it should be small enough to fit in the space, and in addition, strong enough to handle the task.

The area of micromachines covers a wide range of dimensions, from millimeters to nanometers; therefore, different approaches to the development of micromachines are required. While conventional machines can be scaled down into the millimeter range, molecular structures and functions have to be utilized to build machines in the nanometer range. Recent developments in molecular biology and the availability of scanning probe microscopes, such as STM (scanning tunneling microscope), will make building these machines possible in the near future. In the intermediate range, from 0.1 to 100 μm, the fabrication technology for integrated circuits proved to be useful for making gears, springs, motors, and actuators (Gabriel, 1994; Howe et al., 1991; Fan et al., 1988; Meheregany et al., 1988). It is important to classify micromachines according to their size, since the fabrication technology, development principles, and application areas differ between classes.

Table 115.1 compares MEMS manufactured by micromachining and miniaturized machines made by conventional mechanical machining. Unlike miniaturized machines in which the three-dimensional structure, assembled in various shapes, is tightly associated with its function, the limitation of a typical IC-based fabrication process for MEMS only allows us to make planar structures (Fan et al., 1988; Mehregany et al., 1988) and micromotors (Fan et al., 1989; Tai and Muller, 1989), folded structures of thin poly-silicon films (Pister et al., 1992; Sazuki et al., 1992) or a projected image of two-dimensional mask patterns in deep resist (Menz et al., 1991; Guckel et al., 1991). Therefore, it is difficult to realize various functions by only changing the shape

0-8493-8343-9/97/$0.00+$.50

Table 115.1 Comparison Between Micromachined MEMS and Miniaturized Machines

	Micromachined MEMS	Miniaturized machines
Assembly and adjustment	Pre-assembly	Part-by-part
Integration of many elements	Possible	Difficult
Combination with electronics	Integration with the same process	Wire connection
Dimension	2.5 D	3 D

of the machine. However, full use must be made of the advantage that many structures can be obtained simultaneously by pre-assembly and batch processing and that integration with electronic circuits and sensors is possible. Various functions should be realized by using logic circuits with embedded software. We can have many complicated modules in IC-based MEMS, since many micromodules with sensors, actuators, and electronic circuits can be made with exactly the same effort that is required to make just one module.

In spite of these differences, micromachines have a common background. For this reason, a new area of micro science and technology (MSE) should be established to provide the theoretical foundations for the development of micromachines. MSE is the extension of conventional science and technology to the microscopic world. As a result of utilizing a variety of materials and fabrication processes, the micromachines area can benefit from the knowledge offered by a broad range of scientific and engineering disciplines.

The scaling effect is one of the fundamental issues in building micromachines. The friction between sliding surfaces is one of the major problems in rotational micro motors. The reason is that the frictional force obeys an unfavorable scaling law in micro domains. When the characteristic dimension, L, is decreased, the frictional force is proportional to L^2, while the inertial force is proportional to L^3. The frictional force dominates the inertial force and prevents micro gears or rotors from moving smoothly, if moving at all. Friction and tribology in the micro domain are under careful investigation.

115.2 Difficulties in Miniaturization and Proposed Solutions

Medicine, industry, and science are some of the potential application areas for micromachines. This, along with the difficulties in the realization of micromachines, is illustrated in Figure 115.1. There are two major reasons why micromachines are not built and used in large quantities. The difficulty in fabrication of micromachines is one of them. Machining, assembly, and adjustment become more difficult as the machine gets smaller. The other reason is in the control of micromachines. The communication path and the associated cabling occupies more space than

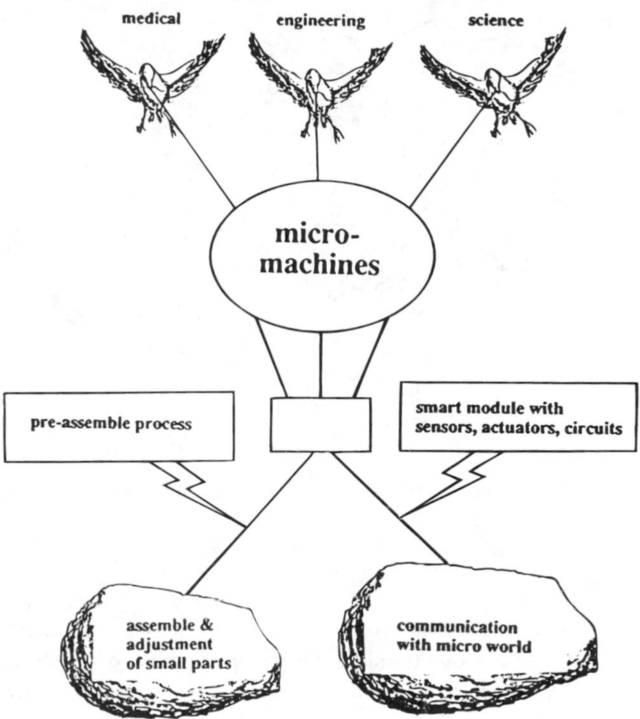

Figure 115.1 Difficulties in miniaturization of micromachines and proposed solutions.

the mechanism itself. This is due to the fact that the number of control signals does not get smaller with the reduction of the physical dimensions of the mechanism. These difficulties in micromachine realization can be overcome by the following technical developments:

1. The difficulty in fabrication of micromachines can be overcome by using the pre-assembly processes based on IC-fabrication (Gabriel, 1984; Howe et al., 1991; Fan et al., 1988; Mehregany, 1988). A complicated micromachine, composed of many parts, can be fabricated simultaneously by using this approach. The assembly and adjustments are minimized or eliminated entirely. The batch fabrication, using deposition, photolithography, and etching allows for mass production of micromachines. The pre-assembly process based on IC fabrication is being improved continuously. Now, the objective is to fabricate truly three-dimensional structures (Fujita and Gabriel, 1991; Menz et al., 1991; Fan et al., 1989).
2. Integration seems to be the answer to the problems involved in the micromechanism control. Micromechanisms and actuators can be integrated with electronic circuits and sensors. The IC-based fabrication technology makes it possible to fabricate complex systems composed of large numbers of micromechanisms. Figure 115.2 depicts an example of a simple-smart module composed of sensors, actuators, and electronic circuits for signal conditioning and logic. The number of control

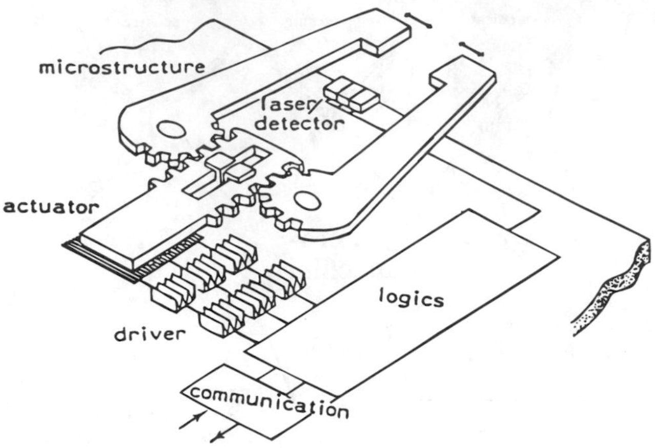

Figure 115.2 An example of a smart module composed of sensors, actuators and circuits.

signals, and as a result connections, between the micromechanism module and its environment can be reduced since most of the information processing and control functions can be executed by the local processors integrated in the module.

The successful fabrication and operation of microactuators and micro mechanical parts by IC-based micromachining technology permitted the production of micro miniature motion-systems (Howe et al., 1991). Although the small size of mechanical components in the system is a very distinctive feature of this emerging technology, it has other, maybe even more attractive, features. The three characteristic features or the three "M"s of the technology are (Gabriel, 1994):

Miniaturization.
Multiplicity.
Microelectronics.

Miniaturization is clearly essential. However, the mere miniaturization of macroscopic machines is not possible because of the scaling effect. Like a swarm of ants carrying food, the cooperative work of many microelements can perform a large task, even when one single device can only produce a small force or perform a simple motion. Multiplicity is the key to successful microsystems. The integration of microelectronics is essential for micro moving elements to cooperate with each other.

To make the development of micromachines and systems possible, the research has to involve a broad range of engineering and science areas, such as materials science, process engineering, device fabrication, system design and control, applications and MSE, etc. Figure 115.3 shows the detailed research issues.

Materials in Micromachining

Silicon is the most commonly used material in micromachining (Petersen, 1982) because the process is well established, (Howe et al., 1991) it has good mechanical properties, and integration with electronics and sensors is possible. Other materials have also been used for specific purposes in micromachining. Table 115.2 summarizes the materials and their useful characteristics. It is important to develop batch fabrication processes for all the materials in the table, so we will reiterate the above mentioned features of micromachining.

Three-Dimensional Micromachining

Microstructures fabricated by surface micromachining are planar in nature and have thicknesses of up to 10 μm in most cases. Some applications require thicker structures or three-dimensional-complicated structures. Modifications of surface micromachining have been attempted.

One technique is to fold up micromachined plates from the substrate to construct a 3-D structure. The plate is released from the substrate and reconnected by hinges (Pister, 1992) or flexible films (Suzuki et al., 1992). Such structures as a cube measuring 300 μm per side and a pair of flaps resembling a butterfly were fabricated. In other trials, overhanging structures were made (Kim et al., 1992). Microscopic tweezers made of polysilicon protrude 400 μm from the edge of a wafer. A single-celled protozoa, a euglena, was held by this microgripper.

Electron beams or laser beams can assist selective growth/solidification/etching of materials (Ikuta and Hirowateri, 1993; Westberg et al., 1991). Three-dimensional structures such as a microhelical spring of 50 μm in diameter or a curved square pipe of 0.1 mm × 0.1 mm × 1 mm were realized. Unfortunately, batch fabrication capability has not yet been proven.

115.3 Microactuators

Because of the scaling consideration (Trimmer, 1990; Pisano, 1989) the electromagnetic force which is most commonly used in macroactuators is not suitable for microactuators. Although some trials were reported on magnetic actuators (Wagner and Benecke, 1991), many microactuators made by micromachining utilize other driving principles such as the electrostatic force. Table 115.3 summarizes recent examples of microactuators. Because of limited space, the table is not all inclusive.

Electrostatic Micromotors

The first row in Table 115.3 lists an electrostatic micromotor with diameters of 60–120 μm by Tai et al. It is called a side-drive type motor, since it utilizes the electrostatic force which acts between the edges of the rotor and the stator. The rotor and stators are made of polysilicon films, 2 μm thick. Stators are placed on a circle and connected in three phases. If the voltage is applied to each phase successively, the rotor rotates in synchronization. The voltage is up to 300 V across the 1–2 μm gap. The torque is estimated to be a few pNm. Rotational speed was reported to be on the order of 500 rpm. The speed is relatively low compared to the theoretical value (Bart and Lang, 1989). The reason is the friction between the rotor and the shaft, although a

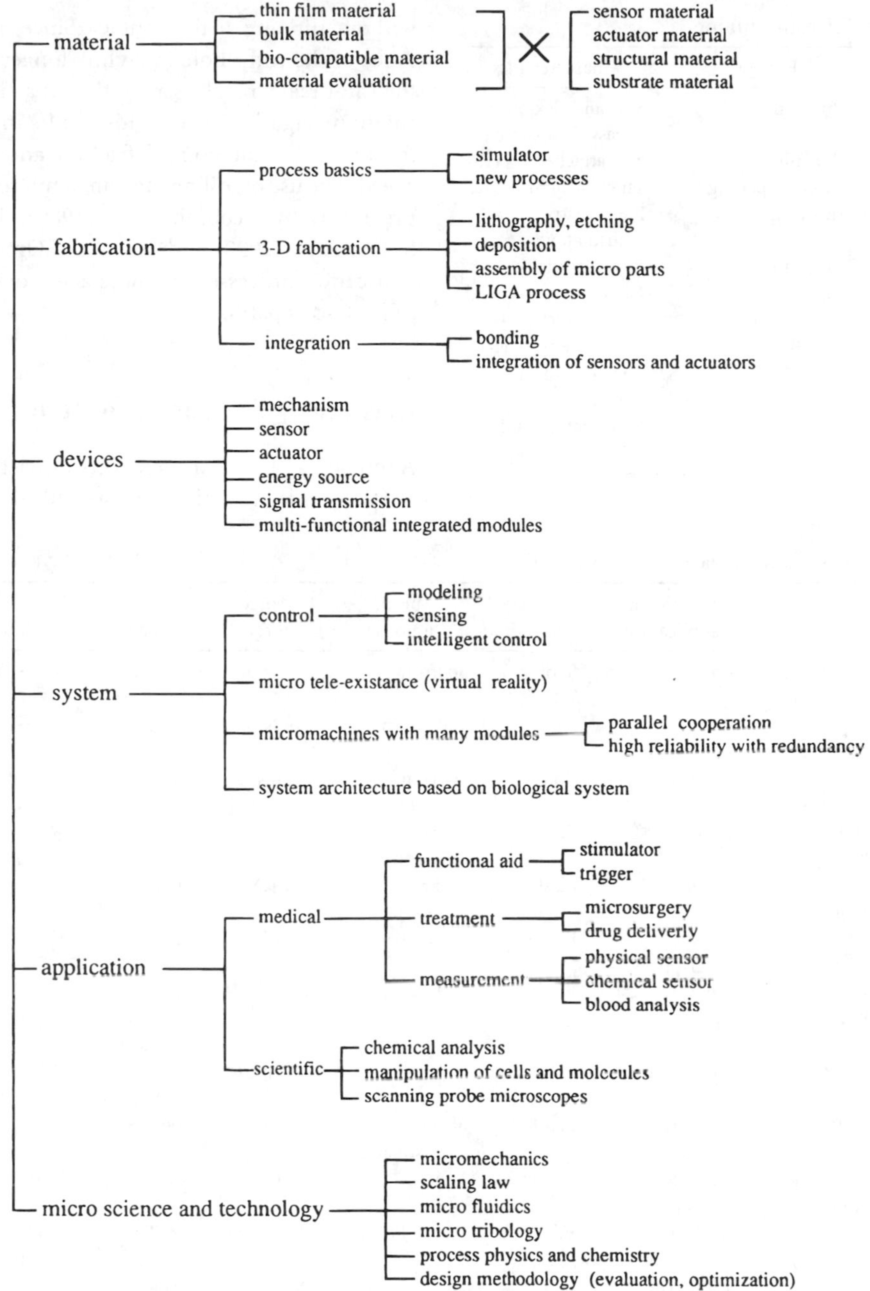

Figure 115.3 Research items in the study of micromachines.

silicon nitride film was deposited on the sliding surface to reduce the friction.

In the second row is the improved version of the side-drive micro motor reported by Mehregany et al. (1990). By improving the design, the fabrication process, and the operating condition of the motor, they achieved rotational speeds of up to 15,000 rpm and continuous operation for more than a week. They reduced the clearance between the rotor and the shaft, formed three dimples under the rotor for both support and electrical contact and operated in nitrogen to avoid oxidation.

Figure 115.4 shows a nickel electrostatic micromotor (Hirano et al., 1995). The rotor, which is only 120 μm in diameter and 7 μm in thickness, can rotate at 10,000 rpm, driven by electrostatic force.

Utilization of Rolling Motion

Even for improved micromotors, friction is a major problem. One solution is to replace the sliding contact at the center with a rolling contact. Mehregany et al. (1990) also made this type of micromotor. As shown schematically in Figure 115.5a, the rotor is a smooth ring whose inner diameter is only a little larger than the shaft. When the voltage is applied sequentially to stators as shown in 115.5b, the rotor rotates eccentrically

Table 115.2 Materials in Micromachining

Material	Usage	Process	Characteristics
Polyimide	Structure	Thin film	Soft and flexible, easy film coating
Tungsten	Structure	Thin film	Not attacked by HF
Ni, Cu, Au	Structure	(Electro) plating	Thick structures
Quartz	Actuation	Anisotropic etching	Piezoelectricity, insulator
ZnO	Actuation	Thin film	Piezoelectricity
PZT	Actuation	Thick film	Large piezoelectricity
TiNi	Actuation	Thin film	Shape memory allay
GaAs	Optics	Thin film	LASER, LED, detector
DLC	Lubrication	Thin film	Low friction and wear

without slipping at the contact. Since the circumferential distance of the rotor hole is slightly longer than that of the shaft, the rotor really revolves a fraction of a circle after one eccentric rotation (Figure 115.5c). This results in two advantages of the motor, e.g., reduction of friction and higher torque at low speed. The use of rolling motion in microactuators was reported previously by Jacobsen et al. (1989), Trimmer et al. (1989), Sakata et al. (1990) and Fujita and Omodaka (1988) although fabrication processes for these actuators were not IC based or fully IC-compatible.

Elastically Supported Actuators

Another way to avoid the effects of friction is with elastic supports. Five electrostatic actuators with elastic supports are shown

Table 115.3 Micro Actuators (fully or partly IC-processed)

Driving principle	Size	Movement, application	Support	Speed, response	Force, torque	Material	Input	Ref. & authors
1. Electrostatic	60 ~ 120 μm (diameter)	Rotation	Sliding	500rpm	a few pNm	poly-Si	60 ~ 400V	Y. C. Tai et al., 1989
2. Electrostatic	100 μm (diameter)	Rotation	Sliding	15000rpm	10pNm	poly-Si	50 ~ 300V	M. Mehregany et al., 1990
3. Electrostatic	100 μm (diameter)	Rotation	Rotation	300rpm	~1nNm	poly-Si	26 ~ 105V	M. Mehregany et al., 1990
4. Electrostatic	5 × 100 × 100 μm^3	10 μm (L. L.)*	Elastic	10 ~ 100kHz (resonance)	2 μN	poly-Si	$40V_{DC}$ + $10V_{AC}$	W. C. Tang et al., 1989
5. Electrostatic	4 × 400 × 400 μm^3	7 μm (L. L.)	Elastic	~3kHz (resonance)	0.8 μN	poly-Si	10V	T. Hirano et al., 1992
6. Electrostatic	2.5 × 60 × 400 μm^3	10 μm (gripper)	Elastic	N.A.	N.A.	poly-Si	20V	C. J. Kim et al., 1992
7. Electrostatic	10 × 500 × 500 μm^3	2 μm (L. L.)	Elastic	N.A.	N.A.	poly-imide metal	200V	R. Mahadevan et al., 1990
8. Electrostatic	0.1 × 0.35 × 0.39 mm^3	on-off valve	Elastic	N.A.	110mmHg	metal, Si_3N_4	30V	T. Ohnstein et al., 1990
9. Electrostatic	4 × 300 × 300 μm^3	5 μm (L. L.)	Elastic	8kHz (resonance)	5 μN	poly-Si	19V	N. Takeshima et al., 1991
10. Piezoelectric	8 × 0.2 mm × 1 mm	7 μm STM scan	Elastic	N.A.	23 μN	ZnO	30V	S. Akamine et al., 1990
11. Piezoelectric	2mm (diameter)	Rotation	Vibration	100–300rpm	25pNm	PZT	4V (100kHz)	K. R. Udagakumar et al., 1991
12. Shape memory alloy	2 × 30 × 2000 μm^3	a few μm	Elastic	20Hz	N.A.	TiNi	2mA, 40V	J. A. Walker et al., 1990
13. Thermal	0.5 × 3 × 3mm^3	45 μm (L. V.)*	Elastic	~5ms	0.6N	Si + liquid	~200mV	M. J. Zdeblik et al., 1987
14. Thermal	0.5 × 8 × 8mm^3	23 μm (L. V.)	Elastic	>1Hz	0.1N	Si	13V	F. C. van de Pol et al., 1989
15. Thermal	6 × 100 × 500 μm^3	74 μm (bending)	Elastic	10Hz (square wave)	N.A.	Si + Au	130mW	W. Riethmüller et al., 1988
16. Thermal	5 × 110 × 500 μm^3	120 μm (bending)	Elastic	8Hz(sinusoidal wave)	N.A.	poly-imide	30mW	M. Ataka et al., 1993
17. Electromagnetic	1.5 × 5.8 × 5.8mm^3	70 μm (L. V.)	Elastic (L. V.)	94Hz (resonance)	450 μN	Au, NdFeB	0.3A	B. Wagner et al., 1991
18. Electromagnetic	0.1 × 10 × 10 mm^3	5mm (L. L.)	Levitation (Meissner effect)	20mm/s	30 μN	YBaCuO, NdFeB	0.3 ~ 0.9A	Y. K. Kim et al., 1990

* L. V.: linear motion in vertical direction, L. L.: linear motion in lateral direction.

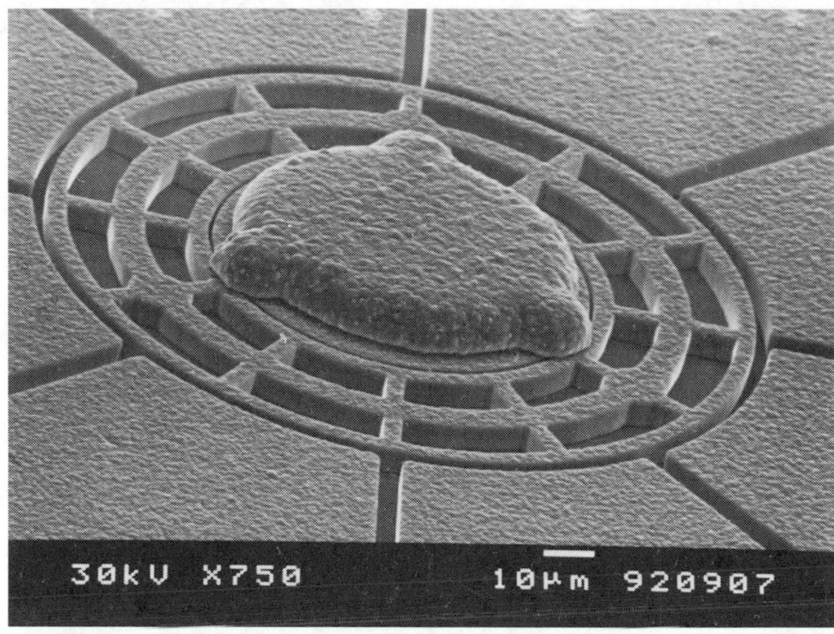

Figure 115.4 Nickel Electrostatic Micromotor (120 μm in diameter, 7 μm in thickness).

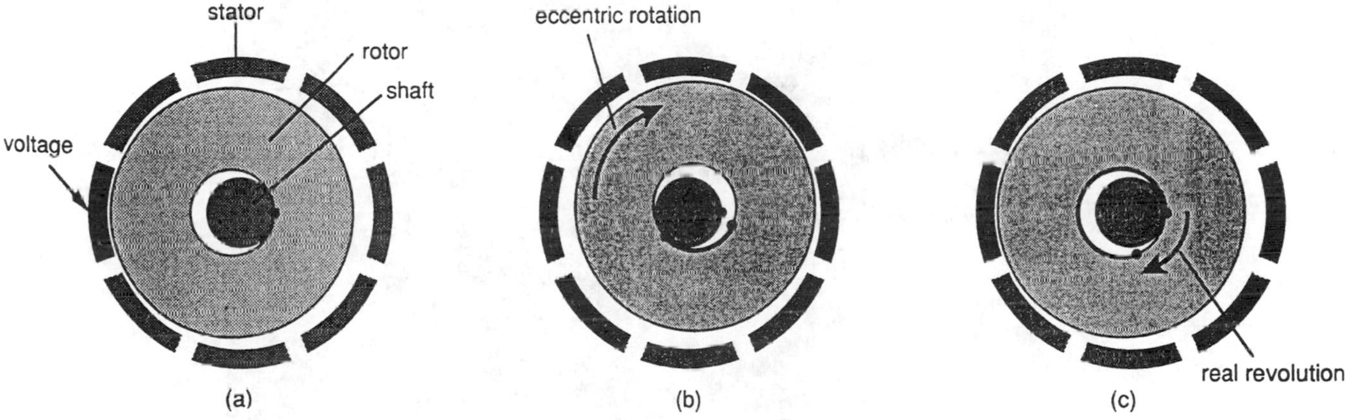

Figure 115.5 Option of a harmonic micromotor.

in the fourth through the eighth rows of Table 115.3. First is an electrostatic resonator by Tang et al. (1989). The resonator is supported by double-fold beams and actuated by comb-like structures. The teeth of the comb, attached to the moving part, overlap those fixed on the substrate. The force to increase the overlapping is generated when voltage is applied between the two combs. An alternating voltage of 10 V with a 40 V DC bias made the suspended part vibrate at resonance. The displacement was 10 μm and the resonant frequency was 18 kHz with 200 μm-long supports.

Furuhata et al. (1991) introduced the oxidation machining technique to obtain sub-micron operational gaps between moving and driving electrodes. The reduced gap enabled them to operate the modified comb-drive actuator with lower voltages that are commonly available in electronic circuits. Hirano et al. (1992) succeeded in obtaining nonresonant deflections of 7 μm with 10 V. The overall shape and the device in operation are shown in Figure 115.6.

Mahadevan et al. (1990) reported a linear actuator made with polyimide. The mover is a polyimide ladder-like structure sandwiched by two driving electrodes. The electrodes are also patterned in stripes which have the same pitch of the mover but are divided into some sections with different phase shifts. The mover is supported by four polyimide beams. Although the mover is not conductive, it is attracted in between the electrodes which make up a parallel-plate-capacitor. The actuator is interesting because it utilizes the force acting on both surfaces of the mover rather than on the edge.

In the eighth row, an electrostatic valve is shown. A plate with one side fixed is driven electrostically and seals an inlet orifice.

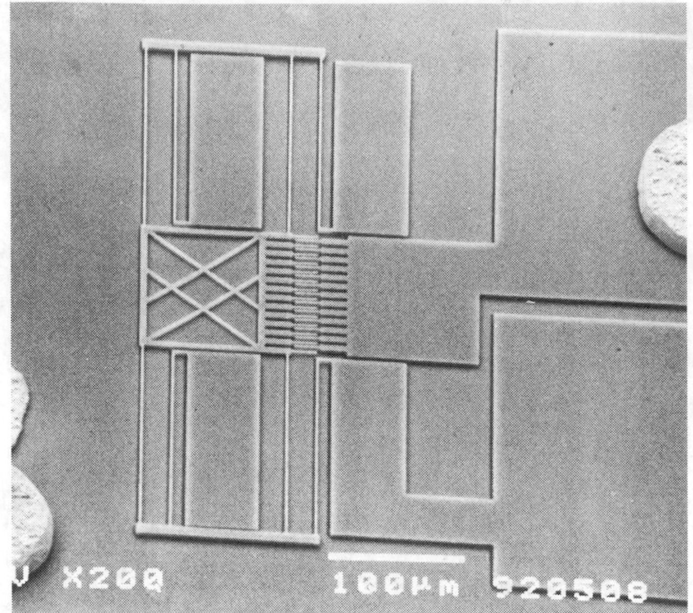

(a)

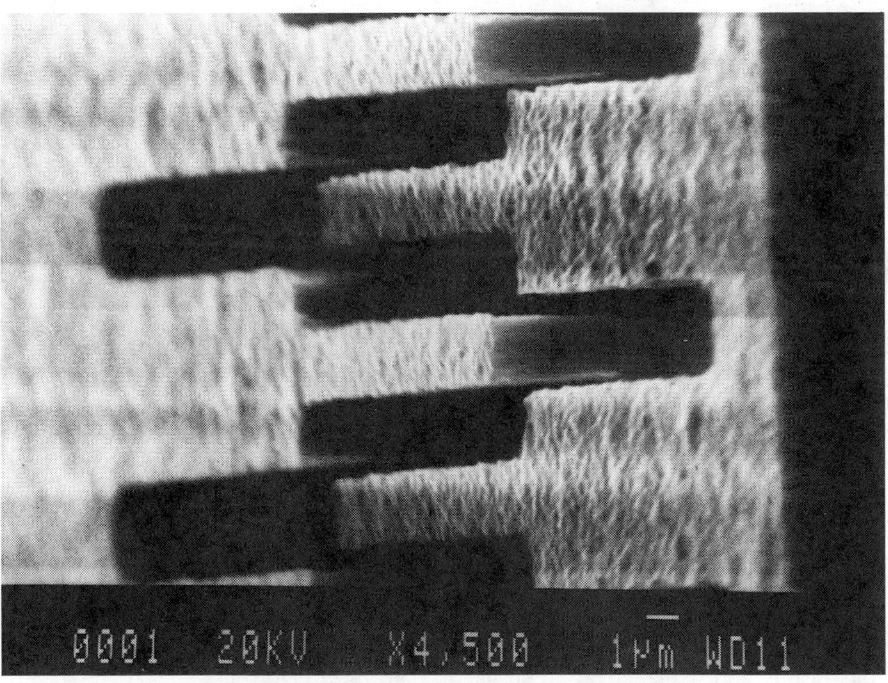

(b)

Figure 115.6 An electrostatic actuator with sub-micron gaps. a. Over all view. A comb-like driver, four positioning and aligment mechanisms, and flexible supports are shown. b. Expanded view of working teeth with 0.5 μm operational gaps. (*Source*: Trimmer, W. S. N. 1990. Micromechanical Systems, Integrated Micro-Motion Systems, F. Harashima, ed., pp. 1–15. Elsevier Science, New York. With permission).

The closure plate is composed of a metal electrode sandwiched by silicon nitride films. The valves are fabricated in a 5 by 5 array, which results in larger flow rate and finer flow control just by closing some of the valves. It was possible to close the valve against pressures of up to 110 mmHg with 30 V applied to the valve.

Other Driving Principles

Microactuators, which utilize other driving principles such as piezoelectric (Udayakumar et al., 1991; Akamine et al., 1990), shape memory alloys (Walker and Gabriel, 1990), thermal expansion (Takashima and Fujita, 1990; Van De Pol et al., 1989; Riethmüller and Benecke, 1988; Zdeblick and Angell, 1987) and electromagnetic (Wagner and Benecke, 1991; Kim et al., 1990) are included in Table 115.3 for comparison. In terms of reducing friction, most of them move elastically with two exceptions. Udayakumar et al. (1991) made the ultrasonic micromotor which utilizes the standing wave to rotate the rotor. A similar trial was made in the linear motion previously by R.M. Moroney et al. (1990). Kim et al. (1990) levitated the permanent-magnet mover by the Meissner effect of the superconducting material.

Each actuator in Table 115.3 has its own advantages and disadvantages. The choice and the optimization should be made according to the requirements of the applications. Generally speaking, the electrostatic actuator is more suitable for performing tasks which can be completed within a chip (positioning of devices/heads/probes, sensors with servo feedback, light deflection, etc.), since it is easily integrated on a chip, easily controlled, and consumes little power. By contrast, the other types of actuators are more robust, produce substantial force, and are suitable for performing external tasks (propulsion, manipulation of objects, etc.).

115.4 Architectures for MEMS: Autonomous Distributed Micromachines

System with Micro Smart Modules

In MEMS, we expect microsystems to perform complicated tasks, such as micromanipulators and self-propelled systems. For example, when a microsystem handles cells, the system must move to the cells by itself and manipulate them. Devices reviewed in the previous section are not strong or complicated enough to complete such tasks. The limitation of the process, which was discussed earlier in this article, must be overcome by the design of microelectromechanical systems (MEMS), which would be completely different from the simple miniaturization of macro machines. As was mentioned above, one of the advantages of MEMS is that many actuators and sensors are supplied with batch processing techniques. Another advantage is that both logic circuits and sensors can be added to the same system. We can expect to have a module which includes sensors, actuators, and logic circuits and has primary information processing and control. Furthermore, many of the modules can be implemented in a small area without assembly. These modules are smart enough to perform elementary control and complex motions with simple input signals. When many modules are arranged on the surface of objects, the surface may be able to perform some functions.

If all the control signals were externally applied to each element of a micro system, wiring would be extremely difficult. Therefore, sensors and controllers will be integrated with actuators because they are necessary to compose a primary servosystem and to reduce the amount of information exchange. The control signal may be given to a group of actuators and that will eliminate the wiring problem in a system with many active elements.

The integration of sensors, actuators, and controllers led to the concept of autonomous distributed micromachines (ADM) as a system architecture suitable for micromachines. An autonomous distributed system is a system which is composed of many smart subsystems called individuals. An individual can gather information with its sensors as well as through communication with neighboring individuals and sometimes with the overall system. It independently determines its behavior based on the information. The way each decides its behavior is to cooperate with each other in order to complete the objectives of the overall system. The ADM are composed of many smart module individuals which are clever enough to control their own actuators and to cooperate with each other. One can find numerous examples of ADM in living organisms. Legs of centipedes and cilia in the respiratory tract are good examples. Motor cells work as the actuator, and interconnected neurons control them. It will be too laborious and inefficient to construct ADM by ordinary machining and assembly. IC-compatible micro fabrication processes (IC-micromachining) are capable of making ADM with many smart modules (Fujita and Gabriel, 1991).

Various applications from a display to an active particle filter could be realized by ADM. In the following, a two-dimensional conveyer system is discussed as an example.

Two-Dimensional Conveyor

The conveyance system is composed of the two-dimensional array of micro modules as shown in Figure 115.7. The module must have actuators and circuits for control and communication. It may have sensors to detect the position and/or weight of an object. All the necessary devices to perform the function can be integrated in modules by IC-micromachining technology. Here, let us consider an in-plane positioner. It is assumed that each module can be addressed by some means, for example by illumination with a laser beam to the optical sensor in the module. The circuit in a module determines the movement of its actuators based on the information from surrounding cells and its sensor signals. The module tells surrounding modules the direction of its own movement and information from its sensors. Such an array can be regarded as a two-dimensional cellular automata with sensors and microactuators.

Actuators in a module must be able to convey the object in four directions ($+x, -x, +y, -y$). The movement of each actuator is coordinated so that the overall pattern of movement matches the task. Patterns for transporting, aligning, positioning, and

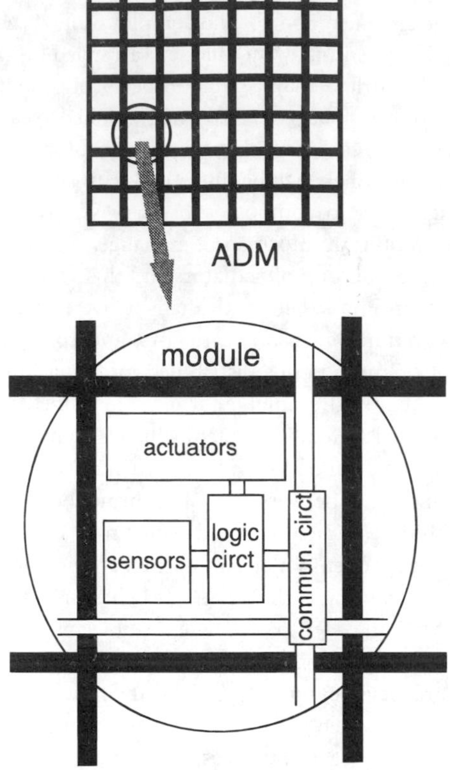

Figure 115.7 The schematic representation of an autonomous distributed micromachine (ADM). (*Source*: *IEEE Trans. Ind. Electronics* 42:449–454. © 1995 IEEE. With permission.)

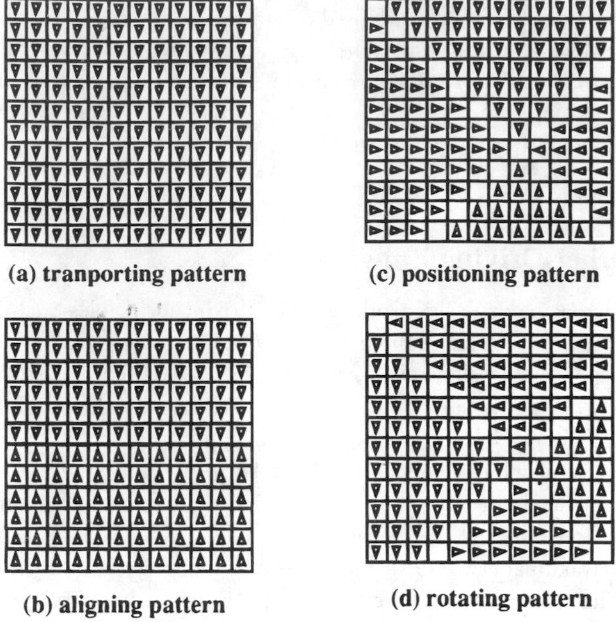

Figure 115.8 Motion patterns of actuators of a planar positioner with many micromodules. (*Source*: *IEEE Trans. Ind. Electronics* 42:449–454. © 1995 IEEE. With permission.)

rotating an object are shown in Figure 115.8. For example, in the positioning pattern, each module activates its actuators in such a way that the direction of movement converges to a desired point in a plane. Note that triangles on the modules indicate the direction of traction. Such patterns can be determined in parallel within the system or triggered by some external signals.

Now, consider an open-loop positioner with two-degrees-of-freedom (lateral and longitudinal). The simplest way to obtain the positioning pattern shown in Figure 115.8c is as follows:

1. A module which is located at the center of the desired position is addressed (Figure 115.9a).
2. The module sends signals to four adjacent modules (Figure 115.9b).
3. When a module receives a signal from one side, the moving direction of the actuators is determined to be toward that side from the opposite side (Figure 115.9c).
4. The module gives signals to three modules which lie in the direction of movement (Figure 115.9d).
5. By repeating 3 and 4, the pattern is generated (Figure 115.9e, 115.9f).
6. Actuators are activated to carry an object to the desired position.

The detection of the object on the positioner is essential when we want to have a closed-loop system. It is assumed that only one flat object is placed on the positioner, and that each module has a sensor which detects the presence of an object above it. The sensor may be a tactile sensor or an optical sensor which detects the shadow of the object under flat illumination. As a whole, the system is like a CCD imager. The edge and the center of gravity can be detected by simple distributed logic. There have been many research works on the distributed processing of image data; these results can be utilized in determining the information processing scheme for ADM.

The positioning task on the conveyor can be completed as follows. Suppose the shape of the object is known. The desired location and orientation of the object is given to the system by teaching it an image of the object when it is at the desired position. The center of gravity of the image is calculated. As was explained in Figure 115.9, a positioning pattern is formed around the center of gravity. When the real object is placed, it is carried to the point. When the center of gravity of the real object comes to the point, the moving pattern changes into a rotational pattern. The match between the given image and the object can be determined by calculating the correlation between them. Repeated use of positioning and rotational patterns leads to the maximum correlation.

Arrayed Actuators

Advantages of Arrayed Actuators

As one of the steps towards ADM, arrayed actuator systems were fabricated and operated, because we believe that the key idea is to coordinate simple motions of many microactuators in order to perform a macroscopic task. Even when each moving step is small, accumulation of many steps covers a large distance.

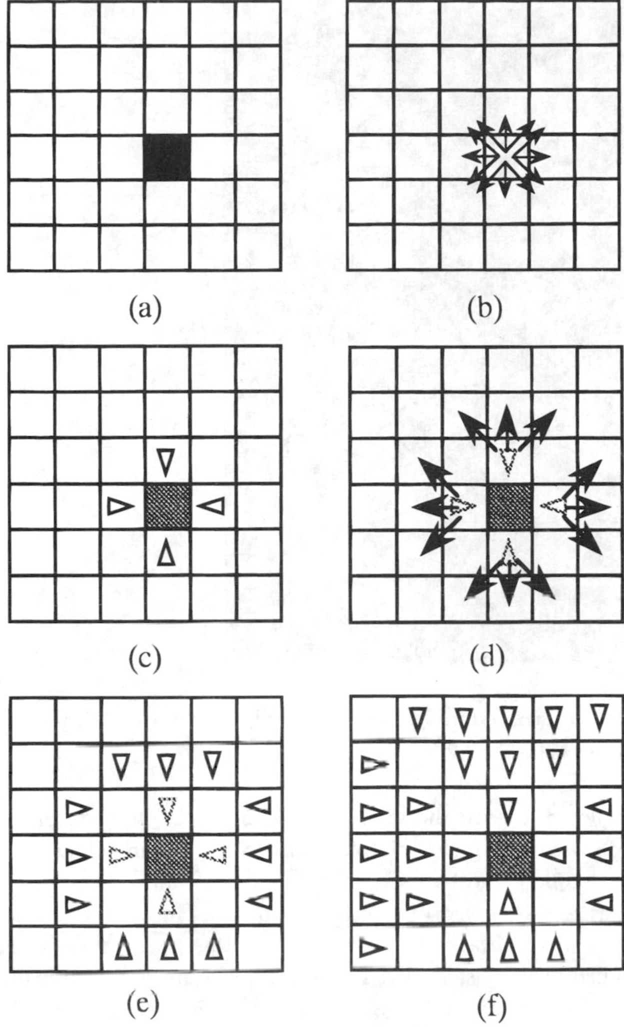

Figure 115.9 Generation of the positioning pattern. a. A module which locates at the center of the desired position is addressed. b. The module sends signals to four adjacent modules. c. When a module receives a signal from one side, the moving direction of its actuator is determined to be toward that side from the opposite side. d. The module gives signals to three modules which lies backward of the moving direction. e. When a module receives a signal from one side, the moving direction of actuators is determined to be toward that side. f. By repeating d. and e., the positioning patter is completed. (*Source*: *IEEE Trans. Ind. Electronics* 42:449–454. © 1995 IEEE. With permission.)

A heavy load may be distributed among many actuators which produce only small force. Flexibility of motion, expandability, and immunity against failure of elements can be achieved. One of the major problems in present microactuators, the problem of friction, can also be solved. Friction in microscale prohibits us from using gears and joints because they waste too much energy. Suspended actuators do not suffer from friction but have a limited range of motion up to a few tens of micrometers. If many such microactuators are arranged in series and parallel (Bobbio et al., 1993; Chen et al., 1993; Minami et al., 1993), the overall structure can produce larger force and displacement and perform more complicated functions than each simple actuator. Because these actuators are driven directly, energy loss associated with transmission of motion is minimal. They can even utilize the friction between themselves and an object to transmit driving force.

To demonstrate the concept, we have fabricated two types of distributed micro motion systems (Fujita and Gabriel, 1991). One is a ciliary motion system (CMS) which mimics the motion and function of cilia in living organisms. Many cantilever actuators vibrate in synchronization and convey objects. As elements of the CMS, thermobimorph cantilever actuators made of polyimide were developed and their motion was experimentally confirmed (Akata et al., 1993). An object was conveyed by the coordinated motion of the CMS composed of many thermobimorph cantilever actuators. The other is an in-plane conveyance system using controlled air flow from many small nozzles on the substrate (Konishi and Fujita, 1994). A plate on the system is levitated and carried by the flow.

Ciliary Motion System

The ciliary motion system mimics the motion of cilia in living organisms (Takeshima and Fujita, 1990). Many cilia vibrate in synchronization to convey objects or fluids. A cilium has two-degrees-of-freedom (rotation and bending). Since the micromachined actuator has only simple motion, two actuator elements are combined to achieve the two-degrees-of-freedom motion. A plate can be carried by the motion sequence of actuators as shown in Figure 115.10. We fabricated an array of microactuators which are thermally driven cantilevers (Akata et al., 1993).

The basic driving principle is based on the thermo-bimorph actuator similar to that reported by Riethmuller et al. (1988). The differences are the material (polyimide) and the simple

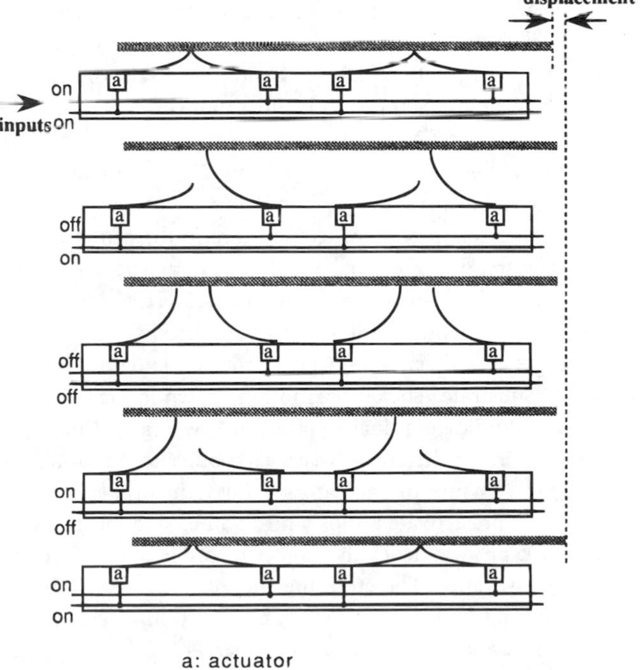

Figure 115.10 The sequence of motion of CMS to convey a plate. (*Source*: Ataka et al. 1993. *IEEE/ASME J. Microelectromechanical Syst.*, 4:146–150. With permission.)

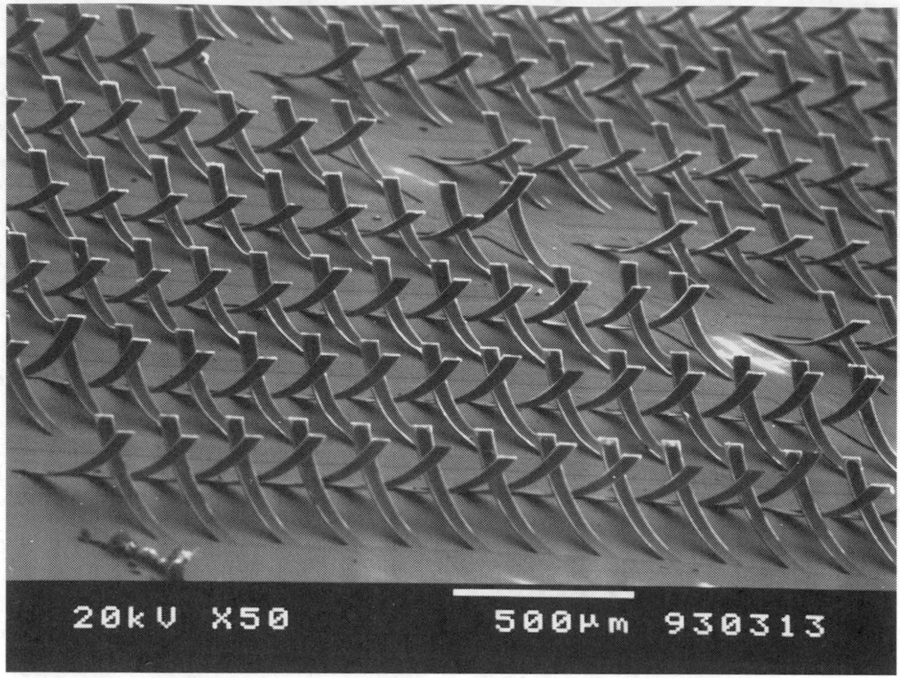

Figure 115.11 SEM photograph of thermal bimorph actuators of CMS. Note that they curl up due to intentionally introduced residual stress. (*Source*: Ataka et al. 1993. *IEEE/ASME J. Microelectromechanical Syst.* 4:146–150. With permission.)

fabrication process. Two layers of polyimide with different thermal expansion coefficients sandwich a metal heater. Aluminum was used as a sacrificial material (Schmidt et al., 1988). Since the polyimide used in the upper layer has a larger thermal expansion coefficient than the lower layer, the residual tensile stress in that layer causes the cantilever to curl. (Note that the tensile stress builds up when the polyimide is cured at elevated temperature and cooled down.) When the current flows in the heater and the temperature rises, the cantilever bends down. The dimensions of the cantilever are: 500 μm in length, 100 μm in width, and 6 μm in thickness. Vertical displacement of 150 μm and horizontal displacement of 80 μm were obtained with 22.5 mA drive current in the heater. The current corresponded to the consumed power of 33 mW and the maximum temperature in the cantilever was 260 C°. The frequency response without any particular cooling was measured. The cut-off frequency was 10 Hz.

Figure 115.11 shows a SEM photograph of CMS which is composed of 512 thermobimorph cantilever actuators on a 1-cm-square substrate. The ciliary motion shown in Figure 115.10 was realized by flowing dual square waves which had the delay of one-quarter wavelength between each other to opposing sets of cantilevers. While the actuators arrayed in one side of the opposing sets are activated, those in the other side must be bent down and kept away from the object in order not to interfere with the conveyance. Therefore the square wave is suitable as a form of driving current for CMS. The direction of conveyance could be reversed by changing the phase of driving voltages applied to opposing sets of actuators.

We operated one half of the CMS shown in Figure 115.11 and observed the conveyance of a load (a silicon piece of 2.6 mm × 1.5 mm × 0.26 mm in size and 2.4 mg in weight). Twenty cantilevers carried the load at the same time. One-half of the CMS was composed of eight modules connected in parallel, and each module had two sets of 16 opposing cantilever actuators in series. It occupied 1 cm × 0.5 cm in area. The input resistance of the system was 250 Ω. During this experiment, the voltage applied to the system was 16V and the current was 65mA; this corresponded to 4 mW power dissipation in each actuator. The experiment was carried out in the ambient air without any cooling equipment.

Conveyor Using Air Flow

Flat objects are levitated and conveyed by controlled air flow in this system (Konishi and Fujita, 1994). Pister et al. (1990) also used a cushion of air to minimize friction and move objects by electrostatic force. We use air flow not only for levitation but also for conveyance. Figure 115.12 shows the concept of one-dimensional conveyance by the directed air flow from micro nozzles. The direction of the flow can be changed from upper-left to lower-right in the figure by closing and opening two air channels of a nozzle as shown in Figure 115.3. An SOI (Silicon On Insulator) wafer is used as a stage of the conveyance system. We made many through-holes in the diaphragm structure using KOH anisotropic etching. The dimension of each hole is about 100 μm × 200 μm. Each hole is covered with a soft layer of polyimide. Two air channels directing opposite sides are made by using the sacrificial layer technique. Two electrodes are placed in the polyimide layer around channels. When the voltage is applied between one of the electrodes and the substrate, the channel is closed (Figure 115.13b). When one nozzle is closed, the air flows through the other nozzle preferentially. The five mask

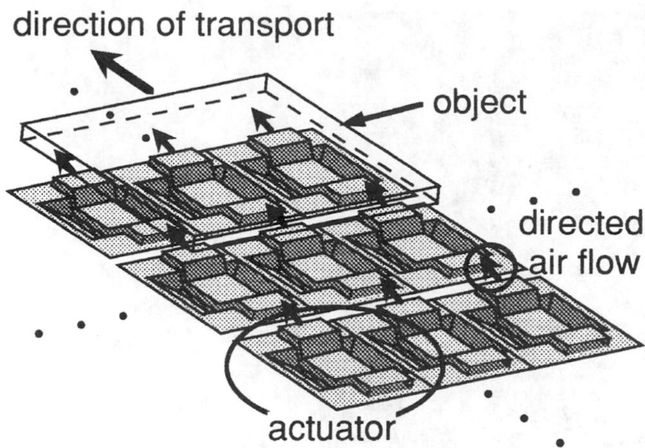

Figure 115.12 Concept of a one-dimensional conveyance system based on controlled air flow (From Konishi and Fujita, 1994. *IEEE/ASME J. Microelectromechanical Syst.*, 3:54–58.).

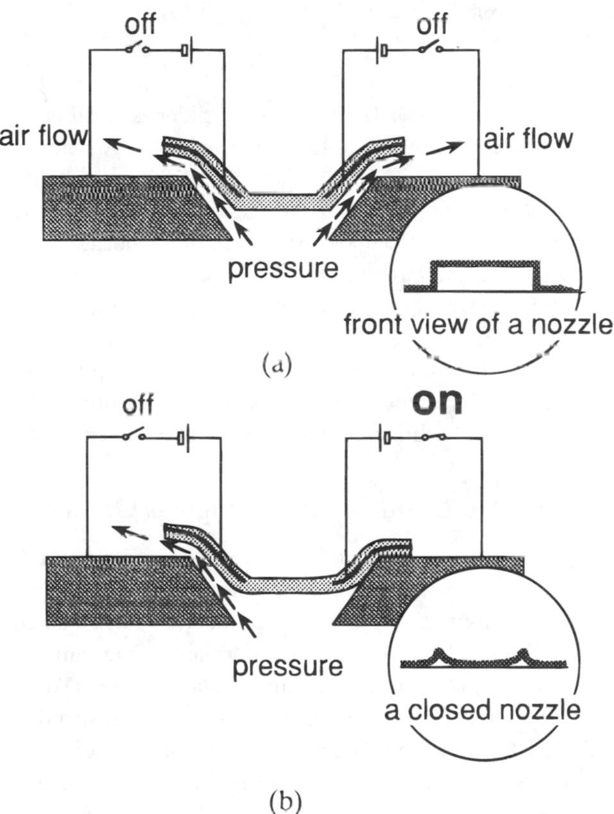

Figure 115.13 Mechanism of the motion of an actuator (cross-section). a. Initial shape. b. When voltage is applied to one electrode to close the channel. (From Konishi and Fujita, 1994. *IEEE/ASME J. Microelectromechanical Syst.*, 3:54–58. With permission.)

process outlined in Konishi and Fujita (1994) was employed. We have made a 7 × 9 array of nozzles in a 2 mm × 3 mm area.

We succeeded in conveying a 1 mm-square and 300 μm-thick piece of silicon to the middle line between two sets of nozzles. The air flow was controlled in such a way that it converged to the middle line from both sides. In other words, nozzles located at the left-hand side of the line blow the air to the right, and ones at the right-hand side blow air to the left. The silicon piece was placed at one edge of the system. It was carried by the flow towards the middle line. When it passed over the line, it was pushed back by the opposing air flow. As a result, it stayed on the middle line as shown in Figure 115.14. The voltage applied to close the channel was 90 V. Air pressure was 2 kPa (0.02 atm.).

115.5 Applications

Figure 115.15 shows some possible applications of microactuators and MEMS. Promising applications of MEMS for the near future will be in optics (Sawada et al., 1991; Nagaoka, 1991; Jerman et al., 1990; Jebens et al., 1989; Petersen, 1977), magnetic and optical heads (Lim et al., 1989), fluidics (Nakagawa et al., 1990; Van De Pol et al., 1990), OA apparatus (Shibata et al., 1987), the handling of cells and macro molecules (Washizu 1990; Fuhr et al., 1991), and microscopy with microprobes (Kobayashi et al., 1992; Yao et al., 1992; Kenny et al., 1991; Tortonese et al., 1991; Akamine et al., 1990) such as STMs (scanning tunneling microscopes) and AFMs (atomic force microscopes). These applications have a common feature in that only very light objects such as mirrors, heads, valves, cells, and microprobes are manipulated and that little physical interaction with the external environment is necessary. One reason is that present microactuators are still primitive and large forces cannot be transmitted to the external world. The other reason is difficulty in packaging. In the following, a few examples are explained.

Optics

Petersen et al. (1977) demonstrated deflecting light beams by small cantilevers driven by electrostatic force in 1977. The dimensions of the cantilever were 100 μm in length, 25 μm in width, and 0.5 μm in thickness. Recently, an optical-fiber switch (Jebens et al., 1989), its aligner (Nagaoaka, 1991), and an adjustable miniature Fabry-Perot interferometer (Jerman et al., 1990) were reported. Sawada et al. (Nakagawa et al., 1990) developed a new integrated optical microencoder. They integrated a U-shaped laser diode with etched mirrors, microlenses, and a photodiode. The size was 0.5 × 0.5 mm^2 square. They claimed a theoretical resolution of 0.01 μm with a 1 μm-pitch grating. Because of its size and the fabrication process, it is possible to integrate the encoder with microactuators, that will result in a micro positioner with very high accuracy.

Fluidics

Good review articles (Nakagawa et al., 1990; Van De Pol et al., 1990) have already been published on micro fluidic systems. Here we discuss only the application of this technology to the ink jet printer. Using silicon micromachining and bonding techniques, Shibata et al. (1987) fabricated micro nozzles and attached a micro heater to each channel. When the pulse current flows in the heater; the ink around the heater turns into a supercritical

Figure 115.14 Photograph showing an object staying at the middle line of the conveyor. (From Konishi and Fujita, 1994. *IEEE/ASME J. Microelectromechanical Syst.*, 3:54–58. With permission.)

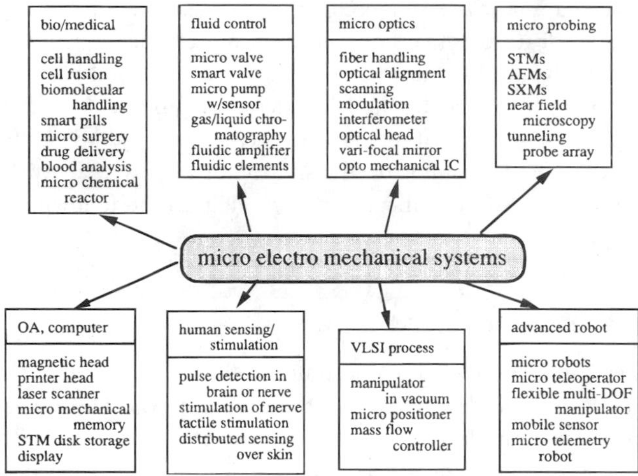

Figure 115.15 Possible applications of micro electro mechanical systems

state and shoots a droplet out from the nozzle. Although there are no moving parts, the heater acts as a microactuator. The printer which utilizes this principle, called a bubble jet printer, has been commercialized and proved to be successful.

Micro Magnetic Head

Micro sliders for read-out can be fabricated by IC-compatible processes. Let us examine the micro system in which the slider is attached to micro flextures and driven by microactuators (Lim et al., 1989, Kogure, 1989). The purposes of the motion are to compensate tracking errors and to avoid crashing. Although large movement such as seeking has to be done by macro structures and actuators, these functions can be miniaturized because of the lighter load. Since the range of movement is limited, the flexible support eliminates friction. Response frequency should be in the order of 10 kHz. If the micro slider is small enough, improved electrostatic actuators will be applicable. Assembly and adjustment are minimized by the pre-assembly capability of micromachining. Small signals associated with the miniaturized head should be amplified by the pre-amplifier located on the same chip. A displacement sensor to detect the gap between the slider and the disk should also be located as close to the slider as possible. The flexture is flexible in driving directions and rigid in undesired directions. The compliance of the flexture should be designed as independently as possible in each direction; it is soft in the moving direction and stiff in other directions.

Electrostatic Handling of Biological Objects

The typical dimensions of biological objects are around 1–10 μm for cells and nanometers in thickness by microns in length for macro molecules. The electric field distribution obtained by microfabricated electrodes can be controlled in the same order of the objects and is suitable for manipulating them (Washizu, 1990; Fuhr et al., 1991). Since the objects are suspended in conductive fluids, the applied voltages are high frequency (more than MHz) alternating voltages. As is the case in the bubble jet printer, the structure does not move but produces a finely determined field around it to actuate the object.

Washizu et al. (1990) developed a cell fusion system using both a micro fluidic system and manipulation with the electric field. Figure 115.16a shows the system. Two types of cells, A and B, are put in the cavities, PA and PB. Each cavity has a piezoelectric pump. The pump pushes the cells into a narrow channel. The channel is so narrow that the cells must proceed one by one to a cell fusion chamber. The cell fusion chamber is shown in Figure 115.16b. Cell-A comes from the left channel and cell-B from the right one. They meet each other at the hole in the wall. Figure 115.16 shows a cell fusion system using a fluid integrated circuit. An electric field is produced through the hole by two electrodes.

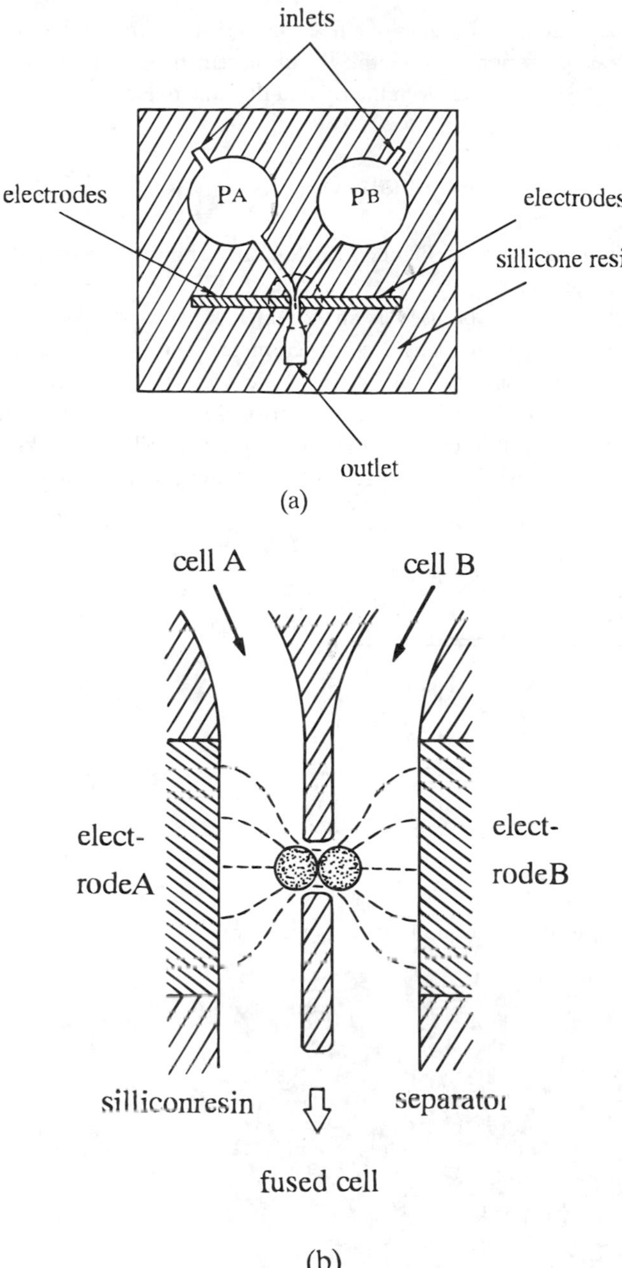

Figure 115.16 A cell fusion system using a fluid integrated circuit. a. The area encircled by a broken line is the cell fusion chamber. PA and PB are piezo-electric micro pumps. b. The field constriction area in the cell fusion chamber.

The cells are attracted to the field constriction area around the hole. They attach together and make a so-called pearl chain. A high but short voltage pulse is applied to the pearl chain from the electrodes in order to fuse the cells. The fused cell is then pushed away by pumps. Based on the same principle, they also made an electric cell conveyer, called a cell shift register, and a cell sorter.

Biological molecules such as DNA or proteins can also be handled by the electric field. For example, the DNA molecule whose normal shape in the water is like a folded string can be stretched by electric fields on the order of 10^6 V/m. The length of DNA molecules is a direct measure of the amount of genetic information in it. Therefore one can determine the amount of information by measuring the length of DNA molecules stretched by the field and stained by fluorescent dyes. Washizu et al. succeeded in orienting DNA molecules along the field. They also align molecules on the microelectrode and cut them at certain lengths by focused UV light. Because the spot size is on the order of 1 μm, it is impossible to cut the molecule with the resolution of base pairs in a DNA molecule held straight by the electric field. They also succeeded in changing the three-dimensional conformation of proteins by the field.

Tunneling Current Detection and Control

Micromachined tunneling units (Kenny et al., 1991; Akamine et al., 1990) and a STM (scanning tunneling microscope) (Yao et al., 1992) have been reported. Some of them require assembly and coarse adjustment of the opposing surface (Kenny et al., 1991; Akamine et al., 1990); while in another the tip is opposed to a protrusion from the substrate (Yao et al., 1992). Kobayashi et al. (1992) reported a lateral tunneling unit (LTU) driven by an electrostatic linear actuator. The lateral configuration of this tunneling unit has the following advantages:

1. Simple surface micromachining process with only one mask.
2. The lateral electrostatic actuator has a large operating range and is easy to control.
3. Integration with other microstructures such as AFM (atomic force microscope) tips.
4. Surfaces of the tip and the opposing wall can be covered by a variety of conductive materials.

The fabrication process of the LTU is a simple one-mask process. We started with a wafer covered with a 2.5 μm-thick oxide layer and a 4 μm-thick polysilicon layer. The polysilicon was patterned by RIE (reactive ion etching) using a nickel mask which was vacuum coated and wet etched. The sacrificial oxide was half etched by straight HF; oxide just beneath the structure remained undissolved. After tungsten was sputtered on to the surface, the structure was released by removing the oxide completely. Relatively large features were not fully undercut and still fixed to the substrate. Figure 115.17 shows the fabricated LTU. The tip and the comb-drive are suspended by four double-folded beams. The distance between the tip and the opposing wall is determined by photolithography to within a few μm; the distance can be covered by the actuator with an applied voltage of 40–100 V. In operation the voltage was gradually increased by the control circuit until the tunneling current (0.1–100 nA) was detected. A small voltage was superimposed to keep the current the same as the reference value.

The result is shown in Figure 115.18. The upper trace is the tunneling current; the lower trace is the applied voltage which followed the reference input of a triangular wave (only the superimposed component is shown). Since this component is much

Figure 115.17 SEM photograph of the LTU

smaller than the DC bias of a few tens of volts, it can be regarded as being proportional to the tip displacement. The current, I, and the distance between the tip and the wall, d, have the following relation:

$$I \sim \exp(-Kd), \text{ where K is a constant.}$$

The upper trace of Figure 115.18 clearly shows the nonlinear dependence.

The LTU can be used as an extremely sensitive displacement sensor, e.g., a very sensitive accelerometer. Another application is the detector for the AFM tip. A piezoresistive cantilever has been proposed for the AFM tip (Tortonese et al., 1991) but the sensitivity is still low. The LTU offers the possibility of making a very sensitive detector. Figure 115.19 shows a SEM photograph

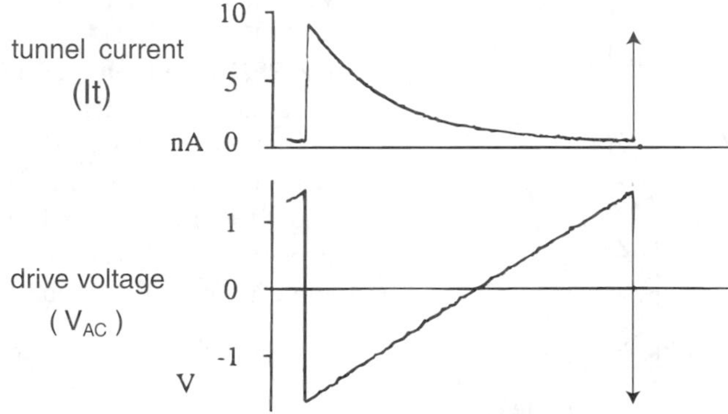

Figure 115.18 The tunneling current (upper trace) and the displacement of the tip (lower trace)

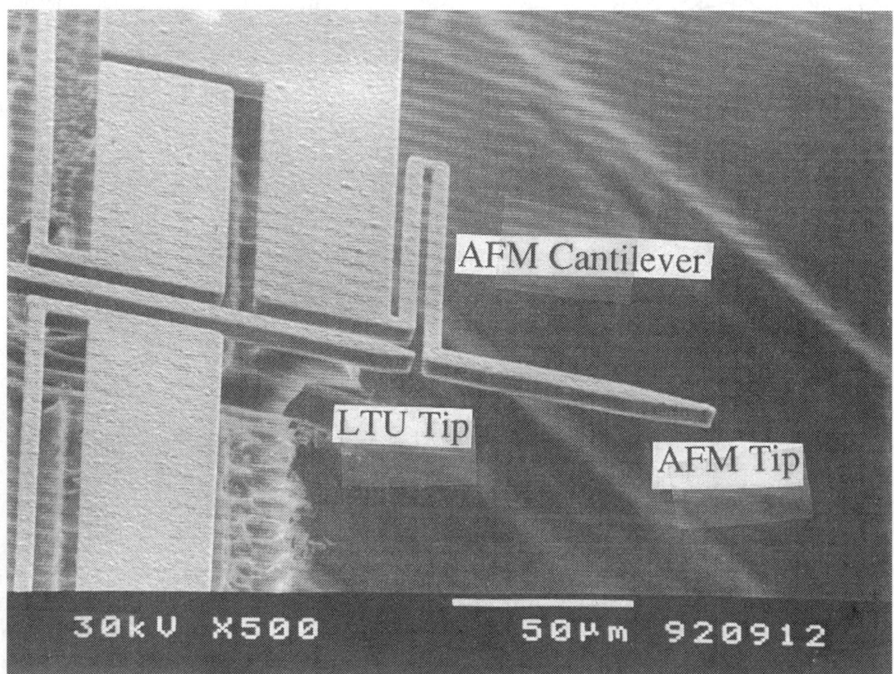

Figure 115.19 SEM photograph of the integrated LTU/AFM tip

of the LTU fabricated together with the suspended sharp tip of the AFM. The integrated LTU/AFM tip allows us to scan the tip on the sample, while the conventional detection scheme using the light lever method only allows the sample to move. We are working towards an experimental verification of the LTU/AFM tip and the development of a process to achieve an overhung (Kim et al., 1992) AFM tip.

115.6 Conclusion

The fabrication technology commonly referred to as IC-compatible micromachining was presented and discussed along with a number of applications. The references quoted in this paper point to the feasibility of both fabrication and operation of these devices. Further development in this area will be driven by specific application areas.

References

Akamine, S., Albrecht, T. R., Zdeblick, M. J., and Quate, C. F., 1990. A planar process for microfabrication of integrated scanning tunneling microscopes, *Sensors and Actuators* A21–A23:964–970.

Ataka, M., Omodaka, A., Takeshima, N., and Fukita, H. 1993. Polyimide bimorph actuators for a ciliary motion system, *IEEE/ASME. J. Microelectromechanical Syst.,* 2(4):146–150.

Bart., S. F., and Lang, J. H. 1989. An analysis of electroquasistatic induction motors, *Sensors and Actuators,* 20:97–106.

Bobbio, S. M., Kellam, M. D., Dudley, B. W., Goodwin-Johnasson, S., Jones, S. K., Jacobson, J. D., Tranjan, F. M., and DuBois, T. D. 1993. Integrated force arrays, *Proc. 6th IEEE Workshop Micro Electro Mechanical Systems,* 149–154, February 7–10, Fort Lauderdale, FL.

Chen, L.-Y., Santos, E. J. P., and MacDonald, N. C. 1993. An isolation technology for joined tungsten MEMS, *Proc. 6th IEEE Workshop Micro Electro Mechanical Systems,* 189–194, February 7–10, Fort Lauderdale, FL.

Fan, L.-S., Tai, Y.-C., and Muller, R. S. 1988. Integrated movable micromechanical structures for sensors and actuators, *IEEE Trans. Electron Devices,* ED-35:724–730.

Fan, L.-S., Tai, Y.-C., and Muller, R. S. 1989. IC-processed electrostatic micromotors, *Sensors and Actuators,* 20:41–48.

Fuhr, G., Hagedorn, R., and Müller, T. 1991. Linear motion of dielectric particles and living cells in microfabricated structures induced by traveling electric fields, *Proc. 4th IEEE Workshop Micro Electro Mechanical Systems,* 259–264, January 30–February 2, Nara, Japan.

Fujita, H. and Gabriel, G. J. 1991. New opportunities for microactuators, *Proc 6th Int. Conf. Solid-State Sensors and Actuators,* 14–20, June 23–27, San Francisco, CA.

Fujita, H. and Omodaka, A. 1988. Fabrication of an electrostatic linear actuator by silicon micromachining, *IEEE Trans. Electron Devices,* ED-38:731–734.

Furuhata, T., Hirano, T., Gabriel, K. J. and Fujita, H. 1992. Sub-micron gaps without submicron lithography, *Proc. IEEE Workshop Micro Electro Mechanical Systems,* 57–62, January 30–February 2, Nara, Japan.

Gabriel, K. J. 1994. Microelectromechanical systems, *J JSME,* 97(905).

Guckel, H., Skrobis, K. J., Christenson, T. R., Klein, J., Han, S., Choi, B., and Lovell, E. G. 1991, Fabrications of assembled micromechanical components via deep x-ray lithography, *Proc. 4th IEEE Workshop Micro Electro Mechanical Systems,* 74–79, January 30–February 2, Nara, Japan.

Hirano, T., Furuhata, T., and Fujita, H. 1995. Dry-released nickel micromotors with low-friction bearing structure, *IEICE Trans. Electron.,* E78-C:132–138.

Hirano, T., Furuhata, T., Gabriel, K. J. and Fujita, H. 1992. Design, fabrication and operation of sub-micron gap comb-drive microactuators, *IEEE/ASME J. Microelectromechanical Syst.,* 1:27–59.

Howe, R. T., Muller, R. S., Gabriel, K. J., and Trimmer, W. S. M. 1991. Silicon micromechanics: sensors and actuators on a chip, *IEEE Spectrum,* June. Pp. 29–35.

Ikuta, K. and Hirowatari, K. 1993. *Proc IEEE Micro Electro Mechanical Systems,* Fort Lauderdale, FL.

Jacobsen, S. C., Price, R. H., Wood, J. E., Rytting, T. H., and Rafaelof, M. 1989. A design overview of an eccentric-motion electrostatic microactuator, *Sensors and Actuators,* 20:1–16.

Jebens, R., Trimmer, W., and Walker, J. 1989. Microactuators for aligning optical fibers, *Sensors and Actuators,* 20:65–73.

Jerman, J. H., Clift, D. J., and Mallinson, S. R. 1990. A miniature Fabry-Perot interferometer with a corrugated diaphragm support, *Tech. Digest IEEE Solid State Sensor and Actuator Workshop,* 140–144, June 4–7, Hilton Head, SC.

Kenny, T. W., Waltman, S. B., Reynolds, J. K., and Kaiser, W. J. 1991. *Appl. Phys. Lett.,* 58:100.

Kim, C. J., Pisano, A. P., and Muller, R. S. 1992. Silicon-process overhanging microgripper, *IEEE/ASME J. Microelectromech. Syst.,* 1:31–36.

Kim, Y.-K., Katsurai, M., and Fujita, H. 1990. Fabrication and testing of a micro superconductive actuator using Miessner effect, *Proc. 3rd IEEE MEMS Workshop,* 61–66, February 11–14, Napa Valley, CA.

Kobayashi, D., Hirano, T., Furuhata, T., and Fujita, H. 1992. An ingegrated lateral tunneling unit, *Proc. 5th IEEE Workshop Micro Electro Mechanical Systems,* 214–219, February 4–7, Travemünde, Germany.

Kogure, K. 1989. Micro-engineering in file memory, *J. JSME,* 92:1056–1060, in Japanese.

Konishi, S. and Fujita, H. 1994. A conveyance system using air flow based on the concept of distributed micro motion systems, *IEEE/ASME J. Microelectromechanical Syst.,* 3:54–58.

Lim, M. G. et al. 1989. Design and fabrication of a linear mocromotor with potential application to magnetic disk file systems, *1989 ASME Winter Annual Meeting,* December, San Francisco, CA.

Mahadevan, R., Mehregany, M., and Gabriel, K. J. 1990. Application of electric microactuators to micromechanics, *Sensors and Actuators* A21–A23:219–225.

Mehregany, M. Gabriel, K. J., and Trimmer, W. S. 1988. Integrated fabrication of polysilicon micro-mechanisms, *IEEE Trans. Electron Devices,* ED-35:719–723.

Mehregany, M., Nagarkar, P., Senturia, S. D. and Lang, J. H. 1990. Operation of microfagricated harmonic and ordinary side-drive motors, *Proc. 3rd IEEE MEMS Workshop,* pp. 1–8, February, Napa Valley, CA.

Menz, W., Bacher, W., Hermening, M., and Michel, A. 1991. The LIGA technique—a novel concept for microstructures and the combination with Si-technologies by injection molding, *Proc. 4th IEEE Workshop Micro Electro Mechanical Systems,* 69–73, January 30–February 2, Nara, Japan.

Minami, M., Kawamura, S., and Esashi, M. 1993. Distributed electrostatic micro actuator (DEMA), Abstract of Late News Papers, *7th Int. Conf. Solid-State Sensors and Actuators,* pp. 2–3, June 7–10, Yokohama, Japan.

Moroney, R. M., White, R. M., and Howe, R. T. 1990. Ultrasonic micromotors: physics and applications, *Proc. 3rd IEEE MEMS Workshop,* 182–187, February 11–14, Napa Valley, CA.

Nagaoka, S., 1991. Micro-magnetic alloy tubes for switching and splicing single-mode fibers, *Proc. 4th IEEE Workshop Micro Electro Mechanical Systems,* 86–91, January 30–February 2, Nara, Japan.

Ohnstain, T., Fukita, T., Ridley, J., and Bonne, U. 1990. Micromachined silicon microvalve, *Proc. 3rd IEEE MEMS Workshop,* pp. 95–98, February, Napa Valley, CA.

Petersen, K. 1982. *Proc. IEEE,* 70:420.

Petersen, K. E. 1977. *Appl. Phys. Let.,* 31(8):521.

Pisano, A. 1989. Resonant-structure micromotors, *Proc. 2nd IEEE Workshop Micro Electro Mechanical Systems,* 44–48, February 20–22, Salt Lake City, UT.

Pister, K., Fearing, R., and How, R. T. 1990. A planar air levitated electrostatic actuator system, *Proc. 3rd IEEE MEMS Workshop,* pp. 67–71, February, Napa Valley, CA.

Pister, K. S. J. 1992. *Dig. IEEE Solid-State Sensor and Actuator Workshop,* p. 136, Hilton Head, SC.

Pister, K. S. J., Judy, M. W., Burgett, S. R., and Fearing, R. S. 1992. Microfabricated hinges, *Sensors and Actuators (A),* 33:249–256.

Riethmüller, W. and Benecke, W. 1988. Thermally excited silicon microactuators, *IEEE Trans. Electron Devices,* ED-35:758–763.

Sakata, M., Hatazawa, Y., Omodaka, A., Kudoh, T., and Fujita, H., 1990. An electrostatic top motor and its characteristics, *Sensors and Actuators,* A21–A23:219–225.

Sawada, R., Tanaka, H., Ohguchi, O., Shimada, J., and Hara, S. 1991. Fabrication of active integrated optical micro-encoder, *Proc. 4th IEEE Workshop Micro Electro Mechanical Systems,* 233–238, January 30–February 2, Nara, Japan.

Schmidt, M., Howe, R. T., Senturia, S., and Haritonidis, J. 1988. Design and calibration of a microfabricated floating-element shear-stress sensor, *IEEE Trans. Electron Devices,* ED-35:750–757.

Shibata, M. et al. 1987. Bubble-jet printing elements, *Proc. 1987 Annual Mtg IEE Japan,* paper S.7-3–4, in Japanese.

Suzuki, K., Shimoyama, I., Miura, H., and Ezura, Y. 1992. Creation of an insect-based microrobot with an external skeleton and elastic joints, *Proc. 5th IEEE Workshop Micro Electro Mechanical Systems,* 190–195, February 4–7, Travemünde, Germany.

Tai, Y.-C. and Muller, R. S. 1989. IC-processed electrostatic synchronous micromotors, *Sensors and Actuators,* 20:49–56.

Takeshima, N. and Fujita, H. 1990a. Design and control of systems with microactuator array, *Proc. IEEE Workshop Advanced Motion Control,* 219–2323, March, Yokohama, Japan.

Takeshima, N., and Fujita, H. 1990b. Design and Control of Systems with Microactuator Array, *Recent Advances in Motion Control,* Ohnishi, K. et al., eds., 125–130, Nikkan Kogyo Shimbun.

Takeshima, N., Gabriel, K. J., Ozaki, M., Takahashi, J., Horiguchi, H., and Fujita, H. 1991. Electrostatic parallelogram actuators, *6th Int. Conf. Solid-State Sensors and Actuators,* 63–66, June 23–27, San Francisco, CA.

Tang, W. C., Nguyen, T.-C. H., Howe, R. T. 1989. Laterally driven polysilicon resonant microstructures, *Sensors and Actuators,* 20:25–32.

Toronese, M. et al. 1991. Atomic force microscope using a piezoresistive cantilever, *Proc. 6th Int. Conf. Solid-State Sensors and Actuators,* 448–451, June 23–27, San Francisco, CA.

Trimmer, W. S. N. 1990. Micromechanical Systems, *Integrated Micro-motion Systems,* Harashima, F., ed., 1–15, Elsvier Science, New York, NY.

Trimmer, W. S. N. and Jebens, R. 1989. Harmonic electrostatic motors, *Sensors and Actuators,* 20:17–24.

Udayakumar, K. R., Bart, S. F., Flynn, A. M., Chen, J., Tavran, L. S., Cross, L. E., Brooks, R. A., and Ehrlich, D. J. 1991. Ferroelectric thin film ultrasonic micromotors, *Proc. 4th IEEE Workshop Micro Electro Mechanical Systems,* 109–113, January 30–February 2, Nara, Japan.

Van De Pol, F. C. M, Wonnink, D. G. J., Elwenspoek, M., and Fluitman, J. H. J. 1989. A thermo-pneumatic actuation principle for a microminiature pump and other micromechanical devices, *Sensors and Actuators,* 17:139–143.

Van De Pol, F. C. M. et al. 1990. Micro Liquid-Handling Devices—A Review, *Micro System Technologies 90,* Riechl, H., ed., p. 793, Springer-Verlag, New York, NY.

Wagner, B. and Benecke 1991, Microfabricated actuator with moving permanent magnet, *Proc. 4th IEEE Workshop Micro Electro Mechanical Systems,* 27–32, January 30–February 2, Nara, Japan.

Walker, J. A., and Gabriel, K. J. 1990. Thin-film processing of TiNi shape memory alloy, *Sensors and Actuators,* A21–A23:243–246.

Washizu, M. 1990. Electrostatic Manipulation of Biological Objects in Microfabricated Structures, *Integrated Micromotion Systems,* Harashima, F., ed., 417–432, Elsevier Science, New York, NY.

Westbérg, H., Boman, M., Johansson, S., and Schweitz, J. 1991. *Proc. 6th Int. Conf. Solid-State Sensors and Actuators,* p. 516, San Francisco, CA.

Yao, J. J., Areny, S. C., and MacDonald, N. C. 1992. Fabrication of high frequency two-dimensional nanoactuators for scanned probe devices, *J. Microelectromechanical Syst.,* 1:14–22.

Zdeblick, M. J., and Angell, J. B. 1987. A microminiature electric-to-fluidic valve, *Tech., Digest, 4th Int. Conf. Solid-State Sensors and Actuators,* 827–829, June 2–5, Tokyo, Japan.

116
Selected Micromachining Fabrication Technologies

A. Bruno Frazier
University of Utah

James Jara-Almonte
SSI Technologies, Inc.

Noel C. MacDonald
Cornell University

M.T.A. Saif
Cornell University

S. A. Miller
Cornell University

Craig R. Friedrich
Louisiana Tech University

Michael J. Vasile
Louisiana Tech University

116.1 Precision Metallic Micro Structures and Micro Molding Technologies 1489
Introduction • Photosensitive Polyimide (PSPI) Process • LIGA-DXRL
116.2 Nanotechnology 1500
Introduction • How Small is a Micrometer? • MEMS Fabrication • MEMS Applications • Conclusions
116.3 Precision Micromachining Technologies 1505
Introduction • Precision Micromachining Techniques • Precision Micromachining Processes

116.1 Precision Metallic Micro Structures and Micro Molding Technologies

A. Bruno Frazier and James Jara-Almonte

Introduction

In addition to the more traditional silicon-based materials used to create micromachined devices (e.g., polysilicon, single crystal silicon, silicon nitride, silicon dioxide), there has been a growing interest in recent years in microstructures fabricated from various elemental metals and metal alloys. Of particular interest to date has been the development of microstructures fabricated using additive metal forming manufacturing processes and micro molding technologies. Some of the elemental metals which have been utilized to realize the metallic microstructures include gold, copper, nickel, iron, silver, and aluminum. The metal alloys that have been utilized thus far include various nickel containing compounds such as nickel-iron, nickel-cobalt, and aluminum containing compounds such as Al-78Zn, LiAlCu, LiAlNi) as well as others on a lesser scale. The choice of metals for the microstructures depends on the application being addressed. For instance, in the case of nickel and nickel containing alloys many times the resulting microstructures are used in magnetic applications (Frazier et al., 1994 Guckel et al., 1994), since nickel and most nickel alloys have desirable magnetic properties. In addition, nickel and some of its alloys are considered good choices for dynamic micro systems due to their superior wear resistant characteristics. In many dynamic micro systems the various components operate in contact with one another, therefore friction and wear properties are of primary interest. In the case of shape memory alloys such as TiNi, the resulting metallic microstructures can be used in micro sensor and actuator applications which utilize this property. Applications to date include precision positioning devices such as fiber optic micro manipulators for fiber switching and maximizing coupling between fibers as well as thermo-sensitive devices and servo actuators (Miyazaki and Nomura, 1994; Ikuta et al., 1994; 1991; Ikuta and Shimizu, 1993; Hirose et al., 1984). In addition to the application areas mentioned above as examples, there are many more applications requiring specific material needs such as bio-compatibility, customized mechanical properties (e.g., residual stress, Young's modulus, hardness, elasticity), and customized electrical properties (e.g., resistivity, high frequency impedance).

Metallic microstructures can be tied to applications through several avenues. First, the microstructures can be used as individual components for applications such as sensors, micro actuators, and miniaturized mechanical systems requiring precision parts.

0-8493-8343-9/97/$0.00+$.50

Second, the microstructures can be components of a more complex micro system containing components of various materials or comprised entirely of metallic components. The micro systems can be manufactured using various system integration methodologies. The most popular methods of producing micro systems are monolithic integration and hybrid packaging. Monolithically integrated systems are those in which all of the micro system components are manufactured on the same substrate using compatible manufacturing processes. Typically these manufacturing technologies are compatible with integrated circuit fabrication and have all the advantages associated with microelectronics technology. An example of a monolithically fabricated system is one in which a sensor is fabricated on the same substrate containing the integrated circuits for signal processing, control and communication. Hybrid packaging technologies have also received a great deal of attention due to the many advantages of the technologies that can be summed up as flexibility. The hybrid technologies that are currently being used include multichip modules, flip-chip bonding, and surface mount packaging. There are many micro systems applications where hybrid technologies are necessary. One of the primary reason for this is process incompatibility between the various technologies used to produce the sensing, micro actuating, and circuit components.

For many of the application areas mentioned above, there is a need for precision metallic microstructures of relatively large thickness (i.e., 10–1000 μm). In general for microstructures over 1–2 microns in thickness, the structure has been realized using additive processing technologies. In most cases, the processing technologies needed to realize the metallic microstructures include two basic processes in addition to other specialized procedures inherent to the particular technologies being used. The two basic processes are a process for creating precision high aspect ratio and/or thick micro molds (typically using polymer materials) and a process for forming the metal into the micro molds. Structures which are relatively thick offer many advantages to micro systems including structural rigidity in actuation systems, high current and flux carrying capability, and the possibility of thick metal sacrificial layers and/or structures for use in the interim stages of a manufacturing process. In addition, processes for the manufacture of thick, high-aspect-ratio (height:width) structures offer the capability of fabricating micromachined devices with compact production of high torque and/or actuation force (Frazier et al., 1991) as well as dominant in-plane (of the substrate) buckling modes for beam type micro structures. Thus, processes for the fabrication of thick, high-aspect-ratio, precision metallic microstructures are of interest.

Several recent micro molding technologies have been developed for the fabrication of thick metallic micromachined devices. Two of the most popular techniques are known as the photosensitive polyimide (PSPI) process (Frazier et al., 1991, 1994) and the LIGA process (Guckel et al., 1994; Becker et al., 1986). Both of these techniques utilize photosensitive polymers as the micro molding material and electroplating to realize the metallic structures. While there are many differences in the two processes, the main differences are in the method of exposing the photosensitive polymers, and the maximum mold thickness available. Polyimides are commercially available materials which are widely used in various aspects of microelectronics. Present applications for polyimide/metal systems include multilevel interconnect technology (e.g., Chakravorty et al., 1989; Milosevic et al., 1988; Moriya et al., 1984) and multiclip packaging (Aclema et al., 1990; Dishon et al., 1989; Rickerl et al., 1987). In addition, UV-exposable, negative-working, photosensitive polyimides which can have spun-on thicknesses in excess of 200 μm in a single coat are now commercially available, thus satisfying the thickness requirement. In addition, many of these systems have extremely sharp sidewalls upon developing, thus allowing the fabrication of relatively high aspect ratio structures. Finally, the additional properties such as compatibility with standard integrated circuit technology (allowing microstructures to be fabricated directly on top of foundry-processed CMOS or other silicon wafers), the option of using the chemically and thermally stable polyimide molding material as an integral part of the micro system (as a dielectric material or structural component), the ability to electroplate in both acidic and alkaline solutions as well as some solvent-based solutions, and the ability to electroplate three dimensionally varying structures using multicoat procedures allows the realization of electroplated microstructures in an inexpensive and manufacturable fashion. Use of photosensitive polyimide in the fabrication of released and non-released micromachined structures made from a variety of metals is discussed in the following section.

One of the most well-known processes for the fabrication of thick and/or high aspect ratio metallic microstructures is the LIGA process (Guckel et al., 1994, 1990; Mehz et al. 1991; Ehrfeld et al., 1987; Becker et al., 1986) (lithography, electroplating, molding). The process, which is detailed later, consists of depositing a thick layer of an X-ray sensitive photoresist, usually based on poly(methyl methacrylate), PMMA, on a metal-coated substrate. The resist is then exposed through an X-ray mask (Menz et al., 1991; Guckel et al., 1989) using a highly collimated and very bright X-ray source, such as a synchrotron. The use of synchrotron-based lithography allows exposure of vertical sidewalls through the thickness of the resist. The resist in the exposed regions is then developed, revealing regions of the underlying metal layer. The photoresist now acts as a 'form' for electroplating-based deposition of a metallic structural material. Once the deposition has been completed, the photoresist is removed, yielding the final electroplated structures. Stationary structures including arrays of pillars and honeycombs have been produced using this method (Ehrfeld et al., 1987; Becker et al., 1986), as well as a variety of sensors (Burbaum et al., 1991; Choi et al., 1991; Mohr et al., 1991a; Guckel et al., 1990b; Ehrfeld et al., 1988). In addition, movable microstructures such as turbines and micro motors have been fabricated using standard sacrificial layer/surface micromachining techniques (Wallrabe et al., 1992; Mohr et al., 1991b, 1990). Analogous methods, such as deep-UV lithography, have been developed for PMMA (Guckel et al., 1990a; Han and Corelli, 1988; Mimura et al., 1978; Lin, 1975), but are limited to single-step maximum thicknesses of approximately 5 μm due to optical absorption of the photoresist. Similar techniques have been developed for conventional thick photoresist materials for inclined microstructures (Beuket et al., 1994). In addition to the use of positive photoresist systems, negative acting photoresist systems have also been used as electroplating molding materials (Mearing, 1986), generally for the printed circuit board industry. Other techniques used to create molds for metallic structures include stenciling used in surface mount technology (Snakamborg, 1983), four level VLSI bipolar metallization designs (Guthrie et al., 1992), photoforming of ultra violet sensitive plastic resins (Takugi and Nakajima, 1993), multi-layer stereo lithography processes for light sensitive plastics (Ikuta et al., 1994; Ikuta and Hirowatari, 1993), laser induced non-planar etching/metallization (Maede et al., 1994), and super-plastic micro forming of metals (Saotome and Inoue, 1994).

Photosensitive Polyimide (PSPI) Process

UV-exposable, negative-working, photosensitive polyimides which can have spun-on thicknesses ranging from 3–150 μm in a single coat depending on the processing conditions are now commercially available, thus allowing the simple fabrication of thick electroplated microstructures. In addition, many of these systems have extremely sharp sidewalls upon developing, thus allowing the fabrication of relatively high aspect ratio structures. The basic process, Figure 116.1, is very similar to ordinary photolithography, with the exception of the large resist thicknesses used. Initially, an electroplating seed layer as well as an adhesion layer (if needed) is deposited on the substrate. This is followed by application of the photosensitive polyimide on top of these layers. The photosensitive polyimide is then soft baked and imaged into the desired pattern using a conventional uv exposure source to form the electroplating mold. Electroplating and (optionally) polyimide stripping are then performed. Electroplated structures of copper, nickel, nickel-iron alloys, gold, silver, aluminum as well as other metals can be realized using this technology.

A typical fabrication process (Frazier and Allen, 1993, 1992) is described in some detail below. Planar silicon, ceramic or compound semiconductor surfaces can be used as the initial substrate upon which a suitable electroplating seed layer of metal is deposited. Photosensitive polyimide (e.g., OCG Microelectronics Probimide 300, 7000 and 7500 series, photo-imageable, thermally-imidizable materials), is then spun on the substrate. The spinning is accomplished in two stages, with a spread stage of 600 rpm for 15 seconds, and a high-speed stage of 1100 rpm for 10 seconds, leading to a film thickness of approximately 40 μm (OCG 349 material). Thinner (or thicker) coats can also be achieved by increasing (decreasing) the speed of the high-speed spin stage. The wafers are then soft baked in a two stage process, 15 minutes at 80°C, then 110°C for 20 minutes to drive off solvent, followed by contact imaging using a standard G-line (436 nm) mask aligner. For a 40 μm thick film, a typical exposure energy of 230 mJ/cm^2 is used. The exposure energy is a linear function of the polyimide thickness for films less than 60 μm. The unexposed areas of the polyimide film are then developed and rinsed, resulting in a polyimide mold through which metal

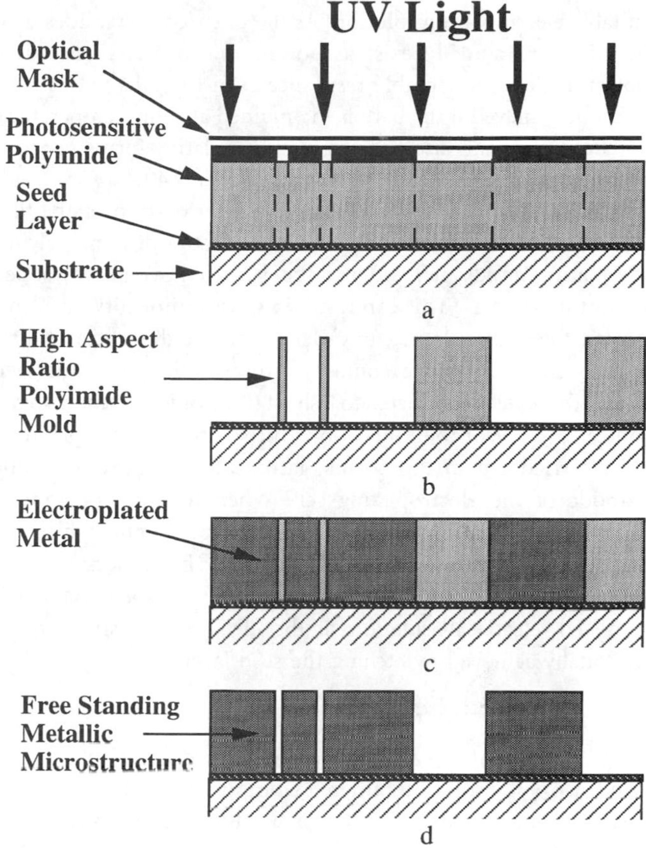

Figure 116.1 The photosensitive polyimide process for the fabrication of high aspect ratio and/or thick metallic microstructures. The process utilizes photosensitive polyimide as a molding material and common cleanroom equipment for processing.

can be electroplated. Either spray or ultrasonic development and rinse have been found to be sufficient for film thicknesses less than 25 μm. Ultrasonic development is needed for films with thicknesses greater than approximately 25 μm. Although no exact measurements of the sidewall profile have been performed, SEM observations indicate nearly vertical sidewalls in both the mold and the electroplated metallic micro structure. Anti-reflection coatings and G-line filters have also been used in standard fashions and have been found to increase the process performance.

The polyimides can be optionally thermally cured (imidized) at this point to achieve increased resistance to solvents and basic solutions. Although the imidization leads to higher resistance of the polyimide to chemical attack, it also results in shrinkage of the film in-plane and orthogonal to the substrate. This shrinkage will substantially decrease the height of the film as well as the compromise the sharpness of the sidewalls. The procedure used in the present studies is: if the polyimide is not to be used as an integral part of the final device, but only as an electroplating mold, do not thermally cure; if the polyimide is to be used as an integral structural part of the final device, then thermal curing is necessary.

To electroplate the microstructures, electrical contact is made to the activated seed layer, and the wafers are immersed in the suitable electroplating solution. As the uncured polyimides used in this work actually exist as polyamic acid esters instead of polyamic acids, acceptable resistance even to baths with pH > 7 can be achieved if the bath is maintained at room temperature. Uncured films can also be used as electroplating forms in acidic plating baths at elevated temperature (typically 40–50°C for acid/copper solutions) with no discernible deterioration of the film patterns. Due to the relatively large aspect ratios of some polyimide plating molds, it is often difficult to remove all entrapped air in the patterns. This can result in gross uniformity problems during the electroplating. In order to remove this entrapped air and to ensure intimate contact of the solution with the seed metal, the wafer is subjected to a short ultrasonic treatment in deionized H_2O prior to immersion in the solution. Electroplating is then carried out in the normal fashion, with the wafer at the cathode of the electroplating cell. When the electroplating is complete, the polyimide is removed. The polyimide, which has not been thermally imidized, can be removed by immersion in hot (70°C) 30 wt% potassium hydroxide solution. If the polyimide is removed, electrical isolation of the electroplated structures is optionally achieved by etching the seed layer.

Nickel Structures

Nickel electroplating was accomplished using a nickel sulfate electrolytic solution buffered with boric acid (Henstock and Spencer-Timms, 1963). The original solution composition used in this study consisted of 200 g/l $NiSO_4 \cdot 6H_2O$, 5 g/l $NiCl_2 \cdot 6H_2O$, 25 g/l H_3BO_3, and 3 g/l saccharin. If nickel-iron alloy (permalloy) is desired, an additional constituent must be added to the electrolytic solution, 8 g/l $FeSO_4 \cdot 7H_2O$. The solution was mixed in a 2000 ml Pyrex beaker and the pH was then adjusted to 2.5 by the addition of H_2SO_4. A 7 cm × 10 cm nickel foil (99.9%) was used as the anode material. The electroplating was carried out with the electrolytic solution at room temperature using a current density of approximately 10 mA/cm^2 corresponding to a plating rate of 0.2–0.3 μm/min. No attempt was made to optimize the above plating conditions.

Using the above concepts and solutions, the micro molds were filled with electroplated nickel. Figure 116.2 shows an electroplated nickel micro gear after the removal of the polyimide mold. As can be seen, an extremely sharp sidewall profile can be obtained. This gear structure is approximately 40 μm in height and 300 μm in diameter with a tooth width of approximately 40 μm.

Figure 116.2 Scanning electron micrograph of a nickel gear structure after removal of the polyimide form, illustrating the extremely sharp sidewall profiles which can be achieved using this process. The gear structure is approximately 40 μm in height and 300 μm in diameter with a tooth width of 40 μm. (*Source*: Burbaum, C., Mohr, J., and Bley, P. 1991. Fabrication of capacitive acceleration sensors by the liga technique. *Sensors and Actuators A*, 25–27:559–563. With permission from Elsevier Science, Amsterdam.)

Copper Structures

The electroplated copper structures were fabricated using both a commercially available acid copper plating solution (LeaRonal Copper Gleam 125S-2) as well as an acid copper solution (Cummings et al., 1982) consisting of 120 g/l $CuSO_4 \cdot 5H_2O$ and 100 g H_2SO_4 per liter of plating solution. A 2000 ml Pyrex beaker was used for both electrolytic copper solutions. During the electroplating process the baths were maintained at 45–50°C, using a current density of approximately 10 mA/cm^2 yielding an approximate deposition rate of 0.2–0.3 μm/min. A 7 cm × 10 cm copper foil was used as the anode material. No attempt was made to optimize the electroplating conditions to minimize surface roughness.

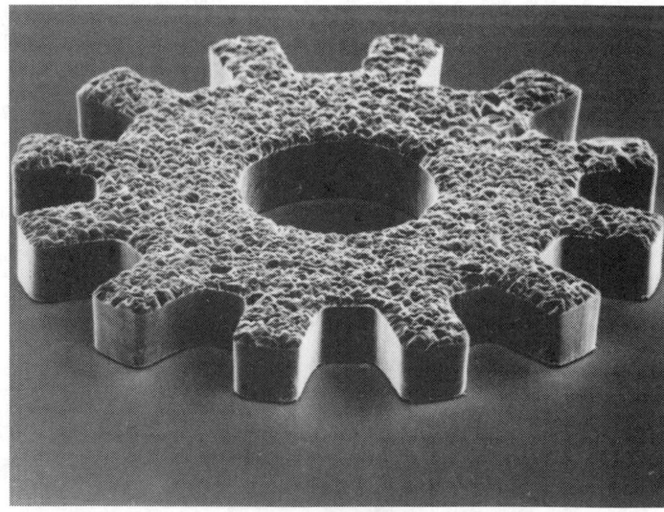

Figure 116.3 Scanning electron micrograph of a copper gear fabricated using the basic process. The gear is approximately 300 μm in diameter and 45 μm in height with a tooth width of 40 μm.

In Figure 116.3 an electroplated copper micro gear is shown. The structure is approximately 45 μm tall and 300 μm in diameter with a tooth width of approximately 40 μm. The surface of the electroplated copper structure is rough due to the fact that the plating process was not optimized. Surface roughness can be reduced by appropriate adjustment of the plating current density and chemical composition of the electrolytic solutions.

Aluminum Structures

Among the electroplated metals that have been demonstrated in micromachining applications, one that shows great potential is aluminum. There are many applications for electroplated aluminum microstructures in which the material properties of aluminum are needed. Electroplated aluminum can be used to develop micromachining processes for high aspect ratio microstructures and to develop micromachining processes to produce controllable thickness gaps between electroplated microstructures (Frazier and Alleen, 1994). One important application of aluminum microstructures is in the fabrication of integrated circuits. Since aluminum is the predominant material used to define the electrical conductors in integrated circuit technology, aluminum microstructures such as high aspect ratio current carrying traces for integrated circuits requiring high input power and/or output power are desirable. In the general case in which conventionally metallized traces are used in combination with electroplated high current carrying traces, using aluminum instead of other metals eliminates the problems associated with intermetallic alloys at junctions between dissimilar metals. In addition, electroplated aluminum can be used to fabricate micro heat sinks. The process required to fabricate these structures from electroplated aluminum is completely compatible with the integrated circuit technology and can be used as a postprocessing step to foundry produce integrated circuits.

To realize micro molded electroplated aluminum microstructures, the micro molding material must have all the desirable properties associated with conventional electroforming materials (e.g., high definition, high aspect ratio molds, simple application and removal processes) as well as have properties desirable for nonconventional electroplating processes (e.g., ability to withstand solvent based electrolytic solutions, chemical resistance to pre-processes required for electroplating). Polyimide materials are shown to be the properties necessary to withstand aluminum electroplating conditions. The polyimide materials that have been used in this work are thermosetting, cross-linked polymers which have characteristics such as high molecular weight and insolubility in ether. The micro molding processes used to create the aluminum microstructures use both photosensitive polyimides and nonphotosensitive polyimides. The nonphotosensitive polyimide process involves the use of plasma or reactive ion etching to produce molds with customized sidewall profiles. This process is covered in detail in reference (Frazier et al., 1994).

The electrolytic solutions used for the deposition of most metals (e.g., copper, gold, nickel, nickel alloys, silver) are water-based solutions. These solutions can either be basic or acidic depending on the metal and the electrolytes used in the bath composition. Aluminum, because it is much more chemically active than hydrogen, probably cannot be electrodeposited from solutions that contain water or any other compound with an acidic hydrogen, for example, acids, alcohols, ammonia, and primary and secondary amines (Harding, 1974). It can be electrodeposited from inorganic and organic fused salt mixtures and from solutions of aluminum compounds in certain organic solvents. The fused salt baths have proven unsuitable for electroforming because of inherent thermal distortion of the deposit due to residual stresses in the films (Schmidt and Hess, 1966). Many other fused salt baths were found to yield only thin or mechanically inferior deposits, and were highly flammable, poisonous, and inconveniently moisture-sensitive (Schmidt and Hess, 1966). For the case of aluminum compounds in organic solvent, the aluminum chloride-lithium aluminum hydride-ethereal solution, originally developed by Conner and Brenner (1958) and Couch and Brenner (1952), yielded satisfactory, low stress deposits. Low volatility, nonflammable derivatives of the aluminum chloride-lithium aluminum hydride-ethereal bath have been developed by replacing part of the ether with a quaternary ammonium salt such as 2-ethoxyethyl trimethylammonium chloride (Begen et al., 1968). Of the several electrodeposition processes that have had some commercial success, only the National Bureau of Standards hydride process has achieved a modest degree of use. Aluminum electroplating has similar commercial applications to conventional electroplated metals such as copper and nickel, but due to the higher cost of the electrolyte (ether vs. water) and the higher initial facility cost (inert atmosphere and safety requirements), it can only compete with conventional processes where the material properties of aluminum are required. The aluminum electroplating solution used in this study was composed of 400 g/l $AlCl_3$ and 15 g/l $LiAlH_4$. Additives can be introduced to the basic solution to reduce grain size and treeing, particularly in thicker deposits. Since the bath contains strong reducing agents and is anhydrous, the electroplating must be carried out in an inert atmosphere. A typical current density used for electroplating is 10–15 mA/cm^2 resulting in an electroplating rate of 0.4 μm/min to 0.8 μm/min.

In the hydride bath, diethyl ether is used as the solvent. For this reason, processes using conventional photoresist materials cannot be utilized for the fabrication of aluminum microstructures. The ether results in swelling and/or decomposition of the polymers used in photoresist systems. Therefore, the photosensitive polyimide process (PSPI) is used as a basis to form the electroplated aluminum microstructures (Frazier and Alhen, 1994).

Figure 116.4 shows an electroplated aluminum gear fabricated using the above process and electroplating bath. The aluminum gear has a thickness of 45 μm, an outer diameter of 250 μm, an inner diameter of 50 μm, and a tooth width of 40 μm. The surface of the microstructure is representative of the grain sizes obtained using the basic aluminum electroplating solution without the addition of additives at a plating rate of 0.4 μm/min.

In addition to the use of aluminum as a material for the fabrication of microstructures, it can also be used as a processing tool to develop fabrication technologies. Two examples of the use of aluminum as a processing tool include the use of electroplated aluminum as a means of producing high aspect ratio metallic microstructures and for producing controllable small gaps in

Figure 116.4 An aluminum electroplated gear structure fabricated using the basic process and the hydride electroplating solution. The gear has an outside diameter of 250 μm, an inside diameter of 50 μm, a height of 45 μm, and a tooth width of 40 μm.

Figure 116.5 An interlocking gear configuration constructed of four copper structures which have been completely released from the substrate. The lower-left and upper-right structures have been turned over revealing the original copper seed layer used for the electroplating process. The gear structures are 300 μm in diameter and 45 μm in height.

metallic microdevices. These processes are outlined in the sections on the extended PSPI process and the controlled gap process.

Surface Micromachined Structures

To achieve electroplated microactuators, provision must be made for release of the electroplated structures or parts of the structures. This release has been achieved in the LIGA process using titanium as a sacrificial layer (Burbaum et al., 1991; Mohr et al., 1991, 1990) as well as special forms of polyimide (Choi et al., 1991; Guckel et al., 1991a/b). Released structures can also be achieved using the polyimide processes using a wide variety of materials as the underlying sacrificial layer. In most applications of this process, the main criterion for determining if a material can be used as a sacrificial layer is simply whether it can be preferentially etched with respect to the electroplated metal. If the sacrificial layer can not be utilized as a seed layer for the electroplated metal, then a thin seed layer can be deposited before the photosensitive polyimide processing.

Fabrication of lifted off micromachined gears can be achieved by applying the basic process to a substrate which contains a chromium release layer underneath of the electroplating seed layer. Figure 116.5 shows four copper structures which have been released and subsequently mechanically positioned into an interlocking gear configuration. The lower-left and upper-right structures have been turned over revealing the original copper seed layer used for the electroplating process. The lateral underetch rate of the chromium sacrificial layer was observed to be approximately 25–35 μm/hr in $HCl:H_2O$ 1:1. Negligible attack on the copper gears was observed during the chromium underetch. The use of this process to create released nickel structures was also achieved.

It has also been possible to achieve selective release (as opposed to blanket release) of electroplated structures fabricated using polyimide electroplated forms. For example, micro motor structures involving electroplated copper and nickel can be achieved by arranging the seed and sacrificial layers such that rotor structures can be selectively released without simultaneously releasing stator structures. Full details of this process have been discussed in Frazier et al. (1991).

Assembled Structures

Assembly of structures fabricated using the photosensitive polyimide process has been realized by using sacrificial layer surface micromachining techniques. Using optical subtraction strategies, gaps in micromachined structures can be greatly reduced. For example, a gear with a hole inner diameter of 64 μm can be fabricated independently of a pin with an outer diameter of 60 μm. These two structures can then be assembled yielding a gear/pin combination with a gap of less than 2 μm. Figure 116.6 shows an assembled electrostatic micro motor fabricated using the PSPI process. In this case, the structure is 50 μm thick with nickel as the structural material for the rotor, stator and pin. The motor was fabricated by forming the rotor on one substrate and the stator/pin on another substrate. The rotor was released from the substrate using a sacrificial layer of physical vapor deposited chromium located under the electroplated structure and mechanically placed on the pin using post assembly techniques. The resulting gap between the rotor and pin in the assembled structure is approximately 2 μm. A similar procedure has been developed for the three dimensional integration of LIGA microstructures (Gockel et al., 1991a).

Figure 116.6 Side-driven electrostatic motor fabricated using post assembly techniques. The design parameters are: number of stator terminals = 12, number of rotor terminals = 8, arc angle for rotor and stator terminals = 38°, pin-rotor gap = 2 μm, and rotor-stator gap = 4 μm. The height of the motor is approximately 50 μm and the diameter of the rotor is 200 μm.

Vertically Integrated Structures

Consider an application where it is desirable to have several projected structures vertically integrated ("stacked") and attached by means of electroplating to form one continuous structure. An example of such an application might be the attachment of a vertical rotor shaft to an electroplated motor to vertically couple mechanical power out of the motor. One way to achieve this vertical integration is to cast a first layer of polyimide, pattern, and electroplate, yielding a metal structure imbedded in the polyimide (as in the basic PSPI process). If the plating is carefully controlled so that the metal structure is approximately coplanar with the resist, a second layer of resist can be spun, a different pattern exposed, and a second metal structure plated to yield a single continuous metal structure with three dimensional variation. The use of photosensitive polyimide in this application is ideal since the photo-crosslinking and/or cure of the first layer of the polyimide induces sufficient solvent stability in the first layer that a second layer can be spun on without dissolving the first. Thus, vertically integrated structures can be achieved using the polyimide process. It should be noted that at all times, lithography is done on surfaces which are nearly planar. In theory, if the second pattern is identical to the first and well-aligned, a continuous projection of the original structure to high aspect ratios can be achieved.

This effect is illustrated by vertically integrated "platform-pin" structures in which an electroplated 'pin' of approximately 1:1 aspect ratio is fabricated on top of a relatively flat and thin electroplated 'platform' using a multilayer polyimide process. The structures described here were all fabricated from nickel, although nickel-on-copper structures have also been fabricated. Figure 116.7 shows a scanning electron micrograph of a vertically integrated "platform-pin" structure. The height of the pin is approximately 45 μm, and the aspect ratio of the pin is approximately one. There appears to be no theoretical limitations on the number of layers which can be applied in this fashion, although no more than two were fabricated in this initial work. The inversion of this structure, i.e., with the "small" (in diameter) component on the bottom and the "large" (in diameter) component on top has also been fabricated using this technique. To realize overhanging structures such as the inverted "platform-pin" structure the complete metal system must be deposited prior to the application of the second polyimide layer.

Extended PSPI Process

Metallic microstructures with aspect ratios as large as 20:1 can be manufactured using an extension of the basic PSPI process. This process differs from other available techniques in that electroplated aluminum micro molds have been used to form the metallic microstructures instead of polymer molding materials. Figure 116.8 outlines the process used to create the high aspect ratio metallic microstructures. The process could be referred to as an inversion process in which the final metallic microstructures are of the same dimensions as the polyimide used as the initial molding material in the process. To outline the extended PSPI

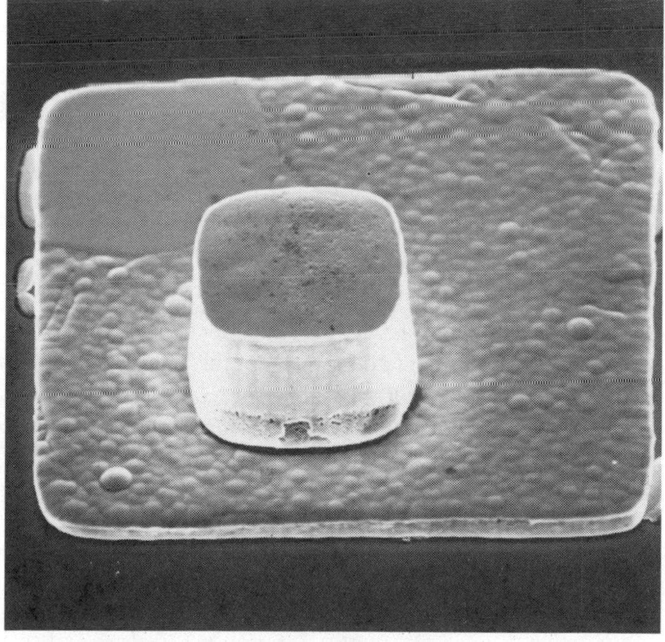

Figure 116.7 A vertically integrated electroplated nickel 'pin-platform' structure fabricated utilizing a multi-coat polyimide procedure. The platform is 120 μm on a side and has a height of 10 μm. The pin is 45 μm in diameter and has a height of 45 μm.

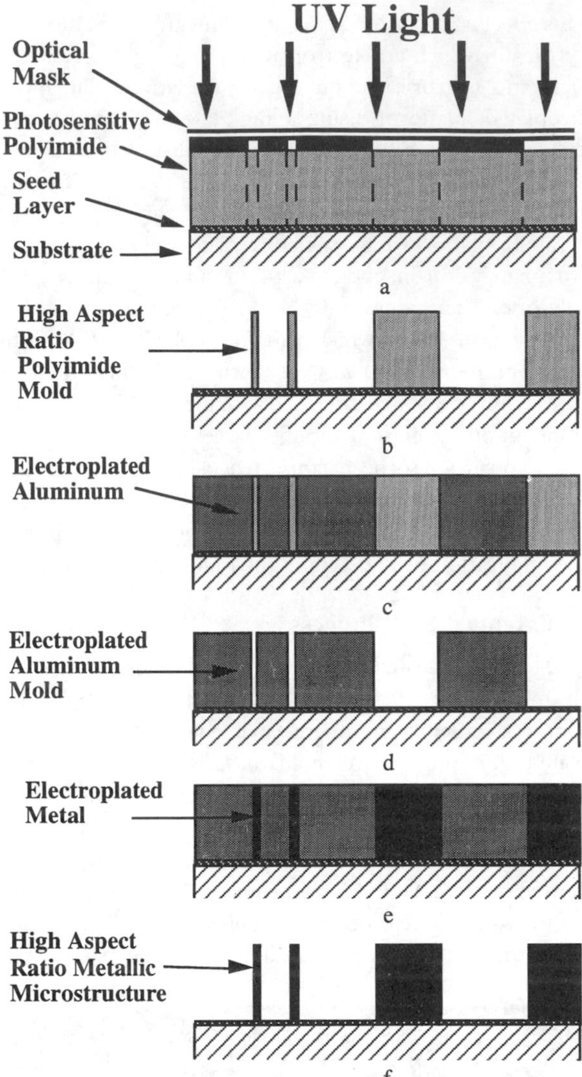

Figure 116.8 The extended PSPI process for the fabrication of high aspect ratio metallic microstructures. Fabrication steps 1a–1d are required for the basic PSPI process. Steps 1e–1f utilize electroplated aluminum as a molding material to form high aspect ratio micro structures.

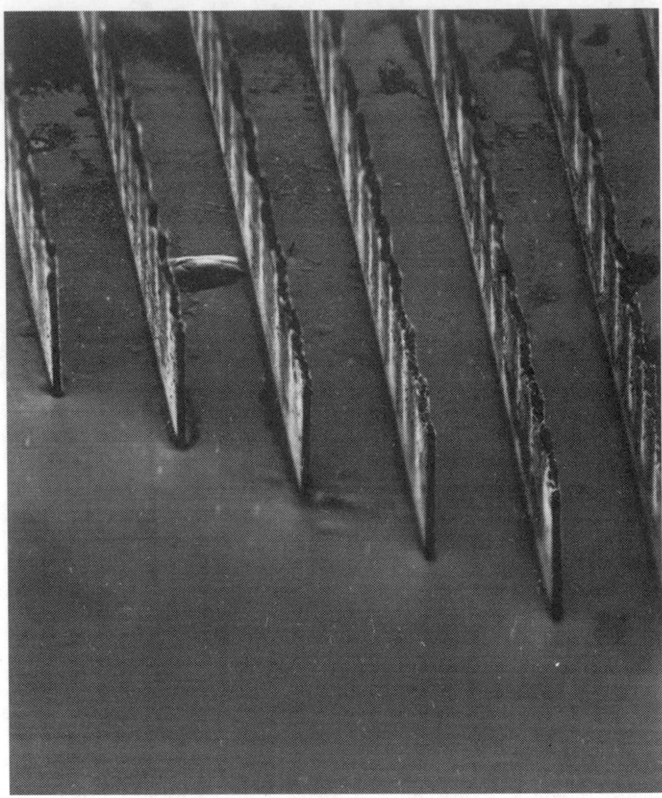

Figure 116.9 An array of high aspect ratio copper beams fabricated using the extended PSPI process. The beams have a height of 42 μm and a width of 2 μm for an aspect ratio of 21:1.

process, fabrication begins with the basic process using photosensitive polyimide as the molding material through which aluminum is electroplated. After the aluminum has been electroplated to the top of the polyimide molds, the polyimide is removed leaving free standing aluminum microstructures (or micro molds in this case). After the polyimide mold is removed from the substrate, the electroplating seed layer is exposed in the regions initially covered by polyimide in the basic PSPI process. At this point in the process, Figure 116.8d, the electroplated aluminum can be used as a mold for further electrodeposition of other metals including copper and nickel. The use of the extended process for the realization of high aspect ratio metallic microstructures is demonstrated in Figure 116.9. In this case, beams of electroplated copper are formed using electroplated aluminum molds. The beams are 42 μm in height and have a width of 2 μm for an aspect ratio of 21:1. Other advantages of this process include the ability to fabricate metallic bimorphs for high current carrying applications, as well as using the electroplated aluminum as a thick sacrificial release layer for surface micromachining applications.

Controlled Gap Process

Electroplated aluminum can be used as a processing tool for the fabrication of controlled thickness gaps between metallic microstructures. In this case, the electroplated aluminum is used as a sacrificial layer between microstructures of metal other than aluminum. The end result is the development of a controllable small gap process for metallic microstructures. Development of a silicon based process for obtaining small gaps has already been reported (Furuhata et al., 1991). Various techniques have been reported including using sacrificial polysilicon, silicon dioxide, and silicon nitride layers. In this process, shown in Figure 116.10, metallic microstructures (other than aluminum) are fabricated using the basic PSPI process. The free standing microstructures are then conformally electroplated with aluminum, thus covering the entire microstructure with a thin layer of aluminum. The thickness of the aluminum "shell" is controlled by the electroplating current density and time duration of the electroplating cycle. The aluminum plating is followed by a second electroplating cycle in which a metal other than aluminum is deposited in the regions surrounding the existing structure. The second electroplating step uses the original seed layer as an electroplating

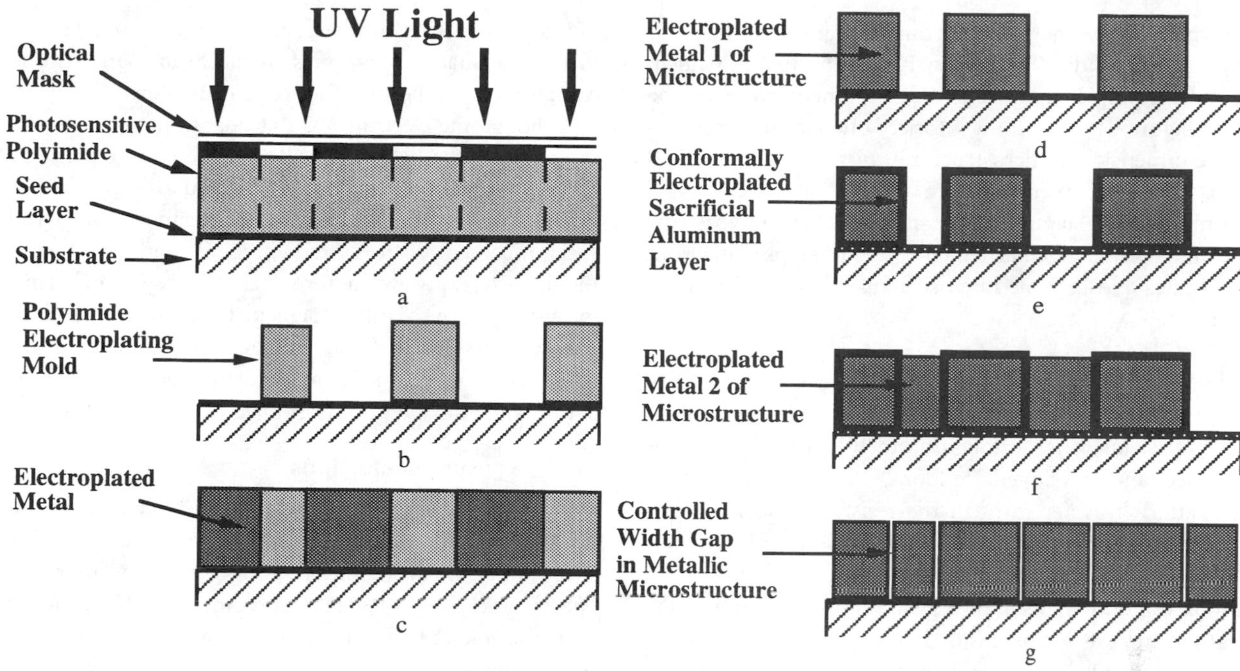

Figure 116.10 The process developed for the realization of controlled gaps between metallic microsystem components. Gaps of < 1 μm to > 10 μm can be realized with sub-micron precision. The process can be extended to create metallic shell structures for packaging applications.

base, and as in the case of the process for the fabrication of high aspect ratio microstructures, the metal deposited during the second cycle does not plate onto the aluminum encompassing the initial microstructure. The second metal is deposited to the desired height with respect to the initial microstructure. Figure 116.11 demonstrates the use of aluminum in the controllable gap process. In this figure, the initial microstructures of copper with the thin film of electroplated aluminum are shown on either side of the metal (copper) deposited second. In the next step the aluminum is removed using a HF solution, leaving a small gap between the copper microstructures. The figure shows a controlled gap of 2 μm between copper structures which are 60 μm and 55 μm in height for a gap aspect ratio of 27.5:1.

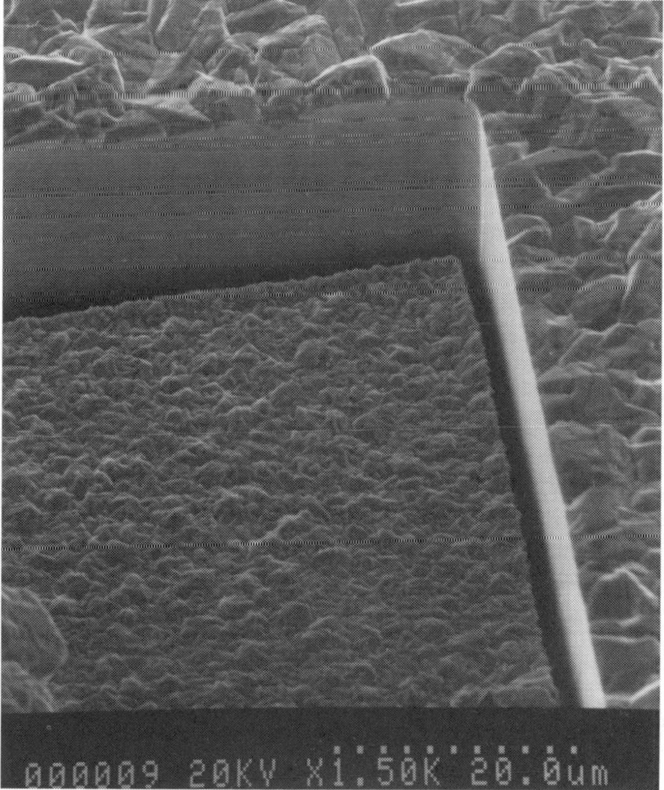

Figure 116.11 A high aspect ratio gap produced between two metallic micro structures using the controlled gap process. The gap is 2 μm and the structural heights are 60 and 55 μm for a gap aspect ratio of 27.5:1.

LIGA-DXRL

The acronym LIGA has its origin in the German names for the three main processes, lithography (*LI*thographie), electroforming (*G*alvanoformung), and molding (*A*bformung) (Krinsky et al., 1993). The commonly used English acronym is DXRL which stands for deep X-ray lithography. As both acronyms imply, the LIGA process begins with deep X-ray lithography using a thick resist layer and highly collimated X-rays. This process is a simple shadow printing process which renders areas of exposed and unexposed resist. This is followed by a removal of the exposed resist with a chemical developer and a backfill of the opened areas with metal by electroforming. Finally, the metal electroformed structure can be removed from the polymer resist and used to replicate microstructures by stamping, injection molding, or reaction injection molding.

The LIGA process sequence need not be followed exactly as

the acronym implies. The sequence can be stopped at any point after exposure depending on the final structure to be attained. For example, multiple exposure and development cycles can be performed before electroforming is carried out to achieve complicated three dimensional microstructures. In some cases the final product is the metal microstructure and it is not necessary to perform molding or stamping. It is also possible to bypass the electroforming step altogether. Much of the developmental work in micro-optics and ceramic microstructures bypasses the electroforming step.

Light Source

Typical resist thicknesses used in LIGA are in the order of several hundred microns, although up to 10 cm thick resist has been exposed and developed (Siddons, 1994). To achieve an accurate pattern transfer through the resist thickness, intense, highly collimated X-rays are needed. Typically, a 1Å to 10Å critical wavelength light source is used for exposures. The light source must also have minimal beam divergence to prevent through the thickness Fresnel diffraction distortions in the resist. For these and other reasons LIGA exposures are carried out with synchrotron radiation.

The light source for LIGA is generated by accelerating electrons or positrons. These are stored in storage rings which bend the beams using electromagnets. The beam is extracted from the bending magnets and directed to the mask and resist system. For example, at the Center for Advanced Microstructures and Devices (CAMD) in Baton Rouge, Louisiana, electrons are accelerated in two LINAC's in series. The electrons are injected into a storage ring where the energy of the electrons is ramped up slowly in an ultra-high-vacuum environment, approximately 10^{-9} Torr, as the electrons travel in a pseudo-circular orbit in the storage ring. The bending magnets, also called dipoles, have a radius of 3 m and have slits on the outside wall to allow extraction of the energy beam.

The emanating light leaves the storage ring bending magnet in a tangential direction. Transformation of the energy from a relativistic reference frame to an inertial reference frame shows that the beam is not fully collimated, but it experiences a natural divergence. The light travels from the bending magnet to the target in a narrow cone. The distribution of light within this cone can be approximated by a Gaussian function and leads to an estimate for the natural divergence of the X-rays, $\sigma_{nat} = mc;s2/E$ (Siddons, 1994), where m is the rest mass of an electron, c is the speed of light, and E is the energy of the electron. Currently, most LIGA efforts are performed at light sources with energies between 1.2 and 2.5 GeV, leading to typical natural divergence of 0.37 to .2 mrad.

The natural divergence is a theoretical quantity that does not take into account the actual path followed by the particles, as the vacuum in the beam line is not perfect. This effect can be incorporated into a term σ_p, yielding a total divergence, $\sigma_{\text{tot}} = \sqrt{\sigma_{\text{nat}}^2 + \sigma_p^2}$. Typical values for the total divergence, assuming small beam sizes, are between 0.3 and 0.5 mrad. This means that at ten meters from the source, an infinitesimal opening in the bending magnet will lead to a Gaussian beam height in the order of 3 to 5 mm. This approximation does not include effects due to beam emitance nor optics in the beam path. The latter is complicated and beyond the scope of this text.

Although the electrons exhibit quantum energy levels in the storage ring, the light received from the storage ring exhibits a continuous spectral distribution. Since the shape of the spectral distribution function is characteristic for most storage rings, the spectral content of the stored energy is represented by a scalar, the characteristic wavelength, $\lambda_c = 18.6/BE^2$, where λ_c is measured in Angstroms (Å), B is the magnetic field intensity of the bending magnet measured in Tesla (T), and E is the energy of the light, measured in GeV. For example, for the DCI storage ring, $\lambda_c =$ 3.4 Å, while for the CAMD storage ring, $\lambda_c =$ 10 Å.

Exposure Conditions

The light is carried through ultra-high-vacuum tubes call beam lines to the exposure's site. It is common to condition the spectrum of the light either by adding optics or inserting filters. For instance, the XRLM1 beam line at CAMD includes an aspheric, gold coated, silica substrate, focusing mirror. This mirror is supposed to focus the Gaussian beam height to 2 mm at the substrate, increasing the incident flux. This increased flux is accompanied by a filtering of some of the intense, short wave X-rays. This mirror can be moved out of the beam path for without-optics exposures. Insertion filters are used to tailor the incident spectrum by removing the low energy radiation. For common LIGA exposures, this low energy radiation does not contribute significantly to dose deposition through the depth of the resist. Instead it contributes to heat generation near the resist surface, and this can lead to undesirable thermal deformations.

LIGA exposure stations are located at the end of the beam lines. The resist-substrate and mask are positioned in the station and held as parallel as possible in a mounting jig. In normal exposures the perpendicularity of the mask-resist assembly to the incident beam will have a significant effect on the quality of the exposures. Exposure station designs are varied; some expose in air, others expose with a helium flush, and the newer designs expose in low-pressure helium environment.

The beam width arriving at the exposure station is a function of both the total divergence, slit opening at the bending magnet, and beam line parameters. Typical beam widths used in LIGA vary from 30 to 80 mm. Thus, the available beam area in LIGA is insufficient to expose an entire silicon wafer area at the same time. Several mechanisms have been tried to overcome this deficiency: step-and-repeat exposures, scanning the mask-resist assembly, scanning the beam. Scanning the mask-resist assembly past the incident beam is easily implemented, and is the more commonly used method by different groups working in LIGA around the world.

In addition to increasing the effective exposure area, scanning can mitigate the deformations caused by temperature gradients at the exposure boundary. The incident light absorbed in the resist generates heat, increasing the local temperature, while the unexposed resist does not experience any heat generation. During scanning exposures, the heat generation will be on and off as a region comes in and out of the beam. This tends to diminish the localized heating effects. In contrast, in a standing exposure, the heat generation in a region is on during the entire exposure, and the steady state temperature reached will be higher than with scanning.

The preferred resist for deep X-ray exposures is a positive tone resist, polymethylmethacrylate (PMMA). PMMA requires a minimum dose of approximately 4 kJ/cm3 to develop out the exposed material, making it a rather insensitive material. The dose below the absorber material, in the shadow, should not exceed 0.1 kJ/cm3 to prevent any noticeable change to the surface of the PMMA. A dose of more than 20 kJ/cm3 can cause the exposed resist to bubble, making this the maximum desirable dose. Thus, the maximum-to-minimum dose ratio of 5 is one of the constraints that determines the desirable spectrum wavelength region and the maximum resist thickness that can be exposed.

Additionally, two independent phenomena are used to select the wavelength region used for LIGA. First, the Fresnel diffraction at the absorber edge will affect the accuracy of the transferred pattern. For monochromatic light, one can estimate the lateral imperfection by assuming that it corresponds to the location of the first maximum of the Fresnel fringe, $\delta = (3/2)\sqrt{\lambda\ z/2}$, where λ is the wavelength z is the distance from the surface of the resist, and is the distance from the absorber edge. Second, the interaction between absorbed X-rays and the resist will release photoelectrons. Some of these electrons will be emitted in a lateral direction leading to imperfections. The lateral penetration, thus the imperfection, can be estimated by assuming that the maximum of the Grün range corresponds to the maximum lateral penetration, $\delta_G = 56\lambda^{-1.75}$ Therefore, the total lateral deviation can be approximated by $\delta_{tot} = \delta + \delta_G$.

The selection of an appropriate wavelength becomes a compromise between minimizing δ and δ_G. δ is minimized at lower wavelengths, more energetic X-rays, while δ_G, is minimized by selecting larger wavelengths, less energetic X-rays. Typically, the range of wavelengths where a trade-off can occur is considered to be $\lambda_c \in [2,10]$ Å. The lower end of this range is favored in LIGA, as this leads to shorter exposure times. The dose is related to the time integral of the power, *P*, arriving at the resist. *P* can be approximated by the relation $P \cong I/E\lambda_c^2$, where *I* is the current. Thus, the dose deposited in the resist is inversely related to the square of the critical wavelength. Shorter wavelengths will yield shorter exposure times.

Working and Intermediate Masks

The intermediate mask is used solely to generate the working mask, while the working mask is used in the LIGA exposures. Intermediate masks have absorber thicknesses between 1 and 3μm, while working masks have absorber thicknesses between 10 μm and 40 μm. There are many ways to generate both working and intermediate masks. The choice of fabrication processes is normally dictated by the desired dimensional accuracy and the minimum feature size, as well as economic considerations.

The intermediate and working masks used in LIGA consist of a membrane held taut by a supporting frame. Absorber structures built on top of the membrane determine the pattern to be transferred. Currently, there is work toward standardization of mask geometries, but a mask standard is not in place. While diamond exhibits ideal characteristics for membrane material, its cost is prohibitive and different membrane materials are used by LIGA groups around the world. Other membrane materials used include silicon, silicon nitride, boron nitride, titanium, beryllium, and polyimide. Titanium foils can exhibit rather large thermal distortions, and should be used with caution. Silicon nitride and boron nitride exhibit lower X-ray stability, and polyimide is used only for developmental work when the life of the mask is limited.

Absorber materials are chosen for their high absorption coefficient in the spectrum of interest in LIGA, $\lambda_c \in [2,10]$ Å. Two metals with high Z that meet this criterion are gold and tungsten. Tungsten absorber structures are generated by anisotropic RIE etching. As gold electroplating is a well established technology, this material is used by most LIGA groups. Electroformed gold absorber structures have low internal stress and fine grain structure. Both of these qualities can be achieved with low-stress gold electroplated gold. Additionally, to maintain uniformity, the gold surface roughness is kept below 5% of the absorber height, and the absorber walls are maintained as vertical as possible. The latter requirement is not difficult to achieve when an intermediate mask is used since the intermediate mask pattern is transferred to the working mask with X-rays.

Resist

Most LIGA groups use PMMA as the resist of choice, both in monomer chain and crosslinked forms. The resist films can be bought from suppliers, as well as produced by casting or pressing. The standard resist has a glass transition temperature, Tg, of approximately 116°C, however, Tg can drop to approximately 75°C depending on the polymerization conditions. PMMA is somewhat insensitive to X-rays, and work to develop more sensitive resists is being carried out at several laboratories.

PMMA is a positive tone resist. The areas exposed to X-rays are damaged by chain scissions. The broken chains can be made soluble in a developer and easily washed away. The scissions process begins with a homolytic scission of the main chain leading to two neighboring radicals. Unfortunately, the two neighboring radicals have a tendency to recombine. The scission-recombination reactions generate heat, and this is used to explain the low sensitivity of PMMA to X-rays. Another hypothesis for the low sensitivity involves the side scissions of the side groups (ester- and a-methyl) that produce gas products.

After exposure, the broken chains are made soluble with different developer formulations, consisting mainly of solvents. The most often reported developer consists of 60% 2-(2-butoxyethoxy)ethanol, 20% tetrahydor-1,4-oxazine, 15% water, and 5% 2-aminoethanol. In general the development process must achieve a balance between the deposited dose in the resist, the time needed to dissolve all cut chains, and the protection of unexposed resist regions from attack by the developer. Furthermore, for small minimum feature sizes and for tight, tall structures, the development process tends to be diffusion limited. The dissolved residues must be removed while at the same time fresh developer must be brought into small regions.

Diffusion limited development can be enhanced with the use

of agitation. The transport velocity and direction can be controlled to a certain extent by the use of megasonic excitation systems. The drawback associated with this method is that if the standing waves in the development bath are setup incorrectly, fine resist structures may be destroyed.

Electroforming

Electroforming occurs in several of the LIGA steps, intermediate and working mask fabrication, and final metallic structures. The masks require gold electroplating. Several LIGA groups use commercially available, low stress, gold electrolytes. For instance, low stress potassium sulfite baths with a Ti/Pt anode are readily available. The gold content of these baths is usually 8–12 gr/lt, and a 0.2–0.4 A/dm^2 current density is used. Care must be used when computing the current density, as the small dimensions of the LIGA structures can cause large variations of the current density.

Typical LIGA structures are rendered in nickel. This metal can be electroplated from several electrolytes. One of these is nickel sulfamate, which can be obtained commercially. Typically, the metal concentration is kept at 105–110 gr/lt, while the current density used is 0.1–2 A/dm^2. Hydrogen yield can be a problem in nickel electroforming, as bubbles can become attached to the surface impeding the migration of fresh electrolyte. In this case, the hydrogen yield can be mitigated by reducing the current density, and surfactants can be added to release the hydrogen bubbles. One drawback of low current densities is long electroplating times, sometimes in the order of days for structures greater than 400 μm high.

Unlike regular electroplating, electroforming of microstructures requires a few additional considerations. Wetting agents are helpful in allowing the electrolyte to reach the tight spaces of high-aspect ratio structures. Electrolytes need to be filtered to maintain cleanliness. Any contaminant in the electrolyte has the potential to impede the flow of electrolyte into tight resist features.

Molding with Nickel Microstructures

Molding refers to the filling of the microstructure cavities with a molten polymer, then separating the solidified plastic parts from the metallic mold. Both of these processes are critical to the successful mass reproduction of the mold features and have been addressed to some extent by different techniques.

Filling of micro cavities is accomplished through injection molding, reaction injection molding, and stamping, also called embossing. In injection molding, polymer pellets are heated past the Tg of the polymer until it has a syrup-like consistency. The molten polymer is then pressurized to several hundred atmospheres mechanically and injected into a network of runners that deliver the polymer to the microstructure mold. Once the mold is filled, it is cooled and the solidified plastic product is released. Reaction injection molding is a similar process that operates at a much lower pressure. Two streams of liquid material are brought separately into the mold where they come in contact and a chemical reaction takes place. The products of this reaction are crosslinked polymers which can then be released from the mold.

Stamping is achieved by heating the polymer sheet past the Tg and bringing the mold insert into the plastic sheet with sufficient force to deform the material, filling all the mold features. This process can be considerably enhanced when it is assisted by a vacuum environment.

Conventional injection molds are made with a slight taper from 87° to 89° to aid release. LIGA molds exhibit nearly vertical walls, requiring careful planning of the polymer microstructure separation from the mold. This is primarily a function of the geometry of the microstructure and mold. The polymer part and nickel mold will exhibit a differential rate of contraction when the mold is cooled. The stress exerted on the polymer structure during release can be substantially increased if a small feature is surrounded by a metal mold; for example, a small plastic column can become trapped as the surrounding nickel mold contracts on it.

116.2 Nanotechnology

Noel C. MacDonald, M. T. A. Saif, and S. A. Miller

Introduction

The evolution of microelectronics has revolutionized electronic computing by reducing the size of the computers and increasing their speed. A computer in the 1950s used thousands of vacuum tubes, weighed several tons, and occupied a large room (Hayes, 1988). Today, an equally capable computer occupies a desktop. It has thousands of transistors assembled in a miniature space. The relatively new technology, named Micro Electro Mechanical Systems (MEMS), is a natural progression of microelectronics. Using the fabrication techniques developed by the microelectronics industry, the size of many conventional *mechanical* systems has been reduced to micrometer-scale dimensions. The reduction in size not only reduces the volume, power consumption, and weight of the mechanical systems, but also produces higher operating performance and allows them to be used in previously inaccessible regions. Because of their similar fabrication processes, MEMS are easily integrated with microelectronics to allow communication with the physical world and render intelligence to the mechanical system. For example, a MEMS pressure sensor records the ambient pressure by the deformation of a thin miniature diaphragm. A microprocessor records the pressure and may activate an actuator if the pressure exceeds a certain critical value. The processor, the diaphragm and the actuator are all on the same chip, designed and fabricated together. As in case of microelectronics, MEMS are batch fabricated. Thousands of components are built at the same time with the same effort as that of one component. The reason is that the fabrication is based on photolithography. To draw an analogy, the effort in taking the picture of one person is same as that of several people—the number of people in the photograph does not matter. Batch fabrication reduces the cost and motivates engineers to shift the emphasis from the design of individual components to the design of complex interconnections between the many components that constitute the system.

How Small is a Micrometer?

An important feature of MEMS is their size, which is on the order of micrometers (μm). A physical intuition for the size scale is obtained from Figure 116.12 which is drawn to scale. For example, the diameter of a human hair varies between 50 to 150 μm. A red blood cell is 7.5 μm and white blood cell is 13 μm in diameter. A MEMS device can be small enough to be engulfed by a white blood cell. However, the overall size of a MEMS device can be large—a few millimeters to a centimeter. The micrometer scale minimum feature size is what makes it a MEMS device.

MEMS Fabrication

The fabrication of MEMS devices is based on the technology that has been developed for microelectronics over the last few decades. The primary ingredients of the MEMS fabrication technology are

1. Lithography.
2. Thin film deposition.
3. Wet and dry etching.
4. Metallization.

Basic Fabrication Ingredients (Wolf and Tauber, 1987)

Lithography. The pattern of the MEMS structure is first designed using a CAD program. The pattern must then be transferred onto the substrate or film being processed. Several methods can be used to achieve this: photolithography, electron beam lithography, or ion beam lithography. The most common method is photolithography. It involves transferring the pattern from a mask to a resist coated substrate—usually single crystal silicon (SCS) wafer. The resist is exposed by shining UV light (or x-rays) through the mask. The pattern is obtained on the resist after development. The resist is then used as a masking material for the subsequent processing step.

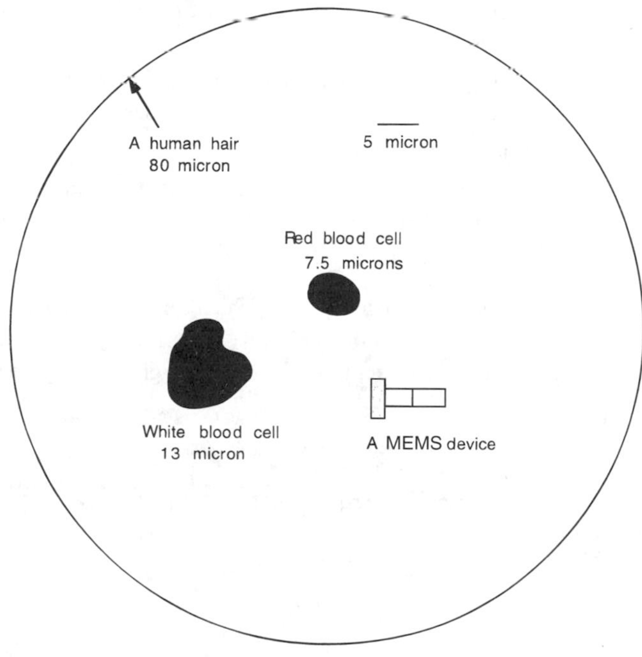

Figure 116.12 How small is a micrometer?

Thin films. Thin films are used extensively in the fabrication of MEMS for a wide variety of reasons, such as to protect a material from etching, to serve as a MEMS structural material, or to insulate one conductor or a semiconductor from another. Thin films are deposited on a substrate by allowing the film to deposit on the substrate from a vapor, by initiating chemical reactions on the surface of the substrate, by evaporation or sputtering, or by other means. The reactions are usually controlled by heating, with or without the presence of a plasma. Films can also be grown by chemically converting the surface layer of the substrate to the required film. The grown films have electrical and mechanical properties superior to those of the corresponding deposited films. A few popular films are silicon dioxide and silicon nitride, used as protective coatings or insulators; polysilicon, used as a structural material; and aluminum, used to form electrodes. Thin films, deposited or grown, often induce considerable stress on MEMS structures due to thermal expansion mismatches between the different films and the substrate. Residual or intrinsic stresses may also develop in the films during their formation. Thermal annealing is generally employed to relieve residual stresses.

Etching. Films or substrates are etched by liquid or gaseous etchants. Liquid etchants are usually highly selective in etching the required material. For example, KOH etches silicon at a much higher rate than silicon dioxide. However, wet etching offers limited control on the dimensions of the etched region. Also it etches isotopically, i.e., etching progresses in all directions unless the etchant has a preferential rate of etching along a crystallographic plane of a crystalline material. Dry etching, involving gaseous etchants, on the other hand, is usually enhanced by a plasma which ionizes the reactant gases and directs them along an applied electric field. Dry etching can be isotropic or anisotropic where etching is along a preferential direction depending on the gases, the power density of the plasma and the etching material.

Metallization. Metallization involves depositing a thin layer of metal to form electrodes on MEMS which might be used as capacitors to generate force or sense motion. The most commonly used metals are aluminum and gold. They can be vapor deposited or sputtered. The former involves heating a metal source in vacuum. The temperature is raised until the metal vaporizes and coats the chamber and hence the substrate. The latter involves bombarding a surface of metal by argon ions. The resulting collisions release metal atoms that deposit on the chamber walls as well as the substrate.

Fabrication Methods (Sze, 1994)

Two distinct methods are used:

Bulk Micromachining. In this method, MEMS structures are made by selectively etching the bulk material of the wafer. The material is usually single crystal silicon (SCS) and a typical etchant is KOH. Figure 116.13 shows the basic steps. One starts with a SCS wafer with its top surface aligned with the (100) direction and a region doped heavily with boron. The wafer is coated with a silicon nitride film. The film is patterned to form the required openings using standard photolithography. The silicon is then etched along preferential directions. For example, in the case of KOH, the (111) plane is etch resistant. The etching stops at the nitride film and the doped region at the bottom of the wafer. The nitride is then removed, and the silicon substrate is bonded with a glass substrate to seal the devices. The process successfully yields diaphragms, cantilever beams and plates as well as V grooves for fiber optic guides. It is suitable for large MEMS devices and when the distance between the devices does not require precise dimensional control. Unfortunately, because etching is at an angle with respect to the vertical, a large portion of the wafer surface has to be sacrificed in this method.

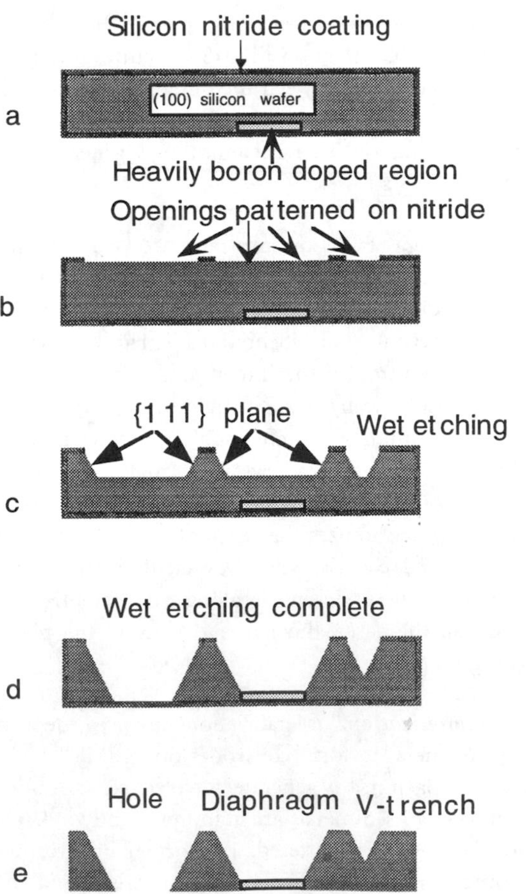

Figure 116.13 Bulk micro machining (Sze, 1994). Steps (a) through (e) show the process sequence. (*Source*: Sze, S. M., ed. 1994. *Semiconductor Sensors*. John Wiley & Sons, NY. With permission.)

Surface Micromachining. This method allows the fabrication of MEMS with micrometer size features and gaps. It essentially involves the deposition and etching of thin films on the surface of a substrate. There are three major processes based on surface micromachining. They are:

Sacrificial Layer Technology. This technique was developed at the University of California at Berkeley. The basic steps are shown in Figure 116.14. They are:

1. The deposition of a silicon nitride film on silicon wafer.
2. The deposition and patterning of a silicon dioxide film on the SCS wafer.
3. The deposition of a polysilicon film on the oxide surface and its subsequent thermal annealing.
4. The patterning of the polysilicon by conventional photolithography.
5. The releasing the polysilicon structures by desolving the oxide film, usually by hydroflouric acid (HF).

The polysilicon structures are anchored to the nitride film. The process is straight forward and easy to implement. But the overall size of the structures is limited by the thickness of the polysilicon film which is about 2 μm. Thus, the structures have low out-of-plane stiffness. Also, during the wet process with HF, the structures may become unstable due to the capillary force (Sze, 1994; Mastrangelo and Hsu, 1993) of the liquid, and they may become attached to the substrate.

LIGA. LIGA is an acronym derived from the German words for lithography, electroforming, and injection molding (*LI*thographie, *G*alvanoformung, *A*bformung). Figure 116.15 shows the basic steps of LIGA. It involves

1. Coating a conducting surface with 50–500 μm of photoresist.
2. Exposing the resist by passing highly collimated x-rays from a synchrotron through an x ray mask.
3. Developing the resist.
4. Filling the the cavity formed in the resist with a metal by electroplating.
5. Removing the resist.

The metal structure then forms a mold which can be used repeatedly as an insert for injection molding. The process results in very high aspect ratio structures which have high out-of-plane

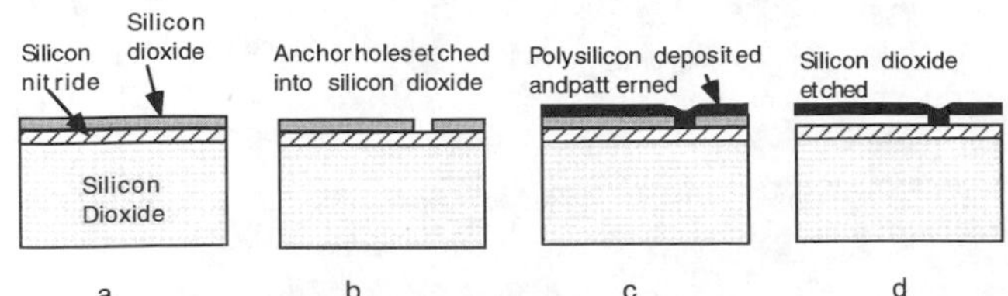

Figure 116.14 Sacrificial layer technology. Steps (a) through (d) show the process sequence.

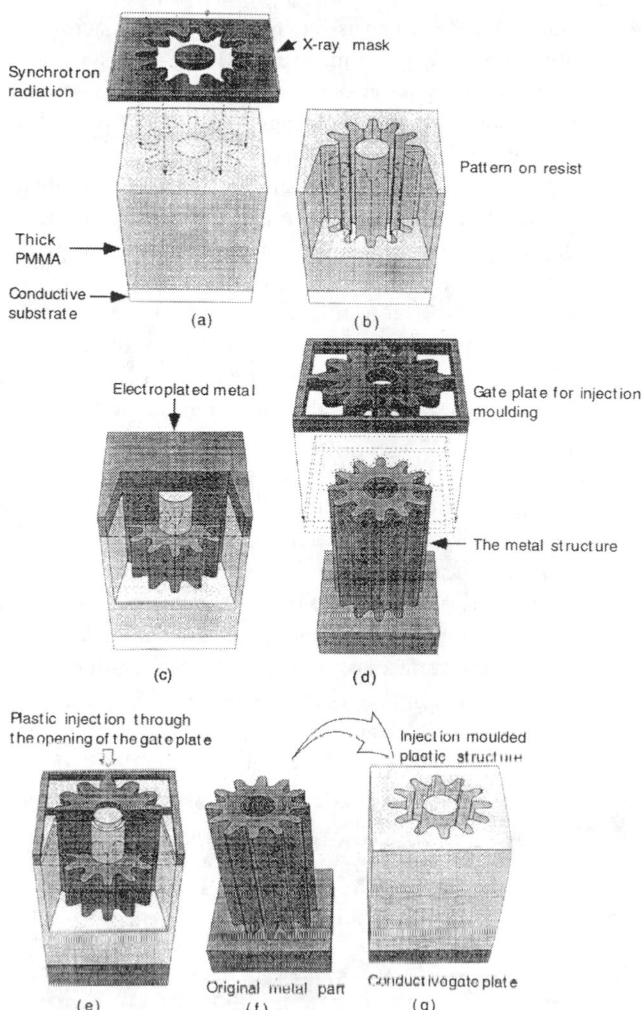

Figure 116.15 The LIGA process (Sze, 1994). Steps (a) through (g) show the process sequence. (*Source*: Sze, S. M., ed. 1994. *Semiconductor Sensors.* John Wiley & Sons, NY. With permission.)

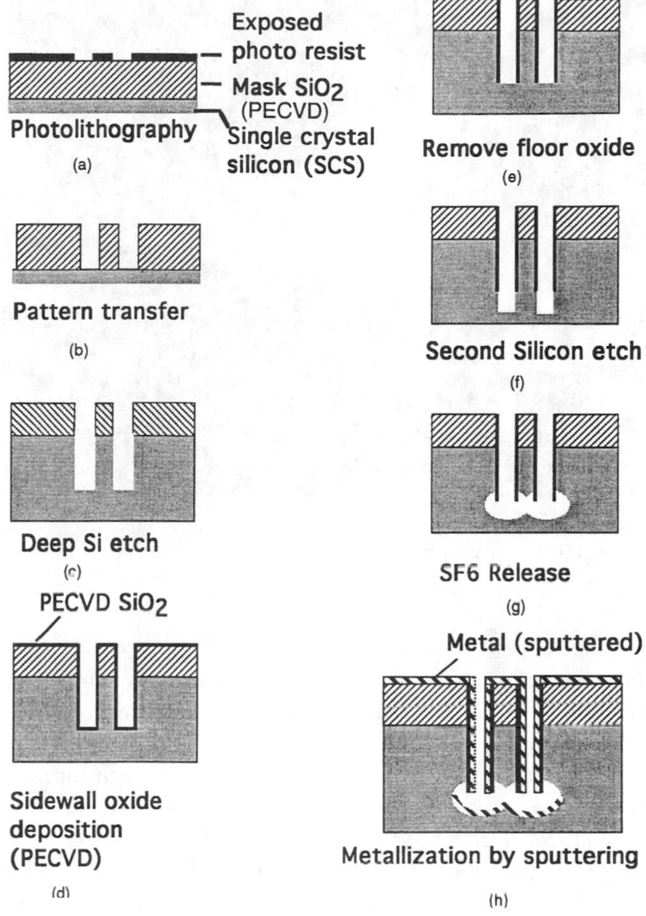

Figure 116.16 A version of Single Crystal Reactive Ion Etching and Metallization (SCREAM) process.

stiffness. Another advantage is that the mold can be used several times. The disadvantages are the initial cost of making the mold using the synchrotron radiation and the difficulty in integrating the structures with other on-chip silicon devices.

Single Crystal Reactive Etching and Metallization (SCREAM). This process was developed at Cornell University. The basic steps of the process are shown in Figure 116.16 (Shaw et al., 1994; Zhang and MacDonald, 1992). The pattern of the device is first formed in photoresist and subsequently transferred to the silicon dioxide film by an anisotropic dry etch. The photoresist is then removed. Next, the silicon is etched anisotropically using a chlorine plasma. Silicon dioxide is deposited or grown to form a conformal coating. The oxide at the bottom of the trenches is then cleared by anisotropic etching. The trench is deepened by another dry etch. Finally, the beams are released by an isotropic dry etch of the silicon. Each beam is held at the end by a support fixed to the substrate or by other beams which are eventually held by supports. Metallization is done on the released structure usually by sputtering. Note that the sidewalls overhang the silicon core because part of the silicon from the beam gets etched out during the release process. This overhang isolates the metal from the silicon.

MEMS Applications

In spite of its infancy, MEMS technology has already been applied to the development of several prototype products. MEMS accelerometers employ small capacitors made of microbeams to sense acceleration. During an impact, the beams move with respect to each other, changing the capacitance. When used in an air bag, this change is monitored by a microprocessor. If the capacitance exceeds a preset limit, the microactuators deploy the air bag. It is important to note that the accelerometer, the processor, and the actuators are all on the same chip. Analog Devices of Massachusetts manufactures these air bag accelerometers and have sold half a million of them to automobile manufacturers (Gabriel, 1995). Another successful application is the pressure sensor. Here, a thin membrane seals the opening of a cavity. The pressure in the cavity is predetermined. The membrane inflates or deflates in response to a pressure difference between the inside and the outside of the cavity. The deformation of the membrane causes

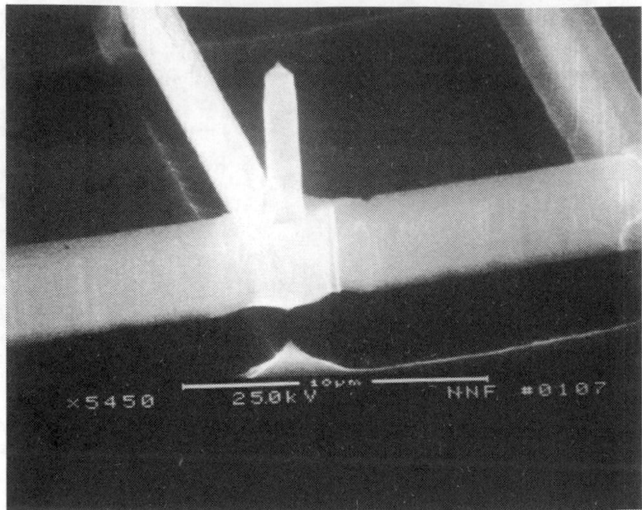

Figure 116.17 A silicon tip for surface scanning or information storage.

a change of resistance across the film. A microprocessor records the change of resistance and hence, the pressure. For a completely different application, Texas Instruments has made micromirrors that can be rotated about an inplane axis. The mirrors are used to display images for television (Gabriel, 1995). Each mirror is only 16 μm × 16 μm in size. It reflects pulses of light onto a screen to form an image pixel. When rotated, the light is deflected away from the screen and the pixel is dark. This approach creates sharper image on large, wall size TV screens. Other applications include Hewlett-Packard's high bandwidth frequency synthesizer and IBM's ink jet nozzle arrays (Petersen, 1982).

MEMS has potential applications in many more areas—the aerospace industry, information storage, robotics, medicine, optics, telecommunications, and materials science. We provide a few examples below.

Aerospace: Research is on going to replace the flaps of conventional aircrafts by independently addressable tiny plates. They will control the air boundary layer structure by tilting with respect to the wing surface. There could be millions of these micro flaps on an aircraft wing. Thus the large conventional flaps and the associated heavy machinery will not be required. This may revolutionize the design concept and architecture of future aircraft (Gabriel, 1995).

Information storage: Small tips with diameters on the order of 10 nm or less have already been used to scan surfaces at the atomic scale. The same tips, held and and maneuvered by a micromechanical actuator, can be used to "pluck" and transport a cluster of atoms from one region of a conducting surface to another. These heaps and valleys of atoms can be read as ones and zeros by a scanning tip. Thus, considerably more data can be stored in a much smaller space than allowed by conventional technology. Figure 116.17 shows such a tip fabricated from single crystal silicon (Xu et al., 1995).

Robotics: MEMS can be used to move and align macroscopic objects by ciliary action. Figure 116.18 shows a grid of beams in teeter totter device which can be twisted about its supporting beams (Bohringer et al., 1995). A substrate may have a large number of these grids that can move a light weight object, such as a sheet of paper, by coordinated motion of the grids. These microrobots may replace the paper transport mechanism of the conventional photocopiers and reduce their size significantly.

Medicine: MEMS can be used as microvalves to inject drugs at timed intervals into the blood stream of patients. The fabrication of miniature laboratories has been proposed for DNA testing and the detection of water borne pathogens.

Motion detection: Research is in progress to develop micromechanical motion detectors (Kenney et al., 1991). These sensors will measure extremely small (sub-nanometer) motions. In addition, the MEMS motion detectors will detect accelerations in the micro-g to nano-g range, that is, one millionth to one billionth the gravitational acceleration at the surface of the Earth. An acceleration as low as 10 micro-g's has been detected by the MEMS accelerometers at Cornell. Figure 116.19 shows an accelerometer made by the SCREAM process (Adams et al., 1995).

Figure 116.18 An actuator for moving light objects.

Figure 116.19 An accelerometer.

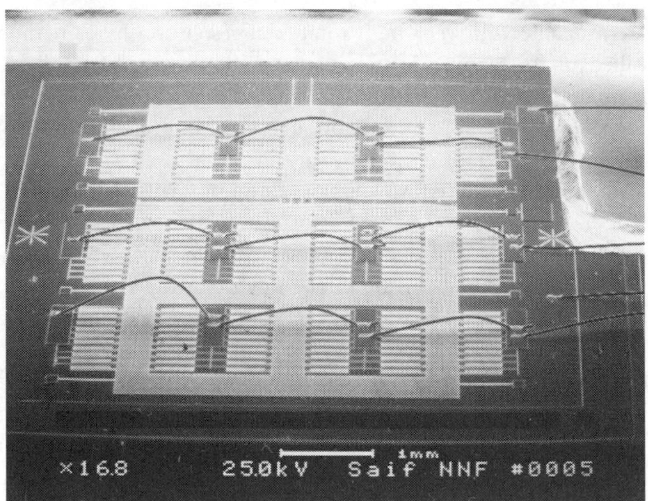

Figure 116.20 (a) A micro loading device and (b) buckled samples.

Materials science: MEMS offer a whole new approach to studying materials. It allows the testing of very small samples and the integration of the sample with the testing machine. The sample and the machine can be designed, patterned, and fabricated at the same time. This avoids the problems of attaching and aligning small samples with a forcing device (Saif and MacDonald, 1995). Furthermore, it allows in situ analysis of the sample, such as transmission electron microscopy of a thin plate under tension. Figure 116.20 shows a microloading machine that applied a compressive force on two beams and to buckle them.

Conclusions

We have rendered a very brief overview of the new and emerging MEMS technology. We have discussed the key ingredients and the primary methods for fabricating MEMS structures. Additionally, we highlighted a few of the current and possible applications of MEMS.

MEMS is a natural progression of microelectronics. Their small size, integrability with microelectronics, batch fabrication, low cost, and high performance make MEMS a potential key technology of the 21st century—a technology that will revolutionize the way we sense and control the physical world around us.

116.3 Precision Micromachining Technologies

Craig R. Friedrich and Michael J. Vasile

Introduction

Precision micromachining has many applications in MEMS in addition to the fundamental practices of precision engineering. Precision engineering practices are used in the design and operation of nearly all MEMS processing equipment because of the small features and demand for small absolute tolerances. However, precision machining, especially at the microscale, can be used in applications which range from the direct production of small parts to the fabrication of masks for lithography applications. Although the focus will be on the tools and processes for fabricating small parts, the same general procedures are used in high precision machining of large parts.

Many definitions of precision engineering have been posed over the years. One quantitative definition is that the tolerance placed on a part or feature is less than or equal to 1 part in 10,000 of the part or feature dimension. In the macroscale region, these tolerances are practical and commonly achieved. At the microscale such a restriction is nearly impossible to achieve. For example, spinnerettes for textile extrusion may require a drilled hole 25 micrometers in diameter. Using this definition of precision engineering would require a tolerance of less than 25 Angstroms. This may not be achievable with direct machining techniques. Micromachining generally requires tolerances small enough to ensure part or system functionality. A second definition of precision engineering is an attitude wherein all effects have an identifiable cause. This is often referred to as determinism and can be applied to both macromachining and micromachining. In practice, there may be insufficient resources to actually identify all the factors which lead to imprecision, however this approach does provide a systematic method for identifying and correcting the factors which are most important.

There is no generally agreed upon dividing point between macromachining and micromachining, however the techniques described in the following sections are generally considered to be in the realm of micromachining. The field of precision macromachining dates to the era wherein accurate measuring instruments and timepieces were needed for navigation. During the industrial revolution, many precision machining companies were formed from the core technology of clock and watch making. In recent times, precision macro-machining was advanced to support the development of nuclear weapons programs (this is because the laws of physics do not accomodate tolerances). However, micromachining with small diamond points dates to the early 20th century when a pantograph was developed which was capable of scribing legible letters 2.5 micrometers in size.

Precision Micromachining Techniques

The equipment and techniques used to achieve precision machining at both the macro and microscales can be separated from the actual machining processes. The techniques include the study of machine tool kinematics, machining error analysis, and error compensation, among others, while the equipment generally includes feedback devices, actuators, and structures. The basic philosophy of precision micromachining is really quite simple. The interface of the material removal process (cutting edge for turning, drilling, milling and energy beam for laser, electrical discharge, and ion beam machining) and the surface of the finished workpiece must be accurately maintained and known.

Unfortunately, the machining process and the environment work against this simple goal.

Common Sources of Errors

Sources of errors in any precision machining operation are as numerous as the number of components in the machine tool itself. Although it is not possible to identify all sources of error for all machining operations, the following are the more commonly encountered ones. Any machine tool will move the workpiece or the tool, or both. Physical movements can take place with a variety of mechanisms including mechanical or fluid bearings, and linear or rotational mechanisms. Linear actuation can take place with devices such as leadscrews driven by stepper or servo motors or linear motor mechanisms.

Linear or rotational mechanisms, such as linear slides and stages or rotary bearings, can be used to move both the cutting tools and the workpiece. Mechanical bearings have the advantages of relatively low cost due to large scale production, high stiffness which maintains dimensional control under high loads from workpiece weight or dynamic loading from workpiece unbalance, and no requirement for a source of pressurized fluid. Disadvantages of mechanical bearings are vibration from the rotation of the mechanical roller elements (low frequency vibration is related to one rotation of the bearing whereas high frequency vibration is related to the rotation of each individual roller element), heat generation, relatively high power consumption due to high mechanical preload to maintain precision, and the need for clean lubricants. Fluid bearings have the advantages of high straightness and flatness, vibrational damping within the fluid film, low force or torque for movement, no heat generation, and no wear. The disadvantages of fluid bearings are low stiffness relative to mechanical bearings (thus requiring very good workpiece balancing in rotational applications), lower load carrying capacity than mechanical bearings, the need for a source of clean pressurized gas, higher cost, and low "crash" tolerance. The choice of a bearing technology for high precision movements depends on a detailed analysis of the application and the level of precision required.

The methods for providing motion are equally as important as the type of bearing used. If a motor can be integrated into a bearing, there is less vibration imposed from pulleys, v-belts, or traction drives. Because fluid bearings are normally used in applications which can not tolerate vibration, the motor armature is integral with the bearing rotor. In large machine tools such as a lathe, the bearing may be driven by a constant or variable speed motor and one or more v-belts through a gearbox which allows for variable spindle speeds. The splice of the v-belts may induce a level of vibration which exceeds the allowable limit. This condition can be lessened by staggering the v-belt splices so that only one splice passes over a drive pulley at any time. This will result in higher frequency vibration but with lower amplitude than if all the splices are aligned. The meshing of gear teeth in a drive system is also a source of vibration and should be avoided.

Linear motion is normally provided by a rotary motor (stepper or servo) and leadscrew or by a linear motor. Stepper motors are available with very high rotational resolution by a method called micro-stepping. Details on this method are widely available from manufacturers. Whether driven by a stepper or servo motor, the slide is still subject to many sources of imprecision. One of the most important is nonlinearity in the leadscrew. Because a leadscrew is machined, it will have a nonuniform pitch along its length. If the motor (a stepper for example) is turned through a constant angle, the distance moved by the stage will depend on the local accuracy of the thread pitch. For the same input to the system, the stage may move different amounts depending on the location along the leadscrew. Fortunately, this nonlinearity can be mapped and compensated. A more expensive, but more accurate, method of compensating for leadscrew nonlinearity is by using a linear feedback device such as a linear encoder or laser interferometer, for example.

Other motion errors arise because of waviness, nonparallelism, and nonflatness of guides and surfaces in which mechanical rolling elements move or on which fluid bearings float. In mechanical slides, these errors arise because of high mechanical preload between the moving and stationary portions of the slide, because of out-of-round rolling elements, nonflat and nonparallel guides, and mechanical hysteresis. Mechanical slides typically have a flatness in the micrometer regime whereas air bearing slides can achieve a flatness in the 0.1 micrometer regime over a large travel.

Other sources of errors in precision machining are coupling between the machining operation, the equipment, and the environment. Effects include vibration and thermal distortion of the workpiece and the machine tool components. Mechanical machining (as opposed to energy beam machining) can cause relatively large forces to be present in the machine structure. Micromachining often uses cutting tools which are small and therefore subject to force-induced deformations. These deformations can lead to both static and dynamic errors. The dynamic error is caused by tool and/orworkpiece vibration which will result in a very rough and periodic surface finish and excessive tool wear or fracture. Vibration can be reduced by increasing the stiffness of the structural loop between the tool tip and the workpiece or by reducing the magnitude of the input forces causing the vibration. A stiffer structure will result in a higher natural frequency for the system and resonance may be avoided. A smaller machining force will also help reduce the amplitude of vibration, in most cases. There are circumstances where an excessively small cut may actually induce vibration because of deformation of the very small cutting edge of a tool and because of the large effective rake angle at which the material is being machined because of the finite edge sharpness of the tool.

As the need for precision increases, thermal effects tend to dominate the machining operation. Because all materials will change size and shape with a change in temperature, thermal effects can be overwhelming. The machining operation itself can produce considerable heat which will cause the workpiece, the cutting tool, and the machine structure to deform. The use of a coolant will help reduce these effects but the coolant temperature must be carefully monitored if the precision is to improve in a predictable manner. Other methods for reducing differential expansion in high precision machine tools is to actually force a coolant throughout the machine structure, particularly the bearings and spindle. The bearing coolant is normally a lubricant with the temperature carefully monitored and controlled by way of a thermal exchange system. Extreme approaches have actually relied upon bathing the entire machine tool in a thin film of temperature controlled oil.

A third source of thermal distortion comes from the air surrounding the machine tool. A precision machine tool should not be placed near any other sources which may conduct, convect, or radiate heat. This is especially true if the machine tool is near an outside wall or a window. Precision machine tools should be placed in interior rooms which are isolated as much as possible. The next ambient source of thermal distortion is the cycling of a heating or cooling system. Most environmental control systems will have a cycling heating or cooling source, even if there is a constant flow of air. This source is usually activated by a thermostat which is an on-off control device. Depending upon the specific conditions, this cycling may take place over a relatively long period of time. Because this temperature change is gradual the thermal mass of the machine tool structure will be able to follow this low frequency cycling with good fidelity. Therefore, it may be better to force the heating or cooling source to cycle more rapidly over a smaller range of deadband temperature. A large thermal mass will not be able to keep up with this high frequency cycling and the expansion or distortion can be minimized.

Yet another source of thermal influence is the machine operator. A person thermally influences a precision machine tool in two ways. First, the exterior of the body has a temperature typically in the range of 24C to 27C. To help control the difference between a machined dimension (made at room temperature or above) and a measured dimension (which is almost always referenced to approximately 20C), most precision machine tools are maintained in a 20C room. The higher body surface temperature of a person will act as both a convective and radiative heat source influencing the machine tool. A second and more substantial effect is the latent heat which a person exhales. In a controlled room, a person will inhale cool, dry air and will exhale very warm and moist air. This warm moisture is far more difficult to remove or cool by an environmental system and will also affect the precision of the machine tool. By both sensible (body surface) and latent heats, a person standing near a precision machine tool has about the same effect as a 150 Watt lightbulb.

Feedback Devices

To monitor or eliminate error sources requires some sort of feedback device to compare the commanded performance with the resulting performance. This performance may be a movement, the isolation of the system from the surroundings, or actual machining of a part. Linear feedback devices include linear encoders and linear variable differential transformers (LVDT). High resolution linear encoders are normally a strip of glass with a grating of closely spaced lines which has been lithographically printed or etched into it. The lines provide a method for periodically interrupting a small spot of light which shines through the glass slide and onto a photodetector. The output of the detector may be wave-shaped to provide a square wave output each time the light is interrupted. By counting the number of waves and by knowing the spacing of the lines, a control system can determine an incremental distance moved. Inaccuracies in a linear encoder include expansion or strain in the glass which will change the spacing of the lines, nonuniformity of line spacing along the glass strip, or a movement to fast for the control system to keep up with resulting in lost counts. Given these sources, a linear stage with an accuracy of +/− 3 micrometers over a 250 millimeter range is typical. With proper electronics, the stage resolution can attain 0.1 micrometers.

An LVDT is a linear transformer with a movable core. The core magnetically couples a primary winding which is provided with an input voltage, with an output winding which indicates the degree of coupling or core movement. The input voltage is normally small (less than 10 Volts rms) but at high frequency (usually in the kiloHertz range). Because the LVDT is essentially an analog device, this requires an analog power supply and voltmeter to maintain high precision. For applications where the LVDT must have very smooth movement or very low contact force, the core can be made to be the moving portion of a linear air bearing. An air bearing LVDT has very high precision because "stiction" is virtually eliminated in the moving core. An air bearing LVDT can attain a resolution of 0.01 micrometers with appropriate control electronics. The contact force can be controlled by regulating the flow of air in the bearing. A contact force less than a milliNewton is possible. Such light contact forces will not allow measurements at high frequency because of the tendency of the core to oscillate in and out of the windings unless the core is rigidly attached to the high frequency source. The sources of error in an LVDT include long term drift of the transformer characteristics and thermal growth due to resistive heating in the windings, contamination (which is minimized in the air bearing type), and a quasi-digital response because of the finite size of the windings in the two transformer halves.

Rotary feedback devices typically include rotary encoders to indicate speed and/or angular position, and capacitive probes to determine concentricity of movement. They have the same general advantages and disadvantages as their linear counterparts. Detailed specifications are readily available from component manufacturers.

Precision Micromachining Processes

Diamond Micromachining

If the desirable characteristics of a cutting tool material were specifically listed, they would include high hardness and high strength, good thermal conductivity, the ability to be polished to a sharp cutting edge, and high elastic stiffness (Young's modulus). Single crystal, natural diamond is such a material. Natural diamond is the hardest known material and has a critical tensile cleavage stress as high as 4 GPa (580,000 psi). At room temperature, diamond has the highest thermal conductivity of any known material (however its high thermal conductivity drops off rapidly both below and above room temperature). Because diamond is a crystalline material, it can be polished to a very sharp edge. A cutting edge radius in the tens-of-nanometers is attainable. The Young's modulus of natural diamond is approximately 1000 GPa (145 million psi) which is nearly five times that of steel. This means that the tool edge will experience minimal deformation under the application of machining forces.

Diamond cutting tools are available in the form of natural crystals, polycrystalline, or synthetic crystals. Polycrystalline diamond is composed of small diamond particles which have been formed with a metal matrix to produce a tool with good hardness and increased toughness because of the metal. However, polycrystalline diamond leaves a surface finish more like grinding because of the particles and therefore is not generally used for demanding applications such as for optics.

Diamond micromachining is carried out with essentially the same equipment as macroscale diamond machining, however the cutting edge dimensions will be smaller to better conform to the required part dimensions. Several small diameter microshafts, which were machined with diamond tools, are shown in Figure 116.21 and 116.22. The 125-micrometer diameter shaft has a root-mean-square surface finish of approximately 25 nanometers with a taper of less than 1 in 200. The 25-micrometer diameter shaft has a length-to-diameter ratio of over four.

Diamond tools with a specific edge shape can be used to contour materials. A section of a micro compact heat exchanger is shown in Figure 116.23. The heat exchanger plate is made from 125-micrometers thick, high conductivity copper foil. The channels are 100-micrometers wide at the bottom and approximately 80-micrometers deep. When the foils are stacked so adjacent foils have the channels perpendicular to each other, a cross-flow heat exchager is formed as shown in Figure 116.24. Because the hot and cold fluids are separated by only 40-micrometers, or less of copper, a high heat transfer can be obtained. Such devices have been demonstrated to have a volumetric heat transfer rate of over 300 megaWatts per cubic meter per Kelvin. At this rate, a one cubic centimeter heat exchanger will provide the same capacity as a conventional automobile radiator. In a similar fashion, diamond machining can be used to enhance the surface heat transfer characteristics of macroscale thermal devices.

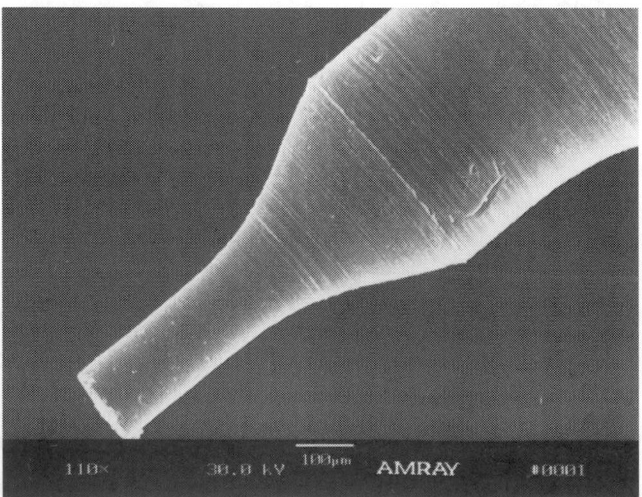

Figure 116.21 125-micrometer (0.005 inch) diameter shaft of aluminum (scale bar = 100 μm).

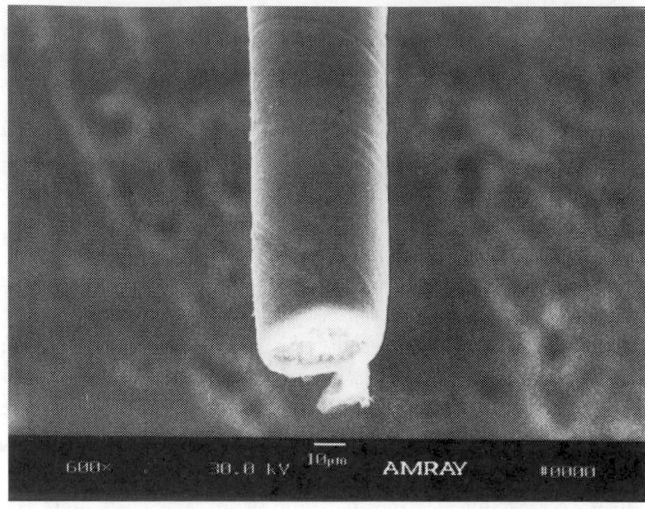

Figure 116.22 25-micrometer (0.001 inch) diameter shaft of aluminum (scale bar = 10 μm).

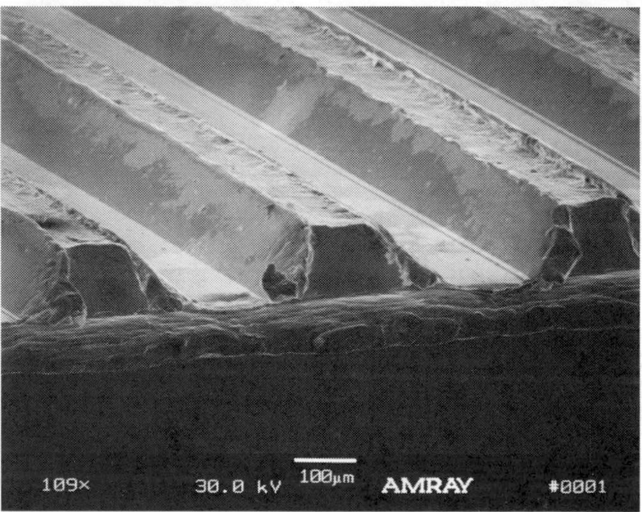

Figure 116.23 Diamond machined fluid micro-channels in copper (scale bar = 100 μm).

Microdrilling

Microdrilling is somewhat similar to conventional drilling with two distinct differences. First, below a diameter of 50 micrometers twist drills are not generally available. Therefore, most microdrilling is done with spade-type drills. Below 50-micrometers in diameter, all microdrilling is done with spade drills. A typical spade drill is shown in Figure 116.25. The second difference is that at small diameter the drill can not be held by conventional means with sufficient precision to eliminate eccentricity when rotated. Microdrills are typically mounted in a high precision mandrel which then rotates in two diamond vee-blocks resulting in four contact points between the outside

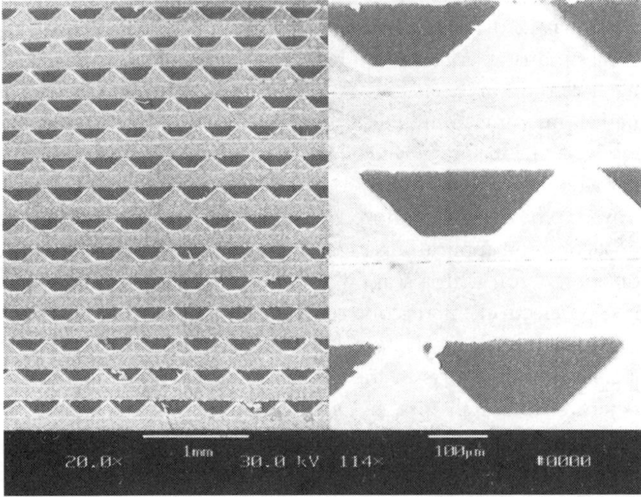

Figure 116.24 Cross section of heat exchanger made by diamond machining.

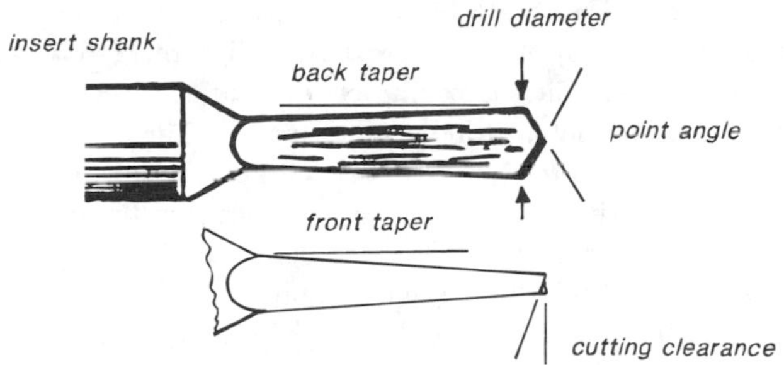

Figure 116.25 Geometry of a micro spade drill.

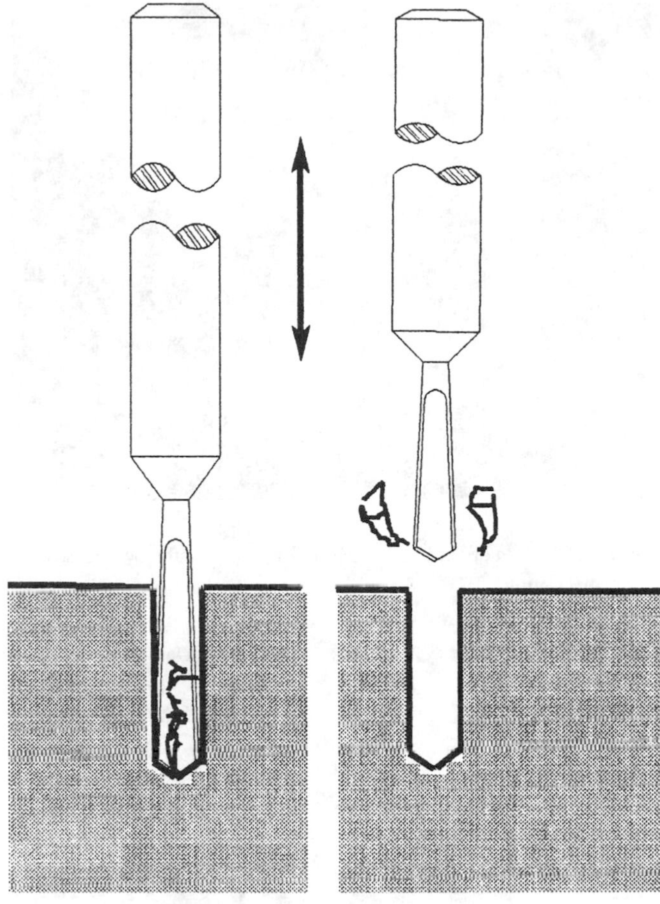

Figure 116.26 Chip removal is aided by "peck drilling" and a thin fluid.

diameter of the mandrel and the drilling machine. The drill-mandrel combination is ground and lapped in a similar vee-block arrangement. The drill is both fabricated and used in the same fashion and the eccentricity is only as much as the variations in the vee-blocks used. Although the mandrel may have some degree of eccentricity, this is virtually eliminated by the vee-block arrangement.

Because a microdrill does not have flutes to help remove chips from the hole being drilled, peck drilling is normally required. This is shown in Figure 116.26. The combination of a light cutting oil and periodic removal of the drill from the hole results in efficient removal of the chips. If the chips remain in the hole, the drilling forces and wall roughness increase. Peck drilling also helps burnish the walls of the hole leaving a very smooth surface. The root-mean-square surface roughness of aluminum, copper, and polymethly methacrylate (PMMA) can be in the tens-of-nanometers. Microdrilling is not normally done at excessive rotational speeds. Actually, a smoother hole will result if the drill speed is in the 4,000 to 6,000 rpm range. At higher rotational speeds, the drill does not cut continuously or cuts extremely thin chips. This can result is severe work hardening which can lead to drill breakage. Under carefully controlled conditions, 25-micrometer microdrills are capable of drilling many hundreds of holes before dulling occurs.

Micromilling

Micromilling is similar to conventional milling except that the machining parameters (tool feed, depth of cut, etc.) must be modified because of the fragility of the milling tool. Small milling tools are being fabricated using the focused ion beam process (see next section). A typical four-fluted tool is shown in Figure 116.27. These tools do operate at higher rotational speeds than do microdrills because recutting of chips in a blind hole is not a problem. Spindle speeds of 10,000 to 20,000 rpm have been used. Because the micromilling tool behaves as a cantilevered beam, the feed of the tool into the material must be kept below the level which will cause tool breakage. In very soft materials, the process leaves behind burring which must be removed with a chemical or electrochemical deburring process.

For micromanufacturing applications, micromilling holds great potential. The process will never be able to rival that of lithography for the production of features on masks or surfaces. However at the larger microscale (100-micrometers sized features), micromilling may have many applications for rapid production of prototype masks and molds for mass fabrication of parts. A simple x-ray lithography mask may take several weeks to have fabricated at a "mask-house" and can cost several thousand dollars, particularly for hard x-rays where the gold absorber layer must be quite thick (15 to 100-micrometers). Given the CAD layout of the features, a similar low resolution mask could be

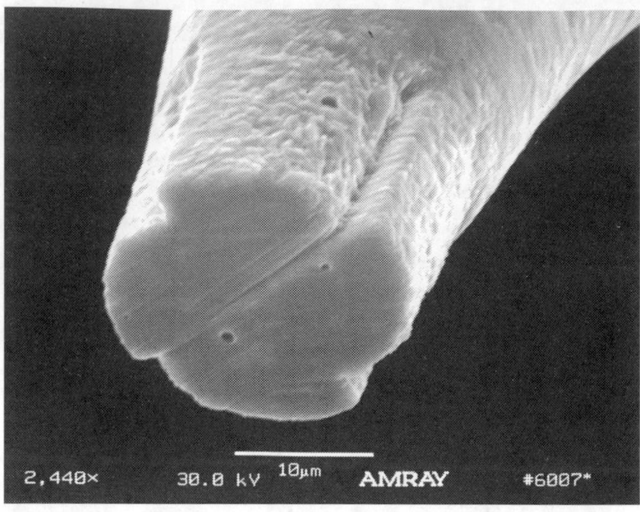

Figure 116.27 Four-fluted micromilling tool made by focused ion beam machining.

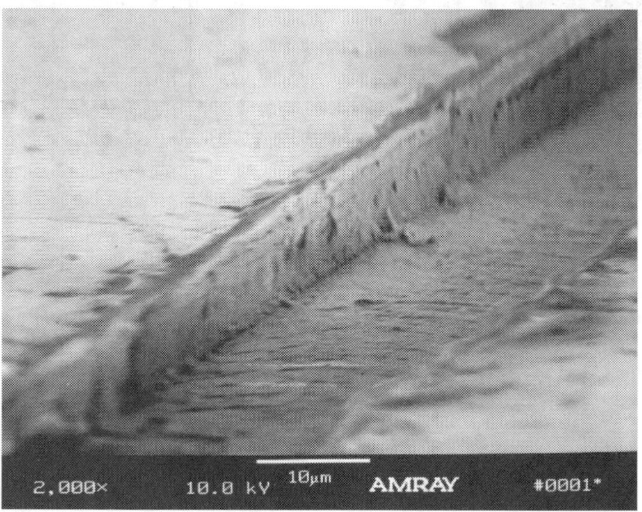

Figure 116.28 Micromilled trench made on "low precisior" milling machine.

milled in less than a day with untended operation of a CNC micromilling machine. A portion of a micromilled trench in PMMA is shown in Figure 116.28. The trench bottom roughness is in the 0.1 micrometer range.

Focused Ion Beam Micromachining

The most precise and the smallest features machined to date have been done using a focused ion beam. Focused ion beam machining is a vacuum process that takes the work pieces sequentially, so there is no advantage of batch production as there is in lithographic based production of microstructures. The material removal process is also relatively slow compared to other linear scan type material removal processes such as laser machining. The main uses of focused ion beam (FIB) machining are therefore in highly specialized areas where the value of the product or prototype warrants the time needed to produce it.

FIB has its origin in the mask repair efforts of the microelectronics industry. A method for material removal with submicrometer resolution was necessary for the masks in use today, hence the FIB micromachining technique was developed, based on the liquid metal ion source which routinely produces 0.1 micrometer diameter ion beams. Other microelectronics applications soon proved to be equally or more valuable than mask repair, and the most common applications today are integrated circuit sectioning for observation with a scanning electron microscope and transmission electron microscope sample preparation.

Microstructure fabrication with FIB is obviously an outgrowth of the research efforts in the microelectronics industry, since it is natural to search for new applications of recently developed tools. FIB has been demonstrated as a micromachining tool for the production of scanning probe microscope tips (AFM or STM tips) and for the production of micromilling tools. Other prototype structures include manipulators and micrometer-sized objects with familiar geometry, such as needles and forks.

Material removal occcurs by sputtering, which is an atomic scale process. Figure 116.29 is an illustration of the material removal process by a focused ion beam. High energy ions (20 to 30 keV) from the source impact the solid and eject surface atoms from the solid by momentum transfer. The efficiency is in the range of three or four solid atoms released for every incident ion. The incident ion beam current is in the range of nanoamperes, which makes the material removal rate approximately 1×10^{-12} cm^3/sec. The ion beam is scanned over a predefined pattern to create the desired shape and the machining time scales with increasing surface area of the scan pattern. The scan pattern is digitally generated and corresponds to features and locations taken from a secondary electron image of the work piece. The secondary electron image is obtained from the ion beam scanning over the work piece in real time.

The time required to remove a volume of material with dimensions 5 μm × 5 μm × 20 μm is approximately 15 minutes (based on the sputter rate of copper). The process is thus confined to making small objects with dimensions up to ten or fifteen micrometers. Faster machining times can be expected when higher current density sources are used. Advantages of this material removal technique are the sub-micron resolution of the dimensions, no substrate heating and no macroscopic debris. Figure 116.30 shows a section through three contacts to an integrated circuit element with dimensional resolution better than

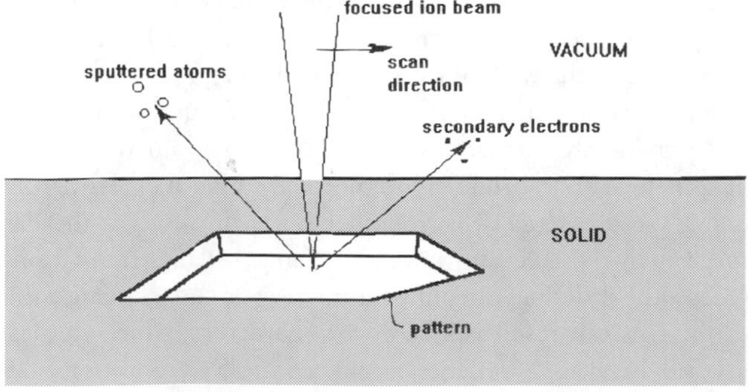

Figure 116.29 Schematic of FIB process.

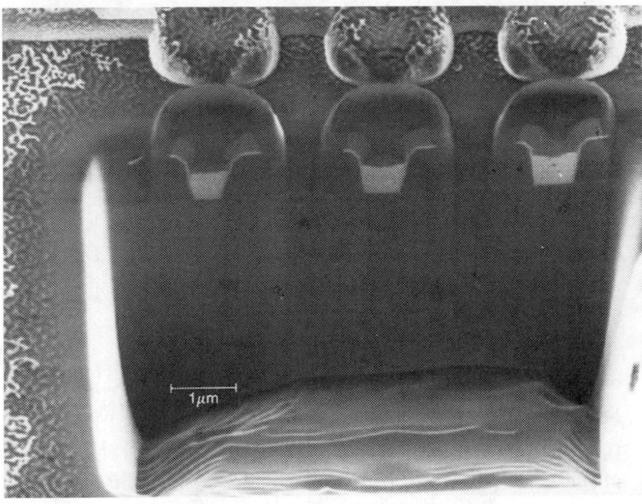

Figure 116.30 Cross section of integrated circuit contacts exposed by FIB.

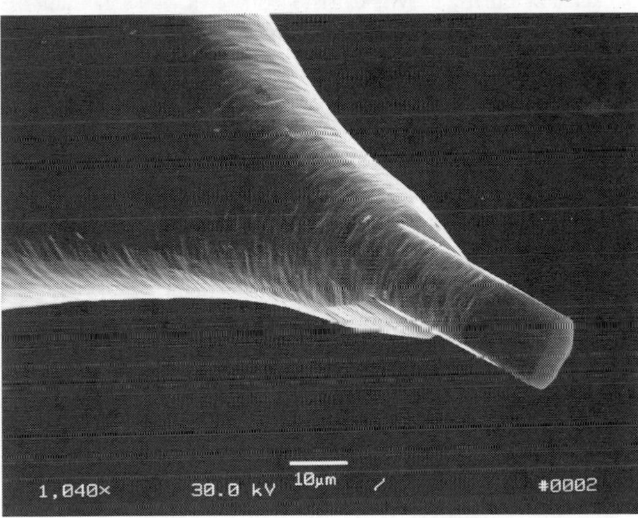

Figure 116.31 Two-fluted micromilling tool with 0.05 µm radius cutting edges.

0.1µm. This does not represent the limit, since smaller diameter focused ion beams are in use.

Figure 116.31 shows an experimental cutting tool made from a high speed steel tool blank by FIB machining. Dimensional resolution is limited by the grain structure or the imperfections in the material, such as microscopic voids or occluded foreign particles. Typical resolution limits for polycrystalline or composite materials is in the range of 0.2 to 0.4 µm.

Laser Micromachining

Laser machining is a well established process for conventional manufacturing. However, at the microscale, the large thermal effects on the workpiece from low frequency or continuous wave lasers can not be tolerated. Laser micromachining is typically performed with excimer lasers which operate in the ultraviolet region and have a relatively high pulse rate (up to 2 kHz). The actual pulse duration is in the nanosecond range with the energy per pulse in the micro-Joule to milli-Joule range. This gives a very high pulse power (megaWatts) but the thermal influence is very short lived. Because workpiece thermal effects such as heating or tempering are time-dependent processes, the nanosecond pulses have minimal effect other than to ablate (vaporize) the workpiece where the pulse is concentrated. By moving the laser beam by mirrors, the workpiece by linear or rotary actuators, or both, complex three dimensional features may be fabricated in any material.

The energy required for ablation is dependent upon the ability of the material to absorb the laser beam and on the material's stability at high temperature. Ablation of metals and ceramics is generally done with a beam fluence in the range of 1 to 20 Joules per square centimeter. Ablative lasers are generally of shorter wavelength and the 248 nm output of the KrF excimer laser is commonly used.

Drawbacks of the process are that the laser beam is not uniformly distributed over its wave front. It has a relatively constant energy distribution in one direction of its cross section, but has a Gaussian distribution in the other direction. This nonhomogeneity can be reduced by filtering the tails of the beam, but this also reduces the total energy into the workpiece and can lead to unwanted diffraction effects. The energy distribution of the beam is also strongly dependent on the cleanliness of the optics and the purity of the gas used in the laser tube.

A second application for laser micromachining is photopolymerization. Similar to optical or x-ray lithographic processes, an excimer laser can be used to expose a photoresist material either by a direct-write process or by a projection mask. The laser energy either forms or breaks the polymer bonds depending on whether the resist is positive or negative in nature. For this application the laser power must be greatly reduced so ablation of the photoresist does not occur. This can be done by lowering the flow of gas into the laser, by using beam directing optics which attenuate the beam strength, or by using variable density filters which also attenuate the beam strength. Polymerization typically uses a beam with a fluence of 1 to 100 milliJoules per square centimeter. Most polymers are more sensitive to longer wavelength ultraviolet light. Therefore, polymerization generally uses the 351nm light available from a XeF laser.

References

Adams, S. G., Bertsch, F. M., Shaw, K. A., Hartwell, P. G., MacDonald, N. C., and Moon, F. C. 1995. Capacitance based tunable micromechanical resonators, *8th Int. Conf. Solid State Sensors and Actuators, and Eurosensors IX*, 438–441, June 25–29, Stockholm, Sweden.

Adema, G. M., Turlik, I., Smith, P. L., and Berry, M. J. 1990. Effects of polymer/metal interaction in thin-film multichip module applications, *Proceedings of the Fourtieth Electronic Components Conference*, 717–726, May, IEEE Electron Devices Society, Las Vegas, Nevada.

Beach, J. G., McGraw, L. D., and Faust, C. D. 1968. *Plating*, 55:936.

Becker, E. W., Ehrfeld, W., Hagmann, P., Maner, A., and Munchmeyer, D. 1986. Fabrication of microstructures with high aspect ratios and great structural heights by synchrotron radiation lithography, galvanoforming, and plastic moulding (LIGA process), *Microelectronic Engineering,* 4:35–56.

Beuret, C., Racine, G., Gobet, J., Luthier R., and de Rooij, N. 1994. Microfabrication of 3D multidirectional inclined structures by UV lithography and electroplating, *Proc. of the IEEE Micro Electro Mechanical Systems Conference,* 81–85, January, Oiso, Japan.

Bohringer, K., Brown, R., Donald, B., Jennings, J., and Rus, D. 1995. Distributed robotic manipulation: experiments in minimalism, *4th Int. Symp. Experimental Robotics,* June, Stanford, CA.

Burbaum C., Mohr, J., and Bley, P. 1991. Fabrication of capacitive acceleration sensors by the liga technique, *Sensors and Actuators A,* 25–27:559–563.

Chakravorty, K. K., Chien, C. P., Cech, J. M., Branson, L. B., Atencio, J. M., White, T. M., Lathrop, L. S., Aker B. W., Tanielian M. H., and Young, P. L. 1989. High density interconnection using photosensitive polyimide and electroplated copper conductor lines, *Proceedings of the Thirty-ninth Electronic Components Conference,* 135–142, May, IEEE Electron Devices Society, Houston, Texas.

Choi, B., Lovell, E. G., Guckel, H., Christenson, T. R., Skrobis, K. J., and Kang, J. W. 1991. Development of pressure transducers utilizing deep x-ray lithography, *Proceedings of the 6th International Conference on Solid-State Sensors and Actuators,* 393–396, IEEE Electron Devices Society, San Francisco, CA.

Couch, D. E. and Brenner, A. 1952. *J. Electrochemical Society,* 96:234.

Connor, J. H. and Brenner, A. J. 1958. *Electrochemical Society,* 103:657.

Cummings, J. P., Jensen, R. J., Kompelien, D. J., and Moravec, T. J. 1982. Technology base for high performance packaging, *Proceedings of the Thirty-second Electronic Components Conference,* 465–470, IEEE Electron Devices Society.

Dishon, G. J., Bobbio, S. M., Tessier, T. G., Ho, Y., and Jewett, R. F. 1989. High rate magnetron RIE of thick polyimide films for advanced computer packaging applications, *Journal of Electronic Materials,* 18:293–299.

Ehrfeld, W., Bley, P., Gotz, F., Hagmann P., Manar A., Mohr, J., Moser, H. O., Munchmeyer, D., Schelb, W., Schmidt, D., and Becker, E. W. 1987. Fabrication of microstructures using the LIGA process. *Proceedings of the 1987 IEEE Micro Robots and Teleoperators Workshop,* TH 02404-8, 9–11, November, IEEE Robotics and Automation Society, Hyannis, MA.

Ehrfeld, W., Gotz, F., Munchmeyer, D., Schelb, W., and Schmidt, D. 1988. Liga process: sensors construction techniques via x-ray lithography, *Proceedings of the IEEE Solid-State Sensor and Actuator Workshop,* 1–4, June, IEEE Electron Devices Society, Hilton Head, SC.

Frazier, A. B. and Allen, M. G. 1992. High aspect ratio electroplated microstructures using a photosensitive polyimide process, *Proceedings of the Micro Electro Mechanical Systems Conference,* 87–92, February, IEEE Robotics and Automation Society, Travemunde, Germany.

Frazier, A. B. and Allen, M. G. 1993. Metallic microstructures fabricated using photosensitive polyimide electroplating molds, *Journal of Microelectromechanical Systems,* 2:87–94.

Frazier, A. B. and Allen, M. G. 1994. Uses of electroplated aluminum in micromachining applications, *IEEE Solid-State Sensor and Actuator Workshop,* 91–96, June, Transducers Research Foundation, Hilton Head, SC.

Frazier, A. B., Babb, J. W., and Allen, M. G. 1991. Design and fabrication of electroplated micromotor structures, *The Winter Ann. Mtg. Am. Soc. of Mech. Eng.* 135–146, December, American Society of Mechanical Engineers, Atlanta, GA.

Frazier, A. B., Ahn, C. H., and Allen, M. G. 1994. Development of micromachined devices using polyimide-based processes, *Sensors and Actuators,* 45:47–55.

Furuhata, T., Hirano, T., Gabriel, K. J., Fujita, H. 1991. *Proceedings of the IEEE Micro Electro Mechanical Systems Conference,* 57–62, January.

Gabriel, K. J. 1995. Engineering microscopic machines, *Scientific American,* 150th Anniversary issue, September, 150–153.

Guckel, H., Burns, D. W., Christenson, T. R., and Tilmans, H. A. C. 1989. Polysilicon x-ray masks, *Microelectronic Engineering,* 9:159–161.

Guckel, H., Christenson, T. R., Skrobis, K. J., Denton, D. D., Choi, B., Lovell, E. G., Lee, J. W., Bajikar, S. S., and Chapman, T. W. 1990a. Deep x-ray and UV lithographies for micromechanics, *Proc. IEEE Solid-State Sensor and Actuator Workshop,* 118–122, June, Hilton Head, SC.

Guckel, H., Christenson, T. R., Skrobis, K. J., Sniegowski, J. J., Kang, J. W., Choi, B., and Lovell, E. G. 1990b. Microstructure sensors, *Proceedings of the 1990 IEEE International Electron Devices Meeting,* 613–616, December, IEEE Electron Devices Society, San Francisco, CA.

Guckel H., Skrobis K. J., Christenson, T. R., Klein, J., Han, S., Choi, B., and Lovell, E. G. 1991a. Fabrication of assembled micromechanical components via deep x-ray lithography, *Proceedings of the 1991 IEEE Micro Electro Mechanical Systems Conference,* 74–79, January, IEEE Robotics and Automation Society, Nara, Japan.

Guckel, H., Skrobis, K. J., Christenson, T. R., Klein, J., Han, S., Choi, B., Lovell, E. G., Chapman, T. W. 1991b. On the application of deep x-ray lithography with sacrificial layers to sensor and actuator construction (the magnetic micromotor with power takeoffs), *Proceedings of the 6th International Conference on Solid-State Sensors and Actuators,* IEEE Electron Devices Society, San Fransisco, CA.

Guckel, H., Christenson, T. R., Earles, T., Klein, J., Zook, J. D., Ohnstein, T., and Karnowski, M. 1994. Laterally driven electromagnetic actuators, *Proc. Solid-State Sensor and Actuator Workshop,* 49–52, June, Hilton Head Island, SC.

Guthrie, W. L., Patrick, W. J., Levine, E., Jones, H. C., Mehter, E. A., Houghton, T. F., Chiu, G. T., Fury, M. A. 1992. A four-level VLSI bipolar metallization design with chemical-mechanical planarization, *IBM J. Res. Develop.,* 36(5):845–857.

Han, C. C. and Corelli, J. C. 1988. Azide-poly(methylmethacrylate) photoresist for ultraviolet lithography, *Journal of Vacuum Science and Technology B,* 6:219–223.

Harding, W. B. 1974. Aluminum, *Modern Electroplating,* Lowenheim, F. A., ed. 63–70, Plainfield, NJ.

Hayes, J. P. 1988. *Computer Architecture and Organization,* second edition, McGraw-Hill, New York, NY.

Henstock, M. E., and Spencer-Timms, E. S. 1963. The composition of thin electrodeposited alloy films with special reference to nickel iron, *Transactions of the Institute of Metal Finishing,* 179–185.

Hirose, S., Ikuta, K., and Umetani, Y. 1984. A new design method of servo-actuators based on the shape memory effect, *Proc. 5th RO.MAN.SY–84 Symp.* Udine, Italy, *Theory and Practice of Robotics and Manipulators,* 339–349. MIT Press, Cambridge, MA.

Ikuta, K. and Hirowatari, K. 1993. Real three dimensional micro fabrication using stereo lithography and metal molding, *Proc. of the IEEE Micro Electro Mechanical Systems Conference,* 42–47, February, Fort Lauderdale, FL.

Ikuta, K., and Shimizu, H. 1993. Two-dimensional mathematical model of shape memory alloy and intelligent SMA-CAD, *Proc. IEEE Micro Electro Mechanical Systems Conf.* 87 92, February, Fort Lauderdale, FL.

Ikuta, K., Tsukamoto, M., and Hirose, S. 1991. Mathematical model and experimental verification of shape memory alloy for designing micro actuator, *Proc. IEEE Micro Electro Mechanical Systems Conf.,* 103–108, February, Nara, Japan.

Ikuta, K., Hayashi, M., and Matsuura, T. 1994. Shape memory alloy thin film fabricated by laser ablation, *Proc. of the IEEE Micro Electro Mechanical Systems Conference,* 355–360, January, Oiso, Japan.

Ikuta, K., Hirowatari, K., and Ogata, T. 1994b. Three dimensional micro integrated fluid system (MIFS) fabricated by stereo lithography, *Proc. of the IEEE Micro Electro Mechanical Systems Conference,* 1–6, January, Oiso, Japan.

Kenny, T. W., Waltman, S. B., Reynolds, J. K., and Kaiser, W. J. 1991. Micromachined silicon tunnel sensor for motion detection, *Appl. Phys. Lett.,* 58:100–102.

Krinsky, S., Perlman, M. L., Watson, R. E. 1994. *Handbook on Synchrotron Radiation,* North-Holland.

Lin, B. J. 1975. Deep UV lithography, *Journal of Vacuum Science and Technology,* 12:1317–1320.

Maeda, S., Minami, K., and Esashi, M. "KrF excimer laser induced selective non-planar metallization", *Proc. of the IEEE Micro Electro Mechanical Systems Conference,* 75–80, January, Oiso, Japan.

Mastrangelo, C. H. and Hsu, C. H. 1993. Mechanical stability and adhesion of microstructures under capillary forces—parts I and IIa. *J. MEMS,* 2(1):33–55.

Mearing, S. G. 1986. Thick liquid photoresist for improving image and plating resolution, *Solid State Technology,* September.

Menz, W., Bacher, W., Harmening, M., and Michel, A. 1991. The liga technique—a novel concept for microstructures and the combination with Si-technologies by injection molding, *Proceedings of the 1991 IEEE Micro Electro Mechanical Systems Conference,* IEEE Robotics and Automation Society, 69–73, January, Nara, Japan.

Milosevic, I., Perret, A., Losert, E., and Schlenkrich, P. 1988. Polyimide enables high lead count TAB, *Semiconductor International,* 28–31, October.

Mimura, Y., Ohkubo, T., Takeuchi, T., and Sekikawa, K. "Deep-UV Photolithography," *Japanese Journal of Applied Physics,* vol. 17, pp. 541–550, March, 1978.

Miyazaki, S. and Nomura, K. 1994. Development of perfect shape memory effect in sputter-deposited Ti-Ni thin films, *Proc. of the IEEE Micro Electro Mechanical Systems Conference,* 176–181, January, Oiso, Japan.

Mohr J., Burbaum C., Bley P., Menz W., and Wallrabe, U. 1990. Movable microstructures manufactured by the liga process as basic elements for microsystems. *Micro System Technologies 90',* 529–537, September, Berlin, Germany.

Mohr J., Anderer, B., and Ehrfeld, W. 1991a. Fabrication of a planar grating spectrograph by deep-etch lithography with synchrotron radiation, *Micro System Technologies 90',* , 529–537, September, Berlin, Germany.

Mohr, J., Bley, P. Burbaum, C., Menz, W., and Wallrabe, U. 1991b. Fabrication of microsensor and microactuator elements by the liga-process, *Proceedings of the 6th International Conference on Solid-State Sensors and Actuators,* 607–609, IEEE Electron Devices Society, San Francisco, CA.

Moriya, K., Ohsaki, T., and Katsura, K. 1984. Photosensitive polyimide dielectric and electroplating conductor, *Proc. 34th Electronic Components Conf.,* 82–87, May , IEEE Electron Devices Society, New Orleans, LA.

Petersen, K. 1982. Silicon as a mechanical material, *Proc. IEEE,* 70(5):420–457.

Rickerl, P. G., Stephanie, J. G., and Slota, P. 1987. Processing of photosensitive polyimides for packaging applications, *IEEE Trans. Components, Hybrids, and Manufacturing Technology,* 690–694, December.

Saif, M. T. A., MacDonald, N. C. 1995. A milli newton micro loading device, *8th Int. Conf. Solid State Sensors and Actuators, and Eurosensors IX,* 60–63, June 25–29, Stockholm, Sweden.

Saotome, Y. and Inoue, A. 1994. Superplastic micro-forming of microstructures, *Proc. of the IEEE Micro Electro Mechanical Systems Conference,* 343–348, January, Oiso, Japan.

Schmidt, F. J. and Hess, I. J. 1966. *Plating,* 53:229.

Snakenborg H. 1983. A new development in electroformed nickel screens, *Proceedings of the Symposium on Electroforming/Deposition Forming,* American Electroplaters Society, March.

Shaw, K. A., Zhang, Z. L., and MacDonald, N. C. 1994. SCREAM I: a single mask, single crystal silicon, reactive etching process for microelectromechanical structures, *Sensors and Actuators,* A-40:63–70.

Siddons, P. 1994. Personal communication, Brookhaven National Laboratory, July.

Sze, S. M., ed., 1994. *Semiconductor Sensors,* John Wiley and Sons, New York, NY.

Takagi, T. and Nakajima, N. 1993. Photoforming applied to fine machining, *Proc. of the IEEE Micro Electro Mechanical Systems Conference,* 173–178, February, Fort Lauderdale, FL.

Visser, C. C. G., Uglow, J. E., Burns, D. W., Wells, G., Redaelli, R., Cerrina, F., and Guckel, H. 1987. A new silicon nitride mask technology for synchrotron radiation x-ray lithography: first results, *Microelectronic Engineering*, 6:299–304.

Wallrabe, U., Bley, P., Krevet, B., Menz, W., and Mohr, J. 1992. Theoretical and experimental results of an electrostatic micromotor with large gear ratio fabricated by the LIGA process, *Proceedings of the 1992 IEEE Micro Electro Mechanical Systems Conference*, 139–140, February, IEEE Robotics and Automation Society Travemunde, Germany.

Wolf, S. and Tauber, R. N. 1987. *Silicon Processing for the VLSI Era*, vol. 1, Lattice Press, Sunset Beach, CA.

Xu, Y., MacDonald, N. C., and Miller S. A. 1995. Integrated micro-scanning tunneling microscope, *Appl. Phys. Lett.*, 67:2305–2307.

Zhang, Z. L. and MacDonald, N. C. 1992. An RIE process for submicron, silicon electromechanical structures, *J. Micromech. Microeng.* 2(1):31–38.

117

Microsensors

117.1 Pressure Sensors and Accelerometers 1515
Terminology • Pressure Sensor Construction • Pressure Sensor Operation • Silicon Accelerometer Construction • Silicon Accelerometer Operation Characteristics • Applications
117.2 Acoustic Wave-Based Chemical Sensors 1519
Introduction • Fundamentals • Chemical Sensors • Sensor Systems • Outlook

Keith O. Warren
Litton Guidance and Control Systems

Antonio J. Ricco
Sandia National Laboratories

117.1 Pressure Sensors and Accelerometers

Keith O. Warren

Pressure sensors and accelerometers constitute the bulk of micro sensors produced for sensing physical non-electromagnetic variables. These silicon microsensors are uniquely positioned to grow rapidly into many new industrial applications due to an important combination of factors:

- Low cost, batch fabrication.
- High accuracy.
- Small size.
- Computer interfacability.

Terminology

Most solid state sensors present the user with a linear transfer function between the sensed variable and an electrical output either analog or digital. The electrical output of a sensor is related to the input variable by a simple linear equation of the form:

$$y = mx + b$$

where y is the electical output in volts (or current, digital word); x is the sensed pressure (or acceleration); m is the scale factor eg. (volts/pascal) (volts/g); b is the bias or offset.

The full scale excursion of the electrical output over the minimum to maximum measured range is often referred to in sensor specifications as the span. Unfortunately, real sensors are often sensitive to changes in other than the variable they were designed to report. Important error sensitivities include temperature sensitivity and mounting sensitivity. The scale factor and bias are each functions of temperature, and the magnitude of this error is listed in the data sheet of commercial sensors. A graphical representation of a typical sensor response is shown in Figure 117.1.

The temperature coefficient of the physical piezoresistance effect that many silicon sensors utilize is negative, resulting in reduced span at higher temperatures.

Pressure Sensor Construction

Silicon pressure sensors are essentially comprised of a thin diaphragm that flexes due to pressure differences across it (see Figure 117.2). If a reference vacuum is sealed on one side of diaphragm it becomes an absolute pressure sensor. The diaphragm contains diffused strain sensitive piezoresistors that convert pressure induced strain into an electrically measurable output.

An alternative construction used in some silicon pressure sensors utilizes a thin silicon diaphragm but relies on the variable capacitance between the flexing diaphragm and a fixed electrode, rather than piezoresistance, as the transduction mode. The devices are typically manufactured by bulk micromachining techniques that form the diaphragm by anisotropic chemical etching.

Since microphotolithographic techniques used by the IC industry are employed to pattern the diaphragm, piezoresistors,

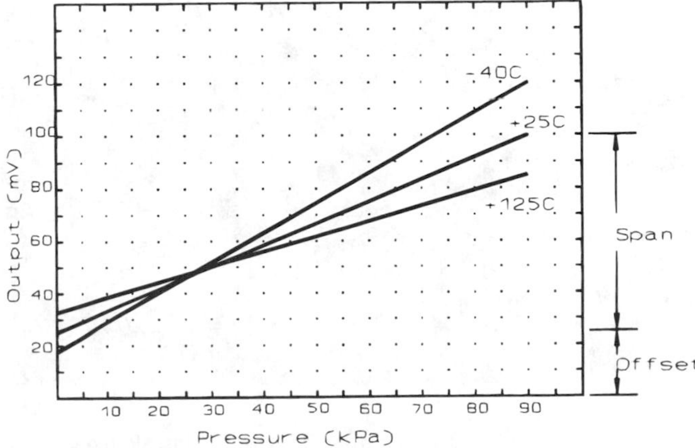

Figure 117.1 Typical sensor output vs. pressure and temperature.

0-8493-8343-9/97/$0.00+$.50

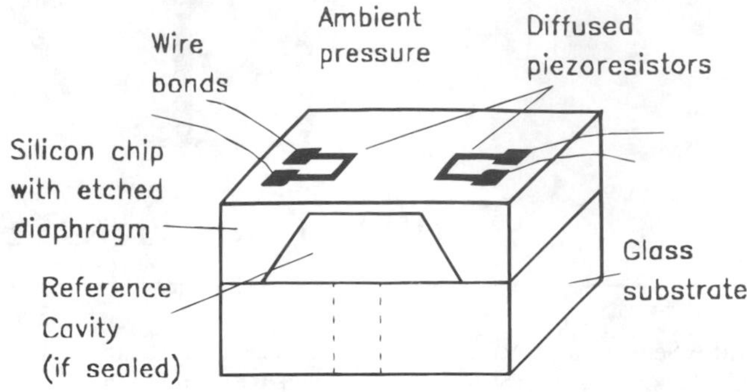

Figure 117.2 Cross-sectional drawing of silicon pressure sensor.

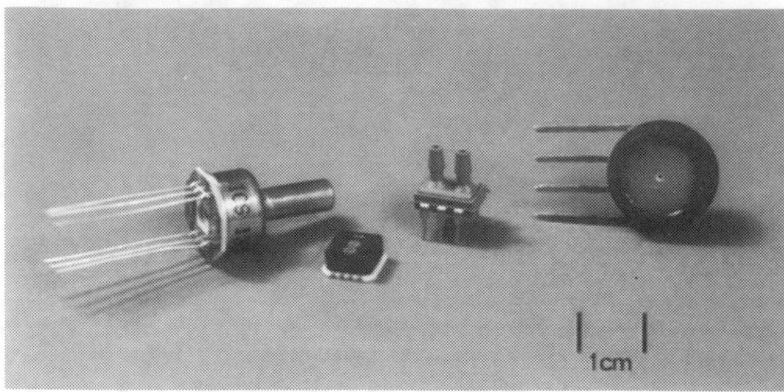

Figure 117.4 A few of the many pressure sensor packages available. Left, two units from IC Sensors; center, minidip Lucas NovaSensor; right, Motorola.

and metallization pads on the sensor, the overall chip size can be relatively small (approximately 1 to 3 mm on a side), with hundreds or thousands of individual units on a 100 mm diameter wafer. The thickness of the silicon wafers used by most manufacturers ranges between 0.3 to 0.5 mm. Typically, the silicon sensor wafers are anodically bonded to a 0.5 mm thick Pyrex glass support wafer before sawing into chips. The Pyrex wafer has a thermal expansion coefficient close to that of silicon and provides a compact, sealed, vacuum reference on one side of the diaphragm if the wafers are bonded in vacuo. Figure 117.3 is a photograph of a variety of unpackaged silicon pressure sensor chips.

The major sensor manufacturers offer a diverse array of low cost packaging styles to house the sensor chip. The package protects the chip and may include molded plastic hose barbs or other pressure interface, as well as easily soldered pins for electrical connection. Several representative package styles are illustrated in the Figure 117.4. Among the popular styles are metal cans with tubing ports, surface mounted leadless ceramic chip carriers, molded epoxy mini-dip with hose barbs, and molded packages for o-ring interface. The stability and reliability of the silicon chip and its low cost package is excellent in applications where the pressure media are benign dry gases such as air, nitrogen, or vacuum. Silicon sensor chips are also commercially available in stainless steel hermetic packages with corregated pressure transparent diaphragms for applications requiring exposure to corrosive gas or fluid media. Exposure to hydrogen gas may cause hydrogen permeation and embrittlement of high nickel alloy stainless steels, requiring care in the selection of a suitable alloy (Swanson, 1993).

Figure 117.3 Unpacked silicon pressure sensor chips. (Photo courtesy Lucas NovaSensor)

Pressure Sensor Operation

Silicon pressure sensors are commercially available with pressure ranges of 2.5 kPa (10″ H_2O) full scale to over 35 MPa (5000 psi) full scale. The small size of the silicon chip and its packaging engenders desirable characteristics such as high frequency response (typically 1 ms), rapid thermal equilibrium and small dead volume, which is important in limited sample volume analysis equipment and for dynamic considerations in feedback control applications.

Piezoresistors on the sensor require a voltage or current excitation and are usually configured as a Wheatstone bridge for first order thermal compensation.

Strain due to pressure change results in a differential output, whereas thermally induced changes are largely cancelled. The bridge output due to full scale pressure excursion is typically 1% to 10% of the supply voltage. However, resistance change of that magnitude can be caused by temperature changes of 2°C to 40°C depending on the device (Bryzek et al., 1990). Further temperature compensation required by most applications can be accomplished using a simple network of external resistors. The compensation resistor values and connection diagram are furnished by the manufacturer along with each device. Some models are supplied with built-in temperature compensation, usually in the form of a ceramic substrate with laser trimmed thick-film resistors assembled into the sensor package. Several manufacturers have incorporated amplifiers along with temperature compensated units to provide high-level output.

Silicon Accelerometer Construction

Silicon accelerometer construction is similar to pressure sensor construction, but is complicated by the necessity of a cantilevered sensing mass subject to measurable displacement in response to applied acceleration. An accelerometer sensing mass is referred to as the proof mass. As depicted conceptually in Figure 117.5, a silicon accelerometer consists of a support frame to which a proof mass is suspended by one or more thin regions that serve as flexures. The flexures, proof mass and frame are precision etched simultaneously along with hundreds of others on a single crystal silicon wafer. Doped piezoresistors may be diffused into the high strain areas of the flexure. The resistance change in the piezoresistors is a linear function of strain, with strain proportional to deflection of the cantilever, due to a force equal to the product of mass and acceleration. This is a spring-mass system with the restoring spring being the thin silicon flexure. For this micromechanical application, silicon is an excellent choice of material due to its extremely elastic nature and no plastic deformation below 600°C (Pearson et al., 1957). Metal springs in contrast can take a set if stressed beyond a narrow elastic range, or may exhibit hysteresis.

Implementation of the silicon suspension spring and proof mass varies among device manufacturers from the single cantilever (Barth et al., 1988) of Figure 117.5, to doubly supported cantilever devices (Terry, 1988) having thin silicon flexures at opposing ends of the proof mass. The improved ruggedness offered by this configuration is offset by the disadvantage of lower sensitivity for a given mass and flexure thickness. Since either type of accelerometer is subject to breakage by excessive shock or vibratory acceleration near the fundamental resonant frequency, overrange protection is provided by one or more capping layers. The small gap formed by the close proximity of the cap and proof mass yields another benefit by suppressing unwanted proof mass resonances using squeeze-film gas damping.

Recently, a few manufacturers have introduced surface micromachined capacitive accelerometers targeted for automotive airbag applications, one of which features on-chip closed loop electronics with high level output to lower the system parts count and ease interfacing (Goodenough, 1991).

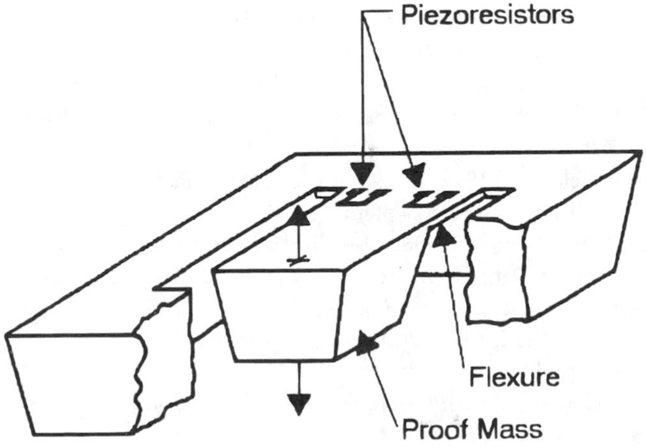

Figure 117.5 Simple single cantilever silicon accelerometer.

Silicon Accelerometer Operation Characteristics

Silicon accelerometers are commercially available for full scale input ranges of +/−2g to +/−200Kg (Quinnel, 1992), with +/−50g devices available from several manufacturers (Link, 1993). Dynamic characteristics must be considered since accelerometers are commonly used to sense shock and vibration, or as feedback sensors in motion control systems. Frequency response, damping factor, noise, and cross-axis sensitivity are usually included on device data sheets. Package mass (weight) can also affect the overall system dynamics if the accelerometer and its housing is an appreciable fraction of the total system mass subject to acceleration. Fortunately, this is seldom a concern due to the miniature size of the chips and low mass of the available packages (typically 1 to 10 grams). Undamped accelerometers exhibit a resonant peak in their frequency response curve which places a practical upper limit on the allowable input frequency band at about 1/3 of the resonant frequency to avoid exciting the resonance, which leads to output saturation or possible device breakage. Shock, random vibration and square wave inputs contain high frequency energy that can excite the accelerometer resonance causing an undesirable ringing in output response. Gas damped accelerometers are desirable for most applications due to the larger usable bandwidth and absence of a resonant peak.

Cross axis sensitivity denotes the accelerometer response to acceleration inputs orthogonal to the designated input axis. Actually this parameter should be termed case axis to true input axis misalignment, since an accelerometer rotated in an acceleration field will show an output changing through positive then crossing over into negative values. The crossover point with zero output while under acceleration means the acceleration is applied exactly orthogonal to the true input axis. The directions may not coincide exactly with the labeled case axes since the proof mass center of mass often lies below the surface of the chip while the flexure attachment lies on the surface of the chip. The line through the proof mass center of mass and intersecting the flexure attachment line at the chip surface is called the pendulous axis and indicates one direction of zero acceleration sensitivity. The true input axis is mutually perpendicular to the pendulous axis and the flexure attachment line.

Increasingly, silicon accelerometers are being considered for use in attitude, heading and reference systems, and for use in inertial navigation systems (Lanco and Geen, 1993; Warren, 1991; Sextant Avionique, 1988). Silicon devices developed for these applications possess much greater dynamic range (140db minimum measurable signal to maximum range) than those designed primarily for airbag crash sensors or general purpose instrumentation. Additionally, navigation quality silicon accelerometers meet much more demanding stability, repeatability, and linearity requirements. The high performance is achieved by using electrostatic force rebalance, rather than a simple spring flexure, to hold the proofmass at a constant null position between two sense/forcer electrodes. The associated closed loop electronics produces an electrostatic force that exactly counters the applied acceleration as required to keep the proofmass from moving. This technique has several benefits:

- Since the proofmass doesn't move, errors due to the flexure spring changing versus temperature are eliminated.
- Vibropendulous rectification is greatly reduced since this is a DC bias error caused by mechanical rectification of vibration as the input axis moves with the proofmass and sees slightly different components of off axis vibration on alternate half cycles.
- Scale factor variations of the proofmass position sensing circuitry have little effect since the loop drives toward null. Bias errors of the null sensing circuitry are divided by the high loop gain.

Further improvements in navigation grade silicon accelerometers performance and stability will be achieved as newer closed-loop silicon devices are fabricated using advanced BESOI and SIMOX technologies (Warren, 1994).

Applications

Automotive applications for silicon pressure sensors have existed since government mileage standards and exhaust emission regulations in the mid-1970s necessitated better control over fuel-air mixtures. Engine computers use input from manifold absolute pressure sensors (MAP) along with known stroke displacement and engine rpm to meter fuel. Several possible automotive applications for silicon pressure sensors, some of which are being actively developed include; barometric pressure, tire pressure, oil pressure, fuel level, turbo boost pressure, air conditioner compressor pressure and diesel engine pressure (Bryzek et al., 1990).

The first major automotive application for silicon accelerometers is the frontal crash sensor for airbag deployment. Another promising application under development is smart vehicle suspension which uses accelerometers in conjunction with pneumatically variable shock absorbers to tailor suspension stiffness and compliance to road conditions. Accelerometers and silicon inertial angular rate sensors may also be used in sophisticated antilock braking system to measure vehicle skid response. These automotive accelerometer uses are expected to be the fastest growing market for smart silicon sensors (MIR, 1992).

In addition to the large and growing volume of silicon sensors consumed by the automotive industry, there are a variety of other industrial applications for silicon microsensors. Silicon pressure sensors find important utility in chemical process control where the basic sensor is often incorporated into a robust housing along with 4–20 ma or digital interface circuitry and sold as a pressure transmitter. HVAC and air handling applications use low cost pressure sensors to measure flows of temperature controlled air for optimum high efficiency building heating or cooling (Tandeske, 1991). Many general flow sensing requirements can be satisfied by measurement of the differential pressure across a known orifice. Tank level sensing of known density liquids can be accomplished by static pressure measurement at the base of the vessel. Industrial safety applications such as explosion suppression systems, commonly use simple pressure switches to monitor volatile reaction chambers and trigger a rapid chamber purge within milliseconds of the beginning of an explosion to avoid destruction of the chamber or plant. Accurate silicon pressure sensors may be required in such an application if the process pressure normally fluctuates within a certain range and a explosive rate of change must be distinguished from the normal expected pressure changes.

Silicon accelerometers may be found in industry monitoring bearings of rotating machinery to detect vibration signatures indicating excessive wear so that preventive maintenance can be performed before a costly unexpected failure. Closed loop motion control systems use silicon accelerometers as feedback sense elements for acceleration or velocity (integrated acceleration). Higher accuracy and stability silicon accelerometers are increasingly being used in low cost inertial measurement units and navigation systems.

References

Barth, P., Pourahmadi, F., Mayer, R., Poydock, J., Petersen, K. 1988. A monolithic silicon accelerometer with integral air damping and overrange protection, *Technical Digest IEEE Solid State Sensors and Actuators Workshop,* p. 35, June 6–9, Hilton Head Island, SC.

Bryzek, J., Petersen, K., Mallon, Jr., J. R., Cristel, L., and Pourahmadi, F. 1990. *Silicon Sensors and Microstructures,* p. 8.1, 2.21, Lucas NovaSensor, Freemont, CA.

Goodenough, F. 1991. Airbags boom when IC accelerometer sees 50G, *Electronic Design,* August 8, 1991.

Lanco, J. and Green, J. 1993. Micromachined inertial sensor development at Northrop, presented at the *Inst. of Navigation Conf.,* June 23, Cambridge, MA.

Link, B. 1993. Field-qualified silicon accelerometers: from 1 milli g to 200,000 *g*, *Sensors,* 10(3):28–33.

Market Intelligence Research Corporation (MIR) 1992. Market Intelligence Research Corporation Forecast, cited in *Design News,* p. 56, February 10.

Pearson, G. L., Read, W. T., and Feldman, W. L. 1957. Deformation and fracture of small silicon crystals, *Acta Metallurgica,* 5:181.

Quinnel, R. A. 1992. Silicon accelerometers tackle cost sensitive applications, *EDN,* September 3, 1992.

Sextant Avionique 1988. *MACSI(r) Silicon Micro Accelerometer,* Crouzet Data Sheet, Ref. 61213 095/809, Sextant Avionique, Division Aerospatial, France.

Swanson, R. M. 1993. Electronic pressure transmitters: the hydrogen problem, *Sensors,* 10(9):33–35.

Tandeske, D. 1991. *Pressure Sensors: Selection and Application,* p. 248, Marcel Dekker, New York, NY.

Terry, S. 1988. A miniature silicon accelerometer with built-in damping, *Technical Digest IEEE Solid State Sensors and Actuators Workshop,* p. 114, June 6–9, Hilton Head Island, SC.

Warren, K. 1991. Electrostatically force-balanced silicon accelerometer, *Navigation, Inst. Navigation,* 38(1):91–99.

Warren, K. 1994. Navigation Grade silicon accelerometers with sacrificially etched SIMOX and BESOI structure, *Technical Digest Solid State Sensors and Actuators Workshop,* June 13–16, Hilton Head Island, SC

117.2 Acoustic Wave-Based Chemical Sensors

Antonio J. Ricco

Introduction

Chemical microsensors have growing roles in many applications, including environmental cleanup and monitoring, industrial process control, automotive and industrial emissions monitoring, aeronautical and space systems, planetary exploration, nonproliferation of weapons, screening for explosives, worker safety, and health care. By integrating a physical transduction platform with a chemically sensitive interface, the chemical microsensor provides some of the functionality of analytical instrumentation, but with vastly reduced cost, size, and power consumption (Hughes et al., 1991). Chemical microsensors utilize many categories of transduction platforms, some of the most popular being based on silicon microelectronic devices, liquid-phase electrochemical methods, ionically and/or electronically conducting ceramic materials, optical waveguides and fibers, thermal transducers, and acoustic wave devices, the last being the subject of this article.

The two platforms predominantly utilized for acoustic wave (AW)-based chemical sensors, the surface acoustic wave (SAW) device and the thickness-shear mode (TSM) resonator, have been extensively explored for this application only in the last 15 years, although seminal experiments date back more than 35 years. In 1959, G. Sauerbrey of the Technical University of Berlin demonstrated the use of a TSM resonator—also known as the "quartz-crystal microbalance" (QCM)—to monitor the thickness of vacuum-deposited metal films (Sauerbrey, 1959), and King demonstrated the first chemical sensor based on the a TSM resonator in 1964 (King, 1964). Meanwhile, the use of interdigital transducers (IDTs) to launch and detect surface acoustic waves on cm-size piezoelectric substrates was demonstrated by R. M. White of the University of California, Berkeley in 1965 (White and Voltmer, 1965), with the first demonstration of chemically sensitive SAW devices in 1979 by H. Wohltjen and R. Dessy of Virginia Polytechnic Institute (Wohltjen and Dessy, 1979). The flexural plate-wave (FPW), shear-transverse wave (STW), leaky SAW (LSAW), and shear-horizontal acoustic plate mode (SH-APM) have recently been added to the acoustic modes used for AW sensors. Over the past ten years, activity in this field has expanded considerably, with perhaps 50 research groups around the world exploring the use of AW devices for gas- and liquid-phase chemical sensing; a few commercial sensors are beginning to address the needs of niche markets as well.

Fundamentals (Martin et al., 1994a,b; Ricco, 1994; Grate et al., 1993; Buttry and Ward, 1992; Wohltjen, 1984)

Acoustic wave devices use piezoelectric crystals, in combination with conductive electrodes, to couple electric fields and mechanical motion, thus exciting and detecting acoustic waves. AWs are sensitive to the mass and mechanical properties of thin, surface-attached films, making them sensitive to chemical species *ab*sorbed or *ad*sorbed by such films and to the physical properties of the films themselves.

The TSM resonator consists of a thinned piezoelectric crystal (typically a disk of AT[1]-cut quartz) with metal electrodes on both faces. Application of an oscillating voltage between these electrodes excites the crystal into a shear-mode mechanical resonance: the two surfaces undergo displacement within their respective planes. The resonant frequency f_0 is equal to $v/2t$, where v is the shear-wave velocity (3.10 km/s for AT-quartz) and t is the crystal thickness. Typically, t ranges from 310 to about 75 μm, yielding resonant frequencies between 5 and 20 MHz. The device functions as a sensor, in either the gas or liquid phase, when incorporated as the frequency-control element of an oscillator circuit (Martin et al., 1994b; Wessendorf, 1993; Buttry and Ward, 1992).

Two SAW device configurations are utilized for chemical sensing. The delay line (Figure 117.6, top) uses one of its two comb-like, photolithographically-defined IDTs to launch a traveling wave along the surface of a piezoelectric substrate, most often ST-cut[1] quartz. SAW resonators utilize one IDT, in combination with a periodic array of ridges (Figure 117.6, bottom), to launch and maintain a standing wave. In both configurations, the second IDT receives the electrical signal associated with the SAW, sending it to external circuitry. In the commonly-used oscillator loop configuration, the signal from the receiving IDT provides the input to an amplifier, the output of which drives the SAW-launching IDT (Figure 117.6). The oscillation frequency f_0 is simply v/d, where v is the SAW velocity (3.16 km/s for ST-quartz) and d is the IDT periodicity. For chemical sensors, d typically ranges from 100 to 6 μm, yielding f_0 in the 30–500 MHz regime.

Perturbation of an AW can affect two of its propagation parameters: velocity and attenuation (the latter manifested as a change in insertion loss for the SAW device, or damping for the TSM resonator). Velocity perturbations are most commonly measured to provide sensor response, changes in frequency being proportional to changes in AW velocity. Because frequency stability of 1 part in 10^8 (e.g., 1 Hz in 100 MHz) is attainable, minute perturbations to wave velocity are measurable. For the SAW, most of the acoustic energy is confined to within one wavelength of the surface, making it very sensitive to surface perturbations. The energy of the TSM is distributed throughout the crystal, so thinner crystals, carrying a greater fraction of the wave energy

[1] The "T" in the designations "AT", "BT", and "ST" stands for temperature; AT and BT were the first two widely recognized temperature-coefficient-optimized cuts for TSM quartz resonators; for the SAW, ST-cut quartz has a minimal temperature coefficient at room temperature.

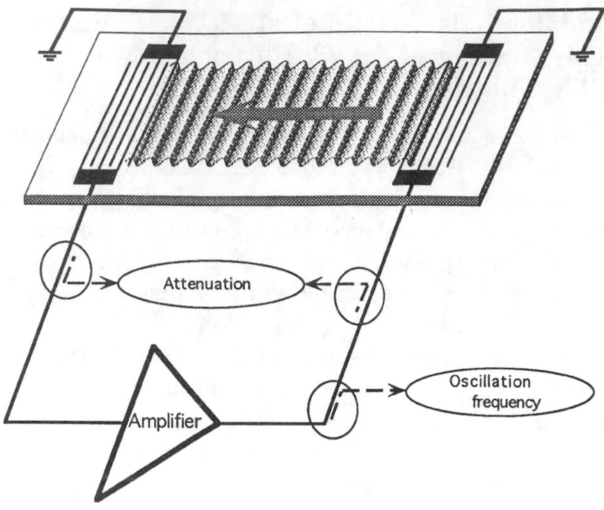

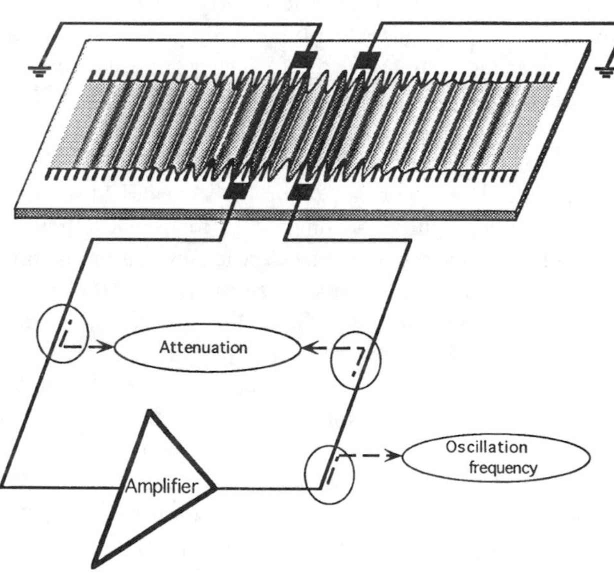

Figure 117.6 Surface acoustic wave (SAW) delay line and associated circuitry (top), showing surface motion resulting from the traveling wave; SAW resonator and associated circuitry (bottom), showing exponentially decaying surface motion associated with the standing wave. IDTs are visible near the left and right ends of the delay line; for the resonator, they are located near the center and are largely obscured by the schematic surface displacement. Sensors are realized by application of a chemically sensitive film to the (shaded) region between the transducers for the delay line and, for the resonator, to the reflector array and the region between transducers.

at the surface, are more sensitive. Thus, higher sensitivity requires smaller t for the TSM resonator and smaller d for the SAW device, the consequence in both cases being higher operating frequencies, more difficulty in device fabrication/handling, and more costly instrumentation.

Change in mass-per-area on the AW device surface is the most-utilized perturbation for sensing applications. Less than 100 pg/cm^2 is detectable by a 100-MHz SAW, while the limit of the TSM resonator is a few ng/cm^2 at 5 MHz. For the TSM resonator, the Sauerbrey Equation (Sauerbrey, 1959) describes the relationship between changes in resonant frequency (Δf) and mass/area ($\Delta(m/A)$):

$$\Delta f = -2\frac{f_0^2}{(\mu_q\rho_q)^{1/2}}\Delta(m/A), \tag{117.1}$$

in which μ_q and ρ_q are the respective shear modulus and density of the quartz substrate and f_0 is the unperturbed resonant frequency.

In addition to mass loading, AW devices respond to a broad range of physical perturbations: changes in the electrical, mechanical, or rheological properties of a thin film or other medium in contact with the device can induce a response. Such multiparameter sensitivity is a double-edged sword: while providing a multitude of physical mechanisms to probe interfacial chemical interactions, it also creates responses to unintended perturbations, with occasionally confusing or misleading results. For a SAW delay line coated with an *acoustically thin* (see below) film having thickness h, complex shear modulus $G\,(= G' + jG'')$, and electrical conductivity σ, changes in oscillation frequency are described by (Martin et al., 1994a; Ricco and Martin, 1991)

$$\Delta f \cong -\kappa c_m f_0^2 \Delta(m/A) + \kappa c_{ve}\frac{f_0^2}{v_0^2}\Delta(hG') - \kappa f_0\frac{K^2}{2}\Delta\left(\frac{(\sigma h)^2}{(\sigma h)^2 + (v_0 C_s)^2}\right). \tag{117.2}$$

Here κ is the fraction of the center-to-center IDT spacing covered by the thin film (κ is often unity for polymer and other electrically insulating films, which can cover the entire device); the other parameters are defined in Table 117.1 along with their values for ST-cut quartz.

Two points regarding Equation 117.2 require comment. First, in the case of a perfectly elastic film—one with*out* viscous loss at the SAW frequency—G' in Equation 117.2 can be replaced by $\mu[(\lambda + \mu)/(\lambda + 2\mu)]$, λ and μ being the Lamé constants of the film. Second, for the viscoelastic term of Equation 117.2 to apply, the constraint that the film be acoustically thin is important and requires explanation. A perfectly elastic film, generally including metals, ceramics, and many glassy polymers, is considered acoustically thin if $h < \Lambda_0(v_f/v_0)^2$, where Λ_0 is the SAW

Table 117.1 Material Constants for X-Propagating, ST-Cut Quartz Substrates for SAW Devices (Ricco and Martin, 1991; Martin et al., 1994a)

Description and symbol	Magnitude and units
Coefficient of mass sensitivity, c_m	1.29 cm^2 g^{-1} MHz^{-1}
Coefficient of viscoelastic sensitivity, c_{ve}	1.55 cm^2 g^{-1} MHz^{-1}
SAW velocity, v_0	3.16 km s^{-1}
Electromechanical coupling coefficient,[a] K^2	0.11%
Capacitance/length,[b] C_a	0.5 pF cm^{-1}

[a] A measure of the piezoelectric strength of the substrate

[b] $C_a = \epsilon_a + \epsilon_o$, the respective permittivities of the substrate and free space

wavelength ($\Lambda_0 \approx d$), and v_f is the velocity of acoustic shear waves *in the thin film* (Martin et al., 1994a; Ricco, 1994); h in the range of 1–3% of the product $\Lambda_0(v_f/v_0)^2$ is a reasonable rule-of-thumb limit for "acoustically thin". If the film is *visco*elastic, then it is acoustically thin if $h < |G|/(f_0 v_0 \rho)$, where ρ is the film density (see Martin et al., 1994a). An acoustically thick film requires a more complex treatment of these effects (Martin et al., 1994a). Note that h need not be particularly large for a film to be acoustically thick: for a 100-MHz ST-quartz SAW device, a 100-nm-thick film with $|G| = 10^8$ dyne/cm^2 (typical for a polymer film in its rubbery state) is acoustically thick. The need to make polymer films acoustically thin (in order to avoid complications when interpreting the response) conflicts with one requirement for rapid sensor response: soft, rubbery polymers are generally far more permeable than hard, glassy films, yielding more rapid response; but glassy polymer films can be made much thicker (hence more sensitive) before becoming acoustically thick.

As mentioned above, perturbation of an AW can affect its attenuation in addition to velocity. While attenuation is seldom measured, it can be quite useful, particularly in the initial phases of system design when candidate sensing materials are being evaluated: unlike most of the other physical parameters to which AW devices are sensitive, mass changes do not affect the wave's attenuation. Monitoring AW attenuation/damping to confirm that it does *not* change therefore verifies that frequency changes can be interpreted as mass loading changes (provided, of course, that extrinsic variables such as temperature, pressure, and electric field do not contribute to the response; in practice, actively minimizing changes in such external perturbations through appropriate system design is necessary to obtain high sensitivity and stability).

Chemical Sensors

The literature of AW device-based chemical sensors has been reviewed in several publications (Janata et al., 1994; Janata, 1992, 1990; Ballantine and Wohltjen, 1989; Nieuwenhuizen and Venema, 1989; Alder and McCallum, 1983). Research focuses on both gas- and liquid-phase analytes and the coatings appropriate for these environments. In the gas phase, the volatile organic compounds (VOCs) are the best-studied analytes, with small-molecule inorganic gases—predominantly CO, CO_2, various NO_x species, NH_3, SO_2, H_2S, and the halogens—in second place. Most of the liquid-phase research has been carried out in aqueous solutions, with dissolved organics, ionic species (particularly heavy metals), and biomolecules all receiving attention. Unfortunately, owing to the surface-normal motion associated with a propagating SAW, these waves are highly attenuated by liquid contact, limiting their use as chemical sensors to the gas phase. One consequence of this limitation is that papers focused on SAW technology invariably deal with gas-phase or thin-film materials characterization applications, while the TSM literature focuses more often, though not exclusively, on the liquid phase. This limitation is also part of the reason for recent interest in "SAW alternatives" such as the SH-APM, STW, LSAW, and FPW, the hope being that such modes will provide the high sensitivity of the SAW in a liquid environment.

To build an AW-based chemical sensor, the device surface must be coated with a chemically sensitive film. For TSM resonators, one or both sides of the disk are coated; for SAW delay lines, the region between IDTs is coated; for SAW resonators, the reflector array is coated. (If the sensing film is not electrically conductive, it can cover SAW IDTs as well.) Film deposition is accomplished using many techniques, including spin, dip, and air-brush casting of polymers and solution-castable oxides; evaporation, sublimation, and sputtering of metals and suitable organic and organometallic compounds; chemical vapor deposition, plasma deposition, and surface chemical derivatization using both organic and inorganic materials; and specialized procedures such as the Langmuir-Blodgett and self-assembling monolayer methods for producing molecularly organized thin films. Regardless of the technique of deposition, film uniformity is important. In addition, if the mass sensitivity of a film-coated device is to follow theory, the sensing film should be acoustically thin (see above), a condition that limits film thickness to between a fraction of one and several microns.

Three very general strategies exist for obtaining selectivity in a chemical microsensor-based system (Ricco et al., 1994):

1. The preparation of a selective chemically sensitive interface, with its own transducer platform, for the species of interest.
2. The use of a chromatographic method to separate analytes from one another according to the time required for each to traverse a medium such as a packed column, causing analytes to impinge one by one on the sensor.
3. The preparation of an array of microsensors, each bearing a chemically distinct coating, but *without* the restriction of perfect selectivity, coupled with the use of mathematical pattern-recognition (PR) techniques to evaluate the response of the array.

When the number of possible species to be analyzed is small and known, or interferences unlikely, technique (1) is optimal; if size, weight, and power consumption are not critical, technique (2) provides laboratory-like versatility; when the number of species is large or unknown, and size, weight, and power must be minimized, the array/PR technique can be the optimal choice.

Regardless of the chosen selectivity strategy, the sensitivity, specificity, reversibility, and response speed of an AW chemical sensor (or system) all depend largely on the properties of the sensing film(s). Therefore, much of the AW sensor research in recent years has focused on chemically sensitive thin films and how they interact with AW devices. Table 117.2 lists some examples of chemically sensitive films and the analytes to which they respond.

Rational selection or design of a chemically selective coating material for a particular analyte is accomplished using knowledge of bulk-phase chemical interactions. For example, the styrene detector (Table 117.2) utilizes a derivative of the first metal/olefin complex ever prepared, Zeise's salt, which has specific affinity

Table 117.2 Chemically Sensitive Films and Analysis for AW-Based Sensors

Film Group	Chemically Sensitive Film	Analyte(s) [Limit of Detection]
Metals	Pd	H_2 [50 ppm]
	Pt	NH_3 [0.5%]
Metal oxides	WO_3	H_2S [10 ppm]
	ZnO	Organic solvents
	t-$PiCl_2$(ethylene)(pyridine)	Styrene [5 ppm]; vinyl acetate [5 ppm] butadiene
Organometallic complexes	CuPc[a]	Cl_2, Br_2, I_2
	H_2Pc, CoPc, CuPc, FePc, MgPc, NiPc, PbPc	NO_2 [500 ppb]
	Cu^{2+}/mercaptourndecanoic acid self-assembled monolayer	DIMP [100 ppb], DMMP[b]
	Co(II) complexes of isonitrilobenzoylacetate and tetra-methylethylenediamine	DIMP
	Triethanolamine	SO_2 [10 ppb]
Organic compounds	Pyridinium tetracyanoquinodimethane	NO_2
	Aminopropyltriethoxysilane	Nitrobenzene and its derivatives
	Histidine hydrochloride	DIMP, malathion, parathion
	Polybutadiene	O_3
	Polyethylene maleate	Cyclopentadiene [200 ppm-min]
Organic polymers[c]	Ethyl cellulose, fluoropolyol, phenoxy resin, poly(amidodixine), poly-1-butadiene, poly(epichlorohydrin), polyethylene, poly-(ethylene maleate), poly(isoprene), poly(methylmethacrylate), polystyrene, poly(vinyl chloride), poly(vinyl stearate), 1,1,1-trifluoroisopropyl methyl siloxane	Benzene, 1-butanol, 2-butanone, 1,2-dichloro-ethane, dichloropentane, *N,N*-dimethylaceta-mide, dimethylphosphite, dodecane, meth-anesulfonyl fluoride, octane, α-pinene oxide, *i*-propylacetate, toluene, triamylphosphite, tributylphosphate, water
	Ethyl cellulose, fluoropolyol, poly(ethylene maleate), poly(vinylpyrrolidone)	Chemical warfare agents and their simulants [30 ppb typical for organophosphonates]
Biomaterials	Goat antibody to human immunoglobulin G	Human immunoglobulin G
	Yeast in alginate gel	Metabolism of glucose

[a] Pc = phthalocyanine: [b] DIMP = diisopropylmethylphosphonate; DMMP = dimethylmethylphosphonate: [c] Because organic polymers generally respond to many compounds, several are often used in an array to provide a distinct response pattern for each analyte. *Sources*: D'Amico et al., 1982; Kepley et al., 1992; Nieuwenhuizen and Venema, 1989; Zhang and Zellers, 1993; Ballantine et al., 1986; Bryant et al., 1983; Costello et al., 1992; Dahint et al., 1994; Fog and Rietz, 1985; Grate, Rose-Pehrrson et al., 1993; Katritzky et al., 1989; Snow and Wohltjen, 1984; Heckl et al., 1990; Martin et al., 1985; Rajakovic et al., 1989; and Smith et al., 1993.

for carbon-carbon double-bond-containing compounds (Zhang and Zellers, 1993). Another example is the detection of hydrogen using a Pd film, which relies on the unique ability of Pd to rapidly and reversibly dissociate and dissolve H_2 at room temperature, with a resulting change in its mechanical properties (D'Amico et al., 1982). An organophosphonate sensor utilizes a surface-immobilized form of Cu^{2+}, a known phosphonate hydrolysis catalyst (Kepley et al., 1992). Figure 117.7 shows the response of this sensor to 500 and 100 ppm-by-volume diisopropylmethylphosphonate (DIMP), a simulant of the chemical warfare agent sarin.

To some degree, there is a tradeoff between specificity and reversibility: the most specific chemical interactions are often the strongest and therefore most difficult to reverse. The notable exception is biological complexes—antibody/antigen interactions and the like—which utilize exquisite recognition of molecular configuration to obtain selectivity, often without irreversible binding. Due to their general frailty in all but carefully controlled environments, as well as difficulties with rigid surface attachment, biocomplexes have yet to be widely applied to AW sensors. The three non-biological examples given above (sensors for styrene, hydrogen, and DIMP) are also exceptions to the specificity/reversibility tradeoff, with the latter two taking their cue from catalysis: to catalyze a chemical reaction, the catalyst must have specific but reversible interactions with the reacting compounds, precisely the characteristics desired for a chemical sensor.

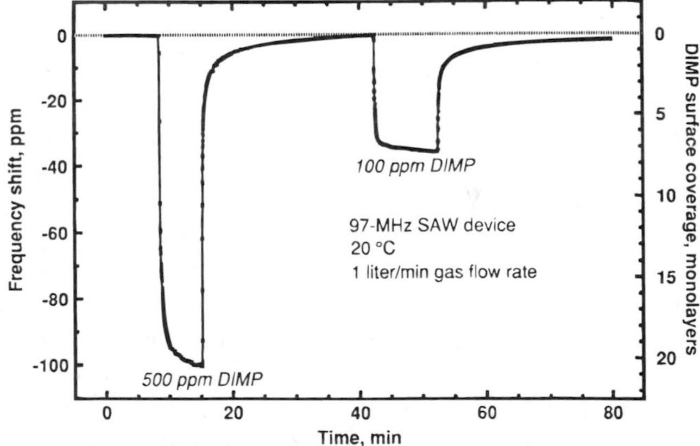

Figure 117.7 Response (frequency shift, in parts per million) of a SAW delay line-based chemical sensor utilizing a self-assembled monolayer of mercaptoundecanoic acid capped with Cu^{2+} ions to selectivity detect diisopropylmethylphosphonate (DIMP, a simulant of the nerve agent sarin) in flowing nitrogen.

While the relatively weak interaction between organic polymers and organic solvents is not highly selective, moderate advances in choosing polymers for detection of a particular solvent have been made through the use of *solubility parameters,* which estimate the affinity of a particular solvent for a given polymer (McGill et al., 1994). The weak nature of polymer/solvent interactions has the advantage that absorption of the solvent is typically rapid and reversible, particularly for polymers in their rubbery state.

Nonspecific adsorption is the bane of the selective AW chemical sensor. Because a response due to mass loading is inevitable, simple physical adsorption of moderate-vapor-pressure compounds can thwart even the most elegant schemes for molecular recognition. The most problematic interferences in this regard are often species with vapor pressures in the milliTorr-to-several-Torr range: volatile enough to produce significant (ppm and above) concentrations in the gas phase, but nonvolatile enough to physically condense readily onto virtually any surface. An additional complication is that the source of such a compound—e.g., a puddle of liquid—does not rapidly evaporate and disperse, as do high-volatility solvents and the like. Thus, none of the materials in Table 117.2 are immune to interference.

Sensor Systems

To perform effectively, AW-based chemical sensor systems often require augmentation(s) relative to the single-sensor, single-film, passive-sampling-of-the-ambient-environment configuration. Perhaps the most basic augmentation is a reference device, which can be either an uncoated device exposed to the same environment as the sensing device, or an identically coated device that is isolated from the chemical species to be detected but exposed to the same "extrinsic environment", i.e., suffering the same fluctuations in temperature, pressure, etc. The former approach is most popular and an obvious answer to the nonspecific adsorption problem, but the latter can be very helpful when, for example, sensitivity of the coating material (as opposed to the sensor platform) to temperature fluctuations is the primary source of noise or drift in the baseline signal. Two fairly complex enhancements, chromatographic separation and the array/PR approach, are mentioned in the previous section. In addition, both sensitivity and limit of detection (LOD) for a given analyte can be enhanced by increasing the area of the surface upon which the chemically sensitive film is formed; for example, the deposition of a porous, high-surface-area thin film can enhance surface area by a factor of 50 or more (Ricco et al., 1989). Preconcentration of the analyte *via* its accumulation on a sorptive column, followed by rapid release upon heating of the column, is a second means to enhance LOD and sensitivity. Baseline drift is a common problem that can be addressed by periodic "rezeroing" of the sensor, accomplished by providing a source of clean gas for reference, or by switching in a scrubbing column (utilizing activated charcoal, for example, in the case of volatile organic analytes), to remove all analytes from the gas stream. For many of the system strategies just described, active sampling is required, with pumps and valves to provide a controlled flow of the analyte stream to the sensor; active temperature control is often necessary as well.

A number of practical AW sensor systems, using one or more of the enhancing strategies listed above, have been developed. Gas-phase systems monitor volatile organics, identify sources of smoke, detect nerve gases or explosives, signal the presence of illegal drugs, and determine particulate sizes. An example of a successful system is the Fuel Dilution Meter developed by Microsensor Systems, Inc.,[2] which uses a polymer-coated 158-MHz SAW delay-line device to monitor the headspace above a volume of lubricant for the presence and concentration of diesel or jet fuel vapors (Jarvis et al., 1994). A schematic diagram and photograph of this system are presented in Figure 117.8. The response of the coated SAW sensor is quite linear over the 0–10% range of lubricant dilution by added fuel. A second application

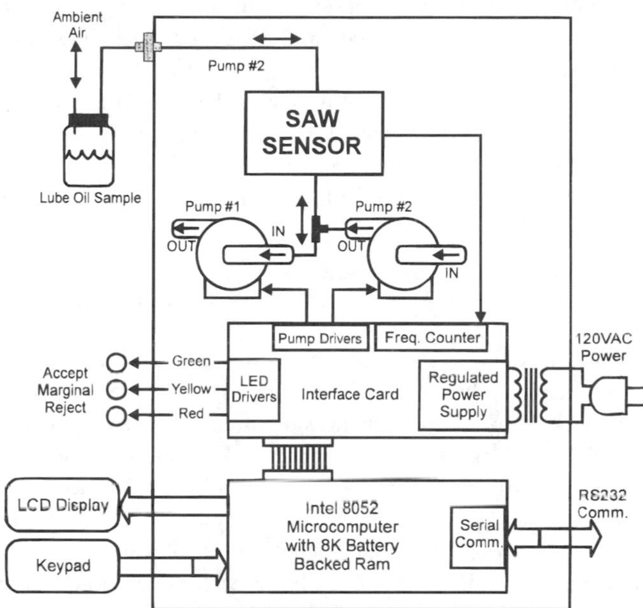

Figure 117.8 Schematic diagram (top) and photograph (bottom) of the Microsensor Systems, Inc. Fuel Dilution Meter.

[2] Microsensor Systems, Inc., 62 Corporate Ct., Bowling Green, KY, 42103; Femtometrics, Inc., 17252 Armstrong Ave., Irvine, CA 92714, Maxtek, Inc., 2908 Oregon Ct., Torrance, CA, 90503; Elchema, Inc., P.O. Box 5067, Potsdam, NY, 13676.

example is the SAW non-volatile residue monitor (Figure 117.9) developed by Femtometrics, Inc.[2] The function of this instrument, which utilizes a 200-MHz SAW resonator and includes a microcontroller and hard disk for data acquisition, is to detect non-volatile airborne contaminants in semiconductor processing facilities.

Liquid-phase systems, most of which are based on the TSM resonator, monitor electrodeposition, measure liquid viscosity, and monitor lubricant breakdown. Maxtek, Inc.[2] markets a commercial system to monitor the electroplating of metals directly onto one electrode of the TSM resonator. Elchema, Inc.[2] sells an Electrochemical Quartz Crystal Microbalance (EQCM) system that includes the instrumentation to carry out a variety of *in-situ* electrochemical experiments on one of the two electrodes of a TSM resonator.

Outlook

Acoustic wave chemical sensors offer considerable promise for the future, particularly for those applications where their exquisite sensitivity to surface mass changes offers a competitive edge that outweighs the costs of making measurements at radio frequencies. Further, the fact that the SAW and other AW devices have been successfully fabricated from polycrystalline thin films of piezoelectric materials, notably ZnO and AIN (Gunshor et al., 1983; Pearce et al., 1981) offers hope of complete integration of this technology onto silicon chips bearing all the necessary electronic circuitry, with resulting economies of mass production.

Despite a fairly thorough understanding of the fundamentals of AW chemical sensors, and with the notable exception of successful niche-market applications like those described in the previous section, they have yet to be widely applied to commercial problems. For example, the *Sensors 1995 Buyers Guide* (*Sensors*, 1994) lists 17 companies in the "Surface Acoustic Wave" category, but this does not equate with high-volume sales of SAW-based chemical sensor systems: the world-wide sales of such systems for 1995 has been estimated at 500 (H. Wohltjen, private communication). For R&D purposes, SAW sensor platforms are available along with measurement systems,[5] in some cases even including coatings tailored to specific sensing problems. Reasons for the lack of commercial proliferation are complex, including regulatory, marketing, and cost factors, as well as technical limitations. In general, however, the entire field of chemical sensors suffers from a common malady: the technology of the physical transduction platforms is far ahead of the development of chemically sensitive, selective interfaces. In particular, interfaces must be developed that not only have interesting and useful interactions with key analytes, but they must be commercially viable in terms of manufacturability, reproducibility, and longevity. Many advances in acoustic wave chemical sensing will come as researchers bring their creativity to bear on chemically selective materials problems.

Figure 117.9 Photograph of the Femtometrics, Inc. non-volatile residue monitor, showing the control unit (including data acquisition and storage) and remote sensor head (including reference and sensing SAW resonators, as well as temperature-control hardware).

Acknowledgments

I gratefully acknowledge innumerable helpful technical discussions during ten years of collaboration with Stephen J. Martin of Sandia National Laboratories (SNL), in particular with regard to the fundamentals of the interaction mechanisms between SAWs and surface films. The technical assistance of A. W. Staton, M. A. Hill, M.-A. Mitchell, and B. L. Wampler, and the graphics expertise of K. Rice, all of SNL, are acknowledged as well. This work was supported by the U.S. DOE under contract DE-AC04-94AL85000.

References

Alder, J. F. and McCallum, J. J. 1983. Piezoelectric crystals for mass and chemical measurements, *The Analyst,* 108: 1169–1189.

Ballantine, Jr., D. S. and Wohltjen, H. 1989. Surface acoustic wave devices for chemical analysis, *Anal. Chem.*, 61:704A–715A.

Ballantine, D. S., Jr., Rose, S. L., Grate, J. W., and Wohltjen, H. 1986.Correlation of surface acoustic wave device coating responses with solubility properties and chemical structure using pattern recognition, *Anal. Chem.* 58, 3058–3066.

Bryant, A., Poirier, M., Riley, D. L., and Vetelino, J. F. 1983. Gas detection using surface acoustic wave delay lines, *Sensors & Actuators* 4, 105–111.

Buttry D. A. and Ward, M. D. 1992. Measurement of interfacial processes at electrode surfaces with the electrochemical quartz crystal microbalance, *Chem. Rev.,* 92:1355–13579.

Costello, B. J., Wang, A. W., and White, R. M. 1992. A flexural-plate-wave microbial sensor, *Technical Digest of the 1992 Solid-State Sensor and Actuator Workshop*, IEEE: New York, pp. 114–117.

Dahint, R., Grunze, M., Josse, F., and Renken, J., 1994. Acoustic plate mode sensor for immunochemical reactions, Anal. Chem., 66, 2888–2892.

D'Amico, A., Palma, A., and Verona, E. 1982. Palladium-surface acoustic wave interaction for hydrogen detection, *Appl. Phys. Lett.*, 41:300–301.

Fog, H. M., and Rietz, B. 1985. Piezoelectric crystal detector for monitoring of ozone in working environment, *Anal. Chem.*, 2634–2638.

Grate, J. W., Rose-Pehrrson, S. L., Venezky, D. L, Klusty, M., and Wohltjen, H. 1993. Smart sensor system for trace organophosphorus and organosulfur vapro detection employing a temperature-controlled array of surface acoustic wave sensors, automated sample preconcentration, and pattern recognition, *Anal. Chem.* 65, 1868–1881.

Grate, J. W., Martin, S. J., and White, R. M. 1993. Acoustic wave microsensors, *Anal. Chem.*, 65:940A-8A:987A–996A.

Gunshor, R. L., Martin, S. J, Pierret, R. F., Surface acoustic wave devices on silicon, *Jap. J. Appl. Phys.* 22, Suppl. 22–1, 37 (1982).

Heckl, W. M., Marassi, F. M., Kallury, K. M. R., Stone, D. C., and Thompson, M. 1990. Surface acoustic wave sensor response and molecular modeling: selective binding of nitrobenzene derivatives to (aminopropyl)triethoxysilane, *Anal. Chem.*, 62, 32–37.

Hughes, R. C., Ricco, A. J., Butler, M. A., and Martin, S. J. 1991. Chemical microsensors, *Science,* 254:74–80.

Janata, J. 1990. Chemical Sensors, Anal. Chem., 62:33R–42R.

Janata, J. 1992. Chemical Sensors, Anal. Chem., 64:196R–218R.

Janata, J., Josowicz, M., and De Vaney, D. M. 1994. Chemical Sensors, *Anal. Chem.*, 66:207R–228R.

Jarvis, N. L., Wohltjen, H., Klusty, M., Gorin, N., Fleck, C., Shay, G., and Smith, A. 1994. Solid-state microsensors for lubricant condition monitoring Part I: fuel dilution meter, *J. Soc. Tribologists and Lubricn. Engineers,* 50:689–693.

Katritzky, A. R., Offerman, R. J., and Wang, Z. 1989. Utilization of pyridinium salts as microsensor coatings, *Langmuir,* 5, 1087–1092 (1989).

Kepley, L. J., Crooks, R. M., and Ricco, A. J. 1992. A selective SAW-based organophosphonate chemical sensor employing a self-assembled, composite monolayer: a new paradigm for sensor design, *Anal. Chem.*, 64:3191–3193.

King, Jr., W. H. 1964. Piezoelectric sorption detector, *Anal. Chem.*, 36:1735–1739.

Martin, S. J., Frye, G. C. and Wessendorf, K. O. 1994b. Sensing liquid properties with thickness-shear mode resonators, *Sensors and Actuators,* 44:209–218.

Martin, S. J., Frye, G. C., and Senturia, S. D. 1994a. Dynamics and response of polymer-coated surface acoustic wave devices: effect of viscoelastic properties and film resonance, *Anal. Chem.*, 66:220–2219.

Martin, S. J., Schweizer, K. S., Schwartz, S. S., and Gunshor, R. L. 1985. Vapor sensing by means of a ZnO-on-Si surface acoustic wave resonator, *Proc. 1984 IEEE Ultrasonics Symp.*, IEEE: New York, pp. 207–213.

McGill, R. A., Abraham, M. H., and Grate, J. W. 1994. Choosing polymer coatings for chemical sensors, *Chemtech,* 24:27–37.

Nieuwenhuizen, M. S. and Venema, A. 1989. Surface acoustic wave chemical sensors, *Sensors and Materials,* 5:261–300.

Pearce, L. G., Gunshor, R. L., Pierret, R. F., Sputtered aluminum nitride on silicon for SAW device applications, *Proc. 1991 IEEE Utrasonics Symp.*, IEEE: New York, pp. 381–383 (1993).

Rajakovic, L., Ghaemmaghami, V., and Thompson, M. 1989. Adsorption on film-free and antibody-coated piezoelectric sensors, *Anal Chem. Acta* 217, 111–121.

Ricco, A. J. 1994. SAW chemical sensors, *The Electrochemical Society Interface,* 3:38–44.

Ricco, A. J. and Martin, S. J. 1991. Thin metal film characterization and chemical sensors: monitoring electronic conductivity, mass loading, and mechanical properties with SAW devices, *Thin Solid Films,* 206:94–101 (1991).

Ricco, A. J. and Martin, S. J. 1994. *Acoustic Wave-Based Sensors, 1995 Yearbook of Science and Technology,* 349–352, McGraw-Hill, New York, NY.

Ricco, A. J., Frye, G. C., and Martin, S. J. 1989. Determination of BET surface areas of porous thin films using surface acoustic wave devices, *Langmuir,* 5:273–276.

Ricco, A. J., Xu, C., Crooks, R. M., and Allred, R. E. 1994. Chemically sensitive interfaces on SAW devices, *Interfacial Design and Chemical Sensing,* Mallouk, T. E. and Harrison, D. J., eds., *ACS Symp. Series,* No. 561, pp. 264–279, American Chemical Society, Washington, DC.

Sauerbrey, Verwendung von Schwingenquarzen zur Wägung dünner Schichten und zur Mikrowägung, *Z. Physik,* 155:206–222.

Sensors 1995 Buyer's Guide, 1994. A. Helmers Publishing, Peterborough, NH.

Smith, D. J., Vetelino, J. F., Falconer, R. S., and Wittman, E. L. 1993. Stability, sensitivity, and selectivity of tungsten trioxide films for sensing applications, *Sensors and Actuators B,* 13–14, 264–268.

Snow, A., Wohltjen, H., Poly(ethylene maleate)-cyclopentadiene: a model reactive polymer-vapor system for evaluation of a SAW microsensor, 1984. *Anal. Chem.* 56, 1411–1416.

Wessendorf, K. O. 1993. The lever oscillator for use in high-resistance resonator applications, *Proc. 1993 IEEE Int. Freq. Control Symp.*, pp. 711–717.

White, M. and Voltmer, F. W. 1965. Direct piezoelectric coupling to surface elastic waves, *Appl. Phys. Lett.*, 7:314–316.

Wohltjen and Dessy, R. 1979. Surface acoustic wave probe for chemical analysis: introduction and instrument description, *Anal. Chem.*, 51:1458–1464.

Wohltjen, 1984. Mechanism of operation and design considerations for surface acoustic wave device vapor sensors, *Sensors and Actuators,* 5:307–325.

Zhang, -Z. and Zellers, E. T. 1993. Coated surface acoustic wave sensor employing a reversible mass-amplifying ligand substitution reaction for real-time measurements of 1,3-butadiene at low-and sub-ppm concentrations, *Anal. Chem.*, 65:1340–1349.

118

Micro Actuators and Energy Supply

118.1 Micro Actuators 1526
Classification of Micro Actuators • Electromagnetic Actuator • Electrostatic Actuator • Piezo Electric Actuator • Giant Magnetostrictive Alloy Actuator • Optical Piezo Electric Actuator • Shape Memory Alloy Actuator • Thermal Actuator • Polymer and Other Actuators

118.2 Energy Supply Methods and Non-Contact Manipulation 1533
Classification of Energy Supply Methods • Internal Supply Methods • External Supply Methods and Non-contact Manipulation

Toshio Fukuda
Nagoya University

Fumihito Arai
Nagoya University

118.1 Micro Actuators

Classification of Micro Actuators

Many different types of micro actuators have been proposed. In designing an actuator, the manner in which to obtain the mechanical energy is the critical point of discussion. The possible energy transformations are shown in Figure 118.1. In Figure 118.1, basic energy is classified as either optical, electrical, thermal, mechanical, and chemical energy. There are many different methods for producing driving power, e.g. an electrostatic motor which converts electric energy into mechanical energy. Several types of micro actuators have been proposed such as an electrostatic actuator, electromagnetic actuator, piezo electric element, GMA (Giant Magnetostrictive Alloy), optical actuator, SMA (Shape Memory Alloy), polymer actuator, pneumatic actuator, and so on. Each actuator has both merits and demerits and several applications have been proposed (Fukuda and Arai, 1993, 1992). The typical micro actuators with examples of their application are introduced here.

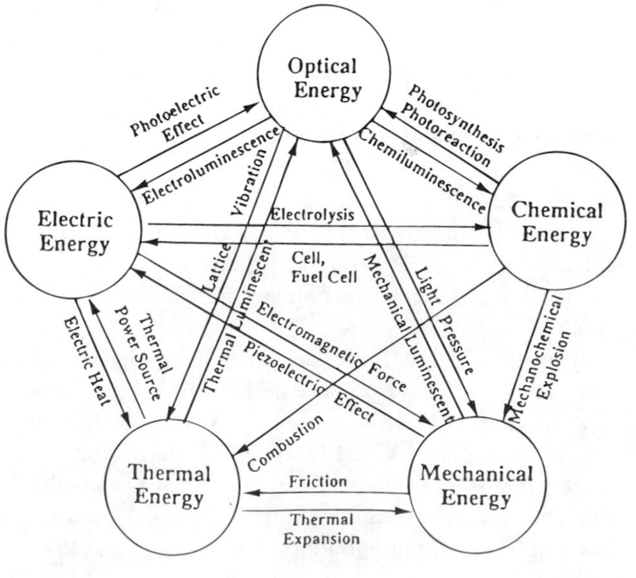

Figure 118.1 Energy transformation between the 5 basic energies.

Electromagnetic Actuator

Electromagnetic actuators are frequently used for miniature sized mechanical systems, such as the focusing control portion of an optical disk drive and a magnetic head positioner for hard disk drives. These types of actuators employ a permanent magnet and the electromagnetic force is produced according to Flemming's law. Figure 118.2 shows the moving coil-type electromagnetic actuator which is used for the focusing system of an optical disk drive. Miniaturization of this actuator is strongly dependent upon the size and power of the permanent magnet.

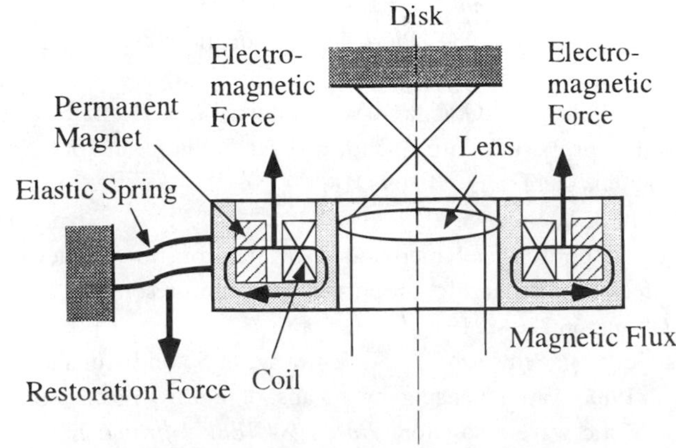

Figure 118.2 Moving coil type electromagnetic actuator used for focusing system of optical disk drive.

0-8493-8343-9/97/$0.00+$.50

Conventional electromagnetic motors are shrinking in size. Hisanaga et al., (1991) developed an electromagnetic motor for a 4.6 mm micro car. Figure 118.3 shows the diagram for an electromagnetic motor (Teshigahara et al., 1992). The motor, which is similar in structure to the stepmotor used in a quartz watch, consists of a small rotating permanent magnet, a small coil and a small core shaft. The SmCo permanent magnet used as a rotor, has a diameter of 1 mm and a thickness of 0.5 mm and is installed on a ZrO_2 rotating shaft. The magnet is 4-pole and magnetized in the radial direction. The intensity of the magnetization is estimated at about 600 Gauss or larger. The coil has a diameter of 1 mm, a length of 2.5 μm and the diameter is wound more than 1000 turns. Electromagnetic stainless steel is used for the core shaft.

Itoh et al. (1991) have developed an electromagnetically actuated small DC motor, that is cylindrical in shape, 3mm in external diameter and 5mm in length. This motor has three thin coils within the motor case. Inside the coils, a rare earth permanent magnet rotates, which is supported by bearings. Moreover, the rotor size is reduced by using a sensorless and brushless DC driving method, which detects the rotor position using the Electro Motive Force (EMF) of the coils. The starting torque of the motor is 2×10^{-5} Nm and the maximum output is 98 mW when the applied voltage is 2 V.

Electrostatic Actuator

When the voltage is applied between two electrodes, the electrostatic force is generated as shown in Figure 118.4. From the scaling law, as the size of the object becomes smaller, the weight decreases in proportion to the cube of the size i.e. ($|L^3|$). The influence of miniaturization on the electrostatic force between the electrodes is not serious ($[L^0]$). An electrostatic actuator is suitable for miniaturization and can easily be miniaturized (Trimmer and Jebens, 1989). In contrast, the electromagnetic actuator, which is often compared with the electrostatic actuator, requires a long cable with enough space to produce a magnetic field, and has a large resistance with a large energy loss, so it is not suitable for miniaturization.

Several kinds of electrostatic actuators have been developed. Electrostatic force can be used in both linear and rotary motors.

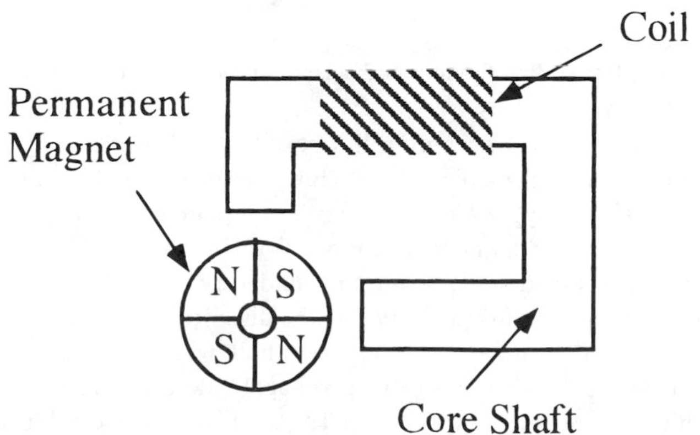

Figure 118.3 Electromagnetic motor (Teshigahara et al., 1992).

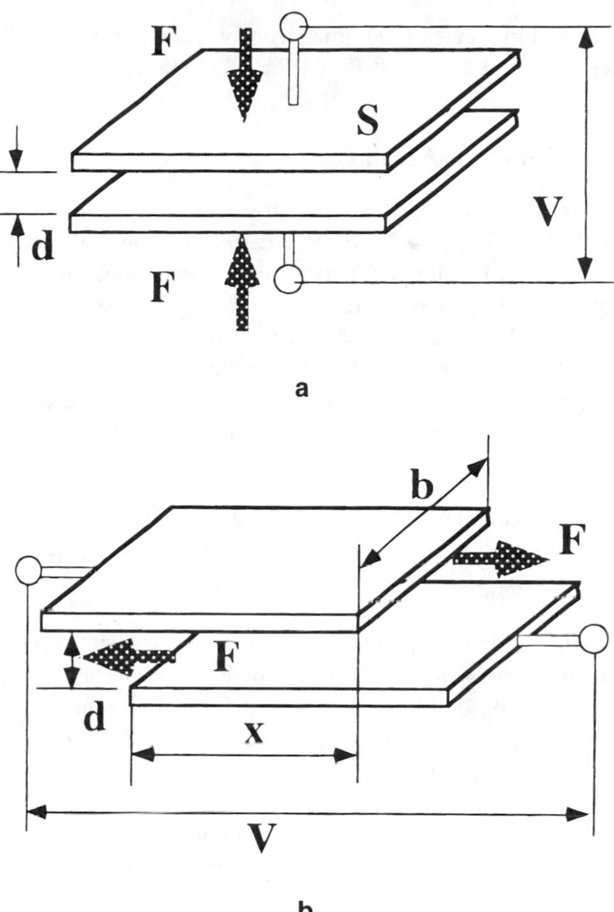

Figure 118.4 Electrostatic force. a. Vertical force. b. Thrust force.

Rotary actuators can be classified as top-drive, side drive, and wobble (harmonic) types. Researchers at UC Berkeley succeeded in rotating the side drive electrostatic micro motor, and a MIT group improved the rotation speed to more than 10,000 rpm and its life-time to more than one week (Mehregany et al., 1990). The main reason for the improvement was said to be the solution of the friction problem.

For the micro actuator, friction is the big problem. The steps used to minimize this problem are considered to be the utilization of

1. Elastic deformation.
2. Rotary motion.
3. Floating device, and so on.

As examples of cases (1), parallel type (Fujita et al., 1988), quad type (Fukuda and Tanaka, 1990), and comb type of electrostatic actuators (Kim et al., 1992) have been developed. Suzuki et al. (1992) have developed electrostatic actuators fabricated from polysilicon and polyimide that can act outside the plane. They are used for the wing joints and muscles of insect-type microrobots. As examples of case (2), a wobble motor (harmonic electrostatic motor), and cylindrical or corn shaped rotary-type have been produced. Examples of case (3), are the utilization of air pressure (Pister et al., 1990), magnetic force and superconductive force using the Meissner effect (Kim et al., 1990).

Piezo Electric Actuator

The Piezo electric actuator has high resolution (on the order of nm) and good response (on the order of kHz), and generates a large force. It is suitable for nano servo-positioning, and it was frequently used as a micro actuator with great success (Hatamura et al., 1990; Higuchi et al., 1990). For example, it is quite popular to use PZT ($Pb(Zr,Ti)O_3$) for the precise positioner of a micro manipulator and STM/AFM (Scanning Tunneling Microscope/Atomic Force Microscope). Its characteristics are summarized as follows.

1. Precise positioning is possible without any clearances.
2. Response speed is high and effective for force operations.

Utilizing these properties, a precise positioner with repetitive control of the quick deformation of PZT has been proposed (Higuchi et al., 1990). A Micro manipulator, with multi-degrees of freedom, has been proposed using a stack of piezo electric elements, and position and force experiments have been conducted to show its effectiveness (Fukuda et al., 1992). Stacked-type piezo electric elements are not suitable for the miniaturization of the total system. Hysteresis and creep phenomena should be properly treated in the control system.

The strain of the piezo electric element is small (0.1%). So continuous actuations of the PZT and sliding mechanism are employed (Higuchi et al., 1990; Ikuta et al., 1994). Ikuta et al. used this mechanism with a brake that is actuated by electromagnetic force. This actuator has been named the cybernetic actuator (Ikuta et al., 1994).

On the other hand, the extension mechanisms are employed in some cases to enlarge the displacement of the PZT. Figure 118.5 shows micro fish that can swim in fluid (Fukuda et al., 1994, 1995). These robots employ the stacked-type PZT with an extension mechanism. For the micro robot in a fluid, the viscosity force is dominant compared with the inertia force. So, the actuator having large output and fast response is suitable. The robot in Figure 118.5a repeats the quick deformation of the PZT to vibrate the double fins symmetrically (150 HZ to 750 Hz) in order to produce a progressive wave as a propulsion force. Because the stacked-type device has little displacement, this robot expands the displacement of the PZT up to 326 times (theoretically) using the hinge extension mechanism made by the electric discharge process. The displacement of the fins are expanded around the resonant frequency and sufficient propulsion force is generated. The swimming speed in the water is 3.7 cm/s. The robot in Figure 118.5b employs a new steering mechanism. The body is made of a magnification mechanism that is designed to enlarge the displacement of PZT by 250 times, geometrically, as shown in Figure 118.5c. This robot moves backward at the first resonant frequency (100 Hz in water) and moves forward at the second resonant frequency (275 Hz in water).

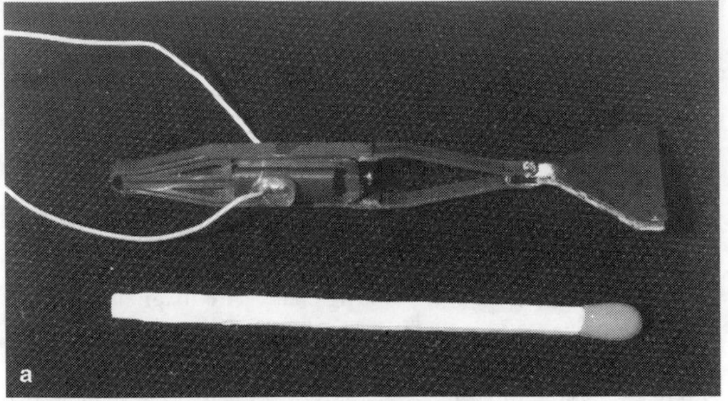

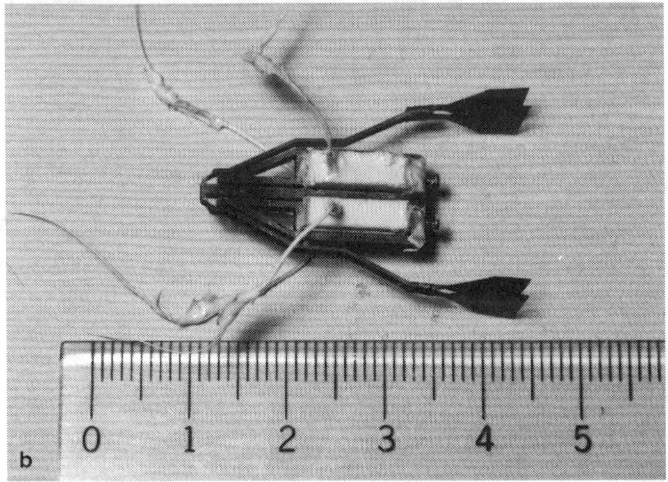

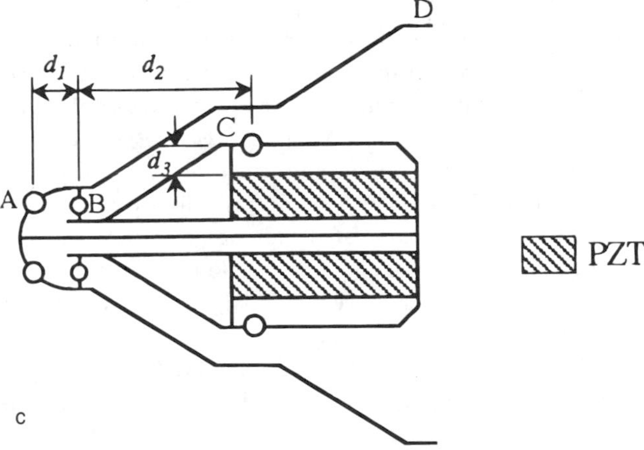

Figure 118.5 Magnification mechanism of prototype II micro fish.

In other examples, a bimorph-type of PZT has been developed for application to mobile robots. Thin films of PZT look promising for future applications of the micro actuator. (Morita et al., 1995) developed an ultrasonic motor based on a PZT thin film. This film is made on a titanium cylinder by the hydrothermal method (Shimomura et al., 1991). The thickness of the PZT film was about 7 μm to 9 μm. A screw bolt was rotated by this actuator (295 rpm at 33 driving voltage) (Morita et al., 1995). Thin films of PZT and PT (Lead Titanate) have been combined with micromechanical structures and MOS integrated circuits for robotic applications (Polla, 1992).

Giant Magnetostrictive Alloy Actuator

Magnetic material that generates strain when a magnetic field is applied is called magnetostrictive material. Study of magnetostrictive material was activated in USA beginning in 1960, and the element (Tb-Dy-Fe alloy) which has large magnetostrictivness has been developed. With the development of crystal growth technology, a large magnetostrictive property is obtained with a comparatively small magnetic field. Giant magnetostrictive alloy (GMA) extends the lines of the magnetic field direction and generates the strain as shown in Figure 118.6. GMA produces large force and large displacement compared with the piezo electric element (about two times, cf: Table 118.1), and the weight per unit stress is small, which can be a great advantage for an actuator (Fukuda et al., 1991). The GMA is driven by the magnetic circuit which controls the outer magnetic field. Yet, the element itself can be used as a cableless actuator, and it was used as a driving actuator for a micro robot (Fukuda et al., 1991). Figure 118.7 shows an in-pipe mobile robot actuated by a GMA (diameter 6 mm, length 4 mm). Maximum speed for this robot is 0.5 mm/s.

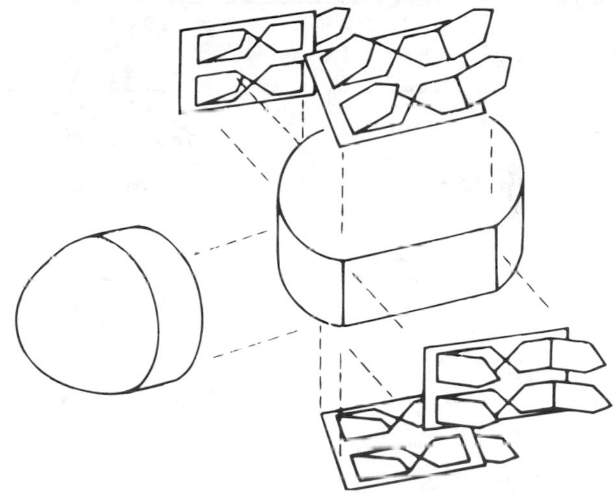

Figure 118.7 Mobile robot in pipe using GMA.

Optical Piezo Electric Actuator

Optical actuators employing light energy for driving are currently the subject of intensive study. Their advantage is that inductive noise is eliminated, electrical insulation is not needed, and non-contact connection is easy. Also, an integrated optical system will be comprised of optical devices and components. For implementation of such an integrated optical system, it is essential that advanced optical actuators be developed.

An optical piezo electric element, which exhibits photostrictive phenomena, has been examined as a new actuator (Uchino et al., 1985; Fukuda et al. 1992). In this actuator, the cable is not a nuisance. As applications, a mobile robot and a relay switched on/off by the light beam have been developed. Optical response characteristics of the optical piezo electric element by UV ray irradiation are characterized by the strain, which results from a combination of the following three different phenomena as shown in Figure 118.8.

1. Photostrictive effect of generating the photostrictive voltage, and the strain caused by the piezo electric effect.
2. Pyroelectric effect of generating pyroelectric current by the temperature difference, and the strain caused by the piezo electric effect.
3. Thermal deformation caused by the applied heat flux.

To improve the response characteristics of the optical piezo electric actuator, a bimorph-type of PLZT (Pb, La) (Zr, Ti)O_3 has been developed. The displacement is increased and the response time of the strain by UV ray irradiation is improved up to about twenty seconds (Fukuda et al., 1992). Moreover, response characteristics are improved by irradiating both sides of the actuator. Response time of the bending motion of this actuator is vastly improved compared with single side irradiation (Fukuda et al., 1994). Figure 118.9 shows the experimental device for the UV ray (365 nm wave length peak with narrow spectral band width) irradiation of the bimorph-type PLZT (La/$PbZrO_3$/$PbTiO_3$:3/52/48). Figure 118.10 shows the displacement at the

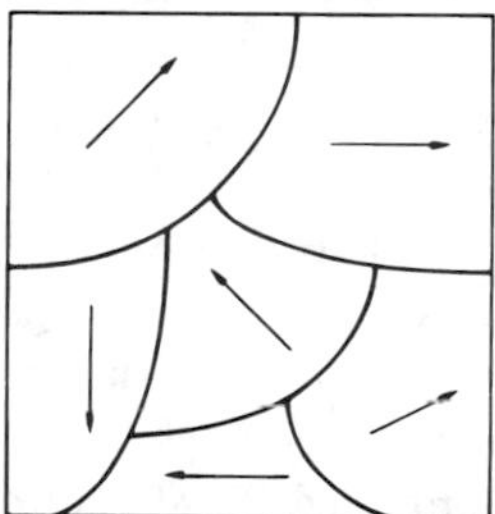

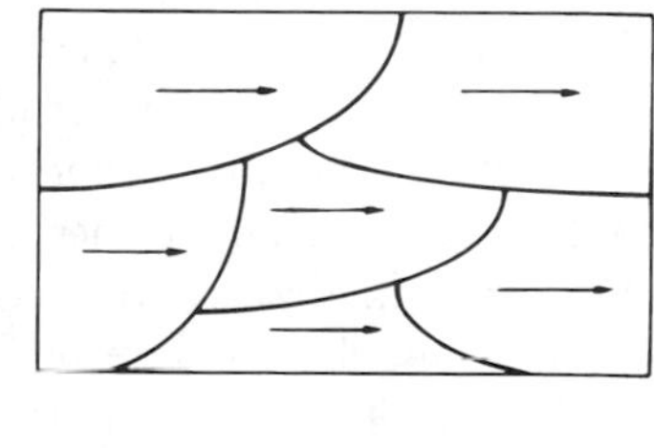

Figure 118.6 Deformation model of GMA.

Table 118.1 Comparison Between GMA and PZT

	GMA	PZT
Energy density [J/m^2]	2250 ~ 36000	670 ~ 950
$\Delta L/L \times 10^6$ (Room temp.)	1650 ~ 2400	670 ~ 950
Response time	ns ~ μs	μs
Weight/power [g/(kgf/mm^2)]	2.0	78

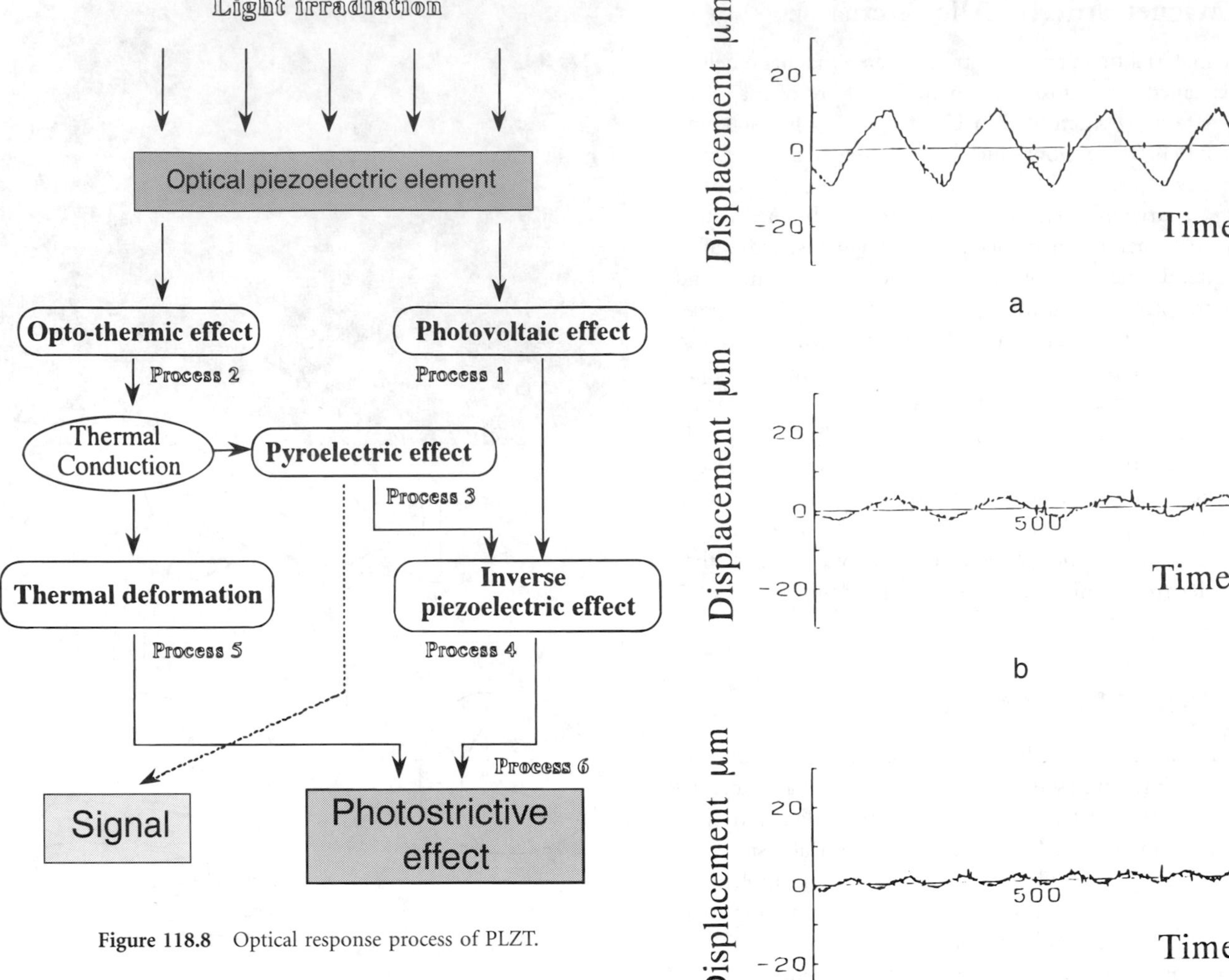

Figure 118.8 Optical response process of PLZT.

Figure 118.10 PLZT response to ON/OFF irradiation.

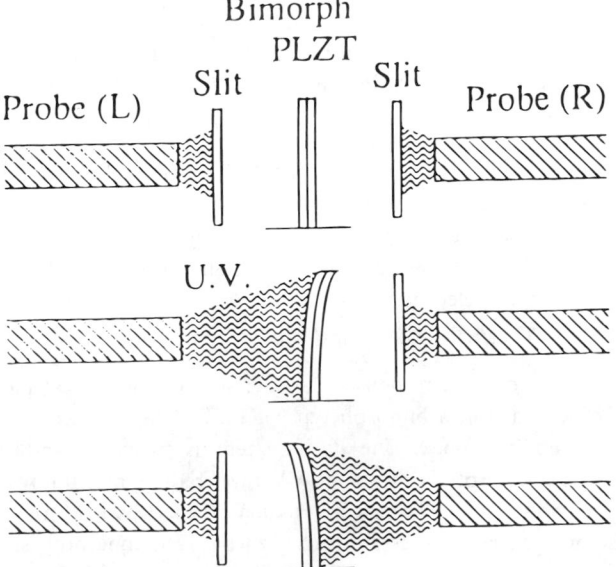

Figure 118.9 UV ray irradiation experiment of PLZT.

tip of the PLZT, when the left and right sides are irradiated alternately at difference frequencies of 1 Hz, 4 Hz, and 8 Hz by the UV intensity of 170 mW/cm^2. Note that the response time of the PLZT is extremely improved.

Hysteresis is a common phenomenon on piezoelectric devices. We examined the hysteresis caused by the accumulation of charges due to a photo electromotive force while the thermal deformation was kept small. For this purpose, the intensity of a light source is set as low as 20 mW/cm^2. We have proposed PWM control for this actuator to change the strength of the UV rays. The UV ray irradiation intensity is almost proportional to the duty ratio. Figure 118.11 shows the hysteresis characteristics of the bimorph PLZT. We can improve the hystersis by double sided irradiation. Based on these improvements, the model of this control system was derived, and the control method for the PLZT proposed. Experimental results of the optical control systems are shown in Figure 118.12.

As applications of this actuator, we have developed an optical micro gripper and an optical mobile robot with non-contact

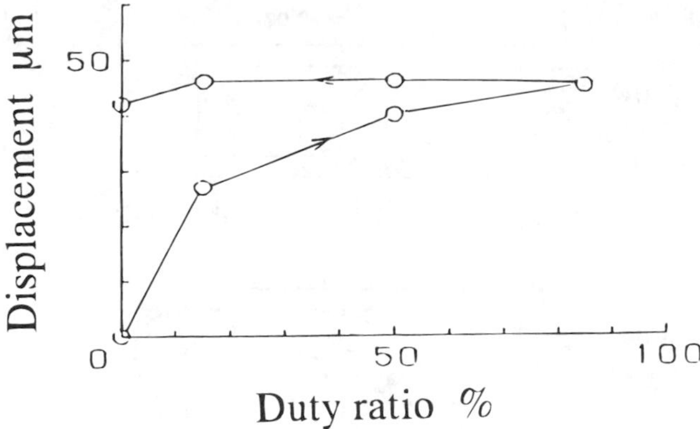

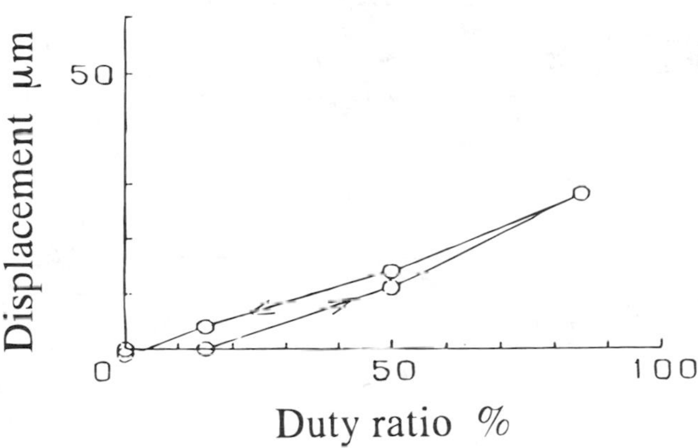

Figure 118.11 Hysteresis characteristics.

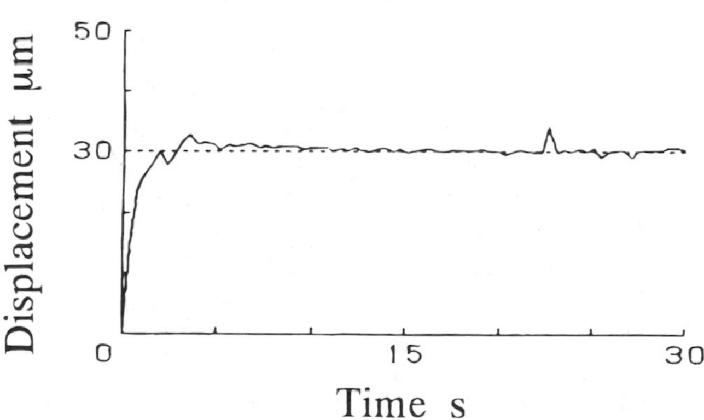

Figure 118.12 Experimental result of optical servoing.

energy transmission. Under consideration is an integrated optical servo control system with energy and information transmission. PLZT has a different optical response for the three different effects stated earlier. So, multi-functional use of the PLZT is expected not only as an actuator but also as the information transmitter.

Shape Memory Alloy Actuator

SMA (shape memory alloy, e.g., NiTi) has a shape memory effect in that it has a unique capability of recovering deformation when heated above a characteristic temperature. The recovery strain is in the range of 6 to 8%. Usually, the maximum strain is set around 2%, so that the shape memory effect will not be partially lost. As a result of miniaturization, heat capacity decreases and heat radiation from the surface increases, which results in an improvement in the response speed. SMA has many application examples such as medical operation tools (Ikuta, 1988; Dario et al., 1991; Fukuda et al., 1993). Figure 118.13 shows an active

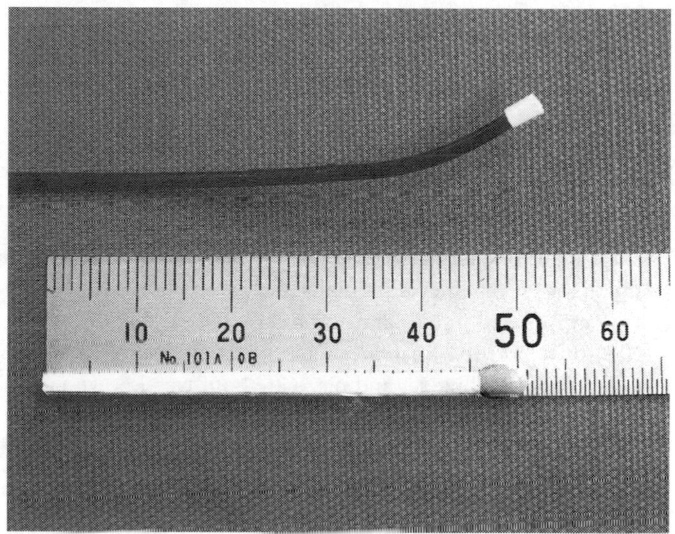

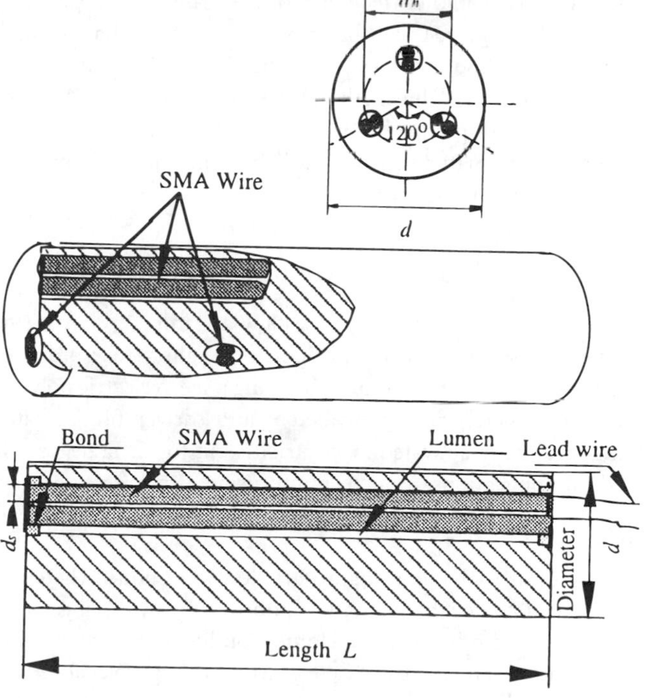

Figure 118.13 Active catheter with multi degrees of freedom for minimal invasive neurosurgery.

catheter with multi degrees of freedom using SMA wires (Fukuda et al., 1993). This catheter is made of serially connected units which contain SMA wires in the lumina, and it has multi degrees of freedom as a result of its serial-parallel structure. This catheter has developed as a medical tool for improving the operability of minimum invasive surgery. The SMA actuator is improved by being integrated with a micro mechanism. A thin film technology realization of the SMA is under development. The cool down speed is improved by miniaturization, and a bimetal-type, which utilizes the difference in thermal expansion, has been proposed.

Thermal Actuator

Thermal capacity is expressed as a product of volume, density, and specific heat. So, if the object is miniaturized, thermal capacity is decreased in proportion to the cube of its relative length. Equivalent heat conductance is also decreased by miniaturization. So, as long as the size of the object becomes small, thermal sensitivity becomes high, thermal response time will be improved and low power consumption will be realized. For these reasons, several types of thermal actuators have been proposed, such as a micro stirling engine, a thermal actuator that uses the pressure of the evaporated liquid, and so on.

Recently, printing quality has been greatly improved by ink jet techniques. Bubble jet and piezo are both well known ink jet methods. The bubble jet method utilizes surface evaporation phenomena. Figure 118.14 illustrates the principle of the bubble jet method. The surface of the heater is heated to around 200°C to 500°C in a couple of micro seconds. This method is suitable for speeding up the printing time if the ink jet nozzle is miniaturized. The recent ink jet frequency of the printer is on the order of a couple of kHz. It can be improved to more than 10 kHz. The piezo method employs a PZT head with a vibrating plate to produce ink jets. This method has some advantages. However, the bubble jet-type ink jet head is simple in its structure and suitable for miniaturization. It will become the basic printing method for the personal printer.

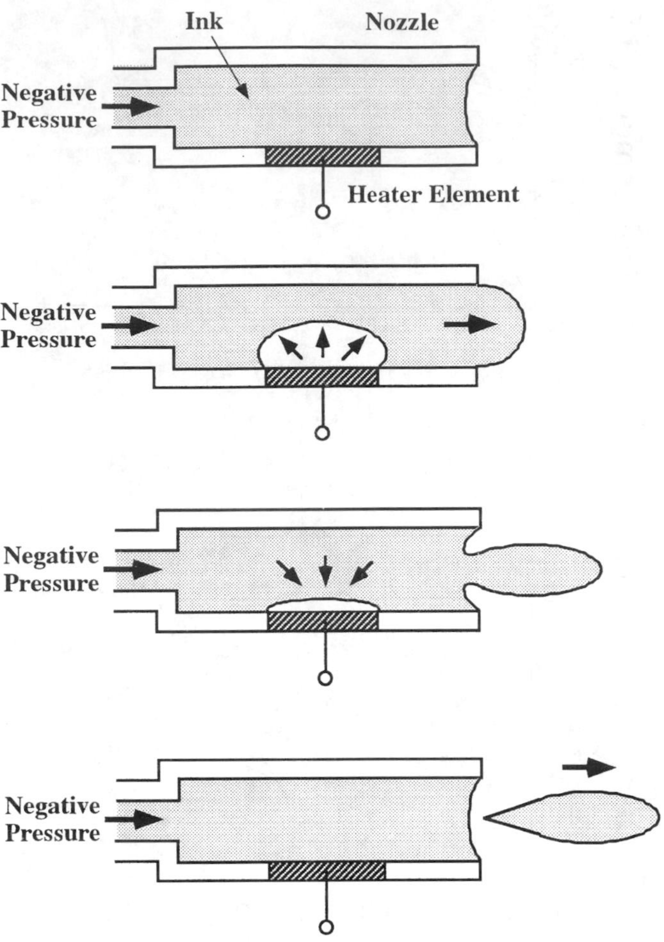

Figure 118.14 Principle of bubble jet method.

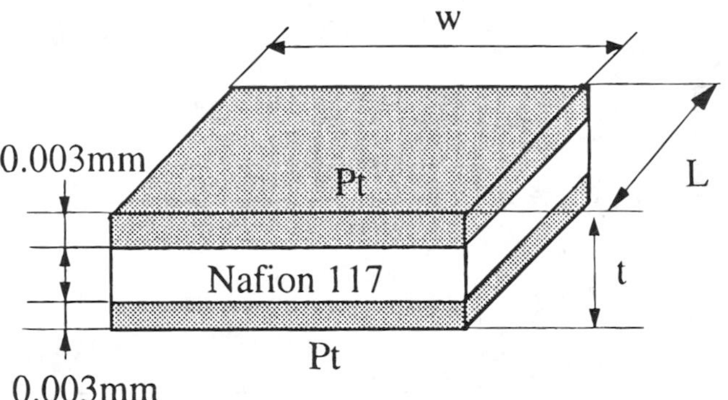

Figure 118.15 Configuration of ICPF.

Polymer and Other Actuators

Polymer actuators are very resistant to impact, force, and moment, easy to process, and light in weight. Their proposed applications include a micro probe using a piezo electric polymer actuator, a micro gripper activated by a PH driven film actuator, a chemical valve activated by electricity and for drug delivery in medical applications, a micro pump using a thermo responsive polymer gel and a water absorbing polymer gel (Hattori et al., 1992).

Most of the polymer actuators are slow to respond. Therefore, a polymer actuator that has fast response with low driving energy is desired. Recently, an ICPF (Ionic Conducting Polymer Film) was developed and studied as a new actuator (Oguro et al., 1993). ICPF is made from a film of perfluorosolfonic acid polymer (Nafion 117, duPont and company) which is chemically plated on both sides with platinum. Figure 118.15 shows the configuration of ICPF. ICPF is driven by low voltage (about 1.5 V) in a wet condition without electrolysis. ICPF is bent on the anode side when the voltage is applied to it. ICPF has been used as the active guide wire for the intravascular neurosurgery (Guo et al., 1995) and is shown in Figure 118.16. The dynamic response of the actuator in a static physiological saline solution (36°C) is shown in Figure 118.17. ICPF has quick response with low driving energy and is superior to the conventional polymer gel actuators.

Other actuators, such as a pneumatic actuator, have been developed (Suzumori et al., 1994) and show promise as a micro actuator.

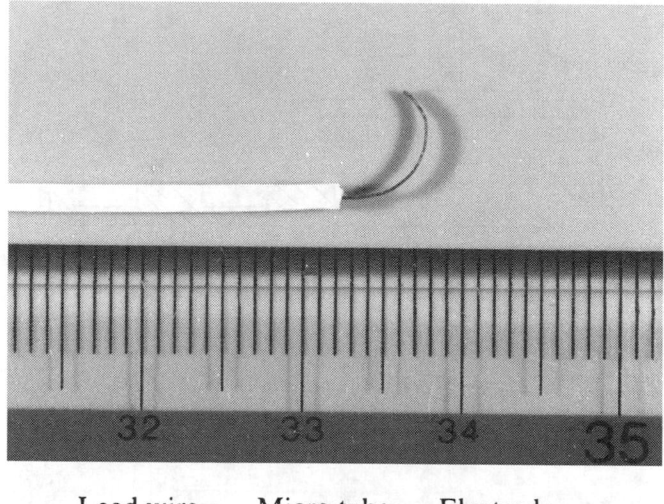

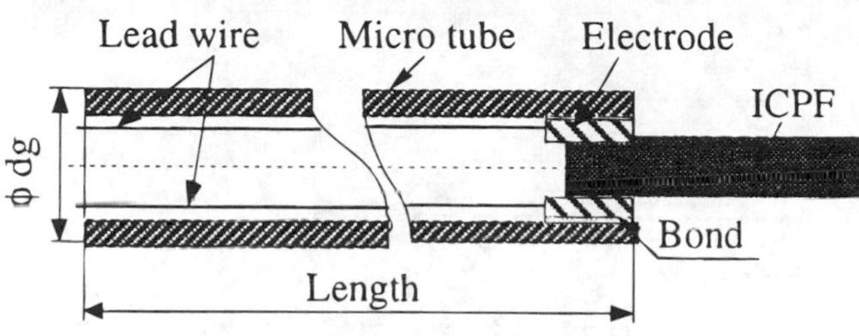

Figure 118.16 Active guide wire employing ICPF.

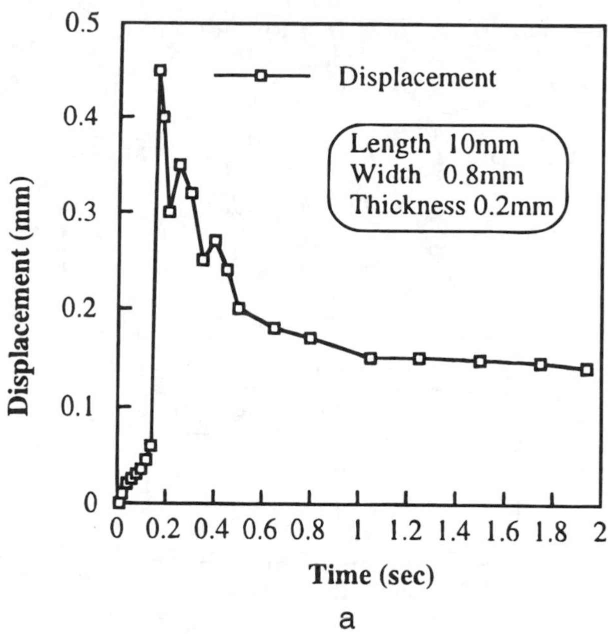

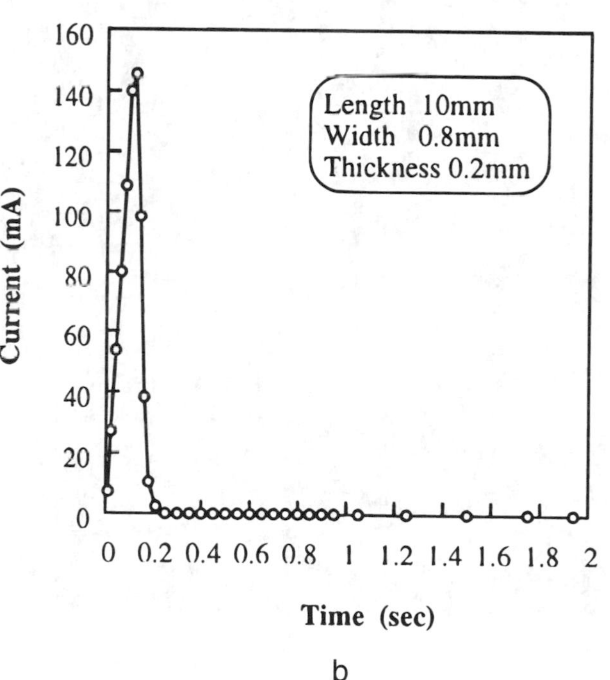

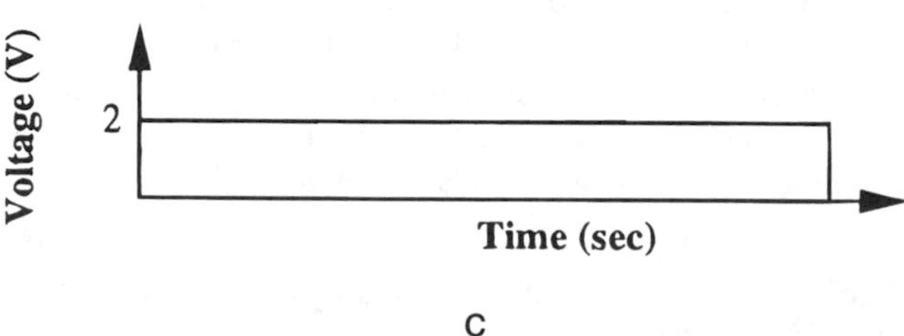

Figure 118.17 Dynamic response of ICPF.

118.2 Energy Supply Methods and Non-contact Manipulation

Classification of Energy Supply Methods

One of the final goals of micro robotics is the realization of an ant-like mobile robot which is small, intelligent, and can perform given tasks. Most of the present micro robots are supplied energy by a cable. However, as the robot becomes smaller, the cable disturbs its motion with much friction. So, the method of energy supply for the micro actuator becomes important. The energy supply methods can be classified as either internal or external. In the former case an internal energy supply source is used. In the latter case, energy is to be supplied to the system without the use of a cable. Non-contact manipulation of the small objects is also discussed.

Internal Supply Methods

In this case, the energy source is contained inside the moving body. Electric energy is frequently used as an internal supply. For this case, a battery and a condenser have been developed. Batteries are good in terms of output and durability, but have difficulty in miniaturization. Recently, a micro lithium battery whose thickness is in the order of microns, electric current density is 60 mA/cm^2, and is rechargeable for 3.6V–1.5V has been developed using thin film technology (Bates et al., 1993). As for condenser types, an autonomous mobile robot 1 cm^3 in volume has been developed in 1992 by Seiko Epson based on the conventional watch production technology. It uses a high capacity condenser 6 mm in diameter, 2 mm in thickness, and 0.33 F electric

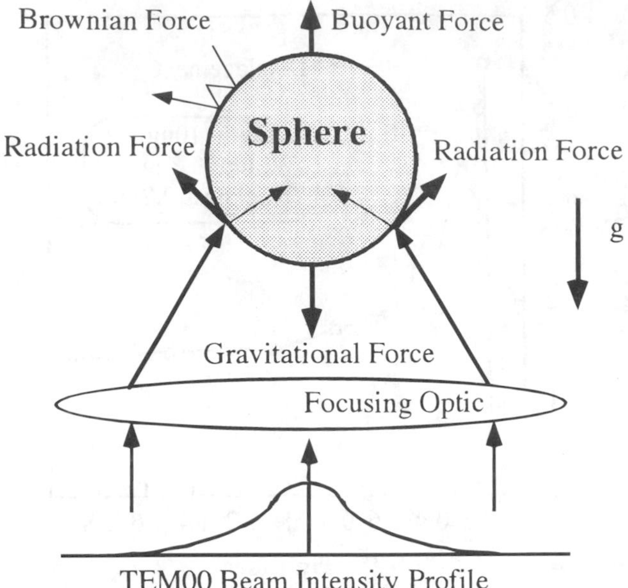

Figure 118.18 Principle of optical trapping (Rambin et al., 1994).

Figure 118.19 Optical mobile robot with non-contact energy transmission on the air table.

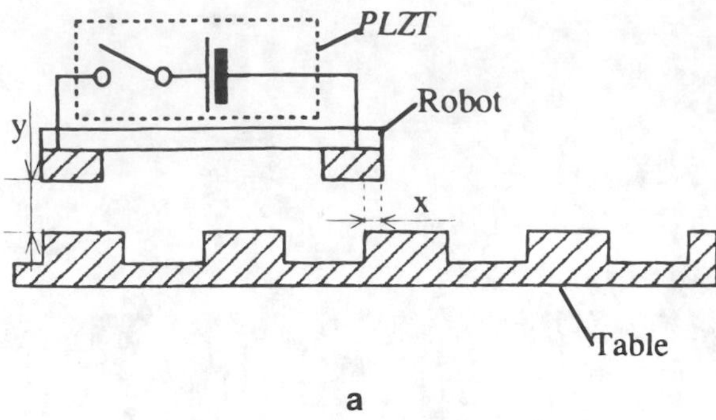

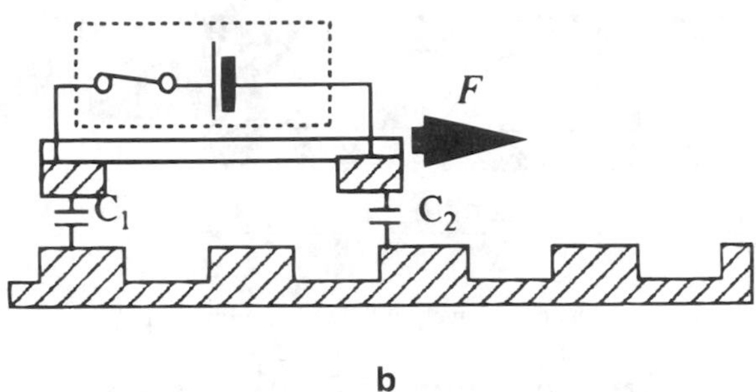

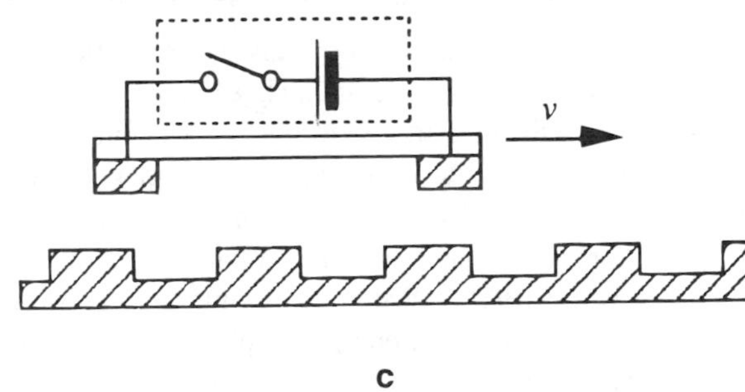

Figure 118.20 Moving principle of the optical mobile robot. (a) Initial condition; (b) light is on; (c) light is off.

capacity as an energy source. The electric capacity of the condenser is small compared with that of the secondary battery. However, this micro robot uses two stepping motors with current control using pulse width modulation, and can move for about 5 minutes after only 30 seconds of charge.

External Supply Methods and Non-contact Manipulation

In this case, energy is supplied to the body from outside. The following types of energy are used as external supply or non-contact manipulation methods.

1. Optical energy.
2. Electromagnetic energy.
3. Super sonic energy.
4. The others.

In the first case, optical energy can be classified as follows:

i) Optical radiation pressure type using a laser beam
ii) Optical energy to strain conversion type using UV ray irradiation and photostrictive phenomena of the element
iii) Optical energy to heat conversion type

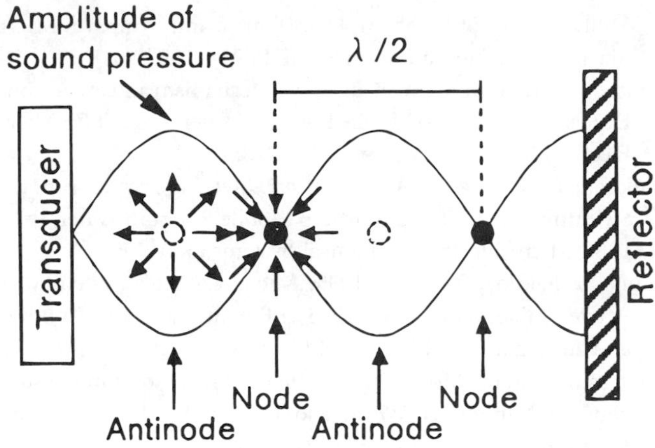

Figure 118.21 Acoustic radiation pressure (Kozuka et al., 1994).

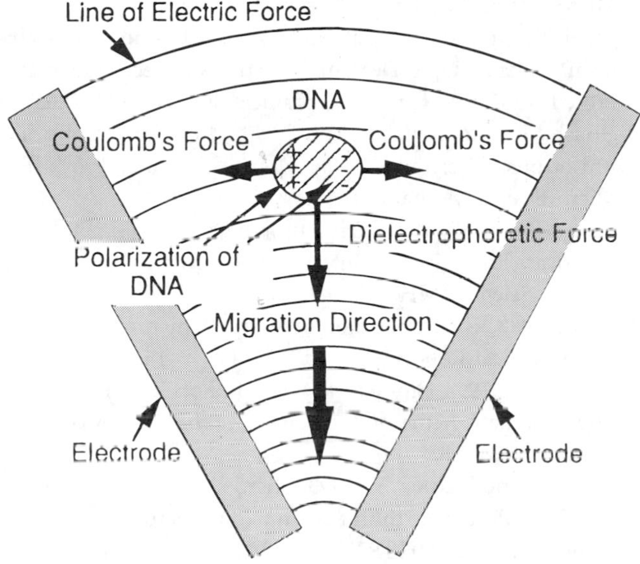

Figure 118.22 Manipulation of DNA molecule and dielectrophoretic force.

As an example of i), the remote operation of a micro object, which is used as tweezers, by a single laser beam has been proposed. The laser has superior properties such as coherency, monochromaticity, focusability, and a short pulse. It has also been demonstrated that optical pressure can be used for noncontact and remote manipulation of micrometer-sized particles (Ashkin et al., 1970). This technology has been used in optical tweezers for trapping and transporting such micro particles as bacteria and microcapsules containing chemical reagents as well as the assembly of micro objects (Masuhara, 1995; Rambin et al., 1994). The basic principle of optical trapping is summarized in Figure 118.18. In this figure, the focused laser beam produces the radiation force for the sphere. The sphere can be trapped by changing the radiation power to balance the external forces acting on it. Masuhara et al. used a 1064 nm TEM_{CO}; mode gausian beam from a CW Nd: YAG laser as a trapping source and focused (1 μm) it into a sample solution. Based on this principle, a micro object can be rotated. Directional control of optical rotation was experimentally demonstrated for artificial SiO_2 micro objects having an anisotropic geometry, which is not bilaterally symmetric but rotationally symmetric in the horizontal cross section (Higurashi et al., 1994).

As an example of ii), an optical piezo electric actuator such as PLZT has been developed as indicated earlier.

As an example of iii), low boiling point liquid material has been used with optical heat conversion material. Moreover, utilization of the pyroelectric effect has been proposed to supply energy from outside. Figure 118.19 shows a moving body on an air table using pyroelectric current which is generated by the temperature difference of the heat applied by UV ray irradiation (Ishihara et al., 1993). PLZT has been employed as an energy transformation method from optical energy to electric energy. Generally, this can be substituted for other pyroelectric elements which can generate pyroelectric current by temperature change. This moving body utilizes electrostatic force as a driving mechanism and the moving principle is shown in Figure 118.20. The field of the moving body is made by the square shaped electrodes which are arranged in squares. Each electrode is 1.5 mm square and placed at intervals of 0.5 mm. The field has many halls (diameter 0.18 mm) placed at intervals of 1 mm, and air is blown to float the moving body. The bottom face of the moving body has several electrodes 1 mm by 1 mm in width and placed at intervals of 4 mm. Each electrode is connected with PLZT. Using UV ray irradiation, thrust force is generated between the bottom face of the moving body and the field, which can be used as the driving force for it. With the air table, friction is considerably reduced and even the weak electrostatic force is enough to move it fast. In an experiment, a moving body moved on the field at a speed of 5 cm/s. Position control of it can be attained by selectively controlling the light beam irradiation.

As an example of (2), micro waves, which have been used for non-contact energy transmission to an airplane and a solar energy generation satellite, have been considered. Sasaki et al. (1993) have proposed a wireless system using microwaves to supply energy to micro robots for performing self-controlled inspections and repairs in thin metal pipes used in such areas as a heat exchanger in electric power generation plants. They expanded the microwave transmission characteristics for pipes of various discontinuous shapes in which transmission loss might occur. They also established, the basic structure for a microwave receiving antenna with a modified monopole and a high conversion efficiency (−0.4 dB). GMA, mentioned earlier can also be considered an example of noncontact energy transmission.

As an example of (3), the radiation pressure of a super sonic wave can be used for both non-contact operation and the driving force for micro objects. It has been shown that alumina particles 16 μm in diameter suspended in water can be trapped and will agglomerate every half wavelength in a standing wave field of 1.75 MHz (Kozuka et al., 1994). Acoustic radiation pressure acting on a micro particle is explained in Figure 118.21. Slight changes in frequency or the distance between transducer and reflector will cause a lateral shift in the column of agglomerated particles. With the orthogonal standing wave field, the agglomeration changes its shape depending on the relative force ratio. By focusing a traveling ultrasound on the trapped particles, we can transport only limited clusters of them and thus demonstrate spatially selective manipulation. Since the acoustic radiation force is different depending on the size, shape, density and compressibility of the particle, characteristically selective manipulation is possible. Concentration and fractionation of small particles in a liquid by ultrasound has also been demonstrated (Yasuda et al., 1995).

As another example, the external force could be obtained through an external medium, e.g. the maintenance pig robot which moves in a pipe line filled with liquid or an electromagnetic field.

A field can be controlled to manipulate a micro object. For example, we manipulated a DNA molecule by electric field control. A DNA molecule moves toward the migration direction as shown in Figure 118.22 using the dielectrophoretic force (Morishima et al., 1995). Selective energy transmission to an elastic object on a vibrating plate has also been proposed.

References

Ashkin, A. 1970. Acceleration and trapping of particles by radiation pressure, *Phys. Rev. Lett.*, 24:156.

Bates, J. B. et al. 1993. Rechargeable solid state lithium microbatteries, *Proc. Micro Electro Mechanical Systems,* 82–86.

Dario, P., Valleggi, R. et al., 1991. A miniature device for medical intracavitary intervention, *Proc. IEEE Micro Electro Mechanical Systems,* 171–175.

Fujita, H. et al., 1988. An integrated micro servosystem, *IEEE Int. Workshop Intelligent Robot and Systems,* 15–20.

Fukuda, T. and Arai, F. 1992. Microrobotics-approach to the realization, *Micro System Technologies 92,* Vde-verlag gmbh, 15–24.

Fukuda, T. and Arai, F. 1993. Microrobotics—on the highway to nanotechnology, *IEEE Ind. Elect. Soc. Newsletter,* 4–5.

Fukuda, T. and Tanaka, T. 1990. Micro Electrostatic Actuator with Three Degrees of Freedom, *Proc. IEEE Micro Electro Mechanical Systems,* 153–158.

Fukuda, T., Hosokai, H. et al., 1991. Giant Magnetostrictive Alloy(GMA) Applications to Micro Mobile Robot as a Micro Actuator without Power Supply Cables, Proc. IEEE Micro Electro Mechanical Systems, 210–215.

Fukuda, T. and Arai, F. 1992. New Actuators for High-Precision Micro Systems, Tzou, H. S. and Fukuda, T. eds., *Precision, Sensors, Actuators and Systems,* 1–37, Kluwer Academic, Norwell, MA.

Fukuda, T., Hattori, S., Arai. F. et al., 1992. Optical Servo System Using Bimorph Optical Piezo-electric Actuator, *Proc. 3rd Int. Symp. Micro Machine and Human Science (MSH'92),* 45–50.

Fukuda, T. et al. 1994. Performance Improvement of Optical Actuator by Double Sides Irradiation, Proc. 20th Int. Conf. on Industrial Electronics, Control and Instrumentation (IECON'94), 3:1472–1477.

Fukuda, T., Guo, S. et al., 1993. Active Catheter System with Multi Degrees of Freedom, Proc. Fourth Int. Symp. on Micro Machine and Human Science (MHS'93), 155–162.

Fukuda, T., Kawamoto, A. et al., 1994. Mechanism and Swimming Experiment of Micro Mobile Robot in Water, Proc. IEEE Micro Electro Mechanical Systems, 273–278.

Fukuda, T., Kawamoto, A. et al., 1995. Steering Mechanism and Swimming Experiment of Micro Mobile Robot in Water, Proc. IEEE Micro Electro Mechanical Systems, 300–305.

Guo, S., Fukuda, T. et al., 1995, Micro Catheter System with Active Guide Wire, Proc. 1995 IEEE Int. Conf. on Robotics and Automation, Vol. 1, 79–84.

Hatamura, Y. and Morishita, H., 1990, Direct Coupling System between Nanometer World and Human World, Proc. IEEE Micro Electro Mechanical Systems, 203–208.

Hattori, S., Fukuda, T. et al., 1992, Structure and Mechanism of Two Types of Micro-Pump Using Polymer Gel, Proc. IEEE Micro Electro Mechanical Systems, 110–115.

Higuchi, T., Yamagata, Y. et al., 1990, Precise Positioning Mechanism Utilizing Rapid Deformations of Piezoelectric Elements, Proc. IEEE Micro Electro Mechanical Systems, 222–226.

Higurashi, E., and Ukita, H. et al., 1994, Rotational Control of Anisotropic Micro-objects by Optical Pressure, Proc. IEEE Micro Electro Mechanical Systems, 291–296.

Hisanaga, M. et al., 1991, Fablication of a 4.8 Millimeter Long Microcar, Proc. Second Int. Symp. on Micro Machine and Human Science (MHS'91), 43–46.

Ikuta, K., 1988, The Application of Micro/Miniature Mechatronics to Medical Robots, Proc. IEEE/IROS, 9–14.

Ikuta, K. et al., 1994, Biomedical Micro Robots Driven by Miniature Cybernetic Actuator, Proc. IEEE Micro Electro Mechanical Systems, 263–268.

Ishihara, H. and Fukuda, T., 1993, Micro Optical Robotic System (MORS), Proc Fourth Int. Symp. on Micro Machine and Human Science (MHS'92), 105–110.

Itoh, T. et al., 1992, Development of Ultra Small DC Motor, Proc. Third Int. Symp. on Micro Machine and Human Science (MHS'93), 27–33.

Kim, Y. K. et al., 1990, Fabrication and Testing of a Micro Superconductive Actuator Using Meissner Effect, Proc. IEEE Micro Electro Mechanical Systems, 61–64.

Kim, C-J. et al, 1992, Silicon-processed Overhanging Microgripper, J. of Microelectromechanical Systems, 1(1):31–36.

Kozuka, T. et al., 1994, Acoustic Manipulation of Micro Objects Using an Ultrasonic Standing Wave, Proc. Fifth Int. Symp. on Micro Machine and Human Science (MHS'94), 83–87.

Masuhara, H., 1995, Microchemistry: Manipulation, Fablication, and Spectroscopy in Small Domains, Proc. IEEE Micro Electro Mechanical Systems, 1–6.

Meheregany, M. et al., 1990, Operation of Microfabricated Harmonic and Ordinary Side-Drive Motors, Proc. Micro Electro Mechanical Systems, 1–8.

Morishima, K., Fukuda, T., Arai, F. et al., 1995, Noncontact Transportation of DNA Molecule by Dilectrophoretic Force, Proc. Sixth Int. Symp. on Micro Machine and Human Science (MHS'95), to be published.

Morita, T. et al., 1995, An Ultrasonic Motor Using Bending Cylindrical Transducer Based on PZT Thin Film, Proc IEEE Micro Electro Mechanical Systems, 49–54.

Oguro, K. et al., 1993, Polymer Film Actuator Driven by a Low Voltage, Proc. Fourth Int. Symp. on Micro Machine and Human Science (MHS'93), 39–40.

Pister, K. S. J. et al., 1990, An Planar Air Levitated Electrostatic Actuator System, Proc IEEE Micro Electro Mechanical Systems, 67–71.

Polla, D. L., 1992, Micromachining of Piezoelectric Microsensors and Microactuators for Robotics Applications, H. S. Tzou and T. Fukuda (eds.), Precision, Sensors, Actuators and Systems, Kluwer Academic Publishers, 139–174.

Rambin, C. L. and Warrington, R. O., 1994, Micro-assembly with a Focused Laser Beam, Proc. IEEE Micro Electro Mechanical Systems, 285–290.

Sasaki, K. et al. Technique of Wireless Energy Service for Microrobots Using Microwave, Proc. Fourth Int. Symp. on Micro Machine and Human Science (MHS'93), 113–117.

Shimomura et al., 1991, Preparation of Lead Zirconate Titanate Thin Film by Hydrothermal Method, Jpn. J. Appl. Phys., 30(9B):2174–2177.

Suzuki et al., 1992, Creation of an Insect-based Microrobot with an External Skeleton and Elastic Joints, Proc. IEEE Micro Electro Mechanical Systems, 190–195.

Suzumori, K. et al., 1994, Microfablication of Integrated FMAs using Stereo Lithography, Proc. IEEE Micro Electro Mechanical Systems, 136–141.

Teshigahara, A. et al., 1992, Fablication of a Shell Body Microcar, Proc. Third Int. Symp. on Micro Machine and Human Science (MHS'92), 137–141.

Trimmer, W. and Jebens, R., 1989, Actuators for Micro Robots, IEEE Int. Conf. on Robotics and Automation, 1547–1552.

Uchino, K. and Aizawa, M., 1985, Photostrictive Actuator Using PLZT Ceramics, Jpn. J. Appl. Phys. 24, Suppl. 42–3, 139–142.

Yasuda, K. et al., 1995, Concentration and Fractionation of Small Particles in Liquid by Ultrasound, Jpn. J. Appl. Phys. Vol. 34, Part 1, No. 5B, 2715–2720.

119

On-Board Power Supply and Remote Driving Mechanisms for Microelectromechanical Systems

119.1 Power Requirements of Microelectromechanical Systems 1538
119.2 On-Board Power Supply: Solar Cell Array 1540
119.3 On-Board Power Supply: Microbattery 1542
119.4 Remote Driving Mechanisms 1544
119.5 Conclusions 1545

Jeong B. Lee
Georgia Institute of Technology

Micromachining is a new technology used to miniaturize electromechanical structures, especially micro scale sensors and micro scale actuators. Since micro scale electromechanical devices are normally built using common microelectronic technologies (integrated circuit fabrication technologies), constructing a microsystem which includes sensors, actuators, data processing, and driving circuitry is feasible on one small die by both integrated batch fabrication and/or hybrid approaches. Due to the intrinsic fabrication advantages of micromachining, low cost and high reliability are expected, contributing to this technology's wide acceptance. Various micromachined devices/systems have been realized, such as accelerometers, pressure sensors, micromotors, micropumps, and microfulidic systems.

Since the power requirements of micromachined devices differ from those of general circuitry, most micromachined devices use an external power supply or power conversion circuitry. Since the size of the micromachined devices ranges from millimeters (mm) to micrometers (μm) in scale, the external power supply is very large in size compared to the micromachined devices as shown in Figure 119.1. In the general case, these large external power supplies are acceptable. However, a miniaturized self-contained power source or a remote power source is desirable in many cases to give more flexibility in the design of micromachined devices. In autonomous applications, such as free moving micro robotic systems, space-based microelectromechanical systems, the self-contained or remote power supply method is essential. In contrast to the evolution of individual micromachined devices and systems, development of the power supplies for micromachined devices has received little attention until recently. Primary areas of research include rechargeable microbatteries, miniaturized solar cell arrays, and energy conversion methods using external magnetic fields. In this section, the power requirements of micromachined devices will be discussed in detail. Then, self-contained on-board power sources and remote driving mechanisms will be described along with applications.

119.1 Power Requirements of Microelectromechanical Systems

The power requirements of micromachined devices depend mainly on the driving principles involved. Several driving principles which are suitable in the micro domain have been used to realize various types of micromachined devices/systems. The most common driving principles include electrostatic drive, magnetic drive, piezoelectric drive, and electrothermal drive. Each drive principle has specific advantages and disadvantages with respect to deflection range, required force, power requirement, environmental durability, and response time. A brief overview of power requirements of each driving principle is shown in Table 119.1.

Electrostatic drive is based on electrostatic forces between the faces of electrodes, as shown in Figure 119.2. If an external voltage V is applied between two electrodes, a potential energy W is stored within the electrodes. The electrostatic forces which act perpendicular to the parallel electrode (F_d) and within the direction of the parallel electrode (F_x) can be represented by

$$F_d = -\frac{\partial W}{\partial d} = -\frac{1}{2}\epsilon_r\epsilon_o\frac{xwV^2}{d^2} \qquad (119.1)$$

0-8493-8343-9/97/$0.00+$.50

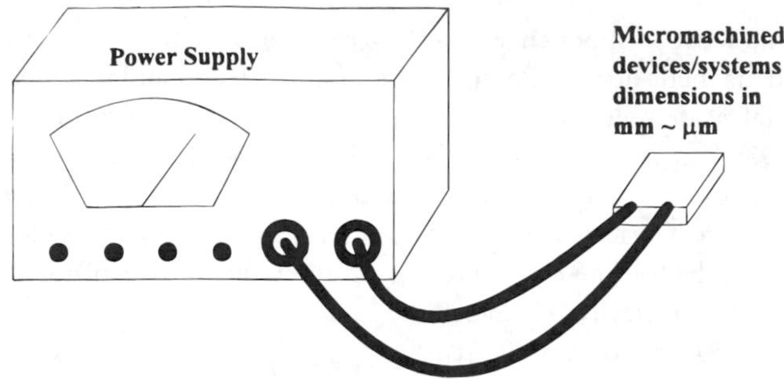

Figure 119.1 A size comparison of common methods of powering micromachined devices.

Table 119.1 Power Requirements of Micromachined Devices (A Brief Overview)

	Voltage [V]	Current
Electrostatic Driving	Tens of volts ~ hundreds of volts	nA ~ μA
Piezoelectric Driving	Tens of volts ~ hundreds of volts	nA ~ μA
Electromagnetic Driving	about 1 V	hundreds of mAs
Electrothermal Driving	A few volts ~ tens of volts	mA ~ tens of mAs

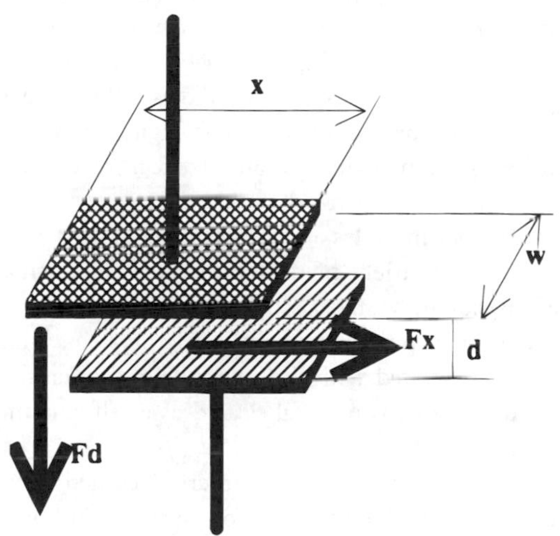

Figure 119.2 Electrostatic forces between parallel electrodes.

$$Fx = -\frac{\partial W}{dx} = \frac{1}{2}\epsilon_r\epsilon_o\frac{wV^2}{d}. \qquad (119.2)$$

Since the force perpendicular to the plates (F_d) decreases with the square of the gap distance, electrostatic forces turn out to be useful for applications where electrode separation (d) is small. Many electrostatic microactuators have been investigated including micromotors (Meheregany et al., 1990; Fan et al., 1989), a comb drive actuator (Tang et al., 1990), and a microvalve (Robertson and Wise, 1994). Both of lateral motion actuators (using the force within the parallel electrode, F_x) (Tang et al., 1990) and vertical motion actuators (using the force perpendicular to the parallel electrode, F_d) (Yamaguchi et al., 1993) have been realized. Some investigations have been made to reduce the driving voltage by reducing the gap between electrodes down to submicron (Hirano et al., 1992) using an oxidation technique, but in general, electrostatically driven devices usually require driving voltages ranging from tens of volts to hundreds of volts and driving currents in the $nA \sim \mu A$ range.

Piezoelectricity is one of the basic material properties of crystals, ceramics, polymers, and liquid crystals. In non-piezoelectric materials, the mechanical and the electrical responses are uncoupled, i.e., the electrical behavior is solely related to electrical input, and the mechanical behavior is solely related to mechanical input. In a piezoelectric material, however, the internal dielectric displacement can be developed not only by the applied electric field, but also by the applied mechanical stress. This displacement is manifested as an internal dielectric polarization or a surface electric charge. Similarly, an applied electric field can cause mechanical deformation of the sample, through its interaction with internal electrical dipoles. Piezoelectricity is widely used in the macroscopic world. Major applications include standard oscillators, ultrasonic and sonar transducers, surface acoustic wave devices, optical modulators, strain gauges and so on. In micromachining, several piezoelectric materials, such as zinc oxide (ZnO) and lead zirconate titanate (PZT) have been successfully prepared in thin films by sputtering and sol-gel deposition techniques. Some organic material, such as polyvinylidene fluoride (PVDF), also exhibit piezoelectricity. Thin film piezoelectric materials have been used to fabricate micromotors (Flynn et al., 1992), micro tips of a scanning tunneling microscope (STM) (Akamine et al., 1990), micropumps (VanLintel et al., 1988), and a micro mobile robot (Fukuda et al., 1994). In general, piezoelectrically driven micromachined devices usually require driving voltages ranging from 10 V_{RMS} to several hundreds V_{RMS} and driving currents from nA to μA range.

The magnetic force is predominant in macro scaled electromechanical devices, but it was not common in the micromachining community until a few years ago. The magnetic forces seemed unsuitable for micro scaled actuators since the magnetic force depends on the volume of the magnet; thus, as the devices became smaller, the resulting magnetic force was unsuitably small due to the scaling effect (Trimmer, 1989). Difficulty in fabrication of the three-dimensional coil winding and the relatively high resistive losses in microfabricated inductive components were also obstacles to realizing micro scaled magnetic devices. Recently, several researchers have successfully fabricated high aspect ratio metallic structures (Guckel et al., 1993; Ahn et al., 1993) which can effectively increase the volume of the magnet to generate suitable magnetic force. As a result, micro fabricated magnetic devices have become popular. Previously, the thickness of micromachined devices was usually less than 5 μm. Device thicknesses have been dramatically increased up to 50 μm to over 100 μm using LIGA (Guckel et al., 1993) or LIGA-like processing (Frazier and Allen, 1993). Magnetic drive is an attractive driving principle if devices need to be operated in a dust-filled environment, in a conducting fluid, and/or in an environment where high driving voltages are unacceptable or unattainable. Representative devices are micromotors (Guckel et al., 1993; Ahn et al., 1993), and a magnetic particle separator which is useful in biomedical applications (Ahn and Allen, 1994). Since the electromagnetic actuation is a current controlled process, these devices usually require driving currents of several hundreds mAs and driving voltages in the range of less than 1 V.

Electrothermal devices use electrically-generated heat as an energy source of actuation. The electrothermal effects can be divided into three different principles: shape memory alloys, electrothermal bimorphs, and thermo-pneumatic actuators. Shape memory alloys are a group of metals that exhibit shape recovery characteristics when heated. These alloys are deformed while below a martensite finish temperature, and they recover their original, undeformed shape when heated above an austenite temperature. During the shape recovery, the alloys produce force and displacement which can be utilized for actuation. The amount of force and displacement depends on the exact geometry of the alloys and the amount of heating. In the macro scale, several shape memory alloy actuators have been investigated for robotic applications. In micromachining, titanium-nickel (TiNi) has been prepared as a thin film, and several actuation demonstration structures (Gabriel et al., 1988) have been fabricated. Devices using this principle usually require driving voltage of tens of volts and driving currents in the mA range.

The electrothermal bimorphs consist of two materials of different coefficients of thermal expansion (CTE) combined in a sandwich structure as shown in Figure 119.3. If the lower layer has a lower CTE than the upper one, the structure will bend upward when heated. The amount of displacement and force can be controlled by heating or cooling the integrated heating resistor. A thermo-pneumatic actuator is similar. It consists of a sealed cavity filled with a thermally expandable medium that can be heated or cooled down, resulting in a pressure change in the

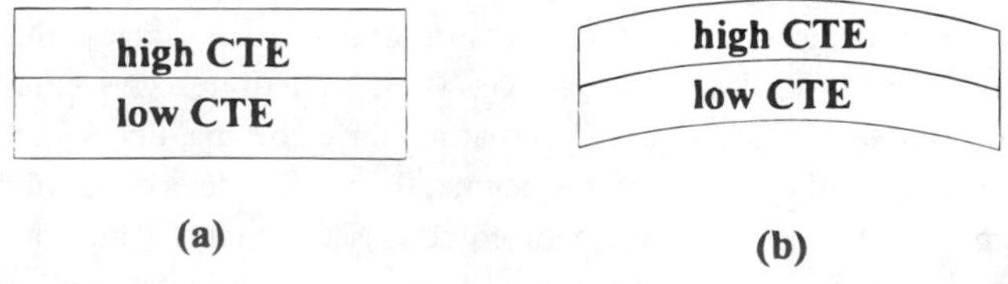

Figure 119.3 Electrothermal bimorph structure (a) normal; (b) when heated.

cavity. Several actuators have been studied using silicon-gold as a sandwich and an integrated heating resistor (Van De Pol et al., 1989). Thermally driven microvalves (Lisec et al., 1994) have also been successfully demonstrated. In general electrothermal bimorph actuators and thermo-pneumatic actuators usually require driving voltages of less than 15 volts and driving currents in the range of tens of mAs.

119.2 On-Board Power Supply: Solar Cell Array

Photovoltaic (PV) conversion of light energy, especially of solar energy, has been well investigated and will be one of the most promising ways of meeting the energy demands of the future in a time when conventional sources of energy, such as fossil fuels, are depleted. Solar cells are operated by converting optical energy (usually the sun) directly into electricity (so called, photovoltaic energy conversion) using the electronic properties of semiconductors. The basic device requirement for the PV energy conversion is an electronic asymmetry which results in a potential barrier. As shown in Figure 119.4, n-type regions have large electron densities, and p-type regions have large hole densities. Under the equilibrium condition, the drift current and the diffusion current balances. When illuminated, excess electron-hole pairs (ehp) are generated by light throughout the cell. The inherent asymmetry of the device physics allows a flow of generated electrons from p-type region to n-type region, and holes flow the opposite direction, giving rise an electrical current which can flow through external electrical loads.

Solar cell power modules are currently being used around the world to power a variety of devices from hand calculators to spacecraft. Beside these macroscopic applications, solar cells are also attractive as a power supply for microsystems, since they can be easily integrated with both circuits and micromachined devices, and therefore can be fabricated as a self-contained on-board power supply.

To be used as power sources for micromachined devices/systems, however, a suitable modification in the traditional design methodology of the solar cell array is required. A major modifications arises from the unique power requirements of micromachined devices/systems. As discussed previously, electrostatic and piezoelectric micromachined devices need voltages in the range of tens of volts ~ hundreds of volts with the current in the range of nA ~ μA. These power requirements are quite different from those traditionally available from solar cells. In case of electrothermal and electromagnetic devices, the power requirements are voltages in the range of 1 V ~ tens of volts with current in the range of mA ~ hundreds of mAs. The possible solar cell designs for micromachined devices can be divided by two categories. One is to generate high voltage with low current and the other is for low voltage with high current. For the high voltage solar cell array, it is necessary that individual solar cells must be able to be connected in series. Additionally, higher open circuit voltage (V_{oc}) of individual cells are desirable. For the high current solar cell, individual cells must deliver as high a short circuit

Table 119.2 Power Requirements of Micromachined Devices (Individual Examples)

Driving Principle	Force, Torque	Material	Power Requirement	Ref.
Electrostatic	a few pNm	poly-Si	60 ~ 400 V/nA ~ μA	Fan, et al. (1989)
Electrostatic	~1 nNm	poly-Si	26 ~ 105 V/nA ~ μA	Meheregany, et al. (1990)
Electrostatic	~80 nN	poly-Si	~10 V/nA ~ μA	Hirano, et al. (1992)
Piezoelectric	23 mN	ZnO	30 V/μA	Flynn, et al. (1992)
Piezoelectric	~0.6 mN	PZT	150 V/μA	Fukuda, et al. (1994)
Electromagnetic	~1 nNm	Ni-Fe	less than 1 V/600 mA	Guckel, et al. (1993)
Electromagnetic	3.3 nNm	Ni-Fe	less than 1 V/500 mA	Ahn, et al. (1993)
Electromagnetic	N/A	Ni-Fe	less than 1 V/500 mA	Ahn, et al. (1994)
Shape Memory Alloy	N/A	TiNi	40 V/2 mA	Gabriel, et al. (1988)
Electrothermal	0.1 N	Si	5 ~ 13 V/tens of mAs	Van De Pol, et al. (1989)

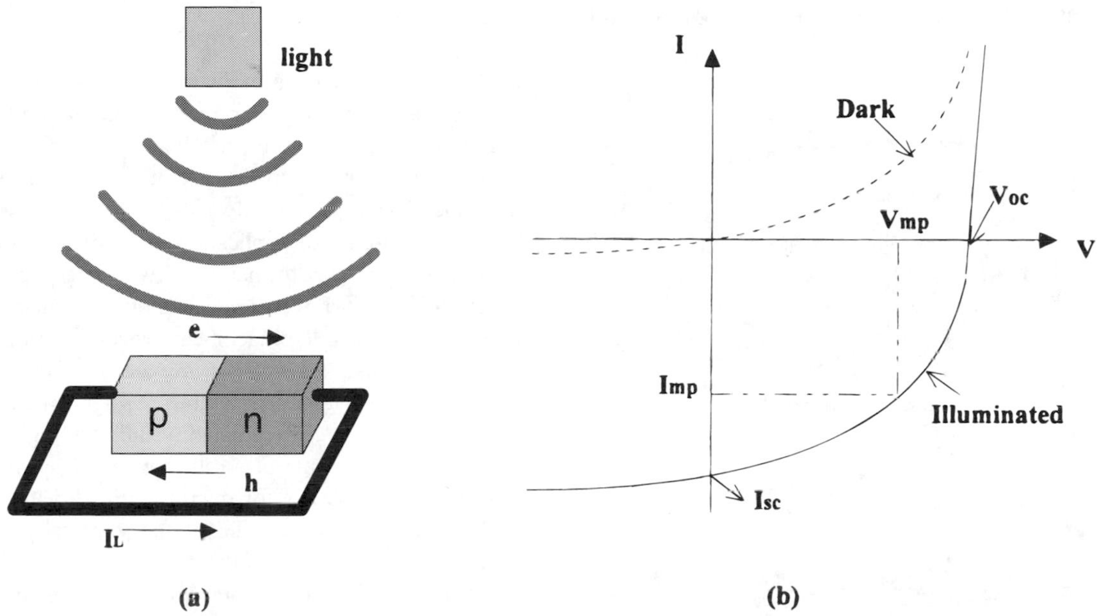

Figure 119.4 (a) p/n junction device; (b) I-V characteristics curve for dark and illuminated conditions.

current density (J_{sc}) as possible. The other main modifications arise from the basic design rules of micromachined devices. Since the micromachined systems are small in size, the solar cell power source must be as compact as possible. Also, when a power source is integrated onto device, the power source should be electrically isolated from the device. To meet these requirements, thin film solar cells are preferable to bulk type solar cells since they can be easily isolated from the device and and since they are vertically compact in size. The thin film solar cells are very attractive in high voltage solar cell applications since the series interconnection between top of one cell to bottom of next cell can be easily obtained. Table 119.3 shows the details of performances of major solar cells in these days.

In the high voltage application, a miniaturized high voltage solar cell array (array area of 1 cm^2) which generates high voltage (array Voc of 150 volts) with low current (short circuit current of 2.8 μA) has been developed (Lee et al., 1995, 1994). The solar cell array has been tested under varying illumination conditions, including Air Mass 1.5 (AM 1.5) condition, incandescent lamps, and fluorescent lamps. The AM 1.5 is a standard solar cell test condition which corresponds to illuminated sunlight on the surface of the earth when the sun is at an inclination of 48.19° relative to overhead. It has been packaged with an electrostatically driven micromachined silicon mirror (Allen et al., 1990) and demonstrated as an on-board power source for micromachined device. The detail view of the solar cell array is shown in Figure 119.5. A chromium (Cr) layer is used as a rear contact of the solar cell. As shown in the Table 119.3, the triple stacked amorphous silicon (a-Si) has the highest V_{oc} with thin film structure. Thus, the triple stacked a-Si p-i-n/p-i-n/p-i-n solar cells which are totally about 1 μm thick are used as active PV layers. A zinc oxide (ZnO) antireflective coating is used to improve the light absorption capability. An indium-tin-oxide (ITO) layer which is optically transparent and electrically conductive is used as both a front contact and an electrical interconnection between cells. Since the variation of light intensity changes the output voltage of the array, it is possible to control the amount of actuation by varying the light intensity. It has been measured that this cell array can generate over 120 volts even under normal room light conditions (Illumination Engineering Society recommends 100 lux for normal residential

Table 119.3 Typical Parameter Summary of Major Photovoltaic Technologies

Solar Cell Description	V_{oc} [V]	J_{sc} [mA/cm^2]	η [%]	Thickness [μm]#	Stability	Cost	Ref.
Single crystal Si PERL cell♣	0.7	41	23	200 ~ 400	Excellent	Very high	Green et al., (1993)
Thin film poly-Si	0.6	33	15.7	~30	Excellent	Medium	Crabb, (1971)
GaAs	1.045	27.6	24.4	2 ~ 10	Excellent	Extremely high	Kazmerski (1989)
Two terminal tandem GaAs	2.403	13.96	27.6	2 ~ 10	Excellent	Extremely high	Kazmerski (1989)
ITO/InP cell	0.313	27.97	18.9	2 ~ 10	Excellent	Extremely high	Kazmerski (1989)
Amorphous Si(a-Si) p/i/n cell	0.33	17.2	9.8	~1	Low	Low	Kazmerski (1989)†
ITO/a-Si/a-Si/steel tandem cell	2.541	6.96	12.4	~1	Fair	Low	Kazmerski (1989)†
Screen printed CdS/CdTe	0.69	31.1	8.1	5 ~ 30	Good	Low	Uda et al. (1982)
CdS/CuInSe$_2$	0.446	35.5	10.3	1 ~ 10	Excellent	Low	Kazmerski (1989)‡

This thickness does not represent the reference cell, but a typical thickness for the technology. From Table 6 in Kazmerski (1989), †From Table 3 in Kazmerski (1989), ‡From Table 2 in Kazmerski (1989), ♣PERL: Passivated Emitter Rear Locally Diffused Cell.

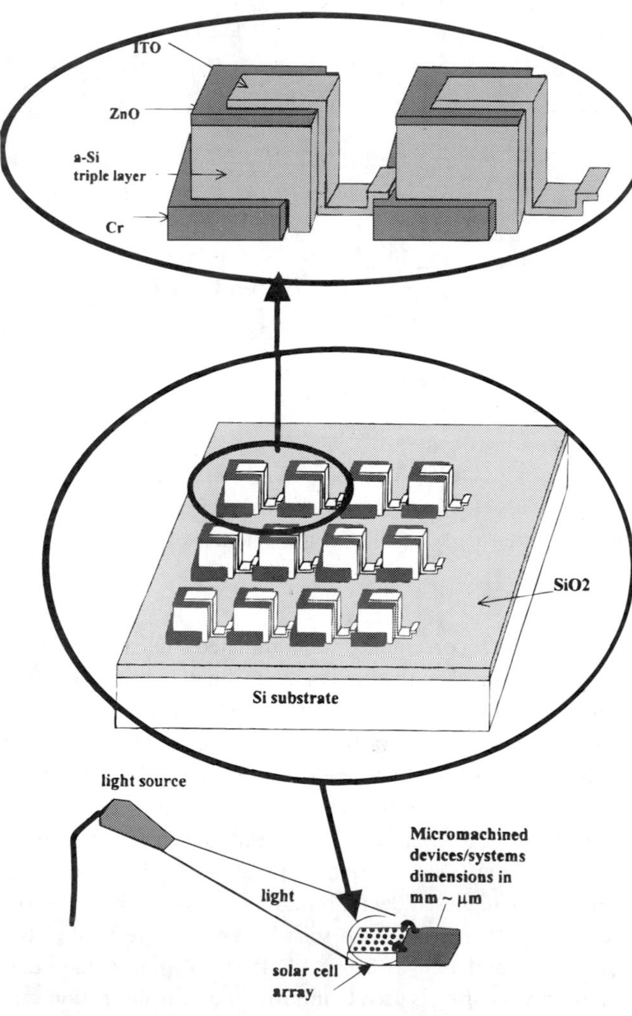

Figure 119.5 A miniaturized solar cell array which interconnects individual cells in series to generate high voltage.

area lighting.). As shown in Figure 119.5, this solar cell array can be used as an on-board power source for autonomous applications, such as movable microrobots. A higher voltage output with suitable current in much smaller size of cell array is expected to be developed in the near future.

In the high current application, individual cells can be interconnected in parallel to increase drive current, but it results in a large solar cell array area which is not suitable as an on-board power source for micromachined devices/systems. Another way of achieving high current is using a high energy density light source, such as a laser. The higher the light intensity, the more electron-hole pairs are generated resulting in higher output current. No high current solar cells have yet been reported as an on-board power source for micromachined devices/systems.

Eventually, it is expected that the solar cell array on-board power source will be one of the major means of supplying power to some microelectronic devices as well as micromachined devices/systems.

119.3 On-Board Power Supply: Microbattery

An electrochemical power source, commonly called a battery, is a device which converts the chemical energy directly into electricity. Batteries are widely used in our everyday life from the small button cells found in electric watches or calculators to lead-acid batteries which are used in automobiles. Batteries can be categorized as non-rechargeable (a so-called primary battery) or rechargeable (a so-called secondary battery). The non-rechargeable battery is used as a portable power source and the rechargeable battery is used as a temporary energy storage and a portable power source. Until recently, conventional batteries used solid electrodes and aqueous electrolytes. Most of modern practical aqueous electrolytes are referred to as dry cells, since the aqueous electrolyte phase has been immobilized by using gelling agents or by incorporation into microporous separators. Common systems for non-rechargeable batteries are based on Zn-MnO$_2$ and Zn-HgO systems, while common rechargeable systems are based on lead-acid or nickel-cadmium. Figure 119.6 shows a common

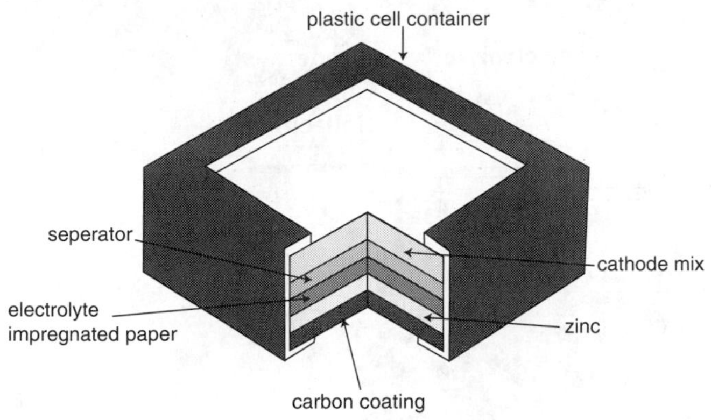

Figure 119.6 A non-rechargeable flat Leclanché cell (Zn-MnO_2 system).

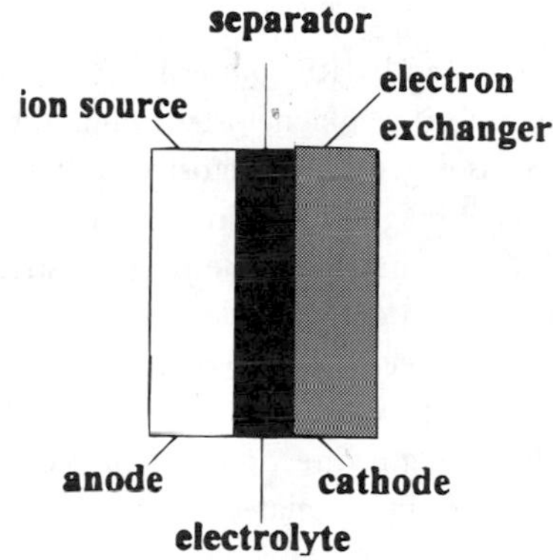

Figure 119.7 Components of solid state batteries.

non-rechargeable flat Leclanché cell (Zn-MnO_2 system) which is basically the same structure as type AA, AAA, C, and D batteries.

As the microelectronic technology has revolutionized the electronics industry, there has been a wide demand for a miniaturized battery systems. Several miniature batteries, using aqueous, non aqueous and solid electrolytes, have been developed as power sources for microelectronic or other miniaturized devices. Solid state batteries consisting of solid electrolytes have advantages in reliability and miniaturization since they have no liquid leakage and can be fabricated using common microelectronic processing, such as thin film evaporation, sputtering, chemical vapor deposition, and molecular beam deposition. They have been used in applications where the reliability and miniaturization is a key factor, for example in implantable electronic instrumentation such as cardiac pacemakers, physiological monitoring/telemetry packages, etc.

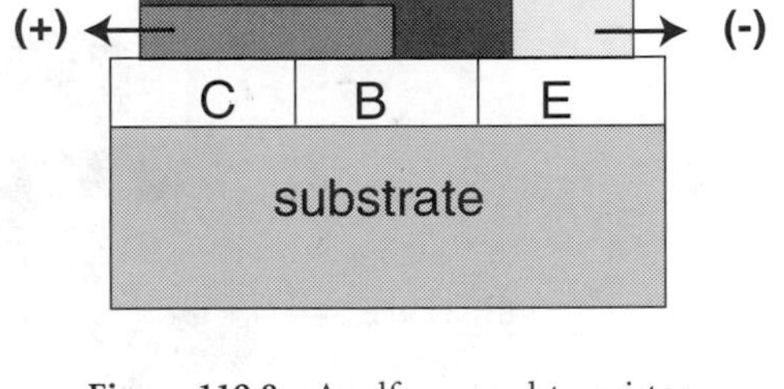

Figure 119.8 A self-powered transistor.

Solid state batteries consist of three components including the ion source (anode), the insulator (electrolyte), and the electron exchanger (cathode) as shown in Figure 119.7. The anode emits positive ions into the separator and delivers electrons to the external load by the oxidation process. The separator should be a good electrical insulator as well as a fast ion conductor. If the electrical insulation is not good enough, electrical leakage will shorten the battery lifetime. The cathode is a mixed ionic-electronic conductor. It accepts electrons from the external load and positive ions through intercalation. The deliverable electrical energy is the electrochemical potential difference between the Fermi level of the metal (anode) and quasi-Fermi level of electron exchanger (cathode). When the battery system is connected to an external load, electrons are extracted from the anode, and positive ions are injected into the separator and diffuse toward the cathode, similar to the transistor operation. Table 119.4 shows several solid state microbattery systems which are a few microns in thickness. Besides these systems, microbatteries using polymer electrolytes are also available. Since the polymer electrolytes are deformable, such microbatteries can be fabricated almost any shape.

As discussed, the miniaturization of batteries, especially solid state microbatteries, is motivated by the demand for miniature batteries which can be used as on-board power sources for microelectronic devices. Using solid state microbattery technology, self-powered microelectronic components or on-board local power sources can be fabricated. The concept of the self-powered microelectronic components is shown in Figure 119.8 (Balkanski and Julien, 1991) (Figure 119.9). Microbatteries also could be useful as on-board power sources for micromachined devices/systems, however, modification of conventional design is required. For the high voltage application, multiple stacked cells or series interconnected array type cells need to be developed. For the high

Table 119.4 Solid State Microbattery Systems

Systems	V_∞ [V]	Current Density [$\mu A/cm^2$]	Ref.
$Li/Li_{3.4}Si_{7.5}P_{0.4}O_4/TIS_1$	2.45	6 ~ 16	Kanehori et al. (1983)
Li/B_2O_3-$0.8Li_2O$-$0.8Li_2SO_4/TiS_aO_y$	2.6	65	Mannier et al. (1988)
Li/B_2O_2-$0.7Li_2O$-$0.5Li_2SO_4/InSe$	2.8	50	Balkamki & Julmer (1991)
$Li/Li_3PO_4/V_2O_5$	1.5 ~ 3.6	60	Butzse et al. (1993)

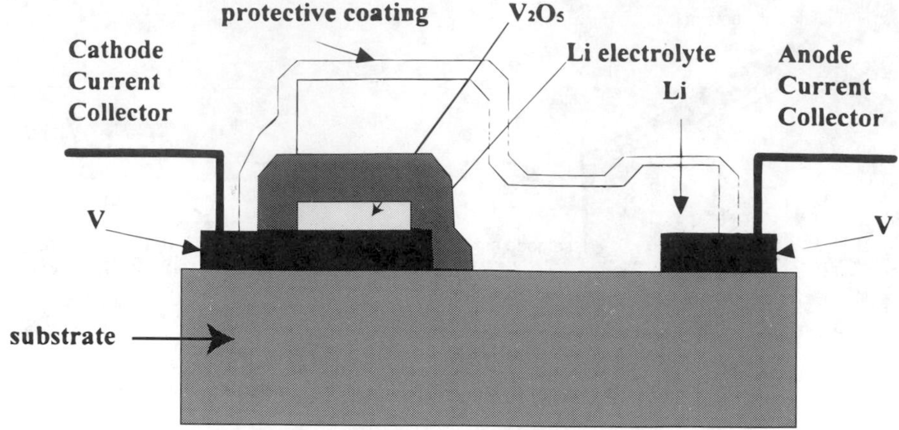

Figure 119.9 A typical example of solid state microbattery (Bates et al., 1993).

current applications, batteries appear inappropriate since the current output is presently unsuitably low. Up to this point, there has been no report on microbattery-powered micromachined devices or systems.

119.4 Remote Driving Mechanisms

Several driving mechanisms based on remote power sources have been investigated in micromachined devices/systems with the goal of realizing a free moving micro device without a power cable. One of the popular sources for supplying energy in remote fashion is the use of an external magnetic field. Both a direct use of external magnetic field (Honda et al., 1994) and an indirect use which converts the magnetic field to other kinds of energy, such as induction heating (Rashidian and Allen, 1993), have been performed. Optical excitation using a high energy density laser source (Hashimoto et al., 1994; Lammerink et al., 1991) has also been investigated.

One of the direct ways to use an external magnetic field as power for micromachined devices is by using magnetostriction. Magnetostriction is a unique material property which can be effectively used in micro scale actuation, since when a magnetostrictive material is in an external magnetic field, it shows a mechanical deformation in certain direction. Recently, Honda et al. (1994) fabricated a magnetostrictive thin film unimorph and bimorph structure using amorphous Terbium-Iron (Tb-Fe) and Samarium-Iron (Sm-Fe) thin films. Both of Tb-Fe and Sm-Fe thin films are prepared by RF sputtering. The amorphous Tb-Fe thin film has a positive magnetostriction and the amorphous Sm-Fe thin film has negative magnetostriction; in addition, both demonstrate a large magnetostriction in low external magnetic filed. A traveling machine using the magnetostrictive thin film bimorph structure with a polyimide interlayer, which is shown in Figure 119.10 has been fabricated and demonstrated. When an alternating magnetic field of 100 Oe at 50 Hz was applied along the machine length direction, it vibrated and traveled at an average speed of approximately 0.5 mm/sec. in the direction indicated.

Rashidian et al. (1993) built a simple microactuator based on a RF heated bimorph structure which indirectly used the external magnetic field as power source. The basic idea behind this device is that if an RF signal is applied to a parallel plate capacitor which consists of a lossy dielectric (in this case, polyvinylidene-fluoride), energy is dissipated as heat. This heat can be used as a source of thermal excitation for an electrothermal actuator, which is the plate capacitor itself. Figure 119.11 shows the concept of this device. As shown in the Figure, it is possible to realize a remote excitation of the device using external magnetic field which forms an inductive coupling of the RF signal.

A high energy density laser light source has been used as a remote excitation of a micro mechanical resonators (Lammerink et al., 1991). As shown in Figure 119.12, the argon (Ar) laser supply an optical power which is absorbed at the upper part of resonator, resulting in a thermal stress. The absorbed optical

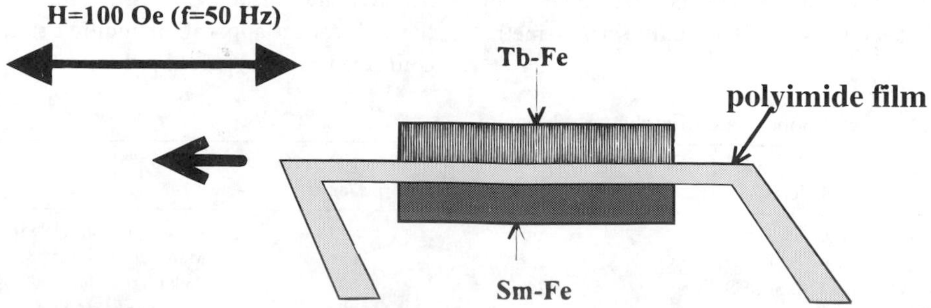

Figure 119.10 A simple travelling machine using a magnetostrictive thin film bimorph.

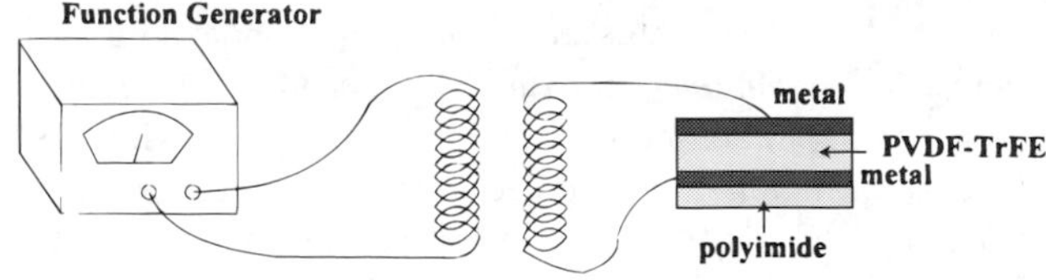

Figure 119.11 Remote operation based on energy coupling method.

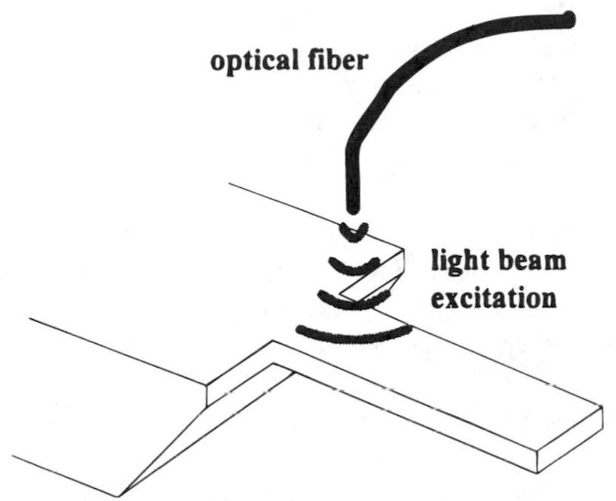

Figure 119.12 Remote operation based on optical excitation.

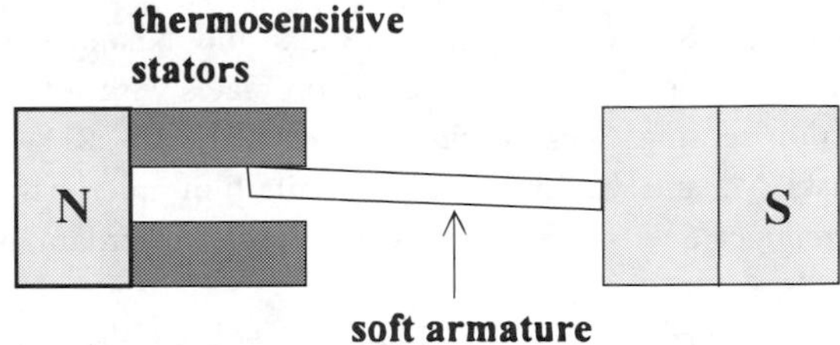

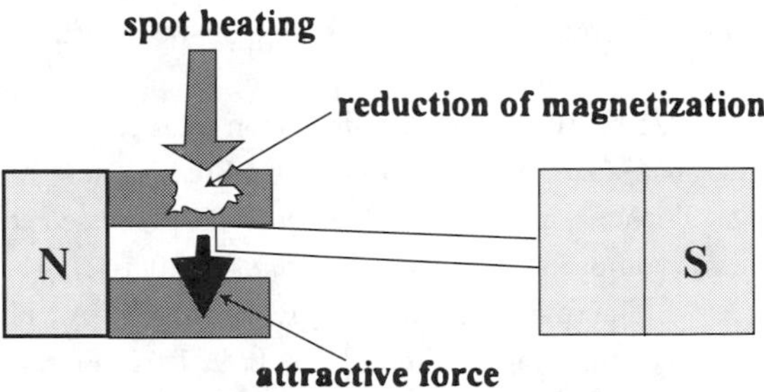

Figure 119.13 Thermally controlled magnetization actuator.

power generates a thermal distribution, which creates a mechanical moment; this mechanical moment is used to actuate the beam.

An interesting example using a high energy density laser as a remote energy source is a thermally controlled magnetization actuator (TCMA) (Hashimoto et al., 1994). Most of the magnetic devices are driven by magnetic fields which are generated by external permanent magnets or currents flowing through coils which are wound around magnets. However, the TCMA consists of permanent magnets, ferromagnetic yokes, and thermosensitive materials, i.e., materials with temperature-dependent magnetic properties. A high energy density laser source as a power supply in order to thermally control the magnetization of the thermosensitive materials which results in changing the reluctance of the magnetic circuits. As shown in Figure 119.13, it consists of permanent magnets, yokes, and an armature made of a soft magnetic materials with low Curie point (Tc). When one of the stators is heated to reduce its magnetization, the force balance between the two stators changes because of the gradient in the magnetic field, and the other stator attracts the armature beam. A remotely located high energy density focused laser beam is used to give spot heating. A nickel (Ni) matrix microrelay has been fabricated using this driving mechanism (Hashimoto et al., 1994).

119.5 Conclusions

External power supplies are widely used as power sources for micromachined devices, however, on-board power sources and remote driving mechanisms can give more flexibility in many cases of the design of micromachined devices. For an autonomous operation, such as free moving micro robots or space based MEMS, on-board power sources or remote driving mechanisms are essential. Miniaturized solar cell arrays have been developed and demonstrated as an on-board power source for high voltage micromachined devices. Up to this point, high current solar cell arrays as power sources for micromachined devices have not been reported. It is expected that, however, a miniaturized high current solar cell array as an on-board power source for micromachined devices will appear using thin film high current PV materials, such as poly-Si thin film solar cell, in a few years. Microbatteries are attractive as on board power sources for micromachined devices as well as microelectronic devices. It is expected that high voltage microbatteries as on-board power sources for micromachined devices will be developed in a few years. The microbattery seemed inappropriate as power source for high current micromachined devices applications. More sophisticated remote driving mechanisms with attractive applications are expected to be developed. Eventually, both on-board self-contained power sources and remote driving mechanisms will be one of the major means of supplying power to micromachined devices and microelectronic devices.

References

Ahn, C. H., Kim, Y. J., and Allen, M. G. 1993. A planar variable reluctance magnetic micromotor with fully integrated stator and coils, *J. Microelectromechanical Systems,* vol. 2(4):165–173.

Ahn, C. H. and Allen, M. G. 1994. A fully integrated micromachined magnetic particle manipulator and separator. *Proc. 7th IEEE Workshop on Micro Electro Mechanical Systems,* Oiso, Japan, 91–96.

Akamine, S., Albrecht, M.J., Zdeblick, and Quate, C. F. 1990. A planar process for microfabrication of integrated scanning tunneling microscopes, *Sensors and Actuators,* A23(1–3): 964–970.

Allen, M. G., Scheidl, M., Smith, R. L., and Nikolich, A. D. 1990. Movable micromachined silicon plates with integrated position sensing, *Sensors and Actuators,* A21(1–3):211–214.

Balkanski M. and C. Julien. 1991. Thin film microbatteries, *Microionics, Solid-state Integrable Batteries,* Amsterdam, Netherlands, 3–39.

Bates, J. B., Gruzalski, G. R., and Luck, C. F. 1993. Rechargeable solid state lithium microbatteries, *Proc. 6th IEEE Workshop on Micro Electro Mechanical Systems,* 82–86 February. Fort Lauderdale, FL.

Crabb, R. 1971. Status report on thin silicon solar cells for flexible arrays, *Solar Cells,* 35–50, Gordon and Breach, New York.

Fan, L. S., Tai, Y. C., and Muller, R. S. 1989. IC-processed electrostatic micromotors, *Sensors and Actuators,* 20(1–2): 41–47.

Flynn, A. M., Tavrow, L. S., Bart, S. F., Brooks, R. A., Ehrlich, D. J., Udayakumar, K. R., and Cross, L. E. 1992. Piezoelectric micromotors for microrobots, *J. Microelectromechanical Systems,* 1(1):44–51.

Frazier, A. B. and Allen, M. G. 1993. Metallic microstructures fabricated using photosensitive polyimide electroplating molds, *J. Microelectromechanical Systems,* 2(2):87–94.

Fukuda, T., Kawamoto, A., Arai, F., and Matsuura, H. 1994. Mechanism and swimming experiment of micro mobile robot in water, *Proc. 7th IEEE Workshop on Micro Electro Mechanical Systems,* 273–278, January, Oiso, Japan.

Gabriel, K. J., Trimmer W. S. N., and Walker, J. A. 1988. A micro rotary actuator using shape memory alloys, *Sensors and Actuators,* 15(1):95–102.

Green, M. A., Wenham, S. R., and Zhao J. 1993. Progress in high efficiency silicon cell and module research, *Proc. 23rd IEEE Photovoltaic Specialists Conference,* 8–13, Louisville, KY.

Guckel, H. Christenson, T. R., Skrobis, K. J. Jung, T. S., Klein, J. Hartojo, K. V., and Widjaja, I. 1993. A first functional current excited planar rotational magnetic micromotor, *Proc. 6th IEEE Workshop on Micro Electro Mechanical Systems,* 7–11, February, Fort Lauderdale, FL.

Hashimoto, E., Tanaka, H., Suzuki, Y., Uenishi, Y., and Watabe, A. 1994. Thermally controlled magnetization actuator (TCMA) using thermosensitive magnetic materials, *Proc. 7th IEEE Workshop on Micro Electro Mechanical Systems,* 108–113, January, Oiso, Japan.

Hirano, T., Furuhata, T., Gabriel, K. J., and Fujita H. 1992. Design, fabrication, and operations of submicron gap comb-drive microactuators, *J. Microelectromechanical Systems,* 1(1):52–59.

Honda, T., Arai, K. I., and Yamaguchi, M. 1994. Fabrication of actuators using magnetostrictive thin films, *Proc. 7th IEEE Workshop on Micro Electro Mechanical Systems,* 51–56, January, Oiso, Japan.

Kanehori, K., Matsumoto, K., Miyauchi, K., and Kudo, T. 1983. *Solid State Ionincs,* 9&10(1445).

Kazmerski, L. L. 1989. Status and assessment of photovaltic Technologies, *International Materials Reviews,* 34(4):185–210.

Lammerink, T. S. J., Elwenspoek, M., and Fluitman, J. H. J. 1991. Optical excitation of micromechanical resonators, *Proc. 4th IEEE Workshop on Micro Electro Mechanical Systems,* 160–165, January, Nara, Japan.

Lee, J. B., Chen, Z., Allen, M. G., Rohatgi, A., and Arya R. 1994. A high voltage solar cell array as an electrostatic MEMS power supply, *Proc. 7th IEEE Workshop on Micro Electro Mechanical Systems,* 331–336, January, Oiso, Japan.

Lee, J. B., Chen, Z., Allen, M. G., and Rohatgi, A. (1995). A miniaturized high voltage solar cell array as an electrostatic MEMS power supply. *J. Microelectromechanical Systems* 4(3):102–108.

Lisec, T., Hoerschelmann, S., Quenzer, H. J., Wagner, B., and Benecke, W. 1994. Thermally driven microvalve with buckling behavior of pneumatic applications, *Proc. 7th IEEE Workshop on Micro Electro Mechanical Systems,* 13–17, January, Oiso, Japan.

Meheregany, M. et al., 1990. Operation of microfabricated harmonic and ordinary side-drive motors, *Proc. 3rd IEEE Workshop on Micro Electro Mechanical Systems,* 1–8, February, Napa Valley, CA.

Meunier, G., Dormoy, R., and Levasseur, A. 1988. French Patent No. 88-14435.

Rashidian B. and Allen, M. G. 1993. Electrothermal microactuators based on dielectric loss heating, *Proc. 6th IEEE Workshop on MIcro Electro Mechanical Systems,* 24–29, February, Fort Lauderdale, FL.

Robertson, J. K. and Wise, K. D. 1994. A nested electrostatically-actuated microvalve for an integrated microflow controller, *Proc. 7th IEEE Workshop on Micro Electro Mechanical Systems,* 7–12, January, Oiso, Japan.

Tang, W. C., Nguyen, T. C., Judy, M. W., and Howe, R. T. 1990. Electrostatic comb-drive of lateral polysilicon resonators, *Sensors and Actuators,* A21(1–3):328–331.

Trimmer, W. S. N. 1989. Microrobots and micromechaical systems, *Sensors and Actuators,* 19(3):267–287.

Uda, H., Matsumoto, H., Komatsu, Y., Nakano, A., and Ikegami, S. 1982. All screen printed CdS/CdTe solar cell, *Proc. 16th IEEE Photovoltaic Specialists Conference,* 801–804.

Van De Pol, F. C. M., Wonnink, D. G. J., Elwenspoek, M., and Fluitman, J. H. J. 1989. A thermopneumatic actuation principle for a microminiature pump and other micromechanical devices, *Sensors and Actuators,* 17(1–2):139–143.

Van Lintel, H. T. G., Van de Pol, F. C. M., and Bouwstra, S. 1988. A piezoelectric micropump based on micromachining of silicon, *Sensors and Actuators,* 15(2):153–167.

Yamaguchi, M., Kawamura, S., Minami, K., and Esashi, M. 1993. Distributed electrostatic micro actuator, *Proc. 6th IEEE Workshop on Micro Electro Mechanical Systems,* 18–23, February, Fort Lauderdale, FL.

120

Si Micromachining in High-Frequency Applications

Linda P. B. Katehi
University of Michigan

Gabriel M. Rebeiz
University of Michigan

Tom M. Weller
University of South Florida

Rhonda F. Drayton
University of Illinois-Chicago

Stephen V. Robertson
University of Michigan

Chen-Yu Chi
QualComm, Inc.

120.1 Introduction 1547

120.2 Applications 1548
Dielectric Membrane Supported Circuits and Antennas • Micromachined Lines for Miniature, Monolithic Packaging

120.3 Fabrication Methodology 1551
Dielectric Membrane Growth • Metallization Pattern Definition • Silicon Micromachining • Backside Metallization • Circuit Assembly

120.4 Membrane Supported Distributed Circuits 1556
Low Pass Filters • Microshield Series Tuning Stubs • Wilkinson Power Divider • Lange Coupler • Resonators on Membrane • Interdigitated Bandpass Filters • Coupled Line Bandpass Filters

120.5 Conformal Micromachined Packaging 1562
Design Approach to Micromachined Circuits • Experimental Validation and Discussion

120.6 Micromachined Lumped Elements 1567
Microwave Measurements: Inductors • Microwave Measurements: Capacitors

120.7 Conclusions 1572

120.1 Introduction

Microwave and millimeter-wave planar-integrated circuits and antennas are the central nervous system in communication radars. In the past decade, the microwave field has been experiencing a technological revolution due to advances in solid-state devices, insulating and semiconducting materials, and circuit analysis techniques. Furthermore, the planarization of guiding structures to transmission lines in microstrip, stripline, or coplanar waveguide form provides great flexibility in design, reduced weight and volume, and compatibility with active devices and radiating elements.

There are, however, drawbacks to these planar geometries. These include frequency dependent mechanisms such as parasitic coupling and radiation, as well as increased ohmic loss, dielectric loss, and dispersion. These effects can seriously deteriorate electrical performance and can lead to costly, time-intensive design cycles (van Deventer et al., 1989; Dunleavy and Katehi, 1988). Suppression of some of these electromagnetic mechanisms, such as parasitic radiation and coupling, has led to improved performance, but requires very sophisticated solutions which add considerably to the weight and cost of the circuits (VandenBerg and Katehi, 1992; Harokopus and Katehi, 1989; Pengelly and Schumacher, 1988). Elimination of other deleterious effects, such as dispersion and ohmic loss, will require fundamentally new approaches to planarization and circuit integration.

In almost all design cycles, system components are typically developed and tested in an open environment. Once performance is tested and design is confirmed, the circuits are mounted into a metal housing which, in most cases introduces multiple parasitic resonances that interfere with the electrical performance of the circuit. In addition, this packaging approach often results in expensive packages that are mostly responsible for the resulting weight and volume of the units. Miniaturized high-frequency circuits with an integrated housing offer lightweight and controllable parasitics, making them appropriate for cellular and mobile communications where system requirements impose strict limits on electrical performance.

Recent advances in semiconductor processing techniques offer a historic opportunity to distinguish the monolithic from the planar character, and provide integration in all of the directions of the three-dimensional space (Weller et al., 1993a,b; Dib et al., 1991). The capability to incorporate one more dimension, and a few more parameters, in the circuit design, can lead to revolutionary shapes and integration schemes. These circuit topologies can reduce ohmic loss and eliminate parasitic radiation or parasitic cavity resonances without affecting the monolithic character of the design circuits. Operating frequencies are thereby extended and performance is optimized.

0-8493-8343-9/97/$0.00+$.50

Silicon miromachining can provide a variety of solutions to the previously mentioned problems resulting in transmission line and array approaches which are characterized by:

- Superior performance.
- Low-weight and volume.
- Easy fabrication.
- Great potential for low cost.

The characteristics shown above may be prioritized according to pre-existing needs, resulting in a number of approaches which use micromachining for millimeter- and sub-millimeter-wave circuit design.

120.2 Applications

The evolution of micromachined circuits and antennas for operation in microwave and millimeter-wave frequencies is still in its infancy. However, presented here is a description of recent accomplishments in this area, with emphasis on the effort performed at the University of Michigan. There are two techniques which have shown promise for use, and which extensively use micromachining to realize novel circuits. The first utilizes dielectric membranes to support transmission line and antenna configurations (Weller et al., 1993a,b; Dib et al., 1991) and emphasizes optimization of circuit performance. The second technique introduces new concepts in packaging such as adaptive or conformal packaging and, in addition to improvement in performance, it emphasizes size/volume/cost reduction (Drayton and Katehi 1982, 1993a,b). The merits of each approach, in relation to electrical performance, fabrication, and compatibility, will be presented, and the impact of the newborn technologies to the state of the art will be discussed.

Dielectric Membrane Supported Circuits and Antennas

Membrane supported radiating elements, such as the integrated-horn antenna, were first developed in 1987 by D.B. Rutledge and G.M. Rebeiz (Rebeiz et al., 1990) and have been further studied and developed at Michigan by G.M. Rebeiz and L.P.B. Katehi. SIS junctions on thin SiN membranes and have been successfully fabricated at MIT by E. Garcia et al. (1993). Furthermore, linearly tapered slot antennas and corner-reflector antennas printed on membranes have been developed by E. Kolberg et al. (Ekstrom et al., 1992) and have shown excellent performance. The fabrication of these radiating structures has led to the establishment of a membrane technology which is reliable, repeatable, and easy to employ. The successful use of membranes for antenna design led to the development of a membrane-supported transmission line, called microshield, which was presented for the first time in the 1991 MTT-S International Microwave Symposium (Dib et al., 1991). The microshield is only one of the possible membrane-supported geometries shown in Figure 120.1. All of these geometries are evolutions of conventional planar lines with one major difference; the substrate material underneath the lines has been removed and a membrane is utilized to support the conductors. Figure 120.1c shows a *membrane coaxial,* which resembles a rectangular coaxial, and is characterized by zero dielectric loss, zero dispersion, zero parasitic radiation while maintaining compatibility to planar monolithic geometries. This propagating structure is completely shielded and can provide passive circuit components with optimum performance. Figure 120.1a shows a *membrane coupled strip* line which very closely resembles the conventional coupled strip line and can provide very efficient antenna feeding networks. The third of the membrane geometries, Figure 120.1b, is the *microshield line* which resembles very closely the conventional coplanar waveguide. This line has zero dispersion, limited parasitic radiation and the capability to suppress the excitation of the unwanted slot mode due to the presence of the folded ground which operates as a continuous air bridge. The membrane line as a transmission medium has created the basis of a new technology, which can provide generic designs appropriate for circuit and antenna applications in the millimeter and submillimeter-wave region.

As shown in Figure 120.1, various configurations of membrane lines may require the use of two or three wafers, to provide a cavity shield on one or both sides. This cavity shield is necessary for some uniplanar geometries, where multiple ground planes or other conductors are in close proximity to the signal line. Figure 120.1a shows a membrane coplanar strip line, made of two high-resistivity <100> Si wafers, in which the electric field is confined to the surface area between the two strips. In this case shielding by the cavity is not required and, depending on the application, the line may operate in a variety of environments. All the membrane line configurations presented so far utilize at least two wafers, and the cavity structures under the signal lines may also be designed to control line characteristics. These membrane monolithic geometries are appropriate for a variety of applications including monolithic antenna and array feeding networks, diode mounting structures for receiver applications, networks for vertical integration, etc.

Membrane supported transmission lines are quasi-planar configurations in which a pure, non-dispersive TEM wave propagates through a two-conductor system embedded in a homogeneous environment. Homogeneity of the environment can be accomplished by using a 1.5-micron-thick dielectric membrane, or a few-micron-thick diaphragm, to support the signal lines, while ground is provided by a metallized micromachined cavity. This cavity is fabricated in Si wafers, using etchants such as KOH (potassium hydroxide) or EDP (ethylene diamine pyrocatechol), and is metallized using evaporation or plating. The signal lines are created by metal deposition on the membranes or diaphragms or can be grafted onto the supporting structure by lift-off techniques. Due to cavity shielding and the pure TEM character of the propagating mode, these lines possess a large single-mode frequency band (DC to $>$ 1 THz), have very low losses and zero signal dispersion. Consequently, circuit components made of these line geometries can provide electrical performance which is superior to those of conventional planar circuits.

The success of membrane-supported circuits relies on the development of thin-film dielectric membranes or diaphragms

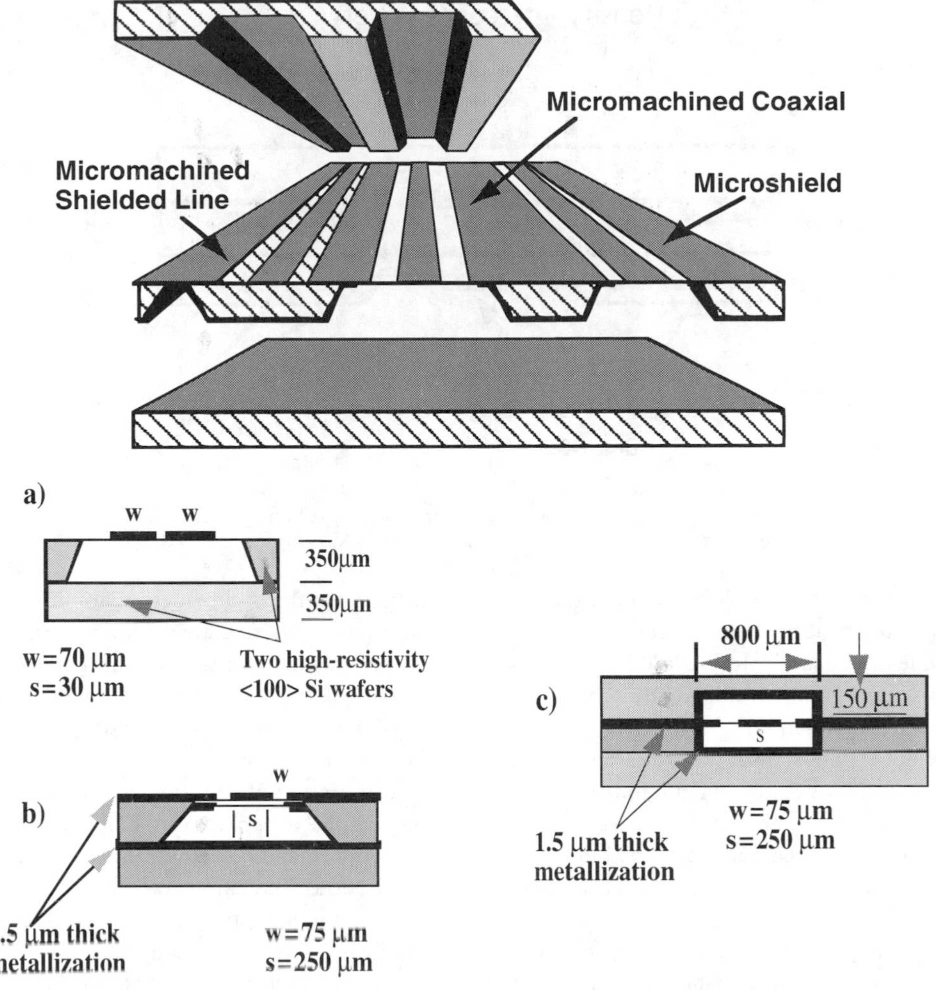

Figure 120.1 Micromachined transmission line geometries.

with good electrical and mechanical properties. These thin-film layers are grown on Si or GaAs wafers, and are used to support the planar conducting strip lines. In view of the previously mentioned performance objectives, the thin films must have low losses at microwave and millimeter-wave frequencies, as well as compatibility with semiconducting and conducting materials. Furthermore, mechanical considerations include reduced sensitivity to applied pressure and temperature variations, along with increased membrane or diaphragm sizes.

For the circuits presented in this chapter, the membranes are in a tri-layer $SiO_2/Si_3N_4/SiO_2$ configuration, with constituent thicknesses of 7000, 3000, and 4000 Å, respectively. The base layer (7000 Å) is grown using thermal oxidation, while the final two layers are formed using a low pressure chemical-vapor-deposition process (LPCVD). The materials used for the development of the dielectric membrane have very low losses from DC ($\rho > 10^{14}$ Ω-cm) well into the THz region. The small size makes the membranes transparent to propagating signals at frequencies as high as 1–3 THz, and provides a near-homogeneous air-filled environment to the propagating electromagnetic wave. Consequently, the wave exhibits very fast wave velocities, zero dispersion, and zero dielectric loss. To date, membrane dimensions as large as 9 mm × 9 mm have been fabricated for use in membrane lines with a yield greater than 95%.

Micromachined Lines for Miniature, Monolithic Packaging

In microwave and millimeter-wave circuit applications, the issue of RF packaging is becoming an important subject to address due to the lack of appropriate packaging configurations for high-frequency circuit design. Several years back, experts in the field of device and component development began to realize that packaging of such components was progressing at a much slower rate than the devices themselves. As a result, many problems observed in device and component performance at these frequencies are being attributed, after diagnostic testing, to the package in which they are housed. Typical problems associated with circuit packages, especially above X-band, include resonances due to the large physical geometry surrounding these circuits, cross-talk caused by parasitic radiation from neighboring circuits, and unwanted excitations that result in power leakage in the form of substrate modes. Many of these issues can be addressed by integrating the package with the circuit monolithically, which

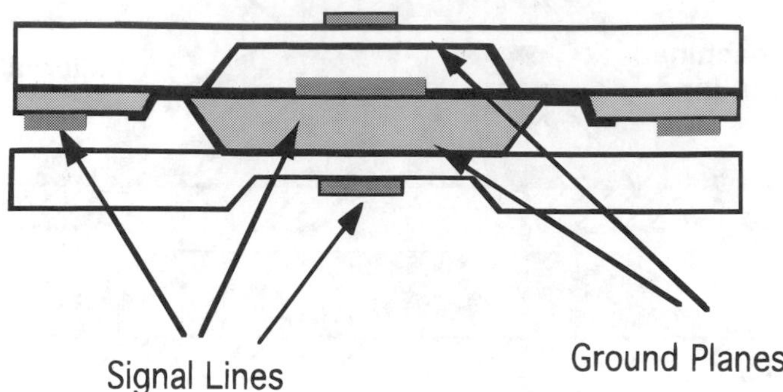

Figure 120.2 High-density micromachined interconnect network.

implies that the package is part of the circuitry and it is designed to meet performance specifications. Other desired attributes of an RF package include the capability to provide protection from hostile environments, appropriate means for heat removal, and mechanical support for components while introducing minimum performance degradation. With state of the art advances in semiconductor processing techniques, silicon micromachining can offer what conventional means have not been able to provide; packages which conform to the circuit geometry, require much less space, and provide superior mechanical, thermal, and electrical performance.

Although micromachining of silicon is a well established technology for sensor and biomedical applications, for the first time this technology is being applied to develop self-packaged circuit components for high-frequency applications. In the past few years, the use of Si micromachining in microwave and millimeter-wave circuit applications has been extensively explored at the University of Michigan, leading to many innovations. Among these have been the capability to develop self-packaged circuit components which have demonstrated superior electrical performance when compared to conventionally developed components. These micromachined circuit components may be of microstrip or coplanar waveguide (CPW) type and they are surrounded by an air-filled cavity in the upper region and a substrate-filled cavity in the lower region. Both cavities are integrated monolithically with the circuits to provide completely shielded geometries which are appropriate for a broad range of applications including high density interconnect networks such as the one shown in Figure 120.2. The use of micromachining has led to the development of a variety of planar circuits with optimum performance (Weller et al., 1993a,b; Drayton and Katehi, 1993, 1992a,b) and has demonstrated the capability to develop novel geometries and integration techniques.

This effort has led to the first demonstration of a high-frequency monolithic conformal package that follows the path of various line geometries, and is sized in such a way that its package resonance exists well above the desired range of operating frequencies. One very important characteristic of this packaging approach is that it provides monolithic self-packaged geometries which can be integrated within more complex planar circuit arrangements. Furthermore, this package can be integrated with any uniplanar technology and can incorporate any type of transmission medium such as microstrip, coplanar waveguide, coaxial line or stripline.

The concept of conformal packaging can be applied to a very broad range of applications. One such application is the development of planar diode mounting structures for detector and mixer applications. To demonstrate the flexibility offered by micromachining, the mounting structure for a K-band detector as shown in Figure 120.3 has been developed with both longitudinal shielding as well as cross cavity shielding. The cross-cavity junction implemented in the shielding allows for the incorporation of commercial diodes or other input paths to the existing circuit. The passive circuit in the mount consists of an input matching network, a diode mounting region, and a lowpass filter output. At the diode placement location, a cross-cavity is formed in both the upper and lower cavity regions and is designed to operate under cut-off as discussed earlier. For the embodiment presented here, the lower cavity dimensions form a substrate-filled waveguide which has the dominant effect on the package resonance.

Figure 120.3 Conformal packaging.

The dimensions of the shield are chosen so that the package resonates at 54 GHz.

120.3 Fabrication Methodology

A detailed description of the processing steps involved in the fabrication of silicon micromachined circuits is presented in this section. The procedures are grouped into five primary categories, which are the growth of the tri-layer dielectric membrane, deposition of the circuit metallization, wet chemical etching (micromachining) of the silicon, backside metallization, and final circuit assembly. The membrane is mainly of importance in the realization of membrane-supported configurations, but is also useful as a temporary masking layer in general micromachining applications. Typically, a single oxide layer is more convenient when only the masking function is required. The silicon wafers must be at least single-side polished on the surface which will be patterned with circuit metallization, and double-side polished wafers are desirable when fine features are to be micromachined on the back side. If the etch patterns consist of relatively large openings and do not require great resolution, the additional expense of double-side polishing can be safely avoided. It is imperative, however, that silicon with a resistivity of at least 1500–2000 Ω-cm be used wherever silicon is exposed to the RF energy. This is done to avoid excessive dielectric loss, and necessitates the use of wafers which are grown using the float zone (FZ) process (Kramer, 1983). For completely metallized ground planes or RF shielding cavities, less expensive, low-resistivity (approximately 10 Ω-cm) wafers are utilized. These are typically grown using the Czochralski (CZ) method (Leadise, 1970).

There are undoubtedly countless numbers of possible variations and extensions of the techniques included here. The intent of this writing is merely to describe a reasonably broad set of procedures which have been found to be reliable for fabrication of the micromachined circuits. For more extensive reviews of silicon processing and micromachining, the reader is referred to Wolf and Tauber, (1986).

Dielectric Membrane Growth

The cornerstone of silicon micromachined, membrane-based transmission line architectures is the three-layer dielectric material on which the conducting lines are supported (see Figure 120.4.) This membrane comprises a thermally grown oxide which is subsequently layered with a silicon nitride and another oxide, both of which are deposited using low pressure chemical vapor deposition (LPCVD). By properly balancing the thickness of each layer, a composite in slight tension is obtained which remains rigid when the silicon substrate is removed, and is robust enough to withstand subsequent processing such as photoresist application and metallization. Furthermore, membranes with dimensions up to 8×8 mm^2 have been fabricated successfully with yields exceeding 95% (Weller et al., 1994a).

The critical issue in achieving flat, large area membranes is determining the correct thicknesses of the constituent layers. It is known that the thermal coefficients of expansion for amorphous SiO_2 and Si_3N_4 are greater and less than that of silicon; The value for silicon is 2.6×10^{-6}/°C, and the values for high-temperature CVD oxide and nitride are 5×10^{-7}/°C and 4–7 $\times 10^{-6}$/°C, respectively (Maissel and Glang, 1970). Therefore, by adjusting the thickness of each layer, a net tensile stress can be left in the membrane when the wafer is cooled to room temperature from the 800–900°C temperatures in the LPCVD furnaces. High-yield membranes are obtained on 10cm diameter, 500–550 micron-thick wafers using values of 7500Å/500Å/4500Å for the thermal oxide, intermediate nitride, and top oxide layer thicknesses, respectively. A direct, experimental technique for determining the appropriate layer thicknesses is outlined in Ling (1993). The specific layer thicknesses given here have been slightly modified with respect to those given in Ling (1993).

Detailed procedures for membrane fabrication are outlined in the following:

- Pre-Furnace Wafer Clean: Prior to the thermal oxide growth, the wafers must be stripped and cleaned of all foreign materials to ensure a high-quality film and to prevent contamination of the furnace. The necessary steps are outlined in Table 120.1.
- Thermal SiO_2 Deposition: The thermal oxide is grown using a dry-wet-dry sequence at a temperature of 1100°C and a pressure of 1 ATM, with the temperature held at 800°C while transferring the wafers in and out of the furnace. The first and last dry oxide layers are dense films which are grown to thicknesses of 5–10 Å using an oxygen flow rate of 3 L/min. The intermediate wet layer is grown using flow rates of 1.7 L/min for O_2 and 2.5 L/min for H_2, yielding a faster growth rate but more porous film in comparison to the dry layers. The time required to grow a 7000 Å-thick oxide is approximately 3–4 hours.
- LPCVD Si_3N_4 and SiO_2 Deposition: Following the thermal oxide growth, the wafers are moved to an LPCVD furnace for deposition of the intermediate nitride and the final oxide. The furnace temperature is approximately 900°C, and the growth rates are around 50 Å/min and 70 Å/min for the nitride and oxide, respectively. Specific parameters for this process are given in Table 120.2.

Metallization Pattern Definition

This section outlines the procedure for depositing the thin-film metallization which constitutes the circuit pattern onto the front side of the wafer (see Figure 120.5). The steps described here are primarily based on the image reversal photoresist technique. In this process, dark features on the circuit mask correspond to areas which will be metallized, and this generally simplifies the alignment procedure since most of the mask is transparent (a clear-field mask). Also, in comparison to alternative methods such as the chlorobenzene process, the image reversal method has been found to yield more ideal metal profiles. Results with the chlorobenzene process demonstrated a phenomena known

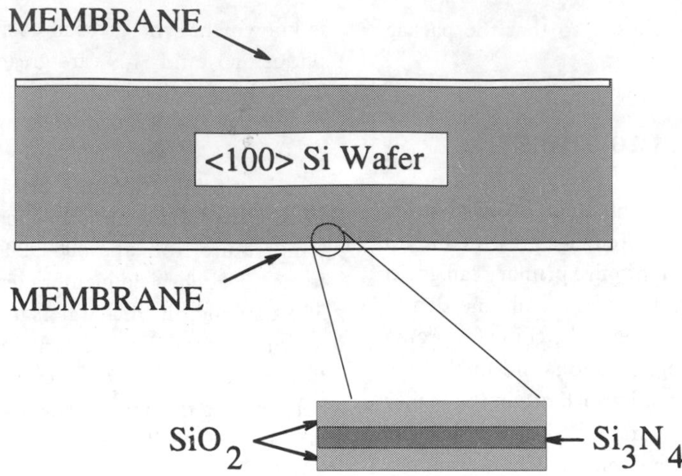

Figure 120.4 The tri-layer dielectric membrane grown on both sides of a silicon wafer.

Table 120.1 Prefurnace Clean Procedure

Solution	Ratio	T°C	Time (Min)
H_2O_2:NH_3OH:H_2O	1:1:15	90	10–20
H_2O quench		24	2
HF:H_2O	1:10 (Note 1)	24	1
H_2O quench		24	2
H_2O_2:HCL:H_2O	4:4:25	90	10–20
H_2O		24	5
H_2O rinse w/N_2 bubbling		24	Note 2
Spin in the rinser/dryer		24	5

Note 1: When cleaning wafers with critical oxide layers, a 1:100 solution can be used instead.
Note 2: The last rinse is continued until the bath resistivity exceeds 13–14 MΩ-cm.

Table 120.2 LPCVD Deposition Parameters. DCS Stands for Dichlorosilane

Dielectric Layer	T°C	Ingredients	Flow Rates (sccm)
Si_3N_4	800	NH_3	160
		DCS	40
SiO_2	900	N2 dilute	290
		N_2O	120
		DCS	60

as "lift-off flags," referring to metal profiles which have sharp, up-turned spikes at the edges of the pattern.

Before starting the procedure, appropriately sized silicon pieces must be prepared. This involves dicing, or simply scribing, a whole wafer into smaller pieces, making sure that the edges follow straight lines along the <110> crystal directions (either parallel or perpendicular to the major flat). If this is not done it will be impossible to correctly align the etch patterns with the crystal structure. This type of misalignment results in poor cavity definition and undercutting of the etch masking layers.

The circuit metallization sequence is as follows:

- Wafer Preparation: For best results, the wafer should be thoroughly cleaned using a 1:1 "piranha" etch (a 1:1 ratio of H_2SO_4:H_2O_2). The duration of the clean should be about 10 minutes, which is the approximate lifetime of the reaction.
- Image Reversal:
  a. Clean the wafer with acetone and IPA. Dehydrate-bake for 3 minutes on a hot-plate at 130°C. Keep desired membrane surface face up.
  b. Spin-apply HMDS (hexamethyldisilazane) adhesion promoter and AZ 5214-E photoresist at 2.5 KRPM. This yields a photoresist thickness of approximately 2.1 μm.

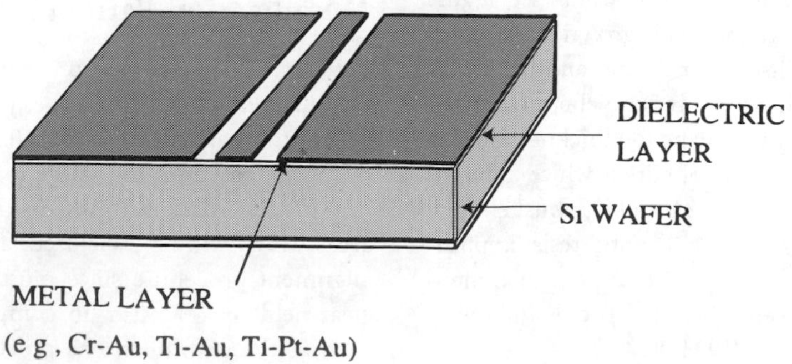

Figure 120.5 Circuit pattern deposited onto the front side of a wafer using thin-film application techniques.

c. Soft-bake 1 minute at 105°C on a hot-plate.
d. Align and expose for 8 seconds at 20 mW/cm^2 ultraviolet light intensity (λ = 405 nm). To achieve success in the subsequent micromachining steps, the mask must be aligned to a straight edge of the wafer, such that the patterns to be etched into the front or back side can be correctly positioned along the <110> crystal directions. This step is made easier if the mask incorporates long, straight alignment marks around the edge of the circuit area. These can be used to align the wafer via Θ-positioning, and if necessary, the pattern can then be translated back to the center of the wafer using *x* and *y* positioning only.
e. Post-bake 1 minute at 130°C on a hot-plate. This step causes the exposed photoresist to "cross-link," and leads to the formation of a thin crust on the upper surface.
f. Flood-expose for 120 seconds at 20 mW/cm^2 UV power density. This long exposure activates the photoresist which was not exposed previously.
g. Develop in AZ327MIF developer for 45–55 seconds. This removes the photoresist which was not exposed during the first alignment, and will also remove some of the photoresist beneath the edges of the cross-linked pattern. It is this undercutting that allows the subsequent lift-off process to succeed.
h. Check pattern definition of the evaporated metal using a microscope.

- Metal Deposition:

a. Descum the wafer using a 36 second, 80 W O_2 plasma etch.
b. Deposit metallization using evaporation techniques. Metal systems such as chrome-gold, titanium-gold, or titanium-platinum-gold are resistant to wet etching using the EDP solution. Alternatively, if thicker layers are necessary to minimize attenuation, copper or silver may be used instead of gold to reduce fabrication costs. Most anisotropic silicon etchants attack aluminum at a substantial rate, and therefore Al should not be used in applications requiring long etch times.
c. Soak in acetone for approximately 6 hours to dissolve the photoresist masking layer and complete the pattern lift-off.

The process described has been verified for line dimensions down to 10 μm, and evaporated metallization thicknesses up to 1.2 μm.

Due to the slow spin-rate for the photoresist deposition, beading will occur around the edges of the silicon piece, leaving up to 5 mm around the circumference which is non-usable. For 1–2 μm features, the beading should be removed using a technique such as that described in Gearhart (1994). Also, for metallization thicknesses up to 0.6 μm, the AZ 5214-E can be spun on at 4.5 KRPM. This provides about 1.4 μm photoresist thickness and is better for feature sizes down to 3–5 μm. Using this method, the first expose and flood expose times should be reduced to 4 seconds and 90 seconds, respectively.

It is important to point out that free-standing membranes of the type described in Dielectric Membrane Growth can withstand the circuit metallization procedure. Therefore, the silicon micromachining can precede the metal deposition if necessary. In this case, photoresist application requires that the wafer should first be attached to a support wafer, separated by small silicon standoffs (see Figure 120.6). The support/standoff/membrane wafer system can be temporarily assembled using a heavy photoresist "glue" such as 1400–37, and baked on a hot plate at 130°C for 1 minute. This assembly will easily survive the photoresist spin and alignment steps, and can be separated during the photoresist development stage to simplify drying. Alternatively, the circuit wafer can be separated during the second hot plate bake by handling it instead of the support wafer. Some type of standoff arrangement is also necessary during the metal evaporation, to prevent membrane failure, since evacuation of the chamber creates a high pressure differential across the membrane.

For applications requiring thicker metallization layers, the evaporation technique can be supplemented with gold electroplating. One method of implementing the plating process begins with the evaporation of Ti-Au-Ti or Cr-Au-Ti seed layers over the entire front (or back) wafer surface, using thicknesses of 500Å/1000Å/500Å, respectively. The photoresist is then applied and patterned using the image reversal technique. Alternatively, a 1400–37 positive photoresist process can be used to provide a thicker photoresist layer. This process requires a dark field mask, and is outlined in steps 1a–1h in the section on Silicon Micromachining. The next step is to remove the exposed Ti in the pattern openings using a 10:1 H_2O:HF etch for 3–10 seconds. The metallization can then be built up using electro-plating, and the Ti layer which remains outside the pattern will help to prevent lateral spreading of the plated Au. Finally, the photoresist is removed with an acetone soak or hot PRS-1000 photoresist stripper. Unwanted seed layer metals are removed using gold and chrome etchants, and 10:1 H_2O:HF for Ti.

Silicon Micromachining

The technologies related to silicon micromachining began to emerge in the 1960s (Lepselter, 1966), and continued to evolve into what is currently a wide array of techniques aimed at manipulating silicon for mechanical purposes. The development of complex micro-electro-mechanical systems (MEMS), sensors,

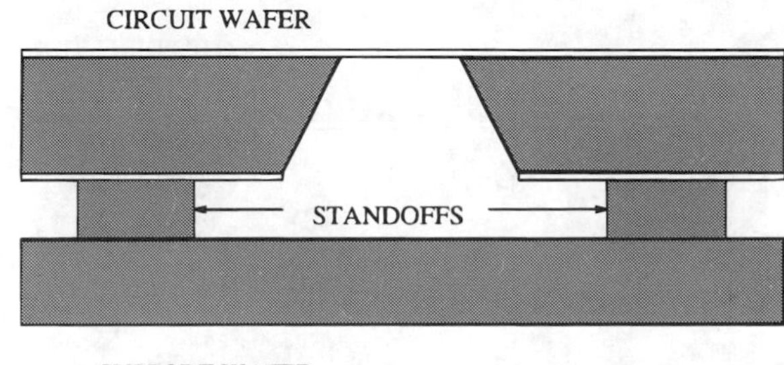

Figure 120.6 Circuit-wafer/standoff/support-wafer assembly used for the photoresist process on free-standing membrane pieces.

and actuators has driven many of the significant advancements in this area, and a large volume of related work has been published in journals on solid state electronics and processing. The microwave community, in which the use of micromachining is relatively new, now has the advantage of drawing upon these established resources to solve problems and develop new concepts for microwave circuits.

The procedure which will be presented here is a very basic approach to bulk silicon micromachining using an anisotropic etchant. It relies on the use of a wet etchant such as KOH (potassium hydroxide) or EDP (Finne and Klein, 1967) (ethylenediamine pyrocatechol) to remove silicon preferentially along certain crystal directions, thereby allowing the formation of well-defined cavity regions. In the work considered here, wafers with a <100> orientation are typically used, and the etch rates are about 50:50:1 for the <100>, <110>, and <111> directions, respectively, using EDP at 100°C (see Figure 120.7). Similar ratios are achievable using KOH, however this is generally not the etchant of choice for dielectric membrane-related work since it etches silicon dioxide much faster than does EDP. The SiO_2 etch rate in EDP is approximately 5 Å/minute, and in KOH the rate can be increased by an order of magnitude. Other masking films for EDP include Si_3N_4,Au, Cr, Ti, Ag, Cu, and Ta (Peterson, 1982).

To describe the complete micromachining procedure, a typical process flow for single-side etching from the back of a wafer is outlined in the following. It is assumed that single-layer oxides or oxide/nitride/oxide layers are in place and will serve as the masking films for the etchant.

- Wafer Preparation:
  a. Clean wafer with acetone and IPA. Dehydrate bake for 3 minutes on a hot-plate at 130°C. Keep desired membrane surface face up.
  b. Spin HMDS adhesion promoter and 1400–37 photoresist at 3.5 KRPM onto the front (desired) membrane surface. This provides a photoresist thickness of approximately 3.5 μm.
  c. Hard-bake 30 minutes at 110C° in an oven. Alternatively, a 1 minute bake at 130°C on a hot plate can be used. This photoresist serves to protect the front side of the wafer during the proceeding masking layer removal steps.
  d. Spin HMDS adhesion promoter and 1400–37 photoresist at 3.5 KRPM onto the back side of the wafer.

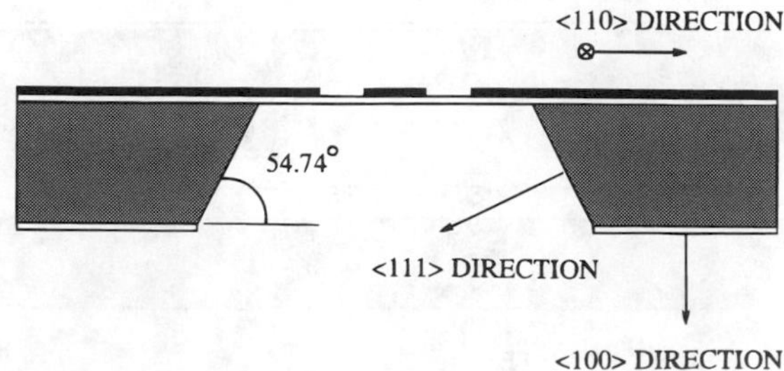

Figure 120.7 Etch profile for a ⟨100⟩ oriented silicon wafer using an anisotropic etchant.

  e. Soft-bake 30 minutes at 90°C in an oven.
  f. Align and expose the patterns to be etched into the silicon using a mask aligner and a dark-field photo mask, at an ultraviolet light power density of 20 mW/cm^2 for 15 seconds. The dark-field mask has openings in the emulsion or chrome which correspond to the areas where photoresist is to be removed, since 1400–37 is a positive resist. The etch pattern must either be aligned to the <110> direction using the wafer edges, or to a pre-aligned front-side metallization layer using infra-red alignment techniques.
  g. Develop in MF319 developer using a 5:1 MF319:H_2O solution for 60 seconds.
  h. Hard bake 30 minutes at 110°C in an oven.

- Masking Layer Removal
  a. Silicon dioxide masking layers can be removed using buffered hydroflouric acid (BHF). The etch rate is approximately 1000 Å/minute at room temperature.
  b. Silicon nitride masking layers are removed with a CF_4 plasma etch. In this process, the etching chamber is pumped down to 75 mTorr and CF_4 and O_2 are then flowed in at about 20 sccm and 0.5 sccm, respectively; the CF_4 flow rate is adjusted until the chamber pressure reaches 250 mTorr. After allowing the system pressure to stabilize, a 100 W RF plasma is ignited which will etch the nitride at a rate of approximately 700 Å/minute. The photoresist masking layer described in the previous steps can safely withstand 11–12 minutes of the plasma etch, but etch durations should be closely monitored if thinner photoresist products (e.g., AZ5214) are employed. When removing the middle nitride layer of the tri-layer composite membrane described in Dielectric Membrane Growth, the integrity of the photoresist mask is critical for the final SiO_2 etch using BHF. The photoresist etch rate in the plasma can be as high as 1500 Å/minute.

- Silicon Etching with EDP
  a. Prepare the EDP solution by combining the following ingredients in a large glass beaker, in the order listed: 96 mL of de-ionized H_2O, 96 gm of catechol, 1.8 gm of pyrazine, and 300 mL of ethylenediamine. (Different quantities of the solution can be mixed by proportionally adjusting the amount of each substance.) Cover the beaker tightly with heavy duty aluminum foil, place on a hot plate, and heat the solution to 110–112°C. It is convenient to use a thermometer probe with feedback to maintain a constant temperature. All mixing, heating, and etching must be done inside a well-ventilated fume hood.
  b. While the EDP is warming up, the final masking layer removal steps should be completed to expose the bare silicon. For example, this might pertain to the BHF etch of the lower thermal oxide layer of the membrane described in Dielectric Membrane Growth. Any native

oxide growth will impede the etching process, and should thus be removed using a quick (30sec) BHF dip. Photoresist can also be stripped off using acetone and IPA, since it will eventually come off in the EDP.

c. Place the wafers in a Teflon holder and place the holder in the heated EDP solution. Be certain to keep the beaker covered. The etch rate is approximately 1.2–1.4 μm/minute, and thus it takes 6.5–7 hours to etch completely through a 500 μm-thick wafer. These results pertain to fairly wide cavities (1–1.2 mm) and include the effects of "notching" (Findler et al., 1982), which refers to the increased etch rate along the cavity edges with respect to the rate in the center of the lower cavity surface. Narrower cavities, particularly those which come down to a point, tend to etch at somewhat higher rates.

d. When the etching is complete, the samples should be rinsed in warm de-ionized water, acetone, and IPA for approximately 15 minutes apiece. The samples should then soak in warm methanol for 6–10 hours to completely remove any EDP residue.

In certain applications it is convenient to use a thin Cr-Au layer to mask the EDP. The metallization can be deposited using the image reversal and lift-off procedure outlined in Metallization Pattern Definition. Also, in situations where different parts of a sample need to be etched to different depths, a multi-step etching procedure can be employed (Drayton et al., 1995; Chi and Rebeiz, 1994). In the first stage the masking layers are removed only from the areas which require the most etching, and the sample is placed in the EDP for a pre determined amount of time. The sample is then removed from the EDP, cleaned with the water/acetone/IPA/methanol sequence (using a shorter duration for the methanol soak), and the masking layer is then removed from the other desired etch regions. The sample is then placed back in the EDP solution to complete the etch.

Backside Metallization

With many of the micromachined circuit geometries, it is necessary to deposit metallization on the backside of the circuit wafer following the silicon etching procedure. In some cases this may simply entail metallizing the entire back surface, which is easily accomplished using evaporation or electro-plating. The process is somewhat more complicated for the membrane-supported microshield line, since the metal must be selectively deposited inside the lower shielding cavity. It is necessary to cover the cavity sidewalls and some portion of the backside of the membrane, to provide an RF short between the upper ground planes and the cavity. To prevent a similar shorting of the transmission line itself, however, the region beneath the slots in the coplanar conducting lines must be shielded from the metallization. This selectivity is achieved with the use of the "shadow" mask illustrated in Figure 120.8. The mask is a silicon wafer that is processed along with the circuit wafer, and has openings etched into it through which the evaporated metal is allowed to pass. The shadow mask is correctly positioned on the backside of the circuit wafer using small alignment cavities that have been etched into each wafer, and temporarily attached using the photoresist adhesion technique described in the next section. The achievable resolution limit of this method is approximately 50 μm. Following the metal evaporation, the shadow mask is removed by soaking the wafers in acetone.

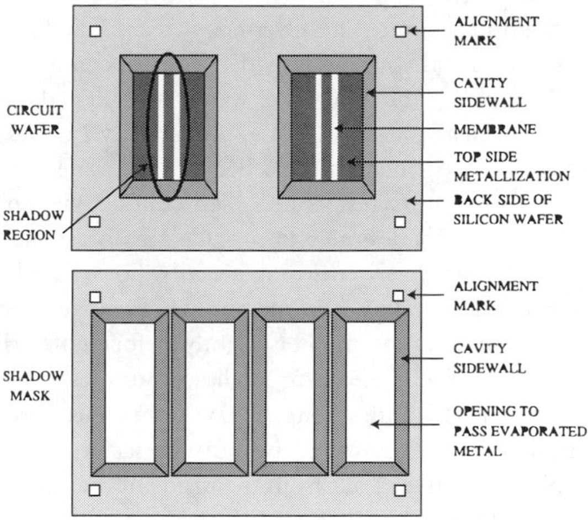

Figure 120.8 The etched shadow mask used for backside Metallization of the wafer.

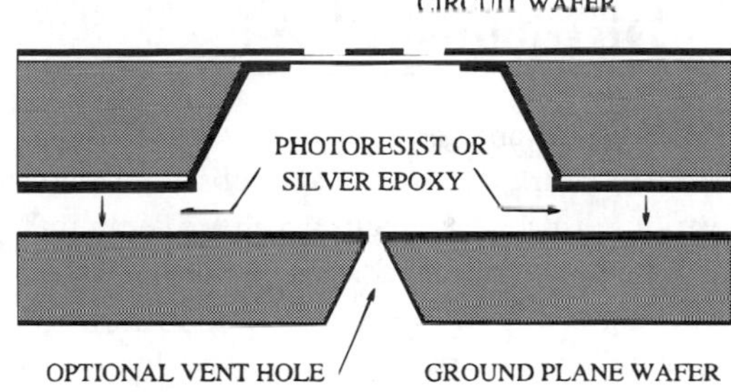

Figure 120.9 Final circuit assembly.

Circuit Assembly

The final phase of the fabrication sequence generally involves the assembly of two to three stacked wafer pieces, such as the ground plane, circuit, and shielding cavity (Figure 120.9). Accurate positioning of each level mandates the incorporation of alignment mechanisms such as patterns in the metallization, and micromachined cavities in the silicon which are etched completely and/or partially through the wafer. These alignment marks should be considered in the first phase of the circuit design.

Two convenient adhesives for bonding the wafers together are photoresist and silver epoxy. The photoresist technique is very useful for temporary assembly, and is common laboratory practice for the purpose of making electrical performance measurements. Using this approach, small drops of a photoresist such as 1400–37 are applied to a wafer, and then two pieces are aligned under a microscope or mask-aligner. The assembled wafers are then baked on a hot-plate at 130°C for approximately 60 seconds to cure the photoresist. The wafers can easily be separated by soaking them in acetone. The silver epoxy technique follows the same basic steps and will also provide electrical contact if necessary. After applying the epoxy, the wafers are aligned and then baked in an oven at 100–120°C for approximately 2 hours, forming an essentially permanent bond between the wafers. The elevated cure temperature results in a high-conductivity contact.

In some cases it is desirable to provide a path for the release of gases when completely enclosing a cavity. For example, when the lower shielding cavity of a membrane-supported geometry is sealed, the membrane may be slightly deformed during the cure of the adhesive. Gases can be allowed to escape either by leaving some part of the contacting wafer surfaces free from photoresist or silver epoxy, or by putting scribe lines in the ground plane which lead out from the inside of the cavity. Alternatively, small air-holes may be etched into the ground plane wafer.

120.4 Membrane-Supported Distributed Circuits

The performance advantages of membrane supported transmission lines can be clearly demonstrated by observing the characteristics of various distributed circuits which are common to planar microwave circuitry and MMICs. The broadband TEM propagation afforded by membrane supported transmission lines permits a significant increase in performance levels for typical planar circuits such as filters, stubs, and power dividers. With conventional substrate supported designs, frequency dependent mechanisms such as radiation loss and dispersion may limit the operating bandwidth of distributed circuits, and may also serve to preclude their use for higher frequency applications. In the following sections, a summary of membrane supported distributed circuits which have been fabricated at the University of Michigan will be presented. In all cases, measurements were made on a vector network analyzer (HP8510) and a Thru-Reflect-Line (TRL) calibration technique was employed to deembed the measurements to the reference planes of the circuits.

Low Pass Filters

Perhaps the simplest example of a low pass filter is the stepped-impedance type of filter which uses short transmission line sections to approximate the effects of a series-L/shunt-C configuration (see Figure 120.10). In this type of filter, series inductances are approximated by short sections of high-impedance transmission line, while low-impedance sections mimic the behavior of shunt capacitors. These transmission line sections may be cascaded to generate the required number of filter poles, and their lengths can be designed to produce either equal ripple (Chebyshev) or maximally flat (Butterworth) filter responses. The benefits of using membrane supported line for low pass filter applications are two-fold: first, the absence of the dielectric substrate ensures low passband insertion loss as a result of reduced dielectric and radiation losses; second, broadband TEM operation preserves the high frequency rejection of the filters, since performance degradation by higher order modes is avoided.

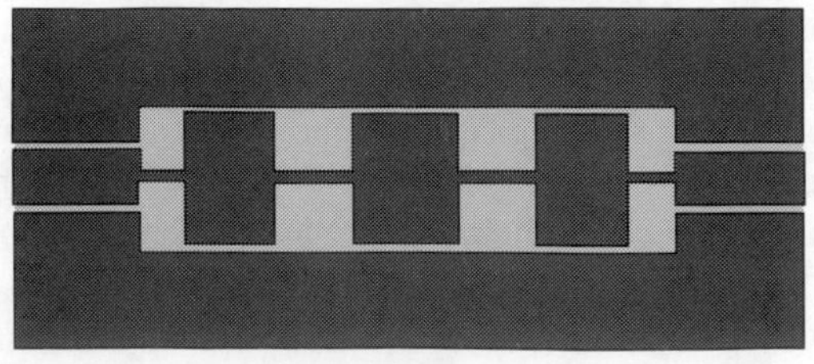

Figure 120.10 Layout of a typical stepped-impedance low pass filter in a CPW type of geometry.

With these benefits in mind, low pass filters have been studied using the microshield line geometry. First, a 5-section, 0.5 dB equal ripple, filter has been designed and measured from 10–40 GHz (Weller et al., 1993c). As shown by the results presented in Figure 120.11, the filter has a cutoff frequency of 26 GHz and a rejection of −25 dB at 40 GHz. The passband insertion loss remains less than 1 dB up to 23 GHz. The measured response agrees very well with theory up to 40 GHz, and the theoretical data for higher frequencies have been plotted to show the filter performance up to 80 GHz.

A microshield line low pass filter for W-band applications has also been designed (Robertson et al., 1994). This filter uses the same stepped-impedance geometry as discussed above, and is based on a 7-section, 0.5 dB equal ripple prototype. Once again, measured results are presented with theory (see Figure 120.12) and excellent agreement is achieved. The W-band filter has a cutoff frequency of 90 GHz, with pass-band insertion loss as low as 0.5 dB and out of band rejection better than 20 dB at 110 GHz. Measured data is presented from 75–110 GHz, which corresponds to the limitations of the test equipment, and theoretical data is shown for the frequency range of 40–140 GHz.

During W-band measurements, data were taken that allowed the extraction of the effective relative dielectric constant ($\epsilon_{r,eff}$) of the microshield line from 75–100 GHz (Marks and Williams, 1993). The graph of $\epsilon_{r,eff}$ in Figure 120.13 illustrates the very

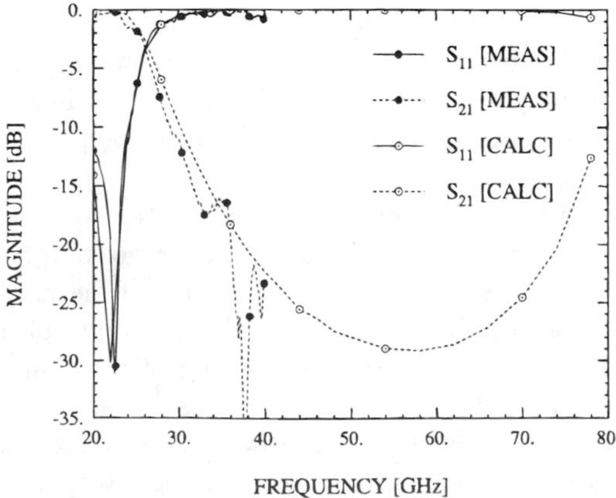

Figure 120.11 Measured S-parameters for a 5-section microshield low-pass filter. Theoretical data is from a method of moments full wave analysis (Pengelly and Schumacher, 1988).

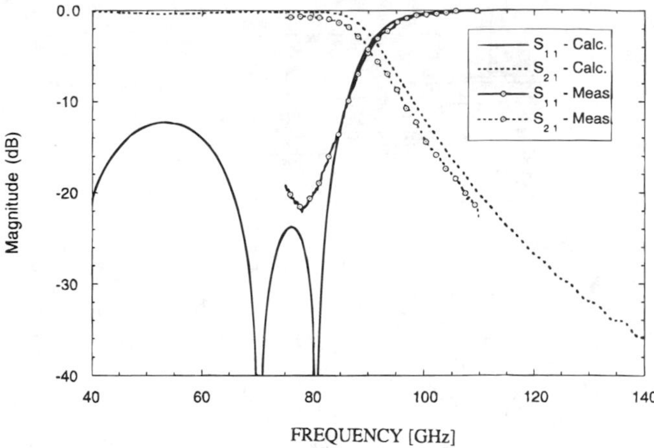

Figure 120.12 Measured and predicted response of a 90 GHz microshield low pass filter. Theoretical data is from a finite-difference time-domain full-wave analysis technique (vanDeventer et al., 1989).

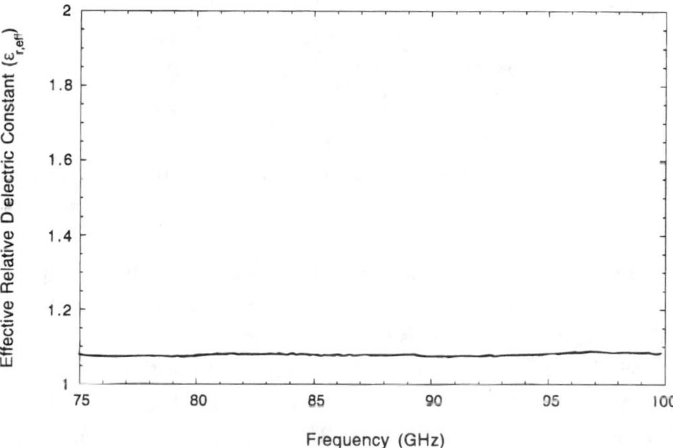

Figure 120.13 Measured effective relative dielectric constant ($\epsilon_{r,eff}$) of a microshield line from 75–100 GHz (Robertson et al., 1994).

minor influence of the membrane on the propagation characteristics of the microshield line. The presence of the dielectrics results in a value of 1.08 for $\epsilon_{r,eff}$, instead of the unity value that would be expected if the signal were propagating entirely in air. Also, very low dispersion is indicated, since the measured $\epsilon_{r,eff}$ remains very nearly constant vs. frequency.

Microshield Series Tuning Stubs

The microshield geometry is also well suited for implementing series type resonant stubs commonly seen in CPW circuits. These types of stubs form the basis for the design of resonant circuits such as filters and matching networks. Generally, stubs which are incorporated into planar transmission lines perform better when they are printed in the center conductor of the line instead of in the ground plane regions on either side of the line. This performance benefit results mainly from reduced parasitic loss caused by radiation from the stub.

Both open-end and short-end series stubs have been fabricated and measured at the University of Michigan (Weller et al., 1995b). The configuration for a typical open-end series stub is shown in Figure 120.14a This stub has a bandpass response when the length is equal to $\lambda_g/4$, and is useful in applications such as DC blocking and LO isolation. The measured response of an open-end series stub is given in Figure 120.14(a) along with theoretical response. The plot also shows the predicted radiation loss, which is below -25 dB throughout the first resonance of the stub. The short-end series stub is very similar to the open-end configuration, as shown in Figure 120.14b. This stub exhibits a band stop characteristic at it's first resonance, as demonstrated by the measured performance.

Wilkinson Power Divider

A Wilkinson power divider was the first microstrip circuit to be adapted to membrane technology (Weller et al., 1994b). Wilkinson power dividers make use of a lumped element resistor which is placed between the two output signal lines and serves to provide an input match at all three ports of the network. In the case of a membrane supported circuit, it is convenient to choose a central impedance of 106 Ω, so 73 Ω matching transformers are included at each port to match the divider to the 50 Ω terminations. The layout of a 33 GHz divider, including the matching sections, is shown in Figure 120.15a; the thin-film resistor is fabricated by evaporation of titanium to a thickness of 400 Å. For measurement purposes, the power divider was mounted on an aluminum fixture to provide the lower ground plane of the circuit and to facilitate connection to other MMIC chips via wire-bonding. The fixture also included an upper shielding surface for the power divider, since it was determined that membrane microstrip has a tendency to radiate due to the lack of high dielectric material to concentrate the electric fields within the transmission line. The *S*-parameter measurements plotted in Figure 120.15 show that the power divider performs well, with better than 15 dB isolation between the two output ports. Analysis of the measured results indicates that the insertion loss for a single divider is approximately 0.2 dB, compared to results of 0.4–0.5 dB for conventional microstrip based circuits (Hamadullah, 1988).

Lange Coupler

Lange (1969) first reported a 3-dB interdigitated microstrip hybrid in 1969. Due to the advantages of broad-band, low-loss and tight-coupling available in his design, the use of the Lange-coupler (Figure 120.16) as a directional coupling scheme has gained popularity among microwave engineers since then. Also due to the perfect symmetry in the design, a Lange-coupler can provide a 90-degree phase difference between the direct port and coupled port over a very wide frequency range (Figure 120.17). Therefore, Lange-couplers have been widely used in balanced amplifiers, single-balanced mixers, image rejection mixers, and phase shifters designed to achieve both phase and amplitude balance. Usually Lange-couplers are designed on a high dielectric constant material such as Duroid or GaAs. At millimeter wave range, the losses from these dielectric materials start to increase and can disturb the performance of the Lange-couplers. However,

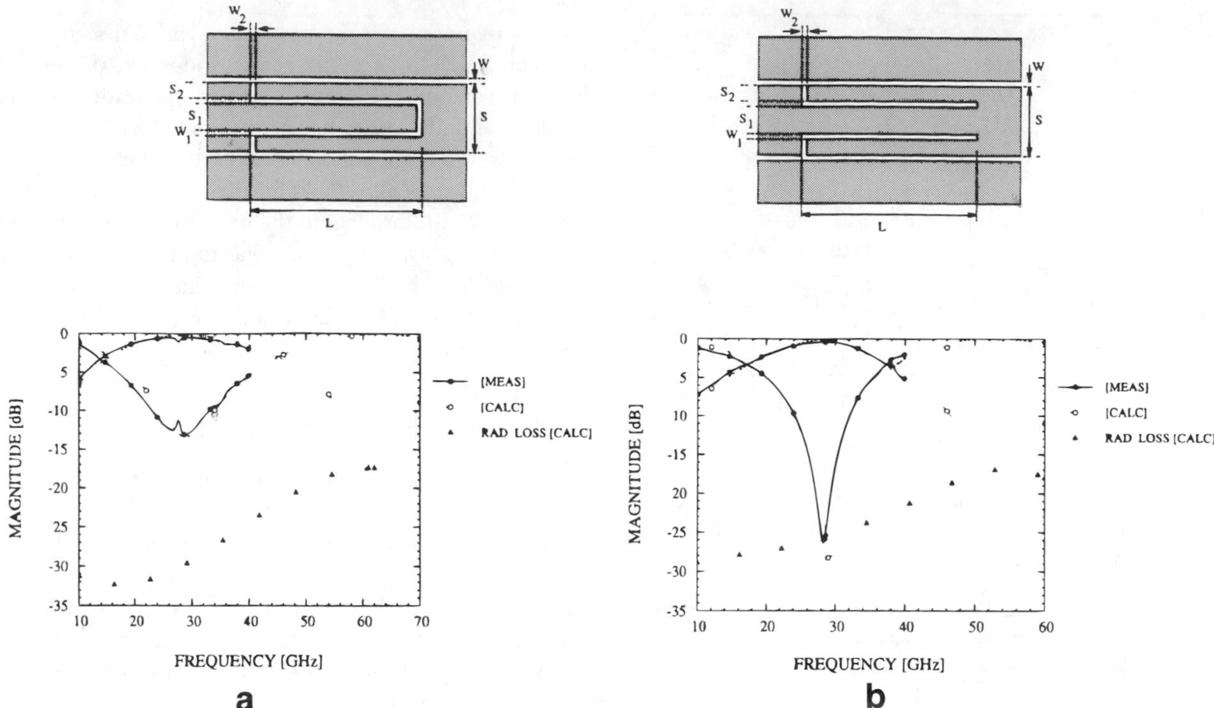

Figure 120.14 Measured S-parameters for open- and short-end series stubs in microshield line: (a) open-end series stub. (b) short-end series stub.

with micromachining technology, Lange-couplers can be built on a thin dielectric membrane to reduce the dielectric loss and improve the performance of the coupler for millimeter wave applications. Based on this concept, a membrane micromachined Lange-coupler has been built on a 350 μm silicon wafer, as shown in Figure 120.18. Grounded coplanar waveguides (GCPW) are used as the input/output feeding structures in the design. Also, six fingers are used to increase the coupling between fingers. The Lange-coupler has been tested from 5 to 20 GHz and the measured results are shown in Figure 120.19a. The coupling bandwidth ranges from 7 GHz to 19 GHz with an isolation better than −18 dB and return loss better than −13 dB. The phase difference between the coupled-port and the direct-port is shown in Figure 120.19b.

Resonators on Membrane

In this section, the performance of stripline and microstrip resonators on thin dielectric membranes will be discussed. The microstrip and stripline resonators are fabricated on a 350 μm-thick high-resistivity silicon wafer using the techniques discussed in Fabrication Methodology. The length of the meander resonator shown in Figure 120.18 is $\lambda_0/2$ at 13.5 GHz and its width is 500 μm. The metallization is gold electroplated to a thickness of 3 μm. The stripline resonator impedance is calculated to be 80.8 Ω and the microstrip resonator impedance is 104 Ω. The resonators are coupled by 150 μm gaps in the 50 Ω feeding transmission line. The measured S_{21} of each resonator is shown in Figure 120.19. The resonant frequencies for the stripline resonator are 13.555 GHz, 27.365 GHz and 39.636 GHz (Figure 120.19a), and are not exactly integer multiples of each other because the gap-coupling capacitance changes with frequency (Chang et al., 1993). Also, the peak S_{21} increases with frequency from −25.8 dB at 13.55 GHz to −10.4 dB at 39.36 GHz also due to the increase in the gap-coupling capacitance. The resonant frequencies for the microstrip resonator are 13.815 GHz and 27.163 GHz. No resonant frequency is seen at 39 GHz due to radiation loss. Table 120.3 shows the measured loaded-Q (Q_L) and the extracted unloaded-Q (Q_u) of these resonators in stripline and microstrip modes.

As is evident in Table 120.3, the stripline Q_s increases as $\sqrt{f}$ with frequency. This is an indication of conductor-loss limited performance and the absence of dielectric and radiation loss mechanisms. On the other hand, the microstrip line resonator suffers from radiation loss and its unloaded-Q decreases with frequency. The 13 GHz Q_u is 235 for the microstrip resonator and is about 15% less than that of the stripline resonator due to a small component of radiation loss. At 27 GHz, the radiation loss component is comparable to the conductor-loss component resulting in $Q_u = 205$ which is around half the value of the stripline resonator at this frequency. At 39 GHz, the radiation loss is very high and no Q-measurements could be done. It is possible to reduce the radiation loss by decreasing the height of the substrate or by increasing the width of the microstrip line.

Interdigitated Bandpass Filters

Micromachining has also been exploited to realize compact interdigitated filters for K-Band and Ka-Band (Chi and Rebeiz,

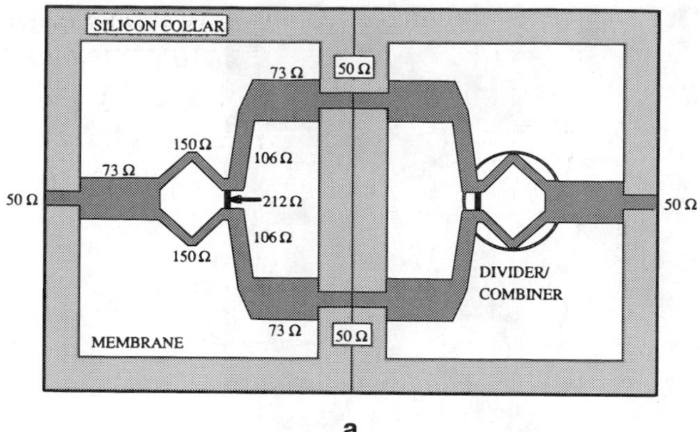

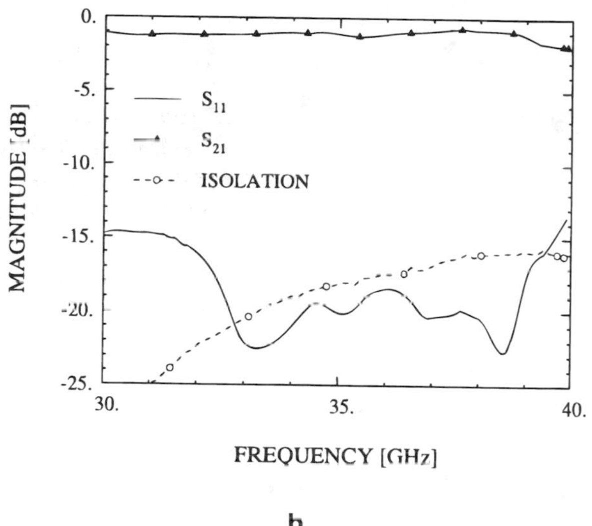

Figure 120.15 (a) Schematic of a membrane microstrip Wilkinson power divider, and (b) Measured S-parameters of back-to-back Wilkinson power dividers fabricated on membrane. Isolation was measured between the two output ports of a single divider.

1994). The interdigitated filter geometry, shown in Figure 110.20(a), uses quarter wavelength resonators similar to the structure discussed above. The resonators are implemented in a stripline configuration and coupled together in a broadside manner. They are suspended on a membrane between two shielding surfaces, but in this case, the shielding enclosure is provided not by a separately machined metal fixture, but by micromachined cavities which are integrated into the filter design. This integrated machined cavity structure, as illustrated by the two-dimensional cross-section in Figure 120.20(b), allows the filter to be fabricated with very small dimensions, allowing its use for high-frequency applications. In addition, the filter design itself is very compact, and can be used to realize filters for a wide range of bandwidths. Another advantage of the interdigitated filter geometry is that the second passband occurs at three times the bandpass frequency.

Synthesis of this filter is not easily accomplished using simple theoretical models, due to the effects of mutual coupling between multiple resonators, including non-adjacent resonators. As an alternative to complex and time consuming full-wave analysis techniques, microwave modeling is used to iteratively design the filter. The microwave model takes the form of a large scale replica of the micromachined filter which operates at 850 MHz. Since a pure TEM mode is excited within the structure, the measurements of the scale model provide a direct correlation to the performance of a miniature filter. And since the scale model can be easily handled and modified, its design can be modified and remeasured with great speed. This allows the design to be optimized before fabrication of the miniaturized interdigitated filter is initiated.

Measurements on the actual micromachined filter are performed through the use of grounded-CPW (GCPW) feed lines which enter the filter cavity through tunnels etched in the top wafer of the cavity assembly. Measured results from 2 to 40 GHz are shown in Figure 120.21, and are compared to scaled measurements of the microwave model performance. The micromachined filter exhibits a return loss better than −15 dB within the passband and a 1.7 dB port-to-port insertion loss at 20.3 GHz. The filter response is predicted very well by the scaled response of the microwave model.

Coupled Line Bandpass Filters

The coupled line bandpass filter is a familiar structure which has found use in many applications that employ microstrip transmission line. It also utilizes the half-wavelength resonator structure discussed earlier, but this type of filter cannot be used at higher frequencies since radiation losses and dispersion associated with the dielectric substrate become so large that filter performance is severely compromised. With membrane supported transmission line technology, however, coupled line bandpass filters which exhibit very high performance have been realized (Robertson et al., 1995). The geometry used for these filters is slightly different from previously discussed microstrip and stripline designs, however, in that it uses micromachining to precisely define the ground plane spacing for the microstrip line. Instead of relying on default wafer thickness to determine shield spacings, this structure employs micromachining to control vertical dimensions which are specified by the electrical performance requirements of the circuits. An example of this type of micromachining is illustrated in Figure 120.22, in which a three wafer assembly comprises a shielded membrane microstrip circuit structure. The middle circuit wafer contains the membrane supported conducting lines of the structure, while the lower ground plane wafer provides the micromachined cavity which acts as a ground to the conducting lines. The third wafer, on top of the assembly, remains completely planar, and is used simply for shielding and enclosure of the circuits. Fabrication of the ground plane wafers uses the 2-sided etching technique discussed in Silicon Micromachining to include windows which allow for on-wafer measurements of the circuits.

On-wafer probing is further facilitated by the use of GCPW probe pads which are required for the ground-signal-ground configuration of the probes. Since the filters are microstrip in

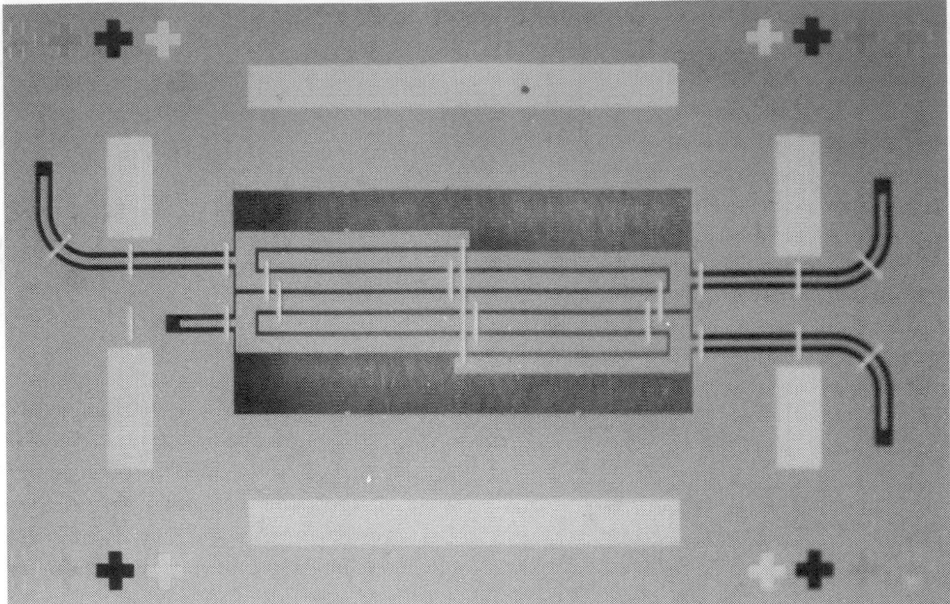

Figure 120.16 Photograph of a Lange coupler fabricated on a thin dielectric membrane. The membrane area appears lighter than the silicon areas. Note the use of air-bridge technology integrated onto the membrane.

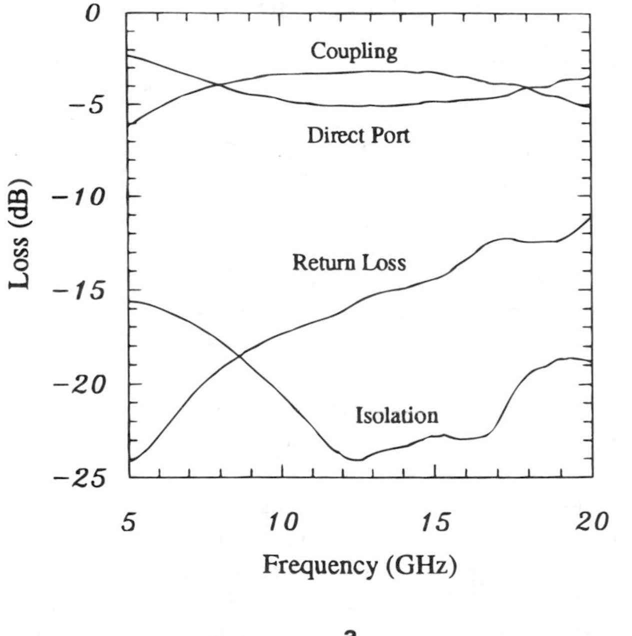

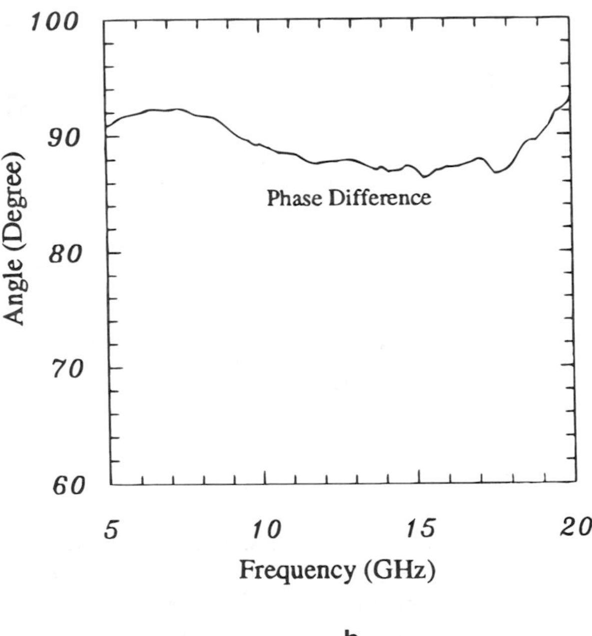

Figure 120.17 Measured performance of a membrane supported Lange coupler. (a) Direct and coupled port transmission, input return loss, and isolation. (b) Phase difference between the two output ports.

nature, a transition which utilizes a Klopfenstein impedance taper is used. This transition serves dual purposes of matching the 50Ω probe impedance to the 90Ω microstrip impedance and rotating the electric fields from their horizontally opposed configuration at the probe tips to a vertical orientation consistent with the dominant mode of microstrip propagation.

Development of bandpass filters for W-band applications is not readily accomplished with conventional quasi-static synthesis techniques available for low frequency filter design. Full-wave analysis techniques must therefore be employed for accurate circuit simulation. These techniques are generally computationally intensive and time consuming, however, so a low frequency modeling approach like that discussed in Interdigitated Bandpass Filters is adopted. Experimental iteration of the circuit design can be easily accomplished with this method, and accurate results may be obtained. Several coupled line bandpass filters have been

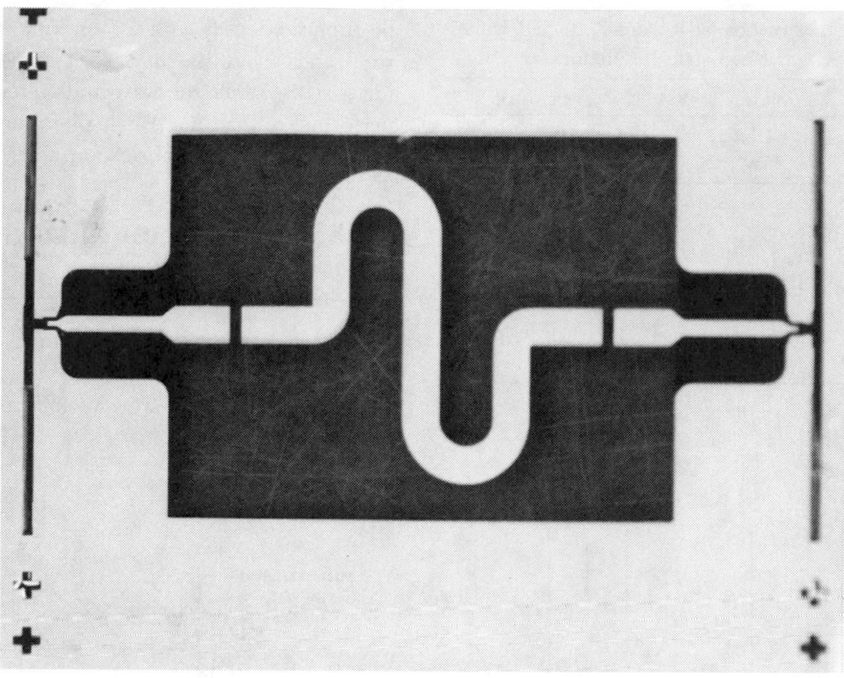

Figure 120.18 Photograph of a resonator structure fabricated on a thin dielectric membrane to study the difference between microstrip and stripline quality factors (Q's).

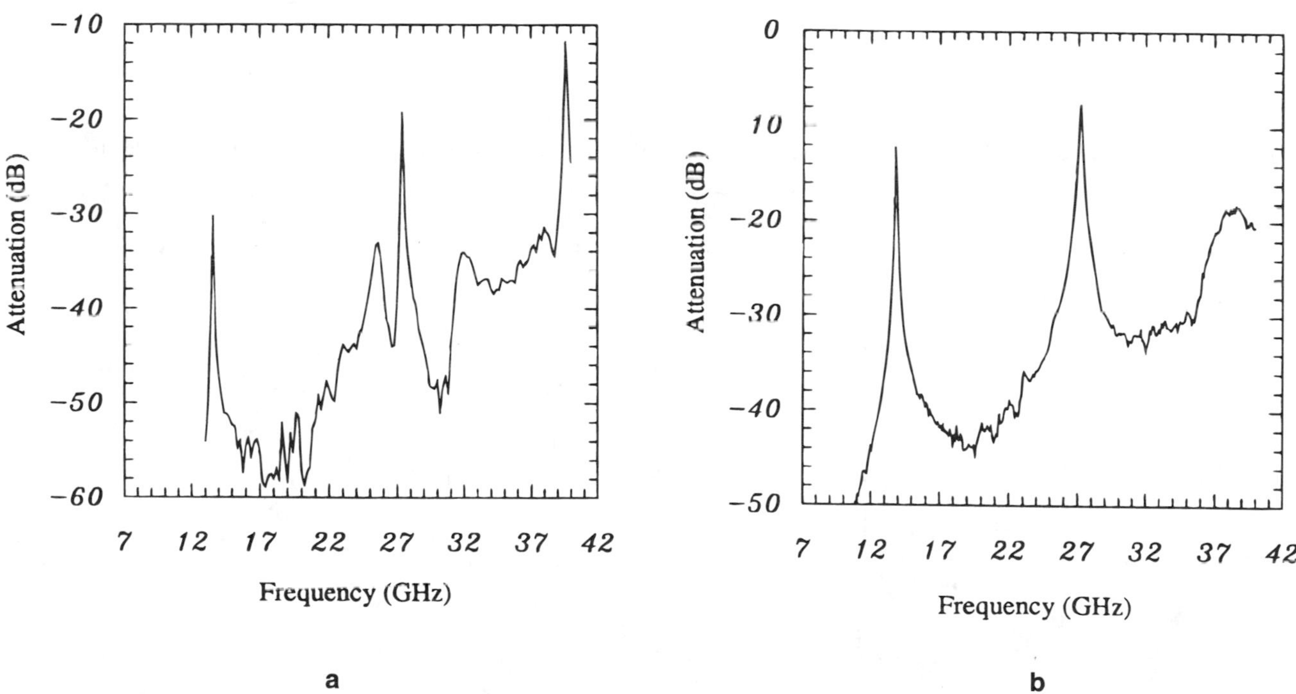

Figure 120.19 Measured insertion loss of (a) stripline and (b) microstrip resonators fabricated on membranes.

developed using a combination of full-wave analysis and modeling techniques. They are all based on equal ripple Chebyshev prototypes, and an example of one of these filters is pictured in Figure 120.23. This photograph shows a narrow-band 5-section filter, including the transitions and probe pads used for on-wafer characterization. The shielding wafer has been removed to view the top surface of the circuit, and the membrane area of the circuit appears dark compared to the lighter gray silicon support rim. Figure 120.24 shows a graph of the measured S-parameters of this filter from 75–110 GHz. The filter performance is characterized by low passband insertion loss, very sharp roll-off, and good out-of-band rejection. This level of performance cannot

Table 120.3 Measured Values of Loaded-Q and Extracted Values of Unloaded-Q for Stripline and Microstrip Resonators.

	Stripline			Microstrip	
	f_{01} (GHz) 13.555	f_{02} (GHz) 27.365	f_{03} (GHz) 39.636	f_{01} (GHz) 13.185	f_{02} (GHz) 27.163
Q_L	258	331	304	155	110
Q_u	272	386	465	234	207
$Q_{u,f}/Q_{u,f_{01}}$	1	1.42	1.71	1	0.88
$\sqrt{f/f_{01}}$	1	1.42	1.71	1	1.40
Skin Depth (μm)	0.676	0.478	0.395	0.676	0.478
α_T (NP/cm)	0.0052	0.0073	0.0111	0.0060	0.0137
R_s (Ω/cm)	0.84	1.18	1.80	1.25	2.85

be duplicated easily with conventional substrate supported structures, and can only be surpassed by expensive rectangular waveguide designs. While the waveguide filters predictably show better performance than the planar filter, the advantages of the planar filter in terms of cost, fabrication, and integrability are significant.

120.5 Conformal Micromachined Packaging

Planar transmission lines such as microstrip, stripline, and coplanar waveguide (CPW) have become conventional structures to use in the design of microwave and millimeter wave circuits due

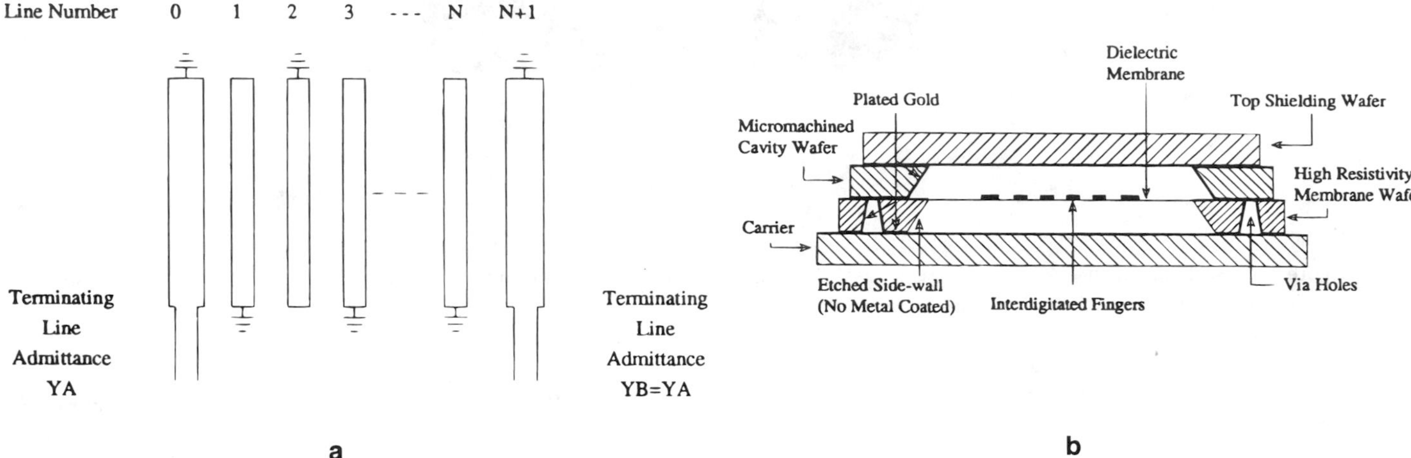

Figure 120.20 Interdigitated filter (a) circuit layout, and (b) two-dimensional geometry used to realize membrane suspended resonators.

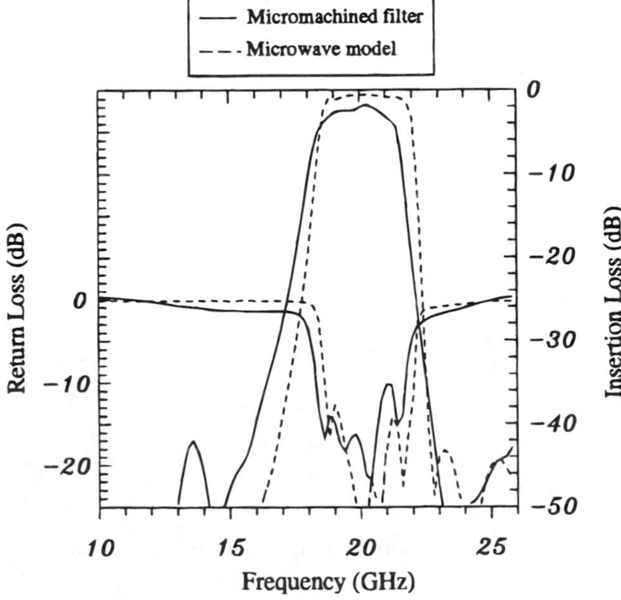

Figure 120.21 Measured response of the interdigitated membrane filter, shown with the scaled measurements of the microwave model.

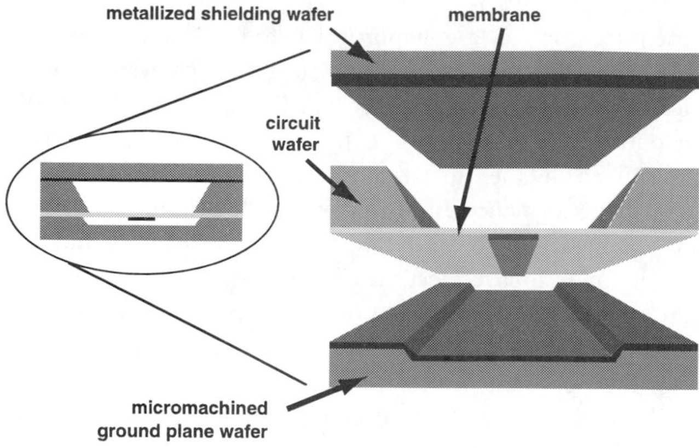

Figure 120.22 Schematic cross-section of a shielded membrane microstrip transmission line. The ground plane wafer and the circuit wafer are micromachined separately and then assembled with the metallized shielding wafer.

to the flexibility provided in fabricating passive components with predefined electrical functions as well as the enhanced ease in mounting active devices. Although microstrip and stripline have been utilized the most in passive circuits, limitations in mounting active devices have made the use of coplanar waveguide more popular since its physical geometry provides inherit advantages. A commonly observed problem in many of these lines, however, is degradation in circuit performance that results when coupling mechanisms associated with parasitics are excited along with radiation effects that arise in dense circuit environments. In order to address the above concerns, the development of a transmission line geometry that offers electrical performance comparable to conventional ones, maintains ease in device mounting, and reduces undesirable coupling effects, is required. This can be obtained with independent shielding of specific circuit components that can be achieved using micromachining techniques.

Since most high frequency circuit applications address development, modeling, fabrication and experimental characterization of systems prior to packaging, the effect of the housing on the electrical performance is very difficult to predict. As a result, the electrical response of many packaged circuits suffers significant performance degradation, mainly attributed to the introduction of unwanted parasitics along with the excitation of multiple shielding resonances resulting from the interaction between the circuit board and metallic housing. To address the issue of proximity coupling and cavity resonances, monolithically integrated cavities can be developed which provide effective shielding to individual components while maintaining an overall geometry that is small enough to avoid the multiple resonance excitation in the range of operating frequencies. From a cost perspective, since conventional housing elements can be rather expensive and impractical to optimize for each generic circuit board, a solution toward cost minimization is easily achieved by using micromachining techniques. Consequently, for system level designs where weight and volume reduction as well as controllable parasitics are critical issues, overall system costs are directly reduced using these techniques.

Since micromachining techniques are well-established in sensor applications, a wealth of information has been discovered on various processing techniques. This has been especially useful in the extension of micromachining techniques from the MEMS arena to high frequency circuit design where emphasis is being placed on the development of self-packaged miniature circuit components used for high frequency systems. In the area of packaging such components, a completely shielded (or self-packaged) micromachined circuit has been developed that is excited by a traditional planar transmission line based on coplanar waveguide. This planar structure is surrounded by an air-filled cavity in the upper region and a substrate-filled one beneath the line as shown in Figure 120.25. To address the specific design issues of shielding and isolation, a variety of micromachining processes have been explored and studied to identify the approach that provides circuits with the best electrical performance at these frequencies. Demonstration of the concept of micromachined circuits for RF applications is given through simple circuit components such as a tuning stub and lowpass filter. These simple components have been developed and their performance has been measured and compared to conventional transmission lines (Drayton and Katehi, 1993c), in this case coplanar waveguide.

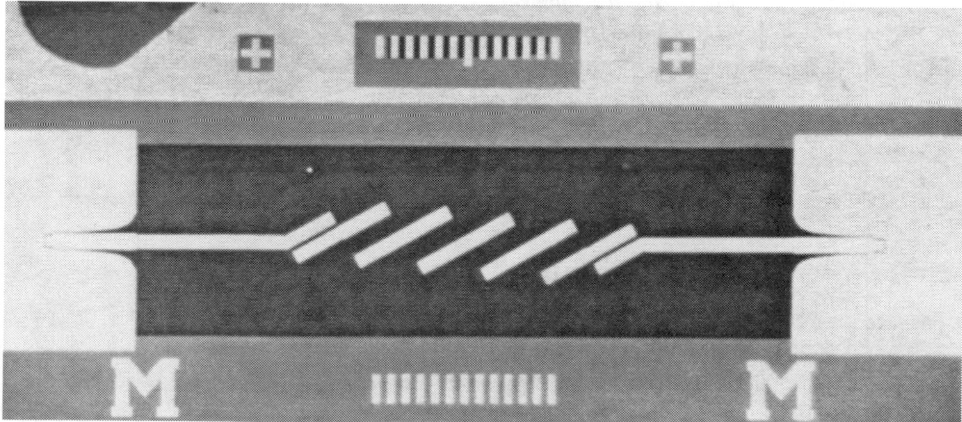

Figure 120.23 Photograph of a 94 GHz coupled line band pass filter. The photo shows the metallized side of the membrane with the micromachined ground plane wafer removed. The darker areas of the picture indicate the thin dielectric membrane.

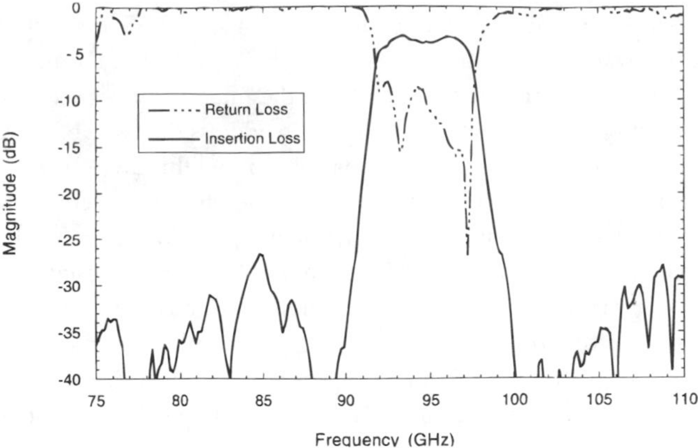

Figure 120.24 Measured response of a 5-section coupled line bandpass filter. The passband insertion loss is 3.4 dB, and the filter has a bandwidth of 6.1% centered at 94.7 GHz.

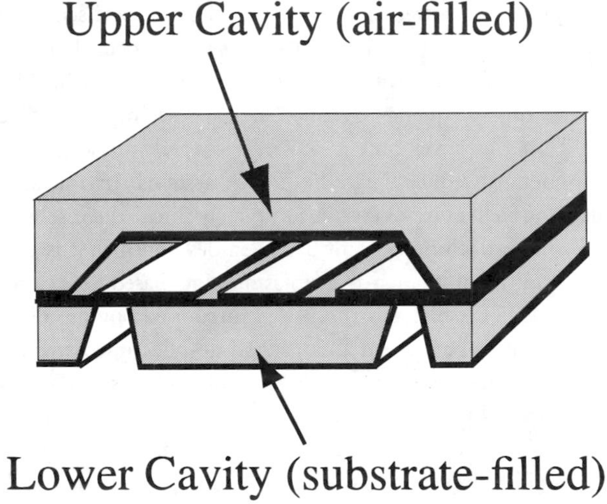

Figure 120.25 Three-dimensional cross section of micromachined lines where the shield and line are integrated monolithically.

The development of micromachined miniature circuit components for high frequency applications has been separated into two parts. The first part concentrates on *circuit development* where the response of a given circuit is determined and the specific circuit parameters are designed to achieve that response. The second part concentrates on *circuit characterization* which includes fabrication, described in Section 120.3, and measurement of the electrical performance of the self-packaged circuits with comparisons to theoretical predictions.

Design Approach for Micromachined Circuits

High frequency self-packaged micromachined circuits are developed in two parts: (a) circuit development and (b) circuit characterization. In part (a), theoretical predictions are made based on very accurate full-wave techniques (Kunz and Luebbers, 1993; Dib and Katehi, 1992; Mei and Fang, 1992; Betz and Mittra, 1992; El-Shandwily and Dib, 1990; Sheen et al., 1990; Zhang and Mei, 1988; Mur, 1981). Since these are complex and computer time intensive, they are inappropriate to use primarily as design tools. Therefore, initial circuit design is best achieved using simpler quasi-static model such as Puff (Wedge et al., 1991). With an entry level circuit design, the specific circuit geometry is determined and then analyzed using full-wave analysis techniques to verify performance. Improvements on the circuit performance are obtained though an iterative method which fine tunes the circuit dimensions to meet the given design requirements. Figure 120.26 shows a brief illustration of the procedures needed for realistic circuit development. In the circuit characterization part once the optimum design has been determined, fabrication is implemented using the procedures presented in Fabrication Methodology, and circuits are then tested to evaluate the overall circuit performance.

The micromachined circuits presented have a shielding environment that has been monolithically integrated into a two-wafer system and is made of cavities in both upper and lower regions. As shown in Figure 120.25, the upper region consists of a metallized air-filled cavity while the lower region has a substrate-filled cavity of high resistivity silicon, $\epsilon_r = 11.7$, which is also metallized on the lower side. To measure the response of these circuits, grounded coplanar waveguide (GCPW) feeding lines are used followed by upper and lower cavities that are grounded through direct contact with the ground planes of the coplanar lines. Shielding is achieved by developing a substrate-filled cavity beneath the line on the lower wafer while an air-filled cavity is formed over the line in the upper wafer. Descriptions of the individual wafer layers used to construct the completely shielded micromachined circuit are discussed in the next paragraph.

The *lower wafer* shown in Figure 120.27 consists of a high resistivity, single-side polished silicon wafer having a dielectric mask of silicon dioxide and a thickness of 350 μm. To develop

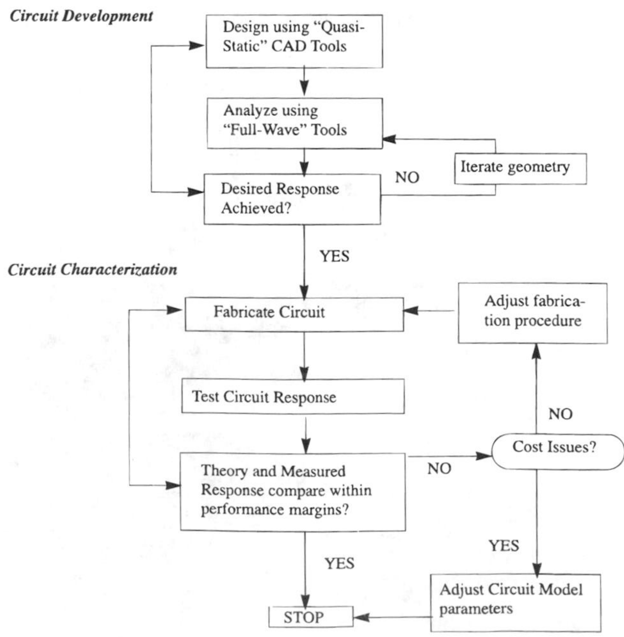

Figure 120.26 Design procedure for micromachined circuits.

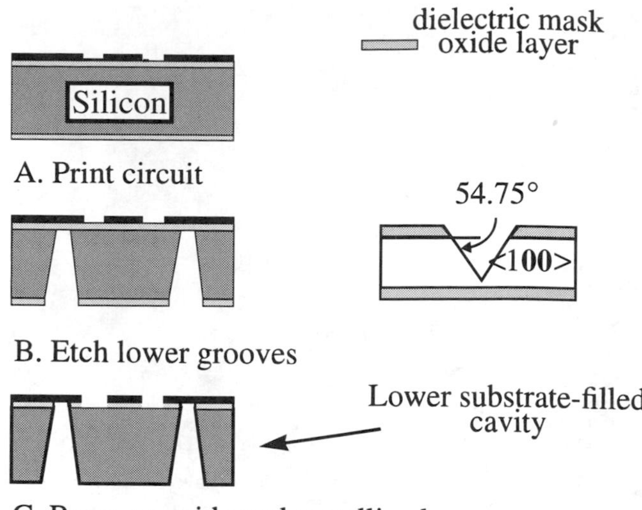

Figure 120.27 Lower Wafer Development. A. Transmission lines are printed on the top surface. B. Lower cavity is formed by etching v-grooves. C. Lower cavity grooves are metallized below the line forming direct contact to the upper ground planes.

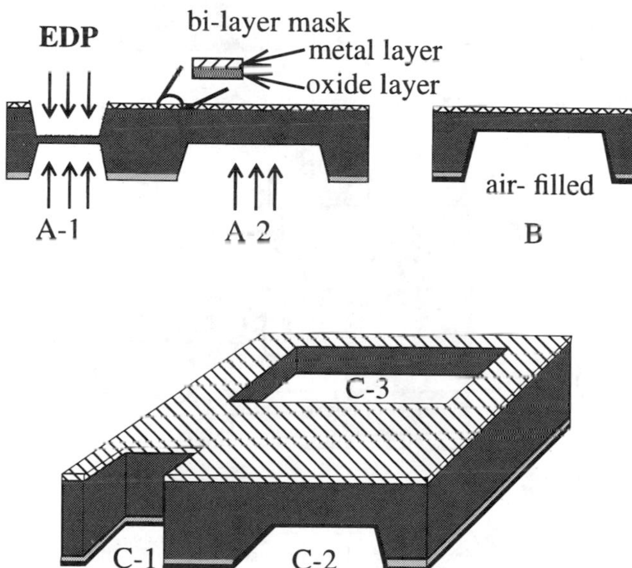

Figure 120.28 Upper Wafer Development. A. Probe windows and alignment marks (A-1) are etched from both sides while the upper cavity (A-2) is etched from one side only. B. The upper cavity is then metallized. C. Finally, the upper wafer sectional view after processing with the alignment marks (C-1), upper cavity (C-2) and the probe window (C-3).

the circuits, the planar lines are printed and electroplated to a desired thickness of 3 μm. With the circuits and alignment marks printed on the upper side, lower cavities are formed below the circuit surface by anisotropically etching the silicon away below the coplanar waveguide grounds to form a substrate filled cavity. Lastly, this cavity is metallized and electro-plated to achieve the desired thickness underneath the coplanar waveguide ground planes.

The *upper wafer*, shown in Figure 120.28, is formed utilizing a double-sided process to define the cavity areas and probe windows required to shield the circuit and allow probing access to the individual components, respectively. In this case, a 500 μm-thick, low resistivity silicon wafer with 7,500 Å of thermally grown oxide on both sides can be used. To allow the sequential etching, this requires at least two masking layers, one must be metal in order to facilitate infrared alignment. This double sided sequential etching process results in mechanically strong wafers when multiple cavities are employed, (Figure 120.30) as a result of the structural beam developed in the probe windows. This is especially useful to increase handling ease during mounting of the upper wafer.

Finally, integration of the upper shield to the lower shielded planar circuits is completed after alignment and attachment of the two wafers via the microscope using regular adhesion methods (see Figure 120.30). In applications where only upper half shielding is required, fabrication is easily achieved by excluding the lower cavity formation.

Experimental Validation and Discussion

To characterize the circuits up to 40 GHz, state of the art measurement systems utilize on-wafer probing techniques to measure planar circuit geometries. Using an HP 8510B Network Analyzer, an Alessi probe station and Cascade Microtech ground-signal-ground probes with a probe pitch of 150 μm accurate characterization is obtained using TRL calibration (Maury et al., 1987; Strid and Gleason, 1989; Ensen and Haer, 1978). As a result, all transitions between the ANA and the newly defined shielded circuit reference plane are taken into account and removed from the circuit response. Simple discontinuities have been implemented to show the realization of conventional circuits in a micromachined configuration. Two elements of interest are the *series open-end stub* and *stepped impedance low-pass filter* which are basic elements to many high frequency circuits such as filters, switches, RF blocks, etc.

The *series open-end tuning stub* has physical dimensions shown in Figure 120.31. Comparison between measurements and full wave analysis results is shown in Figure 120.32. As observed from this figure, the theoretical and experimental results exhibit a shift in the resonant frequency of about 6.9% since the micromachined circuit resonance occurs at 29 GHz compared to the 27 GHz response of the modeled circuit. Although the overall circuit performance is similar, the discrepancy in the resonant frequency can be attributed to the variations between the modeled circuit and the actual one fabricated and measured. The resonant frequency is affected by the fact that the measured line length behaves electrically shorter since there is rounding of the corners and edges of the stub fingers caused during fabrication, which is not accounted for in the model. In addition, metal thickness has been neglected although it has been found to contribute considerably to frequency shifts (Heinrich, 1990). Lastly, the difference in the magnitude between measurement and theory may be attributed to the fact that the theoretical model assumes a lossless system while in reality the circuit has both conductor and dielectric loss.

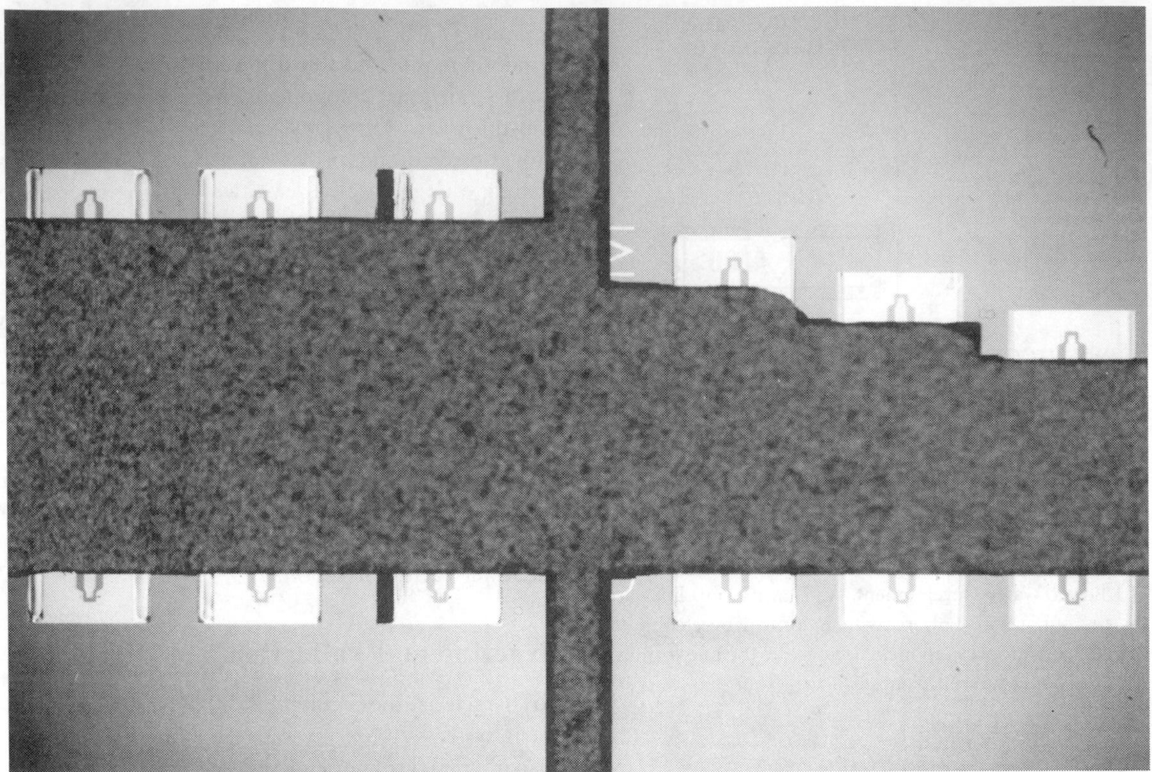

Figure 120.29 Photograph of circuit from top view where the probe windows are shown in relation to the transmission line wafer.

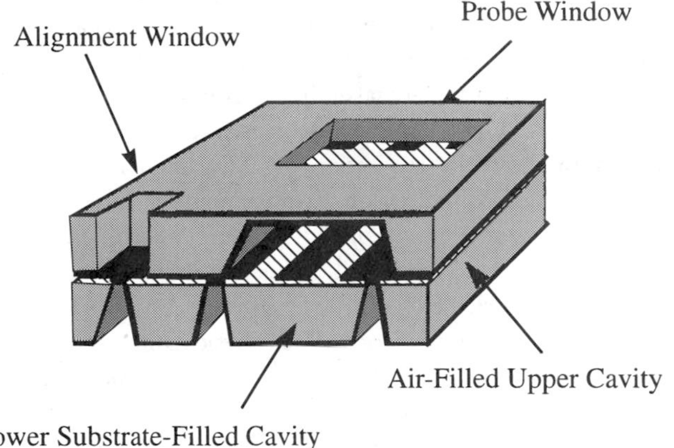

Figure 120.30 Completely shielded micropackaged circuit with lower and upper wafer alignment.

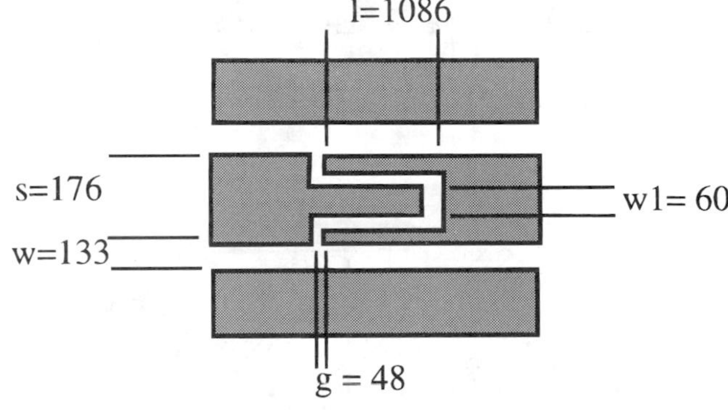

Figure 120.31 Series open-end tuning stub circuit dimensions in microns.

A *five-section stepped-impedance lowpass filter,* designed as shown in Figure 120.33, has high and low impedances of 100Ω and 20Ω respectively and is surrounded by the cavity structure described above. In Figure 120.34, measurements are shown and compared to theoretical results derived from quasi-static models where conductor and dielectric losses are included. Regarding conductor losses, care was taken to incorporate the specific metallization thickness and the appropriate surface resistivity corresponding to the various sections of microstrip line widths (van Dei enter, 1992). To realize 100 and 20 ohm impedance steps, 20 μm and 380 μm wide conductor lines are used with slot widths of 210 μm and 30μm. The total loss in the system, plotted in Figure 120.35, shows good agreement between theory and measured results thus confirming the effectiveness of the micromachined integrated shield and the elimination of loss associated with radiation effects.

To illustrate the effectiveness of the integrated shield, a lowpass filter with a series inductance is shown in Figure 120.36 where a comparison is made between the conventional coplanar waveguide circuit and the micromachined one. The performance of the grounded CPW circuit is affected by the excitation of substrate modes which add constructively and destructively to provide the ripple shown in the high-frequency end of the band. In this case, the effects are observed outside of the lowpass filter band while

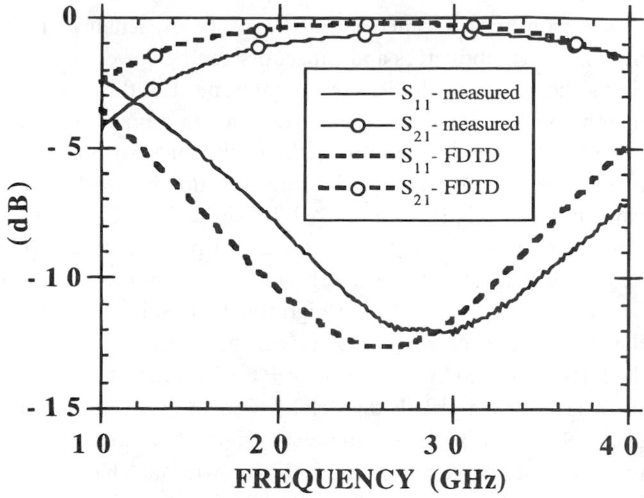

Figure 120.32 Measured vs. FDTD results for the micromachined shielded series open end stub.

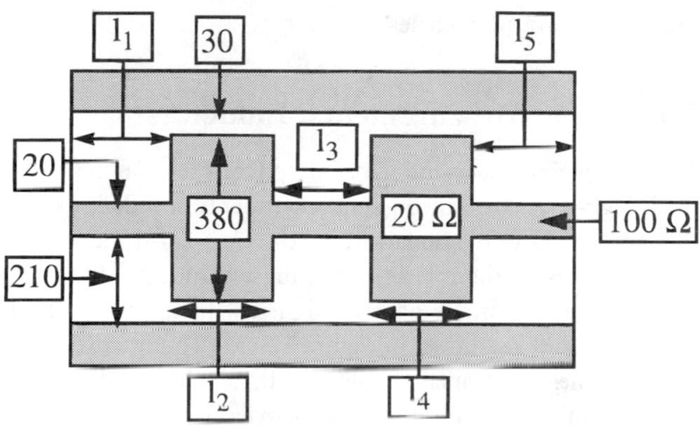

Figure 120.33 Dimensions of a 5-section stepped impedance lowpass filter having low impedance sections of 20 ohms and high impedance sections of 100 ohms.

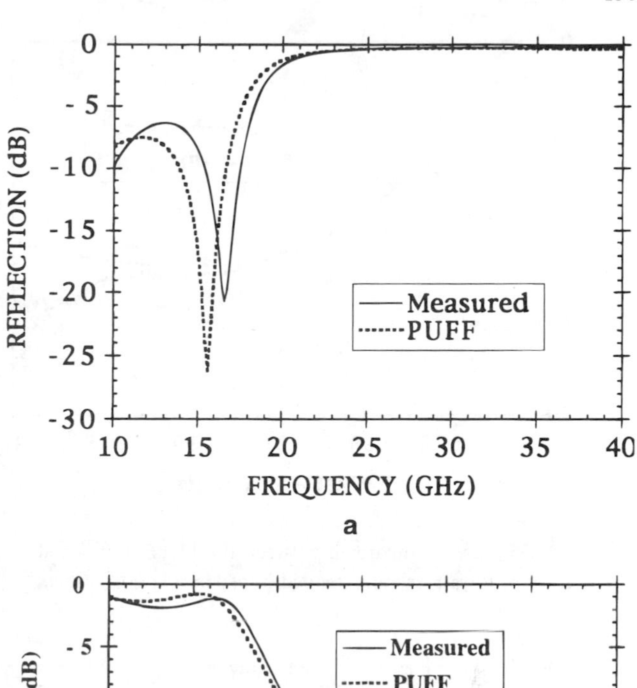

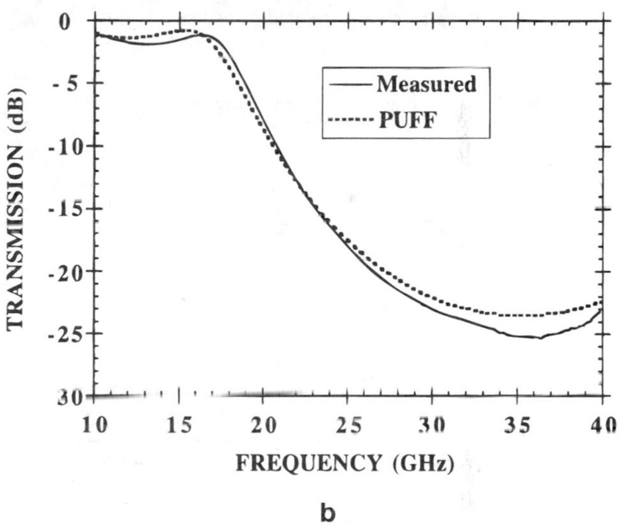

Figure 120.34 Comparison of reflection (a) and transmission (b) coefficient between the PUFF model and measured results for a 5-section stepped impedance lowpass filter.

in other filters, such as bandpass ones, the ripple could occur within the band. By choosing appropriate cavity dimensions in both regions of the micromachined shielded configuration, the effects of substrate modes are eliminated and the electrical response of the circuit in terms of the input reflection and the transmission coefficients are not affected by parasitic radiation but reveal the true characteristics of the circuit component itself.

Simple circuit geometries described above used in conventional planar line designs have been implemented in micromachined form; namely, a series tuning stub and a stepped impedance lowpass filter. The results show that micromachining offers great design flexibility to high frequency applications, while preserving electrical performance. The development of micromachined circuits for miniaturization includes highlights of the design fabrication required followed by measured results compared to quasi-static and full-wave theoretical ones. The resulting data presented prove that monolithic integration of the shield using micromachining techniques allows for the development of circuit components that offer comparable performance to conventional circuits.

120.6 Micromachined Lumped Elements

Planar lumped inductors and capacitors are used in most microwave active and passive integrated circuits as matching elements, bias-chokes and filter components. For frequencies below 12 GHz, these lumped elements are smaller than their transmission line equivalent circuits, and exhibit low-loss and wide bandwidth (Pucel, 1981; Alley, 1970). Recently, planar inductors have been used at 26–40 GHz and their equivalent model has been calculated using a full-wave electromagnetic solution (Abdo-Tuko et al., 1993). Still, the planar inductors show a large parasitic capacitance between the top metal and the ground plane which results in a resonant frequency between 16 and 30 GHz. The planar interdigitated capacitors also suffer from a large parasitic capacitance to ground which affects their performance as true lumped element series capacitors (Esfandiari et al., 1983).

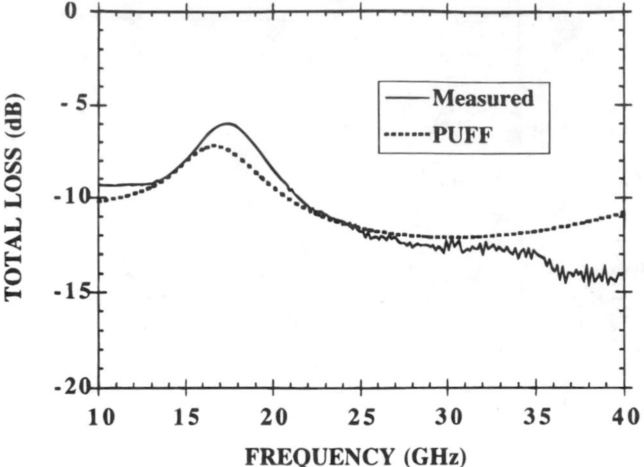

Figure 120.35 Loss comparison between the PUFF model and measured response for a 5-section stepped impedance lowpass filter.

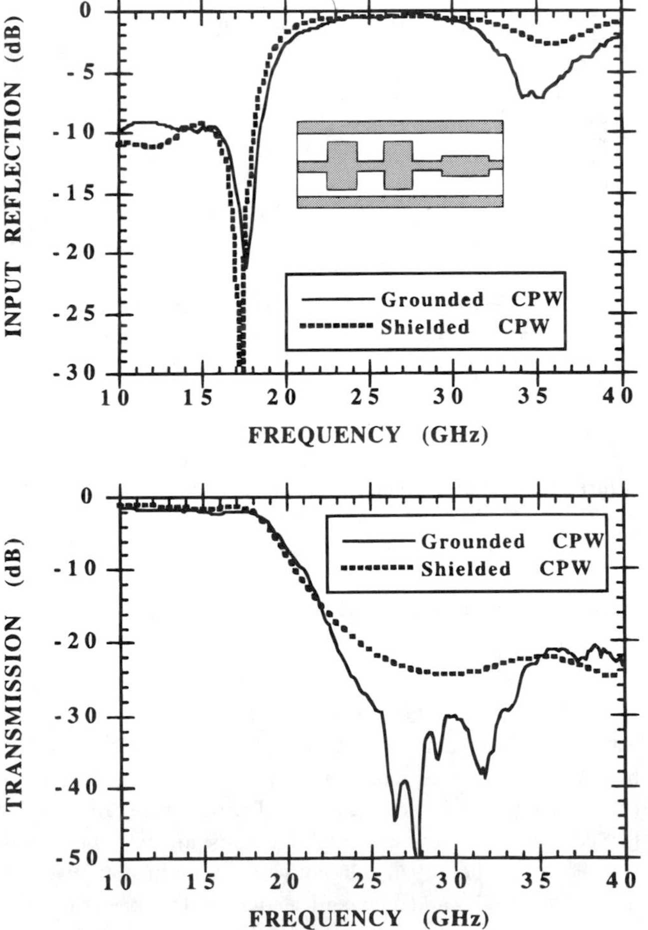

Figure 120.36 Comparison between a conventional CPW and micromachined shielded circuit having a lowpass filter (described above) connected through a half-wave length 50 ohm line to a series inductive line. The 50 ohm line has center conductor (s) and slot widths (w) of 180 and 130 microns, respectively and the inductive section has s and w dimensions of 20 and 210 microns, respectively, with a length of 65 microns.

The problems associated with the parasitic capacitance in planar microstrip inductors and capacitors can be solved by integrating them on a small dielectric membrane. The thin dielectric membrane is defined underneath the lumped element and does not affect the propagation properties of the microstrip line. The membrane is mechanically stable and is compatible with MMIC fabrication techniques as discussed in Fabrication Methodology. The planar inductors and capacitors are suspended in free-space and the quasi-static parasitic capacitance to ground is reduced by a factor of ϵ_r ($\epsilon_r = 11.7$ for high resistivity silicon). Also, for the planar inductor, the quasi-static parasitic capacitance between the lines is reduced by a factor of $(1 + \epsilon_r)/2$ since half the electric fields are in air and half the fields are in the dielectric (Gupta et al., 1981). This reduction in the parasitic capacitance increases the resonant frequency of the inductor without changing the inductance value and the associated series resistance. The application areas of the micro-machined planar inductors and capacitors are in compact phase-shifters, filters, wide band matching networks, and bias circuits for amplifiers, doublers and mixers at millimeter-wave frequencies.

Microwave Measurements: Inductors

Two planar microstrip inductors were fabricated on a 355 μm-thick high-resistivity silicon substrate. Identical inductors using the same masks were also fabricated on a 1.2 μm-thick dielectric membrane using the micro-machining technique outlined previously. The membrane edge is aligned with the physical edge of the inductor as shown in Figure 120.37. The microstrip line is 1 μm-thick electroplated gold and the air-bridge dimensions are 250 μm × 40 μm and it is supported by 2 μm high posts made of electroplated gold. A photo of two completed inductors and their equivalent model are shown in Figure 120.38 and Figure 120.39. The microstrip inductor dimensions are outlined in Table 120.4 and are designed to yield an inductance value of 1.09 nH and 1.69 nH by using the following equation from (Gupta et al., 1981):

$$L_S(nH) = 0.01\, AN^2\pi[\ln(8A/C) + (1/24)(C/A)^2\ln(8A/C + 3.583) - 1/2] \qquad (120.1)$$

where $A = (DO + DI)/4$, $C = (DO - DI)/2$ (see Figure 110.37), A and C are given in "mils" (1 mil = 25.4 μm) and N is the number of turns.

The TRL calibration routine is used to measure the loss of the 50 Ω microstrip line on a high resistivity silicon substrate (2000 Ω-cm) which is attached to the lumped inductor. The microstrip line is 1 μm long on each side of the inductor and exhibits a loss of 0.2 dB/mm from 3 GHz to 20 GHz. This implies that the loss of the microstrip line is dominated by dielectric loss in the substrate. The microstrip line loss is modeled as a matched attenuator (R_1, R_2, R_3). The reference plane for the inductor measurements is defined at the outer limit of the inductor geometries or simultaneously at the edge of the membrane (Figure 120.37).

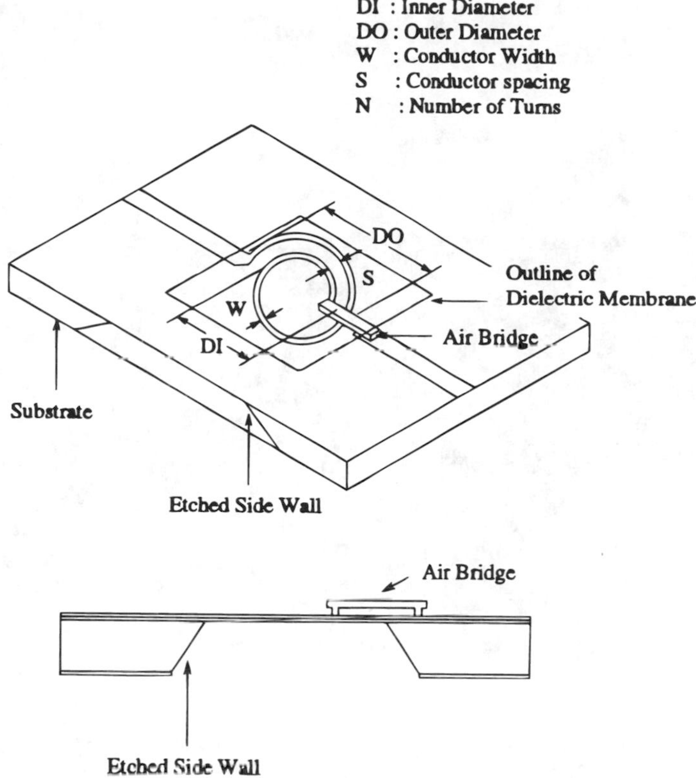

Figure 120.37 Layout of the planar inductor and the membrane outline. The membrane is defined only underneath the lumped element inductor (or capacitor).

First, inductors L_{1S} and L_{2S}, which were built on a high resistivity silicon substrate, are measured from 3 GHz to 20 GHz. Then, the EE_{so}f Touchstone optimization routine is used to fit the equivalent circuit model to the measured S-parameters of these two silicon inductors. The measured and modeled S-parameters of these two inductors are plotted on a Smith chart in Figure 120.40. The equivalent values of L_s, R_s, C_p, C_s for inductors L_{1S} and L_{2S} are summarized in Table 120.5. It is seen that the equivalent inductance agrees quite well with Equation 120.1 and the resonant frequencies are 22 GHz and 17 GHz for a 1.2 nH and a 1.7 nH planar microstrip inductor, respectively.

To predict the behavior of the membrane inductors, the same equivalent circuits used in the silicon inductors are borrowed here. However, a slight modification of the equivalent circuits is required before they can be applied to the membrane inductors. In the equivalent circuit, C_p represents the parasitic capacitance between the spiral inductor and the bottom ground plane. From the quasi-static point of view, the capacitance value of C_p depends on the dielectric constant of the substrate. After the silicon substrate has been etched away, the dielectric material between the spiral inductor and ground plane becomes air only, which means the value of C_p should be reduced by a factor of ϵ_r ($\epsilon_r = 11.7$). In the above equivalent circuit, C_s accounts for the mutual coupling between the inner turns of the spiral inductor. Furthermore, since half of electric field is in the air and the other half is confined in the substrate, the value of C_s need to be reduced by a factor of approximately $(1 + \epsilon_r)/2$ in the membrane inductor cases. The inductance L_s and the resistance R_s are not changed since the membrane and silicon inductors have identical geometries and they are independent of the substrate. The new equivalent circuits for membrane inductors appear and their values are shown in Table 120.5. Good agreement has been achieved between the measured S-parameters for membrane inductors L_{1M}, L_{2M} and the new equivalent circuits derived from L_{1S} and L_{2S} following the procedure discussed above. Both the measured and modeled S-parameters for the membrane inductors are plotted on a Smith chart in Figure 120.41.

The measured and modeled 3 GHz to 20 GHz reactance (X) of the inductors on a thick silicon substrate (L_{1S}, L_{2S}) and the inductors on a thin dielectric membrane (L_{1M}, L_{2M}) is shown in Figure 120.42. It is seen that the measured reactance (X) of the membrane inductors (L_{1M}, L_{2M}) agrees well with the simple equivalent model (3 GHz to 20 GHz). The resonant frequency of the membrane inductors is pushed to around 70 GHz and 50 GHz for a 1.2 nH and a 1.7 nH inductor, respectively. The parasitic capacitances are very low for the micro-machined inductors (C_p, C_s = 2–4 fF) and the membrane inductors can be used as "true" inductors up to 40–60 GHz. The model takes into account only the quasi-static capacitance of the lumped inductor and neglects the transmission-line effects of the pyramidal cavity underneath the lumped inductor and the radiation loss from the air. At mm-wave frequencies, the membrane inductors may result in lower resonant frequencies as predicted above due to non quasi-static effects (high-order modes).

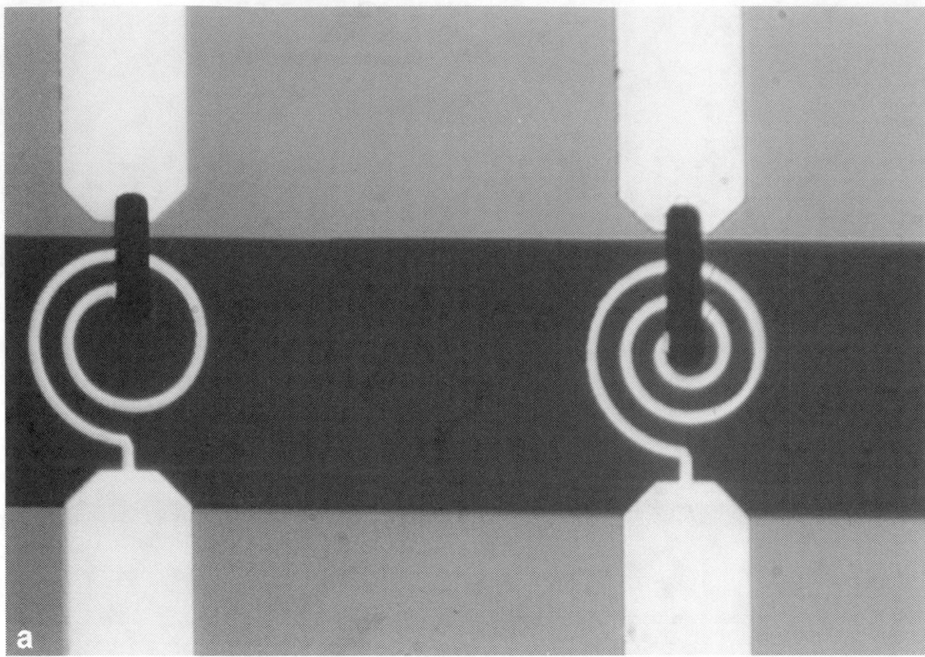

Figure 120.38 (a) A photo of the micromachined inductors L1M(*top left*) and L2M(*top right*) on a small dielectric membrane. (b) A picture of L1M. This picture was taken from the backside of the wafer.

The micromachined inductors behave exactly the same way as the standard planar microstrip inductors at microwave frequencies. Therefore, it is expected that the micro-machined inductor will exhibit a similar Q at microwave frequencies. A small LC series filter composed of a membrane inductor of value 0.9 nH (with a series resistance, $R_s = 1.2$ or 1.3 Ω) and a chip capacitor of value 1.2 pF was fabricated. The chip capacitor is a surface mount MIS type capacitor (Metelics MBIC-1002) and has a very high Q up to 12 GHz, and its effect on the measured Q is neglected at 4.3 GHz. The measured S_{11} of the series LC combination demonstrates a quality factor of $Q = 20$ at 4.3 GHz for the membrane inductor and is close to the expected value from the equation $Q = \omega_0\, L/R_s$ (Figure 120.43). The measurements include the effect of the bond wires used to connect the silicon substrate to the coaxial connectors ($L_{bw} \sim 0.1$ nH). The associated Q of membrane inductors is expected to increase as

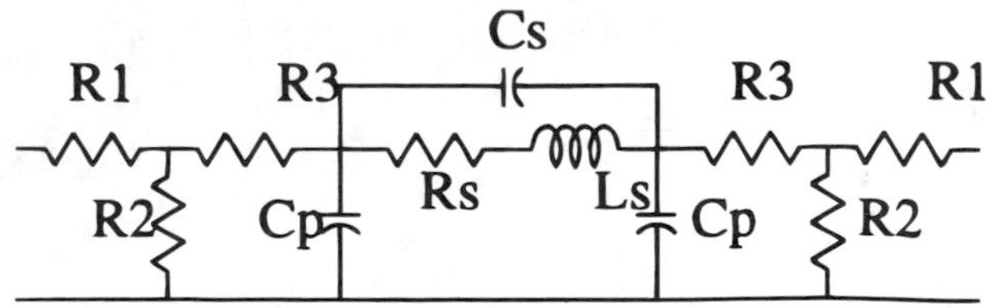

R1 = 0.58Ω R2 = 2171Ω R3 = 0.58Ω

Figure 120.39 The equivalent circuit model of the spiral inductors. A 0.2 dB attenuator is placed at each end to model the loss in the 50 Ω microstrip line on the high resistivity silicon dielectric substrate.

Table 120.4 Physical Dimensions and the Corresponding Calculated Inductance Values for the Spiral Inductors. All Units are in μm (see Figure 120.37).

Components	*DI*	*DO*	*W*	*S*	*N*	$L_s(nH)$
L_{1s}, L_{1M}	254	406.4	25.4	50.8	1.5	1.09
L_{2s}, L_{1M}	101.6	406.4	25.4	50.8	2.5	1.69

Table 120.5 Modeled Values of R_s, L_s, C_s, C_p for Spiral Inductors and Membrane Substrate (see Figure 120.41).

Components	$R_s(\Omega)$	$L_s(nH)$	$C_s(fF)$	$C_p(fF)$
L_{1s}	3	1.2	10	33
L_{1M}	3	1.2	2	2.5
L_{2s}	5	1.7	6	45
L_{2M}	5	1.7	1.2	4

$\sqrt{f}$ with frequency (because the series resistance increases as $\sqrt{f}$ (Gupta et al., 1981), (Daly et al., 1967) to yield a *Q* of 50–60 at 30–40 GHz.

Microwave Measurements: Capacitors

A similar fabrication technique was applied to planar interdigitated capacitors. In this case, the planar capacitors do not suffer from a low resonant frequency but from a relatively large shunt parasitic capacitance to ground (C_p). During modeling the parasitic capacitance to ground is generally included with the interdigitated series capacitance (C_s). However, it is advantageous to eliminate this parasitic capacitance to result in better millimeter-wave filters, phase shifters and matching networks. The micromachined membrane approach reduces the parasitic capacitance by a factor of ϵ_r. However, in this case, it also reduces the interdigitated series capacitance by a factor of $(1 + \epsilon_r)/2$.

Eight finger and four finger interdigitated capacitors were fabricated on a high resistivity silicon substrate and on a membrane. A photo of these two membrane capacitors is shown in Figure 120.44. The capacitor finger is 355 μm long and is 25 μm wide. The gap between the fingers is also 25 μm wide. The measured S_{11} from 7 to 20 GHz for the capacitors are shown in Figure 120.45. It is seen that the membrane interdigitated capacitor (of value around 110 *fF* for the eight finger capacitor and 55 fF for the four finger capacitor) follows the 1-R line on the Smith chart as expected from a capacitor in series with a 50 Ω load. The interdigitated capacitor on the silicon dielectric shows a large shunt capacitance effect due to the parasitic capacitance to ground. The measured S_{11} of the capacitors on the silicon substrate agree very well with published results

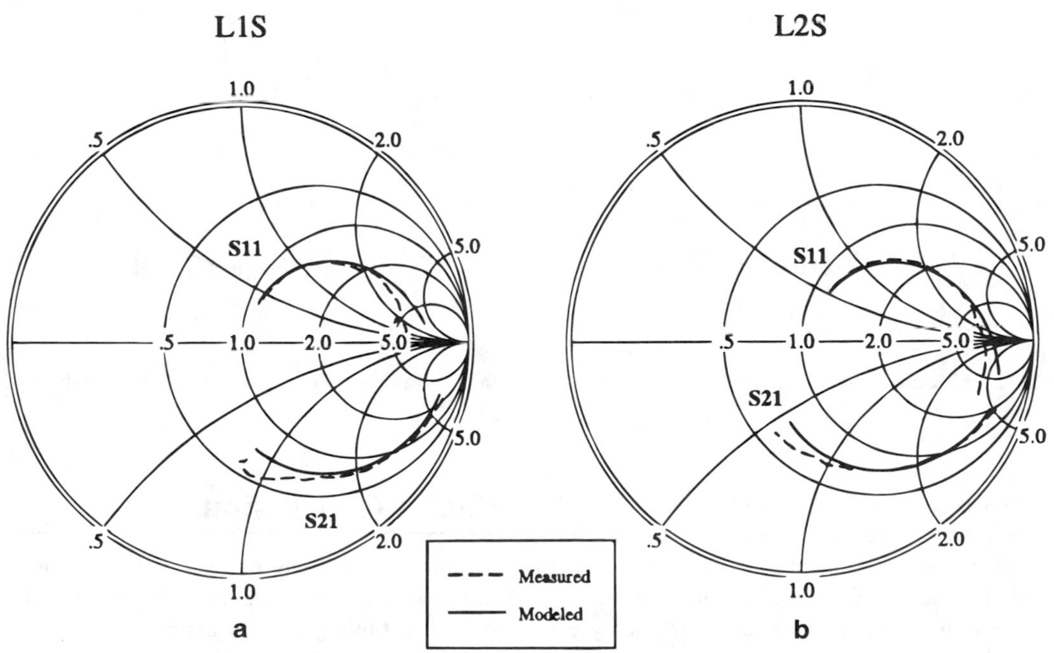

Figure 120.40 Measured and modeled S-parameters of silicon inductors (a) L_{1S} and (b) L_{2S}. Frequency sweep from 3 GHz to 20 GHz.

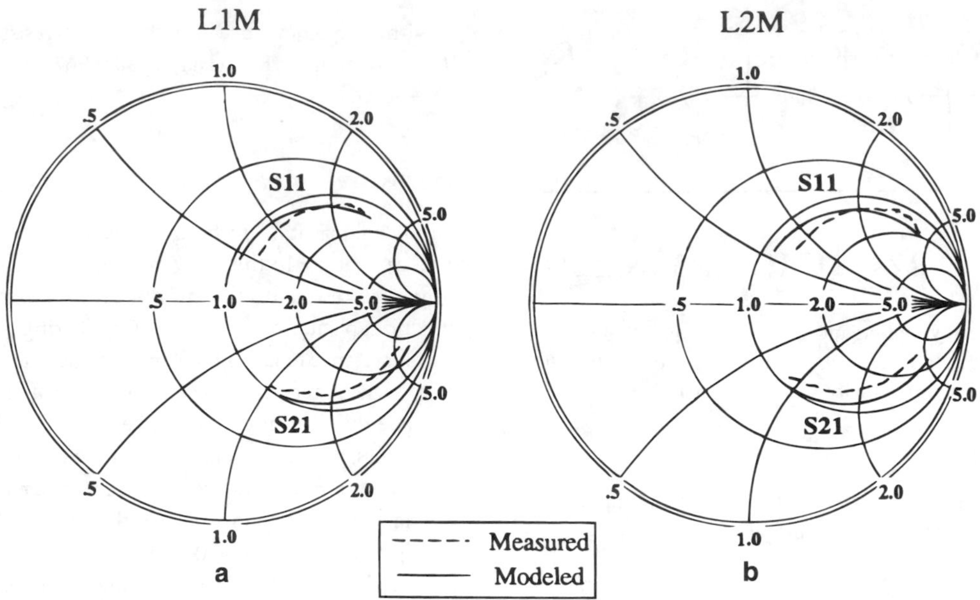

Figure 120.41 Measured and modeled S-parameters of membrane inductors (a) L_{1M} and (b) L_{2M}. Frequency sweeps from 3-GHz to 20 GHz.

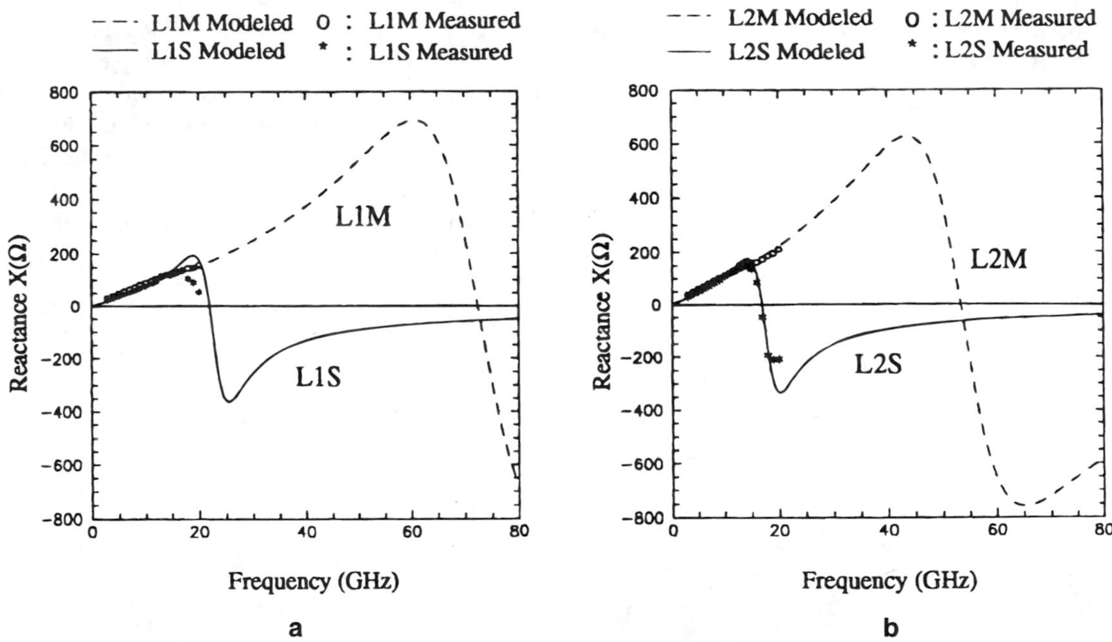

Figure 120.42 Measured and modeled reactance (X) of inductors (a) L_1 and (b) L_2. Frequency sweeps from 3 GHz to 20 GHz.

of similar capacitors on GaAs substrates (Esfandiari, 1983). Compared with dielectric supported capacitors, it is seen that the micro-machined membrane capacitors demonstrate much better performance at microwave frequencies. The quality factor of the membrane capacitors was not measured at microwave frequencies but is expected to be larger than the corresponding capacitors on the silicon substrate due to the absence of the dielectric losses.

120.7 Conclusions

The previous sections have outlined the progress to date on implementing distributed microwave circuit elements such as filters and tuning stubs in membrane supported transmission line geometries. At low frequencies, these membrane supported elements exhibit improved performance over their substrate based counterparts due to the elimination of parasitics and losses

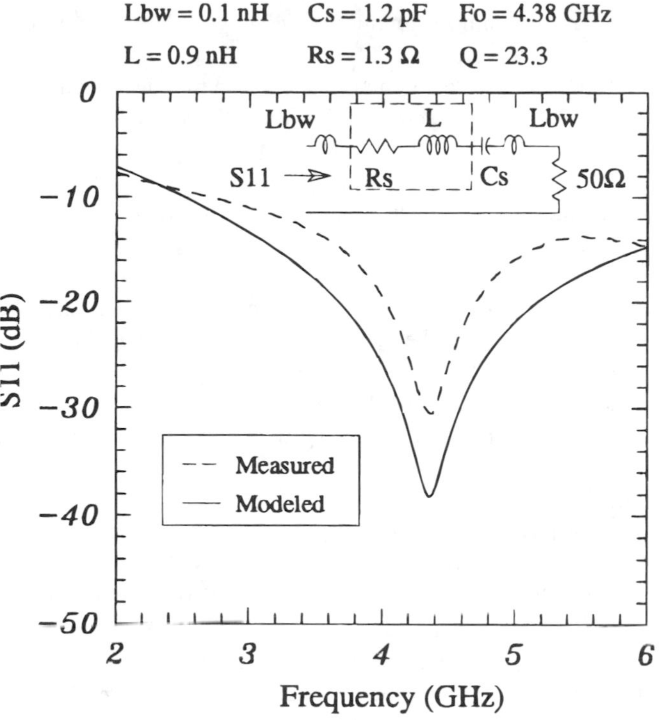

Figure 120.43 Measured and modeled S_{11} of a small LC filter. This filter shows that the 0.9~nH membrane inductor has a quality factor of 23.3 at 4.38 GHz.

associated with the dielectric material. At higher frequencies, problems associated with the substrates make conventional approaches unfeasible, and membrane supported components offer the only planar alternative to costly waveguide-based approaches. Membrane supported transmission lines and circuit components have been shown to perform very well in frequency bands all the way up to W-band (110 GHz). Circuits commonly used in CPW implementations are shown to have superior performance when realized with membrane supported transmission lines like microshield line. Low pass filters and resonant stubs have been measured up to 40 GHz with excellent results, and microshield line low pass filters have been tested as high as 110 GHz. Micromachining has also opened the doors to techniques for fabricating circuits that were previously restricted by cumbersome machining processes. Interdigitated filters were thought to be limited to very low frequencies where they could be manufactured using mechanical techniques, but membrane technology has allowed them to become high performance alternatives at 30 GHz. Planar millimeter-wave microstrip inductors and capacitors have been developed and fabricated on a high-resistivity silicon substrate using micromachining techniques. This micro-machining technique is compatible with via-hole technology in GaAs and InP MMIC processes. The micro machined spiral inductors and interdigitated capacitors are suspended on a thin dielectric membrane to reduce the parasitic capacitance to ground. Since the parasitic capacitance to the ground can be reduced by a factor of ϵ_r, it renders a much higher resonance frequency in this

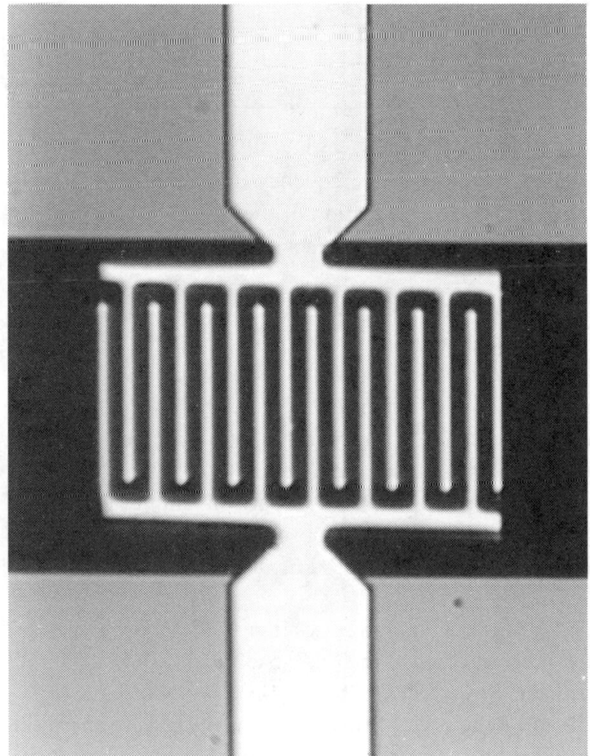

Figure 120.44 A photo of the interdigitated membrane capacitors.

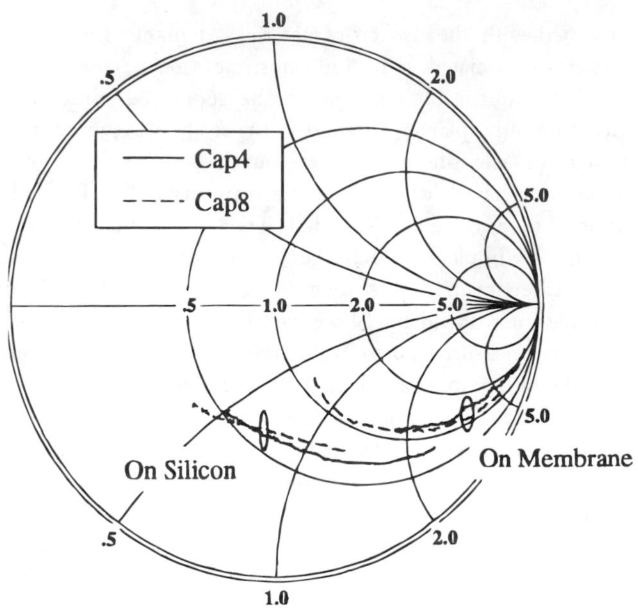

Figure 120.45 Measured S_{11} of eight-finger and four-finger interdigitated capacitors on both membrane and silicon substrate. Frequency starts at 7 GHz and stops at 20 GHz.

membrane structure compared to their Silicon/GaAs counterparts and makes this technology very attractive at millimeter-wave frequencies. This technique can also be applied to lumped elements in coplanar-waveguide transmission lines and micromachined filter design.

References

Abdo-Tuko, M., Naghed, M. and Wolff, I. 1993. Novel 18/36 GHz (M)MIC GaAs FET frequency doublers in CPW-techniques under the consideration of the effects of coplanar discontinuities, *IEEE Trans. Microwave Theory Tech.*, MTT-41(8):1307–1315.

Alley, G. D. 1970. Interdigital capacitors and their application to lumped-element microwave intergrated circuits, *IEEE Trans. Microwave Theory Tech.*, MTT-18:1028–1033.

Altschuler, H. M., and Oliner, A. A. 1960. Discontinuities in the center conductor of symmetric strip transmission line, *IRE Trans. Microwave Theory Tech.*, MTT-8:328–339, May 1960.

Betz V. and Mittra, R. 1992. Comparison and evaluation of boundary conditions for the absorption of guided waves in an FDTD simulation, *IEEE Microwave and Guided Wave Lett.*, 2(12):499–501.

Chang, J. Y., Abidi, A. A. and M. Gaitan, 1993. Large suspended inductors on silicon and their use in a 2-μm CMOS RF amplifier, *IEEE Electron Device Letter,* 14(5):246–248.

Chi, C. Y. and Rebeiz, G. M. 1994a. Planar microwave and millimeter-wave lumped elements and coupled-line filters using micro-machining techniques, *IEEE Trans. Microwave Theory Tech.*, December.

Chi, C. Y. and Rebeiz, G. M. 1994b. Novel low-loss interdigitated filters for microwave and millimeter-wave applications using micromachining techniques, *24th European Microwave Conference Digest,* 2:1357–1359.

Daly, D. A. et al. 1967. Lumped elements in microwave integrated circuits, *IEEE Trans. Microwave Theory Tech.,* MTT-15:713–721.

Dib, N. I. and Katehi, L. P. B. 1992. Impedance calculation for the microshield line, *IEEE Microwave and Guided Wave Letters,* 2(10):406–408.

Dib, N. I., Harokopus, W. P., Katehi, L. P. B., Ling C. C., and Rebeiz, G. M. 1991. A study of a novel planar transmission line, presented at the *1991 IEEE MTT-S Int. Symp.* June, Boston, MA.

Drayton, R. F. and Katehi, L. P. B. 1992. Microwave characterization of microshield lines, *Digest of the 40th ARFTG Conf.,* December, Orlando, FL.

Drayton, R. F. and Katehi, L. P. B. 1993a. Experimental study of micromachined circuits, *Digest of the 1993 Int. Symp. Space Terahertz Technology,* March, Los Angeles, CA.

Drayton, R. F. and Katehi, L. P. B. 1993b. Micromachined circuits for mm-wave applications, *Digest of the 1993 European Microwave Conf.,* September, Madrid, Spain.

Drayton, R. F., Weller, T. M. and Katehi, L. P. 1995. Development of Miniaturized Circuits for High-Frequency Applications using Micromachining Techniques, *Int. J. Microcircuits and Electronic Packaging,* 3rd Quarter.

Dunleavy, L. P. and Katehi, L. P. B. 1988. Shielding effects in microstrip discontinuities, *IEEE Trans. Microwave Theory and Techniques,* MTT-36(12):1767–1774.

Ekstrom, H. A., Gearhart, S. S., Acharya, P. R., Rebeiz, G. M., Kolberg E. L., and Jacobson, S. 1992. 348 GHz endfire slotline antennas on thin dielectric membranes, *IEEE Microwave and Guided Wave Lett.,* 2(9):357–358.

El-Shandwily M. and Dib, N. 1990. Spectral domain analysis of finlines with composite ferrite-dielectric substrate, *Int. J. Electronics,* 68(4):571–583.

Engen, G. and Hoer, C. 1979. Thru-reflect-line: an improved technique for calibrating the six-port automatic network analyzer, *IEEE Trans. Microwave Theory Techn.,* 27(12):987–993.

Esfandiari, R. et al. 1983. Design of interdigitated capacitors and their application to gallium arsenide monolithic filters, *IEEE Trans. Microwave Theory Tech.,* MTT-31(1):57–64.

Findler, G., Muchrow, J., Koch, M., and Munzel, H. 1992. Temporal evolution of silicon surface roughness during anisotropic etching processes, *Micro Electro Mechanical Systems 1992,* February, 62–66.

Finne, R. M. and Klein, D. L. 1967. A water-amine-complexing agent system for etching silicon, *J. Electrochem. Soc.,* 965–970.

Garcia, E., Jacobson, B. R., and Hu, Q. 1993. Fabrication of high-quality superconductor-insulator-superconductor junctions on thin SiN membranes, *Applied Physics Lett.,* 63(7): August.

Gearhart, S. S. 1994. *Integrated Milimeter-Wave and Submillimeter-Wave Antennas and Schottky-Diode Receivers,* Ph.D. Thesis, The University of Michigan.

Gupta, K. C., Garg R., and Chadha, R. 1981. *Computer-Aided Design of Microwave Circuits.,* Artech House, Norwood, MA.

Hamadallah, M. 1988. Microstrip Power Dividers at Mm-Wave Frequencies, *Microwave Journal,* July, pp.116–127.

Harokopus, E. P., and Katehi, L. P. B. 1989. Characterization of microstrip discontinuities on multi-layer dielectric substrates including radiation losses, *IEEE Trans. Microwave Theory and Techniques,* MTT-37(11):2058–2066.

Heinrich, W. 1990. Full-wave analysis of conductor losses on MMIC transmission lines, *IEEE Trans. Microwave Theory Tech.,* 38(10):1468–1472.

Kramer, H. G. 1983. Float-zoning of semiconductor silicon: A perspective, *Solid State Technol.,* 137, January.

Kunz K. and Luebbers, R. 1993. *The Finite Difference Time Domain Method for Electromagnetics,* CRC Press, Boca Raton, FL.

Lange, J. 1969. Interdigitated stripline quadrature hybrid, *IEEE Trans. Microwave Theory Tech.,* MTT-17:1150–1151, December 1969.

Laudise, R. A. 1970. *The Growth of Single Crystals,* Prentice-Hall, Englewood Cliffs, NJ.

Lepselter, M. P. 1966. Beam-lead technology, Bell System Technical Journal, February, 233–254.

Ling, C. C.-S. 1993. *An Integrated 94-GHz Monopulse Tracking Receiver,* Ph.D. Thesis, Radiation Laboratory, University of Michigan.

Maissel, L. I. and Glang, R. 1970. *Handbook of Thin Film Technology,* ch. 6, 29–30, McGraw Hill, New York, NY.

Marks, R. B., Williams, D. F. 1993. NIST De-embedding Software, Program DEEMBED, Revision 4.04.

Maury, M., March, S., and Simpson, G. 1987. LRL calibration of vector automatic network analyzers, *Microwave Journal,* May, 387–391.

Mei K. and Fang, J. 1992. Superabsorbtion-A method to improve absorbing boundary conditions, *IEEE Trans. Antennas and Propagation,* 40(9):1001–1010.

Mur, G. 1981. Absorbing boundary conditions for the finite-difference approximation of the time-domain electromagnetic-field equations, *IEEE Trans. Electromagnagetic Compatibility,* 22(11):377–382.

Pengelly, R. S. and Schumacher, P. 1988. High-performance 20 GHz package for GaAs MIMICs, *Microwave Systems News and Circuit Techniques,* January, 10–19.

Petersen, K. E. 1982. Silicon as a mechanical material, *Proc. IEEE,* 70(5).

Pucel, R. A. 1981. Design considerations for monolithic microwave circuits, *IEEE Trans. Microwave Theory Tech.,* MTT-29(6):513–534.

Rebeiz, G. M., Kasilingam, D. P., Stimson, P. A., Guo, Y., and Rutledge, D. B. 1990. Monolithic millemeter-wave two-dimensional horn imaging arrays, *IEEE Trans. Antennas and Propagation,* AP-28, 1473–1482.

Robertson, S. V., Katehi, L. P. B., and Rebeiz, G. M. 1994. W-Band Microshield Low-Pass Filters, in *1994 IEEE MTT-S Digest,* pp. 625–628.

Robertson, S. V., Katehi, L. P. B., and Rebeiz, G. M. 1995. Micromachined self-packaged W-band band-pass filters, *1995 IEEE MTT-S Digest,* pp. 1543–1546.

Rutledge, D. B., Neikirk D. P., and Kasilingam, D. P. Integrated circuit antennas, *Infrared and Millimeter-Waves,* K. J. Button, ed., 10:1–90, Academic Press, New York.

Rutledge, G. M. 19??. Millimeter-wave and submillimeter-wave antenna structures, *US Patent* 4888, 597.

Sheen, D., Ali, S., Abouzahra, M., and Kong, J. 1990. Finite-difference time-domain method to the analysis of planar microstrip circuits, *IEEE Trans. Microwave Theory Tech.,* 38, 849–857.

Strid, E. W., and Gleason, K. R. 1984. Calibration methods for microwave wafer testing, *1984 IEEE MTT-S Int. Microwave Symp. Digest,* 93–97.

van Deventer, T. E. 1992. *Characterization of Two-Dimensional High Frequency Microstrip and Dielectric Interconnects,* Ph.D dissertation, The University of Michigan. December.

van Deventer, T. E., Katehi, L. P. B., and Cangellaris, A. 1989. An integral equation method for the evaluation of conductor and dielectric losses in high-frequency interconnects, *IEEE Trans. Microwave Theory and Techniques,* MTT-37(11):1964–1972.

VandenBerg, N. L. and Katehi, L. P. B. 1982. Broadband vertical interconnects using slot-coupled shielded microstrip lines, *IEEE Trans. Microwave Theory and Techniques,* MTT-40(1):81–88.

Wedge, S., Compton, R., and Rutledge, D. 1991. *PUFF Computer Aided Design for Microwave Integrated Circuits,* Version 2.0.

Weller, T. M., Katehi, L. P. B., Rebeiz, G. M., Cheng, H. J., and Whitaker, J. F. 1993b. Fabrication and characterization of microshield circuits, *1993 Int. Symp. Space Terahertz Technology,* March, Los Angeles, CA.

Weller, T. M., Rebeiz, G. M., and Katehi, L. P. 1993a. Experimental results on microshield transmission line circuits, *1993 IEEE MTT-S Digest,* pp. 827–830.

Weller, T., Katehi, L., Herman, M., and Wamhof P. 1994. Membrane technology applied to microstrip: A 33 GHz wilkinson power divider, *1994 IEEE MTT-S Int. Microwave Symp. Digest,* 2:911–914.

Weller, T. M., Katehi, L. P. B., and Rebeiz, G. M. 1995a. High performance microshield line components, *IEEE Trans. Microwave Theory Tech.,* 43(3):534–543.

Wolf, S. and Tauber, R. N. 1986. *Silicon Processing for the VLSI Era, Volume 1: Process Technology,* Lattice Press, Sunset Beach, CA.

Zhang X. and Mei, K. 1988. Time-domain finite difference approach to the calculation of the frequency-dependent characteristics of microstrip discontinuities, *IEEE Trans. Microwave Theory Tech.,* 36(12):1775–1781.

121

MEMS Integration—Technical and Economic Considerations

Janusz Bryzek
Intelligent MicroSensor Technology

121.1 Introduction 1576
121.2 Why MEMS Focus on Silicon 1577
Mechanical Performance of Silicon • IC Industry Infrastructure • Dedicated MEMS Developments
121.3 Market Growth Analogy: Transistors, Integrated Circuits and MEMS 1578
121.4 Integrated MEMS Market Overview 1580
121.5 To Integrate or not to Integrate 1582
Pro-Integration Arguments • Counter Integration Arguments • On-Chip Electronics Integration Justification Example
121.6 Mechanical On-Sensor-Chip Integration 1585
121.7 Monolithic or Hybrid 1585
121.8 Case Study: Lucas NovaSensor 1586
On-Sensor-Chip Passive Electronics Integration • On-Sensor-Chip Active Electronics Integration • Summary
121.9 Conclusions 1590

121.1 Introduction

MEMS, Micro-Electro-Mechanical Systems are most often defined as integrated circuits incorporating at least some level of mechanical structures, with the electronic circuitry optional. These mechanical chips are most often fabricated in a batch mode on silicon wafers.

For many, the greatest attractions of such technology was, and still is, the possibility of on-chip electronics integration. Such a view, however, did not find a reflection in practical applications. The largest commercial successes of MEMS are silicon pressure and acceleration sensors, with 1995 volume about 50 and 12 million units respectively. Despite a long history of impressive electronic IC developments, only about 10% of the pressure and 5% of the acceleration sensors have on-sensor-chip integrated active electronics. The key driver behind these applications were the excellent mechanical performance of silicon, and a feasibility of its high-volume batch mode manufacturability, and not circuit integration.

One of the main reasons of this situation is highly diversified and limited size of the sensor market. Sensor technology developments were often undertaken by small companies that did not have semiconductor manufacturing technology. The very few sensor gurus who developed successful commercial sensors thus lacked access to the state-of-the-art IC technology. IC technologists involved in sensor technology, on the other hand, lacked sensor know-how, developing highly integrated devices that were not delivering expected performance. Several semiconductor companies (e.g., Fairchild, National Semiconductor, Burr-Brown) got involved in the silicon pressure sensor technology development early, but later abandoned it as the market development significantly lagged behind the mainstream IC industry. Others restricted their efforts to market niches supporting the higher value added products (e.g., Phillips and Texas Instruments).

Only recently have sensors with on-chip electronics started their journey to the markets. The automotive market with its large volumes became the main technology driver, demanding IC pricing for sensors and justifying—with volume orders—the advanced sensor technology developments. The first on-chip integrated pressure and acceleration sensors addressing the automotive market have reached the production phase.

This chapter overviews the selected technical and economic aspects of the electronics integration for MEMS. It characterizes silicon material properties which make it one of the best materials for mechanical sensors. It overviews the current status of the on-chip integrated sensor market, focusing on pressure and acceleration sensors. It references selected aspects of the transistor and monolithic IC market evolution, contrasting them with the integrated sensors market development characteristics. It presents arguments both in favor and against the on-sensor-chip electronics integration. It finally presents two case studies from NovaSensor (1985 Silicon Valley silicon micromachining start-up

0-8493-8343-9/97/$0.00+$.50

company) development, in which hybrid designs were selected over the single chip solution.

121.2 Why MEMS Focus on Silicon

Silicon is widely used in manufacturing of electronic integrated circuits where only its electrical properties are utilized. Thus on the surface, the most attractive aspect of the silicon micromachining technology should be the potential for the on-MEMS-chip electronics integration. This potential has not yet been greatly utilized so far.

The largest silicon micromachining applications to date, pressure and acceleration sensors, have been enabled primarily by two factors:

- Excellent mechanical performance of silicon enabling it to effectively replace a majority of other sensing technologies.
- The existing infrastructure of the mainstream IC industry, enabling development of products offering an unmatched price to-performance ratio and high volume capabilities.

Micromechanics may be considered as another dimension of the on-sensor-chip integration: mechanical as opposed to electronic. The electronics integration has a very high visibility in the industry. The on-chip integration of mechanical structures, while so far not as visible as electronics, has a much higher potential impact on new applications. Integration of the electronics in sensor applications typically replaces two or more batch manufactured IC chips with just one chip, sensor and signal processing ICs. The penalty is a lower yield, and the benefit is a lower overall cost and increased reliability. Integration of mechanics, however, replaces mechanical components manufactured in a *one-at-a-time* mode with a single batch-manufactured chip. The impact of such a dramatic manufacturing technology change, single-up mode to a typically hundreds-up or thousands up mode, is quite dramatic. It is far greater than the replacement of two IC chips with one, offered by the electronics only integration.

Similarly to electronics, two levels of MEMS-on-a-chip integration are possible:

- Passive elements (e.g., membranes or ink injection nozzles).
- Active, or microactuator, elements (e.g., pressure valves or motors).

Mechanics on-chip integration is feasible due to the same two factors that enabled development of the silicon sensor market: excellent silicon mechanical performance and the available support from the IC industry. The support available from the IC infrastructure for mechanical structures is somewhat limited. MEMS industry has to develop many of their own tools, processes and know-how in many MEMS specific technologies.

Mechanical Performance of Silicon

Silicon and its derivatives (silicon oxide, nitrite, etc.) are some of the best electrically characterized materials in the world. Their mechanical performances importance to MEMS, however, are not well known in the IC industry, and many parameters, such as fracture stress or temperature nonlinearities of stress-strain curves, have yet to be characterized as a function of the used process.

Based on the already known performance obtained in both a direct characterization and indirect measurements (resulting from the devices characterization), silicon can be classified as the best available material for mechanical sensors and many MEMS devices. The major advantages include lack of mechanical and thermal hysteresis, high strength, high sensitivity to stress and excellent long term mechanical stability.

Lack of Mechanical and Thermal Hysteresis

For all mechanical sensors the measure of excellence and simultaneously the performance limit is established by the achievable mechanical and thermal hysteresis and stability. These parameters define the change of the sensor output after, respectively, mechanical (e.g., pressure, temperature cycles and time.) The search for stable low hysteresis materials made older transducer technologies somewhat mystical and loaded with secrets, as classical materials' properties affecting hysteresis were not easily understood and controllable.

For example, one of the high accuracy transducer manufacturers buys tons of the custom steel, machines sample sensors, and tests their long term stability and hysteresis. If the material is bad, the company orders another lot of steel. If the material is good, it will be used for the next several years production.

Silicon delivers good performance on each low-cost wafer, thanks to its extremely pure defect-free crystalline structure.

Mechanical Strength

Silicon strength is comparable to steel, but at a lower density (yielding better dynamic operating devices which depend on the material's weight to strength ratio) and better thermal conductivity, as shown in Table 121.1 below.

High Sensitivity to Stress

Piezoresistive effect in silicon has stress sensitivity two orders of magnitude larger than metal strain gauges. This enables fabrication of the high output devices.

Batch Manufactureability

Capability for manufacturing completed mechanical structures simultaneously on multiple wafers, each carrying multiple

Table 121.1 Comparison of Selected Properties of Steel and Silicon

Parameter	Steel	Silicon	Units
Yield strength	4.2 max	7.0	10^{10} dyne/cm^2
Young's modulus	2.1	1.9	10^{12} dyne/cm^2
Density	7.9	2.3	gram/cm^3
Thermal conductivity	0.97	1.57	W/cm°C

devices, forms the revolutionary aspect of the silicon micromachining technology: batch manufactureability. None of the other sensor technologies has such an advantage. It delivers not only a very high volume manufacturing capacity, but also a very low unit cost.

IC Industry Infrastructure

Besides the excellent mechanical performance, silicon brings significant support from the established mainstream electronic industry, specifically:

- Access to ultra-pure materials.
- Access to advanced semiconductor processes.
- Availability of the high volume packaging technologies.
- Developed high volume batch manufacturing processes for silicon wafers.
- Access to high volume manufacturing equipment.
- Available educated silicon processing technologists.

All these benefits can be "borrowed" for MEMS from the semiconductor industry. Fantastic leverage that MEMS technology enjoys becomes clear when one compares the total R&D funding in MEMS and IC technologies. The total world MEMS R&D budget is on the order of $200M/year, including multiple government funding. The integrated circuit technology spends billions of dollars yearly on R&D in materials, processes, manufacturing equipment and training. Silicon sensors and microstructures benefit from these vast resources. Using materials, processes and equipment developed for mainstream electronic industry significantly lowers the development time and cost for MEMS. Many of the silicon micromachining companies directly utilized the high capacity high volume IC equipment, processes and materials.

Dedicated MEMS Developments

While mainstream electronics brings substantial advantages to MEMS industry, several aspects of its technology are unique and create a substantial technological barrier to entry. These barriers have significantly slowed the development of silicon micromachining markets. Some of the dedicated MEMS technologies that are being developed almost entirely by MEMS industry include:

- Deep (through the wafer) silicon etch.
- Precision dimensional control of vertical silicon structures.
- Ppm level stability of silicon resistors.
- Silicon-silicon and silicon-glass wafer lamination.
- High-volume low-cost pressure/acceleration/temperature testing.
- Chip package stress isolation.
- Packaging enabling mechanical interfacing with electromechanical chips.
- Media isolation for pressure sensors.

While there is a very high level of activities in all these areas at the academic level, lack of production proven results stretches the implementation cycles for more advanced MEMS products to 5 to 15 year timeframe.

121.3 Market Growth Analogy: Transistors, Integrated Circuits and MEMS

The history of semiconductor technology is a mere four decades old. Since its invention, transistor technology has evolved from the discrete components to integrated circuits with millions of transistors on a single silicon chip. The evolution of this market, especially for some of the functional modules such as operational amplifiers, has some parallelism to the largest market for MEMS, silicon sensors. It thus could be expected that other MEMS devices may exhibit similar characteristics.

Initially, operational amplifiers were built using discrete transistors. Eventually, the ceramic based thick-film hybrid technology was used as a packaging foundation. Thick film resistors allowed efficient laser trimming of functional parameters. Many families of hybrid ICs were built providing performance and reliability, but not necessarily offering the lowest cost for the highest volume. The continuous evolution of the analog IC technology gradually allowed it to enter the applications dominated by the analog hybrids, replacing a majority of them. As soon as more advanced performance becomes available in a monolithic form, equivalent hybrid circuits begin losing market share.

Cost, performance and reliability are the strongest drivers in commercial electronics. In some areas, particularly aerospace, weight and size may also be considered important for the application development. The same is true for silicon sensors, as many of MEMS techniques are adapted from the IC industry.

Since the cost is most often the overriding concern for new applications, enabling or disabling them, it is quite interesting to evaluate the initial cost curves for both transistors and ICs (Aaker, 1989; Hadley, 1976) as related technology. Cost learning curves were developed as a function of the cumulative industry volume. For transistors (Figure 121.1), the cost learning curve exhibits two distinctively different phases. Until the cumulative volume reached 10 million units (from 1954 to 1960), yearly price reduction was relatively slow. This period of time can be

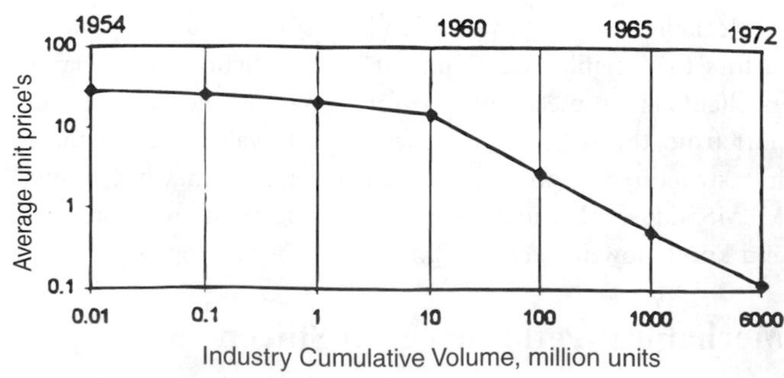

Figure 121.1 Price experience curve for silicon transistors.

characterized as the basic technology and market development. Rapid explosion of the market within the next 12 years increased the cumulative volume about 600 times to 6 billion units, for a compounded growth rate of 70%/year. Simultaneously, the price was reduced 200 times.

Integrated circuits (Figure 121.2) exhibited no early price stability. As the basic market was already initiated by the introduction of transistors, integrated circuits were replacing the functional transistors based assemblies, creating a continuous transition of the market. Cumulative volume has increased two times faster than for transistors. The price experience slope was at 73%, indicating a reduction of price by 27% for each doubling of the shipped volume.

The situation for IC sensors is somewhat different. While the first silicon based pressure sensors were developed in early 1960s, the first mechanically monolithic sensors (a batch fabricated diaphragm integrated with constraint and strain gauges, but without active electronics) were offered in 1967. This sensor integration level could be compared to a single transistor. Its introduction was about 14 years after the introduction of transistors, and 3 years after the introduction of integrated circuits. The cumulative volume of silicon pressure sensors just recently (in 1993) exceeded 100 million units. This matched the 1963 status of transistor industry, or 1968 status of the IC industry. Transistors needed 10 years to reach 100 million units cumulative volume, integrated circuits only 3 years, but pressure sensors required a full 26 years.

A slower market development created a major economic barrier for the implementation of single-chip integrated sensors. Another barrier was created by the evolution of the applications themselves. Out of the annual volume of 45 million silicon pressure sensors manufactured in 1994, only about 25 million require signal conditioning (the rest interfaced with

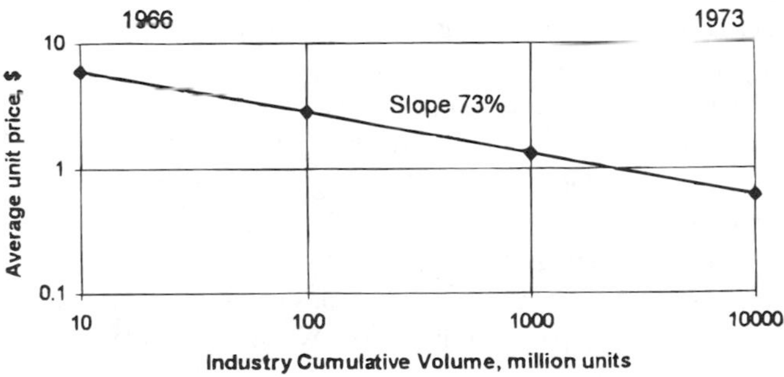

Figure 121.2 Price experience curve for integrated circuits.

Table 121.2 Market Development for Transistors, Integrated Circuits and Silicon Sensors

Item	Transistors	Integrated circuits	Pressure sensors	Acceleration sensors
First products	1953	1965	1960	1967
Cumulative 100 million units	1963	1968	1993	2000?
# years to 100M units	10	3	33	27?
1995 volume, units	Billions	Billions	50 M	12M

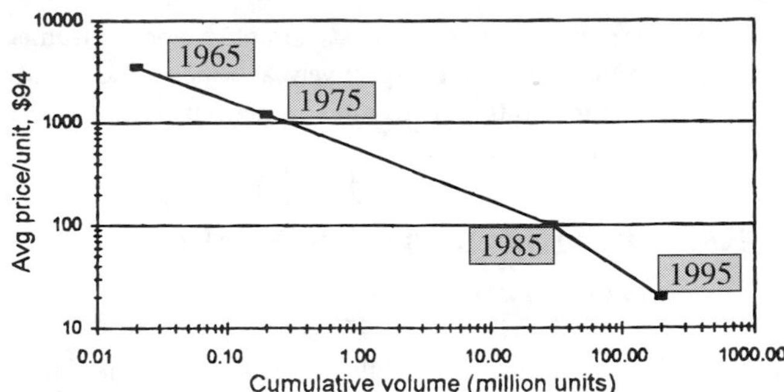

Figure 121.3 Price experience curve for silicon pressure sensors based on author's market research.

the instrumentation designed for low level output sensors, such as hospital monitors).

A large fraction of the signal conditioning sensor market, about 22 million units, represents MAP transducers manufactured by about ten companies. The largest one is GM Delco, shipping over 10 million MAP transducers per year. Even at this volume it was not cost efficient for Delco to develop a single-chip sensor.

Development of the silicon pressure sensor market is illustrated in Figure 121.3. The Price experience curve does not exhibit the initial price holding period visible for transistors. It is more like the price experience curve for integrated circuits. It started however from a significantly higher price, and it required three decades to reach 100 million units which IC's reached in just three years. The average price experience slope through 1985 was about 70%, indicating 30% price drop for each doubling of the cumulative volume. This slope was close to these of transistors and ICs. Around 1985 the price experience curve dropped to about 20% (80% price drop for each doubling of the volume), reflecting the rapidly growing low-cost medical and automotive applications. This slope was significantly steeper than those for transistors and integrated circuits, despite the slower growth of the pressure sensor market.

Development of the silicon acceleration sensor market, the second largest production MEMS device, was even slower than the pressure sensor market. The first sensor was demonstrated in 1967. The cumulative volume of silicon acceleration sensors in 1993 was about 250,000 units. In 1994 the market jumped by about 5 million units. It is forecasted that a cumulative volume of 100 million units will be reached by the year 2000.

It is clear that while roots of the silicon sensor market are the same as for other semiconductor devices, the dynamics of this market is quite different. The dynamic is closer to the development of the market in which MEMS devices are used. These markets include the following main segments:

- Process control.
- Industrial.
- Military/aerospace/commercial aviation.
- Medical.
- Automotive.
- Consumer.

Within these segments, only the automotive and consumer markets, currently estimated respectively at $300M and $50M, have a potential to create growth rates somewhat similar to the IC market growth.

121.4 Integrated MEMS Market Overview

Currently, MEMS market practically consists only of the silicon sensor market. Other MEMS devices exist at either an R&D funding level, or initial commercialization phase (e.g., atomic force microscope tips, display chips or pressure valves).

Integrated sensor market includes two generic levels of electronics integration on sensor chips:

- Passive elements (mainly compensation resistors).
- Active elements (e.g., amplifiers).

Because of limited high volume applications, the on-chip integration did not deliver an expected return-on-investment (ROI) justifying faster industrial implementation of single-chip sensors. The status of the 1994 sensor market is characterized below (Table 121.3).

As shown in Table 121.3, the passive electronics on-sensor chip integration represents only about 14% of the pressure sensor market. Active integration represents about 8% of the pressure sensor market and about 3% of the acceleration sensor market. A limited market share of integrated sensors will likely change in the future, as several large volume applications are visibly entering the automotive market.

Actual implementations can be grouped into several functional categories. These categories can be summarized as follows:

- Passive temperature compensation and calibration.
- Analog electronics for amplification and analog (e.g., laser trimmed resistors), calibration and compensation.
- Analog electronics with digital trimming (e.g., using EEPROM).
- Digital electronics with digital trimming.
- Advanced electronics, e.g., bus output, and/or control algorithm processor (e.g., PID or fuzzy logic), and/or actuator (e.g., power controller).

Table 121.3 Snapshot of the 1994 Sensor Market

Item	Sensor Market [million units]			
	Pressure		Acceleration	
Total market	50	100%	6.0	100%
Total silicon sensor market	45	90%	5.8	97%
Silicon sensors with passive electronics integration (laser trimmed resistors)	7	14%	0.0	0%
Silicon sensors with active electronics integration (amplifiers)	4	8%	0.2	3%

Passive Temperature Compensation and Calibration

Passive compensation and calibration utilizes primarily on-chip laser trimmed thin-film resistors, fuses and anti-fuses to eliminate sensor errors. First production passive temperature compensation was implemented by Kulite Semiconductor, USA, in 1979 (Bryzek et al., 1991). First passive temperature compensation *and* calibration was introduced into production by Motorola, USA, in 1986, for disposable blood pressure sensors (Frank and McCulley, 1985). NovaSensor, USA, introduced a smaller version of such a sensor in 1988 (Bryzek, 1987), but later decided to implement a hybrid version of a transducer, which provided lower cost and higher yields (Bryzek and Gee, 1994). This design is characterized in the section **Case Study** in this Chapter.

Amplification and Analog Calibration and Compensation

Basic signal conditioning for transducers includes amplification, calibration and temperature compensation. With an analog approach, all these functions are accomplished with an analog circuitry: analog amplifier and laser trimmed resistors for calibration and compensation (Figure 121.4).

First silicon pressure sensors with on-chip amplification were developed in late 1970s by IC Transducers (Bryzek, 1980) (first sensor company in Silicon Valley, formed as a spinoff from Fairchild Semiconductor, USA). These integrated sensors delivered such a low yield at that time that they were not practical for commercial applications.

The second major effort was undertaken by Honeywell, USA. This effort delivered an IPT sensor in 1981 (Bryzek, 1981), which integrated 3 bipolar operational amplifiers and 8 thin-film laser trimmable resistors on a single sensor chip. This product delivered excellent performance, meeting requirements of the aerospace applications, but at a cost significantly higher than the alternative implementations. The production program was suspended.

The next major effort to introduce a fully integrated high level output transducer was undertaken by Motorola, USA. It resulted in the 1985 introduction of the integrated sensor with a 3 V output FSO (McCulley et al., 1985). The sensor was based on a bipolar IC technology, SiCr thin resistors and shear gage piezoresistive sensing elements. Technology problems delayed manufacturing of this part until 1993, when Motorola started production sampling.

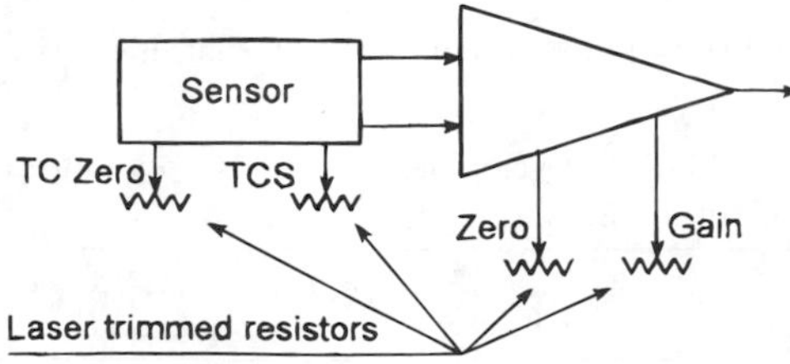

Figure 121.4 Principle of analog signal amplification with analog performance trimming. Typically, temperature errors (TC Zero and TC sensitivity) are compensated at the sensor level, and calibration is performed in the amplifier, all with laser trimmed resistors.

In the meantime, Nippon Denso, Japan, made the decision to introduce a monolithic version of the manifold absolute pressure sensor (MAP) for automotive applications. The product was introduced into production in 1989, and replaced a hybrid version of the MAP manufactured previously. As in the other approaches, sensor design was based on bipolar process amplifiers and thin-film laser trimmed resistors, processing a signal from the piezoresistive bridge. Nippon Denso expected that cost of the fully integrated transducer will be higher than the hybrid for the first several years of production running at about 4 million units/year (Nippon Denso, 1989).

Newest introduction of integrated pressure sensors came in 1993 from Honeywell to selected larger volume customers. This chip was an enhanced descendant of the IPT transducer, with improved manufactureability.

In the acceleration field, Nippon Denso developed the first on-chip integrated sensor in 1989. After several years of struggling with technical problems, the sensor moved into the 1994 model year production *without* the on-chip electronics. Analog Devices introduced a single chip design for crash airbags in 1991. This design was based on a capacitive polysilicon sensor. On-chip electronics included a servo loop maintaining a constant position of the laterally moving polysilicon fingers (Goodenough, 1991). The sensor was implemented in a position of the laterally moving polysilicon fingers (Goodenough, 1991). The sensor was implemented in a surface micromachined technology utilizing 6 masks and the electronics was designed in a BiCMOS technology that required 22 masks, making a total of 26 masks (two masking steps were common). Limited production shipments started in 1993 after an extended debugging cycle.

Chronological industrial efforts in this field are summarized in Table 121.4. As it can be seen, production debugging of the integrated sensor and electronics was a major development theme. It eliminated or significantly delayed production implementation for most of the single chip integrated sensors. All the developments so far have used analog electronics, and only one device used digital trimming technology, with others using analog trimming.

Analog Electronics with Digital Trimming

The major disadvantages of analog trimming are either a low productivity, if potentiometers or discrete resistors are used, or substantial investment in process development and automated laser trimmers, if a sophisticated thin-film technology is used. Analog electronics with digital trimming combines the best of analog signal processing and its lack of signal quantization, with the ease of use and low-cost of digital calibration. It is often called Analog Sensor Signal Processing, or ASSP (Figure 121.5).

This type of processing has already entered the transducer field (Bryzek, 1993). In such approach an amplifier with digitally programmable gain and offset is used to calibrated and compensate the sensor. Individual compensation and calibration constants are stored in a nonvolatile memory. Any type of digital memory can be implemented, and actual smart transducer implementations have used battery backed RAM, EPROM, EEPROM and several modifications of PROMs. D/A converters are usually used to convert a digital code into an analog control signal for either offset or gain adjustment. In this approach the main sensor signal (e.g., pressure) is in an analog domain all the time, without quantization errors. The temperature compensation circuit changes offset and gain in response to temperature, creating an analog correction signal. This correction signal is often quantified, creating temperature error correction quantization error.

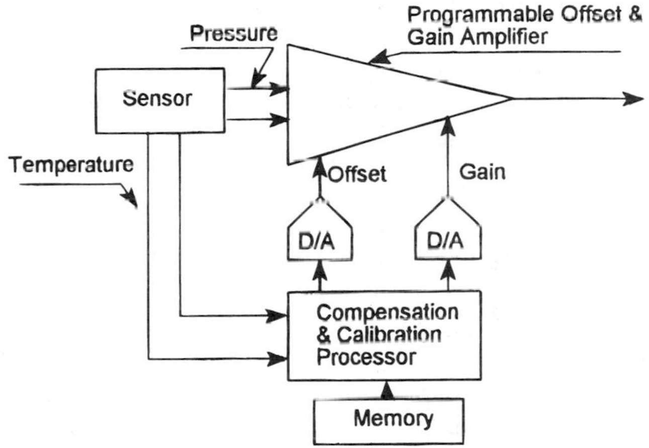

Figure 121.5 Principle of ASSP. Compensation and calibration processor changes offset and gain of the instrumentation amplifier to compensate sensor errors. Digital memory chip replaces analog potentiometers or resistors. D/A converters are used to convert digital correction word to analog domain.

Table 121.4 Chronology of Sensor Developments with Integrated Active Electronics

Company	Sensor type	Electronics	Basic R&D completion	Production
ICT, USA	Pressure	Analog/no trim	1977	Never
Honeywell, USA	Pressure	Analog/analog trim	1981	Never
TDI, USA	Pressure	Analog/analog trim	1985	Never
Motorola	Pressure	Analog/analog trim	1985	1993
Nippon Denso, Japan	Pressure	Analog/analog trim	1988	1990
Honeywell, USA	Pressure	Analog/analog trim	1992	1993
Nippon Denso, Japan	Acceleration	Analog/analog trim	1989	Never
Analog Devices, USA	Acceleration	Analog/analog trim	1991	1994
Bosch, Germany	Pressure	Analog/digital trim	1993	1994

The magnitude of this error is a function of an internal architecture of the compensation and calibration processor. To avoid signal switching between the two temperature compensations signals at the transition temperature, a small temperature hysteresis is often built into the circuit.

There were several developments from leading universities in this field. The most interesting design came from Berlin University (Obermeier et al., 1994) (a single chip sensor), Fraunhofer Institute of Technology (Mokua, 1995) (single chip sensor) and Delft University (Huijsing, 1994) (single chip sensor signal processor). The circuits were implemented in CMOS technology. The first commercial implementation of the analog electronics with digital trimming was introduced in 1994 by Bosch for the automotive MAP transducer (Kress et al., 1995).

Digital Electronics with Digital Trimming

This type of circuitry is often called a DSSP (Digital Sensor Signal Processor). In this configuration, an analog sensor signal is converted to a digital domain with an A/D converter, compensation and calibration is performed in a digital domain, and the compensated and calibrated signal is converted back to an analog domain (Figure 121.6).

This approach was used in the first implementations of smart sensors (Bryzek, 1981). Currently there are several commercial smart sensor designs using this approach, however, none was converted to a single chip solution. The most accurate implementations use 24-bit input A/D (voltage-to-frequency) converters. This configuration gives the best system flexibility, as it is almost entirely based on the implemented software.

More Advanced Electronics

There are several advanced system-on-sensor-chip developments at the academic level. Some already have yielded interesting results. These programs include bus output, control processors and actuators. Some of the leading centers in this field are Delft University, UC California at Berkeley (Yun and Howe, 1983) and the University of Michigan (Najafi, 1992). The most advanced design, University of Michigan's smart sensor with a parallel bus output, was implemented into a real time operation at the University IC Lab.

This emerging high level integration of electronics does not yet have the high volume market pull to accelerate production implementations. It can be expected, however, that within the next decade highly integrated smart sensors will reach the commercial phase.

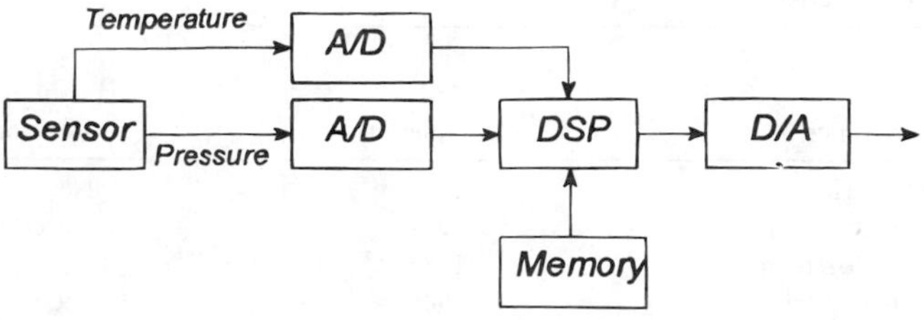

Figure 121.6 Principle of DSSP. Digitized sensor outputs are processed either by a dedicated Digital Signal Processor (DSP) or microcontroller to provide sensor compensation and calibration. Corrected signal is converted to analog domain, if needed, by the D/A converter.

121.5 To Integrate or not to Integrate

There are many compelling reasons behind the drive towards the single chip signal conditioned sensors. Probably the most attractive one is the past track record of the IC industry. Each level of ICs integration improved price to performance ratio in almost every application, having no competition from any other technology. The sensor industry expects the same advantages from the single chip sensors. While a similar progress is still expected, the major encountered problem was a significantly smaller volume of the MEMS industry, currently lagging many orders of magnitude behind integrated circuits. Lower volume makes unit cost higher and allows for a prolonged life of competitive nonsilicon technologies.

The sensor industry often demands from silicon designs performance not needed nor understood in the IC world, such as ppm level silicon resistor stability, low mechanical and thermal hysteresis, stress free processes, etc. On-MEMS-chip electronics integration requires a careful design eliminating fundamental problems brought to mechanical chip by complex electronics: performance degradation and manufacturing logistics.

- **Possible degradation of sensor performance:** Most silicon sensors, being the largest MEMS market, use piezoresistive sensing elements. These elements are sensitive to all parasitic stress. This stress can be (and will be) generated by the additional materials and structures necessary to integrate the IC circuitry on the MEMS chip. Each of the layers has different mechanical and thermal performance, different temperature coefficients of expansion, and is fabricated with a different residual stress. Moreover, the additional IC processes may alter the basic properties of sensing elements, making compensation more difficult. Taking this into account, the design and debugging process for fully integrated sensors is significantly longer than for discrete devices.
- **Logistics of testing and trimming:** In discrete sensor applications, performance of the sensor can be modeled without the influence of the signal conditioning circuitry (Bryzek, 1985). Once the electronics is permanently connected to a MEMS structure, it is more difficult to perform temperature compensation. The logistics of testing and trimming thus will be significantly more complex in comparison to discrete sensors, and must be carefully designed into the circuit and process.

The MEMS industry is still small. Sensor related know-how resided within a limited number of sensor gurus, most outside of companies that could afford an advanced on-chip integration. It is thus not surprising that the learning curves for single chip sensors were painfully slow. The situation is changing, however, and first highly integrated sensors reached the market. It can be expected that more highly integrated MEMS devices will be introduced in coming years. The decision to integrate, however, must be carefully considered, as not all the applications can benefit from it. There are many pro-integration arguments, as there are many against, as discussed below.

Pro-Integration Arguments

Size and Weight

Many of the new applications require that MEMS devices be comparable in size and weight to other integrated components. The smaller the device, the more functional the final product can be made (e.g., a barometer in the wrist watch). Some other applications can even calculate cost savings when the weight is reduced. For example, in modern aerospace systems, the cost of adding weight to an aircraft can be quantified and is not insignificant even for items such as sensors. Increasingly, the same is true for automotive and industrial sensors. Even using small outline (SO) packages, the effect of replacing two individually packaged components with a single integrated component reduces the system weight and space by a factor of at least 2 or 3.

Reliability

On-MEMS-chip integration significantly improves reliability as a result of reduced number of components and interconnections. Other changes that may have noticeable effect are the reduction in electrical power consumption which typically accompanies the integration of multiple components into a single part, resulting in reduction of the operating temperature, thus further increasing reliability.

Cost in Volume

The manufacturing cost in volume is significantly lower for a single chip as compared to a multi-component device. The most obvious reason is that not only has a number of parts and interconnections been removed from the MEMS device, but much of the assembly cost of those parts has disappeared with them. A less obvious benefit is that much of the testing can be carried out while the MEMS devices are still in wafer form. This creates a major challenge for sensors (e.g., pressure or acceleration sensitivity testing), but once solved, facilitates rapid automatic testing borrowing techniques from the IC industry. The ability to weed out the majority of failures at the wafer stage before adding the cost of sensor packaging is a substantial manufacturing cost saver.

Temperature Compensation

Many MEMS devices exhibit cross sensitivity to temperature. To avoid errors due to a temperature gradient between the temperature sensor and MEMS device, the temperature sensor should be a part of MEMS.

Feasibility of Arrays

The integration of several sensors onto a single piece of silicon is not much more of a challenge than the integration of one. Packaging costs are not likely to be significantly more so the main additional cost is only that of silicon area. With a multiplicity of sensors, statistical comparison of several sensors of the same type becomes possible for diagnostics and calibration monitoring. Alternatively, arrays of sensors with multiple cross-sensitivity can be used to calculate outputs free of cross sensitivity (e.g. electronic noses).

Reduced Parasitics

Because of the small dimensions of an integrated sensor, the parasitic impedance of associated interconnects is dramatically reduced and also because of the dimensional stability arising out of the method of manufacture. This is of immense benefit in the design of capacitive sensors where, for example, position is converted to a small capacitance change, often significantly smaller than the interconnection capacitance. The capacitance variation, which constitutes the signal, must frequently be resolved to a few femto-farads and any variation due to parasitics which is greater reduces the sensor accuracy.

Increased RFI Immunity

The small dimensions and short length of the signal paths carrying unamplified signals within integrated MEMS device have an additional benefit. A 1mm connection even in a most demanding automotive requirement of 200V/m electromagnetic field could only acquire a maximum of 200mV of RFI. In real devices the signal would be much smaller due the symmetry and would depend on impedances and geometry. Should these low levels present a problem, then of course it is still a lower cost option to shield a 5 mm cube than, for example, a 50 mm cube.

Counter Integration Arguments

There are many difficulties in integrating sensor and electronics and specific issues which might make integration less desirable for a particular application. These issues are characterized below.

Non-Recurring Engineering Costs

The design (as opposed to manufacture) of fully integrated MEMS is never likely to be cheap. Analogue and digital (mixed mode) simulators will be required for the circuit design as well as tools for modeling physical performance of the new device. Quite often a new process will have to developed and proven. Diversity of applications will make the job more difficult.

Manufacturing Volume

The high NRE cost means that integrated MEMS devices are likely to be available only where that cost can be supported by very high volume applications or where a high cost per sensor can be tolerated over moderate volumes. Outside these scenarios, the only hope for low volume MEMS in the foreseeable future is to share NRE costs with a similar high volume application. Such an initiative was undertaken by ARPA, with funding of North Carolina Microelectronic Center (NCMC). (Markus et al., 1994).

Some sensors manufacturers elected to use an external wafer foundry service to perform the standard IC processes on the wafers, and finish micromachining in-house. This approach creates a problem for smaller volumes (by IC standards), as foundries with advanced and stable processes are hesitant to accept orders for volumes smaller than, e.g., 500 wafers/year.

Media Protection

An accelerometer can be hermetically encapsulated in the medium of the manufacturers choice—even a vacuum. Pressure, flow and chemical sensors must be in intimate contact with the sensed medium, and require reliable media protection to avoid mechanical or chemical destruction. These are problems to which many MEMS devices are inevitably subjected, but not the signal processing circuit. Typically the circuit would be kept well away from any media, especially one which might carry abrasive particles or unknown chemicals. In a case of integrated MEMS, either both device and circuit must be exposed to the medium, or complex arrangements will be required to isolate one part of the silicon surface (with the electronics) from an adjacent part (with MEMS device) exposed to the medium. The former arrangement promises severe technical problems while the latter is likely to be expensive in itself and to require larger silicon chips to accommodate the barrier area.

Combined Yield

For a given semiconductor process the yield falls exponentially as die area increases. In the case of the integrated MEMS device not only will the chips be larger because they contain both the MEMS and the electronics, but also the process will be more complex having more steps of a wider variety. MEMS section of the chip (usually 6 to 12 masking steps) undergoes the unnecessary IC process (9 to 22 masking steps), and vice versa. Usually only the metalization step could be common to both processes. The more complex process will lead to an additional reduction in yield. If both products yield 80%, the combined yield of the larger silicon die will be only 64%. Unless both MEMS device and IC are high yielding, the economic impact of the on-sensor-chip integration may be unacceptable.

It is also possible that even for devices which nominally work, performance will be degraded as the sensor will be subject to circuit processes which it would not normally see, and likewise the circuit would be subject to sensor processes. This will further deduce the yield.

On-Chip Electronics Integration Justification Example

The first company that started volume production of MEMS, a single-chip pressure sensor, is Nippon Denso, Japan. For the first implementation the company selected the automotive market for manifold absolute pressure measurements with large, multimillion units/year, production volume.

The first generation product was a discrete micromachined silicon sensor with hybrid electronics. To justify the on-sensor-chip electronics integration, this company developed a dedicated *cost performance* figure of merit *FOM:* (Tabata, 1995).

$$FOM = \frac{F*T}{A*R*C}$$

where

F = pressure range, Pa
T = compensated temperature range, °C
A = accuracy, % FSO
R = resolution, Pa
C = total cost of all signal conditioning components, assembly and trimming, $

The reference $FOM = 1$ was assigned to their first piezoresistive pressure sensor developed in 1973. The first electronics integration was performed in 1983, where operational amplifiers were integrated on a sensor chip, delivering a high level output. Calibration and compensation was performed by the external circuits. This development increased the cost performance, delivering $FOM = 2$ due to a cost reduction resulting from the integration. In 1989, thin-film resistors were added on the sensor/amplifier chip, providing the self-contained on-chip compensation and calibration. A result yielded $FOM = 8$. In 1991, a capacitive sensor was developed, enabling lowering the pressure range down to 2 kPa with a high resolution. This development increased a figure of merit to 16. Considerations for a further cost reduction brought an interesting perspective on a relative cost of the sensor. In the beginning, the sensor chip, assembly and testing created approximately equal contribution to the transducer cost. The ratio between these cost components started to change with increased volume. Cost of the sensing element decreased, and with the advent of low cost computers the cost of testing also decreased. The dominating cost factor of the transducer was the assembly (Figure 121.7).

The improvement in assembly cost could be achieved from the on-sensor-chip mechanical functions integration. One of the targets for Nippon Denso will be sensor packaging, integrating on-sensor-chip functions such as reference vacuum cavity eliminating a hermetic welding of the package in current package.

The other cost driven future direction foreseen by Nippon

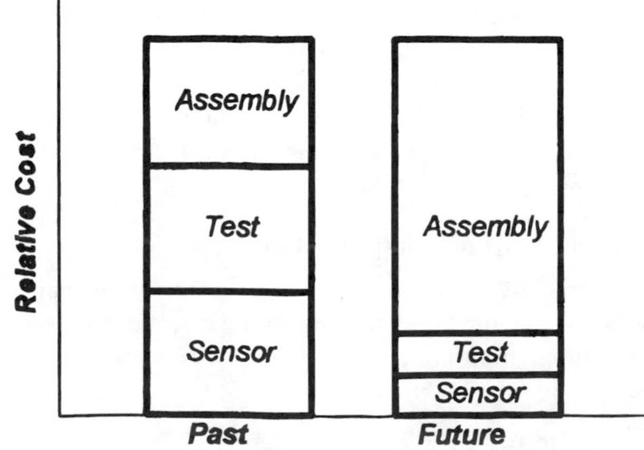

Figure 121.7 Evolution of sensor relative cost. With increased sensor production volume and computerized testing, the assembly cost becomes a dominating cost factor.

Denso will be the integration of actuators, and not microprocessors and sensor communication buses on the sensor chip. Integrated sensor and actuator on a single chip is called *Active Sensor.* A sensor with a servo force-balance loop and on-chip self testing would fit into such a category. The figure of merit increases significantly for such devices, as a result of improvements in accuracy, dynamic range and resolution.

The issue of increased level of mechanical integration will be further discussed in the next section.

121.6 Mechanical On-Sensor-Chip Integration

The integration of mechanical features on MEMS chips provides a potential for substantial benefits to the users, far outweighing the benefits of electronics integration. MEMS devices replace mechanical designs which are manufactured using traditional one-piece-at-a-time technology. The integration of mechanical components on a single chip offers a high-volume low-cost batch mode manufacturing technology, which rewrites the economics for many products.

Several examples of the on-sensor chip mechanical integration (or enhanced micromachining content) include the following:

Acceleration sensors: Two major mechanical functions are currently integrated on the sensor chip: shock protection and selftest. Shock protection allows survival of the sensor during drops on the floor—which generate up to 10,000 g shock—for devices with an operating range of only 2 g. In the classical acceleration sensor designs, this function was accomplished by screw driven mechanical stops.

The selftest generates a mechanical deflection of the seismic mass, simulating the effect of acceleration. It enables testing the integrity of the sensor mechanical and electrical structure in critical applications, such as automotive airbags. Several of the new acceleration sensors have the selftest integrated directly on a chip, while the older generation had been using mechanical hammers to perform the same function.

Pressure sensors: Process control pressure transducers require a single sided pressure overload survival of up to 300 Bar for differential pressure sensors measuring just .07 Bar full scale. Classical designs use expensive mechanical configurations, mostly based on large metal diaphragms activating the valves which seal-off the internal flow channels for the silicone oil surrounding the sensor. This mechanical design protects the sensor, however, at high cost. The newest silicon sensor designs incorporated the overload protection on the sensor chip, dramatically simplifying the transducer design.

Flow controllers: The current generation of products is based on the discrete pressure sensor and valve. There are industrial developments targeting integration of both these components on a single chip, promising a significant reduction of the size and cost.

Microactuators: The current generation of disk drives uses a servo positioner controlling the arm which holds the read/write head. Positioning accuracy is limited to several microns, and response time is limited by the arm's compliance. The integration of the microservo positioner directly at the head location enables an improvement of the positioning accuracy to .1μ, with a simultaneous significant reduction of the response time. These improvements together enable the development of hard drives with 25,000 tpi density, supporting 10 GB/in^2 read-write densities. Such densities were scheduled 10 years from now based on the hard drive industry projections using a classical technology.

These few examples clearly illustrate the potential impact that mechanical on-chip-integration may have on the industry. Mechanical integration could become the fastest growing segment of silicon micromachining, with a larger application base than the electronics integration.

121.7 Monolithic or Hybrid

Hybrid technology is typically based on the ceramic substrate, which is used as a packaging base for monolithic IC chips and other components. The same ceramic substrate is also used as a base for the fabrication of the interconnecting metallization and thick-film resistors, which are typically screen printed in an array form at room temperature, and fired at 850°C. Ceramics have good mechanical properties and offer easy customization at low non-recurring engineering cost (in comparison with custom monolithic ICs). Development cycle times are short, and volume production capabilities are high.

Hybrid integrated circuits have been in use for the last four decades, and are still widely used. This technology offers several advantages for many sensors and other MEMS devices:

- Simplified pressure connectivity, as many pressure port configurations can be directly integrated into the package.
- Capability for a high density packaging of multiple IC chips on a single substrate.
- Access to low-cost laser trimmable resistors.
- Low-cost hermetic feedthroughs from the pressure cavity (vias).
- Broad equipment and process base developed by the electronics industry.
- Capability for handling multiple sensors on a single ceramic plate (snapstrate) through the entire production process, enabling a batch sensor assembly process, equivalent to batch sensor assembly process, equivalent to batch wafer processing.
- High reliability proven in both automotive and aerospace applications.

Similarly to the hybrid analog modules, as long as the production volume is below a critical level, the hybrid design is less

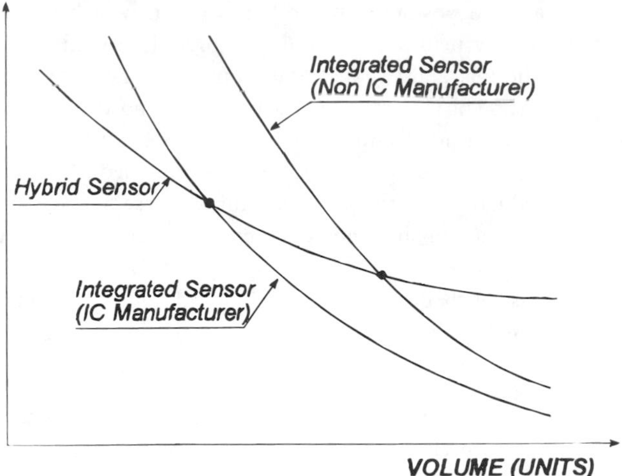

Figure 121.8 Hybrid sensors are less expensive in smaller volume. IC manufacturer will require a smaller volume to match the cost of hybrids with a monolithic solution, as compared to a non IC manufacturer.

expensive (Figure 121.8). This critical volume varies significantly as a function of the existing company infrastructure. For semiconductor companies, the monolithic design delivers lower manufacturing cost at lower volume, typically around 1–3 million units/year. For more traditional transducer and system companies, the critical volume is typically between 5–20 million units/year, as more of the incremental IC infrastructure has to be installed and the production will begin at a lower yield curve and higher unit overhead allocation level.

The wafer fab operation has a high fixed cost of processing wafers. The small volume processing cost (e.g., one wafer/month) is practically the same as the processing cost of several hundred wafers. More importantly, the wafer fab yield increases significantly with increased production volume. Each wafer typically carries between 100 and 10,000 MEMS devices. A high production volume for fully integrated MEMS devices is thus a mandatory requirement to make them cost competitive with the hybrid implementations. Due to a high volume of non-micromachined wafers processed in the IC-manufacturer facility, cost allocation for micromachined wafers is lower, and their yields higher, as compared to the non-IC-manufacturer.

Interestingly, when the IC technology brings new generations of products, hybrid technology can assemble them on a single substrate for even a higher level of integration. The newest trend, Multi-Chip-Modules (MCM), is one of the hottest new hybrid technologies. It is thus expected, that a similar evolution will be visible in the sensor industry, and hybrid sensors will co-exist with monolithic designs for a long time.

121.8 Case Study: Lucas NovaSensor

NovaSensor was founded in 1985 in Silicon Valley, California, with the objective of bringing advanced MEMS devices to production. In 1990, company was acquired by Lucas and the name was changed to Lucas NovaSensor. Within the first years of NovaSensor many new MEMS devices were developed. The major focus was on pressure sensors, as they represented the only developed high volume market at that time.

In two of these developments the on-sensor-chip integration approach had to compete with a hybrid solution, and the outcome is summarized below.

On-Sensor-Chip Passive Electronics Integration

One of the largest markets for pressure sensors is the disposable blood pressure sensor market. It currently consumes yearly about 17 million pressure sensors. As hospital monitors maintain the transducer sensitivity standards initiated decades ago, including 1% calibration of the 5 μV/V/mmHg pressure sensitivity and an option for the AC sensor excitation, sensors in this application can not use active circuits. The only possibility for the on-chip integration is a passive compensation and calibration circuit.

Only silicon piezoresistive sensors are used in this applications. To meet specifications recommended by the US Association for Advancement in Medical Instrumentation (AAMI, 1986), a defacto world standard, a passive signal conditioning circuit must perform the following functions:

- Compensate sensor's initial offset (zero) to a maximum of ±25 mmHg.
- Compensate zero temperature coefficient to a maximum of ±0.3 mmHg/°C.
- Compensate pressure sensitivity to 5 μV/V/mmHg ±1%.
- Compensate temperature coefficient of pressure sensitivity to below .1%/°C.
- Calibrate the output resistance to either 300 or 350 Ω, to enable the shunt calibration feature built into the monitors.
- Provide an output common mode voltage equal to .50 ± .05 of the supply voltage.

The functional circuit delivering all these functions is shown in Figure 121.9. Low impedance sensor bridge is used. Several temperature stable resistors perform compensation and calibration functions, as shown in the drawing. These resistors have to be individually adjusted, or trimmed, for each sensor. The most popular resistor correction technology is laser trimming. Laser trimming can be performed either on thick-film resistors, or on thin-film resistors.

In 1987, NovaSensor started production of a hybrid disposable blood pressure sensor. Silicon sensor die was mounted on a ceramic substrate with laser trimmed thick-film resistors. In the same year, company initiated a development of the next generation of such a sensor based on the on-sensor-chip integrated laser trimmable SiCr resistor technology. The resulting sensor chip P231 was smallest on the market at the time, with a pitch size of 2.5 mm square (Figure 121.10). In addition to basic sensing and calibration, the sensor's layout included several test devices, enabling in-process performance monitoring.

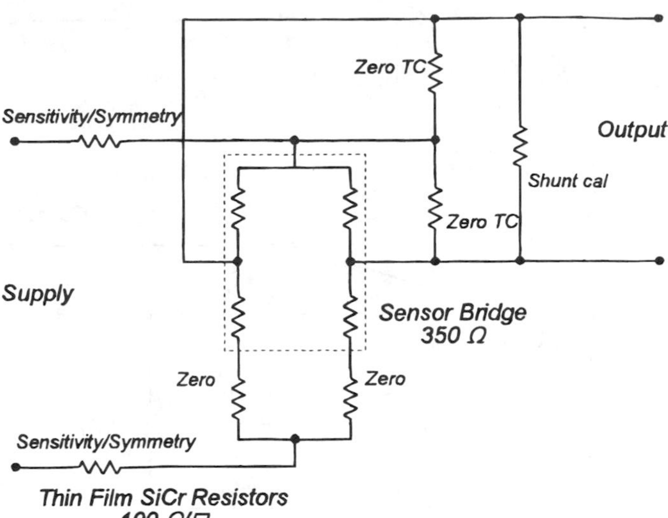

Figure 121.9 Functional circuit used in passive compensation and calibration of disposable blood pressure sensors.

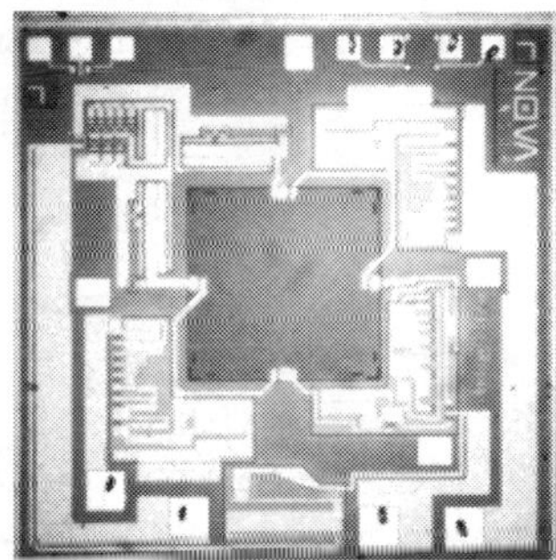

Figure 121.10 Disposable blood pressure sensor chip P231 developed at NovaSensor incorporated SiCr thin-film resistors to perform compensation and calibration of the transducer.

Manufacturing started in 1988, and after shipping approximately 100,000 transducers, production was stopped in 1989. The production cost of the on-chip integrated sensor was consistently higher than the cost of the hybrid version. NovaSensor was a micromachining company, and the cost parity between a hybrid and monolithic implementations was at a significantly higher production volume, than for an established IC manufacturers, such as Motorola.

Development of a new generation hybrid sensor was initiated. This effort resulted in a development of an advanced manufacturing technology based on a batch processing of snapstrates, a ceramic 4.5 × 4.5″ (11.4 × 11.4 cm) plate, each with 120 sensor cells (Figure 121.11). Each cell had an attached sensor die, a plastic pressure port, and thick-film resistors for calibration and compensation. Throughout the entire manufacturing process snapstrates were handled in cassettes carrying between 12 to 30 plates (depending on sensor design). Most of the assembly equipment (such as die attach, wirebond, etc.) was designed to take the cassette and process all the plates automatically. The line achieved a throughput over 1000 units per hour (uph). Due to a very low production cost and appealing design, production

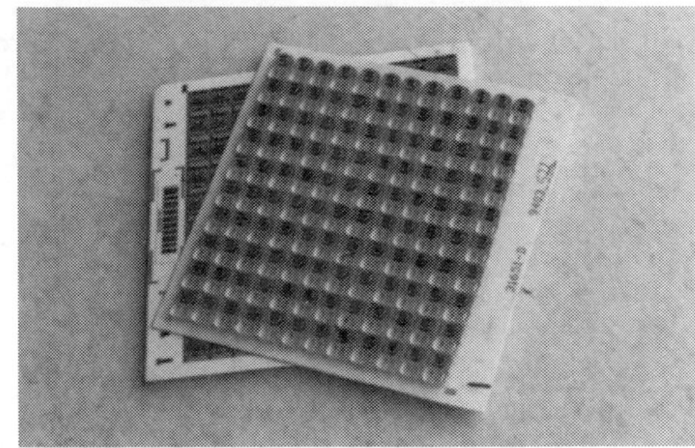

Figure 121.11 Lucas NovaSensor snapstrate technology mirrors batch wafer processing: 120 pressure sensors are located on one ceramic snapstrate, and 12–30 snapstrates are processed in one cassette. One side of the snapstrate includes plastic pressure ports, and the other side laser trimmable thick-film resistors.

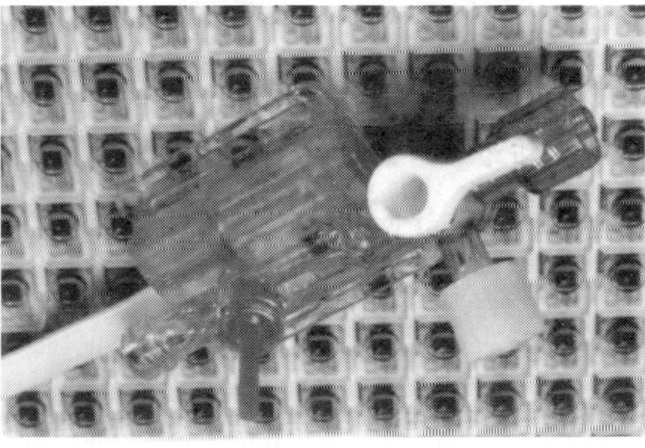

Figure 121.12 Ceramic pressure sensor tile is mounted into a plastic housing for hospital use (Courtesy of Baxter).

volume increased dramatically, forcing an installation of the second production line.

Figure 121.12 shows a disposable blood pressure transducer located on the snapstrates. Plastic housing provides a flow-through channel for a pressure media (saline solution), as well as a fast flush device enabling removal of air bubbles from the system. Ceramic snapstrate carries 120 sensors. One side of the snapstrate has plastic pressure ports, while the other carries thick-film resistors. Connection between both sides of the ceramics is accomplished with a vertical vias.

Snapstrate plates are laser scribed (partially cut). After completing test and assembly operations, ceramic plates are broken, or snapped, along the prescribed lines, yielding compensated and calibrated pressure sensor units. Single-up ceramic sensor assemblies are then mounted in a plastic housing, which is next sterilized and shipped to hospitals.

As a result of batch processing, standard cost of the sensor is very low, enabling a very competitive position on the market against the on-sensor-chip integrated sensors from Motorola.

While batch processing was a new technology for sensors, it was developed quite a long time ago for hybrid integrated circuits. Most of the automated equipment used on Lucas NovaSensor production line was originally designed for the hybrid ICs.

Advantages of hybrid technology in this application can be summarized as follows:

- Low entry cost.
- High volume manufacturing capability resulting from batch manufacturing capability (multiple units on a single plate).
- Fast development cycle.
- Mechanically stable packaging base for the sensor die.
- Very low cost laser trimmable thick-film resistors.
- Inexpensive technology of hermetic interconnections connections (vias).
- Good available equipment and process base.

NovaSensor problems with a single chip sensor technology in this application resulted from the following factors:

- Significant investment in new, for NovaSensor, processes and equipment.
- Long leadtime for a critical equipment delivery (e.g., 9 months for the automated thin-film laser).
- Long process yield debugging cycle and lower yield as compared to a hybrid technology.
- Requirements for a micron level sensor die alignment in the package to enable fast in-package trimming.
- Higher effective die cost as compared to a hybrid approach.

Once the volume of hybrid sensors approached 10 million units/year, Lucas NovaSensor started to consider the reintroduction of the single chip integrated blood pressure sensor.

On-Sensor-Chip Active Electronics Integration

Many applications require a signal conditioned high level output pressure sensor. NovaSensor initiated in 1988 the development of a single chip pressure sensor to support the microprocessor based sensor applications. A bipolar IC process and laser trimmable thin-film SiCr resistors with a bulk micromachined sensor process were selected as a foundation for the sensor.

The functional circuit is shown in Figure 121.13. A traditional two op-amp instrumentation amplifier configuration was selected for the amplification. A proprietary zero temperature compensation technique was used, and a statistical temperature compensation of the FSO was implemented. Calibration of zero was performed within the sensor bridge, and calibration of FSO was achieved via the amplifier's gain control.

A unique output stage circuit design enabled a rail-to-rail output supporting the automotive transducer requirements, providing either a .5 to 4.5 V or .25 to 4.75 V output at single 5 V supply voltage.

Analog laser trimming of thin-film resistors was selected. Special attention has been devoted to the logistics of the test and trim procedures. Discrete sensors allow an easy characterization of their performance. With integrated electronics, the sensor is permanently wired to the electronic circuitry, and many of the standard sensor modeling techniques can not be used.

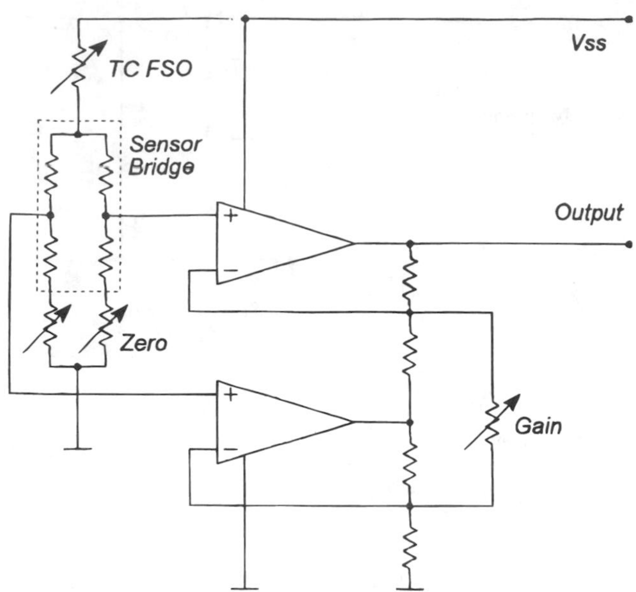

Figure 121.13 Functional circuit of NovaSensor single chip pressure sensor using a bipolar IC process and thin-film SiCr resistors.

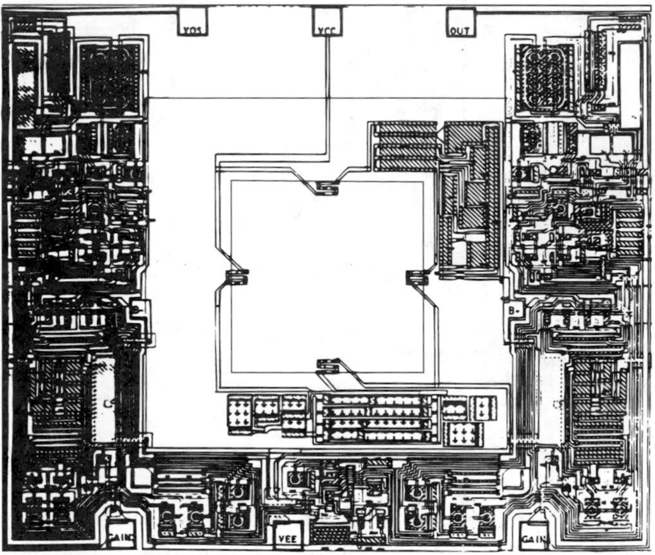

Figure 121.14 NovaSensor single chip pressure sensor was based on a bipolar IC process, SiCr thin-film resistors and bulk piezoresistive sensor process.

Layout of the chip is shown in Figure 121.14. The chip size was 2.5 × 2.5 mm. It was designed as a three terminal device, however, several additional pads were included for the performance modeling. Sensor was designed on a <100> wafer, and performance of bipolar transistors had to be tested and modeled in this orientation. Standard piezoresistor configuration was used.

Facing aggressive market pricing, this sensor was never released to production. Instead, development of an analog hybrid transducer was initiated, and next a development of the smart hybrid transducer.

To address a very low cost high volume application, a simple P592 type sensor was selected for the smart transducer. This sensor was fabricated using silicon fusion bonding that yielded a smallest high volume pressure sensor chip: only 1 × 1 × .6 mm. The chip was processed on 4″ wafers, each carrying 6600 sensors. Chip layout (Figure 121.15) included only the basic sensor bridge and metalization. Silicon fusion bonding allowed incorporation of a silicon constraint, which delivered very good rejection of package stress, thus making possible a direct die attach to mechanically unstable materials, such as printed circuit boards.

Smart signal conditioning ASIC was developed by a Silicon Valley start-up company ISS. This ASIC was developed using DSSP configuration in a low-cost 3 μ CMOS technology. It was based on a digital trimming technology. The circuit incorporated a digitally programmable input amplifier, 10-bit A/D converter, dedicated compensation/calibration DSP processor and 9-bit D/A output.

The most innovative feature of this ASIC was implementation of a proprietary local bus. This bus enabled a parallel connection of multiple sensors in the test oven, and transmission of analog and digital data from each sensor at each test pressure and temperature to the test computer. After calculating correction coefficients, test computer was downloading compensation/calibration directly to transducers' EEPROM. In a consecutive operation, a final test was performed.

To take the advantage of ASIC's bus capabilities, transducer subassembly was designed on a large printed circuit board, called "snapboard" (Figure 121.16). Each snapboard carried multiple transducer assemblies, similarly to the ceramic snapstrate technology or silicon wafers. Four-layer snapboard was terminated with a connector. Several snapboards were inserted into the test oven, enabling very efficient high volume testing. After completing the final test, transducers were separated into individual units, and packaged into the plastic housing.

Incorporation of the manufacturing bus enabled integration of three operations, pre-test, trimming and final test into one manufacturing operation with three automated steps. It translated into a lower manufacturing cost and higher yield, as only one pressure connection to the transducer had to be made on production line.

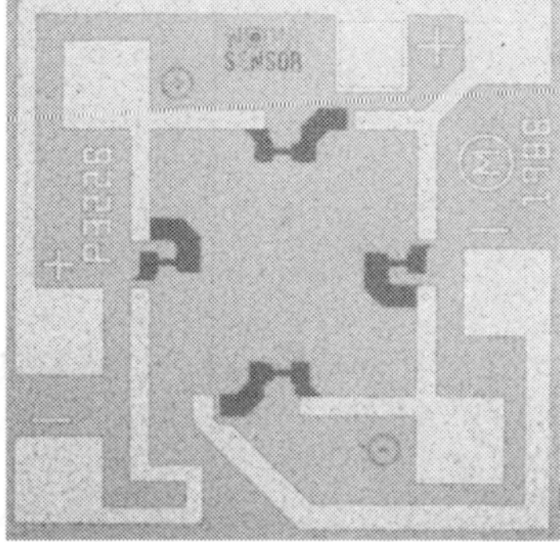

Figure 121.15 Smallest pressure sensor chip for high volume production was only 1 × 1 mm due to silicon fusion bonding.

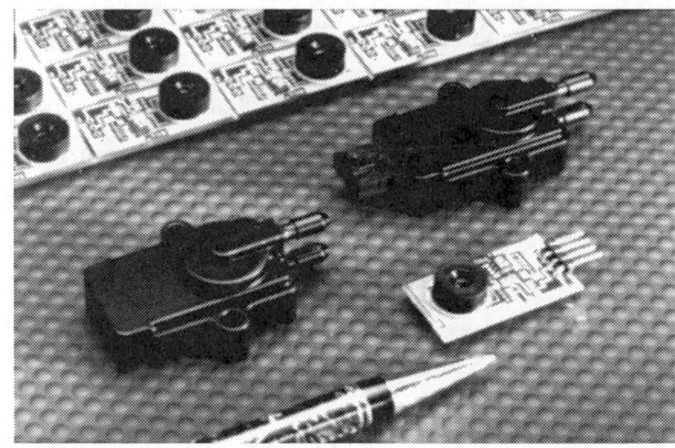

Figure 121.16 Lucas NovaSensor's smart transducer NPS was based on a hybrid design using ASIC signal processor with a local bus enabling batch manufacturing of pressure transducers on snapboard printed circuit boards.

Figure 121.17 High performance Analog Sensor Signal Processor was implemented in a digital CMOS technology (Courtesy of MCA Technology).

Next generation of the Lucas NovaSensor's NPS transducer again rejected the on-sensor-chip electronics integration. Instead, a new generation ASIC was designed. This sensor processor was implemented in ASSP configuration, providing an analog path for a pressure signal with 16 bit performance (Figure 122.17). The most interesting aspect of the design was that a high performance analog circuit implementation was done in a digital 2-metal 2-poly CMOS. Digital CMOS undergoes a continuous feature size shrinkage lowering the unit cost. Since digital processes was separated from a sensor process, it was feasible to cost benefit from the progress in a digital IC arena without redevelopment of the MEMS chip. The development of this chip was performed at another Silicon Valley start-up: MCA Technologies.

Summary

In both discussed examples, hybrid technology was found to be more cost advantageous for Lucas NovaSensor than the monolithic, on-sensor-chip integrated solution. In the case of medical disposable sensors this was true even for a very large volume production, in excess of 5 million units/year. This confirms that a hybrid technology is a very good solution for many silicon micromachining applications, and may be expected to be used in many new products.

It should be noted, that many advantages of hybrid technology result from the special packaging needs of MEMS, which do not exist in electronic only IC business.

121.9 Conclusions

In IC industry, a higher level of integration enabled reduction of the number of components in existing designs, immediately bringing economic benefits. The size of the market was large enough to justify a quick return on investment for a higher level of integration.

MEMS industry is smaller than the IC industry, and much more fragmented. It is missing many high volume applications. For years, there were only two multimillion unit/year applications: disposable blood pressure and manifold automotive pressure. Recently the airbag acceleration sensors emerged joining the multimillion unit/year club. These applications started to attract single chip designs, however their cost was comparable or higher to the hybrid designs.

Sensor market is in a process of very rapid growth. Over the next decade, pressure sensor market is expected to grow from current (1995) 50 million units to about 150 million. Acceleration sensor market will grow from current 12 million to the expected 40 million units. These volumes justify the development of fully integrated devices. The largest market will be the automotive segment, clearly driving the progress in sensor technology. First fully integrated devices have already entered the market place. In the automotive volumes, these sensors should be more cost effective than the hybrid technologies.

While several different IC technologies offer a possibility of the on-sensor-chip integration, the most attractive one seems to be the CMOS technology used for high volume digital integrated circuits. If the sensor circuit implementation is done in this technology (Restegar, 1994), then it could benefit from the continuous transistor size reduction already scheduled for the next decade or two, thus continuously reducing the unit cost.

To take full advantage of the size reduction, new sensor technologies should be explored. These technologies should minimize the number of overall masking steps as well as the resulting chip size. One of the re-emerging options is utilization of the junction based sensors. These type devices (Wlodarski, 1972; Jayaraman et al., 1967; Wartman, 1964) may utilize the same transistors that are used for signal processing, thus potentially eliminating several masking steps unique to piezoresistors. Smaller die sizes will require thinner mechanical structures, which may lead to broader utilization of resonating sensing technology (Petersen et al., 1991).

The question "Integrate or Not to Integrate Sensor Electronics" seems to be clearly evolving into "How to Integrate Sensor and MEMS Electronics for Automotive Applications."

Acknowledgments

Author would like to acknowledge contribution from Gerry Smith, Staff Scientist at Lucas Advanced Engineering Centre, England, to the section on Pros and Cons of integration (Bryzek and Smith, 1994).

References

Aaker, D. A. 1984. *Strategic Market Management,* p. 161, John Wiley & Sons, New York, NY.

AAMI Standard, 1986. *Standard for Interchangeability and Performance of Resistive Bridge Type Blood Pressure Transducers,* AAMI, Arlington, VA.

Bryzek, J. 1980. Personal communication between J. Bryzek and the founder of IC Transducers, Don Lynam.

Bryzek, J. 1981. Personal communication.

Bryzek, J. 1985. Modeling performance of piezoresistive pressure sensors, *Technical Digest 1985 Conf. Solid-State Sensors and Actuators,* June, Philadelphia, PA.

Bryzek, J. 1987. New generation of disposable blood pressure sensors, *Proc. Sensors Expo,* INHCO, Detroit, MI.

Bryzek, J. 1993. Evolution of smart transducers design, *Proc. Sensors Expo West,* March 2–4, San Jose, CA, March 2–4, 1993.

Bryzek, J. and Gee, D. 1994. Low-cost high-volume pressure sensors, *Proc. Sensors Expo 94,* September 18–20, Cleveland, OH.

Bryzek, J. and Smith, G. 1994. On-sensor-chip electronics integration: technical and economic considerations, *Proc. Silicon Sensor Realization Compatible with Microelectronic Circuit Fabrication,* 29–30 September, Nexus, Toulouse, France.

Bryzek, J. et. al., 1991. *Silicon Sensors and Microstructures,* NovaSensor, Silicon Valley, CA.

Frank, R. and McCulley, W. 1985. An update on the integration of silicon pressure sensors, *Professional Program Session Record 27, Wescon/85,* November 19–22, San Francisco, CA.

Goodenough, F. 1991. Airbags boom when IC accelerometer sees 50 G, *Electronic Design,* August 8, 1991.

Hedley, B. 1976. A fundamental approach to strategy development, *Long Range Planning,* December, p. 6.

Huijsing, H. 1994. to J. Bryzek. Personal communication.

Jayaraman, A. et. al. 1967. Effect of hydrostatic pressure on P-N junction characteristics and the pressure variation of the band gap, *J. Appl. Phys.,* 38(11).

Kress, H. J., Marek, J., Mast, M., Schatz, O., Muchow, J. 1995. Integrated pressure sensors with electronic trimming, *Automotive Engineering,* April.

Markus, K. W., Dhuler, V., and Cowen, A. 1994. MEMS Technology Application Center, *Proc. Sensors Expo 94 Conf.* September 20–22, Cleveland, OH.

McCulley, W, et. al. 1985. Fully integrated monolithic pressure sensor with on-chip calibration and compensation, *Proc. OEM Design Conf.* September 9–11, Philadelphia, PA.

Mokwa, W. 1995. Monolithic integrated physical and chemical

sensors in CMOS technology, *Proc. Sensors Expo 95,* May 16–18, Boston, MA.

Najafi, N. 1992. *A Generic Smart Sensing Utilizing A Multi-element Gas Analyzer,* Technical Report No. 209, August, The University of Michigan.

Nippon Denso 1989. to J. Mallon, Jr. Personal communication.

Obermeier, E. et al. Smart pressure sensor with on-chip calibration and compensation capability. *Proc. Sensor Expo West 94,* February 8-10, 1994, Anaheim, CA.

Petersen, K., et. al. 1991. Resonant beam pressure sensor fabricated with silicon fusion bonding, *Digest of Technical Papers, 1991 Int. Conf. Solid-State Sensors and Actuators Transducers '91.*

Rastegar, A. 1994. New Generation of Smart Sensor Signal Conditioning. *Proc. Sensors Expo 94,* September 20–22, Cleveland, OH.

Tabata, O. 1995. MEMS activities in Toyota and Japan, *Proc. Sensors Expo 95,* May 15–16, Boston, MA.

Wlodarski, W. 1972. The possibility of applying the zener diode as a transducer for measuring dynamic pressure, *IEEE Trans. Indu. Electr. and Control Instrumentation,* May, No. 2.

Wortman, J. J., et. al. 1964. Effect of mechanical stress on P-N junction device characteristics, *J. App. Phys.* 35(7).

Yun, W. and Howe, R. 1993. Σ-Δ modulator interfacing with silicon microsensors, *Proc. Sensors Expo West,* March 2–4, San Jose, CA.

122

Multisensor Fusion and Integration for Intelligent Systems

Ren C. Luo
North Carolina State University

Michael G. Kay
North Carolina State University

Kota Takahashi
The University of Electro-Communications

Hiro Yamasaki
Yokogawa Electric Corporation

Kazunori Umeda
Chuo University

Tamio Arai
University of Tokyo

Mark E. Kotanchek
Pennsylvania State University

James P. Helferty
Pennsylvania State University

W. Bosseau Murray
Pennsylvania State University

Charles Palmer
Pennsylvania State University

Zbigniew Korona
Northeastern University

Mieczyslaw M. Kokar
Northeastern University

Ryosuke Masuda
Tokai University

Michio Sasaki
Tokai University

Karl Kluge
University of Michigan

122.1 Introduction 1593

122.2 Issues and Approaches of Multisensor Fusion and Integration 1593
Introduction • The Role of Multisensor Integration and Fusion • Multisensor Integration • Multisensor Fusion • Conclusion

122.3 Audio-Visual Sensor Fusion System for Intelligent Sound Sensing 1609
Introduction • Cue Signal Method • (A + V + K) + A Fusion • Discussion • Conclusions

122.4 Industrial Vision System by Fusing Range Image and Intensity Image 1615
Introduction • Range Image Processing and Intensity Image Processing in Industrial Vision System • Algorithms in Range Image Processing • Development of Vision System with Range Image Sensor and Intensity Image Sensor • Fundamental Experiments of Comparing Range Image and Intensity Image • Bin-Picking of Multiple Cylinders as an Example of the Presented Vision System • Conclusion

122.5 Application of Data Fusion to Neonate Oxygenation Control 1622
Introduction • Physiology and Interactions • Monitor System and Sensors • Proposed Control System • Summary

122.6 Multiresolution Multisensor Target Identification 1627
Introduction • Discrete Wavelet Decomposition • Scale Sequential Identification • Fusion for Target Identification • Multisensor Target Identification Scenario • Results of Experiments • Future Research • Conclusion

122.7 Shaping Control of Plastic Object by Robot Hand with Sensor Fusion Processing 1632
Introduction • Problem of Shaping of Plastic Object • Characteristics of Clay and its Model • Sensor Signal Processing for Shaping • Experiment of Shaping Control • Conclusion

122.8 Multisensor System Integration for Autonomous Navigation Tasks 1639
Selecting Sensors for Autonomous Navigation • The Autonomous Land Vehicle Project • Classes of Integration Techniques • Interaction Between Perception Modules and Choice of Integration Techniques • Lessons and Open Issues

122.9 Future Trends for the Further Development in Multisensor Fusion and Integration 1657
Introduction • Uncertainty and Multisensor Systems • Areas Dependent Upon Advances in Multisensor Fusion and Integration • Future Research Multisensor Fusion and Integration Systems • Conclusion

0-8493-8343-9/97/$0.00+$.50

122.1 Introduction

Ren C. Luo

The synergistic use of multiple sensors by machines and systems is a major factor in enabling some measure of intelligence to be incorporated into their overall operation so that they can interact with and operate in an unstructured environment without the complete control of a human operator. The use of sensors in an intelligent system is an acknowledgment of the fact that it may not be possible or feasible for a system to know *a priori* the state of the outside world to a degree sufficient for its autonomous operation. The reasons a system may lack sufficient knowledge concerning the state of the outside world may be due either to the fact that the system is operating in a totally unknown environment or, while partial knowledge is available and is stored in some form of a world model, it may not be feasible to store large amounts of this knowledge; it may not even be possible in principle to know the state of the world *a priori* since it is dynamically changing and unforeseen events can occur. Sensors allow a system to learn the state of the world as needed and to continuously update its own model of the world. The motivation for using multiple sensors in a system can be considered as the response to the simple question: If a single sensor can increase the capability of a system, would the use of more sensors increase it even further? Over the past decade, a number of researchers have been exploring this question from both a theoretical perspective and by actually building multisensor machines and systems for use in a variety of areas of application. Typical of the applications that can benefit from the use of multiple sensors are automatic target recognition, autonomous mobile robot navigation, industrial tasks like assembly, inspection military command and control for battlefield management, target tracking and medical monitor/control applications.

In this topic area, we will first introduce the role of Multisensor Fusion and Integration (Section 122.2) and the advantages for integrating multiple sensors. The more detailed approaches to different aspects of the multisensor integration problem, the basic integration functions and the common themes among most methods of multisensor integration will then be discussed. In the multisensor fusion section, different methods that have been proposed for fusion of multiple sensors at the signal, pixel, feature and symbol level will be described. Some numerical examples are given to illustrate many of the fusion methods.

A number of case studies and practical applications using multisensor fusion and integration technologies are included. The first example is to extract a target sound signal autonomously from multi-microphone signals corrupted by interference ambient noise, which is presented by Takahasi and Yamasaki (Section 122.3) by using an audio-visual sensor fusion strategy. The second example is to fuse the intensity image and range image of the task environment for detection of target parts and the measurement of pose (position and orientation) of the parts. This sensor fusion approach has provided a flexible vision system for industrial applications such as robot integrated bin-picking and assembly/disassembly tasks. The more detailed information is given by Umeda and Arai (Section 122.4). The third example applies multisensor data fusion concepts to automate inspired neonate oxygen control. Such a control system would monitor the neonate state to facilitate a real-time supply of oxygen appropriate to the patient needs as well as recognize sensor failures and dangerous situations. Thus, the technologies of tactical real-time data fusion are applicable. The more detailed description is given by Kotanchek et al. (Section 122.5). The fourth example is to apply the multisensor fusion concept in Automatic Target Recognition (ATR) problems against targets in natural and man-made clutter environments. The important problem is the fusion of data from multiple sensors (multicolor Infra-Red, visual, microwave, etc.) that provide information in different frequency bands and at different resolutions. The fusion of multi-resolution, multisensor data leads to an increase in the probability of correct identification without a significant increase in the number of computations. The more detailed description is given by Korona and Kokar (Section 122.6). The fifth example combines visual, force, and tactile sensor information for recognizing a situation of dextrous manipulation tasks, for example, shaping of a flexible and a plastic object such as clay to a given shape. This is particularly critical for industrial applications where a robot is used in the fields of chemical, biomedical and food processing industries, due to the need to handle plastic and fragile objects. The more detailed description of this subject is given by Masuda and Sasaki (Section 122.7).

Finally, a concluding section regarding future trends for the further development in multisensor fusion and integration is given by Luo (Section 122.9).

122.2 Issues and Approaches of Multisensor Fusion and Integration

Ren C. Luo and Michael G. Kay

Introduction

There are a number of different means of integrating the information provided by multiple sensors into the operation of a system. The most straightforward approach to multisensor integration is to let the information from each sensor serve as a separate input to the system controller. This approach may be the most appropriate if each sensor is providing information concerning completely different aspects of the environment. The major benefit gained through this approach is the increase in the extent of the environment able to be sensed. The only interaction between the sensors is indirect and based on the individual effect each sensor has on the controller. If there is some degree of overlap between the sensors concerning some aspect of the environment that they are able to sense, it may be possible for a sensor to directly influence the operation of another sensor so that the value of the combined information that the sensors provide is greater than the sum of the value of the information provided by each sensor separately. This synergistic effect from the multisensor integration can be achieved either by using the information from one sensor to provide cues or guide the operation of other sensors, or by actually combining or fusing the information from multiple sensors. The information from the sensors can be fused at a variety of levels of representation depending upon the needs of the system and the degree of similarity between the sensors. The major benefit gained through multisensor fusion is that the system can be provided with information of higher quality concerning, possibly, certain aspects of the environment that can not be directly sensed by any individual sensor operating independently.

A number of books are available that focus on different aspects of multisensor integration and fusion (e.g., Hager, 1990). Abidi and Gonzalez (1992) and Luo and Kay (1995) provide overviews of the general topic of multisensor integration and fusion and present selections of some of fundamental research in the area.

The Role of Multisensor Integration and Fusion

This section describes the role of multisensor integration and fusion in the operation of intelligent machines and systems. The role of multisensor integration and fusion can best be understood with reference to the type of information that the integrated multiple sensors can uniquely provide the system. The potential advantages gained through the synergistic use of this multisensory information can be decomposed into a combination of four fundamental aspects: the redundancy, complementarity, timeliness, and cost of the information. Multisensor integration and the related notion of multisensor fusion are defined and distinguished. The potential advantages in integrating multiple sensors are discussed in terms of four fundamental aspects of the information provided by the sensors and then the possible problems associated with creating a general methodology for multisensor integration and fusion are discussed in terms of the methods used for handling the different sources of error or uncertainty.

Multisensor integration refers to the synergistic use of the information provided by multiple sensory devices to assist in the accomplishment of a task by a system. An additional distinction is made between multisensor integration and the more restricted notion of multisensor fusion. *Multisensor fusion* refers to any stage in the integration process where there is an actual combination (or fusion) of different sources of sensory information into one representational format. The fusion can take place at either the *signal, pixel, feature,* or *symbol* level of representation. The information to be fused may come from multiple sensory devices during a single period of time or from a single sensory device over an extended time period. The distinction between fusion and integration serves to separate the general system-level issues involved in the integration of multiple sensory devices at the architecture and control level, from the more specific mathematical and statistical issues involved in the actual fusion of sensory information.

Potential Advantages in Integrating Multiple Sensors

The purpose of external sensors is to provide a system with useful information concerning some features of interest in the system's environment. The potential advantages in integrating and/or fusing information from multiple sensors are that the information can be obtained more accurately, concerning features that are impossible to perceive with individual sensors, in less time, and at a lesser cost. These advantages correspond, respectively, to the notions of the redundancy, complementarity, timeliness, and cost of the information provided the system:

- *Redundant* information is provided from a group of sensors (or a single sensor over time) when each sensor is perceiving, possibly with a different fidelity, the same features in the environment. The integration or fusion of redundant information can reduce overall uncertainty and thus serve to increase the accuracy with which the features are perceived by the system. Multiple sensors providing redundant information can also serve to increase reliability in the case of sensor error or failure.
- *Complementary* information from multiple sensors allows features in the environment to be perceived that are impossible to perceive using just the information from each individual sensor operating separately. If the features to be perceived are considered dimensions in a space of features, then complementary information is provided when each sensor is only able to provide information concerning a subset of features that form a subspace in the feature space, that is, each sensor can be said to perceive features that are independent of the features perceived by the other sensors; conversely, the dependent features perceived by sensors providing redundant information would form a basis in the feature space.
- *More timely* information, as compared to the speed at which it could be provided by a single sensor, may be provided by multiple sensors due to either the actual speed of operation of each sensor, or the processing parallelism that may be possible to achieve as part of the integration process.
- *Less costly* information, in the context of a system with multiple sensors, is information obtained at a lesser cost when compared to the equivalent information that could be obtained from a single sensor. Unless the information provided by the single sensor is being used for additional functions in the system, the total cost of the single sensor should be compared to the total cost of the integrated multisensor system.

The role of multisensor integration and fusion in the overall operation of a system can be defined as the degree to which each of these four aspects is present in the information provided by the sensors to the system. Redundant information can usually be fused at a lower level of representation compared to complementary information because it can more easily be made commensurate. Complementary information is usually either fused at a symbolic level of representation, or provided directly to different parts of the system without being fused. While in most cases the advantages gained through the use of redundant, complementary, or more timely information in a system are related to technological benefits, in multisensor target tracking fused information is sometimes used in a distributed network of target tracking sensors just to reduce the bandwidth (and cost) required for communication between groups of sensors in the network.

An Object Recognition Example

Figure 122.1 illustrates the distinction between complementary and redundant information in the task of object recognition. Four objects are shown in Figure 122.1a. They are distinguished by the two independent features shape and temperature. Sensors 1 and 2 provide redundant information concerning the shape of an object, and Sensor 3 provides information concerning its temperature. Figures 122.1b and c show hypothetical frequency distributions for both "square" and "round" objects, representing each sensor's historical (i.e., tested) responses to such objects. The bottom axes of both figures represent the range of possible sensor readings. The output values x_1 and x_2 correspond to some numerical "degree of squareness or roundness" of the object as determined by each sensor, respectively. Because Sensors 1 and 2 are not able to detect the temperature of an object, objects *A* and *C* (as well as *B* and *D*) can not be distinguished. The dark portion of the axis in each figure corresponds to the range of output values where there is uncertainty as to the shape of the object being detected. The dashed line in each figure corresponds to the point at which, depending on the output value, objects can be distinguished in terms of a feature. Figure 122.1d is the frequency distribution resulting from the fusion of x_1 and x_2. Without specifying a particular method of fusion, it is usually true that the distribution corresponding to the fusion of redundant information would have less dispersion than its component distributions. Under very general assumptions, a plausibility argument can be made that the relative probability of the fusion process not reducing the uncertainty is zero (Richardson and Marsh, 1988). The uncertainty in Figure 122.1d is shown as approximately half that of Figures 122.1b and c. In Figure 122.1e, complementary information from Sensor 3 concerning the independent feature temperature is fused with the shape information from Sensors 1 and 2 shown in Figure 122.1d. As a result of the fusion of this additional feature, it is now possible to discriminate between all four objects. This increase in discrimination ability is one of the advantages resulting from the fusion of complementary information. As mentioned above, the information resulting from this second fusion could be at a higher representational level (e.g., the result of the first fusion, $x_{1,2}$, may still be a numerical value, while the result of the second, $x_{1,2,3}$, could be a symbol representing one of the four possible objects).

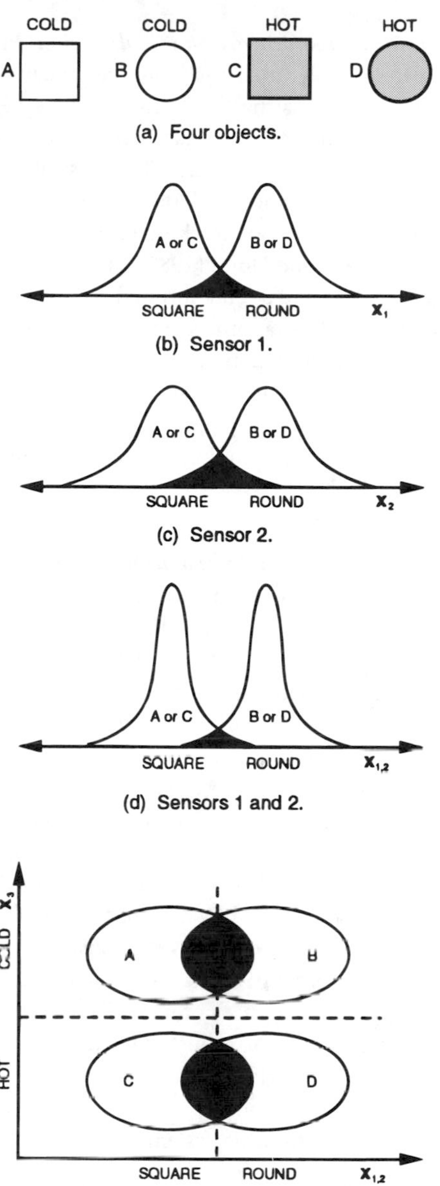

Figure 122.1 The discrimination of four different objects using redundant and complementary information from three sensors. (a) Four objects (*A, B, C,* and *D*) distinguished by the features "shape" (square vs. round) and "temperature" (hot vs. cold). (b) 2-D distributions from Sensor 1 (shape). (c) Sensor 2 (shape). (d) 2-D distributions resulting from fusion of redundant shape information from Sensors 1 and 2. (e) 3-D distributions resulting from fusion of complementary information from Sensors 1 and 2 (shape), and Sensor 3 (temperature). (*Source*: Abidi, M. A. and Gonzalez, R. C., eds. 1992. *Data Fusion in Robotics and Machine Intelligence*, pp. 7–135. Academic Press, Boston. With permission.)

Possible Problems

Many of the possible problems associated with creating a general methodology for multisensor integration and fusion, as well as developing the actual systems that use multiple sensors, center around the methods used for modeling the error or uncertainty in the integration and fusion process, the sensory information, and the operation of the overall system including the sensors. For the potential advantages in integrating multiple sensors to be realized, solutions to these problems will have to be found that are both practical and theoretically sound.

Error in the Integration and Fusion Process. The major problem in integrating and fusing redundant information from multiple sensors is that of "registration"—the determination that the information from each sensor is referring to the same features in the environment. The registration problem is termed the correspondence and data association problem in stereo vision and multitarget tracking research, respectively. Lee and Van Vleet (1988) and Holm (1987) have studied the registration errors between radar and infrared sensors. Lee and Van Vleet have presented an approach that is able to both estimate and minimize the registration error, and Holm has developed a method that is able to autonomously compensate for registration errors in both the total scene as perceived by each sensor ("macroregistration"), and the individual objects in the scene ("microregistration").

Error in Sensory Information. The error in sensory information is usually assumed to be caused by a random noise process that can be adequately modeled as a probability distribution. The noise is usually assumed not to be correlated in space or time (i.e., white), Gaussian, and independent. The major reasons that these assumptions are made is that they enable a variety of fusion techniques to be used that have tractable mathematics and yield useful results in many applications. If the noise is correlated in time (e.g., gyroscope error) it is still sometimes possible to retain the whiteness assumption through the use of a shaping filter (Maybeck, 1979). The Gaussian assumption can only be justified if the noise is caused by a number of small independent sources. In many fusion techniques the consistency of the sensor measurements is increased by first eliminating spurious sensor measurements so that they are not included in the fusion process. Many of the techniques of robust statistics (Huber, 1981) can be used to eliminate spurious measurements. The independence assumption is usually reasonable so long as the noise sources do not originate from within the system.

Error in System Operation. When error occurs during operation due to possible coupling effects between components of a system, it may still be possible to make the assumption that the sensor measurements are independent if the error, after calibration, is incorporated into the system model through the addition of an extra state variable (Maybeck, 1979). In well-known environments the calibration of multiple sensors will usually not be a difficult problem, but when multisensor systems are used in unknown environments, it may not be possible to calibrate the sensors. Possible solutions to this problem may require the creation of detailed knowledge bases for each type of sensor so that a system can autonomously calibrate itself. One other important feature required of any intelligent multisensor system is the ability to recognize and recover from sensor failure.

Multisensor Integration

This section presents approaches to different aspects of the multisensor integration problem discussed in the previous two sections. The basic integration functions are first described and then common themes among most methods of multisensor integration are discussed.

The Basic Integration Functions

Although the process of multisensor integration can take many different forms depending on the particular needs and design of the overall system, certain basic functions are common to most implementations. The diagram shown in Figure 122.2 represents multisensor integration as being a composite of these basic functions. A group of n sensors provide input to the integration process. In order for the data from each sensor to be used for integration it must first be effectively modeled. A *sensor model* represents the uncertainty and error in the data from each sensor and provides a measure of its quality that can be used by the subsequent integration functions. A common assumption is that the uncertainty in the sensory data can be adequately modeled as a Gaussian distribution. After the data from each sensor has been modeled it can be integrated into the operation of the system in accord with three different types of *sensory processing:* fusion, separate operation, and guiding or cueing. The data from Sensors 1 and 2 are shown in the figure as being fused. Prior to its fusion, the data from each sensor must be made commensurate. *Sensor registration* refers to any of the means (e.g., geometrical transformations) used to make the data from each sensor commensurate in both its spatial and temporal dimensions, that is, that the data refer to the same location in the environment over the same period of time. The different types of possible sensor data fusion (i.e., fusion at the signal, pixel, feature, and symbol

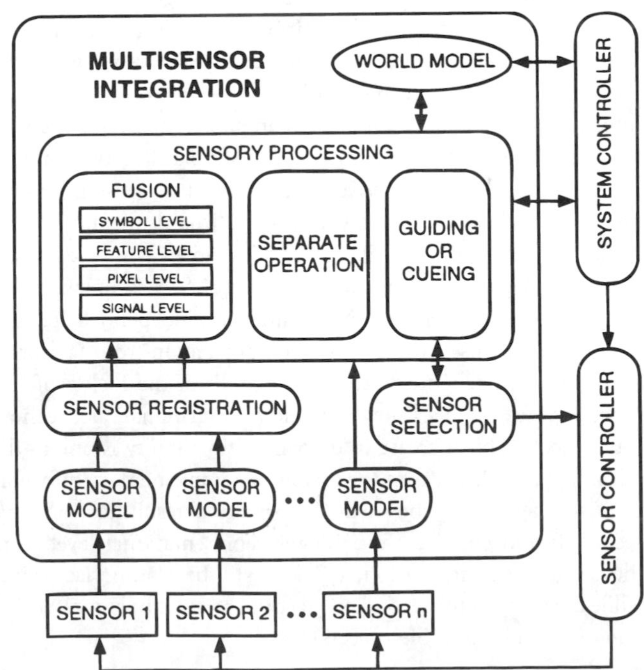

Figure 122.2 Functional diagram of multisensor integration and fusion in the operation of a system. (*Source*: Abidi, M. A. and Gonzalez, R. C., eds. 1992. *Data Fusion in Robotics and Machine Intelligence*, pp. 7–135. Academic Press, Boston. With permission.)

levels) are described in Section 122.4. If the data provided by a sensor is significantly different from that provided by any other sensors in the system, its influence on the operation of the other sensors may be indirect, that is, the *separate operation* of such a sensor will influence the other sensors indirectly through the effects the sensor has on the system controller and the world model. A *guiding or cueing* type of sensory processing refers to the situation where the data from one sensor is used to guide or cue the operation other sensors. A typical example of this type of multisensor integration is found in many robotics applications where visual information is used to guide the operation of a tactile array mounted on the end of a manipulator.

The results of the sensory processing function serve as inputs to the world model. A *world model* is used to store information concerning the state of the environment the system is operating in. A world model can include both a priori information and recently acquired sensory information. High-level reasoning processes can use the world model to make inferences that can be used to direct the subsequent processing of the sensory information and the operation of the system controller. Depending on the needs of a particular application, information stored in the world model can take many different forms, for example, in object recognition tasks the world model might contain just the representations of the objects the system is able to recognize, while in mobile robot navigation tasks the world model might contain the complete representation of the robot's local environment (e.g., the objects in the environment as well as local terrain features). The majority of the research related to the development of multisensor world models has been within the context of the development of suitable high-level representations for multisensor mobile robot navigation and control. The last multisensor integration function, *sensor selection,* refers to any means used to select or allocate the particular group of sensors to be used by the system. The selection process may take place during the initial design of the system or during its actual operation. When selection takes place during operation it can be used to determine the most appropriate sensor or group of sensors to use to guide the operation of other sensors in response to changing environmental or system conditions, for example, sensor failure.

Common Themes in Integration

The frameworks and control structures used for multisensor integration can be distinguished by the degree to which they enable the notions of modularity, hierarchical structures, and adaptability to be efficiently incorporated into the integration process. The means by which multiple sensors are integrated into the operation of an intelligent machine or system are usually a major factor in the overall design of the system. The specific capabilities of the individual sensors and the particular form of the information they provide will have a major influence on the design of the overall architecture of the system. These factors, together with the requirements of the particular tasks the system is meant to perform, make it difficult to define any specific general-purpose methods and techniques that encompass all of the different aspects of multisensor integration. Instead, what has emerged from the work of many researchers is a number of different paradigms, frameworks, and control structures for integration that have proved to be particularly useful in the design of multisensor systems.

Many of the paradigms, frameworks, and control structures used for multisensor integration have been adapted with little or no modification from similar high-level constructs used in systems analysis, computer science, control theory, and AI. In fact, much of multisensor integration research can be viewed as the particular application of a wide range of fundamental systems design principles. Common themes among these constructs that have particular importance for multisensor integration are the notions of modularity, hierarchical structures, and adaptability. In a manner similar to structured programming, *modularity* in the design of the functions needed for integration can reduce the complexity of the overall integration process and can increase its flexibility by allowing many of the integration functions to be designed to be independent of the particular sensors being used; modularity in the operation of the integration functions enables much of the processing to be distributed across the system. The object-oriented programming paradigm and the distributed blackboard control structure are two constructs that are especially useful in promoting modularity for multisensor integration. *Hierarchical structures* are useful in allowing for the efficient representation of the different forms, levels, and resolutions of the information used for sensory processing and control. *Adaptability* in the integration process can be an efficient means of handling the error and uncertainty inherent in the integration of multiple sensors. The use of the artificial neural network formalism allows adaptability to be directly incorporated into the integration process.

A paradigm for multisensor integration is more abstract than a framework, and can be thought of as the inspiration behind the development of more concrete frameworks for integration. A framework, in contrast to a paradigm, typically includes specifications as to the particular form of processing to be used for integration. A single paradigm (e.g., sensory processing using artificial neural networks) may give rise to a variety of different frameworks (e.g., multilayer Perceptrons and associative memories). A number of paradigms and frameworks for integration are described in Luo and Kay (1989), including the influential "logical sensors" paradigm proposed by Henderson and Shilcrat (1984).

Sensor selection is an integration function that can enable a multisensor system to select the most appropriate configuration of sensors (or sensing strategy) from among the sensors available to the system. In order for selection to take place, some type of sensor performance criteria need to be established. In many cases the criteria require that the operation of the sensors be modeled adequately enough so that a cost value can be assigned to measure their performance. Two different approaches to the selection of the type, number, and configuration of sensors to be used in a system can be distinguished: *preselection* during design or initialization, and *real-time selection* in response to changing environmental or system conditions. A number of sensor selection strategies are described in Luo and Kay (1989).

Multisensor Fusion

This section describes different methods that have been proposed for multisensor fusion at the symbol, pixel, feature, and symbol level. Numerical examples are given to illustrate many of the fusion methods.

The fusion of the data or information from multiple sensors or a single sensor over time can take place at different levels of representation (sensory information can be considered data from a sensor that has been given a semantic content through processing and/or the particular context in which it was acquired). As shown in Figure 122.2, a useful categorization is to consider multisensor fusion as taking place at the signal, pixel, feature, and symbol levels of representation. Most of the sensors typically used in practice provide data that can be fused at one or more of these levels. Although the multisensor integration functions of sensor registration and sensor modeling are shown in Figure 122.2 as being separate from multisensor fusion, most of the methods and techniques used for fusion make very strong assumptions, either explicitly or implied, concerning how the data from the different sensors is modeled and to what degree the data is in registration. A fusion method that may be sound in theory can be difficult to apply in practice if the assumed sensor model does not adequately describe the data from a real sensor, for example, the presence of outliers due to sensor failure in an assumed normal distribution of the sensory data can render the fused data useless, or the degree of assumed sensor registration may be impossible to achieve, for example, due to the limited resolution or accuracy of the motors used to control the sensors.

The different levels of multisensor fusion can be used to provide information to a system that can be used for a variety of purposes: Signal-level fusion can be used in real-time applications and can be considered as just an additional step in the overall processing of the signals, pixel-level fusion can be used to improve the performance of many image processing tasks like segmentation, and feature- and symbol-level fusion can be used to provide a system performing an object recognition task with additional features that can be used to increase its recognition capabilities. The different levels can be distinguished by the type of information they provide the system, how the sensory information is modeled, the degree of sensor registration required for fusion, the methods used for fusion, and the means by which the fusion process improves the "quality" of the information provided the system. A comparison of the different levels of fusion is given below and summarized in Table 122.1.

An Example of the Different Fusion Levels

Figure 122.3 provides an example of how the different levels of multisensor fusion can be used in the task of automatic target recognition. In the figure, five sensors are being used by the system to recognize a tank: two millimeter-wave radars (that could be operating at different frequencies), an infrared sensor, a camera providing visual information, and a radio signal detector that can identify characteristic emissions originating from the tank. The complementary characteristics of the information provided by this suite of sensors can enable the system to detect and recognize targets under a variety of different operating conditions, for example, the radars provide range information and their signals are less effected by atmospheric attenuation as compared to the infrared image, while the infrared sensor provides information of greater resolution than the radars and, unlike the camera, is able to operate at night.

The two radars are assumed to be synchronized and coaligned on a platform so that their data is in registration and can be fused at the signal level. The fused signal is shown in the figure as being sent both to the system, where it can be immediately used for the improved detection of targets, and as input to generate a range image of the target. The range image from the radars can then be fused at the pixel level with the intensity image provided by the infrared sensor located on the same platform. In most cases, an element from the range image can only be registered with a neighborhood of pixels from the infrared image because the differences in resolution between the millimeter-wave radars and the infrared sensor. The fused image is sent both to the system, where it can be immediately used to improve target segmentation, and as input to a feature extraction process. The features extracted from the pixel-level fused image can then be fused at the feature level with similar features extracted from visual image provided by the camera. The camera may be located on a different platform because the sensor registration requirements for feature-level fusion are less stringent than those for signal- and pixel-level fusion. The fused features are then sent both to the system, where they can be used to improve the accuracy in the measurement of the orientation or pose of the target, and as input features to an object recognition process. The output of the recognition process is a symbol, with an associated measure of its quality (0.7), indicating the presence of the tank. The symbol can then be fused at the symbol level with a similar symbol derived from the radio signal detector that also indicates the presence of the tank. The fused signal is then sent to the system for the final recognition of the tank. As shown in the figure, the measure of quality of the fused symbol (0.94) is greater than the measures of quality of either of the component symbols and represents the increase in the quality associated with the symbol as a result of the fusion, that is, the increase in the likelihood that the target is a tank.

The transformation from lower to higher levels of representation as the information moves up through the target recognition structure shown in Figure 122.3 is common in most multisensor integration processes. At the lowest level, raw sensory data are transformed into information in the form of a signal. As a result of a series of fusion steps, the signal is transformed into progressively more abstract numeric and symbolic representations. This "signals-to-symbols" phenomenon is also common in computational vision (Fischler and Firschein, 1987) and AI (Chandrasekaran and Goel, 1988).

Signal-Level Methods

Signal-level fusion refers to the combination of signals of a group of sensors to provide a signal that is usually of the same form as the original signals but of greater quality. The signals from the sensors can be modeled as random variables corrupted

Table 122.1 Comparison of Fusion Levels

Characteristics	Signal Level	Pixel Level	Feature Level	Symbol Level
Type of sensory information	Single- or multi-dimensional signals	Multiple images	Features extracted from signals and images	Symbol representing decision
Representation level of information	Low	Low to medium	Medium	High
Model of sensory information	Random variable corrupted by uncorrelated noise	Stochastic process on image or pixels with multidimensional attributes	Non-invariant geometrical form, orientation, position, and temporal extent of features	Symbol with associated uncertainty measure
Degree of registration:				
spatial	High	High	Medium	Low
temporal	High	Medium	Medium	low
Means of registration:				
spatial	Sensor coalignment	Sensor coalignment or shared optics	Geometrical transformations	Spatial attributes of symbol if necessary
temporal	Synchronization or estimation	Synchronization	Synchronization	Temporal attributes of symbol if necessary
Fusion method	Signal estimation	Image estimation or pixel attribute combination	Geometrical and temporal correspondence, and feature attribute combination	Logical and statistical inference
Improvement due to fusion	Reduction in expected variance	Increase in performance of image processing tasks	Reduced processing, increased feature measurement accuracy, and value of additional features	Increase in truth or probability values

(*Source*: Abidi, M. A. and Gonzalez, R. C., eds. 1992. *Data Fusion in Robotics and Machine Intelligence*, pp. 7–135. Academic Press, Boston. With permission.)

by uncorrelated noise, with the fusion process considered as an estimation procedure. As compared to the other types of fusion, signal-level fusion requires the greatest degree of registration between the sensory information. If multiple sensors are used for signal-level fusion their signals must be in temporal as well as spatial registration. If the signals from the sensors are not synchronized they can be put into temporal registration by estimating their values at common points of time. The signals can be registered spatially by having the sensors coaligned on the same platform. Signal-level fusion is usually not feasible if the sensors are distributed on different platforms due to registration difficulties and bandwidth limitations involved in communicating the signals between the platforms. The most common means of measuring the improvement in quality is the reduction in the expected variance of the fused signal (see, e.g., Figure 122.1d). One means of implementing signal level fusion is by taking a weighted average of the composite signals, where the weights are based on the estimated variances of the signals.

Optimal signal-level fusion methods can be developed if certain assumptions concerning the nature of the sensory information are satisfied. The most common assumptions include the use of a measurement model for the information from each sensor that includes a statistically independent additive Gaussian error or noise term (e.g., location data), and an assumption of statistical independence between the error terms for each sensor. Many of the differences in signal-level fusion methods center on the particular techniques (e.g., calibration, thresholding) used for transforming raw sensory data into a form so that the above assumptions become valid and a mathematically tractable fusion method can result. An excellent introduction to the conceptual problems inherent in any signal-level fusion method based on these common assumptions has been provided by Richardson and Marsh (1988). Their paper provides a proof that the inclusion of additional redundant sensory information almost always improves the performance of any signal level fusion method that is based on optimal estimation.

Weighted Average. One of the simplest and most intuitive methods of signal-level fusion is to take a weighted average of redundant information provided by a group of sensors and use this as the fused value. While this method allows for the real-time processing of dynamic low-level data, in most cases the Kalman filter is preferred because it provides a method that is nearly equal in processing requirements and, in contrast to a weighted average, results in an estimate for the fused data that is optimal in a statistical sense.

The weighted average of n sensor measurements x_i with weights $0 < w_i < 1$ is

$$\bar{x} = \sum_{i=1}^{n} w_i x_i$$

where $\Sigma_i w_i = 1$ and $w_i = 0$ if x_i is not within some specified thresholds. The weights can be used to account for the differences in accuracy between sensors and a moving average can be used to fuse together a sequence of measurements from a single sensor so that the more recent measurements are given a greater weight.

Kalman Filter. The Kalman filter (Maybeck, 1979) is used in a number of multisensor systems when it is necessary to fuse dynamic low-level redundant data in real time. The filter

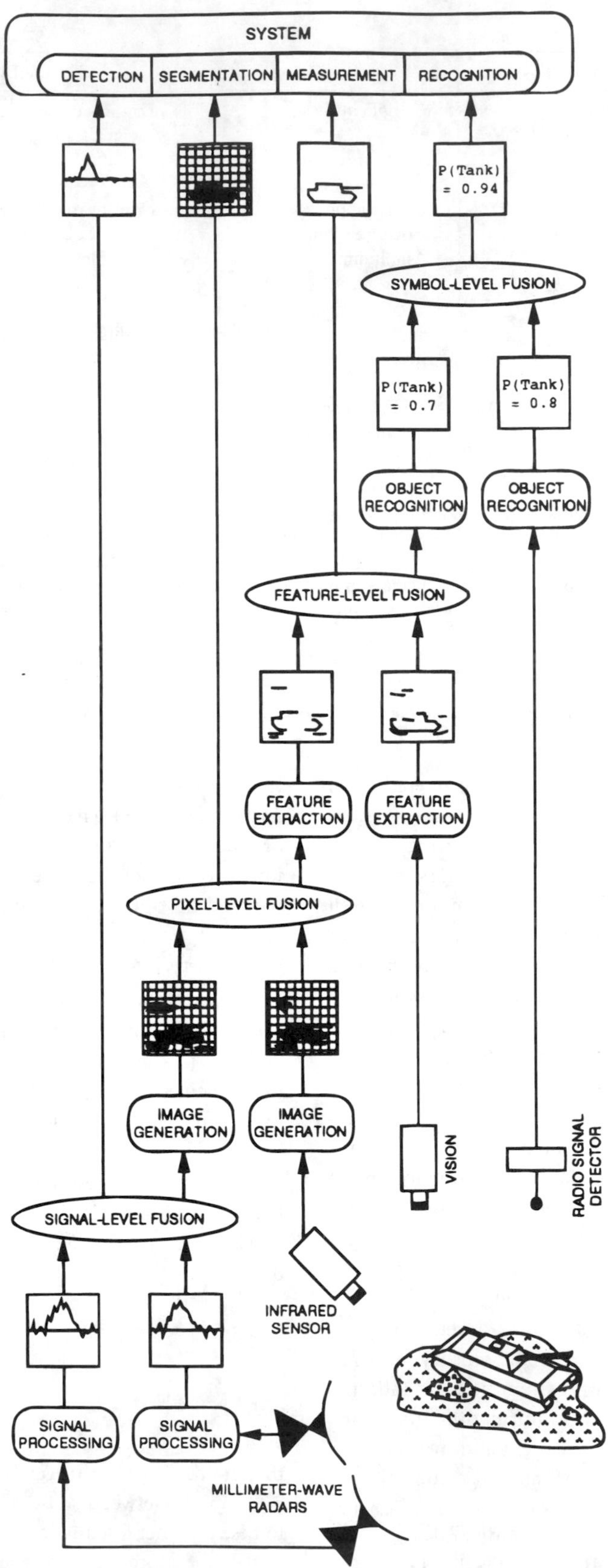

Figure 122.3 Possible uses of signal-, pixel-, feature-, and symbol-level fusion in the automatic recognition of a tank. (*Source*: Abidi, M. A. and Gonzalez, R. C., eds. 1992. *Data Fusion in Robotics and Machine Intelligence*, pp. 7–135. Academic Press, Boston. With permission.)

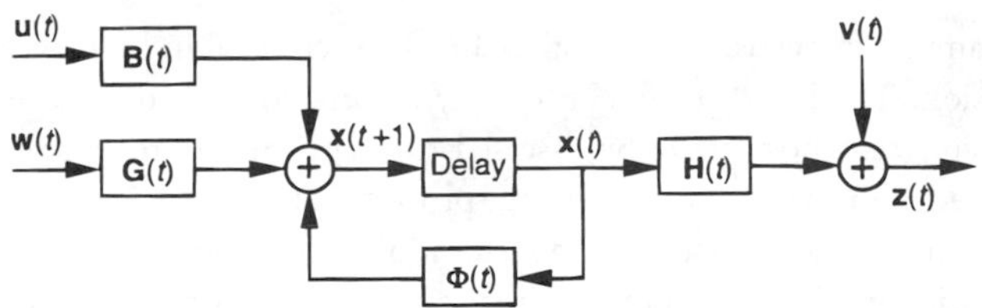

Figure 122.4 Kalman filter block diagram. (*Source*: Abidi, M. A. and Gonzalez, R. C., eds. 1992. *Data Fusion in Robotics and Machine Intelligence*, pp. 7–135. Academic Press, Boston. With permission.)

uses the statistical characteristics of a measurement model to recursively determine estimates for the fused data that are optimal in a statistical sense. If the system can be described with a linear model and both the system and sensor error can be modeled as white Gaussian noise, the Kalman filter will provide unique statistically optimal estimates for the fused data. The recursive nature of the filter makes it appropriate for use in systems without large data storage capabilities. Examples of the use of the filter for multisensor fusion include: object recognition using sequences of images from a sensor, robot navigation, multitarget tracking, inertial navigation, and remote sensing. In some of these applications the "U-D (unit upper triangular and diagonal matrix) covariance factorization filter" or the "extended Kalman filter" is used in place of the conventional Kalman filter if, respectively, numerical instability or the assumption of approximate linearity for the system model presents potential problems. An "adaptive Kalman filter" can be used if the parameters of the filter are not initially known.

The measurements from a group of n sensors can be fused together using a Kalman filter to provide both an estimate of the current state of a system and a prediction of the future state of the system. The state being estimated may, for example, correspond to the current location of a mobile robot, the position and velocity of an object in the environment, features extracted from sensory data (e.g., edges in an image), or to the actual measurements themselves. Given a system represented as a linear discrete Markov process, the "state-space model"

$$\mathbf{x}(t+1) = \mathbf{\Phi}(t)\mathbf{x}(t) + \mathbf{B}(t)\mathbf{u}(t) + \mathbf{G}(t)\mathbf{w}(t$$

and the "measurement model"

$$\mathbf{z}(t) = \mathbf{H}(t\mathbf{x}(t) + \mathbf{v}(t)$$

can be used to describe the system (see Fig. 122.4), where $\mathbf{x} : m$ state vector; $\mathbf{\Phi} : m \infty m$ state transition matrix; $\mathbf{B} : m \infty p$ input transmission matrix; $\mathbf{u} : p$ input vector (e.g., position of sensor platform); $\mathbf{G} : m \infty q$ process noise transmission matrix; $\mathbf{w} : q$ process noise vector; $\mathbf{z} : n$ measurement vector; $\mathbf{H} : n \infty m$ measurement matrix; and $\mathbf{v} : n$ measurement noise vector.

The $\mathbf{w}$ and $\mathbf{v}$ are uncorrelated discrete-time zero-mean white Gaussian noise sequences with covariance kernels

$$E\{\mathbf{w}(t_i)\mathbf{w}^{\mathrm{T}}(t_j)\} = \mathbf{Q}(t_i)\boldsymbol{\delta}_{ij}$$

$$E\{\mathbf{v}(t_i)\mathbf{v}^{\mathrm{T}}(t_j)\} = \mathbf{R}(t_i)\boldsymbol{\delta}_{ij},$$

where $E\{\bullet\}$; denotes the expectation operator and δ_{ij} the Kronecker delta function.

When all of the parameters (the matrices $\mathbf{\Phi}$, **B**, **G**, **H**, **Q**, and **R**) of the models are known, the optimal filtering equations are:

$$(t/t = x(/t-1) + \mathbf{K}(t)[\mathbf{z}(t) - \mathbf{H}(t)x(t/t-1)]$$

$$X(t+1/t) = \mathbf{\Phi}(t)x(t/t) + \mathbf{B}(t)\mathbf{u}(t)$$

where $\mathbf{x}(t|t)$ is the estimate of $\mathbf{x}(t)$ based on the measurements $\{\mathbf{z}(0), \ldots, \mathbf{z}(t)\}$ and $\mathbf{x}(t+1|t)$ is the prediction of $\mathbf{x}(t+1)$ based on the measurements $\{\mathbf{z}(0), \ldots, \mathbf{z}(t)\}$. The $m \infty n$ matrix $\mathbf{K}$ is the "Kalman filter gain" and is defined as

$$\mathbf{K}(t) = \mathbf{P}(t/t-1)[\mathbf{H}^{\mathrm{T}}(t)\mathbf{P}(t/t-1)\mathbf{H}^{\mathrm{T}}(t) + \mathbf{R}(t)]^{-1},$$

where $\mathbf{P}(t|t-1) = E\{(\mathbf{x}(t) - \mathbf{x}(t|t-1))(\mathbf{x}(t) - \mathbf{x}(t|t-1)\}^{T|}$ is the $m \infty m$ conditional covariance matrix of the error in predicting $\mathbf{x}(t)$ and is determined using

$$\mathbf{P}(t+1/t) = \mathbf{\Phi}(t)\mathbf{P}(t/t\mathbf{\Phi}^{\mathrm{T}}(t) + \mathbf{G}(t)\mathbf{Q}(t)\mathbf{G}^{\mathrm{T}}(t,$$

where

$$\mathbf{P}(t/t) = \mathbf{P}(t/t-1) - \mathbf{K}(t(\mathbf{H}(t)\mathbf{P}(t/t-1).$$

The initial conditions for the recursion are given by $\mathbf{x}(0|0) = \mathbf{x}_0$ and $\mathbf{P}(0|0) = \mathbf{P}_0$.

The application of Kalman filtering for multisensor fusion can be illustrated using the object recognition example given in Multisensor Integration. Sensors 1 and 2, S_1 and S_2, provide redundant information relative to each other concerning the shape of the objects to be recognized. The state to be estimated is the shape x of an object and can be assumed to remain constant over time, that is, $x(t) = x$ for all t. The shape measurements z_1 and z_2 from S_1 and S_2, respectively, can be modeled as

$$z_1 = x + v \text{ and } z_2 = x + v_2$$

where v_1 and v_2 are independent zero-mean Gaussian random variables with variances σ_1^2 and σ_2^2, respectively. If the measurements from S_1 and S_2 are available simultaneously, batch processing can be used for fusion, where

$$\mathrm{z} = \begin{bmatrix} z_1 \\ z_2 \end{bmatrix} = \begin{bmatrix} 1 \\ 1 \end{bmatrix} x + \begin{bmatrix} v_{i1} \\ v_2 \end{bmatrix} = \mathbf{H}x + \mathbf{v}.$$

If the measurements are available sequentially, recursive processing can be used to update the estimate of x as new measurements become available. Assuming that the measurement from S_1 is available initially, $\mathbf{x}_0 = \mathbf{x}_0 = z_1$ and $\mathbf{P}_0 = P_0 = \sigma_1^2$ can be considered the a priori information available about x before the receipt of the measurement from S_2. When z_2 becomes available, the optimal estimate of x is given by

$$\begin{aligned}\underline{x} &= \underline{x}_0 + K[z_2 - H\underline{x}_0] \\ &= \underline{x}_0 + P_0 H^{\mathrm{T}}(HP_0 H^{\mathrm{T}} + R)^{-1}[z_2 - H\underline{x}_0] \\ &= z_1 + \sigma_1^2(\sigma_1^2 + \sigma_2^2)^{-1}[z_2 - z_1] \\ &= \frac{\sigma_2^2}{\sigma_1^2 + \sigma_2^2} z_1 + \frac{\sigma_1^2}{\sigma_1^2 + \sigma_2^2} z_2,\end{aligned}$$

where $\mathbf{R} = \sigma_2^2$.

The variances σ_1^2 and σ_2^2 in the estimate of x can be interpreted as providing a means of weighing each measurement z_1 and z^2 so that the measurement with the least variance is given the greatest weight in the fused estimate. The variance of the estimate is $\sigma_1^2\sigma_2^2/(\sigma_1^2 + \sigma_2^2)$, which is less than the variance of either measurement alone. The reduction in variance is shown in Figure 122.1d of Section 2 and represents the reduction in uncertainty due to the fusion of the measurements. x can be further updated as additional measurements become available from either sensor or other sources of information are made available.

Pixel-Level Methods

Pixel-level fusion can be used to increase the information content associated with each pixel in an image formed through a combination of multiple images, for example, the fusion of a range image with a two-dimensional intensity image adds depth information to each pixel in the intensity image that can be useful in the subsequent processing of the image. The different images to be fused can come from a single imaging sensor (e.g., a multispectral camera) or a group of sensors (e.g., stereo cameras). The fused image can be created either through the pixel-by-pixel fusion or through the fusion of associated local neighborhoods of pixels in each of the component images. The images to be fused can be modeled as a realization of a stochastic process defined across the image (e.g., a Markov random field), with the fusion process considered as an estimation procedure, or the information associated with each pixel in a component image can be considered as an additional dimension of the information associated with its corresponding pixel in the fused image (e.g., the two dimensions of depth and intensity associated with each pixel in a fused range-intensity image). Sensor registration is not a problem if either a single sensor is used or multiple sensors are used that provide images of the same resolution and share the same optics and mechanics (e.g., a laser radar operating at the same frequency as an infrared sensor and sharing the same optics and scanning mechanism). If the images to be fused are of different resolution, then a mapping needs to be specified between corresponding regions in the images. The sensors used for pixel-level fusion need to be accurately coaligned so that their images will be in spatial registration. This is usually achieved through locating the sensors on the same platform. The disparity between the locations of the sensors on the platform can be used as an important source of information in the fusion process, for example, to determine a depth value for each pixel in binocular fusion. The improvement in quality associated with pixel-level fusion can most easily be assessed through the improvements noted in the performance of image processing tasks (e.g., segmentation, feature extraction, and restoration) when the fused image is being used as compared to the use of their performance when only the individual component images are used.

The fusion of multisensor data at the pixel level can serve to increase the useful information content of an image so that more reliable segmentation can take place and more discriminating features can be extracted for further processing. Pixel-level fusion can take place at various levels of representation: the fusion of the raw signals from multiple sensors prior to their association with a specific pixel, the fusion of corresponding pixels in multiple registered images to form a composite or fused image, and the use of corresponding pixels or local groups of pixels in multiple registered images for segmentation and pixel-level feature extraction (e.g., an edge image). Fusion at the pixel level is useful in terms of total system processing requirements because use is made of the multisensor data prior to processing-intensive functions like feature matching, and can serve to increase overall performance in tasks like object recognition because the presence of certain substructures like edges in an image from one sensor usually indicates their presence in an image from another sufficiently similar sensor. Duane (1988) has reported better object classification performance using features derived from the pixel-level fusion of TV and forward-looking infrared images as compared to the combined use of features derived independently from each separate image.

In order for pixel-level fusion to be feasible, the data provided by each sensor must be able to be registered and, in most cases, must be sufficiently similar in terms of its resolution and information content. The most obvious candidates for pixel-level fusion include sequences of images from a single sensor and images from a group of identical sensors (e.g., stereo vision). Many of the sensors used for automatic target recognition make extensive use of pixel-level fusion. Although it is possible to use many of the general multisensor fusion methods for pixel-level fusion (e.g., Bayesian estimation) four methods are particularly useful for fusion at the pixel level: logical filters (Ajjimarangsee and Huntsberger, 1988), mathematical morphology (Giardina and Dougherty, 1988), image algebra (Ritter and Wilson, 1987), and simulated annealing (Geman and Geman, 1984; Wolberg and Pavlidis, 1985). What makes these four methods useful for pixel-level fusion is that each method facilitates highly parallel processing because, at most, only a local group of pixels are used to process each pixel, and each method can easily be used to process a wide variety of images from different types of sensors because no problem or sensor specific probability distributions for pixel values are required, thus alleviating the need for either assuming a particular distribution or estimating a distribution through supervised training (only very general assumptions concerning pixel statistics are needed in simulated annealing to characterize the Markov random field used to represent an image).

Feature-Level Methods

Feature-level fusion can be used both to increase the likelihood that a feature extracted from the information provided by a sensor actually corresponds to an important aspect of the environment and as a means of creating additional composite features for use by the system. Features provide for data abstraction. A "primary feature" is created through the attachment of some type of semantic meaning to the results of the processing of some spatial and/or temporal segment of sensory data, while a "composite feature" is created through a combination of existing features. Typical features extracted from an image and used for fusion include edges and regions of similar intensity or depth. When multiple sensors report similar features at the same location in the environment, the likelihood that the features are actually present can be increased and the accuracy with which they are measured can be improved; features that do not receive such support can be as spurious artifacts and eliminated. A feature created as a result of the fusion process may be either a composite of the component features (e.g., an edge that is composed of segments of edges detected by different sensors) or an entirely new type of feature that is composed of the attributes of its component features (e.g., a three-dimensional edge formed through the fusion of corresponding edges in the images provided by stereo cameras). The geometrical form, orientation, and position of a feature, together with its temporal extent, are the most important aspects of the feature that need to be represented so that it can be registered and fused with other features. In some cases, a feature can be made invariant to certain geometrical transformations (e.g., translation and rotation in an image plane) so that all of these aspects do not have to be explicitly represented.

The sensor registration requirements for feature level fusion are less stringent than those for signal- and pixel-level fusion, with the result that the sensors can distributed across different platforms. The geometric transformation of a feature can be used to bring it into registration with other features or with a world model. The improvement in quality associated with feature-level fusion can be measured through the reduction in processing requirements resulting from the elimination of spurious features, the increased accuracy in the measurement of a feature (used, e.g., to determine the pose of an object), and the increase in performance associated with the use of additional features created through fusion (e.g., increased object recognition capabilities).

Feature space mapping (Flachs et al., 1990), Gauss-Markov estimation with constraints (Porrill, 1988), and the extended Kalman filter (Ayache and Faugeras, 1988, 1989) have been used for feature-level fusion. The use of the extended Kalman filter for image fusion allows for the efficient registration of sequences of visual maps so that they can be fused together at the feature level, and enables the overall uncertainty as to the location of objects in the environment to be reduced in the presence of environmental and sensor noise.

Symbol-Level Methods

Symbol-level fusion allows the information from multiple sensors to be effectively used together at the highest level of abstraction. Symbol-level fusion may be the only means by which sensory information can be fused if the sensors are very dissimilar or refer to different regions of the environment. The symbols used for fusion can originate either from the processing of the information provided by the sensors in the system, or through symbolic reasoning processes that may make use of *a priori* information from a world model or sources external to the system (e.g., intelligence reports indicating the likely presence of certain targets in the environment). A symbol derived from sensory information represents a decision that has been made concerning some aspect of the environment (symbol-level fusion is sometimes termed "decision-level fusion"). The decision is usually made by matching features derived from the sensory information to a model. The symbols used for fusion typically have associated with them a measure of the degree to which the sensory information matches the model. A single uncertainty measure is used to represent both the degree of mismatch and any of the inherent uncertainty in the sensory information provided by the sensors. The measure can be used to indicate the relative weight that a particular symbol should be given in the fusion process. Sensor registration is usually not explicitly considered in symbol-level fusion because the spatial and temporal extent of the sensory information upon which a symbol is based has already been explicitly considered in the generation of the symbol, for example, the underlying features upon which a group of symbols are based are already in registration. If the symbols to be fused are not in registration, spatial and temporal attributes can be associated with the symbols and used for their registration.

Different forms of logical and statistical inference can be used for symbol-level fusion (Garvey, 1987; Garvey, Lowrance, and Fischler, 1981). In logical inference the individual symbols to be fused represent terms in logical expressions and the uncertainty measures represent the truth values of the terms. In statistical inference the individual symbols to be fused are represented as conditional probability expressions and their uncertainty measures correspond to the probability measures associated with the expressions. The improvement in quality associated with symbol-level fusion is represented by the increase in the truth or probability values of the symbols created as a result of the inference process.

Bayesian Estimation. Bayesian estimation provides a formalism for multisensor fusion that allows sensory information to be combined according to the rules of probability theory. Uncertainty is represented in terms of conditional probabilities $P(Y|X)$, where $P(Y) = P(Y|X)$ if X remains constant. Each $P(Y|X)$ takes a value between 0 and 1, where 1 represents absolute belief in proposition Y given the information represented by proposition X and 0 represents absolute disbelief. Bayesian estimation is based on the theorem from basic probability theory known as "Bayes' rule":

$$P(Y|X) = \frac{P(X|Y)P(Y)}{P(X)},$$

where $P(Y|X)$, the "posterior probability," represents the belief accorded to the hypothesis Y given the information represented by X. The posterior probability is calculated by multiplying the "prior probability" associated with Y, $P(Y)$, by the "likelihood," $P(X|Y)$, of receiving X given that Y is true. The denominator $P(X)$ is a normalizing constant.

The redundant information from a group of n sensors, S_1 through S_n, can be fused together using the odds and likelihood ratio formulation of Bayes' rule. The information represented by X_i concerning Y from S_i is characterized by $P(X_i|Y)$ and the likelihood $P(X_i|-Y)$ given the negation of Y, or by the "likelihood ratio":

$$L(X_i|Y) = \frac{P(X_i|Y)}{P(X_i|\neg Y)}.$$

Defining the "prior odds" on Y as

$$O(Y) = \frac{P(Y)}{P(\neg Y)},$$

and assuming that the operation of each sensor is independent of the operation of the other sensors in the system, the "posterior odds" on Y given the information $X_1, \ldots, X_n$ from the n sensors are given by the product

$$O(Y|X_i, \ldots, X_n) = O(Y) \prod_{i=1}^{n} L(X_i|Y).$$

The posterior odds are related to the posterior probability by

$$P(Y|X_1, \ldots, X_n) = \frac{O(Y|X_1, \ldots, X_n)}{1 + O(Y|X_1, \ldots, X_n)}.$$

The above formulation can also be used to fuse together a sequence of information from a single sensor provided that the uncertainty of the information can be assumed to be independent over time.

The application of Bayesian estimation to multisensor fusion can be illustrated using the object recognition example in Multisensor Integration. Sensors 1 and 2, S_1 and S_2, provide redundant information relative to each other concerning the shape of the objects to be recognized. Let the propositions S and R represent the hypotheses that the object being sensed is square or round, respectively, and let S_1, R_1, S_2, and R_2 represent the shape indicated in the information provided by S_1 and S_2.

Given the information $P(S_1|S) = 0.82$ and $P(S_2|S) = 0.71$ from S_1 and S_2 concerning the hypothesis S, and assuming that square or round objects are equally likely to be encountered, that is, $P(S) = P(R) = 0.5$, the posterior odds on S given the fusion of the information from both sensors are

$$O(S|S_1, S_2) = \frac{P(S)}{P(\neg S)} \cdot \frac{P(S_1|S)}{P(S_1|\neg S)} \cdot \frac{P(S_2|S)}{P(S_2|\neg S)}$$

$$= \frac{0.5}{0.5} \cdot \frac{0.82}{0.18} \cdot \frac{0.71}{0.29} = 11.15,$$

which corresponds to a posterior probability of

$$P(S|S_1, S_2) = \frac{O(S|S_1, S_2)}{1 + O(S|S_1, S_2)} = \frac{11.15}{1 + 11.15} = 0.92.$$

In a similar manner, given the information $P(R_1|R) = 0.12$ and $P(R_2|R) = 0.14$, the posterior probability accorded the hypothesis R can be determined to be 0.02. The posterior probabilities of both hypotheses do not sum to unity in this example due to an assumed inherent uncertainty in the operation of S_1 and S_2 of 6 and 15 percent, respectively. If, for example, it is known *a priori* that only a third of the objects likely to be encountered are square, the posterior odds on S would be reduced by half and the odds on R would be double.

Dempster-Shafer Evidential Reasoning. Garvey, Lowrance, and Fischler (1981) introduced the possibility of using Dempster-Shafer evidential reasoning for multisensor fusion. The use of evidential reasoning for fusion allows each sensor to contribute information at its own level of detail, for example, one sensor may be able to provide information that can be used to distinguish individual objects, while the information from another sensor may only be able to distinguish classes of objects; the Bayesian approach, in contrast, would not be able to fuse the information from both sensors. Dempster-Shafer evidential reasoning (Zadeh, 1986; Shafer, 1976) is an extension to the Bayesian approach that makes explicit any lack of information concerning a proposition's probability by separating firm belief for the proposition from just its plausibility. In the Bayesian approach all propositions (e.g., objects in the environment) for which there is no information are assigned an equal a priori probability. When additional information from a sensor becomes available and the number of unknown propositions is large relative to the number of known propositions, an intuitively unsatisfying result of the Bayesian approach is that the probabilities of known propositions become unstable. In the Dempster-Shafer approach this is avoided by not assigning unknown propositions an a priori probability (unknown propositions are assigned instead to "ignorance"). Ignorance is reduced (i.e., probabilities are assigned to these propositions) only when supporting information becomes available.

In Dempster-Shafer evidential reasoning the set Θ, termed the "frame of discernment," is composed of mutually exclusive and exhaustive propositions termed "singletons." The level of detail represented by a singleton corresponds to the lowest level of information that is able to be discerned through the fusion of information from a group of sensors or other information sources, for example, a knowledge base. Given n singletons, the power set of Θ, denoted by 2^Θ, contains 2^n elements and is composed of all the subsets of Θ including Θ itself, the empty set ϕ, and each of the singletons. The elements of 2^Θ are termed propositions and each subset is composed of a disjunction of singletons. The set of propositions $\{A_j | A_j \in 2^\Theta\}$ for which a sensor is able to provide direct information are termed its "focal elements." For each sensor S_i, the function

$$m_i: \{A_j | A_j \in 2^{\Theta}\} \varnothing [0, 1],$$

termed a "basic probability assignment," maps a unit of probability mass or belief across the focal elements of S_i subject to the conditions

$$m_i(\phi) = 0$$

and

$$\sum_{A_j \in 2^{\Theta}} m_i(A_j) = 1.$$

Any probability mass not assigned to a proper subset of Θ is included in $m_i(\Theta)$ and is assumed to represent the residual uncertainty of S_i that is distributed in some unknown manner among its focal elements.

A "belief" or "support" function, defined for S_i as

$$bel_i(A) = \sum_{A_j \subseteq A} m_i(A_j),$$

is used to determine the lower probability or minimum likelihood of each proposition A. In a similar manner, "doubt," "plausibility," and "uncertainty" functions are defined as

$$dbt_i(A) = bel_i(A^C),$$

$$pls_i(A) = 1 - dbt_i(A),$$

and

$$u_i(A) = pls_i(A) - bel_i(A).$$

The degree of doubt in A is the degree of belief in the complement of A. The plausibility function determines the upper probability or maximum likelihood of A and represents the mass that is free to move to the belief of A as additional information becomes available. The uncertainty of A represents the mass that has not been assigned for or against belief in A. The Bayesian approach would correspond to the situation where $u_i(A) = 0$ for all $A \in 2^{\Theta}$. The Dempster-Shafer formalism allows for the representation of total ignorance concerning proposition A since $bel(A) = 0$ does not imply $dbt(A) > 0$, even though $dbt(A) = 1$ does imply $bel(A) = 0$. The interval $[bel(A), pls(A)]$ is termed a "belief interval" and represents, by its magnitude, now conclusive the information is for proposition A, for example, total ignorance concerning A is represented as $[0, 1]$, while $[0, 0]$ and $[1, 1]$ represent A as being false and true, respectively.

"Dempster's rule of combination" is used to fuse together the propositions X and Y from the two sensors S_i and S_j.

$$m_{i,j}(A) = \frac{\sum_{X \cap Y = A} m_i(X) m_j(Y)}{1 - \sum_{X \cap Y = \phi} m_i(X) m_j(Y)},$$

whenever $A \quad \phi$, and where $m_{i \oplus j}$ is the orthogonal sum $m_i \quad m_j$ and $X, Y \in 2^{\Theta}$. The denominator is a normalization factor that forces the new masses to sum to unity, and may be viewed as a measure of the degree of conflict or inconsistency in the information provided by S_i and S_j. If the factor is equal to 0 the sensors are completely inconsistent and the orthogonal sum operation is undefined. The combination rule narrows the set of propositions by distributing the total probability mass into smaller and smaller subsets, and can be used to find positive belief for singleton propositions that may be embedded in the complementary information (i.e., focal elements composed of disjunctions of singleton propositions) provided by a group of sensors.

The application of Dempster-Shafer evidential reasoning to multisensor fusion can be illustrated using the object recognition example given previously. Θ is composed of the four singleton propositions A, B, C, and D, corresponding to the four objects to be recognized. Each of the three sensors used to recognize the objects is only able to provide information to distinguish a particular class of objects, for example, square versus round objects. Sensors 1 and 2, S_1 and S_2, provide redundant information relative to each other concerning the shape of the objects, represented as the focal elements A/C (square) and B/D (round). The information from S_1 and S_2 is the same as that used to illustrate Bayesian estimation. Sensor 3, S_3, provides complementary information relative to S_1 and S_2 concerning the temperature of the objects, represented as the focal elements A/B (cold) and C/D (hot).

The mass assignments resulting from the fusion of the information from S_1 and S_2 using Dempster's rule are shown in Table 122.2. The probability mass assigned to each of the focal elements of the sensors reflects the difference in the sensors' accuracy indicated by the frequency distributions shown in Figure 122.1b and 122.1c; for example, given that the object being sensed is most likely square, the greater mass attributed to $m_1(A/C)$ as compared to $m_2(A/C)$ reflects S_1's greater accuracy as compared to S_2. The difference in mass attributed to the object possibly being round reflects the amount of overlap in the distributions for each shape class. The mass attributed to $m(\Theta)$ for each sensor reflects the amount by which the focal element masses have been reduced to account for the inherent uncertainty in the information provided by each sensor. The normalization factor is calculated as 1 minus the sum of the two k's in the table, or $1 - 0.2 = 0.8$. As a result of the fusion, the belief attributed to the object being square has increased from $bel_1(A/C) = 0.82$ and $bel_2(A/C) = 0.71$ to $bel_{1,2}(A/C) = 0.93475$ (the sum of the $m_{1,2}(A/C)$'s in the table). This increase is also indicated by the narrower distribution shown for the fused information in Figure 122.3d.

Table 122.3 shows the mass assignments resulting from the

Table 122.2 Fusion Using Sensors 1 and 2

	$m_2(A \vee C) = 0.71$	$m_2(B \vee D) = 0.14$	$m_2(\Theta) = 0.15$
$m_1(A \vee C) = 0.82$	$m_{1,2}(A \vee C) = 0.72775$	$k = 0.1148$	$m_{1,2}(A \vee C) = 0.15375$
$m_1(B \vee D) = 0.12$	$k = 0.0852$	$m_{1,2}(B \vee D) = 0.021$	$m_1(B \vee D) = 0.0225$
$m_1(\Theta) = 0.06$	$m_{1,2}(A \vee C) = 0.05325$	$m_1(B \vee D) = 0.0105$	$m_{1,2}(\Theta) = 0.01125$

(*Source*: Abidi, M. A. and Gonzalez, R. C., eds. 1992. *Data Fusion in Robotics and Machine Intelligence*, pp. 7–135. Academic Press, Boston. With permission.)

Table 122.3 Fusion Using Sensors 1, 2, and 3

	$m_3(A \vee B) = 0.92$	$m_3(C \vee D) = 0.06$	$m_3(\Theta) = 0.02$
$m_{1,2}(A \vee C) = 0.93475$	$m_{1,2,3}(A) = 0.85997$	$m_{1,2,3}(C) = 0.056085$	$m_{1,2,3}(A \vee C) = 0.018695$
$m_{1,2}(B \vee D) = 0.054$	$m_{1,2,3}(B) = 0.04968$	$m_{1,2,3}(D) = 0.00324$	$m_{1,2,3}(B \vee D) = 0.00108$
$m_{1,2}(\Theta) = 0.01125$	$m_{1,2,3}(A \vee B) = 0.01035$	$m_{1,2,3}(C \vee D) = 0.000675$	$m_{1,2,3}(\Theta) = 0.000225$

(*Source*: Abidi, M. A. and Gonzalez, R. C., eds. 1992. *Data Fusion in Robotics and Machine Intelligence*, pp. 7–135. Academic Press, Boston. With permission.)

fusion of the combined information from sensors 1 and 2, $S_{1,2}$, with the focal elements of S_3. As a result of the fusion, positive belief can be attributed to the individual objects. The most likely object is *A*, as indicated by

$$bel_{1,2,3}(A) = m_{1,2,3}(A) = 0.85997$$

and

$$dbt_{1,2,3}(A) = m_{1,2,3}(B) + m_{1,2,3}(C) + m_{1,2,3}(D) + m_{1,2,3}(B/D) + m_{1,2,3}(C/D) = 0.11076.$$

For this conclusion is quite conclusive as indicated by a small uncertainty and a narrow belief interval for *A:*

$$u_{1,2,3}(A) = pls_{1,2,3}(A) - bel_{1,2,3}(A) = 0.02927,$$

$$[bel_{1,2,3}(A), pls_{1,2,3}(A)] = [0.85997, 0.88924],$$

where the plausibility of *A* is

$$pls_{1,2,3}(A) = 1 - dbt_{1,2,3}(A) = 0.88924.$$

The least likely object is also quite conclusively *D*. If additional information becomes available, for example, that the object was stored inside a refrigerated room, it can easily be combined with the previous evidence to possibly increase the conclusiveness of the recognition process.

Production Rules with Confidence Factors. Production rules can be used to symbolically represent the relation between sensory information and an attribute that can be inferred from the information. Production rules that are not directly based on sensory information can be easily combined with sensory information-based rules as part of an overall high-level reasoning system, for example, expert systems. The use of production rules promotes modularity in the multisensor integration process because additional sensors can be added to the system without requiring the modification of existing rules.

The production rules used for multisensor fusion can be represented as the logical implication of a conclusion *Y* given a premise *X*, denoted as *if X then Y* or $X \varnothing Y$. The premise *X* may be composed of a single proposition or the conjunction, disjunction, or negation of a group of propositions. The inference process can proceed in either a forward or backward chaining manner: in "forward-chaining" inference, a premise is given and its implied conclusions are derived; in "backward-chaining" inference, a proposition is given as a goal to be proven given the known information. In forward-chaining inference, the fusion of sensory information takes place both through the implication of the conclusion of a single rule whose premise is composed of a conjunction or disjunction of information from different sensors, and through the assertion of a conclusion that is common to a group of rules.

Uncertainty is represented in a system using production rules through the association of a "certainty factor" (CF) with each proposition and rule. Each CF is a measure of belief or disbelief and takes a value -1 __ CF __ 1, where CF = 1 corresponds to absolute belief, CF = -1 to absolute disbelief, and CF = 0 to either a lack of information or an equal balance of belief and disbelief concerning a proposition. Uncertainty is propagated through the system using a "certainty factor calculus," for example, the EMYCIN calculus (Buchanan and Shortliffe, 1984).

Each proposition X and its associated CF is denoted as

$$X \text{ cf } (\text{CF}[X]),$$

where CF[*X*] is initially either known or assumed to be equal to 0. Given the set $\leftarrow$ of rules in a system, each rule $r_i \in \leftarrow$ and its associated CF is denoted as

$$r_i\text{: } X \varnothing Y \text{ cf } (\text{CF}_i[X, Y]).$$

The CF of the premise *X* in r_i can be defined as

$$\mathrm{CF}_i[X] = \begin{cases} \mathrm{CF}[X] & \text{if } X = X_1 \\ \min(\mathrm{CF}[X_1], \ldots, \mathrm{CF}[X_n]) & \text{if } X = X_1 \wedge \cdots \wedge X_n \\ \max(\mathrm{CF}[X_1], \ldots, \mathrm{CF}[X_n]) & \text{if } X ;eq X_1 \vee \cdots \vee X_n \\ -\mathrm{CF}[\neg X] & \text{else.} \end{cases}$$

where each X_i is a proposition in X and $-X$ is the negation of X. The CF of the conclusion Y in r_i can be determined using

$$\mathrm{CF}_i[Y] = \begin{cases} -\mathrm{CF}_i[X]\cdot \mathrm{CF}_i[X, Y] & \text{if both CFs} < 0 \\ \mathrm{CF}_i[X]\cdot \mathrm{CF}_i[X, Y] & \text{else.} \end{cases}$$

The $\mathrm{CF}_i[X, Y]$ for r_i can be thought of as the $\mathrm{CF}_i[Y]$ that would result if r_i is invoked and $\mathrm{CF}_i[X] = 1$.

If there is only one rule, r_Y, for which the unknown proposition Y is its conclusion, then $\mathrm{CF}[Y] = \mathrm{CF}_Y[Y]$. If there is more than one rule, then $\mathrm{CF}[Y]$ is determined by fusing together the $\mathrm{CF}_i[Y]$'s of all the r_i for which Y is their conclusion. Let

$$\leftarrow_Y = \{r_i: x \varnothing y \in \leftarrow |\mathrm{CF}_i[X]\_0 \text{ and } y = Y\}$$

be the set of rules with known premises and Y as their conclusion. Given $N = |\leftarrow_Y|$ such rules,

$$\mathrm{CF}[Y] = \mathrm{CF}[Y]_{j=N},$$

where, for every $r_i \in \leftarrow_Y$, $\mathrm{CF}[Y]_0 = 0$ and

$$\mathrm{CF}[Y]_j = \begin{cases} \mathrm{CF}[Y]_{j-1} + \mathrm{CF}_i[Y]\cdot(1 - \mathrm{CF}[Y]_{j-1}) & \text{both CFs} > 0 \\ \mathrm{CF}[Y]_{j-1} + \mathrm{CF}_i[Y]\cdot(1 + \mathrm{CF}[Y]_{j-1}) & \text{both CFs} < 0 \\ \dfrac{\mathrm{CF}[Y]_{j-1} + \mathrm{CF}_i[Y]}{1 - \min(\mathrm{CF}[Y]_{j-1}, \mathrm{CF}_i[Y])} & \text{else,} \end{cases}$$

for $j = 1$ to N.

The application of production rules with certainty factors to multisensor fusion can be illustrated using the object recognition example in Section 122.2. The information from the three sensors S_1, S_2, and S_3 is the same as that used in the illustrations of Bayesian estimation and Dempster-Shafer evidential reasoning. Let S_1 *cf* (0.87) and R_1 *cf* (−0.87) be the known propositions provided by S_1 concerning whether the objects being sensed are either square (S) or round (R), respectively. The two rules

$$r_1\text{: } S_1 \varnothing S \text{ cf } (0.94) \text{ and}$$

$$r_2\text{: } R_1 \varnothing R \text{ cf } (0.94),$$

account for an inherent uncertainty of 6 percent in the information provided by S_1. Using only S_1, the certainty that the object being sensed is square is S *cf* (0.82) and that it is round is R *cf* (−0.82). The information S_2 *cf* (0.84) and R_2 *cf* (−0.84) from S_2, together with the additional rules

$$r_3\text{: } S_2 \varnothing S \text{ cf } (0.85) \text{ and}$$

$$r_4\text{: } R_2 \varnothing R \text{ cf } (0.85)$$

can be fused with the redundant information from S_1 to increase the belief that the object is square to S *cf* (0.9478) and to increase the disbelief that it is round to R *cf* (−0.9478), where $\mathrm{CF}[S] = 0.82 + 0.71(1 - 0.82)$ and $\mathrm{CF}[R] = -0.82 - 0.71(1 - 0.82)$ corresponding to $\leftarrow_S = \{r_i, r_3\}$ and $\leftarrow_R = \{r_2, r_4\}$, respectively.

Let C_3 *cf* (0.94) and H_3 *cf* (−0.94) be the known propositions provided by S_3 concerning whether the objects are either cold (C) or hot (H), respectively. The two rules

$$r_5\text{: } C_3 \varnothing C \text{ cf } (0.98) \text{ and}$$

$$r_6\text{: } H_3 \varnothing H \text{ cf } (0.98)$$

account for the inherent uncertainty in S_3 and can be used together with the additional rules

$$r_7\text{: } S\_C \varnothing A \text{ cf } (1.0),$$

$$r_8\text{: } R\_C \varnothing A \text{ cf } (1.0),$$

$$r_9\text{: } S\_H \varnothing A \text{ cf } (1.0), \text{ and}$$

$$r_{10}\text{: } R\_H \varnothing A \text{ cf } (1.0),$$

to enable the information from S_3 to be fused with the complementary information from S_1 and S_2 to determine the certainty factors associated with the propositions A, B, C, and D, corresponding to the four possible types of objects. Having determined that C *cf* (0.92) and H *cf* (−0.92),

$$\begin{aligned} \mathrm{CF}[A] &= \mathrm{CF}_7[S\_C]\_\ \mathrm{CF}_7[S\_C, A] \\ &= \min(\mathrm{CF}[S]\_\mathrm{CF}[R])\_\ 1.0 \\ &= \min(0.9478, 0.92) = 0.92. \end{aligned}$$

In a similar manner, $\mathrm{CF}[B]$, $\mathrm{CF}[C]$, and $\mathrm{CF}[D]$ can be determined to be −0.9478, −0.92, and −0.9478, respectively.

The definition of a certainty factor calculus to use with production rules for multisensor fusion is ad hoc and will depend upon the particular application for which the system is being used. For example, the results of the object recognition example would more closely resemble the results found using Dempster-Shafer evidential reasoning if the definition of the CF of a conjunction of propositions in the premise of a rule was changed to correspond to the creation of a separate rule for each proposition, for example, $S \varnothing A$ and $C \varnothing A$ instead of $S\_\_C \varnothing A$ in r_7. Using this definition, the resulting CF's for A, B, C, and D would be 0.99, −0.014, 0.014, and −0.99, respectively (where a CF of 0 is assumed to correspond to a probability mass of 0.5).

Conclusion

In addition to multisensor integration and fusion research directed at finding solutions to the problems already mentioned, research in the near future will likely be aimed at developing integration and fusion techniques that will allow multisensory systems to operate in unknown and dynamic environments. The development of sensor modeling and interface standards would accelerate the design of practical multisensor systems. Continued research in the areas of artificial intelligence and neural networks will continue to provide both theoretical and practical insights. AI-based research may prove especially useful in areas like sensor selection, automatic task error detection and recovery, and the development of high-level representations; research based on neural networks may have a large impact in areas like object recognition through the development of distributed representations suitable for the associative recall of multisensory information, and in the development of robust multisensor systems that are able to self-organize and adapt to changing conditions (e.g., sensor failure).

The development of integrated solid-state chips containing multiple sensors has been the focus of much recent research. As current progress in VLSI technology continues, "smart sensors" (Middelhoek and Hoogerwerf, 1985) are being developed that contain many of their low-level signal and fusion processing algorithms in circuits on the same chip as the sensor. In addition to a lower cost, smart sensors provide a better signal-to-noise ratio and abilities for self-testing and calibration. Currently, it is common to supply a multisensor system with just enough sensors for it to complete its assigned tasks; the availability of cheap integrated multisensors may enable some recent ideas concerning "highly redundant sensing" (Brooks, 1990) to be incorporated into the design of intelligent multisensor systems, in some cases, high redundancy may imply the use of up to ten times the number of minimally necessary sensors to provide the system with a greater flexibility and insensitivity to sensor failure.

References

Abidi, M. A. and Gonzalez, R. C., eds., 1992. *Data Fusion in Robotics and Machine Intelligence.* Academic Press, Boston, MA.

Ajjimarangsee, P. and Huntsberger, T. L. 1988. Neural network model for fusion of visible and infrared sensor outputs, Schenker, P. S. ed., *Proc. SPIE,* vol. 1003, *Sensor Fusion: Spatial Reasoning and Scene Interpretation,* 153–160.

Ayache, N. and Faugeras, O. 1988. Building, registrating, and fusing noisy visual maps. *Int. J. Robot. Res.,* 7(6):45–65.

Ayache, N. and Faugeras, O. 1989. Maintaining representations of the environment of a mobile robot. *IEEE Trans. Robot. Automat.,* RA-5(6):804–819.

Brooks, M. 1990. Highly redundant sensing in robotics—Analogies from biology: Distributed sensing and learning, Tou, J. T. and Balchen, J. G., eds., *Highly Redundant Sensing in Robotic Systems,* 35–42, Springer-Verlag, Berlin.

Brooks, R. A. 1986. A Robust layered control system for a mobile robot, *IEEE J. Robotics and Automation,* 2(1):14–23.

Buchanan, B. G. and Shortliffe, E. H., eds. 1984. *Rule-Based Expert Systems: The MYCIN Experiments of the Stanford Heuristic Programming Project.* Addison-Wesley, Reading, MA.

Chandrasekaran, B. and Goel, A. 1988. From numbers to symbols to knowledge structures: Artificial intelligence perspectives on the classification task, *IEEE Trans. Syst., Man Cybern.,* 18(3):415–424.

Duane, G. 1988. Pixel-level sensor fusion for improved object recognition, Weaver, C. W., ed., *Proc. SPIE,* vol. 931, *Sensor Fusion,* 180–185, Orlando, FL.

Fischler, M. A. and Firschein, O. 1987. *Intelligence: the Eye, the Brain and the Computer* (pp. 241–242). Addison-Wesley, Reading, MA.

Flachs, G. M., Jordan, J. B., Beer, C. L., Scott, D. R., and Carlson, J. J. 1990. Feature space mapping for sensor fusion, *J. Robot. Syst.,* 7(3):373–393.

Garvey, T. D. 1987. A survey of AI approaches to the integration of information, Buser, R. G. and Warren, F. B., eds., *Proc. SPIE,* vol. 782, *Infrared Sensors and Sensor Fusion,* 68–82, Orlando, FL.

Garvey, T. D., Lowrance, J. D., and Fischler, M. A. 1981. An inference technique for integrating knowledge from disparate sources, *Proc. 7th Int. Joint Conf. Artificial Intell.,* 319–325, Vancouver, BC, Canada.

Geman, S. and Geman, D. 1984. Stochastic relaxation, Gibbs distributions, and the Bayesian restoration of images, *IEEE Trans. Pattern Anal. Machine Intell.,* PAMI-6(6):721–741.

Giardina, C. R. and Dougherty, E. R. 1988. *Morphological Methods in Image and Signal Processing,* Prentice-Hall, Englewood Cliffs, NJ.

Hager, G. D. 1990. *Task-Directed Sensor Fusion and Planning: A Computational Approach,* Kluwer Academic, Norwell, MA.

Henderson, T. C. and Shilcrat, E. 1984. Logical sensory systems, *J. Robot. Syst.,* 1(2):169–193.

Holm, W. A. 1987. Air-to-ground dual-mode MMW/IR sensor scene registration, Buser, R. G., and Warren, F. B., eds., *Proc. SPIE,* vol. 782, *Infrared Sensors and Sensor Fusion,* 20–27, Orlando, FL.

Huber, P. J. (1981). *Robust Statistics.* Wiley, John & Sons, New York.

Lee, R. H. and Van Vleet, W. B. 1988. Registration error analysis between dissimilar sensors, Weaver, C. W. Ed., *Proc. SPIE,* vol. 931, *Sensor Fusion,* 109–114, Orlando, FL.

Luo, R. C. and Kay, M. G. 1989. Multisensor integration and fusion in intelligent systems, *IEEE Trans. Syst. Man Cybern.,* 19(5):901–931.

Luo, R. C. and Kay, M. G., eds. 1995. *Multisensor Integration and Fusion for Intelligent Machines and Systems.* Ablex, Norwood, NJ.

Maybeck, P. S. 1979. *Stochastic Models, Estimation, and Control,* vol. 1, Academic, New York, NY.

Middelhoek, S. and Hoogerwerf, A. C. 1985. Smart sensors: when and where?, *Sensors and Actuators,* 10:1–8.

Porrill, J. (1988). Optimal combination and constraints for geometrical sensor data, *Int. J. Robot. Res.*, 7(6):66–77.

Richardson, J. M. and Marsh, K. A. 1988. Fusion of multisensor data, *Int. J. Robot. Res.*, 7(6):78–96.

Ritter, G. X. and Wilson, J. N. 1987. The image algebra in a nutshell, *Proc. First Int. Conf. Comp. Vision*, 641–645, London.

Shafer, G. 1976. *A Mathematical Theory of Evidence*, Princeton University Press, Princeton, NJ.

Wolberg, G. and Pavlidis, T. 1985. Restoration of binary images using stochastic relaxation with annealing, *Pattern Recog. Letters*, 3:375–388.

Zadeh, L. A. 1986. A simple view of the Dempster-Shafer theory of evidence and its implication for the rule of combination, *AI Mag.*, 7(3):85–90.

122.3 Audio-Visual Sensor Fusion System for Intelligent Sound Sensing

Kota Takahashi and Hiro Yamasaki

Introduction

The real-time sensor fusion system described here extracts a meaningful sound (target signal), e.g., speech autonomously from multi-microphone signals corrupted by ambient noise interference (disturbance signal), e.g., other speech or background sounds. In short, the system is an intelligent signal receiver which uses sensor fusion techniques.

Many types of intelligent signal receivers with multiple sensors have been proposed recently. The key to such intelligence or autonomy is, we believe, how to distinguish the target signal from a disturbance signal.

Of the many sensing techniques, one of the most popular techniques is Widrow's adaptive noise canceller (Widrow et al., 1975). In this method, "reference sensor" output which does not correlate with a target signal but with interference noise is utilized for adaptation. The system regards a signal which is not included in the reference sensor output as a target signal. In other words, the target signal is defined by zero transfer functions from sound sources to the reference sensor. AMNOR (Kaneda and Ohga, 1986) is another type of adaptive receiver. This system distinguishes the target signal from interference noise by *a priori* knowledge: the position of the target source.

In our system, a target is defined using visual information, or using fused data of audio-visual information and knowledge. That is, we use visual information for adaptation of audio signal processing. Hence, our system can be regarded as an intelligent receiver which uses an audio-visual sensor fusion.

Sensor fusion and integration techniques (Luo and Kay, 1989) are currently an area of intense interest in the field of intelligent sensors. In this area, fusion of different modal sensors is one of the most attractive themes. Although many different modal sensors have been developed, for example, visual sensors, tactile sensors, acoustical sensors, and chemical sensors, if we consider human sensory data, the most typical cross-modality is shown in the fusion of an auditory sensor, i.e., the ear, and a visual sensor, i.e., the eye. The McGurk effect (McDonald and McGurk, 1978; McGurk and McDonald, 1976) can be considered as one proof of the human ability to fuse audio-visual information. In addition, recent neurophysiological studies begin to reveal the mechanism of fusion and the integration of visual and auditory senses (Meredith and Stein, 1986).

However, sensor fusion of audio-visual sensors is not yet the main subject in the area of cross-modal sensor fusion technology today. The only successful theme in this area is speech recognition using a voice and the shape of lips (Aono and Ishikawa, 1991; Petajan et al., 1988).

As mentioned, our audio-visual sensor fusion system distinguishes a target from interference objects using visual information or audio-visual information. The system monitors the environment with a video camera, decides which object is a target, and extracts sound from the target.

The system can be divided into two subsystems: a visual subsystem and an audio subsystem. The audio subsystem extracts a target signal with a digital filter composed of tapped delay lines and adjustable weights (multi-input FIR filter). These weights are adjusted by the cue signal method.

Although the details of the cue signal method will be described later, we briefly describe the theory of the cue signal method here (see Figure 122.5).

For autonomous filter adaptation, it is essential to generate a learning signal internally. Such an internal learning signal can be obtained if we use the cue signal method. An internal learning signal $d(t)$ is formed by the product of the cue signal $\alpha(t)$ and a delayed acoustical signal $\psi(t)$, which contains not only the target signal but also interference noise. The requirement for the cue signal is that the cue signal correlate with the sound level of the target but not with the sound level of the interference noise (details will be described later). Therefore, the sound level of the target must be estimated using other sensors.

Generating the cue signal can be divided into two tasks. One is an automatic segmentation of an image taken by the video camera. The image contains not only the figure of the target but also figures of interference objects. Segmentation is necessary for capture and tracking of the target image only. We call this task V1 task. The other task is estimation of sound power of an object using the segmented image. We call this task the V2 task.

In this paper, two methods for the V1 task are proposed. The first method is generating the cue signal using audio-visual information. We represent this type of fusion as (A + V) + A fusion, whereas a previous method (Yamasaki and Takahashi, 1990) which uses only visual information on the V1 task is written as V + A fusion. The first method, (A + V) + A and the second method, generating the cue signal using audio-visual information and knowledge written as (A + V + K) + A fusion;

This study was performed through special coordination funds of the Science and Technology Agency of the Japanese Government.

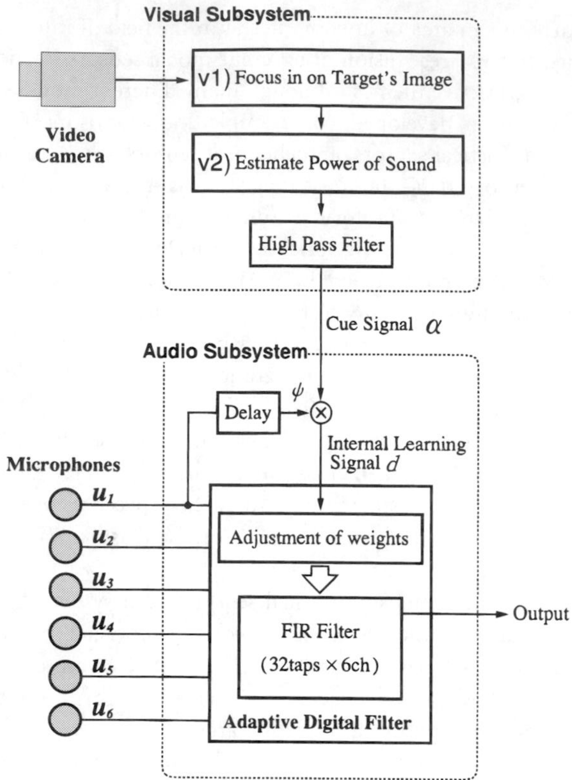

Figure 122.5 Cue signal method.

is described below. In any case, the V1 task and V2 task are performed by the visual subsystem which includes forty DSPs.

Cue Signal Method

To extract the sound of the target, a thirty-two-tap FIR filter is utilized for each microphone. Output of the filter is described as

$$\phi(t) = \sum_{n=1}^{N} f_n y_n(t), \tag{122.1}$$

where f_n are weights of the filter, y_n are values of whole taps, and N is the number of taps ($N = 192$). Let $d(t)$ be a learning signal and $e(t) = \phi(t) - d(t)$ be an error. If the criterion for optimality is to minimize the time average of $e(t)^2$, the optimum solution for the weights can be described as the Wiener solution:

$$p = Rf, \tag{122.2}$$

where $p = (p_n)$ is a correlation (time average) vector between $d(t)$ and $y_n(t)$, $R = (R_{n,n2})$ is a correlation matrix of $y_n(t)$, and $f = (f_n)$ is an optimum weight vector.

As mentioned, the cue signal method is a method for generating an internal learning signal. An internal learning signal $d(t)$ is formed by the product of the cue signal $\alpha(t)$ and a delayed acoustical signal $\psi(t)$.

$$d(t) = \alpha(t)\psi(T). \tag{122.3}$$

Let $\psi_S(t)$ be a target signal component in $\psi(t)$. Under the following four assumptions, the optimum weight vector with this internal learning signal is equal to the optimum weight vector with the learning signal $\psi_S(t)$, which is one of the best-quality learning signals because it does not contain a noise component (Takahashi and Yamasaki, 1990; Yamasaki and Takahashi, 1992). As a result, a target signal can be estimated by the cue signal method.

Assumption 1: Target signal $s(t)$ can be separated into two factors: an envelope signal $a(t)$ and a stationary carrier signal $c(t)$; $s(t) = a(t)\ c(t)$.

Assumption 2: Cue signal $\alpha(t)$ has no correlation with $c(t)$, $\psi(t)$, or $n_j(t)$, where $n_j(t)$ is interference noise of object j. In addition, target signal $s(t)$ has no correlation with $n_j(t)$.

Assumption 3: Cue signal $\alpha(t)$ has a positive correlation with instantaneous target sound power $a(t)^2$, but no correlation with instantaneous power of interference.

Assumption 4: Time average of cue signal $\alpha(t)$ equals zero.

In summary, if we can estimate the target sound power $\alpha(t)^2$, the cue signal method can be used and we can extract the target signal.

(A + V) + A Fusion

V + A Fusion and Its Weak Point

A previous method for generating the cue signal using video images is shown in Figure 122.6, where the video image is represented by intensity $f(x, y, t)$ which is a function of position (x,y) and time t. In Figure 122.6, thick arrows indicate the flow of images $f(x,y,t)$.

This old cue signal generator was based upon a simple audio-visual model: visual stimulus correlates with the sound level of the target. In Figure 122.6, a visual stimulus image is obtained by taking the absolute value of a time differential of an image $f(x,y,t)$.

As has been noted, a task in the visual subsystem can be divided into two tasks: V1 and V2. In the case of Figure 122.6, the V1 task is accomplished by selecting a rectangular area which has the maximum average pixel value in the visual stimulus

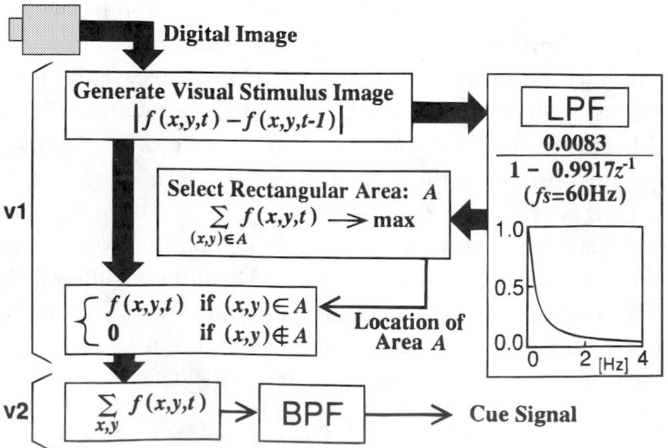

Figure 122.6 Simple method for generating cue signal.

image. The V2 task is accomplished by filtering this average on the time coordinate.

However, this V1 method has the following two problems. First, this method cannot distinguish objects which produce sounds from object which do not produce sounds. Hence, not only the target but also objects without sounds are sometimes selected as a target image. Second, objects with large intensity fluctuation or objects with high contrast induce a large visual stimulus; in addition, conditions of lighting exert a direct effect upon visual stimulus.

In this section, the first problem is solved by introducing an audio signal to the visual subsystem, i.e., (A + V) + A fusion. The second problem is also solved by compensation using spatial difference.

Introducing Audio Signal

Although an audio signal may be introduced and fused in the visual subsystem in various ways, fusion which is based on the same concept as the cue signal method is preferable.

In the cue signal method, an "event" in an audio signal is expressed in terms of an envelope of the sound, and the target sound is distinguished from interference using correlation with the cue signal, which is the "event" of the target sound.

If we introduce the same concept as for the cue signal method in our (A + V) + V fusion, the "event"in an image should be expressed in terms of a visual stimulus, and the target image should be distinguished from other images of silent objects using correlation with the envelope of sounds, which is the "event" of the target sound.

Figure 122.7 shows a block diagram of a cue signal generator based on the above concept. In this method, we assume that only a target induces fluctuation of sound power and fluctuation of the visual stimulus simultaneously. In other words, the generator in Figure 122.7 is designed according to the definition of the target, which is an object with fluctuation of sound power and visual stimulus.

An offset of a microphone signal is eliminated by HPF1, and output of HPF1 is squared to obtain sound power. Next, this signal is smoothed by LPF1 so as to estimate the sum of envelopes of whole sounds. At the end of audio signal processing, the zero-frequency component is eliminated by HPF2.

At the point of (A + V) fusion, the visual stimulus image is multiplied by the output of HPF2. Since this multiplication is executed at each pixel of a visual stimulus image, the output is also digital images. This image is smoothed on the time coordinate by LPF2. The multiplication and smoothing (LPF2) can be regarded as the calculation of a correlation between the visual stimulus image and sound envelope. Output of LPF2 is therefore zero at almost every pixel, and has positive values in the target region. HPF2 is essential because each pixel of the visual stimulus image has a nonnegative value; thus the zero-frequency component in the output of HPF2 has a positive correlation coefficient with the values of these pixels.

To segment the target region, the system selects a rectangular area in which pixels have maximum average values. As a result,

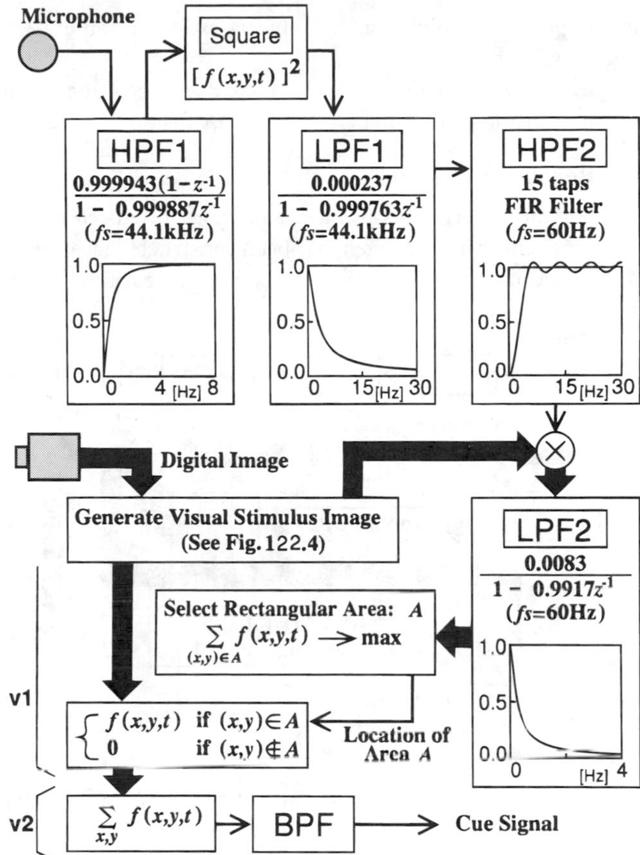

Figure 122.7 Cue signal generator using audio-visual sensor fusion.

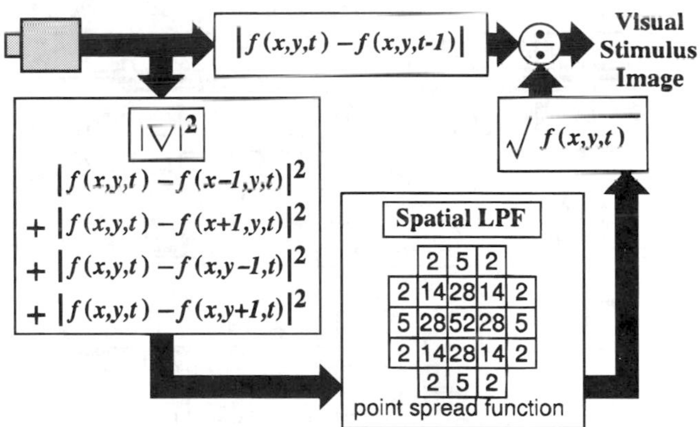

Figure 122.8 Visual stimulus image generator with compensation of spatial difference.

the system can segment the extracted target image using (A + V) fusion.

Compensation Using Spatial Difference

In the old method shown in Figure 122.6, we used the time-difference image as the visual stimulus image. As mentioned, this simple method has problems. Thus, we improve the visual stimulus image generator. Figure 122.8 shows a new generator which compensates time difference using spatial difference. With this compensation, high-contrast objects exert the same effect upon visual stimulus.

Figure 122.9 shows the complete block diagram of the visual subsystem in the case of the (A + V) + A fusion system.

Real-time System

For the experiments using the methods discussed above, a real-time sensor fusion system has been constructed (see Figure 122.10). Seventy DSPs were utilized as general real-time image processors. The system can be divided into two subsystems: an audio subsystem and a visual subsystem.

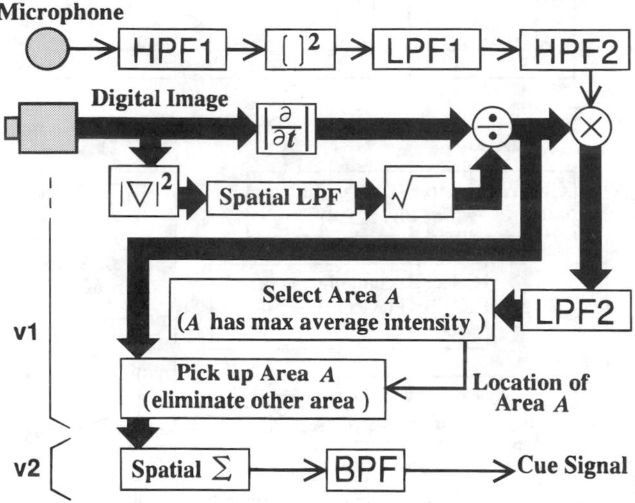

Figure 122.9 The whole block diagram of visual subsystem for (A + V) + A fusion system.

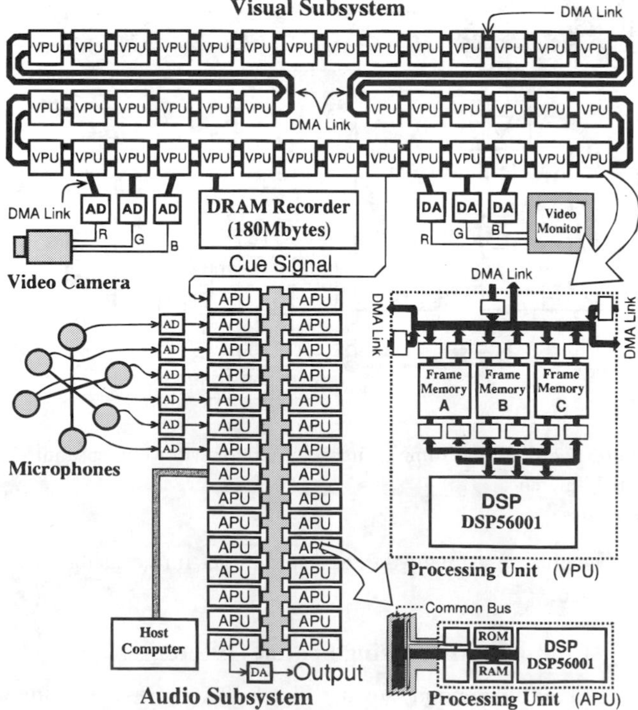

Figure 122.10 Real-time sensor fusion system using 70 DSPs. Visual subsystem is composed of 40 visual processing units (VPU). Audio subsystem is composed of 30 audio processing units (APU).

The audio subsystem is composed of 30 audio processing units (APUs). The APU is a simple processing unit which includes a DSP and local memories. A common bus is utilized as the communication bus between APUs.

The simple processing unit and common bus communication are, we believe, suitable architecture for the audio data property of high sampling rate and small block data.

The visual subsystem is composed of 40 visual processing units (VPUs). A VPU is a complicated processing unit which has three frame memories (A,B,C) connected by DMA links to other frame memories of neighboring VPUs. Each VPU has six DMA links. Since two of them are auxiliary links for the DA converter or AD converter, the other four links can be used for VPU-to-VPU communications. With the four links, not only clockwise communication but also counterclockwise communication is possible. Images can be transferred with one-to-one communication or with broadcast communications through DMA links. DMA communication is highly flexible, so that the algorithm illustrated in Figure 122.9 can be easily implemented.

The complicated processing unit and DMA communication are, we believe, suitable architecture for the visual data property of low sampling rate and large block data.

Experiments of V1 Ability

The V1 ability of the proposed method was compared with that of the old method shown in Figure 122.6.

Figure 122.11 shows the setup for experiments and the results of experiments. In these experiments, objects for sensing were hand clapping of two persons, α and β. In Figure 122.11a, frames

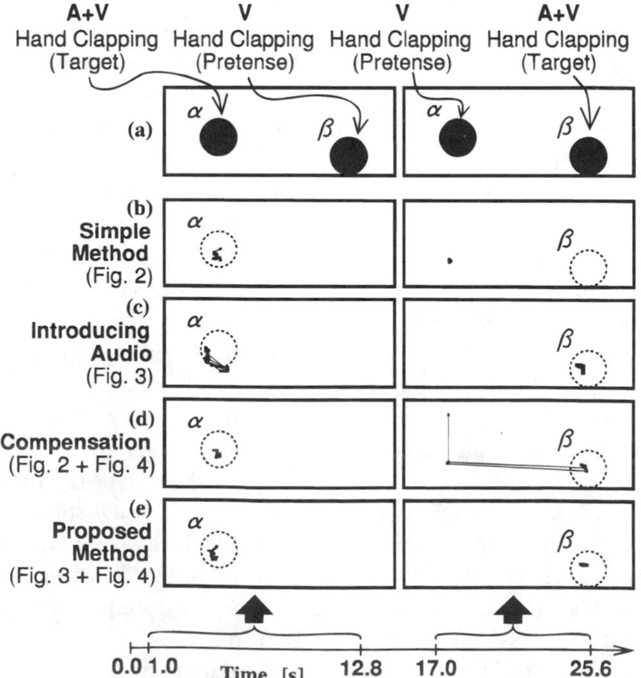

Figure 122.11 Results of experiments. Target is hand clapping; visual disturbance is silent hand clapping.

of digital images of 256 × 48 pixels are represented by large rectangles, locations of hands α and β are represented by black circles.

In the first half (1.0s < *t* < 12.8s) of the experiment, person α was clapping and person β pretended to clap (silent hand clapping). Thus hand clapping α was the target. In the latter half (17.0s < *t* < 25.6s) of the experiment, person β was clapping and person α pretended to clap. Thus hand clapping β was the target in the latter half.

This video image and six microphone outputs were recorded into a DRAM recorder (see Figure 122.10). In order to evaluate the V1 ability of the four methods Figures 122.11b–e), the same audio-visual digital data must be supplied to the four V1 processors. Although real-time processing is possible, the DRAM recorder is essential for evaluating and comparing several methods precisely.

The size of rectangle *A* for choosing the target was fixed: 16 × 16 pixels. Of course, this size should be made changeable in the future.

Figure 122.11b shows the result of the old method, where dotted circles indicate correct target positions. Center positions of rectangle *A* are plotted in the figure. The V1 processor in the old method chose α as the target each time. This can be explained as hand α having higher contrast than hand β.

The result of introducing audio information is shown in Figure 122.11c. Using audio information, the system could distinguish an object which produces sound from an object which does not produce sound. However, the center of *A* sightly drifted because of the high contrast of clothes worn by person α.

Figure 122.11d shows the result of compensation using spatial difference; audio information is not introduced in this experiment. Although the V1 processor chose not only α but also β, the wrong area was chosen in the latter half of the experiment.

Figure 122.11e shows the result of the proposed method; (A + V) + A fusion with compensation. The correct area was chosen each time.

These four results are summarized in Table 122.4. Ratios of correct choice are listed in this table. The results show that precise segmentations were achieved using the proposed method.

(A + V + K) + A Fusion

The cue signal generator described in this section fuses audio information (A), visual information (V),and knowledge(K). This generator was designed to extract voice or speech. Since the (A + V + K) + A fusion system using this generator is complicated, we describe only the outline of the method.

Table 122.4 Effects of Audio-Visual Sensor Fusion and Spatial Difference Compensation

Fig. 7	Method	Figure	Choose Target
(b)	Simple Method	Fig. 2	64%
(c)	Introducing Audio	Fig. 3	74%
(d)	Compensation	Figs. 2 and 4	92%
(e)	Proposed Method	Figs. 3 and 4	96%

Figure 122.12 shows a block diagram of the cue signal generator for the (A + V+ K) + A fusion system. Internal knowledge is stored to find a face in an image. In this figure, internal knowledge is the hue of a human face and an image of a typical face.

Audio information is utilized in different manner from that in the previous section. Locations of sounds are estimated by real-time processing in the visual subsystem.

Visual information and knowledge are fused to find the location of a mouth, which is the target. Since the visual stimulus image is utilized, the system can distinguish a real face from a picture of a face.

V + K information and audio information are fused in the last stage to generate the cue signal. The output signal (TP2) of the V + K fusion block pinpoints the locations of mouths; the output signal (TP3) of the audio block pinpoints the regions of sound sources. If TP2 pinpoints into the regions of TP3, the system assumes that someone is speaking at the moment.

The method shown in Figure 122.12 requires 32 VPUs. Examples of TP2, TP3, and cue signal in Figure 122.12 are shown in Figure 122.13.

Discussion

Although locations estimated from audio information are inaccurate, it does not constitute a problem because we can use visual information which yield accurate locations of objects. Although sound power estimated from visual information are inaccurate, it is not a problem because we can use audio information which yields accurate sound power. These are the advantages of audio-visual sensor fusion shown in Figure 122.12.

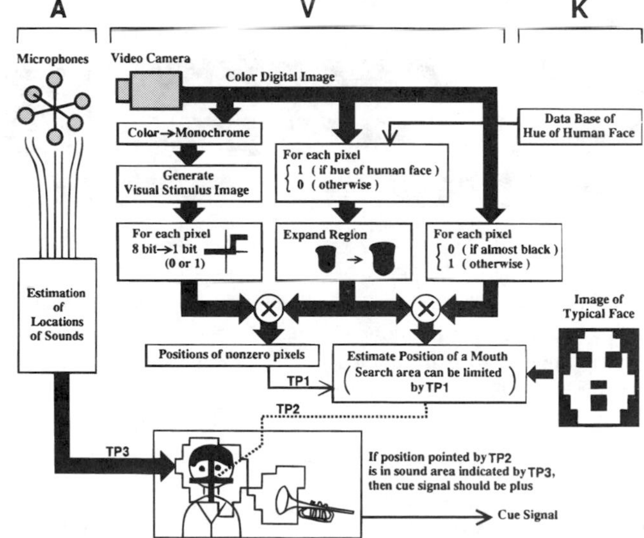

Figure 122.12 A block diagram of cue signal generator for (A + V + K) + A fusion system. Audio information (A) and visual information (K) and internal knowledge (K) are fused to estimate sound power of a target.

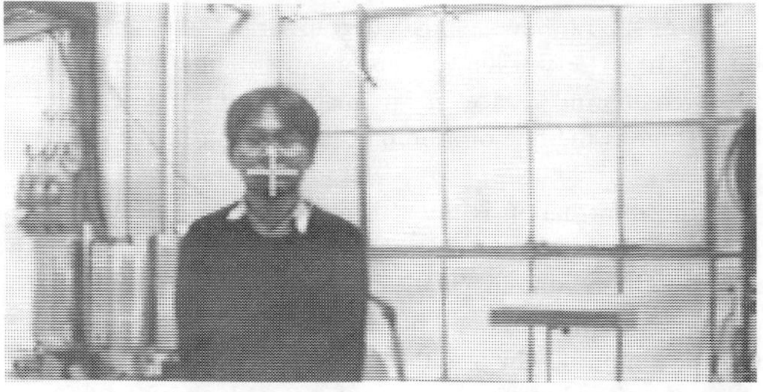

a

b

c

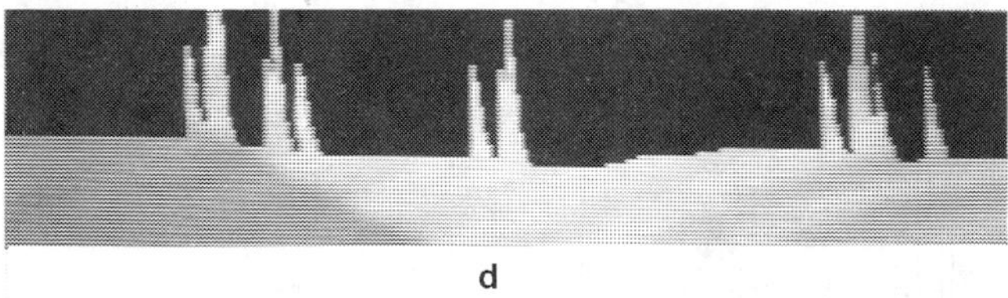

d

Figure 122.13 (a) TP2: video image and internal knowledge were fused so as to estimate the location of a mouth. (b) TP3: six microphone signals were used to estimate the locations of sounds. (c) TP2 and TP3 were fused. (d) Cue signal in the period of 4.3 s.

Conclusions

In this paper, two methods for generating cue signals have been proposed.

The first method, (A + V) + A fusion, generates cue signals using audio-visual sensor fusion. Using audio information in the V1 part, object which produces sounds and object which does not produce sounds can be easily distinguished.

Nevertheless, if interference objects produce sound and simultaneously induce visual stimulus, internal knowledge must be utilized to find the target. The second method described in section IV is one example of such a method. Furthermore, audio information was used to estimate the locations of the sounds, so that audio information and visual information were fused on the pixels of the image.

In conclusion, not only visual information but also audio information and internal knowledge should be used in the cue signal generator. A system using such a cue signal is hierarchical sensor fusion system of audio and visual information.

Acknowledgments

This study was performed through special coordination funds of the Science and Technology Agency of the Japanese Government. The authors wish to thank Dr. Masatoshi Ishikawa and other members of that project for many fruitful discussions on sensor fusion techniques. The authors are grateful to Akira Kimachi, Yoshiko Yamaguchi, Hiroyuki Oohara, and Takeshi Kamazaki for their contributions to the experiments and for building the system.

References

Aono, T. and Ishikawa, M. 1991. Fusion of auditory and visual information by using stochastic process, *34th Japan J. Automatic Control Conf.*, SS6-2:87–88.

Kaneda, Y. and Ohga, J. 1986. Adaptive microphone array system for noise reduction, *IEEE Trans. Acoust., Speech, Signal Processing,* 34(6):1391–1400.

Luo, R. C. and Kay, M. G. 1989. Multisensor integration and fusion in intelligent systems, *IEEE Trans. Syst. Man Cybern.,* 19(5):901–931.

McDonald, J. and McGurk, H. 1978. Visual influence on speech perception process, *Perception and Psychophysics,* 24:253–257.

McGurk, H. and McDonald, J. 1976. Hearing lips and seeing voices, *Nature,* 264:746–748.

Meredith and Stein 1986. Visual, auditory, and somatosensory convergence on cells in superior collilculus results in multisensory integration, *J. Neurophysiol.,* 56:640.

Petajan, B., Bischoff, E. D., and Bodoff, D. 1988. An improved automatic lipreading system to enhance speech recognition, *ACM SIGCH '88,* 19–25.

Takahashi, K. and Yamasaki, H. 1990. Self-adapting multiple microphone system, *Sensors and Actuators,* A21–A23:610–614.

Takahashi, K. and Yamasaki, H. 1992. Real-time sensor fusion system for multiple microphones and video camera, *Proc. 2nd Int. Symp. Measurement and Control in Robotics, (ISMCR '92),* 249–256, Tukuba Science City, Japan.

Widrow, et al. 1975. Adaptive noise canceling: principles and applications, *Proc. IEEE,* 63(12):1692–1716.

Yamasaki, H. and Takahashi, K. 1990. An Intelligent Adaptive Sensing System, *Integrated Micro-Motion Systems,* Harashima, F., ed., 257–277, Elsevier, New York, NY.

Yamasaki, H. and Takahashi, K. 1992. Advanced intelligent sensing system using sensor fusion, *Int. Conf. Ind. Elect., Control, Instrumentation and Automation (IECON '92)*, San Diego, CA.

122.4 Industrial Vision System by Fusing Range Image and Intensity Image

Kazunori Umeda and Tamio Arai

Introduction

Vision systems for industry use intensity image processing, i.e., two dimensional vision system. Today more flexible vision system is needed in many fields. Let us consider recycling as an example. Recycling of industrial products has become very important today when environmental problems are closed up (Dario and Rucci, 1993). Mechanization and automation are indispensable for efficient recycling in a series of processes such as removing screws, disassembling products, cutting them up, and crushing them into small pieces. Among all, the automation of the disassembly process is apparently important with the aid of versatile vision system. Comparing with assembly process, the disassembly process requires more visual feedback because of bad accuracy of positioning and the variety of the products. Considering these backgrounds, more flexible vision system than conventional intensity image processing in two dimensions is needed for industrial use. The fusion of range image and intensity image seems to be effective for the flexible vision system.

This paper proposes a combined vision system of range image and intensity image. Characteristics of both image processings are discussed, and real vision system is composed. First, necessary functions for flexible industrial vision system are shown, and the discussion of characteristics of both image processings is done in the next section. In Algorithms in Range Image Processing, two algorithms in range image processing are presented. The combined system of both a range image and an intensity image is developed in Development of Vision System with Range Image Sensor and Intensity Image Sensor. Fundamental experiments to show the flexibility of both image processings are performed in Fundamental Experiments of Comparing Range Image and Intensity Image. Bin-picking is experimented as a practical industrial example of the composed vision system in Bin-Picking of Multiple Cylinders as an Example of the Presented Vision System. The experiments indicate that the proposed vision system is effective for industrial use.

Range Image Processing and Intensity Image Processing in Industrial Vision System

Necessary Function for Flexible Industrial Vision System

As an example of a target of flexible industrial vision system, let us consider the characteristics of the disassembly process.

- An object is positioned inaccurately: the object to disassemble cannot be positioned accurately as in assembly process, because of its transformation and the wide variety of sizes.
- Transformation, deterioration, and blemishes are added on the product: secondhand goods often have dents and changes on their surfaces such as a flaking-off of painting, rust, and dirt.
- Variety of the objects is large: in production process, the number of kinds of products is limited. On the contrary, in the recycling process, we have to accept various kinds of the products from various manufacturers.
- Geometric models of the products may not be obtained: some products may be too old to obtain their geometric models of production. Ordinary model-based algorithms are difficult to introduce into the disassembly process.

Therefore, the functions of the vision system in the disassembly process are summarized to detect necessary parts from the product, and measure the three dimensional pose of the part.

These two functions must be robust to transformations and blemishes.

These two functions seem to be general for other industrial target like bin-picking. So we suppose the process of the vision system as shown in Figure 122.14.

1. Detection of target parts: necessary target parts are extracted from the product. It consists of two steps as shown in Figure 122.14: (a) segmentation of images, and (b) identification of segmented regions.
2. Measurement of pose of parts: 3D measurement is directly done by means of calculating the distances.

Range image processing

The input technique of the range image (Jain and Jain, 1990) has been studied in two categories: "active methods" and

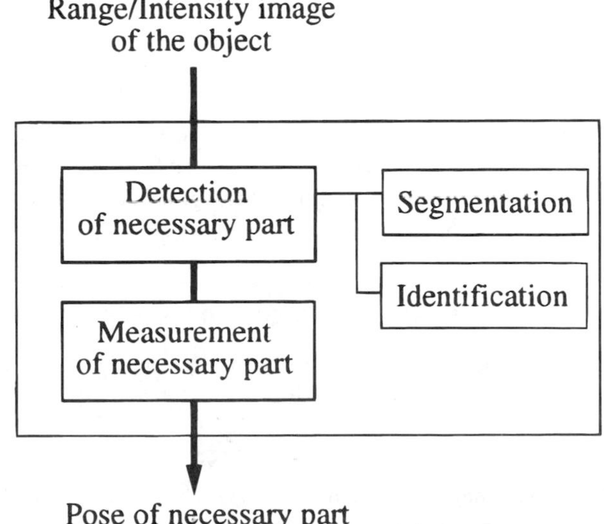

Figure 122.14 Flow chart for recognition of each part.

"passive methods." In passive methods, stereo vision is typical. Triangular measurement using structured light such as slit lights (Shirai, 1972) is one of the most popular active methods. It used to take several seconds or more to acquire a range image, which was the greatest disadvantage of range image processing. In recent years, however, new high-speed sensors (Yamamoto, 1990; Kanade et al., 1990; Sorimachi, 1989; Herbert, 1986) make it shorter, e.g., 1/30 second. But, as compensation for the speed, the number of pixels is restricted to be small. In other words, only a sparse image can be input.

Characteristics of the range image are summarized as below.

- The distance is directly measured.
- The distance is not influenced easily by disturbances.
- Information of the object except shapes is thrown away.
- It is difficult to detect edges of the object because of its sparseness. But it is easy to make a patch of small plane around a measured spot.

Totally the following issue should be realized in range image processing: "basic algorithms to detect and measure simple primitive features such as a plane surface and a cylindrical surface." The detection algorithms are discussed in Algorithms in Range Image Processing and measurement algorithm special for cylinders is shown in Bin-Picking of Multiple Cylinders as an Example of the Presented Vision System.

Intensity Image Processing

Intensity image processing is widely researched and utilized in industrial applications, although the applications are rather simple ones.

Characteristics of intensity image processing are as follows, compared to range image processing.

- Widely used and cheap.
- Standardized.
- The image is dense.
- Two dimensional.
- Several special hardwares have been developed to perform fundamental functions.

Considering the characteristics of two image processing above, *compensation of range image's sparseness with a fast intensity image processing hardware* seems to be useful and practical for an industrial vision system.

Algorithms in Range Image Processing

In this section, segmentation and identification algorithms are defined for range image processing.

Segmentation of a Range Image

Range images and intensity images have to be first divided into several regions for recognition. Various algorithms have been proposed to segment an intensity image. An algorithm to segment a range image is shown here. The segmentation algorithm has to be effective to sparse range images. Hough transformation is studied by Shimizu et al. (1990). The method is suitable for sparse range images, but it is very time-consuming. So we utilize a simple algorithm as shown in Figure 122.15.

1. Segmentation in two dimensional plane: the clustering is made according to distance in a range image as well as to density in an intensity image.
2. Segmentation by discontinuity of distances: adjacent points are classified into two regions when the distances differ from others in more than a threshold.
3. Segmentation by curvature: points are classified more precisely by higher-order features like curvatures or discontinuity of adjacent normal vectors.

Since the discontinuity of distances corresponds directly to the discontinuity of the object in a range image, it can be said that the region can be segmented correctly even by such simple algorithms. It is possible to process the segmentation algorithm at high speed because of its simplicity.

Identification of Four Simple Features in a Range Image

As a range image is assumed to be sparse, it is difficult to deal with a complex object directly. So we focus on simple features. Industrial products usually contain some simple shapes. So it is possible to recognize them, if simple features are detected. We select four simple features: plane, cylinder, cone and sphere. This is because the features are contained in many industrial products and thus important, and they are easy to identify. Extended Gaussian Image (EGI) (Horn, 1986) representation is employed to identify these four features in this paper. EGI expresses an object by distribution of normal vectors of surface planes. Ikeuchi (1981) applied EGI to photometric stereo image. It is an easy and fast operation to make normal vectors from a

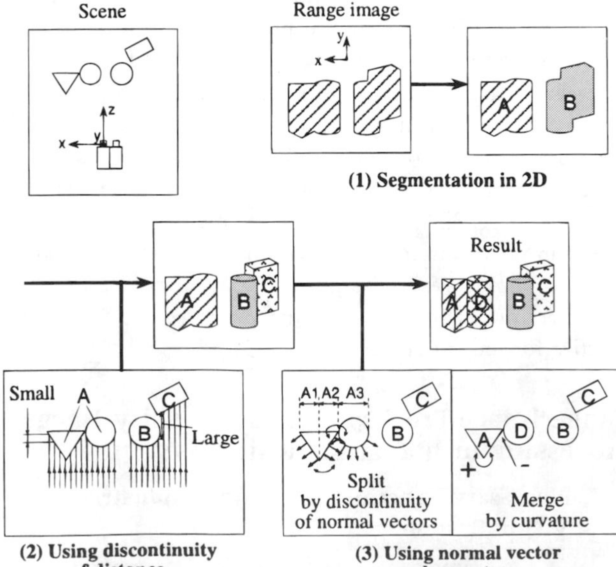

Figure 122.15 Segmentation of a range image.

sparse range image by means of making small patches onto the surface. So EGI is suitable for sparse range image processing, too. Each simple feature corresponds to the following characteristic as shown in Table 122.5. The algorithm to identify these features with EGI is presented in Figure 122.16. As for EGI itself, Kang and Ikeuchi advanced EGI to CEGI(1993) (Complexe EGI), which can deal with not only the orientation of an object but also its translation.

Development of Vision System with Range Image Sensor and Intensity Image Sensor

In this section, the combined vision system with range image sensor and intensity image sensor is shown.

Range Image Sensor

A range image sensor, which was developed by Sorimachi (1989), is used in this paper. It consists of a multiple spotlight projector (approx. 1400 spots), an ordinary 2-dimensional CCD camera (768 × 490 pixels) and a multiple spot detector as shown in Figure 122.17. The spotlight projector uses a halogen lamp (100W) and a metal-evaporated glass plate with many small windows. 8-bit range data (1056 pixels) are obtained as a range image, which is much sparser than an ordinary intensity image ($10^4 \sim 10^6$ pixels). They are calculated by triangulation: 64 pixels

Table 122.5 EGI of Four Simple Features

Feature	Characteristic on EGI
Plane	Point
Cylinder	Circle line passing the center of EGI
Cone	Circle line not passing the center of EGI
Sphere	Sphere

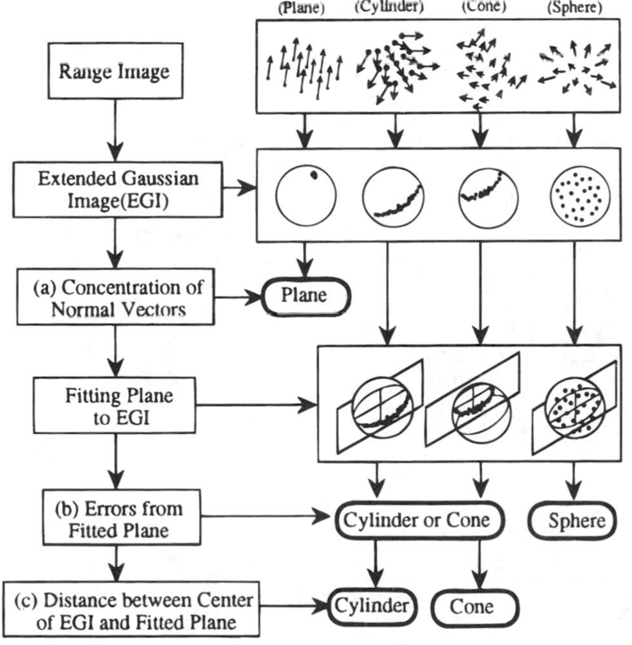

Figure 122.16 Identification of four simple features using EGI.

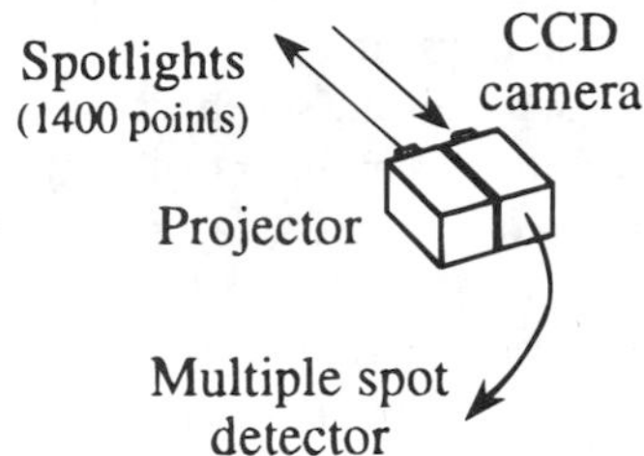

Figure 122.17 Range sensor.

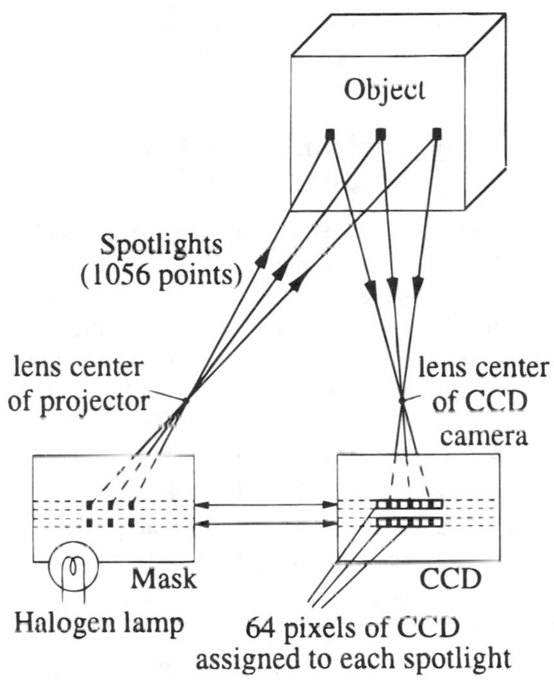

Figure 122.18 Principle of measuring multiple distances of the range sensor.

of CCD are assigned to each spotlight, and the center of gravity of a spotlight image on each "line CCD" is calculated into an 8 bit range datum as illustrated in Figure 122.18. The calculation is done by the multiple spot detector. Ninety-six lines by 11 spots/line makes a frame of 1056 range data at video rate (1/30 sec). Conversion of the 8 bit data to actual distances is performed by a personal computer connected to the range image sensor.

Measurable area is determined according to focal lengths of lenses of both the projector and the CCD camera, and also to the interval between the two lenses. In our experimental system, the measurable distance is 770–860 mm, and the field of view is 14 degrees horizontally by 10 degrees vertically. The resolution in the depth direction is about 0.35 mm.

Matching of different sensors is one of the problems in sensor fusion. Note that the range image sensor can acquire both ordinary intensity images and range images with the same CCD camera. As a result, the problem can be easily avoided in the system.

Configuration of Developed Vision System

Total experimental system is illustrated in Figure 122.19, which consists of a range image processing system and an intensity image processing system.

The range image processing system consists of the range image sensor mentioned above, a monitor to display a range image and a personal computer to execute the data conversion, segmentation, identification, etc.

The intensity image processing system consists of a CCD camera, a commercial image processor and two personal computers to control the image processor. The two computers are connected through RS-232C.

The CCD camera is shared by both systems. The circular fluorescent lamp is turned on for the intensity image, and turned off when a range image is input.

Fundamental Experiments of Comparing Range Image and Intensity Image

To study the robustness of range image processing and intensity image one, fundamental experiments are operated. Two kinds of audio amplifiers are used as an object. One has black finish and the other has golden finish. The process of the experiments is to recognize a large volume dial in front of the amplifier with the two different image systems.

Experimental Conditions

The amplifier is set under the CCD camera as shown in Fig. 122.19. The distance from the lens center of the camera is 830 mm to the black amplifier and 843 mm to the golden one respectively. The diameter of the volume dial is 46.0mm for the black one and 44.5mm for the golden one with the height of 22mm.

Experimental Results

Objects Without Blemishes. Without blemishes, the segmentation of both a range image and an intensity image was successful regardless of the color. In the range image, the dial region can be extracted easily from other regions by the difference of the distances. Even though the dial has the same color as the panel, a fringe made by a hollow on the panel around the dial helped the extraction of the region. Thus, in this experiment, the dial region can be extracted easily in the intensity image, too.

Objects With Blemishes. Blemishes on an object may often cause errors in image processing. Three kinds of tapes: white, black and brown tapes were pasted on/around the dial as typical examples of the blemishes as shown in Figure 122.20. As the color of the amplifier did not influence the experimental results much, results of the black finish amplifier alone are shown principally.

1. Intensity image: binary images of the black amplifier and the golden one are shown in Figure 122.21. For the black one, the dial region is extracted, although the shape of the dial region is chipped at the edge. For the golden one, the dial region cannot be extracted, because the dial region is connected to the panel region.
2. Range image: Figure 122.22 indicates the projected spotlights. A range image is expressed as the network of the neighboring points in Figure 122.23. It is concluded that the influence of blemishes is little on a range image regardless of the color. The influence of blemishes appears not as the change of distances but as lack of some distances in the image.

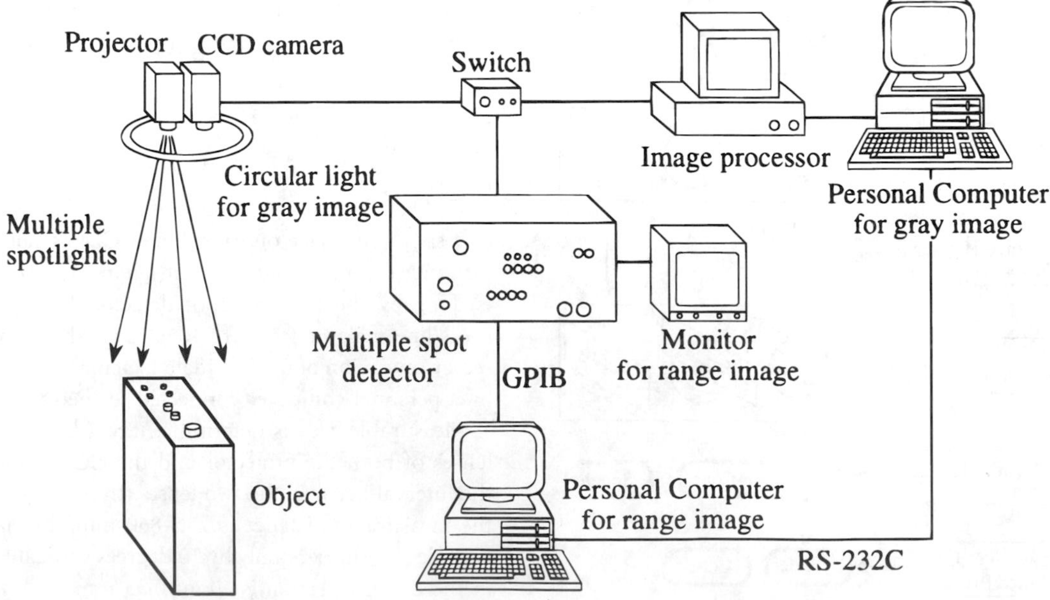

Figure 122.19 Developed vision system.

Figure 122.20 An intensity image of the dial with blemishes.

As a result of the experiments, we conclude that blemishes disturb intensity images strongly but do not disturb range images. In intensity images, the segmentation often fails and the following processes are forced to stop. And even if it succeeds, the measured parameters such as the position of the centroid of the extracted region are inaccurate. On the contrary, range images are robust to blemishes and other changes on the surface of the products. Consequently, the effectiveness of range image for flexible industrial vision system is shown.

Bin-Picking of Multiple Cylinders as an Example of the Presented Vision System

A practical example of the flexible vision system is shown here. The system recognizes one cylinder from among multiple cylinders, measures its pose, and picks it up : a typical bin-picking task. The cylindrical shape is chosen as a typical example of industrial products.

Flow Chart of Bin-picking

As described in Range Image Processing and Intensity Image Processing in Industrial Vision System, the range image cannot be segmented precisely because of its sparseness. This disadvantage can be covered by an intensity image. On the other hand, the bad influence caused by blemishes to an intensity image is compensated by a range image. Combination of both the image systems is desirable, as insisted repeatedly. In this experiment, an intensity image is utilized together with a range image in segmentation process.

Flow chart of the bin-picking process is as follows.

1. Segmentation of an intensity image: Candidate regions, which a robot tries to grasp, are detected and ordered. The top candidate region is chosen. The segmentation is not accurate especially when blemishes are put on the cylinders.

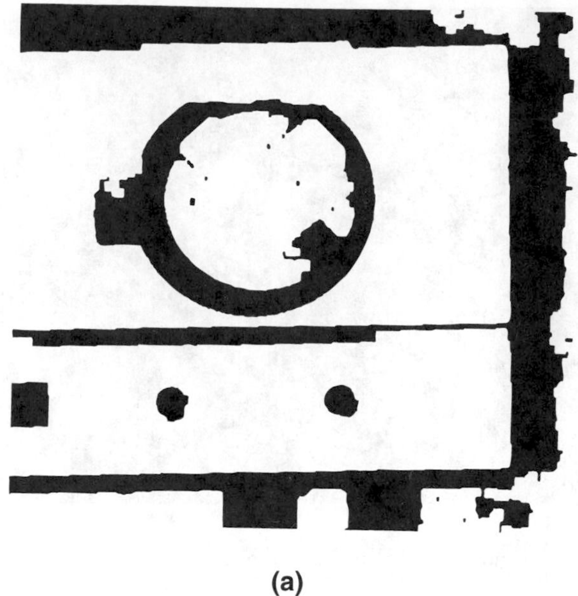

(a)

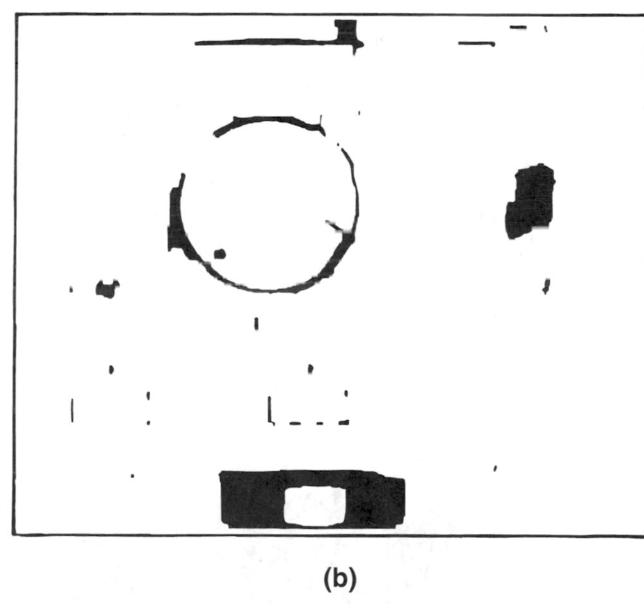

(b)

Figure 122.21 Binary image. (a) Black dial. (b) Gold dial.

2. Segmentation of a range image: A range image around the top candidate region is segmented again as described in Algorithms in Range Image Processing. The largest region is selected as a candidate.
3. Identification of a cylindrical part: The candidate is checked whether it is a cylinder or not by the method in Algorithms in Range Image Processing. If the identification results in 'No,' the next candidate region is selected.

   The next step is only special for cylinders.

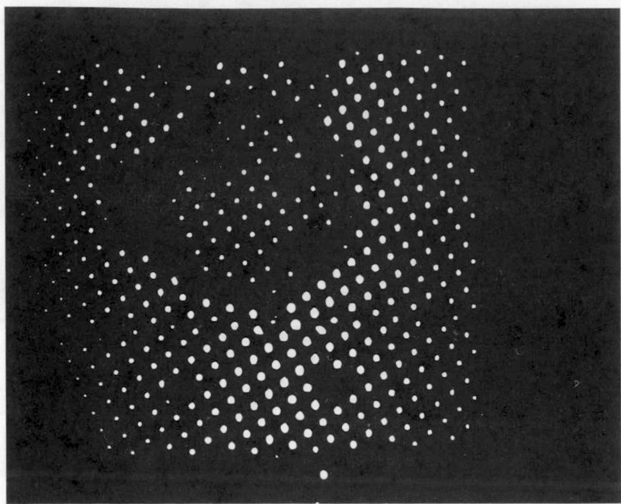

Figure 122.22 Spotlights projected on the amplifier.

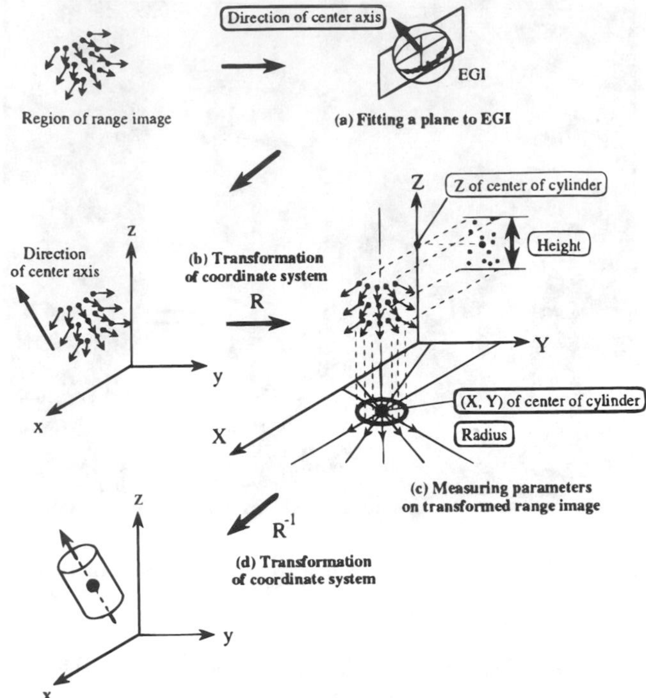

Figure 122.24 Measurement of a cylinder.

4. Measurement of pose: Pose of the selected cylinder is measured as illustrated in Fig.122.24. determine the axis vector of the cylinder and transform the coordinate system to make the axis parallel to the z axis; 2. detect the centroid of the projected circle onto the xy plane; 3. detect the mean value of the z height. The pose is obtained, by transforming reversely, as the coordinates of the center of gravity of the cylinder in original coordinate system.
5. Pick motion: After these image processings succeed, the selected cylinder is picked by the robot.

Experimental Result

The system in Development of Vision System with Range Image Sensor and Intensity Image Sensor was used as the vision system and a robot was installed additionally. The objects were a heap of wooden cylinders of 30 mm diameter and 50 mm height as shown in Figure 122.25. "Bin-picking" here means that the system detects a cylinder in a heap and picks it up from the heap. Figure 122.26 shows the segmentation result obtained in an intensity image. The number indicated in Figure 122.26 represents the priority order.

The results indicate that the success rate of recognition is 90% and that the rate of total bin-picking is 80%. This seems good

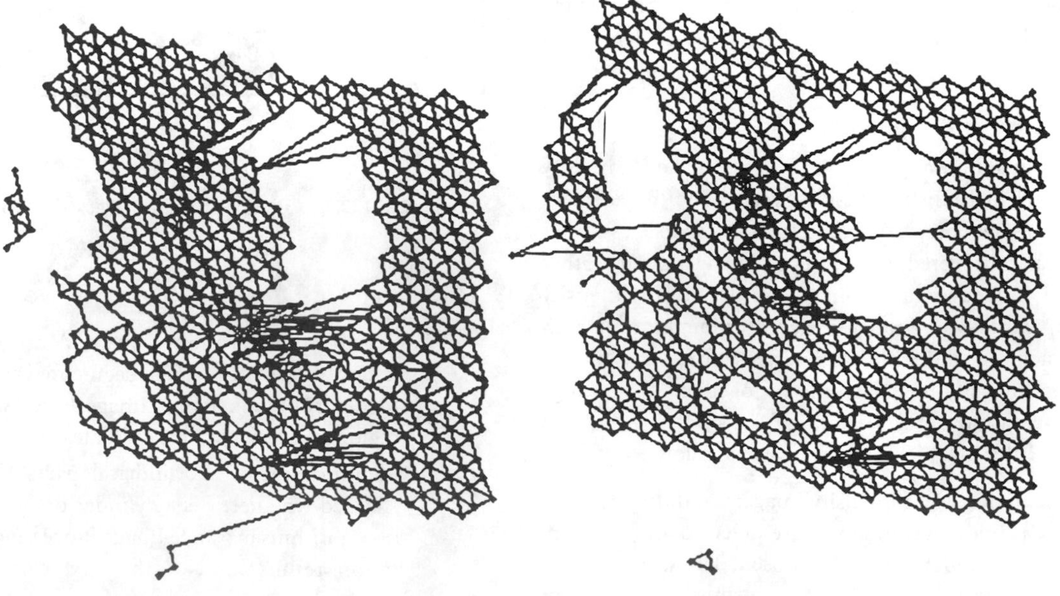

Figure 122.23 A range image in actual distances (left: black dial right: gold dial).

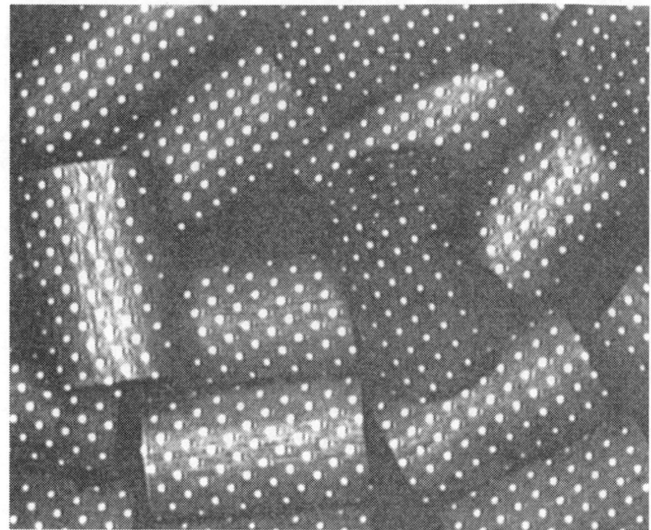

Figure 122.25 An example scene. Multiple cylinders with spotlights projected by range image sensor.

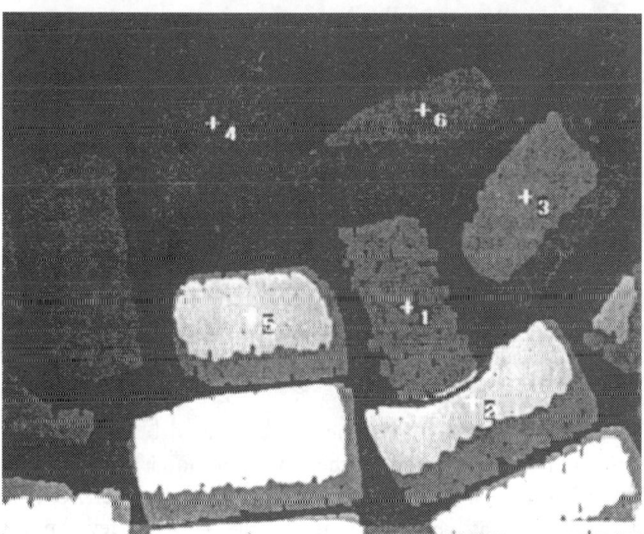

Figure 122.26 Segmentation of an intensity image.

considering short processing time as listed in Table 122.6. The total time in Table 122.6 includes the retrial time on failure of recognition and the communication time between the two image processing systems. In this case, the single range image processing system also worked without the intensity image processing. However, the required time of the combined vision system was shorter than that of the single system.

The experimental view of bin-picking is shown in Figure 122.27.

Table 122.6 Required Time in Bin-Picking an Average

Process	Time
Recognition by an intensity image	2sec
Recognition by a range image	1.3sec
Total time in vision system	6sec

Figure 122.27 Experimental view of bin-picking.

The experiments can be concluded as:

- Range image processing is relatively fast to operate bin-picking.
- Failure rate is still rather high.
- A sparse range image can provide efficient information.
- Segmentation process is well achieved in the combined system. It is faster than single range image processing system.
- The combined system of two image processings can work properly in total.

Although it is mere a simple example of fusing range image and intensity image, the efficiency of the two images' fusion for industrial vision system is indicated in this experiment.

Conclusion

In this chapter, a vision system for industrial use is proposed using the fusion of range image and intensity image. First, functions required in the vision system were analyzed. Second, vision system which can get both range image and intensity image was developed, and experiments were made to verify the system. The following results were derived.

1. A combined system of range image processing and intensity image processing is proposed. It is verified by a bin-picking experiment of multiple cylinders that the system works efficiently.
2. Experiments with some blemishes on products indicate that range image processing is more robust to the changes of environment than intensity image processing.
3. Segmentation algorithm and identification algorithm of four simple features work for a sparse range image.

Acknowledgment

We wish to thank Mr. M. Muto, Mr. K. Kawai, and Mr. Y. Tanaka in Pioneer, Inc. and Mr. K. Sorimachi of Canon, Inc. for supporting experimental apparatus and giving us useful advice.

References

Dario, P. and Rucci, M. 1993. An approach to disassembly problems in robotics, *Proc. IROS '93*, 460–467.

Herbert, M. 1986. Outdoor scene analysis using range data, *Proc. Int. Conf. Robotics and Automation*, 1426–1432.

Horn, B. K. P. 1986. Extended Gaussian images, *IEEE Proc.*, 72(12):1671–1686.

Ikeuchi, K., 1981. Recognition of 3-D objects using the extended Gaussian image, *Proc. 7th IJCAI*, 595–600.

Jain, R. C. and Jain, A. K. 1990. *Analysis and Interpretation of Range Images*, Springer-Verlag, New York, NY.

Kanade, T. Gruss, A., and Carley, L. R. 1990. A VLSI sensor based rangefinding system, *Robotics Research 5th Int. Symp.*, 49–56.

Kang, S. B. and Kkeuchi, K. 1993. The complex EGI: a new representation for 3-D pose determination, *IEEE Trans. PAMI*, 15(7):707–721.

Shimizu, N., Nakazawa, K., Yuta, S., and Kakajima, M. 1990. Cylindrical objects detection using fiber grating vision system, *Int. Symp. Measurement and Control in Robotics*, E1.3.1–7.

Shirai, Y. 1972. Recognition of polyhedrons with a range finder, *Pattern Recognition*, 4:243–250.

Sorimachi, K. 1989. Active range pattern sensor, *J. Robotics and Mechatronics*, 1:269–273.

Yamamoto, M. 1990. Direct estimation of deformable motion parameters from range image sequence, *Proc. ICCV '90*, 460–464.

122.5 Application of Data Fusion to Neonate Oxygenation Control

Mark E. Kotanchek, James P. Helferty, W. Bosseau Murray, and Charles Palmer

Introduction

Although premature infants generally benefit from additional oxygen since one of their most common problems is *Respiratory Distress Syndrome* (RDS) due to underdeveloped lungs, neonates[1] are also vulnerable to oxygen toxicity. One manifestation is Retinopathy[2] of Prematurity (ROP) which may eventually culminate in blindness[3]. Accurate control of the inspired oxygen is required to avoid hypoxemia[4] as well as oxygen toxicity.

Relatively inexpensive non-invasive sensors are widely available to measure the oxygen saturation of the hemoglobin which is related to the partial pressure of oxygen in the blood. Unfortunately, the oxygen uptake and delivery varies dramatically with the neonate's activity so that it is not practical for a human to accurately follow the saturation and adjust the inspired oxygen on the required time scale. As a result, an automated inspired oxygen control system is desirable. Unfortunately, premature infant physiology and problems vary dramatically among infants and over time especially with the very small ones (500–1500 grams) which most require the supplemental oxygen.

Although previous efforts at developing automated systems, (Morozoff et al., 1993; Azhar and Karim, 1991; Dugdale et al., 1988), have been more effective than manual control, these efforts have only exploited a single sensor. Due to the complexity and dynamics of the neonate physiology, *a priori* knowledge about the patient's state as well as expertise supplied by the medical staff and supplemental data supplied by other sensors should be incorporated into the controller. Such a control system would monitor the neonate state to facilitate a real-time supply of oxygen appropriate to the patient needs as well as recognize sensor failures and dangerous situations—thus, the technologies of tactical real-time data fusion are applicable. In the ensuing, we will review the physiology, available sensors, and applicable control system and data fusion concepts which are currently being explored in our research.

Physiology and Interactions

Respiratory Distress Syndrome can be caused by a number of different mechanisms: one of the more common is *Hyaline Membrane Disease* which corresponds to a surfactant deficiency in the lungs so that when the lung alveoli[5] collapse they *stay* collapsed which effectively reduces the surface area available for oxygen transfer to the hemoglobin. Another problem for these neonates is *shunting* wherein the neonate blood flow bypasses ventilated sections of the lung in such a situation, increasing the available oxygen is a moot point; however, pulmonary shunting tends to develop slowly (on the order of 30 minutes) as well as

[1] ne-o-nate\'ne¯-e-na¯t\n: a newborn child; esp: a child less than a month old.

[2] ret-i-nop-a-thy \,ret-en-'a¯p-e-the¯\: any of various noninflammatory disorders of the retina including some that are major causes of blindness

[3] If excess oxygen is present in the blood, the vessels will constrict; however, in neonates the blood vessels overreact and clamp shut—resulting in hypoxic (oxygen deficiency) injury, proliferation of blood vessels, and bleeding. The resulting scar tissue will pull at the retina which eventually detaches to produce blindness.

[4] hyp-ox-emia \,hip¯,a¯k-'se¯-me¯-e, hi¯-,pa¯-\ n: deficient oxygenation of the blood

[5] al-ve-o-lus \al-'ve¯-e-les\ n, pl -li\-,li-, -(,)le¯\: a small cavity or pit: as . . . an air cell of the lungs

dissipate slowly so it does not pose an insurmountable problem for the oxygen control system. Furthermore, neonates susceptible to cardiac shunting can be identified. The latter neonates can have changes with a time course over minutes.

RDS reduces the amount of oxygen available to the infant and can, as a result, produce brain damage or death as well as the other problems associated with hypoxia.[6] In the 1940s, premature infants began to be supplied with supplemental oxygen. Although this reduced the death rate, it was eventually noticed that some of these neonates went blind as they matured[7]—however, without the supplemental oxygen, they may not have had the opportunity to live to become blind.

The mechanisms of ROP and the other manifestations of oxygen toxicity (such as long-term lung damage) are not well understood. Possible mechanisms include the presence of free radicals in the blood stream or phototoxicity due to the continuous bright light commonly present in NICUs. What is understood is that if the partial pressure of oxygen in the arterial blood (P_aO_2) is controlled to be between 80—100 mm Hg (which corresponds to normal atmospheric conditions), ROP can generally be avoided. Unfortunately, P_aO_2 cannot be measured directly in real-time using non-invasive sensors. Fortunately, sensors are available to measure the oxygen saturation of the arterial blood hemoglobin (S_aO_2) which is related to the partial pressure by the nonlinear relationship illustrated in Figure 122.28 (which is derived from an approximation in 1975). Since the mapping from S_aO_2 to P_aO_2 is very sensitive at the desired oxygenation

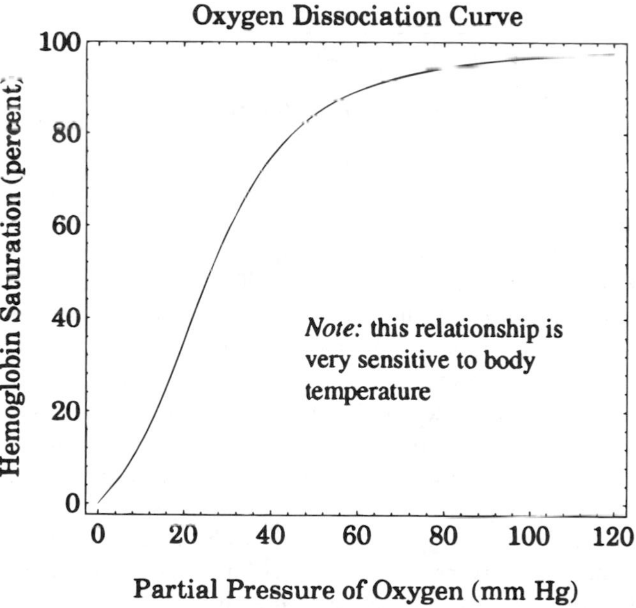

Figure 122.28 Partial pressure of oxygen vs. hemoglobin saturation.

levels as well as subject to large physiologically-induced perturbations, normal procedure is to attempt to maintain a hemoglobin saturation in the 90–95 percent S_aO_2 range to maintain safe operation.[8] (Data fusion of the operating curve characteristics with sensor information concerning temperature, CO_2 tension, pH, etc. may facilitate operating at higher hemoglobin saturation levels.)

From a control system perspective, supplying the neonate with the correct amount of oxygen is a complicated process since the oxygen reaching the body tissues is a function of pulse rate, blood flow, temperature, hemoglobin levels, etc. as well as the controllable *supplied* oxygen.

Monitor System and Sensors

The modern hospital patient is generally surrounded by a suite of invasive and non-invasive sensors providing continuous data about the patient's state. This continuous monitoring is often supplemented by off-line analyses which are useful for calibration although not as part of a real-time control system. The human body is a complicated (non-linear) system with many factors and subtle inter-relationships; as a result it is difficult to model in its full complexity and, therefore, *observability* is an issue as well as *controllability*.

General Sensor Comments

Neonates are fragile and delicate beings so non-invasive sensors are preferred. Thus, although the partial pressure of the oxygen dissolved in the blood may be measured directly from blood samples or continuous measurements made via intra-arterial devices, a less accurate estimate of the partial pressure derived from the non-invasive pulse oxygen analyzer's estimate of hemoglobin oxygenation is preferred. Since prolonged skin contact is a problem for these premature infants, even the non-invasive sensors can pose problems.

In addition to being non-invasive, the ideal sensors should be small and easily affordable to permit continuous monitoring of the infant. One of the goals of neonate care is to prevent brain injury so any tool to detect problems, or potential problems, as well as prevent the evolution of problems, is desired.

As a general comment, it appears that affordable, non-invasive, *bedside* analyzers are needed by the medical community. To illustrate: although a MRI brain scan is non-invasive, the mere act of moving the NICUs[9] neonates can pose health risks.

Patient Status Monitors

A number of companies manufacture sensor suites which monitor and display the patient status. Each system consists of a CRT, speaker, and a chassis into which a variety of modules

[6] hyp-ox-ia \hip-'a¯k-se¯-e, hi¯-'pa¯k-\ n: a deficiency of oxygen reaching the tissues of the body

[7] Retinopathy of prematurity is first clinically evident between the first 4–7 weeks of life in premature infants. Over the following weeks it progresses to retinal detachment (and blindness) in 1% of newborns weighing between 500–1500 grams.

[8] Note that for adults breathing normal air, approximately 97% of the oxygen in the arterial blood is attached to the hemoglobin with the remaining 3% in the form of the dissolved gas which is reflected in the partial pressure. It is this free oxygen which causes the oxygen toxicity.

[9] NICU: *Neonatal Intensive Care Unit*

may be inserted for specific sensing demands. In the Hewlett-Packard (HP) Merlin Monitor™ (which is typical of the genre) each module[10] processes the signal from a single sensor and outputs the results to the CRT if such is selected by the user. The output can consist of a time series e.g., the pulsatile O_2 saturation or the EKG[11] waveforms or a summary statistic such as an estimate of the *averaged* O_2 saturation or the patient heart rate. Operating parameters such as averaging intervals or alarm thresholds may be set by the operator.

Each module is *independent;* while this simplifies the system design and permits "mixing-and-matching" modules as required, information synergy possible due to integrating the information by the full sensor suite must be provided by the attending physician or nurse. A *data fusion* module merging and interpreting the information from a variety of sensors would be a potential boon to the patient care. Such a data fusion system might incorporate aspects of expert systems, fuzzy logic, neural networks, or intelligent control.

Operational Environment

In a typical NICU, the neonate is located on a platform or in an incubator placed in front of a "bench" supporting a plethora of monitors, sensors, speakers, knobs, dials, etc. Each nurse in the NICU is responsible for up to four infants. In this relatively stressful environment the nurse does not have time to closely monitor the oxygen saturation of the neonate's hemoglobin; as a result, the oxygen control is applied at 5–10 (or more) minute intervals. To avoid hypoxemia-induced brain-damage, the nurses tend to shade the saturation levels higher than that ordered by the physician.

The noisy electromagnetic environment of the NICU would have to be handled by any control system. Of course, status displays, alarms, and controls need to be oriented towards the background of the medical rather than that of the control systems communities.

Pulsatile Oxygen Analyzer

Although other invasive sensors are available, the *pulsatile oxygen analyzer* (POA) is attractive as the primary sensor for the oxygenation control due to its non-invasive nature as well as being a relatively inexpensive device. As illustrated in Figure 122.29 the light absorption of oxygenated and deoxygenated hemoglobin is frequency dependent and unique. By pulsing light into the skin, the pulsatile oxygen analyzer measures the absorption at two different frequencies and, thereby, estimates the percent oxygen saturation of the hemoglobin in the blood. For the Merlin monitor, the transmitted frequencies are cycled 250 times per second. This produces a time series similar to that shown in Figure 122.29 (which may optionally be displayed on the monitor) which shows the pulse of oxygenated blood through

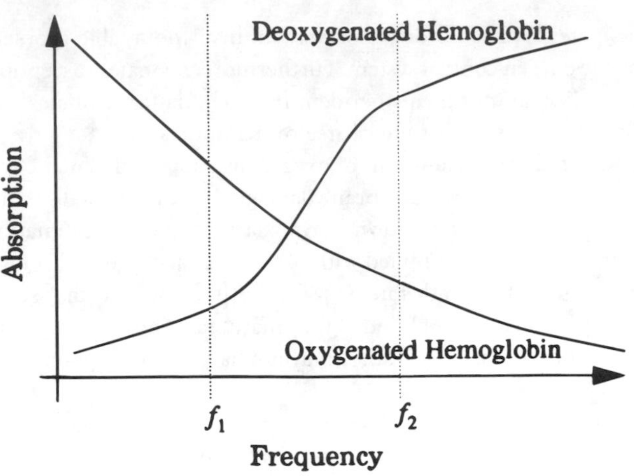

Figure 122.29 Hemoglobin absorption vs. frequency.

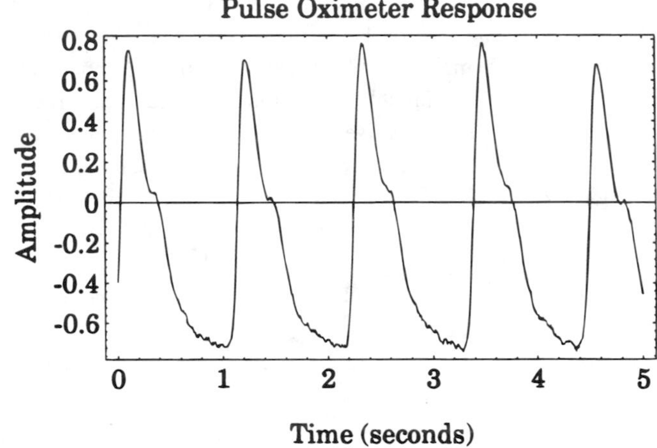

Figure 122.30 Hemoglobin (inverse) absorption vs. time.

the peripheral vascular bed. The algorithms which process the time series to produce estimates of the oxygen saturation are generally proprietary but permit the user to adjust the sampling interval used for the saturation estimate as well as the amplitude scale for the display.

Present pulse oxygen monitor technology does not allow 100% artifact rejection or data validity confirmation. Since patient movement or ambient light can produce erroneous waveforms, pattern recognition technologies may be applicable; random input (achieved by tapping the finger with the sensor attached) resulted in an erroneous saturation estimate of 84 percent. Since the POA waveform is correlated to the patient EKG—albeit with a delay inter-sensor data fusion for data quality assessment is feasible. The correlation of the pulsatile waveform with that of the EKG is illustrated in Figure 122.31.

Supplied Oxygen

The neonate is supplied an air-oxygen mixture where the percent oxygen in the supplied gas is sensed by a paramagnetic oxygen analyzer in the flow tube prior to the neonate. Currently, the gas is supplied at a constant set total flow. Unless the neonate

[10] A module dimension is approximately 1.5 × 5 × 4 inches with a single interface port at the rear.

[11] EKG [G elektrokardiogramm] electrocardiogram, electrocardiograph is abbreviated "ECG" outside of the USA.

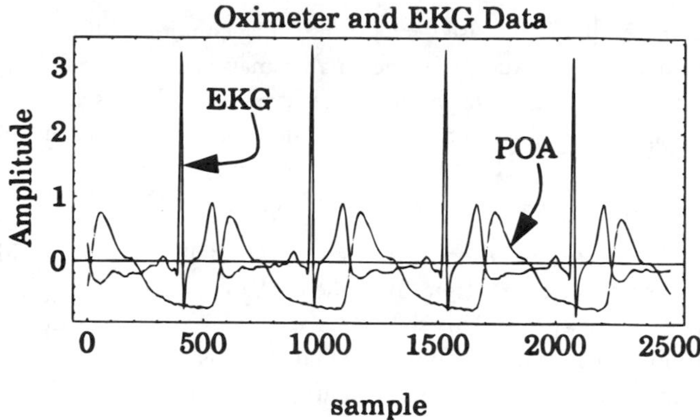

Figure 122.31 PUA waveform vs. EKG waveform.

is in an incubator, additional air is supplied from the environment. The amount of "uncontrolled" air will be determined by the position of the supply tube as well as amount of air inhaled which is a function of activity and position. If the infant is enclosed by an incubator, there will be a "diffusion lag" which must be considered.

Proposed Control System

Figure 122.32 illustrates a proposed control system to monitor the neonate status and adjust the amount of supplied oxygen accordingly. The anticipated operational system would consist of three distinct physical components:

- Gas Flow Controller
- Sensors & Sensor Processing.
- Intelligent Controller.

As with any control system, the plant (neonate) must be well understood; hence, associated with the sensor processing is an implicit effort in neonate and sensor modelling and simulation. An accurate model is required for control system performance analysis as well as design optimization. Note that the model parameters will be a function of the individual patient as well as possibly varying over time for a given patient.

An example of such variation and a possible solution achieved via multi-oximeter data fusion is determining if there is a difference in oxygen saturation in the upper half of the body (blood supplied by the ascending arch of the aorta) compared to the lower half (blood supplied after the ductus arteriosus enters the descending aorta). In cases of pulmonary hypertension which are not infrequent in the NICU deoxygenated blood returning from the body mixes with the oxygenated blood via the ductus arteriosus. This is a peculiarity of the newborn circulation called *persistent fetal circulation.* There are a number of other benefits for multiple oximeter inputs.

Thus, we see that through multi-sensor data fusion, the intelligent controller also offers an improved monitoring and assessment of the patient's state. This additional or more accurate knowledge may then be used by the attending physician and medical staff to improve the treatment plan.

Applicable Technologies

Many signal processing and control system technologies may be applied to biomedical control. The technologies which are most applicable to the current situation are shown in Table 122.7

It is important to realize that although the signal processing associated with MRI and biomedical ultrasound are quite sophisticated, there has been relatively little done in the way of applying signal processing, multi-sensor data fusion, or control system technologies to day-to day patient care. Therefore, fairly simple implementations of the above concepts can be very beneficial.

Control System Components

In this section we briefly describe the system components and associated development issues required for a neonatal oxygenation control system.

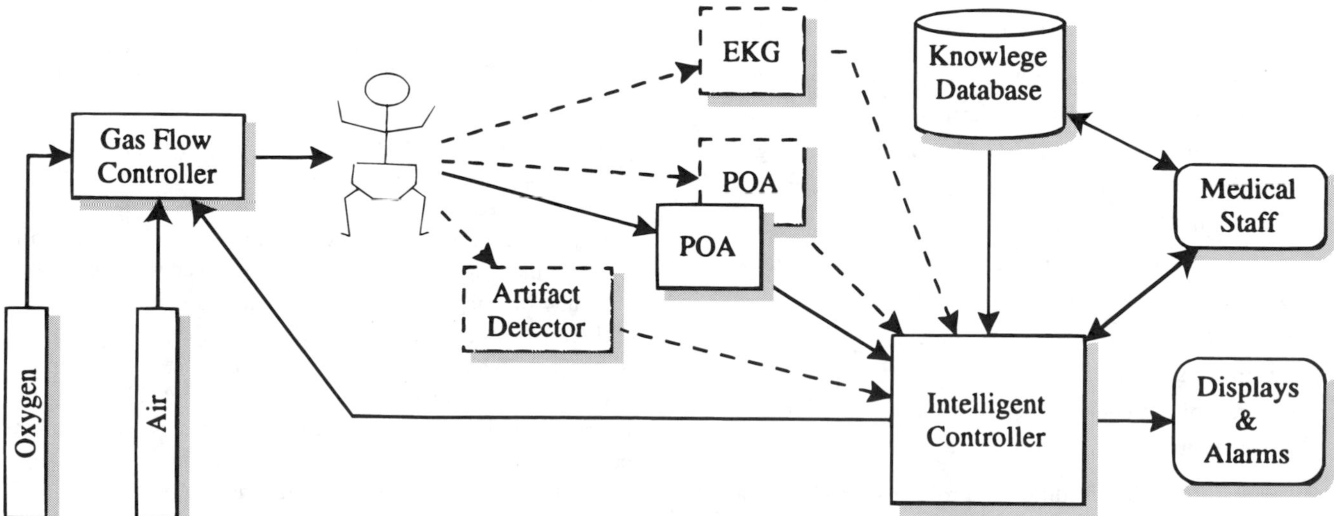

Figure 122.32 Control system block diagram.

Table 122.7 Applicable Technologies

Signal Processing	• Data Fusion
	• Pattern Recognition
	• Matched Filtering
	• Parameter Estimation
Control Systems	• Artificial Intelligence/Expert Systems
	• Fuzzy Logic/Intelligent Control
	• Fault-Tolerant Control
	• Failure Identification
	• Reconfigurable Control

Gas Flow Controller. The gas flow controller is the "autopilot" which will deliver the commanded oxygen flow to the neonate. Envisioned as a classical PID controller along with some safety checking, it would encompass the

- Mixing valve and associated stepper motor or solenoid valve.
- Microprocessor controller and software.
- Pressure reducing valves.
- Rapid response paramagnetic oxygen analyzer.
- Differential transducer (measure flow).
- Fuel cell oxygen analyzer (fail-safe backup).

The control system which is monitoring the neonate would view this device as a "black box." The gas flow controller is functionally distinct from the inspired oxygen controller. Ideally, it would be a general device independent of gas viscosity, pressure, or flow rates.

System Modeling and Sensor Processing. There are two aspects to the system modelling problem: first, the sensor(s) and neonate must be mathematically characterized to permit the implementation of a control system; secondly, the expert knowledge of the physicians and nursing staff should be incorporated into the control system since the complex time-varying nature of a biological system implies more non-linearities and complexity than associated with many mechanical systems. Concerns which must be addressed include:

- Neonate response to supplied oxygen (highly variable).
- Safety issues (defaults and alarms).
- Methods of oxygen supply and supplemental oxygen.
- Sensor characteristics and signal processing.
- Data quality indicators.
- Sensor failure modes and methods of identification.
- Sensor correlation and synergy.
- Expert system database development.
- Control system stability.
- Time constants of sensor and neonate response.
- Sensor and system redundancy.

There are a number of different problems which may afflict a premature infant; each of these can in turn influence the design and operation of the control system. For example, neonates experiencing shunting will not increase their hemoglobin saturation as a result of increased oxygen; such a condition should be recognized. Conversely, some infants may achieve 100% S_aO_2 saturation using atmospheric oxygen levels—which does not pose a risk of ROP. In summary, data fusion and intelligent control are essential aspects of biomedical control systems.

Single-Sensor Controller. The initial prototype would focus on using the output from the POA to control the supplied oxygen percentage. This should be viewed as a relatively simple proof-of-concept prototype oriented towards a robust and safe implementation with extraordinary situations recognized and appropriate alarms generated.

Multi-Sensor Controller. Using the single-sensor controller as a baseline, a multi-sensor system is planned which would perform data corroboration between sensors, do data quality assessments, and utilize the expertise of the medical staff and their *a priori* assessments of the neonate.

The primary patient sensors are anticipated to be the pulse oxygen analyzer (POA) with the EKG used to perform quality assessment of the POA waveforms. This continuous data stream would be processed by the intelligent controller within a context defined by the expert database as well as off-line information gathered through arterial blood gas analysis, blood pressure readings, etc.

The intent here is to develop a framework for future control systems. To illustrate, some current research indicates that some neonate lung problems are due to a lack of surfactants in the lung—so that the lung tissues do not separate properly. Thus, a future system may want to control the introduction of moisture or some other surfactant in the supplied air-oxygen mixture.

Summary

Multi-sensor data fusion coupled with expert knowledge offers a sensor synergy to improve the patient care. Here we have outlined some of the problems associated with neonatal care which inspire the application of control system technologies and real-time data fusion.

There is much to be gained from the synergy possible by making a system responsive to the individual demands of a patient and medical plan. For instance we may lower the ideal oxygen saturation for patients with known intracardiac shunting and raise them for patients with pulmonary hypertension. The possibility of integrating multiple signals is another strength.

There is a multiplicity of monitors in the NICU, each with its own data which the human must interpret and act upon to close the loop in patient care. Data fusion would allow repetitive closing of the loop in routine circumstances by the automated controller with human intervention only required for the seldomly occurring abnormal circumstances. Additionally, multi-sensor integration coupled with an accurate patient model will permit a more accurate assessment of the patient condition—facilitating an improved medical plan and patient care.

Acknowledgments

The authors would like to thank Sorin Vaduva for his assistance in collecting and processing the sensor telemetry used in the control system development.

References

Azhar, N. and Karim, U. 1991. Automatic feedback control of oxygen therapy using pulse oximetry, *Proc. Ann. Conf. Eng. Medicine and Biol.*, 13(4).

Dugdale, R. E. Cameron, R. G., and Lealman, G. T. 1988. Closed-loop control of the partial pressure of arterial oxygen in neonates, *Clin. Phys. Physiol. Meas.*, 9(4):291–305.

Lih, M. M. 1975. *Transport Phenomena in Medicine and Biology,* John Wiley & Sons, New York, NY.

Morozoff, P. E., Evans, R. W., and Smyth, J. A. 1993. Automatic control of blood oxygen saturation in premature infants, *2nd IEEE Conf. Control Applications,* September 13–16, Vancouver, B.C.

122.6 Multiresolution Multisensor Target Identification

Zbigniew Korona and Mieczyslaw M. Kokar

Introduction

The goal of this paper is to investigate a multisensor multiresolution approach to target identification. One sensor often does not provide enough information to build an ATR (Automatic Target Recognition) system that performs effectively against targets in natural and man-made clutter. There are strong indications (Kolodzy, 1993) that the use of several sensors gives target identification better results than the use of one sensor. Distributed filtering is an effective and computationally efficient method for optimal dynamic data fusion for uniform resolution sensors (Hong, 1991). The important problem is the fusion of data from sensors (multicolor IR, visual, microwave, etc.) that provide information in different frequency bands and at different resolutions. Recently, the algorithms for dynamic multiresolution distributed filtering were proposed, (Chon et al., 1994; Hong, 1993), as well. In Hong (1993) the Discrete Wavelet Transform (DWT) is utilized as a bridge linking signals at different resolution levels. However, in most ATR problems, like recognition of ground targets, there is no underlying dynamic model and we cannot apply the idea of multiresolution dynamic filtering. In this work an algorithm for multiresolution multisensor data fusion without a model is proposed. The Inverse Discrete Wavelet Transform (IDWT) is used to fuse data from different resolution levels. We assume that data from all sensors are registered and that signals at different resolution levels are linked by the DWT. In our experiments we use two sensors operating in different frequency bands with differing resolutions. We compare the performance of our multiresolution multisensor identification algorithm with the performance of the scale sequential algorithm (Devaney, 1994) for one sensor.

In Discrete Wavelet Decomposition a short background of the discrete wavelet decomposition is presented. Scale Sequential Identification describes the problem of an M-ary target identification in a sequential procedure that proceeds from low to high scale (resolution levels) in the multiresolution analysis of data space (Devaney, 1994). In a manner analogous to sequential hypothesis testing in time (Wald, 1947). Fusion for Target Identification discusses the problem of sensor/data fusion. It describes two common approaches to fusion. In the first approach data are fused at the levels as dose to sensors as possible (fuse-then-abstract). In the second approach, information is first abstracted and then fused (abstract-then-fuse). Multisensor Target Identification Scenario describes the simulated scenario for target identification and Results of Experiments presents the results of experiments. And finally, Future Research discusses directions for future research.

Discrete Wavelet Decomposition

It is shown in Vaidvanathan (1993) that a discrete signal can be decomposed into multiresolution frequency channels by using the DWT (Figure 122.33). Consider the space $l^2(Z)$ consisting of the square summable sequences $s^j(i)$, $i \in Z$. For a given signal $s^j(i) \in l^2(Z)$ at resolution level j, a lower resolution low frequency signal can be derived by using a lowpass filter with an impulse response $h(i)$ and then by subsampling the output of the lowpass filter by two,

$$s^{j-1}(i) = \sum_n h(2i - n)s^j(n). \tag{122.4}$$

Similarly, a lower resolution high frequency signal can be derived by using a highpass filter with an impulse response $g(i)$ and then by subsampling the output of highpass filtering by two,

$$w_s^{j-1}(i) = \sum_n g(2i - n)s^j(n). \tag{122.5}$$

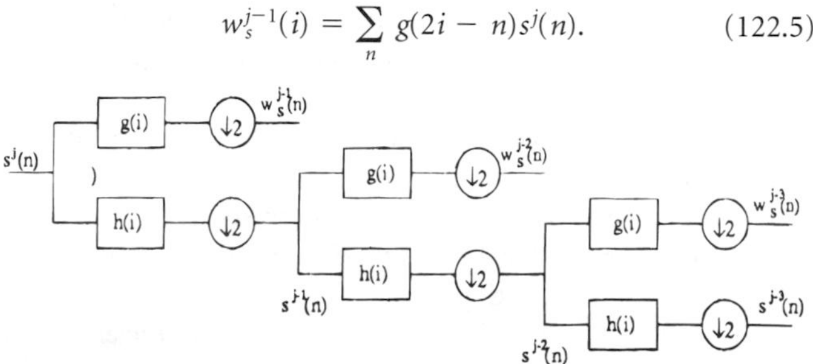

Figure 122.33 Discrete wavelet decomposition.

* This research is partially sponsored by the Advanced Research Projects Agency under Grant F49620-93-1-0490.

The original signal $s^j(i)$ can be recovered from the two lower resolution signals $s^{j-1}(i)$ and $w_s^{j-1}(i)$. However, filters $h(i)$ and $g(i)$ must meet some constraints to guarantee perfect reconstruction: the regularity constraint (Daubechies, 1988) and the orthonormality of the filter impulse responses. In such a case Equations 122.4 and 122.5 can be considered as a decomposition of the signal onto an orthonormal basis. The original signal $s^j(i)$ can be reconstructed by

$$s^j(i) = \sum_n h(2n - i)s^{j-1}(n) + \sum_n g(2n - i)w_s^{j-1}(n). \quad (122.6)$$

In the *discrete wavelet decomposition* mode the output of the lowpass filter $h(i)$ is further decomposed (see Figure 122.33) according to the procedure described above. This decomposition gives "details" (wavelet coefficients $w_s^j(n)$) of the signal as the output of highpass filter $g(i)$ at each resolution level j and the "average signal" (scaling coefficients $s^{jo}(n)$) as the output of the lowpass filter at the lowest resolution level j_0.

Another well known decomposition mode is the *discrete wavepacket decomposition* mode (Dasai and Shazeer, 1991) in which both outputs from the lowpass filter and highpass filters are further uniformly decomposed into low and high frequency signals.

Scale Sequential Identification

In this section we describe the general idea of the scale sequential approach to target identification. More details can be found in Devaney (1994).

We assume that the target signature database is available. The candidate hypotheses for a target ID are defined by

$$H_k\colon y(t) = s_k(t) + N(t), \quad (122.7)$$

where $y(t)$ is the measured signal, $N(t)$ is noise and $s_k(t)$, $k = 0, 1, \ldots, K - 1$, are target signatures. The multiple hypotheses H_k are assumed to be mutually exclusive. The optimal procedure is to test all $k \neq 0$ hypotheses against the null hypothesis H_0: $y(t) = N(t)$ and select that hypothesis that maximizes the log likelihood ratio over this set. However, such a procedure is computationally very intensive. To reduce the number of computations, we decompose the target signatures $s_k(t)$ and the measured signal $y(t)$ into the lower resolution signals using the wavelet decomposition. Computational savings during hypothesis testing are important in the case when we deal with large target signatures databases. They can be large not only because of a large number of targets we want to identify, but also because often there is a need for a large number of signatures corresponding to the same target (for example for different orientations of the target in a case of radar identification).

Let $w_y^j(n)$, $w_s^j(n)$ and $w_N^j(n)$ be the wavelet coefficients of the data $y(t)$, the k'th hypothesis signal $s_k(t)$ and the noise process $N(t)$ at the resolution level j and the time index n. Then, the candidate hypotheses in the wavelet space W^j are defined by

$$H_k\colon w_y^j(n) = w_{s_k}^j(n) + w_N^j(n). \quad (122.8)$$

The wavelet basis is assumed to be orthonormal and the noise to have the Gaussian distribution. These hypotheses are tested by applying a log likelihood ratio approach in the sequential procedure which proceeds from coarse to finer resolution levels. At each resolution level those hypotheses that fail the test are rejected and removed from consideration at higher resolution levels. The sequential decision process is a generalization of Wald's theory of sequential binary hypothesis testing (Wald, 1947). For $\sigma = 1$, the log likelihood ratio becomes

$$L(k) = \log \frac{P_k(y)}{P_0(y)} = \sum_j \sum_n w_{s_k}^j(n) w_y^j(n) - 1/2 \sum_j \sum_n v_{s_k}^j(n)^2. \quad (122.9)$$

In general $j = -\infty$ to $+\infty$. However, in practical applications $j = j_0$ to J, where j_0 is the finest resolution and J is the coarsest resolution. Then Eq. (122.6) becomes

$$L(k) = \sum_{j=j_0}^{J} \sum_n w_{s_k}^j(n) w_y^j(n) - 1/2 \sum_{j=j_0}^{J} \sum_n w_{s_k}^j(n)^2 + L_0(k), \quad (122.10)$$

where $L_0(k)$ is the log likelihood ratio computed at the coarsest resolution:

$$L_0(k) = \sum_n s_k^{jo}(n) y^{jo}(n) - 1/2 \sum_n s_k^{jo}(n)^2, \quad (122.11)$$

and where $y^{jo}(n)$ and $s_k^{jo}(n)$ are the scaling coefficients at the coarsest resolution j_0.

The main reason that we use the wavelet based sequential approach to target identification is due to computational savings. The number of coefficients $w_{s_k}^j(n)$ of each hypothesis signal is $N_j = 2^j$ so that the number of computations for computing the log likelihood ratio decreases geometrically with resolution level. According to Devaney (1994) the number of computations (the number of multiplications and additions) required to test K hypotheses at the finest resolution is

$$N_{fine} = 4K(2^{J_{max}} - 1/4). \quad (122.12)$$

In the wavelet based sequential approach we need only

$$N_{seq} = 4K(2^{j_o} - 1/4 + J_{\max} - j_0) \quad (122.13)$$

computations if 50% of candidate hypotheses are rejected at each resolution level. The computational savings ratio S is

$$S = \frac{N_{fine}}{N^q} = \frac{2^{J_{\max}} - 1/4}{2^{j_o} - 1/4 + J_{\max} - j_0}. \quad (122.14)$$

For example, for $J_{max} = 10$ and $j_0 = 5$ we get $S \approx 38$.

Fusion for Target Identification

Our primary goal in this work is to investigate whether fusing data from two sensors, operating in different frequency bands with differing resolutions, can improve target identification in the presence of noise. The expectation of improved target identification probability based upon the combined information of two different signals over that of just one signal is quite intuitive. It is natural to expect that, through fusing signals from a number of sensors, some complementary information can be obtained that is not obtainable through only one sensor. For example, the radar provides range information and its signal is less affected by atmospheric attenuation as compared to the infrared image, while the infrared sensor provides information of greater resolution and is able to operate at night.

There are two common approaches to perform target identification (classification). In the first one, target is identified by individual sensors separately. Then the information from all sensors is fused to obtain a final decision (abstract-then-fuse). In the second approach, data from all sensors are combined together by so called low level (or sensor level) fusion. Then the target is identified using the fused data (fuse-then-abstract).

Our main objective is to determine how to perform sensor integration and data fusion at levels closer to the sensor. The goal is to increase available information while reducing the amount of processing. By using partially redundant information from the different sensors, robustness of the ATR system is increased. A failure of one sensor does not necessarily result in the failure of the rest of the system. Because of this, less expensive sensors can be used, thus improving the economy of the ATR system.

Multisensor Target Identification Scenario

For the purpose of illustrating the performance of our multisensor identification algorithm we were considering the problem of target identification using two sensors and a database (DBT) of target signatures. Figure 122.34 shows our simulated scenario for target identification. There are two sensors s_1 and s_2 operating in different frequency bands with differing resolutions. In particular, sensor s_1 operates in a high frequency band and its resolution is twice as big as that of sensor s_2 which operates in a low frequency band. We assume that the DBT is created using a high quality sensor operating in the frequency band which completely covers the frequency bands of sensors s_1 and s_2. This high quality sensor has also higher resolution than sensors s_1 and s_2. Additionally, we require that it is possible to decompose the DBT (Figure 122.35) into three new databases DB1, DB2 and DBS such that DB1 consists of the target signatures for sensor s_1 and DB2 consists of the target signatures for sensor s_2. The DBS is the complement signature database such that completeness of the DBT's information is preserved. The DBS database provides data for the middle branch of the scenario presented in Figure 122.34.

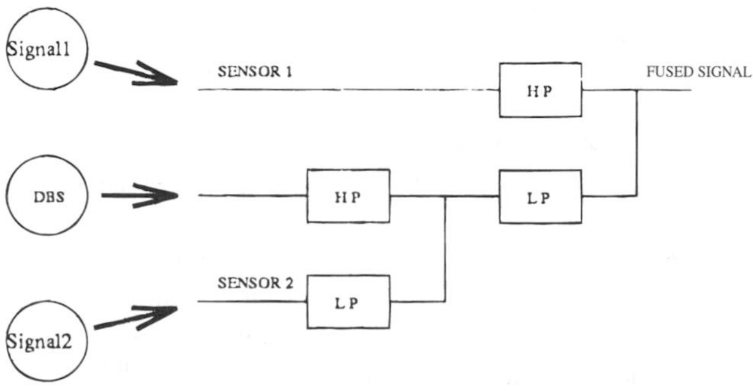

Figure 122.34 Multisensor scenario for target identification.

The M-ary hypothesis testing is performed for both sensors separately. We accept N_1 hypotheses based on data from sensor s_1 and N_2 hypotheses based on data from sensor s_2. The number of hypotheses N_1 and N_2 depends on the way in which we implement the scale sequential hypothesis testing. In general, both N_1 and N_2 are very small. Next, we fuse signatures from the DBS database, which are determined by N_1 and N_2 hypotheses, with the data from sensors s_1 and s_2. This fusion utilizes the perfect reconstruction properties of the IDWT. The reconstructed signal has two advantages. First, it has a higher resolution than each of the individual signals separately. Second, the reconstructed signal is less noisy, because the measurement data are fused with the deterministic signatures. If we assume that the standard deviation of the measurement noise for each sensor is σ then the standard deviation of the fused signal will be less than σ. In particular, for the fused signal in our scenario, the standard deviation σ_f is:

$$\sigma_f = \sqrt{((\sqrt{\sigma^2/2}))^2 + \sigma^2/2} = \sqrt{3}\sigma/2 < \sigma. \quad (122.15)$$

After the fused signal is obtained, we perform hypothesis testing on the fused data. All N_1 and N_2 hypotheses are considered. We directly apply a likelihood ratio test to test these hypotheses. The log likelihood ratio then becomes

$$L(k) = \log \frac{P_k(f)}{P_0(f)} = \sum_i s_k(i)f(i) - 1/2 \sum_i s_k(i)^2, \quad (122.16)$$

where $f(i)$ is the fused signal, $s_k(t)$ are target signatures from the DBT and index i is discrete time. In general, the number of tested hypotheses using the fused signal is not large and the computational overhead is not significant. It could be as few as one hypothesis accepted by each sensor.

Results of Experiments

In our experiments, we performed an M-ary hypothesis testing in a scale sequential procedure for each of the two sensors separately until only one hypothesis was accepted for each sensor. In the case that these two hypotheses differ fusion of the two signals was performed. The final decision were made based on the fused signal using the likelihood ratio test. We formed the DBT by generating random sequences of 256 samples each. The signatures were mapped into the wavelet domain by using the Haar basis.

Figure 122.36–122.38 show a progress of the process of signal identification based on measurement data from sensor s_1 only.

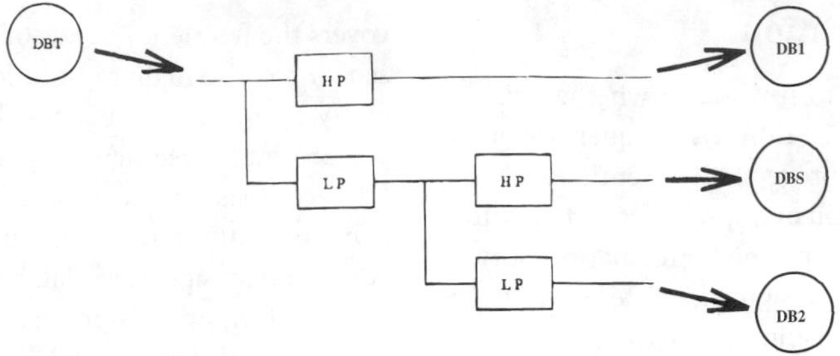

Figure 122.35 Decomposition of the target signatures database (DLT).

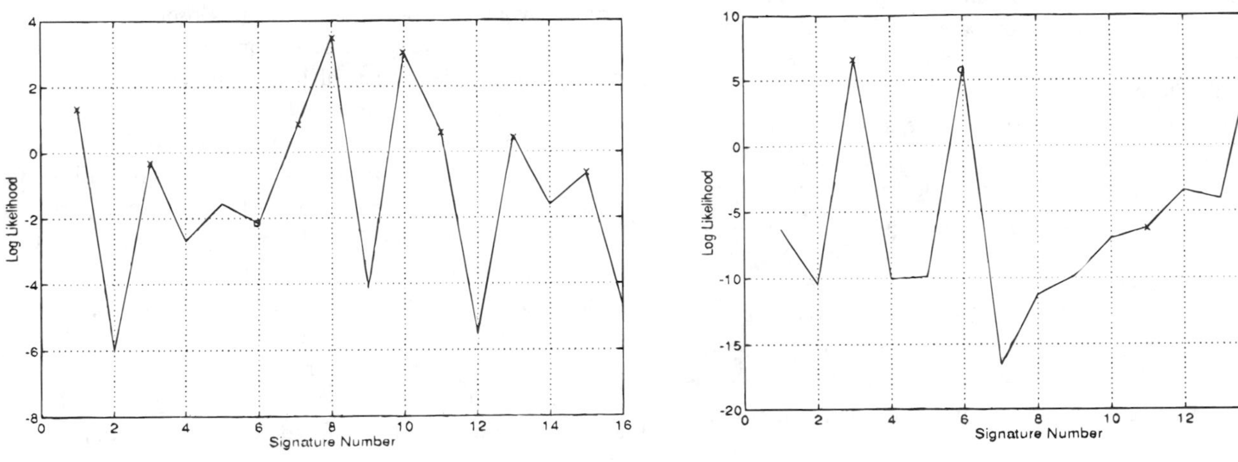

Figure 122.36 Log likelihood for sensor 1 at scale j = 4.

Figure 122.38 Log likelihood for sensor 1 at scale j = 6.

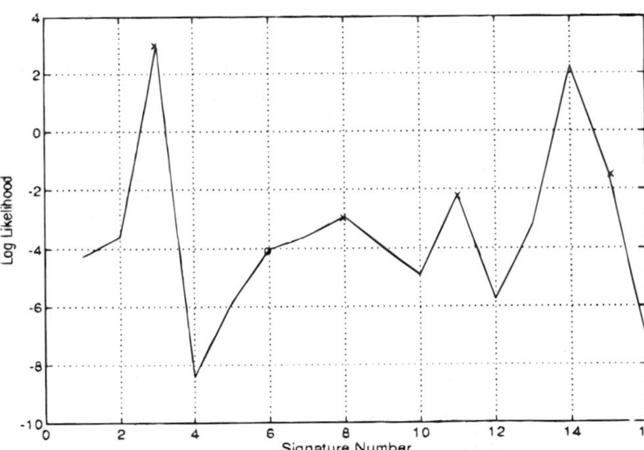

Figure 122.37 Log likelihood for sensor 1 at scale j = 5.

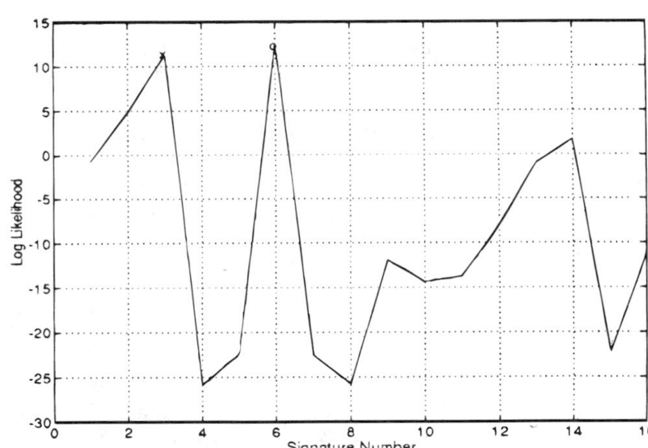

Figure 122.39 Log likelihood for sensor 1 at scale j = 7.

The "x"-es on the diagrams show which hypotheses are accepted at different resolution levels. Originally, we have 16 hypotheses. The small circles indicate the "true" hypothesis. Figure 122.39 shows the final decision for sensors s_1 at the finest scale (hypothesis 3 is accepted in our case). Figures 122.40–122.43 show a progress of the process of signal identification based on measurement data from sensor s_2. The hypothesis 6 is accepted at the finest level. Sensor s_1 works at resolution twice as high as that of sensor s_2. In particular, in our simulation, sensor s_1 was providing 256 measurements and sensor s_2 was providing 128 measurements. Figure 122.44 shows the final (fused) decision. The hypothesis 6 is chosen instead of hypothesis 3 and this decision is correct. In this experiment, the measurements were corrupted by white noise with the standard deviation $\sigma = 2$.

We ran experiments for different levels of noise. Figure 122.45 shows the performance of our algorithm. There are 50 runs for each level of noise. The level of noise is characterized by the standard deviation. The "-." line denotes the number of correct decisions made by sensor s_1 only. The "-" line denotes the number of correct decisions made by sensor s_2 only. The solid line denotes the number of correct decisions based on fused data. Notice that for a lower level of noise all decisions are correct. However, in a

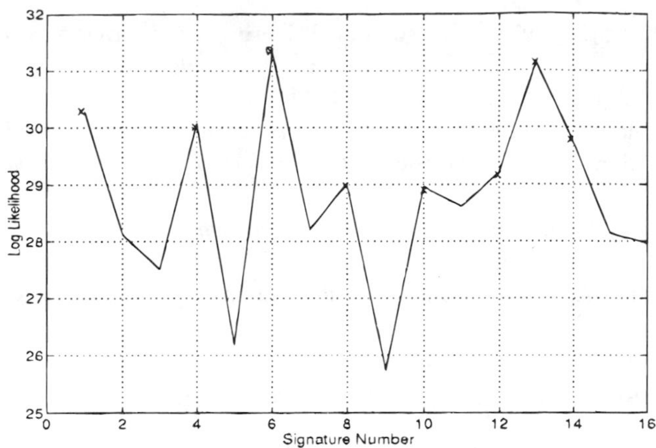

Figure 122.40 Log likelihood for sensor 2 at scale j = 4.

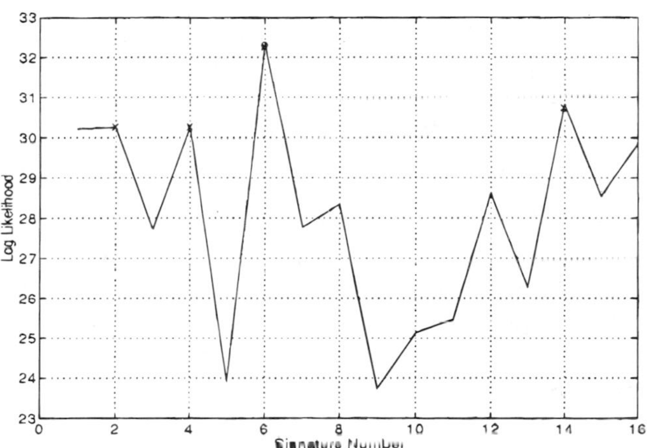

Figure 122.41 Log likelihood for sensor 2 at scale j = 5.

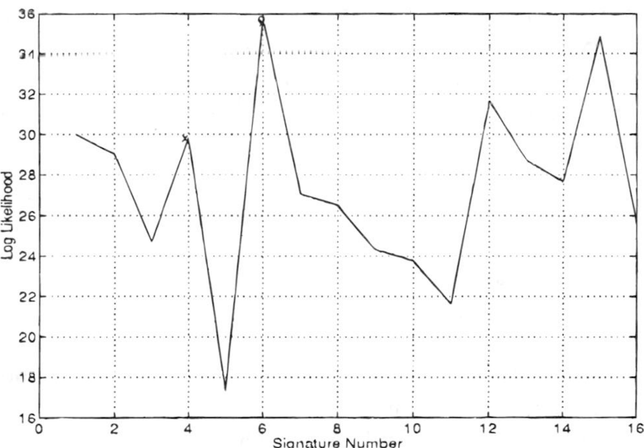

Figure 122.42 Log likelihood for sensor 2 at scale j = 6.

more noisy environment, fused decisions are much more accurate than decisions made by each sensor separately.

Future Research

One of our research goals is to develop a generic sensory data fusion framework. Towards this aim we carry out both theoretical

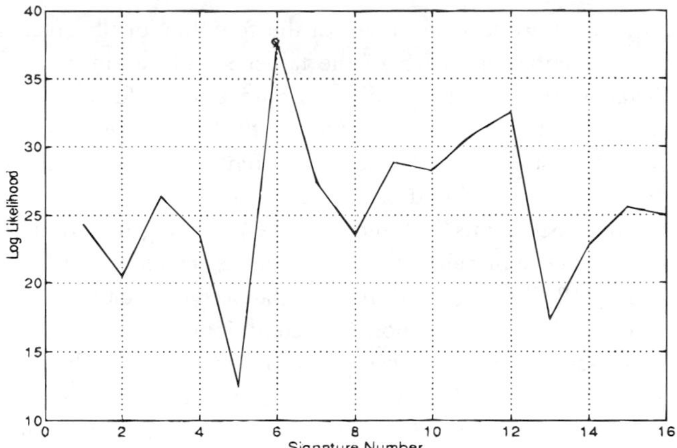

Figure 122.43 Log likelihood for sensor 2 at scale j = 7.

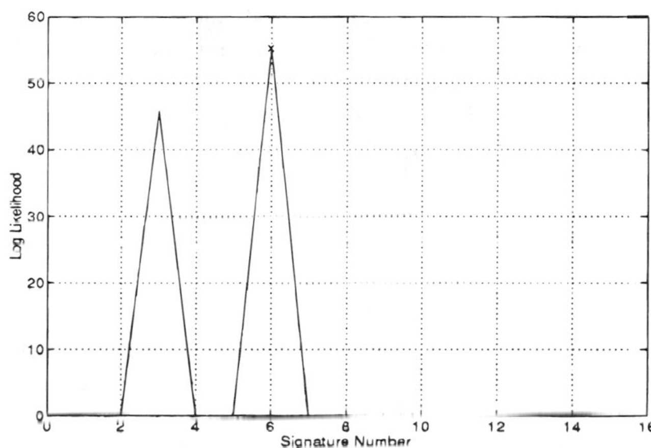

Figure 122.44 Log likelihood for fused signal.

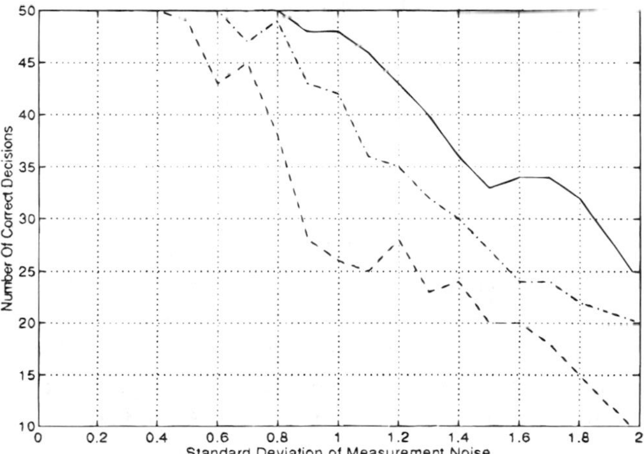

Figure 122.45 Comparison of identification performance between the proposed multisensor algorithm and the algorithm based on one sensor.

investigations and experimental work. In proposing new architectural and algorithmic solutions we try to take advantage of those solutions that are well established in the subject literature, but at the same time we try to both expand the applicability and to improve the performance of our system by adding new features.

In this work we take advantage of the computationally efficient scale sequential approach to the target identification problem (Devaney, 1994). But in order to improve its performance, we extend this approach to multiple multiresolution sensors. The IDWT was used to fuse data at different resolution levels. In particular, we employed the simple Haar wavelets to choose the filter coefficients $h(i)$ and $g(i)$. We are going to investigate performance of our algorithm for other wavelet bases. For example, for FIR filters of width $N + 1$, Daubechies (Kolodey, 1993) has defined a set of smoothness conditions which produce a unique set of coefficients $h(i)$ and $g(i)$, $i = 1, \ldots, N$. The filters defined by these coefficients have better bandpass characteristics than those defined by the Haar wavelets. It was shown in Devaney (1994) that the Daubechies basis has potential to give better results than the Haar basis. Also, we intend to relax gradually the assumption that signals at different resolution levels are linked by the DWT. We are going to investigate the sensor/data fusion problem for unregistered signals. In such a case, in general, it will not be possible to reconstruct the "true" signal. However, we believe, we will be able to obtain an approximation that will lead to the improvement of the performance of the identification process.

Conclusion

An algorithm for multiresolution multisensor target identification using the IDWT is proposed. It is shown that identification decisions using fused data from two sensors are more reliable than using each sensor separately. The significant increase in the number of correct identification decisions (especially for a low signal/noise ratio) is achieved by only a small increase in the number of computations. Another advantage of this method is that we can use two cheap sensors instead of a very expensive one to perform a target identification. The experiments were limited to the multiresolution. The experiments were limited to the multiresolution sensors which are linked by the discrete wavelet transform.

References

Chou, K. C., Willsky, A. S., and Benveniste, A. 1994. Multiscale recursive estimation, data fusion, and regularization, *IEEE Trans. Auto. Control.*, 39(3):464–478.

Daubechies, I. 1988. Orthonormal bases of compact supported wavelets, *Communications on Pure and Applied Mathematics,* 91:909–996.

Desai, M, and Shazeer, D. J. 1991. Acoustic transient analysis using wavelet decomposition, *IEEE Conf. Neural Networks for Ocean Engineering,* Washington, DC.

Devaney, A. J. 1994. Scale Sequential Processing of HRR Radar Data for Automatic Target Identification, *Technical Report, The Center for Electromagnetics Research,* Northeastern University.

Hong, L. 1991. Adaptive distributed filtering in multi-coordinated systems, *IEEE Trans. Aerospace and Electronic Systems,* 28(4):715–724.

Hong, L. 1993. Multiresolutional filtering using wavelet transform, *IEEE Trans. Aerospace and Electronic Systems,* 29(4):1244–1251.

Kolodzy, P. J. 1993. Multidimensional automatic target recognition system evaluation, *The Lincoln Laboratory Journal,* 6(1):117–146.

Vaidyanathan, P. P. 1993. *Multirate Systems and Filter Banks,* Prentice Hall, Englewood Cliffs, NJ.

Wald, A. 1947. *Sequential Analysis,* John Wiley & Sons, New York, NY.

122.7 Shaping Control of Plastic Object by Robot Hand with Sensor Fusion Processing

Ryosuke Masuda and Michio Sasaki

Introduction

Recently, many robots have been used in wider application fields, and then are required to have a dexterous manipulating ability similar to that of humans. For the special-purpose robots which work in service fields and other nonindustrial environments, a dexterous hand system must be designed (Taylor, 1990; Cutkosky, 1989; Mason and Salisbury, 1985). To give dexterity to the robot hand, several kinds of sensors, such as visual and tactile sensors, and a total signal processing system are indispensable (VenKataraman, 1989).

In this chapter, sensor fusion processing is adopted for recognizing a situation of the robot hand, and controlling it to realize the dexterous manipulation tasks (RSJ, 1991; Luo and Kay, 1989). As an example of this task, we discuss shaping a plastic object to a given three-dimensional shape with the special mechanism of a robot hand and the multi-tactile sensors system (Nagahama and Masada, 1991).

As an example of a plastic object, a lump of clay is used in this research. The clay has the easy transformable characteristics according to an external force, and it is difficult to make a mathematical model of the clay because it is vague and time variant. So a simple force feedback control is not adequate to handle the plastic object and then the information of the force and the pressure distribution, and the inner position/velocity are combined to process. In this process, the data fusion technology can be applied and then the state of shaping control can be clearly known. And further, the control signal to the hand actuator can be determined with this processed information. By using the proposed sensor fusion processing, the delicate shaping control system for the robot hand can be accomplished.

In this chapter, first we discuss the mechanism and the sensors which are used to shape a lump of clay to a given shape. And the sensor fusion processing algorism for the shaping control is also mentioned. Then we will show the experimental results of the shaping control and evaluate the proposed method.

Problem of Shaping of Plastic Object

Handling of Plastic Object

Most handling of objects in a robot hand are either rigid or an elastic material in the industrial application. But when a robot is used in the fields of chemical, biomedical, and food industries, it is necessary to handle a plastic and a fragile object. To accomplish such a dexterous handling operation, certain kinds of effector sensors play very important roles as well as the multi-degree of freedom mechanism of the hand.

The purpose of this research is to realize the dexterous handling of a plastic object without damage and define the shaping operation of a lump of clay to a given shape by using the fusion processing of multi-tactile sensors.

Here we propose practical methods of shaping control. The shaping control is tried with the profile position control so far, but in such a case, it is difficult to control both force and position adequately because of the complexity and uncertainty of the characteristics of plastic materials.

In this chapter, we adopt the sensor fusion processing method for the shaping of plastic materials by a robot hand. For this purpose, we use the signal processing system and the robot hand control system, both based on multi-sensor fusion. There are two ways to control the shape of plastic objects: one is a method of rolling an object on the palms which is shown as Figure 122.46a. The other is a fingering method with four fingers on both hands as shown in Figure 122.46b. We call the former method "palm shaping" and the latter "finger shaping."

Construction of Hand for Shaping

Two types of shaping methods are considered, as shown in Figure 122.46. We designed and made the robot hand mechanisms on trial which accomplished the hand tasks like a human hand can do. The structure of the hands and their coordinate systems are shown in Figure 122.47.

Hand (a) has two palms. The upper palm has the functions of pressing down and sliding the object for shaping. The pressing motion forces change in the shape of a plastic object and the sliding motion rolls an object to change its posture. We call this type of hand a "palm hand."

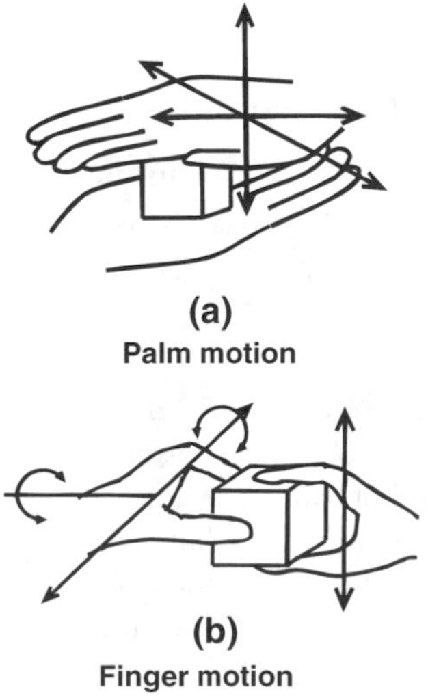

Figure 122.46 State of dextrous handling motion by human hand. (a) Palm motion (b) Finger motion.

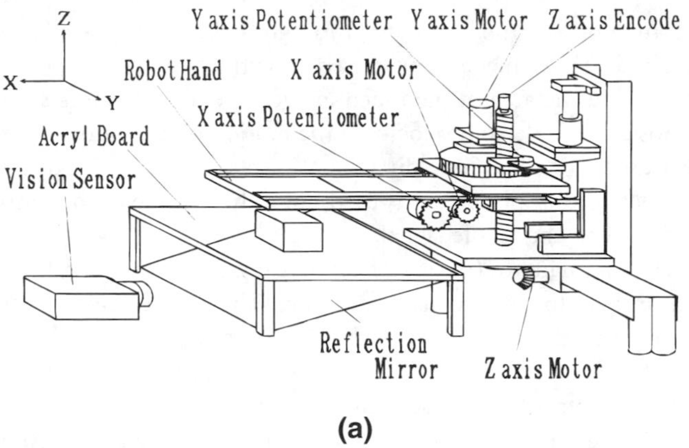

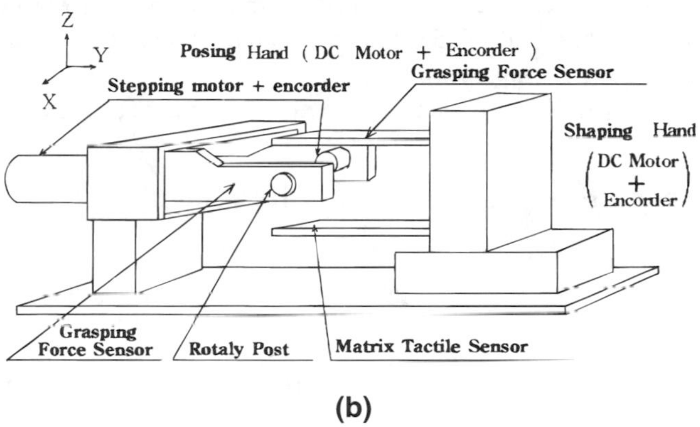

Figure 122.47 Construction of hand system. a. Palm hand system. b. Pair hand system.

Hand (b) has two pairs of fingers. One has the function of open-close on a horizontal plane, which are called "posture fingers," holding the posture of an object. The other pair of fingers on the vertical plane, which are called "shaping fingers," has the pressing and shape-forming functions.

The posture fingers can rotate a grasped object around the X axis by the rotation post. The shaping fingers can rotate itself within the limited angle around the Y axis. We call this type of hand a "pair hand."

Sensor System for Shaping Control

As for the sensor system, we use a conductive rubber tactile sensor which detects a total pressed force on the upper palm, and the tactile image sensor which detects the touch pattern image on the lower palm.

The optical detecting principle is adopted for detecting the touch pattern of the object. This mechanism consists of a transparent board as the lower palm and a light source and a light guide, a plane mirror and a CCD camera, as shown in Figure 122.47a. The contact area of the grasped object is detected as a black and white image of the CCD camera, and it is put into the memory of the computer through a video interface. The detected tactile image is processed to extract the center point, shape, and the area of the touch image. The inner sensors such as position, angle, and velocity of the motion of mechanisms are also used to detect three-directional displacement of the palm. The detected displacement is used not only to feedback but also to fuse with the tactile signal.

The pair hand of (b) has the strain gauges type grasping force sensor and the matrix tactile sensor. The conductive rubber sensor is used for the matrix tactile sensor in which the sandwich structure with the electrode film is adopted. The pressure distribution is detected by the scanning circuit and so the analog force signal of each point is put into the computer through an A/D converter. The force sensors are settled on both the fingers and the matrix tactile sensor is settled on the shaping finger. This hand also has the displacement sensor and rotational angle sensor around the X and Y axis.

The view of the robot hand system is shown in Figure 122.48.

Characteristics of Clay and its Model

Characteristics of Clay

The basic nature of the clay is written as follows (Ogawa, 1991).

1. Composed by fine mud particles.
2. Sticky and plastic.
3. Watery.

The characteristics of clay, especially the plasticity and elasticity, are dependent on two factors: the size of particles and the watery percentage. The clay has sticky characteristics which arise from the smaller size of particles. And the fragile characteristics arise from the larger size of particles. If the water content increases in a lump of clay, the distance between each particle becomes greater, then the fluidity is given, and when the water content decreases in clay, the rigidity appears and the plasticity disappears.

When the robot intended to handle a lump of clay, it was impossible to make a mathematical model properly beforehand because the hypothetic model of grasped clay may change greatly according to the condition of the clay stated above. Needless to say, the percentages of watery content cannot be measured. At the initial state of the task, the characteristics of grasped clay is not known, and although the characteristics are detected before the task starts, it may vary from moment to moment. The shaping of the clay to a given shape is difficult to execute without the model, and so we only make a simple model with the information extracted by multi-sensor processing.

Dynamic Model of Clay

There are several dynamic models of plastic clay indicated so far. Figure 122.49a shows the stress-strain characteristics of the plastic materials, and 122.49b shows the strain rate-stress characteristics of Bingham fluid. The plasticity is the characteristic in which the deformation of the shape remains after removing the applied force, and it is called "permanent set." This permanent set is generated when the strength of the applied force exceeds a certain level. That force is called "yielding force," and is written as $F0$.

The velocity of deformation is represented as follows,

$$D = dZ/dt = (F - F_0)/\eta$$

where η is the constant called "plastic viscosity," and Z is deformation, F is the applied force.

Figure 122.50a shows the visco-elastic model of the elastic material. The rheology model of plastic materials is represented by the parallel connection of a slider and a dumper as shown in Figure 122.50b. (In this figure, the slider represents the fluid phenomenon generates with constant deformation rate, after the force exceeds the yielding force.) This rate is generally proportional to the applied force. This fluid's characteristics differs from that of simple viscose material. The pressed force-fluid characteristics are not linear by the influence of the increase of the cross section. When the force exceeds the elastic limit, the object is shortened to the pressing direction, and is spread to an orthogonal direction. So the pressed area will increase, and then the pressure-fluid curve becomes to 1/2 order curve.

By this fact of characteristics, it is clear that the touched area information is very important for the shaping control.

Knowledge Base for Clay Shaping

From the experience of clay shaping by human fingers, some knowledge of clay shaping is extracted. Examples of the knowledge are listed below.

1. The clay cannot be shaped by the constant force control.
2. The clay cannot be shaped by the position control only.
3. The displacement of the palm and the rotation angle of the posture fingers affect the shaping results.

With the assistance of this kind of knowledge base, the control signal may be determined. The detail of this method is introduced in the following section.

Sensor Signal Processing for Shaping

Extraction of Clay Information

To shape the clay which has the fluid and plastic characteristics stated above, the following data must be extracted before the execution of the task.

Visco-Elastic Characteristics at the Plastic Area. - While holding the object by the posture finger to fix the posture of the object, the robot hand must grasp within the elastic characteristics area by measuring the limit of elastic area. The elastic modulus k is measured by moving a small displacement of the hand ΔZ, step by step, and detecting the increment force Δf, then,

$$k = \Delta f / \Delta Z$$

The viscosity coefficient can be measured by using the rate of

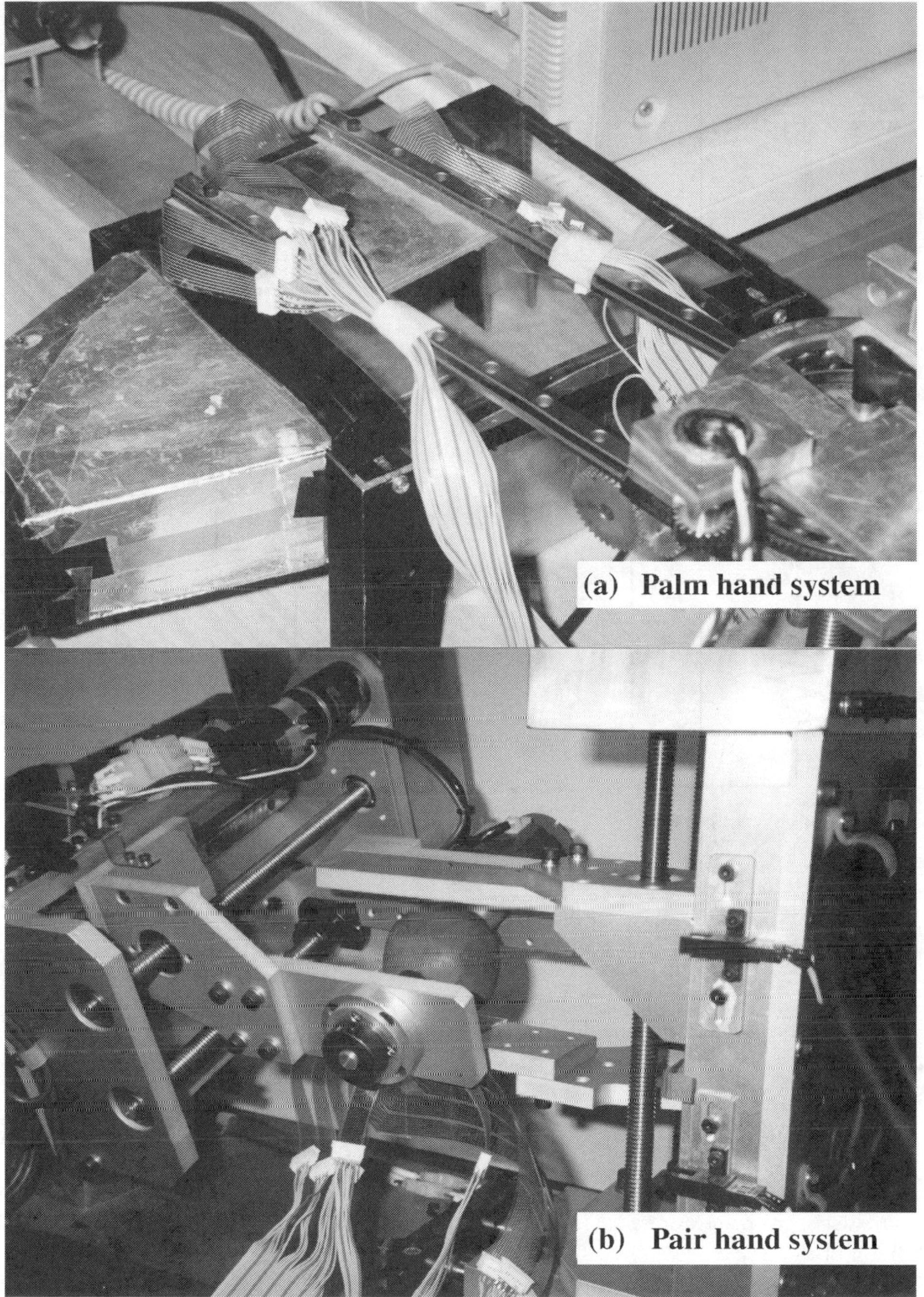

Figure 122.48 View of the hand system for shaping control. a. Palm hand system. b. Pair hand system.

displacement, but it may have much error by this method, and so we do not use this information for the processing.

Judgment of Fluid Generation. Next, the robot hand must find the generation of fluid of the clay. For this purpose, we use the multi-sensor information to judge correctly the yield criterion of pressed force *F*0 and the touch area information.

The fluid generation of clay is judged by the special condition that the change of the deformation becomes large against the increment of pressed force. To use another word, the "creep" generates on the grasped object.

If the state of the clay changes to the plastic from elastic state, the pressed force must be cut down to less than the yielding force.

$$F_0 = \max(F\text{: elastic state})$$

Visco-Elastic Characteristic at the Plastic Area. The main parameter of the visco-elastic characteristics at the plastic area is the plastic viscosity η. It is calculated by using the derivative of displacement by pressing with over yielding force.

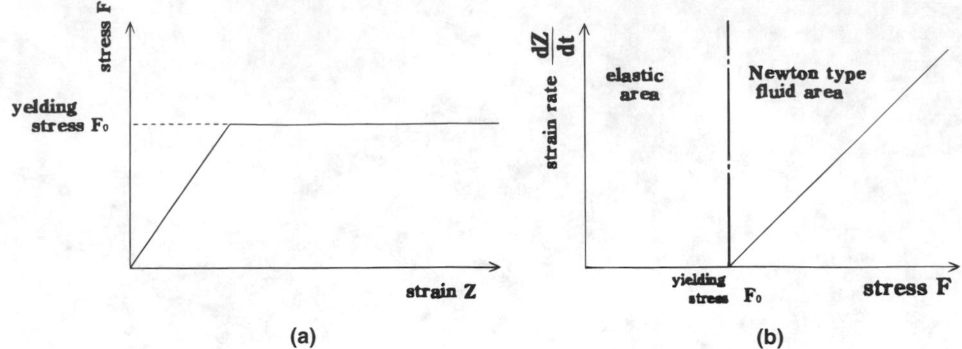

Figure 122.49 Characteristics of a lump of clay. a. Stress—strain Characteristics of plastic materials. b. Stress—strain Rate Characteristics of Bingham Fluid.

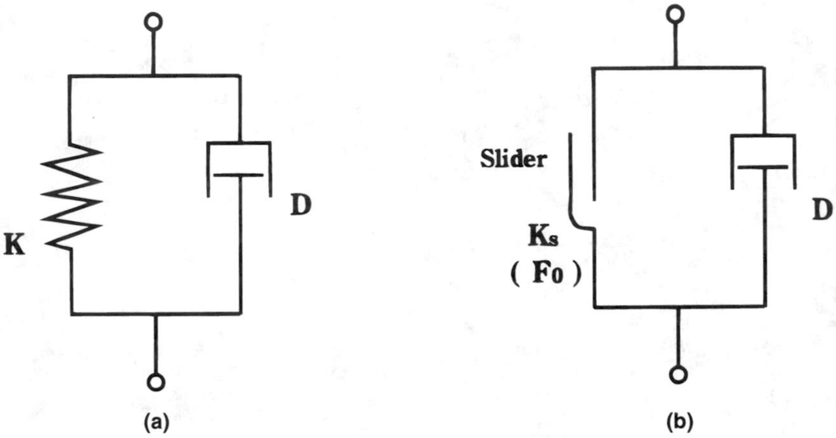

Figure 122.50 Dynamic model of a lump of clay. a. Visco—Elastic Model. b. Visco—Plastic Model.

This parameter can be fused and corrected with the differentiation of touch area ΔA.

$$\eta 1 = fT/\Delta Z$$

$$\eta_2 = fAT/(A\Delta A)$$

where T is the sampling time.

Sensor Information for Shaping

The concept of sensor processing for this shaping is shown in Figure 122.51. This figure shows the flow of the sensor processing and processed information.

In this system, the displacement and the angle of each finger is detected by the inner sensors, and the touch area, shape, and center point of the grasped object is detected by the tactile image sensor or the matrix tactile sensor.

Recognition of Shaping State

The judgment of the shaping state is accomplished using only the three-dimensional shape information. Here, we use the contact area information A by the matrix sensor, displacement of the finger $Z0$ by the position sensor, and the pose information

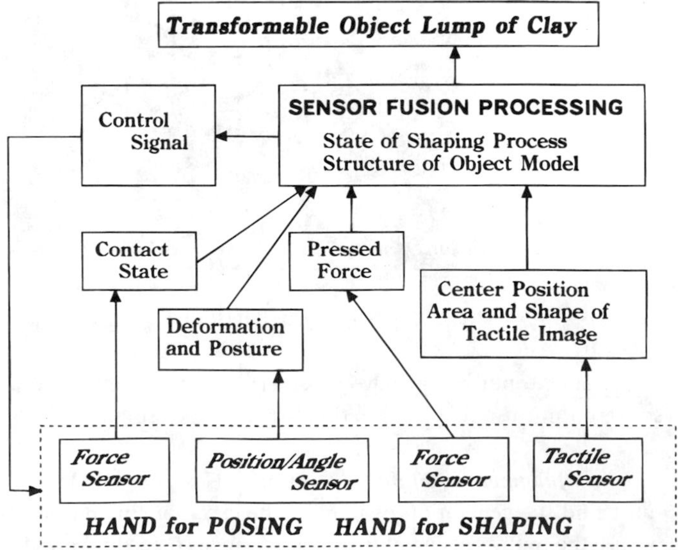

Figure 122.51 Concept of multi-sensor processing system for shaping.

on the X-Y plane by the angle sensor, and they are fused by the simple function to decide the shaping factor.

1. Shaping to sphere is evaluated by the shaping factor *Ss* which is expressed as follows.

$$Ss = (1/A)\pi Z_0 Vi$$

In this equation, the initial volume of clay is *Vi*. There are two case for *Vi*, known and unknown. At the unknown case, the estimated value *Ve* is obtained by multiplying the area *A* with the *z* axis displacement *Z0* at a few step of pressings from the beginning.

$$Ve = (1/n)\sum_j (AjZ_0j) \qquad j = 1 \text{ to } n$$

When the pressing force *F* decreases under the threshold value *Fs*, then it is judged that the shaping is completed.

2. At the inverse shaping where a lump of clay is formed from a ball to a cube, the shaping factor *Sq* is expressed by a touch area *A* and displacement of *z* axis *Z0*.

$$Sq = Z_0 A/Z_0^3 = A/Z_0^2$$

In this case, when the value *V* is known, *V* can be used instead of Z^3. The necessary condition of completion of shaping is when the touch pattern becomes near to the rectangle at all sides, and the force converges within a certain small difference. Final judgement is made when the shaping factor *Sq* becomes near to 1.

Generation of Control Signal

If the shaping state is judged to be insufficient by using the shaping factor, the control commands of the force and the pose must be determined. The algorithm for this control command is defined by using the shaping function in which the multi-sensor information is included.

Function for Sphere Shaping (Forward Shaping). At the palm hand, by pressing the clay at the first stage to generate the small deformation, the system determines the minimum pressing force F_{min}. By the sliding motion with zero pressed force, the maximum and minimum displacement of *Z* is obtained as Z_{max} and Z_{min}. The pressing force *Fs* is determined by the next function.

$$\frac{Fs}{A} = a \cdot \left\{ 1 - \frac{1}{1 + \exp\left(\dfrac{Z - Zq}{b}\right)} \right\}$$

where, *Z*: displacement of the upper hand; *A*: touch area; and *Zq*: [3Zmax + Zmin]/4 (estimated pressed position). The constant *a* and *b* are determined by model parameters and the knowledge base. The forward shaping algorithm is shown in Fig. 122.52.

Function for Cube Shaping (Inverse Shaping Control). This shaping control is accomplished with the pair hand only. It is essentially different from forward shaping. In this control, both the shape recognition process and shape forming process are executed concurrently. The shaping control and the posing control are executed alternatively. The clay inverse shaping algorithm is shown in Figure 122.53.

The control signal of pressing force is determined by the following function.

$$\frac{Fq}{A} = \left\{ \frac{(a + \alpha)}{1 + \exp\left(\dfrac{Sq - 1/2}{b}\right)} + \alpha \right\}$$

In this equation, *Sq* is the shaping factor. The constant *a* and α is determined by the model parameters, and the constant *b* is determined by the knowledge base and task requirement.

The pressing control is made with the force reference calculated by the function with the *Sq* and *A* at the end of the previous sequence. After pressing by the shaping fingers, the rotator of the posing finger is rotated by 90 degrees. Further change the posture of the clay by using the hand rotation mechanism. So when all the face is shaped, this inverse shaping is completed.

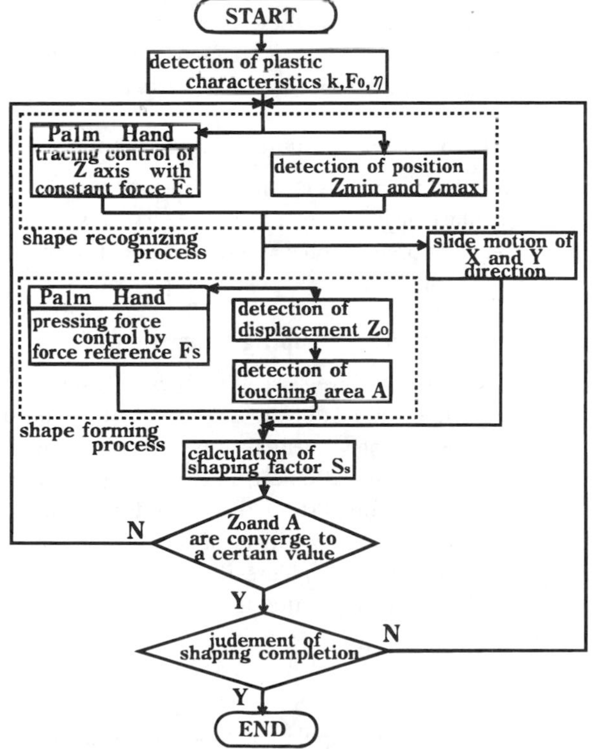

Figure 122.52 Forward shaping control algorithm for the palm hand.

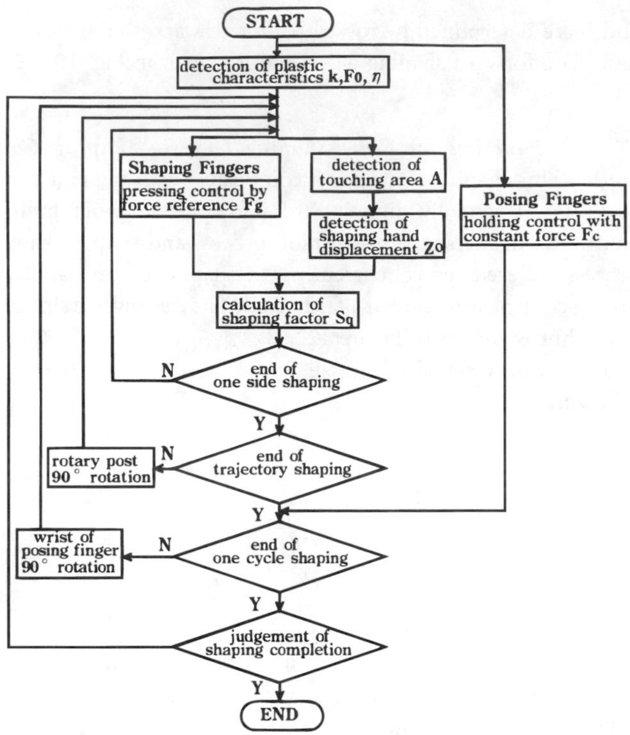

Figure 122.53 Inverse shaping algorithm for the pair hand.

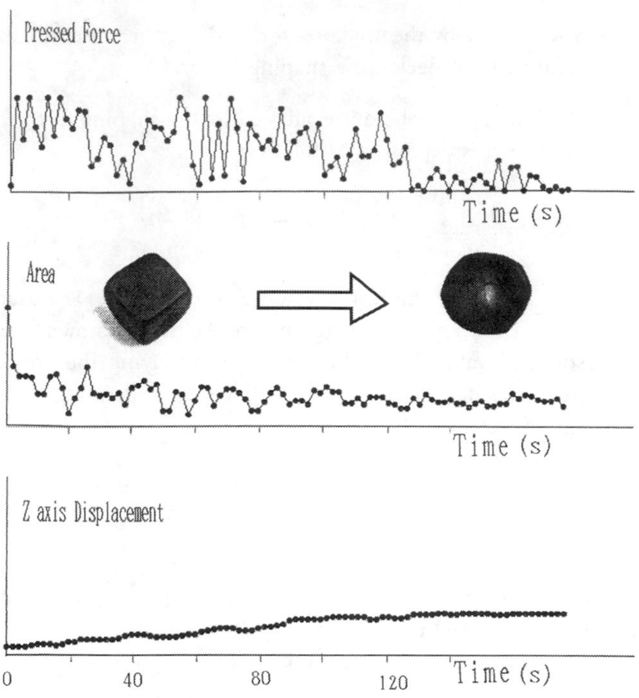

Figure 122.54 Experimental graph of sphere shaping.

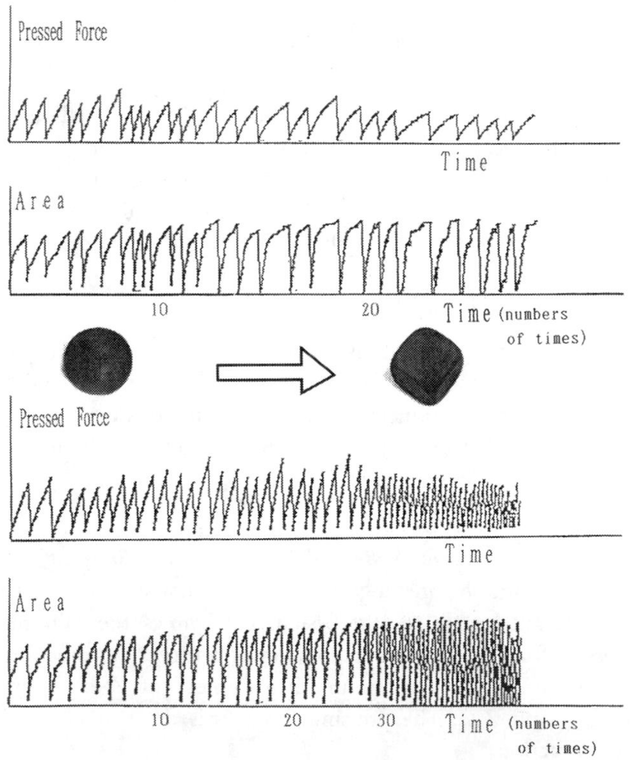

Figure 122.55 Experimental data of cube shaping.

Experiment of Shaping Control

By the hand mechanism shown in Problem of Shaping of Plastic Object and the shaping control algorithm shown in Characteristics of Clay and its Model, the forward and the inverse shaping experiments are executed. In these experiments, the personal computer and on-board computers are used to process the sensor signal and to control the robot hand.

Results of Forward Shaping Experiment

The forward shaping is executed by using the palm hand. The experimental data is shown in Figure 122.54. This data shows the convergence of pressed force and displacement of finger. By this experiment, the nearly full sphere can be made.

Results of Inverse Shaping Experiment

The experiment of inverse shaping is executed by the pair hand. The experimental data of the inverse control is shown in Figure 122.55. We use the sequential control method where first the posture fingers hold the object, next shaping fingers press the object, then open for posture change and press the shape again. By this method, only the data of under minimum of Z displacement is significant in Figure 122.55. When it converges to constant value at every posture, the shaping is completed.

Conclusion

To give the dexterous manipulation function, such as holding the plastic object, to the robot hand, we tried the multi-sensor control. For this purpose, we made up the four-finger pair hand and the three degree of freedom palm hand, and applied the tactile image sensor and the force sensor to these hands. The fusion processing of these sensors is used to recognize the shaping state, and to generate the control signal.

By evaluation of the experimental results, the following important factors are recognized.

1. The three-dimensional finger position and touch area information are very important to estimate the model parameters and to recognize the shaping state.
2. The force, touch area, and finger position must be fused to control the fingers. The model parameters are used to determine the parameters of the functional fusion processing.

The shaping of a plastic object by a robot hand can be accomplished. By the experiments of shaping a lump of clay, the effectiveness of the system is confirmed. Now we are investigating to improve the success rate and the shaping speed, and further shaping to the other complex three-dimensional shapes.

References

Cutkosky, M. R. 1989. On grasp choice, grasp models, and the design of hands for manufacturing tasks, *IEEE Trans. R& A*, 5(3):269–279.

Luo, R. C. and Kay, M. G. 1989. Multisensor integration and fusion in intelligent systems, *IEEE Trans. Syst. Man Cybern.*, 19(5):901–931.

Mason, M. T. and Salisbury, J. K. 1985. *Robot Hands and Mechanics of Manipulation*, MIT Press, Cambridge MA.

Nagahama, K. and Masuda, R. 1991. Shaping of plastic object by the robot hand, *9th Ann Conf. Robotics Society of Japan.*

Ogawa 1991. General Rheology, Sankaido.

Robotics Society of Japan (RSJ) 1991. Report of the research committee on the sensor fusion and integration systems.

Taylor, P. M. 1990, *Sensory Robotics for the Handling of Limb Materials*, Springer-Verlag, New, York, NY.

Venkataraman, S. T. and Iberall, T. 1989. *Dexterous Robot Hands*, Springer-Verlag, New York, NY.

122.8 Multisensor System Integration for Autonomous Navigation Tasks

Karl Kluge

Selecting Sensors for Autonomous Navigation

Autonomous navigation requires perceiving information about a variety of features in the environment around the vehicle. On-road applications require detecting such features as lane markings, other vehicles on the road, traffic signs and signals, and possible obstacles such as pedestrians who may wander into the road. Cross-country applications require recognizing areas of terrain which the vehicle cannot traverse and landmarks verifying the vehicle's position. Three broad approaches are possible to allow an autonomous vehicle to perform the needed perception of the environment: cooperative, single sensor, and multisensor. Both technical and non-technical factors enter into the selection of an appropriate approach to perform a given task.

Cooperative techniques use artificial beacons in the environment to signal the locations of certain features. These beacons are detected by sensors on board autonomous vehicles operating in the environment, simplifying their perception tasks substantially. An example of this type of approach can be seen in the road following work performed in the California PATH project (Shladover, 1992). Ceramic magnets costing about one dollar each are buried in the pavement at one meter intervals along the center of the lane. Four Hall-effect magnetometers are mounted under the front bumper of the test vehicle. The magnitudes of the signals detected by the magnetometers determine the offset of the vehicle from the lane center. This error signal can then be used to steer the vehicle. Cooperative approaches may be a more mature technology at the moment, but suffer from disadvantages as a policy option for applications such as highway automation. Modification of infrastructure is expensive, and can only proceed at a limited pace. The same arguments about modification of the environment also apply in other application areas such as factory automation. In addition, consumers may hesitate to buy expensive optional equipment which can only be used in a limited set of locations.

Single sensor techniques use a single sensor modality to extract all the information needed to perform the task. The VaMoRs group in Munich has pursued this as a research agenda, using monocular intensity images as the sensor modality. They have developed modules to track lane markings and road edges (Dickmanns, 1992, 1988), to detect obstacles in front of the vehicle (Solder and Graefe, 1990), and to detect vehicles approaching from the rear (Efenberger et al., 1992). Using CCD camera data as the only sensor modality has the advantage of using small, relatively inexpensive sensors which can use the existing structure in the environment. This approach suffers from the disadvantage that performing all the necessary perceptual tasks with sufficient robustness using a single sensor is a challenging and unsolved problem. In the case of the road navigation domain, the current fatality rate in the United States corresponds to roughly one death per million vehicle hours. The average number of vehicle-hours per non-fatal accident is still in the tens of thousands (Shladover, 1992). While tremendous advances have been made in the robustness of individual vision modules such as road tracking, the state of the art is still far from being able to guarantee a comparable safety record.

Multisensor approaches use a mixture of sensor modalities in order to simplify the perception tasks the system needs to perform. Different sensors have different strengths and weaknesses. Laser range finders provide a dense depth map of the scene, but do so at a low frame rate and are costly. Sonar and other point range sensors give more sparse data about obstacles in the near environment, but are relatively inexpensive. Infrared cameras provide additional information about object material properties, but high quality, high resolution sensors are still fairly expensive. The TRC HelpMate robot (Skewis et al., 1991) provides an excellent example of a system which uses the complementary strengths of different sensors to achieve impressive performance. Polaroid ultrasound sensors are used in combination with an infrared light stripe projector and camera to detect obstacles in the area in front of the robot. A second CCD camera is mounted looking upward in order to detect ceiling lights. The location of the ceiling lights is used to infer the orientation of the robot with respect to the hallway the robot is traveling along. Only one concession is made to the cooperative approach: elevators in the hospital where the HelpMate is used are equipped with a computer, which the robot communicates with through an infrared transceiver. Doors the robot may need to open are similarly equipped. Information from the different obstacle sensors is pooled into an occupancy grid representation of the area immediately around the robot, and this grid map is then used for path planning purposes.

The main advantage of multisensor approaches is that they simplify parts of the perception problem, and may therefore offer the possibility of reaching commercial levels of reliability earlier than single sensor systems. The main disadvantages of multisensor approaches are that they suffer from potential problems of sensor cost, and raise the problem of integrating the information extracted from the different sensors to intelligently react to the environment around the vehicle. This chapter reviews different architectural approaches to performing this integration. It focuses on work done under the DARPA Autonomous Land Vehicle project during the period 1985 to 1993, using the various perception components which were developed and the integrated systems built from those components to illustrate issues in fusing information from multiple sensors.

The Autonomous Land Vehicle project provided a major impetus to outdoor autonomous navigation work in the United States. Research tended to focus on individual subtasks of the navigation problem once the difficulty of the goals set for the project became clear. The next section reviews the progress made under the ALV project in developing modules to perform individual perception tasks such as tracking the location of the road or navigating through obstacles in the vehicle's neighborhood. This leads into a description of architectural approaches for integrating the results of multiple perception modules which use a mixture of sensor modalities. These approaches can be divided into three classes. *Iconic* techniques simplify scene analysis by registering data from multiple iconic sensor modalities such as color and range images into a fused image. *Symbolic* techniques combine the symbolic results from individual modules into a combined model of the environment. *Behavioral* techniques combine suggested behaviors from different modules to select an appropriate course of action for the vehicle to follow. Examples of all three types of approach will be described. The final section describes major perception modules under investigation at Carnegie Mellon during the period 1985 to 1993, and the ways in which the characteristics of the individual modules interacted with the choice of integration technique. These modules include vision-based road tracking modules, both connectionist and symbolic; obstacle map generation using sonar; generation of terrain elevation maps by range image analysis; and generation of maps and descriptions of discrete obstacles and landmarks in the area near the vehicle using laser range scanner data.

The Autonomous Land Vehicle Project

In 1983 DARPA initiated the ALV (*Autonomous Land Vehicle*) project as part of its Strategic Computing Initiative. A central testbed vehicle was constructed at the Martin Marietta Corporation in Denver, Colorado. The sensors on the vehicle consisted of a color camera, an ERIM laser range scanner, and an inertial navigation system for position estimation. Researchers from a number of institutions developed navigation modules which were tested on the vehicle. These modules included cross-country navigation using laser range data and road following using data from a color camera. The sensor and computing architecture of the vehicle is described in (Hennessy, 1989), along with a description of various demonstrations of capabilities. The project goals and timetable were ambitious (Stefik, 1985). In 1985 the vehicle was supposed to traverse a 20-km route on a paved road at speeds of up to 10 km/hr without performing any obstacle avoidance. In 1987 the vehicle was supposed to autonomously traverse 10 km of open desert at speeds of up to 5 km/hr. In 1988 the vehicle was supposed to traverse a 20 km route on a network of roads using landmarks to aid mission execution, and avoiding obstacles. While these milestones were not fully achieved, solid progress was made in construction of systems to perform the required navigation tasks. This progress is described briefly below.

Cross-Country Navigation

The sensor used for cross-country navigation in the ALV project was the ERIM laser range scanner. The ERIM scanner produces two images per second. Each image consists of 64 rows and 256 columns. Range along the line of sight for each pixel is determined by the phase difference between an emitted amplitude modulated laser beam and the return signal reflected from the nearest object along the line of sight. The range at each pixel is encoded as an eight bit value spanning 0 to 64 feet, giving a range resolution of three inches. The horizontal field of view of the image is 80 degrees, with a 30 degree vertical field of view. Such an active range sensor has several advantages over passive techniques such as stereo. It is not sensitive to ambient lighting conditions, and can operate at night. There is no computationally expensive matching or flow estimation to perform. Finally, the depth map is dense, while many feature-based stereo algorithms produce a sparse depth map. Laser ranging technology suffers from a number of disadvantages, however. At object boundaries part of the laser beam falls on the object and part continues on to the background, giving a mixed range reading. The ERIM scanner's frame rate of two images per second is too low to perform real-time obstacle detection at high speeds. In addition, the slow image scan rate results in distortion in the range images due to vehicle motion during the scan. Besl (1988) reviews this and other active optical range sensing techniques.

Hughes developed a cross country navigation system for the ALV using the ERIM scanner (Olin and Tseng, 1991). Points in range images were backprojected onto a grid on the ground plane to produce an elevation map of the terrain. Interpolation was performed to compensate for uneven coverage due to the sensor geometry. A set of arcs corresponding to a fixed set of steering angles was evaluated to determine how far the vehicle could safely travel along each arc. The criteria used to evaluate traversability were terrain slope, undercarriage clearance, and suspension limits. The safe distance along each arc was combined with a gradient-based global path planner to select the appropriate steering direction.

The cross-country system developed at Carnegie Mellon (Stentz, 1990b) used similar algorithms to convert the ERIM range images into an elevation map of the local terrain, and similar criteria for evaluating the traversability of a path. It was distinguished from the Hughes system described above by the different approach it took in the area of path planning for the vehicle. It treated the problem of path planning as a search in the configuration space of the robot. The search took into account the minimum radius of turn of the vehicle and the constraints on terrain traversability and sensor field of view. It also explicitly modeled the uncertainty in vehicle position due to wheel slippage and sensor error. The planner made sure that a path was not followed beyond the point where uncertainty in vehicle position had accumulated sufficiently to require additional sensing to localize the vehicle with respect to the terrain. The terrain was represented by a quadtree description in order to reduce the cost of searching for an optimal path.

The TraX system developed at SRI (Boblick, 1992) used laser range images to construct a representation of objects in the environment. Rather than use a single representation for object description, a lattice of representations with differing levels of detail was used. The representation chosen for a particular object depended on the quality of the sensor data in which that object appears. The crudest description of an object was as a 2D blob in a range image. A 3D blob gave 3D position and size information and object texture as well as a pointer back to the original 2D blob description. A 3D blob could be elaborated into either a superquadric model of the object or a stick model. When additional data was obtained the description of an object was modified in one of three ways: refinement of existing model parameters, enhancement by the addition of a new property whose value hadn't been estimated previously; or augmentation to a more specific model in the lattice. While TraX used a single sensor modality, the idea of a space of representations at different levels of detail and complexity would extend naturally to data from multiple sensor types.

Road Following

The MARF system developed at the University of Maryland (Waxman, 1987) used edge detection techniques to extract road boundaries and reconstruct the 3D structure of the road surface. Small windows were placed at the predicted locations of the road edges at the bottom of the image. Sobel edge detection and the Hough transform for line detection were used to determine the road edge location in the windows. The system then extrapolated the edges to predict their location higher in the image, and the line extraction process was repeated in those prediction windows. The process was repeated in order to track the road edges to the top of the image. A number of algorithms were constructed for extracting 3D road structure from image data. The most sophisticated algorithm (the "zero-bank" algorithm) (DeMenthon and Davis, 1990) modeled the road as a horizontal segment swept perpendicular to a spine curve which varies in 3D elevation. Global optimization of the result was used to correct for errors in local point matching between the road edges.

The VITS system developed at Martin Marietta (Turk, 1988) used color segmentation rather than edge detection to locate road boundaries. The image was segmented into road and non-road pixels by thresholding a (red minus blue) image. The basic algorithm was extended to include two road classes, sunny and shaded, whose thresholds were found by sampling near the bottom of the image (which was assumed to contain only road). The system assumed that the ground was locally flat, and projected the boundary of the road regions onto the ground to determine the road shape. More sophisticated models of road geometry such as the "hill-and-dale" or "zero-bank" algorithms were rejected because of sensitivity to errors in segmentation and matching of corresponding points on the road edges (Morgenthaler et al., 1990). The VITS system was able to achieve fairly impressive performance, driving at up to 20 km/hr on straight, obstacle-free stretches of road. While Turk (1988) mentions intersection navigation as a criterion for selecting a technique for recovering the road shape, intersection navigation was not implemented. Sacrificing general capability for speed, the restriction to two road color classes limited the robustness of the segmentation.

The Sidewalk II system developed at Carnegie Mellon (Goto et al., 1986) chose a domain which presented a simplified segmentation problem in order to explore issues in map-based navigation. The system used an earlier version of the color classification algorithm which was developed into the SCARF system described below. It fused the color image segmentation with a range image segmentation to distinguish between stairs, an adjacent ramp, and the surrounding grass slope. The system had a map of the geometry of the network of sidewalks it navigated on. Line segments fit to the edges of the extracted road region were matched with expected edges from the map to determine position within an intersection. The fusion technique used to combine the camera and laser range images is described in more detail below.

The SCARF system developed at Carnegie Mellon (Crisman, 1990) used an adaptive color classification technique to classify image pixels into road and non-road classes. Road and non-road regions were modeled by four to eight clusters in RGB color space, and the descriptions of those clusters were updated with each image frame to allow the system to react to changes in illumination and pavement color. SCARF assumed that the road is locally straight, with a known constant width. The classified pixels voted in a Hough scheme to locate the vanishing point and orientation of the road in the image. Once the main road had been found, the pixels on that road were subtracted from the Hough space and further peaks corresponding to intersecting paths were searched for. The combination of the adaptive classification technique with the Hough voting scheme permitted the system to robustly segment a broad variety of roads under a wide range of weather and lighting conditions.

Integrated System Development

Research in complex task domains balances two conflicting tendencies. The first is the centrifugal tendency to focus on individual subtasks as the difficulty of an overall task becomes clear. The second is the centripetal tendency to integrate state-of-the-art modules which perform subtasks within the domain into integrated systems to push the limits of performance on the full task. This conflict manifests itself early in the history of computer vision research. Roberts (1965) constructed a complete object recognition system which performed the entire task from edge detection all the way through to model matching to identify the objects in the scene. Later research fragmented into diverse projects to attack particular subproblems such as edge and region segmentation, object representation schemes, model matching, and model indexing. Similarly, as early efforts revealed the full difficulty of outdoor autonomous navigation many research projects focused on individual navigation subtasks such as road boundary tracking, obstacle avoidance, or terrain mapping for cross-country navigation. As techniques for these problems matured, attention focused again on producing integrated systems which combined the results from different modules.

Asada (1990) constructed a map-based navigation system which incorporated data from both a camera and a range sensor based on structured light. The system had four levels of map. Each sensor had a corresponding physical sensor map, which described sensory data in the sensor's coordinate system. Each sensor's virtual sensor map transformed the sensor data into the vehicle's coordinate system. The virtual sensor maps from different sensors were fused into the local map. These local maps were integrated over time into a global map. Obstacles were extracted from the height data in the laser scanner virtual sensor map. These obstacles were mapped into the camera image coordinates to aid in segmentation and the classification of obstacles as natural or artificial based on image appearance features.

The CODGER system (Stentz, 1990) was developed at Carnegie Mellon to permit the construction of integrated systems using data from multiple sensors. CODGER was a variation on the classical blackboard architecture from artificial intelligence. Lessons learned from CODGER played a key role in the design of later integration architectures at CMU. It is described in detail below, along with the more recent architectures developed to explore different methods of constructing integrated navigation systems.

Classes of Integration Techniques

In the introduction to the chapter, architectures for integrating the results of multiple perception modules were divided into three categories: iconic, symbolic, and behavioral. Iconic techniques which have been investigated include the fusion of color and range image data and the fusion of color and infrared image data. Symbolic architectures include the CODGER "whiteboard" system, the EDDIE communications toolkit and Annotated Map facility, and the ULYSSES system for tactical driving. Behavioral architectures include subsumption architectures, the DAMN architecture (a fine-grained fusion technique for combining suggested courses of action from different modules), and connectionist techniques for combining multiple neural networks such as the MANIAC system. These are described in detail below.

Iconic Fusion Techniques

Camera images provide information about the appearance of objects in a scene in a very direct way, while the geometry of the objects in the scene is encoded in a very indirect fashion. On the other hand, laser range scanner images provide dense depth maps describing the geometry of the scene directly. A number of researchers have registered data from these two types of sensors to simplify the problems of scene segmentation and interpretation.

Morgenthaler et al. (1990) describe an algorithm for fusing range and video images to estimate the 3D structure of the road in front of the ALV testbed vehicle. This was used to estimate ground truth in order to evaluate a number of algorithms for recovering road shape from monocular color images. The image fusion was done using an epipolar plane algorithm. The three points corresponding to the location of a pixel on the image plane, the camera focal point, and the focal point of the ERIM laser range scanner define an epipolar plane which must be searched for the range data point matching a camera image pixel. This plane in the world corresponds to an arc in the range image. This arc is searched for the closest surface it intersects, which corresponds to the point imaged by the camera at the given camera pixel. Road edges detected in the camera image were traced in the fused range/color image to provide the 3D contours of the road edges.

Hebert et al. (1990) describe a similar algorithm to produce fused color/range images. The fused data was used to simplify certain perception problems faced in the Sidewalk II navigation system (Goto et al., 1986). At certain locations in the network of sidewalks the path consisted of a smooth ramp next to a set of stairs. The color segmentation algorithm used, a simple histogram-and-threshold algorithm applied to the blue band of the image, could distinguish the paved surface from the adjacent grass, but could not distinguish the ramp and the stairs. While the stairs could be distinguished from the ramp in the range image, the ramp and adjacent grass had the same slope, and thus could not be distinguished. By combining the results of the two segmentations the system was able to recognize the ramp without any change or improvement to the individual segmentation routines. Similarly, the 3-D shape of a region in the range data (roughly cylindrical and vertical) could be combined directly with color information (brownish) to classify an obstacle as a tree.

The fusion of thermal infrared images with intensity images from a camera is described by Nandhakumar and Aggarwal (1988). Estimates of albedo and the angle between the surface normal and illuminant derived from the intensity image at each pixel are combined with the infrared data to estimate how well the imaged surface conducts heat into its interior. A simple set of rules is used to classify regions based on their intensity and thermal properties. The AIMS system developed by Chu and

Aggarwal (1992) registered thermal IR data with range, reflectivity, and velocity information provided by a laser range finder. Again, the additional information provided by combining the data from the two sensors is used to improve the quality of the scene segmentation produced compared to the results which would be obtained by using only one of the sensors.

Symbolic Fusion Techniques

The iconic fusion techniques just described generate a symbolic description of the environment from a single merged image combining data from multiple types of sensors. Symbolic fusion techniques take symbolic descriptions generated by multiple modules using different sensor modalities and combine them into a unified description of the world.

CODGER. The CODGER system (*COmmunications Database with GEometric Reasoning*) (Stentz, 1990a) was developed at Carnegie Mellon to provide a framework for symbolic fusion of the results of multiple perception modules. CODGER provided facilities for intermodule communication and a central map database for representing information about the environment. It was believed that a central manager for intermodule communication would have the advantages of reducing the number of communication channels in the system and permitting the addition of new modules which could "eavesdrop" on the results of existing modules. Modules which performed parts of the perception tasks necessary for navigation would cooperate through CODGER to execute the desired missions.

The messages passed through CODGER, referred to as *tokens*, consisted of attribute-value pairs. Values could be scalars such as integers, floats, Booleans, strings, or enumerated types; they could also be arrays, pointers to other tokens, or geometric locations in the map stored by CODGER. Modules retrieved tokens of interest by sending a specification to CODGER containing a Boolean expression defining the attribute values needed for a token to match. Primitives to calculate Euclidean distance, polygon intersection, inclusion testing, and minimum bounding rectangles were included to permit specifications involving geometric locations. Module synchronization was performed through three types of query modes. In the first, *immediate non-blocking requests,* CODGER would locate any tokens which matched the specification provided and return them to the calling module. The calling module would then resume execution regardless of whether any matches had been found or not. In the second, *immediate blocking requests,* execution of the calling module was suspended until one or more tokens appeared which matched the given specification. When a match occurred CODGER would send an interrupt to the calling module to resume execution. In the third query mode, *standing requests,* CODGER would generate an interrupt to the calling module every time a new token appeared which matched the request specification, and pass the new token to the module. These query modes allowed both synchronous and asynchronous communication between modules.

In CODGER there were no guaranteed bounds on the latency for a message to be passed back to a module. As a result, integrated systems built using CODGER assumed that the environment was static. Modules using different sensors coordinated action through an abstraction called the driving unit. A driving unit was a patch of ground which the different perception and navigation modules would process in turn. The PILOT module would predict the location of the next driving unit along the road. The road following routine would then use camera data to extract the shape of the road in the predicted driving unit. Once the road location was known, the obstacle detection routine would use the ERIM laser range scanner to check for obstacles in the road. Once the locations of obstacles in a driving unit were known, the PILOT could plan a path through the unit, which the HELM module could then execute. Since CODGER supported multiprocessing systems, different modules could be examining different driving units in front of the vehicle. The PILOT could plan the path through the driving unit ahead of the current one while the obstacle avoidance module scanned the driving unit two ahead, and the road follower detected the road in the driving unit three ahead. This resulted in a pipeline-like system in which modules opportunistically processed a driving unit as soon as the necessary information was available to them (Goto and Shafer, 1990). Figure 122.56 shows the driving pipeline.

EDDIE and the Annotated Map System. In the process of building integrated systems using CODGER it became clear that the architecture had certain limitations. These limitations arose from two main sources: from the bottleneck created by passing all messages through a central module, and from the nature of the map representation used in the system.

When integrated systems were designed, the communication paths were set up in advance. In practice, the only use of the ability to anonymously listen to tokens was for display processes. As a result, the need to pass all messages through CODGER created a bottleneck, reducing system speed.

The map representation was inefficient in several ways. Objects that were entered into CODGER's map were never deleted. Objects were located by providing a coordinate transform between the object's frame of reference and a base frame. The base frame could be the world coordinate system, or another object's frame of reference. As a result, some object's locations were defined by long chains of transforms linking them to a particular landmark. If the location of that landmark was updated, then the change in location had to be propagated through the chains of frame transforms to objects which referred back to it. This resulted in CODGER devoting increasing amounts of time to map updating as a run progressed. Also, the map represented every object at every level of description, rather than permitting partial descriptions. In order to address these problems, a new pair of tools were created to provide communications and map support for integrated navigation systems—the EDDIE communications library and the Annotated Map system.

EDDIE (*Efficient Decentralized Database and Interface Experiment*) (Thorpe, 1991c) provided a set of tools for flexible point-to-point communication between modules in an integrated navigation system. In addition to the communications routines, EDDIE also provided a real-time controller which replaced the

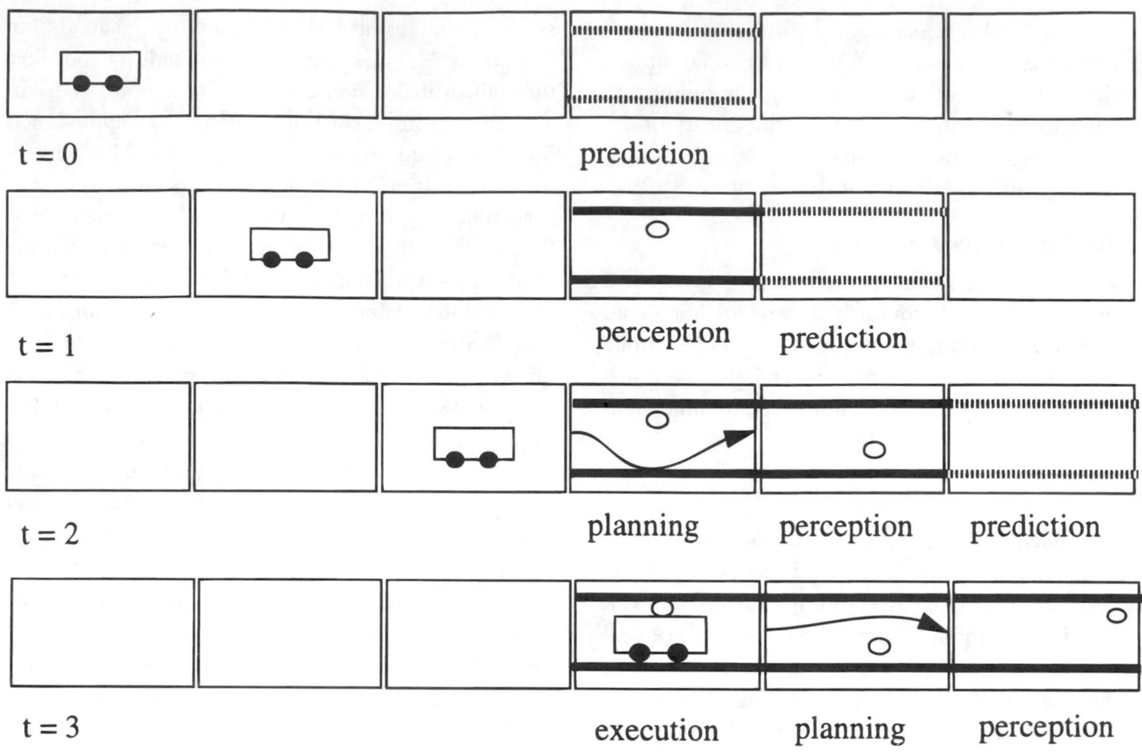

Figure 122.56 Pipelined execution using the driving pipeline.

HELM module and vehicle controller used with CODGER. Switching to point-to-point communications removed the central bottleneck of the CODGER whiteboard.

The map functions provided by the central database in CODGER were handled through the Annotated Map system (Thorpe, 1990). The Annotated Map provided efficient access to geometric information needed to perform navigation scenarios. In addition to storing the location of features in the environment, locations could be annotated with control and object description information. Annotations with control information allowed the system to invoke particular perception routines when specified locations were reached by the vehicle. Annotations with object information contained data about object characteristics which were sent to perception modules which requested them.

Annotations were divided into two types: descriptors and triggers. Descriptors contained information about objects stored in the map, and were passive. They were returned by the Annotated Map in response to specific requests from modules which wished to use them. Modules specified an object type and a polygonal region, and the Annotated Map returned all annotations with that type located inside the specified region. Object types were coarse categories such as landmark, intersection, and road segment. Triggers provided control information, and were active. They resulted in a message being sent to a specified module whenever the vehicle reached a specified location. Locations for triggers could be points or line segments, which could be combined to allow triggers to specify polygonal areas. Typical trigger annotations used in systems constructed with the Annotated Map system included "set speed," "resume vision-based road following," "start landmark matching," "end landmark matching," "stop at landmark," and "start/stop fast obstacle avoidance."

A typical use of the Annotated Map system was to solve the problem of navigating through an intersection represented in a map of the road network. Many existing vision-based road following modules can track road segments between intersections, but cannot navigate through intersections due to limitations of algorithms or sensor field of view. Dead reckoning using encoders on the wheel and steering shafts can be used to drive the vehicle blindly for short periods of time along a pre-planned route through an intersection, but this requires accurate knowledge of the vehicle's location relative to the intersection. Landmark recognition using a laser range finder can determine the location of the vehicle relative to an intersection, but consumes large amounts of computing resources and has a low update rate.

The Annotated Map system provided a framework for combining such modules in order to perform the task of intersection navigation. The human planning a mission would place annotations in the map to invoke the landmark recognition module as the vehicle approached an intersection. The landmark recognition module would request the annotations describing all the landmarks in an area near the estimated vehicle position. The landmark recognition module would match the objects returned by the Annotated Map to the objects detected in sensor data in order to refine the system's estimate of vehicle position. Another annotation would specify the switch to navigation using dead reckoning to travel through the intersection when the vehicle got too close for the vision based road following to proceed. Another annotation on the road which the vehicle turns onto would trigger the switch back to road following using vision once the vehicle was through the intersection. An example of the map annotations near an intersection from a real mission map is shown in Figure 122.57. A trigger annotation, here labeled "Start Intersection," was used to notify the appropriate modules as the vehicle enters the intersection. The small circles to the side of the roads indicate landmarks stored in the map. Another trigger annotation, labeled "End Intersection," was used to notify modules once the vehicle had steered through the intersection and was ready to resume normal road following. The organization of this task information in a geometric framework simplified the planning necessary to execute a mission, and provided a symbolic framework for combining perception modules using different sensor modalities into a system capable of performing complex navigation tasks beyond the capability of any of the modules individually.

Maps were constructed by using perception modules to detect landmarks and other features such as road boundaries while a human drove the vehicle through an area which the vehicle was expected to traverse autonomously in future runs. Many applications involve repeatedly traversing the same or a similar route multiple times. Examples include mail delivery along a particular route, resupply of a military unit stationed at a fixed location, and transportation of materials around mine and construction sites. A facility like the Annotated Map system allows a human to perform the task once, and then incorporate the knowledge necessary to allow autonomous repetition of the task as annotations in the map.

Lessons learned from experience with CODGER led to several key design principles underlying the Annotated Map system. The first was that semantic information contained in annotations was minimized. Other than a small amount of header information needed to perform queries specifying location and object type, all information in a descriptor annotation was stored as a sequence of bytes whose interpretation was left to the requesting module. The second was that global position tracking to determine the location of the vehicle in the map should be performed separately from position tracking for local servoing. The relative error in successive vehicle positions is small when registering data from a small number of closely spaced sensor readings. Large errors appear only as drift accumulates over the course of a mission. As a result, the low level real time controller maintained an estimate of vehicle position based on the encoder and inertial navigation input data, and this estimate was never corrected. Instead, the Annotated Map system maintained a transformation between the controller's estimate of the vehicle location and the location of the vehicle in map coordinates. The third design feature was the centralization of position tracking. By having a central module perform position tracking in the map and allowing the map to trigger particular modules when a specified location is reached, the individual modules were relieved of the task of keeping track of the vehicle location relative to the map model. The fourth design feature was the absence of a central master control module. The map served as a shared scratch pad and alarm clock, but flow of control and data was handled point-to-point between modules, eliminating the bottleneck of a central control mechanism. These features resulted in systems which were faster and easier to maintain than systems built with CODGER.

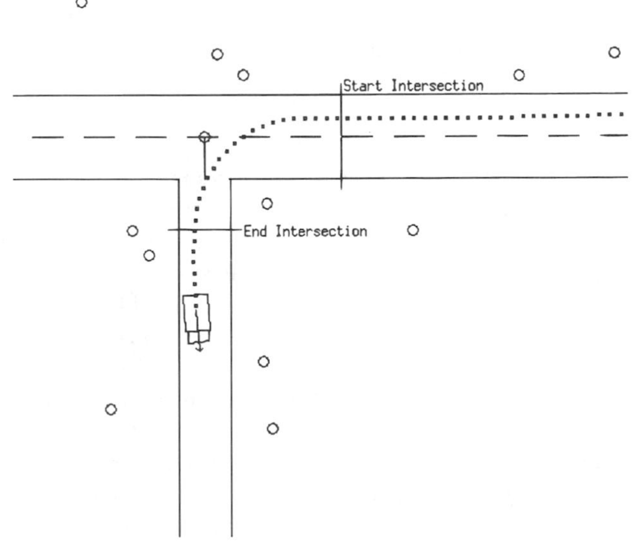

Figure 122.57 Example of an annotated map.

Ulysses. Systems built using CODGER and the driving pipeline technique ran a fixed set of perceptual processing on every road segment. Systems built with EDDIE and the Annotated Map system ran different combinations of perception based on vehicle location in the map. These combinations were set up statically as trigger annotations in the map. General perception for navigation is computationally very expensive, and the context of the vehicle's current situation can be used to significantly reduce the perceptual work load of a system. This was the concept explored in the ULYSSES system (Reece, 1992).

The driving task can be divided into three levels: *strategic, tactical,* and *operational.* Tasks at the strategic level include planning a route to the vehicle's destination or altering the planned course based on traffic conditions along the route. Tasks included at the tactical level are those needed to maintain the vehicle in a safe and legal state given the local environment. Tasks at the operational level provide the perceptual support needed to make tactical decisions and the vehicle control support needed to execute them. ULYSSES modeled decision making at the tactical level. ULYSSES was not implemented on a real vehicle. Instead, it operated within the PHAROS traffic simulator (Reece, 1987) which modeled static aspects of the environment such as lane structure and traffic signs as well as dynamic aspects such as traffic signals and the driving of other vehicles in the road network. Perception was simulated by examining the internal symbolic state of the simulator.

ULYSSES assumed a set of 14 perceptual routines at the operational level. These routines performed isolated perception subtasks such as "find current road," "find crossing cars," "find next car in lane," "mark adjacent lane," and "find next sign." While ULYSSES didn't assume any particular sensor modality for these modules, they are described in a way which implies that they communicate with each other through marking locations in the sensor data. As an example, "mark adjacent lane" might be called to detect the boundaries of the lane to the left of the current lane. These boundaries would be marked in a model of the environment so that a car-finding routine could limit its area of search to that lane. While ULYSSES assumed the markers are placed in a single image, there is no reason they could not be placed in a map of the local environment which each sensor was calibrated with respect to.

ULYSSES maintained five vehicle state variables: acceleration, the current speed limit, whether the vehicle was currently in an intersection, whether the system was waiting for a gap in traffic, and whether to initiate a lane change to the left or right. In each cycle of the system constraints from the task model were used to select an appropriate value for each of these variables. The task model allowed perception to be limited to only those queries needed to determine the values. As an example, acceleration is limited by factors such as road curvature, the distance to the next vehicle in the lane, and the farthest distance the vehicle has seen along the lane. If a particular perception operation could not produce a lower acceleration than the current estimate of the maximum acceleration, then it need not be executed in that cycle. Results of the simulations performed showed that significant reductions in perception are possible by using information about the current local context to limit the perception acts that need to be performed in each decision cycle.

Behavioral Fusion Techniques

The ULYSSES system provided a principled model of the tactical driving task, and implicitly generates a research agenda for perception algorithms based on the specified set of perception routines. Other autonomous navigation tasks may not have as clear an analytic model, and there is a need for architectures which allows the easy combination of perception modules into systems when there is no full model of the task. Behavior-based techniques for system integration allow the incremental improvement of system competence. Individual modules implement particular behaviors. One module might implement an object avoidance behavior, while another implemented a goal-seeking behavior, and a third implemented a road or corridor following behavior. A behavior-based fusion architecture combines the actions suggested by the different modules to select the course of action to follow.

Subsumption. The subsumption method described by Brooks (1986) provides a behavior-based fusion architecture. In a subsumption-based system, the individual behaviors are implemented by finite state machines. Behaviors communicate through output signals which then serve as inputs to other behaviors or actuators. In addition, one behavior can overwrite an input line into another behavior (subsumption) or overwrite an output line from another behavior (inhibition). Such an architecture eliminates the need for any global fusion of data from the individual modules. The subsumption architecture concept suffers from certain limitations, however. Conflicts between behaviors are resolved by the static priorities wired into the subsumption and inhibition signals. Behaviors can only indicate their first choice of possible actions rather than their relative preferences for different options. Finally, signals need to be added into existing lower level behaviors in a subsumption system as modules are added. Hartley and Pipitone (1991) describe these problems in the context of building a subsumption-based controller to land an airplane simulator. They observe that the interaction between high level behaviors added into a subsumption architecture and existing lower level behaviors leads to a loss of modularity. They point out that having behaviors communicate through the state of the world, as suggested by Brooks, doesn't provide an adequate resolution of this problem, as similar states can occur in situations where different responses are appropriate depending on context. They also observed that the strict hierarchy of behaviors imposed by the subsumption architecture was not always appropriate for the task.

DAMN. The DAMN system (*Distributed Architecture for Mobile Navigation*) (Langer et al., 1994; Payton et al., 1990) provides an architecture for behavior-based integration which attempts to address the problems with subsumption architectures described above. Like ULYSSES, DAMN is centered around a number of state variables whose values need to be determined to drive the vehicle (such as speed, steering direction, etc.). Each variable is quantized into a discrete set of alternatives. Steering radii might be quantized into fifteen values representing radii of curvature ranging from a sharp left turn though heading straight to a sharp right turn, while velocity might be quantized into 0.25 m/s increments.

Each perception module generates a value indicating how desirable executing each of the quantized turn radii is. Each perception module has a weight associated with it indicating how important its input is for setting a particular state variable. For example, the contribution of obstacle avoidance should be given more importance than the contribution of a road tracking module in selecting the steering radius. The weighted sum of the contributions of all the perception modules is computed to determine the desirability of choosing each of the quantized values, and the maximum value is selected for execution. This is illustrated in Figure 122.58 below. The circles above the individual behaviors indicate their individual votes for particular actions. Shaded circles represent negative votes, white circles positive votes. The magnitude of the vote is proportional to the size of the circle. The relative weight each behavior has in selecting an action is indicated by the thickness of the arrow connecting its vote for an action with the circles representing the combined votes at top. The votes for each behavior from the individual behaviors are combined in a weighted sum to select the action the vehicle will perform, in this case a soft left turn.

DAMN permits the individual modules to cooperate to select the action the vehicle executes. Adding an additional module involves assigning its input a weight relative to the existing modules and adding its votes into the evaluation of possible actions.

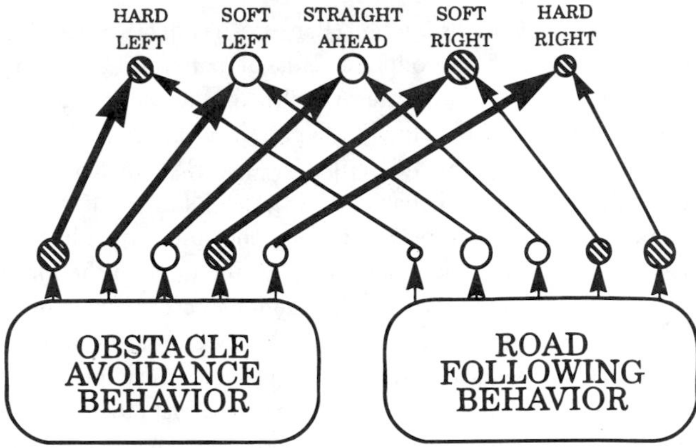

Figure 122.58 Voting of behaviors for actions in the DAMN system.

No communication with or through previously existing modules is needed. DAMN makes no assumptions about the nature of the individual modules, and can be used to construct systems containing a mixture of connectionist and symbolic perception routines. Several behavior-based techniques for combining systems composed entirely of neural networks have been investigated, and are discussed below in the description of the ALVINN connectionist architecture for performing navigation tasks.

Interaction Between Perception Modules and Choice of Integration Techniques

Each of the architectures described above—iconic fusion to assist segmentation, the CODGER whiteboard and Annotated Map systems for symbolic integration of multi-sensor data, the ULYSSES system for task-driven perception, the subsumption and DAMN architectures for behavioral fusion—provides different strengths and weaknesses. Choice of an integration scheme involves constraints which arise from the output of the modules involved as well as those constraints imposed by the requirements of the navigation task to be performed. The next section describes some of the research efforts at Carnegie Mellon exploring modules which perform particular navigation subtasks. Each is described briefly, and the characteristics of the modules which influence the choice of integration techniques used with each are discussed.

ALVINN

The ALVINN system (*Autonomous Land Vehicle In a Neural Network*) (Pomerleau, 1992a) provided a connectionist framework for learning various navigation tasks by observing a human driver. The network architecture was a three layer feed-forward neural net trained using backpropagation. While the description of ALVINN below focuses on camera-based road following, the basic architecture can be adapted to perform a variety of tasks using different sensor modalities.

The input layer of the network was a 30 by 32 "retina." The input layer units encode a reduced resolution version of the camera input. The activation of each input unit was the sum of the blue value of the corresponding pixel and the normalized blue value (blue divided by the pixel intensity). This evened out brightness variations due to shadows and changes in illumination.

Each pixel in the input layer was connected to the four units in the middle hidden layer. These units developed task-specific feature detectors based on the data encountered during the training phase. Each of the hidden units, in turn, was connected to each of the 30 units in the output layer. The units in the output layer represented different steering angles: from a sharp left turn, through heading straight ahead, to a sharp right turn. The network architecture is illustrated in Figure 122.59.

In the training phase, the network was presented with image data from the camera along with the angle of the steering wheel chosen by the human driver. The network was trained to generate a Gaussian pattern of activation centered on the correct steering angle. The range of training instances was extended by transforming the input images and steering responses to simulate situations with more extreme vehicle deviations from the center of the road than would normally be encountered while a human was driving. Structured noise was introduced into randomly chosen images to simulate uncommon events such as guard rails on the side of the road or passing cars. These techniques are described in more detail in Pomerleau (1992a/b). With the use of these techniques the system was able to learn how to track the road after only 5 minutes of observing the human driver perform the task.

Two characteristics of ALVINN are salient when considering the problem of integrating it into a system using other perception

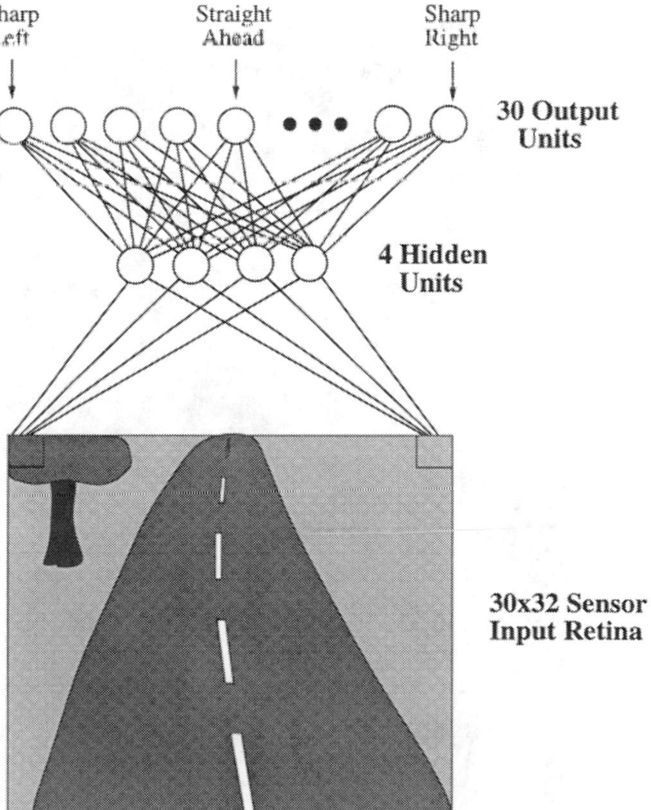

Figure 122.59 The ALVINN architecture.

modules. The first is the lack of an explicit representation of the situation. Only the desired vehicle behavior was represented. Systems which symbolically represent the vehicle's environment and position have to make assumptions about ALVINN's performance. Thorpe et al. (1991a) describe the use of a Kalman filter to integrate position update information from multiple perception modules. Since ALVINN did not produce an explicit representation of the vehicle's offset from the center of the road, the system modeled ALVINN's performance by assuming that it drove the vehicle along the center of the road, with a Gaussian error. The variance of that error was estimated by performing an experiment in which water was dripped from the center of the vehicle's rear bumper, followed by hand measurement of the variation of the water drop locations around the road center.

Individual networks were specialized to perform specific tasks. A general system to perform road following would require separate networks to drive on dirt roads, two lane paved roads, divided highways, etc. Several approaches were tested to select the appropriate network to use in a given situation. The first involved the use of the Annotated Map system. As described above, this system provided the ability to have position-based triggers which invoked particular modules when the vehicle crossed a line or entered a region specified on the map. The annotations in the map statically determined which network should be run at a given place in a mission.

The second approach used the performance of the individual networks to select which network's output should be used to steer the vehicle. This method was called OARE, for *Output Appearance Reliability Estimation* (Pomerleau, 1992a). Since each network was trained to produce a Gaussian pattern of activation around the desired output, the reliability of a network could be estimated based on how closely the pattern of activation of the network's output layer was described by a single Gaussian peak. Figure 122.60 shows the behavior of an ALVINN network trained on an unmarked asphalt bicycle path when presented with three images. Brightness encodes activation for the units in the input retina ("input acts"), hidden layer ("hidden acts"), and output layer ("output acts"). The "target acts" row shows the desired behavior for the net given the input in question. The left example shows a road input image typical of the type of data the network saw during training. The output layer shows a single Gaussian peak of activation at the correct steering direction. The middle example shows an input image where the road branches. The output layer has a bimodal pattern of activation, with the peaks corresponding to the steering directions which would cause the vehicle to turn down the left or right branch. The example on the right shows an input image which is from a different type of road than the one the network was trained on. There is no strong peak in the output layer activation. When the OARE estimate dropped below a threshold value a message was sent to a mapper module. The mapper consulted the route plan and returned a coarse steering direction ("steer left" or "steer right"). The system then used this to select between the peaks in the output layer and steer through the intersection.

The OARE technique selected the output of one of the individual nets as the appropriate steering direction. The MANIAC

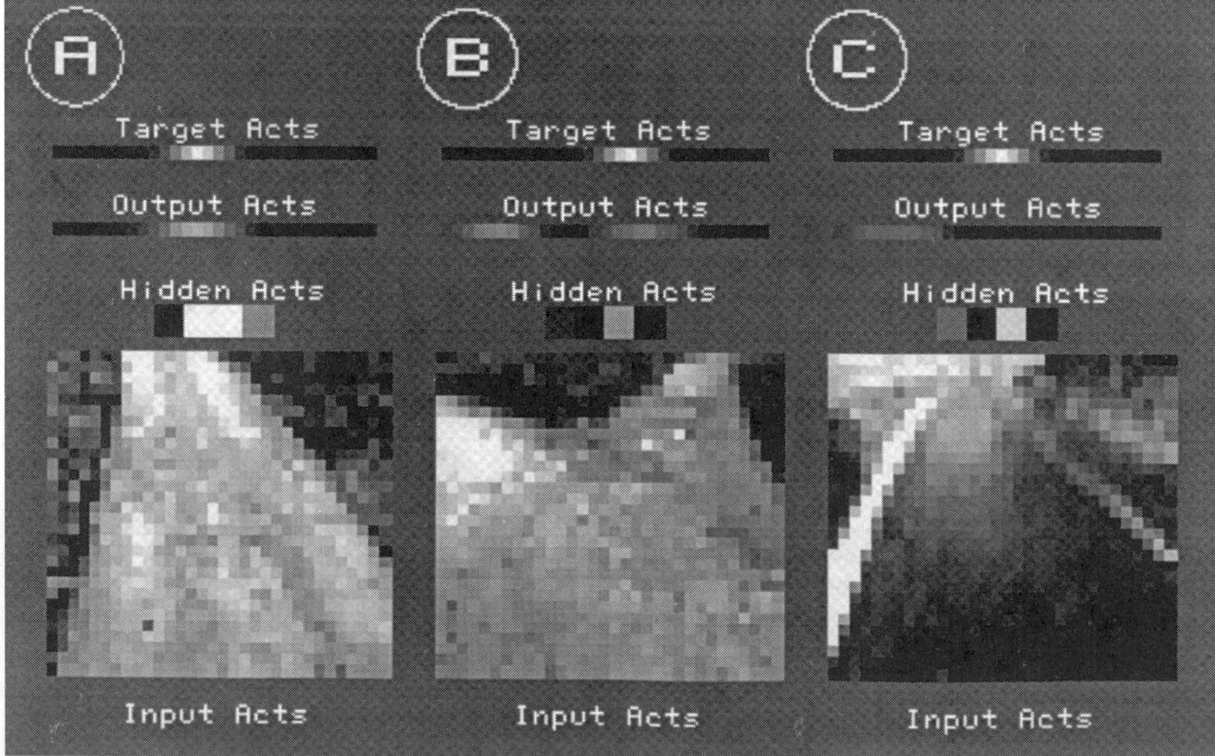

Figure 122.60 Activation for ALVINN's output layer for an image typical of the training data (left), a fork in the road (middle), and a different road (right).

system (Jochem, 1993) was developed to investigate a connectionist method to fuse the results of multiple ALVINN networks. Networks were pretrained to perform individual navigation tasks (road following on an asphalt unmarked road and road following on a two-lane marked road, for instance). The weights between the input and hidden layers of these networks were frozen. The hidden units of these networks formed the input units of a combiner network. This architecture is shown in Figure 122.61. The combiner network was trained off-line using stored examples of the input image/steering direction pairs each of the component networks was trained on. This allowed the combiner network to develop more sophisticated ways of combining the results of the individual networks than simply selecting the output steering direction from a single network. It was hoped that this would allow the combined network to generalize to situations none of the individual networks was trained to handle.

The OARE-based techniques and the MANIAC architecture provided fusion techniques for an integrated system composed of multiple neural networks. ALVINN could also be easily integrated using the DAMN architecture for behavior-based fusion. The activation of the output layer units translated directly into preferences for different steering directions. This provided a behavior-based fusion mechanism for mixed systems combining both neural network modules and traditional symbolic modules.

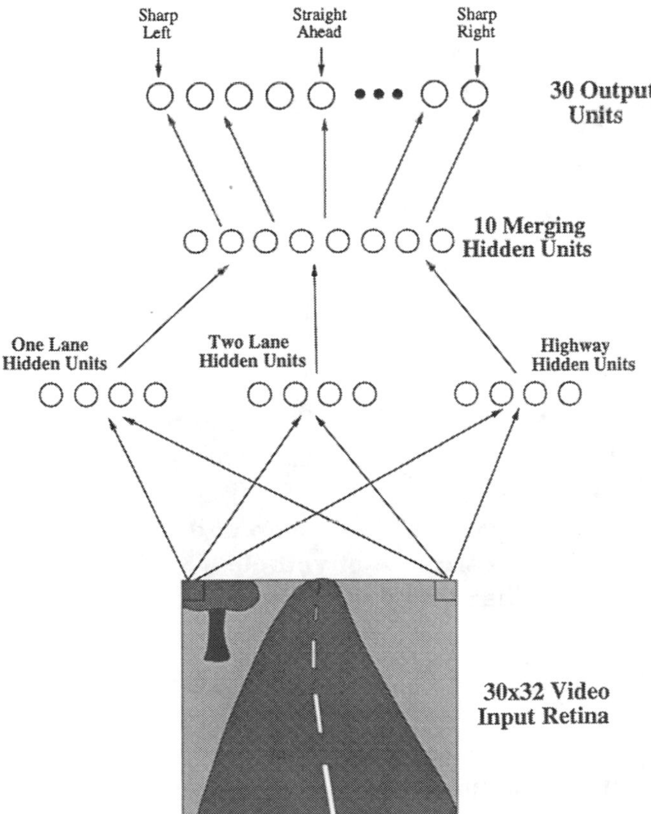

Figure 122.61 The MANIAC architecture for combining neural network perception modules.

YARF

The YARF system (*Yet Another Road Follower*) (Kluge, 1995a,b) used a symbolic model-based approach to vision-based road following. The central data structures in YARF were a global map of the road network and a local map of the environment around the vehicle. The global map provided a partial geometric model of the roads to be traversed. Each road segment was described by a feature cross section indicating the relative positions and appearances of features such as the pavement edges and painted lane markings. The connectivity of the road segments and the relative geometry between segments at an intersection were represented, but there was no global frame of reference relating the positions of different intersections. The global map also served some of the functions of an Annotated Map system, indicating which lane to drive in, which features to track, and where to turn at intersections.

The system used a predict-verify-update model for road tracking. The model of the feature locations and appearances was combined with the estimate of the vehicle's position from the last image and the controller's estimate of the vehicle's motion to predict the location of selected features at a sparse set of locations (five to ten per feature) in the current image. Specialized segmentation techniques were used to detect different types of road features. Each of these techniques was designed to extract stable characteristics of a particular kind of feature under a broad variety of circumstances. As an example, the color of the paint used for yellow stripes appears in roughly the same region of color space for sunlit roads, shadowed roads, and wet roads. This invariance was used to construct a special-purpose segmentation algorithm for detecting yellow painted stripes inside a search window. Whether a particular pixel's color fell in the region containing the yellow paint color could be computed rapidly given a set of planes bounding that region of color space. The pixels in the small search window around the predicted stripe location were classified as "yellow stripe" or "background" and a single pass of thinning was applied to eliminate noise pixels. If there were no pixels classified as "yellow stripe" in the region, then the segmentation routine returned an error value indicating that the feature was not seen. Otherwise, the centroid of the "yellow stripe" pixels was returned as the feature location.

The results of the individual windows in the current image were combined with the results from previous images in a map of the local environment. This local map was used for two purposes. First, the feature locations stored in it from both the current and previous images were used in conjunction with the model of the feature locations to estimate the position of the vehicle relative to the center of the road and the local curvature of the road. Second, the locations where features were expected but not seen were examined for long continuous segments where the features were missing. This information was combined with the knowledge in the global map about which features change or disappear at the next intersection or change in lane structure. If the pattern of missing features corresponded to the expected pattern, then a hypothesis was generated that the vehicle was approaching the next intersection along the route. These hypotheses were then tested using the map knowledge of the geometry of the intersection. If the hypothesis was verified, then YARF switched to the new road segment and tracked it. The YARF processing cycle is illustrated in Figure 122.62.

The design philosophies behind ALVINN and YARF shared a number of features. Both systems were designed to detect changes in the road structure in a data-driven way. YARF did this through reasoning about failures to detect expected features given the context of the map information. ALVINN did this through the OARE technique to detect places where the network output didn't have the desired form. Both sought to replace detailed models of global environment geometry with coarser maps having a more topological flavor. There are important differences which influence the relative ease of integrating the two modules into different fusion architectures, however.

All of the salient information about the task was stored implicitly in an ALVINN network. The hidden units developed an implicit representation of the salient features for the task the network was trained to perform. The output layer implicitly represented the location of the vehicle relative to the road by a steering direction the vehicle should follow.

YARF, on the other hand, had explicit models of the lane structure of the road, local road curvature, and vehicle position relative to the road. Because of this, the output from YARF and an obstacle detection module could be combined in a straightforward manner to answer questions such as "which lane (if any) is the specified obstacle in?" As a result, while both ALVINN and YARF could be easily integrated into a behavior-based fusion architecture such as the DAMN system, YARF was easier to integrate into a symbolic fusion architecture.

The YARF system has been integrated with sonar and laser range obstacle detection modules using the EDDIE toolkit. YARF selected the appropriate path to follow, while the results of obstacle detection were used to set vehicle velocity to avoid collisions. This was demonstrated by having a first year graduate student

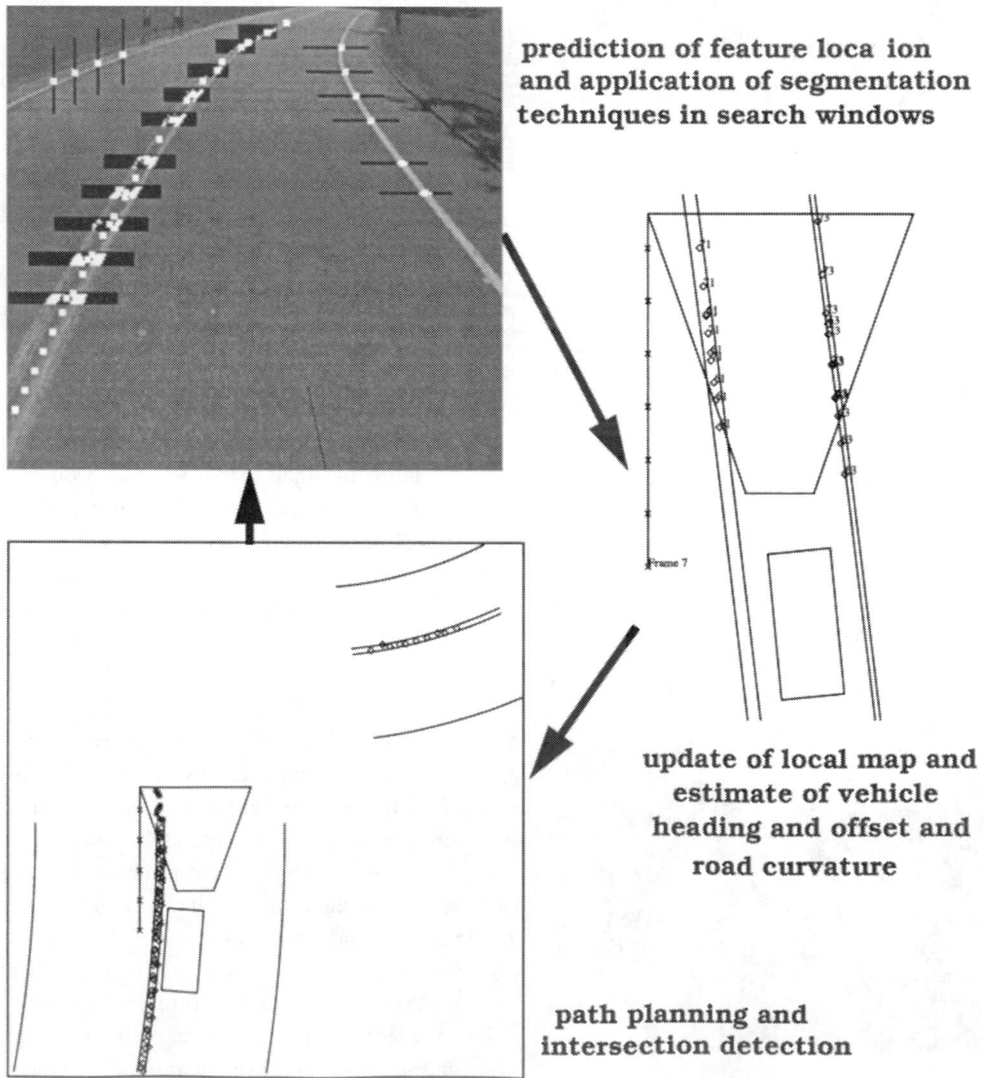

Figure 122.62 Processing cycle in the YARF road following system.

walk in front of the NAVLAB while it was being driven by the system. The vehicle would come to a halt to avoid hitting the pedestrian, then resume driving when he moved out of the vehicle's way. YARF provided most of the perceptual support required by the ULYSSES system for locating lane positions and intersection branches, and one line of future work would integrate the two systems, and would flesh out the remaining perception modules needed to implement the full ULYSSES model of tactical driving.

Sonar-Based Obstacle Detection

The GANESHA system (*Grid-based Approach to Navigation by Evidence Storage and Histogram Analysis*) (Langer, 1994, 1992) used sonar sensors mounted on the front bumper of the NAVLAB vehicle to detect obstacles in the area in front of the robot. Closed type piezoceramic transducers operating at 80 kHz were used. These are robust against moisture and dust in the outdoor environment and are less sensitive to acoustic noise than open type electrostatic transducers. These sensors have an operating range of up to 6 meters, with a beam angle of approximately 5 degrees at a 3 dB fall-off from the major axis of the sensor. Five sensors were mounted to provide good area coverage with reasonable spatial resolution. The sensors were fired simultaneously, and data was collected at a rate of 9 Hz.

In order to fuse data from multiple scans of the sensors, each data point was tagged with the vehicle position at the time of the measurement. Measurements were added into a grid-based map of an area 16.4 meters by 16.4 meters in front and to the sides of the vehicle. Old data points had their positions updated with respect to the map as the vehicle moved, and were deleted when they moved outside the window of the map. Each cell in the grid covered an area of 0.16 square meters. Each grid cell had several variables associated with it: a flag indicating whether an object in the cell (if any) was seen in the current scan or a previous scan; the (x, y) coordinates of the object with respect to the vehicle's coordinate system; and a count indicating the number of times an object was detected in the cell. The (x, y) coordinates are kept to increase the accuracy of the grid update procedure.

When a new set of sensor readings was obtained the existing data points in the grid were updated based on the vehicle controller's estimate of the translation and rotation of the vehicle since the last set of readings. The new data points were then added into the grid. If a new measurement falls into a cell that already had an object in it then the object count for that cell was incremented. Otherwise, a new object was placed in the empty cell. In order to accommodate moving objects and filter out occasional spurious readings the count for a map cell decayed over time.

Another stage of processing detected extended linear obstacles in the local map in order to perform navigation tasks such as driving parallel to a line of parked cars, detecting an open parking space, and parallel parking. Median filtering of the map removed isolated objects due to spurious readings. A piecewise linear fit was then done to the objects in the map by fitting lines to the objects inside a window within the map. Figure 122.63 shows GANESHA find a gap in a row of parked cars.

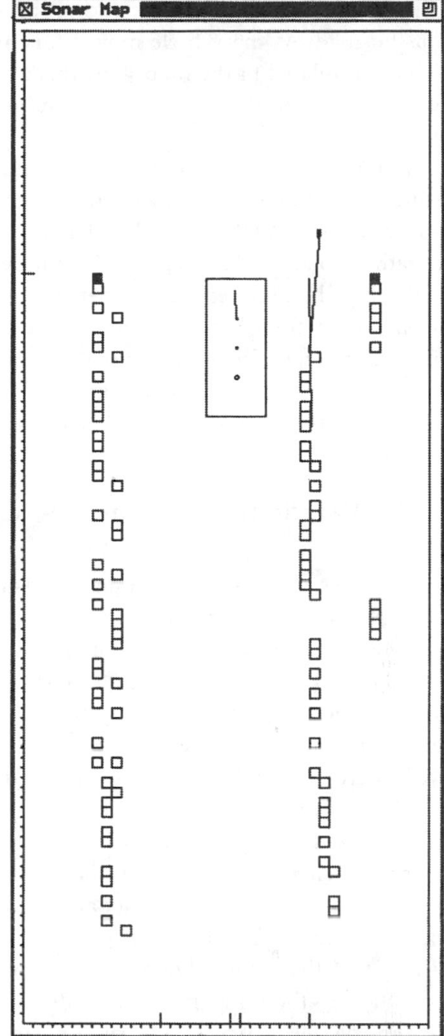

Figure 122.63 Detecting a gap in a row of parked cars using the GANESHA module for sonar-based obstacle detection.

The image on the left shows an open parking space along the right side of the road. As the vehicle moved forward the data points which accumulated in the map show this gap, shown as a solid line to the right of the vehicle in the overhead view on the right.

The grid local map representation permitted integration using a symbolic fusion technique. As described above, GANESHA has been integrated with YARF using the EDDIE system. YARF selected the path to follow. GANESHA set the vehicle speed, and would slow down or halt the vehicle in order to avoid collisions with objects in front of the vehicle. In addition, the system could evaluate the desirability of following various steering directions in order to use a behavior-based fusion technique. This has been done to integrate GANESHA with ALVINN using the DAMN system.

Discrete Obstacle Detection in Laser Range Imagery

Sonar provides a sparse set of range measurements to objects close to the vehicle. Laser range scanners provide a dense set of depth measurements over a greater range of distances. The algorithms applied to the range images depends on the type of navigation task being performed. Navigating along roads or through very gentle terrain with isolated obstacles requires segmentation of discrete obstacles from the surrounding terrain. This section describes one algorithm developed by Hebert for this type of obstacle detection (Thorpe et al., 1991a).

Depth measurements from the range image pixels are projected onto a horizontal grid to construct an elevation map of the terrain in the scanner's field of view. From this grid, an estimate is made of the terrain slope at each cell of the grid by fitting a surface to the range measurements which fall into that cell. Groups of cells which have surface normals far away from the local vertical are segmented out as obstacles. Observations from multiple images are combined in the grid map in order to improve obstacle localization and eliminate spurious objects due to sensor noise and object motion. Objects already in the map have their location predicted in new images. Confirmation of a correct prediction increases the procedure's confidence that the object is real. Failure to observe the object in its predicted location decreases the system's confidence in the object. Each object's location has a mean and covariance matrix associated with it to represent uncertainty in the object position. Additional observations are used to update this uncertainty by applying standard maximum likelihood methods. Figure 122.64 shows a sequence of ERIM images, with objects tracked between frames. Figure 122.65 shows the resulting map that is built up from the data, with each object represented by an ellipse whose size is related to the system's estimate of the covariance of the object location.

These techniques have been used in a number of integrated systems to perform collision avoidance and landmark detection. Systems built using the Annotated Map triggered this module at specified locations to refine the estimate of vehicle position in the map. In addition it could be run continuously while the vehicle was in a selected region to provide collision avoidance. It has been integrated with YARF using the EDDIE system, and with ALVINN using the DAMN and Annotated Map systems.

Constructing Terrain Models From Range Imagery

Cross-country navigation on very rugged terrain requires a more detailed model of the terrain. A grid-based elevation map with a uniform resolution is not an appropriate representation to use as input to path planning. In a system which uses a grid-based representation at a single scale, large regions of easily traversed terrain are represented (and examined) in the same detail as regions which are densely packed with terrain variations. A more efficient representation is to construct a quadtree model of the terrain. Each node contains a description of the minimum and maximum slope and elevation and a measure of the largest elevation discontinuity within the area covered by the node. Large traversable areas with small variation in elevation and slope can be examined by the planner at a high level in the quadtree, reducing the amount of search the planner must perform.

Models extracted from multiple range images are combined to give better accuracy and to provide coverage of previously imaged terrain closer to the vehicle than the bottom of the scanner's field of view. An inertial navigation system provides estimates of vehicle translation in x and y and changes in vehicle roll, pitch and yaw.

Figure 122.66 shows the processing associated with this module. At the top is an ERIM range image. Beneath that is the overhead view of the raw elevation map constructed from the range image. The left image in the third row shows the elevation map after interpolation to compensate for uneven sampling due to sensor geometry. The right image in the third row shows the interpolated elevation map after range shadows (areas occluded by obstacles in the foreground) have been identified. The bottom two rows show features from the quadtree representation of the terrain. The second row from the bottom shows the minimum terrain elevation at decreasing resolution from left to right. The bottom row shows the maximum terrain elevation at decreasing resolutions.

Lessons and Open Issues

The use of multiple sensor modalities is a pragmatic solution to the problem of building systems which achieve useful levels of competence in the short term. In addition, it permits investigation of navigation subtasks dependant on underlying perception capabilities which might not be feasible with a particular sensor modality at present. As an example, research into path planning for cross country navigation requires the support of perception routines capable of building a sufficiently detailed and accurate model of the terrain. Until recently, algorithms to extract depth information from stereo disparities were inadequate to the task. Use of a laser range scanner for this purpose enabled investigation of the issues posed by path planning, research which would

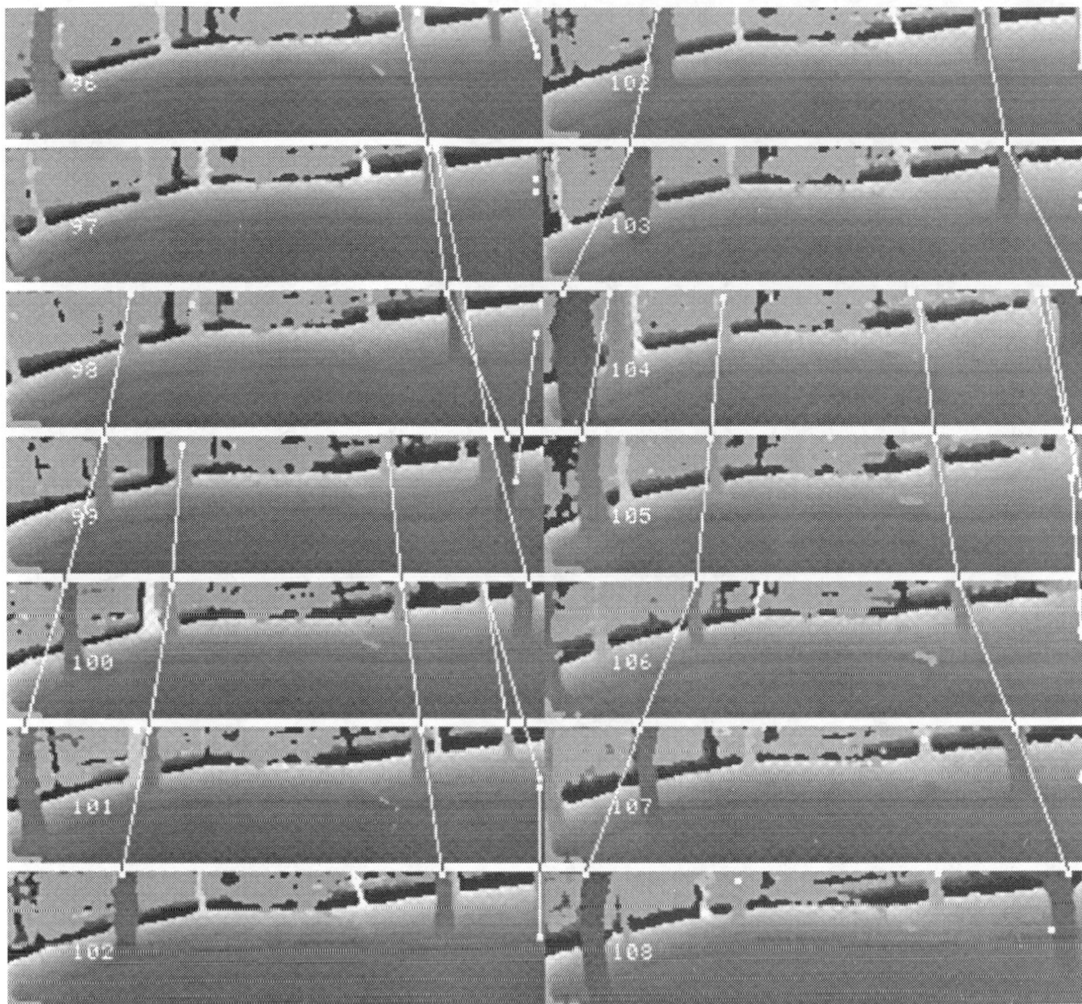

Figure 122.64 Tracking of obstacles over multiple ERIM frames.

otherwise have been blocked by the need for progress in stereo algorithms.

Progress in improving the quality of the results produced by different sensor modalities will require improvements in sensor and computing hardware as well as improvements in the algorithms which process the sensor data. Existing laser range scanners have low image acquisition rates and limited resolution, maximum range, and field of view. Color cameras have inadequate dynamic range in situations with mixed bright sunlight and deep shadow, slow automatic iris response, and limited field of view. The use of active vision to increase sensor coverage of the environment without sacrificing image resolution will require faster hardware and algorithms to provide the necessary processing in the time available.

The choice of an architecture for integrating results from multiple perception modules cannot be made without consideration of the modules to be integrated. There are limits to the symbolic information about the environment which can be inferred from a module which directly produces an action as its output, such as the ALVINN road follower. Similarly, while most modules which produce a symbolic representation of certain features in the environment can use that model to evaluate the desirability of different possible actions, this is not always the case. A module which detects turn lane signs at intersections cannot judge the desirability of different possible lane choices without reference to higher level route information and the results of road tracking.

Systems which perform integration at the symbolic level should specify a minimal representation of objects in the map of the environment, allowing the individual modules to provide free form descriptions agreed upon as needed between modules. It is difficult to foresee all the possible pieces of information it would be useful to include in the map representation, and it is often unnecessary and expensive to describe every object in the map by the same set of features and at the same level of detail. The approach of using free-form strings of bytes as object descriptions was the approach taken to address this issue in the Annotated Map system. The representation space approach taken in the TraX system provided an alternative solution to this problem.

Behavior based approaches lack a mechanism for providing focus of attention. They assume all possibly relevant modules run in parallel and compete or cooperate to drive the vehicle.

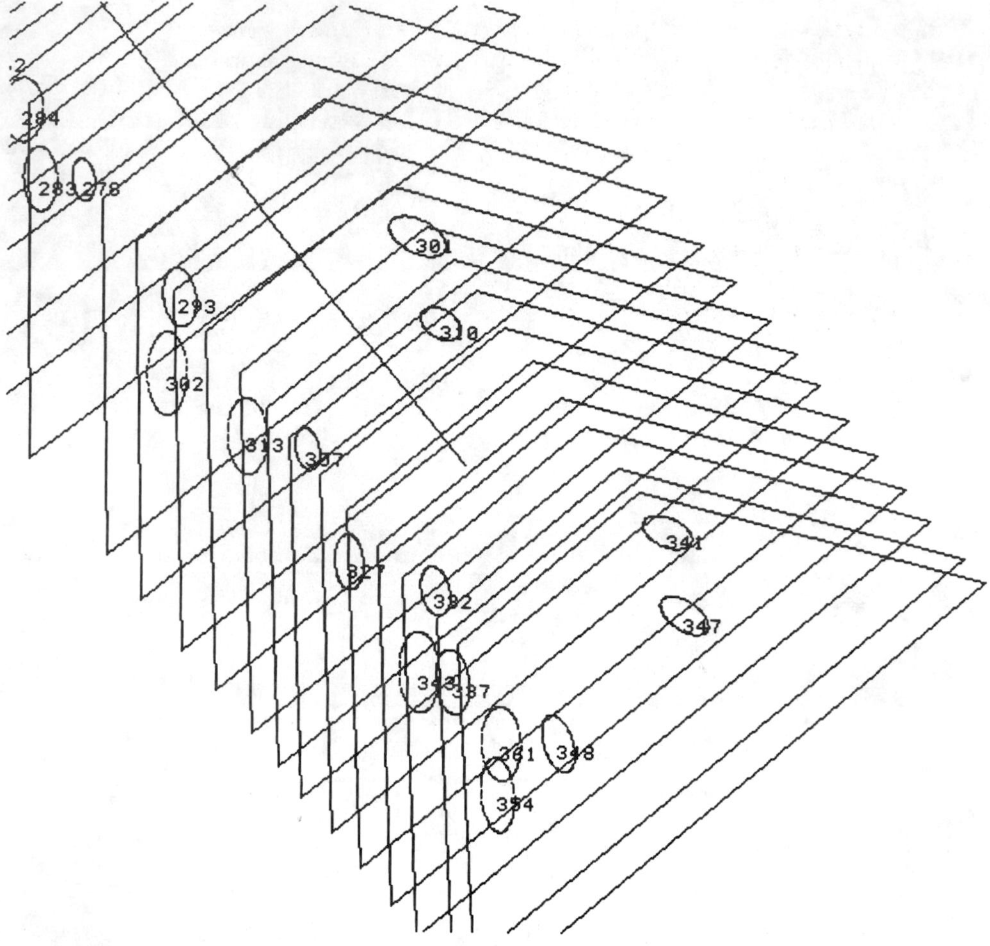

Figure 122.65 Obstacle map built up from multiple ERIM frames.

This limits vehicle speed in all situations, even very "easy" ones, to the safe speed in the most demanding situation the system is designed to cope with. Providing such a mechanism in a manner which doesn't subvert the underlying integration philosophy by introducing a world model is an open research issue.

Integrated systems need to be able to properly handle the distinction between intrinsic physical constraints and socially constructed domain constraints. The minimum braking distance for the vehicle given the current velocity is an example of the first type of constraint. That the vehicle shouldn't cross a solid painted line on the road (particularly a yellow one) when driving in the United States is an example of the second type of constraint. A road following system should violate the lane abstraction when doing so will avoid a collision. The ULYSSES system did not distinguish between the two types of constraints. The DAMN architecture currently solves this problem by weighting obstacle avoidance's preferences more highly than road following's preferences. This requires careful tuning of the module weights in order to avoid the system veering around obstacles in violation of the lane abstraction when the correct solution would be to slow down and stay in the lane. Principled solutions to handle this distinction are needed within all the integration approaches.

Acknowledgments

This research was partly sponsored by ARPA under contracts "Perception for Outdoor Navigation" (contract number DACA76-89-C-0014, monitored by the US Army Topographic Engineering Center) and "Unmanned Ground Vehicle System" (contract number DAAE07-90-C-R059, monitored by TACOM), partly sponsored by the NSF under NSF Contract BCS-9120655, titled "Annotated Maps for Autonomous Underwater Vehicles," and partly supported by the NSF under a grant titled "Massively Parallel Real-Time Computer Vision."

The author wishes to thank the following members of the Unmanned Ground Vehicle project at Carnegie Mellon for contributing comments on the text describing their systems, and figures to illustrate them: Martial Hebert, laser range scanner image segmentation guru (Figure 122.64 and Figure 122.65); Dirk Langer, author of GANESHA (Figure 122.63); Dean Pomerleau, author of ALVINN (Figure 122.47, Figure 122.49, Figure 122.50 and Figure 122.51); Doug Reece, author of ULYSSES; Ken Rosenblatt, author of DAMN (Figure 122.48); Anthony Stentz, author of CODGER (Figure 122.66); and Chuck Thorpe, the author's thesis advisor on the YARF system.

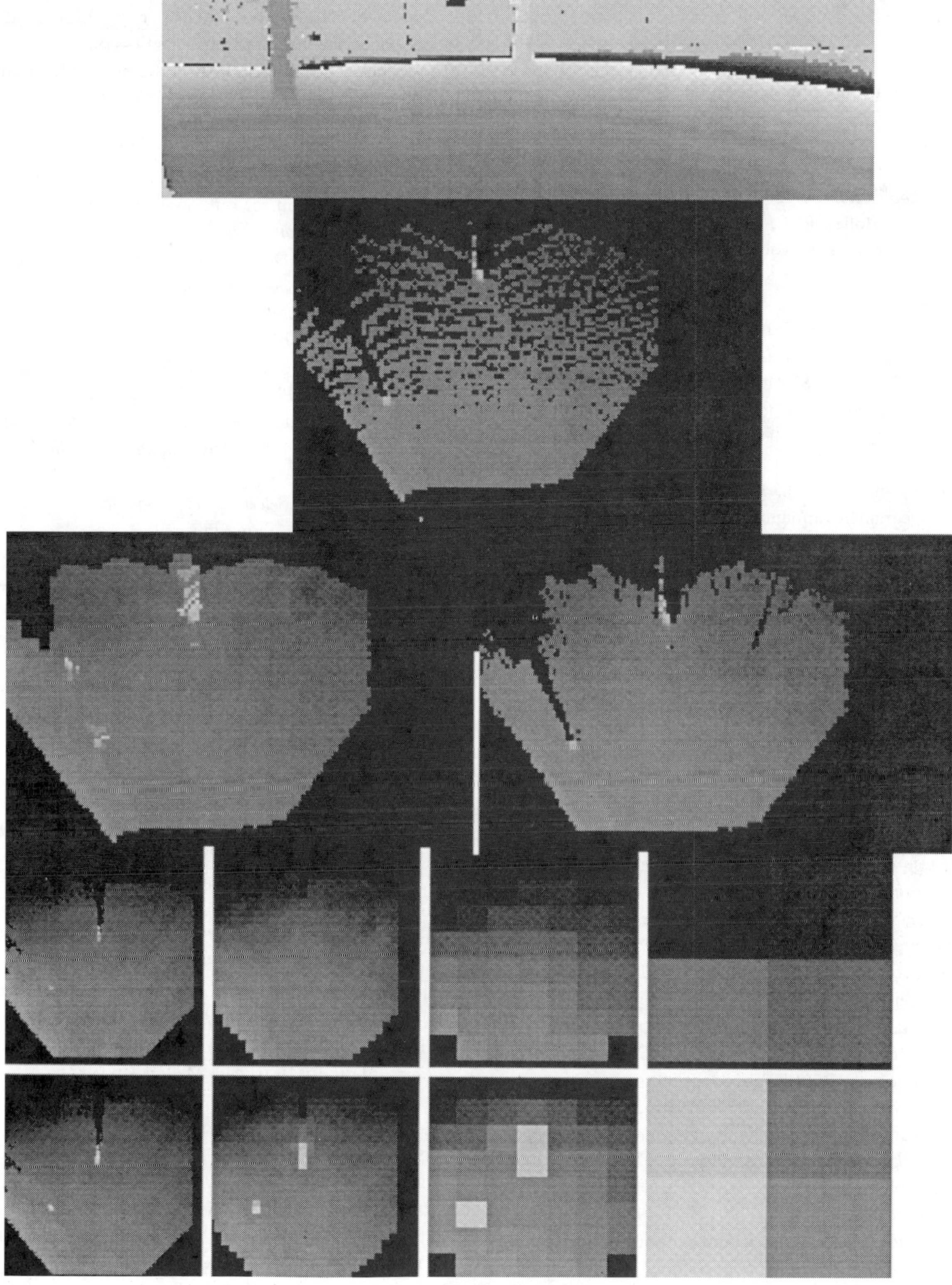

Figure 122.66 Stages in constructing the hierarchical terrain description from ERIM images.

References

Asada, M. 1990. Map Building for a Mobile Robot from Sensory Data, *IEEE Trans. Systems, Man, and Cybernetics,* 20(6):1326–1336.

Besl, P. J. 1988. Active, optical range imaging sensors, *Machine Vision and Applications,* 1:127–152.

Boblick, A. F., and Bolles, R. C. 1992. The representation space paradigm of concurrent evolving object descriptions, *IEEE Trans. Pattern Analysis and Machine Intelligence,* 14(2):145–156.

Chu, C. C., and Aggarwal, J. K. 1992. Image interpretation using multiple sensing modalities, *IEEE Trans. Pattern Analysis and Machine Intelligence,* 14(8):840–847.

Crisman, J. 1990. *Color Vision for the Detection of Unstructured Roads and Intersections,* PhD Thesis, Carnegie-Mellon University.

DeMenthon, D., and Davis, L. 1990. Reconstruction of a road by local image matches and global 3D optimization, *Proc. 1990 IEEE Int. Conf. Robotics and Automation,* May.

Dickmanns, E. D. and Graefe, V. 1988. Applications of dynamic monocular machine vision, *Machine Vision and Applications,* 1:241–261.

Dickmanns, E. D. and Mysliwetz, B. D. 1992. Recursive 3-D road and relative ego-state recognition, *IEEE Trans. Pattern Analysis and Machine Intelligence,* 14:199–213.

Efenberger, W. Ta, Q.-H., Tsinas, L., and Graefe, V. 1992. Automatic recognition of vehicles approaching from behind, *Proc. Intelligent Vehicles '92 Symp.,* June.

Goto, Y., Matsuzaki, K., Kweon, I., and Obatake, T. 1986. CMU sidewalk navigation system: a blackboard-based outdoor navigation system using sensor fusion with colored-range images, *Proc. Fall Joint Computer Conf.,* November.

Goto, Y., Shafer, S. A., and Stentz, A. 1990. The driving pipeline: a driving control scheme for mobile robots, *Vision and Navigation: The Carnegie Mellon Navlab,* ch.10, Thorpe, C. C., ed., Kluwer Academic Publishers, Norwell, MA.

Hennessy, Stephen J., and King, Robert H. Future Mining Technology Spinoffs from the ALV Program. *IEEE Transactions on Industry Applications* 25(2):377–384, 1989.

Hebert, M., Kweon, I., and Kanade, T. 1990. 3-D Vision Techniques for Autonomous Vehicles, *Vision and Navigation: The Carnegie Mellon Navlab,* Ch. 8, Thorpe, C. E., ed., Kluwer Academic Publishers, Norwell, MA.

Jochem, T., Pomerleau, D., and Thorpe, C. 1993. MANIAC: A Next Generation Neurally Based Autonomous Road Follower. *Proceedings, Intelligent Autonomous Systems 3.* February.

Kluge, K. 1993a. *YARF: An Open-Ended Framework for Robot Road Following,* PhD thesis, Carnegie Mellon University, 1993.

Kluge, K., and Thorpe, C. E. 1993b. Intersection detection in the YARF road following system, *Proc. Intelligent Autonomous Systems 3,* February.

Langer, D. and Thorpe, C. 1992. Sonar-based outdoor vehicle navigation and collision avoidance, *Proc. IROS '92.*

Langer, D., Rosenblatt, J. K., and Hebert, M. 1994. An integrated system for autonomous off-road navigation, *Proc. IEEE Int. Conf. Robotics and Automation,* 414–419.

Morgenthaler, D. G., Hennessy, S. J., and DeMenthon, D. 1990. Range-video fusion and comparison of inverse perspective algorithms in static images, *IEEE Trans. Syst., Man, and Cybernetics,* November/December, 20(6):1301–1312.

Nandhakumar, N. and Aggarwal, J. K. 1988. Thermal and visual information fusion for outdoor scene perception, *Proc. IEEE Robotics and Automation.*

Olin, K., and Tseng, D. 1991. Autonomous cross-country navigation, *IEEE Expert,* 6(4):16–29.

Payton, D. W., Rosenblatt, J. K., and Keirsey, D. M. 1990. Plan guided reaction, *IEEE Trans. Syst., Man, and Cybernetics,* 20(6).

Pomerleau, D. A. 1992a. *Neural Network Perception for Mobile Robot Guidance.* PhD thesis, Carnegie Mellon University.

Pomerleau, D. 1992b. Progress in neural network-based vision for autonomous robot driving., *Proc. Intelligent Vehicles '92 Symp.,* June.

Reece, D. A. 1992. *Selective Perception for Robot Driving.* PhD thesis, Carnegie Mellon University.

Reece, D. A. and Shafer, S. 1987. An overview of the PHAROS traffic simulator, *Proc. Int. Conf. Road Safety,* September.

Roberts, L. G. 1965. Machine Perception of Three-Dimensional Solids, *Optical and Electro-Optical Information Processing,* Tippet, J. P., et al., MIT Press, Cambridge, MN.

Shladover, S. E. 1992. The CALIFORNIA PATH program of IVHS research and its approach to vehicle-highway automation, *Proc. Intelligent Vehicles '92 Symp.,* June.

Skewis, T., Evans, J., Lumelsky, V., Krishnamurthy, B., and Barrows, B. 1991. Motion planning for a hospital transport robot, *Proc. 1991 IEEE Int. Conf. Robotics and Automation,* April.

Solder, U. and Graefe, V. 1990. Object detection in real time, *Mobile Robots V.,* November.

Stefik, M. 1985. Strategic computing at DARPA: overview and assessment, *Communications of the ACM,* 28(7):690–704.

Stentz, A. 1990a. The CODGER System for Mobile Robot Navigation, *Vision and Navigation: The Carnegie Mellon Navlab* Ch. 9. Thorpe, C. E., ed., Kluwer Academic Publishers, Norwell, MA.

Stentz, A. 1990b. *The NAVLAB System for Mobile Robot Navigation,* PhD thesis, Carnegie Mellon University.

Thorpe, C., and Gowdy, J. 1990. Annotated maps for autonomous land vehicles, *Proc. DARPA Image Understanding Workshop.* 1990.

Thorpe, C., Amidi, O., Gowdy, J., Hebert, M., and Pomerleau, D. 1991a. Integrating position measurement and image understanding for autonomous vehicle navigation, *Proc. 2nd Int. Workshop High Precision Navigation.*

Thorpe, C., Hebert, M., Kanade, T., and Shafer, S., 1991b. Towards autonomous driving: the CMU navlab, Part I—perception, *IEEE Expert,* 6(4):31–42.

Thorpe, C., Hebert, M., Kanade, T., and Shafer, S. 1991c. Towards autonomous driving: the CMU navlab, Part II—architecture and systems, *IEEE Expert,* 6(4):44–52.

Turk, M. A., Morgenthaler, D. G., Gremban, K. D., and Marra, M. 1988 VITS—a vision system for autonomous land vehicle navigation, *IEEE Trans. Pattern Analysis and Machine Intelligence,* 10(3).

Waxman, A., LeMoigne, J., Davis, L., Srinivasan, B., Kushner, T., Liang, E., and Siddalingaiah, T. 1987. A visual navigation system for autonomous land vehicles, *IEEE J. Robotics and Automation,* RA-3(2):124–141.

122.9 Future Trends for the Further Development in Multisensor Fusion and Integration

Ren C. Luo

Introduction

The need for the operation of intelligent systems in an unstructured, dynamic environment has created a growing demand for the use of distributed, multisensor systems. A robust intelligent agent or decision-support system must have the capability to use multiple disparate and complimentary sensors and to perform situation assessment using information (perhaps from an intelligence group as well as in the form of sensory data) with which there is some degree of uncertainty associated. Sensors for these systems should be chosen such that the data they provide can be correlated, by using methods of fusion and integration, and in such a way that it reduces the uncertainty inherent in the individual sensors (Luo and Kay, 1992).

Various frameworks and control structures have been established for the integration and fusion of multisensor data and the usefulness of each approach is often very application-specific, making it difficult to claim any technique as being general-purpose. Some development background on paradigms and frameworks are presented in (Henderson and Shilcrat, 1984; Henderson et al., 1988a, Luo et al., 1988; Garvey, 1987; Rodger and Browse, 1987; Allen, 1987; Henderson and Grupen, 1990; Thomopoulos, 1990; Pau, 1990). More background on often-used tools for sensor integration, such as artificial neural networks can be found in (Hinton and Anderson, 1989; Pearson and Gelfand, 1988; Priebe and Marchette, 1988; Jakubowicz, 1989). Control structures that have proven to be effective include those such as Bayesian networks (Mitiche et al., 1989; Miltonberger et al., 1989), rule-based systems (Graham and Jones, 1988; Grzymala-Busse, 1991), and distributed blackboards (Johnson et al., 1989; Sikka and Varshney, 1989; Harmon, 1990). Methods for handling the uncertainty that is implicit in all control structures is an area of research that has gained much attention recently.

Uncertainty and Multisensor Systems

Uncertainty implies that the information provided by a sensor or sensor network is imprecise, incomplete, ambiguous, vague or in some way not totally reliable. *Uncertain Reasoning* is the process of forming some opinion or judgment about a situation or an environment (usually as a prelude to the performance of an action) based upon uncertain information. Experience has shown us that practically any attempt to perform some action in the real world requires the use of uncertain reasoning at one or more levels. It is usually impossible-or at least impractical, due to the cost of gathering precise information-to know everything about a given situation. Some methods that have been found to yield favorable results when reasoning under uncertain conditions include: Bayesian belief networks (Pearl, 1986); Certainty Factors (Shortliffe and Buchanan, 1975); the Dempster-Shafer technique for combining evidence; and fuzzy set theory (Zaden, 1965).

Areas Dependent Upon Advances in Multisensor Fusion and Integration

Critical areas for fusion and integration of data supplied by multiple sensors include those in which the amount of raw data required to make a decision tend to overwhelm human operators or decision-makers; especially if loss of life or property hang in the balance. These include applications that require the tracking of multiple targets in real-time, such as: air traffic control (ATC), automatic target recognition (ATR), and intelligent flight control (e.g., jet-fighter pilots).

For example, ATC systems must track multiple targets-perhaps 2000, on the average (Vachar et al., 1992)—in a cluttered environment and most current systems are still using the 1960s-vintage technology of α-β filters (Bar-Shalom, 1992). Multitarget tracking systems must estimate the current state of targets based on uncertain information, and it must calculate the accuracy and credibility (is it really a target, or just clutter?) associated with the state estimate. To do this effectively it must overcome target mode uncertainties (inaccuracies in the sensors due to noise; origin of the measurements is not perfectly certain).

Reduction of the uncertainties associated with origin of measurements is the basis of the research area referred to as *data association.* In a multitarget tracking situation such as ATC, *tracks* (i.e., a set of estimates of the dynamic states of objects) must be assigned to a set of measurements that are generated by each of the sensors such as radar or an optical sensor. This is the same type of problem that occurs in stereo computer vision and is known as the *correspondence problem.* Hence, ongoing research in this area will provide results that will be beneficial across many areas related to multisensor fusion and integration.

Future Research Multisensor Fusion and Integration Systems

Research in multisensor integration and fusion will continue to focus on adaptive systems and on models which allow the highly parallel real-time collection and processing of sensor data. Hybrid systems that accommodate the most useful characteristics of knowledge-based systems, along with neural networks and techniques that focus on effective methods for reasoning under uncertain conditions in establishing a mathematically rigorous foundation for the fusion and integration processes.

Recent advances in the manufacture of silicon chips which are able to integrate sensing devices and signal-processing electronics have opened the world to the development of microsensors on a scale approaching three orders of magnitude smaller than the diameter of a human hair (Middelhoek and Hoogerwarf, 1985). A combination of microsensors and multisensor fusion will make possible a new range of applications. Continuing developments in microsensor technology make it apparent that it will soon be possible to make use of dense populations of highly redundant sensors, in much the same way that they appear in biological systems.

An example of progress in this field is the fringe-effect-capacitive-proximity sensor (*patent pending*) that has recently been developed at the Center for Robotics and Intelligent Machines (CRIM) of North Carolina State University. This sensor is designed and fabricated with solid-state technology and is able to provide a precise measure of proximity in the range from micrometers to millimeters. The use of solid-state technology makes it possible to fabricate these sensors in extremely dense populations and allows the processing circuitry to be located very close to the sensor itself, thus eliminating electronic noise that would be associated with wires and cables.

Research is currently being considered that may make it possible to mount arrays of this sensor inside flexible skin-like layers of Polymide film. This skin-like membrane could then be used in combination with robot grippers that have multi-jointed fingers like the Utah/MIT hand and the Stanford/JPL hand, giving these hands much more sensing ability than is now possible. While the imagination leaps at the possibilities that huge arrays of redundant and complimentary sensors could provide, there are still many problems that must be worked out.

The need for decision support systems in all phases of industry promotes interaction between researchers in multitarget tracking and those in seemingly unrelated fields such as project planning that investigate the problem of resource management and multi-participant cooperation in a dynamic and uncertain environment (Chang, et al., 1993). Project planning and control may offer some insights into the problem of sensor management.

Sensor management becomes critical in systems such as those found in a modern fighter aircraft. Sophisticated airborne-sensor management schemes are necessary to reduce the mental workload of the pilot and to enhance human performance in such a complex system. The *sensor management imperative,* as described in Popoli (1992) is a composite of sensor management roles that are designed to alleviate part of the burden on a fighter pilot by providing much of the feedback between sensor tracking performance and future sensor behavior.

In general, an "intelligent system" will require increased sensory information to ensure effective operation of the system. The effectiveness of the operation should include adaptivity, robustness and ability to handle uncertainty in sensory data. As the technology evolved, we can see more and more attempt to practically fuse multiple sensors in either competitive or complementary manner. Some examples are: ultrasonic range sensors; range image sensor-infrared thermal sensor; intensity vision-infrared thermal sensor; intensity vision-acoustic; vision-tactile sensor; ultrasonic-range image sensor-intensity vision, etc. There are some very good fusion results that have been presented. Certainly, there is a need to continue conducting research in sensor fusion methods incorporate with practical applications, such as autonomous navigation, surveillance, target detection, recognition and tracking, drug interdiction, etc.

Conclusion

Research in the near future will continue to be aimed at developing integration and fusion techniques that will allow multisensory systems to operate in previously unknown and dynamic environments, making use of uncertain or incomplete information. Highly parallel computer architectures will best be able to make use of parallelism that is implicit in systems that use large numbers of sensors. New sensors such as high-redundant microsensor arrays will be incorporated into the design of intelligent multisensor systems.

Multisensor fusion and integration has quickly become a very important field that spreads across many areas of technology and sensor synergism is a key principle in ensuring that maximum information can be extracted in minimal time.

References

Allen, P. K. 1987. A framework for implementing multi-sensor robotic tasks, *AASME Int. Computers in Engr. Conf. and Exhibition,* 303–309.

Bar-Shalom, Y., ed. 1992. *Multitarget-Multisensor Tracking: Applications and Advances,* Artech House, Norwood, MA.

Chang, A.-M., Bailey, J. A. D., et al. 1993. A distributed knowledge-based approach for planning and control projects, *IEEE Trans. Syst. Man and Cyber.,* 23(6):1537–1550.

Chong, C. -Y., Mori, S. et al. 1990. Distributed multitarget multisensor tracking, *Multitarget-Multisensor Tracking: Advanced Applications,* 247–296, Artech House, Norwood, MA.

Graham, I. and Jones, P. L. 1988. *Expert Systems: Knowledge, Uncertainty and Decision,* Chapman and Hall, London.

Grzymala-Busse, J. W. 1991. *Managing Uncertainty in Expert Systems,* Academic Publishers, Boston, MA.

Harmon, S. Y. 1990. Tools for multisensor data fusion in autonomous robots, *Highly Redundant Sensing in Robotic Systems,* 103–125, Springer-Verlag, Berlin.

Hartley, R. and Pipitone, F. 1991. Experiments with the subsumption architecture, *Proc. 1991 IEEE Int. Conf. Robotics and Automation.*

Henderson, T. C. and Grupen, R. 1990. Logical behaviors, *J. Robotic Syst.,* 7(3):309–336.

Henderson, T. C., Weitz, E., et al. 1988. Multisensor knowledge systems: interpreting 3D structure, *Int. J. Robot. Res.,* 7(6):114–137.

Hinton, G. E. and Anderson, J. A., eds. 1989. *Parallel Models of Associative Memory,* Erlbaum, Hillsdale, NJ.

Jakubowicz, O. G. 1989. Multi-layer multi-feature map architecture for situational analysis, *Int. Joint Conf. Neural Networks,* pp. II-23–II-30.

Johnson, D., Shaw, S., et al. 1989. Real-time blackboards for sensor fusion, *Sensor Fusion II,* 61–73.

Luo, R. C. and Kay M. G. 1992. Data Fusion and Sensor Integration: State-of-art 1990's, *Data Fusion in Robotics and Machine Intelligence,* Boston, Academic Press, Boston, MA. (in press).

Luo, R. C., Lin, M., et al. 1988. Dynamic multi-sensor data fusion system for intelligent robots, *J. Robot. and Automat.,* RA-4(4):386–396.

Miltonberger, T., Morgan, D., et al. 1989. Multisensor object recognition from 3D models, *Sensor Fusion: Spatial Reasoning and Scene Interpretation,* 61–169.

Mitiche, A., Lagnaiere, R., et al. 1989. Decision networks for multisensor integration in computer vision, *Sensor Fusion: Spatial Reasoning and Scene Interpretation,* 291–299.

Pau, L. F. 1990. Behavioral knowledge in sensor/data fusion systems, *J. Robotic Syst.,* 7(3):295–308.

Pearl, J. 1986. Fusion, propagation, and structuring in belief networks, *AI,* 29:241–248.

Pearson, J. C., Gelfand, J. J., et al. 1988. Neural network approach to sensory fusion, *SPIE,* Sensor Fusion, 931:103–108.

Popoli, R. 1992. The sensor management imperative, *Multitarget-Multisensor Tracking: Applications and Advances,* 325–392, Artech House, Norwood, MA.

Priebe, C. E. and Marchette, D. J. 1988. Temporal pattern recognition: a network architecture for multisensor fusion, *Intell. Robots Comp. Vision,* 679–685.

Rodger, J. C. and Browse, R. A. 1987. An object-based representation for multisensory robotic perception, *Workshop Spatial Reasoning and Multi-Sensor Fusion,* 13–20.

Shortliffe, E. H. and Buchanan, B. G. 1975. A model of inexact reasoning in medicine, *Math. Biosci.,* 23:351–379.

Sikka, D. I. and Varshney, P. K. 1989. A distributed artificial intelligence approach to object identification and classification, *Sensor Fusion II,* 73–84.

Thomopoulos, S. C. A. 1990. Sensor integration and data fusion, *J. Robotic Syst.* 7(3):37–372, 1990.

Vacher, P., Barret, I., et al. 1992. Design of a tracking algorithm for an advanced ATC system, *Multitarget-Multisensor Tracking: Applications and Advances,* 1–28, Artech House, Norwood, MA.

Zadeh, L. A. 1965. Fuzzy sets, *Infor. Contr.* 8:338–353.

Indexes

Author Index 1663

Subject Index 1669

Author Index

Author	Section	Chapter	Page	Chapter Title
Agehed, K.	X	113	1444	NEBULAS: High Performance Data-Driven Event Building Architectures Based on Asynchronous Self-Routing Packet-Switching Networks
Ang, M. H., Jr.	VI	53	730	Robots
Arai, F.	X	118	1526	Micro Actuators and Energy Supply
Arai, T.	X	122	1592	Multisensor Fusion and Integration for Intelligent Systems
Baginski, T.	I	1	5	Electronics
Barkhordarian, V.	III	10	218	Power MOSFETs
Bejczy, A. K.	VI	53	730	Robots
Belhe, U.	VI	48	663	Production Management Techniques
Benard, M.	X	113	1444	NEBULAS: High Performance Data-Driven Event Building Architectures Based on Asynchronous Self-Routing Packet-Switching Networks
Bencze, J.	III	7	187	Introduction to Power Electronics
Blade, R. A.	X	111	1383	Virtual Reality
Borka, J.	III	13	288	Motor Drives
Boye, A. J.	V	22	447	Modeling for System Control
Boyer, W.	II	6	103	Measurement System Architecture
Brogan, W. L.	V	22	447	Modeling for System Control
Brown, R. H.	III	13	288	Motor Drives
Bryzek, J.	X	121	1576	MEMS Integration—Technical and Economic Considerations
Buczak, A. L.	IX	103	1325	Information Fusion by Fuzzy Set Operations and Genetic Algorithms
Bugajski, D.	V	44	613	μ-Synthesis and Analysis
Calvet, D.	X	113	1444	NEBULAS: High Performance Data-Driven Event Building Architectures Based on Asynchronous Self-Routing Packet-Switching Networks
Carpenter, G. A.	VIII	97	1286	Adaptive Resonance Theory
Carson, J.	VII	73	990	Analog 3-D Neuroprocessor for Fast Frame Focal Plane Image Processing
Caudell, T.	X	111	1383	Virtual Reality
Chi, C.-Y.	X	120	1547	Si Micromachining in High Frequency Applications
Chow, M.	V	39	564	Fuzzy Logic-Based Control
Chow, M.	VIII	83	1096	Fuzzy Systems
Christie, R. W.	IV	18	394	Local Area Networks
Chung, W.	VII	70	951	Supervised Neural Networks for Handwritten Digit Recognition in Industrial Processing
Chung, W.	VII	80	1055	Classifiers: An Overview
Combacau, M.	I	2	22	Digital Control Circuits
Cooper, J. A.	VIII	82	1091	Fuzzy Numbers: The Application of Fuzzy Algebra to Safety and Risk Analysis
Cooper, M. G.	IX	101	1316	Genetic Algorithms
Costa, M.	X	113	1444	NEBULAS: High Performance Data-Driven Event Building Architectures Based on Asynchronous Self-Routing Packet-Switching Networks
Courvoisier, M.	I	2	22	Digital Control Circuits
Dasey, T. J.	VIII	95	1231	Neural Fuzzy Systems in Handwritten Digit Recognition
Daud, T.	VII	73	990	Analog 3-D Neuroprocessor for Fast Frame Focal Plane Image Processing
De Abreu-Garcia, J. A.	V	27	490	The Root Locus Method
DePrycker, M.	X	113	1444	NEBULAS: High Performance Data-Driven Event Building Architectures Based on Asynchronous Self-Routing Packet-Switching Networks
Decotignie, J.-D.	IV	18	394	Local Area Networks
Dempsey, B. J.	IV	20	427	Essential Communications Protocols
Denney, T. S., Jr.	V	38	559	Estimation and Identification
Diy, A.	III	11	229	Insulated Gate Bipolar Transistors
Djidi, K.	X	113	1444	NEBULAS: High Performance Data-Driven Event Building Architectures Based on Asynchronous Self-Routing Packet-Switching Networks

Author	Section	Chapter	Page	Chapter Title
Drayton, R. F.	X	120	1547	Si Micromachining in High-Frequency Applications
Dufey, J.-P.	X	113	1444	NEBULAS: High Performance Data-Driven Event Building Architectures Based on Asynchronous Self-Routing Packet-Switching Networks
Dummermuth, E.	V	41	587	Programmable Logic Control (PLC)
Duong, T. A.	VII	73	990	Analog 3-D Neuroprocessor for Fast Frame Focal Plane Image Processing
Dutta, R.	III	11	229	Insulated Gate Bipolar Transistors
Eide, Å.	VII	75	1014	Radial Basis Function (RBF) Neural Networks
Eide, Å.	VII	76	1019	Hardware Implemented Radial Basis Function (RBF): The IBM Zero Instruction Set Computer
Einolf, C. W., Jr.	II	5	97	Sensors
Enjeti, P.	III	14	349	Main Disturbances
Enns, D.	V	44	613	μ-Synthesis and Analysis
de la Escalera, A.	VI	53	730	Robots
Evers, J.	X	111	1383	Virtual Reality
Fazal, F. A.	VII	72	975	Studies of Pattern Recognition with Self-Learning Layered Neural Networks
Francis, R.	III	11	229	Insulated Gate Bipolar Transistors
Frazier, A. B.	X	116	1489	Selected Micromachining Fabrication Technologies
Friedrich, C. R.	X	116	1489	Selected Micromachining Fabrication Technologies
Fujita, H.	X	115	1472	Micromachines
Fukuda, T.	IX	108	1364	Synthesis of Fuzzy, Artificial Intelligence, Neural Networks, and Genetic Algorithm for Hierarchical Intelligent Control
Fukuda, T.	X	118	1526	Micro Actuators and Energy Supply
Fukushima, K.	VII	71	966	Neocognitron
Gachet, D.	VI	53	730	Robots
Ghorbel, F.	VI	53	730	Robots
Greene, M.	V	28	504	Pole Placement Design
Grossberg, S.	VIII	97	1286	Adaptive Resonance Theory
Gustafsson, L.	X	113	1444	NEBULAS: High Performance Data-Driven Event Building Architectures Based on Asynchronous Self-Routing Packet-Switching Networks
Halász, S.	III	12	244	Conversion
Halász, S.	III	13	288	Motor Drives
Hang, C. C.	V	23	453	Basic Feedback Concept
Harbor, R. D.	V	43	609	Hardware Compensating Networks
Haykin, S.	VII	65	910	Temporal Signal Processing
Hecklesmiller, J.	III	14	349	Main Disturbances
Heinen, J. A.	I	4	73	Signal Processing
Helferty, J. P.	X	122	1592	Multisensor Fusion and Integration for Intelligent Systems
Heragu, S. S.	VI	49	669	Automated Manufacturing System Development Methodology
Heydt, G.	III	14	349	Main Disturbances
Hines, R. M.	IV	17	389	Open Systems Interconnection Basic Reference Model
Hirano, I.	VIII	87	1127	Development of an Intelligent Unmanned Helicopter Based on Fuzzy Systems
Hodel, A. S.	V	21	443	Control System Fundamentals
Hombu, M.	III	13	288	Motor Drives
Hori, T.	III	13	288	Motor Drives
Hsieh, G.-C.	V	34	529	Phase-Locked Loop-Based Control
Huang, K.-L.	VII	64	906	CMAC Neural Networks and Color Correction
Hultberg, S.	X	113	1444	NEBULAS: High Performance Data-Driven Event Building Architectures Based on Asynchronous Self-Routing Packet-Switching Networks
Hung, J. C.	V	25	470	PID Control
Hung, J. C.	V	30	513	Internal Model Control
Hung, J .C.	V	32	522	Dynamic Matrix Control
Hung, J. Y.	V	29	511	The Smith Predictor Technique
Hung, J. Y.	V	37	553	Digital Control
Hung, S. T.	V	42	593	Adaptive Control
Ipsits, I.	III	9	203	Devices and Components
Ishii, R.	VI	52	723	Signal Processing for Factory Production Lines
Jackson, M.	V	44	613	μ-Synthesis and Analysis
Jani, Y.	VIII	86	1116	Fuzzy Logic Control: Basics and Applications
Jani, Y.	VIII	90	1147	Neural Network Learning in Fuzzy Systems
Jara-Almonte, J.	X	116	1489	Selected Micromachining Fabrication Technologies
Jarvis, R.	VI	53	730	Robots
Jiang, B. C.	II	6	103	Measurement System Architecture
Karayiannis, N. B.	VIII	96	1264	Fuzzy Algorithms for Learning Vector Quantization
Karpati, A.	III	12	244	Conversion

Author	Section	Chapter	Page	Chapter Title
Katehi, L. P. B.	X	120	1547	Si Micromachining in High-Frequency Applications
Kay, M. G.	X	122	1592	Multisensor Fusion and Integration for Intelligent Systems
Kaynak, O.	IX	107	1360	Application Techniques: Combining Fuzzy Logic, Artificial Neural Networks, and Probabilistic Reasoning—Soft Computing
Kemeny, S.	VII	73	990	Analog 3-D Neuroprocessor for Fast Frame Focal Plane Image Processing
Kim, C.-J.	X	114	1468	Microelectromechanical Systems (MEMS)
Kleeman, L.	VI	53	730	Robots
Kluge, K.	X	122	1592	Multisensor Fusion and Integration for Intelligent Systems
Kobylecki, K.	X	113	1444	NEBULAS: High Performance Data-Driven Event Building Architectures Based on Asynchronous Self-Routing Packet-Switching Networks
Kohonen, T.	VII	57	835	Strategies and Tactics for the Application of Neural Networks to Industrial Electronics
Kokar, M. M.	X	122	1592	Multisensor Fusion and Integration for Intelligent Systems
Korona, Z.	X	122	1592	Multisensor Fusion and Integration for Intelligent Systems
Kotanchek, M. E.	X	122	1592	Multisensor Fusion and Integration for Intelligent Systems
Kotsu, S.	VIII	87	1127	Development of an Intelligent Unmanned Helicopter Based on Fuzzy Systems
Kurutz, K.	III	9	203	Devices and Components
Kusiak, A.	VI	48	663	Production Management Techniques
Lau, C.	VII	59	858	Neural Networks on a Chip
Layden, D.	III	14	349	Main Disturbances
Lazrak, T.	X	113	1444	NEBULAS: High Performance Data-Driven Event Building Architectures Based on Asynchronous Self-Routing Packet-Switching Networks
Lea, R. N.	VIII	86	1116	Fuzzy Logic Control: Basics and Applications
Lea, R. N.	VIII	90	1147	Neural Network Learning in Fuzzy Systems
Le Dû, P.	X	113	1444	NEBULAS: High Performance Data-Driven Event Building Architectures Based on Asynchronous Self-Routing Packet-Switching Networks
Lee, J. B.	X	119	1538	On-Board Power Supply and Remote Driving Mechanisms for Microelectromechanical Systems
Lee, J. H.	V	31	515	Model Predictive Control
Lee, T. H.	V	23	453	Basic Feedback Concept
Lehr, M.	VII	58	853	The Basic Ideas in Neural Networks
Letheren, M.	X	113	1444	NEBULAS: High Performance Data-Driven Event Building Architectures Based on Asynchronous Self-Routing Packet-Switching Networks
Lim, K.-W.	III	13	288	Motor Drives
Lindblad, T.	VII	75	1014	Radial Basis Function (RBF) Neural Networks
Lindblad, T.	VII	76	1019	Hardware Implemented Radial Basis Function (RBF): The IBM Zero Instruction Set Computer
Lindblad, T.	VIII	89	1143	NeuFuz: A Combined Neural Net/Fuzzy Logic Tool
Lindblad, T.	X	112	1438	Asynchronous Transfer Mode Technology
Lindblad, T.	X	113	1444	NEBULAS: High Performance Data-Driven Event Building Architectures Based on Asynchronous Self-Routing Packet-Switching Networks
Lindsey, C. S.	VII	75	1014	Radial Basis Function (RBF) Neural Networks
Lindsey, C. S.	VII	76	1019	Hardware Implemented Radial Basis Function (RBF): The IBM Zero Instruction Set Computer
Lindsey, C. S.	VIII	89	1143	NeuFuz: A Combined Neural Net/Fuzzy Logic Tool
Lindsey, C. S.	X	113	1444	NEBULAS: High Performance Data-Driven Event Building Architectures Based on Asynchronous Self-Routing Packet-Switching Networks
Lucarelli, C. M.	VI	49	669	Automated Manufacturing System Development Methodology
Luo, R. C.	X	122	1592	Multisensor Fusion and Integration for Intelligent Systems
MacDonald, N. C.	X	116	1489	Selected Micromachining Fabrication Technologies
Magotra, N.	II	6	103	Measurement System Architecture
Manabe, A.	X	113	1444	NEBULAS: High Performance Data-Driven Event Building Architectures Based on Asynchronous Self-Routing Packet-Switching Networks
Mandjavidze, I.	X	113	1444	NEBULAS: High Performance Data-Driven Event Building Architectures Based on Asynchronous Self-Routing Packet-Switching Networks
Mandyam, G.	II	6	103	Measurement System Architecture
Marchette, D. J.	VII	67	923	Wavelets for Pattern Recognition
Marchioro, A.	X	113	1444	NEBULAS: High Performance Data-Driven Event Building Architectures Based on Asynchronous Self-Routing Packet-Switching Networks
Maslowski, W.	III	15	377	Electromagnetic Compatibility for Drives
Masuda, R.	X	122	1592	Multisensor Fusion and Integration for Intelligent Systems
McCoy, W.	II	6	103	Measurement System Architecture
Mica, J. A.	VIII	86	1116	Fuzzy Logic Control: Basics and Applications
Micheli-Tzanakou, E.	VII	61	874	Implementing Neural Networks in Silicon

Author	Section	Chapter	Page	Chapter Title
Micheli-Tzanakou, E.	VII	69	942	Multilayer Perceptrons with ALOPEX and Backpropagation
Micheli-Tzanakou, E.	VII	70	951	Supervised Neural Networks for Handwritten Digit Recognition in Industrial Processing
Micheli-Tzanakou, E.	VII	72	975	Studies of Pattern Recognition with Self-Learning Layered Neural Networks
Micheli-Tzanakou, E.	VII	80	1055	Classifiers: An Overview
Micheli-Tzanakou, E.	VIII	95	1231	Neural Fuzzy Systems in Handwritten Digit Recognition
Miller, S. A.	X	116	1489	Selected Micromachining Fabrication Technologies
Miner, N.	X	111	1383	Virtual Reality
Moffit, C. L.	IV	18	394	Local Area Networks
Moreno, L.	VI	53	730	Robots
Morton, B.	V	44	613	μ-Synthesis and Analysis
Murray, W. B.	X	122	1592	Multisensor Fusion and Integration for Intelligent Systems
Murry, H.	X	111	1383	Virtual Reality
Nagase, H.	III	13	288	Motor Drives
Nagi, R.	VI	47	653	Production Management Architecture
Nagy, I.	III	12	244	Conversion
Nelson, V. P.	I	3	48	Computer Architecture
Niederjohn, R. J.	I	4	73	Signal Processing
Ohnishi, K.	V	33	524	Disturbance Observation-Cancellation Technique
Padgett, M. L.	VII	54	805	Current Applications of Expert Systems in Industrial Electronics
Padgett, M. L.	VII	57	835	Strategies and Tactics for the Application of Neural Networks to Industrial Electronics
Padgett, M. L.	VIII	81	1087	Applications of Fuzzy Systems and Soft Computing in Industrial Electronics
Padgett, M. L.	VIII	84	1103	Fuzzy Hardware
Padgett, M. L.	VIII	85	1112	Fuzzy Modeling Applications: Controls, Visions, Decisions
Padgett, M. L.	VIII	88	1139	Fuzzy and Neural Modeling
Padgett, M. L.	VIII	98	1299	Future Directions for Fuzzy Systems and Soft Computing in Industrial Electronics
Padgett, M. L.	IX	99	1303	Applications of Evolutionary Systems in Industrial Electronics
Padgett, M. L.	IX	100	1307	Evolutionary Computation
Padgett, M. L.	IX	102	1321	Fuzzy Evolutionary and GA Systems
Padgett, M. L.	IX	104	1338	Neural Evolutionary and GA Systems and Applications
Padgett, M. L.	IX	105	1343	Computational Intelligence Applications in Industrial Electronics
Padgett, M. L.	X	111	1383	Virtual Reality
Paillard, C.	X	113	1444	NEBULAS: High Performance Data-Driven Event Building Architectures Based on Asynchronous Self-Routing Packet-Switching Networks
Palmer, C.	X	122	1592	Multisensor Fusion and Integration for Intelligent Systems
Paludetto, M.	I	2	22	Digital Control Circuits
Pancerella, C. M.	IV	16	385	Evolution of Factory Communication
Parr, J.	V	26	474	Bode Diagram Method
Pauwels, B.	X	113	1444	NEBULAS: High Performance Data-Driven Event Building Architectures Based on Asynchronous Self-Routing Packet-Switching Networks
Pedrycz, W.	VIII	94	1207	Fuzzy Pattern Recognition
Pendharkar, S.	III	8	195	Overview: Devices and Components
Petit, G.	X	113	1444	NEBULAS: High Performance Data-Driven Event Building Architectures Based on Asynchronous Self-Routing Packet-Switching Networks
Pettit, R.	II	6	103	Measurement System Architecture
Phillips, C. L.	V	43	609	Hardware Compensating Networks
Pimentel, J. R.	IV	19	417	Manufacturing Automation Protocol (MAP)
Pimentel, J. R.	VI	53	730	Robots
Pleinevaux, P.	IV	18	394	Local Area Networks
Poston, T.	X	111	1383	Virtual Reality
Priebe, C. E.	VII	68	933	Fractals for Pattern Recognition
Proth, J.-M.	VI	47	653	Production Management Architecture
Rahman, M. F.	III	13	288	Motor Drives
Rebeiz, G. M.	X	120	1547	Si Micromachining in High-Frequency Applications
Reed, R. P.	II	6	103	Measurement System Architecture
Reilly, D. L.	VII	77	1025	The RCE Neural Network
Ricco, A. J.	X	117	1515	Microsensors
Riley, G.	VII	55	808	Expert Systems Methodology
Robbi, A. D.	VI	49	669	Automated Manufacturing System Development Methodology
Robertson, S. V.	X	120	1547	Si Micromachining in High-Frequency Applications
Robinson, M.	III	11	229	Insulated Gate Bipolar Transistors
Rogers, G. W.	VII	67	923	Wavelets for Pattern Recognition
Rogers, G. W.	VII	68	933	Fractals for Pattern Recognition

Author	Section	Chapter	Page	Chapter Title
Rogers, S. K.	VII	66	916	Feature Selection for Pattern Recognition Using Multilayer Perceptrons
Roppel, T.	II	6	103	Measurement System Architecture
Rowland, J. R.	V	36	545	Digital Computation
Ruck, D. W.	VII	66	916	Feature Selection for Pattern Recognition Using Multilayer Perceptrons
Rumelhart, D. E.	VII	58	853	The Basic Ideas in Neural Networks
Russell, R. A.	VI	53	730	Robots
Ryerson, D.	II	6	103	Measurement System Architecture
Saeks, R.	VII	62	885	An Avionics Application: MIMD Neural Network Processor
Saif, M. T. A.	X	116	1489	Selected Micromachiningg Fabrication Technologies
Salichs, M. A.	VI	53	730	Robots
Sarkar, D.	IV	20	427	Essential Communications Protocols
Sasaki, M.	X	122	1592	Multisensor Fusion and Integration for Intelligent Systems
Saunders, C.	VII	73	990	Analog 3-D Neuroprocessor for Fast Frame Focal Plane Image Processing
Shelton, R.	VII	54	805	Current Applications of Expert Systems in Industrial Electronics
Shelton, R.	IX	105	1343	Computational Intelligence Applications in Industrial Electronics
Shenai, K.	III	8	195	Overview: Devices and Components
Sheu, B. J.	VII	74	1003	Simulated Annealing, Boltzmann Machine, and Hardware Annealing
Shibata, T.	IX	108	1364	Synthesis of Fuzzy, Artificial Intelligence, Neural Networks, and Genetic Algorithm for Hierarchical Intelligent Control
Sinha, N. K.	V	24	456	Stability Analysis
Sobh, T. M.	VI	50	706	Hybrid Systems and Control
Solka, J. L.	VII	67	923	Wavelets for Pattern Recognition
Solka, J. L.	VII	68	933	Fractals for Pattern Recognition
Solomon, O.	II	6	103	Measurement System Architecture
Specht, D. F.	VII	78	1038	Probabilistic Neural Networks Model
Specht, D. F.	VII	79	1047	General Regression Neural Network Model
Sphicas, P.	X	113	1444	NEBULAS: High Performance Data-Driven Event Building Architectures Based on Asychronous Self-Routing Packet-Switching Networks
Stanislawski, J.	III	14	349	Main Disturbances
Stearns, S. D.	II	6	103	Measurement System Architecture
Steffek, L.	III	14	349	Main Disturbances
Stein, G.	V	44	613	μ-Synthesis and Analysis
Strayer, W. T.	IV	16	385	Evolution of Factory Communication
Sublett, J. W.	IV	18	394	Local Area Networks
Sugeno, M.	VIII	87	1127	Development of an Intelligent Unmanned Helicopter Based on Fuzzy Systems
Sumorok, K.	X	113	1444	NEBULAS: High Performance Data-Driven Event Building Architectures Based on Asynchronous Self-Routing Packet-Switching Networks
Szu, H. H.	IX	109	1369	Advanced Tools for Adaptive Nonlinear Modeling and Control of Power in Large Systems
Tai, Y. C.	X	114	1468	Microelectromechanical Systems (MEMS)
Takagi, H.	X	111	1383	Virtual Reality
Takahashi, K.	X	122	1592	Multisensor Fusion and Integration for Intelligent Systems
Tan, K. K.	V	23	453	Basic Feedback Concept
Tan, L.-Z.	II	6	103	Measurement System Architecture
Tanie, K.	IX	108	1364	Synthesis of Fuzzy, Artificial Intelligence, Neural Networks, and Genetic Algorithm for Hierarchical Intelligent Control
Telfer, B. A.	IX	109	1369	Advanced Tools for Adaptive Nonlinear Modeling and Control of Power in Large Systems
Tenhunen, H.	X	113	1444	NEBULAS: High Performance Data-Driven Event Building Architectures Based on Asynchronous Self-Routing Packet-Switching Networks
Tether, S.	X	113	1444	NEBULAS: High Performance Data-Driven Event Building Architectures Based on Asynchronous Self-Routing Packet-Switching Networks
Thakoor, A.	VII	73	990	Analog 3-D Neuroprocessor for Fast Frame Focal Plane Image Processing
Trent, V.	V	28	504	Pole Placement Design
Trent, V.	V	37	553	Digital Control
Trivedi, M.	III	8	195	Overview: Devices and Components
Tsoukalas, L. H.	VII	56	824	Expert Systems and Their Use in Complex Engineering Systems
Tsoukalas, L. H.	IX	106	1346	Hybrid Artificial Intelligence Systems
Uhrig, R. E.	VII	56	824	Expert Systems and Their Use in Complex Engineering Systems
Uhrig, R. E.	IX	103	1325	Information Fusion by Fuzzy Set Operations and Genetic Algorithms
Uhrig, R. E.	IX	106	1346	Hybrid Artificial Intelligence Systems
Umeda, K.	X	122	1592	Multisensor Fusion and Integration for Intelligent Systems
Upadhyaya, B.	II	6	103	Measurement System Architecture
Utkin, V.	V	35	535	Variable Structure Control Technique

Author	Section	Chapter	Page	Chapter Title
Vasile, M. J.	X	116	1489	Selected Micromachining Fabrication Technologies
Veillette, R. J.	V	27	490	The Root Locus Method
Vemuri, V. R.	IX	99	1303	Applications of Evolutionary Systems in Industrial Electronics
Vemuri, V. R.	IX	101	1316	Genetic Algorithms
Verhille, H.	X	113	1444	NEBULAS: High Performance Data-Driven Event Building Architectures Based on Asynchronous Self-Routing Packet-Switching Networks
Vines, D.	I	1	5	Electronics
Vlacic, L.	VI	46	629	Types of Automated Manufacturing Systems
Walter, P. L.	II	6	103	Measurement System Architecture
Warren, K. O.	X	117	1515	Microsensors
Weaver, A. C.	IV	18	394	Local Area Networks
Weller, T. M.	X	120	1547	Si Micromachining in High-Frequency Applications
Werbos, P. J.	VII	57	835	Strategies and Tactics for the Application of Neural Networks to Industrial Electronics
Werbos, P. J.	VII	63	888	Backpropagation to Neurocontrol
Werbos, P. J.	VIII	91	1157	Neurocontrol and Elastic Fuzzy Logic: Capabilities, Concepts, and Applications
White, C. R.	X	111	1383	Virtual Reality
Widrow, B.	VII	58	853	The Basic Ideas in Neural Networks
Wilhelm, R. G.	VI	51	718	Virtual Manufacturing Environment
Williams, T. J.	VI	46	629	Types of Automated Manufacturing Systems
Winston, H. A.	VIII	87	1127	Development of an Intelligent Unmanned Helicopter Based on Fuzzy Systems
Wolpert, S.	VII	60	867	Commercially Available Artificial Neural Network Chips
Wolpert, S.	VII	61	874	Implementing Neural Networks in Silicon
Wong, W.	VI	46	629	Types of Automated Manufacturing Systems
Wu, T. H.	VII	74	1003	Simulated Annealing, Boltzmann Machine, and Hardware Annealing
Yamasaki, H.	X	122	1592	Multisensor Fusion and Integration for Intelligent Systems
Yee, C.	VI	53	730	Robots
Yen, G. G.	VIII	92	1182	Integrated Health Monitoring and Control in Rotorcraft Machines
Yen, G. G.	VIII	93	1192	Autonomous Neural Control in Flexible Space Structures
Yen, G. G.	IX	110	1372	Application of Model Reference Adaptive Control and Adaptive Time-Delay RBF Networks
Young, B.	III	14	349	Main Disturbances
Zadeh, L. A.	VIII	98	1299	Future Directions for Fuzzy Systems and Soft Computing in Industrial Electronics
Zahner, D. A.	VII	69	942	Multilayer Perceptrons with ALOPEX and Backpropagation
Zhang, D.	V	40	572	Neural Network-Based Control
Zhou, M.	VI	49	669	Automated Manufacturing System Development Methodology
Zurawski, R.	VI	45	625	An Overview of Factory Automation
Zurawski, R.	VI	49	669	Automated Manufacturing System Development Methodology

Subject Index

A

AAC neural network processor
 See Accurate Automation Corp. neural network processor
AAL
 See ATM adaptation layers
ABM
 See Asynchronous balanced mode
Absence of gravity, 764
AC motor (ACM), 289, 290
AC-AC converters, 273, 275
AC-link resonant converter, 272, 274
Accelerometer, 106, 107, 108
Accelerometer gauges, 106, 107
Accelerometers, 1515
Access Control Field, 397, 399
Accessed bit, 66
Accumulator register, 54
Accurate Automation Corp. neural network processor, 885–887
 characteristics, 887
ACM
 See AC motor
Acoustic range sensor, 176–177, 177–178
Acoustic tracking, 1395
Acoustic wave propagation, 739
Acoustic wave-based chemical sensors, 1519–1524
Acoustic based sensing systems
 See Noncontact sensors: acoustic-based
ACR
 See Automatic current regulator
Action selection network, 1148–1149
Action-dependent adaptive critic, 895
Action-dependent heuristic dynamic programming, 893
Action-state evaluation network, 1148–1149
Activation function, 876
Active filters, 363, 365
Active monitor present (AMP), 399, 401
Active sensors, 97, 98, 99, 100
Active-video sensing systems, 177, 178
Actuators, 749
 cables, 757
Acyclic traffic, 407, 409
ADALINE
 See Adaptive linear element model, 942
Adaptive control, 593
Adaptive critic family, 890
Adaptive critics, 893, 894–895, 1173–1175
Adaptive GRNN, 1052–1053
Adaptive linear element model (ADALINE), 942
Adaptive mixtures, 938
Adaptive noise canceller, 1609
Adaptive nonlinear modeling, 1369–1371
Adaptive PNN, 1041–1042
Adaptive resonance theory
 and fuzzy logic, 1288
 applications, 1296–1297
 dynamics, 1288–1290
Adaptive resonance theory (ART), 844, 877, 1286–1297
Adaptive resonance theory classifier, 1056
Adaptive time-delay radial basis function network, 1192, 1194–1195, 1376–1377
Adaptive-network-based fuzzy inference system, 1140
Adaptive-subspace SOM (ASSOM), 842
ADC
 See Analog-to-digital converters
ADD, 57, 58
Address bus, 62
Address latch enable (ALE), 63
ADHDP
 See Action-dependent heuristic dynamic programming, 893
Advanced adaptive critics, 893
Advanced space structures technology research experiments (ASTREX), 1202–1205
Advanced teleoperation (ATOP), 785
AGV
 See Automated guided vehicle
AIC
 See Akaike information criteria
Air Force Office of Scientific Research, 1170
Air gap, 320, 322
Akaike information criteria (AIC), 166, 167
ALCATEL ATM-based event builder demonstrator, 1464
ALE
 See Address latch enable
Algebraic expressions, 23, 24
Algorithmic relation, 1349
Algorithmic state machines (ASM), 32–34
Aliasing, 139–141
All-wheel steering
 See Synchrodrive
ALOPEX, 944–945, 1234–1236
 in very-large-scale integration, 947–949
Alternating current transistor model, 14, 15
Alternative FALVQ algorithms, 1280–1283
ALU
 See Arithmetic and logic unit
ALVINN
 See Autonomous land vehicle in a neural network
American National Standards Institute
 See ANSI
American Production and Inventory Control Society (APICS), 657
American Standard Code for Information Interchange
 See ASCII
AMP
 See Active monitor present
Ampere's law of magnetic force, 751
Amplifier clipping, 114, 115, 116
Analog 3-D neuroprocessors, 990–1002
 six-bit parity problem, 990
Analog ANNs
 See Analog artificial neural networks, 867–869
Analog artificial neural networks (ANNs), 867–869
 advantages, 867–868
 disadvantages, 868
Analog filtering, 105, 106
Analog integrated circuits, 19
Analog multiplexers and multiplexing, 118, 119
Analog trimming, 1581
Analog visual image processor, 867
Analog-to-digital conversion
 See Sampling
Analog-to-digital converters, 139, 140
Analytical modeling, 447
Analytical redundancy, 1184
AND operator, 58, 59
Angle condition, 491
ANN
 See Artificial neural networks
ANN ICs
 See Artificial neural network integrated circuits
ANNA chip, 867
Annular spline, 759
ANSI, 51, 52
ANSI/IEEE Standard 754–1985, 51–52, 52–53
Anthropomorphic robots, 731
Anthropomorphic telemanipulation, 798
Anti-aliasing filter, 92, 93, 108–117, 109–118
 See also Lowpass filter

Anti-aliasing filter distortion, 112–113, 113–114
Anti-lock brake systems, 226, 228
  See also MOSFET: applications
Anticipatory control
  neural fuzzy approaches, 1355–1356
Antiparallel diode, 280, 282
APICS
  See American Production and Inventory Control Society
Application layer, 392, 394
Application level, 841
Application program, 50, 51
Application strategies, 841
Application tasks, 793
Application-specific integrated circuits (ASIC), 41, 42
Approximate reasoning-based intelligent control (ARIC), 1147
Area of influence, 1015
ARIC
  See Approximate reasoning-based intelligent control
Arithmetic and logic unit (ALU), 48, 49
Arithmetic coding, 172, 173
Arithmetic FALVQ 1, 1283
Arithmetic mean, 1280
Arithmetic right shift, 59
ARM
  See Asynchronous response mode
ARQ
  See Automatic repeat request
Arrayed actuators, 1480–1483
  advantages, 1480–1483
Arrival process, 681
ART
  See Adaptive resonance theory (ART)
ART baseline vigilance, 1291
ART choice parameter, 1289
ART field activity vectors, 1292–1293
Artificial intelligence, 824
  hybrid systems, 1346–1357
Artificial neural network integrated circuits (ANN ICs), 868
Artificial neural networks (ANN), 572, 781, 946–947, 1062–1063, 1184–1185, 1236–1240
  advantages, 898
  and fuzzy logic, 1166–1180
  commercial application, 867–872
  functions, 890
  cloning, 890, 894
  optimization, 890
  subsystem, 890
  tracking, 894, 890
  research, 1236–1240
Artificial neuron, 844
Artificial retina, 858
ARTMAP, 1288–1297
ASCII, 53
ASE
  See Associative search element
ASIC
  See Application-specific integrated circuits
ASM
  See Algorithmic state machines
ASR
  See Automatic speed regulator
Assembler, 50
Assembly language, 50
Associative memories, 1063
Associative search element (ASE), 575
ASSOM
  See Adaptive-subspace SOM
ASTREX
  See Advanced space structures technology research experiments
Asymptotic Bode plot, 476
Asymptotic phase plots, 477
Asynchronous balanced mode (ABM), 428, 430
Asynchronous bus, 62
Asynchronous response mode (ARM), 428, 430
Asynchronous state machines, 30
Asynchronous transfer mode (ATM), 412, 414, 1405, 1438–1442
  applications, 1439–1440
AT&T ANNA (analog neural network arithmetic-logic unit), 871
ATLAS AT&T ATM-based real-time demonstrator, 1465
ATLAS detector, 1440
  modeling of, 1453–1454
ATM
  See Asynchronous transfer mode
ATM Adaptation Layers (AAL), 413, 415
ATM cells, 413, 415
ATM data generator, 1463–1464
ATM SONET physical layer board, 1460–1463
ATM switches, 1446–1447
Attention switching, 972
Audio amplifier, 14, 16, 18, 20
Audio-visual sensor-fusion, 1609–1614
  (A+V+K)+A fusion, 1613
Augmented reality, 1411–1414
Auto-zero, 136, 137
Autoassociative memory, 1063
Autocorrelation function, 143, 144
Automated data acquisition, 129, 130
Automated guided vehicle (AGV), 674, 731
Automated manufacturing system design, 677
Automatic control systems
  design, 444
  modeling, 443
Automatic current regulator, 302, 304
Automatic feature extraction, 899
Automatic focusing, 725
Automatic repeat request (ARQ), 428, 430
Automatic speed regulator (ASR), 301, 303
Automatic target recognition (ATR), 1627
Automatic wafer inspection system, 725
Autonomous land vehicle in a neural network (ALVINN), 1647
Autonomous Land Vehicle project, 1640
Autonomous navigation, 1639
Autonomous robots, 1364–1367
Autoregressive models, 451
Axon, 875
Axon hillock, 875

B

Backpropagating utility, 892, 943–944
  convergence of, 944
Backpropagation (BP), 843, 1237, 1349
  See also Multilayer perceptron (MLP)
Backpropagation learning procedure, 856
Backpropagation of utility, 890, 899–900, 1173–1175
Backpropagation through time (BTT), 892
Backtrack, 714
Backward chaining (modus tollens), 809, 827
Backward prediction error, 168, 169
Balance equation, 683
Balanced condition, 145, 146
Balking, 682
Bandpass filters, 79, 80, 362, 364
Bandstop filters, 79, 80
Bandwidth, 61, 62
Barium titanate, 154, 155
Bartlett statistic, 953
Base-collector junction, 17, 18
Base-emitter bias, 18, 19
Base-emitter junction, 17, 18
Baseline drift, 1523
Basic reference model (BRM), 389, 391
Batch learning, 897
Bayes' rule, 828
Bayesian classifiers, 1057–1062
  with multivariate normal populations, 1059–1060
Bayesian optimal classifiers, 1056
  curse of dimensionality, 1056
BCD
  See Binary-coded decimal
Behavioral fusion
  techniques, 1646–1647
Bessel filter, 106, 107, 112–117, 113–118
Biased transistor, 10
BIBO
  See Bounded-input-bounded-output
Bidirectional associative memory, 1195
Bidirectional switches, 280, 282
Big endian, 56
Bilinear approximation, 487
Bilinear transformation, 463
Bill-of-material (BOM), 654, 658
  item, 658
Binary hard-limiter, 876
Binary integers, 52
Binary threshold model, 876
Binary-coded decimal (BCD), 52
Biological neural network (BNN), 835
Biological neuron, 844
Bipolar junction transistor (BJT), 195, 197, 209–210, 211–212
Bipolar transistor, 16–17, 18–19
Bipolar voltage switching, 256, 258
Bird sensing system, 178–179, 179–180
BJT
  See Bipolar junction transistor
BMDO/IST Demonstration Project, 864
BNN
  See Biological neural network
Bode diagram method, 474
Boeing neural information retrieval system, 1287, 1296
Boiling water reactor (BWR), 143, 144
Boltzman's constant, 9
Boltzmann machine, 1004–1005
  characteristics of, 1004–1005

learning algorithm for, 1005
BOM
See Bill-of-material
Boolean algebra, 22, 23
Boolean combinational functions, 35, 36
Boost converter
See Step-up converter
Bootstrapping
for error estimation, 1077–1078
prediction error rates in, 1078–1080
prediction rates from, 1080–1081
Bottleneck operations, 665
Boundary gating, 929–930
See also Pattern recognition
Boundary layer equation, 768
Bounded-input-bounded-output (BIBO), 523
Bounded-input/bounded-output, 1009
Boundedness, 673
BP
See Backpropagation
Breakdown voltage, 201–202, 203–204, 219, 221
Breast cancer detection
and the feature saliency method, 921
Bridge circuits, 145, 146
BRM
See Basic reference model
BTT
See Backpropagation through time
Buck converter
See Step-down converter
Buck-and-boost converter, 254, 256
Buck-boost chopper, 311, 313
BUILDNET routine, 665
Bulk arrivals, 682
Bulk macromachining, 1468–1469, 1502
applications, 1469–1470
Bursting oscillation, 878
Bus-bar charging, 774
Business plan, 654
Busses, 386, 388
Butterworth filter, 80, 81, 112–113, 113–114
BWR
See Boiling water reactor
Byte-addressable, 56

C

C message weight index, 353, 355
Cable frequency response compensation, 119, 120
Cache hit, 65
Cache memory, 48, 49, 64, 65
Cache miss, 65
CAE
See Computer aided environment
CALL (subroutine call), 59, 60
Camcorders, 1428–1430
Cameras, 1427
Cancellation, 526
Canonical correlation analysis, 1055, 1072–1073
Canonical switching cell (CSC), 260, 262
Capacitance, 15, 17
Capacitance bridge, 147, 148
Capacitive filtering, 381, 383
Capacitors, 214, 216
Capacity, 64
Capacity requirement planning (CRP), 654, 658
Carrier sense multiple access with collision detection (CSMA/CD), 395, 397
Cartesian robots, 730
CAS
See Computerised automation system
Cascade backpropagation, 991, 995–999
mathematical model of, 995–997
simulations, 999
Cascade control, 252, 254
Case-based reasoning, 809
Cathode-ray tubes, 49
Cauer filter
See Elliptic filter
CCD
See Charged coupled device
CCD neural processor chip, 859
CD-ROM, 63, 64
Cell decomposition method, 782
Cell membrane, 874
Cell membrane potential, 875
Cellular neural network (CNN), 859
Cellular neural networks, 1007–1013
Central limit theorem, 164, 165
Central processing unit (CPU), 48, 587
Cerebellar model articulation controllers, 906–908
Cerebellar model articulation controllers (CMAC), 843
Certainty factors, 825, 829
Change-source-from-gray-code rule, 817
Character recognition, 1248–1250
Charged coupled device (CCD), 178, 179
Chattering, 540
Chebyshev filter, 80, 81
Chemical microsensors, 1519–1524
Chemical sensors, 1521–1523
Chopper
See DC-DC converter
Chopper control systems
See DC-DC converters
Chunking, 1178
CI
See Computational intelligence
CIE
See Computer integrated enterprise
CIE/CIM
instructional manual, 634
Purdue methodology, 633, 634–638
Ciliary motion system, 1481–1482
CIM
See Computer integrated manufacturing
Circular rotate computer instructions, 59
CISC
See Complex instruction set computers
CISPR
See International Special Committee on Radio Interference
Clamped voltage (CV), 271, 273
Class E converter, 282, 284
Classical adaptive control, 891–892
Classification and regression tree, 1057
Classification, in pattern recognition, 916–917
Classifiers, 1055–1080
Client-server, 406, 408
Client/server instrument control system, 126, 127
CLIPS Object Oriented Language (COOL), 805
CLNP
See Connectionless network protocol
Clock synchronization, 410, 412
Cloning, in neurocontrol, 1158
Closed loop control system, 28, 318, 453
Clustering, 975, 1232–1233, 1245–1248
Clustering techniques, 1043, 1049
CMAC
See Cerebellar model articulation controllers
CMP (compare instruction), 60
CNAPS
See Connected network of adaptive processors, 870–871
CNAPS 1064 Digital Parallel Processor, 870
CNAPS sequencer chip, 870
CNAPS Server II, 871
CNC
See Computerized numerical controllers
CNN
See Cellular neural network
Coaxial cable, 119, 120
COBOL, 50
Cochlear implants, 883–884
Codecs, 1433
Coding, 51
Coefficient of variation, 930
Cohen-Grossberg networks, 889
Collector shorted IGBT, 232, 234
Collector-to-emitter voltage, 10
Collision, 70, 71
Collision avoidance, 1123–1124
Collision-free path, 733
Color vision system, 727
COMFET, 197, 199, 228, 230
Common mode emission, 378, 380
Common mode noise, 106, 107
Common mode rejection ratios, 119, 120
Common-mode voltage, 106, 107
Communication time delay, 793
Communications interface standards
RS-232C, 100, 101
RS-422A, 100, 101
RS-485, 100, 101
Commutation, 186, 188, 246, 248
Commutation overlap, 246, 248
Commutatorless Kraemer control, 312, 313
Compact disk read-only memory
See CD-ROM
Compare instruction
See CMP
Compass-tilt tracking, 1396
Compensation, 609
Compensators
design by root locus method, 495, 497
Competitive learning, 1060–1061
Competitive-learning algorithms, 842
Compiler, 50
Complement coding, 1290
Complementary information, 1326, 1594
Complex conjugation, 356, 358

Complex instruction set computers (CISC), 60, 61
Complex poles, 478
Complex zeros, 478
Compliant control mode, 787
Composite control, 768
Compositional rule of inference, 1100–1101
Compression ratio, 164, 165
Computational intelligence (CI), 835, 1343–1345, 1417–1424
 in aerospace, 1343–1344
 modeling concepts, 1344
Computer aided environment, 42, 43
Computer business oriented language
 See COBOL
Computer graphics, 790
Computer integrated enterprise (CIE), 632
Computer integrated manufacturing (CIM), 632
Computer memory, 48, 49, 55–57, 56–58
Computer modeling
 for asynchronous transfer mode switching networks, 1447–1449
Computer multimedia, 627
Computer software, 50–51
Computer system, 125–127
Computer vision, 724
Computerized automation system (CAS), 632
Computerized numerical controllers (CNC), 403–405
Computers
 mainframe, 48, 49
 microprocessor, 48, 49
 minicomputer, 48, 49
 personal computer, 48, 49
 workstation, 48, 49
Concurrent engineering, 626
Condition code flags, 54, 55
Condition code register, 54, 55
Conductivity modulated field effect transistor
 See COMFET
Conductors, 9
Cone of confusion, 1398
Conical switching fabric, 1453
Connected network of adaptive processors (CNAPS), 870–871
Connectionist neural adaptive processor chip, 861
Connectionless network protocol (CLNP), 429, 431, 432, 434
Conservative limit, 1290
Conservativeness, 673
Constant power operation
 See Field weakening
Contact sensors, 173–174, 174–175
Contents-addressable memory, 1063
Continuous-time filters
 design of, 80, 81
 types of, 79–80, 80–81
Continuous-time signal
 definition, 73, 74
 frequency-domain analysis of, 74–78, 75–79
 time-domain analysis of, 73–74, 74–75
Continuous-time signal processors
 definition, 78, 79
 frequency-domain analysis of, 79, 80
 time-domain analysis of, 78–79, 79–80
Contrast boundaries, 923–924
Contrast enhancement
 See also Histogram equalization, 952
Control structure, 826
Control systems
 architectures, 775–776
Conventional perturbation, 903
 See also Dynamic backpropagation, 903
 See also Forward perturbation, 892
 See also Williams-Zipser method, 903
Converters, 186, 188
Conveyors, 310, 312
 See also Speed sensorless vector control
Convolution, 74, 75, 83, 84
COOL
 See CLIPS Object Oriented Language
Cooper, Leon, 867
Coordinate measuring machine, 175, 176
 See also Three-D measurement: techniques
Copper cable, 118, 119, 120
Copper wire
 types of data transmission via, 119, 120
Corollary of functional determinacy, 158
Corollary of temperature determinacy, 158, 160
Coupled line bandpass filters, 1559–1562
Coupling capacitors, 16, 17
CPU
 See Central processing unit
CPU read cycle, 62
CPU write cycle, 62
CPU-to-device
 See CPU write cycle
CRC
 See Cyclic redundancy check
Credit assignment, 855
Crest factor, 353, 355
Crisp set, 1096–1097
Cross correlation function, 143, 144
Cross-country navigation, 1640
Crossbar switch, 70, 71
Crossover
 and genetic algorithms, 1328
Crosstalk, 119, 120
Crowding selection, 1319
CRP
 See Capacity requirement planning
CSC
 See Canonical switching cell
 See CNAPS sequencer chip, 870
CSI
 See Current-source inverter
CSMA/CD
 See Carrier sense multiple access with collision detection
Cubic spline interpolation, 125, 126
Cue signal method, 1610–1613
Current coupled noise, 120, 121
Current crowding effect, 212, 214
Current feedback control system, 324, 326
Current gain, 10
Current-source inverter (CSI), 263, 265
Curse of dimensionality, 1015, 1056
 See also RBF neural networks, operation and Bayesian optimal classifiers
Custom software system, 128–129, 129–130
Customer-driven manufacturing, 641
Cutoff frequency, 112
CV
 See Clamped voltage
Cyclic redundancy check (CRC), 428, 430
Cycloconverters, 273–275, 275–277

D

D-K iteration, 618
DARPA/Nestor/Intel Ni1000 chip, 1042
DAS
 See Dual attached station
Data acquisition, 137, 138
Data acquisition time, 126, 127
Data bus, 62
Data circuit identification, 390, 392
Data files and organization, 132
Data filtering, 103, 104
Data flow computers, 69, 70
Data manipulation, 130, 131
Data mining
 See Exploratory data analysis
Data processing software, 129, 130
Data recording, 103, 104
Data storage, 129
 magnetic disks, 129
Data strobe, 62
Data transfer acknowledge
 See DTACK
Data transmission path, 103, 104
Data visualization, 838
 See also Neural networks
Datalink layer, 390, 392
Datalink protocols, 427, 429
Datapath, 47, 48, 62
DC brushed motors, 752
DC brushless motors, 752
DC drives, 341, 343
DC motor (DCM), 289, 290
 torque-speed, 289, 291
DC motor control, 253, 255
DC-AC converters, 263, 265
DC-DC converters, 252, 254, 290, 292
DCM
 See DC motor
DCT
 See Discrete cosine transform
Deadlock, 674
 See also Petri nets
Decision functions, 975, 976–977
Decision surface, 975
Decision trees, 809
Decoding, 61, 62
Deconvolution, 119, 120, 122, 123
Dedicated FMS, 649
DEDS
 See Discrete event dynamic systems
Deformation-resistant recognition, 967
Defuzzification, 1101, 1108–1109, 1141
Defuzzification neurons, 1186
Degrees of freedom (DOF), 760
Delay circuit, 725
Delta rule, 943
 See also Least Mean Square algorithm, 943
Dempster's rule of combination, 1605
Dempster-Shafer evidential reasoning, 1604

Denavit-Hartenberg kinematic parameters, 763
Dendrites, 875
Density comparison, 939–940
Destination address, 397, 399
Device-to-CPU
  See CPU read cycle
DHP
  See Dual heuristic programming
Dial indicator method, 174, 175
  See also Three-D measurement: techniques
Dial indicators, 174, 175
Diamond macromachining, 1507–1508
Difference equations, 545
Differential amplifiers, 107–108, 108–109
Differential data recording, 120, 199
Differential mode emission, 378, 380
Diffraction, 739
  See also Target scattering
Digital ANNs, 869–871
  See Digital artificial neural networks, 869–871
Digital artificial neural networks (Digital ANNs), 869–871
Digital circuits, 19
Digital color image, 732
Digital computation, 545
Digital computer, 48
Digital computer instructions
  arithmetic, 57, 58
  control transfer, 58–59, 59–60
  data transfer, 57, 58
  logical, 57–58, 58–59
  process control, 59, 60
  shift/rotate, 58, 59
Digital control, 553
Digital data acquisition system, 105, 106
Digital filtering, 121, 122
Digital neurocomputer, 867
Digital realization, 326, 328
Digital signal processing (DSP), 137, 138
Digital system design, 486
  emulation, 487
  Tustin's method, 487
Digital trimming, 1582
Digitizer, 103, 104
Digitizer noise, 123, 124
Dimensionality reduction, 960
Diode current, 6
Diode equation, 9
Diode forward voltage, 222, 224
Diodes, 5, 193–194, 195–196, 202, 203
Direct converters, 257, 259
Direct current bias transistor model, 13, 15
Direct feedback, 36, 37
Direct feedback loop, 37, 38
Direct inverse approach
  in neural tracking, 891–892
Direct inverse control, 1172–1173
Direct inverse controllers, 890
Direct torque control, 191, 193
Direct volume rendering, 1408
Directional clustering, 1227–1228
Directional sensitivity, 883
Discontinuous mode, 280, 282
Discrete cosine transform (DCT), 170, 171
Discrete event dynamic systems (DEDS), 706
Discrete event manufacturing system, 694
Discrete event simulation, 691, 694
  models, 696–699
  schemes and tools, 699
Discrete event systems, 21, 22, 41, 42
Discrete Fourier series, 83–84, 84–85
Discrete Fourier transforms, 85, 86
Discrete implementation, 509–510
  Ackermann's control formula, 510
Discrete systems control, 21, 22
Discrete wavelet decomposition, 1627–1628
Discrete-time filters
  design of, 90–92, 91–93
  types of, 90, 91
Discrete-time Fourier transforms, 86, 87
Discrete-time signal, 72, 73
Discrete-time signal processors
  definition, 88, 89
  frequency-domain analysis of, 89, 90
  time-domain analysis of, 88–89, 89–90
Discrete-time signals
  definition, 82, 83
  frequency-domain analysis of, 83–84, 84–85
  time-domain analysis of, 83, 84
Discriminant analysis, 961–962
Distortion power, 359, 361
Distributed systems, 387, 389
Disturbance approximation, 526
Disturbance cancellation, 526
Disturbance observation, 526
Disturbance observation-cancellation, 524
Divide-and-conquer technique, 977
DMC
  See Dynamic matrix control
DMOS cells, 230, 232
DOF
  See Degress of freedom
Domain expert, 808
Dot-product, 979
Double conversion UPS, 371, 373
Double diffused power MOSFET
  See DMOS
Double-ended isolated converters, 262, 264
Double-throw switch, 260, 262
Double-way two-pulse bridge connection, 248, 250
Douglas, Rodney, 878
Drain voltage
  See Breakdown voltage
Drift rate, 149, 150
Drive circuitry, 334, 336
DSP
  See Digital signal processing
DTACK, 61, 62
DTC
  See Direct torque control
Dual attached station (DAS), 408, 410
Dual heuristic programming (DHP), 893–894, 895, 1177
Dynamic backpropagation, 903, 1373
Dynamic decay adjustment, 1015
Dynamic headroom, 327, 329
Dynamic matrix control (DMC), 522
Dynamic optimization, in neurocontrol, 1158

E

Echo envelope, 743
Edge detection, 923–924, 952
EDIF
  See Editor files
Editor files, 43
EEPROM, 63, 64
Efficient client/server instrument control system, 126, 127
EFT/B
  See Electric fast transient/burst
Eigenstructure bidirectional associative memory, 1192, 1195–1198
Elastic fuzzy logic, 1088, 1163–1165
Elastic supports, in actuators, 1476–1479
Elasticities, 1164
Elbaum, Charles, 867
Electric drives, 188–189, 190–191
Electric fast transient/burst (EFT/B), 381, 383
Electric fields, 120, 121
Electric power generator, 189, 191
Electrical interference, 119, 120
Electrical lock, 30, 31
Electrical output signals
  analog, 97, 98, 99, 100
  digital, 99, 100
Electrical potential, 9
Electrically trainable ANNs, 868
Electrically-erasable programmable read-only memory
  See EEPROM
Electrocaloric effect, 156, 157
Electroforming, 1500
Electromagnetic actuator, 1526–1527
Electromagnetic compatibility (EMC), 120, 121, 191, 192, 377, 379
Electromagnetic emission, 378, 380
Electromagnetic interference (EMI), 276, 278
Electromagnetic torque, 324, 326
Electromagnetic-based sensing systems
  See Noncontact sensors: electromagnetic-based
Electron irradiation, 232–233, 234–235
Electronic power steering, 226, 228
  See also MOSFET: applications
Electrostatic actuator, 1527–1528
Electrostatic coupling, 120, 121
Electrostatic discharge (ESD), 380, 382
Electrostatic micromotors, 1474–1475
Electrostatic shield
  See Faraday cage
Electrostatic shielding, 120, 121
Electrostatic transducers, 741
Elitist strategy, 1320
Elliptic filter, 80, 81
EMC
  See Electromagnetic compatibility
EMI
  See Electromagnetic interference
Emitter switched thyristor, 198, 200
Empirical modeling, 451
EMSIM, 719
Emulation, 487
  See also Digital system design
End-effectors, 730
End-item, 658

Ending delimiter, 397, 399
Energy barrier, 9
Energy supply methods, 1533–1536
Engineered FMS, 649
Entropy, 163, 164, 165
Entropy coders, 163, 164, 170, 171
EPLD
  See Erasable programmable logic device
EPROM, 63, 64
Equations of motion, 762
Equipment under test (EUT), 381
Equivalent circuits, 302, 304
Equivalent control method, 537
Erasable programmable logic device (EPLD), 41, 42, 107, 108
Erasable programmable read-only memory
  See EPROM
Error backpropagation, 1064–1066
Error correction, 398, 400
Error critic approach, 903
Error-based learning, 1287–1288
ESD
  See Electrostatic discharge
EST
  See Emitter switched thyristor
Etching, 1501
  See also MEMS: fabrication
Ethernet, 70, 71, 386, 388
Ettinghausen effect, 154, 155
Euler's formula, 548
EUT
  See Equipment under test
EVA
  See Extra vehicular activity
Event builder architectures, 1447–1449
  layered structure of, 1455–1457
Event builder demonstrators, 1464–1466
Event building protocols, 1455–1460
Event scheduling simulation, 699
Event-triggered systems, 405, 407
Evolutionary algorithms, 1307
Evolutionary computation (EC), 1307–1315
Evolutionary fuzzy systems, 1322–1323
Evolutionary neural systems, 1340–1341
Evolutionary systems, 1303–1304
  and fuzzy systems, 1303–1304
  and neural systems, 1338–1341
  architectures and connections, 1311–1313
  component functions, 1311
  design, 1307
  functions, 1309–1310
  in industrial electronics, 1303–1304
  modeling concepts, 1303
  models, 1309–1310
  objectives, 1307–1309
  performance measures, 1307–1309
  variable parameters, 1313
Exceptions, 59, 60
Excitation law, 250, 252
Expert systems, 805, 808
  and fuzzy logic systems, 1347
  and neural networks, 1347
  characteristics of, 825
  components of, 825
  definition of, 825
  domain, 808
  heuristics, 808, 824
  implementation issues for, 831
  interface, 825
  legal aspects of, 832
  uses of, 830, 833
ExpertVision
  See Motion Analysis System
Exploratory data analysis, 842
Exponential distribution, 682
Extensible ball bar method, 174–175, 175–176
  See also Three-D measurement: techniques
External energy supply methods, 1534–1536
  See also Non-contact manipulation methods, 1534–1536
Extra vehicular activity (EVA), 796

F

F-15, 893
  advanced adaptive critics in, 893
Factory automation, 625
Factory automation model (FAM), 629
Factory automation systems
  implementations, 641–647
Factory floor communication, 385, 387
Factory network technology, 721
Fail-Dumb syndrome, 837
FALVQ 1 algorithms, 1272–1274
FALVQ 2 algorithms, 1274–1275
FALVQ 3 algorithms, 1275–1277
FAM
  See Factory automation model
Faraday cage, 120, 121
Faraday's Law, 121, 122
Fast Fourier transforms (FFT), 85–86, 86–87, 1023
Fast learning, 1293
Fast recovery diodes, 209, 211
Fast response mode, 1032
Fast-commit, slow-recode option, 1293
Fault condition notification, 390, 392
Fault detection, 1188–1189
Fault detection and identification, 1198–1199
Fault detection, identification and accommodation (FDIA) system, 1187–1189
Fault isolation, 1189
Fault severity estimation, 1189
FBSOA
  See Forward-biased safe operating area
FCFS
  See First come, first served
FCM clustering
  See Fuzzy c-means clustering algorithm
FDDI
  See Fiber distributed data interface
FDDI-II, 411, 413
Feature extraction, 916, 1189, 1232, 1241–1245
Feature Saliency method, 917–921
  applications, 920–921
  automatic selection of features, 919
  breast cancer detection, 921
  computation, 919
  derivation of partial derivatives, 919–920
  feature selection, 917
  scalar form, 919–920
  single hidden layer networks, 920
  vector form, 920
  XOR problem, 920–921
Feature space, 1027
Feature vector, 916
Feature-level multisensor fusion, 1603
Feature-map classifier, 1056
Feed forward networks, 1063, 1064, 1161
  training methods for, 1066
Feedback control systems, 454
Feedback error learning, 891, 1173
Feedback loop, 29, 30
Feldkamp, Lee, 891, 894
FFT
  See Fast Fourier transforms
Fiber distributed data interface (FDDI), 386, 388, 410
Fiber optic cable, 118, 119
Field busses, 403, 405
Field effect transistors, 14
Field programmable gate array (FPGA), 41, 42, 1143
Field programmable logic array (FPLA), 42
Field weakening, 330, 332
Field-oriented control, 329, 331
Fill-result rule, 819
Filter circuit, 269, 271
Filter cutoff frequency, 112
Filter pass-band, 112, 113
Filtering, 109, 110
  See also Signal conditioning and filtering
Finite impulse response (FIR), 89, 90
Finite state machine (FSM), 36–37, 42, 43
Finite-duration discrete-time signals, 82–83, 83–84
Finite-duration impulse-response multilayer perceptrons, 912
FIP, 407
FIR
  See Finite impulse response
FIR filters
  design, 90–91, 91–92
Firing, 809
  See also Rule-based programming
First come, first served (FCFS), 682
First-order-lag-plus-delay (FOLPD), 470
Fitness function, in genetic algorithms, 1317–1318
Fitness proportionate reproduction, 1318–1320
Fitzhugh-Nagumo neuromorphic model, 878
Fixed shunt capacitors, 356, 358
Fixed-frequency control, 266, 268
Fixed-point numbers, 52
FLASH, 63, 64
Flash memory
  See FLASH
Flat belts, 756
Flexible discriminant analysis, 1055
  via optimal scoring, 1075
Flexible joint robots, 766–767
Flexible manufacturing systems (FMS)
  types, 648–650
Flexible multibody systems, 1374–1376
  dynamic modeling of, 1374–1376
Flexible vision system, 1619
Flexspline, 759
Flip-flop feedback, 37, 38, 39, 40

Flip-flop feedback loop, 37, 38
Floating point operations per second
See FLOPS
Floating recording systems, 121, 122
Floating-point numbers, 52
FLOPS, 126, 127
Flow conservation equation
See Balance equation
Flux detection type vector control, 302, 304
Flux linkage, 328, 330
Flux weakening
See Field weakening
FLVQ algorithms, 1269–1270
Flyback converter, 261, 263
FMS
See Flexible manufacturing systems
Focused ion beam micromachining, 1510–1511
Foley effect, 1403
FOLPD
See First-order-lag-plus-delay
Force-reflecting mode, 787
Force/torque sensors, 746
Formula translation
See FORTRAN
FORTRAN, 50
Forward biased diode, 5, 6, 15, 18
Forward chaining (modus ponens), 809, 827
Forward perturbation
See also Conventional perturbation, 892
Forward prediction error, 168, 169
Forward shaping, 1637
Forward-biased safe operating area (FBSOA), 198, 200
Four-quadrant converters, 245, 247
Four-wire Kelvin probe, 146, 147
Fourier series, 75–77
Fourier transforms, 77, 78
FPGA
See Field programmable gate array
FPLA
See Field programmable logic array
Fractal dimension, 933
Fractals, 933–941
Frame check sequence, 397, 399
Frame control field, 397, 399
Frame status field, 397, 399
Free-wheeling diode, 244, 246
Frequency control system, 297, 299
Frequency converter fed squirrel cage induction motor drive, 191, 193
Frequency domain, 72, 73, 74, 75
Frequency response, 105, 106
Frequency shift keying (FSK), 100, 101, 400, 402
Frequency transition zone, 111–112, 112–113
Frustrated internal reflection, 747
FSK
See Frequency shift keying
FSM
See Finite state machine
Fuel dilution meter, 1523
Full-bridge inverters, 264, 266
Full-cycle charging, 774
Full-order controllers, 601
Fused deposition modeling, 627
Fuzzification, 1108
Fuzzy aggregation connectives, 1326–1327
Fuzzy algebra, 1091–1094
Fuzzy algorithm, 1349
Fuzzy algorithms for learning vector quantization (FALVQ), 1270–1272
Fuzzy ART, 1290
algorithms, 1292–1295
stable category learning, 1295
Fuzzy ARTMAP, 1290–1292
algorithm, 1295–1296
Fuzzy c-means (FCM) clustering algorithm, 1245–1248
Fuzzy clustering, 1225, 1233
Fuzzy control, 565, 780, 1101–1102
Fuzzy control modules
in unmanned helicopters, 1131–1133
Fuzzy data processing
in neural networks, 1352–1353
Fuzzy evolutionary systems, 1321–1323
Fuzzy hardware, 1103–1110
architectures and connections, 1108
component functions, 1108–1109
models and functions, 1107
Fuzzy inference, 1108
Fuzzy inference systems, 1140–1141
examples of, 1140–1141
FUZZY ISODATA, 1225–1227
Fuzzy knowledge-based control, 1171–1172
Fuzzy learning vector quantization algorithms, 1268–1269
Fuzzy logic, 1096, 1344
functions
analog, 1108
digital, 1108
in image processing, 1426–1433
camcorders, 1428–1430
cameras, 1427
codecs, 1433
photocopying machines, 1430–1433
television equipment, 1433
in intelligent control systems, 1127–1137
Fuzzy logic control, 1116–1125
Fuzzy logic controllers, 1350–1352
and neural networks, 1352
Fuzzy logic mapping, 1347–1350
Fuzzy logic systems
and expert systems, 1347
Fuzzy modeling, 1112–1114, 1208–1209
engineering approaches, 1112–1114
Fuzzy neural hardware, 1088
Fuzzy neural modeling, 1088
Fuzzy neural networks
learning in, 1216–1218
Fuzzy neural structures, 1213–1218
Fuzzy neurons, 1214–1216
Fuzzy operations, 1098–1100
Fuzzy perceptron, 1220–1221
Fuzzy relation, 1100
Fuzzy set theory, 1096
Fuzzy sets, 1097
Fuzzy systems, 1087–1090
dynamics, 1114
neural network learning in, 1147–1155
system identification in, 1114
Fuzzy-AND neurons, 1185–1186
Fuzzy-based back-propagation network, 1192
Fuzzy-based feedforward neural network, 1185–1187
Fuzzy-genetic fusion
techniques, 1329–1332
fusion from all sensors in one step, 1329–1332
fusion from two sensors in one step, 1329–1332
Fuzzy-OR neurons, 1186
Fuzzy/neural modeling, 1139–1141

G

Gain margin, 480
Gain range, 105, 106
Gain stability with temperature, 105, 106
Gain-control signals, 971
Gain-crossover frequency, 480
Gain-phase relationship, 482
GAL
See Gradient adaptive symmetric lattice
Gamma memory, 910
Gamma model, 911–912
Gas flow controller, 1626
Gate charge, 223, 225
Gate pulse generator (GPG), 290, 292
Gate turn-off (GTO) thyristors, 195, 197
Gauss-Seidel method, 937
Gaussian density function, 142, 143
GDHP
See Globalized dual heuristic programming, 894
Gear backlash, 759
Gear trains, 757
gear shafts, 757
General purpose instrumentation bus (GPIB), 128
General regression neural network, 1047–1053
advantages, 1050
disadvantages, 1050
General-purpose induction motor, 303, 305
General-purpose inverters
applications, 308–310, 310–312
Generalized ARIC, 1147
Generalized learning vector quantization (GLVQ), 1266–1268
Generalized modus ponens (GMP), 1350
Generalized modus tollens (GMT), 1350
Generator excitation systems, 358, 360
Generic tasks, 793
Genetic algorithm operators, 1328
crossover, 1328
mutation, 1328
reproduction, 1328
Genetic algorithms, 1307, 1316–1320
and fuzzy systems, 1357
and neural networks, 1356–1357
optimization problems, 1328–1329
Geometric FALVQ 1, 1282–1283
Geometric mean, 1280
Geometric modeling, 780
Giant magnetostrictive alloy actuator, 1529
Gibb's phenomenon, 77, 90, 91
Global stability in nonlinear systems, 466, 1195
Globalized DHP, 1177

Globalized dual heuristic programming, 894, 895
GLVQ-F algorithms, 1269–1270
Goddard Space Flight Center Remote Manipulator System, 1122
Gold diffusion, 233, 235
GPG
  See Gate pulse generator
GPIB
  See General purpose instrumentation bus
Gradient adaptive symmetric lattice (GAL), 168
Gradient descent, 1003
Grasping task, 710
Gravity compensation, 764
Gravity vector, 763
Gray code representation, 816
Ground loop, 120, 121
GTO
  See Gate turn-off thyristors
Gutschow, Todd, 867

H

Half-bridge inverters, 264, 266
Hall effect, 153, 154
Handwritten character recognition, 969
Handwritten digit recognition, 951–955
Hardware, 845
  See Intelligent electronics applications
Hardware annealing, 1005–1012
  application of, 1009
  on cellular neural networks, 1007–1013
  on Hopfield networks for optimization, 1005–1007
Hardware redundancy, 1184
Harmonic distortion, 351, 353
Harmonic drives, 759
Harmonic FALVQ 1, 1281–1282
Harmonic mean, 1280
Harmonic motion generator (HMG), 789
Harmonics, 250, 252
HART
  See Highway addressable remote transducer
Harvard architecture, 49
HDLC
  See High-level data link control
HDP
  See Heuristic dynamic programming, 893
Head-related transfer functions (HRTFs), 1402
Health monitoring system, 1182–1190
Hebb, Donald, 942
Hecht-Nielsen, Robert, 867
HEMP
  See High-altitude nuclear electromagnetic pulse
Herman-Herring illusion, 882
Heteroassociative memories, 898
Heuristic dynamic programming (HDP), 893, 895
Heuristic knowledge, 824
  See also Expert systems
Heuristics, 808
Hidden units, 855
Hierarchical intelligent control, 1364–1367
Higgs search, 1021
High voltage direct current transmission
  See HVDC
High-altitude nuclear electromagnetic pulse (HEMP), 380, 382
High-fidelity graphics calibration, 791
High-frequency transistor model, 15, 16
High-level data link control (HDLC) transfer modes, 427–430
High-level programming languages, 50
High-Order B_Spline CMAC color correction system, 907
High-voltage diodes, 201–202, 203–204
High voltage direct current (HVDC), 187, 189
High-voltage events, 351, 353
Highpass filters, 79, 80, 138, 139
Highway addressable remote transducer (HART), 101
Histogram equalization
  See also Contrast enhancement, 952
HMG
  See Harmonic motion generator
Hodgkin-Huxley equations, 878
Hoist, 309
  See also Speed sensorless vector control
Hold-out method
  in classifier evaluation, 1077
Hopfield networks, 876, 889, 892
Hopfield, John, 867
Hospital monitoring systems and sensors, 1623–1626
Huffman coders, 171, 172
Huffman model, 29, 30
Human arm-hand system, 786
Human interface, 639
HVDC
  See High voltage direct current
Hybrid ANNs
  See Hybrid artificial neural networks, 871–872
Hybrid artificial neural networks, 871–872
Hybrid connectionist system, 1192
Hybrid integrated circuits, 1585
Hybrid step motor, 333, 335
Hybrid systems, 706
  2-D and 3-D uncertainties, 707–714
Hydraulic actuators, 751
Hypercube field, 1020
Hysteresis
  on piezoelectric devices, 1530
Hysteresis band, 325, 327

I

I/O device interfaces, 66, 67
I/O macrocells, 41–43
IBM ZISCO36, 1019
Iconic fusion
  techniques, 1642–1643
IEEE 802.3 contention bus, 395, 397
IEEE 802.4 token bus, 464
IEEE Standard 519, 352, 354
IEEE Standard 754–1985, 51, 52, 53
IEEE Standard 802.5, 397, 399
IEEE Standard for Binary Floating Point Arithmetic
  See IEEE Standard 754–1985
IGBT, 197–198, 199–200
  applications of, 227–228, 229–230
  comparison with other transistors, 234–235, 236–237
  data sheet parameters of, 235, 237
  design considerations, 230–232, 232–234
  requirement for anti-parallel diode, 234, 236
  structure and operation of, 228–230, 230–232
IGR
  See Insulated gate rectifier
IIR (infinite impulse response), 90
IIR filters
  design, 92–93
ILMI
  See Interim layer management interface
IM
  See Induction motor
Image flow detection, 713
Image motion, 713
Image processing, 713
  and fuzzy logic, 1426–1433
Image subtraction, 725
IMC
  See Internal model control
Immediate operand, 53
Immersive virtual reality, 1410–1411
  software, 1410–1411
Immunity, 380, 382
Impedance, 108, 109
Implication relation, 1349
Implicitly supervised pattern recognition, 1223–1225
Incremental learning, 1185
Incremental learning and forgetting, 1198
Indexing function, 787
Indirect converters, 257, 259, 260, 262
Indirect inverse approach
  in neural tracking, 891–892
Inductances, 214–215, 216–217
Induction motor (IM)
  torque-speed, 294–295, 296–297
Induction motor drives, 342, 344
Inductive coils, 215, 217
Inductive load, 10, 12
Industrial automated systems, 669
  models, 670
Industrial reality, 1414–1417
Industrial vision system, 1615
Inertia link parameters, 763
Inertia matrix
  uniform boundedness of the, 762
Inertial tracking, 1395–1396
Inference engine, 809, 825
Infinite impulse response
  See IIR
Infinite-duration impulse response filters, 913
Information field, 397, 399
Information function architecture, 636
Information fusion, 1325–1336
Input impedance/output impedance, 104, 105
Input neurons, 1185
Input/output devices, 49
Institute for Simulation and Training, 1405
Instruction cycle, 49
Instruction pointer, 54
Instruction set architecture, 49

Instructional manual, CIE/CIM, 634
Instrument Society of America
See ISA
Instrumentation amplifier, 148–149, 149–150
Instrumentation computer system, 126, 127
Instrumentation database, 131, 132
Instrumentation database maintenance software, 131, 132
Insulated gate bipolar transistor
See IGBT
Insulated gate rectifier (IGR), 228, 230
Insulators, 9
Integral of the time multiplied by absolute error (ITAE), 505
Integrated circuit operational amplifier, 147, 148
Integrated circuits, 533
Integrated services digital network (ISDN), 412, 414
Intel 80170, 868–869
Intel 8051 microcontroller, 48, 68
Intel 8086, 48, 53, 54
Intel ETANN 80170, 1019
Intelligent control, 445, 564, 579
Intelligent control systems, 1127–1137
Intelligent electronics applications, 836–846
hardware and software, 845
Intelligent electronics, 1417–1424
Intelligent pattern recognition memory (IPRM), 869–870
Intelligent robotics, 732
Intensity image processing, 1615–1621
Interdigitated bandpass filters, 1558–1559
Interim layer management interface (ILMI), 414, 416
Interior permanent magnet (IPM) motor, 320, 322
Intermediate mask, 1499
Internal energy supply methods, 1533–1534
Internal flow control, 1446
Internal model control (IMC), 513
International Special Committee on Radio Interference (CISPR), 377, 379
Internet message control protocol, 431, 433
Internet protocol (IP), 429, 431
Interphase transformer, 248, 250
Interrupt service routine, 60, 61
Interrupt vector, 60, 61
Inverse discrete wavelet transform, 1627
Inverse shaping control, 1637
Inverted pendulum system, 1118–1122
fuzzy logic approach, 1120–1122
Inverters
See DC-AC converters
IP
See Internet protocol
IPM motor
See Interior permanent magnet motor
IPRM
See Intelligent pattern recognition memory, 869–870
ISA (Instrument Society of America), 96, 97
ISDN
See Integrated services digital network
ISO transport protocol, 436–437, 438–439
Isolate DC-DC converters, 260–261, 262–263
Isolation transformer, 260, 262
ITAE
See Integral of the time multiplied by absolute error

J

Jackknife estimation, 1078
Jackson network, 687–689
Jacobi method, 937
Jam packet, 396, 398
JMPL (Jump and link), 60
Joule effect, 151, 152
Joule heating, 151, 152
JPL ATOP project, 785
JSR (Jump to routine), 60
Jump and link
See JMPL
Jump to routine
See JSR
Jury's stability test, 464
Just-in-time, 666

K

K-fold cross-validation
in classifier evaluation, 1077
K-means clustering, 1043
K-nearest neighbors mode, 1020, 1056, 1062
Kalman filter, 559
Kanban system, 667, 668
rules, 667–668
Karhunen-Loeve expansion, 1241–1242
Karnaugh maps, 22, 23–27, 24–28, 36, 37
Kelvin relations, 152, 153
Kendall-Lee notation, 683
Kharitonov polynomials, 459
Kinematics, 774
See also Mobile robots
Kirchhoff equation, 6, 8, 10
Kirchhoff's current law, 277, 279
Kirchhoff's voltage law, 277, 279
Kirchoffs Current Law, 1006
Knowledge base, 825
Knowledge representation, 826, 1212–1213
rule-based expert system, 826–828
control structure, 826
demons, 826
Kraemer control, 312, 313, 314, 315
Kullback-Leibler distance, 1068–1070
Kullback-Leibler information, 939–941

L

Lag-lead compensator, 485
Lagging, 354, 356
Laminated object manufacturing, 627
LAN
See Local Area Network
Lange coupler, 1557–1558
Laplace transfer function, 116, 117
Laplace transforms, 76–78, 77–79
Large scale integration (LSI), 41, 42
Laser micromachining, 1511
Laser sensors, 777
Laser time-of-flight systems, 735–736
Laser-based sensing systems, 177–178, 178–179
Last come, first served (LCFS), 682
Lateral effect photodiode, 177, 178
Lateral inhibition, 882
Lateral stereopsis, 737
Lattice filters, 167–168, 168–169
LCFS
See Last come, first served
Lead time, 660
Lead zirconate titanate, 154, 155
Leaf nodes, 171, 172
Leakage of energy, 94
Leakage inductance, 270, 272
Leaky integrator, 877, 880
Learning, 979
Learning vector quantization (LVQ), 841, 844, 1043, 1056, 1060–1062, 1264–1284
Least mean square algorithm, 943
Least-squares system identification method, 561
LED
See Light-emitting diode
Level of CI system design, 840
Levinson-Durbin method, 170, 171
LIGA-DXRL, 1497–1500
Light source algorithm, 930
Light-emitting diode (LED), 177, 178
Line commutated converter circuits, 244, 246
Line impedance stabilization networks (LISN), 379, 381
Line regulation, in signal conditioning, 105
Line-frequency diodes, 209, 211
Line-interactive UPS, 372, 374
Linear approximation, 146, 147
Linear discriminant analysis, 1055, 1073
via optimal scoring, 1074–1075
Linear discriminant classifier, 1056
Linear discriminant function, 1221
Linear discriminant score, 1059
Linear integrated circuits
See Analog integrated circuits
Linear motor drives, 753
Linear prediction, 164, 165
Linear time invariant control, 613
Linear time-invariant continuous-time systems, 456
Linear-graded threshold, 876
Linearity, 105
Linearization, 466
Linguistic variables, 1097
Liquid crystal display, 48, 49
LISN
See Line impedance stabilization networks
LISP, 50
Lithium niobate, 154, 155
Lithography, 1501
See also MEMS: fabrication
Little endian, 55, 56
Little's formula, 684
Liveness, 674
Load line
See Thevenin's equivalent circuit
Load line analysis, 5, 7, 8
Load regulation, 104, 105
Load resonant converters, 271, 273
Load transient recovery, 104, 105
Local area network (LAN), 408, 410, 1438
Local features, 928
See also Pattern recognition

Local stability in nonlinear systems, 466
LoFlite, 896
Logic control, 21, 22
Logic expressions, 22, 23
Logic functions, 22, 23, 27, 28
Logic processors, 1218
Logic sensors, 46, 47
Logical address space, 66
Logical link control, 410, 412, 427, 429
Logical shift, 58, 59
Long range resource planning (LRRP), 654
Long-term production planning (LTPP), 654, 655–656
  decisions under the responsibility of participants, 656
  participants, 656
Loop filter, 531
Lorentz force, 153, 154
Lossless reflector, 739
Lossless waveform compression, 163, 164
Lot sizing, 664
Low voltage power disturbances, 352, 354
Low-order controllers, 599
Low-voltage events, 351, 353
Lowpass filters, 79, 80, 138, 139, 1556–1557
LRRP
  See Long range resource planning
LSI
  See Large scale integration
LTPP
  See Long-term production planning
Lucas NovaSensor, 1586–1590
LVDT method, 175–176, 176–177
  See also 3–D measurement: techniques
LVQ
  See Learning vector quantization
Lyapunov's method, 467

M

M64 ETANN, 868
Mach bands, 882
Machine Intelligence Quotient (MIQ), 1299
Machine language, 50
Madaline Rule II, 1065
Magnetic disks, 129
  See also Data storage
Magnetic fields, 120, 121
Magnetic flux, 300, 302
Magnetic materials, 321, 323
Magnetic shielding, 120, 121
Magnetic tracking, 1394–1395
Magnetizing inductance, 270, 272
Magnitude condition, 491
Mahalanobis distance, 1057
Mahowald, Misha, 878
Main electric box (MEB), 794
Man-machine interface
  See Human interface
Manhattan block distance, 1015, 1020
Manipulators, control of, 540
Mantissa, 51–53
Manual/plug-in charging, 774
Manufacturing automation protocol (MAP), 400, 402, 417, 419
  architectures, 418–420, 420–422
  standards, 420–426, 422–428
Manufacturing cell, 650
Manufacturing message specification (MMS), 407, 409, 720
Manufacturing resource planning (MRPII), 664
Manufacturing tasks
  hierarchical model, 629–632
MAP
  See Manufacturing automation protocol
Map field, 1295
Marching cubes method, 1408
MARF autonomous navigation system, 1641
Mark III digital neurocomputer, 867
Mark IV digital neurocomputer, 867
Master plan, 634
  See also CIE/CIM: Purdue methodology
Master production scheduling (MPS), 654, 656–658
Match tracking, 1291
Match-based learning, 1287
Matched filter, 743
Matched pole-zero method, 487
Material requirements planning (MRP), 654, 663
  advantages, 662
  applications, 662
Mathematical models, 447
Matrix converters, 273, 275, 277
Max-iterations-and-gray-code rule, 816
Maximum influence field, 1021
Maximum token rotation time (MTRT), 399, 401
McCullouch and Pitts networks, 877
MCM
  See Multichip module, 870
MCT
  See MOS-controlled thyristor
Mead, Carver, 868, 881, 882
Mealy state machines, 29, 30
Mean time to repair (MTTR), 403, 405
Measurand, 96, 97
Measurement system calibration, 135, 136
Measurement systems, 102–104, 103–105
MEB
  See Main electric box
Mechanical tracking, 1396
Median filter, 928
Median filtering, 952
Medium access control, 410, 412
Medium range production and assembly planning (MRPAP), 654, 657
Medium scale integration, 42
Membership-function neurons, 1185
Memory access time, 64
Memory address bus, 45
Memory addressing
  base/index, 57
  direct, 56
  indirect, 56–57
  program counter relative, 58
Memory cycle time, 64
Memory data field, 44, 45
Memory location, 56
Memory management unit, 66
Memory system organization, 63–64, 64–65
MEMS
  See Microelectromechanical systems
Merge-response rule, 819
Message frame, 401, 403
Metadesign, 836
Metal oxide varistors (MOV), 381, 383
Metal-oxide-semiconductor field effect transistor
  See MOSFET
Metallization, 1501
Metrology, 132–133
Micro molding technologies, 1489–1500
Microactuators, 1474–1479, 1526–1532
  potential applications, 1483
    electrostatic handling of biological objects, 1484
    fluidics, 1483–1484
    micro magnetic head, 1484
    optics, 1483
  types of, 1526
Microbattery, 1542–1544
Microcontroller instruction sets, 68–69
Microcontrollers, 45-47, 49, 50, 66–69
Microdrilling, 1508–1509
Microelectromechanical systems (MEMS), 1468–1470, 1500
  advantages, 1479
  applications, 1503–1505
  economic aspects, 1576–1590
  fabrication, 1501–1503
  integration, 1582–1585
  potential applications, 1483–1484
    fluidics, 1483–1484
    micro magnetic head, 1484
    optics, 1483
  power requirements of, 1538–1540
Microlevel, 840
Micromachined lumped elements, 1567–1572
Micromachines, 1472–1487
  and the scaling effect, 1472–1473
Micromachining
  materials, 1474
Micromachining fabrication, 1489–1500
Micromilling, 1509–1510
Microrobots, 731
Microsensor Systems, Inc., 1523
Microsensors, 1515
Microshield, 1548, 1557
Millions of instructions per second
  See MIPS
MIMO
  See Multi-input-multi-output
Minimax learning rule, 1291
Minimum ECM classifier, 1057–1058
Minimum variance criterion (MVC), 166, 167
Minority carrier lifetime, 232, 234
MIPS, 61, 62
MIPS R4000, 55
Mismatch reset, 1293
Mitsubishi BNU digital ANN IC, 873
MLP
  See Multilayer perceptron
MMS
  See Manufacturing message specification
Mobile robotics, 746
Mobile robots, 541, 773
  applications, 783–784
  control architectures, 775–776
  control of, 780–783

kinematics, 774–775
platforms, 773–774
Mobius transformation
See Bilinear transformation
Modbus plus networking strategy, 641
Model predictive control (MPC), 515, 894
Model reference adaptive control (MRAC), 574
Model reference adaptive system (MRAS), 304, 306
Modeling, 778
Modelling parallelism, 33, 34
Modern filter theory, 105, 106
Modified bit, 65, 66
Modular FMS, 649
Modulus value, 960
Modus ponens
See Forward chaining
Modus tollens
See Backward chaining
Monolithic integrated circuits, 1585
Moore state machines, 29, 30
MOS-controlled thyristor (MCT), 199, 200
MOSFET, 196–197, 198–199, 216–218, 218–220
applications, 225–226, 227–228
dynamic characteristics, 222–225, 224–227
static characteristics, 218–222, 220–224
Mother wavelet, 1023
Motion analysis system, 180, 181
Motion control, 341, 343, 760
Motion tracking, 1383–1384, 1393–1396
Motorola 68000, 48, 53, 54
Motorola 6805 microcontroller, 48, 67, 68
MOV
See Metal oxide varistors
Moving average models, 451
MPC
See Model predictive control
MPS
See Master production scheduling
MRAC
See Model reference adaptive control
MRAS
See Model reference adaptive system
MRP
See Material requirements planning
MRP time-phased record, 663
MRPAP
See Medium range production and assembly planning
MRPII
See Manufacturing resource planning
MSI
See Medium scale integration
MTRT
See Maximum token rotation time
MTTR
See Mean time to repair
Multi chip module (MCM), 1022, 1586
Multi-class optimal classifiers, 1058–1059
Multi-input-multi-output (MIMO), 522
Multi-response regression
and flexible discriminant analysis, 1070–1071
Multi-sensor controller, 1626
Multi-streaming, 891
Multichip module, 870
Multilayer perceptron (MLP), 841, 843, 898
backpropagation, 843
neuron, 843
Multilayer perceptron network, 942–949
Multilayer perceptrons
architecture, 918
See also Feed forward networks
training, 918
Multilayer perceptrons with local feedback, 913
Multiple processor system architectures
collision, 70, 71
multiple-instruction, multiple-data stream, 69, 70
multiple-instruction, single-data stream, 69, 70
single-instruction, multiple-data stream, 69, 70
single-instruction, single-data stream, 68–69, 69–70
Multirobot system, 676
Multisensor fusion, 1325, 1593–1608
potential problems, 1595–1596
Multisensor integration, 1596–1608
advantages, 1594
potential problems, 1595–1596
Multisensor multiresolution
for target identication, 1627–1632
Multistage interconnection network, 1453
Multitasking, 64, 65
Mutation
and genetic algorithms, 1328
MVC
See Minimum variance criterion

N

n-type materials, 9
NAND circuit, 19, 20
Nanotechnology, 1500–1505
Narendra-Parthasarathy temporal neural network model, 910–911
NASP
See National Aerospace Plane
National Aerospace Plane (NASP), 890, 893
National Information Infrastructure (NII), 1405
Navigation of mobile robots, 781
Nearest neighbor classification model, 1221–1222
Nearest neighbor rule, 1062
Nearnst effect, 153, 154
NEBULAS project, 1440–1442
Negative feedback op-amp circuits
inverting, 148, 149
noninverting, 148, 149
Neocognitron, 966–974, 976
and pattern recognition, 976–978
network architecture of, 966
Neonatal oxygenation control system, 1625–1626
Net change processing, 664
Network, 386, 388
Network access, 401, 403
Network connectivity, 387, 389
Network layer, 390, 392
Network performance, 979
Network power factor, 250, 252
Network-to-network interface (NNI), 413, 415
NeuFuz4, 1143–1146, 1356
Neural adaptive control, 1173
Neural adaptive controllers, 890
Neural chips, 1158–1159
Neural evolutionary systems, 1341
Neural fuzzy systems
in handwritten digit recognition, 1248–1256
Neural learning, 840
Neural net clone, 890
Neural network chips
examples, 858–863
Neural network inputs, 1353–1354
fuzzy representations of, 1353–1354
Neural network mapping, 1347
Neural network outputs, 1353–1354
fuzzy representations of, 1353–1354
Neural network processors (NNP), 885–887
Neural network technology
applications of, 864
Neural network-based fuzzy logic decision system, 1354–1355
Neural networks, 853, 1057, 1062–1075, 1344
and expert systems, 1347
and fuzzy logic systems, 1347–1356
architecture, 991
data visualization, 838
design and operation, 991–994
flight control, 895
usage, 1063
Neural systems
and evolutionary systems, 1338–1341
Neural-type cell model, 881
Neurocontrol, 888–900, 1157–1165, 1166–1180
and fuzzy logic, 1166–1180
Neurofuzzy control, 1361
in consumer products, 1361–1362
Neuroidentification, 900–904, 1160
alternative designs for, 901–904
applications, 901
conventional version, 900
definition, 889, 900–901
in diagnostic applications, 901
in image processing, 901
in pattern classification, 901
in prediction problems, 901
in speech recognition, 901
in submarine detection, 901
stochastic version, 900–901
Neurological process
modeling, 881–884
Neuromorphic models, 875–881
Neuron, 843
See also Multilayer perceptron
Neuron detector, 143
Neurons
design, 994
Neurotransmitters, 875
Neutron detector, 144
Next-gray-code rule, 817
NII
See National Information Infrastructure

NI1000 recognition accelerator chip, 863, 1033
Ni1000 Recognition Accelerator IC, 871
Nibbles, 51
Nichols chart, 481
NNI
  See Network-to-Network Interface
NNP
  See Neural network processors
Noise, 105, 106, 119, 120, 454
Non-contact manipulation methods, 1534–1536
Non-real-time learning, 896–897
Non-return-to-zero-inverted (NRZI), 409, 411
Non-uniform mutation, 1328
Noncontact sensors, 176, 177
  acoustic-based, 176, 177
  electromagnetic-based, 181, 182
  optical-based, 177, 178
Nonlinear amplitude response function, 124, 125
Normal density function
  See Gaussian density function
Normal response mode (NRM), 428, 430
NRM
  See Normal response mode
NRZI
  See Non-return-to-zero-inverted
NTC
  See Neural-type cell model
Nuclear power plants
  use of expert systems in, 833
Nyquist contour, 461
Nyquist criterion, 139, 140, 461
Nyquist frequency, 109, 110, 111
Nyquist plot, 461

O

Object oriented design (OOD) software, 129, 130
Object recognition, 1595
Object-oriented models, 698
Observers
  modeling of, 709
Obstacle avoidance, 781
Obstacle detection, 781
OCR IC
  See Optical character recognition IC, 873
Odometry, 731
Odor detectors, 749
Odor discrimination, 749
Off-line learning, 897
Off-line self-tuning, 598
Offset amplifier, 106, 107
Olfactory sensors, 749
On-chip RAM
  See Scratchpad memory
On-line learning, 897
On-line self-tuning, 598
On-line storage, 126, 127
On-sensor chip mechanical integration, 1585
On-sensor-chip active electronics integration, 1588–1589
On-sensor-chip passive electronics integration, 1586–1588
On-state resistance, 220, 222
One-dimensional signal processing, 723
One-quadrant converters, 245, 247
OOD
  See Object oriented design
Op-amp
  See Integrated circuit operational amplifier
Open systems interconnection (OSI), 389, 391
Open-loop control system, 453
Open-loop optimal feedback control
  See Receding horizon control
Operand specifiers, 53–54
Operands, 49, 50
Operating system software, 50, 51, 127, 128
Operation code, 53–54
Operational transconductance amplifier (OTA), 875, 881
Opportunistic-charging, 774
OPT
  See Optimized production technology
Optical character recognition IC, 873
Optical piezo electric actuator, 1529
Optical scanner system, 179, 180
Optical tracking, 1396
Optical-based sensing systems
  See Noncontact sensors: optical-based
Optimal Scaling, 1055
Optimal scoring, 1071
Optimization, 1233–1236
  theory and objectives, 1233–1236
Optimized LVQ1, 1061–1062
Optimized production technology (OPT), 665
Optomechanical sensor array, 746
Orbital operations simulator, 1150
Oscillations
  See Chattering
Oscilloscope, 103, 104
OSI
  See Open Systems Interconnection
OTA
  See Operational transconductance amplifier
Ouput voltage
  filtering of, 268–270
Outages, 351, 353
Output probabilities mode, 1032
Output projection vector, 522
Output voltage
  filtering of, 270–272
Overload recovery, 105, 106
Overloadability, 205–207, 207–209
Overmodulation, 265, 266
Oxidation machining, 1477

P

p-type materials, 9
Page descriptor, 66
  See also Virtual memory management
Page fault, 66
Page table, 66
Paging, 66
Parallel distributed processing, 933
Parallel loaded resonant converter, 282, 284
Parallel resonant circuit, 277, 279
Parallel resonant converter (PRC), 279, 281, 283
Parallel resonant DC-link converter, 272, 274
Parallel simulation, 1452–1453
Parellel inductance, 270, 272
Parks-McClellan algorithm, 91, 92
Partial read-out architecture, 1454–1455
Passive sensors, 97, 98
Passive stereopsis, 737
Passive temperature calibration, 1580
Passive temperature compensation, 1580
Passive-video sensing systems, 179, 180
Passivity-based adaptive control law, 765
Patient status monitors, 1623–1624
Pattern interpretation and recognition application toolkit and environment (PIRATE), 1343
Pattern learning, 897
Pattern recognition, 916–922, 933–941
  applications, 1212–1213
  knowledge representation in, 1212–1213
  role of fuzzy sets in, 1207–1228
  system design, 1240–1245
  theory, 1231–1232
  wavelets for, 923–932
    boundary gating, 929
    local features, 928
PCB
  See Printed circuit board
PDF
  Probability density function
Peak diode recovery, 224, 226
Peak voltage, 5, 8
Peltier effect, 152, 153
Perception, in mobile robots, 776
Perceptron convergence, 1064
Perceptron convergence procedure, 855
Performance verification test, 129–130, 130–131
Periodic continuous-time signals, 73, 74
Periodic convolution, 83, 84
Permanent magnet (PM) excitation, 189, 191
Permanent magnet step motor, 333, 335
Permanent magnet synchronous motor (PMSM), 319, 321
Perturbation analysis, 691
Perturbation approach, 1374
Petri nets, 670
  33–34, 46
  34–35, 47
  description of, 671–672
  methods of analysis of, 674–676
  properties of, 672–674
    deadlock, 674
Phase detector, 530
Phase margin, 480
Phase multiplication, 363, 365
Phase-crossover frequency, 480
Phase-lag compensators, 482
Phase-lead compensators, 482
Phase-locked loop (PLL), 529
Photo transistors, 13, 15
Photocopying machines, 1430–1433
Photosensitive polyimide process, 1491–1497
Physical address space, 65, 66
Physical connection endpoint identifiers, 390, 392
Physical connections in OSI models, 390, 392
Physical layer protocol, 409, 411

Physical medium dependent (PMD), 409, 411
Physical service data units, 390, 392
PID compensators, 485
PID control, 470
PID process controllers, 149, 150
Piezo electric actuator, 1528
Piezoelectric sensors, 747
Piezoelectric transducers, 741
Pin diodes
  See Three-layer diodes
Pipelined recurrent neural network, 913
Pipelining, 69, 70
Pixel, 732
Pixie-level multisensor fusion, 1602
Placement geometry, 733
Planar articulating controls experiment, 1372
Platinum diffusion, 233, 235
PLD
  See Programmable logic devices
PLL
  See Phase-locked loop
Plotting software, 130, 131
PLS
  See Programmable logic sequencer
PM synchronous motor drive, 343, 345
PMD
  See Physical medium dependent
PMSM
  See Permanent magnet synchronous motor
pn junction
  See Diode
Pneumatic actuators, 751, 1532
Point-to-point wiring, 385, 387
Poisson distribution, 682
Polar robotic manipulators, 730
Polaroid ranging module, 742
Polaroid transducer, 741
Pole placement design, 504–510
Polymer actuators, 1532
Polyvinylidene difluoride, 154–155, 155
POP, 58
Porting, 50
Ports, 46, 47
Position control mode, 787
Power conditioning, 367, 370
Power dissipation, 222, 224
Power electronic building block, 1369
Power electronics, 185, 187
Power factor angle, 354, 356
Power quality, 349, 351
Power rectifier, 193, 195
Power semiconductor devices, 193, 195
Power semiconductor diodes, 201, 203
Power supply, 187, 189
PRC
  See Parallel resonant converter
Precision, in information representation, 52
Precision metallic microstructures, 1489–1500
Precision micromachining, 1505–1511
  processes, 1507–1511
Prediction error, 962
Present bit, 65, 66
Presentation layer, 392, 394
Pressure sensors, 1515
Pressurized water reactor (PWR), 143
Primary capital equipment, 648
Primary computer memory, 62, 63
Primary current, 250, 252
Primary pyroelectricity, 155, 156
Primary voltage control system, 296, 298
Primitive flow table, 30, 31
Principal component analysis, 960
Print-header rule, 821
Print-result rule, 821
Printed circuit boards (PCB), 19, 21, 41, 42, 727
Probabilistic neural networks, 1015, 1026, 1038–1046
  advantages of, 1041
  using alternative estimators of f(X), 1041
Probability density function (PDF), 142, 143, 938
Probe-type charging, 774
Process-oriented simulation, 699
Processor coupling
  loosely coupled, 69, 70
  tightly coupled, 69, 70
Processor status register, 55
Producer-distributor-consumer (PDC) model, 406, 407
Production cycle, 657
Production management, 653
PROFIBUS, 407, 409
Program counter, 54
Program proposal, 634
Programmable logic controller, 44–46, 45–47, 587
Programmable logic device architecture
  AND array, 42
  OR array, 42
Programmable logic devices (PLD), 42
Programmable logic sequencer (PLS), 42, 45
Programmable read-only memory
  See PROM
Programming languages, 50, 51
Programming software, 590
PROLOG, 50
PROM, 63, 64
Proton implantation, 233, 235
Prototype cell commitment, 1029–1030
Prototype pattern counts, 1031
Prototype threshold modification, 1030–1031
Prototyping, reverse engineering and, 626–627
Proximeter, 181, 182
Proximity transducer sensing system, 181, 182
Pseudomedian filtering, 1219
PSPICE, 5, 8, 15, 17
Pull system, 666
Pulsatile oxygen analyzer, 1624
Pulse number, 244, 246
Pulse-width modulation (PWM), 263, 265
Pulsed width modulated motor drives, 753
Pumped storage hydroelectric power plants, 317, 319
Punch-through IGBT, 230, 232, 234
Punch-through phenomenon, 202, 204, 219, 221
PUSH, 58
Push system, 666
Push-pull inverters, 264, 266
PWM
  See Pulse-width modulation
PWR
  See Pressurized water reactor
Pyramid Solution Program, 639
Pyroelectric effects, 154, 155
Pyroelectricity, 154, 155

Q

Quadratic discriminant classifier, 1056
Quadratic discriminant score, 1059
Quadratic sigmoid function, 1185
Quality function deployment, 626
Quality loss function, 626
Quantum, 140
Quartz crystal microbalance, 1519
Queuing models, 686–687, 696
Queuing networks, 687
Queuing theory, 681
Quickprop, 1068
Quine-McCluskey method, 27

R

Radial basis function (RBF), 843
Radial basis function neural networks, 1014–1017
  operation of, 1015
    curse of dimensionality, 1015
  training of, 1015–1017
Radiated emissions
  measurements, 380, 382
  mitigation, 380, 382
Radix, 52
RAM (read-write memory), 63, 64
RAM chips
  dynamic, 63, 64
  static, 63, 64
Random access memory
  See RAM
Random FMS, 649
Random noise, 944
Random process, 140, 141
Random signal, 140, 141
Range of numbers, 52
Range image processing, 1615–1621
Range-finding, 735
Rangel, 733
Ranking selection, 1319
Rapid prototyping
  methods, 626
Rate control mode, 787
Raw data, 130, 131
RBF
  See Radial basis function
RBF neural networks
  See Radial basis function neural networks
RCCP
  See Rough cut capacity planning
RCE neural networks, 1025–1036
  advantages, 1026–1027
  applications to pattern recognition, 1032–1033
    character recognition, 1034–1035
    image analysis applications, 1034–1035
  response modes, 1032–1033
  training of, 1027–1032
RCT
  See Reverse-conducting thyristor

RD-31 project, 1445
Reach-through phenomenon, 219–220, 221–222
Reactive power, 352, 354–356, 356–358
Reactive power compensation, 356, 358
Reactive power factor, 355, 358
Read-only memory
  See ROM
Read-write memory
  See RAM
Real-time correction circuit (RTCC), 725
Real-time generalization, 1200
Real-time learning, 896–897
Real-time reconfiguration, 1200
Real-time recurrent learning algorithm, 913
Receding horizon control, 516
Receptivity of state machine, 33
Recognition algorithms, 724
Reconfigurable control, 1199–1202
Reconfiguration, 1199
Rectification, 244, 246
Rectifying circuit, 6
Recurrent neural networks, 913
Reduced instruction set computer (RISC), 47, 48, 54, 55
Redundant information, 1326, 1594
Reference architecture, 634
Reference model
  See Information function architecture
Reflection, 739
Refraction, 875
Refractory period, 875
Regeneration run, 664
Regenerative braking, 293, 295
Region of interest algorithm, 1015
Registers, 47, 48, 54–55
Regularization, 536
Reinforcement learning (RL), 575, 894, 1147
Relaxation methods, 937
Relay ladder diagrams, 41
Relay solenoid, 41
Remote manipulator system, 1122–1123
Rendering hardware, 1409–1410
Repetitive peak reverse voltage, 209
Reproduction
  and genetic algorithms, 1328
Resist, 1499–1500
Resistance temperature detector (RTD), 142, 143
Resistive grid local averaging, 925–928
  advantages and disadvantages, 928
Resistive networks, 925
Resistors, 5, 6, 213–214, 215–216
Resonance, 1293
Resonant converters, 276, 278
Resonant frequency, 114, 115
Resonant switch converters, 271, 273
Respiratory distress syndrome, 1622
Restricted coulomb energy, 872
RET (return), 60
Rete pattern matching algorithm, 810
Return
  See RET
Return from subroutine
  See RTS
Reverse biased diode, 6
Reverse conducting thyristor (RCT), 280, 282
Reverse edges, 740
Reverse recovery phenomenon, 202, 204
Reverse voltage, 202, 204
Reverse-conducting thyristors, 195, 197
Richardson's law, 933
Righi-Leduc effect, 154, 155
Rings, 386, 388
Ripple, 245, 247
Ripple amplitude, 257, 259
RISC
  See Reduced instruction set computer
RL
  See Reinforcement learning
RMS
  See Root mean square
Robot hand system, 1632–1639
Robot platforms, 773
Robot tactile sensing, 745
Robot vision, 732
Robotic manipulator, 730
Robotic manipulators
  Euler-Lagrange equations of, 760–762
Robotic sense of smell, 749
Robots
  qualities and capabilities, 730–732
Robust design, 626
Robust parametric stability, 459
ROM (read-only memory), 63, 64
Root locus, 491
Root locus method, 490
Root mean square (RMS), 353, 355
Root node, 171, 172
Rosenblatt, Frank, 942
Rotary UPS, 372, 374
Rotor, 752
Rotorcraft machines, 1182–1190
Rough cut capacity planning (RCCP), 654
Rough surfaces, 744
Routh table, 458
Routh-Hurwitz criterion, 457, 463
Routing, 431, 433
RS-232C
  See Communications interface standards: RS-232C
RS-422A
  See Communications interface standards: RS-422A
RS-485
  See Communications interface standards: RS-485
RTCC
  See Real-time correction circuit
RTD
  See Resistance temperature detector
RTS (return from subroutine), 60
Rules, 808–809
Rule clusters, 826
Rule interpreter, 827
Rule-based programming, 809
  antecedent, 809
  consequent, 809
  cycle, 809
  firing, 809
Rules of thumb, 808
Runge-Kutta formulas, 550

S

Safety analyses, 1091
Safety grounds, 120, 121
Safety lead time, 664
Safety stock, 664
Sallen-Key filter, 106, 107
Sampling, 72, 73, 80–82, 81–83, 138, 139, 140
Sampling frequency, 81, 82
Sampling period, 81, 82
Sampling theorem, 82, 83
SAP
  See Service access point
SAS
  See Single attached station
Scale sequential approach, 1628
Scaled index, 56, 57
Scaling effect, 1472–1473
SCARF autonomous navigation system, 1641
Scherbius control, 310, 312
Schottky diodes, 209, 211
Scientific data visualization, 1407
Scratchpad memory, 67, 68
SDF
  See Single-degree-of-freedom
SDLC
  See Synchronous data link control
Secondary capital equipment, 648
Secondary computer memory, 63
Secondary pyroelectricity, 155, 156
Secondary resistance control system, 297, 299
Seebeck effect, 153, 154
Seebeck emf, 152, 157–159
Seebeck law, 157, 158
Seebeck source element, 157–158
Seebeck, T.J., 153
Segmentation, 65–66, 66–67
Segmentation, in pattern recognition, 916
Selective attention model, 969–974
  gate signals, 970
  network architecture of, 969–970
Selective laser sintering, 627
Self-commutation, 186, 188
Self-learning layered neural networks, 975–989
Self-organizing feature map, 1061, 1264
Self-organizing map (SOM), 843
Self-tuning control, 607
Self-tuning regulator (STR), 605
SELSPOT, 177, 178
Semiconductor memory, 107, 108
Semiconductors, 6, 9
Semisynchronous bus, 62
Sensitivity approximation, 599
Sensor fusion processing, 1632–1639
Sensor model, 778
Sensors
  classification of, 96, 97
  definition of, 96, 97
Sequence control, 27, 28
Sequence discriminator, 28, 29
Sequencing, 390, 392
Sequential FMS, 649
Sequential system, 35, 36
Series loaded resonant converter, 282, 284
Series resonant converter (SRC), 279, 280, 281, 282

SERVE routine, 665
Service access point (SAP), 410, 412
Service in random order (SIRO), 682
Service process, 682
Servo, 788
Servo drives, 341, 343
Servo system, 303, 305
Servomotor, 560
Session layer, 392, 394
Settling time, 105, 106
Shannon's theorem, 23, 24, 26, 27
Shape memory alloy actuator, 1531–1532
Shells, 808
Shoe box controller, 588
Shunt capacitors, 362, 364
Sidewalk II autonomous navigation system, 1641
Sigmoid potential, 876
Sign-magnitude number format, 51, 52
Signal conditioning and filtering, 105–119, 137, 138
Signal/data transmission, 119–122
  grounding and shielding, 120
Signal integrators, 149, 150
Signal multiplexer, 138, 139
Signal multiplexing, 138, 139
Signal processing, 72, 73
  2-D bar code reading systems, 728
Signal to noise ratio (SNR), 1411
Signal-level multisensor fusion, 1598–1602
Signal-to-noise ratio, 105, 106, 828
Silicon
  mechanical performance of, 1577–1578
Silicon accelerometers
  construction, 1516
  operation characteristics, 1516
Silicon micromachined semiconductor circuits
  fabrication processes, 1551–1556
Silicon micromachining, 1547–1574
Silicon retina, 882–883
Silicon substrate, 13, 21
SIMD numerical array processor (SNAP), 859
SIMD numerical array processor chip, 859
SIMOTRANS HE, 192
Simple boundary gating, 930–931
Simple network management protocol (SNMP), 371, 373, 414, 416
Simulated annealing, 945, 1003–1004
Simulation software, 42, 43
Simultaneous contrast enhancement, 882
Simultaneous recurrent network, 1159, 1161
Sine modulation, 274, 276
Single attached station (SAS), 409, 411
Single instruction multiple data architecture, 947
Single-degree-of-freedom (SDF), 513
Single-ended forward converter, 261, 263
Single-ended hybrid bridge converter, 261, 263
Single-input-single-output (SISO) system, 513
Single-link flexible joint arm, 769
Single-sensor controller, 1626
Single-user operating system, 50, 51
Singular perturbation theory, 540
Singular value decomposition, 1055, 1073
Sinusoid cycle, 6
Sinusoidal voltage, 6
SIRO
  See Service in random order
Six-pulse bridge connection, 248, 250
Six-pulse midpoint connection, 248, 249
Skeletonization, 952
Skewness, 143, 144
Skinlike contact sensors, 746
Skinlike thermal sensors, 748
Slew rate, 105, 106
Sliding mode, 536
Slip frequency type vector control, 301–302, 303–304
Slip power recovery control system, 297, 299, 300, 302
Slip ring wound rotor induction motor, 190, 192
Slip sensors, 749
Slip-power recovery control, 310, 312
Slow reduced order system, 768
SM
  See Synchronous motor
Small scale integration (SSI), 41, 42
Small signal transistor model
  See Alternating current transistor model
Smart gauges, 118, 119
Smart Power[TM], 199, 201
Smith Predictor, 511
Smoothing, 951–952
SMP
  See Standby monitor present
  See Station management protocol
SMPS
  See Switch mode power supply
SMSR
  See Solar maximum satellite repair
SNAP
  See SIMD numerical array processor
SNMP
  See Simple network management protocol
Soft computing, 1087–1090, 1299, 1360–1362
Software, 845
  See Intelligent electronics applications
Software deconvolution, 122–123, 123–124
Software processing, 119, 120
Solar cell array, 1540–1542
Solar maximum satellite repair (SMSR), 794
Solenoids, 751
Solubility parameters, 1522
SOM
  See Self-organizing map
Soma, 875
SONET
  See Synchronous optical network
Sonic digitizing, 626
Sound
  parameters, 1399
Sound content, 1401
Source address, 397, 399
Source current, 104, 105
Spatial locality of reference, 64, 65
Special manufacturing system, 650
Specularity, 739
Speed of sound, 738
Speed sensorless vector control, 303–310
  applications
    hoist, 309
    3-D parking lot, 309
    conveyors, 310
  methods, 304, 306–308, 308–310
SPLIT routine, 665
Square-wave operation, 264, 266
SRC
  See Series resonant converter
SRM
  See Switched reluctance motor
SSI
  See Small scale integration
μ-Stability, 617
Stability, 104, 105
Stability analysis, 456, 1195
Stack, 55
Stack pointer, 55
Stacking stations, 398, 400
Standby power supply, 372, 374
Standby monitor present (SMP), 399, 401
Stars, in network communication, 386, 387
Start-retrieve-result rule, 819
Starting delimiter, 397, 399
State diagram, 28, 29
State estimation, 559
State machine decomposition, 33, 34
State machine evolution
  fundamental mode, 29–33
  pulsed mode, 29–31
State machines, 27–33, 28–34
State observation, 506
State table, 28, 29
State-assignment problem, 35, 36
State-space linear systems, 516
State-transition models, 697
Static pattern recognition, 897
Static VAr compensators, 358, 361
Station management protocol (SMP), 411, 413
Statistical fractals, 933
Stator, 752
Steady-state error analysis, 481
Steel rolling mill drive system, 303, 305
Step motors
  control of, 338–341, 340–342
  drive circuitry for, 334–335, 336–337
  models, 335–338, 337–339
  types and operation of, 331–334, 333–336
Step-down converter, 254, 256
Step-up converter, 254, 256
Step-up/down converter, 254, 256
Stepper motor driver, 755
Stepper motors, 755
Stereo vision, 778
Stereolithography, 627
Sticky neurons, 903
Stochastic process
  See Random process
STR
  See Self-tuning regulator
String encoding, in genetic algorithms, 1317
Striped lighting systems, 736
Structural stability, 1195
Subroutine call
  See CALL

Subtask, 826
SUN SPARC, 47, 48, 53, 54
Superscalar processors, 48, 50
Supervised control, 1171–1172
Supervised controllers, 890
Supervised learning, 888–889, 896–900, 1167–1168, 1218–1223, 1349
  applications of, 897–898
  definition, 889
  for neurocontrol, 1160–1162
    3-layer MLP, 1162–1163
  forms of, 896–897
  implementation of, 1161–1162
Supervised learning systems
  alternative designs for, 898–900
Surface acoustic wave device, 1519
Surface micromachining, 1469
  applications, 1469–1470
Surge voltage, 381, 383
Switch mode power supply (SMPS), 187, 189, 379, 381
Switch peak forward voltage, 285, 287
Switched reluctance motor (SRM), 341, 343
Switched reluctance motor drive
  applications, 348, 350
  construction and operation, 344–345, 346–347
  controls, 347, 349
  equations and equivalent circuits, 345–346, 347–348
  torque-speed, 348, 349
Switches, 19
Switching loss, 230, 232
Symbol-level multisensor fusion, 1603–1607
Symbolic control, 1364
Symbolic fusion
  techniques, 1643–1646
Symmetrical IGBT, 232, 234
Synapse circuits, 993
Synapses, 874
Synaptics, 873
Synchro-resolver, 326, 328
Synchrodrive, 774
Synchronous bus, 61, 62
Synchronous condensers, 358, 360
Synchronous data link control (SDLC), 427, 429
Synchronous motor (SM), 315, 317
Synchronous optical network (SONET), 414, 416
Synchronous state machines, 29, 30
Syncretism, 1161
System identification, 559, 897, 901
System integration, 368, 371
System level, 840
System performance, 640
System program, 49, 50
System topology, 640
μ-Synthesis, 613

T

Tactics for implementation, 841
Tapped delay lines, 1193
Target identification, 1627–1632
Target scattering
  corner, 739
  diffraction, 739
  edges, 739
Target token rotation time (TTRT), 410, 412
Taylor series, 466
TCE
  See Teleoperation configuration editor
TCP
  See Transmission control protocol
TDF
  See Two-degree-of-freedom
TDM
  See Time division multiplexing
TDNN
  See Time-delay neural networks
Telemedicine, 798–799
Teleoperation, 784
Teleoperation configuration editor (TCE), 791
Teleoperators, 784
TELEPERM M automation system, 641
Telephone influence factor, 353, 355
Television equipment, 1433
Temperature, 150–152
Temperature coefficient, 104, 105
Temporal locality of reference, 64, 65
Temporal neural networks
  with hidden states, 912–913
    FIR multilayer perceptrons, 912
    multilayer perceptrons with local feedback, 913
    recurrent neural networks, 913
    time-delay neural network, 912–913
  with observable states, 910–912
    gamma model, 911–912
    Narendra-Parthasarathy model, 910–911
Temporal redundancy, 809
Temporal signal processing, 910–914
Temporal stereopsis, 737
Tensorial pyroelectricity, 155, 156
Tertiary pyroelectricity, 155, 156
THD
  See Total harmonic distortion
Theodolite system, 180, 181
Thermal actuator, 1532
Thermal data, 208, 210
Thermal voltage, 9
Thermally controlled magnetization actuator, 1545
Thermistors, 749
Thermocouple, 97, 98
Thermocouple circuits, 156, 157
Thermoelectric circuit analysis, 158, 159
Thermoelectric effects, 151, 152
Thermomagnetoelectric effects, 153, 154
Thevenin's equivalent circuit, 6
Thickness-shear mode resonator, 1519
  See also Quartz crystal microbalance, 1519
Thin film deposition, 1501
Thin-film disk heads, 725
Three-D measurement
  techniques, 173–182
Three-D triangulation principle, 180, 181
Three-dimensional micromachining, 1474
Three-dimensional parking lot, 309, 311
  See also Speed sensorless vector control
Three-dimensional printing, 627
Three-dimensional signal processing, 724
  computer vision, 724
Three-dimensional sound, 1402
Three-layer diodes, 201, 203
Three-pulse midpoint circuit, 246, 248
Three-pulse midpoint connection, 247, 249
Threshold control, 971
Threshold potential, 875
Threshold voltage, 221, 223
Thresholding, 743
THT
  See Token holding time
Thyristor, 191, 193, 194–197
Thyristor motor, 316, 318
Tichonov's theorem, 768
Time delay, 511
Time division multiplexing (TDM), 412, 414
Time domain, 72, 73, 74
Time-delay neural networks (TDNN), 901–904, 912–913
Time-lagged recurrent network, 891
Time-lagged recurrent networks (TLRN), 891, 901–904, 1161
  real-time adaptation of, 903
Timing belts, 757
TLRN
  See Time-lagged recurrent networks
Toe truncation method, 123, 124
Token bus, 400, 402
Token errors, 398, 400
Token frame, 401, 403
Token holding time (THT), 398, 400
Token ring, 70, 71, 396, 399
Tolerance band control, 266, 268
Torsion oscillations, 542
Total demand distortion, 353, 355
Total harmonic distortion (THD), 353, 355
Total probability of misclassification, 1056
Total pyroelectric effect, 155, 156
Tourmaline, 154, 155
Tournament selection, 1319
Toyota just-in-time system, 666
TPF
  See True power factor
Tracking, in neurocontrol, 1158
Traffic shaping, 1446, 1449
Transconductance, 17, 19, 21, 221, 223
Transducer amplifiers
  performance parameters of, 104–106
Transducer power supplies
  performance parameters of, 104, 105
Transducers, 132–133, 133–134
  nomenclature and modifiers, 97, 98
Transfer batch, 666
Transformer, 247, 249
Transformer leakage reactance, 362, 365
Transient response, 11, 13, 15, 17, 454
Transient transistor model
  See High-frequency transistor model
Transients, 350, 352
Transistor bias, 18, 20
Transistor load line, 8–11
Transistor switch, 9, 11
Transistor terminals
  base, 11, 13
  collector, 11, 13
  emitter, 11, 13
Transistor voltage ratings, 211, 213
Transistor-transistor logic circuit, 18, 20

Transistors, 195, 197
bipolar junction transistors, 8, 10
insulated-gate-bipolar-junction transistors, 8, 10
junction field effect transistors, 8, 10
metal oxide semiconductor field effect transistors, 8, 10
Transmission control protocol (TCP), 435, 437
Transmissions, 756
Transport layer, 391, 393
Trapezoidal modulation, 274, 276
Trigger circuit, 107, 108
Troubleshooting software, 130, 131
True power factor (TPF), 360, 362
Truncation, 892, 903
Truth table, 811
printing, 821
Truth table simplification program, 811
Truth tables, 22, 23, 24
TSP
See Twisted shielded pair
TTRT
See Target token rotation time
Tuned filters, 361, 363
Tustin's method
See Bilinear approximation and Digital system design
Twelve-pulse converter, 248, 250
Twisted shielded pair (TSP), 119, 120
Two's complement number system, 51, 52
Two-degree-of-freedom (TDF), 513
Two-dimensional conveyor, 1479–1480
Two-dimensional signal processing, 724
Two-group regression
and linear discriminant function, 1070
Two pulse bridge connection, 248, 250
Two-pulse midpoint connection, 247, 249
Two quadrant converters, 245, 247

U

UDP
See User datagram protocol
Ultrasonic sensing, 738
Ultrasonic thickness gauges, 748
Ultrasonic transducers, 740
Ultrasound, 738
attenuation of, 738
Ultrasound sensors, 777
Uncertainty analysis, 133, 134
Uncertainty management, 828
UNI
See User-to-network interface
Unidirectional switches, 280, 282
Uninterruptible power supply (UPS), 187, 189, 367, 369
Uninterruptible power system, 352, 354
Unipolar voltage switching, 256, 258
Unitary transforms, 169, 171
Unity-gain buffer, 148, 149
Universal approximation, 1067
Universal codes, 38, 39
UNIX operating system, 127, 128
Unmanned aerial vehicles
design of, 1128–1135
simulations, 1135
Unmanned helicopter
hardware system, 1129–1131
software system, 1131–1135
Unsupervised learning, 967–969, 1063, 1225–1228
UPS
See Uninterruptible power supply
Urgent data, as TCP option, 436, 438
User datagram protocol (UDP), 434–435, 436–437
User-to-network interface (UNI), 413, 415

V

V belts, 757
Variable reluctance step motor, 332–333, 334–335
Variable structure control, 535
VCI
See Virtual circuit identifier
VCO
See Voltage-controlled oscillator
Vector control, 300, 302
Vector quantization, 1056
Vector quantization (VQ), 841
Velocity sensitivity, 883
Very large scale integration (VLSI), 41, 42
Vibration monitoring, 1332
Vibration monitoring system, 1182
Vigilance, 1289
Virtual circuit identifier (VCI), 412, 414
Virtual manufacturing (VM), 718
Virtual manufacturing device (VMD), 720
Virtual measurement, 1355
Virtual memory management, 65–67
memory management unit (MMU), 66
page descriptor, 66
segmentation, 66–67
Virtual path identifier (VPI), 412, 414
Virtual prototyping, 627
Virtual reality, 627, 1383–1388
applications, 1384–1388
in industrial electronics, 1407–1409
Virtual reality systems, 1384
Virtual short-circuit, 148, 149
Virtual sound, 1384, 1397–1404
Virtual tool, 1391
Virtual workbench, 1390–1393
potential applications, 1392–1393
Vision sensors, 778
Visual inspection machine, 727
VLSI
See Very large scale integration
VM
See Virtual manufacturing
VMD
See Virtual manufacturing device
VME-ATM adapter, 1461–1463
Volatility, 62, 63
Voltage cancellation, 268, 270
Voltage divider, 14, 16
Voltage division, 145, 146
Voltage insertion technique, 135, 136
Voltage source inverters (VSI), 263, 265, 287, 289
Voltage transfer function, 15, 17
Voltage-controlled oscillator (VCO), 531, 877
Von Neumann architecture, 48, 49
Von Neumann, Jon, 48, 49
Voxels, 733
VPI
See Virtual path identifier
VSI
See Voltage source inverters
VTTS autonomous navigation system, 1641

W

W-domain, 555
W-plane, 487
WAN
See Wide area network
Ward Leonard control system, 289–290, 291–292
Warm-up time, 104, 105
Waterloo spatial motion analysis and recording techniques
See WATSMART
WATSMART, 177, 178
Wavelet-based segmentation, 923–925
1-D maxima detection wavelet, 924
1-D wavelet transform, 923
2-D wavelet transform, 924–925
Way number, 244, 246
Weight vector, 1292
Weights, 896
Wheatstone bridge, 105, 106, 145, 146
Whisker sensors, 746
White noise, 144, 145
Wide area network (WAN), 126, 127
Widrow's adaptive noise canceller
See also Adaptive noise canceller, 1609
Widrow-Hoff learning procedure, 856
Wiener-Hopf equations, 165, 166
Wilkinson power divider, 1557
Williams-Zipser method, 903
Window grabber, 991
Windowing, 90, 91, 897
Windows
Bartlett, 90, 91
Hamming, 90, 91
Hanning, 91, 92
rectangular, 90–91, 91–92
WIP
See Work-in-process
Wire-bonding system, 724
Withdrawal kanban, 667
Word size, 50, 51
Work-in-process (WIP), 654
Working mask, 1499
Working memory, 808, 827
facts, 808
Write to memory, 64, 65
Writeable bit, 65, 66

X

XOR problem, 920–921
and the Feature Saliency method, 920–921
Xpress transport protocol (XTP), 437, 439
XTP
See Xpress transport protocol

Z

z-Transforms, 86–88, 87–89
Zadeh's principle of incompatibility, 1245
ZCS
  See Zero current switching
ZCS resonant converter, 284, 286
Zernike moment invariants, 957–958
Zernike moments, 955–960
  advantages, 957
  functions of
  See also Zernike moment invariants, 957–958
Zernike moments calculation, 951
Zero current switching (ZCS), 199, 201, 271, 273
Zero padding, 93, 94
Zero stability, 105, 106
Zero voltage switching, 199, 201
Zero voltage switching (ZVS), 271, 273
Zero-order-hold (ZOH), 553
Zeroing, 526
Ziegler-Nichols tuning, 470
ZIP code recognition, 897
ZISCO36 VLSI chip, 1019–1024
  implementation, 1021
ZOH
  See Zero-order-hold
ZOH equivalence method, 510
ZVS
  See Zero voltage switching
ZVS resonant converter, 284, 286